Peterson's Graduate Programs in the Biological Sciences

2007

Book 3

PETERSON'S

A nelnet. COMPANY

About Peterson's, a Nelnet company

Peterson's (www.petersons.com) is a leading provider of education information and advice, with books and online resources focusing on education search, test preparation, and financial aid. Its Web site offers searchable databases and interactive tools for contacting educational institutions, online practice tests and instruction, and planning tools for securing financial aid. Peterson's serves 110 million education consumers annually.

For more information, contact Peterson's, 2000 Lenox Drive, Lawrenceville, NJ 08648; 800-338-3282; or find us on the World Wide Web at www.petersons.com/about.

Editor: Fern A. Oram; Production Editor: Susan W. Dilts; Copy Editors: Bret Bollmann, Michael Haines, Sally Ross, Jill C. Schwartz, Mark D. Snider, Pam Sullivan, Valerie Bolus Vaughan; Research Associate: Amy Weber; Programmer: Phyllis Johnson; Manufacturing Manager: Ray Golaszewski; Composition Manager: Linda M. Williams; Client Relations Representatives: Danielle Groncki, Mimi Kaufman, Karen Mount, Eric Wallace.

ISSN 1088-9434
ISBN-13: 978-0-7689-2159-5
ISBN-10: 0-7689-2159-7

Printed in the United States of America

10 9 8 7 6 5 4 3 2 1   09 08 07

Forty-first Edition

CONTENTS

A Note from the Peterson's Editors

The six volumes of *Peterson's Graduate and Professional Programs*, the only annually updated reference work of its kind, provide wide-ranging information on the graduate and professional programs offered by accredited colleges and universities in the United States, U.S. territories, and Canada and by those institutions outside the United States that are accredited by U.S. accrediting bodies. More than 44,000 individual academic and professional programs at more than 2,000 institutions are listed. *Peterson's Graduate and Professional Programs* have been used for more than forty years by prospective graduate and professional students, placement counselors, faculty advisers, and all others interested in postbaccalaureate education.

Book 1: *Graduate & Professional Programs: An Overview*, contains information on institutions as a whole, while Books 2 through 6 are devoted to specific academic and professional fields.

Book 2: *Graduate Programs in the Humanities, Arts & Social Sciences*

Book 3: *Graduate Programs in the Biological Sciences*

Book 4: *Graduate Programs in the Physical Sciences, Mathematics, Agricultural Sciences, the Environment & Natural Resources*

Book 5: *Graduate Programs in Engineering & Applied Sciences*

Book 6: *Graduate Programs in Business, Education, Health, Information Studies, Law & Social Work*

The books may be used individually or as a set. For example, if you have chosen a field of study but do not know what institution you want to attend or if you have a college or university in mind but have not chosen an academic field of study, it is best to begin with Book 1.

Book 1 presents several directories to help you identify programs of study that might interest you; you can then research those programs further in Books 2 through 6. The *Directory of Graduate and Professional Programs by Field* lists the 476 fields for which there are program directories in Books 2 through 6 and gives the names of those institutions that offer graduate degree programs in each.

For geographical or financial reasons, you may be interested in attending a particular institution and will want to know what it has to offer. You should turn to the *Directory of Institutions and Their Offerings*, which lists the degree programs available at each institution, again, in the 476 academic and professional fields for which Books 2 through 6 have program directories. As in the *Directory of Graduate and Professional Programs by Field*, the level of degrees offered is also indicated.

All books in the series include advice on graduate education, including topics such as admissions tests, financial aid, and accreditation. **The Graduate Adviser** includes two essays and information about accreditation. The first essay, "The Admissions Process," discusses general admission requirements, admission tests, factors to consider when selecting a graduate school or program, when and how to apply, and how admission decisions are made. Special information for international students and tips for minority students are also included. The second essay, "Financial Support," is an overview of the broad range of support available at the graduate level. Fellowships, scholarships, and grants; assistantships and internships; federal and private loan programs, as well as Federal Work-Study; and the GI bill are detailed. This essay concludes with advice on applying for need-based financial aid. "Accreditation and Accrediting Agencies" gives information on accreditation and its purpose and lists first institutional accrediting agencies and then specialized accrediting agencies relevant to each volume's specific fields of study.

With information on more than 44,000 graduate programs in 476 disciplines, *Peterson's Graduate and Professional Programs* give you all the information you need about the programs that are of interest to you in three formats: **Profiles** (capsule summaries of basic information), **Announcements** (information that an institution or program wants to emphasize, written by administrators), and **Close-Ups** (also written by administrators, with more expansive information than the **Profiles**, emphasizing different aspects of their programs). By using these various formats of program information, coupled with **Appendixes** and **Indexes** covering directories and subject areas for all six books, you will find that these guides provide the most comprehensive, accurate, and up-to-date graduate study information available.

Peterson's publishes a full line of resources with information you need to guide you through the graduate admissions process. Peterson's publications can be found at your local bookstore or library—or visit us on the Web at www.petersons.com.

Colleges and universities will be pleased to know that Peterson's helped you in your selection. Admissions staff members are more than happy to answer questions, address specific problems, and help in any way they can. The editors at Peterson's wish you great success in your graduate program search!

THE GRADUATE ADVISER

The Admissions Process

Generalizations about graduate admissions practices are not always helpful because each institution has its own set of guidelines and procedures. Nevertheless, some broad statements can be made about the admissions process that may help you plan your strategy.

General Requirements

Graduate schools and departments have requirements that applicants for admission must meet. Typically, these requirements include undergraduate transcripts (which provide information about undergraduate grade point average and course work applied toward a major), admission test scores, and letters of recommendation. Most graduate programs also ask for an essay or personal statement that describes your personal reasons for seeking graduate study. In some fields, such as art and music, portfolios or auditions may be required in addition to other evidence of talent. Some institutions require that the applicant have an undergraduate degree in the same subject as the intended graduate major.

Most institutions evaluate each applicant on the basis of the applicant's total record, and the weight accorded any given factor varies widely from institution to institution and from program to program.

Admission Tests

The major testing program used in graduate admissions is the Graduate Record Examinations (GRE)® testing program, sponsored by the GRE Board and administered by Educational Testing Service, Princeton, New Jersey.

The Graduate Record Examinations testing program consists of a General Test and eight Subject Tests. The General Test measures verbal reasoning, quantitative reasoning, and analytical writing skills. It is offered as a computer-adaptive test (CAT) in the United States, Canada, and many other countries. In the CAT, the computer determines which question to present next by adjusting to your previous responses. Paper-based General Test administrations are offered in some parts of the world.

The computer-adaptive General Test consists of a 30-minute verbal section, a 45-minute quantitative section, and a 75-minute analytical writing section. In addition, an unidentified verbal or quantitative section that doesn't count toward a score may be included and an identified research section that is not scored may also be included.

The paper-based General Test consists of two 30-minute verbal sections, two 30-minute quantitative sections, and a 75-minute analytical writing section. In addition, an unidentified verbal or quantitative section that doesn't count toward a score may be included.

The Subject Tests measure achievement and assume undergraduate majors or extensive background in the following eight disciplines:

- Biochemistry, Cell and Molecular Biology
- Biology
- Chemistry
- Computer Science
- Literature in English
- Mathematics
- Physics
- Psychology

The Subject Tests are available three times per year as paper-based administrations around the world. Testing time is 2 hours and 50 minutes. You can obtain more information about the GRE tests by visiting the ETS Web site at www.ets.org or consulting the *GRE Information and Registration Bulletin*. The *Bulletin* can be obtained at many undergraduate colleges. You can also download it from the ETS Web site or obtain it by contacting Graduate Record Examinations, Educational Testing Service, PO Box 6000, Princeton, NJ 08541-6000, telephone 1-609-771-7670.

A revised GRE General Test will be introduced in fall 2007 and is designed to increase test validity, provide faculty with better information regarding applicants' performance, address security concerns, increase worldwide access to the test, and make better use of advances in technology and psychometric design. Changes planned to the verbal reasoning section include greater emphasis on higher cognitive skills and less dependence on vocabulary; more text-based materials, such as reading passages; a broader selection of reading passages; emphasis on skills related to graduate work, such as complex reasoning; and expansion of computer-enabled tasks (e.g., clicking on a sentence in a passage to highlight it). In addition, there will be two 40-minute sections rather than one 30-minute section. Changes to the quantitative reasoning section include quantitative reasoning skills that are closer to skills generally used in graduate school, an increase in the proportion of questions involving real-life scenarios and data interpretation, a decrease in the proportion of geometry questions, and better use of technology (e.g., on-screen calculator). In addition, there will be two 40-minute sections rather than one 45-minute section. Changes to the analytical writing section include new, more focused prompts that reduce the possibility of reliance on memorized materials. The issue and argument tasks will each be 30 minutes in length and essay responses will be made available to designated score recipients. It is anticipated that the new score scale range will be 110 to 150, in 1-point increments. Final specification of the score scale will be determined based on data from the initial revised General Test administrations in fall 2007. A concordance table will be available in early January 2008 to assist score users in determining the relationship between old and new verbal and quantitative scores. The score scale for the analytical writing section will continue to be 0 to 6, in half-point increments. An expanded Internet-based testing network will also become available worldwide. Be sure to check the GRE Web site (www.gre.org) for updated information.

If you expect to apply for admission to a program that requires any of the GRE tests, you should select a test date well in advance of the application deadline. Scores on the computer-adaptive General Test are reported within ten to fifteen days; scores on the paper-based General Test and the Subject Tests are reported within six weeks.

Another testing program, the Miller Analogies Test (MAT), is administered at more than 400 Controlled Testing Centers in the United States, Canada, and other countries. The MAT computer-based test is now available. Testing time is 60 minutes. The test consists of 120 partial analogies. You can obtain the *Miller Analogies Test Candidate Information Booklet,* which contains a list of test centers and instructions for taking the test from http://harcourtassessment.com/HAIWEB/Cultures/en-US/dotCom/milleranalogies.com.htm or by calling Harcourt Assessment, Inc., at 1-800-622-3231.

Check the specific requirements of the programs to which you are applying.

Factors Involved in Selecting a Graduate School or Program

Selecting a graduate school and a specific program of study is a complex matter. Quality of the faculty; program and course offerings; the nature, size, and location of the institution; admission requirements; cost; and the availability of financial assistance are among the many factors that affect one's choice of institution. Other considerations are job placement and achievements of the program's graduates and the institution's resources, such as libraries, laboratories, and computer facilities. If you are to make the best possible choice, you need to learn as much as you can about the schools and programs you are considering before you apply. The following steps may help you narrow your choices.

- Talk to alumni of the programs or institutions you are considering to get their impressions of how well they were prepared for work in their fields of study.

- Remember that graduate school requirements change, so be sure to get the most up-to-date information possible.
- Talk to department faculty and the graduate adviser at your undergraduate institution. They often have information about programs of study at other institutions.
- Visit the Web sites of the graduate schools in which you are interested to request a graduate catalog. Contact the department chair in your chosen field of study for additional information about the department and the field.
- Visit as many campuses as possible. Call ahead for an appointment with the graduate adviser in your field of interest and be sure to check out the facilities and talk to students.

When and How to Apply

You should begin the application process at least one year before you expect to begin your graduate study. Find out the application deadline for each institution (many are provided in the **Profile** section of this volume). Go to the institution Web site and find out if you can apply online. If not, request a paper application form. Fill out this form thoroughly and neatly. Assume that the school needs all the information it is requesting and that the admissions officer will be sensitive to the neatness and overall quality of what you submit. Do not supply more information than the school requires.

The institution may ask at least one question that will require a three- or four-paragraph answer. Compose your response on the assumption that the admissions officer is interested in both what you think and how you express yourself. Keep your statement brief and to the point, but, at the same time, include all pertinent information about your past experiences and your educational goals. Individual statements vary greatly in style and content, which helps admissions officers to differentiate among applicants. Many graduate departments give considerable weight to the statement in making their admissions decisions, so be sure to take the time to prepare a thoughtful and concise statement.

If recommendations are a part of the admissions requirements, carefully choose the individuals you ask to write them. It is generally best to ask current or former professors to write the recommendations, provided they are able to attest to your intellectual ability and motivation for doing the work required of a graduate student. It is advisable to provide stamped, preaddressed envelopes to people being asked to submit recommendations on your behalf.

Completed applications, including references and transcripts and admission test scores, should be received at the institution by the specified date.

Be advised that institutions do not usually make admissions decisions until all materials have been received. Enclose a self-addressed postcard with your application, requesting confirmation of receipt. Allow at least 10 days for the return of the postcard before making further inquiries.

If you plan to apply for financial support, it is imperative that you file your application early.

How Admission Decisions Are Made

The program you apply to is directly involved in the admissions process. Although the final decision is usually made by the graduate dean (or an associate) or by the faculty admissions committee, recommendations from faculty members in your intended field are important. At some institutions, an interview is incorporated into the decision process.

A Special Note for International Students

In addition to the steps already described, there are some special considerations for international students who intend to apply for graduate study in the United States. All graduate schools require an indication of competence in English. The purpose of the Test of English as a Foreign Language (TOEFL) is to evaluate the English proficiency of people who are nonnative speakers of English and want to study at colleges and universities where English is the language of instruction. The TOEFL is administered by Educational Testing Service (ETS) under the general direction of a policy board established by the College Board and the Graduate Record Examinations Board.

The TOEFL is administered as a computer-based test and the TOEFL iBT is administered as an Internet-based test throughout most of the world and these are available year-round by appointment only. It is not necessary to have previous computer experience to take the tests. The computer-based test consists of four sections—listening, reading, structure, and writing. Total testing time is approximately 4 hours. The TOEFL iBT consists of four sections—reading, listening, speaking, and writing. Total testing time is approximately 4 hours.

The TOEFL is offered in the paper-based format in areas of the world where computer-based testing is not available. The paper-based TOEFL consists of three sections—listening comprehension, structure and written expression, and reading comprehension. Testing time is approximately 3 hours.

The Test of Written English (TWE) is also given. TWE is a 30-minute essay that measures the examinee's ability to compose in English. Examinees receive a TWE score separate from their TOEFL score. The *Information Bulletin* contains information on local fees and registration procedures.

Additional information and registration materials are available from TOEFL Services, Educational Testing Service, P.O. Box 6151, Princeton, New Jersey 08541-6151. Telephone: 1-609-771-7100. E-mail: toefl@ets.org. World Wide Web: http://www.toefl.org.

International students should apply especially early because of the number of steps required to complete the admissions process. Furthermore, many United States graduate schools have a limited number of spaces for international students, and many more students apply than the schools can accommodate.

International students may find financial assistance from institutions very limited. The U.S. government requires international applicants to submit a certification of support, which is a statement attesting to the applicant's financial resources. In addition, international students *must* have health insurance coverage.

Tips for Minority Students

Indicators of a university's values in terms of diversity are found both in its recruitment programs and its resources directed to student success. Important questions: Does the institution vigorously recruit minorities for its graduate programs? Is there funding available to help with the costs associated with visiting the school? Are minorities represented in the institution's brochures or Web site or on their faculty rolls? What campus-based resources or services (including assistance in locating housing or career counseling and placement) are available? Is funding available to members of underrepresented groups?

At the program level, it is particularly important for minority students to investigate the "climate" of a program under consideration. How many minority students are enrolled and how many have graduated? What opportunities are there to work with diverse faculty and mentors whose research interests match yours? How are conflicts resolved or concerns addressed? How interested are faculty in building strong and supportive relations with students? "Climate" concerns should be addressed by posing questions to various individuals, including faculty members, current students, and alumni.

Information is also available through various organizations, such as the Hispanic Association of Colleges and Universities (HACU), and publications, such as *DIVERSE: Issues in Higher Education* and *Hispanic Outlook* magazine. There are also books devoted to this topic, such as *The Multicultural Student's Guide to Colleges* by Robert Mitchell.

Financial Support

The range of financial support at the graduate level is very broad. The following descriptions will give you a general idea of what you might expect and what will be expected of you as a financial support recipient.

Fellowships, Scholarships, and Grants

These are usually outright awards of a few hundred to many thousands of dollars with no service to the institution required in return. Fellowships and scholarships are usually awarded on the basis of merit and are highly competitive. Grants are made on the basis of financial need or special talent in a field of study. Many fellowships, scholarships, and grants not only cover tuition, fees, and supplies but also include stipends for living expenses with allowances for dependents. However, the terms of each should be examined because some do not permit recipients to supplement their income with outside work. Fellowships, scholarships, and grants may vary in the number of years for which they are awarded.

In addition to the availability of these funds at the university or program level, many excellent fellowship programs are available at the national level and may be applied for before and during enrollment in a graduate program. A listing of many of these programs can be found at the Council of Graduate Schools' Web site: http://www.cgsnet. org/ResourcesForStudents/fellowships.htm.

Assistantships and Internships

Many graduate students receive financial support through assistantships, particularly involving teaching or research duties. It is important to recognize that such appointments should not be simply employment relationships but rather should constitute an integral and important part of a student's graduate education. As such, the appointments should be accompanied by strong faculty mentoring and increasingly responsible apprenticeship experiences. The specific nature of these appointments in a given program should be considered in selecting that graduate program.

TEACHING ASSISTANTSHIPS

These usually provide a salary and full or partial tuition remission and may also provide health benefits. Unlike fellowships, scholarships, and grants, which require no service to the institution, teaching assistantships require recipients to provide the institution with a specific amount of undergraduate teaching, ideally related to the student's field of study. Some teaching assistants are limited to grading papers, compiling bibliographies, taking notes, or monitoring laboratories. At some graduate schools, teaching assistants must carry lighter course loads than regular full-time students.

RESEARCH ASSISTANTSHIPS

These are very similar to teaching assistantships in the manner in which financial assistance is provided. The difference is that recipients are given basic research assignments in their disciplines rather than teaching responsibilities. The work required is normally related to the student's field of study; in most instances, the assistantship supports the student's thesis or dissertation research.

ADMINISTRATIVE INTERNSHIPS

These are similar to assistantships in application of financial assistance funds, but the student is given an assignment on a part-time basis, usually as a special assistant with one of the university's administrative offices. The assignment may not necessarily be directly related to the recipient's discipline.

RESIDENCE HALL AND COUNSELING ASSISTANTSHIPS

These assistantships are frequently assigned to graduate students in psychology, counseling, and social work. Duties can vary from being available in a dean's office for a specific number of hours for consultation with undergraduates to living in campus residences and being responsible for both counseling and administrative tasks or advising student activity groups. Residence hall assistantships often include a room and board allowance and, in some cases, tuition assistance and stipends.

Health Insurance

The availability and affordability of health insurance is an important issue and one that should be considered in an applicant's choice of institution and program. While often included with assistantships and fellowships, this is not always the case and, even if provided, the benefits may be limited. It is important to note that the U.S. government requires international students to have health insurance.

The GI Bill

This provides financial assistance for students who are veterans of the United States armed forces. If you are a veteran, contact your local Veterans Administration office to determine your eligibility and to get full details about benefits. There are a number of programs that offer educational benefits to current military enlistees. Some states have tuition assistance programs for members of the National Guard. Contact the VA office at the college for more information.

Federal Work-Study Program (FWS)

Employment is another way some students finance their graduate studies. The federally funded Federal Work-Study Program provides eligible students with employment opportunities, usually in public and private nonprofit organizations. Federal funds pay up to 75 percent of the wages, with the remainder paid by the employing agency. FWS is available to graduate students who demonstrate financial need. Not all schools have these funds, and some only award them to undergraduates. Each school sets its application deadline and work-study earnings limits. Wages vary and are related to the type of work done. You must file the Free Application for Federal Student Aid (FAFSA) to be eligible for this program.

Loans

Many graduate students borrow to finance their graduate programs when other sources of assistance (which do not have to be repaid) prove insufficient. You should always read and understand the terms of any loan program before submitting your application.

FEDERAL LOANS

Federal Stafford Loans. The Federal Stafford Loan Program offers government-sponsored, low-interest loans to students through a private lender such as a bank, credit union, or savings and loan association.

There are two components of the Federal Stafford Loan program. Under the *subsidized* component of the program, the federal government pays the interest on the loan while you are enrolled in graduate school on at least a half-time basis. Under the *unsubsidized* component of the program, you pay the interest on the loan from the day proceeds are issued. Eligibility for the federal subsidy is based on demonstrated financial need as determined by the financial aid office from the information you provide on the FAFSA. A cosigner is not required, since the loan is not based on creditworthiness.

Although *unsubsidized* Federal Stafford Loans may not be as desirable as *subsidized* Federal Stafford Loans from the student's perspective, they are a useful source of support for those who may not qualify for the subsidized loans or who need additional financial assistance.

Graduate students may borrow up to $18,500 per year through the Stafford Loan Program, up to a cumulative maximum of $138,500, including undergraduate borrowing. This may include up to $8500 in Subsidized Stafford Loans annually, depending on eligibility, up to a cumulative maximum of $65,500, including undergraduate borrowing. The amount of the loan borrowed through the *unsubsidized* Stafford Program equals the total amount of the loan (as much $18,500) minus your eligibility for a Subsidized Stafford Loan (as much as $8500). You may borrow up to the cost of attendance at the school in which you are enrolled or will attend, minus estimated financial assistance from other federal, state, and private sources, up to a maximum of $18,500. The interest rate for the Federal Stafford Loans is fixed at 6.8%.

Two fees may be deducted from the loan proceeds upon disbursement: a guarantee fee of up to 1 percent, which is deposited in an insurance pool to ensure repayment to the lender if the borrower defaults, and a federally mandated 3 percent origination fee, which is used to offset the administrative cost of the Federal Stafford Loan Program. Many lenders do offer reduced-fee or "zero fee" loans.

Under the *subsidized* Federal Stafford Loan Program, repayment begins six months after your last date of enrollment on at least a half-time basis. Under the *unsubsidized* program, repayment of interest begins within thirty days from disbursement of the loan proceeds, and repayment of the principal begins six months after your last enrollment on at least a half-time basis. Some borrowers may choose to defer interest payments while they are in school. The accrued interest is added to the loan balance when the borrower begins repayment. There are several repayment options.

Federal Direct Loans. Some schools participate in the Department of Education's William D. Ford Direct Lending Program instead of the Federal Stafford Loan Program. The two programs are essentially the same except that with the Direct Loans, schools themselves provide the loans with funds from the federal government. Terms and interest rates are virtually the same except that there are a few additional repayment options with Federal Direct Loans.

Federal Perkins Loans. The Federal Perkins Loan is available to students demonstrating financial need and is administered directly by the school. Not all schools have these funds, and some may award them to undergraduates only. Eligibility is determined from the information you provide on the FAFSA. The school will notify you of your eligibility. Eligible graduate students may borrow up to $6000 per year, up to a maximum of $40,000, including undergraduate borrowing (even if your previous Perkins Loans have been repaid). The interest rate for Federal Perkins Loans is 5 percent, and no interest accrues while you remain in school at least half-time. There are no guarantee, loan, or disbursement fees. Repayment begins nine months after your last date of enrollment on at least a half-time basis and may extend over a maximum of ten years with no prepayment penalty.

Deferring Your Federal Loan Repayments. If you borrowed under the Federal Stafford Loan Program or the Federal Perkins Loan Program for previous undergraduate or graduate study, your repayments may be deferred when you return to graduate school, depending on when you borrowed and under which program.

There are other deferment options available if you are temporarily unable to repay your loan. Information about these deferments is provided at your entrance and exit interviews. If you believe you are eligible for a deferment of your loan repayments, you must contact your lender to complete a deferment form. The deferment must be filed prior to the time your repayment is due, and it must be refiled when it expires if you remain eligible for deferment at that time.

SUPPLEMENTAL (PRIVATE) LOANS

Many lending institutions offer supplemental loan programs and other financing plans, such as the ones described here, to students seeking additional assistance in meeting their educational expenses. Some loan programs target all types of graduate students; others are designed specifically for business, law, or medical students. In addition, you can use private loans not specifically designed for education to help finance your graduate degree.

If you are considering borrowing through a supplemental or private loan program, you should carefully consider the terms and be sure to "read the fine print." Check with the program sponsor for the most current terms that will be applicable to the amounts you intend to borrow for graduate study. Most supplemental loan programs for graduate study offer unsubsidized, credit-based loans. In general, a credit-ready borrower is one who has a satisfactory credit history or no credit history at all. A creditworthy borrower generally must pass a credit test to be eligible to borrow or act as a cosigner for the loan funds.

Many supplemental loan programs have minimum and maximum annual loan limits. Some offer amounts equal to the cost of attendance minus any other aid you will receive for graduate study. If you are planning to borrow for several years of graduate study, consider whether there is a cumulative or aggregate limit on the amount you may borrow. Often this cumulative or aggregate limit will include any amounts you borrowed and have not repaid for undergraduate or previous graduate study.

The combination of the annual interest rate, loan fees, and the repayment terms you choose will determine how much you will repay over time. Compare these features in combination before you decide which loan program to use. Some loans offer interest rates that are adjusted monthly, some quarterly, some annually. Some offer interest rates that are lower during the in-school, grace, and deferment periods, and then increase when you begin repayment. Some programs include a loan "origination" fee, which is usually deducted from the principal amount you receive when the loan is disbursed, and must be repaid along with the interest and other principal when you graduate, withdraw from school, or drop below half-time study. Sometimes the loan fees are reduced if you borrow with a qualified cosigner. Some programs allow you to defer interest and/or principal payments while you are enrolled in graduate school. Many programs allow you to capitalize your interest payments; the interest due on your loan is added to the outstanding balance of your loan, so you don't have to repay immediately, but this increases the amount you owe. Other programs allow you to pay the interest as you go, which reduces the amount you later have to repay.

Some examples of supplemental programs follow. The private loan market is very competitive and your financial aid office can help you evaluate these and other programs.

CitiAssist Loans. Offered by Citibank, these no-fee loans help graduate students fill the gap between the financial aid they receive and the money they need for school. Visit www.studentloan.com for more loan information from Citibank.

EXCEL Loan. This program, sponsored by Nellie Mae, is designed for students who are not ready to borrow on their own and wish to borrow with a creditworthy cosigner. Visit www.nelliemae.com for more information.

Key Alternative Loan. This loan can bridge the gap between education costs and traditional funding. Visit www.key.com/html/H-1.3html for more information.

Graduate Access Loan. Sponsored by the Access Group, this is for graduate students enrolled at least half-time. The Web site is www.accessgroup.com.

Signature Student Loan. A loan program for students who are enrolled at least half-time, this is sponsored by Sallie Mae. Visit www.salliemae.com for more information.

Applying for Need-Based Financial Aid

Schools that award federal and institutional financial assistance based on need will require you to complete the FAFSA and, in some cases, an institutional financial aid application.

If you are applying for federal student assistance, you **must** complete the FAFSA. A service of the U.S. Department of Education, it is free to all applicants. Most applicants apply online at www.fafsa.ed.gov. Paper applications are available at the financial aid office of your local college.

After your FAFSA information has been processed, you will receive a Student Aid Report (SAR). If you provided an e-mail address on the

FAFSA, this will be sent to you electronically; otherwise, it will be mailed to your home address.

Follow the instructions on the SAR if you need to correct information reported on your original application. If your situation changes after you file your FAFSA, contact your financial aid officer to discuss amending your information. You can also appeal your financial aid award if you have extenuating circumstances.

If you would like more information on federal student financial aid, visit the FAFSA Web site or download the most recent version of *The Student Guide* at http://studentaid.ed.gov/students/publications/student_guide/index.html. This guide is also available in Spanish.

The U.S. Department of Education also has a toll-free number for questions concerning federal student aid programs. The number is 1-800-4-FED AID (1-800-433-3243). If you are hearing impaired, call toll-free, 1-800-730-8913.

Summary

Remember that these are generalized statements about financial assistance at the graduate level. Because each institution allots its aid differently, you should communicate directly with the school and the specific department of interest to you. It is not unusual, for example, to find that an endowment vested within a specific department supports one or more fellowships. You may fit its requirements and specifications precisely.

Accreditation and Accrediting Agencies

Colleges and universities in the United States, and their individual academic and professional programs, are accredited by nongovernmental agencies concerned with monitoring the quality of education in this country. Agencies with both regional and national jurisdictions grant accreditation to institutions as a whole, while specialized bodies acting on a nationwide basis—often national professional associations—grant accreditation to departments and programs in specific fields.

Institutional and specialized accrediting agencies share the same basic concerns: the purpose an academic unit—whether university or program—has set for itself and how well it fulfills that purpose, the adequacy of its financial and other resources, the quality of its academic offerings, and the level of services it provides. Agencies that grant institutional accreditation take a broader view, of course, and examine university-wide or college-wide services with which a specialized agency may not concern itself.

Both types of agencies follow the same general procedures when considering an application for accreditation. The academic unit prepares a self-evaluation, focusing on the concerns mentioned above and usually including an assessment of both its strengths and weaknesses; a team of representatives of the accrediting body reviews this evaluation, visits the campus, and makes its own report; and finally, the accrediting body makes a decision on the application. Often, even when accreditation is granted, the agency makes a recommendation regarding how the institution or program can improve. All institutions and programs are also reviewed every few years to determine whether they continue to meet established standards; if they do not, they may lose their accreditation.

Accrediting agencies themselves are reviewed and evaluated periodically by the U.S. Department of Education and the Council for Higher Education Accreditation (CHEA). Recognized agencies adhere to certain standards and practices, and their authority in matters of accreditation is widely accepted in the educational community.

This does not mean, however, that accreditation is a simple matter, either for schools wishing to become accredited or for students deciding where to apply. Indeed, in certain fields the very meaning and methods of accreditation are the subject of a good deal of debate. For their part, those applying to graduate school should be aware of the safeguards provided by regional accreditation, especially in terms of degree acceptance and institutional longevity. Beyond this, applicants should understand the role that specialized accreditation plays in their field, as this varies considerably from one discipline to another. In certain professional fields, it is necessary to have graduated from a program that is accredited in order to be eligible for a license to practice, and in some fields the federal government also makes this a hiring requirement. In other disciplines, however, accreditation is not as essential, and there can be excellent programs that are not accredited. In fact, some programs choose not to seek accreditation, although most do.

Institutions and programs that present themselves for accreditation are sometimes granted the status of candidate for accreditation, or what is known as "preaccreditation." This may happen, for example, when an academic unit is too new to have met all the requirements for accreditation. Such status signifies initial recognition and indicates that the school or program in question is working to fulfill all requirements; it does not, however, guarantee that accreditation will be granted.

Institutional Accrediting Agencies—Regional

MIDDLE STATES ASSOCIATION OF COLLEGES AND SCHOOLS
Accredits institutions in Delaware, District of Columbia, Maryland, New Jersey, New York, Pennsylvania, Puerto Rico, and the Virgin Islands.
Jean Avnet Morse, Executive Director
Middle States Commission on Higher Education
3624 Market Street
Philadelphia, Pennsylvania 19104
Telephone: 267-284-5025
Fax: 215-662-5950
E-mail: jmorse@msche.org
World Wide Web: www.msche.org

NEW ENGLAND ASSOCIATION OF SCHOOLS AND COLLEGES
Accredits institutions in Connecticut, Maine, Massachusetts, New Hampshire, Rhode Island, and Vermont.
Barbara E. Brittingham, Interim Director
Commission on Institutions of Higher Education
209 Burlington Road
Bedford, Massachusetts 01730
Telephone: 781-541-5447
Fax: 781-271-0950
E-mail: bbrittingham@neasc.org
World Wide Web: www.neasc.org

NORTH CENTRAL ASSOCIATION OF COLLEGES AND SCHOOLS
Accredits institutions in Arizona, Arkansas, Colorado, Illinois, Indiana, Iowa, Kansas, Michigan, Minnesota, Missouri, Nebraska, New Mexico, North Dakota, Ohio, Oklahoma, South Dakota, West Virginia, Wisconsin, and Wyoming.
Steven D. Crow, Executive Director
The Higher Learning Commission
30 North LaSalle Street, Suite 2400
Chicago, Illinois 60602
Telephone: 312-263-0456
Fax: 312-263-7462
E-mail: scrow@hlcommission.org
World Wide Web: www.ncahigherlearningcommission.org

NORTHWEST COMMISSION ON COLLEGES AND UNIVERSITIES
Accredits institutions in Alaska, Idaho, Montana, Nevada, Oregon, Utah, and Washington.
Sandra E. Elman, President
8060 165th Avenue, NE, Suite 100
Redmond, Washington 98052
Telephone: 425-558-4224
Fax: 425-376-0596
E-mail: selman@nwccu.org
World Wide Web: www.nwccu.org

SOUTHERN ASSOCIATION OF COLLEGES AND SCHOOLS
Accredits institutions in Alabama, Florida, Georgia, Kentucky, Louisiana, Mississippi, North Carolina, South Carolina, Tennessee, Texas, and Virginia.
Belle S. Wheelan, President
Commission on Colleges
1866 Southern Lane
Decatur, Georgia 30033
Telephone: 404-679-4512
Fax: 404-679-4558
E-mail: bwheelan@sacscoc.org
World Wide Web: www.sacscoc.org

WESTERN ASSOCIATION OF SCHOOLS AND COLLEGES
Accredits institutions in California, Guam, and Hawaii.
Ralph A. Wolff, President and Executive Director
The Senior College Commission
985 Atlantic Avenue, Suite 100
Alameda, California 94501
Telephone: 510-748-9001
Fax: 510-748-9797
E-mail: rwolff@wascsenior.org
World Wide Web: www.wascsenior.org/wasc/

Institutional Accrediting Agencies—Other

ACCREDITING COUNCIL FOR INDEPENDENT COLLEGES AND SCHOOLS
Sheryl L. Moody, Executive Director
750 First Street, NE, Suite 980
Washington, DC 20002
Telephone: 202-336-6780
Fax: 202-842-2593
E-mail: smoody@acics.org
World Wide Web: www.acics.org

DISTANCE EDUCATION AND TRAINING COUNCIL
Accrediting Commission
Michael P. Lambert, Executive Director
1601 18th Street, NW
Washington, DC 20009
Telephone: 202-234-5100 Ext. 101
Fax: 202-332-1386
E-mail: detc@detc.org
World Wide Web: www.detc.org

Specialized Accrediting Agencies

[Only Book 1 of *Peterson's Graduate and Professional Programs* Series includes the complete list of specialized accrediting groups recognized by the U.S. Department of Education and the Council on Higher Education Accreditation (CHEA). The lists in Books 2, 3, 4, 5, and 6 are abridged.]

DIETETICS
Beverly E. Mitchell, Director
American Dietetic Association
Commission on Accreditation for Dietetics Education (CADE-ADA)
120 South Riverside Plaza, Suite 2000
Chicago, Illinois 60606
Phone: 312-899-4872
Fax: 312-899-4817
E-mail: bmitchell@eatright.org
Web: www.eatright.org/cade

How to Use These Guides

As you identify the particular programs and institutions that interest you, you can use both Book 1 and the specialized volumes (Books 2–6) to obtain detailed information—Book 1 for information on the institutions overall and Books 2 through 6 for details about the individual graduate units and their degree programs.

Books 2 through 6 are divided into sections that contain one or more directories devoted to programs in a particular field. If you do not find a directory devoted to your field of interest in a specific book, consult *Directories and Subject Areas in Books 2–6* (located at the end of each volume). After you have identified the correct book, consult the *Directories and Subject Areas in This Book* index, which shows (as does the more general directory) what directories cover subjects not specifically named in a directory or section title. This index in Book 2, for example, will tell you that if you are interested in sculpture, you should see the directory entitled Art/Fine Arts. The Art/Fine Arts entry will direct you to the proper page.

Books 2 through 6 have a number of general directories. These directories have entries for the largest unit at an institution granting graduate degrees in that field. For example, the general Engineering and Applied Sciences directory in Book 5 consists of **Profiles** for colleges, schools, and departments of engineering and applied sciences.

General directories are followed by other directories, or sections, that give more detailed information about programs in particular areas of the general field that has been covered. The general Engineering and Applied Sciences directory, in the previous example, is followed by nineteen sections with directories in specific areas of engineering, such as Chemical Engineering, Industrial/Management Engineering, and Mechanical Engineering.

Because of the broad nature of many fields, any system of organization is bound to involve a certain amount of overlap. Environmental studies, for example, is a field whose various aspects are studied in several types of departments and schools. Readers interested in such studies will find information on relevant programs in Book 3 under Ecology and Environmental Biology; in Book 4 under Environmental Management and Policy and Natural Resources; in Book 5 under Energy Management and Policy and Environmental Engineering; and in Book 6 under Environmental and Occupational Health. To help you find all of the programs of interest to you, the introduction to each section of Books 2 through 6 includes, if applicable, a paragraph suggesting other sections and directories with information on related areas of study.

Directory of Institutions with Programs in the Biological Sciences

This directory lists institutions in alphabetical order and includes beneath each name the academic fields in the physical sciences, mathematics, agricultural sciences, the environment, and natural resources in which each institution offers graduate programs. The degree level in each field is also indicated, provided that the institution has supplied that information in response to *Peterson's Annual Survey of Graduate and Professional Institutions*. An *M* indicates that a master's degree program is offered; a *D* indicates that a doctoral degree program is offered; a *P* indicates that the first professional degree is offered; an *O* signifies that other advanced degrees (e.g., certificates or specialist degrees) are offered; and an * (asterisk) indicates that a **Close-Up** and/or **Announcement** is located in this volume. See the index, *Close-Ups and Announcements*, for the specific page number.

Profiles of Academic and Professional Programs in Books 2–6

Each section of **Profiles** has a table of contents that lists the Program Directories, **Announcements**, and **Close-Ups.** Program Directories consist of the **Profiles** of programs in the relevant fields, with **Announcements** following if programs have chosen to include them. **Cross-Discipline Announcements**, if any programs have chosen to submit such entries, and **Close-Ups,** which are more individualized statements, again if programs have chosen to submit them, are also listed.

The **Profiles** found in the 476 directories in Books 2 through 6 provide basic data about the graduate units in capsule form for quick reference. To make these directories as useful as possible, **Profiles** are generally listed for an institution's smallest academic unit within a subject area. In other words, if an institution has a College of Liberal Arts that administers many related programs, the **Profile** for the individual program (e.g., Program in History), not the entire College, appears in the directory.

There are some programs that do not fit into any current directory and are not given individual **Profiles**. The directory structure is reviewed annually in order to keep this number to a minimum and to accommodate major trends in graduate education.

The following outline describes the **Profile** information found in the guides and explains how best to use that information. Any item that does not apply to or was not provided by a graduate unit is omitted from its listing. The format of the **Profiles** is constant, making it easy to compare one institution with another and one program with another. A description of the information in the **Profiles** in Books 2 through 6 follows; the Book 1 **Profile** description is found in that Guide's "How to Use This Guide" article.

Identifying Information. The institution's name, in boldface type, is followed by a complete listing of the administrative structure for that field of study. (For example, University of Akron, Buchtel College of Arts and Sciences, Department of Theoretical and Applied Mathematics, Program in Mathematics.) The last unit listed is the one to which all information in the **Profile** pertains. The institution's city, state, and zip code follow.

Offerings. Each field of study offered by the unit is listed with all postbaccalaureate degrees awarded. Degrees that are not preceded by a specific concentration are awarded in the general field listed in the unit name. Frequently, fields of study are broken down into subspecializations, and those appear following the degrees awarded; for example, "Offerings in secondary education (M.Ed.), including English education, mathematics education, science education." Students enrolled in the M.Ed. program would be able to specialize in any of the three fields mentioned.

Professional Accreditation. Some **Profiles** indicate whether a program is professionally accredited. Because it is possible for a program to receive or lose professional accreditation at any time, students entering fields in which accreditation is important to a career should verify the status of programs by contacting either the chairperson or the appropriate accrediting association.

Jointly Offered Degrees. Explanatory statements concerning programs that are offered in cooperation with other institutions are included in the list of degrees offered. This occurs most commonly on a regional basis (for example, two state universities offering a cooperative Ph.D. in special education) or where the specialized nature of the institutions encourages joint efforts (a J.D./M.B.A. offered by a law school at an institution with no formal business programs and an institution with a business school but lacking a law school), Only programs that are truly cooperative are listed; those involving only limited course work at another institution are not. Interested students should contact the heads of such units for further information.

Part-Time and Evening/Weekend Programs. When information regarding the availability of part-time or evening/weekend study appears in the **Profile**, it means that students are able to earn a degree exclusively through such study.

Postbaccalaureate Distance Learning Degrees. A post-baccalaureate distance learning degree program signifies that course requirements can be fulfilled with minimal or no on-campus study.

Faculty. Figures on the number of faculty members actively involved with graduate students through teaching or research are separated into full- and part-time as well as men and women whenever the information has been supplied.

Students. Figures for the number of students enrolled in graduate and professional programs pertain to the semester of highest enrollment from the 2005–06 academic year. These figures are broken down into full- and part-time and men and women whenever the data have been supplied. Information on the number of matriculated students enrolled in the unit who are members of a minority group or are international students appears here. The average age of the matriculated students is followed by the number of applicants, the percentage accepted, and the number enrolled for fall 2005.

Degrees Awarded. The number of degrees awarded in the calendar year is listed, as is the percentage of students in those degree programs who entered university research/teaching, business/industry, or government service or continued full-time study. Many doctoral programs offer a terminal master's degree if students leave the program after completing only part of the requirements for a doctoral degree; that is indicated here. All degrees are classified into one of four types: master's, doctoral, first professional, and other advanced degrees. A unit may award one or several degrees at a given level; however, the data are only collected by type and may therefore represent several different degree programs.

Median Time to Degree. If provided, information on the median amount of time required to earn the degree for full-time and part-time students is listed here. Also provided is the percentage of students who began their doctoral program in 1997 and received their degree in eight years or less.

Degree Requirements. The information in this section is also broken down by type of degree, and all information for a degree level pertains to all degrees of that type unless otherwise specified. Degree requirements are collected in a simplified form to provide some very basic information on the nature of the program and on foreign language, thesis or dissertation, comprehensive exam, and registration requirements. Many units also provide a short list of additional requirements, such as fieldwork or an internship. No information is listed on the number of courses or credits required for completion or whether a minimum or maximum number of years or semesters is needed. For complete information on graduation requirements, contact the graduate school or program directly.

Entrance Requirements. Entrance requirements are broken down into the four degree levels of master's, doctoral, first professional, and other advanced degrees. Within each level, information may be provided in two basic categories: entrance exams and other requirements. The entrance exams are identified by the standard acronyms used by the testing agencies, unless they are not well known. Other entrance requirements are quite varied, but they often contain an undergraduate or graduate grade point average (GPA). Unless otherwise stated, the GPA is calculated on a 4.0 scale and is listed as a minimum required for admission. Additional exam requirements/recommendations for international students may be listed here. Application deadlines for domestic and international students, the application fee, and whether electronic applications are accepted may be listed here. Note that the deadline should be used for reference only; these dates are subject to change, and students interested in applying should contact the graduate unit directly about application procedures and deadlines.

Expenses. The typical cost of study for the 2005–06 academic year is given in two basic categories: tuition and fees. Cost of study may be quite complex at a graduate institution. There are often sliding scales for part-time study, a different cost for first-year students, and other variables that make it impossible to completely cover the cost of study for each graduate program. To provide the most usable information, figures are given for full-time study for a full year where available and for part-time study in terms of a per-unit rate (per credit, per semester hour, etc.). Occasionally, variances may be noted in tuition and fees for reasons such as the type of program, whether courses are taken during the day or evening, whether courses are at the master's or doctoral level, or other institution-specific reasons. Expenses are usually subject to change; for exact costs at any given time, contact your chosen schools and programs directly. Keep in mind that the tuition of Canadian institutions is usually given in Canadian dollars.

Financial Support. This section contains data on the number of awards administered by the institution and given to graduate students during the 2005–06 academic year. The first figure given represents the total number of students receiving financial support enrolled in that unit. If the unit has provided information on graduate appointments, these are broken down into three major categories: *fellowships* give money to graduate students to cover the cost of study and living expenses and are not based on a work obligation or research commitment, *research assistantships* provide stipends to graduate students for assistance in a formal research project with a faculty member, and *teaching assistantships* provide stipends to graduate students for teaching or for assisting faculty members in teaching undergraduate classes. Within each category, figures are given for the total number of awards, the average yearly amount per award, and whether full or partial tuition reimbursements are awarded. In addition to graduate appointments, the availability of several other financial aid sources is covered in this section. *Tuition waivers* are routinely part of a graduate appointment, but units sometimes waive part or all of a student's tuition even if a graduate appointment is not available. *Federal Work-Study* is made available to students who demonstrate need and meet the federal guidelines; this form of aid normally includes 10 or more hours of work per week in an office of the institution. *Institutionally sponsored loans* are low-interest loans available to graduate students to cover both educational and living expenses. *Career-related internships* or *fieldwork* offer money to students who are participating in a formal off-campus research project or practicum. Grants, scholarships, traineeships, unspecified assistantships, and other awards may also be noted. The availability of financial support to part-time students is also indicated here.

Some programs list the financial aid application deadline and the forms that need to be completed for students to be eligible for financial awards. There are two forms: FAFSA, the Free Application for Federal Student Aid, which is required for federal aid, and the PROFILE®.

Faculty Research. Each unit has the opportunity to list several keyword phrases describing the current research involving faculty members and graduate students. Space limitations prevent the unit from listing complete information on all research programs. The total expenditure for funded research from the previous academic year may also be included.

Unit Head and Application Contact. The head of the graduate program for each unit is listed with academic title and telephone and fax numbers and e-mail address if available. In addition to the unit head, many graduate programs list a separate contact for application and admission information, which follows the listing for the unit head. If no unit head or application contact is given, you should contact the overall institution for information on graduate admissions.

Announcements and Close-Ups

The **Announcements** and **Close-Ups** are supplementary insertions submitted by deans, chairs, and other administrators who wish to offer an additional, more individualized statement to readers. A number of graduate school and program administrators have attached **Announcements** to the end of their **Profile** listings. In them you will find information that an institution or program wants to emphasize. The **Close-Ups** are by their very nature more expansive and flexible than the **Profiles**, and the administrators who have written them may emphasize different aspects of their programs. All of these **Close-Ups** are organized in the same way (with the exception of a few that describe research and training opportunities instead of degree programs), and in each one you will find information on the same basic topics, such as programs of study, research facilities, tuition and fees, financial aid, and application procedures. If an institution or program has submitted a **Close-Up**, a boldface cross-reference appears below its **Profile**. As with the **Announcements**, all of the **Close-Ups** in the guides have been submitted by choice; the absence of an **Announcement** or **Close-Up** does not reflect any type of editorial judgment on the part of Peterson's and their presence in the guides should not be taken as an indication of status, quality, or approval. Statements regarding a university's objectives and accomplishments are a reflection of its own beliefs and are not the opinions of the Peterson's editors.

Cross-Discipline Announcements

In addition to the regular directories that present **Profiles** of programs in each field of study, many sections in Books 2 through 6 contain special notices under the heading **Cross-Discipline Announcements**. Appearing at the end of many **Profile** sections, these **Cross-Discipline Announcements** inform you about programs that you may find of interest described in a different section. A biochemistry department, for example, may place a notice under **Cross-Discipline Announcements** in the Chemistry section (Book 4) to alert chemistry students to that course of study. **Cross-Discipline Announcements**, also written by administrators to highlight their programs, will be helpful to you not only in finding out about programs in fields related to your own but also in locating departments that are actively recruiting students with a specific undergraduate major.

Appendixes

This section contains two appendixes. The first, *Institutional Changes Since the 2006 Edition*, lists institutions that have closed, moved, merged, or changed their name or status since the last edition of the guides. The second, *Abbreviations Used in the Guides*, gives abbreviations of degree names, along with what those abbreviations stand for. These appendixes are identical in all six volumes of *Peterson's Graduate and Professional Programs*.

Indexes

There are three indexes presented here. The first index, *Close-Ups and Announcements*, gives page references for all programs that have chosen to place **Close-Ups** and **Announcements** in this volume. It is arranged alphabetically by institution; within institutions, the arrangement is alphabetical by subject area. It is not an index to all programs in the book's directories of **Profiles**; readers must refer to the directories themselves for **Profile** information on programs that have not submitted the additional, more individualized statements. The second index, *Directories and Subject Areas in Books 2–6*, gives book references for the directories in Books 2-6, for example, "Industrial Design—Book 2," and also includes cross-references for subject area names not used in the directory structure, for example, "Computing Technology (see Computer Science)." The third index, *Directories and Subject Areas in This Book*, gives page references for the directories in this volume and cross-references for subject area names not used in this volume's directory structure.

Data Collection Procedures

The information published in the directories and **Profiles** of all the books is collected through *Peterson's Annual Survey of Graduate and Professional Institutions*. The survey is sent each spring to more than 2,000 institutions offering postbaccalaureate degree programs, including accredited institutions in the United States, U.S. territories, and Canada and those institutions outside the United States that are accredited by U.S. accrediting bodies. Deans and other administrators complete these surveys, providing information on programs in the 476 academic and professional fields covered in the guides as well as overall institutional information. While every effort has been made to ensure the accuracy and completeness of the data, information is sometimes unavailable or changes occur after publication deadlines. All usable information received in time for publication has been included. The omission of any particular item from a directory or **Profile** signifies either that the item is not applicable to the institution or program or that information was not available. **Profiles** of programs scheduled to begin during the 2006–07 academic year cannot, obviously, include statistics on enrollment or, in many cases, the number of faculty members. If no usable data were submitted by an institution, its name, address, and program name appear in order to indicate the availability of graduate work.

Criteria for Inclusion in This Guide

To be included in this guide, an institution must have full accreditation or be a candidate for accreditation (preaccreditation) status by an institutional or specialized accrediting body recognized by the U.S. Department of Education or the Council for Higher Education Accreditation (CHEA). Institutional accrediting bodies, which review each institution as a whole, include the six regional associations of schools and colleges (Middle States, New England, North Central, Northwest, Southern, and Western), each of which is responsible for a specified portion of the United States and its territories. Other institutional accrediting bodies are national in scope and accredit specific kinds of institutions (e.g., Bible colleges, independent colleges, and rabbinical and Talmudic schools). Program registration by the New York State Board of Regents is considered to be the equivalent of institutional accreditation, since the board requires that all programs offered by an institution meet its standards before recognition is granted. A Canadian institution must be chartered and authorized to grant degrees by the provincial government, affiliated with a chartered institution, or accredited by a recognized U.S. accrediting body. This guide also includes institutions outside the United States that are accredited by these U.S. accrediting bodies. There are recognized specialized or professional accrediting bodies in more than fifty different fields, each of which is authorized to accredit institutions or specific programs in its particular field. For specialized institutions that offer programs in one field only, we designate this to be the equivalent of institutional accreditation. A full explanation of the accrediting process and complete information on recognized institutional (regional and national) and specialized accrediting bodies can be found online at www.chea.org or at www.ed.gov/admins/finaid/accred/index.html.

DIRECTORY OF INSTITUTIONS WITH PROGRAMS IN THE BIOLOGICAL SCIENCES

ACADIA UNIVERSITY

Biological and Biomedical
  Sciences—General — M

ADELPHI UNIVERSITY

Biological and Biomedical
  Sciences—General — M

**ALABAMA AGRICULTURAL AND
MECHANICAL UNIVERSITY**

Biological and Biomedical
  Sciences—General — M

ALABAMA STATE UNIVERSITY

Biological and Biomedical
  Sciences—General — M,O

ALBANY MEDICAL COLLEGE

Biological and Biomedical
  Sciences—General — M,D
Cardiovascular Sciences — M,D
Cell Biology — M,D
Immunology — M,D
Microbiology — M,D
Molecular Biology — M,D
Neuroscience — M,D
Pharmacology — M,D

**ALBERT EINSTEIN COLLEGE OF
MEDICINE**

Anatomy — D
Biochemistry — D*
Biological and Biomedical
  Sciences—General — D*
Biophysics — D
Cell Biology — D
Developmental Biology — D
Immunology — D
Microbiology — D
Molecular Biology — D
Molecular Genetics — D
Molecular Pharmacology — D
Neurobiology — D
Pathology — D
Physiology — D

ALCORN STATE UNIVERSITY

Biological and Biomedical
  Sciences—General — M

**ALLIANT INTERNATIONAL
UNIVERSITY–SAN FRANCISCO
BAY**

Pharmacology — M

**AMERICAN HEALTH SCIENCES
UNIVERSITY**

Nutrition — M

AMERICAN UNIVERSITY

Biological and Biomedical
  Sciences—General — M
Biopsychology — M
Neuroscience — D
Toxicology — M,O

**THE AMERICAN UNIVERSITY OF
ATHENS**

Biological and Biomedical
  Sciences—General — M

**AMERICAN UNIVERSITY OF
BEIRUT**

Biochemistry — M
Biological and Biomedical
  Sciences—General — M
Microbiology — M
Neuroscience — M
Nutrition — M
Pharmacology — M
Physiology — M

ANDREWS UNIVERSITY

Biological and Biomedical
  Sciences—General — M
Nutrition — M

ANGELO STATE UNIVERSITY

Biological and Biomedical
  Sciences—General — M

**ANTIOCH NEW ENGLAND
GRADUATE SCHOOL**

Environmental Biology — M

**APPALACHIAN STATE
UNIVERSITY**

Biological and Biomedical
  Sciences—General — M

ARGOSY UNIVERSITY/HAWAI'I

Pharmacology — O

ARIZONA STATE UNIVERSITY

Animal Behavior — M,D
Biochemistry — M,D
Biological and Biomedical
  Sciences—General — M,D
Cell Biology — M,D
Computational Biology — M
Conservation Biology — M,D
Developmental Biology — M,D
Ecology — M,D
Evolutionary Biology — M,D
Genetics — M,D
Microbiology — M,D
Neuroscience — M,D
Physiology — M,D

**ARIZONA STATE UNIVERSITY AT
THE POLYTECHNIC CAMPUS**

Biological and Biomedical
  Sciences—General — M
Nutrition — M

ARKANSAS STATE UNIVERSITY

Biological and Biomedical
  Sciences—General — M,O

**A.T. STILL UNIVERSITY OF
HEALTH SCIENCES**

Biological and Biomedical
  Sciences—General — P,M
Toxicology — P,M

AUBURN UNIVERSITY

Anatomy — M,D
Biological and Biomedical
  Sciences—General — M,D
Botany — M,D
Entomology — M,D
Microbiology — M,D
Nutrition — M,D
Pathobiology — M,D
Pharmacology — M,D
Plant Pathology — M,D
Radiation Biology — M,D
Zoology — M,D

AUSTIN PEAY STATE UNIVERSITY

Biological and Biomedical
  Sciences—General — M
Radiation Biology — M

BALL STATE UNIVERSITY

Biological and Biomedical
  Sciences—General — M,D
Physiology — M

**BARNES-JEWISH COLLEGE OF
NURSING AND ALLIED HEALTH**

Nutrition — M,O

BARRY UNIVERSITY

Anatomy — M

Biological and Biomedical
  Sciences—General — M*

BASTYR UNIVERSITY

Nutrition — M*

BAYLOR COLLEGE OF MEDICINE

Biochemistry — D*
Biological and Biomedical
  Sciences—General — M,D*
Biophysics — D
Cardiovascular Sciences — D*
Cell Biology — D*
Computational Biology — D
Developmental Biology — D*
Genetics — D*
Human Genetics — D
Immunology — D*
Microbiology — D
Molecular Biology — D
Molecular Biophysics — D*
Molecular Medicine — D
Molecular Physiology — D*
Neuroscience — D*
Pathology — D
Pharmacology — D
Structural Biology — D
Virology — D*

BAYLOR UNIVERSITY

Biological and Biomedical
  Sciences—General — M,D
Environmental Biology — M,D
Neuroscience — M,D
Nutrition — M,D

BEMIDJI STATE UNIVERSITY

Biological and Biomedical
  Sciences—General — M

BENEDICTINE UNIVERSITY

Nutrition — M

**BLOOMSBURG UNIVERSITY OF
PENNSYLVANIA**

Biological and Biomedical
  Sciences—General — M

BOISE STATE UNIVERSITY

Biological and Biomedical
  Sciences—General — M

BOSTON COLLEGE

Biochemistry — M,D
Biological and Biomedical
  Sciences—General — M,D*

BOSTON UNIVERSITY

Anatomy — M,D
Biochemistry — M,D*
Biological and Biomedical
  Sciences—General — M,D
Biophysics — M,D*
Biopsychology — M
Cell Biology — M,D*
Genetics — M,D
Genomic Sciences — M,D
Immunology — D*
Microbiology — M,D*
Molecular Biology — M,D
Molecular Medicine — D*
Neurobiology — M,D
Neuroscience — M,D*
Nutrition — M,D*
Pathology — D
Pharmacology — M,D*
Physiology — M,D

**BOWLING GREEN STATE
UNIVERSITY**

Biological and Biomedical
  Sciences—General — M,D*
Nutrition — M

BRADLEY UNIVERSITY

Biological and Biomedical
  Sciences—General — M

BRANDEIS UNIVERSITY

Biochemistry — M,D
Biological and Biomedical
  Sciences—General — M,D,O
Biophysics — M,D
Cell Biology — M,D
Genetics — M,D
Microbiology — M,D
Molecular Biology — M,D
Neurobiology — M,D
Neuroscience — M,D
Structural Biology — M,D

BRIGHAM YOUNG UNIVERSITY

Biochemistry — M,D
Biological and Biomedical
  Sciences—General — M,D
Developmental Biology — M,D
Microbiology — M,D
Molecular Biology — M,D
Neuroscience — M,D
Nutrition — M
Physiology — M,D*

BROCK UNIVERSITY

Biological and Biomedical
  Sciences—General — M,D
Neuroscience — M,D

**BROOKLYN COLLEGE OF THE
CITY UNIVERSITY OF NEW YORK**

Biological and Biomedical
  Sciences—General — M,D
Nutrition — M

BROWN UNIVERSITY

Biochemistry — M,D
Biological and Biomedical
  Sciences—General — M,D
Cancer Biology/Oncology — M,D
Cell Biology — M,D
Developmental Biology — M,D
Ecology — D
Evolutionary Biology — D
Immunology — M,D
Microbiology — M,D
Molecular Biology — M,D*
Molecular Pharmacology — M,D
Neuroscience — D*
Pathobiology — M,D
Pathology — M,D
Physiology — M,D
Toxicology — M,D

BUCKNELL UNIVERSITY

Animal Behavior — M
Biological and Biomedical
  Sciences—General — M

**BUFFALO STATE COLLEGE,
STATE UNIVERSITY OF NEW YORK**

Biological and Biomedical
  Sciences—General — M

**CALIFORNIA INSTITUTE OF
TECHNOLOGY**

Biochemistry — M,D
Biological and Biomedical
  Sciences—General — D*
Biophysics — D
Cell Biology — D
Developmental Biology — D
Genetics — D
Immunology — D
Molecular Biology — D
Molecular Biophysics — M,D
Neurobiology — D
Neuroscience — M,D

CALIFORNIA POLYTECHNIC STATE UNIVERSITY, SAN LUIS OBISPO

Biochemistry	M
Biological and Biomedical Sciences—General	M

CALIFORNIA STATE POLYTECHNIC UNIVERSITY, POMONA

Biological and Biomedical Sciences—General	M
Nutrition	M

CALIFORNIA STATE UNIVERSITY, CHICO

Biological and Biomedical Sciences—General	M
Botany	M
Nutrition	M

CALIFORNIA STATE UNIVERSITY, DOMINGUEZ HILLS

Biological and Biomedical Sciences—General	M,O

CALIFORNIA STATE UNIVERSITY, EAST BAY

Biochemistry	M
Biological and Biomedical Sciences—General	M

CALIFORNIA STATE UNIVERSITY, FRESNO

Biological and Biomedical Sciences—General	M
Ecology	M

CALIFORNIA STATE UNIVERSITY, FULLERTON

Biochemistry	M
Biological and Biomedical Sciences—General	M
Botany	M
Microbiology	M

CALIFORNIA STATE UNIVERSITY, LONG BEACH

Biochemistry	M
Biological and Biomedical Sciences—General	M
Microbiology	M
Nutrition	M

CALIFORNIA STATE UNIVERSITY, LOS ANGELES

Biochemistry	M
Biological and Biomedical Sciences—General	M
Nutrition	M

CALIFORNIA STATE UNIVERSITY, NORTHRIDGE

Biological and Biomedical Sciences—General	M

CALIFORNIA STATE UNIVERSITY, SACRAMENTO

Biological and Biomedical Sciences—General	M

CALIFORNIA STATE UNIVERSITY, SAN BERNARDINO

Biological and Biomedical Sciences—General	M

CALIFORNIA STATE UNIVERSITY, SAN MARCOS

Biological and Biomedical Sciences—General	M

CALIFORNIA STATE UNIVERSITY, STANISLAUS

Marine Biology	M

CALIFORNIA UNIVERSITY OF PENNSYLVANIA

Biological and Biomedical Sciences—General	M

CARLETON UNIVERSITY

Biological and Biomedical Sciences—General	M,D
Neuroscience	M,D

CARNEGIE MELLON UNIVERSITY

Biochemistry	M,D
Biological and Biomedical Sciences—General	M,D*
Biophysics	M,D
Biopsychology	D
Cell Biology	M,D
Computational Biology	M
Developmental Biology	M,D
Genetics	M,D
Molecular Biology	M,D
Neurobiology	M,D
Neuroscience*	

CASE WESTERN RESERVE UNIVERSITY

Anatomy	M,D
Biochemistry	M,D
Biological and Biomedical Sciences—General	M,D*
Biophysics	M,D
Cell Biology	M,D
Developmental Biology	M,D
Genetics	D
Genomic Sciences	D*
Human Genetics	D
Immunology	M,D
Microbiology	D
Molecular Biology	M,D*
Neurobiology	D*
Neuroscience	D*
Nutrition	M,D*
Pathology	M,D
Pharmacology	M,D
Physiology	M,D*
Toxicology	M,D

THE CATHOLIC UNIVERSITY OF AMERICA

Biological and Biomedical Sciences—General	M,D
Cell Biology	M,D
Microbiology	M,D

CENTRAL CONNECTICUT STATE UNIVERSITY

Biological and Biomedical Sciences—General	M,O
Molecular Biology	M

CENTRAL MICHIGAN UNIVERSITY

Biological and Biomedical Sciences—General	M
Conservation Biology	M
Nutrition	M

CENTRAL MISSOURI STATE UNIVERSITY

Biological and Biomedical Sciences—General	M

CENTRAL WASHINGTON UNIVERSITY

Biological and Biomedical Sciences—General	M
Nutrition	M

CHAPMAN UNIVERSITY

Nutrition	M

CHICAGO STATE UNIVERSITY

Biological and Biomedical Sciences—General	M

CITY COLLEGE OF THE CITY UNIVERSITY OF NEW YORK

Biochemistry	M,D
Biological and Biomedical Sciences—General	M,D

CITY OF HOPE NATIONAL MEDICAL CENTER/BECKMAN RESEARCH INSTITUTE

Biological and Biomedical Sciences—General	D*

CLAREMONT GRADUATE UNIVERSITY

Botany	M,D
Computational Biology	M,D

CLARION UNIVERSITY OF PENNSYLVANIA

Biological and Biomedical Sciences—General	M

CLARK ATLANTA UNIVERSITY

Biological and Biomedical Sciences—General	M,D

CLARK UNIVERSITY

Biological and Biomedical Sciences—General	M,D

CLEMSON UNIVERSITY

Biochemistry	M,D*
Biological and Biomedical Sciences—General	M,D*
Biophysics	M,D
Ecology	M,D
Entomology	M,D*
Evolutionary Biology	M,D
Genetics	M,D*
Microbiology	M,D*
Molecular Biology	M,D
Nutrition	M
Plant Biology	M,D
Zoology	M,D

CLEVELAND STATE UNIVERSITY

Biological and Biomedical Sciences—General	M,D
Molecular Medicine	M,D

COLD SPRING HARBOR LABORATORY, WATSON SCHOOL OF BIOLOGICAL SCIENCES

Biological and Biomedical Sciences—General	D*

COLLEGE OF CHARLESTON

Marine Biology	M*

COLLEGE OF SAINT ELIZABETH

Nutrition	M,O

COLLEGE OF STATEN ISLAND OF THE CITY UNIVERSITY OF NEW YORK

Biological and Biomedical Sciences—General	M
Neuroscience	M,D

THE COLLEGE OF WILLIAM AND MARY

Biological and Biomedical Sciences—General	M

COLORADO STATE UNIVERSITY

Biochemistry	M,D
Biological and Biomedical Sciences—General	M,D
Botany	M,D
Cell Biology	M,D*
Ecology	M,D
Entomology	M,D
Genetics	M,D
Immunology	M,D
Microbiology	M,D*
Molecular Biology	M,D
Neuroscience	M,D
Nutrition	M,D
Pathology	M,D
Plant Pathology	M,D
Plant Physiology	M,D
Radiation Biology	M,D
Zoology	M,D

COLORADO STATE UNIVERSITY-PUEBLO

Biochemistry	M
Biological and Biomedical Sciences—General	M

COLUMBIA UNIVERSITY

Anatomy	M,D
Biochemistry	M,D
Biological and Biomedical Sciences—General	M,D*
Biophysics	M,D
Biopsychology	M,D
Cell Biology	M,D
Conservation Biology	M,D,O*
Developmental Biology	M,D
Ecology	D,O
Evolutionary Biology	D,O
Genetics	M,D
Microbiology	M,D
Molecular Biology	M,D
Neurobiology	M,D
Nutrition	M,D*
Pathobiology	M,D
Pathology	M,D
Pharmacology	M,D
Physiology	M,D
Toxicology	M,D

CONCORDIA UNIVERSITY (CANADA)

Biological and Biomedical Sciences—General	M,D,O
Genomic Sciences	M,D,O

CONNECTICUT COLLEGE

Botany	M
Zoology	M

CORNELL UNIVERSITY

Anatomy	M,D
Biochemistry	D
Biological and Biomedical Sciences—General	P,M,D
Biophysics	D
Biopsychology	D
Cell Biology	M,D
Developmental Biology	M,D
Ecology	M,D
Entomology	M,D
Evolutionary Biology	D
Genetics	D
Immunology	P,M,D
Infectious Diseases	M,D
Microbiology	D
Molecular Biology	D
Molecular Medicine	M,D
Neurobiology	M,D*
Nutrition	M,D
Pharmacology	P,M,D*
Physiology	P,M,D
Plant Biology	M,D
Plant Molecular Biology	M,D
Plant Pathology	M,D
Plant Physiology	M,D
Reproductive Biology	M,D
Structural Biology	M,D
Toxicology	M,D

Zoology — P,M,D

CORNELL UNIVERSITY, JOAN AND SANFORD I. WEILL MEDICAL COLLEGE AND GRADUATE SCHOOL OF MEDICAL SCIENCES

Biochemistry — M,D
Biological and Biomedical
  Sciences—General — M,D
Biophysics — D
Cell Biology — M,D
Computational Biology — D
Immunology — M,D
Molecular Biology — M,D
Neuroscience — D
Pharmacology — D
Physiology — D
Structural Biology — M,D
Systems Biology — D

CREIGHTON UNIVERSITY

Biological and Biomedical
  Sciences—General — M,D*
Immunology — M,D
Medical Microbiology — M,D*
Pharmacology — M,D

DALHOUSIE UNIVERSITY

Anatomy — M,D
Biochemistry — M,D
Biological and Biomedical
  Sciences—General — M,D
Biophysics — M,D
Immunology — M,D
Microbiology — M,D
Neurobiology — M,D
Neuroscience — M,D
Pathology — M
Pharmacology — M,D
Physiology — M,D

DARTMOUTH COLLEGE

Biochemistry — D*
Biological and Biomedical
  Sciences—General — D*
Cell Biology — D
Genetics — D*
Immunology — D*
Microbiology — D*
Molecular Biology — D*
Molecular Medicine — D
Neuroscience — D*
Pharmacology — D*
Physiology — D*
Toxicology — D

DELAWARE STATE UNIVERSITY

Biological and Biomedical
  Sciences—General — M

DELTA STATE UNIVERSITY

Biological and Biomedical
  Sciences—General — M

DEPAUL UNIVERSITY

Biochemistry — M
Biological and Biomedical
  Sciences—General — M*

DREXEL UNIVERSITY

Biochemistry — M,D
Biological and Biomedical
  Sciences—General — M,D,O*
Biopsychology — M,D
Cancer Biology/Oncology — M,D
Cell Biology — M,D
Genetics — M,D
Human Genetics — M,D
Immunology — M,D
Microbiology — M,D
Molecular Biology — M,D
Neuroscience — D
Nutrition — M,D
Pathobiology — D
Pharmacology — M,D

DUKE UNIVERSITY

Anatomy — D
Biochemistry — D,O
Biological and Biomedical
  Sciences—General — D,O
Biopsychology — D
Cancer Biology/Oncology — D
Cell Biology — D,O*
Developmental Biology — D,O*
Ecology — M,D,O
Genetics — D*
Immunology — D*
Microbiology — D
Molecular Biology — D,O
Molecular Biophysics — O
Molecular Genetics — D*
Neurobiology — D
Neuroscience — D,O
Pathology — M,D*
Pharmacology — D
Structural Biology — O*
Toxicology — D,O*

DUQUESNE UNIVERSITY

Biochemistry — M,D
Biological and Biomedical
  Sciences—General — M,D*
Pharmacology — M,D
Toxicology — M,D

D'YOUVILLE COLLEGE

Nutrition — M

EAST CAROLINA UNIVERSITY

Anatomy — D
Biochemistry — D
Biological and Biomedical
  Sciences—General — M,D*
Biophysics — M,D
Cell Biology — D
Immunology — D
Microbiology — D
Molecular Biology — M,D
Nutrition — M
Pharmacology — D
Physiology — D

EASTERN ILLINOIS UNIVERSITY

Biological and Biomedical
  Sciences—General — M
Nutrition — M

EASTERN KENTUCKY UNIVERSITY

Biological and Biomedical
  Sciences—General — M
Ecology — M
Nutrition — M

EASTERN MICHIGAN UNIVERSITY

Biological and Biomedical
  Sciences—General — M

EASTERN NEW MEXICO UNIVERSITY

Biological and Biomedical
  Sciences—General — M

EASTERN VIRGINIA MEDICAL SCHOOL

Biological and Biomedical
  Sciences—General — M,D
Reproductive Biology — M

EASTERN WASHINGTON UNIVERSITY

Biological and Biomedical
  Sciences—General — M

EAST STROUDSBURG UNIVERSITY OF PENNSYLVANIA

Biological and Biomedical
  Sciences—General — M

EAST TENNESSEE STATE UNIVERSITY

Anatomy — M,D
Biochemistry — M,D
Biological and Biomedical
  Sciences—General — M,D
Biophysics — M,D
Microbiology — M,D
Nutrition — M
Pharmacology — M,D
Physiology — M,D

EDINBORO UNIVERSITY OF PENNSYLVANIA

Biological and Biomedical
  Sciences—General — M

EMORY UNIVERSITY

Animal Behavior — D
Biochemistry — D
Biological and Biomedical
  Sciences—General — D*
Biophysics — D
Cell Biology — D
Developmental Biology — D
Ecology — D
Evolutionary Biology — D
Genetics — D
Immunology — D
Microbiology — D
Molecular Biology — D
Molecular Genetics — D
Neuroscience — D
Nutrition — M,D
Pharmacology — D

EMPORIA STATE UNIVERSITY

Biological and Biomedical
  Sciences—General — M
Botany — M
Cell Biology — M
Environmental Biology — M
Microbiology — M
Zoology — M

FAIRLEIGH DICKINSON UNIVERSITY, COLLEGE AT FLORHAM

Biological and Biomedical
  Sciences—General — M
Pharmacology — M,O

FAIRLEIGH DICKINSON UNIVERSITY, METROPOLITAN CAMPUS

Biological and Biomedical
  Sciences—General — M

FAYETTEVILLE STATE UNIVERSITY

Biological and Biomedical
  Sciences—General — M

THE FEINSTEIN INSTITUTE FOR MEDICAL RESEARCH

Molecular Medicine — D

FISK UNIVERSITY

Biological and Biomedical
  Sciences—General — M

FITCHBURG STATE COLLEGE

Biological and Biomedical
  Sciences—General — M

FLORIDA AGRICULTURAL AND MECHANICAL UNIVERSITY

Biological and Biomedical
  Sciences—General — M
Entomology — M
Pharmacology — M,D
Toxicology — M,D

EAST TENNESSEE STATE UNIVERSITY columns continued →

FLORIDA ATLANTIC UNIVERSITY

Biochemistry — M,D
Biological and Biomedical
  Sciences—General — M,D
Neuroscience — D

FLORIDA INSTITUTE OF TECHNOLOGY

Biological and Biomedical
  Sciences—General — M,D
Cell Biology — M,D*
Ecology — M
Marine Biology — M*
Molecular Biology — M,D

FLORIDA INTERNATIONAL UNIVERSITY

Biological and Biomedical
  Sciences—General — M,D
Nutrition — M,D*

FLORIDA STATE UNIVERSITY

Biochemistry — M,D
Biological and Biomedical
  Sciences—General — M,D
Cell Biology — M,D
Computational Biology — D
Developmental Biology — M,D
Ecology — M,D
Evolutionary Biology — M,D
Genetics — M,D
Immunology — M,D
Marine Biology — M,D
Microbiology — M,D
Molecular Biology — M,D
Molecular Biophysics — D
Neuroscience — D
Nutrition — M,D
Physiology — M,D
Plant Biology — M,D
Structural Biology — D

FORDHAM UNIVERSITY

Biological and Biomedical
  Sciences—General — M,D*

FORT HAYS STATE UNIVERSITY

Biological and Biomedical
  Sciences—General — M

FRAMINGHAM STATE COLLEGE

Nutrition — M

FROSTBURG STATE UNIVERSITY

Biological and Biomedical
  Sciences—General — M
Conservation Biology — M
Ecology — M

GEORGE MASON UNIVERSITY

Biological and Biomedical
  Sciences—General — M,D
Cell Biology — M,D,O
Ecology — M,D
Evolutionary Biology — M,D,O
Microbiology — M,D,O
Molecular Biology — M,D,O
Neuroscience — M,D,O

GEORGETOWN UNIVERSITY

Biochemistry — M,D
Biological and Biomedical
  Sciences—General — M,D*
Biophysics — M,D
Cancer Biology/Oncology —
Cell Biology — D
Immunology — M,D
Infectious Diseases — M,D
Microbiology — M,D*
Molecular Biology — D
Neuroscience — D
Pathology — M,D
Pharmacology — D
Physiology — M,D
Radiation Biology — M

THE GEORGE WASHINGTON UNIVERSITY

Biochemistry	M,D*
Biological and Biomedical Sciences—General	M,D*
Genetics	M,D
Genomic Sciences	M*
Immunology	D*
Infectious Diseases	M
Microbiology	M,D
Molecular Biology	M,D
Molecular Medicine	D*
Neuroscience	D
Pharmacology	D

GEORGIA COLLEGE & STATE UNIVERSITY

Biological and Biomedical Sciences—General	M

GEORGIA INSTITUTE OF TECHNOLOGY

Biochemistry	M,D
Biological and Biomedical Sciences—General	M,D*
Physiology	M

GEORGIAN COURT UNIVERSITY

Biological and Biomedical Sciences—General	M,O

GEORGIA SOUTHERN UNIVERSITY

Biological and Biomedical Sciences—General	M

GEORGIA STATE UNIVERSITY

Biochemistry	M,D
Biological and Biomedical Sciences—General	M,D*
Cell Biology	M,D
Environmental Biology	M,D
Microbiology	M,D
Molecular Genetics	M,D
Neurobiology	M,D*
Nutrition	M,O
Physiology	M,D

GOUCHER COLLEGE

Biological and Biomedical Sciences—General	O

GOVERNORS STATE UNIVERSITY

Environmental Biology	M

GRADUATE SCHOOL AND UNIVERSITY CENTER OF THE CITY UNIVERSITY OF NEW YORK

Biochemistry	D
Biological and Biomedical Sciences—General	D
Biopsychology	D
Neuroscience	D

GRAND VALLEY STATE UNIVERSITY

Biological and Biomedical Sciences—General	M

HAMPTON UNIVERSITY

Biological and Biomedical Sciences—General	M

HARVARD UNIVERSITY

Biochemistry	D
Biological and Biomedical Sciences—General	M,D,O*
Biophysics	D*
Biopsychology	D
Cell Biology	D
Evolutionary Biology	D*
Genetics	D
Genomic Sciences	D
Immunology	D

Infectious Diseases	D
Microbiology	D
Molecular Biology	D*
Molecular Genetics	D
Molecular Pharmacology	D
Neurobiology	D
Neuroscience	D*
Nutrition	D
Pathology	D
Physiology	M,D
Structural Biology	D
Systems Biology	D*

HERITAGE UNIVERSITY

Biological and Biomedical Sciences—General	M

HOFSTRA UNIVERSITY

Biological and Biomedical Sciences—General	M*

HOOD COLLEGE

Biological and Biomedical Sciences—General	M
Environmental Biology	M

HOWARD UNIVERSITY

Anatomy	M,D
Biochemistry	M,D*
Biological and Biomedical Sciences—General	M,D*
Biophysics	D
Biopsychology	M,D
Genetics	M,D
Human Genetics	M,D
Microbiology	D
Molecular Biology	M,D
Nutrition	M,D
Pharmacology	M,D
Physiology	D

HUMBOLDT STATE UNIVERSITY

Biological and Biomedical Sciences—General	M*

HUNTER COLLEGE OF THE CITY UNIVERSITY OF NEW YORK

Biochemistry	M
Biological and Biomedical Sciences—General	M,D
Biopsychology	M
Genetics	

ICR GRADUATE SCHOOL

Biological and Biomedical Sciences—General	M

IDAHO STATE UNIVERSITY

Biological and Biomedical Sciences—General	M,D
Medical Microbiology	M,D
Microbiology	M,D
Nutrition	M,O
Pharmacology	M,D

ILLINOIS INSTITUTE OF TECHNOLOGY

Biochemistry	M,D
Biological and Biomedical Sciences—General	M,D
Cell Biology	M,D
Microbiology	M,D
Molecular Biology	M,D
Molecular Biophysics	M,D

ILLINOIS STATE UNIVERSITY

Biological and Biomedical Sciences—General	M,D*
Botany	M,D
Ecology	M,D
Genetics	M,D
Microbiology	M,D
Physiology	M,D
Zoology	M,D

IMMACULATA UNIVERSITY

Nutrition	M

INDIANA STATE UNIVERSITY

Biological and Biomedical Sciences—General	M,D*
Ecology	M,D
Microbiology	M,D
Nutrition	M
Physiology	M,D

INDIANA UNIVERSITY BLOOMINGTON

Anatomy	M,D
Animal Behavior	O
Biochemistry	M,D
Biological and Biomedical Sciences—General	M,D*
Cell Biology	M,D
Developmental Biology	D
Ecology	M,D
Evolutionary Biology	M,D
Genetics	D
Microbiology	M,D
Molecular Biology	D
Neuroscience	D
Nutrition	M,D
Pharmacology	M,D
Physiology	M,D
Plant Biology	M,D
Zoology	M,D

INDIANA UNIVERSITY OF PENNSYLVANIA

Biological and Biomedical Sciences—General	M
Nutrition	M

INDIANA UNIVERSITY–PURDUE UNIVERSITY FORT WAYNE

Biological and Biomedical Sciences—General	M

INDIANA UNIVERSITY–PURDUE UNIVERSITY INDIANAPOLIS

Anatomy	M,D
Biochemistry	D*
Biological and Biomedical Sciences—General	M,D*
Biophysics	M,D
Biopsychology	M,D
Cell Biology	M,D
Immunology	M,D
Microbiology	M,D*
Molecular Biology	D
Molecular Genetics	M,D*
Neurobiology	M,D
Nutrition	M
Pathology	M,D
Pharmacology	M,D
Physiology	M,D
Toxicology	M,D

IOWA STATE UNIVERSITY OF SCIENCE AND TECHNOLOGY

Biochemistry	M,D*
Biological and Biomedical Sciences—General	M,D
Biophysics	M,D
Cell Biology	M,D
Computational Biology	M,D
Developmental Biology	M,D
Ecology	M,D
Entomology	M,D
Evolutionary Biology	M,D
Genetics	M,D*
Immunology	M,D
Microbiology	M,D
Molecular Biology	M,D*
Neuroscience	M,D*
Nutrition	M,D
Pathology	M,D
Plant Pathology	M,D
Plant Physiology	M,D
Structural Biology	M,D

IMMACULATA UNIVERSITY

Toxicology	M,D

JACKSON STATE UNIVERSITY

Biological and Biomedical Sciences—General	M,D

JACKSONVILLE STATE UNIVERSITY

Biological and Biomedical Sciences—General	M

JAMES MADISON UNIVERSITY

Biological and Biomedical Sciences—General	M

JOHN CARROLL UNIVERSITY

Biological and Biomedical Sciences—General	M

THE JOHNS HOPKINS UNIVERSITY

Anatomy	D
Biochemistry	M,D
Biological and Biomedical Sciences—General	M,D*
Biophysics	M,D
Cell Biology	D*
Developmental Biology	D
Evolutionary Biology	D
Genetics	M,D
Human Genetics	D*
Immunology	M,D*
Infectious Diseases	M,D
Microbiology	M,D
Molecular Biology	M,D
Molecular Medicine	D
Neuroscience	D
Nutrition	M,D
Pathobiology	D
Pathology	D
Pharmacology	D*
Physiology	M,D
Toxicology	M,D

KANSAS STATE UNIVERSITY

Anatomy	M,D
Biochemistry	M,D
Biological and Biomedical Sciences—General	M,D
Entomology	M,D
Genetics	M,D
Microbiology	D
Nutrition	M,D
Pathobiology	M,D
Physiology	M,D
Plant Pathology	M,D

KECK GRADUATE INSTITUTE OF APPLIED LIFE SCIENCES

Biological and Biomedical Sciences—General	M

KENT STATE UNIVERSITY

Biochemistry	M,D
Biological and Biomedical Sciences—General	M,D*
Botany	M
Cell Biology	M,D
Ecology	M,D
Molecular Biology	M,D
Neuroscience	M,D
Nutrition	M
Pharmacology	M,D
Physiology	M,D

LAKEHEAD UNIVERSITY

Biological and Biomedical Sciences—General	M

LAMAR UNIVERSITY

Biological and Biomedical Sciences—General	M

LAURENTIAN UNIVERSITY

Biochemistry	M
Biological and Biomedical Sciences—General	M

LEHIGH UNIVERSITY

Biochemistry	D
Biological and Biomedical Sciences—General	D
Molecular Biology	D

LEHMAN COLLEGE OF THE CITY UNIVERSITY OF NEW YORK

Biological and Biomedical Sciences—General	M
Nutrition	M

LESLEY UNIVERSITY

Ecology	M,D,O

LOMA LINDA UNIVERSITY

Anatomy	M,D
Biochemistry	M,D
Biological and Biomedical Sciences—General	M,D
Microbiology	M,D
Nutrition	M,D
Pathology	M,D
Pharmacology	M,D
Physiology	M,D

LONG ISLAND UNIVERSITY, BROOKLYN CAMPUS

Biological and Biomedical Sciences—General	M
Pharmacology	M,D
Toxicology	M,D

LONG ISLAND UNIVERSITY, C.W. POST CAMPUS

Biological and Biomedical Sciences—General	M
Cardiovascular Sciences	M,O
Immunology	M
Microbiology	M
Nutrition	M,O

LOUISIANA STATE UNIVERSITY AND AGRICULTURAL AND MECHANICAL COLLEGE

Biochemistry	M,D
Biological and Biomedical Sciences—General	M,D*
Biopsychology	M,D
Entomology	M,D
Plant Pathology	M,D
Toxicology	M

LOUISIANA STATE UNIVERSITY HEALTH SCIENCES CENTER

Anatomy	M,D
Biological and Biomedical Sciences—General	M,D
Cell Biology	M,D
Developmental Biology	M,D
Human Genetics	M,D
Immunology	M,D
Microbiology	M,D
Neurobiology	M,D
Neuroscience	D*
Parasitology	M,D
Pathology	M,D
Pharmacology	M,D
Physiology	M,D

LOUISIANA STATE UNIVERSITY HEALTH SCIENCES CENTER AT SHREVEPORT

Anatomy	M,D
Biochemistry	M,D
Cell Biology	M,D
Immunology	M
Microbiology	M
Molecular Biology	M,D
Pharmacology	D

Physiology	M,D

LOUISIANA TECH UNIVERSITY

Biological and Biomedical Sciences—General	M
Nutrition	M

LOYOLA UNIVERSITY CHICAGO

Anatomy	M,D
Biochemistry	M,D*
Biological and Biomedical Sciences—General	M
Cell Biology	M,D*
Immunology	M,D
Infectious Diseases	M
Microbiology	M,D*
Molecular Biology	D
Molecular Physiology	M,D*
Neurobiology	M,D
Neuroscience	M,D
Pharmacology	M,D*
Virology	M,D

MAHARISHI UNIVERSITY OF MANAGEMENT

Physiology	M,D

MARQUETTE UNIVERSITY

Biological and Biomedical Sciences—General	M,D
Cell Biology	M,D
Developmental Biology	M,D
Ecology	M,D
Evolutionary Biology	M,D
Genetics	M,D
Microbiology	M,D
Molecular Biology	M,D
Neurobiology	M,D
Physiology	M,D

MARSHALL UNIVERSITY

Biological and Biomedical Sciences—General	M,D

MARYWOOD UNIVERSITY

Nutrition	M*

MASSACHUSETTS COLLEGE OF PHARMACY AND HEALTH SCIENCES

Pharmacology	M,D

MASSACHUSETTS INSTITUTE OF TECHNOLOGY

Biochemistry	D
Biological and Biomedical Sciences—General	P,M,D*
Computational Biology	D
Environmental Biology	M,D,O
Molecular Pathology	M,D
Molecular Pharmacology	M,D
Molecular Toxicology	M,D
Neuroscience	D*
Systems Biology	D
Toxicology	M,D

MAYO GRADUATE SCHOOL

Biochemistry	D*
Biological and Biomedical Sciences—General	D*
Cancer Biology/Oncology	D
Cell Biology	D
Genetics	D
Immunology	D*
Molecular Biology	D
Molecular Pharmacology	D*
Neuroscience	D
Structural Biology	D
Virology	D*

MCGILL UNIVERSITY

Anatomy	M,D
Biochemistry	M,D
Biological and Biomedical Sciences—General	M,D*

Cell Biology	M,D
Entomology	M,D
Human Genetics	M,D
Immunology	M,D
Microbiology	M,D
Neuroscience	M,D
Nutrition	M,D
Parasitology	M,D,O
Pathology	M,D
Pharmacology	M,D
Physiology	M,D

MCMASTER UNIVERSITY

Biochemistry	M,D
Biological and Biomedical Sciences—General	M,D
Cancer Biology/Oncology	M,D
Cardiovascular Sciences	M,D
Cell Biology	M,D
Genetics	M,D
Immunology	M,D
Molecular Biology	M,D
Neuroscience	M,D
Pharmacology	M,D
Physiology	M,D
Virology	M,D

MCNEESE STATE UNIVERSITY

Biological and Biomedical Sciences—General	M

MEDICAL COLLEGE OF GEORGIA

Anatomy	D
Biochemistry	D
Biological and Biomedical Sciences—General	M,D*
Cardiovascular Sciences	M,D
Cell Biology	D
Molecular Biology	D
Molecular Medicine	D
Neuroscience	D
Pharmacology	D
Physiology	D

MEDICAL COLLEGE OF WISCONSIN

Biochemistry	M,D
Biological and Biomedical Sciences—General	M,D*
Biophysics	D*
Cell Biology	M,D
Developmental Biology	M,D
Microbiology	M,D
Molecular Genetics	M,D
Pathology	M,D
Pharmacology	M,D
Physiology	M,D*
Toxicology	M,D

MEDICAL UNIVERSITY OF SOUTH CAROLINA

Anatomy	P,D
Biochemistry	P,M,D
Biological and Biomedical Sciences—General	P,M,D
Cell Biology	P,D
Immunology	P,M,D
Microbiology	P,M,D*
Molecular Biology	P,M,D
Molecular Pharmacology	P,M,D
Neuroscience	P,M,D
Pathobiology	P,D
Pathology	P,M,D

MEHARRY MEDICAL COLLEGE

Biochemistry	D
Biological and Biomedical Sciences—General	D
Microbiology	D
Pharmacology	M,D
Physiology	D

MEMORIAL UNIVERSITY OF NEWFOUNDLAND

Biochemistry	M,D
Biological and Biomedical Sciences—General	M,D,O

Biopsychology	M,D
Human Genetics	M,D
Marine Biology	M,D

MEREDITH COLLEGE

Nutrition	M

MIAMI UNIVERSITY

Biochemistry	M,D
Botany	M,D*
Microbiology	M,D
Plant Biology	M,D
Zoology	M,D*

MICHIGAN STATE UNIVERSITY

Biochemistry	M,D*
Biological and Biomedical Sciences—General	M,D
Cell Biology	M,D
Ecology	D*
Entomology	M,D
Evolutionary Biology	D
Genetics	M,D
Microbiology	M,D*
Molecular Biology	M,D
Molecular Genetics	M,D
Neuroscience	M,D
Nutrition	M,D
Pathobiology	M,D
Pathology	M,D
Pharmacology	M,D
Physiology	M,D
Plant Biology	M,D
Plant Pathology	M,D
Toxicology	M,D
Zoology	M,D

MICHIGAN TECHNOLOGICAL UNIVERSITY

Biological and Biomedical Sciences—General	M,D
Ecology	M
Plant Molecular Biology	M,D

MIDDLE TENNESSEE STATE UNIVERSITY

Biological and Biomedical Sciences—General	M
Nutrition	M

MIDWESTERN STATE UNIVERSITY

Biological and Biomedical Sciences—General	M

MIDWESTERN UNIVERSITY, DOWNERS GROVE CAMPUS

Biological and Biomedical Sciences—General	M*

MIDWESTERN UNIVERSITY, GLENDALE CAMPUS

Biological and Biomedical Sciences—General	M
Cardiovascular Sciences	M

MILLERSVILLE UNIVERSITY OF PENNSYLVANIA

Biological and Biomedical Sciences—General	M

MILLS COLLEGE

Biological and Biomedical Sciences—General	O

MILWAUKEE SCHOOL OF ENGINEERING

Cardiovascular Sciences	M

MINNESOTA STATE UNIVERSITY MANKATO

Biological and Biomedical Sciences—General	M
Ecology	M

MISSISSIPPI COLLEGE

Biological and Biomedical
  Sciences—General   M

MISSISSIPPI STATE UNIVERSITY

Biochemistry   M,D
Biological and Biomedical
  Sciences—General   M,D
Entomology   M,D
Molecular Biology   M,D
Nutrition   M,D
Plant Pathology   M,D

MISSOURI STATE UNIVERSITY

Biological and Biomedical
  Sciences—General   M
Cell Biology   M
Molecular Biology   M

MONTANA STATE UNIVERSITY

Biochemistry   M,D
Biological and Biomedical
  Sciences—General   M,D
Ecology   M,D
Environmental Biology   M,D
Microbiology   M,D
Molecular Biology   M,D
Neuroscience   M,D
Plant Pathology   M,D
Zoology   M,D

MONTCLAIR STATE UNIVERSITY

Biochemistry   M
Biological and Biomedical
  Sciences—General   M,O
Molecular Biology   M,O
Nutrition   M,O

MOREHEAD STATE UNIVERSITY

Biological and Biomedical
  Sciences—General   M

MOREHOUSE SCHOOL OF MEDICINE

Biological and Biomedical
  Sciences—General   D*

MORGAN STATE UNIVERSITY

Biological and Biomedical
  Sciences—General   M,D
Environmental Biology   D*

MOUNT ALLISON UNIVERSITY

Biological and Biomedical
  Sciences—General   M

MOUNT MARY COLLEGE

Nutrition   M

MOUNT SAINT VINCENT UNIVERSITY

Nutrition   M

MOUNT SINAI SCHOOL OF MEDICINE OF NEW YORK UNIVERSITY

Biological and Biomedical
  Sciences—General   M,D*
Biophysics   M,D
Cell Biology   M,D
Genetics   M,D
Genomic Sciences   M,D
Microbiology   M,D
Molecular Biology   M,D
Neuroscience   M,D
Pathology   M,D
Structural Biology   M,D

MURRAY STATE UNIVERSITY

Biological and Biomedical
  Sciences—General   M,D
Marine Biology   M

NEW JERSEY INSTITUTE OF TECHNOLOGY

Biological and Biomedical
  Sciences—General   M,D
Computational Biology   M

NEW MEXICO HIGHLANDS UNIVERSITY

Biological and Biomedical
  Sciences—General   M

NEW MEXICO INSTITUTE OF MINING AND TECHNOLOGY

Biochemistry   M,D
Biological and Biomedical
  Sciences—General   M

NEW MEXICO STATE UNIVERSITY

Biochemistry   M,D
Biological and Biomedical
  Sciences—General   M,D
Entomology   M
Molecular Biology   M,D
Plant Pathology   M

NEW YORK INSTITUTE OF TECHNOLOGY

Nutrition   M

NEW YORK MEDICAL COLLEGE

Anatomy   M,D
Biochemistry   M,D
Biological and Biomedical
  Sciences—General   M,D*
Cell Biology   M,D*
Immunology   M,D
Microbiology   M,D
Molecular Biology   M,D
Neuroscience   M,D
Pathology   M,D*
Pharmacology   M,D
Physiology   M,D*

NEW YORK UNIVERSITY

Biological and Biomedical
  Sciences—General   M,D*
Cancer Biology/Oncology   M,D
Cell Biology   D
Computational Biology   D*
Developmental Biology   D
Genetics   M,D
Immunology   M,D*
Microbiology   M,D*
Molecular Biology   M,D
Molecular Genetics   M,D
Molecular Pharmacology   D
Molecular Toxicology   M,D
Neurobiology   M,D
Neuroscience   D*
Nutrition   M,D
Parasitology   D
Pharmacology   D
Physiology   D
Plant Biology   M,D
Reproductive Biology   D
Structural Biology   D*
Toxicology   M,D

NICHOLLS STATE UNIVERSITY

Environmental Biology   M
Marine Biology   M

NORTH CAROLINA AGRICULTURAL AND TECHNICAL STATE UNIVERSITY

Biological and Biomedical
  Sciences—General   M
Nutrition   M

NORTH CAROLINA CENTRAL UNIVERSITY

Biological and Biomedical
  Sciences—General   M

NORTH CAROLINA STATE UNIVERSITY

Biochemistry   M,D*
Biological and Biomedical
  Sciences—General   M,D
Botany   M,D
Cell Biology   M,D
Ecology   M,D
Entomology   M,D
Genetics   M,D
Genomic Sciences   M,D*
Immunology   M,D
Microbiology   M,D
Molecular Toxicology   M,D*
Nutrition   M,D
Pathology   M,D
Pharmacology   M,D
Physiology   M,D
Plant Pathology   M,D
Toxicology   M,D
Zoology   M,D

NORTH DAKOTA STATE UNIVERSITY

Biochemistry   M,D
Biological and Biomedical
  Sciences—General   M,D
Botany   M,D
Cell Biology   M,D
Conservation Biology   M,D
Ecology   M,D
Entomology   M,D
Genomic Sciences   M,D
Microbiology   M,D
Molecular Biology   M,D
Molecular Pathology   M,D
Nutrition   M
Pathology   M,D
Plant Pathology   M,D
Zoology   M,D

NORTHEASTERN ILLINOIS UNIVERSITY

Biological and Biomedical
  Sciences—General   M

NORTHEASTERN UNIVERSITY

Biochemistry   M,D
Biological and Biomedical
  Sciences—General   M,D*
Cardiovascular Sciences   M
Marine Biology   M,D
Pharmacology   M,D*
Toxicology   M

NORTHERN ARIZONA UNIVERSITY

Biochemistry   M
Biological and Biomedical
  Sciences—General   M,D
Ecology   M,O

NORTHERN ILLINOIS UNIVERSITY

Biochemistry   M,D
Biological and Biomedical
  Sciences—General   M,D
Nutrition   M

NORTHERN MICHIGAN UNIVERSITY

Biochemistry   M
Biological and Biomedical
  Sciences—General   M

NORTHWESTERN UNIVERSITY

Biochemistry   D
Biological and Biomedical
  Sciences—General   D*
Biophysics   D
Biopsychology   D
Cancer Biology/Oncology   D
Cell Biology   D
Computational Biology   M
Developmental Biology   D
Evolutionary Biology   D
Genetics   D

Immunology (North Carolina State University continued)

Immunology   D
Microbiology   D
Molecular Biology   D
Neurobiology   M,D
Neuroscience   D*
Pharmacology   M
Physiology   M
Reproductive Biology   D
Structural Biology   D
Toxicology   D

NORTHWEST MISSOURI STATE UNIVERSITY

Biological and Biomedical
  Sciences—General   M

NOTRE DAME DE NAMUR UNIVERSITY

Biological and Biomedical
  Sciences—General   O

NOVA SCOTIA AGRICULTURAL COLLEGE

Botany   M
Ecology   M
Environmental Biology   M
Physiology   M
Plant Pathology   M
Plant Physiology   M

NOVA SOUTHEASTERN UNIVERSITY

Biological and Biomedical
  Sciences—General   M
Marine Biology   M,D
Pharmacology   M

OAKLAND UNIVERSITY

Biological and Biomedical
  Sciences—General   M
Cell Biology   M

OCCIDENTAL COLLEGE

Biological and Biomedical
  Sciences—General   M

OGI SCHOOL OF SCIENCE & ENGINEERING AT OREGON HEALTH & SCIENCE UNIVERSITY

Biochemistry   M,D
Molecular Biology   M,D

THE OHIO STATE UNIVERSITY

Anatomy   M,D
Biochemistry   M,D
Biological and Biomedical
  Sciences—General   M,D
Biophysics   M,D
Biopsychology   M,D
Cell Biology   M,D
Developmental Biology   M,D
Ecology   M,D
Entomology   M,D
Evolutionary Biology   M,D
Genetics   M,D*
Immunology   M,D
Microbiology   M,D
Molecular Biology   M,D
Molecular Genetics   M,D
Neuroscience   D
Nutrition   M,D
Pathobiology   M,D
Pathology   M
Pharmacology   M,D
Physiology   M,D
Plant Biology   M,D
Plant Pathology   M,D
Toxicology   M,D
Virology   M,D

OHIO UNIVERSITY

Biochemistry   M,D
Biological and Biomedical
  Sciences—General   M,D*

Cell Biology — M,D
Ecology — M,D
Environmental Biology — M,D
Evolutionary Biology — M,D
Microbiology — M,D
Molecular Biology — M,D*
Neuroscience — M,D
Nutrition — M
Physiology — M,D
Plant Biology — M,D

OKLAHOMA STATE UNIVERSITY

Biochemistry — M,D*
Botany — M,D
Conservation Biology — M,D
Entomology — D
Microbiology — M,D
Molecular Biology — M,D
Molecular Genetics — M,D
Nutrition — M,D
Plant Pathology — D
Zoology — M,D

OKLAHOMA STATE UNIVERSITY CENTER FOR HEALTH SCIENCES

Biological and Biomedical
 Sciences—General — M,D
Molecular Biology — M
Toxicology — M

OLD DOMINION UNIVERSITY

Biochemistry — M
Biological and Biomedical
 Sciences—General — M,D
Ecology — D

OREGON HEALTH & SCIENCE UNIVERSITY

Biochemistry — D
Biological and Biomedical
 Sciences—General — M,D,O
Biopsychology — M,D
Cell Biology — D
Developmental Biology — D
Genetics — D
Immunology — D
Microbiology — D
Molecular Biology — D
Neuroscience — M,D*
Pharmacology — D
Physiology — D

OREGON STATE UNIVERSITY

Biochemistry — M,D
Biophysics — M,D
Botany — M,D
Cell Biology — M,D
Genetics — M,D
Microbiology — M,D
Molecular Biology — M,D
Molecular Toxicology — M,D
Nutrition — M,D
Pathology — M
Plant Pathology — M,D
Plant Physiology — M,D
Toxicology — M,D
Zoology — M,D

PACIFIC GRADUATE SCHOOL OF PSYCHOLOGY

Biopsychology — D

PALMER COLLEGE OF CHIROPRACTIC

Anatomy — M

THE PENNSYLVANIA STATE UNIVERSITY MILTON S. HERSHEY MEDICAL CENTER

Anatomy — M,D
Biochemistry — M,D
Biological and Biomedical
 Sciences—General — M,D*
Cell Biology — M,D
Genetics — M,D
Immunology — M,D

Microbiology — M,D
Molecular Biology — M,D
Neuroscience — M,D
Pharmacology — M,D
Physiology — M,D
Virology — M,D

THE PENNSYLVANIA STATE UNIVERSITY UNIVERSITY PARK CAMPUS

Biochemistry — M,D*
Biological and Biomedical
 Sciences—General — M,D*
Biopsychology — M,D*
Cell Biology — M,D*
Developmental Biology — M,D
Ecology — M,D
Entomology — M,D
Evolutionary Biology — M,D
Genetics — M,D
Microbiology — M,D
Molecular Biology — M,D
Molecular Medicine — M,D
Neuroscience — M,D
Nutrition — M,D
Pathobiology — M,D
Physiology — M,D
Plant Pathology — M,D
Plant Physiology — M,D

PHILADELPHIA COLLEGE OF OSTEOPATHIC MEDICINE

Biological and Biomedical
 Sciences—General — M,O

PITTSBURG STATE UNIVERSITY

Biological and Biomedical
 Sciences—General — M

PONCE SCHOOL OF MEDICINE

Biological and Biomedical
 Sciences—General — D

PORTLAND STATE UNIVERSITY

Biological and Biomedical
 Sciences—General — M,D

PRAIRIE VIEW A&M UNIVERSITY

Biological and Biomedical
 Sciences—General — M

PRESCOTT COLLEGE

Ecology — M

PRINCETON UNIVERSITY

Biological and Biomedical
 Sciences—General — D
Biophysics — D
Ecology — D*
Evolutionary Biology — D
Molecular Biology — D*
Neuroscience — D

PURDUE UNIVERSITY

Anatomy — M,D
Biochemistry — M,D
Biological and Biomedical
 Sciences—General — M,D
Biophysics — M,D
Botany — M,D
Cell Biology — M,D
Developmental Biology — M,D
Ecology — M,D
Entomology — M,D
Evolutionary Biology — M,D
Genetics — M,D
Immunology — M,D
Infectious Diseases — M,D
Microbiology — M,D
Molecular Biology — M,D
Molecular Pharmacology — M,D
Neurobiology — M,D
Nutrition — M,D
Parasitology — M,D
Pathobiology — M,D
Pathology — M,D

Pharmacology — M,D
Physiology — M,D
Plant Pathology — M,D
Plant Physiology — M,D
Toxicology — M,D
Virology — M,D

PURDUE UNIVERSITY CALUMET

Biological and Biomedical
 Sciences—General — M

QUEENS COLLEGE OF THE CITY UNIVERSITY OF NEW YORK

Biochemistry — M
Biological and Biomedical
 Sciences—General — M

QUEEN'S UNIVERSITY AT KINGSTON

Anatomy — M,D
Biochemistry — M,D
Biological and Biomedical
 Sciences—General — M,D
Cell Biology — M,D
Immunology — M,D
Microbiology — M,D
Pathology — M,D
Pharmacology — M,D
Physiology — M,D
Toxicology — M,D

QUINNIPIAC UNIVERSITY

Biological and Biomedical
 Sciences—General — M
Cell Biology — M
Microbiology — M
Molecular Biology — M*
Pathology — M

RENSSELAER POLYTECHNIC INSTITUTE

Biochemistry — M,D
Biological and Biomedical
 Sciences—General — M,D*
Biophysics — M,D
Cell Biology — M,D
Developmental Biology — M,D
Microbiology — M,D
Molecular Biology — M,D

RHODE ISLAND COLLEGE

Biological and Biomedical
 Sciences—General — M

RICE UNIVERSITY

Biochemistry — M,D*
Cell Biology — M,D
Ecology — M,D
Evolutionary Biology — M,D

ROCHESTER INSTITUTE OF TECHNOLOGY

Biological and Biomedical
 Sciences—General — M

THE ROCKEFELLER UNIVERSITY

Biological and Biomedical
 Sciences—General — D*

ROSALIND FRANKLIN UNIVERSITY OF MEDICINE AND SCIENCE

Anatomy — M,D
Biochemistry — M,D
Biological and Biomedical
 Sciences—General — M,D*
Cell Biology — M,D
Immunology — M,D
Medical Microbiology — M,D
Microbiology — M,D
Molecular Pharmacology — M,D
Neuroscience — D*
Nutrition — M
Pathology — M,D
Physiology — M,D*

RUSH UNIVERSITY

Anatomy — M,D
Biochemistry — D
Cell Biology — M,D
Immunology — M,D
Microbiology — M,D
Neuroscience — M,D
Nutrition — M
Pharmacology — M,D
Physiology — D
Virology — M,D

RUTGERS, THE STATE UNIVERSITY OF NEW JERSEY, CAMDEN

Biological and Biomedical
 Sciences—General — M

RUTGERS, THE STATE UNIVERSITY OF NEW JERSEY, NEWARK

Biochemistry — M,D
Biological and Biomedical
 Sciences—General — M,D*
Biopsychology — D
Computational Biology — M
Neuroscience — D

RUTGERS, THE STATE UNIVERSITY OF NEW JERSEY, NEW BRUNSWICK/PISCATAWAY

Biochemistry — M,D
Biological and Biomedical
 Sciences—General — D
Biopsychology — D
Cell Biology — M,D
Developmental Biology — M,D
Ecology — M,D
Entomology — M,D
Environmental Biology — M,D
Evolutionary Biology — M,D
Genetics — M,D
Immunology — M,D
Marine Biology — M,D
Medical Microbiology — M,D
Microbiology — M,D
Molecular Biology — M,D*
Molecular Genetics — M,D
Molecular Pharmacology — D
Neurobiology — D
Neuroscience — D
Nutrition — M,D
Physiology — D
Plant Biology — M,D
Plant Molecular Biology — M,D
Plant Pathology — M,D
Plant Physiology — M,D
Toxicology — M,D
Virology — M,D

SAGE GRADUATE SCHOOL

Nutrition — M

ST. CLOUD STATE UNIVERSITY

Biological and Biomedical
 Sciences—General — M

SAINT FRANCIS UNIVERSITY

Biological and Biomedical
 Sciences—General — M

ST. FRANCIS XAVIER UNIVERSITY

Biological and Biomedical
 Sciences—General — M

ST. JOHN'S UNIVERSITY (NY)

Biological and Biomedical
 Sciences—General — M,D*
Toxicology — M

SAINT JOSEPH COLLEGE

Biological and Biomedical
 Sciences—General — M
Cell Biology — M
Molecular Biology — M

SAINT JOSEPH'S UNIVERSITY

Biological and Biomedical Sciences—General	M

SAINT LOUIS UNIVERSITY

Anatomy	M,D
Biochemistry	D
Biological and Biomedical Sciences—General	M,D*
Immunology	D
Microbiology	D
Molecular Biology	D
Nutrition	D
Pathology	D
Pharmacology	D
Physiology	D

SALISBURY UNIVERSITY

Physiology	M

SAM HOUSTON STATE UNIVERSITY

Biological and Biomedical Sciences—General	M

SAN DIEGO STATE UNIVERSITY

Biological and Biomedical Sciences—General	M,D*
Cell Biology	M,D
Ecology	M,D*
Microbiology	M
Molecular Biology	M,D*
Nutrition	M
Toxicology	M,D

SAN FRANCISCO STATE UNIVERSITY

Biochemistry	M
Biological and Biomedical Sciences—General	M
Cell Biology	M
Conservation Biology	M
Ecology	M
Marine Biology	M
Microbiology	M
Molecular Biology	M
Physiology	M

SAN JOSE STATE UNIVERSITY

Biological and Biomedical Sciences—General	M
Microbiology	M
Molecular Biology	M
Nutrition	M
Physiology	M

SARAH LAWRENCE COLLEGE

Human Genetics	M

THE SCRIPPS RESEARCH INSTITUTE

Biochemistry	D
Biological and Biomedical Sciences—General	D
Biophysics	D

SETON HALL UNIVERSITY

Biochemistry	M,D
Biological and Biomedical Sciences—General	M,D
Microbiology	M,D
Molecular Biology	M,D*
Neuroscience	M

SHIPPENSBURG UNIVERSITY OF PENNSYLVANIA

Biological and Biomedical Sciences—General	M

SIMMONS COLLEGE

Nutrition	M,O

SIMON FRASER UNIVERSITY

Biochemistry	M,D
Biological and Biomedical Sciences—General	M,D,O
Biophysics	M,D
Entomology	M,D,O
Molecular Biology	M,D
Toxicology	M,D,O

SMITH COLLEGE

Biological and Biomedical Sciences—General	M,D

SONOMA STATE UNIVERSITY

Biological and Biomedical Sciences—General	M
Environmental Biology	M

SOUTH CAROLINA STATE UNIVERSITY

Nutrition	M

SOUTH DAKOTA STATE UNIVERSITY

Biochemistry	M,D
Biological and Biomedical Sciences—General	M,D
Entomology	M
Microbiology	M
Plant Pathology	M

SOUTHEASTERN LOUISIANA UNIVERSITY

Biological and Biomedical Sciences—General	M

SOUTHEAST MISSOURI STATE UNIVERSITY

Biological and Biomedical Sciences—General	M
Nutrition	M

SOUTHERN CONNECTICUT STATE UNIVERSITY

Biological and Biomedical Sciences—General	M

SOUTHERN ILLINOIS UNIVERSITY CARBONDALE

Biochemistry	M,D
Biological and Biomedical Sciences—General	M,D
Microbiology	M,D
Molecular Biology	M,D*
Nutrition	M
Pharmacology	M,D*
Physiology	M,D*
Plant Biology	M,D*
Zoology	M,D*

SOUTHERN ILLINOIS UNIVERSITY EDWARDSVILLE

Biological and Biomedical Sciences—General	M

SOUTHERN METHODIST UNIVERSITY

Biological and Biomedical Sciences—General	M,D*

SOUTHERN UNIVERSITY AND AGRICULTURAL AND MECHANICAL COLLEGE

Biochemistry	M
Biological and Biomedical Sciences—General	M

SOUTHWESTERN OKLAHOMA STATE UNIVERSITY

Microbiology	M

STANFORD UNIVERSITY

Biochemistry	D*
Biological and Biomedical Sciences—General	M,D
Biophysics	D
Cancer Biology/Oncology	D
Developmental Biology	D
Genetics	D
Immunology	D
Microbiology	D*
Molecular Pharmacology	D
Neuroscience	D
Physiology	D
Structural Biology	D

STATE UNIVERSITY OF NEW YORK AT BINGHAMTON

Biological and Biomedical Sciences—General	M,D
Biopsychology	M,D

STATE UNIVERSITY OF NEW YORK AT BUFFALO

Anatomy	M,D
Biochemistry	M,D
Biological and Biomedical Sciences—General	M,D*
Biophysics	M,D
Cancer Biology/Oncology	D
Cell Biology	D
Immunology	M,D
Microbiology	M,D
Molecular Biology	D
Molecular Pharmacology	D*
Neuroscience	M,D
Nutrition	M,D
Pathology	M,D
Pharmacology	M,D
Physiology	M,D
Structural Biology	M,D
Toxicology	M,D

STATE UNIVERSITY OF NEW YORK AT NEW PALTZ

Biological and Biomedical Sciences—General	M

STATE UNIVERSITY OF NEW YORK COLLEGE AT BROCKPORT

Biological and Biomedical Sciences—General	M

STATE UNIVERSITY OF NEW YORK COLLEGE AT ONEONTA

Biological and Biomedical Sciences—General	M

STATE UNIVERSITY OF NEW YORK COLLEGE OF ENVIRONMENTAL SCIENCE AND FORESTRY

Biochemistry	M,D
Conservation Biology	M,D
Ecology	M,D
Entomology	M,D
Environmental Biology	M,D
Plant Pathology	M,D

STATE UNIVERSITY OF NEW YORK DOWNSTATE MEDICAL CENTER

Biological and Biomedical Sciences—General	M,D*
Cell Biology	D
Molecular Biology	D
Neuroscience	D

STATE UNIVERSITY OF NEW YORK, FREDONIA

Biological and Biomedical Sciences—General	M

STATE UNIVERSITY OF NEW YORK UPSTATE MEDICAL UNIVERSITY

Anatomy	M,D

BIOCHEMISTRY

Biochemistry	M,D
Biological and Biomedical Sciences—General	M,D*
Cell Biology	M,D
Developmental Biology	D
Immunology	M,D
Microbiology	M,D
Molecular Biology	M,D
Neuroscience	D
Pharmacology	M,D
Physiology	M,D

STEPHEN F. AUSTIN STATE UNIVERSITY

Biological and Biomedical Sciences—General	M

STEVENS INSTITUTE OF TECHNOLOGY

Biochemistry	M,D,O

STONY BROOK UNIVERSITY, STATE UNIVERSITY OF NEW YORK

Anatomy	D*
Biochemistry	D*
Biological and Biomedical Sciences—General	D*
Biophysics	D*
Biopsychology	D*
Cell Biology	M,D
Developmental Biology	M,D*
Ecology	D*
Evolutionary Biology	D
Genetics	D*
Immunology	M,D
Microbiology	D*
Molecular Biology	M,D*
Molecular Genetics	D*
Molecular Physiology	D*
Neuroscience	D*
Pathology	M,D
Pharmacology	D*
Physiology	D*
Structural Biology	D

SUL ROSS STATE UNIVERSITY

Biological and Biomedical Sciences—General	M

SYRACUSE UNIVERSITY

Biochemistry	D
Biological and Biomedical Sciences—General	M,D*
Biophysics	D
Neuroscience	M,D
Nutrition	M
Structural Biology	D

TARLETON STATE UNIVERSITY

Biological and Biomedical Sciences—General	M

TEACHERS COLLEGE COLUMBIA UNIVERSITY

Neuroscience	M,D
Nutrition	M,D

TEMPLE UNIVERSITY

Anatomy	M,D
Biochemistry	M,D*
Biological and Biomedical Sciences—General	M,D*
Cell Biology	M,D
Genetics	D
Immunology	M,D
Microbiology	M,D*
Molecular Biology	D
Neuroscience	M,D
Pathology	D
Pharmacology	M,D
Physiology	D

TENNESSEE STATE UNIVERSITY

Biological and Biomedical
Sciences—General — M,D*

TENNESSEE TECHNOLOGICAL UNIVERSITY

Biological and Biomedical
Sciences—General — M
Environmental Biology — M

TEXAS A&M INTERNATIONAL UNIVERSITY

Biological and Biomedical
Sciences—General — M

TEXAS A&M UNIVERSITY

Anatomy — M,D
Biochemistry — M,D
Biological and Biomedical
Sciences—General — M,D
Biophysics — M,D
Biopsychology — M,D
Botany — M,D
Cell Biology — D
Entomology — M,D
Genetics — M,D
Microbiology — M,D
Molecular Biology — D
Neuroscience — M,D*
Nutrition — M,D
Parasitology — M,D
Pathobiology — M,D
Pathology — M,D
Physiology — M,D
Plant Biology — M,D
Plant Pathology — M,D
Toxicology — M,D
Zoology — M,D

TEXAS A&M UNIVERSITY–COMMERCE

Biological and Biomedical
Sciences—General — M

TEXAS A&M UNIVERSITY–CORPUS CHRISTI

Biological and Biomedical
Sciences—General — M

TEXAS A&M UNIVERSITY–KINGSVILLE

Biological and Biomedical
Sciences—General — M

TEXAS A&M UNIVERSITY SYSTEM HEALTH SCIENCE CENTER

Biochemistry — D
Biological and Biomedical
Sciences—General — M,D*
Cardiovascular Sciences — D
Cell Biology — D
Immunology — D
Medical Microbiology — D
Microbiology — D
Molecular Biology — D
Neuroscience — D*
Pathology — D
Pharmacology — D
Structural Biology — D
Toxicology — D
Virology — D

TEXAS CHRISTIAN UNIVERSITY

Biological and Biomedical
Sciences—General — M
Ecology — M

TEXAS SOUTHERN UNIVERSITY

Biological and Biomedical
Sciences—General — M
Nutrition — M
Toxicology — M,D

TEXAS STATE UNIVERSITY-SAN MARCOS

Biochemistry — M
Biological and Biomedical
Sciences—General — M
Conservation Biology — M
Marine Biology — M

TEXAS TECH UNIVERSITY

Biological and Biomedical
Sciences—General — M,D
Entomology — M,D
Microbiology — M,D
Nutrition — M,D
Toxicology — M,D
Zoology — M,D

TEXAS TECH UNIVERSITY HEALTH SCIENCES CENTER

Biochemistry — M,D
Biological and Biomedical
Sciences—General — M,D
Cell Biology — M,D
Medical Microbiology — M,D
Molecular Genetics — M,D
Molecular Pathology — M
Neuroscience — M,D
Pharmacology — M,D
Physiology — M,D

TEXAS WOMAN'S UNIVERSITY

Biological and Biomedical
Sciences—General — M,D
Molecular Biology — M,D
Nutrition — M,D

THOMAS JEFFERSON UNIVERSITY

Biochemistry — M,D*
Biological and Biomedical
Sciences—General — M,D,O*
Cell Biology — D
Developmental Biology — M,D
Genetics — D*
Immunology — D*
Microbiology — M,D
Molecular Biology — D
Molecular Pharmacology — D
Neuroscience — D
Pathology — D
Pharmacology — M*
Physiology — D
Structural Biology — D

TOURO COLLEGE

Biological and Biomedical
Sciences—General — M

TOWSON UNIVERSITY

Biological and Biomedical
Sciences—General — M

TRENT UNIVERSITY

Biological and Biomedical
Sciences—General — M,D

TROPICAL AGRICULTURE RESEARCH AND HIGHER EDUCATION CENTER

Conservation Biology — M,D

TRUMAN STATE UNIVERSITY

Biological and Biomedical
Sciences—General — M

TUFTS UNIVERSITY

Biochemistry — D
Biological and Biomedical
Sciences—General — M,D*
Cell Biology — D
Developmental Biology — D
Genetics — D
Immunology — D
Microbiology — D
Molecular Biology — D
Molecular Physiology — D
Neuroscience — D
Nutrition — M,D
Pharmacology — D
Physiology — D

TULANE UNIVERSITY

Biochemistry — M,D*
Biological and Biomedical
Sciences—General — M,D,O
Cell Biology — M,D
Human Genetics — M,D
Immunology — M,D
Microbiology — M,D
Molecular Biology — M,D
Neuroscience — M,D
Nutrition — M
Parasitology — M,D,O
Pharmacology — M,D
Physiology — M,D
Structural Biology — M,D

TUSKEGEE UNIVERSITY

Biological and Biomedical
Sciences—General — M
Nutrition — M

UNIFORMED SERVICES UNIVERSITY OF THE HEALTH SCIENCES

Biological and Biomedical
Sciences—General — M,D*
Cell Biology — D
Immunology — D
Infectious Diseases — D*
Microbiology — D*
Molecular Biology — D*
Neuroscience — D*
Pathobiology — D
Pathology — D
Zoology — M,D

UNIVERSITÉ DE MONCTON

Biochemistry — M
Biological and Biomedical
Sciences—General — M
Nutrition — M

UNIVERSITÉ DE MONTRÉAL

Biochemistry — M,D,O
Biological and Biomedical
Sciences—General — M,D
Biophysics — M,D
Cancer Biology/Oncology — O
Cell Biology — M,D
Genetics — O
Immunology — M,D
Infectious Diseases — O
Microbiology — M,D,O
Molecular Biology — M,D
Neuroscience — M,D,O
Nutrition — M,D
Pathology — M,D
Pharmacology — M,D
Physiology — M,D
Toxicology — O
Virology — D

UNIVERSITÉ DE SHERBROOKE

Biochemistry — M,D
Biological and Biomedical
Sciences—General — M,D,O
Biophysics — M,D
Cell Biology — M,D
Immunology — M,D
Microbiology — M,D
Pharmacology — M,D
Physiology — M,D
Radiation Biology — M,D

UNIVERSITÉ DU QUÉBEC À CHICOUTIMI

Genetics — M

UNIVERSITÉ DU QUÉBEC À MONTRÉAL

Biological and Biomedical
Sciences—General — M,D

UNIVERSITÉ DU QUÉBEC À TROIS-RIVIÈRES

Biophysics — M,D

UNIVERSITÉ DU QUÉBEC, INSTITUT NATIONAL DE LA RECHERCHE SCIENTIFIQUE

Biological and Biomedical
Sciences—General — M,D
Immunology — M,D
Medical Microbiology — M,D
Microbiology — M,D
Virology — M,D

UNIVERSITÉ LAVAL

Anatomy — M,D,O
Biochemistry — M,D,O
Biological and Biomedical
Sciences—General — M,D,O
Cancer Biology/Oncology — O
Cardiovascular Sciences — O
Cell Biology — M,D
Immunology — M,D
Infectious Diseases — O
Microbiology — M,D
Molecular Biology — M,D
Neurobiology — M,D
Nutrition — M,D
Pathology — O
Physiology — M,D
Plant Biology — M,D

UNIVERSITY AT ALBANY, STATE UNIVERSITY OF NEW YORK

Biochemistry — M,D
Biological and Biomedical
Sciences—General — M,D*
Biopsychology — M,D,O
Cell Biology — M,D
Conservation Biology — M
Developmental Biology — M,D
Ecology — M,D
Evolutionary Biology — M,D
Genetics — M,D
Immunology — M,D
Molecular Biology — M,D
Neurobiology — M,D
Neuroscience — M,D
Pathology — M,D
Structural Biology — M,D
Toxicology — M,D*

THE UNIVERSITY OF AKRON

Biological and Biomedical
Sciences—General — M
Nutrition — M

THE UNIVERSITY OF ALABAMA

Biological and Biomedical
Sciences—General — M,D
Nutrition — M

THE UNIVERSITY OF ALABAMA AT BIRMINGHAM

Biochemistry — D*
Biological and Biomedical
Sciences—General — M,D*
Biophysics — M,D
Cell Biology — M,D*
Genetics — D
Microbiology — D*
Molecular Biology — M,D
Molecular Genetics — M,D
Molecular Physiology — M,D
Neurobiology — D*
Neuroscience — M,D*
Nutrition — M,D,O
Pathology — D
Pharmacology — D
Physiology — M,D
Toxicology — M,D

THE UNIVERSITY OF ALABAMA IN HUNTSVILLE

Biological and Biomedical Sciences—General — M

UNIVERSITY OF ALASKA ANCHORAGE

Biological and Biomedical Sciences—General — M
Nutrition — O

UNIVERSITY OF ALASKA FAIRBANKS

Biochemistry — M,D
Biological and Biomedical Sciences—General — M,D
Botany — M,D
Marine Biology — M,D
Zoology — M,D

UNIVERSITY OF ALBERTA

Biochemistry — M,D
Biological and Biomedical Sciences—General — M,D,O
Cancer Biology/Oncology — M,D
Cell Biology — M,D
Conservation Biology — M,D
Ecology — M,D
Environmental Biology — M,D
Evolutionary Biology — M,D
Genetics — M,D
Immunology — M,D
Medical Microbiology — M,D
Microbiology — M,D
Molecular Biology — M,D
Neuroscience — M,D
Pathology — M,D
Pharmacology — M,D
Physiology — M,D
Plant Biology — M,D

THE UNIVERSITY OF ARIZONA

Anatomy — D
Biochemistry — M,D
Biological and Biomedical Sciences—General — M,D
Cancer Biology/Oncology — D
Cell Biology — M,D
Ecology — M,D
Entomology — M,D
Evolutionary Biology — M,D
Genetics — M,D
Immunology — M,D
Microbiology — M,D
Molecular Biology — M,D
Neuroscience — D
Nutrition — M,D
Pathobiology — M,D
Pharmacology — M,D,O
Physiology — D
Plant Pathology — M,D
Toxicology — M,D,O

UNIVERSITY OF ARKANSAS

Biological and Biomedical Sciences—General — M,D
Cell Biology — M,D
Entomology — M,D
Molecular Biology — M,D
Plant Pathology — M

UNIVERSITY OF ARKANSAS AT LITTLE ROCK

Biological and Biomedical Sciences—General — M

UNIVERSITY OF ARKANSAS FOR MEDICAL SCIENCES

Anatomy — M,D
Biochemistry — M,D
Biological and Biomedical Sciences—General — M,D*
Biophysics — M,D
Immunology — M,D

MICROBIOLOGY (UNIVERSITY OF ARKANSAS FOR MEDICAL SCIENCES)

Microbiology — M,D
Molecular Biology — M,D
Neurobiology — M,D*
Nutrition — M
Pathology — M
Pharmacology — M,D*
Physiology — M,D
Toxicology — M,D

UNIVERSITY OF BRIDGEPORT

Nutrition — M

THE UNIVERSITY OF BRITISH COLUMBIA

Anatomy — M,D
Biochemistry — M,D
Biological and Biomedical Sciences—General — M,D
Biopsychology — M,D
Botany — M,D
Cell Biology — M,D
Genetics — M,D
Immunology — M,D
Microbiology — M,D
Molecular Biology — M,D
Neuroscience — M,D
Nutrition — M,D
Pathology — M,D
Pharmacology — M,D
Physiology — M,D
Reproductive Biology — M,D
Zoology — M,D

UNIVERSITY OF CALGARY

Biochemistry — M,D
Biological and Biomedical Sciences—General — M,D
Cancer Biology/Oncology — M,D
Cardiovascular Sciences — M,D
Immunology — M,D
Infectious Diseases — M,D
Microbiology — M,D
Molecular Biology — M,D
Neuroscience — M,D

UNIVERSITY OF CALIFORNIA, BERKELEY

Biochemistry — M,D
Biological and Biomedical Sciences—General — D
Biophysics — D
Cell Biology — D
Immunology — D
Infectious Diseases — M,D
Microbiology — D
Molecular Biology — D
Molecular Toxicology — D
Neuroscience — D*
Nutrition — M,D*
Physiology — M,D
Plant Biology — D

UNIVERSITY OF CALIFORNIA, DAVIS

Animal Behavior — M,D
Biochemistry — M,D
Biophysics — M,D
Cell Biology — M,D
Developmental Biology — M,D
Ecology — M,D
Entomology — M,D
Evolutionary Biology — D
Genetics — M,D
Immunology — M,D
Microbiology — M,D*
Molecular Biology — M,D
Neuroscience — D*
Nutrition — M,D
Pathology — M,D
Pharmacology — M,D
Physiology — M,D
Plant Biology — M,D
Plant Pathology — M,D
Toxicology — M,D
Zoology — M

UNIVERSITY OF CALIFORNIA, IRVINE

Anatomy — M,D
Biochemistry — M,D
Biological and Biomedical Sciences—General — M,D
Biophysics — D
Cell Biology — M,D
Developmental Biology — M,D
Ecology — M,D
Evolutionary Biology — M,D
Genetics — D
Microbiology — M,D
Molecular Biology — M,D
Molecular Genetics — M,D
Neurobiology — M,D
Pharmacology — M,D*
Physiology — D
Toxicology — M,D

UNIVERSITY OF CALIFORNIA, LOS ANGELES

Anatomy — D
Biochemistry — M,D
Biological and Biomedical Sciences—General — M,D
Cell Biology — M,D*
Developmental Biology — M,D
Human Genetics — M,D
Immunology — M,D
Microbiology — M,D
Molecular Biology — M,D
Molecular Genetics — M,D
Molecular Toxicology — D*
Neurobiology — D
Neuroscience — M,D
Pathology — M,D
Pharmacology — D*
Physiology — M,D
Plant Molecular Biology — M,D

UNIVERSITY OF CALIFORNIA, RIVERSIDE

Biochemistry — M,D
Biological and Biomedical Sciences—General — M,D
Botany — M,D
Cell Biology — M,D
Developmental Biology — M,D
Entomology — M,D
Evolutionary Biology — M,D
Genetics — D
Genomic Sciences — D
Microbiology — M,D
Molecular Biology — M,D
Molecular Genetics — D
Neuroscience — D
Plant Biology — M,D
Plant Pathology — M,D
Toxicology — M,D

UNIVERSITY OF CALIFORNIA, SAN DIEGO

Biochemistry — M,D*
Biological and Biomedical Sciences—General — M,D*
Biophysics — M,D
Cancer Biology/Oncology — D
Cardiovascular Sciences — D
Cell Biology — D*
Developmental Biology — D
Ecology — D
Evolutionary Biology — D*
Genetics — D
Immunology — D
Marine Biology — M,D
Microbiology — D
Molecular Biology — D
Molecular Pathology — D*
Neurobiology — D*
Neuroscience — D*
Pharmacology — D*
Physiology — D
Plant Biology — D
Plant Molecular Biology — D
Structural Biology — D

Systems Biology — D*
Virology — D

UNIVERSITY OF CALIFORNIA, SAN FRANCISCO

Anatomy — D
Biochemistry — D
Biological and Biomedical Sciences—General — D
Biophysics — D
Cell Biology — D
Developmental Biology — D
Genetics — D
Genomic Sciences — D
Immunology — D
Microbiology — D
Molecular Biology — D
Neuroscience — D
Pathology — D
Pharmacology — D
Physiology — D

UNIVERSITY OF CALIFORNIA, SANTA BARBARA

Biochemistry — M,D
Biophysics — M,D
Cell Biology — M,D
Developmental Biology — M,D
Ecology — M,D
Evolutionary Biology — M,D
Marine Biology — M,D
Molecular Biology — M,D

UNIVERSITY OF CALIFORNIA, SANTA CRUZ

Biochemistry — M,D
Cell Biology — M,D
Ecology — M,D
Environmental Biology — M,D
Evolutionary Biology — M,D
Molecular Biology — M,D
Toxicology — M,D

UNIVERSITY OF CENTRAL ARKANSAS

Biological and Biomedical Sciences—General — M

UNIVERSITY OF CENTRAL FLORIDA

Biological and Biomedical Sciences—General — M,D,O
Conservation Biology — M,D,O
Microbiology — M
Molecular Biology — M,D

UNIVERSITY OF CENTRAL OKLAHOMA

Biological and Biomedical Sciences—General — M
Nutrition — M

UNIVERSITY OF CHICAGO

Anatomy — D
Biochemistry — D
Biological and Biomedical Sciences—General — P,M,D
Cancer Biology/Oncology — D
Cell Biology — D
Developmental Biology — D
Ecology — D
Evolutionary Biology — D
Genetics — D
Human Genetics — D
Immunology — D
Microbiology — D
Molecular Biology — D
Molecular Genetics — D
Molecular Physiology — D
Neurobiology — D
Neuroscience — D
Nutrition — D
Pathology — D
Pharmacology — D
Physiology — D

Zoology	D

UNIVERSITY OF CINCINNATI

Biochemistry	M,D
Biological and Biomedical Sciences—General	M,D*
Biophysics	D
Cell Biology	D
Developmental Biology	D*
Genomic Sciences	M,D
Immunology	M,D*
Microbiology	M,D
Molecular Biology	M,D
Molecular Genetics	M,D
Molecular Medicine	D
Molecular Toxicology	M,D
Neuroscience	D
Nutrition	M
Pathobiology	D
Pathology	D
Pharmacology	D
Physiology	D
Toxicology	D*

UNIVERSITY OF COLORADO AT BOULDER

Animal Behavior	M,D
Biochemistry	M,D
Cell Biology	M,D
Developmental Biology	M,D
Ecology	M,D
Evolutionary Biology	M,D
Genetics	M,D
Marine Biology	M,D
Microbiology	M,D
Molecular Biology	M,D
Neurobiology	M,D
Physiology	M,D

UNIVERSITY OF COLORADO AT DENVER AND HEALTH SCIENCES CENTER

Biochemistry	D
Biological and Biomedical Sciences—General	D
Biophysics	D
Cancer Biology/Oncology	D
Cell Biology	D
Developmental Biology	D
Genetics	D
Immunology	D*
Microbiology	D*
Molecular Biology	D*
Neuroscience	D*
Pharmacology	D
Physiology	D
Toxicology	P,D

UNIVERSITY OF COLORADO AT DENVER AND HEALTH SCIENCES CENTER—DOWNTOWN DENVER CAMPUS

Biological and Biomedical Sciences—General	M

UNIVERSITY OF CONNECTICUT

Biochemistry	M,D
Biological and Biomedical Sciences—General	M,D
Biophysics	M,D
Biopsychology	M,D
Botany	M,D
Cell Biology	M,D*
Developmental Biology	M,D
Ecology	M,D
Entomology	M,D
Genetics	M,D
Genomic Sciences	M
Microbiology	M,D
Molecular Biology	M
Neurobiology	M,D
Neuroscience	M,D
Nutrition	M,D
Pathobiology	M,D
Pharmacology	M,D*
Physiology	M,D*
Plant Biology	M,D

Plant Molecular Biology	M,D
Structural Biology	M,D
Toxicology	M,D
Zoology	M,D

UNIVERSITY OF CONNECTICUT HEALTH CENTER

Biochemistry	D
Biological and Biomedical Sciences—General	D*
Cell Biology	D*
Developmental Biology	D
Genetics	D*
Immunology	D*
Molecular Biology	D*
Molecular Pharmacology	D*
Neuroscience	D*

UNIVERSITY OF DAYTON

Biological and Biomedical Sciences—General	M,D*

UNIVERSITY OF DELAWARE

Biochemistry	M,D
Biological and Biomedical Sciences—General	M,D
Cancer Biology/Oncology	M,D
Cell Biology	M,D
Developmental Biology	M,D
Ecology	M,D
Entomology	M,D
Evolutionary Biology	M,D
Genetics	M,D
Microbiology	M,D
Molecular Biology	M,D
Neuroscience	D
Nutrition	M
Physiology	M,D

UNIVERSITY OF DENVER

Biological and Biomedical Sciences—General	M,D

UNIVERSITY OF DETROIT MERCY

Biochemistry	M

UNIVERSITY OF FLORIDA

Anatomy	D
Biochemistry	M,D
Biological and Biomedical Sciences—General	D*
Botany	M,D
Cell Biology	M,D
Ecology	M,D
Entomology	M,D
Genetics	D
Genomic Sciences	D
Immunology	D
Microbiology	M,D
Molecular Biology	M,D
Molecular Genetics	M,D
Neuroscience	M,D
Nutrition	M,D
Pathology	D
Pharmacology	M,D
Physiology	M,D
Plant Biology	M,D
Plant Molecular Biology	M,D
Plant Pathology	M,D
Toxicology	M,D,O
Zoology	M,D

UNIVERSITY OF GEORGIA

Anatomy	M
Biochemistry	M,D
Biological and Biomedical Sciences—General	D
Cell Biology	M,D
Ecology	M,D
Entomology	M,D
Genetics	M,D
Infectious Diseases	M,D
Medical Microbiology	M,D
Microbiology	M,D*
Molecular Biology	M,D
Neuroscience	D
Nutrition	M,D

Pathology	M,D
Pharmacology	M,D
Physiology	M,D
Plant Biology	M,D
Plant Pathology	M,D
Toxicology	M,D

UNIVERSITY OF GUAM

Biological and Biomedical Sciences—General	M
Marine Biology	M

UNIVERSITY OF GUELPH

Anatomy	M,D
Biochemistry	M,D
Biological and Biomedical Sciences—General	M,D
Biophysics	M,D
Botany	M,D
Cell Biology	M,D
Ecology	M,D
Entomology	M,D
Environmental Biology	M,D
Evolutionary Biology	M,D
Immunology	M,D,O
Infectious Diseases	M,D,O
Microbiology	M,D
Molecular Biology	M,D
Molecular Genetics	M,D
Neuroscience	M,D,O
Nutrition	M,D
Pathology	M,D,O
Pharmacology	M,D
Physiology	M,D
Plant Pathology	M,D
Toxicology	M,D
Zoology	M,D

UNIVERSITY OF HARTFORD

Biological and Biomedical Sciences—General	M
Neuroscience	M*

UNIVERSITY OF HAWAII AT MANOA

Anatomy	M,D
Biological and Biomedical Sciences—General	M,D
Botany	M,D
Conservation Biology	M,D
Ecology	M,D
Entomology	M,D
Evolutionary Biology	M,D
Genetics	M,D
Marine Biology	M,D
Medical Microbiology	M,D
Microbiology	M,D
Molecular Biology	M,D
Nutrition	M
Physiology	M,D
Plant Pathology	M,D
Reproductive Biology	M,D
Zoology	M,D

UNIVERSITY OF HOUSTON

Biochemistry	M,D
Biological and Biomedical Sciences—General	M,D
Pharmacology	P,M,D

UNIVERSITY OF HOUSTON–CLEAR LAKE

Biological and Biomedical Sciences—General	M

UNIVERSITY OF IDAHO

Biochemistry	M,D
Biological and Biomedical Sciences—General	M,D
Computational Biology	M,D
Entomology	M,D
Microbiology	M,D
Molecular Biology	M,D

Pathology	M,D
Pharmacology	M,D
Physiology	M,D
Plant Biology	M,D
Plant Pathology	M,D
Toxicology	M,D

UNIVERSITY OF ILLINOIS AT CHICAGO

Anatomy	M,D
Biochemistry	M,D
Biological and Biomedical Sciences—General	M,D
Biophysics	M,D
Cell Biology	M,D
Developmental Biology	M,D
Ecology	M,D
Evolutionary Biology	M,D
Genetics	M,D
Immunology	D
Microbiology	D*
Molecular Biology	M,D
Molecular Genetics	D
Neurobiology	M,D*
Nutrition	M,D
Pharmacology	D*
Physiology	M,D
Plant Biology	M,D

UNIVERSITY OF ILLINOIS AT SPRINGFIELD

Biological and Biomedical Sciences—General	M

UNIVERSITY OF ILLINOIS AT URBANA–CHAMPAIGN

Biochemistry	M,D*
Biological and Biomedical Sciences—General	M,D
Biophysics	D*
Biopsychology	M,D
Cell Biology	D
Computational Biology	D
Developmental Biology	D
Ecology	M,D
Entomology	M,D*
Evolutionary Biology	M,D
Microbiology	M,D*
Molecular Biology	
Neuroscience	D
Nutrition	M,D
Pathobiology	M,D
Physiology	M,D*
Plant Biology	M,D
Zoology	M,D

UNIVERSITY OF INDIANAPOLIS

Biological and Biomedical Sciences—General	M

THE UNIVERSITY OF IOWA

Anatomy	D
Bacteriology	M,D
Biochemistry	M,D*
Biological and Biomedical Sciences—General	M,D*
Biophysics	M,D
Cell Biology	M,D
Evolutionary Biology	M,D
Genetics	M,D,O*
Immunology	M,D
Microbiology	M,D*
Molecular Biology	D*
Neurobiology	M,D
Neuroscience	D*
Pathology	M
Pharmacology	M,D
Physiology	M,D
Plant Biology	M,D
Radiation Biology	M,D*
Virology	M,D

UNIVERSITY OF KANSAS

Anatomy	M,D
Biochemistry	M,D*
Biological and Biomedical Sciences—General	M,D*
Biophysics	M,D
Botany	M,D
Cell Biology	M,D
Developmental Biology	M,D
Ecology	M,D
Entomology	M,D
Evolutionary Biology	M,D

Immunology — D
Microbiology — M,D
Molecular Biology — M,D
Molecular Genetics — D
Neuroscience — M,D
Nutrition — M,O
Pathology — M,D
Pharmacology — M,D
Physiology — M,D
Toxicology — M,D

UNIVERSITY OF KENTUCKY

Anatomy — D
Biochemistry — D
Biological and Biomedical
  Sciences—General — M,D
Entomology — M,D
Microbiology — D*
Neurobiology — D
Nutrition — M,D*
Pharmacology — D
Physiology — M,D
Plant Pathology — M,D
Plant Physiology — D
Toxicology — M,D*

UNIVERSITY OF LETHBRIDGE

Biochemistry — M,D
Biological and Biomedical
  Sciences—General — M,D
Molecular Biology — M,D
Neuroscience — M,D

UNIVERSITY OF LOUISIANA AT LAFAYETTE

Biological and Biomedical
  Sciences—General — M,D
Environmental Biology — M,D
Evolutionary Biology — M,D

UNIVERSITY OF LOUISIANA AT MONROE

Biological and Biomedical
  Sciences—General — M

UNIVERSITY OF LOUISVILLE

Anatomy — M,D
Biochemistry — M,D
Biological and Biomedical
  Sciences—General — M*
Biophysics — M,D
Environmental Biology — D
Immunology — M,D
Microbiology — M,D
Molecular Biology — M,D
Neurobiology — M,D
Pharmacology — M,D
Physiology — M,D
Toxicology — M,D

UNIVERSITY OF MAINE

Biochemistry — M,D
Biological and Biomedical
  Sciences—General — D
Botany — M
Ecology — M,D
Entomology — M
Marine Biology — M,D
Microbiology — M,D
Molecular Biology — M,D
Nutrition — M,D
Plant Biology — M,D
Plant Pathology — M
Zoology — M,D

UNIVERSITY OF MANITOBA

Anatomy — M,D
Biochemistry — M,D
Biological and Biomedical
  Sciences—General — M,D
Botany — M,D
Entomology — M,D
Human Genetics — M,D
Immunology — M,D
Medical Microbiology — M,D*

Microbiology — M,D
Nutrition — M,D
Pathology — M
Pharmacology — M,D
Physiology — M,D
Zoology — M,D

UNIVERSITY OF MARYLAND

Anatomy — M,D
Biochemistry — D*
Biological and Biomedical
  Sciences—General — M,D
Cardiovascular Sciences
Cell Biology — M,D
Human Genetics — M,D
Immunology — M,D
Microbiology — M,D*
Molecular Biology — D
Molecular Medicine — D
Neurobiology — M,D
Neuroscience — M,D
Pathology — M,D
Pharmacology — M,D
Physiology — M,D
Toxicology — M,D

UNIVERSITY OF MARYLAND, BALTIMORE COUNTY

Biochemistry — D*
Biological and Biomedical
  Sciences—General — M,D
Cell Biology — D
Molecular Biology — M,D
Neuroscience — D

UNIVERSITY OF MARYLAND, COLLEGE PARK

Biochemistry — M,D
Biological and Biomedical
  Sciences—General — M,D
Cell Biology — M,D
Conservation Biology — M
Ecology — M,D
Entomology — M,D
Evolutionary Biology — M,D
Molecular Biology — D
Molecular Genetics — M,D
Neuroscience — M,D*
Nutrition — M,D
Plant Biology — M,D

UNIVERSITY OF MARYLAND EASTERN SHORE

Toxicology — M,D

UNIVERSITY OF MASSACHUSETTS AMHERST

Biochemistry — M,D
Biological and Biomedical
  Sciences—General — M,D
Cell Biology — D
Developmental Biology — M,D
Entomology — M,D
Environmental Biology — M,D
Evolutionary Biology — M,D
Microbiology — M,D*
Molecular Biology — D*
Neuroscience — M,D
Nutrition — M,D
Plant Biology — M,D*
Plant Molecular Biology — M,D
Plant Physiology — M,D

UNIVERSITY OF MASSACHUSETTS BOSTON

Biological and Biomedical
  Sciences—General — M
Cell Biology — D
Environmental Biology — D
Molecular Biology — D

UNIVERSITY OF MASSACHUSETTS DARTMOUTH

Biological and Biomedical
  Sciences—General — M

Marine Biology — M

UNIVERSITY OF MASSACHUSETTS LOWELL

Biochemistry — M,D
Biological and Biomedical
  Sciences—General — M,D

UNIVERSITY OF MASSACHUSETTS WORCESTER

Biochemistry — D
Biological and Biomedical
  Sciences—General — D*
Cancer Biology/Oncology — D
Cell Biology — D*
Immunology — D*
Microbiology — D
Molecular Genetics — D*
Molecular Pharmacology — D
Neuroscience — D
Physiology — D*
Virology — D

UNIVERSITY OF MEDICINE AND DENTISTRY OF NEW JERSEY

Biochemistry — M,D
Biological and Biomedical
  Sciences—General — M,D,O
Cancer Biology/Oncology — D
Cardiovascular Sciences — M,D
Cell Biology — M,D
Immunology — M,D
Microbiology — M,D*
Molecular Biology — M,D*
Molecular Genetics — M,D
Molecular Medicine — D
Molecular Pathology — D
Molecular Pharmacology — M,D
Neuroscience — M,D
Nutrition — M,D,O
Pathology — D
Pharmacology — D,O
Physiology — M,D

UNIVERSITY OF MEMPHIS

Biological and Biomedical
  Sciences—General — M,D
Nutrition — M

UNIVERSITY OF MIAMI

Biochemistry — D
Biological and Biomedical
  Sciences—General — M,D*
Biophysics — D
Cancer Biology/Oncology — D*
Cell Biology — D
Developmental Biology — D
Ecology — M,D
Evolutionary Biology — M,D
Genetics — M,D
Immunology — D
Marine Biology — M,D
Microbiology — D*
Molecular Biology — D
Neuroscience — M,D*
Pharmacology — D*
Physiology — D

UNIVERSITY OF MICHIGAN

Biochemistry — D
Biological and Biomedical
  Sciences—General — M,D*
Biophysics — D
Biopsychology — D
Cell Biology — M,D*
Developmental Biology — M,D
Ecology — M,D
Evolutionary Biology — M,D
Genetics
Human Genetics — M,D
Immunology — D
Microbiology — D
Molecular Biology — M,D
Molecular Biophysics
Neuroscience — D

Nutrition — M
Pathology — D*
Pharmacology — D
Physiology — D
Toxicology — M,D

UNIVERSITY OF MICHIGAN–FLINT

Biological and Biomedical
  Sciences—General — M

UNIVERSITY OF MINNESOTA, DULUTH

Biochemistry — M,D
Biological and Biomedical
  Sciences—General — M
Biophysics — M,D
Immunology — M,D
Medical Microbiology — M,D
Molecular Biology — M,D
Pharmacology — M,D
Physiology — M,D
Toxicology — M,D

UNIVERSITY OF MINNESOTA, TWIN CITIES CAMPUS

Animal Behavior — M,D
Biochemistry — D
Biological and Biomedical
  Sciences—General — M,D
Biophysics — M,D
Biopsychology — D
Cell Biology — M,D
Conservation Biology — M,D
Developmental Biology — M,D
Ecology — M,D
Entomology — M,D
Evolutionary Biology — M,D
Genetics — M,D
Infectious Diseases — M,D
Microbiology — D*
Molecular Biology — M,D
Neuroscience — M,D
Nutrition — M,D
Pharmacology — M,D
Physiology — M,D
Plant Biology — M,D
Plant Pathology — M,D
Toxicology — M,D

UNIVERSITY OF MISSISSIPPI

Biological and Biomedical
  Sciences—General — M,D
Pharmacology — M,D
Toxicology — M,D

UNIVERSITY OF MISSISSIPPI MEDICAL CENTER

Anatomy — M,D
Biochemistry — M,D*
Biological and Biomedical
  Sciences—General — M,D*
Biophysics — M,D
Microbiology — M,D
Pathology — M,D
Pharmacology — M,D*
Physiology — M,D
Toxicology — M,D

UNIVERSITY OF MISSOURI–COLUMBIA

Biochemistry — M,D
Biological and Biomedical
  Sciences—General — M,D
Biopsychology — M,D
Cell Biology — M,D
Ecology — M,D
Entomology — M,D
Evolutionary Biology — M,D
Genetics — D
Immunology — M,D
Microbiology — M,D
Molecular Biology — M,D
Neurobiology — M,D
Nutrition — M,D*
Pathobiology — M,D

Pharmacology	M,D
Physiology	M,D
Plant Biology	M,D
Plant Pathology	M,D

UNIVERSITY OF MISSOURI–KANSAS CITY

Biochemistry	D
Biological and Biomedical Sciences—General	M,D
Biophysics	D
Cell Biology	D*
Molecular Biology	D*

UNIVERSITY OF MISSOURI–ROLLA

Biological and Biomedical Sciences—General	M
Environmental Biology	M

UNIVERSITY OF MISSOURI–ST. LOUIS

Animal Behavior	M,D,O
Biochemistry	M,D,O
Biological and Biomedical Sciences—General	M,D,O
Botany	M,D,O
Cell Biology	M,D,O
Conservation Biology	M,D,O
Developmental Biology	M,D,O
Ecology	M,D,O
Evolutionary Biology	M,D,O
Genetics	M,D,O
Molecular Biology	M,D,O
Physiology	M,D,O

THE UNIVERSITY OF MONTANA

Animal Behavior	M,D,O
Biochemistry	M,D
Biological and Biomedical Sciences—General	M,D
Ecology	M,D
Microbiology	M,D
Pharmacology	M,D
Toxicology	M,D
Zoology	M,D

UNIVERSITY OF NEBRASKA AT KEARNEY

Biological and Biomedical Sciences—General	M

UNIVERSITY OF NEBRASKA AT OMAHA

Biological and Biomedical Sciences—General	M
Biopsychology	M,D,O

UNIVERSITY OF NEBRASKA–LINCOLN

Biochemistry	M,D
Biological and Biomedical Sciences—General	M,D
Entomology	M,D
Nutrition	M,D
Toxicology	M,D

UNIVERSITY OF NEBRASKA MEDICAL CENTER

Anatomy	M,D
Biochemistry	M,D
Biological and Biomedical Sciences—General	M,D*
Cancer Biology/Oncology	M,D
Cell Biology	M,D
Microbiology	M,D
Molecular Biology	M,D
Neuroscience	M,D
Nutrition	O
Pathology	M,D
Pharmacology	M,D
Physiology	M,D
Toxicology	M,D

UNIVERSITY OF NEVADA, LAS VEGAS

Biochemistry	M,D
Biological and Biomedical Sciences—General	M,D

UNIVERSITY OF NEVADA, RENO

Biochemistry	M,D*
Biological and Biomedical Sciences—General	M,D
Cell Biology	M,D
Conservation Biology	D
Ecology	D
Evolutionary Biology	D
Molecular Biology	M,D
Molecular Pharmacology	M,D
Nutrition	M
Physiology	M,D

UNIVERSITY OF NEW BRUNSWICK FREDERICTON

Biological and Biomedical Sciences—General	M,D

UNIVERSITY OF NEW BRUNSWICK SAINT JOHN

Biological and Biomedical Sciences—General	M,D

UNIVERSITY OF NEW ENGLAND

Biological and Biomedical Sciences—General	M

UNIVERSITY OF NEW HAMPSHIRE

Biochemistry	M,D
Genetics	M,D
Microbiology	M,D
Molecular Biology	M,D
Nutrition	M,D
Plant Biology	M,D
Zoology	M,D

UNIVERSITY OF NEW HAVEN

Cell Biology	M*
Molecular Biology	M
Nutrition	M

UNIVERSITY OF NEW MEXICO

Biochemistry	M,D
Biological and Biomedical Sciences—General	M,D
Biophysics	M,D
Cell Biology	M,D
Genetics	M,D
Microbiology	M,D
Molecular Biology	M,D
Neuroscience	M,D
Nutrition	M
Pathology	M,D
Physiology	M,D
Toxicology	M,D

UNIVERSITY OF NEW ORLEANS

Biological and Biomedical Sciences—General	M,D

THE UNIVERSITY OF NORTH CAROLINA AT CHAPEL HILL

Biochemistry	M,D
Biological and Biomedical Sciences—General	M,D*
Biophysics	M,D
Botany	M,D*
Cancer Biology/Oncology	
Cell Biology	D*
Developmental Biology	M,D
Ecology	M,D
Evolutionary Biology	M,D
Genetics	M,D
Immunology	M,D
Microbiology	M,D
Molecular Biology	M,D
Molecular Physiology	D
Neurobiology	D*
Nutrition	M,D

Pathology	D
Pharmacology	D
Toxicology	M,D*

THE UNIVERSITY OF NORTH CAROLINA AT CHARLOTTE

Biological and Biomedical Sciences—General	M,D

THE UNIVERSITY OF NORTH CAROLINA AT GREENSBORO

Biochemistry	M
Biological and Biomedical Sciences—General	M
Nutrition	M,D

THE UNIVERSITY OF NORTH CAROLINA WILMINGTON

Biological and Biomedical Sciences—General	M,D
Marine Biology	M,D*

UNIVERSITY OF NORTH DAKOTA

Anatomy	M,D
Biochemistry	M,D
Biological and Biomedical Sciences—General	M,D
Botany	M,D
Ecology	M,D
Entomology	M,D
Environmental Biology	M,D
Genetics	M,D
Immunology	M,D
Microbiology	M,D
Pharmacology	M,D
Physiology	M,D
Zoology	M,D

UNIVERSITY OF NORTHERN COLORADO

Biological and Biomedical Sciences—General	M,D

UNIVERSITY OF NORTHERN IOWA

Biological and Biomedical Sciences—General	M

UNIVERSITY OF NORTH FLORIDA

Biological and Biomedical Sciences—General	M
Nutrition	M,O

UNIVERSITY OF NORTH TEXAS

Biochemistry	M,D
Biological and Biomedical Sciences—General	M,D
Molecular Biology	M,D

UNIVERSITY OF NORTH TEXAS HEALTH SCIENCE CENTER AT FORT WORTH

Anatomy	M,D
Biochemistry	M,D
Biological and Biomedical Sciences—General	M,D*
Genetics	M,D
Immunology	M,D
Microbiology	M,D
Molecular Biology	M,D
Pharmacology	M,D
Physiology	M,D

UNIVERSITY OF NOTRE DAME

Biochemistry	M,D
Biological and Biomedical Sciences—General	M,D*
Cell Biology	M,D
Ecology	M,D
Evolutionary Biology	M,D
Genetics	M,D
Molecular Biology	M,D
Parasitology	M,D
Physiology	M,D

UNIVERSITY OF OKLAHOMA

Biochemistry	M,D
Botany	M,D
Microbiology	M,D
Zoology	M,D*

UNIVERSITY OF OKLAHOMA HEALTH SCIENCES CENTER

Biochemistry	M,D
Biological and Biomedical Sciences—General	M,D
Biopsychology	M,D
Cell Biology	M,D
Immunology	M,D
Microbiology	M,D
Molecular Biology	M,D
Neuroscience	M,D
Nutrition	M
Pathology	D
Physiology	M,D
Radiation Biology	M,D

UNIVERSITY OF OREGON

Biochemistry	M,D
Biological and Biomedical Sciences—General	M,D
Biopsychology	M,D
Ecology	M,D
Evolutionary Biology	M,D
Genetics	M,D
Marine Biology	M,D
Molecular Biology	M,D
Neuroscience	M,D
Physiology	M,D

UNIVERSITY OF OTTAWA

Biochemistry	M,D
Biological and Biomedical Sciences—General	M,D
Cell Biology	M,D
Immunology	M,D
Microbiology	M,D
Molecular Biology	M,D

UNIVERSITY OF PENNSYLVANIA

Biochemistry	D
Biological and Biomedical Sciences—General	M,D
Cancer Biology/Oncology	D
Cell Biology	D
Computational Biology	D
Developmental Biology	D
Ecology	D
Evolutionary Biology	D
Genetics	D
Genomic Sciences	D
Immunology	D*
Microbiology	D
Molecular Biology	D
Molecular Biophysics	D
Neurobiology	D
Neuroscience	D
Parasitology	D
Pharmacology	D
Physiology	D
Plant Biology	D
Structural Biology	D
Virology	D

UNIVERSITY OF PITTSBURGH

Biochemistry	M,D
Biological and Biomedical Sciences—General	M,D*
Cell Biology	M,D
Computational Biology	D
Developmental Biology	D*
Ecology	M,D*
Evolutionary Biology	M,D
Human Genetics	M,D
Immunology	M,D
Infectious Diseases	M,D
Microbiology	M,D*
Molecular Biology	D*
Molecular Biophysics	D
Molecular Genetics	M,D
Molecular Pathology	M,D
Molecular Pharmacology	D

Molecular Physiology	M,D
Molecular Toxicology	M,D,O
Neurobiology	M,D
Neuroscience	D*
Nutrition	M
Pathology	M,D
Structural Biology	D
Virology	M,D

UNIVERSITY OF PRINCE EDWARD ISLAND

Anatomy	M,D
Bacteriology	M,D
Biological and Biomedical Sciences—General	M
Immunology	M,D
Parasitology	M,D
Pathology	M,D
Pharmacology	M,D
Physiology	M,D
Toxicology	M,D
Virology	M,D

UNIVERSITY OF PUERTO RICO, MAYAGÜEZ CAMPUS

Biological and Biomedical Sciences—General	M

UNIVERSITY OF PUERTO RICO, MEDICAL SCIENCES CAMPUS

Anatomy	M,D
Biochemistry	M,D
Biological and Biomedical Sciences—General	M,D*
Microbiology	M,D
Nutrition	M,D,O
Pharmacology	M,D
Physiology	M,D
Toxicology	M,D
Zoology	M,D

UNIVERSITY OF PUERTO RICO, RÍO PIEDRAS

Biological and Biomedical Sciences—General	M,D
Nutrition	M

UNIVERSITY OF REGINA

Biochemistry	M,D
Biological and Biomedical Sciences—General	M,D

UNIVERSITY OF RHODE ISLAND

Biochemistry	M,D
Biological and Biomedical Sciences—General	M,D
Cell Biology	M,D
Entomology	M,D
Microbiology	M,D
Molecular Biology	M,D
Molecular Genetics	M,D
Nutrition	M,D
Pharmacology	M,D
Plant Pathology	M,D
Toxicology	M,D

UNIVERSITY OF RICHMOND

Biological and Biomedical Sciences—General	M

UNIVERSITY OF ROCHESTER

Anatomy	M,D
Biochemistry	M,D
Biological and Biomedical Sciences—General	M,D*
Biophysics	M,D
Computational Biology	M,D
Genetics	M,D
Immunology	M,D
Microbiology	M,D
Molecular Medicine	
Neurobiology	M,D
Neuroscience	M,D
Pathobiology	

Pathology	M,D*
Pharmacology	M,D
Physiology	M,D
Toxicology	M,D

UNIVERSITY OF ST. MICHAEL'S COLLEGE

Ecology	P,M,D,O

UNIVERSITY OF SAN FRANCISCO

Biological and Biomedical Sciences—General	M

UNIVERSITY OF SASKATCHEWAN

Anatomy	M,D
Biochemistry	M,D,O
Biological and Biomedical Sciences—General	M,D,O
Cell Biology	M,D
Microbiology	M,D
Pathology	M,D
Pharmacology	M,D
Physiology	M,D
Reproductive Biology	M,D
Toxicology	M,D,O

THE UNIVERSITY OF SCRANTON

Biochemistry	M

UNIVERSITY OF SOUTH ALABAMA

Biochemistry	D
Biological and Biomedical Sciences—General	M,D*
Cell Biology	D
Immunology	D
Microbiology	D
Molecular Biology	D
Neuroscience	D
Pharmacology	D
Physiology	D
Toxicology	M

UNIVERSITY OF SOUTH CAROLINA

Biochemistry	M,D
Biological and Biomedical Sciences—General	M,D,O*
Cell Biology	M,D
Developmental Biology	M,D
Ecology	M,D*
Evolutionary Biology	M,D
Molecular Biology	M,D*

THE UNIVERSITY OF SOUTH DAKOTA

Biological and Biomedical Sciences—General	M,D
Cardiovascular Sciences	M,D
Cell Biology	M,D
Immunology	M,D
Microbiology	M,D
Molecular Biology	M,D
Neuroscience	M,D
Pharmacology	M,D
Physiology	M,D

UNIVERSITY OF SOUTHERN CALIFORNIA

Biochemistry	M,D
Biological and Biomedical Sciences—General	M,D*
Biophysics	M,D
Cell Biology	M,D
Computational Biology	D
Genetics	M,D
Immunology	M,D
Marine Biology	D
Microbiology	M,D
Molecular Biology	M,D*
Molecular Pharmacology	M,D*
Neurobiology	M,D
Neuroscience	D*
Nutrition	M
Pathobiology	M,D*

Pathology	M,D
Pharmacology	M,D
Physiology	M,D
Toxicology	M,D

UNIVERSITY OF SOUTHERN MAINE

Biological and Biomedical Sciences—General	M
Immunology	M
Molecular Biology	M

UNIVERSITY OF SOUTHERN MISSISSIPPI

Biochemistry	M,D
Biological and Biomedical Sciences—General	M,D*
Environmental Biology	M,D
Marine Biology	M,D
Microbiology	M,D
Molecular Biology	M,D
Nutrition	M,D

UNIVERSITY OF SOUTH FLORIDA

Anatomy	M,D
Biochemistry	M,D
Biological and Biomedical Sciences—General	M,D
Biophysics	M,D
Botany	M,D
Cancer Biology/Oncology	D*
Ecology	M,D
Immunology	M,D
Medical Microbiology	M,D
Microbiology	M,D
Molecular Biology	M,D
Pathology	M,D
Pharmacology	M,D
Physiology	M,D
Zoology	M,D

THE UNIVERSITY OF TENNESSEE

Anatomy	M,D
Animal Behavior	M,D
Biochemistry	M,D
Biological and Biomedical Sciences—General	M,D
Ecology	M,D
Entomology	M,D*
Evolutionary Biology	M,D
Genetics	M,D
Genomic Sciences	M,D
Microbiology	M,D
Nutrition	M
Physiology	M,D
Plant Pathology	M,D
Plant Physiology	M,D

THE UNIVERSITY OF TENNESSEE AT MARTIN

Nutrition	M

THE UNIVERSITY OF TENNESSEE HEALTH SCIENCE CENTER

Anatomy	D
Bacteriology	D
Biochemistry	D
Biological and Biomedical Sciences—General	M,D*
Biophysics	M,D
Cell Biology	D
Immunology	D
Microbiology	D
Molecular Biology	D
Neurobiology	D*
Pathology	M,D
Pharmacology	M,D
Physiology	M,D
Structural Biology	D
Virology	D

THE UNIVERSITY OF TENNESSEE– OAK RIDGE NATIONAL LABORATORY GRADUATE SCHOOL OF GENOME SCIENCE AND TECHNOLOGY

Biological and Biomedical Sciences—General	M,D
Genomic Sciences	M,D*

THE UNIVERSITY OF TEXAS AT ARLINGTON

Biological and Biomedical Sciences—General	M,D

THE UNIVERSITY OF TEXAS AT AUSTIN

Animal Behavior	D
Biochemistry	M,D
Biological and Biomedical Sciences—General	M,D
Biopsychology	M,D
Cell Biology	D
Developmental Biology	D
Ecology	D
Evolutionary Biology	D
Genetics	D
Immunology	D
Microbiology	M,D*
Molecular Biology	D*
Neurobiology	M,D
Neuroscience	M,D*
Nutrition	M,D
Plant Biology	M,D

THE UNIVERSITY OF TEXAS AT BROWNSVILLE

Biological and Biomedical Sciences—General	M

THE UNIVERSITY OF TEXAS AT DALLAS

Biological and Biomedical Sciences—General	M,D
Cell Biology	M,D
Molecular Biology	M,D*
Neuroscience	M,D

THE UNIVERSITY OF TEXAS AT EL PASO

Biological and Biomedical Sciences—General	M,D

THE UNIVERSITY OF TEXAS AT SAN ANTONIO

Biological and Biomedical Sciences—General	M,D*
Cell Biology	D
Molecular Biology	D
Neurobiology	D

THE UNIVERSITY OF TEXAS AT TYLER

Biological and Biomedical Sciences—General	M

THE UNIVERSITY OF TEXAS HEALTH SCIENCE CENTER AT HOUSTON

Biochemistry	M,D
Biological and Biomedical Sciences—General	M,D*
Cancer Biology/Oncology	M,D
Cell Biology	M,D
Developmental Biology	M,D
Genetics	M,D
Human Genetics	M,D
Immunology	M,D
Microbiology	M,D
Molecular Biology	M,D
Molecular Genetics	M,D
Molecular Pathology	M,D
Neuroscience	M,D
Toxicology	M,D
Virology	M,D

THE UNIVERSITY OF TEXAS HEALTH SCIENCE CENTER AT SAN ANTONIO

Biochemistry	M,D
Biological and Biomedical Sciences—General	P,M,D,O
Cell Biology	D
Immunology	M,D
Microbiology	M,D*
Molecular Medicine	M,D*
Pharmacology	D
Physiology	M,D
Structural Biology	D

THE UNIVERSITY OF TEXAS MEDICAL BRANCH

Bacteriology	D
Biochemistry	D
Biological and Biomedical Sciences—General	M,D
Biophysics	D
Cell Biology	D
Computational Biology	D
Genetics	D
Immunology	M,D
Infectious Diseases	D*
Microbiology	M,D*
Molecular Biophysics	M,D
Neuroscience	D
Pathology	D*
Pharmacology	M,D
Physiology	M,D
Structural Biology	D
Toxicology	D
Virology	D

THE UNIVERSITY OF TEXAS OF THE PERMIAN BASIN

Biological and Biomedical Sciences—General	M

THE UNIVERSITY OF TEXAS–PAN AMERICAN

Biological and Biomedical Sciences—General	M

THE UNIVERSITY OF TEXAS SOUTHWESTERN MEDICAL CENTER AT DALLAS

Biochemistry	D
Biological and Biomedical Sciences—General	M,D*
Cell Biology	D
Developmental Biology	D
Genetics	D
Immunology	D
Microbiology	D
Molecular Biophysics	D
Neuroscience	D
Pharmacology	D
Radiation Biology	M,D

UNIVERSITY OF THE INCARNATE WORD

Biological and Biomedical Sciences—General	M
Nutrition	M,O

UNIVERSITY OF THE PACIFIC

Biological and Biomedical Sciences—General	M

UNIVERSITY OF THE SCIENCES IN PHILADELPHIA

Biochemistry	M,D
Cell Biology	M
Pharmacology	M,D
Toxicology	M,D

THE UNIVERSITY OF TOLEDO

Anatomy	M
Biochemistry	M,D*
Biological and Biomedical Sciences—General	M,D*
Cell Biology	D
Ecology	M,D*
Genetics	M
Microbiology	M
Molecular Biology	M,D
Neuroscience	M,D
Pathology	M
Pharmacology	M
Physiology	M

UNIVERSITY OF TORONTO

Biochemistry	M,D
Biological and Biomedical Sciences—General	M,D,O
Biophysics	M,D
Botany	M,D
Genetics	M,D
Immunology	M,D
Nutrition	M,D
Pathobiology	M,D
Pharmacology	M,D
Physiology	M,D
Zoology	M,D

UNIVERSITY OF TULSA

Biological and Biomedical Sciences—General	M,D

UNIVERSITY OF UTAH

Anatomy	M,D
Biochemistry	M,D*
Biological and Biomedical Sciences—General	M,D*
Cancer Biology/Oncology	M,D
Cell Biology	
Ecology	M,D
Evolutionary Biology	M,D
Genetics	M,D
Human Genetics	M,D
Immunology	
Microbiology	
Molecular Biology	D*
Neurobiology	M,D
Neuroscience	D*
Nutrition	M
Pathology	M,D
Pharmacology	M,D
Physiology	D
Toxicology	M,D

UNIVERSITY OF VERMONT

Anatomy	D
Biochemistry	M,D*
Biological and Biomedical Sciences—General	M,D*
Biophysics	M,D
Botany	M,D
Cell Biology	M,D*
Microbiology	M,D
Molecular Biology	M,D
Molecular Genetics	M,D*
Molecular Physiology	M,D
Neurobiology	D
Neuroscience	D
Nutrition	M
Pathology	M
Pharmacology	M,D

UNIVERSITY OF VICTORIA

Biochemistry	M,D
Biological and Biomedical Sciences—General	M,D
Microbiology	M,D

UNIVERSITY OF VIRGINIA

Bacteriology	D
Biochemistry	D
Biological and Biomedical Sciences—General	M,D
Biophysics	M,D
Cardiovascular Sciences	
Cell Biology	D
Immunology	D*
Microbiology	D*
Molecular Genetics	D
Molecular Medicine	D
Molecular Physiology	M,D
Neuroscience	D

Pharmacology	D
Physiology	D
Systems Biology	D
Virology	D

UNIVERSITY OF WASHINGTON

Bacteriology	M,D
Biochemistry	D
Biological and Biomedical Sciences—General	M,D
Biophysics	D
Botany	M,D
Cell Biology	D*
Genetics	M,D
Genomic Sciences	D
Immunology	D
Microbiology	D
Molecular Biology	D
Molecular Medicine	M,D
Neurobiology	D
Neuroscience	D
Nutrition	M,D
Parasitology	M,D
Pathobiology	M,D*
Pathology	M,D
Pharmacology	M,D*
Physiology	D
Structural Biology	D
Toxicology	M,D
Zoology	D

UNIVERSITY OF WATERLOO

Biological and Biomedical Sciences—General	M,D

THE UNIVERSITY OF WESTERN ONTARIO

Anatomy	M,D
Biochemistry	M,D
Biological and Biomedical Sciences—General	M,D
Biophysics	M,D
Cell Biology	M,D
Immunology	M,D
Microbiology	M,D
Molecular Biology	M,D
Neuroscience	M,D
Pathology	M,D
Physiology	M,D
Plant Biology	M,D
Zoology	M,D

UNIVERSITY OF WEST FLORIDA

Biochemistry	M
Biological and Biomedical Sciences—General	M
Environmental Biology	M

UNIVERSITY OF WEST GEORGIA

Biological and Biomedical Sciences—General	M

UNIVERSITY OF WINDSOR

Biochemistry	M,D
Biological and Biomedical Sciences—General	M,D
Biopsychology	M,D

UNIVERSITY OF WISCONSIN–EAU CLAIRE

Biological and Biomedical Sciences—General	M

UNIVERSITY OF WISCONSIN–LA CROSSE

Biological and Biomedical Sciences—General	M
Cell Biology	M
Medical Microbiology	M
Microbiology	M
Molecular Biology	M
Physiology	M

UNIVERSITY OF WISCONSIN–MADISON

Anatomy	M,D

Bacteriology	M
Biochemistry	M,D*
Biological and Biomedical Sciences—General	M,D
Biophysics	D
Biopsychology	D
Botany	M,D
Cancer Biology/Oncology	D
Cell Biology	M,D*
Conservation Biology	M
Ecology	M,D
Entomology	M,D
Environmental Biology	M,D
Genetics	M,D*
Medical Microbiology	D
Microbiology	D*
Molecular Biology	M,D
Neurobiology	D
Neuroscience	M,D
Nutrition	M,D
Pathology	D*
Pharmacology	M,D*
Physiology	M,D
Plant Pathology	M,D
Toxicology	M,D
Zoology	M,D

UNIVERSITY OF WISCONSIN–MILWAUKEE

Biological and Biomedical Sciences—General	M,D

UNIVERSITY OF WISCONSIN–OSHKOSH

Biological and Biomedical Sciences—General	M
Botany	M
Microbiology	M
Zoology	M

UNIVERSITY OF WISCONSIN–PARKSIDE

Molecular Biology	M

UNIVERSITY OF WISCONSIN–STEVENS POINT

Nutrition	M

UNIVERSITY OF WISCONSIN–STOUT

Nutrition	M

UNIVERSITY OF WYOMING

Botany	M,D
Entomology	M,D
Molecular Biology	M,D
Neuroscience	D
Nutrition	M
Pathobiology	M
Physiology	M,D
Reproductive Biology	M,D
Zoology	M,D

UTAH STATE UNIVERSITY

Biochemistry	M,D
Biological and Biomedical Sciences—General	M,D
Ecology	M,D
Microbiology	M,D
Molecular Biology	M,D
Nutrition	M,D
Toxicology	M,D

VANDERBILT UNIVERSITY

Biochemistry	M,D
Biological and Biomedical Sciences—General	M,D*
Biophysics	M,D
Cancer Biology/Oncology	M,D
Cell Biology	M,D
Immunology	M,D
Microbiology	M,D
Molecular Biology	M,D
Molecular Biophysics	
Molecular Physiology	M,D
Neuroscience	D

Pathology D
Pharmacology D
Toxicology M

VILLANOVA UNIVERSITY

Biological and Biomedical
  Sciences—General M*

VIRGINIA COMMONWEALTH UNIVERSITY

Anatomy M,D,O*
Biochemistry M,D,O*
Biological and Biomedical
  Sciences—General M,D,O*
Genetics M,D,O
Human Genetics M,D,O
Immunology M,D,O
Microbiology M,D,O
Molecular Biology M,D,O
Molecular Biophysics M,D,O
Neuroscience M,D,O*
Pathology M,D*
Pharmacology M,D,O*
Physiology M,D,O
Toxicology M,D,O

VIRGINIA POLYTECHNIC INSTITUTE AND STATE UNIVERSITY

Biochemistry M,D*
Biological and Biomedical
  Sciences—General M,D
Botany M,D
Computational Biology D
Developmental Biology M,D
Ecology M,D
Entomology M,D
Evolutionary Biology M,D
Genetics M,D
Microbiology M,D
Nutrition M,D
Plant Pathology M,D
Plant Physiology M,D
Zoology M,D

VIRGINIA STATE UNIVERSITY

Biological and Biomedical
  Sciences—General M

WAGNER COLLEGE

Biological and Biomedical
  Sciences—General M
Microbiology M

WAKE FOREST UNIVERSITY

Anatomy D
Biochemistry D*
Biological and Biomedical
  Sciences—General M,D
Cancer Biology/Oncology D*
Genomic Sciences D
Human Genetics D
Immunology D
Microbiology D*
Molecular Biology D
Molecular Genetics D*
Molecular Medicine M,D*
Neurobiology D
Neuroscience D
Pathobiology M,D*
Pharmacology D
Physiology D

WALLA WALLA COLLEGE

Biological and Biomedical
  Sciences—General M

WASHINGTON STATE UNIVERSITY

Biochemistry M,D
Biological and Biomedical
  Sciences—General M
Biophysics M,D
Botany M,D
Cell Biology M,D
Entomology M,D
Genetics M,D
Microbiology M,D
Molecular Biology M,D
Neuroscience M,D
Nutrition M,D
Pathology M,D
Pharmacology M,D
Plant Molecular Biology M,D
Plant Pathology M,D
Toxicology M,D
Zoology M,D

WASHINGTON STATE UNIVERSITY SPOKANE

Physiology M,O

WASHINGTON STATE UNIVERSITY TRI-CITIES

Biological and Biomedical
  Sciences—General M
Toxicology M,D

WASHINGTON UNIVERSITY IN ST. LOUIS

Biochemistry D
Biological and Biomedical
  Sciences—General D
Cell Biology D
Computational Biology D
Developmental Biology D
Ecology D
Environmental Biology D
Evolutionary Biology D
Genetics M,D,O
Immunology D
Microbiology D
Molecular Biology D
Molecular Biophysics D
Molecular Genetics D
Neuroscience D
Plant Biology D

WAYNE STATE UNIVERSITY

Anatomy M,D
Biochemistry M,D
Biological and Biomedical
  Sciences—General M,D
Biopsychology M
Cancer Biology/Oncology M,D*
Genetics M,D*
Immunology M,D
Microbiology M,D
Molecular Biology M,D*
Neurobiology M,D
Neuroscience M,D
Nutrition M,D
Pathology M,D
Pharmacology P,M,D
Physiology M,D
Toxicology M,D,O*

WESLEYAN UNIVERSITY

Biochemistry M,D*
Biological and Biomedical
  Sciences—General D*
Cell Biology D
Developmental Biology D
Evolutionary Biology D

Genetics D
Molecular Biology D*
Neurobiology D
Physiology D

WEST CHESTER UNIVERSITY OF PENNSYLVANIA

Biological and Biomedical
  Sciences—General M

WESTERN CAROLINA UNIVERSITY

Biological and Biomedical
  Sciences—General M

WESTERN CONNECTICUT STATE UNIVERSITY

Biological and Biomedical
  Sciences—General M

WESTERN ILLINOIS UNIVERSITY

Biological and Biomedical
  Sciences—General M,O
Marine Biology M,O
Zoology M,O

WESTERN KENTUCKY UNIVERSITY

Biological and Biomedical
  Sciences—General M

WESTERN MICHIGAN UNIVERSITY

Biological and Biomedical
  Sciences—General M,D

WESTERN WASHINGTON UNIVERSITY

Biological and Biomedical
  Sciences—General M

WEST TEXAS A&M UNIVERSITY

Biological and Biomedical
  Sciences—General M

WEST VIRGINIA UNIVERSITY

Biochemistry M,D*
Biological and Biomedical
  Sciences—General M,D*
Cancer Biology/Oncology M,D*
Cell Biology M,D
Developmental Biology M,D
Entomology M
Environmental Biology M,D
Evolutionary Biology M,D
Genetics M,D*
Human Genetics M,D
Immunology M,D
Microbiology M,D*
Molecular Biology M,D
Neuroscience M,D
Nutrition M
Physiology M,D*
Plant Pathology M
Reproductive Biology M,D
Toxicology M,D

WICHITA STATE UNIVERSITY

Biological and Biomedical
  Sciences—General M*

WILLIAM PATERSON UNIVERSITY OF NEW JERSEY

Biological and Biomedical
  Sciences—General M*
Ecology M

Genetics D
Molecular Biology D*
Neurobiology D
Physiology D

WEST CHESTER UNIVERSITY OF PENNSYLVANIA

Genetics D
Molecular Biology D*
Neurobiology D
Physiology D

Molecular Biology M
Physiology M

WINTHROP UNIVERSITY

Biological and Biomedical
  Sciences—General M
Nutrition M

WOODS HOLE OCEANOGRAPHIC INSTITUTION

Marine Biology M,D,O

WORCESTER POLYTECHNIC INSTITUTE

Biochemistry M,D
Biological and Biomedical
  Sciences—General M,D*

WRIGHT STATE UNIVERSITY

Anatomy M
Biochemistry M
Biological and Biomedical
  Sciences—General M,D*
Biophysics M
Immunology M
Microbiology M
Molecular Biology M
Pharmacology M
Physiology M
Toxicology M

YALE UNIVERSITY

Biochemistry M,D
Biological and Biomedical
  Sciences—General D*
Biophysics M,D
Cancer Biology/Oncology D
Cell Biology D
Computational Biology D
Developmental Biology D
Ecology D*
Evolutionary Biology D
Genetics D
Genomic Sciences D
Immunology D
Infectious Diseases D
Microbiology D
Molecular Biology D
Molecular Biophysics D
Molecular Medicine D
Molecular Pathology D
Molecular Physiology D
Neurobiology D
Neuroscience D
Parasitology D*
Pathobiology D
Pathology D
Pharmacology D
Physiology D
Plant Biology D*
Structural Biology D
Virology D

YORK UNIVERSITY

Biological and Biomedical
  Sciences—General M,D

YOUNGSTOWN STATE UNIVERSITY

Biological and Biomedical
  Sciences—General M

ACADEMIC PROGRAMS
IN THE BIOLOGICAL SCIENCES

Section 1
Biological and Biomedical Sciences

This section contains a directory of institutions offering graduate work in biological and biomedical sciences, followed by in-depth entries submitted by institutions that chose to prepare detailed program descriptions. Additional information about programs listed in the directory but not augmented by an in-depth entry may be obtained by writing directly to the dean of a graduate school or chair of a department at the address given in the directory.

Programs in fields related to the biological and biomedical sciences may be found throughout this book as well as in Book 2 in Psychology and Counseling and Sociology, Anthropology, and Archaeology. In Book 4, see Chemistry, Marine Sciences and Oceanography, and Mathematical Sciences; in Book 5, see Agricultural Engineering and Bioengineering, Biomedical Engineering and Biotechnology, Civil and Environmental Engineering, Management of Engineering and Technology, and Ocean Engineering; and in Book 6, see Allied Health, Chiropractic, Dentistry and Dental Sciences, Medicine, Nursing, Optometry and Vision Sciences, Pharmacy and Pharmaceutical Sciences, Public Health, and Veterinary Medicine and Sciences.

CONTENTS

Biological and Biomedical Sciences—General

Acadia University, Faculty of Pure and Applied Science, Department of Biology, Wolfville, NS B4P 2R6, Canada. Offers M Sc. *Faculty:* 16 full-time (3 women), 4 part-time/adjunct (0 women). *Students:* 21 full-time (9 women), 12 part-time (6 women); includes 1 minority (Asian American or Pacific Islander), 1 international. 15 applicants, 53% accepted, 7 enrolled. In 2005, 11 degrees awarded. *Degree requirements:* For master's, thesis, comprehensive exam, registration. *Entrance requirements:* For master's, minimum B- average in last 2 years of major. Additional exam requirements/recommendations for international students: Required—TOEFL (minimum score 580 paper-based; 237 computer-based). *Application deadline:* For fall admission, 2/1 for domestic students. Application fee: $50. *Financial support:* In 2005–06, 10 research assistantships (averaging $5,000 per year), 6 teaching assistantships (averaging $8,000 per year) were awarded; scholarships/grants also available. Financial award application deadline: 2/1. *Faculty research:* Respiration physiology, estuaries and fisheries, limnology, plant biology, conservation biology. *Unit head:* Dr. Dan Toews, Head, 902-585-1162. *Application contact:* Nancy Roscoe-Huntley, Administrative Secretary, 902-585-1695, Fax: 902-585-1059, E-mail: nancy.roscoe-huntley@acadiau.ca.

Adelphi University, Graduate School of Arts and Sciences, Department of Biology, Garden City, NY 11530-0701. Offers MS. Part-time and evening/weekend programs available. *Students:* 3 full-time (0 women), 24 part-time (16 women); includes 8 minority (2 African Americans, 3 Asian Americans or Pacific Islanders, 3 Hispanic Americans), 3 international. Average age 26. In 2005, 9 degrees awarded. *Degree requirements:* For master's, thesis or alternative. *Entrance requirements:* For master's, 3 letters of recommendation. Additional exam requirements/recommendations for international students: Required—TOEFL (minimum score 550 paper-based; 213 computer-based). *Application deadline:* Applications are processed on a rolling basis. Application fee: $50. *Expenses:* Tuition: Full-time $21,150; part-time $650 per credit. Required fees: $550; $450 per year. Tuition and fees vary according to degree level, campus/location and program. *Financial support:* In 2005–06, 9 research assistantships with full and partial tuition reimbursements (averaging $2,895 per year) were awarded; teaching assistantships, career-related internships or fieldwork, Federal Work-Study, institutionally sponsored loans, and unspecified assistantships also available. Financial award application deadline: 2/15; financial award applicants required to submit FAFSA. *Faculty research:* Plant-animal interactions, physiology (plant, cornea), reproductive behavior, topics in evolution, fish biology. *Unit head:* Dr. James Dooley, Chairperson, 516-877-4200, E-mail: dooley@adelphi.edu. *Application contact:* Christine Murphy, Director of Admissions, 516-877-3050, Fax: 516-877-3039, E-mail: admissions@adelphi.edu.

Alabama Agricultural and Mechanical University, School of Graduate Studies, School of Arts and Sciences, Department of Natural and Physical Sciences, Area in Biology, Huntsville, AL 35811. Offers MS. Part-time and evening/weekend programs available. *Degree requirements:* For master's, thesis, comprehensive exam. *Entrance requirements:* For master's, GRE General Test. Electronic applications accepted. *Faculty research:* Radiation and chemical mutagenesis, human cytogenetics, microbial biotechnology, microbial metabolism, environmental toxicology.

Alabama State University, School of Graduate Studies, College of Arts and Sciences, Department of Biology, Montgomery, AL 36101-0271. Offers biology (M Ed, MS); biology education (Ed S). Part-time programs available. *Degree requirements:* For master's, one foreign language, thesis, comprehensive exam; for Ed S, thesis. *Entrance requirements:* For master's, GRE, GRE Subject Test, graduate writing competency test; for Ed S, graduate writing competency test, GRE, MAT. Additional exam requirements/recommendations for international students: Required—TOEFL (minimum score 500 paper-based; 173 computer-based). *Faculty research:* Salmonella pseudomonas, cancer cells.

Albany Medical College, Graduate Programs in the Biological Sciences, Albany, NY 12208-3479. Offers MS, PhD. Part-time programs available. *Faculty:* 80 full-time (24 women), 15 part-time/adjunct (6 women). *Students:* 185 full-time (126 women), 3 part-time (1 woman); includes 17 minority (7 African Americans, 7 Asian Americans or Pacific Islanders, 3 Hispanic Americans), 23 international. Average age 28. 112 applicants, 55% accepted, 34 enrolled. In 2005, 31 master's, 11 doctorates awarded. Terminal master's awarded for partial completion of doctoral program. *Degree requirements:* For master's, thesis; for doctorate, thesis/dissertation, oral qualifying exam, written preliminary exam, 1 published paper review. *Entrance requirements:* For master's and doctorate, GRE General Test, letters of recommendation. Additional exam requirements/recommendations for international students: Required—TOEFL. *Application deadline:* For fall admission, 3/15 for domestic students. Applications are processed on a rolling basis. Application fee: $0 ($60 for international students). *Financial support:* In 2005–06, 70 research assistantships with full tuition reimbursements (averaging $18,000 per year) were awarded; Federal Work-Study, scholarships/grants, and tuition waivers (full) also available. Financial award applicants required to submit FAFSA. *Unit head:* Dr. Henry S. Pohl, Senior Associate Dean for Education Programs, 518-262-5253, Fax: 518-262-5183. *Application contact:* Jean M. Cornwell, Admissions Coordinator, 518-262-5253, Fax: 518-262-5183, E-mail: graduatestudies@mail.amc.edu.

Albert Einstein College of Medicine, Medical Scientist Training Program, Bronx, NY 10461. Offers MD/PhD. Students must apply through the Albert Einstein College of Medicine.

Albert Einstein College of Medicine, Sue Golding Graduate Division of Medical Sciences, Bronx, NY 10461. Offers PhD, MD/PhD. *Degree requirements:* For doctorate, thesis/dissertation. *Entrance requirements:* For doctorate, GRE General Test. Additional exam requirements/recommendations for international students: Required—TOEFL.

See Close-Up on page 81.

Alcorn State University, School of Graduate Studies, School of Arts and Sciences, Department of Biology, Alcorn State, MS 39096-7500. Offers MS. *Students:* 11 full-time (4 women), 12 part-time (8 women); includes 16 minority (all African Americans), 5 international. In 2005, 1 degree awarded. *Application deadline:* For fall admission, 7/15 for domestic students; for spring admission, 11/25 for domestic students. Applications are processed on a rolling basis. Application fee: $0 ($10 for international students). *Unit head:* Dr. Bettaiya Rajanna, Chairperson, 601-877-6236.

American University, College of Arts and Sciences, Department of Biology, Program in Biology, Washington, DC 20016-8001. Offers MA, MS. Part-time programs available. *Students:* 9 full-time (7 women), 5 part-time (3 women). Average age 24. In 2005, 5 degrees awarded. *Degree requirements:* For master's, directed literature research. *Entrance requirements:* For master's, GRE General Test, GRE Subject Test, minimum GPA of 3.0. Additional exam requirements/recommendations for international students: Required—TOEFL. *Application deadline:* For fall admission, 2/1 for domestic students; for spring admission, 10/1 for domestic students. Application fee: $50. *Expenses:* Tuition: Full-time $17,802; part-time $989 per credit. Required fees: $380. *Financial support:* Fellowships, research assistantships, teaching assistantships, career-related internships or fieldwork, Federal Work-Study, and institutionally sponsored loans available. Financial award application deadline: 2/1.

The American University of Athens, The School of Graduate Studies, Athens, Greece. Offers biomedical sciences (MS); business (MBA); business communication (MA); computer sciences (MS); engineering and applied sciences (MS); politics and policy making (MA); systems engineering (MS); telecommunications (MS). *Faculty:* 15 full-time (2 women), 13 part-time/adjunct (4 women). *Students:* 20 full-time (2 women), 8 part-time, 10 international. *Entrance requirements:* Additional exam requirements/recommendations for international students: Required—TOEFL (minimum score 550 paper-based; 213 computer-based). Application fee: 100 euros. *Faculty research:* Nanotechnology, environmental sciences, rock mechanics, human skin studies, Monte Carlo algorithms and software. *Unit head:* Dr. Rita Roussos,

Director of the School of Graduate Studies, 302-725-9301-3, Fax: 302-10-7259304, E-mail: rroussos@aua.edu.

American University of Beirut, Graduate Programs, Faculty of Arts and Sciences, Beirut, Lebanon. Offers anthropology (MA); Arabic language and literature (MA); archaeology (MA); biology (MS); business administration (MBA); chemistry (MS); computer science (MS); economics (MA); education (MA); English language (MA); English literature (MA); environmental policy planning (MSES); finance and banking (MFB); financial economics (MFE); geology (MS); history (MA); mathematics (MS); Middle Eastern studies (MA); philosophy (MA); physics (MS); political studies (MA); psychology (MA); public administration (MA); sociology (MA). *Degree requirements:* For master's, one foreign language, thesis (for some programs), comprehensive exam, registration. *Entrance requirements:* For master's, GRE, letter of recommendation.

Andrews University, School of Graduate Studies, College of Arts and Sciences, Department of Biology, Berrien Springs, MI 49104. Offers MAT, MS. *Faculty:* 7 full-time (0 women). *Students:* 6 full-time (2 women), 3 part-time (1 woman); includes 3 minority (1 African American, 1 Asian American or Pacific Islander, 1 Hispanic American), 3 international. Average age 27. 2 applicants. In 2005, 2 degrees awarded. *Degree requirements:* For master's, thesis, comprehensive exam. *Entrance requirements:* For master's, GRE Subject Test. *Application deadline:* Applications are processed on a rolling basis. Application fee: $43. *Financial support:* Fellowships, research assistantships, teaching assistantships, career-related internships or fieldwork, Federal Work-Study, and institutionally sponsored loans available. Financial award application deadline: 3/15. *Unit head:* Dr. David A. Steen, Chairman, 269-471-3243. *Application contact:* Carolyn Hurst, Supervisor of Graduate Admission, 800-253-2874, Fax: 269-471-3228, E-mail: graduate@andrews.edu.

Angelo State University, College of Graduate Studies, College of Sciences, Department of Biology, San Angelo, TX 76909. Offers MS. Part-time and evening/weekend programs available. *Faculty:* 12 full-time (3 women). *Students:* 8 full-time (6 women), 6 part-time (2 women); includes 2 minority (both Hispanic Americans) Average age 31. 4 applicants, 75% accepted, 2 enrolled. In 2005, 6 degrees awarded. *Degree requirements:* For master's, thesis optional. *Entrance requirements:* For master's, GRE General Test. Additional exam requirements/recommendations for international students: Required—TOEFL or IELT. *Application deadline:* For fall admission, 7/15 priority date for domestic students, 6/15 priority date for international students; for spring admission, 12/8 for domestic students, 11/1 for international students. Applications are processed on a rolling basis. Application fee: $25 ($50 for international students). Electronic applications accepted. *Expenses:* Tuition, state resident: part-time $126 per credit. Tuition, nonresident: full-time $7,236; part-time $402 per credit. Required fees: $844; $94 per credit. One-time fee: $25. Tuition and fees vary according to course load. *Financial support:* In 2005–06, 11 students received support, including 1 research assistantship, 2 teaching assistantships (averaging $10,251 per year); career-related internships or fieldwork, Federal Work-Study, scholarships/grants, and unspecified assistantships also available. Support available to part-time students. Financial award application deadline: 3/1. *Faculty research:* Texas poppy-mallow project, Chisos hedgehog cactus, skunks, reptiles, amphibians, and rodents, seed germination, mammals. *Unit head:* Dr. Kelly McCoy, Head, 325-942-2189 Ext. 246, E-mail: kelly.mccoy@angelo.edu. *Application contact:* Dr. Bonnie B. Amos, Graduate Advisor, 325-942-2189 Ext. 256, E-mail: bonnie.amos@angelo.edu.

Appalachian State University, Cratis D. Williams Graduate School, College of Arts and Sciences, Department of Biology, Boone, NC 28608. Offers MS. Part-time programs available. *Faculty:* 25 full-time (6 women). *Students:* 26 full-time (10 women), 10 part-time (4 women), 1 international. 22 applicants, 59% accepted, 10 enrolled. In 2005, 6 degrees awarded. *Degree requirements:* For master's, one foreign language, thesis, comprehensive exam, registration. *Entrance requirements:* For master's, GRE General Test. Additional exam requirements/recommendations for international students: Required—TOEFL (minimum score 570 paper-based; 230 computer-based). *Application deadline:* For fall admission, 7/1 priority date for domestic students, 1/1 priority date for international students; for spring admission, 11/1 for domestic students, 6/1 for international students. Application fee: $45. *Expenses:* Tuition, state resident: full-time $2,593. Tuition, nonresident: full-time $12,176. Required fees: $1,726. *Financial support:* In 2005–06, 2 fellowships (averaging $1,000 per year), research assistantships (averaging $8,000 per year), 40 teaching assistantships (averaging $8,105 per year) were awarded; career-related internships or fieldwork, Federal Work-Study, scholarships/grants, and unspecified assistantships also available. Support available to part-time students. Financial award application deadline: 7/1; financial award applicants required to submit FAFSA. *Faculty research:* Aquatic and terrestrial ecology, animal and plant physiology, behavior and systematics, immunology and cell biology, molecular biology and microbiology. *Unit head:* Dr. Steven Seagle, Chairman, 828-262-3025. *Application contact:* Dr. Gary Walker, Graduate Coordinator, 828-262-3025, E-mail: walkergl@appstate.edu.

Arizona State University, Division of Graduate Studies, College of Liberal Arts and Sciences, Department of Biology, Tempe, AZ 85287. Offers behavior (MS, PhD); biology (MNS); biology education (MS, PhD); cell and developmental biology (MS, PhD); computational, statistical, and mathematical biology (MS, PhD); conservation (MS, PhD); ecology (MS, PhD); evolution (MS, PhD); genetics (MS, PhD); history and philosophy of biology (MS, PhD); molecular and cellular biology (MS, PhD); neuroscience (MS, PhD); physiology (MS, PhD). Terminal master's awarded for partial completion of doctoral program. *Degree requirements:* For master's, thesis; for doctorate, thesis/dissertation, oral exam, comprehensive exam. *Entrance requirements:* For master's, GRE General Test; for doctorate, GRE General Test, GRE Subject Test. Additional exam requirements/recommendations for international students: Required—TOEFL (minimum score 600 paper-based; 250 computer-based); Recommended—TSE. Electronic applications accepted. *Faculty research:* Animal behavior, comparative endocrinology, invertebrate neurophysiology, urban ecology, signal transduction.

Arizona State University, Division of Graduate Studies, College of Liberal Arts and Sciences, Division of Natural Sciences and Mathematics, School of Life Sciences, Tempe, AZ 85287. Offers biology (MNS); cell and developmental biology (MS, PhD); microbiology (MNS, MS, PhD). *Accreditation:* NAACLS. *Degree requirements:* For master's, thesis optional for MNS, required for MS; for doctorate, one foreign language, thesis/dissertation. *Entrance requirements:* For master's and doctorate, GRE.

Arizona State University at the Polytechnic Campus, East College, Department of Applied Biological Sciences, Mesa, AZ 85212. Offers MS. Part-time programs available. *Degree requirements:* For master's, thesis, oral defense. *Entrance requirements:* For master's, GRE General Test or MAT. Additional exam requirements/recommendations for international students: Required—TOEFL (minimum score 550 paper-based; 213 computer-based); Recommended—TWE, TSE. Electronic applications accepted. *Faculty research:* Ecological restoration, wildlife ecology, urban horticulture, geographic information systems, riparian ecology.

Arkansas State University, Graduate School, College of Sciences and Mathematics, Department of Biological Sciences, Jonesboro, State University, AR 72467. Offers biological sciences (MA); biology (MS); biology education (MSE, SCCT). Part-time programs available. *Faculty:* 13 full-time (4 women). *Students:* 12 full-time (5 women), 21 part-time (9 women); includes 2 minority (both African Americans) Average age 27. 17 applicants, 100% accepted, 10 enrolled. In 2005, 13 degrees awarded. *Degree requirements:* For master's, thesis (for some programs), comprehensive exam. *Entrance requirements:* For master's, GRE General Test, appropriate bachelor's degree, letters of reference; for SCCT, GRE General Test or MAT, interview, master's degree, letters of reference. Additional exam requirements/recommendations for international students: Required—TOEFL (minimum score 213 computer-based). *Application deadline:* For fall admission, 7/1 for domestic students; for spring admission, 11/15 priority date for domestic students. Applications are processed on a rolling basis. Application fee: $15

Biological and Biomedical Sciences—General

Arkansas State University (continued)
($25 for international students). Electronic applications accepted. *Expenses:* Tuition, state resident: full-time $3,232; part-time $180 per hour. Tuition, nonresident: full-time $8,164; part-time $454 per hour. Required fees: $716; $37 per hour. $25 per semester. Tuition and fees vary according to course load and program. *Financial support:* Teaching assistantships, scholarships/grants and unspecified assistantships available. Financial award application deadline: 7/1; financial award applicants required to submit FAFSA. *Unit head:* Dr. Aldemaro Romero, Chair, 870-972-3082, Fax: 870-972-2638, E-mail: aromero@astate.edu.

A.T. Still University of Health Sciences, Kirksville College of Osteopathic Medicine, Kirksville, MO 63501. Offers biomedical sciences (MS); osteopathic medicine (DO). *Accreditation:* AOsA. *Faculty:* 57 full-time (15 women), 23 part-time/adjunct (2 women). *Students:* 680 full-time (269 women), 24 part-time (9 women); includes 108 minority (5 African Americans, 5 American Indian/Alaska Native, 86 Asian Americans or Pacific Islanders, 12 Hispanic Americans), 13 international. Average age 27. 2,600 applicants, 182 enrolled. In 2005, 150 first professional degrees awarded. *Degree requirements:* For DO, level I and 2 CE comprehensive osteopathic medical Licensing Exam and COMLEX PE; for master's, thesis. *Entrance requirements:* For DO, MCAT, minimum undergraduate GPA of 2.5 (cumulative and science) or 90 semester hours with minimum GPA of 3.5 (cumulative and science) and minimum MCAT of 28; for master's, GRE, MCAT, or DAT, minimum undergraduate GPA of 2.5 (cumulative and science). *Application deadline:* For fall admission, 2/15 for domestic students, 2/15 for international students. Applications are processed on a rolling basis. Application fee: $60. Electronic applications accepted. *Expenses:* Tuition: Full-time $20,740. Required fees: $1,000. Tuition and fees vary according to degree level, program and student level. *Financial support:* In 2005–06, 637 students received support, including 8 fellowships with full tuition reimbursements available (averaging $9,000 per year); career-related internships or fieldwork, Federal Work-Study, institutionally sponsored loans, and scholarships/grants also available. Financial award application deadline: 5/1; financial award applicants required to submit FAFSA. *Faculty research:* Osteopathic manipulation and pneumonia in the elderly, basis of hypocholestrolemic drug-induced cataracts, membrane structure in cataracts, role of caveolin in the ocular lens, cell membrane. Total annual research expenditures: $670,095. *Unit head:* Dr. Philip C. Slocum, Vice President for Medical Affairs and Dean, 660-626-2354, Fax: 660-626-2080, E-mail: pslocum@atsu.edu. *Application contact:* Donna Sparks, Associate Director for Admissions, 660-626-2237, Fax: 660-626-2969, E-mail: admissions@atsu.edu.

Auburn University, College of Veterinary Medicine and Graduate School, Graduate Programs in Veterinary Medicine, Auburn University, AL 36849. Offers biomedical sciences (MS, PhD), including anatomy, physiology and pharmacology (MS), biomedical sciences (PhD), clinical sciences (MS), large animal surgery and medicine (MS), pathobiology (MS), radiology (MS), small animal surgery and medicine (MS). Part-time programs available. *Faculty:* 76 full-time (24 women). *Students:* 11 full-time (8 women), 33 part-time (20 women); includes 3 minority (all Hispanic Americans), 14 international. 41 applicants, 37% accepted, 11 enrolled. In 2005, 9 master's, 3 doctorates awarded. *Degree requirements:* For doctorate, thesis/dissertation. *Entrance requirements:* For master's, GRE General Test; for doctorate, GRE General Test, GRE Subject Test. *Application deadline:* For fall admission, 7/7 for domestic students; for spring admission, 11/24 for domestic students. Applications are processed on a rolling basis. Application fee: $25 ($50 for international students). Electronic applications accepted. *Financial support:* Research assistantships, teaching assistantships, Federal Work-Study available. Support available to part-time students. Financial award application deadline: 3/15. *Application contact:* Dr. Stephen L. McFarland, Acting Dean of the Graduate School, 334-844-4700.

Auburn University, Graduate School, College of Sciences and Mathematics, Department of Biological Sciences, Auburn University, AL 36849. Offers botany (MS, PhD); microbiology (MS, PhD); zoology (MS, PhD). *Faculty:* 28 full-time (5 women). *Students:* 44 full-time (23 women), 49 part-time (27 women); includes 8 minority (5 African Americans, 3 Hispanic Americans), 17 international. 65 applicants, 51% accepted, 22 enrolled. In 2005, 11 master's, 1 doctorate awarded. *Entrance requirements:* For master's and doctorate, GRE General Test. Additional exam requirements/recommendations for international students: Required—TOEFL. *Application deadline:* For fall admission, 7/7 for domestic students; for spring admission, 11/24 for domestic students. Electronic applications accepted. *Financial support:* Research assistantships, teaching assistantships available. *Unit head:* Dr. James M Barbaree, Chair, 334-844-1647, Fax: 334-844-1645. *Application contact:* Dr. Stephen L. McFarland, Acting Dean of the Graduate School, 334-844-4700.

Austin Peay State University, Graduate School, College of Science and Mathematics, Department of Biology, Clarksville, TN 37044-0001. Offers biology (MS); clinical laboratory science (MS); radiologic science (MS). Part-time programs available. *Faculty:* 12 full-time (3 women). *Students:* 4 full-time (2 women), 17 part-time (12 women); includes 2 minority (1 African American, 1 Hispanic American), 4 international. Average age 31. In 2005, 4 degrees awarded. *Degree requirements:* For master's, thesis optional. *Entrance requirements:* For master's, GRE General Test, 3 letters of recommendation. Additional exam requirements/recommendations for international students: Required—TOEFL (minimum score 500 paper-based; 173 computer-based). *Application deadline:* For fall admission, 7/31 for domestic students; for spring admission, 12/17 priority date for domestic students. Applications are processed on a rolling basis. Application fee: $25. *Expenses:* Tuition, state resident: full-time $4,936; part-time $340 per credit hour. Tuition, nonresident: full-time $14,248; part-time $744 per credit hour. Required fees: $957. Part-time tuition and fees vary according to course load. *Financial support:* In 2005–06, research assistantships (averaging $9,250 per year), teaching assistantships with partial tuition reimbursements (averaging $6,900 per year) were awarded; career-related internships or fieldwork, Federal Work-Study, institutionally sponsored loans, scholarships/grants, and unspecified assistantships also available. Support available to part-time students. Financial award application deadline: 4/1. *Faculty research:* Nonpaint source pollution, amphibian biomonitoring, aquatic toxicology, biological indicators of water quality, taxonomy. *Unit head:* Dr. Don Dailey, Interim Chair, 931-221-7781, E-mail: daileyd@apsu.edu.

Ball State University, Graduate School, College of Sciences and Humanities, Department of Biology, Muncie, IN 47306-1099. Offers biology (MA, MAE, MS); biology education (Ed D). *Accreditation:* NCATE (one or more programs are accredited). *Faculty:* 22. *Students:* 27 full-time (18 women), 26 part-time (11 women); includes 3 minority (2 Asian Americans or Pacific Islanders, 1 Hispanic American), 9 international. Average age 24. 33 applicants, 76% accepted, 15 enrolled. In 2005, 18 master's, 4 doctorates awarded. *Degree requirements:* For doctorate, thesis/dissertation. *Entrance requirements:* For master's, GRE General Test; for doctorate, GRE General Test, minimum graduate GPA of 3.2. *Application fee:* $25 ($35 for international students). *Expenses:* Tuition, state resident: full-time $6,246. Tuition, nonresident: full-time $16,006. *Financial support:* In 2005–06, research assistantships with full tuition reimbursements (averaging $19,153 per year), 40 teaching assistantships with full tuition reimbursements (averaging $8,663 per year) were awarded; career-related internships or fieldwork also available. Financial award application deadline: 3/1. *Faculty research:* Aquatics and fisheries, tumors, water and air pollution, developmental biology and genetics. *Unit head:* Dr. Carl E. Warnes, Chairman, 765-285-8820, Fax: 765-285-8804, E-mail: cwarnes@bsu.edu.

Barry University, School of Natural and Health Sciences, Program in Biology and Biomedical Sciences, Miami Shores, FL 33161-6695. Offers biology (MS); biomedical sciences (MS). Part-time and evening/weekend programs available. *Students:* 147. *Degree requirements:* For master's, thesis (for some programs), comprehensive exam. *Entrance requirements:* For master's, GRE General Test or Florida Teacher's Certification Exam (biology); GRE General Test, MCAT, or DAT (biomedical sciences). *Application deadline:* Applications are processed on a rolling basis. Application fee: $30. Electronic applications accepted. *Expenses:* Tuition: Full-time $12,330; part-time $685 per credit. *Financial support:* Application deadline: 5/1; *Faculty research:* Genetics, immunology, anthropology. *Unit head:* Dr. Ralph Laudan, Associ-

ate Dean, 305-899-3229, Fax: 305-899-3225, E-mail: rlaudan@mail.barry.edu. *Application contact:* Dr. Jocelyn Goulet, Director, Health Services Admissions Operation, 305-899-3541, Fax: 305-899-3232, E-mail: jgoulet@mail.barry.edu.

See Close-Up on page 83.

Baylor College of Medicine, Graduate School of Biomedical Sciences, Houston, TX 77030-3498. Offers MS, PhD, MD/PhD. *Faculty:* 457 full-time (109 women). *Students:* 551 full-time (254 women); includes 106 minority (18 African Americans, 2 American Indian/Alaska Native, 48 Asian Americans or Pacific Islanders, 38 Hispanic Americans), 213 international. Average age 28. 969 applicants, 25% accepted, 105 enrolled. In 2005, 7 master's, 53 doctorates awarded. Terminal master's awarded for partial completion of doctoral program. *Median time to degree:* Of those who began their doctoral program in fall 1997, 75% received their degree in 8 years or less. *Degree requirements:* For master's, thesis, registration; for doctorate, thesis/dissertation, public defense. *Entrance requirements:* For doctorate, GRE General Test, GRE Subject Test (strongly recommended), minimum GPA 3.0. Additional exam requirements/recommendations for international students: Required—TOEFL. *Application deadline:* For fall admission, 2/1 for domestic students. Applications are processed on a rolling basis. Application fee: $30. Electronic applications accepted. *Expenses:* Tuition: Full-time $8,200. Full-time tuition and fees vary according to program. *Financial support:* In 2005–06, 208 fellowships (averaging $23,000 per year), 328 research assistantships (averaging $23,000 per year), 1 teaching assistantship were awarded; career-related internships or fieldwork, Federal Work-Study, institutionally sponsored loans, health care benefits, tuition waivers (full and partial), and stipends also available. Financial award applicants required to submit FAFSA. *Faculty research:* Cell and molecular biology of cardiac muscle, structural biophysics, gene expression and regulation, human genomes, viruses. *Unit head:* Dr. William R. Brinkley, Dean of Graduate Sciences, 713-798-5263, Fax: 713-798-6325, E-mail: brinkley@bcm.tmc.edu. *Application contact:* Donna Otwell, Director of Administrative Operations, 713-798-4029, Fax: 713-798-6325, E-mail: dotwell@bcm.edu.

See Close-Up on page 85.

Baylor University, Graduate School, College of Arts and Sciences, Department of Biology, Waco, TX 76798. Offers biology (MA, MS, PhD); environmental biology (MS); limnology (MSL). Part-time programs available. *Faculty:* 13 full-time (3 women). *Students:* 21 full-time (6 women), 16 part-time (8 women); includes 2 minority (1 African American, 1 Hispanic American), 7 international. In 2005, 4 master's, 2 doctorates awarded. *Degree requirements:* For master's, thesis (for some programs); for doctorate, thesis/dissertation. *Entrance requirements:* For master's and doctorate, GRE General Test. *Application deadline:* For fall admission, 1/31 for domestic students. Applications are processed on a rolling basis. Application fee: $25. *Financial support:* Teaching assistantships, career-related internships or fieldwork, Federal Work-Study, institutionally sponsored loans, and tuition waivers (full and partial) available. Support available to part-time students. Financial award application deadline: 2/28. *Faculty research:* Terrestrial ecology, aquatic ecology, genetics. *Unit head:* Dr. Ken Wilkins, Graduate Program Director, 254-710-2911, Fax: 254-710-2969, E-mail: ken_wilkins@baylor.edu. *Application contact:* Sandy Tighe, Administrative Assistant, 254-710-2911, Fax: 254-710-2969, E-mail: sandy_tighe@baylor.edu.

Baylor University, Graduate School, Institute of Biomedical Studies, Waco, TX 76798. Offers MS, PhD. *Students:* 23 full-time (8 women); includes 3 minority (all Asian Americans or Pacific Islanders), 12 international. In 2005, 4 master's, 4 doctorates awarded. *Entrance requirements:* For master's and doctorate, GRE General Test. *Application deadline:* Applications are processed on a rolling basis. Application fee: $25. *Financial support:* Research assistantships, teaching assistantships available. *Unit head:* Dr. Robert Kane, Interim Director, 254-710-2514, Fax: 254-710-3878, E-mail: robert_kane@baylor.edu. *Application contact:* Suzanne Keener, Administrative Assistant, 254-710-3588, Fax: 254-710-3870.

Bemidji State University, School of Graduate Studies, College of Social and Natural Sciences, Department of Biology, Bemidji, MN 56601-2699. Offers MS. Part-time programs available. *Faculty:* 14 part-time/adjunct (5 women). *Students:* 3 full-time (2 women), 5 part-time (2 women). Average age 34. 3 applicants, 67% accepted. In 2005, 3 degrees awarded. *Degree requirements:* For master's, thesis or alternative, departmental qualifying exam. *Entrance requirements:* For master's, GRE General Test. Additional exam requirements/recommendations for international students: Required—TOEFL. *Application deadline:* For fall admission, 5/1 for domestic students. Applications are processed on a rolling basis. Application fee: $20. Electronic applications accepted. *Expenses:* Tuition, state resident: full-time $4,716. Required fees: $384. One-time fee: $20 full-time. *Financial support:* In 2005–06, teaching assistantships with partial tuition reimbursements (averaging $8,000 per year); career-related internships or fieldwork, Federal Work-Study, scholarships/grants, and unspecified assistantships also available. Support available to part-time students. Financial award application deadline: 5/1. *Unit head:* Dr. Patrick Guilfoile, Chair, 218-755-2800, Fax: 218-755-4107.

Bloomsburg University of Pennsylvania, School of Graduate Studies, College of Science and Technology, Department of Biological and Allied Health Sciences, Program in Biology, Bloomsburg, PA 17815-1301. Offers MS. *Faculty:* 17 full-time (5 women). *Students:* 1 (woman) full-time, 5 part-time (2 women), 1 international. Average age 30. 5 applicants, 100% accepted, 2 enrolled. In 2005, 5 degrees awarded. *Degree requirements:* For master's, thesis or alternative. *Entrance requirements:* For master's, GRE General Test, minimum GPA of 3.0, 2 letters of recommendation. Additional exam requirements/recommendations for international students: Required—TOEFL. *Application deadline:* Applications are processed on a rolling basis. Application fee: $30. Electronic applications accepted. *Expenses:* Tuition, state resident: full-time $5,888; part-time $327 per credit. Tuition, nonresident: full-time $9,422; part-time $523 per credit. Required fees: $1,359. *Financial support:* Unspecified assistantships available. *Unit head:* Dr. Carl Hansen, Coordinator, 570-389-4580, Fax: 570-389-3028, E-mail: chansen@bloomu.edu.

Boise State University, Graduate College, College of Arts and Sciences, Department of Biology, Boise, ID 83725-0399. Offers biology (MA, MS); raptor biology (MS). Part-time programs available. *Degree requirements:* For master's, thesis. *Entrance requirements:* For master's, GRE General Test, minimum GPA of 3.0. Electronic applications accepted. *Faculty research:* Soil and stream microbial ecology, avian ecology.

Boston College, Graduate School of Arts and Sciences, Department of Biology, Chestnut Hill, MA 02467-3800. Offers biochemistry (MS, PhD); biology (MS, PhD). *Students:* 4 full-time (all women), 44 part-time (24 women); includes 7 minority (1 African American, 6 Asian Americans or Pacific Islanders), 6 international. 118 applicants, 28% accepted, 12 enrolled. In 2005, 4 master's, 2 doctorates awarded. Terminal master's awarded for partial completion of doctoral program. *Degree requirements:* For master's and doctorate, thesis/dissertation. *Entrance requirements:* For master's and doctorate, GRE General Test, GRE Subject Test. Additional exam requirements/recommendations for international students: Required—TOEFL (minimum score 550 paper-based; 213 computer-based). *Application deadline:* For fall admission, 1/15 for domestic students. Application fee: $70. Electronic applications accepted. *Financial support:* Fellowships with full tuition reimbursements, research assistantships with full tuition reimbursements, teaching assistantships with full tuition reimbursements, Federal Work-Study and scholarships/grants available. Support available to part-time students. Financial award application deadline: 3/1; financial award applicants required to submit FAFSA. *Faculty research:* DNA replication in mammalian cells, control of the cell cycle, immunology, plant genetics. *Unit head:* Dr. Marc Muskavitch, Chairperson, 617-552-3540, E-mail: marc.muskavitch@bc.edu. *Application contact:* Dr. Daniel Kirschner, Graduate Program Director, 617-552-3540, E-mail: daniel.kirschner@bc.edu.

See Close-Up on page 87.

Boston University, Graduate School of Arts and Sciences, Department of Biology, Boston, MA 02215. Offers MA, PhD. *Students:* 105 full-time (60 women), 4 part-time (1 woman); includes

Biological and Biomedical Sciences—General

8 minority (3 African Americans, 5 Asian Americans or Pacific Islanders), 20 international. Average age 30. 179 applicants, 28% accepted, 25 enrolled. In 2005, 13 master's, 8 doctorates awarded. Terminal master's awarded for partial completion of doctoral program. *Degree requirements:* For master's, one foreign language, thesis (for some programs); registration; for doctorate, one foreign language, thesis/dissertation, comprehensive exam, registration. *Entrance requirements:* For master's and doctorate, GRE General Test, GRE Subject Test, 3 letters of recommendation. Additional exam requirements/recommendations for international students: Required—TOEFL (minimum score 600 paper-based; 250 computer-based). *Application deadline:* For fall admission, 1/1 for domestic students, 1/1 for international students. Application fee: $60. *Expenses:* Tuition: Full-time $31,530; part-time $985 per credit. Required fees: $316; $40 per semester. Tuition and fees vary according to course level and program. *Financial support:* In 2005–06, 110 students received support, including 7 fellowships with full tuition reimbursements available (averaging $16,500 per year), 62 research assistantships with full tuition reimbursements available (averaging $16,000 per year), 38 teaching assistantships with full tuition reimbursements available (averaging $16,000 per year); Federal Work-Study, institutionally sponsored loans, scholarships/grants, and traineeships also available. Financial award application deadline: 1/1; financial award applicants required to submit FAFSA. *Unit head:* Dr. Geoffrey M. Cooper, Chairman, 617-353-3856, Fax: 617-353-6340, E-mail: gmcooper@bu.edu. *Application contact:* Ruth Greene, Academic Administrator, 617-353-2432, Fax: 617-353-6340, E-mail: greene@bu.edu.

Boston University, School of Medicine, Division of Graduate Medical Sciences, Boston, MA 02215. Offers MA, MS, PhD, MBA/MA, MD/MA, MD/PhD, MPH/MA. Part-time programs available. *Faculty:* 80 full-time (20 women), 134 part-time/adjunct (19 women). *Students:* 471 full-time (229 women), 26 part-time (14 women); includes 128 minority (19 African Americans, 82 Asian Americans or Pacific Islanders, 27 Hispanic Americans), 77 international. Average age 27. In 2005, 75 master's, 11 doctorates awarded. Terminal master's awarded for partial completion of doctoral program. *Degree requirements:* For master's, qualifying exam; for doctorate, thesis/dissertation, qualifying exam. *Entrance requirements:* Additional exam requirements/recommendations for international students: Required—TOEFL. *Application deadline:* For spring admission, 10/15 priority date for domestic students. Electronic applications accepted. *Expenses: Contact institution.* Tuition and fees vary according to course level and program. *Financial support:* In 2005–06, 38 fellowships with tuition reimbursements, 121 research assistantships with tuition reimbursements, 6 teaching assistantships with tuition reimbursements were awarded; Federal Work-Study, scholarships/grants, and traineeships also available. *Unit head:* Dr. Carl Franzblau, Associate Dean, 617-638-5120, Fax: 617-638-4842, E-mail: medsci@bu.edu. *Application contact:* Michelle Hall, Assistant Director of Admissions, 617-638-5121, Fax: 617-638-5740, E-mail: natashah@bu.edu.

Bowling Green State University, Graduate College, College of Arts and Sciences, Department of Biological Sciences, Bowling Green, OH 43403. Offers MAT, MS, PhD. Part-time programs available. *Faculty:* 27 full-time (7 women), 27 part-time/adjunct (7 women). *Students:* 83 full-time (45 women), 17 part-time (6 women); includes 2 African Americans, 4 Asian Americans or Pacific Islanders, 35 international. Average age 29. 68 applicants, 53% accepted, 24 enrolled. In 2005, 15 master's, 4 doctorates awarded. *Degree requirements:* For master's, thesis or alternative; for doctorate, thesis/dissertation, comprehensive exam. *Entrance requirements:* For master's and doctorate, GRE General Test. Additional exam requirements/recommendations for international students: Required—TOEFL. *Application deadline:* For fall admission, 2/1 for domestic students. Application fee: $30. Electronic applications accepted. *Financial support:* In 2005–06, 10 research assistantships with full tuition reimbursements (averaging $13,391 per year), 71 teaching assistantships with full tuition reimbursements (averaging $11,946 per year) were awarded; Federal Work-Study and unspecified assistantships also available. *Faculty research:* Aquatic ecology, helminth energetics, endocrinology and neurophysiology, nitrogen fixation, photosynthesis. *Unit head:* Dr. Scott Rogers, Chair, 419-372-2332. *Application contact:* Dr. Stan Smith, Graduate Coordinator, 419-372-8259.

See Close-Up on page 89.

Bradley University, Graduate School, College of Liberal Arts and Sciences, Department of Biology, Peoria, IL 61625-0002. Offers MS. Part-time programs available. *Students:* 4 full-time (3 women), 2 part-time (1 woman). 8 applicants, 63% accepted, 2 enrolled. *Degree requirements:* For master's, thesis, comprehensive exam. *Entrance requirements:* For master's, GRE General Test, 2 letters of recommendation. Additional exam requirements/recommendations for international students: Required—TOEFL (minimum score 550 paper-based; 213 computer-based). *Application deadline:* For fall admission, 5/15 priority date for domestic students, 5/15 priority date for international students; for spring admission, 10/15 priority date for domestic students, 10/15 priority date for international students. Applications are processed on a rolling basis. Application fee: $40 ($50 for international students). *Financial support:* In 2005–06, 1 research assistantship with full and partial tuition reimbursement (averaging $5,060 per year) was awarded; scholarships/grants, tuition waivers (partial), and unspecified assistantships also available. Financial award application deadline: 4/1. *Unit head:* Dr. Erich Stabenau, Chairperson, 309-677-3012, E-mail: eks@bradley.edu. *Application contact:* Dr. Keith Johnson, Graduate Coordinator, 309-677-3015, E-mail: kjohnso@bradley.edu.

Brandeis University, Graduate School of Arts and Sciences, Post-Baccalaureate Premedical Program, Waltham, MA 02454-9110. Offers Certificate. *Students:* 9; includes 1 African American, 1 Asian American or Pacific Islander. Average age 26. 62 applicants, 63% accepted, 9 enrolled. *Entrance requirements:* For degree, GRE, SAT, resumé with paid and volunteer work relevant to field of medicine, letters of recommendation. Additional exam requirements/recommendations for international students: Required—TOEFL (minimum score 600 paper-based; 250 computer-based). *Application deadline:* For fall admission, 5/1 for domestic students. Applications are processed on a rolling basis. Application fee: $55. Electronic applications accepted. *Financial support:* Application deadline: 4/15; *Unit head:* Kate Fukawa-Connelly, Assistant Dean for Health Professions Advising, 781-736-3470, Fax: 781-736-3469, E-mail: kfconnelly@brandeis.edu.

Brandeis University, Graduate School of Arts and Sciences, Programs in Life Sciences, Waltham, MA 02454-9110. Offers biochemistry (MS, PhD); biophysics and structural biology (MS, PhD); molecular and cell biology (MS, PhD), including genetics (PhD), microbiology (PhD), molecular and cell biology, molecular biology (PhD); neuroscience (MS, PhD). Part-time programs available. *Faculty:* 56 full-time (15 women). *Students:* 156 full-time (67 women). 473 applicants, 16% accepted. In 2005, 21 master's, 20 doctorates awarded. Terminal master's awarded for partial completion of doctoral program. *Degree requirements:* For master's, thesis/dissertation, registration; for doctorate, thesis/dissertation, comprehensive exam, registration. *Entrance requirements:* For doctorate, GRE General Test. Additional exam requirements/recommendations for international students: Required—TOEFL (minimum score 600 paper-based; 250 computer-based). *Application deadline:* Applications are processed on a rolling basis. Application fee: $55. Electronic applications accepted. *Financial support:* Fellowships, research assistantships, teaching assistantships, career-related internships or fieldwork, scholarships/grants, and tuition waivers (full and partial) available. Support available to part-time students. Financial award application deadline: 4/15; financial award applicants required to submit CSS PROFILE or FAFSA. *Application contact:* Margaret Haley, Assistant Dean, Graduate Admissions, 781-736-3406, Fax: 781-736-3412, E-mail: haley@brandeis.edu.

Brigham Young University, Graduate Studies, College of Biological and Agricultural Sciences, Department of Integrative Biology, Provo, UT 84602-1001. Offers biological science education (MS); integrative biology (MS, PhD); wildlife and wildlands conservation (MS, PhD). *Faculty:* 35 full-time (3 women). *Students:* 19 full-time (8 women), 24 part-time (7 women); includes 2 minority (1 Asian American or Pacific Islander, 1 Hispanic American). Average age 27. 31 applicants, 74% accepted, 17 enrolled. In 2005, 13 master's awarded. *Median time to degree:* Of those who began their doctoral program in fall 1997, 100% received their degree in 8 years or less. *Degree requirements:* For master's and doctorate, thesis/dissertation,

comprehensive exam, registration. *Entrance requirements:* For master's and doctorate, GRE General Test, minimum GPA of 3.0 for last 60 credit hours of course work. Additional exam requirements/recommendations for international students: Required—TOEFL (minimum score 550 paper-based; 213 computer-based), GRE. *Application deadline:* For fall admission, 1/31 for domestic students, 1/31 for international students. Application fee: $50. Electronic applications accepted. *Financial support:* In 2005–06, 56 students received support, including 1 fellowship with full and partial tuition reimbursement available (averaging $5,500 per year), 22 research assistantships with full and partial tuition reimbursements available (averaging $12,000 per year), 33 teaching assistantships with full and partial tuition reimbursements available (averaging $12,000 per year); career-related internships or fieldwork, institutionally sponsored loans, scholarships/grants, tuition waivers (full and partial), and unspecified assistantships also available. Financial award application deadline: 3/1. *Faculty research:* Systematics, bioinformatics, conservation. Total annual research expenditures: $268,851. *Unit head:* Dr. Larry L. St. Clair, Chair, 801-422-2582, Fax: 801-422-0090, E-mail: larry_stclair@byu.edu. *Application contact:* Nancy P. Heiss, Graduate Secretary, 801-422-2010, Fax: 801-422-0090, E-mail: nancy_heiss@byu.edu.

Brock University, Graduate Studies, Faculty of Mathematics and Science, Program in Biological Sciences, St. Catharines, ON L2S 3A1, Canada. Offers biology (M Sc, PhD). Part-time programs available. *Faculty:* 18 full-time (5 women), 6 part-time/adjunct (3 women). *Students:* 34 full-time (20 women), 1 part-time, 9 international. 28 applicants, 43% accepted. In 2005, 5 degrees awarded. *Degree requirements:* For master's and doctorate, thesis/dissertation. *Entrance requirements:* For master's, honors B Sc in biology, minimum undergraduate GPA of 3.0. Additional exam requirements/recommendations for international students: Required—TOEFL. *Application deadline:* Applications are processed on a rolling basis. Application fee: $75. Electronic applications accepted. *Financial support:* Fellowships, research assistantships, teaching assistantships, career-related internships or fieldwork, scholarships/grants, and unspecified assistantships available. Support available to part-time students. *Faculty research:* Viticulture, neurobiology, ecology, molecular biology, molecular genetics. *Unit head:* Dr. Miriam Richards, Graduate Program Director, 905-688-5550 Ext. 3956, Fax: 905-688-1855, E-mail: miriam@brocku.ca. *Application contact:* Dr. Miriam Richards, Graduate Program Director, 905-688-5550 Ext. 3956, Fax: 905-688-1855, E-mail: miriam@brocku.ca.

Brooklyn College of the City University of New York, Division of Graduate Studies, Department of Biology, Brooklyn, NY 11210-2889. Offers applied biology (MA); biology (MA, PhD). The department offers courses at Brooklyn College that are creditable toward the CUNY doctoral degree. *Students:* 10 applicants, 100% accepted, 0 enrolled. In 2005, 1 degree awarded. *Degree requirements:* For master's, one foreign language, thesis or alternative. *Entrance requirements:* For master's, GRE General Test, GRE Subject Test, minimum GPA of 3.0, 2 letters of recommendation. Additional exam requirements/recommendations for international students: Required—TOEFL. *Application deadline:* For fall admission, 3/1 for domestic students, 2/1 for international students; for spring admission, 11/1 for domestic students, 10/1 for international students. Application fee: $125. *Expenses:* Tuition, state resident: full-time $6,400; part-time $270 per credit. Tuition, nonresident: full-time $12,000; part-time $500 per credit. Required fees: $118 per term. *Financial support:* Federal Work-Study, institutionally sponsored loans, and scholarships/grants available. Support available to part-time students. Financial award application deadline: 5/1; financial award applicants required to submit FAFSA. *Faculty research:* Evolutionary biology, molecular biology of development, cell biology, comparative endocrinology, ecology. *Unit head:* Dr. John Blamire, Chairperson, 718-951-5396, E-mail: jblamire@brooklyn.cuny.edu. *Application contact:* Marianne Booufall-Tynan, Director of Admissions, 718-951-5902, E-mail: adminqry@brooklyn.cuny.edu.

Brown University, Graduate School, Division of Biology and Medicine, Providence, RI 02912. Offers M Med Sc, MA, MPH, MS, Sc M, PhD, MD/PhD. Part-time programs available. Terminal master's awarded for partial completion of doctoral program. *Degree requirements:* For doctorate, thesis/dissertation. *Entrance requirements:* For master's and doctorate, GRE General Test. Additional exam requirements/recommendations for international students: Required—TOEFL. Electronic applications accepted.

Bucknell University, Graduate Studies, College of Arts and Sciences, Department of Biology, Lewisburg, PA 17837. Offers MA, MS. Part-time programs available. *Degree requirements:* For master's, thesis. *Entrance requirements:* For master's, GRE General Test, GRE Subject Test, minimum GPA of 2.8. Additional exam requirements/recommendations for international students: Required—TOEFL.

Buffalo State College, State University of New York, Graduate Studies and Research, Faculty of Natural and Social Sciences, Department of Biology, Buffalo, NY 14222-1095. Offers biology (MA); secondary education (MS Ed), including biology. Evening/weekend programs available. *Degree requirements:* For master's, thesis (for some programs), project. *Entrance requirements:* For master's, minimum GPA of 2.75. Additional exam requirements/recommendations for international students: Required—TOEFL (minimum score 550 paper-based; 213 computer-based).

California Institute of Technology, Division of Biology, Pasadena, CA 91125-0001. Offers biochemistry and molecular biophysics (PhD); cell biology and biophysics (PhD); developmental biology (PhD); genetics (PhD); immunology (PhD); molecular biology (PhD); neurobiology (PhD). *Faculty:* 38 full-time (8 women). *Students:* 72 full-time (33 women); includes 6 Hispanic Americans. 200 applicants, 15% accepted, 6 enrolled. In 2005, 20 doctorates awarded. *Degree requirements:* For doctorate, thesis/dissertation, qualifying exam. *Entrance requirements:* For doctorate, GRE General Test. Additional exam requirements/recommendations for international students: Required—TOEFL. *Application deadline:* For fall admission, 1/1 for domestic students, 1/1 for international students. Application fee: $80. Electronic applications accepted. *Financial support:* In 2005–06, fellowships with full tuition reimbursements (averaging $21,195 per year), teaching assistantships with full tuition reimbursements (averaging $4,305 per year) were awarded; research assistantships with full tuition reimbursements, institutionally sponsored loans also available. Financial award application deadline: 1/1. *Faculty research:* Molecular genetics of differentiation and development, structure of biological macromolecules, molecular and integrative neurobiology. *Unit head:* Elliot Meyerowitz, Chairman, 626-395-4951, Fax: 626-683-3343, E-mail: biograd@cco.caltech.edu. *Application contact:* Elizabeth Ayala, Graduate Program Coordinator, 626-395-4497, Fax: 626-683-3343, E-mail: biograd@cco.caltech.edu.

See Close-Up on page 91.

California Polytechnic State University, San Luis Obispo, College of Science and Mathematics, Department of Biological Sciences, San Luis Obispo, CA 93407. Offers MS. *Faculty:* 8 full-time (0 women), 3 part-time/adjunct (1 woman). *Students:* 10 full-time (5 women), 8 part-time (2 women). 26 applicants, 54% accepted, 5 enrolled. In 2005, 7 degrees awarded. *Degree requirements:* For master's, thesis (for some programs), comprehensive exam (for some programs). *Entrance requirements:* For master's, GRE General Test, minimum GPA of 3.0 in last 90 quarter units. Additional exam requirements/recommendations for international students: Required—TOEFL, TWE. *Application deadline:* For fall admission, 6/1 for domestic students, 11/30 for international students. For winter admission, 8/1 for domestic students; for spring admission, 12/1 for domestic students. Applications are processed on a rolling basis. Application fee: $55. Electronic applications accepted. *Expenses:* Tuition, nonresident: part-time $226 per unit. Required fees: $1,063 per unit. *Financial support:* Teaching assistantships, career-related internships or fieldwork and Federal Work-Study available. Support available to part-time students. Financial award application deadline: 3/2; financial award applicants required to submit FAFSA. *Faculty research:* Ancient fossil DNA, restoration ecology microbe biodiversity indices, biological inventories. *Unit head:* Dennis F. Frey, Graduate Coordinator, 805-756-2802, Fax: 805-756-1419, E-mail: dfrey@calpoly.edu. *Application contact:* Dennis F. Frey, Graduate Coordinator, 805-756-2802, Fax: 805-756-1419, E-mail: dfrey@calpoly.edu.

California State Polytechnic University, Pomona, Academic Affairs, College of Science, Program in Biological Sciences, Pomona, CA 91768-2557. Offers MS. Part-time programs avail-

Biological and Biomedical Sciences—General

California State Polytechnic University, Pomona (continued)
able. *Students:* 38 full-time (21 women), 37 part-time (18 women); includes 33 minority (1 African American, 19 Asian Americans or Pacific Islanders, 13 Hispanic Americans), 6 international. Average age 28. 42 applicants, 69% accepted, 21 enrolled. In 2005, 17 degrees awarded. *Degree requirements:* For master's, thesis. *Entrance requirements:* For master's, GRE General Test. *Application deadline:* For fall admission, 5/1 for domestic students. For winter admission, 10/15 for domestic students; for spring admission, 1/20 for domestic students. Applications are processed on a rolling basis. Application fee: $55. Electronic applications accepted. *Expenses:* Tuition, nonresident: full-time $9,021. Required fees: $3,597. *Financial support:* Career-related internships or fieldwork, Federal Work-Study, and institutionally sponsored loans available. Support available to part-time students. Financial award application deadline: 3/2; financial award applicants required to submit FAFSA. *Unit head:* Dr. David J. Moriarty, Graduate Coordinator, 909-869-4055, E-mail: djmoriarty@csupomona.edu.

California State University, Chico, Graduate School, College of Natural Sciences, Department of Biological Sciences, Program in Biological Sciences, Chico, CA 95929-0515. Offers MS. *Degree requirements:* For master's, thesis, oral exam. *Entrance requirements:* For master's, GRE General Test, 2 letters of recommendation. Additional exam requirements/recommendations for international students: Required—TOEFL (minimum score 550 paper-based; 213 computer-based). Electronic applications accepted.

California State University, Dominguez Hills, College of Natural and Behavioral Science, Department of Biology and Bioinformatics, Carson, CA 90747-0001. Offers biology (MA); human cytogenic technology (Certificate). Part-time and evening/weekend programs available. *Degree requirements:* For master's, thesis (for some programs). *Entrance requirements:* For master's, GRE General Test, GRE Subject Test, minimum GPA of 2.5.

California State University, East Bay, Academic Programs and Graduate Studies, College of Science, Department of Biological Sciences, Hayward, CA 94542-3000. Offers biological sciences (MS); marine sciences (MS). Part-time programs available. *Students:* 24. 24 applicants, 50% accepted. *Degree requirements:* For master's, thesis. *Entrance requirements:* For master's, GRE Subject Test, minimum GPA of 3.0 in field, 2.75 overall. Additional exam requirements/recommendations for international students: Required—TOEFL (minimum score 550 paper-based; 213 computer-based). *Application deadline:* For fall admission, 5/31 for domestic students, 4/30 for international students. For winter admission, 9/30 for domestic students. Applications are processed on a rolling basis. Application fee: $55. Electronic applications accepted. *Financial support:* Career-related internships or fieldwork, Federal Work-Study, and institutionally sponsored loans available. Support available to part-time students. Financial award application deadline: 3/2. *Unit head:* Dr. Donald Gailey, Chair, 510-885-4763, E-mail: donald.gailey@csueastlay.edu. *Application contact:* Deborah Baker, Associate Director, 510-885-3286, Fax: 510-885-4777, E-mail: deborah.baker@csueastbay.edu.

California State University, Fresno, Division of Graduate Studies, College of Science and Mathematics, Department of Biology, Fresno, CA 93740-8027. Offers MA. Part-time and evening/weekend programs available. *Degree requirements:* For master's, thesis. *Entrance requirements:* For master's, GRE General Test, GRE Subject Test, minimum GPA of 2.5 in last 60 units. Additional exam requirements/recommendations for international students: Required—TOEFL. Electronic applications accepted. *Faculty research:* Genome neuroscience, ecology conflict resolution, biomechanics, cell death, vibrio cholerae.

California State University, Fullerton, Graduate Studies, College of Natural Science and Mathematics, Department of Biological Science, Fullerton, CA 92834-9480. Offers biological science (MS); botany (MS); microbiology (MS). Part-time programs available. *Students:* 18 full-time (10 women), 41 part-time (26 women); includes 23 minority (3 African Americans, 9 Asian Americans or Pacific Islanders, 11 Hispanic Americans), 3 international. Average age 27. 61 applicants, 39% accepted, 19 enrolled. In 2005, 10 degrees awarded. *Degree requirements:* For master's, thesis. *Entrance requirements:* For master's, DAT, GRE General Test and GRE Subject Test, or MCAT, minimum GPA of 3.0 in biology. Application fee: $55. *Expenses:* Tuition, nonresident: part-time $339 per unit. *Financial support:* Teaching assistantships, career-related internships or fieldwork, Federal Work-Study, institutionally sponsored loans, and scholarships/grants available. Support available to part-time students. Financial award application deadline: 3/1. *Faculty research:* Glycosidase release and the block to polyspermy in ascidian eggs. *Unit head:* Dr. Robert Koch, Chair, 714-278-3614. *Application contact:* Dr. Michael Horn, Adviser, 714-278-3707.

California State University, Long Beach, Graduate Studies, College of Natural Sciences and Mathematics, Department of Biological Sciences, Long Beach, CA 90840. Offers microbiology (MPH, MS), including medical technology, microbiology (MS), nurse epidemiology. Part-time programs available. *Faculty:* 18 full-time (2 women). *Students:* 9 full-time (8 women), 42 part-time (24 women); includes 14 minority (2 American Indian/Alaska Native, 4 Asian Americans or Pacific Islanders, 8 Hispanic Americans), 1 international. Average age 29. 68 applicants, 37% accepted, 15 enrolled. In 2005, 6 degrees awarded. *Entrance requirements:* For master's, GRE Subject Test, minimum GPA of 3.0. *Application deadline:* For fall admission, 7/1 for domestic students; for spring admission, 12/1 for domestic students. Applications are processed on a rolling basis. Application fee: $55. Electronic applications accepted. *Expenses:* Tuition, nonresident: part-time $339 per semester hour. *Financial support:* Teaching assistantships, Federal Work-Study, institutionally sponsored loans, scholarships/grants, traineeships, and unspecified assistantships available. Financial award application deadline: 3/1. *Unit head:* Dr. Editte Gharakhanian, Chair, 562-985-8878, Fax: 562-985-8878, E-mail: eghara@csulb.edu. *Application contact:* Dr. Judith A Brusslan, Graduate Advisor, 562-985-4806, Fax: 562-985-8878, E-mail: bruss@csulb.edu.

California State University, Los Angeles, Graduate Studies, College of Natural and Social Sciences, Department of Biological Sciences, Los Angeles, CA 90032-8530. Offers biology (MS). Part-time and evening/weekend programs available. *Faculty:* 3 full-time (1 woman), 1 part-time/adjunct (0 women). *Students:* 90 full-time (61 women), 48 part-time (27 women); includes 105 minority (14 African Americans, 50 Asian Americans or Pacific Islanders, 41 Hispanic Americans). In 2005, 13 degrees awarded. *Degree requirements:* For master's, comprehensive exam or thesis. *Entrance requirements:* Additional exam requirements/recommendations for international students: Required—TOEFL. *Application deadline:* For fall admission, 6/30 for domestic students; for spring admission, 2/1 for domestic students. Applications are processed on a rolling basis. Application fee: $55. *Financial support:* Federal Work-Study available. Support available to part-time students. Financial award application deadline: 3/1. *Faculty research:* Ecology, environmental biology, cell and molecular biology, physiology, medical microbiology. *Unit head:* Dr. Nancy McQueen, Acting Chair, 323-343-2050, Fax: 323-343-6451.

California State University, Northridge, Graduate Studies, College of Science and Mathematics, Department of Biology, Northridge, CA 91330. Offers biology (MS); genetic counseling (MS). *Degree requirements:* For master's, comprehensive exam or thesis. *Entrance requirements:* For master's, GRE Subject Test. Additional exam requirements/recommendations for international students: Required—TOEFL. *Faculty research:* Cell adhesion, cancer research, fishery research.

California State University, Sacramento, Graduate Studies, College of Natural Sciences and Mathematics, Department of Biological Sciences, Sacramento, CA 95819-6048. Offers biological sciences (MA, MS); immunohematology (MS); marine science (MS). Part-time programs available. *Students:* 22 full-time (11 women), 53 part-time (37 women); includes 16 minority (3 African Americans, 7 Asian Americans or Pacific Islanders, 6 Hispanic Americans), 4 international. Average age 30. 79 applicants, 42% accepted, 24 enrolled. *Degree requirements:* For master's, thesis, writing proficiency exam. *Entrance requirements:* For master's, bachelor's degree in biology or equivalent, minimum GPA of 3.0 in biology, minimum overall GPA of 2.75 during last 2 years of course work. Additional exam requirements/recommendations for inter-

national students: Required—TOEFL. *Application deadline:* Applications are processed on a rolling basis. Application fee: $55. Electronic applications accepted. *Expenses:* Tuition, nonresident: part-time $339 per unit. Required fees: $276 per semester hour. *Financial support:* Research assistantships, teaching assistantships, career-related internships or fieldwork and Federal Work-Study available. Support available to part-time students. Financial award application deadline: 3/1. *Unit head:* Dr. Nick Ewing, Chair, 916-278-6535, Fax: 916-278-6993.

California State University, San Bernardino, Graduate Studies, College of Natural Sciences, Department of Biology, San Bernardino, CA 92407-2397. Offers MS. Part-time programs available. *Faculty:* 14 full-time, 18 part-time/adjunct (0 women). *Students:* 5 full-time (1 woman), 12 part-time (9 women); includes 5 minority (1 African American, 3 Asian Americans or Pacific Islanders, 1 Hispanic American), 1 international. Average age 29. 13 applicants, 46% accepted, 6 enrolled. In 2005, 1 degree awarded. *Degree requirements:* For master's, thesis or alternative. *Entrance requirements:* For master's, minimum GPA of 3.0. *Application deadline:* For fall admission, 8/31 for domestic students. Application fee: $55. *Expenses:* Tuition, nonresident: full-time $8,136; part-time $226 per unit. Required fees: $3,884. *Financial support:* Fellowships, research assistantships, teaching assistantships, career-related internships or fieldwork available. *Faculty research:* Ecology, molecular biology, physiology, cell biology, neurobiology. *Unit head:* Dr. David M. Polcyn, Chair, 909-537-5313, Fax: 909-537-7038, E-mail: dpolcyn@csusb.edu.

California State University, San Marcos, College of Arts and Sciences, Program in Biological Sciences, San Marcos, CA 92096-0001. Offers MS. Part-time programs available. *Faculty:* 12 full-time (6 women), 7 part-time/adjunct (2 women). *Students:* 19 full-time (15 women), 9 part-time (6 women); includes 2 minority (1 African American, 1 Hispanic American), 1 international. Average age 32. 18 applicants, 56% accepted. In 2005, 1 degree awarded. *Degree requirements:* For master's, thesis. *Entrance requirements:* For master's, GRE Subject Test, minimum GPA of 2.7 in mathematics and science or minimum GPA of 3.0 in the last 35 units of mathematics and science. *Application deadline:* 3/15 for domestic students; for spring admission, 10/15 priority date for domestic students. Application fee: $55. *Expenses:* Tuition, nonresident: part-time $339 per unit. Required fees: $1,171 per term. Tuition and fees vary according to course load. *Financial support:* Fellowships with full tuition reimbursements, research assistantships, teaching assistantships available. *Faculty research:* Gene regulation of lifestates, carbon cycling, genetic markers of viral infection, neurobiology. *Unit head:* Denise Garcia, Chair, 760-750-4132, E-mail: dgarcia@csusm.edu. *Application contact:* Catalina Huggins, Administrative Coordinator, 760-750-4103, E-mail: chuggins@csusm.edu.

California University of Pennsylvania, School of Graduate Studies, School of Science and Technology, Department of Biological and Environmental Sciences, California, PA 15419-1394. Offers biology (M Ed, MS). Part-time and evening/weekend programs available. *Faculty:* 5 full-time (1 woman). *Students:* 1 (woman) full-time, 2 part-time (both women). Average age 25. In 2005, 3 degrees awarded. *Degree requirements:* For master's, thesis, comprehensive exam. *Entrance requirements:* For master's, GRE General Test, minimum GPA of 2.5, teaching certificate. Additional exam requirements/recommendations for international students: Required—TOEFL. *Application deadline:* Applications are processed on a rolling basis. Application fee: $25. Electronic applications accepted. *Financial support:* Tuition waivers (full) and unspecified assistantships available. *Unit head:* Dr. David Argent, Coordinator, 724-938-1529, E-mail: argent@cup.edu.

Carleton University, Faculty of Graduate Studies, Faculty of Science, Department of Biology, Ottawa, ON K1S 5B6, Canada. Offers M Sc, PhD. *Degree requirements:* For master's, thesis/dissertation, seminar; for doctorate, thesis/dissertation, seminar, comprehensive exam. *Entrance requirements:* For master's, honors degree in science; for doctorate, M Sc. Additional exam requirements/recommendations for international students: Required—TOEFL. *Application deadline:* Applications are processed on a rolling basis. Application fee: $75 Canadian dollars. *Financial support:* Fellowships, research assistantships, teaching assistantships, institutionally sponsored loans, scholarships/grants, and unspecified assistantships available. *Faculty research:* Biochemical, structural, and genetic regulation in cells; behavioral ecology; insect taxonomy; physiology of cells. *Unit head:* James Cheetham, Chair, 613-520-2600 Ext. 3867, Fax: 613-520-2569, E-mail: cns@carleton.ca. *Application contact:* Lenore Fahrig, Supervisor of Graduate Studies, 613-520-2600 Ext. 3515, Fax: 613-520-2569, E-mail: cns@carleton.ca.

Carnegie Mellon University, Mellon College of Science, Department of Biological Sciences, Pittsburgh, PA 15213-3891. Offers biochemistry (PhD); biophysics (PhD); cell biology (PhD); computational biology (MS, PhD); developmental biology (PhD); genetics (PhD); molecular biology (PhD); neurobiology (PhD). *Degree requirements:* For doctorate, thesis/dissertation, comprehensive exam. *Entrance requirements:* For doctorate, GRE General Test, GRE Subject Test, interview. Electronic applications accepted. *Faculty research:* Genetic structure, function, and regulation; protein structure and function; biological membranes; biological spectroscopy.

See Close-Up on page 93.

Case Western Reserve University, School of Graduate Studies, Department of Biology, Cleveland, OH 44106. Offers MS, PhD. Part-time programs available. Terminal master's awarded for partial completion of doctoral program. *Degree requirements:* For master's, thesis or alternative; for doctorate, thesis/dissertation. *Entrance requirements:* For master's and doctorate, GRE General Test and GRE Subject Test or MCAT. Additional exam requirements/recommendations for international students: Required—TOEFL. *Faculty research:* Cellular, developmental, and molecular biology; genetics; genetic engineering; biotechnology; ecology.

See Close-Up on page 95.

Case Western Reserve University, School of Medicine, Biomedical Sciences Training Program, Cleveland, OH 44106. Offers PhD. *Degree requirements:* For doctorate, thesis/dissertation. *Entrance requirements:* For doctorate, GRE General Test. Additional exam requirements/recommendations for international students: Required—TOEFL. Electronic applications accepted. *Faculty research:* Biochemistry, molecular biology, immunology, genetics, neurosciences.

See Close-Up on page 95.

Case Western Reserve University, School of Medicine, Medical Scientist Training Program, Cleveland, OH 44106. Offers MD/PhD. *Students:* 12 full-time (3 women); includes 2 minority (both African Americans) Average age 23. 304 applicants, 16% accepted, 12 enrolled. *Median time to degree:* Of those who began their doctoral program in fall 1997, 100% received their degree in 8 years or less. *Application deadline:* For fall admission, 12/15 for domestic students. Applications are processed on a rolling basis. Application fee: $25. Electronic applications accepted. *Financial support:* In 2005–06, 90 students received support, including 12 fellowships with full tuition reimbursements (averaging $23,000 per year); traineeships and MSTP Grant also available. *Unit head:* Clifford V Harding, Professor and Director, 216-368-5059, Fax: 216-368-5295, E-mail: cvh3@case.edu. *Application contact:* Donna Mcilwain, Department Assistant, 216-368-3404, Fax: 216-368-5295, E-mail: djh5@case.edu.

The Catholic University of America, School of Arts and Sciences, Department of Biology, Washington, DC 20064. Offers cell and microbial biology (MS, PhD), including cell biology, microbiology; clinical laboratory science (MS, PhD). Part-time programs available. *Faculty:* 8 full-time (3 women), 1 (woman) part-time/adjunct. *Students:* 6 full-time (2 women), 20 part-time (11 women); includes 3 African Americans, 1 Asian American or Pacific Islander, 13 international. Average age 32. 30 applicants, 50% accepted, 3 enrolled. Terminal master's awarded for partial completion of doctoral program. *Degree requirements:* For master's, thesis or alternative, comprehensive exam; for doctorate, thesis/dissertation, comprehensive exam. *Entrance requirements:* For master's and doctorate, GRE General Test, GRE Subject Test, 3 letters of recommendation. Additional exam requirements/recommendations for international students: Required—TOEFL (minimum score 580 paper-based; 237 computer-based). *Application deadline:* For fall admission, 2/1 for domestic students; for spring admission, 11/15 priority date for domestic students. Applications are processed on a rolling basis. Application fee: $55. Electronic applications accepted. *Expenses:* Tuition: Full-time $24,800; part-time $940 per

Biological and Biomedical Sciences—General

credit. Required fees: $1,090; $285 per term. Part-time tuition and fees vary according to course load and program. *Financial support:* Fellowships, research assistantships, teaching assistantships, career-related internships or fieldwork, scholarships/grants, tuition waivers (full and partial), and unspecified assistantships available. Support available to part-time students. Financial award application deadline: 2/1; financial award applicants required to submit FAFSA. *Faculty research:* Cell differentiation, regulation of cell growth, drug resistance, gene cloning and sequencing, developmental biology and neurobiology. *Unit head:* Dr. Venigalla Rao, Chair, 202-319-5267, Fax: 202-319-5721, E-mail: rao@cua.edu.

Central Connecticut State University, School of Graduate Studies, School of Arts and Sciences, Department of Biology, New Britain, CT 06050-4010. Offers anesthesia (MS); biological sciences (MA, MS), including ecology and environmental sciences (MA), general biology (MA); general health (Certificate). Part-time and evening/weekend programs available. *Faculty:* 11 full-time (4 women), 6 part-time/adjunct (4 women). *Students:* 72 full-time (47 women), 29 part-time (14 women); includes 9 minority (3 African Americans, 1 American Indian/Alaska Native, 3 Asian Americans or Pacific Islanders, 2 Hispanic Americans), 2 international. Average age 33. 26 applicants, 54% accepted, 9 enrolled. In 2005, 38 degrees awarded. *Degree requirements:* For master's, thesis or alternative, comprehensive exam; for Certificate, qualifying exam. *Entrance requirements:* For master's, minimum GPA of 2.7. Additional exam requirements/recommendations for international students: Required—TOEFL. *Application deadline:* For fall admission, 7/1 for domestic students; for spring admission, 12/1 for domestic students. Applications are processed on a rolling basis. Application fee: $50. Electronic applications accepted. *Expenses:* Tuition, area resident: full-time $3,780. Tuition, state resident: full-time $5,670; part-time $362 per credit. Tuition, nonresident: full-time $10,530; part-time $362 per credit. Required fees: $3,064. One-time fee: $62 part-time. Tuition and fees vary according to degree level and program. *Financial support:* In 2005–06, 4 students received support, including 1 research assistantship; career-related internships or fieldwork, Federal Work-Study, scholarships/grants, and unspecified assistantships also available. Support available to part-time students. Financial award application deadline: 3/1; financial award applicants required to submit FAFSA. *Faculty research:* Environmental science, anesthesia, health sciences, zoology, animal behavior. *Unit head:* Dr. Ruth Rollins, Chair, 860-832-2645.

Central Michigan University, College of Graduate Studies, College of Science and Technology, Department of Biology, Mount Pleasant, MI 48859. Offers biology (MS); conservation biology (MS). *Faculty:* 25 full-time (7 women). *Students:* 19 full-time (10 women), 31 part-time (13 women). Average age 27. In 2005, 18 degrees awarded. *Degree requirements:* For master's, thesis or alternative, registration. *Entrance requirements:* For master's, bachelor's degree in biology, minimum GPA of 3.0. *Application deadline:* Applications are processed on a rolling basis. Application fee: $35 ($45 for international students). *Expenses:* Tuition, area resident: Part-time $325 per credit hour. Tuition, state resident: part-time $603 per credit hour. Tuition and fees vary according to degree level and reciprocity agreements. *Financial support:* In 2005–06, 2 fellowships with tuition reimbursements, 11 research assistantships with tuition reimbursements, 26 teaching assistantships with tuition reimbursements were awarded; career-related internships or fieldwork and Federal Work-Study also available. Financial award application deadline: 3/7. *Faculty research:* Vertebrates, morphology and taxonomy of aquatic plants, molecular biology and genetics, microbials and invertebrate ecology. *Unit head:* Dr. Claudia B. Douglass, Chairperson, 989-774-3227, Fax: 989-774-3462, E-mail: doug1cb@cmich.edu. *Application contact:* Dr. A. Scott McNaught, Graduate Program Coordinator, 989-774-1335, Fax: 989-774-3462, E-mail: scott.mcnaught@cmich.edu.

Central Missouri State University, The Graduate School, College of Arts and Sciences, Department of Biology and Earth Science, Warrensburg, MO 64093. Offers biology (MS). Part-time programs available. *Faculty:* 12 full-time (3 women), 1 part-time/adjunct (0 women). *Students:* 4 full-time (2 women), 18 part-time (9 women); includes 1 minority (Hispanic American) Average age 33. 13 applicants, 69% accepted, 10 enrolled. In 2005, 6 degrees awarded. *Degree requirements:* For master's, oral exam, thesis or research report. *Entrance requirements:* For master's, GRE Subject Test, 30 hours of course work in biology, minimum undergraduate GPA of 2.5. Additional exam requirements/recommendations for international students: Required—TOEFL (minimum score 500 paper-based; 173 computer-based). *Application deadline:* For fall admission, 6/1 priority date for domestic students, 5/1 priority date for international students; for spring admission, 10/1 priority date for domestic students, 10/1 priority date for international students. Applications are processed on a rolling basis. Application fee: $30 ($50 for international students). *Expenses:* Tuition, state resident: full-time $5,160; part-time $215 per credit hour. Tuition, nonresident: full-time $10,320; part-time $430 per credit hour. Required fees: $336; $14 per credit hour. *Financial support:* In 2005–06, 8 teaching assistantships (averaging $3,750 per year) were awarded; Federal Work-Study, scholarships/grants, unspecified assistantships, and laboratory assistantships also available. Support available to part-time students. Financial award application deadline: 3/1; financial award applicants required to submit FAFSA. *Unit head:* Dr. John Gole, Chair, 660-543-4933, Fax: 660-543-8006, E-mail: gole@cmsu1.cmsu.edu.

Central Washington University, Graduate Studies, Research and Continuing Education, College of the Sciences, Department of Biological Sciences, Ellensburg, WA 98926. Offers MS. Part-time programs available. *Faculty:* 17 full-time (5 women). *Students:* 7 full-time (3 women), 2 part-time (1 woman); includes 2 minority (1 American Indian/Alaska Native, 1 Asian American or Pacific Islander). 8 applicants, 75% accepted, 6 enrolled. In 2005, 4 degrees awarded. *Degree requirements:* For master's, thesis or alternative. *Entrance requirements:* For master's, GRE General Test, minimum GPA of 3.0. Additional exam requirements/recommendations for international students: Required—TOEFL (minimum score 550 paper-based; 213 computer-based). *Application deadline:* For fall admission, 4/1 for domestic students. For winter admission, 10/1 for domestic students; for spring admission, 1/1 for domestic students. Applications are processed on a rolling basis. Application fee: $50. *Expenses:* Tuition, state resident: full-time $5,904; part-time $197 per credit. Tuition, nonresident: full-time $12,960; part-time $432 per credit. Required fees: $623. Tuition and fees vary according to degree level. *Financial support:* In 2005–06, 7 teaching assistantships with partial tuition reimbursements (averaging $8,100 per year) were awarded; research assistantships with partial tuition reimbursements, Federal Work-Study also available. Financial award application deadline: 3/1; financial award applicants required to submit FAFSA. *Unit head:* Dr. David Darda, Chair, 509-963-2731. *Application contact:* Justine Eason, Admissions Program Coordinator, 509-963-3103, Fax: 509-963-1799, E-mail: masters@cwu.edu.

Chicago State University, School of Graduate and Professional Studies, College of Arts and Sciences, Department of Biological Sciences, Chicago, IL 60628. Offers MS. Part-time and evening/weekend programs available. *Degree requirements:* For master's, thesis. *Entrance requirements:* For master's, minimum GPA of 2.75, 15 credit hours in biological sciences. *Faculty research:* Molecular genetics of gene complexes, mammalian immune cell function, genetics of agriculturally important microbes, environmental toxicology, neuromuscular physiology.

City College of the City University of New York, Graduate School, College of Liberal Arts and Science, Division of Science, Department of Biology, New York, NY 10031-9198. Offers MA, PhD. Part-time programs available. *Students:* 5 full-time (4 women), 27 part-time (18 women); includes 20 minority (7 African Americans, 9 Asian Americans or Pacific Islanders, 4 Hispanic Americans), 6 international. 31 applicants, 45% accepted, 12 enrolled. In 2005, 6 degrees awarded. Terminal master's awarded for partial completion of doctoral program. *Degree requirements:* For master's, thesis or alternative; for doctorate, one foreign language, thesis/dissertation, teaching experience. *Entrance requirements:* For master's and doctorate, GRE General Test. Additional exam requirements/recommendations for international students: Required—TOEFL (minimum score 500 paper-based; 213 computer-based). *Application deadline:* For fall admission, 5/1 for domestic students; for spring admission, 11/15 for domestic students. Application fee: $125. *Financial support:* Fellowships, research assistantships, teaching assistantships, career-related internships or fieldwork and scholarships/grants available. *Faculty research:*

Animal behavior, ecology, genetics, neurobiology, molecular biology. *Unit head:* Jane Gallagher, Chair, 212-650-6584, E-mail: janegall@sci.ccny.cuny.edu. *Application contact:* Ralph Zuzulo, Graduate Adviser, 212-650-6800.

City of Hope National Medical Center/Beckman Research Institute, City of Hope Graduate School of Biological Sciences, Duarte, CA 91010. Offers PhD. *Faculty:* 57 full-time (12 women). *Students:* 51 full-time (27 women). Average age 24. 135 applicants, 16% accepted, 14 enrolled. In 2005, 39 degrees awarded. *Median time to degree:* Of those who began their doctoral program in fall 1997, 87% received their degree in 8 years or less. *Degree requirements:* For doctorate, thesis/dissertation, comprehensive exam. *Entrance requirements:* For doctorate, GRE General Test, GRE Subject Test, 2 years of course work in chemistry (general and organic); 1 year course work in each biochemistry, general biology, and general physics; 2 semesters of course work in mathematics; significant research laboratory experience. Additional exam requirements/recommendations for international students: Required—TOEFL (minimum score 595 paper-based). *Application deadline:* For spring admission, 2/1 priority date for domestic students, 2/1 priority date for international students. Application fee: $0. All students receive a fellowship. *Financial support:* In 2005–06, 6 fellowships (averaging $26,000 per year) were awarded; health care benefits and tuition waivers (full) also available. *Faculty research:* DNA damage and repair, protein structure, cancer biology, T cells and immunology, RNA splicing and binding. Total annual research expenditures: $65 million. *Unit head:* Dr. John J. Rossi, Dean, 626-256-8775, Fax: 626-301-8105, E-mail: jrossi@coh.org. *Application contact:* Dr. Steven J. Novak, Associate Dean for Administration, 626-256-8775, Fax: 626-301-8105, E-mail: snovak@coh.org.

See Close-Up on page 97.

Clarion University of Pennsylvania, College of Graduate Studies, College of Arts and Sciences, Department of Biology, Clarion, PA 16214. Offers MS. *Degree requirements:* For master's, thesis or alternative. *Entrance requirements:* For master's, GRE General Test, minimum QPA of 2.75. Additional exam requirements/recommendations for international students: Required—TOEFL (minimum score 600 paper-based; 250 computer-based).

Clark Atlanta University, School of Arts and Sciences, Department of Biology, Atlanta, GA 30314. Offers MS, PhD. Part-time programs available. Terminal master's awarded for partial completion of doctoral program. *Degree requirements:* For master's, one foreign language, thesis; for doctorate, 2 foreign languages, thesis/dissertation. *Entrance requirements:* For master's, GRE General Test, minimum GPA of 2.5; for doctorate, GRE General Test, minimum graduate GPA of 3.0. *Faculty research:* Regulation of amino-DNA, cellular regulations.

Clark University, Graduate School, Department of Biology, Worcester, MA 01610-1477. Offers MA, PhD. *Faculty:* 9 full-time (3 women), 2 part-time/adjunct (1 woman). *Students:* 24 full-time (12 women), 1 part-time; includes 2 minority (1 Asian American or Pacific Islander, 1 Hispanic American), 9 international. Average age 31. 24 applicants, 50% accepted, 9 enrolled. In 2005, 3 master's, 1 doctorate awarded. *Degree requirements:* For master's and doctorate, thesis/dissertation. *Entrance requirements:* For master's and doctorate, GRE General Test. Additional exam requirements/recommendations for international students: Required—TOEFL. *Application deadline:* For fall admission, 2/15 for domestic students. Applications are processed on a rolling basis. Application fee: $50. Electronic applications accepted. *Expenses:* Tuition: Full-time $29,300. Required fees: $30. *Financial support:* In 2005–06, fellowships with full tuition reimbursements (averaging $18,750 per year), 4 research assistantships with full tuition reimbursements (averaging $18,750 per year), 12 teaching assistantships with full tuition reimbursements (averaging $18,750 per year) were awarded; scholarships/grants and tuition waivers (full and partial) also available. *Faculty research:* Nitrogen assimilation in marine algae, phylogenetic relationships, fungal tree of life, ancestral plasticity, drosophi genetic analysis, cytokinesis proteins. Total annual research expenditures: $606,000. *Unit head:* Dr. Susan Foster, Chair, 508-793-7173. *Application contact:* Paula Kupstas, Department Secretary, 528-793-7173, Fax: 528-793-8861, E-mail: biology@clarku.edu.

Clemson University, Graduate School, College of Agriculture, Forestry and Life Sciences, Department of Biological Sciences, Program in Biological Sciences, Clemson, SC 29634. Offers MS, PhD. *Students:* 21 full-time (14 women), 4 part-time (3 women); includes 1 minority (Hispanic American), 3 international. 12 applicants, 42% accepted, 5 enrolled. In 2005, 1 master's, 1 doctorate awarded. *Degree requirements:* For master's, thesis optional; for doctorate, thesis/dissertation, comprehensive exam. *Entrance requirements:* For master's and doctorate, GRE General Test. Additional exam requirements/recommendations for international students: Required—TOEFL. *Application deadline:* For fall admission, 6/1 for domestic students, 4/15 for international students. Application fee: $50. *Financial support:* Research assistantships, teaching assistantships available. Financial award application deadline: 3/15; financial award applicants required to submit FAFSA. *Application contact:* Information Contact, 864-656-3587, Fax: 864-656-0435.

See Close-Ups on pages 99 and 101.

Cleveland State University, College of Graduate Studies, College of Science, Department of Biological, Geological, and Environmental Sciences, Cleveland, OH 44115. Offers biology (MS); environmental science (MS); molecular medicine (PhD); regulatory biology (PhD). Part-time programs available. *Faculty:* 19 full-time (3 women), 34 part-time/adjunct (7 women). *Students:* 54 full-time (34 women), 34 part-time (17 women); includes 8 minority (6 African Americans, 2 Asian Americans or Pacific Islanders), 32 international. Average age 30. 34 applicants, 65% accepted, 18 enrolled. In 2005, 1 master's, 3 doctorates awarded. Terminal master's awarded for partial completion of doctoral program. *Median time to degree:* Of those who began their doctoral program in fall 1997, 100% received their degree in 8 years or less. *Degree requirements:* For master's, thesis (for some programs); for doctorate, thesis/dissertation, comprehensive exam. *Entrance requirements:* For master's and doctorate, GRE General Test, 2 letters of recommendation. Additional exam requirements/recommendations for international students: Required—TOEFL (minimum score 525 paper-based; 197 computer-based); Recommended—TSE. *Application deadline:* For fall admission, 4/1 priority date for domestic students, 4/1 priority date for international students; for spring admission, 12/1 priority date for domestic students. Applications are processed on a rolling basis. Application fee: $30. Electronic applications accepted. *Expenses:* Tuition, state resident: full-time $10,700. Tuition, nonresident: full-time $14,628. Tuition and fees vary according to program. *Financial support:* In 2005–06, 29 students received support, including research assistantships with full and partial tuition reimbursements available (averaging $16,500 per year), teaching assistantships with full and partial tuition reimbursements available (averaging $16,500 per year); institutionally sponsored loans and unspecified assistantships also available. *Faculty research:* Molecular and cell biology, immunology. *Unit head:* Dr. Michael Gates, Chairperson, 216-687-3917, Fax: 216-687-6972, E-mail: m.gates@csuohio.edu. *Application contact:* Dr. Jeffrey Dean, Graduate Program Director, 216-687-2440, Fax: 216-687-6972, E-mail: gpd.bges@csuohio.edu.

Cold Spring Harbor Laboratory, Watson School of Biological Sciences, Graduate Program, Cold Spring Harbor, NY 11724. Offers biological sciences (PhD). *Degree requirements:* For doctorate, thesis/dissertation, lab rotations, teaching experience, qualifying exam, post-doctoral proposals. *Entrance requirements:* For doctorate, GRE General Test, GRE Subject Test. Additional exam requirements/recommendations for international students: Required—TOEFL. *Faculty research:* Genetics, molecular, cellular and structural biology, neurobiology, cancer, plant biology.

See Close-Up on page 103.

College of Staten Island of the City University of New York, Graduate Programs, Program in Biology, Staten Island, NY 10314-6600. Offers MS. Part-time programs available. *Faculty:* 2 full-time (1 woman). *Students:* Average age 29. 5 applicants, 100% accepted, 3 enrolled. In 2005, 1 degree awarded. *Degree requirements:* For master's, thesis. *Entrance requirements:* For master's, GRE General Test, GRE Subject Test (biology), minimum GPA of 3.0 in science

Biological and Biomedical Sciences—General

College of Staten Island of the City University of New York (continued)
and math, 2.75 overall, bachelor's degree in biology, 2 letters of recommendation. Additional exam requirements/recommendations for international students: Required—TOEFL (minimum score 550 paper-based; 213 computer-based). *Application deadline:* Applications are processed on a rolling basis. Application fee: $125. *Expenses:* Tuition, state resident: full-time $6,400; part-time $270 per credit. Tuition, nonresident: part-time $500 per credit. Required fees: $328; $101 per semester. *Financial support:* Application deadline: 3/15; *Faculty research:* Nanotechnology at the biology-chemistry interface, reassessment of species boundaries in endemic Arkansas salamanders of plethodon ouachitar using molecular phylogeographic techniques, prenatal cocaine exposure induces AMPA receptor dysfunction, dispersal and population dynamics of the New England American oystercatcher, importance of metal storage in prey and digestion in predators to metal trophic transfer in estuarine food chains. Total annual research expenditures: $226,200. *Unit head:* Dr. Elena McCoy, Coordinator, 718-982-3862, Fax: 718-982-3852, E-mail: biologymasters@mail.csi.cuny.edu. *Application contact:* Emmanuel Esperance, Deputy Director of Office of Recruitment and Admissions, 718-982-2190, Fax: 718-982-2500, E-mail: admissions@mail.csi.cuny.edu.

The College of William and Mary, Faculty of Arts and Sciences, Department of Biology, Williamsburg, VA 23187-8795. Offers MS. *Faculty:* 26 full-time (10 women), 1 part-time/adjunct (0 women). *Students:* 18 full-time (13 women), 1 (woman) part-time; includes 1 minority (Asian American or Pacific Islander) Average age 25. 32 applicants, 59% accepted, 9 enrolled. In 2005, 9 degrees awarded. *Degree requirements:* For master's, thesis (for some programs), comprehensive exam. *Entrance requirements:* For master's, GRE Subject Test, GRE General Test, minimum GPA of 3.0. Additional exam requirements/recommendations for international students: Required—TOEFL. *Application deadline:* For fall admission, 1/15 for domestic students. Applications are processed on a rolling basis. Application fee: $30. *Expenses:* Tuition, state resident: full-time $5,828; part-time $245 per credit. Tuition, nonresident: full-time $17,980; part-time $685 per credit. Required fees: $3,051. Tuition and fees vary according to program. *Financial support:* In 2005–06, 7 research assistantships with full tuition reimbursements (averaging $9,000 per year), 15 teaching assistantships with full tuition reimbursements (averaging $9,000 per year) were awarded; Federal Work-Study, institutionally sponsored loans, and unspecified assistantships also available. Financial award application deadline: 3/1; financial award applicants required to submit FAFSA. *Faculty research:* Cellular and molecular biology, genetics, ecology, organismic biology, physiology. Total annual research expenditures: $1.6 million. *Unit head:* Dr. Paul D. Heideman, Chair, 757-221-2207, Fax: 757-221-6483, E-mail: pdheid@wm.edu. *Application contact:* Dr. John P. Swaddle, Graduate Director, 757-221-2231, Fax: 757-221-6483, E-mail: pswad@wm.edu.

Colorado State University, College of Veterinary Medicine and Biomedical Sciences, Department of Biomedical Sciences, Fort Collins, CO 80523-0015. Offers MS, PhD. *Faculty:* 25 full-time (6 women), 2 part-time/adjunct (0 women). *Students:* 59 full-time (31 women), 27 part-time (16 women); includes 6 minority (1 American Indian/Alaska Native, 2 Asian Americans or Pacific Islanders, 3 Hispanic Americans), 4 international. Average age 28. 76 applicants, 53% accepted, 39 enrolled. In 2005, 34 master's, 7 doctorates awarded. *Degree requirements:* For master's, thesis (for some programs); for doctorate, thesis/dissertation. *Entrance requirements:* For master's and doctorate, GRE General Test, GRE Subject Test. Additional exam requirements/recommendations for international students: Required—TOEFL (minimum score 650 paper-based; 280 computer-based). *Application deadline:* For fall admission, 4/1 priority date for domestic students, 4/1 priority date for international students. Applications are processed on a rolling basis. Application fee: $50. Electronic applications accepted. *Expenses:* Tuition, state resident: full-time $3,690; part-time $205 per credit. Tuition, nonresident: full-time $14,958; part-time $831 per credit. Required fees: $1,061. *Financial support:* In 2005–06, 7 fellowships with full tuition reimbursements (averaging $18,426 per year), 37 research assistantships with full and partial tuition reimbursements (averaging $16,704 per year), 11 teaching assistantships with full tuition reimbursements (averaging $16,655 per year) were awarded; Federal Work-Study and traineeships also available. Financial award application deadline: 4/1. *Faculty research:* Structural biology, integrative neuroscience, developmental neurobiology, reproductive physiology, equine reproduction. Total annual research expenditures: $8.8 million. *Unit head:* Dr. Barbara M. Sanborn, Chair, 970-491-7842, Fax: 970-491-7569, E-mail: barbara.sanborn@colostate.edu. *Application contact:* Alice M. Alexander, Graduate Education Coordinator, 970-491-6199, Fax: 970-491-7569, E-mail: alice.alexander@colostate.edu.

Colorado State University, Graduate School, College of Natural Sciences, Department of Biology, Fort Collins, CO 80523-0015. Offers botany (MS, PhD); zoology (MS, PhD). Part-time programs available. *Faculty:* 22 full-time (6 women). *Students:* 18 full-time (11 women), 21 part-time (11 women); includes 4 minority (2 American Indian/Alaska Native, 1 Asian American or Pacific Islander, 1 Hispanic American), 7 international. Average age 28. 43 applicants, 33% accepted, 9 enrolled. In 2005, 5 master's, 4 doctorates awarded. Terminal master's awarded for partial completion of doctoral program. *Degree requirements:* For master's, thesis (for some programs), comprehensive exam; for doctorate, thesis/dissertation, comprehensive exam. *Entrance requirements:* For master's and doctorate, GRE General Test, minimum GPA of 3.0. Additional exam requirements/recommendations for international students: Required—TOEFL. *Application deadline:* For fall admission, 1/15 priority date for domestic students, 1/15 priority date for international students; for spring admission, 11/1 for domestic students, 11/1 for international students. Applications are processed on a rolling basis. Application fee: $50. Electronic applications accepted. *Expenses:* Tuition, state resident: full-time $3,690; part-time $205 per credit. Tuition, nonresident: full-time $14,958; part-time $831 per credit. Required fees: $1,061. *Financial support:* In 2005–06, 127 students received support, including 6 fellowships with full tuition reimbursements available (averaging $22,500 per year), 27 research assistantships with full tuition reimbursements available (averaging $13,000 per year), 94 teaching assistantships with full tuition reimbursements available (averaging $12,888 per year); career-related internships or fieldwork, Federal Work-Study, institutionally sponsored loans, and traineeships also available. Financial award application deadline: 2/15. *Faculty research:* Aquatic and terrestrial ecology, cell biology and genetics, plant/animal physiology, developmental biology, evolutionary biology. Total annual research expenditures: $3.6 million. *Unit head:* Daniel R. Bush, Chair, 970-491-7011, Fax: 970-491-0649. *Application contact:* Dorothy Ramirez, Graduate Coordinator, 970-491-1923, Fax: 970-491-0649, E-mail: dorothy.ramirez@colostate.edu.

Colorado State University-Pueblo, College of Science and Mathematics, Pueblo, CO 81001-4901. Offers applied natural science (MS), including biochemistry, biology, chemistry. Part-time and evening/weekend programs available. *Faculty:* 15 full-time (6 women). *Students:* 14 full-time (9 women), 5 part-time (1 woman); includes 2 minority (both Hispanic Americans) 8 applicants, 100% accepted, 8 enrolled. In 2005, 9 degrees awarded. *Degree requirements:* For master's, thesis (for some programs), internship report (if non-thesis), comprehensive exam (for some programs), registration. *Entrance requirements:* For master's, GRE General Test, minimum GPA of 3.0. Additional exam requirements/recommendations for international students: Required—TOEFL (minimum score 500 paper-based; 173 computer-based). *Application deadline:* For fall admission, 6/15 priority date for domestic students, 6/15 priority date for international students; for spring admission, 10/15 priority date for domestic students, 10/15 priority date for international students. Applications are processed on a rolling basis. Application fee: $35. *Expenses:* Tuition, state resident: full-time $2,177; part-time $121 per credit hour. Tuition, nonresident: full-time $10,157; part-time $564 per credit hour. Required fees: $490; $41 per credit hour. *Financial support:* In 2005–06, 9 students received support, including 1 fellowship (averaging $1,000 per year), 3 teaching assistantships with partial tuition reimbursements available (averaging $9,000 per year); research assistantships, career-related internships or fieldwork, scholarships/grants, and unspecified assistantships also available. Financial award application deadline: 6/1; financial award applicants required to submit FAFSA. *Faculty research:* Fungal cell walls, molecular biology, bioactive materials synthesis, forensic chemistry, atomic force microscopy-surface chemistry. Total annual research expenditures: $400,000. *Unit head:* Dr. Kristina Proctor, Dean, 719-549-2340, Fax: 719-549-2732, E-mail: kristina.

proctor@colostate-pueblo.edu. *Application contact:* Dr. Melvin Druelinger, Director, MSANS Program, 719-549-2325, Fax: 719-549-2071, E-mail: mel.druelinger@colostate-pueblo.edu.

Columbia University, College of Physicians and Surgeons and Graduate School of Arts and Sciences, Graduate School of Arts and Sciences at the College of Physicians and Surgeons, New York, NY 10032. Offers M Phil, MA, PhD, MD/PhD. Only candidates for the PhD are admitted. Terminal master's awarded for partial completion of doctoral program. *Degree requirements:* For doctorate, thesis/dissertation. *Entrance requirements:* For master's and doctorate, GRE General Test. Additional exam requirements/recommendations for international students: Required—TOEFL. Expenses: Contact institution. Tuition and fees vary according to course level, course load, campus/location and program. *Faculty research:* Molecular and cellular biology, neurobiology, biochemistry, genetics and developmental biology, immunology.

See Close-Up on page 105.

Columbia University, Graduate School of Arts and Sciences, Division of Natural Sciences, Department of Biological Sciences, New York, NY 10027. Offers M Phil, MA, PhD, MD/PhD. *Faculty:* 25 full-time (34 women); includes 6 minority (1 African American, 1 American Indian/Alaska Native, 3 Asian Americans or Pacific Islanders, 1 Hispanic American), 42 international. Average age 28. 125 applicants, 24% accepted. In 2005, 28 master's, 15 doctorates awarded. *Degree requirements:* For master's, teaching experience, written exam; for doctorate, thesis/dissertation. *Entrance requirements:* For master's and doctorate, GRE General Test, GRE Subject Test. Additional exam requirements/recommendations for international students: Required—TOEFL. Application fee: $75. *Expenses:* Tuition: Full-time $31,448. Tuition and fees vary according to course level, course load, campus/location and program. *Financial support:* Fellowships, teaching assistantships, Federal Work-Study, and institutionally sponsored loans available. Support available to part-time students. Financial award application deadline: 1/5; financial award applicants required to submit FAFSA. *Unit head:* Michael Sheetz, Chair, 212-854-4857, Fax: 212-865-8246, E-mail: ms2001@columbia.edu.

See Close-Up on page 107.

Concordia University, School of Graduate Studies, Faculty of Arts and Science, Department of Biology, Montréal, QC H3G 1M8, Canada. Offers biology (M Sc, PhD); biotechnology and genomics (Diploma). *Students:* 68 full-time (39 women). In 2005, 10 master's, 1 doctorate, 4 other advanced degrees awarded. *Degree requirements:* For master's, thesis; for doctorate, thesis/dissertation, pedagogical training. *Entrance requirements:* For master's, honors degree in biology; for doctorate, M Sc in life science. *Application deadline:* For fall admission, 1/15 for domestic students. For winter admission, 8/31 for domestic students. Application fee: $50. *Expenses:* Tuition, state resident: full-time $834; part-time $334 per term. Tuition, nonresident: full-time $2,200; part-time $880 per term. Required fees: $680 per term. Tuition and fees vary according to degree level and program. *Faculty research:* Cell biology, animal physiology, ecology, microbiology/molecular biology, plant physiology/biochemistry and biotechnology. *Unit head:* Dr. Luc Varin, Chair, 514-848-2424 Ext. 3390, Fax: 514-848-2881. *Application contact:* Dr. Paul Widden, Director, 514-848-2424 Ext. 3413, Fax: 514-848-2881.

Cornell University, College of Veterinary Medicine, Ithaca, NY 14853-0001. Offers comparative biomedical science (PhD); immunology (MS, PhD); pharmacology (PhD); physiology (PhD); veterinary medicine (DVM); zoology (MS, PhD). Accreditation: AVMA. *Faculty:* 155 full-time (53 women). *Students:* 334 full-time (267 women); includes 71 minority (25 African Americans, 2 American Indian/Alaska Native, 18 Asian Americans or Pacific Islanders, 26 Hispanic Americans), 2 international. Average age 26. 871 applicants, 11% accepted, 78 enrolled. In 2005, 81 first professional degrees, 15 doctorates awarded. *Degree requirements:* For first-professional, thesis or alternative, on-site clinical training. *Entrance requirements:* GRE General Test or MCAT, undergraduate pre-medical science program, animal or veterinary experience, letter of recommendation. *Application deadline:* For fall admission, 10/1 for domestic students, 10/1 for international students. Application fee: $40. Electronic applications accepted. *Expenses:* Contact institution. *Financial support:* In 2005–06, 303 students received support, including 30 fellowships (averaging $24,854 per year), 88 research assistantships with tuition reimbursements available (averaging $24,854 per year); Federal Work-Study, institutionally sponsored loans, scholarships/grants, and unspecified assistantships also available. Financial award application deadline: 2/1; financial award applicants required to submit CSS PROFILE or FAFSA. *Faculty research:* Extensive biomedical research, comparative cancer, food safety. Total annual research expenditures: $49.3 million. *Unit head:* Dr. Donald F. Smith, Dean, 607-253-3771. *Application contact:* Jennifer A Mailey, Director of Admissions, 607-253-3700, Fax: 607-253-3709, E-mail: vet_admissions@cornell.edu.

Cornell University, Graduate School, Graduate Fields of Comparative Biomedical Sciences, Field of Comparative Biomedical Sciences, Ithaca, NY 14853-0001. Offers cellular and molecular medicine (MS; PhD); developmental and reproductive biology (MS, PhD); infectious diseases (MS, PhD); population medicine and epidemiology (MS); population medicine and epidemiology sciences (PhD); structural and functional biology (MS, PhD). *Faculty:* 135 full-time (38 women). *Students:* 41 full-time (23 women); includes 4 minority (1 African American, 3 Asian Americans or Pacific Islanders), 19 international. 58 applicants, 60% accepted, 33 enrolled. In 2005, 4 degrees awarded. *Degree requirements:* For master's, thesis/dissertation; for doctorate, thesis/dissertation, comprehensive exam. *Entrance requirements:* For master's and doctorate, GRE General Test, 2 letters of recommendation. Additional exam requirements/recommendations for international students: Required—TOEFL (minimum score 550 paper-based; 213 computer-based). *Application deadline:* For fall admission, 12/15 for domestic students. Application fee: $60. Electronic applications accepted. *Financial support:* In 2005–06, 41 students received support, including 13 fellowships with full tuition reimbursements available, 28 research assistantships with full tuition reimbursements available; teaching assistantships with full tuition reimbursements available, institutionally sponsored loans, scholarships/grants, health care benefits, tuition waivers (full and partial), and unspecified assistantships also available. Financial award applicants required to submit FAFSA. *Faculty research:* Receptors and signal transduction, viral and bacterial infectious diseases, tumor metastasis, clinical sciences/nutritional disease, development/neurologic disorders. *Unit head:* Director of Graduate Studies, 607-253-3276, Fax: 607-253-3756. *Application contact:* Graduate Field Assistant, 607-253-3276, Fax: 607-253-3756, E-mail: graduate_edcvm@cornell.edu.

Cornell University, Joan and Sanford I. Weill Medical College and Graduate School of Medical Sciences, Weill Graduate School of Medical Sciences, New York, NY 10021. Offers MS, PhD, MD/PhD. *Faculty:* 233 full-time (61 women). *Students:* 328 full-time (181 women); includes 47 minority (10 African Americans, 24 Asian Americans or Pacific Islanders, 13 Hispanic Americans), 132 international. Average age 22. 403 applicants. In 2005, 11 master's, 40 doctorates awarded. *Median time to degree:* Of those who began their doctoral program in fall 1997, 100% received their degree in 8 years or less. *Degree requirements:* For doctorate, thesis/dissertation, final exam. *Entrance requirements:* For doctorate, GRE General Test, GRE Subject Test. Additional exam requirements/recommendations for international students: Required—TOEFL. Application fee: $60. Electronic applications accepted. *Expenses:* Contact institution. *Financial support:* In 2005–06, 4 fellowships (averaging $19,968 per year) were awarded; scholarships/grants, health care benefits, tuition waivers (full), and stipends also available. *Unit head:* Dr. David P. Hajjar, Dean, 212-746-6900, E-mail: dphajjar@med.cornell.edu.

Cornell University, Joan and Sanford I. Weill Medical College and Graduate School of Medical Sciences, Weill Medical College, Tri-Institutional MD/PhD Program, New York, NY 10021-4896. Offers MD/PhD. Offered through the Tri-Institutional Program with The Rockefeller University and Sloan-Kettering Institute. *Faculty:* 249 full-time (49 women). *Students:* 108 full-time (39 women); includes 37 minority (11 African Americans, 15 Asian Americans or Pacific Islanders, 11 Hispanic Americans), 1 international. 392 applicants, 11% accepted, 11 enrolled. *Median time to degree:* Of those who began their doctoral program in fall 1997, 75% received their degree in 8 years or less. *Application deadline:* For fall admission, 10/15 for domestic students. Application fee: $0. Electronic applications accepted. *Expenses:* Contact institution. *Financial support:* In 2005–06, 108 students received support, including 108 fellow-

Biological and Biomedical Sciences—General

ships with tuition reimbursements available (averaging $24,833 per year); health care benefits, tuition waivers (full), and stipend and research supplement also available. *Faculty research:* Neuroscience, pharmacology, immunology, structural biology, genetics. *Unit head:* Dr. Olaf S. Andersen, Director, 212-746-6023, Fax: 212-746-8678, E-mail: mdphd@med.cornell.edu. *Application contact:* Ruth Gotian, Manager, 212-746-6023, Fax: 212-746-8678, E-mail: mdphd@med.cornell.edu.

Creighton University, School of Medicine and Graduate School, Graduate Programs in Medicine, Department of Biomedical Sciences, Omaha, NE 68178-0001. Offers MS, PhD, MD/PhD. Terminal master's awarded for partial completion of doctoral program. *Degree requirements:* For master's and doctorate, thesis/dissertation. *Entrance requirements:* For master's and doctorate, GRE General Test. Additional exam requirements/recommendations for international students: Required—TOEFL. Electronic applications accepted. *Faculty research:* Molecular biology and gene transfection.

See Close-Up on page 109.

Dalhousie University, Faculty of Graduate Studies, College of Arts and Science, Faculty of Science, Department of Biology, Halifax, NS B3H 4R2, Canada. Offers M Sc, PhD. Part-time programs available. *Degree requirements:* For master's and doctorate, thesis/dissertation. *Entrance requirements:* Additional exam requirements/recommendations for international students: Required—TOEFL. Electronic applications accepted. *Faculty research:* Marine biology, ecology, animal physiology, plant physiology, microbiology (cell, molecular, genetics, development).

Dalhousie University, Faculty of Graduate Studies and Faculty of Medicine, Graduate Programs in Medicine, Halifax, NS B3H 4R2, Canada. Offers M Sc, PhD, MD/PhD. Part-time programs available. *Degree requirements:* For master's and doctorate, thesis/dissertation. *Entrance requirements:* Additional exam requirements/recommendations for international students: Required—TOEFL. Expenses: Contact institution.

Dartmouth College, School of Arts and Sciences, Department of Biological Sciences, Hanover, NH 03755. Offers biology (PhD). *Faculty:* 19 full-time (7 women), 4 part-time/adjunct (2 women). *Students:* 69 full-time (31 women); includes 1 minority (Asian American or Pacific Islander), 25 international. Average age 27. 363 applicants, 20% accepted, 35 enrolled. In 2005, 7 doctorates awarded. *Degree requirements:* For doctorate, thesis/dissertation, teaching experience. *Entrance requirements:* For doctorate, GRE General Test, GRE Subject Test. Additional exam requirements/recommendations for international students: Required—TOEFL. *Application deadline:* For fall admission, 2/1 for domestic students. Application fee: $40. Electronic applications accepted. *Expenses:* Tuition: Full-time $31,770. *Financial support:* In 2005–06, 67 students received support, including fellowships with full tuition reimbursements available (averaging $21,000 per year), research assistantships with full tuition reimbursements available (averaging $21,000 per year); Federal Work-Study, institutionally sponsored loans, scholarships/grants, traineeships, tuition waivers (full), and unspecified assistantships also available. Financial award applicants required to submit FAFSA. *Faculty research:* Population, community and ecosystem ecology; development (Coelegans, Drosophila and Arabidopsis); circadian rhythms and plant nutrition (Arabidopsis); cell motility and the cytoskeleton (squid and sea urchins). Total annual research expenditures: $4.6 million. *Unit head:* Dr. Mark McPeek, Chair, 603-646-2378. *Application contact:* Amy F. Layne, Administrative Assistant, 603-646-3847, Fax: 603-646-1347, E-mail: amy.f.layne@dartmouth.edu.

See Close-Up on page 111.

Delaware State University, Graduate Programs, Department of Biology, Dover, DE 19901-2277. Offers biology (MS); biology education (MS). Part-time and evening/weekend programs available. *Degree requirements:* For master's, thesis (for some programs). *Entrance requirements:* For master's, GRE, minimum GPA of 3.0 in major, 2.75 overall. Electronic applications accepted. *Faculty research:* Cell biology, immunology, microbiology, genetics, ecology.

Delta State University, Graduate Programs, College of Arts and Sciences, Department of Biological and Physical Sciences, Cleveland, MS 38733-0001. Offers MSNS. Part-time programs available. *Degree requirements:* For master's, research project or thesis. *Entrance requirements:* For master's, GRE General Test. *Application deadline:* For fall admission, 8/1 for domestic students; for spring admission, 12/1 priority date for domestic students. Applications are processed on a rolling basis. Application fee: $0. *Expenses:* Tuition, state resident: full-time $3,762; part-time $205 per hour. Tuition, nonresident: full-time $8,702; part-time $490 per hour. *Financial support:* Research assistantships, career-related internships or fieldwork, Federal Work-Study, and institutionally sponsored loans available. Support available to part time students. Financial award application deadline: 6/1. *Unit head:* Dr. John E. Tiftickjian, Chair, 662-846-4240, Fax: 662-846-4798, E-mail: jtift@deltastate.edu.

DePaul University, College of Liberal Arts and Sciences, Department of Biological Sciences, Chicago, IL 60604-2287. Offers MA, MS. *Faculty:* 10 full-time (5 women), 6 part-time/adjunct (1 woman). *Students:* 11 full-time (4 women), 1 part-time; includes 1 minority (Hispanic American), 1 international. Average age 25. 18 applicants, 50% accepted. *Degree requirements:* For master's, thesis (for some programs), oral exam. *Entrance requirements:* For master's, GRE or MCAT, minimum GPA of 2.7. Additional exam requirements/recommendations for international students: Required—TOEFL (minimum score 590 paper-based; 243 computer-based). *Application deadline:* For fall admission, 3/31 priority date for domestic students, 3/31 priority date for international students. For winter admission, 7/31 for domestic students; for spring admission, 11/30 for domestic students. Applications are processed on a rolling basis. Application fee: $25. Electronic applications accepted. *Financial support:* In 2005–06, 10 teaching assistantships with full tuition reimbursements (averaging $9,500 per year) were awarded; Federal Work-Study, institutionally sponsored loans, scholarships/grants, and tuition waivers (full and partial) also available. Support available to part-time students. Financial award application deadline: 4/1. *Faculty research:* Blood oxygen transport in vertebrates, cell motility, detoxification in plant cells, molecular biology of fungi, B-lymphocyte development. Total annual research expenditures: $118,725. *Unit head:* Dr. Stanley Cohn, Chair, 773-325-7595, Fax: 773-325-7596, E-mail: schon@depaul.edu. *Application contact:* Dr. Margaret Silliker, Director of Graduate Admissions, 773-325-2194, E-mail: msillike@depaul.edu.

See Close-Up on page 113.

Drexel University, College of Arts and Sciences, Department of Bioscience and Biotechnology, Philadelphia, PA 19104-2875. Offers biological science (MS, PhD); nutrition and food sciences (MS, PhD), including food science (MS), nutrition science (PhD). Part-time programs available. *Degree requirements:* For doctorate, thesis/dissertation. *Entrance requirements:* For master's and doctorate, GRE General Test. Additional exam requirements/recommendations for international students: Required—TOEFL. Electronic applications accepted. *Faculty research:* Genetic engineering, physiological ecology.

Drexel University, College of Medicine, Biomedical Graduate Programs, Philadelphia, PA 19104-2875. Offers MBS, MLAS, MMS, MS, PhD, Certificate, MD/PhD. Part-time programs available. Terminal master's awarded for partial completion of doctoral program. *Degree requirements:* For master's, comprehensive exam; for doctorate, thesis/dissertation, qualifying exam. *Entrance requirements:* For master's and doctorate, GRE General Test. Additional exam requirements/recommendations for international students: Required—TOEFL. Electronic applications accepted. Expenses: Contact institution.

See Close-Ups on pages 115 and 117.

Drexel University, College of Medicine, MD/PhD Program, Philadelphia, PA 19104-2875. Offers MD/PhD. Electronic applications accepted.

Drexel University, School of Biomedical Engineering, Science and Health Systems, Program in Biomedical Science, Philadelphia, PA 19104-2875. Offers MS, PhD. *Degree requirements:*

For master's, thesis (for some programs); for doctorate, thesis/dissertation. Electronic applications accepted.

Duke University, Graduate School, Department of Biological and Biologically Inspired Materials, Durham, NC 27708-0586. Offers PhD, Certificate. *Faculty:* 35 full-time. *Students:* 1 full-time (0 women). 5 applicants, 0% accepted, 0 enrolled. *Entrance requirements:* Additional exam requirements/recommendations for international students: Required—IELTS (preferred) or TOEFL. *Application deadline:* For fall admission, 12/31 for domestic students. Application fee: $75. Electronic applications accepted. *Unit head:* Robert Clark, Director, 919-660-5435, Fax: 919-660-5409, E-mail: rws6@duke.edu.

Duke University, Graduate School, Department of Biology, Durham, NC 27708. Offers PhD. *Faculty:* 45 full-time. *Students:* 79 full-time (47 women); includes 4 minority (3 Asian Americans or Pacific Islanders, 1 Hispanic American), 24 international. 100 applicants, 21% accepted, 13 enrolled. In 2005, 18 doctorates awarded. *Degree requirements:* For doctorate, one foreign language, thesis/dissertation. *Entrance requirements:* For doctorate, GRE General Test, GRE Subject Test (recommended). Additional exam requirements/recommendations for international students: Required—IELT (preferred) or TOEFL. *Application deadline:* For fall admission, 12/31 for domestic students, 12/31 for international students. Application fee: $75. Electronic applications accepted. *Financial support:* Fellowships, research assistantships, teaching assistantships, Federal Work-Study available. Financial award application deadline: 12/31. *Unit head:* John Willis, Contact, 919-684-3649, Fax: 919-660-7293, E-mail: aslzoo@duke.edu.

Duquesne University, Bayer School of Natural and Environmental Sciences, Department of Biological Sciences, Pittsburgh, PA 15282-0001. Offers biology (MS, PhD). Part-time programs available. *Faculty:* 19 full-time (9 women), 1 part-time/adjunct (0 women). *Students:* 39 full-time (24 women), 4 part-time (1 woman); includes 3 minority (1 African American, 2 American Indian/Alaska Native), 6 international. Average age 26. 40 applicants, 68% accepted, 15 enrolled. In 2005, 11 degrees awarded. Terminal master's awarded for partial completion of doctoral program. *Degree requirements:* For master's, thesis (for some programs), comprehensive exam (for some programs); registration; for doctorate, thesis/dissertation, comprehensive exam, registration. *Entrance requirements:* For master's, GRE General Test, BS in biological sciences or related field; for doctorate, GRE General Test, MS in biological sciences or related field. Additional exam requirements/recommendations for international students: Required—TOEFL. *Application deadline:* For fall admission, 3/1 priority date for domestic students, 2/1 priority date for international students; for spring admission, 10/1 priority date for domestic students, 9/1 priority date for international students. Applications are processed on a rolling basis. Application fee: $0. *Expenses: Contact institution.* Tuition and fees vary according to degree level and program. *Financial support:* In 2005–06, 1 fellowship with full tuition reimbursement (averaging $15,080 per year), 2 research assistantships with full tuition reimbursements (averaging $19,000 per year), 18 teaching assistantships with full tuition reimbursements (averaging $14,310 per year) were awarded; Federal Work-Study, scholarships/grants, tuition waivers (partial), and unspecified assistantships also available. Financial award application deadline: 5/1; financial award applicants required to submit FAFSA. *Faculty research:* Cell and developmental biology, molecular biology and genetics, evolution, ecology, physiology. Total annual research expenditures: $717,033. *Unit head:* Dr. Joseph R. McCormick, Chair, 412-396-4775, Fax: 412-396-5907, E-mail: mccormick@duq.edu. *Application contact:* Mary Ann Quinn, Assistant to the Dean Graduate Affairs, 412-396-6339, Fax: 412-396-4881, E-mail: gradinfo@duq.edu.

See Close-Up on page 119.

East Carolina University, Brody School of Medicine and Graduate School, Graduate Programs in Medicine, Greenville, NC 27858-4353. Offers MPH, PhD. *Students:* 60 full-time (43 women), 55 part-time (35 women); includes 26 minority (21 African Americans, 5 Asian Americans or Pacific Islanders), 12 international. Average age 27. 98 applicants, 49% accepted. In 2005, 7 master's, 11 doctorates awarded. *Median time to degree:* Of those who began their doctoral program in fall 1997, 70% received their degree in 8 years or less. *Degree requirements:* For doctorate, thesis/dissertation, comprehensive exam, registration. *Entrance requirements:* For doctorate, GRE General Test. Additional exam requirements/recommendations for international students: Required—TOEFL. *Application deadline:* Applications are processed on a rolling basis. Application fee: $50. *Expenses:* Tuition, state resident: full-time $2,516. Tuition, nonresident: full-time $12,832. *Financial support:* Fellowships available. Financial award application deadline: 6/1. *Unit head:* Dr. John Lehman, Associate Dean for Research and Graduate Studies, 252-744-9346, Fax: 252-744-3260, E-mail: lehmanj@ecu.edu. *Application contact:* Dean of Graduate School, 252-328-6012, Fax: 252-328-6071, E-mail: gradschool@ecu.edu.

Announcement: Doctoral education and training programs in diverse contemporary research areas designed to be consistent with the rapid advances in the biomedical sciences. High-quality programs with small class sizes, personalized instruction, ready access to faculty members, frequent contact with a wide variety of health science professionals, and opportunity for interdisciplinary research.

See Close-Up on page 123.

East Carolina University, Brody School of Medicine, Interdisciplinary Program in Biological Sciences, Greenville, NC 27858-4353. Offers PhD. *Faculty:* 5 full-time (0 women). *Students:* 10 full-time (4 women), 7 part-time (1 woman); includes 1 minority (Asian American or Pacific Islander), 4 international. Average age 28. 10 applicants, 50% accepted. In 2005, 1 degree awarded. *Median time to degree:* Of those who began their doctoral program in fall 1997, 100% received their degree in 8 years or less. *Degree requirements:* For doctorate, thesis/dissertation, comprehensive exam, registration. *Entrance requirements:* For doctorate, GRE General Test, bachelor's degree in biological chemistry or physical science. Additional exam requirements/recommendations for international students: Required—TOEFL. *Application deadline:* For fall admission, 6/1 for domestic students. Applications are processed on a rolling basis. Application fee: $50. *Expenses:* Tuition, state resident: full-time $2,516. Tuition, nonresident: full-time $12,832. *Financial support:* In 2005–06, 10 fellowships (averaging $21,500 per year) were awarded Financial award application deadline: 6/1. *Faculty research:* Immunochemistry and allergens, immunological disorders, cell biology of tumors, membrane antigens, microphage biology. *Unit head:* Dr. Gerhard Kalmus, Program Director, 252-744-2803, Fax: 252-744-3616, E-mail: kalmusa@ecu.edu. *Application contact:* Dr. Gerhard Kalmus, Program Director, 252-744-2803, Fax: 252-744-3616, E-mail: kalmusa@ecu.edu.

East Carolina University, Graduate School, Thomas Harriot College of Arts and Sciences, Department of Biology, Greenville, NC 27858-4353. Offers biology (MS); molecular biology/biotechnology (MS). Part-time programs available. *Faculty:* 38 full-time (10 women). *Students:* 24 full-time (9 women), 40 part-time (15 women); includes 6 minority (4 African Americans, 1 Asian American or Pacific Islander, 1 Hispanic American), 2 international. Average age 27. 24 applicants, 13% accepted, 9 enrolled. In 2005, 28 degrees awarded. *Degree requirements:* For master's, one foreign language, thesis, comprehensive exam. *Entrance requirements:* For master's, GRE General Test, GRE Subject Test. Additional exam requirements/recommendations for international students: Required—TOEFL. *Application deadline:* For fall admission, 6/1 for domestic students; for spring admission, 10/15 for domestic students. Applications are processed on a rolling basis. Application fee: $50. *Expenses:* Tuition, state resident: full-time $2,516. Tuition, nonresident: full-time $12,832. *Financial support:* Fellowships with partial tuition reimbursements, research assistantships with partial tuition reimbursements, teaching assistantships with partial tuition reimbursements, career-related internships or fieldwork, Federal Work-Study, scholarships/grants, and unspecified assistantships available. Support available to part-time students. Financial award application deadline: 6/1. *Faculty research:* Biochemistry, microbiology, cell biology. *Unit head:* Dr. Ronald Newton, Chair, 252-328-6204, Fax: 252-328-4178, E-mail: newtonro@ecu.edu. *Application contact:* Interim Dean of Graduate School, 252-328-6012, Fax: 252-328-6071, E-mail: gradschool@ecu.edu.

See Close-Up on page 121.

Biological and Biomedical Sciences—General

Eastern Illinois University, Graduate School, College of Sciences, Department of Biological Sciences, Charleston, IL 61920-3099. Offers MS. In 2005, 1 degree awarded. *Degree requirements:* For master's, exam. *Application deadline:* For fall admission, 7/31 for domestic students. Applications are processed on a rolling basis. Application fee: $30. *Expenses:* Tuition, state resident: part-time $150 per credit hour. Tuition, nonresident: part-time $452 per credit hour. Required fees: $738. *Financial support:* In 2005–06, 15 research assistantships with tuition reimbursements (averaging $7,200 per year), 5 teaching assistantships with tuition reimbursements (averaging $7,200 per year) were awarded; career-related internships or fieldwork also available. *Unit head:* Dr. Andrew Methven, Chair, 217-581-3011, Fax: 217-581-7141, E-mail: cfasm@eiu.edu. *Application contact:* Dr. Charles Costa, Coordinator, 217-581-2520, E-mail: cfcjc@eiu.edu.

Eastern Kentucky University, The Graduate School, College of Arts and Sciences, Department of Biological Sciences, Richmond, KY 40475-3102. Offers biological sciences (MS); ecology (MS). Part-time programs available. *Degree requirements:* For master's, thesis. *Entrance requirements:* For master's, GRE General Test, minimum GPA of 2.5. *Faculty research:* Systematics, ecology, and biodiversity; animal behavior; protein structure and molecular genetics; biomonitoring and aquatic toxicology; pathogenesis of microbes and parasites.

Eastern Michigan University, Graduate School, College of Arts and Sciences, Department of Biology, Ypsilanti, MI 48197. Offers MS. Evening/weekend programs available. *Faculty:* 20 full-time (6 women). *Students:* 8 full-time (3 women), 32 part-time (17 women); includes 2 minority (both Asian Americans or Pacific Islanders), 8 international. Average age 28. In 2005, 14 degrees awarded. *Entrance requirements:* For master's, GRE General Test, GRE Subject Test. Additional exam requirements/recommendations for international students: Required—TOEFL. *Application deadline:* For fall admission, 5/15 priority date for domestic students, 5/1 priority date for international students. For winter admission, 10/15 for domestic students; for spring admission, 3/15 for domestic students. Applications are processed on a rolling basis. Application fee: $35. *Expenses:* Tuition, state resident: full-time $7,838; part-time $327 per credit hour. Tuition, nonresident: full-time $15,770; part-time $657 per credit hour. Required fees: $33 per credit hour. $40 per term. Tuition and fees vary according to course level, course load and degree level. *Financial support:* In 2005–06, fellowships (averaging $4,000 per year); research assistantships with full tuition reimbursements, teaching assistantships with full tuition reimbursements, Federal Work-Study and scholarships/grants also available. Support available to part-time students. Financial award applicants required to submit FAFSA. *Unit head:* Dr. Tamara Greco, Interim Head, 734-487-4242.

Eastern New Mexico University, Graduate School, College of Liberal Arts and Sciences, Department of Biology, Portales, NM 88130. Offers MS. Part-time programs available. *Faculty:* 8 full-time (0 women). *Students:* 4 full-time (3 women), 12 part-time (8 women); includes 1 minority (African American), 5 international. Average age 30. 21 applicants, 33% accepted. In 2005, 4 degrees awarded. *Degree requirements:* For master's, one foreign language, thesis optional. *Entrance requirements:* For master's, minimum GPA of 2.5. *Application deadline:* For fall admission, 8/20 for domestic students. Applications are processed on a rolling basis. Application fee: $10. Electronic applications accepted. *Expenses:* Tuition, state resident: full-time $2,316; part-time $97 per credit hour. Tuition, nonresident: full-time $7,872; part-time $328 per credit hour. Required fees: $33 per credit hour. *Financial support:* In 2005–06, 1 fellowship (averaging $9,450 per year), 3 research assistantships (averaging $7,700 per year), 9 teaching assistantships (averaging $7,700 per year) were awarded; Federal Work-Study also available. Support available to part-time students. Financial award application deadline: 3/1. *Unit head:* Dr. Darren Pollack, Graduate Coordinator, 505-562-2862, E-mail: darren.pollack@enmu.edu.

Eastern Virginia Medical School, Doctoral Program in Biomedical Sciences, Norfolk, VA 23501-1980. Offers PhD, MD/PhD. *Faculty:* 58. *Students:* 32 full-time (22 women); includes 2 African Americans, 1 Asian American or Pacific Islander, 1 Hispanic American, 14 international. 34 applicants, 15% accepted, 25 enrolled. In 2005, 2 degrees awarded. *Degree requirements:* For doctorate, thesis/dissertation. *Entrance requirements:* For doctorate, GRE General Test. Additional exam requirements/recommendations for international students: Required—TOEFL. *Application deadline:* For fall admission, 2/1 for domestic students. Applications are processed on a rolling basis. Application fee: $30. *Expenses:* Contact institution. *Financial support:* Research assistantships, career-related internships or fieldwork available. *Faculty research:* Reproductive physiology, cancer, molecular biology and immunology, cytogenetics and development, systems biology. Total annual research expenditures: $14 million. *Unit head:* Dr. William J. Wasilenko, Director, 757-446-8480, Fax: 757-446-8449, E-mail: wasilewj@evms.edu. *Application contact:* Toni Dorn, Administrator, 757-446-8480, Fax: 757-446-8449, E-mail: dornma@evms.edu.

Eastern Virginia Medical School, Master's Program in Biomedical Sciences, Norfolk, VA 23501-1980. Offers MS. *Faculty:* 25. *Students:* 18 full-time (5 women); includes 3 minority (all Asian Americans or Pacific Islanders) 170 applicants, 20% accepted, 18 enrolled. In 2005, 19 degrees awarded. *Application deadline:* For fall admission, 4/1 for domestic students. Application fee: $50. *Expenses:* Contact institution. *Financial support:* Federal Work-Study and institutionally sponsored loans available. *Unit head:* Dr. Donald Meyer, Director, 757-446-8480, Fax: 757-446-8449, E-mail: meyerdc@evms.edu. *Application contact:* Toni Dorn, Administrator, 757-446-8480, Fax: 757-446-8449, E-mail: dornma@evms.edu.

Eastern Virginia Medical School, Master's Program in Biomedical Sciences (Clinical Embryology and Andrology), Norfolk, VA 23501-1980. Offers MS. *Faculty:* 16. *Students:* 41 full-time (28 women); includes 13 minority (3 African Americans, 7 Asian Americans or Pacific Islanders, 3 Hispanic Americans). 46 applicants, 50% accepted, 23 enrolled. In 2005, 20 degrees awarded. *Degree requirements:* For winter admission, 2/15 for domestic students. Applications are processed on a rolling basis. Application fee: $50. *Expenses:* Contact institution. *Financial support:* In 2005–06, 10 students received support. *Unit head:* Dr. Jacob Mayer, Director, 757-446-5049, Fax: 757-446-5905. *Application contact:* Nancy Garcia, Administrator, 757-446-8935, Fax: 757-446-5905, E-mail: garcianw@evms.edu.

Eastern Virginia Medical School, Master's Program in Biomedical Sciences (Research Track), Norfolk, VA 23501-1980. Offers MS. *Faculty:* 57. *Students:* 5 full-time (all women); includes 1 Asian American or Pacific Islander, 1 international. 5 applicants, 60% accepted, 4 enrolled. In 2005, 2 degrees awarded. *Application deadline:* For spring admission, 4/1 for domestic students. Applications are processed on a rolling basis. Application fee: $50. *Expenses:* Contact institution. *Financial support:* In 2005–06, 4 students received support. *Unit head:* Dr. William J. Wasilenko, Director, 757-446-8480, Fax: 757-446-8449, E-mail: wasilewj@evms.edu. *Application contact:* Toni Dorn, Administrator, 757-446-8480, Fax: 757-446-8449, E-mail: dornma@evms.edu.

Eastern Washington University, Graduate Studies, College of Science, Mathematics and Technology, Department of Biology, Cheney, WA 99004-2431. Offers MS. *Degree requirements:* For master's, thesis, comprehensive exam. *Entrance requirements:* For master's, GRE General Test, minimum GPA of 3.0. *Faculty research:* Ecology of Eastern Washington Scablands, Columbia River fisheries, biotechnology applied to vaccines, role of mycorrhiza in plant nutrition, exercise and estrous cycles.

East Stroudsburg University of Pennsylvania, Graduate School, School of Arts and Sciences, Department of Biology, East Stroudsburg, PA 18301-2999. Offers biology (M Ed, MS). Part-time and evening/weekend programs available. *Degree requirements:* For master's, thesis or alternative, comprehensive exam. *Entrance requirements:* For master's, GRE, undergraduate major in life sciences, 2 semesters in chemistry, 3 letters of recommendation. Additional exam requirements/recommendations for international students: Required—TOEFL (minimum score 550 paper-based; 213 computer-based).

East Tennessee State University, James H. Quillen College of Medicine, Biomedical Science Graduate Program, Johnson City, TN 37614. Offers anatomy (MS, PhD); biochemistry

(MS, PhD); biophysics (MS, PhD); microbiology (MS, PhD); pharmacology (MS, PhD); physiology (MS, PhD). Part-time programs available. *Faculty:* 49 full-time (12 women), 1 (woman) part-time/adjunct. *Students:* 30 full-time (19 women), 5 part-time (4 women); includes 2 minority (1 African American, 1 Asian American or Pacific Islander), 11 international. Average age 31. 78 applicants, 13% accepted, 9 enrolled. In 2005, 1 master's, 4 doctorates awarded. Terminal master's awarded for partial completion of doctoral program. *Degree requirements:* For master's, one foreign language, thesis, comprehensive qualifying exam; for doctorate, 2 foreign languages, thesis/dissertation. *Entrance requirements:* For master's, GRE General Test, minimum GPA of 3.0, bachelor's degree in biological or related science; for doctorate, GRE General Test, GRE Subject Test. Additional exam requirements/recommendations for international students: Required—TOEFL (minimum score 550 paper-based; 213 computer-based). *Application deadline:* For fall admission, 3/15 for domestic students; for spring admission, 3/1 for domestic students. Application fee: $25 ($35 for international students). *Expenses:* Contact institution. *Financial support:* In 2005–06, 7 research assistantships with full tuition reimbursements (averaging $15,000 per year) were awarded; teaching assistantships with full tuition reimbursements, career-related internships or fieldwork, Federal Work-Study, institutionally sponsored loans, scholarships/grants, and tuition waivers (full) also available. Financial award application deadline: 7/1; financial award applicants required to submit FAFSA. Total annual research expenditures: $2.1 million. *Unit head:* Dr. Mitchell E. Robinson, Assistant Dean, Director, 423-439-4658, E-mail: robinson@etsu.edu.

East Tennessee State University, School of Graduate Studies, College of Arts and Sciences, Department of Biological Sciences, Johnson City, TN 37614. Offers biology (MS); microbiology (MS). *Faculty:* 12 full-time (0 women). *Students:* 18 full-time (11 women), 1 (woman) part-time; includes 1 minority (African American), 3 international. Average age 27. 22 applicants, 73% accepted, 8 enrolled. In 2005, 6 degrees awarded. *Degree requirements:* For master's, thesis or alternative, comprehensive exam. *Entrance requirements:* For master's, GRE General Test or GRE Subject Test, minimum GPA of 3.0. Additional exam requirements/recommendations for international students: Required—TOEFL (minimum score 550 paper-based; 213 computer-based). *Application deadline:* For fall admission, 7/15 for domestic students; for spring admission, 11/1 for domestic students. Applications are processed on a rolling basis. Application fee: $25 ($35 for international students). *Expenses:* Tuition, nonresident: full-time $9,312; part-time $404 per hour. Required fees: $261 per hour. *Financial support:* In 2005–06, 1 research assistantship with full tuition reimbursement (averaging $6,000 per year), 15 teaching assistantships with full tuition reimbursements (averaging $6,000 per year) were awarded; career-related internships or fieldwork and institutionally sponsored loans also available. Financial award application deadline: 7/1; financial award applicants required to submit FAFSA. *Faculty research:* Vertebrate natural history, mutation rates in fruit flies, regulation of plant secondary metabolism, plant biochemistry, timekeeping in honeybees, gene expression in diapausing flies. Total annual research expenditures: $226,807. *Unit head:* Dr. Dan M. Johnson, Chair, 423-439-4329, Fax: 423-439-5958, E-mail: johnsodm@etsu.edu.

Edinboro University of Pennsylvania, Graduate Studies and Research, School of Science, Management and Technology, Department of Biology and Health Services, Edinboro, PA 16444. Offers biology (MS). Part-time and evening/weekend programs available. *Faculty:* 7 full-time (2 women). *Students:* 6 full-time (3 women), 1 (woman) part-time; includes 1 minority (Hispanic American) Average age 29. In 2005, 2 degrees awarded. *Degree requirements:* For master's, thesis or alternative, competency exam. *Entrance requirements:* For master's, GRE or MAT, minimum QPA of 2.5. *Application deadline:* Applications are processed on a rolling basis. Application fee: $25. Electronic applications accepted. *Expenses:* Tuition, state resident: full-time $5,888; part-time $327 per credit. Tuition, nonresident: full-time $9,422; part-time $523 per credit. Required fees: $1,839; $40 per credit. *Financial support:* In 2005–06, 4 research assistantships with full and partial tuition reimbursements (averaging $3,850 per year) were awarded; Federal Work-Study, scholarships/grants, and unspecified assistantships also available. Support available to part-time students. Financial award application deadline: 3/15; financial award applicants required to submit FAFSA. *Faculty research:* Microbiology, molecular biology, zoology, botany, ecology. *Unit head:* Dr. Martin J. Mitchell, Chairperson, 814-732-2500, Fax: 814-732-2792. *Application contact:* Dr. Mary Margaret Bevevino, Dean of Graduate Studies and Research, 814-732-2856, Fax: 814-732-2611, E-mail: mbevevino@edinboro.edu.

Emory University, Graduate School of Arts and Sciences, Division of Biological and Biomedical Sciences, Atlanta, GA 30322-1100. Offers PhD. *Students:* 311 full-time (69 women). *Students:* 389 full-time (254 women); includes 76 minority (39 African Americans, 2 American Indian/Alaska Native, 24 Asian Americans or Pacific Islanders, 11 Hispanic Americans), 54 international. Average age 27. 601 applicants, 28% accepted, 93 enrolled. In 2005, 40 doctorates awarded. *Median time to degree:* Of those who began their doctoral program in fall 1997, 100% received their degree in 8 years or less. *Degree requirements:* For doctorate, thesis/dissertation, comprehensive exam, registration. *Entrance requirements:* For doctorate, GRE General Test, minimum GPA of 3.0 in science course work. Additional exam requirements/recommendations for international students: Required—TOEFL. *Application deadline:* For fall admission, 1/3 for domestic students, 1/3 for international students. Application fee: $50. Electronic applications accepted. *Expenses:* Contact institution. *Financial support:* In 2005–06, 165 students received support, including 165 fellowships with full tuition reimbursements available (averaging $23,000 per year); institutionally sponsored loans, health care benefits, and tuition waivers (full) also available. *Faculty research:* Biochemistry and genetics, immunology and microbiology, neuroscience and pharmacology, nutrition, population biology and ecology. Total annual research expenditures: $154.6 million. *Unit head:* Dr. Keith Wilkinson, Acting Director, 404-727-2545, Fax: 404-727-3322. *Application contact:* Kathy Smith, Director of Recruitment and Admissions, 404-727-2545, Fax: 404-727-3322, E-mail: kathy.smith@emory.edu.

See Close-Up on page 125.

Emporia State University, School of Graduate Studies, College of Liberal Arts and Sciences, Department of Biological Sciences, Emporia, KS 66801-5087. Offers botany (MS); environmental biology (MS); general biology (MS); microbial and cellular biology (MS); zoology (MS). Part-time programs available. *Faculty:* 13 full-time (2 women), 4 part-time/adjunct (2 women). *Students:* 3 full-time (1 woman), 17 part-time (7 women), 3 international. 10 applicants, 90% accepted, 7 enrolled. In 2005, 5 degrees awarded. *Degree requirements:* For master's, comprehensive exam or thesis. *Entrance requirements:* For master's, GRE, appropriate undergraduate degree, interview, letters of reference. Additional exam requirements/recommendations for international students: Required—TOEFL (minimum score 450 paper-based; 133 computer-based). *Application deadline:* For fall admission, 8/15 for domestic students. Applications are processed on a rolling basis. Application fee: $30 ($75 for international students). Electronic applications accepted. *Expenses:* Tuition, state resident: full-time $2,890; part-time $132 per credit. Tuition, nonresident: full-time $9,258; part-time $422 per credit. Required fees: $626; $41 per credit. Tuition and fees vary according to degree level. *Financial support:* In 2005–06, 4 research assistantships with full tuition reimbursements (averaging $6,492 per year), 9 teaching assistantships with full tuition reimbursements (averaging $6,492 per year) were awarded; fellowships, career-related internships or fieldwork, Federal Work-Study, institutionally sponsored loans, health care benefits, and unspecified assistantships also available. Financial award application deadline: 3/15; financial award applicants required to submit FAFSA. *Faculty research:* Fisheries, range, and wildlife management; aquatic, plant, grassland, vertebrate, and invertebrate ecology; mammalian and plant systematics, taxonomy, and evolution; immunology, virology, and molecular biology. *Unit head:* Dr. J. Richard Schrock, Chair, 620-341-5311, Fax: 620-341-5607, E-mail: jschrock@emporia.edu. *Application contact:* Dr. Derek Zelmer, Graduate Coordinator, 620-341-5623, Fax: 620-341-5607, E-mail: dzelmer@emporia.edu.

Fairleigh Dickinson University, College at Florham, Maxwell Becton College of Arts and Sciences, Department of Biological and Allied Health Sciences, Madison, NJ 07940-1099. Offers biology (MS). *Students:* 3 full-time (all women), 11 part-time (8 women), 1 international. Average age 26. 10 applicants, 90% accepted, 5 enrolled. In 2005, 5 degrees awarded.

Biological and Biomedical Sciences—General

Entrance requirements: For master's, GRE General Test. *Application deadline:* Applications are processed on a rolling basis. Application fee: $40. *Unit head:* Dr. William Fordham, Acting Chair, 973-443-8779, Fax: 973-443-8766, E-mail: fordham@fdu.edu.

Fairleigh Dickinson University, Metropolitan Campus, University College: Arts, Sciences, and Professional Studies, School of Natural Sciences, Program in Biology, Teaneck, NJ 07666-1914. Offers MS. *Students:* 2 full-time (1 woman), 5 part-time (2 women), 2 international. Average age 34. 20 applicants, 70% accepted, 4 enrolled. In 2005, 5 degrees awarded. *Application deadline:* Applications are processed on a rolling basis. Application fee: $40. *Unit head:* Dr. Irwin Isquith, Director, School of Natural Sciences, 201-692-2395, Fax: 201-692-7349, E-mail: isquith@fdu.edu.

Fayetteville State University, Graduate School, Department of Natural Sciences, Fayetteville, NC 28301-4298. Offers biology (MS). Part-time and evening/weekend programs available. *Faculty:* 5 full-time (1 woman). *Students:* 1 (woman) full-time, 8 part-time (6 women); includes 3 minority (all African Americans) *Degree requirements:* For master's, thesis, internship, comprehensive exam. *Entrance requirements:* For master's, GRE General Test. *Application deadline:* For fall admission, 7/1 for domestic students; for spring admission, 12/1 for domestic students. Applications are processed on a rolling basis. Application fee: $25. Electronic applications accepted. *Expenses:* Tuition, state resident: full-time $1,918; part-time $202 per credit. Tuition, nonresident: full-time $11,508; part-time $1,401 per credit. Required fees: $975. *Faculty research:* Mechanism of antibiotic resistance, genetic disruption of anticrobial agents, electro kinetic chromatography; nanotechnology, catalytic activity and substrate specificity of enzymes, control of soil borne pathogens. *Unit head:* Dr. Ronald Johnston, Chairperson, 910-672-1691, E-mail: rjohnston@uncfsu.edu.

Fisk University, Graduate Programs, Department of Biology, Nashville, TN 37208-3051. Offers MA. Part-time programs available. *Degree requirements:* For master's, thesis, comprehensive exam. *Entrance requirements:* For master's, GRE. *Faculty research:* Real space, topographical imaging, serotonin receptors in rats, enzyme assays, developmental biology.

Fitchburg State College, Division of Graduate and Continuing Education, Programs in Biology and Teaching Biology (Secondary Level), Fitchburg, MA 01420-2697. Offers MA, MAT. *Accreditation:* NCATE. Part-time and evening/weekend programs available. *Students:* 3 full-time (all women), 7 part-time (4 women). Average age 36. 3 applicants, 100% accepted, 3 enrolled. In 2005, 5 degrees awarded. *Entrance requirements:* For master's, GRE General Test, letters of recommendation, resumé. Additional exam requirements/recommendations for international students: Required—TOEFL (minimum score 550 paper-based; 213 computer-based). *Application deadline:* Applications are processed on a rolling basis. Application fee: $25 ($50 for international students). *Expenses:* Tuition, state resident: part-time $150 per credit. Tuition, nonresident: part-time $150 per credit. Required fees: $85 per credit. *Financial support:* In 2005–06, research assistantships with partial tuition reimbursements (averaging $5,500 per year); Federal Work-Study, scholarships/grants, and unspecified assistantships also available. Support available to part-time students. Financial award application deadline: 3/1; financial award applicants required to submit FAFSA. *Unit head:* Dr. Christopher Cratsley, Chair, 978-665-3617, Fax: 978-665-3658, E-mail: gce@fsc.edu. *Application contact:* Director of Admissions, 978-665-3144, Fax: 978-665-4540, E-mail: admissions@fsc.edu.

Florida Agricultural and Mechanical University, Division of Graduate Studies, Research, and Continuing Education, College of Arts and Sciences, Department of Biology, Tallahassee, FL 32307-3200. Offers MS. *Degree requirements:* For master's, thesis, comprehensive exam. *Entrance requirements:* For master's, GRE General Test, minimum GPA of 3.0. Additional exam requirements/recommendations for international students: Required—TOEFL (minimum score 550 paper-based).

Florida Atlantic University, Charles E. Schmidt College of Science, Department of Biological Sciences, Boca Raton, FL 33431-0991. Offers MS, MST. Part-time programs available. *Faculty:* 21 full-time (4 women), 3 part-time/adjunct (2 women). *Students:* 90 full-time (62 women), 19 part-time (13 women); includes 26 minority (1 African American, 9 Asian Americans or Pacific Islanders, 16 Hispanic Americans), 18 international. Average age 29. 67 applicants, 64% accepted, 36 enrolled. In 2005, 22 degrees awarded. *Degree requirements:* For master's, thesis (for some programs). *Entrance requirements:* For master's, GRE General Test, minimum GPA of 3.0. Additional exam requirements/recommendations for international students: Required—TOEFL. *Application deadline:* For fall admission, 3/15 for domestic students, 2/15 for international students; for spring admission, 11/1 for domestic students, 8/15 for international students. Application fee: $30. *Expenses:* Tuition, state resident: full-time $4,394; part-time $244 per credit. Tuition, nonresident: full-time $16,441; part-time $912 per credit. *Financial support:* In 2005–06, 3 research assistantships, 42 teaching assistantships with tuition reimbursements were awarded; fellowships, career-related internships or fieldwork and Federal Work-Study also available. *Faculty research:* Ecology of the Everglades, molecular biology and biotechnology, marine biology. Total annual research expenditures: $410,000. *Unit head:* Dr. Guritno Roesijadi, Chair, 561-297-3320, Fax: 561-297-2749, E-mail: groesija@fau.edu. *Application contact:* Dr. Ramasoan Narayana, Graduate Coordinator, 561-297-2018, Fax: 561-297-2749, E-mail: rnarayanan@fau.edu.

Florida Atlantic University, Charles E. Schmidt College of Science, Department of Biomedical Science, Boca Raton, FL 33431-0991. Offers MS, PhD. *Faculty:* 30 full-time (9 women). *Students:* 33 full-time (18 women), 6 part-time (4 women); includes 18 minority (6 African Americans, 4 Asian Americans or Pacific Islanders, 8 Hispanic Americans), 5 international. Average age 26. 33 applicants, 70% accepted, 22 enrolled. In 2005, 10 degrees awarded. *Degree requirements:* For master's, thesis (for some programs). *Entrance requirements:* For master's, GRE, minimum GPA of 3.0. *Application deadline:* For fall admission, 6/1 for domestic students, 2/15 for international students; for spring admission, 11/1 for domestic students, 8/15 for international students. Application fee: $30. *Expenses:* Tuition, state resident: full-time $4,394; part-time $244 per credit. Tuition, nonresident: full-time $16,441; part-time $912 per credit. *Financial support:* Research assistantships available. *Unit head:* Dr. Dwight Warren, Chair, 561-297-2219, E-mail: dwarren@fau.edu. *Application contact:* Kathy Jurewicz, Program Assistant, 561-297-2216, E-mail: kjurewicz@fau.edu.

Florida Institute of Technology, Graduate Programs, College of Science, Department of Biological Sciences, Melbourne, FL 32901-6975. Offers biological sciences (PhD); biotechnology (MS); cell and molecular biology (MS, PhD); ecology (MS); marine biology (MS). Part-time programs available. *Faculty:* 14 full-time (1 woman). *Students:* 46 full-time (22 women), 11 part-time (8 women); includes 3 minority (2 Asian Americans or Pacific Islanders, 1 Hispanic American), 13 international. Average age 29. 130 applicants, 24% accepted, 10 enrolled. In 2005, 8 master's, 1 doctorate awarded. *Degree requirements:* For master's, thesis, graduate thesis seminar; for doctorate, thesis/dissertation, dissertations seminar, publications, comprehensive exam, registration. *Entrance requirements:* For master's, GRE General Test, 3 letters of recommendation, minimum GPA of 3.0, resumé; for doctorate, GRE General Test, GRE Subject Test, resumé, 3 letters of recommendation, minimum GPA of 3.2. Additional exam requirements/recommendations for international students: Required—TOEFL (minimum score 550 paper-based; 213 computer-based). *Application deadline:* Applications are processed on a rolling basis. Application fee: $50. Electronic applications accepted. *Expenses:* Tuition: Part-time $825 per credit. *Financial support:* In 2005–06, 33 students received support, including 15 research assistantships with full and partial tuition reimbursements available (averaging $12,840 per year), 18 teaching assistantships with full and partial tuition reimbursements available (averaging $7,911 per year); career-related internships or fieldwork and tuition remissions also available. Financial award application deadline: 3/1; financial award applicants required to submit FAFSA. *Faculty research:* Initiation of protein synthesis in eukaryotic cells, fixation of radioactive carbon, changes in DNA molecule, endangered or threatened avian and mammalian species, hydroacoustics and feeding preference of the West Indian manatee. Total annual research expenditures: $962,328. *Unit head:* Dr. Gary N. Wells, Head, 321-674-8034,

Fax: 321-674-7238, E-mail: gwells@fit.edu. *Application contact:* Carolyn P. Farrior, Director of Graduate Admissions, 321-674-7118, Fax: 321-723-9468, E-mail: cfarrior@fit.edu.

Florida International University, College of Arts and Sciences, Department of Biology, Miami, FL 33199. Offers MS, PhD. *Faculty:* 38 full-time (12 women). *Students:* 86 full-time (44 women), 26 part-time (11 women); includes 26 minority (3 Asian Americans or Pacific Islanders, 23 Hispanic Americans), 23 international. 74 applicants, 20% accepted, 13 enrolled. In 2005, 13 master's, 4 doctorates awarded. *Degree requirements:* For master's and doctorate, one foreign language, thesis/dissertation. *Entrance requirements:* For master's, GRE General Test, 2 letters of recommendation; for doctorate, GRE General Test, 3 letters of recommendation. Additional exam requirements/recommendations for international students: Required—TOEFL. Application fee: $25. *Expenses:* Tuition, state resident: full-time $4,294; part-time $239 per credit. Tuition, nonresident: full-time $15,641; part-time $869 per credit. Required fees: $252; $126 per term. Tuition and fees vary according to program. *Unit head:* Dr. James Fourqrean, Chairperson, 305-348-4084, Fax: 305-348-4096, E-mail: jim.fourqrean@fiu.edu.

Florida State University, College of Medicine, Division of Research and Graduate Programs, Tallahassee, FL 32306. Offers PhD. *Faculty:* 19 full-time (5 women), 3 part-time/adjunct (all women). *Students:* 9 full-time (6 women); includes 1 African American, 3 Asian Americans or Pacific Islanders, 4 international. 26 applicants, 23% accepted, 4 enrolled. *Degree requirements:* For doctorate, thesis/dissertation, comprehensive exam. *Entrance requirements:* For doctorate, GRE. Additional exam requirements/recommendations for international students: Required—TOEFL (minimum score 600 paper-based). *Application deadline:* For fall admission, 3/1 for domestic students, 3/1 for international students. Applications are processed on a rolling basis. Application fee: $30. Electronic applications accepted. *Financial support:* In 2005–06, 4 students received support, including 5 fellowships with full tuition reimbursements available (averaging $21,000 per year), 4 research assistantships with full tuition reimbursements available (averaging $21,000 per year); scholarships/grants also available. Financial award application deadline: 7/1; financial award applicants required to submit FAFSA. *Faculty research:* Neuroscience, development, cell cycle, molecular basis of disease. Total annual research expenditures: $3.4 million. *Unit head:* Dr. Myra M. Hurt, Associate Dean for Graduate Programs, 850-644-8935, Fax: 850-645-7153, E-mail: myra.hurt@med.fsu.edu. *Application contact:* Jeremy Farris, Program Assistant for Graduate Programs, 850-645-6420, Fax: 850-645-7153, E-mail: jeremy.farris@med.fsu.edu.

Florida State University, Graduate Studies, College of Arts and Sciences, Department of Biological Science, Tallahassee, FL 32306. Offers cell biology (MS, PhD); developmental biology (MS, PhD); ecology (MS, PhD); evolutionary biology (MS, PhD); genetics (MS, PhD); immunology (MS, PhD); marine biology (MS, PhD); microbiology (MS, PhD); molecular biology (MS, PhD); neuroscience (PhD); plant sciences (MS, PhD). *Faculty:* 44 full-time (6 women). *Students:* 99 full-time (44 women); includes 8 minority (2 African Americans, 4 Asian Americans or Pacific Islanders, 2 Hispanic Americans), 27 international. 142 applicants, 22% accepted, 17 enrolled. In 2005, 4 master's, 7 doctorates awarded. *Degree requirements:* For master's and doctorate, thesis/dissertation, teaching experience seminar presentations, comprehensive exam, registration. *Entrance requirements:* For master's, GRE General Test (minimum 1100: U-500, 2-500), minimum upper division GPA of 3.0; for doctorate, GRE General Test (minimum 1100: U-500, Q-500), minimum upper division GPA of 3.0. Additional exam requirements/recommendations for international students: Required—TOEFL (minimum score 600 paper-based; 250 computer-based), IB-100. *Application deadline:* For fall admission, 1/15 for domestic students, 12/1 for international students; for spring admission, 10/15 for domestic students, 9/1 for international students. Application fee: $30. Electronic applications accepted. *Financial support:* In 2005–06, 98 students received support, including 11 fellowships with full tuition reimbursements available (averaging $19,000 per year), 36 research assistantships with full tuition reimbursements available (averaging $19,000 per year), 51 teaching assistantships with full tuition reimbursements available (averaging $17,600 per year); traineeships and unspecified assistantships also available. Financial award application deadline: 1/15; financial award applicants required to submit FAFSA. *Unit head:* Dr. George W. Bates, Professor and Associate Chairman, 850-644-3023, Fax: 850-644-9829, E-mail: bates@bio.fsu.edu. *Application contact:* Judy Bowers, Coordinator, Graduate Affairs, 850-644-3023, Fax: 850-644-9829, E-mail: gradinfo@bio.fsu.edu.

Florida State University, Graduate Studies, College of Arts and Sciences, Department of Mathematics, Tallahassee, FL 32306. Offers applied mathematics (MS, PhD); biomedical mathematics (MS, PhD); financial mathematics (MS, PhD); pure mathematics (MS, PhD). Part-time programs available. *Faculty:* 41 full-time (4 women), 10 part-time/adjunct (6 women). *Students:* 117 full-time (27 women), 5 part-time; includes 20 minority (3 African Americans, 1 American Indian/Alaska Native, 7 Asian Americans or Pacific Islanders, 9 Hispanic Americans), 56 international. Average age 26. 317 applicants, 66% accepted, 41 enrolled. In 2005, 31 master's, 7 doctorates awarded. Terminal master's awarded for partial completion of doctoral program. *Degree requirements:* For master's, thesis optional; for doctorate, thesis/dissertation, preliminary exam. *Entrance requirements:* For master's and doctorate, GRE General Test, minimum GPA of 3.0, 4-year bachelor's degree. Additional exam requirements/recommendations for international students: Required—TOEFL (minimum score 550 paper-based; 213 computer-based). *Application deadline:* For fall admission, 3/1 priority date for domestic students, 1/1 priority date for international students; for spring admission, 8/15 priority date for domestic students, 5/1 priority date for international students. Applications are processed on a rolling basis. Application fee: $30. Electronic applications accepted. *Financial support:* In 2005–06, 100 students received support, including 2 fellowships with full tuition reimbursements available (averaging $18,000 per year), 12 research assistantships with full tuition reimbursements available (averaging $16,000 per year), 79 teaching assistantships with full tuition reimbursements available (averaging $16,000 per year); career-related internships or fieldwork, scholarships/grants, and unspecified assistantships also available. Financial award application deadline: 3/1. *Faculty research:* Low-dimensional manifolds, algebra geometry, fluid dynamics, financial mathematics, biomedical mathematics. *Unit head:* Dr. Philip L Bowers, Chairperson, 850-645-3338, Fax: 850-644-4053, E-mail: bowers@math.fsu.edu. *Application contact:* Dr. Eric P. Klassen, Associate Chair for Graduate Studies, 850-644-2202, Fax: 850-644-4053, E-mail: klassen@math.fsu.edu.

Fordham University, Graduate School of Arts and Sciences, Department of Biological Sciences, New York, NY 10458. Offers MS, PhD. Part-time and evening/weekend programs available. *Faculty:* 19 full-time (2 women). *Students:* 33 full-time (21 women), 6 part-time (5 women); includes 4 minority (3 Asian Americans or Pacific Islanders, 1 Hispanic American), 9 international. Average age 27. 39 applicants, 54% accepted, 9 enrolled. In 2005, 4 master's, 3 doctorates awarded. Terminal master's awarded for partial completion of doctoral program. *Median time to degree:* Of those who began their doctoral program in fall 1997, 33% received their degree in 8 years or less. *Degree requirements:* For master's, one foreign language, comprehensive exam; for doctorate, one foreign language, thesis/dissertation, comprehensive exam. *Entrance requirements:* For master's and doctorate, GRE General Test, GRE Subject Test (recommended). Additional exam requirements/recommendations for international students: Required—TOEFL (minimum score 550 paper-based; 213 computer-based). *Application deadline:* For fall admission, 1/4 for domestic students; for spring admission, 11/1 for domestic students. Application fee: $85. Electronic applications accepted. *Expenses:* Tuition: Part-time $875 per credit. Required fees: $239 per term. Tuition and fees vary according to course load. *Financial support:* In 2005–06, 38 students received support, including 6 fellowships with full and partial tuition reimbursements available (averaging $23,167 per year), 24 research assistantships with full and partial tuition reimbursements available (averaging $23,000 per year), 6 teaching assistantships with full and partial tuition reimbursements available (averaging $9,850 per year); Federal Work-Study, institutionally sponsored loans, scholarships/grants, tuition waivers (full and partial), and unspecified assistantships also available. Support available to part-time students. Financial award application deadline: 1/4; financial award applicants required to submit FAFSA. *Faculty research:* Analysis of tumor suppressor genes, chromatic biology and epigenetics, insect ecology, ecology and physiology of mycorrhizal associations, molecular biology of familial Dysautonomia. Total annual research expenditures:

Biological and Biomedical Sciences—General

Fordham University (continued)
$1.1 million. *Unit head:* Dr. Robert Ross, Chair, 718-817-3640, Fax: 718-817-3645, E-mail: rross@fordham.edu. *Application contact:* Charlene Dundie, Director of Graduate Admissions, 718-817-4420, Fax: 718-817-3566, E-mail: dundie@fordham.edu.

See Close-Up on page 127.

Fort Hays State University, Graduate School, College of Health and Life Sciences, Department of Biological Sciences, Program in Biology, Hays, KS 67601-4099. Offers MS. Part-time programs available. *Faculty:* 14 full-time (2 women). *Students:* 18 full-time (6 women), 4 part-time (1 woman); includes 1 Hispanic American. 12 applicants, 50% accepted. In 2005, 8 degrees awarded. *Degree requirements:* For master's, thesis optional. *Entrance requirements:* Additional exam requirements/recommendations for international students: Required—TOEFL (minimum score 550 paper-based; 213 computer-based). *Application deadline:* For fall admission, 7/1 for domestic students. Applications are processed on a rolling basis. Application fee: $30 ($35 for international students). Electronic applications accepted. *Expenses:* Tuition, state resident: part-time $141 per credit. Tuition, nonresident: part-time $371 per credit. *Financial support:* In 2005–06, 7 teaching assistantships (averaging $5,000 per year) were awarded; research assistantships, tuition waivers (full) also available. *Unit head:* Dr. Elmer Finck, Chair, Department of Biological Sciences, 785-628-4214, E-mail: efinck@fhsu.edu.

Frostburg State University, Graduate School, College of Liberal Arts and Sciences, Department of Biology, Frostburg, MD 21532-1099. Offers applied ecology and conservation biology (MS); fisheries and wildlife management (MS). Part-time and evening/weekend programs available. *Faculty:* 7 full-time (1 woman), 7 part-time/adjunct (0 women). *Students:* 15 full-time (10 women), 15 part-time (8 women). Average age 28. 18 applicants, 33% accepted, 5 enrolled. In 2005, 10 degrees awarded. *Degree requirements:* For master's, thesis. *Entrance requirements:* For master's, GRE General Test, resumé. *Application deadline:* For fall admission, 7/15 for domestic students. Applications are processed on a rolling basis. Application fee: $30. Electronic applications accepted. *Expenses:* Tuition, state resident: full-time $5,292; part-time $294 per credit hour. Tuition, nonresident: full-time $6,066; part-time $337 per credit hour. Required fees: $67; $67 per credit hour. $9 per term. One-time fee: $30 full-time. *Financial support:* In 2005–06, 15 research assistantships with full tuition reimbursements (averaging $5,000 per year) were awarded; career-related internships or fieldwork and Federal Work-Study also available. Financial award application deadline: 4/1; financial award applicants required to submit FAFSA. *Faculty research:* Molecular and morphological evolution, ecology and behavior of birds, binecology of forest nematodes and associated insects, conservation genetics of amphibians and fishes, biology of endangered species. *Unit head:* Dr. William Seddon, Chair, 301-687-4166. *Application contact:* Patricia C. Spiker, Director, Graduate Services, 301-687-7053, Fax: 301-687-4597, E-mail: pspiker@frostburg.edu.

George Mason University, College of Science, Department of Biology, Fairfax, VA 22030. Offers biology (MS, PhD), including bioinformatics (MS), ecology, systematics and evolution (MS), interpretive biology (MS), molecular and microbiology (PhD), molecular, microbial, and cellular biology (MS), organismal biology (MS). Part-time programs available. *Faculty:* 17 full-time (11 women), 24 part-time/adjunct (11 women). *Students:* 45 full-time (21 women), 200 part-time (105 women); includes 42 minority (16 African Americans, 1 American Indian/Alaska Native, 12 Asian Americans or Pacific Islanders, 13 Hispanic Americans), 11 international. Average age 33. 190 applicants, 46% accepted, 55 enrolled. In 2005, 14 degrees awarded. Terminal master's awarded for partial completion of doctoral program. *Degree requirements:* For master's, thesis or alternative; for doctorate, thesis/dissertation, internship. *Entrance requirements:* For master's, GRE General Test, GRE Subject Test, bachelor's degree in biology or equivalent; for doctorate, GRE General Test, GRE Subject Test. *Application deadline:* For fall admission, 5/1 for domestic students; for spring admission, 11/1 for domestic students. Electronic applications accepted. *Expenses:* Tuition, state resident: full-time $5,244; part-time $219 per credit. Tuition, nonresident: full-time $15,636; part-time $651 per credit. Required fees: $1,524; $65 per credit. *Financial support:* Fellowships, research assistantships, teaching assistantships available. Support available to part-time students. Financial award application deadline: 3/1; financial award applicants required to submit FAFSA. *Faculty research:* Animal behavior, animal physiology. *Unit head:* Dr. Vikas E. Chandhoke, Director, 703-993-2674, Fax: 703-993-1046, E-mail: biologygrad@gmu.edu.

Georgetown University, Graduate School of Arts and Sciences, Department of Biology, Washington, DC 20057. Offers MS, PhD. Terminal master's awarded for partial completion of doctoral program. *Degree requirements:* For master's and doctorate, thesis/dissertation, comprehensive exam. *Entrance requirements:* For master's and doctorate, GRE General Test, GRE Subject Test (biology). Additional exam requirements/recommendations for international students: Required—TOEFL (minimum score 550 paper-based; 213 computer-based). Electronic applications accepted. *Faculty research:* Parasitology, ecology, evaluation and behavior, neuroscience and development, cell and molecular biology, immunology.

See Close-Up on page 129.

Georgetown University, Graduate School of Arts and Sciences, Programs in Biomedical Sciences, Washington, DC 20057. Offers MS, PhD, MD/PhD, MS/PhD. *Entrance requirements:* For doctorate, GRE General Test. Additional exam requirements/recommendations for international students: Required—TOEFL.

Georgetown University, National Institutes of Health Sponsored Programs, Programs in Biomedical Sciences, Washington, DC 20057. Offers MS, PhD, MD/PhD, MS/PhD. *Faculty:* 252 full-time (73 women), 14 part-time/adjunct (3 women). *Students:* 236 full-time (92 women), 52 part-time (24 women); includes 63 minority (4 African Americans, 1 American Indian/Alaska Native, 51 Asian Americans or Pacific Islanders, 7 Hispanic Americans), 43 international. In 2005, 84 master's, 10 doctorates awarded. *Entrance requirements:* For doctorate, GRE General Test. Additional exam requirements/recommendations for international students: Required—TOEFL. *Application fee:* $50 ($55 for international students). *Financial support:* Career-related internships or fieldwork available. *Application contact:* Graduate School Admissions Office, 202-687-5568.

See Close-Up on page 131.

The George Washington University, Columbian College of Arts and Sciences, Department of Biological Sciences, Washington, DC 20052. Offers MS, PhD. Part-time and evening/weekend programs available. *Faculty:* 20 full-time (6 women), 10 part-time/adjunct (1 woman). *Students:* 25 full-time (14 women), 6 part-time (4 women); includes 2 minority (1 African American, 1 Asian American or Pacific Islander), 9 international. 59 applicants, 22% accepted, 7 enrolled. In 2005, 1 master's, 4 doctorates awarded. Terminal master's awarded for partial completion of doctoral program. *Degree requirements:* For master's, comprehensive exam; for doctorate, thesis/dissertation, general exam. *Entrance requirements:* For master's and doctorate, GRE General Test, minimum GPA of 3.0. Additional exam requirements/recommendations for international students: Required—TOEFL (minimum score 550 paper-based; 213 computer-based). *Application deadline:* For fall admission, 1/2 priority date for domestic students, 1/2 priority date for international students; for spring admission, 10/1 priority date for domestic students, 10/1 priority date for international students. Applications are processed on a rolling basis. Application fee: $60. Electronic applications accepted. *Financial support:* In 2005–06, 20 students received support, including fellowships with full tuition reimbursements available (averaging $18,000 per year), teaching assistantships with full tuition reimbursements available (averaging $18,000 per year); Federal Work-Study also available. Financial award application deadline: 1/2. *Faculty research:* Systematics, evolution, ecology, developmental biology, cell/molecular biology. Total annual research expenditures: $900,000. *Unit head:* Dr. James M Clark, Chair, 202-994-6090, Fax: 202-994-6100. *Application contact:* Dr. John R. Burns, Professor, 202-994-7149, Fax: 202-994-6100, E-mail: jrburns@gwu.edu.

See Close-Up on page 133.

The George Washington University, Columbian College of Arts and Sciences, Department of Environmental Studies, Program in Hominid Paleobiology, Washington, DC 20052. Offers MS, PhD. Part-time and evening/weekend programs available. Terminal master's awarded for partial completion of doctoral program. *Degree requirements:* For master's, thesis, comprehensive exam; for doctorate, thesis/dissertation, general exam. *Entrance requirements:* For master's, GRE General Test, bachelor's degree in field, minimum GPA of 3.0; for doctorate, GRE General Test, minimum GPA of 3.0. Additional exam requirements/recommendations for international students: Required—TOEFL (minimum score 550 paper-based; 213 computer-based). Electronic applications accepted.

The George Washington University, Columbian College of Arts and Sciences, Institute for Biomedical Sciences, Washington, DC 20052. Offers biochemistry and molecular biology (PhD); genetics (MS, PhD); microbiology and immunology (PhD); molecular medicine (PhD); neuroscience (PhD); pharmacology (PhD). Part-time and evening/weekend programs available. *Degree requirements:* For doctorate, thesis/dissertation. *Entrance requirements:* For doctorate, GRE General Test, minimum GPA of 3.0. Additional exam requirements/recommendations for international students: Required—TOEFL (minimum score 600 paper-based; 250 computer-based). Electronic applications accepted.

See Close-Up on page 135.

The George Washington University, National Institutes of Health Sponsored Programs, National Institutes of Health Graduate Program in Biomedical Sciences, Washington, DC 20052. Offers MS, PhD. *Unit head:* Dr. Diana Johnson, Director, 202-994-7120, Fax: 202-994-6100.

See Close-Up on page 137.

Georgia College & State University, Graduate School, School of Liberal Arts and Sciences, Department of Biology, Milledgeville, GA 31061. Offers MS. Part-time programs available. *Faculty:* 15 full-time (3 women). *Students:* 13 full-time (5 women), 8 part-time (3 women); includes 5 minority (4 African Americans, 1 Asian American or Pacific Islander). 21 applicants, 81% accepted, 6 enrolled. In 2005, 10 degrees awarded. *Degree requirements:* For master's, thesis optional. *Entrance requirements:* For master's, GRE, 30 hours undergraduate course work in biological science. Additional exam requirements/recommendations for international students: Required—TOEFL. *Application deadline:* For fall admission, 7/1 for domestic students. Applications are processed on a rolling basis. Application fee: $25. Electronic applications accepted. *Financial support:* In 2005–06, 15 research assistantships with tuition reimbursements were awarded; career-related internships or fieldwork, Federal Work-Study, and unspecified assistantships also available. Support available to part-time students. Financial award application deadline: 3/1; financial award applicants required to submit FAFSA. *Faculty research:* Vertebrate collecting and monitoring, paleontologic expedition. *Unit head:* Dr. William Wall, Chair, 478-445-0818, E-mail: bill.wall@gcsu.edu. *Application contact:* Dr. Harold Reed, Graduate Coordinator, 478-445-0815, E-mail: harold.reed@gcsu.edu.

Georgia Institute of Technology, Graduate Studies and Research, College of Sciences, School of Biology, Atlanta, GA 30332-0001. Offers applied biology (MS, PhD); bioinformatics (MS, PhD); biology (MS). Part-time programs available. Terminal master's awarded for partial completion of doctoral program. *Degree requirements:* For master's, thesis; for doctorate, thesis/dissertation, qualifying exam. *Entrance requirements:* For master's, GRE General Test, minimum GPA of 2.9; for doctorate, GRE General Test, minimum GPA of 3.0. Additional exam requirements/recommendations for international students: Required—TOEFL. Electronic applications accepted. *Faculty research:* Microbiology, molecular and cell biology, ecology.

Georgian Court University, School of Sciences and Mathematics, Lakewood, NJ 08701-2697. Offers biology (MS); counseling psychology (MA); holistic health (Certificate); holistic health studies (MA); mathematics (MA); professional counselor (Certificate); school psychology (Certificate). Part-time and evening/weekend programs available. *Faculty:* 22 full-time (15 women), 7 part-time/adjunct (4 women). *Students:* 30 full-time (25 women), 175 part-time (154 women); includes 10 minority (3 African Americans, 1 American Indian/Alaska Native, 3 Asian Americans or Pacific Islanders, 3 Hispanic Americans), 1 international. Average age 37. 126 applicants, 74% accepted, 76 enrolled. In 2005, 52 degrees awarded. *Degree requirements:* For master's, thesis (for some programs), comprehensive exam (for some programs). *Entrance requirements:* For master's, GRE General Test, GRE Subject Test in biology (MS), 3 letters of recommendation. Additional exam requirements/recommendations for international students: Required—TOEFL (minimum score 550 paper-based; 213 computer-based). *Application deadline:* For fall admission, 8/1 priority date for domestic students, 4/1 priority date for international students; for spring admission, 1/1 priority date for domestic students, 7/1 priority date for international students. Applications are processed on a rolling basis. Application fee: $40. Electronic applications accepted. *Expenses:* Tuition: Part-time $566 per credit. *Financial support:* Scholarships/grants, health care benefits, and unspecified assistantships available. Financial award application deadline: 4/15; financial award applicants required to submit FAFSA. *Unit head:* Dr. Linda James, Dean, 732-987-2617. *Application contact:* Eugene Soltys, Director of Graduate Admissions, 732-987-2760 Ext. 2760, Fax: 732-987-2000, E-mail: admissions@georgian.edu.

Georgia Southern University, Jack N. Averitt College of Graduate Studies, Allen E. Paulson College of Science and Technology, Department of Biology, Statesboro, GA 30460. Offers MS. Part-time programs available. *Students:* 31 full-time (18 women), 12 part-time (7 women); includes 4 minority (1 African American, 2 Asian Americans or Pacific Islanders, 1 Hispanic American), 1 international. Average age 26. 16 applicants, 75% accepted, 9 enrolled. In 2005, 9 degrees awarded. *Degree requirements:* For master's, thesis, terminal exam. *Entrance requirements:* For master's, GRE General Test, minimum GPA of 2.8, BS in biology, 2 letters of reference. Additional exam requirements/recommendations for international students: Required—TOEFL (minimum score 550 paper-based; 216 computer-based). *Application deadline:* For fall admission, 3/1 priority date for domestic students, 3/1 priority date for international students; for spring admission, 10/1 priority date for domestic students, 10/1 priority date for international students. Applications are processed on a rolling basis. Application fee: $50. Electronic applications accepted. *Expenses:* Tuition, state resident: full-time $2,926; part-time $122 per semester hour. Tuition, nonresident: full-time $11,704; part-time $488 per semester hour. Required fees: $1,024. *Financial support:* In 2005–06, 35 students received support, including research assistantships with partial tuition reimbursements available (averaging $5,500 per year), teaching assistantships with partial tuition reimbursements available (averaging $5,500 per year); career-related internships or fieldwork, Federal Work-Study, scholarships/grants, and unspecified assistantships also available. Support available to part-time students. Financial award application deadline: 4/15; financial award applicants required to submit FAFSA. *Faculty research:* Cellular and molecular biology, conservation biology, ecology and genetics, marine invertebrates. Total annual research expenditures: $615,362. *Unit head:* Dr. Stephen Vives, Chair, 912-681-5487, Fax: 912-681-0845, E-mail: svives@georgiasouthern.edu. *Application contact:* 912-681-5384, Fax: 912-681-0740, E-mail: gradschool@georgiasouthern.edu.

Georgia State University, College of Arts and Sciences, Department of Biology, Atlanta, GA 30303-3083. Offers applied and environmental microbiology (MS, PhD); cell biology and physiology (MS, PhD); molecular genetics and biochemistry (MS, PhD); neurobiology and behavior (MS, PhD). Part-time and evening/weekend programs available. *Degree requirements:* For master's, thesis or alternative, exam; for doctorate, thesis/dissertation, exam. *Entrance requirements:* For master's and doctorate, GRE General Test. Additional exam requirements/recommendations for international students: Required—TOEFL. Electronic applications accepted. *Expenses:* Tuition, state resident: full-time $4,368; part-time $182 per semester hour. Tuition, nonresident: full-time $8,732; part-time $728 per semester hour. Required fees: $46 per semester hour. *Faculty research:* Physiological biochemistry, gene expression, molecular virology, microbial ecology, integration in neural systems.

Goucher College, Program in Post-Baccalaureate Premedical Studies, Baltimore, MD 21204-2794. Offers Certificate. *Faculty:* 6 full-time (2 women), 2 part-time/adjunct (1 woman). *Students:*

Biological and Biomedical Sciences—General

30 full-time (16 women); includes 4 minority (1 African American, 3 Asian Americans or Pacific Islanders). Average age 26. *Application deadline:* Applications are processed on a rolling basis. Application fee: $50. *Expenses: Contact institution. Financial support:* Institutionally sponsored loans and scholarships/grants available. Financial award application deadline: 3/1; financial award applicants required to submit FAFSA. *Unit head:* Liza Thompson, Director, 800-414-3437, Fax: 410-337-6461, E-mail: lthompso@goucher.edu.

Graduate School and University Center of the City University of New York, Graduate Studies, Program in Biology, New York, NY 10016-4039. Offers PhD. *Faculty:* 78 full-time (24 women). *Students:* 182 full-time (105 women), 3 part-time (all women); includes 37 minority (12 African Americans, 1 American Indian/Alaska Native, 7 Asian Americans or Pacific Islanders, 17 Hispanic Americans), 55 international. Average age 32. 115 applicants, 52% accepted, 33 enrolled. In 2005, 18 degrees awarded. *Degree requirements:* For doctorate, thesis/dissertation, teaching experience. *Entrance requirements:* For doctorate, GRE General Test. Additional exam requirements/recommendations for international students: Required—TOEFL. *Application deadline:* For fall admission, 2/1 for domestic students. Application fee: $125. Electronic applications accepted. *Financial support:* In 2005–06, 89 students received support, including 79 fellowships, 1 research assistantship, 1 teaching assistantship; career-related internships or fieldwork, Federal Work-Study, institutionally sponsored loans, and tuition waivers (full and partial) also available. Financial award application deadline: 2/1; financial award applicants required to submit FAFSA. *Unit head:* Dr. Richard L. Chappell, Executive Officer, 212-817-8101, Fax: 212-817-1504, E-mail: rchappell@gc.cuny.edu. *Application contact:* Executive Officer, 212-817-8100, Fax: 212-817-1504, E-mail: biology@gc.cuny.edu.

Grand Valley State University, College of Liberal Arts and Sciences, Biology Department, Allendale, MI 49401-9403. Offers MS. *Faculty:* 11 full-time (4 women), 5 part-time/adjunct (1 woman). *Students:* 18 full-time (9 women), 10 part-time (9 women); includes 1 minority (Hispanic American) Average age 29. 16 applicants, 69% accepted, 11 enrolled. In 2005, 3 degrees awarded. *Entrance requirements:* Additional exam requirements/recommendations for international students: Required—TOEFL. Application fee: $30. *Expenses:* Tuition, state resident: full-time $5,364; part-time $298 per credit. Tuition, nonresident: full-time $10,800; part-time $600 per credit. *Financial support:* In 2005–06, 17 research assistantships with full and partial tuition reimbursements (averaging $8,000 per year) were awarded *Faculty research:* Natural resources conservation biology, aquatic sciences, terrestrial ecology, behavioral biology. *Unit head:* Dr. Shaily Menon, Director, 616-331-2586, E-mail: menons@gvsu.edu. *Application contact:* Dr. Mark Luttenton, Graduate Program Director, 616-331-2505, E-mail: Luttentm@gvsu.edu.

Grand Valley State University, College of Liberal Arts and Sciences, Department of Biomedical and Health Sciences, Allendale, MI 49401-9403. Offers health science (MHS, MS). Part-time programs available. *Faculty:* 7 full-time (1 woman). *Students:* 9 full-time (8 women), 2 part-time (1 woman); includes 2 minority (1 African American, 1 Asian American or Pacific Islander). Average age 26. 17 applicants, 71% accepted, 6 enrolled. In 2005, 1 degree awarded. *Degree requirements:* For master's, thesis, qualifying exam. *Entrance requirements:* For master's, GRE General Test, minimum GPA of 3.0, 3 letters of recommendation. Additional exam requirements/recommendations for international students: Required—TOEFL. *Application deadline:* For fall admission, 2/1 priority date for domestic students, 2/1 priority date for international students. Applications are processed on a rolling basis. Application fee: $30. Electronic applications accepted. *Expenses:* Tuition, state resident: full-time $5,364; part-time $298 per credit. Tuition, nonresident: full-time $10,800; part-time $600 per credit. *Financial support:* In 2005–06, 2 students received support. Scholarships/grants and unspecified assistantships available. *Unit head:* Dr. Anthony Nieuwkoop, Head, 616-331-3318, Fax: 616-331-2090, E-mail: nieuwot@gvsu.edu. *Application contact:* Dr. Debra Burg, Director, 616-331-3721, Fax: 616-331-2090, E-mail: burgd@gvsu.edu.

Hampton University, Graduate College, Department of Biological Sciences, Hampton, VA 23668. Offers MA, MS. Part-time and evening/weekend programs available. *Degree requirements:* For master's, thesis optional. *Entrance requirements:* For master's, GRE General Test. *Faculty research:* Marine ecology, microbial and chemical pollution, pesticide problems.

Harvard University, Extension School, Cambridge, MA 02138-3722. Offers applied sciences (CAS); biotechnology (ALM); educational technologies (ALM); English for graduate and professional studies (DGP); environmental management (ALM, CEM); information technology (ALM); journalism (ALM); liberal arts (ALM); management (CM); mathematics for teaching (ALM); museum studies (ALM); premedical studies (Diploma); publication and communication (CPC). Part-time and evening/weekend programs available. *Faculty:* 236 part-time/adjunct. *Students:* 101 full-time (56 women), 564 part-time (278 women); includes 167 minority (35 African Americans, 1 American Indian/Alaska Native, 84 Asian Americans or Pacific Islanders, 47 Hispanic Americans). Average age 36. In 2005, 112 master's, 184 Diplomas awarded. *Degree requirements:* For master's, thesis. *Entrance requirements:* For master's, 3 completed graduate courses with grade of B or higher. Additional exam requirements/recommendations for international students: Required—TOEFL (minimum score 600 paper-based; 250 computer-based), TWE (minimum score 5). *Application deadline:* Applications are processed on a rolling basis. Application fee: $75. *Expenses: Contact institution.* Full-time tuition and fees vary according to program and student level. *Financial support:* In 2005–06, 268 students received support. Scholarships/grants available. Support available to part-time students. Financial award application deadline: 8/6; financial award applicants required to submit FAFSA. *Unit head:* Michael Shinagel, Dean. *Application contact:* Program Director, 617-495-4024, Fax: 617-495-9176.

Harvard University, Graduate School of Arts and Sciences, Department of Organismic and Evolutionary Biology, Cambridge, MA 02138. Offers biology (PhD). *Students:* 57 full-time (22 women). 98 applicants, 13% accepted. In 2005, 11 doctorates awarded. *Degree requirements:* For doctorate, 2 foreign languages. *Entrance requirements:* For doctorate, GRE General Test, GRE Subject Test (recommended), 7 courses in biology, chemistry, physics, mathematics, computer science, or geology. Additional exam requirements/recommendations for international students: Required—TOEFL. *Application deadline:* For fall admission, 12/15 for domestic students. Application fee: $60. *Expenses:* Tuition: Full-time $28,752. Full-time tuition and fees vary according to program and student level. *Financial support:* Fellowships, research assistantships, teaching assistantships, career-related internships or fieldwork, Federal Work-Study, and institutionally sponsored loans available. Financial award application deadline: 12/30. *Unit head:* Betsey Cogswell, Administrator, 617-495-5497, Fax: 617-495-5264. *Application contact:* Departmental Office, 617-495-2305.

Harvard University, Graduate School of Arts and Sciences, Division of Medical Sciences, Boston, MA 02115. Offers biological chemistry and molecular pharmacology (PhD); cell biology (PhD); genetics (PhD); microbiology and molecular genetics (PhD); pathology (PhD), including experimental pathology. *Students:* 433 full-time (210 women). In 2005, 83 doctorates awarded. *Degree requirements:* For doctorate, thesis/dissertation. *Entrance requirements:* For doctorate, GRE General Test, GRE Subject Test. Additional exam requirements/recommendations for international students: Required—TOEFL. Application fee: $60. *Expenses:* Tuition: Full-time $28,752. Full-time tuition and fees vary according to program and student level. *Financial support:* Fellowships, research assistantships, teaching assistantships, institutionally sponsored loans and tuition waivers (full) available. Financial award application deadline: 1/1. *Unit head:* Administrator, 617-432-2029. *Application contact:* Administrator, 617-432-2029.

Harvard University, Graduate School of Arts and Sciences, Program in Biological Sciences in Public Health (BPH), Boston, MA 02115. Offers PhD. *Degree requirements:* For doctorate, thesis/dissertation, qualifying exam. *Entrance requirements:* For doctorate, GRE General Test, GRE Subject Test. Additional exam requirements/recommendations for international students: Required—TOEFL. *Application deadline:* For fall admission, 12/15 for domestic students. Application fee: $60. *Expenses:* Tuition: Full-time $28,752. Full-time tuition and fees vary according to program and student level. *Financial support:* Fellowships, research assistantships, teaching assistantships, institutionally sponsored loans and tuition waivers (full) available.

Financial award application deadline: 1/1. *Faculty research:* Nutrition biochemistry, molecular and cellular toxicology, cardiovascular disease, cancer biology, tropical public health, environmental health physiology. *Unit head:* Ruth Kenworthy, Administrator, 617-432-2932, Fax: 617-432-0433, E-mail: kenworthy@cvlab.harvard.edu. *Application contact:* Ruth Kenworthy, Administrator, 617-432-2932, Fax: 617-432-0433, E-mail: kenworthy@cvlab.harvard.edu.

See Close-Up on page 139.

Harvard University, Medical School and Graduate School of Arts and Sciences, Division of Health Sciences and Technology, Biomedical Enterprise Program, Cambridge, MA 02138. Offers SM. *Students:* 22 full-time (4 women). Average age 31. 44 applicants, 27% accepted, 12 enrolled. In 2005, 3 degrees awarded. *Degree requirements:* For master's, thesis. *Entrance requirements:* For master's, GMAT, bachelor's degree in engineering or sciences, work experience in biomedical business. *Application deadline:* For fall admission, 12/15 for domestic students, 12/15 for international students. Application fee: $70. Electronic applications accepted. *Financial support:* In 2005–06, 4 students received support, including 4 research assistantships with partial tuition reimbursements available (averaging $19,493 per year); teaching assistantships with partial tuition reimbursements available, institutionally sponsored loans, health care benefits, and unspecified assistantships also available. *Faculty research:* Entrepreneurship; technology strategy management; organizational strategies and models; epidemiology and biostatistics; biomedical research from the molecular to the whole-organism level. *Application contact:* Catherine Modica, Admissions Coordinator, 617-253-2307, Fax: 617-253-6692, E-mail: cmodica@mit.edu.

Harvard University, Medical School and Graduate School of Arts and Sciences, Division of Medical Sciences, Boston, MA 02115. Offers PhD, MD/PhD. *Degree requirements:* For doctorate, thesis/dissertation, qualifying exam. *Entrance requirements:* For doctorate, GRE General Test, GRE Subject Test. Additional exam requirements/recommendations for international students: Required—TOEFL. Expenses: Contact institution. Full-time tuition and fees vary according to program and student level.

Heritage University, Graduate Programs in Education, Program in Professional Development, Toppenish, WA 98948-9599. Offers bilingual education/ESL (M Ed); biology (M Ed); English and literature (M Ed); reading/literacy (M Ed); special education (M Ed). Part-time and evening/weekend programs available. *Degree requirements:* For master's, thesis (for some programs), comprehensive exam (for some programs), registration.

Hofstra University, College of Liberal Arts and Sciences, Division of Natural Sciences, Mathematics, Engineering, and Computer Science, Department of Biology, Hempstead, NY 11549. Offers MA, MS. Part-time programs available. *Faculty:* 7 full-time (4 women), 1 part-time/adjunct (0 women). *Students:* 12 full-time (6 women), 8 part-time (5 women); includes 2 minority (1 Asian American or Pacific Islander, 1 Hispanic American), 1 international. Average age 28. 53 applicants, 72% accepted, 21 enrolled. In 2005, 16 degrees awarded. *Degree requirements:* For master's, thesis, registration. *Entrance requirements:* For master's, GRE General Test, bachelor's degree in biology or equivalent, 2 letters of recommendation. Additional exam requirements/recommendations for international students: Required—TOEFL (minimum score 550 paper-based; 213 computer-based). *Application deadline:* Applications are processed on a rolling basis. Application fee: $60. Electronic applications accepted. *Expenses:* Tuition: Full-time $12,060; part-time $670 per credit. Required fees: $930; $155 per term. Tuition and fees vary according to course load and program. *Financial support:* In 2005–06, 12 students received support, including 3 fellowships with full tuition reimbursements available (averaging $5,331 per year); research assistantships, scholarships/grants also available. Financial award applicants required to submit FAFSA. *Faculty research:* Temperature sex determination in reptiles, the role of genes in determining fate of reproductive cells in celegans, chemosensory functions in lobsters, relationship b/w hormones, bird behavior and environmental conditions, natural history of deep-sea elasmobranes. Total annual research expenditures: $500,000. *Unit head:* Dr. Peter C. Daniel, Program Director, 516-463-6718, Fax: 516-463-5112, E-mail: biopcd@hofstra.edu. *Application contact:* Carol Drummer, Dean of Graduate Admissions, 516-463-4876, Fax: 516-463-4664, E-mail: gradstudent@hofstra.edu.

See Close-Up on page 141.

Hood College, Graduate School, Program in Biomedical Science, Frederick, MD 21701-8575. Offers MS. Part-time and evening/weekend programs available. *Faculty:* 4 full-time (2 women), 4 part-time/adjunct (2 women). *Students:* 4 full-time (all women), 89 part-time (62 women); includes 13 minority (9 African Americans, 3 Asian Americans or Pacific Islanders, 1 Hispanic American), 3 international. Average age 31. In 2005, 13 master's awarded. *Degree requirements:* For master's, thesis or alternative. *Entrance requirements:* For master's, bachelor's degree in biology; minimum GPA of 2.5; undergraduate course work in cell biology, chemistry, organic chemistry, and genetics. Additional exam requirements/recommendations for international students: Required—TOEFL. *Application deadline:* Applications are processed on a rolling basis. Application fee: $35. Electronic applications accepted. *Expenses:* Tuition: Full-time $6,300; part-time $350 per credit. Required fees: $40. *Financial support:* Tuition waivers (partial) available. Financial award applicants required to submit FAFSA. *Unit head:* Dr. Craig Laufer, Director, 301-696-3656, Fax: 301-696-3597. *Application contact:* Lori White Drega, Associate Dean of Graduate School, 301-696-3811, Fax: 301-696-3597, E-mail: gofurther@hood.edu.

Howard University, Graduate School of Arts and Sciences, Department of Biology, Washington, DC 20059-0002. Offers MS, PhD. Part-time programs available. *Degree requirements:* For master's and doctorate, thesis/dissertation, qualifying exams. *Entrance requirements:* For master's and doctorate, GRE General Test, minimum GPA of 3.0. Additional exam requirements/recommendations for international students: Required—TOEFL. Electronic applications accepted. *Faculty research:* Physiology, molecular biology, cell biology, microbiology, environmental biology.

See Close-Up on page 143.

Humboldt State University, Graduate Studies, College of Natural Resources and Sciences, Department of Biological Sciences, Arcata, CA 95521-8299. Offers MA. *Students:* 33 full-time (20 women), 24 part-time (16 women); includes 40 minority (all Hispanic Americans) Average age 31. 61 applicants, 52% accepted, 20 enrolled. In 2005, 6 degrees awarded. *Degree requirements:* For master's, project or thesis. *Entrance requirements:* For master's, GRE General Test, appropriate bachelor's degree, minimum GPA of 2.5. Additional exam requirements/recommendations for international students: Required—TOEFL (minimum score 500 paper-based; 173 computer-based). *Application deadline:* For fall admission, 2/1 for domestic students, 2/1 for international students. Applications are processed on a rolling basis. Application fee: $55. *Expenses:* Tuition, nonresident: part-time $339 per unit. Required fees: $3,102. *Financial support:* Application deadline: 3/1; *Faculty research:* Plant ecology, DNA sequencing, invertebrates. *Unit head:* Dr. Michael Mesler, Coordinator, 707-826-3674, E-mail: mm1@humboldt.edu.

See Close-Up on page 145.

Hunter College of the City University of New York, Graduate School, School of Arts and Sciences, Department of Biological Sciences, New York, NY 10021-5085. Offers MA, PhD. Part-time programs available. *Faculty:* 19 full-time (9 women), 3 part-time/adjunct (1 woman). *Students:* Average age 30. 19 applicants, 68% accepted, 8 enrolled. In 2005, 14 degrees awarded. Terminal master's awarded for partial completion of doctoral program. *Degree requirements:* For master's, one foreign language. *Entrance requirements:* For master's, GRE, 1 year of course work in organic chemistry (including laboratory), college physics, calculus; undergraduate major in biology, botany, physiology, zoology, chemistry or physics. Additional exam requirements/recommendations for international students: Required—TOEFL. *Application deadline:* For fall admission, 4/1 for domestic students, 2/1 for international students; for spring admission, 11/1 for domestic students, 9/1 for international students. Application fee: $125. *Expenses:* Tuition, state resident: full-time $6,400; part-time $270 per credit. Tuition, nonresident:

Biological and Biomedical Sciences—General

Hunter College of the City University of New York (continued)
part-time $500 per credit. Required fees: $50 per term. Part-time tuition and fees vary according to course load and program. *Financial support:* Fellowships, research assistantships, teaching assistantships, scholarships/grants and tuition waivers (partial) available. Support available to part-time students. *Faculty research:* Analysis of prokaryotic and eukaryotic DNA, protein structure, mammalian DNA replication, oncogene expression, neuroscience. *Unit head:* Dr. Shirley Raps, Chairperson, 212-772-5293. *Application contact:* William Zlata, Director for Graduate Admissions, 212-772-4482, Fax: 212-650-3336, E-mail: admissions@hunter.cuny.edu.

Hunter College of the City University of New York, Hunter Center for Gene Structure and Function, New York, NY 10021-5085.

ICR Graduate School, Graduate Programs, Santee, CA 92071. Offers astro/geophysics (MS); biology (MS); geology (MS); science education (MS). Part-time programs available. *Faculty:* 5 full-time (1 woman), 1 part-time/adjunct (0 women). *Students:* 6 full-time (2 women), 6 part-time (3 women). Average age 45. In 2005, 4 degrees awarded. *Degree requirements:* For master's, thesis (for some programs), comprehensive exam (for some programs). *Entrance requirements:* For master's, minimum undergraduate GPA of 3.0, bachelor's degree in science or science education. *Application deadline:* Applications are processed on a rolling basis. Application fee: $30. *Expenses:* Tuition: Part-time $150 per semester hour. *Faculty research:* Age of the earth, limits of variation, catastrophe, optimum methods for teaching. Total annual research expenditures: $200,000. *Unit head:* , Dr. Kenneth B. Cumming, Dean, 619-448-0900, Fax: 619-448-3469. *Application contact:* Dr. Jack Kriege, Registrar, 619-448-0900 Ext. 6016, Fax: 619-448-3469, E-mail: jkriege@icr.org.

Idaho State University, Office of Graduate Studies, College of Arts and Sciences, Department of Biological Sciences, Pocatello, ID 83209. Offers biology (MNS, MS, DA, PhD); clinical laboratory science (MS); microbiology (MS). *Accreditation:* NAACLS. *Degree requirements:* For master's, one foreign language, thesis, comprehensive exam, registration (for some programs); for doctorate, 2 foreign languages, thesis/dissertation, comprehensive exam, registration. *Entrance requirements:* For master's, GRE General Test, minimum GPA of 3.0 in all upper division classes; for doctorate, GRE General Test, GRE Subject Test, minimum GPA of 3.0 in all upper division classes. Additional exam requirements/recommendations for international students: Required—TOEFL (minimum score 550 paper-based; 213 computer-based). *Faculty research:* Ecology and evolutionary biology, plant and animal physiology, plant and animal developmental biology, immunology, molecular biology.

Illinois Institute of Technology, Graduate College, College of Science and Letters, Department of Biological, Chemical and Physical Sciences, Biology Division, Chicago, IL 60616-3793. Offers biochemistry (MS); biology (MBS, PhD); biotechnology (MS); cell biology (MS); microbiology (MS); molecular biochemistry and biophysics (MS, PhD). Part-time and evening/weekend programs available. Postbaccalaureate distance learning degree programs offered (no on-campus study). Terminal master's awarded for partial completion of doctoral program. *Degree requirements:* For master's, thesis (for some programs), comprehensive exam; for doctorate, thesis/dissertation, comprehensive exam. *Entrance requirements:* For master's and doctorate, GRE General Test, minimum undergraduate GPA of 3.0. Additional exam requirements/recommendations for international students: Required—TOEFL (minimum score 550 paper-based; 213 computer-based). Electronic applications accepted. *Faculty research:* Protein crystallography, small angle x-ray differation of muscle, spectroscopy of multidomain proteins, structure and function of cell cycle proteins, development of anticancer drugs.

Illinois State University, Graduate School, College of Arts and Sciences, Department of Biological Sciences, Normal, IL 61790-2200. Offers biological sciences (MS); biology (PhD); biotechnology (MS); botany (PhD); ecology (PhD); genetics (PhD); microbiology (PhD); physiology (PhD); zoology. (PhD). Part-time programs available. *Faculty:* 27 full-time (6 women). *Students:* 36 full-time (28 women), 25 part-time (17 women); includes 1 minority (Asian American or Pacific Islander), 23 international. 80 applicants, 21% accepted. In 2005, 14 master's, 5 doctorates awarded. *Degree requirements:* For master's, thesis or alternative; for doctorate, variable foreign language requirement, thesis/dissertation, 2 terms of residency. *Entrance requirements:* For master's, GRE General Test, minimum GPA of 2.6 in last 60 hours of course work; for doctorate, GRE General Test. *Application deadline:* Applications are processed on a rolling basis. Application fee: $30. *Expenses:* Tuition, state resident: full-time $3,060; part-time $170 per credit hour. Tuition, nonresident: full-time $6,390; part-time $355 per credit hour. Required fees: $1,411; $47 per credit hour. *Financial support:* In 2005–06, 22 research assistantships (averaging $13,909 per year), 38 teaching assistantships (averaging $12,617 per year) were awarded; Federal Work-Study, tuition waivers (full), and unspecified assistantships also available. Financial award application deadline: 4/1. *Faculty research:* CRUI: osmoregulation in euryhaline fish: physiology, ecology and molecular biology; the PRISM project: enhancing science and math education; cell structure—functions of Na pump assembly. *Unit head:* Dr. Hou Tak Takucheung, Acting Chairperson, 309-438-3669. *Application contact:* Derek A. McCracken, Graduate Adviser, 309-438-3664.

See Close-Up on page 147.

Indiana State University, School of Graduate Studies, College of Arts and Sciences, Department of Life Sciences, Terre Haute, IN 47809-1401. Offers ecology (PhD); life sciences (MS); microbiology (PhD); physiology (PhD); science education (MS); sports medicine (PhD). *Faculty:* 24 full-time (8 women), 12 part-time/adjunct (2 women). *Students:* 51 full-time (24 women), 24 part-time (10 women); includes 3 minority (all Asian Americans or Pacific Islanders), 11 international. Average age 26. 45 applicants, 73% accepted, 29 enrolled. In 2005, 9 master's, 2 doctorates awarded. *Degree requirements:* For master's, thesis (for some programs); for doctorate, thesis/dissertation, comprehensive exam. *Entrance requirements:* For master's and doctorate, GRE General Test. *Application deadline:* For fall admission, 7/1 for domestic students; for spring admission, 11/1 priority date for domestic students. Applications are processed on a rolling basis. Application fee: $35. Electronic applications accepted. *Expenses:* Tuition, state resident: full-time $6,288; part-time $262 per credit hour. Tuition, nonresident: full-time $12,504; part-time $521 per credit hour. *Financial support:* In 2005–06, 26 teaching assistantships with partial tuition reimbursements (averaging $8,005 per year) were awarded; research assistantships with partial tuition reimbursements, Federal Work-Study, institutionally sponsored loans, and tuition waivers (partial) also available. Financial award application deadline: 3/1; financial award applicants required to submit FAFSA. *Unit head:* Dr. Swapan Ghosh, Interim Chairperson, 812-237-2400.

See Close-Up on page 149.

Indiana University Bloomington, Graduate School, College of Arts and Sciences, Department of Biology, Bloomington, IN 47405-7000. Offers biology teaching (MAT); evolution, ecology, and behavior (MA, PhD); genetics (PhD); microbiology (MA, PhD); molecular, cellular, and developmental biology (PhD); plant sciences (MA, PhD); zoology (MA, PhD). PhD offered through the University Graduate School. Part-time programs available. *Faculty:* 42 full-time (8 women), 21 part-time/adjunct (6 women). *Students:* 203 full-time (107 women); includes 21 minority (9 African Americans, 2 Asian Americans or Pacific Islanders, 10 Hispanic Americans), 29 international. In 2005, 6 master's, 8 doctorates awarded. Terminal master's awarded for partial completion of doctoral program. *Degree requirements:* For master's and doctorate, thesis/dissertation, oral defense. *Entrance requirements:* For master's and doctorate, GRE General Test. Additional exam requirements/recommendations for international students: Required—TOEFL. *Application deadline:* For fall admission, 1/5 for domestic students; for spring admission, 9/1 priority date for domestic students. Applications are processed on a rolling basis. Application fee: $45. Electronic applications accepted. *Expenses:* Tuition, state resident: full-time $5,437; part-time $227 per credit hour. Tuition, nonresident: full-time $15,836; part-time $660 per credit hour. Required fees: $821. Tuition (and fees vary according to campus/location and program. *Financial support:* In 2005–06, 187 students received support, including fellowships with tuition reimbursements available (averaging $19,500 per year),

research assistantships with tuition reimbursements available (averaging $19,500 per year), teaching assistantships with tuition reimbursements available (averaging $18,000 per year); scholarships/grants and tuition waivers (full) also available. *Unit head:* Dr. Elizabeth C. Raff, Chair, 812-855-5522. *Application contact:* Gretchen Clearwater, Adviser for Graduate Affairs, 812-855-1861, Fax: 812-855-6705, E-mail: biograd@bio.indiana.edu.

See Close-Up on page 151.

Indiana University Bloomington, Medical Sciences Program, Bloomington, IN 47405-7000. Offers anatomy and cell biology (MA, PhD); pharmacology (MS, PhD); physiology (MA, PhD). *Students:* 11 full-time (2 women), 9 part-time (5 women); includes 5 minority (1 African American, 4 Asian Americans or Pacific Islanders), 2 international. Average age 29. *Entrance requirements:* For master's, GRE, minimum GPA of 3.0; for doctorate, GRE. Additional exam requirements/recommendations for international students: Required—TOEFL. *Application deadline:* For fall admission, 1/15 for domestic students; for spring admission, 9/1 for domestic students. Application fee: $45 ($55 for international students). *Expenses:* Tuition, state resident: full-time $5,437; part-time $227 per credit hour. Tuition, nonresident: full-time $15,836; part-time $660 per credit hour. Required fees: $821. Tuition and fees vary according to campus/location and program. *Financial support:* Fellowships available. *Unit head:* Dr. John B. Watkins, Assistant Dean/Director, 812-855-0616. *Application contact:* Kimberly Bunch, Director of Graduate Admissions, 812-855-1119, E-mail: kbunch@indiana.edu.

Indiana University of Pennsylvania, Graduate School and Research, College of Natural Sciences and Mathematics, Department of Biology, Program in Biology, Indiana, PA 15705-1087. Offers MS. *Degree requirements:* For master's, thesis optional. *Entrance requirements:* For master's, 2 letters of recommendation. Additional exam requirements/recommendations for international students: Required—TOEFL.

Indiana University–Purdue University Fort Wayne, School of Arts and Sciences, Department of Biological Sciences, Fort Wayne, IN 46805-1499. Offers biology (MS). Part-time and evening/weekend programs available. *Faculty:* 18 full-time (2 women). *Students:* 16 full-time (9 women), 10 part-time (7 women); includes 2 minority (both Asian Americans or Pacific Islanders), 1 international. Average age 27. 19 applicants, 100% accepted, 13 enrolled. In 2005, 5 degrees awarded. *Degree requirements:* For master's, thesis optional. *Entrance requirements:* For master's, GRE General Test, minimum GPA of 2.8, major or minor in biology. Additional exam requirements/recommendations for international students: Required—TOEFL (minimum score 600 paper-based; 260 computer-based). *Application deadline:* For fall admission, 4/1 for domestic students; for spring admission, 12/1 for domestic students. Applications are processed on a rolling basis. Application fee: $55. *Expenses:* Tuition, state resident: full-time $4,023; part-time $232 per credit. Tuition, nonresident: full-time $9,182; part-time $503 per credit. Required fees: $383. Tuition and fees vary according to course load. *Financial support:* In 2005–06, 1 research assistantship with partial tuition reimbursement (averaging $11,700 per year), 8 teaching assistantships with partial tuition reimbursements (averaging $11,700 per year) were awarded; scholarships/grants and unspecified assistantships also available. Support available to part-time students. Financial award application deadline: 3/1; financial award applicants required to submit FAFSA. *Faculty research:* Behavior, aquatic and physiological ecology, molecular biology and immunology, botany, conservation. Total annual research expenditures: $459,778. *Unit head:* Dr. Bruce Kingsbury, Chairperson, 260-481-6305, Fax: 260-481-6087, E-mail: kingsbur@ipfw.edu. *Application contact:* Dr. George Mourad, Director of Graduate Programs, 260-481-5704, Fax: 260-481-6087, E-mail: mourad@ipfw.edu.

Indiana University–Purdue University Indianapolis, School of Science, Department of Biology, Indianapolis, IN 46202-2896. Offers MS, PhD. Part-time and evening/weekend programs available. *Faculty:* 7 full-time (2 women). *Students:* 49 full-time (25 women), 24 part-time (9 women); includes 10 minority (4 African Americans, 3 Asian Americans or Pacific Islanders, 3 Hispanic Americans), 8 international. Average age 26. In 2005, 45 degrees awarded. Terminal master's awarded for partial completion of doctoral program. *Degree requirements:* For master's, thesis (for some programs); for doctorate, thesis/dissertation. *Entrance requirements:* For master's and doctorate, GRE General Test. *Application deadline:* For fall admission, 6/1 for domestic students. Application fee: $50 ($60 for international students). *Expenses:* Tuition, state resident: full-time $5,159; part-time $215 per credit hour. Tuition, nonresident: full-time $14,890; part-time $620 per credit hour. Required fees: $614. Tuition and fees vary according to campus/location and program. *Financial support:* Fellowships with partial tuition reimbursements, research assistantships with partial tuition reimbursements, teaching assistantships with partial tuition reimbursements, career-related internships or fieldwork available. Financial award application deadline: 4/1. *Faculty research:* Cell and model membranes, cell and molecular biology, immunology, oncology, developmental biology. *Unit head:* Dr. N. Douglas Lees, Chair, 317-274-0588, Fax: 317-274-2846.

See Close-Up on page 153.

Iowa State University of Science and Technology, College of Veterinary Medicine and Graduate College, Graduate Programs in Veterinary Medicine, Department of Biomedical Sciences, Ames, IA 50011. Offers MS, PhD. *Faculty:* 21 full-time, 1 part-time/adjunct. *Students:* 17 full-time (10 women); includes 2 minority (1 African American, 1 Asian American or Pacific Islander), 12 international. 16 applicants, 6% accepted, 1 enrolled. In 2005, 3 master's, 1 doctorate awarded. *Degree requirements:* For master's, thesis or alternative; for doctorate, thesis/dissertation. *Entrance requirements:* For master's and doctorate, GRE General Test (strongly recommended). Additional exam requirements/recommendations for international students: Required—TOEFL, TOEFL (paper score 530; computer score 197) or IELTS (score 6). *Application deadline:* For fall admission, 1/1 priority date for domestic students, 1/1 priority date for international students; for spring admission, 9/1 priority date for domestic students, 9/1 priority date for international students. Application fee: $30 ($70 for international students). Electronic applications accepted. *Financial support:* In 2005–06, 11 research assistantships with partial tuition reimbursements (averaging $15,305 per year), 3 teaching assistantships with partial tuition reimbursements (averaging $15,000 per year) were awarded; career-related internships or fieldwork, scholarships/grants, health care benefits, and unspecified assistantships also available. *Faculty research:* Cerebellar research, endocrine physiology, memory and learning and associated diseases ion-channels and dry resistance, glia-neuron signaling, neurobiology of pain. *Unit head:* Dr. Richard J. Martin, Chair, 515-294-2440, Fax: 515-294-2315, E-mail: biomedsci@iastate.edu. *Application contact:* Dr. Walter H. Hsu, Director of Graduate Education, 515-294-6864, E-mail: whsu@iastate.edu.

Jackson State University, Graduate School, School of Science and Technology, Department of Biology, Jackson, MS 39217. Offers biology education (MST); environmental science (MS, PhD). Part-time and evening/weekend programs available. *Degree requirements:* For master's, thesis (alternative accepted for MST); for doctorate, thesis/dissertation, comprehensive exam. *Entrance requirements:* For master's, GRE General Test; for doctorate, MAT. Additional exam requirements/recommendations for international students: Required—TOEFL. *Faculty research:* Comparative studies on the carbohydrate composition of marine macroalgae, host-parasite relationship between the spruce budworm and entomepathogen fungus.

Jacksonville State University, College of Graduate Studies and Continuing Education, College of Arts and Sciences, Department of Biology, Jacksonville, AL 36265-1602. Offers MS. *Faculty:* 12 full-time (1 woman). *Students:* 7 full-time (4 women), 29 part-time (17 women); includes 7 minority (5 African Americans, 1 American Indian/Alaska Native, 1 Asian American or Pacific Islander). In 2005, 13 degrees awarded. *Degree requirements:* For master's, thesis optional. *Entrance requirements:* For master's, GRE General Test or MAT. *Application deadline:* Applications are processed on a rolling basis. Application fee: $20. *Expenses:* Tuition, state resident: full-time $4,848; part-time $202 per credit hour. Tuition, nonresident: full-time $9,696; part-time $404 per credit hour. One-time fee: $20 full-time. *Financial support:* In 2005–06, 11 teaching assistantships were awarded Support available to part-time students. Financial award application deadline: 4/1; financial award applicants required to submit FAFSA.

Biological and Biomedical Sciences—General

Unit head: Dr. Frank Romano, Head, 256-782-5038. *Application contact:* 256-782-5329, Fax: 256-782-5321, E-mail: graduate@jsu.edu.

James Madison University, College of Graduate and Professional Programs, College of Science and Mathematics, Department of Biology, Harrisonburg, VA 22807. Offers MS. Part-time programs available. *Faculty:* 15 full-time (4 women), 2 part-time/adjunct (1 woman). *Students:* 10 full-time (6 women), 2 part-time, 1 international. Average age 28. In 2005, 5 degrees awarded. *Degree requirements:* For master's, thesis. *Entrance requirements:* For master's, GRE General Test, GRE Subject Test. Additional exam requirements/recommendations for international students: Required—TOEFL. *Application deadline:* For fall admission, 2/15 for domestic students. Applications are processed on a rolling basis. Application fee: $55. Electronic applications accepted. *Expenses:* Tuition, state resident: full-time $5,904; part-time $246 per credit hour. Tuition, nonresident: full-time $16,824; part-time $701 per credit hour. *Financial support:* In 2005–06, 9 students received support, including 1 research assistantship with full tuition reimbursement available (averaging $6,691 per year); teaching assistantships, Federal Work-Study and unspecified assistantships also available. Financial award application deadline: 3/1; financial award applicants required to submit FAFSA. *Faculty research:* Evolutionary ecology, gene regulation, microbial ecology, plant development, biomechanics. *Unit head:* Dr. Louise M. Temple-Rosebrook, Academic Unit Head, 540-568-6225.

John Carroll University, Graduate School, Department of Biology, University Heights, OH 44118-4581. Offers MA, MS. Part-time programs available. *Degree requirements:* For master's, essay or thesis. *Entrance requirements:* For master's, undergraduate major in biology, 1 semester of biochemistry. *Faculty research:* Algal ecology, systematics, molecular genetics, neurophysiology, behavioral ecology.

The Johns Hopkins University, National Institutes of Health Sponsored Programs, Department of Biology, Baltimore, MD 21218-2699. Offers biochemistry (PhD); biophysics (PhD); cell biology (PhD); developmental biology (PhD); genetic biology (PhD); molecular biology (PhD). *Faculty:* 25 full-time (4 women). *Students:* 126 full-time (72 women); includes 36 minority (3 African Americans, 1 American Indian/Alaska Native, 21 Asian Americans or Pacific Islanders, 11 Hispanic Americans), 19 international. 282 applicants, 26% accepted, 36 enrolled. In 2005, 15 degrees awarded. *Median time to degree:* Of those who began their doctoral program in fall 1997, 81.2% received their degree in 8 years or less. *Degree requirements:* For doctorate, thesis/dissertation, comprehensive exam, registration. *Entrance requirements:* For doctorate, GRE General Test. Additional exam requirements/recommendations for international students: Required—TOEFL (minimum score 600 paper-based; 250 computer-based), TWE, TSE. *Application deadline:* For fall admission, 12/15 for domestic students. Application fee: $60. *Expenses:* Tuition: Full-time $30,960. Tuition and fees vary according to degree level and program. *Financial support:* In 2005–06, 24 fellowships (averaging $23,000 per year), 93 research assistantships (averaging $23,000 per year), 22 teaching assistantships (averaging $23,000 per year) were awarded; Federal Work-Study, institutionally sponsored loans, scholarships/grants, traineeships, health care benefits, tuition waivers (partial), and unspecified assistantships also available. Financial award application deadline: 4/15; financial award applicants required to submit FAFSA. *Faculty research:* Protein and nucleic acid biochemistry and biophysical chemistry, molecular biology and development. Total annual research expenditures: $11.2 million. *Unit head:* Dr. Allen Shearn, Chair, 410-516-4693, Fax: 410-516-5213, E-mail: bio_cals@jhu.edu. *Application contact:* Joan Miller, Academic Affairs Manager, 410-516-5502, Fax: 410-516-5213, E-mail: joan@jhu.edu.

See Close-Up on page 155.

The Johns Hopkins University, School of Medicine, Graduate Programs in Medicine, Baltimore, MD 21218-2699. Offers MA, MS, PhD, MD/PhD. *Faculty:* 246 full-time (75 women), 32 part-time/adjunct (9 women). *Students:* 743 full-time; includes 155 minority (40 African Americans, 1 American Indian/Alaska Native, 96 Asian Americans or Pacific Islanders, 18 Hispanic Americans), 262 international. 1,207 applicants, 21% accepted, 137 enrolled. In 2005, 7 master's, 94 doctorates awarded. *Degree requirements:* For doctorate, thesis/dissertation. *Entrance requirements:* Additional exam requirements/recommendations for international students: Required—TOEFL. *Application deadline:* For fall admission, 1/10 priority date for domestic students, 1/10 priority date for international students. Applications are processed on a rolling basis. Application fee: $75. Electronic applications accepted. *Expenses:* Contact institution. Tuition and fees vary according to degree level and program. *Financial support:* In 2005–06, fellowships with full tuition reimbursements (averaging $23,000 per year); research assistantships, teaching assistantships with tuition reimbursements, career-related internships or fieldwork, Federal Work-Study, institutionally sponsored loans, and tuition waivers (full) also available. Financial award applicants required to submit FAFSA. *Unit head:* Dr. William B. Guggino, Director, 410-955-7166, Fax: 410-614-8331.

The Johns Hopkins University, Zanvyl Krieger School of Arts and Sciences, Department of Biology, Baltimore, MD 21218-2699. Offers PhD. *Faculty:* 25 full-time (5 women), 21 part-time/adjunct (5 women). *Students:* 130 full-time (68 women); includes 34 minority (4 African Americans, 1 American Indian/Alaska Native, 15 Asian Americans or Pacific Islanders, 14 Hispanic Americans), 21 international. Average age 23. 221 applicants, 27% accepted, 32 enrolled. In 2005, 12 doctorates awarded. Terminal master's awarded for partial completion of doctoral program. *Median time to degree:* Of those who began their doctoral program in fall 1997, 100% received their degree in 8 years or less. *Degree requirements:* For doctorate, thesis/dissertation, comprehensive exam, registration. *Entrance requirements:* For doctorate, GRE General Test, GRE Subject Test. Additional exam requirements/recommendations for international students: Required—TOEFL (minimum score 600 paper-based; 250 computer-based), IELT, TWE, TSE. *Application deadline:* For fall admission, 12/15 for domestic students, 12/15 for international students. Application fee: $60. Electronic applications accepted. *Expenses:* Tuition: Full-time $30,960. Tuition and fees vary according to degree level and program. *Financial support:* In 2005–06, 5 fellowships with tuition reimbursements (averaging $2,000 per year), 64 research assistantships with tuition reimbursements (averaging $25,200 per year), 35 teaching assistantships with tuition reimbursements (averaging $15,000 per year) were awarded; Federal Work-Study, institutionally sponsored loans, scholarships/grants, traineeships, health care benefits, tuition waivers (full), and unspecified assistantships also available. Financial award application deadline: 4/15; financial award applicants required to submit FAFSA. *Faculty research:* Cell biology, molecular biology and development, biochemistry, developmental biology, biophysics. Total annual research expenditures: $9.9 million. *Unit head:* Dr. Allen Shearn, Chair, 410-516-4693, Fax: 410-516-5213, E-mail: bio_cals@jhu.edu. *Application contact:* Joan Miller, Academic Affairs Manager, 410-516-5502, Fax: 410-516-5213, E-mail: joan@jhu.edu.

Kansas State University, College of Veterinary Medicine and Graduate School, Graduate Programs in Veterinary Medicine, Department of Anatomy and Physiology, Manhattan, KS 66506. Offers biomedical science (MS); physiology (PhD). *Faculty:* 14 full-time (3 women), 1 part-time/adjunct (0 women). *Students:* 16 full-time (5 women), 7 part-time (4 women); includes 1 minority (Hispanic American), 12 international. Average age 30. 4 applicants, 100% accepted, 4 enrolled. In 2005, 2 degrees awarded. Terminal master's awarded for partial completion of doctoral program. *Degree requirements:* For master's, thesis; for doctorate, one foreign language, thesis/dissertation. *Entrance requirements:* Additional exam requirements/recommendations for international students: Required—TOEFL (minimum score 550 paper-based; 213 computer-based). *Application deadline:* For fall admission, 2/1 for domestic students. Applications are processed on a rolling basis. Application fee: $30 ($55 for international students). *Expenses:* Tuition, state resident: full-time $5,160; part-time $215 per credit hour. Tuition, nonresident: full-time $12,816; part-time $534 per credit hour. Required fees: $564. *Financial support:* In 2005–06, 12 students received support, including teaching assistantships (averaging $21,325 per year) were awarded; teaching assistantships with partial tuition reimbursements, Federal Work-Study, institutionally sponsored loans, and scholarships/grants also available. Financial award application deadline: 3/1. *Faculty research:* Cardiovascular and pulmonary, immunophysiology, neuroscience, pharmacology, epithelial. Total annual research expenditures: $4 million. *Unit head:* Frank Blecha,

Head, 785-532-2741, Fax: 785-532-4557, E-mail: blecha@vet.ksu.edu. *Application contact:* Chris Ross, Director, 785-532-4507, Fax: 785-432-4557, E-mail: ross@vet.ksu.edu.

Kansas State University, College of Veterinary Medicine and Graduate School, Graduate Programs in Veterinary Medicine, Department of Clinical Sciences, Manhattan, KS 66506. Offers biomedical science (MS). *Faculty:* 17 full-time (5 women), 9 part-time/adjunct (3 women). *Students:* 8 full-time (3 women), 7 part-time (4 women); includes 1 minority (American Indian/Alaska Native), 2 international. Average age 27. 14 applicants, 100% accepted, 14 enrolled. In 2005, 4 degrees awarded. *Degree requirements:* For master's, thesis. *Entrance requirements:* For master's, GRE, DVM. Additional exam requirements/recommendations for international students: Required—TOEFL (minimum score 550 paper-based; 213 computer-based). *Application deadline:* For fall admission, 2/1 for domestic students; for spring admission, 10/1 for domestic students. Applications are processed on a rolling basis. Application fee: $30 ($55 for international students). Electronic applications accepted. *Expenses:* Tuition, state resident: full-time $5,160; part-time $215 per credit hour. Tuition, nonresident: full-time $12,816; part-time $534 per credit hour. Required fees: $564. *Financial support:* In 2005–06, 2 students received support, including 13 research assistantships (averaging $16,687 per year); teaching assistantships, institutionally sponsored loans and scholarships/grants also available. Financial award application deadline: 3/1; financial award applicants required to submit FAFSA. *Faculty research:* Clinical trials, equine gastrointestinal ulceration, leptospirosis, food animal pharmacology, equine immunology, diabetes. Total annual research expenditures: $560,276. *Unit head:* Greg Grauer, Head, 785-532-4890, E-mail: ggrauer@vet.ksu.edu. *Application contact:* Mike Sanderson, Director, 785-532-4264, E-mail: sandersn@vet.ksu.edu.

Kansas State University, College of Veterinary Medicine and Graduate School, Graduate Programs in Veterinary Medicine, Department of Diagnostic Medicine/Pathobiology, Manhattan, KS 66506. Offers biomedical science (MS); diagnostic medicine/pathobiology (PhD). *Faculty:* 25 full-time (4 women), 6 part-time/adjunct (2 women). *Students:* 20 full-time (9 women), 6 part-time (4 women), 14 international. Average age 30. 6 applicants, 0% accepted, 0 enrolled. In 2005, 4 master's, 3 doctorates awarded. Terminal master's awarded for partial completion of doctoral program. *Degree requirements:* For master's, thesis. *Entrance requirements:* For master's and doctorate, interviews. Additional exam requirements/recommendations for international students: Required—TOEFL (minimum score 550 paper-based; 213 computer-based). *Application deadline:* For fall admission, 2/1 for domestic students; for spring admission, 10/1 for domestic students. Applications are processed on a rolling basis. Application fee: $30 ($55 for international students). *Expenses:* Tuition, state resident: full-time $5,160; part-time $215 per credit hour. Tuition, nonresident: full-time $12,816; part-time $534 per credit hour. Required fees: $564. *Financial support:* In 2005–06, 20 research assistantships (averaging $16,257 per year) were awarded; teaching assistantships, Federal Work-Study, institutionally sponsored loans, and scholarships/grants also available. Financial award application deadline: 3/1; financial award applicants required to submit FAFSA. *Faculty research:* Infectious disease of animals, food safety and security, epidemiology and public health, toxicology, and pathology. Total annual research expenditures: $2.8 million. *Unit head:* M. M. Chengappa, Head, 785-532-4403, E-mail: chengapa@vet.ksu.edu. *Application contact:* T. G. Nagaraja, Director, 785-532-1214, E-mail: tnagaraj@ksu.edu.

Kansas State University, Graduate School, College of Arts and Sciences, Division of Biology, Manhattan, KS 66506. Offers biology (MS, PhD); microbiology (PhD). *Faculty:* 38 full-time (11 women), 14 part-time/adjunct (5 women). *Students:* 60 full-time (31 women), 9 part-time (3 women); includes 3 minority (2 Asian Americans or Pacific Islanders, 1 Hispanic American), 20 international. 123 applicants, 11% accepted, 14 enrolled. In 2005, 5 master's, 1 doctorate awarded. Terminal master's awarded for partial completion of doctoral program. *Degree requirements:* For master's and doctorate, thesis/dissertation. *Entrance requirements:* For master's, GRE General Test, minimum undergraduate GPA of 3.0; for doctorate, GRE General Test, minimum GPA of 3.0. Additional exam requirements/recommendations for international students: Required—TOEFL (minimum score 550 paper-based; 213 computer-based). *Application deadline:* For fall admission, 1/15 priority date for domestic students, 1/15 priority date for international students; for spring admission, 8/1 for domestic students, 8/1 for international students. Applications are processed on a rolling basis. Application fee: $30 ($55 for international students). Electronic applications accepted. *Expenses:* Tuition, state resident: full-time $5,160; part-time $215 per credit hour. Tuition, nonresident: full-time $12,816; part-time $534 per credit hour. Required fees: $564. *Financial support:* In 2005–06, 39 research assistantships (averaging $17,488 per year), 17 teaching assistantships with full tuition reimbursements (averaging $16,487 per year) were awarded; fellowships, institutionally sponsored loans and scholarships/grants also available. Support available to part-time students. Financial award application deadline: 3/1; financial award applicants required to submit FAFSA. *Faculty research:* Ecology, genetics, developmental biology, microbiology, cell biology. Total annual research expenditures: $9.1 million. *Unit head:* Brian S. Spooner, Director, 785-532-6615, Fax: 785-532-6653, E-mail: spoon1@ksu.edu. *Application contact:* S. Keith Chapes, Director, 785-532-6795, Fax: 785-532-6653, E-mail: skcbiol@ksu.edu.

Keck Graduate Institute of Applied Life Sciences, Bioscience Program, Claremont, CA 91711. Offers MBS. *Faculty:* 18 full-time (6 women). *Students:* 66 full-time (19 women). *Degree requirements:* For master's, project. *Entrance requirements:* For master's, GRE General Test or MCAT. Additional exam requirements/recommendations for international students: Required—TOEFL. *Application deadline:* Applications are processed on a rolling basis. Application fee: $60. Electronic applications accepted. *Expenses:* Tuition: Full-time $35,050. *Financial support:* In 2005–06, 66 fellowships were awarded; career-related internships or fieldwork, institutionally sponsored loans, and scholarships/grants also available. *Faculty research:* Computational biology, drug discovery and development, molecular and cellular biology, biomedical engineering, biomaterials and tissue engineering. *Unit head:* Dr. Sheldon M. Schuster, President, 909-607-7855, Fax: 909-607-8086. *Application contact:* Katherine A. Jernberg, Dean of Admissions and Student Services, 909-607-8208, Fax: 909-607-8086, E-mail: katherine_jernberg@kgi.edu.

Kent State University, College of Arts and Sciences, Department of Biological Sciences, Kent, OH 44242-0001. Offers botany (MS); ecology (MS, PhD); physiology (MS, PhD). *Degree requirements:* For master's and doctorate, thesis/dissertation. *Entrance requirements:* For master's, GRE General Test, minimum GPA of 3.0; for doctorate, GRE General Test, minimum GPA of 3.25. Additional exam requirements/recommendations for international students: Required—TOEFL (minimum score 600 paper-based; 257 computer-based). Electronic applications accepted.

See Close-Up on page 157.

Kent State University, Program in Biological Anthropology, Kent, OH 44242-0001. Offers PhD. Offered in cooperation with Northeastern Ohio Universities College of Medicine. *Degree requirements:* For doctorate, thesis/dissertation. *Entrance requirements:* For doctorate, GRE General Test. *Faculty research:* Human evolution, paleodemography, orofacial anatomy, osteology, primate behavior.

Kent State University, School of Biomedical Sciences, Kent, OH 44242-0001. Offers MS, PhD. Offered in cooperation with Northeastern Ohio Universities College of Medicine. Terminal master's awarded for partial completion of doctoral program. *Degree requirements:* For master's and doctorate, thesis/dissertation. *Entrance requirements:* For master's and doctorate, GRE General Test.

See Close-Up on page 159.

Lakehead University, Graduate Studies, Faculty of Social Sciences and Humanities, Department of Biology, Thunder Bay, ON P7B 5E1, Canada. Offers M Sc. Part-time and evening/weekend programs available. *Degree requirements:* For master's, thesis, department seminar, oral examination. *Entrance requirements:* For master's, minimum B average. Additional exam requirements/recommendations for international students: Required—TOEFL. *Faculty research:*

Biological and Biomedical Sciences—General

Lakehead University (continued)
Systematics and biogeography, wildlife parasitology, plant physiology and biochemistry, plant ecology, fishery biology.

Lamar University, College of Graduate Studies, College of Arts and Sciences, Department of Biology, Beaumont, TX 77710. Offers MS. Part-time and evening/weekend programs available. *Faculty:* 8 full-time (1 woman). *Students:* 1 full-time (0 women), 6 part-time (3 women). Average age 34. 13 applicants, 23% accepted, 0 enrolled. *Degree requirements:* For master's, thesis. *Entrance requirements:* For master's, GRE General Test, minimum GPA of 2.5 in last 60 hours of undergraduate course work. Additional exam requirements/recommendations for international students: Required—TOEFL. *Application deadline:* For fall admission, 8/1 for domestic students; for spring admission, 12/1 for domestic students. Applications are processed on a rolling basis. Application fee: $25 ($50 for international students). *Expenses:* Tuition, state resident: part-time $137 per semester hour. Tuition, nonresident: part-time $413 per semester hour. Required fees: $102 per semester hour. Tuition and fees vary according to course load. *Financial support:* In 2005–06, 3 teaching assistantships (averaging $6,200 per year) were awarded Financial award application deadline: 4/1. *Faculty research:* Physiology, ichthyology, microbiology, limnology, behavior. *Unit head:* Dr. Michael E. Warren, Chair, 409-880-8262, Fax: 409-880-1827. *Application contact:* Dr. R. C. Harrel, Graduate Adviser, 409-880-8255, Fax: 409-880-1827.

Laurentian University, School of Graduate Studies and Research, Programme in Biology, Sudbury, ON P3E 2C6, Canada. Offers M Sc. Part-time programs available. *Degree requirements:* For master's, thesis. *Entrance requirements:* For master's, honors degree with second class or better. *Faculty research:* Recovery of acid-stressed lakes, effects of climate change, origin and maintenance of biocomplexity, radionuclide dynamics, cytogenetic studies of plants.

Lehigh University, College of Arts and Sciences, Department of Biological Sciences, Bethlehem, PA 18015-3094. Offers biochemistry (PhD); integrative biology (PhD); molecular biology (PhD). Postbaccalaureate distance learning degree programs offered (no on-campus study). *Faculty:* 22 full-time (9 women). *Students:* 31 full-time (21 women), 51 part-time (29 women); includes 2 minority (1 Asian American or Pacific Islander, 1 Hispanic American), 6 international. 65 applicants, 29% accepted. In 2005, 3 doctorates awarded. *Median time to degree:* Of those who began their doctoral program in fall 1997, 100% received their degree in 8 years or less. *Degree requirements:* For doctorate, thesis/dissertation, comprehensive exam. *Entrance requirements:* For doctorate, GRE General Test. Additional exam requirements/recommendations for international students: Required—TOEFL, TSE. *Application deadline:* For fall admission, 1/15 for domestic students. Applications are processed on a rolling basis. Application fee: $50. Electronic applications accepted. *Financial support:* In 2005–06, 30 students received support, including 4 fellowships with tuition reimbursements available (averaging $20,000 per year), 6 research assistantships with tuition reimbursements available (averaging $20,000 per year), 16 teaching assistantships with tuition reimbursements available (averaging $20,000 per year); scholarships/grants, tuition waivers (full and partial), and unspecified assistantships also available. Financial award application deadline: 1/31. *Faculty research:* Gene expression, cytoskeleton and cell structure, cell cycle and growth regulation, neuroscience, animal behavior. *Unit head:* Dr. Neal G. Simon, Chairperson, 610-758-3680, Fax: 610-758-4004. *Application contact:* Dr. Michael R. Kuchka, Graduate Coordinator, 610-758-3687, Fax: 610-758-4004, E-mail: mrk5@lehigh.edu.

Lehman College of the City University of New York, Division of Natural and Social Sciences, Department of Biological Sciences, Program in Biology, Bronx, NY 10468-1589. Offers MA.

Loma Linda University, School of Science and Technology, Department of Biological and Earth Sciences, Loma Linda, CA 92350. Offers MS, PhD. *Faculty:* 8 full-time, 7 part-time/adjunct. *Students:* 15 full-time (7 women), 11 part-time (2 women); includes 5 minority (1 American Indian/Alaska Native, 1 Asian American or Pacific Islander, 3 Hispanic Americans, 1 international. *Unit head:* Dr. Ron L. Carter, Coordinator, 909-824-4530.

Long Island University, Brooklyn Campus, Richard L. Conolly College of Liberal Arts and Sciences, Department of Biology, Brooklyn, NY 11201-8423. Offers MS. Part-time and evening/weekend programs available. *Degree requirements:* For master's, thesis or alternative. *Entrance requirements:* For master's, 2 letters of recommendation. Additional exam requirements/recommendations for international students: Required—TOEFL (minimum score 500 paper-based; 173 computer-based). Electronic applications accepted.

Long Island University, C.W. Post Campus, College of Liberal Arts and Sciences, Department of Biology, Brookville, NY 11548-1300. Offers biology (MS); biology education (MS). Part-time and evening/weekend programs available. *Degree requirements:* For master's, thesis optional. *Entrance requirements:* For master's, GRE General Test, minimum GPA of 2.75 in major. Electronic applications accepted. *Faculty research:* Immunology, molecular biology, systematics, behavioral ecology, microbiology.

Long Island University, C.W. Post Campus, School of Health Professions and Nursing, Department of Biomedical Sciences, Program in Medical Biology, Brookville, NY 11548-1300. Offers hematology (MS); immunology (MS); medical biology (MS); medical chemistry (MS); medical microbiology (MS). Part-time and evening/weekend programs available. *Degree requirements:* For master's, thesis. *Entrance requirements:* For master's, minimum GPA of 2.75 in major. Electronic applications accepted. *Faculty research:* Hematopoiesis, growth factors in cancer, interleukins in allergy, PCR techniques.

Louisiana State University and Agricultural and Mechanical College, College of Basic Sciences, Department of Biological Sciences, Baton Rouge, LA 70803. Offers biochemistry (MS, PhD); biological science (MS, PhD). Part-time programs available. *Faculty:* 79 full-time (13 women). *Students:* 145 full-time (74 women), 8 part-time (5 women); includes 18 minority (15 African Americans, 2 Asian Americans or Pacific Islanders, 1 Hispanic American), 51 international. Average age 29. 185 applicants, 22% accepted, 145 enrolled. In 2005, 3 master's, 17 doctorates awarded. Terminal master's awarded for partial completion of doctoral program. *Degree requirements:* For doctorate, thesis/dissertation. *Entrance requirements:* For master's and doctorate, GRE General Test, minimum GPA of 3.0. Additional exam requirements/recommendations for international students: Required—TOEFL (minimum score 550 paper-based; 213 computer-based). *Application deadline:* For fall admission, 5/15 for domestic students, 5/15 for international students; for spring admission, 10/15 for domestic students, 10/15 for international students. Applications are processed on a rolling basis. Application fee: $25. Electronic applications accepted. *Financial support:* In 2005–06, 17 fellowships with full and partial tuition reimbursements (averaging $18,054 per year), 48 research assistantships with full and partial tuition reimbursements (averaging $21,414 per year), 74 teaching assistantships with full and partial tuition reimbursements (averaging $18,451 per year) were awarded; Federal Work-Study, institutionally sponsored loans, and unspecified assistantships also available. Support available to part-time students. Financial award applicants required to submit FAFSA. Total annual research expenditures: $8.1 million. *Unit head:* Dr. Terry Bricker, Chair, 225-578-2601, Fax: 225-578-2597, E-mail: btbric@lsu.edu. *Application contact:* Dr. Thomas S. Moore, Associate Chairman, 225-578-1556, Fax: 225-578-7299, E-mail: gradoff@lsu.edu.

See Close-Up on page 161.

Louisiana State University Health Sciences Center, School of Graduate Studies in New Orleans, New Orleans, LA 70112-2223. Offers MPH, MS, PhD, MD/PhD. Part-time and evening/weekend programs available. Terminal master's awarded for partial completion of doctoral program. *Degree requirements:* For master's and doctorate, thesis/dissertation. *Entrance requirements:* For master's and doctorate, GRE General Test. Additional exam requirements/recommendations for international students: Required—TOEFL.

Louisiana State University Health Sciences Center at Shreveport, Department of Biochemistry and Molecular Biology, Shreveport, LA 71130-3932. Offers MS, PhD, MD/PhD. *Faculty:* 16 full-time (3 women), 5 part-time/adjunct (0 women). *Students:* 38 full-time (15 women); includes 14 minority (3 African Americans, 11 Asian Americans or Pacific Islanders). Average age 26. 50 applicants, 10% accepted, 5 enrolled. In 2005, 2 master's, 1 doctorate awarded. *Median time to degree:* Of those who began their doctoral program in fall 1997, 100% received their degree in 8 years or less. *Degree requirements:* For master's and doctorate, thesis/dissertation. *Entrance requirements:* For master's and doctorate, GRE General Test. Additional exam requirements/recommendations for international students: Required—TOEFL. *Application deadline:* For fall admission, 4/15 for domestic students. Applications are processed on a rolling basis. Application fee: $30. *Financial support:* In 2005–06, 38 students received support, including research assistantships (averaging $18,000 per year); institutionally sponsored loans also available. *Faculty research:* Metabolite transport, regulation of translation and transcription, prokaryotic molecular genetics, cell matrix biochemistry, yeast molecular genetics, oncogenes. Total annual research expenditures: $160,000. *Unit head:* Dr. Robert E. Rhoads, Head, 318-675-5160, Fax: 318-675-5180. *Application contact:* Dr. Robert L. Smith, Associate Professor and Director of Graduate Studies, 318-675-5168, Fax: 318-675-5180, E-mail: rsmith2@lsuhsc.edu.

See Close-Up on page 163.

Louisiana Tech University, Graduate School, College of Applied and Natural Sciences, School of Biological Sciences, Ruston, LA 71272. Offers MS. Part-time programs available. *Degree requirements:* For master's, thesis or alternative. *Entrance requirements:* For master's, GRE General Test, GRE Subject Test. *Faculty research:* Genetics, animal biology, plant biology, physiology biocontrol.

Loyola University Chicago, Graduate School, Department of Biology, Chicago, IL 60611-2196. Offers MA, MS. *Faculty:* 19 full-time (5 women), 7 part-time/adjunct (2 women). *Students:* 74 full-time (40 women), 5 part-time (2 women); includes 14 minority (11 Asian Americans or Pacific Islanders, 3 Hispanic Americans), 4 international. Average age 26. 25 applicants, 48% accepted, 7 enrolled. In 2005, 51 degrees awarded. *Degree requirements:* For master's, thesis (for some programs), registration. *Entrance requirements:* For master's, GRE General Test, 3 letters of recommendation. Additional exam requirements/recommendations for international students: Required—TOEFL. *Application deadline:* For fall admission, 7/1 for domestic students; for spring admission, 12/1 for domestic students. Applications are processed on a rolling basis. Application fee: $40. Electronic applications accepted. *Expenses:* Tuition: Full-time $11,610; part-time $645 per credit. Required fees: $55 per semester. *Financial support:* In 2005–06, 7 students received support, including 7 fellowships with full tuition reimbursements available (averaging $16,000 per year); Federal Work-Study and institutionally sponsored loans also available. Financial award application deadline: 2/1; financial award applicants required to submit FAFSA. *Faculty research:* Molecular biology and genetics, biochemistry, cell biology and physiology, population biology and aquatic ecology. Total annual research expenditures: $2.5 million. *Application contact:* Dr. Terry Grande, Graduate Program Director, 773-508-3649, Fax: 773-508-3646, E-mail: tgrande@luc.edu.

Marquette University, Graduate School, College of Arts and Sciences, Department of Biology, Milwaukee, WI 53201-1881. Offers cell biology (MS, PhD); developmental biology (MS, PhD); ecology (MS, PhD); endocrinology (MS, PhD); evolutionary biology (MS, PhD); genetics (MS, PhD); microbiology (MS, PhD); molecular biology (MS, PhD); muscle and exercise physiology (MS, PhD); neurobiology (MS, PhD); reproductive physiology (MS, PhD). Terminal master's awarded for partial completion of doctoral program. *Degree requirements:* For master's, thesis, 1 year of teaching experience or equivalent, comprehensive exam; for doctorate, thesis/dissertation, 1 year of teaching experience or equivalent, qualifying exam. *Entrance requirements:* For master's and doctorate, GRE General Test, GRE Subject Test. Additional exam requirements/recommendations for international students: Required—TOEFL. *Faculty research:* Microbial and invertebrate ecology, evolution of gene function, DNA methylation, DNA arrangement.

Marshall University, Academic Affairs Division, Graduate College, College of Science, Department of Biological Science, Huntington, WV 25755. Offers MA. *Faculty:* 13 full-time (3 women), 1 (woman) part-time/adjunct. *Students:* 35 full-time (15 women), 4 part-time (3 women); includes 6 minority (2 African Americans, 3 Asian Americans or Pacific Islanders, 1 Hispanic American), 5 international. Average age 27. In 2005, 18 degrees awarded. *Degree requirements:* For master's, thesis (for some programs). *Entrance requirements:* For master's, GRE General Test, GRE Subject Test. *Financial support:* Career-related internships or fieldwork available. *Unit head:* Dr. Charles Somerville, Chairperson, 304-696-2424, E-mail: somervil@marshall.edu. *Application contact:* Information Contact, 304-746-1900, Fax: 304-746-1902, E-mail: services@marshall.edu.

Marshall University, Joan C. Edwards School of Medicine and Graduate College, Program in Biomedical Sciences, Huntington, WV 25755. Offers MS, PhD. *Faculty:* 27 full-time (7 women), 2 part-time/adjunct (0 women). *Students:* 30 full-time (15 women), 2 part-time (both women); includes 2 minority (1 African American, 1 Asian American or Pacific Islander), 4 international. Average age 26. 30 applicants, 67% accepted, 16 enrolled. In 2005, 2 master's, 4 doctorates awarded. Terminal master's awarded for partial completion of doctoral program. *Degree requirements:* For master's, thesis optional; for doctorate, thesis/dissertation, written and oral qualifying exams. *Entrance requirements:* For master's, GRE General Test or MCAT (medical science only), 1 year of course work in biology, physics, chemistry, and organic chemistry and associated labs; for doctorate, GRE General Test, 1 year of course work in biology, physics, chemistry, and organic chemistry and associated labs. Additional exam requirements/recommendations for international students: Required—TOEFL (minimum score 525 paper-based; 216 computer-based). *Application deadline:* Applications are processed on a rolling basis. Application fee: $30 ($40 for international students). *Financial support:* In 2005–06, research assistantships with tuition reimbursements (averaging $21,700 per year; career-related internships or fieldwork, Federal Work-Study, and institutionally sponsored loans also available. Support available to part-time students. Financial award application deadline: 5/1; financial award applicants required to submit FAFSA. *Faculty research:* Neurosciences, cardiopulmonary science, molecular biology, toxicology, endocrinology. *Unit head:* Dr. Richard M. Niles, Associate Dean for Research and Graduate Education, 304-696-7323, Fax: 304-696-7171, E-mail: niles@marshall.edu. *Application contact:* Dr. Vernon E. Reichenbecher, Director of Graduate Studies, 304-696-7327, Fax: 304-696-7171, E-mail: reichenb@marshall.edu.

Massachusetts Institute of Technology, School of Science, Department of Biology, Cambridge, MA 02139-4307. Offers biological oceanography (PhD); biology (PhD). *Faculty:* 52 full-time (11 women). *Students:* 247 full-time (129 women); includes 50 minority (3 African Americans, 1 American Indian/Alaska Native, 34 Asian Americans or Pacific Islanders, 12 Hispanic Americans), 31 international. Average age 27. 550 applicants, 17% accepted, 29 enrolled. In 2005, 35 doctorates awarded. *Degree requirements:* For doctorate, thesis/dissertation, comprehensive exam. *Entrance requirements:* For doctorate, GRE General Test. Additional exam requirements/recommendations for international students: Required—TOEFL (minimum score 577 paper-based; 233 computer-based). *Application deadline:* For fall admission, 12/15 for domestic students, 12/15 for international students. Application fee: $70. Electronic applications accepted. *Expenses:* Tuition: Full-time $32,100. Required fees: $200. Part-time tuition and fees vary according to course load. *Financial support:* In 2005–06, 214 students received support, including 107 fellowships with tuition reimbursements available (averaging $26,164 per year), 114 research assistantships with tuition reimbursements available (averaging $26,424 per year); teaching assistantships, Federal Work-Study, institutionally sponsored loans, scholarships/grants, traineeships, health care benefits, and unspecified assistantships also available. *Faculty research:* DNA recombination, replication, and repair; transcription and gene regulation; signal transduction; cell cycle; neuronal cell fate. Total annual research expenditures: $95.4 million.

Unit head: Prof. Chris Kaiser, Department Head, 617-253-4701, Fax: 617-253-8699. *Application contact:* Biology Education Office, 617-253-3717, Fax: 617-258-9329, E-mail: gradbio@mit.edu.

See Close-Up on page 165.

Massachusetts Institute of Technology, Whitaker College of Health Sciences and Technology, Harvard-MIT Division of Health Sciences and Technology, Biomedical Enterprise Program, Cambridge, MA 02139-4307. Offers SM. *Students:* 22 full-time (4 women). Average age 31. 44 applicants, 27% accepted, 12 enrolled. In 2005, 3 degrees awarded. *Degree requirements:* For master's, thesis. *Entrance requirements:* For master's, GMAT, bachelor's degree in engineering or science, work experience in biomedical business. *Application deadline:* For fall admission, 12/15 for domestic students, 12/15 for international students. Application fee: $70. *Expenses:* Tuition: Full-time $32,100. Required fees: $200. Part-time tuition and fees vary according to course load. *Financial support:* In 2005–06, 4 students received support, including 4 research assistantships with partial tuition reimbursements available (averaging $19,493 per year); teaching assistantships with partial tuition reimbursements available, institutionally sponsored loans, health care benefits, and unspecified assistantships also available. Financial award application deadline: 1/15; financial award applicants required to submit FAFSA. *Faculty research:* Entrepeneurship, technology strategy management, organizational strategies and models, epidemiology and biostatics, biomedical research from the molecular to the whole organism level. *Application contact:* Catherine A. Modica, Admissions Coordinator, 617-253-2307, Fax: 617-253-6692, E-mail: cmodica@mit.edu.

Massachusetts Institute of Technology, Whitaker College of Health Sciences and Technology, Harvard-MIT Division of Health Sciences and Technology, Program in Medical Sciences, Cambridge, MA 02139-4307. Offers MD, MD/MS, MD/PhD. *Students:* 192 full-time (59 women). Average age 26. 575 applicants, 8% accepted, 30 enrolled. In 2005, 32 degrees awarded. *Degree requirements:* For first-professional, thesis. *Entrance requirements:* MCAT. *Application deadline:* For fall admission, 11/15 for domestic students. Application fee: $85. *Expenses: Contact institution.* Part-time tuition and fees vary according to course load. *Financial support:* In 2005–06, 60 students received support, including 48 research assistantships with full and partial tuition reimbursements available (averaging $7,277 per year), 12 teaching assistantships with full and partial tuition reimbursements available (averaging $4,314 per year); fellowships with full and partial tuition reimbursements available, institutionally sponsored loans, scholarships/grants, and unspecified assistantships also available. Financial award application deadline: 1/15. *Unit head:* Dr. Joseph Bonventre, Director, 617-432-1738. *Application contact:* Dr. David Earl Cohen, Director of MD Admissions, 617-726-5576.

Mayo Graduate School, Graduate Programs in Biomedical Sciences, Rochester, MN 55905. Offers PhD, MD/PhD. *Degree requirements:* For doctorate, oral defense of dissertation, qualifying oral and written exam. *Entrance requirements:* For doctorate, GRE, 1 year of chemistry, biology, calculus, and physics. Additional exam requirements/recommendations for international students: Required—TOEFL. Electronic applications accepted.

See Close-Up on page 167.

McGill University, Faculty of Graduate and Postdoctoral Studies, Faculty of Medicine, Department of Medicine, Montréal, QC H3A 2T5, Canada. Offers experimental medicine (M Sc, PhD), including bioethics (M Sc), experimental medicine.

McGill University, Faculty of Graduate and Postdoctoral Studies, Faculty of Medicine, Program in Experimental Medicine, Montréal, QC H3A 2T5, Canada. Offers bioethics (M Sc); experimental medicine (M Sc, PhD). Terminal master's awarded for partial completion of doctoral program. *Degree requirements:* For master's and doctorate, thesis/dissertation. *Entrance requirements:* For doctorate, B Sc or M Sc in medical field or MD, minimum GPA of 3.4. Additional exam requirements/recommendations for international students: Required—TOEFL.

McGill University, Faculty of Graduate and Postdoctoral Studies, Faculty of Science, Department of Biology, Montréal, QC H3A 2T5, Canada. Offers biology (M Sc, PhD); neo-tropical environment (M Sc, PhD). Terminal master's awarded for partial completion of doctoral program. *Degree requirements:* For master's, thesis, registration; for doctorate, thesis/dissertation, residency, research seminar, comprehensive exam, registration. *Entrance requirements:* For master's and doctorate, minimum GPA of 3.0. Additional exam requirements/recommendations for international students: Required—TOEFL (minimum score 550 paper-based; 213 computer-based), IELT (minimum score 7). Electronic applications accepted. *Faculty research:* Evolution; ecology, biodiversity and conservation; bioinformatics, molecular biology and genetics; cell and developmental biology; neurobiology.

Announcement: Program emphasis for the MSc and PhD degrees is on development of the intellectual and technical skills necessary for independent research. The department is very well equipped for graduate training and research in all areas, and its resources are greatly extended by affiliation with other research organizations, institutes, and hospitals.

McMaster University, Faculty of Health Sciences and School of Graduate Studies, Program in Medical Sciences, Hamilton, ON L8S 4M2, Canada. Offers cell biology and metabolism (M Sc, PhD); hemostasis, thromboembolism, and atherosclerosis (M Sc, PhD); molecular biology, genetics, and cancer (M Sc, PhD); molecular immunology, virology, and inflammation (M Sc, PhD); neurosciences and behavioral sciences (M Sc, PhD); physiology/pharmacology (M Sc, PhD). *Students:* 186 full-time, 10 part-time. In 2005, 24 master's, 11 doctorates awarded. *Degree requirements:* For master's, thesis/dissertation; for doctorate, thesis/dissertation, comprehensive exam. *Entrance requirements:* For master's, honors B Sc, B+ average in related field; for doctorate, M Sc, minimum B+ average, students with proven research experience and an A average may be admitted with a B Sc degree. Additional exam requirements/recommendations for international students: Required—TOEFL (minimum score 580 paper-based; 237 computer-based). *Application deadline:* For fall admission, 9/30 for domestic students. For winter admission, 3/31 for domestic students. Applications are processed on a rolling basis. Application fee: $85. *Financial support:* Teaching assistantships available. *Application contact:* Dr. Carl Richards, Associate Dean, 905-525-9140 Ext. 22983, Fax: 905-546-1129.

McMaster University, School of Graduate Studies, Faculty of Science, Department of Biochemistry and Biomedical Sciences, Hamilton, ON L8S 4M2, Canada. Offers M Sc, PhD. Terminal master's awarded for partial completion of doctoral program. *Degree requirements:* For master's, thesis/dissertation; for doctorate, thesis/dissertation, comprehensive exam. *Entrance requirements:* For master's and doctorate, minimum B+ average. Additional exam requirements/recommendations for international students: Required—TOEFL (minimum score 550 paper-based; 213 computer-based). *Faculty research:* Molecular and cell biology, biomolecular structure and function, molecular pharmacology and toxicology.

McMaster University, School of Graduate Studies, Faculty of Science, Department of Biology, Hamilton, ON L8S 4M2, Canada. Offers M Sc, PhD. Part-time programs available. *Degree requirements:* For master's, thesis/dissertation; for doctorate, thesis/dissertation, comprehensive exam. *Entrance requirements:* Additional exam requirements/recommendations for international students: Required—TOEFL (minimum score 550 paper-based; 213 computer-based).

McNeese State University, Graduate School, College of Science, Department of Biological and Environmental Sciences, Lake Charles, LA 70609. Offers environmental and chemical sciences (MS). Evening/weekend programs available. *Faculty:* 10 full-time (1 woman). *Students:* 27 full-time (14 women), 8 part-time (5 women); includes 11 minority (9 African Americans, 2 Hispanic Americans), 9 international. In 2005, 7 degrees awarded. *Degree requirements:* For master's, thesis or alternative, comprehensive exam. *Entrance requirements:* For master's, GRE General Test. *Application deadline:* For fall admission, 7/15 for domestic students. Applications are processed on a rolling basis. Application fee: $20 ($30 for international students).

Expenses: Tuition, area resident: Part-time $193 per hour. Tuition, state resident: full-time $2,226. Required fees: $862; $106 per hour. Tuition and fees vary according to course load. *Financial support:* Application deadline: 5/1. *Unit head:* Dr. Mark L. Wygoda, Head, 337-475-5674, Fax: 337-475-5677, E-mail: mwygoda@mcneese.edu. *Application contact:* Dr. Harold Stevenson, Coordinator, 337-475-5663, Fax: 337-475-5677, E-mail: hstevens@mcneese.edu.

Medical College of Georgia, School of Graduate Studies, Augusta, GA 30912. Offers MN, MS, MSN, DNP, PhD. Part-time programs available. Postbaccalaureate distance learning degree programs offered (no on-campus study). *Faculty:* 181 full-time (54 women), 2 part-time/adjunct (1 woman). *Students:* 191 full-time (116 women), 71 part-time (52 women); includes 38 minority (26 African Americans, 1 American Indian/Alaska Native, 6 Asian Americans or Pacific Islanders, 5 Hispanic Americans), 55 international. Average age 32. 123 applicants, 47% accepted, 27 enrolled. In 2005, 43 master's, 14 doctorates awarded. Terminal master's awarded for partial completion of doctoral program. *Degree requirements:* For doctorate, thesis/dissertation. *Entrance requirements:* For master's, GRE, MAT; for doctorate, GRE General Test, MCAT. Additional exam requirements/recommendations for international students: Required—TOEFL. *Application deadline:* For fall admission, 6/30 for domestic students, 1/15 for international students. Applications are processed on a rolling basis. Application fee: $30. *Financial support:* In 2005–06, 112 students received support, including 6 fellowships with partial tuition reimbursements available (averaging $25,500 per year), 105 research assistantships with partial tuition reimbursements available (averaging $22,500 per year), 1 teaching assistantship with partial tuition reimbursement available (averaging $22,500 per year); career-related internships or fieldwork, Federal Work-Study, institutionally sponsored loans, scholarships/grants, traineeships, and unspecified assistantships also available. Support available to part-time students. Financial award application deadline: 5/31; financial award applicants required to submit FAFSA. *Faculty research:* Cancer, cardiovascular biology, neurosciences, inflammation/infection, diabetes. *Unit head:* Dr. Gretchen B. Caughman, Dean, 706-721-3278, Fax: 706-721-6829, E-mail: gcaughma@mail.mcg.edu. *Application contact:* Carol S. Nobles, Director of Student Recruitment and Admissions, 706-721-2725, Fax: 706-721-7279, E-mail: cnobles@mail.mcg.edu.

See Close-Ups on pages 169 and 171.

Medical College of Wisconsin, Graduate School of Biomedical Sciences, Milwaukee, WI 53226-0509. Offers MA, MPH, MS, PhD, MD/MA, MD/MS, MD/PhD. Part-time and evening/weekend programs available. Postbaccalaureate distance learning degree programs offered (minimal on-campus study). *Degree requirements:* For master's, thesis/dissertation, registration; for doctorate, thesis/dissertation, comprehensive exam, registration. *Entrance requirements:* For master's and doctorate, GRE General Test. Additional exam requirements/recommendations for international students: Required—TOEFL. *Faculty research:* Bioterrorism, genomics, SIDS.

See Close-Up on page 173.

Medical University of South Carolina, College of Graduate Studies, Charleston, SC 29425-0002. Offers Pharm D, MBS, MCR, MS, PhD, DMD/PhD, MD/PhD. *Faculty:* 390 full-time (122 women). *Students:* 208 full-time (114 women), 56 part-time (36 women); includes 33 minority (14 African Americans, 1 American Indian/Alaska Native, 11 Asian Americans or Pacific Islanders, 7 Hispanic Americans), 32 international. Average age 28. 355 applicants, 33% accepted, 78 enrolled. In 2005, 22 master's, 39 doctorates awarded. Terminal master's awarded for partial completion of doctoral program. *Median time to degree:* Of those who began their doctoral program in fall 1997, 93% received their degree in 8 years or less. *Degree requirements:* For master's, thesis, research seminar; for doctorate, thesis/dissertation, teaching and research seminar, oral and written exams. *Entrance requirements:* For doctorate, GRE General Test, interview. Additional exam requirements/recommendations for international students: Required—TOEFL (minimum score 600 paper-based; 250 computer-based). *Application deadline:* For fall admission, 1/15 priority date for domestic students, 1/15 priority date for international students). Applications are processed on a rolling basis. Application fee: $0 ($75 for international students). Electronic applications accepted. *Expenses: Contact institution. Financial support:* In 2005–06, fellowships with partial tuition reimbursements (averaging $21,000 per year); Federal Work-Study and scholarships/grants also available. Support available to part-time students. Financial award application deadline: 3/1; financial award applicants required to submit FAFSA. *Faculty research:* Lipid biochemistry, central nervous system injuries, cell signaling and cancer biology, neural injury and repair, aging, drug delivery systems. *Unit head:* Dr. Perry V. Halushka, Dean, 843-792-3012, Fax: 843-792-6590, E-mail: halushpv@musc.edu. *Application contact:* Cheryl Brown, Assistant Dean for Admissions, 800-792-3391, Fax: 843-792-6590, E-mail: cookla@musc.edu.

Medical University of South Carolina, College of Health Professions, Department of Clinical Services, Program in Cytology and Biosciences, Charleston, SC 29425-0002. Offers MS. *Faculty:* 1 (woman) full-time. *Students:* 25 full-time (18 women); includes 9 minority (5 African Americans, 3 Asian Americans or Pacific Islanders, 1 Hispanic American). Average age 27. 23 applicants, 65% accepted, 12 enrolled. In 2005, 7 degrees awarded. *Degree requirements:* For master's, didactic and clinical curriculum. *Entrance requirements:* For master's, GRE, minimum GPA of 3.0, 3 letters of reference. Additional exam requirements/recommendations for international students: Required—TOEFL (minimum score 600 paper-based; 250 computer-based). *Application deadline:* For fall admission, 6/1 priority date for domestic students, 6/1 priority date for international students. Application fee: $75. Electronic applications accepted. *Financial support:* Federal Work-Study and scholarships/grants available. Support available to part-time students. Financial award application deadline: 3/15; financial award applicants required to submit FAFSA. *Unit head:* Prof. Karen B. Geils, Program Director, 843-792-4013, Fax: 843-792-3383, E-mail: brinkerk@musc.edu. *Application contact:* Jennifer R. Bailey, Director of Student Affairs, 843-792-3326, Fax: 843-792-0253, E-mail: baileyje@musc.edu.

Meharry Medical College, School of Graduate Studies, Division of Biomedical Sciences, Nashville, TN 37208-9989. Offers PhD. *Degree requirements:* For doctorate, thesis/dissertation, comprehensive exam. *Entrance requirements:* For doctorate, GRE General Test, GRE Subject Test. *Faculty research:* Molecular mechanisms of biological systems and their relationship to human diseases, regulatory biological and cellular structure and function, genetic regulation of growth and cellular metabolisms.

Memorial University of Newfoundland, Faculty of Medicine and School of Graduate Studies, Graduate Programs in Medicine, St. John's, NL A1C 5S7, Canada. Offers M Sc, PhD, Diploma, MD/PhD. Part-time programs available. *Degree requirements:* For master's, thesis; for doctorate, thesis/dissertation, oral defense of thesis, comprehensive exam. *Entrance requirements:* For master's, MD or B Sc; for doctorate, MD or M Sc; for Diploma, bachelor's degree in health-related field. Additional exam requirements/recommendations for international students: Required—TOEFL (minimum score 550 paper-based; 213 computer-based). Electronic applications accepted. Students attend 2 or 3 semesters per year. *Expenses:* Tuition, nonresident: full-time $1,466; part-time $733 per semester. International tuition: $1,906 full-time. Tuition and fees vary according to degree level and program. *Faculty research:* Human genetics, immunology, molecular biology, stroke, cancer.

Memorial University of Newfoundland, School of Graduate Studies, Department of Biology, St. John's, NL A1C 5S7, Canada. Offers biology (M Sc, PhD); marine biology (M Sc, PhD). Part-time programs available. *Students:* 74 full-time (46 women), 6 part-time (2 women), 25 international. 37 applicants, 22% accepted, 6 enrolled. In 2005, 15 master's, 5 doctorates awarded. *Degree requirements:* For master's, thesis; for doctorate, thesis/dissertation, oral defense of thesis, comprehensive exam. *Entrance requirements:* For master's, honors degree (minimum 2nd class standing) in related field. *Application deadline:* Applications are processed on a rolling basis. Application fee: $40 Canadian dollars. Electronic applications accepted. Students attend 2 or 3 semesters per year. *Expenses:* Tuition, nonresident: full-time $1,466; part-time $733 per semester. International tuition: $1,906 full-time. Tuition and fees vary according to degree level and program. *Financial support:* Fellowships, research assistantships, teaching assistantships, institutionally sponsored loans available. *Faculty*

Biological and Biomedical Sciences—General

Memorial University of Newfoundland (continued)
research: Northern flora and fauna, especially cold ocean and boreal environments. *Unit head:* Dr. Chris Parrish, Chair, 709-737-3225, Fax: 709-737-3018, E-mail: pmarino@mun.ca. *Application contact:* Dr. Paul Snelgrove, Graduate Officer, 709-737-3440, E-mail: graduate.biology@mun.ca.

Michigan State University, College of Human Medicine and The Graduate School, Graduate Programs in Human Medicine, East Lansing, MI 48824. Offers biochemistry and molecular biology (MS, PhD); bioethics, humanities, and society (MA); epidemiology (MS, PhD); microbiology (MS); microbiology and molecular genetics (PhD); pharmacology and toxicology (MS, PhD); physiology (MS, PhD). *Students:* 57 full-time (35 women), 22 part-time (17 women); includes 11 minority (6 African Americans, 5 Asian Americans or Pacific Islanders), 32 international. Average age 30. *Entrance requirements:* Additional exam requirements/recommendations for international students: Required—TOEFL (minimum score 550 paper-based; 213 computer-based). *Expenses:* Tuition, state resident: part-time $330 per credit hour. Tuition, nonresident: part-time $685 per credit hour. Tuition and fees vary according to program. *Financial support:* Institutionally sponsored loans available. *Unit head:* Dr. Nigel S. Paneth, Associate Dean, Research and Graduate Studies, 517-432-4789, E-mail: paneth@msu.edu.

Michigan State University, College of Osteopathic Medicine and The Graduate School, Graduate Studies in Osteopathic Medicine, East Lansing, MI 48824. Offers biochemistry and molecular biology (MS, PhD); microbiology (MS); microbiology and molecular genetics (MS, PhD); pharmacology and toxicology (MS, PhD), including pharmacology and toxicology, pharmacology and toxicology-environmental toxicology (PhD); physiology (MS, PhD). *Students:* 2 full-time (1 woman), 1 international. Average age 27. *Expenses:* Tuition, state resident: part-time $330 per credit hour. Tuition, nonresident: part-time $685 per credit hour. Tuition and fees vary according to program. *Unit head:* Dr. Veronica M. Maher, Associate Dean, Graduate Studies, 517-353-7785, Fax: 517-353-9004, E-mail: maher@msu.edu. *Application contact:* Kathie Schafer, Director of Admissions, 517-353-7740, Fax: 517-355-3296, E-mail: comadm@com.msu.edu.

Michigan State University, College of Veterinary Medicine and The Graduate School, Graduate Program in Veterinary Medicine, Program in Comparative Medicine and Integrative Biology, East Lansing, MI 48824. Offers MS, PhD. *Students:* 21 full-time (11 women), 5 part-time (all women); includes 2 minority (both African Americans), 14 international. Average age 29. 16 applicants, 38% accepted. In 2005, 2 degrees awarded. *Degree requirements:* For master's, oral defense of thesis or research paper; for doctorate, thesis/dissertation, defense of dissertation, comprehensive exam. *Entrance requirements:* For master's, GRE General Test, minimum GPA of 3.0, bachelor's degree or higher in life sciences or related field, 3 letters of recommendation; for doctorate, GRE General Test, minimum GPA of 3.0 bachelor's degree or higher in life sciences or related field, 3 letters of recommendation. Additional exam requirements/recommendations for international students: Required—TOEFL (minimum score 550 paper-based; 213 computer-based). *Application deadline:* For fall admission, 12/27 for domestic students. Application fee: $50. Electronic applications accepted. *Expenses:* Tuition, state resident: part-time $330 per credit hour. Tuition, nonresident: part-time $685 per credit hour. Tuition and fees vary according to program. *Financial support:* In 2005–06, 21 fellowships with tuition reimbursements (averaging $8,093 per year), 11 research assistantships with tuition reimbursements (averaging $14,513 per year) were awarded; scholarships/grants, health care benefits, and unspecified assistantships also available. *Faculty research:* Human and animal health and well-being; whole-animal systems- transmission, differentiation and the pathobiology of disease. *Unit head:* Dr. Vilma Yuzbasiyan-Gurkan, Program Director, 517-355-6463 Ext. 1562, Fax: 517-353-8957, E-mail: yuzbasiyan@cvm.msu.edu. *Application contact:* Dr. Victoria Hoelzer-Maddox, Administrative Assistant, 517-353-3118, Fax: 517-432-1037, E-mail: hoelzer-maddox@cvm.msu.edu.

Michigan Technological University, Graduate School, College of Sciences and Arts, Department of Biological Sciences, Houghton, MI 49931-1295. Offers MS, PhD. Part-time programs available. *Faculty:* 15 full-time (6 women), 7 part-time/adjunct (0 women). *Students:* 15 full-time (7 women), 4 part-time (2 women), 6 international. Average age 28. 43 applicants, 30% accepted, 6 enrolled. In 2005, 5 master's, 1 doctorate awarded. Terminal master's awarded for partial completion of doctoral program. *Median time to degree:* Of those who began their doctoral program in fall 1997, 60% received their degree in 8 years or less. *Degree requirements:* For master's, thesis/dissertation, comprehensive exam (for some programs), registration; for doctorate, thesis/dissertation, comprehensive exam, registration. *Entrance requirements:* For master's and doctorate, GRE. Additional exam requirements/recommendations for international students: Required—TOEFL (minimum score 550 paper-based; 213 computer-based). *Application deadline:* For fall admission, 3/1 for domestic students; for spring admission, 11/1 priority date for domestic students. Applications are processed on a rolling basis. Application fee: $40 ($45 for international students). Electronic applications accepted. *Expenses:* Tuition, nonresident: full-time $11,232; part-time $468 per credit. Required fees: $754; $377 per semester. Full-time tuition and fees vary according to course load, degree level and program. *Financial support:* In 2005–06, 16 students received support, including fellowships with full tuition reimbursements available (averaging $9,542 per year), 2 research assistantships with full tuition reimbursements available (averaging $9,542 per year), 11 teaching assistantships with full tuition reimbursements available (averaging $9,542 per year); career-related internships or fieldwork, Federal Work-Study, scholarships/grants, health care benefits, tuition waivers (partial), unspecified assistantships, and co-op also available. Financial award applicants required to submit FAFSA. *Faculty research:* Aquatic ecology, biological control, predator-prey interactions, environmental microbiology, microbial and plant biochemistry, genomics and bioinformatics. Total annual research expenditures: $274,972. *Unit head:* Dr. John H. Adler, Chair, 906-487-2025, Fax: 906-487-3167, E-mail: jhadler@mtu.edu. *Application contact:* Dr. Donald F. Lueking, Director of Graduate Studies, 906-487-2027, Fax: 906-487-3167, E-mail: drluekin@mtu.edu.

Middle Tennessee State University, College of Graduate Studies, College of Basic and Applied Sciences, Department of Biology, Murfreesboro, TN 37132. Offers MS. Part-time and evening/weekend programs available. Postbaccalaureate distance learning degree programs offered. *Degree requirements:* For master's, one foreign language, thesis, comprehensive exam. *Entrance requirements:* For master's, GRE or MAT. Additional exam requirements/recommendations for international students: Required—TOEFL (minimum score 525 paper-based; 195 computer-based). Electronic applications accepted.

Midwestern State University, Graduate Studies, College of Science and Mathematics, Program in Biology, Wichita Falls, TX 76308. Offers MS. Part-time and evening/weekend programs available. *Degree requirements:* For master's, thesis, comprehensive exam. *Entrance requirements:* For master's, GRE General Test, MAT. Additional exam requirements/recommendations for international students: Required—TOEFL (minimum score 550 paper-based; 213 computer-based). Electronic applications accepted. *Faculty research:* Ecology and systematics of spiders and mammals, plant physiology and molecular biology, Drosophila genetics.

Midwestern University, Downers Grove Campus, College of Health Sciences, Illinois Campus, Program in Biomedical Sciences, Downers Grove, IL 60515-1235. Offers MBS. Part-time programs available. *Students:* 34 full-time (23 women), 1 part-time; includes 9 minority (3 African Americans, 6 Asian Americans or Pacific Islanders). Average age 24. 80 applicants, 76% accepted, 28 enrolled. In 2005, 8 degrees awarded. *Entrance requirements:* For master's, GRE General Test, MCAT or PCAT, 2 letters of recommendation. *Application deadline:* Applications are processed on a rolling basis. Application fee: $50. *Expenses:* Tuition: full-time $22,579; part-time $47 per credit hour. Required fees: $412. Full-time tuition and fees vary according to degree level and program. *Unit head:* Dr. Fred D. Romano, Director, 630-515-6392, E-mail: froman@midwestern.edu. *Application contact:* Mark Clancy, Director of Admissions, 630-515-7200, Fax: 630-971-6086, E-mail: admissil@midwestern.edu.

Announcement: Midwestern University is committed to educating the health-care team of the new century. The University operates campuses in Downers Grove, IL, and in Glendale, AZ. The Biomedical Science Program offers Master of Biomedical Sciences (MBS) degrees at both Downers Grove and Glendale. The program is designed for students interested in biotechnology, biomedical research, and the pharmaceutical industry. Successful students are also competitive candidates for admission to postbaccalaureate professional schools in the health sciences. Contact: Office of Admissions, 800-458-6253 (Downers Grove), 888-247-9277 (Glendale); e-mail: admissil@midwestern.edu (Downers Grove), admissaz@midwestern.edu (Glendale). Web site: http://www.midwestern.edu.

Midwestern University, Glendale Campus, College of Health Sciences, Arizona Campus, Program in Biomedical Sciences, Glendale, AZ 85308. Offers MBS. *Students:* 26 full-time (16 women), 3 part-time (1 woman); includes 13 minority (3 African Americans, 6 Asian Americans or Pacific Islanders, 4 Hispanic Americans). Average age 27. 77 applicants, 74% accepted, 43 enrolled. In 2005, 8 degrees awarded. Application fee: $50. *Expenses:* Contact institution. *Unit head:* Dr. William P. Baker, Director, 623-572-3622. *Application contact:* James Walters, Director of Admissions, 888-247-9277, Fax: 623-572-3229, E-mail: admissaz@midwestern.edu.

Millersville University of Pennsylvania, Graduate School, School of Science and Mathematics, Department of Biology, Millersville, PA 17551-0302. Offers MS. Part-time and evening/weekend programs available. *Faculty:* 19 full-time (7 women), 3 part-time/adjunct (all women). *Students:* 2 full-time (both women), 1 (woman) part-time. Average age 27. 3 applicants, 100% accepted, 1 enrolled. In 2005, 3 degrees awarded. *Degree requirements:* For master's, thesis optional. *Entrance requirements:* For master's, GRE General Test, GRE Subject Test (biology), minimum undergraduate GPA of 2.75. Additional exam requirements/recommendations for international students: Required—TOEFL (minimum score 500 paper-based; 183 computer-based). *Application deadline:* For fall admission, 3/1 for domestic students; for spring admission, 10/1 priority date for domestic students. Applications are processed on a rolling basis. Application fee: $35. *Expenses:* Tuition, state resident: full-time $5,888; part-time $327 per credit. Tuition, nonresident: full-time $9,422; part-time $523 per credit. Tuition: $1,216; $60 per credit. Tuition and fees vary according to course load. *Financial support:* In 2005–06, 3 students received support, including 3 research assistantships with full tuition reimbursements available (averaging $4,000 per year); Federal Work-Study, institutionally sponsored loans, and unspecified assistantships also available. Support available to part-time students. Financial award application deadline: 3/15; financial award applicants required to submit FAFSA. *Unit head:* Dr. Joel B. Piperberg, Chair, 717-872-3273, Fax: 717-872-3905, E-mail: joel.piperberg@millersville.edu. *Application contact:* Dr. Victor S. DeSantis, Dean of Graduate Studies, 717-872-3099, Fax: 717-871-2022, E-mail: victor.desantis@millersville.edu.

Mills College, Graduate Studies, Program in Pre-Med, Oakland, CA 94613-1000. Offers Certificate. Part-time programs available. *Faculty:* 5 full-time (2 women), 4 part-time/adjunct (all women). *Students:* 42 full-time (40 women), 25 part-time (all women); includes 13 minority (2 African Americans, 7 Asian Americans or Pacific Islanders, 4 Hispanic Americans), 1 international. Average age 27. 118 applicants, 73% accepted, 45 enrolled. In 2005, 33 degrees awarded. *Entrance requirements:* For degree, GRE General Test, bachelor's degree in a non-science area. Additional exam requirements/recommendations for international students: Required—TOEFL. *Application deadline:* For fall admission, 2/1 for domestic students, 2/1 for international students. Applications are processed on a rolling basis. Application fee: $50. Electronic applications accepted. *Expenses:* Tuition: Full-time $20,350. Required fees: $2,748. *Financial support:* In 2005–06, 12 students received support, including 20 fellowships with partial tuition reimbursements available (averaging $2,000 per year), 6 teaching assistantships with partial tuition reimbursements available (averaging $4,290 per year); institutionally sponsored loans and scholarships/grants also available. Support available to part-time students. Financial award application deadline: 2/1; financial award applicants required to submit FAFSA. *Faculty research:* Bacterial and viral genetics, microbiology, lipid biochemistry, inorganic nitrogen chemistry. *Unit head:* Dr. John Brabson, Co-Director, 510-430-2203, Fax: 510-430-3314, E-mail: gradstudies@mills.edu. *Application contact:* Randy McGlauthing, Director of Graduate Admissions, 510-430-2355, Fax: 510-430-2159, E-mail: rmcglauth@mills.edu.

Minnesota State University Mankato, College of Graduate Studies, College of Science, Engineering and Technology, Department of Biological Sciences, Mankato, MN 56001. Offers biology (MS); biology education (MS); environmental science (MS), including ecology, economic and political systems, human ecosystems, physical science, technology. Part-time programs available. *Students:* 14 full-time (7 women), 17 part-time (10 women). Average age 31. In 2005, 2 degrees awarded. *Degree requirements:* For master's, one foreign language, thesis or alternative, comprehensive exam. *Entrance requirements:* For master's, minimum GPA of 3.0 during previous 2 years of course work. Additional exam requirements/recommendations for international students: Required—TOEFL. *Application deadline:* For fall admission, 7/1 for domestic students; for spring admission, 11/1 for domestic students. Applications are processed on a rolling basis. Application fee: $40. Electronic applications accepted. *Expenses:* Tuition, state resident: part-time $243 per credit. Tuition, nonresident: part-time $400 per credit. Required fees: $30 per credit. *Financial support:* Fellowships, research assistantships with full tuition reimbursements, teaching assistantships with full tuition reimbursements, career-related internships or fieldwork, Federal Work-Study, institutionally sponsored loans, and unspecified assistantships available. Support available to part-time students. Financial award application deadline: 3/15; financial award applicants required to submit FAFSA. *Faculty research:* Limnology, enzyme analysis, membrane engineering, converters. *Unit head:* Dr. Gregg Marg, Chairperson, 507-389-2786. *Application contact:* 507-389-2321, E-mail: grad@mnsu.edu.

Mississippi College, Graduate School, College of Arts and Sciences, Program in Combined Sciences, Major in Biology, Clinton, MS 39058. Offers MCS. *Degree requirements:* For master's, comprehensive exam. *Entrance requirements:* For master's, GRE General Test, minimum GPA of 2.5.

Mississippi State University, College of Arts and Sciences, Department of Biological Sciences, Mississippi State, MS 39762. Offers MS, PhD. *Faculty:* 17 full-time (5 women), 1 part-time/adjunct (0 women). *Students:* 29 full-time (13 women), 8 part-time (4 women); includes 6 minority (5 African Americans, 1 Asian American or Pacific Islander), 9 international. Average age 27. 45 applicants, 24% accepted, 8 enrolled. In 2005, 8 master's, 2 doctorates awarded. Terminal master's awarded for partial completion of doctoral program. *Degree requirements:* For master's and doctorate, one foreign language, thesis/dissertation, comprehensive oral or written exam. *Entrance requirements:* For master's and doctorate, GRE General Test. Additional exam requirements/recommendations for international students: Required—TOEFL. *Application deadline:* For fall admission, 7/1 for domestic students; for spring admission, 11/1 for domestic students. Applications are processed on a rolling basis. Application fee: $30. *Expenses:* Tuition, state resident: full-time $4,312; part-time $240 per hour. Tuition, nonresident: full-time $9,772; part-time $543 per hour. International tuition: $10,102 full-time. Tuition and fees vary according to course load. *Financial support:* In 2005–06, 29 teaching assistantships with full tuition reimbursements (averaging $11,434 per year) were awarded; research assistantships, Federal Work-Study, institutionally sponsored loans, scholarships/grants, and unspecified assistantships also available. Financial award applicants required to submit FAFSA. *Faculty research:* Botany, zoology, microbiology, ecology. Total annual research expenditures: $5.7 million. *Unit head:* Dr. Narasiah Gavini, Head, 662-325-3483, Fax: 662-325-7939, E-mail: gavini@biology.msstate.edu. *Application contact:* Philip G. Bonfanti, Director of Admissions, 662-325-4104, Fax: 662-325-8872, E-mail: admit@msstate.edu.

Missouri State University, Graduate College, College of Natural and Applied Sciences, Department of Biology, Springfield, MO 65804-0094. Offers biology (MNAS, MS); secondary education (MS Ed), including biology. *Faculty:* 16 full-time (3 women), 1 (woman) part-time/adjunct. *Students:* 22 full-time (12 women), 10 part-time (4 women); includes 1 minority (American Indian/Alaska Native), 1 international. Average age 27. 18 applicants, 72% accepted, 11 enrolled. In 2005, 12 degrees awarded. *Degree requirements:* For master's, thesis or alternative, comprehensive exam. *Entrance requirements:* For master's, GRE (MS and MNAS), 24 hours of course work in biology, minimum GPA of 3.0 (MS), 9-12 teacher certification (MS Ed), 3.0 GPA (MNAS). Additional exam requirements/recommendations for international students: Required—TOEFL (minimum score 550 paper-based; 213 computer-based), IELT (minimum score 6). *Application deadline:* For fall admission, 7/20 for domestic students; for spring admission, 12/20 priority date for domestic students. Applications are processed on a rolling basis. Application fee: $30. Electronic applications accepted. *Expenses:* Tuition, state resident: full-time $3,402; part-time $189 per credit. Tuition, nonresident: full-time $6,804; part-time $378 per credit. Required fees: $207 per semester. Part-time tuition and fees vary according to course level, course load and program. *Financial support:* In 2005–06, 1 fellowship (averaging $8,575 per year), 5 research assistantships with full tuition reimbursements (averaging $8,750 per year), 22 teaching assistantships with full tuition reimbursements (averaging $7,605 per year) were awarded; Federal Work-Study, scholarships/grants, and unspecified assistantships also available. Financial award application deadline: 3/31; financial award applicants required to submit FAFSA. *Faculty research:* Field biology, organismal biology, microbiology. *Unit head:* Dr. S. Alicia Mathis, Head, 417-836-5126, Fax: 417-836-6934, E-mail: biology@missouristate.edu. *Application contact:* Dr. Thomas Tomasi, Graduate Director, 417-836-5169, Fax: 417-836-6934, E-mail: tomtomasi@missouristate.edu.

Montana State University, College of Graduate Studies, College of Letters and Science, Department of Ecology, Bozeman, MT 59717. Offers biological sciences (MS, PhD); fish and wildlife biology (PhD); fish and wildlife management (MS); land rehabilitation (intercollege) (MS). Part-time programs available. *Faculty:* 15 full-time (4 women). *Students:* 4 full-time (3 women), 61 part-time (23 women); includes 1 minority (American Indian/Alaska Native), 1 international. Average age 31. 8 applicants, 75% accepted, 6 enrolled. In 2005, 9 master's, 4 doctorates awarded. *Degree requirements:* For master's, thesis (for some programs), comprehensive exam, registration; for doctorate, thesis/dissertation, comprehensive exam, registration. *Entrance requirements:* For master's and doctorate, GRE General Test. Additional exam requirements/recommendations for international students: Required—TOEFL (minimum score 550 paper-based; 213 computer-based). *Application deadline:* For fall admission, 7/15 priority date for domestic students, 5/15 priority date for international students; for spring admission, 12/1 priority date for domestic students, 10/1 priority date for international students. Applications are processed on a rolling basis. Application fee: $30. Electronic applications accepted. *Expenses:* Tuition, state resident: full-time $4,132. Tuition, nonresident: full-time $1,132. *Financial support:* In 2005–06, 3 fellowships with full tuition reimbursements (averaging $21,300 per year), 45 research assistantships with full and partial tuition reimbursements (averaging $10,830 per year), 22 teaching assistantships with full and partial tuition reimbursements (averaging $10,268 per year) were awarded; career-related internships or fieldwork, Federal Work-Study, scholarships/grants, health care benefits, tuition waivers, and unspecified assistantships also available. Financial award application deadline: 3/1; financial award applicants required to submit FAFSA. *Faculty research:* Population dynamics, landscape ecology, evolution and genetics, ecological modeling, aquatic ecosystems. Total annual research expenditures: $2.1 million. *Unit head:* Dr. David Roberts, Department Head, 406-994-4548, Fax: 406-994-3190, E-mail: droberts@montana.edu.

Montclair State University, The Graduate School, College of Science and Mathematics, Department of Biology and Molecular Biology, Montclair, NJ 07043-1624. Offers biology (MS), including biology science education, molecular biology; molecular biology (Certificate). Part-time and evening/weekend programs available. *Faculty:* 19 full-time (7 women), 19 part-time/adjunct (9 women). *Students:* 14 full-time (9 women), 61 part-time (36 women); includes 20 minority (10 African Americans, 6 Asian Americans or Pacific Islanders, 4 Hispanic Americans), 3 international. 52 applicants, 62% accepted, 23 enrolled. In 2005, 19 master's, 4 other advanced degrees awarded. *Degree requirements:* For master's, thesis or alternative, comprehensive exam. *Entrance requirements:* For master's, GRE General Test, 24 credits of course work in undergraduate biology, 2 letters of recommendation, teaching certificate (biology sciences education concentration). Additional exam requirements/recommendations for international students: Required—TOEFL (minimum score 83 computer-based). *Application deadline:* Applications are processed on a rolling basis. Application fee: $60. Electronic applications accepted. *Expenses:* Tuition, state resident: part-time $409 per credit. Tuition, nonresident: part-time $604 per credit. Required fees: $56 per credit. Tuition and fees vary according to course load, degree level and program. *Financial support:* In 2005–06, 7 research assistantships with full tuition reimbursements (averaging $7,000 per year) were awarded; Federal Work-Study, scholarships/grants, and unspecified assistantships also available. Support available to part-time students. Financial award application deadline: 3/1; financial award applicants required to submit FAFSA. *Faculty research:* Cells, algea blooms, scallops, NJ bays, Barnegat Bay. Total annual research expenditures: $48,000. *Unit head:* Dr. Scott Kight, Chairperson, 973-655-7047. *Application contact:* Dr. Reginald Halaby, Adviser, 973-655-4397, E-mail: halabyr@mail.montclair.edu.

Morehead State University, Graduate Programs, College of Science and Technology, Department of Biological and Environmental Sciences, Morehead, KY 40351. Offers biology (MS); regional analysis and public policy (MS). Part-time programs available. *Faculty:* 14 full-time (2 women), 2 part-time/adjunct (1 woman). *Students:* 9 full-time (4 women), 3 part-time (all women), 2 international. Average age 25. 16 applicants, 94% accepted. In 2005, 7 degrees awarded. *Degree requirements:* For master's, oral and written final exams, thesis optional. *Entrance requirements:* For master's, GRE General Test, minimum GPA of 3.0 in biology, 2.5 overall; undergraduate major/minor in biology, environmental science, or equivalent. Additional exam requirements/recommendations for international students: Required—TOEFL (minimum score 525 paper-based; 197 computer-based). *Application deadline:* For fall admission, 8/1 priority date for domestic students, 8/1 priority date for international students; for spring admission, 12/1 priority date for domestic students, 12/1 priority date for international students. Applications are processed on a rolling basis. Application fee: $0 ($55 for international students). Electronic applications accepted. *Financial support:* In 2005–06, 2 research assistantships (averaging $6,000 per year) were awarded; career-related internships or fieldwork and Federal Work-Study also available. Financial award application deadline: 4/1; financial award applicants required to submit FAFSA. *Faculty research:* Atherosclerosis, RNA evolution, cancer biology, water quality/ecology, immunoparasitology. *Unit head:* Dr. David Magrane, Chair, 606-783-2944, E-mail: d.magrane@moreheadstate.edu. *Application contact:* Betty R. Cowsert, Graduate Admissions/Records Manager, 606-783-2039, Fax: 606-783-5061, E-mail: b.cowsert@moreheadstate.edu.

Morehouse School of Medicine, Program in Biomedical Sciences, Atlanta, GA 30310-1495. Offers PhD. *Faculty:* 19 full-time (13 women). *Students:* 26 full-time (17 women); all minorities (all African Americans) Average age 26. 24 applicants, 38% accepted, 7 enrolled. In 2005, 3 degrees awarded. *Degree requirements:* For doctorate, thesis/dissertation. *Entrance requirements:* For doctorate, GRE General Test. Additional exam requirements/recommendations for international students: Required—TOEFL (minimum score 550 paper-based; 200 computer-based). *Application deadline:* For fall admission, 2/1 for domestic students, 2/1 for international students. Application fee: $35. Electronic applications accepted. *Expenses:* Contact institution. Full-time tuition and fees vary according to degree level, program and student level. *Financial support:* In 2005–06, 14 fellowships with full and partial tuition reimbursements (averaging $14,000 per year) were awarded; career-related internships or fieldwork, institutionally sponsored loans, scholarships/grants, traineeships, health care benefits, and tuition waivers (full) also available. Financial award application deadline: 5/1; financial award applicants required to submit FAFSA. Total annual research expenditures: $5.8 million. *Unit head:* Dr. Douglas Paulsen, Director, 404-752-1559. *Application contact:* Dr. Sterling Roaf, Director of Admissions, 404-752-1650, Fax: 404-752-1512, E-mail: phdadmissions@msm.edu.

See Close-Up on page 175.

Morgan State University, School of Graduate Studies, School of Computer, Mathematical, and Natural Sciences, Department of Biology, Baltimore, MD 21251. Offers bio-environmental science (PhD). *Expenses:* Tuition, state resident: part-time $272 per credit. Tuition, nonresident: part-time $478 per credit. Required fees: $58 per credit. *Application contact:* Dr. James E. Waller, Admissions and Program Officer, 443-885-3185, Fax: 443-885-8226, E-mail: jwaller@moac.morgan.edu.

Morgan State University, School of Graduate Studies, School of Computer, Mathematical, and Natural Sciences, Interdisciplinary Program in Science, Baltimore, MD 21251. Offers biology (MS); chemistry (MS); physics (MS). *Students:* 9 (2 women); includes 4 minority (all African Americans) 4 international. In 2005, 5 degrees awarded. *Degree requirements:* For master's, thesis, oral defense of thesis, comprehensive exam. *Entrance requirements:* For master's, GRE General Test, minimum GPA of 2.5. *Application deadline:* For fall admission, 2/1 for domestic students; for spring admission, 10/1 priority date for domestic students. Applications are processed on a rolling basis. Application fee: $0. *Expenses:* Tuition, state resident: part-time $272 per credit. Tuition, nonresident: part-time $478 per credit. Required fees: $58 per credit. *Financial support:* Fellowships, research assistantships, career-related internships or fieldwork, Federal Work-Study, institutionally sponsored loans, scholarships/grants, health care benefits, and unspecified assistantships available. Support available to part-time students. *Unit head:* Dr. Juarine Stewart, Dean, 443-885-4515, Fax: 443-885-8215. *Application contact:* Dr. James E. Waller, Admissions and Program Officer, 443-885-3185, Fax: 443-885-8226, E-mail: jwaller@moae.morgan.edu.

Mount Allison University, Faculty of Science, Department of Biology, Sackville, NB E4L 1E4, Canada. Offers M Sc. *Degree requirements:* For master's, thesis. *Entrance requirements:* For master's, honors degree. *Faculty research:* Ecology, evolution, physiology, behavior, biochemistry.

Mount Sinai School of Medicine of New York University, Graduate School of Biological Sciences, New York, NY 10029-6504. Offers biophysics, structural biology and biomathematics (PhD); community medicine (MPH); genetic counseling (MS); genetics and genomic sciences (PhD); mechanisms of disease and therapy (PhD); microbiology (PhD); molecular, cellular, biochemical and developmental sciences (PhD); neurosciences (PhD). *Students:* 218 full-time (109 women). 4,208 applicants, 7% accepted, 117 enrolled.Terminal master's awarded for partial completion of doctoral program. *Degree requirements:* For master's, registration; for doctorate, thesis/dissertation, registration. *Entrance requirements:* For doctorate, GRE General Test, GRE Subject Test, 3 years of college pre-med course work. Additional exam requirements/recommendations for international students: Required—TOEFL. *Application deadline:* For fall admission, 1/15 for domestic students. Application fee: $75. Electronic applications accepted. *Expenses:* Tuition: Full-time $33,250. Required fees: $1,600. Full-time tuition and fees vary according to degree level, program and reciprocity agreements. *Financial support:* In 2005–06, fellowships with full tuition reimbursements (averaging $26,000 per year), research assistantships with full tuition reimbursements (averaging $26,000 per year) were awarded; Federal Work-Study, institutionally sponsored loans, scholarships/grants, health care benefits, and unspecified assistantships also available. Financial award application deadline: 4/5; financial award applicants required to submit FAFSA. *Faculty research:* Cancer, gene therapy, minimally invasive surgery, cardiac translational research. Total annual research expenditures: $162.2 million. *Unit head:* Dr. Diomedes Logothetis, Dean, 212-241-6546, Fax: 212-241-0651, E-mail: diomedes.logothetis@mssm.edu. *Application contact:* Lily Recanati, Manager, 212-241-3267, Fax: 212-241-0651, E-mail: lily.recantati@mssm.edu.

Announcement: The Graduate School of Biological Sciences at Mount Sinai School of Medicine offers PhD and MSTP (MD/PhD) training in more than 200 active research laboratories. Students entering the program are free to choose from 6 multidisciplinary training areas. The research and course work emphasize individualized approaches to foster the training goals of the student. The Mount Sinai campus in New York City offers the advantages of a world-class cultural center in which students benefit from recreational activities as well as the larger scientific community of New York City. Stipends, tuition, health insurance, and housing are offered to all students.

See Close-Up on page 177.

Murray State University, College of Science, Engineering and Technology, Department of Biological Sciences, Murray, KY 42071-0009. Offers MAT, MS, PhD. Part-time programs available. *Degree requirements:* For master's, thesis (for some programs). *Entrance requirements:* For master's, GRE General Test. Additional exam requirements/recommendations for international students: Required—TOEFL.

New Jersey Institute of Technology, Office of Graduate Studies, College of Science and Liberal Arts, Department of Life Sciences and Biology, Program in Biology, Newark, NJ 07102. Offers MS, PhD. Part-time and evening/weekend programs available. *Faculty:* 5 full-time (1 woman). *Students:* 4 full-time (2 women), 1 (woman) part-time; includes 3 minority (1 African American, 2 Asian Americans or Pacific Islanders), 1 international. Average age 32. 27 applicants, 19% accepted, 4 enrolled. In 2005, 1 degree awarded. *Entrance requirements:* For master's, GRE General Test. Additional exam requirements/recommendations for international students: Required—TOEFL (minimum score 550 paper-based; 213 computer-based). *Application deadline:* For fall admission, 6/5 for domestic students; for spring admission, 10/15 for domestic students. Applications are processed on a rolling basis. Application fee: $60. Electronic applications accepted. *Expenses:* Tuition, state resident: full-time $9,620; part-time $520 per credit. Tuition, nonresident: full-time $13,542; part-time $715 per credit. Required fees: $78; $54 per credit. $78 per year. Tuition and fees vary according to course load. *Financial support:* Fellowships with full and partial tuition reimbursements, research assistantships with full and partial tuition reimbursements, teaching assistantships with full and partial tuition reimbursements, career-related internships or fieldwork, Federal Work-Study, institutionally sponsored loans, and unspecified assistantships available. Financial award application deadline: 3/15. *Faculty research:* Realistic building codes, optimization of training programs, effect of physical and mental fatigue of training. *Application contact:* Kathryn Kelly, Director of Admissions, 973-596-3300, Fax: 973-596-3461, E-mail: admissions@njit.edu.

New Mexico Highlands University, Graduate Studies, College of Arts and Sciences, Department of Natural Sciences, Las Vegas, NM 87701. Offers applied chemistry (MS); biology (MS); environmental science and management (MS). Part-time programs available. *Faculty:* 11 full-time (4 women), 7 part-time/adjunct (2 women). *Students:* 15 full-time (6 women), 6 part-time (3 women); includes 4 minority (1 American Indian/Alaska Native, 3 Hispanic Americans), 13 international. Average age 29. 6 applicants, 100% accepted, 6 enrolled. In 2005, 8 degrees awarded. *Degree requirements:* For master's, thesis, comprehensive exam, registration. *Entrance requirements:* For master's, minimum undergraduate GPA of 3.0. Additional exam requirements/recommendations for international students: Required—TOEFL (minimum score 540 paper-based; 190 computer-based). *Application deadline:* For fall admission, 8/1 for domestic students. Applications are processed on a rolling basis. Application fee: $15. *Expenses:* Tuition, state resident: full-time $2,280; part-time $101 per credit. Tuition, nonresident: full-time $3,420; part-time $151 per credit. One-time fee: $20 full-time. *Financial support:* In 2005–06, 4 students received support, including 13 teaching assistantships (averaging $11,500 per year); research assistantships with full and partial tuition reimbursements available, Federal Work-Study, institutionally sponsored loans, scholarships/grants, and unspecified assistantships also available. Support available to part-time students. Financial award application deadline: 3/1. *Unit head:* Dr. Merritt Helvenston, Chair, 505-454-3263, Fax: 505-454-3103, E-mail: merritt@nmhu.edu. *Application contact:* Diane Trujillo, Administrative Assistant Graduate Studies, 505-454-3266, Fax: 505-454-3558, E-mail: dtrujillo@nmhu.edu.

New Mexico Institute of Mining and Technology, Graduate Studies, Department of Biology, Socorro, NM 87801. Offers MS. Part-time programs available. *Degree requirements:* For master's, thesis. *Entrance requirements:* For master's, GRE General Test. Additional exam requirements/recommendations for international students: Required—TOEFL (minimum score 540 paper-based; 207 computer-based). Electronic applications accepted. *Faculty research:* Molecular biology, evolution and evolutionary ecology, immunology, endocrinology.

Biological and Biomedical Sciences—General

New Mexico State University, Graduate School, College of Arts and Sciences, Department of Biology, Las Cruces, NM 88003-8001. Offers MS, PhD. Part-time programs available. *Faculty:* 15 full-time (3 women), 8 part-time/adjunct (4 women). *Students:* 55 full-time (31 women), 19 part-time (9 women); includes 19 minority (2 African Americans, 1 American Indian/Alaska Native, 2 Asian Americans or Pacific Islanders, 14 Hispanic Americans), 18 international. Average age 31. 36 applicants, 69% accepted, 14 enrolled. In 2005, 7 master's, 2 doctorates awarded. *Degree requirements:* For master's, thesis (for some programs) for doctorate, thesis/dissertation, comprehensive exam. *Entrance requirements:* Additional exam requirements/recommendations for international students: Required—TOEFL. *Application deadline:* For fall admission, 1/15 for domestic students; for spring admission, 10/5 priority date for domestic students. Applications are processed on a rolling basis. Application fee: $30 ($50 for international students). Electronic applications accepted. *Expenses:* Tuition, state resident: full-time $3,156; part-time $175 per credit. Tuition, nonresident: full-time $12,510; part-time $565 per credit. Required fees: $1,050. *Financial support:* In 2005–06, 5 fellowships, 16 research assistantships, 26 teaching assistantships were awarded; Federal Work-Study also available. Support available to part-time students. Financial award application deadline: 1/15. *Faculty research:* Microbiology, cell and organismal physiology, ecology and ethology, evolution, genetics, developmental biology. *Unit head:* Dr. Daniel J. Howard, Head, 505-646-3611, Fax: 505-646-5665, E-mail: dahoward@nmsu.edu.

New York Medical College, Graduate School of Basic Medical Sciences, Valhalla, NY 10595-1691. Offers MS, PhD, MD/PhD. Part-time and evening/weekend programs available. Terminal master's awarded for partial completion of doctoral program. *Degree requirements:* For master's, thesis/dissertation; for doctorate, thesis/dissertation, comprehensive exam. *Entrance requirements:* For master's and doctorate, GRE General Test. Additional exam requirements/recommendations for international students: Required—TOEFL.

See Close-Up on page 179.

New York University, Graduate School of Arts and Science, Department of Biology, New York, NY 10012-1019. Offers biology (PhD); biomedical journalism (MS); cancer and molecular biology (PhD); computational biology (PhD); computers in biological research (MS); developmental genetics (PhD); general biology (MS); immunology and microbiology (PhD); molecular genetics (PhD); neurobiology (PhD); oral biology (MS); plant biology (PhD); recombinant DNA technology (MS). Part-time programs available. *Faculty:* 24 full-time (5 women), 8 part-time/adjunct. *Students:* 104 full-time (52 women), 41 part-time (23 women); includes 28 minority (2 African Americans, 20 Asian Americans or Pacific Islanders, 6 Hispanic Americans), 47 international. Average age 27. 349 applicants, 56% accepted, 39 enrolled. In 2005, 59 master's, 4 doctorates awarded. Terminal master's awarded for partial completion of doctoral program. *Degree requirements:* For master's, thesis or alternative, qualifying paper; for doctorate, thesis/dissertation, comprehensive exam. *Entrance requirements:* For master's, GRE General Test; for doctorate, GRE General Test, GRE Subject Test. Additional exam requirements/recommendations for international students: Required—TOEFL. *Application deadline:* For fall admission, 1/4 for domestic students. Application fee: $80. *Financial support:* Fellowships with tuition reimbursements, research assistantships with tuition reimbursements, teaching assistantships with tuition reimbursements, career-related internships or fieldwork, Federal Work-Study, institutionally sponsored loans, scholarships/grants, health care benefits, and unspecified assistantships available. Financial award application deadline: 1/4; financial award applicants required to submit FAFSA. *Faculty research:* Genomics, molecular and cell biology, development and molecular genetics, molecular evolution of plants and animals. *Unit head:* Gloria Coruzzi, Chairman, 212-998-8200, Fax: 212-995-4015, E-mail: biology@nyu.edu. *Application contact:* Stephen Small, Director of Graduate Studies, 212-998-8200, Fax: 212-995-4015, E-mail: biology@nyu.edu.

New York University, Graduate School of Arts and Science, Department of Environmental Medicine, New York, NY 10012-1019. Offers environmental health sciences (MS, PhD), including biostatistics (PhD), environmental hygiene (MS), epidemiology (PhD), ergonomics and biomechanics (PhD), exposure assessment and health effects (PhD), molecular toxicology/carcinogenesis (PhD), toxicology. Part-time programs available. *Faculty:* 26 full-time (7 women). *Students:* 42 full-time (33 women), 21 part-time (10 women); includes 11 minority (2 African Americans, 6 Asian Americans or Pacific Islanders, 3 Hispanic Americans), 21 international. Average age 32. 56 applicants, 38% accepted, 9 enrolled. In 2005, 7 master's, 6 doctorates awarded. Terminal master's awarded for partial completion of doctoral program. *Degree requirements:* For master's, thesis or alternative; for doctorate, one foreign language, thesis/dissertation, oral and written exams. *Entrance requirements:* For master's and doctorate, GRE General Test, GRE Subject Test, minimum GPA of 3.0; bachelor's degree in biological, physical, or engineering science. Additional exam requirements/recommendations for international students: Required—TOEFL. *Application deadline:* For fall admission, 11/14 for domestic students. Application fee: $80. *Financial support:* Fellowships with tuition reimbursements, teaching assistantships with tuition reimbursements, career-related internships or fieldwork, Federal Work-Study, institutionally sponsored loans, and health care benefits available. Financial award application deadline: 2/1; financial award applicants required to submit FAFSA. *Unit head:* Dr. Max Costa, Chair, 845-731-3661, Fax: 845-351-3317, E-mail: ehs@env.med.nyu.edu. *Application contact:* Dr. Jerome J Solomon, Director of Graduate Studies, 845-731-3661, Fax: 845-351-3317, E-mail: ehs@env.med.nyu.edu.

See Close-Up on page 183.

New York University, School of Medicine and Graduate School of Arts and Science, Medical Scientist Training Program, New York, NY 10012-1019. Offers MD/MS, MD/PhD. Students must be accepted by both the School of Medicine and the Graduate School of Arts and Science. *Faculty:* 163 full-time (32 women). *Students:* 75 full-time (27 women); includes 21 minority (4 African Americans, 15 Asian Americans or Pacific Islanders, 2 Hispanic Americans). Average age 25. 340 applicants, 10% accepted, 9 enrolled. *Application deadline:* For fall admission, 10/15 for domestic students. Application fee: $100. Electronic applications accepted. *Expenses:* Contact institution. *Financial support:* In 2005–06, fellowships with full tuition reimbursements (averaging $25,000 per year), research assistantships with full tuition reimbursements (averaging $25,000 per year) were awarded; teaching assistantships, health care benefits and unspecified assistantships also available. *Faculty research:* Neurosciences, cell biology and molecular genetics, structural biology, microbial pathogenesis and host defense. *Unit head:* Dr. Rodney E. Ulane, Director, 212-263-2149, Fax: 212-263-3766, E-mail: rodney.ulane@med.nyu.edu. *Application contact:* Cindy D. Meador, Academic Coordinator, 212-263-3767, E-mail: cindy.meador@med.nyu.edu.

See Close-Up on page 181.

North Carolina Agricultural and Technical State University, Graduate School, College of Arts and Sciences, Department of Biology, Greensboro, NC 27411. Offers MS. Part-time and evening/weekend programs available. *Degree requirements:* For master's, thesis (for some programs), qualifying exam, comprehensive exam. *Entrance requirements:* For master's, GRE General Test, minimum GPA of 2.6. *Faculty research:* Physical ecology, cytochemistry, botany, parasitology, microbiology.

North Carolina Central University, Division of Academic Affairs, College of Arts and Sciences, Department of Biology, Durham, NC 27707-3129. Offers MS. *Degree requirements:* For master's, one foreign language, thesis, comprehensive exam. *Entrance requirements:* For master's, GRE, minimum GPA of 3.0 in major, 2.5 overall. Additional exam requirements/recommendations for international students: Required—TOEFL.

North Carolina State University, College of Veterinary Medicine, Program in Comparative Biomedical Sciences, Raleigh, NC 27695. Offers cell biology and morphology (MS, PhD); epidemiology and population medicine (MS, PhD); immunology (MS, PhD); microbiology and immunology (MS, PhD); pathology (MS, PhD); pharmacology (MS, PhD); specialized veterinary medicine (MS). Part-time programs available. *Degree requirements:* For master's and doctorate, thesis/dissertation. *Entrance requirements:* For master's and doctorate, GRE General

Test. Additional exam requirements/recommendations for international students: Required—TOEFL (minimum score 550 paper-based; 213 computer-based). Electronic applications accepted. *Expenses:* Contact institution. *Faculty research:* Infectious diseases, cell biology, pharmacology and toxicology, genomics, pathology and population medicine.

North Carolina State University, Graduate School, College of Agriculture and Life Sciences, Raleigh, NC 27695. Offers M Tox, MAEE, MB, MBAE, MFG, MFM, MFS, MG, MMB, MN, MP, MS, MZS, PhD. Part-time programs available. Electronic applications accepted.

North Dakota State University, The Graduate School, College of Science and Mathematics, Department of Biological Sciences, Fargo, ND 58105. Offers biological sciences (MS); botany (MS, PhD); cellular and molecular biology (PhD); environmental and conservation sciences (MS, PhD); genomics (MS, PhD); natural resource management (MS, PhD); zoology (MS, PhD). *Faculty:* 20. *Students:* 30 full-time (10 women), 5 part-time (1 woman); includes 1 minority (Asian American or Pacific Islander), 3 international. Average age 24. 14 applicants, 43% accepted. In 2005, 4 master's awarded. *Degree requirements:* For master's and doctorate, thesis/dissertation. *Entrance requirements:* For master's and doctorate, GRE General Test. Additional exam requirements/recommendations for international students: Required—TOEFL. *Application deadline:* For fall admission, 3/15 for domestic students; for spring admission, 10/30 priority date for domestic students. Applications are processed on a rolling basis. Application fee: $45 ($60 for international students). Electronic applications accepted. *Financial support:* In 2005–06, 3 fellowships with full tuition reimbursements (averaging $15,000 per year), 9 research assistantships with full tuition reimbursements (averaging $14,400 per year), 19 teaching assistantships with full tuition reimbursements (averaging $9,550 per year) were awarded; career-related internships or fieldwork, Federal Work-Study, institutionally sponsored loans, scholarships/grants, tuition waivers (full), and unspecified assistantships also available. Support available to part-time students. Financial award application deadline: 4/15; financial award applicants required to submit FAFSA. *Faculty research:* Comparative endocrinology, physiology, behavioral ecology, plant cell biology, aquatic biology. Total annual research expenditures: $675,000. *Unit head:* Dr. William J. Bleier, Chair, 701-231-7087, Fax: 701-231-7149, E-mail: william.bleier@ndsu.nodak.edu.

Northeastern Illinois University, Graduate College, College of Arts and Sciences, Department of Biology, Program in Biology, Chicago, IL 60625-4699. Offers MS. Part-time and evening/weekend programs available. *Degree requirements:* For master's, thesis optional. *Entrance requirements:* For master's, minimum GPA of 2.75. *Faculty research:* Paleoecology and freshwater biology; protein biosynthesis and targeting; microbial growth and physiology; molecular biology of antibody production; reptilian neurobiology.

Northeastern University, College of Arts and Sciences, Department of Biology, Boston, MA 02115-5096. Offers bioinformatics (PMS); biology (MS, PhD); biotechnology (MS); marine biology (MS). Part-time programs available. *Faculty:* 26 full-time (10 women), 2 part-time/adjunct. *Students:* 79 full-time (43 women), 11 part-time (9 women); includes 7 minority (1 African American, 5 Asian Americans or Pacific Islanders, 1 Hispanic American), 22 international. Average age 28. 148 applicants, 25% accepted. In 2005, 20 master's, 3 doctorates awarded. Terminal master's awarded for partial completion of doctoral program. *Degree requirements:* For master's, thesis; for doctorate, thesis/dissertation, qualifying exam. *Entrance requirements:* For master's, GRE General Test, GRE Subject Test; for doctorate, GRE General Test. Additional exam requirements/recommendations for international students: Required—TOEFL (minimum score 250 computer-based). *Application deadline:* For fall admission, 1/15 for domestic students, 2/1 for international students. Application fee: $50. *Financial support:* In 2005–06, 34 teaching assistantships with tuition reimbursements (averaging $16,475 per year) were awarded; fellowships with tuition reimbursements, research assistantships with tuition reimbursements, career-related internships or fieldwork, Federal Work-Study, and tuition waivers (full and partial) also available. Financial award application deadline: 3/1; financial award applicants required to submit FAFSA. *Faculty research:* Biochemistry, cell and systems physiology, ecology, marine sciences, molecular biology. *Unit head:* Dr. Susan Powers-Lee, Chair, 617-373-2260, Fax: 617-373-3724, E-mail: gradbio@neu.edu. *Application contact:* Janeen Greene, Administrative Assistant, 617-373-2262, Fax: 617-373-3724, E-mail: gradbio@neu.edu.

See Close-Up on page 185.

Northern Arizona University, Graduate College, College of Engineering and Natural Science, Department of Biological Sciences, Flagstaff, AZ 86011. Offers biology (MS, PhD); biology education (MAT). *Degree requirements:* For master's, MAT final, MS thesis, oral exam; for doctorate, thesis/dissertation. *Entrance requirements:* For master's, GRE General Test, GRE Subject Test; for doctorate, GRE General Test. Electronic applications accepted. *Faculty research:* Genetic levels of trophic levels, plant hybrid zones, insect biodiversity, natural history and cognition of wild jays.

Northern Illinois University, Graduate School, College of Liberal Arts and Sciences, Department of Biological Sciences, De Kalb, IL 60115-2854. Offers MS, PhD. Part-time programs available. *Faculty:* 30 full-time (6 women), 7 part-time/adjunct (1 woman). *Students:* 48 full-time (24 women), 20 part-time (15 women); includes 8 minority (2 American Indian/Alaska Native, 5 Asian Americans or Pacific Islanders, 1 Hispanic American), 13 international. Average age 30. 75 applicants, 33% accepted, 12 enrolled. In 2005, 11 master's, 1 doctorate awarded. Terminal master's awarded for partial completion of doctoral program. *Degree requirements:* For master's, thesis optional; for doctorate, thesis/dissertation, candidacy exam, dissertation defense. *Entrance requirements:* For master's, GRE General Test, bachelor's degree in related field, minimum GPA of 2.75; for doctorate, GRE General Test, bachelor's or master's degree in related field, minimum undergraduate GPA of 2.75, minimum graduate GPA of 3.2. Additional exam requirements/recommendations for international students: Required—TOEFL (minimum score 550 paper-based; 213 computer-based). *Application deadline:* For fall admission, 6/1 for domestic students, 5/1 for international students; for spring admission, 11/1 for domestic students, 10/1 for international students. Applications are processed on a rolling basis. Application fee: $30. Electronic applications accepted. *Expenses:* Tuition, state resident: full-time $4,565; part-time $191 per credit hour. Tuition, nonresident: full-time $9,129; part-time $382 per credit hour. *Financial support:* In 2005–06, 11 research assistantships with full tuition reimbursements, 35 teaching assistantships with full tuition reimbursements were awarded; fellowships with full tuition reimbursements, career-related internships or fieldwork, Federal Work-Study, scholarships/grants, tuition waivers (full), and unspecified assistantships also available. Support available to part-time students. Financial award applicants required to submit FAFSA. *Faculty research:* Plant molecular biology, neurosecretory control, ethnobotony, organellar genomes, carbon metabolism. *Unit head:* Dr. Michael Parrish, Chair, 815-753-1753, Fax: 815-753-0461, E-mail: mparrish@niu.edu. *Application contact:* Dr. Carl von Ende, Director of Graduate Studies, 815-753-7826.

Northern Michigan University, College of Graduate Studies, College of Arts and Sciences, Department of Biology, Marquette, MI 49855-5301. Offers MS. Part-time programs available. Postbaccalaureate distance learning degree programs offered (minimal on-campus study). *Degree requirements:* For master's, thesis or alternative. *Entrance requirements:* For master's, GRE, minimum GPA of 3.0. *Faculty research:* Molecular genetics of sex-linked genes, biology of protozoan parasites, wildlife ecology, organochlorines in the environment, insect development.

Northwestern University, The Graduate School and Judd A. and Marjorie Weinberg College of Arts and Sciences, Interdepartmental Biological Sciences Program (IBiS), Evanston, IL 60208. Offers biochemistry, molecular biology, and cell biology (PhD), including biochemistry, cell and molecular biology, molecular biophysics, structural biology; biotechnology (PhD); cell and molecular biology (PhD); developmental biology and genetics (PhD); hormone action and signal transduction (PhD); neuroscience (PhD); structural biology, biochemistry, and biophysics (PhD). Participants in the Interdepartmental Biological Sciences Program include the Departments of Biochemistry, Molecular Biology, and Cell Biology; Chemistry; Neurobiology and Physiology; Chemical Engineering; Civil Engineering; and Evanston Hospital. *Entrance requirements:* For doctorate, GRE General Test. Additional exam requirements/recommendations for international students: Required—TOEFL (minimum score 600 paper-based),

Biological and Biomedical Sciences—General

TSE(minimum score 50). Electronic applications accepted. *Faculty research:* Developmental genetics, gene regulation, DNA-protein interactions, biological clocks, bioremediation.

Announcement: The IBiS Graduate Program is focused on one mission: to develop its PhD students into independent, creative research scientists, teachers, and professionals. The IBiS Graduate Program provides excellent research opportunities in an environment that develops and prepares the whole scientist. Central to this development is a rigorous curriculum and a number of life science career development programs that prepare the student for a variety of career options.

See Close-Up on page 187.

Northwestern University, Northwestern University Feinberg School of Medicine, Combined MD/PhD Medical Scientist Training Program, Chicago, IL 60611. Offers MD/PhD. Application must be made to both The Graduate School and the Medical School. *Accreditation:* LCME/AMA. *Students:* 76 full-time (31 women); includes 27 minority (2 African Americans, 2 American Indian/Alaska Native, 20 Asian Americans or Pacific Islanders, 3 Hispanic Americans). Average age 25. 458 applicants, 7% accepted, 12 enrolled. *Application deadline:* For fall admission, 10/15 for domestic students. Applications are processed on a rolling basis. Application fee: $70. Electronic applications accepted. *Financial support:* In 2005–06, 16 fellowships with full tuition reimbursements (averaging $24,500 per year) were awarded *Faculty research:* Cardiovascular epidemiology, cancer epidemiology, nutritional interventions for the prevention of cardiovascular disease and cancer, women's health, outcomes research. *Unit head:* David M. Engman, Director, 312-503-1288, E-mail: d-engman@northwestern.edu. *Application contact:* Dr. Sandra Lee, Associate Director, 312-503-5232, Fax: 312-908-5253, E-mail: mstp@northwestern.edu.

Northwestern University, Northwestern University Feinberg School of Medicine and Interdepartmental Degree Programs, Integrated Graduate Programs in the Life Sciences, Chicago, IL 60611. Offers cancer biology (PhD); cell biology (PhD); developmental biology (PhD); evolutionary biology (PhD); immunology and microbial pathogenesis (PhD); molecular biology and genetics (PhD); neurobiology (PhD); pharmacology and toxicology (PhD); structural biology and biochemistry (PhD). *Degree requirements:* For doctorate, thesis/dissertation, written and oral qualifying exams, comprehensive exam. *Entrance requirements:* For doctorate, GRE General Test. Additional exam requirements/recommendations for international students: Required—TOEFL (minimum score 600 paper-based; 250 computer-based). Electronic applications accepted.

See Close-Up on page 189.

Northwest Missouri State University, Graduate School, College of Arts and Sciences, Department of Biology, Maryville, MO 64468-6001. Offers MS. Part-time programs available. *Faculty:* 8 full-time (2 women). *Students:* 4 full-time (3 women). 6 applicants, 17% accepted, 1 enrolled. *Degree requirements:* For master's, thesis, comprehensive exam. *Entrance requirements:* For master's, GRE General Test, minimum GPA of 3.0 in last 60 hours or 2.75 overall, writing sample. Additional exam requirements/recommendations for international students: Required—TOEFL (minimum score 550 paper-based; 213 computer-based). *Application deadline:* For fall admission, 7/1 for domestic students, 7/1 for international students; for spring admission, 11/15 for domestic students, 11/15 for international students. Applications are processed on a rolling basis. Application fee: $0 ($50 for international students). *Expenses:* Tuition, state resident: full-time $2,077; part-time $231 per credit hour. Tuition, nonresident: full-time $3,650; part-time $406 per credit hour. Required fees: $105 per term. Tuition and fees vary according to campus/location and reciprocity agreements. *Financial support:* In 2005–06, research assistantships (averaging $5,500 per year), 2 teaching assistantships with full tuition reimbursements (averaging $5,500 per year) were awarded; tutorial assistantships also available. Financial award application deadline: 3/1; financial award applicants required to submit FAFSA. *Unit head:* Dr. Gregg Dieringer, Chairperson, 660-562-1812. *Application contact:* Dr. Frances Shipley, Dean of Graduate School, 660-562-1145, Fax: 660-562-1096, E-mail: gradsch@nwmissouri.edu.

Notre Dame de Namur University, Division of Academic Affairs, School of Sciences, Department of Natural Sciences, Belmont, CA 94002-1908. Offers premedical studies (Certificate). *Expenses:* Tuition: Full-time $11,790; part-time $655 per unit. Required fees: $30 per semester hour. *Unit head:* Dr. Neil Marshall, Chair, 650-508-3554. *Application contact:* Barbara Sterner, Assistant Director of Graduate Admissions, 650-508-3600, Fax: 650-508-3426, E-mail: grad.admit@ndnu.edu.

Nova Southeastern University, Health Professions Division, College of Allied Health and Nursing, Department of Physician Assistant Studies, Fort Lauderdale, FL 33314-7796. Offers medical science/physician assistant (MMS). Students enter program as undergraduates. *Faculty:* 15 full-time (4 women), 1 part-time/adjunct (0 women). *Students:* 198 full-time (152 women), 15 part-time (11 women); includes 45 minority (9 African Americans, 1 American Indian/Alaska Native, 15 Asian Americans or Pacific Islanders, 20 Hispanic Americans). Average age 27. 586 applicants, 20% accepted, 83 enrolled. In 2005, 75 degrees awarded. *Entrance requirements:* For master's, GRE, minimum GPA of 2.7. *Application deadline:* For winter admission, 12/31 for domestic students. Applications are processed on a rolling basis. Application fee: $120. Electronic applications accepted. *Expenses:* Contact institution. *Financial support:* In 2005–06, 85 students received support. *Unit head:* Bill Marquardt, Chair and Program Director, 954-262-1252, E-mail: marquardt@nsu.nova.edu. *Application contact:* Judy Dickman, Admissions Counselor, 954-262-1109, E-mail: dickman@nsu.nova.edu.

Nova Southeastern University, Health Professions Division, College of Medical Sciences, Fort Lauderdale, FL 33314-7796. Offers biomedical sciences (MBS). *Faculty:* 30 full-time (10 women), 1 (woman) part-time/adjunct. *Students:* 28 full-time (14 women), 26 part-time (17 women); includes 12 minority (5 African Americans, 2 Asian Americans or Pacific Islanders, 5 Hispanic Americans), 1 international. Average age 27. 108 applicants, 23% accepted. In 2005, 36 degrees awarded. *Degree requirements:* For master's, thesis. *Entrance requirements:* For master's, minimum GPA of 2.5. *Application deadline:* For spring admission, 4/15 for domestic students. Applications are processed on a rolling basis. Application fee: $50. *Expenses:* Contact institution. *Financial support:* Applicants required to submit FAFSA. *Faculty research:* Neurophysiology, mucosal immunology, allergies involving the lungs, cardiovascular physiology parasitology. Total annual research expenditures: $125,000. *Unit head:* Dr. Harold E. Laubach, Dean, 954-262-1303, Fax: 954-262-1802, E-mail: harold@nsu.nova.edu. *Application contact:* Doreen Palmer, Admissions Counselor, 954-262-1111, Fax: 954-262-2282, E-mail: medinfo@nsu.nova.edu.

Oakland University, Graduate Study and Lifelong Learning, College of Arts and Sciences, Department of Biological Sciences, Rochester, MI 48309-4401. Offers biological sciences (MA, MS); cellular biology of aging (MS). *Faculty:* 7 full-time (1 woman), 4 (woman) part-time/adjunct. *Students:* 10 full-time (5 women), 15 part-time (10 women), 1 international. Average age 26. 11 applicants, 73% accepted, 9 enrolled. In 2005, 7 degrees awarded. *Degree requirements:* For master's, thesis. *Entrance requirements:* For master's, GRE Subject Test, GRE General Test, minimum GPA of 3.0 for unconditional admission. Additional exam requirements/recommendations for international students: Required—TOEFL (minimum score 550 paper-based; 213 computer-based). *Application deadline:* For fall admission, 7/15 priority date for domestic students, 5/1 priority date for international students. For winter admission, 12/1 for domestic students; for spring admission, 3/15 for domestic students. Applications are processed on a rolling basis. Application fee: $30. Electronic applications accepted. *Expenses:* Contact institution. *Financial support:* Federal Work-Study, institutionally sponsored loans, and tuition waivers (full) available. Financial award application deadline: 3/1; financial award applicants required to submit FAFSA. *Faculty research:* Mechanism producing rhythmic beating in cilia and flagella, biosynthesis in Rhodobacter spaeroides, promotor escape by RNA polyerase II, biochemical characterization of carbofuron hydroxylase. Total annual research expenditures: $433,454. *Unit head:* Dr. Arik Dvir, Chair, 248-370-3375, Fax: 248-370-4225.

Application contact: Dr. Keith Berven, Coordinator, 248-370-3581, Fax: 248-370-4225, E-mail: berven@oakland.edu.

Occidental College, Graduate Studies, Department of Biology, Los Angeles, CA 90041-3314. Offers MA. Part-time programs available. *Degree requirements:* For master's, thesis, final exam. *Entrance requirements:* For master's, GRE General Test, GRE Subject Test, minimum GPA of 3.0. Additional exam requirements/recommendations for international students: Required—TOEFL (minimum score 625 paper-based; 263 computer-based). Expenses: Contact institution.

The Ohio State University, College of Medicine and Public Health and Graduate School, Graduate Programs in the Basic Medical Sciences, Integrated Biomedical Science Graduate Program, Columbus, OH 43210. Offers immunology (MS, PhD); medical genetics (MS, PhD); molecular virology (MS, PhD); pharmacology (MS, PhD). *Degree requirements:* For doctorate, thesis/dissertation. *Entrance requirements:* For master's, GRE General Test; for doctorate, GRE. Additional exam requirements/recommendations for international students: Required—TOEFL (minimum score 600 paper-based; 250 computer-based), TSE. Electronic applications accepted.

The Ohio State University, Graduate School, College of Biological Sciences, Columbus, OH 43210. Offers MS, PhD. Part-time programs available. *Degree requirements:* For doctorate, thesis/dissertation. *Entrance requirements:* For master's and doctorate, GRE General Test, GRE Subject Test in biology or biochemistry (recommended). Additional exam requirements/recommendations for international students: Required—TOEFL (minimum score 600 paper-based; 250 computer-based), TSE. Electronic applications accepted.

Ohio University, Graduate Studies, College of Arts and Sciences, Department of Biological Sciences, Athens, OH 45701-2979. Offers biological sciences (MS, PhD); cell biology and physiology (MS, PhD); ecology and evolutionary biology (MS, PhD); exercise physiology and muscle biology (MS, PhD); microbiology (MS, PhD); neuroscience (MS, PhD). *Faculty:* 51 full-time (17 women), 6 part-time/adjunct (1 woman). *Students:* 88 full-time (40 women), 1 part-time; includes 1 minority (Hispanic American), 41 international. Average age 24. 50 applicants, 24% accepted, 10 enrolled. In 2005, 9 master's, 12 doctorates awarded. *Median time to degree:* Of those who began their doctoral program in fall 1997, 90% received their degree in 8 years or less. *Degree requirements:* For master's, thesis, 1 quarter of teaching experience; for doctorate, thesis/dissertation, 2 quarters of teaching experience, comprehensive exam. *Entrance requirements:* For master's and doctorate, GRE General Test. Additional exam requirements/recommendations for international students: Required—TOEFL (minimum score 620 paper-based; 260 computer-based). *Application deadline:* For fall admission, 1/15 for domestic students, 1/15 for international students. Application fee: $45. Electronic applications accepted. *Financial support:* In 2005–06, 87 students received support, including 2 fellowships with full tuition reimbursements available (averaging $15,000 per year), 10 research assistantships with full tuition reimbursements available (averaging $15,500 per year), 75 teaching assistantships with full tuition reimbursements available (averaging $15,500 per year); Federal Work-Study, institutionally sponsored loans, and tuition waivers (full) also available. Financial award application deadline: 1/15. *Faculty research:* Ecology and evolutionary biology, exercise physiology and muscle biology, neurobiology, cell biology, physiology. Total annual research expenditures: $2.8 million. *Unit head:* Dr. Ralph DiCaprio, Chair, 740-593-2290, Fax: 740-593-0300, E-mail: dicaprir@ohio.edu. *Application contact:* Dr. Donald B. Miles, Graduate Chair, 740-593-2317, Fax: 740-593-0300, E-mail: milesd@ohio.edu.

See Close-Up on page 191.

Oklahoma State University Center for Health Sciences, Program in Biomedical Sciences, Tulsa, OK 74107-1898. Offers MS, PhD, DO/PhD. *Faculty:* 27 full-time (5 women), 15 part-time/adjunct (4 women). *Students:* 7 full-time (3 women), 5 part-time (1 woman). Average age 31. In 2005, 1 degree awarded. *Degree requirements:* For master's, thesis/dissertation; for doctorate, thesis/dissertation, comprehensive exam. *Entrance requirements:* For master's, GRE General Test; for doctorate, GRE General Test. Additional exam requirements/recommendations for international students: Required—TOEFL. *Application deadline:* Applications are processed on a rolling basis. Application fee: $40 ($75 for international students). *Expenses:* Tuition, state resident: full-time $16,045. Tuition, nonresident: full-time $31,265. Tuition and fees vary according to program. *Financial support:* In 2005–06, 3 research assistantships with partial tuition reimbursements (averaging $17,000 per year) were awarded; scholarships/grants and tuition waivers (partial) also available. Financial award application deadline: 3/31; financial award applicants required to submit FAFSA. *Faculty research:* Neuroscience, cell biology, cell signaling, infectious disease, parasitology, virology. Total annual research expenditures: $900,000. *Unit head:* Dr. Earl L. Blewett, Director, 918-561-8405, Fax: 918-561-8276. *Application contact:* Leah Haines, Associate Director of Admissions and Student Records, 800-677-1972, Fax: 918-561-8243, E-mail: ldhaines@chs.okstate.edu.

Old Dominion University, College of Sciences, Program in Biology, Norfolk, VA 23529. Offers MS. Part-time programs available. *Faculty:* 20 full-time (2 women), 23 part-time/adjunct (2 women). *Students:* 12 full-time (9 women), 26 part-time (14 women), 1 international. Average age 26. 31 applicants, 71% accepted, 18 enrolled. In 2005, 12 degrees awarded. *Degree requirements:* For master's, thesis optional. *Entrance requirements:* For master's, GRE General Test, MCAT, minimum GPA of 3.0 in major, 2.7 overall, faculty advisor. Additional exam requirements/recommendations for international students: Required—TOEFL (minimum score 550 paper-based; 213 computer-based). *Application deadline:* For fall admission, 2/1 priority date for domestic students, 2/1 priority date for international students. For winter admission, 6/1 for domestic students; for spring admission, 10/1 for domestic students. Application fee: $40. Electronic applications accepted. *Expenses:* Tuition, state resident: part-time $263 per credit hour. Tuition, nonresident: part-time $661 per credit hour. Required fees: $39 per semester. Part-time tuition and fees vary according to campus/location. *Financial support:* In 2005–06, 2 fellowships (averaging $6,575 per year), 10 research assistantships with partial tuition reimbursements (averaging $15,000 per year), 6 teaching assistantships with partial tuition reimbursements (averaging $15,000 per year) were awarded; career-related internships or fieldwork and scholarships/grants also available. Support available to part-time students. Financial award application deadline: 2/1; financial award applicants required to submit FAFSA. *Faculty research:* Wetland ecology, systematics and ecology of vertebrates, marine biology, molecular and cellular microbiology, physiological and reproductive biology. Total annual research expenditures: $2 million. *Unit head:* Dr. Wayne L. Hynes, Graduate Program Director, 757-683-3613, Fax: 757-683-5283, E-mail: biolgpd@odu.edu.

Old Dominion University, College of Sciences, Program in Biomedical Sciences, Norfolk, VA 23529. Offers PhD. *Faculty:* 27 full-time (5 women). *Students:* 43 full-time (32 women), 18 part-time (6 women); includes 13 minority (6 African Americans, 2 American Indian/Alaska Native, 3 Asian Americans or Pacific Islanders, 2 Hispanic Americans), 23 international. Average age 30. 27 applicants, 41% accepted, 7 enrolled. In 2005, 5 degrees awarded. *Degree requirements:* For doctorate, thesis/dissertation, comprehensive exam. *Entrance requirements:* For doctorate, GRE General Test, minimum GPA of 3.0. Additional exam requirements/recommendations for international students: Required—TOEFL (minimum score 213 computer-based). *Application deadline:* For fall admission, 2/15 priority date for domestic students, 2/15 priority date for international students. Application fee: $40. Electronic applications accepted. *Expenses:* Tuition, state resident: part-time $263 per credit hour. Tuition, nonresident: part-time $661 per credit hour. Required fees: $39 per semester. Part-time tuition and fees vary according to campus/location. *Financial support:* In 2005–06, 38 students received support, including 2 fellowships with full tuition reimbursements available (averaging $18,000 per year), 2 research assistantships with full tuition reimbursements available (averaging $18,000 per year), 4 teaching assistantships with full tuition reimbursements available (averaging $15,000 per year); career-related internships or fieldwork, scholarships/grants, tuition waivers (partial), and unspecified assistantships also available. Support available to part-time students. Financial award application deadline: 2/15; financial award applicants required to submit FAFSA. *Faculty research:* Systems biology and biophysics, pure and

Biological and Biomedical Sciences—General

Old Dominion University (continued)
applied biomedical sciences, biological chemistry, clinical chemistry, cell biology and molecular pathogenesis. Total annual research expenditures: $3.7 million. *Unit head:* Dr. R. James Swanson, Graduate Program Director, 757-683-3614, Fax: 757-683-5283, E-mail: bimdgpd@odu.edu.

Old Dominion University, Darden College of Education, Programs in Secondary Education, Norfolk, VA 23529. Offers biology (MS Ed); chemistry (MS Ed); English (MS Ed); instructional technology (MS Ed); library science (MS Ed); secondary education (MS Ed). *Accreditation:* NCATE. Part-time and evening/weekend programs available. Postbaccalaureate distance learning degree programs offered (minimal on-campus study). *Faculty:* 28 full-time (11 women). *Students:* 50 full-time (37 women), 149 part-time (93 women); includes 24 minority (15 African Americans, 2 American Indian/Alaska Native, 2 Asian Americans or Pacific Islanders, 5 Hispanic Americans), 2 international. Average age 37. 44 applicants, 95% accepted. In 2005, 119 degrees awarded. *Degree requirements:* For master's, thesis optional. *Entrance requirements:* For master's, GRE General Test, or MAT, PRAXIS I for master's with licensure, minimum GPA of 2.8, teaching certificate. Additional exam requirements/recommendations for international students: Required—TOEFL. *Application deadline:* Applications are processed on a rolling basis. Application fee: $40. Electronic applications accepted. *Expenses:* Tuition, state resident: part-time $263 per credit hour. Tuition, nonresident: part-time $661 per credit hour. Required fees: $39 per semester. Part-time tuition and fees vary according to campus/location. *Financial support:* In 2005–06, 58 students received support, including 2 research assistantships with tuition reimbursements available ($6,777 per year), 3 teaching assistantships with tuition reimbursements available (averaging $5,333 per year); fellowships, career-related internships or fieldwork, Federal Work-Study, institutionally sponsored loans, scholarships/grants, and tuition waivers (partial) also available. Support available to part-time students. Financial award application deadline: 2/15; financial award applicants required to submit FAFSA. *Faculty research:* Mathematics retraining, writing project for teachers, geography teaching, reading. *Unit head:* Dr. Robert Lucking, Graduate Program Director, 757-683-5545, Fax: 757-683-5862, E-mail: ecisgpd@odu.edu.

Oregon Health & Science University, School of Medicine, Graduate Programs in Medicine, Portland, OR 97239-3098. Offers MS, PhD, Certificate, MD/PhD. Part-time programs available. Terminal master's awarded for partial completion of doctoral program. *Degree requirements:* For master's, thesis, capstone experience; for doctorate, thesis/dissertation, qualifying exam. *Entrance requirements:* For master's and doctorate, GRE General Test. Expenses: Contact institution.

Oregon Health & Science University, School of Medicine, Integrative Biomedical Sciences Program, Portland, OR 97239-3098. Offers PhD, MD/PhD. *Degree requirements:* For doctorate, thesis/dissertation. *Entrance requirements:* For doctorate, GRE General Test. Additional exam requirements/recommendations for international students: Required—TOEFL. *Faculty research:* Physiology, geophysics, pharmacology, electrophysiology, molecular/cell biology.

The Pennsylvania State University Milton S. Hershey Medical Center, Graduate School Programs in the Biomedical Sciences, Hershey, PA 17033-2360. Offers MS, PhD, MD/PhD, PhD/MBA. *Students:* 727 applicants, 13% accepted, 44 enrolled. In 2005, 19 master's, 27 doctorates awarded. Terminal master's awarded for partial completion of doctoral program. *Degree requirements:* For master's, thesis or alternative, registration; for doctorate, thesis/dissertation, oral exam, comprehensive exam, registration. *Entrance requirements:* For doctorate, GRE, minimum GPA of 3.0. Additional exam requirements/recommendations for international students: Required—TOEFL (minimum score 560 paper-based; 220 computer-based). *Application deadline:* For fall admission, 2/1 priority date for domestic students, 2/1 priority date for international students. Applications are processed on a rolling basis. Application fee: $45. Electronic applications accepted. *Expenses:* Contact institution. *Financial support:* Fellowships with full tuition reimbursements, research assistantships with full tuition reimbursements, teaching assistantships with tuition reimbursements, scholarships/grants, health care benefits, tuition waivers (full), and unspecified assistantships available. Financial award applicants required to submit FAFSA. *Unit head:* Dr. Michael F. Verderame, Associate Dean for Graduate Studies, 717-531-3892, Fax: 717-531-4139, E-mail: grad-hmc@psu.edu. *Application contact:* Kathleen M. Simon, Administrative Assistant, 717-531-8892, Fax: 717-531-4139, E-mail: grad-hmc@psu.edu.

See Close-Up on page 195.

The Pennsylvania State University University Park Campus, Graduate School, College of Earth and Mineral Sciences, Department of Geosciences, State College, University Park, PA 16802-1503. Offers astrobiology (PhD); earth sciences (M Ed); geosciences (MS, PhD). *Students:* 98 full-time (41 women), 5 part-time (1 woman); includes 7 minority (1 African American, 2 American Indian/Alaska Native, 1 Asian American or Pacific Islander, 3 Hispanic Americans), 19 international. *Entrance requirements:* For master's and doctorate, GRE General Test. Additional exam requirements/recommendations for international students: Required—TOEFL. *Expenses:* Tuition, state resident: full-time $12,518; part-time $522 per credit. Tuition, nonresident: full-time $23,004; part-time $959 per credit. Required fees: $484. Tuition and fees vary according to course load, campus/location and program. *Unit head:* Dr. Timothy J. Bralower, Professor of Geosciences, 814-863-8177, Fax: 814-863-7823.

The Pennsylvania State University University Park Campus, Graduate School, Eberly College of Science, Department of Biology, State College, University Park, PA 16802-1503. Offers biology (MS, PhD); molecular evolutionary biology (MS, PhD). *Students:* 51 full-time (24 women), 1 part-time; includes 5 minority (2 African Americans, 2 Asian Americans or Pacific Islanders, 1 Hispanic American), 30 international. *Entrance requirements:* For master's and doctorate, GRE General Test. Application fee: $45. *Expenses:* Tuition, state resident: full-time $12,518; part-time $522 per credit. Tuition, nonresident: full-time $23,004; part-time $959 per credit. Required fees: $484. Tuition and fees vary according to course load, campus/location and program. *Financial support:* Fellowships, research assistantships, teaching assistantships available. *Unit head:* Dr. Douglas R. Cavener, Head, 814-865-5497, Fax: 814-865-9131, E-mail: drc9@psu.edu. *Application contact:* Dr. Douglas R. Cavener, Head, 814-865-5497, Fax: 814-865-9131, E-mail: drc9@psu.edu.

See Close-Up on page 193.

The Pennsylvania State University University Park Campus, Graduate School, Intercollege Graduate Programs, Intercollege Graduate Program in Integrative Biosciences, State College, University Park, PA 16802-1503. Offers integrative biosciences (PhD), including biomolecular transport dynamics, cell and developmental biology, cellular and molecular mechanisms of toxicity, chemical biology, ecological and molecular plant physiology, immunobiology, molecular medicine, neuroscience, nutrition science. *Students:* 110 full-time (58 women), 2 part-time (both women); includes 6 minority (3 African Americans, 3 Asian Americans or Pacific Islanders), 61 international. *Entrance requirements:* For master's and doctorate, GRE General Test. Application fee: $45. *Expenses:* Tuition, state resident: full-time $12,518; part-time $522 per credit. Tuition, nonresident: full-time $23,004; part-time $959 per credit. Required fees: $484. Tuition and fees vary according to course load, campus/location and program. *Financial support:* Fellowships available. *Unit head:* Dr. Richard J. Frisque, Co-Director, 814-863-3523, Fax: 814-863-1357, E-mail: rjf6@psu.edu.

See Close-Up on page 197.

Philadelphia College of Osteopathic Medicine, Graduate and Professional Programs, Program in Biomedical Sciences, Philadelphia, PA 19131-1694. Offers MS, Certificate. *Faculty:* 28 full-time (14 women), 5 part-time/adjunct (3 women). *Students:* 91 full-time (50 women); includes 34 minority (23 African Americans, 9 Asian Americans or Pacific Islanders, 2 Hispanic Americans). Average age 26. 249 applicants, 38% accepted, 64 enrolled. In 2005, 16 degrees awarded. *Degree requirements:* For master's, thesis, registration. *Entrance requirements:* For master's, GRE or MCAT, minimum GPA of 3.0, course work in biology, chemistry,

English, physics. *Application deadline:* For fall admission, 7/15 for domestic students. Applications are processed on a rolling basis. Application fee: $50. *Faculty research:* Developmental biology, cytokine and inflammation, neurobiology of aging, pain mechanisms, cell death. Total annual research expenditures: $244,095. *Unit head:* Dr. Ruth D. Thornton, Chair, 215-871-6440, Fax: 215-871-6865, E-mail: rutht@pcom.edu. *Application contact:* Carol A. Fox, Associate Vice President for Enrollment Management, 215-871-6700, Fax: 215-871-6719, E-mail: carolf@pcom.edu.

Pittsburg State University, Graduate School, College of Arts and Sciences, Department of Biology, Pittsburg, KS 66762. Offers MS. *Students:* 9. *Degree requirements:* For master's, thesis or alternative. *Entrance requirements:* For master's, GRE ($60 for international students). *Expenses:* Tuition, state resident: full-time $2,015; part-time $170 per credit hour. Tuition, nonresident: full-time $4,953; part-time $415 per credit hour. Tuition and fees vary according to course load, campus/location and program. *Financial support:* Research assistantships, teaching assistantships, career-related internships or fieldwork and Federal Work-Study available. *Unit head:* Dr. James Triplett, Chairperson, 620-235-4730. *Application contact:* Marvene Darraugh, Administrative Officer, 620-235-4220, Fax: 620-235-4219, E-mail: mdarraug@pittstate.edu.

Ponce School of Medicine, Program in Biomedical Sciences, Ponce, PR 00732-7004. Offers PhD. *Faculty:* 51 full-time (23 women), 18 part-time/adjunct (8 women). *Students:* 29 full-time (21 women); all minorities (all Hispanic Americans) Average age 29. 20 applicants, 50% accepted, 9 enrolled. In 2005, 3 degrees awarded. *Degree requirements:* For doctorate, one foreign language, thesis/dissertation, comprehensive exam, registration. *Entrance requirements:* For doctorate, GRE General Test, proficiency in Spanish and English, minimum overall GPA of 2.75, 3 letters of recommendation, minimum 35 credits in science. *Application deadline:* For fall admission, 3/15 for domestic students, 3/15 for international students. Application fee: $100. *Expenses:* Tuition: Part-time $245 per credit. Required fees: $1,779 per year. Full-time tuition and fees vary according to degree level and program. *Financial support:* In 2005–06, 3 fellowships with full tuition reimbursements (averaging $6,560 per year), 16 research assistantships with full tuition reimbursements (averaging $9,308 per year) were awarded; scholarships/grants also available. Financial award application deadline: 5/30; financial award applicants required to submit FAFSA. *Unit head:* Dr. José Torres, Associate Dean for Graduate Studies and Research, 787-840-2158, E-mail: jtorres@psm.edu.

Portland State University, Graduate Studies, College of Liberal Arts and Sciences, Department of Biology, Portland, OR 97207-0751. Offers MA, MS, PhD. *Faculty:* 19 full-time (6 women), 4 part-time/adjunct (1 woman). *Students:* 28 full-time (14 women), 7 part-time (3 women); includes 3 minority (1 African American, 2 Asian Americans or Pacific Islanders), 1 international. Average age 29. 18 applicants, 83% accepted, 10 enrolled. In 2005, 10 degrees awarded. *Degree requirements:* For master's, one foreign language, thesis; for doctorate, thesis/dissertation. *Entrance requirements:* For master's, GRE General Test, GRE Subject Test, minimum GPA of 3.0 in upper-division course work or 2.75 overall, 2 letters of reference; for doctorate, GRE General Test, GRE Subject Test, minimum GPA of 3.5 in science. Additional exam requirements/recommendations for international students: Required—TOEFL (minimum score 550 paper-based; 213 computer-based). *Application deadline:* For fall admission, 4/1 for domestic students, 3/1 for international students. For winter admission, 9/1 for domestic students; for spring admission, 11/1 for domestic students. Applications are processed on a rolling basis. Application fee: $50. *Expenses:* Tuition, state resident: full-time $6,648; part-time $231 per credit. Tuition, nonresident: full-time $11,319; part-time $231 per credit. Required fees: $686; $67 per credit. *Financial support:* In 2005–06, 3 research assistantships with full tuition reimbursements (averaging $16,403 per year), 4 teaching assistantships with full tuition reimbursements (averaging $14,608 per year) were awarded; Federal Work-Study, scholarships/grants, tuition waivers (partial), and unspecified assistantships also available. Support available to part-time students. Financial award application deadline: 3/1; financial award applicants required to submit FAFSA. *Faculty research:* Genetic diversity and natural population, vertebrate temperature regulation, water balance and sensory physiology, trace elements and aquatic ecology, molecular genetics. Total annual research expenditures: $3.3 million. *Unit head:* Dr. Stanley Hillman, Chair, 503-725-3851.

Prairie View A&M University, Graduate School, College of Arts and Sciences, Department of Biology, Prairie View, TX 77446-0519. Offers MS. *Students:* 1 (woman) full-time; minority (African American) Average age 37. 1 applicant, 100% accepted, 1 enrolled. *Degree requirements:* For master's, thesis, comprehensive exam. *Entrance requirements:* For master's, GRE General Test, BS in biology or equivalent. Additional exam requirements/recommendations for international students: Required—TOEFL. *Application deadline:* Applications are processed on a rolling basis. Application fee: $25. *Expenses:* Tuition, state resident: full-time $1,440; part-time $80 per credit. Tuition, nonresident: full-time $6,444; part-time $358 per credit. *Financial support:* Federal Work-Study available. Financial award application deadline: 4/1; financial award applicants required to submit FAFSA. *Faculty research:* Geonomics, hypertension, control of gene express, proteins, kigands that interact with hormone receptos. Total annual research expenditures: $10,000. *Unit head:* Dr. Harriet E. Howard-Lee, Interim Head, 936-857-3911, Fax: 936-857-4944, E-mail: george_brown@pvamu.edu. *Application contact:* Dr. Seab A. Smith, Associate Professor, 936-857-3911, Fax: 936-857-4944, E-mail: seab_smith@pvamu.edu.

Princeton University, Graduate School, Department of Ecology and Evolutionary Biology, Princeton, NJ 08544-1019. Offers biology (PhD); neuroscience (PhD). *Degree requirements:* For doctorate, thesis/dissertation. *Entrance requirements:* For doctorate, GRE General Test, GRE Subject Test. Additional exam requirements/recommendations for international students: Required—TOEFL (minimum score 600 paper-based; 250 computer-based). Electronic applications accepted.

See Close-Up on page 705.

Purdue University, Graduate School, PULSe—Purdue University Life Sciences Program, West Lafayette, IN 47907. Offers PhD. *Students:* 143 full-time (86 women); includes 3 African Americans, 7 Asian Americans or Pacific Islanders, 5 Hispanic Americans, 88 international. 352 applicants, 21% accepted, 21 enrolled. *Entrance requirements:* For doctorate, GRE. Additional exam requirements/recommendations for international students: Required—TOEFL. *Application deadline:* For fall admission, 2/1 for domestic students, 2/1 for international students. Applications are processed on a rolling basis. Application fee: $55. Electronic applications accepted. *Unit head:* Dr. Colleen Gabauer, Director. *Application contact:* Elizabeth Ann Chandler, Coordinator, 765-494-1634, Fax: 765-496-1475.

Purdue University, Graduate School, School of Science, Department of Biological Sciences, West Lafayette, IN 47907. Offers biochemistry (PhD); biophysics (PhD); cell and developmental biology (PhD); ecology, evolutionary and population biology (MS, PhD), including ecology, evolutionary biology, population biology; genetics (MS, PhD); microbiology (MS, PhD); molecular biology (PhD); neurobiology (MS, PhD); plant physiology (PhD). *Faculty:* 47 full-time (9 women), 4 part-time/adjunct (1 woman). *Students:* 97 full-time (53 women), 8 part-time (4 women); includes 13 minority (3 African Americans, 1 American Indian/Alaska Native, 4 Asian Americans or Pacific Islanders, 5 Hispanic Americans), 50 international. Average age 28. 168 applicants, 29% accepted, 23 enrolled. In 2005, 18 master's, 9 doctorates awarded. Terminal master's awarded for partial completion of doctoral program. *Degree requirements:* For master's, thesis (for some programs); for doctorate, thesis/dissertation, seminars, teaching experience. *Entrance requirements:* For master's and doctorate, GRE General Test. Additional exam requirements/recommendations for international students: Required—TOEFL, TSE. *Application deadline:* For fall admission, 2/15 for domestic students, 1/31 for international students. Applications are processed on a rolling basis. Application fee: $55. Electronic applications accepted. *Financial support:* In 2005–06, 15 fellowships, 60 research assistantships, 53 teaching assistantships were awarded. Support available to part-time students. Financial award application deadline: 2/15; financial award applicants required to submit FAFSA. *Unit head:* Dr. Richard J Kuhn, Head, 765-494-4407. *Application contact:* Nancy Konopka, Graduate Studies Office Manager, 765-494-8142, Fax: 765-494-0876, E-mail: njk@bilbo.bio.purdue.edu.

Biological and Biomedical Sciences—General

Purdue University Calumet, Graduate School, School of Engineering, Mathematics, and Science, Department of Biological Sciences, Hammond, IN 46323-2094. Offers biology (MS); biology teaching (MS); biotechnology (MS). *Entrance requirements:* For master's, GRE. Additional exam requirements/recommendations for international students: Required—TOEFL. Electronic applications accepted. *Faculty research:* Cell biology, molecular biology, genetics, microbiology, neurophysiology.

Queens College of the City University of New York, Division of Graduate Studies, Mathematics and Natural Sciences Division, Department of Biology, Flushing, NY 11367-1597. Offers MA. Part-time and evening/weekend programs available. *Faculty:* 18 full-time (6 women). *Students:* 28 applicants, 93% accepted, 21 enrolled. In 2005, 9 degrees awarded. *Degree requirements:* For master's, thesis or alternative, qualifying exam, comprehensive exam. *Entrance requirements:* For master's, minimum GPA of 3.0. Additional exam requirements/recommendations for international students: Required—TOEFL. *Application deadline:* For fall admission, 4/1 for domestic students; for spring admission, 11/1 for domestic students. Applications are processed on a rolling basis. Application fee: $125. *Expenses:* Tuition, state resident: part-time $270 per credit. Tuition, nonresident: part-time $500 per credit. Required fees: $112 per year. *Financial support:* Career-related internships or fieldwork, Federal Work-Study, institutionally sponsored loans, tuition waivers (partial), and unspecified assistantships available. Support available to part-time students. Financial award application deadline: 4/1; financial award applicants required to submit FAFSA. *Faculty research:* Cell biology, evolutionary biology, environmental biology, microbiology. *Unit head:* Dr. Corrine Michels, Chairperson, 718-997-3400, E-mail: corinne_michels@qc.edu. *Application contact:* Dr. Jeanne Szalay, Graduate Adviser, 718-997-3400, E-mail: jeanne_szalay@qc.edu.

Queen's University at Kingston, School of Graduate Studies and Research, Faculty of Arts and Sciences, Department of Biology, Kingston, ON K7L 3N6, Canada. Offers M Sc, PhD. Part-time programs available. *Degree requirements:* For master's, thesis/dissertation; for doctorate, thesis/dissertation, comprehensive exam. *Entrance requirements:* Additional exam requirements/recommendations for international students: Required—TOEFL. *Faculty research:* Limnology, plant morphogenesis, nitrogen fixation, cell cycle, genetics.

Quinnipiac University, School of Health Sciences, Programs in Medical Laboratory Sciences, Hamden, CT 06518-1940. Offers biomedical sciences (MHS); laboratory management (MHS); microbiology (MHS). *Accreditation:* NAACLS. Part-time and evening/weekend programs available. *Faculty:* 2 full-time (0 women), 3 part-time/adjunct (2 women). *Students:* 21 full-time (15 women), 26 part-time (20 women); includes 6 minority (3 African Americans, 2 Asian Americans or Pacific Islanders, 1 Hispanic American), 2 international. Average age 29. 31 applicants, 77% accepted, 17 enrolled. In 2005, 10 degrees awarded. *Degree requirements:* For master's, thesis optional. *Entrance requirements:* For master's, minimum GPA of 2.5; bachelor's degree in biological, medical, or health sciences. Additional exam requirements/recommendations for international students: Required—TOEFL (minimum score 575 paper-based; 233 computer-based). *Application deadline:* For fall admission, 7/30 priority date for domestic students, 5/30 priority date for international students; for spring admission, 12/15 priority date for domestic students, 10/15 priority date for international students. Applications are processed on a rolling basis. Application fee: $45. Electronic applications accepted. *Expenses:* Tuition: Part-time $570 per credit. *Financial support:* Tuition waivers (partial) and unspecified assistantships available. Support available to part-time students. Financial award application deadline: 4/15; financial award applicants required to submit FAFSA. *Faculty research:* Microbial physiology, fermentation technology. *Unit head:* Dr. Kenneth Kaloustian, Director, 203-582-8676, Fax: 203-582-3443, E-mail: ken.kaloustian@quinnipiac.edu. *Application contact:* Louise Howe, Associate Director of Graduate Admissions, 800-462-1944, Fax: 203-582-3443, E-mail: graduate@quinnipiac.edu.

Rensselaer Polytechnic Institute, Graduate School, School of Science, Department of Biology, Troy, NY 12180-3590. Offers biochemistry (MS, PhD); biophysics (MS, PhD); cell biology (MS, PhD); developmental biology (MS, PhD); microbiology (MS, PhD); molecular biology (MS, PhD). Part-time programs available. Terminal master's awarded for partial completion of doctoral program. *Degree requirements:* For master's and doctorate, thesis/dissertation, comprehensive exam, registration. *Entrance requirements:* For master's and doctorate, GRE General Test. Additional exam requirements/recommendations for international students: Required—TOEFL. Electronic applications accepted. *Expenses:* Tuition: Full-time $31,000; part-time $1,320 per credit. Required fees: $1,623. *Faculty research:* Bioinformatics, molecular biology/biochemistry, cell and tissue biology, environment, ecology.

See Close-Up on page 199.

Rhode Island College, School of Graduate Studies, Faculty of Arts and Sciences, Department of Biology, Providence, RI 02908-1991. Offers MA. *Faculty:* 12 full-time (4 women). *Students:* Average age 29. *Degree requirements:* For master's, thesis (for some programs). *Entrance requirements:* For master's, GRE General Test and GRE Subject Test or MAT. *Application deadline:* For fall admission, 4/1 for domestic students. Applications are processed on a rolling basis. Application fee: $50. *Expenses:* Tuition, state resident: part-time $227 per credit hour. Tuition, nonresident: part-time $475 per credit hour. Required fees: $14 per credit hour. *Financial support:* Career-related internships or fieldwork available. Financial award application deadline: 4/1. *Unit head:* Dr. Edythe Anthony, Chair, 401-456-8010.

Rochester Institute of Technology, Graduate Enrollment Services, College of Science, Department of Biological Sciences, Rochester, NY 14623-5603. Offers MS. *Students:* 24 full-time (14 women), 2 part-time (both women); includes 5 minority (1 African American, 3 Asian Americans or Pacific Islanders, 1 Hispanic American), 7 international. 33 applicants, 42% accepted, 5 enrolled. In 2005, 6 degrees awarded. *Expenses:* Tuition: Full-time $25,392; part-time $713 per credit. Required fees: $183; $61 per term. *Unit head:* Douglas Merrill, Director, 585-475-2496, E-mail: dpmsbi@rit.edu.

The Rockefeller University, Program in Biomedical Sciences, New York, NY 10021-6399. Offers PhD, MD/PhD. *Faculty:* 117 full-time (31 women), 186 part-time/adjunct (45 women). *Students:* 200 full-time (93 women); includes 39 minority (10 African Americans, 1 American Indian/Alaska Native, 23 Asian Americans or Pacific Islanders, 5 Hispanic Americans), 77 international. Average age 25. 690 applicants, 10% accepted, 26 enrolled. In 2005, 22 degrees awarded. *Degree requirements:* For doctorate, thesis/dissertation. *Entrance requirements:* Additional exam requirements/recommendations for international students: Required—TOEFL. *Application deadline:* For fall admission, 1/1 for domestic students, 1/1 for international students. For winter admission, 12/1 for domestic students. Application fee: $75. Electronic applications accepted. *Financial support:* In 2005–06, 200 fellowships with full tuition reimbursements (averaging $25,500 per year) were awarded; institutionally sponsored loans, scholarships/grants, traineeships, and health care benefits also available. *Unit head:* Dr. Sidney Strickland, Dean of Graduate Studies, 212-327-8086, Fax: 212-327-8505, E-mail: phd@rockefeller.edu. *Application contact:* Kristen Cullen, Admissions and Records Administrator, 212-327-8088, Fax: 212-327-8505, E-mail: cullenk@rockefeller.edu.

See Close-Up on page 201.

Rosalind Franklin University of Medicine and Science, Integrated Bioscience Program, North Chicago, IL 60064-3095.

Rosalind Franklin University of Medicine and Science, School of Graduate and Postdoctoral Studies, North Chicago, IL 60064-3095. Offers MS, PhD, MD/MS, MD/PhD. Part-time programs available. *Degree requirements:* For doctorate, thesis/dissertation. *Entrance requirements:* For master's and doctorate, GRE General Test. Additional exam requirements/recommendations for international students: Required—TOEFL, TWE. *Faculty research:* Extracellular matrix, nutrition and mood, neuropsychopharmacology, membrane transport, brain metabolism.

See Close-Up on page 203.

Rutgers, The State University of New Jersey, Camden, Graduate School of Arts and Sciences, Program in Biology, Camden, NJ 08102-1401. Offers MS. Part-time and evening/weekend programs available. *Degree requirements:* For master's, thesis (for some programs), comprehensive exam (for some programs), registration. *Entrance requirements:* For master's, GRE General Test, GRE Subject Test (recommended). Additional exam requirements/recommendations for international students: Required—TOEFL. Electronic applications accepted. *Faculty research:* Neurobiology, biochemistry, ecology, developmental biology, biological signaling mechanisms.

Rutgers, The State University of New Jersey, Newark, Graduate School, Program in Biology, Newark, NJ 07102. Offers MS, PhD. Part-time and evening/weekend programs available. *Faculty:* 27 full-time (7 women), 4 part-time/adjunct (1 woman). *Students:* 40 full-time (22 women), 49 part-time (32 women); includes 36 minority (7 African Americans, 22 Asian Americans or Pacific Islanders, 7 Hispanic Americans). 119 applicants, 63% accepted, 34 enrolled. In 2005, 22 master's, 2 doctorates awarded. Terminal master's awarded for partial completion of doctoral program. *Degree requirements:* For master's, thesis optional; for doctorate, thesis/dissertation, qualifying exam. *Entrance requirements:* For master's, GRE General Test, minimum undergraduate B average; for doctorate, GRE General Test, GRE Subject Test, minimum B average. *Application deadline:* For fall admission, 2/15 for domestic students; for spring admission, 12/1 for domestic students. Applications are processed on a rolling basis. Application fee: $50. Electronic applications accepted. *Expenses:* Tuition, state resident: full-time $10,440; part-time $435 per credit. Tuition, nonresident: full-time $15,520; part-time $637 per credit. *Financial support:* In 2005–06, 36 students received support, including 3 fellowships with partial tuition reimbursements available (averaging $18,000 per year), 24 teaching assistantships with full tuition reimbursements available (averaging $16,988 per year); Federal Work-Study, tuition waivers (full and partial), and unspecified assistantships also available. Support available to part-time students. Financial award application deadline: 3/1. *Faculty research:* Cell-cytoskeletal elements, development and regeneration in the nervous system, cellular trafficking, environmental stressors and their impact on development, opportunistic parasitic infections in AIDS. *Unit head:* Dr. Ed Bonder, Program Director, 973-353-1047, Fax: 973-353-5518, E-mail: ebonder@andromeda.rutgers.edu. *Application contact:* Dr. Ed Bonder, Program Director, 973-353-1047, Fax: 973-353-5518, E-mail: ebonder@andromeda.rutgers.edu.

See Close-Up on page 205.

Rutgers, The State University of New Jersey, Newark, Graduate School, Program in Computational Biology, Newark, NJ 07102. Offers MS. In 2005, 1 degree awarded. *Entrance requirements:* For master's, GRE, minimum undergraduate B average. Additional exam requirements/recommendations for international students: Required—TOEFL. *Application deadline:* For fall admission, 6/1 for domestic students, 6/1 for international students; for spring admission, 12/1 for domestic students, 12/1 for international students. Application fee: $50. *Expenses:* Tuition, state resident: full-time $10,440; part-time $435 per credit. Tuition, nonresident: full-time $15,520; part-time $637 per credit.

Rutgers, The State University of New Jersey, New Brunswick/Piscataway, Graduate School, BioMaps Institute for Quantitative Biology (Biology at the Inter-face of the Mathematical and Physical Sciences), New Brunswick, NJ 08901-1281. Offers PhD. *Faculty:* 15 full-time (4 women). *Students:* 17 full-time (6 women); includes 2 minority (both Asian Americans or Pacific Islanders), 9 international. Average age 27. 32 applicants, 31% accepted, 6 enrolled. *Degree requirements:* For doctorate, thesis/dissertation, comprehensive exam, registration. *Entrance requirements:* For doctorate, GRE. Additional exam requirements/recommendations for international students: Required—TOEFL. *Application deadline:* For winter admission, 2/15 for domestic students. Applications are processed on a rolling basis. Application fee: $50. Electronic applications accepted. *Expenses:* Tuition, state resident: full-time $10,440; part-time $435 per credit. Tuition, nonresident: full-time $15,520; part-time $647 per credit. Required fees: $129 per credit. Tuition and fees vary according to program. *Financial support:* In 2005–06, 16 students received support, including 9 fellowships with full tuition reimbursements available (averaging $24,000 per year), 3 research assistantships with full tuition reimbursements available (averaging $20,000 per year), 4 teaching assistantships with full tuition reimbursements available (averaging $20,000 per year); traineeships, health care benefits, and unspecified assistantships also available. Financial award application deadline: 6/30; financial award applicants required to submit FAFSA. *Faculty research:* Protein folding; nucleic acid structure; systems biology; transcriptional regulation; signal transduction. *Unit head:* Dr. Wilma Olsen, Co-Director, 732-445-3993, Fax: 732-445-5958, E-mail: olson@rutchem.rutgers.edu. *Application contact:* Dr. Paul H. Ehrlich, Administrative and Associate Graduate Program Director, 732-445-8377, Fax: 732-445-5958, E-mail: pehrlich@biomaps.rutgers.edu.

St. Cloud State University, School of Graduate Studies, College of Science and Engineering, Department of Biological Sciences, St. Cloud, MN 56301-4498. Offers MA, MS. *Faculty:* 23 full-time (7 women). *Students:* 18 full-time (11 women), 3 part-time (all women); includes 3 minority (2 Asian Americans or Pacific Islanders, 1 Hispanic American), 4 international. 11 applicants, 45% accepted. In 2005, 5 degrees awarded. *Degree requirements:* For master's, thesis or alternative, comprehensive exam (for some programs). *Entrance requirements:* For master's, GRE General Test, minimum GPA of 2.75. Additional exam requirements/recommendations for international students: Recommended—TOEFL (minimum score 550 paper-based), IELT (minimum score 7). *Application deadline:* For fall admission, 6/1 priority date for domestic students, 4/1 priority date for international students; for spring admission, 10/1 priority date for domestic students, 8/1 priority date for international students. Applications are processed on a rolling basis. Application fee: $35. Electronic applications accepted. *Expenses:* Tuition, state resident: part-time $277 per credit. Tuition, nonresident: part-time $379 per credit. Required fees: $23 per credit. Tuition and fees vary according to course load and reciprocity agreements. *Financial support:* Federal Work-Study, scholarships/grants, and unspecified assistantships available. Financial award application deadline: 3/1. *Unit head:* Dr. Timothy Schuh, Chairperson, 320-308-2036, Fax: 320-308-4166, E-mail: tjschuh@stcloudstate.edu. *Application contact:* Linda Lou Krueger, School of Graduate Studies, 320-308-2113, Fax: 320-308-5371, E-mail: lekrueger@stcloudstate.edu.

Saint Francis University, Department of Physician Assistant Sciences, Medical Science Program, Loretto, PA 15940-0600. Offers MMS. Part-time programs available. Postbaccalaureate distance learning degree programs offered (no on-campus study). *Degree requirements:* For master's, thesis or alternative. Electronic applications accepted. Expenses: Contact institution. *Faculty research:* Health care policy, physician assistant practice roles, health promotion/disease prevention, public health epidemiology.

St. Francis Xavier University, Graduate Studies, Department of Biology, Antigonish, NS B2G 2W5, Canada. Offers M Sc. *Faculty:* 5 full-time (1 woman). *Students:* 3 applicants, 67% accepted. *Degree requirements:* For master's, thesis, registration. *Entrance requirements:* For master's, minimum undergraduate B average, undergraduate major in biology or related area. Additional exam requirements/recommendations for international students: Required—TOEFL (minimum score 580 paper-based; 236 computer-based). *Application deadline:* For fall admission, 9/1 for domestic students. Applications are processed on a rolling basis. Application fee: $40. *Financial support:* Scholarships/grants and tuition waivers (partial) available. *Faculty research:* Cellular, whole organism, and population levels; marine photosynthesis; biophysical mechanisms; aquatic biology. Total annual research expenditures: $350,000. *Unit head:* Dr. David J. Garbary, Professor, 902-867-2164, Fax: 902-867-2389, E-mail: dgarbary@stfx.ca. *Application contact:* 902-867-2219, Fax: 902-867-2329, E-mail: admit@stfx.ca.

St. John's University, St. John's College of Liberal Arts and Sciences, Department of Biological Sciences, Queens, NY 11439. Offers MS, PhD. Part-time and evening/weekend programs available. *Faculty:* 16 full-time (9 women), 16 part-time/adjunct (5 women). *Students:* 25 full-time (18 women), 16 part-time (8 women); includes 13 minority (4 African Americans, 7 Asian Americans or Pacific Islanders, 2 Hispanic Americans), 20 international. Average age 29. 49 applicants, 61% accepted, 8 enrolled. In 2005, 4 doctorates awarded. *Median time to*

Biological and Biomedical Sciences—General

St. John's University (continued)
degree: Of those who began their doctoral program in fall 1997, 100% received their degree in 8 years or less. *Degree requirements:* For master's, residency, thesis optional; for doctorate, thesis/dissertation, residency, comprehensive exam. *Entrance requirements:* For master's, GRE General Test, GRE Subject Test, minimum GPA of 3.0, 2 letters of recommendation; for doctorate, GRE General Test, GRE Subject Test, minimum GPA of 3.0 (undergraduate), 3.5 (graduate); 2 letters of recommendation. Additional exam requirements/recommendations for international students: Required—TOEFL (minimum score 500 paper-based; 173 computer-based). *Application deadline:* For fall admission, 5/1 priority date for domestic students, 5/1 priority date for international students; for spring admission, 11/1 priority date for domestic students, 11/1 priority date for international students. Applications are processed on a rolling basis. Application fee: $40. Electronic applications accepted. *Expenses:* Tuition: Full-time $17,520; part-time $730 per credit. Required fees: $125 per term. Tuition and fees vary according to program. *Financial support:* Fellowships, research assistantships, scholarships/grants available. Support available to part-time students. Financial award application deadline: 3/1; financial award applicants required to submit FAFSA. *Faculty research:* Regulation of gene transcription, molecular control of development in yeast, physiology of aging, cellular signal transduction. *Unit head:* Dr. Jay Zimmerman, Chair, 718-990-1679, E-mail: zimmermj@stjohns.edu. *Application contact:* Matthew Whelan, Director, Office of Admissions, 718-990-2000, Fax: 718-990-2096, E-mail: admissions@stjohns.edu.

See Close-Up on page 207.

Saint Joseph College, Graduate Division, Department of Biology, West Hartford, CT 06117-2700. Offers biology (MS), including general biology, molecular and cellular biology; biology/chemistry (MS). MS biology (including general biology; molecular and cellular biology) offered online only. Part-time and evening/weekend programs available. Postbaccalaureate distance learning degree programs offered (no on-campus study). *Faculty:* 4 full-time (2 women), 5 part-time/adjunct (3 women). *Students:* Average age 34. 34 applicants, 100% accepted, 11 enrolled. In 2005, 13 degrees awarded. *Degree requirements:* For master's, thesis or alternative, comprehensive exam. *Entrance requirements:* For master's, 2 letters of recommendation. *Application deadline:* Applications are processed on a rolling basis. Application fee: $50. Electronic applications accepted. *Expenses:* Tuition: Part-time $540 per credit. Required fees: $25 per credit. *Financial support:* Career-related internships or fieldwork, health care benefits, and unspecified assistantships available. Support available to part-time students. Financial award application deadline: 7/15; financial award applicants required to submit FAFSA. *Faculty research:* Neurology, cardiology, immunology, mircobiology. *Unit head:* Dr. Charles Morgan, Chair, 860-231-5335, E-mail: cmorgan@sjc.edu.

Saint Joseph College, Graduate Division, Department of Chemistry, West Hartford, CT 06117-2700. Offers biology/chemistry (MS); chemistry (MS). Part-time and evening/weekend programs available. *Faculty:* 5 full-time (1 woman). *Students:* 1 (woman) full-time, 4 part-time (3 women); includes 2 minority (1 Asian American or Pacific Islander, 1 Hispanic American). Average age 30. 4 applicants, 100% accepted, 2 enrolled. In 2005, 3 degrees awarded. *Degree requirements:* For master's, thesis optional. *Entrance requirements:* For master's, 2 letters of recommendation. *Application deadline:* Applications are processed on a rolling basis. Application fee: $50. Electronic applications accepted. *Expenses:* Tuition: Part-time $540 per credit. Required fees: $25 per credit. *Financial support:* Career-related internships or fieldwork, health care benefits, and unspecified assistantships available. Support available to part-time students. Financial award application deadline: 7/15; financial award applicants required to submit FAFSA. *Faculty research:* Regulation of cancer cells, selective oxidation of organic concept mapping in general chemistry. *Unit head:* Dr. Peter Markow, Chair, 860-231-5240, Fax: 860-233-5695, E-mail: pmarkow@sjc.edu.

Saint Joseph's University, College of Arts and Sciences, Department of Biology, Philadelphia, PA 19131-1395. Offers MA, MS. *Faculty:* 6 full-time (1 woman). *Students:* 4 full-time (3 women), 14 part-time (10 women); includes 2 minority (1 Asian American or Pacific Islander, 1 Hispanic American), 5 international. In 2005, 7 degrees awarded. *Entrance requirements:* For master's, GRE, 2 letters of recommendation. *Application deadline:* For fall admission, 7/15 for domestic students; for spring admission, 11/15 for domestic students. Application fee: $35. *Expenses:* Tuition: Part-time $692 per credit. Tuition and fees vary according to program. *Unit head:* Dr. Karen Snetselaar, Director, 610-660-1820.

Saint Louis University, Graduate School, College of Arts and Sciences and Graduate School, Department of Biology, St. Louis, MO 63103-2097. Offers MS, MS-R, PhD. Part-time programs available. *Faculty:* 24 full-time (5 women). *Students:* 38 full-time (22 women), 3 part-time (1 woman); includes 4 minority (1 African American, 2 Asian Americans or Pacific Islanders, 1 Hispanic American), 8 international. Average age 27. 21 applicants, 48% accepted, 6 enrolled. In 2005, 6 master's, 7 doctorates awarded. *Degree requirements:* For master's, thesis (for some programs), comprehensive exam; for doctorate, thesis/dissertation, preliminary exams. *Entrance requirements:* For master's and doctorate, GRE General Test or MCAT, letters of recommendation, resumé. Additional exam requirements/recommendations for international students: Required—TOEFL (minimum score 550 paper-based; 213 computer-based). *Application deadline:* For fall admission, 6/1 for domestic students, 6/1 for international students; for spring admission, 11/1 for domestic students, 11/1 for international students. Applications are processed on a rolling basis. Application fee: $40. *Expenses:* Tuition: Part-time $760 per credit hour. Required fees: $55 per semester. *Financial support:* In 2005–06, 39 students received support, including 14 research assistantships with tuition reimbursements available (averaging $16,000 per year), 14 teaching assistantships with tuition reimbursements available (averaging $15,500 per year); health care benefits and tuition waivers (partial) also available. Financial award application deadline: 6/1; financial award applicants required to submit FAFSA. *Faculty research:* Molecular systematics, pathogen-host interactions, developmental regulation. *Unit head:* Dr. Richard L. Mayden, Chairperson, 314-977-3494, Fax: 314-977-3658, E-mail: maydenrl@slu.edu. *Application contact:* Gary Behrman, Associate Dean of the Graduate School, 314-977-3827, E-mail: behrmang@slu.edu.

Announcement: The department offers dissertation or thesis research options in cellular and molecular regulation, ecology, systematics, and evolution. Research includes population biology, genetics and ecology, systematics, biogeography, speciation, signal transduction, host-parasite interactions, cell and molecular biology, microbiology, neurobiology, and other areas. Reis Biological Station is available for field research. Students may do research at the St. Louis Zoo or Missouri Botanical Garden. Research and teaching assistantships are available. Contact Dr. Joe Leverich, Graduate Program Director (telephone: 314-977-3900, fax: 314-977-3658, e-mail: leverich@slu.edu, Web site: http://bio.slu.edu).

See Close-Up on page 209.

Saint Louis University, Graduate School and School of Medicine, Graduate Program in Biomedical Sciences, St. Louis, MO 63103-2097. Offers PhD. *Faculty:* 139 full-time (58 women), 5 part-time/adjunct (2 women). *Students:* 81 full-time (48 women), 5 part-time (3 women); includes 6 minority (2 African Americans, 2 Asian Americans or Pacific Islanders, 2 Hispanic Americans), 13 international. Average age 27. 46 applicants, 83% accepted, 28 enrolled. In 2005, 11 degrees awarded. *Entrance requirements:* Additional exam requirements/recommendations for international students: Required—TOEFL (minimum score 550 paper-based; 213 computer-based). *Application deadline:* For fall admission, 7/1 for domestic students, 7/1 for international students; for spring admission, 11/1 for domestic students, 11/1 for international students. Applications are processed on a rolling basis. Application fee: $40. *Expenses:* Tuition: Part-time $760 per credit hour. Required fees: $55 per semester. *Financial support:* In 2005–06, 57 students received support, including 10 research assistantships (averaging $15,950 per year), 2 teaching assistantships (averaging $15,500 per year); fellowships, Federal Work-Study, health care benefits, and tuition waivers (partial) also available. Support available to part-time students. Financial award application deadline: 6/1; financial

award applicants required to submit FAFSA. *Application contact:* Gary Behrman, Associate Dean of the Graduate School, 314-977-3827, E-mail: behrmang@slu.edu.

Sam Houston State University, College of Arts and Sciences, Department of Biological Sciences, Huntsville, TX 77341. Offers biology (MA, MS). Part-time programs available. *Faculty:* 12 full-time (3 women). *Students:* 4 full-time (1 woman), 5 part-time (4 women); includes 2 minority (both Hispanic Americans) Average age 26. In 2005, 1 degree awarded. *Degree requirements:* For master's, thesis (for some programs). *Entrance requirements:* For master's, GRE General Test. Additional exam requirements/recommendations for international students: Required—TOEFL (minimum score 550 paper-based; 213 computer-based). *Application deadline:* For fall admission, 8/1 for domestic students; for spring admission, 12/1 for domestic students. Application fee: $20. *Financial support:* Research assistantships, teaching assistantships available. Financial award application deadline: 5/31; financial award applicants required to submit FAFSA. *Unit head:* Dr. Monte Thies, Chair, 936-294-1538, Fax: 936-294-3940, E-mail: bio_mlt@shsu.edu. *Application contact:* Dr. Diane Neudorf, Advisor, 936-294-1548.

San Diego State University, Graduate and Research Affairs, College of Sciences, Department of Biological Sciences, San Diego, CA 92182. Offers biology (MA, MS), including ecology (MS), molecular biology (MS), physiology (MS), systematics/evolution (MS); biostatistics and biometry (PhD); cell and molecular biology (PhD); ecology (MS, PhD); microbiology (MS). *Students:* 73 full-time (36 women), 100 part-time (53 women); includes 33 minority (2 African Americans, 16 Asian Americans or Pacific Islanders, 15 Hispanic Americans), 33 international. Average age 26. 228 applicants, 23% accepted, 32 enrolled. In 2005, 27 master's, 9 doctorates awarded. Terminal master's awarded for partial completion of doctoral program. *Degree requirements:* For master's and doctorate, thesis/dissertation. *Entrance requirements:* For master's, GRE General Test, GRE Subject Test, resumé or curriculum vitae, 2 letters of recommendation. Additional exam requirements/recommendations for international students: Required—TOEFL. *Application deadline:* For fall admission, 5/1 for domestic students; for spring admission, 11/1 for domestic students, 10/1 for international students. Applications are processed on a rolling basis. Application fee: $55. Electronic applications accepted. *Financial support:* In 2005–06, 116 teaching assistantships were awarded; fellowships, research assistantships, career-related internships or fieldwork and unspecified assistantships also available. Financial award applicants required to submit FAFSA. Total annual research expenditures: $9.9 million. *Unit head:* Christopher Glembotski, Chair, 619-594-6767, Fax: 619-594-5676. *Application contact:* Terry Frey, Graduate Coordinator, 619-594-6756, Fax: 619-594-5676, E-mail: gradcoor@sciences.sdsu.edu.

Announcement: The department offers MS degrees in biology, earned by thesis, with concentrations in ecology, microbiology, molecular biology, physiology, and evolutionary biology. It also has PhD programs in ecology (jointly with University of California, Davis) and in cell/molecular biology (jointly with University of California, San Diego). Teaching assistantships pay approximately $10,600 per year; stipends for PhD students are about $15,200 (ecology) and $17,200 (cell/molecular biology). For applications and additional information, contact Graduate Coordinator, Biology Department, San Diego State University, San Diego, CA 92182-4614 (telephone: 619-594-6919, fax: 619-594-5676, e-mail: gradcoor@sciences.sdsu.edu).

San Francisco State University, Division of Graduate Studies, College of Science and Engineering, Department of Biology, San Francisco, CA 94132-1722. Offers cell and molecular biology (MA); conservation biology (MA); ecology and systematic biology (MA); marine biology (MA); microbiology (MA); physiology and behavioral biology (MA). *Entrance requirements:* For master's, minimum GPA of 2.5 in last 60 units.

San Jose State University, Graduate Studies and Research, College of Science, Department of Biological Sciences, San Jose, CA 95192-0001. Offers biological sciences (MA, MS); molecular biology and microbiology (MS); organismal biology, conservation and ecology (MS); physiology (MS). Part-time programs available. *Students:* 44 full-time (35 women), 33 part-time (24 women); includes 39 minority (33 Asian Americans or Pacific Islanders, 6 Hispanic Americans), 12 international. Average age 31. 140 applicants, 40% accepted, 29 enrolled. In 2005, 39 degrees awarded. *Entrance requirements:* For master's, GRE. *Application deadline:* For fall admission, 6/29 for domestic students; for spring admission, 11/30 for domestic students. Applications are processed on a rolling basis. Application fee: $59. Electronic applications accepted. *Expenses:* Tuition, nonresident: part-time $339 per unit. Required fees: $1,286 per semester. Tuition and fees vary according to course load and degree level. *Financial support:* In 2005–06, 13 teaching assistantships were awarded; Federal Work-Study also available. Financial award applicants required to submit FAFSA. *Faculty research:* Systemic physiology, molecular genetics, SEM studies, toxicology, large mammal ecology. *Unit head:* Dr. Sally Veregge, Chair, 408-924-4900, Fax: 408-924-4840. *Application contact:* Dr. Howard Shellhammer, Graduate Coordinator, 408-924-4897.

The Scripps Research Institute, Kellogg School of Science and Technology, Program in Biology and Biophysical Sciences, La Jolla, CA 92037. Offers biology (PhD); biophysics (PhD). *Faculty:* 99 full-time (29 women). *Students:* 96 full-time (40 women). 299 applicants, 17% accepted, 12 enrolled. In 2005, 10 degrees awarded. *Degree requirements:* For doctorate, thesis/dissertation. *Entrance requirements:* For doctorate, GRE General Test, GRE Subject Test. Additional exam requirements/recommendations for international students: Required—TOEFL. *Application deadline:* For fall admission, 1/1 for domestic students, 1/1 for international students. Application fee: $0. *Expenses:* Tuition: Full-time $5,000. *Financial support:* Institutionally sponsored loans and stipends available. *Faculty research:* Biocatalysis and enzyme engineering, molecular structure and function, neurosciences, immunology, plant biology. *Unit head:* Dr. Stephen P. Mayfield, Associate Dean, 784-9848. *Application contact:* Marylyn Rinaldi, Administrative Director, 858-784-8469, Fax: 858-784-2802, E-mail: mrinaldi@scripps.edu.

Seton Hall University, College of Arts and Sciences, Department of Biological Sciences, South Orange, NJ 07079-2697. Offers biology (MS); microbiology (MS); molecular bioscience (PhD). Part-time and evening/weekend programs available. *Students:* 17 full-time (14 women), 35 part-time (18 women). Average age 29. 33 applicants, 70% accepted, 14 enrolled. In 2005, 8 degrees awarded. *Degree requirements:* For master's, research paper or thesis, seminar. *Application deadline:* For fall admission, 7/1 priority date for domestic students, 7/1 priority date for international students; for spring admission, 11/1 priority date for domestic students, 11/1 priority date for international students. Applications are processed on a rolling basis. Application fee: $50. Electronic applications accepted. *Financial support:* Teaching assistantships, career-related internships or fieldwork and Federal Work-Study available. *Faculty research:* Neurobiology, genetics, immunology, molecular biology, cellular physiology, toxicology, microbiology, bioinformatics. *Unit head:* Dr. Carolyn Bentivegna, Chair, 973-761-9044, Fax: 973-761-9596, E-mail: bentivca@shu.edu. *Application contact:* Dr. Carroll D. Rawn, Director of Graduate Studies, 973-761-9054, Fax: 973-761-9596, E-mail: rawncarr@shu.edu.

See Close-Up on page 593.

Shippensburg University of Pennsylvania, School of Graduate Studies, College of Arts and Sciences, Department of Biology, Shippensburg, PA 17257-2299. Offers MS. Part-time and evening/weekend programs available. *Faculty:* 10 full-time (3 women). *Students:* 12 full-time (7 women), 17 part-time (9 women), 1 international. Average age 28. 15 applicants, 80% accepted, 8 enrolled. In 2005, 7 degrees awarded. *Degree requirements:* For master's, thesis optional. *Entrance requirements:* For master's, 33 credits of course work in biology, minimum 4 courses/labs in chemistry. Additional exam requirements/recommendations for international students: Required—TOEFL (minimum score 560 paper-based; 220 computer-based). *Application deadline:* Applications are processed on a rolling basis. Application fee: $30. Electronic applications accepted. *Expenses:* Tuition, state resident: full-time $5,888; part-time $327 per credit. Tuition, nonresident: full-time $9,422; part-time $523 per credit. Required fees: $854; $27 per credit. *Financial support:* In 2005–06, 9 research assistantships with full tuition reimbursements (averaging $2,575 per year) were awarded; career-related

Biological and Biomedical Sciences—General

internships or fieldwork, scholarships/grants, and unspecified assistantships also available. Support available to part-time students. Financial award application deadline: 3/1; financial award applicants required to submit FAFSA. *Unit head:* Dr. Gregory Paulson, Chairperson, 717-477-1401, Fax: 77-477-4064, E-mail: gspaul@ship.edu. *Application contact:* Renee Payne, Associate Dean of Graduate Admissions, 717-477-1231, Fax: 717-477-4016, E-mail: rmpayn@ship.edu.

Simon Fraser University, Graduate Studies, Faculty of Science, Department of Biological Sciences, Burnaby, BC V5A 1S6, Canada. Offers biological sciences (M Sc, PhD); environmental toxicology (Diploma), including food and drug toxicology, industrial toxicology; pest management (MPM). *Degree requirements:* For masters, doctorate, and Diploma, thesis/dissertation. *Entrance requirements:* For master's and Diploma, minimum GPA of 3.0; for doctorate, minimum GPA of 3.5. Additional exam requirements/recommendations for international students: Required—TOEFL or IELTS. Electronic applications accepted. *Faculty research:* Molecular biology, marine biology, ecology, wildlife biology, endocrinology.

Smith College, Graduate Programs, Department of Biological Sciences, Northampton, MA 01063. Offers MA, MAT, PhD. Part-time programs available. *Faculty:* 14 full-time (5 women), 4 part-time/adjunct (1 woman). *Students:* 1 full-time (0 women), 8 part-time (6 women); includes 1 minority (Asian American or Pacific Islander), 2 international. Average age 30. 6 applicants, 67% accepted, 4 enrolled. In 2005, 4 master's awarded. *Degree requirements:* For master's, one foreign language, thesis (for some programs); for doctorate, 2 foreign languages, thesis/dissertation. *Entrance requirements:* For master's and doctorate, GRE General Test, GRE Subject Test. Additional exam requirements/recommendations for international students: Required—TOEFL. *Application deadline:* For fall admission, 4/1 for domestic students, 1/15 for international students; for spring admission, 12/1 for domestic students. Application fee: $60. *Expenses:* Tuition: Full-time $30,520; part-time $955 per credit. *Financial support:* In 2005–06, 9 students received support, including 7 teaching assistantships with full tuition reimbursements available (averaging $10,780 per year); institutionally sponsored loans and scholarships/grants also available. Support available to part-time students. Financial award application deadline: 1/15; financial award applicants required to submit CSS PROFILE or FAFSA. *Unit head:* Richard Briggs, Chair, 413-585-3823, E-mail: rbriggs@smith.edu. *Application contact:* Rob P. Dorit, Graduate Student Advisor, 413-585-3638, E-mail: rdorit@smith.edu.

Sonoma State University, School of Science and Technology, Department of Biology, Rohnert Park, CA 94928-3609. Offers environmental biology (MA); general biology (MA). Part-time programs available. *Faculty:* 8 full-time (2 women), 7 part-time/adjunct (4 women). *Students:* 12 full-time (8 women), 1 international. Average age 33. 14 applicants, 36% accepted, 3 enrolled. In 2005, 2 degrees awarded. *Degree requirements:* For master's, thesis or alternative, oral exam. *Entrance requirements:* For master's, GRE General Test, GRE Subject Test, minimum GPA of 3.0. *Application deadline:* For fall admission, 11/30 for domestic students. Applications are processed on a rolling basis. Application fee: $55. *Expenses:* Tuition, nonresident: part-time $339 per unit. Required fees: $2,099 per semester. *Financial support:* In 2005–06, 2 research assistantships, 9 teaching assistantships were awarded; career-related internships or fieldwork and Federal Work-Study also available. Financial award application deadline: 3/2. *Faculty research:* Molecular biology, genetics, riparian and wetland ecology, plant ecology, feeding mechanisms of invertebrates, ichthyology, microbiology. Total annual research expenditures: $32,000. *Unit head:* Dr. James Christmann, Chair, 707-664-2189, E-mail: james.christmann@sonoma.edu. *Application contact:* John Hopkirk, Graduate Adviser, 707-664-2180.

South Dakota State University, Graduate School, College of Agriculture and Biological Sciences, Department of Biology/Microbiology, Brookings, SD 57007. Offers biology (MS); microbiology (MS). *Degree requirements:* For master's, thesis, oral exam. *Entrance requirements:* For master's, GRE. Additional exam requirements/recommendations for international students: Required—TOEFL. *Faculty research:* Plant tissue culture, molecular biology studies of metabolic regulation in plants, mechanisms of mammalian gene expression, aquatic-wetland ecosystem ecology, stress-induced immunosuppression on parasite-induced pathology.

South Dakota State University, Graduate School, College of Agriculture and Biological Sciences, Program in Biological Sciences, Brookings, SD 57007. Offers PhD. *Degree requirements:* For doctorate, thesis/dissertation, preliminary oral and written exams. *Entrance requirements:* Additional exam requirements/recommendations for international students: Required—TOEFL.

Southeastern Louisiana University, College of Science and Technology, Department of Biological Sciences, Hammond, LA 70402. Offers biology (MS). Part-time programs available. *Faculty:* 15 full-time (3 women). *Students:* 18 full-time (7 women), 19 part-time (13 women); includes 4 minority (2 African Americans, 2 Hispanic Americans), 2 international. Average age 27. 14 applicants, 100% accepted, 12 enrolled. In 2005, 11 degrees awarded. *Degree requirements:* For master's, thesis, oral and written exams. *Entrance requirements:* For master's, GRE General Test, minimum GPA of 3.0, 30 undergraduate hours in biology, 3 letters of reference. Additional exam requirements/recommendations for international students: Required—TOEFL (minimum score 500 paper-based; 173 computer-based). *Application deadline:* For fall admission, 7/15 priority date for domestic students, 6/1 priority date for international students; for spring admission, 12/1 priority date for domestic students, 10/1 priority date for international students. Applications are processed on a rolling basis. Application fee: $20 ($30 for international students). Electronic applications accepted. *Expenses:* Tuition, state resident: full-time $3,131; part-time $174 per credit. Tuition, nonresident: full-time $7,127; part-time $396 per credit. *Financial support:* In 2005–06, 2 fellowships with full tuition reimbursements (averaging $2,450 per year), 13 research assistantships with full tuition reimbursements (averaging $2,200 per year), 8 teaching assistantships with full tuition reimbursements (averaging $2,200 per year) were awarded; Federal Work-Study, institutionally sponsored loans, scholarships/grants, and unspecified assistantships also available. Support available to part-time students. Financial award application deadline: 5/1; financial award applicants required to submit FAFSA. *Faculty research:* Immunology, molecular biology, parasitology, microbiology, environmental biology. Total annual research expenditures: $1.5 million. *Unit head:* Dr. David Sever, Department Head, 985-549-3740, Fax: 985-549-3851, E-mail: dsever@selu.edu. *Application contact:* Sandra Meyers, Graduate Admissions Analyst, 985-549-2066, Fax: 985-549-5632, E-mail: admissions@selu.edu.

Southeast Missouri State University, School of Graduate Studies, Department of Biology, Cape Girardeau, MO 63701-4799. Offers MNS. Part-time programs available. *Faculty:* 16 full-time (4 women). *Students:* 7 full-time (2 women), 23 part-time (13 women); includes 2 minority (both African Americans) Average age 27. 6 applicants, 100% accepted. In 2005, 4 degrees awarded. *Degree requirements:* For master's, thesis or alternative. *Entrance requirements:* For master's, GRE General Test, minimum undergraduate GPA of 2.75, minimum of 30 hours of undergraduate course work in science and mathematics. Additional exam requirements/recommendations for international students: Required—TOEFL (minimum score 550 paper-based; 213 computer-based). *Application deadline:* For fall admission, 8/1 for domestic students, 4/1 for international students; for spring admission, 11/21 for domestic students, 9/1 for international students. Applications are processed on a rolling basis. Application fee: $20 ($100 for international students). Electronic applications accepted. *Expenses:* Tuition, state resident: part-time $186 per hour. Tuition, nonresident: part-time $339 per hour. Required fees: $114; $13 per hour. Tuition and fees vary according to course load, degree level and campus/location. *Financial support:* In 2005–06, 24 students received support, including 18 teaching assistantships with full tuition reimbursements available (averaging $6,600 per year); fellowships, unspecified assistantships also available. Financial award applicants required to submit FAFSA. *Faculty research:* Wildlife biology, molecular genetics and cytogenetics, plant physiology. Total annual research expenditures: $160,286. *Unit head:* Dr. William Eddleman, Chairperson, 573-651-2171, Fax: 573-651-2223, E-mail: weddleman@semo.edu. *Application contact:* Marsha L. Arant, Senior Administrative Assistant, Office of Graduate Studies, 573-651-2192, Fax: 573-651-2001, E-mail: marant@semo.edu.

Southern Connecticut State University, School of Graduate Studies, School of Arts and Sciences, Department of Biology, New Haven, CT 06515-1355. Offers biology (MS); biology for nurse anesthetists (MS). Part-time and evening/weekend programs available. *Degree requirements:* For master's, thesis optional. *Entrance requirements:* For master's, previous course work in biology, chemistry, and mathematics; interview. Electronic applications accepted.

Southern Illinois University Carbondale, Graduate School, College of Science, Biological Sciences Program, Carbondale, IL 62901-4701. Offers MS. *Students:* Average age 25. 9 applicants, 0 enrolled. *Degree requirements:* For master's, thesis or alternative. *Entrance requirements:* For master's, GRE General Test, minimum GPA of 2.7. Additional exam requirements/recommendations for international students: Required—TOEFL. *Application deadline:* Applications are processed on a rolling basis. Application fee: $0. *Financial support:* In 2005–06, 3 students received support; fellowships with full tuition reimbursements available, research assistantships with full tuition reimbursements available, teaching assistantships with full tuition reimbursements available, Federal Work-Study, institutionally sponsored loans, and tuition waivers (full) available. Support available to part-time students. *Faculty research:* Molecular mechanisms of mutagenesis, reproductive endocrinology, avian energetics and nutrition, developmental plant physiology. *Unit head:* Brooks Burr, Director, 618-453-4112.

Southern Illinois University Carbondale, Graduate School, Graduate Program in Medicine, Carbondale, IL 62901-4701. Offers molecular, cellular and systemic physiology (MS); pharmacology (MS, PhD); physiology (MS, PhD). *Faculty:* 31 full-time (5 women). *Students:* 24 full-time (7 women), 16 part-time (7 women); includes 5 minority (3 African Americans, 1 Asian American or Pacific Islander, 1 Hispanic American), 22 international. 48 applicants, 33% accepted, 10 enrolled. In 2005, 3 master's, 3 doctorates awarded. Terminal master's awarded for partial completion of doctoral program. *Degree requirements:* For master's and doctorate, thesis/dissertation. *Entrance requirements:* For master's, minimum GPA of 3.0; for doctorate, minimum GPA of 3.25. Additional exam requirements/recommendations for international students: Required—TOEFL. Application fee: $0. *Financial support:* In 2005–06, 27 students received support, including 12 fellowships with full tuition reimbursements available, 1 research assistantship with full tuition reimbursement available, 10 teaching assistantships with full tuition reimbursements available; institutionally sponsored loans and tuition waivers (full) also available. *Faculty research:* Cardiovascular physiology, neurophysiology of hearing. *Application contact:* Graduate Program Committee, 618-536-5513.

Southern Illinois University Edwardsville, Graduate Studies and Research, College of Arts and Sciences, Department of Biological Sciences, Program in Biology, Edwardsville, IL 62026-0001. Offers MA, MS. *Expenses:* Tuition, state resident: part-time $190 per semester hour. Tuition, nonresident: part-time $380 per semester hour. Tuition and fees vary according to course load, reciprocity agreements and student level. *Unit head:* Dr. Kurt Schulz, Director, 618-650-3005.

Southern Methodist University, Dedman College, Department of Biological Sciences, Dallas, TX 75275. Offers MA, MS, PhD. *Faculty:* 9 full-time (3 women). *Students:* 16 full-time (8 women); includes 1 minority (Asian American or Pacific Islander), 10 international. Average age 29. In 2005, 1 degree awarded. Terminal master's awarded for partial completion of doctoral program. *Degree requirements:* For master's, thesis (for some programs), thesis (MS), oral exam; for doctorate, thesis/dissertation, qualifying exam. *Entrance requirements:* For master's and doctorate, GRE General Test, minimum GPA of 3.0. Additional exam requirements/recommendations for international students: Required—TOEFL (minimum score 550 paper-based; 217 computer-based). *Application deadline:* For fall admission, 2/1 for domestic students; for spring admission, 11/30 priority date for domestic students. Applications are processed on a rolling basis. Application fee: $60. *Financial support:* In 2005–06, 7 research assistantships with full tuition reimbursements (averaging $19,200 per year), 7 teaching assistantships with full tuition reimbursements (averaging $19,200 per year) were awarded; Federal Work-Study and tuition waivers (partial) also available. Financial award applicants required to submit FAFSA. *Faculty research:* Free radicals and aging, protein structure, chromatin structure, signal processes, retroviral pathogenesis. Total annual research expenditures: $2 million. *Unit head:* Larry Ruben, Head, 214-768-2321, Fax: 214-768-3955. *Application contact:* Dr. Steven Vik, Graduate Adviser, 214-768-4228, Fax: 214-768-3955, E-mail: svik@mail.smu.edu.

See Close-Up on page 211.

Southern University and Agricultural and Mechanical College, Graduate School, College of Sciences, Department of Biology, Baton Rouge, LA 70813. Offers MS. *Faculty:* 12 full-time (4 women). *Students:* 19 full-time (14 women), 9 part-time (all women); includes 23 minority (22 African Americans, 1 Asian American or Pacific Islander). Average age 26. 19 applicants, 84% accepted, 13 enrolled. In 2005, 8 degrees awarded. *Degree requirements:* For master's, thesis, comprehensive exam. *Entrance requirements:* For master's, GRE General Test. Additional exam requirements/recommendations for international students: Required—TOEFL (minimum score 525 paper-based; 193 computer-based). *Application deadline:* For fall admission, 4/15 priority date for domestic students, 4/15 priority date for international students; for spring admission, 11/1 priority date for domestic students, 11/1 priority date for international students. Applications are processed on a rolling basis. Application fee: $25. *Financial support:* In 2005–06, 4 teaching assistantships (averaging $7,000 per year) were awarded; research assistantships Financial award application deadline: 4/15; financial award applicants required to submit FAFSA. *Faculty research:* Toxicology, neuroendocrinology, mycotoxin, virology. *Unit head:* Dr. Willis Jacob, Chair, 225-771-5210, Fax: 225-771-5386, E-mail: whjacob@att.net.

Stanford University, School of Humanities and Sciences, Department of Biological Sciences, Stanford, CA 94305-9991. Offers MS, PhD. Terminal master's awarded for partial completion of doctoral program. *Degree requirements:* For doctorate, thesis/dissertation, oral exam. *Entrance requirements:* For master's, GRE General Test; for doctorate, GRE General Test, GRE Subject Test. Additional exam requirements/recommendations for international students: Required—TOEFL. Electronic applications accepted.

Stanford University, School of Medicine, Graduate Programs in Medicine, Stanford, CA 94305-9991. Offers MS, PhD. Terminal master's awarded for partial completion of doctoral program. *Degree requirements:* For master's and doctorate, thesis/dissertation. *Entrance requirements:* For master's, GRE General Test or MCAT. Additional exam requirements/recommendations for international students: Required—TOEFL. Electronic applications accepted.

State University of New York at Binghamton, Graduate School, School of Arts and Sciences, Department of Biological Sciences, Binghamton, NY 13902-6000. Offers MA, PhD. Terminal master's awarded for partial completion of doctoral program. *Degree requirements:* For master's, thesis, oral exam, seminar presentation; for doctorate, thesis/dissertation, comprehensive exam. *Entrance requirements:* For master's and doctorate, GRE General Test, GRE Subject Test. Additional exam requirements/recommendations for international students: Required—TOEFL. Electronic applications accepted.

State University of New York at Buffalo, Graduate School, College of Arts and Sciences, Department of Biological Sciences, Buffalo, NY 14260. Offers MA, MS, PhD. *Faculty:* 26 full-time (5 women), 3 part-time/adjunct (1 woman). *Students:* 55 full-time (27 women), 1 part-time; includes 1 minority (Asian American or Pacific Islander), 20 international. Average age 24. 169 applicants, 25% accepted, 24 enrolled. In 2005, 16 master's, 5 doctorates awarded. Terminal master's awarded for partial completion of doctoral program. *Median time to degree:* Of those who began their doctoral program in fall 1997, 75% received their degree in 8 years or less. *Degree requirements:* For master's, research rotation, seminar, written exam; for doctorate, thesis/dissertation, oral candidacy exam, research, seminar, comprehensive exam, registration. *Entrance requirements:* For master's and doctorate, GRE General Test, 2 semesters of course work in calculus, course work in chemistry through organic chemistry, strong biology background. Additional exam requirements/recommendations for international students: Required—TOEFL (minimum score 600 paper-based; 240 computer-based). *Application*

Biological and Biomedical Sciences—General

State University of New York at Buffalo (continued)
deadline: For fall admission, 2/1 priority date for domestic students, 2/1 priority date for international students; for spring admission, 11/1 for domestic students, 11/1 for international students. Applications are processed on a rolling basis. Application fee: $35. Electronic applications accepted. *Financial support:* In 2005–06, 5 fellowships with full tuition reimbursements (averaging $7,500 per year), 16 research assistantships with full tuition reimbursements (averaging $20,000 per year), 33 teaching assistantships with full tuition reimbursements (averaging $20,000 per year) were awarded; scholarships/grants and health care benefits also available. Financial award application deadline: 2/15; financial award applicants required to submit FAFSA. *Faculty research:* Osmotic regulation via ion channels, control of gene expression in prokaryotes and eukaryotes, signal transduction, molecular evolution in marine and terrestrial systems, biochemistry and molecular genetics of protein-nucleic acid complexes. Total annual research expenditures: $8 million. *Unit head:* Dr. Gerald Koudelka, Chairman, 716-645-2363 Ext. 158, Fax: 716-645-2975, E-mail: koudelka@buffalo.edu. *Application contact:* Dr. Stephen Free, Director of Graduate Studies, 716-645-2363 Ext. 149, Fax: 716-645-2975, E-mail: free@buffalo.edu.

See Close-Up on page 213.

State University of New York at Buffalo, Graduate School, Graduate Programs in Cancer Research and Biomedical Sciences at Roswell Park Cancer Institute, Buffalo, NY 14260. Offers MS, PhD. Part-time programs available. *Faculty:* 129 part-time/adjunct (33 women). *Students:* 138 full-time (75 women), 33 part-time (20 women); includes 26 minority (7 African Americans, 12 Asian Americans or Pacific Islanders, 7 Hispanic Americans), 56 international. Average age 24. 318 applicants, 35% accepted, 50 enrolled. In 2005, 29 master's, 20 doctorates awarded. Terminal master's awarded for partial completion of doctoral program. *Median time to degree:* Of those who began their doctoral program in fall 1997, 90% received their degree in 8 years or less. *Degree requirements:* For master's and doctorate, thesis/dissertation. *Entrance requirements:* For master's, GRE General Test or MCAT; for doctorate, GRE General Test. Additional exam requirements/recommendations for international students: Required—TOEFL (minimum score 600 paper-based; 250 computer-based), TWE. *Application deadline:* For fall admission, 2/1 priority date for domestic students, 2/1 priority date for international students. Applications are processed on a rolling basis. Application fee: $35. Electronic applications accepted. *Financial support:* In 2005–06, 23 fellowships with full tuition reimbursements (averaging $21,000 per year), 100 research assistantships with full tuition reimbursements (averaging $21,000 per year), 1 teaching assistantship with full tuition reimbursement (averaging $8,500 per year) were awarded; Federal Work-Study, institutionally sponsored loans, and unspecified assistantships also available. Financial award application deadline: 2/1; financial award applicants required to submit FAFSA. *Faculty research:* Basic and biomedical cancer research, cell and molecular biology, biophysics, biochemistry, pharmacology. Total annual research expenditures: $60 million. *Unit head:* Dr. Arthur M. Michalek, Dean, 716-845-2339, Fax: 716-845-8178. *Application contact:* Craig R. Johnson, Director of Admissions, 716-845-2339, Fax: 716-845-8178, E-mail: craig.johnson@roswellpark.edu.

State University of New York at Buffalo, Graduate School, School of Medicine and Biomedical Sciences, Graduate Programs in Medicine and Biomedical Sciences, Buffalo, NY 14260. Offers MA, MS, PhD, MD/PhD. Degrees awarded through participating departments. *Faculty:* 101 full-time (23 women), 22 part-time/adjunct (6 women). *Students:* 153 full-time (79 women), 16 part-time (5 women); includes 11 minority (1 American Indian/Alaska Native, 8 Asian Americans or Pacific Islanders, 2 Hispanic Americans), 64 international. Average age 23. 470 applicants, 23% accepted, 49 enrolled. In 2005, 16 master's, 26 doctorates awarded. Terminal master's awarded for partial completion of doctoral program. *Median time to degree:* Of those who began their doctoral program in fall 1997, 98% received their degree in 8 years or less. *Degree requirements:* For master's, thesis (for some programs), comprehensive exam (for some programs), registration; for doctorate, thesis/dissertation, comprehensive exam, registration. *Entrance requirements:* For master's, GRE General Test or MCAT; for doctorate, GRE General Test, GRE Subject Test (recommended), 3 letters of recommendation. Additional exam requirements/recommendations for international students: Required—TOEFL (minimum score 600 paper-based; 250 computer-based). *Application deadline:* For fall admission, 2/1 priority date for domestic students, 2/1 priority date for international students. Applications are processed on a rolling basis. Application fee: $35. Electronic applications accepted. *Expenses: Contact* institution. *Financial support:* In 2005–06, 4 fellowships with full tuition reimbursements (averaging $25,000 per year), 30 research assistantships with full tuition reimbursements (averaging $21,000 per year), 31 teaching assistantships with full tuition reimbursements (averaging $2,000 per year) were awarded; career-related internships or fieldwork, Federal Work-Study, institutionally sponsored loans, scholarships/grants, traineeships, health care benefits, and unspecified assistantships also available. Financial award application deadline: 2/1; financial award applicants required to submit FAFSA. *Faculty research:* Neuroscience; molecular, cell, and structural biology; microbial pathogenesis; cardiopulmonary physiology; biochemistry, biotechnology and clinical laboratory science. Total annual research expenditures: $117.3 million. *Unit head:* Dr. Suzanne G. Laychock, Senior Associate Dean for Research and Graduate Studies, 716-829-3398, Fax: 716-829-2437, E-mail: laychock@acsu.buffalo.edu. *Application contact:* Amy J. Sierocki, Office Manager, 716-898-8585, E-mail: akuzdale@buffalo.edu.

See Close-Up on page 217.

State University of New York at Buffalo, Graduate School, School of Medicine and Biomedical Sciences, Interdisciplinary Graduate Program in Biomedical Sciences, Buffalo, NY 14214. Offers PhD. PhD awarded through participating departments. *Students:* 18 full-time (9 women), 6 international. Average age 23. 374 applicants, 15% accepted, 18 enrolled. *Degree requirements:* For doctorate, thesis/dissertation, comprehensive exam, registration. *Entrance requirements:* For doctorate, GRE General Test, GRE Subject Test (recommended), 3 letters of recommendation. Additional exam requirements/recommendations for international students: Required—TOEFL (minimum score 600 paper-based; 250 computer-based). *Application deadline:* For fall admission, 2/1 priority date for domestic students, 2/1 priority date for international students. Applications are processed on a rolling basis. Application fee: $35. Electronic applications accepted. *Financial support:* In 2005–06, 18 students received support, including 3 fellowships with full tuition reimbursements available (averaging $25,000 per year), 17 teaching assistantships with full tuition reimbursements available (averaging $21,000 per year); Federal Work-Study, scholarships/grants, traineeships, health care benefits, and unspecified assistantships also available. Financial award application deadline: 2/1; financial award applicants required to submit FAFSA. *Faculty research:* Molecular, cell and structural biology, pharmacology and toxicology, neurosciences, microbiology, pathogenesis and disease. Total annual research expenditures: $117.3 million. *Unit head:* Dr. Richard A. Rabin, Director, 716-829-3398, Fax: 716-829-2437, E-mail: smbs-gradprog@buffalo.edu. *Application contact:* Amy J. Sierocki, Staff Associate, 716-829-3398, Fax: 716-829-2437, E-mail: ehayden@buffalo.edu.

See Close-Up on page 215.

State University of New York at New Paltz, Graduate School, Faculty of Liberal Arts and Sciences, Department of Biology, New Paltz, NY 12561. Offers MA, MAT, MS Ed. *Faculty:* 7 full-time (4 women), 5 part-time/adjunct (0 women). *Students:* 6 full-time (2 women), 11 part-time (4 women). Average age 32. In 2005, 1 degree awarded. *Degree requirements:* For master's, thesis (for some programs), comprehensive exam. *Entrance requirements:* For master's, GRE General Test, GRE Subject Test, minimum GPA of 3.0. Additional exam requirements/recommendations for international students: Required—TOEFL (minimum score 550 paper-based; 213 computer-based). *Application deadline:* For fall admission, 5/15 priority date for domestic students, 3/1 priority date for international students; for spring admission, 11/15 for domestic students, 10/1 for international students. Application fee: $50. *Expenses:* Tuition, state resident: full-time $6,900; part-time $288 per credit hour. Tuition, nonresident: full-time $10,920; part-time $455 per credit hour. Required fees: $27 per credit. $130

per semester. *Financial support:* In 2005–06, 2 research assistantships with partial tuition reimbursements (averaging $5,000 per year), 3 teaching assistantships with partial tuition reimbursements (averaging $5,000 per year) were awarded; Federal Work-Study and institutionally sponsored loans also available. *Unit head:* Dr. Hon Hing Ho, Chairman, 845-257-3770. *Application contact:* Dr. Thomas Nolen, Coordinator, 845-257-3738, E-mail: nolent@newpaltz.edu.

State University of New York College at Brockport, School of Letters and Sciences, Department of Biological Sciences, Brockport, NY 14420-2997. Offers MS. Part-time programs available. *Students:* 5 full-time (4 women), 11 part-time (4 women); includes 1 minority (Hispanic American) 12 applicants; 83% accepted, 6 enrolled. In 2005, 3 degrees awarded. *Degree requirements:* For master's, thesis or alternative, comprehensive exam. *Entrance requirements:* For master's, GRE General Test; GRE Subject test in biology, or biochemistry, cell and molecular biology, letters of recommendation, minimum GPA of 3.0, scientific writing sample. Additional exam requirements/recommendations for international students: Required—TOEFL (minimum score 550 paper-based; 213 computer-based). *Application deadline:* For fall admission, 7/15 for domestic students, 7/15 for international students; for spring admission, 11/15 for international students, 11/15 for international students. Application fee: $50. *Expenses:* Tuition, state resident: full-time $6,900; part-time $288 per credit. Tuition, nonresident: full-time $10,920; part-time $455 per credit. Required fees: $685; $28 per credit. *Financial support:* In 2005–06, 3 research assistantships with tuition reimbursements, 7 teaching assistantships with tuition reimbursements (averaging $6,000 per year) were awarded; career-related internships or fieldwork, Federal Work-Study, scholarships/grants, and unspecified assistantships also available. Support available to part-time students. Financial award application deadline: 3/15; financial award applicants required to submit FAFSA. *Faculty research:* Ecology and environmental science, physiology, cellular molecular biology, microbiology, plant and animal biology. *Unit head:* Dr. Stuart Tsubota, Chairperson, 585-395-2193, Fax: 585-395-2741, E-mail: stsubota@brockport.edu. *Application contact:* Dr. Adam Rich, Graduate Program Director, 585-395-5740, Fax: 585-395-2741, E-mail: arich@brockport.edu.

State University of New York College at Oneonta, Graduate Studies, Department of Biology, Oneonta, NY 13820-4015. Offers MA. Part-time and evening/weekend programs available. *Students:* 5. *Degree requirements:* For master's, comprehensive exam. *Entrance requirements:* For master's, GRE General Test, GRE Subject Test. *Application deadline:* For fall admission, 3/25 for domestic students; for spring admission, 10/1 priority date for domestic students. Applications are processed on a rolling basis. Application fee: $50. *Expenses:* Tuition, state resident: full-time $6,900. Tuition, nonresident: full-time $10,920. *Unit head:* Dr. William Pietraface, Chair, 607-436-3703, Fax: 607-436-3646, E-mail: pietrawj@oneonta.edu.

State University of New York Downstate Medical Center, School of Graduate Studies, Brooklyn, NY 11203-2098. Offers MS, PhD, MD/PhD. *Faculty:* 119 full-time (29 women), 18 part-time/adjunct (8 women). *Students:* 71 full-time (32 women); includes 2 African Americans, 8 Asian Americans or Pacific Islanders, 1 Hispanic American, 41 international. Average age 30. 81 applicants, 23% accepted, 11 enrolled. In 2005, 1 master's, 10 doctorates awarded. *Degree requirements:* For doctorate, thesis/dissertation, registration. *Entrance requirements:* For doctorate, GRE. *Application deadline:* For fall admission, 7/1 for domestic students. Application fee: $35. *Financial support:* In 2005–06, 70 students received support, including 70 teaching assistantships with tuition reimbursements available (averaging $23,500 per year); fellowships, research assistantships, career-related internships or fieldwork, Federal Work-Study, health care benefits, and tuition waivers (full and partial) also available. *Faculty research:* Cellular and molecular neurobiology, role of oncogenes in early cardiogenesis, mechanism of gene regulation, cardiovascular physiology, yeast molecular genetics. *Unit head:* Dr. Susan Schwartz-Giblin, Dean, 718-270-1155. *Application contact:* Denise Sheares, Admissions Officer, 718-270-2738, Fax: 718-270-3378, E-mail: dsheares@downstate.edu.

See Close-Up on page 219.

State University of New York, Fredonia, Graduate Studies, Department of Biology, Fredonia, NY 14063-1136. Offers MS, MS Ed. Part-time and evening/weekend programs available. *Faculty:* 8 full-time (2 women). *Students:* 7 full-time (3 women), 4 part-time (2 women). Average age 25. In 2005, 8 degrees awarded. *Degree requirements:* For master's, thesis optional. *Application deadline:* For fall admission, 8/5 for domestic students; for spring admission, 12/1 for domestic students. Application fee: $50. *Expenses:* Tuition, state resident: full-time $6,912; part-time $288 per credit hour. Tuition, nonresident: full-time $10,920; part-time $455 per credit hour. Required fees: $543; $45 per credit hour. *Financial support:* In 2005–06, 3 teaching assistantships with partial tuition reimbursements (averaging $5,477 per year) were awarded; research assistantships, tuition waivers (full and partial) also available. Support available to part-time students. Financial award application deadline: 3/15. *Unit head:* Dr. Ted Lee, Chairman, 716-673-2282, E-mail: teodore.lee@fredonia.edu.

State University of New York Upstate Medical University, College of Graduate Studies, Syracuse, NY 13210-2334. Offers MS, PhD, MD/PhD. Part-time programs available. *Faculty:* 59 full-time (12 women), 3 part-time/adjunct (1 woman). *Students:* 123 (60 women); includes 22 minority (3 African Americans, 11 Asian Americans or Pacific Islanders, 8 Hispanic Americans) 33 international. Average age 27. 189 applicants, 33% accepted, 23 enrolled. In 2005, 7 master's, 9 doctorates awarded. Terminal master's awarded for partial completion of doctoral program. *Degree requirements:* For master's, thesis/dissertation; for doctorate, thesis/dissertation, comprehensive exam. *Entrance requirements:* For master's and doctorate, GRE General Test, GRE Subject Test, interview. Additional exam requirements/recommendations for international students: Required—TOEFL. *Application deadline:* For fall admission, 1/15 priority date for domestic students, 1/15 priority date for international students. Applications are processed on a rolling basis. Application fee: $40. *Financial support:* In 2005–06, fellowships (averaging $15,300 per year), research assistantships (averaging $15,300 per year) were awarded; Federal Work-Study, institutionally sponsored loans, and scholarships/grants also available. Support available to part-time students. Financial award application deadline: 4/15; financial award applicants required to submit FAFSA. *Faculty research:* Cancer research, cardiovascular disease, microbiology and virology, neuroscience, developmental and structural biology. *Unit head:* Dr. Maxwell M. Mozell, Dean, 315-464-4538, Fax: 315-464-4544. *Application contact:* Therese A. Brown, Information Contact, 315-464-4541, Fax: 315-464-4544, E-mail: gradstud@upstate.edu.

See Close-Up on page 221.

Stephen F. Austin State University, Graduate School, College of Sciences and Mathematics, Department of Biology, Nacogdoches, TX 75962. Offers MS. *Faculty:* 16 full-time (1 woman), 4 part-time/adjunct (1 woman). *Students:* 19 full-time (11 women), 9 part-time (5 women); includes 5 minority (3 African Americans, 1 Asian American or Pacific Islander, 1 Hispanic American), 3 international. 12 applicants, 100% accepted. In 2005, 10 degrees awarded. *Degree requirements:* For master's, thesis optional. *Entrance requirements:* For master's, GRE General Test, minimum GPA of 2.8 in last 60 hours, 2.5 overall. Additional exam requirements/recommendations for international students; Required—TOEFL. *Application deadline:* For fall admission, 8/1 for domestic students; for spring admission, 12/15 for domestic students. Applications are processed on a rolling basis. Application fee: $0 ($50 for international students). *Expenses:* Tuition, state resident: full-time $2,628; part-time $146 per credit hour. Tuition, nonresident: full-time $7,596; part-time $422 per credit hour. Required fees: $900; $170. *Financial support:* In 2005–06, 13 teaching assistantships (averaging $9,500 per year) were awarded; Federal Work-Study and unspecified assistantships also available. Financial award application deadline: 3/1. *Unit head:* Dr. Don A. Hay, Chair, 936-468-3601, E-mail: dhay@sfasu.edu.

Stony Brook University, State University of New York, Graduate School, College of Arts and Sciences, Program in Biological and Biomedical Sciences, Stony Brook, NY 11794. Offers PhD. *Expenses:* Tuition, state resident: full-time $6,900; part-time $288 per credit.

Tuition, nonresident: full-time $10,920; part-time $455 per credit. Required fees: $704. *Unit head:* Dr. James V. Staros, Dean, College of Arts and Sciences, 631-632-6999, Fax: 631-632-6900.

Stony Brook University, State University of New York, Health Sciences Center, School of Medicine and Graduate School, Graduate Programs in Medicine, Stony Brook, NY 11794. Offers PhD. *Students:* 101 full-time (57 women); includes 17 minority (8 African Americans, 6 Asian Americans or Pacific Islanders, 3 Hispanic Americans), 30 international. 135 applicants, 26% accepted. In 2005, 21 degrees awarded. *Degree requirements:* For doctorate, thesis/dissertation, exam. *Entrance requirements:* For doctorate, GRE General Test. Additional exam requirements/recommendations for international students: Required—TOEFL. *Application deadline:* For fall admission, 1/15 for domestic students. Application fee: $50. Electronic applications accepted. *Expenses: Contact institution. Financial support:* Fellowships, research assistantships, teaching assistantships, career-related internships or fieldwork and Federal Work-Study available. Financial award application deadline: 3/15. Total annual research expenditures: $51.9 million. *Application contact:* Dr. William Jungers, Chairman, Committee on Admissions, 631-444-2113, Fax: 631-444-6032, E-mail: admissions@dean.som.sunysb.edu.

Stony Brook University, State University of New York, Health Sciences Center, School of Medicine, Medical Scientist Training Program, Stony Brook, NY 11794. Offers MD/PhD. *Application deadline:* For fall admission, 1/15 for domestic students. *Expenses:* Tuition, state resident: full-time $6,900; part-time $288 per credit. Tuition, nonresident: full-time $10,920; part-time $455 per credit. Required fees: $704. *Financial support:* Tuition waivers (full) available. *Application contact:* Dr. William Jungers, Chairman, Committee on Admissions, 631-444-2113, Fax: 631-444-6032, E-mail: admissions@dean.som.sunysb.edu.

Announcement: The purpose of the Medical Scientist Training Program (MSTP) is to train academic medical scientists for both research and teaching in medical schools and research institutions. Students in this program will be equipped to study major medical problems at the basic level and, at the same time, recognize the clinical significance of their pursuits and discoveries. Interested candidates should complete the MSTP Application of Interest and submit it along with the general application materials for the School of Medicine. MSTP applicants may be invited for additional interviews with the MSTP Committee on Admissions. The MSTP is a fully funded program. The M.D. with Recognition in Research is a partially funded program of supervised research completed within the 4 years of medical training. No further application is required at this time. Students indicate their interest in this program and are eligible in the spring of their first year.

See Close-Up on page 223.

Sul Ross State University, School of Arts and Sciences, Department of Biology, Alpine, TX 79832. Offers MS. Part-time programs available. *Degree requirements:* For master's, thesis optional. *Entrance requirements:* For master's, GRE General Test, minimum GPA of 2.5 in last 60 hours of undergraduate work. *Faculty research:* Plant-animal interaction, Chihuahuan desert biology, insect biological control, plant and animal systematics, wildlife biology.

Syracuse University, Graduate School, College of Arts and Sciences, Department of Biology, Syracuse, NY 13244. Offers biology (MS, PhD); structural biology, biochemistry and biophysics (PhD). *Students:* 50 full-time (22 women), 5 part-time (4 women). 107 applicants, 12% accepted, 8 enrolled.Terminal master's awarded for partial completion of doctoral program. *Degree requirements:* For master's and doctorate, thesis/dissertation. *Entrance requirements:* For master's and doctorate, GRE General Test, GRE Subject Test. Additional exam requirements/recommendations for international students: Required—TOEFL. *Application deadline:* For fall admission, 1/10 for domestic students. Applications are processed on a rolling basis. Application fee: $65. Electronic applications accepted. *Financial support:* Fellowships with full tuition reimbursements, research assistantships with full and partial tuition reimbursements, teaching assistantships with full tuition reimbursements, tuition waivers (partial) available. *Faculty research:* Cell signaling, plant ecosystem ecology, aquatic ecology, genetics and molecular biology of color vision, ion transport by cell membranes. *Unit head:* Dr. John M Russell, Chairperson, 315-443-3962. *Application contact:* Evelyn Lott, Information Contact, 315-443-9154, Fax: 315-443-2012, E-mail: ealott@syr.edu.

See Close-Up on page 225.

Tarleton State University, College of Graduate Studies, College of Science and Technology, Department of Biological Sciences, Stephenville, TX 76402. Offers biology (MS). Part-time and evening/weekend programs available. *Faculty:* 7 full-time (0 women), 3 part-time/adjunct (0 women). *Students:* 7 full-time (3 women), 10 part-time (6 women); includes 1 minority (Asian American or Pacific Islander) Average age 36. In 2005, 4 degrees awarded. *Degree requirements:* For master's, thesis (for some programs), comprehensive exam. *Entrance requirements:* For master's, GRE General Test, minimum GPA of 3.0. Additional exam requirements/recommendations for international students: Required—TOEFL (minimum score 550 paper-based; 220 computer-based). *Application deadline:* For fall admission, 8/5 for domestic students; for spring admission, 12/1 for domestic students. Applications are processed on a rolling basis. Application fee: $25 ($75 for international students). *Financial support:* In 2005–06, 1 research assistantship (averaging $12,000 per year), 8 teaching assistantships (averaging $12,000 per year) were awarded; career-related internships or fieldwork and Federal Work-Study also available. Support available to part-time students. Financial award application deadline: 5/1; financial award applicants required to submit FAFSA. *Unit head:* Dr. John Calahan, Head, 254-968-9159.

Temple University, Graduate School, College of Science and Technology, Department of Biology, Philadelphia, PA 19122-6096. Offers MS, PhD. *Faculty:* 19 full-time (5 women). *Students:* 2 full-time (both women), 26 part-time (11 women); includes 6 minority (all Asian Americans or Pacific Islanders), 10 international. 40 applicants, 33% accepted, 7 enrolled. In 2005, 4 master's, 2 doctorates awarded. Terminal master's awarded for partial completion of doctoral program. *Degree requirements:* For master's and doctorate, thesis/dissertation. *Entrance requirements:* For master's and doctorate, GRE General Test, minimum GPA of 3.0 Additional exam requirements/recommendations for international students: Required—TOEFL (minimum score 600 paper-based; 250 computer-based). *Application deadline:* For fall admission, 4/15 for domestic students, 12/15 for international students; for spring admission, 11/15 for domestic students, 8/1 for international students. Applications are processed on a rolling basis. Application fee: $50. *Expenses:* Tuition, state resident: full-time $8,694; part-time $483 per credit. Tuition, nonresident: full-time $12,672; part-time $704 per credit. Required fees: $500; $122 per semester. Tuition and fees vary according to course level, campus/location and program. *Financial support:* Fellowships, research assistantships, teaching assistantships, Federal Work-Study and tuition waivers (full) available. Financial award application deadline: 1/15; financial award applicants required to submit FAFSA. *Faculty research:* Membrane proteins, genetics, molecular biology, neuroscience, aquatic biology. Total annual research expenditures: $1 million. *Unit head:* Dr. Shohreh Amini, Chair, 215-204-0604, Fax: 215-204-6646, E-mail: shohreh.amini@temple.edu.

See Close-Up on page 227.

Temple University, Health Sciences Center, School of Medicine and Graduate School, Graduate Programs in Medicine, Philadelphia, PA 19122-6096. Offers MS, PhD, MD/PhD. *Faculty:* 77 full-time (13 women). *Students:* 63 full-time (33 women), 78 part-time (38 women); includes 23 minority (9 African Americans, 11 Asian Americans or Pacific Islanders, 3 Hispanic Americans), 53 international. 162 applicants, 34% accepted, 28 enrolled. In 2005, 1 master's, 19 doctorates awarded. Terminal master's awarded for partial completion of doctoral program. *Degree requirements:* For master's, thesis; for doctorate, thesis/dissertation, research seminars. *Entrance requirements:* For master's and doctorate, GRE General Test. Additional exam requirements/recommendations for international students: Required—TOEFL. Application fee: $50. Electronic applications accepted. *Expenses: Contact institution.* Tuition and fees vary according to course level, campus/location and program. *Financial support:* Fellowships, research assistantships,

career-related internships or fieldwork, Federal Work-Study, institutionally sponsored loans, scholarships/grants, and tuition waivers (full and partial) available. Support available to part-time students. Financial award application deadline: 1/15; financial award applicants required to submit FAFSA. *Faculty research:* Molecular biology and biochemistry; cardiovascular, renal, and neurophysiological pharmacology; reproductive and developmental biology; immunology and microbiology; cancer research. Total annual research expenditures: $8.3 million. *Unit head:* Dr. Barrie Ashby, Associate Dean for Graduate Studies, 215-707-3252, Fax: 215-707-4725, E-mail: ashbar@temple.edu.

Tennessee State University, Graduate School, College of Arts and Sciences, Department of Biological Sciences, Nashville, TN 37209-1561. Offers MS, PhD. *Faculty:* 8 full-time (3 women). *Students:* 36 full-time (25 women), 1 (woman) part-time; includes 26 minority (24 African Americans, 2 Asian Americans or Pacific Islanders), 4 international. Average age 27. 30 applicants, 70% accepted. In 2005, 11 degrees awarded. *Degree requirements:* For master's and doctorate, thesis/dissertation. *Entrance requirements:* For master's, GRE General Test, GRE Subject Test, minimum GPA of 2.5. *Application deadline:* Applications are processed on a rolling basis. Application fee: $15. *Financial support:* Unspecified assistantships available. Support available to part-time students. Financial award application deadline: 5/1. *Faculty research:* Soybean tissue culture, neurochemistry, microbial genetics. *Unit head:* Dr. Terrence Johnson, Head, 615-963-5748.

See Close-Up on page 229.

Tennessee Technological University, Graduate School, College of Arts and Sciences, Department of Biology, Cookeville, TN 38505. Offers environmental biology (MS); fish, game, and wildlife management (MS). Part-time programs available. *Faculty:* 22 full-time (2 women). *Students:* 22 full-time (8 women), 17 part-time (7 women); includes 3 minority (1 African American, 2 Asian Americans or Pacific Islanders). Average age 25. 14 applicants, 50% accepted, 7 enrolled. In 2005, 9 degrees awarded. *Degree requirements:* For master's, thesis. *Entrance requirements:* For master's, GRE General Test. Additional exam requirements/recommendations for international students: Required—TOEFL. *Application deadline:* For fall admission, 3/1 for domestic students; for spring admission, 8/1 for domestic students. Application fee: $25 ($30 for international students). *Expenses:* Tuition, state resident: full-time $8,421; part-time $307 per hour. Tuition, nonresident: full-time $22,389; part-time $711 per hour. *Financial support:* In 2005–06, 22 research assistantships (averaging $9,000 per year), 9 teaching assistantships (averaging $7,500 per year) were awarded. Financial award application deadline: 4/1. *Faculty research:* Aquatics, environmental studies. *Unit head:* Dr. Daniel Combs, Interim Chairperson, 931-372-3134, Fax: 931-372-6257, E-mail: dcombs@tntech.edu. *Application contact:* Dr. Francis O. Otuonye, Associate Vice President for Research and Graduate Studies, 931-372-3233, Fax: 931-372-3497, E-mail: fotuonye@tntech.edu.

Texas A&M International University, Office of Graduate Studies and Research, College of Arts and Sciences, Department of Biology and Chemistry, Laredo, TX 78041-1900. Offers biology (MS). *Expenses:* Tuition, state resident: full-time $1,580. Tuition, nonresident: full-time $5,432. Required fees: $3,808. *Application contact:* Rosie Espinoza, Director of Admissions, 956-326-2200, Fax: 956-326-2199, E-mail: enroll@tamiu.edu.

Texas A&M University, College of Science, Department of Biology, College Station, TX 77843. Offers biology (MS, PhD); botany (MS, PhD); microbiology (MS, PhD); molecular and cell biology (PhD); neuroscience (MS, PhD); zoology (MS, PhD). *Faculty:* 25 full-time (4 women). *Students:* 87 full-time (46 women), 10 part-time (4 women); includes 11 minority (1 African American, 1 American Indian/Alaska Native, 4 Asian Americans or Pacific Islanders, 5 Hispanic Americans), 29 international. Average age 28. 101 applicants, 35% accepted, 29 enrolled. In 2005, 2 master's, 12 doctorates awarded. *Degree requirements:* For master's, thesis or alternative, registration; for doctorate, thesis/dissertation, comprehensive exam, registration. *Entrance requirements:* For master's and doctorate, GRE General Test. Additional exam requirements/recommendations for international students: Required—TOEFL. *Application deadline:* For fall admission, 1/15 for domestic students. Applications are processed on a rolling basis. Application fee: $50 ($75 for international students). Electronic applications accepted. *Expenses:* Tuition, state resident: full-time $4,488; part-time $187 per credit hour. Tuition, nonresident: full-time $11,112; part-time $463 per credit hour. Required fees: $1,974. *Financial support:* Fellowships, research assistantships, teaching assistantships available. Financial award application deadline: 4/1; financial award applicants required to submit FAFSA. *Unit head:* Dr. Paul E. Hardin, Chair, 979-845-0184, Fax: 979-845-2891. *Application contact:* Graduate Advisor, 979-845-7755.

Texas A&M University–Commerce, Graduate School, College of Arts and Sciences, Department of Biological and Earth Sciences, Commerce, TX 75429-3011. Offers M Ed, MS. *Faculty:* 7 full-time (2 women), 2 part-time/adjunct (0 women). *Students:* 10 full-time (5 women), 13 part-time (10 women); includes 7 minority (2 African Americans, 4 Asian Americans or Pacific Islanders, 1 Hispanic American), 3 international. Average age 36. In 2005, 3 degrees awarded. *Degree requirements:* For master's, thesis (for some programs), comprehensive exam. *Entrance requirements:* For master's, GRE General Test. *Application deadline:* For fall admission, 6/1 for domestic students; for spring admission, 11/1 priority date for domestic students. Applications are processed on a rolling basis. Application fee: $0 ($25 for international students). Electronic applications accepted. *Financial support:* In 2005–06, research assistantships (averaging $7,875 per year), teaching assistantships (averaging $7,875 per year) were awarded; Federal Work-Study, institutionally sponsored loans, and scholarships/grants also available. Financial award application deadline: 5/1; financial award applicants required to submit FAFSA. *Faculty research:* Microbiology, botany, environmental science, birds. Total annual research expenditures: $3,000. *Unit head:* Dr. Don R. Lee, Interim Head, 903-886-5378, Fax: 903-886-5997. *Application contact:* Tammi Thompson, Graduate Admissions Adviser, 843-886-5167, Fax: 843-886-5165, E-mail: tammi_thompson@tamu-commerce.edu.

Texas A&M University–Corpus Christi, Graduate Studies and Research, College of Science and Technology, Program in Sciences, Corpus Christi, TX 78412-5503. Offers biology (MS); environmental sciences (MS); mariculture (MS). Part-time and evening/weekend programs available. *Degree requirements:* For master's, thesis (for some programs), comprehensive exam, registration. *Entrance requirements:* For master's, GRE General Test. Additional exam requirements/recommendations for international students: Required—TOEFL. Electronic applications accepted.

Texas A&M University–Kingsville, College of Graduate Studies, College of Arts and Sciences, Department of Biology, Kingsville, TX 78363. Offers MS. Part-time programs available. *Degree requirements:* For master's, thesis or alternative, comprehensive exam. *Entrance requirements:* For master's, GRE General Test, minimum GPA of 3.0. Additional exam requirements/recommendations for international students: Required—TOEFL. *Faculty research:* Venom physiology, monoclonal research with venom, shore bird ecology, metabolism of foreign amino acids.

Texas A&M University System Health Science Center, Baylor College of Dentistry, Graduate Division, Department of Biomedical Sciences, College Station, TX 77840. Offers MS, PhD. Part-time programs available. Terminal master's awarded for partial completion of doctoral program. *Degree requirements:* For master's and doctorate, thesis/dissertation. *Entrance requirements:* For master's, GRE General Test; for doctorate, GRE General Test, DDS or DMD. Additional exam requirements/recommendations for international students: Required—TOEFL. *Faculty research:* Craniofacial biology, aging, neuroscience, physiology, molecular/cellular biology.

Texas A&M University System Health Science Center, Graduate School of Biomedical Sciences, Department of Medical Physiology, College Station, TX 77840. Offers PhD. *Degree requirements:* For doctorate, thesis/dissertation. *Entrance requirements:* For doctorate, GRE General Test. *Faculty research:* Cardiovascular physiology, vascular cell and molecular biology.

Biological and Biomedical Sciences—General

Texas A&M University System Health Science Center, Institute of Biosciences and Technology, College Station, Houston, TX 77030-3303. Offers medical sciences (PhD). Degree awarded by the Graduate School for Biomedical Sciences. *Degree requirements:* For doctorate, thesis/dissertation. *Entrance requirements:* For doctorate, GRE General Test. Additional exam requirements/recommendations for international students: Required—TOEFL, TWE. Expenses: Contact institution. *Faculty research:* Cancer biology, DNA structure, extracellular matrix biology, development, birth defects.

See Close-Up on page 231.

Texas Christian University, College of Science and Engineering, Department of Biology, Fort Worth, TX 76129-0002. Offers biology (MA, MS); environmental sciences (MS), including earth sciences, ecology. Part-time and evening/weekend programs available. *Entrance requirements:* For master's, GRE General Test, GRE Subject Test. Additional exam requirements/recommendations for international students: Required—TOEFL. *Application deadline:* For fall admission, 3/1 for domestic students; for spring admission, 12/1 for domestic students. Applications are processed on a rolling basis. Application fee: $0. *Expenses:* Tuition: Part-time $740 per credit hour. *Financial support:* Unspecified assistantships available. Financial award application deadline: 3/1. *Unit head:* Dr. Ray Drenner, Chairperson, 817-257-7165. *Application contact:* Dr. Bonnie Melhart, Associate Dean, College of Science and Engineering, E-mail: b.melhart@tcu.edu.

Texas Southern University, Graduate School, School of Science and Technology, Department of Biology, Houston, TX 77004-4584. Offers MS. Part-time and evening/weekend programs available. *Faculty:* 6 full-time (2 women), 1 part-time/adjunct (0 women). *Students:* 15 full-time (10 women), 10 part-time (5 women); 23 African Americans, 1 Asian American or Pacific Islander, 1 Hispanic American. Average age 31. 3 applicants, 100% accepted, 2 enrolled. In 2005, 4 degrees awarded. *Degree requirements:* For master's, one foreign language, thesis, comprehensive exam. *Entrance requirements:* For master's, GRE General Test, minimum GPA of 2.5. Additional exam requirements/recommendations for international students: Required—TOEFL. *Application deadline:* For fall admission, 7/15 for domestic students. Applications are processed on a rolling basis. Application fee: $35 ($75 for international students). *Expenses:* Tuition, state resident: full-time $1,728; part-time $96 per credit hour. Tuition, nonresident: full-time $6,174; part-time $343 per credit hour. Tuition and fees vary according to course load and degree level. *Financial support:* Fellowships, teaching assistantships, career-related internships or fieldwork, Federal Work-Study, and institutionally sponsored loans available. Financial award application deadline: 5/1. *Faculty research:* Microbiology, cell and molecular biology, biochemistry, biochemical virology, biophysics. *Unit head:* Dr. Debabrate Ghosh, Chairman, 713-313-1032. *Application contact:* Shirley Harris, Information Contact, 713-313-7838.

Texas State University-San Marcos, Graduate School, College of Science, Department of Biology, Program in Biology, San Marcos, TX 78666. Offers M Ed, MA, MS. *Students:* 23 full-time (12 women), 14 part-time (10 women); includes 6 minority (1 American Indian/Alaska Native, 3 Asian Americans or Pacific Islanders, 2 Hispanic Americans), 1 international. Average age 30. 17 applicants, 88% accepted, 9 enrolled. In 2005, 21 degrees awarded. *Entrance requirements:* For master's, GRE General Test, minimum GPA of 2.75 in last 60 hours of undergraduate work. Additional exam requirements/recommendations for international students: Required—TOEFL. *Application deadline:* For fall admission, 6/15 for domestic students, 6/1 for international students; for spring admission, 10/15 for domestic students, 10/1 for international students. Applications are processed on a rolling basis. Application fee: $40 ($90 for international students). *Expenses:* Tuition, area resident: Part-time $116 per credit. Tuition, state resident: full-time $3,168; part-time $176 per credit. Tuition, nonresident: full-time $8,136; part-time $452 per credit. Required fees: $1,112; $74 per credit. Full-time tuition and fees vary according to course load. *Financial support:* In 2005–06, 28 students received support; research assistantships, teaching assistantships available. *Unit head:* Dr. David Lemker, Graduate Advisor, 512-245-3364, E-mail: dl10@txstate.edu.

Texas State University-San Marcos, Graduate School, Interdisciplinary Studies Program in Biology, San Marcos, TX 78666. Offers MSIS. *Students:* 4 applicants, 25% accepted. *Degree requirements:* For master's, comprehensive exam. *Application deadline:* For fall admission, 6/15 for domestic students; for spring admission, 10/15 priority date for domestic students. Applications are processed on a rolling basis. Application fee: $40 ($90 for international students). *Expenses:* Tuition, area resident: Part-time $116 per credit. Tuition, state resident: full-time $3,168; part-time $176 per credit. Tuition, nonresident: full-time $8,136; part-time $452 per credit. Required fees: $1,112; $74 per credit. Full-time tuition and fees vary according to course load. *Financial support:* Application deadline: 4/1; *Unit head:* Dr. David Lemker, Graduate Advisor, 512-245-3364, E-mail: dl10@txstate.edu.

Texas Tech University, Graduate School, College of Arts and Sciences, Department of Biological Sciences, Lubbock, TX 79409. Offers biological informatics (MS); biology (MS, PhD); microbiology (MS); zoology (MS, PhD). Part-time programs available. *Faculty:* 29 full-time (4 women). *Students:* 101 full-time (47 women), 9 part-time (2 women); includes 7 minority (2 Asian Americans or Pacific Islanders, 5 Hispanic Americans), 42 international. Average age 30. 74 applicants, 50% accepted, 16 enrolled. In 2005, 8 master's, 5 doctorates awarded. *Degree requirements:* For master's, thesis (for some programs); for doctorate, thesis/dissertation. *Entrance requirements:* For master's and doctorate, GRE General Test. Additional exam requirements/recommendations for international students: Required—TOEFL (minimum score 550 paper-based; 213 computer-based). *Application deadline:* Applications are processed on a rolling basis. Application fee: $50 ($60 for international students). Electronic applications accepted. *Expenses:* Tuition, state resident: full-time $4,296. Tuition, nonresident: full-time $10,920. Required fees: $1,992. Tuition and fees vary according to program. *Financial support:* In 2005–06, 51 students received support, including 22 research assistantships with partial tuition reimbursements available (averaging $13,849 per year), 74 teaching assistantships with partial tuition reimbursements available (averaging $13,899 per year); career-related internships or fieldwork, Federal Work-Study, and institutionally sponsored loans also available. Support available to part-time students. Financial award application deadline: 4/15; financial award applicants required to submit FAFSA. *Faculty research:* Genome organization and evolution, plant stress response, climate change on arid ecosystems, cell signaling hormone regulation. Total annual research expenditures: $3.1 million. *Unit head:* Dr. John C. Zak, Chair, 806-742-2715, Fax: 806-742-2963, E-mail: john.zak@ttu.edu. *Application contact:* Dr. Randall M. Jeter, Graduate Adviser, 806-742-2710, Fax: 806-742-2963, E-mail: randall.jeter@ttu.edu.

Texas Tech University Health Sciences Center, Graduate School of Biomedical Sciences, Lubbock, TX 79430-0002. Offers MS, PhD, MD/PhD, MS/PhD. *Faculty:* 94 full-time (21 women), 10 part-time/adjunct (2 women). *Students:* 79 full-time (32 women), 1 (woman) part-time; includes 9 minority (1 African American, 1 American Indian/Alaska Native, 5 Asian Americans or Pacific Islanders, 2 Hispanic Americans), 34 international. Average age 29. 131 applicants, 12% accepted, 11 enrolled. In 2005, 8 master's, 4 doctorates awarded. Terminal master's awarded for partial completion of doctoral program. *Degree requirements:* For master's and doctorate, thesis/dissertation. *Entrance requirements:* For master's and doctorate, GRE General Test, minimum GPA of 3.0. Additional exam requirements/recommendations for international students: Required—TOEFL (minimum score 550 paper-based; 213 computer-based). *Application deadline:* For fall admission, 5/15 priority date for domestic students, 4/15 priority date for international students; for spring admission, 11/15 priority date for domestic students, 10/15 priority date for international students. Applications are processed on a rolling basis. Application fee: $45. Electronic applications accepted. *Financial support:* In 2005–06, 6 fellowships with partial tuition reimbursements (averaging $20,500 per year), 43 research assistantships with full and partial tuition reimbursements (averaging $20,500 per year) were awarded; institutionally sponsored loans and scholarships/grants also available. Financial award applicants required to submit FAFSA. *Faculty research:* Genetics of neurological disorders, hemodynamics to prevent DVT, toxin A synthesis, DA neurons, peroxidases. Total annual research expenditures:

$6.5 million. *Unit head:* Dr. Barbara C. Pence, Associate Dean, 806-743-2556, Fax: 806-743-2656, E-mail: acagsbs@ttuhsc.edu. *Application contact:* Pamela Johnson, Director of Graduate Programs, 806-743-2556, Fax: 806-743-2656, E-mail: pamela.johnson@ttuhsc.edu.

Texas Woman's University, Graduate School, College of Arts and Sciences, Department of Biology, Denton, TX 76201. Offers biology (MS); biology teaching (MS); molecular biology (PhD). Part-time programs available. *Students:* 23 full-time (15 women), 13 part-time (12 women); includes 9 minority (6 African Americans, 1 Asian American or Pacific Islander, 2 Hispanic Americans), 19 international. Average age 31. In 2005, 8 master's, 4 doctorates awarded. Terminal master's awarded for partial completion of doctoral program. *Degree requirements:* For master's, thesis (for some programs), comprehensive exam; for doctorate, thesis/dissertation, residency, comprehensive exam. *Entrance requirements:* For master's and doctorate, GRE General Test, 3 letters of reference. Additional exam requirements/recommendations for international students: Required—TOEFL (minimum score 550 paper-based; 213 computer-based). *Application deadline:* Applications are processed on a rolling basis. Application fee: $30 ($50 for international students). Electronic applications accepted. *Expenses:* Tuition, state resident: full-time $5,868; part-time $163 per credit. Tuition, nonresident: full-time $15,948; part-time $443 per credit. Required fees: $20 per credit. $152 per term. *Financial support:* In 2005–06, 8 research assistantships (averaging $10,706 per year), 1 teaching assistantship (averaging $10,706 per year) were awarded; career-related internships or fieldwork, Federal Work-Study, institutionally sponsored loans, scholarships/grants, traineeships, health care benefits, and unspecified assistantships also available. Support available to part-time students. Financial award application deadline: 3/1; financial award applicants required to submit FAFSA. *Faculty research:* Plants and plant viruses, plasmia DNA in bacteria, transcriptional regulation of RNA, hormone actions, serotonin modulation of female reproductive behavior. *Unit head:* Dr. Sarah McIntire, Chair, Mail: 940-898-2351, Fax: 940-898-2382, E-mail: smcintire@mail.twu.edu. *Application contact:* Samuel Wheeler, Coordinator of Graduate Admissions, 940-898-3188, Fax: 940-898-3081, E-mail: wheelersr@twu.edu.

Thomas Jefferson University, Jefferson College of Graduate Studies, Philadelphia, PA 19107. Offers MS, DPT, PhD, Certificate, MD/PhD. Part-time and evening/weekend programs available. Postbaccalaureate distance learning degree programs offered (no on-campus study). *Students:* 346 full-time (264 women), 301 part-time (225 women); includes 115 minority (60 African Americans, 45 Asian Americans or Pacific Islanders, 10 Hispanic Americans), 36 international. Average age 29. 896 applicants, 47% accepted, 243 enrolled. In 2005, 166 master's, 13 doctorates awarded. Terminal master's awarded for partial completion of doctoral program. *Degree requirements:* For master's, thesis (for some programs), registration; for doctorate, thesis/dissertation, comprehensive exam, registration. *Entrance requirements:* For master's, GRE; for doctorate, GRE, minimum GPA of 3.2. Additional exam requirements/recommendations for international students: Required—TOEFL (minimum score 213 computer-based). *Application deadline:* For fall admission, 3/1 priority date for domestic students, 3/1 priority date for international students. Applications are processed on a rolling basis. Application fee: $50. Electronic applications accepted. *Expenses:* Tuition: Full-time $14,894; part-time $800 per credit. *Financial support:* In 2005–06, 251 students received support, including 116 fellowships with full tuition reimbursements available; research assistantships, Federal Work-Study, institutionally sponsored loans, scholarships/grants, and traineeships also available. Support available to part-time students. Financial award application deadline: 5/1; financial award applicants required to submit FAFSA. *Faculty research:* Developmental biology, immunology, genetics, oncology, molecular biology. *Unit head:* Dr. James H. Keen, Dean, 215-503-8982, Fax: 215-503-6690, E-mail: james.keen@jefferson.edu. *Application contact:* Jessie F. Pervall, Director of Admissions, 215-503-0155, Fax: 215-503-9920, E-mail: jessie.pervall@jefferson.edu.

See Close-Up on page 233.

Touro College, Barry Z. Levine School of Health Sciences, Biomedical Sciences Program, New York, NY 10010. Offers MS, MD/MS. Professional course work completed at the Technion Faculty of Medicine, Israel. *Degree requirements:* For master's, comprehensive exam or project. *Entrance requirements:* For master's, GRE Subject Test or MCAT. Expenses: Contact institution.

Towson University, Graduate School, Program in Biology, Towson, MD 21252-0001. Offers MS. Part-time and evening/weekend programs available. *Faculty:* 22 full-time (8 women). *Students:* 59. 20 applicants, 75% accepted, 9 enrolled. In 2005, 16 degrees awarded. *Degree requirements:* For master's, exam, thesis optional. *Entrance requirements:* For master's, GRE General Test (thesis students), minimum GPA of 3.0, 24 credits in related course work. Additional exam requirements/recommendations for international students: Required—TOEFL. *Application deadline:* Applications are processed on a rolling basis. Application fee: $40. Electronic applications accepted. *Financial support:* In 2005–06, 4 research assistantships with full tuition reimbursements (averaging $11,000 per year), 12 teaching assistantships with full tuition reimbursements (averaging $11,000 per year) were awarded; career-related internships or fieldwork, Federal Work-Study, and unspecified assistantships also available. Support available to part-time students. Financial award application deadline: 4/1; financial award applicants required to submit FAFSA. *Faculty research:* Microbiology, molecular biology, ecology, physiology, conservation biology. *Unit head:* Dr. Gail E. Gasparich, Graduate Program Co-Director, 410-704-4515, Fax: 410-704-2405, E-mail: ggasparich@towson.edu. *Application contact:* 410-704-2501, Fax: 410-704-4675, E-mail: grads@towson.edu.

Trent University, Graduate Studies, Program in Applications of Modeling in the Natural and Social Sciences, Peterborough, ON K9J 7B8, Canada. Offers applications of modeling in the natural and social sciences (MA); biology (M Sc, PhD); chemistry (M Sc); computer studies (M Sc); geography (M Sc, PhD); physics (M Sc). Part-time programs available. *Degree requirements:* For master's, thesis. *Entrance requirements:* For master's, honours degree. *Faculty research:* Computation of heat transfer, atmospheric physics, statistical mechanics, stress and coping, evolutionary ecology.

Trent University, Graduate Studies, Program in Watershed Ecosystems and Program in Applications of Modeling in the Natural and Social Sciences, Department of Biology, Peterborough, ON K9J 7B8, Canada. Offers M Sc, PhD. Part-time programs available. *Degree requirements:* For master's and doctorate, thesis/dissertation. *Entrance requirements:* For master's, honours degree; for doctorate, master's degree. *Faculty research:* Aquatic and behavioral ecology, hydrology and limnology, human impact on ecosystems, behavioral ecology of birds, ecology of fish.

Truman State University, Graduate School, Division of Science, Program in Biology, Kirksville, MO 63501-4221. Offers MS. *Students:* 6 full-time. 3 applicants, 67% accepted. In 2005, 1 degree awarded. *Degree requirements:* For master's, thesis, comprehensive exam. *Entrance requirements:* For master's, GRE General Test, minimum GPA of 3.0. Additional exam requirements/recommendations for international students: Required—TOEFL (minimum score 550 paper-based; 213 computer-based). *Application deadline:* For fall admission, 6/15 for domestic students; for spring admission, 11/1 for domestic students. Applications are processed on a rolling basis. Application fee: $0. Electronic applications accepted. *Expenses:* Tuition, state resident: full-time $4,572; part-time $254 per hour. Tuition, nonresident: full-time $7,812; part-time $434 per hour. Required fees: $36 per term. *Financial support:* In 2005–06, research assistantships with tuition reimbursements (averaging $8,000 per year), teaching assistantships with tuition reimbursements (averaging $8,000 per year) were awarded; career-related internships or fieldwork and Federal Work-Study also available. Financial award application deadline: 5/1; financial award applicants required to submit FAFSA. *Unit head:* Dr. Scott Burt, Director, 660-785-7133. *Application contact:* Crista Chappell, Graduate Office Secretary, 660-785-4109, Fax: 660-785-7460, E-mail: gradinfo@truman.edu.

Tufts University, Cummings School of Veterinary Medicine, Program in Comparative Biomedical Sciences, Medford, MA 02155. Offers PhD, DVM/MS. *Faculty:* 13 full-time. *Students:* 10 full-time (9 women); includes 3 minority (all Asian Americans or Pacific Islanders) Average age

Biological and Biomedical Sciences—General

26. 3 applicants, 100% accepted, 3 enrolled. *Degree requirements:* For doctorate, thesis/dissertation. *Entrance requirements:* For doctorate, GRE General Test. Additional exam requirements/recommendations for international students: Required—TOEFL. *Application deadline:* For fall admission, 2/1 for domestic students, 2/1 for international students. Application fee: $75. Electronic applications accepted. *Financial support:* In 2005–06, 5 students received support, including fellowships with full tuition reimbursements available (averaging $20,000 per year), teaching assistantships with full tuition reimbursements available (averaging $5,000 per year) Financial award application deadline: 3/15. *Faculty research:* Infectious disease, reproductive biology. *Unit head:* Dr. Arthur Donohue-Rolfe, Head, 508-887-7919, E-mail: arthur.donohue-rolfe@tufts.edu. *Application contact:* Rebecca Russo, Director of Admissions, 508-839-7920, Fax: 508-839-2953, E-mail: rebecca.russo@tufts.edu.

Tufts University, Graduate School of Arts and Sciences, Department of Biology, Medford, MA 02155. Offers MS, PhD. Part-time programs available. *Faculty:* 12 full-time, 1 part-time/adjunct (0 women). *Students:* 47 (32 women); includes 4 minority (3 Asian Americans or Pacific Islanders, 1 Hispanic American) 13 international. 151 applicants, 17% accepted, 14 enrolled. In 2005, 14 master's, 1 doctorate awarded. Terminal master's awarded for partial completion of doctoral program. *Degree requirements:* For master's, thesis (for some programs); for doctorate, thesis/dissertation. *Entrance requirements:* For master's and doctorate, GRE General Test. Additional exam requirements/recommendations for international students: Required—TOEFL (minimum score 550 paper-based; 213 computer-based). *Application deadline:* For fall admission, 1/15 for domestic students, 12/30 for international students; for spring admission, 10/15 for domestic students, 9/15 for international students. Applications are processed on a rolling basis. Application fee: $65. Electronic applications accepted. *Expenses:* Tuition: Full-time $32,360. Tuition and fees vary according to program. *Financial support:* Research assistantships with full and partial tuition reimbursements, teaching assistantships with full and partial tuition reimbursements, Federal Work-Study, scholarships/grants, and tuition waivers (partial) available. Financial award application deadline: 1/15; financial award applicants required to submit FAFSA. *Unit head:* Dr. Harry Bernheim, Chair, 617-627-3195. *Application contact:* Dr. Juliet A. Fuhrman, Information Contact, 617-627-3195.

See Close-Up on page 235.

Tufts University, Sackler School of Graduate Biomedical Sciences, Boston, MA 02155. Offers MS, PhD, DVM/PhD, MD/PhD. *Faculty:* 155 full-time (41 women). *Students:* 250 full-time (160 women), 2 part-time (1 woman); includes 41 minority (3 African Americans, 27 Asian Americans or Pacific Islanders, 11 Hispanic Americans), 62 international. Average age 28. 696 applicants, 16% accepted, 41 enrolled. In 2005, 9 master's, 31 doctorates awarded. *Degree requirements:* For doctorate, thesis/dissertation. *Entrance requirements:* For doctorate, GRE General Test, 3 letters of reference. Additional exam requirements/recommendations for international students: Required—TOEFL. *Application deadline:* For fall admission, 1/15 priority date for domestic students, 1/15 priority date for international students. Applications are processed on a rolling basis. Application fee: $65. Electronic applications accepted. *Expenses: Contact institution.* *Financial support:* In 2005–06, 228 students received support, including 228 research assistantships with full tuition reimbursements available (averaging $29,000 per year); scholarships/grants, health care benefits, and tuition waivers (full and partial) also available. Financial award application deadline: 1/15. *Faculty research:* Cell biology, molecular biology, biochemistry, genetics, immunology. *Unit head:* Naomi Rosenberg, Dean, 617-636-6767, Fax: 617-636-0375, E-mail: naomi.rosenberg@tufts.edu. *Application contact:* 617-636-6767, Fax: 617-636-0375, E-mail: sackler-school@tufts.edu.

Tulane University, Graduate School, Department of Ecology and Evolutionary Biology, New Orleans, LA 70118-5669. Offers MS, PhD. Terminal master's awarded for partial completion of doctoral program. *Degree requirements:* For master's, thesis or alternative; for doctorate, thesis/dissertation. *Entrance requirements:* For master's, GRE General Test, minimum B average in undergraduate course work; for doctorate, GRE General Test. Additional exam requirements/recommendations for international students: Required—TOEFL; Recommended—TSE. Electronic applications accepted. *Faculty research:* Ichthyology, plant systematics, crustacean endocrinology, ecotoxicology, ornithology.

Tulane University, School of Medicine and Graduate School, Graduate Programs in Medicine, New Orleans, LA 70118-5669. Offers MS, PhD, Diploma, MD/MS, MD/PhD. *Degree requirements:* For doctorate, thesis/dissertation. *Entrance requirements:* For master's, GRE General Test, minimum B average in undergraduate course work; for doctorate, GRE General Test. Additional exam requirements/recommendations for international students: Required—TOEFL or TSE. Expenses: Contact institution.

Tuskegee University, Graduate Programs, College of Agricultural, Environmental and Natural Sciences, Department of Biology, Tuskegee, AL 36088. Offers MS. *Faculty:* 12 full-time (3 women). *Students:* 15 full-time (10 women), 2 part-time (both women); includes 14 minority (all African Americans), 2 international. Average age 26. In 2005, 5 degrees awarded *Degree requirements:* For master's, thesis. *Entrance requirements:* For master's, GRE General Test, GRE Subject Test. Additional exam requirements/recommendations for international students: Required—TOEFL (minimum score 500 paper-based; 173 computer-based). *Application deadline:* For fall admission, 7/15 for domestic students. Applications are processed on a rolling basis. Application fee: $25 ($35 for international students). *Expenses:* Tuition: Full-time $12,400. Required fees: $300; $490 per credit. *Financial support:* Fellowships, teaching assistantships, Federal Work-Study and institutionally sponsored loans available. Support available to part-time students. Financial award application deadline: 4/15. *Unit head:* Dr. Roberta Troy, Head, 334-727-8829.

Uniformed Services University of the Health Sciences, School of Medicine, Programs in Biomedical Sciences, Bethesda, MD 20814-4799. Offers emerging infectious diseases (PhD); medical and clinical psychology (PhD), including clinical psychology, medical psychology; medical history (MMH); microbiology and immunology (PhD); molecular and cell biology (PhD); neuroscience (PhD); pathology (PhD), including comparative pathology, molecular pathobiology; preventive medicine and biometrics (MPH, MSPH, MTMH, Dr PH, PhD), including environmental health science (PhD), medical zoology (PhD), public health (MPH, MSPH, Dr PH), tropical medicine and hygiene (MTMH). *Faculty:* 319 full-time (100 women), 3,978 part-time/adjunct (798 women). *Students:* 164 full-time (90 women), 3 part-time (1 woman); includes 36 minority (14 African Americans, 3 American Indian/Alaska Native, 16 Asian Americans or Pacific Islanders, 5 Hispanic Americans), 9 international. Average age 28. 256 applicants, 39% accepted, 66 enrolled. In 2005, 46 master's, 13 doctorates awarded. Terminal master's awarded for partial completion of doctoral program. *Median time to degree:* Of those who began their doctoral program in fall 1997, 100% received their degree in 8 years or less. *Degree requirements:* For master's, thesis or alternative, comprehensive exam; for doctorate, thesis/dissertation, qualifying exam, comprehensive exam. *Entrance requirements:* For master's, GRE General Test; for doctorate, GRE General Test, minimum GPA of 3.0. Additional exam requirements/recommendations for international students: Required—TOEFL. *Application deadline:* For fall admission, 1/15 for domestic students. Applications are processed on a rolling basis. Application fee: $0. *Financial support:* In 2005–06, fellowships with full tuition reimbursements (averaging $23,000 per year), research assistantships with full tuition reimbursements (averaging $23,000 per year) were awarded; career-related internships or fieldwork and tuition waivers (full) also available. *Unit head:* Dr. Eleanor S. Metcalf, Associate Dean, 301-295-1104, E-mail: emetcalf@usuhs.mil. *Application contact:* Janet M. Anastasi, Graduate Program Coordinator, 301-295-9474, Fax: 301-295-6772, E-mail: janastasi@usuhs.mil.

See Close-Up on page 237.

Université de Moncton, Faculty of Science, Department of Biology, Moncton, NB E1A 3E9, Canada. Offers M Sc. *Degree requirements:* For master's, one foreign language, thesis. *Entrance requirements:* For master's, minimum GPA of 3.0. Electronic applications accepted. *Faculty research:* Terrestrial ecology, aquatic ecology, marine biology, aquaculture, ethology, biotechnology.

Université de Montréal, Faculty of Graduate Studies, Faculty of Arts and Sciences, Department of Biological Sciences, Montréal, QC H3C 3J7, Canada. Offers M Sc, PhD. Part-time programs available. *Faculty:* 35 full-time (10 women), 6 part-time/adjunct (1 woman). *Students:* 82 full-time (51 women), 1 part-time. 47 applicants, 32% accepted, 15 enrolled. In 2005, 52 master's, 17 doctorates awarded. *Degree requirements:* For master's, thesis; for doctorate, thesis/dissertation, general exam. *Entrance requirements:* For doctorate, MS in biology or related field. *Application deadline:* For fall and spring admission, 2/1. For winter admission, 11/1 for domestic students. Application fee: $30. Electronic applications accepted. *Financial support:* Fellowships, research assistantships, teaching assistantships available. Support available to part-time students. Financial award application deadline: 9/1. *Faculty research:* Fresh water ecology, plant biotechnology, neurobiology, genetics, cell physiology. *Unit head:* Thérèse Cabana, Chairman, 514-343-6878, Fax: 514-343-2293. *Application contact:* François-Joseph Lapointe, Professor, 514-343-7999, Fax: 514-343-2293.

Université de Montréal, Faculty of Medicine and Faculty of Graduate Studies, Graduate Programs in Medicine, Programs in Biomedical Sciences, Montréal, QC H3C 3J7, Canada. Offers M Sc, PhD. *Students:* 325 full-time (199 women), 24 part-time (13 women). 145 applicants, 48% accepted, 65 enrolled. *Degree requirements:* For master's, thesis; for doctorate, thesis/dissertation, general exam. *Entrance requirements:* For master's and doctorate, proficiency in French, knowledge of English. *Application deadline:* For fall and spring admission, 2/1. For winter admission, 11/1 for domestic students. Application fee: $30. Electronic applications accepted. *Unit head:* Daniel Lajeunesse, Director, 514-343-5778, Fax: 514-343-5751. *Application contact:* Denise Varennes, Information Contact, 514-343-6111 Ext. 4134, E-mail: denise.varennes@umontreal.ca.

Université de Sherbrooke, Faculty of Medicine and Health Sciences, Graduate Programs in Medicine, Sherbrooke, QC J1K 2R1, Canada. Offers M Sc, PhD. Part-time programs available. *Students:* 162 full-time (94 women), 143 part-time (79 women). 106 applicants, 49% accepted, 46 enrolled. In 2005, 41 master's, 19 doctorates awarded. Terminal master's awarded for partial completion of doctoral program. *Degree requirements:* For master's and doctorate, thesis/dissertation. *Application deadline:* For fall admission, 6/30 for domestic students. For winter admission, 10/31 for domestic students; for spring admission, 2/28 for domestic students. Application fee: $50. Electronic applications accepted. *Financial support:* Fellowships, research assistantships, tuition waivers (full) available. *Unit head:* Dr. Claude Asselin, Vice Dean for Graduate Studies, 819-564-5276, E-mail: claude.asselin@usherbrooke.ca.

Université de Sherbrooke, Faculty of Sciences, Department of Biology, Sherbrooke, QC J1K 2R1, Canada. Offers M Sc, PhD, Diploma. *Faculty:* 26 full-time (4 women), 7 part-time/adjunct (0 women). *Students:* 88 full-time (49 women). Average age 23. 40 applicants, 25% accepted. In 2005, 10 master's, 3 doctorates awarded. *Degree requirements:* For master's, thesis/dissertation; for doctorate, thesis/dissertation, comprehensive exam. *Entrance requirements:* For doctorate, master's degree. *Application deadline:* For fall admission, 6/30 for domestic students. Applications are processed on a rolling basis. Application fee: $50. Electronic applications accepted. *Financial support:* Fellowships, research assistantships, teaching assistantships available. *Faculty research:* Microbiology, ecology, molecular biology, cell biology, biotechnology. *Unit head:* Dr. Gabriel Girard, Chairman, 819-821-8000 Ext. 3030, Fax: 819-821-8200.

Université du Québec à Montréal, Graduate Programs, Program in Biology, Montréal, QC H3C 3P8, Canada. Offers M Sc, PhD. Part-time programs available. *Degree requirements:* For master's and doctorate, thesis/dissertation. *Entrance requirements:* For master's, appropriate bachelor's degree or equivalent, proficiency in French; for doctorate, appropriate master's degree or equivalent, proficiency in French.

Université du Québec, Institut National de la Recherche Scientifique, Graduate Programs, Research Center—INRS—Institut Armand-Frappier—Human Health, Québec, QC G1K 9A9, Canada. Offers applied microbiology (M Sc); biology (PhD); experimental health sciences (M Sc); virology and immunology (M Sc, PhD). Programs given in French. Part-time programs available. *Faculty:* 46. *Students:* 170 full-time (100 women), 25 international. Average age 28. In 2005, 24 master's, 5 doctorates awarded. *Degree requirements:* For doctorate, thesis/dissertation. *Entrance requirements:* For master's and doctorate, appropriate bachelor's degree, proficiency in French. *Application deadline:* For fall admission, 3/30 for domestic students, 3/30 for international students. For winter admission, 11/1 for domestic students. Application fee: $30 Canadian dollars. *Financial support:* Fellowships, research assistantships, teaching assistantships available. *Faculty research:* Immunity, infection and cancer; toxicology and environmental biotechnology; molecular pharmacochemistry *Unit head:* Pierre Talbot, Director, 450-681-5010 Ext. 4406, E-mail: pierre.talbot@iaf.inrs.ca. *Application contact:* Michel Barbeau, Registrar, 418-654-2518, Fax: 418-654-3858, E-mail: michel.barbeau@adm.inrs.ca.

Université Laval, Faculty of Medicine, Graduate Programs in Medicine, Québec, QC G1K 7P4, Canada. Offers M Sc, PhD, Diploma. *Degree requirements:* For doctorate, thesis/dissertation, comprehensive exam. *Entrance requirements:* For doctorate, knowledge of French, comprehension of written English; for Diploma, knowledge of French. Electronic applications accepted.

Université Laval, Faculty of Sciences and Engineering, Department of Biology, Programs in Biology, Québec, QC G1K 7P4, Canada. Offers M Sc, PhD. Terminal master's awarded for partial completion of doctoral program. *Degree requirements:* For master's, thesis/dissertation; for doctorate, thesis/dissertation, comprehensive exam. *Entrance requirements:* For master's and doctorate, knowledge of French and English. Electronic applications accepted.

University at Albany, State University of New York, College of Arts and Sciences, Department of Biological Sciences, Albany, NY 12222-0001. Offers biodiversity, conservation, and policy (MS); ecology, evolution, and behavior (MS, PhD); forensic molecular biology (MS); molecular, cellular, developmental, and neural biology (MS, PhD). *Students:* 52 full-time (28 women), 26 part-time (11 women). Average age 27. 135 applicants, 27% accepted, 21 enrolled. In 2005, 13 master's, 5 doctorates awarded. *Degree requirements:* For master's, one foreign language; for doctorate, one foreign language, thesis/dissertation. *Entrance requirements:* For master's and doctorate, GRE General Test. Additional exam requirements/recommendations for international students: Required—TOEFL (minimum score 550 paper-based; 213 computer-based). *Application deadline:* For fall admission, 2/15 priority date for domestic students, 5/1 priority date for international students; for spring admission, 11/1 for domestic students, 11/1 for international students. Applications are processed on a rolling basis. Application fee: $60. Electronic applications accepted. *Financial support:* Fellowships, research assistantships, teaching assistantships, unspecified assistantships and minority assistantships available. Financial award application deadline: 5/1. *Faculty research:* Interferon, neural development, RNA self-splicing, behavioral ecology, DNA repair enzymes. *Unit head:* Dr. Albert Millis, Chair, 518-442-4300.

University at Albany, State University of New York, School of Public Health, Department of Biomedical Sciences, Albany, NY 12222-0001. Offers biochemistry, molecular biology, and genetics (MS, PhD); cell and molecular structure (MS, PhD); immunobiology and immunochemistry (MS, PhD); molecular pathogenesis (MS, PhD); neuroscience (MS, PhD). *Students:* 37 full-time (23 women), 17 part-time (11 women). Average age 28. In 2005, 3 master's, 9 doctorates awarded. *Degree requirements:* For master's and doctorate, thesis/dissertation. *Entrance requirements:* For master's and doctorate, GRE General Test, GRE Subject Test. Additional exam requirements/recommendations for international students: Required—TOEFL (minimum score 550 paper-based; 213 computer-based). *Application deadline:* For fall admission, 1/1 for domestic students, 1/1 for international students; for spring admission, 10/30 for domestic students, 11/1 for international students. Applications are processed on a rolling basis. Application fee: $60. Electronic applications accepted. *Financial support:* Fellowships, research assistantships available. Financial award application deadline: 2/1. *Faculty research:* Geno

Biological and Biomedical Sciences—General

University at Albany, State University of New York (continued)
expression; RNA processing; membrane transport; immune response regulation; etiology of AIDS, Lyme disease, epilepsy. *Unit head:* Dr. James Dias, Chair, 518-474-2662.

See Close-Up on page 239.

The University of Akron, Graduate School, Buchtel College of Arts and Sciences, Department of Biology, Akron, OH 44325. Offers MS. Part-time programs available. *Faculty:* 18 full-time (3 women), 3 part-time/adjunct (1 woman). *Students:* 32 full-time (19 women), 3 part-time (2 women); includes 4 minority (1 Asian American or Pacific Islander, 3 Hispanic Americans), 1 international. Average age 29. 15 applicants, 47% accepted, 7 enrolled. In 2005, 13 degrees awarded. *Degree requirements:* For master's, oral defense of thesis, oral exam, seminars, thesis optional. *Entrance requirements:* For master's, GRE, minimum overall GPA of 2.75, minimum GPA of 3.0 in biology. Additional exam requirements/recommendations for international students: Required—TOEFL (minimum score 550 paper-based; 213 computer-based), Michigan English Language Assessment Battery. *Application deadline:* For fall admission, 8/15 for domestic students. Applications are processed on a rolling basis. Application fee: $30 ($40 for international students). Electronic applications accepted. *Expenses:* Tuition, state resident: full-time $5,816; part-time $323 per credit. Tuition, nonresident: full-time $9,976; part-time $554 per credit. Required fees: $794; $43 per credit. $12 per term. Tuition and fees vary according to course load, degree level and program. *Financial support:* In 2005–06, 17 research assistantships with full tuition reimbursements, 27 teaching assistantships with full tuition reimbursements were awarded; scholarships/grants and tuition waivers (full) also available. *Faculty research:* Genetics, immunology, molecular biology, physiology, virology. Total annual research expenditures: $1.3 million. *Unit head:* Dr. Richard Londraville, Interim Chair, 330-972-7155, E-mail: londraville@uakron.edu. *Application contact:* Dr. Monte Turner, Director of Graduate Studies, 330-972-7129, E-mail: meturner@uakron.edu.

The University of Alabama, Graduate School, College of Arts and Sciences, Department of Biological Sciences, Tuscaloosa, AL 35487. Offers MS, PhD. *Faculty:* 27 full-time (8 women). *Students:* 59 full-time (27 women), 7 part-time (3 women); includes 8 minority (4 African Americans, 1 American Indian/Alaska Native, 3 Hispanic Americans), 14 international. Average age 26. 104 applicants, 25% accepted, 17 enrolled. In 2005, 5 master's, 5 doctorates awarded. Terminal master's awarded for partial completion of doctoral program. *Median time to degree:* Of those who began their doctoral program in fall 1997, 56% received their degree in 8 years or less. *Degree requirements:* For master's, preliminary written exam, thesis optional; for doctorate, thesis/dissertation, preliminary written and oral exams. *Entrance requirements:* For master's and doctorate, GRE General Test, minimum GPA of 3.0. *Application deadline:* For fall admission, 7/6 for domestic students. Applications are processed on a rolling basis. Application fee: $25. Electronic applications accepted. *Expenses:* Tuition, state resident: full-time $4,864; part-time $382 per hour. Tuition, nonresident: full-time $13,516; part-time $796 per hour. *Financial support:* In 2005–06, 42 students received support, including 9 fellowships with tuition reimbursements available (averaging $11,000 per year), 32 teaching assistantships (averaging $11,000 per year); research assistantships with tuition reimbursements available, Federal Work-Study and institutionally sponsored loans also available. Support available to part-time students. Financial award application deadline: 8/14; financial award applicants required to submit FAFSA. *Faculty research:* Developmental genetics, limnology, taxonomy, microbiology. Total annual research expenditures: $2.1 million. *Unit head:* Dr. Martha J. Powell, Chair, 205-348-5960, Fax: 205-348-1786, E-mail: mpowell@biology.as.ua.edu. *Application contact:* Dr. Keller F. Suberkropp, Graduate Director, 205-348-1795, Fax: 205-348-1403, E-mail: ksuberkp@biology.as.ua.edu.

The University of Alabama at Birmingham, School of Natural Sciences and Mathematics, Department of Biology, Birmingham, AL 35294. Offers MS, PhD. *Students:* 36 full-time (17 women), 10 part-time (6 women); includes 11 minority (5 African Americans, 1 American Indian/Alaska Native, 5 Asian Americans or Pacific Islanders), 8 international. 84 applicants, 40% accepted. In 2005, 9 master's, 4 doctorates awarded. Terminal master's awarded for partial completion of doctoral program. *Degree requirements:* For master's and doctorate, thesis/dissertation. *Entrance requirements:* For master's and doctorate, GRE General Test, previous course work in biology, calculus, organic chemistry, and physics. Additional exam requirements/recommendations for international students: Required—TOEFL. *Application deadline:* Applications are processed on a rolling basis. Application fee: $35 ($60 for international students). Electronic applications accepted. *Expenses:* Tuition, state resident: part-time $170 per credit hour. Tuition, nonresident: full-time $4,612; part-time $425 per credit hour. International tuition: $10,732 full-time. Required fees: $11 per credit hour. $124 per term. Tuition and fees vary according to course load, degree level and program. *Financial support:* In 2005–06, 22 students received support, including 3 fellowships with full tuition reimbursements available (averaging $14,000 per year), 19 teaching assistantships with full tuition reimbursements available (averaging $14,000 per year); research assistantships, career-related internships or fieldwork, Federal Work-Study, institutionally sponsored loans, and tuition waivers (full) also available. Support available to part-time students. *Faculty research:* Invertebrate physiology, marine biology, environmental biology. *Unit head:* Dr. Ken R. Marion, Chair, 205-934-3582, Fax: 205-975-6097, E-mail: kmarion@uab.edu.

See Close-Up on page 241.

The University of Alabama in Huntsville, School of Graduate Studies, College of Science, Department of Biological Sciences, Huntsville, AL 35899. Offers MS. Part-time and evening/weekend programs available. *Faculty:* 12 full-time (3 women), 1 part-time/adjunct (0 women). *Students:* 23 full-time (18 women), 7 part-time (all women); includes 8 minority (6 African Americans, 1 Asian American or Pacific Islander, 1 Hispanic American), 1 international. Average age 29. 14 applicants, 71% accepted, 8 enrolled. In 2005, 7 degrees awarded. *Degree requirements:* For master's, thesis or alternative, oral and written exams, comprehensive exam, registration. *Entrance requirements:* For master's, GRE General Test, previous course work in biochemistry and organic chemistry, minimum GPA of 3.0. Additional exam requirements/recommendations for international students: Required—TOEFL (minimum score 550 paper-based; 213 computer-based). *Application deadline:* For fall admission, 5/30 priority date for domestic students, 2/30 priority date for international students; for spring admission, 10/10 priority date for domestic students, 7/10 priority date for international students. Applications are processed on a rolling basis. Application fee: $40. *Expenses:* Tuition, state resident: full-time $5,866; part-time $244 per credit hour. Tuition, nonresident: full-time $12,060; part-time $500 per credit hour. Tuition and fees vary according to course load. *Financial support:* In 2005–06, 20 students received support, including 2 fellowships with full and partial tuition reimbursements available (averaging $10,800 per year), 6 research assistantships with full and partial tuition reimbursements available (averaging $18,357 per year), 12 teaching assistantships with full and partial tuition reimbursements available (averaging $8,213 per year); career-related internships or fieldwork, Federal Work-Study, institutionally sponsored loans, scholarships/grants, health care benefits, tuition waivers (full and partial), and unspecified assistantships also available. Support available to part-time students. Financial award application deadline: 4/1; financial award applicants required to submit FAFSA. *Faculty research:* Cellular and developmental biology, reproductive physiology, immunology, genetics and molecular biology, microbiology. Total annual research expenditures: $997,228. *Unit head:* Dr. Gopi K. Podila, Chair, 256-824-6263, Fax: 256-824-6305, E-mail: podilag@email.uah.edu.

University of Alaska Anchorage, College of Arts and Sciences, Department of Biological Sciences, Anchorage, AK 99508-8060. Offers MS. Part-time programs available. *Students:* 10 full-time (7 women), 8 part-time (4 women); includes 1 minority (Asian American or Pacific Islander), 1 international. 15 applicants, 13% accepted. In 2005, 7 degrees awarded. *Degree requirements:* For master's, thesis. *Entrance requirements:* For master's, GRE General Test, GRE Subject Test, bachelor's degree in biology, chemistry or equivalent science. Additional exam requirements/recommendations for international students: Required—TOEFL (minimum score 550 paper-based; 213 computer-based). *Application deadline:* For fall admission, 7/1 priority date for domestic students, 7/1 priority date for international students; for spring

admission, 11/1 priority date for domestic students, 11/1 priority date for international students. Applications are processed on a rolling basis. Application fee: $45. *Financial support:* In 2005–06, 2 research assistantships with full tuition reimbursements were awarded; teaching assistantships with full tuition reimbursements, Federal Work-Study and traineeships also available. Support available to part-time students. Financial award application deadline: 4/1; financial award applicants required to submit FAFSA. *Faculty research:* Taxonomy and vegetative analysis in Alaskan ecosystems, fish environment and seafood, biochemistry, arctic ecology, vertebrate ecology. *Unit head:* Dr. Garry Davies, Chair, 907-786-4765, Fax: 907-786-4607. *Application contact:* Cheryl Wright, Program Manager, 907-786-1558.

University of Alaska Fairbanks, College of Natural Sciences and Mathematics, Department of Biology and Wildlife, Fairbanks, AK 99775-7520. Offers biological sciences (MS, PhD), including biology, botany, zoology; biology (MAT); wildlife biology (MS, PhD). Part-time programs available. *Faculty:* 29 full-time (7 women), 2 part-time/adjunct (1 woman). *Students:* 89 full-time (53 women), 24 part-time (15 women); includes 6 minority (1 African American, 4 Asian Americans or Pacific Islanders, 1 Hispanic American), 11 international. Average age 30. 74 applicants, 53% accepted, 17 enrolled. In 2005, 14 master's, 4 doctorates awarded. Terminal master's awarded for partial completion of doctoral program. *Degree requirements:* For master's and doctorate, thesis/dissertation, comprehensive exam, registration. *Entrance requirements:* For master's and doctorate, GRE General Test, GRE Subject Test. Additional exam requirements/recommendations for international students: Required—TOEFL (minimum score 550 paper-based; 213 computer-based); Recommended—TWE, TSE. *Application deadline:* For fall admission, 6/1 for domestic students, 3/1 for international students; for spring admission, 12/1 for domestic students, 9/1 for international students. Applications are processed on a rolling basis. Application fee: $50. Electronic applications accepted. *Expenses:* Tuition, state resident: full-time $4,392; part-time $244 per credit. Tuition, nonresident: full-time $8,964; part-time $498 per credit. Required fees: $800; $5 per credit. $48 per contact hour. Tuition and fees vary according to course level, course load, campus/location and reciprocity agreements. *Financial support:* In 2005–06, 36 research assistantships with tuition reimbursements (averaging $9,207 per year), 23 teaching assistantships with tuition reimbursements (averaging $5,358 per year) were awarded; fellowships with tuition reimbursements, career-related internships or fieldwork, Federal Work-Study, and scholarships/grants also available. Financial award application deadline: 2/1; financial award applicants required to submit FAFSA. *Faculty research:* Plant-herbivore interactions, plant metabolic defenses, insect manufacture of glycerol, ice nucleators, structure and functions of arctic and subarctic freshwater ecosystems. *Unit head:* Dr. Kent E. Schubegerle, Chair, 907-474-7671, Fax: 907-474-6716, E-mail: fybio@uaf.edu.

University of Alberta, Faculty of Graduate Studies and Research, Department of Biological Sciences, Edmonton, AB T6G 2E1, Canada. Offers environmental biology and ecology (M Sc, PhD); microbiology and biotechnology (M Sc, PhD); molecular biology and genetics (M Sc, PhD); physiology and cell biology (M Sc, PhD); plant biology (M Sc, PhD); systematics and evolution (M Sc, PhD). *Faculty:* 72 full-time (15 women), 15 part-time/adjunct (4 women). *Students:* 238 full-time (117 women), 32 part-time (15 women), 31 international. 206 applicants, 42% accepted. In 2005, 29 master's, 31 doctorates awarded. Terminal master's awarded for partial completion of doctoral program. *Degree requirements:* For master's and doctorate, thesis/dissertation, registration. *Entrance requirements:* Additional exam requirements/recommendations for international students: Required—TOEFL. *Application deadline:* For fall admission, 3/1 for domestic students. Applications are processed on a rolling basis. Application fee: $0. Tuition and fees charges are reported in Canadian dollars. *Expenses:* Tuition, state resident: part-time $562 Canadian dollars per term. Tuition, nonresident: full-time $3,375 Canadian dollars. Required fees: $573 Canadian dollars; $84 Canadian dollars per term. *Financial support:* In 2005–06, 4 research assistantships with partial tuition reimbursements (averaging $12,000 per year), 103 teaching assistantships with partial tuition reimbursements (averaging $12,300 per year) were awarded; career-related internships or fieldwork and scholarships/grants also available. *Unit head:* Laura Frost, Chair, 780-492-1904. *Application contact:* Dr. John P. Chang, Associate Chair for Graduate Studies, 780-492-1257, Fax: 780-492-9457, E-mail: bio.grad.coordinator@ualberta.ca.

University of Alberta, Faculty of Medicine and Dentistry and Faculty of Graduate Studies and Research, Graduate Programs in Medicine, Edmonton, AB T6G 2E1, Canada. Offers M Sc, MPH, PhD, Postgraduate Diploma, MD/PhD. Part-time programs available. Terminal master's awarded for partial completion of doctoral program. *Degree requirements:* For doctorate, thesis/dissertation. Tuition and fees charges are reported in Canadian dollars. *Expenses:* Tuition, state resident: part-time $562 Canadian dollars per term. Tuition, nonresident: full-time $3,375 Canadian dollars. Required fees: $573 Canadian dollars; $84 Canadian dollars per term. *Faculty research:* Basic, clinical, and applied biomedicine.

The University of Arizona, College of Medicine, Graduate Programs in Medicine, Tucson, AZ 85721. Offers MPH, MS, PhD, MD/PhD. Part-time programs available. Terminal master's awarded for partial completion of doctoral program. *Degree requirements:* For doctorate, thesis/dissertation. *Entrance requirements:* For master's and doctorate, GRE General Test. Expenses: Contact institution.

University of Arkansas, Graduate School, J. William Fulbright College of Arts and Sciences, Department of Biological Sciences, Fayetteville, AR 72701-1201. Offers biology (MA, MS, PhD). *Students:* 38 full-time (18 women), 13 part-time (5 women); includes 1 minority (African American), 6 international. 48 applicants, 23% accepted. In 2005, 5 master's, 10 doctorates awarded. *Degree requirements:* For doctorate, one foreign language, thesis/dissertation. *Entrance requirements:* For master's and doctorate, GRE Subject Test. Application fee: $40 ($50 for international students). *Financial support:* In 2005–06, 2 fellowships with tuition reimbursements, 17 research assistantships, 22 teaching assistantships were awarded; career-related internships or fieldwork and Federal Work-Study also available. Support available to part-time students. Financial award application deadline: 4/1; financial award applicants required to submit FAFSA. *Unit head:* Dr. Kimberly Smith, Departmental Chairperson, 479-575-3251, Fax: 479-575-4010, E-mail: kgsmith@uark.edu. *Application contact:* Dr. William Etges, Graduate Coordinator, 479-575-6358, Fax: 479-575-4010, E-mail: wetges@uark.edu.

University of Arkansas at Little Rock, Graduate School, College of Science and Mathematics, Program in Biology, Little Rock, AR 72204-1099. Offers MS.

University of Arkansas for Medical Sciences, College of Medicine and Graduate School, Graduate Programs in Medicine, Little Rock, AR 72205-7199. Offers MS, PhD, MD/PhD. *Students:* 109 full-time, 20 part-time. *Degree requirements:* For doctorate, thesis/dissertation. *Entrance requirements:* For master's and doctorate, GRE General Test. Application fee: $0. *Expenses:* Contact institution. *Financial support:* Fellowships, research assistantships, teaching assistantships, scholarships/grants and unspecified assistantships available. Support available to part-time students. *Unit head:* Dr. Robert E. McGehee, Dean, Graduate School, 501-686-5454. *Application contact:* Dr. Kristen Sterba, Coordinator and Recruiter for Graduate Studies, 501-526-7396, E-mail: kmsterba@uams.edu.

See Close-Up on page 243.

The University of British Columbia, Faculty of Medicine and Faculty of Graduate Studies, Graduate Programs in Medicine, Vancouver, BC V6T 1Z1, Canada. Offers M Sc, MH Sc, MHA, MOT, MPT, MRSc, PhD, MD/PhD. Part-time programs available. *Expenses:* Contact institution. *Financial support:* Fellowships, research assistantships, teaching assistantships, career-related internships or fieldwork, Federal Work-Study, and institutionally sponsored loans available. Support available to part-time students. *Application contact:* Dr. J. Carter, Associate Dean of Admissions, 604-822-4482.

University of Calgary, Faculty of Graduate Studies, Faculty of Science, Department of Biological Sciences, Calgary, AB T2N 1N4, Canada. Offers M Sc, PhD. Part-time programs available. *Faculty:* 44 full-time (4 women), 3 part-time/adjunct (2 women). *Students:* 126 full-

Biological and Biomedical Sciences—General

time (59 women). 65 applicants, 35% accepted. In 2005, 14 master's, 6 doctorates awarded. *Degree requirements:* For master's, thesis, registration; for doctorate, thesis/dissertation, candidacy exam. *Entrance requirements:* Additional exam requirements/recommendations for international students: Required—TOEFL. *Application deadline:* For fall admission, 6/1 priority date for domestic students, 5/1 priority date for international students. For winter admission, 10/1 for domestic students; for spring admission, 3/1 for domestic students. Applications are processed on a rolling basis. Application fee: $60. Electronic applications accepted. *Financial support:* In 2005–06, 81 students received support, including 30 research assistantships (averaging $4,000 per year), 51 teaching assistantships (averaging $12,422 per year) *Faculty research:* Biochemistry; cellular, molecular, and microbial biology; botany; ecology; zoology. *Unit head:* Dr. Jeffrey L. Goldberg, Head, 403-220-5260, Fax: 403-289-9311. *Application contact:* K. A. Barron, Graduate and Scholarship Administrator, 403-220-6623, Fax: 403-289-9311, E-mail: kbarron@ucalgary.ca.

University of Calgary, Faculty of Medicine and Faculty of Graduate Studies, Department of Medical Science, Calgary, AB T2N 1N4, Canada. Offers cancer biology (M Sc, PhD); immunology (M Sc, PhD); joint injury and arthritis research (M Sc, PhD); medical education (M Sc, PhD); medical science (M Sc, PhD); mountain medicine and high altitude physiology (M Sc). *Faculty:* 114 full-time (17 women), 5 part-time/adjunct (0 women). *Students:* 121 full-time (69 women), 1 part-time. 68 applicants, 29% accepted, 19 enrolled. In 2005, 19 master's, 7 doctorates awarded. *Median time to degree:* Of those who began their doctoral program in fall 1997, 100% received their degree in 8 years or less. *Degree requirements:* For master's, thesis; for doctorate, thesis/dissertation, candidacy exam. *Entrance requirements:* For master's, minimum undergraduate GPA of 3.2; for doctorate, minimum graduate GPA of 3.2. Additional exam requirements/recommendations for international students: Required—TOEFL (minimum score 600 paper-based; 250 computer-based). *Application deadline:* For fall admission, 6/15 priority date for domestic students, 5/15 priority date for international students. For winter admission, 10/15 for domestic students; for spring admission, 3/15 for domestic students. Applications are processed on a rolling basis. Application fee: $100 ($130 for international students). Electronic applications accepted. *Financial support:* In 2005–06, 30 students received support, including 22 research assistantships, 2 teaching assistantships; scholarships/grants and tuition waivers (partial) also available. *Faculty research:* Cancer biology, immunology, joint injury and arthritis, medical education, population genomics. *Unit head:* Dr. Francine Smith, Graduate Coordinator, 403-220-6852, Fax: 403-210-8109, E-mail: fsmith@ucalgary.ca. *Application contact:* Christine Szefer, Graduate Program Administrator, 403-220-6852, Fax: 403-210-8109, E-mail: cszefer@ucalgary.ca.

University of California, Berkeley, Graduate Division, College of Letters and Science, Department of Integrative Biology, Berkeley, CA 94720-1500. Offers PhD. *Degree requirements:* For doctorate, thesis/dissertation, oral qualifying exam. *Entrance requirements:* For doctorate, GRE General Test, GRE Subject Test. Additional exam requirements/recommendations for international students: Required—TOEFL. *Faculty research:* Morphology, physiology, development of plants and animals, behavior, ecology.

University of California, Irvine, Office of Graduate Studies, School of Biological Sciences, Irvine, CA 92697. Offers MS, PhD, MD/PhD. *Degree requirements:* For doctorate, thesis/dissertation. *Entrance requirements:* For master's and doctorate, GRE General Test, GRE Subject Test, minimum GPA of 3.0. Additional exam requirements/recommendations for international students: Required—TOEFL (minimum score 550 paper-based; 213 computer-based), TSE. Electronic applications accepted. *Faculty research:* Molecular biology and biochemistry, developmental and cell biology, physiology and biophysics, neurosciences, ecology and evolutionary biology.

University of California, Los Angeles, Graduate Division, College of Letters and Science, Department of Organismic Biology, Ecology and Evolution, Los Angeles, CA 90095. Offers biology (MA, PhD); plant molecular biology (PhD). *Degree requirements:* For master's, comprehensive exam or thesis; for doctorate, thesis/dissertation, oral and written qualifying exams. *Entrance requirements:* For master's, GRE General Test, GRE Subject Test (biology), minimum GPA of 3.0; for doctorate, GRE General Test, GRE Subject Test (biology), minimum undergraduate GPA of 3.0. Electronic applications accepted. *Faculty research:* Molecular, cell, and developmental biology; interactive biology; organisms and populations.

University of California, Los Angeles, School of Medicine and Graduate Division, Graduate Programs in Medicine, Los Angeles, CA 90095. Offers MA, MS, PhD, MD/PhD. Terminal master's awarded for partial completion of doctoral program. *Degree requirements:* For doctorate, thesis/dissertation, qualifying exams. *Entrance requirements:* For master's, GRE General Test.

University of California, Riverside, Graduate Division, Department of Biology, Riverside, CA 92521-0102. Offers biology (MS, PhD); evolution, ecology and organismal biology (MS, PhD). Department also affiliated with following interdepartmental graduate programs: Cell, Molecular, and Developmental Biology; Evolution and Ecology; Genetics.. *Faculty:* 22 full-time (5 women), 1 (woman) part-time; includes 3 minority (1 Asian American or Pacific Islander, 2 Hispanic Americans), 6 international. Average age 30. In 2005, 6 master's, 3 doctorates awarded. Terminal master's awarded for partial completion of doctoral program. *Degree requirements:* For master's, oral defense of thesis; for doctorate, thesis/dissertation, 3 quarters of teaching experience, qualifying exams. *Entrance requirements:* For master's and doctorate, GRE General Test, minimum GPA of 3.2. Additional exam requirements/recommendations for international students: Required—TOEFL (minimum score 550 paper-based; 213 computer-based); Recommended—TSE (minimum score 50). *Application deadline:* For fall admission, 5/1 for domestic students, 2/1 for international students. For winter admission, 9/1 for domestic students; for spring admission, 12/1 for domestic students. Applications are processed on a rolling basis. Application fee: $60 ($75 for international students). Electronic applications accepted. *Expenses:* Tuition, nonresident: full-time $14,694. Required fees: $9,009. Full-time tuition and fees vary according to program. *Financial support:* In 2005–06, research assistantships (averaging $14,000 per year), teaching assistantships with tuition reimbursements (averaging $15,000 per year) were awarded; fellowships, career-related internships or fieldwork, Federal Work-Study, institutionally sponsored loans, and tuition waivers (full and partial) also available. Financial award application deadline: 1/5; financial award applicants required to submit FAFSA. *Faculty research:* Molecular genetics, neurophysiology, evolutionary biology, physiology and organismal biology, signal transduction. *Unit head:* Dr. Richard Cardullo, Chair, 951-827-5901, Fax: 951-827-4286. *Application contact:* Zina Romero, Graduate Program Assistant, 800-735-0717, Fax: 951-827-5913, E-mail: biograd@ucr.edu.

University of California, Riverside, Graduate Division, Program in Biomedical Sciences, Riverside, CA 92521-0102. Offers PhD, MD/PhD. *Faculty:* 25 full-time (7 women). *Students:* 16 full-time (10 women); includes 4 minority (1 African American, 3 Asian Americans or Pacific Islanders), 8 international. Average age 29. In 2005, 1 degree awarded. *Median time to degree:* Of those who began their doctoral program in fall 1997, 100% received their degree in 8 years or less. *Degree requirements:* For doctorate, thesis/dissertation, qualifying exams. *Entrance requirements:* For doctorate, GRE General Test, minimum GPA of 3.2. Additional exam requirements/recommendations for international students: Required—TOEFL (minimum score 550 paper-based; 213 computer-based); Recommended—TSE (minimum score 50). *Application deadline:* For fall admission, 5/1 for domestic students, 2/1 for international students. For winter admission, 9/1 for domestic students; for spring admission, 12/1 for domestic students. Applications are processed on a rolling basis. Application fee: $60 ($75 for international students). Electronic applications accepted. *Expenses:* Tuition, nonresident: full-time $14,694. Required fees: $9,009. Full-time tuition and fees vary according to program. *Financial support:* In 2005–06, research assistantships (averaging $14,000 per year), teaching assistantships (averaging $15,000 per year) were awarded; fellowships, scholarships/grants also available. Financial award application deadline: 2/1; financial award applicants required to submit FAFSA. *Faculty research:* Regulation of cell proliferation; signal transduction in endocrine,

nervous, reproductive, and immune tissues; human genetic disorders; microbiology of human pathogens. *Unit head:* Dr. Craig V. Byus, Interim Dean and Program Director, 951-827-5705, Fax: 951-827-5504, E-mail: mstema@ucrac1.ucr.edu. *Application contact:* Kathy Redd, Graduate Program Assistant, 800-735-0717, Fax: 951-827-5517, E-mail: bmpasst@ucr.edu.

University of California, San Diego, Graduate Studies and Research, Division of Biology, La Jolla, CA 92093. Offers biochemistry (PhD); biology (MS); cell and developmental biology (PhD); computational neurobiology (PhD); ecology, behavior, and evolution (PhD); genetics and molecular biology (PhD); immunology, virology, and cancer biology (PhD); molecular and cellular biology (PhD); neurobiology (PhD); plant molecular biology (PhD); plant systems biology (PhD); signal transduction (PhD). Offered in association with the Salk Institute. *Degree requirements:* For doctorate, thesis/dissertation, qualifying exam. *Entrance requirements:* For doctorate, GRE General Test, pre-application available in September. Additional exam requirements/recommendations for international students: Required—TOEFL. Electronic applications accepted.

See Close-Up on page 245.

University of California, San Diego, School of Medicine and Graduate Studies and Research, Graduate Studies in Biomedical Sciences, La Jolla, CA 92093-0685. Offers molecular cell biology (PhD); pharmacology (PhD); physiology (PhD); regulatory biology (PhD). *Degree requirements:* For doctorate, thesis/dissertation, qualifying exam. *Entrance requirements:* For doctorate, GRE General Test. Additional exam requirements/recommendations for international students: Required—TOEFL. Electronic applications accepted. *Faculty research:* Molecular and cellular biology, molecular and cellular pharmacology, cell and organ physiology.

See Close-Up on page 247.

University of California, San Diego, School of Medicine, Medical Scientist Training Program, La Jolla, CA 92093. Offers MD/PhD.

University of California, San Francisco, Graduate Division, Biomedical Sciences Graduate Group, San Francisco, CA 94143. Offers anatomy (PhD); endocrinology (PhD); experimental pathology (PhD); physiology (PhD). *Degree requirements:* For doctorate, thesis/dissertation. *Entrance requirements:* For doctorate, GRE General Test.

University of Central Arkansas, Graduate School, College of Natural Sciences and Math, Department of Biological Science, Conway, AR 72035-0001. Offers MS. Part-time programs available. *Faculty:* 18 full-time (4 women), 2 part-time/adjunct (both women). *Students:* 18 full-time (4 women), 92 part-time (73 women); includes 14 African Americans, 5 international. 33 applicants, 82% accepted, 27 enrolled. *Degree requirements:* For master's, thesis optional. *Entrance requirements:* For master's, GRE General Test, minimum GPA of 2.7. Additional exam requirements/recommendations for international students: Required—TOEFL (minimum score 550 paper-based; 213 computer-based). *Application deadline:* For fall admission, 3/1 for domestic students; for spring admission, 10/1 priority date for domestic students. Applications are processed on a rolling basis. Application fee: $25 ($40 for international students). *Expenses:* Tuition, state resident: part-time $190 per hour. Tuition, nonresident: part-time $380 per hour. Required fees: $31 per hour. $88 per term. Tuition and fees vary according to course load and program. *Financial support:* In 2005–06, research assistantships with partial tuition reimbursements (averaging $8,500 per year), 21 teaching assistantships with partial tuition reimbursements (averaging $8,500 per year) were awarded; unspecified assistantships also available. Financial award application deadline: 2/15; financial award applicants required to submit FAFSA. *Faculty research:* Cell apoptosis, cortical neurons, pedal 3 neurons. *Unit head:* Dr. Steven Runge, Chairperson, 501-450-3146, Fax: 501-450-5914, E-mail: srunge@uca.edu. *Application contact:* Brenda Herring, Admissions Assistant, 501-450-3124, Fax: 501-450-5678, E-mail: bherring@uca.edu.

University of Central Florida, Burnett College of Biomedical Sciences, Orlando, FL 32816. Offers MS, PhD. *Faculty:* 17 full-time (6 women), 4 part-time/adjunct (2 women). *Students:* 57 full-time (31 women), 8 part-time (3 women); includes 6 minority (1 African American, 3 Asian Americans or Pacific Islanders, 2 Hispanic Americans), 24 international. 67 applicants, 51% accepted, 25 enrolled. In 2005, 7 master's, 2 doctorates awarded. *Expenses:* Tuition, state resident: full-time $5,788. Tuition, nonresident: full-time $21,927. Required fees: $241 per credit hour. *Financial support:* In 2005–06, 52 research assistantships (averaging $8,200 per year), 28 teaching assistantships (averaging $4,500 per year) were awarded. *Unit head:* Dr. Pappachan E. Kolattukudy, Dean, 407-823-1206, Fax: 407-823-0956, E-mail: pk@mail.ucf.edu

University of Central Florida, College of Sciences, Department of Biology, Orlando, FL 32816. Offers biology (MS); conservation biology (PhD, Certificate). Part-time and evening/weekend programs available. *Faculty:* 19 full-time (6 women), 6 part-time/adjunct (3 women). *Students:* 43 full-time (34 women), 45 part-time (27 women); includes 6 minority (1 Asian American or Pacific Islander, 5 Hispanic Americans), 4 international. Average age 29. 86 applicants, 59% accepted, 35 enrolled. In 2005, 19 degrees awarded. *Degree requirements:* For master's, thesis or alternative, biology field exam, comprehensive exam. *Entrance requirements:* For master's, GRE General Test, minimum GPA of 3.0 in last 60 hours. Additional exam requirements/recommendations for international students: Required—TOEFL. *Application deadline:* For fall admission, 3/1 for domestic students; for spring admission, 10/15 for domestic students. Electronic applications accepted. *Expenses:* Tuition, state resident: full-time $5,788. Tuition, nonresident: full-time $21,927. Required fees: $241 per credit hour. *Financial support:* In 2005–06, 18 fellowships with partial tuition reimbursements (averaging $1,903 per year), 30 research assistantships with partial tuition reimbursements (averaging $4,600 per year), 26 teaching assistantships with partial tuition reimbursements (averaging $10,000 per year) were awarded; career-related internships or fieldwork, Federal Work-Study, institutionally sponsored loans, tuition waivers (partial), and unspecified assistantships also available. Financial award application deadline: 3/1; financial award applicants required to submit FAFSA. *Unit head:* Dr. David Borst, Chair, 407-823-2976, Fax: 407-823-5769, E-mail: dborst@mail.ucf.edu. *Application contact:* Dr. John F. Weishampel, Coordinator, 407-823-2141, Fax: 407-823-5769, E-mail: jweisham@mail.ucf.edu.

University of Central Oklahoma, College of Graduate Studies and Research, College of Mathematics and Science, Department of Biology, Edmond, OK 73034-5209. Offers MS. Part-time programs available. *Faculty:* 14 full-time (4 women), 6 part-time/adjunct (2 women). *Students:* 4 full-time (2 women), 4 part-time (all women), 1 international. Average age 25. 5 applicants, 80% accepted. In 2005, 2 degrees awarded. *Degree requirements:* For master's, thesis. *Entrance requirements:* For master's, GRE General Test, GRE Subject Test (biology). Additional exam requirements/recommendations for international students: Required—TOEFL (minimum score 550 paper-based; 213 computer-based). *Application deadline:* Applications are processed on a rolling basis. Application fee: $25. Electronic applications accepted. *Expenses:* Tuition, state resident: full-time $2,988; part-time $125 per credit hour. Tuition, nonresident: full-time $4,728; part-time $197 per credit hour. Required fees: $716; $16 per credit hour. *Financial support:* Federal Work-Study and unspecified assistantships available. Financial award application deadline: 3/31; financial award applicants required to submit FAFSA. *Faculty research:* Environmental (*legionella*), aquatic biology (ecological), mammalogy field studies, microbiology, genetics. *Unit head:* Dr. Jenna Hellack, Chairperson, 405-974-5773, Fax: 405-974-3824.

University of Chicago, Division of the Biological Sciences, Chicago, IL 60637-1513. Offers MD, MS, PhD, MD/PhD. *Faculty:* 458 full-time (99 women), 18 part-time/adjunct (8 women). *Students:* 437 full-time (220 women); includes 88 minority (19 African Americans, 1 American Indian/Alaska Native, 49 Asian Americans or Pacific Islanders, 19 Hispanic Americans), 78 international. Average age 27. 915 applicants, 22% accepted, 80 enrolled. In 2005, 25 master's, 63 doctorates awarded. *Degree requirements:* For doctorate, thesis/dissertation, registration. *Entrance requirements:* For doctorate, GRE General Test. Additional exam requirements/

Biological and Biomedical Sciences—General

University of Chicago (continued)
recommendations for international students: Required—TOEFL. *Application deadline:* For fall admission, 12/28 priority date for domestic students, 12/28 priority date for international students. Application fee: $55. Electronic applications accepted. *Financial support:* In 2005–06, 333 students received support, including fellowships with full tuition reimbursements available (averaging $26,301 per year), research assistantships with full tuition reimbursements available (averaging $26,301 per year); institutionally sponsored loans, scholarships/grants, traineeships, and health care benefits also available. Financial award applicants required to submit FAFSA. *Unit head:* Dr. James Madara, Dean, 773-702-9000. *Application contact:* Parag M. Shah, Administrator, Graduate Affairs, 773-702-5853, Fax: 773-834-1618, E-mail: pshah@bsd.uchicago.edu.

University of Cincinnati, Division of Research and Advanced Studies, College of Medicine, Biomedical Sciences Flex Option Program, Cincinnati, OH 45221. Offers PhD. *Degree requirements:* For doctorate, qualifying exam. *Entrance requirements:* For doctorate, GRE, 2 letters of recommendation, essay/personal statement. Additional exam requirements/recommendations for international students: Required—TOEFL. Electronic applications accepted. *Faculty research:* Environmental health, developmental biology, cell and molecular biology, immunobiology, molecular genetics.

See Close-Up on page 249.

University of Cincinnati, Division of Research and Advanced Studies, College of Medicine, Graduate Programs in Biomedical Sciences, Cincinnati, OH 45221. Offers MS, PhD. Terminal master's awarded for partial completion of doctoral program. *Degree requirements:* For master's, thesis; for doctorate, thesis/dissertation, qualifying exam. *Entrance requirements:* For master's and doctorate, GRE General Test. Additional exam requirements/recommendations for international students: Required—TOEFL. Electronic applications accepted. Expenses: Contact institution.

University of Cincinnati, Division of Research and Advanced Studies, College of Medicine, Physician Scientist Training Program, Cincinnati, OH 45267. Offers MD/PhD. Electronic applications accepted.

University of Cincinnati, Division of Research and Advanced Studies, McMicken College of Arts and Sciences, Department of Biological Sciences, Cincinnati, OH 45221-0006. Offers MS, PhD. Part-time programs available. Terminal master's awarded for partial completion of doctoral program. *Degree requirements:* For master's, thesis/dissertation, registration; for doctorate, thesis/dissertation, comprehensive exam, registration. *Entrance requirements:* For master's and doctorate, GRE General Test, GRE Subject Test, BS in biology, chemistry, or equivalent. Additional exam requirements/recommendations for international students: Required—TOEFL; Recommended—TSE. Electronic applications accepted. *Faculty research:* Physiology and development, cell and molecular, ecology and evolutionary.

University of Colorado at Denver and Health Sciences Center, Graduate School, Program in Biomedical Sciences, Denver, CO 80262. Offers PhD, PhD/MD. *Students:* 286 full-time (159 women), 25 part-time (18 women); includes 29 minority (4 African Americans, 9 Asian Americans or Pacific Islanders, 16 Hispanic Americans), 39 international. 415 applicants, 16% accepted, 65 enrolled. In 2005, 34 doctorates awarded. Terminal master's awarded for partial completion of doctoral program. *Degree requirements:* For doctorate, thesis/dissertation. *Entrance requirements:* Additional exam requirements/recommendations for international students: Required—TOEFL (minimum score 550 paper-based; 213 computer-based). Application fee: $50. *Expenses:* Contact institution. Tuition and fees vary according to degree level and program. *Financial support:* Fellowships, research assistantships, teaching assistantships, career-related internships or fieldwork, Federal Work-Study, institutionally sponsored loans, and traineeships available. Support available to part-time students. Financial award applicants required to submit FAFSA. *Unit head:* Dr. Steven Anderson, Director, 303-724-3278, E-mail: steve.anderson@uchsc.edu. *Application contact:* Julie Westerdahl, Program Administrator, 303-724-3278, E-mail: julie.westerdahl@uchsc.edu.

University of Colorado at Denver and Health Sciences Center—Downtown Denver Campus, College of Liberal Arts and Sciences, Department of Biology, Denver, CO 80217-3364. Offers MS. Part-time programs available. *Faculty:* 12 full-time (6 women). *Students:* 8 full-time (6 women), 24 part-time (15 women); includes 5 minority (3 African Americans, 2 Asian Americans or Pacific Islanders), 2 international. Average age 31. 23 applicants, 43% accepted, 9 enrolled. In 2005, 9 degrees awarded. *Degree requirements:* For master's, thesis or alternative, comprehensive exam. *Entrance requirements:* For master's, GRE General Test, minimum GPA of 3.0. Additional exam requirements/recommendations for international students: Required—TOEFL (minimum score 525 paper-based; 197 computer-based). *Application deadline:* For fall admission, 4/15 for domestic students; for spring admission, 10/15 for domestic students. Applications are processed on a rolling basis. Application fee: $50 ($75 for international students). Electronic applications accepted. *Expenses:* Tuition, state resident: part-time $325 per credit hour. Tuition, nonresident: part-time $1,077 per credit hour. Required fees: $145 per credit hour. One-time fee: $115 part-time. Tuition and fees vary according to course level and program. *Financial support:* Research assistantships, teaching assistantships, Federal Work-Study available. Financial award application deadline: 4/1; financial award applicants required to submit FAFSA. *Unit head:* Dr. Gerald Audesirk, Chair, 303-556-2593, Fax: 303-556-4352, E-mail: gerald.audesirk@cudenver.edu.

University of Connecticut, Graduate School, College of Liberal Arts and Sciences, Department of Molecular and Cell Biology, Storrs, CT 06269. Offers applied genomics (MS, PSM); biobehavioral science (PhD); biochemistry (MS, PhD); biophysics and structural biology (MS, PhD); biotechnology (MS); cell and developmental biology (MS, PhD); genetics, genomics, and bioinformatics (MS), including genetics (MS, PhD); genetics, genomics, and bioinformation (PhD), including genetics (MS, PhD); microbial systems analysis (MS, PSM); microbiology (MS, PhD); plant cell and molecular biology (MS, PhD). *Faculty:* 64 full-time (13 women). *Students:* 124 full-time (58 women), 15 part-time (8 women); includes 17 minority (5 African Americans, 1 American Indian/Alaska Native, 9 Asian Americans or Pacific Islanders, 2 Hispanic Americans), 36 international. Average age 27. 285 applicants, 28% accepted, 55 enrolled. In 2005, 23 master's, 10 doctorates awarded. Terminal master's awarded for partial completion of doctoral program. *Degree requirements:* For master's, comprehensive exam; for doctorate, thesis/dissertation. *Entrance requirements:* For master's and doctorate, GRE General Test, GRE Subject Test. Additional exam requirements/recommendations for international students: Required—TOEFL (minimum score 550 paper-based; 213 computer-based). *Application deadline:* For fall admission, 2/1 priority date for domestic students, 2/1 priority date for international students; for spring admission, 11/1 for domestic students, 10/1 for international students. Applications are processed on a rolling basis. Application fee: $55. Electronic applications accepted. *Expenses:* Tuition, state resident: part-time $444 per credit hour. Tuition, nonresident: part-time $1,154 per credit hour. Tuition and fees vary according to course load. *Financial support:* In 2005–06, 35 research assistantships with full tuition reimbursements, 59 teaching assistantships with full tuition reimbursements were awarded; fellowships, Federal Work-Study, scholarships/grants, health care benefits, and unspecified assistantships also available. Financial award application deadline: 2/1; financial award applicants required to submit FAFSA. *Unit head:* Philip L. Yeagle, Head, 860-486-4329, Fax: 860-486-4331, E-mail: yeagle@uconnvm.uconn.edu. *Application contact:* Anne St. Onje, Graduate Secretary, 860-486-4314, Fax: 860-486-3943, E-mail: ann.st_onje@uconn.edu.

University of Connecticut, Graduate School, University of Connecticut Health Center, Field of Biomedical Science, Storrs, CT 06269. Offers PhD. *Faculty:* 140 full-time (33 women). *Students:* 167 full-time (97 women), 23 part-time (11 women); includes 13 minority (3 African Americans, 1 American Indian/Alaska Native, 7 Asian Americans or Pacific Islanders, 2 Hispanic Americans), 82 international. Average age 28. 319 applicants, 16% accepted, 51 enrolled. In 2005, 20 degrees awarded. *Degree requirements:* For doctorate, thesis/dissertation. *Entrance requirements:* For doctorate, GRE General Test, GRE Subject Test. Additional exam

requirements/recommendations for international students: Required—TOEFL (minimum score 550 paper-based; 213 computer-based). *Application deadline:* For fall admission, 2/1 priority date for domestic students, 2/1 priority date for international students; for spring admission, 11/1 for domestic students, 10/1 for international students. Applications are processed on a rolling basis. Application fee: $55. Electronic applications accepted. *Expenses:* Tuition, state resident: part-time $444 per credit hour. Tuition, nonresident: part-time $1,154 per credit hour. Tuition and fees vary according to course load. *Financial support:* In 2005–06, 145 research assistantships with full tuition reimbursements were awarded; fellowships, Federal Work-Study, scholarships/grants, health care benefits, and unspecified assistantships also available. Financial award application deadline: 2/1; financial award applicants required to submit FAFSA. *Application contact:* Tricia Avolt, Graduate Coordinator, 860-679-4306, Fax: 860-679-1899, E-mail: robertson@nso2.uchc.edu.

University of Connecticut Health Center, Graduate School and School of Medicine, Combined Degree Program in Biomedical Sciences, Farmington, CT 06030. Offers MD/PhD. *Entrance requirements:* Additional exam requirements/recommendations for international students: Required—TOEFL (minimum score 550 paper-based; 213 computer-based). Expenses: Contact institution.

University of Connecticut Health Center, Graduate School, Programs in Biomedical Sciences, Farmington, CT 06030. Offers PhD, DMD/PhD, MD/PhD, PhD/Certificate. Part-time and evening/weekend programs available. *Degree requirements:* For doctorate, thesis/dissertation, comprehensive exam, registration. *Entrance requirements:* For doctorate, GRE General Test. Additional exam requirements/recommendations for international students: Required—TOEFL (minimum score 600 paper-based; 250 computer-based). Electronic applications accepted.

See Close-Up on page 251.

University of Dayton, Graduate School, College of Arts and Sciences, Department of Biology, Dayton, OH 45469-1300. Offers MS, PhD. *Faculty:* 15 full-time (4 women). *Students:* 14 full-time (8 women), 1 part-time; includes 1 minority (Asian American or Pacific Islander), 2 international. Average age 24. 31 applicants, 13% accepted, 4 enrolled. In 2005, 1 degree awarded. Terminal master's awarded for partial completion of doctoral program. *Degree requirements:* For master's and doctorate, thesis/dissertation, comprehensive exam, registration. *Entrance requirements:* For master's and doctorate, GRE General Test, GRE Subject Test, minimum undergraduate GPA of 3.0. Additional exam requirements/recommendations for international students: Required—TOEFL (minimum score 550 paper-based; 213 computer-based). *Application deadline:* For fall admission, 3/15 priority date for domestic students, 3/1 priority date for international students. Applications are processed on a rolling basis. Application fee: $0. Electronic applications accepted. *Expenses:* Tuition: Part-time $567 per credit hour. Required fees: $25 per term. Tuition and fees vary according to degree level and program. *Financial support:* In 2005–06, 3 research assistantships with full and partial tuition reimbursements (averaging $17,574 per year), 16 teaching assistantships with full and partial tuition reimbursements (averaging $11,017 per year) were awarded; institutionally sponsored loans, health care benefits, and unspecified assistantships also available. Financial award application deadline: 3/15; financial award applicants required to submit FAFSA. *Faculty research:* Plant and animal physiology; cell, molecular, and developmental biology; animal behavior and ecology; community ecology and environmental biology; genetics and microbiology. Total annual research expenditures: $500,000. *Unit head:* Dr. John J. Rowe, Chair, 937-229-2521, Fax: 937-229-2021. *Application contact:* E. Eavers.

See Close-Up on page 253.

University of Delaware, College of Arts and Sciences, Department of Biological Sciences, Newark, DE 19716. Offers biotechnology (MS); cancer biology (MS, PhD); cell and extra-cellular matrix biology (MS, PhD); cell and systems physiology (MS, PhD); developmental biology (MS, PhD); ecology and evolution (MS, PhD); microbiology (MS, PhD); molecular biology and genetics (MS, PhD). *Faculty:* 39 full-time (11 women). *Students:* 64 full-time (46 women), 2 part-time; includes 7 minority (4 African Americans, 2 Asian Americans or Pacific Islanders, 1 Hispanic American), 18 international. Average age 26. 113 applicants, 31% accepted, 19 enrolled. In 2005, 5 master's, 4 doctorates awarded. Terminal master's awarded for partial completion of doctoral program. *Median time to degree:* Of those who began their doctoral program in fall 1997, 100% received their degree in 8 years or less. *Degree requirements:* For master's, thesis/dissertation, preliminary exam; for doctorate, thesis/dissertation, preliminary exam, comprehensive exam. *Entrance requirements:* For master's and doctorate, GRE General Test. Additional exam requirements/recommendations for international students: Required—TOEFL (minimum score 600 paper-based; 250 computer-based); Recommended—TWE, TSE. *Application deadline:* For fall admission, 4/15 for domestic students, 1/15 for international students; for spring admission, 10/1 for domestic students. Applications are processed on a rolling basis. Application fee: $60. Electronic applications accepted. *Financial support:* In 2005–06, 26 students received support, including fellowships with full tuition reimbursements available (averaging $19,000 per year), 19 research assistantships with full tuition reimbursements available (averaging $19,000 per year), 26 teaching assistantships with full tuition reimbursements available (averaging $19,000 per year); tuition waivers (partial) also available. Financial award application deadline: 4/15. *Faculty research:* Microorganisms, bone, cancer metastasis, developmental biology, cell biology, DNA. Total annual research expenditures: $8.3 million. *Unit head:* Dr. Daniel D. Carson, Chair, 302-831-6977, Fax: 302-831-2281, E-mail: dcarson@udel.edu. *Application contact:* Dr. Melinda K. Duncan, Graduate Coordinator, 302-831-1841, Fax: 302-831-2281, E-mail: danders@udel.edu.

University of Denver, Faculty of Natural Sciences and Mathematics, Department of Biological Sciences, Denver, CO 80208. Offers MS, PhD. Part-time programs available. *Faculty:* 14 full-time (3 women). *Students:* 2 full-time (both women), 18 part-time (10 women), 4 international. 23 applicants, 30% accepted. In 2005, 4 master's, 5 doctorates awarded. Terminal master's awarded for partial completion of doctoral program. *Degree requirements:* For master's, thesis; for doctorate, one foreign language, thesis/dissertation. *Entrance requirements:* For master's and doctorate, GRE General Test. Additional exam requirements/recommendations for international students: Required—TOEFL. *Application deadline:* For fall admission, 3/1 for domestic students. Applications are processed on a rolling basis. Application fee: $45. *Expenses:* Tuition: Full-time $27,756; part-time $771 per credit. Required fees: $174. *Financial support:* In 2005–06, 1 research assistantship with full and partial tuition reimbursement (averaging $13,300 per year), 18 teaching assistantships with full and partial tuition reimbursements (averaging $13,392 per year) were awarded; Federal Work-Study and institutionally sponsored loans also available. Support available to part-time students. Financial award application deadline: 3/1; financial award applicants required to submit FAFSA. *Faculty research:* Molecular biology, cell biology, neurobiology, ecology, molecular evolution. Total annual research expenditures: $996,000. *Unit head:* 303-871-3661, E-mail: tquinn@du.edu.

University of Florida, College of Medicine and Graduate School, Interdisciplinary Program in Biomedical Sciences, Gainesville, FL 32611. Offers PhD, JD/MS, JD/PhD, MBA/MS, MBA/PhD, MS/M Ed. *Faculty:* 247. *Students:* 234 full-time (130 women); includes 54 minority (10 African Americans, 39 Asian Americans or Pacific Islanders, 5 Hispanic Americans). 267 applicants, 18% accepted. In 2005, 43 doctorates awarded. *Degree requirements:* For doctorate, thesis/dissertation. *Entrance requirements:* For doctorate, GRE General Test, minimum GPA of 3.0. Additional exam requirements/recommendations for international students: Required—TOEFL. *Application deadline:* For fall admission, 2/15 for domestic students. Application fee: $30. Electronic applications accepted. *Expenses: Contact institution.* Tuition and fees vary according to program. *Financial support:* In 2005–06, research assistantships with full tuition reimbursements (averaging $21,500 per year); fellowships with full tuition reimbursements, teaching assistantships, institutionally sponsored loans, traineeships, health care benefits, and unspecified assistantships also available. *Unit head:* Dr. Wayne McCormack, Associate Dean of Graduate Education, 352-392-7413, Fax: 352-846-3466, E-mail: mccormac@pathology.ufl.edu.

See Close-Up on page 255.

University of Georgia, Graduate School, Biomedical and Health Sciences Institute, Athens, GA 30602. Offers neuroscience (PhD). *Students:* 1 applicant, 100% accepted, 1 enrolled. *Application contact:* Dr. Jan Sandor, Director of Graduate Admissions, 706-542-1787, Fax: 706-542-9480, E-mail: gradadm@uga.edu.

University of Guam, Graduate School and Research, College of Arts and Sciences, Program in Biology, Mangilao, GU 96923. Offers tropical marine biology (MS). *Degree requirements:* For master's, comprehensive exam. *Entrance requirements:* For master's, GRE General Test, GRE Subject Test. Additional exam requirements/recommendations for international students: Required—TOEFL. *Faculty research:* Maintenance and ecology of coral reefs.

University of Guelph, Graduate Program Services, College of Biological Science, Guelph, ON N1G 2W1, Canada. Offers M Sc, PhD. Part-time programs available. *Faculty:* 97 full-time (19 women). *Students:* 303 full-time (153 women), 12 part-time (9 women). 162 applicants, 35% accepted. In 2005, 75 master's, 17 doctorates awarded. *Degree requirements:* For doctorate, thesis/dissertation. *Application deadline:* Applications are processed on a rolling basis. Application fee: $75. *Financial support:* In 2005–06, 80 students received support; research assistantships, teaching assistantships, scholarships/grants available. *Unit head:* Prof. Michael James Emes, Dean, 519-824-4120 Ext. 56102, Fax: 519-767-2044, E-mail: memes@uoguelph.ca. *Application contact:* Laurie Winn, Graduate Admissions Secretary, 519-824-4320 Ext. 52730, Fax: 519-767-1656, E-mail: lwinn@uoguelph.ca.

University of Hartford, College of Arts and Sciences, Department of Biology, West Hartford, CT 06117-1599. Offers biology (MS); neuroscience (MS). Part-time and evening/weekend programs available. *Faculty:* 2 full-time (1 woman), 2 part-time/adjunct (1 woman). *Students:* 10 full-time (6 women), 6 part-time (5 women); includes 1 minority (Hispanic American), 7 international. Average age 29. 17 applicants, 65% accepted. In 2005, 5 degrees awarded. *Degree requirements:* For master's, oral exams, thesis optional. *Entrance requirements:* For master's, GRE or MCAT. Additional exam requirements/recommendations for international students: Required—TOEFL (minimum score 550 paper-based; 213 computer-based). *Application deadline:* Applications are processed on a rolling basis. Application fee: $40 ($55 for international students). Electronic applications accepted. *Expenses:* Tuition: Part-time $515 per credit. Required fees: $200 per term. Tuition and fees vary according to program. *Financial support:* Research assistantships, teaching assistantships, Federal Work-Study and tuition waivers (partial) available. Support available to part-time students. Financial award application deadline: 6/1; financial award applicants required to submit FAFSA. *Faculty research:* Neurobiology of aging, central actions of neural steroids, neuroendocrine control of reproduction, retinopathies in sharks, plasticity in the central nervous system. Total annual research expenditures: $15,900. *Unit head:* Dr. Martin Cohen, Chairman, 860-768-5372, Fax: 860-768-5002. *Application contact:* Reneé Murphy, Assistant Director of Graduate Admissions, 860-768-4371, Fax: 860-768-5160, E-mail: rmurphy@hartford.edu.

University of Hawaii at Manoa, John A. Burns School of Medicine and Graduate Division, Graduate Programs in Biomedical Sciences, Honolulu, HI 96822. Offers MS, PhD. Part-time programs available. *Students:* Average age 31. *Degree requirements:* For doctorate, thesis/dissertation. Application fee: $25 ($50 for international students). *Expenses:* Contact institution. Tuition and fees vary according to program. *Financial support:* Fellowships, research assistantships, teaching assistantships, career-related internships or fieldwork, Federal Work-Study, institutionally sponsored loans, and tuition waivers (full and partial) available. Support available to part-time students. *Application contact:* Rosanne Harrigan, 808-692-0904.

University of Houston, College of Natural Sciences and Mathematics, Department of Biology and Biochemistry, Houston, TX 77204. Offers biochemistry (MA, MS, PhD); biology (MA, MS, PhD). *Faculty:* 24 full-time (4 women), 2 part-time/adjunct (1 woman). *Students:* 107 full-time (46 women), 6 part-time (2 women); includes 9 minority (2 African Americans, 1 American Indian/Alaska Native, 4 Asian Americans or Pacific Islanders, 2 Hispanic Americans), 80 international. Average age 28. 37 applicants, 76% accepted, 21 enrolled. In 2005, 8 master's, 11 doctorates awarded. Terminal master's awarded for partial completion of doctoral program. *Degree requirements:* For master's, thesis (for some programs); for doctorate, thesis/dissertation, comprehensive exam. *Entrance requirements:* For master's and doctorate, GRE General Test. Additional exam requirements/recommendations for international students: Required—TSE and SPEAK Test. *Application deadline:* For fall admission, 4/7 for domestic students; for spring admission, 10/2 priority date for domestic students. Applications are processed on a rolling basis. Application fee: $0 ($75 for international students). *Financial support:* In 2005–06, research assistantships with full tuition reimbursements (averaging $14,300 per year), 62 teaching assistantships with full tuition reimbursements (averaging $14,300 per year) were awarded; fellowships with full tuition reimbursements, career-related internships or fieldwork, Federal Work-Study, institutionally sponsored loans, scholarships/grants, health care benefits, and unspecified assistantships also available. Support available to part-time students. Financial award application deadline: 3/10. *Faculty research:* Evolutionary biology, neuroscience, infectious diseases, circadian rhythm, protein structure. *Unit head:* Dr. Stuart Dryer, Chairman, 713-743-2666, Fax: 713-743-2632, E-mail: sdryer@uh.edu. *Application contact:* Elizabeth Bullock, Graduate Adviser and Program Coordinator, 713-743-2633, Fax: 713-743-2899, E-mail: ebullock@uh.edu.

University of Houston–Clear Lake, School of Science and Computer Engineering, Program in Biological Sciences, Houston, TX 77058-1098. Offers MS. Part-time and evening/weekend programs available. *Entrance requirements:* For master's, GRE General Test. Additional exam requirements/recommendations for international students: Required—TOEFL (minimum score 550 paper-based; 213 computer-based).

University of Idaho, College of Graduate Studies, College of Science, Department of Biological Sciences, Moscow, ID 83844-2282. Offers bioinformatics and computational biology (MS, PhD); biological sciences (M Nat Sci). *Students:* 22 full-time (10 women), 4 part-time (3 women), 6 international. Average age 28. In 2005, 2 master's, 2 doctorates awarded. *Degree requirements:* For doctorate, one foreign language, thesis/dissertation. *Entrance requirements:* For master's, GRE, minimum GPA of 2.8; for doctorate, GRE, minimum undergraduate GPA of 2.8, 3.0 graduate. *Application deadline:* For fall admission, 8/1 for domestic students; for spring admission, 12/15 for domestic students. Application fee: $55 ($60 for international students). *Expenses:* Tuition, state resident: full-time $4,508. Tuition, nonresident: full-time $8,770; part-time $130 per credit. Required fees: $217 per credit. *Financial support:* Research assistantships, teaching assistantships available. Financial award application deadline: 2/15. *Unit head:* Larry J. Forney, Chair, 208-885-6280.

University of Illinois at Chicago, College of Medicine and Graduate College, Graduate Programs in Medicine, Chicago, IL 60607-7128. Offers anatomy and cell biology (MS, PhD); biochemistry and molecular biology (MS, PhD); genetics (PhD), including molecular genetics; health professions education (MHPE); microbiology and immunology (PhD); pharmacology (PhD); physiology and biophysics (MS, PhD); surgery (MS). Part-time programs available. Terminal master's awarded for partial completion of doctoral program. *Degree requirements:* For master's and doctorate, thesis/dissertation. *Entrance requirements:* For master's and doctorate, GRE General Test. Expenses: Contact institution.

University of Illinois at Chicago, Graduate College, College of Liberal Arts and Sciences, Department of Biological Sciences, Chicago, IL 60607-7128. Offers cell and developmental biology (PhD); ecology and evolution (MS, DA, PhD); genetics and development (PhD); molecular biology (MS, PhD); neurobiology (MS, PhD); plant biology (MS, DA, PhD). *Degree requirements:* For master's, thesis; for doctorate, thesis/dissertation, preliminary exam. *Entrance requirements:* For master's and doctorate, GRE General Test, GRE Subject Test, previous

course work in physics, calculus, and organic chemistry; minimum GPA of 2.75. Additional exam requirements/recommendations for international students: Required—TOEFL. Electronic applications accepted.

University of Illinois at Springfield, Graduate Programs, College of Liberal Arts and Sciences, Program in Biology, Springfield, IL 62703-5407. Offers MS. Part-time and evening/weekend programs available. *Faculty:* 6 full-time (3 women). *Students:* 11 full-time (6 women), 12 part-time (7 women); includes 2 minority (both African Americans) Average age 27. 19 applicants, 84% accepted, 10 enrolled. In 2005, 7 degrees awarded. *Degree requirements:* For master's, thesis or alternative, registration. *Entrance requirements:* For master's, GRE General Test, GRE Subject Test in biology, bachelor's degree with courses in biology, microbiology, genetics, botany, vertebrate biology and/or ecology; minimum undergraduate GPA of 3.0. Additional exam requirements/recommendations for international students: Required—TOEFL (minimum score 550 paper-based; 213 computer-based). *Application deadline:* Applications are processed on a rolling basis. Application fee: $50 ($60 for international students). Electronic applications accepted. *Expenses:* Tuition, state resident: full-time $4,726; part-time $163 per credit hour. Tuition, nonresident: full-time $14,178; part-time $490 per credit hour. Required fees: $1,382; $582 per term. *Financial support:* In 2005–06, fellowships with full tuition reimbursements (averaging $7,650 per year), research assistantships with full tuition reimbursements (averaging $7,200 per year), teaching assistantships with full tuition reimbursements (averaging $7,200 per year) were awarded; career-related internships or fieldwork, Federal Work-Study, scholarships/grants, health care benefits, and unspecified assistantships also available. Support available to part-time students. Financial award application deadline: 11/15; financial award applicants required to submit FAFSA. *Faculty research:* Growth of fresh water algae, reproductive biology, microbial ecology, plant systematics. *Unit head:* Dr. Gary Butler, Program Administrator, 217-206-7340, Fax: 217-206-6217, E-mail: butter.gary@uis.edu.

University of Illinois at Urbana–Champaign, Graduate College, College of Liberal Arts and Sciences, School of Chemical Sciences, Champaign, IL 61820. Offers MS, PhD. *Faculty:* 53 full-time (6 women), 2 part-time/adjunct (1 woman). *Students:* 411 full-time (136 women), 4 part-time (1 woman); includes 36 minority (7 African Americans, 25 Asian Americans or Pacific Islanders, 4 Hispanic Americans), 113 international. 629 applicants, 22% accepted, 87 enrolled. In 2005, 34 master's, 47 doctorates awarded. *Degree requirements:* For doctorate, thesis/dissertation. *Entrance requirements:* For master's, minimum GPA of 3.0. *Application deadline:* Applications are processed on a rolling basis. Application fee: $50 ($60 for international students). Electronic applications accepted. *Expenses:* Contact institution. *Financial support:* In 2005–06, 116 fellowships, 251 research assistantships, 267 teaching assistantships were awarded; career-related internships or fieldwork, Federal Work-Study, institutionally sponsored loans, and tuition waivers (full and partial) also available. Financial award application deadline: 2/15. *Unit head:* Thomas B. Rauchfuss, Director, 217-333-5070, Fax: 217-333-3120, E-mail: rauchfuz@uiuc.edu. *Application contact:* Cheryl Kappes, Administrative Aide, 217-333-5070, Fax: 217-333-3120, E-mail: dambache@uiuc.edu.

University of Illinois at Urbana–Champaign, Graduate College, College of Liberal Arts and Sciences, School of Integrative Biology, Champaign, IL 61820. Offers MS, PhD. *Faculty:* 32 full-time (8 women). *Students:* 123 full-time (58 women), 17 part-time (10 women); includes 14 minority (2 African Americans, 8 Asian Americans or Pacific Islanders, 4 Hispanic Americans), 17 international. 223 applicants, 17% accepted, 32 enrolled. In 2005, 20 master's, 12 doctorates awarded. Application fee: $50 ($60 for international students). *Financial support:* In 2005–06, 9 fellowships, 68 research assistantships, 55 teaching assistantships were awarded. *Unit head:* Fred Delcomyn, Director, 217-333-3044, Fax: 217-244-1224, E-mail: delcomyn@uiuc.edu. *Application contact:* Carol Hall, Secretary, 217-333-8208, Fax: 217-244-1224, E-mail: cahall@uiuc.edu.

University of Indianapolis, Graduate Programs, College of Arts and Sciences, Department of Biology, Indianapolis, IN 46227-3697. Offers human biology (MS). Part-time and evening/weekend programs available. *Faculty:* 9 full-time. *Students:* 6 full-time (4 women), 4 part-time (2 women). Average age 27. In 2005, 5 degrees awarded. *Degree requirements:* For master's, thesis. *Entrance requirements:* For master's, GRE Subject Test, minimum GPA of 3.0. *Application deadline:* Applications are processed on a rolling basis. Application fee: $50. *Financial support:* Federal Work-Study available. Financial award application deadline: 5/1; financial award applicants required to submit FAFSA. *Unit head:* Dr. L. Mark Harrison, Chairperson, 317-788-3282, E-mail: harrison@uindy.edu. *Application contact:* Dr. Daniel Briere, Dean, 317-788-3277, Fax: 317-788-3480, E-mail: dbriere@uindy.edu.

The University of Iowa, Graduate College, College of Liberal Arts and Sciences, Department of Biological Sciences, Iowa City, IA 52242-1316.

See Close-Up on page 257.

The University of Iowa, Graduate College, Program in Translational Biomedicine, Iowa City, IA 52242-1316. Offers MS, PhD. *Students:* 6 full-time (2 women), 1 international. 3 applicants, 100% accepted, 3 enrolled. In 2005, 1 degree awarded. *Degree requirements:* For master's, exam, thesis optional; for doctorate, thesis/dissertation, comprehensive exam, registration. *Entrance requirements:* For master's and doctorate, GRE General Test, minimum GPA of 3.0. Additional exam requirements/recommendations for international students: Required—TOEFL (minimum score 550 paper-based; 213 computer-based). *Application deadline:* Applications are processed on a rolling basis. Application fee: $60 ($85 for international students). Electronic applications accepted. *Expenses:* Tuition, state resident: part-time $1,882 per term. Tuition, nonresident: full-time $17,338; part-time $4,907 per term. Tuition and fees vary according to course load and program. *Financial support:* In 2005–06, 1 research assistantship with partial tuition reimbursement was awarded; teaching assistantships with partial tuition reimbursements Financial award applicants required to submit FAFSA. *Unit head:* , Gary Hunninghake, Director, 319-356-4187.

The University of Iowa, Roy J. and Lucille A. Carver College of Medicine and Graduate College, Biosciences Program, Iowa City, IA 52242-1316. Offers PhD. *Faculty:* 241 full-time. *Students:* 21 full-time (10 women). 71 applicants, 52% accepted, 21 enrolled. *Degree requirements:* For doctorate, thesis/dissertation. *Entrance requirements:* For doctorate, GRE General Test, minimum GPA of 3.0. Additional exam requirements/recommendations for international students: Required—TOEFL (minimum score 600 paper-based; 250 computer-based). *Application deadline:* For fall admission, 1/15 priority date for domestic students, 1/15 priority date for international students. Applications are processed on a rolling basis. Application fee: $60 ($85 for international students). Electronic applications accepted. *Expenses:* Contact institution. Tuition and fees vary according to course load and program. *Financial support:* In 2005–06, 21 research assistantships with full tuition reimbursements (averaging $22,000 per year) were awarded; fellowships, teaching assistantships *Unit head:* Dr. Andrew F. Russo, Director, 319-335-7872, Fax: 319-335-7656, E-mail: andrew-russo@uiowa.edu. *Application contact:* Jodi M. Graff, Program Associate, 319-335-8305, E-mail: jodi-graff@uiowa.edu.

The University of Iowa, Roy J. and Lucille A. Carver College of Medicine and Graduate College, Graduate Programs in Medicine, Iowa City, IA 52242-1316. Offers MA, MPAS, MS, DPT, PhD, JD/MHA, MBA/MHA, MD/JD, MD/PhD, MHA/MA, MHA/MS, MPH/MHA, MS/MA, MS/MS. Part-time programs available. In 2005, 115 master's, 57 doctorates awarded. *Degree requirements:* For doctorate, thesis/dissertation. Application fee: $30 ($50 for international students). Electronic applications accepted. *Expenses:* Contact institution. Tuition and fees vary according to course load and program. *Financial support:* Fellowships, research assistantships, teaching assistantships, career-related internships or fieldwork, Federal Work-Study, institutionally sponsored loans, and tuition waivers (full and partial) available. Support available to part-time students. Financial award applicants required to submit FAFSA.

The University of Iowa, Roy J. and Lucille A. Carver College of Medicine and Graduate College, Medical Scientist Training Program, Iowa City, IA 52242-1316. Offers MD/PhD.

Biological and Biomedical Sciences—General

The University of Iowa (continued)
Faculty: 130 full-time (30 women), 5 part-time/adjunct (2 women). *Students:* 63 full-time (21 women); includes 10 minority (1 African American, 1 American Indian/Alaska Native, 7 Asian Americans or Pacific Islanders, 1 Hispanic American). Average age 24. 125 applicants, 21% accepted, 9 enrolled. *Application deadline:* For fall admission, 12/15 for domestic students. Applications are processed on a rolling basis. Application fee: $50. Electronic applications accepted. *Expenses:* Tuition, state resident: part-time $1,882 per term. Tuition, nonresident: full-time $17,338; part-time $4,907 per term. Tuition and fees vary according to course load and program. *Financial support:* In 2005–06, 24 fellowships with full tuition reimbursements (averaging $22,000 per year), 34 research assistantships with full tuition reimbursements (averaging $22,000 per year) were awarded; scholarships/grants and traineeships also available. Total annual research expenditures: $750,000. *Unit head:* Dr. Pamela Geyer, Director, 319-335-6953, Fax: 319-335-7656, E-mail: mstp@uiowa.edu. *Application contact:* Leslie Arnold, Program Associate—MSTP, 319-335-8304, Fax: 319-335-7656, E-mail: mstp@uiowa.edu.

The University of Iowa, Roy J. and Lucille A. Carver College of Medicine, Program in Translational Biomedical Research, Iowa City, IA 52242-1316. Offers MS, PhD. *Accreditation:* NAACLS. *Students:* 6 full-time (3 women); includes 1 minority (Hispanic American) Average age 34. 1 applicant, 0% accepted. In 2005, 1 degree awarded. *Degree requirements:* For master's, thesis optional; for doctorate, thesis/dissertation, comprehensive exam. *Entrance requirements:* For master's and doctorate, GRE General Test, minimum GPA of 3.0. Additional exam requirements/recommendations for international students: Required—TOEFL (minimum score 550 paper-based; 213 computer-based). *Application deadline:* For fall admission, 1/15 for domestic students. Applications are processed on a rolling basis. Application fee: $30 ($50 for international students). Electronic applications accepted. *Expenses:* Tuition, state resident: part-time $1,882 per term. Tuition, nonresident: full-time $17,338; part-time $4,907 per term. Tuition and fees vary according to course load and program. *Financial support:* In 2005–06, 3 fellowships with full tuition reimbursements (averaging $40,000 per year), 1 research assistantship with full tuition reimbursement were awarded. *Faculty research:* Microbial organisms, advanced CT-scan technology, cellular mechanisms, gene discovery and expression. *Unit head:* Dr. Gary W. Hunninghake, Director, 319-356-4187, Fax: 319-356-8101. *Application contact:* Emily L. Avgenackis, Program Associate, 319-384-6381, Fax: 319-353-6406, E-mail: emily-avgenackis@uiowa.edu.

University of Kansas, Graduate School, College of Liberal Arts and Sciences, Division of Biological Sciences, Lawrence, KS 66045. Offers MA, PhD. *Faculty:* 76. *Students:* 115 full-time (63 women), 26 part-time (12 women); includes 5 minority (1 American Indian/Alaska Native, 1 Asian American or Pacific Islander, 3 Hispanic Americans), 46 international. Average age 29. 167 applicants, 23% accepted. In 2005, 15 master's, 9 doctorates awarded. Terminal master's awarded for partial completion of doctoral program. *Degree requirements:* For master's and doctorate, thesis/dissertation, comprehensive exam. *Entrance requirements:* For master's and doctorate, GRE General Test. Additional exam requirements/recommendations for international students: Required—TOEFL. *Application deadline:* For fall admission, 1/10 priority date for domestic students, 1/10 priority date for international students. Application fee: $55 ($60 for international students). Electronic applications accepted. *Expenses:* Tuition, state resident: full-time $4,859. Tuition, nonresident: full-time $12,000. Required fees: $589. Tuition and fees vary according to program. *Financial support:* Fellowships, research assistantships with partial tuition reimbursements, teaching assistantships with full and partial tuition reimbursements, career-related internships or fieldwork, Federal Work-Study, and institutionally sponsored loans available. Support available to part-time students. Financial award application deadline: 3/1; financial award applicants required to submit FAFSA. *Faculty research:* Ecology, evolutionary biology, genetics, developmental biology, cell biology. *Unit head:* James A. Orr, Chair, 785-864-4301, Fax: 785-864-5321, E-mail: jorr@ku.edu.

University of Kansas, Graduate Studies Medical Center, Interdisciplinary Graduate Program in Biomedical Sciences, Kansas City, KS 66160-7836. Offers MA, MPH, MS, PhD, MD/MPH, MD/MS, MD/PhD. Part-time and evening/weekend programs available. *Students:* 19 full-time (13 women), 1 (woman) part-time; includes 3 African Americans, 3 Asian Americans or Pacific Islanders, 1 Hispanic American, 1 international. Average age 25.Terminal master's awarded for partial completion of doctoral program. *Degree requirements:* For master's, thesis/dissertation; for doctorate, thesis/dissertation, comprehensive exam. *Entrance requirements:* For master's and doctorate, GRE. Additional exam requirements/recommendations for international students: Required—TOEFL, TSE. *Application deadline:* For fall admission, 1/15 for domestic students. Applications are processed on a rolling basis. Application fee: $0. Electronic applications accepted. *Expenses:* Tuition, state resident: full-time $4,859. Tuition, nonresident: full-time $12,000. Required fees: $589. Tuition and fees vary according to program. *Financial support:* Fellowships with tuition reimbursements, research assistantships with partial tuition reimbursements, teaching assistantships with full and partial tuition reimbursements, Federal Work-Study, institutionally sponsored loans, traineeships, and unspecified assistantships available. Support available to part-time students. Financial award application deadline: 3/30; financial award applicants required to submit FAFSA. *Faculty research:* Cardiovascular biology, neurosciences, signal transduction and cancer biology, molecular biology and genetics, and developmental biology. *Unit head:* Dr. Michael P. Sarras, Director, 913-588-2039, Fax: 913-588-2711, E-mail: igpbs@kumc.edu. *Application contact:* Lori Reeks, Coordinator, 913-588-2719, Fax: 913-588-2711, E-mail: lreeks@kumc.edu.

See Close-Up on page 259.

University of Kentucky, Graduate School, Graduate School Programs from the College of Arts and Sciences, Program in Biology, Lexington, KY 40506-0032. Offers MS, PhD. *Faculty:* 25 full-time (5 women). *Students:* 49 full-time (22 women), 5 part-time; includes 3 minority (1 African American, 2 Hispanic Americans), 21 international. Average age 28. 148 applicants, 22% accepted, 17 enrolled. In 2005, 7 master's, 5 doctorates awarded. *Median time to degree:* Of those who began their doctoral program in fall 1997, 78.7% received their degree in 8 years or less. *Degree requirements:* For master's, thesis optional; for doctorate, thesis/dissertation, comprehensive exam. *Entrance requirements:* For master's, GRE General Test, minimum undergraduate GPA of 2.5; for doctorate, GRE General Test, minimum graduate GPA of 3.0. Additional exam requirements/recommendations for international students: Required—TOEFL (minimum score 550 paper-based; 213 computer-based). *Application deadline:* For fall admission, 7/17 priority date for domestic students, 2/1 priority date for international students; for spring admission, 12/13 priority date for domestic students, 6/15 priority date for international students. Applications are processed on a rolling basis. Application fee: $40 ($55 for international students). Electronic applications accepted. *Expenses:* Tuition, state resident: full-time $6,308; part-time $331 per credit hour. Tuition, nonresident: full-time $13,968; part-time $756 per credit hour. Tuition and fees vary according to course load, degree level and program. *Financial support:* In 2005–06, 46 students received support, including 1 fellowship with full tuition reimbursement available, 9 research assistantships with full tuition reimbursements available (averaging $7,500 per year), 36 teaching assistantships with full tuition reimbursements available (averaging $14,400 per year); Federal Work-Study, institutionally sponsored loans, scholarships/grants, traineeships, health care benefits, tuition waivers (partial), and unspecified assistantships also available. Support available to part-time students. Financial award application deadline: 3/15. *Faculty research:* General biology, microbiology, *Drosophila* molecular genetics, molecular virology, multiple loci inheritance. *Unit head:* Dr. Peter Mirabito, Director of Graduate Studies, 859-257-7642, Fax: 859-257-1717, E-mail: pmmira00@uky.edu. *Application contact:* Dr. Brian Jackson, Senior Associate Dean, 859-257-8176, Fax: 859-323-1928.

University of Kentucky, Graduate School, Graduate School Programs from the College of Medicine, Lexington, KY 40506-0032. Offers MS, PhD, MD/PhD. *Faculty:* 155 full-time (33 women), 4 part-time/adjunct (0 women). *Students:* 215 full-time (119 women), 22 part-time (15 women); includes 18 minority (6 African Americans, 7 Asian Americans or Pacific Islanders, 5 Hispanic Americans), 84 international. Average age 28. 389 applicants, 45% accepted, 147 enrolled. In 2005, 11 master's, 24 doctorates awarded. *Median time to degree:* Of those who

began their doctoral program in fall 1997, 89.6% received their degree in 8 years or less. *Degree requirements:* For master's, thesis (for some programs), comprehensive exam; for doctorate, thesis/dissertation, comprehensive exam. *Entrance requirements:* For master's, GRE General test, minimum undergraduate GPA of 2.5; for doctorate, GRE General test, minimum undergraduate GPA of 3.0. Additional exam requirements/recommendations for international students: Required—TOEFL (minimum score 550 paper-based; 213 computer-based). *Application deadline:* For fall admission, 7/17 priority date for domestic students, 2/1 priority date for international students; for spring admission, 12/13 priority date for domestic students, 6/15 priority date for international students. Applications are processed on a rolling basis. Application fee: $40 ($55 for international students). Electronic applications accepted. *Expenses:* Tuition, state resident: full-time $6,308; part-time $331 per credit hour. Tuition, nonresident: full-time $13,968; part-time $756 per credit hour. Tuition and fees vary according to course load, degree level and program. *Financial support:* In 2005–06, 32 fellowships with full tuition reimbursements (averaging $4,589 per year), 202 research assistantships with full tuition reimbursements (averaging $21,000 per year), 9 teaching assistantships with full tuition reimbursements (averaging $20,000 per year) were awarded; Federal Work-Study, scholarships/grants, traineeships, health care benefits, tuition waivers (partial), and unspecified assistantships also available. Support available to part-time students. Financial award application deadline: 3/15; financial award applicants required to submit FAFSA. *Unit head:* Dr. E. Wilson, Dean of College of Medicine. *Application contact:* Dr. Brian Jackson, Senior Associate Dean, 859-257-8176, Fax: 859-323-1928.

University of Lethbridge, School of Graduate Studies, Lethbridge, AB T1K 3M4, Canada. Offers accounting (MScM); addictions counseling (M Sc); agricultural biotechnology (M Sc); agricultural studies (M Sc, MA); anthropology (MA); archaeology (MA); art (MA); biochemistry (M Sc); biological sciences (M Sc); biomolecular science (PhD); biosystems and biodiversity (PhD); Canadian studies (MA); chemistry (M Sc); computer science (M Sc); computer science and geographical information science (M Sc); counseling psychology (M Ed); dramatic arts (MA); earth, space, and physical science (PhD); economics (MA); educational leadership (M Ed); English (MA); environmental science (M Sc); evolution and behavior (PhD); exercise science (M Sc); finance (MScM); French (MA); French/German (MA); French/Spanish (MA); general education (M Ed); general management (MScM); geography (M Sc, MA); German (MA); health sciences (M Sc, MA); history (MA); human resource management and labour relations (MScM); individualized multidisciplinary (M Sc, MA); information systems (MScM); international management (MScM); kinesiology (M Sc, MA); management (M Sc, MA); marketing (MScM); mathematics (M Sc); music (MA); Native American studies (MA); neuroscience (M Sc, PhD); new media (MA); nursing (M Sc); philosophy (MA); physics (M Sc); policy and strategy (MScM); political science (MA); psychology (M Sc, MA); religious studies (MA); sociology (MA); theoretical and computational science (PhD); urban and regional studies (MA). Part-time and evening/weekend programs available. *Faculty:* 250. *Students:* 193 full-time, 145 part-time. 35 applicants, 100% accepted, 35 enrolled. In 2005, 40 degrees awarded. *Degree requirements:* For doctorate, thesis/dissertation, comprehensive exam. *Entrance requirements:* For master's, GMAT (M Sc management), bachelor's degree in related field, minimum GPA of 3.0 during previous 20 graded semester courses, 2 years teaching or related experience (M Ed); for doctorate, master's degree, minimum graduate GPA of 3.5. Additional exam requirements/recommendations for international students: Required—TOEFL. Application fee: $60 Canadian dollars. *Expenses:* Tuition, nonresident: part-time $531 per course. Required fees: $83 per year. Tuition and fees vary according to degree level and program. *Financial support:* Fellowships, research assistantships, teaching assistantships, scholarships/grants, health care benefits, and unspecified assistantships available. *Faculty research:* Movement and brain plasticity, gibberellin physiology, photosynthesis, carbon cycling, molecular properties of main-group ring components. *Unit head:* Dr. Shamsul Alam, Dean, 403-329-2121, Fax: 403-329-2097, E-mail: inquiries@uleth.ca. *Application contact:* Kathy Schrage, Administrative Assistant, Office of the Academic Vice President, 403-329-2121, Fax: 403-329-2097, E-mail: inquiries@uleth.ca.

University of Louisiana at Lafayette, Graduate School, College of Sciences, Department of Biology, Lafayette, LA 70504. Offers biology (MS); environmental and evolutionary biology (PhD). *Faculty:* 20 full-time (4 women), 2 part-time/adjunct (0 women). *Students:* 58 full-time (34 women), 12 part-time (10 women); includes 3 minority (2 African Americans, 1 Asian American or Pacific Islander), 23 international. Average age 29. 40 applicants, 38% accepted, 8 enrolled. In 2005, 7 master's, 3 doctorates awarded. Terminal master's awarded for partial completion of doctoral program. *Degree requirements:* For master's, thesis, registration; for doctorate, 2 foreign languages, thesis/dissertation, comprehensive exam, registration. *Entrance requirements:* For master's, GRE General Test, minimum GPA of 2.75; for doctorate, GRE General Test, GRE Subject Test, minimum GPA of 3.0. Additional exam requirements/recommendations for international students: Required—TOEFL (minimum score 550 paper-based; 213 computer-based). *Application deadline:* For fall admission, 5/15 for domestic students, 5/15 for international students; for spring admission, 10/1 for domestic students, 10/1 for international students. Applications are processed on a rolling basis. Application fee: $20 ($30 for international students). Electronic applications accepted. *Expenses:* Tuition, state resident: full-time $3,330; part-time $93 per credit hour. Tuition, nonresident: full-time $9,510; part-time $350 per credit hour. International tuition: $9,646 full-time. *Financial support:* In 2005–06, 17 fellowships with full tuition reimbursements (averaging $14,289 per year), 9 research assistantships with full tuition reimbursements (averaging $8,024 per year), 17 teaching assistantships with full tuition reimbursements (averaging $12,044 per year) were awarded; Federal Work-Study, institutionally sponsored loans, and unspecified assistantships also available. Financial award application deadline: 5/1. *Faculty research:* Structure and ultrastructure, system biology, ecology, processes, environmental physiology. *Unit head:* Dr. Darryl L. Felder, Head, 337-482-6748, Fax: 337-482-5834, E-mail: dlf4517@louisana.edu. *Application contact:* Dr. Paul Leberg, Graduate Coordinator, 337-482-6750, Fax: 337-482-5834, E-mail: leberg@louisiana.edu.

University of Louisiana at Monroe, Graduate Studies and Research, College of Arts and Sciences, Department of Biology, Monroe, LA 71209-0001. Offers MS. *Degree requirements:* For master's, thesis. *Entrance requirements:* For master's, GRE General Test, minimum GPA of 2.8 overall or 3.0 during last 21 hours of biology. *Faculty research:* Fish systematics and zoogeography, taxonomy and distribution of Louisiana plants, aquatic biology, secondary succession, microbial ecology.

University of Louisville, Graduate School, College of Arts and Sciences, Department of Biology, Program in Biology, Louisville, KY 40292-0001. Offers MS. *Students:* 25 full-time (15 women), 19 part-time (11 women); includes 4 minority (3 African Americans, 1 Asian American or Pacific Islander), 5 international. Average age 30. In 2005, 9 degrees awarded. *Degree requirements:* For master's, thesis (for some programs). *Entrance requirements:* For master's, GRE General Test. *Application deadline:* Applications are processed on a rolling basis. Application fee: $50. *Expenses:* Tuition, state resident: full-time $6,006; part-time $334 per credit hour. Tuition, nonresident: full-time $16,554; part-time $920 per credit hour. Tuition and fees vary according to course load, degree level and program. *Application contact:* Dr. Joseph M. Steffen, Director of Graduate Studies, 502-852-6771, Fax: 502-852-0725, E-mail: joe.steffen@louisville.edu.

See Close-Up on page 261.

University of Maine, Graduate School, College of Natural Sciences, Forestry, and Agriculture, Department of Biological Sciences, Program in Biological Sciences, Orono, ME 04469. Offers PhD. *Faculty:* 34. *Students:* 7 full-time (2 women), 5 part-time (2 women), 1 international. Average age 33. 10 applicants, 10% accepted, 1 enrolled. In 2005, 1 doctorate awarded. *Degree requirements:* For doctorate, thesis/dissertation. *Entrance requirements:* For doctorate, GRE General Test. Additional exam requirements/recommendations for international students: Required—TOEFL. *Application deadline:* For fall admission, 2/1 for domestic students. Applications are processed on a rolling basis. Application fee: $50. Electronic applications accepted. *Financial support:* Research assistantships with tuition reimbursements, teach-

Biological and Biomedical Sciences—General

ing assistantships with tuition reimbursements available. Financial award application deadline: 3/1. *Unit head:* Dr. Stelbs Tavanteis, Coordinator, 207-581-2986. *Application contact:* Scott G. Delcourt, Associate Dean of the Graduate School, 207-581-3219, Fax: 207-581-3232, E-mail: graduate@maine.edu.

University of Manitoba, Faculty of Graduate Studies, Faculty of Science, Biological Sciences Division, Winnipeg, MB R3T 2N2, Canada. Offers botany (M Sc, PhD); microbiology (M Sc, PhD); zoology (M Sc, PhD).

University of Manitoba, Faculty of Medicine and Faculty of Graduate Studies, Graduate Programs in Medicine, Winnipeg, MB R3T 2N2, Canada. Offers M Sc, PhD, MD/PhD. *Accreditation:* LCME/AMA. Part-time programs available. Expenses: Contact institution.

University of Maryland, School of Medicine, Graduate Program in Life Sciences, Baltimore, MD 21201. Offers biochemistry (MS, PhD); epidemiology (MS, PhD); gerontology (PhD); microbiology (PhD); molecular and cell biology (MS); molecular medicine (PhD); neuroscience (MS, PhD); pharmacology (MS); physiology (MS); rehabilitation sciences (PhD); toxicology (MS, PhD). *Faculty:* 245 full-time (52 women). *Students:* 268 full-time (165 women), 46 part-time (30 women); includes 43 minority (25 African Americans, 13 Asian Americans or Pacific Islanders, 5 Hispanic Americans), 86 international. 435 applicants, 26% accepted, 61 enrolled. In 2005, 18 master's, 46 doctorates awarded. *Median time to degree:* Of those who began their doctoral program in fall 1997, 99% received their degree in 8 years or less. *Degree requirements:* For master's, registration; for doctorate, thesis/dissertation, lab rotations, comprehensive exam, registration. *Entrance requirements:* For master's and doctorate, GRE or MCAT. Additional exam requirements/recommendations for international students: Required—TOEFL (minimum score 550 paper-based; 213 computer-based). *Application deadline:* For winter admission, 1/15 for domestic students. Applications are processed on a rolling basis. Application fee: $50. Electronic applications accepted. *Expenses:* Tuition, state resident: full-time $8,079; part-time $409 per credit hour. Tuition, nonresident: full-time $18,384; part-time $731 per credit hour. Required fees: $695; $10 per credit hour. Tuition and fees vary according to degree level and program. *Financial support:* In 2005–06, 30 fellowships with full tuition reimbursements (averaging $23,000 per year), 22 research assistantships with full tuition reimbursements (averaging $23,000 per year) were awarded; health care benefits also available. *Faculty research:* Cancer, reproduction, neuroscience, cardiovascular, immunology. *Unit head:* Dr. Margaret Merril McCarthy, Assistant Dean for Graduate Studies, 410-706-2655, Fax: 410-706-8341, E-mail: mmcarthy@umaryland.edu.

University of Maryland, Baltimore County, Graduate School, College of Natural and Mathematical Sciences, Department of Biological Sciences, Baltimore, MD 21250. Offers applied molecular biology (MS); biological sciences (MS, PhD); marine-estuarine-environmental sciences (MS, PhD); molecular and cell biology (PhD); neurosciences and cognitive sciences (PhD). Part-time programs available. *Faculty:* 36 full-time (12 women), 3 part-time/adjunct (1 woman). *Students:* 94 full-time (45 women); includes 42 minority (13 African Americans, 26 Asian Americans or Pacific Islanders, 3 Hispanic Americans). 145 applicants, 48% accepted, 30 enrolled. In 2005, 11 master's, 12 doctorates awarded. *Degree requirements:* For doctorate, thesis/dissertation. *Entrance requirements:* For master's and doctorate, GRE General Test, minimum GPA of 3.0. Additional exam requirements/recommendations for international students: Required—TOEFL. *Application deadline:* For fall admission, 2/1 for domestic students, 1/1 for international students. Applications are processed on a rolling basis. Application fee: $50. Electronic applications accepted. *Expenses:* Tuition, state resident: part-time $395 per credit. Tuition, nonresident: part-time $652 per credit. Required fees: $82 per credit. Tuition and fees vary according to course load, program and reciprocity agreements. *Financial support:* In 2005–06, 85 students received support, including 6 fellowships with tuition reimbursements available (averaging $22,500 per year), 18 research assistantships with tuition reimbursements available (averaging $21,500 per year), 37 teaching assistantships with tuition reimbursements available (averaging $20,500 per year); career-related internships or fieldwork and tuition waivers (partial) also available. *Faculty research:* Molecular genetics, neurobiology, metabolism. *Unit head:* Dr. Lasse Lindahl, Chairman, 410-455-2261, Fax: 410-455-3875. *Application contact:* Dr. David M. Eisenmann, Director, Graduate Program, 410-455-3669, Fax: 410-455-3875, E-mail: biograd@umbc.edu.

University of Maryland, College Park, Graduate Studies, College of Chemical and Life Sciences, Department of Biology, Program in Biology, College Park, MD 20742. Offers MS, PhD. Part-time and evening/weekend programs available. *Students:* 42 full-time (19 women), 2 part-time (1 woman); includes 4 minority (3 African Americans, 1 Asian American or Pacific Islander), 9 international. 35 applicants, 34% accepted, 7 enrolled. In 2005, 8 master's, 5 doctorates awarded. Terminal master's awarded for partial completion of doctoral program. *Degree requirements:* For master's, thesis optional; for doctorate, thesis/dissertation, oral exam. *Entrance requirements:* For master's and doctorate, GRE General Test, GRE Subject Test, minimum GPA of 3.0, 3 letters of recommendation. Additional exam requirements/recommendations for international students: Required—TOEFL. *Application deadline:* For fall admission, 2/1 for domestic students, 2/1 for international students; for spring admission, 10/1 for domestic students, 6/1 for international students. Applications are processed on a rolling basis. Application fee: $60. Electronic applications accepted. *Financial support:* In 2005–06, 11 fellowships with full tuition reimbursements (averaging $8,438 per year) were awarded; research assistantships with tuition reimbursements, teaching assistantships with tuition reimbursements Financial award application deadline: 2/1; financial award applicants required to submit FAFSA. *Unit head:* Dr. Irwin Forseth, Associate Chair, 301-405-1629, Fax: 301-314-9358, E-mail: if2@umail.umd.edu. *Application contact:* Dean of Graduate School, 301-405-4190, Fax: 301-314-9305.

University of Maryland, College Park, Graduate Studies, College of Chemical and Life Sciences, Program in Life Sciences, College Park, MD 20742. Offers MLS. *Students:* 1 (woman) full-time, 133 part-time (83 women); includes 14 minority (10 African Americans, 3 Asian Americans or Pacific Islanders, 1 Hispanic American), 1 international. 57 applicants, 95% accepted, 33 enrolled. In 2005, 15 degrees awarded. *Degree requirements:* For master's, scholarly paper. *Entrance requirements:* For master's, 1 year of teaching experience, letters of recommendation. *Application deadline:* For fall admission, 5/1 priority date for domestic students, 2/1 priority date for international students; for spring admission, 10/1 priority date for domestic students, 6/1 priority date for international students. Applications are processed on a rolling basis. Application fee: $50. Electronic applications accepted. *Financial support:* Fellowships with tuition reimbursements, research assistantships with tuition reimbursements, teaching assistantships with tuition reimbursements, Federal Work-Study and scholarships/grants available. Support available to part-time students. Financial award applicants required to submit FAFSA. *Faculty research:* Genetic engineering, gene therapy, ecology, biocomplexity. *Unit head:* Dr. Paul Mazzocchi, Director, 301-405-4265, E-mail: pmazzocc@deans.umd.edu. *Application contact:* Dean of Graduate School, 301-405-4190, Fax: 301-314-9305.

University of Massachusetts Amherst, Graduate School, College of Natural Resources and the Environment, Program in Animal Biotechnology and Biomedical Sciences, Amherst, MA 01003. Offers mammalian and avian biology (MS, PhD). Part-time programs available. *Faculty:* 18 full-time (7 women). *Students:* 19 full-time (10 women), 1 part-time; includes 4 minority (2 African Americans, 2 Hispanic Americans), 5 international. Average age 28. 22 applicants, 50% accepted, 6 enrolled. In 2005, 4 master's, 6 doctorates awarded. Terminal master's awarded for partial completion of doctoral program. *Degree requirements:* For master's, thesis or alternative; for doctorate, thesis/dissertation. *Entrance requirements:* For master's and doctorate, GRE General Test. Additional exam requirements/recommendations for international students: Required—TOEFL (minimum score 530 paper-based; 197 computer-based). *Application deadline:* For fall admission, 2/1 priority date for domestic students, 2/1 priority date for international students; for spring admission, 10/1 for domestic students, 10/1 for international students. Applications are processed on a rolling basis. Application fee: $40 ($65 for international students). Electronic applications accepted. *Expenses:* Tuition, state resident: part-time $110 per credit. Tuition, nonresident: part-time $414 per credit. Required

fees: $2,824 per term. One-time fee: $250 part-time. Full-time tuition and fees vary according to course load, campus/location, program and reciprocity agreements. *Financial support:* In 2005–06, research assistantships with full tuition reimbursements (averaging $11,772 per year), teaching assistantships with full tuition reimbursements (averaging $11,259 per year) were awarded; fellowships with full tuition reimbursements, career-related internships or fieldwork, Federal Work-Study, scholarships/grants, traineeships, and unspecified assistantships also available. Support available to part-time students. Financial award application deadline: 2/1. *Unit head:* Dr. Sam Black, Director, 413-545-2312, Fax: 413-545-6326.

University of Massachusetts Boston, Office of Graduate Studies and Research, College of Science and Mathematics, Program in Biology, Boston, MA 02125-3393. Offers MS. Part-time and evening/weekend programs available. *Degree requirements:* For master's, thesis, oral exams. *Entrance requirements:* For master's, GRE General Test, GRE Subject Test, minimum GPA of 2.75. *Faculty research:* Microbial ecology, population and conservation genetics energetics of insect locomotion, science education, evolution and ecology of marine invertebrates.

University of Massachusetts Boston, Office of Graduate Studies and Research, College of Science and Mathematics, Program in Biotechnology and Biomedical Science, Boston, MA 02125-3393. Offers MS. Part-time and evening/weekend programs available. *Degree requirements:* For master's, oral exams, thesis optional. *Entrance requirements:* For master's, GRE General Test, GRE Subject Test, minimum GPA of 2.75, 3.0 in science and math. *Faculty research:* Evolutionary and molecular immunology, molecular genetics, tissue culture, computerized laboratory technology.

University of Massachusetts Dartmouth, Graduate School, College of Arts and Sciences, Department of Biology, North Dartmouth, MA 02747-2300. Offers biology (MS); marine biology (MS). Part-time programs available. *Faculty:* 17 full-time (6 women), 4 part-time/adjunct (3 women). *Students:* 7 full-time (4 women), 7 part-time (4 women), 1 international. Average age 27. 18 applicants, 33% accepted, 6 enrolled. In 2005, 2 degrees awarded. *Degree requirements:* For master's, thesis. *Entrance requirements:* For master's, GRE General Test, GRE Subject Test. Additional exam requirements/recommendations for international students: Required—TOEFL (minimum score 500 paper-based). *Application deadline:* For fall admission, 5/7 for domestic students, 3/7 for international students; for spring admission, 11/15 priority date for domestic students, 9/15 priority date for international students. Application fee: $35 ($55 for international students). Electronic applications accepted. *Expenses:* Tuition, state resident: full-time $2,071; part-time $86 per credit. Tuition, nonresident: full-time $8,099; part-time $337 per credit. Required fees: $7,282; $393 per credit. *Financial support:* In 2005–06, 7 teaching assistantships with full tuition reimbursements (averaging $10,340 per year) were awarded; research assistantships with full tuition reimbursements, Federal Work-Study and unspecified assistantships also available. Support available to part-time students. Financial award application deadline: 3/1; financial award applicants required to submit FAFSA. *Faculty research:* Marine life and quality analysis, environmental impact to cranberry soil, oceanography, aquaculture. Total annual research expenditures: $190,000. *Unit head:* Dr. Nancy O'Connor, Director, 508-999-8217, Fax: 508-999-8196, E-mail: noconnor@unmassd.edu. *Application contact:* Carol Novo, Graduate Admissions Officer, 508-999-8604, Fax: 508-999-8183, E-mail: graduate@umassd.edu.

University of Massachusetts Lowell, Graduate School, College of Arts and Sciences, Department of Biological Sciences, Lowell, MA 01854-2881. Offers biochemistry (PhD); biological sciences (MS); biotechnology (MS). Part-time programs available. *Degree requirements:* For master's and doctorate, thesis/dissertation. *Entrance requirements:* For master's and doctorate, GRE General Test. Electronic applications accepted.

University of Massachusetts Worcester, Graduate School of Biomedical Sciences, Interdisciplinary Graduate Program in the Biomedical Sciences, Worcester, MA 01655-0115. Offers PhD. *Faculty:* 41 full-time (10 women). *Entrance requirements:* For doctorate, GRE General Test. Additional exam requirements/recommendations for international students: Required—TOEFL (minimum score 600 paper-based; 250 computer-based). *Application deadline:* For fall admission, 12/15 for domestic students, 12/15 for international students. Applications are processed on a rolling basis. Application fee: $25 ($50 for international students). *Expenses:* Tuition, state resident: full-time $2,640. Tuition, nonresident: full-time $9,856. Required fees: $5,685. *Financial support:* In 2005–06, research assistantships (averaging $25,235 per year) *Unit head:* Dr. William Theurkauf, Director, 508-856-4900, E-mail: william.theurkauf@umassmed.edu. *Application contact:* Michael Cole, Director of Admissions and Recruitment, 508-856-4779, Fax: 508-856-3659, E-mail: michael.cole@umassmed.edu.

See Close-Up on page 265.

University of Massachusetts Worcester, Graduate School of Biomedical Sciences, Program in Biomedical Sciences, Worcester, MA 01655-0115. Offers MD/PhD. *Students:* 32 full-time (15 women). *Application deadline:* For fall admission, 11/1 for domestic students. Applications are processed on a rolling basis. Application fee: $75. *Expenses:* Tuition, state resident: full-time $2,640. Tuition, nonresident: full-time $9,856. Required fees: $5,685. *Financial support:* In 2005–06, research assistantships with full tuition reimbursements (averaging $25,235 per year) *Unit head:* Dr. Elliot Androphy, Director, 508-856-6602.

See Close-Up on page 263.

University of Medicine and Dentistry of New Jersey, Graduate School of Biomedical Sciences, Graduate Programs in Biomedical Sciences–Newark, Newark, NJ 07107. Offers biochemistry and molecular biology (MS, PhD); biomedical sciences (interdisciplinary) (PhD); cell biology and molecular medicine (PhD); experimental pathology (PhD); integrative neuroscience (PhD); microbiology and molecular genetics (PhD); pharmacological sciences (Certificate); pharmacology and physiology (PhD). *Students:* 331 full-time (192 women), 3 part-time (2 women); includes 131 minority (46 African Americans, 54 Asian Americans or Pacific Islanders, 31 Hispanic Americans), 67 international. Average age 27. 451 applicants, 51% accepted, 112 enrolled. In 2005, 28 master's, 19 doctorates awarded. Terminal master's awarded for partial completion of doctoral program. *Degree requirements:* For master's, thesis; for doctorate, thesis/dissertation, qualifying exam. *Entrance requirements:* For master's and doctorate, GRE General Test. Additional exam requirements/recommendations for international students: Required—TOEFL. *Application deadline:* Applications are processed on a rolling basis. Application fee: $40. *Financial support:* Fellowships, research assistantships, teaching assistantships, career-related internships or fieldwork, Federal Work-Study, institutionally sponsored loans, and tuition waivers (full and partial) available. Financial award application deadline: 5/1. *Unit head:* Dr. Nicholas A. Ingoglia, Associate Dean, Graduate School, 973-972-4776, Fax: 973-972-7148, E-mail: ingoglia@umdnj.edu. *Application contact:* Dr. Henry E. Brezenoff, Acting Dean, 973-972-5333, Fax: 973-972-7148, E-mail: hbrezeno@umdnj.edu.

University of Medicine and Dentistry of New Jersey, Graduate School of Biomedical Sciences, Graduate Programs in Biomedical Sciences–Piscataway, Piscataway, NJ 08854-5635. Offers biochemistry and molecular biology (MS, PhD); biomedical engineering (MS, PhD); cellular and molecular pharmacology (MS, PhD); environmental sciences/exposure assessment (PhD); molecular genetics, microbiology and immunology (MS, PhD); neuroscience (MS, PhD); physiology and integrative biology (MS, PhD). *Students:* 535 full-time (290 women); includes 112 minority (17 African Americans, 1 American Indian/Alaska Native, 61 Asian Americans or Pacific Islanders, 33 Hispanic Americans), 230 international. Average age 28. In 2005, 15 master's, 36 doctorates awarded. Terminal master's awarded for partial completion of doctoral program. *Degree requirements:* For master's, thesis, internship, comprehensive exam; for doctorate, thesis/dissertation, oral and written qualifying exams. *Entrance requirements:* For master's and doctorate, GRE General Test. Additional exam requirements/recommendations for international students: Required—TOEFL. *Application deadline:* For fall admission, 1/5 for domestic students. Applications are processed on a rolling basis. Application fee: $40. *Financial support:* Fellowships, research assistantships, teaching assistantships, career-related internships or fieldwork, Federal Work-Study, institutionally sponsored loans, traineeships, and tuition waivers (full and partial) available. Financial

Biological and Biomedical Sciences—General

University of Medicine and Dentistry of New Jersey (continued)
award application deadline: 5/1. *Unit head:* Dr. Michael J. Leibowitz, Associate Dean, Graduate School, 732-235-5016, Fax: 732-235-4720, E-mail: gsbspisc@umdnj.edu.

See Close-Up on page 267.

University of Medicine and Dentistry of New Jersey, Graduate School of Biomedical Sciences, Graduate Programs in Biomedical Sciences–Stratford, Stratford, NJ 08084-5634. Offers biomedical sciences (MBS, MS); cell and molecular biology (MS, PhD). *Students:* 53 full-time (27 women); includes 22 minority (7 African Americans, 1 American Indian/Alaska Native, 11 Asian Americans or Pacific Islanders, 3 Hispanic Americans), 7 international. Average age 25. 53 applicants. In 2005, 1 master's, 1 doctorate awarded. *Degree requirements:* For master's, thesis; for doctorate, thesis/dissertation, qualifying exam. *Entrance requirements:* For master's and doctorate, GRE General Test. Additional exam requirements/recommendations for international students: Required—TOEFL. *Application deadline:* For fall admission, 2/1 for domestic students; for spring admission, 10/1 for domestic students. Applications are processed on a rolling basis. Application fee: $40. *Financial support:* Fellowships, Federal Work-Study available. Financial award application deadline: 5/1. *Unit head:* Dr. Carl E. Hock, Associate Dean, Graduate School, 856-566-6282, Fax: 856-566-6195, E-mail: hock@umdnj.edu.

University of Memphis, Graduate School, College of Arts and Sciences, Department of Biology, Memphis, TN 38152. Offers MS, PhD. *Faculty:* 25 full-time (2 women). *Students:* 41 full-time (24 women), 19 part-time (8 women); includes 7 minority (5 African Americans, 2 Asian Americans or Pacific Islanders), 18 international. Average age 29. 73 applicants, 48% accepted. In 2005, 6 master's, 2 doctorates awarded. *Degree requirements:* For master's, thesis or alternative, comprehensive exam; for doctorate, thesis/dissertation, comprehensive exam. *Entrance requirements:* For master's, GRE General Test, GRE Subject Test, minimum GPA of 2.5; for doctorate, GRE General Test, GRE Subject Test, master's degree. *Application deadline:* For fall admission, 8/1 for domestic students; for spring admission, 12/1 for domestic students. Applications are processed on a rolling basis. Application fee: $25 ($50 for international students). *Financial support:* In 2005–06, 22 students received support, including 1 fellowship, 4 research assistantships, 17 teaching assistantships with full tuition reimbursements available (averaging $11,000 per year) Financial award application deadline: 2/1. *Faculty research:* Evolution and ecology, physiology and behavior, signal transduction, protein trafficking, receptor-mediated carcinogenesis. Total annual research expenditures: $1.9 million. *Unit head:* Dr. Melvin L. Beck, Chairman, 901-678-2970, Fax: 901-678-4746, E-mail: mbeck@memphis.edu. *Application contact:* Dr. Stephan Schoech, Information Contact, 901-678-2327, Fax: 901-678-4746, E-mail: sschoech@memphis.edu.

University of Miami, Graduate School, College of Arts and Sciences, Department of Biology, Coral Gables, FL 33124. Offers biology (MS, PhD); genetics and evolution (MS, PhD). *Faculty:* 21 full-time (3 women), 10 part-time/adjunct (3 women). *Students:* 49 full-time (26 women); includes 2 minority (both Asian Americans or Pacific Islanders), 16 international. Average age 30. 39 applicants, 36% accepted, 10 enrolled. In 2005, 6 degrees awarded. Terminal master's awarded for partial completion of doctoral program. *Median time to degree:* Of those who began their doctoral program in fall 1997, 100% received their degree in 8 years or less. *Degree requirements:* For master's, thesis (for some programs), comprehensive exam (for some programs); for doctorate, thesis/dissertation, oral and written qualifying exam. *Entrance requirements:* For master's and doctorate, GRE General Test, 3 letters of recommendation, research papers. Additional exam requirements/recommendations for international students: Required—TOEFL (minimum score 550 paper-based; 213 computer-based), TSE. *Application deadline:* For fall admission, 1/1 for domestic students, 1/1 for international students. Application fee: $50. Electronic applications accepted. *Financial support:* In 2005–06, 49 students received support, including 9 fellowships with full tuition reimbursements available (averaging $18,000 per year), 15 research assistantships with full tuition reimbursements available (averaging $16,500 per year), 20 teaching assistantships with full tuition reimbursements available (averaging $16,500 per year); career-related internships or fieldwork, Federal Work-Study, institutionally sponsored loans, scholarships/grants, health care benefits, and unspecified assistantships also available. Financial award application deadline: 3/1; financial award applicants required to submit FAFSA. *Faculty research:* Neuroscience to ethology; plants, vertebrates and mycorrhizae; phylogenies, life histories and species interactions; molecular biology, gene expression and populations; cells, auditory neurons and vertebrate locomotion. Total annual research expenditures: $946,278. *Unit head:* Dr. Leonel O. Sternberg, Director of Graduate Studies, 305-284-6436, Fax: 305-284-3039, E-mail: leo@bio.miami.edu. *Application contact:* Beth E. Goad, Graduate Coordinator, 305-284-5116, Fax: 305-284-3039, E-mail: bgoad@bio.miami.edu.

Announcement: Faculty members have active research programs in ecology, behavior, neuroethology, conservation, and evolutionary and tropical biology. Proximity to natural communities in southern Florida and to the Neotropics facilitates field studies. Special aspects include a Distinguished Visiting Professor Program, support for pilot research, and scientific meeting attendance for graduate students. Research opportunities in collaborative studies involve the National Biological Service, Fairchild Tropical Gardens, and the Smithsonian Tropical Research Institute (STRI) in Panama, among others. The University is a founding member of the Organization for Tropical Studies (OTS) and is actively involved in its program. University, McLamore, and Maytag Fellowships are available for outstanding students. Teaching assistantships (9 months) offered to incoming students have stipends of $16,500.

See Close-Up on page 269.

University of Miami, Graduate School, College of Arts and Sciences, Program in Tropical Biology, Ecology, and Behavior, Coral Gables, FL 33124. Offers MS, PhD. Terminal master's awarded for partial completion of doctoral program. *Degree requirements:* For master's, thesis optional; for doctorate, thesis/dissertation, oral and written qualifying exam. *Entrance requirements:* For master's and doctorate, GRE General Test, GRE Subject Test. *Application deadline:* For fall admission, 2/1 for domestic students. Applications are processed on a rolling basis. Application fee: $50. *Financial support:* Fellowships, research assistantships, teaching assistantships, career-related internships or fieldwork and institutionally sponsored loans available. Financial award application deadline: 2/15. *Faculty research:* Behavioral ecology, plant-animal and plant-environmental interactions and coevolution, biogeography, conservation biology, genetics, *Unit head:* Dr. Leonel O. Sternberg, Director of Graduate Studies, 305-284-5116, Fax: 305-284-3039, E-mail: leo@bio.miami.edu.

Announcement: Faculty members in tropical biology, ecology, and behavior program have active research programs in the Neotropics and Paleotropics. Proximity to natural communities in southern Florida and Neotropics facilitates field studies. Tropical biology fellowship, as well as funds to attend Organization for Tropical Studies (OTS) and other field courses, awarded annually. Maytag, McLamore, and University Fellowships, currently $18,000, also available.

University of Miami, Graduate School, Miller School of Medicine, Graduate Programs in Medicine, Interdisciplinary Program in Biomedical Studies, Coral Gables, FL 33124. Offers PhD. *Faculty:* 136 full-time (36 women). *Students:* 3 full-time (1 woman), 1 international. Average age 26. 39 applicants, 21% accepted, 3 enrolled. *Degree requirements:* For doctorate, thesis/dissertation. *Entrance requirements:* For doctorate, GRE General Test. Additional exam requirements/recommendations for international students: Required—TOEFL (minimum score 600 paper-based; 250 computer-based). *Application deadline:* For fall admission, 7/1 for domestic students; for spring admission, 11/1 for domestic students. Applications are processed on a rolling basis. Application fee: $50. Electronic applications accepted. *Financial support:* In 2005–06, fellowships with full tuition reimbursements (averaging $22,000 per year); research assistantships *Faculty research:* Cell and molecular biology, neuroscience, microbiology, immunology, pharmacology. *Unit head:* Dr. Richard J. Bookman, Associate Dean for Research and Graduate Studies, 305-243-1094, Fax: 305-243-3593, E-mail: biomedgrad@miami.edu.

Application contact: Dr. Claudia M. Ochatt, Administrator, 305-243-6278, Fax: 305-243-3593, E-mail: cochatt@med.miami.edu.

University of Michigan, Horace H. Rackham School of Graduate Studies, Cellular Biotechnology Program, Ann Arbor, MI 48109.

University of Michigan, Medical School and Horace H. Rackham School of Graduate Studies, Medical Scientist Training Program, Ann Arbor, MI 48109. Offers MD/PhD. *Accreditation:* LCME/AMA. *Students:* 47 full-time (13 women); includes 13 minority (3 African Americans, 8 Asian Americans or Pacific Islanders, 2 Hispanic Americans). *Application deadline:* For fall admission, 10/15 for domestic students. Applications are processed on a rolling basis. Application fee: $60. Electronic applications accepted. *Expenses:* Tuition, state resident: full-time $14,082; part-time $894 per credit hour. Tuition, nonresident: full-time $28,500; part-time $1,675 per credit hour. Required fees: $189; $189 per unit. *Financial support:* In 2005–06, 46 students received support, including 46 fellowships with full tuition reimbursements available (averaging $23,500 per year); scholarships/grants, traineeships, and health care benefits also available. *Unit head:* , Dr. Ronald J. Koenig, Director, 734-764-6176, Fax: 734-764-8180, E-mail: rkoenig@umich.edu. *Application contact:* Laurie Koivupalo, Program Secretary, 734-764-6176, Fax: 734-764-8180, E-mail: lkoivupl@umich.edu.

University of Michigan, Medical School and Horace H. Rackham School of Graduate Studies, Program in Biomedical Sciences (PIBS), Ann Arbor, MI 48109. Offers MS, PhD. *Students:* 80 full-time (47 women); includes 23 minority (7 African Americans, 2 American Indian/Alaska Native, 9 Asian Americans or Pacific Islanders, 5 Hispanic Americans), 10 international. Average age 25. 675 applicants, 31% accepted, 78 enrolled. *Degree requirements:* For doctorate, oral defense of dissertation, preliminary exam. *Entrance requirements:* For doctorate, GRE General Test, research experience. Additional exam requirements/recommendations for international students: Required—TOEFL (minimum score 650 paper-based; 250 computer-based). *Application deadline:* For fall admission, 12/31 for domestic students, 12/31 for international students. Application fee: $65 ($75 for international students). Electronic applications accepted. *Expenses:* Tuition, state resident: full-time $14,082; part-time $894 per credit hour. Tuition, nonresident: full-time $28,500; part-time $1,675 per credit hour. Required fees: $189; $189 per unit. *Financial support:* In 2005–06, 76 students received support, including 76 fellowships with full tuition reimbursements available (averaging $23,500 per year); scholarships/grants, health care benefits, tuition waivers (full), and unspecified assistantships also available. Financial award application deadline: 12/31. *Faculty research:* Genetics, cellular and molecular biology, microbial pathogenesis, cancer biology, neuroscience. *Unit head:* Dr. David R. Engelke, PIBS Director/Professor of Biological Chemistry/Assistant Dean for Research and Graduate Studies, 734-615-7005, Fax: 734-647-7022, E-mail: engelke@umich.edu. *Application contact:* Tiffany L. Porties, Recruitment Coordinator, 734-647-7005, Fax: 734-647-7022, E-mail: pibs@umich.edu.

See Close-Up on page 271.

University of Michigan–Flint, College of Arts and Sciences, Flint, MI 48502-1950. Offers American culture (MLS); biology (MS); computer and information systems (MS); public administration (MPA); social sciences (MA). Part-time programs available. *Faculty:* 10 full-time (2 women). *Students:* 2 full-time (both women), 42 part-time (14 women); includes 3 minority (all African Americans) Average age 33. 46 applicants, 80% accepted, 29 enrolled. In 2005, 2 degrees awarded. *Degree requirements:* For master's, thesis optional. *Entrance requirements:* Additional exam requirements/recommendations for international students: Required—TOEFL (minimum score 550 paper-based; 220 computer-based), IELT (minimum score 7). *Application deadline:* For fall admission, 8/1 for domestic students; for winter admission, 11/15 for domestic students; for spring admission, 3/15 for domestic students. Applications are processed on a rolling basis. Application fee: $55. Electronic applications accepted. *Expenses:* Contact institution. Full-time tuition and fees vary according to degree level and program. *Financial support:* In 2005–06, 19 students received support; fellowships, Federal Work-Study and scholarships/grants available. Support available to part-time students. Financial award applicants required to submit FAFSA. *Faculty research:* Household hazardous waste, e. coli, invasive aquatic species, small heterafullerene molecules, contour-derived elevation and terrain slope. *Unit head:* Dr. D. J. Trela, Dean, 810-762-3234, Fax: 810-762-3006, E-mail: djtrela@umflint.edu. *Application contact:* Bradley T. Maki, Director of Graduate Admissions, 810-762-3171, Fax: 810-766-6789, E-mail: bmaki@umflint.edu.

University of Minnesota, Duluth, Graduate School, College of Science and Engineering, Department of Biology, Duluth, MN 55812-2496. Offers MS. Part-time programs available. *Faculty:* 43 full-time (6 women), 2 part-time/adjunct (1 woman). *Students:* 30 full-time (15 women), 1 (woman) part-time; includes 1 minority (Asian American or Pacific Islander) Average age 27. 30 applicants, 37% accepted, 11 enrolled. In 2005, 6 degrees awarded *Degree requirements:* For master's, seminar. *Entrance requirements:* For master's, GRE General Test, minimum GPA of 3.0. Additional exam requirements/recommendations for international students: Required—TOEFL (minimum score 550 paper-based; 213 computer-based). *Application deadline:* For fall admission, 7/15 for domestic students; for spring admission, 11/15 for domestic students. Applications are processed on a rolling basis. Application fee: $55 ($75 for international students). *Financial support:* In 2005–06, research assistantships with full tuition reimbursements (averaging $13,500 per year), teaching assistantships with full tuition reimbursements (averaging $12,394 per year) were awarded; fellowships with full tuition reimbursements, career-related internships or fieldwork, Federal Work-Study, institutionally sponsored loans, traineeships, and tuition waivers (full and partial) also available. Support available to part-time students. Financial award application deadline: 3/15; financial award applicants required to submit FAFSA. *Faculty research:* Cell biology, developmental biology, forest ecology, freshwater ecology, landscape ecology. Total annual research expenditures: $2.5 million. *Unit head:* Dr. John Pastor, Director of Graduate Studies, 218-726-6262, Fax: 218-726-8142, E-mail: biograd@d.umn.edu. *Application contact:* 218-726-7523.

University of Minnesota, Twin Cities Campus, Graduate School, College of Biological Sciences, Biological Science Program, Minneapolis, MN 55455-0213. Offers MBS. Part-time and evening/weekend programs available. *Entrance requirements:* For master's, 2 years of work experience. Electronic applications accepted. *Expenses:* Contact institution. Full-time tuition and fees vary according to class time, course load, program and reciprocity agreements.

University of Minnesota, Twin Cities Campus, Medical School and Graduate School, Graduate Programs in Medicine, Minneapolis, MN 55455-0213. Offers MA, MS, PhD. Part-time and evening/weekend programs available. *Expenses:* Contact institution. Full-time tuition and fees vary according to class time, course load, program and reciprocity agreements.

University of Mississippi, Graduate School, College of Liberal Arts, Department of Biology, Oxford, University, MS 38677. Offers MS, PhD. *Faculty:* 21 full-time (6 women), 3 part-time/adjunct (all women). *Students:* 26 full-time (12 women), 1 (woman) part-time; includes 2 minority (both African Americans), 4 international. 30 applicants, 20% accepted, 2 enrolled. In 2005, 1 master's, 3 doctorates awarded. *Degree requirements:* For master's and doctorate, thesis/dissertation. *Entrance requirements:* For master's and doctorate, GRE General Test, GRE Subject Test, minimum GPA of 3.0. Additional exam requirements/recommendations for international students: Required—TOEFL. *Application deadline:* For fall admission, 4/1 for domestic students; for spring admission, 10/1 for domestic students. Applications are processed on a rolling basis. Application fee: $25. Electronic applications accepted. *Expenses:* Tuition, state resident: full-time $4,320; part-time $240 per credit hour. Tuition, nonresident: full-time $9,744; part-time $301 per credit hour. Tuition and fees vary according to program. *Financial support:* Research assistantships, teaching assistantships, scholarships/grants available. Financial award application deadline: 3/1; financial award applicants required to submit FAFSA.

Faculty research: Freshwater biology, including ecology and evolutionary biology; environmental and applied biology. *Unit head:* Dr. Murray W. Nabors, Chair, 662-915-7203, Fax: 662-915-5144.

University of Mississippi Medical Center, School of Graduate Studies in the Health Sciences, Jackson, MS 39216-4505. Offers MS, MSN, PhD, MD/PhD. *Faculty:* 98 full-time (26 women), 2 part-time/adjunct (0 women). *Students:* 209 full-time (139 women), 121 part-time (98 women); includes 21 minority (3 African Americans, 15 Asian Americans or Pacific Islanders, 3 Hispanic Americans). Average age 28. 172 applicants, 54% accepted, 68 enrolled. In 2005, 31 master's, 15 doctorates awarded. Terminal master's awarded for partial completion of doctoral program. *Degree requirements:* For master's, thesis; for doctorate, thesis/dissertation, first authored publication. *Application deadline:* Applications are processed on a rolling basis. Application fee: $10. *Financial support:* In 2005–06, 71 students received support, including 71 research assistantships (averaging $16,234 per year) Financial award application deadline: 3/15; financial award applicants required to submit FAFSA. *Faculty research:* Immunology; protein chemistry and biosynthesis; cardiovascular, renal, and endocrine physiology; rehabilitation therapy on immune system/hypothalamic/adrenal axis interaction. Total annual research expenditures: $11.3 million. *Unit head:* Dr. I. K. Ho, Dean, 601-984-1600, Fax: 601-984-1637, E-mail: iho@pharmacology.umsmed.edu. *Application contact:* Barbara Westerfield, Director, Student Records and Registrar, 601-984-1080, Fax: 601-984-1079, E-mail: bwesterfield@registrar.umsmed.edu.

See Close-Up on page 273.

University of Missouri–Columbia, Graduate School, College of Arts and Sciences, Division of Biological Sciences, Columbia, MO 65211. Offers cellular, molecular and developmental biology (MA, PhD); evolutionary biology and ecology (MA, PhD); neurobiology and behavior (MA, PhD). *Faculty:* 43 full-time (12 women). *Students:* 48 full-time (26 women), 28 part-time (10 women); includes 7 minority (2 African Americans, 3 Asian Americans or Pacific Islanders, 2 Hispanic Americans), 20 international. In 2005, 3 master's, 8 doctorates awarded. Terminal master's awarded for partial completion of doctoral program. *Degree requirements:* For master's, thesis/dissertation; for doctorate, thesis/dissertation, comprehensive exam. *Entrance requirements:* For master's and doctorate, GRE General Test, minimum GPA of 3.0. *Application deadline:* For fall admission, 1/15 for domestic students. Applications are processed on a rolling basis. Application fee: $45 ($60 for international students). *Financial support:* Fellowships, research assistantships, teaching assistantships, institutionally sponsored loans available. *Unit head:* Dr. Ray Semlitsch, Director of Graduate Studies, 573-884-6396, E-mail: semlitschr@missouri.edu. *Application contact:* Nila Emerich, Application Contact, 800-553-5698.

University of Missouri–Columbia, School of Medicine and Graduate School, Graduate Programs in Medicine, Columbia, MO 65211. Offers MS, MSPH, PhD. Part-time programs available. *Faculty:* 62 full-time (15 women), 6 part-time/adjunct (3 women). *Students:* 116 full-time (66 women), 59 part-time (29 women); includes 25 minority (17 African Americans, 4 Asian Americans or Pacific Islanders, 4 Hispanic Americans), 39 international. In 2005, 6 master's, 7 doctorates awarded. *Degree requirements:* For doctorate, thesis/dissertation. *Entrance requirements:* For master's and doctorate, GRE General Test, minimum GPA of 3.0. *Application deadline:* Applications are processed on a rolling basis. Application fee: $45 ($60 for international students). *Expenses: Contact institution. Financial support:* Fellowships, research assistantships, teaching assistantships, career-related internships or fieldwork and institutionally sponsored loans available. *Application contact:* Dr. William Altemeier, Associate Dean for Students, 573-882-3490, E-mail: altemeirw@missouri.edu.

University of Missouri–Kansas City, School of Biological Sciences, Kansas City, MO 64110-2499. Offers biology (MA); cell biology and biophysics (PhD); cellular and molecular biology (MS); molecular biology and biochemistry (PhD). Part-time and evening/weekend programs available. *Faculty:* 37 full-time (8 women), 6 part-time/adjunct (3 women). *Students:* 14 full-time (6 women), 31 part-time (24 women); includes 4 minority (1 African American, 2 American Indian/Alaska Native, 1 Asian American or Pacific Islander), 2 international. Average age 33. 35 applicants, 51% accepted, 15 enrolled. In 2005, 13 degrees awarded. *Degree requirements:* For doctorate, thesis/dissertation, comprehensive exam. *Entrance requirements:* For master's, GRE, minimum GPA of 3.0; for doctorate, GRE General Test. Additional exam requirements/recommendations for international students: Required—TOEFL. *Application deadline:* Applications are processed on a rolling basis. Application fee: $35 ($50 for international students). *Expenses:* Tuition, state resident: full-time $4,738; part-time $263 per credit hour. Tuition, nonresident: full-time $12,235; part-time $679 per credit hour. Required fees: $582. Tuition and fees vary according to course load, program and student level. *Financial support:* In 2005–06, 27 research assistantships with full tuition reimbursements (averaging $22,000 per year), 11 teaching assistantships with full tuition reimbursements (averaging $22,000 per year) were awarded; Federal Work-Study, institutionally sponsored loans, scholarships/grants, tuition waivers (full and partial), and unspecified assistantships also available. Support available to part-time students. *Faculty research:* Structural biology, molecular genetics. Total annual research expenditures: $4.5 million. *Unit head:* Dr. Lawrence A. Dreyfus, Dean, 816-235-5246, Fax: 816-235-5158, E-mail: dreyfusl@umkc.edu. *Application contact:* Laura Batenic, Information Contact, 816-235-2352, Fax: 816-235-5158, E-mail: batenicl@umkc.edu.

University of Missouri–Rolla, Graduate School, College of Arts and Sciences, Department of Biological Sciences, Rolla, MO 65409-0910. Offers MS.

University of Missouri–St. Louis, College of Arts and Sciences, Department of Biology, St. Louis, MO 63121. Offers biology (MS, PhD), including animal behavior (MS), biochemistry (MS), biotechnology (MS), conservation biology (MS), development (MS), ecology (MS), environmental studies (PhD), evolution (MS), genetics (MS), molecular/cellular biology (MS), physiology (MS), plant systematics, population biology (MS), tropical biology (MS); biotechnology (Certificate); tropical biology and conservation (Certificate). Part-time programs available. *Faculty:* 50. *Students:* 26 full-time (15 women), 101 part-time (51 women); includes 13 minority (5 African Americans, 6 Asian Americans or Pacific Islanders, 2 Hispanic Americans), 41 international. Average age 32. In 2005, 22 master's, 2 doctorates awarded. *Degree requirements:* For master's, thesis or alternative; for doctorate, one foreign language, thesis/dissertation, 1 semester of teaching experience. *Entrance requirements:* For doctorate, GRE General Test. *Application deadline:* For spring admission, 12/1 priority date for domestic students. Applications are processed on a rolling basis. Application fee: $35 ($40 for international students). Electronic applications accepted. *Expenses:* Tuition, state resident: part-time $263 per credit hour. Tuition, nonresident: part-time $680 per credit hour. Required fees: $53 per credit hour. Tuition and fees vary according to program. *Financial support:* In 2005–06, 11 fellowships with full tuition reimbursements (averaging $30,000 per year), 15 research assistantships with full and partial tuition reimbursements (averaging $16,000 per year), 22 teaching assistantships with full and partial tuition reimbursements (averaging $16,000 per year) were awarded; career-related internships or fieldwork and Federal Work-Study also available. Support available to part-time students. Financial award application deadline: 2/1. *Faculty research:* Molecular biology, microbial genetics. *Unit head:* Zuleyma Tang-Martinez, Director of Graduate Studies, 314-516-6498, Fax: 314-516-6233, E-mail: zuleyma@umsl.edu. *Application contact:* 314-516-5458, Fax: 314-516-5310, E-mail: gradadm@umsl.edu.

The University of Montana, Graduate School, College of Arts and Sciences, Division of Biological Sciences, Missoula, MT 59812-0002. Offers biochemistry and microbiology (MS, PhD), including biochemistry (MS), integrative microbiology and biochemistry (PhD), microbial ecology, microbiology (MS); organismal biology and ecology (MS, PhD). *Faculty:* 39 full-time (4 women). *Students:* 51 full-time (23 women), 15 part-time (4 women); includes 3 minority (2 American Indian/Alaska Native, 1 Asian American or Pacific Islander), 16 international. 70 applicants, 30% accepted, 10 enrolled. In 2005, 4 master's, 6 doctorates awarded. Terminal master's awarded for partial completion of doctoral program. *Degree requirements:* For master's and doctorate, thesis/dissertation. *Entrance requirements:* For master's and doctorate, GRE General Test. Additional exam requirements/recommendations for international students: Required—TOEFL. *Application deadline:* For fall admission, 2/1 for domestic students. Applica-

tions are processed on a rolling basis. Application fee: $45. *Expenses:* Tuition, state resident: part-time $267 per credit. Tuition, nonresident: part-time $665 per credit. Part-time tuition and fees vary according to course load and degree level. *Financial support:* In 2005–06, research assistantships with full tuition reimbursements (averaging $14,000 per year), teaching assistantships with full tuition reimbursements (averaging $14,000 per year) were awarded; Federal Work-Study and unspecified assistantships also available. Financial award application deadline: 3/1; financial award applicants required to submit FAFSA. *Faculty research:* Biochemistry/microbiology, organismal biology, ecology. Total annual research expenditures: $6.1 million. *Unit head:* Erick P. Greene, Acting Associate Dean, 406-243-5122. *Application contact:* Janean Clark, Graduate Programs Secretary, 406-243-5222, Fax: 406-243-4184, E-mail: jmclark@selway.umt.edu.

University of Nebraska at Kearney, College of Graduate Study, College of Natural and Social Sciences, Department of Biology, Kearney, NE 68849-0001. Offers biology (MS); science education (MS Ed). *Accreditation:* NCATE (one or more programs are accredited). Part-time and evening/weekend programs available. *Degree requirements:* For master's, thesis optional. *Entrance requirements:* For master's, GRE General Test. Additional exam requirements/recommendations for international students: Required—TOEFL (minimum score 550 paper-based; 213 computer-based). Electronic applications accepted. *Faculty research:* Pollution injury, molecular biology-viral gene expression, prairie range condition modeling, evolution of symbiotic nitrogen fixation.

University of Nebraska at Omaha, Graduate Studies and Research, College of Arts and Sciences, Department of Biology, Omaha, NE 68182. Offers MS. Part-time programs available. *Faculty:* 25 full-time (7 women). *Students:* 6 full-time (5 women), 25 part-time (14 women); includes 1 minority (Hispanic American), 4 international. Average age 29. 18 applicants, 56% accepted, 8 enrolled. In 2005, 9 degrees awarded. *Degree requirements:* For master's, thesis (for some programs), comprehensive exam (for some programs). *Entrance requirements:* For master's, GRE General Test, minimum GPA of 3.0. Additional exam requirements/recommendations for international students: Required—TOEFL (minimum score 500 paper-based; 173 computer-based). *Application deadline:* For fall admission, 3/1 for domestic students; for spring admission, 10/15 priority date for domestic students. Applications are processed on a rolling basis. Application fee: $45. Electronic applications accepted. *Expenses:* Tuition, state resident: part-time $172 per credit. Tuition, nonresident: part-time $452 per credit. Part-time tuition and fees vary according to campus/location. *Financial support:* In 2005–06, 24 students received support; fellowships, research assistantships with tuition reimbursements available, teaching assistantships with tuition reimbursements available, Federal Work-Study, institutionally sponsored loans, scholarships/grants, tuition waivers (partial), and unspecified assistantships available. Support available to part-time students. Financial award application deadline: 3/1; financial award applicants required to submit FAFSA. *Unit head:* Dr. William Tapprich, Chairperson, 402-554-2641.

University of Nebraska–Lincoln, Graduate College, College of Agricultural Sciences and Natural Resources, Department of Veterinary and Biomedical Sciences, Lincoln, NE 68588. Offers MS, PhD. Postbaccalaureate distance learning degree programs offered (minimal on-campus study). *Degree requirements:* For master's, thesis optional; for doctorate, thesis/dissertation, comprehensive exam. *Entrance requirements:* For master's, GRE General Test; for doctorate, GRE General Test, MCAT, or VCAT. Additional exam requirements/recommendations for international students: Required—TOEFL (minimum score 550 paper-based; 213 computer-based). Electronic applications accepted. *Faculty research:* Virology, immunobiology, molecular biology, mycotoxins, ocular degeneration.

University of Nebraska–Lincoln, Graduate College, College of Arts and Sciences, School of Biological Sciences, Lincoln, NE 68588. Offers MA, MS, PhD. *Degree requirements:* For master's, thesis optional; for doctorate, thesis/dissertation, comprehensive exam. *Entrance requirements:* For master's and doctorate, GRE General Test. Additional exam requirements/recommendations for international students: Required—TOEFL (minimum score 550 paper-based; 213 computer-based). Electronic applications accepted. *Faculty research:* Behavior, botany, and zoology; ecology and evolutionary biology; genetics; cellular and molecular biology; microbiology.

University of Nebraska Medical Center, Graduate Studies, Biomedical Research Training Program, Omaha, NE 68198.

See Close-Up on page 275.

University of Nebraska Medical Center, Graduate Studies, Medical Sciences Interdepartmental Area, Omaha, NE 68198. Offers MS, PhD. Part-time programs available. *Faculty:* 315. *Students:* 16 full-time (9 women), 27 part-time (14 women); includes 2 minority (1 African American, 1 Hispanic American), 17 international. 18 applicants, 33% accepted, 4 enrolled. In 2005, 4 master's, 9 doctorates awarded. Terminal master's awarded for partial completion of doctoral program. *Median time to degree:* Of those who began their doctoral program in fall 1997, 100% received their degree in 8 years or less. *Degree requirements:* For master's and doctorate, thesis/dissertation, comprehensive exam, registration. *Entrance requirements:* For master's and doctorate, GRE General Test. Additional exam requirements/recommendations for international students: Required—TOEFL. *Application deadline:* Applications are processed on a rolling basis. Application fee: $40. *Expenses:* Tuition, area resident: Part-time $200 per hour. Tuition, nonresident: part-time $538 per hour. Required fees: $308; $59 per term. *Financial support:* In 2005–06, 6 research assistantships with full tuition reimbursements (averaging $20,000 per year) were awarded; fellowships, teaching assistantships, institutionally sponsored loans also available. Support available to part-time students. Financial award application deadline: 3/1. *Faculty research:* Molecular genetics, oral biology, veterinary pathology, newborn medicine, immunology. *Unit head:* , Dr. M. Patricia Leuschen, Graduate Committee Chair, 402-559-9291, Fax: 402-559-9333, E-mail: pleusche@unmc.edu.

University of Nevada, Las Vegas, Graduate College, College of Science, Department of Biological Sciences, Las Vegas, NV 89154-9900. Offers MS, PhD. Part-time programs available. *Faculty:* 33 full-time (8 women), 9 part-time/adjunct (3 women). *Students:* 40 full-time (17 women), 20 part-time (11 women); includes 3 minority (all Asian Americans or Pacific Islanders), 8 international. 40 applicants, 38% accepted, 13 enrolled. In 2005, 2 master's, 2 doctorates awarded. *Degree requirements:* For master's, thesis, oral exam; for doctorate, one foreign language, thesis/dissertation, comprehensive exam. *Entrance requirements:* For master's, GRE General Test, minimum GPA of 3.0; for doctorate, GRE General Test, GRE Subject Test, minimum GPA of 3.0. Additional exam requirements/recommendations for international students: Required—TOEFL (minimum score 550 paper-based; 213 computer-based). *Application deadline:* For fall admission, 6/15 for domestic students, 5/1 for international students; for spring admission, 10/31 for domestic students, 10/1 for international students. Application fee: $60 ($75 for international students). Electronic applications accepted. *Expenses:* Tuition, state resident: part-time $150 per credit. Tuition, nonresident: part-time $315 per credit. Tuition and fees vary according to course load, program and reciprocity agreements. *Financial support:* In 2005–06, 26 teaching assistantships with partial tuition reimbursements (averaging $11,000 per year) were awarded; research assistantships with full tuition reimbursements, career-related internships or fieldwork, Federal Work-Study, institutionally sponsored loans, scholarships/grants, health care benefits, and unspecified assistantships also available. Support available to part-time students. Financial award application deadline: 3/1. *Unit head:* Dr. Carl Reiber, Chair, 702-895-3399. *Application contact:* Graduate College Admissions Evaluator, 702-895-3320, Fax: 702-895-4180, E-mail: gradcollege@unlv.edu.

University of Nevada, Reno, Graduate School, College of Science, Department of Biology, Reno, NV 89557. Offers MS. *Faculty:* 17 full-time (5 women). *Students:* 4 full-time (2 women), 13 part-time (10 women). Average age 30. 5 applicants, 40% accepted, 0 enrolled. In 2005, 2 degrees awarded. *Degree requirements:* For master's, thesis optional. *Entrance requirements:* For master's, GRE General Test, minimum GPA of 2.75. Additional exam requirements/recommendations for international students: Required—TOEFL. *Application deadline:* For fall

Biological and Biomedical Sciences—General

University of Nevada, Reno (continued)
admission, 3/1 for domestic students; for spring admission, 11/1 for domestic students. Applications are processed on a rolling basis. Application fee: $60 ($95 for international students). *Expenses:* Tuition, area resident: Full-time $2,767; part-time $923 per semester. Tuition, state resident: full-time $5,733; part-time $1,911 per semester. Tuition, nonresident: full-time $12,679; part-time $1,911 per semester. International tuition: $13,878 full-time. Required fees: $404; $202 per term. One-time fee: $90. *Financial support:* In 2005–06, 18 research assistantships, 18 teaching assistantships were awarded; Federal Work-Study and institutionally sponsored loans also available. Financial award application deadline: 3/1. *Faculty research:* Gene expression, stress protein genes, secretory proteins, conservation biology, behavioral ecology. *Unit head:* Dr. Steve Vander Wall, Graduate Program Director, 775-784-6583.

University of Nevada, Reno, School of Medicine and Graduate School, Graduate Programs in Medicine, Reno, NV 89557. Offers MS, PhD. *Faculty:* 5. *Students:* 72 full-time (52 women), 32 part-time (18 women); includes 11 minority (1 African American, 6 Asian Americans or Pacific Islanders, 4 Hispanic Americans), 29 international. Average age 28. 60 applicants, 80% accepted, 33 enrolled. In 2005, 21 master's, 8 doctorates awarded. *Degree requirements:* For doctorate, thesis/dissertation. *Entrance requirements:* For master's, GRE General Test, minimum GPA of 2.75; for doctorate, GRE General Test, minimum GPA of 3.0, *Application deadline:* For fall admission, 3/1 for domestic students. Applications are processed on a rolling basis. Application fee: $60 ($95 for international students). *Expenses:* Tuition, area resident: Full-time $2,767; part-time $923 per semester. Tuition, state resident: full-time $5,733; part-time $1,911 per semester. Tuition, nonresident: full-time $12,679; part-time $1,911 per semester. International tuition: $13,878 full-time. Required fees: $404; $202 per term. One-time fee: $90. *Financial support:* Fellowships, research assistantships, teaching assistantships, Federal Work-Study available. Support available to part-time students. Financial award application deadline: 3/1.

University of New Brunswick Fredericton, School of Graduate Studies, Faculty of Science, Department of Biology, Fredericton, NB E3B 5A3, Canada. Offers M Sc, PhD. Part-time programs available. *Degree requirements:* For master's and doctorate, thesis/dissertation. *Entrance requirements:* For master's and doctorate, minimum GPA of 3.0. Additional exam requirements/recommendations for international students: Required—TOEFL, TWE. Electronic applications accepted. *Faculty research:* Molecular biology, wildlife ecology, aquaculture, systematics.

University of New Brunswick Saint John, Faculty of Science, Applied Science and Engineering, Saint John, NB E2L 4L5, Canada. Offers biology (M Sc, PhD). Part-time programs available. *Degree requirements:* For master's and doctorate, thesis/dissertation. *Entrance requirements:* Additional exam requirements/recommendations for international students: Required—TOEFL.

University of New England, College of Arts and Sciences, Programs in Professional Science, Biddeford, ME 04005-9526. Offers applied biosciences (MS); marine sciences (MS). *Faculty:* 5 full-time (4 women). *Students:* 5 full-time (3 women); includes 1 minority (Asian American or Pacific Islander), 2 international. Average age 25. 5 applicants, 100% accepted, 5 enrolled. Application fee: $40. *Expenses:* Contact institution. *Unit head:* Lawrence Fritz, Chair, Department of Biological Sciences, E-mail: lfritz@une.edu. *Application contact:* Robert Pecchia, Associate Dean of Admissions, 207-283-0171 Ext. 2297, Fax: 207-602-5900, E-mail: admissions@une.edu.

University of New Mexico, Graduate School, College of Arts and Sciences, Department of Biology, Albuquerque, NM 87131-2039. Offers MS, PhD. *Faculty:* 53 full-time (15 women), 9 part-time/adjunct (4 women). *Students:* 83 full-time (48 women), 18 part-time (11 women); includes 18 minority (4 American Indian/Alaska Native, 2 Asian Americans or Pacific Islanders, 12 Hispanic Americans), 7 international. Average age 32. 66 applicants, 20% accepted, 13 enrolled. In 2005, 10 master's, 17 doctorates awarded. *Degree requirements:* For master's, thesis optional; for doctorate, thesis/dissertation, comprehensive exam. *Entrance requirements:* For master's and doctorate, GRE General Test, GRE Subject Test, minimum GPA of 3.2, letters of recommendation. Additional exam requirements/recommendations for international students: Required—TOEFL (minimum score 550 paper-based; 213 computer-based), Internet Based TOEFL-Minimum Score: 79. *Application deadline:* For fall admission, 1/15 for domestic students. Electronic applications accepted. *Expenses:* Tuition, state resident: full-time $5,676. Tuition, nonresident: full-time $14,974; part-time $238 per credit hour. Required fees: $385 per term. Tuition and fees vary according to course load and program. *Financial support:* In 2005–06, 7 fellowships (averaging $7,213 per year), 37 research assistantships (averaging $15,220 per year), 86 teaching assistantships (averaging $15,220 per year) were awarded; Federal Work-Study, scholarships/grants, health care benefits, tuition waivers (full), and unspecified assistantships also available. Financial award application deadline: 3/1; financial award applicants required to submit FAFSA. *Faculty research:* Genomics, molecular genetics, immunology, physiological ecology, evolution. Total annual research expenditures: $11 million. *Unit head:* Dr. Eric Loker, Chair, 505-277-5508, Fax: 505-277-0304, E-mail: esloker@unm.edu. *Application contact:* Vivian J. Kent, Graduate Coordinator, 505-277-3411, Fax: 505-277-0304, E-mail: vkent@unm.edu.

University of New Mexico, School of Medicine, Biomedical Sciences Graduate Program, Albuquerque, NM 87131-5196. Offers biochemistry and molecular biology (MS, PhD); cell biology and physiology (MS, PhD); molecular genetics and microbiology (MS, PhD); neuroscience (MS, PhD); pathology (MS, PhD); toxicology (MS, PhD). Part-time programs available. Terminal master's awarded for partial completion of doctoral program. *Degree requirements:* For master's, thesis/dissertation; for doctorate, thesis/dissertation, comprehensive exam. *Entrance requirements:* For master's and doctorate, GRE General Test, minimum undergraduate GPA of 3.0. Additional exam requirements/recommendations for international students: Required—TOEFL. Electronic applications accepted. *Expenses:* Tuition, state resident: full-time $5,676. Tuition, nonresident: full-time $14,974; part-time $238 per credit hour. Required fees: $385 per term. Tuition and fees vary according to course load and program. *Faculty research:* Signal transduction, infectious disease, biology of cancer, structural biology, neuroscience.

University of New Orleans, Graduate School, College of Sciences, Department of Biological Sciences, New Orleans, LA 70148. Offers MS, PhD. *Faculty:* 8 full-time (1 woman), 2 part-time/adjunct (both women). *Students:* 34 full-time (25 women), 11 part-time (9 women); includes 2 minority (both African Americans), 4 international. Average age 29. 45 applicants, 44% accepted, 8 enrolled. In 2005, 7 degrees awarded. *Degree requirements:* For master's, one foreign language, thesis. *Entrance requirements:* For master's, GRE General Test. Additional exam requirements/recommendations for international students: Required—TOEFL (minimum score 550 paper-based; 213 computer-based). *Application deadline:* For fall admission, 7/1 priority date for domestic students; for spring admission, 11/15 priority date for domestic students, 10/1 priority date for international students. Applications are processed on a rolling basis. Application fee: $20. Electronic applications accepted. *Financial support:* Application deadline: 5/15; *Faculty research:* Biochemistry, genetics, vertebrate and invertebrate systematics and ecology, cell and mammalian physiology, morphology. *Unit head:* Dr. Steven G. Johnson, Chairperson, 504-280-6741, Fax: 504-280-6121, E-mail: sgjohnso@uno.edu. *Application contact:* Dr. Kathleen Burt-Utley, Graduate Coordinator, 504-280-7058, Fax: 504-280-6121, E-mail: kburtutl@uno.edu.

The University of North Carolina at Chapel Hill, Graduate School, College of Arts and Sciences, Department of Biology, Chapel Hill, NC 27599. Offers botany (MA, MS, PhD); cell biology, development, and physiology (MA, MS, PhD); cell motility and cytoskeleton (PhD); ecology and behavior (MA, MS, PhD); genetics and molecular biology (MA, MS, PhD); morphology, systematics, and evolution (MA, MS, PhD). Terminal master's awarded for partial completion of doctoral program. *Degree requirements:* For master's, thesis (for some programs), comprehensive exam; for doctorate, thesis/dissertation, comprehensive exam. *Entrance*

requirements: For master's, GRE General Test, GRE Subject Test, 2 semesters of calculus or statistics, 2 semesters of physics, organic chemistry, 3 semesters of biology; for doctorate, GRE General Test, GRE Subject Test, 2 semesters calculus or statistics, 2 semesters physics, organic chemistry, 3 semesters of biology. Additional exam requirements/recommendations for international students: Required—TOEFL (minimum score 550 paper-based; 213 computer-based). Electronic applications accepted. *Faculty research:* Gene expression, biomechanics, yeast genetics, plant ecology, plant molecular biology.

See Close-Up on page 277.

The University of North Carolina at Chapel Hill, School of Medicine and Graduate School, Graduate Programs in Medicine, Chapel Hill, NC 27599. Offers allied health sciences (MPT, MS, Au D, DPT, PhD), including human movement science (MS, PhD), occupational science (MS, PhD), physical therapy (MPT, MS, DPT), rehabilitation counseling and psychology (MS); speech and hearing sciences (MS, Au D, PhD); biochemistry and biophysics (MS, PhD); biomedical engineering (MS, PhD); cell and developmental biology (PhD); cell and molecular physiology (PhD); genetics and molecular biology (MS, PhD); microbiology and immunology (MS, PhD), including immunology, microbiology; neurobiology (PhD); pathology and laboratory medicine (PhD), including experimental pathology; pharmacology (PhD). Postbaccalaureate distance learning degree programs offered. *Faculty:* 429 full-time (149 women), 125 part-time/adjunct (20 women). *Students:* 712 full-time (430 women), 34 part-time (25 women); includes 91 minority (38 African Americans, 5 American Indian/Alaska Native, 41 Asian Americans or Pacific Islanders, 7 Hispanic Americans), 76 international. In 2005, 73 master's, 62 doctorates awarded. Terminal master's awarded for partial completion of doctoral program. *Degree requirements:* For master's, comprehensive exam; for doctorate, thesis/dissertation. *Application deadline:* Applications are processed on a rolling basis. Application fee: $65. Electronic applications accepted. *Expenses:* Contact institution. *Financial support:* In 2005–06, 77 fellowships with full and partial tuition reimbursements, 309 research assistantships with full tuition reimbursements, 23 teaching assistantships with full tuition reimbursements were awarded; career-related internships or fieldwork, Federal Work-Study, institutionally sponsored loans, traineeships, tuition waivers (full and partial), and unspecified assistantships also available. Support available to part-time students. Financial award applicants required to submit FAFSA. *Unit head:* Dr. William I. Roper, Dean, 919-966-4161, Fax: 919-966-6354.

The University of North Carolina at Charlotte, Graduate School, College of Arts and Sciences, Department of Biology, Charlotte, NC 28223-0001. Offers MA, MS, PhD. Part-time and evening/weekend programs available. *Faculty:* 19 full-time (4 women), 3 part-time/adjunct (1 woman). *Students:* 17 full-time (10 women), 34 part-time (24 women); includes 11 minority (4 African Americans, 5 Asian Americans or Pacific Islanders, 2 Hispanic Americans), 10 international. Average age 29. 29 applicants, 52% accepted, 13 enrolled. In 2005, 8 master's, 3 doctorates awarded. *Degree requirements:* For master's and doctorate, thesis/dissertation. *Entrance requirements:* For master's, GRE General Test, minimum GPA of 3.0 in undergraduate major, 2.75 overall; for doctorate, GRE General Test, minimum GPA of 3.5 in biology, 3.0 in chemistry, 3.0 in math, 3.0 overall. Additional exam requirements/recommendations for international students: Required—TOEFL (minimum score 557 paper-based; 220 computer-based). *Application deadline:* For fall admission, 7/15 for domestic students, 5/1 for international students; for spring admission, 11/15 for domestic students, 10/1 for international students. Applications are processed on a rolling basis. Application fee: $55. Electronic applications accepted. *Expenses:* Tuition, state resident: full-time $2,504; part-time $157 per credit. Tuition, nonresident: full-time $12,711; part-time $794 per credit. Required fees: $1,424; $89 per credit. Tuition and fees vary according to course load and program. *Financial support:* In 2005–06, 1 fellowship (averaging $15,000 per year), 25 research assistantships (averaging $9,096 per year), 57 teaching assistantships (averaging $6,184 per year) were awarded; career-related internships or fieldwork, Federal Work-Study, institutionally sponsored loans, scholarships/grants, and unspecified assistantships also available. Support available to part-time students. Financial award application deadline: 4/1; financial award applicants required to submit FAFSA. *Faculty research:* Liver blood flow in response to stress/injury, host response to bacterial and viral infection, mechanisms of cancer development and spread, stress responses in marine organisms as a measure of environmental change. *Unit head:* Dr. Michael C Hudson, Chair, 704-687-8694, Fax: 704-687-3128, E-mail: mchudson@emial.uncc.edu. *Application contact:* Kathy B. Giddings, Director of Graduate Admissions, 704-687-3366, Fax: 704-687-3279, E-mail: gradadm@email.uncc.edu.

The University of North Carolina at Greensboro, Graduate School, College of Arts and Sciences, Department of Biology, Greensboro, NC 27412-5001. Offers MS. *Students:* 9 full-time, 40 part-time. *Degree requirements:* For master's, thesis. *Entrance requirements:* For master's, GRE General Test, GRE Subject Test. Additional exam requirements/recommendations for international students: Required—TOEFL. *Application deadline:* For fall admission, 3/1 for domestic students; for spring admission, 11/1 for domestic students. Applications are processed on a rolling basis. *Expenses:* Tuition, state resident: part-time $302 per credit hour. Tuition, nonresident: part-time $1,683 per credit hour. Required fees: $51 per credit hour. Tuition and fees vary according to course load and program. *Financial support:* Research assistantships with full tuition reimbursements, teaching assistantships with full tuition reimbursements, career-related internships or fieldwork, scholarships/grants, traineeships, and unspecified assistantships available. *Faculty research:* Environmental biology, biochemistry, animal ecology, vertebrate reproduction. *Unit head:* Dr. John J. Lepri, Head, 336-334-5391, Fax: 336-334-5839, E-mail: jjlepri@uncg.edu. *Application contact:* Michelle Harkleroad, Director of Graduate Admissions, 336-334-4884, Fax: 336-334-4424, E-mail: mbharkle@uncg.edu.

The University of North Carolina Wilmington, College of Arts and Sciences, Department of Biological Sciences, Wilmington, NC 28403-3297. Offers biology (MS); marine biology (MS, PhD). Part-time programs available. *Faculty:* 28 full-time (5 women). *Students:* 4 full-time (2 women), 53 part-time (31 women); includes 4 minority (2 African Americans, 2 Hispanic Americans), 2 international. Average age 29. 92 applicants, 15% accepted, 14 enrolled. In 2005, 13 degrees awarded. *Degree requirements:* For master's, thesis/dissertation, comprehensive exam; for doctorate, thesis/dissertation. *Entrance requirements:* For master's, GRE General Test, GRE Subject Test, minimum B average in undergraduate major. *Application deadline:* For fall admission, 3/15 for domestic students. Applications are processed on a rolling basis. Application fee: $45. *Financial support:* In 2005–06, 31 teaching assistantships were awarded; career-related internships or fieldwork and Federal Work-Study also available. Support available to part-time students. Financial award application deadline: 3/15. *Faculty research:* Marine responses, estuaries studies, biotechnology, underwater research, acid rain. *Unit head:* Dr. Martin H. Posey, Chairman, 910-962-3470, E-mail: poseym@uncw.edu. *Application contact:* Dr. Robert D. Roer, Dean, Graduate School, 910-962-4117, Fax: 910-962-3787, E-mail: roer@uncw.edu.

See Close-Up on page 815.

University of North Dakota, Graduate School, College of Arts and Sciences, Department of Biology, Grand Forks, ND 58202. Offers botany (MS, PhD); ecology (MS, PhD); entomology (MS, PhD); environmental biology (MS, PhD); fisheries/wildlife (MS, PhD); genetics (MS, PhD); zoology (MS, PhD). *Faculty:* 16 full-time (3 women). *Students:* 15 applicants, 13% accepted, 2 enrolled. In 2005, 5 degrees awarded. Terminal master's awarded for partial completion of doctoral program. *Degree requirements:* For master's, thesis/dissertation, final exam; for doctorate, thesis/dissertation, final exam, comprehensive exam. *Entrance requirements:* For master's, GRE General Test, GRE Subject Test, minimum GPA of 3.0; for doctorate, GRE General Test, GRE Subject Test, minimum GPA of 3.5. Additional exam requirements/recommendations for international students: Required—TOEFL (minimum score 550 paper-based; 213 computer-based). *Application deadline:* For fall admission, 10/1 for domestic students, 10/1 for international students. Application fee: $35. Electronic applications accepted. *Financial support:* In 2005–06, 8 research assistantships with full tuition reimbursements (averaging $11,375 per year), 13 teaching assistantships with full tuition reimbursements (averaging $10,813 per year) were awarded; fellowships, Federal Work-Study, institutionally sponsored loans, scholarships/grants, and tuition waivers (full and partial) also available.

Biological and Biomedical Sciences—General

Support available to part-time students. Financial award application deadline: 3/15; financial award applicants required to submit FAFSA. *Faculty research:* Population biology, wildlife ecology, RNA processing, hormonal control of behavior. *Unit head:* Dr. Richard Sweitzel, Graduate Director, 701-777-4676, Fax: 701-777-2623, E-mail: richard_sweitzel@und.nodak.edu.

University of North Dakota, School of Medicine and Graduate School, Graduate Programs in Medicine, Grand Forks, ND 58202. Offers MOT, MPAS, MPT, MS, DPT, PhD, MD/PhD. Postbaccalaureate distance learning degree programs offered (minimal on-campus study). *Faculty:* 70 full-time (29 women), 4 part-time/adjunct (all women). *Students:* 142 full-time (102 women), 148 part-time (83 women). 126 applicants, 56% accepted, 66 enrolled. In 2005, 77 master's, 30 doctorates awarded. *Degree requirements:* For doctorate, thesis/dissertation, final exam, comprehensive exam. *Entrance requirements:* For master's and doctorate, minimum GPA of 3.0. Additional exam requirements/recommendations for international students: Required—TOEFL. *Application deadline:* For fall admission, 3/1 for domestic students. Application fee: $35. Electronic applications accepted. *Expenses: Contact institution. Financial support:* In 2005–06, 43 students received support, including 22 research assistantships with full tuition reimbursements available (averaging $13,997 per year), 22 teaching assistantships with full tuition reimbursements available (averaging $13,997 per year); fellowships, Federal Work-Study, institutionally sponsored loans, scholarships/grants, and tuition waivers (full and partial) also available. Support available to part-time students. Financial award application deadline: 3/15; financial award applicants required to submit FAFSA. Total annual research expenditures: $4.7 million. *Application contact:* Brenda Halle, Admissions Specialist, 701-777-2947, Fax: 701-777-3619, E-mail: brendahalle@mail.und.edu.

University of Northern Colorado, Graduate School, College of Natural and Health Sciences, School of Biological Sciences, Greeley, CO 80639. Offers biological education (PhD); biological sciences (MS). Part-time programs available. *Faculty:* 14 full-time (5 women). *Students:* 23 full-time (12 women), 4 part-time (3 women). Average age 29. 17 applicants, 100% accepted, 8 enrolled. In 2005, 5 master's, 1 doctorate awarded. *Degree requirements:* For master's, comprehensive exam; for doctorate, thesis/dissertation, comprehensive exam. *Entrance requirements:* For master's and doctorate, GRE General Test, 3 letters of recommendation. *Application deadline:* Applications are processed on a rolling basis. Application fee: $50 ($60 for international students). Electronic applications accepted. *Expenses:* Tuition, state resident: full-time $4,968; part-time $207 per credit hour. Tuition, nonresident: full-time $14,688; part-time $612 per credit hour. Required fees: $645; $32 per credit hour. *Financial support:* In 2005–06, 24 students received support, including 8 fellowships (averaging $959 per year), research assistantships (averaging $17,950 per year), 12 teaching assistantships (averaging $14,360 per year); unspecified assistantships also available. Financial award application deadline: 3/1; financial award applicants required to submit FAFSA. *Unit head:* Dr. Catherine Gardiner, Director, 970-351-2921, Fax: 970-351-2335.

University of Northern Iowa, Graduate College, College of Natural Sciences, Department of Biology, Cedar Falls, IA 50614. Offers MA, MS. Part-time programs available. *Faculty:* 26 full-time (10 women). *Students:* 8 full-time (3 women), 1 part-time. 8 applicants, 75% accepted, 2 enrolled. In 2005, 4 degrees awarded. *Degree requirements:* For master's, thesis or alternative, comprehensive exam (for some programs). *Entrance requirements:* Additional exam requirements/recommendations for international students: Required—TOEFL (minimum score 500 paper-based; 180 computer-based). *Application deadline:* For fall admission, 8/1 for domestic students. Applications are processed on a rolling basis. Application fee: $30 ($50 for international students). Electronic applications accepted. *Expenses:* Tuition, state resident: full-time $5,708. Tuition, nonresident: full-time $13,532. Required fees: $712. *Financial support:* Scholarships/grants available. Financial award application deadline: 2/1. *Unit head:* Dr. Barbara A. Hetrick, Head, 319-273-2456, Fax: 319-273-7125, E-mail: barbara.hetrick@uni.edu.

University of North Florida, College of Arts and Sciences, Department of Biology, Jacksonville, FL 32224-2645. Offers MA, MS. Part-time programs available. *Faculty:* 13 full-time (2 women). *Students:* 13 full-time (8 women), 7 part-time (4 women); includes 2 minority (1 African American, 1 Hispanic American). Average age 28. 27 applicants, 41% accepted, 7 enrolled. In 2005, 1 degree awarded. *Degree requirements:* For master's, thesis (for some programs). *Entrance requirements:* For master's, GRE General Test, minimum GPA of 3.0 in last 60 hours, letters of recommendation. Additional exam requirements/recommendations for international students: Required—TOEFL (minimum score 570 paper-based). *Application deadline:* For fall admission, 6/1 for domestic students, 3/1 for international students; for spring admission, 7/1 for domestic students, 7/1 for international students. Application fee: $30. *Expenses:* Tuition, state resident: full-time $4,391; part-time $244 per semester hour. Tuition, nonresident: full-time $15,036; part-time $835 per semester hour. Required fees: $789, $44 per semester hour. *Financial support:* In 2005–06, 12 teaching assistantships (averaging $6,289 per year) were awarded. Financial award application deadline: 4/1; financial award applicants required to submit FAFSA. Total annual research expenditures: $770,633. *Unit head:* Dr. Joe Butler, Chair, 904-620-2830, Fax: 904-620-3885, E-mail: jbutler@unf.edu. *Application contact:* Dr. Anthony Rossi, Graduate Coordinator, 904-620-2830, E-mail: arossi@unf.edu.

University of North Texas, Robert B. Toulouse School of Graduate Studies, College of Arts and Sciences, Department of Biological Sciences, Denton, TX 76203. Offers biochemistry (MS, PhD); biology (MA, MS, PhD); environmental science (MS, PhD); molecular biology (MA, MS, PhD). *Faculty:* 30 full-time (3 women). *Students:* 105 full-time (58 women), 59 part-time (34 women); includes 29 minority (6 African Americans, 1 American Indian/Alaska Native, 15 Asian Americans or Pacific Islanders, 7 Hispanic Americans), 31 international. Average age 32. 115 applicants, 40% accepted, 23 enrolled. In 2005, 20 master's, 11 doctorates awarded. Terminal master's awarded for partial completion of doctoral program. *Degree requirements:* For master's, oral defense of thesis; for doctorate, one foreign language, thesis/dissertation, comprehensive exam. *Entrance requirements:* For master's and doctorate, GRE General Test. *Application deadline:* For fall admission, 7/1 for domestic students; for spring admission, 11/1 for domestic students. Application fee: $50 ($75 for international students). *Expenses:* Tuition, state resident: full-time $3,258; part-time $181 per semester hour. Tuition, nonresident: full-time $8,226; part-time $451 per semester hour. Required fees: $1,219; $68 per semester hour. *Financial support:* Fellowships, research assistantships, teaching assistantships, career-related internships or fieldwork, Federal Work-Study, and institutionally sponsored loans available. Support available to part-time students. *Faculty research:* Network neuroscience, toxicology with earthworm model, aquatic toxicology, biodegradation of toxic chemicals, physiology with roundworm parasite. *Unit head:* Dr. Arthur Goven, Chair, 940-565-3590, Fax: 940-565-3821, E-mail: goven@unt.edu. *Application contact:* Beth Chlapek, Graduate Adviser, 940-565-3627, Fax: 940-565-3821, E-mail: bsloan@unt.edu.

University of North Texas Health Science Center at Fort Worth, Graduate School of Biomedical Sciences, Fort Worth, TX 76107-2699. Offers anatomy and cell biology (MS, PhD); biochemistry and molecular biology (MS, PhD); biomedical sciences (MS, PhD); biotechnology (MS); forensic genetics (MS); integrative physiology (MS, PhD); medical science (MS); microbiology and immunology (MS, PhD); pharmacology (MS, PhD); science education (MS). *Faculty:* 57 full-time (9 women), 2 part-time/adjunct (0 women). *Students:* 177 full-time (98 women), 42 part-time (33 women); includes 61 minority (15 African Americans, 3 American Indian/Alaska Native, 27 Asian Americans or Pacific Islanders, 16 Hispanic Americans), 53 international. Average age 28. 237 applicants, 62% accepted, 91 enrolled. In 2005, 37 master's, 15 doctorates awarded. Terminal master's awarded for partial completion of doctoral program. *Degree requirements:* For master's and doctorate, thesis/dissertation. *Entrance requirements:* For master's and doctorate, GRE General Test. Additional exam requirements/recommendations for international students: Required—TOEFL. *Application deadline:* For fall admission, 5/1 for domestic students. Application fee: $25 ($50 for international students). *Expenses: Contact institution. Financial support:* In 2005–06, 80 research assistantships (averaging $16,000 per year) were awarded; fellowships, teaching assistantships, career-related internships or fieldwork, Federal Work-Study, institutionally sponsored loans, scholarships/grants, and traineeships

also available. Support available to part-time students. Financial award application deadline: 4/1; financial award applicants required to submit FAFSA. *Faculty research:* Alzheimer's disease, aging, eye diseases, cancer, cardiovascular disease. Total annual research expenditures: $21 million. *Unit head:* Dr. Thomas Yorio, Dean, 817-735-2560, Fax: 817-735-0243, E-mail: yoriot@hsc.unt.edu. *Application contact:* Carla Lee, Director of Graduate Admissions and Services, 817-735-2560, Fax: 817-735-0243, E-mail: gsbs@hsc.unt.edu.

See Close-Up on page 279.

University of Notre Dame, Graduate School, College of Science, Department of Biological Sciences, Notre Dame, IN 46556. Offers aquatic ecology, evolution and environmental biology (MS, PhD); cellular and molecular biology (MS, PhD); genetics (MS, PhD); physiology (MS, PhD); vector biology and parasitology (MS, PhD). *Faculty:* 34 full-time (8 women), 3 part-time/adjunct (0 women). *Students:* 124 full-time (55 women); includes 11 minority (1 African American, 6 Asian Americans or Pacific Islanders, 4 Hispanic Americans), 39 international. 95 applicants, 34% accepted, 22 enrolled. In 2005, 4 master's, 11 doctorates awarded. Terminal master's awarded for partial completion of doctoral program. *Median time to degree:* Of those who began their doctoral program in fall 1997, 61% received their degree in 8 years or less. *Degree requirements:* For master's and doctorate, thesis/dissertation, comprehensive exam. *Entrance requirements:* For master's and doctorate, GRE General Test. Additional exam requirements/recommendations for international students: Required—TOEFL. *Application deadline:* For fall admission, 2/1 for domestic students; for spring admission, 11/1 for domestic students. Applications are processed on a rolling basis. Application fee: $50. Electronic applications accepted. *Financial support:* In 2005–06, 124 students received support, including 24 fellowships with full tuition reimbursements available (averaging $22,000 per year), 47 research assistantships with full tuition reimbursements available (averaging $15,250 per year), 45 teaching assistantships with full tuition reimbursements available (averaging $16,000 per year); traineeships and tuition waivers (full) also available. Financial award application deadline: 2/1. *Faculty research:* Tropical disease, molecular genetics, neurobiology, evolutionary biology, aquatic biology. Total annual research expenditures: $15.3 million. *Unit head:* Dr. Gary A. Lamberti, Director of Graduate Studies, 574-631-6552, Fax: 574-631-7413, E-mail: biology.biosadm.1@nd.edu. *Application contact:* Dr. Terrence J. Akai, Director of Graduate Admissions, 574-631-7706, Fax: 574-631-4183, E-mail: gradad@nd.edu.

See Close-Up on page 281.

University of Oklahoma Health Sciences Center, College of Medicine and Graduate College, Graduate Programs in Medicine, Oklahoma City, OK 73190. Offers biochemistry and molecular biology (MS, PhD), including biochemistry, molecular biology; cell biology (MS, PhD); medical sciences (MS); microbiology and immunology (MS, PhD), including immunology, microbiology; neuroscience (MS, PhD); pathology (PhD); physiology (MS, PhD); psychiatry and behavioral sciences (MS, PhD), including biological psychology; radiological sciences (MS, PhD), including medical radiation physics. Part-time programs available. Terminal master's awarded for partial completion of doctoral program. *Degree requirements:* For doctorate, thesis/dissertation. *Entrance requirements:* For doctorate, GRE General Test, 3 letters of recommendation. Additional exam requirements/recommendations for international students: Required—TOEFL. *Expenses:* Contact institution. *Faculty research:* Behavior and drugs, structure and function of endothelium, genetics and behavior, gene structure and function, action of antibiotics.

University of Oregon, Graduate School, College of Arts and Sciences, Department of Biology, Eugene, OR 97403. Offers ecology and evolution (MA, MS, PhD); marine biology (MA, MS, PhD); molecular, cellular and genetic biology (PhD); neuroscience and development (PhD). *Faculty:* 42 full-time (14 women), 5 part-time/adjunct (2 women). *Students:* 92 full-time (43 women), 2 part-time (both women); includes 9 minority (2 African Americans, 1 American Indian/Alaska Native, 4 Asian Americans or Pacific Islanders, 2 Hispanic Americans), 6 international. 18 applicants, 83% accepted. In 2005, 11 master's, 7 doctorates awarded. Terminal master's awarded for partial completion of doctoral program. *Degree requirements:* For master's, thesis (for some programs); for doctorate, thesis/dissertation. *Entrance requirements:* For master's and doctorate, GRE General Test, minimum GPA of 3.2. Additional exam requirements/recommendations for international students: Required—TOEFL. *Application deadline:* For fall admission, 12/15 for domestic students. *Financial support:* In 2005–06, 36 teaching assistantships were awarded; research assistantships, Federal Work-Study, institutionally sponsored loans, and scholarships/grants also available. Financial award application deadline: 2/1. *Faculty research:* Developmental neurobiology; evolution, population biology, and quantitative genetics; regulation of gene expression; biochemistry of marine organisms. *Unit head:* George Sprague, Head, 541-340-6051, Fax: 541-346-6056. *Application contact:* Lynne Romans, Admissions Contact, 541-346-4252, Fax: 541-346-6056, E-mail: lromans@uoregon.edu.

University of Ottawa, Faculty of Graduate and Postdoctoral Studies, Faculty of Science, Ottawa-Carleton Institute of Biology, Ottawa, ON K1N 6N5, Canada. Offers M Sc, PhD. Part-time programs available. *Faculty:* 29 full-time (4 women), 6 part-time/adjunct (2 women). *Students:* 87 full-time, 12 part-time. 68 applicants, 51% accepted, 25 enrolled. In 2005, 15 master's, 7 doctorates awarded. *Degree requirements:* For master's, thesis/dissertation, seminar; for doctorate, thesis/dissertation, seminar, comprehensive exam. *Entrance requirements:* For master's, honors B Sc degree or equivalent, minimum B average; for doctorate, honors B Sc with minimum B+ average or M Sc with minimum B+ average. *Application deadline:* For fall admission, 3/1 priority date for domestic students, 2/15 priority date for international students. For winter admission, 11/15 for domestic students; for spring admission, 4/1 for domestic students. Applications are processed on a rolling basis. Application fee: $75. Electronic applications accepted. *Expenses:* Tuition: Part-time $260 per credit. Tuition and fees vary according to course load and program. *Financial support:* Fellowships, research assistantships with full tuition reimbursements, teaching assistantships with full tuition reimbursements, career-related internships or fieldwork, Federal Work-Study, scholarships/grants, traineeships, tuition waivers (full and partial), and unspecified assistantships available. Financial award application deadline: 2/15. *Faculty research:* Physiology/biochemistry, cellular and molecular biology, ecology, behavior and systematics. *Unit head:* Dr. Steve Perry, Chair, 613-520-2600 Ext. 3873. *Application contact:* Lise Maisonneuve, Graduate Studies Administrator, 613-562-5800 Ext. 6335, Fax: 613-562-5486, E-mail: lise@science.uottawa.ca.

University of Pennsylvania, School of Arts and Sciences, Graduate Group in Biology, Philadelphia, PA 19104. Offers PhD. *Degree requirements:* For doctorate, thesis/dissertation. *Entrance requirements:* For doctorate, GRE General Test, GRE Subject Test. Additional exam requirements/recommendations for international students: Required—TOEFL. Electronic applications accepted.

University of Pennsylvania, School of Medicine, Biomedical Graduate Studies, Philadelphia, PA 19104. Offers MS, PhD, DMD/PhD, MD/PhD, VMD/PhD. Part-time programs available. *Faculty:* 592. *Students:* 673 full-time (338 women); includes 150 minority (21 African Americans, 4 American Indian/Alaska Native, 92 Asian Americans or Pacific Islanders, 33 Hispanic Americans), 73 international. 1,035 applicants, 25% accepted, 89 enrolled. In 2005, 12 master's, 80 doctorates awarded. Terminal master's awarded for partial completion of doctoral program. *Degree requirements:* For master's, comprehensive exam; for doctorate, thesis/dissertation. *Entrance requirements:* For master's and doctorate, GRE General Test. Additional exam requirements/recommendations for international students: Required—TOEFL. *Application deadline:* For fall admission, 12/15 priority date for domestic students, 12/1 priority date for international students. Applications are processed on a rolling basis. Application fee: $70. Electronic applications accepted. *Expenses: Contact institution. Financial support:* In 2005–06, 643 students received support; fellowships, research assistantships, scholarships/grants, traineeships, and unspecified assistantships available. *Unit head:* Dr. Susan R. Ross, Director, 215-898-1030. *Application contact:* Michelle Romolini, Admissions Coordinator, 215-898-1030, Fax: 215-898-2671, E-mail: romolini@mail.med.upenn.edu.

Biological and Biomedical Sciences—General

University of Pittsburgh, School of Arts and Sciences, Department of Biological Sciences, Pittsburgh, PA 15260. Offers ecology and evolution (MS, PhD); molecular, cellular, and developmental biology (PhD). *Faculty:* 29 full-time (9 women), 4 part-time/adjunct (1 woman). *Students:* 61 full-time (36 women); includes 4 minority (2 African Americans, 2 Hispanic Americans), 10 international. Average age 23. 249 applicants, 2% accepted, 4 enrolled. In 2005, 1 master's, 9 doctorates awarded. *Median time to degree:* Of those who began their doctoral program in fall 1997, 100% received their degree in 8 years or less. *Degree requirements:* For master's and doctorate, thesis/dissertation, comprehensive exam, registration. *Entrance requirements:* For master's and doctorate, GRE General Test, GRE Subject Test. Additional exam requirements/recommendations for international students: Required—TOEFL (minimum score 550 paper-based; 213 computer-based). *Application deadline:* For fall admission, 1/15 priority date for domestic students, 12/15 priority date for international students. Applications are processed on a rolling basis. Application fee: $0 ($40 for international students). Electronic applications accepted. *Expenses:* Tuition, state resident: full-time $13,194; part-time $537 per credit. Tuition, nonresident: full-time $25,012; part-time $1,026 per credit. Required fees: $700; $164 per term. Tuition and fees vary according to campus/location and program. *Financial support:* In 2005–06, 36 fellowships with full tuition reimbursements, 97 research assistantships with full tuition reimbursements (averaging $20,332 per year), 34 teaching assistantships with full tuition reimbursements (averaging $20,332 per year) were awarded; Federal Work-Study, scholarships/grants, traineeships, health care benefits, and tuition waivers (full) also available. *Faculty research:* Molecular biology, cell biology, molecular biophysics, developmental biology, ecology and evolution. Total annual research expenditures: $7.1 million. *Unit head:* Dr. Graham F. Hatfull, Chairman, 412-624-4350, Fax: 412-624-4759, E-mail: gfh@pitt.edu. *Application contact:* Cathleen M. Barr, Graduate Administrator, 412-624-4268, Fax: 412-624-4759, E-mail: cbarr@pitt.edu.

University of Pittsburgh, School of Medicine, Graduate Programs in Medicine, Interdisciplinary Biomedical Sciences Program, Pittsburgh, PA 15260. Offers PhD. *Faculty:* 306 full-time (69 women). *Students:* 43 full-time (31 women); includes 6 minority (3 African Americans, 1 Asian American or Pacific Islander, 2 Hispanic Americans), 12 international. Average age 28. 415 applicants, 22% accepted, 42 enrolled. *Degree requirements:* For doctorate, thesis/dissertation, comprehensive exam, registration. *Entrance requirements:* For doctorate, GRE General Test, GRE Subject Test, minimum QPA of 3.0. Additional exam requirements/recommendations for international students: Required—TOEFL (minimum score 600 paper-based; 250 computer-based), IELT (minimum score 7). *Application deadline:* For fall admission, 12/15 priority date for domestic students, 12/15 priority date for international students. Application fee: $40. Electronic applications accepted. *Expenses:* Tuition, state resident: full-time $13,194; part-time $537 per credit. Tuition, nonresident: full-time $25,012; part-time $1,026 per credit. Required fees: $700; $164 per term. Tuition and fees vary according to campus/location and program. *Financial support:* In 2005–06, 48 research assistantships with full tuition reimbursements (averaging $21,500 per year) were awarded; teaching assistantships, institutionally sponsored loans, scholarships/grants, traineeships, and unspecified assistantships also available. *Faculty research:* Biochemistry and molecular genetics, cell biology and molecular physiology, cellular and molecular pathology, immunology, molecular pharmacology. *Application contact:* Graduate Studies Administrator, 412-648-8957, Fax: 412-648-1077, E-mail: gradstudies@medschool.pitt.edu.

See Close-Up on page 283.

University of Prince Edward Island, Faculty of Science, Charlottetown, PE C1A 4P3, Canada. Offers biology (M Sc); chemistry (M Sc). *Degree requirements:* For master's, thesis. *Entrance requirements:* Additional exam requirements/recommendations for international students: Required—TOEFL (minimum score 550 paper-based; 213 computer-based), Canadian Academic English Language Assessment, Michigan English Language Assessment Battery, Canadian Test of English for Scholars and Trainees. Tuition charges are reported in Canadian dollars. *Expenses:* Tuition, area resident: Full-time $3,816 Canadian dollars. International tuition: $6,356 Canadian dollars full-time. Part-time tuition and fees vary according to course level, degree level, campus/location, program and student level. *Faculty research:* Ecology and wildlife biology, molecular, genetics and biotechnology, organametallic, bio-organic, supra-molecular and synthetic organic chemistry, neurobiology and stoke materials science.

University of Puerto Rico, Mayagüez Campus, Graduate Studies, College of Arts and Sciences, Department of Biology, Mayagüez, PR 00681-9000. Offers MS. Part-time programs available. *Faculty:* 26. *Students:* 40 full-time (19 women), 58 part-time (44 women); includes 64 minority (all Hispanic Americans), 34 international. 57 applicants, 25% accepted, 13 enrolled. In 2005, 6 degrees awarded. *Degree requirements:* For master's, one foreign language, thesis, comprehensive exam. *Entrance requirements:* For master's, GRE. *Application deadline:* For fall admission, 2/15 for domestic students, 2/15 for international students; for spring admission, 9/15 for domestic students, 9/15 for international students. Applications are processed on a rolling basis. Application fee: $20. *Expenses:* Tuition, state resident: part-time $100 per credit. International tuition: $4,655 full-time. Part-time tuition and fees vary according to course level and course load. *Financial support:* In 2005–06, 74 students received support, including fellowships (averaging $1,200 per year), 25 research assistantships (averaging $1,500 per year), 49 teaching assistantships (averaging $927 per year); Federal Work-Study and institutionally sponsored loans also available. *Faculty research:* Herpetology, immunology, microbiology, immunology, botany. Total annual research expenditures: $48,264. *Unit head:* Dr. Lucy Williams, Director, 787-265-3837.

University of Puerto Rico, Medical Sciences Campus, School of Medicine, Division of Graduate Studies, San Juan, PR 00936-5067. Offers MS, PhD. *Faculty:* 52 full-time (22 women), 10 part-time/adjunct (3 women). *Students:* 78 full-time (56 women); includes 77 Hispanic Americans, 1 international. Average age 23. 52 applicants, 40% accepted, 16 enrolled. In 2005, 4 master's, 6 doctorates awarded. *Terminal master's awarded for partial completion of doctoral program. *Degree requirements:* For master's, one foreign language, thesis/dissertation, registration; for doctorate, one foreign language, thesis/dissertation, comprehensive exam, registration. *Entrance requirements:* For master's and doctorate, GRE General Test, GRE Subject Test, interview, 3 letters of recommendation, minimum GPA of 3.0. *Application deadline:* For fall admission, 9/15 for domestic students, 9/15 for international students; for spring admission, 2/15 for domestic students, 2/15 for international students. Application fee: $15. *Expenses:* Contact institution. Tuition and fees vary according to class time, degree level and program. *Financial support:* Fellowships, research assistantships, teaching assistantships, career-related internships or fieldwork, Federal Work-Study, institutionally sponsored loans, and tuition waivers (full and partial) available. Support available to part-time students. Financial award application deadline: 4/30. *Unit head:* Dr. Walter I. Silva, Associate Dean for Biomedical Sciences and Director Graduate Studies, 787-758-2525 Ext. 1831, Fax: 787-767-8693, E-mail: wsilva@rcm.upr.edu. *Application contact:* Julia M. Prado-Otero, Administrator Graduate Program, 787-758-2525 Ext. 7017, Fax: 787-767-8693, E-mail: jprado@rcm.upr.edu.

See Close-Up on page 285.

University of Puerto Rico, Río Piedras, College of Natural Sciences, Department of Biology, San Juan, PR 00931-3300. Offers MS, PhD. Part-time and evening/weekend programs available. *Students:* 86 full-time (54 women), 28 part-time (14 women); includes 113 minority (1 Asian American or Pacific Islander, 112 Hispanic Americans). Average age 27. In 2005, 9 master's, 12 doctorates awarded. *Degree requirements:* For master's and doctorate, one foreign language, thesis/dissertation, comprehensive exam. *Entrance requirements:* For master's, GRE Subject Test, EXADEP, interview, minimum GPA of 3.0, letter of recommendation; for doctorate, GRE Subject Test, interview, master's degree, minimum GPA of 3.0, letter of recommendation. *Application deadline:* For fall admission, 2/1 for domestic students, 2/1 for international students. Application fee: $17. *Expenses:* Tuition, state resident: part-time $100 per credit. Tuition, nonresident: part-time $294 per credit. Required fees: $72 per term. *Financial support:* Fellowships, research assistantships, teaching assistantships, Federal Work-Study, institutionally sponsored loans, and tuition waivers (partial) available. Financial award applica-

tion deadline: 5/31. *Faculty research:* Environmental, poblational and systematic biology. *Unit head:* Dr. Paul Bayman, Coordinator, 787-764-0000 Ext. 3551.

University of Regina, Faculty of Graduate Studies and Research, Faculty of Science, Department of Biology, Regina, SK S4S 0A2, Canada. Offers M Sc, PhD. *Faculty:* 12 full-time (3 women), 6 part-time/adjunct (0 women). *Students:* 19 full-time (14 women), 1 (woman) part-time. 12 applicants, 42% accepted. In 2005, 4 master's awarded. *Degree requirements:* For master's, thesis/dissertation, registration; for doctorate, thesis/dissertation, comprehensive exam, registration. *Entrance requirements:* Additional exam requirements/recommendations for international students: Required—TOEFL (minimum score 580 paper-based; 237 computer-based). *Application deadline:* Applications are processed on a rolling basis. Application fee: $60 ($100 for international students). *Financial support:* In 2005–06, 3 fellowships (averaging $14,886 per year), 2 research assistantships (averaging $12,750 per year), 2 teaching assistantships (averaging $13,501 per year) were awarded; scholarships/grants also available. Financial award application deadline: 6/15. *Faculty research:* Moss developmental regulation, orthopteran population genetics, microbial toxin synthesis, fish endocrinology, terrestrial ecology. *Unit head:* Dr. William Chapco, Head, 306-585-4145, Fax: 306-585-4894, E-mail: william.chapco@uregina.ca. *Application contact:* Dr. Christopher Yost, Assistant Professor, 306-585-5223, Fax: 306-585-4894, E-mail: christopher.yost@uregina.ca.

University of Rhode Island, Graduate School, College of Arts and Sciences, Department of Biological Sciences, Kingston, RI 02881. Offers MS, PhD. *Expenses:* Tuition, state resident: full-time $5,522; part-time $307 per credit. Tuition, nonresident: full-time $15,992; part-time $888 per credit. Required fees: $1,786; $73 per credit. One-time fee: $80 part-time. *Unit head:* Dr. Marian Goldsmith, Chair, 401-874-2637.

University of Richmond, Graduate School of Arts and Sciences, Department of Biology, Richmond, University of Richmond, VA 23173. Offers MS, JD/MS. *Faculty:* 12 full-time (7 women). In 2005, 2 degrees awarded. *Degree requirements:* For master's, thesis, registration. *Entrance requirements:* For master's, GRE General Test, GRE Subject Test, undergraduate major in biology or related area. Additional exam requirements/recommendations for international students: Required—TOEFL. *Application deadline:* For fall admission, 11/15 for domestic students; for spring admission, 11/15 for domestic students. Application fee: $30. *Expenses:* Tuition: Full-time $26,300; part-time $1,320 per hour. Tuition and fees vary according to course load. *Financial support:* Research assistantships, teaching assistantships, Federal Work-Study, institutionally sponsored loans, tuition waivers (partial), unspecified assistantships, and tuition awards available. Financial award application deadline: 3/15; financial award applicants required to submit FAFSA. *Faculty research:* DNA repair to gene regulation in microorganisms, phylogenetic relationships among microhylid subfamilies. *Unit head:* Dr. John Hayden, Coordinator, 804-289-8232. *Application contact:* Suzanne V. Blyer, Program Coordinator, 804-289-8417, Fax: 804-289-8818, E-mail: asgrad@richmond.edu.

University of Rochester, The College, Arts and Sciences, Department of Biology, Rochester, NY 14627-0250. Offers MS, PhD. Terminal master's awarded for partial completion of doctoral program. *Degree requirements:* For doctorate, thesis/dissertation, qualifying exam. *Entrance requirements:* For master's and doctorate, GRE General Test. Additional exam requirements/recommendations for international students: Required—TOEFL.

See Close-Up on page 289.

University of Rochester, School of Medicine and Dentistry, Graduate Programs in Medicine and Dentistry, Rochester, NY 14627-0250. Offers MA, MPH, MS, PhD, MBA/MPH, MBA/MS, MD/MPH, MD/MS, MD/PhD, MPH/MS, MPH/PhD. Part-time programs available. *Degree requirements:* For doctorate, thesis/dissertation, qualifying exam. *Entrance requirements:* For master's and doctorate, GRE General Test. Additional exam requirements/recommendations for international students: Required—TOEFL. Electronic applications accepted.

See Close-Up on page 287.

University of Rochester, School of Medicine and Dentistry, Integrative Biomedical Science Program, Rochester, NY 14627.

University of San Francisco, College of Arts and Sciences, Department of Biology, San Francisco, CA 94117-1080. Offers MS. *Faculty:* 9 full-time (6 women). *Students:* 3 full-time (1 woman), 3 part-time (1 woman); includes 2 minority (1 Asian American or Pacific Islander, 1 Hispanic American). Average age 25. 28 applicants, 32% accepted, 3 enrolled. In 2005, 3 degrees awarded. *Degree requirements:* For master's, thesis. *Entrance requirements:* For master's, GRE General Test, GRE Subject Test, BS in biology or the equivalent. *Application deadline:* For fall admission, 4/15 for domestic students; for spring admission, 10/15 for domestic students. Application fee: $55 ($65 for international students). *Expenses:* Tuition: Part-time $925 per unit. Tuition and fees vary according to degree level, campus/location and program. *Financial support:* In 2005–06, 6 students received support; teaching assistantships, career-related internships or fieldwork, Federal Work-Study, institutionally sponsored loans, and tuition waivers available. Financial award application deadline: 3/2; financial award applicants required to submit FAFSA. *Unit head:* , Dr. John T. Sullivan, Chair, 415-422-6755, Fax: 415-422-6363.

University of Saskatchewan, College of Graduate Studies and Research, College of Arts and Sciences, Department of Biology, Saskatoon, SK S7N 5A2, Canada. Offers M Sc, PhD, Diploma. *Degree requirements:* For master's, thesis (for some programs), registration; for doctorate, thesis/dissertation, registration. *Entrance requirements:* Additional exam requirements/recommendations for international students: Required—TOEFL.

University of Saskatchewan, Western College of Veterinary Medicine and College of Graduate Studies and Research, Graduate Programs in Veterinary Medicine, Department of Veterinary Biomedical Sciences, Saskatoon, SK S7N 5A2, Canada. Offers veterinary anatomy (M Sc); veterinary biomedical sciences (M Vet Sc); veterinary physiological sciences (M Sc, PhD). *Faculty:* 15 full-time (5 women). *Students:* 36 full-time (19 women); includes 3 minority (all African Americans) 6 applicants, 33% accepted. In 2005, 5 master's, 1 doctorate awarded. *Degree requirements:* For master's and doctorate, thesis/dissertation. *Faculty research:* Toxicology, animal reproduction, pharmacology, chloride channels, pulmonary pathobiology. *Unit head:* Dr. Barry Blakley, Head, 306-966-7349, Fax: 306-966-7376, E-mail: barry.blakley@usask.ca.

University of South Alabama, College of Medicine and Graduate School, Program in Basic Medical Sciences, Mobile, AL 36688-0002. Offers biochemistry and molecular biology (PhD); cell biology and neuroscience (PhD); microbiology and immunology (PhD); pharmacology (PhD); physiology (PhD). *Faculty:* 61 full-time (7 women). *Students:* 39 full-time (26 women), 3 part-time (2 women); includes 1 minority (American Indian/Alaska Native), 7 international. In 2005, 4 degrees awarded. *Degree requirements:* For doctorate, thesis/dissertation. *Application deadline:* For fall admission, 4/1 for domestic students. Applications are processed on a rolling basis. Application fee: $25. *Expenses:* Contact institution. *Financial support:* Fellowships, research assistantships, institutionally sponsored loans available. Financial award application deadline: 4/1. *Faculty research:* Microcirculation, molecular biology, cell biology, growth control. *Unit head:* Lanette Flagge, Academic Advisor, 251-460-6153.

See Close-Up on page 291.

University of South Alabama, Graduate School, College of Arts and Sciences, Department of Biological Sciences, Mobile, AL 36688-0002. Offers MS. Part-time programs available. *Faculty:* 10 full-time (2 women). *Students:* 12 full-time (7 women), 7 part-time (4 women), 2 international. 18 applicants, 39% accepted, 3 enrolled. In 2005, 4 degrees awarded. *Degree requirements:* For master's, one foreign language, comprehensive exam. *Entrance requirements:* For master's, GRE Subject Test, minimum GPA of 3.0. *Application deadline:* For fall admission, 9/1 for domestic students. Applications are processed on a rolling basis. Application fee: $25. *Expenses:* Tuition, state resident: full-time $4,008. Tuition, nonresident: full-time $8,016.

Required fees: $692. *Financial support:* Fellowships, research assistantships, teaching assistantships available. Support available to part-time students. Financial award application deadline: 4/1. *Faculty research:* Aquatic and marine biology, molecular biochemistry, plant and animal taxonomy. *Unit head:* Dr. John Freeman, Chair, 251-460-6331.

University of South Carolina, The Graduate School, College of Science and Mathematics, Department of Biological Sciences, Columbia, SC 29208. Offers biology (MS, PhD); biology education (IMA, MAT); ecology, evolution and organismal biology (MS, PhD); molecular, cellular, and developmental biology (MS, PhD). IMA and MAT offered in cooperation with the College of Education. Terminal master's awarded for partial completion of doctoral program. *Degree requirements:* For master's, one foreign language, thesis (for some programs); for doctorate, one foreign language, thesis/dissertation. *Entrance requirements:* For master's and doctorate, GRE General Test, minimum GPA of 3.0 in science. Electronic applications accepted. *Faculty research:* Marine ecology, population and evolutionary biology, molecular biology and genetics, development.

University of South Carolina, School of Medicine and The Graduate School, Graduate Programs in Medicine, Columbia, SC 29208. Offers biomedical science (MBS, PhD); genetic counseling (MS); nurse anesthesia (MNA); rehabilitation counseling (MRC, Certificate), including psychiatric rehabilitation (Certificate), rehabilitation counseling (MRC). Terminal master's awarded for partial completion of doctoral program. *Degree requirements:* For master's, thesis (for some programs), practicum, comprehensive exam; for doctorate, thesis/dissertation, comprehensive exam. *Entrance requirements:* For master's, doctorate, and Certificate, GRE General Test. Electronic applications accepted. Expenses: Contact institution. *Faculty research:* Cardiovascular diseases, oncology, neuroscience, psychiatric rehabilitation, genetics.

University of South Carolina, School of Medicine and The Graduate School, Graduate Programs in Medicine, Graduate Program in Biomedical Science, Doctoral Program in Biomedical Science, Columbia, SC 29208. Offers PhD. *Degree requirements:* For doctorate, thesis/dissertation, comprehensive exam. *Entrance requirements:* For doctorate, GRE General Test. Electronic applications accepted. *Faculty research:* Cancer, neuroscience, cardiovascular, reproductive, immunology.

See Close-Up on page 293.

University of South Carolina, School of Medicine and The Graduate School, Graduate Programs in Medicine, Graduate Program in Biomedical Science, Master's Program in Biomedical Science, Columbia, SC 29208. Offers MBS. *Degree requirements:* For master's, thesis, comprehensive exam. *Entrance requirements:* For master's, GRE General Test. Electronic applications accepted. *Faculty research:* Cardiovascular diseases, oncology, reproductive biology, neuroscience, microbiology.

See Close-Up on page 293.

The University of South Dakota, Graduate School, College of Arts and Sciences, Department of Biology, Vermillion, SD 57069-2390. Offers MA, MNS, MS, PhD. *Faculty:* 11 full-time (4 women), 2 part-time/adjunct (both women). *Students:* 27 (12 women); includes 1 minority (African American) In 2005, 12 master's, 3 doctorates awarded. *Degree requirements:* For master's, thesis (for some programs), comprehensive exam (for some programs); for doctorate, thesis/dissertation, comprehensive exam. *Entrance requirements:* For master's, GRE Subject Test and GRE General Test, minimum GPA of 2.7; for doctorate, GRE General Test, GRE Subject Test, minimum GPA of 2.7. Additional exam requirements/recommendations for international students: Required—TOEFL (minimum score 550 paper-based; 213 computer-based), IBT 70. *Application deadline:* Applications are processed on a rolling basis. Application fee: $35. Electronic applications accepted. *Expenses:* Tuition, state resident: part-time $116 per credit hour. Tuition, nonresident: part-time $341 per credit hour. Required fees: $85 per credit hour. Tuition and fees vary according to course load, program and reciprocity agreements. *Financial support:* In 2005–06, 1 fellowship with partial tuition reimbursement (averaging $17,000 per year), 1 research assistantship with partial tuition reimbursement (averaging $12,000 per year), 18 teaching assistantships with partial tuition reimbursements (averaging $12,000 per year) were awarded; Federal Work-Study and unspecified assistantships also available. Support available to part-time students. Financial award applicants required to submit FAFSA. *Faculty research:* Evolutionary and ecological informatics, neuroscience, stress physiology. *Unit head:* Dr. David Swanson, Chair, 605-677-5211, Fax: 605-677-6557, E-mail: dlswanso@usd.edu. *Application contact:* Dr. Paula Mabee, Graduate Coordinator, 605-677-5211, Fax: 605-677-6557, E-mail: pmabee@usd.edu.

The University of South Dakota, School of Medicine and Health Sciences and Graduate School, Biomedical Sciences Graduate Program, Vermillion, SD 57069-2390. Offers cardiovascular research (MA, PhD); cellular and molecular biology (MA, PhD); molecular microbiology and immunology (MA, PhD); neuroscience (MA, PhD); physiology and pharmacology (MA, PhD). Part-time programs available. *Faculty:* 35 full-time (8 women). *Students:* 36 full-time (22 women), 21 international. Average age 28. 73 applicants, 33% accepted, 17 enrolled. In 2005, 3 master's, 3 doctorates awarded. Terminal master's awarded for partial completion of doctoral program. *Degree requirements:* For master's, thesis/dissertation, registration; for doctorate, thesis/dissertation, comprehensive exam, registration. *Entrance requirements:* For master's and doctorate, GRE General Test, minimum GPA of 3.0. Additional exam requirements/recommendations for international students: Required—TOEFL (minimum score 550 paper-based; 213 computer-based). *Application deadline:* For fall admission, 4/15 priority date for domestic students, 4/15 priority date for international students. Applications are processed on a rolling basis. Application fee: $35. *Expenses:* Contact institution. Tuition and fees vary according to course load, program and reciprocity agreements. *Financial support:* In 2005–06, 36 students received support, including 29 fellowships with partial tuition reimbursements available (averaging $20,772 per year), 7 research assistantships with full and partial tuition reimbursements available (averaging $10,386 per year); teaching assistantships, Federal Work-Study also available. Financial award applicants required to submit FAFSA. *Faculty research:* Molecular biology, microbiology, neuroscience, cellular biology, physiology. Total annual research expenditures: $7.4 million. *Unit head:* Dr. Steven B. Waller, Associate Dean of Basic Biomedical Sciences, 605-677-5157, Fax: 605-677-6381, E-mail: swaller@usd.edu. *Application contact:* Stacie Marie Peitz, Program Assistant, 605-677-5254, Fax: 605-677-6381, E-mail: biomed@usd.edu.

University of Southern California, Graduate School, College of Letters, Arts and Sciences, Department of Biological Sciences, Los Angeles, CA 90089. Offers marine environmental biology (PhD); molecular and computational biology (PhD). *Degree requirements:* For doctorate, thesis/dissertation. *Entrance requirements:* For doctorate, GRE General Test. Additional exam requirements/recommendations for international students: Required—TOEFL. *Expenses:* Tuition: Full-time $25,416; part-time $1,059 per unit. Required fees: $484; $484 per year. Tuition and fees vary according to course load and program.

University of Southern California, Keck School of Medicine and Graduate School, Graduate Programs in Medicine, Los Angeles, CA 90089. Offers MPAP, MPH, MS, PhD, MD/PhD. *Faculty:* 258 full-time (85 women), 20 part-time/adjunct (7 women). *Students:* 578 full-time (370 women), 11 part-time (8 women); includes 245 minority (19 African Americans, 1 American Indian/Alaska Native, 129 Asian Americans or Pacific Islanders, 96 Hispanic Americans), 174 international. Average age 26. 1,522 applicants, 31% accepted, 194 enrolled. In 2005, 121 master's, 36 doctorates awarded. Terminal master's awarded for partial completion of doctoral program. *Degree requirements:* For doctorate, thesis/dissertation. *Entrance requirements:* For master's, GRE General Test, minimum GPA of 3.0; for doctorate, GRE General Test. Additional exam requirements/recommendations for international students: Required—TOEFL. Application fee: $65 ($75 for international students). Electronic applications accepted. *Expenses:* Tuition: Full-time $25,416; part-time $1,059 per unit. Required fees: $484; $484 per year. Tuition and fees vary according to course load and program. *Financial support:* In 2005–06, 409 students received support, including 20 fellowships with full tuition reimbursements available (averaging $23,499 per year), 193 research assistantships with full tuition reimburse-

ments available (averaging $23,100 per year), 52 teaching assistantships with full tuition reimbursements available (averaging $23,100 per year); career-related internships or fieldwork, Federal Work-Study, institutionally sponsored loans, scholarships/grants, traineeships, and tuition waivers (full and partial) also available. Support available to part-time students. Financial award application deadline: 2/1. Total annual research expenditures: $25.9 million. *Unit head:* Dr. Francis S. Markland, Associate Dean for Research, 323-442-1607, Fax: 323-442-1610. *Application contact:* Oralia Gonzales, Administrative Services Manager, 323-442-1607, Fax: 323-442-1610, E-mail: oraliago@hsc.usc.edu.

See Close-Up on page 295.

University of Southern Maine, College of Arts and Science, Portland, ME 04104-9300. Offers American and New England studies (MA); biology (MS); creative writing (MFA); social work (MSW); statistics (MS). Part-time and evening/weekend programs available. Post-baccalaureate distance learning degree programs offered (minimal on-campus study). *Degree requirements:* For master's, thesis optional. *Entrance requirements:* For master's, GRE General Test or MAT. Additional exam requirements/recommendations for international students: Required—TOEFL. Electronic applications accepted.

University of Southern Mississippi, Graduate School, College of Science and Technology, Department of Biological Sciences, Hattiesburg, MS 39406-0001. Offers environmental biology (MS, PhD); marine biology (MS, PhD); microbiology (MS, PhD); molecular biology (MS, PhD). *Degree requirements:* For master's and doctorate, thesis/dissertation, comprehensive exam. *Entrance requirements:* For master's, GRE General Test, minimum GPA of 3.0; for doctorate, GRE General Test, minimum GPA of 3.5. Additional exam requirements/recommendations for international students: Required—TOEFL.

See Close-Up on page 297.

University of South Florida, College of Graduate Studies, College of Arts and Sciences, Department of Biology, Tampa, FL 33620-9951. Offers biology (PhD); botany (MS); ecology (PhD); microbiology (MS); physiology (PhD); zoology (MS). Part-time programs available. *Faculty:* 21. *Students:* 54 full-time (32 women), 26 part-time (17 women); includes 8 minority (2 African Americans, 1 American Indian/Alaska Native, 1 Asian American or Pacific Islander, 4 Hispanic Americans), 14 international. 76 applicants, 34% accepted, 13 enrolled. In 2005, 3 master's, 3 doctorates awarded. *Degree requirements:* For master's, thesis (for some programs), graduate seminar in biology; for doctorate, 2 foreign languages, thesis/dissertation, essay of research interest, comprehensive exam. *Entrance requirements:* For master's, GRE General Test, minimum undergraduate GPA of 3.0 in last 60 hours of course work; for doctorate, GRE General Test, GRE Subject Test in biology, minimum undergraduate GPA of 3.0 in last 60 hours of course work. Additional exam requirements/recommendations for international students: Required—TOEFL (minimum score 570 paper-based), TSE (minimum score 50). *Application deadline:* For fall admission, 2/1 priority date for domestic students, 3/1 priority date for international students; for spring admission, 10/1 for domestic students, 8/1 for international students. Application fee: $30. Electronic applications accepted. *Financial support:* Fellowships with full tuition reimbursements, research assistantships with full tuition reimbursements, teaching assistantships with full tuition reimbursements, Federal Work-Study and unspecified assistantships available. Financial award application deadline: 6/30. *Unit head:* Sydney Pierce, Chairperson, 813-974-3250, Fax: 813-974-3263. *Application contact:* Christine Smith, Graduate Advisor, 813-974-4747, Fax: 813-974-3263, E-mail: csmith2@chuma1.cas.usf.edu.

University of South Florida, College of Medicine and College of Graduate Studies, Graduate Programs in Medical Sciences, Tampa, FL 33620-9951. Offers anatomy (PhD); biochemistry and molecular biology (MS, PhD), including biochemistry and molecular biology (PhD); bioinformatics and computational biology (MS); medical microbiology and immunology (PhD); pathology (PhD); pharmacology and therapeutics (PhD), including medical sciences; physiology and biophysics (PhD). *Students:* 108 full-time (53 women), 34 part-time (26 women); includes 31 minority (9 African Americans, 9 Asian Americans or Pacific Islanders, 13 Hispanic Americans), 30 international. 117 applicants, 99% accepted, 116 enrolled. In 2005, 9 master's, 4 doctorates awarded. *Degree requirements:* For doctorate, thesis/dissertation. *Entrance requirements:* For doctorate, GRE General Test, minimum GPA of 3.0. Application fee: $30. *Expenses:* Contact institution. *Financial support:* Institutionally sponsored loans and scholarships/grants available. Financial award application deadline: 4/1; financial award applicants required to submit FAFSA. *Unit head:* Dr. Joseph J. Krzanowski, Associate Dean for Research and Graduate Affairs, 813-974-4181, Fax: 813-974-4317, E-mail: jkrzanow@com1.med.usf.edu.

The University of Tennessee, Graduate School, College of Arts and Sciences, Program in Life Sciences, Knoxville, TN 37996. Offers genome science and technology (MS, PhD); plant physiology and genetics (MS, PhD). *Degree requirements:* For doctorate, one foreign language, thesis/dissertation. *Entrance requirements:* For master's and doctorate, GRE General Test, minimum GPA of 2.7. Additional exam requirements/recommendations for international students: Required—TOEFL. Electronic applications accepted.

The University of Tennessee, Graduate School, Intercollegiate Programs, Program in Comparative and Experimental Medicine, Knoxville, TN 37996. Offers MS, PhD. *Degree requirements:* For master's and doctorate, thesis/dissertation. *Entrance requirements:* For master's and doctorate, GRE General Test, minimum GPA of 2.7. Additional exam requirements/recommendations for international students: Required—TOEFL. Electronic applications accepted.

The University of Tennessee Health Science Center, College of Graduate Health Sciences, Integrated Program in Biomedical Sciences, Memphis, TN 38163-0002. Offers MS, PhD. *Faculty:* 220 full-time (20 women), 30 part-time/adjunct (5 women). *Students:* 71 full-time (35 women); includes 6 minority (5 African Americans, 1 Hispanic American), 30 international. Average age 26. 251 applicants, 17% accepted, 23 enrolled. In 2005, 1 degree awarded. Terminal master's awarded for partial completion of doctoral program. *Degree requirements:* For master's, thesis/dissertation, registration; for doctorate, thesis/dissertation, comprehensive exam, registration. *Entrance requirements:* For doctorate, GRE, minimum GPA of 3.0, 3 letters of recommendation. Additional exam requirements/recommendations for international students: Required—TOEFL (minimum score 600 paper-based; 213 computer-based). *Application deadline:* For fall admission, 3/1 for domestic students, 3/1 for international students. Applications are processed on a rolling basis. Application fee: $0. Electronic applications accepted. *Financial support:* In 2005–06, 23 students received support, including 23 research assistantships with full tuition reimbursements available (averaging $20,000 per year); scholarships/grants and tuition waivers (full) also available. *Faculty research:* Molecular biology, physiology, pharmacology, pathology, neuroscience. Total annual research expenditures: $50 million. *Unit head:* Dr. Patrick Ryan, Program Director, 901-448-8764, Fax: 901-448-6958, E-mail: pryan@utmem.edu. *Application contact:* Janie VanProoijen, Assistant to Director, 901-448-7030, Fax: 901-448-6958, E-mail: jvanprooijen@utmem.edu.

See Close-Up on page 299.

The University of Tennessee–Oak Ridge National Laboratory Graduate School of Genome Science and Technology, Graduate Program, Oak Ridge, TN 37830-8026. Offers life sciences (MS, PhD). *Faculty:* 1 (woman) full-time, 87 part-time/adjunct (19 women). *Students:* 43 full-time (20 women), 2 part-time (1 woman); includes 4 minority (2 African Americans, 2 Asian Americans or Pacific Islanders), 21 international. Average age 30. 49 applicants, 33% accepted, 9 enrolled. In 2005, 3 master's, 3 doctorates awarded. *Median time to degree:* Of those who began their doctoral program in fall 1997, 98% received their degree in 8 years or less. *Degree requirements:* For master's, thesis/dissertation, registration; for doctorate, thesis/dissertation, comprehensive exam, registration. *Entrance requirements:* For master's and doctorate, GRE General Test. Additional exam requirements/recommendations for international students: Required—TOEFL (minimum score 550 paper-based; 213 computer-based). *Application deadline:* For fall admission, 1/15 priority date for domestic students, 2/1 priority date for international students. Applications are processed on a rolling basis. Application fee: $35.

Biological and Biomedical Sciences—General

The University of Tennessee–Oak Ridge National Laboratory Graduate School of Genome Science and Technology (continued)
Electronic applications accepted. *Expenses:* Tuition, state resident: part-time $296 per hour. Tuition, nonresident: part-time $895 per hour. *Financial support:* In 2005–06, 25 students received support, including 9 research assistantships with full tuition reimbursements available (averaging $18,000 per year); fellowships, institutionally sponsored loans, health care benefits, tuition waivers (full), and unspecified assistantships also available. Financial award application deadline: 3/31. *Faculty research:* Genetics/genomics, structural biology/proteomics, computational biology/bioinformatics, bioanalytical technologies. *Unit head:* Dr. Cynthia B Peterson, Director, 865-974-4083, Fax: 965-974-0361, E-mail: cbpeters@utk.edu. *Application contact:* Kay Gardner, Program Resource Specialist, 865-574-1227, Fax: 865-574-4812, E-mail: gardnerk@utk.edu.

See Close-Up on page 799.

The University of Texas at Arlington, Graduate School, College of Science, Department of Biology, Arlington, TX 76019. Offers biology (MS); quantitative biology (PhD). Part-time and evening/weekend programs available. *Faculty:* 11 full-time (3 women), 2 part-time/adjunct (1 woman). *Students:* 51 full-time (30 women), 25 part-time (17 women); includes 12 minority (2 African Americans, 4 Asian Americans or Pacific Islanders, 6 Hispanic Americans), 14 international. 47 applicants, 66% accepted, 19 enrolled. In 2005, 13 master's, 3 doctorates awarded. *Degree requirements:* For master's, thesis, oral defense of thesis; for doctorate, thesis/dissertation, oral defense of dissertation, comprehensive exam, registration. *Entrance requirements:* For master's and doctorate, GRE General Test. Additional exam requirements/recommendations for international students: Required—TOEFL (minimum score 550 paper-based; 213 computer-based), TSE. *Application deadline:* For fall admission, 6/16 for domestic students. Applications are processed on a rolling basis. Application fee: $35 ($50 for international students). *Expenses:* Tuition, state resident: full-time $3,350. Tuition, nonresident: full-time $8,318. International tuition: $8,448 full-time. Required fees: $1,277. Full-time tuition and fees vary according to course level and program. *Financial support:* In 2005–06, 4 fellowships (averaging $1,000 per year), 4 research assistantships (averaging $15,500 per year), 26 teaching assistantships (averaging $15,500 per year) were awarded; Federal Work-Study and institutionally sponsored loans also available. Financial award application deadline: 6/1; financial award applicants required to submit FAFSA. *Unit head:* Dr. Johnathan Campbell, Chair, 817-272-2871, Fax: 817-272-2855, E-mail: campbell@exchange.uta.edu. *Application contact:* Dr. Daniel R. Formanowicz, Graduate Adviser, 817-272-2871, Fax: 817-272-2855, E-mail: formanow@uta.edu.

The University of Texas at Austin, Graduate School, College of Natural Sciences, School of Biological Sciences, Austin, TX 78712-1111. Offers MA, PhD. *Entrance requirements:* For master's and doctorate, GRE General Test. Electronic applications accepted.

The University of Texas at Brownsville, Graduate Studies, College of Science, Mathematics and Technology, Brownsville, TX 78520-4991. Offers biological sciences (MS, MSIS); mathematics (MS); physics (MS). Part-time and evening/weekend programs available. *Faculty:* 51 full-time (10 women). *Students:* 34 (16 women); includes 24 minority (all Hispanic Americans) 7 applicants. In 2005, 9 degrees awarded. *Degree requirements:* For master's, thesis optional. *Entrance requirements:* For master's, GRE General Test. Additional exam requirements/recommendations for international students: Required—TOEFL. *Application deadline:* For fall admission, 7/1 for domestic students; for spring admission, 12/1 priority date for domestic students. Applications are processed on a rolling basis. Application fee: $30. *Financial support:* Federal Work-Study, scholarships/grants, and tuition waivers (partial) available. Support available to part-time students. Financial award application deadline: 4/3; financial award applicants required to submit FAFSA. *Faculty research:* Fish, insects, barrier islands, algae, curlits. *Unit head:* Terry Jay Phillips, Interim Dean, 956-882-6701, Fax: 956-882-8988. *Application contact:* Irma C. Hernandez, Information Contact, 956-882-7787, Fax: 956-882-7279, E-mail: irma.c.hernandez@utb.edu.

The University of Texas at Dallas, School of Natural Sciences and Mathematics, Program in Biology, Richardson, TX 75083-0688. Offers bioinformatics and computational biology (MS); biotechnology (MS); molecular and cell biology (MS, PhD). Part-time and evening/weekend programs available. *Faculty:* 15 full-time (2 women). *Students:* 64 full-time (40 women), 18 part-time (14 women); includes 18 minority (1 African American, 15 Asian Americans or Pacific Islanders, 2 Hispanic Americans), 46 international. Average age 28. 133 applicants, 71% accepted, 39 enrolled. In 2005, 10 master's, 7 doctorates awarded. *Degree requirements:* For master's, thesis optional; for doctorate, thesis/dissertation, publishable paper. *Entrance requirements:* For master's and doctorate, GRE General Test. Additional exam requirements/recommendations for international students: Required—TOEFL (minimum score 550 paper-based; 213 computer-based). *Application deadline:* For fall admission, 7/15 for domestic students; for spring admission, 11/15 for domestic students. Applications are processed on a rolling basis. Application fee: $50 ($100 for international students). Electronic applications accepted. *Expenses:* Tuition, state resident: full-time $5,450; part-time $303 per credit. Tuition, nonresident: full-time $12,648; part-time $703 per credit. Tuition and fees vary according to program. *Financial support:* In 2005–06, 17 research assistantships with tuition reimbursements (averaging $14,125 per year), 16 teaching assistantships with tuition reimbursements (averaging $11,972 per year) were awarded; fellowships, career-related internships or fieldwork, Federal Work-Study, institutionally sponsored loans, and scholarships/grants also available. Support available to part-time students. Financial award application deadline: 4/30; financial award applicants required to submit FAFSA. *Faculty research:* DNA replication, regulation of gene expression, subcellular organelles, physical chemistry of macromolecules, damage and repair of cellular DNA. Total annual research expenditures: $4.1 million. *Unit head:* Dr. Donald Gray, Head, 972-883-2513, Fax: 972-883-2502, E-mail: dongray@utdallas.edu. *Application contact:* Dr. Ernest Hannig, Graduate Advisor, 972-883-2505, Fax: 972-883-2409, E-mail: hannig@utdallas.edu.

See Close-Up on page 647.

The University of Texas at El Paso, Graduate School, College of Science, Department of Biological Sciences, El Paso, TX 79968-0001. Offers bioinformatics (MS); biological science (MS, PhD); environmental science and engineering (PhD). Part-time and evening/weekend programs available. *Degree requirements:* For master's, thesis. *Entrance requirements:* For master's, GRE General Test, minimum GPA of 3.0; for doctorate, GRE General Test. Additional exam requirements/recommendations for international students: Required—TOEFL. Electronic applications accepted.

The University of Texas at San Antonio, College of Sciences, Department of Biology, San Antonio, TX 78249-0617. Offers biology (PhD), including cell and molecular biology, neurobiology; biology and biotechnology (MS), including biology, biotechnology. *Degree requirements:* For master's and doctorate, comprehensive exam, registration. *Entrance requirements:* For master's, GRE General Test, minimum GPA of 3.0; for doctorate, GRE General Test, minimum GPA of 3.3. Additional exam requirements/recommendations for international students: Required—TOEFL (minimum score 500 paper-based; 173 computer-based). Electronic applications accepted.

See Close-Up on page 301.

The University of Texas at Tyler, College of Arts and Sciences, Department of Biology, Tyler, TX 75799-0001. Offers biology (MAT, MS); interdisciplinary studies (MSIS). *Faculty:* 9 full-time (4 women), 4 part-time/adjunct (2 women). *Students:* 10 full-time (4 women), 3 part-time (2 women); includes 1 minority (Asian American or Pacific Islander) 5 applicants, 100% accepted, 0 enrolled. In 2005, 4 degrees awarded. *Degree requirements:* For master's, thesis, oral qualifying exam, thesis defense, comprehensive exam. *Entrance requirements:* For master's, GRE General Test, GRE Subject Test, bachelor's degree in biology or equivalent. *Application deadline:* Applications are processed on a rolling basis. Application fee: $0. Electronic applications accepted. *Expenses:* Tuition, state resident: part-time $321. Tuition, nonresident: part-time $597. Tuition and fees vary according to course load. *Financial support:* In 2005–06, 7 students received support, including 1 research assistantship (averaging $9,200 per year), 9 teaching assistantships (averaging $9,200 per year); scholarships/grants also available. Financial award application deadline: 7/1; financial award applicants required to submit FAFSA. *Faculty research:* Phenotypic plasticity and heritability of life history traits, invertebrate ecology and genetics, systematics and phylogenetics of reptiles, hibernation physiology in turtles, landscape ecology, host-microbe interaction; outer membrane proteins in bacteria. Total annual research expenditures: $100,000. *Unit head:* Dr. Neil Ford, Department Graduate Coordinator, 903-566-7249, E-mail: nford@uttyler.edu. *Application contact:* Carol A. Hodge, Office of Graduate Studies, 903-566-5642, Fax: 903-566-7068, E-mail: chodge@mail.uttyl.edu.

The University of Texas Health Science Center at Houston, Graduate School of Biomedical Sciences, Houston, TX 77225-0036. Offers MS, PhD, MD/PhD. *Faculty:* 549 full-time (135 women). *Students:* 545 full-time (315 women); includes 123 minority (23 African Americans, 2 American Indian/Alaska Native, 46 Asian Americans or Pacific Islanders, 52 Hispanic Americans), 186 international. Average age 26. 601 applicants, 41% accepted, 160 enrolled. In 2005, 23 master's, 63 doctorates awarded. Terminal master's awarded for partial completion of doctoral program. *Degree requirements:* For master's and doctorate, thesis/dissertation. *Entrance requirements:* For master's and doctorate, GRE General Test. Additional exam requirements/recommendations for international students: Required—TOEFL, TWE. *Application deadline:* For fall admission, 1/15 priority date for domestic students, 12/15 priority date for international students; for spring admission, 11/1 priority date for domestic students. Applications are processed on a rolling basis. Application fee: $10. Electronic applications accepted. *Financial support:* Fellowships with full tuition reimbursements, research assistantships with full tuition reimbursements, teaching assistantships, institutionally sponsored loans, scholarships/grants, and health care benefits available. Financial award application deadline: 1/15. *Unit head:* Dr. George M. Stancel, Dean, 713-500-9880, Fax: 713-500-9877, E-mail: george.m.stancel@uth.tmc.edu. *Application contact:* Dr. Victoria P. Knutson, Assistant Dean of Admissions, 713-500-9860, Fax: 713-500-9877, E-mail: victoria.p.knutson@uth.tmc.edu.

See Close-Up on page 303.

The University of Texas Health Science Center at San Antonio, Graduate School of Biomedical Sciences, San Antonio, TX 78229-3900. Offers Pharm D, MS, MSN, PhD, Certificate. Part-time and evening/weekend programs available. *Entrance requirements:* GRE General Test. Electronic applications accepted. Expenses: Contact institution.

The University of Texas Medical Branch, Graduate School of Biomedical Sciences, Galveston, TX 77555. Offers MA, MMS, MPH, MS, PhD, JD/PhD, MD/MA, MD/PhD. Part-time programs available. *Faculty:* 321 full-time (83 women). *Students:* 330 (174 women); includes 55 minority (13 African Americans, 2 American Indian/Alaska Native, 16 Asian Americans or Pacific Islanders, 24 Hispanic Americans) 83 international. Average age 31. In 2005, 16 master's, 36 doctorates awarded. *Degree requirements:* For doctorate, thesis/dissertation. *Entrance requirements:* For master's and doctorate, GRE General Test. Additional exam requirements/recommendations for international students: Required—TOEFL (minimum score 550 paper-based; 213 computer-based). *Application deadline:* Applications are processed on a rolling basis. Application fee: $30 ($75 for international students). Electronic applications accepted. *Expenses:* Tuition, state resident: full-time $8,350; part-time $90 per credit hour. Tuition, nonresident: full-time $21,450; part-time $366 per credit hour. Required fees: $1,027; $11 per credit hour. $60 per term. *Financial support:* In 2005–06, fellowships (averaging $23,000 per year), research assistantships with full tuition reimbursements (averaging $23,000 per year) were awarded; teaching assistantships, career-related internships or fieldwork, Federal Work-Study, institutionally sponsored loans, scholarships/grants, traineeships, health care benefits, and unspecified assistantships also available. Support available to part-time students. Financial award applicants required to submit FAFSA. *Unit head:* Dr. Cary W. Cooper, Dean, 409-772-2665, Fax: 409-747-0772, E-mail: ccooper@utmb.edu. *Application contact:* Dr. Dorian H. Coppenhavor, Assistant Dean for Student Affairs, 409-772-2665, Fax: 409-747-0772, E-mail: dcoppen@utmb.edu.

The University of Texas of the Permian Basin, Office of Graduate Studies, College of Arts and Sciences, Department of Sciences and Mathematics, Program in Biology, Odessa, TX 79762-0001. Offers MS. Part-time and evening/weekend programs available. *Degree requirements:* For master's, thesis or alternative, comprehensive exam, registration. *Entrance requirements:* For master's, GRE General Test. Additional exam requirements/recommendations for international students: Required—TOEFL (minimum score 550 paper-based; 213 computer-based).

The University of Texas–Pan American, College of Science and Engineering, Department of Biology, Edinburg, TX 78541-2999. Offers MS. Part-time and evening/weekend programs available. *Degree requirements:* For master's, comprehensive exam. *Entrance requirements:* For master's, GRE General Test, minimum GPA of 2.75 in biology. *Expenses:* Tuition, state resident: full-time $2,268; part-time $68 per credit hour. Tuition, nonresident: full-time $7,236; part-time $370 per credit hour. Required fees: $488. *Faculty research:* Flora and fauna of South Padre Island, plant taxonomy of Rio Grande Valley.

The University of Texas Southwestern Medical Center at Dallas, Southwestern Graduate School of Biomedical Sciences, Division of Basic Science, Dallas, TX 75390. Offers biological chemistry (PhD); cell regulation (PhD); genetics and development (PhD); immunology (PhD); integrative biology (PhD); molecular biophysics (PhD); molecular microbiology (PhD); neuroscience (PhD). *Faculty:* 239 full-time (52 women), 10 part-time/adjunct (0 women). *Students:* 477 full-time (236 women), 10 part-time (2 women); includes 88 minority (8 African Americans, 1 American Indian/Alaska Native, 49 Asian Americans or Pacific Islanders, 30 Hispanic Americans), 166 international. Average age 28. 979 applicants, 16% accepted, 76 enrolled. In 2005, 57 doctorates awarded. *Degree requirements:* For doctorate, thesis/dissertation, qualifying exam. *Entrance requirements:* For doctorate, GRE, minimum GPA of 3.0. Additional exam requirements/recommendations for international students: Required—TOEFL. *Application deadline:* For fall admission, 1/5 for domestic students. Applications are processed on a rolling basis. Application fee: $0. Electronic applications accepted. *Expenses:* Tuition, state resident: full-time $6,550; part-time $50 per credit hour. Tuition, nonresident: full-time $19,650; part-time $326 per credit hour. Required fees: $42 per credit hour. Tuition and fees vary according to degree level and program. *Financial support:* Fellowships, research assistantships, institutionally sponsored loans and traineeships available. *Unit head:* Dr. Mike Roth, Associate Dean, 214-648-3276, Fax: 214-648-0320, E-mail: michael.roth@utsouthwestern.edu. *Application contact:* Dr. Nancy E. Street, Associate Dean, 214-648-6708, Fax: 214-648-2102, E-mail: nancy.street@utsouthwestern.edu.

See Close-Up on page 305.

The University of Texas Southwestern Medical Center at Dallas, Southwestern Graduate School of Biomedical Sciences, Division of Clinical Science, Clinical Science Program, Dallas, TX 75390. Offers MCS. Part-time programs available. *Students:* 2 full-time (1 woman), 48 part-time (22 women); includes 18 minority (2 African Americans, 1 American Indian/Alaska Native, 9 Asian Americans or Pacific Islanders, 6 Hispanic Americans), 4 international. 52 applicants, 100% accepted, 50 enrolled. *Degree requirements:* For master's, 1 year clinical research project. *Entrance requirements:* For master's, graduate degree in biomedical science. *Application deadline:* For spring admission, 12/16 for domestic students. Applications are processed on a rolling basis. Electronic applications accepted. *Expenses:* Tuition, state resident: full-time $6,550; part-time $50 per credit hour. Tuition, nonresident: full-time $19,650; part-time $326 per credit hour. Required fees: $42 per credit hour. Tuition and fees vary according to degree level and program. *Unit head:* Dr. Milton Packer, Chair, 214-648-0491, Fax: 214-648-6417, E-mail: milton.packer@utsouthwestern.edu. *Application contact:* Dena Wheaton, Program Coordinator, 214-648-2410, Fax: 214-648-3978, E-mail: dena.wheaton@utsouthwestern.edu.

Biological and Biomedical Sciences—General

The University of Texas Southwestern Medical Center at Dallas, Southwestern Graduate School of Biomedical Sciences, Medical Scientist Training Program, Dallas, TX 75390. Offers PhD, MD/PhD. *Students:* Average age 24. 35 applicants, 43% accepted, 15 enrolled. *Application deadline:* For fall admission, 11/1 for domestic students. Application fee: $0. Electronic applications accepted. *Expenses:* Tuition, state resident: full-time $6,550; part-time $50 per credit hour. Tuition, nonresident: full-time $19,650; part-time $326 per credit hour. Required fees: $42 per credit hour. Tuition and fees vary according to degree level and program. *Financial support:* Application deadline: 3/1. *Unit head:* Dr. Dennis McKearin, Associate Dean, 214-648-2057, Fax: 214-648-2814, E-mail: dennis.mckearin@utsouthwestern.edu. *Application contact:* Robin Downing, Education Coordinator, 214-648-6764, Fax: 214-648-2814, E-mail: robin.downing@utsouthwestern.edu.

See Close-Up on page 307.

The University of Texas Southwestern Medical Center at Dallas, Southwestern Graduate School of Biomedical Sciences, Program in Integrative Biology, Dallas, TX 75390. Offers PhD. *Faculty:* 62 full-time (13 women), 1 part-time/adjunct (0 women). *Students:* 51 full-time (26 women); includes 8 minority (1 African American, 5 Asian Americans or Pacific Islanders, 2 Hispanic Americans), 19 international. Average age 29. In 2005, 6 doctorates awarded. *Degree requirements:* For doctorate, thesis/dissertation, qualifying exam. *Entrance requirements:* For doctorate, GRE General Test, minimum GPA of 3.0. *Application deadline:* For fall admission, 1/5 for domestic students. Application fee: $0. Electronic applications accepted. *Expenses:* Tuition, state resident: full-time $6,550; part-time $50 per credit hour. Tuition, nonresident: full-time $19,650; part-time $326 per credit hour. Required fees: $42 per credit hour. Tuition and fees vary according to degree level and program. *Financial support:* Fellowships, research assistantships, institutionally sponsored loans and traineeships available. *Faculty research:* Muscle physiology, ion transport in secretory cells, nuclear hormone receptors, contractile protein phosphorylation, cardiovascular homeostasis. *Unit head:* Dr. Yi Liu, Chair, 214-645-6033, Fax: 214-645-6050, E-mail: yi.liu@utsouthwestern.edu. *Application contact:* Dr. Nancy E. Street, Associate Dean, 214-648-6708, Fax: 214-648-2102, E-mail: nancy.street@utsouthwestern.edu.

University of the Incarnate Word, School of Graduate Studies and Research, School of Mathematics, Sciences, and Engineering, Program in Biology, San Antonio, TX 78209-6397. Offers MA, MS. Part-time and evening/weekend programs available. *Students:* 4 full-time (2 women), 27 part-time (21 women); includes 15 minority (2 African Americans, 13 Hispanic Americans). Average age 26. In 2005, 5 degrees awarded. *Degree requirements:* For master's, thesis optional. *Entrance requirements:* For master's, GRE General Test, minimum GPA of 3.0. Additional exam requirements/recommendations for international students: Required—TOEFL. *Application deadline:* For fall admission, 8/15 for domestic students; for spring admission, 12/31 for domestic students. Applications are processed on a rolling basis. Application fee: $20. *Expenses:* Tuition: Full-time $9,810; part-time $545 per credit hour. Required fees: $828; $46 per credit. One-time fee: $30. *Financial support:* Research assistantships, teaching assistantships, Federal Work-Study and Federal Loans available. Financial award application deadline: 9/12. *Faculty research:* Mammalogy, zoogeography, cell biology, physical chemistry, molecular genetics. *Unit head:* Dr. Bonnie McCormick, Chair, 210-829-3831, Fax: 210-829-3153, E-mail: mccormic@universe.uiwtx.edu. *Application contact:* Andrea Cyterski-Acosta, Dean of Enrollment, 210-829-6005, Fax: 210-829-3921, E-mail: cyterski@uiwtx.edu.

University of the Pacific, College of the Pacific, Department of Biological Sciences, Stockton, CA 95211-0197. Offers MS. *Faculty:* 13 full-time (3 women). *Students:* Average age 25. 20 applicants, 40% accepted, 8 enrolled. In 2005, 3 degrees awarded. *Degree requirements:* For master's, thesis. *Entrance requirements:* For master's, GRE General Test, GRE Subject Test. Additional exam requirements/recommendations for international students: Required—TOEFL (minimum score 475 paper-based; 150 computer-based). *Application deadline:* For fall admission, 3/1 for domestic students; for spring admission, 10/1 priority date for domestic students. Applications are processed on a rolling basis. Application fee: $75. *Expenses:* Tuition: Full-time $25,658. Required fees: $430; $886 per unit. Tuition and fees vary according to course load. *Financial support:* In 2005–06, 22 teaching assistantships were awarded; institutionally sponsored loans also available. Support available to part-time students. Financial award application deadline: 3/1; financial award applicants required to submit FAFSA. *Unit head:* Dr. Gregg Jongeward, Chairman, 209-946-2181.

The University of Toledo, College of Graduate Studies, College of Arts and Sciences, Department of Biological Sciences, Program in Biology, Toledo, OH 43606-3390. Offers MS, PhD. *Expenses:* Tuition, state resident: full-time $6,623; part-time $308 per credit hour. Tuition, nonresident: full-time $13,232; part-time $735 per credit hour.

See Close-Up on page 309.

The University of Toledo, Graduate School, College of Arts and Sciences, Department of Earth, Ecological and Environmental Sciences, Toledo, OH 43606-3390. Offers biology (ecology track) (MS, PhD); geology (MS), including earth surface processes, general geology. Part-time programs available. *Faculty:* 28. *Students:* 8 full-time (6 women), 2 part-time (both women). Average age 31. 6 applicants, 83% accepted, 3 enrolled. In 2005, 2 degrees awarded. *Degree requirements:* For master's, thesis. *Entrance requirements:* For master's, GRE General Test. Additional exam requirements/recommendations for international students: Required—TOEFL. *Application deadline:* For fall admission, 8/1 for domestic students. Applications are processed on a rolling basis. Application fee: $45. Electronic applications accepted. *Expenses:* Tuition, area resident: Part-time $308 per credit hour. Tuition, state resident: full-time $3,312. Tuition, nonresident: full-time $6,616; part-time $735 per credit hour. *Financial support:* In 2005–06, 2 research assistantships (averaging $14,000 per year), 30 teaching assistantships (averaging $11,724 per year) were awarded; Federal Work-Study, institutionally sponsored loans, and tuition waivers (full) also available. Support available to part-time students. Financial award application deadline: 4/1; financial award applicants required to submit FAFSA. *Faculty research:* Environmental geochemistry, geophysics, petrology and mineralogy, paleontology, geohydrology. *Unit head:* Dr. Michael Phillips, Chair, 419-530-4572, Fax: 419-530-4421, E-mail: michael.phillips@utoledo.edu. *Application contact:* Johan Gottgens, 419-530-8451, E-mail: john.gottgens@utoledo.edu.

See Close-Up on page 715.

The University of Toledo, College of Graduate Studies, Program in Molecular and Cellular Biology, Toledo, OH 43606-3390. Offers PhD, MD/PhD. *Degree requirements:* For doctorate, thesis/dissertation, qualifying exam. *Entrance requirements:* For doctorate, GRE General Test, minimum undergraduate GPA of 3.0. *Expenses:* Tuition, state resident: full-time $6,623; part-time $308 per credit hour. Tuition, nonresident: full-time $13,232; part-time $735 per credit hour.

See Close-Up on page 311.

University of Toronto, School of Graduate Studies, Life Sciences Division, Toronto, ON M5S 1A1, Canada. Offers M Sc, M Sc BMC, M Sc F, MA, MFC, MH Sc, MN, PhD, Certificate, Diploma, M Sc/PhD, MBA/MN, MD/PhD. Part-time programs available. *Degree requirements:* For doctorate, thesis/dissertation.

University of Tulsa, Graduate School, College of Engineering and Natural Sciences, Department of Biological Sciences, Tulsa, OK 74104-3189. Offers MS, MTA, PhD. Part-time programs available. *Faculty:* 12 full-time (4 women). *Students:* 14 full-time (11 women); includes 1 minority (American Indian/Alaska Native), 7 international. Average age 29. 20 applicants, 35% accepted, 2 enrolled. In 2005, 3 master's awarded. *Degree requirements:* For master's, thesis, oral exams; for doctorate, thesis/dissertation, comprehensive exam. *Entrance requirements:* For master's and doctorate, GRE General Test. Additional exam requirements/recommendations for international students: Required—TOEFL (minimum score 550 paper-based; 213 computer-based), IELT (minimum score 6). *Application deadline:* Applications are processed on a rolling basis. Application fee: $30. Electronic applications accepted. *Expenses:*

Tuition: Full-time $12,132; part-time $674 per credit hour. Required fees: $60; $3 per credit hour. *Financial support:* In 2005–06, 13 students received support, including 1 fellowship with full tuition reimbursement available (averaging $30,000 per year), 1 research assistantship with full and partial tuition reimbursement available (averaging $10,500 per year), 9 teaching assistantships with full tuition reimbursements available (averaging $14,500 per year); career-related internships or fieldwork, Federal Work-Study, scholarships/grants, tuition waivers (full and partial), and unspecified assistantships also available. Support available to part-time students. Financial award application deadline: 2/1; financial award applicants required to submit FAFSA. *Faculty research:* Biomedical research, wetland conservation, environmental biology, biochemical analysis, animal behavior. Total annual research expenditures: $1.1 million. *Unit head:* Dr. Glen E. Collier, Chairperson, 918-631-2758, Fax: 918-631-2762, E-mail: glen-collier@utulsa.edu. *Application contact:* Dr. Kenton S. Miller, Adviser, 918-631-3065, Fax: 918-631-2762, E-mail: grad@utulsa.edu.

University of Utah, The Graduate School, College of Science, Department of Biology, Salt Lake City, UT 84112-1107. Offers biology (M Phil); ecology and evolutionary biology (MS, PhD); genetics (MS, PhD); molecular biology (PhD). Part-time programs available. *Faculty:* 42 full-time (8 women), 1 part-time/adjunct (0 women). *Students:* 55 full-time (30 women), 15 part-time (2 women); includes 4 minority (3 Asian Americans or Pacific Islanders, 1 Hispanic American), 19 international. Average age 29. 26 applicants, 54% accepted, 11 enrolled. In 2005, 3 master's, 6 doctorates awarded. Terminal master's awarded for partial completion of doctoral program. *Median time to degree:* Of those who began their doctoral program in fall 1997, 20% received their degree in 8 years or less. *Degree requirements:* For master's and doctorate, thesis/dissertation. *Entrance requirements:* For master's and doctorate, GRE General Test, minimum GPA of 3.0. Additional exam requirements/recommendations for international students: Required—TOEFL (minimum score 500 paper-based; 173 computer-based). *Application deadline:* For fall admission, 1/13 for domestic students, 1/13 for international students. Application fee: $45 ($65 for international students). *Expenses:* Tuition, state resident: full-time $2,932; part-time $369 per credit. Tuition, nonresident: full-time $10,350; part-time $1,302 per credit. Required fees: $516 per term. Tuition and fees vary according to course load and program. *Financial support:* In 2005–06, 2 fellowships with full tuition reimbursements (averaging $30,000 per year), 33 research assistantships with full tuition reimbursements (averaging $22,000 per year), 33 teaching assistantships with full tuition reimbursements (averaging $15,000 per year) were awarded; career-related internships or fieldwork, scholarships/grants, traineeships, and health care benefits also available. Financial award application deadline: 2/15; financial award applicants required to submit FAFSA. *Faculty research:* Behavioral ecology, cellular neurobiology, DNA replication, ecological genetics, herpetology. Total annual research expenditures: $11.4 million. *Unit head:* David R. Wolstenholme, Chair, 801-581-6517, Fax: 801-581-4668, E-mail: wolstenholme@bioscience.utah.edu. *Application contact:* Shannon Nielsen, Administrative Program Coordinator, 801-581-5636, Fax: 801-581-4668, E-mail: shannon.nielsen@bioscience.utah.edu.

See Close-Up on page 313.

University of Utah, School of Medicine and The Graduate School, Graduate Programs in Medicine, Salt Lake City, UT 84112-1107. Offers M Phil, M Stat, MPAS, MPH, MS, MSPH, PhD. Part-time programs available. Electronic applications accepted. *Expenses:* Tuition, state resident: full-time $2,932; part-time $369 per credit. Tuition, nonresident: full-time $10,350; part-time $1,302 per credit. Required fees: $516 per term. Tuition and fees vary according to course load and program.

University of Vermont, College of Medicine and Graduate College, Graduate Programs in Medicine, Burlington, VT 05405. Offers anatomy and neurobiology (PhD); biochemistry (MS, PhD); microbiology and molecular genetics (MS, PhD); molecular physiology and biophysics (MS, PhD); pathology (MS); pharmacology (MS, PhD). *Students:* 73 (37 women); includes 3 minority (1 African American, 1 Asian American or Pacific Islander, 1 Hispanic American) 26 international. 130 applicants, 27% accepted, 16 enrolled. In 2005, 2 master's, 8 doctorates awarded. *Degree requirements:* For master's and doctorate, thesis/dissertation. *Entrance requirements:* For master's and doctorate, GRE General Test. Additional exam requirements/recommendations for international students: Required—TOEFL (minimum score 550 paper-based; 213 computer-based). *Application deadline:* For fall admission, 4/1 for domestic students. Applications are processed on a rolling basis. Application fee: $40. Electronic applications accepted. *Expenses:* Tuition, area resident: Part-time $410 per credit hour. Tuition, nonresident: part-time $1,034 per credit hour. *Financial support:* Fellowships, research assistantships, teaching assistantships, traineeships and analytical assistantships available. Financial award application deadline: 3/1.

See Close-Up on page 315.

University of Vermont, Graduate College, College of Arts and Sciences, Department of Biology, Burlington, VT 05405. Offers biology (MS, PhD); biology education (MAT, MST). *Faculty:* 17. *Students:* 31 (15 women) 16 international. 35 applicants, 26% accepted, 8 enrolled. In 2005, 2 master's, 2 doctorates awarded. *Degree requirements:* For master's and doctorate, thesis/dissertation. *Entrance requirements:* For master's and doctorate, GRE General Test. Additional exam requirements/recommendations for international students: Required—TOEFL (minimum score 550 paper-based; 213 computer-based). *Application deadline:* For fall admission, 4/1 for domestic students. Applications are processed on a rolling basis. Application fee: $40. Electronic applications accepted. *Expenses:* Tuition, area resident: Part-time $410 per credit hour. Tuition, nonresident: part-time $1,034 per credit hour. *Financial support:* Fellowships, research assistantships, teaching assistantships available. *Unit head:* Dr. Judith Van Houten, Chairperson, 802-656-2922. *Application contact:* N. Gotelli, Coordinator, 802-656-2922.

University of Victoria, Faculty of Graduate Studies, Faculty of Science, Department of Biology, Victoria, BC V8W 2Y2, Canada. Offers M Sc, PhD. *Faculty:* 23 full-time (6 women), 15 part-time/adjunct (1 woman). *Students:* Average age 26. 92 applicants, 17% accepted, 16 enrolled. In 2005, 7 master's, 3 doctorates awarded. *Median time to degree:* Of those who began their doctoral program in fall 1997, 98% received their degree in 8 years or less. *Degree requirements:* For master's, thesis, seminar; for doctorate, thesis/dissertation, seminar, candidacy exam. *Entrance requirements:* For master's and doctorate, GRE General Test, minimum B+ average in previous 2 years of biology course work. Additional exam requirements/recommendations for international students: Required—TOEFL (minimum score 575 paper-based; 233 computer-based), IELT (minimum score 7). *Application deadline:* For fall admission, 5/31 for domestic students, 12/15 for international students. Applications are processed on a rolling basis. Application fee: $75 ($125 for international students). Electronic applications accepted. Tuition charges are reported in Canadian dollars. *Expenses:* Tuition, nonresident: full-time $4,492 Canadian dollars; part-time $749 Canadian dollars per term. International tuition: $5,346 Canadian dollars full-time. Tuition and fees vary according to course load, campus/location and program. *Financial support:* In 2005–06, 2 fellowships (averaging $12,200 per year), 65 teaching assistantships (averaging $6,500 per year) were awarded; research assistantships, career-related internships or fieldwork, institutionally sponsored loans, scholarships/grants, unspecified assistantships and stipends, awards also available. Financial award application deadline: 2/15. *Faculty research:* Neurobiology of vertebrates and invertebrates, physiology, reproduction and tissue culture of forest trees, evolution and ecology, cell and molecular biology, molecular biology of environmental health. Total annual research expenditures: $4 million. *Unit head:* Dr. Will Hintz, Chair, 250-721-7091, Fax: 250-721-7120, E-mail: biochair@uvic.ca. *Application contact:* Dr. Brad Anholt, Graduate Adviser, 250-721-7094, Fax: 250-721-7120, E-mail: gradsec@uvvm.uvic.ca.

University of Virginia, College and Graduate School of Arts and Sciences, Department of Biology, Charlottesville, VA 22903. Offers MA, MS, PhD. *Faculty:* 32 full-time (7 women), 1 part-time/adjunct (0 women). *Students:* 48 full-time (35 women); includes 2 minority (1 Asian American or Pacific Islander, 1 Hispanic American), 16 international. Average age 27. In 2005, 4 master's, 6 doctorates awarded. *Degree requirements:* For master's, thesis; for doctorate,

Biological and Biomedical Sciences—General

University of Virginia *(continued)*
one foreign language, thesis/dissertation. *Entrance requirements:* For master's and doctorate, GRE General Test, GRE Subject Test. *Application fee:* $40. Electronic applications accepted. *Expenses:* Tuition, state resident: full-time $7,731. Tuition, nonresident: full-time $18,672. Required fees: $1,479. Full-time tuition and fees vary according to degree level and program. *Financial support:* Applicants required to submit FAFSA. *Unit head:* Douglas Taylor, Chair, 434-982-5474, Fax: 434-982-5626, E-mail: drt3b@virginia.edu. *Application contact:* Peter C. Brunjes, Associate Dean for Graduate Programs and Research, 434-924-7184, Fax: 434-924-6737, E-mail: grad-a-s@virginia.edu.

University of Virginia, School of Medicine, Department of Molecular Physiology and Biological Physics, Program in Biological and Physical Sciences, Charlottesville, VA 22903. Offers MS. *Students:* 11 full-time (6 women); includes 1 minority (Asian American or Pacific Islander) Average age 26. In 2005, 21 degrees awarded. *Entrance requirements:* For master's, GRE General Test. Additional exam requirements/recommendations for international students: Required—TOEFL. *Application deadline:* Applications are processed on a rolling basis. *Application fee:* $60. Electronic applications accepted. *Expenses:* Tuition, state resident: full-time $7,731. Tuition, nonresident: full-time $18,672. Required fees: $1,479. Full-time tuition and fees vary according to degree level and program. *Financial support:* Fellowships, research assistantships available. Financial award applicants required to submit FAFSA. *Application contact:* Peter C. Brunjes, Associate Dean for Graduate Programs and Research, 434-924-7184, Fax: 434-924-6737, E-mail: grad-a-s@virginia.edu.

University of Washington, School of Medicine and Graduate School, Graduate Programs in Medicine, Seattle, WA 98195. Offers MOT, MS, MSE, DPT, PhD. Part-time programs available. *Degree requirements:* For doctorate, thesis/dissertation. *Entrance requirements:* For doctorate, GRE. Electronic applications accepted. Expenses: Contact institution.

University of Waterloo, Graduate Studies, Faculty of Science, Department of Biology, Waterloo, ON N2L 3G1, Canada. Offers M Sc, PhD. Part-time programs available. *Faculty:* 33 full-time (9 women), 59 part-time/adjunct (13 women). *Students:* 102 full-time (59 women), 14 part-time (6 women). 86 applicants, 15% accepted, 12 enrolled. In 2005, 21 master's, 3 doctorates awarded. *Degree requirements:* For master's, thesis/dissertation, graduate seminar; for doctorate, thesis/dissertation, graduate seminar, comprehensive exam. *Entrance requirements:* For master's, honors degree, minimum B average; for doctorate, master's degree, minimum B average. Additional exam requirements/recommendations for international students: Required—TOEFL, TWE. *Application deadline:* For fall admission, 8/1 for domestic students. Applications are processed on a rolling basis. Application fee: $75 Canadian dollars. Electronic applications accepted. *Financial support:* Research assistantships, teaching assistantships, career-related internships or fieldwork available. *Faculty research:* Biosystematics, ecology and limnology, molecular and cellular biology, biochemistry, physiology. *Unit head:* Dr. B. R. Glick, Chair, 519-888-4567 Ext. 5208, Fax: 519-746-0614. *Application contact:* J. D. Lehman, Graduate Coordinator, 519-888-4567 Ext. 6392, Fax: 519-746-0614, E-mail: gradbio@sciborg.uwaterloo.ca.

The University of Western Ontario, Faculty of Graduate Studies, Biosciences Division, London, ON N6A 5B8, Canada. Offers M Cl Sc, M Sc, MA, MPT, PhD, MD/M Sc, MD/PhD. Part-time programs available. Postbaccalaureate distance learning degree programs offered.

University of West Florida, College of Arts and Sciences: Sciences, Division of Life and Health Sciences, Department of Biology, Pensacola, FL 32514-5750. Offers biological chemistry (MS); biology (MS); biology education (MST); coastal zone studies (MS); environmental biology (MS). *Accreditation:* NCATE. *Faculty:* 16 full-time (5 women), 2 part-time/adjunct (9 women). *Students:* 12 full-time (10 women), 31 part-time (19 women); includes 4 minority (1 African American, 2 Asian Americans or Pacific Islanders, 1 Hispanic American), 2 international. Average age 28. 17 applicants, 71% accepted, 8 enrolled. In 2005, 4 degrees awarded. *Degree requirements:* For master's, thesis. *Entrance requirements:* For master's, GRE General Test. Additional exam requirements/recommendations for international students: Required—TOEFL (minimum score 550 paper-based; 213 computer-based). *Application deadline:* For fall admission, 6/1 for domestic students, 5/15 for international students; for spring admission, 11/1 for domestic students, 10/1 for international students. Applications are processed on a rolling basis. Application fee: $30. *Expenses:* Tuition, state resident: full-time $5,833; part-time $243 per credit hour. Tuition, nonresident: full-time $21,204; part-time $884 per credit hour. Tuition and fees vary according to campus/location. *Financial support:* In 2005–06, 20 students received support, including 5 research assistantships with partial tuition reimbursements available (averaging $5,000 per year), 11 teaching assistantships with partial tuition reimbursements available (averaging $8,000 per year) Financial award application deadline: 4/15; financial award applicants required to submit FAFSA.

University of West Georgia, Graduate School, College of Arts and Sciences, Department of Biology, Carrollton, GA 30118. Offers MS. Part-time programs available. *Faculty:* 10 full-time (3 women), 1 part-time/adjunct (0 women). *Students:* 8 full-time (5 women), 1 (woman) part-time, 1 international. 17 applicants, 76% accepted, 10 enrolled. In 2005, 4 degrees awarded. *Degree requirements:* For master's, thesis (for some programs), comprehensive exam (for some programs), registration. *Entrance requirements:* For master's, GRE General Test, minimum GPA of 2.5, undergraduate degree in biology. Additional exam requirements/recommendations for international students: Required—TOEFL. *Application deadline:* For fall admission, 8/1 priority date for domestic students, 6/6 priority date for international students; for spring admission, 12/18 for domestic students, 10/3 for international students. Application fee: $20. *Expenses:* Tuition, state resident: full-time $2,196; part-time $122 per semester hour. Tuition, nonresident: full-time $8,784; part-time $488 per semester hour. Required fees: $228; $25 per semester hour. $114 per semester. *Financial support:* In 2005–06, 10 research assistantships with full tuition reimbursements (averaging $6,000 per year), 8 teaching assistantships with full tuition reimbursements (averaging $8,000 per year) were awarded; career-related internships or fieldwork, scholarships/grants, and unspecified assistantships also available. Support available to part-time students. Financial award application deadline: 8/1; financial award applicants required to submit FAFSA. *Faculty research:* Molecular systematics, animal physiology, marine ecology, plant ecology. Total annual research expenditures: $200,000. *Unit head:* Dr. Henry G. Zot, Chair, 678-839-6547, Fax: 678-839-6548, E-mail: hzot@westga.edu. *Application contact:* Dr. Jack O. Jenkins, Dean, Graduate School, 678-839-6419, Fax: 678-839-5949, E-mail: jjenkins@westga.edu.

University of Windsor, Faculty of Graduate Studies and Research, Faculty of Science, Department of Biological Sciences, Windsor, ON N9B 3P4, Canada. Offers M Sc, PhD. Part-time programs available. *Faculty:* 18 full-time (3 women), 5 part-time/adjunct (0 women). *Students:* 46 full-time (20 women), 2 part-time (1 woman). 34 applicants, 38% accepted. In 2005, 12 master's, 4 doctorates awarded. *Degree requirements:* For master's, thesis/dissertation; for doctorate, thesis/dissertation, comprehensive exam. *Entrance requirements:* For master's and doctorate, minimum B average. Additional exam requirements/recommendations for international students: Required—TOEFL (minimum score 560 paper-based; 220 computer-based). *Application deadline:* For fall admission, 7/1 for domestic students. For winter admission, 11/1 for domestic students; for spring admission, 3/1 for domestic students. Applications are processed on a rolling basis. Application fee: $55. Electronic applications accepted. *Financial support:* In 2005–06, 41 teaching assistantships (averaging $8,956 per year) were awarded; research assistantships, Federal Work-Study, scholarships/grants, tuition waivers (full and partial), unspecified assistantships, and bursaries also available. Financial award application deadline: 2/15. *Faculty research:* Great Lakes Institute: aquatic ecotoxicology, regulation and development of the olfactory system, mating system evolution, signal transduction, aquatic ecology. *Unit head:* Dr. William Crosby, Head, 519-253-3000 Ext. 2697, Fax: 519-971-3609, E-mail: bcrosby@uwindsor.ca. *Application contact:* Biological Sciences, 519-253-3000 Ext. 2697, Fax: 519-971-3609, E-mail: biosci@uwindsor.ca.

University of Wisconsin–Eau Claire, College of Arts and Sciences, Program in Biology, Eau Claire, WI 54702-4004. Offers MS. *Faculty:* 17 full-time (6 women). *Students:* 2 full-time (1 woman), 2 part-time (both women). Average age 30. 3 applicants, 67% accepted, 2 enrolled. In 2005, 2 degrees awarded. *Degree requirements:* For master's, thesis, comprehensive exam. *Entrance requirements:* For master's, bachelor's degree in biology or related field, minimum GPA of 3.0. *Application deadline:* For fall admission, 7/1 for domestic students; for spring admission, 12/1 for domestic students. Applications are processed on a rolling basis. Application fee: $45. *Expenses:* Tuition, state resident: full-time $6,223; part-time $345 per credit. Tuition, nonresident: full-time $16,833; part-time $935 per credit. Tuition and fees vary according to program and reciprocity agreements. *Financial support:* In 2005–06, 3 students received support, including teaching assistantships (averaging $7,780 per year); Federal Work-Study also available. Financial award applicants required to submit FAFSA. *Unit head:* Dr. Paula Kleintjes, Chair, 715-836-4166, Fax: 715-836-5089, E-mail: kleintpk@uwec.edu.

University of Wisconsin–La Crosse, Office of University Graduate Studies, College of Science and Health, Department of Biology, La Crosse, WI 54601-3742. Offers aquatic sciences (MS); biology (MS); cellular and molecular biology (MS); clinical microbiology (MS); microbiology (MS); nurse anesthesia (MS); physiology (MS). *Accreditation:* AANA/CANAEP. Part-time programs available. *Faculty:* 18 full-time (5 women), 1 part-time/adjunct (0 women). *Students:* 23 full-time (10 women), 52 part-time (25 women); includes 5 minority (1 American Indian/Alaska Native, 3 Asian Americans or Pacific Islanders, 1 Hispanic American), 3 international. Average age 26. 61 applicants, 44% accepted, 23 enrolled. In 2005, 14 degrees awarded. *Degree requirements:* For master's, thesis, comprehensive exam, registration. *Entrance requirements:* For master's, GRE General Test, minimum GPA of 2.85. Additional exam requirements/recommendations for international students: Required—TOEFL (minimum score 550 paper-based; 213 computer-based). *Application deadline:* For fall admission, 3/1 for domestic students. Applications are processed on a rolling basis. Application fee: $45. Electronic applications accepted. *Expenses:* Tuition, state resident: part-time $354 per credit. Tuition, nonresident: part-time $943 per credit. Tuition and fees vary according to course load, program and reciprocity agreements. *Financial support:* In 2005–06, 10 students received support, including 4 research assistantships with partial tuition reimbursements available (averaging $10,000 per year), 10 teaching assistantships with partial tuition reimbursements available (averaging $9,600 per year); career-related internships or fieldwork, Federal Work-Study, health care benefits, unspecified assistantships, and grant-funded positions also available. Support available to part-time students. Financial award application deadline: 3/15; financial award applicants required to submit FAFSA. *Faculty research:* Cell and molecular biology, physiology, environmental sciences, mycology, biomedical general. Total annual research expenditures: $700,000. *Unit head:* Dr. Tom Volk, Program Director, 608-785-6972, Fax: 608-785-6959, E-mail: volk.thom@uwlax.edu. *Application contact:* Kathryn Kiefer, Associate Director of Admissions, 608-785-8939, E-mail: admissions@uwlax.edu.

University of Wisconsin–Madison, Medical School and Graduate School, Graduate Programs in Medicine, Madison, WI 53706-1380. Offers biomolecular chemistry (MS, PhD); cancer biology (PhD); genetics and medical genetics (MS, PhD), including genetics (PhD), medical genetics (MS); medical physics (MS, PhD), including health physics (MS), medical physics; microbiology (PhD); molecular and cellular pharmacology (PhD); pathology and laboratory medicine (PhD); physiology (PhD); population health (MPH, MS, PhD). Part-time programs available. Postbaccalaureate distance learning degree programs offered (minimal on-campus study). Terminal master's awarded for partial completion of doctoral program. Application fee: $45. Electronic applications accepted. *Expenses:* Contact institution. *Financial support:* Fellowships with full tuition reimbursements, research assistantships with full tuition reimbursements, teaching assistantships with full tuition reimbursements, scholarships/grants, traineeships, and tuition waivers (full) available. *Unit head:* Dr. Paul M. DeLuca, Associate Dean of Research and Graduate Studies, 608-265-0524, Fax: 608-265-0522, E-mail: pmdeluca@facstaff.wisc.edu.

University of Wisconsin–Madison, Medical School, Medical Scientist Training Program, Madison, WI 53705-2221. Offers MD/PhD. *Accreditation:* LCME/AMA. *Faculty:* 380 full-time (80 women). *Students:* 58 full-time (25 women); includes 11 minority (5 African Americans, 1 American Indian/Alaska Native, 4 Asian Americans or Pacific Islanders, 1 Hispanic American). 161 applicants, 17% accepted, 10 enrolled. *Median time to degree:* Of those who began their doctoral program in fall 1997, 100% received their degree in 8 years or less. *Application deadline:* For fall admission, 12/16 for domestic students. Applications are processed on a rolling basis. Application fee: $45. Electronic applications accepted. *Financial support:* In 2005–06, fellowships with full tuition reimbursements (averaging $20,000 per year), research assistantships with full tuition reimbursements (averaging $20,000 per year) were awarded; traineeships and health care benefits also available. *Unit head:* Dr. Deane Mosher, Director, 608-262-1576, Fax: 608-263-4969, E-mail: dfmosher@wisc.edu. *Application contact:* Paul Cook, Program Administrator, 608-262-6321, Fax: 608-262-4226, E-mail: pscook@wisc.edu.

University of Wisconsin–Madison, School of Veterinary Medicine, Department of Animal Health and Biomedical Sciences, Madison, WI 53706-1380. Offers comparative biosciences (MS, PhD), including anatomy, biochemistry, cellular and molecular biology, environmental toxicology, neurosciences, pharmacology, physiology. Part-time programs available. Terminal master's awarded for partial completion of doctoral program. *Degree requirements:* For master's, thesis or alternative, registration; for doctorate, thesis/dissertation, comprehensive exam, registration. *Entrance requirements:* For master's and doctorate, GRE. Additional exam requirements/recommendations for international students: Required—TOEFL. Electronic applications accepted. *Faculty research:* Infectious disease, neuroscience, genomics physiology, pharmacology and toxicology.

University of Wisconsin–Milwaukee, Graduate School, College of Health Sciences, Interdepartmental Program in Health Sciences, Milwaukee, WI 53201-0413. Offers PhD. *Students:* 4 full-time (2 women), 2 part-time (both women); includes 1 minority (Asian American or Pacific Islander), 1 international. 6 applicants, 33% accepted, 1 enrolled. *Expenses:* Tuition, area resident: Part-time $716 per credit. Tuition, state resident: part-time $776 per credit. Tuition, nonresident: part-time $1,614 per credit. Required fees: $229 per term. Tuition and fees vary according to course load and program. *Unit head:* Cynthia Hasbrook, Representative, 414-229-5677, E-mail: cah@uwm.edu.

University of Wisconsin–Milwaukee, Graduate School, College of Letters and Sciences, Department of Biological Sciences, Milwaukee, WI 53201-0413. Offers MS, PhD. *Faculty:* 31 full-time (7 women). *Students:* 50 full-time (22 women), 26 part-time (17 women); includes 6 minority (3 African Americans, 3 Hispanic Americans), 17 international. 89 applicants, 28% accepted, 18 enrolled. In 2005, 13 master's, 5 doctorates awarded. *Degree requirements:* For master's, thesis; for doctorate, thesis/dissertation, 1 foreign language or data analysis proficiency. *Entrance requirements:* For master's and doctorate, GRE General Test. *Application deadline:* For fall admission, 3/1 for domestic students. Applications are processed on a rolling basis. Application fee: $45 ($75 for international students). *Expenses:* Tuition, area resident: Part-time $716 per credit. Tuition, state resident: part-time $776 per credit. Tuition, nonresident: part-time $1,614 per credit. Required fees: $229 per term. Tuition and fees vary according to course load and program. *Financial support:* In 2005–06, 2 research assistantships, 66 teaching assistantships were awarded; fellowships, career-related internships or fieldwork and unspecified assistantships also available. Support available to part-time students. Financial award application deadline: 4/15. *Unit head:* Steve Forst, Representative, 414-229-6373, Fax: 414-229-3926, E-mail: sforst@uwm.edu.

University of Wisconsin–Oshkosh, The School of Graduate Studies, College of Letters and Science, Department of Biology and Microbiology, Oshkosh, WI 54901. Offers biology (MS), including botany, microbiology, zoology. *Degree requirements:* For master's, thesis, comprehensive exam, registration. *Entrance requirements:* For master's, GRE General Test, minimum GPA of

Biological and Biomedical Sciences—General

3.0, BS in biology. Additional exam requirements/recommendations for international students: Required—TOEFL (minimum score 550 paper-based; 213 computer-based). Electronic applications accepted.

Utah State University, School of Graduate Studies, College of Science, Department of Biology, Logan, UT 84322. Offers biology (MS, PhD); ecology (MS, PhD). Part-time programs available. *Faculty:* 39 full-time (7 women). *Students:* 139 full-time (73 women), 16 part-time (9 women); includes 4 minority (2 African Americans, 1 Asian American or Pacific Islander, 1 Hispanic American), 42 international. Average age 26. 44 applicants, 61% accepted, 21 enrolled. In 2005, 8 master's, 3 doctorates awarded. *Degree requirements:* For master's and doctorate, thesis/dissertation. *Entrance requirements:* For master's and doctorate, GRE General Test, minimum GPA of 3.0. Additional exam requirements/recommendations for international students: Required—TOEFL (minimum score 575 paper-based). *Application deadline:* For fall admission, 6/15 for domestic students; for spring admission, 10/15 for domestic students. Applications are processed on a rolling basis. Application fee: $50 ($60 for international students). *Financial support:* In 2005–06, 3 fellowships with partial tuition reimbursements (averaging $20,000 per year), research assistantships with partial tuition reimbursements (averaging $16,000 per year), 44 teaching assistantships with partial tuition reimbursements (averaging $11,275 per year) were awarded; career-related internships or fieldwork, Federal Work-Study, and institutionally sponsored loans also available. Support available to part-time students. Financial award application deadline: 2/15. *Faculty research:* Plant, insect, microbial, and animal biology. *Unit head:* Dr. Jon Y. Takemoto, Head, 435-797-1909, E-mail: jon@biology.usu.edu. *Application contact:* Nancy Kay Harrison, Coordinator of Graduate Studies, 435-797-1770, Fax: 435-797-1575, E-mail: nancykay@biology.usu.edu.

Vanderbilt University, Graduate School, Department of Biological Sciences, Nashville, TN 37240-1001. Offers MS, PhD, MD/PhD. *Faculty:* 38 full-time (5 women). *Students:* 49 full-time (20 women); includes 3 minority (all African Americans), 20 international. 161 applicants, 6% accepted, 6 enrolled. In 2005, 4 doctorates awarded. *Degree requirements:* For doctorate, thesis/dissertation, final and qualifying exams. *Entrance requirements:* For master's and doctorate, GRE General Test. *Application deadline:* For fall admission, 1/15 for domestic students, 1/15 for international students. Application fee: $0. Electronic applications accepted. *Expenses:* Tuition: Part-time $1,283 per semester hour. Required fees: $2,202; $1,101 per semester. One-time fee: $30. Tuition and fees vary according to course load, program and student level. *Financial support:* Fellowships with full and partial tuition reimbursements, research assistantships with full tuition reimbursements, teaching assistantships with full tuition reimbursements, Federal Work-Study, institutionally sponsored loans, traineeships, and health care benefits available. Financial award application deadline: 1/15. *Unit head:* Dr. Charles K. Singleton, Chair, 615-322-2008, Fax: 615-343-6707. *Application contact:* Doug McMahon, Director of Graduate Studies, 615-322-2008, Fax: 615-343-6707, E-mail: douglas.g.mcmahon@vanderbilt.edu.

See Close-Up on page 317.

Vanderbilt University, School of Medicine, Interdisciplinary Graduate Program in Biomedical and Biological Sciences, Nashville, TN 37240-1001. Offers PhD. First-year students in biomedical sciences enter the program. Degrees are awarded through participating departments of biochemistry, biological sciences, cancer biology, cell and developmental biology, human genetics, microbiology, and immunology, molecular physiology and biophysics, neuroscience, cellular and molecular pathology, and pharmacology. *Degree requirements:* For doctorate, thesis/dissertation, final and qualifying exams, dissertation defense, comprehensive exam, registration. *Entrance requirements:* For doctorate, GRE General Test. Electronic applications accepted. *Expenses:* Tuition: Part-time $1,283 per semester hour. Required fees: $2,202; $1,101 per semester. One-time fee: $30. Tuition and fees vary according to course load, program and student level. *Faculty research:* Genetics; immunology; neurobiology; cell and developmental biology; signal transduction.

Vanderbilt University, School of Medicine and Graduate School, Medical Scientist Training Program, Nashville, TN 37240-1001. Offers MD/PhD. Electronic applications accepted. Expenses: Contact institution. One-time fee: $30. Tuition and fees vary according to course load, program and student level.

Villanova University, Graduate School of Liberal Arts and Sciences, Department of Biology, Villanova, PA 19085-1699. Offers MA, MS. Part-time and evening/weekend programs available. *Faculty:* 7 full-time (2 women), 1 part-time/adjunct (0 women). *Students:* 11 full-time (4 women), 34 part-time (21 women); includes 3 minority (1 African American, 1 Asian American or Pacific Islander, 1 Hispanic American), 3 international. Average age 26. 41 applicants, 63% accepted. In 2005, 7 degrees awarded. *Degree requirements:* For master's, thesis (for some programs), comprehensive exam (for some programs). *Entrance requirements:* For master's, GRE General Test, GRE Subject Test, minimum GPA of 3.0. Additional exam requirements/recommendations for international students: Required—TOEFL. *Application deadline:* For fall admission, 8/1 priority date for domestic students, 8/1 priority date for international students; for spring admission, 12/1 for domestic students, 12/1 for international students. Application fee: $50. Electronic applications accepted. *Expenses:* Contact institution. Tuition and fees vary according to program and student level. *Financial support:* Research assistantships with tuition reimbursements, teaching assistantships with tuition reimbursements, Federal Work-Study and scholarships/grants available. Support available to part-time students. Financial award applicants required to submit FAFSA. *Unit head:* Dr. Russel Gardner, Chair, 610-519-4830.

See Close-Up on page 319.

Virginia Commonwealth University, Graduate School, College of Humanities and Sciences, Department of Biology, Doctoral Program in Integrative Life Sciences, Richmond, VA 23284-9005. Offers PhD. *Students:* 9 full-time (5 women). 35 applicants, 26% accepted. *Entrance requirements:* For doctorate, GRE, minimum GPA of 3.0 in last 60 credits of undergraduate work or in graduate degree, 3 Letters of recommendation. Additional exam requirements/recommendations for international students: Required—TOEFL. Application fee: $50. *Expenses:* Tuition, state resident: full-time $3,185; part-time $405 per credit. Tuition, nonresident: full-time $7,952; part-time $940 per credit. Required fees: $751 per semester hour. Tuition and fees vary according to course load and program. *Unit head:* Dr. Leonard A. Smock, Chair, 804-828-1562, Fax: 804-828-0503, E-mail: lasmock@vcu.edu. *Application contact:* 804-828-6916, Fax: 804-828-6949.

Virginia Commonwealth University, Medical College of Virginia-Professional Programs, School of Medicine Graduate Programs, Richmond, VA 23284-9005. Offers MPH, MS, PhD, CBHS, MD/MPH, MD/PhD. Part-time programs available. *Students:* 735 applicants, 51% accepted. In 2005, 51 master's, 29 doctorates, 41 other advanced degrees awarded. Terminal master's awarded for partial completion of doctoral program. *Degree requirements:* For doctorate, thesis/dissertation, comprehensive oral and written exams. *Entrance requirements:* For doctorate, GRE General Test, MCAT. Application fee: $50. *Expenses:* Tuition, state resident: full-time $6,268; part-time $405 per credit. Tuition, nonresident: full-time $15,904; part-time $940 per credit. Required fees: $751 per semester hour. Tuition and fees vary according to course load and program. *Financial support:* Fellowships, research assistantships, teaching assistantships, career-related internships or fieldwork, Federal Work-Study, institutionally sponsored loans, and tuition waivers (full) available. *Application contact:* Dr. Jan F. Chlebowski, Associate Dean for Graduate Education, 804-828-1023, Fax: 804-828-1473, E-mail: jfchlebo@vcu.edu.

See Close-Up on page 321.

Virginia Polytechnic Institute and State University, Graduate School, College of Science, Department of Biological Sciences, Blacksburg, VA 24061. Offers botany (MS, PhD); ecology and evolutionary biology (MS, PhD); genetics and developmental biology (MS, PhD); microbiology (MS, PhD); zoology (MS, PhD). *Faculty:* 38 full-time (9 women). *Students:* 70 full-time (30 women), 4 part-time (1 woman); includes 6 minority (2 African Americans, 1 American Indian/Alaska Native, 2 Asian Americans or Pacific Islanders, 1 Hispanic American), 12 international. Average age 27. 81 applicants, 22% accepted, 14 enrolled. In 2005, 9 master's, 8 doctorates awarded. *Entrance requirements:* For master's and doctorate, GRE General Test. Additional exam requirements/recommendations for international students: Required—TOEFL (minimum score 550 paper-based; 213 computer-based). *Application deadline:* Applications are processed on a rolling basis. Application fee: $45. Electronic applications accepted. *Expenses:* Tuition, state resident: full-time $6,558; part-time $364 per credit. Tuition, nonresident: full-time $11,296; part-time $628 per credit. Required fees: $1,419; $468 per credit. $234 per term. *Financial support:* In 2005–06, 28 research assistantships with full tuition reimbursements (averaging $16,689 per year), 37 teaching assistantships with full tuition reimbursements (averaging $13,902 per year) were awarded; career-related internships or fieldwork, Federal Work-Study, scholarships/grants, and unspecified assistantships also available. *Faculty research:* Freshwater ecology, cell cycle regulation, behavioral ecology, motor proteins. *Unit head:* Dr. Bob Jones, Chairman, 540-231-9514, Fax: 540-231-9307, E-mail: rhjones@vt.edu. *Application contact:* Sue Rasmussen, Graduate Secretary, 540-231-8929, Fax: 540-231-9307, E-mail: sueras@vt.edu.

Virginia State University, School of Graduate Studies, Research, and Outreach, School of Engineering, Science and Technology, Department of Life Sciences, Petersburg, VA 23806-0001. Offers biology (MS). *Degree requirements:* For master's, one foreign language, thesis. *Entrance requirements:* For master's, GRE General Test. *Faculty research:* Schwann cell cultures, selection of apios as an alternative crop, systematic botany, flowers of three species of wild ginger.

Wagner College, Division of Graduate Studies, Department of Biological Sciences, Staten Island, NY 10301-4495. Offers advanced physician assistant studies (MS); microbiology (MS). Part-time and evening/weekend programs available. *Students:* 19 full-time (12 women), 12 part-time (9 women); includes 11 minority (5 African Americans, 5 Asian Americans or Pacific Islanders, 1 Hispanic American). 20 applicants, 95% accepted, 16 enrolled. In 2005, 18 degrees awarded. *Degree requirements:* For master's, comprehensive exam or thesis. *Entrance requirements:* For master's, minimum GPA of 2.5, proficiency in statistics, undergraduate major in science. Additional exam requirements/recommendations for international students: Required—TOEFL. *Application deadline:* For fall admission, 8/1 priority date for domestic students, 6/30 priority date for international students; for spring admission, 12/10 for domestic students, 11/15 for international students. Applications are processed on a rolling basis. Application fee: $50 ($85 for international students). *Expenses:* Tuition: Full-time $14,760; part-time $820 per credit. *Financial support:* Traineeships, tuition waivers (partial), unspecified assistantships, and alumni fellowships available. *Unit head:* Dr. Donald Stearns, Chair, 718-390-3197, E-mail: dstearns@wagner.edu. *Application contact:* Kristina Muller, Admissions Office—Senior Associate Director, 718-390-3411, Fax: 718-390-3105, E-mail: kmuller@wagner.edu.

Wake Forest University, Graduate School, Department of Biology, Winston-Salem, NC 27109. Offers MS, PhD. Part-time programs available. *Faculty:* 21 full-time (5 women), 3 part-time/adjunct (1 woman). *Students:* 38 full-time (25 women); includes 2 minority (1 African American, 1 Asian American or Pacific Islander), 7 international. Average age 30. 59 applicants, 24% accepted, 13 enrolled. In 2005, 7 master's, 3 doctorates awarded. *Median time to degree:* Of those who began their doctoral program in fall 1997, 100% received their degree in 8 years or less. *Degree requirements:* For master's, one foreign language, thesis, registration; for doctorate, 2 foreign languages, thesis/dissertation, comprehensive exam, registration. *Entrance requirements:* For master's and doctorate, GRE General Test. Additional exam requirements/recommendations for international students: Required—TOEFL (minimum score 213 computer-based). *Application deadline:* For fall admission, 1/15 for domestic students, 1/15 for international students. Application fee: $45. Electronic applications accepted. *Financial support:* In 2005–06, 33 students received support, including 1 fellowship with full tuition reimbursement available (averaging $18,850 per year), 7 research assistantships with full tuition reimbursements available (averaging $15,350 per year), 25 teaching assistantships with full tuition reimbursements available (averaging $15,350 per year); scholarships/grants and tuition waivers (full and partial) also available. Support available to part-time students. Financial award application deadline: 1/15; financial award applicants required to submit FAFSA. *Faculty research:* Cell biology, ecology, parasitology, immunology. *Unit head:* Dr. Brian Tague, Director, 336-758-5774, Fax: 336-758-6008, E-mail: taguebw@wfu.edu.

Wake Forest University, School of Medicine and Graduate School, Graduate Programs in Medicine, Winston-Salem, NC 27109. Offers MS, PhD, MD/PhD. *Degree requirements:* For master's and doctorate, thesis/dissertation. *Entrance requirements:* For master's and doctorate, GRE General Test. Additional exam requirements/recommendations for international students: Required—TOEFL. Electronic applications accepted. Expenses: Contact institution. *Faculty research:* Atherosclerosis, cardiovascular physiology, pharmacology, neuroanatomy, endocrinology.

Walla Walla College, Graduate School, Department of Biological Sciences, College Place, WA 99324-1198. Offers biology (MS). Part-time programs available. *Faculty:* 5 full-time (1 woman), 1 part-time/adjunct (0 women). *Students:* 6 full-time (3 women), 3 part-time (2 women), 1 international. Average age 28. 8 applicants, 63% accepted, 5 enrolled. In 2005, 1 degree awarded. *Degree requirements:* For master's, thesis, marine station experience. *Entrance requirements:* For master's, GRE General Test, GRE Subject Test, minimum GPA of 2.75. *Application deadline:* For fall admission, 4/1 for domestic students. Applications are processed on a rolling basis. Application fee: $50. Electronic applications accepted. *Expenses:* Tuition: Full-time $19,071; part-time $489 per credit. *Financial support:* In 2005–06, 9 students received support, including 9 teaching assistantships with full tuition reimbursements available (averaging $10,704 per year); Federal Work-Study also available. Financial award application deadline: 4/1; financial award applicants required to submit FAFSA. *Faculty research:* Marine biology, plant development, neurobiology, animal physiology, behavior. *Unit head:* Dr. Scott H. Ligman, Chair, 509-527-2603, E-mail: ligmsc@wwc.edu. *Application contact:* Dr. Joe G. Galusha, Dean of Graduate Studies, 509-527-2421, Fax: 509-527-2237, E-mail: galujo@wwc.edu.

Washington State University, Graduate School, College of Sciences, School of Biological Sciences, Program in Biology, Pullman, WA 99164. Offers MS. *Faculty:* 33. *Students:* 2 full-time (both women). 13 applicants, 8% accepted, 1 enrolled. *Degree requirements:* For master's, thesis, registration. *Entrance requirements:* For master's, GRE, minimum GPA of 3.0, 3 letters of recommendation. Additional exam requirements/recommendations for international students: Required—TOEFL. *Application deadline:* For fall and spring admission, 1/15. Application fee: $35. *Expenses:* Tuition, state resident: full-time $6,295; part-time $336 per credit. Tuition, nonresident: full-time $15,949; part-time $819 per credit. Required fees: $933. Part-time tuition and fees vary according to campus/location and program. *Financial support:* In 2005–06, 4 students received support, including 1 research assistantship with tuition reimbursement available (averaging $13,635 per year), 1 teaching assistantship with tuition reimbursement available (averaging $13,635 per year) *Faculty research:* Inter-&intra-cellular signaling in plant reproduction, biodiversity. *Application contact:* Graduate Coordinator, 509-335-1666, Fax: 509-335-3184, E-mail: sbs@wsu.edu.

Washington State University Tri-Cities, Graduate Programs, Program in Biology, Richland, WA 99352-1671. Offers MS. *Faculty:* 20. *Students:* 1 (woman) full-time, 1 (woman) part-time. 2 applicants, 50% accepted, 0 enrolled. *Degree requirements:* For master's, special project, thesis optional. *Entrance requirements:* For master's, GRE, minimum GPA of 3.0, 3 letters of recommendation. Additional exam requirements/recommendations for international students: Required—TOEFL. *Application deadline:* For fall admission, 1/15 priority date for domestic students, 3/1 priority date for international students; for spring admission, 9/15 priority date for domestic students, 7/1 priority date for international students. Application fee: $35. *Expenses:* Tuition, state resident: full-time $6,295; part-time $336 per credit. Tuition, nonresident: full-time $15,949; part-time $819 per credit. Required fees: $429. Full-time tuition and fees vary

Biological and Biomedical Sciences—General

Washington State University Tri-Cities (continued)

according to campus/location and program. Part-time tuition and fees vary according to course load and program. *Financial support:* In 2005–06, 2 students received support. *Application contact:* 509-372-7250, Fax: 509-372-7100.

Washington University in St. Louis, Graduate School of Arts and Sciences, Division of Biology and Biomedical Sciences, St. Louis, MO 63130-4899. Offers biochemistry (PhD); chemical biology (PhD); computational biology (PhD); developmental biology (PhD); evolution, ecology and population biology (PhD), including ecology, environmental biology, evolutionary biology, genetics; immunology (PhD); molecular biophysics (PhD); molecular cell biology (PhD); molecular genetics (PhD); molecular microbiology and microbial pathogenesis (PhD); neurosciences (PhD); plant biology (PhD). *Degree requirements:* For doctorate, thesis/dissertation. *Entrance requirements:* For doctorate, GRE General Test, GRE Subject Test. Electronic applications accepted.

Wayne State University, Graduate School, College of Liberal Arts and Sciences, Department of Biological Sciences, Detroit, MI 48202. Offers biological sciences (MA, MS, PhD); molecular biotechnology (MS). *Faculty:* 13 full-time (2 women). *Students:* 67 full-time (40 women), 22 part-time (14 women); includes 17 minority (6 African Americans, 11 Asian Americans or Pacific Islanders), 45 international. Average age 29. 86 applicants, 37% accepted, 17 enrolled. In 2005, 10 master's, 4 doctorates awarded. Terminal master's awarded for partial completion of doctoral program. *Degree requirements:* For master's, thesis (for some programs); for doctorate, thesis/dissertation. *Entrance requirements:* For master's, GRE General Test, minimum GPA of 3.0; for doctorate, GRE General Test, GRE Subject Test, minimum GPA of 3.2. Additional exam requirements/recommendations for international students: Required—TOEFL; Recommended—TWE (minimum score 6). *Application deadline:* For fall admission, 7/1 for domestic students, 6/1 for international students. Applications are processed on a rolling basis. Application fee: $30 ($50 for international students). Electronic applications accepted. *Expenses:* Tuition, state resident: part-time $338 per credit hour. Tuition, nonresident: part-time $746 per credit hour. Required fees: $24 per credit hour. Full-time tuition and fees vary according to program. *Financial support:* In 2005–06, 7 research assistantships with tuition reimbursements (averaging $16,304 per year), 45 teaching assistantships with tuition reimbursements (averaging $14,997 per year) were awarded; fellowships with tuition reimbursements, Federal Work-Study and institutionally sponsored loans also available. *Faculty research:* Cell and developmental biology, neurobiology, molecular biology and biotechnology, evolutionary biology, ecology. Total annual research expenditures: $2.3 million. *Unit head:* James D. Tucker, Chair, 313-577-2783, Fax: 313-577-6891, E-mail: ao1754@wayne.edu. *Application contact:* John Lopes, Graduate Director, 313-993-7816, Fax: 313-577-6891, E-mail: jlopes@sun.science.wayne.edu.

Wesleyan University, Graduate Programs, Department of Biology, Middletown, CT 06459-0260. Offers cell biology (PhD); comparative physiology (PhD); developmental biology (PhD); genetics (PhD); neurophysiology (PhD); population biology (PhD). *Faculty:* 12 full-time (3 women). *Students:* 29 full-time (15 women), 10 international. Average age 26. 131 applicants. In 2005, 2 doctorates awarded. *Degree requirements:* For doctorate, one foreign language, thesis/dissertation. *Entrance requirements:* For doctorate, GRE Subject Test. *Application deadline:* For fall admission, 2/15 for domestic students. Applications are processed on a rolling basis. Application fee: $0. *Expenses:* Tuition: Full-time $24,732. One-time fee: $20 full-time. *Financial support:* Research assistantships, teaching assistantships, stipends available. *Faculty research:* Microbial population genetics, genetic basis of evolutionary adaptation, genetic regulation of differentiation and pattern formation in *drosophila*. *Unit head:* Dr. Michael Weir, Chairman, 860-685-2402, E-mail: mweir@wesleyan.edu. *Application contact:* Marjorie Fitzgibbons, Information Contact, 860-685-2157, E-mail: mfitzgibbons@wesleyan.edu.

See Close-Up on page 323.

West Chester University of Pennsylvania, Graduate Studies, College of Arts and Sciences, Department of Biology, West Chester, PA 19383. Offers MS. Part-time and evening/weekend programs available. *Students:* 4 full-time (2 women), 22 part-time (12 women); includes 1 minority (Asian American or Pacific Islander) Average age 32. 21 applicants, 90% accepted, 7 enrolled. In 2005, 1 degree awarded. *Degree requirements:* For master's, thesis, comprehensive exam. *Entrance requirements:* For master's, GRE General Test, GRE Subject Test. *Application deadline:* For fall admission, 4/15 for domestic students; for spring admission, 10/15 for domestic students. Applications are processed on a rolling basis. Application fee: $35. *Expenses:* Tuition, state resident: full-time $5,988; part-time $327 per credit. Tuition, nonresident: full-time $9,422; part-time $523 per credit. Required fees: $54 per semester. *Financial support:* In 2005–06, 2 research assistantships with full tuition reimbursements (averaging $5,000 per year) were awarded; unspecified assistantships also available. Support available to part-time students. Financial award application deadline: 2/15; financial award applicants required to submit FAFSA. *Faculty research:* Cell physiology of insect ovarian follicles, field inventory of reptiles and amphibians. *Unit head:* Dr. Jack Waber, Chair, 610-436-2926, E-mail: jwaber@wcupa.edu. *Application contact:* Dr. Giovanni Casotti, Graduate Coordinator, 610-436-2856, E-mail: giovanni@bio.wcupa.edu.

Western Carolina University, Graduate School, College of Arts and Sciences, Department of Biology, Cullowhee, NC 28723. Offers biology (MAT, MS); comprehensive education-biology (MA Ed). Part-time and evening/weekend programs available. *Degree requirements:* For master's, thesis, comprehensive exam. *Entrance requirements:* For master's, GRE General Test. Additional exam requirements/recommendations for international students: Required—TOEFL (minimum score 550 paper-based; 213 computer-based).

Western Connecticut State University, Division of Graduate Studies, School of Arts and Sciences, Department of Biological and Environmental Sciences, Danbury, CT 06810-6885. Offers MA. Part-time and evening/weekend programs available. *Degree requirements:* For master's, comprehensive exam or thesis. *Entrance requirements:* For master's, minimum GPA of 2.5.

Western Illinois University, School of Graduate Studies, College of Arts and Sciences, Department of Biological Sciences, Macomb, IL 61455-1390. Offers biological sciences (MS); zoo and aquarium studies (Certificate). Part-time programs available. *Students:* 43 full-time (27 women), 20 part-time (16 women); includes 6 minority (2 African Americans, 1 American Indian/Alaska Native, 1 Asian American or Pacific Islander, 2 Hispanic Americans), 3 international. Average age 26. 38 applicants, 79% accepted. In 2005, 14 master's, 6 other advanced degrees awarded. *Degree requirements:* For master's, thesis or alternative. *Entrance requirements:* Additional exam requirements/recommendations for international students: Required—TOEFL (minimum score 550 paper-based; 213 computer-based). *Application deadline:* Applications are processed on a rolling basis. Application fee: $30. Electronic applications accepted. *Expenses:* Tuition, state resident: full-time $3,599; part-time $200 per semester hour. Tuition, nonresident: full-time $7,198; part-time $400 per semester hour. Required fees: $890; $49 per semester hour. Tuition and fees vary according to campus/location. *Financial support:* In 2005–06, 21 students received support, including 7 research assistantships with full tuition reimbursements available (averaging $6,288 per year), 14 teaching assistantships (averaging $6,288 per year) Financial award applicants required to submit FAFSA. *Unit head:* Dr. Richard V. Anderson, Chairperson, 309-298-2408. *Application contact:* Dr. Barbara Baily, Director of Graduate Studies/Associate Provost, 309-298-1806, Fax: 309-298-2345, E-mail: grad-office@wiu.edu.

Western Kentucky University, Graduate Studies, Ogden College of Science and Engineering, Department of Biology, Bowling Green, KY 42101-3576. Offers biology (MA Ed, MS). *Faculty:* 7 full-time (2 women). *Students:* 16 full-time (10 women), 8 part-time (5 women); includes 1 minority (African American), 10 international. Average age 28. 10 applicants, 80% accepted, 3 enrolled. In 2005, 13 degrees awarded. *Degree requirements:* For master's, research tool, thesis optional. *Entrance requirements:* For master's, GRE General Test, minimum GPA of 2.75. Additional exam requirements/recommendations for international students:

Required—TOEFL (minimum score 555 paper-based; 213 computer-based). *Application deadline:* For fall admission, 7/1 priority date for domestic students, 5/15 priority date for international students; for spring admission, 11/1 for domestic students, 9/15 for international students. Applications are processed on a rolling basis. Application fee: $35. *Expenses:* Tuition, state resident: full-time $5,816; part-time $299 per credit hour. Tuition, nonresident: full-time $6,356; part-time $326 per credit hour. *Financial support:* In 2005–06, 16 students received support, including 3 research assistantships with partial tuition reimbursements available (averaging $9,000 per year), 13 teaching assistantships with partial tuition reimbursements available (averaging $9,000 per year); Federal Work-Study, institutionally sponsored loans, tuition waivers (partial), unspecified assistantships, and service awards also available. Support available to part-time students. Financial award application deadline: 4/1; financial award applicants required to submit FAFSA. *Faculty research:* Phytoremediation, culturing of salt water organisms, PCR-based standards, biological monitoring (water) bioremediation, genetic diversity. Total annual research expenditures: $81,235. *Unit head:* Dr. Richard G Bowker, Head, 270-745-3696, Fax: 270-745-6856, E-mail: richard.bowker@wku.edu.

Western Michigan University, Graduate College, College of Arts and Sciences, Department of Biological Sciences, Kalamazoo, MI 49008-5202. Offers biological sciences (MS, PhD); molecular biotechnology (MS). *Degree requirements:* For master's and doctorate, thesis/dissertation, oral exam. *Entrance requirements:* For master's and doctorate, GRE General Test.

Western Washington University, Graduate School, College of Sciences and Technology, Department of Biology, Bellingham, WA 98225-5996. Offers MS. Part-time programs available. *Faculty:* 18. *Students:* 8 full-time (7 women), 14 part-time (11 women); includes 2 minority (1 African American, 1 Asian American or Pacific Islander), 1 international. 18 applicants, 61% accepted, 6 enrolled. In 2005, 8 degrees awarded. *Degree requirements:* For master's, thesis. *Entrance requirements:* For master's, GRE General Test, GRE Subject Test (biology), minimum GPA of 3.0 in last 60 semester hours or last 90 quarter hours. Additional exam requirements/recommendations for international students: Required—TOEFL (minimum score 567 paper-based; 227 computer-based). *Application deadline:* For fall admission, 2/1 for domestic students. For winter admission, 10/1 for domestic students; for spring admission, 2/1 for domestic students. Applications are processed on a rolling basis. Application fee: $50. *Expenses:* Tuition, state resident: Part-time $188 per credit. Tuition, state resident: full-time $5,628; part-time $539 per credit. Tuition, nonresident: full-time $16,176. Required fees: $624. *Financial support:* In 2005–06, 2 research assistantships with partial tuition reimbursements (averaging $10,170 per year), 12 teaching assistantships with partial tuition reimbursements (averaging $10,629 per year) were awarded; Federal Work-Study, institutionally sponsored loans, scholarships/grants, tuition waivers (partial), and unspecified assistantships also available. Support available to part-time students. Financial award application deadline: 2/15; financial award applicants required to submit FAFSA. *Faculty research:* Cell biology, evolutionary developmental biology, organismal biology, ecology and evolutionary biology, marine biology. *Unit head:* Dr. Joanne Otto, Chair, 360-650-4044. *Application contact:* Dr. Merrill Peterson, Graduate Program Advisor, 360-650-3636.

West Texas A&M University, College of Agriculture, Nursing, and Natural Sciences, Department of Life, Earth, and Environmental Sciences, Program in Biology, Canyon, TX 79016-0001. Offers MS. Part-time programs available. *Degree requirements:* For master's, thesis optional. *Entrance requirements:* For master's, GRE General Test. Additional exam requirements/recommendations for international students: Required—TOEFL (minimum score 550 paper-based). Electronic applications accepted. *Faculty research:* Aeroallegen concentration, scorpions, kangaroo mice, seed anatomy with light and scanning electron microscope.

West Virginia University, Eberly College of Arts and Sciences, Department of Biology, Morgantown, WV 26506. Offers cell and molecular biology (MS, PhD); environmental and evolutionary biology (MS, PhD); integrative organismal biology (PhD); integrative organismal biology (MS). *Faculty:* 18 full-time (3 women), 4 part-time/adjunct (all women). *Students:* 29 full-time (20 women), 5 part-time (3 women); includes 1 minority (Asian American or Pacific Islander), 9 international. Average age 26. 50 applicants, 10% accepted. In 2005, 1 master's, 2 doctorates awarded. Terminal master's awarded for partial completion of doctoral program. *Degree requirements:* For master's, thesis, final exam; for doctorate, thesis/dissertation, preliminary and final exams. *Entrance requirements:* For master's, GRE General Test, GRE Subject Test, minimum GPA of 3.0; for doctorate, GRE General Test, minimum GPA of 3.0. Additional exam requirements/recommendations for international students: Required—TOEFL. *Application deadline:* For fall admission, 4/1 for domestic students; for spring admission, 10/1 for domestic students. Applications are processed on a rolling basis. Application fee: $45. *Expenses:* Tuition, state resident: full-time $4,582; part-time $258 per credit hour. Tuition, nonresident: full-time $13,820; part-time $741 per credit hour. *Financial support:* In 2005–06, 4 research assistantships, 22 teaching assistantships were awarded; Federal Work-Study and institutionally sponsored loans also available. Financial award application deadline: 4/1; financial award applicants required to submit FAFSA. *Faculty research:* Environmental biology, genetic engineering, developmental biology, global change, biodiversity. *Unit head:* Dr. Jonathan Cumming, Chair, 304-293-5201 Ext. 2508, Fax: 304-293-6363, E-mail: jonathan.cumming@mail.wvu.edu. *Application contact:* Dr. William T. Peterjohn, Director of Graduate Studies, 304-293-5201 Ext. 2510, Fax: 304-293-6363, E-mail: william.peterjohn@mail.wvu.edu.

See Close-Up on page 325.

West Virginia University, School of Medicine, Graduate Programs at the Health Science Center, Morgantown, WV 26506. Offers MS, PhD, MD/PhD. Part-time and evening/weekend programs available. Postbaccalaureate distance learning degree programs offered (minimal on-campus study). *Students:* 115 full-time (63 women), 45 part-time (28 women). Average age 29. In 2005, 47 master's, 6 doctorates awarded. Application fee: $45. *Expenses:* Contact institution. *Financial support:* In 2005–06, 42 research assistantships, 41 teaching assistantships were awarded; fellowships, career-related internships or fieldwork, Federal Work-Study, institutionally sponsored loans, tuition waivers (full and partial), and graduate administrative assistantships also available. Financial award applicants required to submit FAFSA. *Unit head:* Dr. Thomas Saba, Interim Associate Dean/Graduate Coordinator, 304-293-4011, Fax: 304-293-7038, E-mail: ccraig@hsc.wvu.edu. *Application contact:* Claire Noel, Graduate Adviser and Assistant Director, 304-293-7116, Fax: 304-293-7038, E-mail: claire.noel@hsc.wvu.edu.

See Close-Up on page 327.

Wichita State University, Graduate School, Fairmount College of Liberal Arts and Sciences, Department of Biological Sciences, Wichita, KS 67260. Offers MS. Part-time programs available. *Degree requirements:* For master's, variable foreign language requirement, comprehensive exam. *Entrance requirements:* For master's, GRE Subject Test. Additional exam requirements/recommendations for international students: Required—TOEFL. Electronic applications accepted. *Faculty research:* Molecular biology, environmental science, reproductive endocrinology, cancer, plant ecology.

See Close-Up on page 329.

William Paterson University of New Jersey, College of Science and Health, Department of Biology, General Biology Program, Wayne, NJ 07470-8420. Offers general biology (MA); limnology and terrestrial ecology (MA); molecular biology (MA); physiology (MA). Part-time and evening/weekend programs available. *Students:* 2 full-time (1 woman), 13 part-time (9 women); includes 4 Hispanic Americans. In 2005, 4 degrees awarded. *Degree requirements:* For master's, independent study or thesis. *Entrance requirements:* For master's, GRE General Test, minimum GPA of 2.75. *Application deadline:* Applications are processed on a rolling basis. Application fee: $50. Electronic applications accepted. *Expenses:* Tuition, state resident: full-time $476. Tuition, nonresident: full-time $717. *Financial support:* Research assistantships, career-related internships or fieldwork and unspecified assistantships available. Financial award application deadline: 4/1; financial award applicants required to submit FAFSA.

Application contact: Danielle Liautaud, Assistant Director, 973-720-3579, Fax: 973-720-2035, E-mail: liautaudd@wpunj.edu.

See Close-Up on page 331.

Winthrop University, College of Arts and Sciences, Department of Biology, Rock Hill, SC 29733. Offers MS. Part-time programs available. *Degree requirements:* For master's, thesis optional. *Entrance requirements:* For master's, GRE General Test, minimum GPA of 3.0. Electronic applications accepted.

Worcester Polytechnic Institute, Graduate Studies and Enrollment, Department of Biology and Biotechnology, Worcester, MA 01609-2280. Offers biology (MS); biotechnology (MS, PhD). Part-time and evening/weekend programs available. *Faculty:* 12 full-time (5 women). *Students:* 17 full-time (10 women), 2 part-time (1 woman); includes 1 minority (Asian American or Pacific Islander), 2 international. 48 applicants, 31% accepted, 8 enrolled. In 2005, 2 degrees awarded. *Degree requirements:* For master's, thesis; for doctorate, thesis/dissertation, qualifying exam, comprehensive exam. *Entrance requirements:* For master's and doctorate, GRE General Test, 3 letters of recommendation. Additional exam requirements/recommendations for international students: Required—TOEFL (minimum score 550 paper-based; 213 computer-based). *Application deadline:* For fall admission, 1/15 for domestic students; for spring admission, 10/15 priority date for domestic students. Applications are processed on a rolling basis. Application fee: $70. Electronic applications accepted. *Expenses:* Tuition: Part-time $997 per credit hour. *Financial support:* In 2005–06, 19 students received support, including 2 fellowships with full tuition reimbursements available, 3 research assistantships with full and partial tuition reimbursements available, 11 teaching assistantships with full and partial tuition reimbursements available; career-related internships or fieldwork, institutionally sponsored loans, scholarships/grants, and unspecified assistantships also available. Financial award application deadline: 1/15. *Faculty research:* Molecular biology, plant physiology, immunology and neurobiology, biomedication, genetic engineering. Total annual research expenditures: $544,457. *Unit head:* Dr. Eric Overstrom, Head, 508-831-5538, Fax: 508-831-5936, E-mail: ewo@wpi.edu. *Application contact:* Dr. Samuel M Politz, Graduate Coordinator, 508-831-5028, Fax: 508-831-5936, E-mail: spolitz@wpi.edu.

See Close-Up on page 333.

Wright State University, School of Graduate Studies, College of Science and Mathematics, Department of Biological Sciences, Dayton, OH 45435. Offers biological sciences (MS); environmental sciences (MS). *Degree requirements:* For master's, thesis optional. *Entrance requirements:* Additional exam requirements/recommendations for international students: Required—TOEFL.

Wright State University, School of Graduate Studies, College of Science and Mathematics and School of Medicine, Program in Biomedical Sciences, Dayton, OH 45435. Offers PhD. *Degree requirements:* For doctorate, thesis/dissertation. *Entrance requirements:* Additional exam requirements/recommendations for international students: Required—TOEFL.

Announcement: This interdisciplinary doctoral program in biomedical sciences is designed to give the in-depth training required to solve complex research problems of contemporary life sciences. Opportunities and support are available in molecular biology, biochemistry, cell biology, physiology, chemistry, structural biology, applied biomedical computation, immunology, neuroscience, applied/predictive toxicology, and epidemiology.

See Close-Up on page 335.

Yale University, School of Medicine and Graduate School of Arts and Sciences, Combined Program in Biological and Biomedical Sciences (BBS), New Haven, CT 06520. Offers PhD, MD/PhD. *Faculty:* 280 full-time. *Students:* 77 full-time. 825 applicants, 26% accepted. *Degree requirements:* For doctorate, thesis/dissertation. *Entrance requirements:* For doctorate, GRE General Test. Additional exam requirements/recommendations for international students: Required—TOEFL. *Application deadline:* For fall admission, 12/8 for domestic students, 12/8 for international students. Application fee: $85. Electronic applications accepted. *Expenses:* Contact institution. *Financial support:* In 2005–06, fellowships with full tuition reimbursements (averaging $26,000 per year), research assistantships with full tuition reimbursements (averaging $26,000 per year), teaching assistantships with full tuition reimbursements (averaging $26,000 per year) were awarded. *Unit head:* , Dr. Lynn Cooley, Director, 203-785-5067, E-mail: bbs@yale.edu. *Application contact:* Dr. John Alvaro, Administrative Director, 203-785-3735, Fax: 203-785-3734, E-mail: bbs@yale.edu.

See Close-Up on page 337.

York University, Faculty of Graduate Studies, Faculty of Pure and Applied Science, Program in Biology, Toronto, ON M3J 1P3, Canada. Offers M Sc, PhD. Part-time and evening/weekend programs available. *Faculty:* 52 full-time (11 women), 7 part-time/adjunct (3 women). *Students:* 94 full-time (58 women), 11 part-time (6 women). 112 applicants, 30% accepted, 34 enrolled. In 2005, 18 master's, 7 doctorates awarded. *Degree requirements:* For master's, thesis or alternative, registration; for doctorate, thesis/dissertation, preliminary exam, comprehensive exam, registration. *Application deadline:* Applications are processed on a rolling basis. Application fee: $80. Electronic applications accepted. *Expenses:* Tuition, state resident: part-time $798 per term. Tuition, nonresident: full-time $3,190. International tuition: $7,515 full-time. Required fees: $217. Tuition and fees vary according to program. *Financial support:* In 2005–06, fellowships (averaging $12,137 per year), research assistantships (averaging $3,426 per year), teaching assistantships (averaging $9,616 per year) were awarded; career-related internships or fieldwork and fee bursaries also available. *Unit head:* Joel Shore, Director, 416-736-2100 Ext. 22342.

Youngstown State University, Graduate School, College of Arts and Sciences, Department of Biological Sciences, Youngstown, OH 44555-0001. Offers MS. Part-time programs available. *Degree requirements:* For master's, thesis, oral review, comprehensive exam. *Entrance requirements:* For master's, GRE General Test, minimum GPA of 2.7. Additional exam requirements/recommendations for international students: Required—TOEFL. *Faculty research:* Cell biology, neurophysiology, molecular biology, neurobiology, gene regulation.

Cross-Discipline Announcements

Florida Institute of Technology, Graduate Programs, College of Engineering, Department of Marine and Environmental Systems, Program in Oceanography, Melbourne, FL 32901-6975.

Program awards the MS and PhD in biological oceanography, chemical oceanography, geological oceanography, and physical oceanography and the MS in coastal zone management. Part of the Department of Marine and Environmental Systems in the College of Engineering. Visit the Web site at http://www.fit.edu/AcadRes/dmes/.

Massachusetts Institute of Technology, School of Engineering, Biological Engineering Division, Cambridge, MA 02139-4307.

Program provides opportunities for study and research at the interface of biology and engineering leading to specialization in bioengineering and applied biosciences. The areas include understanding how biological systems operate, especially when perturbed by genetic, chemical, or materials interventions or subjected to pathogens or toxins, and designing innovative biology-based technologies in diagnostics, therapeutics, materials, and devices for application to human health and diseases, as well as other societal problems and opportunities.

Oregon Health & Science University, School of Medicine, Graduate Programs in Medicine, Department of Behavioral Neuroscience, Portland, OR 97239-3098.

The department offers an interdisciplinary graduate program consisting of basic science training in behavioral neuroscience with specialization in such areas as physiological psychology, behavioral and molecular genetics, behavioral pharmacology, neuroendocrinology, and biological bases of addiction. Students with degrees in biological sciences, psychology, or neuroscience are encouraged to apply. Visit http://www.ohsu.edu/behneuro/.

Princeton University, Graduate School, Department of Molecular Biology, Princeton, NJ 08544-1019.

Graduate studies in the Department of Molecular Biology at Princeton emphasize training in research and encourage students to apply molecular, biochemical, structural, genetic, and computational approaches to biological problems. Faculty members and students pursue research in a wide variety of areas of biochemistry and cell biology, biological dynamics and computational biology, biomedicine and societal issues, biophysics and structural biology, developmental biology, genetics and genomics, microbiology, neurobiology, oncology, and virology.

University of Minnesota, Twin Cities Campus, Medical School and Graduate School, Graduate Programs in Medicine and Institute of Technology, Program in Microbial Engineering, Minneapolis, MN 55455-0213.

The microbial engineering Master of Science program, coordinated by the BioTechnology Institute, involves faculty members and facilities from 12 departments and 6 institutes in cross-disciplinary training and research. The microbial engineering degree fulfills the minor requirement for graduates who choose to pursue a PhD in related fields.

University of Pittsburgh, School of Arts and Sciences, Department of Biological Sciences, Program in Molecular, Cellular, and Developmental Biology, Pittsburgh, PA 15260.

In the department's graduate program in molecular, cellular, and developmental biology, faculty research groups use prokaryotic and eukaryotic experimental systems and molecular and genetic approaches to understand the structure and function of genes and proteins, macromolecular interactions, cell-specific gene regulation, regulation of cell proliferation, and embryogenesis.

University of Southern California, School of Pharmacy and Graduate School, Graduate Programs in Pharmacy, Graduate Program in Pharmaceutical Sciences, Los Angeles, CA 90089.

The graduate program, staffed by 12 full-time faculty members, emphasizes research in drug delivery and targeting, utilizing medicinal chemistry, computational chemistry, cell biology, molecular pharmacology and biology, pharmaceutics, pharmacokinetics, and pharmacodynamics. Teaching and research assistantships are awarded to highly qualified applicants in chemistry, biology, or other sciences.

University of Wisconsin–Madison, Graduate School, Training Program in Biotechnology, Madison, WI 53706-1380.

The University of Wisconsin–Madison offers a predoctoral training program in biotechnology. Trainees receive a PhD in their major field, for example, biology and biomedical sciences, while receiving extensive cross-disciplinary training through the minor degree. Trainees participate in industrial internships and a weekly student seminar series with other program participants. These experiences reinforce the cross-disciplinary nature of the program. Students choose a major and minor professor from a list of more than 130 faculty members in 40 different departments conducting research related to biotechnology.

Washington University in St. Louis, School of Medicine, Graduate Programs in Medicine, Program in Occupational Therapy, St. Louis, MO 63130-4899.

Professional entry MS and OTD (offers 3-2 program with 26 affiliated colleges). Excellent application of biological and social sciences to rewarding professional career. Prepares for practice in occupational therapy, both in rehabilitation and community settings. Program addresses the health, work, physical, cognitive, and social factors that affect the lives of persons with disabilities.

ALBERT EINSTEIN COLLEGE OF MEDICINE OF YESHIVA UNIVERSITY

Sue Golding Graduate Division

Program of Study

The Einstein Graduate Programs in Biomedical Sciences offer interdisciplinary graduate studies centrally administered through the Sue Golding Graduate Division, with Ph.D. degrees granted through the ten basic science departments (Anatomy and Structural Biology, Biochemistry, Cell Biology, Developmental and Molecular Biology, Microbiology and Immunology, Molecular Genetics, Molecular Pharmacology, Neuroscience, Pathology, and Physiology and Biophysics), as well as a new Ph.D. in clinical investigation. Students develop an individualized program of graduate study utilizing the total faculty and resources of the Albert Einstein College of Medicine. There is no master's degree program, although students receive an M.S. degree while completing Ph.D. course and qualifying exam requirements.

Prospective students apply directly to the Graduate Division, not to a specific department. Once admitted, the student spends the first year taking foundation graduate courses and carrying out three different laboratory rotations. This allows the student to experience the breadth of scientific inquiry and to identify an exciting area of research before committing to a particular laboratory. Once a thesis laboratory is chosen, the student becomes a member of the appropriate academic department. By the end of two years of study, most students will have passed the qualifying examination and made significant progress on a thesis project. Students also take part in frequent informal and formal research seminars, journal clubs, and meetings of special-interest research groups, which help develop critical learning skills. In "Works-in-Progress" sessions, students present their own work to obtain feedback and develop presentation skills. There is a rich schedule of national and international invited speakers, offering students outstanding opportunities to hear about a broad range of cutting-edge research projects in the biomedical sciences.

Research Facilities

Students in the graduate program work in laboratories and offices in the Leo Forchheimer Medical Sciences Building, the Ullmann Research Center, the Chanin Institute for Cancer Research, the Rose F. Kennedy Center, the Rachel Golding Pavilion, the Belfer Educational Center for Health Sciences, and the Caroline and Joseph S. Gruss Magnetic Resonance Research Center. A new building, the Price Center for Genetic and Translational Medicine, is currently under construction and will house forty new laboratories and broaden the areas of investigation on the campus. Laboratories are equipped with state-of-the-art instrumentation required for modern research in biological sciences. Specialized facilities of the College include a hybridoma facility; a mass spectrometer; oligonucleotide synthesis and DNA sequencing facilities; flow cytometry facilities; a transgenic mouse and stem cell facility; an image analysis facility; an analytical ultrastructure center; a cytogenetics facility; a genetic and physical mapping facility; a cDNA microarray facility; a proteomics facility; the new MMRC building, which provides access to unparalleled imagery of the interior of the human body and creates the most detailed images ever seen of the anatomy and physiology of living organisms; an animal institute; a bioinstrumentation center; a new Innovation Laboratory for noninvasive visualization of cells; and a library that contains an extensive collection of current periodicals.

Financial Aid

The graduate program is supported by training grants, individual investigator-initiated research grants, and institutional support. The stipend for students entering in fall 2006 was $26,000 per year. In addition, tuition and fees, including individual health insurance, are provided.

Cost of Study

Academic costs are waived for all graduate students. In addition, all graduate students receive an annual stipend.

Living and Housing Costs

The residence complex of the Albert Einstein College of Medicine provides unfurnished apartments located in three 27-story towers directly across the street from the medical school. The average monthly rental is $336 for a shared one-bedroom apartment. A cafeteria in the Mazer Building is open to all students.

Student Group

Approximately 370 students are enrolled in the Sue Golding Graduate Division at the Albert Einstein College of Medicine, including 65 M.D./Ph.D. students in the Ph.D. phase and 735 students in the medical school. An active Graduate Student Council elects representatives to all school committees, plans community service and social events, and provides a cohesive student forum. Multitudes of student-run clubs provide additional social, cultural, and volunteer opportunities.

Location

The Albert Einstein College of Medicine is located in a pleasant residential community at the northeastern edge of the Bronx. There is easy access to Manhattan as well as to beaches and parks by public transportation or car. New York City offers unlimited opportunities for residents to pursue a wide range of scientific and cultural interests.

The College

The Albert Einstein College of Medicine is a privately endowed, coeducational, and nondenominational constituent college of Yeshiva University. Since the enrollment of its first class in 1955, it has gained international stature as a center of research and teaching in the medical sciences. The Sue Golding Graduate Division was established in 1957 to provide advanced study and research training leading to the Ph.D. The Graduate Division granted its first Ph.D. in 1961 and, since then, more than 1,000 Ph.D. degrees have been conferred. The student body is drawn from undergraduate schools throughout the United States and from many other countries. The formal Ph.D. program is supplemented by postdoctoral research training, so that graduate students have ample opportunity for interaction with biomedical scientists at all career levels.

Applying

The prerequisite for admission is a bachelor's degree or evidence of an equivalent education from a college or university of recognized standing. Undergraduate preparation should include course work in chemistry, organic chemistry, calculus, physics, and biology. The GRE General Test is required. International students whose native language is not English must also submit their scores on the Test of English as a Foreign Language (TOEFL).

The Einstein Graduate Programs in the Biomedical Sciences utilize an online application available at http://www.aecom.yu.edu/phd. The admissions decision is made after careful consideration of transcripts, GRE scores, letters of recommendation, personal statement describing specific research/career goals, and an on-campus interview with faculty members.

Applications must be completed by January 15 to guarantee consideration for fall enrollment.

Correspondence and Information

Graduate Admissions
Graduate Division of Biomedical Sciences
Albert Einstein College of Medicine
1300 Morris Park Avenue (Belfer 201)
Bronx, New York 10461-1602
Phone: 718-430-2345
Fax: 718-430-8655
E-mail: phd@aecom.yu.edu
Web site: http://www.aecom.yu.edu/phd

Albert Einstein College of Medicine of Yeshiva University

GRADUATE PROGRAM DEPARTMENTS AND ASSOCIATED FACULTY

Anatomy and Structural Biology
J. S. Condeelis, Ph.D., Co-Chairman: molecular basis of chemotaxis and invasion. R. H. Singer, Ph.D., Co-Chairman: cellular molecular biology; spatial organization of specific nucleic acid sequences. D. Cox, Ph.D.: macrophage phagocytosis and motility. A. M. Cuervo, M.D./Ph.D.: lysosomes and aging. D. Frenz, Ph.D.: development of inner ear. R. E. Hirsch, Ph.D.: structural biology of hemoglobins. U. T. Meier, Ph.D.: nucleocytoplasmic transport. B. Ovryn, Ph.D.: biophotonics applied to elucidating the cellular basis of human disease. B. H. Satir, Ph.D.: signal transduction events in secretion and membrane fusion. P. Satir, Ph.D., Distinguished University Professor: cell biology of microtubule-based motility. J. Segall, Ph.D.: molecular and genetic analysis of amoeboid chemotaxis. E. Snapp, Ph.D.: organization and dynamics of endoplasmic reticulum proteins and membranes. A. W. Wolkoff, M.D.: receptor-mediated endocytosis in hepatocytes. M. Symons, Ph.D.: signal transduction through Rho family proteins.

Biochemistry
V. L. Schramm, Ph.D., Chairman: enzymes; transition states and inhibitors. E. Abel-Santos, Ph.D.: hydroput screening for antibiotics. S. Almo, Ph.D.: X-ray crystallography. R. Angeletti, Ph.D.: functional proteomics; mass spectrometry. J. Blanchard, Ph.D.: mechanisms of enzymes. M. Brenowitz, Ph.D.: protein–nucleic acid interactions. A. R. Bresnick, Ph.D.: regulation of the actin cytoskeleton. R. Briehl, M.D.: biophysical chemistry of proteins. R. Callender, Ph.D.: enzymes; protein folding; spectroscopy. M. Charron, Ph.D.: glucose transporters and glucagon receptors. D. Cohen, M.D., Ph.D.: cholesterol; phospholipids; lipid transfer proteins. S. Englard, Ph.D.: metabolic regulation. A. Fiser, Ph.D.: bioinformatics and computational biology; structural genomics; proteomics. M. Girvin, Ph.D.: membrane protein NMR. V. B. Hatcher, Ph.D.: bacterial–endothelial cell interactions. D. Lawrence, Ph.D.: bioorganic chemistry. T. Leyh, Ph.D.: enzymology of sulfate activation. I. Listowsky, Ph.D.: protein-ligand interactions. M. Makman, M.D., Ph.D.: neurotransmitter and hormone action. S. Roderick, Ph.D.: X-ray crystallography. S. Schwartz, Ph.D.: biophysical systems; quantum and statistical mechanics. H. Steinman, Ph.D.: bacterial pathogenesis and physiology. I. Willis, Ph.D.: eukaryotic gene transcription.

Cell Biology
A. I. Skoultchi, Ph.D., Chairman: gene expression; chromatin structure; terminal differentiation; cancer biology; chromatin; differentiation and development; knock-out mice. B. K. Birshtein, Ph.D.: immunoglobulin gene rearrangement and expression. E. Bouhassira, Ph.D.: human embryonic stem cells, epigenetics, gene expression, and silencing. W. Edelmann, Ph.D.: mouse models to study the role of DNA mismatch repair genes in DNA repair, cancer, and meiosis. D. Fyodorov, Ph.D.: chromosome assembly and dynamics; biochemistry of ATP-dependent chromatin remodeling factors; *Drosophila melanogaster*. M. Kielian, Ph.D.: virus-membrane fusion; virus assembly; novel antiviral therapies; membrane protein structure and function. R. N. Kitsis, M.D.: fundamental mechanisms of apoptosis and their application to heart disease and cancer. S. G. Nathenson, M.D.: structural and molecular basis for the regulation of T cells through costimulatory molecules. C. Query, M.D., Ph.D.: mechanism and modulation of spliceosome function. M. D. Scharff, M.D.: mutation and switching of immunoglobulin genes and use of monoclonal antibodies in infections. P. Scherer, Ph.D.: role of the adipocyte in energy homeostasis and inflammation. C. L. Schildkraut, Ph.D.: replication of viral DNA, oncogenes, and immunoglobulin genes; role of nuclear localization in gene expression. D. A. Shafritz, M.D.: liver-specific gene expression; hepatic stem cells; liver cell transplantation and gene therapy. P. Stanley, Ph.D.: functions of mammalian glycans in development, cancer, and notch signal transduction. J. R. Warner, Ph.D.: regulation of the biosynthesis and assembly of a molecular machine, the ribosome. H. Ye, Ph.D.: oncogenes in normal and malignant hematopoiesis.

Developmental and Molecular Biology
E. R. Stanley, Ph.D., Chairman: growth factors and signaling in development and disease. R. H. Angeletti, Ph.D.: proteomics of liver cancer and infectious disease; mass spectrometry. Y. Chen, Ph.D.: morphogen signaling in vertebrate development. T. Evans, Ph.D.: molecular regulation of embryonic blood and heart development. T. Graf, Ph.D.: transcription factor regulation of hematopoietic stem cell. R. Kuliawat, Ph.D.: protein targeting to secretory granules and melanosomes. U. Maitra, Ph.D.: translation initiation and ribosome biogenesis in eukaryotic cells. A. Melnick, M.D.: cancer epigenomics: the mechanism of action of transcriptional oncoproteins and transcription therapy. T. Michaeli, Ph.D.: signal transduction pathways regulating cellular proliferation, differentiation, and physiology. H. Nguyen, Ph.D.: genetic pathways regulating muscle development in *Drosophila* and vertebrates. J. W. Pollard, Ph.D.: growth factor and steroid hormone regulation of mammalian development; macrophages and breast cancer. R. Ruggieri, Ph.D.: stress-activated protein kinases in DNA damage response, cell motility, and differentiation. D. Shields, Ph.D.: protein sorting in the secretory pathway. N. Sibinga, M.D.: molecular mechanisms underlying vascular obstructive disease. D. Wilson, Ph.D.: herpes simplex virus envelope assembly; protein trafficking and membrane biology. L. Zhu, M.D., Ph.D.: the retinoblastoma protein family of growth suppressors and cell-cycle control.

Microbiology and Immunology
M. S. Horwitz, M.D., Chairman: adenovirus immunoregulatory proteins and molecular pathogenesis. K. Auborn, Ph.D.: chemoprevention of lupus, papillomavirus infections, and hormone-enhanced cancers. J. Brojatsch, Ph.D.: mechanism and regulation of anthrax-toxin-mediated cell killing. R. D. Burk, M.D.: pathogenesis of human papillomavirus/cervix cancer and human tumor-suppressor gene functions. A. Casadevall, M.D., Ph.D.: molecular genetics and immune responses to *C. neoformans*. J. Chan, M.D.: host defense and pathogenesis in tuberculosis. E. Dadachova, Ph.D.: radioimmunotherapy and targeted radiation therapy of cancer and infectious diseases. L. D'Adamio, M.D., Ph.D.: regulation of programmed cell death by presenilins and amyloid precursor protein. T. DiLorenzo, Ph.D.: beta-cell antigens in type 1 diabetes. T. Dragic, Ph.D.: mechanisms of viral entry, using the human immunodeficiency virus and the hepatitis C virus as models. M. Feldmesser, M.D.: pathogenesis and host response to *Aspergillus fumigatus*. D. Fidock, Ph.D.: *Plasmodium falciparum* malaria: mechanisms of drug resistance and pathogenesis, drug discovery. H. Goldstein, M.D.: mouse models of the human immune system to study gene therapy and HIV infection. J. A. Hardin, M.D.: regulatory mechanisms for immune responses to chromatin antigens in patients with systemic lupus erythematosus. W. R. Jacobs Jr., Ph.D., Howard Hughes Medical Institute: molecular genetics of pathogenic mycobacteria. K. Kim, M.D.: pathogenesis of *Toxoplasma gondii* infection; malaria chemotherapy. S. G. Nathenson, M.D.: structural basis for antigen presentation to T-cell receptors. E. Peeva, M.D.: breakdown of tolerance; immunomodulatory effects of sex hormones; SLE. L. Pirofski, M.D.: human immunity to encapsulated organisms. S. A. Porcelli, M.D.: mechanisms of antigen processing and presentation. V. R. Prasad, Ph.D.: mechanisms of drug resistance in HIV; structure-function studies of HIV-1 RT and human telomerase. C. Putterman, M.D.: tolerance and autoimmunity; anti-DNA antibodies; systemic lupus erythematosus. C. Rogler, Ph.D.: hepatocellular carcinoma and hepatitis B viruses. B. Steinberg, Ph.D.: interactions between human papillomaviruses and their target epithelial cells, with major focus on laryngeal papillomas and altered signal transduction. K. Tracey, M.D.: pathogenic basis of systemic inflammation; identification of therapeutic molecular targets and pathways.

Molecular Genetics
J. Lenz, Ph.D., Interim Chair: molecular genetics of viral oncogenesis. N. Baker, Ph.D.: development and growth of *Drosophila*. N. Barzilai, M.D.: searching for longevity genes in humans. G. Childs, Ph.D.: functional genomics of murine development and disease. A. Cvekl, Ph.D.: transcriptional regulation in mammalian ocular development and disease. S. Emmons, Ph.D.: genetic control of development and behavior in *Caenorhabditis*. J. Greally, M.D., Ph.D.: epigenetic regulation of the mammalian genome. G. V. Kalpana, Ph.D.: role of chromatin remodeling proteins in HIV-1 replication and cancer: development of therapeutic interventions. S. Mani, M.D.: molecular basis for the variation in antitumor drug response. B. Morrow, Ph.D.: human developmental disorders. H. M. Nitowsky, M.D.: human biochemical and molecular genetics. L. Ozelius, Ph.D.: genetics of movement disorders. G. Prelich, Ph.D.: transcriptional regulatory mechanisms. J. Roy-Chowdhury, M.B.B.S., M.R.C.P.: liver-directed gene therapy and hepatocyte transplantation. N. Roy-Chowdhury, Ph.D.: molecular mechanisms of inherited disorders of bilirubin glucuronidation. N. Schreiber-Agus, Ph.D.:

transcriptional regulation and cancer pathogenesis. T. Weber, M.D.: hereditary predisposition to colorectal cancer.

Molecular Pharmacology
S. B. Horwitz, Ph.D., Co-Chair: mechanisms of action and resistance to the microtubule stabilizing agents, Taxol, the epothilones, and discodermolide. C. S. Rubin, Ph.D., Co-Chair: cyclic AMP and protein phosphorylation in signal transduction. J. M. Backer, M.D.: signal transduction by phosphoinositide 3'-kinases. C. F. Brewer, Ph.D.: protein-carbohydrate interactions in signal transduction. N. Carrasco, M.D.: molecular characterization and clinical applications of the sodium/iodide symporter (NIS) and other transporters. C.-W. Chow: Ph.D.: signal transduction and gene regulation by transcription factor NFAT. L. D. Fricker, Ph.D.: neuropeptides and neuropeptide processing enzymes. I. D. Goldman, M.D.: cloning and characterization of membrane transporters; mechanisms of action, resistance, and transport of new-generation antifolates. R. Gorlick, M.D.: mechanisms of drug resistance in leukemia and osteosarcoma. G. S. Kroog, M.D.: bombesin receptor signal transduction. M. P. Lisanti, M.D., Ph.D.: role of caveolae organelles and caveolin proteins in signaling, oncogenic transformation, and muscular dystrophy. T. V. McDonald, M.D.: ion channel regulation in disease. G. A. Orr, Ph.D.: proteomics of the cytoskeleton and associated signaling scaffolds. L. Rossetti, M.D.: biochemical and molecular mechanisms of nutrient sensing. C. A. Stein, M.D., Ph.D.: mechanisms of action of antisense oligonucleotides.

Neuroscience
D. S. Faber, Ph.D., University Chair: synaptic transmission and neural plasticity. C. Abrams, M.D., Ph.D.: mechanisms of connexin diseases, peripheral nerve diseases. M. H. Akabas, M.D., Ph.D.: ion channel structure-function. J. C. Arezzo, Ph.D., physiologic models of neurotoxic and neurogenetic diseases. T. Bargiello, Ph.D.: molecular neurogenetics. M. V. L. Bennett, D.Phil.: excitatory amino acid receptors, gap junctions. L. Brown, Ph.D.: basal ganglia. F. Bukauskas, Ph.D.: biophysics of gap junctions. R. Carroll, Ph.D.: synaptic plasticity; glutamate receptor regulation. P. Castillo, Ph.D.: synaptic plasticity in the mammalian brain. P. Davies, Ph.D.: neurobiology of Alzheimer's disease. K. Dobrenis, Ph.D.: treatment of diseases affecting the CNS; microglial cell biology. A. M. Etgen, Ph.D.: hormone-neurotransmitter interactions; reproductive neuroendocrinology. L. Fricker, Ph.D.: neuropeptides and neuropeptide-processing enzymes. A. Galanopoulou, M.D., Ph.D.: molecular aspect on neonatal seizures. G. Haddad, M.D.: Neuronal susceptibility to hypoxic injury, vertebrate and invertebrate. D. H. Hall, Ph.D.: *C. elegans* ultrastructure and synapse formation. J. Hebert, Ph.D.: genetics of forebrain development. E. L. Hertzberg, Ph.D.: gap-junction communication. N. Hiroi, Ph.D.: genetic basis of addiction and motor disorders. Z. Kaprielian, Ph.D.: pattern formation and axon guidance in developing spinal cord and brain. K. Khodakhah, Ph.D.: calcium regulation of excitability. J. Larocca, Ph.D.: regulation of myelin biogenesis. M. Mehler, M.D.: CNS development. S. Moshe, M.D.: developmental epilepsy. S. Nawy, Ph.D.: synaptic transmission and modulation; retinal function. A. Peinado, Ph.D.: development of cerebral cortex. A. Pereda, M.D., Ph.D.: synaptic plasticity; electrical synapses. D. Pettit, Ph.D.: synaptic imaging and physiology. C. S. Raine, Ph.D., D.Sc.: CNS; autoimmunity; multiple sclerosis; demyelination; neuroimmunology. D. M. Rosenbaum, M.D.: molecular mechanisms of ischemic neuronal injury neuroprotection. C. Schroeder, Ph.D.: cognitive neuroscience; attention-multisensory integration. D. Spray, Ph.D.: gap junctions; single channels. M. Steinschneider, M.D., Ph.D.: auditory cortex, speech music. E. Sussman, Ph.D.: auditory perception. I. Vathy, Ph.D.: prenatal drug exposure, CNS development. L. Velisek, M.D.: Ph.D.: prenatal brain damage, brain metabolism, and seizures. J. Veliskova, M.D., Ph.D.: epilepsy and cell death; hormones in epilepsy. V. K. Verselis, Ph.D.: ion channel structure-function. S. U. Walkley, Ph.D.: cellular neurobiology and experimental neuropathology. R. S. Zukin, Ph.D.: excitatory amino acid receptors.

Pathology
M. Prystowsky, M.D., Ph.D., Chairman: cytokine regulation of inflammation. T. Belbin, Ph.D.: molecular tumor classification; epigenomic status of human malignancies. A. Bergman, Ph.D.: evolutionary systems biology; gene networks. J. Berman, Ph.D.: endothelial cell molecular biology; atherosclerosis; neuroimmunology. C. Brosnan, Ph.D.: inflammation in the central nervous system. B. Cannella, Ph.D.: inflammation of the central nervous system. L. Cannizzaro, Ph.D.: mapping of cancer gene loci. P. Davies, Ph.D.: neurobiology of dementia. S. Factor, M.D.: cardiovascular pathology. R. Hazan, Ph.D.: breast cancer; metastasis; cell adhesion molecules; growth factor receptors. Z. Kaprielian, Ph.D.: developmental neuroscience; axon guidance. S. Lee, M.D.: neuroimmunology; AIDS; neuropathology. J. Locker, M.D., Ph.D.: transcription factors; developmental gene regulation. F. Macian, M.D., Ph.D.: molecular basis of immune tolerance. P. Novikoff, Ph.D.: liver carcinogenesis; receptor-mediated endocytosis. A. Orlofsky, Ph.D.: host defense; innate immunity. C. Raine, Ph.D.: neuroimmunology; multiple sclerosis; CNS autoimmunity. M. Sadofsky, M.D., Ph.D.: mechanism and regulation of V(D)J recombination. L. Santambrogio, M.D., Ph.D.: immunobiology of dendritic cells and microglial cells. B. Shafit-Zagardo, Ph.D.: molecular neurobiology; developmental neuropathology. H. B. Tanowitz, M.D.: Chagas' disease; cardiomyopathy. B. Terman, Ph.D.: tumor angiogenesis; growth factors; signal transduction. K. Weidenheim, M.D.: neuropathology; neuropathology of autism. L. Weiss, M.D., M.P.H.: *Toxoplasma gondii*; microsporidiosis. M. Wittner, M.D., Ph.D.: Chagas' disease and pathogenesis of cardiomyopathy; microsporidiosis.

Physiology
D. L. Rousseau, Ph.D., Chairman: elucidation of the molecular basis of heme protein function by resonance Raman spectroscopy; molecular mechanisms of nitric oxide synthase and terminal oxidase; time-resolved studies of protein folding. S. A. Acharya, Ph.D.: semisynthesis of proteins through reverse proteolysis; hemoglobin chemistry, structure, and function; biophysical chemistry of polymerization of sickle-cell hemoglobin; blood substitutes. P. Aisen, M.D.: biochemistry of iron metabolism; spectroscopy of iron-bearing proteins; mechanisms of iron binding and release; interactions of iron-bearing proteins with their specific cell-membrane receptors. M. H. Akabas, M.D./Ph.D.: ion channels in synaptic transmission; structure-function studies of the GABA-A receptor ion channels; mechanisms of general anesthetic action. R. W. Briehl, M.D.: biophysical chemistry of proteins; physical chemistry and structure in the polymerization and gelation of sickle-cell hemoglobin. A. Finkelstein, Ph.D.: mechanisms of transport across cell membranes; voltage-dependent channels in lipid bilayer membranes; reconstitution of biological transport systems in these membranes; protein translocation associated with channel gating. J. M. Friedman, M.D./Ph.D.: time-resolved laser spectroscopy of proteins; comparative biophysics of hemoglobin; conformational dynamics, disorder, and reactivity in hemoglobins; blood substitutes; Ca binding proteins; encapsulation of proteins in sol-gels. G. J. Gerfen, Ph.D.: structural determination of paramagnetic intermediates in enzymatic catalysis; adenosylcobalamin-dependent enzymes; advanced EPR techniques, including pulsed and high-frequency EPR. R. K. Gupta, Ph.D.: NMR studies of intracellular metal ions and oxidative stress. H. Hetherington, Ph.D.: in vivo NMR studies of the metabolic mechanisms underlying epilepsy. L. A. Jelicks, Ph.D.: magnetic resonance imaging and spectroscopy of small-animal models of cardiac disease. R. L. Nagel, M.D.: hemoglobin hemoglobinopathies; gene therapy for sickle cell anemia; physiology of transgenic mice with hemoglobinopathies; physiology, genetic defects, and parasitosis of red cells. J. Peisach, Ph.D.: paramagnetic metalloproteins, including those containing Cu, Fe, Co, and Mn: elucidation of the metal-binding site structures in these molecules from hyperfine interactions obtained by electron spin echo methods. V. Schuster, M.D.: molecular mechanisms and regulation of epithelial anion transport; prostaglandin metabolism. S. D. Schwartz, Ph.D.: theoretical studies of biophysical systems; quantum mechanics; statistical mechanics; enzyme mechanism. D. J. Sharp, Ph.D.: molecular mechanisms of mitosis; roles of microtubule-based motor proteins in chromosome movement and segregation; design and characterization of antitumor agents. S. Slatin, Ph.D.: molecular mechanisms of voltage-dependent channels in lipid bilayer membranes. H. J. Sosa, Ph.D.: structure of motor proteins; kinesin; cryo-electron microscopy; light microscopy; fluorescence spectroscopy. J. Tardiff, M.D./Ph.D.: molecular physiology of thin filament-related cardiomyopathies; functional effects of cardiac troponin T mutations. S.-R. Yeh, Ph.D.: protein function and dynamics; protein folding; fast kinetics with continuous-flow and stopped-flow techniques; laser spectroscopy.

BARRY UNIVERSITY

Programs in Biology and Biomedical Sciences

Programs of Study

Barry University's School of Natural and Health Sciences provides high-quality education in the life sciences, enabling students to become competent, thoughtful, ethical, and compassionate natural and biomedical scientists and health professionals. This is accomplished within a caring environment that is supportive of the religious dimension of the University.

The M.S. in Biomedical Science Program offers three program options, each requiring the completion of 36 semester hours of course work with a minimum 3.0 GPA. The accelerated one-year program and the eighteen-month/two-year program are designed for students who have completed their undergraduate premed preparation and want to enhance their qualifications for entrance into medical school. These students must register for and pass a qualifying examination. The Industrial Track/Research program is designed for students who are currently employed in the health-care industry and want to advance in their chosen field, those who wish to enter the biomedical fields in industry or teaching, and those who are preparing for Ph.D. programs. These students must complete a significant research project and research paper in lieu of a comprehensive exam.

The Master of Science (M.S.) in Biology Program offers secondary school biology teachers the opportunity to continue their life science studies in greater depth, become current in their field, and obtain the credentials required for career advancement. The curriculum covers core areas in biology, the practical application of this information in the high school setting, and the relationship of biology and technology. Students must complete 30 semester hours of course work, including at least 3 hours of research or internship.

Research Facilities

In 1995, the Natural and Health Sciences Building was built to expand the number of laboratory facilities for student instruction and faculty and student research. The building has thirteen faculty offices and research labs, as well as classrooms, seminar rooms, and program offices.

The Monsignor William Barry Memorial Library contains more than 713,270 items, including 2,600 periodical titles, 5,000 audiovisual items, and 150 electronic databases. Among its special collections is the Catholic American collection. In addition, the Barry Library Information Services System Web (BLISSWeb) serves as the computerized online catalog of the library's holdings and can be searched from computers in the library, computer labs, residence halls, and offices across campus as well as home and from off-campus sites.

Financial Aid

The Stafford Student Loan Program enables students to borrow up to $8500 in subsidized loans or $18,500 in unsubsidized loans. Recipients must be enrolled on at least a half-time basis and show satisfactory academic progress. Other loans may be available in amounts up to $20,000. The state of Florida provides grants to qualified students, including the Jose Marti Scholarship Challenge Grant, Seminole/Miccosukee Indian Scholarships, and Scholarships for Children of Deceased or Disabled Veterans. The University also receives NIH grants, including the Minority Access to Research Centers grant and the Minority Biomedical Research Support grant. Scholarships from other sources may be available.

Cost of Study

In the 2006–07 academic year, graduate tuition is $725 per academic credit.

Living and Housing Costs

For the 2006–07 academic year, full-time, degree-seeking graduate students are eligible to live in housing administered by Residential Life. Graduate students who apply for housing are offered a space at the Best Western Inn-on-the-Bay, a local hotel at which the University has accommodated students for several years. Barry University does not offer married and/or family housing; all units are for single students only. All spaces offered to graduate students are private rooms with private baths, and no meal plan is attached; the cost is $4925 per semester. Barry University does not provide transportation service to or from the hotel site, which is located 5.2 miles from the University. Hotel assignments are for first-year graduate students only, and are not renewable after May 2007. Graduate students may apply to reside in a residence hall for summer 2007 prior to moving off campus for the remainder of their tenure at Barry. There is a variety of meal plans available for purchase for those students who wish to utilize the campus dining hall. Off-campus apartments typically range from $600 to more than $1000 per month.

Student Group

Approximately 150 students are enrolled in the program each year. Although students come from a wide variety of social and academic backgrounds, the majority enter the program with B.S. degrees and significant course work in general biology, general chemistry, organic chemistry, or physics. A grade of C or better is required for all of the courses listed.

Student Outcomes

In June 2005, 95 percent of the programs' graduates were accepted to medical and dental school. These schools include Emory University, Mayo Medical Center, Mount Sinai Medical School, Tulane University, and Robert Wood Johnson Medical School. Some students were also accepted into optometry, podiatry, and veterinary programs.

Location

Located in Miami Shores, Florida, Barry University is situated between the cities of Miami and Fort Lauderdale. This location gives the students the benefits of a small suburb yet allows easy access to major metropolitan areas. Golf, tennis, swimming, soccer, scuba diving, waterskiing, sailing, and horseback riding are available throughout the year. The Florida Grand Opera, Miami City Ballet, and the Coconut Grove Playhouse provide a full season of highly acclaimed performances. Miami also offers professional baseball, basketball, and football teams. Among the numerous recreational and ecological features of the Miami area are the Florida Keys, the Everglades, national parks, and marine and state parks.

The University

Since Barry University was established in 1940, its administration has strived to develop high-quality academic programs that serve the needs of both the students and the local community. Today, Barry offers more than sixty traditional undergraduate programs and more than fifty graduate programs to approximately 7,000 students representing forty-nine states and eighty countries. In the 2005 edition of *U.S. News & World Report*'s America's Best Colleges, Barry's academic quality was ranked as one of the strongest in South Florida.

Applying

Prospective students are required to submit an application for admission; an undergraduate transcript showing a GPA of 3.0 or higher; acceptable GRE, MCAT, or FTCE scores; two letters of recommendation; a personal statement describing academic and professional goals; and a $30 application fee. There are no deadlines to apply for admission; however, applicants are encouraged to apply at least thirty days prior to the beginning of the anticipated start term.

Correspondence and Information

Health Sciences Admissions
Barry University
11300 NE 2nd Avenue
Miami Shores, Florida 33161-6695
Phone: 305-899-3379
      800-756-6000 Ext. 3379 (toll-free)
E-mail: healthsciences@mail.barry.edu
Web site: http://www.barry.edu/snhs

Barry University

THE FACULTY AND THEIR RESEARCH

Michael Bill, Professor; Ed.D., Nova Southeastern. Human anatomy and physiology; histology; gross anatomy.

Jacqueline Chang, Professor; Ph.D., Princeton.

Gil Ellis, Professor; Ed.D., Nova Southeastern. Physiology and histology; cardiovascular physiology, with an emphasis on cardiac-electro-physiology.

Angela Fickel, Instructor; M.S., Barry.

Sr. John Karen Frei, Professor and Dean; Ph.D., Miami (Florida). Orchid seed, protocorm, and seedling growth response on varying orchid substrates including tree bark.

Elizabeth T. Hays, Professor; Ph.D., Maryland. Caffeine and theophylline action on vertebrate skeletal muscle physiology.

Christoph Hengartner, Professor; Ph.D., MIT. Regulation of yeast RNA polymerase II holoenzyme transcription; PRV gene expression in epithelial and neuronal cells.

Xiao Tang Hu, Professor; Ph.D., Hunan Medical. Growth factor–mediated cell-cycle control and apoptosis in human hematopoietic cells.

Ana M. Jimenez, Professor; Ph.D., Miami (Florida). Mouse auditory function as a model for effects of advanced age and noise exposure on the human ear.

Ralph Laudan, Professor and Program Director; Ph.D., Rutgers. Immune response of fish to parasitic infections; immune response association with stress and insulin-dependent diabetes.

Peter Lin, Professor; Ph.D., Johns Hopkins. Reproductive biology in fish and amphibians; elucidation of molecular and biological mechanisms by which gonadotropins, gonadal peptides, and steroids control episodic reproductive activities in vertebrates.

Nicholas Lutfi, Assistant Professor of Anatomy; D.P.M., Barry.

Sylvia Maciá, Professor; Ph.D., Miami (Florida). Feasibility of using laboratory-reared seagrass seedlings for replanting of damaged and unvegetated shallow-water seagrass communities.

Jeremy R. Montague, Professor; Ph.D., Syracuse. Spatial dispersions and behavior of sea urchins; polyp growth of soft coral species; biometrical analysis of dinosaur and fish fossils.

Laura Mudd, Professor; Ph.D., Florida. Regulation of growth, differentiation, and survival of neurons from distinct regions of the mammalian brain, which are particularly susceptible to neurodegenerative disorders.

Gerhild Packert, Professor; Ph.D., South Florida. Fruit fly *Drosophila melanogaster;* genetics and developmental biology.

Flona Redway, Professor; Ph.D., Cambridge. In vitro development of ornamentals; effects of synthetic rotenoids on plant growth.

Allen F. Sanborn, Professor; Ph.D., Illinois at Urbana-Champaign. Mechanisms that permit habitat utilization and communication in insects of the superfamily *Cicadoidea.*

Graham Shaw, Professor; Ph.D., Aston (England). Role of *Helicobacter pylori* in peptic ulcer relapse; impact of technology on education.

Leticia Vega, Professor; Ph.D., MIT. How telomerase contributes to genomic integrity.

BAYLOR COLLEGE OF MEDICINE

Graduate School of Biomedical Sciences

Programs of Study

The Graduate School of Biomedical Sciences at Baylor College of Medicine offers the Ph.D. degree in eight basic science areas: biochemistry and molecular biology, immunology, molecular and cellular biology, molecular and human genetics, molecular physiology and biophysics, molecular virology and microbiology, neuroscience (with a special track in neurobiology of disease), and pharmacology. Interdepartmental programs in cardiovascular sciences, cell and molecular biology (with a special track in the biology of aging), developmental biology, structural and computational biology and molecular biophysics, and translational biology and molecular medicine are also available. A program that allows students to obtain both the Ph.D. and M.D. degrees in a seven-to eight-year period is also available. Students with multidisciplinary research interests find the environment at Baylor to be particularly advantageous, as interdepartmental interactions are encouraged. The graduate programs share the common goal of developing an understanding of fundamental biological processes.

Under a reciprocal arrangement, students enrolled in the Graduate School may take degree-related graduate courses for credit at Texas A&M University, Rice University, the University of Houston, and the University of Texas Health Science Center at Houston's Graduate School of Biomedical Sciences.

Each student is directed by an advisory committee and the director of their program. A flexible curriculum is available that consists of interdisciplinary courses offered by the Graduate School and specialized courses offered by any of Baylor's graduate programs. Course requirements are generally completed in the first year. Qualifying exams and dissertation research are performed in subsequent years.

Research Facilities

Baylor College of Medicine is one of several biomedical institutions within the Texas Medical Center. The Graduate School faculty utilizes state-of-the-art research facilities housed in numerous research centers. More than 1 million square feet of space is devoted to the research and teaching activities of the 1,700 full-time faculty members, who conduct more than $400-million worth of sponsored research annually. The Graduate School executive offices, as well as classrooms, a student lounge, and additional research space for the College, are located in the sixteen-story Albert B. Alkek Graduate School of Biomedical Sciences Building. The Houston Academy of Medicine–Texas Medical Center Library, a modern facility with the largest professional staff of any medical library in the nation, is available to all students.

Financial Aid

Costs for Ph.D. students are normally covered by scholarships awarded by the College. Tuition scholarships are available to all students. The stipend for all Ph.D. students is $23,000 for the 2006–07 academic year. Health insurance is provided for the students; additional financial aid is also available. Requests for information regarding financial aid should be addressed to the Financial Aid Office, Baylor College of Medicine, Houston, Texas 77030.

Cost of Study

Tuition for the 2006–07 academic year is $8200. Students pay a one-time matriculation fee of $25, a one-time graduation fee of $140 during the fourth year, and a student fee of $150 for first-year students and $20 for subsequent years. Students on temporary visas also pay an annual international services fee of $75 for an F-1 visa or $100 for a J-1 visa.

Living and Housing Costs

One residence hall, located adjacent to the College, offers moderately priced, fully furnished rooms and apartments. Privately owned furnished and unfurnished apartments and houses, offered at a wide range of rents, are available nearby. The cost of living in Houston is less than in most major U.S. cities.

Student Group

The 534 Ph.D. students enrolled for the 2005–06 academic year represented a variety of states and countries; 48 percent of those enrolled were women.

Student Outcomes

Graduates typically go on to postdoctoral research appointments followed by careers in academics, medicine, and industry. Career options also include science education and government. A Career Resource Center is available.

Location

Houston is a young and dynamic city with nearly 4 million inhabitants. Extensive and affordable cultural and recreational facilities are available: theaters, a symphony orchestra, opera, art museums, a natural history museum, a planetarium, and professional sports arenas. The metropolitan area includes several large university campuses, the Lyndon B. Johnson NASA Manned Spacecraft Center installation, public parks, beaches for fishing and swimming along the coast, and sports facilities of every kind.

The College and The School

Baylor College of Medicine, an independent, private, and fully accredited biomedical university, is located in the heart of the Texas Medical Center, one of the largest medical centers in the world. The Medical Center, adjacent to residential areas, covers more than 675 acres and has 100 permanent buildings, 20,165 students, and 51,273 employees. Baylor is recognized as a leader in biomedical and basic science research and ranks first in the nation for research expenditures in the biological sciences by universities and colleges. The Graduate School is committed to excellence in graduate training. There is a high degree of interdisciplinary cooperation, not only among the faculty members in basic science areas but also with clinical investigators in the College and associated institutions in the Texas Medical Center. Ongoing research programs being carried out by productive and widely recognized investigators in both the basic sciences and the clinical faculty, coupled with the favorable faculty-student ratio, permit students to be directly involved in and contribute to significant research projects.

Applying

The school year begins in August and ends in July; it consists of five terms, each eight weeks long. Applicants must hold a bachelor's degree or a more advanced degree or must be in the final stages of a program leading to a bachelor's degree or the equivalent at the time of application. Undergraduate course suggestions include a year each of general biology, physics, mathematics, chemistry, and organic chemistry, all with a B average or better. The TOEFL is required of all students who have not earned a degree at a university in which the primary language is English. Official transcripts from each college or university attended, letters of evaluation from at least 3 professors, and scores from the Graduate Record Examinations (General Test and Subject Test recommended in biology, chemistry, or biochemistry, cell and molecular biology) must be mailed directly to the School rather than by the applicant. Medical College Admission Test scores are acceptable for dual M.D./Ph.D. degree candidates; admission to this program must go through the Medical School Admissions Office. January 1 is the target date for completed applications; early application is encouraged. The electronic application can be accessed online at http://www.bcm.tmc.edu/gradschool/gs-apply.html.

Correspondence and Information

Graduate School of Biomedical Sciences
Baylor College of Medicine
One Baylor Plaza
Mail Stop: BCM215
Houston, Texas 77030-3498
Phone: 713-798-3312
    888-550-9288 (toll-free)
Fax: 713-798-6325
E-mail: gradappboss@bcm.edu
Web site: http://www.bcm.edu/gradschool/

Baylor College of Medicine

DIRECTORS OF GRADUATE STUDIES AND RESEARCH PROGRAMS

William R. Brinkley, Ph.D., Senior Vice President for Graduate Sciences and Dean of the Graduate School. 713-798-4028; Fax: 713-798-6325 (E-mail: eldridge@bcm.edu)

Biochemistry and Molecular Biology. Adam Kuspa, Ph.D., Director of Graduate Studies. 713-798-4527 (E-mail: biochemistry@bcm.edu)
Structural biophysics; protein design and engineering; molecular genetics and gene regulation; developmental biology and molecular embryology; neurobiochemistry; molecular immunology; macromolecular assembly and recognition; cell-cycle regulation; genomics; developmental biology and cellular regulation; membrane and lipid biochemistry; neurobiochemistry and signal transduction.

Cardiovascular Sciences. Alan R. Burns, Ph.D., Director of Graduate Studies. 713-798-4977 (E-mail: carolar@bcm.edu)
An interdepartmental program addressing all aspects of the cardiovascular system. Cell and molecular biology of cardiac muscle, vascular smooth muscle, and endothelium, including cellular membranes and contractile proteins; control of gene expression in cardiac development and hypertrophy; control of vascular cell proliferation; control of cardiac function and blood pressure; role of various adhesion molecules in the inflammatory aspects of cardiovascular pathologies; myocardial ischemia, metabolism, and oxygen-free radicals; ultrastructural analysis of striated and smooth muscle; mechanisms of atherogenesis; molecular biology of lipoproteins; physical chemistry of lipids; NMR spectroscopy; in vivo instrumentation monitoring of cardiovascular parameters in experimental and clinical settings; clinical correlations of basic disease mechanisms and models of cardiovascular diseases; functional cardiovascular genomics; stem-cell biology.

Cell and Molecular Biology. Thomas A. Cooper, M.D., Director of Graduate Studies. 713-798-6557 (E-mail: cmbprog@bcm.edu; Web site: http://bcm.edu/cmb)
More than 95 participating faculty members from eleven different departments provide students with a diverse set of choices for their Ph.D. research. The range of research interests includes molecular mechanisms of inherited diseases, cancer and cell-cycle regulation, biology of aging, human gene therapy, signal transduction and membrane biology, the human genome project, functional genomics, structural and computational biology, gene expression and regulation, developmental biology, molecular virology, medical microbiology, and immunology. This breadth of topics is complemented by a small class size, allowing hands-on training and education.

Developmental Biology. Hugo Bellen, D.V.M., Ph.D., Director of Graduate Studies. 713-798-7696 (E-mail: cat@bcm.edu; Web site: http://bcm.edu/db/)
The Interinstitutional and Interdepartmental Program in Developmental Biology provides a wide spectrum of exciting research possibilities and a broad cross-disciplinary training. The field of developmental biology is integrative. In order to understand how a single cell develops into a complex organism, the program uses molecular biology, cell biology, biochemistry, imaging, genetics, and genomics. Studies of organisms—as diverse as social molds, worms, flies, frogs, chickens, fish, mice, and humans—are conducted using a wide variety of approaches, instruments, and techniques of modern biological research. Members of the Program in Developmental Biology study basic biological mechanisms of direct and fundamental relevance to human development and disease. This allows the faculty members to unravel the principles and mechanisms that guide embryonic development, the differentiation of adult cell types, regeneration, and aging. The faculty members' major research interests are neurobiology, cancer biology, cell death, neurodegenerative and other human diseases, stem cell biology, gene therapy, reproductive development, oogenesis, muscle, heart, kidney, bone, skin, limb and eye development, cell lineage specification, X chromosome dosage compensation, and plant differentiation.

Immunology. Dorothy E. Lewis, Ph.D., Director of Graduate Studies. 713-798-6054 (E-mail: dlewis@bcm.edu; Web site: http://www.bcm.edu/immuno)
The Department of Immunology offers students the opportunity to conduct cutting edge research covering a broad range of basic and translational immunology. Research programs include cancer immunology and inflammatory mechanisms; gene therapy and genetic vaccines; immune-related diseases, including asthma, autoimmunity, AIDS, and inherited immunodeficiency; lymphocyte activation; signal transduction and transcriptional regulation; development of T and B lymphocytes and dendritic cells; antigen presentation and biochemistry of MHC molecules.

Molecular and Cellular Biology. JoAnne S. Richards, Ph.D., Director of Graduate Studies. 713-798-4598 (E-mail: mcbgrad@bcm.edu)
The Department of Molecular and Cellular Biology is internationally acclaimed for its outstanding research and graduate program. Faculty research interests are diverse and include aging, cancer biology (breast, prostate and skin), cell-cycle regulation, chromatin and transcription factors, cellular signaling, developmental biology, diabetes and molecular metabolism, molecular biology of nuclear receptors and coactivators, molecular genetics and knockout mice, gene therapy, molecular neurobiology, reproductive biology, and stem cell biology.

Molecular and Human Genetics. Gad Shaulsky, Ph.D., Director of Graduate Studies. 713-798-5056 (E-mail: genetics-gradprgm@bcm.edu; Web site: http://www.bcm.edu/molgen)
Genetic approaches to understanding human disease; genome structure and maintenance; genetic instability; microbial genetics, including bacterial, phage, yeast, and *Dictyostelium* molecular genetics; gene regulation; mouse and *Drosophila* developmental genetics; cytogenetics; gene therapy; genetic technologies; functional genomics; informatics.

Molecular Physiology and Biophysics. Eric Klann, Ph.D., Director of Graduate Studies. 713-798-5630 (E-mail: molphys@bcm.edu; Web site: http://public.bcm.edu/physio/)
Structure and function of ion channels and transport proteins; signal transduction; synaptic plasticity; cell-cycle control; reactive oxygen species; neuronal morphology; drug and gene delivery; biosensors for genetic diagnosis; knockout and transgenic mice; muscle and cardiovascular function; learning and memory; cancer; small animal in vivo neuronal tract tracing utilizing manganese enhanced magnetic resonance imaging (MEMRI); confocal microscopy; recombinant DNA technology.

Molecular Virology and Microbiology. Robert F. Ramig, Ph.D., Director of Graduate Studies. 713-798-4830 (E-mail: rramig@bcm.edu; Web site: http://www.bcm.edu/molvir/)
Molecular microbiology, including molecular genetic and biochemical approaches to virulence of pathogenic microorganisms; physical and genetic basis for enzyme-substrate recognition; topoisomerase and DNA topology; oncogenes and cancer; molecular biology of virus replication; viral genetics; transcriptional and translational regulation in viruses; viral infectious diseases, with emphasis on the gastrointestinal, respiratory, and blood-transmitted infections; vaccine development; development and application of viral assays; animal models and antiviral development.

Neuroscience. J. David Sweatt, Ph.D., Director of Graduate Studies. 713-798-7270 or 800-367-9149 (toll-free) (E-mail: nsc_ask@bcm.edu; Web site: http://neuro.neusc.bcm.edu)
Molecular and developmental neurobiology; neurobiology of disease; neuroanatomy; neurophysiology; neural systems analysis; biophysics; imaging; computer-assisted neural modeling.

Pharmacology. Pui-Kwong Chan, Ph.D., Director of Graduate Studies. 713-798-7902 (E-mail: pchan@bcm.edu)
Cancer chemotherapy; drug development; nuclear proteins and the organization of nucleoli; neuropharmacology; gene expression of detoxification enzymes.

Structural and Computational Biology and Molecular Biophysics. Wah Chiu, Ph.D., Director of Graduate Studies; Timothy G. Palzkill, Ph.D., Co-Director. 713-798-5197 (E-mail: scb@bcm.edu)
Proteomics; genomics; bioinformatics; X-ray crystallography; NMR; electron cryomicroscopy; optical microscopy; molecular spectroscopy; MRI; molecular dynamics; protein folding; prediction of macromolecule structure and function; medical informatics; computational and theoretical neuroscience; drug and inhibitor design; ion channel biophysics; biomedical applications of grid computing.

Translational Biology and Molecular Medicine. David P. Huston, M.D. and Mary K. Estes, Ph.D., Co-Directors of Graduate Studies. 713-798-1077 (E-mail: pramirez@bcm.tmc.edu; Web site: http://www.bcm.edu/tbmm)
TBMM is an interdepartmental program designed to train a new generation of Ph.D. or M.D./Ph.D. biomedical researchers whose primary goals and training are oriented toward human health research. More than 100 faculty members from all of the basic and clinical science departments and clinical research centers provide students with a broad spectrum of research opportunities related to every aspect of molecular medicine and human diseases. Unique aspects of the program include dual mentoring by a basic scientist and a clinical scientist, novel courses created to teach molecular mechanisms and clinical aspects of human diseases in an integrated fashion, seminars and journal clubs targeted to disease mechanisms and treatment, clinical rotations relevant to research projects, and training in translating bench research into clinical application.

BOSTON COLLEGE

Biology Department

Programs of Study	The Department offers programs of study leading to the M.S. and Ph.D. degrees in biology. Basic areas of study include biochemistry, cellular and developmental biology, genetics, cell cycle, vector biology, neurobiology, bioinformatics, and structural biology. The M.S. program provides a research-oriented experience through course work and a research-based thesis project performed under the guidance of a graduate research faculty member. The Ph.D. degree provides a more in-depth training experience. Core course work is provided in cell biology, biochemistry, molecular biology, and genetics. Advanced electives are available in all areas of faculty expertise. Seminar courses provide students with ongoing training in critical thinking and oral presentation of scientific data. Research experience is provided by working in close cooperation with faculty members, postdoctoral fellows, and senior students in a collaborative, supportive environment. In cooperation with the School of Education, the Master of Science in Teaching (M.S.T.) degree in biology is also offered.
Research Facilities	The Biology Department, with a wing that opened in 2000, occupies more than 70,000 square feet of research space in Higgins Hall. Faculty laboratories have state-of-the-art equipment. Shared facilities include several tissue-culture rooms; common equipment rooms; TEM, fluorescence, and confocal microscopes; X-ray diffraction and a capillary DNA sequencer; machine and electronic workshops; and state-of-the-art computers for online data analysis, production of publication-quality figures, and bioinformatic research and analysis. The university science library subscribes to more than 600 scientific journals. Access to libraries of institutions in the greater Boston area is available through consortium arrangements.
Financial Aid	Graduate assistantships (teaching and research based) are available with full tuition remission. Stipends are $25,000 per calendar year.
Cost of Study	In 2006–07, the tuition and fees for a full-time student are $1040 per credit, 100 percent of which is covered by tuition remission for students receiving financial aid.
Living and Housing Costs	The Housing Office provides an extensive list of off-campus housing options. Most graduate students rent rooms or apartments near Chestnut Hill; many biology students share apartments with other students in the program. Average monthly expenses (rent, food, utilities) are $800 for students.
Student Group	The enrollment at Boston College is 14,500, including 4,200 students enrolled in the various graduate schools. There are 46 graduate students in the Department, of whom 43 are in the Ph.D. program and 3 are in the M.S. program. In addition, 2 students are enrolled in the M.S.T. program. The graduate students are geographically and ethnically diverse. More than half are women. Approximately 90 percent of the students receive some form of financial aid.
Location	Boston College is located in the Chestnut Hill section of Newton, an attractive residential area about 6 miles from the heart of Boston, with easy access to the city by public transportation. The Boston area, with its numerous educational and biomedical research institutions, offers countless outstanding seminars, lectures, colloquia, and concerts throughout the year. A wide variety of cultural and recreational opportunities can be found close to the campus.
The College	Founded in Massachusetts in 1863, Boston College currently includes the Graduate School of Arts and Sciences and graduate schools of law, social work, management, nursing, and education. Its expanding campus is graced with many attractive Gothic buildings. Boston College has a strong tradition of academic excellence and service to the community.
Applying	Preference is given to completed applications received prior to January 15. This deadline is especially important for those seeking financial aid. Later applications may be considered until June 15 if space is available. Admission is granted on the basis of academic background and demonstrated aptitude in biology and related disciplines. A year of organic chemistry, physics, and mathematics and a solid background in biology are highly recommended for admission. Scores on the Graduate Record Examinations General Test and the Subject Test in biology are recommended but not required.
Correspondence and Information	Professor Charles Hoffman Director, Graduate Program Biology Department Higgins Hall Boston College Chestnut Hill, Massachusetts 02467-3961 Phone: 617-552-3540 E-mail: gradbio@bc.edu Web site: http://www.bc.edu/schools/cas/biology

Boston College

THE GRADUATE RESEARCH FACULTY

Anthony T. Annunziato, Professor; Ph.D., Massachusetts Amherst, 1979. Biochemistry/molecular biology; DNA replication and nucleosome assembly in mammalian cells.

David R. Burgess, Professor; Ph.D., California, Davis, 1974. Spatial and temporal regulation of cytokinesis; role of the actin- and microtubule-based cytoskeletons in early development.

Thomas C. Chiles, Professor and Chairman of Biology; Ph.D., Florida, 1988. Cell biology, signal transduction; cell-cycle control, gene regulation in mature B lymphocytes.

Jeffrey Chuang, Assistant Professor; Ph.D., MIT, 2001. Computational approaches to comparative genomics, gene regulation, and molecular evolution.

Peter G. Clote, Professor; Ph.D., Duke, 1979. Algorithms and mathematical modeling in computational biology: genomic motif detection, protein folding on lattice models, RNA secondary structure, functional genomics via gene expression profile.

Kathleen Dunn, Associate Professor; Ph.D., North Carolina at Chapel Hill, 1982. Plant molecular biology; cloning and characterization of genes induced during alfalfa nodulation.

Marc-Jan Gubbels, Assistant Professor; Ph.D., Utrecht (Netherlands), 2000. Genetics and cell biology of the apicomplexan parasite *Toxoplasma gondii*.

Laura E. Hake, Associate Professor; Ph.D., Tufts, 1992. Molecular control of early development in *Xenopus;* protein degradation; RNA-protein interactions; translational regulation during gametogenesis.

Charles Hoffman, Professor; Ph.D., Tufts, 1986. Signal transduction and transcriptional regulation in fission yeast; analysis of PKA and MAPK signal pathways in nutrient monitoring.

Daniel Kirschner, Professor; Ph.D., Harvard, 1972. Structural biochemistry.

Gabor T. Marth, Assistant Professor; D.Sc., Washington (St. Louis), 1994. DNA polymorphism discovery and analysis; genomic and algorithmic approaches to population genetics; long-term human demography, haplotype structure, and medical genetics.

Junona Moroianu, Associate Professor; Ph.D., Rockefeller, 1996. Cell biology; molecular mechanisms of nucleocytoplasmic transport of cellular and viral macromolecules in mammalian cells.

Marc A. T. Muskavitch, Professor and DeLuca Chair; Ph.D., Stanford, 1981. Developmental biology: intercellular signaling and cell-fate specification in *Drosophila;* host-parasite interactions in *Anopheles.*

Clare M. O'Connor, Associate Professor; Ph.D., Purdue, 1977. Cellular biochemistry.

Janet L. Paluh, Assistant Professor; Ph.D., Stanford, 1996. Chromosome and microtubule dynamics in fission yeast; role of mitotic klps, MAPs, and MTOC function in chromosome segregation.

William H. Petri, Associate Professor and Associate Chairman; Ph.D., Berkeley, 1972. Molecular, developmental, and genetic aspects of development in *Drosophila.*

Donald J. Plocke, Associate Professor; Ph.D., MIT, 1961. Molecular biology; structure and function of phytochelatins; metal ions as regulators in microbial systems; metal ion detoxification mechanisms.

Thomas N. Seyfried, Professor; Ph.D., Illinois, 1976. Neurogenetics: use of genetics and neurochemistry in neural membrane function and developmental neurobiology.

Mohammed Shahabuddin, Assistant Professor; Ph.D., Edinburgh (Scotland), 1990. Cell biology, signal transduction, and gene expression in pathogen transmission by insects.

Anne Stellwagen, Assistant Professor; Ph.D., California, San Francisco, 1997. Chromosome dynamics in budding yeast; telomeres, telomerase, and DNA repair.

Stephen R. Wicks, Assistant Professor; Ph.D., British Columbia, 1996. Molecular genetics of chemosensory processing in the nematode *C. elegans.*

Higgins Hall, home of the Biology Department.

BOWLING GREEN STATE UNIVERSITY

Department of Biological Sciences

Programs of Study

The Department of Biological Sciences at Bowling Green State University offers graduate training for professional careers in both applied and fundamental areas of biology. Programs are available leading to the degrees of Doctor of Philosophy, Master of Science, and Master of Arts in Teaching. Major areas of specialization include aquatic ecology, conservation biology and genetics, entomology and parasitology, microbiology, biochemistry and molecular biology, neuroscience and behavior, and plant science. The research interests of the individual faculty members are listed on the reverse of this page.

Completion of the Ph.D. program, which consists of formal and informal course work as well as dissertation research, usually takes four years after completion of the master's degree or five years after the completion of the Bachelor of Science degree. Master's programs, with both thesis and nonthesis options, take about two years to complete. The degree plan is selected by the student with the assistance of a faculty adviser and advisory committee; the program of study is flexible to allow students to pursue their individual professional goals. In addition, graduate students are expected to gain teaching experience as assistants in laboratory and lecture courses.

The Department of Biological Sciences also offers interdisciplinary research programs. Members of the department interact with faculty members in the Departments of Chemistry, Geology, and Psychology in forming cooperative programs to meet the specialized needs and goals of graduate students.

Research Facilities

The Department of Biological Sciences is housed in the Life Sciences Building, a modern five-story, 120,000-square-foot research and teaching complex. The forty laboratories housed in the building are equipped with state-of-the-art instrumentation. Included are scanning and transmission electron microscopes, X-ray analyzers, ultracentrifuges, scanning and dual-beam spectrophotometers, PCR machines, high-performance liquid chromatography equipment, gamma and liquid scintillation counters, environmental control chambers, darkrooms, cold rooms, biochemistry and tissue culture labs, and a biohazard work area. Other departmental facilities include 80-acre and 100-acre forest preserves; a greenhouse complex with areas for the propagation and breeding of temperate, tropical, subtropical, and desert plants; and a herbarium containing more than 20,000 specimens. A 9,000-square-foot animal facility has thirty breeding and maintenance rooms and a P3 area. An excellent science library is located adjacent to the Life Sciences Building.

Financial Aid

Graduate assistantship and fellowship stipends ranged from $9311 to $14,076 (plus $2328 to $4687 for the summer) plus a waiver of tuition and general fees ($15,573 in-state and $26,535 out-of-state, including the summer) for a half-time appointment during 2006–07.

Predoctoral nonservice fellowships are available to Ph.D. students, normally during the last year of study. These awards provide a stipend ($17,595 for 2006–07) and a waiver of instructional, nonresident, and general fees.

Renewal of assistantships and fellowships is contingent upon satisfactory completion of assignments and progress toward the degree. Research assistantships are also available through research grants to individual faculty members.

Cost of Study

Tuition and general fees in 2006–07 were $15,573 for Ohio residents and $26,535 for nonresidents and included the summer. Book costs were estimated at $700 per year.

Living and Housing Costs

The University does not provide housing for graduate students. However, there are numerous apartments and rooms for rent near the campus at costs comparable to those in other midwestern cities of similar size.

Student Group

There were 95 graduate students (40 pursuing master's degrees and 55 pursuing Ph.D. degrees) in the Department of Biological Sciences for 2006–07. A diverse group, they were nearly equally divided between men and women and included representatives of a dozen or more states and several countries. The majority of the department's students receive some form of financial aid from the University.

The University enrolls 18,000 students, including 2,400 graduate students.

Student Outcomes

Recent Ph.D. recipients have accepted postdoctoral positions at academic institutions, including Harvard, University of California at Berkeley, University of Michigan, and Yale; or have accepted positions in government, at institutions such as the Centers for Disease Control and the Environmental Protection Agency; or have accepted positions in industry. Recent master's degree recipients have gone to graduate or medical schools, such as Ohio State, Purdue, Tufts, and University of Michigan; or have accepted positions in industry with such companies as Amgen and Wyeth Laboratories.

Location

Bowling Green is a northwestern Ohio community of approximately 28,000 residents. It is located 16 miles south of Toledo; Detroit, Cleveland, and Columbus are all within a 100-mile radius. The community offers numerous recreational and social programs that supplement the activities offered by the University.

The University

Bowling Green, a state-assisted university, was founded in 1910. The attractive campus occupies 1,250 acres and has more than 100 buildings. The atmosphere is friendly, open, and reasonably informal. Graduate faculty members in the Department of Biological Sciences and throughout the University are committed to excellence in both research and teaching.

Applying

Requirements for admission include a baccalaureate degree, normally with a major in one of the sciences; GRE scores on the verbal, quantitative, and analytical portions of the General Test; favorable grade point averages in all subjects and in the major; and three letters of recommendation.

Applicants can apply online at https://www.applyweb.com/apply/bgsug. Applicants are encouraged to correspond directly with faculty members in their areas of interest. Bowling Green State University is an Equal Opportunity/Affirmative Action employer. Women and members of minorities are encouraged to apply.

Correspondence and Information

Graduate Committee Chair
Department of Biological Sciences
Bowling Green State University
Bowling Green, Ohio 43403-0212

Fax: 419-372-2024
E-mail: ldevenn@bgnet.bgsu.edu
Web site: http://www.bgsu.edu/departments/biology/

Bowling Green State University

THE FACULTY AND THEIR RESEARCH

Verner P. Bingman, Professor (Psychology); Ph.D., SUNY at Albany, 1981. Neuroethology of learning and memory; navigation in birds.

Juan L. Bouzat, Associate Professor; Ph.D., Illinois at Urbana-Champaign, 1998. Molecular ecology; conservation and population genetics; molecular evolution.

George S. Bullerjahn, Professor; Ph.D., Virginia, 1984. Microbial physiology; regulation of stress-induced functions in cyanobacteria.

Sheryl L. Coombs, Professor; Ph.D., Hawaii, 1980. Mechanosensory processing by auditory and lateral line systems of aquatic vertebrates; neuroethology.

Donald W. Deters, Assistant Professor; Ph.D., California, Irvine, 1972. Microbiology; mitochondrial biogenesis; molecular biology of yeasts.

Carmen F. Fioravanti, Professor; Ph.D., UCLA, 1973. Comparative biochemistry; anaerobic energetics of the parasitic helminths; mitochondrial transhydrogenase systems; electron transport mechanisms and isoprene biosynthesis.

Michael E. Geusz, Associate Professor; Ph.D., Vanderbilt, 1990. Circadian rhythms in mammalian and molluscan neurons and control of intracellular calcium.

John S. Graham, Professor; Ph.D., Washington State, 1984. Plant molecular biology and biochemistry; regulation of gene expression; plant-pest interactions; proteinases and proteinase inhibitors; hormone physiology.

Carol A. Heckman, Professor and Director of Electron Microscopy Center; Ph.D., Massachusetts Amherst, 1972. Cell biology; role of ruffling in growth control; image analysis; molecular biology of transforming proteins.

Robert Huber, Associate Professor; Ph.D., Texas Tech, 1989. Neurochemistry of aggression.

Roudabeh J. Jamasbi, Professor; Ph.D., Arkansas, 1974. Biology and immunology of digestive and respiratory tract carcinomas; generation of monoclonal antibodies against carcinomas.

Ray A. Larsen, Assistant Professor; Ph.D., Montana, 1986. Bacterial membrane energetics.

Rex L. Lowe, Professor; Ph.D., Iowa State, 1970. Aquatic ecology; community ecology of benthic algae.

R. Michael McKay, Associate Professor; Ph.D., McGill, 1992. Aquatic microbial ecology; photoplankton-trace metal interactions; inorganic carbon acquisition and assimilation; phycotoxins.

Lee A. Meserve, Distinguished Teaching Professor; Ph.D., Rutgers, 1972. Mammalian endocrinology; altered thyroid status in development and aging.

Helen Michaels, Associate Professor; Ph.D., Illinois at Urbana-Champaign, 1986. Molecular evolution; chloroplast DNA and angiosperm phylogeny; evolution and ecology of plant mating systems; ecological genetics; molecular systematics; plant ecology.

Jeffrey G. Miner, Associate Professor; Ph.D., Ohio State, 1990. Aquatic community ecology; abiotic effects on predator-prey interactions; ecology of fishes.

Paul A. Moore, Professor; Ph.D., Boston University, 1990. Marine chemical ecology; sensory ecology; physiology and behavior of marine and aquatic organisms.

Paul F. Morris, Associate Professor; Ph.D., Queen's at Kingston, 1987. Molecular plant-microbe interactions; regulation of chemoattraction in *Phytophthora* sp.

Kevin Pang, Associate Professor (Psychology); Ph.D., Colorado, 1986. Neural basis of learning, memory, and attention; role of basal forebrain system in forms of learning, memory, and attention.

Daniel M. Pavuk, Lecturer; Ph.D., Ohio State, 1990. Terrestrial ecology; insect ecology; parasitoid biology; plant-insect interactions; agroecology.

C. Lee Rockett, Professor; Ph.D., LSU, 1971. Ecology of soil mites; behavior and control of mosquitoes.

Scott O. Rogers, Professor and Chair; Ph.D., Washington (Seattle), 1987. Molecular biology, evolution, introns; ancient DNA, DNA preservation, biotechnology; ancient ice, plants, microbes.

Karen V. Root, Assistant Professor; Ph.D., Florida Tech, 1996. Conservation biology and population ecology.

Stan L. Smith, Professor; Ph.D., Northwestern, 1978. Insect endocrinology and biochemistry; ecdysteroid metabolism and cytochrome P-450 systems; effects of allelochemicals on insect development and reproduction.

Tami C. Steveson, Assistant Professor; Ph.D., Denver, 1994. Protein biochemistry; molecular neurobiology; cell biology; protein trafficking.

Eileen M. Underwood, Associate Professor; Ph.D., Indiana, 1979. Developmental and molecular genetics of oogenesis and early embryogenesis of *Drosophila;* maternal-effect mutations.

Moira J. van Staaden, Associate Professor; Ph.D., Texas Tech, 1989. Ethology and evolution.

Daniel D. Wiegmann, Associate Professor; Ph.D., Wisconsin–Madison, 1993. Behavioral ecology; reproductive biology of fishes; animal decision making.

Ronny C. Woodruff, Distinguished Research Professor; Ph.D., Utah State, 1972. Mutagenesis and genetics of transposable DNA elements in *Drosophila;* genetics of natural populations; role of mutation in molecular evolution.

CALIFORNIA INSTITUTE OF TECHNOLOGY

Division of Biology

Program of Study

There are two principal research foci in the Division of Biology at Caltech.

The goals of research in molecular, cellular, and developmental biology and genetics are to understand the molecular mechanisms that regulate cellular proliferation, communication, and function and the mechanisms of differentiation and pattern formation that form the diverse and functionally integrated cell types in multicellular organisms. This area includes the development of the many different cell types and connections found in the nervous system.

Research in integrative neurobiology involves the analysis of the interactions of collections of neurons in behavior, both in simple insect nervous systems and in the more complex vertebrate systems. Integrative neurobiology is closely associated with computational analysis and modeling as carried out in the Computation and Neural Systems Option.

Emphasis in graduate study is on individual research under the guidance of faculty members. Normally, research is initiated during the first year of graduate study by three periods of rotation, each approximately one quarter in duration, in three different research laboratories. By the end of the first year, students will have chosen a laboratory for their thesis work. Course requirements are minimal and are limited to those that contribute to the major program of the student. Major work may be pursued in any of the seven areas: biophysics and cell biology, developmental biology, genetics, immunology, integrative neurobiology, molecular biology and biochemistry, and molecular and cellular neurobiology. Any one of these majors may be combined with biotechnology as part of a dual major. Students are admitted only for study toward the Ph.D. degree. Five to seven years are usually needed to complete the Ph.D. requirements.

Research Facilities

Caltech's Division of Biology includes the Kerckhoff, Alles, Church, Beckman, and Braun laboratories of biology; the Beckman Institute; Broad Center; and the Kerckhoff Marine Laboratory at Corona del Mar. All equipment pertinent to modern biology is available to graduate students. A complete and modern biology library is housed along with the related chemistry library in the nine-story Robert Andrews Millikan Memorial Library.

Financial Aid

Students admitted to graduate standing in biology are provided with financial support adequate to meet their normal expenses. They receive a full tuition grant and, in addition, an NIH traineeship, a fellowship, a part-time teaching assistantship, or a research assistantship. The annual stipend for 2005–06 was $25,500. There are job opportunities for spouses in the Pasadena area. Loan programs are available.

Cost of Study

Tuition costs are covered by grants, as noted above, as are the fees required for participation in the health program. Funds are provided for research expenses and travel to scientific meetings.

Living and Housing Costs

A single student, or a married student whose spouse has a job, can live simply but comfortably on his or her graduate student income. University housing is available for all entering graduate students. Caltech is an informal place, and clothing costs can be kept low.

Student Group

The Division of Biology has, at any one time, approximately 90 graduate students and approximately 150 postdoctoral fellows, in addition to the 39 professors of biology. About 45 percent of the graduate students are women.

Student Outcomes

Most of the Division's graduates pursue postdoctoral research prior to joining university faculties as assistant professors. Recent examples include postdoctoral positions with the MIT Center for Genome Research and the University of California, San Francisco and a faculty position at Yonsei University in Korea.

Location

Caltech is situated in Pasadena, a city of approximately 125,000 people about 15 miles from the center of downtown Los Angeles. It is located midway between the mountains (for skiing in the winter and mountain climbing in the summer) and the sea (for swimming, surfing, and boating). Caltech offers extensive opportunities for those interested in swimming, tennis, track, and all related recreational activities. Caltech's relationship with the surrounding community is harmonious and peaceful.

The Institute

Caltech, including its Division of Biology, is an uncommon institution by virtue of its small size (about 900 undergraduate students, 1,100 graduate students, and 275 professors), its high quality, and its leadership in scientific education and research. The Division of Biology is young, having been founded in 1928, and has been able to adapt to changing circumstances in the ensuing years. Both Caltech and the Division of Biology have remarkable histories of continued important contributions in scientific research and scholarship.

Applying

To apply for admission, the applicant should write to the Dean of Graduate Studies at the address below, stating an intention to apply for graduate study in biology. Application forms and related information will be mailed. There is an application fee of $50. Additional information about the research and training programs of the Division of Biology should be obtained by writing to the graduate secretary of the Division of Biology.

Applicants are normally expected to meet the following minimal requirements: mathematics through calculus and elementary differential equations; at least one year of college physics; chemistry, including organic chemistry; and elementary biology. Applications are particularly encouraged from students who have had additional course work in basic sciences and biology equivalent to the undergraduate biology option at Caltech and who have demonstrated laboratory ability and research motivation in undergraduate research projects or employment. Students may be admitted on the basis of advanced course work in psychology or other relevant subjects in lieu of some of the basic science courses normally required.

Students are usually admitted either at the beginning of the academic year in the fall or at the beginning of the summer term, July 1. Completed application materials should be received by January 1; early submission is recommended to allow ample time for a visit to the campus. Scores on the General Test of the Graduate Record Examinations should be submitted, and, if possible, the test should be taken by the November date preceding application.

Correspondence and Information

For research and training information:
Graduate Secretary
Division of Biology 156-29
California Institute of Technology
Pasadena, California 91125
Phone: 626-395-4497

For application information:
Dean of Graduate Studies 02-31
California Institute of Technology
Pasadena, California 91125

California Institute of Technology

THE FACULTY AND THEIR RESEARCH

Cellular Biology and Biophysics
Raymond J. Deshaies. Genetic and biochemical dissection of cell-cycle regulation in yeast.
William Dunphy. Biochemistry of cell-cycle control in *Xenopus* embryos.
Michael Elowitz. In vivo modeling: a synthetic approach to regulatory networks.
Scott E. Fraser. Developmental neurobiology; optical methods.
Jean-Paul Revel. Morphology, biochemistry, and evolution of plasma membrane specializations, with particular emphasis on gap junctions.
Alexander J. Varshavsky. Mechanics, functions, and applications of intracellular proteolysis; the ubiquitin system and the N-end rule pathway.

Molecular Biology, Biochemistry, Developmental Biology, Genetics, and Immunology
Giuseppe Attardi. The mitochondrial genetic system in animal cells; selective amplification of nuclear genes in human cell lines.
David Baltimore. Immunology and molecular biology.
Pamela J. Bjorkman. Crystal structures of cell-surface proteins involved in the immune response.
Marianne Bronner-Fraser. Analysis of neural crest cell migration and differentiation in the developing nervous system using embryological, cell biological, and molecular approaches.
Judith L. Campbell. Regulation of DNA replication in the yeast cell cycle: genetic and biochemical analysis.
David C. Chan. Biochemical and structural analysis of Fzo protein.
Eric H. Davidson. Molecular biology of early development; regulation of lineage-specific gene activation in sea urchin embryos; control of gene expression in eukaryotes; DNA sequence organization and its evolution.
Bruce Hay. Molecular mechanisms of *Drosophila* development; genetics; signal transduction; cell death.
Grant Jensen. Study of structure of large protein complexes and their arrangement within living cells by cryoelectron microscopy.
Stephen L. Mayo. Protein design; protein structure/stability correlations; protein NMR.
Elliot M. Meyerowitz. Molecular developmental genetics; molecular analysis of flower development in *Arabidopsis*; signal transduction in plants; plant evolution.
Ellen Rothenberg. Development of lymphocytes; changes in gene expression during T-cell maturation; lymphocyte activation pathways.
Melvin Simon. Mechanisms of signal transduction; molecular mechanisms in development; mouse genetics.
Angelike Stathopoulos. Gene regulatory network controlling gastrulation in *Drosophila;* control of gene expression; mechanisms of signal transduction.
Paul Sternberg. Molecular genetics of nematode development, behavior, and evolution; signal transduction; tumor suppressor genes.
James H. Strauss. Replication and evolution of alphaviruses and flaviviruses and of other RNA viruses.
Barbara Wold. Molecular genetic studies of the structure, function, and regulation of mammalian cell-surface receptors.

Neurobiology
John M. Allman. Physiological basis of perception; organization and evolution of the primate visual system.
Richard A. Andersen. Cortical neurophysiology, visuospatial perception, and visual-motor integration.
David Anderson. Cell fate determination in vertebrate neurogenesis.
Seymour Benzer. Neurogenetics of *Drosophila* nervous system development and behavior.
Mary B. Kennedy. Molecular mechanisms of central nervous system synaptic plasticity.
Christof Koch. Biophysics; computational neuroscience; computational vision.
Masakazu Konishi. Neuroethology; behavioral, neurophysiological, and neuroanatomical studies of avian auditory and vocal systems.
Gilles Laurent. Olfactory and mechanosensory processing in insects; synaptic transmission in the avian brain.
Henry A. Lester. Chemical transmission at synapses and within cells; gating of membrane channels; molecular approaches.
Paul H. Patterson. Regulation of neuronal phenotype by cytokines; interactions between the nervous and immune systems; mapping of position in the mammalian nervous system.
Erin Schuman. Mechanisms of synaptic modification; role of NO as a retrograde second messenger.
Shinsuke Shimojo. Visual psychophysics and behavioral studies of sensory-motor functions.
Athanassios Siapas. Learning and memory formation across distributed networks of neurons; multitetrode electrophysiological recordings from freely behaving rodents.
Kai Zinn. Molecular genetic studies of insect nervous system development; signal transduction in vertebrate olfaction.

CARNEGIE MELLON UNIVERSITY

Department of Biological Sciences

Programs of Study

The Department of Biological Sciences at Carnegie Mellon University offers a research-oriented program leading to the Ph.D. degree. Major research activities encompass biochemistry, biophysics, cell biology, developmental biology, genetics, molecular biology, neurobiology, and computational biology, with a continuing special emphasis on interdisciplinary projects. In collaboration with a faculty committee, students construct unique course programs according to their needs and professional goals. To qualify for the Ph.D., a graduate student must take a core course in biochemistry, cell biology, and either molecular biology or neuroscience; pass a qualifying examination; write and defend a research proposal; submit a thesis based on the student's research; and pass a final oral examination. Research is carried out under the guidance of a faculty member and a research advising committee.

In addition, the Department of Biological Sciences offers a three- to four-semester program leading to the M.S. in computational biology for promising students who plan careers in this field. The focus of the program is the development of practical skills required for success in computational biology. There is a low student-faculty ratio, ensuring that each student receives individual attention and advising; a unique program of research and course work is designed to address the needs and goals of each student. More information is available at http://www.cmu.edu/bio/graduate.

One of the most attractive features of the Department is the opportunity for close interaction among graduate students and faculty members. In addition to formal teaching and research programs, the Department offers a seminar series and weekly journal club meetings. The seminar series invites scientists from the United States and abroad to discuss their research with faculty members and graduate students. At journal club meetings, students and faculty members present topical and controversial papers from current literature.

Research Facilities

The Department of Biological Sciences is based in the Mellon Institute at Carnegie Mellon. On site are modern research and teaching laboratories, sophisticated computing facilities, the biology and chemistry library, an electron microscopy facility, two animal facilities, lecture and seminar rooms, instrument and chemical storage areas, glassblowing and machine shop services, a drafting facility, and office space. In addition to the Department of Biological Sciences, the eight-story Mellon Institute building houses the University's Department of Chemistry, the Pittsburgh NMR Center, the Center for the Neural Basis of Cognition, and the Molecular Biosensor and Imaging Center.

Financial Aid

Each graduate student accepted into the full-time Ph.D. program automatically receives financial support, which includes a full tuition remission valued at approximately $31,800 (2006–07), a stipend valued at approximately $22,600 (2006–07), and a health insurance allowance of $870 (2006–07). All Ph.D. candidates are required to perform teaching service that is unrelated to financial support. A limited number of fellowships are available for students in the M.S. in computational biology program.

Cost of Study

Tuition for the M.S. in computational biology program is approximately $31,800 (2006–07).

Living and Housing Costs

University housing for graduate students is not available, but off-campus housing is plentiful and affordable. The Carnegie Mellon Housing Office provides further information at http://www.housing.cmu.edu.

Student Group

Graduate enrollment at Carnegie Mellon University totals more than 4,000 and includes students from all parts of the United States and many other countries. In 2006, the Department of Biological Sciences had 50 Ph.D. and M.S. graduate students and 10 postdoctoral fellows.

Student Outcomes

Most of the students who graduated from the Department of Biological Sciences in the last three years have obtained postdoctoral positions. They are currently at the University of California in San Francisco, Northwestern University, University of Chicago, Albert Einstein College of Medicine at Yeshiva University, Brown University, Cornell University, and Carnegie Mellon, among others.

Location

Pittsburgh, a large metropolitan area of more than 2 million people, is the headquarters for many of the nation's largest corporations. There is a large concentration of research laboratories, biotechnology companies, hospitals, and universities in the area. Carnegie Mellon is located in Oakland, the educational center of the city, and borders on attractive residential areas and Schenley Park, the largest of Pittsburgh's many parks. The campus is close to the many cultural and sports activities of the city and only 4 miles from the downtown business district.

The University

Carnegie Mellon, a private institution devoted to liberal professional education, was established in 1900 as the Carnegie Technical School with a gift from Andrew Carnegie. In 1912, the name of the school was changed to Carnegie Institute of Technology. Mellon Institute was founded in 1913 by A. W. Mellon and R. B. Mellon for general research in the sciences and for cooperation with industry in sponsored research and engineering projects. The two institutions merged in 1967 to become Carnegie Mellon University. The University has a total enrollment of about 9,665 students and approximately 1,125 teaching faculty members.

Applying

The Department of Biological Sciences seeks promising students interested in careers in biological research. Students with strong backgrounds in all areas of biological sciences, chemistry, mathematics, or physics who have graduated from a recognized four-year college, university, or institute of technology are considered for entry to the graduate program. The deadline for applying to the Ph.D. program is January 1.

For the M.S. program, highest priority is given to those who apply by February 15, and the final deadline for acceptance into the M.S. program is May 1.

Applicants are required to provide official reports of scores on the General Test of the Graduate Record Examinations, transcripts from all college-level institutions attended, and three letters from professional references. International students are required to submit official TOEFL scores. Prospective applicants should visit the Web site listed in the Correspondence and Information section to complete the online application.

Correspondence and Information

Graduate Programs Office
Department of Biological Sciences
Carnegie Mellon University
4400 Fifth Avenue
Pittsburgh, Pennsylvania 15213
Phone: 412-268-3012
Web site: http://www.cmu.edu/bio

Carnegie Mellon University

THE FACULTY AND THEIR RESEARCH

Eric T. Ahrens, Assistant Professor; Ph.D., UCLA. Biological imaging; advancing the state of the art of high-resolution MRI and using these techniques to visualize development, connectivity, function, and pathology of the vertebrate nervous system.

Alison L. Barth, Assistant Professor; Ph.D., Berkeley. Identifying the molecules and pathways involved in developmental and adult plasticity, using in vivo manipulations to induce changes in synaptic strength as well as whole-cell electrophysiological recordings.

Peter B. Berget, Associate Professor; Ph.D., Minnesota. Functional proteomics and genomics; gene and protein discovery and genome annotation in mammalian cells, using CD-tagging; constructing modified transposon and retroviral CD-cassette delivery vectors to be used in both DNA library and cell tagging experiments.

William E. Brown, Professor; Ph.D., Minnesota. Protein structure-function relationships; proteins and small-molecular-weight mediators that play important roles in airway injury and response; structure and function characterization of such molecules and their relationship to physiological responses such as inflammation, hypersensitivity, and epithelial cell loss.

Justin Crowley, Assistant Professor; Ph.D., Duke. Formation of neural circuitry; development of neural processing modules in primary visual cortex; combination of physiological and anatomical techniques used to explore both structure and function of neural circuitry in the developing brain.

Dannie Durand, Associate Professor; Ph.D., Columbia. Computational molecular biology; use of computational approaches to study the role of gene duplication in the acquisition of new gene function and the evolution of vertebrate genomes.

Charles A. Ettensohn, Professor; Ph.D., Yale. Cell migration and cell adhesion during development; understanding morphogenesis in developing multicellular animals; morphogenesis of the primary mesenchyme cells and cell-cell interactions that regulate the choice of cell fates during embryogenesis in the sea urchin embryo.

David D. Hackney, Professor; Ph.D., Berkeley. Enzyme mechanisms, regulation, and structure; the three main enzyme systems involved in biological energy transductions: kinesin ATPase, myosin ATPase, and ATP synthesizing complex of the inner mitochondrial membrane.

Chien Ho, Professor; Ph.D., Yale. Correlating the structure-function relationships in biological systems, particularly allosteric proteins, using hemoglobin as a model; membrane-associated proteins and enzymes; structure-function relationship in cell membranes; applications of NMR imaging and in vivo spectroscopy to investigate cellular structures and functions of living systems.

Jeffrey O. Hollinger, Professor and Director, Center for Bone Tissue Engineering; D.D.S., Ph.D., Maryland. Developing and designing tissue-engineered therapies to regenerate bone; matrices to deliver cells and soluble signaling molecules to bone-deficient sites; basic and translational efforts in molecular, cell, and developmental biology and in polymers and surgical research.

Jonathan W. Jarvik, Associate Professor; Ph.D., MIT. Functional proteomics and genomics; CD-tagging, a technique that allows simultaneous tagging of genes, transcripts, and proteins in a single biochemical event; CD-tagging mammalian cells with epitope and GFP tags; identification of tagged genes; observation of tagged proteins in live cells; purification of tagged proteins.

Elizabeth W. Jones, Professor and Department Head; Ph.D., Washington (Seattle). Roles of intracellular proteinases; activities and genesis of the lysosome-like vacuole and enclosed hydrolases; study of the yeast *Saccharomyces cerevisiae*, which is amenable to genetic, biochemical, molecular biological, and cell biological analysis.

Frederick Lanni, Associate Professor; Ph.D., Harvard. Biophysical aspects of the cytoskeleton and cell motility; biophysics of cell motility; microscopy research and development; use of high-resolution light microscopy in the study of motile function in live cells and in reconstituted model systems.

Tina Lee, Assistant Professor; Ph.D., California, San Francisco. Membrane trafficking, organelle structure, and dynamics; complementary in vivo and in vitro approaches to study the mechanisms by which changes in cell physiology regulate trafficking steps within the secretory pathway of mammalian cells.

Adam D. Linstedt, Associate Professor; Ph.D., California, San Francisco. Regulation of organelle assembly; mechanisms regulating the assembly of the Golgi complex; role of giantin in the reassembly of the Golgi after cell division.

A. Javier López, Associate Professor; Ph.D., Duke. Developmental biology; posttranscriptional regulation of gene function; structure and function of protein family encoded by the Ubx locus of *Drosophila*, which specifies the pathway of differentiation followed by a particular group of segments in the insect body.

Brooke McCartney, Assistant Professor; Ph.D., Duke. Mechanisms of signal transduction and cytoskeletal organization during *Drosophila* development; genetic, cell biological, developmental, and biochemical techniques used to understand intersections between signal transduction and cytoskeletal organization in *Drosophila*, using the adenomatous polyposis coli (APC) family of tumor suppressors as a model.

William R. McClure, Professor Emeritus; Ph.D., Wisconsin–Madison. Mechanism and regulation of *Escherichia coli* DNA-dependent RNA polymerase; effect of DNA sequence and structure on promoter function; interaction of protein activators and repressors with RNA polymerase during initiation of RNA synthesis.

Jonathan S. Minden, Associate Professor; Ph.D., Yeshiva (Einstein). Developmental biology; molecular basis of pattern formation; use of a wide variety of approaches, from classical genetics to state-of-the-art computer-assisted fluorescence microscopy, to investigate pattern formation in *Drosophila*.

Robert F. Murphy, Professor; Ph.D., Caltech. Investigation of mechanisms and pathways of receptor-mediated endocytosis and protein localization using fluorescence techniques, especially multiparameter flow cytometry, and computational biology.

John F. Nagle, Professor; Ph.D., Yale. Phase transitions in biomembranes and molecular interactions in lipid bilayer; molecular mechanisms for proton transport through membranes in connection with bioenergetic processes.

Gordon S. Rule, Professor; Ph.D., Carnegie Mellon. NMR studies of protein structure dynamics; enzyme-substrate, protein-lipid, antibody-antigen, and protein–nucleic acid interactions.

Nathan N. Urban, Assistant Professor; Ph.D., Pittsburgh. How circuitry of the olfactory bulb transforms the spatially segregated, rate-coded, and combinatorial glomerular odor representation into one that is more spatially homogenous and more sparse and its dependence on the precise timing of mitral cell spikes; transformation implementation in the circuitry of the bulb.

James F. Williams, Professor; Ph.D., Toronto. Defining the organization of the adenovirus genome in terms of location and function of genes controlling virus development during infection; requirement for adenogene products in oncogenic transformation of rodent cells.

John L. Woolford Jr., Professor; Ph.D., Duke. Mechanism of ribosome assembly; genes necessary for messenger RNA splicing in yeast; utilizing *Saccharomyces cerevisiae*, due to the ease with which it can be manipulated for biochemical, genetic, molecular biological, and cell biological experiments.

CASE WESTERN RESERVE UNIVERSITY

Biomedical Sciences Training Program

Programs of Study	The Biomedical Sciences Training Program (BSTP) offers unique opportunities for Ph.D. students in the biomedical sciences. Case School of Medicine (SOM) is an outstanding research-intensive medical school that is strongly committed to graduate education and research. With more than 200 faculty members who can serve as mentors, the BSTP gives students maximum flexibility in selecting from over a dozen Ph.D. programs described in the section on the faculty and their research. Students seeking the combined Ph.D./M.D. degree may perform their thesis research under the aegis of the Medical Scientist Training Program. BSTP faculty members are located in the CWRU School of Medicine and at affiliated research institutions that include the University Hospitals of Cleveland, the Cleveland Clinic Foundation, the MetroHealth Medical Center, and the Cleveland VA Medical Center. These institutions provide a highly interactive environment for teaching and research.

BSTP students spend their first semester doing laboratory rotations with prospective thesis mentors. During the first year, students choose their advisers and join one of the graduate programs. They then earn their degrees by completing the requirements of the individual Ph.D. programs. First-year students also take a comprehensive course in cell and molecular biology. This course provides a broad exposure to modern biology and a foundation for advanced courses.

After the first year, the emphasis is on thesis research. Students interact closely with their advisers as they develop their skills in planning, performing, and interpreting experiments. Students also take advanced courses devoted to current discoveries. In addition, students present seminars and participate in journal clubs. Travel support is provided for presentation of research results at national meetings. The BSTP is committed to bringing the brightest students from a broad range of backgrounds to tackle the biomedical challenges of the future.

Research Facilities Excellent facilities enhance the training environment. Modern facilities are available for the biomedical, genetic, and behavioral analyses of bacteria, fruit flies, frogs, and mammals, including a state-of-the-art transgenic mouse facility and proteomic and microarray cores. Imaging facilities for confocal, two-photon electron microscopic analyses are available. Centralized facilities are available for amino acid and DNA sequencing, oligonucleotide synthesis, electron microscopy, X-ray crystallography, mass spectrometry, Raman spectrometry, EPR, NMR, and flow cytometry. The Wolstein biomedical research facility, which opened in 2004, significantly increases research laboratories at Case, and new development of the West Quad is anticipated.

Financial Aid Ph.D. students are appointed fellows and receive an annual stipend during their graduate training; the stipend is $23,000 for 2006–07. Full tuition and health insurance are provided for all students.

Cost of Study The cost of study, including tuition, is covered for all students throughout their training.

Living and Housing Costs Many students live off campus in nearby apartments and houses, and the Case Housing Office maintains postings for reasonably priced housing in the vicinity of the University. Residential neighborhoods are adjacent to the campus.

Student Group There are 9,447 students enrolled in the University; of this number, 2,179 are enrolled in the graduate school, 3,293 in the professional schools, and 3,975 are undergraduates. About 60 students enroll in the Ph.D. programs at Case SOM each year, and a total of 350 doctoral graduate students are registered throughout the biomedical sciences at Case.

Location Cleveland is an exceptional city in which to work, live, and learn, with some 2 million people in a cosmopolitan community rated among the most livable in the United States. Noted favorably were its health-care institutions, the arts, and the four-season climate. Cleveland is home to thirty Fortune 500 companies that are characteristic of its healthy, diversified economy. Downtown Cleveland offers a range of attractions to rival any city in the nation. Visitors marvel at assets, which include the Cleveland Museum of Art, Rock and Roll Hall of Fame and Museum, Great Lakes Science Center, and first-rate professional sports teams. Nightlife features a variety of theater and dance options in Playhouse Square as well as the popular riverfront clubs of the Flats and numerous restaurants and bistros offering live music. Numerous winter and summer recreational facilities are situated in the surrounding area.

The University Since its founding in 1843, Case SOM has been a vanguard of progress in research and education. Case was recently named "the most wired" university in the nation, reflecting the adoption of electronic-learning strategies. As a major research institution, Case enjoys outstanding programs in genomics, cancer biology, the study of the brain, and the biology of infections as well as emerging programs in imaging research and nanotechnology. The University has outstanding sports facilities and a variety of student organizations.

Applying Prerequisites for entrance into the program are organic chemistry and mathematics through calculus. A full course in biochemistry and molecular biology is recommended. Applications should be submitted by January 15, but late applications are accepted. Overseas applicants should apply early. Online applications are preferred (http://www.case.edu/med/BSTP/), but applications may also be submitted in hard copy. Scores from the Graduate Record Examinations (GRE) are required.

Admission to the Ph.D. program is determined by a committee representing all participating programs. Admitted students are free to join any graduate training program.

Correspondence and Information
Debbie Noureddine
School of Medicine, TG1
Case Western Reserve University
10900 Euclid Avenue
Cleveland, Ohio 44106-4934

Phone: 216-368-3347
E-mail: bstp@case.edu
Web site: http://www.case.edu/med/BSTP/
       http://casemed.case.edu/gradprog

Case Western Reserve University

THE FACULTY AND THEIR RESEARCH

The BSTP has a dozen member graduate programs with more than 200 faculty members. A complete listing of individual graduate programs and roster of faculty members and their research can be accessed via the Web at http://www.case.edu/med/BSTP and searched by research interest at http://casemed.case.edu/gradprog.

Anatomy
The anatomy program provides multidisciplinary training with strengths in biological anthropology, cell injury (responses to ischemia and toxins), and developmental neurobiology (mammalian brain stem respiratory centers).

Biochemistry
Biochemistry research extends from the study of how genes are transcribed to the structure and function of the proteins encoded by these genes. Increasingly, biochemistry has become the common language for translating the advances in molecular genetics into cellular and molecular terms. Biochemistry program faculty members have research interests in the areas of structural biology, regulation of gene expression, proteins and enzymes, and metabolic regulation and gene therapy.

Cell Biology
The cell biology program is focused on understanding the relationship between cellular organization and cell function. This problem is addressed by studying intact cells, subcellular organelles, and macromolecular assemblies. Areas of research in the cell biology program include cell growth and its control, cell-cell recognition and signaling, cytoskeleton and extracellular matrix, protein secretion and endocytosis of membrane components, organelle genesis, and functions of the nuclear envelope.

Molecular and Cellular Basis of Disease and Immunology
Experimental pathology is concerned with the origins, manifestations, and mechanisms of disease; with homeostatic and protective mechanisms of the body; and with therapy. The program has a commitment to the areas of immunology, cancer biology, tissue injury and healing, biomaterials biocompatibility, and aging and Alzheimer's disease.

Molecular Biology
The molecular biology program emphasizes investigations of the structure-function relationships of genes, RNA, and proteins in important model systems such as viruses, bacteria, yeast, and mammalian cells. Research and training strengths include regulation of gene expression, signal transduction, cell division, molecular genetics, and the structure, function, and enzymology of RNA.

Molecular, Developmental, and Human Genetics
The combination of classical and molecular genetics represents one of the most powerful sets of tools for probing the mechanisms underlying biological processes. The program emphasizes the use of modern genetic and molecular approaches to understand fundamental biological processes in *Drosophila, C. elegans,* mouse, and human systems. Studies include chromosome structure and function, the genetic basis of development, and the molecular basis of human disease. The training environment benefits from a close association between basic scientists and medical scientists in the Center for Human Genetics.

Molecular Virology
Viruses are the pathogenic agents of diseases ranging from the common cold to AIDS and cancer. Viruses also provide outstanding model systems for studying biological processes such as cellular proliferation, transcription, translation, splicing, and DNA replication. Furthermore, because viruses introduce genetic material into cells as part of their life cycle, they are being developed as vectors for gene therapy. Faculty members study all these aspects of viruses and their hosts. Research strengths include viral replication, virus-cell interactions, mechanisms of interferon action, viral oncogenesis, and the construction of viral vectors for gene therapy.

Neurosciences
Understanding the brain is one of the great frontiers of biology. Important problems include how the nervous system processes information and controls behavior, how it is assembled, and how it malfunctions in neurological and psychiatric disorders. Research strengths in the program include cellular and developmental neurobiology, with an emphasis on synaptic function and plasticity; development of the nervous system; information processing by simple neurocircuits; and pharmacology of neurotransmitters and receptors. The basis of neurological and psychiatric diseases such as Alzheimer's, schizophrenia, and depression are also major research areas.

Nutritional Sciences
The program has both fundamental and applied components. The fundamental component studies the metabolism of nutrients in the framework of metabolic regulation. Research focuses on developing nutrient therapy in newborn infants, investigating metabolism in humans by stable isotope technology, noninvasive procedures for studying nutrient metabolism, developing transgenic animal models for metabolic disorders, developing techniques to introduce chimeric genes into animals to correct metabolic defects, and characterizing genes of metabolic interest. Applied nutritional sciences include dietetics, epidemiology, education, and population counseling.

Pharmacological Sciences
The goal of pharmacology is to discover, devise, and develop agents that are useful in the treatment of disease. This is best accomplished when complex cellular functions and mechanisms of drug action are understood at the molecular level. The training program is broadly defined to meet these difficult challenges. Research opportunities are concerned with the interaction of chemical signals with biological systems. Areas of research strength include intracellular signaling, cancer biology, transcriptional regulation, apoptosis, structural biology, bacterial pathogenesis, neuronal signaling and development, cytokines, drug metabolism, functional genomics, and endocrinology.

CITY OF HOPE NATIONAL MEDICAL CENTER / BECKMAN RESEARCH INSTITUTE

City of Hope Graduate School of Biological Sciences

Programs of Study

The mission of the City of Hope Graduate School of Biological Sciences is to train students to be outstanding research scientists in chemical, molecular, and cellular biology. Graduates of this program, awarded the degree of Doctor of Philosophy in biological sciences, are equipped to address fundamental questions in the life sciences and biomedicine for careers in academia, industry, and government. The time spent in the program is devoted to full-time study and research. During the first year, the student completes the core curriculum and three laboratory rotations (ten to twelve weeks each). The core curriculum contains biochemistry, molecular biology, cell biology, and biostatistics/bioinformatics. One Advanced Topics course is taken during spring of the first year. After the first year, the student prepares and orally defends a research proposal based on an original topic not related to previous work conducted by the student. An additional Advanced Topics course is required after the first year and students are required to take a literature-based journal club every year after the first year. Students also participate in workshops on scientific communication and on the responsible conduct of research. After successfully completing the core curriculum and research proposal, students concentrate the majority of their time on their individual dissertation laboratory research project. The written thesis/dissertation must be presented by the student for examination by 4 members of the City of Hope staff and 1 qualified member from an outside institution.

Research Facilities

City of Hope is a premier medical center, one of forty National Cancer Institute–designated Comprehensive Cancer Centers. Its Beckman Research Institute launched the biotech industry by creating the first human recombinant gene products, insulin and growth hormone, which are now used by millions of people worldwide. State-of-the-art facilities include mass spectrometry, NMR, molecular modeling, cell sorting, DNA sequencing, molecular pathology, scanning and transmission electron microscopy, confocal microscopy, and molecular imaging. The Lee Graff Medical and Scientific Library allows access to the latest biomedical information via its journal and book collection, document delivery, interlibrary loans, and searches of online databases.

Financial Aid

All students in the Graduate School receive a fellowship of $26,000 per year as well as health and dental insurance.

Cost of Study

There are no tuition charges. A registration fee of $50 per semester ($150 per year) is the student's only financial obligation to City of Hope.

Living and Housing Costs

The School has limited, low-cost housing available. Additional housing is available within the immediate area at a cost of $600 to $900 per month.

Student Group

The Graduate School faculty consists of 69 of City of Hope's investigators. Fifty-one graduate students were working toward the Ph.D. degree in biological sciences in 2004–05.

Student Outcomes

Graduates have gone on to work as postdoctoral fellows at California Institute of Technology; Harvard University; Scripps Research Institute; Stanford University; University of California, Los Angeles; University of California, San Diego; University of California, Irvine; University of Missouri; University of Southern California; and Washington University in St. Louis. Graduates have also found positions with Wyeth-Ayerst Research; Allergan, Inc.; and the U.S. Biodefense and its subsidiary, Stem Cell Research Institute of California, Inc.

Location

City of Hope is located 25 miles northeast of downtown Los Angeles, minutes away from Pasadena and close to beaches, mountains, and many recreational and cultural activities. Many academic institutions, such as California Institute of Technology, UCLA, and USC, are less than an hour's drive away.

The Medical Center and The Institute

City of Hope was founded in 1913, initially as a tuberculosis sanatorium. Research programs were initiated in 1951, and, in 1983, the Beckman Research Institute of City of Hope was established with support from the Arnold and Mabel Beckman Foundation. The Institute comprises basic science research groups within the Divisions of Biology, Immunology, Molecular Medicine, and Neurosciences, among others.

Applying

The deadline for application is February 1 for classes starting in September. Applying early is advisable. Candidates must take the Graduate Record Examinations, including both the General Test and an appropriate Subject Test. For further information and an application, students should contact the School at the address listed in this description.

Correspondence and Information

Tom LeBon, Ph.D.
City of Hope Graduate School of Biological Sciences
1450 East Duarte Road
Duarte, California 91010-3000
Phone: 626-301-8293
Fax: 626-301-8105
E-mail: tlebon@coh.org
Web site: http://gradschool.coh.org

City of Hope National Medical Center/Beckman Research Institute

THE FACULTY AND THEIR RESEARCH

Karen S. Aboody. Neural stem cells and cancer.
Adam M. Bailis. Molecular mechanisms of genome stability; mechanisms of recombination.
Ganesaratnam Balendiran. Structure of nucleic acid and nucleic acid–binding proteins.
Michael E. Barish. Development and regulation of potassium currents in the central nervous system.
William P. Bennett. Breast cancer chemoprevention by tamoxifen and mutation spectrum analysis of iatrogenic cancers.
Ravi Bhatia. Abnormal beta 1 integrin-mediated microenvironmental regulation of hematopoiesis in chronic myelogenous leukemia; pathogenesis of myelodysplastic syndrome.
Chauncey W. Bowers. Extracellular signaling.
Edouard M. Cantin. Pathogenesis of herpes simplex virus in the nervous system.
Saswati Chatterjee. Adeno-associated virus vectors for gene therapy.
Shiuan Chen. Role of aromatase in breast cancer; molecular basis of catalytic differences between isoforms of DT-diaphorase.
Wenyong Chen. Epigenetics; cancer and aging.
Yuan Chen. Structural studies of proteins involved in transcription regulation; protein-protein recognition in the ubiquitination pathway.
Fong-Fong Chu. Role of GPX-GI in colon carcinogenesis.
Don J. Diamond. Vaccine research and development.
Richard W. Ermel. Graduate training in laboratory animal medicine; applied animal research.
Barry Marc Forman. Endocrinology of orphan nuclear receptors.
Rajesh Gaur. Mechanisms of protein RNA interactions.
Carlotta A. Glackin. Transcription factors in osteogenesis.
Gerald P. Holmquist. Mutation mechanisms, chromosome structure and evolution.
Kazuo Ikeda. Synaptic transmission mechanisms.
Keiichi Itakura. Function of the human cytomegalovirus (HCMV) matrix attachment region (MAR); function and regulation of the Hoxa-13 gene in differentiation and liver carcinogenesis.
Michael Jensen. T-lymphocyte gene therapy MicM.
Markus Kalkum. Mass spectrometry of peptides and proteins.
Susan Kovats. MHC class II–mediated antigen presentation.
Theodore G. Krontiris. Genetic instability and gene-gene interactions leading to cancer.
Terry D. Lee. Mass spectral analysis of peptides and proteins.
Ren-Jang Lin. Structure and mechanisms of RNA splicing.
Chih-Pin Liu. TCR/CD3 signaling and T-cell development; TCR and MHC-antigen interactions in autoimmune disease.
Qiang Lu. Formation of neuronal circuits.
Jeffrey R. Mann. Role of epigenetics in mammalian development.
Marcia M. Miller. Immunobiology of the avian major histocompatibility complex.
Rama Natarajan. Cellular and molecular mechanisms of diabetes-induced accelerated cardiovascular disease.
Edward M. Newman. Biochemistry and pharmacology of antineoplastic drugs.
Timothy R. O'Connor. Function of DNA repair proteins and repair mechanisms.
Gerd P. Pfeifer. Mechanisms of carcinogenesis.
Arthur D. Riggs. Gene regulation, chromatin structure and mechanisms of heritable epigenetic changes.
John J. Rossi. RNA processing and therapeutic RNA applications.
Paul M. Salvaterra. Neural gene regulation.
Binghui Shen. DNA replication/repair nucleases in mutagenesis and tumorigenesis.
Yanhong Shi. Nuclear receptors in neural stem cells and adult neurogenesis.
Chu-Chih Shih. Creation and maintenance of long-term hematopoietic stem cells; SCID-hu mouse system.
John E. Shively. CEA gene family; protein chemistry; anti-CEA antibodies for tumor imaging and therapy.
Judith Singer-Sam. Monoallelic expression in the central nervous system.
Steven S. Smith. DNA methyltransferase, terminal transferase and tumor biology.
Steve S. Sommer. Spontaneous mutagenesis in humans; genetic predisposition to multifactorial disease.
John Termini. Endogenous oxidative damage of nucleic acids and mutations.
Toshifumi Tomoda. Role of axonal trafficking in axon/synapse formation.
James E. Vaughn. Development of the mammalian nervous system.
Jeffrey N. Weitzel. Clinical cancer genetics; ovarian carcinogenesis.
Jiing-Kuan Yee. Development of lentiviral vectors for human gene therapy.
Yun Yen. Resistance to chemotherapeutic agents.
John A. Zaia. Experimental AIDS therapy.
Defu Zeng. Regulatory T cells in transplantation tolerance.

Associate Faculty

Daniela Castanotto. Gene therapy vector development.
Steven Esworthy. Role of GPX-GI in colon carcinogenesis.
Steve Flanagan. Transcription factors in osteogenesis.
Stephen Forman. Hematopoietic stem cell biology and bone marrow transplantation.
Kathleen Hill. Molecular biology of metagenesis.
Thomas LeBon. Proteins involved in diabetes and stem cell biochemistry.
Andrew Raubitschek. Radioimmunotherapy.
Mark Sherman. Molecular modeling.
Stella Tommasi. Mechanisms of carcinogenesis.
Robert Whitson. Transcription factors involved in metabolism and development.
Jeffrey Wong. Radiation research.
Mary Young. Protein mass spectrometry sequencing.

CLEMSON UNIVERSITY

Master of Science in Biological Sciences

Program of Study

The Master of Science in Biological Sciences Program at Clemson University is offered through the Department of Biological Sciences. The biological sciences program encompasses a wide variety of disciplines, including animal behavior, biomechanics and functional morphology, cell biology, conservation biology, developmental biology, ecology, evolution, immunology, physiology, population genetics, and systematics. The overall goals of the program are to develop scientists with strong interdisciplinary skills in research design, critical thinking, and communication in the biological sciences as well as expertise in a specific research area.

The M.S. degree program is usually completed within three years and requires 24 semester credit hours of formal course work and 6 semester credit hours of thesis research (a nonthesis option is available). The first year emphasizes general course work and research topic and proposal development. The second and third years focus on research and thesis preparation and defense. The M.S. in Biological Sciences Program prepares students for careers as biologists in private industry, consulting, state and federal agencies, and junior college instruction.

Research Facilities

The Department of Biological Sciences houses research instrumentation and maintains affiliations with several campuswide facilities. Imaging laboratories contain confocal and scanning and transmission electron microscopes. Additional electron microscopes are available through the University's microscopy facility off campus. The Clemson University DNA-sequencing facility provides ABI and LiCor instrumentation for DNA sequencing and fragment analysis. A real-time PCR machine, fluorimeter plate reader, and spectrophotometers are also available. Instron 8874 and 1321 biaxial mechanical testing systems are available through the Department of Bioengineering. A CT scanning facility is also available through the Greenville Hospital System. Comparative osteological and alcohol-preserved specimens as well as more than 104,000 herbarium specimens are available in the Campbell Museum of Natural History.

Research infrastructure for field studies includes access to field sites through the South Carolina Botanical Garden and the Clemson University Experimental Forest. Greenhouse space is available in the new Biotechnology Greenhouses, and flow-through systems and artificial stream channels are available in the University's Aquatic Research Facility.

Financial Aid

Graduate teaching assistantships are available through the Department of Biological Sciences to support teaching in undergraduate laboratories. These assistantships may be renewed annually as long as a student is making satisfactory progress toward the degree and are limited to three years for M.S. candidates. The twelve-month stipends are currently $12,000 for M.S. students. Graduate research assistantships are also available through extramurally funded research programs with individual faculty members within the department. A limited number of graduate teaching assistantships are currently available for the upcoming academic year.

Cost of Study

Tuition for 2006–07 is $4643 per semester for in-state students and $9255 per semester for nonresidents. Off-campus rates are $535 per hour for in-state students and $918 per hour for nonresidents. Graduate assistants pay a flat fee of $1079 per semester and $348 per summer session. Graduate fellows pay South Carolina resident fees.

Living and Housing Costs

On-campus housing is available; for information, students should visit http://www.housing.clemson.edu. The cost of living in Clemson is quite low compared to the national average; students who choose to live off campus typically spend $350 to $450 per month for rent, depending on location, amenities, roommates, etc.

Student Group

The program has approximately 10 students. Seventy-three percent are women, 91 percent attend on a full-time basis, and 9 percent are international students.

Student Outcomes

Students who matriculate in the M.S. program seek advanced training through entrance to professional schools or Ph.D. programs or work as professionals in industrial or public utilities positions or health professions or in teaching positions at public, private, or technical schools.

Location

Clemson is a small, beautiful college town near the Blue Ridge Mountains and Lake Hartwell in upstate South Carolina. The Upstate region is one of the country's fastest-growing areas and is an important part of the I-85 corridor, a multistate area along Interstate 85 that runs from metropolitan Atlanta to Richmond, Virginia, and encompasses Charlotte, North Carolina, and North Carolina's Research Triangle. Atlanta and Charlotte are each a 2-hour drive away. Many financial institutions and other industries have national headquarters for a major presence in the Upstate, including Wachovia, Bank of America, BMW, Bon Secours St. Francis Health System, Bosch North America, Bowater, Charter Communications, Ernst & Young, Fluor Corporation, IBM, Microsoft, Michelin of North America, and many others.

The University

Clemson is classified by the Carnegie Foundation as Doctoral/Research University–Extensive, a category comprising less than 4 percent of all universities in America. The University's mission is to fulfill the covenant between its founder and the people of South Carolina to establish a "high seminary of learning" through its responsibilities of teaching, research, and extended public service. The University has identified eight areas of academic emphasis that create collaborations that, in turn, help fulfill the University's mission.

Applying

The biological sciences program seeks students with strong backgrounds in the biological sciences whose interests complement existing emphasis areas in evolution and ecology, cell and developmental biology, and comparative organismal biology. Applicants may apply on the Web at http://www.grad.clemson.edu/p_apply.html. Applications, with a $50 nonrefundable fee, should be received no later than five weeks prior to registration. Every required item in support of the application must be on file by that date. Students are advised to contact the department for the deadlines of the program of proposed study.

Correspondence and Information

Dr. Margaret B. Ptacek
Graduate Programs Coordinator
Department of Biological Sciences
132 Long Hall
Clemson University
Clemson, South Carolina 29634-0314
Phone: 864-656-2328
Fax: 864-656-0435
E-mail: mptacek@clemson.edu
Web site: http://www.clemson.edu/biosci/graduate/

Clemson University

THE FACULTY AND THEIR RESEARCH

Albert G. Abbott, Professor; Ph.D., Brown, 1980. Structural and functional genomic approaches to understanding gene and genome structure and evolution in eukaryotic organisms; genetic engineering of biobased materials. (E-mail: aalbert@clemson.edu)

Robert E. Ballard, Professor; Ph.D., Iowa, 1975. Biosystematics and chemotaxonomy of flowering plants. (E-mail: ballard@clemson.edu) .

Richard W. Blob, Assistant Professor; Ph.D., Chicago, 1998. Biomechanics and evolution of animal function; animal locomotion; comparative vertebrate anatomy, physiology, and functional morphology; herpetology; vertebrate paleontology. (E-mail: rblob@clemson.edu)

N. Dwight Camper, Professor; Ph.D., North Carolina State, 1966. Tissue and cell culture of medicinal plants; isolation and analysis of active ingredients of medicinal plants; selected bioassays of biological activity of medicinal plants, including antibacterial, antineoplastic, mutagenesis, estrogen binding activity, and antifungal bioassays; activity and mode of action of plant growth regulating chemicals. (E-mail: dcamper@clemson.edu)

Michael J. Childress, Assistant Professor; Ph.D., Florida State, 1995. Behavioral ecology; marine ecology; comparative sociobiology; invertebrate zoology; animal behavior; communication; evolutionary biology. (E-mail: mchildr@clemson.edu)

James M. Colacino, Associate Professor; Ph.D., SUNY at Buffalo, 1973. Comparative respiratory and circulatory physiology; invertebrate hemoglobin function; mathematical models of physiological systems. (E-mail: jmclc@clemson.edu)

Saara J. DeWalt, Assistant Professor; Ph.D., LSU, 2003. Population ecology and genetics of invasive plants; community ecology of woody plants, with emphasis on lianas (woody vines); tropical ecology. (E-mail: saarad@clemson.edu)

Yuqing Dong, Research Assistant Professor; Ph.D., Peking, 1999. Identifying and functionally characterizing the evolutionarily conserved genetic determinants important for longevity, using *C. elegans;* specifically, determining the transcriptional cofactors of DAF-16 and exploring the potential longevity pathways in *C. elegans.* (E-mail: ydong@clemson.edu)

Gene W. Eidson, Adjunct Assistant Professor; Ph.D., Clemson, 1990. Use of constructed wetlands for wastewater treatment; ecological restoration of wetlands, swamps, bottomland hardwood forests, and sandhill streams; remediation of pit lakes. Current projects involve restoration of Phinizy Swamp, Augusta, Georgia; restoration of Kennecott-Ridgeway Gold Mine, Ridgeway, South Carolina; and optimization of a 450-acre constructed wetland for advanced wastewater treatment, Augusta, Georgia. (E-mail: scientryst@aol.com)

Linda J. Gahan, Research Associate Professor; Ph.D., Illinois, 1968. Understanding mechanisms of Bt resistance in Lepidoptera (moths), most specifically the tobacco budworm, which is an agricultural pest of cotton, tobacco, tomatoes, and other crops. Bt is a naturally occurring pesticide produced by the bacterium, *Bacillus thuringensis,* and it is used to control insect pests of important agricultural crops. (E-mail: glinda@clemson.edu)

Vincent S. Gallicchio, Professor and Associate Vice President for Research; Ph.D., NYU, 1976. Experimental drug therapeutics for AIDS and cancer, with a focus on compounds that inhibit ribonucleotide reductase and antioxidants derived from natural food products; nonpsychiatric clinical uses of lithium, in particular, for Alzheimer's disease and viral-induced pathogens; experimental hematopoiesis, with particular interest in stem cell biology and role of the stroma/microenvironment in these processes; international education pertaining to biomedical laboratory science. (E-mail: vsgall@clemson.edu)

John J. Hains, Associate Professor; Ph.D., Clemson, 1987. Limnology; aquatic biology and ecology; river, lake, and reservoir ecosystems; watershed processes. (E-mail: jhains@clemson.edu)

Bradley Hersh, Assistant Professor; Ph.D., MIT, 2002. Development and evolution of animal shapes, with emphasis on insect wings; molecular mechanisms of gene regulation during *Drosophila* development; regulation of target genes by Hox proteins.

J. Jeffrey Isely, Professor and Leader of the South Carolina Cooperative Fish and Wildlife Research Unit; Ph.D., Texas A&M, 1984. Restoration ecology of endangered species, primarily shortnose and Gulf sturgeon and robust redhorse; behavioral ecology of fish, primarily movement in relation to natural and anthropogenic environmental changes and age and growth of fish, specializing in the evaluation of otolith microstructure. (E-mail: jisely@clemson.edu)

Patrick G. R. Jodice, Assistant Professor; Ph.D., Oregon State, 1999. Wildlife ecology; conservation biology; physiological ecology; ornithology; ecological energetics, foraging ecology, diet and nutrition, and avian diving behavior; various avian taxa, especially seabirds and shorebirds, although mammals and reptiles are also studied. (E-mail: pjodice@clemson.edu)

Alan R. Johnson, Assistant Professor; Ph.D., Tennessee, 1988. Application of computer simulation models to ecology, environmental toxicology, and risk assessment, specifically, the dynamics of fragmented populations (metapopulations), species invasions, and biogeochemical cycling of metals. (E-mail: alanj@clemson.edu)

Stephen J. Klaine, Professor; Ph.D., Rice, 1982. Fate and effects of contaminants in the environment, specifically contaminants that migrate from various land uses into aquatic ecosystems and their effects on aquatic plants and animals; contaminant effects on fish, aquatic invertebrates, plants, and algae; toxicity of metals and pesticides; development of strategies that facilitate the coexistence of economically sound land use with good environmental quality. (E-mail: sklaine@clemson.edu)

William R. Marcotte Jr., Associate Professor; Ph.D., Virginia, 1987. Plant molecular and developmental biology; plant genetic mechanisms, including hormone-mediated and posttranscriptional regulation of gene expression; physiology of desiccation-related proteins. (E-mail: marcotw@clemson.edu)

Peter B. Marko, Assistant Professor; Ph.D., California, Davis, 1997. Molecular population biology, biogeography, and conservation genetics. (E-mail: pmarko@clemson.edu)

Amy L. Moran, Assistant Professor; Ph.D., Oregon, 1997. Ecology and evolution of marine organisms; physiological and morphological adaptations of early life-history stages to varying environments. (E-mail: moran@clemson.edu)

Andrew S. Mount, Lecturer; Ph.D., Clemson, 1999. Cellular and molecular biology of biomineralization in mollusks, using the Eastern oyster *Crassostrea virginica* as a model: understanding how the organism nucleates calcium carbonate crystals, the mantle as a shell-forming organ, the role of collagen, and the investigation of the role of the immune system in shell formation; developing novel functional genomic approaches that will enable transcriptome analysis of specific cell types related to the secretion of organic matrix proteins. (E-mail: mount@clemson.edu)

Kimberly S. Paul, Assistant Professor; Ph.D., Princeton, 1998. Parasite-host adaptation in African trypanosomes; cell biology, biochemistry, and molecular biology of fatty acid metabolism; environmental sensing and regulation of lipid uptake and metabolism; lipid trafficking; mitochondrial biology. (E-mail: kpaul@clemson.edu)

Edward B. Pivorun, Professor; Ph.D., Minnesota, 1973. Comparative physiology of vertebrates, with emphasis on the physiology of daily torpor and hibernation in mammals; role of neuropeptides and neuroendocrines in modulating the physiological state of daily torpor; population dynamics of small mammals in the Great Smoky Mountains National Park. (E-mail: ebpvr@clemson.edu)

Margaret B. Ptacek, Associate Professor; Ph.D., Missouri–Columbia, 1991. Speciation; animal behavior and mating systems; population divergence in fishes; conservation genetics. (E-mail: mptacekk@clemson.edu)

Thomas E. Schwedler, Professor; Ph.D., Auburn, 1980. Infection kinetics of bacterial diseases of fish; effects of therapeutic agents on immune function in fish; health assessment and system design for ornamental fish. (E-mail: tschwdl@clemson.edu)

Barbara J. Speziale, Associate Professor; Ph.D., Clemson, 1985. Aquatic ecology and limnology research; education and extension outreach, K–12 youth development. (E-mail: bjspz@clemson.edu)

Timothy P. Spira, Professor; Ph.D., Berkeley, 1983. Ecology, evolution, and conservation biology of plants, with emphasis on plant-animal interactions. (E-mail: stimoth@clemson.edu)

Lesly A. Temesvari, Associate Professor; Ph.D., Windsor, 1987. Molecular and cellular mechanisms that govern the biogenesis and function of endosomes and lysosomes; cellular and molecular biological approaches used to investigate the role of several small molecular weight Rab GTPases in endosomal and lysosomal membrane and protein trafficking and in pathogenicity of the protozoan parasite, *Entamoeba histolytica.* (E-mail: ltemesv@clemson.edu)

David W. Tonkyn, Associate Professor; Ph.D., Princeton, 1985. Population and community ecology and conservation biology. (E-mail: tdavid@clemson.edu)

Peter Van den Hurk, Assistant Professor; Ph.D., William and Mary, 1998. Toxicology of environmental pollutants in aquatic ecosystems; effects of mixtures of contaminants on enzyme systems responsible for the detoxification of pollutants; fish species as relevant models and effect indicators for contaminated field situations; metabolism of toxicants in liver cells and intestinal subcellular fractions; cytochromes P-450, sulfotransferase, UDP-glucuronosyltransferase and glutathione-S-transferase. (E-mail: pvdurk@clemson.edu)

Alfred P. Wheeler, Professor and Department Chair; Ph.D., Duke, 1975. Cell and invertebrate physiology; whole organism, cellular, and molecular approaches to mechanisms of mineralization, especially the function of organic matrix molecules; development of biodegradable peptide polymers for industrial applications. (E-mail: wheeler@clemson.edu)

CLEMSON UNIVERSITY

Ph.D. in Biological Sciences

Program of Study	The Ph.D. in Biological Sciences Program is offered through the Department of Biological Sciences. The biological sciences program encompasses a wide variety of disciplines, including animal behavior, biomechanics and functional morphology, cell biology, conservation biology, developmental biology, ecology, evolution, immunology, physiology, population genetics, and systematics. The overall goals of the program are to develop scientists with strong interdisciplinary skills in research design, critical thinking, and communication in the biological sciences as well as expertise in a specific research area.
	The Ph.D. degree program is usually completed within five years and requires 18 semester credit hours of research. In addition, specific graduate courses may be recommended to develop breadth of knowledge in biological sciences as well as the specific area of research concentration. The program requires an approved plan of study (second year), successful completion of the qualifying exam for candidate status (third year), and research and dissertation preparation and defense (second through fifth years). The Ph.D. in Biological Sciences Program prepares students for careers as research scientists in academic institutions and with state, federal, and private agencies.
Research Facilities	The Department of Biological Sciences houses research instrumentation and maintains affiliations with several campuswide facilities. Imaging laboratories contain confocal and scanning and transmission electron microscopes. Additional electron microscopes are available through the University's microscopy facility off campus. The Clemson University DNA-sequencing facility provides ABI and LiCor instrumentation for DNA sequencing and fragment analysis. A real-time PCR machine, fluorimeter plate reader, and spectrophotometers are also available. Instron 8874 and 1321 biaxial mechanical testing systems are available through the Department of Bioengineering. A CT scanning facility is also available through the Greenville Hospital System. Comparative osteological and alcohol-preserved specimens as well as more than 104,000 herbarium specimens are available in the Campbell Museum of Natural History.
	Research infrastructure for field studies includes access to field sites through the South Carolina Botanical Garden and the Clemson University Experimental Forest. Greenhouse space is available in the new Biotechnology Greenhouses, and flow-through systems and artificial stream channels are available in the University's Aquatic Research Facility.
Financial Aid	Graduate teaching assistantships are available through the Department of Biological Sciences to support teaching in undergraduate laboratories. These assistantships may be renewed annually as long as a student is making satisfactory progress toward the degree and are limited to five years for Ph.D. candidates. The twelve-month stipends are currently $15,000 for students in the Ph.D. program. Graduate research assistantships are also available through extramurally funded research programs with individual faculty members within the department. A limited number of graduate teaching assistantships are currently available for the upcoming academic year.
Cost of Study	Tuition for 2006–07 is $4643 per semester for in-state students and $9255 per semester for nonresidents. Off-campus rates are $535 per hour for in-state students and $918 per hour for nonresidents. Graduate assistants pay a flat fee of $1079 per semester and $348 per summer session. Graduate fellows pay South Carolina resident fees.
Living and Housing Costs	On-campus housing is available; for information, students should visit http://www.housing.clemson.edu. The cost of living in Clemson is quite low compared to the national average; students who choose to live off campus typically spend $350 to $450 per month for rent, depending on location, amenities, roommates, etc.
Student Group	The program has approximately 10 students. Seventy-three percent are women, 91 percent attend on a full-time basis, and 9 percent are international students.
Student Outcomes	Students who matriculate in the Ph.D. program seek postdoctoral and university faculty appointments or high-level research positions in government or industry.
Location	Clemson is a small, beautiful college town near the Blue Ridge Mountains and Lake Hartwell in upstate South Carolina. The Upstate region is one of the country's fastest-growing areas and is an important part of the I-85 corridor, a multistate area along Interstate 85 that runs from metropolitan Atlanta to Richmond, Virginia, and encompasses Charlotte, North Carolina, and North Carolina's Research Triangle. Atlanta and Charlotte are each a 2-hour drive away. Many financial institutions and other industries have national headquarters for a major presence in the Upstate, including Wachovia, Bank of America, BMW, Bon Secours St. Francis Health System, Bosch North America, Bowater, Charter Communications, Ernst & Young, Fluor Corporation, IBM, Microsoft, Michelin of North America, and many others.
The University	Clemson is classified by the Carnegie Foundation as Doctoral/Research University–Extensive, a category comprising less than 4 percent of all universities in America. The University's mission is to fulfill the covenant between its founder and the people of South Carolina to establish a "high seminary of learning" through its responsibilities of teaching, research, and extended public service. The University has identified eight areas of academic emphasis that create collaborations that, in turn, help fulfill the University's mission.
Applying	The biological sciences program seeks students with strong backgrounds in the biological sciences whose interests complement existing emphasis areas in evolution and ecology, cell and developmental biology, and comparative organismal biology. Applicants may apply on the Web at http://www.grad.clemson.edu/p_apply.html. Applications, with a $50 nonrefundable fee, should be received no later than five weeks prior to registration. Every required item in support of the application must be on file by that date. Students are advised to contact the department for the deadlines of the program of proposed study.
Correspondence and Information	Dr. Margaret B. Ptacek Graduate Programs Coordinator Department of Biological Sciences 132 Long Hall Clemson University Clemson, South Carolina 29634-0314 Phone: 864-656-2328 Fax: 864-656-0435 E-mail: mptacek@clemson.edu Web site: http://www.clemson.edu/biosci/graduate/

Clemson University

THE FACULTY AND THEIR RESEARCH

Albert G. Abbott, Professor; Ph.D., Brown, 1980. Structural and functional genomic approaches to understanding gene and genome structure and evolution in eukaryotic organisms; genetic engineering of biobased materials. (E-mail: aalbert@clemson.edu)

Robert E. Ballard, Professor; Ph.D., Iowa, 1975. Biosystematics and chemotaxonomy of flowering plants. (E-mail: ballard@clemson.edu)

Richard W. Blob, Assistant Professor; Ph.D., Chicago, 1998. Biomechanics and evolution of animal function; animal locomotion; comparative vertebrate anatomy, physiology, and functional morphology; herpetology; vertebrate paleontology. (E-mail: rblob@clemson.edu)

N. Dwight Camper, Professor; Ph.D., North Carolina State, 1966. Tissue and cell culture of medicinal plants; isolation and analysis of active ingredients of medicinal plants; selected bioassays of biological activity of medicinal plants, including antibacterial, antineoplastic, mutagenesis, estrogen binding activity, and antifungal bioassays; activity and mode of action of plant growth regulating chemicals. (E-mail: dcamper@clemson.edu)

Michael J. Childress, Assistant Professor; Ph.D., Florida State, 1995. Behavioral ecology; marine ecology; comparative sociobiology; invertebrate zoology; animal behavior; communication; evolutionary biology. (E-mail: mchildr@clemson.edu)

James M. Colacino, Associate Professor; Ph.D., SUNY at Buffalo, 1973. Comparative respiratory and circulatory physiology; invertebrate hemoglobin function; mathematical models of physiological systems. (E-mail: jmclc@clemson.edu)

Saara J. DeWalt, Assistant Professor; Ph.D., LSU, 2003. Population ecology and genetics of invasive plants; community ecology of woody plants, with emphasis on lianas (woody vines); tropical ecology. (E-mail: saarad@clemson.edu)

Yuqing Dong, Research Assistant Professor; Ph.D., Peking, 1999. Identifying and functionally characterizing the evolutionarily conserved genetic determinants important for longevity, using *C. elegans;* specifically, determining the transcriptional cofactors of DAF-16 and exploring the potential longevity pathways in *C. elegans.* (E-mail: ydong@clemson.edu)

Gene W. Eidson, Adjunct Assistant Professor; Ph.D., Clemson, 1990. Use of constructed wetlands for wastewater treatment; ecological restoration of wetlands, swamps, bottomland hardwood forests, and sandhill streams; remediation of pit lakes. Current projects include restoration of Phinizy Swamp, Augusta, Georgia; restoration of Kennecott-Ridgeway Gold Mine, Ridgeway, South Carolina; and optimization of a 450-acre constructed wetland for advanced wastewater treatment, Augusta, Georgia. (E-mail: scientryst@aol.com)

Linda J. Gahan, Research Associate Professor; Ph.D., Illinois, 1968. Understanding mechanisms of Bt resistance in Lepidoptera (moths), most specifically the tobacco budworm, which is an agricultural pest of cotton, tobacco, tomatoes, and other crops. Bt is a naturally occurring pesticide produced by the bacterium, *Bacillus thuringensis,* and it is used to control insect pests of important agricultural crops. (E-mail: glinda@clemson.edu)

Vincent S. Gallicchio, Professor and Associate Vice President for Research; Ph.D., NYU, 1976. Experimental drug therapeutics for AIDS and cancer, with a focus on compounds that inhibit ribonucleotide reductase and antioxidants derived from natural food products; nonpsychiatric clinical uses of lithium, in particular, for Alzheimer's disease and viral-induced pathogens; experimental hematopoiesis, with particular interest in stem cell biology and role of the stroma/microenvironment in these processes; international education pertaining to biomedical laboratory science. (E-mail: vsgall@clemson.edu)

John J. Hains, Associate Professor; Ph.D., Clemson, 1987. Limnology; aquatic biology and ecology; river, lake, and reservoir ecosystems; watershed processes. (E-mail: jhains@clemson.edu)

Bradley Hersh, Assistant Professor; Ph.D., MIT, 2002. Development and evolution of animal shapes, with emphasis on insect wings; molecular mechanisms of gene regulation during *Drosophila* development; regulation of target genes by Hox proteins.

J. Jeffrey Isely, Professor and Leader of the South Carolina Cooperative Fish and Wildlife Research Unit; Ph.D., Texas A&M, 1984. Restoration ecology of endangered species, primarily shortnose and Gulf sturgeon and robust redhorse; behavioral ecology of fish, primarily movement in relation to natural and anthropogenic environmental changes and age and growth of fish, specializing in the evaluation of otolith microstructure. (E-mail: jisely@clemson.edu)

Patrick G. R. Jodice, Assistant Professor; Ph.D., Oregon State, 1999. Wildlife ecology; conservation biology; physiological ecology; ornithology; ecological energetics, foraging ecology, diet and nutrition, and avian diving behavior; various avian taxa, especially seabirds and shorebirds, although mammals and reptiles are also studied. (E-mail: pjodice@clemson.edu)

Alan R. Johnson, Assistant Professor; Ph.D., Tennessee, 1988. Application of computer simulation models to ecology, environmental toxicology, and risk assessment, specifically, the dynamics of fragmented populations (metapopulations), species invasions, and biogeochemical cycling of metals. (E-mail: alanj@clemson.edu)

Stephen J. Klaine, Professor; Ph.D., Rice, 1982. Fate and effects of contaminants in the environment, specifically contaminants that migrate from various land uses into aquatic ecosystems and their effects on aquatic plants and animals; contaminant effects on fish, aquatic invertebrates, plants, and algae; toxicity of metals and pesticides; development of strategies that facilitate the coexistence of economically sound land use with good environmental quality. (E-mail: sklaine@clemson.edu)

William R. Marcotte Jr., Associate Professor; Ph.D., Virginia, 1987. Plant molecular and developmental biology; plant genetic mechanisms, including hormone-mediated and posttranscriptional regulation of gene expression; physiology of desiccation-related proteins. (E-mail: marcotw@clemson.edu)

Peter B. Marko, Assistant Professor; Ph.D., California, Davis, 1997. Molecular population biology, biogeography, and conservation genetics. (E-mail: pmarko@clemson.edu)

Amy L. Moran, Assistant Professor; Ph.D., Oregon, 1997. Ecology and evolution of marine organisms; physiological and morphological adaptations of early life-history stages to varying environments. (E-mail: moran@clemson.edu)

Andrew S. Mount, Lecturer; Ph.D., Clemson, 1999. Cellular and molecular biology of biomineralization in mollusks, using the Eastern oyster *Crassostrea virginica* as a model: understanding how the organism nucleates calcium carbonate crystals, the mantle as a shell-forming organ, the role of collagen, and the investigation of the role of the immune system in shell formation; developing novel functional genomic approaches that will enable transcriptome analysis of specific cell types related to the secretion of organic matrix proteins. (E-mail: mount@clemson.edu)

Kimberly S. Paul, Assistant Professor; Ph.D., Princeton, 1998. Parasite-host adaptation in African trypanosomes; cell biology, biochemistry, and molecular biology of fatty acid metabolism; environmental sensing and regulation of lipid uptake and metabolism; lipid trafficking; mitochondrial biology. (E-mail: kpaul@clemson.edu)

Edward B. Pivorun, Professor; Ph.D., Minnesota, 1973. Comparative physiology of vertebrates, with emphasis on the physiology of daily torpor and hibernation in mammals; role of neuropeptides and neuroendocrines in modulating the physiological state of daily torpor; population dynamics of small mammals in the Great Smoky Mountains National Park. (E-mail: ebpvr@clemson.edu)

Margaret B. Ptacek, Associate Professor; Ph.D., Missouri–Columbia, 1991. Speciation; animal behavior and mating systems; population divergence in fishes; conservation genetics. (E-mail: mptacekk@clemson.edu)

Thomas E. Schwedler, Professor; Ph.D., Auburn, 1980. Infection kinetics of bacterial diseases of fish; effects of therapeutic agents on immune function in fish; health assessment and system design for ornamental fish. (E-mail: tschwdl@clemson.edu)

Barbara J. Speziale, Associate Professor; Ph.D., Clemson, 1985. Aquatic ecology and limnology research; education and extension outreach, K–12 youth development. (E-mail: bjspz@clemson.edu)

Timothy P. Spira, Professor; Ph.D., Berkeley, 1983. Ecology, evolution, and conservation biology of plants, with emphasis on plant-animal interactions. (E-mail: stimoth@clemson.edu)

Lesly A. Temesvari, Associate Professor; Ph.D., Windsor, 1987. Molecular and cellular mechanisms that govern the biogenesis and function of endosomes and lysosomes; cellular and molecular biological approaches used to investigate the role of several small molecular weight Rab GTPases in endosomal and lysosomal membrane and protein trafficking and in pathogenicity of the protozoan parasite, *Entamoeba histolytica.* (E-mail: ltemesv@clemson.edu)

David W. Tonkyn, Associate Professor; Ph.D., Princeton, 1985. Population and community ecology and conservation biology. (E-mail: tdavid@clemson.edu)

Peter Van den Hurk, Assistant Professor; Ph.D., William and Mary, 1998. Toxicology of environmental pollutants in aquatic ecosystems; effects of mixtures of contaminants on enzyme systems responsible for the detoxification of pollutants; fish species as relevant models and effect indicators for contaminated field situations; metabolism of toxicants in liver cells and intestinal subcellular fractions; cytochromes P-450, sulfotransferase, UDP-glucuronosyltransferase and glutathione-S-transferase. (E-mail: pvdurk@clemson.edu)

Alfred P. Wheeler, Professor and Department Chair; Ph.D., Duke, 1975. Cell and invertebrate physiology; whole organism, cellular, and molecular approaches to mechanisms of mineralization, especially the function of organic matrix molecules; development of biodegradable peptide polymers for industrial applications. (E-mail: wheeler@clemson.edu)

COLD SPRING HARBOR LABORATORY

Watson School of Biological Sciences

Program of Study	The Watson School of Biological Sciences at Cold Spring Harbor Laboratory (CSHL) offers an accredited graduate training program, which leads to the Ph.D. degree, to a select group of self-motivated students of outstanding ability and intellect. The curriculum takes advantage of the unique and flexible environment of CSHL and includes the following innovative features: approximately four years from matriculation to Ph.D. degree award, broad representation of the biological sciences, a first year with course work and laboratory rotations in separate phases, emphasis on the principles of scientific reasoning and logic, continued advanced course instruction throughout the graduate curriculum, and two-tier mentoring.

The program provides an exciting and intensive educational experience. The curriculum is designed to train self-reliant students who, under their own guidance, can acquire and assimilate the knowledge that their research or career demands require. The course work is varied, involving core courses, focused topic courses, and CSHL postgraduate courses.

The current fields of research expertise of CSHL faculty members are genetics, cellular and molecular biology, structural biology, developmental biology, virology, protein chemistry, cell-cycle regulation, plant genetics, electrophysiology, behavior, imaging, bioinformatics, computational neurobiology, and genomics. The laboratories of all CSHL research faculty members are available to students in the program.

Requirements for the award of the Ph.D. degree are successful completion of all course work, laboratory rotations, teaching (at the Laboratory's Dolan DNA Learning Center), the Ph.D. qualifying exam, thesis research and postdoctoral proposals, and defense of a written thesis that describes original research. The program aims to train future leaders in the biological sciences.

Research Facilities

Cold Spring Harbor Laboratory has state-of-the-art facilities for research in genetics; molecular, cellular, and structural biology; neuroscience; cancer; plant biology; and bioinformatics. As a National Cancer Institute–designated Cancer Center, there is an extensive set of shared resources. There are several libraries and one archive on campus. Library services, such as database searching and reference and interlibrary loan services, are available. An information technology department provides campuswide support of computing.

Financial Aid

The Watson School of Biological Sciences supports each student with an annual stipend, health benefits, affordable housing, subsidized food, and funds for tuition and research costs. To enhance their careers, students are encouraged to seek independent funding through predoctoral fellowships from, for example, the National Science Foundation.

Cost of Study

The Watson School of Biological Sciences provides full remission of all tuition fees for all accepted students. The School also supports the stipend and research costs of each student for four years.

Living and Housing Costs

The Laboratory provides affordable housing to all graduate students through a network of on-site and off-site housing. Single graduate students are offered single rooms in shared houses with house-cleaning services; married students are housed in apartments. First-year students of the Watson School are offered housing in the Townsend Knight House, a newly renovated house from 1810 that is located on the shore of Cold Spring Harbor opposite the Laboratory.

Student Group

The class size is approximately 8 to 10 students per year. The entering class of 2006 included students from the United States, Israel, Canada, France, and Italy. The Laboratory aims to produce graduates in the biological sciences who are likely to become leaders in science and society.

Location

The Laboratory is located on the wooded north shore of Long Island, 35 miles east of Manhattan in New York City, and offers many amenities, both cultural and recreational. Recreational activities at CSHL include a fitness room, tennis and volleyball courts, a private beach, sailboats and windsurfers, and many quiet back roads for running or walking. Students may also attend classical music performances and art exhibitions sponsored by the Laboratory for scientists and the neighboring community.

The Laboratory

Since its inception in 1890, CSHL has been involved in higher education and is today a world leader in biology education. The CSHL Press publishes internationally recognized books and journals. The Dolan DNA Learning Center educates students and teachers about the world of DNA. The Undergraduate Research Program, started in 1959, hosts exceptional undergraduates from around the world for a summer research experience. CSHL is also involved in education at the highest levels through a postgraduate program of twenty-five advanced courses in biology and many large and small international conferences. These meetings and courses attract 8,000 scientists annually to the Laboratory. The Laboratory has also been involved in graduate education leading to the Ph.D. degree for more than twenty-five years, particularly through shared graduate programs with Stony Brook University.

Applying

Applicants must have received a baccalaureate degree from an accredited university or college prior to matriculation. Admission is based on the perceived ability of the applicant to excel in this doctoral program, without regard to gender, race, color, ethnic origin, sexual orientation, disability, or marital status. Suitable applicants are assessed on the basis of their academic record, recommendations from their mentors, and an on-site interview. Students should ensure that the school receives all application materials (transcripts, examination scores, letters of recommendation, etc.) no later than December 15 for the following fall term. Early application is advisable. Further information about the School and the application procedure may be requested by mail or obtained from the Web site at http://www.cshl.edu/gradschool.

Correspondence and Information

Dr. Lilian Gann, Dean
Watson School of Biological Sciences
Cold Spring Harbor Laboratory
P.O. Box 100
One Bungtown Road
Cold Spring Harbor, New York 11724
Phone: 516-367-6890
Fax: 516-367-6919
E-mail: gradschool@cshl.edu
Web site: http://www.cshl.edu/gradschool

Cold Spring Harbor Laboratory

THE FACULTY AND THEIR RESEARCH

Research Faculty

Carlos D. Brody, Associate Professor; Ph.D., Caltech, 1998. Computational neuroscience; psychophysics of short-term memory; neural data analysis; computation with spiking neural networks.

Dmitri Chklovskii, Associate Professor; Ph.D., MIT, 1994. Theoretical neuroscience; principles of brain design.

Hollis Cline, Professor; Ph.D., Berkeley, 1985. Neuronal development; experience-dependent plasticity; visual system; *Xenopus* in vivo imaging; electrophysiology; synaptogenesis.

Josh Dubnau, Assistant Professor; Ph.D., Columbia, 1995. Learning; memory; genetics; behavior.

Grigori Enikolopov, Associate Professor; Ph.D., USSR Academy of Sciences (Moscow), 1978. Signal transduction in neurons; development; gene expression; nitric oxide.

Gregory Hannon, Professor; Ph.D., Case Western Reserve, 1992. Growth control in mammalian cells.

Eli Hatchwell, Investigator; M.D., Cambridge, 1985; Ph.D., Oxford, 1995. Sporadic human genetic disease.

Tatsuya Hirano, Professor; Ph.D., Kyoto, 1989. Chromosome structure and function; cell-cycle control; *Xenopus* cell-free system.

Z. Josh Huang, Associate Professor; Ph.D., Brandeis, 1994. Neuroscience; experience-dependent development and plasticity of the neocortex; mouse genetics.

David Jackson, Associate Professor; Ph.D., East Anglia (England), 1991. Plant development; genetics; cell-to-cell mRNA and protein trafficking.

Leemor Joshua-Tor, Professor; Ph.D., Weizmann (Israel), 1991. Structural biology; X-ray crystallography; molecular recognition; transcription; proteases.

Alexei Koulakov, Assistant Professor; Ph.D., Minnesota, 1998. Theoretical neurobiology; quantitative principles of cortical design.

Adrian R. Krainer, Professor; Ph.D., Harvard, 1986. Posttranscriptional regulation of gene expression; pre-mRNA splicing mechanisms; alternative splicing; RNA-protein interactions; cell-free systems.

Yuri Lazebnik, Professor; Ph.D., St. Petersburg State, 1986. Apoptosis; caspases; cancer chemotherapy; proteases.

Scott Lowe, Professor; Ph.D., MIT, 1994. Modulation of apoptosis, chemosensitivity, and senescence by oncogenes and tumor-suppressor genes.

Robert Lucito, Assistant Professor; Ph.D., NYU, 1993. Genomic analysis of cancer.

Wolfgang Lukowitz, Assistant Professor; Ph.D., Tübingen (Germany), 1996. Genetics; development; plant biology; gene expression; signal transduction; plants.

Zachary Mainen, Associate Professor; Ph.D., California, San Diego, 1995. Computational neuroscience; olfactory system; synaptic transmission and plasticity; learning and memory.

Roberto Malinow, Professor, M.D., NYU, 1984; Ph.D., Berkeley, 1986. Neuroscience; learning and memory; central synaptic transmission and plasticity.

Robert Martienssen, Professor; Ph.D., Cambridge, 1986. Plant genetics; transposons; development; gene regulation; DNA methylation.

W. Richard McCombie, Professor; Ph.D., Michigan, 1982. Genome structure; DNA sequencing; computational molecular biology; Human Genome Project.

Alea A. Mills, Assistant Professor; Ph.D., California, Irvine, 1997. Functional genomics; tumorigenesis; development.

Partha P. Mitra, Professor; Ph.D., Harvard, 1993. Neuroinformatics; theoretical engineering; animal communications; neural prostheses; brain imaging; developmental linguistics.

Vivek Mittal, Assistant Professor; Ph.D., Nehru (New Delhi), 1988. Functional genomics of cancer genes; mouse models of human cancer; DNA chip technology.

Senthil K. Muthuswamy, Assistant Professor; Ph.D., McMaster, 1995. Understanding cancer initiation using 3-dimensional epithelial structures.

Michael P. Myers, Assistant Professor; Ph.D., Case Western Reserve, 1996. Protein complexes in signal transduction.

Andrew Neuwald, Associate Professor; Ph.D., Iowa, 1987. Classification and modeling of protein domains; predicting protein structure and function.

Scott Powers, Associate Professor; Ph.D., Columbia, 1983. Cancer gene discovery; cancer diagnostics and therapeutics; cancer biology.

Jonathan Sebat, Assistant Professor; Ph.D., Idaho, 2002. Copy number variation; segmental duplication; genetics; neurogenetics; ROMA; microarray.

Jacek Skowronski, Associate Professor; M.D., 1980, Ph.D., 1981, Lodz (Poland). HIV pathogenesis; *nef* gene; signal transduction; protein sorting; animal models.

David L. Spector, Professor; Ph.D., Rutgers, 1980. Cell biology; nuclear structure; microscopy; pre-mRNA splicing.

Lincoln Stein, Professor; M.D., Ph.D., Harvard, 1989. Genome informatics; bioinformatics; mapping; software; World Wide Web.

Arne Stenlund, Associate Professor; Ph.D., Uppsala (Sweden), 1984. Papillomavirus; cancer; DNA replication.

Bruce Stillman, President and CEO; Ph.D., Australian National, 1979. DNA replication; chromatin assembly; biochemistry; yeast genetics; cancer; cell cycle.

Karel Svoboda, Professor and Assistant Investigator, Howard Hughes Medical Institute; Ph.D., Harvard, 1994. Neocortical circuits and their plasticity.

William Tansey, Professor; Ph.D., Sydney, 1991. Oncogene regulation; transcription; protein destruction.

Marja Timmermans, Associate Professor; Ph.D., Rutgers, 1996. Plant development; axis specification; homeobox genes; stem cell function.

Nicholas Tonks, Professor; Ph.D., Dundee (Scotland), 1985. Posttranslational modification; phosphorylation; phosphatases; signal transduction; protein structure and function.

Tim Tully, Professor; Ph.D., Illinois, 1981. Neuroscience; learning and memory; *Drosophila* genetics.

Linda Van Aelst, Associate Professor; Ph.D., Leuven (Belgium), 1991. Signal transduction; Ras and Rac proteins; tumorigenesis; metastasis.

Michael Wigler, Professor; Ph.D., Columbia, 1978. Cancer; genomics; oncogenes; signal transduction; Ras; yeast genetics.

Rui-Ming Xu, Professor; Ph.D., Brandeis, 1990. X-ray crystallography; protein-RNA interactions; gene expression; cell cycle.

Anthony Zador, Associate Professor; M.D., 1994, Ph.D., 1994, Yale. Computational neuroscience; synaptic plasticity; auditory processing; cortical circuitry.

Michael Zhang, Professor; Ph.D., Rutgers, 1987. Computational genomics; nucleic acid pattern recognition; gene expression.

Yi Zhong, Professor; Ph.D., Iowa, 1991. Neurophysiology; *Drosophila;* learning and memory; neurofibromatosis; signal transduction.

Non-Research Faculty

James D. Watson, Chancellor; Ph.D., Indiana, 1950.

Alexander A. F. Gann, Editorial Director of Coldspring Harbor Laboratory Press; Ph.D., Edinburgh, 1989.

Terri Grodzicker, Assistant Director for Academic Affairs; Ph.D., Columbia, 1969.

John R. Inglis, Executive Director of Cold Spring Harbor Laboratory Press; Ph.D., Edinburgh, 1976.

David A. Micklos, Executive Director of the DNA Learning Center; M.A., Maryland, 1982.

David J. Stewart, Director of Meetings and Courses; Ph.D., Cambridge, 1988.

Jan A. Witkowski, Executive Director of the Banbury Center; Ph.D., London, 1972.

COLUMBIA UNIVERSITY

Graduate School of Arts and Sciences
The Coordinated Doctoral Program in Basic Sciences
at the College of Physicians and Surgeons

Programs of Study	The Coordinated Doctoral Program in Basic Sciences at the College of Physicians and Surgeons of Columbia University offers ten Ph.D. programs in basic sciences—Biochemistry and Molecular Biophysics; Biomedical Informatics; Cell Biology and Pathobiology; Genetics and Development; Human Nutrition; the Integrated Program in Cellular, Molecular, and Biophysical Studies; Microbiology; Neurology and Behavior; Pharmacology; and Physiology and Cellular Biophysics.
	The Coordinated Doctoral Program allows students to become part of the exciting and highly interactive research community of faculty members, graduate students, and postdoctoral fellows at the College of Physicians and Surgeons. Courses are concentrated in the first two years. The first-year curriculum emphasizes research rotations in three different laboratories, including the option of doing at least one rotation in a basic science laboratory of any Ph.D. program. In this way, students experience a range of research topics and potential Ph.D. mentors. At the beginning of the second year, students choose their mentors and begin their research for the Ph.D. degree. While students are likely to choose laboratories within the Ph.D. program that they initially entered, the Coordinated Doctoral Program makes it possible for a student to arrange to work with a mentor in any basic science Ph.D. program. The basic sciences faculty is committed to providing a complete graduate education that emphasizes scholarship, intellectual challenge, intimate guidance, independence, and sophisticated training in research. Ph.D. graduates from the College of Physicians and Surgeons have consistently gone on to become world leaders in biomedical research.
	In addition to the Ph.D. program, a terminal M.A. degree program is available in biomedical informatics. All other programs only admit candidates who are interested in obtaining the Ph.D. degree. M.A. and M.Phil. degrees are awarded to students in the Ph.D. program who have successfully completed their courses and qualifying examinations.
Research Facilities	The College's 170,000 square feet of facilities include extensively equipped and well-funded laboratories that contain the modern instrumentation necessary for state-of-the-art research in the biomedical sciences.
Financial Aid	All students in the Coordinated Doctoral Program receive full support for required tuition, student health services, and hospitalization insurance. Students also receive a generous stipend for their personal use that begins at registration and normally continues throughout the period of graduate study. This stipend is $27,336 for the 2006–07 academic year. Both international students and U.S. citizens are eligible for this support.
Cost of Study	Tuition and fees are fully covered by the predoctoral fellowship for all doctoral students, domestic and international, at the College of Physicians and Surgeons. For the 2006–07 academic year, this level of support is $35,515.
Living and Housing Costs	University housing is available on the Health Sciences Campus of Columbia University. Accommodations include University Residence Halls, which consist of furnished 2- or 4-person suites, and Institutional Real Estate Apartments, which include studios and one-, two-, and three-bedroom apartments. Students receive a free membership to the Bard Athletic Center, which features a swimming pool, squash courts, a gymnasium, a sauna, and exercise equipment. A full-time trainer is on staff, and the center is accessible to the handicapped. Programs offered include aerobics, weight training, and swim lessons.
Student Group	There are currently 349 predoctoral students enrolled in the various basic science Ph.D. programs at the College of Physicians and Surgeons. Fifty-six percent are women, and 29 percent are international students. There are an additional 35 M.D./Ph.D. students in basic science labs.
Location	The College of Physicians and Surgeons of Columbia University is located on the Health Sciences Campus at 168th Street and Broadway in upper Manhattan. The complex includes the Columbia–Presbyterian Medical Center and its subdivisions, the Audubon Biomedical Science and Technology Park, and the New York State Psychiatric Institute. New York's world-renowned cultural activities are all easily accessible by public transportation, as are sporting events and other recreational opportunities.
The University	Columbia University, a privately supported institution, is one of the world's leading educational and research centers. Founded by charter as King's College in 1754, it is one of the oldest universities in the country.
	The Health Sciences Campus of Columbia University includes the College of Physicians and Surgeons, the School of Nursing, the School of Public Health, and the School of Dental and Oral Surgery. In addition, the Comprehensive Cancer Center, the Center for Molecular Recognition, the Center for Neurobiology and Behavior, the Center for Reproductive Sciences, the Howard Hughes Institute for Neurobiology, the Howard Hughes Institute for Structural Biology, the Institute for Cancer Research, and the Institute of Human Nutrition are also based on the Health Sciences Campus. Classrooms and laboratory facilities for graduate school programs are located in the College of Physicians and Surgeons, the William Black Medical Research Building, the Julius and Armand Hammer Health Sciences Center, the Research Annex of the New York State Psychiatric Institute, and the recently completed buildings of the Audubon Biomedical Science and Technology Park, which house basic, clinical, and biotechnology research laboratories, including the Genome Center and the Center for Disease Prevention.
Applying	The General Test of the Graduate Record Examinations is required. Requirements for Subject Tests vary among the specific Ph.D. programs. International applicants whose native language is not English are required to take the Test of English as a Foreign Language (TOEFL); applicants with a TOEFL score below 250 (computer-based test) or 600 (paper-based test) are reviewed with caution. Completed applications and all supporting material should be submitted by the beginning of January for admission to the fall semester. Applications received after January are reviewed until the incoming class is filled. Admission is usually in the fall; only under special circumstances and with prearrangement with the accepting doctoral program are applications accepted for admission to the spring term. Applications and information about the specific Ph.D. programs within the Coordinated Doctoral Program in Basic Sciences may be obtained from the Office of Graduate Affairs or from the appropriate Graduate Program Director.
Correspondence and Information	Office of Graduate Affairs Room 205 Columbia University Medical Center 701 West 168th Street New York, New York 10032 Phone: 212-305-8058 Fax: 212-305-1031 E-mail: BiomedicalSciences@columbia.edu Web site: http://cpmcnet.columbia.edu/dept/gsas/

GRADUATE PROGRAM DIRECTORS

Biochemistry: Dr. Oliver Hobert (telephone: 212-305-3885; e-mail: rh2021@columbia.edu)

Biomedical Informatics: Dr. Stephen Johnson (telephone: 212-305-1858; e-mail: sbj2@columbia.edu)

Cell Biology and Pathobiology: Dr. Ronald Liem (telephone: 212-305-4078; e-mail: rkl2@columbia.edu)

Cellular, Molecular, and Biophysical Studies: Dr. Ronald Liem (telephone: 212-305-4078; e-mail: rkl2@columbia.edu); Dr. Lorraine Symington (telephone: 212-305-4793; e-mail: lss5@columbia.edu)

Genetics and Development: Dr. Virginia Papaioannou (telephone: 212-305-4753; e-mail: vep1@columbia.edu)

Microbiology: Dr. Howard Shuman (telephone: 212-305-6913; e-mail: has7@columbia.edu)

Neurobiology and Behavior: Dr. John Koester (telephone: 212-543-5239; e-mail: jdk3@columbia.edu); Dr. Darcy Kelley (telephone: 212-854-5108; e-mail: dbk3@columbia.edu)

Nutrition: Dr. Debra Wolgemuth (telephone: 212-305-7900; e-mail: djw3@columbia.edu)

Pharmacology: Dr. Daniel Goldberg (telephone: 212-305-1673; e-mail: dig5@columbia.edu)

Physiology: Dr. Amy MacDermott (telephone: 212-305-3889; e-mail: abm1@columbia.edu)

RESEARCH TOPICS

Axonal transport, glycoproteins, regeneration, X-ray crystallography, exocytosis, RAB3 GTP-binding proteins, germ cells, muscle contraction, differentiation, molecular developmental genetics, temporal and spatial gene regulation, DNA repair, cell-cycle checkpoint control, fission yeast, enteric nervous system, cellular asymmetry, cyclic nucleotides, development, cell biology, molecular biology, neuronal cytoskeleton, breast cancer susceptibility genes, cell adhesion, integrin receptors, growth cone–target cell interactions, bioengineering analysis, growth systems, endocrine systems, biogenic amines, protein transport, mitochondria, organelle motility, synaptic physiology, neuronal development, developmental biology, neurosecretory cells, estrogen, growth factors, neural development, muscle cell fate, myogenic gene regulation, embryogenesis, molecular genetics, olfactory perception, regulation, gene expression, B lymphocyte, T lymphocyte, B cell, T cell, structure, function, retrovirus, genomes, *Trichinella*, secretory proteins, gene regulation, protein folding, *E. coli* phages, temperate bacteriophages, oncogenesis, thyroid cells, membrane proteins, cell-cell interactions, signal transduction, cell fate choice, *C. elegans*, structural biology, macromolecules, diffraction, computational methods, RNA processing, synaptogenesis, neural circuitry, computer, nucleic acids, molecular mechanisms, neuronal differentiation, axon guidance, cell recognition vertebrates, nervous system, learning, memory, invertebrates, receptors, enzyme, hydrogenase, pattern formation, homeotic genes, nuclear magnetic resonance, dynamics, protein-ligand interactions, protein-receptor complexes, RNA structure, RNA reactivity, RNA folding, ribozyme tertiary structure, enzymatic activity, gene sequence, nucleotide sequence, DNA, genome project, 3-D protein structure, chromosome–nucleic acid biosynthesis, eukaryotic cells, prokaryotic cells, bacteria, virus, fungi, yeast, transcriptional activators, repressors, transformation, Src, nonreceptor tyrosine kinases, nuclear organization, DNA methyltransferases, epigenetic effects, cytokinesis, fission yeast, retroviral replication, DNA replication, RNA replication, microtubule, cytokines, mammalian, inflammation, immune response, pharmacological studies, cell differentiation, Wnt, notch, signaling, glycosylphosphatidylinositol (GPI) anchor, *Drosophila melanogaster*, meiosis, cardiac-autonomic interactions, second messengers, aorta; epithelial cell, smooth muscle, chemistry, ATP-driven transporters, host-pathogen interactions, pathogenesis, infectious disease, polymorphonuclear mononuclear phagocytes, PMN, platelets, endothelial cells (EC), human herpes, papillomavirus, infected cells, early embryonic development, pattern formation, gametogenesis, mitosis, hematopoietic stem cell, gene transfer, globin gene expression, transcriptional control, protein kinase pathways, skin, hair, pathophysiology, inductive interactions, organogenesis, murine models, cancer, molecular pathogenesis, lymphoma, leukemia, lymphoid tissue, growth factors, tumorigenesis, parental imprinting, psychiatric genetics, behavioral genetics, behavior, human genetics, gene mapping, multigenic inheritance, vertebrate limb, genetic linkage, preimplantation stages; T-box genes, stem-cell technology, targeted mutagenesis, psychiatric disorders, genetic recombination, genomic stability, functional genomics, neurological, neuromuscular disorders, mitochondrial genetics, cellular interactions, neurochemistry, food intake, metabolic rate, cytogenetics, congenital malformations, miscarriages, genetic toxicology, user interfaces, outpatient applications, practice guidelines, prevention, computer-based medical records, nuclear cardiology, medical concept representation, clinical decision support, medical knowledge, decision making, medicine, clinical information systems, efficacy, utility, security, confidentiality, methods, standards, data interchange, health care, bioinformatics, medical language processing, medical vocabulary, knowledge representation, statistical methods, automated decision support, data acquisition, mobile computing, expert systems, electronic medical records, database design, clinical anesthesiology, medical informatics, navigation; computer-based curriculum development, networking; security, systems architecture, distributed computing systems, human-computer interfaces, image processing, knowledge encoding, decision support systems, dental informatics, training, World Wide Web, information technologies, education, electronic oral health records, cell division, actin ring, biochemical approaches, DNA replication, partition, segregation, broad host range, bacterial plasmids, conjugation, antibiotic resistance, genetic tools, *Actinobacillus actinomycetemcomitans*, periodontal disease, adherence, biofilm, reverse transcriptase, in vitro mutagenesis, targeted gene disruption, transcription initiation, transcription termination, HIV-1, poliovirus, immunoglobulin class-switch, JAK-STAT, active transport; intracellular growth, *Legionella pneumophila*, phagosome, DNA recombination, two-component signal transduction, bacterial pathogenesis, viral promoters, *Salmonella*, morphological changes, nerve cells, cognition, perception, cognitive neuroscience, cell patterning, axon guidance, olfactory transduction, visually orienting behavior, primates, neurotropin, central nervous system, monoaminergic synapses, limb movement, axon injury, glial lineage, attention, vision, eye movements, associative, nonassociative learning, serotonin receptors, physiological, pathological states, psychophysics, electrophysiology, ion channels, modulation by neurohormones, sexual differentiation, neurons, synaptic connections, segmental differentiation, synaptic transmission, spinal cord dorsal horn, metacognition, neuropsychology, computational, psychophysical, visual processing, mechanisms, drug action, voltage-gated ion channels, microenvironments, plasticity, dopamine-receptor, serotonin-receptor, pre-mRNA processing, psychotic diseases, deficiency, knockout mice, neurobiology, circadian rhythms, neuroendocrinology, segmental organization, *Drosophila* eye, neurotrophin family, growth factors, biophysical studies, cerebral cortex, obesity, clinical trial, quantitative genetics, statistical, research methodology, lipid, lipoprotein, metabolism, retinoids, vitamin A, mutant mouse models, food, nutrition, policy, law, lipid emulsion, free fatty acids, ion transport, nutrients, diarrheal diseases, cystic fibrosis, virus infectivity; infant immunity, varicella-zoster virus, chicken pox, shingles, plasma lipoprotein, apoprotein B, secretion, hepatocytes, dietary regulation, lipolytic enzymes, atherosclerosis, energy expenditures, radiation, eating disorders, parasitology, environmental sciences, pregnancy, lactation, psychosocial factors, birth weight, pediatrics, molecular epidemiology, risk assessment, carbohydrate, diabetes mellitus, chemical carcinogens, calcium metabolism, hypertriglyceridemia, transgenic mouse, vascular cell biology, blood vessel, thrombosis, extracellular, intracellular sterol transport, cholesterol esterification, ACAT reaction, macrophage, endocytic pathways, cholesteryl ester, cellular responses, Kaposi's sarcoma, tumor suppression, neuron migration, lamination, CNS, cell lineages, neurodegenerative disease, cortical development, extracellular matrices, basement membranes, synapse, infectious agents, human host, cytogenetics, apoptosis, programmed cell death, head, neck tumors, aging, major histocompatibility complex, Alzheimer's disease, cervical epithelial cells, transformation, hormone action, cardiac electrophysiology, arrhythmias, reproduction, primates, phospholipid, organelle movement, environmental lead exposure, GTPases, immunity, antiarrhythmic drug action, G-protein–coupled receptors, pharmacological specificity, neurohormones, calcium release channels, cardiac autonomic responsiveness, membrane biophysics, antisense oligodeoxynucleotides, myocardial infarction, ion conduction, epithelial transport; ion pumps, biopolymers, water channels, corneal physiology, phosphatidylinositol, mononuclear leukocytes, coagulation.

COLUMBIA UNIVERSITY

Graduate School of Arts and Sciences
Department of Biological Sciences

Program of Study

The Department offers training, leading to the Ph.D. degree, in cellular, molecular, computational, developmental, and structural biology; genetics; molecular biophysics; and neurobiology. The graduate program provides each student with a solid background in contemporary biology and an in-depth knowledge of one or more of the above areas. The specific nature and scheduling of courses taken during the first two graduate years are determined by the student's consultation with the graduate student adviser, taking into account the background and specific research interests of the student. During the first year, all students take an intensive core course that provides a solid background in structural biology, cell biology, genetics, neurobiology, molecular biology, developmental biology, immunology, and bioinformatics.

Beginning in the first year, graduate students attend advanced seminar courses, including the preresearch seminar, which is a forum for faculty-student research discussion. Important components of graduate education include the ability to analyze critically the contemporary research literature and to present such analyses effectively through oral and written presentations. Students acquire training in these skills through participation in advanced-level seminars and journal clubs and presentation and defense of original research proposals during the second year of graduate study.

Beginning in the first year of graduate work, students also engage in research training. Students may choose laboratories in the Department of Biological Sciences on Columbia's main Morningside Heights campus or in about twenty-five laboratories at Columbia's Health Sciences campus. To inform incoming students of research opportunities, faculty members discuss ongoing research projects with them in the preresearch seminar held in the autumn term of the first year. All students are strongly encouraged to participate in ongoing research in up to three different laboratories during the first year. The choice of a dissertation sponsor is made after consultation between the student and potential faculty advisers, and intensive research begins following the spring term of the student's first year. Each student is assigned a Ph.D. advisory committee made up of the student's sponsor and 2 other faculty members.

Research Facilities

The Department of Biological Sciences is located in the modern Sherman Fairchild Center for the Life Sciences. The building provides 50,000 square feet of laboratory space for the Department's laboratories, as well as extensive shared instrument facilities. The latter include automated DNA synthesis and sequencing; fluorescence and digital microscopy; analytical and preparative biochemistry; FACS analysis and microinjection facilities; and housing and care of research animals, including transgenic mice. A library, designed for ready access to the collections, is housed in the same building. It includes the latest equipment for database searching, an extensive microfilm collection, and audiovisual facilities.

Financial Aid

All accepted students receive generous stipends, complete tuition exemption, and medical insurance. Special fellowships with larger stipends are also available to members of minority groups.

Cost of Study

Tuition and fees are paid for all graduate students accepted into the Department.

Living and Housing Costs

Most students live in University-owned, subsidized apartments or dormitories within walking distance of the laboratories. In addition, the campus is easily reached by public transportation from all areas of the city.

Student Group

There are about 100 graduate students and 60 postdoctoral fellows in the Department.

Location

New York is the cultural center of the country and offers unrivaled opportunities for attending concerts, operas, plays, and sporting events, for visiting outstanding museums, and for varied, affordable dining. Many excellent beaches, ski slopes, and state and national parks are within reasonable driving distance.

The University and The Department

Columbia was established as King's College in 1754 and has grown into one of the major universities of the world. The Department is located on the beautiful main campus in Morningside Heights, which combines the advantages of an urban setting and a peaceful college-town atmosphere.

Applying

Undergraduate training in one of the natural or physical sciences is required. It is desirable for students to have had at least one year of calculus, as well as courses in organic and physical chemistry, physics, genetics, biochemistry, and cell biology. Any deficiencies may be made up while in graduate school. The Graduate Record Examinations, including the Subject Test in biology, chemistry, or physics, is also required, as is the Test of English as a Foreign Language for international applicants whose native language is not English. Completed applications should be returned by January 3 for admission to the fall semester. Application forms and additional information can be obtained from the Department's Web site.

Columbia University is an Equal Opportunity/Affirmative Action institution.

Correspondence and Information

Graduate Student Adviser
Department of Biological Sciences
600 Fairchild
Columbia University
1212 Amsterdam Avenue, Mail Code 2402
New York, New York 10027
Phone: 212-854-4581
Fax: 212-865-8246
E-mail: biology@columbia.edu
Web site: http://www.columbia.edu/cu/biology/

Columbia University

THE FACULTY AND THEIR RESEARCH

Walter J. Bock, Professor; Ph.D., Harvard, 1959. General evolutionary theory; evolutionary and functional morphology; morphology and classification of birds; history and philosophy of evolutionary biology.

Ronald Breslow, Adjunct Professor; Ph.D., Harvard, 1956. Aromaticity and antiaromaticity; biochemical model systems; biomimetic synthetic methods; small ring compounds; organic reaction mechanisms; organic electrochemistry.

J. Chloë Bulinski, Professor; Ph.D., Wisconsin, 1980. Microtubule dynamics and function during the cell cycle and cell differentiation.

Harmen Bussemaker, Assistant Professor; Ph.D., Utrecht (Netherlands), 1995. Bioinformatics research aimed at understanding how regulatory proteins control chromatin structure and gene expression, using a combined analysis of complete genome sequences and DNA microarray or SAGE data.

Martin Chalfie, Professor; Ph.D., Harvard, 1976; Member, National Academy of Sciences. Developmental genetics of identified nerve cells in *Caenorhabditis elegans;* genetic analysis of cell differentiation, mechanosensory transduction, synapse specification, and aging.

Lawrence A. Chasin, Professor; Ph.D., MIT, 1967. Molecular genetics of pre-mRNA processing; molecular recognition of RNA splice sites in long transcripts.

Julio Fernandez, Professor; Ph.D., Berkeley, 1982. Study of the cellular events that lead to the release of histamine or catecholamine-containing secretory granules from single, isolated mast cells or chromaffin cells; analysis of single protein elasticity by atomic force microscopy (AFM).

Stuart Firestein, Professor; Ph.D., Berkeley, 1988. Cellular and molecular physiology of transduction; coding and neuronal regeneration in the vertebrate olfactory system.

John F. Hunt, Associate Professor; Ph.D., Yale, 1993. Structural genomics and biophysical studies of the molecular mechanism of transmembrane transport.

Daniel D. Kalderon, Professor; Ph.D., London, 1984. Molecular mechanisms of cellular interactions mediated by cAMP-dependent protein kinase (PKA) in *Drosophila;* roles of PKA in hedgehog signaling and in generating anterior/posterior polarity in oocytes.

Darcy B. Kelley, Professor and Howard Hughes Medical Institute Professor; Ph.D., Rockefeller, 1975. Sexual differentiation of the nervous system; molecular analyses of androgen-regulated development in neurons and muscle; neuroethology of vocal communication; evolution of the nuclear receptor family.

James L. Manley, Professor; Ph.D., SUNY at Stony Brook, 1976. Regulation of mRNA synthesis in animal cells; biochemical and genetic analysis of mechanisms and control of mRNA transcription, splicing, and polyadenylation; developmental control of gene expression.

Ann McDermott, Adjunct Professor; Ph.D., Berkeley, 1987. Solid-state NMR of enzyme active sites and model systems.

Elizabeth Miller, Assistant Professor; Ph.D., La Trobe (Australia), 1999. Protein folding, assembly, and the regulation of intracellular protein transport.

James Mohler, Adjunct Professor; Ph.D., MIT, 1982. Genetic control of pattern formation during *Drosophila* development.

Robert E. Pollack, Professor; Ph.D., Brandeis, 1966. Critical analysis of issues involving molecular biology and religion.

Carol L. Prives, Professor; Ph.D., McGill, 1968; Member, National Institute of Medicine. Structure and function of the p53 tumor suppressor protein and p53 family members; studies on cell cycle and apoptosis; stress-activated signaling and control of proteolysis.

Ron Prywes, Professor; Ph.D., MIT, 1984. Normal and cancerous mechanisms of regulation of cellular proliferation and gene expression; signal transduction and activation of transcription factors; activation of transcription by the ER stress/unfolded protein response.

Michael P. Sheetz, Professor; Ph.D., Caltech, 1972. Motility studies of cells and microtubule motor proteins, with an emphasis on the force-dependent interactions relevant to transformed cells and neuron pathfinding, using laser tweezers.

Brent Stockwell, Assistant Professor; Ph.D., Harvard, 1997. Diagramming disease networks with chemical and biological tools.

Liang Tong, Professor; Ph.D., Berkeley, 1989. Structural biology of fatty acid metabolism and pre-mRNA processing; structure-based drug design.

Alexander A. Tzagoloff, Alan H. Kempner Professor of Biological Sciences; Ph.D., Columbia, 1962. Energy-coupling mechanisms; structure of membrane enzymes; biogenesis of mitochondria; genetics of mitochondria in yeast.

Lili Yamasaki, Associate Professor; Ph.D., Texas Health Science Center at San Antonio, 1991. Role of E2F/DP transcription factors in growth control by the retinoblastoma tumor suppressor protein (pRB) in vivo.

Jian Yang, Associate Professor; Ph.D., Washington (Seattle), 1991. Structure and function of ion channels; molecular mechanisms of ion channel regulation and localization.

Rafael Yuste, Associate Professor and Howard Hughes Medical Institute Investigator; M.D., Madrid, 1987; Ph.D., Rockefeller, 1992. Dendritic integration and cortical microcircuitry.

Additional Faculty Sponsors for Ph.D. Research

Richard Axel, Biochemistry; Howard Hughes Medical Institute Investigator and Nobel Laureate in Physiology or Medicine 2004. Central and peripheral organization of the olfactory system.

Kathryn Calame, Microbiology/Immunology. Lymphocyte gene regulation.

Marian Carlson, Genetics/Development. Regulation of gene expression in yeast; mechanisms of transcriptional control.

Virginia Cornish, Chemistry. Development of in vivo selection strategies for evolving proteins with novel catalytic properties.

Frank Costantini, Genetics/Development. Molecular genetics of mammalian development.

Jean Gautier, Genetics/Development. Cell cycle and cell death during early development.

Steve Goff, Biochemistry; Howard Hughes Medical Institute Investigator. Retroviral replication.

Lloyd Greene, Pathology. Mechanisms of neuronal differentiation and degeneration and their regulation by external growth factors.

Iva Greenwald, Biochemistry; Howard Hughes Medical Institute Investigator. Development and cell-cell interactions.

Tulle Hazelrigg, Biological Sciences. mRNA localization in *Drosophila* oocytes.

René Hen, Neurobiology and Behavior. Serotonin receptors and behavior.

Oliver Hobert, Biochemistry and Molecular Biophysics; Howard Hughes Medical Institute Investigator. Nervous system development and function.

Wayne Hendrickson, Biochemistry; Howard Hughes Medical Institute Investigator. Macromolecular structure; X-ray crystallography.

David Hirsh, Biochemistry. Molecular genetics of development.

Tom Jessell, Biochemistry; Howard Hughes Medical Institute Investigator. Molecular mechanisms of neural differentiation.

Laura Johnston, Genetics/Development. Control of growth and cell division during development.

Eric Kandel, Physiology; Howard Hughes Medical Institute Investigator and Nobel Laureate in Physiology or Medicine 2000. Cell and molecular mechanisms of associative and nonassociative learning.

Arthur Karlin, Biochemistry. Molecular mechanisms of receptor function.

Richard Mann, Biochemistry. Transcriptional control.

Aaron Mitchell, Microbiology. Regulatory pathways that govern cellular differentiation in the two yeasts: *Saccharomyces cerevisiae* and *Candida albicans.*

Art Palmer, Biochemistry. Biomolecular dynamics, structure, and function; NMR spectroscopy.

Virginia Papaioannou, Genetics/Development. Genetic control of mammalian development in the peri-implantation period.

Rodney Rothstein, Genetics/Development. Yeast genetics; mechanisms of genetic recombination; control of genome stability; functional genomics.

Andrey Rzhetsky, Medical Informatics. Comparative genomics; gene networks; genomic informatics; neural networks.

Chris Schindler, Microbiology/Medicine. JAK-STAT signaling and immune response.

Steve Siegelbaum, Pharmacology; Howard Hughes Medical Institute Investigator. Molecular studies of ion channel structure and function; synaptic transmission and plasticity in the mammalian brain.

Gary Struhl, Genetics/Development; Howard Hughes Medical Institute Investigator. Developmental genetics.

Lorraine Symington, Microbiology. Homologous recombination in the yeast *Saccharomyces cerevisiae.*

Richard Vallee, Pathology. Motor proteins in axonal transport, brain developmental disease, and synaptic function.

CREIGHTON UNIVERSITY

School of Medicine
Department of Biomedical Sciences

Programs of Study

The Department of Biomedical Sciences in the School of Medicine at Creighton University offers Master of Science (M.S.) and Doctor of Philosophy (Ph.D.) degrees. These flexible programs foster a multidisciplinary approach that uses research, course work, and facilities to cater to a student's needs and research interests. Research training is available in biochemistry, bioorganic chemistry, bioinformatics physiology, cell and developmental biology, molecular biology, neurobiology, and proteomics. In the first semester, students join the research laboratories of their major advisers. All students must complete independent research and either a master's thesis or a doctoral dissertation. Foundation courses include the fundamentals of biochemistry, cell and molecular biology, physiology, human neuroanatomy, and cytochemistry and histochemistry. Students may also register for didactic courses of the graduate programs in pharmacology and medical microbiology, including the receptor pharmacology course.

The master's program in clinical anatomy is primarily designed for those who wish to continue their professional careers as teachers of clinical anatomy and is offered in the Department of Biomedical Sciences with participation by the Departments of Radiology, Surgery, and Pathology. This eighteen-month program prepares students with the necessary skills and experiences to teach clinically relevant anatomy in any of the health sciences. The curriculum includes human gross anatomy and neuroanatomy, pathology, surgery, radiology, and embryology. Students have opportunities to dissect the entire human body, to attend autopsies and surgeries, and to participate in case-based discussions of regional anatomy. A portion of the curriculum is devoted to lecture techniques, clinical correlations, and computer-aided instruction and to the proper and safe preparation and use of preserved and fresh tissue for anatomical demonstration. Students must begin the program in August with the study of human gross anatomy and take classes during the summer session of their first year. Students graduate in December.

Research Facilities

The Molecular Biology Research Core Facility provides both services and instrumentation for all investigators who use molecular biology techniques in their research. Services provided include DNA sequencing and oligonucleotide synthesis. In addition, the facility offers some of the most precise technology and equipment available, including high-speed centrifugation and ultracentrifugation, phosphorimaging for both radioactive and chemiluminescent samples, lyophilizing, autoradiographic film processing, UV/visible gel documentation, spectrophotometry and GCG-DNA, and protein-sequence analysis computer programs.

The Department contains the Nebraska Center for Cell Biology offers the Zeiss LSM 510 META NLO system, an LSM 510 META confocal scanning system with three lasers (Ar 458/477/488/514 nm, Green HeNe 543 nm, Red HeNe 633 nm), a META scanning module with two single-channel detectors and one multichannel detector, a Coherent Chameleon XR (705–980 nm) for multiphoton excitation, two nondescanned detectors for multiphoton work, a Becker & Hickl spectroscopy system for fluorescence correlation spectroscopy (FCS) and time-correlated single-photon counting (TCSPC), a cooled CCD video camera and image acquisition software, two optical stretchers with IPG Photonics 10 W 1060-nm IR fiber lasers, and a binocular dissection microscope.

The genome sequencing projects are providing vast quantities of new information that is invaluable for identifying and characterizing gene products. However, primary structure is only part of the complete structural characterization required for understanding mechanisms of protein functions; the 3-D structures must also be known. For initial structure-function studies, fragments/domains of proteins can be efficiently synthesized using solid phase peptide synthesis. These fragments could be structural, functional domains, or epitopes for raising antibodies. The selection of fragments/domains is done using the Bioinformatics Facility, which is specialized in 3-D structural characterizations of polypeptides. This collaboration provides a unique, knowledge-based design, synthesis, and characterization of polypeptides.

The Bioinformatics Facility also has an 18-node AlphaPowered cluster, a dual AMD Athlon (1.33 GHz) file server (105 Gbyte raid) and 40-node dual AMD Athlon (1.33 GHz) cluster with both fast Internet and gigabit connections, and a 6-node dual AMD Opteron (2.2 GHz) cluster. A dual R14000 (600 MHz) CPU SGI Octane 2 workstation is used for visualization.

The Structural Proteomics Facility is currently equipped with a Jasco J-810 electronic circular dichroism (ECD) spectrometer connected to a stopped-flow apparatus, a BioTools vibrational circular dichroism (VCD) spectrophotometer, and a Perking-Elmer HPLC apparatus connected to a PE SCIEX API 150 EX quadruple mass spectrometer (LC-MS). The Proteomics Facility also has an ACT Apogee and ABI SYNREGY peptide synthesizers.

The 34,000-square-foot Health Sciences Library and Learning Resources Center has more than 210,000 volumes and access to 1,600 serials as well as more than 600 study seats, 174 laptop-access ports, and wireless network access. Medical School facilities available to students include a fluorescence-activated cell sorter and transmission and scanner electron microscopes.

For more information, applicants are encouraged to visit the Graduate Program Web site at http://www.biomedsci.creighton.edu.

Financial Aid

All students accepted into the M.S. and Ph.D. programs receive tuition remission. Students accepted into the Ph.D. program are provided yearly stipends ($19,802 in 2005–06). Subject to satisfactory performance, stipends are given for three years if the student enters the program with a master's degree and for four years if the student enters the program with a bachelor's degree. Stipends come from research grants, doctoral training programs, and a limited number of departmental fellowships. Teaching duties are not required for graduate students to receive a stipend. With the help of their advisers, students are encouraged to seek predoctoral funding from local and national funding agencies.

Cost of Study

Tuition is $36,426 per year, and the University fee is $764. However, all graduate students in the biomedical sciences program receive tuition remission.

Living and Housing Costs

In 2004–05, typical annual on-campus room and board charges ranged between $7200 and $7530, depending on the accommodations and meal plan.

Student Group

In 2004–05, there were 5 students enrolled in the anatomy master's degree program, and 8 students were enrolled in the M.S. in biological sciences program as well as 34 Ph.D. candidates.

Location

Located on the western bank of the Missouri River, Omaha has been an important agricultural and transportation center since its establishment in the late 1850s. The five-county, 2,500-square-mile Omaha metropolitan area has a population of more than 775,000, and there are more than 18,000 businesses, which provide students with endless job and internship opportunities. Attractions, big-time sports, top-notch restaurants, and fascinating historic sites make Omaha a tourist destination. Local bands often play on the Creighton campus in addition to their performances at the nightspots around the city.

The University and The Department

Creighton University is committed to being the outstanding comprehensive Jesuit university in the United States. The Creighton Jesuit education has always been focused upon the development of the total person. Creighton educates men and women to be leaders in their careers and professions and to be of service to society. The Department of Biomedical Sciences in the Creighton University School of Medicine was established in 1988 when the Departments of Anatomy, Biochemistry, and Physiology merged. The mission of the Department is to provide education and opportunities for research in the sciences basic to medicine.

Applying

All applicants must have a bachelor's degree or the equivalent, preferably with satisfactory completion of course work in a biological, chemical, or physical science; a GPA of at least 3.0 overall; and GRE scores in the 50th percentile for the quantitative and verbal parts of the exam. The Graduate School requires all students whose native language is not English to score 550 or better on the Test of English as a Foreign Language (TOEFL) unless they can demonstrate proficiency in some other way.

Applicants should send to the Dean of the Graduate School the completed application form; the nonrefundable $40 application fee; official transcripts of all college work attempted, sent directly from the undergraduate institutions attended; GRE scores; three letters of recommendation; and TOEFL scores, if applicable.

Correspondence and Information

Richard F. Murphy, Director of Graduate Studies
Department of Biomedical Sciences
School of Medicine
Creighton University
2500 California Plaza
Omaha, Nebraska 68178
Phone: 402-280-2918
Fax: 402-280-2690
E-mail: barrym@creighton.edu
Web site: http://www.biomedsci.creighton.edu/

Creighton University

THE FACULTY AND THEIR RESEARCH

Primary Faculty

Devendra K. Agrawal, Professor; Ph.D., Lucknow (India), 1978; Ph.D., McMaster, 1984. Immunomodulation of inflammatory diseases of the lung and blood vessels.

Donald R. Babin, Professor; Ph.D., New Brunswick, 1962. Peptide chemistry.

Kirk W. Beisel, Professor; Ph.D., Rutgers, 1978. Molecular delineation of developmental and functional mechanisms of the inner ear.

Dale R. Bergren, Associate Professor; Ph.D., North Dakota, 1976. Pulmonary physiology and pharmacology.

Philip R. Brauer, Associate Professor; Ph.D., Medical College of Wisconsin, 1985. Cell-cell interactions in development.

Laura L. Bruce, Associate Professor; Ph.D., Georgetown, 1982. Neuronal development and comparative anatomy.

Arthur F. Fishkin, Associate Professor; Ph.D., Iowa, 1957. Glycoprotein biochemistry.

Bernd Fritzsch, Professor; Ph.D., Darmstadt, 1978. Development and survival of inner ear sensory neurons.

Henry Gale, Assistant Professor; Ph.D., Illinois, 1966. Muscle mechanics.

Richard J. Hallworth, Associate Professor; Ph.D., Baylor College of Medicine, 1983. Biophysics and biomechanics of cochlear outer hair cell motility.

Laura A. Hansen, Assistant Professor; Ph.D., North Carolina, 1993. Signal transduction in carcinogenesis and skin biology.

David Zhi-Zhou He, Associate Professor; Ph.D., Shanghai Institute of Physiology, 1990. Development, biophysics of somatic motility, and mechanoelectrical (forward) transduction in the outer hair cells.

Joseph A. Knezetic, Associate Professor; Ph.D., Cincinnati, 1986. Transcriptional regulation of hemoglobin genes.

Sándor Lovas, Associate Professor; Ph.D., Szeged (Hungary), 1985. Computational chemistry, structure-activity relationships in peptides.

Robert B. Mackin, Associate Professor; Ph.D., Emory, 1987. Regulation and specificity of enzymes involved in peptide hormone biosynthesis.

Barbara J. McLaughlin, Professor; Ph.D., Stanford, 1972. Cell biological mechanisms underlying the pathogenesis of retinal and corneal disease.

Rita A. Meyer, Assistant Professor; Ph.D., Chicago, 1981. Gap junctions and cell adhesion in cell growth control and differentiation.

Richard F. Murphy, Professor and Chair; Ph.D., National (Ireland), 1968. Structure-function relationships in regulatory peptides.

David H. Nichols, Associate Professor; Ph.D., Oregon, 1975. Cell biology of development.

Eric B. Patterson, Assistant Professor; Ph.D., Meharry Medical College, 1984. Molecular biology of collagen synthesis.

David H. Petzel, Professor; Ph.D., Illinois, 1982. Comparative physiology of ion transport.

Thomas E. Pisarri, Assistant Professor; Ph.D., Wisconsin, 1983. Sensory innervation of airways and heart.

Thomas H. Quinn, Professor and Director of the M.S. in Anatomy Program; Ph.D., Nebraska Medical Center, 1981. Clinical anatomy, comparative morphology.

Roger D. Reidelberger, Professor; Ph.D., California, Davis, 1981. Control of appetite and digestion.

D. David Smith, Associate Professor; Ph.D., Edinburgh, 1986. Polypeptide synthesis.

Garrett A. Soukup, Assistant Professor; Ph.D., Nebraska Medical Center, 1997. Nucleic acid structure, function, and molecular recognition.

John A. Yee, Professor; Ph.D., Utah, 1974. Control of bone growth.

Secondary Faculty

Robert J. Anderson, Professor; M.D., Northwestern, 1973. Role of sulfate conjugation in hormone, neurotransmitter, and drug metabolism.

John M. Bertoni, Professor; M.D., Michigan, 1971; Ph.D., Michigan, 1979. Neurochemistry of development and degenerative disorders.

Dominic E. Cosgrove, Staff Scientist III; Ph.D., Nebraska Medical Center, 1989. Animal models for human disease; gene-targeting and gene therapy.

Diane M. Cullen, Associate Professor; Ph.D., Wisconsin, 1989. Bone response to mechanical forces.

Joseph G. Dulka, Assistant Professor; Ph.D., Alberta, 1989. Sensory neurobiology; neuroendocrinology; neuroanatomy; hormone-induced neural plasticity; physiological basis of behavior.

Gleb R. Haynatzki, Assistant Professor; Ph.D., Ohridski (Bulgaria), 1989; Ph.D., California, Santa Barbara, 1995. Biostatistics; applied probability; cancer and AIDS epidemiology; carcinogenesis modeling.

Martin R. Hulce, Associate Professor; Ph.D., John Hopkins, 1983. New synthetic methods.

William B. Jeffries, Associate Professor; Ph.D., University of the Sciences in Philadelphia, 1985. Human essential hypertension.

Walter Jesteadt, Director of Research; Ph.D., Pittsburgh, 1971.

Philip Kelley, Staff Scientist II; Ph.D., Washington (St. Louis), 1979.

William J. Kimberling, Professor; Ph.D., Indiana, 1967. Positional cloning characterization of genes that cause hearing loss.

Anthony E. Kincaid, Assistant Professor; Ph.D., Michigan 1991. Functional anatomy of the basal ganglia and animal models for movement disorders.

JoAnn McGee, Staff Scientist; Ph.D., Illinois, 1989. Neurobiology of the developing CNS.

Barbara J. Morley, Professor; Ph.D., Maine, 1973. Neurochemistry.

Michael G. Nichols, Assistant Professor; Ph.D., Rochester, 1996. Optical biophysics.

Edward J. Walsh, Coordinator/Staff Scientist; Ph.D., Creighton, 1983. Development of hearing.

Adjunct Faculty

Dominique Crapon de Caprona, Adjunct Associate Professor; Ph.D., Geneva, 1977. Quantitative ethology of behavioral deficits and compensation in vestibular-impaired mice exposed to balance tasks.

DARTMOUTH COLLEGE

Department of Biological Sciences
Graduate Program in Ecology and Evolutionary Biology
Graduate Program in Molecular and Cellular Biology

Programs of Study

The Department of Biological Sciences at Dartmouth College offers two graduate programs leading to the Ph.D. degree: the Graduate Program in Ecology and Evolutionary Biology and the Graduate Program in Molecular and Cellular Biology. Each program emphasizes independent research to prepare students for careers in academic institutions, government agencies, and industry. Five years are usually required for completion of course work and research leading to the Ph.D. In addition to offering two graduate programs, the Department also participates in the M.D./Ph.D. program, based in the Dartmouth Medical School. Many biologists located in other departments in the College as well as in the Dartmouth Medical School and in the Thayer School of Engineering interact with members of the Department of Biological Sciences, contribute to graduate teaching, and are available for consultation on graduate student research.

Faculty members and students in the Graduate Program in Ecology and Evolutionary Biology conduct basic and applied research on topics in population, community, and ecosystem ecology pertaining to aquatic and terrestrial systems. Dartmouth's location provides easy access to a great variety of natural habitats, including several extensive College-owned areas. Entering students begin thesis research in a lab of their choice and take a series of courses chosen in consultation with members of their advisory committee.

The Graduate Program in Molecular and Cellular Biology is offered in conjunction with faculty members from the Departments of Biochemistry, Chemistry, Genetics, and Microbiology and Immunology. Research by biology faculty members emphasizes model systems, with various labs pursuing studies in development *(C. elegans, Drosophila,* and *Arabidopsis),* circadian rhythms and plant nutrition *(Arabidopsis),* organelle assembly (yeast), cell motility, and the cytoskeleton (squid and sea urchins). During the first year, students attend a comprehensive two-quarter course in biochemistry and cell and molecular biology. A series of three 1-term research rotations in individual faculty members' labs trains students in research techniques and allows students to select a thesis adviser from among the program faculty members by the end of their first year.

Research Facilities

The Department occupies the Charles Gilman Life Sciences Laboratory, which is connected to the Dartmouth Medical School and the Dana Biomedical Library. Facilities permit study at all levels of organization and include a rooftop greenhouse and controlled environment chambers. Specialized instrumentation is housed in the Rippel Electron Microscope Facility, the Robert D. Allen Video Light Microscopy Facility, and the Molecular Genetics Center (with DNA and peptide synthesis and sequencing capabilities). All laboratories are directly linked to the Kiewit Computation Center.

Financial Aid

Research and teaching assistantships, plus federal fellowships and traineeships, provide support for most students. In 2006–07, stipends for students in the Graduate Program in Ecology and Evolutionary Biology are approximately $21,600 for twelve months. In addition, the Department covers the cost of health insurance. For students in the Graduate Program in Molecular and Cellular Biology, stipends are $23,500 for twelve months, plus health insurance coverage.

Cost of Study

Tuition of approximately $44,395 for the 2006–07 academic year is paid for all graduate students by the various fellowship and assistantship awards.

Living and Housing Costs

The College assists graduate students in arranging for appropriate housing, either in College facilities or in privately owned accommodations in the Hanover area. College-owned apartments are available at various rents for married graduate students.

Student Group

There are more than 40 students in Ph.D. programs who are working in labs in the Department of Biological Sciences. In addition, there are approximately 100 Ph.D. students in the Graduate Program in Molecular and Cellular Biology who are working in other departments. Nearly one quarter of the students are married; about one half are women.

Location

Located midway between Boston and Montreal, Hanover is the quintessential New England college town, situated in the Connecticut River Valley, which forms the border between New Hampshire and Vermont. The College supports much of the nonacademic life of the area, operating major film, music, and drama programs throughout the year, a large number of hiking trails and cabins, and facilities for cross-country and downhill skiing. It also provides the usual athletic opportunities through its swimming pools, gymnasiums, tennis courts, skating rink, and other facilities.

The College

Dartmouth College was founded as a liberal arts college in 1769 but has had students in professional degree programs since its medical school opened in 1797. Its present undergraduate population numbers 4,250. The three professional schools (medicine, engineering, and business administration) enroll approximately 800 additional students. Approximately 400 men and women are now enrolled in the arts and sciences graduate division. The smallest of the Ivy League institutions, Dartmouth has a long-standing tradition of close student-faculty ties.

Applying

Applicants should be prepared in biology, chemistry, physics, and mathematics, although some deficiencies can be made up during the first year of graduate study. Applications are encouraged from women and men of diverse backgrounds.

Applications for admission to the Graduate Program in Molecular and Cellular Biology should be received by January 7 for admission in September. For the Graduate Program in Ecology and Evolutionary Biology, applications should be received by February 1. All applicants must submit scores from the General Test of the Graduate Record Examinations (GRE). If possible, applicants to the ecology program should submit scores from the Subject Test in biology.

It is the long-standing policy of Dartmouth College to actively support equality of opportunity for all persons regardless of race or ethnic background. No student will be denied admission or be otherwise discriminated against because of race, color, sex, religion, handicap, or national or ethnic origin.

For more information, students should visit the Department's Web site at http://www.dartmouth.edu/~biology.

Correspondence and Information

Chair, Graduate Admissions Committee
Ecology and Evolutionary Biology
Department of Biological Sciences
Dartmouth College
6044 Gilman
Hanover, New Hampshire 03755-3576
E-mail: ecology@mac.dartmouth.edu

Chair, Graduate Admissions Committee
Program in Molecular and Cellular Biology at Dartmouth
7560 Remsen Building
Room 239
Dartmouth College
Hanover, New Hampshire 03755-3842
E-mail: mcb@dartmouth.edu

Dartmouth College

THE FACULTY AND THEIR RESEARCH

For a complete listing of MCB Program faculty members, students should see the description under Dartmouth, Molecular and Cellular Biology.

Matthew P. Ayres, Associate Professor; Ph.D., 1991. Terrestrial ecology; plant-herbivore interactions, nutritional ecology, plant defenses, temperature responses, population dynamics, and distribution limits.

Edward M. Berger, Professor; Ph.D., 1969. Genetics and molecular biology: ecdysterone action in *Drosophila,* mechanisms of transcriptional regulation.

Sharon E. Bickel, Associate Professor; Ph.D., 1991. Molecular and genetic analysis of chromosome behavior in *Drosophila* meiosis.

Douglas T. Bolger, Adjunct Associate Professor; Ph.D., 1991. Conservation biology: effects of habitat fragmentation on animal populations, landscape ecology.

Ryan Calsbeek, Assistant Professor; Ph.D., 2001. Ecology and evolution: importance of ecology to natural and sexual selection in wild lizard populations.

Diane L. Church, Research Assistant Professor; Ph.D., 1988. Genetics, developmental biology, and molecular biology: control of germ line development in the nematode *Caenorhabditis elegans.*

Kathryn L. Cottingham, Associate Professor; Ph.D., 1996. Aquatic community and ecosystem ecology; quantitative ecology and biostatistics; environmental monitoring and ecological indicators.

Patrick J. Dolph, Associate Professor; Ph.D., 1989. Genetic and molecular analysis of phototransduction in *Drosophila melanogaster.*

Albert J. Erives, Assistant Professor; Ph.D., 1999. Gene regulation in the evolution and development of metazoan systems.

Michael W. Fanger, Adjunct Professor; Ph.D., 1967. Immunology and hybridomas; tumor-associate antigens; tumor cell biology; regulation and expression of secretory immunity.

Carol L. Folt, Professor; Ph.D., 1982. Aquatic ecology: competition and feeding strategies in zooplankton communities.

Andrew J. Friedland, Adjunct Professor; Ph.D., 1985. Elemental cycling in forested ecosystems: quantification of cycling rates, residence times, and elemental pools, using chemical mass balance, stable isotopes, and some micrometeorological techniques.

Tillman U. Gerngross, Adjunct Assistant Professor; Ph.D., 1992. Enzymology of microbial polymer formation; fermentation process development; development of protein expression systems; metabolic engineering of industrial microorganisms.

Amy Gladfelter, Assistant Professor; Ph.D., 2001. Cell biology and genetics: cell-cycle control in multinucleated cells, evolution of the cell cycle, morphogenesis.

Robert H. Gross, Associate Professor; Ph.D., 1974. Computational molecular biology.

Mary Lou Guerinot, Professor; Ph.D., 1979. Genetics and molecular biology: genetic regulation of the nitrogen-fixing symbiosis between rhizobia and legumes; iron regulation of gene expression in plants *(Arabidopsis thaliana)* and bacteria.

Rebecca E. Irwin, Assistant Professor; Ph.D., 2000. Population and community ecology; evolutionary ecology; plant-animal interactions; mutualisms; plant mating systems; invasive species.

Thomas Jack, Associate Professor; Ph.D., 1990. Molecular genetics of flower development in *Arabidopsis.*

Eric J. Lambie, Associate Professor; Ph.D., 1987. Developmental genetics of gonadal development in *Caenorhabditis elegans.*

Lee R. Lynd, Adjunct Associate Professor; D.E., 1987. Biochemical engineering, microbiology, and molecular biology applied to conversion of plant biomass into fuels and chemicals; sustainable resource utilization.

C. Robertson McClung, Professor; Ph.D., 1986. Genetics and molecular biology of circadian rhythms in *Arabidopsis.*

Mark A. McPeek, Professor; Ph.D., 1989. Macroevolutionary ecology; evolution of community structure; phylogeography; phenotypic evolution.

David R. Peart, Professor; Ph.D., 1982. Plant ecology: plant populations, communities and succession.

Kevin J. Peterson, Associate Professor; Ph.D., 1996. Origin and early evolution of animal body plans.

G. Eric Schaller, Associate Professor; Ph.D., 1990. Molecular, biochemical, and genetic analysis of hormone signaling pathways in *Arabidopsis thaliana.*

Roger D. Sloboda, Professor; Ph.D., 1974. Cell biology: microtubule assembly and biochemistry, mitotic apparatus assembly and function, intracellular particle motility.

Elizabeth F. Smith, Associate Professor; Ph.D., 1992. Molecular, genetic, and biochemical analysis of eukaryotic flagellar motility and assembly in *Chlamydomonas reinhardtii.*

Richard S. Stemberger, Research Associate Professor; Ph.D, 1982. Zooplankton ecology, systematics, and community responses to environmental disturbance.

Samuel J. Vélez, Associate Professor; Ph.D., 1974. Neurobiology: development of neuronal connections, nerve regeneration.

Ross A. Virginia, Adjunct Professor; Ph.D., 1980. Terrestrial ecosystem ecology; nutrient cycling and soil ecology, especially in deserts.

Lee A. Witters, Professor; M.D., 1969. Metabolism and enzyme regulation: hormonal and substrate regulation of protein kinases in mammalian cells; role of protein phosphorylation in cell metabolism and adaptation to cellular stress.

DEPAUL UNIVERSITY

Department of Biological Sciences

Program of Study	The Department of Biological Sciences offers graduate studies leading to the M.S. (thesis) and M.A. (final project) degrees. M.S. students are required to complete a minimum of 52 quarter hours of credit, up to 16 of which may include research credits. They are also expected to take at least 8 quarter hours of courses in the three core areas of study that include ecology, evolution, and population biology; genetics, cell, and molecular biology; and physiology and neurobiology. Students who select the M.A. option are required to complete a minimum of 50 quarter hours of credit with 12 quarter hours of study in each of the three core areas mentioned previously.

The graduate program emphasizes contemporary biology, with major areas of instruction and research in cell biology, comparative vertebrate physiology, plant biochemistry, developmental biology, genetics, immunobiology, microbiology, molecular biology, neurobiology, aquatic biology, and ecology. Selection of a program of study and research is based on the needs and aims of each student, determined in consultation with the student's faculty advisory committee. Students may direct their studies toward a professional career in teaching and/or research at a variety of levels. |

Research Facilities

Research facilities are available in the McGowan Biological and Environmental Sciences Center and include laboratories, offices, reading/conference and computer rooms, animal rooms, greenhouses, sterile transfer rooms, equipment rooms, light- and temperature-controlled incubators, cell-culture facilities, and low-temperature rooms. Standard instrumentation for research includes electrophoresis equipment; ultracentrifuges and preparative centrifuges; ultralow-temperature freezers; electrophysiological equipment for intracellular and extracellular recordings; an imaging system; multichannel scintillation counter; spectrophotometers; HPLC; DNA thermal cycler and automated DNA sequencer; chromatographic equipment; electronic recording equipment; osmometers and ion analysis equipment; rotary, sliding, and cryostatic microtomes; phase-contrast, polarization, video, and fluorescence microscopes; and both microcomputers and mainframe computers. Equipment for collecting and analyzing cells from freshwater and marine habitats is also available. Arrangements can also be made for equipment sharing with other Chicago institutions.

Financial Aid

Student support during graduate study is provided through laboratory teaching assistantships. Stipends for graduate assistantships were available at a level of $9000 per academic year (ten months) in 2005–06 and also included full tuition remission. Low-interest loans are also available to students.

Cost of Study

Graduate student tuition was $427 per quarter hour in 2005–06. The initial application fee is $25; fees for laboratory courses range from $20 to $50.

Living and Housing Costs

Student dormitories, centrally located on campus, as well as moderately priced private housing within easy commuting distance of the campus, are available. Costs, especially housing, are substantially lower than those found in most other major cities. Help is available in finding good off-campus housing.

Student Group

Approximately 23,000 students attend DePaul University's schools and colleges. Graduate enrollment in liberal arts and sciences was more than 1,500 in 2005–06. The majority of students are from the greater Chicago area, but nearly all sections of the United States as well as several other countries are also represented, especially in the graduate programs.

Location

A great number of cultural and recreational activities are available in Chicago, frequently on or near DePaul's several campuses. These include opera; ballet; concerts; serious and light theater; instrumental and vocal music recitals; folk singing; documentary and other film showings; art and sculpture exhibits; museum displays; lectures; tours of areas that are of historic, architectural, or ethnic interest; zoo, aquarium, and planetarium visits; dining at many good restaurants in the area; nightclub entertainment; park and beach picnicking or bicycling; Lake Michigan swimming, sailing, waterskiing, and fishing; and many amateur and professional sports.

The University

Founded in 1898 on Judeo-Christian principles, DePaul University, the largest Catholic university in the United States, is an urban, coeducational institution of higher learning whose students, faculty members, and administrators are racially, ethnically, and religiously diverse, being selected solely on the basis of individual ability and personal commitment. The Loop campus in downtown Chicago houses the main administrative offices as well as the Schools of Commerce, Law, Computer Science, and New Learning. The Lincoln Park campus on the Near North lakefront of Chicago houses the College of Liberal Arts and Sciences, which includes all undergraduate and graduate science facilities.

Applying

In order to receive full consideration, applicants are encouraged to apply in the early spring for fall enrollment. This is especially the case for individuals seeking financial assistance. Completed application forms and all supporting credentials should be received by the Director of Graduate Studies at least eight weeks prior to the date of first enrollment. Supporting information usually includes scores on the General Test of the Graduate Record Examinations, three letters of recommendation, and a one- to two-page statement of purpose. Applications can be made online at https://robin.depaul.edu/onlineapps/webapp/laspass.asp.

Correspondence and Information

For information about the program and financial support:

Director of Graduate Studies
Department of Biological Sciences
DePaul University
2325 North Clifton
Chicago, Illinois 60614
Phone: 773-325-7595
     773-325-8000
Web site: http://condor.depaul.edu/~biology/

Requests for application forms should be made to:

Graduate Information Office
DePaul University
1 East Jackson Boulevard
Chicago, Illinois 60604
E-mail: admitdpu@depaul.edu
Web site: http://www.depaul.edu

DePaul University

THE FACULTY AND THEIR RESEARCH

S. A. Cohn, Chairman; Ph.D., Colorado. Cell biology; cell movements; physiological ecology of diatoms.
J. S. Brooke, Ph.D., Western Ontario. Microbiology; cell surfaces.
J. V. Dean, Ph.D., Illinois. Plant biochemistry; membrane transport; secondary product metabolism.
P. E. Funk, Ph.D., Loyola Chicago. Immunology; microbiology; B-lymphocyte development.
D. A. Kozlowski, Ph.D., Texas at Austin. Neurobiology; neural injury and repair.
E. E. LeClair, Ph.D., Chicago. Embryology; cartilage morphogenesis; lung development.
L. A. Maginniss, Ph.D., Hawaii. Comparative physiology; blood-oxygen transport.
J. F. Masken, Ph.D., Colorado State. Physiology; toxicology; pharmacology and endocrinology.
D. A. Meritt, Ph.D., Illinois at Chicago. Biology and management of neotropical mammals in captivity.
T. T. Rajah, Ph.D., Osmania (India). Environmental estrogens; cell signaling and cancer.
K. Shimada, Ph.D., Illinois at Chicago. Paleontology; geology; fossil sharks.
M. E. Silliker, Ph.D., Berkeley. Molecular biology; genetics; mycology.
T. C. Sparkes, Ph.D., Kentucky. Behavior and population ecology of aquatic organisms.

William G. McGowan Biological and Environmental Sciences Center.

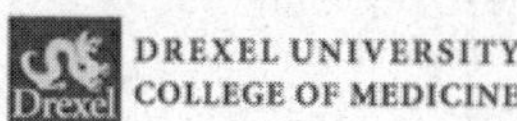

DREXEL UNIVERSITY

College of Medicine
Biomedical Graduate Studies

Program of Study	Drexel University College of Medicine (Drexel Med), formerly MCP Hahnemann University, has been educating students in biomedical sciences for more than 150 years and is committed to developing excellence in leadership, education, and training. At Drexel Med, education is the top priority, and the College is committed to preparing students for careers as academic scientists or successful professionals. The College believes that to be an effective and productive member of the scientific community, one must be trained with an extensive scientific knowledge. Drexel Med graduates leave with expertise in their own fields as well as an understanding of other disciplines—education and training that best prepares students to meet the health-care needs of today and tomorrow.
	Master's and doctoral programs are offered in the following disciplines: biochemistry, microbiology and immunology, molecular and cell biology and genetics, molecular pathobiology, neuroscience, and pharmacology and physiology. A combined M.D./Ph.D. program is also available for a small pool of highly qualified applicants.
Research Facilities	Such a diverse and prolific research environment serves as an exciting backdrop for a multitude of opportunities in graduate education and research training, which have been further fortified by MCP Hahnemann's merger with Drexel University in July 2002. This merger with Drexel, a preeminent computing and engineering university, has further expanded opportunities and has culminated in the development of many new undergraduate and graduate programs interweaving medicine with biomedical engineering, robotics, computing, education, and business. Drexel Med believes that innovation is at the interface of disciplines, and this merger affords students the opportunity to discover and invent.
Financial Aid	Stipends, currently at $23,500 plus health insurance coverage, may be awarded to qualified doctoral students in each of the graduate programs. Qualified doctoral students also receive tuition remission.
	Although no tuition remission or stipend waivers are currently available to students in the master's programs, the Office of University Student Financial Affairs (http://www.drexel.edu/provost/finaid/mc/pg/) strives to help all students find the financial resources they need to attend the Drexel University College of Medicine. Many work-study opportunities exist in the College, which provides substantial work experiences and financial support to qualified and deserving students. Students must complete and submit the Free Application for Federal Student Aid (FAFSA) by May 1 to determine if they are eligible to receive Federal Work-Study aid.
Cost of Study	The full-time master's student tuition rate for the 2006–07 academic year is $18,200 per year. Part-time or nonmatriculating student tuition is $1010 per credit hour. Additional fees include a $100 nonrefundable deposit, a general student fee of $135, and a student activity fee of $55 per semester. All particulars relating to tuition, fees, and financial aid are subject to change.
Living and Housing Costs	The Center City campus residence hall has been undergoing renovations. In order to help students with off-campus housing, the Office of Residential Life has prepared an apartment listing of properties in the vicinity of the University and some detailed information regarding off-campus housing options. For more information regarding off-campus housing, students should contact the Housing and Student Life Programs, at 215-762-1400, or via e-mail at studentlife.cchc@drexel.edu.
Student Group	The relatively small size of the program facilitates informal and informative exchanges between students and faculty members. Students consider the accessibility of the faculty one of the strengths of the program.
Location	Located in the vibrant city of Philadelphia, Drexel Med students have access to more than 143,000 businesses, corporations, and firms; forty-four major pharmaceutical companies; and the third-largest concentration of research institutions in the U.S. In addition to the plethora of opportunities for career growth, Philadelphia has much to offer culturally, recreationally, and historically. For avid travelers, Philadelphia is situated 1 to 3 hours from New York City, Atlantic City and other New Jersey shore points, the Pocono Mountains, and Washington, D.C.
The College	Drexel College of Medicine has three campuses within the city of Philadelphia. The main medical school campus, Queen Lane, is located in the East Falls section of Philadelphia. This is the location of all of the educational programs for the first two years of the medical school curriculum as well as the home for the Departments of Microbiology and Immunology and of Neurobiology and Anatomy.
	The Center City campus is home to the Departments of Biochemistry, Pharmacology and Physiology, and Pathology. It is also the site for the premedical programs. At this location, Drexel Med also has another major teaching hospital, Hahnemann University Hospital, operated by Tenet Healthcare Corporation. The main campus, located near the 30th Street train station, is the primary home of undergraduate education at Drexel University.
Applying	The Drexel University College of Medicine has a rolling admissions policy, which means that complete applications are reviewed as they are received. Applicants are therefore advised to apply early, as decisions to accept or deny admission may be made before the official deadlines. The application deadline is February 1 for all programs except the M.D./Ph.D. program, which is November 1.
	Applicants must hold a baccalaureate degree from an accredited institution and present evidence of their ability to pursue graduate work, as exemplified by high scholarship achievement, high aptitude scores, and strong recommendations. Previous research experience is highly regarded.
	Satisfactory scores on the Graduate Record Examinations (GRE) General Test are required. Individual programs generally request personal interviews. Certain requirements may be waived in unusual circumstances. International applicants whose primary language is not English must demonstrate competence in English as indicated by the Test of English as a Foreign Language (TOEFL).
Correspondence and Information	Office of Biomedical Graduate Studies Drexel University College of Medicine 2900 Queen Lane, Suite 239A Philadelphia, Pennsylvania 19129 Phone: 215-991-8570 E-mail: biograd@drexel.edu Web site: http://www.drexelmed.edu/biograd

Drexel University

DIRECTORS OF GRADUATE PROGRAMS AND AREAS OF RESEARCH

Biochemistry
Under the direction of Patrick J. Loll, Ph.D., Professor of Biochemistry and Program Director, this is a challenging and broad-based graduate program of research and course work leading to the master's or doctoral degree. The aim of the biochemistry program is to train scientists to identify, address, and solve biomedical problems at the molecular level. The themes of molecular structure, molecular mechanisms, and molecular regulation are recurrent throughout the diverse research areas represented by the biochemistry faculty.

M.D./Ph.D. Program
Jane Azizkhan-Clifford, Professor and Chair of the Department of Biochemistry, is the also the Director of the M.D./Ph.D. Program. Each year, a small number of highly qualified applicants with prior research experience are accepted into the program, designed to train students for careers in academic medicine that include both research and clinical practice.

Microbiology and Immunology
The Microbiology and Immunology Program, directed by Lawrence Bergman, Ph.D., Professor of Microbiology and Immunology, consists of faculty members from four different departments. The faculty members in this program have diverse research interests ranging from studying the cellular and molecular pathogenesis of infectious agents to the effect of aging on the immune function. In the first year, students spend most of their time completing required courses in the core curriculum and completing the research laboratory rotation requirements.

Molecular and Cell Biology and Genetics
Under the leadership of Joseph Nickels, Ph.D., Assistant Professor of Biochemistry, more than 60 investigators from the various departments within the University have formed an interdisciplinary program under the broad title of Molecular and Cell Biology and Genetics. Areas of intense research focus include cancer biology, regulation of gene expression, cell-cycle regulation, cell signaling, immunobiology, and pathogenic microbiology.

Molecular Pathobiology
This program provides students with a comprehensive education in contemporary knowledge of pathophysiological mechanisms and prepares them for careers in research and teaching in academic and corporate institutions. Greg Johannes, Ph.D., serves as the Program Director and leads the research team with topics related to neurodegeneration, tumor metastasis, angiogenesis, tumor suppressor genes, oncogenes, and mechanisms of host defense and disease.

Neuroscience
Peter Baas, Ph.D., Professor of Neurobiology and Anatomy, directs the Neuroscience Program. One of the few programs of its kind in the area, it gives students an opportunity to gain interdisciplinary research training. Participating faculty includes members of the Departments of Neurobiology and Anatomy, Pathology and Laboratory Medicine, Pharmacology and Physiology, and Neurology. Current research emphasizes the basic process underlying the organization and functioning of the nervous system and incorporates approaches ranging from molecular biology to systems neurobiology.

Pharmacology and Physiology
A degree in the Pharmacology and Physiology Program, directed by Robert Moreland, Ph.D., requires independent research under the direction of faculty members in the department, who are engaged in highly active research programs involving molecular, cellular, and behavioral approaches to experimental pharmacology and physiology in a strongly collaborative environment. Pharmacology examines and characterizes the action of drugs in humans and animals. It emphasizes the therapeutic responses of drugs, their mechanisms of action, the fate of drugs in the body, potential adverse reactions, and drug-drug interactions. Physiology considers processes that control and regulate the functioning of systems within an intact organism. Basic physiological processes underlie all fields in biomedical science. Understanding and exploiting the specific actions of drugs can also furnish a way to probe physiological and biochemical processes in both normal and pathological circumstances.

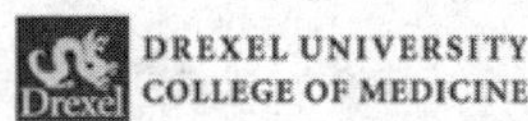

DREXEL UNIVERSITY

College of Medicine
Post-Baccalaureate Pre-Professional Education Programs
Medical Science Programs

Programs of Study

Drexel University College of Medicine (DUCOM), formerly MCP Hahnemann University, has been educating students in the biomedical sciences for more than 150 years and is committed to developing excellence in leadership, education, and training. The Division of Post-Baccalaureate Pre-Professional Education is dedicated to providing exceptional educational opportunities that combine rigorous academic learning, practical experience, and the use of advanced technologies. DUCOM is committed to educating students to make an immediate contribution to their organizations and communities.

Preprofessional education programs are offered in the following disciplines: Master of Science in clinical research and organizational management (CROM), Master of Science in pathologist's assistant studies (Path A), Master of Forensic Science (M.F.S.), and Master of Laboratory Animal Science (M.L.A.S.). The medical science programs offer one-year certificates in interdepartmental medical science (IMS); medical science preparation (MSP), and veterinary medical science (VMS), as well as the Drexel Pathway to Medical School (DPMS) and special master's programs, including the Master in Medical Science (M.M.S.) and Master of Biological Science (M.B.S.). A two-year part-time evening program for career changers who wish to apply to health professional schools is offered, the Post-Baccalaureate Premedical (PMED) program.

Research Facilities

A diverse and prolific research environment serves as an exciting backdrop for a multitude of opportunities in graduate education and research training, which have been further fortified by the merger with Drexel University in 2002. This merger with Drexel, a preeminent computing and engineering university, has further expanded the University's opportunities, interweaving medicine with biomedical engineering, robotics, computing, education, and business. DUCOM believes that innovation is at the interface of disciplines, and this merger affords students the opportunity to discover and invent.

Financial Aid

Although currently no tuition remission or stipend waivers are available to students in the master's programs, the Office of University Student Financial Affairs (http://www.drexel.edu/provost/finaid/mc/pg/) strives to help all students find the financial resources they need to attend the Drexel University College of Medicine. The College offers many work-study opportunities that provide substantial work experiences and financial support to qualified and deserving students.

Cost of Study

The tuition rate for the 2006–07 academic year ranges from $15,000 to $23,000. Part-time or nonmatriculating student tuition is approximately $775 per credit hour. All particulars relating to tuition, fees, and financial aid are subject to change.

Living and Housing Costs

In order to help students with off-campus housing, the Office of Residential Life has prepared an apartment listing of properties in the vicinity of the University and some detailed information regarding off-campus housing options. For more information regarding off-campus housing, students should contact the Housing and Student Life Programs at 215-762-1400 or via e-mail at studentlife.cchc@drexel.edu.

Student Group

In 2006–07, 473 students matriculated in the Post-Baccalaureate Pre-Professional Education programs, along with postdoctoral fellows and medical and allied health students. The relatively small size of many of the programs facilitates informal and informative exchanges between students and faculty members. Even in the larger programs, students consider the accessibility of the faculty and staff one of the strengths of the programs.

Location

Located in the vibrant city of Philadelphia, Drexel medical students have accessibility to more than 143,000 businesses, corporations, and firms; forty-four major pharmaceutical companies; and the third-largest concentration of research institutions in the U.S. In addition to the plethora of opportunities for career growth, Philadelphia has much to offer culturally, recreationally, and historically. For avid travelers, Philadelphia is situated 1 to 3 hours from New York City, Atlantic City and other New Jersey shore points, the Pocono Mountains, and Washington, D.C.

The College

Drexel College of Medicine has three campuses within the city of Philadelphia. The main medical school campus (Queen Lane) is located in the East Falls section of Philadelphia. This is the location of all the educational programs for the first two years of the medical school curriculum, as well as the home of the Departments of Microbiology and Immunology and of Neurobiology and Anatomy.

The Center City campus is home to the Departments of Biochemistry, Pharmacology and Physiology, and Pathology. It is also the site for the premedical programs. At this location, Drexel College of Medicine also has one of its major teaching hospitals, Hahnemann University Hospital, operated by Tenet Healthcare Corporation.

The main campus, located near the 30th Street train station, is the primary home of undergraduate education at Drexel University.

Applying

The application process requires that applicants submit a completed application form, two letters of recommendation in the sciences, college transcripts, and MCAT, GRE, or Miller Analogies Test scores, depending upon the program of interest. Applications are available online at http://www.drexelmed.edu/PostBaccalaureatePremedPrograms/ApplicationsResources/DownloadIMSMSPDPMSMFS/tabid/681/Default.aspx or by contacting the Post-Baccalaureate Pre-Professional Education Programs address. Applications are due July 13 for August matriculation, August 15 for the PMED Program, and September 9 for the M.F.S.

To determine eligibility for the Federal Work-Study Program, students should complete and submit the Free Application for Federal Student Aid (FAFSA) by May 1.

Correspondence and Information

Post Baccalaureate Pre-Professional Education Programs
Drexel University College of Medicine
245 North 15th Street, Mail Stop 344
Philadelphia, Pennsylvania 19102
Phone: 215-762-4692
E-mail: imsinfo@drexel.edu
Web site: http://www.drexelmed.edu/ims

Drexel University

GRADUATE PROGRAMS AND PROGRAM DIRECTORS

Master of Science in Clinical Research and Organizational Management
Under the direction of William Hirschhorn, this two-year, lockstep online program was designed after extensive consultation with individuals involved in clinical research and research administration to meet the particular educational and professional needs of this increasingly regulated yet very important field of investigation for future leaders of the research enterprise. Students are selected and courses are conducted in such a manner as to maximize the opportunities for learning from each other via the Web as well as from a distinguished faculty, which is composed of outstanding teachers, researchers, administrators, and practitioners.

Master of Science in Pathologist's Assistant Studies
Under the direction of James W. Moore, M.H.S., the Master of Science in pathologist's assistant studies program is a two-year, full-time, lock-step program beginning in June of each year. The program is designed for students who want to pursue an advanced career as a physician extender in pathology and laboratory medicine. The first year is the didactic portion of the program, supplemented by pathology laboratory exposure. The second year is composed of several hospital-based rotations offering progressively responsible experience in autopsy and surgical pathology. These rotations are supplemented with formal classroom education. Upon completion of the degree, students will be able to gain employment in community hospitals; academic centers, such as medical schools and university hospitals; private pathology laboratories; medical research centers; government hospitals; and medical examiner offices.

Master of Laboratory Animal Science
The M.L.A.S. program, directed by Julian E. Mesina, D.V.M., Ph.D., is a two-year program, which can be completed full- or part-time, designed for students who want to pursue advanced careers in laboratory animal science or laboratory animal facility management. Most classes are scheduled in late afternoon and early evening to accommodate working professionals. The M.L.A.S. degree is also a proven enhancement to the veterinary school application.

Master of Forensic Science
Detective William R. Welsh Jr., M.S., and Fredric N. Hellman, M.D., M.B.A., are the Co-Directors for the M.F.S. program. The M.F.S. program launched in September 2005 and is designed to allow the student exposure to both the intricacies of problem solving as well as an exposure to the real-world application of the related disciplines within the field of forensic science. The curriculum is designed to provide the student with a solid foundation within the forensic sciences, while at the same time encouraging growth and leadership in new and emerging applications within the field. A collaborative network of municipal agencies, private enterprise, and allied professional programs within the University has been built to prepare professionals who can confront the forensic challenges of the new millennium.

Veterinary Medical Science
Under the leadership of Julian E. Mesina, D.V.M., Ph.D., Program Director, the Veterinary Medical Science (VMS) Preparatory Program (the only one in the nation) is a one-year postbaccalaureate program designed to strengthen skills in the basic sciences and enhance the application to veterinary school. After successful completion of the VMS program, students have the option to enroll in the Master of Laboratory Animal Science degree program.

Medical Science Programs
Gerald Soslau, Ph.D., Professor of Biochemistry and Molecular Biology and Senior Associate Dean of Post-Baccalaureate Pre-Professional Education, is also the Director for the Interdepartmental Medical Science (IMS), Medical Science Preparatory (MSP), Master of Biological Science (M.B.S.), Master of Medical Science (M.M.S.), Drexel Pathways to Medical School (DPMS), and Evening Post-Baccalaureate Pre-Medical (PMED) programs. IMS is a graduate program in which students take six medical school courses and are graded relative to the performance of the College's first-year medical school students. MSP is a graduate program with course work at the graduate and postbaccalaureate levels for preparation to improve GPA and MCAT scores. The IMS and MSP programs are one-year certificate programs with the option to return for a second year to complete a master's degree (M.M.S. or M.B.S., respectively). The DPMS program is a graduate early assurance program for entry into Drexel University College of Medicine. The DPMS program is dedicated to disadvantaged and underrepresented students. The PMED program is an undergraduate program for career changers interested in gaining admittance into a health professional school.

DUQUESNE UNIVERSITY

Department of Biological Sciences

Programs of Study

The Department offers the Master of Science (M.S.) and the Doctor of Philosophy (Ph.D.) degrees in biological sciences. Both thesis and nonthesis M.S. options are available. There are three main areas of concentration: cell and molecular biology, cell and systems physiology, and microbiology. Subdisciplines within these concentration areas include evolutionary biology, prokaryotic and eukaryotic molecular genetics, reproductive biology, microbial physiology, ecology, neurobiology, developmental biology, and signal transduction. For full-time students, the emphasis is on experimental research and a publishable thesis. Individualized programs are prepared to complement the students' background and career plans. The M.S. thesis option entails 24 credits of course work, 6 credits of thesis research, and 2 one-credit seminar courses. The M.S. nonthesis option requires 32 credits of course work, 3 of which may be in research. The curriculum is designed to allow most full-time M.S. students to complete the program in two years. The Ph.D. program entails a total of 44 to 56 credits (depending on the student's background) for students entering with a bachelor's degree. Twelve of these credits are awarded for successful completion of the Ph.D. dissertation. The Ph.D. Committee determines credit requirements for students entering the Ph.D. program with the M.S. degree.

Research Facilities

The Department occupies about 48,000 square feet in the Mellon Hall of Science, a Bauhaus-style building designed by Mies van der Rohe. Facilities include two DNA sequencers, darkrooms, a computer lab, a cell and tissue culture lab, an incubator room, biotechnology lab, a light microscopy and imaging lab, and an electron microscopy suite. The equipment inventory includes ultracentrifuges, spectrophotometers, incubators, HPLC, gas analyzers, electrophoresis units, DNA thermal cyclers, DNA sequencers, a phosphoimager, fraction collectors, fluorescent microscopes with computer image enhancement features, a recently purchased confocal laser scanning microscope, and computer-based physiology labs. Equipment is maintained by 3 full-time electronics technicians in a renovated instrument shop. Adjacent to Mellon Hall is Bayer Learning Center, a multimedia classroom building housing the USDA-approved 5,700-square-foot animal-care facility.

Financial Aid

Most full-time students enrolled in the M.S. thesis option and Ph.D. programs finance their graduate studies by working as teaching assistants. M.S. teaching assistantships carry a stipend of $14,740 for twelve months plus a full waiver of tuition; teaching duties require approximately 15 hours per week. Ph.D. program teaching assistantships carry an annual stipend of $19,415 plus a full waiver of tuition. The Ph.D. stipend is available to students transferring from the Department's M.S. program after successful completion of the qualifier exam (taken near the end of the second year of study) and to students who enter the Ph.D. program with the M.S. degree. Individual faculty members support highly qualified M.S. and Ph.D. students with research assistantships.

Cost of Study

In 2006–07, graduate tuition is $710 per credit.

Living and Housing Costs

Most graduate students live in off-campus apartments in the city or surrounding towns. Students typically pay between $400 and $500 per month for a studio/one-bedroom apartment within walking distance of Duquesne; housing in many surrounding neighborhoods is less expensive and is accessible by public transportation. The cost of clothing, food, and housing in Pittsburgh tends to be slightly less than cities of comparable size.

Student Group

The University enrolls more than 6,000 undergraduate and 4,000 graduate students. Typically, there are approximately 40 graduate students in biology.

Location

Allegheny County has a population of more than 1.5 million residents; nearly one third of these people live in the city of Pittsburgh. Altitude varies from 710 to 1,370 feet above sea level, and average temperatures range from 33°F in January to 75°F in July. An academic community of seven colleges and universities enhances the intellectual life of the area, and the many businesses, industries, and services provide economic strength and extensive employment opportunities. In addition, Metropolitan Pittsburgh offers a wide range of cultural and recreational activities, among them the Pittsburgh Symphony Orchestra, Pittsburgh Ballet Theater, Duquesne Tamburitzans, International Poetry Forum, Three Rivers Arts Festival, Pittsburgh Public Theater, Pittsburgh Opera Center (headquartered at Duquesne University), Art Institute, Carnegie Art Museum, Carnegie Library, Carnegie Museum of Natural History, Carnegie Science Center, Allegheny Observatory, Andy Warhol Museum, Phipps Conservatory, National Aviary, and the Pittsburgh Zoo. The Steelers, the Pirates, and the Penguins are the local professional sports teams. The city has an extensive system of parks with areas for biking, running, and walking. Also in the area are freshwater beaches, golf courses, and areas for camping.

The University

Duquesne University, founded in 1878, is a private Catholic coeducational university. The 40-acre campus is located on a quiet promontory overlooking the historical Monongahela River at the edge of the famed "Golden Triangle" of downtown Pittsburgh. Its students and more than 400 faculty members are organized in nine academic units: the College of Liberal Arts, the School of Natural and Environmental Sciences, the School of Pharmacy, the School of Nursing, the School of Law, the School of Business and Administration, the School of Education, the School of Music, and the School of Health Sciences. These academic units are supported by the University library of more than 276,700 volumes and nearly 2,000 periodicals, along with extensive holdings at the University of Pittsburgh, a University press, a computer center, and a University radio station affiliated with National Public Radio (WDUQ). Basketball is the major intercollegiate sport (Duquesne is a member of the Atlantic Ten), and the football team is a member of the Metro Atlantic Athletic Conference. The Tamburitzans and the Red Masquers are well-established ethnic dance and theatrical groups, respectively.

Applying

It is recommended that students who wish to be considered for assistantships submit completed application forms, three letters of recommendation, and transcripts of all academic work by February 15. Assistantships are normally awarded by April 15 for the following academic year. To be considered for admission, students should have a bachelor's or master's degree from an accredited institution with a major in biology or a related science with a foundation in biology and a strong background in chemistry, physics, and mathematics. Students must take the GRE exam and include scores with their application forms. In addition, international students must have a minimum TOEFL score of 600 on the paper-based test or 250 on the computer-based test.

Correspondence and Information

Graduate Programs
Bayer School of Natural and Environmental Sciences
Duquesne University
100 Mellon Hall
Pittsburgh, Pennsylvania 15282

Phone: 412-396-4900
E-mail: biology@duq.edu
Web site: http://www.science.duq.edu/biology/

Duquesne University

THE FACULTY AND THEIR RESEARCH

Mary Alleman, Associate Professor; Ph.D., Berkeley, 1985. Plant genome structure; regulation of gene expression during development.

Philip E. Auron, Professor and Chair; Ph.D., Penn State, 1980. Structure-function relationships in proteins and nucleic acids; signal transduction and subsequent effects on gene induction and cell morphology; cytokine biology.

Peter A. Castric, Professor; Ph.D., Montana State, 1969. *Pseudomonas aeruginosa:* HCN biosynthesis, production of pili; molecular biology of virulence factors.

Richard Elinson, Professor and Chair; Ph.D., Yale, 1970. Developmental biology; origin of the embryo body plan; development and evolution.

Michael Jensen-Seaman, Assistant Professor; Ph.D., Yale, 2000. Molecular evolution and population genetics of primates; genome evolution; bioinformatics.

David J. Lampe, Associate Professor; Ph.D., Illinois, 1992. Molecular biology and evolution of transposable elements; use of transposons as genetic tools.

Joseph R. McCormick, Associate Professor; Ph.D., Rochester, 1989. Genetics of bacterial cell division; developmental regulation of gene expression.

Jana Patton-Vogt, Assistant Professor; Ph.D., Kentucky, 1992. Molecular genetics of phospholipid metabolism in *Saccharomyces cerevisiae.*

John A. Pollock, Associate Professor; Ph.D., Syracuse, 1984. Multigene analysis of neural fate determination; signal transduction to gene regulation in the developing fly eye.

Brady A. Porter, Assistant Professor; Ph.D., Ohio State, 1999. Systematics; population genetics; phylogeography and molecular parentage on North American freshwater fishes.

Kyle W. Selcer, Associate Professor; Ph.D., Texas Tech, 1986. Environmental; comparative reproduction; roles of estrogens and androgens in reproduction and cancer.

John F. Stolz, Professor; Ph.D., Boston, 1984. Microbial ecology and evolution; environmental and applied microbiology; biochemistry and ecophysiology of metal-reducing bacteria and phototrophic bacteria.

Nancy J. Trun, Assistant Professor; Ph.D., Princeton, 1988. Microbial genetics; the genetics, physiology, and biochemistry of chromosome folding.

Sarah Woodley, Assistant Professor; Ph.D., Arizona State, 1999. Behavioral neuroendocrinology; pheromonal regulation of endocrine physiology and reproductive behavior.

EAST CAROLINA UNIVERSITY

Graduate Programs in Biology and Molecular Biology / Biotechnology

Programs of Study

The Department of Biology offers programs leading to the Master of Science degree in biology or in molecular biology/biotechnology. The M.S. degree requires a thesis and 30 semester hours of course work. It is designed for those who plan to pursue the Ph.D. degree as well as for those who are planning careers in health-related, environmental, and technical areas.

Faculty members from the Department of Biology may act as mentors for students applying to interdisciplinary Ph.D. programs in coastal resources management and the interdisciplinary doctoral program in the biological sciences.

Students choose a major area of specialization in which a broad spectrum of courses is available: coastal ecology and environmental biology; biochemistry, evolution, cell biology, and developmental biology; or molecular biology and biotechnology. Each student works closely with a thesis adviser in developing and executing research. Faculty members of the medical school may also serve as thesis advisers for biology students.

To meet minimum requirements for the M.S. degree, students must complete course work, pass a comprehensive written exam, and write and orally defend a thesis based on original research.

Research Facilities

For students interested in fieldwork on coastal ecosystems, there are salt marshes, estuaries, lakes, swamp forests, and pocosins within short driving distance of the campus. Field stations on the Pamlico River estuary, operated by the University, and other state-owned coastal facilities are available for work in the field.

For students oriented toward laboratory research, the following are available: a small-animal facility well equipped for physiological and biochemical studies, radioactive-isotope detectors, sterile rooms for microbiology and tissue culture, transmission and scanning electron microscopes, gas chromatograph and electrophoresis units, and hybridoma, DNA sequencing, and protein purification labs. In addition, equipment is available for the molecular biology and biotechnology curriculum.

The University's computer has numerous terminals on campus and a remote batch–teleprocessing link to the Triangle Universities Computation Center in Research Triangle Park, North Carolina, where a wide variety of statistical software is available. The library receives most of the major biological journals and has many related monographs, proceedings of symposia, and government documents.

Financial Aid

The department offers teaching assistantships to many graduate students. Full teaching loads normally include three laboratories and service to the department totaling 20 hours per week; in 2005–06, stipends were $8500 for nine months. Limited summer stipends and out-of-state tuition waivers are available.

Cost of Study

During 2005–06, tuition and all fees for full-time students (taking 9 or more semester hours of courses) were estimated to be $2005 per semester for North Carolina residents and $7160 per semester for nonresidents. Costs of books vary considerably, depending on the nature of course work.

Living and Housing Costs

It is estimated that in 2005–06 the average student, who is a North Carolina resident, incurred necessary expenses of approximately $4620 for room and board during a two-semester academic year. Off-campus housing rates typically ranged from $350 to $500 per month for one- or two-bedroom apartments.

Student Group

Approximately 100 graduate students and 350 undergraduate majors are enrolled in the Department of Biology. The total enrollment of the University is approximately 23,000, including more than 4,000 in graduate programs. Most graduate students are North Carolina residents; most out-of-state students are from Atlantic coastal states.

Student Outcomes

Recent graduates have found employment in industry, academic and research institutions, and government. Several have continued their studies at professional schools or in doctoral programs.

Location

East Carolina University is located in Greenville, North Carolina, a rapidly growing city of about 68,600 people. Agriculture, forestry, and fishing have traditionally provided the economic base for the region, but a number of industries are opening plants in the area.

Mild temperatures and proximity to beaches make eastern North Carolina attractive for outdoor recreation. The campus is about 1½ hours from the Atlantic Coast and a 2-hour drive from Raleigh, the state capital.

The University

East Carolina University is the third-largest of the senior colleges in the University of North Carolina system. The main campus, approximately 340 acres within the city of Greenville, is convenient to the downtown and shopping areas of the city. The University provides services to the region through its continuing education activities and institutes. Music and the arts have a long tradition in the University and the community. The University has cooperative exchange programs with the National University in Costa Rica and Acadia University in Canada.

Applying

Admission requires a bachelor's degree from an accredited institution; acceptance is based on scores on the General Test of the Graduate Record Examinations, undergraduate records, and three letters of recommendation. All applicants are required to pay a $50 nondeductible, nonrefundable application fee. Information about graduate study in biology and teaching assistantships is furnished on request by the Biology Graduate Director, listed in the Correspondence and Information section of this description. Application for admission must be sent to the Dean of the Graduate School, also listed in the correspondence section.

Correspondence and Information

Department of Biology
East Carolina University
Greenville, North Carolina 27858-
    4353
Web site: http://www.ecu.edu/biology/

Dr. G. W. Kalmus
Biology Graduate Director
Phone: 252-328-6722
E-mail: kalmusg@mail.ecu.edu

Dean of the Graduate School
Phone: 252-328-6012
Web site: http://www.ecu.edu/
    gradschool

East Carolina University

THE FACULTY AND THEIR RESEARCH

Charles E. Bland, Professor; Ph.D., North Carolina at Chapel Hill, 1969. Mycology.
Jason E. Bond, Associate Professor; Ph.D., Virginia Tech, 1999. Arachnid and myriapod systematics and evolution.
Mark M. Brinson, Professor; Ph.D., Florida, 1973. Forested wetland, salt-marsh, and estuarine ecology; nutrient cycling.
Anthony A. Capehart, Associate Professor; Ph.D., Wake Forest, 1991. Developmental biology; cell interactions.
David R. Chalcraft, Assistant Professor; Ph.D., Illinois at Urbana-Champaign, 2002. Ecology and evolutionary biology.
Robert R. Christian, Professor; Ph.D., Georgia, 1976. Microbial and estuarine ecology.
Lisa M. Clough, Associate Professor; Ph.D., SUNY at Stony Brook, 1993. Marine benthic ecology; biogeochemistry.
John T. Conoley, Visiting Assistant Professor; Ph.D., North Carolina State, 2005. Educational technology.
Hal J. Daniel III, Professor Emeritus; Ph.D., Southern Mississippi, 1969. Animal behavior; comparative anatomy.
Mary A. Farwell, Associate Professor; Ph.D., Berkeley, 1989. Molecular biology; biochemistry; biotechnology.
Thomas L. Feldbush, Professor; Ph.D., Ohio State, 1996. Immunology.
Alexandros G. Georgakilas, Assistant Professor; Ph.D., Athens, 1992. Radiation damage to DNA.
Carol Goodwillie, Assistant Professor; Ph.D., Washington (Seattle), 1997. Plant ecology and evolution; evolutionary genetics.
Paul W. Hager, Visiting Assistant Professor; Ph.D., Berkeley, 1984. Biochemistry.
Jinling Huang, Assistant Professor; Ph.D., Georgia, 2000. Computational biology and bioinformatics.
Claudia L. Jolls, Associate Professor; Ph.D., Colorado, 1980. Evolutionary plant ecology.
Elizabeth A. Jones, Visiting Assistant Professor; Ph.D., East Carolina, 1996. Pharmacology.
Gerhard W. Kalmus, Professor and Director of Graduate Studies; Ph.D., Rutgers, 1977. Developmental biology; cell and tissue culture; differentiation.
Alfred C. Lamb III, Professor; Ph.D., Georgia, 1986. Herpetology; population genetics; evolution.
Joseph J. Luczkovich, Associate Professor; Ph.D., Florida State, 1987. Marine ecology.
Thomas J. McConnell, Professor; Ph.D., Florida, 1986. Molecular immunology.
Susan B. McRae, Visiting Assistant Professor; Ph.D., Cambridge, 1994. Ornithology.
Ronald J. Newton, Professor and Chairman; Ph.D., Texas A&M, 1972. Botany.
Anthony S. Overton, Assistant Professor; Ph.D., Maryland, 2003. Fisheries science.
Cindy Putnam-Evans, Associate Professor; Ph.D., Georgia, 1988. Biochemistry; cell regulation.
William H. Queen, Professor and Director, Institute for Coastal and Marine Resources; Ph.D., Duke, 1967. Marsh ecology.
Enrique Reyes, Associate Professor; Ph.D., LSU, 1992. Marine sciences.
Roger E. Robbins, Visiting Assistant Professor; Ph.D., Duke, 1986. Zoology.
Roger A. Rulifson, Professor; Ph.D., North Carolina State, 1980. Fisheries biology.
Jean-Luc Scemama, Assistant Professor; Ph.D., Toulouse (France), 1984. Molecular pharmacology and biotechnology.
Margit Schmidt, Visiting Assistant Professor; Ph.D., Heidelberg, 1986. Molecular biology.
Charles A. Singhas, Assistant Professor and Director of Undergraduate Studies; Ph.D., Virginia, 1975. Endocrinology; reproductive physiology.
Donald W. Stanley, Associate Professor; Ph.D., North Carolina State, 1974. Estuarine productivity; nutrient cycling.
Edmund J. Stellwag, Associate Professor; Ph.D., Virginia Commonwealth, 1978. Microbiology; bacterial physiology; molecular genetics.
John W. Stiller, Assistant Professor; Ph.D., Washington (Seattle), 1995. Phycology; molecular biology; evolution.
Kyle G. Summers, Associate Professor; Ph.D., Michigan, 1990. Behavioral ecology; evolution.
Leonard F. Sutton, Visiting Assistant Professor; Ph.D., North Carolina State, 2002. Physiology and anatomy.
Wei Tang, Visiting Assistant Professor; Ph.D., Chinese Academy of Sciences (Beijing), 1996. Forest biotechnology.
Heather D. Vance-Chalcraft, Visiting Assistant Professor; Ph.D., Illinois, 2003. Ecology.
Terry L. West, Associate Professor; Ph.D., Duke, 1982. Marine invertebrate zoology and ecology.
Alan R. White, Professor and Dean, Harriot College of Arts and Sciences; Ph.D., North Carolina at Chapel Hill, 1981. Cell wall polysaccharide biosynthesis.
Nancy M. White, Associate Professor; Ph.D., North Carolina State, 1996. Landscape ecology and water quality.
Yong Zhu, Assistant Professor; Ph.D., Tokyo, (Japan), 1991. Comparative animal physiology and molecular endocrinology.

Adjunct Faculty

The Department of Biology has adjunct faculty members from various units at East Carolina University, such as the Institute for Coastal and Marine Resources and departments in the School of Medicine, plus faculty from other institutions in the United States and abroad. These faculty members expand opportunities for course offerings and thesis research.

EAST CAROLINA UNIVERSITY

Brody School of Medicine
Doctoral Programs in the Biomedical Sciences

Programs of Study	The Departments of Anatomy and Cell Biology, Biochemistry and Molecular Biology, Microbiology and Immunology, Pharmacology and Toxicology, and Physiology and the Interdisciplinary Doctoral Program in Biological Sciences offer Ph.D. programs for highly qualified persons. The programs include a core curriculum of course work designed to meet the particular interests, aptitudes, and needs of individual students. Each candidate is encouraged to acquire a broad understanding of human biology as well as in-depth knowledge of a specific area. All of the programs are highly research oriented.
	The typical program of study consists of a minimum of 58 semester hours in graduate-level courses combined with individualized instruction in current laboratory research techniques. Each student may work with several faculty members before selecting a thesis preceptor. By the end of the second year, after approval of a dissertation research proposal and a successful doctoral candidacy examination, the student is admitted to candidacy for the degree. Thesis research, along with some specialized graduate-level courses, occupies most of the third and fourth years of training. Committee acceptance of a written doctoral thesis and successful defense of the thesis are the final degree requirements. Publication of thesis work in refereed journals is expected. The individual programs are intended mainly as preparation for careers in biomedical research, but some emphasis is also placed on the development of teaching and administrative skills.
Research Facilities	Each department is well equipped with instrumentation for current preparative, analytical, and bioelectric recording techniques. Extensive computer facilities with networking are available. Appropriate facilities accommodate tissue culture and virological studies and the handling of pathogenic and recombinant agents. Access to specialized instrumentation is provided by the Core Laboratories System and the Biotechnology Program. Such instrumentation includes units for flow cytometry, oligonucleotide synthesis, phosphorimaging analysis, peptide synthesis, and the sequencing of protein and DNA. Facilities exist for the preparation of monoclonal antibodies; databases for molecular biology and genetics are also available.
	All basic science departments, the Health Sciences Library, and excellent animal-care facilities are located in the Brody Medical Sciences Building, the research and education center of the Brody School of Medicine. Additional research laboratories and facilities are separately housed in the Biotechnology Center and the Life Sciences Building. Many investigators at East Carolina University have cooperative working arrangements with other major research facilities throughout the state and region. Financial support for research is broad based, coming from federal and state governments, private foundations, corporations, and local sources.
Financial Aid	Students awarded a full fellowship also receive total tuition remission for fall and spring semesters as well as reimbursement of up to $1000 for medical insurance. The demands of the programs, which include summer work, usually preclude a student's holding part-time employment outside the department in which he or she is pursuing the Ph.D. Spouses usually have no difficulty securing employment in the area.
Cost of Study	Tuition and fees for North Carolina residents are $4484 for 2006–07; this figure is subject to change. Out-of-state students qualifying for graduate assistantships may pay resident tuition. Books and supplies cost about $500 per year.
Living and Housing Costs	Unmarried students living in privately owned off-campus quarters should estimate a minimum of $13,000 per year for housing, food, and other essential living expenses. A variety of apartments and other rooming facilities are available in the Greenville area.
Student Group	There are approximately 65 predoctoral students in various stages of training. These students come from many different geographical areas of the United States and abroad. The small enrollment provides an unusual opportunity for individualized curricular plans and easy access to faculty and facilities.
Location	The University is located in Greenville (population 65,000), 85 miles east of Raleigh, the state capital, and 80 miles from the ocean. Greenville is a rapidly expanding city with a broadly structured and balanced financial base consisting of the University and its medical center, agriculture, light industry, and small businesses. The University, local institutions of worship, and social organizations provide a variety of scholarly, cultural, and recreational pastimes. Eastern North Carolina has a relatively mild winter climate, proximity to the seacoast, and excellent recreational opportunities.
The University and The School	East Carolina University is a comprehensive institution with a total enrollment of approximately 17,500 students. Curriculums encompass the arts, sciences, education, and technology. The Brody School of Medicine was established as a four-year school in 1976 and in 1981 graduated its first class of physicians. The Ph.D. programs in the basic biomedical sciences were initiated in 1979. The School of Medicine and the graduate programs in the biomedical sciences are in a phase of dynamic growth and have attracted outstanding faculty members from all over the United States. Full-time members of the graduate faculty in the basic biomedical sciences number 96.
Applying	Application forms are available from the departments offering the Ph.D. degree, the office listed in this Close-Up, or the office of the Dean of the Graduate School. Applications can also be downloaded from the East Carolina University Web site. General requirements for admission include a baccalaureate degree from an accredited institution, creditable scores on the General Test of the Graduate Record Examinations, references from academic preceptors, and a personal interview when feasible. Undergraduate backgrounds strong in the sciences and mathematics are preferred. More information about specific departmental requirements and preferences can be obtained from the particular departments. Applications for the fall semester should be submitted no later than June 1. Because decisions are usually made in April or May, early applications are encouraged. Interested students may obtain information from the appropriate departmental chairperson, listed under the Faculty and Their Research section, or from the address listed in this Close-Up.
Correspondence and Information	Associate Dean for Research and Graduate Studies School of Medicine East Carolina University Greenville, North Carolina 27834 Web site: http://www.ecu.edu

East Carolina University

THE FACULTY AND THEIR RESEARCH

Department of Anatomy and Cell Biology

Professors Cheryl B. Knudson (Chairman), Jack E. Brinn, Hubert W. Burden, Ronald W. Dudek, Donald J. Fletcher, Warren Knudson, Thomas M. Louis, Randell H. Renegar, David M. Terrian. Associate Professor Ann O. Sperry. Assistant Professors Yan-hua Chen, Qun Lu. Professor Emeritus Irvin E. Lawrence.

Molecular mechanisms of Alzheimer dementia, epileptic seizure, and mental retardation using presenilin mouse models; molecular mechanisms of prostate cancer using a variety of models that mimic acquisition of androgen-independence and metastasis; molecular mechanisms of embryological development identifying the role of microtubule-associated proteins (MAPS) and the cytoskeleton; effects of the peripheral autonomic nervous system and altered gravity on female mammalian reproductive function; regulation of tissue-specific gene expression in placental trophoblast cells using promotor-reporter vector constructs and a rat choriocarcinoma cell line, signaling pathways involved in tight junction assembly and disassembly, development of distance learning curriculum for the health sciences.

Department of Biochemistry and Molecular Biology

Professors Phillip H. Pekala, (Interim Chairman), Joseph M. Chalovich, Joseph G. Cory, Ronald S. Johnson, George J. Kasperek, Richard H. L. Marks. Assistant Professors Brett D. Keiper, Ruth A. Schwalbe, Brian M. Shewchuk. Professor Emeritus Sam N. Pennington.

Metabolic regulation—lipogenesis and protein turnover; cellular differentiation and control of gene expression; nucleotide metabolism in tumor cells; regulation of muscle contraction and cell motility; muscle cytoskeleton; transcriptional specificity of RNA polymerase; protein-DNA interactions; translational control during animal development; molecular physiology and biophysics of potassium channels; drug resistance in tumor cells; cancer cell biology.

Department of Microbiology and Immunology

Professors C. Jeffrey Smith (Chairman), James A. McCubrey, R. Martin Roop. Associate Professors Paul L. Fletcher, Richard A. Franklin, Mark D. Mannie, Everett Pesci. Assistant Professors Shaw Akula, Fred E. Bertrand, James P. Coleman, Isabelle Lemasson, Daniel W. Martin, Rachel L. Roper. Professors Emeritus Robert S. Fulghum, A. Mason Smith.

Physiological and molecular genetic studies of metabolic regulation in bacteria; genetic regulation and evolution of bacterial antibiotic resistance; genetic analysis of bacterial oxidative stress response; molecular mechanisms of bacterial pathogenesis; analysis of cell to cell signaling in bacteria; production of monoclonal antibodies; interaction of human intestinal flora with the host; cloning, sequencing, and expression of viral genes; transformation of cells by viral and cellular oncogenes; regulation of hematopoietic cell growth and differentiation; experimental allergic encephalomyelitis; immune system involvement in multiple sclerosis; structure and function analysis of animal virus genes; receptor-mediated intracellular signaling pathways in lymphocytes; B- and T-cell development; receptor-mediated stimulus-secretion mechanisms; DNA vaccines to viral antigens; induction of drug resistance in breast and prostate cells; regulation of bacterial virulence factors; poxvirus pathogenesis; signaling and structural components of Kaposi's sarcoma–associated herpesvirus entry; SARS virulence genes.

Department of Pharmacology and Toxicology

Professors David A. Taylor (Chairman), Abdel A. Abdel-Rahman, Donald W. Barnes, M. Saeed Dar, James E. Gibson, Brian A. McMillen. Assistant Professors Tatyana Nikalova, Ken Soderstrom, Rukiyah Van Dross. Research Faculty Mona M. McConnaughey. Professors Emeritus Alphonse J. Ingenito, Robert D. Myers, Wallace R. Wooles.

CNS/behavioral pharmacology; cardiovascular pharmacology; influence of ethanol and cannabinoid intoxication on motor coordination; functional interaction between nicotine and ethanol/cannabinoid-induced ataxia, glutamate/nitric oxide/cGMP signaling; brain catecholamine turnover, adenosine, and inhibitory substances; cell signaling and second messengers; isolation, characterization, and functional roles of neurotransmitter receptors; mechanisms of action of novel antianxiety drugs; development of drugs to block craving for abused substances; nicotine dependence and brain opioids; influence of nicotine and cannabinoids on brain development; role of bioflavinoids in prevention of skin cancer, cellular mechanisms of cancer development; mechanisms of ethanol-induced hypertension; actions of antihypertensive drugs on central and reflex cardiovascular control; molecular pharmacology; molecular biology of ion channels; role of potassium channels in disease; immunoactive drugs and microsomal drug-metabolizing enzymes; psychostimulants; toxicology; molecular toxicology; mechanisms of cellular toxicity; mechanisms of cellular adaptation.

Department of Physiology

Professors Robert M. Lust (Chairman), Robert G. Carroll, G. Lynis Dohm, S. Gregory Iams, Edward M. Lieberman, Richard H. Ray, Edward R. Seidel, Michael R. Van Scott. Associate Professors Jian Ding, Christopher Wingard, Laxmansa C. Katwa, Alexander Murashov. Joint Faculty Ron Cortright, Scott Gordon, Robert C. Hickner. Professors Emeritus David L. Beckman, William H. Waugh.

Cardiovascular, cellular molecular, gastrointestinal, and respiratory physiology; neuroscience; bioenergetics; pain; inflammation in the CNS; electrophysiology of neurons; neuron-glial interactions; circadian rhythm; arteriosclerosis; congestive heart failure; hemorrhagic shock; hypertension; myocardial ischemia; coronary blood flow; lipid metabolism; peptide hormones; vascular remodeling; epithelial growth and differentiation; polyamines; asthma, bronchial hyperreactivity, airway smooth muscle, airway remodeling, and pulmonary inflammation; diabetes, exercise, insulin signaling, and obesity; environmental cardiology.

Interdisciplinary Doctoral Program in Biological Sciences

Professors Gerhard W. Kalmus (Program Director), Libero J. Bartolotti, Arthur P. Bode, Charles E. Boklage, Paul Bolin Jr., Joseph M. Chalovich, Robert Christian, John D. Christie, Joseph G. Cory, Paul Gemperline, Carl Haisch, Charles A. Hodson, Donald R. Hoffman (Graduate Director for Biomedical Science Track), Ronald S. Johnson, Alfred C. Lamb III, John M. Lehman, Thomas J. McConnell, James A. McCubrey, Larry W. Means, Ronald J. Newton, Phillip H. Pekala, Roy M. Roop, Roger Rulifson, Andrew L. Sargent, George Sigounas, C. Jeffrey Smith, Paul H. Strausbauch, David M. Terrian, Kathryn M. Verbanac. Associate Professors William E. Allen, Jason E. Bond (Graduate Director for Biology Track), Lisa Clough, Larry Dobbs, Mary A. Farwell, Paul L. Fletcher Jr., Richard A. Franklin, L. Robert Hanrahan, Anne Kellogg, Prabhaker Khazanie, Brian Love, Mark D. Mannie, Cindy Putnam-Evans, Art Rodriguez (Graduate Director for Chemistry Track), Edmund J. Stellwag, Yu Yang. Assistant Professors Shaw M. Akula, Fred E. Bertrand III, Colin S. Burns, Anthony Capehart, David Chalcraft, James P. Coleman, Allison S. Danell, D. Erik Everhart, Alexander Georakilas, Carol Goodwillie, Anthony Hayford, Michael R. Hoane, Daren H. Kaiser, Brett D. Keiper, Yumin Li, Kwang Hun Lim, Andrew T. Morehead, Anthony Overton, Everett C. Pesci, Lorita M. Rebellato-deVente, Jonathan M. Reed, Timothy Romack, Rachel L. Roper, Jean-Luc Scemama, Margit Schmidt, Ruth A. Schwalbe, Charles A. Singhas, John W. Stiller, Kyle Summers, Yong Zhu.

Surgical pathology; experimental hematology; clinical microbiology; toxicology; oncology; electron microscopy; transplantation; genetics; nephrology; muscle; molecular biology; neuroscience; diabetes, adipocytes; virology; molecular immunology; molecular bacteriology; physical chemistry; organic chemistry; quantum biochemistry; NMR protein interactions; organometallic chemistry; analytical chemistry; molecular botany; molecular phylogeny; developmental biology; signaling; teratology; herpetology; molecular endocrinology; radiation biology; ecology; fisheries science; marine ecology; microbiology ecology; evolutionary systematics and genetics; bioorganic chemistry; bioanalytical chemistry; biophysical chemistry.

EMORY UNIVERSITY

Graduate School of Arts and Sciences
Graduate Division of Biological and Biomedical Sciences

Programs of Study

The Graduate Division of Biological and Biomedical Sciences is composed of eight interdisciplinary programs, each leading to the degree of Doctor of Philosophy. Each training program focuses on a major research area in contemporary biology or medicine and is based upon the realization that an interdisciplinary approach is essential not only for the solution of research problems but also for successful competition in modern biology and medicine. Each program seeks to provide both a broad background in molecular, cellular, and integrative biology and in-depth training in the concepts and methods of at least two biological/biomedical research disciplines. Research training and the acquisition of teaching and communication skills are stressed.

Students enroll with either a bachelor's or a master's degree, and the time to degree for the Ph.D. is approximately 5.5 years. Training in the Graduate Division is interdisciplinary, and students typically perform three rotations in their first year before affiliating with a faculty member for their dissertation research. The training students receive prepares them for jobs in many different career areas, including faculty and postdoctoral positions at top research universities, positions in prestigious institutions such as the U.S. Centers for Disease Control and Prevention and the National Institutes of Health, and positions in government and the pharmaceutical industry.

The Division offers eight interdisciplinary programs of study: the Program in Biochemistry, Cell and Developmental Biology; the Program in Genetics and Molecular Biology; the Program in Immunology and Molecular Pathogenesis; the Program in Microbiology and Molecular Genetics; the Program in Molecular and Systems Pharmacology; the Program in Neuroscience; the Program in Nutrition and Health Sciences; and the Program in Population Biology, Ecology, and Evolution. The Division provides students with the opportunity to work with world-renowned researchers who are located on or adjacent to the Emory campus. The 311 faculty members in the eight programs are drawn from the School of Medicine, the School of Public Health, the College of Arts and Sciences, Winship Cancer Institute, and Woodruff Health Sciences Center, as well as from the U.S. Centers for Disease Control and Prevention (CDC), the American Cancer Society, and the Yerkes National Primate Research Center.

Research Facilities

Emory University is one of the major biological research and medical referral centers in the Southeast. The state-of-the-art instrumentation that is needed to study virtually any aspect of modern biology or medicine is contained within the laboratories of the Division training faculty or in the centralized research facilities of participating departments and centers. Excellent research facilities are available, including the Biomolecular Computing Resource Facility, Transgenic Mouse Facility, Microchemical Facility, and Vaccine Research Center. Additional facilities for high-resolution structural biology, proteomics, microscopy, DNA array analysis, and the production of monoclonal antibodies are also housed on the Emory campus. Complete computer facilities are readily accessible. The Health Sciences Library and the Woodruff Library for Advanced Studies house most literature resources, along with up-to-date information retrieval facilities and online journal subscriptions.

Financial Aid

Division students receive a tuition scholarship, health insurance, and stipend support. For the 2006–07 academic year (September through August), tuition awards are $29,800 and Division stipends are $23,500

Cost of Study

Tuition and fees for the 2006–07 year are $31,586, and the cost of books and supplies averages $600 per year. A tuition scholarship is provided for Division students.

Living and Housing Costs

University housing is available for graduate and professional students at a monthly cost of about $1000 for an unfurnished one-bedroom apartment that includes all utilities except long-distance telephone service and certain cable television channels and provides convenient access to the Student Activity and Academic Center. Most students live off campus in areas that are attractive and conveniently located, where a one-bedroom apartment rents for $500 to $600 per month and a two-bedroom apartment rents for approximately $150 per month more. The University provides assistance in locating both on- and off-campus housing. In addition, the cost of living in the Emory area is lower than that of most university cities.

Student Group

Emory University has an enrollment of approximately 10,800 students, more than 4,950 of whom are enrolled in its nine graduate and professional schools. The students are drawn from all areas of the country and abroad; they come from large research-oriented universities and from small liberal arts colleges. The Division currently has nearly 400 students in various stages of graduate training, and last year they were primary or coauthors on more than 200 research papers or abstracts.

Location

Metropolitan Atlanta, populated by about 4 million people and widely considered the cultural and industrial center of the Southeast, offers a host of diverse opportunities. The Jimmy Carter Presidential Center and Library and the Martin Luther King, Jr. Center for Nonviolent Social Change are both internationally known educational service facilities. Culturally, Atlanta offers the Atlanta Ballet, the Atlanta Symphony Orchestra, the Alliance Theatre Company, and the High Museum of Art. The Atlanta Botanical Gardens, Zoo Atlanta, and the new Georgia Aquarium, which is considered one of the finest in the world, are also available in the city. Atlanta has four major professional sports teams and is easily accessible by a convenient bus and rail system. The Appalachian Mountains are located just north of the city, and Atlantic and Gulf Coast beaches are accessible by car.

The University

Emory University is a privately controlled, coeducational institution. Founded in Oxford, Georgia, in 1836, the college moved to the Druid Hills area of Atlanta in 1915, at which time the medical school was established. Since that time, Emory has grown into an internationally recognized teaching, research, and service center. The major components of the University are Emory College and Oxford College; the Graduate School of Arts and Sciences; the Schools of Business Administration, Law, and Theology; and the Woodruff Health Sciences Center, which includes the Schools of Medicine, Public Health, Nursing, the University hospital and other major hospitals, the Division of Allied Health Professions, and the Yerkes National Primate Research Center. University affiliates include the U.S. Centers for Disease Control and Prevention and Georgia Institute of Technology. Emory is one of the finest settings for advanced training in the biological and biomedical sciences in the Southeast.

Trends in NIH funding rank Emory among the fastest-growing medical centers in the United States. Nationally, Emory ranks fourth in the number of individual NIH fellowships and nineteenth in terms of both total NIH research dollars and institutional training-grant dollars. Emory has continued its rapid growth in research funding for a total of more than $351 million in sponsored research funding for 2004.

Applying

Students normally matriculate at the beginning of the fall semester. Applications should be completed by January 3 to be considered for fall enrollment; application information is available on the Division's Web site. Prospective applicants should plan to take the General Test of the Graduate Record Examinations; official score reports must be transmitted to the University from ETS by the application deadline and must be less than five years old. International applicants are also required to submit TOEFL scores that are no more than two years old. The minimum requirements for admission are a bachelor's degree and a grade of B or better in science courses. Minority students are encouraged to apply.

Correspondence and Information

Recruitment and Admissions
Graduate Division of Biological and Biomedical Sciences
Emory University
1462 Clifton Road, Suite 314
Atlanta, Georgia 30322

Phone: 404-727-2545
E-mail: gdbbs@emory.edu
Web site: http://www.biomed.emory.edu

Emory University

MAJOR AREAS OF FACULTY RESEARCH

The faculty members of each program and their particular areas of research are described in detail in separate program listings elsewhere in *Peterson's Graduate Programs in the Biological Sciences* and on the Web at http://www.petersons.com. The following overview is provided for the convenience of the reader.

Program in Biochemistry, Cell and Developmental Biology. This program has 61 faculty members from fifteen departments. Major areas of faculty research include cell and developmental biology, signal transduction, molecular biology, enzymes and cofactors, receptor and ion channel function, stem cells, and cancer biology.

Program in Genetics and Molecular Biology. This program has 51 faculty members from thirteen departments. In this diverse program, a plethora of model systems and technologies are used to explore and elucidate the molecular and biological principles governing the genetics of inherited human disorders, mechanisms of inheritance and epigenetics, mechanisms of genomic and chromosomal stability, replication, recombination and repair, gene expression and chromatin structure, cancer genetics and pathways, and the genetic control of development and sex determination.

Program in Immunology and Molecular Pathogenesis. This program has 47 faculty members from seven departments and the Emory Vaccine Center. Areas of research include T-cell-mediated immune responses; viral immunology; immunogenetics; transplant tolerance; molecular virology of HIV, herpesviruses, and influenza; vaccine design; T-cell activation and development; mucosal immunology; autoimmunity; and innate immunity.

Program in Microbiology and Molecular Genetics. This program has 26 faculty members from five departments and the U.S. Centers for Disease Control and Prevention. Areas of research interest include molecular biology of viruses and bacterial pathogens, bacterial genetics and physiology, microbial development, molecular biology of DNA recombination and transposition, and mechanisms of bacterial and viral pathogenesis.

Program in Molecular and Systems Pharmacology. This program has 45 faculty members from twelve University and School of Medicine departments. This program offers broad training in the biomedical sciences for students interested in learning how the drugs of today work and how the novel therapeutics of tomorrow can be developed. The curriculum emphasizes fundamental pharmacological principles and understanding the mechanisms of action for current therapeutics. Particular research strengths of this program include neuropharmacology, cancer biology, AIDS research, DNA repair, signal transduction, cardiovascular pharmacology, toxicology, and chemical biology.

Program in Neuroscience. This program has 104 faculty members from twenty-one departments and centers, including both basic science and clinical research areas. This multidisciplinary program covers broad areas of cutting-edge neuroscience research that includes cellular, molecular, and developmental neuroscience; neurological and psychiatric disease; behavioral neuroscience; motor control and movement science; systems neuroscience; computational neuroscience; and neuropharmacology. In addition to the wealth of research resources and a host of highly accomplished faculty members, the cornerstone of the Emory Program in Neuroscience is its strong sense of academic community and a structure that encourages a collaborative, multidisciplinary approach to solving challenging and exciting research questions.

Program in Nutrition and Health Sciences. This program has 47 faculty members from fourteen departments across the University and from the American Cancer Society and the U.S. Centers for Disease Control and Prevention. Faculty research spans basic (gene-nutrient regulation of metabolism), clinical (predictive health, metabolic ward studies), and population (epidemiology, surveillance, interventions with both U.S. domestic and international foci) nutrition.

Program in Population Biology, Ecology, and Evolution. This program has 27 faculty members from eleven departments. Major areas of faculty research include disease ecology, evolution of infectious disease, molecular evolution, evolutionary genetics and genomics, evolution and ecology of the immune system, evolution of genetic diseases in humans, population dynamics, and mathematical biology.

FORDHAM UNIVERSITY

Graduate School of Arts and Sciences
Department of Biological Sciences

Programs of Study

The Department of Biological Sciences at Fordham University offers programs of study leading to the M.S. and Ph.D. degrees. Requirements for the M.S. are 30 credits beyond the B.S. or B.A., knowledge of a foreign language or computer language, and a comprehensive examination. A student may elect to earn 6 credits by writing a thesis. Opportunities for part-time students are also available. Requirements for the Ph.D. are 30 credits beyond the M.S., completion of a program of study and research as recommended by a committee, knowledge of a foreign or computer language, a comprehensive examination, and a dissertation based on original research. Upon registration, students are assigned an interim adviser. Comprehensive exams are usually taken before the end of the second year.

Areas of specialization are cell and molecular biology and ecology. Research areas in cell and molecular biology include immunomodulators and cancer immunology; eukaryotic gene expression; ion channel physiology; genetic basis of aging; genetic toxicology; cytogenetic and molecular analysis of chromosomes; spermatogenesis and early development; cellular differentiation; regeneration in invertebrates; neuronal differentiation; electron microscopy and trace-element analysis; neurophysiology; and growth factors. Research areas in ecology include population and community ecology; conservation biology; microbial ecology and physiological ecology of plants, terrestrial insects, and aquatic organisms; and thermal biology. Emphases include insect-parasitoid interactions, paleoecology, trace-element biogeochemistry, freshwater ecology, systematics and evolution of mammals and insects, behavioral ecology of mammals, and extinction dynamics. Opportunities are available for collaborative research projects at the nearby Wildlife Conservation Society (Bronx Zoo) and New York Botanical Gardens.

Research Facilities

The Department of Biological Sciences is housed in a renovated 27,000-square-foot building containing sixteen modern laboratories and a central animal-care facility. The laboratories are equipped for broad-based research employing modern technologies. Items of particular interest for graduate research are tissue culture facilities, microorganism culture facilities, transmission and scanning electron microscopes, fluorescence microscopes, microinjection equipment, ultracentrifuges, a fluorometer, high-performance liquid chromatography equipment, scintillation counters, a video densitometer, DNA sequencing equipment, and an oligonucleotide synthesizer. The department also maintains the Louis Calder Conservation and Ecology Center, a research, teaching, and conference facility comprising 113 wooded acres, a 10-acre lake, wetlands, research labs, a greenhouse, animal facilities, a conference center, and student housing. Research at the center concentrates on applied environmental sciences, basic ecological research (with emphasis on aquatic studies), and conservation biology. The University maintains a computing center and a separate science library.

Financial Aid

Financial support in the form of assistantships and teaching fellowships is available for full-time graduate students. Tuition remission scholarships are usually held jointly with these awards. In 2005–06, stipends ranged from $17,000 to $25,000. Opportunities for support in the form of research assistantships are also available.

Cost of Study

For 2005–06, tuition was $875 per credit. The general activities fee was $159 per semester. The technology fee was $80 per semester.

Living and Housing Costs

On-campus housing is available for unmarried students at approximately $8000 per twelve-month lease (2005–06). Most students live in off-campus housing near the University or elsewhere in the New York City area. Rents vary widely.

Student Group

The total enrollment of Fordham University is approximately 12,000 students. The Graduate School of Arts and Sciences has an enrollment of approximately 1,150 students. Sixty graduate students are enrolled in the Department of Biological Sciences. These students have diverse ethnic, cultural, and religious backgrounds. Most graduates of the program are engaged in teaching, research, or both.

Location

The department is housed in Larkin Hall on the Rose Hill campus, which is in the North Bronx, adjacent to the New York Botanical Gardens and the Wildlife Conservation Society (Bronx Zoo). The 85-acre wooded and picturesque campus is conveniently near public transportation and parkways. The Louis Calder Conservation and Ecology Center is located 25 miles north of the campus in suburban Armonk, New York. Opportunities for cultural and social enrichment in New York are virtually unlimited.

The University and The Department

Fordham University, founded in 1841, consists of ten colleges and schools. The Graduate School of Arts and Sciences offers a wide range of master's and doctoral degree programs and participates in the New York City Doctoral Consortium.

The Department of Biological Sciences maintains an atmosphere that is demanding, yet friendly and human. It sponsors weekly seminars by outside speakers, and students have many opportunities to participate in journal clubs and other forms of scientific discussion. The Louis Calder Conservation and Ecology Center, with its conference center, provides an outstanding resource for urban ecological research and departmental events.

Applying

Applicants must hold a bachelor's degree in the biological sciences or the equivalent, have a minimum undergraduate GPA of 3.0 (on a 4.0 scale), submit GRE scores (including the Subject Test in biology), and arrange for official transcripts and two letters of recommendation to be sent to the University. TOEFL scores are required of most international students. Early application is encouraged. The deadlines are April 3, 2006, for the fall term and November 1, 2005, for the spring term. Applications for financial aid must be received by January 4, 2006. The fee for a paper application is $85; the fee for an online application is $65.

Correspondence and Information

For applications and further information, students should contact:

Director of Admissions
Graduate School of Arts and Sciences
Fordham University
Bronx, New York 10458

Phone: 718-817-4415
E-mail: fuga@fordham.edu
Web site: http://www.fordham.edu/gsas/

Specific questions can be addressed to:

Chairperson
Department of Biological Sciences
Fordham University
Bronx, New York 10458

Phone: 718-817-3640

Fordham University

THE FACULTY AND THEIR RESEARCH

Sergio Abreu, Ph.D., Connecticut. Biochemical, physiological, and morphological basis of pattern formation; metabolic analysis of early embryonic differentiation events; modeling of cellular and biochemical systems.

David Burney, Ph.D., Duke. Environmental changes and extinction in tropical areas, including Africa, Madagascar, and Hawaii; global trends in climate and ecological disturbance; fossil pollen analysis and other paleoecological techniques; biodiversity conservation.

George Dale, Ph.D., CUNY Graduate Center. Behavioral and population ecology of fishes; coral reef community ecology; the eye of fishes.

Craig Frank, Ph.D., California, Irvine. Integrative mammalian ecology; foraging; nutrition; plant-herbivore relationships; thermal biology.

Masaaki Hamaguchi, M.D., Ph.D., Tokyo. Molecular biology and cancer biology; analysis of tumor suppressor genes for breast and lung cancers.

Melissa Henriksen, Ph.D., Pennsylvania. Molecular mechanisms of signal transduction and transcription by STATs; chromatin biology and epigenetics.

Gerard Iwantsch, Ph.D., Penn State. Physiological and developmental interactions of the insect and its insect parasites.

Raj Kandpal, Ph.D., Bangalore (India). Molecular genetics and functional genomics.

Levente Kapás, M.D., Szeged (Hungary). Physiology and biochemistry of sleep regulation; characterization of endogenous sleep-promoting substances; the role of the endocrine system in the regulation of vigilance.

Gail Langellotto, Ph.D., Maryland. Insect ecology; trophic interactions and terrestrial community structure; pollinator conservation in developed landscapes; stable isotopes as a tool in ecological studies.

James Lewis, Ph.D., Duke. Plant physiological and community ecology, focusing on mechanisms regulating responses to climate change, invasive species, and pollution.

Asit Mukherjee, Ph.D., Utah. Molecular-cytogenetic and molecular studies on human cellular aging; chemical-induced genotoxicity and human cellular aging; human-mammalian cytogenetics and cell genetics.

Gordon Plague, Ph.D., Georgia. Ecological genetics; genomic interactions between bacterial endosymbionts and their insect hosts; speciation; evolution and ecology of aquatic invertebrates.

Robert Ross, Chair; Ph.D., Cornell. Role of cellular phenotype and the proto-oncogene N-myc in human neuroblastoma cell malignancy and differentiation; regulation of N-myc amplification and expression.

Berish Rubin, Ph.D., CUNY Graduate Center. Biology and biochemistry of immunomodulators, with primary emphasis on the study of their mechanisms of action and their efficacy as anticancer agents.

Daniel Sullivan, S.J., Ph.D., Berkeley. Biological control of insect pests; behavior of microwasps attacking aphids; tri-trophic ecology of aphid parasitoids and hyperparasitoids; international agriculture in Africa (IITA Nigeria), South America (CIAT Colombia), and India (ICRISAT).

William Thornhill, Ph.D., California, Santa Barbara. Molecular and cellular neurobiology; gene expression and regulation of ion channel expression, trafficking, and function in membranes.

Amy Tuininga, Ph.D., Rutgers. Community structure and ecosystem function; response of biodiversity and nutrient cycling to disturbances such as atmospheric pollution, fire, or invasive species; fungal symbioses and pathogenesis.

John Wehr, Ph.D., Durham (England). Ecology of algae in stream and river food webs; role of allochthonous carbon and nutrients in freshwater ecosystems; links between biodiversity and ecosystem services; population biology and biogeography of freshwater brown algae.

GEORGETOWN UNIVERSITY

Department of Biology

Programs of Study

The Department of Biology of Georgetown University offers graduate studies leading to a Ph.D. degree; on rare occasions, students are accepted into a program directed toward attaining an M.S. degree. The Department represents a wide range of biological disciplines. The aim of the program is to prepare scientists for diverse careers in research and teaching.

During the first two years of study, students take formal course work planned with a faculty adviser. Core courses include ecology and evolutionary biology; molecular, cellular, and developmental biology; biostatistics; and teaching biology. By the end of the first year, students select a thesis adviser and undertake advanced study and research in any of the areas represented by faculty research (listed in the Faculty and Their Research section). In general, Departmental research programs fall into the areas of animal behavior; biochemistry; cell, molecular, and developmental biology; ecology and evolutionary biology; immunology; neurobiology; and parasitology. Prospective students should contact individual faculty members concerning research opportunities and topics.

Because Georgetown University is a member of the Consortium of Universities of the Washington Metropolitan Area, the facilities and courses of all the Washington, D.C., area universities are available. In addition, many faculty members have collaborative research arrangements with scientists at organizations such as the National Institutes of Health, the U.S. Department of Agriculture, and the Smithsonian Institution.

Students must complete the general University requirements for examinations and serve as teaching fellows in undergraduate courses.

Research Facilities

The Department of Biology occupies approximately 14,000 square feet of teaching and research space in the Reiss Science Center and 2,000 square feet of space devoted to evolutionary biology on Observatory Hill. The Department is well equipped with research instrumentation, including shared computing and image analysis equipment, HPLCs, scintillation and gamma counters, high-speed centrifuges and ultracentrifuges, fluorescence and video microscopes, laser-scanning confocal and transmission electron microscopes, electrophysiology facilities, PCR thermal cyclers, DNA sequencing facilities, tissue culture facilities, and a greenhouse. Facilities are available in the Reiss Science Center chemistry department and the adjacent Medical Center campus for peptide sequencing and synthesis, FACS flow cytometry, confocal microscopy, and animal care.

Libraries on the main campus contain more than 117,000 bound volumes in the sciences, receive more than 700 journals, and maintain electronic search and document retrieval systems. In addition, there are extensive holdings in the Medical Center library. These facilities are supplemented by access to the Library of Congress, the National Archives, the National Library of Medicine, the National Agricultural Library, and the libraries of the Consortium of Universities of the Washington Metropolitan Area.

Financial Aid

Departmental teaching fellowships covering full tuition and providing a stipend are available to Ph.D. candidates. University fellowships are available on a competitive basis. Several tuition scholarships and research assistantships are also available. Recent students have been successful at obtaining support from national foundations. Tuition and stipend support is not available to M.S. candidates. Current tuition and stipend amounts are listed on the Department of Biology's Web site at http://bioserver.georgetown.edu.

Cost of Study

Tuition for Ph.D. candidates is covered by financial aid. Tuition was $27,528 in 2004–05.

Living and Housing Costs

Off-campus housing is available within walking distance and throughout the metropolitan area. University-operated bus service provides regular connections to the Washington public bus and subway systems, facilitating transport to a variety of housing options.

Student Group

There are approximately 25 graduate students in the Department of Biology and 550 graduate students in the sciences, including the Medical Center. There are a total of 2,800 graduate students in the University.

Location

The nation's capital has a population of about 600,000, with about 4 million in the greater metropolitan area. Cultural attractions abound in the form of art, music, museums, and theater, and numerous parks are located throughout the city. The climate is moderate. The Blue Ridge Mountains lie 1 hour by car to the west, Chesapeake Bay is about 1 hour to the east, and the Atlantic Ocean is 3 hours to the east.

The University

Georgetown University was founded in 1789 as a Jesuit educational institution, and about 60 percent of the undergraduate students are Catholic. The University is made up of the College of Arts and Sciences, the Graduate School, the Law Center, and the Schools of Medicine, Nursing and Health Studies, and Business Administration. The total enrollment is approximately 12,000, with most students on the main campus where the Reiss Science Center is located. The Law Center is located near the federal government offices in downtown Washington.

Applying

Applicants must apply by January 3 to be considered for assistantship or fellowship support and fall semester admission. However, students may apply at any time to begin their studies in June, August, or January. Applicants are strongly encouraged to discuss research opportunities with faculty members prior to applying for admission. Graduate Record Examinations—both the General Test and the Subject Test in biology or in biochemistry, cell and molecular biology—are recommended, and it is suggested that these examinations be taken no later than October of the preceding year. Applicants whose native language is not English and who will not have a degree from an English-speaking institution prior to matriculation are required to take the TOEFL; a TOEFL score of 600 or greater (250 or greater on the computer exam) is preferred, and a score of at least 550 (213 on the computer exam) is required.

Correspondence and Information

Chair
Committee on Graduate Students and Studies
Department of Biology–Reiss 406
Georgetown University
Washington, D.C. 20057-1229

Phone: 202-687-6247
Fax: 202-687-5662
E-mail: biology@georgetown.edu
Web site: http://bioserver.georgetown.edu

Georgetown University

THE FACULTY AND THEIR RESEARCH

Peter Armbruster, Assistant Professor; Ph.D., Oregon, 1997. Ecological genetics; quantitative and molecular population genetics; conservation biology; invasive species biology.

Edward M. Barrows, Professor; Ph.D., Kansas, 1975. Biodiversity, ecology, behavior, and evolution emphasizing insects; scientific communication.

Elena Silva Casey, Assistant Professor; Ph.D., Stanford, 1996. Developmental biology: neural induction and patterning in the amphibian *Xenopus laevis* and the ascidian *Ciona intestinalis*.

George B. Chapman, Professor; Ph.D., Princeton, 1953. Cell biology: electron microscopy of viruses, bacteria, protozoa, and selected plant, invertebrate, and vertebrate cells and tissues.

Douglas A. Eagles, Associate Professor; Ph.D., Massachusetts, 1972. Neurobiology: role of ketogenic diets in reducing seizures and effects upon neuronal activity.

Heidi G. Elmendorf, Assistant Professor; Ph.D., Stanford, 1993. Cellular and molecular parasitology; unique aspects of the cytoskeleton and gene expression in *Giardia lamblia*.

Matthew B. Hamilton, Assistant Professor; Ph.D., Brown, 1995. Evolutionary genetics; plant population genetics; conservation genetics.

Ellen J. Henderson, Professor and Chair; Ph.D., Purdue, 1971. Glycobiology: genetic, biochemical, and cellular analysis of developmental control of protein glycosylation and glycoconjugate roles in intercellular cohesion.

Janet Mann, Associate Professor; Ph.D., Michigan, 1991. Behavioral ecology; cetology; primatology; ethology; development.

Joseph H. Neale, Professor; Ph.D., Georgetown, 1970. Cell and molecular biology of the peptide neurotransmitter N-acetylaspartylglutamate and metabotropic glutamate receptors; development of drug therapies for neurodegeneration, neurotoxicity, and chronic pain.

Ronda J. Rolfes, Assistant Professor; Ph.D., Purdue, 1990. Molecular genetics: regulation of gene expression in response to nutrient conditions and signal transduction pathways.

Anne G. Rosenwald, Assistant Professor; Ph.D., Johns Hopkins, 1989. Molecular cell biology: regulation of membrane traffic; signal transduction.

Steven M. Singer, Assistant Professor; Ph.D., Stanford, 1993. Immunology and molecular biology: regulation of mucosal immune responses to the protozoan parasite *Giardia lamblia*; analysis of *Giardia lamblia* transcription and genome organization.

Philip Sze, Associate Professor; Ph.D., Cornell, 1972. Phycology: aquatic and marine biology.

Diane W. Taylor, Professor; Ph.D., Hawaii, 1975. Immunology: infectious disease immunity with emphasis on the immunoparasitology of malaria.

Martha R. Weiss, Assistant Professor; Ph.D., Berkeley, 1992. Evolutionary ecology: plant-insect interactions; plant reproductive biology and pollination ecology; insect behavior.

Adjunct and Research Faculty

Tomasz Bzdega, Associate Research Professor; Ph.D., Warsaw, 1985. Molecular neurobiology of the peptide neurotransmitter N-acetylaspartylglutamate.

Rose Leke, Adjunct Associate Professor; Ph.D., Montreal. Immunology: infectious disease immunity, with emphasis on the immunoparasitology of malaria.

Marcela Parra, Assistant Research Professor; Ph.D., Georgetown, 1991. Immunoparasitology: malarial parasites and immune responses.

Isabella A. Quakyi, Adjunct Associate Research Professor; Ph.D., London, 1980. Immunoparasitology: characterization of human malaria parasite antigens for vaccine development; cellular mechanisms of malaria immunity; immunoepidemiological studies in populations exposed to malaria; immune dysregulation in malaria and concomitant measles.

Joshua P. Rosenthal, Adjunct Research Professor; Ph.D., Berkeley, 1993. Biodiversity and conservation; plant-insect interactions.

Stephen D. Schiffer, Adjunct Professor; D.V.M., Michigan State, 1976. Comparative anatomy and animal models of human disease.

Barbara Wroblewska, Associate Research Professor; Ph.D., Warsaw, 1979. Molecular biology of metabotropic glutamate receptors; role of N-acetylaspartylglutamate in neurotransmission.

GEORGETOWN UNIVERSITY / NATIONAL INSTITUTES OF HEALTH

Graduate Program in Biomedical Sciences

Program of Study	The Georgetown University/National Institutes of Health (GU-NIH) Graduate Partnership Program (GPP) in Biomedical Sciences is designed to allow graduate students to explore all areas of biomedical research while being exposed to the expertise of both research institutions. Ph.D. programs at GU currently participating in the partnership include: Biochemistry and Molecular Biology, Cell Biology, Microbiology and Immunology, Pharmacology, Physiology and Biophysics, and Tumor Biology.

A key feature of the GU-NIH GPP is that instead of choosing a single laboratory in which to conduct his or her thesis research, the student selects an ongoing collaborative project between Georgetown University and NIH laboratories. A student may also establish a new collaboration between a GU and NIH scientist based on the student's interests and those of the mentors involved.

The first year of graduate training focuses on a core curriculum that enhances and expands the base of the student's scientific knowledge. Core courses include the following areas: biochemistry and molecular biology, biostatistics, cell biology, microbiology and immunology, physiology and biophysics, and ethics. Each student begins his or her research training by performing three laboratory rotations at the NIH and at Georgetown University, with at least one rotation to be performed at each institution. At the conclusion of the first year, each student must select a department to focus their didactic training and a collaborative project for their dissertation research, which by definition, requires that the student conduct research on both campuses. Additional courses specific to the degree program may be required during the second year of training. During periods of research at the NIH, students are expected to continue to attend seminars, journal clubs, etc. of the degree-granting program at Georgetown.

Research Facilities
As the federal government's primary agency for the support and conduct of biomedical research, the NIH's many institutes and centers employ approximately 1,200 tenured or tenure-track investigators and nearly 3,700 postdoctoral scientists with medical, dental, or graduate degrees. The NIH intramural research program, located on a 300-acre campus in Bethesda, Maryland, 10 miles from downtown Washington, D.C., has been the scene of many exciting scientific advances. Four Nobel Laureates made their prize-winning discoveries in NIH laboratories.

Basic research at NIH in the biomedical sciences is complemented by an active clinical research program. NIH's unique 250-bed research hospital and laboratory complex, the Warren Grant Magnuson Clinical Center, allows NIH's intramural scientists to bring research closer to the bedside and clinic. The NIH campus is also home to the National Library of Medicine, the world's largest medical library, and NIH-GU students have free online access to hundreds of full-text research journals. Georgetown University is located in the Georgetown area of the District of Columbia and is adjacent to many of the attractions of the city. The medical campus houses the basic science departments, which are currently training more than 350 M.S. and Ph.D. students as well as several hundred postdoctoral fellows.

Financial Aid
Students admitted to the program receive a full tuition award supported by the Georgetown University and the National Institutes of Health as well as a Predoctoral Intramural Research Training Award (Pre-IRTA) from the NIH. The current Pre-IRTA stipend is $24,800 for entering students.

Cost of Study
Students admitted into the GU-NIH Program in Biomedical Sciences receive funding from NIH and GU for stipend, tuition, and medical benefits.

Living and Housing Costs
The cost of living in the Washington, D.C., area is similar to other major cities in the United States. Various types of accommodations are available along the Metro (subway) lines that run between the University and the NIH. Near the NIH and GU there are various houses, apartments, and rooms in private residences available for rent. For information on available housing near NIH, applicants should visit the Web sites of local newspapers or contact the Graduate Partnerships Program office at the Web site listed in the Correspondence and Information section or the GU Office of Off-campus Housing.

Student Group
There are approximately 5,800 students at Georgetown University; 550 are graduate students studying the sciences, including those enrolled in the Medical Center. Students selected to participate in the GU-NIH program are members of the larger group of biomedical graduate students. The NIH has more than 4,000 trainees at the postbaccalaureate, graduate, and postdoctoral level, including over 400 graduate students from more than 100 universities. While at the NIH, graduate students enjoy services and activities sponsored by the Graduate Partnerships Program that build a strong graduate student community, similar to that of a university campus.

Location
Georgetown University is located in the Georgetown area of Washington, D.C. The National Institutes of Health is in a northwest suburban area of the metropolitan region, approximately 10 miles from the University. The cultural and academic life of Washington, D.C., is rich, and there are pleasant outdoor experiences year-round.

The University and The National Institutes of Health
Georgetown University, founded in 1789, is located near the many cultural and educational offerings of the entire metropolitan area. The GU program includes internationally regarded faculty members who are drawn from several academic departments, each opening the opportunity to develop a collaborative project between an NIH Investigator and a University professor.

The NIH has a long history of independent and creative training of scientists and physicians. Many NIH-trained scientists have received international recognition for their work. The NIH environment is rich in scientific exchange and provides opportunities for a broad biomedical research experience. Many graduate students have received their graduate training at the NIH through arrangements between the NIH and universities.

Applying
To apply, prospective students must be U.S. citizens or noncitizen nationals of the United States or must have been lawfully admitted for permanent residence (i.e., possession of a valid Alien Registration Receipt Card or some other verification of such status) and hold a bachelor's degree from an accredited university. The student's academic record, GRE scores, and prior research should reflect strong academic ability and evidence of superior promise. An entering class typically includes a variety of different undergraduate majors in the biological, chemical, or physical sciences. Information on submitting an application to the collaborative GU-NIH program may be found at http://gpp.nih.gov. Applications to Georgetown University should be completed and received by the posted deadline. Early submissions are encouraged.

Correspondence and Information

Graduate Partnerships Program
Building 2, Room 2E06
National Institutes of Health–DHHS
2 Center Drive
Bethesda, Maryland 20892-0234
Phone: 301-594-9605
Fax: 301-594-9606
E-mail: gpp@nih.gov
Web site: http://gpp.nih.gov

NIH Partnership Directors
Mark Cookson, Ph.D.
Phone: 301-451-3870
E-mail: cookson@mail.nih.gov

Allison McBride, Ph.D.
Phone: 301-496-1370
E-mail: amcbride@niaid.nih.gov

Thomas G. Sherman, Ph.D.
Director, Biomedical Sciences Program
233 Basic Science Building
Georgetown University Medical Center
3900 Reservoir Road NW
Washington, D.C. 20057-1460
Phone: 202-687-7044
Fax: 202-687-7407
E-mail: shermant@georgetown.edu

Georgetown University/National Institutes of Health

THE FACULTY AND THEIR RESEARCH

NIH Investigators
Presently, there are more than 500 NIH Investigators who participate in or would like to participate in the training of graduate students. Prospective students should refer to the Web site at http://gpp.nih.gov/researchers/ for details about specific researchers. Investigators listed on this Web site are available for dissertation research at the National Institutes of Health.

Georgetown University Professors
The Biomedical Sciences Program encompasses more than 140 faculty investigators, whose research interests include interdisciplinary studies in neuroscience and oncology/tumor biology as well as basic science research in biochemistry and molecular biology, cell biology, microbiology and immunology, pathology, pharmacology, and physiology and biophysics. Faculty members also participate in this program from main campus departments such as biology, chemistry, linguistics, nursing, and psychology as well as from clinical departments at the medical center, such as pediatrics, otolaryngology, medicine, psychiatry, neurology, neurosurgery, radiation medicine, and molecular and human genetics.

Collaborative Dissertation Research Projects
Students are required to engage in dissertation research that involves a collaboration with at least one GU faculty member and one NIH scientist. Students may create their own or join an established collaboration. A list of mentors with ongoing or desired collaborations can be found at http://gumc.georgetown.edu/departments/mcgrad/biomed/bspnihcolab.html.

The Dale and Betty Bumpers Vaccine Research Center (VRC), one of more than thirty major research buildings on the NIH campus, was established to facilitate research in vaccine development. The VRC is dedicated to improving global human health through the rigorous pursuit of effective vaccines for human diseases.

The National Human Genome Research Institute (NHGRI) led the Human Genome Project for the National Institutes of Health, which culminated in the completion of the full human genome sequence in April 2003. Now NHGRI moves forward into the genomic era with research aimed at improving human health and fighting disease.

THE GEORGE WASHINGTON UNIVERSITY

Department of Biological Sciences

Programs of Study

The Department of Biological Sciences offers programs leading to the Master of Science or Doctor of Philosophy. Both full-time and part-time students are accepted.

Each master's degree program is designed by the student in consultation with a departmental adviser. The master's degree with thesis requires 24 course credits, plus a thesis (6 credits) based on original research; the nonthesis master's degree requires completion of 36 course credits. Both require satisfactory completion of a final comprehensive examination.

Each doctoral program is formulated by the student in consultation with an advisory committee. A total of 72 credits beyond the bachelor's degree (including a dissertation) is required for the Ph.D. Course work is designed to prepare the student for general examinations in three or more fields relevant to his or her major area of interest. Upon completion of these examinations, the student undertakes dissertation research under the guidance of a dissertation director. The final examination is an oral defense of the dissertation.

Graduate research in the Department generally focuses on one of three areas: cell/molecular/developmental biology, systematics and evolution, or ecology (http://www.gwu.edu/~biology/faculty). Graduate research opportunities in cell/molecular/developmental biology cover a wide variety of topics, but most are linked through common interests in protein-protein and receptor-ligand interactors. Graduate research in systematics and evolution includes comparative studies of many different kinds of organisms and is enhanced by a formal agreement with the Smithsonian's National Museum of Natural History. Graduate research in ecology involves the use of field sampling, experimentation, and laboratory analyses to develop an understanding of the ecological factors shaping natural populations and communities of plants and animals. Students and faculty members interact regularly with scientists in the GW Institute for Biomedical Sciences, the Smithsonian's National Museum of Natural History and National Zoo, the National Institutes of Health, the National Park Service, the Food and Drug Administration, and the Environmental Protection Agency.

Research Facilities

The University has excellent facilities for research in the biological sciences, including electron and confocal microscopy, DNA sequencing, fossil preparation, digital imaging, and microcomputing. Animal-care facilities are available in the Department, and vehicles are available for field research. In addition, some students may use facilities at the local institutions listed above. Accessible library facilities include the Library of Congress, National Library of Medicine, National Agricultural Library, and the several libraries of the Consortium of Universities of the Washington Metropolitan Area.

Financial Aid

Ten graduate teaching fellowships and ten graduate research fellowships, all with tuition waivers, are available in the Department. Funding for graduate students working in specific research areas is often available from grants.

Cost of Study

Tuition in 2005–06 was $924 per semester hour. There are additional fees for certain laboratory courses, the University Center, computer use, graduation, and thesis binding.

Living and Housing Costs

The cost of living in Washington is comparable to that of other metropolitan areas. Most graduate students live off-campus, and the Metro system provides safe, clean transportation throughout the area. Complete information is available from the director of housing (http://www.och.gwu.edu).

Student Group

The George Washington University has approximately 17,000 students from all parts of the United States and 120 other countries. The Columbian College of Arts and Sciences includes 1,500 M.S. and Ph.D. candidates; various other schools of the University enroll more than 7,500 additional full- and part-time graduate students.

Student Outcomes

M.S. graduates are employed in the fields of biotechnology industry, basic research, law, teaching, information management, and medicine. Ph.D. recipients go on to postdoctoral research, university faculty positions, government and industry laboratories, and museum curatorships.

Location

The George Washington University is located in downtown Washington, D.C., near the White House, the World Bank, the Corcoran Gallery of Art, the Department of State, the National Academy of Sciences, the Kennedy Center for the Performing Arts, the Smithsonian Institution, and numerous professional organizations.

The University

The George Washington University, chartered by Congress in 1821, is private and nonsectarian. The campus includes the School of Medicine and Health Sciences, School of Public Health and Health Services, School of Engineering and Applied Science, Elliot School of International Affairs, School of Business and Public Management, GWU Law School, and Graduate School of Education and Human Development. The campus is a mixture of new buildings, traditional town houses, and older classroom buildings.

Applying

Application forms may be obtained from the Columbian School of Arts and Sciences (http://www.gwu.edu/~csas/). Students who complete applications for admission in the fall semester by January 2 are notified of the results in April; students who complete applications after January 2 are notified in July. For admission in the spring semester, applications must be completed by October 1. Completed applications must include transcripts, letters of recommendation, and GRE scores. Students are strongly encouraged to communicate with a potential adviser before submitting an application.

For application information, students should contact the graduate admissions office (telephone: 202-994-6210, e-mail: csasgrad@gwu.edu) or visit the Web page (http://www.gwu.edu/~csas/adminfo.html).

Correspondence and Information

Dr. John R. Burns
Department of Biological Sciences
The George Washington University
Washington, D.C. 20052
Phone: 202-994-7149
Fax: 202-994-6100
E-mail: jrburns@gwu.edu
Web site: http://www.gwu.edu/~biology

The George Washington University

THE FACULTY AND THEIR RESEARCH

Full-Time Faculty

Marc W. Allard, Associate Professor; Ph.D., Harvard. Molecular systematics; evolution and phylogenetics of the Mammalia.

Ken M. Brown, Professor; Ph.D., Michigan State. Developmental biology; neurohormones in early embryogenesis; developmental toxicology.

John R. Burns, Professor; Ph.D., Massachusetts. Histology; reproductive biology and comparative morphology of fishes.

Sheri A. Church, Assistant Professor; Ph.D., Virginia. Molecular evolution and bioinformatics; genetics of speciation and diversification.

James M. Clark, Associate Professor; Ph.D., Chicago. Paleontology and systematics of tetrapods; field collection of dinosaur-age fossils.

Robert P. Donaldson, Professor; Ph.D., Michigan State. Cell and molecular biology; plant peroxisomal membranes; electron transport; peroxisomal targeting sequence receptor; antioxidant metabolism.

Patrick S. Herendeen, Associate Professor; Ph.D., Indiana. Plant systematics and paleontology: systematics and history of the family Leguminosae; systematics and paleontology of early flowering plants; anatomy and evolution of angiosperm wood.

L. Patricia Hernández, Assistant Professor; Ph.D., Harvard. Evolutionary developmental biology of vertebrates; pattern formation in fishes.

Gustavo Hormiga, Associate Professor; Ph.D., Maryland. Systematics and evolutionary biology of spiders (orb weavers and their relatives).

Diana E. Johnson, Associate Professor; Ph.D., Chicago. Population genetics; speciation; evolution of gene families.

Robert E. Knowlton, Professor; Ph.D., North Carolina. Invertebrate zoology; marine biology; physiological ecology and morphogenesis of decapod crustacean larvae under natural and experimental conditions.

John T. Lill, Assistant Professor; Ph.D., Missouri–St. Louis. Plant-herbivore interactions; insect community ecology; life history evolution.

Diana L. Lipscomb, Professor; Ph.D., Maryland. Evolution; evolution of protozoa; origin of multicellular organisms; systematic theory.

Henry Merchant, Associate Professor; Ph.D., Rutgers. Population, community, and ecosystem ecology; energetics; biology of urban areas.

David W. Morris, Assistant Professor; Ph.D., Leeds (England). Molecular genetics; genetic manipulation of microorganisms for methanol and ethanol fuel production from organic waste; molecular biology of medicinal plants.

Randall K. Packer, Professor; Ph.D., Penn State. Kidney function; hypertension.

L. Courtney Smith, Associate Professor; Ph.D., UCLA. Molecular evolution of the deuterostome immune system.

Frank J. Turano, Associate Professor; Ph.D., Miami (Ohio). Plant signaling networks; plant biochemistry and molecular biology.

Elizabeth F. Wells, Associate Professor; Ph.D., North Carolina. Plant ecology of disturbed sites; conservation biology; floristics of the mid-Atlantic states; alien plant species.

Adjunct Faculty

Matthew Carrano, Adjunct Assistant Professor and Curator of Dinosauria, Smithsonian Institution; Ph.D. Chicago. Dinosaur phylogeny, functional morphology, and paleoecology.

Jonathan Coddington, Adjunct Professor and Curator of Arachnids and Myriapods, Smithsonian Institution; Ph.D., Harvard. Systematics and evolutionary biology of spiders.

Kevin De Queiroz, Adjunct Associate Professor, Smithsonian Institution; Ph.D., Berkeley. Vertebrate systematics and evolution.

Kristian Fauchald, Adjunct Professor; Ph.D., USC. Morphology, systematics, and biology of polychaete annelids; marine benthic ecology.

Benjamin R. Fisher, Adjunct Associate Professor, Covance Laboratories; Ph.D., George Washington. Developmental toxicology.

Vicki Funk, Adjunct Professor, Smithsonian Institution; Ph.D., Ohio State. Systematics of flowering plants, especially Asteraceae; conservation biology.

Peter L. Goering, Adjunct Associate Professor and Research Toxicologist, U.S. Food and Drug Administration; Ph.D., Kansas. Toxic responses of liver and kidney; toxicology of metals.

W. John Kress, Adjunct Professor, Smithsonian Institution; Ph.D., Duke. Systematics of flowering plants, especially tropical monocotyledons; conservation biology.

Patrick Nolan, Adjunct Associate Professor and Primary Examiner, U.S. Patent and Trademark Office; Ph.D., George Washington. Allergy and autoimmunity.

Lynne R. Parenti, Adjunct Professor and Curator of Fishes, Smithsonian Institution; Ph.D., CUNY Graduate Center. Systematics; comparative anatomy and biogeography of atherinomorph and gobioid fishes.

Hans-Dieter Sues, Adjunct Professor and Associate Vice President for Research and Collections, National Museum of Natural History; Ph.D., Harvard. Systematics and evolution of fossil reptiles.

F. Chris Thompson, Adjunct Professor and Research Entomologist, U.S.D.A.; Ph.D., Massachusetts. Dipteran systematics.

Stanley H. Weitzman, Adjunct Professor and Curator of Fishes, Smithsonian Institution; Ph.D., Stanford. Systematics and evolution of South American freshwater fishes.

SELECTED PUBLICATIONS OF THE FACULTY

A more complete list of publications is available at the individual faculty Web pages.

Allard, M. W., et al. Characterization of human control region sequences of the African American SWGDAM forensic mtDNA data set. *Forensic Sci. Int.* 148(2-3):169–79, 2005.

Papaconstantinou, A. D., et al. **(K. M. Brown).** Regulation of uterine hsp90α, hsp72, and HSF-1 transcription in B6C3F1 mice by β-estradiol and bisphenol A: Involvement of the estrogen receptor and protein kinase. *C. Toxicol. Lett.* 144:257–270, 2003.

Burns, J. R., and S. H. Weitzman. Insemination in ostariophysan fishes. In *Viviparous Fishes,* pp. 107–34, eds. H. J. Grier and M. C. Uribe. *New Life Publications,* 2005.

Church, S. A. Molecular phylogenetics of *Houstonia* (Rubiaceae): Descending aneuploidy and breeding system evolution in the radiation of the lineage across North America. *Mol. Phylog. Evol.* 27:223–38, 2003.

Clark, J. M., M. Norell, and P. Makovicky. Cladistic approaches to the relationships of birds to other theropods. In *Mesozoic Birds: Above the Heads of Dinosaurs,* chap. 2, eds. L. Chiappe and L. Witmer. University of California Press, 2002.

Donaldson, R. P. Peroxisomal membrane enzymes. In *Plant Peroxisomes,* chap. 8, pp. 259–78, eds. A. Baker and I. A. Graham. Kluwer Academic Publishers, 2002.

Herendeen, P. S., A. Bruneau, and G. P. Lewis. Phylogenetic relationships in caesalpinioid legumes: A preliminary analysis based on morphological and molecular data. In *Advances in Legume Systematics,* part 10, pp. 37–62, eds. B. B. Klitgaard and A. Bruneau. Kew: Royal Botanic Gardens, 2003.

Hernández, L. P., S. E. Patterson, and S. H. Devoto. The development of muscle fiber type identity in zebrafish cranial muscles. *Anat. Embryol.* 209:323–34, 2005.

Hormiga, G., M. A. Arnedo, and R. Gillespie. Speciation on a conveyor belt: Sequential colonization of the Hawaiian Islands by *Orsonwelles* spiders (Araneae, Linyphiidae). *Systematic Biol.* 52(1):70–88, 2003.

Knowlton, R. E., and C. J. Vargo. The larval morphology of *Palaemon floridanus* Chace, 1942 (Decapoda, Palaemonidae) compared with other species of *Palaemon* and *Palaemonetes. Crustaceana* 77:683–715, 2004.

Lill, J. T., and R. J. Marquis. Leaf ties as colonization sites for forest arthropods: An experimental study. *Ecol. Entomol.* 29:300–8, 2004.

Lipscomb, D. L., N. Platnick, and Q. Wheeler. The intellectual content of taxonomy. *Trends Ecol. Evol.* 18:65–6, 2003.

Merchant, H. C., R. N. Khan, and **R. E. Knowlton.** The effect of macrophytic cover on survival of *Palaemonetes pugio* and *P. vulgaris* (grass shrimp) in the presence of predatory *Fundulus heteroclitus* (killfish). *Contrib. Zool.* 70:61–71, 2001.

Lin, B., **D. W. Morris,** and J. Y. Chou. Hepatocyte nuclear factor 1α is an accessory factor required for activation of G-6-P gene transcription by glucocorticoids. *DNA Cell Biol.* 17:967–74, 1998.

Beutler, K. T., et al. **(R. K. Packer).** Long-term regulation of EnaC expression in kidney by angiotensin II. *Hypertension* 41:1143–50, 2003.

Nair, S. V., et al. **L. C. Smith.** Macroarray analysis of coelomocyte gene expression in response to LPS in the sea urchin, *Strongylocentrotus purpuratus:* Identification of unexpected immune diversity in an invertebrate. *Physiol. Genomics* 22:33–47, 2005.

Kang, J., and **F. J. Turano.** The putative glutamate receptor 1.1 (AtGLR1.1) functions as a regulator of carbon and nitrogen metabolism in *Arabidopsis thaliana. Proc. Natl. Acad. Sci. U.S.A.* 100:6872–7, 2003.

Shetler, S. G., S. S. Orli, **E. F. Wells,** and M. Beyersdorfer. *Checklist of the Vascular Plants of Plummers Island, Maryland,* 165 pp. *Bull. Biol. Soc. Washington,* in press.

THE GEORGE WASHINGTON UNIVERSITY

Institute for Biomedical Sciences
Interdisciplinary Doctoral Programs in the Biomedical Sciences

Programs of Study

The George Washington University's (GW) interdisciplinary doctoral programs in the biomedical sciences are organized within its Institute for Biomedical Sciences. Students are prepared for careers in research and teaching through an integrated program taught by faculty members from GW's Columbian College of Arts and Sciences and School of Medicine and Heath Sciences, including scientists from the Children's Research Institute of Children's National Medical Center, the National Institutes of Health, and the Institute for Genomic Research.

A core curriculum during the first year of graduate training covers macromolecular interactions of proteins and nucleic acids, cell biology, developmental biology, immunology, neurobiology, and metabolism. Laboratory rotations enable students to become more familiar with research techniques, scientific communication, and potential dissertation research projects. After the program area is selected, the student develops a plan of study consistent with that field and begins work on a dissertation research topic under the guidance of an Institute faculty mentor.

Admission to the Institute for Biomedical Sciences is the mechanism whereby students can ultimately become affiliated with the following Ph.D. programs: biochemistry and molecular genetics, microbiology and immunology, and molecular medicine.

Research Facilities

Extensive research facilities are available in faculty laboratories in GW's Medical Center and Columbian College of Arts and Sciences, the Children's Research Institute as well as through the GW/NIH Graduate Partner Program in selected laboratories of adjunct faculty members at the National Institutes of Health. The University's Gelman Library and Himmelfarb Health Sciences Library are available to graduate students, and there are numerous government agency and other research libraries in the Washington area, including the National Library of Medicine.

Financial Aid

Institute Fellowships are available on a competitive basis. They carry a $23,000-per-year stipend and 24-credit-hour tuition for the first two years (support during the second year is dependent on a satisfactory record during the first year). Subsequently, students are supported by extramural fellowships, scholarships, or research grants to the laboratory in which they are doing their thesis research project. Student tuition and fees are provided in full.

The George Washington University offers one Presidential Merit Fellowship in the Biomedical Sciences that is available on a competitive basis. It carries a $24,000-per-year stipend and 24-credit-hour tuition for two years.

Cost of Study

The cost of tuition for the 2006–07 academic year is $970 per credit hour and the cost of student association fees is $1 per credit hour.

Living and Housing Costs

Cost of living varies widely according to the type of accommodations and the area in which the student chooses to live. University housing is not generally available.

Student Group

The Institute for Biomedical Sciences admits 10–15 students per year. During the first year, the students participate in the core curriculum and a dissertation adviser and program are chosen at the end of the first academic year. These affiliations are important for the students' subsequent studies and research.

Location

The George Washington University is located in downtown Washington. In the immediately adjacent areas are the Corcoran Gallery of Art, the John F. Kennedy Center for the Performing Arts, and many other governmental and cultural institutions. The University's proximity to the National Institutes of Health, the National Library of Medicine, the Library of Congress, and the Food and Drug Administration are of particular interest to the students in the health sciences.

The University

The George Washington University is both a national and international institution, educating students from fifty states and 150 countries in a climate of academic accomplishment and scholarly integrity. More than 10,000 graduate students, along with almost 7,000 undergraduate students, interact with a notable faculty members and a myriad of cultural, governmental, corporate, and research institutions in the vibrant atmosphere of one of the world's most beautiful and influential cities.

The main GW campus is located five blocks from the White House in a historic area of Washington, D.C., known as Foggy Bottom. In addition to the extensive offerings of the on-campus libraries, students may have access through their research committee to a multitude of information and research resources, including the Library of Congress, the Smithsonian Institution, and the National Institutes of Health and other associations, nonprofit organizations, and institutions.

Ready transportation to the city's resources is available through Metro, Washington's extensive rapid transit system. With a station located on campus, Metro makes the University easily accessible from many parts of the city and the Virginia and Maryland suburbs. The Children's Research Institute of Children's Hospital and the National Institutes of Health are also near Metro stations.

Applying

Ph.D. graduate students are admitted to the Institute after a review of their qualifications and an interview. The Admissions Committee seeks students with broad interests and enthusiasm for further in-depth study in the biomedical sciences. It is important that applicants have course work that prepares them for a rigorous core curriculum in modern molecular and cellular biology. Normally, a minimum of a B average is required (3.0 on a 4.0 scale). All applicants must submit GRE General Test scores, three letters of recommendation, transcripts from all institutions attended, and a statement of purpose. International applicants should note that a minimum TOEFL score of 600 (250 on the computer-based test) is required. Applications are accepted for the fall semester only. The deadline for application materials is January 4.

Correspondence and Information

For information:
Institute for Biomedical Sciences
Ross Hall 605
The George Washington University
2300 Eye Street, NW
Washington, D.C. 20037

Phone: 202-994-2179
Fax: 202-994-0967
E-mail: gwibs@gwu.edu
Web site: http://www.gwumc.edu/ibs/

For application forms and catalogs:
Columbian College of Arts and Sciences
The Office of Graduate Studies
Phillips Hall, Room 107
The George Washington University
Washington, D.C. 20052

Phone: 202-994-6210
Fax: 202-994-6213
E-mail: askccas@gwu.edu
Web site: http://columbian.gwu.edu

The George Washington University

THE FACULTY AND THEIR RESEARCH

Julia A. Albright, Ph.D. Immunology of parasitic diseases.

Marc Allard, Ph.D. Molecular systematics and evolution.

Mahnaz Badamchian, Ph.D. Characterization of thymosins.

J. Martyn Bailey, Ph.D., D.Sc. Prostaglandins and corticosteroids.

Mark Batshaw, Ph.D. Inborn errors of urea synthesis-gene therapy.

James F. Battey, M.D. Structure, function, and regulation of G-protein coupled receptors.

Michael Bell, M.D.

Patricia Berg, Ph.D. Role of BP1 and other homeobox genes in the progression of breast cancer and prostate cancer.

Jeff Bethony, Ph.D.

Bernard Bouscarel, Ph.D. Role of bile acids in hepatocellular signal transduction.

John N. Brady, Ph.D. Viral regulatory proteins.

Beda Brichacek, Ph.D.

Kenneth Brown, Ph.D. Role of neurotransmitters and hormones in cell differentiation.

Michael Bukrinsky, Ph.D. Mechanisms of HIV-1 replication and pathogenesis, including virus-cell interactions at the step of entry, transport of viral genome into the nucleus, virus-induced cell-cycle perturbations, cellular innate anti-HIV responses, and impairment of cholesterol metabolism in HIV-infected macrophages.

Susan Ceryak, Ph.D. Carcinogenesis: Survival signaling after genotoxic insult.

Yi-Wen Chen, Ph.D.

Vincent A. Chiappinelli, Ph.D. Properties of neuronal nicotinic receptors.

Anne Chiaramello, Ph.D. Gene regulatory networks and neurodegenerative disorders.

Janice Y. Chou, Ph.D. Molecular genetics of human inborn errors of metabolism.

Philip H. Cogen, M.D., Ph.D. Molecular markers in brain tumors.

Anamaris Colberg-Poley, Ph.D. Cytomegalovirus pathogenicity.

Stephanie Constant, Ph.D. Regulation of inflammatory responses and study of immune evasion mechanisms used by parasites.

George Davis, Ph.D.

Edward C. DeFabo, Ph.D. UV effects of cellular immunity.

Louis DePalma, M.D. Gene expression in activated lymphocytes.

Benjamin F. Dickens, Ph.D. Mechanisms of myocardial injury.

Robert P. Donaldson, Ph.D. Glyoxysomal membrane electron transport.

Stephen Dopkins, Ph.D. Mechanisms of memory.

Clare M. Fraser, Ph.D. Genome sequencing and analysis; evolution of gene families and species.

Robert J. Freishtat, M.D., Ph.D.

Sydney W. Fu, M.D., Ph.D.

Linda L. Gallo, Ph.D. Cholesterol metabolism.

Vittorio Gallo, Ph.D.

Elodie Ghedin, Ph.D. Parasite genomics *(Brugia malayi, Trypanosoma cruzi);* development of tools for global detection and identification of emerging viral pathogens.

David Goldman, M.D. Characterization of genetic behavioral differences.

Allan L. Goldstein, Ph.D. Chemical and biological properties of the thymosins.

Gordon Hager, Ph.D. Transcriptional regulation and chromatin structure.

Tim G. Hales, Ph.D. Molecular mechanism of action of general anesthetics and opioids.

Yetrib Hathout, Ph.D.

John Hawdon, Ph.D. Hookworm development and infection; hookworm population genetics.

Robert G. Hawley, Ph.D.

Tarik F. Haydar, Ph.D. Cellular and molecular studies of neural stem cells during development of cerebral cortex.

L. Patricia Hernandez, Ph.D.

Eric Hoffman, Ph.D. Gene discovery; diagnostic and therapeutics of neurological disorders.

Peter J. Hotez, M.D., Ph.D. Vaccine development for parasitic and tropical diseases; Human Hookworm Vaccine Initiative.

Valerie W. Hu, Ph.D. Gap junction regulation.

Arthur Hurwitz, Ph.D.

Steven Jacobson, Ph.D. Pathogenesis of the human disorder, multiple sclerosis.

Diana Johnson, Ph.D. Population genetics and molecular evolution.

Fatah Kashanchi, Ph.D. Genomics and proteomics of human viruses; mutation analysis of HIV-1 and HTLV-1; use of cell-cycle inhibitor to study viral infections; molecular mechanisms of pathogenesis related to HIV-1, HTLV-1, and HHV-8.

Jonathan R. Keller, Ph.D. Molecular and cellular biology of hematopoiesis.

Katherine A. Kennedy, Ph.D. Molecular actions of antitumor drugs.

Ewen F. Kirkness, Ph.D. cDNA libraries; receptors; ion-channels; recombinant expression; structure; serotonin; GABA.

Jay H. Kramer, Ph.D. Mechanisms of myocardial injury.

Andrei Komarov, Ph.D.

Janette Krum, Ph.D. Astroflial and vascular mechanisms in brain repair; roles of Vascular Endothelial Growth Factor (VEGF) in CNS repair and development.

Ajit Kumar, Ph.D. RNA-protein interactions.

Stephan Ladisch, M.D. Metabolism and biological functions of tumor gangliosides.

Patricia S. Latham, M.D. Monocyte gene regulation and cytokine response.

Norman H. Lee, Ph.D. Studying mRNA regulation and global patterns of gene expression with DNA microarrays.

David Leitenberg, M.D., Ph.D. Regulation of T-lymphocyte activation and development.

Craig Linebaugh, Ph.D. Neurologic speech and language disorders.

Bai Lu, Ph.D. Neurotrophic regulation of synapse development and plasticity.

Tobey MacDonald, M.D. Molecular biology of childhood brain tumors.

Rose G. Mage, Ph.D. Molecular immunogenetics.

I. Tong Mak, Ph.D. Oxidative endothelial-cell injury.

H. George Mandel, Ph.D. Cancer drug metabolism.

Timothy A. McCaffrey, Ph.D. Cardiovascular disease: genomics and stem cells.

Ronald D. McKay, Ph.D. Brain development from stem cells to synapses.

David Mendelowitz, Ph.D. Electrophysiology of central cardiorespiratory neurons; anesthetics; nicotine receptors; viral tracing.

Susana Mendez, D.V.M., Ph.D. Parasite immunology and vaccine development against parasitic diseases.

Carl Merril, M.D. Methods to increase sensitivity of protein detection; use of bacterial viruses for treatment of bacterial infections.

Sally A. Moody, Ph.D. Molecular determination of neuronal phenotypes.

Terry W. Moody, Ph.D. Growth factor receptors in cancer cells.

Bernard Moss, Ph.D. Molecular characterization of vaccinia virus.

Kanneboyina Nagaraju, Ph.D.

Jenae Neiderhiser, Ph.D. Behavioral development and behavioral genetics.

Sergei Nekhai, Ph.D.

William C. Nierman, Ph.D. Genomics-based functional analysis of microbial pathogenicity.

Nancy Noben-Trauth, Ph.D. Focus on immune regulation by cytokines using models of parasitic infections and inflammatory bowel disease.

Frances P. Noonan, Ph.D. UV-radiation carcinogenesis.

Stephen J. O'Brien, Ph.D. Population and evolutionary genetics.

Travis O'Brien, Ph.D.

Randall K. Packer, Ph.D. Renal ion and acid-base balance.

Steven R. Patierno, Ph.D. Molecular mechanisms of carcinogenesis and metastasis.

David C. Perry, Ph.D. Neuronal apoptosis; nicotinic receptors.

Scott Peterson, Ph.D.

Kenna D. Peusner, Ph.D. Central vestibular neural circuit development.

John Philbeck, Ph.D. Human visual space perception and navigation.

Ligia Pinto, Ph.D.

Sasa Radoja, Ph.D.

David Reiss, M.D. Family process research; behavioral genetics.

Marcos Rojkind, M.D., Ph.D.

Brian Rood, M.D.

Mary C. Rose, Ph.D. Mucin glycoproteins, inflammatory mediators, and airway diseases.

Jeffrey M. Rosenstein, Ph.D. Vascular changes in brain tumors.

Lawrence A. Rothblat, Ph.D. Psychobiology of memory.

Thomas D. Sargent, Ph.D.; PP Coordinator at NIH. Control of tissue differentiation.

Narine Sarvazyan, Ph.D. Cellular origins of arrhythmias.

Jeffrey Schlom, Ph.D. Tumor immunology.

Gary L. Simon, M.D., Ph.D. Treatment of HIV infection; metabolic complications of antiretroviral therapy; sepsis and septic shock.

L. Courtney Smith, Ph.D. Evolution of innate immunity.

Steven J. Soldin, Ph.D. Transplant immunology.

Eva M. Sorenson, Ph.D. Localization and function of neuronal nicotinic receptors.

Mary Ann Stepp, Ph.D. Cell-cell and cell-substrate interactions.

Gerald V. Stokes, Ph.D. Mechanisms of chlamydia infections.

Yan A. Su, M.D. Ph.D.

Margaret Sutherland, Ph.D. Transgenics; glutamate transporter; electrophysiology; embryonic stem cells, amyotrophic lateral sclerosis; epilepsy; neuronal death; apoptosis.

Mendel Tuchman, M.D. Nitrogen metabolism and its disorders.

Jack Vanderhoek, Ph.D. Eicosanoid metabolism.

Stanisku Vukmanovic, M.D., Ph.D.

Ray Walsh, Ph.D.

William B. Weglicki, M.D. Mechanisms of injury of myocardial membranes and cells.

Thomas E. Wellems, M.D., Ph.D.

Linda L. Werling, Ph.D. Regulation of cerebral catecholamine release; nicotinic, sigma, and PCP receptors.

Alexander Wlodawer, Ph.D. Structure-function relationship of proteins by the method of X-ray crystallography; particular targets include proteases, kinases, lectins, and cytokines and their receptors.

THE GEORGE WASHINGTON UNIVERSITY / NATIONAL INSTITUTES OF HEALTH

Graduate Program in Biomedical Sciences

Program of Study

The George Washington University/National Institutes of Health (GWU/NIH) Graduate Program in Biomedical Sciences offers Ph.D. degrees in three areas: biochemistry and molecular genetics, microbiology and immunology, and molecular medicine. Within molecular medicine, the student may specialize in neuroscience, oncology, or pharmacology. The graduate program utilizes the expertise of the graduate faculty at the George Washington University and the research expertise of NIH scientists. This partnership promises students the unique opportunity to exploit the rich network of collaborations between scientists working at both sites.

During the first year of study, students are enrolled in a core curriculum in biomedical sciences that includes the study of proteins and macromolecules, nucleic acids, cell biology, information processing, and related topics. They then proceed to core curricula in the three Ph.D. degree areas designed to introduce them to the specialty field in which they may pursue the Ph.D. At the end of the first year of study, students choose their dissertation mentors (one at NIH and one at GWU) and individual Ph.D. program, and plan a program of study including courses specific to their discipline with the program adviser. Additional seminars and special lectures are available at both campuses. Students in the NIH partnership program complete laboratory rotations at the NIH and GWU with scientists working on collaborative projects or may choose to forge new collaborative relationships between partner labs. By the end of the second year, the student should have completed all course work and be fully engaged in dissertation research.

Research Facilities

As the federal government's primary agency for the support and conduct of biomedical research, the NIH's many institutes and centers employ approximately 1,200 tenured or tenure-track investigators and more than 3,700 postdoctoral scientists with medical, dental, or graduate degrees. The NIH intramural research program, located on a 300-acre campus in Bethesda, Maryland, 10 miles from downtown Washington, D.C., has been the scene of many exciting scientific advances. Four Nobel Laureates made their prizewinning discoveries in NIH laboratories on topics that range from breaking the genetic code to discoveries concerning the chemical transmitters in nerve terminals, including the mechanisms for their storage, release, and deactivation.

Basic research at NIH in the biomedical sciences is complemented by an active clinical research program. NIH's unique 250-bed research hospital and laboratory complex, the Warren Grant Magnuson Clinical Center, allows NIH's intramural scientists to bring research closer to the bedside and clinic. The NIH campus is also home to the National Library of Medicine, the world's largest medical library, and NIH/GWU students have free online access to hundreds of full-text research journals.

Financial Aid

Students admitted to the program receive a full-tuition award supported by the George Washington University and the National Institutes of Health, as well as a full-time Predoctoral Intramural Research Training Award (Pre-IRTA) from the NIH. The current Pre-IRTA stipend is $24,800 for entering students, with a possible merit increase each year thereafter, based on the student's performance.

Cost of Study

Students admitted into the GWU/NIH Graduate Program in Biomedical Sciences receive funding from NIH and GWU for stipend, tuition, and medical benefits.

Living and Housing Costs

The cost of living in the Washington, D.C., area is similar to that of other major cities in the United States. The Office of Campus Life at the George Washington University provides information on off- and on-campus housing and also hosts apartment-hunting weekends during the summer. Various types of accommodations are available along the Metro (subway) lines that run between the University and the NIH. Near NIH, there are various houses, apartments, and rooms in private residences available for rent. For information on available housing near NIH, applicants should visit the Web sites of local newspapers or contact the Graduate Partnerships Program office at the NIH Web site listed in the contact section.

Student Group

The total student body of GWU includes more than 10,200 full-time students and approximately 8,700 part-time students. Students selected to participate in the GWU/NIH program are actually part of a larger group of graduate students in the GWU Program in Biomedical Sciences, where enrollment during recent academic years has been approximately 80 students. There are more than 400 graduate students at NIH from more than 100 universities. While at NIH, graduate students enjoy services and activities sponsored by the Graduate Partnerships Program that build a strong graduate student community, similar to that of a university campus.

Location

The George Washington University is located in the Foggy Bottom area of Washington, D.C., just a few blocks west of the White House. The NIH is in a northwest suburban area of the metropolitan region, approximately 10 miles from the University. The two institutions are easily accessible by the Metro system. The cultural and academic life of Washington, D.C., is rich, and there are pleasant outdoor activities available year-round.

The University and The National Institutes of Health

Scientists at the National Institutes of Health have played a major role in the development of the understanding of genetics, from the breaking of the genetic code in the 1960s to the first successful use of gene therapy to cure a genetic disease in the 1990s. Many graduate students through the years have enjoyed the richness of the unparalleled facilities and research opportunities at NIH through arrangements between the NIH and local universities such as GWU.

The George Washington University, chartered by Congress in 1821, benefits from the cultural and educational offerings of the entire metropolitan area, where cultural, professional, and recreational opportunities abound. The GWU program is composed of members drawn from several departments, Children's National Medical Center, the Institute for Genomic Research, government agencies including the NIH, and independent laboratories. The University holds regional accreditation from the Middle States Association of Colleges and Schools.

Applying

To apply, prospective students must be U.S. citizens or noncitizen nationals of the United States or must have been lawfully admitted for permanent residence (i.e., possess a valid Alien Registration Card or some other verification of such status), with a bachelor's degree from an accredited university. The student's academic record, GRE scores, and prior research should reflect strong academic ability and evidence of superior promise. An entering class typically includes a variety of different undergraduate majors in the biological, chemical, or physical sciences. Information on submitting an application to the collaborative GWU/NIH program or joining an NIH laboratory for dissertation research may be found at http://gpp.nih.gov. Applications to the George Washington University should be completed and received by the posted deadline. Early submissions are encouraged.

Correspondence and Information

Graduate Partnerships Program
Building 2, Room 2E06
National Institutes of Health–DHHS
2 Center Drive
Bethesda, Maryland 20892-0234
Phone: 301-594-9605
Fax: 301-594-9606
E-mail: gpp@nih.gov
Web site: http://gpp.nih.gov

Dr. Bill Pavan
NIH Partnership Director
Phone: 301-496-7584
E-mail: bpavan@mail.nih.gov

Dr. Linda Werling
Partnership Director
Institute for Biomedical Sciences
The George Washington University
Washington, D.C. 20052
Phone: 202-994-2142
Fax: 202-994-0967
E-mail: msdmib@gwumc.edu
Web site: http://www.gwumc.edu/ibs/

The George Washington University/National Institutes of Health

THE FACULTY AND THEIR RESEARCH

Listed below are the currently established collaborations between George Washington University and NIH investigators. Students choose one of these groups of collaborators to work with for their dissertation or establish new collaborations between GWU and NIH scientists.

GWU: Michael Bukrinsky, M.D., Ph.D., Professor and Vice Chair, Microbiology, Immunology, and Tropical Medicine, and Professor of Biochemistry and Molecular Biology.
NIH: Leonid Margolis, Ph.D., Chief, Section of Intercellular Interactions, Deputy Director, NASA/NIH Center for Three-Dimensional Tissue Culture, NICHD.

GWU: Susan Ceryak, Ph.D., Research Assistant Professor of Pharmacology and Physiology.
NIH: Phillip Dennis, M.D., Ph.D., Head, Lung Cancer Biology Section, Cancer Therapeutics Branch, NCI.

GWU: Anamaris M. Colberg-Poley, Ph.D., Professor of Pediatrics, Biochemistry and Molecular Biology, Children's Research Institute.
NIH: Jennifer Lippincott-Schwartz, Ph.D., Head, Section on Organelle Biology, Cell and Molecular Biology Branch, NICHD.

GWU: Tim Hales, Ph.D., Professor of Pharmacology and Physiology.
NIH: David Lovinger, Ph.D., Chief, Laboratory of Integrative Neuroscience, NIAAA.

GWU: Eric Hoffman, Ph.D., Professor of Genetics, Director and Senior Investigator, Department of Genetic Medicine, Children's National Medical Center.
NIH: Howard A. Fine, M.D., Chief, Neuro-Oncology Branch, NCI/NINDS.
NIH: Kenneth Fischbeck, M.D., Chief, Neurogenetics Branch, NINDS.
NIH: Vittorio Sartorelli, M.D., Group Leader, Muscle Gene Expression Group, Laboratory of Muscle Biology, NIAMS.
NIH: Colin L. Stewart, D.Phil., Chief, Laboratory of Cancer and Developmental Biology, NCI.

GWU: Peter Hotez, M.D. Ph.D., Professor and Chair, Department of Microbiology, Immunology, and Tropical Medicine.
NIH: Allan Saul, Ph.D., and Louis Miller, M.D., Co-Chiefs, Malaria Vaccine Development Branch, NIAID.

GWU: Fatah Kashanchi, Ph.D., Associate Professor of Biochemistry and Molecular Biology and Co-Director, Keck Institute for Proteomics Technology.
NIH: John Brady, Ph.D., Head, Virus Tumor Biology Section, Laboratory of Cellular Oncology, NCI.
NIH: Steve Jacobson, Ph.D., Head, Viral Immunology Section, NINDS.
NIH: Kuan Jeang, M.D., Ph.D., Head, Molecular Virology Section, Laboratory of Molecular Microbiology, NIAID.
NIH: Ulrich Siebenlist, Ph.D., Head, Immune Activation Section, Laboratory of Immunoregulation, NIAID.

GWU: Ajit Kumar, Ph.D, Professor of Biochemistry, Molecular Biology and of Genetics.
NIH: David Symer, M.D., Ph.D., Laboratory of Immunobiology, NCI.

GWU: Tim McCaffrey, Ph.D., Professor of Biochemistry and Molecular Biology, Director, Catherine Birch McCormick Genomics Center.
NIH: Lawrence Brody, Ph.D., Head, Molecular Pathogenesis Section, NHGRI.

GWU: Susana Mendez D.V.M., Ph.D., Assistant Research Professor, Microbiology and Tropical Medicine.
NIH: Yasmine Belkaid, Ph.D., Laboratory of Mucosal Immunology, NIAID.

GWU: Nancy Noben-Trauth, Ph.D., Professor of Immunology.
NIH: David Sacks, Ph.D., Head, Intracellular Parasite Biology Section, Laboratory of Parasitic Diseases, NIAID.

GWU: Randall K. Packer, Ph.D., Professor of Biology.
NIH: Mark A. Knepper, M.D., Ph.D., Chief, Laboratory of Kidney and Electrolyte Metabolism, NHLBI.

GWU: Mary Rose, Ph.D. Associate Professor, Department of Pediatrics and of Biochemistry and Molecular Biology, and Senior Investigator, Center for Genetic Medicine Research, Children's National Medical Center.
NIH: Sonia Jakowlew, Ph.D., Head, Experimental Biochemistry Section, Cell and Cancer Biology Branch, NCI.

The Dale and Betty Bumpers Vaccine Research Center (VRC), one of more than thirty major research buildings on the NIH campus, was established to facilitate research in vaccine development.

The National Human Genome Research Institute (NHGRI) led the Human Genome Project for the NIH, which culminated in the completion of the full human genome sequence in April 2003.

HARVARD UNIVERSITY

School of Public Health
Program in Biological Sciences in Public Health

Program of Study	The Biological Sciences in Public Health (BPH) program, leading to the Ph.D. degree, is located at the Harvard School of Public Health. This program is offered through the faculty of the Graduate School of Arts and Sciences of Harvard University. The program trains a cadre of leaders who, while possessing expertise in the individual fields of biological research, also possess a broad interdisciplinary knowledge of epidemiology and biostatistics. The program trains research scientists who are interested in the following areas of cellular and molecular biology: bioengineering; molecular mechanisms of adaptive responses to stress; molecular and cellular toxicology; radiobiology; nutritional biochemistry; genetic and molecular mechanisms of chronic diseases such as obesity, diabetes, and cancer; cell-environment interactions; toxicology; cancer; pulmonary inflammation; immunology; infectious diseases: protozoa, helminths, viruses, and bacteria; and genetic approaches to disease mechanisms. Students apply cutting-edge technology to the solution of worldwide problems, with a focus toward better treatment and prevention of human diseases. It has become increasingly evident that progress in disease prevention is optimally promoted by a close interaction between epidemiologists and laboratory scientists, where laboratory discoveries and epidemiological observations interact in an iterative manner to advance research in both fields.

The program offers a firm foundation in biomedical sciences, epidemiology, and biostatistics. The faculty includes approximately 60 members from the School of Public Health as well as faculty members from other Harvard-affiliated institutions. Specific courses supplement this core as dictated by individual research interests. To realistically evaluate their research interests, the suitability of a laboratory, and a potential thesis adviser, students engage in three laboratory rotations. A qualifying exam must be passed prior to beginning thesis work, and the thesis must be defended prior to the granting of the Ph.D. degree.

Applicants generally have a bachelor's degree and demonstrated competence in organic and biological chemistry, general biology, physics, and calculus.

It is believed that there will be a substantial growth in career opportunities for biomedical scientists not only in academic and world-health organizations but in the biotechnology and pharmaceutical industries as well.

Research Facilities Founded in 1922, the School of Public Health is one of the newer schools at Harvard. Students have access to all facilities used by the participating programs. These modern facilities are housed in fourteen floors of three buildings devoted to laboratory research. Included among these facilities is a state-of-the-art cardiovascular research laboratory that was formally dedicated in 1992. Students have access to Countway Library, one of the most complete biomedical research collections in the country.

Financial Aid Students are awarded fellowships that cover tuition, health fees, and a stipend. For the 2005–06 academic year, the stipend was $27,000 ($2250 per month). Applicants are encouraged to apply for support from extramural agencies. Enrolled students who receive competitively funded and externally sponsored fellowships may be eligible to receive an educational allowance. International applicants are urged to seek financial support from their national governments and fellowship agencies. Special scholarships for qualified international students are available for students from Cypress/geographical region, with interests in environmental health; Nigeria; and developing central African countries.

Cost of Study All students enrolled in the program are awarded fellowships for tuition and health insurance as well as a stipend. Tuition and health fees in 2005–06 were $31,280 for the first and second years and $10,002 for subsequent years. These costs are usually covered by institutional training grants and sponsored research grants.

Living and Housing Costs Many students live near the School of Public Health, in Boston or in Brookline. Others live in Cambridge and use the free shuttle-bus service. For information on housing in Cambridge, students should write the Graduate School of Arts and Sciences, Lehman Hall, Building B2, Harvard University, Cambridge, Massachusetts 02138; for housing in the medical area, they should write to Vanderbilt Hall, Harvard Medical School, 107 Avenue Louis Pasteur, Boston, Massachusetts 02115.

Student Group More than 50 graduate students are enrolled in the program. In 2005–06, there were approximately 60 affiliated faculty members.

Location The Harvard School of Public Health is located in the midst of other Harvard-affiliated health institutions, including Harvard Medical School, the School of Dental Medicine, Countway Library, hospitals, and scientific research institutes, all concentrated in Boston's Longwood Medical Area. Metropolitan Boston is a stimulating center of cultural and academic activity, with more than forty colleges and universities. It is within easy access to magnificent recreational areas in the city, in the mountains, and at the seashore.

Applying Applications specifying interest in the Biological Sciences in Public Health Program (BPH/8500) must be returned to the Graduate School of Arts and Sciences by December 8. Scores from the Graduate Record Examinations (GRE) General Test are required; GRE Subject Test scores are optional (institutional code: R-3451). To ensure that scores arrive in time to be considered, these tests should be taken by November. International applicants need a minimum score of 600 on the TOEFL in order to demonstrate adequate English proficiency. The 2006–07 application booklet has comprehensive updated information. The deadline for applying to the Ph.D. program is December 8. Graduate School of Arts and Sciences (GSAS) application forms must be used when applying to the Ph.D. Program in Biological Sciences in Public Health.

Online application submissions are encouraged. The Graduate School of Arts and Sciences online application form can be found at http://www.gsas.harvard.edu/4-forms/applyonline.html. Hard copies of the Graduate School of Arts and Sciences application form can be requested at http://www.gsas.harvard.edu/4-forms/application.html. Students who have specific questions should contact the Program Administrator, listed in this In-Depth Description.

Correspondence and Information
Mrs. Ruth Kenworthy, Program Administrator
Division of Biological Sciences
Harvard School of Public Health
665 Huntington Avenue, Building 1-1312
Harvard University
Boston, Massachusetts 02115
Phone: 617-432-2932
Fax: 617-432-0433
E-mail: bph@hsph.harvard.edu
Web site: http://www.hsph.harvard.edu/bph/

Harvard University

THE FACULTY AND THEIR RESEARCH

The following is a representative sample, not a complete listing, of the faculty.

Robert B. Banzett, Associate Professor of Medicine (HMS) and Associate Professor of Physiology (HSPH). Perception of afferent information from the respiratory system.

Barry R. Bloom, Dean of the Faculty and Professor of Immunology and Infectious Diseases (HSPH). Study of pathogenesis and protection in tuberculosis and development of vaccines.

Joseph D. Brain, Cecil K. and Philip Drinker Professor of Environmental Physiology (HSPH). Function and structure of pulmonary and hepatic macrophages; responses to inhaled gases and particles.

Barbara Burleigh, Associate Professor of Immunology and Infectious Diseases (HSPH). Studies of the molecular basis of host-cell invasion, signaling, and differentiation by the human pathogen *Trypanosoma cruzi.*

Hannia Campos, Associate Professor of Nutrition (HSPH). Gene-environment interactions in human lipoprotein metabolism.

David C. Christiani, Professor of Occupational Medicine and Epidemiology, Department of Environmental Health (MPH). Assessment of the impact of workplace pollutants on health.

John R. David, Richard Pearson Strong Professor Emeritus of Tropical Public Health (HSPH). Biology of cytokines and control of parasitic infections.

Bruce Demple, Professor of Toxicology (HSPH). Cellular defenses against oxygen radicals.

Manoj Duraisingh, Assistant Professor of Immunology and Infectious Diseases (HSPH). Molecular mechanisms underlying the pathogenesis of human malaria.

Raymond Erikson, American Cancer Society Professor of Cellular and Developmental Biology (HU). Protein phosphorylation and gene expression in normal and transformed cells.

Mryon E. Essex, Mary Woodard Lasker Professor of Health Sciences (HSPH). Study of human and primate T-lymphotrophic retroviruses including agents that cause AIDS.

Jeffrey J. Fredberg, Professor of Bioengineering and Physiology (HSPH). Identification of the mechanical basis of airway and lung parenchymal function at the levels of organ, tissue, cell, and protein.

Laurie H. Glimcher, Irene Heinz Given Professor of Immunology (HSPH) and Professor of Medicine (HMS). Genetic regulation of immune response; role of class II major histocompatability complex proteins in T-lymphocyte activation.

Marcia B. Goldberg, Associate Professor of Medicine (HMS). Host-pathogen interactions of *Shigella.*

Beatriz Gonzalez-Flecha, Assistant Professor of Molecular Biology (HSPH). Characterization of the pathways of oxidant-dependent promotion of cell growth in lung epithelial cells.

Michael J. Grusby, Professor of Molecular Immunology (HSPH) and Associate Professor of Medicine (HMS). In vivo models of immune deficiency by homologous recombination in ES cells.

Donald A. Harn Jr., Professor of Tropical Public Health (HSPH). Immune regulation; immunoparasitology of B- and T-cell subsets; immunology and molecular biology of *Schistosome mansoni.*

Woodland Hastings, Paul C. Mangelsdorf Professor of Natural Sciences (HU). Molecular mechanism of cellular circadian regulation.

I-Cheng Ho, Assistant Professor of Medicine (HMS) and Assistant Professor of Immunology and Infectious Diseases (HSPH). Differentiation and activation of helper T cells.

Gökhan S. Hotamisligil, James Stevens Simmons Professor of Genetics and Metabolism (HSPH). Signaling mechanisms of peptide hormones; genetic and molecular basis of obesity and diabetes.

Howard Hu, Associate Professor of Occupational Medicine (HSPH) and Associate Professor of Medicine (HMS). Metals toxicity and gene-metal interactions.

David J. Hunter, Professor of Epidemiology (HSPH). Cancer epidemiology; molecular epidemiology.

Phyllis J. Kanki, Professor of Pathobiology (HSPH). Study of epidemiologic and biological characteristics of HIV viruses in Africa.

Karl T. Kelsy, Professor of Cancer Biology and Environmental Health (HSPH). Study of workplace mutagen and carcinogen exposure.

David M. Knipe, Professor of Microbiology and Molecular Genetics (HMS). Mechanisms by which herpes simplex virus (HSV) undergoes a productive infection in epithelial cells; host immune response to viral infection; use of mutant strains of HSV as a herpes vaccine and as an AIDS vaccine vector.

Lester Kobzik, Associate Professor of Environmental Health (HSPH) and Assistant Professor of Pathology (HMS). Lung macrophage differentiation and function; flow cytometry applications for respiratory cell biology.

Roberto Kolter, Professor of Microbiology and Molecular Genetics (HMS). Molecular biology of bacterial interactions; peptide production and release; growth phase regulation of gene expression.

Igor Kramnik, Assistant Professor of Immunology (HSPH). Genetic dissection of mechanisms of host susceptibility to tuberculosis.

Chih-Hao Lee, Assistant Professor of Genetics and Complex Diseases (HSPH). Nuclear lipid receptors as therapeutic targets of metabolic diseases.

Tun-Hou Lee, Professor of Virology (HSPH). Human and related primate retroviruses.

Marc Lipsitch, Associate Professor of Epidemiology (HSPH). Theoretical, statistical, and experimental approaches to population biology; epidemiology of infectious diseases.

John B. Little, James Stevens Simmons Professor Emeritus of Radiobiology (HSPH). Radiation mutagenesis and carcinogenesis; genetic instability.

Brendan D. Manning, Assistant Professor of Genetics and Complex Diseases (HSPH). Signaling pathways underlying tumorigenesis and metabolic diseases.

Joseph P. Mizgerd, Associate Professor of Physiology and Cell Biology (HSPH). Regulation of acute inflammatory responses by intercellular and intracellular signaling molecules.

Karl Munger, Associate Professor of Pathology (HMS). Human papillomaviruses (HPVs): the cause of hyperplastic skin lesions.

Heather Nelson, Assistant Professor of Environmental Epidemiology (HSPH). Integrating genetic susceptibility and tumor profiling into the epidemiologic study of skin cancer to build new etiologic models and improve risk estimates for disease.

Bjorn R. Olsen, Hersey Professor of Cell Biology (HMS). Molecular studies of skeletal and vascular morphogenesis.

Eric J. Rubin, Assistant Professor of Immunology and Infectious Diseases (HSPH). Virulence factors of mycobacteria; acquisition of virulence determinants of *Vibrio cholerae*; generalized transposon mutagenesis systems for bacteria.

Frank M. Sacks, Professor of Nutrition (HSPH) and Associate Professor of Medicine (HMS). Human lipoprotein metabolism: biochemistry and metabolic modeling.

Stephanie A. Shore, Senior Lecturer of Physiology (HSPH). Physiological and pharmacological aspects of bronchoconstriction.

Eric S. Silverman, Assistant Professor of Environmental Health (HSPH) and Associate Professor of Medicine (HMS). Molecular and genetic determinant of asthma.

Thomas J. Smith, Professor and Director of Industrial Hygiene (HSPH). Environmental exposures for studies in health effects and investigation of the relationship between environmental exposure and internal dose.

Joseph Sodroski, Professor of Pathology (HMS) and Professor of Cancer Biology (HSPH). Human immunodeficiency virus envelope glycoproteins and vaccine development; human immunodeficiency virus vectors; pathogenesis of human retroviruses.

Bruce M. Spiegelman, Professor of Biological Chemistry and Molecular Pharmacology (HMS). Regulation of gene expression in mammalian cell differentiation; adipose cell and tissue development; nuclear hormone receptor and basic-helix-loop-helix families of transcription factors.

Armen H. Tashjian Jr., Professor Emeritus of Toxicology (HSPH) and Professor of Biological Chemistry and Molecular Pharmacology (HMS). Molecular mechanisms of control of the biosynthesis, secretion, and action of polypeptide hormones; signal transduction mechanisms.

Ning Wang, Associate Professor of Physiology and Cell Biology (HSPH). Mechanical mechanisms of cytoskeleton and its regulatory role in cell growth and migration.

Marianne Wessling-Resnick, Professor of Nutritional Biochemistry (HSPH). Regulation of the cellular uptake of macromolecular nutrients.

Walter C. Willett, Professor of Epidemiology and Nutrition (HSPH) and Professor of Medicine (HMS). Relations of dietary factors to the occurrence of human disease.

Dyann F. Wirth, Professor of Tropical Public Health (HSPH). Molecular genetic analysis of gene expression, transsplicing, and homologous recombination in *Leishmania enrietti.*

Dieter Wolf, James Stevens Simmons Associate Professor of Molecular Oncology (HSPH). Control and regulation of DNA replication and control of normal and abnormal cell growth.

Xiping Xu, Associate Professor of Environmental Health (HSPH) and Associate Professor of Medicine and Epidemiology (HMS). Genetic dissection of complex diseases such as asthma, MI, diabetes, and colon cancer.

Zhi-Min Yuan, James Stevens Simmons Associate Professor of Radiobiology (HSPH). Delineation of biochemical and molecular basis of stress-induced responses.

HOFSTRA UNIVERSITY

College of Liberal Arts and Sciences
Department of Biology

Programs of Study

The Department of Biology offers an M.A./M.S. degree in biology, with concentrations in marine and freshwater biology and cell and molecular biology. Students in the M.A. in biology program must complete 30 credits of graduate course work and 3 credits toward a written essay and its defense, for a total of 33 credits. Students in the M.S. in biology program must complete 24 credits of graduate course work and 6 credits of laboratory/field research leading to a thesis and its defense, for a total of 30 credits.

The marine and freshwater biology concentration focuses on aspects of marine ecology, evolution, physiology, and behavior. Students may undertake research projects in the Long Island area or at the Marine Laboratory in Jamaica, West Indies. The cell and molecular biology concentration focuses on aspects of cell biology and molecular genetics pertinent to biotechnology. Course work emphasizes both current knowledge and laboratory techniques in the field

Research Facilities

The Department of Biology is located in Gittleson Hall and contains scanning electron microscopes, a greenhouse, animal facilities, tissue culture facilities, a high-speed imaging apparatus, digital particle image velocimetry (DPIV) instrumentation, and several research laboratories for cellular and molecular genetic techniques and fluorescent in situ hybridization. The department oversees the Hofstra University Marine Laboratory in Jamaica, West Indies. Hofstra University houses a world-class library system that contains more than 1.4 million volumes and electronic/print access to most of the leading journals in biology. Students have access to the academic computing facility and may obtain individual or group instruction. The Hofstra Career Center provides advice and support for all graduate students seeking employment and the University Advisement Office assists students with their applications to professional schools and Ph.D. programs.

Financial Aid

Most graduate students depend, in part, on student loans to finance their graduate careers. The Department of Biology offers a limited number of Hofstra University Graduate Scholarships (full- and part-time) on a competitive basis to matriculated graduate students. In addition, students interested in pursuing a research interest in conservation biology or ecology relevant to Long Island habitats may apply for the Donald Axinn Fellowship in Ecology and Conservation, which provides full tuition remission. Teaching assistantships and instructor positions in undergraduate laboratories are available for qualified graduate students. Graduate assistantships, with varying stipend and tuition remission benefits, are available from the Office of Student Employment. The Office of Residential Life offers RA positions for campus dormitories.

Cost of Study

The application fee for all graduate programs is $60. Tuition for full-time graduate students is typically $6660 per semester or $740 per credit hour. University fees are approximately $300 per semester.

Living and Housing Costs

Residence hall fees range from $2625 to $5625, depending on level of occupancy. Students who prefer an off-campus apartment can find more information in the Office of Residential Life.

Student Group

Approximately 10 to 15 students are accepted into the graduate programs each year, with a total of about 50 part-time and full-time students in any one year. At least half of the students come from the New York-New Jersey-Connecticut region. The remaining half of incoming students are from other regions in the United States and from abroad. No preference is given to students from New York; applications are evaluated on academic merit only. In general, student distribution is equal among the various programs.

Student Outcomes

Graduates with a master's degree pursue professional degrees in the health sciences, continue graduate work at the doctoral level, and find employment as lab technicians for hospitals and companies and as marine and wildlife specialists for governmental and nongovernmental organizations.

Location

Hofstra University is located in Hempstead, Long Island, about 25 miles east of Manhattan and in the center of Long Island. The University has easy access to three international airports, suburban and major railroad terminals, and bus lines. Long Island and New York City provide continuous access to world famous museums, libraries, research facilities, theater, and sporting events.

The University

Hofstra University was founded in 1935 as a private, nonsectarian, coeducational university. The University includes eight colleges and schools on campus. The total enrollment for the University is approximately 13,000, with 1,246 faculty members. The campus is a recognized member of the American Association of Botanical Gardens and Arboreta and has 130 buildings on 240 acres.

Applying

Applicants for admission must hold a bachelor's degree in biology or its equivalent from a college or university of recognized standing. In addition, they are expected to have completed courses in general and organic chemistry, physics, and mathematics and have a minimum GPA of 3.0. Students with deficiencies in prerequisites may be accepted conditionally but may be required to make up deficiencies within their first year of graduate work. Applicants should submit transcripts, GRE General Test scores, two letters of recommendation, and a personal statement indicating research interests and career goals. Applicants should submit all materials to the Graduate Admissions Office. Acceptance decisions are made by the Department of Biology on a rolling basis. Students interested in competitive fellowships should apply by June 30 to be considered for an award.

Correspondence and Information

Dr. Peter Daniel, Graduate Director or
Dr. Robert Seagull, Chair
Department of Biology
114 Hofstra University
Hempstead, New York 11549-1140
Phone: 516-463-5518
Fax: 516-463-5112
E-mail: biology@hofstra.edu
Web site: http://www.hofstra.edu/Academics/Graduate/
Programs/GP_BIO

Office of Graduate Admissions
106 Memorial Hall
Hofstra University
Hempstead, New York 11549
Phone: 516-463-4723
866-GRADHOF(toll-free)
E-mail: gradstudent@hofstra.edu
Web site: http://www.hofstra.edu/Academics/Graduate

Hofstra University

THE FACULTY AND THEIR RESEARCH

Russell Burke, Associate Professor, Ph.D. Evolutionary ecology; conservation biology; herpetology. Dr. Burke's work centers around the interface between the applied field of conservation biology and the typically nonapplied fields of ecology and evolution. He uses basic natural history, food habits studies, and manipulative field experiments to better understand animal ecology and test evolutionary models. These findings are then used as a basis for suggesting management actions for conservation. Most often, he studies reptiles and the predator species associated with them, especially raccoons.

Beverly Clendening, Associate Professor, Ph.D. Developmental biology of *Drosophila*. Research in Dr. Clendening's laboratory is centered broadly around late development and the adult function of mesodermal tissues in *Drosophila*. Current projects include the isolation and characterization of a gene that produces a muscular dystrophy–like syndrome when mutated and the characterization of two genes that are important for normal fertility. Her lab uses behavioral assays, classic genetic and molecular techniques, and specialized *Drosophila* transposon technology. A secondary interest is the molecular basis of temperature-dependent sex determination in turtles.

Peter Daniel, Associate Professor, Ph.D. Sensory biology of lobsters. Dr. Daniel is interested in how chemosensory input mediates behavior. His model organism for studying chemosensory biology is the lobster, because the chemical senses play a paramount role in informing the lobster about relevant features in the environment. In particular, he studies behaviors that aid in detection, recognition, and orientation toward food sources. He is ultimately interested in how the sensory system detects and processes information leading to these behaviors. Techniques employed in the lab include motion analysis and electrophysiology.

Julie Heath, Assistant Professor, Ph.D. Physiological ecology of birds. Dr. Heath is interested in animal physiology and ecology, particularly, the relationships among hormones, bird behavior, and environmental conditions. By understanding how external conditions trigger changes in hormone concentrations and, subsequently, changes in behavior, insight is gained into the evolutionary ecology of organisms. She has examined these relationships in a variety of bird species, including falcons, warblers, and water birds. Current, as well as future, research in the Long Island area focuses on hormones of young birds as they fledge from the nest, effects of toxins on parental behavior, and how isotopes are transferred in the food chain. Dr. Heath welcomes students to her lab who are interested in conducting research on animal physiology and ecology.

Maureen Krause, Assistant Professor, Ph.D. Molecular evolution and biochemistry of marine bivalves; functional genomics. Dr. Krause is interested in the link between genetic variation and protein function from an evolutionary perspective. Most of her work focuses on marine organisms, because they display such amazing adaptations to their environment. Projects in her lab include a molecular evolutionary study of an enzyme found in bay scallops, a conservation genetics study of green sea turtles, and work to develop a new method for characterizing gene expression at the genome-wide level (in collaboration with Brookhaven National Laboratory).

John Morrissey, Associate Professor, Ph.D. Marine biology; shark taxonomy and behavior. Dr. Morrissey is the Director of the Hofstra University Marine Laboratory. His research interests concern the natural history of deep-sea elasmobranchs, specifically those inhabiting the Cayman Trench on the north coast of Jamaica. His students are involved with various studies of deep-sea sharks, including describing new species, reproduction, feeding ecology, age and growth, visual adaptations, and others. In addition, one student is studying the movements of sand tiger sharks on the east coast of the United States using satellite telemetry.

Dorothy Pumo, Professor, Ph.D. Molecular genetics; evolutionary biology of bats. Dr. Pumo's students study the evolution of molecules and also try to use molecules (proteins and DNA) to improve their understanding of animal evolution and development. Most recent work has dealt with mammals, especially bats. Her lab amplifies, clones, and sequences specific regions of DNA from a variety of animals. Computer analyses (bioinformatics) of the DNA and protein sequences are subsequently performed to examine hypotheses about animal relationships.

Christopher Sanford, Associate Professor, Ph.D. Functional morphology in fishes. Dr. Sanford is interested in the functional design of lower vertebrates, specifically fishes. His research centers on every aspect of prey capture, from prey acquisition to prey processing. This work uses an integrative approach designed to cover many biological disciplines such as biomechanics, physiology, evolution, and ecology. He is interested in both theoretical and experimental approaches to understanding the design of vertebrates, including fluid dynamics, biomechanics, and origin of new behaviors.

Robert Seagull, Professor, Ph.D. Cytoskeleton; cell biology of cotton. Dr. Seagull's research is aimed at developing an understanding of the biological mechanisms that regulate cotton fiber growth and development. The long-term goal of his research is to develop strategies to alter fiber growth in ways that improve the textile traits of cotton. He uses a multifaceted approach to his study of cotton fiber and has used an array of experimental techniques, such as light and electron microscopy, cell culture, pharmacological experiments, physiological experiments, cell fractionation, protein purification, immunolocalization, and more.

Laura Vallier, Assistant Professor, Ph.D. Gene expression and development of *C. elegans.* Dr. Vallier uses genetic analysis to investigate the genes involved in cell-fate decisions during development in the soil nematode *C. elegans.* Her research involves understanding the development of the gonadal sheath, a tissue that is made of five pairs of cells surrounding the proximal half of each gonadal arm that is required for gametogenesis, spermatogenesis, meiotic progression, and ovulation. Currently, her studies focus on the identification and characterization of the genes required to specify the cell fate of the gonadal sheath cells using RNAi.

Joanne Willey, Associate Professor, Ph.D. Gene regulation and morphogenesis in bacteria. The *Streptomyces* are a group of filamentous soil bacteria that are a rich source of antibiotics and other medically and industrially important compounds. Her lab has isolated several surfactants that function as biological wetting agents from several *Streptomyces* species. She is interested in how these compounds facilitate bacterial growth and how they might be useful to humans.

Jason Williams, Assistant Professor, Ph.D. Biology of marine invertebrates. Research in Dr. Williams's lab focuses on the systematics and biology of marine polychaete worms and crustaceans. Currently he is investigating the associates of hermit crabs, particularly shell-burrowing polychaete worms and isopod parasites. A variety of marine invertebrates live as commensals or parasites of hermit crabs, and his studies attempt to more accurately define the interactions in this unique marine system. Research in the Indo-West Pacific has led to the description of new polychaete and isopod species as well as the first report of egg predation by polydorid worms on host hermit crabs.

HOWARD UNIVERSITY

Department of Biology

Programs of Study	The Department of Biology offers graduate programs leading to the Master of Science (M.S.) and Doctor of Philosophy (Ph.D.) degrees. The department offers two major areas of concentration: cell and molecular biology (CMB) and ecological, environmental, and systematic biology (EESB). Training in CMB offers specialization in molecular biology, molecular genetics, biochemical genetics, developmental biology, animal physiology, and microbiology. Training in EESB offers specialization in basic and applied ecology, environmental and evolutionary studies of various organisms, ecological studies of adaptive strategies, and modeling atmospheric and weathering interactions. Although students may elect programs leading to M.S. or Ph.D. degrees, the Ph.D. program is the major focus of the graduate program. The department also participates in the M.D./Ph.D. degree program jointly offered by the Graduate School and College of Medicine.
Research Facilities	The Department of Biology has facilities for graduate-level teaching and research. Instructional facilities include a SMART classroom outfitted with Internet access and computerized laboratories. Modern instrumentation for research and instruction includes phase-contrast, fluorescence, interference-contrast, and electron microscopes; scintillation and gamma counters; ultracentrifuges; spectrophotometers; an HPLC; DNA sequencing apparatus; PCR and electrophoresis apparatus; laminar flow hoods; fermentors; a cold room; an animal facility; environmental chambers; sonicators; lyophilizers; an herbarium; and a greenhouse. Research laboratories are equipped with computers and the necessary software for data analyses. Other equipment accessible to students through collaborative effort of faculty members includes laser desorption/ionization of time-of-flight mass spectrometry, a gas chromatography mass spectrometer, a nuclear and magnetic resonance spectrometer in the department of chemistry, and flow cytometry and confocal microscopy in the College of Medicine.
Financial Aid	Financial support for graduate students is available through a variety of graduate and research assistantships, fellowships, traineeships, tuition scholarships, loans, college work-study, and Howard University Student Employment Program work-study. Graduate and teaching assistantships are generally provided from the University budget and other sources, which typically provide a stipend of $10,000 to $13,000 per academic year, plus tuition for part-time teaching or other departmental activities (15 hours per week). Research assistantships are generally provided by research grants secured by faculty members from private and public funding sources. Students receive stipends per academic year that range from $10,000 to $27,000 plus tuition for part-time research assistance (12–20 hours per week). Tuition scholarships provided by University funds, gifts, endowments, and public and private sources cover the cost of tuition and are available to full-time graduate students who are in good academic standing.
Cost of Study	Full-time students enrolled for 9 to 15 hours pay tuition of approximately $6635 per semester. Part-time students pay tuition of $737 per credit hour. The general fees are $402.50. Book and supply costs vary. All fees are subject to change.
Living and Housing Costs	Students may elect to live in any of the University housing, which includes modern apartments ranging from studios to three-bedroom apartments. Off-campus housing is also available. Housing information may be obtained from the Office of Residence Life.
Student Group	Howard University has a total student body of approximately 11,000 students from every state in the U.S. and every continent. About 2,000 of these are graduate students representing thirty-five states and fifty-nine countries worldwide. This diversity offers excellent opportunities for multicultural experiences and stimulating scholarly exchanges among students.
Location	The location of Howard University in Washington, D.C., the nation's capital, provides students ready access to major national research centers. These include the National Institutes of Health, the Food and Drug Administration, the Department of Agriculture, the Smithsonian Institution, the National Library of Medicine, and the Library of Congress.
The University	Founded in 1867, Howard University is a comprehensive, private, predominantly African-American institution. The University has sixteen schools and colleges. The Graduate School of Arts and Sciences is internationally renowned as a center for research and academic excellence. It offers twenty-three Ph.D. and thirty-two master's programs.
Applying	The Admissions Application should be received by April 1 for the fall semester, November 1 for the spring semester, and March 15 for the summer session. The Application for Graduate Study and the $45 fee are valid for two continuous semesters only. Applicants applying to graduate programs must have the equivalent of a four-year baccalaureate degree from an accredited institution. Regular admission may be granted to a student having a minimum 3.0 grade point average (on a 4.0 scale), appropriate training in biology, a minimum of one year of college-level mathematics and physics, and two years of chemistry. Applicants are required to submit three letters of recommendation, a statement of interest, transcripts, and the most recent Graduate Record Examinations scores. A student who shows significant academic promise but does not meet all of the requirements for regular admission may be accepted provisionally. Such a student is not classified as a regular graduate student until that student satisfies the requirements of provisional admission specified by the Graduate Studies Committee. These should be met during the first year of residence. A provisional student must obtain a 3.0 average during the first year of full-time study. Students with at least a bachelor's degree may be admitted to either the Ph.D. or M.S. program. Official certificates, university transcripts, and/or mark sheets must be sent directly from institutions abroad to the address in accordance with the program of interest. If the documents are not in English, they must be accompanied by official translation. All applicants from countries where English is not the official language must score satisfactorily on the Test of English as a Foreign Language (TOEFL). International applicants requesting an I-20 form (Certificate of Eligibility, International Certificate of Eligibility) must submit adequate documents for financial resources for the duration of the program. The I-20 is issued after the student is admitted to the University, has paid the necessary $150 enrollment fee, and has submitted the Statement of Financial Resources Form with the correct amount of funds, signed by the sponsor(s).
Correspondence and Information	Graduate School Admission Office Howard University Fourth and College Street NW Washington, D.C. 20059 Phone: 202-806-7793 Web site: http://www.gs.howard.edu

Howard University

THE FACULTY AND THEIR RESEARCH

Franklin R. Ampy, Professor; Ph.D., Oregon State, 1962. Biostatistics and genetics: mutagenic effects of environmental carcinogens. (e-mail: fampy@howard.edu; telephone: 202-806-6952)

Winston A. Anderson, Professor; Ph.D., Brown, 1966. Cellular biology and reproductive endocrinology; mechanism of hormone action in normal and neoplastic breast and uterine cell growth. (e-mail: wanderson@howard.edu; telephone: 202-806-6950)

Theodore A. Bremner, Associate Professor; Ph.D., Howard, 1972. Biochemical genetics: oxidant regulation of gene expression in myeloid differentiation and neoplasia. (e-mail: tbremner@howard.edu; telephone: 202-806-6957)

Priscila Chaverri, Assistant Professor; Ph.D., Penn State, 2003. Mycology: biodiversity, systematics, and phylogenetics/evolution of fungi important to agriculture and tropical forest conservation. (e-mail: pchaverri@howard.edu; telephone 202-806-6933)

Richard M. Duffield, Professor; Ph.D., Georgia, 1976. Entomology: terrestrial and freshwater populations; insect pheromones in behavior. (e-mail: rduffield@howard.edu; telephone: 202-806-6127)

Sisir K. Dutta, Professor Emeritus; Ph.D., Kansas State, 1960. Molecular genetics: genes that confer ability to biodegrade hazardous chemicals; bioremediation of soils; effects of electromagnetic radiation on gene expression. (e-mail: sdutta@howard.edu; telephone: 202-806-6942)

Atanu Duttaroy, Assistant Professor; Ph.D., Calcutta, 1989. Molecular and cellular biology: aging; prevention and repair of cellular damage due to superoxide radicals. (e-mail: aduttaroy@howard.edu; telephone: 202-806-7939)

William R. Eckberg, Professor and Chairman; Ph.D., Michigan State, 1975. Cellular and developmental biology: control of egg activation at fertilization and control of cell division; mechanisms regulating calcium waves in living cells; role of protein kinases and phosphatases in development; egg organization in the control of early development. (e-mail: weckberg@howard.edu; telephone: 202-806-6933)

Broderick E. Eribo, Associate Professor; Ph.D., Wayne State, 1987. General microbiology: characterization of food and environmental bacteria. (e-mail: beribo@howard.edu; telephone: 202-806-6937)

Jack S. Frankel, Professor; Ph.D., Rutgers, 1976. Genetics: ontogenetics of teleostean isozyme systems; population and evolutionary genetics of *Sceloporus*. (e-mail: jfrankel@howard.edu; telephone: 202-806-6959)

Abner B. Lall, Professor; Ph.D., Maryland, 1971. Neuroscience: neural processing of visual information in insects and fishes. (e-mail: alall@howard.edu; telephone: 202-806-6797)

Clarence M. Lee, Professor; Ph.D., Howard, 1969. Immunoparasitology: immunomodulation in animal models using various parasites. (e-mail: cmlee@howard.edu; telephone: 202-806-9733)

George A. Middendorf III, Associate Professor; Ph.D., Tennessee, Knoxville, 1979. Ecology: behavior and ecology of reptiles and amphibians; population biology of spiny lizards (genus *Sceloporus*); host-parasite relationships between lizards and chiggers. (e-mail: gmiddendorf@howard.edu; telephone: 202-806-7289)

Karen E. Nelson, Associate Professor; Ph.D., Cornell, 1996. Microbial genomics: lateral gene transfer, evolution, metagenomics, and microbial physiology. (e-mail: kenelson@howard.edu; telephone: 202-806-6933)

Raymond L. Petersen, Professor; Ph.D., Rutgers, 1976. Environmental science/biomonitoring/plant stress: air pollution and salt tolerance. (e-mail: rpetersen@howard.edu; telephone: 202-806-6943)

David Schwartzman, Professor; Ph.D., Brown, 1971. Biogeochemistry: geomicrobiology; isotope geology. (e-mail: dws@scs.howard.edu; telephone: 202-806-6926)

Hemayet Ullah, Assistant Professor, Ph.D., North Carolina at Chapel Hill, 2002. Plant cellular signal transduction and G proteins; stress physiology; hormone response. (e-mail: hullah@howard.edu; telephone: 202-806-6958)

HUMBOLDT STATE UNIVERSITY

Department of Biological Sciences

Program of Study

The Department of Biological Sciences offers the Master of Arts (M.A.) in biology. Students can choose to pursue investigations with an emphasis in the laboratory, forests, marine habitats, or a variety of other settings. Students can study living processes at all levels, from cellular and molecular to ecosystems. Students begin their program by taking both required courses and courses specifically in their area of thesis research. A total of 30 units at the upper-division and graduate levels are required to complete the program. Most students complete their degree in two years, but completion in one year is possible under exceptional circumstances. Some areas of study may require a longer period of enrollment, depending on the complexity of study and data availability. A written thesis and oral thesis defense is also required. With a strong emphasis now placed on the training of secondary school teachers in California, the M.A. can be combined with a California teaching credential for increased depth of learning and job preparation that allows graduates to compete for jobs on both the national and international levels.

Each graduate student accepted into the program automatically becomes a member of the Biology Graduate Student Association (BGSA), which actively participates with the Department in sponsoring a lecture series featuring widely respected university teachers and researchers from throughout the country. Another highlight of the year is the annual mini-symposium, in which students and members of the faculty and staff share their research interests with informative presentations. A spirit of both social and intellectual interaction is fostered each semester by a traditional potluck barbecue, where students and members of the faculty and staff join together for an afternoon of relaxation and conversation.

Research Facilities

The Department offers facilities and faculty expertise that allow master's students to complete their programs in almost any area of biology. Laboratories on the main campus are equipped for biotechnology, scanning and transmission electron microscopy, mammalogy, genetic analysis (including DNA sequencing), and computer modeling. Field opportunities include the marine laboratory, located 14 miles from the main campus; a University forest; and a 90-foot oceangoing research vessel, R/V *Coral Sea*. Humboldt State's splendid greenhouse contains plant specimens from more than 175 families—one of the most diverse collections in California. Individual rooms, ranging from a desert room to a fern room, offer students a unique opportunity to study the world's plant life in one setting.

The library's collection includes 557,776 volumes, 2,128 print and 5,622 electronic subscriptions to scholarly and popular periodicals, and extensive holdings of microforms and other material. The library provides round-the-clock access to 160 index, reference, and full-text databases. Virtually all of these resources are accessible from the library home page. The library's Digital Literacy Closet provides facilities and assistance to students who want to explore, evaluate, and apply multimedia technologies to their scholarly endeavors. Within the library, students have access to more than fifty computer workstations for study and research. These computers provide access not only to local information resources and library materials but also to those of other institutions within the California State University (CSU) system and beyond via the Internet.

Financial Aid

California residents may apply for federal, state, and local aid by completing the Free Application for Federal Student Aid (FAFSA) forms. There are a limited number of teaching associate positions and a fee-waiver program for 1 or 2 qualified out-of-state students available per semester, depending on funding. A number of work-study positions are also available.

Cost of Study

California residents pay $3752 in tuition; out-of-state students pay $10,470. Books, materials, and other expenses total approximately $3000.

Living and Housing Costs

Humboldt's facilities, located in a spectacular natural setting, consist of five different residence-hall living areas. Each is unique and provides various options for individual styles and personal preferences. The cost for room and board ranges from $4300 to $9000, depending on the type of room and meal plan. Off-campus choices in Arcata include apartments, houses, and rooms in distinctive and historic Victorian homes. Bus service in Arcata is available free of charge for students.

Student Group

Of the 53 master's students in the Department of Biological Sciences, 28 are full-time and 20 are women.

Location

Humboldt State University is located in the town of Arcata, 100 miles south of the Oregon border and 300 miles north of San Francisco, on the northern coast of California between scenic redwoods and the Pacific Ocean. In Arcata and nearby Eureka, students find a casual and safe environment that is a hub of North Coast music, theater, and art. The University's innovative CenterArts program, cited as a "model in the West" by the National Endowment for the Arts, brings culturally diverse talent to the campus, with more than fifty events staged annually. The gentle coastal climate produces winter rains offset by comfortable sunny days. More extreme temperatures are within a half-hour drive inland. There, students can relax or float down a river in the hot summer sun or enjoy winter snow activities. Skiing, rock climbing, hiking, sailing, kayaking, and surfing are just some of the activities to experience in the areas surrounding Humboldt. The University is surrounded by relatively untouched biological habitats that provide a wealth of challenging opportunities for study at the master's level by the aspiring biologist, botanist, or zoologist. These habitats are home to hundreds of different animals, plants, and fungi, many of which are unique to northwestern California.

The University

The essence of graduate education at Humboldt State University is academic quality and faculty involvement at an affordable price. Students who attend Humboldt become part of a university that has achieved national recognition. *U.S. News & World Report* consistently ranks Humboldt among the top 10 percent of regional colleges and universities in the nation. Nearly 90 percent of Humboldt's master's graduates are either employed in a job in their fields or enrolled in a doctoral program. Graduate students have access to distinctive research facilities that provide experience needed for life beyond the classroom.

Applying

Applicants should have a bachelor's degree with a major in biology, botany, zoology, or a related subject area; an overall undergraduate grade point average of at least 2.5 for the last 60 semester units or 90 quarter units of credit; and acceptance by an approved faculty member who agrees to act as the major adviser. Applicants must submit the completed CSU application (available online at http://www.csu.mentor.com), the $55 nonrefundable application fee, and official transcripts of all undergraduate work to the Office of Research and Graduate Studies. Applicants must send three letters of recommendation, GRE test scores, and a statement of objectives to the Department of Biological Sciences. The deadline is February 1.

Correspondence and Information

Dr. Michael R. Messler
Graduate Coordinator
Department of Biological Sciences
Humboldt State University
1 Harpst Street
Arcata, California 95521-8299
Phone: 707-826-3245
Fax: 707-826-3201
E-mail: mrm1@humboldt.edu
Web site: http://www.humboldt.edu/~biosci/gradprogram.html

Humboldt State University

THE FACULTY AND THEIR RESEARCH

Full-Time

Brian S. Arbogast, Assistant Professor and Curator of Mammals; Ph.D., Wake Forest. Molecular systematics, population genetics, and phylogeography of mammals. Historical demography and genetic structure of sister species: Deer mice (*Peromyscus* spp.) in the North American temperate rainforest. *Mol. Ecol.* 12:711–24, 2003 (with Zheng and Kenagy).

Milton J. Boyd, Professor and Department Chair; Ph.D., California, Davis. Community ecology of salt marshes, coastal bays, intertidal rocky and sand beach environments, and nearshore benthos.

Michael A. Camann, Associate Professor; Ph.D., Georgia. Effects of land use and land management on insect community structure and on regional ecosystem processes, especially nutrient dynamics. Acari and Collembola at Black Mountain Experimental Forest: An interim report on community structure and prescribed fire effects. *USDA Forest Serv. PSW Res. Station Rep.,* 2001 (with Lamoncha and Plant).

Sean F. Craig, Assistant Professor; Ph.D., SUNY Stony Brook. Population dynamics, ecology, and life-history strategies of colonial marine organisms. Isolation and characterization of microsatellites in the bryozoan *Crisia denticulata. Mol. Ecol. Notes* 100(1):1–2, 2001 (with D'Amato et al.).

Megan Donahue, Assistant Professor; Ph.D., California, Davis. Spatial population dynamics; integration of theory and data; influence of habitat selection on dynamics; marine community ecology; parasite-mediated interactions. Size-dependent competition in a gregarious porcelain crab *Petrolisthes cinctipes* (Anomura: Porcellanidae). *Mar. Ecol. Prog. Ser.* 267:196–207, 2004.

P. Dawn Goley, Associate Professor; Ph.D., California, Santa Cruz. Behavioral ecology of marine mammals.

Terry W. Henkel, Assistant Professor; Ph.D., Duke. Ecology and systematics of neotropical macromycetes and the role of mycorrhizae in structuring forest communities. Mast fruiting and seedling survival of the ectomycorrhizal *Dicymbe corymbosa* (Caesalpiniaceae) in Guyana. *New Phytol.* 167:543–56, 2005 (with Mayor and Woolley).

Erik S. Jules, Associate Professor; Ph.D., Michigan. Spread of invasive organisms; effects of habitat fragmentation on plant populations and community interactions; environmental history. Spread of an invasive pathogen over a variable landscape: A nonnative root rot on Port Orford cedar. *Ecology* 83:3167–81, 2002 (with Kaufmann, Ritts, and Carroll).

Casey R. Lu, Associate Professor; Ph.D., Michigan. Heavy-metal stress in plants as related to phytoremediation; plant biotechnology; improving science education. *Creating a Sustainable Future: Living in Harmony with the Earth,* New Delhi: Research Book Centre, 2002 (with Kaufman et al.).

Sharyn Marks, Associate Professor; Ph.D., Berkeley. Herpetology; conservation biology; evolutionary developmental biology. Metamorphosis and evolution of feeding behavior in salamanders of the family Plethodontidae. *Zoological J. Linnean Soc.* 134:375–400, 2002 (with Deban).

Michael R. Messler, Professor; Ph.D., Michigan. Pollination biology and reproductive biology of pteridophytes (ferns). The radiation of Calochortus: Generalist flowers moving through a mosaic of potential pollinators. *Oikos* 89:209–22, 2002 (with Dilley and Wilson).

Edward C. Metz, Assistant Professor; Ph.D., Hawaii. Species specificity of fertilization among marine invertebrates as a model for the molecular evolution of reproductive recognition processes.

Bruce A. O'Gara, Assistant Professor; Ph.D., Iowa State. Neurobiological and pharmacological control of feeding in the medicinal leech; effects of environmental toxicants on the nervous system of the aquatic oligochaete worm *Lumbriculus variegatus.*

John O. Reiss, Assistant Professor; Ph.D., Harvard. Evolution and morphology of amphibian metamorphosis.

Rollin C. Richmond, President; Ph.D., Rockefeller. Evolutionary genetics of *Drosophila.*

Frank J. Shaughnessy, Associate Professor; Ph.D., British Columbia. Population and community ecology of seaweeds and seagrasses, with an emphasis on pursuing studies relevant to marine resource managers. *Non-indigenous Marine Species of Humboldt Bay, California.* A report to the California Department of Fish and Game, 110 pp., 2002 (with Boyd and Mulligan).

Patricia L. Siering, Associate Professor; Ph.D., Cornell. Microbial ecology and physiology, especially pertaining to biogeochemistry and life in extreme environments. Investigation of environmental and microbial diversity in high-temperature, low-pH geothermal features at Lassen Volcanic National Park. *101st Gen. Meeting Am. Soc. Microbiol.,* Abst. N-197, May 19–23, 2001. American Society for Microbiology, Washington, D.C. (with Wilson).

Stephen C. Sillett, Associate Professor; Ph.D., Oregon State. Forest canopy biology, especially epiphytes of the canopy in temperate and tropical forests. The limits to tree height. *Nature* 428:851–4, 2004 (with Koch, Jennings, and Davis).

Joe Szewczak, Assistant Professor; Ph.D., Brown. Comparative physiology and physiological ecology, especially of bats. Open-flow plethysmography with pressure-decay compensation. *Respiration Physiol.* 134:57–67, 2003 (with Powell).

Mihai Tomescu, Assistant Professor; Ph.D., Ohio. Plant morphology, anatomy, and paleobotany. Probing the seasonality signal in pollen spectra of Eneolithic coprolites (Harsova-tell, Constanta County, southeast Romania). *Cult. Civilization Lower Danube* 22:207–21, 2005.

Jacob P. Varkey, Professor; Ph.D., Illinois State. Genetics of sperm development in the free-living nematode *C. elegans.*

Jeffrey W. White, Assistant Professor; Ph.D., Michigan State. Biogeography and evolution of rare plants, using GIS applications and phylogeny reconstruction; ecological studies and monitoring of rare plants. Student attitudes surveyed in an introductory environmental resources engineering course. *Proc. Frontiers Educ. 2004 Conference* (with Espinoza, Eschenbach, and Cashman).

Mark S. Wilson, Assistant Professor; Ph.D., Cornell. Using molecular genetics to address environmental questions, particularly those concerning microbial ecology and diversity. Horizontal transfer of phn-Ac dioxygenase genes within one of two phenotypically and genotypically distinctive naphthalene-degrading guilds from adjacent soil environments. *Appl. Environ. Microbiol.* 69:2172–81, 2003 (with Herrick, Jeon, Hinman, and Madsen).

Half-Time

James P. Smith Jr., Professor Emeritus. Systematics of grasses; biology of poisonous plants; occult botany.

Part-Time

Jessica Edwards, Lecturer. Bacterial chemotaxis and bioenergetics; elucidating the bacterial mechanism for sensing chemical and energetic stimulants in their environment.

Michael King, Collection Manager, Vertebrate Museum; M.A., Humboldt State. Vertebrate morphology; vertebrate natural history; environmental science.

Leslie J. VanderMolen, Lecturer; M.A., Humboldt State.

ILLINOIS STATE UNIVERSITY

Department of Biological Sciences

Programs of Study

The Department offers the M.S. (thesis option only) and Ph.D. degrees in biological sciences. The degree programs offer students the opportunity to combine broad training in the biological sciences with specialization in an area of particular interest, preparing them for careers in teaching and research in both academic and nonacademic positions. Areas of specialization include animal behavior, biochemistry, botany, cell biology, developmental biology, ecology, genetics, immunology, microbiology, molecular biology, physiology, systematics, ultrastructure, and zoology.

The M.S. degree requires 30 credit hours. It is expected that most students will complete an M.S. in less than three years. M.S. students may elect to pursue a sequence in biotechnology, a program to prepare students for careers in genetic engineering and the biotechnology industry, or conservation biology, a discipline linking ecology, genetics, and systematics to applied problems in biodiversity preservation and ecosystem function, or behavior, ecology, evolution, and systematics, a sequence that provides students with training in whole-organism biology (Ph.D. students may also elect to pursue a similar sequence). Students pursuing the Ph.D. may enter the doctoral program either after first obtaining a master's degree or directly following completion of the baccalaureate. Ph.D. students planning a career in academia may request to enter the Department's Ph.D. Scholar/Educator program. Those admitted into the program receive not only the customary training in research and scholarship but also specialized training and supervision in classroom instruction.

Research Facilities

The Department of Biological Sciences is housed in modern facilities that are well equipped for the wide variety of research opportunities available at Illinois State University. Instrumentation includes both transmission and scanning electron microscopes, a confocal microscope, an automated DNA sequencer, a DNA microarray system, a real-time PCR, a fluorescence-activated cell sorter, epifluorescence microscopes, patch clamp apparatus, networked computers, liquid scintillation counters, recording spectrophotometers, electrophoresis equipment, gamma counters, ultracentrifuges, and gas and high-pressure liquid chromatographs. Additional facilities include a computer lab, a herbarium, a natural history museum, greenhouses, animal rooms, temperature- and humidity-controlled areas, and darkroom facilities. Outdoor areas near campus provide a variety of habitats for fieldwork, and the Department maintains vehicles and equipment for use in these areas. Milner Library, with collections totaling more than 1.5 million items, is adjacent to the biological sciences facility.

Financial Aid

Teaching and research assistantships are available to students with a minimum 3.0 GPA or a combined verbal and quantitative GRE score of more than 1000. Students holding assistantships normally work 6 to 8 hours per week. Tuition costs are waived for students holding an assistantship. In 2005–06, nine-month stipends for M.S. students were $1020 per month and for Ph.D. students, $1620 per month.

Cost of Study

Although teaching and research assistants receive tuition waivers, they pay general fees, which include insurance. In 2005–06, insurance and fees for full-time graduate students were $565 per semester for 9 credit hours.

Living and Housing Costs

University apartments are available to families and single graduate students; costs ranged from $334 to $457 per month in 2005–06. Off-campus housing is readily available.

Student Group

Approximately 100 students are enrolled in the Department's M.S. and Ph.D. programs. They come from a wide variety of colleges and universities in the United States and abroad. Many graduates of the master's program have gone on to professional schools; most of the others are employed in research. Doctoral graduates have gone on to postdoctoral study or positions in academia, government, or private industry.

Location

Illinois State University is located in the central Illinois community of Normal-Bloomington, with a population of about 100,000. The community provides many of the attractions offered in larger metropolitan areas but retains a rural atmosphere. The twin cities are easily accessible by car, bus, train, or plane.

The University and The Department

Abraham Lincoln drafted the documents establishing Illinois State Normal University, which was founded in 1857 as the first public institution of higher education in Illinois. The University began to offer graduate work in several different departments in 1943. In the 1960s, when programs were first offered in the liberal arts as well as in teacher education and when doctoral-level curricula were introduced, the institution was renamed Illinois State University.

The Department of Biological Sciences has a strong commitment to excellence in all aspects of the graduate program. Graduate classes are small enough to permit extensive faculty-student interaction.

Applying

Students who have completed work for a bachelor's degree from an accredited institution may apply for admission to the Graduate School. Students must file an application and arrange for scores on the General Test of the Graduate Record Examinations and official transcripts to be sent to the Office of Admissions. Additional materials are required by the Department and students should visit the Department's Web site for further information. All application materials should be on file by February 1 for full consideration for admission the following fall semester. For the M.S. program, applications received after February 1 may also be considered, but all applicants are strongly urged to have their applications complete by February 1.

Correspondence and Information

Biology Graduate Program Office
Department of Biological Sciences
Illinois State University
Normal, Illinois 61790-4120

Phone: 309-438-3664
E-mail: cawinch@ilstu.edu
Web site: http://www.bio.ilstu.edu

Illinois State University

THE FACULTY AND THEIR RESEARCH

Roger C. Anderson, Professor; Ph.D., Wisconsin–Madison, 1968. Plant ecology; phytosociology; grassland fire ecology; mycorrhizae.

Joseph E. Armstrong, Professor; Ph.D., Miami (Ohio), 1975. Floral biology; floral morphogenesis, function, and evolution; beetle pollination in rain forest communities; biology of hemiparasitic plants in prairies.

Victoria A. Borowicz, Adjunct Assistant Professor; Ph.D., Penn State, 1986. Mutualistic and antagonistic species interactions in plants.

David W. Borst Jr., Professor; Ph.D., UCLA, 1973. Hormonal regulation of development in anthropods.

Rachel M. Bowden, Assistant Professor; Ph.D., Indiana, 2001. Maternal effects on temperature-dependent sex determination in reptiles.

Diane Byers, Associate Professor; Ph.D., Rutgers, 1993. Evolution and adaptation in small populations of plants; population genetics.

Angelo Capparella, Associate Professor; Ph.D., LSU, 1987. Avian systematics, biogeography, and conservation biology; Neotropical and Nearctic birds.

Joseph M. Casto, Adjunct Assistant Professor; Ph.D., Johns Hopkins, 2001. Neuroendocrinology; integrative animal physiology; stroke; chronic pain.

Hou T. Cheung, Professor and Chair; Ph.D., Wisconsin–Madison, 1977. Cellular and tumor immunology.

Martha Cook, Associate Professor; Ph.D., Wisconsin–Madison, 1996. Evolution of plants from green algae; microscopy.

Kevin A. Edwards, Assistant Professor; Ph.D., Duke, 1996. Cytoskeletal and signal transduction proteins of the fruit fly, *Drosophila melanogaster*.

Anne-Katrin Eggert, Adjunct Assistant Professor; Ph.D., Bielefeld (Germany), 1990. Behavioral ecology; sexual selection; parental care; insects.

Paul A. Garris, Associate Professor; Ph.D., Indiana, 1990. Neurobiology; endocrinology; biosensors.

Craig Gatto, Associate Professor; Ph.D., Missouri–Columbia, 1994. Structure and function of the sodium pump (Na,K-ATPase).

Christopher D. Horvath, Associate Professor; Ph.D., Duke, 1992. Philosophy of biology.

Radheshyam K. Jayasawal, Professor; Ph.D., Purdue, 1985. Microbial genetics; plant molecular biology; molecular basis of host-pathogen interaction.

Marjorie Jones, Professor; Ph.D., Texas Health Science Center at San Antonio, 1982. Enzymes involved in heme synthesis.

Steven Juliano, Professor; Ph.D., Penn State, 1985. Community ecology; predator-prey systems, competition and other species interactions; reproductive tactics; insect ecology; biostatistics.

Alan J. Katz, Professor; Ph.D., Ohio State, 1974. Genetic toxicology of *Drosophila*; induced somatic mutagenesis and recombination; biostatistics.

Jeffery M. Kramer, Adjunct Assistant Professor; Ph.D., Illinois at Urbana-Champaign, 2001. Neurobiology; molecular and integrative physiology; chronic pain.

Sabine S. Loew, Associate Professor; Ph.D., SUNY at Stony Brook, 1992. Molecular conservation genetics; behavioral ecology.

Cynthia Moore, Associate Professor; Ph.D., Temple, 1978. Developmental genetics of zebra fish.

Wade Nichols, Associate Professor; Ph.D., Iowa, 1992. Bacterial pathogenesis of the respiratory tract.

Anthony J. Otsuka, Professor; Ph.D., California, San Diego, 1979. Molecular genetics; biotin operon in *E. coli* and developmental genetics in *C. elegans*.

William L. Perry, Assistant Professor; Ph.D., Notre Dame, 1998. Ecological genetics and community dynamics of invading species.

Robert L. Preston, Professor; Ph.D., California, Irvine, 1970. Membrane physiology; transport of nonelectrolytes; osmoregulation; membrane toxicology.

David A. Rubin, Assistant Professor; Ph.D., Denver, 1994. Molecular evolution of calcium regulatory hormones and receptors.

Scott K. Sakaluk, Professor; Ph.D., Toronto, 1986. Behavioral ecology; sexual selection; communication; insects.

John C. Sedbrook, Assistant Professor; Ph.D., Wisconsin–Madison, 1997. Developmental genetics of *Arabidopsis*.

Charles F. Thompson, Professor; Ph.D., Indiana, 1971. Population and evolutionary ecology of birds.

Laura A. Vogel, Associate Professor; Ph.D., Medical College of Ohio, 1995. Regulation of the immune response.

Sharon L. Weldon, Assistant Professor; Ph.D., California, San Diego, 1984. Structure-function of proteins; protein and mRNA stability.

Douglas Whitman, Professor; Ph.D., Berkeley, 1982. Insect behavior, ecology, and physiology; insect-plant interactions and predator-prey interactions.

Brian J. Wilkinson, Professor; Ph.D., Sheffield (England), 1971. Microbial physiology and biochemistry; resistance of *Staphylococcus aureus* to methicillin and vancomycin; bacterial stress biology.

David L. Williams, Associate Professor; Ph.D., Illinois at Urbana-Champaign, 1990. Parasitology; biochemistry and molecular biology of *S. mansoni*.

INDIANA STATE UNIVERSITY

Department of Life Sciences

Programs of Study	The Department of Life Sciences at Indiana State University (ISU) offers graduate programs that lead to the Master of Science (thesis or nonthesis) and Doctor of Philosophy degrees in the areas of cellular and molecular biology; microbiology; molecular ecology, evolution, and bioinformatics; physiology; science education; and sports medicine. The Department offers students a general background in biological science plus advanced training in selected research areas that prepare them for careers in research and teaching. Master's degree requirements include a total of 32 semester hours of graduate work, with a minimum of 18 semester hours in the area of specialization. Doctoral degree requirements include 83 semester hours of graduate work (which may include 32 hours from other graduate study), a dissertation, research tools, and two semesters of participation in the teaching program. Students typically are accepted to work with a particular adviser. Students have the opportunity to participate in the Department's research programs in a variety of laboratories. The student selects a committee that plans an academic program and provides guidance in the research program, which students are encouraged to begin as early as possible. Research strengths of faculty members include bacteriology, behavioral ecology, botany, cellular physiology, community ecology, conservation biology, cytology, curricular and pedagogical development, developmental biology, ecosystem biology, endocrinology, evolutionary biology, exercise physiology, hematology, immunology, landscape biology, microbiology, molecular biology, molecular evolution, molecular ecology, molecular genetics, mycology, neurobiology, physiological ecology, plant physiology, population genetics, phylogenetics, restoration biology, soil ecology, and systematics.
Research Facilities	The Department occupies two floors of the Science Building and the top floor of the Student Services Building and has use of an adjacent animal-research building. Additional research facilities are available in the Departments of Athletic Training, Chemistry, and Physics and in the Indiana University School of Medicine Terre Haute Center, which adjoins the Science Building. The departmental research laboratories include facilities for tissue and cell culture, cell fusion and hybridoma production, recombinant DNA, molecular biology, and automated DNA sequencing. Controlled-environment rooms, large greenhouses, personal and mainframe computers, research-animal facilities, and shop facilities are also available. In addition, the University has supercomputing facilities for high-end computational research. The University maintains 55-acre and 28-acre field study areas and owns/manages two nature preserves of 80 and 8 acres, respectively. The University also maintains a 92-acre multipurpose field station near campus. Diverse terrestrial and aquatic habitats and biotas are available for study in nearby parks, wildlife refuges, and accessible private lands. The Department facilitates interdepartmental collaborations with the Department of Chemistry and the Department of Geography, Geology, and Anthropology.
Financial Aid	Graduate assistantships and University fellowships are available to qualified students. Twelve-month stipends for 2005–06 were approximately $9900 for students pursuing the Master of Science (with thesis) degree and $10,900 for doctoral students (increasing to $12,000 when the student advances to candidacy). Students with appointments work part-time as teaching assistants. Tuition scholarships are awarded to all graduate students who hold teaching assistantships. Tuition assistance is also available on a competitive basis for Master of Science (nonthesis) students. Research assistantships are available through individual faculty research grants.
Cost of Study	Tuition and fees for the 2005–06 academic year were $262 per semester hour for in-state students and $521 per semester hour for out-of-state students during regular semesters. Competitively awarded scholarship funds for tuition assistance are available.
Living and Housing Costs	Room and board are provided for single students in on-campus residence halls at a cost of $2648 to $3073 per semester. University apartments are also available near campus for single or married students and their families for $526 to $625 per month, including utilities. Low-cost housing is available in the surrounding community.
Student Group	The University's enrollment of 11,700 includes 1,900 graduate students. The graduate students in the Department of Life Sciences are a nationally and internationally diverse group, with about an equal number of men and women.
Student Outcomes	Graduates of the program pursue a variety of professional paths, finding employment in academia, industry, and government.
Location	ISU adjoins the central business district of Terre Haute. The city, with a population of more than 70,000, is located on the banks of the Wabash River in west-central Indiana. Cultural activities include amateur and professional theatrical productions, symphonies, and art exhibitions. Numerous city parks offer an active schedule of athletic programs. Excellent county and state parks and outdoor recreational areas for camping and boating are within easy driving distance.
The University	Indiana State University has grown during its more than 140-year history from Indiana State Normal School to Indiana State Teachers College and Indiana State College to full university status.
Applying	Requirements for admission include 32 semester hours in biology, chemistry through organic chemistry, two semesters of physics, and mathematics through calculus or statistics. Students with an undergraduate major in another physical science may apply up to 8 semester hours of advanced work in their major as part of the 32 semester hours in biology. Graduate Record Examinations General Test scores must be submitted. TOEFL scores are also required of students from countries in which English is not the native language. Applications are accepted throughout the year for admission in August, January, or June. For financial support, application should be made by February 1 for admission in September. For additional information about graduate courses as well as the graduate student handbook and research profiles of the Department's professors, applicants should consult the Web site listed in the Correspondence and Information section.
Correspondence and Information	Admissions Department of Life Sciences Indiana State University Terre Haute, Indiana 47809 Phone: 812-237-3880 Fax: 812-237-3378 E-mail: admission@biology.indstate.edu Web site: http://biology.indstate.edu

Indiana State University

THE FACULTY AND THEIR RESEARCH

H. Kathleen Dannelly, Associate Professor of Life Sciences; Ph.D., Arizona State. Microbiology: control mechanisms of bacterial metabolic pathways, specifically, *Euglena mutablis* and *Helicobactor pylori;* neuroimmune interactions; adaptive mechanisms of acid-tolerant organisms. (LSDANNEL@isugw. indstate.edu)

*Taihung Duong, Associate Professor of Life Sciences and of Anatomy and Cell Biology; Ph.D., UCLA. Neuroanatomy: brain aging and neurodegenerative diseases. (TDUONG@medicine.indstate.edu)

**Jeffrey E. Edwards, Adjunct Associate Professor of Life Sciences; Ph.D., Indiana. Exercise physiology: human energy balance and expenditure. (j-edwards@indstate.edu)

*Roy W. Geib, Professor of Life Sciences, Alvin S. Levine Professor of Microbiology and Immunology and of Pathology and Laboratory Medicine, and Assistant Dean and Director, Indiana University School of Medicine–Terre Haute Center; Ph.D., Texas Health Science Center at Dallas. Virology and immunology; complementary/integrative therapies for chronic diseases. (RGEIB@medicine.indstate.edu)

‡Swapan K. Ghosh, Professor of Life Sciences and Interim Chair; Ph.D., Calcutta. Immunology: tumor immunology; recombinant vaccine development; immunoregulation; development of sensitive immunoassays for medical application; environmental factors in autoimmunity. (LSGHOSH@isugw.indstate. edu)

Rusty A. Gonser, Assistant Professor of Life Sciences; Ph.D., SUNY at Albany. Conservation genetics; behavioral, evolutionary, and molecular ecology. (LSGONSER@isugw.indstate.edu)

James P. Hughes, Professor of Life Sciences; Ph.D., Berkeley. Molecular and cellular endocrinology: mechanism of hormone and growth factor signal transduction. (LSHUGHS@isugw.indstate.edu)

*Mary T. Johnson, Associate Professor of Life Sciences, Microbiology, and Immunology; Ph.D., Indiana State. Pulsed field therapy and inflammation; stress and immune system function; cell culture and proteomics. (johnsomt@iupui.edu)

*Michael W. King, Professor of Life Sciences, Biochemistry, and Molecular Biology; Ph.D., California, Riverside. Molecular and developmental biology: regenerative biology; nervous system development. (mking@medicine.indstate.edu)

Mary Ann L. McLean, Assistant Professor of Life Sciences; Ph.D., Calgary. Soil microbial community and ecosystem ecology: disturbance impacts on soil diversity and processes. (LSMCLEAN@isugw.indstate.edu)

*Margaret M. Moga, Associate Professor of Life Sciences and of Anatomy and Cell Biology; Ph.D., Loyola Chicago. Anatomy and neurobiology: biology of acupuncture; bioenergy therapies; complementary and alternative medicine. (mmoga@medicine.indstate.edu)

Timothy J. Mulkey, Associate Professor of Life Sciences; Ph.D., Ohio State. Plant physiology: hormones and calcium in the control of plant growth, development, and tropic responses. (LSMULKY@isugw.indstate.edu)

†Stanley S. Shimer, Professor of Life Sciences and Science Education and Coordinator, Center for Science Education; Ph.D., Indiana. Earth/space science; biology; physical science; science education. (S-SHIMER@indstate.edu)

**Catherine L. Stemmans, Adjunct Assistant Professor of Life Sciences; Ph.D., Southern Mississippi. Clinical instruction: pedagogy. (CAT@indstate.edu)

†Marcella Stevens, Assistant Professor of Life Sciences and Coordinator, Clinical Laboratory Science; Ph.D., Capella. Clinical laboratory science and science education; hematology and stem cell research. (MSTEVENS@indstate.edu)

Gary W. Stuart, Associate Professor of Life Sciences; Ph.D., Washington (Seattle). Molecular genetics: molecular systematics; recombination mechanisms; vertebrate transgenesis. (LSSTUAR@isugw.indstate.edu)

Elaina M. Tuttle, Assistant Professor of Life Sciences; Ph.D., SUNY at Albany. Behavioral and molecular ecology: evolution of life history strategies; reproductive trade-offs; population genetics; evolutionary biology; ecological genetics. (LSTUTTLE@isugw.indstate.edu)

*Gabi Nindl Waite, Assistant Professor of Life Sciences and of Cellular and Integrative Physiology; Ph.D., Stuttgart. Medical, cellular, and integrative physiology; pathophysiology; biophysical interactions with cell systems; treatment of inflammatory diseases; development of therapeutic devices. (GNINDL@medicine.indstate.edu)

---

*Joint appointment with the Terre Haute Center for Medical Education, a division of the Indiana University School of Medicine
**Regular appointment in the Department of Athletic Training of Indiana State University
†Joint appointment with the Center for Science Education
‡Adjunct appointment with the Terre Haute Center for Medical Education, a division of the Indiana University School of Medicine

INDIANA UNIVERSITY BLOOMINGTON

Department of Biology

Programs of Study	Graduate programs are offered in genetics (Ph.D.); microbiology (M.A. and Ph.D.); molecular, cellular, and developmental biology (Ph.D.); plant sciences (M.A. and Ph.D.); zoology (M.A. and Ph.D.); and evolution, ecology, and behavior (Ph.D.). In addition, there is a Master of Arts in Teaching (M.A.T.) program. These programs were developed as a result of the merger of the Departments of Microbiology, Plant Sciences, and Zoology into a single Department of Biology. The programs are flexible so that students can design a course of study that can keep pace with the ever-changing and rapidly expanding subdisciplines of biology. Each program emphasizes the interdisciplinary nature of modern biology, and faculty members often participate in administering the degree requirements of students in several different programs.

Normally, the first year is devoted primarily to course work. A research sponsor is selected before the end of the first year, and research is begun soon thereafter. A qualifying examination is administered before the beginning of the third year. Most students require an additional two to three years to complete their Ph.D. thesis research.

The requirement for the M.A. degree is completion of 30 credit hours of graduate courses, including some independent work. Individual programs may include independent research, an oral examination, or a thesis.

Research Facilities
Jordan Hall and Myers Hall house complete facilities for research in cell, molecular, and developmental biology; microbiology; genetics; evolution; ecology; and organismal biology; and the Center for Genomics and Bioinformatics. The department includes laboratories specifically designed for research in molecular, evolutionary, and developmental genetics; microbiology; and molecular biology. The Biology Library, an audiovisual center, animal and microbial cell-culture units, and equipment centers with facilities for electron microscopy, centrifugation, automated DNA sequencing, isotope counting, and other techniques are available to all students. Work in ecology and field biology is done on an 11,300-acre reservoir, Lake Monroe, on the University-owned farm in the Hoosier National Forest, and in local woodlands.

Financial Aid
University fellowships, research assistantships, and NIH and NSF training grants in the areas of genetics, molecular and cellular biology, and animal behavior are available to qualified students. Stipends range from $19,500 to $22,000 per year. Associate instructorships ($16,250 for ten months plus summer support of approximately $3360) are also available. All forms of support include tuition and health insurance, except for allocated fees (approximately $700 per semester). All qualified Ph.D. candidates are supported by one of these forms of aid. Indiana University continues its commitment to the achievement of equal opportunities for all.

Cost of Study
Tuition in 2005–06 was $212 per credit hour for Indiana residents and $618 per credit hour for nonresidents. Tuition is paid by the department except for allocated fees (approximately $600 per semester).

Living and Housing Costs
Housing and dining facilities are provided at Eigenmann Hall. University apartments are available for married students. The estimated cost of rooms for the 2006–07 academic year is $10,473 for the eight-month academic year (the rate may vary depending on the size of the room). Apartment rents in Bloomington range from $400 per month for an unfurnished one-bedroom apartment to $800 per month for an unfurnished three-bedroom apartment. Application forms and complete information about accommodations (including up-to-date rates) may be obtained by writing to the Halls of Residence at Indiana University.

Student Group
About 32,000 students attend Indiana University Bloomington. Of the 8,000 graduate and professional students, about 180 are in the biological sciences. Approximately half of the students are women and/or members of minority groups.

Location
Indiana University Bloomington is located within the city of Bloomington, which has a population of about 61,000. The 1,850-acre campus is situated in the wooded hills of southern Indiana, 8 miles from Lake Monroe (with 100 miles of shoreline, it is the largest lake in the state), 18 miles from Nashville (a popular tourist and vacation town in Brown County), and 55 miles from Indianapolis. Bloomington is also within 100 miles of Louisville and Cincinnati, and it is within 250 miles of Chicago and St. Louis. In addition, Bloomington is within easy driving distance of four state parks, two state forests, and Lake Lemon.

The University and The Department
Founded in 1820, Indiana University Bloomington is the principal liberal arts institution in the state. The Department of Biology was formed several years ago by the merger of the Departments of Microbiology, Plant Sciences, and Zoology, each of which had maintained excellent national and international reputations. By merging, the departments gained the flexibility to add graduate programs in rapidly expanding subdisciplines while continuing to offer top-quality education in more traditional aspects of biology.

Applying
Students are admitted into one of the various programs on the basis of GRE scores on the General and Subject Tests, undergraduate performance, and three letters of recommendation. TOEFL scores are required for international students whose native language is not English.

Students may apply directly to the Department of Biology or electronically at http://www.bio.indiana.edu/~grdschl/infoadm.html. All applications for admission should be received by January 5 (December 1 for international applicants); application for financial support is made on the same form as application for admission.

Correspondence and Information
Graduate Admissions
Department of Biology
Indiana University Bloomington
Bloomington, Indiana 47405-3700

Phone: 812-855-1861
E-mail: biograd@bio.indiana.edu
Web site: http://www.bio.indiana.edu

Indiana University Bloomington

THE FACULTY AND THEIR RESEARCH

Justen Andrews: Gene regulatory networks controlling sex in arthropods.
Carl Bauer: Oxygen and light regulation of gene expression; biosynthesis of heme and chlorophyll; prokaryotic development.
James Bever: Ecology and evolution of plants and fungi.
Edmond D. Brodie III: Evolutionary biology; predator-prey coevolution; maternal effects; indirect genetic effects and social behavior.
Yves Brun: Cell-cycle control, cell division, and cell differentiation in bacteria; bacterial adhesion.
Lingling Chen: Structural studies of protein-protein interactions in GroEL-mediated protein folding process and host-pathogen communications.
Peter Cherbas: *Drosophila* development and genomics; nuclear receptors.
Keith Clay: Microbial interactions; ecology; symbiosis; disease.
Lynda Delph: Evolutionary ecology; plant reproductive biology.
Greg Demas: Behavioral endocrinology; neuroendocrine-immune interactions; biological rhythms; seasonality; aggression.
Thomas Donahue: Genetic, molecular, and biochemical analysis of translation initiation in yeast.
James Drummond: Mechanistic studies of DNA mismatch and lesion processing.
Joseph Duffy: Developmental genetics, pattern formation, and evolution of signal transduction in *Drosophila*.
Viola Ellison: Human chromosome duplication and maintenance of genome integrity.
Mark Estelle: Molecular genetics of hormone action in *Arabidopsis*.
Wayne Forrester: Mechanisms of directed cell migration in *C. elegans*.
Patricia Foster: Mutagenesis, DNA replication, and recombination.
Clay Fuqua: Multicellular interactions of bacteria.
Matthew Hahn: Computational and evolutionary genomics; evolution of transcriptional regulation; molecular population genetics.
Roger Hangarter: Plant physiology: environmental sensory-response and plant development.
Richard Hardy: Genome functions of RNA viruses and the roles of *trans*-acting factors.
Laura Hurley: Neuromodulation of the auditory system; modulator-induced plasticity in neural circuits underlying behavior.
Roger Innes: Molecular genetics of plant-pathogen interactions.
Daniel Kearns: Bacterial multicellular behavior.
David Kehoe: Light-regulated signal transduction in cyanobacteria.
Ellen Ketterson: Avian biology; avian migration; mating systems and parental care; hormones and behavior; physiological mechanisms underlying trade-offs in life histories; using hormones to explore adaptation; dominance and aggression; population dynamics during the nonbreeding season.
Justin Kumar: Compound eye development in the fruit fly, *Drosophila melanogaster.*
Curt Lively: Population biology; predatory-prey and host-parasitic interactions.
Michael Lynch: Evolution of molecules, genome structure, and phenotypes.
Emília Martins: Evolution of complex behavioral traits; phylogenies and the comparative method.
Scott Michaels: Molecular genetics of flowering time regulation in *Arabidopsis* and other species.
Armin Moczek: Evolutionary developmental biology; phenotypic plasticity; morphological and behavioral diversity in arthropods.
Leonie Moyle: Genetics of speciation and adaptation; comparative genomics; evolutionary ecology; plant reproduction.
Tuli Mukhopadhyay: Structure and assembly of enveloped RNA viruses.
Jeffrey D. Palmer: Molecular evolution: lateral transfer of mitochondrial genes to the nucleus and between organisms; evolution of mutation rates; molecular phylogeny.
Heather Reynolds: Plant community ecology; plant-microbe interactions.
Loren Rieseberg: Plant evolutionary genetics; systematics; reproductive biology; speciation; conservation genetics.
William Saxton: Cell and developmental biology; molecular motors and cytoplasmic motility.
Peggy Schultz: Ecology of plants and mycorrhizal fungi; restoration ecology.
Sidney Shaw: Microtubule dynamics and organization in acentriolar *Arabidopsis* cells.
Troy Smith: Neural and hormonal control of reproductive and communication behavior; ion channels and membrane excitability; motor pattern generation.
Susan Strome: Germ cell development in *C. elegans.*
Michael Wade: Evolution in metapopulations; genetic basis of speciation in Tribolium; epistasis; evolutionary genetics of maternal effects; sexual selection; coevolution of arthropod hosts and Wolbachia endosymbionts.
Maxine Watson: Plant development ecology.
Malcolm E. Winkler: Physiology, pathogenesis, molecular genetics, stress responses, and genomics of the gram-positive human respiratory pathogenic bacterium, *Streptococcus pneumoniae.*
Joel Ybe: X-ray crystallographic and biochemical studies of membrane vesicle protein coats.
Miriam Zolan: Meiosis and DNA repair.

Adjunct Faculty

David Daleke, Department of Medical Sciences: Structure and organization of biological membranes; phospholipid transporters; effect of diabetes on membrane structure.
Richard DiMarchi: Relationship of protein structure and function; novel methods of drug delivery.
Elizabeth Housworth, Department of Mathematics: Statistical genetics.
Fredrika Kaestle, Department of Anthropology: Ancient DNA; human genetics; population genetics; phylogenetics; prehistoric population movement; Americas; Siberia; Pacific.
Vicky Meretsky, School of Public and Environmental Affairs: Conservation biology.
Joseph Near, Department of Medical Sciences: Understanding various biochemical processes involved in chemical neurotransmission; catecholamine and serotonin transport systems in storage vesicles and the presynaptic plasma membrane.
Anton Neff, Department of Medical Sciences: Developmental biology.
Kenneth Nephew, Department of Medical Sciences: Nuclear receptors; hormone-dependent cancers.
Martha Oakley, Department of Chemistry: Biochemistry and bioorganic chemistry.
David Parkhurst, School of Public and Environmental Affairs: Applied mathematical modeling; physiological plant ecology; risk and decision analysis.
Flynn Picardal, School of Public and Environmental Affairs: Environmental microbiology; microbial geochemistry, and bioremediation.
Henry Prange, Department of Medical Sciences: Comparative physiology of respiration; temperature regulation.
Anne Prieto, Department of Psychology: Function and signaling of receptor tyrosine kinases and their ligands in the developing and mature nervous system.
Christine Quirk, Medical Sciences Program: Deciphering the role of p8, an HMG-like protein, in development, tumorigenesis, and malignancy.
Dale Sengelaub, Department of Psychology: Neural development; neural plasticity; hormones and behavior.
Martin J. Stone, Department of Chemistry: Structural molecular biology.
Roderick Suthers, Department of Medical Sciences: Neural and physiological bases of acoustic behavior.
William Timberlake, Department of Psychology: Environmental regulation of photosynthetic carbon assimilation in *Chlamydomonas reinhardtii.*
Thomas J. Tolbert, Department of Biemistry: Glycoprotein synthesis; protein chemistry/proteomics.
Claire E. Walczak, Department of Medical Sciences: Mechanisms of mitotic spindle assembly and chromosome segregation; regulation of microtubule dynamics during interphase and mitosis.
Meredith West, Department of Psychology: Development of communicative behavior.

INDIANA UNIVERSITY–PURDUE UNIVERSITY INDIANAPOLIS

Department of Biology

Programs of Study	The Department of Biology offers a wide variety of research experiences. Students may work with faculty members on research projects in immune system genetics, endocrinology, DNA repair, spinal cord regeneration, yeast gene cloning and antifungal drug studies, biology teaching methods, membrane biochemistry and biophysics, oncology, renal physiology, molecular development, plant molecular biology, plant cell biology, and regenerative biology and medicine. The Ph.D. is pursued through a program based in Indianapolis. Written qualifying examinations are administered at the end of the first year of study. By the end of the second year, students must pass the qualifying exam, available in immunobiology, biochemistry and molecular biology, cell and developmental biology, and membrane biology. Before graduation, students are required to write a dissertation and defend it before an advisory committee. Students earning an M.S. thesis degree take a minimum of 9 credit hours of graduate-level course work. The remainder of the 30 credit hours of registration is taken as thesis research and a seminar, a 1-credit-hour registration during which the student presents the results of the research prior to the thesis defense. The nonthesis M.S. degree includes 27 hours of course work, an independent project, and a seminar. Up to 9 hours of course work can be taken in a secondary area chosen to complement students' career objectives. Examples of secondary areas include, but are not limited to, chemistry, mathematics, public affairs, business, law, statistics, computer science, education, and health administration. The preprofessional nonthesis M.S. degree is designed for students who want to enhance their credentials for application to a professional school; 30 hours must be completed in two semesters. Applicants must have a minimum of 16 credit hours toward a biology major and meet professional school requirements in chemistry and physics.
Research Facilities	The Department is located in a modern research building with well-equipped laboratories. The research laboratories of the Indiana University Schools of Medicine and Dentistry are located in adjacent buildings. The Department is also home to administrative offices and some laboratories of the Indiana University Center for Regenerative Biology and Medicine. The IUPUI campus has state-of-the-art core facilities for life science research, including support for bioinformatics, gene knockouts, transgenics, DNA microarrays, proteomics, DNA and protein sequencing, and rapid SNP analysis.
Financial Aid	Ph.D. students are eligible to apply for University or departmental fellowships, teaching assistantships, or research assistantships, for an annual stipend of $20,000. For M.S. thesis students, teaching assistantships are available at a calendar-year stipend of $16,000. Tuition remission, which pays most of the tuition costs, and payment of student health insurance premiums are included in the support package. Research assistantships (RA) are sometimes available through individual faculty members who hold external funds for this purpose. RA support for M.S. students includes tuition remission and health insurance in addition to a stipend of $15,000.
Cost of Study	In the 2005–06 academic year, tuition was $194.10 per credit for in-state residents and $560.15 per credit for out-of-state residents.
Living and Housing Costs	Apartments are $795 per month for one bedroom, $681 per student per month for two bedrooms, and $567 per student per month for four bedrooms. The cost of off-campus housing varies upward from $450. Communications fees, including telephone and cable TV service, range from $30 to $65 per month, and electricity charges range from $58 to $76 per month.
Student Group	Indiana University–Purdue University Indianapolis (IUPUI) has more than 7,900 students in its graduate programs.
Student Outcomes	Students graduating with Ph.D. degrees have been successful in obtaining postdoctoral research experience and have gone on to secure faculty positions at accredited institutions of higher learning or research positions in industry or government. Students graduating with M.S. thesis degrees are highly successful in obtaining employment on campus or in local industry. Indianapolis is home to several large pharmaceutical and biotechnology companies that have provided high-paying jobs for many IUPUI graduates.
Location	The campus is only a few blocks from downtown Indianapolis and adjacent to White River State Park. Indianapolis attracts more than 18 million visitors a year and is easily accessible from a wide variety of locations. Indianapolis has an enriching array of arts and attractions, including the Indiana Repertory Theatre, Indianapolis Museum of Art, Indianapolis Symphony Orchestra, Indianapolis Colts, Indianapolis 500 race, and Indiana Pacers.
The University	Indiana University–Purdue University Indianapolis is an urban research university created in 1969 as a partnership between Indiana and Purdue Universities. IUPUI grants degrees in some 185 programs and offers the broadest range of academic programs of any campus in Indiana. It ranks among the top fifteen nationwide in the number of first professional degrees it confers and among the top five in the number of health-related degrees. IUPUI is the home campus for statewide programs in medicine, dentistry, nursing, allied health, and social work.
Applying	Applications can be made online at the Department's Web site. The application deadline for admission and financial aid for the Ph.D. program is March 1. Applications for the M.S. thesis degree (with financial support) should be made by June 1 for fall admission or November 1 for spring admission. Applications for nonthesis M.S. programs should be made by August 1 or December 1 for fall or spring admission, respectively. A nonrefundable application fee of $45 is required of all applicants. Applicants should send official GRE scores and transcripts directly to Jennifer Howard at the Department.
Correspondence and Information	Jennifer Howard Department of Biology SL 306, 723 West Michigan Street Indiana University–Purdue University Indianapolis Indianapolis, Indiana 46202-5191 Phone: 317-274-0577 Fax: 317-274-2846 E-mail: biograd@iupui.edu Web site: http://www.biology.iupui.edu/

Indiana University–Purdue University Indianapolis

THE FACULTY AND THEIR RESEARCH

Resident Faculty

Ruth Allen, Ph.D. Graft-versus-host disease arising after bone marrow transplantation; autoimmune syndromes; effect of genetic variation on immune function.

Martin Bard, Ph.D. Sterol biosynthesis and regulation in the yeasts *Saccharomyces cerevisiae* and *Candida albicans;* genetic analysis and biochemistry of the sterol enzyme complex.

Teri Belecky-Adams, Ph.D. Mechanisms of optic cup development and regeneration.

Bonnie Blazer-Yost, Ph.D. Hormonal regulation of renal ion transport; electrophysiological, biochemical, and proteomic studies of salt and water homeostasis.

Ellen Chernoff, Ph.D. Spinal cord and limb regeneration: amphibian development and regeneration, stem cell properties, and patterning gene expression.

James Clack, Ph.D. Electrophysiology of vertebrate photoreceptors.

Dring Crowell, Ph.D. Regulation of plant growth and development by isoprenoid compounds; isoprenylation and post-isoprenylation processing of proteins in *Arabidopsis thaliana.*

Pamela Crowell, Ph.D. Chemoprevention and chemotherapy of pancreatic cancer with dietary phytochemicals; expression and function of oncogenic PRL phosphatases in human cancer.

N. Douglas Lees, Ph.D. Ergosterol biosynthesis and regulation in *Saccharomyces cerevisiae* and *Candida albicans;* identification of new targets for antifungal drug development.

Anna Malkova, Ph.D. Mechanisms of DNA repair and recombination; connections between recombination and replication; mechanisms leading to gross chromosomal rearrangements similar to those leading to cancer.

Kathleen Marrs, Ph.D. Biology education; using Just-in-Time Teaching and the Web to identify student misconceptions in undergraduate biology; using the Web and interactive lectures to increase student learning and motivation.

Stephen Randall, Ph.D. The role of calcium and ion-binding proteins in cold and other abiotic stress responses in plants.

Simon Rhodes, Ph.D. Gene regulation during endocrine organ development; molecular aspects of hormone actions on bone; nervous system regeneration in amphibians.

William Stillwell, Ph.D. Membrane biophysics; role of omega-3 fatty acids in controlling membrane structure and function.

David Stocum, Ph.D. Cellular and molecular processes that specify axial patterns of tissue differentiation in developing systems; molecular pathways that characterize formation of a regeneration blastema in regenerating and nonregenerating amphibian limbs.

Mark Terrell, Ed.D. Teaching innovations in anatomical science; applying the science of learning to the learning of science.

Xianzhong Wang, Ph.D. Photosynthetic and respiratory responses to global environmental changes by plants; effects of elevated atmospheric CO_2 on male and female individuals of dioecious species.

John Watson, Ph.D. Photomorphogenesis and signal transduction in plants; phototropins, the blue-light photoreceptors for phototropism; light- and auxin-regulated protein kinases.

Adjunct Faculty

Charles Barman, Ed.D. Science and environmental education: student's science concepts, teaching models, and conceptual understanding.

Subba Chintalacharuvu, Ph.D. Immunology: autoimmune diseases and the role of T and NK T cells in acute inflammation.

Keith Dunker, Ph.D. Protein order and disorder.

Mark Heiman, Ph.D. Neuroendocrine regulation of energy balance; pharmaceutical approach to fat attrition.

Gary Krishnan, Ph.D. Characterization of osteogenic, myogenic, and adipogenic factors that regulate the cell fate of pluripotent mesenchymal stem cells.

John McIntyre, Ph.D. Identification and regulation of human autoantibodies; specificities, mechanisms, and associated pathologies.

Joseph Petolino, Ph.D. Transgenic production of crop plants; gene expression; plant cell and tissue culture.

John Schild, Ph.D. Ionic channel dynamics underlying the neural coding of cardiovascular sensory afferents and information transmission across the first synapse in the medial nucleus tractus solitarius.

Stephanie Sen, Ph.D. Development of insect hormone inhibitors; biomimetic polyene cyclizations; antibody-assisted chemical transformations.

Rafat Siddiqui, Ph.D. Molecular mechanisms of signal transduction in inflammatory diseases.

Rosamund Smith, D.Phil. Discovery and development of protein and antibody therapeutics as human pharmaceuticals.

Edward Srour, Ph.D. Characterization and biology of human hematopoietic stem cells.

Chris Vlahos, Ph.D. Identification and role of kinases in cardiovascular function and disease.

Frank Witzmann, Ph.D. Differential protein expression analysis using proteomics technologies in the assessment of chemical toxicity; various paradigms, including in vivo and in vitro models of environmental toxins in liver, lung, and kidney and ethanol in the brain.

Steve Zuckerman, Ph.D. Regulation of macrophage activation in inflammation and immunology; cytokine biology and evaluation of changes in macrophage surface antigens/receptors.

JOHNS HOPKINS UNIVERSITY / NATIONAL INSTITUTES OF HEALTH

Graduate Program in Cell, Molecular, and Developmental Biology and Biophysics

Program of Study

To take advantage of the unique intellectual and scientific resources available at the National Institutes of Health (NIH), a cooperative graduate program was established between the Department of Biology at Johns Hopkins University (JHU) and the NIH through the NIH Graduate Partnerships Program (GPP). The goal of the program is to combine the educational excellence of Johns Hopkins in the biological sciences with the tremendous variety of research possibilities available at the NIH, bringing students the benefits of working with leading researchers in the biomedical life sciences. The Graduate Program in Cell, Molecular, and Developmental Biology and Biophysics at JHU offers training in the areas of biochemistry, biophysics, cell biology, developmental biology, genetics, and molecular biology. Students can enter the graduate program from a variety of backgrounds, including biology, chemistry, and physics.

The curriculum of the program includes a set of four core courses taken by all students during their first year at JHU: advanced molecular biology, graduate biophysical chemistry, advanced cell biology, and advanced developmental biology and genetics. During the first year, students do a total of four laboratory rotations between JHU and NIH. At the end of the first year, students choose their research advisers at NIH and begin their dissertation research. At the end of the second year, students take examinations that qualify them to continue toward a Ph.D. Dissertation research is done in the second through fifth years within the NIH laboratories with frequent visits to JHU.

In addition to formal course work, the students actively participate in a weekly seminar series with faculty members and invited speakers and personally conduct their own journal club. While at NIH, students present their research in a weekly research seminar series and are invited weekly to participate in a luncheon for nationally renowned speakers invited to the NIH Lecture Series.

Research Facilities

The NIH is the world's premier biomedical research location. As the federal government's primary agency for the support and conduct of biomedical research, the NIH's many institutes and centers employ nearly 1,200 tenured or tenure-track investigators and as many as 3,700 postdoctoral scientists with either medical, dental, or graduate degrees. The NIH intramural research program, located on a 300-acre campus in Bethesda, Maryland, 10 miles from downtown Washington, D.C., has been the scene of many exciting scientific advances. It includes more than thirty biomedical research buildings that house a broad spectrum of biomedical and related scientific research. Four Nobel Laureates made their prize-winning discoveries in NIH laboratories and more than 100 received training at NIH. Basic research in the biomedical sciences at the NIH is complemented by an active clinical research program at the unique 250-bed research hospital and laboratory complex, the Warren Grant Magnuson Clinical Center. The NIH campus is also home to the National Library of Medicine, the world's largest medical library.

The laboratories and teaching space of the Johns Hopkins Department of Biology are located in Seeley G. Mudd Hall. The Milton S. Eisenhower Library, which has a superb integrated science collection, is nearby. All equipment and instrumentation relevant to contemporary biology are available to graduate students through the facilities of the Homewood Campus, the Carnegie Institution, the School of Medicine, and the School of Hygiene and Public Health.

Financial Aid

All JHU/NIH graduate students are supported through NIH Intramural Research Training Awards and by Johns Hopkins and receive support for stipend, tuition, and medical insurance throughout their years of training. Stipends for first-year students are $24,800 (2006–07) and increase yearly based on the student's performance.

Cost of Study

Students' tuition is funded during their studies and dissertation research.

Living and Housing Costs

Apartments, houses, and rooms in private residences are available for rent near the NIH campus. For information on available housing, prospective students should visit the Web sites of local newspapers or contact the GPP Web site at http://gpp.nih.gov. Near JHU in the Baltimore area, rates for rooms and apartments vary from $500 to $750 per month. The Housing Office on campus assists students in finding rooms and apartments.

Student Group

Johns Hopkins has approximately 3,300 undergraduate and 1,300 graduate students. Up to 5 new students are admitted each year into the collaborative JHU/NIH program and are members of a larger class totaling nearly 20 students who enter the biology department each year. The biology department has about 120 graduate students and 50 postdoctoral fellows. While at the NIH, students join more than 400 other graduate students from more than 100 universities who are doing their research in NIH laboratories. The Graduate Partnerships Program office at the NIH sponsors graduate student services and activities similar to those at a university to ensure student success and create a strong graduate student community.

Location

The 300-acre campus of NIH in Bethesda, Maryland, is close to Washington, D.C., affording a spectacular cultural and community environment as well as pleasant outdoor activities all year round. Johns Hopkins University is in Baltimore, Maryland, just a short distance north of the NIH campus. Maryland is a small state bounded by the Atlantic Ocean and the Allegheny and Appalachian mountains, providing the opportunity for pleasurable outdoor activities. The cultural and academic environments of Baltimore and Washington also enhance opportunities for many recreational experiences.

The University and The National Institutes of Health

The concept of graduate study in America originated with the founding of Johns Hopkins in 1876. The objective of the University has been to enrich young men and women, not merely through acquired knowledge but also in the spirit of inquiry, through independent and creative research. The NIH has a long history of independent and creative training of scientists and physicians. Many NIH-trained scientists have received international recognition for their work. The NIH environment is rich in scientific exchange, and it provides opportunities for a broad biomedical research experience. Many graduate students have received their graduate training at the NIH through arrangements between the NIH and universities such as Johns Hopkins.

Applying

To apply, prospective students must be U.S. citizens or noncitizen nationals of the United States or must have been lawfully admitted for permanent residence (i.e., possess a valid Alien Registration Receipt Card or some other verification of such status). An entering class typically includes a variety of undergraduate majors in the biological, chemical, or physical sciences. Information on submitting an application to the JHU/NIH Graduate Partnerships Program in biology or joining an NIH laboratory for dissertation research can be found on the Web at http://gpp.nih.gov. Applications to Johns Hopkins University should be completed and received by the posted deadline.

Correspondence and Information

Graduate Partnerships Program
Building 2, Room 2E06
National Institutes of Health–DHHS
2 Center Drive
Bethesda, Maryland 20892-0234
Phone: 301-594-9605
Fax: 301-594-9606
E-mail: gpp@nih.gov
Web site: http://gpp.nih.gov

Dr. Susan Gottesman, Partnerships
 Director
Phone: 301-496-3524
E-mail: gottesms@mail.nih.gov

Partnerships Co-Directors:
Dr. Beverly Wendland
Dr. Mark Van Doren
Department of Biology
Seeley G. Mudd Hall
Johns Hopkins University
Baltimore, Maryland 21218
Phone: 410-516-0460 (Wendland)
 410-516-4717 (Van Doren)
E-mail: bwendland@jhu.edu
 vandoren@jhu.edu
Web site: http://www.bio.jhu.edu

Johns Hopkins University/National Institutes of Health

THE FACULTY AND THEIR RESEARCH

NIH Investigators

Currently, there are more than 500 NIH investigators who participate in or would like to lead the training of graduate students. Given the broad biological basis of the Johns Hopkins program, virtually all of these investigators are potential dissertation mentors. Prospective students should refer to the Web site at http://gpp.nih.gov/Researchers/ for details about participating NIH scientists. To assist students in identification of a dissertation adviser at NIH, Dr. Susan Gottesman, the NIH Partnerships Director, works with individual students and refers them to other leading NIH investigators to help match them with potential mentors for laboratory rotations.

JHU Professors

Karen Beemon, Professor and Chair. Retroviral RNA processing and transport; avian leukosis virus tumorigenesis.

Maurice J. Bessman, Professor. Metabolism of nucleotides and nucleic acids; proteomics.

Ludwig Brand, Professor. Conformation and activity of biological macromolecules with emphasis on applications of fluorescence techniques.

Victor Corces, Professor. Role of chromatin structure in the control of gene expression; study of insect retroviruses and their mechanisms of infectivity and insertional mutagenesis.

Kyle Cunningham, Professor. Calcium signaling and mechanisms of gene expression in yeast.

Michael Edidin, Professor. Membrane organization and dynamics; immunology.

Ernesto Freire, Professor. Structure-based thermodynamics of molecular recognition and function.

Samer Hattar, Assistant Professor. Light reception for nonimage detection: role of rods, cones, and the new photoreceptors (melanopsin-containing retinal ganglion cells).

Edward Hedgecock, Professor. Developmental genetics of the nervous system of *Caenorhabditis elegans.*

Blake Hill, Assistant Professor. Protein design, protein folding, and structure; NMR spectroscopy.

Andy Hoyt, Professor. Yeast chromosome segregation, with emphasis on mitotic motor proteins; cell-cycle regulation.

Ru Chih Huang, Professor. Gene regulation, molecular virology, retrotransposition, and chromosomal structure and function.

Rejji Kuruvilla, Assistant Professor. Control of neuronal development by target-derived neurotrophins.

Yuan Lee, Professor. Glycoproteins, glycolipids, and cell-surface substances.

Richard McCarty, Professor. Structure, mechanism, and regulation of the chloroplast ATP synthase; chloroplast metabolite transport.

Evangelos Moudrianakis, Professor. Assembly and dynamics of nucleoproteins and chromosomes; bacterial and chloroplast bioenergetics.

Peter Privalov, Professor. Energetics of protein structure and mechanism of protein folding and stabilization; microcalorimetry.

Saul Roseman, Professor. Carbohydrate metabolism and transport.

Joel Schildbach, Associate Professor. Structural biology of bacterial conjugation.

Robert Schleif, Professor. Protein-DNA interactions and regulation of gene activity.

Trina Schroer, Professor. Microtubule-based motor enzymes and their roles in membrane traffic.

Allen Shearn, Professor and Vice Chair. Normal and tumorous imaginal disk development studied in mutants of *Drosophila.*

Mark Van Doren, Assistant Professor. Primordial germ-cell development and migration.

Beverly Wendland, Associate Professor. Molecular mechanisms of endocytosis in yeast and mammalian cells.

Haiqing Zhao, Assistant Professor. Function and development of olfactory sensory neurons.

The Dale and Betty Bumpers Vaccine Research Center (VRC), one of more than thirty major research buildings on the NIH campus, was established to facilitate research in vaccine development. The VRC is dedicated to improving global human health through the rigorous pursuit of effective vaccines for human diseases.

The National Human Genome Research Institute (NHGRI) led the Human Genome Project for the National Institutes of Health, which culminated in the completion of the full human genome sequence in April 2003. Now NHGRI moves forward into the genomic era with research aimed at improving human health and fighting disease.

KENT STATE UNIVERSITY

Department of Biological Sciences

Programs of Study	Kent State University's Department of Biological Sciences offers graduate study leading to M.S. and Ph.D. degrees, as well as an M.A. program for teachers. Faculty research areas include aquatic ecology; evolutionary, population, and systematic biology; microbiology; reproductive physiology/endocrinology; cell and molecular biology; and neurobiology and behavior. Faculty members in aquatic ecology participate with others in geology, geography, and chemistry through the interdisciplinary Water Resources Research Institute. Students working with faculty members in the Department major in biological sciences, with a concentration in ecology or physiology, or in biomedical sciences, which offers several concentrations, including cell and molecular biology and neurobiology (http://www.kent.edu/biomedical).

A program of study is developed for each student based on his or her background and interests. The M.S. program, including the thesis, normally takes about two years to complete. Course work is emphasized in the first year, and closely supervised research and writing is emphasized in the second. In the Ph.D. program, courses selected by the student and a guidance committee are usually completed in three or four semesters and are followed by a candidacy examination. Emphasis is placed upon conducting significant, original research.

Graduate research deals with both applied and basic problems, and the relationship between student and adviser is a close one. In the doctoral program, particularly, the student becomes a junior colleague, which, together with the teaching experience provided by the University, improves the student's opportunities for securing a desirable position after graduation. Graduates of the program are employed in universities, industry, and government.

Research Facilities

The Department of Biological Sciences is housed in Cunningham Hall, which includes a state-of-the art research wing. This research wing is an $8.5-million, 40,000-square-foot annex of Cunningham Hall. The annex contains nineteen research and support labs, a tissue-culture facility, RIA laboratory, a chemical storage room, a chemical- and biohazardous-waste room, a seminar room, and a classroom. The annex features several shared laboratories, including an aquatic ecology support lab, a genomics lab, and a confocal microscope and imaging laboratory. Resources available to faculty members and students include comprehensive suites of equipment for genomics, proteomics, and imaging, including the 3-D immersive development facility and classroom.

In addition, the Department houses an extensive herbarium and the Herrick Conservatory, and several properties are available for research. The Department operates a unique on-campus experimental wetland, the Art and Margaret Herrick Memorial Aquatic Ecology Research Facility (AERF), which features ten independent wetland cells that are amenable to experimentation.

Financial Aid

Major awards available for the academic year include graduate assistantships, research assistantships, and teaching fellowships (with stipends for the 2004–05 academic year of $16,000 for doctoral students); University-awarded nonservice doctoral fellowships; and other awards sponsored by NIH, NSF, private foundations, and industry. All graduate appointees receive a full waiver of tuition and fees.

Cost of Study

Tuition and fees for full-time graduate students in 2004–05 were $3990 per semester. Out-of-state students were assessed an additional $3506 per semester.

Living and Housing Costs

Single rooms in dormitories designated for upperclass students and graduate students rent for $2015 to $2700 per semester. Apartments for both single and married students rent for $635 to $665 per month. Students may also find off-campus housing, which varies widely in quality and cost. Additional information about housing can be found online at http://www.res.kent.edu.

Student Group

Currently, there are about 23,000 students enrolled at the main campus of Kent State University, of whom approximately 5,000 are graduate students.

Location

Kent, a city of 30,000, is located 35 miles southeast of Cleveland and 12 miles east of Akron and has both the advantages of a semirural setting and the benefits of proximity to metropolitan areas, with their cultural attractions and academic centers. There are several theater groups at the University and in the community. Blossom Music Center, the summer home of the Cleveland Orchestra and site of Kent State's cooperative programs in art, music, and theater, is 15 miles away.

The University

Established in 1910, Kent State University is one of Ohio's largest state universities. The main campus consists of 820 acres of wooded hillsides, an airport, an eighteen-hole golf course, and approximately 105 buildings valued at more than $200 million. Bachelor's, master's, and doctoral degrees are offered in more than thirty subject areas, and the faculty numbers 800.

The Northeastern Ohio Universities College of Medicine, a consortium of Kent State, the University of Akron, and Youngstown State University, is 10 miles east of Kent. The faculty of the medical college offers opportunities for graduate study through programs given under the auspices of Kent State's School of Biomedical Sciences.

Applying

Students may be admitted to begin graduate work during any semester, but graduate assistantships are normally awarded in March and April for the following fall; therefore, application materials should be completed by January 1. Scores from the General Test of the Graduate Record Examinations (GRE) are required.

Correspondence and Information

Dr. Ferenc A. de Szalay
Coordinator of Graduate Studies
Department of Biological Sciences
Kent State University
P.O. Box 5190
Kent, Ohio 44242-0001
Phone: 330-672-2819
E-mail: bscigrad@kent.edu
Web site: http://www.kent.edu/biology/graduateprograms/index.cfm

Kent State University

THE FACULTY AND THEIR RESEARCH

Aquatic Ecology

Robert E. Carlson, Ph.D., Minnesota. Empirical and geospatial modeling; lake classification; nutrient and food web dynamics in lakes and urban sedimentation basins.

Ferenc A. de Szalay, Ph.D., Berkeley. Wetland ecology, with focus on trophic structure of macroinvertebrate communities; effects of flood-pulsing on invertebrates and plants in riparian wetlands; unionid mussel ecology.

Robert T. Heath, Ph.D., USC. Food web dynamics in the Great Lakes and coastal wetlands; phosphorus and carbon dynamics in aquatic ecosystems; planktonic microbial ecology; land-use impact on planktonic community development in Lake Erie; composition and significance of dissolved organic matter in controlling planktonic community production.

Mark W. Kershner, Ph.D., Ohio State. Community ecology of lakes and streams; ecological simulation modeling; predator-prey interactions; fish ecology; invading species.

Laura G. Leff, Ph.D., Georgia. Stream ecology; microbial and molecular ecology; bacteria and dissolved organic matter; biofilms; microbial biodiversity.

Evolutionary, Population, and Systematic Ecology

Todd A. Blackledge (Adjunct), Ph.D., Ohio State. Behavioral ecology of spiders; evolutionary ecology of spider webs; biomechanics of silk.

Andrea L. Case, Ph.D., Toronto. Evolutionary and ecological genetics of plant breeding systems; genetics of cytoplasmic male sterility; cytonuclear coevolution; evolution of gender and sexual dimorphism; plant-pollinator interactions.

Volodymyr Dvornyk, Ph.D., Moscow Pedagogical State. Evolution of circadian systems; population genetics; functional genomics; genetic epidemiology; genetics of complex traits.

W. Randolph Hoeh, Ph.D., Michigan. Bivalve systematics, freshwater mussel (Unionoida) evolution, and ecology; molecular systematics; mating system evolution; evolution of doubly uniparental inheritance of mitochondrial DNA.

Patrick Lorch (Adjunct), Ph.D., Toronto. Behavioral and evolutionary ecology of insect mating behavior, group formation and movement, and sexual selection.

Randall J. Mitchell (Adjunct), Ph.D., California, Riverside. Plant-pollinator interactions; evolutionary ecology; foraging behavior of nectar feeders; conservation biology; effects of population size and habitat fragmentation.

Peter H. Niewiarowski (Adjunct), Ph.D., Pennsylvania. Evolution of life history traits; population dynamics; physiological and biophysical ecology of reptiles and amphibians.

Helen Piontkivska, Ph.D., Penn State. Molecular evolution of genes and genomes; large-scale analyses of genomic sequences; adaptive evolution of host-parasite interactions.

Oscar Rocha, Ph.D., Pennsylvania. Plant ecology; plant reproductive ecology; ecological genetics and conservation biology.

Stephen C. Weeks (Adjunct), Ph.D., Rutgers. Evolution of mating systems, especially maintenance of sexual reproduction; aquatic ecology of freshwater crustaceans; behavioral ecology; evolution of life history traits.

Microbiology

Robert T. Heath, Ph.D., USC. Biochemical limnology; phosphorus dynamics in aquatic ecosystems; planktonic biochemistry and physiology ecology; land-use impact on planktonic community development in Lake Erie; composition and significance of dissolved organic matter in controlling planktonic community production; biochemical indicators of eutrophication; Great Lakes coastal wetlands.

Laura G. Leff, Ph.D., Georgia. Stream ecology; microbial and molecular ecology; bacteria and dissolved organic matter; biofilms; microbial biodiversity.

Helen Piontkivska, Ph.D., Penn State. Molecular evolution of genes and genomes; large-scale analyses of genomic sequences; adaptive evolution of host-parasite interactions.

Christopher J. Woolverton, Ph.D., West Virginia. Infectious disease biosensors; microbial informatics; antimicrobial drug/device discovery; in vitro and in vivo models of infectious disease.

Reproductive Physiology and Endocrinology

P. Bagavandoss, Ph.D., Michigan. Ovarian physiology; action of cannabinoids on ovarian surface epithelium and granulosa cells.

James L. Blank, Ph.D., Indiana. Male reproductive physiology; physiological ecology; neuroendocrinology; reproductive toxicology.

Brent C. Bruot, Ph.D., Oklahoma State. Cell physiology of gonads; endocrine and paracrine regulation of steroid hormone synthesis; immunoendocrine regulation of steroidogenesis.

Dean E. Dluzen, Ph.D., IIT. Neuroendocrinology; effects of gonadal steroid hormones upon neuroprotection and memory.

Gail C. Fraizer, Ph.D., Berkeley. Molecular and cellular biology; tumor-suppressor genes and cancer; mechanism of tumorigenesis; urological malignancies.

J. David Glass, Ph.D., Wesleyan. Molecular and cellular neuroscience; neural regulation of the circadian biological clock; neurochemical processes regulating behavioral and metabolic rhythms in mammals; cellular and behavioral plasticity of the adult brain.

Douglas W. Kline, Ph.D., California, Davis. Gamete physiology; sperm-egg interaction; in vitro fertilization of mammalian eggs; oocyte maturation and the regulation of egg activation.

Jennifer L. Marcinkiewicz, Ph.D., Illinois at Urbana-Champaign. Ovarian physiology; ovarian follicle development; oocyte–somatic cell interactions; effects of cytokines on ovarian function; effects of endocrine disrupters on female reproduction.

Eric M. Mintz, Ph.D., California, Santa Cruz. Behavioral neuroscience; neurochemical regulation of circadian rhythms; sex differences in circadian rhythms.

John R. D. Stalvey, Ph.D., USC. Molecular endocrinology; cell physiology of the gonads and adrenal gland; interaction between hormonal and genetic factors regulating steroid hormone synthesis; ovarian cancer.

Sean L. Veney, Ph.D., USC. Behavioral neuroendocrinology; sexual differentiation of the brain; avian syrinx development and anatomy; neuroethology of birdsong.

Srinivasan Vijayaraghavan, Ph.D., New Delhi. Sperm physiology and biochemistry; regulation of sperm motility and fertility; signal transduction; molecular mechanisms of protein kinase and protein phosphatase action.

Cell and Molecular Biology

P. Bagavandoss, Ph.D., Michigan. Ovarian physiology; action of cannabinoids on ovarian surface epithelium and granulosa cells.

Brent C. Bruot, Ph.D., Oklahoma State. Cell physiology of gonads; endocrine and paracrine regulation of steroid hormone synthesis; immunoendocrine regulation of steroidogenesis.

Andrea L. Case, Ph.D., Toronto. Evolutionary and ecological genetics of plant breeding systems; genetics of cytoplasmic male sterility; cytonuclear coevolution; evolution of gender and sexual dimorphism; plant-pollinator interactions.

Robert V. Dorman, Ph.D., Ohio State. Biochemistry of vascular diseases; brain trauma and multiple sclerosis.

Gail C. Fraizer, Ph.D., Berkeley. Molecular and cellular biology; tumor-suppressor genes and cancer; mechanism of tumorigenesis; urological malignancies.

Robert T. Heath, Ph.D., USC. Biochemical limnology; phosphorus dynamics in aquatic ecosystems; planktonic biochemistry and physiology ecology; land-use impact on planktonic community development in Lake Erie; composition and significance of dissolved organic matter in controlling planktonic community production; biochemical indicators of eutrophication; Great Lakes coastal wetlands.

W. Randolph Hoeh, Ph.D., Michigan. Bivalve systematics, freshwater mussel (Unionoida) evolution, and ecology; molecular systematics; mating system evolution; evolution of doubly uniparental inheritance of mitochondrial DNA.

Douglas W. Kline, Ph.D., California, Davis. Gamete physiology; sperm-egg interaction; in vitro fertilization of mammalian eggs; oocyte maturation and the regulation of egg activation.

Laura G. Leff, Ph.D., Georgia. Stream ecology; microbial and molecular ecology; bacteria and dissolved organic matter; biofilms; microbial biodiversity.

Jennifer L. Marcinkiewicz, Ph.D., Illinois at Urbana-Champaign. Ovarian physiology; ovarian follicle development; oocyte–somatic cell interactions; effects of cytokines on ovarian function; effects of endocrine disrupters on female reproduction.

Eric M. Mintz, Ph.D., California, Santa Cruz. Behavioral neuroscience; neurochemical regulation of circadian rhythms; sex differences in circadian rhythms.

Helen Piontkivska, Ph.D., Penn State. Molecular evolution of genes and genomes; large-scale analyses of genomic sequences; adaptive evolution of host-parasite interactions.

Oscar Rocha, Ph.D., Pennsylvania. Plant ecology; plant reproductive ecology; ecological genetics and conservation biology.

Ronald L. Salisbury (Adjunct), Ph.D., Medical College of Virginia. Steroid hormone action in the central nervous system; brain sexual differentiation; neuroendocrinology of gonadotropin secretion; sex steroids and hypertension.

John R. D. Stalvey, Ph.D., USC. Molecular endocrinology; cell physiology of the gonads and adrenal gland; interaction between hormonal and genetic factors regulating steroid hormone synthesis; ovarian cancer.

Sean L. Veney, Ph.D., Virginia. Behavioral neuroendocrinology; sexual differentiation of the brain; avian syrinx development and anatomy; neuroethology of birdsong.

Srinivasan Vijayaraghavan, Ph.D., New Delhi. Sperm physiology and biochemistry; regulation of sperm motility and fertility; signal transduction; molecular mechanisms of protein kinase and protein phosphatase action.

Christopher J. Woolverton, Ph.D., West Virginia. Infectious disease biosensors; microbial informatics; antimicrobial drug/device discovery; in vitro and in vivo models of infectious disease.

Neurobiology and Behavior

Todd A. Blackledge (Adjunct), Ph.D., Ohio State. Behavioral ecology of spiders; evolutionary ecology of spider webs; biomechanics of silk.

James L. Blank, Ph.D., Indiana. Male reproductive physiology; physiological ecology; neuroendocrinology; reproductive toxicology.

Robert V. Dorman, Ph.D., Ohio State. Biochemistry of vascular diseases; brain trauma and multiple sclerosis.

J. David Glass, Ph.D., Wesleyan. Molecular and cellular neuroscience; neural regulation of the circadian biological clock; neurochemical processes regulating behavioral and metabolic rhythms in mammals; cellular and behavioral plasticity of the adult brain.

Eric M. Mintz, Ph.D., California, Santa Cruz. Behavioral neuroscience; neurochemical regulation of circadian rhythms; sex differences in circadian rhythms.

Ronald L. Salisbury (Adjunct), Ph.D., Medical College of Virginia. Steroid hormone action in the central nervous system; brain sexual differentiation; neuroendocrinology of gonadotropin secretion; sex steroids and hypertension.

Sean L. Veney, Ph.D., Virginia. Behavioral neuroendocrinology; sexual differentiation of the brain; avian syrinx development and anatomy; neuroethology of birdsong.

KENT STATE UNIVERSITY

School of Biomedical Sciences

Program of Study	The School of Biomedical Sciences at Kent State University offers an interdisciplinary approach to graduate education in biomedical research areas. The program involves more than 170 nationally recognized faculty members associated with three universities, a medical school, and various clinical institutions. This is the largest interinstitutional graduate program in the state and, as such, provides exceptional opportunities for advanced graduate education. Graduate degrees are offered in five program areas: biological anthropology, cell and molecular biology, neuroscience, pharmacology, and physiology. All students follow a core curriculum that includes cell biology, biochemistry, and statistical analyses. Course work within each program is both focused and flexible, allowing students to gain in-depth understanding of their chosen field.
Research Facilities	The research facilities available to graduate students are modern and extensive due to the involvement of multiple departments and institutions in the School. These include fully accredited animal research facilities; clinical facilities in local hospitals; superb computational facilities; extensive library holdings, including research journals and reference materials; and the most modern research equipment and space available.
Financial Aid	Financial support during the academic year may be available in the form of graduate assistantships, research assistantships, or teaching fellowships. Beginning in academic year 2006–07, the annual stipend award for precandidate doctoral students is $18,000; for postcandidate doctoral students, the award is $19,000. In addition, University-sponsored nonservice fellowships are available, as are awards from government agencies, private foundations, and industry. All recipients of graduate assistantships receive a full waiver of tuition and fees.
Cost of Study	Tuition and fees for a full-time graduate student for 2006–07 is $4350 per semester, with an additional $4110 assessed for out-of-state students.
Living and Housing Costs	The cost of living is quite reasonable in the Kent area, and housing facilities are available on and off the campus. Single graduate students may reside in on-campus dormitories, whereas students and families may choose from among many available apartments. Information on housing can be found at http://www.res.kent.edu.
Student Group	Kent State is the second-largest university in Ohio and enrolls about 35,000 students, including approximately 5,000 graduate students. The Graduate Student Senate is active on campus, as are organizations for international students.
Location	Kent is a city of about 30,000 located on the Cuyahoga River 33 miles from Cleveland and 11 miles from Akron. The main campus is situated in a peaceful suburban setting, and artistic, athletic, social, and recreational activities abound on or near the campus. In addition, the nearby cities offer all the advantages of a major metropolis, including museums, theaters, and professional sports teams. The University also has links with nearby Blossom Music Center, the summer home of the renowned Cleveland Orchestra. In addition, the rolling hills, rivers, and spring-fed lakes make Kent a popular site for numerous outdoor activities in all seasons.
The University and The School	Kent State University was established in 1910, and the main campus covers 820 wooded acres and includes about 150 buildings. The University also manages a golf course, an airport, and the newly constructed Wellness Center, which is free to registered students. Bachelor's, master's, and doctoral degrees are offered in more than thirty subject areas, and the faculty numbers 800. More information on the University may be obtained at http://www.kent.edu. The School of Biomedical Sciences was founded in 1987 to enhance graduate education in the biomedical sciences. Although the graduate program is housed solely in the School, faculty members from several universities and clinical institutions participate in this educational program. The Northeastern Ohio Universities College of Medicine participates in a medical education consortium with Kent State University, the University of Akron, and Youngstown State University. The basic science campus of this medical school is located near the Kent campus, and its clinical facilities include eighteen local teaching hospitals. Other participating institutions are the Cleveland Clinic, the Oak Clinic, the University of Akron, Youngstown State University, Akron General Medical Center, NASA, and Summa Health System. This diversity offers a truly unique educational opportunity for the graduate students, and 95 percent finish with the Ph.D. degree.
Applying	Students may apply at any time and be admitted during any semester; however, financial aid is normally distributed on an academic-year-basis beginning in the fall semester. Therefore, application materials completed by January 1 provide the greatest potential for funding for the following fall. A report from the General Test of the Graduate Record Examinations is required. Applications can be completed online at http://www.kent.edu/RAGS.
Correspondence and Information	Dr. Robert V. Dorman, Director School of Biomedical Sciences Kent State University Kent, Ohio 44242 Phone: 330-672-2263 E-mail: rdorman@kent.edu Web site: http://www.kent.edu/biomedical

Kent State University

THE FACULTY AND THEIR RESEARCH

Biological Anthropology
J. L. Blank, Ph.D. Primate reproduction.
C. O. Lovejoy, Ph.D. Human evolution.
R. S. Meindl, Ph.D. Demographic anthropology.
M. A. Norconk, Ph.D. Primate ecology in neotropics.
J. G. M. Thewissen, Ph.D. Mammalian evolution; whales.
M. C. Verstraete, Ph.D. Human motion; gait.
C. J. Vinyard, Ph.D. Evolution of primate head.
S. C. Ward, Ph.D. Miocene hominoids; dental evolution.

Cellular and Molecular Biology
C. F. Ansevin, M.D. Neuromuscular disorders.
D. Asch, Ph.D. Gene control and transcription.
N. E. Brasch, Ph.D. Bioinorganic/medicinal chemistry.
B. C. Bruot, Ph.D. Reproductive endocrinology.
J. Y.-L. Chiang, Ph.D. Gene expression; P450.
M. J. Cismowski, Ph.D. GPCR signal integration.
C. R. Cooper, Ph.D. Molecular fungal pathogenesis.
B. Datta, Ph.D. Regulation of protein synthesis.
A. DeLucia, Ph.D. DNA tumor viruses.
K. Doane, Ph.D. Receptor matrix interactions.
J. J. Docherty, Ph.D. Novel antiviral agents.
R. V. Dorman, Ph.D. Neurodegenerative diseases.
G. C. Fraizer, Ph.D. Mechanisms of tumorigenesis.
E. J. Freeman, Ph.D. Neurodegenerative diseases.
J. A. Fulton, Ph.D. Wound healing; tissue engineering.
A. Gericke, Dr.rer.nat. Lipid-mediated protein functions.
J. D. Glass, Ph.D. Regulation of the biological clock.
R. B. Gregory, Ph.D. Proteomics methods development.
J. P. Hardwick, Ph.D. Molecular biology; P-450.
R. T. Heath, Ph.D. Biodiversity; bacterial assemblages.
W. Horton, Ph.D. Cartilage biology; gene expression.
D. King, Ph.D. Skeletal biology; gene therapy.
D. W. Kline, Ph.D. Gamete physiology.
P. H. Koo, Ph.D. Neurodegeneration; a-2-macroglobulin.
W. J. Landis, Ph.D. Vertebrate mineral formation.
L. G. Leff, Ph.D. Molecular environmental microbiology.
H. Lorimer, Ph.D. Mitochondrial DNA replication in yeast.
W. P. Lynch, Ph.D. Mechanisms of neurodegeneration.
H. Mao, Ph.D. Molecular biophysics.
J. L. Marcinkiewicz, Ph.D. Ovarian physiology.
J. McDonough, Ph.D. Neurodegenerative diseases.
A. Milsted, Ph.D. Genetics of hypertension.
E. M. Mintz, Ph.D. Behavioral neurobiology.
R. E. Papka, Ph.D. Neurochemical anatomy.
H. Piontkivska, Ph.D. Comparative and evolutionary genomics.
K. S. Rosenthal, Ph.D. HSV; pathobiology.
M. A. Russell, Ph.D. Intermediate filament protein synemin.
L. M. Siperko, Ph.D. Tissue mineralization.
J. R. D. Stalvey, Ph.D. Molecular endocrinology.
D. Stroup, Ph.D. Eukaryotic gene expression.
C.-C. Tsai, Ph.D. Drug–nucleic acid interactions.
S. Veney, Ph.D. Behavioral neuroendocrinology.
S. Vijayaraghavan, Ph.D. Sperm physiology.
G. R. Walker, Ph.D. Muscular disease; immunity.
J. M. Walro, Ph.D. Neuromuscular biology.
F. G. Walz, Ph.D. Enzymology; mechanisms.
C. J. Woolverton, Ph.D. Infection; biosensors; antibiotics.
J. Yun, Ph.D. Signaling via adrenergic receptors.

Cellular and Molecular Biology/Cleveland Clinic Foundation
J. C. Adams, Ph.D. Cell–extracellular matrix adhesion.
A. Almasan, Ph.D. Apoptosis; cell-cycle control.
S. Apte, Ph.D. Biological role of ADAMTS proteases.
G. Casey, Ph.D. Genetic analysis of human malignancies.
G. M. Chisolm, Ph.D. Lipoproteins and cell function.
J. W. Crabb, Ph.D. Proteomics, biochemistry of vision.
D. Damron, Ph.D. Anesthesia; diabetic cardiomyopathy.
K. T. Dayie, Ph.D. Regulators of gene expression.
P. E. DiCorleto, Ph.D. Endothelial cell gene expression.
J. A. DiDonato, Ph.D. Cytokine signal transduction.
S. C. Erzurum, M.D. Airway biology and inflammation.
P. Fox, Ph.D. Translational control of gene expression.
L. M. Graham, M.D. Prosthetic vascular grafts.
A. V. Gudkov, D.Sci., Ph.D. p53, interferons; TNF.
S. J. Haque, Ph.D. Cytokine-mediated cell signaling.
J. G. Hollyfield, Ph.D. Retinal degenerative disease.
J. L. Hoover-Plow, Ph.D. Pathogenesis-lipoprotein(a).
P. H. Howe, Ph.D. Transforming growth factor beta.
D. W. Jacobsen, Ph.D. Homocysteine; vitamin B12.
D. Janigro, Ph.D. Blood-brain barrier; human disease.
G. K. Koski, Ph.D. Cancer immunobiology.
A. C. Larner, M.D., Ph.D. Actions of interferons.
X. Li, Ph.D. Signaling in immunity.
M. Luciano, Ph.D. Diagnosis and treatment of hydrocephalus.
W. B. Macklin, Ph.D. Regulation of brain development.
R. E. Morton, Ph.D. Lipid and lipoprotein metabolism.
R. A. Padgett, Ph.D. RNA splicing in vivo and in vitro.

P. E. Pellett, Ph.D. Herpes virus molecular biology.
M. Penn, Ph.D. Myocardial regeneration.
D. Perez, Ph.D. Function of adrenergic receptors.
G. Plautz, Ph.D. T-cell immunotherapy.
M. Perin, Ph.D. Molecular biology of the synapse.
J. Qin, Ph.D. Structural biology; biomolecular NMR.
M. E. Quinones-Mateu, Ph.D. HIV evolution; resistance.
R. Rackley, M.D. Urothelial cell biology; bladder cancer.
M. L. Ruehr, Ph.D. PKA signaling; cardiac myocytes.
R. H. Silverman, Ph.D. Interferon; RNase L; prostate.
N. Sizemore, Ph.D. Inflammation and cancer.
G. R. Stark, Ph.D. p53, interferons, NFkB.
D. J. Stuehr, Ph.D. Biochemistry of nitric oxide.
T. Subramanian, M.D. Experimental therapeutics.
B. D. Trapp, Ph.D. Myelination; demyelination.
D. R. Van Wagoner, Ph.D. Cardiovascular physiology.
Q. Wang, Ph.D. Cardiac genetics; development.
B. R. G. Williams, Ph.D. Innate immunity regulation.
A. Wolfman, Ph.D. Ras dependent signal transduction.
Y. Xu, Ph.D. Cancer development; lysolipids; receptors.
T. Yi, Ph.D. Hematopoiesis and leukemogenesis.

Neuroscience
C. F. Ansevin, M.D. Neuromuscular disorders.
J. L. Blank, Ph.D. Endocrine disruptors; toxicology.
D. E. Dluzen, Ph.D. Estrogen; neuroprotection.
R. V. Dorman, Ph.D. Neurodegenerative disorders; MS.
S. B. Fountain, Ph.D. Biopsychology; cognition.
E. J. Freeman, Ph.D. Neurodegenerative diseases; MS.
A. V. Galazyuk, Ph.D. Neuroscience of hearing.
D. P. Gans, Ph.D. Auditory neurophysiology.
J. D. Glass, Ph.D. Molecular and cellular neuroscience.
R. L. Joynes, Ph.D. Spinal cord injury; plasticity.
P. H. Koo, Ph.D. Neurodegeneration; a-2-macroglobulin.
Q. Liu, Ph.D. Vertebrate visual system; development.
J. A. Lovell, Ph.D. Chronic pain; aging; antioxidants.
W. P. Lynch, Ph.D. Mechanisms of neurodegeneration.
Y. Lu, Ph.D. Neurotransmitter systems.
J. McDonough, Ph.D. Neurodegeneration; mitochondrial genes; MS.
E. M. Mintz, Ph.D. Behavior; circadian rhythms.
R. E. Papka, Ph.D. Neurochemical anatomy.
D. C. Riccio, Ph.D. Memory processes; amnesia.
B. R. Schofield, Ph.D. Functional anatomy.
M. A. Simmons, Ph.D. Mechanisms of drug actions.
S. L. Veney, Ph.D. Behavioral neuroendocrinology.
G. R. Walker, Ph.D. Muscular disease; immunity.
J. M. Walro, Ph.D. Developmental neurobiology.
J. J. Wenstrup, Ph.D. Auditory neuroscience.

Pharmacology
D. D. Allen, Ph.D. Nicotine addiction.
S. Basu, Ph.D. Molecular basis of RNA structure and function.
J. Y.-L. Chiang, Ph.D. Gene expression; P450; cloning.
M. J. Cismowski, Ph.D. GPCR; cardiac hypertrophy.
D. E. Dluzen, Ph.D. Steroid hormones; neuroprotection.
J. G. Meszaros, Ph.D. Mechanisms of cardiac remodeling.
M. A. Simmons, Ph.D. Neuropharmacology.
J. Yun, Ph.D. GPCR signaling.

Physiology
P. Bagavandoss, Ph.D. Ovarian physiology.
J. L. Blank, Ph.D. Reproductive physiology; disruptors.
B. C. Bruot, Ph.D. Reproductive endocrinology.
D. E. Dluzen, Ph.D. Neuroendocrinology; steroids.
R. V. Dorman, Ph.D. Neurodegenerative diseases; MS.
D. E. Ely, Ph.D. Hypertension; cardiovascular physiology.
H. G. Folkesson, Ph.D. Lung development, RDS.
E. J. Freeman, Ph.D. Neurodegenerative diseases; MS.
J. D. Glass, Ph.D. Regulation of the biological clock.
E. L. Glickman, Ph.D. Metabolic responses to cold.
R. T. Heath, Ph.D. Biodiversity; bacterial assemblages.
M. I. Kalinski, Ph.D. Biochemical adaptation to exercise.
D. W. Kline, Ph.D. Gamete physiology; oocytes.
R. L. Londraville, Ph.D. Physiology and molecular biology of fish.
J. L. Marcinkiewicz, Ph.D. Ovarian physiology.
M. B. Maron, Ph.D. Pulmonary edema; alveoli.
J. G. Meszaros, Ph.D. Mechanisms of cardiac remodeling.
A. Milsted, Ph.D. Genetics of hypertension.
E. M. Mintz, Ph.D. Behavioral neuroscience; circadian rhythms.
G. D. Niehaus, Ph.D. Detection of microbes; pathology.
C. F. Pilati, Ph.D. Cardiac contractility; autonomic.
R. L. Salisbury, Ph.D. Steroid hormones; CNS.
S. P. Schmidt, Ph.D. Vascular grafts; wound healing.
J. R. D. Stalvey, Ph.D. Molecular endocrinology.
S. Vijayaraghavan, Ph.D. Sperm physiology.
J. M. Walro, Ph.D. Neural regulation of cardiac function.
C. J. Woolverton, Ph.D. Infection; biosensors; antibiotics.

LOUISIANA STATE UNIVERSITY

Department of Biological Sciences

Programs of Study

The Department of Biological Sciences at Louisiana State University offers programs leading to the M.S. and Ph.D. degrees in biochemistry and biological sciences.

Research Facilities

Research areas emphasized in the department are biochemistry and molecular biology; cell, developmental, and integrative biology; and systematics, ecology, and evolutionary biology. The facilities in the department include well-equipped modern laboratories as well as NMR analysis, mass spectrometry (bio-ion plasma desorption time-of-flight mass spectrometer), automated DNA sequencers, and confocal and electron microscopy. Research support is also provided by the Macromolecular Computing Analysis Facility. Software for molecular graphics and modeling, as well as the GCG programs, is available. The department also has access to the Internet via the campuswide fiber-optic network.

The department has shared equipment, including ultracentrifuges, scintillation counters, ultralow freezers, tissue culture facilities, biohazard hoods, autoclaves, and darkrooms. The department also has access to the research collections of the Museum of Natural Science and the LSU Herbarium, the Plant Growth Facility, an Aquatic Facility, and Sea Grant. Several faculty members collaborate with faculty members from the Departments of Animal Science, Chemistry, Civil and Environmental Engineering, Chemical Engineering, Plant Pathology and Crop Physiology, and Oceanography and Coastal Science, as well as the Pennington Biomedical Research Center, the Institute for Environmental Studies, and the School of Veterinary Medicine. Louisiana's Universities Marine Consortium (LUMCON) provides support for marine/estuarine research and maintains two research vessels.

Financial Aid

Teaching and research assistantships are available. Stipends are provided on a twelve-month basis and are competitive. In 2005–06, stipends ranged from $15,250 to $25,000 for Ph.D. students. Fellowships in the amount of $25,000 per year plus tuition fees are available to outstanding candidates. Students are also encouraged to seek fellowships from granting agencies, such as the National Science Foundation and other private and public agencies.

Cost of Study

Full tuition waivers exist for graduate assistants on University support. Fees for graduate assistants on University support were approximately $720 per semester in 2005–06 for both state residents and nonresidents. Diploma and thesis-binding fees total $60 for the master's degree and $105 for the doctorate.

Living and Housing Costs

The 2005–06 room rates for University housing ranged from approximately $450 to $1000 per semester for each occupant. Unfurnished University apartments for married students rented for approximately $200 to $300 per month. Off-campus apartments rented for $350 to $550 per month (two bedrooms, unfurnished).

Student Group

The on-campus enrollment for the 2005 fall term was about 31,560; approximately 5,165 were graduate and professional students. Graduate students come from all over the United States and from more than forty other countries. The Department of Biological Sciences currently has 145 Ph.D. students, 7 M.S. students, and 21 postdoctoral fellows.

Location

Baton Rouge, with a population of more than 300,000, is the capital of Louisiana and is located on the Mississippi River. The semitropical climate permits outdoor activities the year round; hunting, fishing, and boating opportunities are excellent. South Louisiana has a rich cultural heritage and is internationally renowned for its music and cuisine. The Baton Rouge Symphony, Little Theatre, Fine Arts Center, the LSU Union, and the new LSU Museum of Art offer cultural experiences of high quality. Antebellum homes and other historic sites are nearby. LSU has a rich tradition in

The University

The 1,944-acre campus at Baton Rouge is the largest of the University's six major campuses. The University awarded the first advanced degree in 1869. LSU is one of only nine universities in the country designated as both a land-grant and sea-grant institution and one of about eighty universities rated as a Research I institution.

Applying

Inquiries concerning graduate assistantships should be addressed to the Department of Biological Sciences nine months before expected enrollment; later requests are considered if positions are available. All applicants are required to take the General Test of the Graduate Record Examinations. Applicants must have a bachelor's degree from an accredited institution. Applicants from other countries must submit scores on the Test of English as a Foreign Language (TOEFL). There is a $25 application fee. Louisiana State University adheres to the principle of equal educational and employment opportunity without regard to race, sex, color, creed, or national origin. This policy extends to all programs and activities supported by the University. An online application procedure is available at http://gradlsu.gs.lsu.edu.

Correspondence and Information

Associate Chair for Graduate Studies
Department of Biological Sciences
107 Life Sciences Building
Louisiana State University
Baton Rouge, Louisiana 70803
E-mail: gradoff@lsu.edu
Web site: http://www.biology.lsu.edu

Louisiana State University

THE GRADUATE FACULTY AND THEIR RESEARCH

Fareed Aboula-ela, Assistant Professor; Ph.D., California, Berkeley, 1988. NMR and RNA structure and small-molecule RNA interactions. (E-mail: faboul@lsu.edu)

Christopher Austin, Adjunct Assistant Professor; Ph.D., Texas at Austin. Herpetology. (E-mail: ccaustin@lsu.edu)

Sue G. Bartlett, Associate Professor; Ph.D., Duke, 1978. Chloroplast biogenesis, especially the expression and assembly of chloroplast proteins. (E-mail: sbartle@lsu.edu)

J. R. Battista, Professor; Ph.D., Wayne State, 1986. Molecular biology; genetic and biochemical characterization of mutagenesis and DNA repair in *E. coli* and *Deinococcus radiodurans.* (E-mail: jbattis@lsu.edu)

M. A. Batzer, Andrew C. Pereboom Alumni Departmental Professor; Ph.D., LSU, 1988. Comparative genomics; computational biology; molecular genetics; human evolution; forensics; architecture of the human genome. (E-mail: mbatzer@lsu.edu)

M. Blackwell, Boyd Professor; Ph.D., Texas at Austin, 1973. Mycology, fungal systematics, and evolution; fungal life histories; arthropod-dispersal of fungi. (E-mail: mblackwell@lsu.edu)

T. M. Bricker, Morehouse Family Professor; Ph.D., Miami (Florida), 1981. Physiology, biochemistry, and molecular biology of photosynthesis; protein-protein interactions of membrane proteins. (E-mail: btbric@lsu.edu)

Kenneth M. Brown, Professor; Ph.D., Iowa, 1976. Ecology of oysters and their predators; conservation biology of freshwater mussels; community and population ecology of freshwater snails. (E-mail: kmbrown@lsu.edu)

Richard C. Bruch, Associate Professor; Ph.D., Delaware, 1982. Signal transduction in peripheral olfactory neurons. (E-mail: rbruch@lsu.edu)

Robert Brumfield, Adjunct Assistant Professor; Ph.D., Maryland, College Park, 1999. Genetic resources and ornithology. (E-mail: brumfld@lsu.edu)

John T. Caprio, Kent Professor; Ph.D., Florida State, 1976. Olfaction and taste in fish. (E-mail: jcap@lsu.edu)

Kevin R. Carman, Professor; Ph.D., Florida State, 1989. Microbial ecology; benthic ecology; biological oceanography. (E-mail: zocarm@lsu.edu)

Stephania Cormier, Assistant Professor; Ph.D., LSU Medical Center, 1997. Immunopathology; lung cell and molecular biology; regulation of gene and protein expression; neonatal immunity. (E-mail: scormier@lsu.edu)

J. T. Cronin, Associate Professor; Ph.D., Florida State, 1991. Population and community ecology; plant-herbivore–natural enemy interactions; ecological genetics and biological pest management. (E-mail: jcronin@lsu.edu)

Michael J. Dagg, Adjunct Professor; Ph.D., Washington (Seattle), 1975. LUMCON, Cocodrie, Louisiana. Biological oceanography. (E-mail: mdagg@lsu.edu)

Patrick DiMario, Associate Professor; Ph.D., Indiana, 1985. Molecular cytology of the nucleolus; proteins that interact with nascent ribosomal RNAs. (E-mail: pdimari@lsu.edu)

Huangen Ding, Associate Professor; Ph.D., Pennsylvania, 1995. Regulatory function and metabolism of iron sulfur proteins during oxidative stress. (E-mail: hding@lsu.edu)

William T. Doerrler, Assistant Professor; Ph.D., Texas Southwestern Medical Center at Dallas, 1999. Membrane biogenesis in *E. coli.* (E-mail: wdoerr@lsu.edu)

David Donze, Assistant Professor; Ph.D., Alabama at Birmingham, 1996. Chromatin structure and gene expression. (E-mail: ddonze@lsu.edu)

J. Michael Fitzsimons, Adjunct Professor; Ph.D., Michigan, 1970. Systematic icthyology. (E-mail: fitzsimons@lsu.edu)

John W. Fleeger, Kent Professor; Ph.D., South Carolina, 1977. Marine ecology. (E-mail: zoflee@lsu.edu)

David W. Foltz, Professor; Ph.D., Michigan, 1979. Population genetics. (E-mail: dfoltz@lsu.edu)

Fernando Galvez, Assistant Professor; Ph.D., McMaster, 2000. Integrative fish biology; environmental physiology; aquatic toxicology. (E-mail: galvezf@lsu.edu)

Evanna Gleason, Associate Professor; Ph.D., California, Davis, 1990. Neurobiology. (E-mail: egleaso@lsu.edu)

Anne Grove, Associate Professor; Ph.D., Copenhagen, 1990. Protein–nucleic acid interactions and sequence-dependent DNA bendability, analyzed in the context of the yeast RNA polymerase III. (E-mail: agrove@lsu.edu)

Mark S. Hafner, Professor; Ph.D., Berkeley, 1979. Molecular systematics and mammalogy. (E-mail: namark@lsu.edu)

Hollie Hale-Donze, Assistant Professor; Ph.D., Alabama at Birmingham, 1997. Infection and immunity. (E-mail: hhaled1@lsu.edu)

S. C. Hand, Thompson Distinguished Professor; Ph.D., Oregon State, 1980. Molecular and integrative physiology of animals; mechanisms of dormancy in invertebrates, adaption of animals to low oxygen environment. (E-mail: shand@lsu.edu)

Kyle Harms, Assistant Professor; Ph.D., Princeton, 1997. Population and community ecology; tropical ecosystem ecology; evolutionary ecology of plants and their interactions with other organisms. (E-mail: kharms@lsu.edu)

Craig M. Hart, Assistant Professor; Ph.D., Cornell, 1988. Chromosome organization; chromatin structure and gene expression. (E-mail: chart4@lsu.edu)

Michael Hellberg, Associate Professor; Ph.D., California, Davis, 1993. Population genetics and speciation of marine organisms. (E-mail: mhellbe@lsu.edu)

Dominique G. Homberger, Professor; Ph.D., Zurich (Switzerland), 1976. Functional and evolutionary morphology. (E-mail: zodhomb@lsu.edu)

Naohiro Kato, Assistant Professor; Ph.D., Hiroshima (Japan), 1998. In situ visualization of genomic function dynamics. (E-mail: kato@lsu.edu)

Joomyeong Kim, Associate Professor; Ph.D., LSU Medical Center, 1995. Mammalian genomic imprinting; genome evolution and function. (E-mail: jkim@lsu.edu)

Roger A. Laine, Professor; Ph.D., Rice, 1970. Biosynthesis and function of cell-surface complex carbohydrates, glycoproteins, and glycolipids. (E-mail: rlaine@usa.net)

John C. Larkin, Associate Professor; Ph.D., Carnegie-Mellon, 1985. Plant developmental genetics. (E-mail: jlarkin@lsu.edu)

Yong-Hwan Lee, Assistant Professor; Ph.D., SUNY at Stony Brook, 1995. Structure/function studies of the metabolic regulator proteins and their regulatory mechanisms. (E-mail: yhlee@lsu.edu)

Vincent LiCata, Associate Professor; Ph.D., John Hopkins, 1991. Energetic and structural relationships in proteins: role of water and other components in protein function and protein-DNA interactions. (E-mail: licata@lsu.edu)

David J. Longstreth, Associate Professor; Ph.D., Duke, 1976. Physiological ecology. (E-mail: btlong@lsu.edu)

John W. Lynn, Kent Professor; Ph.D., California, Davis, 1981. Reproductive physiology. (E-mail: zolynn@lsu.edu)

Thomas S. Moore Jr., Professor; Ph.D., Indiana, 1970. Plant lipid biochemistry. (E-mail: btmoor@lsu.edu)

James V. Moroney, Streva Memorial LSU Alumni Association Professorship; Ph.D., Cornell, 1982. Plant cell biology. (E-mail: btmoro@lsu.edu)

Marcia Newcomer, Professor; Ph.D., Rice, 1979. Protein crystallography. (E-mail: newcomer@lsu.edu)

G. S. Pettis, Associate Professor; Ph.D., Missouri–Columbia, 1989. Molecular biology; genetic and biochemical analysis of conjugation in *Streptomyces.* (E-mail: gpettis@lsu.edu)

William J. Platt, Professor; Ph.D., Cornell. 1972. Plant population biology/ecology. (E-mail: btplat@lsu.edu)

Kirsten Prufer, Assistant Professor; Ph.D., Friedrich-Schiller University Jena (Germany), 1995. Cell biology of nuclear receptors; trafficking; nuclear import and export; molecular endocrinology; biochemistry; molecular biology. (E-mail: kprufer@lsu.edu)

F. A. Rainey, Associate Professor; D.Phil., Waikato (New Zealand), 1992. Molecular systematics and ecology; determination of prokaryotic diversity using culturing and nonculturing methods; enviratoxonomy. (E-mail: frainey@lsu.edu)

J. Van Remsen Jr., Adjunct Professor; Ph.D., Berkeley, 1978. Ornithology. (E-mail: najames@lsu.edu)

Frederick H. Sheldon, Adjunct Associate Professor; Ph.D., Yale, 1986. Molecular systematics and ornithology. (E-mail: fsheld@lsu.edu)

Ding S. Shih, Associate Professor; Ph.D., Virginia Tech, 1969. Viral gene expression and its regulation; application of biotechnology methods for agricultural and medical purposes. (E-mail: dshih@lsu.edu)

Joseph F. Siebenaller, Professor; Ph.D., California, San Diego (Scripps), 1978. Biochemical adaptation to the marine environment. (E-mail: zojose@lsu.edu)

Harold Silverman, Professor; Ph.D., Ohio, 1977. Freshwater mussel anatomy and physiology; muscle structure and function. (E-mail: cxsilv@lsu.edu)

Jacqueline Stephens, Professor; Ph.D., East Carolina, 1992. Diabetes research; transcription factors; biochemistry; molecular biology; cell biology. (E-mail: jsteph1@lsu.edu)

Richard Stevens, Assistant Professor; Ph.D., Texas Tech, 2002. Community ecology; macroecology; biogeography. (E-mail: rstevens@lsu.edu)

William B. Stickle Jr., Professor; Ph.D., Saskatchewan, 1970. Marine environmental physiology. (E-mail: zostic@lsu.edu)

Kurt Svoboda, Assistant Professor; Ph.D., SUNY at Stony Brook, 1996. Neurobiology and behavior. (E-mail: ksvobo1@lsu.edu)

Lowell E. Urbatsch, Professor; Ph.D., Georgia, 1970. Vascular plant systematics. (E-mail: leu@lsu.edu)

Grover L. Waldrop, Associate Professor; Ph.D., SUNY at Buffalo, 1988. Kinetic and chemical mechanisms of enzymatic reactions; structure and function of biotin-dependent carboxylases. (E-mail: gwaldro@lsu.edu)

Andrew Whitehead, Assistant Professor; Ph.D., California, Davis, 2003. Environmental genomics; population genomics; stress biology; ecotoxicology. (E-mail: andreww@lsu.edu)

G. Bruce Williamson, Professor; Ph.D., Indiana, 1975. Ecology. (E-mail: btwill@lsu.edu)

E. William Wischusen, Associate Professor; Ph.D., Cornell, 1990. Vertebrate ecology. (E-mail: ewischu@lsu.edu)

Tin-Wein Yu, Assistant Professor; Ph.D., East Anglia, (England), 1994. Functional genetics of microbial metabolites and structural diversity of natural products. (E-mail: yu@lsu.edu)

Wayne Zhou, Associate Professor; Ph.D., Oregon State, 1991. Structure and function of protein tyrosine phosphatases; structure and function of the phox (PX) domain-containing proteins. (E-mail: zhouw@lsu.edu)

LOUISIANA STATE UNIVERSITY
HEALTH SCIENCES CENTER AT SHREVEPORT

School of Graduate Studies

Programs of Study	The School of Graduate Studies offers programs leading to the M.S. and Ph.D. degrees in biochemistry, cellular biology, microbiology, pharmacology, and physiology. A combined M.D./Ph.D. degree is also offered to exceptional students enrolled in the School of Medicine. The Department of Biochemistry and Molecular Biology emphasizes research training in molecular biology, molecular genetics, protein biochemistry, protein synthesis, nucleic acid biochemistry, and membrane biochemistry. A separate track within the department provides training in biotechnology. The Department of Cellular Biology and Anatomy offers courses and research training in the modern disciplines of cancer biology, including cellular immunology and neurobiology. The Department of Microbiology and Immunology offers a program of courses and laboratory research training in cell and molecular biology, immunology, virology, bacteriology, and molecular mechanisms of infectious diseases. The Department of Pharmacology, Toxicology and Neuroscience emphasizes broad training and research in pharmacology, with specialization in neuropharmacology, behavioral pharmacology, neuroimmunology, and toxicology. The Department of Molecular and Cellular Physiology offers training in cardiovascular, gastrointestinal, renal, and cell physiology.
Research Facilities	The library that serves the Health Sciences Center occupies 39,000 square feet over three floors. This includes small group teaching rooms, teleconferencing facilities, and two state-of-the-art computer labs. The library houses a collection of more than 190,000 print volumes and offers more than 2,357 journals in either print or electronic format. It also provides access to sixty-nine databases and 291 electronic books. Reference services include interlibrary loans and a computerized bibliographic service, utilizing the National Library of Medicine's MEDLARS database. The Academic Computing Center provides use of mainframe, minicomputer, and microcomputer facilities. The research facilities are located in the Medical School and adjacent Biomedical Research Institute. The Department of Biochemistry and Molecular Biology has more than 14,000 square feet of space, with laboratories equipped for state-of-the-art research in molecular biology and biochemistry. Equipment includes a phosphorimager, preparative and analytical ultracentrifuges, DNA sequencing equipment, several HPLCs, computers, spectrophotometers, a spectrofluorometer, and scintillation counters. The Department of Cellular Biology and Anatomy occupies more than 20,000 square feet of modern work space. Complete facilities are available for performing a wide range of biological techniques, including transmission electron microscopy; bright-field, dark-field, and fluorescence imaging; autoradiography; tissue culture; histochemistry; immunocytochemistry; chromatography; HPLC; electrophoresis; and neurophysiology. The Department of Microbiology and Immunology has more than 16,000 square feet of research and teaching space made up of laboratories, common equipment rooms (housing centrifuges, freezers, scintillation counters, etc.), and special-function suites. Among these are facilities for HPLC, fluorescence microscopy, monoclonal antibody production, cell culture and virus cultivation, recombinant DNA and radiation work, a germfree mouse colony, flow cytometry, and a darkroom complex. The Department of Pharmacology, Toxicology and Neuroscience has 12,000 square feet of space and equipment for biochemical, behavioral, molecular, and physiological studies. Programs emphasize neurochemical, behavioral, and neuroendocrine approaches to neuropharmacology. Specialized facilities include laboratories for operant and other behaviors, brain imaging, cell culture, and neurochemistry. The Department of Molecular and Cellular Physiology occupies 12,000 square feet of space that is fully equipped for research with cellular, organ, and whole animal models of the cardiovascular system, gastrointestinal tract, and kidneys. Complete facilities are available for intravital video microscopy, confocal microscopy and image analysis, cell culture, subcellular fractionation, antibody production, immunocytochemistry, protein and lipid analysis, chromatography, and computer data acquisition and analysis. The Research Core Facility (RCF), located in the Biomedical Research Institute, provides shared instrumentation for confocal microscopy, automated cellular imaging, fluorescence microscopy, DNA-array analysis, flow cytometry, laser capture microdissection, mass spectrometry, and real-time PCR. The Animal Imaging Facility (part of the RCF) provides microPET, microCT, and in vivo chemiluminescence and fluorescence imaging capabilities.
Financial Aid	Financial support in the form of stipends and waiver of tuition fees is available. These stipends are awarded on a competitive basis.
Cost of Study	Tuition is waived for all qualified, full-time Ph.D. students.
Living and Housing Costs	Furnished and unfurnished apartments and houses, offered at a wide range of rents, are available throughout the city. A listing is maintained by the Office of Student Affairs at the Louisiana State University Health Sciences Center at Shreveport.
Student Group	There are about 100 students enrolled in the graduate programs at the Louisiana State University Health Science Center at Shreveport.
Location	Shreveport is located in northwest Louisiana, where climate and geography combine to offer year-round recreational opportunities. The metropolitan area has a population of about 350,000. Shreveport is a major center for health and medical services. LSU-Shreveport, Centenary College, and branches of several universities; the R. W. Norton Art Gallery; the Shreveport Symphony and Opera Company; minor-league baseball (Shreveport Sports); ice hockey (Bossier-Shreveport Mudbugs); and numerous theater, dance, and music groups contribute to the area's educational and cultural advantages.
The School and The Center	In addition to the School of Graduate Studies, the Health Sciences Center at Shreveport also includes the Schools of Medicine and Allied Health Professions. The Health Sciences Center and the University Hospital have a student body of about 600 and a faculty of about 250.
Applying	Admission to the graduate school is gained through admission to a departmental program. Requirements include a baccalaureate degree from an accredited undergraduate institution with an acceptable GPA, an acceptable score on the GRE, and satisfactory standing at the school most recently attended. A TOEFL score of 550 or better is required of international students. Acceptable scores for the GRE and the required GPA are defined by individual departments. All departments encourage applicants to come for an interview whenever possible, and an interview may be a requirement for admission. The Dean for Graduate Studies (Sandra C. Roerig, Ph.D., 318-675-7877, sroeri@lsuhsc.edu) provides general information. The following representatives should be contacted for admission requirements and application procedures for specific departments: Biochemistry—Lucy Robinson, Ph.D. (318-675-5164, lrobin@lsuhsc.edu); Cellular Biology—Stephen Pruett, Ph.D. (318-675-4386, spruet@lsuhsc.edu); Microbiology—Dennis O'Callaghan, Ph.D. (318-675-5750, docall@lsuhsc.edu); Pharmacology—Heather Kleiner, Ph.D. (318-675-8559, hklein@lsuhsc.edu); and Physiology—Steven Alexander, Ph.D. (318-675-4151, jalexa@lsuhsc.edu).
Correspondence and Information	Office of Graduate Studies Louisiana State University Health Sciences Center at Shreveport P.O. Box 33932 Shreveport, Louisiana 71130-3932 Web site: http://www.sh.lsuhsc.edu/gradschool/

Louisiana State University Health Sciences Center at Shreveport

THE FACULTY AND THEIR RESEARCH

Biochemistry and Molecular Biology

Eric J. Aamodt, Ph.D., Associate Professor. Molecular developmental biology; establishment of gene-expression patterns during *C. elegans* development.

Jun Chung, Ph.D., Assistant Professor. Integrin-matrix regulation of oncogene expression.

John L. Clifford, Ph.D., Associate Professor. Molecular mechanism of skin carcinogenesis, retinoid and interferon signaling.

Nathan Davis, Ph.D., Associate Professor. Transcriptional regulation of oncogene expression.

Arrigo De Benedetti, Ph.D., Associate Professor. Gene expression and growth deregulation in mammalian cells.

Eric A. First, Ph.D., Associate Professor. Catalytic mechanisms of tRNA synthetases.

Tony Giordano, Ph.D., Associate Professor. Small molecule regulation of gene expression; mRNA translational inhibition.

Sidney R. Grimes, Ph.D., Research Professor. Regulation of histone gene expression.

David S. Gross, Ph.D., Professor. Transcriptional response to stress; silenced chromatin; mechanism of p53 transactivation; yeast genetics.

Shile Huang, Ph.D., Assistant Professor. IGF-I/mTOR signaling in tumorigenesis and metastasis.

Sushil K. Jain, Ph.D., Professor. Red blood cells; free radicals and nutrition involvement in the disease process.

Nadejda Korneeva, Ph.D., Research Assistant Professor. Regulation of protein synthesis in eukaryotic cells; interaction between translational initiation factors and RNA; enzymatic activity of initiation factors.

Nancy J. Leidenheimer, Ph.D., Associate Professor. Molecular biology of GABA receptors.

Neal Mathias, Ph.D., Assistant Professor. Assembly, mechanism of action, and regulation of SCE ubiquitin ligases in yeast.

Shari Meyers, Ph.D., Associate Professor. Regulation of myeloid cell differentiation; molecular mechanism of transformation of hematopoietic cells.

Brent C. Reed, Ph.D., Associate Professor. Regulation of insulin receptor and glucose transporter expression and function.

Robert E. Rhoads, Ph.D., Professor and Head. Protein synthesis in eukaryotic cells; messenger RNA structure and function; regulation of cell growth; viral pathogenesis involving translation.

Lucy C. Robinson, Ph.D., Associate Professor. Protein kinase cascades and growth control; yeast genetics.

Sergey Slepenkov, Ph.D., Assistant Research Professor. Nucleic acid-protein interactions: enzymology of RNA and protein unfolding.

Robert L. Smith, Ph.D., Associate Professor. Studies of interactions among protein domains of fibronectin and collagen.

Kelly Tatchell, Ph.D., Professor. Regulation of protein phosphorylation; yeast genetics and molecular biology.

B. Jill Williams, Ph.D., Associate Professor. Mechanisms of cancer formation in the prostate.

Stephan N. Witt, Ph.D., Associate Professor. Kinetic and mechanistic studies of the 70-kDa family of heat shock proteins; site-specific mutagenesis; protein folding.

Cellular Biology and Anatomy

John A. Beal, Ph.D., Professor. Structure connectivity and development of sensory neurons of spinal cord.

Anping Chen, Ph.D., Assistant Professor. Fundamental mechanisms that govern the interaction of tumors with the immune system.

Kurt Doege, Ph.D., Associate Professor. Transcriptional mechanisms that regulate cartilage-specific gene expression.

Wei-Ming Duan, M.D., Ph.D., Assistant Professor. Molecular and cellular mechanisms responsible for neurodegenerative disorders, especially for Parkinson's disease, and gene therapy and cell replacement therapy with stem cells for this disease.

Jonathan Glass, M.D., Professor. Iron uptake in erythroid cells.

Kathryn Hamilton, Ph.D., Associate Professor. Structure and function of chemical sensory systems.

Pamela Hebert, Ph.D., Instructor. Immuno (toxico) logical effects of stress on natural killer cells and their ability to reject tumor metastasis.

Guillermo Herrera, M.D., Professor. Pathogenesis of light chain–mediated glomerular injury.

Daniel Keppler, Ph.D., Assistant Professor. The role of eystatin M in breast tumor progression.

David S. Knight, Ph.D., Associate Professor. Hypothalamus in stress and obesity.

Andrew A. Marino, Ph.D., Professor. Biocompatibility of implantable materials; bioelectricity.

Michael J. Mathis, Ph.D., Associate Professor. BRCA1 locus involvement in sporadic breast and ovarian cancer.

Kevin J. McCarthy, Ph.D., Professor. Cell/molecular biology of basement membranes in the development and pathology of kidney disease.

Joe E. Penny, Ph.D., Associate Professor. Inner-ear structure, function, and response to ototoxic drugs, noise, and infection.

Stephen B. Pruett, Ph.D., Professor. Neuroendocrine mediators and immunomodulation by ethanol and other chemicals.

Leonard L. Seelig Jr., Ph.D., Professor and Head. Maternal-fetal immunobiology.

Omar Skalli, Ph.D., Associate Professor. Glial filament associated proteins in human astrocytomas.

William M. Steven, Ph.D., Associate Professor. Alcohol and maternal/neonatal immune responses.

Qian-Jin Zhang, Ph.D., Assistant Professor. Fundamental mechanisms that govern the interaction of tumors with the immune system.

Li-Ru Zhao, M.D., Ph.D., Assistant Professor. Molecular and cellular mechanisms underlying stroke; neurogenesis from stem cells in the brain after brain ischemia.

Microbiology and Immunology

James A. Cardelli, Ph.D., Professor. Phagocytosis of helicobacter and brucella; breast cancer metastasis.

Robert P. Chervenak, Ph.D., Professor. T-lymphocyte development.

Lindsey M. Hutt-Fletcher, Ph.D., Professor. Role of Epstein-Barr virus glycoproteins in replication and pathogenesis.

Stephanie Karst, Ph.D., Assistant Professor. Pathogenesis of noroviruses.

William B. Klimstra, Ph.D., Assistant Professor. Molecular pathogenesis of arbovirus infection.

David McGee, Ph.D., Associate Professor. *Helicobacter pylori* host-pathogen interactions.

Martin I. Muggeridge, Ph.D., Associate Professor. Molecular biology of herpesvirus glycoproteins and membrane fusion.

Dennis J. O'Callaghan, Ph.D., Boyd Professor and Head. Molecular biology of herpesvirus replication and viral genes essential for pathogenesis and virulence.

Kenneth M. Peterson, Ph.D., Associate Professor. Molecular pathogenesis and intestinal colonization by *Vibrio cholerae* and *cholerae* vaccine development.

Kate D. Ryman, Ph.D., Assistant Professor. Host response to arbovirus infection.

Martin Sapp, Ph.D., Associate Professor. Virus cell interactions, pathogenesis of human papillomaviruses.

Rona S. Scott, Ph.D., Instructor. Epstein-Barr virus and its role in lymphomagenesis.

Daniel W. Shelver, Ph.D., Assistant Professor. Molecular pathogenesis of group B *Streptococcus*.

John W. Sixbey, M.D., Professor. EBV persistence; pathobiology of EBV replication in cancer.

Patrick M. Smith, Ph.D., Instructor. Role of herpesvirus glycoproteins in severe inflammatory disease.

John Staczek, Ph.D., Professor. Vectors for *Pseudomonas* vaccines.

Traci Testerman, Ph.D., Assistant Professor. *Helicobacter pylori* physiology and pathogenesis.

Scott Tibbets, Ph.D., Assistant Professor. Pathogenesis of oncogenic herpesviruses.

R. Michael Wolcott, Ph.D., Professor. B-lymphocyte development.

Andrew D. Yurochko, Ph.D., Associate Professor. Cell activation by human cytomegalovirus.

Molecular and Cellular Physiology

J. Steven Alexander, Ph.D., Associate Professor. Endothelial biology; leukocyte-endothelial interactions; inflammatory bowel disease; junctional proteins, vascular inflammation.

Tak Yee Aw, Ph.D., Professor. Oxidative stress, antioxidants, and cell death; redox regulation of cell turnover, apoptosis, and proliferation; ischemia-reperfusion injury; hypercholesterolemia.

Shayne Barlow, D.V.M., Ph.D., Assistant Professor. Mouse genetics and comparative pathology.

D. Neil Granger, Ph.D., Boyd Professor and Head. Inflammation; leukocyte-endothelial cell adhesion; ischemia-reperfusion injury; hypercholesterolemia.

Matthew B. Grisham, Ph.D., Professor. Inflammation, free radical biology, gene regulation.

Norman R. Harris, Ph.D., Associate Professor. Physiology of microvascular transport; alterations of microcirculation during inflammation and cardiovascular disease.

Lynn Harrison, Ph.D., Associate Professor. DNA repair mechanisms and molecular physiology.

Robert D. Specian, Ph.D., Professor. Intestinal goblet cell function; mucosal injury and repair; epithelial differentiation in colitis.

Tomas C. Welbourne, Ph.D., Professor. Regulation mechanism in glutamine/glutamate/acid base metabolism.

Pharmacology, Toxicology and Neuroscience

Tammy R. Dugas, Ph.D., Assistant Professor. Vascular toxicology; gender-specific differences in toxicity; peroxidase-mediated metabolism of xenobiotics.

Adrian J. Dunn, Ph.D., Boyd Professor. Behavioral neurochemistry; role of CRF and norepinephrine in stress; cytokine effects on the brain.

Donard S. Dwyer, Ph.D., Associate Professor. Neuroimmunology; mechanisms of nuclear translocation.

Nicholas E. Goeders, Ph.D., Professor and Head. Neurobiology of reinforcement; behavioral neuroscience; autoradiography.

Anita Kablinger, M.D., Associate Professor. Psychopharmacology; clinical trials; gender issues; prodromal schizophrenia; bipolar disorder.

Ronald L. Klein, Ph.D., Assistant Professor. Viral vector transfer for gene therapy of neurodegenerative disorders.

Heather E. Kleiner, Ph.D., Assistant Professor. Chemoprevention of cancer; chemical carcinogenesis; xenobiotic biotransformation.

Kenneth E. McMartin, Ph.D., Professor. Toxicology of alcohols; biochemical toxicology; regulation of folate metabolism.

James C. Patterson, M.D., Ph.D., Assistant Professor. Cognitive neuroscience; functional neuroimaging (PET, fMRI); schizophrenia; Alzheimer's dementia.

Sandra C. Roerig, Ph.D., Professor. Opioid pharmacology; G-proteins; opioid-adrenergic interactions; inflammatory pain; nitric oxide.

Lisa M. Schrott, Ph.D., Assistant Professor. Drugs of abuse; neural-immune interactions; development disorders; learning and memory.

Artur H. Swiergiel, Ph.D., Research Assistant Professor. Behavioral responses in stress; role of environmental stimuli and immune factors in behavior.

James H. Zavecz, Ph.D., Associate Professor. Cardiovascular pharmacology; excitation-contraction coupling.

MASSACHUSETTS INSTITUTE OF TECHNOLOGY

Department of Biology

Programs of Study

The Department of Biology offers graduate study leading to the Doctor of Philosophy degree. The course of study includes graduate courses in the intellectual foundations of modern biology, as well as advanced courses in specialized topics. Research in the Department's laboratories represents a wide range of areas: biochemistry, biophysical chemistry and molecular structure, cell biology, developmental biology, genetics/microbiology, immunology, and neurobiology.

Biochemistry involves the study of the chemical properties of proteins, nucleic acids, protein–nucleic acid complexes, carbohydrates, and complex lipids. Specific areas of study include the chemistry of oncogenes, the mechanism of RNA splicing, analysis of cytoskeletal proteins, the mechanism of ion pumps and photoreceptors, and the role of complex carbohydrates in cell-surface function and protein compartmentalization.

In biophysical chemistry and molecular structure, studies are made of the principles that underlie the folding, stability, molecular design, and assembly of proteins and nucleic acids. Specific areas of concentration include the mechanism of protein folding; detailed analysis of conformation, using techniques such as X-ray crystallography and NMR; genetic strategies for enhancing the stability, ligand affinity, and catalytic efficiency of proteins and enzymes; protein–nucleic acid recognition; and antigen-antibody interactions.

Cell biology involves the molecular biological, genetic, and phenotypic analyses of eukaryotic cells. Specific areas of research include the organization, expression, and regulation of eukaryotic genomes; the structure and function of membranes and cytoskeletons; the molecular basis of cellular structure, organization, proliferation, and movement; the differentiation and functions of specialized cell types; and the molecular basis of disease.

Developmental biology embraces study of the cellular, genetic, and molecular mechanisms responsible for generating the diversity of cell types that arise during development and controlling the ways in which cells interact to produce organ systems and whole organisms. These problems are studied using animals (e.g., mice, *C. elegans, Drosophila,* zebra fish) and plants. Specific topics of interest include the regulation of gene expression, cell-cell interactions, cell lineages, cell migrations, and the mechanism of sex determination.

Genetics and microbiology studies include classical and molecular genetic analyses of fundamental problems in bacteria, bacteriophage, viruses, yeast, *Drosophila,* and plants. Areas of specific interest include protein secretion, DNA transposition protein turnover, DNA synthesis and repair, and mechanisms of genetic recombination and electron transport in mitochondria. More complex problems under study are cellular responses to stress, mechanisms of aging, plant-bacterial interactions, high-resolution structure-function studies of proteins and tRNAs, and the control circuits regulating gene expression.

Immunology is the study of the genetic, cellular, and molecular mechanisms underlying the sensitivity and specificity of the immune system. The immunologists in the Department study the chemistry of antigen-antibody and antigen T-cell receptor interactions, using the tools of molecular biology as well as classical immunological approaches and the role of idiotypic and cellular interactions in the regulation of the immune system, as studied by organ culture, hybridoma technology, and the behavior of transgenic mice.

Neurobiology emphasizes molecular, cell-biological, developmental, and genetic approaches to basic problems in the field. Areas of research interest include the molecular determinants of neuronal diversity and shape of cell-adhesive, cell-inductive, and synaptic interactions and the genetic and molecular determinants of cell lineages, memory storage, and sensory transduction.

Research Facilities

A state-of-the-art biology building opened in 1994. Other Department faculty members have their laboratories in the Center for Cancer Research, the Whitaker Building, and the Whitehead Institute for Biomedical Research. The Science Library contains 300,000 volumes and 1,800 journals for reference and provides a congenial atmosphere for study.

Financial Aid

Funds are available to support qualified graduate students as predoctoral fellows or research assistants. In 2006–07, the twelve-month stipend is $27,400 and includes medical extended hospital insurance.

Cost of Study

Students supported by the Department have all tuition remitted. Tuition is $33,400 for the 2006–07 academic year.

Living and Housing Costs

The Institute has a number of on-campus dormitory rooms and apartments for single and married graduate students. Rooms for single students average $7054 per academic year. Rents for single-student apartments are between $641 and $1333 per month depending upon the building. Student family housing averages $1193 per month. A student housing office is available to assist those who choose to live off campus in finding suitable apartments and/or roommates. Off-campus rents vary widely, but most students pay between $575 and $900 per month in shared apartments, excluding heat and utilities.

Student Group

Graduate School enrollment now stands at 6,140 students, who represent all states and many countries. In the 2006–07 academic year, there are 227 graduate students in the Department of Biology.

Student Outcomes

Nearly all of MIT's doctoral students go on to postdoctoral training in a wide range of fields and institutions all over the world. MIT has followed its students over a longer period as well. Most take up research positions in academics or industry. Others have taken primarily teaching positions. A small group work in alternative careers, including law, business, and public health.

Location

MIT's 125-acre campus extends for more than a mile along the Cambridge side of the Charles River basin, facing the city of Boston. Many cultural and recreational opportunities are easily accessible.

The Institute

The Massachusetts Institute of Technology is an independent, coeducational, endowed institution committed by charter and plan to the extension of knowledge through teaching and research. It is organized into five academic schools—Architecture and Planning, Engineering, Humanities and Social Science, Management, and Science—and a number of interdisciplinary groups. There are 10,206 students, including 3,550 women and 4,066 undergraduates. There are 992 faculty members and a total teaching staff of more than 1,500 members, not including teaching assistants and instructors.

Applying

Applications should be postmarked by December 15 for matriculation in the following September.

Correspondence and Information

Educational Administrator
Department of Biology, Room 68-120
Massachusetts Institute of Technology
Cambridge, Massachusetts 02139
E-mail: gradbio@mit.edu
Web site: http://web.mit.edu/biology/www/

Massachusetts Institute of Technology

THE FACULTY

Angelika Amon, Linda and Howard Stern Career Development Associate Professor of Biology and Investigator, HHMI; Ph.D., Vienna.
Tania A. Baker, Professor of Biology and Associate Investigator, HHMI; Ph.D., Stanford.
David P. Bartel, Helen and Irwin Sizer Professor of Biology; Member, Whitehead Institute; and Investigator, HHMI; Ph.D., Harvard.
Stephen P. Bell, Professor of Biology; Assistant Investigator, HHMI; and Chair of the Graduate Committee; Ph.D., Berkeley.
Chris B. Burge, Associate Professor of Biology; Ph.D., Stanford.
Jianzhu Chen, Professor of Immunology and Co-Chair of the Undergraduate Committee; Ph.D., Stanford.
Sallie W. Chisholm, Professor of Civil and Environmental Engineering and Professor of Biology; Ph.D., SUNY at Albany.
Martha Constantine-Paton, Professor of Neurobiology; Ph.D., Cornell.
Herman N. Eisen, Professor Emeritus and Senior Lecturer; M.D., NYU.
Gerald R. Fink, American Cancer Society Professor of Genetics; Ph.D., Yale.
Paul A. Garrity, Whitehead Career Development Assistant Professor of Biology; Ph.D., Caltech.
Frank B. Gertler, Professor of Biology; Ph.D., Wisconsin–Madison.
Alan D. Grossman, Professor of Biology; Ph.D., Wisconsin–Madison.
Leonard P. Guarente, Novartis Professor of Biology; Ph.D., Harvard.
Michael Hemann, Assistant Professor of Biology; Ph.D., Johns Hopkins.
Nancy H. Hopkins, Amgen, Inc., Professor of Biology; Ph.D., Harvard.
H. Robert Horvitz, Professor of Biology and Investigator, HHMI; Ph.D., Harvard.
David E. Housman, Ludwig Professor of Biology; Ph.D., Brandeis.
Richard O. Hynes, Daniel K. Ludwig Professor for Cancer Research and Investigator, HHMI; Ph.D., MIT.
Barbara Imperiali, Ellen Swallow Richards Professor of Chemistry; Ph.D., MIT.
Vernon M. Ingram, John and Dorothy Wilson Professor of Biology; Ph.D., London.
Tyler Jacks, Professor of Biology; Investigator, HHMI; and Director, Center for Cancer Research; Ph.D., California, San Francisco.
Rudolf Jaenisch, Professor of Biology and Member, Whitehead Institute; M.D., Munich.
Chris A. Kaiser, Professor of Biology and Head of Department; Ph.D., MIT.
Amy Keating, Assistant Professor of Biology; Ph.D., UCLA.
Gobind Khorana, Alfred P. Sloan Professor Emeritus of Biology and Chemistry and Senior Lecturer; Ph.D., Liverpool (England).
Dennis Kim, Assistant Professor of Biology; M.D./Ph.D., Harvard.
Jonathan A. King, Professor of Molecular Biology; Ph.D., Caltech.
Monty Krieger, Whitehead Professor of Biology, Professor of Molecular Genetics, and Charles F. Hopewell Faculty Fellow; Ph.D., Caltech.
Eric S. Lander, Professor of Biology; Member, Whitehead Institute; and Director of Broad Institute; D.Phil., Oxford.
Douglas Lauffenburger, Professor of Biological Engineering and Biology; Ph.D., Minnesota.
Jacqueline Lees, Professor of Biology and Associate Director, Center for Cancer Research; Ph.D., London.
Susan Lindquist, Professor of Biology and Member, Whitehead Institute for Biomedical Research; Ph.D., Harvard.
Troy Littleton, Associate Professor of Biology and Picower Institute for Learning and Memory; M.D./Ph.D., Baylor College of Medicine.
Harvey Lodish, Professor of Biology, Professor of Bioengineering, and Member, Whitehead Institute; Ph.D., Rockefeller.
Carlos Lois, Assistant Professor of Neurobiology; Ph.D., Rockefeller.
Boris Magasanik, Jacques Monod Professor of Microbiology; Ph.D., Columbia.
Paul T. Matsudaira, Professor of Biology, Professor of Bioengineering, and Member, Whitehead Institute; Ph.D., Dartmouth.
Elly Nedivi, Associate Professor Brain and Cognitive Sciences and Biology; Ph.D., Stanford.
Terry L. Orr-Weaver, Professor of Biology and Member, Whitehead Institute; Ph.D., Harvard.
David C. Page, Professor of Biology; Investigator, HHMI; and Director, Whitehead Institute; M.D., Harvard.
Mary-Lou Pardue, Boris Magasanik Professor of Biology; Ph.D., Yale.
Hidde Ploegh, Professor of Biology and Member, Whitehead Institute; Ph.D., Harvard.
William G. Quinn, Professor of Neurobiology; Ph.D., Princeton.
Uttam L. RajBhandary, Lester Wolfe Professor of Molecular Biology and Associate Head of Department; Ph.D., Durham (England).
Peter Reddien, Assistant Professor of Biology and Associate Member, Whitehead Institute; Ph.D., MIT.
Aviv Regev, Assistant Professor in Biology and Member of the Broad Institute; Ph.D., Tel Aviv.
Alexander Rich, William Thompson Sedgwick Professor of Biophysics; M.D., Harvard.
David Sabatini, Associate Professor of Biology and Associate Member, Whitehead Institute; M.D./Ph.D., Johns Hopkins.
Leona Samson, Professor of Biological Engineering and Director, Center for Environmental Health Sciences; Ph.D., London.
Robert T. Sauer, Salvador E. Luria Professor of Biology; Ph.D., Harvard.
Thomas Schwartz, Assistant Professor of Biology; Ph.D., Free University of Berlin.
Phillip A. Sharp, Institute Professor of Biology; Ph.D., Illinois at Chicago.
Morgan Sheng, Menicon Professor of Neuroscience and Associate Investigator, HHMI; Ph.D., Harvard.
Anthony J. Sinskey, Professor of Microbiology; Sc.D., MIT.
Hazel L. Sive, Professor of Biology and Co-Chair of the Undergraduate Committee; Ph.D., Rockefeller.
Frank Solomon, Professor of Biology; Ph.D., Brandeis.
Lisa A. Steiner, Professor of Immunology; M.D., Yale.
JoAnne Stubbe, Novartis Professor of Chemistry and Professor of Biology; Ph.D., Berkeley.
Susumu Tonegawa, Picower Professor of Biology and Neuroscience; Investigator, Howard Hughes Medical Institute; and Director, Picower Institute for Learning and Memory; Ph.D., California, San Diego.
Graham C. Walker, American Cancer Society Professor of Biology; Ph.D., Illinois at Urbana-Champaign.
Robert A. Weinberg, Daniel K. Ludwig Professor for Cancer Research, American Cancer Society Professor of Biology, and Member, Whitehead Institute; Ph.D., MIT.
Matthew Wilson, Professor, Department of Brain and Cognitive Sciences; Ph.D., Caltech.
Michael B. Yaffe, Associate Professor of Biology; M.D./Ph.D., Case Western Reserve.
Richard A. Young, Professor of Biology and Member, Whitehead Institute; Ph.D., Yale.

MAYO GRADUATE SCHOOL

Graduate Program in Biomedical Sciences

Program of Study

Mayo Graduate School, an integral part of the internationally recognized Mayo Clinic, offers a truly unique environment for Ph.D. and M.D./Ph.D. training in the biomedical sciences. Students start with a one-year core curriculum that covers the scientific principles and methodologies common to the specialized disciplines within biomedical research. During the second year, students complete a focused curriculum in one of six areas of specialization: biochemistry and molecular biology (with subtracks in biochemistry and structural biology, cancer biology, and cell biology and genetics), biomedical engineering (with subtracks in biomechanics, biomedical imaging, molecular biophysics, and physiology), immunology, molecular neuroscience, molecular pharmacology and experimental therapeutics, or virology and gene therapy. Thesis advisers can be chosen from a full spectrum of research faculty members, giving students optimum flexibility in creating their own curriculum. Mayo Graduate School's particular focus is on educating future independent research investigators, especially those who want to study questions directly applicable to medically related problems. Mayo offers an outstanding environment in which to work and learn, emphasizing teamwork and mutual respect among all students and staff.

Further details concerning the individual programs of study that are listed in the above paragraph are provided in separate descriptions.

Research Facilities

Each area of specialization is well equipped with state-of-the-art facilities for research and education in the life sciences. The campus in Rochester contains more than 965,000 square feet of space devoted to research and related activities. Research opportunities in molecular neuroscience and pharmacology are also available at the campus in Jacksonville, Florida, and opportunities in biochemistry, molecular biology, and cancer biology are available at the Scottsdale, Arizona, campus. A full-time faculty of 100 scientists and an annual budget of $296 million, with more than half coming from NIH and other extramural sources, support current research. Mayo Medical Library contains 413,000 volumes and subscribes to 4,300 medical and scientific journals.

Financial Aid

Full-time graduate students are provided with Mayo Graduate School predoctoral research fellowships. They provide a yearly stipend ($23,600 in 2006–07) in addition to medical insurance and other benefits. Receipt of fellowship stipends does not obligate students to work as laboratory assistants or as teaching assistants. Since stipends are provided by the Graduate School, students are not constrained in their choice of research advisers by funding limitations.

Cost of Study

Tuition costs for course work taken at Mayo's Rochester, Minnesota; Jacksonville, Florida; and Scottsdale, Arizona, campuses are provided in addition to the yearly stipend. Students pay no tuition or ancillary fees.

Living and Housing Costs

Living costs in Rochester are comparable to those in cities of similar size within the Upper Midwest and generally lower than those in urban areas of the East and West Coasts. A single student can live comfortably in Rochester on the stipend provided.

Student Group

Approximately 25 students are admitted each year into the Ph.D. in biomedical sciences program and 6 into the M.D./Ph.D. program. Enrollment in other Mayo educational programs includes 1,100 clinical residents and fellows and 300 postdoctoral research fellows.

Student Outcomes

Approximately three fourths of students enrolled in Mayo Ph.D. programs graduate with a Ph.D. degree. Students take a little more than five years to complete a Ph.D. Most students have authored several publications by the time they graduate.

Location

The city of Rochester combines the best of two worlds—the warmth and friendliness of a small town with the bustling commerce, entertainment, and conveniences of a metropolis. Lectures, symphony concerts, art exhibits, and a civic theater contribute to a cosmopolitan atmosphere, unusual for a city this size. The city and its environs provide an extensive, four-season calendar of recreation. Attractions in Minneapolis and St. Paul, a major metropolitan area, just 80 miles north of Rochester, include professional sports and a variety of cultural opportunities.

The Graduate School

The Mayo Graduate School is part of the Mayo Clinic College of Medicine, which includes Mayo Medical School, Mayo School of Graduate Medical Education, Mayo School of Health Sciences, and Mayo School of Continuing Medical Education. Mayo Graduate School is accredited by the Higher Learning Commission of the North Central Association of Colleges and Schools.

Applying

The general requirements for admission to the Mayo Graduate School Ph.D. program include a bachelor's degree from an accredited college or university and scores on the GRE General Test, or equivalent, indicating strong academic ability. Specific course prerequisites include 2 years of college chemistry (including organic), 1 year of calculus, 1 year of biology, and 1 year of physics, with evidence of superior performance in all courses. A foundation course in biochemistry or molecular biology is strongly recommended. Certain program tracks have additional requirements such as course work in physical chemistry and advanced mathematics. Official transcripts from schools attended, three letters of recommendation, and a summary of the student's scientific interests and career goals are also required. Interviews are required. Applications to the M.D./Ph.D. program are initiated via the AMCAS system and must be submitted by November 1. Applications for the Ph.D. program should be submitted by December 15, and there is a $30 application fee. Mayo Clinic College of Medicine is an affirmative action and equal opportunity educator and employer.

Correspondence and Information

Graduate Program in Biomedical Sciences
Mayo Graduate School
200 First Street, SW
Rochester, Minnesota 55905
Phone: 507-538-1160
E-mail: phd.training@mayo.edu
Web site: http://www.mayo.edu/mgs/

Mayo Graduate School

THE FACULTY

Biochemistry and Molecular Biology: Z. Bajzer, M. E. Bolander, T. P. Burghardt, R. Cattaneo, G. W. Dewald, F. J. Couch, N. L. Eberhardt, D. N. Fass, M. J. Federspiel, S. J. Gendler, G.J. Gores, J. P. Grande, P. C. Harris, B. F. Horazdovsky, G. Isaya, R. G. Janknecht, R. B. Jenkins, D. J. Katzmann, R. Kumar, N. F. LaRusso, J. J. Lee, N. A. Lee, E. B. Leof, A. H. Limper, J. C. Loftus, S. I. Macura, L. J. Maher III, D. J. McCormick, C. T. McMurray, M. A. McNiven, G. Mer, L. J. Miller, J. C. Morris, M. J. Oursler, W. G. Owen, R. E. Pagano, R. Patel, L. R. Pease, J. F. Poduslo, F. G. Prendergast, M. Ramirez-Alvarado, J. R. Riordan, S. J. Russell, J. L. Salisbury, R. D. Simari, D. F. Smith, D. I. Smith, T. C. Spelsberg, E. E. Strehler, A. H. Tang, S. N. Thibodeau, D. J. Tindall, D. O. Toft, R. T. Turner, R. Urrutia, S. Vuk-Pavlovic, C. J. Wetmore, E. D. Wieben, C. Y. F. Young, Z. Zhang.

Biomedical Engineering: K. An, Z. Bajzer, M. Belohlavek, T. P. Burghardt, J. C. Burnett, N. W. Chbat, R. L. Ehman, B. J. Erickson, G. Farrugia, M. Fatemi, J. P. Felmlee, B. K. Gilbert, J. F. Greenleaf, M. G. Herman, R. D. Hubmayr, C. R. Jack, M. J. Joyner, K. R. Kaufman, R. W. Kline, S. I. Macura, A. Manduca, V. M. Miller, M. K. O'Connor, Y. P. Pang, S. J. Riederer, E. L. Ritman, R. A. Robb, G. C. Sieck, S. M. Sine, R. J. Vetter, M. J. Yaszemski.

Immunology: D. D. Billadeau, R. J. Bram, M. Cascalho, C. S. David, G. J. Gleich, K. E. Hedin, D. F. Jelinek, H. Kita, E. D. Kwon, P. J. Leibson, V. A. Lennon, D. J. McKean, L. R. Pease, J. L. Platt, E. M. Poeschla, M. Rodriguez, , P. J. Wettstein.

Molecular Neuroscience: A. J. Bieber, W. S. Brimijoin, D. W. Dickson, C. B. Eckman, M. J. Farrer, T. E. Golde, M. L. Hutton, G. Isaya, C. D. James, R. B. Jenkins, J. J. Lee, V. A. Lennon, M. McKinney, C. T. McMurray, M. A. McNiven, M. M. Nicolle, R. E. Pagano, Y. P. Pang, J. F. Poduslo, G. A. Poland, C. Raffel, E. Richelson, R. A. Robb, M. Rodriguez, T. L. Rosenberry, C. Shin, G. C. Sieck, S. M. Sine, V. K. Somers, E. E. Strehler, R. Urrutia, R. M. Weinshilboum, A. J. Windebank, S. G. Younkin.

Molecular Pharmacology and Experimental Therapeutics: M. J. Ackerman, M. M. Ames, W. S. Brimijoin, J. Chen, A. P. Fields, L. M. Karnitz, Z. S. Katusic, S. H. Kaufmann, M. McKinney, C. T. McMurray, L. J. Miller, Y. P. Pang, G. A. Poland, F. G. Prendergast, E. Richelson, C. Shin, A. Terzic, R. M. Weinshilboum, S. H. Yen, S. G. Younkin.

Cancer Biology: M. M. Ames, D. D. Billadeau, M. Cascalho, J. Chen, F. J. Couch, C. S. David, G. W. Dewald, R. L. Ehman, M. J. Federspiel, S. J. Gendler, K. E. Hedin, C. D. James, R. Janknecht, D. F. Jelinek, R. B. Jenkins, L. M. Karnitz, S. H. Kaufmann, P. J. Leibson, V. A. Lennon, W. L. Lingle, W. Liu, R. V. Lloyd, J. A. Lust, L. J. Maher III, M. A. McNiven, D. Mukhopadhyay, D. J. O'Kane, Y. P. Pang, M. R. Pittelkow, J. L. Platt, F. G. Prendergast, C. Raffel, S. J. Russell, J. L. Salisbury, S. P. Scully, V. Shah, D. I. Smith, T. C. Spelsberg, E. E. Strehler, D. J. Tindall, D. O. Toft, R. Urrutia, J. M. van Deursen, R. G. Vile, R. M. Weinshilboum, P. J. Wettstein, A. J. Windebank, C. Y. F. Young.

Virology and Gene Therapy: A. D. Bradley, N. M. Caplice, R. B. Cattaneo, M. J. Federspiel, Z. S. Katusic, C. G. A. McGregor, J. C. Morris III, E. M. Poeschia, G. A. Poland, M. Rodriguez, S. J. Russell, R. D. Simari, R. G. Vile.

MEDICAL COLLEGE OF GEORGIA

School of Graduate Studies

Programs of Study	The School of Graduate Studies, part of Georgia's health science center known as the Medical College of Georgia (MCG), offers graduate programs in biomedical sciences, biostatistics, allied health sciences, and nursing. The School of Graduate Studies grants the following degrees: Ph.D. in biomedical science (with programs in cellular biology and anatomy, biochemistry and molecular biology, molecular medicine, neuroscience, pharmacology, physiology, and vascular biology); Ph.D. in nursing; Doctor of Nursing Practice (D.N.P.) in nursing; Master of Nursing (nursing anesthesia, family nurse practitioner, nursing, pediatric nurse practitioner, and RN-M.N./M.S.N.); M.S. in Nursing (adult/critical care and clinical nurse specialist); Ph.D. and M.S. in oral biology; M.S. in biostatistics; M.S. in allied health sciences (concentrations in dental hygiene, diagnostic medical sonography, medical technology, nuclear medicine technology, occupational therapy, physician assistant studies, radiation therapy, and respiratory therapy); M.S. in medical illustration; and Master of Public Health (M.P.H.) in health informatics. The Doctor of Physical Therapy (D.P.T.) and the M.H.S. in occupational therapy programs are offered through the School of Allied Health Sciences. A combined M.D./Ph.D. program and a combined D.M.D./Ph.D. are also available.
Research Facilities	MCG has a comprehensive, interdisciplinary biomedical sciences research program, with laboratory and clinical research in a number of areas, including cancer biology, infection, and inflammation and neurological and cardiovascular diseases. In addition to classic basic science and clinical departments that conduct excellent teaching and research, MCG also has several interdisciplinary centers and institutes with state-of-the-art research programs and equipment. The facilities to support biomedical research activities have been significantly increased, modernized, and renovated in recent years. Phase one of the Interdisciplinary Research Building was completed in the mid-1990s to house the Institute of Molecular Medicine and Genetics (IMMAG), founded to promote research excellence in the basic biomedical and clinical sciences. Phase two of the Interdisciplinary Research Building, a 94,000-square-foot structure completed in 2004, houses the researchers, laboratories, and equipment that advance research in areas such as diabetes, hypertension, biotechnology, and genomic medicine. The Carl T. Sanders Research and Education Building provides laboratories and additional classrooms designed to optimize educational and research opportunities. MCG has developed a number of state-of-the-art core facilities to ensure that investigators have access to cutting-edge technologies essential for leading biomedical research, including facilities and/or resources for cell imaging, genomics, proteomics, flow cytometry, biotelemetry, biostatistics, bioinformatics, transgenic mouse, and transgenic zebrafish. There are more than 300,000 square feet of dedicated research space on campus. Currently under construction are a 183,000-square-foot new Health Sciences Building, the future home of the Schools of Allied Health Sciences and Nursing, and a 160,000-square-foot Cancer Research Center that will house research in areas such as cellular therapeutics and immunotherapy, molecular chaperones, and breast, gynecologic, gastrointestinal, oral, and respiratory tract cancer. The Robert B. Greenblatt, M.D., Library contains more than 170,000 bound volumes and 9,000 unique title journals. It is highly computerized, with access to other University System of Georgia library collections, database searching, and interlibrary loans.
Financial Aid	For 2006–07, several means of financial aid are available. Eligible students in the Ph.D. biomedical sciences graduate program are offered a research assistantship package (funded through the School of Graduate Studies and/or research grants) that includes stipend support in the amount of $23,000 per year (before taxes). Biomedical science Ph.D. students with stipend support are eligible for reduced tuition at a cost of $25 per semester. Information on other financial aid options is available at http://www.mcg.edu/students/finaid/studentguide.pdf.
Cost of Study	Tuition for full-time students enrolled in the School of Graduate Studies in non-nursing programs for the 2005–06 academic year (per semester) was $2184 for Georgia residents and $8732 for nonresidents. Tuition for full-time students enrolled in School of Graduate Studies programs in nursing for the 2005–06 academic year (per semester) was $2724 for Georgia residents and $9232 for nonresidents. All students are required to pay an additional $75 technology fee per semester. Biomedical science Ph.D. students with graduate research assistantships are eligible for reduced tuition at a cost of $25 per semester. Students with 6 or more hours per semester pay health, activity, and miscellaneous fees ranging from $240 to $293 per semester, depending upon the specific program, program location, and semester enrolled. Nonresident graduate students may qualify for resident fees through the tuition waiver program administered by the School. A complete listing of tuition and fees for each program may be found at http://www.mcg.edu/students/Tuition_and_fees/gradfees.htm. Tuition rates and fees may vary and are subject to change for 2006–07.
Living and Housing Costs	The Office of Student Housing (http://www.mcg.edu/students/affairs/housing/) manages on-campus housing for approximately 220 students. Air-conditioned dormitory rooms are available for single or double occupancy. One- and two-bedroom apartments for married students are also available. Dining accommodations are offered on campus at the various units of the Food Service Division. An off-campus housing office (http://www.mcg.edu/students/affairs/housing/offcampus.htm) helps students locate rooms and apartments in the vicinity.
Student Group	There are about 265 graduate students at MCG. Students come from institutions in the U.S. and recognized international institutions. Graduates speak highly of their MCG education, citing factors such as small class sizes, extensive faculty support, in-depth hands-on training, a friendly environment, and cultivation of highly marketable skills.
Location	A variety of cultural, sports, and other recreational opportunities are available in the Augusta area (population 492,000). The cultural attractions of Augusta include the Morris Museum of Art, Augusta Ballet, Augusta Symphony, Augusta Opera, Augusta Players, and several concert and film series. The mean temperature is 64 degrees. Atlanta and Savannah, Georgia; Charleston, South Carolina; and Asheville, North Carolina, are all within a 3-hour drive of Augusta. For more information on Augusta, prospective students should visit http://www.augustaga.org/.
The Medical College	The Medical College of Georgia, chartered in 1828, has become a major academic health center and health sciences research university committed to being one of the nation's premier academic health centers for outstanding education, public service, and leading-edge research and scholarship. The institution, consisting of the Schools of Graduate Studies, Allied Health Sciences, Dentistry, Medicine, and Nursing and the hospital and clinics, offers a broad range of health-care and biomedical research training opportunities. The School of Graduate Studies granted its first degree in 1953.
Applying	Deadlines for completed applications for 2006 admission are specific for each program. A full listing of programs and deadlines can be found at http://www.mcg.edu/students/gradpgms/gradapply.pdf. Programs begin in the fall, unless otherwise noted. The deadlines for completed 2006 applications for graduate programs are November 1 for nursing anesthesia M.N. and nursing R.N.-M.N./M.S.N. (spring semester), December 31 for biomedical sciences Ph.D. early application, January 15 for biomedical sciences Ph.D. and oral biology Ph.D., January 31 for medical illustration M.S. (preliminary portfolio/form), March 1 for biostatistics, March 15 for nursing RN-M.N./M.S.N (summer semester), April 15 for D.N.P. (summer semester) and clinical nurse leader M.S.N. (summer semester), May 1 for physician assistant studies M.S., June 1 for nursing Ph.D. and nursing M.N. and M.S.N, and July 1 for other allied-health M.S. programs, health informatics M.P.H., nursing RN-M.S.N., RN-M.N, and oral biology M.S. A completed application form, three letters of recommendation, two official transcripts from each college or university attended, and a $30 application fee are required for most graduate programs. In many cases, appropriate entrance exam test scores (e.g., GRE, TOEFL) are required. Students should complete the application as early as possible, particularly if financial aid is requested. Candidates should check the application form to confirm deadlines and with individual programs for specific and/or additional admission requirements (http://www.mcg.edu/Admissions/index.htm and http://www.mcg.edu/catalog/gencatalog/AdmissIndex.htm). Printed applications are available at http://www.mcg.edu/students/GradPgms/apply.htm. To apply online, applicants should visit https://www.applyweb.com/apply/mcg/menu.html. To request an application or additional information, prospective students should contact the Office of Academic Admissions (706-721-2725, gradadm@mail.mcg.edu, http://www.mcg.edu/students/gradpgms/). All application materials should be sent to the Office of Academic Admissions, AA-170, Medical College of Georgia, Augusta, Georgia 30912-7310. Students applying to the combined M.D./Ph.D. program have specific requirements and should apply through the School of Medicine (stdadmin@mail.mcg.edu or einscho@mail.mcg.edu).
Correspondence and Information	Patricia L. Cameron, Ph.D., Associate Dean Director of the STAR Programs for Undergraduate Research School of Graduate Studies Medical College of Georgia, CJ 2201 1120 15th Street Augusta, Georgia 30912-1500 Phone: 706-721-3279 E-mail: gradstudies@mail.mcg.edu Web site: http://www.mcg.edu/GradStudies/index.html

Medical College of Georgia

FACULTY HEADS AND AREAS OF RESEARCH

BIOMEDICAL SCIENCES

Ph.D. Program. The School grants the Ph.D. degree with programs in the following seven areas: biochemistry and molecular biology (http://www.mcg.edu/GradStudies/gbiochem.htm), cellular biology and anatomy (http://www.mcg.edu/GradStudies/gcba.htm), molecular medicine (http://www.mcg.edu/gradmm/), neuroscience (http://www.mcg.edu/neuroscience/), pharmacology (http://www.mcg.edu/GradStudies/gpharmtox.htm), physiology (http://www.mcg.edu/GradStudies/gphyendo.htm), and vascular biology (http://www.mcg.edu/GradStudies/vascularbio.htm). Patricia L. Cameron, Ph.D., Associate Dean, School of Graduate Studies (biomed@mail.mcg.edu).

Department of Biochemistry and Molecular Biology. Areas of research interest are focused on important human health issues, including drug abuse and its effect on the developing fetus, drug delivery, membrane transporters and their relevance to human disease, pathogenic bacteria and factors affecting their virulence, the regulation of inflammation and cancer by cyclic nucleotide–dependent protein kinases, the control of globin gene expression, corneal wound healing, and RNA degradation. Vadivel Ganapathy, Ph.D., Regents' Professor and Chair (vganapat@mail.mcg.edu); Darren D. Browning, Ph.D., Assistant Professor and Program Director (dbrowning@mail.mcg.edu).

Department of Cellular Biology and Anatomy. The Cell Biology and Anatomy Program provides training in developmental biology, cell biology, and neuroscience. Students investigate biomedical problems related to the brain, eye and visual system, development and fate of the neural crest cells, development and regulation of the body axes, development of organs of special sense, neurodegeneration, cell wounding and repair, reparative potential of stem cells, cell death and dysfunction of cellular signaling, cancer biology, and bone formation. Sally Atherton, Ph.D., Professor and Chair (satherton@mail.mcg.edu); Zheng Dong, Ph.D., Associate Professor and Program Director (zdong@mail.mcg.edu).

Department of Pharmacology. Pharmacology is a wide-ranging discipline encompassing chemistry, molecular and cellular biology, physiology, and behavior. Faculty research interests include receptor-associated proteins, cell signaling via G-protein–coupled receptors, protein kinases and ion channels, and cognitive function in aged primates. R. William Caldwell, Ph.D., Professor and Chair (wcaldwel@mail.mcg.edu); Richard E. White, Ph.D., Professor and Program Director (rwhite@mail.mcg.edu).

Department of Physiology. Physiology integrates concepts and principles from all areas of medical science to understand function, from the molecular to the whole-animal level. Faculty members have expertise in cardiovascular and renal physiology, hypertension, behavioral neuroscience and genetics, and adrenal endocrinology and genetics, with experimental techniques ranging from chronic animal studies to state-of-the-art genomic and proteomic work. R. Clinton Webb, Ph.D., Professor and Chair (cwebb@mail.mcg.edu); Michael Brands, Ph.D., Professor and Program Director (mbrands@mail.mcg.edu).

Institute for Molecular Medicine and Genetics. The Molecular Medicine Graduate Program is based in an interdisciplinary research institute rather than an academic department. Research interests are focused in developmental neurobiology, cancer biology and gene regulation, molecular immunology, biotechnology and genomic medicine, regenerative medicine and molecular chaperones, and radiobiology. Robert Yu, Med.Sc.D., Professor and Institute Director (ryu@mail.mcg.edu); Wendy Bollag, Ph.D., Associate Professor and Program Director (wbollag@mail.mcg.edu).

Neuroscience. The Graduate Program in Neuroscience is an interdisciplinary program that provides students with an understanding of the disorders of the nervous system by integrating study of basic molecular-, cellular-, and systems-level neuroscience. Faculty members mentor students in research on drug abuse, memory, vision, hearing, ion channels, neuronal migration, growth of dendrites, development, synapse formation, regeneration, puberty, epilepsy, stroke, posttraumatic stress disorder, schizophrenia, and Parkinson's and Alzheimer's diseases. Deborah L. Lewis, Ph.D., Professor and Program Director (dlewis@mcg.edu).

Vascular Biology. Vascular biology is concerned with understanding vascular function at the molecular, cellular, organ, and whole-body levels and with designing tools for diagnosis, treatment, and prevention of vascular diseases, such as atherosclerosis, diabetes, hypertension, and ischemia/reperfusion. Active research programs include investigations on the roles of nitric oxide, renin-angiotensin, endothelin, reactive oxygen/nitrogen species, and growth factors in the pathophysiology of the coronary, pulmonary, renal, and retinal vascular beds. John Catravas, Ph.D., Regents' Professor and Director (jcatrava@mail.mcg.edu); David Pollock, Ph.D., Professor and Program Director (dpollock@mail.mcg.edu).

Department of Oral Biology and Maxillofacial Pathology. The department offers Master of Science and doctoral degrees (http://www.mcg.edu/GradStudies/goralbio.htm). Biomedical engineering and molecular approaches are used to study orthodontic tooth movement, regulation of cranial suture closure, dentin dynamics, fluoride metabolism, hormonal effects on oral health, temporomandibular joint histomorphology, chemopreventive agents and biomarkers of oral cancer, salivary biology, and biocompatibility of dental materials. George S. Schuster, D.D.S., Ph.D., Professor and Chair (gschuste@mail.mcg.edu); Jill Lewis, Associate Professor and Program Director (jillewis@mail.mcg.edu).

Department of Biostatistics. The main objective of the M.S. program in biostatistics is to produce M.S.-level biostatisticians able to consult, at the staff level, with medical and health researchers. The M.S. program gives an intensive exposure to the wide range of ideas, methodology, and techniques needed to perform as a staff-level statistician in the areas of designing and collecting data from experiments, observational studies, and clinical trials and analysing and reporting such data, with the aim of extracting information from these data stored in various forms. The application focus is on medicine, nursing, allied health, laboratory science, and other biological disciplines. James K. Dias, Ph.D., Associate Professor and Academic Program Director (jdias@mcg.edu).

GRADUATE PROGRAMS IN ALLIED HEALTH SCIENCES

Master of Health Education, Master of Science. As of January 1, 2005, no new students are being enrolled into this degree program until further notice; several of the disciplines originally served by this degree have moved to the master's entry-level program. The Master of Science is a research degree program for allied health and related professionals. The M.S. program is interdisciplinary and research oriented and requires completion of a thesis. The School of Allied Health Sciences offers concentrations in biomedical and radiological technologies, dental hygiene, health informatics, medical illustration, occupational therapy, physician assistant studies, and respiratory therapy. Kent Guion, M.D., Associate Dean for Academic Affairs (wguion@mcg.edu; http://www.mcg.edu/SAH/index.html or http://www.mcg.edu/GradStudies/Index.html).

Department of Medical Illustration. The M.S. in medical illustration program is one of only five in the nation. The curriculum includes basic medical science courses (gross anatomy, histology, embryology, neuroanatomy, pathology, and orientation to surgery); training in the materials, techniques, and uses of medical illustration; planning the production of instructional media; and experience in preparing scientific information for presentation through publication and audiovisual methods of communication. Steven J. Harrison, M.S., Associate Professor, Chair, and Program Director (sharriso@mail.mcg.edu); Andrew Swift, Chairman, Medical Illustration Admissions Committee (medart@mail.mcg.edu).

Department of Dental Hygiene. Marie Collins, M.S., Assistant Professor and Chair (mcollins@mail.mcg.edu).

Department of Health Informatics. The Master of Public Health in Health Informatics (iMPH) is a new program that produces public-health professionals with skills and competencies that contribute to the management, analysis, and dissemination of personal and population health-related information. This program is a comprehensive and integrated educational experience encompassing public health, health informatics, and health-administration themes. The curriculum is delivered using a hybrid approach, with approximately three quarters of the courses being Internet based. Miguel A. Zuniga, M.D., Associate Professor and Program Director (http://www.mcg.edu/sah/DHI/MPH/).

Department of Biomedical and Radiological Technologies. Elizabeth Kenimer Leibach, Ed.D., Assistant Professor, Chair, and Program Director (ekenimer@mail.mcg.edu).
Department of Occupational Therapy. Kathy Bradley, Ed.D., Associate Professor, Chair, and Program Director (kbradley@mail.mcg.edu).
Department of Physical Therapy. Douglas Keskula, Ph.D., Associate Professor, Chair, and Program Director (dkeskula@mail.mcg.edu).
Physician Assistant Department. Bonnie Dadig, Ed.D., PA-C, Associate Professor, Chair, and Program Director (bdadig@mail.mcg.edu).
Department of Respiratory Therapy. Randy Baker, Ph.D., RRT, Associate Professor, Chair, and Program Director (rbaker@mail.mcg.edu).

GRADUATE PROGRAMS IN NURSING

The focus of the nursing graduate programs is to prepare nurses as outstanding health-care leaders for the state of Georgia and beyond. The School of Nursing faculty is dedicated to the integration of teaching, research, and practice, ensuring that every graduate is prepared to reach the farthest ends of their career goals. Students find a variety of highest-quality, innovative, educational programs to meet their needs. Students should visit the MCG School of Nursing Web site at http://www.mcg.edu/SON/, or the School of Graduate Studies Web site at http://www.mcg.edu/GracStudies/index.html, or contact Georgia Narsavage, Ph.D., APRN, BC, Professor and Associate Dean for Academic Affairs (gnarsavage@mail.mcg.edu)

Adult Nursing (Critical-Care Clinical Nurse Specialist). The Master of Science degree in adult nursing prepares nurses for advanced nursing practice as clinical nurse specialists, with an emphasis on acute care and critical care related to health promotion and management of health problems of adults. Students have the opportunity to formulate, implement, and test a conceptual framework for nursing practice in selected adult populations in primary, secondary, and tertiary settings. Full- and part-time study is available, and the curriculum provides a foundation for doctoral study.

Family Nurse Practitioner (FNP). The FNP program prepares advanced-practice nurses to provide primary health care to clients, families, and communities. This program is offered on MCG's main campus in Augusta and in the Georgia cities of Athens and Columbus via on-site faculty members and distance learning and interactive Web instruction. Clinical assignments are made within the geographic location of the student's home. Full- and part-time study is available, and there is a post-master's option for students who have already obtained a master's degree in nursing.

Pediatric Nurse Practitioner (PNP). The PNP program prepares advanced-practice nurses to provide primary health care to children and young adults. The program's core courses and the didactic portion of the clinical courses are offered on MCG's main campus in Augusta and in the Georgia cities of Athens and Columbus via on-site faculty members and distance learning and interactive Web instruction. Clinical assignments are made within the geographic location of the student's home. Full-time and part-time study is available, and there is a post-master's option for students who already obtained a master's degree in nursing.

Nursing Anesthesia (CRNA). This program is a twenty-eight-month course of didactic and clinical study leading to a Master of Nursing degree. The program follows a two-phase model: an initial eight-month period (two semesters) of didactic instruction is followed by twenty months (five semesters) of clinical and didactic instruction. Students are eligible to take the National Certification Examination required for advanced practice as a nurse anesthetist in the state of Georgia. Participation in the Nursing Anesthesia Program requires a full-time commitment.

Nursing RN to M.N./M.S.N. This program provides diploma and associate-degree nurses with a streamlined curriculum to obtain a master's degree in nursing. Specific tracks, depending on enrollment, include M.N. degrees in family nurse practitioner and pediatric nurse practitioner and M.S.N. degrees in clinical nurse specialist in adults and critical care. Full- and part-time study is available.

Doctor of Nursing Practice (DNP). This program educates advanced nurse clinicians for expert practice in leadership and clinical roles. The program enables nurses to become expert nurse leaders and collaborators in solving health-care problems in their institutions. This degree encourages nurses to stay in health-care practice and contribute to issues faced in the field. This program is offered on MCG's main campus in Augusta and in the Georgia cities of Athens and Columbus via on-site faculty members and distance learning and interactive Web instruction. The program can be completed on a full- or part-time basis.

Doctor of Philosophy Program in Nursing (Ph.D.). The program prepares nurses to provide leadership in nursing education, practice, and research. The program emphasizes interdisciplinary development and dissemination through biobehavioral nursing research in the following areas of faculty research focus: health outcomes, health disparities and health-provider performance, and patient safety. The program can be completed on a full- or part-time basis.

MEDICAL COLLEGE OF GEORGIA

University System of Georgia M.D./Ph.D. Program

Program of Study	The Medical College of Georgia (MCG), in conjunction with the University of Georgia at Athens (UGA), Georgia Institute of Technology, and Georgia State University, offers a program leading to a combined M.D./Ph.D. degree. The program is designed to train physician-scientists as both excellent clinicians and critically trained scientists. This program is directed toward those select individuals focused on preparation for careers as biomedical investigators. Students complete all of the normal requirements of the Medical College of Georgia for the M.D. degree. The Ph.D. degree can be earned in any of the degree-granting biomedical science departments of the Medical College of Georgia, Georgia Institute of Technology, the University of Georgia, or Georgia State University.

The M.D./Ph.D. program normally requires seven to eight years of study. Students entering the program complete the standard two-year preclinical program at the Medical College of Georgia. Many of the preclinical medical school courses also earn graduate credit for the M.D./Ph.D. students. During the summer between the first and second years, students perform laboratory rotations at one of the four research campuses. Choices for laboratory rotations are made by students in consultation with program advisers. Following the preclinical years, students enter graduate training at one of the four graduate campus sites. Students are required to complete all of the normal Ph.D. requirements, including preliminary exams, thesis, and defense. Following the completion of graduate studies, students complete the clinical requirements for their M.D. degree. Throughout the program years, students participate in special M.D./Ph.D. seminars, including a series of bimonthly sessions with graduates of dual-degree programs from around the country.

Research Facilities State-of-the-art research facilities are available at all four research campuses of the M.D./Ph.D. program. In addition, a number of core facilities aid in important key technologies, including DNA sequencing facilities at MCG and UGA, peptide synthesis and sequencing facilities at MCG and UGA, a monoclonal antibody facility at UGA, a transgenic and knock-out mouse facility associated with the Institute for Molecular Medicine and Genetics at MCG, a biotelemetry and small-animal behavior core at MCG, a transgenic zebrafish core at MCG, proteomic and mass spectroscopy cores at MCG and UGA, and a fluorescence imaging facility at MCG. The libraries of the four research campuses are connected through the World Wide Web and share information on periodicals and publications.

Financial Aid Funding for the M.D./Ph.D. program comes from the state of Georgia and other institutional funds. Students in the M.D./Ph.D. program receive a tuition waiver. The stipend for students beginning July 1, 2005, was $22,500, and stipend support is maintained throughout the program. No unfunded students are supported.

Cost of Study All academic costs are covered by the M.D./Ph.D. program except for a $25-per-semester filing fee. Students are responsible for optional comprehensive health insurance. Microscope rental is currently $60 per year. The expense of supplies, instruments, and books in the first year is estimated at $2030, with lower costs in subsequent years.

Living and Housing Costs Both on-campus and off-campus housing options are available. The current estimated on-campus room and board (with utilities and phone) are $1580 per month; estimated off-campus room and board (with utilities and phone) are $1785 per month.

Student Group The M.D./Ph.D. program admits up to 4 students per year out of a class of approximately 180 medical students. There are currently 23 students enrolled in the program at various levels of study.

Location The Augusta metropolitan area, with a population of more than 400,000, is the second-largest metropolitan area in Georgia. Augusta is located between the Atlantic Ocean and the Appalachian Mountains, with easy access to a number of outdoor activities, including hiking, sailing, and river rafting. Cultural life in Augusta includes a professional civic symphony, an opera company, and a ballet company. Both Atlanta and Savannah are 2½-hour drives from Augusta. The University of Georgia is located in Athens, which lies approximately halfway between Augusta and Atlanta. Both the Georgia Institute of Technology and Georgia State University are located in Atlanta.

The College Founded in 1828, the Medical College of Georgia is the eleventh-oldest continuously operating medical education institution in the United States and is third-oldest medical school in the Southeast. The Medical College is the health sciences university of the University System of Georgia. The Medical College of Georgia includes a 540-bed teaching hospital and a regional Children's Medical Center.

Applying Students applying to the M.D./Ph.D. program require previous substantive research experience. Interested students should apply to the Medical College of Georgia through the American Medical College Application Service (AMCAS) and check the appropriate box on the form. MCAT scores are required for application, but no GRE scores are needed. Applicants are not required to be Georgia residents and applications from nonresidents are encouraged. However, program entry is only available to citizens or permanent residents of the United States. Qualified candidates are interviewed by 2 members of the M.D./Ph.D. committee in addition to their regular medical school interviews.

Correspondence and Information
Director
M.D./Ph.D. Program
Medical College of Georgia
CB2803
1120 Fifteenth Street
Augusta, Georgia 30912

Phone: 706-721-6306
Fax: 706-721-8727
E-mail: mdphd@mcg.edu
Web site: http://www.mcg.edu/som/mdphd

Medical College of Georgia

THE FACULTY AND THEIR RESEARCH

Georgia Institute of Technology
Main contact:
> The Wallace H. Coulter Department of Biomedical Engineering
> Georgia Tech and Emory University
> 1639 Pierce Drive, Suite 2001
> Atlanta, Georgia 30322-4600
> Director: Stephen Deweerth, Ph.D.
> Telephone: 404-894-7063

Department involved in biomedical and health-related research:
> Biomedical Engineering Department: http://www.bme.gatech.edu/academics/grad_prog.html; Biomedical Engineering faculty list: http://www.bme.gatech.edu/facultystaff/faculty.php

Georgia State University
Main contact:
> College of Arts and Sciences
> P.O. Box 4038
> Atlanta, Georgia 30302-4038
> Associate Dean: Charles Derby, Ph.D.
> Telephone: 404-651-1800

Departments involved in biomedical and health-related research:
> Biology programs: http://biology.gsu.edu/graduate/index.html; Biology faculty list: http://biology.gsu.edu/people/faculty/alphabetical.cfm
> Chemistry programs: http://chemistry.gsu.edu/School/grad-info.php; Chemistry faculty list: http://chemistry.gsu.edu/faculty/
> Psychology programs: http://www2.gsu.edu/~wwwpsy/ClinProg.htm; Psychology faculty list: http://www2.gsu.edu/~wwwpsy/FacDirectory.htm

Medical College of Georgia
Main contact:
> School of Graduate Studies
> 1120 15th Street, Room CJ-2201
> Augusta, Georgia 30912
> Associate Dean: Gretchen Caughman, Ph.D.
> Telephone: 706-721-3278
> Web site: http://www.mcg.edu/GradStudies/dprograms.htm

Departments involved in biomedical and health-related research:
> Biochemistry: http://www.mcg.edu/bmb
> Cellular Biology and Anatomy: http://www.mcg.edu/som/cba
> Molecular Medicine: http://www.mcg.edu/institutes/immag
> Pharmacology: http://www.mcg.edu/som/phmtox/faculty.html
> Physiology: http://www.mcg.edu/som/phy/faculty.htm
> Vascular Biology: http://www.mcg.edu/centers/vbc/index.html

University of Georgia
Main contact:
> University of Georgia Biomedical and Health Sciences Institute
> 220 Veterinary Medicine
> Athens, Georgia 30602-7394
> Director: Harry A. Dailey, Ph.D.
> Telephone: 706-542-5922
> Web site: http://www.biomed.uga.edu

Colleges and departments involved in biomedical and health-related research:
> University of Georgia BHSI Directory: http://www.biomed.uga.edu/directory.html
> Franklin College of Arts and Sciences: http://www.franklin.uga.edu
> College of Agricultural and Environmental Sciences: http://www.caes.uga.edu
> College of Education: http://www.coe.uga.edu
> College of Family and Consumer Sciences: http://www.fcs.uga.edu
> College of Pharmacy: http://www.rx.uga.edu/main/home/httpd/html/index.html
> College of Public Health: http://www.publichealth.uga.edu
> College of Veterinary Medicine: http://www.vet.uga.edu

MEDICAL COLLEGE OF WISCONSIN

Graduate School of Biomedical Sciences
Interdisciplinary Program in Biomedical Sciences

Program of Study

The Interdisciplinary Program in Biomedical Sciences at the Graduate School of Biomedical Sciences, Medical College of Wisconsin (MCW) is a multidisciplinary doctoral training program that is designed to prepare students for a research career in cellular, molecular, and integrative biology. Areas of concentration within the program include molecular biology and genetics, cell biology, developmental biology, immunology and microbial pathogenesis, virology, neurobiology, enzymology, pharmacology and toxicology, structural biology, free radical biology, and biochemistry. Students are admitted only to a course of study leading to a Ph.D.

The integrated program is designed to provide graduate students with the broadest possible range of research training opportunities by permitting them to select a research adviser from among all graduate faculty members in the program, regardless of departmental affiliation. Participating faculty members are drawn from the Departments of Biochemistry; Biophysics; Cell Biology, Neurobiology, and Anatomy; Microbiology and Molecular Genetics; and Pharmacology and Toxicology.

During the first two semesters, Ph.D. trainees participate in a core curriculum that integrates molecular, cellular, and systems biology classes. Courses include lectures and problem-solving and discussion sessions that focus on current research problems. Through faculty member interaction and laboratory rotations, students select a dissertation adviser from the large and diverse graduate faculty and, at the end of the first year, matriculate into one department for advanced training. Advancement to Ph.D. candidacy occurs after successful completion of a qualifying examination and acceptance of a research plan.

Detailed information regarding each participating program can be obtained via the MCW Web page (http://www.mcw.edu/gradschool).

Research Facilities

The participating faculty members occupy modern laboratories that are well equipped for research in their areas of interest. There are campuswide facilities, such as a transgenic mouse facility, a state-of-the-art biotechnology center with the latest nucleic acid and protein sequencing and synthesizing instrumentation. Hardware and software components for protein structure/function, X-ray crystallography, and protein interaction analysis are available. A core microscopy facility includes transmission electron, scanning electron, and confocal laser microscopes. Image-processing facilities provide computer enhancement and analysis, allowing three-dimensional analysis of subcellular structures.

Financial Aid

All eligible students are provided with a tuition scholarship and a competitive stipend. The 2005–06 stipend was $22,973 per annum. Health insurance is provided.

Cost of Study

Tuition is covered for all students in the Ph.D. program.

Living and Housing Costs

A wide variety of affordable living accommodations are available near the campus.

Student Group

More than 550 graduate students are enrolled at the College. The Medical College also enrolls approximately 800 medical students, some of whom study for the combined M.D./Ph.D. degree. A large M.P.H. program is also a part of the MCW education mission.

Student Outcomes

The majority of graduates from the participating programs accept postdoctoral fellowships. Other students have taken advantage of opportunities to go directly into teaching positions at the assistant professor level and into positions in biomedical-related industries.

Location

Milwaukee, which provides some 14,000 acres of parks and areas of urban development, is located on Lake Michigan, approximately 90 miles from Chicago. Students enjoy Wisconsin's many inland lakes, state parks, and national forests; these provide a wealth of year-round outdoor opportunities. The campus of the Medical College is located in a safe, attractive suburban area that is minutes from downtown Milwaukee.

The College

The College was established in 1913 as the Marquette University School of Medicine. It was reorganized in 1967 as an independent institution and was renamed the Medical College of Wisconsin in 1970. The faculty includes approximately 900 full-time members. The College has attracted more than $115 million a year in research and training funding and ranks in the top third of U.S. medical schools in NIH funding.

Applying

Applicants should have completed 8 semester hours each of biology, general chemistry, and physics as well as courses in college-level mathematics and statistics. Some experience in laboratory research, as might be obtained from employment or participation in undergraduate science projects, is highly desirable. The General Test of the Graduate Record Examinations (GRE) is required. Applicants must have a minimum cumulative undergraduate GPA of 3.0 (on a 4.0 scale). For students whose native language is not English, the Test of English as a Foreign Language (TOEFL) is required.

Correspondence and Information

Interdisciplinary Program in Biomedical Sciences
Graduate School of the Biomedical Sciences
Medical College of Wisconsin
8701 Watertown Plank Road
Milwaukee, Wisconsin 53226
Phone: 414-456-8218
Fax: 414-456-6555
E-mail: gradschool@mcw.edu
Web site: http://www.mcw.edu/biomed

Medical College of Wisconsin

THE FACULTY AND THEIR RESEARCH

Biochemistry

John E. Baker, Professor; Ph.D. Role of nitric oxide and K_{ATP} channels in myocardial ischemic reperfusion injury. **Nancy M. Dahms,** Professor; Ph.D. Role of carbohydrate signals and receptors in the targeting of lysosomal enzymes to lysosomes. **Robert Deschenes,** Professor and Chairman; Ph.D. Eukaryotic signal transduction; *ras* oncogene structure-function; subcellular localization of protein complexes. **Albert W. Girotti,** Professor; Ph.D. Membrane-damaging effects of activated oxygen species; mechanism of antineoplastic photosensitizing agents. **George J. Giudice,** Professor; Ph.D. Epidermal cell biology; molecular biology of desmosomes and hemidesmosomes. **Owen W. Griffith,** Professor; Ph.D. Study of enzyme mechanisms; amino acid metabolism; nitric oxide biology; cancer chemotherapy. **Vaughn E. Jackson,** Associate Professor, Ph.D. DNA replication and transcription through nucleosomes; regulation of gene expression. **Jung-Ja P. Kim,** Professor; Ph.D. Structure and function of enzymes; protein–nucleic acid interactions. **Ravi P. Misra,** Associate Professor; Ph.D. Regulation of gene expression during early development; molecular mechanisms of cardiac hypertrophy. **Richard L. Sabina,** Associate Professor; Ph.D. N-domain diversity of AMP deaminase isoforms; regulation of human AMP deaminase gene family. **Bellur Seetharam,** Professor; Ph.D. Molecular and cell biology of cobalamin binding and transport proteins. **Sally S. Twining,** Professor; Ph.D. Proteases and protease inhibitors in eye diseases. **Brian F. Volkman,** Associate Professor; Ph.D. Structural biology of signaling proteins and NMR spectroscopy in structural genomics.

Biophysics

William E. Antholine, Associate Professor; Ph.D. Uptake studies of metal antitumor agents; development of radiosensitizing agents. **Brian Bennett,** Assistant Professor; Ph.D. Metalloenzymes in cancer; AIDS; nerve gas detoxification and toxic insult. **Jimmy B. Feix,** Professor; Ph.D. Electron spin resonance studies of membrane proteins and peptide-membrane interactions. **Neil Hogg,** Associate Professor; Ph.D. Biological chemistry of nitric oxide and its oxidation products in pathophysiology. **Balaraman Kalyanaraman,** Professor and Chairman; Ph.D. Free-radical metabolites in the biological system. **Candice S. Klug,** Assistant Professor; Ph.D. Protein structure and functional dynamics studies using site-directed spin labeling EPR spectroscopy. **Witold K. Subczynski,** Assistant Professor; Ph.D. Spin label studies on membrane dynamics and organization; spin-label oximetry and NO-metry. **Jeannette Vasquez Vivar,** Assistant Professor; Ph.D. Mechanisms regulating superoxide and nitric oxide formation from nitric oxide synthase.

Cell Biology, Neurobiology, and Anatomy

Joseph C. Besharse, Professor and Chairman; Ph.D. Cellular and molecular basis of circadian rhythmicity; retinal photoreceptors. **Janice M. Burke,** Professor; Ph.D. Morphogenesis and cellular aging of the retinal pigment epithelium (RPE). **Lisa A. Cirillo,** Assistant Professor; Ph.D. Role of chromatin structure and its modification on transcriptional regulation of genes in liver. **Maria J. Crowe,** Assistant Professor; Ph.D. Behavioral, cellular, and molecular analysis of spinal cord injury and traumatic brain injury. **Stephen A. Duncan,** Associate Professor; D.Phil. Molecular mechanisms underlying mammalian development. **Carol Everson,** Associate Professor; Ph.D. Physical health effects of sleep deprivation, with a focus on neuroendocrine and immune systems. **Claudia S. Huettner,** Assistant Adjunct Professor; Ph.D. In vivo targeting of hematopoietic stem cells; mouse models of human leukemias. **Brian A. Link,** Assistant Professor; Ph.D. Genetics of ocular development and maintenance in zebrafish. **John W. Lough,** Professor; Ph.D. Cell and molecular biology of early heart development. **Alan N. Mayer,** Assistant Professor; M.D., Ph.D. Molecular genetics of digestive organ development. **Jay Neitz,** Professor; Ph.D. Biological basis of visual perception; molecular biology and genetics of the retina. **Maureen Neitz,** Professor; Ph.D. Biological basis of visual perception; molecular biology and genetics of the retina. **Danny A. Riley,** Professor; Ph.D. Spaceflight unloading of skeletal muscle, vibration injury of arteries and nerves, and muscular dystrophy. **Elena V. Semina,** Assistant Professor; Ph.D. Discovery of genes underlying complex developmental disorders, with an emphasis on developmental glaucoma. **D. J. Sidjanin,** Assistant Professor; Ph.D. Molecular genetics of hereditary eye diseases. **Fritz Sieber,** Professor; Ph.D. Photochemotherapy; development of merocyanine dyes as fluorescent probes. **Maya Sieber-Blum,** Professor; Ph.D. Mechanisms of differentiation of neural crest cells into neurons, smooth-muscle cells, and pigment cells. **Cheryl L. Stucky,** Assistant Professor; Ph.D. Cellular and molecular mechanisms underlying pain. **Margaret Wong-Riley,** Professor; Ph.D. Metabolic and neurochemical plasticity in the adult primate visual system.

Microbiology and Molecular Genetics

Joseph T. Barbieri, Professor; Ph.D. Mechanism of action of bacterial toxins and their interaction with eukaryotic cells. **Bonnie N. Dittel,** Assistant Adjunct Professor; Ph.D. Immune regulation of inflammation in immunity and autoimmunity. **William R. Drobyski,** Professor; M.D. Immunobiology of bone marrow transplantation; graft versus host disease; autoimmunity and graft versus leukemia reactivity. **Michael B. Dwinell,** Assistant Professor; Ph.D. Pathogenesis of mucosal immunity signaling; function of chemokines and chemokine receptors. **Dara W. Frank,** Professor; Ph.D. Molecular analysis of type III toxin synthesis and translocation by *Pseudomonas aeruginosa.* **Jack Gorski,** Associate Adjunct Professor; Ph.D. Molecular immunology; immune system memory; thymic maturation and function of MHC class II. **William J. Grossman,** Assistant Professor; M.D., Ph.D. Immune regulation: role of the perforin/granzyme pathway in human and murine regulatory T-cell function. **Amy W. Hudson,** Assistant Professor; Ph.D. Viral immunoevasion strategies: disruption of MHC presentation by human herpesviruses. **Ming Lei,** Associate Professor; Ph.D. Regulation of DNA replication initiation in eukaryotes. **Roy M. Long,** Associate Professor; Ph.D. Intracellular transport and localization of mRNA in *Saccharomyces cerevisiae.* **Subramaniam Malarkannan,** Assistant Professor, Ph.D. Minor histocompatibility antigens; CD8+ T cells in transplant rejection and tumor clearance. **Mark T. McNally,** Associate Professor; Ph.D. RNA processing in Rous sarcoma virus; role of *cis*- and *trans*-acting factors in splicing control. **Debra K. Newman,** Assistant Adjunct Professor; Ph.D. Structural and functional aspects of T-cell receptor-mediated antigen recognition. **Rimas J. Orentas,** Associate Professor; Ph.D. Epstein-Barr virus; neuroblastomas; bone marrow transplantation and tumor immunology. **Nita H. Salzman,** Assistant Professor; M.D., Ph.D. Role of defensins in innate mucosal immunity. **Andrey Sorokin,** Associate Professor; Ph.D. Mechanisms and consequences of signal transduction: endothelin-mediated signaling through small GTPases; cyclooxygenase-2 and the prevention of apoptosis. **Jerry L. Taylor,** Professor; Ph.D. Regulation of herpesvirus gene expression by interferons. **Paula Traktman,** Professor and Chairman; Ph.D. Molecular, genetic, and biochemical analysis of vaccinia virus. **Robert L. Truitt,** Professor; Ph.D. Experimental bone marrow transplantation; GVL/GVH reactivity transplant tolerance; immunoregulation. **Demin Wang,** Assistant Adjunct Professor; Ph.D. Function of signal transduction pathways of cytokine and B-cell receptors. **Calvin B. Williams,** Associate Professor; M.D., Ph.D. T-cell tolerance; thymocyte selection and maintenance of the peripheral repertoire. **Li Wu,** Assistant Professor; Ph.D. HIV interactions with immune cells; HIV transmission mechanisms. **Thomas C. Zahrt,** Assistant Professor; Ph.D. Host-pathogen interactions of *Mycobacterium tuberculosis.*

Pharmacology and Toxicology

John A. Auchampach, Associate Professor; Ph.D. Cardiovascular pharmacology; adenosine receptors; myocardial ischemia. **Alan S. Bloom,** Professor; Ph.D. Biochemical neuropharmacology of aging and dementia; drugs of abuse; functional MRI. **William B. Campbell,** Professor and Chairman; Ph.D. Cardiovascular pharmacology; vascular tone and adrenal steroidogenesis. **Garrett J. Gross,** Professor; Ph.D. Cardiovascular pharmacology; ischemia-reperfusion inquiry. **David D. Gutterman,** Professor; M.D. Cardiovascular pharmacology; vascular biology. **Cecilia J. Hillard,** Professor; Ph.D. Biochemical neuropharmacology of drugs of abuse. **Ronald N. Hines,** Professor; Ph.D. Molecular toxicology; genes involved in xenobiotic metabolism; cytochromes P450. **Judy Kersten,** Professor; M.D. Cardiovascular pharmacology; regulation of coronary blood flow; diabetes. **Wai-Meng Kwok,** Associate Professor; Ph.D. Cardiovascular pharmacology; ATP-regulated potassium channels; anesthetics. **Sang H. Lee,** Assistant Professor; Ph.D. Neuropharmacology; memory; glutamate receptors. **Pin-Lan Li,** Professor; M.D., Ph.D. Regulation of ion channels in vascular smooth muscle by drugs and endogenous compounds. **Marilyn P. Merker,** Professor; Ph.D. Metabolic functions of normal and injured pulmonary endothelium in the intact lung. **Charles R. Myers,** Professor; Ph.D. Molecular toxicology; toxicology of metals; biotransformation of xenobiotics. **Peter J. Newman,** Professor; Ph.D. Cell and molecular biology of human platelet and endothelial cell adhesion molecules. **Kasem Nithipatikom,** Associate Professor; Ph.D. Spectroscopic methods and trace analyses of drugs and endogenous hormones; cancer biology. **Sandra L. Pfister,** Associate Professor; Ph.D. Cardiovascular pharmacology; role of eicosanoids in the regulation of vascular tone. **Kirkwood A. Pritchard Jr.,** Professor; Ph.D. Biochemical pharmacology; mechanism of nitric oxide synthase regulation. **David C. Warltier,** Professor; M.D., Ph.D. Cardiovascular physiology and pharmacology; myocardial ischemia and infarction. **Carol L. Williams,** Associate Professor; Ph.D. Cardiovascular pharmacology; GTPases in vascular smooth-muscle-cell growth; cancer biology.

MOREHOUSE SCHOOL OF MEDICINE

Ph.D. in the Biomedical Sciences

Program of Study

The Ph.D. program in the biomedical sciences is designed to develop independent investigators who are capable of assuming leadership roles in academic and corporate biomedical research. A track for obtaining combined M.D./Ph.D. degrees is available. Students may study with graduate faculty members from a variety of basic science and clinical departments. Available areas of focus within and across these disciplines include AIDS and infectious disease, cancer, cardiovascular disease, cell biology, developmental biology, immunity, microbiology, molecular biology, neuroscience, pathology, pharmacology, physiology, space medicine and life sciences, stroke, toxicology, and vision research.

The first year of graduate study is devoted primarily to core courses, which include course sequences in graduate cell biology and fundamentals of professional science. Students complete advanced graduate (elective) courses and begin dissertation research in their second year. At least four years of full-time study beyond the baccalaureate degree and 3½ years in residence at Morehouse School of Medicine (MSM) are required to complete the Ph.D. program.

Research Facilities

Morehouse School of Medicine currently occupies three buildings on the main campus in which essentially all biomedical research and research-related activities take place. The Basic Medical Sciences Building (BMSB) and the attached Medical Education Building (MEB) provide more than 35,000 square feet of research and research support space. This includes sixty-five individual, shared-use, and core facilities laboratories. The 35,000-square-foot Multi-Disciplinary Research Center (MRC) currently houses the Neuroscience Institute and the Clinical Research Center. The Center for Laboratory Animal Resources, which is located in the BMSB, occupies an additional 8,500 square feet, and the Medical Library, which is located in the MEB, comprises 10,000 square feet. A research wing on the MEB was completed in 2000 and provides an additional 35,000 square feet of research space.

Faculty investigators currently have an average of approximately 400 square feet of individual laboratory space. In addition, many core and shared-use, state-of-the-art research technology facilities and major instrument laboratories are available to all researchers. The core support facilities and individual investigator laboratories are extremely well equipped to support biomedical research. Institutional facilities include flow cytometry/FACS, monoclonal antibody preparation, nucleic acid sequencing and DNA synthesis, an Affymetrix and Agilent microarray scanner and GeneNet/GeneSpring software, SELDI and mass-spectrometry proteomics instruments, protein/peptide purification (HPLC) and two-dimensional gel electrophoresis, and imaging and image analysis facilities for brightfield, fluorescence, laser-dissection, and confocal microscopy and scanning and transmission electron microscopy. Scientific imaging and graphics preparation services are provided by the Division of Information.

Additional collaborative research opportunities are available through existing links with the Centers for Disease Control and Prevention, the Cleveland Clinic, the Environmental Protection Agency, the National Aeronautics and Space Administration, the National Institutes of Health, and other universities.

Financial Aid

The program provides yearly stipends for all of its graduate students. Stipends are a minimum of $20,000 per year. In subsequent years, stipends derive from a variety of funding sources. To continue stipend support, a student must maintain satisfactory progress in the program for which the stipend was awarded, must devote full time to study or research in the biomedical sciences, and must not engage in gainful employment outside the program.

Cost of Study

Tuition waivers or reimbursements are provided for all full-time students in good academic standing unless the student has extramural grant support to cover these expenses.

Living and Housing Costs

University housing is not currently available. Rental units are available throughout the Atlanta metropolitan area at a reasonable cost.

Student Group

Morehouse School of Medicine is a health sciences institution comprising four postbaccalaureate programs (M.D., M.P.H., M.S. in clinical research, and Ph.D.), with a total student population approaching 300.

Location

Morehouse School of Medicine is a member of the Atlanta University Center, a consortium of five independent institutions of higher education (Clark Atlanta University, the Interdenominational Theological Center, Morehouse College, Morehouse School of Medicine, and Spelman College). Together, the center's institutions constitute the largest predominantly black private educational complex in the world. Other major educational and research institutions in Atlanta include the Centers for Disease Control and Prevention, Emory University, Georgia Institute of Technology, and Georgia State University. In addition to being a center for higher education, Atlanta offers a variety of cultural opportunities, including the Atlanta Symphony Orchestra, the Atlanta Ballet, the Alliance Theatre, and several smaller theater companies. The High Museum of Art is an architectural masterpiece that houses an extensive collection of its own and hosts several first-rate touring collections each year. Atlanta is also known for its world-famous centers honoring Dr. Martin Luther King Jr. and former president Jimmy Carter. Sports and entertainment facilities include the Atlanta Braves' Turner Field, the Georgia Dome, broadcasting's CNN Center, the Fox Theater, the Phillips Arena, and many jazz and blues clubs. It is not surprising that many of America's best-known and most respected African-American businesspeople, politicians, and professionals call Atlanta home, making this city one of the premier centers of African-American culture in the country.

The School

The institution was established in 1975 to address the shortage of minority physicians and related problems in medically underserved communities. Beginning as a two-year preclinical program in 1978, Morehouse School of Medicine was approved to become a four-year medical school in 1981 and granted its first M.D. degrees in 1985. The School was accredited in 1992 by the Southern Association of Colleges and Schools to award the Ph.D. degree, and its first graduates finished in 1998.

Applying

Applications must be submitted by February 1 for consideration for admission and stipend support in July and by October 1 for admission and stipend support in January. Bachelor's degrees, with strong performance in science courses, are expected. GRE General Test scores are required, and scores on the Subject Test in chemistry or biology are recommended. International applicants are required to submit TOEFL scores, and third-party verification of academic records and references may be required. Application materials and information can be obtained through the Assistant Director of Admissions in the Office of Admissions and Student Affairs via phone (404-752-1650) or e-mail (phdadmissions@msm.edu).

Correspondence and Information

Office of Graduate Education in the Biomedical Sciences Program
Morehouse School of Medicine
720 Westview Drive
Atlanta, Georgia 30310-1495
Phone: 404-752-1580
Fax: 404-752-1064

Morehouse School of Medicine

THE FACULTY AND THEIR RESEARCH

Mukaila Akinbami, Ph.D., Missouri–Columbia. Integrated physiology of transgenic animal models; role of high blood pressure on vascular function; gene expression profiles and stroke. akin@msm.edu

Leonard M. Anderson, Ph.D., Northwestern. Cardiovascular genomics; vascular smooth-muscle-cell fate determination from stem cells. landerson@msm.edu

Methode Bacanamwo, Ph.D., Illinois. Chromatin remodeling and epigenetic mechanisms in the regulation of vascular gene expression in health and disease. mbacanmw@msm.edu

Mohamed A. Bayorh, Ph.D., Howard. Cardiovascular, neurochemical, and signal transduction pathways involved in the actions of polyunsaturated fatty acids, vasoactive substances, and drugs of abuse. bayorh@msm.edu

Jorge A. Benitez, Ph.D., CENIC (Cuba). Development of bacterial vaccines. jbenitez@msm.edu

Vincent C. Bond, Ph.D., Penn State. DNA virology; mammalian cell biology. bond@msm.edu

L. DiAnne Bradford, Ph.D., Georgia Tech. Psychopharmacology; predicting clinical efficacy and safety across patient populations. bradford@msm.edu

Eugene Chen, Ph.D., Western Ontario. Role of nuclear receptors in vascular function. echen@msu.edu

Teh-Ching Chu, Ph.D., Louisville. Receptor pharmacology; medical acupuncture; herbal medicine. tc@msm.edu

Margaret Colden-Stanfield, Ph.D., Texas Medical Branch. Cardiovascular pharmacology; membrane biophysics; cellular physiology; immunology; ion channel activation in vascular endothelial cells and leukocytes; role in inflammatory response. stanfiel@msm.edu

Hector DeLeon, Ph.D., McGill. Molecular relationships between recruited inflammatory cells and resident cells in injured tissues; DNA microarray analysis and gene-targeting technology to identify regulatory inflammatory molecules as therapeutic targets for cardiovascular diseases. hdeleon@msm.edu

Kamla Dutt, Ph.D., Punjab (India). Culture methodology for studies of retinitis pigmentosa; retinal and pigment epithelial interactions; HIV in ocular tissues; normal skin and benign nevus growth. duttk@msm.edu

Francis Eko, Ph.D., Vienna. Vaccine development for *Chlamydia*. feko@msm.edu

Byron Ford, Ph.D., Meharry. Cellular and molecular mechanisms involved in the pathophysiology of atherosclerosis and stroke. bford@msm.edu

Minerva Garcia-Barrio, Ph.D., Salamanca (Spain). Role of bHLH transcription factors as modulators of vascular remodeling. garbarm@msm.edu

Gary Gibbons, M.D., Harvard. Regulation of vascular remodeling. ggibbons@msm.edu

Sandra A. Harris-Hooker, Ph.D., Atlanta. Loss of endothelial cells and smooth-muscle cells in the pathogenesis of atherosclerosis; in vitro blood vessel modeling. hooker@msm.edu

Jacqueline Hibbert, Ph.D., West Indies. Metabolic responses to disease; stable isotope tracer analysis of disease effects on protein and energy nutritional requirements. hibberj@msm.edu

Joseph Igietseme, Ph.D., Georgetown. Investigation of the cellular, molecular, and biochemical mechanisms of chlamydial immunity and immunopathogenesis. igiets@msm.edu

Ward Kirlin, Ph.D., Emory. Chemical carcinogenesis and toxicology; molecular regulation of induction pathways involved in activation and detoxification of carcinogens. kirlin@msm.edu

Brenda J. Klement, Ph.D., Kansas State. Endochondral bone formation and skeletal tissue changes that occur in microgravity conditions. klement@msm.edu

Gordon J. Leitch, Ph.D., Chicago. Pathophysiology of diarrheal diseases and host-enteroparasite relationships. leitch@msm.edu

James Lillard, Ph.D., Kentucky. Role of chemokines in immunity, inflammation, and cancer. lillard@msm.edu

Woo-Kuen Lo, Ph.D., Wayne State. Ultrastructure and cell biology of the eye; intercellular junctions; cell membrane and cytoskeleton of the lens. lowk@msm.edu

Deborah A. Lyn, Ph.D., West Indies. Molecular biomarkers in tumor development and cardiovascular hypertension; sites of cellular HIV-1 integration and disease progression. lyn@msm.edu

Peter MacLeish, Ph.D., Harvard. Understanding the functional organization of the vertebrate retina. macleip@msm.edu

David R. Mann, Ph.D., Rutgers. Reproductive endocrinology: stress and testicular steroidogenesis, sexual and behavioral differentiation in primates, postmenopausal osteoporosis; immune-endocrine interactions. mann@msm.edu

Julian Menter, Ph.D., George Washington. Dermatology, photobiology, and photochemistry; physical organic and physical biochemistry. menteri@msm.edu

Gale Newman, Ph.D., LSU. Host responses to *Helicobacter pylori*, the causative agent of ulcers, toward vaccine development; HIV infection of human macrophages. newmang@msm.edu

John W. Patrickson, Ph.D., Howard. Chronobiology; neural mechanisms in the generation of circadian rhythms. patricks@msm.edu

Douglas F. Paulsen, Ph.D., Wake Forest. Skeletal pattern formation during embryonic development; microgravity effects on the musculoskeletal system. paulsen@msm.edu

Silvia S. Pierangeli, Ph.D., Louisville. Antiphospholipid antibodies; systemic lupus erythematosus. pierans@msm.edu

Michael D. Powell, Ph.D., Texas at Dallas. Role of cellular factors in the regulation of HIV-1 reverse transcription. powellm@msm.edu

Veena N. Rao, Ph.D., Osmania (India). Molecular and functional dissection of ELK-1 and BRCA-1 tumor-suppressor genes; role in cell growth; differentiation, signal transduction, and apoptosis in cancers. vrao@msm.edu

E. Shyam P. Reddy, Ph.D., Andhra (India). Functional role of ets, fusion onco-proteins, and tumor suppressors in leukemias, lymphomas, and sarcomas; function-based therapeutic approaches to human cancer. ereddy@msm.edu

Gary L. Sanford, Ph.D., Brown. Modulation of lung growth, maturation, and function; remodeling of the pulmonary vasculature; cellular and molecular studies of the role of a soluble lectin in these processes. sanford@msm.edu

Qing Song, M.D., Peking; Ph.D., South Carolina. Molecular mechanisms of genetic susceptibility to cardiovascular disease, obesity, and diabetes. qsong@msm.edu

Rajagopala Sridaran, Ph.D., University of Health Sciences (Chicago). Reproductive endocrinology; gravity during pregnancy; molecular and cellular antifertility action of gonadotropin-releasing hormone (GnRH); role of luteal GnRH in corpus luteum demise and parturition. sridaran@msm.edu

Jonathan Stiles, Ph.D., Salford (England). Molecular and cellular biology of *Trypanosoma-*, *Plasmodium-*, and *Trichomonas*-induced pathogenesis. stiles@msm.edu

Myrtle Thierry-Palmer, Ph.D., Wisconsin–Madison. Vitamins D and K metabolism and function. theirrm@msm.edu

Kelwyn H. Thomas, Ph.D., California, San Diego. Gene regulatory mechanisms involved in cellular differentiation; germ cell development in mouse testis. thomask@msm.edu

Winston Thompson, Ph.D., Rutgers. Cell and reproductive biology; molecular mechanisms of ovarian follicular development and cyst formation. thompsw@msm.edu

Gianluca Tosini, Ph.D., Bristol (England). Interactions between retinal and hypothalamic circadian clocks. tosinig@msm.edu

Wenli Wang, M.D., Ph.D., Beijing. Role of Notch signaling pathway components in vascular smooth-muscle-cell growth, hypertrophy, and apoptosis. wwenli@msm.edu

Joseph A. C. Whittaker, Ph.D., Howard. Basal ganglia anatomy and physiology; electrophysiology; excitotoxic mechanisms; brain stem neural circuits in cardiorespiratory control. whittaj@msm.edu

Evan F. Williams, Ph.D., Howard. Role of nucleoside transporters in cardiovascular function; ocular purinergic systems. evan@msm.edu

Lawrence E. Wineski, Ph.D., Illinois. Neural organization of craniofacial musculature; microgravity effects on the musculoskeletal system. wineski@msm.edu

Qinling Yang, Ph.D., Washington State. Molecular cardiovascular disease mechanisms. qyang@msm.edu

Xuebiao Yao, Ph.D., Berkeley. Mechanisms of mitotic chromosome segregation; establishment and maintenance of cell polarity. xyao@msm.edu

MOUNT SINAI SCHOOL OF MEDICINE OF NEW YORK UNIVERSITY

Graduate School of Biological Sciences

Programs of Study

The Graduate School of Biological Sciences of Mount Sinai School of Medicine offers graduate education in diverse, cutting-edge areas of biomedically important basic sciences in the Ph.D. and M.D./Ph.D. programs. Participating laboratories reflect a special commitment to developing and translating new findings in state-of-the-art basic biomedical sciences to problems of human disease and therapy. The Graduate School has adopted a model for its predoctoral training programs that reflects the multidisciplinary nature of contemporary biomedical sciences. Ph.D. and M.D./Ph.D. students are encouraged to pursue a course of study that focuses on rigorous mastery of the art, craft, and conceptual framework for an important research problem.

All students enter without a formal commitment to a particular training area. They pursue a series of rotations through diverse research laboratories before formally choosing a Ph.D. research mentor and one of six multidisciplinary training areas (MTAs). The MTAs were designed to facilitate interdepartmental interdisciplinary interactions and include biophysics, structural biology, and biomathematics (BSBB); genetics and genomic sciences (GGS); mechanisms of disease and therapy (MDT); molecular, cellular, biochemical, and developmental sciences (MCBDS); microbiology (MIC); and neurosciences (NEU). Within each of these broad categories, there is flexibility for students to explore virtually every realm of biomedical sciences, including cancer biology, endocrinology, enzymology, gene expression–transcription, gene therapy, immunobiology, mathematical modeling, molecular endocrinology, organellar biology–trafficking, pathobiology, peptide hormones, pharmacology, physical biochemistry, rational drug design, receptors, signal transduction–cell cycle, structural biology, systems physiology, transport channels/metabolism, and virology and oncogenes.

Research Facilities

Modern research facilities are housed primarily in the Annenberg and East Buildings. Throughout their graduate careers at Mount Sinai, students take advantage of the extraordinarily equipped laboratories, including core facilities for special biophysical, molecular biological, and immunological techniques. In addition, there is a superb library that offers ongoing opportunities for the enhancement of computer skills.

Financial Aid

All students are supported by a common stipend ($26,500 for 2006–07). The fellowship also covers the cost of tuition and provides a comprehensive health insurance package.

Living and Housing Costs

Subsidized housing is available for all students at a wide range of costs. Most single students reside in Aron Hall, which is across the street from the East Building.

Student Group

The 254 graduate students (178 Ph.D. and 76 M.D./Ph.D. students) are drawn from a national and international pool of applicants. About half of the students are women, and applicants from all racial and ethnic groups are welcome.

Student Outcomes

Mount Sinai's Ph.D. and M.D./Ph.D. students have enjoyed much professional success following completion of their training. Graduates have assumed academic positions at premier institutions in this country and abroad and have begun to assume leadership positions on their campuses and in national advisory groups. They are contributing to cutting-edge areas of experimental science. Mount Sinai's Ph.D. and M.D./Ph.D. alumni are at institutions such as Barnard; Case Western Reserve; Columbia; Cornell Medical School; the City University of New York; Duke; Einstein; Harvard; McGill; National Yang-Min University (Taiwan); New Jersey Institute of Technology; Northwestern; New York Medical College; New York University; New York University Medical School; Rutgers; Sloan-Kettering; the State University of New York Upstate Medical Center; Tulane; the Universities of California, Colorado, Connecticut, Indiana, Massachusetts, Minnesota, North Carolina, Pennsylvania, Pittsburgh, Puerto Rico School of Medicine, Texas, and Washington; Williams; and Yale. A number of graduates are at such institutions as the Fox Chase Cancer Institute and the National Institutes of Health. Still others have pursued their careers in industrial settings, some working on very basic research efforts and others conducting research or supervising programs that have a more applied theme. They have been employed at companies such as Advanced Tissue Sciences; Alliance Pharmaceuticals; Amgen; Astra; Bristol-Myers Squibb; DuPont-Merck; Eli Lilly; Enzon, Inc.; Immunomedia; Incstar; Metpath; Novartis; Oncogene Science, Inc.; OrthoBiotech; OrthoLogic; Procter & Gamble; Schering-Plough; Sephcor; Smith-Kline Beecham Pharmaceuticals; Trophix Pharmaceuticals; and Wyeth-Ayerst Pharmaceuticals.

Location

The Mount Sinai Medical Center occupies a four-block area on Fifth Avenue in Manhattan, across from Central Park. A Recreation Office provides students with extraordinary access to the entertainment and cultural opportunities of New York City's exciting urban environment. Tickets to concerts, the theater, and the opera are available, and major museums are in the neighborhood. Facilities for tennis, jogging, and other sports are available across the street in Central Park. Universal equipment and basketball facilities for students are located in the Residence Hall. Students also have access to the outstanding facilities of the nearby 92nd Street Y.

The Graduate School

An informal collegial spirit pervades the institution, providing easy interaction between graduate students and the faculty and between graduate students and other trainee groups. Students enjoy journal clubs and seminars, travel regularly to scientific meetings, and use the services of other scientific institutions in New York. Graduate students enjoy an environment in which basic research, research applied to fundamental problems in human biology, and more clinical research and clinical applications enrich one another. The predoctoral education prepares students to take creative and sophisticated approaches to new scientific problems in a variety of career settings.

Applying

Applicants to the Ph.D. program should have a strong science background. Prior to matriculation, students generally have completed one year of differential and integral calculus, one year of physics, chemistry through second-semester organic chemistry, one year of biology, and one semester of biochemistry. It is desirable for applicants to have taken some advanced science course work. Students with an interest in areas such as biomathematics and neuroscience may have different backgrounds appropriate for graduate work in those fields. The GRE General Test is required of all applicants. Submitting results of GRE Subject Tests is strongly recommended. The TOEFL is required for students for whom English is not the first language and who did not complete a four-year undergraduate degree at an English-speaking institution.

Aside from the quality of undergraduate performance and the indications of general aptitude that are derived from standardized test scores, the most important criterion for acceptance into the Ph.D. program is evidence that the applicant has those characteristics that lead to an independent and productive research career. Thus, the applicant's exposure to research is of great interest to the Admissions Committee. Those applicants who are under the most serious consideration for acceptance are invited for an interview. Students who have or will have completed B.A. or B.S. programs may obtain application materials online at http://www.mssm.edu/gradschool. The deadline for the receipt of all application materials is January 15, but early submission of applications is highly recommended for full competition for fellowships. All application materials must be sent to the graduate school at the address in the contact information section.

Correspondence and Information

Admissions Officer
Graduate School of Biological Sciences
Mount Sinai School of Medicine
One Gustave L. Levy Place, Box 1022
New York, New York 10029-6574

Phone: 212-241-6546
Fax: 212-241-0651
E-mail: grads@mssm.edu
Web site: http://www.mssm.edu/gradschool

Mount Sinai School of Medicine of New York University

REPRESENTATIVE FACULTY AND THEIR RESEARCH

Most faculty members are associated with more than one training area. The primary affiliation is provided.

Biophysics, Structural Biology, and Biomathematics (BSBB)
A. Aggarwal: Structural biology; DNA-binding proteins.
C. Bancroft: Transcriptional signals.
C. Bodian: Statistical problems in research.
K. Borden: RING proteins.
Z. Fayad: Magnetic resonance imaging; cardiovascular.
D. Logothetis: G protein; K^+ channels.
S. Maayani: Membrane receptors and effectors.
J. Mandeli: Statistical design and analysis of experiments.
M. Max: Circadian rhythms.
M. Mezel: Monte Carlo computer simulations.
R. Osman: Receptors; molecular biophysics; DNA.
S. Rackovsky: Protein folding.
R. Sanchez: Bioinformatics.
L. Sirovich: Image analysis; fluid dynamics.
A. Sornborger: Detecting signals in imaging data.
S. Wallenstein: Clinical trials; experimental design; epidemiology.
S. Wearne: Mathematical neuroscience; computational biology.
M. Zhou: Protein NMR spectroscopy; structural biology.

Genetics and Genomic Sciences (GGS)
D. Bishop: Biochemical/molecular genetics.
J. Chen: Genetic susceptibility and gene interactions in human diseases.
R. Desnick: Biochemical/molecular genetics.
G. Diaz: Molecular genetics of inherited metabolic disease.
B. Gelb: Molecular cardiology.
K. Hirschhorn: Cytogenetics.
Y. Ioannou: Molecular genetics.
J. Martingetti: Molecular biology; cancer genetics; tumor suppressor genes.
E. Schuchman: Molecular genetics.
M. Walsh: Modulation of chromatin structure and function.
P. Warburton: Chromosome structure.
J. Wetmur: DNA recombination and repair.

Molecular, Cellular, Biochemical, and Developmental Sciences (MCBDS)
D. Bechhofer: Regulation of bacterial gene expression.
J. Bieker: Transcription; gene expression.
A. Caplan: Molecular chaperones; signal transduction.
A. Cederbaum: Ethanol-induced oxidative stress and cytotoxicity.
A. Chan: Ras-related oncogenes in human cancer.
P. Cortes: Antigen receptor gene.
M. Diverse-Pierluissi: Receptor-mediated signals.
D. Felsenfeld: Receptor function.
J. Hirsch: G-protein–mediated signaling.
S. Kohtz: Growth/differentiation of cardiomyoblasts.
R. Krauss: Cancer; oncogene; cell cycle.
J. Laitman: Vocal tract development.
B. Laurent: Chromosome dynamics during the processes of chromosome segregation; maintenance of genome integrity; gene transcription.
A. Levine: Steroid hormone and growth factor regulation of prostate cell development and carcinogenesis; prostate cancer-bone interactions.
Z. Ma: Phospholipases in insulin secretion/beta-cell apoptosis.
S. Masur: Regulation of cells and the extracellular matrix.
M. Mlodzik: Cellular interactions and signaling pathways.
M. O'Connell: Cell-cycle regulation.
Z. Pan: Cycle controls.
J. Reidenberg: Anatomy of respiratory/vocal tract and skull.
L. Satlin: Transport; ion channels.
P. Shaw: Cysteine proteinase inhibitor genes.
R. Taneja: Signaling mechanisms in cellular differentiation.
A. Ting: Immunobiology; signal transduction.
M. Walsh: Cell cycle; transcriptional control.
L. Wang: Oncogenes; signal transduction.
P. Wassarman: Fertilization in mammals.
T. Weber: Development of nonviral gene therapy vectors.
D. Weinstein: Neural patterning.

Mechanisms of Disease and Therapy (MDT)
S. Aaronson: Molecular genetics of cancer.
M. Abreu: Toll-like receptor signaling in the intestine; inflammatory bowel disease.
M. Babyatsky: Nervous system and the immune response.
M. Baron: Mammalian blood and vascular development.
A. Branch: RNA biochemistry and therapeutics.
J. Bromberg: Molecular and cellular transplantation immunobiology.
D. Burstein: Cell differentiation.
O. Candia: Epithelial transport; water channels.
S. Chen: cancer immunology.
C. Cunningham-Rundles: Antibody production.
T. Davies: autoimmune thyroid disease.
E. Flatow: Rotator cuff tendon failure; shoulder mechanisms.
P. Frenette: Adhesion molecules in hematopoiesis and sickle cell disease.

S. Friedman: Gene expression during liver injury.
S. Ghaffari: Molecular mechanisms of regulation of hematopoietic stem and progenitor cell fate.
J. Gil: Image analysis in pathology.
R. Gordon: Fibrosis atherosclerosis.
M. Grace: Enzymology and gene expression.
S. Hall: Gene therapy for prostate cancer.
B. Herold: Herpes virus.
R. Iyengar: Signal transduction; G protein; drug design.
K. Jepsen: Skeletal biology.
M. Klotman: Molecular pathogenesis and therapy of HIV-I.
P. Klotman: Gene therapy; HIV-associated nephropathy.
D. LeRoith: Role of insulin and insulin-like growth factors in pathological states, including diabetes, growth, bone disorders, and cancer.
S. Lira: Chemokines; leukocyte trafficking; angiogenesis.
R. Majeska: Bone-cell biology.
J. Manfredi: p53 tumor-suppressor activity.
L. Mayer: T- and B-cell activation; antigen presentation.
R. Mira-y-Lopez: Aberrant vitamin A signaling in breast cancer.
L. Ossowski: Cancer metastasis; tumor dormancy.
G. Randolph: Antigen-presenting dendritic cells.
H. Sampson: Immunopathogenic mechanisms.
M. Schaffler: Biomechanics; skeletal biology.
A. Schecter: Mechanisms of atherosclerosis and arterial injury.
M. Skobe: Tumor metastasis; angiogenesis and lymphangiogenesis.
H. Snoeck: Biology of hematopoiesis.
F. Suchy: Transporters for bile acid.
Y. Suzuki: Asbestos related diseases.
N. Tulchin: Mammary carcinogenesis.
J. Unkeless: Fc receptors and signal transduction.
S. Woo: Gene therapy.
M. Zaidi: Intracellular calcium homeostasis.
D. Zhang: Cancer biology/signal transduction.
K. Zier: Immunotherapy; immunosuppression.

Microbiology (MIC)
C. Basler: Influenza virus and hemorrhagic fever virus.
J. Blaho: Regulation of HSV replication.
C. Bona: Immunology/molecular immunology.
A. Garcia-Sastre: Influenza virus vectors.
T. Krulwich: Tetracycline and Na^+ transporters; oxidative phosphorylation.
M. Linden: Gene therapy.
T. Moran: Immunology; virus infection.
S. Morgello: HIV infection in the brain.
P. Palese: Molecular and genetic analysis of RNA viruses.
B. Pogo: Viral oncology.
D. Tortorella: Viral subversion of the immune system.

Neurosciences (NEU)
C. Alberini: Molecular mechanisms of long-term memory.
D. Benson: Synaptogenesis in CNS.
A. Bergemann: Receptor tyrosine kinase.
V. Brezina: Neurophysiological systems.
J. Buxbaum: Molecular basis of neuropsychiatric disorders.
B. Cohen: Neurophysiology.
E. Cropper: Neural basis of behavioral plasticity.
S. Dracheva: Role of serotonic 2C receptor mRNA editing in psychiatric conditions (suicide, drug addiction); molecular abnormalities in the brain of schizophrenia patients.
G. Elder: Gene regulation of neurodegenerative diseases.
V. Haroutunian: Alzheimer's disease.
P. Hof: Cortical connectivity; dementia.
G. Holstein: Neurobiology of vestibular system.
G. Huntley: Cerebral cortex.
E. Kaplan: Neurophysiology.
E. Landau: Electrophysiology.
R. Margolskee: Sensory transduction; transgenics; molecular neurobiology.
A. McInnis: Genetics of neurobehavioral disorders.
C. Mobbs: Age-correlated pathologies.
S. Moore: Vstibular control of gaze and posture.
J. Morrison: Neurodegeneration; aging.
G. Pasinetti: Molecular neurogerontology.
I. Prohovnik: Functional neuroimaging; fMRI.
P. Rapp: Learning and memory.
N. Robakis: Alzheimer's disease.
S. Salton: Molecular neurobiology.
S. Sealfon: Signal transduction.
M. Shapiro: Memory and brain.
P. Shashidharan: Dystonia.
K. Weiss: Neurobiology of behavior.

NEW YORK MEDICAL COLLEGE

Graduate School of Basic Medical Sciences

Programs of Study

The Graduate School of Basic Medical Sciences (GSBMS) of New York Medical College offers programs leading to the M.S. and Ph.D. degrees in biochemistry and molecular biology, cell biology, experimental pathology, microbiology and immunology, pharmacology, and physiology plus an interdisciplinary M.S. program. There are specialized M.S. tracks in environmental science and health and toxicological pathology. The full-time faculty of 89 basic medical scientists, related through joint research ventures with a core of distinguished clinical research faculty members and their programs, offers a special opportunity to those with the requisite training and stamina.

Ph.D. degrees are awarded in six basic medical sciences. During the first year, students undertake an interdisciplinary core curriculum of courses and rotate through laboratories throughout the Graduate School. After this first year, students choose their major discipline and dissertation sponsor, complete the remaining didactic requirements in the chosen discipline, and begin intensive research training. Formal course work is usually completed within approximately two years, after which the student completes the qualifying exam, forms a dissertation advisory committee, presents a formal thesis proposal, and devotes his or her primary effort to the dissertation research project.

The M.S. requires completion of 30 credits. Two M.S. degree sequences are available: (1) a research program consisting of 25 didactic and up to 5 research credits and a research thesis or (2) a program consisting of 30 didactic credits and a scholarly literature review. The M.S. degree is earned full- or part-time in evening classes. The interdisciplinary M.S. program is particularly suitable for students wishing to prepare for a career in medicine, dentistry, or other health professions.

The Department of Cell Biology offers training in cell biology and neuroscience leading to careers in academia and industry. The research utilizes recombinant DNA, protein sequencing and mass spectroscopy, electron microscopy, tissue culture, fluorescent digital image analysis, confocal microscopy, electrophysiology and phosphorimaging, subcellular fractionation, and biochemistry.

The Department of Biochemistry and Molecular Biology provides students with a solid foundation in the concepts and applications of modern biochemistry and molecular biology. Areas of research include protein structure and function, enzyme reaction mechanisms, mechanisms of hormone action and cell signaling, enzymology, mechanisms of DNA replication and repair, cell-cycle regulation, molecular biology of cancer cells, and molecular neurobiology.

The Department of Pathology offers a vigorous multidisciplinary milieu for training in experimental pathology. The programs focus on the comprehensive study of pathogenic mechanisms of human disease. Areas of interest in the department include apoptosis, cell signaling and gene activation in cardiovascular disease, flow cytometry and cell-cycle analysis, tumor cell biology and immunology, biochemical toxicology, carcinogenesis, tissue engineering, tuberculosis and other chronic infectious diseases, free-radical pathobiology, aging, hypersensitivity, cytokine and growth factor analysis, and environmentally induced pathology.

In the Department of Microbiology and Immunology, the student acquires a broad acquaintance with microbiology, molecular biology, and immunology as well as depth in an elective field. Areas available for thesis research include molecular biology of tumor cells and cell growth factors, bacterial genetics, pathogenesis of infectious disease, monoclonal antibody synthesis, immune function in AIDS, structure and function of influenza virus antigens, molecular virology, and the biochemistry and genetics of emerging bacterial pathogens.

The Department of Pharmacology emphasizes training in research methods for examining of the action of drugs at the systemic, cellular, and subcellular levels. Areas of research include renal and corneal cytochrome P-450 arachidonic acid metabolism, patch-clamp analysis of electrolyte transport, GC mass spectroscopy of lipid metabolites, vasoactive hormones in hypertension and integrative cardiovascular function, neuroendocrinology and prohormone processing, pathophysiologic factors in stroke and renal vascular disease, and gene-based therapy for cardiovascular disease.

The Department of Physiology provides students with an understanding of the function of the body's cells and organ systems and the mechanisms for regulation of these functions. Research opportunities include cellular neurophysiology, neural and endocrine control of the heart and circulation, microcirculation, the physiology of gene expression, heart failure, cardiac metabolism, and the physiological effects of oxygen metabolites.

Research Facilities

The College has an extensive laboratory complex in the basic medical and clinical sciences. The Basic Sciences Building houses the medical sciences library, which maintains 200,000 volumes, major journals, and a computer-based retrieval system. There are also a fully accredited comparative medicine facility, a well-equipped and staffed instrumentation shop, a variety of classrooms, a bookstore, a cafeteria, and student lounges.

Financial Aid

Federal and state loan programs are available for M.S. students. Ph.D. students receive tuition remission, medical insurance, and combinations of College fellowships and research assistantships. Inquiries regarding College support should be directed to the Graduate School Admissions Office. The Office of Financial Aid should be consulted for information on federal and state programs.

Cost of Study

In 2006–07, tuition is $645 per credit, or $10,320 annually, for a full-time master's student taking at least 8 credits per semester. Annual Ph.D. tuition is $14,160 before candidacy (first two years) and $4000 after candidacy. Fees range between $30 and $330 per year depending upon options chosen. Comprehensive medical insurance is available for individual ($2100 annually) or family ($5100 annually) coverage.

Living and Housing Costs

A limited number of rooms and apartments are available for graduate students in on-campus College housing. On-campus housing costs range from $490 to $600 per month for suite-style apartments to $1315 per month for a three-bedroom apartment (families with children). Private off-campus accommodations are also available. Students should contact the Director of Housing, Candy Hack, Administration Building (telephone: 914-594-4832), well in advance of arrival in order to make housing arrangements.

Student Group

The total College enrollment is 1,407. There are 53 Ph.D. and 102 M.S. students in the Graduate School of Basic Medical Sciences.

Location

The College campus is located in the Westchester Medical Center campus, 5 miles from White Plains and 28 miles north of New York City.

The College

New York Medical College, one of the largest medical schools in the country, was established in 1860. Graduate education at the College began informally in 1910, graduate degrees were offered as early as 1938, and a graduate division was established in 1963.

Applying

Applications for admission may be submitted at any time during the year. For optimal review of credentials and consideration for financial aid and housing, however, applications for fall enrollment into Ph.D. programs should be received by February 1. Specific program requirements are published in the Graduate School bulletin and are also available on the College Web site. Students may apply online at the College Web site or download a blank application. Paper applications may also be obtained from the Graduate School Admissions Office. M.S. and Ph.D. applicants must submit GRE General Test scores. International students are required to submit results of the TOEFL. Undergraduate transcripts and two letters of recommendation from teachers or scientists personally familiar with the applicant must be submitted directly by the school or recommenders separately.

Correspondence and Information

Francis L. Belloni, Ph.D., Dean
Graduate School of Basic Medical Sciences
Basic Sciences Building, Room A41
New York Medical College
Valhalla, New York 10595
Web site: http://www.nymc.edu/gsbms/

New York Medical College

THE GRADUATE FACULTY AND THEIR RESEARCH

Biochemistry and Molecular Biology. E. Y. C. Lee, Ph.D., Professor and Chairman: enzymology, structure-function relationships, and regulation of ser/thr protein phosphatases. D. N. Frick, Ph.D., Associate Professor: molecular mechanisms of hepatitis C virus replication and drug resistance; protein expression and purification; structure-based rational antiviral drug design. M. I. Horowitz, Ph.D., Professor: interaction of glucose with histones and membrane lipids; properties and characterization of sulfotransferases; nutritional biochemistry. M. Y. W. Lee, Ph.D., Professor: DNA replication, polymerases, and repair; cell-cycle regulation. S. C. Olson, Ph.D., Associate Professor: signal transduction; regulation of phospholipase D pathway by protein kinase C and G proteins. E. L. Sabban, Ph.D., Professor: molecular neurobiology; molecular mechanisms of stress; cloning and regulation of gene expression for catecholamine-synthesizing enzymes and neuropeptides. Y. C. Tse-Dinh, Ph.D., Professor: protein-DNA interactions; topoisomerase structure and function; gene regulation and DNA supercoiling. B. I. Weinstein, Ph.D., Professor and Graduate Program Director: biochemistry of steroid action; metabolism and biologic activity of cortisol; enzyme deficiencies in glaucoma. J. M. Wu, Ph.D., Professor: regulation of gene expression in leukemic and prostate cancer cells; cell-cycle control; chemoprevention by fenretinide and resveratrol. Z. Zhang, Ph.D., Assistant Professor: X-ray crystallography; stem-cell factor; quinine reductase2.

Cell Biology and Anatomy. J. D. Etlinger, Ph.D., Professor and Chairman: skeletal muscle growth and atrophy; intracellular proteolysis in erythroid and muscle cells; role of proteasomes and ubiquitin; spinal cord injury. A. B. Drakontides, Ph.D., Professor Emerita: pathogenesis of early and late changes at the neuromuscular junction and skeletal muscle induced by chemical irritants. V. A. Fried, Ph.D., Professor and Graduate Program Director: ubiquitin and cellular regulation; cytoskeletal structure and functions. F. Hannan, Ph.D., Assistant Professor: *Drosophila melanogaster;* neurofibromastosis; learning a memory; Res; adonglyl cyclose; expression profiles. J. Kang, M.D., Ph.D., Associate Professor: astrocyte-mediated modulation of inhibitory synaptic transmission; interplay between excitatory and inhibitory synapses; properties of gap junction, K+, and GABA-A channels. K. M. Lerea, Ph.D., Associate Professor: mechanisms of signal transduction; role of protein ser/thr kinases and phosphatases in integrin functions and platelet activation. S. A. Newman, M.D., Professor: physical and molecular mechanisms of development and evolution; pattern formation in the vertebrate limb; collagen assembly. R. Rozental, M.D., Ph.D., Associate Professor: role of connexins in nervous system development and dysfunction in ischemia and perinatal seizures. T. Sato, M.D., Associate Professor: regulation of calcium in normal and dystrophic skeletal muscles. P. B. Sehgal, M.D., Ph.D., Professor: interleukin-6; p53; gene expression; signal transduction (STAT3). S. C. Sharma, Ph.D., Professor: genetic approaches to regeneration of adult CNS neurons. A. D. Springer, Ph.D., Professor: engineering models of retinal development; optic nerve regeneration. P. K. Stanton, Ph.D., Professor: neuronal plasticity; long-term depression and potentiation of synaptic strength; synaptic functional changes in epilepsy; mechanisms of ischemia-induced delayed neuronal death. G. Suarez, M.D., Research Associate Professor: nonenzymatic protein glycation; sorbitol pathway; cell senescence; diabetic complications; self-assembly of collages and lens crystallins. R. J. Zeman, Ph.D., Associate Professor: β_2-adrenoceptors in musculoskeletal growth; mechanisms of spinal cord injury; regulation of intracellular calcium.

Experimental Pathology. P. M. Chander, M.B.B.S., Professor: pathogenesis of renal and vascular damage in stroke-prone spontaneously hypertensive rats; pathogenesis of HIV-associated nephropathy. W. Dai, Ph.D., Associate Professor: cell-cycle regulation. Z. Darzynkiewicz, M.D., Ph.D., Professor: development of new methods of cell analysis using flow cytometry; analysis of cell-cycle specificity of antitumor drugs. H. P. Godfrey, M.D., Ph.D., Professor and Graduate Program Director (Ph.D. programs): mechanisms of pathogenesis in tuberculosis; biomedical mechanisms of delayed hypersensitivity, chronic inflammation, and infectious disease. M. I. Iatropoulos, M.D., Research Professor: comparative mechanisms of toxicity and carcinogenesis. A. M. Jeffrey, Ph.D., Research Professor: toxicology and chronic carcinogenesis. M. Jhanwar-Uniyal, Ph.D., Research Associate Professor: signal transduction, BRCA, p53, cancer, central nervous system in obesity. A. Kumar, Ph.D., Professor: role of renin-angiotensin system in hypertension and atherosclerosis. J. H.-C. Lin, Ph.D., Research Assistant Professor: molecular mechanisms governing endothelial cell dysfunctions during atherogenesis. P. A. Lucas, Ph.D., Research Associate Professor: wound healing and tissue engineering. M. R. Melamed, M.D., Professor: flow and static cytometry of human cancer cells; cytochemical, immunochemical, and in situ nucleic acid reactions for diagnostic and prognostic purposes. F. H. Moy, Ph.D., Associate Professor of Clinical Pathology and Graduate Program Director (M.S. programs): biostatistics and epidemiology, methodology, and applications in environmetrics and risk assessment. F. Traganos, Ph.D., Professor: mechanisms of cell-cycle progression (checkpoints) and cell death (apoptosis) in cell cultures and clinical models. J. H. Weisburger, Ph.D., M.D. (hon.), Research Professor: mechanisms of toxicity and carcinogenicity; mechanisms and role of promoters in major human cancers; role of nutrition in human carcinogenesis; rational means of prevention of cancer, coronary heart disease, and stroke. G. M. Williams, M.D., Professor: mechanisms of carcinogenesis; metabolic and genetic effects of chemical carcinogens. R. E. Zachrau, M.D., Professor: spontaneous and induced tumor-specific, cell-mediated immunity in human breast cancer and its role in development of systemic metastasis and second primary cancers of breast and nonbreast origin.

Microbiology and Immunology. I. S. Schwartz, Ph.D., Professor and Chairman: molecular pathogenesis of Lyme disease and other emerging bacterial pathogens; functional genomics. M. E. Aguero-Rosenfeld, M.D., Associate Professor: pathogenesis and diagnosis of Lyme disease and human granulocytic ehrlichiosis. A. Banerjee, Ph.D., Assistant Professor: cloning, characterization, and pathogenesis of phase-variable genes in *Neisseria;* role of bacterial and mycobacterial surface glycans. R. Banerjee, Ph.D., Assistant Professor: molecular virology and molecular oncology. D. Bessen, Ph.D., Professor: molecular pathogenesis, epidemiology, and evolutionary biology of group A *Streptococcus* (GAS); role of GAS infection in pediatric neuropsychiatric disorders. D. Bucher, Ph.D., Associate Professor: structure, function, and immunochemistry of viral antigens. F. Cabello, M.D., Professor: microbial genetics; infectious disease; recombinant DNA. J. Geliebter, Ph.D., Associate Professor: immunology and molecular biology of prostate cancer. C. V. Hamby, Ph.D., Associate Professor: molecular biology and immunology of human tumors. E. D. Kilbourne, M.D., Professor Emeritus: virology; viral genetics; influenza vaccines. B. Safai, M.D., Professor: pathophysiology and pathogenesis of skin diseases. R. K. Tiwari, Ph.D., Associate Professor and Graduate Program Director: tumor immunology and chemoprevention; cellular immunology; immune dysregulation in disease. F. E. Wassermann, Ph.D., Professor Emeritus: virus genetics; epidemiology, bioethics.

Pharmacology. J. C. McGiff, M.D., Professor and Chairman: neural and hormonal control of circulation and renal function. N. G. Abraham, Ph.D., Professor: gene transfer and gene therapy in the cardiovascular system and CD34+ cells. M. Balazy, Ph.D., Professor: biochemistry of arachidonic acid metabolism. M. A. Carroll, Ph.D., Professor: renal cytochrome P-450 metabolites of arachidonic acid. C. Conaway, Ph.D., Adjunct Assistant Professor: mechanisms of cancer chemoprevention; metabolism pharmacokinetics. N. R. Ferreri, Ph.D., Professor: cytokine production and function in the kidney and vascular smooth muscle. M. S. Goligorsky, M.D., Ph.D., Professor: basic mechanisms of endothelial dysfunction, its prevention and reversal; translation of bench findings to clinical physiology and pharmacology. K. Gronert, Ph.D., Assistant Professor: eicosanoid; inflammation; wound healing; lipaxtons. M. A. Inchiosa Jr., Ph.D., Professor: biochemical pharmacology of muscle. A. Nasjletti, M.D., Professor and Graduate Program Co-director: hormonal mediators of blood pressure regulation. C. A. Powers, Ph.D., Associate Professor: neuroendocrinology. J. Quilley, Ph.D., Associate Professor: Interactions of vasoactive hormones and eicosanoids in vascular regulation in diabetes and hypertension. M. L. Schwartzman, Ph.D., Professor and Graduate Program Co-director: cytochrome P-450 metabolism of arachidonic acid in inflammation and hypertension. C. J. Smith, Ph.D., Adjunct Associate Professor: mechanisms of altered hormone-dependent activation/expression of cardiovascular cyclic nucleotide phosphodiesterases, protein kinase C, and immediate-early gene expression during heart failure, vascular injury, and diabetes. C. T. Stier, Ph.D., Associate Professor and M.S. Program Director: pharmacological protection against vascular damage and stroke. W. Wang, M.D., Professor: regulation of renal electrolytes transport.

Physiology. G. Kaley, Ph.D., Professor and Chairman: control of blood pressure and blood flow. P. Anversa, M.D., Professor of Medicine: cardiac hypertrophy and aging; myocardial cell apoptosis. F. L. Belloni, Ph.D., Professor: cardiovascular control; vascular and cardiac actions of adenosine. John G. Edwards, Ph.D., Assistant Professor: physiological control of gene transcription; regulation of transcription factors; cardiac hypertrophy; exercise biochemistry and overload alterations of the myocardial phenotype. T. H. Hintze, Ph.D., Professor: cardiovascular functions in chronically instrumented animals. A. Huang, M.D., Ph.D., Assistant Professor of Physiology: role of estrogens in vascular function. A. Koller, M.D., Professor: regulation of blood flow in the microcirculation. C. S. Leonard, Ph.D., Professor: modulation of mesopontine cholinergic nervous and neocortical interneurons; mammalian oculomotor system in CNS. E. M. Levee, D.V.M., Assistant Professor: comparative medicine. N. Levine, Ph.D., Professor: fluid and electrolyte secretion in the male reproductive system. E. J. Messina, Ph.D., Professor: microvascular control and regulation of smooth-muscle reactivity. S. S. Passo, Ph.D., Professor: neuroendocrine control of blood pressure. Fabio A. Recchia, M.D., Associate Professor: control of myocardial metabolism; nitric oxide; heart failure; cardiac mechanics and efficiency; coronary circulation. W. N. Ross, Ph.D., Professor: regional properties of neurons. D. Sun, M.D., Ph.D., Associate Professor: role of endothelial stress on coronary arteriolar function. C. I. Thompson, Ph.D., Associate Professor and Graduate Program Director: renal hemodynamics and GFR control. M. S. Wolin, Ph.D., Professor: vascular regulation via cyclic GMP, metabolites, and oxygen tension.

NEW YORK UNIVERSITY

School of Medicine
Sackler Institute of Graduate Biomedical Sciences

Programs of Study

The Sackler Institute of Graduate Biomedical Sciences at the New York University School of Medicine is a division of New York University's Graduate School of Arts and Sciences, which offers graduate training programs in the basic medical sciences leading to the Ph.D. degree and, in coordination with the Medical Scientist Training Program, a combined M.D./Ph.D. program. The Institute encompasses the basic medical science departments at the School of Medicine that offer interdisciplinary training programs in biomedical imaging, cellular and molecular biology, computational biology, developmental genetics, medical and molecular parasitology, microbiology, molecular oncology and immunology, molecular pharmacology and signal transduction, neuroscience and physiology, pathobiology, and structural biology. Each program is individually administered with its own requirements. Students in most programs complete their doctoral training in five years. There are no terminal master's degree programs offered at the Sackler Institute.

When applying for admission to the Sackler Institute, students have the option of either applying directly to individual training programs or entering the "Open Program." This latter option allows students the opportunity to perform research rotations during their first academic year in any laboratory of a member of the graduate faculty in the Sackler Institute, regardless of their departmental or program affiliation. Students then select a thesis adviser and program affiliation by the end of their first academic year. This is accomplished with the help of a Graduate Advisory Committee, exposure to all research possibilities through a series of faculty seminars, and participation in elective courses in the various disciplines.

Research Facilities

The New York University Medical Center is one of the largest teaching and research complexes in the country. Research laboratories are located in the Basic Medical Science Building, Tisch and Bellevue Hospitals, the Public Health Research Institute, the Center for Biomedical Imaging, the Skirball Institute of Biomolecular Medicine, and the newly completed Smilow Research Center. The participating faculty members occupy modern research laboratories that are equipped for research in their areas of interest. There are centralized departmental facilities as well as Medical Center–wide resources available through the Kaplan Cancer Center, the GCRC, and the Skirball Institute. These facilities and resources include the latest nucleic acid and protein sequencing and synthesizing instrumentation; structural biology laboratories with state-of-the-art X-ray crystallography equipment; fMRI and mass spectrometry facilities; P-2 and P-3 laboratories for studying highly infectious agents; modern microscopy equipment such as laser confocal microscopes, transmission and scanning electron microscopes, and image-processing facilities; and major animal-care centers, which include a transgenic mouse unit as well as a monoclonal antibody production laboratory. Excellent libraries and state-of-the-art computer facilities (ranging from individual personal computers to access to mainframe computers and minicomputers) are available to all students.

Financial Aid

All graduate students are supported by either assistantships or traineeships, which carry stipends of $27,000 for the 2007–08 academic year, in addition to all tuition fees and health insurance costs. Financial support is provided for the duration of study. No teaching or laboratory assisting is required for the receipt of financial aid. Low-interest housing loans of $1500 per year are also available for qualified students, as are loans for the purchase of personal computers.

Cost of Study

The cost of study is usually met by the financial aid awards, as described above.

Living and Housing Costs

University housing in dormitory facilities and apartment complexes within the immediate vicinity of the Medical Center is guaranteed for all incoming graduate students. Depending on the type of accommodation, rents range from $710 to $840 per month for shared apartments. Married student housing ranges from $1120 to $1790 per month. The services of the Medical Center's housing office are also available to students who have special needs or wish to live off campus.

Student Group

There are approximately 230 Ph.D. and 80 M.D./Ph.D. candidates in the Institute. On average, 35–45 new students matriculate each year. Students are drawn from a pool of highly qualified national and international applicants. About 15 percent of the students are from underrepresented minority groups, 55 percent are women, and 25–30 percent are international students. In addition to the graduate students, there are 675 M.D. candidates enrolled in the School of Medicine.

Location

The New York University Medical Center, where the Sackler Institute is located, is in the heart of midtown Manhattan, on First Avenue between 23rd and 38th Streets. The location provides easy access to the unparalleled myriad of cultural, social, and educational opportunities of New York City, with its world-class museums, libraries, theater, dance, music, and sporting events. To facilitate access to these activities, the Medical School has established a student activities office, which supplies information and discount tickets to many of these events.

The University

New York University, which was founded in 1831, is the largest private university in the country, with an enrollment of more than 50,000 students. The University includes thirteen colleges and schools at five major centers in Manhattan. The main campus is located around Washington Square in the historic Greenwich Village area of the city. University-sponsored bus service connects the main campus to the Medical Center. The Medical Center is composed of the School of Medicine and affiliated hospitals and research institutes and has a history of more than 155 years of training nationally and internationally recognized physicians and scientists.

Applying

Students are admitted to a Ph.D. program or, for those seeking a combined medical career, an M.D./Ph.D. program. Applicants for admission must have at least a bachelor's degree or its equivalent from a college or university of recognized standing and have a strong background in the biological, chemical, and physical sciences. Applicants must supply letters of recommendation and a personal statement describing their goals for graduate study, research interests and experiences, and reasons for applying for graduate study. The GRE General Test is required; Subject Tests are not required, though they are recommended. For international students from non-English-speaking countries, the TOEFL is required. Evaluation for admission to the different programs is carried out by individual program Admissions Committees and is based on previous academic achievement, research ability, letters of recommendation, and personal interviews. Applicants are encouraged to submit application forms and all supporting material by December 18. Applications received after this date are considered at the discretion of the Admissions Committee of the desired program. Students can obtain application materials, bulletins, and additional information from the Institute's Web site or directly from the Institute's offices.

Correspondence and Information

Matthew Cipriano
Admissions Coordinator
Sackler Institute of Graduate Biomedical Sciences
New York University School of Medicine
550 First Avenue
New York, New York 10016

Phone: 212-263-5648
Fax: 212-263-7600
E-mail: sackler-info@med.nyu.edu
Web site: http://www.med.nyu.edu/sackler

New York University

THE FACULTY

Students can do their thesis research in the laboratories of more than 165 faculty members who have appointments in the basic science departments at the School of Medicine and who are members of the Sackler Institute, as well as selected members of other NYU divisions, such as the Departments of Biology and Chemistry, the Courant Institute, and the Center for Neural Sciences. The names of the faculty members are listed in this volume under the individual graduate programs: Biomedical Imaging, Cellular and Molecular Biology, Computational Biology, Developmental Genetics, Medical and Molecular Parasitology, Microbiology, Molecular Oncology and Immunology, Molecular Pharmacology, Neuroscience and Physiology, Pathobiology, and Structural Biology (in coordination with the NIH). Below are the names of the program directors, graduate advisers, and coordinators or administrators for each program.

Biomedical Imaging (phone: 212-263-3308; e-mail: qun.chen@med.nyu.edu)
Director: Daniel H. Turnbull, Ph.D.
Graduate Advisor: Qun Chen, Ph.D.

Cellular and Molecular Biology (phone: 212-263-5360; e-mail: CMB.Program@med.nyu.edu)
Director: Daniel Rifkin, Ph.D.
Graduate Advisors: Lynette Wilson, Ph.D., Department of Cell Biology; Ed Ziff, Ph.D., Department of Biochemistry.

Computational Biology (phone: 212-263-6337; e-mail: cardot01@nyu.edu)
Director: Tamar Schlick, Ph.D.
Graduate Advisor: Timothy Cardozo, M.D., Ph.D.

Developmental Genetics (phone: 212-263-7290; e-mail: treisman@saturn.med.nyu.edu)
Director: Ruth Lehmann, Ph.D.
Graduate Advisor: Jessica Treisman, Ph.D.

Medical and Molecular Parasitology (phone: 212-263-8160; e-mail: eichid01@popmail.med.nyu.edu)
Director: Karen Day, Ph.D.
Graduate Advisor: Daniel Eichinger, Ph.D.

Microbiology (phone: 212-263-7662; e-mail: garabm01@popmail.med.nyu.edu; phone: 212-263-0415; e-mail: mohri01@popmail.med.nyu.edu)
Director: Claudio Basilico, M.D.
Graduate Advisors: Michael Garabedian, Ph.D.; Ian Mohr, Ph.D.

Molecular Oncology and Immunology (phone: 212-263-8192; e-mail: levyd01@popmail.med.nyu.edu; phone: 212-263-2540; e-mail: smithsu@saturn.med.nyu.edu)
Director: Angel Pellicer, M.D.
Graduate Advisors: David Levy, Ph.D.; Susan Smith, Ph.D.

Molecular Pharmacology and Signal Transduction (phone: 212-263-5963; e-mail: bache02@popmail.med.nyu.edu)
Director: Herbert H. Samuels, M.D.
Graduate Advisor: Erika Bach, Ph.D.

Neuroscience and Physiology (phone: 212-263-5770; e-mail: blooms01@popmail.med.nyu.edu)
Director: Rodolfo R. Llinas, M.D., Ph.D.
Graduate Advisor: Stewart Bloomfield, Ph.D.

Pathobiology (phone: 212-263-6827; e-mail: loomic01@popmail.med.nyu.edu)
Director: David B. Roth, M.D., Ph.D.
Graduate Advisor: Cindy Loomis, M.D., Ph.D.

Structural Biology (phone: 212-263-8634; e-mail: wang@saturn.med.nyu.edu)
NYU Director: David L. Stokes, Ph.D.
NIH Director: Nico Tjandra, Ph.D.
NYU Graduate Advisor: Da Nang Wang, Ph.D.

Medical Sciences Training Program (M.D./Ph.D.) (phone: 212-263-5649; e-mail: cindy.meador@med.nyu.edu)
Director: Rodney Ulane, Ph.D.
Administrative Officer: Cindy Meador.

Joan and Joel Smilow Research Center.

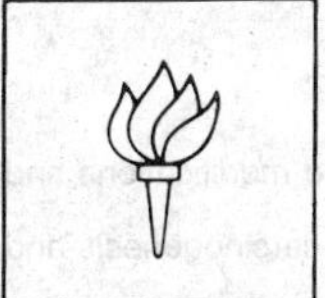

NEW YORK UNIVERSITY

School of Medicine and the Graduate School of Arts and Sciences
Department of Environmental Medicine
Graduate Program in Environmental Health Sciences

Programs of Study

The Department of Environmental Medicine at New York University (NYU) is a diverse and dynamic research-oriented facility that is dedicated to the study of all aspects of environmental health science, with emphasis on such major health problems as cancer, respiratory illness, and cardiovascular disease. The department offers a strong teaching and research program leading to both the M.S. and Ph.D. degrees through the Graduate School of Arts and Sciences of NYU.

The M.S. is offered in the Graduate Program in Environmental Health Sciences, with areas of study in environmental hygiene and environmental toxicology. Areas of study toward the Ph.D. are offered in biostatistics, epidemiology, ergonomics and biomechanics, exposure assessment and health effects, molecular toxicology/carcinogenesis, and toxicology.

During the first two years, Ph.D. students devote their time to course work and to research training via laboratory rotations. Comprehensive exams are usually taken at the end of the second year. The final two to four years are spent conducting independent research under the supervision of a thesis adviser chosen by mutual consent of adviser and student. Weekly departmental seminars by visiting scientists as well as student seminar groups and presentations contribute to widening the students' outlook and keeping them abreast of the advancing state of knowledge in the environmental health sciences. Because the student-faculty ratio is kept low, substantial interaction with faculty members is possible.

Research Facilities

Graduate student research is carried out in the department's research laboratories, which are well equipped with a variety of up-to-date, specialized scientific equipment to support the approximately ninety ongoing research projects.

A specialized environmental health library is located within the department, and the NYU Medical Center library provides a large resource of biomedical references. Computerized bibliographic aids are available. Computer facilities are available in-house for scientific and word processing use. Tie-ins are also available to the NYU computer network; free computer time is made available for student research.

Financial Aid

Successful Ph.D. and M.S. applicants qualify for financial aid plus tuition and fees (ergonomics and biomechanics candidates are not included). In 2006–07, Ph.D. students receive a stipend of $26,000 per year; M.S. students receive a stipend of $13,000 per year.

Cost of Study

Tuition for 2006–07 is $1080 per credit plus fees, but these costs are waived for students who receive financial aid.

Living and Housing Costs

The cost of living varies according to individual requirements and location. NYU apartments are available near the department's labs in Tuxedo, New York, and at NYU in Manhattan.

Student Group

Because of the diversity of the program, graduate students are accepted from a variety of backgrounds, including the biological, physical, and social sciences; mathematics; and engineering. Fifty-four graduate students are enrolled in the training program. Recent doctoral graduates have taken academic (40 percent), corporate (40 percent), and government (20 percent) scientific positions. A number have received postdoctoral fellowships at leading universities.

Location

NYU is in the heart of New York City. The many cultural centers of Manhattan are only minutes away by bus or subway. The Department of Environmental Medicine has a major facility located at Sterling Forest, a 20,000-acre reserve about 45 miles north of New York City. This location offers the dual benefits of rural/suburban living and proximity to New York City. Most courses are offered in Manhattan, and transportation between locations is provided.

The University and The Department

New York University, which was founded in 1831, is a large, private university with impressive scientific, scholarly, and cultural resources. The Department of Environmental Medicine has been designated by the National Institute of Environmental Health Sciences as a university center of excellence in teaching and research. It is also supported by the National Cancer Institute as a specialized center for research on environmental cancer.

Applying

Applications and supporting documents are due by December 15 for consideration for admission and financial aid. It is necessary to submit an NYU Graduate School application form, transcripts, three letters of recommendation, and scores on the GRE General Test. International students must submit scores on the Test of English as a Foreign Language (TOEFL). It is also helpful to arrange to visit the Department of Environmental Medicine and meet with several members of the faculty.

Correspondence and Information

Graduate Coordinator
Department of Environmental Medicine
New York University School of Medicine
57 Old Forge Road
Tuxedo, New York 10987
Phone: 845-731-3661
Fax: 845-351-3317
E-mail: ehs@env.med.nyu.edu
Web site: http://www.med.nyu.edu/environmental/graduate/

New York University

THE FACULTY AND THEIR RESEARCH

Max Costa, Ph.D., Chair, Department of Environmental Medicine. Metal carcinogenesis and toxicology; DNA-protein interactions; DNA damage; histone modifications and epigenetic mechanism of carcinogenesis.

Jerome J. Solomon, Ph.D., Director of Graduate Studies. DNA-carcinogen interaction; biological consequences of DNA adducts; mass spectrometry in carcinogenesis and environmental research.

Ilana Belitskaya-Levy, Ph.D. High-dimensional data analysis; algorithms for missing data analysis; EM algorithm; cluster analysis; developing statistical methods for analyzing large data arising in genomics and molecular biology; DNA microarrays; flow cytometry; statistical design and analysis of clinical trials; data mining; teaching.

Maarten C. Bosland, D.V.Sc., Ph.D. Hormonal carcinogenesis; prostate cancer chemoprevention; prostate and breast cancer; endocrine disruption; experimental pathology.

Fredric J. Burns, Ph.D. Cancer prevention and multiple stages in radiation carcinogenesis; patched gene and DNA repair genes in cancer susceptibility; arsenic cocarcinogenesis; DNA repair and proliferation.

Lung Chi Chen, Ph.D. Inhalation toxicology; exposure-response relationships; air pollution.

Beverly S. Cohen, Ph.D. Measurement of personal exposures to airborne toxicants; dosimetry of inhaled pollutant gases and aerosols; airborne radioactivity.

Mitchell D. Cohen, Ph.D. Pulmonary immunotoxicology of inhaled pollutants; effects of inhaled pollutants on lung/lung immune-cell iron homeostasis; modulation of cytokine biochemistry by metals and complex mixtures; pulmonary/immunotoxicology of World Trade Center dusts.

Norman Cohen, Ph.D. Radiobiology, radiochemistry, and radiophysics of the biokinetics, metabolism, and toxicology of heavy-metal, bone-seeking radionuclides in human and nonhuman primates; environmental ecology of radionuclides, bioassay, internal dosimetry, and assessment of low-level radiation exposures using measurements of internal radioactive contaminants; chelation and other therapeutic techniques for reducing health risks associated with systemic deposits of radioactive materials.

Mark S. Condon, Ph.D. Stromal-epithelial interactions in carcinogenesis; in vitro and animal models of prostate cancer progression and metastasis.

Hugh L. Evans, Ph.D. Neurotoxicology.

Emerich Fiala, Ph.D. Mechanisms of chemical carcinogenesis and cancer chemoprevention.

Krystyna Frenkel, Ph.D. Carcinogenesis and chemoprevention; role of endogenous oxidative stress in cancer and aging; contribution of inflammatory cytokines to carcinogenesis; effects of radiation-, metal-, and chemical-induced free radicals and their interactions with DNA on cancer development; biomarkers of cancer risk.

George Friedman-Jiménez, M.D. Occupational and clinical epidemiology; epidemiology of radiation and cancer; epidemiology of asthma; epidemiologic methods; urban populations.

Judith D. Goldberg, D.Sc. Design and analysis of clinical trials; survival analysis; disease screening and misclassification; analysis of observational data; statistical genomics.

David Goldsheyder, M.A., M.S. Biomechanics; workplace design; workstation modification; ergonomics.

Terry Gordon, Ph.D. Genetic susceptibility of lung disease produced by environmental and occupational agents.

Albert F. Gunnison, Ph.D. Molecular mechanisms and toxicology of pulmonary inflammation; DNA microarray technology; reproductive toxicology.

Manny Halpern, Ph.D. Ergonomics; workplace intervention; injury prevention methodology; job analysis.

Naomi H. Harley, Ph.D. Dosimetry of internally deposited radionuclides; measurement of radiation and radioactivity; risk modeling of radiation carcinogenesis.

Maire S. A. Heikkinen, Ph.D. Measurement of ultrafine and nanometer aerosols; development of instrumentation for collection and analysis of acidic, radioactive, and biological particles.

Chuanshu Huang, Ph.D. Signal transduction in tumor promotion and prevention; molecular mechanism of carcinogenesis caused by ultraviolet radiation, metal compounds, and smoking.

Xi Huang, Ph.D. Implication of iron and oxidative stress in human diseases.

Kazuhiko Ito, Ph.D. Human health effects of air pollution and risk analysis.

Rudolph J. Jaeger, Ph.D. Inhalation toxicology; aerosol science; plastics toxicology and the toxicology of their monomers; combustion products; tobacco smoke toxicology; pulmonary pathophysiology; liver toxicity and pathophysiology; effects of lead and heavy metals on the developing nervous system.

Catherine B. Klein, Ph.D. Mammalian mutagenesis; epigenetic gene control; DNA methylation; oxidants; metals; estrogens; molecular cytogenetics.

Karen Koenig, Ph.D. Epidemiology of coronary heart disease and cancer; epidemiologic methods.

Morton Lippmann, Ph.D. Inhalation toxicology; aerosol science and physiology; occupational and environmental hygiene; air pollution.

Angela Lis, M.A. Occupational musculoskeletal disorders; low back pain; prevention of injury; prevention of disability; biomechanics; ergonomics.

Mengling Liu, Ph.D. Analysis of longitudinal data with informative censoring; survival analysis; semiparametric inference; analysis for quality-of-life data.

Michael Marmor, Ph.D. Epidemiology and prevention of HIV/AIDS, tuberculosis, and other infectious diseases; clinical trials of HIV vaccines and nonvaccine interventions; environmental, occupational, and ophthalmologic epidemiology.

Jomol P. Mathew, Ph.D. Design, management, and analysis of large biological/clinical databases; data mining; design and analysis of clinical trials; statistical modeling of biological data; design and development of statistical computing tools; bioinformatics, computational biology, GIS, and spatial statistics.

Assieh Melikian, Ph.D. Mechanisms of environmental carcinogenesis; cancer chemoprevention; biomarkers; molecular epidemiology.

Arthur Nádas, Ph.D. Mathematical statistics; biostatistics; mathematical biology; statistical design of HIV immunotypes, with the goal of a broadly effective polyvalent vaccine for HIV; experimental design and analysis using microarrays and gene chips; statistical analysis of telemetry data; mathematical modeling of spontaneous mutagenesis; rapid multivariate diagnostic tests for tuberculosis; pattern recognition using dynamic programming, hidden Markov modeling, and neural networks.

Bhagavathi A Narayanan, Ph.D. Prostate and colon cancer chemoprevention; nonsteroidal anti-inflammatory drugs; genomic and proteomic approaches; potential molecular targets; biomarkers.

Narayanan K. Narayanan, Ph.D. Chemopreventive proteomics; Omega-3 polyunsaturated fatty acid against prostate cancer; proteomic profiling of differentiation-inducing proteins.

Margareta Nordin, Med.Dr.Sci. Occupational musculoskeletal disorders, low back pain; prevention of injury; prevention of disability; motor control; biomechanics; ergonomics.

Qingshan Qu, M.D. Pulmonary toxicology; biomarker application and risk assessment.

William N. Rom, M.D., M.P.H. Environmental and occupational lung diseases; molecular mechanisms of lung cancer; tuberculosis (TB)/AIDS; interferon-gamma therapy for TB and TB vaccine and immune response; environmental policy, wilderness preservation, and global warming.

Toby G. Rossman, Ph.D. Spontaneous mutagenesis; genotoxicity of metal compounds; mechanisms of resistance to metals; arsenic carcinogenicity.

Nirmal Roy, Ph.D. Molecular biology of the aromatic hydrocarbon receptor pathway; DNA lesions and mutations induced by xenobiotic compounds.

Yongzhao Shao, Ph.D. Genetic linkage/association analysis; genetic epidemiology; statistical inference; design of experiments; likelihood theory; mixture models.

Ali Sheikhzadeh, Ph.D. Occupational biomechanics; biomechanical modeling and testing; electromyography; ergonomic product evaluation.

Roy E. Shore, Ph.D., Dr.P.H. Environmental and genetic epidemiology of cancer; radiation epidemiology; epidemiologic methods.

Ock Soon Sohn, Ph.D. Mechanism of chemical carcinogenesis and chemoprevention; in vivo and in vitro metabolism of xenobiotics; pharmacokinetics; analytical chemistry.

Bernard G. Steinetz, Ph.D. Environment of the newborn: possible role of milk-borne hormones in phenotypic expression of inherited hip dysplasia; role of hormones in protection against environmental carcinogen–induced breast cancer afforded by early pregnancy; role of hormones in the modulation of insulin resistance of pregnancy.

Ting-Chung Suen, Ph.D. Oncogenes and tumor-suppressor genes; breast cancer; transcriptional regulation of gene expression; effects of carcinogens on gene expression; gene chips and microarrays.

Moon-shong Tang, Ph.D. Carcinogenesis and mutagenesis; DNA damage; DNA repair.

Kam-Meng Tchou-Wong, Ph.D. p53 pathways in metal- and carcinogen-induced lung cancer; Wnt signaling pathways in lung fibrosis and cancer; chemoprevention of lung carcinogenesis; infection and ethnic disparities in diabetes risk and cardiovascular diseases.

George D. Thurston, D.Sc. Human health effects of inhaled air pollutants; asthma; aerosol science; acidic air pollution; air pollution meteorology and modeling; risk analysis.

Paolo G. Toniolo, M.D. Cancer epidemiology; role of endogenous hormones in the etiology of chronic diseases; influence of diet on endogenous hormones in health and disease; health consequences of human exposure to hormonally active agents in the environment.

Chi-hong Tseng, Ph.D. Survival analysis; measurement error models; design of clinical trials.

Sherri Weiser, Ph.D. Biopsychosocial models; low back pain; personality and health; occupational stress.

Isaac Wirgin, Ph.D. Molecular biology of carcinogenesis; cancer in aquatic organisms; population genetics and molecular evolution.

Judy Xiong, Ph.D. Occupational hygiene; environmental chemistry; aerosol science.

Anne Zeleniuch-Jacquotte, M.D. Cancer epidemiology; methods in epidemiology and clinical trials.

Judith T. Zelikoff, Ph.D. Immunotoxicology; development of immune biomarkers and alternative animal models for immunotoxicological studies; effects of inhaled pollutants on host resistance and pulmonary immune defense mechanisms; metal-induced immunotoxicity.

NORTHEASTERN UNIVERSITY

Department of Biology

Programs of Study	The Department of Biology at Northeastern University offers programs of study leading to the Ph.D. and M.S. in biology. A new professional M.S. in marine biology was added in April 2005. In addition, the department participates in two interdisciplinary professional M.S. programs: the M.S. in bioinformatics and computational molecular biology and the M.S. in biotechnology. The biology Ph.D. and M.S. programs provide a broad background knowledge base and an in-depth study of a specialized area of biology. The programs emphasize close interaction between graduate students and faculty members in developing the intellectual and experimental skills required for creative independent research. Research interests of the faculty members range from biochemistry and molecular and cell biology to organismal, ecological, and evolutionary biology. Communication and shared research among the laboratories are encouraged. Prior to choosing a thesis/dissertation adviser, students generally do laboratory rotations.

Ph.D. and master's students participate in seminars and graduate biology courses as well as dissertation and thesis research. There are no course credit hour requirements for the doctoral student who holds a master's degree. The breadth of course offerings allows selection of courses appropriate for the student's specialized needs. Doctoral students are required to have at least one year of experience as teaching assistants in a laboratory course. When available, teaching assistantships are also provided to master's students. Part-time M.S. study is facilitated by the availability of some evening graduate-level courses and by the option of a literature M.S. thesis, where the student critically analyzes original research literature. Both Ph.D. and M.S. students may elect to take some graduate courses in other divisions of the University, such as engineering, pharmacy, and business.

Research Facilities

In addition to well-equipped research laboratories, facilities include microarray analysis, a bioinformatics computer cluster, an electron microscopy center, a research museum for vertebrate study, controlled-environment rooms, cell-culture facilities, and a wide variety of preparative and analytical instruments, including FACS, real-time PCR, phosphoimager, confocal microscopes, electrophysiological devices, and equipment for image analysis. Facilities of the Marine Science Center, located nearby at Nahant, include an ocean-going research vessel, a running seawater system, and a robotics laboratory. The large, technologically sophisticated University library contains extensive print, database, and media collections and has access to other major research collections, including the Boston Library Consortium. The library, major software packages, and the Internet are accessible through laboratory and office computers.

Financial Aid

Graduate research and teaching assistantships, which provide stipends of up to $25,536, tuition remission, and partial coverage of Northeastern University student health insurance, are available to full-time students. Research assistantships are provided by research grants to individual faculty members and departmental research funds. Internships with biotechnology companies constitute an additional option for doctoral students.

Northeastern University awards need-based financial aid to graduate students through federal loans and work-study programs. The University also offers minority fellowships and Martin Luther King Jr. Scholarships.

Cost of Study

The cost of tuition for the 2006–07 academic year in the Graduate School of Arts and Sciences is $930 per semester hour of credit. Other charges required of all full-time students are health insurance and the Student Center fee.

Living and Housing Costs

A wide variety of off-campus living accommodations is available in the vicinity of the University and in the greater Boston area. Northeastern University is near the heart of Boston's efficient public transportation system, which provides access to the entire area. For more information about on- and off-campus housing, students should refer to http://www.housing.neu.edu.

Student Group

Graduate students in the Department of Biology are part of a large, diverse, and dynamic academic community that includes students and faculty members from related disciplines within the University, such as chemistry, pharmacy, psychology, physics, and electrical engineering as well as from the many other academic, medical, and research institutions in the Boston area. Seminars at other institutions are open to Northeastern students, and the University belongs to the Boston Area Graduate Student Association. In 2005–06, there were 45 Ph.D. students, 15 full-time M.S. students, and 6 part-time M.S. students in the Department of Biology. About 62 percent of the graduate students in the Department were women, and about 33 percent were international students.

Location

Boston, the state capital, offers a broad spectrum of academic, cultural, and recreational opportunities. The campus is located in the Back Bay area of Boston, within walking distance of Symphony Hall, the Museum of Fine Arts, and Fenway Park, home of the Red Sox. The greater Boston area is accessible by the "T," Boston's efficient subway system. Within a 15-minute ride are the Boston Common, the Museum of Science, the New England Aquarium, Cambridge, and downtown Boston. Outdoor recreational activities are enhanced by Boston's proximity to both the seacoast and the mountains of New England. Boston is a mixture of Colonial tradition and modern America, and it is home to people of every intellectual, political, economic, racial, ethnic, and religious background. It is a place where the past is appreciated, the present enjoyed, and the future anticipated.

The University

Founded in 1898, Northeastern University is a privately endowed, nonsectarian institution of higher learning. It offers a variety of curricula through seven undergraduate colleges, nine graduate and professional schools, several institutes, two part-time undergraduate divisions, and a number of continuing education programs. In fall 2005, approximately 24,000 students were enrolled at Northeastern University, including about 4,800 graduate and professional students. Northeastern University is a Carnegie Research II university.

Applying

Applications must be received by January 15 in order to receive full consideration for financial aid; the forms can be obtained online at http://www.biology.neu.edu/graduate_programs.html. Applicants must have earned a baccalaureate degree and have adequate undergraduate training in biology, chemistry, mathematics, and physics. Graduate Record Examinations (GRE) scores must be submitted; the GRE Subject Test is not required. International students must demonstrate proficiency in writing and speaking English with a minimum TOEFL score of 250 on the computer-based test (the equivalent of 100 on the Internet-based test or 600 on the paper-based test).

Correspondence and Information

Department of Biology
134 Mugar Hall
Northeastern University
Boston, Massachusetts 02115
E-mail: gradbio@neu.edu
Web site: http://www.biology.neu.edu/graduate_programs.html

Northeastern University

THE FACULTY AND THEIR RESEARCH

Ahmed Abdelal, Professor and Provost; Ph.D., California, Davis. Microbial physiology and biochemistry.
Joseph L. Ayers Jr., Associate Professor; Ph.D., California, Santa Cruz. Neurophysiology and behavior.
Kostia Bergman, Associate Professor; Ph.D., Caltech. Signal transduction in bacteria.
Donald P. Cheney, Associate Professor; Ph.D., South Florida. Biotechnology and evolutionary ecology of marine plants.
Erin Cram, Assistant Professor; Ph.D., Berkeley. Molecular and cell biology.
Frederick C. Davis, Professor; Ph.D., Texas at Austin. Neurobiology; circadian rhythms.
H. William Detrich, Professor; Ph.D., Yale. Biochemistry and molecular biology of marine fishes.
Slava Epstein, Associate Professor; Ph.D., Institute of Oceanology (Moscow). Microbial evolution and ecology.
Veronica Godoy-Carter, Assistant Professor; Ph.D., Tufts. Molecular biology and microbiology.
Valentin Ilyin, Assistant Professor; Ph.D., Institute of Crystallography (Moscow). Theoretical molecular biology and bioinformatics.
Edward Jarroll, Professor and Director, Marine Science Center; Ph.D., West Virginia. Parasitology; biochemistry.
Gwilym S. Jones, Professor; Ph.D., Indiana State. Mammalogy; vertebrate systematics and ecology.
Kim Lewis, Professor; Ph.D., Moscow. Molecular microbiology; drug discovery.
James M. Manning, Professor; Ph.D., Tufts. Protein structure-function.
Richard L. Marsh, Professor; Ph.D., Michigan. Biology of muscle and locomotion.
Charles A. M. Meszoely, Professor; Ph.D., Boston University. Parasitology; vertebrate paleontology; herpetology.
Donald O'Malley, Associate Professor; Ph.D., Harvard. Cellular and systems neurobiology.
Jacqueline Piret, Associate Professor and Graduate Coordinator; Ph.D., MIT. Molecular and industrial microbiology.
Susan G. Powers-Lee, Professor and Chair; Ph.D., Berkeley. Protein structure and function.
Nathan W. Riser, Professor Emeritus; Ph.D., Stanford. Invertebrate biology.
Rebeca B. Rosengaus, Assistant Professor; Ph.D., Boston University. Behavioral ecology and insect sociobiology.
Daniel C. Scheirer, Associate Professor; Ph.D., Penn State. Plant molecular biology and morphology.
Michail Sitkovsky, Professor; Ph.D., Moscow State. Biochemistry and immunopharmacology.
Wendy A. Smith, Associate Professor; Ph.D., Duke. Cellular endocrinology.
Phyllis R. Strauss, Professor; Ph.D., Rockefeller. DNA repair mechanisms.
Geoffrey Trussell, Assistant Professor; Ph.D., William and Mary. Evolutionary and community marine ecology.
Steven Vollmer, Assistant Professor; Ph.D., Harvard. Marine speciation and population genetics.
Carol M. Warner, Professor; Ph.D., UCLA. Immunology; molecular biology; developmental biology.

Behrakis Health Sciences Center.

NORTHWESTERN UNIVERSITY

Interdepartmental Biological Sciences Program
Evanston Campus and Evanston Hospital

Programs of Study

The Interdepartmental Biological Sciences Program at Northwestern University leads to the degree of Doctor of Philosophy in biological sciences in one of seven areas of concentration: biotechnology, cell and molecular biology, developmental biology and genetics, hormone action and signal transduction, neuroscience, cancer biology, or structural biology, biochemistry, and biophysics. These areas represent the strengths of the laboratories of nearly 75 faculty members in six academic departments on the Evanston campus and nearby Evanston Hospital.

Graduate study in the doctoral program has four significant components: research, course work, teaching, and independent learning through seminars and interaction with other scientists. The primary mission of the doctoral program is to encourage the development of independent, creative research scientists and teachers in several disciplines. The program is deeply committed to helping students prepare for the career options that await them after completion of their degrees. New students are introduced to research opportunities in the program through three laboratory rotations, an annual retreat, and individual consultation. At the end of the first year, students select a faculty adviser and initiate research projects that form the basis of the dissertation. Each student takes three of seven core courses, three electives, and two primary literature-based special-topics courses.

Research Facilities

The University and Evanston Hospital provide extensive support facilities for research programs. These include the recently established Facility for Automated High Throughput Analysis, a 600-MHz NMR facility; the Analytical Services Laboratory, which includes EPR, ENDOR, and Raman spectrometers; the Biotechnology Research Services Laboratory; a biological X-ray facility for protein crystallography and computer graphics; the Keck Biophysics Facility; and the Cell Culture and Analysis Laboratory. Other facilities include those for recombinant DNA technology and DNA sequencing, development of transgenic animals, synthesis of oligonucleotides and peptides, and protein purification. Microscope facilities include modern electron microscopes for darkfield EM, STEM, SEM, and diffraction modes; fluorescence microscopes; confocal microscopes; and an atomic-force microscope.

Financial Aid

Every student can expect to receive financial support for the entire period of graduate study, provided he or she remains in residence and makes satisfactory progress in meeting degree requirements. The 2005–06 stipend was $24,000 plus $500 for moving costs for the calendar year, exclusive of tuition. A health insurance and dental benefit is also provided.

Cost of Study

All appointments carry full remission of tuition and fees. Tuition amounted to $39,921 based on the 2004–05 academic year.

Living and Housing Costs

The University offers on-campus housing for graduate students in Englehart Hall. Both furnished and unfurnished units are available. The 2004–05 rates for one- or two-bedroom apartments ranged from $1050 to $1150. Most students find satisfactory accommodations in apartments in the vicinity of the campus and in nearby Chicago. For more information, students should visit http://www.northwestern.edu/studentaffairs/.

Student Group

The participating laboratories currently have 115 graduate students who come from colleges throughout the United States and several other countries. In addition, there are 80 postdoctoral fellows who are actively engaged in research. In 2004, the undergraduate student body numbered 7,669, and the graduate student body totaled 5,791. More information is available at http://www.northwestern.edu/about/facts/.

Location

The departments are located on the 170-acre Evanston campus, which stretches for a mile along the western shore of Lake Michigan, as well as at close-by Evanston Hospital. The University is 12 miles north of Chicago and is home to some of the country's finest museums, libraries, art centers, and concert halls. On the eastern edge of the campus are fine beaches which provide opportunities for sailing, swimming, and other water sports.

The University

Northwestern University was founded in 1851 and is one of the nation's largest private universities. Graduate programs are conducted on both the Evanston and Chicago campuses. The Evanston campus includes the Weinberg College of Arts and Sciences; the Robert R. McCormick School of Engineering and Applied Science; the Schools of Education, Journalism, Music, and Communication; and the Kellogg Graduate School of Management. The Medical School and School of Law are located on the Chicago campus.

To keep pace with rapid advances in the life sciences, the University has a continuing commitment to develop life sciences programs of excellence. These programs are generously supported by the University and the Searle Leadership Fund. A life sciences building was completed in 1992 and houses interdisciplinary programs in cell and molecular biology as well as the Rice Institute for Biomedical Research. The Center for Structural Biology, completed in 1996, houses the Keck Biophysics Facility. The new Pancoe–ENH Life Sciences Pavilion was completed in 2003 and houses basic scientists and clinical investigators to facilitate translational research.

Northwestern University is committed to increasing the number of underrepresented minority students seeking the Ph.D. degree and welcomes applications for admission and financial aid from qualified minority students.

Applying

The program admits students at the beginning of the fall session. Interested students should refer to the program's Web site (listed below) for the official deadline for the receipt of applications. The General Test of the Graduate Record Examinations (GRE) is required. International students are required to pass the Test of English as a Foreign Language (TOEFL) with a minimum score of 600 before admission and earn a score of at least 50 on the Test of Spoken English (TSE) before the end of the first year of study. More information about application requirements can be found at http://www.biochem.northwestern.edu/ibis/stud/appProc.htm#jump.

Correspondence and Information

Chairperson, Graduate Admissions
Interdepartmental Biological Sciences Program (IBiS)
2-100 Hogan Hall
Northwestern University
2205 Tech Drive
Evanston, Illinois 60208-3500
Phone: 847-491-4301
      800-545-1761 (toll-free)
E-mail: ibis@northwestern.edu
Web site: http://www.biochem.northwestern.edu/ibis

Northwestern University

THE FACULTY AND THEIR RESEARCH

Ravi Allada, Ph.D., Michigan. The molecular basis of circadian rhythms and sleep.
Luis A. Nunes Amaral, Ph.D., Boston University. Complex networks in nature; system-level modeling of biological processes.
Guillermo Ameer, Ph.D., MIT. Bioartificial organ systems; cell delivery and transportation; tissue engineering.
Hamid Band, Ph.D., All-India Institute of Medical Sciences; M.D., Medical College of Srinagar (India). Role of proto-oncogenes as negative regulators of signals.
Vimla Band, Ph.D., All-India Institute of Medical Sciences. Delineating molecular basis of early steps of human breast cancer.
Annelise Barron, Ph.D., Berkeley. The design, synthesis, and structural characterization of nonnatural sequence-specific heteropolymers designed to adopt peptidomimetic folded structures.
Joseph Bass, M.D., Ph.D., Medical College of Pennsylvania. Cell biology of diabetes mellitus.
Greg Beitel, Ph.D., MIT. Genetic control of epithelial tube size in the *Drosophila* tracheal system.
Rhonda Brand, Ph.D., Michigan. Chemical absorption through the skin.
Michael Caplan, M.D., Chicago. Platelet activating factor; necrotizing enterocolitis and inflammation.
Richard Carthew, Ph.D., MIT. Cell differentiation and morphogenesis in *Drosophila* development.
Peter Dallos, Ph.D., Northwestern. Biophysics and physiology of the auditory system; bioacoustics; sensory neurobiology.
Goberdhan P. Dimri, Ph.D., Jawaharlal Nehru (New Delhi). Molecular mechanisms of cellular senescense; oncogenesis in human cells.
Andrew Dudley, Ph.D., Harvard. Uncovering the mechanisms that govern morphogenesis.
David Ferster, Ph.D., Harvard. Receptive field properties and synaptic organization of visual cortex.
Heike Fölsch, Ph.D., Munich. Establishment and maintenance of cell polarity.
Christopher Froelich, M.D., Loyola. NK cells in innate and adaptive host defenses; lymphocyte serine proteases: cytotoxic and regulatory properties.
Richard F. Gaber, Ph.D., Wisconsin–Madison. Molecular genetics of ion transport and glucose transport in yeast.
Qingshen Gao, M.D., Shandong Medical University (China). Dissecting the BRCA2 tumor suppressor pathway.
Hilary Arnold Godwin, Ph.D., Stanford. Fluorescent zinc probes based on metal-induced peptide folding.
Erwin Goldberg, Ph.D., Iowa. Biochemistry and molecular biology of mammalian spermatogenesis; immunoinfertility and contraceptive vaccine development.
Vassily Hatzimanikatis, Ph.D., Caltech. Functional genomics; mathematical and computational biotechnology/bioinformatics.
Linda Hicke, Ph.D., Berkeley. Down regulation of signal transducing receptors.
Emmet Hirsch, M.D., Northwestern. Molecular pathogenesis of infection-induced preterm birth.
Brian M. Hoffman, Ph.D., Caltech. Metallobiochemistry; bioinorganic chemistry; electron transfer at long distance; EPR; ENDOR.
Robert A. Holmgren, Ph.D., Harvard. Specification of cell fate in *Drosophila* development.
Curt M. Horvath, Ph.D., Northwestern. Signal transduction and gene regulation in innate immune responses to cancer and viruses.
Theodore S. Jardetzky, Ph.D., Basel (Switzerland). Crystallography and biochemistry of immune system proteins; histocompatibility antigens; superantigens; IgE Fc receptor.
Tamas Jilling, M.D., Pecs (Hungary). Pathogenesis of necrotizing enterocolitis.
Jian-Ping Jin, M.D., Fourth Military Medical (China); Ph.D., Iowa. Gene regulation and structure-function relationships of contractile and cytoskeleton proteins.
Karen L. Kaul, M.D., Ph.D., Northwestern. Role of p53 in breast cancer; molecular diagnostic laboratory.
David M. Kelso, Ph.D., Northwestern. Medical instrumentation; biosensors; kinetics of antibody and DNA binding reactions in solution and on solid phases; pharmacokinetics; optimization of drug administration.
William L. Klein, Ph.D., UCLA. Molecular development and plasticity of CNS synapses; growth cones and nascent synapses; Alzheimer's disease.
Carole LaBonne, Ph.D., Harvard. Development of the vertebrate neural crest.
Robert A. Lamb, Ph.D., Sc.D., Cambridge. Synthesis and intracellular transport of integral membrane proteins; molecular and cellular biology of enveloped negative strand RNA viruses; protein engineering.
Jon E. Levine, Ph.D., Illinois. Neuroendocrinology; molecular, cellular, and integrative physiology of neuropeptidergic neurons and neuropeptide receptors.
Robert A. Linsenmeier, Ph.D., Northwestern. Microenvironment of the mammalian retina, including ionic balance and transport of nutrients, especially oxygen; electrophysiology of the mammalian retina.
Daniel I. H. Linzer, Ph.D., Princeton. Regulation of mammalian cell growth and differentiation; expression and action of protein hormones.
Howard L. Lipton, M.D., Nebraska. Molecular pathogenesis of persistent virus infections.
Paul A. Loach, Ph.D., Yale. Structure and function relationships in supramolecular systems, especially photoreceptor complexes of photosynthetic organisms; mechanisms of electron and proton transport.
Robert C. MacDonald, Ph.D., UCLA. Membrane biochemistry and biophysics.
Andreas Matouschek, Ph.D., Cambridge. Mechanisms of protein folding and transport.
Kelly E. Mayo, Ph.D., Washington (Seattle). Gene regulation in the mammalian neuroendocrine system.
Thomas J. Meade, Ph.D., Ohio State. Bioinorganic chemistry; biological molecular imaging.
William Miller, Ph.D., Berkeley. Biochemical engineering; mammalian cell culture applications in biotechnology, medicine, and tissue engineering.
Chad A. Mirkin, Ph.D., Penn State. Nanotechnology.
Alfonso Mondragón, Ph.D., Cambridge. Crystallography; protein–nucleic acid interactions; topoisomerases.
Richard I. Morimoto, Ph.D., Chicago. Transcriptional regulation and function of heat shock proteins and molecular chaperones.
Thomas V. O'Halloran, Ph.D., Columbia. Metal-responsive gene regulation; protein-DNA interactions; heavy-metal toxicity.
Eleftherios Papoutsakis, Ph.D., Purdue. Tissue engineering of the bone marrow for transplantation therapies; cell culture engineering; genetic engineering of anaerobic clostridia.
Lawrence H. Pinto, Ph.D., Northwestern. Genetic basis for visual function; ion channels of viruses; information processing in the retina.
Ishwar Radhakrishnan, Ph.D., Columbia. Structural studies of protein-protein interactions and enzyme action in transcription regulation.
Indira Raman, Ph.D., Wisconsin–Madison. Ionic mechanisms of neuronal excitability.
Amy Rosenzweig, Ph.D., Massachusetts Amherst. Structure and biochemistry of metalloenzymes.
Aryeh Routtenberg, Ph.D., Michigan. Molecular basis for brain information storage.
Mark Segraves, Ph.D., Pennsylvania. Neuronal mechanisms underlying oculomotor behavior; contributions of cerebral cortex to eye movements.
Lonnie Shea, Ph.D., Michigan. Tissue engineering and gene therapy.
Richard Silverman, Ph.D., Harvard. Bioorganic, medicinal, and enzyme chemistry.
Erik J. Sontheimer, Ph.D., Yale. Ribonucleoproteins and eukaryotic gene expression.
Nelson Spruston, Ph.D., Baylor College of Medicine. Synaptic integration in the central nervous system.
Joseph S. Takahashi, Ph.D., Oregon. Molecular neurobiology and genetics of circadian clocks.
Fred W. Turek, Ph.D., Stanford. Photoperiodic control of neuroendocrine-gonadal activity; circadian rhythmicity.
Olke C. Uhlenbeck, Ph.D., Harvard. RNA biochemistry.
Thomas A. Victor, M.D., Ph.D., Illinois at Chicago. Study of the effect of dietary lipids on the genesis and progression of human breast cancer.
San Ming Wang, M.D., Swiss Institute for Experimental Cancer Research (Switzerland). Genome sciences.
Xiaozhong (Alec) Wang, Ph.D., NYU. Genetic analysis of protocadherin diversity in the central nervous system.
Eric Weiss, Ph.D., Colorado at Boulder. Spatially and temporally coordinated cell growth and division.
Jonathan Widom, Ph.D., Stanford. Biophysical chemistry; structure and function of chromosomes.
Teresa K. Woodruff, Ph.D., Northwestern. Mechanisms of ovarian follicle growth and development; hormone and growth-factor interactions with intracellular signaling cascades.
Catherine S. Woolley, Ph.D., Rockefeller. Hormonal regulation of structural and functional plasticity of brain circuitry.
Tai T. Wu, Ph.D., Harvard. Structural and functional relationships of proteins and DNAs, in particular those pertaining to immunoglobulins.
Alice M. Wyrwicz, Ph.D., Illinois at Chicago. Biomedical applications of NMR spectroscopy and imaging.

NORTHWESTERN UNIVERSITY

Chicago Campus–Medical School
Integrated Graduate Program in the Life Sciences

Program of Study

The Integrated Graduate Program in the Life Sciences (IGP) is a multidisciplinary doctoral training program designed to prepare students for research careers in cellular, molecular, and integrative biology. Areas of concentration include biochemistry, cancer biology, cell biology, developmental biology, drug discovery and chemical biology, evolutionary biology, immunology and microbial pathogenesis, molecular biology and genetics, neurobiology, pharmacology and toxicology, and structural biology.

The IGP is designed to provide graduate students with a broad range of research training opportunities by permitting them to select a research adviser from among all faculty members in the IGP, regardless of departmental affiliation. The diversity of scientific interests among the faculty members on campus creates a highly stimulating environment for graduate training and research. Participating faculty members are drawn from the Departments of Cell and Molecular Biology, Microbiology-Immunology, Pathology, Molecular Pharmacology and Biological Chemistry, and Physiology and from several clinical departments. Students take core courses in biochemistry, cell biology, and molecular biology during the first year. Each area of specialization requires two to three additional courses. Students round out their course requirements with two to three electives. Students begin research rotations during their first quarter of study. Typically, the student selects a thesis adviser before the end of the summer quarter of the first year. At the end of the winter quarter of the second year, the student must pass a written and oral qualifying examination in order to be admitted to candidacy for the Ph.D. Completion of the doctoral program requires approximately 5¼ years. A substantial portion of this time is devoted to thesis research under the guidance of a faculty member.

Research Facilities

The participating faculty members occupy modern laboratories that are well equipped for research in their areas of interest. Each department has central facilities that are shared by all its research groups, such as culture rooms and dishwashing services, common equipment rooms that house major instrumentation, and computing facilities. There are also campuswide and University-wide facilities, such as transgenic mouse facilities; a state-of-the-art biotechnology center with nucleic acid and protein sequencing and synthesizing instrumentation, digital detection instruments, and chip array technology; a modern microscopy center, including a laser confocal microscope and transmission and scanning electron microscopes; a modern image-processing facility; animal-care facilities; machine shops; and electronics services. Numerous departmental reading rooms contain specialized journals and books, while the Galter Health Sciences Library contains more than 300,000 bound volumes and receives more than 2,500 biological sciences serials.

Financial Aid

Typically, all entering students are provided full tuition support and a stipend for living expenses. The stipend level for 2006–07 is $24,500. New students are also provided with a one-time moving allowance of $500. Stipend and tuition are provided for the entire period of residence for students in good standing. Entering students are initially supported for six quarters (eighteen months) by predoctoral University fellowships. In subsequent years, tuition and stipends are provided from faculty research grants, outside scholarships, teaching assistantships, training grants, or special graduate fellowships.

Cost of Study

Tuition is typically included as part of the fellowship award and subsequent support. Student group health insurance benefits are provided in full.

Living and Housing Costs

Abundant and reasonably priced off-campus housing is located within convenient walking or commuting distance of the campus. Campus parking for students commuting more than 2 miles is available.

Student Group

Currently, there are approximately 150 predoctoral IGP students and more than 100 postdoctoral fellows on the Chicago campus. The student body is derived from widely distributed areas of the United States and other countries. IGP students share courses and labs with members of other life science programs, such as the Northwestern University Institute for Neuroscience (NUIN) and the Interdepartmental Biological Science program (IBiS). Several students enrolled in the combined M.D./Ph.D. degree program belong to the IGP.

Student Outcomes

Most Ph.D. graduates continue training in postdoctoral positions, primarily in preparation for careers in teaching and research at universities and in industry. Of 23 recent Ph.D. graduates, 18 entered postdoctoral training, 2 sought additional degrees, 2 were employed by law or biotech firms, and 1 took a college faculty position.

Location

The Medical School is located on the shores of Lake Michigan in the midst of a large medical and scientific complex. Chicago has substantial cultural, recreational, commercial, and sports activities. Many of these are within walking or a short commuting distance of the downtown Chicago campus. On campus, the Lake Shore Center residence has a pool, squash and racquetball courts, a basketball court, a weight room, and a gymnasium, as well as a pub and student lounges.

The University

Northwestern University is a privately supported, independent, nondenominational, and coeducational institution founded in 1851 by people determined to create "a university of the highest order of excellence." The undergraduate schools are located on the Evanston campus, about 13 miles north of the Chicago Campus. Among the 2,300 full-time students enrolled in The Graduate School at Northwestern, about 250 are on the Chicago campus. There are also about 1,500 full-time students on the Chicago campus in the medical, business, and law schools.

Applying

Students are admitted only to a course of study leading to a Ph.D. Applicants for admission should have a strong undergraduate background in the basic biological, chemical, and physical sciences and a compelling dedication to a research career in basic biomedical sciences. Applications must include a statement describing goals for graduate study, research interest and experience, and reasons for applying for graduate study. The GRE General Test is required. The TOEFL is also required for international applicants, with a minimum score of 600 on the paper-based test or 250 on the computer-based test. All students enter the graduate program at the beginning of the fall quarter. Evaluation for admission is based on previous academic achievement, GRE scores, and letters of recommendation. Applications should be submitted to the Graduate School by December 31.

Correspondence and Information

Dr. Steven Anderson
Northwest University Feinberg School of Medicine
Integrated Graduate Program, Ward 12-373
303 East Chicago Avenue
Chicago, Illinois 60611

Phone: 800-255-4166 (toll-free)
E-mail: sja314@northwestern.edu
Web site: http://www.feinberg.northwestern.edu/igp

Northwestern University

THE FACULTY AND THEIR RESEARCH

Michael Abecassis, M.D. Cytomegalovirus latency and reactivation.

Sara Ahlgren, Ph.D. Craniofacial growth and development.

Wayne Anderson, Ph.D. Crystallographic studies of protein interactions.

Hamid Band, M.D., Ph.D. Cbl regulation of tyrosine kinases.

Vimla Band, Ph.D. Human breast and cervical cancer.

Terry Barrett, M.D. Oral tolerance; mucosal immune system.

James Bartles, Ph.D. Plasma membrane differentiation; cell polarity.

Daniel Batlle, M.D. Ion transport in renal tubular cells.

Raymond Bergan, M.D. Human prostate cells.

Lester Binder, Ph.D. Microtubules and neurodegenerative disease.

Martha Bohn, Ph.D. Neurotrophic factors and gene therapies for neurodegenerative diseases.

Melissa Brown, Ph.D. Mast cells in autoimmunity.

Irina Budunova, M.D., Ph.D. Intercellular communication in carcinogenesis.

Navdeep Chandel, Ph.D. Tumor cell responses to oxygen deprivation.

Anjen Chenn, M.D., Ph.D. Regulation of cell fate and proliferation in CNS.

Rex Chisholm, Ph.D. Cell motility; morphogenesis; myosin; dynein; molecular motors and disease.

Nicholas Cianciotto, Ph.D. Pathogenesis of *Legionella pneumophila;* Legionnaires' disease.

Vincent Cryns, M.D. Apoptosis and caspase-mediated proteolysis.

Marian Dagosto, Ph.D. Primate evolution: functional morphology, systematics, paleontology.

Syamal Datta, M.D. Autoimmunity and regulation of B-cell function.

David Dean, Ph.D. Nuclear transport of plasmids and gene therapy vectors.

Robert Decker, Ph.D. Protein turnover and intercellular junctions in organogenesis.

Christine DiDonato, Ph.D. Molecular basis of spinal muscular atrophy.

Margarita Dubocovich, Ph.D. Monoamine receptors in the retina, brain, and other organs.

Elizabeth Eklund, M.D. Molecular biology of late myeloid differentiation.

David Engman, M.D., Ph.D. Autoimmunity, heat shock proteins, parasites, heart disease.

Adriana Ferreira, M.D., Ph.D. Neurite elongation in central neurons.

Douglas Freymann, Ph.D. X-ray crystallography of GTPase; signal recognition particle.

Karla Fullner-Satchell, Ph.D. *Vibrio cholerae* RTX toxin and host-pathogen interactions.

Jaime García-Añoveros, Ph.D. Genetic dissection of touch in *Caenorhabditis elegans.*

Robert Goldman, Ph.D. Cell motility; structure and function of cytoplasmic filament systems.

Cara Gottardi, Ph.D. Catenins and cadherins in cell adhesion.

Kathleen Green, Ph.D. Analysis of desmosomes and keratins.

Adrian Gross, M.D. Structure and function of ion channels.

Jaime Grutzendler, M.D. Synaptic stability in health and disease.

Kasturi Haldar, Ph.D. Protozoan and bacterial entry and development in mammalian cells.

Alan Hauser, M.D., Ph.D. Type III secretion in *Pseudomonas.*

Laura Herzing, Ph.D. Epigenetics of neurodevelopmental disorders.

Sui Huang, Ph.D. Structure-function of perinucleolar compartment.

Philip Iannaccone, M.D., D.Phil. Tumor clonality; preimplantation; reproductive toxicity.

Kouichi Iwasaki, Ph.D. Molecular biology and genetics of a rhythmic behavior.

J. Larry Jameson, M.D., Ph.D. Nuclear receptors; gene transcription; cell signaling; genetic diseases.

Jonathan Jones, Ph.D. Assembly of intercellular junctions; development of epithelial tissues.

Geoffrey Kansas, Ph.D. Selections in leukocyte adhesion and migration.

William Karpus, Ph.D. Pathogenesis and regulation of autoimmune encephalomyelitis (EAE).

Dixon Kaufman, M.D., Ph.D. Immunobiology of islet transplant rejection.

Byung Kim, Ph.D. Antigen processing; T-cell activation; viral immunity; viral disease.

David Klumpp, Ph.D. Molecular pathogenesis of urinary tract infections.

Jhumku Kohtz, Ph.D. Sonic hedgehog in telencephalic development.

Peter Kopp, M.D. Molecular basis of thyroid and endocrine disease.

James Kramer, Ph.D. Genetics and development in *C. elegans.*

Lou Laimins, Ph.D. In vitro studies of papillomavirus infection.

Chung Lee, Ph.D. Role of hormones in tumor growth.

Jonathan Leis, Ph.D. Retrovirus replication, reverse transcription, integration.

Honglin Li, Ph.D. Caspases and Bcl-2 family members in apoptosis.

Liming Li, Ph.D. Structural properties of prion proteins.

Jon Lomasney, M.D. Alpha receptors in cardiac myocytes.

Richard Longnecker, Ph.D. Latent infection by Epstein-Barr virus.

William Lowe Jr., M.D. Transcriptional regulation of insulin-like growth factor expression.

Thomas McGarry, M.D., Ph.D. Control of cell division during development.

Richard Miller, Ph.D. Calcium channels in nerve communication and neurodegenerative disease.

Stephen Miller, Ph.D. Immunoregulation of T-cell–mediated immune responses.

Bernard Mirkin, Ph.D. Growth and differentiation of neural ectoderm-derived cells.

Jill Morris, Ph.D. Molecular basis of schizophrenia.

Toshio Narahashi, Ph.D. Electrophysiology and pharmacology of nerve and synaptic membrane diseases.

Puneet Opal, M.D., Ph.D. Cellular basis of neurodegeneration.

Boris Pasche, M.D., Ph.D. Transforming growth-factor beta in cancer development.

Leonidas Platanias, M.D., Ph.D. Signal transduction in malignant cells.

Richard Pope, M.D. Immune mechanisms in rheumatoid arthritis.

Yi Rao, Ph.D. Axon guidance and neuronal migration.

Matthew Ravosa, Ph.D. Functional anatomy of the primate cranium.

Janardan Reddy, M.D. Carcinogenesis and peroxisome proliferators.

Steven Rosen, M.D. Immune mechanisms in carcinomas.

Vijay Sarthy, Ph.D. Retinal development and degeneration.

Richard Scarpulla, Ph.D. Gene regulation; transcription factors; intracellular signaling; mitochondrial disease.

Anthony Schaeffer, M.D. Molecular pathogenesis of urinary tract infections.

H. William Schnaper, M.D. Cell-matrix interactions; extracellular matrix proteases.

H. Steven Seifert, Ph.D. Mechanisms of pilus antigenic variation in *Neisseria gonorrhoeae.*

Brian Shea, Ph.D. Evolutionary morphology of primates.

Teepu Siddique, M.D. Genetics of human motor neuron disorders.

Eugene Silinsky, Ph.D. Synaptic transmission; ATP; electrophysiology.

Hans-Georg Simon, Ph.D. Pattern formation; limb development; limb regeneration.

Greg Smith, Ph.D. Herpesvirus transport in the nervous system.

Patricia Spear, Ph.D. Molecular mechanism of herpesvirus infection.

M. Sharon Stack, Ph.D. Regulation of matrix proteases.

Paul Stein, Ph.D. Src family kinases in lymphoid cells.

Paula Stern, Ph.D. Mechanisms of bone formation and resorption.

D. James Surmeier, Ph.D. Modulation of neuronal excitability.

Francis Szele, Ph.D. Neurogenesis in the adult subventricular zone.

Jacob Sznajder, M.D. Regulation of Na,K-ATPase in the alveolar epithelium.

Bayar Thimmapaya, Ph.D. Transcription factors; genetics; cancer; metastasis; adenovirus; E1A.

Jacek Topczewski, Ph.D. Morphogenic processes in zebrafish development.

Warren Tourtellotte, M.D., Ph.D. *Egr* transcription factors in the central nervous system.

Margrit Urbanek, Ph.D. Susceptibility genes for complex diseases.

Linda Van Eldik, Ph.D. Calcium-modulated proteins in cell function.

Robert Vasser, Ph.D. Molecular basis of Alzheimer's disease.

Tom Volpe, Ph.D. Heterochromatin formation in higher eukaryotes.

Olga Volpert, Ph.D. Mechanisms of angiogenesis inhibitors.

Mark Wainwright, M.D., Ph.D. Mechanisms of acute brain injury.

David Walterhouse, M.D. Function/regulation of the oncogene GLI.

Zhou Wang, Ph.D. Apoptosis and cell proliferation in the prostate.

D. Martin Watterson, Ph.D. Calcium in signal transduction and as a regulator of homeostasis in eukaryotes.

Susan Winandy, Ph.D. Transcriptional regulation of T-cell development.

Gayle Woloschak, Ph.D. Radiosensitivity and motor neuron disease.

Nabeel Yaseen, M.D., Ph.D. Nuclear transport and leukemia.

Yijun Zhu, M.D. Estrogen receptor coactivators and breast cancer development.

OHIO UNIVERSITY

College of Arts and Sciences
Department of Biological Sciences

Programs of Study	The graduate program offers master's and doctoral degrees in biological sciences and microbiology. Graduate education is conducted in five research focus groups. The cell biology and physiology group employs molecular, cellular, and systems approaches to study animal and plant function. The ecology and evolutionary biology group integrates research on ecology, functional morphology, phylogeny, genetics, and life history of natural populations and model organisms to study evolutionary patterns, processes, and mechanisms. The exercise physiology and muscle biology group focuses on effects of exercise, nutrition, gender, and aging on human performance, muscle histology, muscle physiology, and reproductive endocrinology. The microbiology group applies molecular biological techniques to theoretical and practical problems in bacteriology, virology, parasitology, and immunology. The neurobiology group emphasizes systems and computational neurobiology.

Master's students complete 45 quarter system credit hours for the degree, including a minimum of 30 hours of formal courses, including seminars. They perform one quarter of supervised teaching, carry out original research, and write a thesis, which entails thesis proposal and thesis defense examinations. Students normally complete the program in 2.25 years. A nonthesis master's degree is offered for elementary or secondary school teachers. Doctoral students complete 135 hours of graduate work; of these, 45 are in formal courses or seminars. They also complete at least two quarters of supervised teaching, a candidacy examination by their ninth quarter, and a dissertation proposal examination by their thirteenth quarter. Average time for completion of the doctoral degree is 5.5 years.

Research Facilities

In addition to standard laboratory equipment used in biological laboratories, special research equipment available to graduate students includes an electron microscopy facility; a quantitative microscopy and imaging facility, which houses a laser scanning confocal microscope with Silicon Graphics Indigo workstations and a Neurolucida 3-D reconstruction and analysis system; exercise physiology facilities with metabolic carts, a Biodex isokinetic dynamometer, and an echocardiograph; a hybridoma facility; and a 15,000-specimen vertebrate collection. Facilities for mass spectrophotometry; NMR; computer-assisted DNA, RNA, and protein sequence analysis; and housing for animals are also available on campus. Library holdings include 1.1 million volumes with approximately 1,000 periodicals on biological and biomedical science topics.

Financial Aid

Program students are supported by teaching or research assistantships or by fellowships. In 2005–06, a teaching assistantship provided an annual stipend of $14,000 for master's students and $17,000 for doctoral students and a waiver of tuition. The Department's Graduate Committee may, without solicitation and at their discretion, make additional awards in recognition of a student's academic excellence in the program. In addition, any student may apply for small Houk Awards for partial support of research expenses.

Cost of Study

In 2005–06, tuition and fees were $2977 per quarter for residents of Ohio and $5641 per quarter for nonresidents. Except for general and recreational fees ($516 per quarter), tuition and fees are waived for students awarded teaching or research assistantships or fellowships.

Living and Housing Costs

Unmarried students may live in graduate dormitory housing, which cost approximately $3210 per quarter for room and board for 2005–06. Married students may live in one the University's apartment buildings, where monthly rents are approximately $700. Houses and apartments in the community are also available with one-bedroom apartments ranging from $350 to $650.

Student Group

The enrollment on the Athens campus is 19,000, including a graduate enrollment of 2,600, of whom 80 are enrolled in the Biological Sciences Graduate Program. In the current group of students, 40 percent are women, 39 percent are international, and 3 percent are part time. An undergraduate GPA of at least 3.2 is required for admission to the program. A TOEFL score of at least 620 is required for the admission of students whose native language is not English. All students must score in the fiftieth percentile or above on the General Test of the GRE.

Student Outcomes

Sixty percent of the graduates of the master's program have continued their education in graduate or medical school and 40 percent have sought employment. All doctoral graduates in the last five years have obtained employment in their area within one year of completing their doctoral degrees. Of these doctoral graduates, 32 percent obtained employment in industry and 68 percent are employed in academic positions.

Location

Ohio University is located in Athens, which has a population of 21,000. Athens is located 75 miles southeast of Columbus. The community and University offer a broad range of cultural opportunities, including art, dance, music, and theater. The rural setting in the scenic Appalachian foothills offers many outdoor recreational activities, including four attractive state parks within a short drive.

The University

Founded in 1804, Ohio University is the oldest institution of higher education in the old Northwest Territory. It is accredited by the North Central Association of Colleges and Schools. Over the past two decades, a major emphasis on excellence in research has brought the University recognition by the Carnegie Foundation for the Advancement of Teaching as being in the RU/H: Research Universities (high research activity) category. Professional programs include a School of Physical Therapy and a College of Osteopathic Medicine. Ohio University operates on the quarter system.

Applying

An application form and a more extensive description of faculty interests may be obtained from the Department of Biological Sciences address, listed in this description. Students must submit applications by January 15 for entrance in the following fall quarter. In addition to the application form, applicants must submit transcripts, GRE General Test scores, TOEFL scores (if the student's native language is not English), three letters of recommendation, a letter of research intent, and the names of 3 faculty members they would like to review their file.

Correspondence and Information

Graduate Secretary
Department of Biological Sciences
Irvine Hall
Ohio University
Athens, Ohio 45701-2979
Phone: 740-593-2334
E-mail: nilsen@ohio.edu
Web site: http://www.biosci.ohiou.edu/grad

Ohio University

THE GRADUATE FACULTY AND THEIR RESEARCH

Cell Biology and Physiology
Mark Berryman, Assistant Professor; Ph.D. Molecular aspects of membrane-cytoskeletal attachments in epithelial cells.
Mary Chamberlin, Associate Professor; Ph.D. Developmental changes in insect epithelial physiology; mitochondrial metabolism in ectotherms.
Elizabeth Crockett, Assistant Professor; Ph.D. Physiological/biochemical adaptations to varying or extreme physical environments.
Janet Duerr, Assistant Professor; Ph.D. Using *C. elegans* to examine regulation of neuronal gene expression, neurotransmitter trafficking, and behavior.
Frank Horodyski, Associate Professor; Ph.D. Hormonal control of insect development and metamorphosis.
Sharon Inman, Assistant Professor; Ph.D. Renal microcirculation and ischemia/reperfusion injury in kidney transplantation.
Kelly Johnson, Assistant Professor; Ph.D. Insect physiology; evolution of feeding specialization; chemical ecology.
Richard E. Klabunde, Associate Professor; Ph.D. Cardiac and vascular function in septic shock.
John Kopchick, Professor; Ph.D. Molecular mechanism(s) of growth hormone action.
Yang V. Li, Assistant Professor; Ph.D. Cell-to-cell communication in the central nervous system; how the brain modifies its function and structure through experiences.
Anne Loucks, Professor; Ph.D. Effects of diet and exercise on endocrine regulation of the human reproductive system.
Felicia V. Nowak, Associate Professor; Ph.D., M.D. Regulation and mechanism of action of neuropeptides in brain development and function.
Robert Rakowski, Professor; Ph.D. Ion channels and electrogenic pumps.
Allan Showalter, Professor; Ph.D. Molecular biology and biochemistry of plant cell wall proteins and halophytes.
Soichi Tanda, Assistant Professor; Ph.D. Genetics; molecular and developmental biology.

Ecology and Evolutionary Biology
Audrone Biknevicius, Associate Professor; Ph.D. Biodynamics and biometrics of vertebrate feeding and locomotor systems.
Elizabeth Crockett, Assistant Professor; Ph.D. Physiological/biochemical adaptations to varying or extreme physical environments.
Kim Cuddington, Assistant Professor; Ph.D. Theoretical population ecology.
Warren Currie, Assistant Professor; Ph.D. Biological oceanography; aquatic ecology.
R. Patrick Hassett, Assistant Professor; Ph.D. Biological oceanographic/zooplankton ecology and physiology.
Frank Horodyski, Associate Professor; Ph.D. Hormonal control of insect development and metamorphosis.
Kelly Johnson, Assistant Professor; Ph.D. Insect physiology; evolution of feeding specialization; chemical ecology.
Donald Miles, Associate Professor; Ph.D. Morphological and ecological correlates of locomotion; life history evolution in squamate reptiles.
Molly R. Morris, Assistant Professor; Ph.D. Behavioral ecology; sexual selection and evolution of communication in fishes.
Patrick O'Connor, Assistant Professor; Ph.D. Respiratory evolution in birds and other archosaurs (e.g., dinosaurs); evolution and biogeographic history of Gondwanan vertebrates during the Cretaceous period.
Steve Reilly, Associate Professor; Ph.D. Ecological and functional morphology, morphometrics, ontogeny, heterochrony, vertebrate metamorphosis, systematics.
Willem Roosenberg, Associate Professor; Ph.D. Ectotherm life history evolution; ecology of environmental sex determination; conservation biology.
Nancy Stevens, Assistant Professor; Ph.D. Biomechanical underpinnings of arboreality in primates and other vertebrates; evolutionary and biogeographic history of Tertiary vertebrates from continental Africa and Madagascar.
Soichi Tanda, Assistant Professor; Ph.D. Genetics; molecular and developmental biology.
Matthew White, Associate Professor; Ph.D. Population and evolutionary genetics; biochemical systematics; conservation biology.
Lawrence Witmer, Associate Professor; Ph.D. Craniofacial ontogeny, functional morphology, and paleontology of archosaurs.

Exercise Physiology and Muscle Biology
Joseph Eastman, Professor; Ph.D. Evolutionary morphylogy of Antarctic fish.
Robert Hikida, Professor; Ph.D. Adaptive and nucleocytoplasmic responses of human skeletal muscles to changes in activity.
John Howell, Associate Professor; Ph.D. Skeletal muscle physiology at both the cellular and organ system levels.
Richard Klabunde, Associate Professor; Ph.D. Cardiac and vascular function in septic shock.
Anne Loucks, Professor; Ph.D. Effects of diet and exercise on endocrine regulation of the human reproductive system.
Robert Staron, Associate Professor; Ph.D. Adaptation of skeletal muscle fibers in response to various workloads.

Microbiology
Bonita Biegalke, Associate Professor; Ph.D. Molecular biology of herpes viruses; determinants of pathogenicity.
Xiao-Zhuo Chen, Assistant Professor; Ph.D. Molecular and cellular biology of cancer and cancer gene therapy.
Karen Coschigano, Assistant Professor; Ph.D. Elucidation of genes involved in the development of or protection from kidney damage.
Peter Coschigano, Assistant Professor; Ph.D. Molecular and genetic analysis of anaerobic biodegradation of toxic compounds.
Kenneth Goodrum, Associate Professor; Ph.D. Cytokine responses to Group B streptococci.
Mario Grijalva, Assistant Professor; Ph.D. Molecular diagnosis, immunology, and epidemiology of tropical diseases.
Donald Holzschu, Assistant Professor; Ph.D. Retroviruses; mechanisms of retroviral induction of cell proliferation and tumorigenesis.
Calvin James, Associate Professor; Ph.D. Aspects of signal transduction pathways relevant to both viral and cellular gene regulation.
John Kopchick, Professor; Ph.D. Molecular mechanism(s) of growth hormone action.
Edwin Rowland, Associate Professor; Ph.D. Immunobiology of *Trypanosoma cruzi* infection, including disease pathogenesis.
Tomohiko Sugiyama, Assistant Professor; Ph.D. Biochemistry and molecular biology of meiotic recombination and DNA double-strand-break repair.

Neurobiology
Robert Colvin, Associate Professor; Ph.D. Calcium transport across plasma membranes; neurobiology of aging.
Ralph DiCaprio, Professor; Ph.D. Cellular neurophysiology; neuronal basis of behavior; organization and generation of rhythmic motor patterns.
Janet Duerr, Assistant Professor; Ph.D. Using *C. elegans* to examine regulation of neuronal gene expression, neurotransmitter trafficking, and behavior.
Robert Hikida, Professor; Ph.D. Adaptive and nucleocytoplasmic responses of human skeletal muscles to changes in activity.
William Holmes, Associate Professor; Ph.D. Mathematical and computational models; long-term potentiation; dendritic spines.
Scott Hooper, Associate Professor; Ph.D. Central and peripheral mechanisms underlying rhythmic motor pattern production.
John Howell, Associate Professor; Ph.D. Skeletal muscle physiology at both the cellular and organ system levels.
Dae-Woo Lee, Assistant Professor; Ph.D. Degeneration of dopamine neurons in Parkinson's disease; molecular and cellular mechanisms of synaptic plasticity.
Felicia V. Nowak, Associate Professor; Ph.D., M.D. Regulation and mechanism of action of neuropeptides in brain development and function.
Ellengene Peterson, Professor; Ph.D. Vestibular neurobiology; sensorimotor control of head movement.
Robert Rakowski, Professor; Ph.D. Ion channels and electrogenic pumps.
Michael Rowe, Professor; Ph.D. Image analysis and spatial sampling in the vertebrate retina; development and evolution of vision in vertebrates.

THE PENNSYLVANIA STATE UNIVERSITY

Department of Biology

Programs of Study	Graduate study in biology at Penn State offers a wide variety of choices for work toward the M.S. and Ph.D. degrees.

Programs of Study

Graduate study in biology at Penn State offers a wide variety of choices for work toward the M.S. and Ph.D. degrees.

The Department has active research in bioinformatics, ecology, evolution, disease dynamics, genetics, cell biology, developmental biology, molecular biology, neuroscience, and physiology. More specific descriptions of research areas are found listed after faculty names. The Department also houses the Institute for Molecular Evolution and Genetics (IMEG). For detailed information on IMEG, students should refer to the Web site (http://imeg.psu.edu/).

Ph.D. students are encouraged to gain at least one year of teaching experience as assistants in laboratory courses during their program. The program emphasizes close interaction between graduate students and faculty members in developing the intellectual and experimental skills required to create independent research. Students may carry out laboratory rotations before making a final choice of a thesis adviser. Since the doctoral dissertation is based on original research, students are encouraged to participate in laboratory and research activities under the direction of a faculty adviser as early as possible in their training.

In addition to carrying out the research necessary to produce a thesis, students are required to participate in seminars and in graduate biology courses. There are very few uniform course requirements across the Department, and the breadth of course offerings allows each student, in consultation with a faculty adviser, to select courses appropriate for his or her own specialized training.

Although the time needed to complete the requirements for a graduate degree varies, normally a Ph.D. program is expected to take no longer than five years. Students who elect to complete an M.S. degree are expected to finish in two years. All students are expected to begin their research during the first year.

In addition to a written thesis, the components of a doctoral program include a candidacy exam (administered in the second year of study), a comprehensive examination (taken when course work is completed), and a final oral defense of the thesis. Students completing an M.S. degree must also defend their thesis.

Research Facilities

Modern, well-equipped research facilities are located in Mueller, North Frear, and Wartik Laboratories, as well as in the recently completed Life Sciences Building. Additional resources include animal-care facilities, the Biotechnology Institute, an electron microscopy laboratory, a Linus computational cluster, confocal laser microscopes, a state-of-the-art greenhouse, transgenic capability, and a herbarium of more than 150,000 plant specimens. Field studies are possible in local areas ranging from forests, upland bogs, streams, and lakes to cultivated farmland. Field projects are also conducted by faculty members and students in tropical America, Africa, the Caribbean, and the deep sea. As a member of the Organization for Tropical Studies, Penn State offers access to courses and research facilities in Costa Rica. For more information on field projects, students should access the Department Web site.

Financial Aid

Students admitted into the Department were awarded graduate assistantships of at least $14,445 for two semesters and $2889 for summer support plus a waiver of tuition for the 2005–06 academic year. The Eberly College of Science Braddock Awards provides financial supplements in amounts of $2000 to $16,000 to outstanding students selected by the Department. A limited number of Graduate School fellowships, which pay a stipend of $15,500 and cover tuition, are also available on a competitive basis. Research assistantships are often available through individual faculty members.

Cost of Study

Tuition is covered for all students awarded teaching and research assistantships and Graduate School fellowships.

Living and Housing Costs

Housing is available both on campus and off campus. In 2005–06, on-campus housing rates for single graduate students began at $545 per month (utilities included). A meal plan is extra. One-, two-, and three-bedroom apartments are also available for married graduate students and families, starting at $725 per month (utilities included). The cost of off-campus housing varies, depending on the type of accommodations desired.

Student Group

The University Park campus has a total enrollment of more than 40,000, including nearly 7,000 graduate students. There are more than 800 undergraduate biology majors and about 110 graduate students in the Department.

Location

The University Park campus is adjacent to the town of State College in scenic central Pennsylvania. The area is characterized by a series of long ridges and valleys, and much of the surrounding land is state forest, providing opportunities for boating, camping, fishing, hiking, hunting, skiing, and swimming. Residential areas range from a small, convenient urban center and various suburban neighborhoods to the surrounding rural countryside. The county's population totals about 130,000. Travel to and from the region is made convenient by the proximity of Interstates 80 and 99 and the University Park airport (SCE), only 5 miles from campus.

The University

The Pennsylvania State University, founded in 1855, is the land-grant university of Pennsylvania. The University has more than 4,500 faculty members and offers degrees in about 160 baccalaureate and 150 graduate programs. Penn State regularly ranks among the nation's top fifteen public research universities, with yearly research expenditures of more than $500 million. The University operates on a fifteen-week semester system, with two 6-week summer sessions.

Applying

Although there is no fixed deadline for fall applications, the fullest consideration for fellowships and Braddock Supplements is given to applications received by January 31. Scores on the General Test of the Graduate Record Examinations (GRE) are required and the Subject Test is recommended. International students whose native language is not English must also submit TOEFL scores. Transcripts, three letters of recommendation, and a letter indicating the applicant's specific area of research interest and objectives are needed to complete the application. Students lacking some undergraduate courses, e.g., chemistry, calculus, and physics, are required to remedy these deficiencies soon after arrival.

Formal application is required online at http://www.bio.psu.edu/home/degrees/gradprog/apply. Students may also submit a free online preapplication for evaluation at http://www2.bio.psu.edu/biology/preapplication.html before completing the formal application.

Correspondence and Information

Graduate Program Secretary
Department of Biology
208 Mueller Laboratory
The Pennsylvania State University
University Park, Pennsylvania 16802

Phone: 814-863-7034
E-mail: gradinfo@email.bio.psu.edu
Web site: http://www.bio.psu.edu

The Pennsylvania State University

THE FACULTY AND THEIR RESEARCH

H. Akashi, Assistant Professor; Ph.D., Chicago, 1996. Molecular evolution and genome evolution; molecular adaptation and weak selection. Metabolic efficiency and amino acid composition in the proteomes of *Escherichia coli* and *Bacillus subtilis*. *Proc. Natl. Acad. Sci. U.S.A.* 99:3695, 2002 (with Gojobori).

S. M. Assmann, Waller Professor of Plant Biology; Ph.D., Stanford, 1986. Plant signal transduction; second-messenger regulation of ion channels; molecular biology of G proteins; plant RNA biology. Plants: The latest model system for G-protein research. *EMBO Rep.* 5:572–8, 2004 (with Jones).

O. N. Bjørnstad, Associate Professor; Ph.D., Oslo, 1997. Population ecology; ecology of infectious disease. Evolution and emergence of *Bordetella* in humans. *Trends Microbiol.* 13:355–9, 2005 (with Harvill).

D. M. Braun, Assistant Professor; Ph.D., Missouri, 1997. Plant genetics, signal transduction, and development. Utility and distribution of conserved noncoding sequences (CNSs) in the grasses. *Proc. Natl. Acad. Sci. U.S.A.* 99:6147–51, 2002 (with Kaplinsky et al.).

D. R. Cavener, Professor and Department Head; Ph.D., Georgia, 1980. Genetic regulation of secretory functions in endocrine and exocrine pancreas and the skeletal system using mouse genetic models of human disease, including diabetes, growth retardation, and skeletal dysplasias. Activating transcription factor 3 is integral to the eukaryotic initiation factor 2 kinase stress response. *Mol. Cell. Biol.* 24:1365–77, 2004 (with Jiang et al.).

G. Chen, Assistant Professor; Ph.D., Shanghai Institute of Physiology, 1993. Molecular and cellular mechanisms of synaptogenesis and synaptic plasticity. Cyclothiazide induces robust epileptiform activity in rat hippocampal neurons both in vitro and in vivo. *J. Physiol.* 571:605–8, 2006 (with Qi et al.).

D. J. Cosgrove, Professor and Eberly Chair of Biology; Ph.D., Stanford, 1980. Plant growth and development; molecular biology and physiology of cell wall enlargement; loosening of plant cell walls by expansins; cell-wall biochemistry. Growth of the plant cell wall. *Nature Rev. Mol. Cell Biol.* 6:850–61, 2005.

R. J. Cyr, Professor; Ph.D., California, Irvine, 1986. Plant cell and developmental biology. The cortical microtubule array: From dynamics to organization. *Plant Cell* 16:2546–52, 2004 (with Dixit).

C. W. dePamphilis, Associate Professor; Ph.D., Georgia, 1988. Molecular evolution, molecular systematics, and evolutionary genomics of plants. Widespread genome duplications throughout the history of flowering plants. *Genome Res.*, in press (with Cui et al.).

N. V. Fedoroff, Willaman Professor of Life Sciences and Evan Pugh Professor; Ph.D., Rockefeller, 1972. Stress and hormone signaling in plants, in particular, the role of heterotrimeric G proteins; miRNA regulation of gene expression; regulation of transcript stability in plant stress responses. Different signaling and cell death roles of heterotrimeric G protein alpha and beta subunits in the *Arabidopsis* oxidative stress response to ozone. *Plant Cell* 17:1–14, 2005.

C. R. Fisher, Professor; Ph.D., California, Santa Barbara, 1985. Physiology and ecology of hydrothermal-vent and cold-seep fauna. High hydrogen sulfide demand of long-lived vestimentiferan tube worm aggregations modifies the chemical environment at deep-sea hydrocarbon seeps. *Ecol. Lett.* 6:212–19, 2003 (with Cordes et al.).

S. Gilroy, Associate Professor; Ph.D., Edinburgh, 1987. Hormone perception and signal transduction in plant cells; regulation of secretion and gravitropism. *Arabidopsis* H+-PPase AVP1 regulates auxin-mediated organ development. *Science* 310:121–5, 2005 (with Li et al.).

B. T. Grenfell, Alumni Professor of the Biological Sciences; Ph.D., York, 1980. Disease dynamics; population biology; quantitative biology. Integrating life history and cross-immunity into the evolutionary dynamics of pathogens. *Proc. Royal. Soc. B–Biol. Sci.* 273:409–16, 2006 (with Restif).

M. J. Guiltinan, Professor; Ph.D., California, Irvine, 1986. Developmental and cell biology; control of plant gene expression; starch biosynthesis in maize; genetic engineering of horticultural crops. Maize starch branching enzyme (SBE) isoforms and amylopectin structure: In the absence of SBEIIb, the further absence of SBEIa leads to increased branching. *Plant Physiol.* 136:3515–23, 2004 (with Yao and Thompson).

S. B. Hedges, Professor; Ph.D., Maryland, 1988. Evolutionary biology, evolutionary genomics, and astrobiology. The origin and evolution of model organisms. *Nature Rev. Genet.* 3:838–49, 2002.

E. C. Holmes, Professor; Ph.D., Cambridge, 1990. Molecular evolution; virology; disease dynamics. The population genetics and evolutionary epidemiology of RNA viruses. *Nat. Rev. Microbiol.* 2:279–87, 2004 (with Moya and González-Candelas).

P. J. Hudson, Willaman Professor; Ph.D., Oxford, 1979. Population dynamics of wildlife diseases. Large shifts in pathogen virulence relate to host population structure. *Science* 303:842–4, 2004 (with Boots and Sasaki).

Z. C. Lai, Associate Professor; Ph.D., Yeshiva (Einstein), 1988. Developmental neurobiology and cancer biology. Control of cell proliferation and apoptosis by Mob as tumor suppressor, Mats. *Cell* 120:675–85, 2005.

A. Liu, Assistant Professor; Ph.D., NYU, 2000. Biogenesis and function of cilia in signal transduction and mammalian embryonic development. Intraflagellar transport proteins regulate both the activator and repressor functions of Gli transcription factors. *Development* 132:3103–11, 2005.

S. J. Liu, Assistant Professor; Ph.D., Rochester, 1990. Mechanisms underlying activity-dependent changes in synaptic transmission and in membrane excitability. Subunit interaction with PICK and GRIP controls Ca^{2+}-permeability of AMPARs at cerebellar synapses. *Nature Neurosci.* 8:768–75, 2005 (with Cull-Candy).

B. Lüscher, Associate Professor; Ph.D., Zurich, 1983. Molecular, cellular, and systems neuroscience; synaptogenesis and plasticity of GABAergic inhibitory synapses; receptor trafficking; genetic analysis of the neural substrate of anxiety. Distinct gamma2 subunit domains mediate clustering and synaptic function of postsynaptic GABA-A receptors and gephyrin. *J. Neurosci.* 25:594–603, 2005 (with Alldred et al.).

H. Ma, Professor; Ph.D., MIT, 1988. Molecular genetics of flower development, anther differentiation, meiosis, and pollen development; evolutionary genomics of flower development; evolutionary and bioinformatic studies of recombination genes. The *Arabidopsis RAD51* gene is dispensable for vegetative growth but required for meiosis. *Proc. Natl. Acad. Sci. U.S.A.* 101:10596–601, 204 (with Li et al.).

W. Makalowski, Associate Professor; Ph.D., Poznan (Poland), 1991. Comparative genomics of eukaryotic genomes. Not junk after all. *Science* 300:1246–7, 2003.

K. Makova, Assistant Professor; Ph.D., Texas Tech, 1999. Molecular evolution; population genetics; evolutionary genomics. Strong male-driven evolution of DNA sequences in humans and apes. *Nature* 41:624–6, 2002 (with Li).

J. H. Marden, Professor; Ph.D., Vermont, 1988. Integrative, ecological, and evolutionary physiology. A candidate locus for variation in dispersal rate in a butterfly metapopulation. *Proc. Royal Soc. B* 272:2449–56, 2005 (with Haag et al.).

P. McSteen, Assistant Professor; Ph.D., East Anglia (England), 1996. Plant development genetics and signal transduction. *barren inflorescence2* regulates axillary meristem development in the maize inflorescence. *Development* 128:2881–91, 2001 (with Hake).

W. Miller, Professor; Ph.D., Washington (Seattle), 1969. Algorithms and software for molecular biology. Galaxy: A platform for interactive large-scale genome analysis. *Genome Res.* 15:1451–5, 2005 (with Giardine et al.).

M. Nei, Evan Pugh Professor; Ph.D., Kyoto (Japan), 1959. Molecular evolutionary genetics. *Molecular Evolution and Phylogenetics*, Oxford University Press, 2000 (with Kumar).

R. W. Ordway, Associate Professor and Chair, Genetics Graduate Program; Ph.D., Massachusetts Medical Center, 1990. Cellular and molecular mechanisms of chemical synaptic transmission. Active zone localization of presynaptic calcium channels encoded by the cacophony locus of *Drosophila*. *J. Neurosci.* 24:282–5, 2004 (with Kawasaki et al.).

R. L. Patterson, Assistant Professor; Ph.D., Maryland, Baltimore, 2000. Molecular and cellular neuroscience; calcium signaling; computational biology. Phospholipase C-g: diverse roles in receptor-mediated calcium signaling. *Trends Biochem. Sci.* 30:688–97, 2005.

E. S. Post, Associate Professor; Ph.D., Alaska Fairbanks, 1995. Climate change ecology; community dynamics; analytical and experimental approaches. Ecological responses to recent climate change. *Nature* 416:389–95, 2002.

S. W. Schaeffer, Associate Professor; Ph.D., Georgia, 1985. Molecular population genetics and genomics. Comparative genome sequencing of *Drosophila pseudoobscura*: Chromosomal, gene and cis-element evolution. *Genome Res.* 15:1–18, 2005.

K. Shea, Assistant Professor; Ph.D., London, 1994. Invasion ecology; applications of ecological theory. Community ecology theory as a framework for biological invasions. *Trends Ecol. Evol.* 17(4):170–6, 2002 (with Chesson).

A. G. Stephenson, Professor; Ph.D., Michigan, 1978. Ecology and evolution of plant reproduction. Interrelationships among inbreeding, herbivory, and disease on reproduction in a wild gourd. *Ecology* 85:3023–34, 2004 (with Leyshon et al.).

G. H. Thomas, Associate Professor; Ph.D., Edinburgh, 1984. Role of the actin cytoskeleton in cell polarity and the development of *Drosophila melanogaster*. Brush border spectrin is required for early endosome recycling in *Drosophila*. *J. Cell. Sci.* 119:1361–70, 2006 (with Phillips).

C. Uhl, Professor; Ph.D., Michigan State, 1980. Plant ecology; human ecology; environmental impacts; sustainability assessments. Leaf demography and phenology in Amazonian rain forest: A census of 40,000 leaves of twenty-three tree species. *Ecol. Monogr.* 74:3–23, 2004 (with Reich et al.).

A. Walker, Evan Pugh Professor; Ph.D., London, 1967. Primate and human evolution; functional anatomy. The Lothagam hominids. In *Lothagam: The Dawn of Humanity in Eastern Africa*, pp. 249–56, eds. J. Harris and M. Leakey. New York: Columbia University Press, 2003.

K. M. Weiss, Evan Pugh Professor; Ph.D., Michigan, 1972. Genetics and evolution of complex traits; human genetic variation; developmental genetics of vertebrate dentition. The phenogenetic logic of life. *Nature Rev. Genet.* 6:2–11, 2005.

M. D. Whim, Assistant Professor; Ph.D., Cambridge, 1988. Regulation of neuropeptide secretion. FMRF amide tagging: How an ionotropic receptor can be used to measure peptide secretion. *Methods* 33:295–301, 2004 (with Moss).

THE PENNSYLVANIA STATE UNIVERSITY

College of Medicine
Milton S. Hershey Medical Center
Graduate Programs in the Biomedical Sciences

Programs of Study	Anatomy Program: A multidisciplinary program deadline to an M.D. and/or Ph.D. degree. Comprehensive training in anatomical sciences and research in areas related to neurosciences, cellular biology, diabetes, cancer, vision, and pediatrics, with a special emphasis on translational research. The master's program is especially directed toward students in need of establishing credentials for further professional training (e.g., medical school) Biochemistry and Molecular Biology Program: A departmental program with research opportunities in structure and function of macromolecules, carcinogenesis and cancer prevention, and molecular genetics and genomics. Specialization includes X-ray crystallography of protein complexes, protein folding and protein-ligand interactions, metalloproteinase and multimeric protease regulation, protein degradation and quality control, glycoconjugates and malaria pathogenesis, glycosphingo lipid interactions with viruses and toxins, tobacco and environmental carcinogenesis, DNA replication and repair, gene regulation, chromatin and chromosome structure and function, molecular genetics of anesthesia, and RNA processing and subcellular distribution of RNAs and proteins. Bioengineering Program: An intercollege program with research opportunities in artificial organs; cardiovascular fluid dynamics; biomaterials research, including blood-polymer interactions, surface modification/characterization, and scanning probe microscopy; controlled drug delivery via biodegradable and smart biomaterials, polymer-protein, polymer-cell, and protein-cell interactions are also of interest along with orthopedic biomechanics, implants, tissue adaptation, mechanotransduction, and large-scale finite-element modeling. Cell and Molecular Biology Program: An interdepartmental program with a cross-section of research opportunities in nine departments within the College. Research includes membrane structure, receptors, and modulators; extracellular matrix; organelle assembly, structure, and function; cell division, differentiation, and regulation of gene expression; gene mapping and recombinant DNA; molecular interface between cells and human viruses; intracellular events relevant to the immune response; cell-cell communication; cancer; and tissue engineering. Genetics: An intercollege program designed to prepare graduates for rapidly expanding opportunities in genetics at academic institutions, biotechnology and pharmaceutical companies, private research institutes, and governmental research laboratories. Training encompasses the spectrum of genetics, from basic to molecular and genomic sciences to human diseases using a variety of model systems including yeast, zebrafish, mammalian viruses, and mice as well as human populations. Health Evaluation Sciences: A program designed to prepare for clinical investigations and health services research, two related fields in which shortages of trained investigators have been identified. Research opportunities for physicians, nurses, epidemiologists, and health services and policy experts are rapidly expanding into a wide variety of settings, including academic health centers, hospitals, health insurance and pharmaceutical industries, and government. A two-year M.S. program teaches population-based disciplines that enable planning, execution, analysis, and dissemination of health-care research, as well as evaluation and improvement of health-care practices. Integrative Biosciences: An intercollege, intercampus program that develops conceptual connections leading to innovative, interdisciplinary research. Options for study include bioinformatics and genomics, cell and developmental biology, chemical biology, immunobiology, molecular medicine, and molecular toxicology. Students matriculate at either the College of Medicine or the University Park campus. M.D./Ph.D. Program: This program offers students interested in careers in academic medicine and research to obtain the training in clinical and basic sciences. The dual-degree program provides a firm foundation in medicine plus research experience essential to pursue a career in academic medicine and translational research. Microbiology and Immunology Program: A departmental program with research opportunities to explore interactions of viruses with host cells and animal model systems, with emphasis on adenoviruses, hepatitis viruses, herpesviruses, papillomaviruses, papovaviruses, parvoviruses, and retroviruses. Individual programs center on virus replication and assembly, neurovirulence, pathogenesis, latency and reactivation, oncogenesis, and the role of the cellular immune response in these processes. Viral systems serve as models for eukaryotic gene regulation, signal transduction, and the human immune response. Research programs are maintained in eukaryotic cellular differentiation, tumor cell biology, and cancer immunology. Neuroscience Programs: An intercollege program with broad representation in neuroscience research including areas of cellular and molecular neuroscience: growth factors, cell lineage, oxidative damage and neurotoxicology, aging, and tumor cell growth. Research in systems and behavioral neuroscience includes information processing in sensory systems, taste and mechanisms of reward, pain, and autonomic regulation. Pharmacology Program: A departmental program with research in cellular and molecular pharmacology, focused on identification of molecular targets for drug development in cancer and cardiovascular and neurological diseases. Areas of specialization include signal transduction and ion channel–receptor structure-function analyses. Moreover, the department possesses state-of-the-art drug discovery, molecular biology, and proteomics facilities. A dual Ph.D./M.B.A. program is available for interested students. Physiology Program: An interdisciplinary graduate program with research focused on mechanisms regulating cell function, metabolism, growth, and development under both physiological and pathophysiological conditions. Areas of specialization include DNA repair and carcinogenesis, transcriptional and translational control of gene expression, proteosomal and lysosomal mediated proteolysis, cell-cell and cell–extracellular matrix interactions, genetic factors in disease susceptibility, intracellular trafficking of proteins and vesicles, ion channels and cardiovascular function, signal transduction, metabolic dysregulation, telomerase and cell senescence, cancer, diabetes, and obesity.
Research Facilities	Core facilities and shared research support include an MRI/MRS facility; whole animal luminescent imaging; a BL3 containment facility; confocal, deconvolution, fluorescence, and transmission electron microscopy; laser capture microdissection; histology services; a transgenic/gene knock-out and rederivation facility; flow cytometric analysis and cell sorting; custom peptide and DNA synthesis; automated peptide and DNA sequencing; proteomics/MALDI-TOF-TOF and hybrid ion trap mass spectroscopy; 1D and 2D LC and Gel separation systems; a viral vector core; molecular modeling and docking software; detection/analysis systems for chemiluminescent, radiolabeled, and fluorescent samples; an X-ray crystallographyc facility; an organic synthesis facility; and quantitative real-time PCR and microarray analysis facility. Consulting centers for biostatistics, informatics, and epidemiology are available
Financial Aid	Students receive graduate assistantships of at least $20,772 per year, plus tuition waiver, or fellowships from various sources. Financial support allows full-time effort in graduate studies.
Cost of Study	Students are responsible for costs of books, student activity fees, and 20 percent of the Penn State Graduate Assistant and Fellow Health Insurance Policy (MEGA) premium, if selected.
Living and Housing Costs	For housing information, students should contact the Housing Office, PSU/COM, P.O. Box 850, Hershey, Pennsylvania 17033 (telephone: 717-531-8210).
Student Group	Students have outstanding records of academic achievement, with exceptional GRE performance. Students make important contributions to scientific literature during their research training and intellectual development. Graduates hold postdoctoral positions at prestigious institutions and establish careers in academia, industry, and government.
Location	The College of Medicine is in Hershey, a community of approximately 20,000, an outstanding environment for research and scholarly achievement with easy access to major attractions and metropolitan areas.
The College	The 550-acre College campus includes Medical Sciences and Biomedical Research Buildings, a University Hospital, Animal Research Facilities, apartments, and fitness and day-care centers.
Applying	Qualified students with undergraduate preparation in the biological, biochemical, or physical sciences; an overall GPA of 3.0 or higher; competitive GRE scores; three letters of recommendation; and a personal essay summarizing background and professional goals are considered for admission. The most qualified are accepted on a space-available basis.
Correspondence and Information	Office of Graduate Student Affairs The Pennsylvania State University College of Medicine H170 500 University Drive Hershey, Pennsylvania 17033-2390 Phone: 717-531-8892 Fax: 717-531-4139 E-mail: grad-hmc@psu.edu Web site: http://www.hmc.psu.edu/gsa

The Pennsylvania State University

THE FACULTY AND THEIR RESEARCH

S. F. Abcouwer. Growth factor and cytokine expression in retinal diseases and cancer. K. D. Alloway. Neural circuits and the coordination of neuronal activity. S. A. Amin. Polycyclic aromatic hydrocarbons and tobacco-specific nitrosamines. D. A. Antonetti. Vascular permeability in diabetes and cancer. A. J. Barber. Neurodegeneration in diabetic retinopathy. R. Bascom. Inhalation toxicology; occupational respiratory disorders, lung cancer imaging. C. Berlin. Pediatric pharmacology; lactation; nutrition. M. Bewley. X-ray crystallography; protein complexes. E. Bixler. Normal sleep: patterns and mechanisms. J. S. Bond. Metalloproteases; meprins; cancer; diabetes; kidney disease. R. H. Bonneau. Neuroendocrine effects on immunity to herpes simplex virus (HSV). J. J. Botti. Exercise during pregnancy; perinatal infection; immunologic diseases in pregnancy. S. K. Bronson. Stem cells; osteoblast; genetics. C. R. Brown. Protein trafficking and degradation. K. K. Burkhart. Stroke: cytoskeletal changes and signal transduction. V. A. Canfield. Differentiation and development in gastric mucosa. L. Carrel. Genetic, epigenetic, and genomic control of gene regulation on the mammalian inactive X-chromosome. G. Chase. Lifetime risk analysis; statistical issues in genetic counseling. V. Chau. Enzymology of protein ubiquitination and ubiquitin-mediated proteolysis. J. Y. Cheng. Excitation—contraction coupling; myocardial infarction, exercise training; phospholemman. K. C. Cheng. Genomic instability; cell differentiation and cancer. H.-L. Chiang. Lysosomal protein targeting and degradation. V. M. Chinchilli. Biostatistics; multivariate analysis; repeated measurements analysis; clinical trials; cross-over designs. M. R. Chinoy. Pulmonary hypoplasia; retinoic acid in lung morphogenesis/alveolar formation; MMPs and vascular abnormalities. M. J. Chorney. Structure of MHC; hemochromatosis; quantitative trait loci. N. D. Christensen. Molecular analysis of papillomavirus immunity. G. A. Clawson. Early changes in carcinogenesis; therapeutic ribozymes and antisense. C. M. Collins. Electromagnetic fields; magnetic resonance imaging; numerical calculations; temperature. J. R. Connor. Iron regulation; cell stress; aging; myelin; CNS; brain tumors. R. N. Cooney. Sepsis; cell signaling; metabolism. R. J. Courtney. Proteins and glycoproteins of herpes simplex virus. T. J. Craig. Asthma; COPD; rhinitis; sleep. R. C. Craven. Retrovirus replication and assembly. L. M. Demers. Breast cancer; metastatic bone disease. H. J. Donahue. Gap junctions; mechanotransduction; metastasis; bone and cartilage cell biology; tissue engineering, regenerative medicine. B. R. Dworkin. Learning and physiological regulation. K. A. Eckert. Mechanism of human cell mutagenesis. K. El-Bayoumy. Mechanisms of tobacco and environmental carcinogenesis; role of nutrition in cancer prevention. J. Ellis. Molecular neuropharmacology; regulation of acetylcholine receptors. P. J. Eslinger. Functional brain imaging of cognition; Alzheimer's disease; concussion. L. A. Evey. Neuroanatomy; neurophysiology; sensory coding; neural convergence; anesthetics; implementation and efficacy of teaching technologies. A. G. Ewing. Cellular and developmental neuroscience. E. Eyster. Hemophilia, HIV/AIDS, and hepatitis. J. Fang. Neurobiology of sleep; neural and immune interactions; interactions between sleep and memory. D. J. Feith. Polyamines; ornithine decarboxylase; carcinogenesis; mouse models; skin cancer. J. Flanagan. Pathway of protein folding as it occurs in vitro and in vivo. J. Floros. Surfactant protein genes in lung health and diseases. R. A. Frost. Mechanisms of growth factor signaling and resistance. R. A. Gabbay. Improving diabetes health-care delivery; non-invasive glucose monitoring and insulin delivery. T. W. Gardner. Diabetic retinopathy; insulin receptor signaling; apoptosis. I. C. Gilchrist. Platelet receptors and arteriosclerosis therapy. C. D. Gowda. Malaria parasite-glycobiology cytoadherence; immune regulation. J. W. Griffith. Veterinary/comparative pathology of laboratory animals, aging, and cancer. S. A. Grigoryev. Chromatin structure and epigenetic chromosomal markers in cancer. P. S. Grigson. Neural basis of reward comparison. E. Gunther. Breast cancer and treatment. A. Hajnal. Neural mechanisms of altered food preferences and meal-size control in obesity, type-2 diabetes, and in response to weight reduction. C. S. Hollenbeak. Health economics; Bayesian statistics; pharmacoeconomics. C. J. Holliman. Trauma; prehospital care; clinical emergency medicine. J. Hu. Replication and pathogenesis of hepatitis B and C viruses. R. B. Irby. Src in colon tumor progression and metastasis; LGL leukemia. H. C. Isom. Liver pathophysiology: injury, hemochromatosis, and HBV. L. S. Jefferson. Signaling pathways and effector mechanisms involved in the translational control of gene expression. B. C. Jones. Neurogenetics; complex traits analysis of brain iron. K. D. Karpa. Probiotics; medication therapeutic management; drug-disease state management. M. Katzman. Retroviral integration; integrase; AIDS pathogenesis. G. L. Kauffman. Regulation of gastrointestinal function by brain pesticides. R. L. Keil. Mechanisms of volatile anesthetic action; amino acid regulation of translation. M. Kester. Kinase cascades and lipid-derived second messengers. J. K. Kim. Role of obesity and inflammation in type 2 diabetes and its complications. S. R. Kimball. Hormonal and nutritional regulation of gene expression; mRNA translation; cell signaling. T. S. King. Biostatistics; measures of agreement; analysis of categorical data. K. Kjerulff. Women's health; health services research; health psychology. D. W. Knutson. Clinical studies of hypertension and renal dialysis. K. Koch. Gastric dysrhythmias; nausea; electrogastrography. W. Koltun. Genetics of inflammatory bowel disease; immunology of gut inflammation. C. H. Lang. Infection/stress/trauma; IGF-I; cytokines; metabolism. K. F. LaNoue. Neurotransmitters; mechanisms of type II diabetes. P. Lazarus. Tobacco carcinogenesis; carcinogen metabolism; molecular epidemiology and cancer susceptibility; p53 tumor suppressor gene. R. S. Legro. Polycystic ovary syndrome; anovulation; insulin resistance; infertility; androgen excess. E. J. Lengerich. Epidemiology of chronic disease and cancer. U. A. Leuenberger. Neural control of circulation; sleep apnea. R. Levenson. Molecular neurobiology; dopamine receptors and their role in schizophrenia; inner ear development in zebrafish. D. Liao. Cardiovascular disease; environmental pollution; population studies. H.-M. Lin. Longitudinal analysis; categorical data analysis; sleep; organ transplantation. W. Liu. Linkage and association analysis; statistical genetics. T. A. Lloyd. Lifestyle determinants of health. T. P. Loughran. LGL leukemia. T. L. Lowe. Drug delivery; polymer-protein-cell interactions. K. Lucas. EBV virus; bone marrow transplant. R. F. Lundy. Behavioral neuroscience; sensory processing in the brain. C. J. Lynch. Nutrient signaling to adipose tissue; obesity. G. I. Makhatadze. Molecular basis of protein folding and macromolecular recognition; prostate cancer. A. Manni. Cellular mechanisms controlling breast cancer development, progression, and proliferation. G. Matters. Metalloproteases in cancer cell progression, invasion, and metastasis. D. T. Mauger. Bioequivalence; biopharmaceutical models; measurement error; clinical traits. J. M. McAllister. Human ovary; steroidogenesis; gene expression. D. E. McCloskey. Cancer therapeutics; polyamines; mechanism of drug action. M. McEchron. Neural networks of learning and memory; neurophysiology of aging and learning. T. J. McGarrity. GI cancers; genetics; carcinogenesis. P. J. McLaughlin. Growth factors; receptors; cardiovascular development; cancer. C. Meyers. Papillomaviruses replication and oncogenesis. B. Miller. Calcium channels in signal transduction of erythropoietin. R. J. Milner. Molecular neurobiology: gene expression in glia cells. K. M. Mulder. TGFβ; growth factors; cell signaling; cancer drug discovery. J. Muscat. Molecular epidemiology of cancer; susceptibility to tobacco smoke. Y.-C. Ng. Na-pump; heart failure and hypertrophy; skeletal muscle aging. C. Niyibizi. Stem cell; cell therapy; gene therapy; skeletal disease; connective tissue biochemistry; regenerative medicine. C. Norbury. Mechanisms of antigen processing and presentation in vivo during induction of antiviral T-cell responses. R. Norgren. Neural bases of motivation: taste, hunger, and thirst. R. W. Ordway. Cellular and molecular mechanism of synaptic transmission. F. K. Orkin. Clinical epidemiology; outcomes research. A. Ouyang. Brain-gut axis in functional bowel disease. C. Palmer. Hypotic-ischemic brain injury to developing brain: iron, free radicals, pharmacoprotection, neutrophils, nitric oxide. L. J. Parent. Intracellular trafficking of retroviral proteins and RNA. I. M. Paul. Newborn healthcare and health outcomes; obesity prevention; cough asthma. A. E. Pegg. Polyamine metabolism and role in neoplastic growth; DNA repair. X. Peng. Transgenic animal models of tumorigenesis, sleep and memory regulation, and cardiovascular diseases. T. J. Perrault. Multisensory integration. B. Z. Peterson. Modulation of neuronal and cardiac ion channels. D. S. Phelps. Surfactant regulation of lung immune cell function. M. D. Planas-Silva. Estrogen signaling and cell-cycle control in breast cancer. T. C. Pritchard. Central organization of taste. Z. Qian. Environmental epidemiology; exposure assessment; asthma. S. R. Rannels. Vitamin K in lung; lung development; lung cancer. C. A. Ray. Neural control of circulation; exercise. T. Ritty. Extracellular matrix proteins; matrix protein receptors; matrix metalloproteinases. G. P. Robertson. Molecular dissection of melanoma development. I. J. Ropson. Folding and function of β-sheet proteins. G. Rosenberg. Mechanical circulatory support. V. J. Ruiz-Valasco. Signaling mechanisms between ion channels, G proteins, and receptors. L. Sandirasegarane. Regulation of Akt (protein Kinase B) by nitric oxide in vascular smooth muscle cells. M. Saunders. Biomechanics; bone remodeling; osteolysis.. T. Schell. Cancer imunotherapy; CD8+T-cell tolerance. C.-L. Schengrund. Glyosphingolipids; HIV; botulinum toxin. I. U. Scott. Diabetic retinopathy; age-related macular degeneration; clinical/outcomes/epidemiologic research in ophthalmology. M. Shaffer. Clustered and longitudinal data; missing data; Bayesian statistics. L. M. Shantz. Signal transduction pathways and regulation of ornithine decarboxylase in cancer. C. A. Siedlecki. Blood-biomaterial interactions; probe microscopy; applications of nanotechnology to biomaterials. I. Simpson. Role of nutrient transporters in cerebral metabolism. L. I. Sinoway. Neural control of circulation. J. P. Smith. Peptides and growth of pancreatic cancer. M. B. Smith. In vivo NMR imaging and spectroscopy. A. J. Snyder. Mechanical circulatory support; interventional devices; biomaterials. W. W. Souba. Amino acid treatment; glutamine metabolism; glutamine during catabolic states; system N transporter. D. J. Spector. Adenovirus; human cytomegalovirus; gene regulation. T. Spratt. Mechanism of catalysis and fidelity of DNA replication and DNA repair. T. Subramanian. Neural transplantation; gene therapy; basal ganglia neurophysiology; pathobiology of neurodegeneration. J. Y. Summy-Long. Neurobiology of cytokines on dendritic release; neural-glial plasticity of the magnocellular system. S.-C. Sun. Signal transduction in innate immunity; lymphocyte activation and retroviral oncogenesis. R. B. Tenser. Viral infection of the nervous system; multiple sclerosis. M. J. Tevethia. Transforming mechanisms of DNA viruses. S. S. Tevethia. Cytotoxic T-cell vaccines against viruses and tumors. D. Thiboutot. Factors regulating human sebum proliferation in acne. M. E. Truckenmiller. Mechanisms of antigen processing and presentation; dendritic cell function; neuroendocrine effects. A. Undar. Cardiopulmonary bypass; pediatric heart pumps; artificial organs. T. C. Vary. Regulation of protein metabolism in alcoholic cardiomyopathy, inflammation, and sepsis. M. F. Verderame. Breast cancer; tyrosine kinase signal transduction; organotypic cultures. K. E. Vrana. Molecular neuroscience; stem cells; proteomics; substance abuse. C. Weisman. Health services research; women's health; quality of health care. W. J. Weiss. Implantable circulatory support devices; Doppler ultrasound. J. Weisz. Chronic oxidative stress; breast carcinogenesis; cancer initiating stem cells; cytochemistry; imaging. J. W. Wills. Protein targeting; molecular biology of virus assembly and budding. R. P. Wilson. Anesthesia and pain management of laboratory anmals; wound healing. R. T. Wilson. Occupational exposure; VDR polymorphisms; renal cell cancer; childhood leukemia. Q. Yang. Magnetic resonance imaging and functional magnetic resonance imaging. J. You. Role of mechanical signals in regulating molecular and cellular processes. J. K. Yun. Sphingolipid biology; molecular characterization of sphingosine kinases; role of sphingolipid metabolic enzymes in diseases. I. S. Zagon. Peptides and receptors in cancer, cell renewal, wound healing, and development. J. Zhu. Mechanisms of cellular aging; telomerase regularion; cancer.

THE PENNSYLVANIA STATE UNIVERSITY

The Huck Institutes of the Life Sciences
Graduate Degree Programs

Programs of Study	Graduate education in the Huck Institutes of the Life Sciences is interdisciplinary, intercollegiate, flexible, and team oriented, providing in-depth training in multiple aspects of the life sciences as well as exposing students to a variety of scientific investigation approaches in several disciplines. Students can choose from six Intercollege Graduate Degree Programs (IGDP): Ecology, Genetics, Integrative Biosciences, Neuroscience, Physiology, and Plant Biology. Within Integrative Biosciences are various research specialties (called options), including bioinformatics and genomics, focusing on computational, evolutionary, and functional analysis of genomes; cell and developmental biology, studying the control and coordination of gene expression; chemical biology, studying structures and functions of macromolecules using physical/chemical and molecular biological techniques; immunobiology, exploring the interaction among cells and molecules contributing to innate and adaptive reactivity to foreign antigenic challenge and the breakdown in normal regulation leading to pathogenic states of self-reactivity and infection; molecular medicine, investigating the pathophysiological basis of human disease in the areas of cancer biology, infectious disease, metabolic regulation, and molecular and human genetics; and molecular toxicology, studying biomedical and physiological manifestations resulting from potentially noxious agents in the environment at the molecular and cellular levels in biological organisms. It is anticipated that many of these options will become independent graduate degree programs in the near future.

The Huck Institutes has nearly 300 faculty members, who are affiliated with thirty-eight departments housed in the Colleges of Agricultural Sciences, Earth and Mineral Sciences, Engineering, Health and Human Development, and Liberal Arts; the Eberly College of Science at the University Park campus; and the College of Medicine at the Hershey Medical Center campus. Dual faculty mentors, laboratory rotations, and graduate internships are incorporated into the program. Both master's and doctoral degrees are awarded.

Research Facilities

The Shared Technology Facilities exist to facilitate and support research of Penn State's faculty members and students. The following facilities are available: the DNA Microarray Facility—custom DNA microarray preparation, hybridization, and scanning; the Nucleic Acid Facility—DNA synthesis and sequencing; the Center for Computational Genomics—equipped with both PC and Macintosh computers running a full complement of molecular biology and modeling software packages, a Silicon Graphics Indigo Extreme computing workstation, and Sun SPARCstations; the Hybridoma and Cell Culture Laboratory—isolation and production of monoclonal antibodies of interest to researchers in the life sciences, cell culture, cell freezing, storage, and media preparation; the Electron Microscope Facility—a multiuser service lab for ultrastructural characterization of biological and biomaterial specimens through transmission and scanning electron microscopy, energy-dispersive X-ray spectroscopy, and selected area electron diffraction; the Center for Quantitative Cell Analysis—instrumentation for the quantitative analysis and automated separation of individual cells from large populations; the Nuclear Magnetic Resonance Facility—360-, 400-, 500-, and 600-MHz spectrometers for biological studies; the X-Ray Crystallography Facility—rotating anode diffractometer, image plate detectors, and Silicon Graphics computers for data analysis; the Proteomics and Mass Spectrometry Core Facility—research in characterization of proteins, peptides, lipids, nucleic acids, bioconjugates, metabolites, and other substances; the Macromolecular Core Facility—N-terminal protein sequencing, peptide synthesis, amino acid analysis, oligonucleotide synthesis, and radiometric and densitometric analysis; the Cell Science Core Facility—flow cytometry and imaging; the Molecular Genetics Core Facility—DNA sequencing, genotyping, and DNA computing; the Transgenic Mouse Core Facility—transgenic and targeted-gene-disruption mouse models for use in biomedical research; and the Bioinformatics Consulting Center—support in computational biology, bioinformatics, and biostatistics, including genomic EST analysis, primer design, gene expression analysis (both Affymetrix and spotted arrays), and classical statistical genetics.

Financial Aid

Most students who are admitted to a Huck Institutes graduate program receive financial support in the form of a graduate assistantship for the first two semesters (plus summer session) from the Huck Institutes. Depending on the graduate program, a total of five years of support is guaranteed to doctoral students who remain in good standing.

Cost of Study

Students who are awarded a graduate assistantship receive a stipend and a tuition waiver. Students are responsible for books, thesis costs, taxes on stipends, and partial payment of health benefits.

Living and Housing Costs

The cost of living at the University Park campus and surrounding areas is moderate. The University offers various graduate on-campus apartment accommodations between $575 and $965 per month (including utilities) for single and married students. Students may purchase meal plans costing between $1435 and $1815 per semester. Off-campus housing in private homes and apartment complexes provides living accommodations at a wide range of prices. Housing at the Hershey Medical Center campus is available on campus for single and married students. A typical two-bedroom apartment currently rents for $828 per month, including utilities. Convenient off-campus housing is also available.

Student Group

There were 71 Huck Institutes graduate students accepted for 2006. They joined the 320 Huck graduate students who are currently pursuing their degrees. A minimum of forty-four Huck Institutes graduate assistantships are anticipated for fall 2007.

Student Outcomes

Graduates enter academic institutions; research laboratories; pharmaceutical, biotechnology, and agricultural companies; government and regulatory laboratories; federal agencies; and science-based companies and organizations.

Location

Home of the University Park campus, State College, Pennsylvania (population 100,000), is the major cultural center of central Pennsylvania. Within a 3-hour drive of Baltimore and Pittsburgh and a 4-hour drive of New York City, Philadelphia, and Washington, D.C., the town has a collegiate atmosphere. There is bus and air commuting service to five major cities in the mid-Atlantic area. The University and the community sponsor cultural, athletic, professional, and scholarly events. There are excellent recreational opportunities on campus and in the surrounding countryside. The Hershey Medical Center campus is 15 minutes from Harrisburg, Pennsylvania's state capital; 2 hours from the University Park campus, Philadelphia, Baltimore, and Washington, D.C.; and 3 hours from New York City. Nearby tourist attractions include Hersheypark, Hershey Museum, and Hershey Gardens.

The Institutes

The Huck Institutes of the Life Sciences represent more than 300 scientists in seven colleges on the University Park and Hershey Medical Center campuses. They are working together for quality and innovation in research and education in the life sciences. Graduate education integrates research, teaching, theory, and practice using the most contemporary techniques, equipment, and electronic communications. Huck Institutes graduate students are free to explore the wide variety of education, training, and career options available to today's scientists.

Applying

General admission requirements include appropriate course work in the life sciences, a junior/senior undergraduate GPA of at least 3.0 on a 4.0 scale (some programs require a higher GPA), and test scores from three GRE General Tests. While the biology GRE Subject Test is required for ecology, Subject Tests in biology, chemistry, biochemistry, or cell and molecular biology are encouraged but not required for the other Huck graduate programs.

Correspondence and Information

Information about a particular graduate program may be obtained by contacting the program chair, visiting the Huck Institutes' Web site, or calling, writing, or e-mailing:

Admissions Committee, Huck Institutes of the Life Sciences
101 Life Sciences Building
The Pennsylvania State University
University Park, Pennsylvania 16802-5807

Phone: 866-774-2467 (toll-free in the U.S.)
Fax: 814-863-1357
E-mail: gradinfo@huck.psu.edu
Web site: http://www.HUCK.psu.edu/GradEd/

The Pennsylvania State University

THE HUCK INSTITUTES GRADUATE DEGREE PROGRAMS AND THEIR CHAIRS

ECOLOGY GRADUATE PROGRAM
Ecology is a highly interdisciplinary cross-college program that educates scientists about the interactions among organisms and the environment from molecular to ecosystem levels of organization. The program fosters a learning environment that is centered on ecological theory and application.

David Mortensen, Professor of Weed Ecology; Ph.D., North Carolina State. Phone: 814-865-1906; e-mail: dam37@psu.edu.

GENETICS GRADUATE PROGRAM
Genetics is a remarkable field in which classical concepts combine with cutting-edge approaches to reveal the mysteries of biology. Training encompasses the full spectrum of genetics from the molecular to the population levels in a wide range of model systems. Beginning and advanced graduate-level courses are taught by active research faculty members in their own areas of specialization.

Sarah Bronson, Associate Professor of Cellular and Molecular Physiology; Ph.D., Washington (St. Louis). Phone: 717-531-5194; e-mail: skb8@psu.edu.
Richard Ordway, Associate Professor of Biology; Ph.D., Massachusetts. Phone: 814-863-5693; e-mail: rwo4@psu.edu.

INTEGRATIVE BIOSCIENCES (IBIOS) GRADUATE PROGRAM
Peter Hudson, Willaman Professor of Biology; D.Phil., Oxford. Phone: 814-865-0522; e-mail: pjh18@psu.edu.

IBIOS Program Research Options and Their Chairs

Bioinformatics and Genomics Option
Recent advances in sequencing and high-throughput expression technology are generating unprecedented amounts of genomic data. Computational and statistical challenges arise in processing, storing, and analyzing these data. At the same time, evolutionary concepts underlie nearly all aspects of genome analysis. This IBIOS program option brings together statistical, computational, evolutionary, and biological expertise to address current analytical problems in genome research.

John Carlson, Associate Professor of Molecular Genetics and Director, Schatz Center for Tree Molecular Genetics; Ph.D., Illinois. Phone: 814-863-9164; e-mail: jec16@psu.edu.
Keith Cheng, Associate Professor of Pathology and Adjunct Professor of Biochemistry and Molecular Biology; M.D., NYU; Ph.D., Washington (Seattle). Phone: 717-531-5635; e-mail: kcc2@psu.edu.

Cell and Developmental Biology Option
The ability to study and manipulate genes at the molecular level allows unprecedented approaches to fundamental questions in cell and developmental biology. These issues include the mechanisms of regulation of cellular function and how differentiating organisms make decisions about cell fate, gene expression, and patterning.

Hong Ma, Professor of Biology; Ph.D., MIT. Phone: 814-863-8082; e-mail: hxm16@psu.edu.

Chemical Biology Option
Fundamental and technical advances in chemistry, biochemistry, and molecular biology provide unprecedented opportunities to probe the mysteries of life at the molecular level. Questions can now be framed in terms of the physical properties that are responsible for the structure-function relationships of DNA, RNA, and proteins, including enzymatic catalysis, mechanisms of gene expression, and the mechanism of protein folding.

George Makhatadze, Professor of Biochemistry and Molecular Biology; Ph.D., Institute of Protein Research, Moscow. Phone: 717-531-0712; e-mail: makhatadze@psu.edu.

Immunobiology Option
Immunobiology focuses on the fundamental concepts underlying the immune network. Emphasis is also placed on clinical perspectives residing within the realms of transplantation, autoimmunity, hypersensitivity, infection, and cancer. An appreciation of immunological components (cytokines, antibodies, and lymphocytes) that are currently translatable into therapeutic modalities is underscored.

Andrew Henderson, Assistant Professor of Veterinary Science; Ph.D., California, Riverside. Phone: 814-863-0340; e-mail: ajh6@psu.edu.
Neil Christensen, Associate Professor of Pathology and Microbiology and Immunology; Ph.D., Auckland (New Zealand). Phone: 717-531-6185; e-mail: ndc1@psu.edu.

Molecular Medicine Option
The Molecular Medicine option is centered on the pathophysiological basis of human disease. Research opportunities range from molecular biology to the physiology of cell and organ function. Specific research areas include cancer, metabolic regulation, infectious disease, and human genetics.

Avery August, Associate Professor of Immunology; Ph.D., Cornell. Phone: 814-863-3539; e-mail: axa45a@psu.edu.
Craig Meyers, Professor of Microbiology and Immunology; Ph.D., UCLA. Phone: 717-531-6240; e-mail: cmeyers@psu.edu.

Molecular Toxicology Option
Graduate training encompasses studies at the molecular, cellular, and organ levels as well as the intact animal. The research emphasis is mechanistic in nature, with two major scientific themes: transgenic models of toxicity and molecular model systems of toxicology. The diversity of interests represented in this IBIOS option provides its predoctoral trainees with a unique interdisciplinary background in toxicology that ultimately permits them to address toxicological problems in comprehensive and innovative ways.

Gary H. Perdew, Professor of Molecular Toxicology; Ph.D., Oregon State. Phone: 814-865-0400; e-mail: ghp2@psu.edu.

NEUROSCIENCE GRADUATE PROGRAM
The Neuroscience Graduate Program emphasizes interdisciplinary approaches at molecular, cellular, systems, and behavioral levels to understand the structure and function of the nervous system. Specific research areas include molecular neurobiology and neurogenetics, neural development, systems and computational neuroscience, behavioral neuroscience, brain metabolism, aging, and neurological diseases.

Byron Jones, Professor of Neuroscience and Anatomy; Ph.D., Arizona. Phone: 814-863-0167; e-mail: bcj1@psu.edu.
Robert J. Milner, Professor and Chair of Neuroscience and Anatomy; Ph.D., Rockefeller. Phone: 717-531-8652; e-mail: rjm11@psu.edu.

PHYSIOLOGY GRADUATE PROGRAM
Research training in physiology helps prepare students for generating and using information gathered at all levels of inquiry, from the gene to the whole body. Students can select their thesis topics from research projects in endocrinology, nutrition, neurophysiology, respiration, inflammation and immunology, reproduction, protein synthesis and regulation, muscular and skeletal physiology, cardiovascular physiology, and thermoregulation.

Leonard Jefferson, Evan Pugh Professor and Chair, Cell and Molecular Biology; Ph.D., Vanderbilt. Phone: 717-531-8567; e-mail: lgg3@psu.edu.

PLANT BIOLOGY GRADUATE PROGRAM
Plants respond dynamically to environmental stimuli, both natural and anthropogenic, modifying traits in ways that profoundly impact both natural and managed ecosystems, including agriculture. This program seeks to educate students who can integrate molecular, physiological, and ecological approaches to studying plant development and responses to the environment.

Teh-hui Kao, Professor of Biochemistry and Molecular Biology; Ph.D., Yale. Phone: 814-863-1042; e-mail: txk3@psu.edu.

RENSSELAER POLYTECHNIC INSTITUTE

Department of Biology

Programs of Study

It is an exciting time to be in Rensselaer's Department of Biology. Throughout Rensselaer, there is a shift in emphasis toward initiatives in biotechnology, integrative systems biology, and cell and tissue engineering. The Department is growing and has recently added several new faculty members and expanded its programs. Faculty members and students engage in cutting-edge research in state-of-the-art laboratories and facilities, including the brand new $100-million Center for Biotechnology and Interdisciplinary Studies with state-of-the-art equipment for cellular, biochemical, and biophysical approaches to life science research. With excellent leadership, an outstanding faculty, dedicated staff, bright students eager to explore, and modern facilities, the Department is at the forefront of groundbreaking research efforts in this vitally important field.

Rensselaer's Department of Biology offers three graduate-level degree programs: a Ph.D. in biochemistry and biophysics, an M.S. in bioinformatics, and a Ph.D. in biology. The M.S. degree programs require 30 credit hours beyond the bachelor's degree. The Ph.D. requires 90 credit hours beyond the bachelor's degree.

Research Facilities

The Department maintains extensive research and instructional laboratories that house special and unique equipment developed for specific studies as well as extensive analytical and optical instrumentation, minicomputers, and microcomputers. The Biochemistry and Molecular Biology Teaching Laboratory includes a walk-in cold room and is equipped with facilities for modern molecular biology, gene expression, DNA sequencing, protein purification, plate reader–based screening assays, and state-of-the-art separation technology. The Bioinformatics Laboratory is equipped with twenty Silicon Graphics workstations, distance delivery apparatus, and audiovisual equipment. It is connected via a high-speed network to local Challenger and Origin 2000 servers and provides local access to high-speed RAID drives for searches of the most generally important biological sequence and structure databases. Available software includes some of the most powerful and generally used searching, sequence alignment, and analysis software, accessible through GCG seqlab and other formats, as well as molecular visualization, modeling, and dynamics packages from Molecular Simulations. Available software includes Vector NTI and Tripos packages.

Several research areas involve participation and cooperation with other departments, including the Departments of Biomedical Engineering, Chemistry and Chemical Biology, Earth and Environmental Science, and Economics, and a number of Rensselaer's major interdisciplinary and independent, collaborative facilities, including the new $100-million Center for Biotechnology and Interdisciplinary Studies, Darrin Fresh Water Institute, the Wadsworth Center, the Ordway Research Institute, and the Genomics Institute.

Rensselaer's Center for Biotechnology and Interdisciplinary Studies is a $100-million facility that provides a platform for collaboration among many diverse academic and research disciplines to enhance discovery and encourage innovation. Key areas of emphasis include drug discovery and development, regenerative medicine, and functional materials and devices, all of which build upon the fundamental disciplines of systems biology, biocatalysis, computational biology, and tissue engineering. The core research facilities within the center contain laboratories for molecular biology, analytical biochemistry, microbiology, imaging, histology, tissue and cell culture, proteomics, and scientific computing and visualization. The center contains an 800-MHz nuclear magnetic resonance (NMR) spectrometer and the computing and visualization infrastructure needed to model molecular structure at the atomic level.

The Darrin Fresh Water Institute (DFWI) conducts research that increases public awareness of environmental issues and contributes to answers for tough questions concerning the protection of land, water, and air. The DFWI is well known for its all-encompassing study of fresh water systems and ecological processes. This institute provides Rensselaer students and faculty members as well as visiting scientists the opportunity to study a number of ecosystems and to conduct research on important environmental problems. There are facilities on the RPI campus, at the lakeside Adirondack field station, and at a remote monitoring station.

Wadsworth Center is the most comprehensive state health laboratory in the country. Research is conducted through four divisions: environmental disease prevention, genetic disorders, infectious disease, and molecular medicine. Ordway Research Institute, Inc., is a not-for-profit, freestanding corporation with specific research themes—cancer, genomics/pharmacogenetics, emerging infections and host defense, and neural and vascular biology—and a mission to translate basic science observations into therapeutics. Core research facilities are an infrastructure of laboratories and equipment for flow cytometry, hollow fiber research, microscopy, pharmacokinetics and pharmacodynamics modeling, signal transduction, and target and drug discovery. The Genomics Institute, an Ordway Research Institute center of research, is a collaboration between the Wadsworth Center and Albany Medical College. This institute targets research for better health by the discovery of genetic approaches for the understanding of mammalian development and disease in order to design new treatment answers.

Financial Aid

Financial aid is available in the forms of teaching and research assistantships and fellowships, which include tuition scholarships and stipends. Rensselaer assistantships and university, corporate, or national fellowships fund many of Rensselaer's full-time graduate students. Outstanding students may qualify for university-sponsored Rensselaer Graduate Fellowship Awards, which carry a minimum stipend of $20,000 and a full-tuition and fees scholarship. All fellowship awards are calendar-year awards for full-time graduate students. Summer support is also available in many departments. Low-interest, deferred-repayment graduate loans are available to U.S. citizens with demonstrated need.

Cost of Study

Full-time graduate tuition for the 2006–07 academic year is $32,600. Other costs (estimated living expenses, insurance, etc.) are projected to be about $12,400. Therefore, the cost of attendance for full-time graduate study is approximately $45,000. Part-time study and cohort programs are priced differently. Students should contact Rensselaer for specific cost information related to the program they wish to study.

Living and Housing Costs

Graduate students at Rensselaer may choose from a variety of housing options. On campus, students can select one of the many residence halls, and there are abundant options off campus as well, many within easy walking distance.

Student Group

There are 1,234 graduate students, of whom 30 percent are women, 90 percent are full-time, and 69 percent study at the doctoral level.

Student Outcomes

Rensselaer's graduate students are hired in a variety of industries and sectors of the economy and by private and public organizations, the government, and institutions of higher education. Starting salaries average $63,262 for master's degree recipients.

Location

Located just 10 miles northeast of Albany, New York State's capital city, Rensselaer's historic 275-acre campus sits on a hill overlooking the city of Troy, New York, and the Hudson River. The area offers a relaxed lifestyle with many cultural and recreational opportunities, with easy access to both the high-energy metropolitan centers of the Northeast—such as Boston, New York City, and Montreal, Canada—and the quiet beauty of the neighboring Adirondack Mountains.

The Institute

Recognized as a leader in interactive learning and interdisciplinary research, Rensselaer continues a tradition of excellence and technological innovation dating back to 1824. More than 100 graduate programs in more than fifty disciplines attract top students, researchers, and professors. The discovery of new scientific concepts and technologies, especially in emerging interdisciplinary fields, is the lifeblood of Rensselaer's culture and a core goal for the faculty, staff, and students. Fueled by significant support from government, industry, and private donors, Rensselaer provides a world-class education in an environment tailored to the individual.

Applying

The admission deadline for the fall semester is January 1. Basic admission requirements are the submission of a completed application form (available online), the required application fee ($75), a statement of background and goals, official transcripts, official scores on the GRE General Test, TOEFL or IELTS scores (if applicable), and two recommendations. It is recommended that applicants also submit scores for the GRE Subject Test in Biology.

Correspondence and Information

Sharon Simmons, Admissions Coordinator
Department of Biology
1W14 Jonsson-Rowland Science Center
Rensselaer Polytechnic Institute
110 8th Street
Troy, New York 12180
Phone: 518-276-2808
E-mail: simmos2@rpi.edu
Web site: http://j2ee.rpi.edu/biology/

Rensselaer Polytechnic Institute

THE FACULTY AND THEIR RESEARCH

Blanca Barquera, Assistant Professor; Ph.D., National Autonomous University of Mexico. Bacterial physiology; biochemistry/biophysics; bioenergetics; sodium metabolism.

Donna L. Bedard, Research Professor; Ph.D., Chicago. Molecular environmental biology; molecular microbial ecology.

Charles Boylen, Professor; Ph.D., Wisconsin. Physiological responses of microorganisms and aquatic angiosperms to environmental perturbation.

Christopher Bystroff, Associate Professor; Ph.D., California, San Diego. Folding of proteins via database modeling and simulation.

Lenore S. Clesceri, Associate Professor Emeritus and Acting Program Director for IGERT, Division of Graduate Education, NSF; Ph.D., Wisconsin. Natural polymer degradation and transformations of synthetic organics.

Joyce J. Diwan, Professor; Ph.D., Illinois at Chicago. Development of an integrated package of computer-based learning tools for teaching biochemistry of metabolism in a studio format.

Jonathan Dordick, Professor; Ph.D., MIT. Biocatalysis in nonaqueous media; molecular bioprocessing; combinatorial biocatalysis; nanobiotechnology.

Henry L. Ehrlich, Professor Emeritus; Ph.D., Wisconsin–Madison. Bacterial oxidation of Mn(II) and reduction of Mn(IV), in particular as it applies to the development and fate of marine ferromanganese concretions and the possible bacterial origin of Mn(IV) oxide in calcareous deposits along the western shore of the Dead Sea.

Russell J. Ferland, Assistant Professor; Ph.D., Rochester. Joubert syndrome; neuroscience; genetics; neurodevelopment; learning and memory.

Fern Finger, Assistant Professor; Ph.D., Yale. Septins; cytoskeleton; cellular and axonal migration; organogenesis; *C. elegans.*

Angel E. Garcia, Professor; Ph.D.. Cornell. Mathematical modeling and computational analysis of problems in cellular and molecular biology; theoretical and computational aspects of biomolecular dynamics.

Michael H. Hanna, Associate Professor; Ph.D., Illinois at Urbana-Champaign. Microbiology, molecular biology: directed gene/protein evolution.

Jane F. Koretz, Professor and Acting Chair; Ph.D., Chicago. Characterization of human crystalline lens development and aging; dynamics of human visual focusing and presbyopia.

Lee Ligon, Assistant Professor; Ph.D., Virginia. Cytoskeletal organization and dynamics; microtubule cytoskeletal structure and function.

Robert J. Linhardt, Professor; Ph.D., Johns Hopkins. Glycoprotein, proteoglycans, and other glycoconjugates are prepared by fermentation using recombinant technology, extraction from tissues, or chemical and enzymatic synthesis to determine their structure and study their biological activities for application to new drug development.

Bradford Lister, Research Professor and Director, Anderson Center for Innovation in Undergraduate Education; Ph.D., Princeton. Ecology; undergraduate education.

Carl N. McDaniel, Professor; Ph.D., Wesleyan. Environmental sciences; developmental biology; interface between biology and economics.

Joel Morgan, Research Assistant Professor; Ph.D., Caltech. Molecular mechanism of energy transduction in biological systems.

Sandra A. Nierzwicki-Bauer, Professor and Director, Darrin Fresh Water Institute; Ph.D., New Hampshire. Microbiology; molecular biology.

Andrea Page-McCaw, Assistant Professor; Ph.D., MIT. Analyzing the in vivo functions of MMPs through mutant analysis and by identifying interacting proteins.

Patrick Page-McCaw, Assistant Professor; Ph.D., MIT. Organization, regulation, and function of the neural circuits that coordinate the startle response.

Robert E. Palazzo, Professor and Director, Center for Biotechnology and Interdisciplinary Studies; Ph.D., Wayne State. Cellular and molecular biology of the centrosome.

Robert H. Parsons, Associate Professor; Ph.D., Oregon State. Physiology; molecular biology.

Charles J. Pfau, Professor Emeritus; Ph.D., Indiana. Virology; immunology.

George E. Plopper, Assistant Professor; Ph.D., Harvard. Biochemical control of tumor cell migration; stem cell differentiation.

Harry Roy, Professor; Ph.D., Johns Hopkins. Development of muscle cells; sea urchin rRNA synthesis; biochemistry of ATP synthesis in chloroplasts and structure and biogenesis of the chloroplast enzyme ribulose bisphosphate carboxylase (Rubisco), particularly regarding its dependence on the action of molecular chaperones.

Douglas M. Swank, Assistant Professor; Ph.D., Pennsylvania. How muscle is designed to perform an amazingly wide variety of tasks including locomotion, pumping blood, and sound production.

Chunyu Wang, Assistant Professor; Ph.D., Cornell; M.D., Peking Union Medical College. Application of nuclear magnetic resonance (NMR) spectroscopy to study problems in neuroscience and aging, e.g. Alzheimer's disease.

Michael Zuker, Professor; Ph.D., MIT. Development of algorithms to predict RNA and DNA secondary structure by free energy minimization using empirically derived thermodynamic parameters.

RESEARCH GROUPS

Biochemistry and Biophysics: Several biology faculty members, all of whom are members of the interdepartmental Center for Biophysics, have research projects in the areas of biochemistry and biophysics. Many of these projects have goals that are both fundamental and applied. One project is examining changes in patterns of gene expression associated with adaptation to different conditions by *Vibrio cholera*, the organism responsible for the disease cholera. Aging of the human crystalline lens is being studied, including loss of the ability to focus (accommodate). Experiments at a molecular level are characterizing alpha-crystallin, a small heat shock protein that is the major protein of the mammalian lens. The role of chaperones in assembly of the photosynthetic enzyme Ribulose Bisphosphate Carboxylase/Oxygenase (RuBisCO) is also being examined. The bioinformatics studies of nitric oxide synthase (NOS) are being combined with spectroscopic studies (e.g., electron paramagnetic resonance and circular dichroism), thermodynamic and kinetics experiments, and site-directed mutagenesis to elucidate mechanisms of catalysis and regulation of this enzyme. Research in one laboratory has goals relating to enzyme technology. Projects include the study of enzymatic catalysis under extreme conditions and the enzymatic synthesis of polymeric materials. Biochemical research in one laboratory focuses on heparan sulfate proteoglycans. Kinetics and thermodynamics of interactions of carbohydrates with proteins are being studied using biophysical approaches such as calorimetry, spectroscopy, X-ray crystallography, and molecular modeling. The internal structure of mitochondria is being characterized using electron tomography and computer modeling. The structure of the outer mitochondrial membrane channel VDAC is also being determined using cryoelectron microscopy with 2-D crystals. Faculty: Barquera, Dordick, Koretz, Linhardt, Roy.

Bioinformatics and Computational Biology: Statistical models and molecular simulations are being applied to predict protein structures and protein folding pathways. A hidden Markov model for sequence-structure correlations has been developed, along with methods for predicting inter-residue contacts and helix propensities of short peptides. Collaborative research by 2 faculty members involves protein structure predictions based on gene sequence alignments. Results of these studies, along with experiments utilizing approaches of biochemistry and molecular genetics, have advanced understanding of the structural basis of the function and regulation of isoforms of the enzyme nitric oxide synthase (NOS). Algorithms to predict RNA and DNA secondary structure are being developed by free energy minimization, using empirically derived thermodynamic parameters. Computed partition functions, for systems containing two molecules that can fold as well as hybridize with each other, allow prediction of melting curves. The Bioinformatics Center, a joint initiative of Rensselaer and the Wadsworth Center of the New York State Department of Health, provides additional opportunities for collaborative research. Faculty: Bystroff, Garcia, Zuker.

Biotechnology: Several projects being carried out by biology faculty members, alone or in collaboration with members of other departments, are aimed at developing new technology or useful products. Many patents and patent applications have resulted from this research. New methods for site-directed mutagenesis and combinatorial chimeragenesis have been developed. These are being used in collaborative research aimed at directed evolution of small heat shock proteins. In another project, a database of naturally occurring plasmids has been developed. Biocatalysts with unique activities and selectivities are being generated, and biomolecules are being incorporated into nanostructures and composites. Other research is using combinatorial and high-throughput biocatalysis for drug discovery. Biochips containing heparan sulfates are being prepared for screening of the mouse glycome. Carbohydrates equivalent to tumor antigens are being synthesized for potential use in cancer vaccines. Faculty: Dordick, Hanna, Linhardt.

Cell Biology and Cell Signaling: The centrosome, the organelle that directs formation and organization of the cellular microtubule network, is being characterized using isolated and reconstituted centrosomes. Some experiments are aimed at elucidating genetic and biochemical mechanisms that control replication of the centrosome in relation to cell-cycle events. How intracellular signaling pathways are modulated in response to interaction of cells with the extracellular matrix is being examined in two biology laboratories. Differentiation relating to interaction of mesenchymal stem cells with extracellular matrix proteins is being studied, along with regulation of migration of breast cancer cells across lung endothelium. Another project is defining the role of plasma membrane integrins in mediating cell responses to changes in the extracellular microenvironment. Experiments are aimed at elucidating roles of matrix constituents in regulating tissue maintenance and repair, as well as tumor progression. Faculty: Finger, Ligon, Palazzo, Plopper.

Educational Innovation: Several faculty members are engaged in developing computer-based courseware and formats for studio teaching. Many of these educational innovations have been evaluated for student perceptions and learning outcomes. Papers describing these innovations, and/or the learning materials themselves, have been successfully subjected to peer review. A resource of expertise is the Anderson Center for Innovation in Undergraduate Education, directed by a biology faculty member. Web-based materials for the studio-format teaching of courses in biochemistry, human physiology, genetics, and ecology have been developed and tested. Four Biology faculty members are jointly developing computer-based materials for teaching, in a unique modified studio format, multiple sections of a high-enrollment course in introductory biology. Undergraduate laboratory courses have been revised to emphasize teaching students how to design, carry out, and report results of modern life science experiments. Faculty: Diwan, Edick, Hanna, Lister, McDaniel, Roy.

Microbial Ecology, Geomicrobiology, and Environmental Biology: Five members of the Biology Department work in areas of microbial ecology, geomicrobiology, and environmental biology. In addition, other faculty members use microbes as model systems or for gene expression. The Darrin Fresh Water Institute, administered by a member of the Department, has modern facilities for research relating to aquatic biology and the environment. An interdepartmental Environmental Science degree program has fostered collaborative interactions, particularly with members of the Earth and Environmental Science and Economics Departments at Rensselaer. Tools of molecular biology are being used to identify bacteria of ancient lineage in subsurface sediments, to characterize aquatic phytoplankton communities, and to detect the presence of zebra mussel veligers in water samples. Other studies focus on the identification and molecular genetics of cyanobacteria that grow in symbiotic association with the aquatic fern Azolla. One laboratory is identifying and characterizing microorganisms in aquatic sediments that have enzymatic capabilities for oxidizing or dechlorinating PCBs. A goal is to use such organisms or enzymes derived from them to degrade contaminant PCBs. Another project is focusing on the molecular ecology of microbial communities in hot springs. Physiological responses of microorganisms and aquatic angiosperms to environmental stress, such as that arising from acid rain, are being studied. Ecological alteration of regional lakes by introduction of exotic invasive plant species is also being examined. The work of one faculty member deals with broad issues of ecology and the environment, such as environmental effects on biodiversity and challenges to sustainability. Faculty: Bedard, Boylen, Ehrlich, McDaniel, Nierzwicki-Bauer.

Molecular Genetics and Developmental Biology: Four members of the Biology Department, with additional collaborators, are involved in research relating to molecular genetics and developmental biology. Several model organisms are being studied using an array of modern molecular and cytological tools. Experiments, utilizing the nematode worm *C. elegans* as a model system, are elucidating developmental functions of septins, GTP-binding proteins involved in cytokinesis that are implicated in human cancers and neurodegenerative diseases. Directed evolution of small heat shock proteins (chaperones) is a goal of one project that involves collaborative interactions of several faculty members. Using in vitro recombination, chimeric genes are being assembled from fragments of genes for naturally occurring heat shock proteins. These are expressed and assayed in *E. coli*. Developmental roles of matrix metalloproteases are being studied in the fruit fly *Drosophila melanogaster*. These proteases, which have important signaling roles, are increased in inflammatory diseases and cancer. Another model system, the larval zebra fish, is being for studies of developmental genetics and neuroscience. Neural circuits involved in the highly conserved startle response are being characterized, and genes involved in habituation to startle identified. Mammalian genes that influence development, differentiation, and behavior are being examined using mice as a model system. Mouse behaviors are being correlated with specific genes, and mouse models of human behavioral conditions and diseases are being created using transgenic and gene knockout approaches. Faculty: Finger, Hanna, A. Page-McCaw, P. Page-McCaw.

Neurobiology and Behavior: Three members of the Biology Department, with additional collaborators, are involved in neurobiological research. Several different model systems are being used for studies ranging from the cell biology of nervous system development to elucidating the molecular bases of animal behavior. The functions of septin-family GTPases in axonal migration during nervous system development are being studied in the nematode worm *C. elegans*. Septins are evolutionarily conserved proteins implicated in Parkinson's and other neurodegenerative diseases. Experimental approaches include in vivo studies of nervous system development and locomotor behavior as well as cell biological studies of axonal migration using primary cultures of embryonic worm neurons. The startle response, a highly conserved behavior, is being studied in another model system, the larval zebra fish. Approaches include quantitative assays of startle behavior, characterization of the neural circuits involved in the startle response, and identification of genes required for the naive startle response and for habituation to startle. Genetic polymorphisms that influence learning and memory, anxiety and fearfulness, and left-right laterality in the brain are being studied in the mouse. Mouse behaviors are being correlated with specific genes, and mouse models of human behavioral conditions and diseases are being created using transgenic and gene knockout approaches. Faculty: Ferland, Finger, P. Page-McCaw.

THE ROCKEFELLER UNIVERSITY

Graduate Programs

Programs of Study	Graduate education leading to the Ph.D. is offered to outstanding students regarded as potential leaders in their scientific fields. The University's research covers a wide range of biomedical and related sciences, including biochemistry, structural biology, biophysics, and chemistry; molecular, cell, and developmental biology; medical sciences and human genetics; immunology and microbiology; neurosciences; and bioinformatics, biophysics, and computational neuroscience, as summarized by the faculty list in this description. Students work closely with a faculty of active scientists and are encouraged to learn through a combination of course work, tutorial guidance, and apprenticeship in research laboratories. Graduate Fellows spend the first two years engaged in a flexible combination of courses geared toward academic qualification while conducting research in laboratories pertaining to their area of scientific interest. They choose a laboratory for thesis research by the end of the first year and devote their remaining time to pursuit of significant experimental or theoretical research, culminating in a dissertation and thesis defense. Students can spend full time in research; there are no teaching or other service obligations.

The faculties of the Rockefeller University, Weill Medical College of Cornell University, the Weill Graduate School of Medical Sciences of Cornell University, and Sloan-Kettering Institute collaborate in offering a combined M.D./Ph.D. program in the biomedical sciences to about 90 students. This program, conducted on the adjacent campuses of these three institutions in New York City, normally requires six or seven years of study and leads to an M.D. degree conferred by Cornell University and a Ph.D. degree conferred by either the Rockefeller University or the Weill Graduate School of Cornell University, depending upon the organizational affiliation of the student's adviser.

Research Facilities The University and its affiliate Howard Hughes Medical Institute maintain a full range of laboratories and services for the research activities of the professional staff and students. Facilities include clinical and animal research centers on campus, a library, computing services, a field research center in Dutchess County, the Aaron Diamond AIDS Research Center (ADARC), as well as new centers for human genetics, studies in physics and biology, biochemistry and structural biology, immunology and immune diseases, sensory neuroscience, and Alzheimer's disease research.

Financial Aid Each student accepted into the Ph.D. program receives a stipend ($26,750 in 2006–07) that is adequate to meet all living expenses. Students also receive an annual budget of $2500 that can be used for travel, books and journals, computer purchases, and lab supplies.

Cost of Study Full remission of all tuition and fees is provided by the University for all accepted students.

Living and Housing Costs On-campus housing is available for all students at subsidized rates. The stipend is designed to cover the cost of food, housing, and other basic living expenses. Students may elect to live off campus, but rents in the vicinity are very high.

Student Group There are 200 graduate students, of whom 154 are enrolled in the Ph.D. program and 46 in the Ph.D. phase of the combined M.D./Ph.D. program. It is the policy of the Rockefeller University to support equality of educational opportunity. No individual is denied admission to the University or otherwise discriminated against with respect to any program of the University because of creed, color, national or ethnic origin, race, sex, or disability.

Student Outcomes Graduates of the Rockefeller University have excelled in their professions. Two graduates have been awarded the Nobel Prize, and 20 graduates are members of the National Academy of Sciences. Most Ph.D. graduates move to postdoctoral positions at academic and research centers and subsequently have careers in academics, biotechnology, and the pharmaceutical industry. A few have pursued careers in medicine, law, and business. Almost all M.D./Ph.D. graduates first complete residencies in medical specialties, and most become medical scientists at major academic and medical research centers.

Location The University is situated between 62nd and 68th streets in Manhattan, overlooking the East River. Despite its central metropolitan location, the 15-acre campus has a distinctive nonurban character, featuring gardens, picnic areas, fountains, and a tennis court. In addition to administrative and residential buildings, there are seven large laboratory buildings and a forty-bed hospital that serves as a clinical research center. Immediate neighbors are the New York Hospital, the Weill Medical College of Cornell University, Memorial Hospital, and the Sloan-Kettering Institute for Cancer Research. The wide range of institutions in New York City affords unlimited opportunities in research specialties, library facilities, and cultural resources.

The University The Rockefeller University is dedicated to benefiting humankind through scientific research and its application. Founded in 1901 by John D. Rockefeller as the Rockefeller Institute for Medical Research, it rapidly became a source of major scientific innovation in treating and preventing human disease. Since 1954, the institute has extended its function by offering graduate work at the doctoral level to a select group of qualified students.

Laboratories, rather than departments, are the fundamental units of the University. The absence of departmental barriers between laboratories encourages interdisciplinary, problem-oriented approaches to research and facilitates intellectual interaction and collaboration. The collegial atmosphere fosters independence and initiative in students. In addition to the 200 doctoral students, there are 375 postdoctoral associates and fellows and a faculty of 71 full, associate, and assistant professors on campus who head laboratories.

Applying Applications for the M.D./Ph.D. program must be completed by October 16; those for the Ph.D. program must be completed by December 8. Applicants are required to submit a personal statement describing research experience and goals as well as reasons for pursuing graduate study at the Rockefeller University. Also required are official transcripts and at least three letters of recommendation. Official GRE General Test scores are required and Subject Test scores are highly recommended for admission to the Ph.D. program. MCAT scores are required for the M.D./Ph.D. program. Further information about each program and details on application procedures may be obtained from the programs' respective Web sites. This information is also available on the University Web site, from which application forms and instructions can be downloaded.

Correspondence and Information

For the Ph.D. program:
Office of Graduate Studies
The Rockefeller University
1230 York Avenue
New York, New York 10021-6399

Phone: 212-327-8086
E-mail: phd@rockefeller.edu
Web site: http://www.rockefeller.edu

For the M.D./Ph.D. program:
Tri-Institutional M.D./Ph.D. Program
Weill Medical College of Cornell University
1300 York Avenue, Room D-115
New York, New York 10021-4896

Phone: 212-746-6023
   888-U2-MD-PHD (toll-free)
E-mail: mdphd@mail.med.cornell.edu
Web site: http://www.med.cornell.edu/mdphd

The Rockefeller University

LABORATORY HEADS AND THEIR RESEARCH

C. David Allis, Ph.D. (Histone Modifications and Chromatin Biology). Enzymology and function of covalent histone modifications; histone code and epigenetic regulation.

Cori Bargmann, Ph.D. (Neuroscience). Genetic analysis of olfactory behavior and neural development.

Günter Blobel, M.D., Ph.D. (Cell Biology). Protein translocation across membranes; macromolecular traffic into and out of the nucleus.

Sean Brady, Ph.D. (Genetically Encoded Small Molecules). Structure and function of genetically encoded small molecules.

Jan L. Breslow, M.D. (Biochemical Genetics and Metabolism). Identifying the genes that control atherosclerosis susceptibility.

Brian T. Chait, D.Phil. (Mass Spectrometry and Gaseous Ion Chemistry). Protein mass spectrometry.

Nam-Hai Chua, Ph.D. (Plant Molecular Biology). Gene regulation and signal transduction in plants.

Joel Cohen, Ph.D., Dr.P.H. (Populations). Population dynamics; ecology; epidemiology.

Barry Coller, M.D. (Clinical Hematology). Biochemistry of platelet disorders; study of heritable coagulopathies.

Frederick P. Cross, Ph.D. (Molecular Genetics). Cell-cycle control in budding yeast.

George A. M. Cross, Ph.D. (Molecular Parasitology). Regulation of gene and surface glycoprotein expression in trypanosomes.

James E. Darnell Jr., M.D. (Molecular Cell Biology). Signal transduction and gene control in mammalian differentiation.

Robert B. Darnell, M.D., Ph.D. (Molecular Neuro-Oncology). Neuro-oncology and autoimmunity; molecular neurobiology.

Seth Darst, Ph.D. (Molecular Biophysics). Protein crystallography and electron microscopy of macromolecular assemblies.

Titia de Lange, Ph.D. (Cell Biology and Genetics). Chromosome function in vertebrates.

Madhav Dhodapkar, M.D. (Cancer Biology). Dendritic cell-based immunotherapy of cancer and viral diseases.

Mitchell J. Feigenbaum, Ph.D. (Mathematical Physics).

Vincent A. Fischetti, Ph.D. (Bacterial Pathogenesis). Pathogenesis of streptococcal diseases and mucosal vaccine development.

Jeffrey M. Friedman, M.D., Ph.D. (Molecular Genetics). Genes controlling food intake and body weight; mouse genetics.

Elaine Fuchs, Ph.D. (Mammalian Cell Biology and Development). Molecular mechanisms underlying the coordination of proliferation, transcription, and cell adhesion in tissue morphogenesis and in cancer.

Hinonori Funabiki, Ph.D. (Chromosome and Cell Biology). Mechanisms controlling accurate chromosome segregation during the cell division cycle.

David C. Gadsby, Ph.D. (Cardiac and Membrane Physiology). Mechanism and function of ion pumps and channels.

Ulrike Gaul, Ph.D. (Developmental Neurogenetics). Axon pathfinding and target recognition in the developing *Drosophila* visual system.

Charles D. Gilbert, M.D., Ph.D. (Neurobiology). Visual spatial integration and cortical dynamics.

Konstantin A. Goulianos, Ph.D. (Experimental High-Energy Physics).

Paul Greengard, Ph.D. (Molecular and Cellular Neuroscience). Roll of phosphoproteins in signal transduction in the developing and adult nervous system.

Howard C. Hang, Ph.D. (Chemical Biology and Microbial Pathogenesis). Chemical tools for studying posttranslational modifications in living cells.

Mary E. Hatten, Ph.D. (Developmental Neurobiology). Control of CNS neuronal specification and migration during vertebrate brain development.

Nathaniel Heintz, Ph.D. (Molecular Biology). Cell-cycle regulation; molecular neurobiology; mammalian neurogenetics.

Ali Hemmati-Brivanlou, Ph.D. (Molecular Embryology). Molecular embryology of vertebrates.

David D. Ho, M.D. (Dynamics of HIV/SIV Replication). Kinetics of CD4 lymphocyte turnover; determinants of disease progression; therapy of HIV infection.

A. James Hudspeth, M.D., Ph.D. (Sensory Neuroscience). Transduction and synaptic signaling by hair cells of the inner ear.

Tarun Kapoor, Ph.D. (Chemistry and Cell Biology). Small molecule probes of cellular processes.

Maria Karayiorgou, M.D. (Human Neurogenetics). Genetics and neurobiology of schizophrenia and obsessive-compulsive disorder.

Bruce W. Knight Jr. (Biophysics). Neurophysiology and applied mathematics.

M. Magda Konarska, Ph.D. (Molecular Biology and Biochemistry). Splicing of mRNA precursors and replication of hepatitis delta virus.

Mary Jeanne Kreek, M.D. (Neuroscience). Neurobiology and molecular genetics of addictive diseases; endogenous opioid system.

James G. Krueger, M.D., Ph.D. (Investigative Dermatology). Cutaneous pathobiology.

Stanislas Leibler, Ph.D. (Physics and Mathematical Biology). Analysis of biological networks.

Albert J. Libchaber, Ph.D. (Experimental Condensed-Matter Physics).

Roderick MacKinnon, M.D. (Molecular Neurobiology and Biophysics). Structure and function of ion channels and associated regulatory proteins.

Marcelo Magnasco, Ph.D. (Mathematical Physics). Stochastic processes in biology systems.

Bruce S. McEwen, Ph.D. (Neuroendocrinology). Hormonal regulation of neural plasticity.

John McKinney, Ph.D. (Infectious Diseases). Mechanisms of pathogenesis and protection in tuberculosis.

Peter Mombaerts, M.D., Ph.D. (Vertebrate Developmental Neurogenetics). Olfaction.

Tom W. Muir, Ph.D. (Synthetic Protein Chemistry). Combinatorial protein chemistry.

Christian Munz, Ph.D. (Viral Immunobiology). Immune control of the persistent human tumorvirus, Epstein-Barr virus.

Fernando Nottebohm, Ph.D. (Animal Behavior). Animal communication; mechanisms of learning, memory duration, and brain repair.

Michel C. Nussenzweig, M.D., Ph.D. (Molecular Immunology). Molecular basis of B-cell development.

Michael O'Donnell, Ph.D. (DNA Replication). Underlying principles of DNA replication in the human and *E. coli* systems.

Jürg Ott, Ph.D. (Statistical Genetics). Developing, implementing, and applying statistical methods of human genetic mapping.

F. Nina Papavasiliou, Ph.D. (Molecular Immunology). Molecular mechanisms of lymphocyte diversity.

Donald W. Pfaff, Ph.D. (Neurobiology and Behavior). Gene expression in brain; hormone action; brain control of behavior.

Jeffrey V. Ravetch, M.D., Ph.D. (Molecular Genetics and Immunology). Genetics of the humoral immune response; genetic variation in malaria parasite.

George N. Reeke Jr., Ph.D. (Biological Modeling). Theoretical models of brain functions; protein structure.

Charles Rice, Ph.D. (Virology). Molecular genetics of animal RNA viruses (alphaviruses and flaviviruses, in particular hepatitis C virus); replication and pathogenesis.

Robert G. Roeder, Ph.D. (Biochemistry and Molecular Biology). Transcriptional regulatory mechanisms in animal cells.

Michael P. Rout, Ph.D. (Structural Cell Biology). Nucleocytoplasmic transport; nuclear pore complex structure, function, and assembly.

Thomas P. Sakmar, M.D. (Molecular Biology and Biochemistry). Biochemistry and molecular biology of transmembrane signal transduction and visual phototransduction.

Shai Shaham, Ph.D. (Cancer Biology). Programmed cell death in the nematode *Caenorhabditis elegans.*

Eric Siggia, Ph.D. (Theoretical Condensed-Matter Physics). Statistical physics and dynamical systems to cellular biophysics and bioinformatics.

Sanford M. Simon, Ph.D. (Cellular Biophysics). Protein biogenesis, membrane protein assembly, tumorigenesis, and drug resistance.

C. Erec Stebbins, Ph.D. (Structural Microbiology). Structural studies of bacterial virulence factors and their host cell targets.

Ralph M. Steinman, M.D. (Cellular Physiology and Immunology). Antigen presenting cell function for initiating immune responses in health and disease, especially HIV-1 infection.

Hermann Steller, Ph.D. Molecular biology of apoptosis and cancer biology.

Sidney Strickland, Ph.D. (Neurobiology and Genetics). Genetics of neuronal function and dysfunction; genetics of early development.

Alexander Tarakhovsky, M.D., Ph.D. (Immunology). Mechanisms of the dynamic tuning of antigen receptor-mediated signaling in lymphocytes.

Alexander Tomasz, Ph.D. (Microbiology). Mechanisms of antibiotic resistance and virulence in bacteria.

Thomas Tuschl, Ph.D. (Chemistry). Regulation of gene expression by double-stranded RNA in humans.

Leslie Vosshall, Ph.D. (Sensory Neuroscience). Molecular genetics of olfaction in *Drosophila melanogaster.*

Milton H. Werner, Ph.D. (Molecular Biophysics). Communication mechanisms governing gene regulation and cell death.

Michael W. Young, Ph.D. (Genetics). Genes controlling behavior and development in *Drosophila;* molecular control of circadian rhythms.

ROSALIND FRANKLIN UNIVERSITY
OF MEDICINE AND SCIENCE
School of Graduate and Postdoctoral Studies

Programs of Study	The School of Graduate and Postdoctoral Studies offers programs of study and research leading to the Ph.D. and M.S. degrees in anatomy, biochemistry, experimental and clinical psychology, microbiology and immunology, neuroscience, pathology, pharmacology, and physiology and biophysics. Close relationships between the departments in course work and in research are characteristic of the graduate school.

Areas of research specialization include anatomy (neuroanatomical and neurophysiological studies of central and peripheral nervous systems and ultrastructure of the conduction system of the heart), biochemistry (antibody biosynthesis, cytochemistry, biochemical genetics, metabolic regulation, cancer, development, immunochemistry, and cell membranes and nuclear structure), microbiology (leukemia, immunotherapy, virology, amebiasis and host-parasite relations, environmental hazards, ecology, and molecular biology of microbes), neuroscience (mechanisms of neuronal death and repair, neuronal regeneration and transplantation, neurogenesis and stem cells, neuronal receptors, drug abuse, and drug-induced neuroplasticity), pathology (clinical microbiology, clinical chemistry, immunochemistry, cancer biology, cytogenetics, and hematology), pharmacology (neuropharmacology, autonomic and developmental pharmacology, growth factors and hormones, and control of gene expression), and physiology and biophysics (cardiac metabolism, cardiovascular dynamics, biophysics, neurophysiology, endocrinology, and reproductive and pathological physiology).

Combined M.D./M.S., M.D./Ph.D., and D.P.M/Ph.D. programs are offered as well. For these programs, students must be accepted by the Chicago Medical School, the School of Graduate and Postdoctoral Studies, and the Scholl College of Podiatric Medicine.

Research Facilities Well-equipped and modern research laboratories are available for graduate students. The institutional library has extensive holdings in the medical and related health sciences; the scientific libraries of neighboring institutions are also available for use by graduate students. The general facilities include fully equipped animal quarters.

Financial Aid The University offers a number of fellowships, which are awarded on a competitive basis. In 2004–05, these provided stipends that ranged up to $21,000 per year, and some included a waiver of tuition fees. In addition, a limited number of fellowships provided by private foundations and research assistantships funded by grants are available.

Cost of Study Tuition for 2004–05 was $5118.67 per quarter for three terms for a normal course load of 12 to 16 units.

Living and Housing Costs The University has moderately priced, on-campus housing facilities.

Student Group During the 2004–05 academic year, the University had a total enrollment of approximately 1,690 students in its four schools: the Medical School, School of Graduate and Postdoctoral Studies, undergraduate College of Health Professions, and Dr. William M. Scholl College of Podiatric Medicine.

Student Outcomes Most of the recent graduates who received the Ph.D. degree in the basic medical sciences are in postdoctoral positions at universities across the nation. Some examples of permanent employment after postdoctoral training include positions as university faculty members, positions in the pharmaceutical industry, or government positions. Combined M.D./Ph.D. and D.P.M/Ph.D. graduates continue their training as physician scientists. Virtually all recent graduates in the health-related sciences (clinical psychology) entered their chosen professional careers.

Location The School of Graduate and Postdoctoral Studies is part of the University complex, which is located in North Chicago, adjacent to the grounds of the North Chicago Veterans Affairs Hospital Medical Center. The Northshore suburbs offer a variety of cultural and recreational opportunities.

The School The School of Graduate and Postdoctoral Studies was established in 1968 as a further development of the Chicago Medical School, which was founded in 1912. The graduate program is directed toward the education of students who plan careers in teaching and research in the basic medical and health-related sciences.

Applying Candidates for admission must have a bachelor's degree or its equivalent from an accredited college or university. Applicants are selected on the basis of previous academic work; preparation in the proposed field of graduate study, as determined by the graduate faculty in that field; grade point average; scores on the Graduate Record Examinations (GRE); and recommendations from persons involved in the student's previous educational experience. The nonrefundable application fee is $25.

All applications for the basic science programs must be received by February 1. The deadlines for the programs in psychology and applied physiology are December 31 and June 1, respectively. Fellowships are awarded on a competitive basis. Early application is advised for applicants seeking fellowships.

Correspondence and Information Office of Admissions
School of Graduate and Postdoctoral Studies
Rosalind Franklin University of Medicine and Science
3333 Green Bay Road
North Chicago, Illinois 60064
Phone: 847-578-3209
E-mail: grad.admissions@rosalindfranklin.edu
Web site: http://www.rosalindfranklin.edu

Rosalind Franklin University of Medicine and Science

FACULTY HEADS

Michael Sarras, Ph.D., Dean, School of Graduate and Postdoctoral Studies, and Vice President for Research.
Biochemistry and Molecular Biology: Kenneth Neet, Ph.D., Chair.
Cell Biology and Anatomy: William Frost, Ph.D., Chair.
Cellular and Molecular Pharmacology: Gloria Meredith, Ph.D., Chair.
Clinical Psychology: Michael Seidenberg, Ph.D., Chair.
Microbiology and Immunology: Bala Chandran, Ph.D., Chair.
Neuroscience: Marina Wolf, Ph.D., Chair.
Pathology: Arthur Schneider, M.D., Chair.
Physiology and Biophysics: Robert Bridges, Ph.D., Chair.

RUTGERS, THE STATE UNIVERSITY OF NEW JERSEY, NEWARK

Programs in Biology

Programs of Study

The Department of Biological Sciences provides students with advanced knowledge of both plant and animal biology and microbiology. In recent years, it has increased both the span of research and support for scientific investigations. The department is involved in a number of innovative research initiatives, including collaborations with other Rutgers-Newark science departments, the Center for Molecular and Behavioral Neuroscience, the New Jersey Institute of Technology; and the University of Medicine and Dentistry of New Jersey.

Students earning the Master of Science in biology may choose one of two options. The thesis option requires 24 course credits and at least 6 research credits, as well as a thesis based on an experimental laboratory or field project. The nonthesis option requires 30 credits of course work; students must also pass a written comprehensive examination and write an extended report on a topic to be determined by the student and his or her major adviser. The degree also includes a research component that students meet by either writing a thesis on an experimental laboratory or field project or a research paper based on current literature in the field.

The Ph.D. in biology is divided into three tracks: cell/molecular/biochemical; ecology/evolution; and computational biology. Each track requires a minimum of 36 credits of course work and 36 credits of research, as well as at least two rotations in research laboratories during the first two years, which require a written report with a public oral presentation. At the completion of the two rotations, students must take and pass an oral candidacy examination and, with a research mentor, begin to develop and execute an independent thesis project.

Research Facilities

The Biology Learning Center contains self-paced laboratories to allow students flexibility in scheduling their lab work. The labs are designed to take about 4 hours, but some students prefer to work faster. Discussion classes held in the center cover new topics exploring the latest developments in science. Students who are visiting the center for the first time are required to bring a lab manual and a biology kit. The Rutgers library system ranks among the top university research libraries in the nation, with twenty-six libraries holding 6.4 million bound volumes. The John Cotton Dana Library on the Newark campus has a collection of more than 600,000 volumes, including approximately 300,000 books, 100,000 bound periodicals, and 200,000 federal and state publications; as well as more than 600,000 pieces of microform and 15,000 audiovisual items. The Dana Library currently subscribes to about 1,500 periodicals in print and has access to several hundred electronic books and more than 25,000 electronic journals.

Financial Aid

The Graduate School offers teaching and graduate assistantships to Ph.D. students with strong records of scholastic achievement. Assistantships include remission of tuition and student fees and have an annual value of $32,000 per year. New Jersey residents with backgrounds of historic poverty and early educational disadvantage may apply for Educational Opportunity Fund grants of up to $2500 per year. The Federal Work-Study Program awards up to $5000 through part-time employment. Students may borrow under a number of loan programs, including the Federal Perkins Loan, the William D. Ford Federal Direct Student Loan, and the New Jersey College Loan to Assist State Students.

Cost of Study

Tuition for full-time study (12 credits) in 2005–06 was $4834 per term for New Jersey residents and $7186 for nonresidents. Tuition for part-time study was $403 per credit for state residents and $598.80 per credit for nonresidents. Annual fees were about $1885 for commuters and $2220 for students living on campus.

Living and Housing Costs

Students living on campus pay $6134 for a nine-month single-unit lease or $5614 for a nine-month double-unit lease; twelve-month leases cost $7365 and $6763, respectively. A variety of meal plans are available, either in block plans for the entire semester or in weekly plans of ten to nineteen meals per week, at $1600 to $1700 per semester. Off-campus housing is available for $400 to $1000 per month, depending on size and location.

Student Group

Of the 543 ethnically diverse students enrolled in the program, two thirds are from New Jersey and 54 percent are women. More than two thirds are enrolled part-time, and they enter the program with undergraduate degrees in one of the biological sciences.

Student Outcomes

Recent departmental graduates have gone on to postdoctoral fellowships at such outstanding research institutions as Harvard, University of California at Los Angeles, Washington University in St. Louis, Case Western Reserve, Vanderbilt, the National Institutes of Health, and Yale. They have achieved academic and industrial positions throughout the country, from California to Washington, D.C.

Location

Newark, the third-oldest city in the United States, has something for everyone. Arts lovers can enjoy shows at the New Jersey Performing Arts Center or artwork at the Newark Museum. Sports enthusiasts can work out at Branch Brook Park, play golf at Weequahic Park, or attend a game at Riverfront Stadium. Residents can also take the PATH train to New York City or Hoboken.

The University

Founded in 1766, Rutgers University is the eighth-oldest institution of higher education in the nation and a member of the prestigious Association of American Universities. It offers more than 100 bachelor's, 100 master's, and seventy doctoral and professional degree programs. The University has more than 130 specialized research centers that bring in more than $250 million in research grants annually. In addition, its student body is highly diverse, coming from almost every state and more than 130 other countries.

Applying

Applicants are required to submit an application form, three letters of recommendation, undergraduate and graduate transcripts, and Graduate Record Examinations (GRE) scores of ranking in the 50 percentile or better on each of the General Examinations. The Subject Test in biology is recommended. Completed applications should be submitted by February 15 for admission to the Ph.D. program the following fall. M.S. applications are accepted until July 15. Late applications are accepted only if space permits.

Correspondence and Information

Department of Biological Sciences
Rutgers, The State University of New Jersey, Newark
101 Warren Street
Newark, New Jersey 07102
Phone: 973-353-5347
Fax: 973-353-5518
E-mail: biosci@newark.rutgers.edu
Web site: http://newarkbiosci.rutgers.edu

Rutgers, The State University of New Jersey, Newark

THE FACULTY AND THEIR RESEARCH

Jonathan Adams, Assistant Professor; Ph.D., Marseille, 1995. Broad-scale patterns and processes in ecology, including biogeography, the global carbon cycle, and climate-vegetation feedbacks.

Nihal Altan-Bonnet, Assistant Professor; Ph.D., Rockefeller, 1998. Biogenesis of organelles specialized for viral RNA replication; epigenetic regulation of protein activity in the cell via the Golgi apparatus; endoplasmic reticulum/spindle dynamics during mitosis

Edward M. Bonder, Associate Professor, Chair, and Graduate Program Director; Ph.D., Pennsylvania, 1983. Molecular mechanisms of nonmuscle-cell motility, primarily actin microfilament-based movements.

Ann Cali, Professor; Ph.D., Ohio State. Ultrastructure, development, and pathology of microsporidian parasites.

John Crow, Associate Professor; Ph.D., Washington State. Plant ecology; general horticulture; Alaskan wetlands.

Harvey H. Feder, Professor and Associate Provost; Ph.D., Oregon. Development and adult function of neuroendocrine mechanisms underlying the expression of reproductive behaviors.

Gerald D. Frenkel, Professor; Ph.D., Harvard, 1971. Effects of selenium compounds on tumors in vivo and tumor cells in vitro; prevention of drug resistance in cancer.

Wilma Friedman, Associate Professor; Ph.D., Rockefeller, 1986. Cellular mechanisms of cytokine and neurotrophin actions on CNS neurons and glia.

Doina Ganea, Professor; Ph.D., Illinois at Chicago, 1985. Molecular mechanisms of immunoglobulin gene rearrangement; neuroimmunology.

Lion F. Gardiner, Associate Professor; Ph.D., Rhode Island.

Jorge Golowasch, Associate Professor; Ph.D., Brandeis, 1990. Cellular mechanisms of activity-dependent regulation of ionic currents, neuronal excitability, and neural network activity.

Erik Hamerlynck, Assistant Professor; Ph.D., Kansas State, 1995. Plant ecophysiology; responses of plants to biotic and abiotic factors affecting species distribution and ecosystem dynamics.

Claus Holzapfel, Assistant Professor; Ph.D., Göttingen (Germany), 1993. Community ecology; plant population and community processes, with special interest in role of facilitation and competition in desert systems.

G. Miller Jonakait, Professor; Ph.D., Cornell. Developmental neurobiology; factors controlling neurotransmitter development and function during embryogenesis; neuroimmunology.

David Kafkewitz, Professor; Ph.D., Cornell, 1969. Physiology and metabolism of anaerobic prokaryotes; asparagine and glutamine functions in animal cells.

Andrew E. Kasper Jr., Associate Professor; Ph.D., Connecticut, 1970. Evolution of early land plants; morphology, anatomy, and systematics of Devonian plant fossils.

Haesun Kim, Assistant Professor; Ph.D., Cincinnati, 1996. Cellular and molecular biology of myelinating glial cells: mechanism of signal transduction involved in axon–Schwann cell interaction; molecular mechanisms of cell-fate determination in developing peripheral nervous system.

Edward G. Kirby, Professor and Dean, Faculty of Arts and Sciences. Ph.D., Florida. Plant development and tissue culture; forest biotechnology.

John M. Maiello, Associate Professor and Chair, Prehealth Advisory Committee; Ph.D., Rutgers. Fungal morphology and physiology.

Douglas W. Morrison, Associate Professor and Coordinator, Undergraduate Studies; Ph.D., Cornell, 1975. Ecological and neurobiological studies of social and foraging behavior of birds.

Farzan Nadim, Associate Professor; Ph.D., Boston University, 1994. Electrophysiology of neurons and networks; computational and mathematical modeling of biological systems.

Gareth J. Russell, Assistant Professor; Ph.D., Tennessee, 1996. Computational ecology.

Judith S. Weis, Professor; Ph.D., NYU. Effects of environmental factors, including pollutants, on growth, development, behavior, life history, and interactions of marine organisms, particularly fish and crustaceans; ecotoxicology; estuarine and salt marsh ecology and ecotoxicology.

ST. JOHN'S UNIVERSITY

Department of Biological Sciences

Programs of Study	The Department of Biological Sciences offers the M.S. and Ph.D. degrees in biology as well as the M.S. with a specialization in biotechnology. The primary concentration of research is in molecular and cell biology, but there is also active research in aging, developmental biology, ecology, microbiology, and neuroendocrinology. A major research theme is signal transduction mechanisms.

In both the M.S. and Ph.D. programs, students begin their studies with a core curriculum in biochemistry and cell and molecular biology. A variety of upper-level courses are offered in these areas as well as on subjects that reflect the research interests of the faculty. These courses are augmented by weekly seminars, which provide the opportunity to interact with distinguished scientists from other institutions. M.S. students take a comprehensive examination in the last semester of graduate studies. Although master's students may choose a nonthesis option, they are encouraged to work in faculty laboratories to gain research experience, for which they receive course credit.

Doctoral students take both a qualifying exam, following the completion of the core curriculum, and a comprehensive exam, covering their area of expertise, upon satisfactory completion of a research proposal. In addition, doctoral students rotate through faculty labs in order to acquaint themselves with ongoing research projects and to gain experience in a variety of techniques. Recognizing the importance of participation in national meetings for the intellectual and professional development of the student, the Department provides some support for travel.

Research Facilities

The Department of Biological Sciences is located on two floors of a science building which is shared with the chemistry and physics departments, as well as the College of Pharmacy and Allied Health Professions. Many of the research labs were recently renovated. A central facility contains centrifuges, ultracentrifuges, and scintillation counters. All equipment essential for modern molecular and cell biology is available within the Department, including apparatus for oligonucleotide and peptide synthesis, fluorescence and confocal microscopy and image analysis, FPLC, and PCR. Both scanning and transmission electron microscope facilities are located in the building. A fully accredited animal facility is housed in the science building. Personal computers and software for word processing, statistics, and graphics, as well as programs for current literature review and analysis of gene sequences, are available in the Department for student use. Educational and research opportunities are augmented by cooperation with other institutions in the New York City area.

Financial Aid

Teaching assistantships, providing a stipend and tuition remission, are available to qualified doctoral and master's degree students. Assistantships require 9–12 contact hours per week. Doctoral fellowships, some requiring no service to the University, are awarded on a competitive basis based on GRE scores, grades, and research productivity. Women doctoral students are eligible for the prestigious Clare Boothe Luce fellowships, awarded on the basis of academic merit. Further, highly qualified students are eligible for GAANN fellowships, supported by the U.S. Department of Education, and for IMSD fellowships, funded by the National Institutes of Health. Numerous loan programs and specifically designated award programs are also available. Support from research grants is sometimes available.

Cost of Study

For 2005–06, tuition was $730 per credit hour. There was a general fee of $75 per semester. A few courses may require a $60 lab fee, and there are various minor fees for placement examinations.

Living and Housing Costs

Housing is available nearby. The cost of living is similar to that in the suburbs of most metropolitan areas. Dormitories are available. Excellent public transportation is reasonably priced.

Student Group

St. John's has a multicultural student body of approximately 18,000, including more than 1,200 students in the Graduate Division of Arts and Sciences. The master's and doctoral programs in biological sciences have about 40 students each. Many master's degree students maintain employment while continuing their studies. Most doctoral students are supported by teaching assistantships or fellowships. Graduates go on to pursue careers in academia, the biotechnology and pharmaceutical industries, and government.

Location

The Queens campus of St. John's University, comprising 99 acres, is located in a pleasant residential area offering both a suburban environment and proximity to Manhattan and Long Island. There are several public parks nearby that have excellent athletic and picnic facilities as well as walking trails. The University is easily accessible by both automobile and public transportation.

The University

St. John's University was founded by the Vincentian community in 1870. The Graduate Division of Arts and Sciences, established in 1913, oversees the graduate programs in biology. The Master of Science and Doctor of Philosophy programs are fully accredited by the Middle States Association of Colleges and Schools and by New York State.

Applying

Applications for all programs are reviewed on a rolling admissions basis but should be submitted by March 1 for the fall semester. Applicants for the master's program must have undergraduate training in biology with a minimum 3.0 cumulative average, including a 3.0 index in biology. For the Ph.D. program, applicants should have a cumulative index of 3.5, with at least a 3.0 in biology, and satisfactory GRE scores on both the General and Subject Tests.

Correspondence and Information

For additional information:
Jay A. Zimmerman, Ph.D., Chairman
Department of Biological Sciences
St. John's University
8000 Utopia Parkway
Jamaica, New York 11439
Phone: 718-990-6288
E-mail: zimmermj@stjohns.edu

For applications:
Graduate Admissions
Newman Hall
St. John's University
8000 Utopia Parkway
Jamaica, New York 11439
Phone: 718-990-2000

St. John's University

THE FACULTY AND THEIR RESEARCH

Cell Biology and Biochemistry
Diana Bartelt, Associate Professor; Ph.D., CUNY, Hunter, 1982. Calmodulin-regulated protein kinases; cellular signal transduction.
Dipak Haldar, Professor; Ph.D., London, 1966. Mitochondria: structure, synthetic processes, and biogenesis.
Yue J. Lin, Associate Professor; Ph.D., Ohio State, 1976. Cytology and cytogenetics; spontaneous and induced abnormalities in chromosomes; sister chromatid exchange.
Laura Schramm, Assistant Professor; Ph.D., Stony Brook, SUNY, 2001. Regulation of mechanisms of eukaryotic gene expression.
Louis D. Trombetta, Professor (joint appointment with the College of Pharmacy and Allied Health Professions); Ph.D., Fordham, 1974. Electron microscopy; neurotoxicology; experimental pathology and biochemistry; effects of drugs and toxic substances on the central nervous system.
Ales Vancura, Associate Professor; Ph.D., Prague Institute of Chemical Technology, 1989. Cellular signal transduction by lipid and protein phosphorylation; phosphatidylinositol kinases; mechanisms of intracellular protein targeting and localization.

Ecology and Physiological Ecology
Frank Cantelmo, Associate Professor; Ph.D., CUNY, City College, 1978. Physiological ecology; physiology of the hard clam *Mercenaria mercenaria* in relation to depuration.
Richard Stalter, Professor; Ph.D., South Carolina, 1968. Physiological ecology; coastal plant community ecology; plant ecology of Indian shell rings.

Molecular Biology and Microbiology
Timothy H. Carter, Professor; Ph.D., Princeton, 1972. Regulation of mammalian gene expression; biochemical action of tumor promoters and protein kinases.
Anne M. Dranginis, Associate Professor; Ph.D., Michigan, 1982. Molecular control of developmental programs; determination of yeast cell types.
Irvin N. Hirshfield, Associate Professor; Ph.D., Pittsburgh, 1966. Molecular response of bacteria to environmental stresses, with emphasis on acidity and temperature; global regulatory mechanisms in bacteria.
Ivana Vancurova, Associate Professor; Ph.D., Czechoslovak Academy of Sciences, 1989. Molecular mechanisms of inflammation.

Physiology and Developmental Biology
Chris Bazinet, Associate Professor; Ph.D., MIT, 1986. Molecular genetics and assembly of protein structures; role of clathrin in development of multicellular organisms.
Jaya Haldar, Professor; Ph.D., London, 1966. Neurophysiology; regulation and neural effects of posterior pituitary hormones.
Richard A. Lockshin, Professor; Ph.D., Harvard, 1963. Developmental cell physiology, gerontology; mechanisms of cell death and homeostatic down-regulation; mechanisms of proteolysis; use of computers in video monitoring of behavior.
Jay Zimmerman, Associate Professor and Chair; Ph.D., Rutgers, 1975. Physiology of aging, myocardial responses to anoxia and ischemia during senescence; age-related susceptibility to chemical carcinogens and activation of oncogenes.

SAINT LOUIS UNIVERSITY

Department of Biology

Programs of Study

The Department of Biology offers programs leading to the M.S., M.S. (Research), and Ph.D. degrees in two areas of concentration: ecology, evolution, and systematics and cellular and molecular regulation. Candidates for the M.S. degree must complete a minimum of 30 semester hours of approved course work; no formal thesis is required, but research experience in a faculty research laboratory is recommended. Candidates for the M.S. (R) degree must complete a minimum of 24 semester hours of course work approved for their area of concentration in addition to 6 hours of thesis research. The course program for the Ph.D. degree requires 60 semester hours, including 12 hours of dissertation research. Each area of concentration has specific prerequisites and requirements for course work. Previous graduate work done at other institutions is evaluated individually, but a minimum of 24 credit hours plus the dissertation must be completed at Saint Louis University (SLU). Requirements for the M.S. and M.S. (R) degrees should be completed within two years of full-time study. The Ph.D. program should be completed in four to five years. Students completing graduate degrees in biology are qualified for advanced research programs and professional opportunities in research and higher education.

Research Facilities

Teaching and research facilities for the Department of Biology are housed in Macelwane Hall on the main campus and at the Reis Biological Station in the Missouri Ozarks. Many of the labs in Macelwane have recently received substantial renovation, and state-of-the-art equipment is available both in the Department and at the Health Sciences Center. Research within the Department is supplemented by collaborations with researchers at the Missouri Botanical Garden and the St. Louis Zoo, offering some students the opportunity to conduct specialized research at these institutions.

The Department of Biology operates the Reis Biological Station near Steelville, Missouri, in Crawford County, on 225 acres of upland, oak-hickory forest within the 1.5-million-acre Mark Twain Forest. The Reis Biological Station has a variety of aquatic and terrestrial communities located on-site or nearby. These include caves, streams, rivers, reservoirs, glades, fens, and forests. There are well-equipped research and teaching laboratories, kitchens, dormitories, and housing for visiting scientists.

Financial Aid

Many students are supported by graduate teaching or research assistantships. Some students receive fellowship support. Several teaching and research assistantships are available annually on a competitive basis. Each assistantship provides a stipend for eleven months ($15,500 for 2005–06), tuition remission for 9 hours each semester and 3 hours during the summer, and medical benefits. Graduate fellowships also are available from governmental and private sources. The Graduate School sponsors Presidential Fellowships and several fellowships for minority students, all on a competitive basis. Students not supported by assistantships or fellowships may be eligible for other financial aid. Student support in 2005–06 included fourteen teaching assistantships, twenty-one research assistantships, and three fellowships. Evidence of research accomplishment and potential is highly valued in review of applications for admission and assistantships.

Cost of Study

Tuition remission is provided for all students receiving assistantships and fellowships. The Graduate School tuition rate for 2005–06 was $760 per credit hour.

Living and Housing Costs

St. Louis is one of the most affordable metropolitan areas in the country. Various residence halls and apartments are available on campus. Most graduate students attending Saint Louis University live off campus. The majority of students live in nearby apartments, with rent for a one-bedroom apartment typically starting at $400 per month. Shared housing is often a more economical option that some students prefer.

Student Group

The University enrolls more than 11,000 full- and part-time students, including more than 500 internationals from nearly eighty countries. During 2003–04, the Department of Biology enrolled 16 full-time master's and 25 full-time Ph.D. students in addition to several part-time students.

Location

Located near the heart of the metropolitan area, midtown St. Louis is also home to the St. Louis Symphony's Powell Hall. The Metrolink transit line leads directly from SLU to Lambert Airport and to downtown attractions, including the homes of the baseball Cardinals, the football Rams, and the hockey Blues, as well as the landmark Gateway Arch and Laclede's Landing. Nearby attractions include Forest Park, with the St. Louis Zoo, and the St. Louis Art Museum.

The University and The Department

Saint Louis University, a private university under Catholic and Jesuit auspices, traces its history to the foundation of the Saint Louis Academy in 1818. The Society of Jesus took over the direction of the school in 1827. The college received its charter as Saint Louis University in 1832, becoming the first university established west of the Mississippi. The University settled at its present site on Grand Boulevard in 1888. The Carnegie Foundation for the Advancement of Teaching classifies Saint Louis University as a Doctoral/Research University–Extensive institution.

Applying

Applicants for all programs should have a bachelor's degree that includes at least 18 credit hours of upper-division biology course work, 8 hours of upper-division chemistry (typically two semesters of organic chemistry), two semesters of physics, and a semester of calculus. Certain deficiencies can be made up during residency in the graduate program. Applications must include transcripts, three letters of reference, GRE General Test scores, and a personal statement. Requests for application packets should be made directly to the Dean of the Graduate School, 3634 Lindell Boulevard, St. Louis, Missouri 63108 (telephone: 314-977-2240), or can be obtained from the Graduate School's Web site (http://www.slu.edu/graduate/apply_now.html). Requests for information may also be made via the Graduate School's Web site at http://www.slu.edu/graduate/request_form.html.

Correspondence and Information

Dr. Joe Leverich, Graduate Program Director
Department of Biology
Saint Louis University
3507 Laclede Avenue
St. Louis, Missouri 63103
Phone: 314-977-3900
Fax: 314-977-3658
E-mail: leverich@slu.edu
Web site: http://bio.slu.edu

Saint Louis University

THE FACULTY AND THEIR RESEARCH

FULL-TIME FACULTY

Robert D. Aldridge, Professor; Ph.D., New Mexico. Reproductive biology of reptiles, with a specialty in snakes: especially, morphology and histology of the reproductive system, pheromone communication, and the environmental and endocrine control of reproduction, functions, and sources of gonadotropins and sex steroids.

Nevin Aspinwall, Professor and Director, Reis Biological Station; Ph.D., British Columbia. Evolutionary biology, particularly at the population level; hybridization phenomena in freshwater fishes; strength and nature of isolating mechanisms that maintain the genetic integrity of closely related syntopic species.

Jan Barber, Assistant Professor; Ph.D., Texas at Austin. Evolutionary relationships among flowering plants and how those relationships illustrate biological processes; research utilizing molecular phylogenetic methods to reconstruct the evolutionary history of groups of plants.

Peter Bernhardt, Professor; Ph.D., Melbourne (Australia). Floral evolution within America's tallgrass prairies; the woodlands and rainforests of Australia; the shrublands of southern Africa; pollinator-flower interactions and pollen-stigma receptivity of the orchid, lily, and iris families; relictual dicots and Proteaceae (macadamia nut family).

Barrie P. Bode, Associate Professor; Ph.D., Florida. Cancer and molecular physiology; glutamine transporter expression in liver cancer; amino acid–dependent regulation of glutamine transporter SN1 in liver and muscle.

Gerardo R. Camilo, Associate Professor; Ph.D., Texas Tech. Ecology of population and community dynamics as these respond to variability in the environment, currently involving a long-term study of different scales and intensities of logging in relation to the spatial and temporal structure of forests in the Ozarks.

Steven J. Dina, Associate Professor and Associate Dean, College of Arts and Sciences; Ph.D., Utah.

Brian Downes, Assistant Professor; Ph.D., Purdue. Peptide tags, ubiquitin pathway, ubiquitin protein ligase 3 (UPL3) in *Arabidopsis thaliana;* analysis of ubiquitin-like peptide tags.

Jon Fisher, Assistant Professor; Ph.D., Washington (St. Louis). Applied physiology for skeletal muscle glucose metabolism.

Joseph C. Fortier, S.J., Assistant Professor; Ph.D., Wyoming. Evolution, systematics, biological diversity, biogeography, and host relationships of parasitoid wasps, especially those in the family Braconidae.

Jack Kennell, Associate Professor; Ph.D., Florida. Molecular fossils or contemporary genetic elements that are ancient in origin, specifically, plasmids that replicate by reverse transcription in the mitochondria of filamentous fungi; nuclear-mitochondrial interactions: how mitochondria communicate with the nucleus in eukaryotic cells, using fungi as a model system.

Joe Leverich, Professor and Coordinator, Graduate Program; Ph.D., Texas at Austin. Evolutionary biology of plant populations; plant reproductive ecology; fitness measures; organization of genetic variation in plant populations.

Richard L. Mayden, Professor and Chair; Ph.D., Kansas. Systematics, ecology, and biogeography of freshwater fish species, specifically, species and ecosystems in North America, Russia, Turkmenistan, Kazakhstan, Brazil, and Mexico.

Shawn E. Nordell, Assistant Professor; Ph.D., New Mexico. How animals make decisions from both a proximate and evolutionary level, examining several types of behavior (mate choice, schooling in fish, and foraging); what information individuals use to make decisions and how they use information once they have acquired it (i.e., decision rules).

John G. Severson, Professor and Coordinator, Undergraduate Programs; Ph.D., British Columbia. Mechanism of plant growth stimulation by petroleum-derived naphthenic acids.

Susan Spencer, Assistant Professor; Ph.D., Washington (St. Louis). Identification of new components of the EGF-receptor signaling pathway; R8 photoreceptor neurons in *Drosophila.*

William S. Stark, Professor; Ph.D., Wisconsin–Madison. Visual receptor function; visual sensitivity to ultraviolet light, rhodopsin, and vitamin A's diverse effects in visual receptors; use of mutants and knockouts to develop a mouse model for ceroid lipofuscinosis.

Thomas Valone, Associate Professor; Ph.D., Arizona. Community ecology; conservation biology; effects of climate change on ecosystems; effects of disturbances on diversity; foraging behavior, involving work on birds, mammals, ants, and plants.

Yuqi Wang, Assistant Professor; Ph.D., Creighton. Mechanism and regulation of signaling initiated by G-protein–coupled receptors (GPCR); GPCR signaling pathways in *Saccharomyces cervisiae.*

Robert M. Wood, Associate Professor; Ph.D., Alabama. Reconstructing phylogenetic/evolutionary relationships among North American freshwater fishes, primarily focusing on two groups of stream-dwelling fishes, darters and minnows.

EMERITUS FACULTY

Dorothy J. Feir, Ph.D., Wisconsin.
Judith Z. Medoff, Ph.D., Brandeis.
Fr. Raymond H. Reis, S.J., Ph.D., Saint Louis.
Raymond R. Walsh, Ph.D., Cornell.

ASSOCIATED FACULTY

Preprofessional Health Studies
Donald O. Schreiweis, Associate Professor and Director, Preprofessional Health Studies; Ph.D., Washington State.

Missouri Botanical Garden
Thomas B. Croat, Adjunct Associate Professor; Ph.D., Kansas.
Peter Goldblatt, Adjunct Associate Professor; Ph.D., Cape Town (South Africa).
Peter H. Raven, Adjunct Professor; Ph.D., UCLA.
P. Mick Richardson, Adjunct Professor; Ph.D., London.

St. Louis Zoological Park
Cheryl S. Asa, Adjunct Professor; Ph.D., Wisconsin–Madison.

SOUTHERN METHODIST UNIVERSITY

Department of Biological Sciences

Programs of Study

The Department of Biological Sciences offers graduate programs leading to the M.S., M.A., and Ph.D. degrees, but concentration is on the doctoral program. Research programs are directed toward developing an understanding of biological systems at the molecular, cellular, and organismal levels. Major areas of emphasis are molecular biology, cell biology, genetics, biochemistry, and parasitology. Furthermore, it is the aim of the Department to prepare students for careers in teaching and research.

The Ph.D. program includes academic course work and research training that may be completed in approximately five years. After passing a preliminary examination that includes the defense of a research proposal, Ph.D. candidates devote the major part of their academic effort to the research program and preparation of a dissertation. The M.S. degree program includes 6 semester hours of thesis work and may be completed in about two years. The M.A. degree program includes a project in lieu of a thesis and requires eighteen months to two years of study.

Research Facilities

In 2002, the Department moved into the Dedman Life Sciences Building, a state-of-the-art research facility. Currently, the Department has laboratories equipped for research in each of the areas of emphasis. New equipment includes a confocal microscope, a flow cytometer, electron spin resonance instrumentation, and a computer workstation for protein modeling. Facilities also include RT-PCR, scintillation systems, ultracentrifuges, HPLC systems, cold and warm rooms, tissue culture hoods, spectrophotometers, fluorometers, and general equipment to support laboratory investigations.

Financial Aid

Teaching assistantships are awarded on a competitive basis to eligible candidates. In 2006–07, these assistantships provide stipends of $14,850 for the academic year. Summer research stipends increase total stipends to $19,800.

Cost of Study

In 2006–07, tuition and fees are $1196 per semester hour. Students admitted to the Ph.D. program usually receive departmental support to cover these costs.

Living and Housing Costs

On-campus housing for single students is available for $5085 for the 2006–07 academic year. University-owned apartments have a variety of different accommodations and are available from $436 to $900 per month. There are three married/family housing facilities on campus with costs that range from $4110 to $4890 per year. A variety of services are offered in the Hughes-Trigg Student Center, such as camping rental, a post office, and a food court.

Student Group

Graduate students in the Department of Biological Sciences have a wide range of educational and geographic backgrounds; students come from countries such as France, Ukraine, India, and China, as well as the United States. In 2006–07, 17 graduate students are working toward advanced degrees in the Department. Approximately 2,075 students were enrolled in graduate programs in the sciences, humanities, business, arts, and engineering. Total University enrollment is approximately 11,000 students.

Location

The SMU campus is located within the city of University Park, approximately 5 miles from downtown Dallas. Residents are offered a broad spectrum of cultural and recreational opportunities, such as the Dallas Symphony, the Dallas Ballet, and the Dallas Theater Center, as well as numerous other musical, dance, and theater groups. Within the immediate vicinity of campus, on Greenville Avenue, students experience the flavor of Dallas through many restaurants, cafes, and other entertainment spots located there. Furthermore, every year Dallas is host to the Texas State Fair, which presents a wide variety of exhibits and amusements to the public.

The University

Founded in 1911, Southern Methodist University is a private coeducational institution located on a 164-acre campus that is noted for its beauty. Graduate studies in Dedman College include master's programs in seventeen fields and doctoral programs in twelve. Other divisions of SMU are the School of Engineering and Applied Sciences, Cox School of Business, Meadows School of the Arts, the Dedman School of Law, and Perkins School of Theology.

Applying

Applications, transcripts, and three letters of recommendation should be submitted by February 1 for candidates to be fully considered for financial aid. Those who apply after February 1 are notified about admission and considered for any uncommitted financial aid on a rolling basis, but no later than August 1 for the fall semester. Scores on the General Test of the Graduate Record Examinations are required.

For additional information about the Department of Biological Sciences, prospective students may write to any member of the Department's faculty or to the graduate committee at the address below.

Correspondence and Information

Graduate Committee
Department of Biological Sciences
Southern Methodist University
Dallas, Texas 75275-0376
Phone: 214-SMU-2730 (768-2730)

Dr. Steven B. Vik
Department of Biological Sciences
113 Dedman Life Sciences Building
Southern Methodist University
Dallas, Texas 75275-0376
Phone: 214-SMU-2730 (768-2730)
E-mail: svik@mail.smu.edu
Web site: http://www.smu.edu/biology

Southern Methodist University

THE FACULTY AND THEIR RESEARCH

Christine E. Buchanan, Professor; Ph.D., Chicago, 1973. Molecular biology, microbiology, biochemistry; characterization of penicillin-binding protein activities during growth and sporulation of *Bacillus subtilis.*

Robert Harrod, Assistant Professor; Ph.D., University of Maryland, Baltimore County, 1996. Transcriptional regulation; HIV-1 and HTLV-1 trans-activator interactions.

Richard S. Jones, Professor; Ph.D., Wesleyan, 1984. Developmental genetics and mechanisms of gene regulation in *Drosophila.*

William C. Orr, Professor; Ph.D., Wayne State, 1982. Molecular genetics of aging and oxidative stress in eukaryotes, especially *Drosophila* spp.

Larry S. Ruben, Professor and Chair; Ph.D., Minnesota, 1979. Molecular parasitology; biochemistry and molecular biology of calcium-dependent regulatory pathways in African trypanosomes.

John E. Ubelaker, Professor; Ph.D., Colorado State, 1967. Helminth morphology, ecology, and fine structure; cellular morphology of cestodes; reproduction, especially gametogenesis, in helminths; host-parasite relationships of *Angiostrongylus costaricensis* and *Trichinella* spp.; the inflammatory response directed against helminths.

Steven B. Vik, Professor; Ph.D., Oregon, 1980. Protein biochemistry; assembly, structure, and function of membrane-bound enzymes.

Pia D. Vogel, Assistant Professor; Ph.D., Kaiserslautern (Germany), 1987. Structure-function analysis of enzymes; protein-protein interactions; electron spin resonance.

James Waddle, Assistant Professor; Ph.D., Washington (St. Louis), 1993. Genetic analysis of insulin signaling and longevity in *C. elegans;* biogenesis and function of the intestinal lumen.

STATE UNIVERSITY OF NEW YORK AT BUFFALO

Department of Biological Sciences

Programs of Study	The Department of Biological Sciences offers programs of study leading to the Ph.D., M.S., and M.A. degrees, with specializations in molecular biology, genetics, biological chemistry, cell biology, developmental biology, plant science, physiology (cellular, comparative, and endocrine), bioinformatics, neuroscience (molecular, cellular, systems, and behavioral), skeletal muscle, and sensory transduction. The Ph.D. and M.S. programs are designed to give the student a comprehensive understanding of living organisms and the background and experience required for carrying out independent research. While course work is required, the emphasis of these programs is on research training. A faculty adviser, assigned to each entering student, is responsible for advising the student on the appropriate course work for the first year of study. By the end of the second semester, Ph.D. students select a research adviser, and by the fourth semester of study they must pass an examination that includes the preparation of an original research proposal and an oral exam covering the proposal and areas of biology pertinent to their studies. A weekly seminar program with distinguished national and international speakers provides students with the latest information on current research problems.
Research Facilities	The department is exceptionally well equipped for research and teaching. The research laboratories are housed in a modern building, with such adjunct facilities as instrument shops, animal quarters, and a greenhouse. Major instrumentation and equipment are available, including such items as a 750-MHz NMR spectrometer, fluorescence and electron microscopes, and a phosphorimagers. The department has extensive computer resources. Programs for protein and nucleic acid sequence analysis and the molecular modeling and molecular dynamics simulations of macromolecules are accessible both through high-quality graphics terminals and desktop microcomputers. Mainframe computer access is provided to all graduate students.
Financial Aid	Most graduate students qualify for financial assistance in the form of fellowships, traineeships, and research or teaching assistantships. For 2007–08, Ph.D. appointments provide stipends that range from $20,000 to $28,000 per calendar year, plus full tuition costs. In addition, health-care insurance covering medical, dental, and vision services is provided for all stipend-supported students and their dependents.
Cost of Study	Graduate tuition in 2006–07 is $288 per credit, or $3450 per semester for 12 or more credits, for in-state students and $455 per credit, or $5640 per semester, for out-of-state students. Fees are additional. Most students in the department are covered by a full-tuition scholarship.
Living and Housing Costs	There are moderately priced ($350 to $650 per month) apartments on and in the vicinity of the campus. Assistance in finding campus housing can be obtained from the University Housing Office at 716-829-2224. Off-campus housing can be found at http://www.subboard.com/sbi-och.
Student Group	Total enrollment at the University exceeds 27,000, with total enrollment of part- and full-time graduate and professional students in excess of 9,000. The Department of Biological Sciences enrolls about 60 full-time graduate students, who come from around the world.
Student Outcomes	Graduates of the doctoral program are routinely successful in securing excellent postdoctoral and professional positions. Recent graduates have accepted positions at many outstanding institutions, including NIH, Princeton University, Yale University, NIEHS, Harvard University, and Scripps Institute.
Location	The State University of New York at Buffalo is located in the second-largest city in New York State. The city is conveniently located near outstanding boating, swimming, camping, and skiing areas. Only a short drive away are the cities of Toronto and Niagara Falls and the world-famous Stratford (Canada) Shakespeare Festival. The cultural attractions of Buffalo include the Albright-Knox Art Gallery, the Buffalo Philharmonic, science and historical museums, several permanent professional theater companies, and numerous amateur theater groups. Buffalo also has several professional sports teams, including football, hockey, and minor-league Triple-A baseball.
The University and The Department	Founded in 1846 as the University of Buffalo, the State University of New York at Buffalo is today the largest single unit and the most comprehensive graduate center in the SUNY system. Graduate enrollment, faculty recruitment, research, and public service programs receive significant University support. The pace of these achievements has been greatly accelerated by the University's multimillion-dollar building program, which provides the necessary physical backdrop for enriched and expanded services in education. There are currently 26 departmental faculty members, 6 of whom have been recruited in the last two years. Faculty members regularly collaborate with colleagues in many departments, as well as those of the Roswell Park Cancer Institute and the State University of New York at Buffalo Health Sciences faculty. Funded research in the department is currently more than $4 million annually. The research interests of the faculty members are listed in the Faculty and Their Research section.
Applying	To be considered for special University fellowships that supplement the regular assistantships, applications must be received prior to February 1. All other applications for admission to the program and for financial assistance should be filed by February 15. Applications should be submitted online at http://www.gradmit.buffalo.edu. Appointments for the upcoming academic year are made beginning March 15. All students are expected to be proficient in the use of spoken and written English. Results of the TOEFL are required for all international applicants.
Correspondence and Information	Director of Graduate Studies Department of Biological Sciences, 109 Cooke Hall University at Buffalo, The State University of New York Buffalo, New York 14260-1300 Phone: 716-645-2363 Fax: 716-645-2975 E-mail: ub-biosci@buffalo.edu Web site: http://www.biology.buffalo.edu

State University of New York at Buffalo

THE FACULTY AND THEIR RESEARCH

Biological Chemistry, Molecular Biology, and Genetics

Richard R. Almon, Professor; Ph.D., Illinois at Urbana-Champaign. Molecular mechanisms of hormone action; quantitative models that describe the integrated influence of neural, hormonal, mechanical, and nutritional factors in control of muscle protein mass.

James O. Berry, Associate Professor; Ph.D., Iowa State. Molecular biology of gene expression in plants.

Jeremy A. Bruenn, Professor; Ph.D., Berkeley. Fungal viruses; replication and transcription of yeast and *Ustilago maydis* viral dsRNAs; synthesis of killer toxins.

Debra DuBois, Research Assistant Professor; Ph.D., SUNY at Buffalo. Molecular mechanisms of muscle atrophy/hypertrophy.

Paul Gollnick, Professor; Ph.D., Iowa State. Protein-RNA interactions involved in regulating the *trp* operon in *Bacillus*.

Kiong Ho, Assistant Professor; Ph.D., Cornell. Eukaryotic gene expression; RNA processing and repair.

Margaret Hollingsworth, Associate Professor; Ph.D., Colorado. RNA-protein complexes involved in chloroplast gene expression.

Gerald B. Koudelka, Professor and Chair; Ph.D., SUNY at Buffalo. DNA-protein interactions at the molecular level; regulation of transcription.

Randall Shortridge, Associate Professor; Ph.D., North Texas. The study of visual transduction in *Drosophila* and phosphatidylinositol-specific phospholipase C-mediated signaling mechanisms, using a combination of genetic, biochemical, molecular, and physiological approaches.

Grayson Snyder, Associate Professor; Ph.D., Harvard. Physical chemistry of disulfide exchange reactions; protein folding and protein engineering; structure of the AIDS virus envelope protein.

Michael Yu, Assistant Professor; Ph.D., UCLA. Arginine methylation; chromatin structure and gene regulation in *Saccharomyces cerevisiae*.

Cell and Developmental Biology

Ronald Berezney, Professor; Ph.D., Purdue. Cell biology; proteins of the nucleus; role of the nuclear matrix in DNA replication; major proteins that constitute the nuclear matrix structure.

Paul J. Cullen, Assistant Professor; Ph.D., Washington (St. Louis). Signal transduction and cell polarity.

Stephen J. Free, Professor; Ph.D., Stanford. Regulation of gene expression in lower eukaryotes; structure and regulation of genes for extracellular hydrolases in *Neurospora*.

Todd Hennessey, Professor; Ph.D., Wisconsin–Madison. Excitable membranes of *Paramecium;* cellular mechanisms involved in controlling membrane excitability and ciliary beating, using the unicellular eukaryote *Paramecium* as a simple model system.

James R. LaFountain, Professor; Ph.D., SUNY at Albany. Cell biology; physiology of mitosis, meiosis, and cell motility; chromosome movements that occur during mitosis and meiosis of behavior and locomotion.

Cellular, Comparative, and Endocrine Physiology

Mary A. Bisson, Professor; Ph.D., Duke. Plant physiology; water balance; membrane transport.

Denise M. Ferkey, Assistant Professor; Ph.D., Washington (Seattle). Chemosensation and signal transduction in *C. elegans*.

Charles R. Fourtner, Professor; Ph.D., Michigan State. Neurobiology; invertebrate neurophysiology; role of the nervous system in initiating and controlling behavior in invertebrates, particularly the arthropods.

Christopher Loretz, Associate Professor; Ph.D., UCLA. Physiological mechanisms of osmoregulation; characterization of membrane and transepithelial electrolyte transport processes in vertebrates.

Kathryn F. Medler, Assistant Professor; Ph.D., Louisiana State. Physiological regulation of cell signaling mechanisms in chemical sensory systems.

Scott Medler, Research Assistant Professor; Ph.D., Louisiana State. Cellular and molecular physiology of muscles.

Brian A. Pierchala, Assistant Professor; Ph.D., Johns Hopkins. Role of trophic factors in development of neurons and the nervous system.

Matthew Xu-Friedman, Assistant Professor; Ph.D., Cornell. Synaptic physiology and neuronal computation in the auditory brainstem.

Ecology and Evolution

Clyde F. Herreid, Distinguished Teaching Professor; Ph.D., Penn State. Physiological ecology; comparative physiology; animal behavior; energetics and hormonal regulation.

Derek J. Taylor, Associate Professor; Ph.D., Guelph. Evolutionary biology; molecular systematics; ecology of freshwater invertebrates.

STATE UNIVERSITY OF NEW YORK AT BUFFALO

School of Medicine and Biomedical Sciences
Interdisciplinary Graduate Program in Biomedical Sciences (IGPBS)

Program of Study

The Interdisciplinary Graduate Program in Biomedical Sciences (IGPBS) allows students to choose from ten doctoral degree programs involving the Departments of Biochemistry, Microbiology and Immunology, Oral Biology, Pathology and Anatomy, Pharmacology and Toxicology, Physiology and Biophysics, and Structural Biology and the Neuroscience Program. This program provides students with a unique ability to obtain a broad-based scientific foundation as they fully explore an exceptionally diverse choice of research areas. Students have the opportunity to study and collaborate with faculty members involved in cutting-edge research encompassing virtually all aspects of biomedical sciences. While furnishing the requisite scientific background, the program also has the flexibility to meet the individual needs and interests of the students, whether they plan to enter careers in academia, industry, or government. Students should visit http://www.smbs.buffalo.edu for more information. During the first year of study, students are enrolled in a core curriculum designed to provide fundamental scientific concepts and knowledge, which includes biochemistry, molecular biology, and cell biology. Elective courses allow each student the opportunity to pursue their own individual interests. A series of laboratory rotations of the student's choosing provides both research experience and the opportunity to explore different research areas and interact with potential mentors prior to making any commitment. At the end of the first year, the student chooses a Ph.D. mentor, matriculates into the mentor's department, and ultimately completes the degree requirements within that department. Students may expect to complete research and course work leading to the Ph.D. in four to five years.

Research Facilities

The School of Medicine and Biomedical Sciences is located on the South Campus and has developed a number of multidisciplinary research focus centers, which combine the expertise of basic science and clinician-scientist faculty members. The Buffalo Center of Excellence in Bioinformatics merges high-end technology with expertise in genomics and proteomics. Core facilities associated with the University include those for protein and DNA sequencing, oligonucleotide synthesis, laser confocal microscopy, electron microscopy, transgenic animals, DNA microarray and gene chips, NMR, animal MRI, MALDI-TOF, high-throughput and molecular targeting, and X-ray crystallography. The Hauptman-Woodward Medical Research Institute, directed by 1985 Nobel laureate Dr. Herbert A. Hauptman, houses the Department of Structural Biology. The Health Sciences Library has extensive journal and book holdings. Students have desktop computer access to bibliographic databases, full-text journals, clinical manuals and textbooks, drug information resources, and Web communication tools.

Financial Aid

Stipends are provided through the School of Medicine and Biomedical Sciences. In fall 2005, stipends ranged from $21,000 to $24,000 and include health-care benefits.

Cost of Study

Full tuition scholarships are available for all students admitted to this program. Tuition and fees for 2005–06 were approximately $4045 per semester for in-state and $5845 per semester for out-of-state students. Fees are not included in tuition scholarship.

Living and Housing Costs

Moderately priced off-campus housing is available in the city and suburbs near the South Campus. A separate graduate student residence hall is available, as are new graduate student apartments at the North Campus. The cost of living in Buffalo is much lower than that in many comparable urban areas, and the quality of life in Buffalo for both single and married students is described as a place "Where a small-town smile meets a big city heart."

Student Group

Approximately 27,000 students, including 8,000 graduate students, are enrolled at the University. The School of Medicine and Biomedical Sciences enrolls approximately 175 students in its graduate programs.

Student Outcomes

Graduates of the biomedical sciences program bring a broad-based academic experience to their postdoctoral pursuits. Their training in contemporary methods and technology, complemented by a firm foundation in the basic sciences, gives them a competitive edge. Whether they are attracted to positions in academia, industry, or government, they will be poised to contribute to the overall mission of their employer.

Location

The School of Medicine and Biomedical Sciences and other health science programs are located on the State University of New York at Buffalo (UB) South Campus, which is in the northeast corner of the city of Buffalo and borders the suburbs of Amherst, Cheektowaga, Williamsville, and Kenmore. The city is located at the eastern end of Lake Erie, a few miles south of the Canadian border. The University is approximately 1 hour by plane from Boston, Chicago, New York City, and Philadelphia. The western New York area provides excellent year-round recreational opportunities. The Buffalo Philharmonic Orchestra and world-renowned Albright-Knox Art Gallery are just two of the main attractions for music and art lovers. The downtown Buffalo theater district is home to the Studio Arena Theatre and the Shea's Buffalo. Buffalo is also close to such cultural centers as Toronto, the Chautauqua Institution, the Stratford Shakespeare Festival in Canada, the Shaw Festival at Niagara-on-the-Lake, and Artpark in Lewiston. Niagara Falls is a short drive away. Sports enthusiasts enjoy NHL and NFL teams as well as Triple A baseball. The University's sports teams participate at the NCAA Division I level.

The University

The University was established in 1856 and incorporated into the New York State System in 1962. It is enriched by the presence of more than 2,000 international students from 100 countries. The State University of New York at Buffalo is committed to nurturing its diverse cultural, racial, and ethnic student population. In addition to the work of individual faculty members and the investigations of research units within departments, UB supports more than fifty interdisciplinary organized research units. UB's libraries, the largest in the SUNY system and fourth largest among all university library systems in New York State, provide support for research and scholarship.

Applying

Applicants for admission should have a strong undergraduate background with a dedication to a career in the biomedical sciences. Candidates are evaluated on the basis of their undergraduate curriculum, grade point average, three letters of recommendation, and scores on the General Test of the Graduate Record Examinations (GRE). For international applicants, the TOEFL is also required. Similarly, research experience is desirable but not required. Applicants who wish to be considered for fellowships or scholarships in addition to the stipend must submit the online application by February 1. Applications are accepted and reviewed through April.

Correspondence and Information

Director, Interdisciplinary Graduate Program in Biomedical Sciences
School of Medicine and Biomedical Sciences
128 BIOED Building
State University of New York at Buffalo
3435 Main Street
Buffalo, New York 14214-3013
Phone: 716-829-3398
Fax: 716-829-2437
E-mail: smbs-gradprog@buffalo.edu
Web site: http://www.smbs.buffalo.edu/rbe/igpbs

State University of New York at Buffalo

DEPARTMENTAL CHAIRPERSONS AND DIRECTORS OF GRADUATE STUDIES

Individual faculty member listings may be found at http://www.smbs.buffalo.edu/departments. All the addresses given below are in the School of Medicine and Biomedical Sciences, 3435 Main Street, Buffalo, New York 14214-3013.

Senior Associate Dean, Research and Biomedical Education
Dr. Suzanne Laychock, 128 BIOED Building (telephone: 716-829-3398, fax: 716-829-2437).

Director, Interdisciplinary Graduate Program in Biomedical Sciences
Dr. Richard A. Rabin, 128 BIOED Building (telephone: 716-829-3398, fax: 716-829-2437, e-mail: rarabin@buffalo.edu).

M.D./Ph.D. Program
Director: Dr. Paul Knight, 128 BIOED Building (telephone: 716-829-2172, fax: 716-829-2437, e-mail: mdphd@buffalo.edu).

STATE UNIVERSITY OF NEW YORK AT BUFFALO

School of Medicine and Biomedical Sciences
Medical Scientist Training Program

Program of Study	The Medical Scientist Training Program (MSTP) of the School of Medicine and Biomedical Sciences at the State University of New York at Buffalo provides highly talented students with an opportunity to combine intensive scientific training with medical school experience. The MSTP curriculum is designed to integrate both a rigorous clinical and research experience throughout the program and to provide the student with the skills needed for an academic career path. The primary purpose of the MSTP is to develop physician scientists who will become leaders in basic, translational, or clinical research and in clinical medicine in medical schools and teaching hospitals. This program is directed toward helping the student prepare for a career in which they excel as both an investigator and a clinician educator. The resources available for graduate training are extensive and diversified, making the School uniquely suited for training MSTP students. MSTP students are provided with a vast opportunity to choose from a number of research programs and schools in which to complete their Ph.D. thesis. Opportunities for research include not only those available within the Medical School but also the graduate departments in the School of Pharmacy, the Department of Oral Biology within the School of Dental Medicine, and the Departments of Biology and Psychology as well as the graduate programs at Roswell Park Cancer Institute. Research interests are considerably diverse and include translational medicine, bioinformatics, nanomedicine, and molecular and structural biology. The breadth of the research available for students and the high quality of the faculty members throughout the University attest to the strength of the MSTP at SUNY at Buffalo. Specific information about faculty members and their research interests within several of the departments may be found at http://www.smbs.buffalo.edu. Information may also be obtained by visiting department Web sites at http://www.buffalo.edu or by writing to the departments directly.
Research Facilities	The facilities and resources available for graduate and clinical training are extensive and diverse, providing the MSTP students with an extraordinary opportunity to choose from a number of research programs and schools for their doctoral thesis. The School of Medicine and Biomedical Sciences has developed a number of multidisciplinary research focus centers that combine the expertise of basic science and clinician-scientist faculty members. The centers are housed in a multidisciplinary research facility. Core facilities associated with the University include those for protein and DNA sequencing, oligonucleotide synthesis, laser confocal microscopy, electron microscopy, transgenic animals, DNA and protein microarray and gene chips, NMR, animal MRI, MALDI-TOF, high-throughput and molecular targeting, fluorescent cell sorting (FACS), and X-ray crystallography. Roswell Park Cancer Institute is a leading research center in the field of cancer. The Hauptman-Woodward Medical Institute houses the Department of Structural Biology and is located in the medical corridor near the Cancer Institute. The Buffalo Center of Excellence in Bioinformatics merges high-end technology with expertise in genomics and proteomics. The Health Sciences Library has extensive journal and book holdings. Students have computer access to online bibliographic databases, full-text journals, clinical manuals and textbooks, drug information resources, and communication tools such as e-mail.
Financial Aid	All eligible students enrolled in the MSTP receive a tuition scholarship while enrolled in the program. In addition, MSTP students receive fellowships provided by the School of Medicine and Biomedical Sciences, the Graduate School, and individual graduate departments. The current minimum level of fellowships during the medical school phase of the student's training is $20,000 per year. During the graduate years, fellowship levels are set and paid for by the graduate departments, and the stipend range is $21,000–$24,000 annually.
Cost of Study	Although tuition is paid for, students are responsible for paying fees. The 2006–07 fees are $1270 per year. During the first two years in the program, there is an additional microscope use fee of $150.
Living and Housing Costs	The University is located in a residential suburb of Buffalo, and it is easy for students to find moderately priced accommodations. University apartments are available, and the University Housing Bureau is very helpful in finding accommodations for both staff and students. A rapid transit system connects the Health Sciences Complex with downtown Buffalo, Roswell Park Cancer Institute, the Buffalo Center of Excellence in Bioinformatics, and Hauptman-Woodward Medical Institute.
Student Group	Approximately 585 medical students and 175 graduate students are enrolled in the medical school. The MSTP currently has approximately 30 students enrolled in its program.
Student Outcomes	Upon graduation from the MSTP, students routinely enter research-oriented residency programs in a variety of specialties at eminent academic and clinical institutions. Professional opportunities for the MSTP graduates are excellent and encompass a breadth of areas, from clinical medicine to research-oriented positions. Past students have distinguished themselves by obtaining faculty positions at leading medical schools throughout the country.
Location	The School of Medicine and Biomedical Sciences and other health science programs are located on the UB South Campus, which is in the northeast corner of the city of Buffalo and borders the suburbs of Amherst, Cheektowaga, Williamsville, and Kenmore. The city is located at the eastern end of Lake Erie, several miles south of the Canadian border. The University is approximately 1 hour by plane from Boston, New York City, and Philadelphia. The western New York area provides excellent year-round recreational opportunities. The Buffalo Philharmonic Orchestra and world-renowned Albright-Knox Art Gallery are just two of the main attractions for music and art lovers. The downtown Buffalo theater district is home to the Studio Arena Theatre and the Shea's Buffalo. Buffalo is also close to such cultural centers as Toronto, the Chautauqua Institution, the Stratford Shakespeare Festival in Canada, the Shaw Festival at Niagara-on-the-Lake, and Artpark in Lewiston. Niagara Falls is a short drive away. Sports enthusiasts enjoy NHL and NFL teams as well as AAA baseball. The University's sports teams participate at the NCAA Division I level.
The University	The State University of New York at Buffalo started as a private institution, the University of Buffalo, when a group of local physicians decided to open a university and a medical school in Buffalo in 1846. Around the turn of the century, expansion began, and some sixty years later the University had grown into a small but diversified institution. In particular, a College of Arts and Sciences was established in 1915 and a graduate division in 1939. The University of Buffalo was incorporated into the state system in 1962 with the intention of making Buffalo its foremost graduate center. As predicted, the merger caused an unprecedented growth, with student registration jumping from nearly 6,000 to more than 27,000. Currently, the University has more than 3,800 full-time, part-time, and volunteer faculty members, encompassing 175 graduate and professional programs that enroll approximately 8,500 students.
Applying	Students applying for admission to the MSTP at SUNY at Buffalo's School of Medicine and Biomedical Sciences should follow the standard procedure for application to the Medical School, and they should request a supplemental MSTP application form. Evaluation of applications is based on consideration of the candidate's undergraduate credentials, MCAT scores, statement of career objectives, prior research experience, and letters of support.
Correspondence and Information	Dr. Paul Knight, MSTP Director Office of Research and Graduate Studies 128 BIOED Building School of Medicine and Biomedical Sciences State University of New York at Buffalo Buffalo, New York 14214 Phone: 716-829-3398 E-mail: smbs-mdphd@buffalo.edu Web site: http://www.smbs.buffalo.edu/rbe/mstp

State University of New York at Buffalo

FACULTY HEADS OF PROGRAMS AFFILIATED WITH THE MSTP

The area code for all faculty telephone numbers listed below is 716.

Senior Associate Dean for Research and Biomedical Education: Dr. Suzanne G. Laychock, 829-3398.

SCHOOL OF MEDICINE (http://www.smbs.buffalo.edu/)
Biochemistry: Dr. Kenneth Blumenthal, Chairperson, 829-2727.
Microbiology and Immunology: Dr. John Hay, Chairperson, 829-2907.
Neurosciences: Dr. Malcolm Slaughter, Director, 829-3240.
Pathology and Anatomical Sciences: Dr. Francisco Velazquez, Chairperson, 626-7200 Ext. 8027.
Pharmacology and Toxicology: Dr. Ronald Rubin, Chairperson, 829-2800.
Physiology and Biophysics: Dr. Harold Strauss, Chairperson, 829-2738.
Structural Biology: Dr. George DeTitta, Chairperson, 856-9600.

ROSWELL PARK CANCER INSTITUTE (http://www.roswellpark.org/)
Cancer Pathology and Prevention: Dr. Clement Ip, Director, 845-3063.
Cellular and Molecular Biology/Cancer Genetics: Dr. John Yates, Chairperson, 845-8964.
Cellular and Molecular Biophysics: Dr. John Subjeck, Chairperson, 845-3135.
Immunology: Dr. Soldano Ferrone, Chairperson, 845-5758.
Molecular Pharmacology and Cancer Therapeutics: Dr. Michael Braitten, Chairperson, 845-8223.

SCHOOL OF DENTAL MEDICINE (http://www.sdm.buffalo.edu/)
Oral Biology: Dr. Frank Scannapieco, Chairperson, 829-3373.

SCHOOL OF PHARMACY (http://www.pharmacy.buffalo.edu/)
Center for Drug Discovery and Experimental Therapeutics: Dr. Robert Straubinger, 645-2842 Ext. 243.
Pharmaceutics: Dr. Marilyn Morris, Director of Graduate Studies, 645-2842.

SCHOOL OF PUBLIC HEALTH (http://sphhp.buffalo.edu/)
Social and Preventive Medicine and Division of Biostatistics: Dr. Maurizio Trevisan, Chairperson, 829-2975.

COLLEGE OF ARTS AND SCIENCES (http://www.cas.buffalo.edu/)
Biological Sciences: Dr. Mary Bisson, Chairperson, 645-2363.
Communicative Disorders: Dr. Elaine Stathopoulos, Chair, 829-2797.
Psychology: Dr. Jack Meacham, Chairperson, 645-3650 Ext. 203.

BUFFALO CENTER OF EXCELLENCE IN BIOINFORMATICS (http://www.bioinformatics.buffalo.edu/coeb/content/)
Director: Dr. Jeffrey Skolnick, 849-6711.

STATE UNIVERSITY OF NEW YORK
DOWNSTATE MEDICAL CENTER
School of Graduate Studies

Programs of Study

The School of Graduate Studies of the State University of New York (SUNY) Downstate Medical Center grants three biomedical science Doctor of Philosophy (Ph.D.) degrees: neural and behavioral science, molecular and cellular biology, and biomedical engineering. All programs offer training leading to a combined M.D./Ph.D. The Ph.D. in biomedical engineering is granted jointly by SUNY Downstate and Polytechnic University.

The chief objective of the biomedical science Ph.D. programs is to educate students to become investigators and teachers in the biomedical sciences. Each student's academic program includes a thorough foundation in current biomedical sciences, as well as in-depth studies in a particular area of research interest. Most students complete their formal course requirements within the first two years of study, after which their efforts are focused primarily on their research, leading eventually to a formal thesis proposal and thesis defense. Students select a thesis adviser after completion of two research rotations in their chosen area. Doctoral programs sponsor seminar series of invited, prestigious speakers. All graduate students are enrolled as student members in the New York Academy of Sciences, which sponsors events such as lectures and seminars. Through a recent affiliation, students may be eligible to do rotations at the MRC Centre for Synaptic Plasticity at the University of Bristol, England.

In the last five years, 94 percent of graduates have continued on to postdoctoral positions at excellent academic institutions. Also, the growing Advanced Biotechnology Park, located adjacent to the campus, provides students with opportunities to do rotations in an industrial setting. As part of a strategic alliance with Polytechnic University for fostering research and education in biomedicine and bioengineering, numerous research collaborations with investigators from the neighboring campus are flourishing. In addition, faculty members from the School of Graduate Studies provide the biomedical sciences portion of Polytechnic University's M.S. program in biomedical engineering. Areas of combined faculty expertise include molecular genetics, biomaterials, drug delivery systems, bioimaging, and neuroengineering. Downstate's joint Ph.D. program with Polytechinic in biomedical engineering was mentioned above, in the first paragraph of this description.

Research Facilities

Extensive research facilities are available in the laboratories of individual faculty members at SUNY Downstate. There are also excellent institutional core facilities for DNA synthesis and purification, electron microscopy, peptide analysis, mass spectrometry, cell imaging and confocal microscopy, a transgenic mouse facility, a microarray laboratory, a central animal facility, a computer center, a scientific instrumentation center, and a central photographic and illustration service.

The Medical Research Library of Brooklyn occupies three floors of the modern Health Science Education Building. It has extensive computer-related resources for information retrieval and subscribes to 1,846 periodicals. Through the library's Web site, students have 24-hour access to the online catalog, electronic journals, and databases. The circulating collection holds 46,072 books and 260,613 bound journals, making it one of the largest medical libraries in the U.S.

Financial Aid

Most graduate students are supported through a comprehensive program of financial aid based on support from institutional sources and extramural grant funds. Students entering Ph.D. and M.D./Ph.D. programs in fall 2006 receive twelve-month stipends, including discretionary funds totaling $25,000.

Cost of Study

In 2006–07, tuition for full-time graduate students is $6900 per year for New York State residents and $10,920 per year for nonresidents. Student fees are $220 per year. All matriculating students are eligible for tuition waiver fellowships.

Living and Housing Costs

Accommodations for single and married students are available in the campus residence halls. All student rooms and study lounges have been wired for fast Internet access. The minimum yearly cost for a single student in 2006–07 is $5936. Students may live in off-campus housing near the School or elsewhere in the metropolitan area. Off-campus rental costs vary widely.

Student Group

Total student enrollment at the SUNY Downstate Medical Center in fall 2005 was 1,426, with 83 students in the School of Graduate Studies, 790 students in the College of Medicine (including the M.P.H. program), 325 students in the College of Nursing, and 228 students in the College of Health Related Professions. In fall 2005, 46 percent of the students in the School of Graduate Studies were women.

Location

New York City is one of the world's leading educational and cultural centers. The School is located only minutes away from the Brooklyn Academy of Music, Brooklyn Botanic Garden, and Brooklyn Museum; a short distance by public transportation from the superb cultural facilities of lower Manhattan; and within an hour by public transportation of Manhattan's world-renowned concert halls, fine art museums, and Broadway theaters. Student discount tickets are available for many plays and musical performances. There are excellent recreational facilities in the New York metropolitan area and extensive opportunities for skiing, hiking, and camping within a 2-hour drive.

The Medical Center

In addition to the School of Graduate Studies, SUNY Downstate Medical Center comprises the Colleges of Medicine, Health Related Professions, Nursing, and the University Hospital of Brooklyn. King's County Hospital Center is located just across the main avenue from SUNY Downstate. Also, SUNY Downstate has affiliations with many research institutes and teaching hospitals in Brooklyn and Staten Island. In the Basic Science Building, where most research laboratories are located, there is a graduate student lounge with computers.

The Health Science Education Building includes modern classrooms and teaching laboratories. The Student Center, adjacent to the dormitories, has a swimming pool and squash courts as well as other athletic and meeting facilities. An extensive program of extracurricular courses and other organized activities is offered at the Student Center.

Applying

Students are selected by a Schoolwide admissions committee and are admitted on a rolling basis, usually for the fall semester. Applicants are required to submit official transcripts from all colleges and universities attended, Graduate Record Examinations scores, and two letters of recommendation. M.D./Ph.D. applicants must submit separate applications to both the College of Medicine and the School of Graduate Studies. Preference is given to competitive applicants who submit completed applications prior to March 1. Review of Ph.D. applications begins in December.

Applicants are selected based on their qualifications, without regard to race, color, sex, or national origin. Qualities such as demonstrated interest in research, intellectual curiosity, perceptivity, and ability to reason, as well as enthusiastic letters of recommendation and a record of high academic achievement, all play an important role in the selection process.

Correspondence and Information

Denise Sheares, Director of Admissions
School of Graduate Studies
SUNY Downstate Medical Center
450 Clarkson Avenue, Box 41P
Brooklyn, New York 11203-2098

Phone: 718-270-2738
Fax: 718-270-3378
E-mail: denise.sheares@downstate.edu
Web site: http://www.downstate.edu/grad

State University of New York Downstate Medical Center

THE FACULTY AND THEIR RESEARCH

Vahé E. Amassian, M.B. Magnetic or electrical transcranial stimulation of humans and animal models in analyzing various functions.
Randall L. Barbour, Ph.D. Optical tomographic imaging methods for the evaluation of tissue function.
Olcay A. Batuman, M.D. Molecular level studies of vascular endothelial functions in pathogenesis of human diseases.
Peter J. Bergold, Ph.D. Analysis of the pathophysiology of neuronal disorders.
Paulette Bernd, Ph.D. Nerve-growth factors and their receptors; developmental neurobiology; cardiac development.
Stacy W. Blain, Ph.D. Cell-cycle progression, focusing on the cyclin-cdks and their inhibitors, the Cip/Kips and Ink4s.
Martin Bluth, M.D., Ph.D. ICAM, selectin in leukocytes; PBMC in inflammatory diseases; IgE in autoimmunity, transplantation, and viral infections.
Ivan Bodis-Wollner, M.D., D.Sc. Relationship between high-frequency changes in the EEG and saccadic eye movements, voluntarily or cued.
Mohamed Boutjdir, Ph.D. Autonomic regulation of native and heterologously expressed ion channels; intracellular signaling; arrhythmias.
John K. Chapin, Ph.D. Technologies for recording from large populations of neurons in sensory and motor areas of brain; neurorobotics.
Brahim Chaqour, Ph.D. Mechanotransduction mechanisms and regulation and function of mechanosensitive genes (e.g., Cyr61, CTGF, Nov).
William J. Chirico, Ph.D. Growth factors: biogenesis, angiogenesis, and cancer; chaperone-dependent protein folding and translocation.
Eva B. Cramer, Ph.D. Early inflammatory response in intestinal and urinary tract infections; mucosal immunity.
Howard A. Crystal, M.D. Clinicopathologic and epidemiologic studies of memory and cognition in normal aging and dementia.
Diana L. Dow-Edwards, Ph.D. Developmental toxicity of AZT and cocaine; drug abuse, brain imaging, and behavior.
Helen G. Durkin, Ph.D. Immune responses in HIV-1 disease; regulation of allergic/IgE responses.
André A. Fenton, Ph.D. Studies of spatial cognition, memory, and hippocampal function, using integrated methods of research.
Miriam H. Feuerman, Ph.D. Identification of genes regulating gene expression during liver regeneration; susceptibility to tumorigenesis.
Steven E. Fox, Ph.D. Hippocampal EEG rhythms and location-specific firing of cells.
Robert F. Furchgott, Ph.D. Mechanisms of relaxation of vascular smooth muscle by endogenous and exogenous nitric oxide and by light.
Gregory G. Gick, Ph.D. Molecular mechanisms underlying regulation of mammalian Na, K-ATPase subunit gene expression.
Alan R. Gintzler, Ph.D. Biochemistry of addiction/narcotic tolerance; G-protein cascades; gender-dependent regulation of pain.
Mimi N. Halpern, Ph.D. Nasal chemical senses: transduction, coding, development, degeneration/regeneration, and functional analysis.
Christopher U. Hellen, D.Phil. Roles of eukaryotic initiation factors (eIFs) in translation of viral and cellular mRNAs.
Ellen Hsu, Ph.D. Molecular mechanisms in generation of antibody diversity; DNA rearrangement and somatic hypermutation.
M. Mahmood Hussain, Ph.D. Protein-protein interactions and molecular mechanisms of intestinal lipoprotein assembly.
Xian-Cheng Jiang, Ph.D. Creation and development of mouse models for study of relation between lipid metabolism and heart disease.
Ira S. Kass, Ph.D. Mechanisms of anoxic damage to brain.
Elizabeth Kornecki, Ph.D. Molecular mechanisms of platelet/endothelial cell adhesion underlying inflammatory thrombosis.
John L. Kubie, Ph.D. Rat hippocampus: navigation, learning, and memory.
John A. Lewis, Ph.D. Inhibition of interferon-induced signal transduction by viruses; regulation of gene expression by interferons.
Douglas S. F. Ling, Ph.D. Cortical inhibitory circuits and regulation of excitatory transmission; epilepsy and synaptic plasticity.
William W. Lytton, M.D. Computer modeling of neurons and neural networks; applications to epilepsy, stroke, learning, and memory.
Maureen V. McLeod, Ph.D. Signal transduction mechanisms that regulate growth and development; nuclear import and export.
Lisa R. Merlin, M.D. Role of metabotropic glutamate receptors in epilepsy; network properties of hippocampus; synaptic plasticity.
Hillary B. Michelson, Ph.D. Functional connectivity, morphology, and maturation of hippocampal inhibitory circuitry; epilepsy.
Josef Michl, M.D. Host defense against infection and cancer; molecular carcinogenesis in novel disease models.
Donald R. Mills, Ph.D. Translational regulation of gene expression in prokaryotes.
Suzanne S. Mirra, M.D. Neuropathology of Alzheimer's disease, other neurodegenerative disorders, and vascular dementia.
Foroozan Mokhtarian, M.P.H., Ph.D. Experimental autoimmune encephalomyelitis used to test potential therapeutic agents for MS.
Robert U. Muller, Ph.D. Molecular, synaptic, and network foundations of spatial memory and learning in rodents; hippocampus.
Allen J. Norin, Ph.D. Role of the Haymaker gene product in cancer and normal cell physiology.
Maja Nowakowski, Ph.D. Macrophage immune functions and NO production in infections; lung immunity in infection and inflammation.
George K. Ojakian, Ph.D. Integrin signaling pathways and regulation of epithelial tubule formation.
Camilo A. Parada, Ph.D. Mechanism(s) by which the HIV-1-encoded Tat protein enhances HIV-1 transcription.
Nicholas J. Penington, Ph.D. Neuropharmacology of 5-HT neurons: patch clamp studies, signal transduction, and ion channel modulation.
Katherine L. Perkins, Ph.D. Synaptic transmission in the hippocampus, primarily GABAergic transmission; depolarizing GABA response.
Tatyana Pestova, D.Sc. Mechanism of initiation of eukaryotic protein synthesis.
Matthew R. Pincus, M.D., Ph.D. Oncogenesis; oncoprotein structure; mitogenic signal transduction; design of anticancer agents.
Christopher A. J. Roman, Ph.D. Regulation of B-cell development; B-cell immunodeficiencies; cancer.
Leonard A. Rosenblum, Ph.D. Primate behavioral and brain development; psychopathology; psychopharmacology; psychoimmunology.
Julie I. Rushbrook, Ph.D. Expression and modification of developmental myosins; allosterism of NAD⁺-isocitrate dehydrogenase.
Todd C. Sacktor, M.D. Protein kinase C isozymes; PKMζ; long-term potentiation; long-term depression; learning and memory.
Frank R. Scalia, Ph.D. Target recognition during development and regeneration of the visual pathway.
M. A. Q. Siddiqui, Ph.D. Gene regulation in muscle development and disease; signal transduction pathways; transcriptional adaptation.
Sheryl S. Smith, Ph.D. Neurosteroid effects on GABA-A receptor plasticity; hippocampal physiology; anxiety and epilepsy.
Armin Stelzer, M.D., Ph.D. Hippocampus; synaptic plasticity; role of synaptic inhibition; regulation of GABA receptor.
Mark G. Stewart, M.D., Ph.D. Physiology and pathophysiology of limbic neurons and limbic neuron circuits.
Alfred Stracher, Ph.D. Molecular basis of neuromuscular and neurodegenerative disorders, role of calpain; design of protease inhibitors.
Gladys N. Teitelman, Ph.D. Pancreatic islet cell differentiation; isolation of insulin precursor cells, signals that control their maturation.
Henri Tiedge, Ph.D. Neuronal gene expression; RNA transport; dendritic protein synthesis; synaptic plasticity; Alzheimer's disease.
Roger D. Traub, M.D. Studies of how large populations of neurons generate collective behaviors; electrophysiology; computer modeling.
Mario Vassalle, M.D. Electrophysiology of the heart: automaticity, ionic mechanisms and control in relation to impulse formation.
Michael A. Wagner, Ph.D. Molecular mechanisms underlying vertebrate neurogenesis; tumor-suppressor genes; cardiogenesis.
Dalton Wang, Ph.D. Chemosignal transduction in vomeronasal system; protein chemistry; chemoattractive compounds.
Keith Williams, Ph.D. Structure, function, pharmacology, and regulation of glutamate receptors.
Robert K. S. Wong, Ph.D. Calcium and intercellular signaling and circuit organization of the hippocampus.
Michael E. Zenilman, M.D. Gastrointestinal and pancreatic surgery and molecular physiology; laparoscopic surgery; aging.

STATE UNIVERSITY OF NEW YORK
UPSTATE MEDICAL UNIVERSITY

College of Graduate Studies

Programs of Study

The College of Graduate Studies at SUNY Upstate Medical University educates students to be research scientists at the Ph.D. or master's level, preparing them for careers in academic medical centers, biomedical research institutes, the biotechnology industry, and government agencies. Ph.D. degrees are offered by the Departments of Cell and Developmental Biology, Biochemistry and Molecular Biology, Microbiology and Immunology, Neuroscience and Physiology, and Pharmacology. Under the administration of the Department of Neuroscience and Physiology, there is also an interdepartmental program offering a Ph.D. degree in neuroscience. Master's degrees are offered through the Departments of Cell and Developmental Biology, Biochemistry and Molecular Biology, Microbiology and Immunology, and Neuroscience and Physiology. To provide a maximum choice in selecting a research specialization, the College gives students a full year to make the crucial decision as to the topic and mentor for their thesis research. During the first year, students rotate through three labs, choosing from a variety of well-funded labs in leading-edge research areas. To help with the rotation selection, faculty members from each department program take an afternoon early in the semester to present the research carried on in the program. All first-year students also take a cross-departmental core curriculum designed to provide a broad-based education in the up-to-date fundamentals of research in basic biomedical sciences. After successful completion of the lab rotations and a first-year core curriculum taught by faculty members from throughout the College, students select their thesis mentor and transition into the appropriate degree-granting department or program. They then proceed to fulfill the course requirements specific to that program and conduct their thesis research. The thesis research gives students the technical and critical-thinking skills needed to carry on independent scientific experimentation. Participation in a vigorous research program facilitates the transition from student to professional colleague. Research at SUNY Upstate Medical University spans many disciplines, from basic science using model systems to translational and clinical research. Some examples of the major research areas are cancer research, cardiovascular disease, cell signaling, developmental biology, functional genomics, gene expression, immunology, membrane biology and biophysics, microbiology and virology, musculoskeletal science, neuroscience, sensory biology, and structural biology. The University has a high proportion of young, vigorous faculty members and a low student-faculty ratio. Consequently, students can expect a high degree of faculty involvement in their graduate training. The Ph.D. program, including research, didactic course work, and successful defense of a dissertation, is intended to be completed in four or five years.

Research Facilities

In addition to a plethora of specialized equipment found in individual research labs, a variety of state-of-the-art instruments are available as shared resources. These include ultracentrifuges; spectrophotometers; amino acid analyzers; cell and organ culture facilities; gas chromatograph–mass spectrometers; high-pressure liquid chromatography apparatuses; a fluorescence-activated cell sorter; light, confocal, and electron microscopes; an image-analysis facility; an automated DNA sequencer; equipment for microarray analysis; and a 30-parallel-processor supercomputer. The Institute for Human Performance and the Cardiovascular Research Institute offer additional state-of-the-art research facilities. There are also full research support services, including laboratory animal facilities; network access to the SeqWeb suite of software; a computer-age medical library with 1,450 serial titles; electronics and machine shops; and photographic and computer services.

Financial Aid

All accepted Ph.D. students are fully supported throughout their education by tuition waivers and a stipend ($20,388 per year). Support comes from graduate assistantships, departmental assistantships, and NIH and NSF grants.

Cost of Study

Stipends and tuition waivers are available for all students accepted into the Ph.D. program. Student fees, which include a health service fee, were $466 for the 2005–06 academic year. Tuition and fees for master's students for the 2005–06 academic year were $6900 for in-state students and $10,920 for out-of-state students. Costs are subject to change.

Living and Housing Costs

On-campus housing is available in a ten-story apartment building. These apartments ranged from $3585.61 (standard room, double occupancy) to $7488.18 (married/family accommodations, one-bedroom apartment) for the academic year 2005–06. Many graduate students rent houses or apartments within a mile of the campus and bicycle or walk to and from campus during much of the year. The cost of living in the Syracuse area is low.

Student Group

There are 123 graduate students in the biomedical sciences (50 percent women; 100 percent full-time) and approximately 600 medical students, 200 nursing students, and 200 students in the health professions enrolled at Upstate Medical University. Twenty-five percent of the graduate students come from Canada, Europe, and Asia. Syracuse University and the SUNY College of Environmental Science are located within a mile of the University, resulting in a population of approximately 23,000 students in the immediate area.

Location

Syracuse is the center of a medium-sized metropolitan area located in the scenic center of New York State. The area offers excellent boating, hiking, biking, and skiing in the nearby Finger Lakes region, the Adirondack and the Catskill Mountains, and Lake Ontario. Cultural activities include a professional symphony orchestra and opera company, chamber music groups, several top-notch music festivals (classical, blues, and jazz), and a repertory theater, as well as art and history museums. The area also offers many excellent school districts. Syracuse University's top-level collegiate sporting events are a major Syracuse recreational activity. Syracuse is easily reached by air, rail, and auto.

The University

In addition to the College of Graduate Studies, SUNY Upstate Medical University includes three other colleges. These are the Colleges of Medicine, Nursing, and Health Professions. Upstate Medical University, formerly known as SUNY Health Science Center at Syracuse, is one of four medical centers in the SUNY system. It was established in 1834 as the medical department of Geneva College and had the distinction of graduating Elizabeth Blackwell, the first woman to receive an M.D. in the United States. The University is located close to downtown Syracuse and is immediately adjacent to (but not affiliated with) the campus of Syracuse University. The Campus Activities Building houses a swimming pool, a sauna bath, a gymnasium, squash courts, a handball/paddleball court, a weightlifting area with a Universal Gym and a full Nautilus room, billiards, table tennis, a television room, a bookstore, a snack bar, and a lounge. Conference rooms are also available for student use.

Applying

The College of Graduate Studies at SUNY Upstate does not have an application deadline; however, the Admissions Committee begins reviewing applications in December and continues until all positions are filled, which can be as early as the beginning of April. The State University of New York requires a $40 application fee. Minimum requirements are a bachelor's degree or its equivalent and course work that includes biology, mathematics (preferably through calculus), physics, and chemistry (organic and inorganic). GRE General Test scores are required, and scores from the Subject Test in chemistry or biology are recommended. International applicants must provide clear evidence of English proficiency (including speaking) by taking the Test of English as a Foreign Language (TOEFL), which, as of September 2005, also tests the ability to speak English.

Correspondence and Information

Office of Graduate Studies
State University of New York Upstate Medical University
750 East Adams Street
Syracuse, New York 13210
Phone: 315-464-4538
Fax: 315-464-4544
E-mail: gradstud@upstate.edu
Web site: http://www.upstate.edu/grad/

State University of New York Upstate Medical University

THE FACULTY AND THEIR RESEARCH

Matthew J. Allen, Ph.D., Adjunct Assistant Professor. Tumor-bone cell signaling and the cellular/molecular basis of implant loosening.

David C. Amberg, Ph.D., Associate Professor. Regulation of actin dynamics and analysis of genomic influences on actin function.

Justus Anumonwo, Ph.D., Assistant Professor. Biophysical and molecular bases of cardiac impulse generation and propagation.

Robert B. Barlow, Ph.D., Adjunct Professor. Neural basis of visual behavior; computational models of neural coding; circadian and metabolic modulation of human visual sensitivity.

Omer Berenfeld, Ph.D., Assistant Professor. Biophysics of impulse propagation and mechanisms of cardiac arrhythmias.

Scott D. Blystone, Ph.D., Associate Professor. Extracellular matrix regulation of the leukocyte inflammatory phenotype.

Blair Calancie, Ph.D., Professor. CNS plasticity after trauma; intraoperative electrophysiology.

David Cameron, Ph.D., Assistant Professor. Regeneration and development of the retina.

Enrico Camporesi, M.D., Adjunct Professor. Cardiac and respiratory responses to exercise in extreme environments and during postoperative recovery.

Gregory Canute, M.D., Assistant Professor. Brain tumor research.

Howard Chang, Ph.D., M.D., Assistant Professor. Pathology and anatomy of dementia-related brain; spinal cord injury.

Gino Cingolani, Ph.D., Assistant Professor. X-ray crystallography of large viral DNA-pumping enzymes and structural cell biology of nucleocytoplasmic transport.

Richard L. Cross, Ph.D., Professor and Chair. Mechanisms of mitochondrial oxidation phosphorylation; biological rotary motors; single molecule measurement.

Michael H. Cynamon, M.D., Professor. Antituberculosis activity of pyrazinamide.

Timothy A. Damron, M.D., Professor. Radioprotectant strategies for protecting the pediatric growth plate.

Mario Delmar, M.D., Ph.D., Professor. Cellular and subcellular bases of cardiac rhythm disturbances; regulation of gap-junction channels.

Joseph Domachowske, M.D., Adjunct Associate Professor. Pneumovirus pathogenesis.

Gerold Feuer, Ph.D., Associate Professor. HTLV pathogenesis in the SCID-hu mouse model; lentivirus-based gene therapy vectors; KSHV/HHV-8.

Eileen A. Friedman, Ph.D., Adjunct Professor. Cell signaling in colon cancer progression and in muscle development.

Jerrie Gavalchin, Ph.D., Professor. Regulation of pathogenic antibody production in autoimmune glomerulonephritis; cell-surface receptors for retroviruses.

David M. Gilbert, Ph.D., Professor. Regulation of DNA replication during the cell cycle and development; nuclear dynamics and genome plasticity/embryonic stem cells.

Charles J. Hodge, M.D., Professor. Mechanisms of cortical plasticity and cortical reorganization after injury.

James Holsapple, M.D., Associate Professor. Visual association cortex.

Huaiyu Hu, Ph.D., Assistant Professor. Mechanisms of neuronal migration and axonal growth in the developing brain.

Ying Huang, M.D., Ph.D., Associate Professor. Oncogenic signaling in cellular transformation and apoptosis; tumor suppressor genes; regulation of microtubule stability by RASSF1A tumor suppressor.

Charles B. C. Hwang, Ph.D., Associate Professor. DNA replication of herpes viruses.

José Jalife, M.D., Professor, Chair, and Director, Institute for Cardiovascular Research. Cardiac electrophysiology and mechanisms of arrhythmias.

Burk Jubelt, M.D., Professor and Chair. CNS acute and chronic poliovirus and enterovirus infections.

Patricia M. Kane, Ph.D., Professor. Mechanisms and regulation of cellular pH control; V-type ATPases.

Wendy Kates, Ph.D., Associate Professor. Anatomic and functional imaging investigations of neurodevelopment in individuals with genetic or psychiatric disorders.

Grant Kelley, M.D., Associate Professor. Elucidating the regulation of PLC-epsilon and its role in glucose signaling and endothelial cell function in diabetes.

Dilip Kittur, M.D., Sc.D., Professor. Endothelial cell function; hemirenal transplant.

Barry Knox, Ph.D., Professor. Light transduction and retinal development; regulatory genes involved in phototransduction.

James Listman, M.D., Adjunct Assistant Professor. Cytomegalovirus and transplantation.

Stewart N. Loh, Ph.D., Associate Professor. Mechanism and kinetics of protein folding; structure and function of the p53 tumor suppressor; design of proteins with new or enhanced functions; protein-based molecular switches.

John J. Lucas, Ph.D., Professor. Glycobiology; complex glycoconjugate structure and function; parasite cysteine proteases.

Kenneth Mann, Ph.D., Professor. Mechanical and biological factors in total joint replacement.

Paul Massa, Ph.D., Adjunct Professor. Genetic regulation of glial cell differentiation.

James S. McCasland, Ph.D., Professor. Cortical plasticity; development of somatotopic representations in cortex.

Michael M. Meguid, M.D., Adjunct Professor. Neurophysiological regulation of food intake.

Frank Middleton, Ph.D., Assistant Professor. Molecular basis of cortical–basal ganglia and cortical-cerebellar circuit and dysfunction in neurological and psychiatric disease.

Michael Miller, Ph.D., Professor and Chair. Factors that regulate the proliferation and survival/death of neurons (and their precursors) in the developing central nervous system; models of fetal alcohol syndrome and autism.

David R. Mitchell, Ph.D., Professor. Dynein ATPase function in flagellar motility.

Jenifer Moffat, Ph.D., Associate Professor. Varicella zoster pathogenesis.

Sandra Mooney, Ph.D., Assistant Professor. Cell death and survival in the developing brain; mechanisms of ethanol toxicity; models of fetal alcohol syndrome and autism.

Brad Motter, Ph.D., Research Associate Professor. Visual neurophysiology; attention and search behavior.

Maxwell M. Mozell, Ph.D., Professor. Physiology of olfactory discrimination.

Eric Olson, Ph.D., Assistant Professor. Cellular and molecular mechanisms of cerebral cortex development.

Andras Perl, M.D., Ph.D., Adjunct Professor. Apoptosis; endogenous retroviruses; transaldolase; autoimmunity; cancer.

Arkadii Pertsov, Ph.D., Professor. Biophysical mechanisms of cardiac arrhythmias; fluorescence imaging; bioinformatics.

Rosemary Rochford, Ph.D., Associate Professor. Etiology of viral-associated malignancies; gammaherpesvirus pathogenesis.

Mark E. Schmitt, Ph.D., Associate Professor. Ribonucleoprotein assembly and biogenesis; mitochondrial RNA import; mRNA degradation; cell-cycle control.

M. Saeed Sheikh, M.D., Ph.D., Associate Professor. Apoptotic signal transduction and cancer biology.

Edward J. Shillitoe, B.D.S., Ph.D., Professor and Chair. Gene therapy for cancer.

Allen E. Silverstone, Ph.D., Professor. How dioxins and estrogens and estrogenic compounds affect the immune system.

Joseph A. Spadaro, Ph.D., Professor. Electromagnetic and mechanical regulation of bone physiology; skeletal growth and bone density.

Dennis J. Stelzner, Ph.D., Professor and Interim Chair. CNS regeneration, spinal cord injury research, and cortical plasticity after fetal ethanol exposure.

Nikolaus Szeverenyi, Ph.D., Professor. Magnetic resonance imaging; image analysis.

Steven M. Taffet, Ph.D., Professor. Regulation of intercellular communication in the heart; gene expression during macrophage activation.

Arthur Tatum, M.D., Associate Professor. Monoclonal antibody-mediated demyelination; molecular and cell biology of neural cell adhesion molecules.

Daniel Ts'o, Ph.D., Associate Professor. Neuronal mechanisms of visual perception, studied through physiological, anatomical, and functional imaging techniques.

Christopher E. Turner, Ph.D., Professor. Signaling to the cytoskeleton during cell adhesion and cell motility.

Mary Lou Vallano, Ph.D., Professor. Neuronal survival and development.

Richard D. Veenstra, Ph.D., Associate Professor. Regulation of connexin-specific gap junctions; gap-junction channel biophysics.

Karen L. Vikstrom, Ph.D., Research Assistant Professor. Molecular analysis of heart muscle disease. (Collaboration with Dr. José Jalife's group.)

Brent Vogt, Ph.D., Professor. Structure, functions, and pathologies of cingulate cortex.

Stephan Wilkens, Ph.D., Associate Professor. Structure and mechanism of membrane-bound transport proteins.

Richard J. H. Wojcikiewicz, Ph.D., Associate Professor. Intracellular signaling via InsP$_3$ receptors and the ubiquitin/proteasome pathway.

Steven Youngentob, Ph.D., Associate Professor. Olfactory neural plasticity; olfactory signal transduction; peripheral mechanisms of odorant quality coding.

STONY BROOK UNIVERSITY, STATE UNIVERSITY OF NEW YORK

School of Medicine with
Cold Spring Harbor Laboratory and Brookhaven National Laboratory
Medical Scientist Training Program

Program of Study	The School of Medicine admits approximately 100 medical students and 6 Medical Scientist Training Program (M.S.T.P.) students annually. The purpose of the M.S.T.P. is to train academic medical scientists for both research and teaching in medical schools and research institutions. The graduates of this program are equipped to study major medical problems at the basic level and, at the same time, to recognize the clinical significance of their pursuits and discoveries. The program has diversity and flexibility that permits the design of individualized educational opportunities to meet the needs of students with a variety of interests, educational backgrounds, and career goals. This diversity is fulfilled by the combined faculties of the Division of Biological Sciences and the School of Medicine, in conjunction with the Cold Spring Harbor Laboratory and Brookhaven National Laboratory. M.S.T.P. students typically take seven to eight years to complete the dual-degree program. Because the research requirements for this combined-degree program are not fulfilled in a time-dependent manner but rather based upon accomplishments in the laboratory, some variability in the time it takes to complete the program is anticipated.
Research Facilities	The primary training facility is Stony Brook University, State University of New York. Secondary facilities are the Cold Spring Harbor Laboratory and the Brookhaven National Laboratory. At Stony Brook, faculty members are drawn from three departments of the College of Arts and Sciences (A&S) and five departments of the Health Sciences Center (HSC), which includes the dental school. The three A&S departments as well as the Department of Microbiology of the HSC are housed in the Life Sciences Building, a seven-story structure with excellent facilities and equipment. The other HSC departments are situated nearby in the Health Sciences Center. The recently constructed 60,000-square-foot Center for Molecular Medicine (CMM) houses interdepartmental groups of faculty focused on structural biology, infectious disease, developmental genetics, and molecular neurobiology. The CMM is connected by a bridge to the Life Sciences Building, fostering connections to the groups housed there, and is designed to facilitate interlaboratory mixing and collaboration. Each floor of the CMM includes a small library; computer rooms for writing, data management, database retrieval, and graphics; both small and large conference rooms for discussions and meetings; and a kitchen/lunch room where trainees can interact in an informal setting. The laboratories have been designed with an open architecture. The functional focus of this design is enhancement of the interaction of personnel from various laboratories. A number of new research facilities are under construction, including a center for yeast genetics, a new building to house the rapidly growing program in bioengineering, and a Cancer Center. The Cold Spring Harbor Laboratory and Brookhaven National Laboratory are modern research facilities that provide special environments for trainees. The medical school and associated life sciences faculty at Stony Brook rank in the top 5 percent nationally for research productivity.
Financial Aid	Students admitted to the M.S.T.P. are covered for medical and graduate school tuition during their time in the program and receive a stipend. The stipend for year one of the program in 2006–07 is $25,000. Students who successfully compete for research fellowships in the later years receive a 10 percent supplement beyond that (i.e., to $27,500).
Cost of Study	Students in the program receive a full tuition waiver. All fees, including the cost of health insurance, are also paid. The expense of supplies, equipment, and books is approximately $2100 for the first year, with correspondingly lower expenses in succeeding years.
Living and Housing Costs	For 2006–07, a single student living on campus requires from $13,000 to $15,000 to cover normal living expenses for twelve months. All medical/graduate students, including married couples, can be accommodated in the residence halls. Garden apartments provide housing for nearly 1,000 students and their families. Off-campus apartments and houses can be found in the vicinity of the University. Information about housing can be obtained from the University's Off-Campus Housing Office or from the Health Sciences Center Office of Student Services.
Student Group	The total enrollment at the University is more than 20,000, including 7,000 graduate students and 100 new medical students per year. The M.S.T.P. has 40 to 50 students at various levels of training and representing a diverse pool recruited from across the nation regardless of geographical origin or New York State residency.
Location	Stony Brook is located in a region of coves, beaches, and small historic villages on the North Shore of Long Island, approximately 50 miles east of New York City. Traveling to the city is easy since the local commuter train (the Long Island Railroad) makes a stop right on campus. The area has retained its distinctive New England flavor and combines the charm of a rural setting minutes from the waters of Long Island Sound with proximity to the cultural, scientific, and industrial resources of the nation's largest city. The Cold Spring Harbor Laboratory is about 25 miles west, and Brookhaven National Laboratory about 18 miles east of Stony Brook.
The University	Stony Brook University was founded in 1957 at Oyster Bay, Long Island. In 1960 it was designated the State University's fourth University Center with a mandate to develop undergraduate and graduate programs in the humanities, sciences, social sciences, and engineering. In 1962, it moved to its present location at Stony Brook, where it has grown to a campus with eighty-one buildings on 1,100 acres, including the nineteen-story University Hospital. Beyond its research eminence, the University is well known for its biannual film festivals and the extensive series of musical and cultural events staged each year in its performance houses, the Staller and Wang Centers. The Cold Spring Harbor Laboratory, under its former director, Dr. J. D. Watson, and now under the current director, Dr. B. W. Stillman, has grown into one of the world's most important independent centers for biological research. The Brookhaven National Laboratory, one of the major government research centers in the country, is managed by Stony Brook and Battelle Memorial Institute and carries out important research in all areas of science and engineering.
Applying	Applicants to Stony Brook's School of Medicine may apply to the M.S.T.P. by submitting the appropriate application form provided with the medical school secondary application. Applications to the M.S.T.P. are screened by the Steering Committee, and applicants of interest are invited for a formal interview with members of the M.S.T.P. Committee and the Medical School Admissions Committee on the same day.
Correspondence and Information	Director, M.S.T.P. School of Medicine, Health Sciences Center Stony Brook University, State University of New York Stony Brook, New York 11794-8651 Phone: 631-444-3219 Fax: 631-444-6229 E-mail: carron@pharm.sunysb.edu Web site: http://www.pharm.sunysb.edu/mstp

Stony Brook University, State University of New York

TRAINING PROGRAMS

The faculty members of each program and their particular areas of research are described in detail in separate program listings elsewhere in Peterson's Graduate and Professional Programs series. The following overview is provided for the convenience of the reader. More information about each of these programs can be obtained by writing to them directly or online (http://www.pharm.sunysb.edu/mstp/gradprograms.htm). Correspondence should be addressed to: Director, (specific graduate program), Stony Brook University, State University of New York, Stony Brook, New York 11794.

Anatomical Sciences.
Biochemistry and Structural Biology.
Biomedical Engineering.
Cellular and Molecular Pathology.
Chemistry.
Genetics.
Molecular and Cellular Biology.
Molecular and Cellular Pharmacology.
Molecular Genetics and Microbiology.
Neuroscience.
Physics and Astronomy.
Physiology and Biophysics.
Psychology.
Sociology.

Major Areas of Faculty Research Interest

AIDS, including molecular biology, virology, pharmacology, and epidemiology.
Biomedical engineering, including stem cell therapy, nanotechnology, biomaterials/biomechanics, biosensors, bionanotechnology, medical instrumentation and imaging, tissue engineering, and biomedical modeling.
Cancer biology.
Cell structure and function, including organelle biogenesis, cell cycle, differentiation, transformation, and extracellular matrix.
Cellular immunology.
Computer graphics and molecular modeling.
Developmental biology and genetics.
Endocrinology, including programs in diabetes and obesity.
Enzymology.
Genetic toxicology.
Glycobiology.
Lyme disease, bacterial pathogenicity.
Mammalian vectors and gene therapy.
Membrane biology and biophysics, including ion channels, ion transport, excitability, and cell-cell communication.
Molecular biology and molecular genetics, including DNA replication, transcription, regulation of gene expression, protein synthesis and transport, oncogenes, molecular immunology, and hematology.
Molecular cardiology.
Molecular virology, with emphasis on SV40, adenovirus, adeno-associated virus, poliovirus, retroviruses, HIV, and poxvirus.
Neurobiology, cellular, molecular, and systems.
Neuropharmacology.
Signal transduction.
Structural biology and physical biochemistry of proteins and nucleic acids.
Vertebrate morphology, including systematics, paleoprimatology, and paleoanthropology.
Vision research.

SYRACUSE UNIVERSITY

Department of Biology

Programs of Study

Graduate programs leading to either the M.S. or Ph.D. degree are offered in a wide variety of disciplines within the areas of biochemistry, molecular biology, cell biology, ecology, and evolution. Both degrees emphasize individual research, supplemented by training in formal course work and seminars. Students' programs are individually designed, in consultation with their research adviser and a research committee, to reflect their particular background and interests. Students may rotate through several research laboratories to help them select a research topic and adviser. Courses may be taken in other departments of the University or in the State University of New York Health Science Center or College of Environmental Sciences and Forestry, both of which are adjacent to the campus. Experts from these institutions may also serve on the student's research committee. Because of the various formal and informal collaborative training and research programs, an exceptionally rich variety of stimulating opportunities is available for graduate research and training.

The time needed to complete the programs is approximately three years for the M.S. and five years for the Ph.D. Qualification for the Ph.D. degree requires the student to pass an oral defense of a selected research topic. The successful completion of a student's degree requires the writing and oral defense of a thesis (M.S.) or dissertation (Ph.D.) based on an independent research project.

An excellent seminar program featuring visiting biologists provides ongoing information about current research at other institutions. The biology department sponsors an annual conference at which graduate students have an opportunity to present their work.

Research Facilities

Research facilities currently include an AAALAC-accredited animal facility and extensive facilities and instrumentation for carrying out most kinds of modern biological research at the molecular, cellular, organismal, and population levels. Extensive library holdings and computing facilities are readily accessible for student and faculty use. Construction is underway for the new life sciences complex. The 210,000-square-foot building, the University's largest, most ambitious construction project, will bring the biology, chemistry, and biochemistry departments under one roof for the first time in the University's history. The complex is set to open in fall 2008.

Financial Aid

Virtually all departmental graduate students are supported by full tuition scholarships, as well as teaching assistantships, research assistantships, or fellowships that pay competitive stipends. These include fellowship awards, guaranteed for a three-year period, to the most outstanding students. Advanced students may also compete for University funds that are distributed annually and can be used to facilitate their research. Supplemental summer fellowships are available for exceptionally well qualified students in the Ph.D. program.

Cost of Study

Tuition for graduate students is estimated at $940 per credit hour for the 2006–07 academic year.

Living and Housing Costs

Graduate student housing is available on and off campus. Numerous apartments and rooms located in the University area are within walking distance to campus or are accessible by the free campus bus service.

Student Group

The department currently has approximately 40 graduate students, mostly from the eastern United States, but many other states and several countries are also represented. An active departmental Graduate Student Organization sponsors various social and athletic events.

Student Outcomes

Departmental graduates who wish to pursue academic careers commonly obtain postdoctoral fellowships at the leading universities and research centers nationwide. Departmental alumni hold positions in academic institutions, government and clinical laboratories, and the private sector.

Location

Syracuse has approximately 200,000 inhabitants, and the population of the metropolitan area exceeds half a million. Located in the center of New York State in one of the most beautiful regions of the country, Syracuse has most of the cultural advantages of a large city, and recreational facilities for all seasons are within easy reach. A major symphony orchestra; numerous other musical groups performing chamber music, opera, and choral music; a professional repertory theater; the ballet company; and two art museums provide a full schedule of events year-round, featuring both local and visiting artists. The University's 50,000-seat Carrier Dome is used for University events, such as football and basketball games, and also for occasional concerts. Residents enjoy the seasonal variation, with mild springs, warm summers, crisp falls with spectacular colors, and cold winters with much snow. Within less than an hour's drive there are numerous opportunities for exceptional recreational activities, such as downhill and cross-country skiing, camping, fishing, boating, and hiking. Syracuse has an easily accessible airport that is within an hour's air travel of the major cities of the Northeast. There are also excellent interstate highways and a railway connection.

The University

Syracuse University, a nonsectarian, fully accredited, private institution, was founded in 1870. It consists of thirteen schools and colleges. On the main campus there are 11,455 full-time undergraduates, 3,256 full-time graduate students, and 864 full-time faculty members. Both the State University of New York Upstate Medical University and State University of New York College of Environmental Science and Forestry border the campus.

Applying

For fall admission and consideration of financial support, preference is given to completed applications received by February 1. Later applications and spring admissions are considered on a space available basis. All applicants are required to provide a personal statement, transcripts from college-level institutions attended, three letters of recommendation, and scores on the General Test of the Graduate Record Examinations. TOEFL scores are required of all applicants whose native language is not English. Applications from women and members of minority groups are strongly encouraged. Prospective students are encouraged to correspond directly with faculty members in their areas of interest.

Correspondence and Information

Graduate Program Secretary
Department of Biology
Syracuse University
108 College Place, 122 Lyman Hall
Syracuse, New York 13244-1220

Phone: 315-443-9154
Fax: 315-443-2156
E-mail: biology@syr.edu
Web site: http://biology.syr.edu

Syracuse University

THE FACULTY AND THEIR RESEARCH

R. Craig Albertson, Assistant Professor; Ph.D., New Hampshire, 2002. Genetic and developmental basis of craniofacial evolution.

David M. Althoff, Assistant Research Professor; Ph.D., Washington State, 1998. Species interactions; insect community ecology; molecular ecology.

John M. Belote, Professor; Ph.D., North Carolina at Chapel Hill, 1979. Genetics and molecular biology of *Drosophila* proteasomes. The proteasome is a major component of the cellular machinery that is responsible for the selective breakdown of unwanted proteins. The goal is to understand the details of the proteasome's structure, regulation, and function and to elucidate its roles during the development of a multicellular organism.

Brian R. Calvi, Associate Professor; Ph.D., Harvard, 1993. Cell cycle control of DNA replication and genome stability.

Samuel H. P. Chan, Professor; Ph.D., Rochester, 1970. Bioenergetics, with particular interest in the mechanism by which subcellular organelles such as mitochondria convert nutrients into ATP for cellular activities. The lab is studying membrane-bound enzyme complexes, including cytochrome oxidase, adenine nucleotide translocase, and ATPase from normal and pathological tissues.

Michael S. Cosgrove, Assistant Professor; Ph.D., Syracuse, 1998. Structural biochemistry of proteins involved in the epigenetic regulation of chromatin structure.

Scott E. Erdman, Associate Professor; Ph.D., California, Davis, 1994. Cellular and molecular mechanisms in eukaryotes that underlie cell shape change and cell adhesion in response to extracellular signals; functional genomic studies of mechanisms pertaining to lipid and membrane homeostasis in eukaryotes. These questions are being approached in the model system, baker's yeast, through studies using a variety of biochemical, genetic, and light microscopy techniques.

Douglas A. Frank, Associate Professor; Ph.D., Syracuse, 1990. Plant and ecosystem ecology. Research examines native ungulate regulation of grassland ecosystems. Scope includes grazer effects on grassland species composition and animal-plant-microbe energy and nutrient webs.

Jason Fridley, Assistant Professor; Ph.D., North Carolina, 2002. Plant ecology.

Anthony Garza, Assistant Professor; Ph.D., Texas A&M, 1995. Biofilm formation; bacterial development; stress resistance in bacteria.

F. Reed Hainsworth, Professor; Ph.D., Pennsylvania, 1968. Comparative animal physiology and behavior. The research focuses on testing predictions of models of organism performance concerned with foraging, temperature regulation, movement, and their interactions.

Richard L. Hallberg, Professor; Ph.D., Johns Hopkins, 1968. The role of phosphatases in stress regulation. Translational thermotolerance. The role of molecular chaperones in mitochondrial biogenesis.

Eleanor M. Maine, Associate Professor; Ph.D., Princeton, 1984. Genetic and epigenetic control of development in the nematode, *Caenorhabditis elegans*. A variety of genetic and molecular approaches are used to study how germ cell fate is regulated, particularly focusing on inductive cell-signaling and epigenetic mechanisms that regulate germ cell proliferation and differentiation of gametes.

Melissa E. Pepling, Assistant Professor; Ph.D., SUNY at Stony Brook, 1995. Genetics and cell biology of early oocyte development; immunocytochemistry coupled with confocal microscopy used to analyze sterile mouse mutants that disrupt the early development of oocytes.

Scott Pitnick, Associate Professor; Ph.D., Arizona State, 1992. Reproductive behavior, morphology, and physiology; evolution of reproductive and life history strategies; diversification, and speciation.

Ramesh Raina, Associate Professor; Ph.D., Jawaharlal Nehru (New Delhi), 1991. Genetic and molecular mechanisms regulating plant-pest interactions; functional genomics of cell signaling in plants.

Mark E. Ritchie, Professor; Ph.D., Michigan, 1987. Plant-animal interactions and community ecology. Emphasis on how insect and mammalian herbivores interact with grassland plant communities and ecosystems and how the species diversity of these groups changes across space and continental environmental gradients. Research employs both mathematical, field experimental, and data synthesis methods in Minnesota, Utah, and South Africa.

John M. Russell, Professor; Ph.D., Utah, 1971. Ionic homeostasis and cell biology; causes and roles of effects of human cytomegalovirus on host cell ionic transport processes; basic mechanisms, regulation, and protein trafficking of the coupled Na,K,Cl cotransporter.

William T. Starmer, Professor; Ph.D., Arizona, 1972. Ecological genetics of microorganisms and insects. The lab investigates the reasons for general and specific associations of yeasts, *Drosophila,* and their host plants; the coadaptation of yeasts and their insect vectors; and the community organization of yeast, including widespread geographic studies of "killer" yeasts.

J. Albert C. Uy, Assistant Professor; Ph.D., Maryland, College Park, 2000. Behavioral ecology; sexual selection; animal communication and signal evolution; evolution of premating isolation.

Roy D. Welch, Assistant Professor; Ph.D., Wisconsin–Madison, 1997. Behavioral genetics of self-organization in bacterial biofilms.

Larry L. Wolf, Professor; Ph.D., Berkeley, 1966. Behavior and ecology. Behavioral investigations, including studies of mating systems, foraging, aggression, and life history traits, are aimed at elucidating proximate and ultimate environmental influences and their interactions with the characteristics of the organism. The group also investigates the role of behavior in community organization.

TEMPLE UNIVERSITY
of the Commonwealth System of Higher Education

College of Science and Technology
Department of Biology

Programs of Study

The department offers Ph.D. and M.S. programs in biology. On the basis of faculty research interests, the department is divided into three areas: molecular biology, genetics, and biochemistry; physiology, neurobiology, and behavior; and cell and developmental biology. Students take courses in all three areas but specialize in one of them.

Students admitted to the Ph.D. program usually require five years of full-time study to complete degree requirements, which include satisfactory completion of a core curriculum (a course in biochemistry, a laboratory preparation course, and one course in each of the three areas described above) and participation in five graduate seminars. In addition, all Ph.D. candidates must teach undergraduate laboratory sections for at least two semesters. During the first two years, students are expected to complete their core requirements and embark on a laboratory research project leading to a Ph.D. dissertation. In the spring of the second year, each student takes a preliminary examination, in which he or she prepares a written description of the research project and orally defends the project before members of the appropriate area committee. Students who pass the preliminary exam are formally admitted to Ph.D. candidacy. The Ph.D. is awarded when the dissertation has been approved by the thesis committee after a public defense.

Students admitted to the M.S. program normally take two years of full-time study to complete degree requirements, which include satisfactory completion of the core curriculum described above plus one graduate seminar. A minimum of 24 credit hours are required for the M.S. degree. In addition, students must do laboratory research that leads to a master's thesis. The M.S. is awarded when the research has been defended orally before members of the appropriate area committee.

Research Facilities

The department is housed in the Biology–Life Sciences Building on Temple's main campus. Research laboratories occupy the top two floors of the building. Specialized equipment available for graduate student research includes transmission and scanning electron microscopes, a confocal microscope, ultracentrifuges, HPLCs, scintillation counters, and computers for molecular modeling. Laptop personal computers are available for student use. The departmental library subscribes to 250 current journals and fifty-eight continuations and provides a network system for accessing electronic databases. Darkrooms, animal-care facilities, seawater tank rooms, a machine shop, and greenhouses are also located in the building. The Biology Life Sciences Building also houses the Center for Neurovirology and Cancer Biology, a 20,000-square-foot state-of-the-art biomedical research facility where investigators study the molecular pathogenesis of viral- and nonviral-induced disorders of the central nervous system.

Financial Aid

The department attempts to provide financial support for all qualified graduate students, although first-year international students are not supported by the department unless they can demonstrate an excellent command of English. Most students are supported by teaching assistantships. For students who entered in 2005–06, teaching and research assistantships carried a nine-month academic-year stipend of $14,535 and tuition remission. Summer stipends are available for students in the Ph.D. and M.S. programs who are doing laboratory research.

The Graduate School awards Presidential Fellowships (annual stipend of $20,000 in 2006–07) and University Fellowships (annual stipend of $18,000 in 2003–04) on a competitive basis. All Graduate School fellowships also carry a tuition remission. A limited number of fellowships for minority candidates are available. In addition, some graduate students are supported by grants held by faculty members.

Cost of Study

Temple is a state-related university. In 2003–04, tuition was $402 per credit for Pennsylvania residents and $582 per credit for nonresidents.

Living and Housing Costs

Furnished and unfurnished University-sponsored apartments are available in the vicinity of the campus. Rent for a one-bedroom unfurnished apartment was $543 per month in 2002–03. Rentals off campus may be more expensive. The University Housing Office provides a list of currently available housing.

Student Group

There are 32 graduate students in the department, a majority of whom are working toward the Ph.D. degree.

Location

Philadelphia is the sixth-largest city in the country, with a metropolitan population of more than 5 million. The city is famous for its restaurants, parks, and history. In addition, it offers a wide variety of cultural attractions, including art museums, a world-class symphony orchestra, chamber music, dance, theater, art film houses, and jazz clubs. Philadelphia is home to professional teams in baseball, football, ice hockey, basketball, and lacrosse. Within a 2- to 3-hour drive, one can ski, hike, or fish in the Pocono Mountains, swim in the Atlantic Ocean, or canoe in the streams that wind through the Pine Barrens in south-central New Jersey.

The University

Temple University has a total enrollment of approximately 30,000 students. It has developed with the ideal of "educational opportunity for the able and deserving student of limited means." With a rich heritage of social purpose, Temple seeks to provide education of high quality without regard to race, creed, or station in life. Affiliation with the System of Higher Education of the Commonwealth of Pennsylvania enhances Temple's role as a public institution.

Applying

Although there are no strict course requirements, applicants should have a background in modern biology, preferably including genetics and cell biology. Full-year courses in general and organic chemistry, physics, and calculus are strongly recommended. In addition, applicants must take the General Test of the Graduate Record Examinations; international applicants whose native language is not English should take the TOEFL.

Application forms may be obtained from the address given in this In-Depth Description. For September admission, the deadline is February 1 for applicants who wish to be considered for Graduate School fellowships; applicants who wish to be considered for other types of financial aid must submit their applications by April 1. For January admission, the deadline is November 1.

Applicants are evaluated on the basis of their academic record, letters of recommendation, and their scores on the GRE. Notification of admission is made as soon as the application is evaluated.

Correspondence and Information

Chairperson, Graduate Committee
Department of Biology
Temple University
Philadelphia, Pennsylvania 19122
Phone: 215-204-8854

Temple University

THE FACULTY AND THEIR RESEARCH

Molecular Biology, Genetics, and Biochemistry

Shohreh Amini, Professor; Ph.D., Pennsylvania, 1983. Molecular pathogenesis of HIV-1–induced neurological disease and gene delivery to the central nervous system (CNS).

Frank N. Chang, Professor; Ph.D., Wisconsin, 1968. Studies on protein-protein interactions and functional proteomics; identification and function of protein allergens associated with asthma.

Antonio Giordano, Professor; M.D./Ph.D., Naples (Italy), 1990. The family of retinoblastoma proteins and cyclin-dependent kinases in the cell cycle and cancer.

Kamel Khalili, Professor; Ph.D., Pennsylvania, 1983. Molecular biology and genetics of CNS diseases, with emphasis on viral-induced cancer and neurodegeneration.

Karen Palter, Associate Professor; Ph.D., Princeton, 1978. Genetics and molecular biology of *Drosophila* sialic acid pathway genes.

Harry P. Rappaport, Professor; Ph.D., Yale, 1956. Synthetic complementary base pairs in DNA and RNA.

Jay Rappaport, Professor; Ph.D., Pennsylvania, 1986. HIV vaccine development and pathogenesis of HIV in CNS.

Richard B. Waring, Associate Professor; Ph.D., Essex (England), 1978. RNA processing and catalytic RNA.

Physiology, Neurobiology, Behavior, and Ecology

Edward R. Gruberg, Professor; Ph.D., Illinois, 1969. Electrophysiology; anatomy; behavior; vertebrate vision.

Ee Lin Lim, Assistant Professor; Ph.D., MIT, 1997. Molecular detection and taxonomy of aquatic microorganisms; diversity and ecology of protists.

Stuart E. Neff, Professor; Ph.D., Cornell, 1960. Aquatic biology; insect behavior.

Shepherd K. Roberts, Professor; Ph.D., Princeton, 1959. Circadian rhythms.

Robert W. Sanders, Associate Professor; Ph.D., Georgia, 1988. Aquatic ecology; ecology of protists; plankton; microbiology.

Jacqueline Tanaka, Associate Professor; Ph.D., Illinois, 1981. Ion channels; retina; protein structure-function.

Cell and Developmental Biology

Karen Palter, Associate Professor; Ph.D., Princeton, 1978. Genetics and molecular biology of *Drosophila* sialic acid and pathway genes.

Joel B. Sheffield, Professor; Ph.D., Chicago, 1969. Cell-cell interactions in retinal development.

Richard W. Weisenberg, Professor; Ph.D., Chicago, 1968. Microtubules and axonal transport.

RECENT REPRESENTATIVE FACULTY PUBLICATIONS

Chang, F. N., et al. A simple method for assigning multiple immunogens to their protein on a two-dimensional blot and its applications to asthma-causing allergens. *Electrophoresis* 22:2098–102, 2001.

Chang, F. N. A new broad-spectrum protease inhibitor from the entomopathogenic bacterium *Photorhabus luminescens. Microbiology* 146:3141–7, 2000.

Macaluso, M., et al. **(A. Giordano).** pRb2/p130-E2F4/5-HDAC1-SUV39H1-p300 and pRb2/p130-E2F4/5-HDAC1-SUV39H1-DNMT1 multimolecular complexes mediate the transcription of estrogen-receptor-α in breast cancer. *Oncogene* 22:3511–17, 2003.

Gruberg, E. R., and E. A. Dudkin. Nucleus isthmi enhances calcium influx into optic nerve fiber terminals in *Rana pipiens. Brain Res.* 969:44–52, 2003.

Chipitsyna G., B. E. Sawaya, **K. Khalili,** and **S. Amini.** Cooperativity between Rad51 and C/EBP family transcription factors modulates basal and Tat-induced activation of the HIV-1 LTR in astrocytes. *J. Cell Physiol.* 207(3):605–13, 2006.

Darbinian, N., M. K. White, and **K. Khalili.** Regulation of the Pur-alpha promoter by E2F-1. *J. Cell Biochem.,* 2006.

White M. K., T. S. Gorrill, and **K. Khalili.** Reciprocal transactivation between HIV-1 and other human viruses. *Virology* 352(1):1–13, 2006.

Khalili, K., and M. K. White. Human demyelinating disease and the polyomavirus JCV. *Mult. Scler.* 12(2):133–42, 2006.

Lim, E. L., A. V. Tonnita, W. G. Thilly, and M. F. Polz. Combination of competitive quantitative PCR and constant denaturant capillary electrophoresis for high resolution detection and enumeration of microbial cells. *Appl. Environ. Microbiol.* 67(9):3897–903, 2001.

Neff, S. E., and E. J. Bacon. Bottom fauna in Doe Valley Lake, Meade County, Kentucky. *Trans. Ky. Acad. Sci.* 4393–4:158–67, 1982.

Palter, K., et al. Expression of a functional *Drosophila melanogaster* N-acetylneuraminic acid (Neu5Ac) phosphate synthase gene: Evidence for endogenous sialic acid biosynthetic ability in insects. *Glycobiology* 12(1):1–11, 2002.

Darbinian-Sarkissian, N., et al. **(J. Rappaport, K. Khalili,** and **S. Amini).** Dysregulation of NGF-signaling and Egr-1 expression by Tat in neuronal cell culture. *J. Cell Physiol.* 208(3):506–15, 2006.

Abraham S., et al. **(J. Rappaport, K. Khalili,** and **S. Amini).** Cooperative interaction of C/EBP beta and Tat modulates MCP-1 gene transcription in astrocytes. *J. Neuroimmunol.* 160(1–2):219-27, 2005.

Rappaport, H. P. Synthesis of 9-(2-deoxy-β-D-ribo furanosyl) purine-2-thione. *Nucleosides Nucleotides* 19(4), 2000.

Sanders, R. W., et al. Nutrient acquisition and population growth of a mixotrophic alga in axenic and bacterized cultures. *Microbial Ecol.* 42:513–23, 2001.

Claudio P. P., et al. **(J. B. Sheffield, A. Giordano, K. Khalili,** and **S. Amini).** Cdk9 phosphorylates p53 on serine 392 independently of CKII. *J. Cell Physiol.* 208(3):602–12, 2006.

Sheffield, J. B. Creating pseudocolor images using ImageJ. *Microscopy Today* 14:52, 2006. (Invited Paper)

Zhou, J., et al. **(J. Sheffield).** Cloning and characterization of angiocidin, a tumor cell binding protein for thrombospondin-1. *J. Cell Biochem.* 92:125–46, 2004.

Sheffield, J. B., et al. The use of Avidin as a probe for the distribution of mitochondrial carboxylases in developing chick retina. *J. Histochem. Cytochem.* 46(2):177–82, 1998.

Tanaka, J. Structure-function study of photoreceptor cyclic nucleotide-gated channels. *Biophys. J.* 78:2321–33, 2000.

Waring, R. B., and W. J. Geese. A comprehensive characterization of a group IB intron and its encoded maturase reveals that protein-assisted splicing requires an almost intact intron RNA. *J. Mol. Biol.* 308:609–22, 2001.

Weisenberg, R. B. Microtubule gelation-contraction: Essential components and relation to slow axonal transport. *Science* 238:1119–22, 1987.

TENNESSEE STATE UNIVERSITY

Department of Biological Sciences

Programs of Study	The Department of Biological Sciences at Tennessee State University (TSU) offers graduate training to prepare scholars for the pursuit of research careers in academia, government, and industry and to improve the level of competency of high school and college teachers. The program leads to the degrees of Master of Science in biology and Doctor of Philosophy in biological science. The major areas of specialization are neurophysiology, agricultural biotechnology, and cell and molecular biology. The research interests of individual faculty members are listed in a separate section.

The curriculum of the Ph.D. program, which consists of formal course work and dissertation research, is designed to be completed in four years. The master's degree program takes two years to complete and can be achieved with or without thesis research. Students entering this degree program are encouraged to select the thesis option. The master's degree with thesis consists of a total of 36 semester hours of course work, including the thesis (4 semester hours). The nonthesis degree requires 39 semester hours of course work plus a nonthesis project.

Students select their program of study with the assistance of a faculty adviser and advisory committee. The program has flexibility and consists of a core of required courses and electives, allowing students to pursue their individual professional goals. In addition, graduate students are expected to gain teaching experience as assistants in the Department of Biological Sciences laboratory and lecture courses.

Research Facilities

The Department of Biological Sciences' graduate program is housed in two buildings located on the main campus of TSU—Harned and McCord Halls. These buildings are equipped with modern scientific facilities and equipment, such as electron, confocal, and fluorescent microscopes; tissue culture facilities; ultra- and high-speed centrifuges; research-grade spectrophotometers; high-performance liquid chromatographs; dark and cold rooms; PCR and electroporation machines; and an automated DNA sequencing system. The research faculty members enjoy fully functional research laboratories within which their graduate students conduct their research projects. Also available to graduate students is a landscape office with individual cubicles and minor storage space.

Financial Aid

Financial aid for doctoral and master's degree students is available through graduate research and teaching assistantships. The assistantships for doctoral students provide tuition and fees plus a stipend of $18,000 per year. Assistantships for master's degree students provide $11,205, including tuition and fees.

Cost of Study

Tuition and general fees for the 2005–06 academic year were $5672 for Tennessee residents and $14,984 for nonresidents.

Living and Housing Costs

The University does not provide housing for graduate students; however, some students have applied for and received housing in undergraduate dormitories. Most of the graduate students seek housing in the Nashville metropolitan area, where there are numerous apartments, rooms, and houses for rent or purchase. Apartment rent is comparable to that in Southeastern cities of similar size and ranges from $475 per month to $750 per month for a fashionable facility. The extremities of the Nashville metropolitan area are within 15 to 20 minutes of the TSU campus.

Student Group

There are approximately 1,800 students enrolled in various graduate programs at TSU, 34 who are seeking degrees in the programs of the Department of Biological Sciences. The group is rather diverse and consists of 64 percent women, 52 percent minority (African American, Hispanic, and Native American), and 13 percent international students. The majority of the department's students receive some form of financial aid from the University, and many of the master's-degree students work part-time in the Nashville metropolitan area.

Location

Tennessee State University is located in Nashville, Tennessee, known as Music City USA, a major metropolitan area in middle Tennessee that has a population of more than 1 million people. In addition to being a preeminent educational center for the region, major industries include tourism, printing and publishing, music production, finance, insurance, automobile manufacturing, and health-care services. Thus, the community offers numerous recreational and social programs that supplement the artistic, cultural, recreational, and scientific activities offered by the various universities in the area.

The University

Tennessee State University is a comprehensive 1890 land-grant institution founded in 1912. The University provides programming in agriculture, allied health, arts and sciences, business, education, engineering and technology, home economics, human services, nursing, and public administration. There are two campuses—the main campus, which occupies 450 acres, and a modern four-story teaching, research, and conference complex, which is located in downtown Nashville.

Applying

Admission to the graduate program in biological sciences requires a baccalaureate degree, normally with a major in the sciences, and acceptable Graduate Record Examinations (GRE) scores (verbal and quantitative). In addition, the student must meet all other requirements as specified by the School of Graduate Studies. The department recommends that all students take the GRE Subject Test, and all students seeking admission into the doctoral program must submit three letters of recommendation and a personal statement discussing academic and career goals, interests, and related experiences in the sciences. For online applications, students should visit the University's Web site at http://www.tnstate.edu/, click Admissions, and then click Apply Online. Students may then click on the link to the Graduate Studies Web site, where a Graduate School Online Application link is available.

Correspondence and Information

Graduate Admissions Committee
Department of Biological Sciences
Tennessee State University
P.O. Box 9536
3500 John Merritt Boulevard
Nashville, Tennessee 37209-1561
Web site: http://www.tnstate.edu

Tennessee State University

THE FACULTY AND THEIR RESEARCH

BIOLOGICAL SCIENCES FACULTY

Mary Ann Asson-Batres, Associate Professor; Ph.D. (biochemistry), Oregon Health Sciences. Role of vitamin A in neurogenesis.

M. Ann Blackshear, Associate Professor; Ph.D. (pharmacology), Meharry Medical College. The role of brain dopamine receptors in the behavioral effects of psychostimulant and psychoactive drugs and eating behavior.

Carolyn B. Caudle, Associate Professor; Ph.D. (microbiology), Meharry Medical College. Development of biotechnical methods for the control of *Salmonellae*.

Anthony O. Ejiofor, Associate Professor; Ph.D. (microbiology), Nigeria. Phylogeny and community structure of prokaryotes in extreme environments and dynamics of groundwater pollution.

Philip F. Ganter, Associate Professor; Ph.D. (ecology), North Carolina. The evolution and ecology of natural yeast communities.

Michael T. Ivy, Associate Professor; Ph.D. (physiology and biophysics), Illinois at Chicago. Cholinergic neurochemistry in the area of integral membrane protein traffic and the role of trafficking in the physiological regulation of the choline cotransporter.

Terrance L. Johnson, Department Chair and Professor; Ph.D. (microbiology), North Texas. Understanding the interactions between microbes and the biotic and abiotic components of various polluted environments.

Prem S. Kahlon, Professor; Ph.D. (plant breeding and genetics), LSU. Plant tissue culture and evaluation of plant germplasm for biological stress.

Elaine Martin, Assistant Professor; Ph.D. (secondary education–biology), Alabama at Birmingham. Distance education; evaluation and assessment of student achievement in science.

Elbert L. Myles Jr., Professor; Ph.D. (genetics), Arizona. Analysis of growth inhibition by secondary compounds and regeneration of medicinal plants.

Robert F. Newkirk, Professor; Ph.D. (mammalian physiology), Colorado State. Physiological regulation of the choline cotransporter and molecular mechanisms that undergird its trafficking into and out of the nerve terminal membrane.

John T. Robinson Jr., Assistant Professor; Ph.D. (cell and developmental biology), North Carolina at Chapel Hill. Developmental function and regulation of cytoskeletal proteins and associated transport complexes in *Drosophila*.

Xiaofie Wang, Assistant Professor; Ph.D. (molecular and genome giology), Hong Kong. Application of the genomic approach in the studies of circadian physiology and lipid homeostasis.

Benny Washington Jr., Associate Professor; Ph.D. (biochemistry), Atlanta. Physiological role of histamine in cardiovascular regulation and cardiovascular response to acute and chronic ethanol exposure.

AGRICULTURAL SCIENCES FACULTY

Desh R. Duseja, Professor; Ph.D. (soil chemistry and biometrology), Utah State. Soil/water-pesticide interactions/movements and soil/water quality.

Constantine L. Fenderson, Professor and Interim Dean, School of Agriculture and Consumer Sciences; Ph.D. (animal nutrition), Michigan State. Nutrient metabolism and utilization in ruminants.

Makonnen W. Lema, Associate Professor; Ph.D. (animal nutrition), Oklahoma State. Nutrient utilization, nutritional energetics and nutrient partitioning in livestock as influenced by genetic modification, ambient temperature distress and nutrient consumption level.

Surendra P. Singh, Professor; Ph.D. (agricultural economics and statistics), Penn State. Resource economics and regulations.

THE TEXAS A&M UNIVERSITY SYSTEM HEALTH SCIENCE CENTER

Institute of Biosciences and Technology

Program of Study

The Institute of Biosciences and Technology (IBT) offers a program of study that leads to a Ph.D. degree in biomedical sciences. Incoming students learn the guiding principles of how to conduct biomedical research. An individually tailored curriculum is designed to ensure that each student acquires the necessary theoretical background and appropriate knowledge and skills. Through frequent interaction with fellow students, postdoctoral fellows, and faculty members, the students share in the excitement of making scientific discoveries and learn to design and develop successful research programs. The development of the student is closely monitored during the course of the program.

The IBT is organized around focused centers, which include the Center for Cancer Biology and Nutrition (CCBN), Center for Environmental and Genetic Medicine (CEGM), Center for Extracellular Matrix Biology (CEMB), Center for Genome Research (CGR), Center for Molecular Development and Diseases, and the W. M. Keck Center for Informatics. Areas of faculty research expertise within these centers include protein structure and function; RNA and DNA structure and function; molecular basis of genetic disease; structure-function relationships between growth factors and their receptors; elucidation and regulation of signal transduction cascades; transcription factors and growth and development; environmental and genetic impact on fetal growth and development; biosynthesis, function, and organization of the extracellular matrix; developmental and genetic aspects of congenital heart disease; and vaccine development against pathogenic bacteria.

Research groups at the IBT utilize a variety of experimental approaches and systems for solving questions of biomedical importance. These include protein engineering, X-ray crystallography, NMR spectroscopy, enzymology, molecular biology, microscopy, molecular genetics, conventional and conditional gene knockouts in the mouse and the development of transgenic mouse lines, mouse developmental biology, human genetics, and yeast genetics.

Research Facilities

The IBT occupies an eleven-floor research building, which was completed in 1991, in the Texas Medical Center in Houston. The IBT is a modern research facility with state-of-the-art equipment resources, in addition to spacious laboratories and transgenic facilities. The IBT is a member of the Texas Medical Center, along with Baylor College of Medicine, the University of Texas–Houston Health Science Center, and the M. D. Anderson Cancer Center. The IBT shares the Texas Medical Center Library, which is centrally located within the Texas Medical Center.

Financial Aid

Graduate assistantships are available, currently with stipends of $22,000 per year.

Cost of Study

The total cost of tuition and fees is approximately $4400 per fiscal year. Tuition payments are made by the IBT, as well as other insurance options that are available to graduate students.

Living and Housing Costs

The Texas Medical Center is adjacent to excellent neighborhoods that offer affordable housing opportunities. A typical one-bedroom apartment or shared apartment rents for between $500 and $800 per month.

Student Group

The IBT has approximately 50 Ph.D students, including students from various departments on the Texas A&M College Station campus, the University of Texas–Houston Health Science Center Graduate School of Biomedical Sciences (GSBS), and international institutions. The IBT graduate students are part of the Texas A&M University System Health Science Center Graduate School of Biomedical Sciences and interact with fellow students within the GSBS program at College Station, Baylor College of Dentistry in Dallas, and the School of Rural and Public Health in College Station through yearly graduate student symposia and other organized group events. An umbrella Graduate Student Organization governs the IBT's graduate student population.

Location

Houston, the country's fourth-largest city, is located close to the Texas gulf coast and has a tropical climate most of the year. Houston is a cosmopolitan city with superb cultural assets, including world-class theater, opera, symphony, and ballet and wonderful and varied museums. Sports fans can attend major league baseball (Houston Astros), basketball (Houston Rockets and the Comets), American football (Houston Texans), or ice hockey (Houston Aeros) games. Houston is a major venue for concerts in all areas of music and is proud to offer outstanding restaurants reflecting the city's cultural and ethnic diversity. Houston is also the home of the annual Houston Rodeo and Livestock Show, one of the largest of its kind in the U.S. Houston's proximity to Galveston and the Gulf of Mexico (45 miles) makes the beach quite accessible, as is ocean fishing. Other outdoor opportunities within Texas include the cypress swamps and Big Thicket State Park of eastern Texas, the hill country of central Texas, dramatic canyons, and the high desert and mountains of Big Bend National Park.

The University and The Institute

Texas A&M University, founded in 1876, is the oldest public institution of higher learning in the state. The Texas A&M University campus in College Station, 90 miles north of Houston, has roughly 50,000 undergraduates and more than 7,000 graduate students. IBT, with the College of Medicine in College Station and the Baylor Dental School in Dallas, is part of the Texas A&M University System Health Science Center.

Applying

Applications for entry in the fall semester are requested to arrive no later than February 1. Typically, students are notified of their acceptance shortly after March 1 and are expected to respond by April 1 at the latest. Students applying to the IBT should have a strong undergraduate background in biology, biochemistry, chemistry, mathematics, and/or molecular biology. Strong letters of recommendation indicating academic excellence, personal maturity, and exceptional motivation and interest in the experimental sciences are an important part of the application. In addition, previous research experience is considered favorably. IBT also requires GRE General Test scores, in addition to TOEFL/TWE scores for all international applicants.

Correspondence and Information

Graduate Program Director
Institute of Biosciences and Technology
The Texas A&M University System Health Science Center
2121 West Holcombe Boulevard
Houston, Texas 77030-3303
Phone: 713-677-7612
Fax: 713-677-7725
E-mail: jbender@ibt.tamhsc.edu
Web site: http://www.ibt.tamhsc.edu/

The Texas A&M University System Health Science Center

THE FACULTY AND THEIR RESEARCH

Brad A. Amendt, Associate Professor; Ph.D., Iowa, 1994. Molecular and biochemical mechanisms of gene expression during embryogenesis; molecular basis of genetic disorders and homeobox genes.

M. Gabriela Bowden, Research Assistant Professor; Ph.D., Texas Health Science Center at Houston, 1999. Microbial proteins; structure, function, and role in pathogenesis.

Richard H. Finnell, Professor; Ph.D., Oregon, 1980. Gene-environment interactions mediating susceptibility to birth defects.

Weiman He, Assistant Professor; Ph.D., Maryland, Baltimore, 1997. Nuclear hormone receptor PPARγ and its role in the occurrence and treatment of obesity, diabetes, atherosclerosis, and osteoporosis.

Magnus Höök, Professor; Ph.D., Uppsala (Sweden), 1974. Extracellular matrix biology; cell surface receptors; microbial pathogenesis; protein-protein interactions.

Mingyao Liu, Associate Professor; Ph.D., Maryland College Park, 1992. GTPases; G-protein–coupled receptors; molecular basis of signal transduction; mechanism of human stem cell proliferation and differentiation.

James F. Martin, Associate Professor; M.D., 1986, Ph.D., 1995, Texas–Houston Health Science Center. Homeobox genes; vertebrate embryogenesis; mouse genetics; birth defects.

Wallace L. McKeehan, Professor; Ph.D., Texas at Austin, 1970. Cytokine-receptor interactions, apoptosis, and tumor suppression in cell and mouse models of prostate and liver cancer.

Laura E. Mitchell, Associate Professor; Ph.D., Yale, 1991. Genetic epidemiology of birth defects; maternal genetic effects; statistical genetics.

Vladimir Potaman, Research Assistant Professor; Ph.D., Moscow Institute of Physics and Technology, 1981. DNA structure and interaction with ligands; structural effects in DNA replication, repair, and transcription.

Stephen H. Safe, Distinguished Professor; D.Phil., Oxford, 1966. Mechanism of estrogen-induced gene expression; mechanisms of Ah receptor–estrogen receptor crosstalk; development of drugs for treating multiple cancers; PPARγ agonists; endocrine disruptors.

Robert Schwartz, Professor and Interim Director; Ph.D., Pennsylvania, 1972. Molecular development and disease related to gene networks and drug discovery.

Robert Y.-L. Tsai, Assistant Professor, M.D., National Taiwan, 1988; Ph.D., Johns Hopkins, 1996. Molecular mechanism controlling stem cell and cancer cell self-renewal: implications in tissue regeneration, normal aging, and tumor recurrence.

Fen Wang, Assistant Professor; Ph.D., Clarkson, 1994. Structure and function of fibroblast growth factors and receptors in prostate cancer, birth defects, and neuron regeneration; mouse molecular genetics; cooperation between extopic FGFR1 and depression of FGFR2 signaling in induction of high-grade prostatic intraepithelial neoplasia in mouse prostate.

Robert D. Wells, Professor; Ph.D., Pittsburgh, 1964. DNA structure, triplet repeats, and human hereditary neuromuscular disease.

Yi Xu, Research Assistant Professor; Ph.D., Texas Health Science Center at Houston, 1998. Pathogenic bacteria surface proteins; roles in pathogenesis and host interactions.

The Joint Faculty and Their Research

Bharat B. Aggarwal, Ph.D., Professor, Department of Experimental Therapeutics, University of Texas M. D. Anderson Cancer Center. Cytokines; chemoprevention; cell signaling; NF-kappaB; apoptosis; natural products.

Susan Carmichael, Ph.D., Birth Defects Monitoring Program, Berkeley, California.

Benoit de Crombrugghe, M.D., Professor and Chair, Department of Molecular Medicine, University of Texas M. D. Anderson Cancer Center. Cell fate determination and cell differentiation in bone and cartilage.

Rena N. D'Souza, Ph.D., Professor and Director of Research, University of Texas Health Science Center at Houston, Dental Branch. Molecular and genetic determinants of patterning and matrix formation during craniofacial, oral, and dental development.

K. C. Donnelly, Ph.D., Associate Professor, Veterinary Anatomy and Public Health, Texas A&M University. Exposure assessment and risk assessment of complex environmental mixtures.

Xin-Hua Feng, Ph.D., Associate Professor; Departments of Molecular and Cellular Biology and Surgery, Baylor College of Medicine. TGF-ß/SMAD signaling in cell-growth control; tissue differentiation and tumorigenesis.

Stanley Glasser, Ph.D., Adjunct Professor and Professor Emeritus, Cell Biology, Baylor College of Medicine. Embryo-endometrial interactions; in vitro models; environmental disruptors.

Jordan Gutterman, M.D., Professor, Molecular Therapeutics, M.D. Anderson Cancer Center. Identification and characterization of novel plant compounds that regulate the growth of tumor cells and endothelial cells.

Mikio Kan, M.D., Ph.D., Director of the Board, Central Research Laboratories, Zeria Pharmaceutical Company, Ltd. A role of heparan sulfate for interaction of FGF and FGF receptor.

Richard Mayne, Ph.D., Professor, Department of Cell Biology, University of Alabama at Birmingham. Structure, function, and pathophysiology of extracellular matrix.

David McQuillan, Ph.D., Director of Research, LifeCell Corporation, Branchburg, New Jersey.

John C. Pérez, Ph.D., Regents Professor and Director, Natural Toxins Research Center (NTRC), Texas A&M University–Kingsville. Disintegrins found in snake venoms and their role in blocking normal function of integrins on cells.

Bernhard Rupp, Ph.D., Dr.habil. CEO, q.e.d. life science discoveries, inc., Livermore, CA

James C. Sacchettini, Ph.D., Professor, Department of Biochemistry and Physics, Texas A&M University. Structure and function analysis of proteins using X-ray crystallography; protein-lipid interactions; rational drug design on tuberculosis; enzymes in nucleotide metabolism.

Richard E. Wainerdi, Ph.D., President and CEO, Texas Medical Center; PE. Radioanalytical and microanalytical chemistry.

Jianming Xu, Ph.D., Assistant Professor, Department of Molecular and Cellular Biology, Baylor College of Medicine. Use of transgenic and knockout mouse models in combination with cell-culture systems to study the normal and pathological roles and molecular mechanisms of several oncogenic transcriptional coregulators in breast and prostate cancers and in normal development and organ function.

THOMAS JEFFERSON UNIVERSITY

Jefferson College of Graduate Studies

Programs of Study

Jefferson College of Graduate Studies offers graduate programs leading to the degree of Doctor of Philosophy in biochemistry and molecular biology, cell and developmental biology, genetics, immunology and microbial pathogenesis, molecular pharmacology and structural biology, neuroscience, physiology, and tissue engineering and regenerative medicine. The College also offers graduate programs leading to the degree of Master of Science in biomedical chemistry, cell and developmental biology, microbiology, pharmacology, and public health. Ph.D. students generally complete their program courses during the first two years and a comprehensive examination follows. The focus is on doctoral thesis research for the remainder of the program. Research culminates in a scholarly dissertation that is defended during a final examination.

Jefferson Medical College and Jefferson College of Graduate Studies jointly sponsor a combined M.D./Ph.D. program for students who wish to pursue a career in academic medicine and medical research. During the first two years, students complete the basic medical preclinical curriculum in Jefferson Medical College. The next three years are spent in the Graduate College in pursuit of the Ph.D. degree. The final two years involve completion of medical studies in Jefferson Medical College.

Research Facilities

The Bluemle Life Sciences Building, with 157,000 square feet of laboratory space, serves as the primary research facility for molecular biology and genetics, molecular virology, microbiology, and immunology. Other basic science departments are housed in spacious, modern facilities. Farber Institute for Neurosciences is housed in the Jefferson Hospital for Neuroscience building and Jefferson Alumni Hall. Students have access to modern research equipment for molecular analysis of gene expression and functional aspects of the immune system, sophisticated studies of cell physiology, and in-depth studies of embryonic development and all aspects of drug metabolism.

Financial Aid

Financial support is available to the majority of full-time Ph.D. students in the form of University fellowships, training grants, and research assistantships. In 2006–07, students granted full fellowship support receive funds for payment of tuition and a stipend of $24,500. Also available to students demonstrating financial need are Title IV funds, including those from the Federal Family Education Loan Programs (federally subsidized and unsubsidized Stafford student loans), Federal Perkins Loan, and Federal Work-Study programs. University loan programs are also available to qualifying students.

Cost of Study

Tuition and fees for the 2006–07 academic year are $15,640 for full-time Ph.D. students. Tuition for M.S. basic science students is $790 per credit.

Living and Housing Costs

On-campus student housing includes four residence facilities that provide apartment and dormitory accommodations. Expenses for housing range from $318 per month for a shared room in the Martin Building to $917 per month for an apartment in the Barringer Building and $869 per month in the Orlowitz Building.

Student Group

Currently, the University enrolls about 2,500 students. The College of Graduate Studies enrolls approximately 600 students, about half of whom are women.

Location

Thomas Jefferson University is centrally located in Philadelphia within walking distance of many places of cultural interest, including concert halls, theaters, museums, art galleries, and historic landmarks. There are numerous intercollegiate and professional athletic events. Convenient bus and subway lines connect the University with other local universities and colleges and with several outstanding libraries. The proximity of the New Jersey shore and the Pennsylvania mountains offers year-round recreational opportunities, and New York City and Washington, D.C., are each just two hours away.

The University

The rapidly expanding Thomas Jefferson University is an academic health center emphasizing the biological sciences. Its origin dates back to 1824, with the founding of Jefferson Medical College and includes, in addition to the Jefferson Medical College, the Jefferson College of Graduate Studies, the Jefferson College of Health Professions, the University Hospital, and various affiliated hospitals and institutions. The Bluemle Life Sciences Building houses the Kimmel Cancer Center. The Jefferson Hospital for Neurosciences houses the Farber Institute for Neurosciences. Jefferson Alumni Hall houses, in addition to departmental laboratories, the Alcohol Research Center, the Ischemia-Shock Research Institute, and the Jefferson Institute of Molecular Medicine. Student activities offices and facilities (i.e., swimming pool, exercise rooms, gymnasium, and handball/squash court), student lounges, and JCGS administrative offices are also located in Jefferson Alumni Hall.

Applying

Applications to Ph.D. programs should be submitted by March 1 for optimal consideration for admission and fellowship support. Applications to M.S. programs are evaluated on a rolling basis. Graduate Record Examinations (GRE) scores, three letters of recommendation, academic transcripts, and a $50 application fee are required of all applicants. While GRE Subject Test scores are appreciated and can enhance an application, they are not required for admission. For those whose native language is not English, scores on the Test of English as a Foreign Language (TOEFL) are required. Campus visits and tours of respective departments are accommodated upon request.

Correspondence and Information

Office of Admissions
Jefferson College of Graduate Studies
Thomas Jefferson University
1020 Locust Street, M-60
Philadelphia, Pennsylvania 19107-6799
Phone: 215-503-4400
Fax: 215-503-9920
E-mail: jcgs-info@jefferson.edu
Web site: http://www.jefferson.edu/cgs

Thomas Jefferson University

THE FACULTY AND THEIR RESEARCH

Biochemistry and Molecular Biology–Ph.D.
Research faculty: Arthur Allen, Emad S. Alnemri, Carol L. Beck, Jeff Benovic (Department Chair), George C. Brainard, David Capuzzi, Mon-Li Chu, Edgar Davidson, John I. Farber, Sam Gandy, Barry J. Goldstein, Gerald B. Grunwald, Noreen J. Hickok, Shi-Ying Ho, Jan B. Hoek, Ya-Ming Hou, James B. Jaynes, Sergio Jimenez, Erica Johnson, Hideko Kaji, James H. Keen, Michael P. King, Jose Martinez, Alexander Mazo, W. Edward Mercer (Program Director), Urich Rodeck, Peter Ronner, Michael Root, Barbara Schick, Matthias J. Schnell, Stephanie Schulz, Charles P. Scott, Sandor Shapiro, Thomas Tulenko, Jouni Uitto, Scott Waldman, Philip B. Wedegaertner, David A. Wenger, Eric Wickstrom, Charlene J. Williams, John C. Williams, Kevin J. Williams, Edward P. Winter, Kyonggeun Yoon, Allen R. Zeiger.

Research interests: regulation of DNA replication; RNA metabolism; mechanisms of cell synthesis; steroid and protein hormone receptors; mitochondrial DNA mutations; molecular biology of protein degradation; molecular biology–gene cloning; base sequencing of genes; gene expression in cell cultures; apoptosis and signal transduction; molecular biology of human diseases.

Cell and Developmental Biology
Research faculty: Eleni Anni, Robert Barsotti, David Birk, Alan Cahill, Ronald A. Coss, Manuel L. Covarrubias, Leonard M. Eisenman, John S. Ellingson, John L. Farber, Mark Feitelson, Bruce A. Fenderson, Giancarlo Ghiselli, Gregory E. Gonye, Fred Gorstein, Gerald B. Grunwald, Gyorgy Hajnoczky, David Herrick, Jan B. Hoek, Richard Horn, Lorraine Iacovitti, Renato Iozzo, Nathan Janes, James B. Jaynes, Suresh K. Joseph, Hideko Kaji, Boris N. Kholodenko, Devendra M. Kochhar, Edward. B. Lankford, Alexander Mazo, Peter A. McCue, A. Sue Menko, Pamela A. Norton, John G. Pastorino, Nancy J. Philp, Biddanda C. Ponnappa, Emanuel Rubin, Raphael Rubin, James D. San Antonio, Barbara P. Schick, Jay S. Schneider, James Schwaber, David S. Strayer, Christopher Stubbs, Theodore F. Taraschi (Director), Jouni J. Uitto, Elisabeth J. Van Bockstaele, Philip Wedegaertner, Edward P. Winter.

Research interests: alcohol-induced liver diseases; bioinformatics; calcium signaling; cell cycle and apoptosis; calcium metabolism; cell-cell and cell matrix adhesion; cellular differentiation; cellular effects of alcohol; cellular signaling; chemical carcinogenesis; contractility; cytoskeleton; developmental aspects of oncology; developmental neurobiology; DNA repair in malaria parasites; extracellular matrix structure and function; eye development; functional genomics; gene therapy; hepatitis virus and hepatocellular carcinoma; ion channel structure and function; mechanisms of anesthetic action; mechanisms of cell injury; membrane biology; mitochondria and cell death; neurodegeneration and pathophysiology of Parkinsonism; protein trafficking and organelle biogenesis in malaria-infected erythrocytes; regulation of cellular growth; stem cells; structure-function relationships in biological membranes; systems biology.

Genetics–Ph.D.
Research faculty: David Abraham, Emad S. Alnemri, Raffaele Baffa, Renato Baserga, Jeffrey L. Benovic, Bruce M. Boman, Arthur M. Buchberg Bruno Calabretta, Marcella Devoto, Ya-Ming Hou, James B. Jaynes, Erica Johnson, James H. Keen, Carlisle Landel, Alexander Mazo, Steven McKenzie, W. Edward Mercer, Diane Merry, Jay L. Rothstein, Linda D. Siracusa (Program Director), Saul Surrey, Jouni J. Uitto, Scott Waldman, Ya Wang, Charlene J. Williams, Edward P. Winter.

Research interests: functional genomics; genetics of cancer susceptibility; molecular genetic analysis of normal human genome; cytogenetics of aneuploidy syndromes; genetics of immune system; molecular genetics of animal models of human disease; molecular genetics of hematopoietic neoplasias and solid tumors; genetic analysis of G-1 phase and molecular mechanisms of altered growth regulation by oncogenes and tumor suppressor genes; transcriptional regulation and control of gene expression; mechanisms of ionizing and nonionizing radiation damage to cells.

Immunology and Microbial Pathogenesis–Ph.D.
Research faculty: David Abraham, Kishore R. Alugupalli, Melvin Bosma, Arthur M. Bunchberg, Bruno Calabretta, Catherine E. Calkins, Bernhard Dietzschold, Laurence C. Eisenlohr, Mark A Feitelson, Neal Flomenberg, Phyllis Flomenberg, Richard R. Hardy, D. Craig Hooper, Donald L. Jungkind, Tim Manser, W. Edward Mercer, Glen F. Rall, Ulrich Rodeck, Michael J. Root, Jay L. Rothstein (Program Director), Matthias Schnell, Charles P. Scott, Linda D. Siracusa, Alagarsamy Srinivasan, David S. Strayer, Yuri Sykulev, Theodore F. Taraschi, John L. Wagner, Scott A. Waldman, David West, Hui Zhang, Jianke Zhang.

Research interests: antigen presentation; antiviral agents; autoimmunity; B-cell development; cancer immunology; cell biology of malaria; cell growth regulation and differentiation; cellular immunology; cell activation and signal transduction; chemical and antigenic structure of virus particles; cytokines; developmental immunology; immunochemistry; immunogenetics; immunoparasitology; immunoregulation; latent virus infections; microbial immunology and pathogenesis; molecular immunology; neuroimmunology; neurovirology; regulation of viral gene expression; reproductive immunology; transplantation immunology; viral oncogenes; viral replication; virus-cell interactions.

Molecular Pharmacology and Structural Biology
Research faculty: Emad S. Alnemri, Renato Baserga, Jeffrey L. Benovic (Program Director), Bruce Boman, George C. Brainard, John L. Farber, Barry J. Goldstein, Jan B. Hoek, Ya-Ming Hou, James H. Keen, Walter Koch, Jose Martinez, Ulrich Rodeck, Michael J. Root, Charles P. Scott, Scott A. Waldman, Philip Wedegaertner, Eric Wickstrom, John Williams, Edward P. Winter.

Research interests: biochemistry and cell, molecular, and structural biology of cell-surface and intracellular receptors; structural biology and molecular modeling of protein-protein and nucleic acid interactions; molecular mechanisms of membrane sorting and intracellular organization; neuropharmacology; signal transduction; clinical pharmacology.

Neuroscience
Research faculty: Jeffrey Benovic, George C. Brainard, Manuel Covarrubias, Leonard Eisenman, Michelle Ehrlich, Jayasri Das Sarma, Samuel Gandy (Director of Farber Institute for Neurosciences), Gregory Gonye, Gerald Grunwald, Larry Harshyne, Richard Horn, Lorraine Iacovitti, Hector Lopez, Jeffrey Mason, Sue Menko, Diane Merry, Michael Oshinsky, Michelle Page, Raymond Regan, A. M. Rostami, Jay Schneider, James Schwaber, Stephen D. Silberstein, Joseph I. Tracy, Elisabeth Van Bockstaele (Program Director), Stephen Weinstein, David A. Wenger, Ji-Fang Zhang.

Research interests: neurodegenerative disorders, neuropharmacology, neuronal signaling, protein-protein interactions, structure and function of ion channels, affective disorders, neurodevelopment, neuroopthalmology, addiction and motivated behaviors.

Physiology
Research faculty: Robert J. Barsotti, Carol L. Beck, Thomas M. Butler, Manuel L. Covarrubias, Andrea Eckhart, Masumi Eto, John T. Flynn, Richard Horn, Walter J. Koch, Edward B. Lankford, A. Sue Menko, Rosario Scalia (Program Director), Marion J. Siegman (Chairperson), Marilyn J. Woolkalis, Donna S. Woulfe, Ji-Fang Zhang.

Research interests: energetics and regulation of smooth-muscle contraction; ultrastructural analysis of proteins; regulation of cross-bridge function in smooth and cardiac muscle; integrative cardiovascular physiology; cardiovascular gene therapy; physiology of the endothelium; thrombosis and hemostasis; the role of endothelial cells in innate immunity; physiology of leukocytes-endothelium interactions in vivo; diabetic vasculopathies; the role of integrins in the regulation of cell differentiation; changes in G-protein–coupled receptor signaling in hypertension; signal transduction mediated by protein phosphorylation; structure-function studies of voltage-gated ion channels; molecular regulation of the function of cloned voltage-sensitive potassium channels; molecular genetic basis of ion channel diseases; skeletal muscle voltage-gated chloride channels; voltage-gated calcium channels; regulation of calcium channels by second messengers.

Tissue Engineering and Regenerative Medicine–Ph.D.
Research Faculty: Chris Adams, Robert Akins Jr, Mon-Li Chu, Keith Danielson, Noreen Hickok, William J. Hozack, Masahiro Iwamoto, Motomi Enomoto-Iwamoto, Jeffrey I. Joseph, Eiki Koyama, Sue A. Menko, Maurizio Pacifici, Makarand V. Risbud, Richard H. Rothman, James D. San Antonio, Irving Shapiro (Program Director), Peter F. Sharkey, Vickram Srinvas, Marla Steinbeck, Jouni Uitto, Eric Wickstrom, Charlene J. Williams.

Research interests: cell and tissue engineering of bone, cartilage, pancreas, and blood vessels; bioreactors; biomaterials; tissue regeneration; cell-cell and cell-matrix interactions; angiogenesis; osteogenesis; chondrogenesis; genetic engineering; cell-biomaterial interaction; implant design; cellular and functional imaging; antisense technology.

Basic Science M.S. Programs
Program Directors: Carol L. Beck, Dennis M. Gross, Gerald B. Grunwald.

The programs give students sound theoretical and practical foundations in four important areas of the basic biomedical sciences. Students can specialize within a discipline by selecting from the track options within each program. M.S. basic program offerings are biomedical chemistry, cell and developmental biology , microbiology, and pharmacology.

Public Health–M.S. Programs
Research faculty: James Diamond, Jennifer Lofland, David B. Nash (Codirector), James D. Plumb, Richard Wender (Codirector).

Research interests: public-health biostatistics; epidemiology of diabetes; molecular microbiology uses in infectious diseases; health-services administration; health-education programs in immunization; behavioral science–tools to improve community well-being; environmental science–toxic substances in the schools.

TUFTS UNIVERSITY

Department of Biology

Programs of Study

The Department of Biology at Tufts University offers a program of advanced study and thesis research leading to the degrees of Doctor of Philosophy (Ph.D.) and Master of Science (M.S.) in biology and biotechnology. A master's degree can also be obtained for course work, without research. The program is arranged into six concentrations: cell physiology; conservation and the environment; developmental biology; ecology, behavior, and evolution; genetics and molecular biology; and neurobiology and animal behavior. Each concentration has its own adviser and a core of faculty researchers. Befitting the integrative nature of biology, the department encourages interdisciplinary research and collaborations between academic fields.

The department holds regular seminars presented by invited scientists to keep students and faculty members abreast of exciting research going on elsewhere. Students may take advantage of similar seminars at the many research institutions in the Boston area, for which notices are posted regularly. In addition to invited speakers, each of the departmental faculty members and postdoctoral fellows presents an annual research seminar.

Research Facilities

The Department of Biology is located in modern buildings on the Medford campus of Tufts University. State-of-the-art research equipment and facilities are available for the specialties represented by faculty members. Most members of the faculty support their research and that of their students with grants awarded by various federal agencies and private foundations. Tufts is part of the Internet2 consortium, and the University's central computer facilities are excellent. The department has high-speed network access throughout the buildings and has dedicated computers, scanners, and printers for graduate students. A comfortable departmental reading room holds current issues of major biological journals, while the main library building contains back issues. In addition, Tufts' Health Sciences Library and the libraries of many other Boston-area universities and teaching hospitals are accessible to Department of Biology graduate students, and the extensive electronic journal resources can be accessed through students' accounts both on and off campus.

Financial Aid

In 2005–06, nine-month academic year and departmental summer-supplement teaching assistantships carried a stipend of approximately $18,000 for entering thesis master's candidates and $22,000 for entering doctoral candidates, as well as a full-tuition scholarship. Stipends may increase for 2006–07. Some students beyond the first year are supported as research assistants. Research thesis students in both doctoral and master's programs are supported by teaching assistantships.

Cost of Study

Tuition for the academic year 2006–07 is estimated at approximately $29,500. A tuition scholarship is granted to all students on University assistantships and to students supported by private scholarships.

Living and Housing Costs

The cost of living in the Boston area is high. Most graduate students live in rooms or apartments in Cambridge, Somerville, Medford, and Arlington. Living expenses for a student sharing an apartment and doing his or her own cooking average nearly $800 per month. The University has limited on-campus housing for incoming graduate students.

Student Group

The department admits 5 to 10 new students each year. There are approximately 30 graduate students who are currently working toward biology degrees. The University is coeducational and nonsectarian. On the Medford campus, there are more than 1,300 graduate students in all fields of study and approximately 4,900 undergraduates.

Location

The Department of Biology is located on the Medford campus of Tufts, 6 miles north of downtown Boston and 15 minutes by public transportation from Harvard Square in Cambridge. The Boston area is justifiably famous for its active and varied intellectual and artistic life. As part of the greater Boston community, Tufts provides an environment where education assumes a broader context than one's own specialty. The richness of Boston's music, art, theater, and sports life is complemented by the beauty of the historic New England countryside.

The University

Tufts University is a private coeducational institution founded in 1852. Its liberal arts, sciences, and engineering colleges are located in Medford, on a hill overlooking Boston and Cambridge. These colleges and the Tufts University School of Medicine, in downtown Boston, offer comprehensive coverage of biological areas of research. In addition to taking advantage of course offerings at the medical school, registered students may enroll in one graduate course each semester at Boston College, Boston University, or Brandeis University. Students may also participate in research collaborations with scientists at other area institutions.

Applying

Prospective students are encouraged to apply by January 15 for admission for the following academic year. The deadline for international students is December 15. Applications should be made through the Graduate School and should include a recent transcript, three letters of recommendation, and a statement of purpose. Applicants should also submit scores from the General Test of the Graduate Record Examinations. The Subject Test in biology or biochemistry is recommended but not required. International students whose native language is not English must submit TOEFL scores. Prospective students are encouraged to visit the department to consult with those faculty members in whose work they are interested.

Correspondence and Information

For application materials:
Graduate School of Arts and Sciences
Tufts University
Ballou Hall
Medford, Massachusetts 02155
Phone: 617-627-3395
Web site: http://ase.tufts.edu/graduate/

For specific information about the department:
Graduate Program Assistant
Department of Biology
Tufts University
Medford, Massachusetts 02155
Phone: 617-627-3195
Web site: http://ase.tufts.edu/biology/

Tufts University

THE FACULTY AND THEIR RESEARCH

Astier M. Almedom, Assistant Professor and Henry R. Luce Professor of Science and Humanitarianism; D.Phil., Oxford. Humanitarian policies regarding famine; public health policies.

Harry A. Bernheim, Associate Professor and Department Chair; Ph.D., Michigan. Immunology.

Frances S. Chew, Professor; Ph.D., Yale. Ecology.

David E. Cochrane, Professor; Ph.D., Vermont. Physiology of the mast cell, with emphasis on the role of peptides in regulating mast cell function in inflammation and allergy.

George S. Ellmore, Associate Professor; Ph.D., Berkeley. Experimental plant anatomy and morphology; protein transport and targeting in seeds and storage organs.

Susan G. Ernst, Professor; Ph.D., Massachusetts. Developmental biology; regulation of gene expression during early development and cell differentiation; molecular basis of mitochondrial diseases.

Ross S. Feldberg, Associate Professor; Ph.D., Michigan. Biochemistry.

Catherine Freudenreich, Assistant Professor; Ph.D., Duke. Yeast genetics and molecular biology; yeast as a model system for human triplet repeat diseases; chromosome structure and fragility.

Juliet A. Fuhrman, Associate Professor and Graduate Program Advisor; Ph.D., Johns Hopkins. Immunology and infectious disease; filarial parasites and parasite-host-vector relationships; molecular and biochemical basis of parasite virulence.

Sara M. Lewis, Associate Professor; Ph.D., Duke. Evolutionary ecology; tropical community ecology; reproductive behavior and sexual selection in natural populations.

Kelly M. McLaughlin, Assistant Professor; Ph.D., Massachusetts. Molecular development (organogenesis); molecular events leading to cell differentiation and patterning during organ formation; signaling pathways in pronephric nephrogenesis (embryonic kidney).

Mitch McVey, Assistant Professor; Ph.D., MIT. Molecular genetics; mechanisms of DNA repair and recombination in *Drosophila*.

Colin M. Orians, Associate Professor; Ph.D., Penn State. Ecology; tropical ecology; chemical ecology; plant-herbivore-environment interactions; effects of hybridization on plant resistance.

Jan A. Pechenik, Professor; Ph.D., Rhode Island. Invertebrate zoology; physiological ecology; reproduction and development of marine invertebrates, with emphasis on larval growth, differentiation, and metamorphosis.

J. Michael Reed, Associate Professor; Ph.D., North Carolina State. Conservation biology; ornithology; dispersal; effects of behavior on population biology.

L. Michael Romero, Associate Professor; Ph.D., Stanford. Physiology; endocrinology; glucocorticoid release in response to stress.

Eli Siegel, Professor; Ph.D., Rutgers. Microbial genetics.

Philip Starks, Assistant Professor; Ph.D., Cornell. Behavioral ecology; recognition systems; evolution of eusociality; cooperation and conflict in social animals.

Barry A. Trimmer, Professor; Ph.D., Cambridge. Neurobiology; neurotransmitters and receptors in insects; neural control of locomotion.

UNIFORMED SERVICES UNIVERSITY OF THE HEALTH SCIENCES
Programs in Biomedical Sciences

Programs of Study	The Graduate Programs of the School of Medicine offer training leading to master's degrees and doctoral degrees in the biomedical sciences and public health. The Doctor of Philosophy, Doctor of Public Health, Master of Science, Master of Public Health, and Master of Tropical Medicine and Hygiene degrees are offered. A physician/scientist's program (M.D./Ph.D.) is also offered. Detailed information on specific graduate programs is listed in the Graduate Programs in the Basic Health Sciences section.
Research Facilities	Graduate training programs are conducted in state-of-the-art research facilities on the Uniformed Services University (USU) campus. Individual laboratories and core facilities are well equipped with the instrumentation required for modern biomedical research. The University is located in proximity to and has collaborations with major research facilities, including the National Institutes of Health, the National Library of Medicine, Walter Reed Army Institute of Research, the Armed Forces Radiology Research Institute, and numerous biotechnology companies.
Financial Aid	The University offers stipends on a competitive basis to civilian doctoral students who are U.S. citizens or resident international students. For academic year 2006–07, USU-supported stipends are $24,000 for entering students and $25,000 after advancement to candidacy. Outstanding applicants may be nominated for Dean's Special Fellowships, which support a stipend of $29,000. Grant-supported stipends may also be available for international students and other students with specific mentors.
Cost of Study	USU is a federal institution. Tuition and fees are waived for civilian students. Active-duty military personnel accepted to study at USU must have the consent and sponsorship of their parent service and incur a service obligation at the completion of their studies. Civilian students do not incur a service obligation to the United States government after the completion of their graduate training program.
Living and Housing Costs	Living costs in the greater Washington, D.C., area are comparable to other major metropolitan areas. Every effort is made to assist students in making satisfactory living arrangements since on-campus housing is not available.
Student Group	During the past academic year, approximately 165 graduate students were enrolled full-time. The total number of graduate degrees awarded in the last five years is 260. Nearly 700 students are enrolled in the medical degree (M.D.) program. In addition, the Graduate School of Nursing also offers M.S. and Ph.D. degrees in nursing.
Location	USU is located in Bethesda, Maryland, an immediately adjacent northern suburb of Washington, D.C. The University is located in a parklike setting on the grounds of the National Naval Medical Center. Wooded areas and jogging and biking trails surround the University. A nearby Metro subway station provides convenient access to downtown Washington and surrounding areas. The D.C. metropolitan area has more than 3 million people and contains seven major universities (including three other medical schools), numerous colleges, and internationally renowned research facilities. Cultural and recreational activities are plentiful, with a major park system, recreational facilities, museums, theaters, symphonies, opera, and major-league sports teams. The Chesapeake Bay, Atlantic Ocean, and the Shenandoah Mountains are easily accessible.
The University	The Uniformed Services University of the Health Sciences was established by Congress in 1972 to provide a comprehensive education in medicine to those who demonstrate potential for careers as Medical Corps officers in the uniformed services. Civilian graduate programs in the basic medical sciences are an essential part of the academic environment at the University. Military officers are also eligible for admission with consent of their service. USU subscribes fully to the policy of equal educational opportunity and accepts students on a competitive basis without regard to race, color, sex, age, or creed.
Applying	Applicants must have completed a bachelor's degree from an accredited academic institution prior to enrollment. Applicants must send official transcripts of all college-level courses taken and recent GRE scores (within the last two years) to the Office of Graduate Education. GRE Subject Tests are not required. Applicants must also arrange for three letters of recommendation from individuals familiar with their academic work to be sent to the University. An online application form is available at http://cim.usuhs.mil/geo/ (click on Prospective Student). Completed applications should be received before January 15 for matriculation in late August. There is no application fee.
Correspondence and Information	Associate Dean for Graduate Education Uniformed Services University 4301 Jones Bridge Road Bethesda, Maryland 20814-4799 Phone: 301-295-3913 800-772-1747 (toll-free) E-mail: graduateprogram@usuhs.mil Web site: http://www.usuhs.mil (click on Graduate Education)

Uniformed Services University of the Health Sciences

GRADUATE PROGRAMS IN THE BASIC HEALTH SCIENCES

INTERDISCIPLINARY PH.D. PROGRAMS

Emerging Infectious Diseases. Christopher Broder, Ph.D., Program Director. This interdisciplinary Ph.D. program provides the scientific community with broadly trained, outstanding scientists that can contribute significantly to the increasingly complex field of infectious disease mechanisms and pathogenesis. In a rigorous academic environment, students learn to ask well-informed questions, develop the skills to answer those questions at the bench, expand their capacity to think creatively and broadly, and acquire the skills necessary to communicate their ideas. The EID Program has a core interdisciplinary curriculum, three academic tracks that permit the development of areas of expertise, advanced graduate courses, and broadly based research programs. Areas of research emphasize the cellular and molecular basis of microbial pathogenesis, immunology, the pathology of infectious diseases, virology, bacteriology, and parasitology, as well as the epidemiology of infectious diseases caused by these pathogens. Strong research programs are available within each of these areas, and emphasize modern methods in molecular biology, cell biology, and interdisciplinary experimental strategies. Prospective students should contact the Program Director by e-mail at cbroder@usuhs.mil or visit the Web at http://www.usuhs.mil/eid.

Molecular and Cell Biology. Jeff Harmon, Ph.D., Program Director. The interdisciplinary Ph.D. program in molecular and cell biology offers instruction and research training in areas that address fundamental questions in contemporary biology. Individual faculty members employ model systems ranging from simple prokaryotes to complex transgenic mammals to pursue a broad spectrum of interests that includes genomics, functional proteomics, molecular mechanisms of signal transduction, cell-cell signaling, regulation of cell growth and division, developmental biology, and regulation of gene expression. A rigorous academic environment, combined with access to state-of the-art equipment and a low student to faculty ratio, provides students with the opportunity to pursue an individualized training program. Applicants should have a strong background in biology or chemistry. Prospective students should contact the Program Director by e-mail at jharmon@usuhs.mil or visit the Web at http://www.usuhs.mil/mcb.

Neuroscience. Regina Armstrong, Ph.D., Program Director. The interdisciplinary Ph.D. program in neuroscience offers extensive training opportunities in diverse areas of neuroscience. Interdisciplinary instruction is provided in the development, structure, function, and pathology of the nervous system. Research areas include cellular, molecular, physiological, and pharmacological approaches to analysis of nervous system development, regeneration, and plasticity. Neurobiological and behavioral research includes adaptive responses of the nervous system to stress, injury, and a changing environment. Prospective students should have a strong background in biological, chemical, and/or behavioral sciences. Prospective students should contact the Program Director by e-mail at rarmstrong@usuhs.mil or visit the Web at http://www.usuhs.mil/nes/home.html.

Physician/Scientist Program (M.D./Ph.D.). The M.D./Ph.D. program at USU was established to train outstanding, dedicated military officers as independent physician/scientists to carry out both clinical investigations and biomedical research in the basic sciences. The program combines a rigorous basic science graduate curriculum with outstanding clinical training and special integrated M.D./Ph.D. activities that qualify students for careers in academic medicine, biomedical, and clinical research, as well as clinical practice. The curriculum combines and integrates the requirements for both the M.D. and Ph.D. degrees. The M.D./Ph.D. program consists of three phases to be completed in seven to eight years. Prospective students should contact the Graduate Education Office by e-mail at graduateprograms@usuhs.mil or visit the Web at http://cim.usuhs.mil/geo/mdphd.htm.

DEPARTMENTALLY BASED PH.D., DR.P.H., AND MASTER'S DEGREE PROGRAMS

Department of Medical and Clinical Psychology. Tracy Sbrocco, Ph.D., Program Director. Doctoral programs and research in this area emphasize the application of psychology to behavioral medicine and to clinical psychology. Study in applied areas on the interface of health, psychology, and behavior, and in basic areas of psychology, is offered. An American Psychological Association-accredited clinical psychology Ph.D. program is offered to selected members of the Uniformed Services. Prospective students should contact the Program Director by e-mail at tsbrocco@usuhs.mil or visit the Web at http://www.usuhs.mil/mps/Psychology/index.html.

Department of Pathology. Radha Maheshwari, Ph.D., Program Director. Programs of research and study focus on molecular approaches to pathology of human disease. They include genetics, carcinogenesis, immunology, and cell biology. Prospective students should contact the Program Director by e-mail at rmaheshwari@usuhs.mil or visit the Web at http://wwwpath.usuf2.usuhs.mil/default.html.

Department of Preventive Medicine and Biometrics. David Cruess, Ph.D., Program Director. Graduate programs in public health are offered at the master's and doctoral levels. The Master of Public Health (M.P.H.), Master of Tropical Medicine and Hygiene (M.T.M.H.), and the Master of Science in Public Health (M.S.P.H.) programs are designed primarily for uniformed officers with experience in a health-related field. The Doctorate in Public Health (Dr.P.H.) program prepares individuals for leadership roles in research, teaching, or policy development in the field of public health. In addition, two Ph.D. programs are offered: Medical Zoology, for students with a master's degree in entomology or parasitology, who wish to pursue further study in field-oriented medical parasitology or vector biology; and Environmental Health Sciences, with a focus on military-relevant research, particularly in the area of exposure assessment. Prospective students should contact the Program Director by e-mail at dcruess@usuhs.mil or visit the Web at http://www.usuhs.mil/pmb/PMBintropage.html.

Additional departments contributing to the teaching and research training of doctoral students include Anatomy, Physiology and Genetics, Biochemistry, Microbiology and Immunology, Pharmacology, and Physiology and some clinical departments (e.g., Medicine, Neurology, and Psychiatry).

UNIVERSITY AT ALBANY,
STATE UNIVERSITY OF NEW YORK

School of Public Health
Department of Biomedical Sciences

Program of Study

The department offers broadly based research training in the areas of biodefense and emerging infectious disease, molecular genetics, structural and cell biology, immunology and infectious diseases, and neuroscience. Students are admitted for study toward the Ph.D. or M.S. degree. Training is individualized, with course selection based on the area of specialization and on the background and interests of each student. During the first year, projects are undertaken in the laboratories of several faculty members. At the end of the laboratory rotation period, the student selects a mentor for thesis research. Emphasis is placed on informal instruction and interaction between students and faculty members in the laboratory, seminars, colloquia, and journal clubs.

Research Facilities

Most of the faculty members in the Department of Biomedical Sciences are research scientists at the Wadsworth Center, the central laboratory complex of the New York State Department of Health. The Wadsworth Center has two main sites; one is located at the Empire State Plaza in downtown Albany, and the other is at the nearby David Axelrod Institute, which is adjacent to the Albany Medical Center, the home base of several other faculty members. The Wadsworth laboratories are spacious modern facilities that are among the most technologically advanced and comprehensive laboratory complexes in the country. Faculty and student research at the Wadsworth Center is supported by outstanding core facilities in biochemistry, genetics, imaging, immunology, molecular structure, mass spectrometry, and NMR, as well as by national centers in biological imaging (NIH/NCRR), genetics (CDC), and nanobiotechnology (NSF). Facilities include high-resolution cryoelectron microscopes, advanced computer graphics, X-ray crystallography, biological mass spectroscopy, and multidimensional NMR. The comprehensive molecular genetic and biochemistry core facilities include peptide and oligonucleotide synthesis, protein and DNA sequencing, and transgenic mouse facilities.

Financial Aid

Graduate assistantships and tuition scholarships are awarded to all incoming doctoral students. Assistantship stipends range from $20,000 to $24,000. Other forms of financial support include research assistantships from faculty research grants and merit-based University fellowships. Outstanding candidates may apply for a fellowship in biodefense and emerging infectious disease, which includes a competitive stipend, tuition, health insurance, and travel to scientific meetings. These fellowships are awarded to students who wish to receive a broad training in biodefense, emerging infectious disease, and epidemiology.

Cost of Study

Full-time tuition in 2005–06 was $3450 per semester for New York State residents and $5640 per semester for out-of-state and international students. University fees are approximately $500 per semester.

Living and Housing Costs

In addition to student residence halls, off-campus apartments are available at rents ranging from $400 to $600 per month.

Student Group

Of the 17,000 students at the State University of New York at Albany, 4,500 are graduate students. They come from all parts of the United States and more than forty countries.

Location

Albany, the capital of the state of New York, offers a large choice of cultural attractions ranging from music, dance, and theater, to horseracing and other sports. In the nearby Adirondack and Catskill Mountains and in the Berkshires, students find an unlimited variety of outdoor activities including hiking, camping, canoeing, skiing, and climbing. Other attractions include Lake George, Saratoga Springs, Tanglewood (in Massachusetts), and the Olympic Sports Complex in Lake Placid. New York City, Boston, Montreal, Vermont, and Atlantic Ocean beaches are all within a few hours' drive. Albany has an airport, bus and train services, and an extensive public transportation system.

The University

The University at Albany is the senior campus and one of four University Centers of the sixty-four campus State University of New York system. The main campus is housed in a modern complex occupying a 400-acre site at the western edge of Albany.

Applying

College transcripts; Graduate Record Examinations scores on the General Test; three references from people who are familiar with the applicant's academic qualifications; a statement of the applicant's educational and professional aims; and a $60 application fee must be submitted to the Office of Graduate Admissions. A Subject Test in biology, biochemistry, chemistry, or physics is recommended. International applicants must submit a minimum TOEFL score of 600. The application deadline for the Ph.D. program is January 1; the deadline for the M.S. program is April 1.

Correspondence and Information

Department of Biomedical Sciences
Wadsworth Center
University at Albany
P.O. Box 509
Albany, New York 12201-0509

Phone: 518-473-7553
Fax: 518-473-8520
E-mail: bmsdept@wadsworth.org
Web site: http://www.wadsworth.org/sph/bms

University at Albany, State University of New York

THE FACULTY AND THEIR RESEARCH

James A. Dias, Chair; Ph.D., Washington State, 1979. Structure and function of gonadotropins and their receptors.

Terence Wagenknecht, Associate Chair; Ph.D., Minnesota, 1977. Structures of macromolecular complexes.

Rajendra Kumar Agrawal, Ph.D., Banaras Hindu (India), 1993. Structure and function of macromolecular assemblies.

Julio A. Aguirre-Ghiso, Ph.D., Buenos Aires, 1997. Mechanisms of invasion and metastasis; cancer dormancy; mitogen and stress-activated signal transduction; oncogenomics.

David Anders, Ph.D., Kansas State, 1983. Molecular biology and pathogenesis of human herpesviruses.

Thomas J. Begley, Ph.D., SUNY at Albany, 1999. Systems toxicology; mechanisms of DNA repair; damage signaling.

Marlene Belfort, Ph.D., California, Irvine, 1972. Gene expression; RNA splicing; intron mobility; molecular evolution; inteins in biotechnology.

Kristen Bernard, D.V.M., California, Davis, 1989; Ph.D., Wisconsin–Madison, 1995. Pathogenesis of arboviruses.

Valerie Bolivar, Ph.D., Dalhousie. Genetics of complex behaviors (e.g., learning and memory, anxiety) in the mouse; development of mouse models of autism.

Samuel Bowser, Ph.D., SUNY at Albany, 1984. Cell biology; cytoskeleton and cell motility; physiological ecology of rhizopod protists; molecular evolution.

April Burch, Ph.D., Arizona, 2000. Virus assembly and the host-pathogen interface.

Michele Caggana, Sc.D., Harvard, 1991. Population-based gene frequency studies, newborn screening, clinical testing, nanobiotechnology.

David Carpenter, M.D., Harvard, 1964. Electrophysiology of synaptic transmission.

Sudha Chaturvedi, Ph.D., Delhi (India), 1990. Fungal and parasite pathogeneses; protein transport and secretion mechanisms.

Vishnu Chaturvedi, Ph.D., Delhi (India), 1988. Unique metabolism of fungal pathogens; fungal-human phagocyte interactions.

Xiang Yang Chen, Ph.D., Hong Kong, 1990. Spinal reflex conditioning.

Nick M. Cirino, Ph.D., Case Western Reserve, 1995. Pathogen characterization, protein biochemistry, molecular biology, assay development.

Douglas S. Conklin, Ph.D., Wisconsin–Madison, 1992. Functional genomics of cancer; RNAi-based genetic technology; cell cycle physiology.

Jan Conn, Ph.D., Toronto, 1987. Evolutionary genetics of mosquitoes that transmit infectious diseases.

Joan Curcio, Ph.D., George Washington, 1987. Retrotransposons in yeast *S. cerevisiae*: mechanisms and regulation of retrotransposition.

Keith Derbyshire, Ph.D., Edinburgh, 1983. Mobile DNA; transposons; conjugation; molecular genetics of mycobacteria.

Victoria Derbyshire, Ph.D., Yale, 1990. Structure and function of intron- and intein-encoded endonucleases.

Xinxin Ding, Ph.D., Michigan, 1988. Molecular toxicology; pharmacogenetics; gene regulation and biological function of cytochrome P-450 monooxygenases; transgenic/knockout mouse models of human diseases.

Gregory D. Ebel, Sc.D., Harvard, 2000. Ecology and evolution of arboviruses.

Christina Egan, Ph.D., Albany Medical College, 1997. Diagnostic assay development for the detection of biothreat agents.

Michael Fasullo, Ph.D. DNA recombination; DNA repair; yeast genetics; cell-cycle checkpoints; molecular biology.

Joachim Frank, Ph.D., Munich Technical, 1970. 3-D cryoelectron microscopy of macromolecules; mechanisms of translation.

Robert Glaser, Ph.D., Cornell, 1989. Chromatin structure and function in *Drosophila*; palmitoyl-protein thioesterases and Batten disease.

Todd Gray, Ph.D., Michigan, 1993. Epigenetics and genomic imprinting in mammals; genetics and function of the MAKORIN zinc-finger protein family.

Steven Hanes, Ph.D., Brown, 1988. Homeobox genes, protein–DNA interaction; prolyl-isomerases, cell-cycle control.

Charles Hauer, Ph.D., Virginia, 1991. Proteomics; protein modifications; biological mass spectometry.

Bruce J. Herron, Ph.D., SUNY at Albany, 1999. Functional analysis of the mouse genome; investigation of defects in mammalian organogenesis.

Joachim Jaeger, D.Phil., Basel (Switzerland), 1991. Structure/function studies of viral replication complexes using X-ray crystallography, solution scattering, imaging techniques, mutagenesis, and kinetics.

John Kaplan, Ph.D., Albany Medical College, 1975. Interactions of phagocytes, platelets, and endocytic cells.

Janet Keithly, Ph.D., Iowa State, 1968. Parasite metabolism and drug design using biochemical and molecular techniques.

Alexey Khodjakov, Ph.D., Moscow, 1990. Cell biology: Structure and functions of centrosomes in vertebrates.

Michael Koonce, Ph.D., Berkeley, 1987. Molecular biology of the cytoskeleton; molecular motors.

Laura D. Kramer, Ph.D., Cornell, 1974. Epidemiology and ecology of arthropod-borne viruses.

David Lawrence, Ph.D., Boston College, 1971. Biochemistry of lymphocytes; host resistance; immunotoxicology.

William Lee, Ph.D., Johns Hopkins, 1987. Cellular immunology; generation of immunologic memory; lymphocyte interactions.

David LeMaster, Ph.D., Yale, 1980. Protein sturcture and dynamics; enzyme catalysis; NMR spectroscopy.

Hongmin Li, Ph.D., Institute of Biophysics (Beijing), 1995. Crystal structure of macromolecular complexes in bacteria or viral infection and host response.

Ronald Limberger, Ph.D., West Virginia, 1984. Genetics and structural analysis of treponemal motility genes.

Carmen Mannella, Ph.D., Pennsylvania, 1974. Mitochondrial structure and metabolite transport.

Nicholas Mantis, Ph.D., Cornell, 1994. Mucosal immunity and vaccine development to enteropathogenic bacteria and toxins.

Paul Masters, Ph.D., Brandeis, 1981. Molecular virology of coronaviruses; assembly of viral structural proteins and mechanism of RNA synthesis.

Joseph Mazurkiewicz, Ph.D., Colorado, 1973. Cutaneous expression of neuropeptides and their regulation; relationships of cutaneous innervation to injury and pain.

Kathleen McDonough, Ph.D., Stanford, 1990. Microbial pathogenesis; host-parasite interactions; TB and plague.

Bruce McEwen, Ph.D., Cornell, 1982. Use of cutting-edge technology in light and electron microscopy to study the structure and function of vertebrate kinetochores and axonemes.

James McSharry, Ph.D., Virginia, 1970. Pharmacodynamics and pharmacokinetics of antiviral compounds.

Anne Messer, Ph.D., Oregon, 1972. Gene therapy of brain disorders.

Randall Morse, Ph.D., Caltech, 1981. Transcription and chromatin in yeast; roles of the histone amino termini in transcriptional regulation.

Donal Murphy, Ph.D., Michigan, 1974. Expression and function of major histocompatibility complex (MHC) genes and their products.

Kimberlee A. Musser, Ph.D., Albany Medical College, 1998. Molecular analysis of bacterial agents; bioterrorism preparedness.

Dilip Nag, Ph.D., Calcutta, 1983. Genetic recombination; yeast genetics and meiosis.

Kenneth Pass, Ph.D., Medical University of South Carolina, 1976. Analysis of hemoglobin variants.

Janice D. Pata, Ph.D., Colorado at Boulder, 1994. Structure-function studies of specialized cellular and viral polymerases.

Haydeh Payami, Ph.D., Berkeley, 1985. Genetics of aging and neurodegenerative diseases.

Robert Rej, Ph.D., Albany Medical College, 1976. New biochemical tests for disease diagnosis.

Conly L. Rieder, Ph.D., Oregon, 1977. Mitosis and the physiology of cell division; centrosome and kinetochore function.

Thomas Ryan, Ph.D., SUNY at Stony Brook, 1970. Proteases; protease inhibitors; blood clotting; fibrinolysis.

Max Salfinger, M.D., Basel (Switzerland), 1977. Clinical mycobacteriology; tuberculosis.

Erasmus Schneider, Ph.D., Bern (Switzerland), 1983. New mechanisms of multidrug resistance in cancer cells; cooperation between drug metabolism and drug transport.

Richard F. Seegal, Ph.D., Georgia, 1972. Developmental neurotoxicity of PCBs and heavy metals; neuroimmune interactions and Parkinson's disease.

Stewart Sell, M.D., Pittsburgh, 1960. Stem cells in tissue renewal and carcinogenesis; chemical hepatocarcinogenesis.

William Shain, Ph.D., Temple, 1972. Brain response to injury and disease; neuroimmunology.

Jason R. E. Shepard, Ph.D., Tufts, 2004. Biotechnology; biodefense diagnostics; fiber optics; biosensor development.

Pei-Yong Shi, Ph.D., Georgia State, 1995. Molecular biology and biochemistry of flavivirus replication.

Abigail Snyder-Keller, Ph.D., Pittsburgh, 1983. Developmental neuroanatomy and neurotoxicology; effects of prenatal drug exposure.

Derek Symula, Ph.D., Wisconsin–Madison, 1995. Chemical mutagenesis in mouse embryonic stem cells; microarray-based transcriptional profiling to discover and characterize genes involved in heart disease.

Harry Taber, Ph.D., Rochester, 1963. Bacterial gene expression and antibiotic resistence.

Scott A. Tenenbaum, Ph.D., Tulane, 1994. Posttranscriptional gene expression as mediated by RNA-binding proteins; viral-host interactions; ribonomics.

Robert Trimble, Ph.D., RPI, 1969. Structural and biochemical studies of glycoprotein glycan biosynthesis and processing.

James Turner, Ph.D., SUNY at Buffalo, 1973. 3-D light microscopy; quantitative 3-D image analysis; nanobiotechnology.

Anne Walsh, Ph.D., Albany Medical College, 1990. Immunogenetics; interindividual genetic variation; forensic DNA testing.

David E. Wentworth, Ph.D., Wisconsin, 1996. Molecular determinants of interspecies transmission and pathogenesis of influenza and corona viruses.

Gary Winslow, Ph.D., Colorado, 1989. Immunity to intracellular bacteria; superantigen recognition by T cells.

Jonathan R. Wolpaw, M.D., Case Western Reserve, 1970. Neuronal and synaptic substrates of memory; brain-computer interfaces.

Susan J. Wong, Ph.D., Saskatchewan, 1980. Clinical laboratory immunology; vector-borne infectious diseases.

Adjunct Faculty

M. Behr, Ph.D.; J. Carp, Ph.D.; S. Chittur, Ph.D.; J. Figge, M.D.; T. Gemmill, Ph.D.; A. Habura, Ph.D.; G. Hernandez, Ph.D.; J. Izard, Ph.D.; E. Jacobs, Ph.D.; R. Keller, Ph.D.; Q. Lin, Ph.D.; J. Linden, M.D., M.P.H.; A. Macario, M.D.; L. Mayerhofer, Ph.D.; A. Mikhailov, Ph.D., M.D.; S. Philpott, Ph.D.; I. Salkin, Ph.D.; A. Schneider, Ph.D.; N. Strominger, Ph.D.; S. Tieman, Ph.D.; R. Webster, Ph.D.; A. Willey, Ph.D.; W. Wolfgang, Ph.D.; D. Woodland, Ph.D.

THE UNIVERSITY OF ALABAMA AT BIRMINGHAM

School of Natural Sciences and Mathematics
Department of Biology

Programs of Study

The Department of Biology offers programs of study leading to the M.S. and Ph.D. degrees. Graduate students may specialize in research activities at all levels of biological organization, with emphases on ecophysiology, cellular and molecular biology, endocrinology, and ecology of aquatic organisms, or on models related to human disease. The aim of the Department is to provide a broad background and a field of specialty that prepare the student for a professional career in research and/or teaching.

Two types of master's programs are available. A student may choose a research-based program that requires, in addition to a thesis, a minimum of 24 hours of committee-approved course work. The nonresearch plan requires a minimum of 30 hours of approved course work and a thesis incorporating a review and analysis of a topic of current or historical interest in biology. Either plan of study can be completed in approximately two years.

No specific number of courses is required for the Ph.D. Programs are individually designed to meet the needs of the student and to fulfill the aims of the Department. However, a dissertation embodying the results and analysis of an original experimental investigation is required.

Seminars and teaching experience are part of the training program for both the M.S. and Ph.D. degrees. To qualify for candidacy, the student in the master's program must take either a written or an oral comprehensive examination. The Ph.D. student must take both written and oral examinations. The final examination for all candidates consists of an oral defense of the research thesis or comprehensive review paper.

Research Facilities

Well-equipped research laboratories for the Department are located in Campbell Hall. Facilities are available for vertebrates and invertebrates, including marine and freshwater forms, and for botanical specimens. The University operates a farm suitable for field studies. For students interested in marine biology, the University is a member of the Marine Environmental Science Program at Dauphin Island near Mobile, Alabama. The Medical Center library and the University College library have extensive holdings in biological and related sciences.

Financial Aid

Teaching assistantships, graduate assistantships, and fellowships are available. Stipends are awarded on a yearly basis; for 2006–07, they are $14,500 for the master's program and $19,000 for the doctoral program. Tuition and other fees are paid for all students who are awarded stipends. Fellowships can require teaching on a regular basis, and assistantships typically require teaching a maximum of 8 contact hours per week. Students not receiving stipends may teach laboratory sections on a fee-for-service basis.

Cost of Study

Graduate tuition for in-state students was $200 per credit hour in 2005–06. Out-of-state students were charged $300 per credit hour. Tuition and fees are paid for stipend recipients.

Living and Housing Costs

The cost of living in Birmingham is slightly lower than the average for major American cities. Many reasonably priced apartments are available near campus, and some University apartments are available.

Student Group

The total enrollment at the University of Alabama at Birmingham is approximately 17,000; 11,400 are undergraduates and more than 5,700 are in graduate and professional school programs. The Department of Biology averages about 45 graduate students in M.S. or Ph.D. programs.

Student Outcomes

Recent graduates have assumed professorships in departments such as biology, zoology, immunology, or marine science at various academic institutions. Some have chosen a career in the medical or dental profession. Other positions assumed recently by graduates include staff scientists at NASA, the Army Corps of Engineers, and marine research laboratories. Also, different environmental consulting companies and biotechnology companies have employed graduates as technicians, staff scientists, or laboratory directors.

Location

The University of Alabama at Birmingham is a comprehensive urban university situated on a campus that occupies an 80-block area in the southern section of Birmingham. Many cultural resources are available, including the Alys Robinson Stephens Performing Arts Center, museums, the Jimmy Morgan Zoo, and the Botanical Gardens. Recreational opportunities include athletic events and a variety of outdoor activities at nearby lakes and parks or along the Gulf Coast, which is 5 hours away by car. The city has a mild climate throughout the year.

The University

The University of Alabama at Birmingham has forty-six master's and thirty-three doctoral programs. Students benefit from the active research programs that attract $600 million in research funds each year, making the University one of the highest-ranked institutions in receipt of federal research support.

Applying

Application forms, the *Bulletin of the Graduate School,* and other information can be obtained from the Dean of the Graduate School, the University of Alabama at Birmingham, UAB Station, Birmingham, Alabama 35294-1150. For admission in good standing, students should have a baccalaureate degree in biology or a related field, an overall B average in undergraduate courses, and a satisfactory score on the General Test of the Graduate Record Examinations or an equivalent test. It is also desirable that entering students have completed two years of chemistry (including a year of organic chemistry), a year of physics, and mathematics through calculus. A statement of career objectives, three letters of evaluation, and an official copy of transcripts should be included with the application.

Correspondence and Information

Dr. Stephen A. Watts, Program Director for Biology
Department of Biology
CH 375
The University of Alabama at Birmingham
1530 3rd Avenue South
Birmingham, Alabama 35294-1170
Phone: 205-934-2045
Fax: 205-975-6097
E-mail: sawatts@uab.edu
Web site: http://www.uab.edu/uabbio/

The University of Alabama at Birmingham

THE FACULTY AND THEIR RESEARCH

Charles D. Amsler, Professor; Ph.D., California, Santa Barbara. Phycology and chemical ecology.
Robert A. Angus, Professor; Ph.D., Connecticut. Aquatic ecology and toxicology.
Richard B. Aronson, Adjunct Professor; Ph.D., Harvard. Marine ecology and paleoecology.
Asim K. Bej, Professor; Ph.D., Louisville. Molecular genetics and microbial ecology.
Larry R. Boots, Adjunct Assistant Professor; Ph.D., Ohio State. Reproductive endocrinology.
George F. Crozier, Adjunct Professor; Ph.D., California, San Diego. Marine vertebrate physiology.
Joseph J. Gauthier, Associate Professor; Ph.D., UCLA. Microbial physiology.
Vithal K. Ghanta, Professor; Ph.D., Southern Illinois at Carbondale. Immunology.
David T. Jenkins, Associate Professor; Ph.D., Tennessee. *Basidiomycete* taxonomy.
Daniel D. Jones, Professor Emeritus; Ph.D., Michigan State. Plant physiology; microbial ecology.
David W. Kraus, Associate Professor; Ph.D., Clemson. Comparative invertebrate physiology.
Ken R. Marion, Professor; Ph.D., Washington (St. Louis). Vertebrate ecology.
James B. McClintock, Professor; Ph.D., South Florida. Invertebrate biology; marine chemical ecology.
Timothy Nagy, Adjunct Associate Professor; Ph.D., Utah. Nutritional physiology.
Robert W. Thacker, Associate Professor; Ph.D., Michigan. Marine and freshwater ecology.
Trygve Tollefsbol, Associate Professor; Ph.D., North Texas; D.O., North Texas Health Science at Fort Worth. Molecular biology; telomerase and
    DNA methylation.
R. Douglas Watson, Professor; Ph.D., Iowa. Developmental endocrinology.
Stephen A. Watts, Professor; Ph.D., South Florida. Physiology and nutrition of aquatic invertebrates; aquaculture.
Thane Wibbels, Associate Professor; Ph.D., Texas A&M. Comparative reproductive physiology of vertebrates.

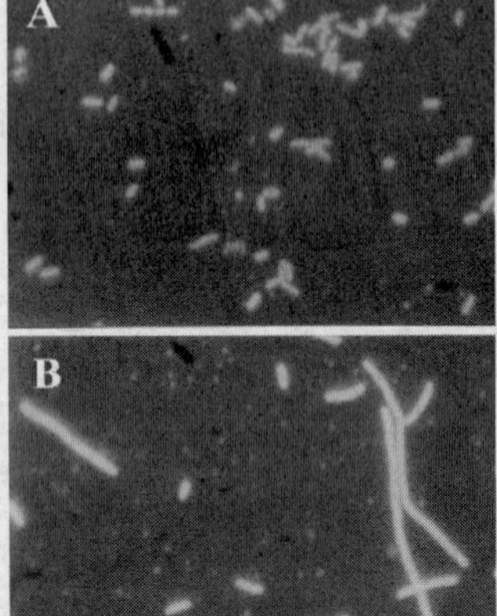

Salmonella—How expression of selected genes and their regulation enable microbes to survive in hostile environments is one area of prokaryotic research. For example, note the influence of cold temperature on morphology of *Salmonella typhimurium* LT2 grown for 78 hours at 37°C (panel A) or 10°C (panel B).

Population dynamics, reproductive ecology, and conservation of several species, including sea turtles, are studied.

Studies of marine chemical ecology are conducted in the Gulf of Mexico, the Atlantic, and, as seen here, beneath the sea ice in Antarctica.

UNIVERSITY OF ARKANSAS FOR MEDICAL SCIENCES
Graduate School

Programs of Study	The Graduate School offers graduate study leading to a Master of Science degree with major work in any one of the following fields: biochemistry and molecular biology, interdisciplinary biomedical sciences, interdisciplinary toxicology, microbiology and immunology, neurobiology and developmental sciences, occupational and environmental health, pathology, pharmacology, and physiology and biophysics. Work is offered leading to the degree of Doctor of Philosophy in biochemistry and molecular biology, interdisciplinary biomedical sciences, interdisciplinary toxicology, microbiology and immunology, neurobiology and developmental sciences, pharmacology, and physiology and biophysics. A Master of Science degree in pharmaceutical sciences is offered through the College of Pharmacy. Master of Science degrees in clinical nutrition, communicative disorders, and genetic counseling are offered through the College of Health Related Professions, and Master of Nursing Science and Doctor of Philosophy programs are offered through the College of Nursing. Each program has its own requirements; details may be obtained from the appropriate program director listed in the faculty section. In general, it takes a student three or four years after receipt of a master's degree to complete the course work, qualifying examinations, research, and dissertation necessary for the Ph.D. degree. Approximately two years are required for the M.S. or the M.N.Sc. degree.
Research Facilities	Research laboratories are located in the Barton Institute of Medical Research, the Shorey building, the Education II building, the Education III building, two biomedical research buildings, the Harvey and Bernice Jones Eye Institute, the Arkansas Cancer Research Center, the Jackson T. Stephens Spine and Neurosciences Institute, and the Donald W. Reynolds Center on Aging. Additional research facilities are the McClellan Veteran's Administration Hospital and the National Center for Toxicological Research. The campus is also a member of the Council of Sponsoring Institutions of Oak Ridge Associated Universities. Experimental animals are housed in a modern vivarium staffed by 2 full-time veterinarians and their assistants. The library has a collection of 177,842 volumes and subscribes to 1,927 journals.
Financial Aid	The campus participates in the full range of federal financial aid programs. Graduate assistantships are awarded through the individual programs, and, for nursing students, Professional Nurse Traineeship Funds are available. Students are also provided limited funding for individual research projects and travel to national meetings.
Cost of Study	Arkansas residents who enrolled for 9 or fewer hours paid tuition of $240 per semester hour in 2005–06. The nonresident rate was $514 per hour. Tuition for full-time students was $2400 for in-state students and $5140 for out-of-state students.
Living and Housing Costs	There is a residence hall on campus. Rooms, apartments, and houses are available in the immediate vicinity of the campus. A cafeteria is available on campus in the University Hospital. Single students should allow about $20,000 per year for general living expenses.
Student Group	In 2005–06, there were 2,328 students enrolled at the University of Arkansas for Medical Sciences: 581 in the College of Medicine, 360 in the College of Pharmacy, 317 in the College of Nursing, 547 in the College of Health Related Professions, 128 in the College of Public Health, and 395 in the Graduate School.
Location	Little Rock, the capital and largest city in Arkansas, is located in the geographic center of the state and is served by major airlines and the interstate highway system. The population of the metropolitan area exceeds 540,000, and a wide range of recreational and cultural activities are available.
The University	The University's Medical Sciences Campus includes the Colleges of Medicine, Pharmacy, Nursing, Health Related Professions, and Public Health and the Graduate School. A 450-bed teaching hospital and various area health education centers across the state are provided to further research facilities for students attending the University Graduate School.
Applying	For application forms and further information about graduate work at the University of Arkansas for Medical Sciences, students should write to the UAMS address or visit the UAMS home page.
	International applicants, including resident aliens whose native language is not English and who do not have a master's degree from a regionally accredited U.S. graduate school, are required to submit a minimum score of 550 on the paper-based version or 213 on the computer-based version of the Test of English as a Foreign Language (TOEFL) before any action is taken on their applications. All international applicants are also required to submit Graduate Record Examinations (GRE) scores.
Correspondence and Information	Graduate School University of Arkansas for Medical Sciences 4301 West Markham Street, Slot 601 Little Rock, Arkansas 72205 E-mail: graduateschooluams@uams.edu Web site: http://www.uams.edu/gradschool

University of Arkansas for Medical Sciences

FACULTY HEADS AND PROGRAM AREAS

Biochemistry and Molecular Biology. Alan D. Elbein, Ph.D., Professor and Chairman. The departmental faculty comprises more than 30 full-time and joint appointees. Research programs include protein structure; glycolipids and glycoproteins; NMR imaging; neurochemistry; biochemistry of hormone receptors; chorionic gonadotrophin; lipid, carbohydrate, and amino acid metabolism; mechanism of oxidative phosphorylation; mechanisms underlying oncogene action, mutagenesis, teratogenesis, carcinogenesis, and detoxification of carcinogens; nucleic acid biochemistry; synthesis and uses of photoaffinity analogues; regulation of gene expression in *E. coli,* yeast, tetrahymena, human muscle cells, and other mammalian cells; DNA methylation; development and aging; structure and function of genes encoding the major histocompatibility complex, ATPases, multidrug resistance transporters, liver redox enzymes, and receptors for LDL and adenosine; identification of genes determining life span; and mechanisms of recombination in human cells.

Interdisciplinary Biomedical Sciences. William D. Wessinger, Ph.D., Associate Professor and Director. The interdisciplinary biomedical sciences (IBS) graduate program offers a flexible array of options for Ph.D. training. Students take a common core curriculum in the first year that provides the foundation for subsequent training in interdisciplinary research tracks. All graduate faculty members in the College of Medicine may participate in the IBS program. By the end of the first year, students choose a doctoral adviser and a research track in the IBS program. Interdisciplinary tracks include aging biology, cancer biology, cell biology, cellular and molecular immunology/immunopathology, and infectious disease and pathogenesis.

Interdisciplinary Toxicology. Jack A. Hinson, Ph.D., Professor and Director, Division of Toxicology, Department of Pharmacology and Toxicology. Research activity centers on the evaluation of the health effects of chemical toxicants. Emphasis is placed on mechanisms whereby chemical toxicants exert their effects. Course work covers the areas of biochemistry, biometry, immunology, pathology, pharmacology, physiology, toxicology, laboratory instrumentation, and laboratory animal techniques. Twenty-four adjunct faculty members are located at the National Center for Toxicological Research (NCTR). Research at NCTR emphasizes the areas of carcinogenesis, mutagenesis, teratogenesis, neurobehavioral toxicology, defense and repair mechanisms, and general toxicology.

Microbiology and Immunology. Roger G. Rank, Ph.D., Professor and Chairman. Faculty research interests in microbiology include bacterial bioremediation, the genetics and pathogenesis of mycobacteria, the molecular pathogenesis of *Staphylococcus aureus* in musculoskeletal infections, the epidemiology of *Giardia lamblia,* and the population dynamics of *Clinostomum marginatum.* Faculty research in virology stresses molecular pathogenesis of infection with papillomavirus, parvoviruses, respiratory syncytial virus, Epstein-Barr virus, varicella-zoster virus, herpes simplex viruses, and picornaviruses. Related topics include responses in viral lung diseases, T-cell immunotherapy of virus-induced tumors, cancer virology, gene therapy, viral diagnostics, antiviral therapy, and vaccines. Faculty interests in immunology include the immune response to tumor cells, age-associated loss of immune function, immunotoxicity from drugs of abuse, the cellular and molecular mechanisms of tolerance induction, and the immunopathogenesis of chlamydial infections.

Neurobiology and Developmental Sciences. Gwen V. Childs, Ph.D., Professor and Chairman. Research interests within the department include cell kinetic analyses, chronobiology, chronotherapy of neoplasms, drug-susceptibility rhythms, developmental neurobiology, neurotrophic factors, neurotransplantation and regeneration of the nervous system, physiology and anatomy of brain stem–spinal locomotor systems, basal ganglia and locomotion, biology of astrocytes, neural cells in culture, molecular cytology, cryobiology and cryobanking, male infertility, electron microscopy, and pattern formation and morphogenesis in regenerating and developing limbs.

Occupational and Environmental Health. Jay Gandy, Ph.D., Professor and Director. The Master of Science program in occupational and environmental health is a cooperative program of the University of Arkansas for Medical Sciences, the University of Arkansas at Little Rock, and the National Center for Toxicological Research. The program trains students and health professionals in industrial hygiene. The program emphasizes occupational hazards related to exposure to the chemicals, infectious diseases, and other substances encountered in hospitals, research laboratories, and industrial environments. The master's thesis project may be either a field experience or a laboratory research project. The 15 full-time and adjunct faculty members conduct research in diverse areas, including industrial toxicology, agricultural toxicology, health physics, noise evaluation and control, air-sampling techniques, and risk assessment.

Pathology. Bruce R. Smoller, M.D., Chairman. Faculty members have M.D. and/or Ph.D. degrees. The Department of Pathology is committed to excellence in teaching, service, and research. Research activities cover both experimental and clinical areas, including theory, application, and techniques in clinical pathology, electron microscopy, histochemistry, cytogenetics, neurotoxicology and neuropathology, heavy-metal toxicology, endocrine pathology, biochemical pathology, musculoskeletal pathology, carcinogenesis, teratology, immunopathology, and advanced techniques in anatomic pathology. The Department of Pathology offers graduate training leading to an M.S. degree in experimental pathology and is also an integral part of the Interdisciplinary Toxicology Graduate Program. Faculty members and students in the pathology department can use the fully staffed research facilities located at the Medical School, the Arkansas Children's Hospital, and the Veterans Administration Hospitals.

Pharmacology. Nancy Rusch, Ph.D., Professor and Chairman, Department of Pharmacology and Interdisciplinary Toxicology. Areas of research interest include neuroscience, substance abuse, behavioral pharmacology and toxicology, autonomic and cardiovascular pharmacology, the pharmacology of aging, the electrophysiology of cardiac muscle, general anesthetics, immunopharmacology, biochemical pharmacology, renal pharmacology, and drug metabolism and disposition. Collaborative arrangements with other departments in the College of Medicine and the College of Pharmacy and with other institutions, such as the National Center for Toxicological Research and the Veterans Administration Hospitals, support the training program.

Physiology and Biophysics. Michael L. Jennings, Ph.D., Professor and Chairman. The Department of Physiology and Biophysics has 35 faculty members (12 primary and 23 joint faculty with primary appointments in other departments). Faculty research interests cover a wide range of fields within the overall theme of regulation of biological function at the level of the gene, protein, cell organ, or whole animal. Specific research areas include protein secretion and targeting, receptor regulation, calcium signaling, ion transport, retinal development, endocrinology, brain metabolism, myelination, exercise science, renal pathophysiology, cardiovascular physiology, space biology, respiratory physiology, gastrointestinal physiology, thermal physiology, musculoskeletal biology, and tumor biology.

College of Health Related Professions. Ronald H. Winters, Ph.D., Dean.

Thomas Guyette, Ph.D., Chair of the Audiology and Speech Pathology Program. The graduate communicative disorders training program holds Education and Service Board certification through the American Speech-Language-Hearing Association in audiology and speech-language pathology. Students may select an emphasis in audiology or in speech-language. Specialized pediatric and geriatric aural rehabilitation training is available. This training program is a consortium of the University of Arkansas for Medical Sciences and the University of Arkansas at Little Rock and uniquely combines the clinical resources of UAMS with the academic offerings of UALR. Students must successfully complete an applied thesis or research project. Exceptional supporting facilities include the Audiology and Speech Pathology Department at the Little Rock Veterans Administration Hospitals and other clinics and agencies in the Little Rock urban area.

Reza Hakkak, Ph.D., Chair and Director of the Clinical Nutrition Program. The Master of Science degree program in clinical nutrition is designed to prepare graduates to practice at an advanced or specialized level. Areas of emphasis are geriatric nutrition, pediatric nutrition, and nutrition and wellness/disease prevention.

Shannon N. Barringer, M.S., C.G.C., Chair of the Genetic Counseling Program.

College of Nursing. Linda C. Hodges, Ed.D., Dean. The Master of Nursing Science program provides opportunities for study in the following areas of advanced nursing practice: adult nurse practitioner/clinical nurse specialist–case manager, family nurse practitioner, gerontology nurse practitioner, nursing administration, pediatric nurse practitioner/clinical nurse specialist–case manager, and women's health nurse practitioner. The doctoral program prepares graduates to advance the art and science of nursing through research and scholarship. Faculty research focuses on management of individual and family responses to acute and chronic heart problems across the life span as well as health services research.

College of Pharmacy. L. D. Milne, Ph.D., Dean. Research activity is conducted in the following fields: medicinal chemistry, pharmaceutics, pharmacology, hospital pharmacy, and radiological health. All research laboratories are on the campus.

 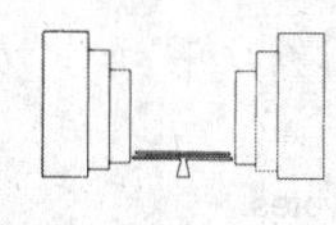

UNIVERSITY OF CALIFORNIA, SAN DIEGO

Division of Biological Sciences in Association with the Salk Institute for Biological Studies

Program of Study	The University of California, San Diego (UCSD) Division of Biological Sciences, in association with the Salk Institute for Biological Studies, offers a program of graduate studies leading to the Ph.D. degree in biology. The program provides an extraordinary setting for highly motivated students interested in training with distinguished faculty members. The major areas of faculty research are biochemistry and biophysics; bioinformatics; cell and developmental biology; computational neurobiology; ecology, behavior, and evolution; genetics and molecular biology; immunology, virology, and cancer biology; neurobiology; plant molecular biology; plant systems biology; and signal transduction. During the first year, students complete four to six 6-week research internships (rotations) in laboratories that the student determines to be of interest. To promote a diversity of research experiences and help initiate new students into the community of advanced graduate study and research, students are encouraged to pursue rotations in at least one lab not directly related to their thesis research interests. Following the rotation period, students select a faculty adviser and lab for their thesis research. First-year students are also expected to enroll in a graduate biology course sequence. Specific course sequences are determined by the student's area of specialization. The first year concludes with the administration of a qualifying exam that is given in conjunction with the first-year course sequence. Subsequent years are comprised primarily of thesis-oriented research (the thesis proposal is presented at the end of the second year), teaching assistant responsibilities (one quarter per year for three years), and meetings with the 5-member Thesis Committee until the thesis defense, typically after 5.5 years.
Research Facilities	The laboratories of the Division of Biological Sciences and the Salk Institute are fully equipped for modern biological research. Students may pursue a thesis with any participating faculty member in the Division or at the Salk Institute. Students may also be given permission to complete thesis research in the facilities of other science departments on campus, including the Department of Chemistry and Biochemistry, the School of Medicine, the Scripps Institution of Oceanography, and the UCSD Moores Cancer Center. Students may, in some cases, pursue their research in the labs of adjunct faculty members from non-UCSD facilities, such as the Burnham Institute, Scripps Research Institute, the La Jolla Institute for Allergy and Immunology, and the Center for the Reproduction of Endangered Species at the San Diego Zoo.
Financial Aid	All Ph.D. students in good standing receive full financial support (tuition, fees, and health insurance) for 5½ years. In addition, students receive an annual stipend. Currently, the stipend level is $24,500.
Cost of Study	Cost of study is detailed in the Financial Aid section of this In-Depth Description.
Living and Housing Costs	Limited affordable on-campus housing is available for both single and married graduate students. There are also several off-campus housing options for graduate students in neighborhoods near or easily accessible to campus. The housing and dining office (http://hds.ucsd.edu/housing/) maintains current information on rental options.
Student Group	In fall 2005, there were 20,980 undergraduate and 5,160 graduate (including medical) students at UCSD. There are currently approximately 250 graduate students from across the United States and from several other countries in the Division of Biological Sciences.
Location	The University and the Salk Institute are located on the San Diego coast, overlooking the Pacific Ocean. The climate is conducive to year-round outdoor activities. The area shares a rich heritage with its neighbor, Mexico, and supports many cultural events. The scientific community includes the UCSD School of Medicine, the Scripps Institution of Oceanography, the Scripps Research Institute, the San Diego Supercomputer Center, the La Jolla Institute for Allergy and Immunology, the Burnham Institute, and hundreds of biotechnology research firms.
The University	UCSD has rapidly established itself as one of the top institutions in the country for higher education and research. This success has been driven by the University's dynamic science faculty and graduate programs. As evidence, UCSD ranks seventh in National Academy of Sciences membership, the National Science Foundation ranks UCSD fifth in federal R&D expenditures, and, in 2001, the *Nature Yearbook of Science and Technology* described UCSD as one of the ten most powerful research universities in the United States.
Applying	Inquiries concerning application procedures should be directed to the Biology Graduate Admissions Coordinator. Applications are accepted for fall enrollment only and due in mid-December. GRE General Test scores are required. GRE Subject Test scores are recommended. International applicants must first complete a preapplication available for download from the Biology Graduate Admissions Web site. If the preapplication is deemed competitive, the applicant is invited to complete the official online application. TOEFL scores are also required for international applicants whose first language is not English. Competitive applicants must demonstrate strong performance in the physical sciences and mathematics. Research experience is also considered desirable. The program encourages applications from members of underrepresented groups.
Correspondence and Information	Biology Graduate Admissions Coordinator Division of Biological Sciences, 0348 University of California, San Diego 9500 Gilman Drive La Jolla, California 92093-0348 Phone: 858-534-3835 Fax: 858-534-4459 E-mail: gradprog@biology.ucsd.edu Web site: http://www-biology.ucsd.edu/

University of California, San Diego

THE FACULTY AND THEIR RESEARCH

UNIVERSITY OF CALIFORNIA, SAN DIEGO

Eduardo Macagno, Dean: cellular and molecular mechanisms underlying the specification of neuronal identity and the generation of neuronal arbors.

Peter Andolfatto: population genetics; adaptation; demography and genome evolution; *Drosophila* and *Lepidopteran* genomics.

Raffi Aroian: relationship between cellular architecture (cytoskeleton) and cellular function.

Doris Bachtrog: population genetics, molecular evolution and comparative genomics; evolution of sex.

Tim Baker: cryoelectron microscopy; three-dimensional computer reconstruction of macromolecules.

Darwin K. Berg: molecular and cellular neurobiology; regulation of synaptic components on neurons.

Ethan Bier: *Drosophila* neural development.

Lisa Boulanger: how proteins of the immune system participate in normal brain development and synaptic plasticity.

Steven Briggs: embryonic stem cell fate; plant resistance to infectious disease.

Stuart Brody: biochemical genetics of circadian rhythms.

Lin Chao: evolutionary processes in microbes; bacteria and their viruses.

Maarten J. Chrispeels: plant molecular and cell biology; regulation of gene expression and protein targeting.

Nigel M. Crawford: plant molecular biology; control of gene expression and ion transport.

Michael David: mechanisms by which interferons and other cytokines activate the transcription of early response genes.

Russell F. Doolittle: evolution of protein structure/function.

Robert Dutnall: regulation of chromatin structure and function.

Scott Emr: phosphoinositide signaling and the control of membrane traffic in eukaryotic cells.

Daniel Feldman: neurophysiology; development and plasticity of sensory cortical maps.

Marla B. Feller: mechanisms and function of spontaneous activity in the developing mammalian visual system.

Richard A. Firtel: mechanisms regulating gene expression and cell differentiation during development.

Douglass Forbes: nuclear transport in vitro; nuclear pore structure.

E. Peter Geiduschek: biochemical and genetic approaches to control of gene activity.

Anirvan Ghosh: molecular neurobiology; mechanisms of cortical development and plasticity.

Partho Ghosh: mechanisms of bacterial and protozoan pathogenesis; host response against infectious microbes.

Ananda Goldrath: generation and maintenance of T-cell immunity.

Randy Hampton: genetics and molecular biology of protein degradation in yeast.

Stephen M. Hedrick: immune recognition and development.

Donald R. Helinski: regulation of plasmid DNA replication and plasmid DNA partitioning.

Hopi Hoekstra: evolution and population genetics; the genetic basis of adaptation; genomics.

David Holway: ecological basis of invasive species; Argentine ants; fire ants—California.

John Huelsenbeck: molecular evolution; phylogenetics; genetics of adaptation.

Colin Jamora: cell biology of morphogenesis; molecular mechanisms of organ development in mice.

Walter Jetz: ecological patterns at the level of the individual, population and community.

Randall Johnson: regulation of response to hypoxia and other microenvironmental stresses in vivo.

Tracy Johnson: mechanisms of pre-mRNA splicing; mRNA synthesis and processing.

James T. Kadonaga: transcriptional regulation in eukaryotes.

Amy Kiger: cellular morphogenesis; functional genomic analysis of cell shape changes in *Drosophila* development.

Joshua Kohn: plant population biology; mating system evolution; ecological genetics.

William B. Kristan: neurobiology of simple nervous systems.

Mitchell Kronenberg: presentation of peptide and nonpeptide antigens to T lymphocytes and mucosal immunity.

Russell Lande: modeling the static and dynamic properties of quantitative characters to derive laws that will be useful in the analysis of phenotypic evolution.

William F. Loomis Jr.: developmental biology of *Dictyostelium*.

Vivek Malhotra: organization of the Golgi membranes; regulation of vesicular transport.

Karen Marchetti: evolution and behavior; species divergence in color patterns and color vision; host-parasite interactions.

William McGinnis: molecular genetics of early *Drosophila* embryogenesis.

Mauricio Montal: molecular basis of electrical excitability and signal transduction in neural membranes.

Cornelis Murre: helix-loop-helix proteins in B-cell development and oncogenesis.

John Newport: nuclear structure, regulation of the cell cycle, and early development.

James Nieh: evolution of animal language: symbolic communication in highly social bees; sensory physiology and mechanisms of multimodal communication; bioacoustics; cyborg bee project.

Maho Niwa: signal transduction pathways that regulate the functional capacity of organelles.

Amy Pasquinelli: genetic and molecular approaches for understanding how tiny regulatory RNAs control gene expression.

Gentry Patrick: neurobiology; the ubiquitin-proteasome system (UPS), one of the major cellular pathways controlling protein turnover in mammalian cells.

Lorraine Pillus: chromosome structure and function; genetics and epigenetics of transcription; position effects and transcriptional repression.

Joe Pogliano: DNA segregation in prokaryotes.

Kit Pogliano: how the progeny of a single cell division assume differing fates.

James W. Posakony: pattern formation and gene expression in *Drosophila* development.

Paul A. Price: analysis of protein structure and physiological function; posttranslational protein modification.

Pamela Reinagel: neural coding of visual information by thalamic neurons; temporal and population coding, natural scenes, and visual behavior.

Michael G. Rosenfeld: transcriptional regulation of development in the endocrine and neural systems.

Kaustuv Roy: physical and biotic controls on the distribution and diversity of species in benthic marine ecosystems.

Oliver Ryder: genetics of endangered species; zoo biology.

Milton H. Saier: transmembrane transport and the regulation of cellular physiology and transcription.

Massimo Scanziani: developmental neurobiology.

William R. Schafer: molecular basis of behavior in the nematode *Caenorhabditis elegans*.

Immo E. Scheffler: somatic-cell genetics and molecular biology.

Robert J. Schmidt: plant molecular biology; mechanisms of tissue-specific gene expression in maize.

Julian I. Schroeder: plant cell and molecular biology; signal transduction and ion channels in higher plant cells.

Terrence J. Sejnowski: computational models of the nervous system; mechanisms for synaptic plasticity.

Laurie Smith: molecular genetic analysis of plant cell division and morphogenesis.

Nicholas C. Spitzer: control of expression of neuronal phenotypes.

Suresh Subramani: organelle homeostasis; peroxisome function and biogenesis; peroxisomal disorders; peroxisomal protein import; peroxisome turnover by autophagy.

David Traver: vertebrate blood cell development; biology of hematopoietic stem and progenitor cells.

Jing W. Wang: systems neurobiology; representation and processing of olfactory information in simple olfactory system.

Carl Ware: immune regulation by TNF-related cytokines and viral evasion strategies.

Steven A. Wasserman: molecular genetic analysis of patterning, cell-cycle control, and signal transduction in *Drosophila* development.

Christopher J. Wills: molecular evolution; genetic variation at DNA level.

David S. Woodruff: conservation and evolutionary biology; speciation; genetics of endangered species.

Yang Xu: lymphocyte development and signaling pathways in response to DNA damage.

Michael P. Yaffe: membrane and organelle biogenesis; intracellular movement of macromolecules.

Martin F. Yanofsky: molecular and genetic analyses of flower development.

Yunde Zhao: plant biology, molecular mechanisms of auxin homeostasis, and signal transduction.

Charles S. Zuker: molecular neurobiology; signal transduction.

SALK INSTITUTE

Senyon Choe: X-ray crystallography of biological macromolecules.

Joanne Chory: genes that control light-regulated gene expression in *Arabidopsis*.

Andrew Dillin: pathways that regulate the aging process.

Joseph Ecker: molecular genetics of *Arabidopsis*.

Beverly Emerson: transcriptional regulation in differentiation and disease.

Ron Evans: molecular analysis of neuroendocrine function.

Fred Gage: regeneration in the adult mammalian nervous system.

Martyn Goulding: homeobox genes in neural development.

Martin Hetzer: organization and assembly of the nuclear envelope during development and cell proliferation.

Tony Hunter: role of protein phosphorylation in oncogenesis and cell-cycle regulation.

Katherine Jones: HIV transcriptional regulation in vitro.

Leanne Jones: mechanisms controlling stem cell behavior.

Jan Karlseder: structure and function of mammalian telomeres.

Chris Kintner: molecular biology of neurogenesis in amphibian embryos.

Nathaniel Landau: chemokine receptors and HIV accessory genes.

Kuo-Fen Lee: study of neural development and function using gene targeting.

Jeffrey Long: understanding apical/basal polarity in plants.

Marc Montminy: characterization of the structure, regulation of nuclear factors which stimulate gene expression.

Joseph Noel: stereochemical principles governing biological processes; emphasis on signal transduction.

Dennis O'Leary: development of the vertebrate nervous system.

Satchin Panda: molecular mechanism of the biological clock in a mouse model system.

Samuel Pfaff: molecular mechanisms that control vertebrate motor axon pathfinding.

Bart Sefton: cellular transformation, signal transduction, and the regulation of cell proliferation.

Paul Slesinger: "molecular gates" that regulate the movement of ions in and out of a cell.

John Thomas: molecular genetics of neuronal development in *Drosophila*.

Jim Umen: control of cell size and organelle DNA inheritance.

Inder Verma: oncogenes, proto-oncogenes, and gene transfer.

Geoff Wahl: genome remodeling in mammalian cells.

Lei Wang: strategies for molecular evolution and molecular imaging.

Matthew Weitzman: virus replication and the development of viral vectors for gene therapy.

John Young: mechanisms of retroviral and anthrax toxin entry into cells.

UNIVERSITY OF CALIFORNIA, SAN DIEGO

School of Medicine
Graduate Studies in Biomedical Sciences

Program of Study

The graduate program in biomedical sciences based at the School of Medicine of the University of California, San Diego (UCSD), provides training at the cutting-edge of basic and interdisciplinary biosciences. Advanced training tracks in molecular cell biology, molecular pharmacology, physiology, genetics, microbiology/immunology, and molecular pathology are offered as well as cross-track focus areas in bioinformatics, cancer biology, chemical biology, developmental biology, glycobiology, reproductive endocrinology, neurobiology, and structural and chemical biology. The cross-disciplinary design of the program is supported by faculty members from multiple departments who have in common a commitment to Ph.D. training in the biomedical sciences. The program's strengths lie in the multiple quantitative technologies utilized and developed by the basic investigators to study fundamental problems in molecular, cellular, and integrative biology. Students choose the research area most compatible with their interests.

The program has developed as distinguished faculty members were recruited from the School of Medicine, Salk Institute, UCSD Cancer Center, and the Howard Hughes Medical Institute for Molecular Biology. Collaborative efforts within this diverse group of faculty members provide research opportunities that traverse multiple traditional disciplines and permit insightful scientific advances.

First-year students take a core curriculum that covers both fundamental aspects and current advances in biomedical research. Laboratory rotations and 15 units of advanced electives complete the course requirements. A second exam at the end of the third year permits the student to enlist input into a thesis project from a selected doctoral committee. After choosing a dissertation adviser, students write and defend a theoretical research proposal as a qualifying examination. Students generally receive their Ph.D. at the end of the fifth year.

Research Facilities

The program's research laboratories are located in the Biomedical Science, Leichtag Center for Molecular Genetics, and Cell and Molecular Medicine Buildings on the Health Sciences campus in La Jolla, adjacent to the Biomedical Library. The Skaggs School of Pharmacy and the Pharmaceutical Sciences (SSPPS) and the Departments of Biology and Chemistry also house several of the program's faculty members. Within walking distance are the San Diego Supercomputer Center, the Scripps Institution of Oceanography, the Salk Institute, and the Veterans Administration Medical Center. Additional laboratories are located in the Medical and Clinical Teaching Facilities, the Moores Cancer Center, and at the UCSD Medical Center at Hillcrest.

Financial Aid

All students in good standing in the program receive full financial support (tuition, fees, health insurance, and a living stipend) throughout their Ph.D. training. Sources of support are NIH training grants, University fellowships, and research grants.

Cost of Study

All graduate students receive full financial support.

Living and Housing Costs

In 2005–06, limited on-campus housing was available from $550 per month for a single to $1250 per month for a family. A wide range of off-campus housing is available in San Diego and in neighboring communities within a 15-minute drive of the campus and the ocean. The off-campus housing office maintains up-to-date information on rental listings.

Student Group

In 2005–06, there were 3,629 graduate and 19,763 undergraduate students at UCSD. Students in the basic biomedical sciences program were selected from outstanding colleges throughout the nation and abroad and include recent college graduates as well as those having substantial postcollege research experience. Students currently enrolled in the program total 160. Graduates routinely obtain postdoctoral positions in excellent laboratories and move on to challenging careers in academia, research institutes, and industry.

Location

The 1,300-acre UCSD campus lies in La Jolla, on the northern edge of San Diego, atop high bluffs overlooking the Pacific Ocean. The intellectual activities on campus are enriched by the proximity of the Salk Institute, the Scripps Clinic and Research Foundation, and a burgeoning group of companies engaged in biotechnological research and development. San Diego offers a year-round mild Mediterranean climate, miles of shoreline, and ready access to mountains, deserts, and Mexico. Cultural pleasures include the Old Globe Theatre, La Jolla Playhouse, the San Diego Symphony and Opera, and a wide variety of fine art galleries.

The University

In its forty-six years of existence, UCSD has progressed rapidly to become one of the premier research institutions in the United States. Built on the strong scientific base first established by the Scripps Institution of Oceanography, UCSD's research support is second among American universities. The scientific environment is full of excitement, with a national supercomputer center, Howard Hughes Medical Institute, Ludwig Institute for Cancer Research, the Skaggs School of Pharmacy and Pharmaceutical Sciences, the Rebecca and John Moores UCSD Cancer Center, and several other new research buildings on campus.

Applying

Applications are encouraged from outstanding students who have majored in any laboratory science or in mathematics. Desirable features include a minimum GPA of 3.25 and GRE scores above the 80th percentile on the verbal, quantitative, and analytical sections of the General Test (required) and on the Subject Test in biology, chemistry, or biochemistry and cell and molecular biology (recommended). Undergraduate prerequisites include organic chemistry, biochemistry, physical chemistry, calculus, and biology; some exposure to cell or molecular biology and mammalian physiology is strongly advised. Prior research experience is desirable. The pre-application is required before the application is considered. Applications, considered for the fall quarter only, are due December 1; earlier submission is advantageous.

Students can initiate the application process by visiting http://biomedsci.ucsd.edu.

Correspondence and Information

Graduate Studies in Biomedical Sciences
School of Medicine, 0685
University of California, San Diego
9500 Gilman Drive
La Jolla, California 92093-0685
Phone: 858-534-3982
Web site: http://biomedsci.ucsd.edu

University of California, San Diego

THE FACULTY AND THEIR RESEARCH

Joseph Adams, Associate Professor; Ph.D., Pennsylvania. Protein kinases; enzyme catalysis and regulation.

Katerina Agassaglou, Assistant Professor; Ph.D., Athens (Greece). Multiple sclerosis; extracellular matrix; fibrin; regeneration.

Kim E. Barrett, Professor; Ph.D., University College (London). Role of mast cell in allergic and inflammatory diseases.

Timothy Bigby, Professor; M.D., Baylor. Airway inflammation and lipid mediator pathways involved in diseases of the airway.

Roland C. Blantz, Professor; M.D., Johns Hopkins. Basic renal physiology; filtration resorption.

Anthony Wynshaw Boris, Professor; M.D., Ph.D., Case Western. Mouse cancer genetics.

Gerry R. Boss, Professor; M.D., California, Irvine. Signal transduction and gene expression.

Philip E. Bourne, Professor; Ph.D., Flinders (Australia). Bioinformatics, with emphasis on structural bioinformatics.

Joan Heller Brown, Professor; Ph.D., Yeshiva (Einstein). Neurotransmitters; phosphoinositide turnover; protein kinases.

Laurence L. Brunton, Professor; Ph.D., Virginia. Cyclic nucleotide metabolism; regulatory actions of protein kinases.

John Carethers, Professor; M.D., Wayne State. Colorectal cancer.

Dennis A. Carson, Professor; M.D., Columbia. Immunology of autoimmune, immunodeficiency, and neoplastic diseases.

Webster K. Cavenee, Professor; Ph.D., Kansas. Human cancer genetics; tumor suppression; oncogene function.

Ju Chen, Associate Professor; Ph.D., Indiana. Molecular mechanisms of heart and skeletal muscle diseases.

Shu Chien, Professor; M.D., National Taiwan; Ph.D., Columbia. Molecular basis of blood cell biophysics; endothelial transport.

Mario Chojkier, Professor; M.D., Buenos Aires. Regulation of collagen gene transcription and tissue fibrogenesis.

Jerold Chun, Adjunct Professor; M.D., Ph.D., Stanford. Cerebral cortex development.

Don W. Cleveland, Professor; Ph.D., Princeton. Microtubule motors and chromosome movement.

Edward A. Dennis, Professor; Ph.D., Harvard. Phospholipase regulation and mechanism; prostaglandin generation.

Arshad Desai, Assistant Professor; Ph.D., California, San Francisco. Spatial segregation of the replicated genome during cell division.

Nazneen Dewji, Associate Adjunct Professor; Ph.D., London. Molecular and cellular basis of Alzheimer's disease.

Wolfgang H. Dillmann, Professor; M.D., Munich. T_3 effects on SR Ca^{2+}; ATPase gene; function of heat shock proteins in the heart.

Jack Dixon, Professor and Dean of Scientific Affairs; Ph.D., California, Santa Barbara. Roles of proteins in cellular response to molecular signals.

Daniel Donoghue, Professor; Ph.D., MIT. Growth factors; cell cycle regulation; autocrine; $cdc2$/cyclin transformation; FGFR.

Steven Dowdy, Professor; Ph.D., California, Irvine. Regulation of G1 cell cycle progression in cancer.

Mark H. Ellisman, Professor; Ph.D., Colorado at Boulder. Cellular neurobiology and ultrastructural plasticity of nervous systems.

Scott D. Emr, Professor; Ph.D., Harvard. Cell biology of intracellular protein sorting and organelle biogenesis.

Jeffrey D. Esko, Professor; Ph.D., Wisconsin–Madison. Molecular and genetic analysis of glycosylation.

Ronald M. Evans, Adjunct Professor; Ph.D., UCLA. Molecular genetics of steroid, thyroid, and retinoid receptors.

Marilyn G. Farquhar, Professor; Ph.D., Berkeley. Signaling in intracellular membrane traffic; cell and molecular basis.

James R. Feramisco, Professor; Ph.D., California, Davis. Role of protogenes and oncogenes in cell growth and differentiation.

Theodore Friedmann, Professor; M.D., Pennsylvania. Human gene therapy; viral vectors; genome mapping.

Xiang-Dong Fu, Professor; Ph.D., Case Western Reserve. RNA processing and regulation.

Richard Gallo, Professor; M.D., Ph.D., Rochester. Mechanisms of innate defense by antimicrobial peptides and proteoglycans.

Christopher K. Glass, Professor; M.D./Ph.D., California, San Diego. Transcriptional control of macrophage development.

Joseph Gleeson, Associate Professor; M.D., Chicago. Molecular genetics of brain development.

Larry S. B. Goldstein, Professor; Ph.D., Washington (Seattle). Intracellular modality; cell division; cytoskeleton; neurobiology.

John Guatelli, Professor; M.D., California, San Diego. Molecular virology of HIV-1 replication, emphasizing the function of the Nef protein.

Bruce Hamilton, Professor; Ph.D., Caltech. Genetics and functional genomics.

Stephen Hedrick, Professor; Ph.D., California, Irvine. Immune recognition and development.

Michael Hogan, Adjunct Professor; Ph.D., Tennessee, Knoxville. Exercise physiology; skeletal muscle metabolism; peripheral oxygen delivery.

Stephen B. Howell, Professor; M.D., Harvard. Pharmacology of anticancer drugs; chemotherapeutic agents.

Paul A. Insel, Professor; M.D., Michigan. Signal transduction by catecholamines, ATP, and G-proteins.

Wolfgang G. Junger, Adjunct Professor; Ph.D., Vienna (Austria). Cellular immune response to trauma, hemorrhage, and burns.

Martin F. Kagnoff, Professor; M.D., Harvard. Immunoglobulin isotype regulation; HLA class II D region genes.

Michael Karin, Professor; Ph.D., UCLA. Regulation of gene transcription; cell type specific gene expression.

Kenneth Kaushansky, Professor; M.D., Chicago. Gene expression; growth control; hematopoesis; megakaryocyte development; molecular biology; stem cell transplantation.

John Kelsoe, Professor; M.D., Alabama at Birmingham. Identification of genes which predispose to psychiatric illnesses.

Thomas J. Kipps, Professor; M.D., Ph.D., Harvard. Human cell physiology, signal transduction, and gene expression.

Richard D. Kolodner, Professor; Ph.D., California, Irvine. Genetics and biochemistry of DNA repair and recombination.

Elizabeth A. Komives, Professor; Ph.D., California, San Francisco. Protein-protein interactions.

Ronald Kuczenski, Professor; Ph.D., Michigan State. Regulation and pharmacology of brain transmitters.

Mark Lawson, Assistant Professor; Ph.D., California, Irvine. Molecular mechanisms of hormone action in the pituitary.

Hyam L. Leffert, Professor; M.D., Yeshiva (Einstein). Growth processes of hepatocytes; ion fluxes and gene expression; gene therapy.

Fred Levine, Professor; Ph.D., Washington. Gene transfer/gene therapy for diabetes.

Richard Lieber, Professor; Ph.D., California, Davis. Structure-function in skeletal muscle and plasticity.

Jamey Marth, Professor; Ph.D., Washington (Seattle). Molecular and developmental biology of vertebrate oligosaccharides.

J. Andrew McCammon, Professor; Ph.D., Harvard. Computer-aided molecular and drug design; theory of molecular recognition.

Pamela Mellon, Professor; Ph.D., Berkeley. Developmental and hormonal regulation of gene expression.

Marc R. Montminy, Adjunct Professor; M.D., Ph.D., Tufts. CREB and cyclic AMP responsive transcription.

Alexandra Newton, Professor; Ph.D., Stanford. Signal transduction involving protein phosphorylation; protein kinase C.

Sanjay Nigam, Professor; M.D., Pennsylvania. Identifying key growth factors and genes.

Victor Nizet, Associate Professor; M.D., Stanford. Molecular bacterial pathogenesis and innate immunity.

Daniel T. O'Connor, Professor; M.D., California, Davis. Hypertension; catecholamine storage vesicles.

Karen Oegema, Assistant Professor; Ph.D., California, San Francisco. Structural rearrangements of the cytoskeleton required for cell division.

Jerrold M. Olefsky, Professor; M.D., Illinois. Molecular and cellular mechanisms of insulin action; pathophysiology of diabetes mellitus.

Renate Pilz, Professor; M.D., California, San Diego. Signal transduction by cyclic nucleotide-dependent protein kinases.

Frank L. Powell, Professor; Ph.D., California, Davis. Respiratory physiology; CO_2 sensitivity; acclimatization.

Morton P. Printz, Professor; Ph.D., Pittsburgh. Genetics of stress and hypertension; brain renin-angiotensin system.

Oswald Quehenberger, Professor; Graz (Austria). Mechanisms of monocyte recruitment and activation.

Eyal Raz, Professor; M.D., Hebrew (Jerusalem). Immunostimulatory DNA sequences.

Bing Ren, Assistant Professor; Ph.D., Harvard. Apoptosis; β lymphocyte development.

Douglas D. Richman, Professor; M.D., Stanford. HIV: Replication and pathogenesis.

Geof Rosenfeld, Professor; M.D., Rochester. Molecular biology of neuroendocrine gene expression.

Geert Schmid-Schoenbein, Professor; Ph.D., California, San Diego. Microcirculation; ischemia; hypertension; rheology.

Shunichi Shimasaki, Professor; Ph.D., Hiroshima (Japan). Molecular and cellular biology of ovarian function.

Gregg Silverman, Professor; M.D., Rutgers. Autoimmune disease.

Stephen Spector, Professor; M.D., Tufts. HIV: Pathogenesis and treatment of congenital and perinatal infections; viral diseases.

Charles F. Stevens, Professor; M.D., Yale; Ph.D., Rockefeller. Synaptic transmission.

Palmer Taylor, Professor; Ph.D., Wisconsin. Molecular structure and regulation of enzymes and receptors.

Susan S. Taylor, Professor; Ph.D., Johns Hopkins. Biochemistry: Structure and function of protein kinases.

Lynn Ten Eyck, Adjunct Professor; Ph.D., Princeton. Protein interactions; crystallography computational chemistry.

Roger Y. Tsien, Professor; Ph.D., Cambridge. Cell biology of intracellular signaling; molecular engineering.

Robert H. Tukey, Professor; Ph.D., Iowa. Use of recombinant DNA to characterize rabbit and human cytochrome P-450.

Eric Turner, Professor; M.D., Ph.D., Washington. Basis mechanisms of brain development.

Wylie W. Vale, Professor; Ph.D., Baylor. Neuroendocrine hormones and growth factors and their receptors.

Ajit P. Varki, Professor; M.D., Christian Medical College, Vellore (India). Glycobiology: Biochemistry and molecular biology.

Francisco Villareal, Adjunct Professor; M.D., Ph.D., California, San Diego. Pathophysiology of cardiac hypertrophy and remodeling.

Judith Varner, Associate Professor; Ph.D., Basel (Switzerland). Angiogenesis; metastasis.

Peter D. Wagner, Professor; M.D., Sydney. Pulmonary physiology; oxygen transport and utilization; exercising skeletal muscle.

Nicholas J. G. Webster, Professor; Ph.D., Stanford. Transcriptional regulation of insulin receptor gene.

David S. Williams, Adjunct Professor; Ph.D., Australian National. Photoreceptor cells; retinal structure and signal transduction.

Joseph L. Witztum, Professor; M.D., Washington (St. Louis). Role of oxidation of LDL.

Virgil Woods, Professor; M.D., California, San Francisco. Functional proteomics.

Anthony Wynshaw-Boris, Professor; M.D., Ph.D., Case Western. Mouse cancer genetics.

Tony L. Yaksh, Professor; Ph.D., Purdue. Role of spinal receptors in modulating sensory and autonomic transmission.

Jason Yuan, Professor; M.D., Ph.D., Peking Union. Vascular physiology and pathophysiology.

Maurizio Zanetti, Professor; M.D., Padova (Italy). Structure-function of immunoglobulins.

Huilin Zhou, Assistant Professor; Ph.D., Stanford. Cell cycle regulation; signal transduction.

UNIVERSITY OF CINCINNATI

College of Medicine
Biomedical Sciences Flex Option Program

Programs of Study	Graduate programs in biomedical sciences at the University of Cincinnati (UC) College of Medicine are designed to educate and train graduate students for careers in biomedical research, including academia, industry, government, and private enterprise. The graduate programs provide a collegial atmosphere that fosters top-quality research combined with in-depth course work, seminars, and journal clubs. In this environment, students learn to apply the latest scientific technology to the most exciting areas of biomedical research. The graduate programs are supported by faculty research that has direct bearing on questions of human biology and disease, such as human genetics and cancer. In addition, extensive use of research facilities, such as functional genomics, transgenic technology in model organisms, and imaging, is a hallmark of many individual research programs, with exceptional access to state-of-the-art technical and core facilities.

Although it may vary somewhat depending on the program, during the first year, most students take a series of core curriculum courses and carry out laboratory rotations with 2 or 3 faculty members. At the end of this year, each student selects a thesis adviser. During the second year, students begin their thesis research and attend advanced courses or special topics seminars. By the end of the second year, students take the qualifying examination, which is in the form of a written proposal and an oral exam. The last two to three years are devoted exclusively to research, which includes presenting at national meetings and publishing findings in peer-reviewed journals.

The Biomedical Sciences Flex Option Program is designed to offer first-year doctoral students maximum exposure to the range of research available at the UC Academic Health Center and maximum flexibility in the choice of graduate programs in the College of Medicine, prior to the selection of a graduate program and area of dissertation research. Students who gain acceptance into graduate study via the Flex Option join the doctoral program of their choice in their second year. It is understood that acceptance of any student into the Flex Option guarantees acceptance of that student into any of the ten Ph.D. programs in the College of Medicine, as long as the student is in good academic standing. (M.D., M.D./Ph.D., M.S., and Sc.D. programs are not available to Flex students.)

Flex students are advised in the Office of Research and Graduate Education and enroll in the core course sequence along with other first-year graduate students in the College of Medicine. Students may select from more than 200 research faculty members in twelve doctoral programs in the College of Medicine and enter their chosen program as a second-year student in that program.

Students admitted in the fall are expected to start laboratory rotations during the preceding summer. This introduces the student to the research areas available so that he or she is ready to focus on core courses in the fall quarter. The student should be able to complete one or two lab rotations during the summer quarter. By the end of the first year of study, as a student prepares to enter a program, he or she is referred to potential mentors.

Research Facilities
The University of Cincinnati has fifteen libraries, including the Chemistry and Biology Library and the Academic Information Technology and Libraries (AIT&L), which houses the Health Sciences Library and the Cincinnati Medical Heritage Center. AIT&L also has 4,300 journals in paper and electronic form, approximately 292,000 volumes, and 326 bibliographic databases. The University's faculty members are making stunning advancements in biotechnology; human and environmental genetics; cardiovascular, cardiopulmonary, and cancer biology; neuroscience; developmental biology; immunology; and microbiology as well as many other disciplines. Core research facilities support researchers across disciplines and include neuroimaging (high-field MRI), BSL3 biocontainment, LAMS, biostatistics, DNA sequencing, transgenics, gene targeting/gene knockout, mouse phenotyping, genomics/gene array, proteomics and bioinformatics, structural biology, GCRC, and health policy/outcomes research. Sponsored program awards (grant funding) at the UC Medical Center exceeded $280 million in 2005.

Financial Aid
Every doctoral student in the Biomedical Sciences Flex Option is awarded a scholarship for 100 percent tuition and fees, plus a generous annual stipend ($20,500 in 2005) and paid health insurance. Competitive fellowships, such as the Albert Yates Fellowship, are available and can raise the stipend amount.

Cost of Study
In the 2005–06 academic year, full-time tuition, including fees, was $3599 per quarter for Ohio residents and $6634 for nonresidents.

Living and Housing Costs
Estimated living expenses per year are about $14,080. Students living off-campus can expect to pay $300–$700 per month for a one-bedroom apartment or $500–$900 per month for a two-bedroom unit.

Student Group
Approximately 7 new students enroll each year into the Biomedical Sciences Flex Option Program, for a total enrollment of 7 out of 70 applicants. In fall 2005, successful candidates had an average undergraduate GPA of 3.65 and a cumulative GRE average of 1363 (out of 1600) for the quantitative and verbal scoring and an average analytical score of 4.5 (out of 6.0).

Student Outcomes
Career options after graduation include academic teaching/research, with best positions gained following postdoctoral specialization training; research as a staff member of a public or private laboratory where postgraduate training and research is the major focus; and private industry, such as pharmaceutical firms where new drug design, discovery, development, testing, and screening procedures are frequently combined with high-level basic research. Most graduates prefer to enhance their professional expertise by securing postdoctoral specialized training before accepting such positions.

Location
The campus is located in Cincinnati, nestled on the banks of the Ohio River. The downtown area offers upscale shopping, inexpensive parking, and a variety of restaurants and museums. Sports fans can watch the Reds, Bengals, or Cyclones in action. The Arnoff Center for the Arts is the home of dance, theater, and Broadway shows. The cultural life of the city includes the Cincinnati Ballet, the Cincinnati Symphony Orchestra, the Cincinnati Pops Orchestra, the Contemporary Arts Center, the Cincinnati Art Museum, the Cincinnati History Museum, the Museum of Natural History and Science, the Newport Aquarium, Krohn Conservatory, and the world-famous Cincinnati Zoo and Botanical Garden.

The College
The College of Medicine was established in 1819 and is considered the oldest medical college west of the Allegheny Mountains. The College's exceptional alumni and current and past faculty members have made considerable contributions to medicine and to the medical sciences. For example, the College developed the first medical laser laboratory in the country, the first oral polio vaccine, the drug Benadryl, and the first residency program in emergency medicine, as well as numerous medical devices.

The College's twelve graduate programs had 725 applications and matriculated 114 students in 2005, of whom 29 were international and 69 were from Ohio. There were 98 students who graduated (24 M.S. and 53 Ph.D.), bringing the total student body to 434, of whom 47 percent are women, 25 percent are international, and 5 percent are from U.S. underrepresented minority groups.

Applying
The application must include the following: a completed online application form, official notification of GRE scores, official transcripts from every undergraduate institution attended, an essay stating the applicant's research and career goals and reasons for choosing the program, letters of recommendation from 2 people qualified to assess the applicant's potential for success, and a $40 application fee. TOEFL scores are required of students whose first language is not English. Successful applicants are invited to the campus for interviews with three small groups of faculty members. The deadline to apply is January 1. Applications received after that date are not guaranteed full consideration.

Correspondence and Information
Laura Hildreth, Assistant Dean
University of Cincinnati College of Medicine
P.O. Box 670555
Cincinnati, Ohio 45267-0555

Phone: 513-558-6791
Fax: 513-558-2850
E-mail: hildrele@uc.edu
Web site: http://med.uc.edu/GradEd/GradFlex.cfm

University of Cincinnati

THE FACULTY AND THEIR RESEARCH

Flex students may choose from more than 200 faculty members who are potential mentors for graduate students in the College of Medicine. The following list is composed of the Graduate Program Directors in the various Ph.D. programs available to Flex students.

Robert F. Highsmith, Professor of Molecular Physiology, Flex Graduate Student Adviser, Associate Dean, and Director, Office of Research and Graduate Education; Ph.D. (physiology and biophysics), Cincinnati, 1972. Intercellular signaling within blood vessel wall of coronary arteries; using cell culture as a model, how products of endothelial cells and peripheral blood mononuclear cells may influence the vascular smooth muscle cell in overall regulation of vasomotor tone and blood flow.

Robert Brackenbury, Professor and Director, Graduate Program in Cell and Cancer Biology; Ph.D. (biology), Brandeis, 1976. How cell-cell interactions regulate cell movement.

C. Ralph Buncher, Professor and Director, Environmental Health Graduate Programs; Sc.D. (biostatistics/epidemiology), Harvard, 1967. Effects of exposure to lead; clinical trials, especially of pharmaceutical products; evaluating imaging systems; cancer studies, especially with relation to Cincinnati and Ohio; studies of effects of radiation.

David Butler, Professor and Director, Biomedical Engineering; Ph.D. (engineering mechanics), Michigan State, 1976. Tissue-engineered natural tendon repair and meniscus transplantation.

Iain Cartwright, Associate Professor and Director, Molecular Genetics, Biochemistry, and Microbiology Program; Ph.D. (biochemistry), Warwick (England), 1965. Investigating genetic, physiological, and organismal responses to toxic heavy metals (e.g., arsenic) in the environment, including the basis for differential susceptibility.

Gary A. Gudelsky, Professor and Director, College of Pharmacy and Neuroscience Program; Ph.D. (pharmacology), Michigan State, 1977. Neuropharmacology; systems/behavioral approaches to the study of drugs of abuse (methamphetamine, MDMA) and antipsychotic agents.

James Herman, Professor and Director, Neuroscience Program; Ph.D. (neurobiology and anatomy), Rochester, 1987. Neurocircuit regulation of hormonal and behavioral stress responses; neurobiology of anxiety and depression.

Nelson Horseman, Professor and Director, Systems Biology and Physiology Program; Ph.D. (physiology), LSU, 1978. Molecular endocrinology; lactation; breast cancer; stress-trauma interactions.

Wallace Ip, Professor and Director of Admissions and Recruitment, Cell and Cancer Biology; Ph.D. (cell biology), IIT, 1977. Cell biology of tumor formation and metastasis; cell motility; cytoskeleton.

Simon Newman, Professor and Director, Pathobiology and Molecular Medicine; Ph.D. (immunology), Alabama at Birmingham, 1978. Innate immunity to fungal pathogens, particularly *Histoplasma capsulatum*.

Robert M. Rapoport, Associate Professor and Director, Molecular, Cellular, and Biochemical Pharmacology; Ph.D. (pharmacology), UCLA, 1980. Regulation of the cerebral vasculature, with an emphasis on elucidating mechanisms underlying the vasospasm following subarachnoid hemorrhage, hypocapnia, and cocaine.

Marsha Wills-Karp, Professor and Director, Immunobiology Program; Ph.D., California, Santa Barbara, 1986. Immunogenetics of allergic disorders such as asthma, with an emphasis on elucidating the role of the innate and adaptive immune systems in disease pathogenesis.

Chris Wylie, William Schubert Professor and Director, Division of Developmental Biology, Cincinnati Children's Hospital Research Foundation; Ph.D., University College, London, 1971. Molecular and genetic analysis of early vertebrate morphogenesis, including cell signaling, cell specification, cell migration, and control of the cytoskeleton, using *Xenopus* and mouse model systems.

Katherine Yutzey, Associate Professor and Director, Molecular and Developmental Biology; Ph.D. (biology), Purdue, 1992. Molecular regulation of embryonic heart development; experimental embryological, genetic, and molecular approaches used for mechanistic analyses of early cardiac lineage determination and heart morphogenesis.

UNIVERSITY OF CONNECTICUT HEALTH CENTER

Graduate Programs in Biomedical Sciences

Programs of Study

Work leading to the Ph.D. degree in biomedical sciences and master's degrees in dental sciences and public health is offered through Graduate School faculty members associated with the Schools of Medicine and Dental Medicine at the University of Connecticut Health Center in Farmington. A combined-degree program with the School of Medicine offers an M.D./Ph.D. degree to qualified students interested in academic medicine and research. In addition, the Schools of Medicine and Dental Medicine, in conjunction with the Public Health Program, offer a combined program leading to the M.D./M.P.H. or D.M.D./M.P.H. The School of Dental Medicine offers a D.M.D./Ph.D. and a Combined Certificate Training Ph.D. program for students with advanced dental degrees. Ph.D. students apply to the Integrated Admissions Mode, which offers a first year of study in the basic science curriculum prior to the selection of an area of concentration in which to pursue the Ph.D. thesis work.

Research Facilities

The program offices and laboratories are part of the University of Connecticut Health Center. A wide range of general and specialized equipment and expertise in the biological, biochemical, and biophysical sciences is available. Students have access to all facilities and equipment necessary for the pursuit of their research programs. In addition, major institutional resources include central small-animal facilities and a library that contains approximately 140,000 volumes and 450 CAI programs and subscribes to more than 3,200 current periodicals.

Financial Aid

Support for doctoral students engaged in full-time degree programs at the Health Center is provided on a competitive basis. Graduate research assistantships for 2006–07 provide a stipend of $26,000 per year, which includes a waiver of tuition/University fees for the fall and spring semesters and a student health insurance plan. While financial aid is offered competitively, the Health Center makes every possible effort to address the financial needs of all students during their period of training.

Cost of Study

For 2006–07, tuition is $3996 per semester ($7992 per year) for full-time students who are Connecticut residents and $10,368 per semester ($20,772 per year) for full-time out-of-state residents. General University fees are added to the cost of tuition for students who do not receive a tuition waiver. These costs are usually met by traineeships or research assistantships for doctoral students.

Living and Housing Costs

There is a wide range of affordable housing options in the greater Hartford area within easy commuting distance of the campus, including an extensive complex that is adjacent to the Health Center. Costs range from $600 to $800 per month for a one-bedroom unit; 2 or more students sharing an apartment usually pay less. University housing is not available at the Health Center.

Student Group

The facilities in Farmington are used by approximately 550 students in the Schools of Medicine and Dental Medicine, 400 graduate students in the Ph.D. and master's programs, and numerous postdoctoral fellows.

Location

The Health Center is located in the historic town of Farmington, Connecticut. Set in the beautiful New England countryside on a hill overlooking the Farmington Valley, it is close to ski areas, hiking trails, and facilities for boating, fishing, and swimming. Connecticut's capital city of Hartford, 7 miles east of Farmington, is the center of an urban region of approximately 800,000 people. The beaches of the Long Island Sound are about 50 minutes away to the south, and the beautiful Berkshires are a short drive to the northwest. New York City and Boston can be reached within 2½ hours by car. Hartford is the home of the acclaimed Hartford Stage Company, TheatreWorks, the Hartford Symphony and Chamber orchestras, two ballet companies, an opera company, the Wadsworth Atheneum (the oldest public art museum in the nation), the Mark Twain house, the Hartford Civic Center, and many other interesting cultural and recreational facilities. The area is also home to several branches of the University of Connecticut, Trinity College, and the University of Hartford, which includes the Hartt School of Music. Bradley International Airport (about 20 minutes from campus) serves the Hartford/Springfield area with frequent airline connections to major cities in this country and abroad. Frequent bus and rail service is also available from Hartford.

The Health Center

The 200-acre Health Center campus at Farmington houses a division of the University of Connecticut Graduate School, as well as the School of Medicine and Dental Medicine. The campus also includes the John Dempsey Hospital, associated clinics, and extensive medical research facilities, all in a centralized facility with more than 1 million square feet of floor space. The Health Center's newest research addition, the Academic Research Building, was opened in 1999. This impressive eleven-story structure provides 170,000 square feet of state-of-the-art laboratory space. The faculty at the center includes more than 260 full-time members. The institution has a strong commitment to graduate study within an environment that promotes social and intellectual interaction among the various educational programs. Graduate students are represented on various administrative committees concerned with curricular affairs, and the Graduate Student Organization (GSO) represents graduate students' needs and concerns to the faculty and administration, in addition to fostering social contact among graduate students in the Health Center.

Applying

Applications for admission should be submitted on standard forms obtained from the Graduate Admissions Office or the Web site and should be filed together with transcripts, three letters of recommendation, a personal statement, and recent results from the General Test of the Graduate Record Examinations. International students must take the Test of English as a Foreign Language (TOEFL) to satisfy Graduate School requirements. The deadline for completed applications and receipt of all supplemental materials is December 15. In accordance with the laws of the state of Connecticut and of the United States, the University of Connecticut Health Center does not discriminate against any person in its educational and employment activities on the grounds of race, color, creed, national origin, sex, age, or physical disability.

Correspondence and Information

Graduate Programs in Biomedical Sciences
Graduate Admissions Office, MC 3906
University of Connecticut Health Center
Farmington, Connecticut 06030-3906

Phone: 860-679-2175
E-mail: robertson @nso2.uchc.edu
Web site: http://grad.uchc.edu

University of Connecticut Health Center

FACULTY AND RESEARCH AREAS

The Health Center's graduate faculty of more than 100 members is drawn from both the basic and clinical departments of the Schools of Medicine and Dental Medicine.

Cell Biology. This interdisciplinary program offers the student the opportunity to bring modern molecular and physical techniques to bear on problems in cell biology. Emphasis is placed on cellular physiology, with special attention to membrane and surface phenomena of normal and neoplastic cells, cell differentiation, signal transduction, protein and lipid biosynthesis and trafficking, and cellular regulation. Linda Shapiro, Assistant Professor of Cell Biology and Program Director; Kevin Claffey, Associate Professor of Cell Biology and Associate Program Director.

Genetics and Developmental Biology. This program emphasizes cellular and molecular bases of differentiation and development and includes opportunities in molecular human genetics. Research opportunities are available in cell and tissue interactions, regulation of gene expression, humoral regulation of gene expression in differentiation, molecular aspects of connective tissue development, physiology of fertilization, molecular mechanisms of sperm-egg binding and fusion, and molecular genetics and biology of connective tissue disorders. William Mohler, Assistant Professor of Genetics and Developmental Biology and Program Director.

Immunology. The central focus of this program is to train the student to become an independent investigator and educator who will provide research and educational contributions to basic, applied, or clinical immunology. This goal is achieved through lectures, seminars, research presentations, and a concentration on laboratory research in immunochemistry, signal transduction, cellular immunology, molecular immunology, molecular immunoparasitology, lymphocyte development, and immunology of the eye. Lynn Puddington, Associate Professor of Immunology and Program Director; Anthony Vella, Assistant Professor of Immunology and Associate Program Director.

Molecular Biology and Biochemistry. Research in this program is directed toward explaining biological phenomena at the molecular level. The program includes four major areas of concentration and research: relation of the structure of macromolecules to their function, biosynthesis of macromolecules, biochemical genetics, and assembly of macromolecules into complex cellular structures. Henry Furneaux, Associate Professor of Molecular, Microbial, and Structural Biology and Program Director; Stephen M. King, Professor of Biochemistry and Associate Program Director.

Neuroscience. This interdepartmental program encompasses experimental approaches ranging from the molecular to the systems level, including areas of concentration in cellular, molecular, and developmental neurobiology; neuroanatomy; neurophysiology; neurochemistry; neuroendocrinology; neuroimmunology; neuropharmacology; and neuropathology. Eric Levine, Associate Professor of Pharmacology and Program Director.

Skeletal, Craniofacial and Oral Biology. Topics of basic biological research include matrix biochemistry, bone cell growth and differentiation, regulation of gene expression during limb development, regulation of cell migration and aggregation, molecular and cellular biology of inflammation and repair, tissue interactions in craniofacial development, radiation-induced DNA damage and repair, neural structure and function in gustatory and visual systems, and control of voluntary movements in speech production and mastication. William Upholt, Professor of Oral Rehabilitation, Biomaterials and Skeletal Development, Center for Regenerative Medicine and Skeletal Development, and Program Director.

Cellular and Molecular Pharmacology. Research in this interdisciplinary program is concerned with cellular and molecular mechanisms of signal transduction underlying drug action in several areas of pharmacology concentration, including that of the central and peripheral nervous systems, cardiovascular and pulmonary systems, endocrine-reproductive system, immunopharmacology, chemotherapy and hepatic biotransformation of endogenous substances, and xenobiotics. Joel Pachter, Professor of Pharmacology and Program Director.

Combined M.D./Ph.D. Program. This program is designed for students interested in careers in medical research and academic medicine. It enables students to acquire competence in both the basic science and clinical aspects of their chosen fields. The program allows a student to combine the curricula of two schools in a way that meets the specific degree requirements of each, and yet it allows the completion of both in a period less than that needed if the two curricula were taken in sequence. Entry into the program is limited to a small number of unusually well qualified students who are either currently enrolled in the medical school or who have been accepted into the first-year class. Barbara Kream, Professor of Medicine and of Genetics and Developmental Biology and Program Director.

Combined D.M.D./Ph.D. Program. This program is designed for students interested in careers in dental research and academic dental medicine. It enables students to acquire competence in both the basic science and clinical aspects of their chosen fields. The program allows a student to combine the curricula of two schools in a way that meets the specific degree requirements of each, and yet it allows the completion of both in a period less than that needed if the two curricula were taken in sequence. Entry into the program is limited to a small number of unusually well qualified students who are either currently enrolled in the dental school or who have been accepted into the first-year class. Alan Lurie, Professor of Oral Health and Diagnostic Sciences; Head, Division of Oral and Maxillofacial Radiology; and Program Director.

Combined M.D./M.P.H. or D.M.D./M.P.H. Program. A joint-degree program leading to the Master of Public Health in addition to the Doctor of Medicine or the Doctor of Dental Medicine is sponsored by the Graduate Program in Public Health and the Schools of Medicine and Dental Medicine. The joint-degree program has been developed to prepare future physicians and dentists to deal creatively with the rapidly changing environment of medicine and health care. It is possible to complete the degree requirements for both programs during the four years of medical or dental school. David Gregorio, Professor of Community Medicine and Health Care and Program Director.

Dental Science. The Master of Dental Science degree program is an interdepartmental program whose primary objective is to provide instruction in dental science that will enhance the student's ability to instruct and undertake research in dental schools. This program provides an opportunity for cooperative study and research between dentistry, the basic sciences, and allied health fields. Both M.Dent.Sc. and oral biology Ph.D. students may combine their work in these programs with advanced clinical training in endodontics, orthodontics, oral pathology, pedodontics, periodontics, oral medicine, oral radiology, and oral and maxillofacial surgery. Monty MacNeil, Associate Dean for Academic Affairs, oversees this program. In addition, an intercampus program in dental materials, under the direction of Jon Goldberg, is included in the School of Dental Medicine.

Public Health. This multidisciplinary master's program, accredited by the Council for Education in Public Health, is based in the Department of Community Medicine and Health Care. It offers a core curriculum in epidemiology, biostatistics, health administration, environmental health, the sociomedical sciences, health law, and electives in these and related areas. David Gregorio, Professor of Community Medicine and Health Care and Program Director.

UNIVERSITY OF DAYTON

Department of Biology

Programs of Study	The Department of Biology offers programs leading to the Master of Science and the Doctor of Philosophy degrees. The degrees are in biology, but each program is tailored to the student's own interests and career plans. Specialization is accomplished by selection of courses, choice of thesis or dissertation topic, and participation in weekly seminars in the area of interest. The specific program is determined after consultation between the student and the advisory committee. The department offers two major areas of specialization: environmental and ecological sciences and basic biomedical sciences. Special fields within these areas are indicated by the research interests of the faculty, listed on the reverse of this page.

Each entering student enrolls in a two-semester core course in bioinstrumentation. After the first semester, the student chooses a major professor who forms a committee that works closely with the student throughout the graduate training to plan a program tailored to his or her needs and interests. This may include some courses in other departments, such as chemistry, physics, mathematics, and computer sciences. Occasionally, a student may be encouraged to spend time at a biological station or an off-campus research laboratory. Clinical studies may be taken by arrangement with local hospitals. The University is a member of the Southwestern Ohio Council for Higher Education; course work taken at other member institutions may be applied toward the degree. Although attendance on a full-time basis is required, a student is permitted to take two courses prior to commencing full-time study.

Twenty-four credit hours of course work and a research thesis are required for the M.S. degree. Because the Ph.D. is centered on development of professional competence, there are no set course requirements; preliminary examinations and periodic meetings determine the extent of course work and progress in the dissertation.

The department maintains an active program of guest lecturers and visiting biologists. Classes are small, and students have a close working relationship with the faculty. The University calendar consists of two trimesters and two 6-week spring and summer sessions. The fall term usually begins in late August.

Research Facilities

Fully equipped laboratories are maintained for graduate training and advanced research in the special areas indicated in the list of faculty members. Specialized equipment includes standard molecular biology equipment, ultracentrifuges, recording spectrophotometers, liquid scintillation systems, gamma counters, a $^{14}CO_2$ labeling system, gas chromatographs, electrophoretic apparatus, electron microscopes, growth and environmental chambers, electrophysiological apparatus, fraction collectors, and the usual variety of centrifuges, light microscopes, and specialized optical and photographic equipment. The on-campus location of the University of Dayton Research Institute provides excellent electronic instrumentation and supporting facilities for glassblowing, computer usage, graphics, and the design of special equipment.

Financial Aid

Qualified applicants are eligible for assistance in the form of fellowships, traineeships, or research or teaching assistantships. Stipends for academic-year appointments range from $16,881 to $17,902 in 2000–07. Financial aid in the form of competitive fellowships is available during the summer. All appointments are exempt from tuition during both the academic year and the summer session.

Cost of Study

For 2006–07, the cost of graduate courses is $567 per credit hour for the M.S. and $644 per credit hour for the Ph.D., plus a $25 University fee each semester.

Living and Housing Costs

Graduate students arrange for their own housing, although the Housing Office will provide assistance upon request. The cost varies considerably depending on the type of accommodations. Apartment rents average $425 per month.

Student Group

Seventeen students are currently enrolled in the graduate program in biology. Total full-time enrollment throughout the University is more than 3,300 graduate students and 6,000 undergraduates.

Location

Dayton, Ohio, is the center of a metropolitan area of more than 850,000 people. The climate is generally mild, with a tendency to be humid. The city offers many cultural attractions, including the Dayton Art Institute, the Museum of Natural History, the Dayton Opera Association, the Dayton Civic Ballet, the Dayton Philharmonic Orchestra, and several drama groups. Professional baseball, football, and hockey are available in nearby Cincinnati, an hour's drive by I-75. The Air Force Museum is located on the grounds of the nearby Wright-Patterson Air Force Base.

The University

The University of Dayton is a medium-sized, private coeducational school located in the heart of the Midwest. Although the origins of the institution date from 1850, it became known as the University of Dayton in 1920. It attracts its student body from the local community, the state of Ohio and other Midwestern and Eastern states, and a number of other lands. Its four schools and one college offer a broad range of programs in more than forty departments of instruction.

Applying

For consideration for full admission to the biology M.S. or Ph.D. degree program, applicants should submit the following information: the graduate application form (students may submit at no charge online at http://gradadmission.udayton.edu); a personal statement indicating specific areas of research interest, any prior research experience, and professional goals; an official college transcript; three letters of recommendation; and current scores on the General Test of the Graduate Record Examinations. International students must submit TOEFL scores. Applicants seeking financial aid should apply before March 1 for fall term and Oct. 15 for winter term.

Correspondence and Information

Jayne B. Robinson, Chairman
Department of Biology
University of Dayton
300 College Park Avenue
Dayton, Ohio 45469-2320
Phone: 937-229-2521
E-mail: graduate.biology@notes.udayton.edu
Web site: http://biology.udayton.edu

Shirley J. Wright, Director
Biology Graduate Program
University of Dayton
300 College Park Avenue
Dayton, Ohio 45469-2320
Phone: 937-229-2857
E-mail: graduate.biology@notes.udayton.edu
Web site: Web site: http://biology.udayton.edu

University of Dayton

THE GRADUATE FACULTY AND THEIR RESEARCH

Randall J. Breitwisch, Associate Professor; Ph.D., Miami (Florida), 1987. Behavioral ecology; ethology.
Albert J. Burky, Professor; Ph.D., Syracuse, 1969. Invertebrate physiology; physiological ecology.
Carl F. Friese, Associate Professor; Ph.D., Utah State, 1991. Community and microbial ecology.
Sudhindra R. Gadagkar, Assistant Professor; Ph.D., Dalhousie, 1997. Computational biology.
Donald R. Geiger, Professor; Ph.D., Ohio State, 1963. Plant physiology; radiation biology.
Marie-Claude Hofmann, Associate Professor; Ph.D., Lausanne (Switzerland), 1988. Cell biology.
Yi-ling Hong, Assistant Professor; Ph.D., Kentucky, 1997. Molecular biology.
Robert J. Kearns, Professor; Ph.D., Washington State, 1978. Medical microbiology; immunology.
Carissa M. Krane, Assistant Professor; Ph.D., Washington (St. Louis), 1996. Molecular physiology.
Mark G. Nielsen, Assistant Professor; Ph.D., Stanford, 1994. Genetics.
Jayne B. Robinson, Professor and Chairman; Ph.D., Ohio State, 1991. Environmental microbiology.
John J. Rowe, Professor; Ph.D., Kansas, 1975. Microbial physiology; nitrate reduction.
Panagiotis Tsonis, Professor; Ph.D., Nagoya (Japan), 1983. Molecular biology; developmental biology.
P. Kelly Williams, Professor; Ph.D., Indiana, 1973. Ecology; vertebrate zoology.
Shirley J. Wright, Associate Professor; Ph.D., Iowa, 1987. Developmental biology; cell biology.

UNIVERSITY OF FLORIDA

College of Medicine
Interdisciplinary Program in Biomedical Sciences

Programs of Study	The College of Medicine at the University of Florida (UF) offers graduate training in biomedical research leading to a Ph.D. degree through the Interdisciplinary Program (IDP) in Biomedical Sciences. The goal of the Interdisciplinary Program is to prepare students for a diversity of careers in research and teaching in academic and commercial settings. The program provides a modern, comprehensive graduate education in biomedical science while providing both maximum program flexibility as well as appropriate specialization for graduate students. It represents a cooperative effort among six interdisciplinary graduate programs with a total membership of more than 250 faculty members. The six programs are Biochemistry and Molecular Biology, Genetics, Immunology and Microbiology, Molecular Cell Biology, Neuroscience, and Physiology and Pharmacology. During the first year of study, incoming graduate students undertake a common, comprehensive interdisciplinary core curriculum of classroom study developed in a cooperative effort by all of the graduate programs in the College of Medicine. In addition, students select from any of the College of Medicine faculty members for participation in several laboratory rotations conducted throughout the first year, concurrent with the core curriculum. By the end of their first year, students select a laboratory in which to conduct their dissertation research and, once again, they may choose from any of the graduate faculty members in the College of Medicine. At that time, students also elect an affiliation with any one of the six graduate programs for their advanced training. Formal selection of a graduate program and mentor is made after completion of the core curriculum to maximize flexibility and facilitate an informed decision; however, students may make an informal commitment to a program or lab at any time. Through the individual graduate programs, students have access to advanced courses and seminars in their chosen specialty, which can be undertaken along with their dissertation laboratory research.
Research Facilities	The College of Medicine houses state-of-the-art research facilities maintained by the Interdisciplinary Center for Biomedical Research (ICBR), the McKnight Brain Institute, the Clinical Research Center, several other University of Florida Research Centers, and individual research laboratories. Together these facilities provide services for DNA and protein synthesis and sequencing, hybridoma production, confocal and electron microscopy, NMR spectroscopy, computing and molecular modeling, flow cytometry, transgenic mouse production, and gene therapy vector construction. The University of Florida libraries, including the Health Center library, form the largest information resource system in the state of Florida, and support up-to-date computer-based bibliographic retrieval services.
Financial Aid	Students accepted into the Interdisciplinary Program receive a stipend of $22,000 per annum. Students also receive a waiver that covers tuition.
Cost of Study	Graduate research assistants pay fees of approximately $1000 per year. Books and supplies cost about $1600.
Living and Housing Costs	Married student housing is available in six apartment villages operated by the University; dormitories for single students are available in limited quantities. The cost of housing for a single graduate student is approximately $5000 per year. Most graduate students live in abundant and comfortable off-campus housing that is near the University; rents vary but are low on a national scale.
Student Group	Enrollment at the University of Florida is both culturally and geographically diverse and numbers about 50,000, including about 10,000 graduate students. There are about 300 graduate students in the College of Medicine.
Location	Situated in north-central Florida, midway between the Atlantic Ocean and the Gulf of Mexico, Gainesville is a nationally recognized academic and research center, and was rated the number one place to live in the U.S. in a 1995 *Money* magazine survey. There are about 200,000 residents in the Gainesville metropolitan area, excluding University of Florida students. The climate is moderate, permitting outdoor activities the year around. The University and the Gainesville community also host numerous cultural events, including music, dance, theater, lectures, and art exhibits. Many of the men's and women's athletic programs are considered to be in the top ten programs in the country.
The University and The College	The University of Florida, located on 2,000 acres, is among the nation's leading research universities as categorized by the Carnegie Commission on Higher Education. UF is a member of the Association of American Universities, the nation's most prestigious higher education organization. UF is also one of the nation's top three universities in the breadth of academic programs offered on a single campus. It has twenty colleges and schools and 100 interdisciplinary research and education centers, bureaus, and institutes. The College of Medicine, opened in 1956, has become a nationally recognized leader in medical education and research.
Applying	The Interdisciplinary Program in Biomedical Sciences seeks promising students with undergraduate training in chemistry, biology, psychology, or related disciplines in the life sciences. Applicants are selected on the basis of previous academic work, research experience, GRE General Test scores, letters of recommendation, and a personal interview. Students are admitted for the fall semester, which begins in late August. Early application is recommended and should be completed by February 1.
Correspondence and Information	Interdisciplinary Program in Biomedical Sciences Associate Dean for Graduate Education University of Florida College of Medicine P.O. Box 100215 Gainesville, Florida 32610-0215 Phone: 352-392-5461 E-mail: idp@ufl.edu Web site: http://idp.med.ufl.edu

University of Florida

INTERDEPARTMENTAL GRADUATE PROGRAMS AND DIRECTORS

The College of Medicine at the University of Florida is located in Gainesville, Florida. For program information, interested students should address the individuals below with the name of the graduate program or center and the box number. The zip code for all addresses is 32610.

GRADUATE PROGRAMS

Biochemistry and Molecular Biology (P.O. Box 100245): Dr. Susan Frost (352-392-3207).
Genetics (P.O. Box 100266): Dr. Henry V. Baker (352-392-0680).
Immunology and Microbiology (P.O. Box 100275): Dr. Laurence Morel (352-392-3790).
Molecular Cell Biology (P.O. Box 100235): Dr. Phyllis LuValle (352-392-6121).
Neuroscience (P.O. Box 100244): Dr. Susan L. Semple-Rowland (352-392-3598).
Physiology/Pharmacology (P.O. Box 100267): Dr. Jeffrey Harrison (352-392-3227).

RESEARCH CENTERS AND DIRECTORS

Alcohol Research (P.O. Box 100244): Dr. Don W. Walker (352-392-4219).
Biotechnology Research, Interdisciplinary Center for (P.O. Box 101580): Dr. Rob Ferl, Director (352-273-8030).
Brain Institute (P.O. Box 100244): Dr. Dennis Steindler (352-392-0490).
Cancer Center (P.O. Box 100232): Dr. W. Stratford May (352-846-1145).
Clinical Research (P.O. Box 100322): Dr. Peter W. Stacpoole (352-395-0032).
Environmental Human Toxicology (P.O. Box 110885): Dr. Stephen M. Roberts (352-392-4700 Ext. 5500).
Forensic Medicine, Maples Center for (P.O. Box 100275): Dr. Bruce A. Goldberger (352-265-0680 Ext. 72001).
Gene Therapy (P.O. Box 100266): Dr. Nicholas Muzyczka (352-392-8541).
Genetics Institute (P.O. Box 100266): Dr. Kenneth I. Berns (352-846-2739).
Hypertension (P.O. Box 100274): Dr. Mohan Raizada (352-392-9299).
Immunology and Transplantation (P.O. Box 100275): Dr. Mark Atkinson (352-392-9719).
Mammalian Genetics (P.O. Box 100266): Dr. Thomas P. Yang (352-392-3054).
Neurobiological Sciences (P.O. Box 100244): Dr. Charles J. Vierck (352-392-6555).
Neurobiology of Aging (P.O. Box 100487): Dr. Edwin M. Meyer (352-392-6364).
Pain Research (P.O. Box 100444) Dr. Robert P. Yezierski (352-392-3032).
Periodontal Disease Research (P.O. Box 100434): Dr. William McArthur (352-392-4377).
Smell & Taste, Center for (P.O. Box 100015): Dr. Barry Ache (904-461-4034).
Structural Biology (P.O. Box 100245): Dr. Thomas H. Mareci (352-392-3375).
Toxicology Specialization (P.O. Box 100267): Dr. Kathleen T. Shiverick (352-392-3545).
Vision Science (P.O. Box 100266): Dr. William W. Hauswirth (352-392-0679).
The Whitney Laboratory for Marine Bioscience (6505 Ocean Shore Drive, St. Augustine, Florida 32080): Dr. Peter A. V. Anderson (904-461-4000).
Women's Health Research Center (P.O. Box 100171): Dr. May Ann Burg (352-392-9000 Ext. 9000).

THE UNIVERSITY OF IOWA

Department of Biological Sciences

Programs of Study	The graduate program in the Department of Biological Sciences provides students with the training they need to excel in state-of-the-art research and effective teaching careers. The Department has four core areas: cell and developmental biology, evolution, genetics, and neurobiology. Encompassed within these core areas are 32 faculty members employing a wide range of experimental approaches and model systems, spanning unicellular organisms, plants, and vertebrates. Entering graduate students are offered a broad range of course work, including specialized seminars in which they learn to critique, discuss, and present scientific results. During their first year, Ph.D. students undertake research rotations in three laboratories of their choice, one of which they join to pursue thesis research. A comprehensive examination, including an exercise in grant writing, is given in the second or third year. At the successful completion of graduate research, the student writes and orally defends a dissertation, leading to the Ph.D. degree. The Ph.D. program ordinarily takes from four to six years; the master's degree program with thesis takes two to three years.
Research Facilities	Each student has office and laboratory space in a modern, comfortable, well-equipped building. Major equipment of all types required for modern biological research is available to graduate students. A Departmental library is available to graduate students 24 hours a day. Students also have access to the $3.5-million Health Sciences Library. A forest preserve within 30 minutes of the campus is available for experimental projects. The Iowa Lakeside Laboratory is a biological station featuring a variety of natural habitats. Available on campus are a number of modern facilities designed to assist students' research efforts, including facilities for DNA microarray analysis; a molecular biology facility, with capabilities that include synthesizing oligopeptides, sequencing oligopeptides, and providing enzymes; a mass spectrometry/NMR facility; a hybridoma facility that provides monoclonal antibodies; a fermentor facility for large cultures; and two electron microscopy labs, including one in the Department of Biological Sciences. The Department also maintains its own computer facility, in addition to the main campus networks, a confocal microscopy-imaging facility, and its own DNA core facility.
Financial Aid	It is the policy of the Department to offer financial support to all its graduate students. Teaching and research assistantships that average $22,500 per year, plus full-tuition scholarships, combine for a total compensation package exceeding $30,000 per twelve months. An attractive health insurance plan is offered. University fellowships and traineeships are also available, including those for interdisciplinary programs in cell and molecular biology, genetics, and neurobiology. Students are encouraged to apply for competitive national fellowships. Additional fellowships are available for members of minority groups.
Cost of Study	Tuition for the 2005–06 academic year (fall, spring, and summer semesters combined) was $6344. Doctoral students were each provided with full-tuition scholarships. Students receiving assistantships gain in-state/resident status. Fees are subject to change.
Living and Housing Costs	Privately owned housing and University housing are both available in various price ranges. Applicants are sent details on University housing and an appropriate selection of current classified ads. A car is not essential, as both the city and University provide extensive bus service.
Student Group	The approximately 55 graduate students in the Department come from a variety of institutions, including major state universities, private universities, private colleges, and universities outside the United States. A few arrive with M.S. degrees, and a few have degrees in specialties other than biology, zoology, or botany (for example, biochemistry, biopsychology, physics, and science education). All current students are receiving financial aid from fellowships, traineeships, or teaching or research assistantships. Graduates who previously received their training in the Department include numerous distinguished researchers and teachers.
Location	Iowa City, with a population of more than 60,000, is situated in rolling country, with rivers, lakes, and woodlands as well as remnants of tall-grass prairie. The air is clean, and the atmosphere is friendly. Extensive cultural activities are available in Iowa City, including the programs in Hancher Auditorium, Clapp Recital Hall, and the Mabie Theater. Chicago is 240 miles away, and transportation is readily available. State, county, and city parks are attractive, uncrowded, and varied.
The University and The Department	The University of Iowa is internationally known for its Writers' Workshop, innovative programs in the arts, and a variety of strong departments in the humanities, social sciences, and natural sciences, of which the Department of Biological Sciences is one. The Department has built upon traditional strengths and added new faculty members of diverse interests as well as new space and equipment. The Department operates an NIH program project in cell motility and a Center for Comparative Genomics. The Department faculty members participate in three NIH predoctoral training grants. The faculty is active, well funded, productive, and gregarious. The Department regards its group of graduate students as a central element in its commitment to research and teaching.
Applying	The Department encourages applications from talented and qualified students in biology and other fields; superior students are sought regardless of undergraduate training. Ordinarily, calculus, physics, organic chemistry or biochemistry, and basic courses in biology should be completed while the student is an undergraduate. Students accepted in previous years have received scores averaging above 1200 on the combined verbal and quantitative portions of the GRE General Test, and a GPA of at least 3.0 is expected. Exceptions to these standards are made when background or special talents dictate. Special attention is paid to undergraduate research experience. Students are admitted primarily for the fall semester (August) but, in special cases, may be admitted at other times. Students notified of financial support by April 1 must accept or decline by April 15. Further information may be obtained by calling or writing to the Department's address or to any member of the faculty.
Correspondence and Information	Phil Ecklund Graduate Admissions Committee Department of Biological Sciences, Room 142 Biology Building The University of Iowa Iowa City, Iowa 52242-1342 Phone: 319-335-1092 Fax: 319-335-1069 E-mail: biology-admissions@uiowa.edu Web site: http://www.uiowa.edu/ http://www.biology.uiowa.edu/ (for the Department of Biological Sciences)

The University of Iowa

THE FACULTY AND THEIR RESEARCH

Debashish Bhattacharya, Ph.D., Simon Fraser, 1988. Evolutionary biology.
* Chi-Lien Cheng, Ph.D., Connecticut, 1982. Plant molecular biology: genetic and molecular analysis of gene regulation in *Arabidopsis.*
Josep M. Comeron, Ph.D., Barcelona, 1997. Population genetics; molecular evolution; evolutionary genomics.
Michael E. Dailey, Ph.D., Washington (St. Louis), 1990. Neural development; mechanisms of synapse formation and plasticity; cell migration; confocal time-lapse imaging.
Jeffrey Denburg, Ph.D., Johns Hopkins, 1970. Developmental neurobiology: molecular basis of cell recognition and other cellular events occurring in developing and regenerating nervous systems.
Daniel F. Eberl, Ph.D., Guelph, 1991. Neurobiology and genetics: molecular mechanisms of hearing and auditory behavior in *Drosophila.*
Jan Fassler, Ph.D., Purdue, 1983. Genetic identification and characterization of transcription factors in yeast.
Joseph Frankel, Ph.D., Yale, 1960. Pattern formation: genetic and phenotypic analysis of the organization of cell structures on the cell surface of ciliated protozoa.
Steven H. Green, Ph.D., Caltech, 1982. Molecular and developmental neurobiology: mechanism of action of nerve growth factor; neuronal regulation of cardiac development.
Gary Gussin, Ph.D., Harvard, 1966. Molecular genetics: bacteriophage genetics; control of transcription.
* Stephen D. Hendrix, Ph.D., Berkeley, 1969. Plant-herbivore interactions; plant demography; succession.
Lilach Hadany, Ph.D., Tel-Aviv, 2002. Evolution of recombination, sexual reproduction, dispersal patterns, mutation rates, and complex adaptation.
Diana G. Horton, Ph.D., Alberta, 1981. Systematics of bryophytes; peatland ecology.
Douglas W. Houston, Ph.D., Miami (Florida), 1999. Maternal transcription factor and signaling pathways in vertebrate development.
Erin E. Irish, Ph.D., Indiana, 1984. Developmental genetics, molecular biology of plant development, floral determination, and sex determination in maize.
Alan Kay, Ph.D., Cambridge, 1984. Biophysics of neuronal ion channels: mathematical modeling of neurons and networks.
Jack Lilien, Ph.D., Chicago, 1967. Developmental and molecular biology; molecular mechanisms affecting the function of adhesion molecules.
Jim Jung-Ching Lin, Ph.D., Connecticut, 1979. Cell and molecular biology: structure and function of protein components of the cytoskeleton; expression and organization of tropomyosin genes in normal and transformed cells; protein isoform switches during heart development.
John M. Logsdon Jr., Ph.D., Indiana, 1995. Molecular evolutionary biology: genome evolution; origin and evolution of meiosis; molecular phylogeny of eukaryotes.
Robert Malone, Ph.D., Oregon, 1976. Molecular biology and genetics of recombination and DNA repair in eukaryotes (yeast); meiosis as a developmental pathway.
Linda Maxson, Ph.D., San Diego State, 1973. Systematics; molecular evolution.
Bryant F. McAllister, Ph.D., Rochester, 1996. Evolutionary genetics and genome evolution.
John R. Menninger, Ph.D., Harvard, 1964. Cell and molecular biology; information transfer and control in biological systems; accuracy of protein synthesis; ribosome editing; protein biosynthesis; mechanism of inhibition of antibiotics; cellular aging.
* Jonathan E. Poulton, D.Phil., Oxford, 1974. Plant biochemistry; secondary plant products.
* Jeff T. Schabilion, Ph.D., Kansas, 1969. Paleobotany: paleoecology, morphology, development, and evolution of coal-age plants and Cretaceous Cycadeoids.
* Ming-Che Shih, Ph.D., Iowa, 1983. Plant molecular biology: regulation of gene expression in *Arabidopsis thaliana;* chloroplast molecular evolution.
Diane C. Slusarski, Ph.D., Northwestern, 1993. Developmental biology: analysis of signal transduction mechanisms involved in pattern formation in the vertebrate embryo (zebrafish).
David R. Soll, Ph.D., Wisconsin–Madison, 1970. Cell and molecular biology: regulation of gene expression during differentiation and dedifferentiation in *Dictyostelium discoideum;* cell motility and chemotaxis; high-frequency switching and genomic rearrangements in *Candida.*
Barbara Stay, Ph.D., Radcliffe, 1953. Insect reproduction and development physiology, endocrinology, fine structure.
Christopher Stipp, Ph.D., MIT, 1996. Cell surface receptors for extracellular matrix; cell motility.
* WeiYeh Wang, Ph.D., Missouri, 1972. Genetics, biochemistry, and molecular biology of chlorophyll biosynthesis in *Chlamydomonas.*
Joshua A. Weiner, Ph.D., San Diego, 1999. Neuronal cell interactions and development.
Chun-Fang Wu, Ph.D., Purdue, 1976. Neurobiology: cellular neurophysiology in *Drosophila;* genetic control of neural function and development.

Emeritus Faculty
Richard G. Baker, Professor of Geology; Ph.D., Colorado, 1969. Palynology: plant macrofossils; Quaternary paleoecology.
* Wayne R. Carlson, Ph.D., Indiana, 1968. Genome structure and cytogenetics of maize.
* Robert W. Cruden, Ph.D., Berkeley, 1967. Pollination ecology: reproductive biology of flowering plants; systematics.
* Robert W. Embree, Ph.D., Berkeley, 1962. Mycology: taxonomy and ecology of zygomycetes.
Joseph P. Hegmann, Ph.D., Illinois, 1968. Quantitative genetics.
Richard G. Kessel, Ph.D., Iowa, 1959. Cytology and developmental biology: germ-cell differentiation; structure, function, and morphogenesis of cellular organelles; mechanisms of secretion; scanning electron microscopy of tissues and organs; microvascular casts; structure and function of annulate lamellae and nuclear envelope.
Jerry J. Kollros, Ph.D., Chicago, 1942. Experimental embryology: size-control mechanisms in the nervous system; skin differentiation; integrative mechanisms in amphibian metamorphosis; regeneration.
* Thomas E. Melchert, Ph.D., Texas at Austin, 1963. Taxonomy; biochemical systematics and cytotaxonomy.
Roger D. Milkman, Ph.D., Harvard, 1956. Population genetics and molecular evolution: nucleotide sequence variation in wild strains of *Escherichia coli;* selection.
* Richard D. Sjölund, Ph.D., California, Davis, 1968. Cell biology: regulation of phloem differentiation in plant tissue culture; membrane transport; ultrastructure; video microscopy.
Eugene Spaziani, Ph.D., UCLA, 1956. Endocrinology: organismal, cellular, molecular, steroidogenesis; mechanisms of peptide hormone action; reproductive physiology; hormonal control of molting and growth.
Norman E. Williams, Ph.D., UCLA, 1958. Protozoan development control of synthesis and assembly of proteins involved in cortical differentiation in ciliates: electron microscopy of cortical development.

Joint Appointment
Jeffrey C. Murray, Professor of Pediatrics: M.D., Tufts, 1978. Human genetics.

**Plant sciences faculty member.*

UNIVERSITY OF KANSAS MEDICAL CENTER

Interdisciplinary Graduate Program in Biomedical Sciences

Programs of Study

The Interdisciplinary Graduate Program in Biomedical Sciences (IGPBS) leads to a Ph.D. or M.D./Ph.D. degree in various biomedical fields. Research at University of Kansas Medical Center (KUMC) encompasses a broad spectrum, including investigations of the underlying mechanisms of protein structure and function; fundamental research in pharmacology and toxicology; viral, microbial, molecular, cellular, developmental, reproductive, neuronal, immunological, renal, and general physiological biology; and clinically related studies focusing on a broad range of human diseases, such as cancer and AIDS.

At KUMC, students entering IGPBS have time to receive an education in the most current areas of the biomedical sciences before they select a laboratory in which to conduct their graduate research program. During the first year of the IGPBS, students participate in a state-of-the-art and highly integrated core curriculum that involves faculty members from all the basic science departments and some clinical departments.

In addition to courses that provide the fundamental principles essential for an understanding of the biomedical sciences, students in the first year of their studies are provided with an introduction to the practical aspects of research, such as the use of biostatistics, biographies, bioethics, the appropriate use of animals in research, and procedures for human studies research; and a laboratory resource course that teaches the principles and procedures involved in the use of cutting-edge laboratory techniques.

In the latter part of the first year of study, students also take advanced reading courses and laboratory rotations with KUMC's research faculty. This provides each student with the tie to evaluate the various research programs at KUMC before they select the research program that is best for them. Students select faculty mentors based on their research interests. Degrees are granted from the Departments of Anatomy and Cell Biology; Biochemistry and Molecular Biology; Microbiology, Molecular Genetics and Immunology; Pathology and Laboratory Medicine; Molecular and Integrative Physiology; Pharmacology, Toxicology and Therapeutics; and Neuroscience and the Training Program in Environmental Toxicology.

Research Facilities

Extensive research facilities are located throughout the main 60-acre campus include the Lied Research Building. Other research support facilities include computer services, a molecular biotechnology center, biometry support, and electron microscopy and confocal microscopy support services.

Financial Aid

Teaching and research assistantships are available. Students admitted into the IGPBS are awarded $21,700 in financial support and given a tuition waiver. Student travel awards are also available as students progress through the program. Scholarships for M.D./Ph.D. students are awarded annually on a competitive basis.

Cost of Study

Tuition is currently estimated at $179 per credit hour for state residents and $459 per credit hour for nonresidents. However, students accepted into the IGPBS receive a tuition waiver upon meeting enrollment requirements. Students are responsible for campus and library fees, estimated at $312 per academic year.

Living and Housing Costs

There are a multitude of options available to KUMC students near the campus. Current living expenses are between $450 and $800 per month.

Student Group

Twenty-five percent of the students enrolled in the IGPBS are from international locales. The age range of all students falls between 22 and 32 years.

Student Outcomes

Upon graduation from KUMC, students can expect to obtain a position in the biotechnology, academic, or governmental career fields.

Location

The University of Kansas Medical Center is located at 39th and Rainbow Boulevard in Kansas City, Kansas. It is on the border of Kansas and Missouri, with quick access to Westport, the Plaza, the Nelson-Atkins Museum of Art, and the Kansas City Art Institute.

The Graduate Program

The Interdisciplinary Graduate Program in Biomedical Sciences is an educational program within the School of Medicine at the University of Kansas Medical Center. It consists of eight degree-granting departments or programs. The IGPBS is made up of the Departments of Anatomy and Cell Biology; Biochemistry and Molecular Biology; Microbiology, Molecular Genetics and Immunology; Pathology and Laboratory Medicine; Molecular and Integrative Physiology; Pharmacology, Toxicology and Therapeutics; and Neuroscience and the Training Program in Environmental Toxicology.

Applying

Students who are interested in applying to the IGPBS should request an application packet. Students may also apply online via the information listed. Applications must be received by January 15.

Correspondence and Information

Director
Interdisciplinary Graduate Program in Biomedical Sciences
5009 Wescoe, MS3025
University of Kansas Medical Center
3901 Rainbow Boulevard
Kansas City, Kansas 66160-7836
Phone: 913-588-2719
      800-408-2039 (toll-free)
Fax: 913-588-2711
E-mail: igpbs@kumc.edu
Web site: http://www.kumc.edu/igpbs

University of Kansas Medical Center

AREAS OF RESEARCH

Faculty members at KUMC have fifteen areas of research emphasis:

Cell and developmental biology
Molecular and cellular immunology
Molecular and cellular biophysics
Molecular toxicology and environmental health
Neurosciences
Renal biology
Signal transduction and cancer biology
Cardiovascular biology
Molecular pathogenesis of infectious diseases
Molecular biology and genetics
Molecular virology
Pharmacological sciences
Proteomics
Reproductive biology
Structural biology

UNIVERSITY OF LOUISVILLE

Department of Biology

Programs of Study	The Department of Biology at the University of Louisville offers degree programs leading to the Ph.D. and the M.S. (thesis and nonthesis). A wide range of courses in the broader area of biology are taught. Current areas of strength are focused in two divisions: Molecular, Cellular, and Developmental Biology and Evolution, Ecology, and Behavior. Courses are offered in a variety of areas, including microbial ecology, genetics and evolution, population genetics and evolutionary medicine, plant biochemistry, behavioral ecology, developmental neural biology, environmental and exercise physiology, and community, population, and ecosystem ecology. Several faculty members are also associated with the interdisciplinary Center for Genetics and Molecular Medicine. A new Center for Evolutionary Medicine, for which several faculty members have already been hired, is planned for the near future. The M.S. programs are sufficiently flexible to meet any career goal. Both the M.S. and Ph.D. programs require that all students complete a flexible core program. Research for the doctoral dissertation or master's thesis may be conducted under the supervision of any member of the Department who holds membership in the graduate faculty. A cooperative Ph.D. program exists with the Department of Biology at Murray State University. Students seeking the Ph.D. degree often possess a master's degree or its equivalent; however, students may enter the program with only a baccalaureate degree. Generally, the first year or two is spent in course work and completion of the candidacy exams. Research is begun in the second year, and the dissertation completed in the final year.
Research Facilities	The Department of Biology occupies approximately 50,000 square feet of research, teaching, and office space in the Life Sciences Building and the new Belknap Research Building, located on the University's Belknap Campus. Special on-site research facilities and equipment include a greenhouse, low-speed and ultracentrifuges, spectrophotometers, gas chromatographs, sterile transfer hoods, a phosphorimager, an automated DNA sequencer, digital photomicroscopy equipment, scintillation counters, animal care facilities, controlled environment chambers, extensive and well-curated animal and plant research collections, field vehicles, and boats. The Environmental Analysis Laboratory houses an HPLC, gamma counter for dating sediments, atomic absorption spectrometer, mass spectrometer for stable isotope analysis, C and N analyzer, and automated equipment for water quality analyses. The Ohio River Experimental Station is a facility for conducting experiments on river biota and is located within minutes of campus, on the Ohio River. Faculty members and students also conduct research at the University's 200-acre Horner Wildlife Sanctuary and on the 14,000 acres of the private Bernheim Forest, located just south of Louisville.
Financial Aid	Teaching assistantships and University fellowships are available competitively; research assistantships are available through grants and contracts held by individual faculty members. Teaching assistantships and fellowships provide a stipend ($20,000 for twelve months), tuition remission, and University-paid health insurance.
Cost of Study	Tuition rates are available from the Bursar at 502-852-7325 or via the World Wide Web at http://www.louisville.edu/admin/bursar/burshome.html.
Living and Housing Costs	Housing costs are available from the Bursar at 502-852-7325 or via the World Wide Web at http://www.louisville.edu/student/life/housing/rates.html.
Student Group	There are approximately 60 students in the graduate programs. More than half receive aid through teaching assistantships, research assistantships, and University fellowships.
Location	The Department of Biology is located on the University's main campus, which occupies 160 acres about 2 miles south of downtown. The Louisville metropolitan area, the sixteenth-largest city in the United States with a current population of approximately 1 million, is a center for activities in the arts, education, and sports. Louisville has consistently been rated as one of the most desirable areas of the country in which to live and has a very moderate cost of living for a metropolitan area.
The University	The University of Louisville was founded in 1798 and, prior to its entry into the Kentucky state system, was the oldest municipal university in the nation. At present, there are more than 20,000 students in a broad range of undergraduate and graduate programs. The University is one of the two major research universities in the commonwealth and has been designated as the primary center for urban-related programs.
Applying	Applications must be complete by July 1 for fall admission or December 1 for spring admission. To ensure competitiveness for financial support, applications for fall should be complete by February 1, and applications for spring should be complete by October 1.
Correspondence and Information	Dr. Joseph M. Steffen Associate Professor, Director of Graduate Studies Department of Biology University of Louisville Louisville, Kentucky 40292 Phone: 502-852-6771 Fax: 502-852-0725 E-mail: joe.steffen@louisville.edu Web site: http://www.louisville.edu/a-s/biology

University of Louisville

THE FACULTY AND THEIR RESEARCH

Ronald M. Atlas, Professor and Dean of the Graduate School; Ph.D., Rutgers, 1972. Microbial ecology and physiology; hydrocarbon metabolism; microbial abatement of pollution; polar microbiology. (r.atlas@louisville.edu)

Margaret M. Carreiro, Associate Professor; Ph.D., Rhode Island, 1989. Nutrient cycling; ecosystem ecology and microbial ecology of terrestrial habitats, particularly those in urban and suburban landscapes. (m.carreiro@louisville.edu)

Gary A. Cobbs, Professor; Ph.D., California, Riverside, 1975. Biostatistics; population genetics; protein polymorphism; evolutionary theory; *Drosophila* genetics. (gacobb01@gwise.louisville.edu)

Cynthia Corbitt, Assistant Professor; Ph.D., Alaska Fairbanks, 1997. Behavioral neuroendocrinology; environmental signaling and the central nervous system. (cynthia.corbitt@louisville.edu)

Lee A. Dugatkin, Professor; Ph.D., SUNY at Binghamton, 1991. Behavioral and evolutionary ecology; evolution of cooperative and altruistic behavior; cultural transmission and female mate choice; evolutionary psychology. (laduga01@athena.louisville.edu)

Daniel E. Dykhuizen, Professor; Ph.D., Chicago, 1970. Microbial population biology; evolution of infectious disease; emerging pathogens; population genetics and evolution. (dandykclife.bio.sunysb.edu)

Perri K. Eason, Associate Professor and Curator, Monroe Terrestrial Vertebrate Collections; Ph.D., California, Davis, 1991. Behavioral ecology and conservation biology of terrestrial vertebrates and insects. (pkeaso01@gwise.louisville.edu)

Paul W. Ewald, Professor; Ph.D., Washington (Seattle), 1980. Evolutionary ecology of parasitism; evolutionary medicine; agonistic behavior; pollination biology. (p0ewal01@gwise.louisville.edu)

Ronald D. Fell, Professor and Chair; Ph.D., Iowa State, 1977. Applied physiology, nutrition, muscle metabolism, and exercise physiology. (rfell@louisville.edu)

Jay M. Gulledge, Assistant Professor; Ph.D., Alaska Fairbanks, 1996. Microbial ecology and physiology; biogeochemistry; soil trace gas and nutrient cycling; global change biology; molecular community analysis. (jay.gulledge@louisville.edu)

Jeffrey D. Jack, Associate Professor; Ph.D., Dartmouth, 1993. Aquatic ecology; zooplankton ecology in large rivers; stream restoration and bioassessment. (jeff.jack@louisville.edu)

Awdhesh Kalia, Assistant Professor; Ph.D., All India Institute of Medicine, 2000. Pathogen epidemiology and evolution; evolution of bacterial pathogenesis. (a0kali02@gwise.louisville.edu)

Arnold J. Karpoff, Associate Professor; Ph.D., Oregon, 1969. Plant anatomy; plant growth and development; plant tissue culture. (ajkarp01@gwise.louisville.edu)

Martin G. Klotz, Associate Professor; Ph.D., Jena (Germany), 1986. Molecular genetics and evolution of bacterial autotrophic nitrification; oxidative stress tolerance of fluorescent pseudomonads; plant-bacterial interactions. (martin.klotz@louisville.edu)

William D. Pearson, Professor; Curator, Clay Fish Collection; Ph.D., Utah State, 1970. Fisheries biology; ichthyology; aquatic ecology; biospeleology. (wdpear01@gwise.louisville.edu)

Michael H. Perlin, Professor; Ph.D., Chicago, 1983. Molecular genetics; structure and function of the gene; molecular biology of drug action; gene mapping and splicing; fungal phytopathogens; host-pathogen interaction. (mhperl01@gwise.louisville.edu)

Susan K. Remold, Assistant Professor; Ph.D., Cornell, 1998. Use of microbial systems to explore the genetic architecture of complex traits and the distribution of organisms in space and hosts. (susanna.remold@yale.edu)

David J. Schultz, Assistant Professor; Ph.D., Penn State, 1996. Molecular genetics, biotechnology, and biochemistry; production of unusual lipids and secondary metabolites in plants; trichome-mediated pest resistance in *Pelargonium xhortorum* (geranium). (david.schultz@louisville.edu)

Joseph M. Steffen, Associate Professor; Ph.D., New Mexico, 1980. Cellular and molecular physiology; environmental physiology; endocrine and growth factor effects on muscle. (joe.steffen@louisville.edu)

Eric V. Wong, Assistant Professor; Ph.D., Case Western Reserve, 1996. Developmental neurobiology; axon guidance in visual system development; inhibitory proteins in central nervous system regeneration. (evwong@gwise.louisville.edu)

UNIVERSITY OF MASSACHUSETTS WORCESTER

Medical School
Graduate School of Biomedical Sciences

Program of Study	The University of Massachusetts Worcester Graduate School of Biomedical Sciences offers programs leading to the Ph.D. degree in biomedical sciences. The Graduate School is composed of three divisions: Basic and Biomedical Sciences, Biomedical Engineering, and Clinical and Population Health Research. The Basic and Biomedical Sciences division includes specialization in biochemistry and molecular pharmacology, cell biology, cellular and molecular physiology, immunology and virology, molecular genetics and microbiology, and neuroscience; an interdisciplinary graduate program is also offered. All of the divisions share the common objective of educating students to become biomedical scientists/educators who are equipped to conduct research with direct relevance to human disease. The University of Massachusetts Medical School is one of the fastest-growing biomedical research institutes in the nation and received more than $175 in research funding in 2005. First-year students in the Basic and Biomedical Sciences division take an integrated interdisciplinary core curriculum and participate in laboratory rotations. At the start of the second year, students select an area of specialization and a research adviser and take the qualifying examination. Advanced course requirements and thesis research occupy the remainder of the program. The final examination is the defense of the research thesis. Students in the Biomedical Engineering division take a mixture of biomedical courses at the Medical School and engineering courses at Worcester Polytechnic Institute (WPI) in addition to completing laboratory rotations during their first and second years. Students in the Clinical and Population Health Research division complete a core curriculum based on appropriate statistical, epidemiologic, and research methods, as well as develop an understanding of the contextual issues of health and health care.
Research Facilities	The Medical School opened in 1970 and has been generously endowed with office, classroom, and laboratory space. Facilities and equipment are available in all modern areas of biomedical research. Most recently, a ten-story, 360,000-square-foot research building opened in 2001. Major facilities include a large experimental animal-care facility, a virus laboratory, a cell and tissue culture facility, a protein chemistry laboratory, a Diabetes Research Center, and electronics and machine shops. New technologies include three-dimensional analysis of proteins and other large molecules through X-ray crystallography and nuclear magnetic resonance spectroscopy, a facility for transgenic animals—animals that receive foreign genes, isolated and transferred from other animals or humans—to test the consequences of gene function on the human body, and a Drug Development Program to assist in the development of products by pharmaceutical and biotechnology companies.
Financial Aid	Beginning full-time students are offered a graduate assistantship (in 2005–06 the annual stipend was $25,235). Research assistantships paying $25,235 are available to students starting thesis research. All students are awarded assistantships. Tuition is waived for both graduate and research assistants.
Cost of Study	All students are supported with assistantships and tuition waivers. Student health, dental, and disability insurance is provided by the School.
Living and Housing Costs	Although no on-campus housing is provided, many apartments and rooms are available in the Worcester area. The large number of students in the area makes sharing of apartments feasible at costs that fall within the budget dictated by the assistantship stipend; local public transportation is available and quite economical.
Student Group	There are currently 398 doctoral students from throughout the United States and other countries enrolled. This includes 34 students enrolled in the combined M.D./Ph.D. program.
Student Outcomes	Graduates pursue postdoctoral positions at outstanding institutions and eventually take staff positions in industry or government or faculty positions in academic institutions throughout the United States and abroad. The program held its first graduation in 1984 and has more than 275 alumni.
Location	The third-largest city in New England, Worcester is 40 miles west of Boston. Worcester and the surrounding towns offer varied living conditions and a wide range of educational and cultural activities. Located in the city are numerous other educational institutions, including Clark University; Holy Cross, Assumption, and Worcester State Colleges; and Worcester Polytechnic Institute.
The University and The Medical School	The University of Massachusetts has five campuses: the flagship campus at Amherst; the Medical School in Worcester; and campuses in Boston, Dartmouth, and Lowell. The Medical School opened in 1970 with 16 students; there are now 100 medical students in each class. The Medical School joins regional institutions and the University of Massachusetts at Amherst in the training of other health-care professionals. The Graduate School of Nursing at the Medical School opened in 1985.
Applying	Each candidate for admission to the Basic and Biomedical Sciences division is expected to have a bachelor's degree in one of the physical or biological sciences. Senior undergraduates may be admitted pending successful completion of their baccalaureate programs. While no minimum grade point average is required, it is expected that applicants will have demonstrated strong performance in their science undergraduate studies. In addition, applicants are expected to have at least one year of research experience in a laboratory setting. Applicants to the Biomedical Engineering division are expected to have an undergraduate degree or a strong background in mathematics, physics, or engineering. Applicants to the Clinical and Population Health Research division are expected to have a Master's degree and completed introductory work in biostatistics and epidemiology. Strong applicants may be admitted conditionally prior to completing such course work. Applicants in each division must take the General Test of the Graduate Record Examinations. Three letters of recommendation, original transcripts, and an application fee ($25 for Massachusetts residents and $50 for nonresidents), which is subject to change, are required. The deadline for fall admission is December 15 for the Basic and Biomedical Sciences and Biomedical Engineering divisions and February 1 for the Clinical and Population Health Research division. The University of Massachusetts Medical School is an Equal Opportunity/Affirmative Action employer. The University of Massachusetts Medical School recruits members of minority groups and women candidates, considers all applicants, and, once students are accepted, ensures that they are not discriminated against in any area. The graduate catalog and application forms may be accessed at the Web site listed in this In-Depth Description. Potential applicants are welcome to contact graduate directors (listed in this description) with specific questions about departmental programs.
Correspondence and Information	Director of Recruitment and Admission Graduate School of Biomedical Sciences University of Massachusetts Worcester Worcester, Massachusetts 01655-0116 Phone: 508-856-4135 E-mail: gsbs@umassmed.edu Web site: http://www.umassmed.edu/gsbs

University of Massachusetts Worcester

GRADUATE REPRESENTATIVES AND AREAS OF SPECIALIZATION

Program in Biochemistry and Molecular Pharmacology
C. Robert Matthews, Ph.D., Chair; Nick Rhind, Ph.D., Graduate Director. Research areas in molecular mechanisms of protein folding, gene expression and regulation, DNA repair, cell signaling and cell cycle regulation, developmental biology, protein trafficking, transport across biological membranes and membrane biophysics, neurobiology, structural and chemical biology, and viral molecular biology.

Program in Biomedical Engineering and Medical Physics
Glenn Gaudette, Ph.D., Program and Graduate Director (Department of Surgery). This joint program between the Graduate School of Biomedical Sciences and Worcester Polytechnic Institute employs the advanced technical knowledge and expertise of engineering and medical faculty members to provide students with the knowledge and skills to apply engineering principles to medically related problems. The key areas of research focus are biosensors and instrumentation, medical imaging, and tissue engineering.

Program in Cell Biology
Gary S. Stein, Ph.D., Chair; Anthony Imbalzano, Ph.D., Graduate Director. Cellular, molecular, and genetic approaches are taken to address structure-function relationships associated with cell growth and differentiation, fertilization, development, cell signaling, gene expression, cell motility, and DNA repair. State-of-the-art techniques are used to explore the underlying cellular bases of diseases, including cancer, muscular dystrophy, rheumatoid arthritis, diabetes, osteoporosis, skeletal abnormalities, Huntington's and other neurodegenerative diseases, and central nervous system and immune function defects.

Program in Cellular and Molecular Physiology
H. Maurice Goodman, Ph.D., Chair; Jack Leonard, Ph.D., Graduate Director. Research utilizes molecular, biophysical, and cellular techniques to investigate cell function, with emphasis on signal transduction processes, cell growth, and proliferation. The cellular actions of neural transmitters, hormones, and sensation are also being investigated. In addition, studies are being performed to characterize the properties of membrane ionic channels and transporters. This research focuses on elucidating the functions of normal cells that constitute the tissues of the body in order to understand perturbations caused by aging and diseases that afflict mankind.

Program in Clinical and Population Health Research
Carole Upshur, Ed.D., Associate Dean. The program emphasizes clinical and health services research skills and provides students with the tools needed to conduct research on health-care access, screening, treatment, quality, and outcomes.

Program in Immunology/Virology
Alan Rothman, M.D., Program Director; Trudy Morrison, Ph.D., and Janet Stavnezer, Ph.D., Graduate Advisors. Research in an interdepartmental program in immunology and virology studying lymphocyte development, differentiation and function, immune effector mechanisms, signal transduction in lymphocytes, autoimmune disorders, antigen presentation, and immune responses to viral and bacterial pathogens.

Program in Molecular Genetics and Microbiology
Allan Jacobson, Ph.D., Chair; Richard Baker, Ph.D., Graduate Director. Research interests of the program faculty members include molecular biology, molecular and cellular immunology, genetics, virology, bacteriology, microbial pathogenesis, neurobiology, and development biology.

Program in Neuroscience
Steven Treistman, Ph.D., Program Director; David Weaver, Ph.D., Graduate Director. Research in this interdepartmental program stresses molecular approaches to nervous system function and dysfunction. Faculty members interests span the spectrum from molecular biology and biophysics to behavior, and investigators utilize a wide variety of techniques.

Interdisciplinary Graduate Program in Biomedical Sciences
William Theurkauf, Ph.D., Program Director; Craig Peterson, Ph.D., Program Co-Director. The program encourages interdisciplinary approaches to complex problems of biomedical significance and draws on faculty members from all of the basic and biomedical science departments and several clinical departments within the Medical School. In order to facilitate rapid initiation of full-time thesis research, the program has a streamlined and flexible graduate curriculum that is tailored to the specific needs of individual students.

The student-faculty relationship is highlighted by a shared passion for scientific exploration.

The Lazare Research Building at the University of Massachusetts Medical School.

Students benefit from the first-class facilities of the Medical School campus.

UNIVERSITY OF MASSACHUSETTS WORCESTER

Interdisciplinary Graduate Program

Program of Study	The Interdisciplinary Graduate Program (IGP) at the University of Massachusetts Medical School (UMMS) in Worcester offers a program of study and research leading to a Ph.D. degree in biomedical sciences. Much of modern biomedical research is interdisciplinary in nature, utilizing a combination of genetic, molecular, biochemical, and cellular approaches to address fundamental problems in complex systems. The IGP supports and encourages graduate student training in broad-based interdisciplinary approaches to problems of biomedical significance. The program is characterized by a streamlined and flexible graduate curriculum that is tailored to the specific needs of individual students and encourages rapid initiation of full-time thesis research. The IGP currently includes more than 100 UMMS faculty members, representing all of the basic science and several clinical departments.

A major goal of the IGP is to provide a curriculum that allows students to become fully engaged in thesis research as early as possible, usually by the end of the first year. In order to ensure a solid academic background and experience in a variety of experimental approaches, all first-year students take a full-year core course covering biochemistry and molecular and cellular biology and perform a series of laboratory rotations. Optimally, rotations are completed and a thesis laboratory selected by the summer of a student's first year. Advanced course work, journal clubs, and other enrichment activities beyond the graduate core course are tailored to the requirements of an individual student and are determined by the faculty adviser and student. A minimum of two advanced topics courses are required.

After a thesis laboratory is selected, students begin preparation for the qualifying exam (composed of a written research proposal and oral defense), which can be based on the proposed thesis topic or on an independent subject. In consultation with the adviser, the student selects a Qualifying Committee, which generally becomes the Thesis Advisory Committee. Students meet with the Research Advisory Committee at least once a year. The Ph.D. degree is awarded following completion and oral defense of a written thesis.

Research Facilities
The program's extensive state-of-the-art research facilities are housed in the main Medical School building, the adjacent Biotechnology Park, and the Lazare Research Building. Specialized research cores include facilities for gene chip analysis, mass spectroscopy, X-ray crystallography, transgenics, DNA sequencing and synthesis, analytical ultracentrifugation, proteomics, and biomedical imaging.

Financial Aid
In 2005–06, beginning full-time students were offered a graduate assistantship at an annual stipend of $25,235. Research assistantships paying $25,235 were available to students starting thesis research. Tuition is waived for both graduate and research assistants. The University provides health and hospitalization insurance as well as a disability insurance plan to all students.

Cost of Study
Tuition is waived for all graduate students.

Living and Housing Costs
Apartments and rooms are available in the Worcester area and nearby communities. The large number of students in the area makes the sharing of apartments feasible, at costs that can be borne by students receiving financial aid. Local public transportation is also available.

Student Group
There are approximately 400 graduate students enrolled in the Graduate School of Biomedical Sciences; more than 100 are students enrolled in the IGP Ph.D. program.

Location
Located in the heart of the commonwealth, Worcester is the third-largest city in New England and is 40 miles west of Boston. Worcester and the surrounding towns offer a wide range of educational, cultural, and recreational activities. Skiing and hiking areas and numerous ponds and rivers are located in the Worcester region, as are many other educational institutions, including Worcester Polytechnic Institute, Assumption College, Clark University, and Holy Cross College. The beaches of Cape Cod and Rhode Island, as well as the ski areas of Vermont, New Hampshire, and Maine, are within easy driving distance of the campus.

The Medical School
The Medical School is an integral part of the state university system, which consists of four other campuses in Lowell, Dartmouth, Amherst, and Boston.

Applying
The IGP accepts graduate students through the Graduate School of Biomedical Sciences. Applicants should have a bachelor's degree in a physical or biological science. Specific course requirements include a year each of calculus, organic chemistry, physics, and biology. Applicants must take the General Test of the Graduate Record Examinations. No GRE Subject Tests are required. Three letters of recommendation and an application fee ($25 for in-state students, $50 for out-of-state students) are required. Applications should be received by January 2 for admission the following fall.

The University of Massachusetts Medical School is an Equal Opportunity/Affirmative Action employer. The University of Massachusetts Medical School recruits minority and women candidates, considers all applicants, and, once students are accepted, ensures that they are not discriminated against in any area.

The graduate catalog and application forms may be obtained from the address in this In-Depth Description. Additional information about the IGP is available at http://www.umassmed.edu/igp/. Specific questions can be directed to the IGP Administrator, Darla Cavanaugh (telephone: 508-856-6877; e-mail: igp@umassmed.edu) or the Program Director, William Theurkauf (telephone: 508-856-4900; e-mail: william.theurkauf@umassmed.edu).

Correspondence and Information
Dean, Graduate School of Biomedical Sciences
University of Massachusetts Medical School
55 Lake Avenue, North
Worcester, Massachusetts 01655-0116
Phone: 508-856-4135
Web site: http://www.umassmed.edu/gsbs

University of Massachusetts Worcester

THE FACULTY AND THEIR RESEARCH

Usha Acharya, Ph.D., Assistant Professor. Molecular genetics of lipid metabolism and signaling.

Schahram Akbarian, M.D., Ph.D. Assistant Professor. Noradrenergic neuron function in neurotrophin mutant mice.

Elliot Androphy, M.D., Professor. Molecular biology of papillomaviruses.

Zheng-Zheng Bao, Ph.D., Assistant Professor. Molecular mechanisms in vertebrate development.

Leslie Berg, Ph.D., Professor. T-lymphocyte development and activation.

Rita Bortell, Ph.D., Associate Professor. Pathogenesis of autoimmune diabetes.

Victor Boyartchuk, Ph.D., Assistant Professor. Studies of genetic factors controlling susceptibility to infectious diseases using animal model systems.

Michael Brodsky, Ph.D., Assistant Professor. *Drosophila* p53 and DNA damage–induced apoptosis.

Vivian Budnik, Ph.D., Professor. Molecular mechanisms of synapse assembly and plasticity.

Sharon Cantor, Ph.D., Professor. Hereditary breast cancer.

Lucio Castilla, Ph.D., Assistant Professor. Genetics of leukemia; mouse models.

Francis Chan, Ph.D., Assistant Professor. Analysis of TNF receptor signaling in healthy and diseased immune systems.

Jason Chen, Ph.D., Assistant Professor. Proliferation and apoptosis of cells expressing papillomavirus oncogenes.

Paul Clapham, Ph.D., Associate Professor. HIV receptors and cellular tropisms.

Silvia Corvera, M.D., Professor. Receptor trafficking.

Michael Czech, Ph.D., Professor. Signal transduction by the insulin and insulin-like growth factor receptor tyrosine kinases.

Roger Davis, Ph.D., Professor. Signal transduction mechanisms.

Job Dekker, Ph.D., Assistant Professor. Epigenetic control of chromosome activity.

Ronald Desrosiers, Ph.D., Adjunct Professor. Viral pathogenesis.

Stephen Doxsey, Ph.D., Associate Professor. Regulation of mitotic spindle assembly.

Patrick Emery, Ph.D., Assistant Professor. Circadian rhythms and photoreception in *Drosophila*.

Robert Finberg, M.D., Professor. Host-microbial interactions.

Katherine Fitzgerald, Ph.D., Assistant Professor. Innate immune signaling.

Harvey Florman, Ph.D., Professor. Signal transduction mechanisms.

Paul Gardner, Ph.D., Associate Professor. Molecular studies of neurotransmitter receptor gene expression.

Rachel Gerstein, Ph.D., Assistant Professor. Developmental regulation of lymphocyte development and V(D)J recombination.

Reid Gilmore, Ph.D., Professor. Molecular mechanism of secretory protein translocation.

Edward Ginns, M.D., Ph.D., Professor. Molecular basis of psychiatric and neurologic disorders.

Douglas T. Golenbock, M.D., Professor. Toll receptors; pathophysiology of sepsis and pelvic inflammatory disease.

Heinrich Gottlinger, M.D., Ph.D., Professor. Molecular biology of HIV-1.

Michael Green, M.D., Ph.D., Professor. Regulation of gene expression in eukaryotic cells.

Dale Greiner, Ph.D., Professor. Transplantation tolerance and autoimmune diabetes.

Steven Grossman, M.D., Ph.D., Assistant Professor. Ubiquitin/proteasome system and cancer.

Kirsten Hagstrom, Ph.D., Assistant Professor. Chromosome structure and segregation during cell division.

Lawrence Hayward, M.D., Ph.D., Associate Professor. Hyperkalemic periodic paralysis: Na channelopathy.

Shuk-Mei Ho, Ph.D., Professor. Hormonal carcinogenesis in the ovary, prostate, and breast.

Mitsuo Ikebe, Ph.D., Professor. Function and regulation of motor proteins in cellular processes.

Anthony Imbalzano, Ph.D., Associate Professor. Chromatin structure and regulation of gene expression, cell cycle, cell differentiation, and development.

Tony Ip, Ph.D., Associate Professor. Gene regulatory molecules in innate immunity and development.

Allan Jacobson, Ph.D., Professor. Cytoplasmic aspects of the posttranscriptional regulation of gene expression.

Stephen Jones, Ph.D., Assistant Professor. Genes in development and tumorigenesis.

Paul Kaufman, Ph.D., Associate Professor. Assembly and function of eukaryotic chromosomes.

Michelle Kelliher, Ph.D., Assistant Professor. Role of human leukemia-associated transcriptional factors in development of cancer.

Kendall Knight, Ph.D., Associate Professor. Genetic recombination: functional organization of multidomain enzymes and multiprotein complexes.

William Kobertz, Ph.D., Assistant Professor. Structure, function, and modulation of ion channels.

Timothy Kowalik, Ph.D., Assistant Professor. Cell activation, apoptosis, and DNA viruses.

Stephen Lambert, Ph.D., Assistant Professor. Membrane-cytoskeleton interactions in axonal function.

David Lambright, Ph.D., Associate Professor. Structural studies of intracellular signaling and vesicle trafficking.

Nathan Lawson, Ph.D., Assistant Professor. Determining the signals responsible for blood vessel development using zebrafish.

Tzumin Lee, M.D., Ph.D., Associate Professor. Neural circuitry formation and plasticity in *Drosophila* brain.

John Leong, M.D., Ph.D., Associate Professor. Microbial pathogens and the mammalian host.

Brian Lewis, Ph.D., Assistant Professor. Molecular genetics of pancreatic and liver cancers.

Hong-Sheng Li, Ph.D., Assistant Professor. Neuronal signal transduction and degeneration in the fly eye.

Egil Lien, Ph.D., Assistant Professor. Mechanisms for microbial activation and evasion of innate immune responses via toll-like receptors.

Elizabeth Luna, Ph.D., Professor. Membrane skeleton dynamics; motility; adhesion; signal transduction.

Jianyuan Luo, Ph.D., Assistant Professor. P53 acetylation/deacetylation pathway and cancer.

Katherine Luzuriaga, M.D., Professor. Viral pathogenesis and immunopathogenesis of persistent viral infections.

Martin Marinus, Ph.D., Professor. Recombinational DNA repair.

Dannel McCollum, Ph.D., Assistant Professor. Cell cycle, signaling, and cytokinesis.

Haley Melikian, Ph.D., Assistant Professor. Cocaine and antidepressant-sensitive monoamine transporters.

Craig Mello, Ph.D., Professor. Regulation of gene expression during early embryogenesis in *C. elegans*.

Stephen Miller, Ph.D., Assistant Professor. Cellular processes by GTPases.

John Mordes, M.D., Professor. Diabetes.

Mary Munson, Ph.D., Assistant Professor. Regulation of vesicle targeting and fusion.

Gregory Pazour, Ph.D., Assistant Professor. Role of the intraflagellar transport (IFT) proteins in eukaryotic ciliary assembly.

Thoru Pederson, Ph.D., Professor. RNA processing, RNA-protein interactions, and RNA traffic in eukaryotic cells.

Craig Peterson, Ph.D., Professor. Role of chromatin in gene expression and DNA repair.

Peter Pryciak, Ph.D., Assistant Professor. Signal transduction and cell polarity control.

Steven Reppert, M.D., Professor. Molecular clocks.

Nick Rhind, Ph.D., Assistant Professor. Checkpoint regulation of the fission yeast cell cycle.

Joel Richter, Ph.D., Professor. Translational control in early development; translational control in the mammalian central nervous system.

Ann Rittenhouse, Ph.D., Associate Professor. Calcium channels and neuronal plasticity.

Alonzo Ross, Ph.D., Professor. PTEN; CNS stem cells; NGF signal transduction.

Aldo Rossini, M.D., Professor. Mechanisms of tolerance applied to the pathogenesis of autoimmune insulin-dependent diabetes mellitus and its cure.

Charles Sagerstrom, Ph.D., Assistant Professor. Zebrafish developmental neurobiology.

Celia Schiffer, Ph.D., Associate Professor. Structural basis for molecular recognition in HIV protease.

Neal Silverman, Ph.D., Assistant Professor. Signal transduction during the insect immune response.

Ravindra Singh, Ph.D., Assistant Professor. Use of evolutionary selection to understand alternative splicing and the protein function.

Greenfield Sluder, Ph.D., Professor. Mitosis.

Merav Socolovsky, M.D., Ph.D., Assistant Professor. Molecular mechanisms regulating the homeostasis of hematopoietic progenitors.

Janet Stavnezer, Ph.D., Professor. Molecular genetics of the immunoglobulin heavy-chain switch.

Lawrence Stern, Ph.D., Associate Professor. Molecular recognition in the immune system.

Mario Stevenson, Ph.D., Professor. AIDS pathogenesis.

John Sullivan, M.D., Professor. Relationship between HIV-1 pathogenesis and the human immune system.

William Theurkauf, Ph.D., Professor. Cell division control and tumor-suppressor function.

Donald Tipper, Ph.D., Professor. Yeast molecular genetics.

Heidi Tissenbaum, Ph.D., Assistant Professor. Molecular mechanisms of aging in *C. elegans*.

Steven Treistman, Ph.D., Professor. Mechanism of action of alcohol and opiate drugs.

Fumihiko Urano, M.D., Ph.D., Assistant Professor. Molecular pathogenesis of protein conformational diseases.

Bert van den Berg, Ph.D., Assistant Professor. Structure and function of membrane transport proteins.

Scott Waddell, Ph.D., Assistant Professor. *Drosophila* olfactory memory.

Marion Walhout, Ph.D., Assistant Professor. Mapping transcription regulatory circuits in the nematode *C. elegans*.

Yong-Xu Wang, Ph.D., Assistant Professor. Transcriptional control of energy metabolism and metabolic diseases by the nuclear receptor PPAR subfamily.

Yu-Li Wang, Ph.D., Professor. Mechanism of cell migration and cell division.

Doyle Ward, Ph.D., Assistant Professor. Type IV secretion and bacterial virulence.

David Weaver, Ph.D., Associate Professor. Molecular physiology of circadian rhythms.

George Witman, Ph.D., Professor. Cilia, flagella, and the molecular motors.

Scot Wolfe, Ph.D., Assistant Professor. Creating novel transcription factors for targeted gene regulation.

Lan Xu, Ph.D., Assistant Professor. TGFβ signal transduction.

Zuoshang Xu, Ph.D., Associate Professor. Structure, function, and degeneration of spinal motor neurons.

Phillip Zamore, Ph.D., Associate Professor. Dissecting the RNAi and stRNA pathways.

Maria Zapp, Ph.D., Assistant Professor. Regulation of retroviral and cellular mRNA.

Jianhua Zhou, Ph.D., Assistant Professor. Molecular pathogenesis of neurodegenerative diseases.

UNIVERSITY OF MEDICINE AND DENTISTRY OF NEW JERSEY

Graduate School of Biomedical Sciences
at Robert Wood Johnson Medical School, Piscataway Division

Programs of Study

The graduate programs in cellular and molecular pharmacology; biochemistry and molecular biology; molecular genetics, microbiology, and immunology; and physiology and integrative biology, together with corresponding programs based at Rutgers University, constitute the programs in the molecular biosciences. These programs afford maximum flexibility by integrating more than 200 faculty members from the nine Ph.D. programs in the two universities, which share admissions and the first-year curriculum in the molecular biosciences. Incoming graduate students also complete three or four 8-week research rotations in different laboratories to broaden their perspective and help them to select a thesis adviser. A written qualifying examination and defense of a proposition paper are required. After the first year, each student chooses a specific graduate program, dependent on research interest, and follows the curriculum of that program. The Ph.D. degree requires completion and successful defense of a thesis based on original research.

The graduate programs in biomedical engineering, neuroscience, and exposure assessment (environmental sciences), together with the corresponding programs at Rutgers University, also offer the Ph.D. After completing courses and passing the qualifying examination, each student selects a thesis adviser from the either university. The Ph.D. degree is awarded based on completion and successful defense of a thesis based on original research.

Research Facilities

The extensive research facilities of both universities provide opportunities for studying virtually all aspects of biomedical sciences. Research facilities of participating faculty members are located in approximately fifteen academic and clinical departments and research institutes at Rutgers and UMDNJ–Robert Wood Johnson Medical School. Research institutes and centers include the Waksman Institute, Cancer Institute of New Jersey, Center for Advanced Biotechnology and Medicine, Center for Agricultural Molecular Biotechnology, Center for Alcohol Studies, Child Health Institute, Gerontological Institute, and Environmental and Occupational Health Sciences Institute. A major computer facility, the Center for Computer and Information Services, provides powerful computing capabilities for advanced research. Microcomputers are available to students in most laboratories. The library system has holdings of more than 3 million bound volumes and 1 million documents; major scientific holdings are located at the Library of Science and Medicine and departmental and research institute libraries.

Financial Aid

All incoming Ph.D. graduate students are provided with financial support, including a stipend, tuition remission, and health insurance. Multiple NIH-sponsored training grants and the Initiative for Minority Student Development and Bridge to the Doctoral Degree Awards are available for students with outstanding credentials.

Cost of Study

For 2005–06, tuition for New Jersey residents was $5556 per academic year for full-time enrollment. Tuition for nonresident students was $8049 per academic year for full-time enrollment. Student fees are modest.

Living and Housing Costs

For the academic year 2005–06, on-campus housing was available for $6912 per person. Residence halls for men and women are available. Family housing and meal plans at dining halls are also available. Information on housing in the New Brunswick/Piscataway area is available through the Department of Housing (telephone: 732-445-2215).

Student Group

Students are admitted through the programs in the molecular biosciences and in other programs of Rutgers University and UMDNJ, which annually admit 70 students. These students interact with more than 500 students in the biomedical sciences at the two universities that share the Piscataway campus.

Student Outcomes

Most graduates obtain postdoctoral research positions in laboratories in academia, government laboratories, or industry and subsequently are employed in similar institutions. New Jersey is the home of many pharmaceutical, chemical, and biotechnology research laboratories.

Location

The campus of UMDNJ–Robert Wood Johnson Medical School is located in Piscataway, New Jersey, on the Busch campus of Rutgers University. It is about 45 miles south of New York City and is within easy access of highways and mass transit. The Medical School is convenient to New York City's cultural life, the many recreational beach areas of the New Jersey shoreline, and the camping areas of the state parks.

The University and The School

The University of Medicine and Dentistry of New Jersey is a state institution under the administration of a Board of Trustees appointed by the governor. It includes the New Jersey medical and dental schools, the School of Nursing, and the School of Health Related Professions, all located at the Newark campus; the Robert Wood Johnson Medical School in Piscataway and Camden; the School of Public Health, with campuses in Piscataway, Newark, and Stratford/Camden; and the New Jersey School of Osteopathic Medicine in Stratford. The Graduate School of Biomedical Sciences comprises three divisions on the Newark, Piscataway, and Stratford campuses. The Piscataway campus is located on the Busch campus of Rutgers, The State University of New Jersey.

Applying

A bachelor's degree or its equivalent, with a strong record of academic achievement, is required for admission. The most appropriate preparation is an undergraduate concentration in either biology (molecular and cellular biology and microbiology) or chemistry (organic, analytical, and physical chemistry) and should include courses in calculus and physics. Selection is made on the basis of previous academic work, letters of recommendation, and scores on the Graduate Record Examinations (the required General Test and the optional Subject Test in biochemistry, cell and molecular biology; biology; or chemistry). TOEFL scores are required of students for whom English is not the native language.

Correspondence and Information

Admissions
Graduate School of Biomedical Sciences
Robert Wood Johnson Medical School
University of Medicine and Dentistry of New Jersey
675 Hoes Lane
Piscataway, New Jersey 08854-5635
Phone: 732-235-5016
E-mail: gsbspisc@umdnj.edu
Web site: http://www2.umdnj.edu/~gsbspisc

University of Medicine and Dentistry of New Jersey

THE FACULTY

BIOCHEMISTRY AND MOLECULAR BIOLOGY
Professor M. Inouye (Department Chair).
Associate Professor K. Madura (Graduate Director).

Professors
C. Abate-Shen, E. Arnold, S. Brill, B. Brodsky, K. Chada, M. Driscoll, R. Ebright, M. R. Gartenberg, C. Gélinas, M. Hampsey, J. Lenard, P. Lobel, F. Matsumura, J. W. Messing, G. Montelione, S. Patel, D. Reinberg, M. Roth, A. Shatkin, A. Stock, P. Yurchenco.

Associate Professors
C. Carr, P. Casaccia-Bonnefil, R. Habas, S. C. Hsu, S. Inouye, K. Madura, S. Shiff, V. Studitsky, L. D. Vales, Y. H. Wang.

Adjunct Professors
A. Gabriel, J. Laskin, P. Manowitz, M. Rosenberg, S. Shiff, W. Wadsworth, N. Walworth.

BIOMEDICAL ENGINEERING GRADUATE PROGRAM
Associate Professor T. Shinbrot (Graduate Director).
Professor N. A. Langrana (Department Chair).

Professors
H. Alexander, E. Arnold, L. Axel, H. Berman, K. Breslauer, Y. Chabal, K. Ciuffreda, B. Coleman, S. Danforth, D. Denhardt, G. M. Drzewiecki, S. Dunn, R. Ebright, S. J. England, S. Garofalini, M. Grumet, N. Guzelsu, J. Kedem, J. B. Kohn, J. B. Kostis, E. Kowler, C. Kulikowski, N. A. Langrana, R. M. Lehman, M. Lewis, J. K.-J. Li, Y. Lu, T. Madey, D. Metaxas, E. Micheli-Tzanakou, G. Montelione, F. Muzzio, W. Olson, T. V. Papathomas, J. R. Parsons, H. Pederson, R. Riman, A. J. Salkind, J. L. Semmlow, A. Shatkin, L. Shepp, G. K. Shoane, F. H. Silver, J. Tischfield, G. S. Tzanakos, Y. Vardi, W. Warren, H. Weiss, J. Wilder, W. Young, M. Zimmerman.

Associate Professors
H. M. Buettner, G. C. Burdea, W. Craelius, M. G. Dunn, D. Foran, C. Gatt, P. Georgopoulos, P. V. Moghe, J. A. Neubauer, R. Nowakowski, J. L. Ricci, P. Sinko, K. Uhrich.

Assistant Professors
Y. Androulakis, G. Atlas, B. Biswal, N. Boustany, M. Bouzit, L. Cai, J. Carbeck, Z. Guo, R. D. Harten, S. Kim, A. Madabhushi, A. Mann, D. Mavroidis, G. Nackman, P. O'Connor, C. Roth, D. Schreiber, P. Yim, D. Zhang.

Instructors
C. Glass, K. Scarbrough.

CELLULAR AND MOLECULAR PHARMACOLOGY
Professor L. Liu (Department Chair).
Professor M. R. Gartenberg (Graduate Director).

Professors
J. Bertino, G. Brewer, J. D. Chen, A. Conney, M. R. Gartenberg, C. Gélinas, W. Hait, B. A. Kamen, F. C. Kauffman, A.-N. Tong Kong, T. Kong, J. D. Laskin, J. Lenard, L. Liu, P. Lobel, J. Ma, R. Nagele, N. C. Partridge, L. A. Pohorecky, E. Rubin, K. Scotto, Y. Shi, T. Shimamura, N. C. Walworth, W. Welsh, D. J. Wolff, C. S. Yang, P. D. Yurchenco.

Associate Professors
D. Banerjee, D. Pilch, J. Parness, A. G. Ryazanov, P. Sonsalla, V. Studitsky, G. Werlen, J. Williams, G. You, X. F. S. Zheng, R. Zhou.

Assistant Professors
X. Chen, P. Cole, B. L. Firestein, J. D. Fondell, S. (Victor) Jin, F. Liu, R. D. McKinnon, S. Meiners, J. Ramos, L. Runnels, F. Sesti.

Adjunct Professors
E. Abali, S. Wimalawansa.

ENVIRONMENTAL SCIENCES–EXPOSURE ASSESSMENT
C. Weisel (Director).
C. G. Uchrin (Co-Director).

Professors
M. Gallo, P. Georgopoulos, J. Y. Hong, H. Kipen, P. J. Lioy, R. Tate III, C. Weisel.

Associate Professors
M. Robson, S. Shalat, B. Turpin, J. Zhang.

Assistant Professors
Z.-H. Fan, G. Mainelis.

MOLECULAR GENETICS, MICROBIOLOGY, AND IMMUNOLOGY
Professor S. Pestka (Department Chair).
Professor J. P. Dougherty (Graduate Director).

Professors
C. Abate-Shen, E. Arnold, K. J. Breslauer, G. Brewer, K. M. Das, D. T. Denhardt, J. Dougherty, D. T. Dubin, M. R. Gartenberg, C. Gélinas, A. B. Gottlieb, M. Hampsey, M. Inouye, W. G. Johnson, T. G. Kinzy, J. B. Kohn, E. C. Lattime, M. J. Leibowitz, J. Lenard, L. F. Liu, P. Lobel, J. Ma, J. W. Messing, W. R. Moyle, N. C. Partridge, S. Peltz, S. Pestka, J. R. Pintar, A. Rabson, D. Reinberg, M. Reiss, Y. Ron, A. Sahota, A. J. Shatkin, M. Shen, Y. Shi, A. Stock, V. Stollar, D. A. Winkelmann, P. D. Yurchenco, X. F. S. Zheng.

Associate Professors
J. D. Dinman, D. J. Foran, A. Gabriel, C. W. Hewitt, J. A. Langer, H. Li, R. D. McKinnon, J. Millonig, R. J. Strair, N. Walworth, G. Werlen, N. A. Woychik, J. Zheng.

Assistant Professors
G. F. Arnold, P. Casaccia-Bonnefil, P. R. Copeland, S. Corbett, H. Fan, R. Foty, F. J. Germino, B. Haimovich, S. G. Kramer, M. P. Matise, J. W. Ramos, M. Soto, M. Xiang.

Adjunct Professor
S. Stein.

NEUROSCIENCE
Professor C. F. Dreyfus (Interim Department Chair).
Professor M. D. Egger (Interim Co-Director, Graduate Program).
Professor J. Pintar (Interim Co-Director, Graduate Program).

Professors
C. Abate-Shen, T. M. Casey, K. Y. Chen, K. V. Chin, E. DiCicco-Bloom, C. F. Dreyfus, M. Driscoll, M. D. Egger, M. Grumet, R. Hart, S. Hitchcock-DeGregori, A. Hyndman, W. Johnson, B. Kamen, J. J.-K. Li, R. W. Padgett, C. H. Page, J. Pintar, J. Rabii, D. J. Riley, D. Sarkar, M. Schachner, M. M. Shen, T. Shimamura, T. Shors, J. M. Stern, G. Wagner, M. West, W. Young, R. Zhou.

Associate Professors
J. P. Advis, S. B. Auerbach, D. Cowen, R. Davis, I. Edery, J. M. Fagan, N. Hayes, S.-C. Hsu, J. V. Martin, R. McKinon, J. H. Millonig, J. Neubauer, R. S. Nowakowski, T. Otto, T. Papathomas, M. R. Plummer, W. Saidel, D. K. Sarkar, P. Sonsalla, M. V. K. Sukhdeo, D. Vicario, W. Wadsworth, K. Wu, M. Xiang, G. Zeevalk, J. Zheng.

Assistant Professors
P. Casaccia-Bonnefil, D. Crockett, H. Fan, B. L. Firestein, A. Kusnecov, M. P. Matise, C. G. Rongo, G. Shumyatsky.

PHYSIOLOGY AND INTEGRATIVE BIOLOGY
Professor N. C. Partridge (Department Chair).
Professor J. Ma (Graduate Director).

Professors
J. P. Advis, A. Conney, D. Denhardt, C. F. Dreyfus, M. Gallo, J. Kedem, J. Lenard, P. Lobel, J. Ma, G. Merrill, J. Neubauer, N. C. Partridge, J. Pintar, D. Sarkar, P. Scholz, Y. Shi, H. R. Weiss, W. J. Welsh, S. Wimalawansa, E. Zambraski.

Associate Professors
M. Driscoll, L. Grandison, G. Lambert, K. McKeever, R. McKinnon, T. Otto, J. Parness, S. Shapses, A. K. Sinha, N. R. Stevenson, M. Watford, R. Zachow, R. Zhou.

Assistant Professors
M. Brotto, D. Cowen, H. Fan, J. Fondell, B. Grant, E. Jacinto, F. Liu, Z. Pan, H. Ruan, L. Runnels, F. Sesti, C.-C. Tsai.

UNIVERSITY OF MIAMI

Department of Biology

Programs of Study	The department offers both Ph.D. and M.S. degree programs. Strengths of the program include ecology, genetics, behavior, neuroethology, evolutionary biology, and tropical biology. Students with a bachelor's degree may apply directly to either program, but terminal M.S. programs are not encouraged. Requirements for the Ph.D. are 60 graduate credits, with a minimal average grade of B and no grade below C; a formal written proposal; an oral defense; and later a dissertation and an oral defense. The distribution of the 60 credits among formal courses, independent studies, and dissertation research depends on the student's background and is worked out with the student's dissertation committee. Requirements for the M.S. are a research proposal, 26 graduate credits, a thesis, and oral defense of thesis. Course work may include courses offered at the Rosenstiel School of Marine and Atmospheric Science and at the University of Miami School of Medicine, if deemed appropriate by the student's committee. Any number of independent studies can be arranged, subject to approval by the student's committee. Students are required to select a research topic that associates them with one or more members of the regular graduate faculty listed on the reverse of this page. The University is a charter member of the Organization for Tropical Studies (OTS). Several of the department's faculty members participate in OTS field courses. The department also maintains close ties with the Smithsonian Tropical Research Institute. The Distinguished Visiting Professor Program has brought to the department such outstanding scientists as Drs. F. Bazzaz, J. Brockman, D. Brooks, H. Caswell, E. L. Charnov, N. Davies, L. Ehrman, J. S. Farris, D. J. Futuyma, L. E. Gilbert, R. K. Koehn, J. R. Krebs, M. F. Mickevich, N. Nadkarni, G. Nelson, G. Northcutt, R. Pulliam, P. J. Regal, D. Reznick, R. E. Ricklefs, D. E. Rosen, M. Rosenzweig, J. Roughgarden, M. J. Ryan, D. Simberloff, J. Travis, E. O. Willis, and L. Wolf.
Research Facilities	The department provides a full range of facilities in most major areas of biology. Modern equipment is available for molecular, cellular, electrophysiological, organismal, and field studies. There are several computer labs with state-of-the-art Pentium II and Macintosh Power PC systems and full e-mail and Internet capabilities. These resources can be accessed from personal machines. A full-time technical support staff maintains the departmental servers and network while assisting all end-user needs. There is an extensive live tropical tree collection in the John C. Gifford Arboretum. The department is situated on three floors of the Cox Science Building, along with the Departments of Chemistry and Geology. It is across the street from the Richter Library, which houses more than 8,900 journals and more than 1 million volumes, and adjacent to the Ungar Computer Center. Field stations of the Organization for Tropical Studies in Costa Rica and the Smithsonian Tropical Research Institute in Panama may be used for research. The Fairchild Tropical Garden is 10 minutes from campus, and Everglades National Park and Biscayne National Marine Park are also nearby.
Financial Aid	Teaching assistantships (TAs) are available on a competitive basis. Current stipends are $16,500 for first-year students, plus a tuition waiver for 18 credits per year. Students may compete for Maytag, University, McLamore, and Tropical Biology Fellowships ($18,000 per year). Research assistantships are occasionally available from individual faculty grants. It is the department's policy to support all Ph.D. students. The department funds students to attend field courses. Up to $1500 per student is available for those presenting papers at scientific meetings.
Cost of Study	Tuition ($1208 per credit in 2005–06) is waived for all students with TAs and fellowships. The University fees per semester include a $62 University fee, $35 activity fee, $112 Wellness Center fee (optional), and $30 athletic fee (optional).
Living and Housing Costs	Students may live in University housing accommodations that range from shared rooms to apartments. The costs range from $2920 to $3010 per semester. Off-campus housing is available from approximately $650 per month upward. The cost of living in Miami is equivalent to that of other metropolitan areas.
Student Group	The University of Miami has an enrollment of 14,000 students, with approximately 2,900 graduate students. The department usually has about 50 graduate students.
Location	The University of Miami is located in Coral Gables. The Department of Biology is situated on the main campus in a subtropical environment, 30 minutes from the Everglades and 10 minutes from the Atlantic Ocean. Local cultural facilities include an opera, art and science museums, three theaters that offer Broadway plays, and various smaller theaters. Local tourist and recreational attractions include professional football, basketball, baseball, and hockey teams; Parrot Jungle; Monkey Jungle; the SeaQuarium; the Metro Zoo; and various marinas, parks, and paths for running or biking. Areas for fishing and diving in Biscayne Bay and the Florida Keys are within 10 to 90 minutes of the campus. Everglades National Park, where hiking, canoeing, and camping can be enjoyed, is less than 1 hour from campus.
The University	The University was founded in 1925 as a private, coeducational institution and is now composed of fourteen colleges and schools. The Coral Gables campus is a 260-acre tract containing 120 buildings and is located 15 minutes from both the University's Rosenstiel School of Marine and Atmospheric Sciences (on Virginia Key) and School of Medicine (in Miami).
Applying	Application materials for admission may be obtained from the department's director of graduate studies. All completed application forms, including letters of application, must be mailed directly to the director of graduate studies of the biology department by January 1. Before submitting the formal application, students are required to contact one or more appropriate members of the graduate faculty, stating their career goals and research interests. Following tentative acceptance of a student and his or her research topic, an evaluation of the complete file is made by the Graduate Admissions and Advisory Committee, which notifies the Graduate School or International Admissions regarding admission. Recommendations from at least three academic referees should address the student's motivation, ability to conceptualize and deal quantitatively with biological problems, and research potential. Evidence of research capability should be included. Scores on the GRE General Test and Subject Test in biology are essential for TA or fellowship consideration and recommended for all students. International students must submit a TOEFL score of at least 550 (paper-based) or at least 212 (computer-based). Decisions on admission and awards are announced by April 15.
Correspondence and Information	Leonel Sternberg Director of Graduate Studies Department of Biology University of Miami P.O. Box 249118 Coral Gables, Florida 33124 Phone: 305-284-3973 Fax: 305-284-3039 E-mail: l.sternberg@miami.edu Web site: http://fig.cox.miami.edu

University of Miami

THE FACULTY AND THEIR RESEARCH

Theodore H. Fleming, Ph.D., Michigan, 1969. Population and community ecology; co-evolution between tropical and desert plants and vertebrate visitors; foraging strategies of plant-visiting bats. *Ecol. Monogr.* 71:511–30, 2001.

Michael S. Gaines, Ph.D., Indiana, 1970. Effect of habitat fragmentation and patchiness on the dynamics and genetic structure of small mammal populations. Effects of tree island size and hydroperiod on the population dynamics of small mammals in the Everglades. In *Tree Islands of the Everglades*, pp. 429–44, eds. F. H. Sklar and A. van der Valk. Dordrecht: Kluwer Academic Publishers, 2002 (with Diffendorfer, Sasso, and Beck).

Luis Glaser, Ph.D., Washington (St. Louis), 1956. Biological chemistry of neuron-neuron and neuron–Schwann cell interactions. *J. Cell Biol.* 101:744–54, 1985.

Guillermo Goldstein, Ph.D., Washington, 1981. Ecophysiology of tropical trees. *Plant Cell Environ.* 21:397–406, 1998 (with Andrade et al.).

Steven M. Green, Ph.D., Rockefeller, 1973. Ethology and behavioral ecology, particularly animal communication and social organization, especially of primates and fish; conservation biology. *NOAA Tech. Report NMFS* 151, 2001.

Thomas J. Herbert, Ph.D., Johns Hopkins, 1970. Plant ecophysiology, leaf movement, canopy structure. *Photo. Eng. Remote Sensing* 61:89–92, 1995.

Carol C. Horvitz, Ph.D., Northwestern, 1980. Population dynamics of tropical plants: matrix models, stochastic demography, plant-animal interactions, dispersal, invasion. The many growth rates and elasticities of populations in random environments. *Am. Naturalist* 162:489–502, 2003 (with Tuljapurkar and Pascarella).

David P. Janos, Ph.D., Michigan, 1975. Tropical botany; role of mycorrhizae in tropical forests and agroecosystems; evolutionary ecology of mutuality associations. In *Fungi and Environmental Change,* vol. 20, pp. 129–62, eds. J. C. Frankland, N. Magan, and G. M. Gadd. Cambridge: Cambridge University Press, 1996.

Elli Kohen, M.D., Istanbul, 1954; Ph.D., Karolinska (Stockholm), 1973. Microspectrofluorometry. In *Advances in Optical and Electron Microscopy*, vol. 14, pp. 121–211, eds. T. Malvey and C. Sheppard. San Diego: Academic Press, 1994.

Julian C. Lee, Ph.D., Kansas, 1977. Vertebrate ecology, biogeography, and systematics, particularly of amphibians and reptiles. *Field Guide to the Amphibians and Reptiles of the Maya World*. New York: Cornell University Press, 2000.

Zhongmin (John) Lu, Ph.D., Loyola Chicago, 1995. Auditory neurobiology, neural mechanisms of hearing in fish, and Florida red tides and hearing. *J. Comp. Physiol. A* 187:453–65, 2001.

Peter Luykx, Ph.D., Berkeley, 1964. Contractile vacuoles. In *Vacuolar Compartments, Annual Plant Reviews,* pp. 43–70, eds. D. G. Robinson and J. C. Rogers. Sheffield, England: Sheffield Academic Press, 2000.

Charles H. Mallery, Ph.D., Georgia, 1970. Developmental biochemistry and physiology; macromolecular syntheses during root cell development; mineral ion absorption and transport; cation/anion-stimulated ATPase properties in teleosts. *Plant Cell Physiol.* 25:1123–31, 1985.

James O'Reilly, Ph.D., Northern Arizona, 1998. Physiology and biomechanics of vertebrate locomotion and feeding. *Am. Zoologist* 40:123–35, 2000 (with Summers and Ritter).

Matthew D. Potts, Ph.D., Harvard, 2001. Habitat heterogeneity and niche structure of trees in two tropical rain forests. *Oecologia* 129:446–74, 2004.

Jeffrey S. Prince, Ph.D., Cornell, 1971. Ecology of marine algae; environmental induction of reproduction; ultrastructure. *J. Exp. Biol.* 201:1595–1613, 1998.

Kathleen Sullivan Sealey, Ph.D., California, San Diego (Scripps Institution), 1982. Evaluating the use of roving diver and transect surveys to assess coral reef fish assemblages off southeastern Hispaniola. *Coral Reefs* 21:216–23, 2002.

William A. Searcy, Ph.D., Washington (Seattle), 1977. Behavioral ecology, especially animal communication, avian mating systems. *Am. Naturalist* 159:221–30, 2002.

Leonel Sternberg, Ph.D., California, Riverside, 1978. Uptake of water by lateral roots of small trees and saplings in an Amazonian tropical forest. *Plant Soil* 238:151–8, 2002.

Richard R. Tokarz, Ph.D., Colorado, 1977. Hormonal regulation of reproductive behavior in nonmammalian vertebrates. *Copeia* 2003(3):502–11, 2003 (with Patterson and McMann).

Keith D. Waddington, Ph.D., Kansas, 1977. Behavioral ecology; pollination biology; foraging behavior and communication. *J. Comp. Physiol.* 187:293–301, 2001.

Barbara A. Whitlock, Ph.D., Harvard, 2000. Plant evolution; molecular systematics; molecular evolution. *Syst. Botany* 26:420–37, 2001.

David L. Wilson, Ph.D., Chicago, 1969. Biology of aging: biology of mind and consciousness. *Mech. Aging Dev.* 113:101–16, 2000.

Research Faculty

Donald L. DeAngelis, Ph.D., Yale, 1972. Population, community, and ecosystem theory; mathematical and computer modeling. (U.S. Geological Survey). *Am. Naturalist* 159: 231–44, 2002 (with Holland and Bronstein).

Linda Farmer, Ph.D., Miami (Florida), 1977. Physiological ecology of marine organisms. Osmotic and ionic regulation. In *Biology of Crustacea,* vol. V. New York: Academic Press, 1982 (with Mantel and Farmer).

Faculty with Joint Appointments

Robert K. Cowen, Ph.D., California, San Diego, 1985; Maytag Professor of Ichthyology, Marine Biology and Fisheries, Rosenstiel School of Marine and Atmospheric Sciences. Ecology and early life history of marine fishes; population connectivity of marine populations. *Science* 287:857–9, 2000.

Adjunct Faculty

Patti J. Anderson, Ph.D., Florida, 1998. Plant demography and population ecology; economic botany; natural forest management, especially in tropical forests; ethnobotany; conservation and sustainable development. (Fairchild Tropical Garden)

Judith Lee Bronstein, Ph.D., Michigan, 1986. Evolutionary ecology of mutualisms; plant-insect interactions, especially obligate relationships. (Arizona)

Jack B. Fisher, Ph.D., California, Davis, 1969. Plant structure and function; mycorrhizae of native plants. (Fairchild Tropical Garden)

Javier Francisco-Ortega, Ph.D., Birmingham (England), 1992. Origin and evolution of plants endemic to islands. (Fairchild Tropical Garden)

Gerald F. Guala, Ph.D., Florida, 1998. Systematics; biogeography of tropical and subtropical bamboos and other grasses. (Fairchild Tropical Garden)

Martine Hossaert-McKey, Ph.D., Pau France, 1988. Evolutionary biology of nursery pollination mutualisms. (CNRS, Montpellier)

Julia Kornegay, Ph.D., Cornell, 1985. Genetic improvement of the common bean, using molecular and conventional approaches; horticultural, research, and education missions of botanical gardens. (Fairchild Tropical Garden)

Carl E. Lewis, Ph.D., Cornell, 2000. Molecular biology, systematics, and evolution of palms. *Plant Syst. Evol.* 236:1–17, 2002 (with Doyle). (Fairchild Tropical Garden)

Doyle McKey, Ph.D., Michigan, 1979. Evolutionary ecology of ant-plant mutualisms; evolutionary approaches in ethnobiology. (CNRS, Montpellier)

John J. Pipoly III, Ph.D., CUNY, New York Botanical Garden, 1986. Systematics of Myrsinaceae, Clusiaceae; anatomy; tree architecture; tropical forest inventory; conservation and sustainable development; ICAD design. (Fairchild Tropical Garden)

Joel C. Trexler, Ph.D., Florida State, 1986. Trade-offs and resource allocation; genetic and ecological constraints; spatial ecology and migration; management and restoration at the Everglades. (Florida International)

Scott Zona, Ph.D., Claremont and Rancho Santa Ana Botanical Garden, 1989. Taxonomy, systematics, anatomy, and phytochemistry of tropical plants, especially palms. (Fairchild Tropical Garden)

UNIVERSITY OF MICHIGAN

Graduate Program in Biomedical Sciences

Program of Study	The Graduate Program in Biomedical Sciences (PIBS) at the University of Michigan is made up of thirteen different Ph.D. degree–granting programs in the Horace H. Rackham Graduate School. These programs include nanotechnology; stem cell research; bioinformatics; biological chemistry; biophysics; cell and developmental biology; cellular and molecular biology; human genetics; immunology; microbiology and immunology; molecular, cellular, and developmental biology; molecular and integrative physiology; neuroscience; pathology; and pharmacology. PIBS offers flexibility in the choice of any of the participating Ph.D. programs while retaining the small, focused environment of each of the individual training programs. Entering students can immediately begin training in any of the thirteen participating programs or take a course of study that is compatible with several programs. In the meantime, students engage in research rotations in two to four laboratories of their choice out of more than 300 graduate research mentors who are available in the thirteen programs. PIBS has been set up so that students have until the end of their first year to commit to a research laboratory and training program. Course work and qualifying examinations are specific to each Ph.D. program. Many research areas are represented in the forty-one research training programs, including behavior and cognition; biological chemistry; bioinformatics; biophysics; biotechnology; cancer biology; cell biology; cellular and molecular immunology; developmental biology; drug metabolism; electrophysiology; enzyme mechanism; gene expression; gene therapy; genetics; genetics of disease; hormone action; immunology; inflammation; information processing in sensory systems; integrative biology; learning and memory systems; metabolic regulation; microbial pathogenesis; molecular basis of aging; molecular, cellular, and developmental biology; molecular evolution; molecular membrane biology; neurobiology; neuroscience; organogenesis; pharmacology; physiology; plant biology; protein trafficking; receptor interactions; reproductive biology; signal transduction; structural biology; virology; and vision research. Students are being prepared for roles as scholars in biomedical research in academic institutions and the private sector and teaching.
Research Facilities	The University of Michigan has one of the most vibrant research communities in the United States. For nearly two decades, the Medical School has ranked in the top ten schools nationally in funding by the National Institutes of Health. The Biomedical Research Core Facilities are conveniently located for PIBS faculty members and students. They provide state-of-the-art technical support for research laboratories, including the Biochemical Nuclear Magnetic Resonance Core, Biopolymers Cores: Protein and Carbohydrate Structure Facility and Restriction Enzyme Store, the Biosafety Level 3 Core for culture and analysis of pathogenic organisms, the DNA Sequencing Core, the Flow Cytometry Core, and the Transgenic Animal Model Core. Additional core facilities on the Medical School campus include the Cell Biology Laboratories, the Hybridoma Core, the National Gene Vector Core, the Organogenesis Morphology Core, and the Vector and Cell Culture Core.
Financial Aid	The major emphasis of graduate training in the participating programs is on laboratory research. To expedite progress toward completing Ph.D. requirements, students are supported throughout their graduate studies. Tuition, stipend ($23,500 for the 2005–06 academic year), and health benefits are paid from a combination of individual fellowships, predoctoral training grant fund, institutional funds, and research grants. Many students are teaching assistants for one term for the purpose of gaining teaching experience.
Cost of Study	Fellowships that cover stipend, tuition, and fees are offered to incoming students.
Living and Housing Costs	Most graduate students live in apartments or rooms in private homes. A limited number of rooms with board are available in three international houses. The cost of a one-bedroom unfurnished apartment that is close to campus is about $695 or more per month. A three-bedroom unfurnished house can be rented for approximately $1260 per month. University-operated apartment developments are available for married students. These units range in price from $645 per month for a one-bedroom efficiency to $1064 per month for a three-bedroom furnished apartment.
Student Group	The diverse student body of the University of Michigan is composed of approximately 38,000 students. About 60 percent of the students are engaged in undergraduate studies, and 40 percent are in graduate or professional schools. The thirteen participating departments and programs in PIBS matriculate a class of approximately 65 graduate students each fall.
Location	The University of Michigan is located in Ann Arbor, a small cosmopolitan city in the southeast corner of Michigan that is approximately 1 hour west of the greater Detroit metropolitan area and only a 20-minute drive from the Detroit Metropolitan International Airport. With a population of approximately 125,000, Ann Arbor combines the congenial atmosphere and ease of living of a small city with the wide range of cultural opportunities that are typically found in a much larger community. The proximity of the University of Michigan to the city of Ann Arbor provides easy access to these activities. More than 140 parks in the city and surrounding areas offer many outdoor activities, including boating on the Huron River, biking, in-line skating, hiking, cross-country skiing, and skating. Forested areas along inland lakes and streams offer canoeing, fishing, and camping opportunities. In addition, Ann Arbor is only a few hours from the sandy beaches of three Great Lakes—Michigan, Superior, and Huron. The combination of cultural activities, outdoor and sporting activities, excellent restaurants, and a lively community make Ann Arbor an exceptional place for graduate studies.
The University	The University of Michigan, chartered in 1817, is a state-supported institution with a national and international reputation for excellence in scholarship. The 2,500-acre University of Michigan campus in Ann Arbor includes the Medical School, Main Campus, and North Campus areas that are each readily accessible on foot or by shuttle bus. The intellectual, cultural, and environmental components of the University of Michigan work together to provide a comfortable home to thousands of students, postdoctoral trainees, and faculty and staff members.
Applying	A bachelor's degree is required for admission to the program. It is recommended that entering students have a strong background in physical and biological sciences, including laboratory experience. Students may be admitted with deficiencies, but they are required to correct them early, in consultation with their specific program of study. The General Test of the Graduate Record Examinations is required for all graduate students. It is recommended that PIBS applicants also take one of the science Subject Tests. The deadline for application is December 31.
Correspondence and Information	Program in Biomedical Sciences 2960 Taubman Medical Library, Box 0619 University of Michigan 1150 West Medical Center Drive Ann Arbor, Michigan 48109-0619 Phone: 734-647-7005 877-294-0120 (toll-free) Fax: 734-647-7022 E-mail: pibs@umich.edu Web site: http://www.med.umich.edu/pibs/

University of Michigan

THE FACULTY AND THEIR RESEARCH

Listings of the more than 300 faculty members and their research can be accessed online at http://www.med.umich.edu/pibs/.

Bioinformatics: David States, M.D., Ph.D., Professor of Human Genetics and Director, Bioinformatics Training Program. The Graduate Program in Bioinformatics merges molecular biology and genetics with advanced computer science technology to understand the complex web of interactions linking the individual components of a living cell to the integrated behavior of the entire organism. Computational biology and bioinformatics are opening new windows into disease processes and are leading to novel diagnostic and treatment strategies.

Biological Chemistry: William L. Smith, Ph.D., Minor J. Coon Professor of Biochemistry and Department Chair. The graduate program that leads to the Ph.D. degree in biological chemistry is designed to provide outstanding training toward careers in research and scholarship in the areas of biochemistry; molecular, cellular, and developmental biology; and related areas. This program exposes students to a broad spectrum of current research through innovative course structures and seminars. Students have the opportunity to pursue dissertation research in any of approximately fifty laboratories in a department that provides an intellectually stimulating and interactive environment.

Biophysics: James Penner-Hahn, Ph.D., Professor of Chemistry and Chair, Biophysics Research Division. The Biophysics Graduate Program prepares students to pursue independent research at the interface of biology, chemistry, and physics. Faculty members and students engage in fundamental and applied studies of proteins, nucleic acids, membranes, and complex cellular systems using powerful theoretical and experimental tools of structural biology, biochemistry, molecular biology, spectroscopy, and microscopy.

Cell and Developmental Biology: James Douglas Engel, Ph.D., Professor of Cell and Developmental Biology and Chair. The mission of the Graduate Program in Cell and Developmental Biology is to train Ph.D. students to conduct modern research that combines molecules and morphology in understanding the mechanisms that underlie cellular and developmental processes.

Cellular and Molecular Biology: Jessica Schwartz, Ph.D., Professor of Physiology and Director, Cellular and Molecular Biology Training Program. The Graduate Program in Cellular and Molecular Biology provides interdisciplinary Ph.D. training that is individually tailored for each student. Trainees choose research opportunities from among the more than 80 faculty members who represent more than twenty departments throughout the University, including Clinical and Basic Biomedical Sciences, Biology, Chemistry, Biophysics, and others. Training faculty members are leaders in their fields and provide the highest quality research training in fields such as cancer biology, cellular architecture, developmental biology, gene expression, gene therapy, macromolecular structure, molecular genetics, molecular basis of disease, signal transduction, and virology.

Human Genetics: Sally A. Camper, Ph.D., Professor of Human Genetics and Internal Medicine, James V. Neel Professor of Human Genetics, and Chair, Department of Human Genetics. The Graduate Program in Human Genetics offers theoretical and practical training to prepare Ph.D. scientists for careers in genetics, including genomics, developmental genetics, molecular basis of human genetic disease, regulation of gene expression, gene therapy, cancer genetics, genetic epidemiology, and population genetics. The educational program includes laboratory and statistical research, state-of-the-art molecular genetics, and consideration of clinical applications and associated ethical issues. There are more than 20 training faculty members.

Immunology: D. Keith Bishop, Ph.D., Associate Professor of Surgery and Microbiology and Immunology and Director, Immunology Training Program. This interdepartmental program offers formal course work and research training leading to the award of the Ph.D. in immunology. Faculty research interests include both basic immunology and the translation of research findings to clinical use, and focus on T- and B-cell development, autoimmunity, transplantation and tumor immunity, antigen presentation, immunogenetics, immunopathology, and immune senescence.

Molecular and Integrative Physiology: John A. Williams, M.D., Ph.D., Professor of Physiology and Chair. The Graduate Program in Molecular and Integrative Physiology is designed to provide an optimal atmosphere for education in molecular, cellular, and systems physiology. Ongoing research covers a broad spectrum of research areas, including muscle, cardiovascular, renal, gastrointestinal tract, and neuroendocrine system physiology and disease; aging; signal transduction; molecular genetics; normal and neoplastic growth and development; gene therapy; sensory function; and ion-channel structure and function.

Molecular, Cellular, and Developmental Biology: Richard Hume, Ph.D., Professor and Chair, Molecular, Cellular, and Developmental Biology. The Graduate Program in Molecular, Cellular, and Developmental Biology (MCDB) is intended for training researchers who are capable of carrying out independent and distinguished scholarly activities and thus contributing to the body of knowledge in their selected fields of specialization. The faculty members share technical approaches, such as recombinant DNA, genetics, biochemistry, and specialized imaging. They also share a common intellectual approach that emphasizes mechanistic and experimental strategies to investigate a diverse set of biological problems using a variety of model systems, including prokaryotes, yeast, *C. elegans, Drosophila, Arabidopsis, Xenopus,* zebrafish, chicks, and mammals.

Microbiology and Immunology: Harry L. T. Mobley, Ph.D., Frederick G. Novy Collegiate Professor and Chair, Department of Microbiology and Immunology. The Department of Microbiology and Immunology offers a wide range of research and course work opportunities that lead to the Ph.D. degree. Research in the department emphasizes the molecular and cellular aspects of microbial biology and pathogenesis, immunology, and tumor virology. Students have an opportunity to choose a mentor for their Ph.D. research projects from among approximately 25 investigators in the department whose approaches combine the most modern experimental and theoretical techniques.

Neuroscience: Peter F. Hitchcock, Ph.D., Professor of Cell and Developmental Biology and Director of the Neuroscience Program. This interdisciplinary program leads to a Ph.D. degree in neuroscience. Students select a mentor from approximately 70 program faculty members throughout the University. Research opportunities are diverse and include studies of animal behavior, neural imaging, electrophysiology, neuropharmacology, molecular neurobiology, genetics, and human disease.

Pathology: Jay Hess, M.D., Ph.D., Carl V. Weller Professor of Pathology and Chair, Department of Pathology. The Department of Pathology Ph.D. Program offers training in molecular and cellular mechanisms of disease, with emphasis on tissue injury and repair, inflammation, aging, tumor biology, apoptosis, and the biology and pathobiology of cytokines, adhesion, molecules, and extracellular matrix.

Pharmacology: Paul F. Hollenberg, Ph.D., Maurice H. Seevers Collegiate Professor of Pharmacology and Chair. The Graduate Program in Pharmacology provides students with a multifaceted approach to the study of drug action. The major programs of the department include neuropharmacology, signal transduction, autonomic pharmacology, psychopharmacology, addiction research, cardiovascular pharmacology, cancer pharmacology, endocrine pharmacology, drug metabolism, pharmacokinetics, toxicology, medicinal chemistry, and pharmacogenetics.

UNIVERSITY OF MISSISSIPPI MEDICAL CENTER

Graduate Studies

Programs of Study

The degree of Doctor of Philosophy is offered by the University of Mississippi Medical Center in biomaterials science, the biomedical sciences (programs in the Departments of Anatomy, Biochemistry, Microbiology, Pathology, Pharmacology and Toxicology, Physiology and Biophysics, and Preventive Medicine), the clinical health sciences, and nursing science. The purpose of the program is to prepare students for careers as scientists with high levels of competence in research and teaching. The general requirements for the doctorate include advanced course work that is sufficient to provide a basis for continuing study and demonstration of the ability to independently perform original and meritorious research.

Research Facilities

All the facilities of the Medical Center are available to graduate students. These include well-equipped classrooms, the Verner Holmes Learning Resources Center, and animal facilities with a full-time veterinary staff.

Financial Aid

Students are encouraged to apply for predoctoral fellowships from various sources. In addition, a limited number of research assistantships are available. Federally insured loans for graduate students and Federal Perkins Loans are also available. Some departments also have NIH training grant funds for the support of graduate training.

Cost of Study

In 2005–06, full-time students who were Mississippi residents paid tuition of $1077 for each quarter. Part-time students pay a fee that is prorated according to the number of quarter hours registered. In addition to resident tuition, nonresidents also paid a nonresident fee for each quarter ($1225 in 2005–06). Nonresident part-time students pay prorated tuition and nonresident fees. Nonresident students who receive assistantships are exempt from nonresident fees.

Living and Housing Costs

On-campus housing is not available for graduate students attending the University; however, there are numerous apartments and small houses available for rent within walking distance of the campus. The overall cost of living in Jackson is lower than the average cost across the United States. Stipends are adequate to cover modest living expenses. For additional information, applicants should contact the Student Apartment Manager, 2590 North State Street, Jackson, Mississippi 39216.

Student Group

In 2005–06, there were 327 students in the Graduate Programs in Health Sciences, 400 in the School of Medicine, 119 in the School of Dentistry, 237 in the School of Nursing, and 394 in the School of Health Related Professions. Residents, interns, fellows, and students in certificate programs brought the total Medical Center enrollment to 2,027.

Location

The city of Jackson, the economic center of the area, is located along the Pearl River, which divides Hinds and Madison counties from Rankin County. Jackson is about 40 miles east of the Mississippi River and 100 miles west of the Alabama state line at the intersection of Interstates 55 and 20. Situated approximately halfway between Memphis, Tennessee, to the north and New Orleans, Louisiana, to the south, it is sometimes called the "Crossroads of the South." Jackson is the heart of a thriving three-county metropolitan area that had 395,396 inhabitants at the time of the 1990 census.

The Medical Center

The University of Mississippi Medical Center, situated on a 164-acre campus in the heart of Jackson, is the nucleus of a complex that has more than tripled in size since 1955. The campus includes the School of Medicine, the School of Nursing, the School of Graduate Studies in the Health Sciences, the School of Dentistry, the School of Health Related Professions, and University Hospital, the teaching hospital for all of the programs. The main campus of the University of Mississippi is located at University, Mississippi, 165 miles from Jackson.

Applying

Applications are available from the Office of Student Services and Records at the address listed in the Correspondence and Information section of this Close-Up. The application fee cannot be waived. For information regarding specific programs, interested students should contact that program's Director of Graduate Studies.

Correspondence and Information

Office of Student Services and Records
University of Mississippi Medical Center
2500 North State Street
Jackson, Mississippi 39216-4505
Web site: http://umc.edu/

University of Mississippi Medical Center

THE FACULTY AND THEIR RESEARCH

Department of Anatomy. (601-984-1640) Duane E. Haines, Ph.D., Chairperson; James C. Lynch, Ph.D., Director of Graduate Studies.

The graduate program provides a broad background in biomedical science and expertise in a selected area of research. It helps the students to develop the skills and insights necessary to become effective teachers and independent investigators. Research opportunities include somatosensory, visual, and vestibular systems neurobiology; motor systems neurobiology; and cell, developmental, and cardiovascular biology. The department has excellent research facilities for light and electron microscopy and image analysis, tissue culture, intracellular and extracellular electrophysiology, behavioral testing, and biochemical analysis.

Department of Biochemistry. (601-984-1500) Mark O. J. Olson, Ph.D., Chairperson; Drazen Raucher, Ph.D., Director of Graduate Studies.

The graduate program provides a strong foundation in biochemical, molecular, and cell biological methods by focusing on problems of individual interest. Areas of research emphasis include regulation of gene expression, function of nucleolar proteins, nuclear organization and disease, mechanism of drug resistance, design and application of macromolecular drug carrier systems, three-dimensional structural analysis of multiprotein complexes, mechanism of electron and proton transfer through proteins, biophysical analysis of drug-DNA and drug-microtubule interactions, assembly of membrane protein complexes, free radical enzymology, and microtubule polymerization.

Department of Clinical Health Sciences. (601-984-6333) William J. Rudman, Ph.D., Co-Director of Graduate Studies; Hamed Benghuzzi, Ph.D., Co-Director.

The graduate program in clinical health sciences makes available to qualified students an organized educational framework that offers preparation for outcomes research in clinically significant areas requiring a strong background knowledge in principles of clinical health sciences. The program of study and research at the M.S. and Ph.D. levels leads students to a better understanding of health information technology, health-care systems, and new technologies in clinical laboratory science as they relate to clinical practice.

Department of Microbiology. (601-984-1700) Richard J. O'Callaghan, Ph.D., Chairperson and Co-Director of Graduate Studies; V. Gregory Chinchar, Ph.D., Co-Director.

Programs of study and research employ cellular, molecular, and biochemical approaches to address fundamental and applied questions in immunology and microbiology. The range of scientific disciplines includes bacteriology, animal virology, microbial physiology and genetics, parasitology, immunochemistry, and immunobiology.

Department of Nursing. (601-984-6220) Kaye Bender, Ph.D., RN, FAAN, Dean.

The doctoral program provides a strong foundation in theoretical and methodological content that is essential for the scholarly investigation of health-related problems that are encountered in the practice of nursing. The School of Nursing houses one of the few basic science laboratories in the southeastern United States. Scientists in this lab focus on research related to the effect of drug combinations on cancer cells, cell differentiation and migration, and proliferation with a variety of cell types. This includes facs, facs analysis, and protein biochemistry. Other nurse researchers are involved in the Jackson Heart Study, one of the few research projects with an emphasis on both qualitative and quantitative methods for understanding the multifaceted aspects of hypertension in African-American populations.

Department of Pathology. (601-984-1565) Julius M. Cruse, M.D., Ph.D., Director of Graduate Studies.

Major emphasis on immunopathology, with active research programs in AIDS, especially regarding gene identification related to AIDS susceptibility and disease progression; transplantation immunology in which mechanisms of rejection are actively evaluated based on antibody identification and cellular reactivity; and neuroendocrine immune system interactions in spinal-cord injury patients. There is also active research on diseases of the kidney, including pathological changes underlying hypertension, and the progression of renal disease in high-risk population groups; prostate cancer; and on the relationship of human papilloma viral types to cervical cancer. The department has always had a strong research interest in autoimmune diseases. In addition, there is also research on hypertension, obesity, and renal disease.

Department of Pharmacology and Toxicology. (601-984-1600) Ing K. Ho, Ph.D., Chairperson; Jerry M. Farley, Ph.D., Director of Graduate Studies.

Areas of research in the program include biochemical pharmacology, immunopharmacology, endocrinology, electrophysiology, cardiovascular pharmacology, molecular pharmacology and toxicology, neurotoxicology, neuropharmacology, and drugs of abuse.

Department of Physiology and Biophysics. (601-984-1801) John E. Hall, Ph.D., Chairperson; Thomas E. Lohmeier, Ph.D., Director of Graduate Studies.

Research in the department focuses on mechanisms for the control of the cardiovascular, renal, and neuroendocrine systems by integrating results from studies of genetic, molecular, and cellular physiology with data from studies at the tissue, organ, and whole-animal levels. Faculty members have expertise in each of these broad areas, and specific areas of interest include mechanisms underlying hypertension and cardiovascular disease caused by obesity, intracellular messenger systems in vascular smooth muscle, genetic control of the renin-angiotensin system, mechanisms controlling blood vessel growth, and vascular endothelial function and its role in the pathophysiology of diabetes, pregnancy-induced hypertension, and postmenopausal hypertension. Integration of this information from virtually all levels of circulatory control is facilitated by faculty members with expertise in control system analyses with mathematical modeling. The result is a comprehensive approach to the study of the physiology and pathophysiology of the circulatory system.

Department of Preventive Medicine. (601-984-1935) Edward F. Meydrech, Ph.D., Chairperson and Director of Graduate Studies.

Graduate divisions within the department emphasize biostatistics and epidemiology. The M.S. and Ph.D. programs provide broad experience in all aspects of these two disciplines.

The University of Mississippi Medical Center offers equal opportunity in employment and education, M/F/D/V.

UNIVERSITY OF NEBRASKA
MEDICAL CENTER
Biomedical Research Training Program

Program of Study	The Biomedical Research Training Program (BRTP) is a first-year graduate program that offers students the opportunity to explore a variety of research laboratories and academic programs before deciding on a Ph.D.-granting department. Students rotate through any of the more than 100 research laboratories affiliated with the five basic science departments: Biochemistry and Molecular Biology; Genetics, Cell Biology and Anatomy; Pathology and Microbiology; Pharmacology and Experimental Neuroscience; and Cellular and Integrative Physiology and the Cancer Research Graduate Program. This includes nearly all of the University of Nebraska Medical Center (UNMC) research laboratories, since most research faculty members in other departments and institutes at UNMC have adjunct appointments in a basic science department. During the first year, students take a strong, integrated curriculum that forms the foundation for any of the Ph.D.-granting graduate programs. The core curriculum consists of four courses: macromolecular structure and function, molecular cell biology, the cell and gene regulation, and cell signaling. The curriculum, along with research rotations in four laboratories, provides the first-year student with a multidisciplinary foundation that is important for success in modern biomedical research. Before the end of the first year, students select a Ph.D.-granting department and their thesis research mentor. After the first year, students complete the curriculum of the selected department. Additional course requirements depend upon the interests and goals of the student and are defined on an individual basis by the student's Supervisory Committee. A comprehensive examination is required for admission to Ph.D. candidacy. After the second year, students spend most of their time performing the Ph.D. thesis research project. The Ph.D. is awarded upon the successful completion of the thesis research dissertation.
Research Facilities	UNMC is located on a 60-acre campus in the center of Omaha, Nebraska. UNMC's research laboratories are housed in ten buildings, with more than 374,000 square feet of laboratory space. The Durham Research Center, a state-of-the-art, 280,000-square-foot facility that adds another 124,000 square feet of modern research laboratory space, was recently completed. UNMC has twenty research core facilities that provide state-of-the-art research equipment and expertise, such as the Molecular Biology Core Facility (DNA synthesis/sequencing), Protein Structure Core Facility (protein sequencing, peptide analysis, and amino acid analysis), mass spectrometry (chemical identification), Flow Cytometry and Cell Sorting Facility, Confocal Microscopy Facility, Monoclonal Antibody Facility (both monoclonal and polyclonal antibody production), and new facilities for Biacore real-time kinetic analysis, gene knockouts, and DNA microarray technology. In addition, the various departments, institutes, and investigators have a full complement of the current equipment used in modern biological and biomedical research.
Financial Aid	All students are supported throughout their first year ($21,000 for the 2005–06 academic year). Students are supported in subsequent years by a variety of fellowships, assistantships, and grants.
Cost of Study	Tuition is waived for BRTP students.
Living and Housing Costs	UNMC has on-campus housing, and there are a variety of apartment complexes found within walking distance of the campus. Within a short driving distance there are a wide variety of housing options available in a range of prices; the average cost of monthly rent is $250 to $550. The BRTP office is available to assist with housing information.
Student Group	There are more than 2,800 students at UNMC receiving health sciences–related training (such as M.D., pharmacy, nursing, and allied health). As of fall 2005, 181 students were pursuing the Ph.D. in the various research laboratories; more than 50 percent were women.
Student Outcomes	Former students who have received their Ph.D. degrees at UNMC currently hold academic positions at a variety of universities and have positions at a variety of biotechnology and pharmaceutical companies.
Location	The Omaha metropolitan area has approximately 801,000 citizens. Omaha has been rated as one of the top ten U.S. cities for quality of living due to low living expenses and excellent educational, recreational, and cultural opportunities (Omaha Symphony, Opera Omaha, and the College Baseball World Series). Omaha is also known for its variety and quality of restaurants and live music venues.
The Medical Center	The University of Nebraska Medical Center is the health sciences training university for the state of Nebraska. The Medical Center consists of the School of Allied Health and the Colleges of Dentistry, Medicine, Nursing, and Pharmacy. The 60-acre UNMC campus is located in the center of Omaha, within a 5-minute drive of downtown Omaha and the historic Old Market (restaurant and entertainment) district.
Applying	Admission to the program requires a bachelor's degree from an accredited undergraduate institution, with a minimum 3.0 GPA. Students should have successfully completed undergraduate courses in biology, calculus, physics, and chemistry, including organic chemistry. The Graduate Record Examinations, transcripts, and three letters of recommendation are also required. International applicants are encouraged to apply to one of the five basic science departments and to indicate their interest in participating in the BRTP.
Correspondence and Information	BRTP Coordinator Biomedical Research Training Program University of Nebraska Medical Center 985825 Nebraska Medical Center Omaha, Nebraska 68198-5825 Phone: 402-559-3362 800-517-3362 (toll-free) Fax: 402-559-3363 E-mail: brtp@unmc.edu Web site: http://www.unmc.edu/brtp

University of Nebraska Medical Center

FACULTY RESEARCH

The Biomedical Research Training Program includes more than eighty UNMC research laboratories. The general research interests of BRTP faculty members include those indicated by the department designations: Biochemistry and Molecular Biology, Cell Biology and Anatomy, Pathology and Microbiology, Pharmacology, and Physiology and Biophysics. More specifically, faculty research interests target one or more of the following eight focus areas.

Cancer Biology: Despite a century of unprecedented progress in the understanding and treatment of human diseases, cancer remains one of the most complex and challenging problems of medical science—complex because its secrets are intricately embedded within the fundamental processes of life; challenging because it demands insight and creativity from the scientists who seek to understand its mysteries. At UNMC, there are thirty-nine research laboratories focused upon cancer research. Many of the faculty members are associated with the Eppley Institute for Research in Cancer and Allied Diseases. Areas of particular strength include focus groups working on the cancers of prostate, breast, pancreas, and blood.

Cardiovascular Biology: The research interests of the 21 faculty members in the cardiovascular biology area can be categorized into four general, but overlapping, themes: cardiovascular development, cell transport and signaling, microvascular function, and neurohumoral control. Ongoing studies in each area examine clinically relevant issues at all levels of biological activity, from the molecular and cellular to the coordinated integration of separate organ systems. They include gene expression during cardiac development; viral pathogenesis of the myocardium; regulation of cell receptors, ion channels, and second-messenger systems; control of microcirculatory vessels; pathophysiological alterations in capillary permeability; and the reflex and humoral control of cardiovascular, pulmonary, and renal function.

Cell Signaling: In the past several years, the area of cell signaling has very rapidly expanded into being a major field in its own right. Recent results indicate that there is a very large number of proteins whose function is to modulate various aspects of cell signaling and thereby regulate cell function. These range from the literally hundreds of cell-surface receptors that mediate cell-to-cell signals to a wide array of second-messenger and to transducer/adapter proteins that convert activated receptors into an intracellular biochemical/electrical signal. Understanding cell signaling is critical to the research activities of the 34 UNMC faculty members and has implications for all of the other focus areas. At UNMC, cell signaling plays a central role in research related to cell proliferations in development and cancer, neuronal cell signaling, cardiovascular cell function, immune cell regulation, and gene regulation. These studies use a wide variety of techniques, from molecular biology and gene knockouts to electrophysiology.

Developmental Biology: One of the greatest challenges facing biomedical research is to fully understand the cellular and molecular mechanisms of the process of development. The application of the powerful tools of genetics, combined with the techniques of modern cellular and molecular biology, has led to an explosion of interest in the study of development. A major emphasis of the cell and developmental biology area at UNMC is on the genes that control developmental events and interactions between developmentally important genes and their environment. Both transgenic and knockout mouse technologies are routinely used. Physiological systems studied at UNMC include the hematopoietic and immune systems, the skeletal system, the cardiovascular system, and the nervous system.

Genetics: Molecular genetics at UNMC is strongly represented by 32 faculty members from multiple departments, including several faculty members in the Center for Human Molecular Genetics. Major areas of research interest are the genetics of human disorders and cancer, the genetic control of embryonic development, and gene-environment interactions. Genetic engineering (transgenic and knockout technologies) and functional genomics (DNA microarray) strategies are being used to address gene function and regulation. Positional cloning and linkage studies complement the efforts to understand genetic predisposition and susceptibility for developmental disorders and common diseases.

Immunology: UNMC researchers (approximately 30) focus on a variety of immunology topics: autoimmune and inflammatory diseases, oncology, hematology, infectious diseases, and immunorelated neurologic dysfunction. Research on the mechanism(s) of stem cell mobilization and immunologic reconstitution following transplantation is a major focus. The use of cytokines for adjuvant activity in vaccines and the use of viruses as vectors for vaccines and cytokines are research interests as well. Researchers are examining neural immunologic interactions in neurodegenerative diseases. Other research interests include T-cell receptor activation, the effects of alcohol on host defense mechanisms, the assembly of tumor-specific antigens, the psychoneural endocrine regulation of the immune system, and acute-phase proteins and their role in inflammatory processes.

Molecular Biology: Molecular biological approaches are used by the majority of UNMC research faculty members. These techniques allow analysis of protein structure and function; determination of how and which genes are turned on and off during many physiologic, developmental, pathologic, and oncogenic processes; and understanding of the day-to-day workings of living cells. Some laboratories use these approaches to elucidate the molecular mechanisms that control gene expression and cellular responses to external signals and stimuli. State-of-the-art facilities are available for DNA microinjection, DNA microarrays, growth of mouse embryonic stem cells, and the preparation, propagation, and maintenance of transgenic mice.

Neuroscience: At UNMC, neuroscience researchers (thirty research laboratories) are interested in three major areas of neuroscience: (1) Brain development and stem cells. Several faculty members at UNMC currently use transgenic techniques to uncover fundamental mechanisms of brain development and embryology. (2) How neurons communicate: UNMC investigators use a variety of electrophysiological, imaging, and biochemical techniques to examine how various signaling agents (neurotransmitters, hormones, growth factors) modulate synaptic transmission. (3) How neurodegenerative diseases are protected and damaged by the immune system, the focus of the Center for Neurovirology and Neurodegenerative Diseases.

Omaha downtown.

UNIVERSITY OF NORTH CAROLINA AT CHAPEL HILL
Department of Biology

Programs of Study

The Department of Biology offers the Ph.D. degree with specializations in botany; genetics and molecular biology; cell biology, development, and physiology; morphology, systematics, and evolution; and ecology and behavior. Although the Department offers M.A. and M.S. degrees, it does not encourage students who are seeking a terminal master's degree to enroll.

Each student's program of study is supervised by his or her special committee and is tailored to the student's needs and aspirations. The master's degree requires 30 hours of credit. Requirements for the Ph.D. do not include a specific number of credits.

Research Facilities

The Department is housed in three adjacent modern laboratory buildings with a full range of facilities for research in fields ranging from molecular to environmental biology, including environment rooms, greenhouses, an extensive herbarium, three electron microscopes, ultracentrifugation and radioisotope equipment, and modern computer facilities. Students also have access to the facilities of the Institute of Marine Sciences and the research vessel *Cape Hatteras* of the Duke/UNC Oceanographic Consortium. A botanical garden, a behavioral research station, and extensive and varied tracts of land in their natural states (owned by the University) are available for research only a few miles from campus. Proximity to two other major universities, Duke and North Carolina State, and to the nationally known Research Triangle Park provides exceptional and diverse opportunities for graduate training and research. The University is a member of the Organization for Tropical Studies, which provides research and training opportunities in Costa Rica.

Financial Aid

Teaching assistantships in the amount of $14,300 to $17,750 for the academic year are available from the Department. Most doctoral fellowships include tuition and fees. University research fellowships are available from the Graduate School on recommendation of the Department. In 2005–06, research fellowships paid $21,000 plus tuition per academic year for up to five years. Research assistantships are available through faculty members; for information concerning these assistantships, students should write directly to the faculty member whose research is in their area of interest.

Cost of Study

Tuition and fees for the 2005–06 academic year were $5012 for North Carolina residents and $19,010 for nonresidents.

Living and Housing Costs

The University owns residence halls for single students and apartments for married students. Room costs in 2006–07 average $3960 per academic year in the residence halls for double occupancy. Rents for apartments for married students start at $5900 per year. Off-campus housing is more expensive.

Student Group

Total enrollment is more than 26,300 students, and more than one third of the students are in graduate or professional training. The Department of Biology has approximately 90 graduate students, almost equally divided among the five programs of study.

Location

Chapel Hill is a university town in the Piedmont region, which is located nearly 180 miles from the coast and 170 miles from the mountains. The surrounding area is largely wooded with a mixture of pine and deciduous trees. The universities and cities in the Research Triangle area provide Chapel Hill with a stimulating mixture of the arts, culture, recreation, and science.

The University

The University of North Carolina at Chapel Hill was chartered in 1789 and is the oldest state university in the nation. The University has one of the Southeast's most highly ranked and comprehensive graduate schools.

Applying

Application forms for admission and for graduate appointments must be obtained from the Graduate School. Applications, accompanied by credentials and scores on the General Test and the Subject Test in biology of the Graduate Record Examinations (GRE), must be submitted by January 1 if the applicant wishes to be considered for all forms of financial aid. For online applications, students should visit the University's Web site at http://www.unc.edu or the Graduate School Web site at http://gradschool.unc.edu. For additional information about the Department of Biology, students should visit the Web site listed in the Correspondence and Information section. Information concerning opportunities in specific fields of research may be obtained by contacting the appropriate faculty members.

Correspondence and Information

For department information:
Graduate Admissions
CB #3280, Coker Hall
University of North Carolina
Chapel Hill, North Carolina 27599
E-mail: jnichols@bio.unc.edu
Web site: http://www.bio.unc.edu/

For applications:
Graduate School
CB #4010, Bynum Hall
University of North Carolina
Chapel Hill, North Carolina 27599
E-mail: gradinfo@unc.edu

University of North Carolina at Chapel Hill

THE FACULTY AND THEIR RESEARCH

S. C. Ahmed, Assistant Professor; Ph.D., Iowa State. Telomeres; DNA damage and germline immortality.
A. S. Baldwin, Professor; Ph.D., Virginia. Class I MHC complex gene expression and HIV gene expression.
V. L. Bautch, Professor; Ph.D., Illinois at Chicago. Molecular biology of blood vessel formation in transgenic mice.
K. Bloom, Professor; Ph.D., Purdue. Structural organization of eukaryotic chromosomes.
W. E. Bollenbacher, Professor; Ph.D., UCLA. Neuroendocrinology; developmental neurobiology.
C. L. Burch, Assistant Professor; Ph.D., California, San Diego. Virus experimental evolution.
J. Carter, Professor of Geology and Biology; Ph.D., Yale. Paleoecology; invertebrate paleontology.
G. P. Copenhaver, Assistant Professor; Ph.D., Princeton. Plant genome biology; recombination; centromeres.
J. L. Dangl, J. N. Couch Professor; Ph.D., Stanford. Genetic and molecular analysis of disease resistance.
R. J. Duronio, Associate Professor; Ph.D., Washington (St. Louis). Cell-cycle control during *Drosophila* development.
J. A. Feduccia, Heninger Professor; Ph.D., Michigan. Evolutionary biology; avian evolution, paleontology, and systematics.
P. G. Gensel, Professor; Ph.D., Connecticut. Paleobotany (palynology); plant morphology, evolution.
L. I. Gilbert, Kenan Professor; Ph.D., Cornell. Insect endocrinology, molecular genetics, biochemistry, and physiology.
R. P. Goldstein, Assistant Professor; Ph.D., Texas. Anterior/posterior polarity at the cellular and organismal levels during development.
A. K. Harris, Professor; Ph.D., Yale. Movement of tissue cells in animal development and cancer invasiveness.
A. M. Jones, Professor; Ph.D., Illinois. Plant cell biology; light- and hormone-regulated growth.
C. D. Jones, Assistant Professor; Ph.D., Rochester. Evolutionary genetics and genomics.
J. J. Kieber, Associate Professor; Ph.D., MIT. Molecular genetic analysis of plant hormone signal transduction.
W. M. Kier, Professor; Ph.D., Duke. Functional morphology of invertebrates; biomechanics.
J. G. Kingsolver, Kenan Professor; Ph.D., Stanford. Evolutionary ecology and physiological ecology.
J. D. Lieb, Assistant Professor; Ph.D., Berkeley. Understanding genomewide, protein-DNA interactions.
S. Liljegren, Assistant Professor; Ph.D., California, San Diego. Molecular genetic analysis of flower development.
K. J. Lohmann, L. G. Hoggard Distinguished Professor; Ph.D., Washington (Seattle). Neurobiology; physiological mechanisms for sensing magnetic fields.
G. P. Maroni, Associate Professor; Ph.D., Wisconsin–Madison. Molecular genetics and genetic regulation in *Drosophila melanogaster*.
W. F. Marzluff, Professor of Biochemistry and Biology; Ph.D., Duke. Control of gene expression in animal cells.
S. W. Matson, Professor and Chairman; Ph.D., Rochester. Molecular biology; biochemistry of DNA replication and repair.
A. G. Matthysse, Professor; Ph.D., Harvard. Plant molecular biology; bacterial plant pathogens; bacterial genetics.
C. E. Mitchell, Assistant Professor; Ph.D., Minnesota. Host-pathogen interactions at the community level; conservation biology.
R. K. Peet, Professor; Ph.D., Cornell. Ecology: plant communities and populations.
M. A. Peifer, Professor; Ph.D., Harvard. Developmental genetics.
C. H. Peterson, Professor of Marine Science and Biology; Ph.D., California, Santa Barbara. Population and community ecology in marine environments.
T. D. Petes, Professor; Ph.D., Washington (Seattle). Yeast genetics; chromosome structure, recombination, and replication.
D. W. Pfennig, Associate Professor; Ph.D., Texas at Austin. Behavioral ecology and evolutionary biology.
K. Pfennig, Assistant Professor of Biology and Carolina Environmental Program; Ph.D., Illinois at Urbana–Champaign. Ecology, behavior, and evolution.
R. D. Podolsky, Assistant Professor; Ph.D., Washington (Seattle). Ecological and evolutionary responses of organisms to environmental variation.
J. R. Pringle, Kenan Professor; Ph.D., Harvard. Cell cycle; cytoskeleton and cellular morphogenesis; genome organization; yeast genetics.
P. J. Pukkila, Associate Professor; Ph.D., Yale. Molecular mechanisms in genetic recombination and meiosis.
J. W. Reed, Associate Professor; Ph.D., MIT. Light signal transduction in plants.
S. R. Reice, Associate Professor; Ph.D., Michigan State. Experimental stream ecology; community and ecosystems analysis.
S. L. Rogers, Assistant Professor; Ph.D., Illinois. Cytoskeletal dynamics; cellular morphology.
E. D. Salmon, Professor; Ph.D., Pennsylvania. Cell biology; cell motility; microtubules; mechanisms of mitosis and cell division.
L. L. Searles, Associate Professor; Ph.D., Cornell. Molecular biology; eukaryotic gene regulation; transposable elements.
J. J. Sekelsky, Assistant Professor; Ph.D., Harvard. Meiotic recombination and DNA repair in *Drosophila*.
M. R. Servedio, Assistant Professor; Ph.D., Texas at Austin. Models in speciation and behavioral ecology.
K. Sockman, Assistant Professor; Ph.D., Washington State. Neuroendocrine control of reproductive flexibility.
D. W. Stafford, Professor; Ph.D., Miami (Florida). Recombinant DNA; DNA synthesis; DNA sequencing; gene structure.
T. J. Vision, Assistant Professor; Ph.D., Princeton. Evolutionary and computational genetics.
P. S. White, Professor; Ph.D., Dartmouth. Plant ecology; conservation ecology.
R. H. Wiley, Professor; Ph.D., Rockefeller. Animal behavior, especially communication and social organization.

Associated Faculty
F. L. Conlon, Adjunct Assistant Professor; Ph.D., Columbia. *Xenopus*; mesoderm; heart; T-box genes.
S. Crews, Adjunct Associate Professor; Ph.D., Caltech. Molecular genetics of nervous system development.
B. H. Judd, Adjunct Professor; Ph.D., Caltech. Cytogenetics and regulation.
M. A. Resnick, Adjunct Professor; Ph.D., Berkeley. Yeast genetics.
R. E. Wyatt, Adjunct Professor; Ph.D., Duke. Ecology and evolution of plant reproduction.

UNIVERSITY OF NORTH TEXAS
HEALTH SCIENCE CENTER AT FORT WORTH
Graduate School of Biomedical Sciences

Program of Study

A student interested in graduate work in biomedical sciences at UNT Health Science Center may pursue either the M.S. or the Ph.D. degree. Doctoral degrees culminate with a specialization in cell biology and genetics, biochemistry and molecular biology, biomedical sciences, health psychology, microbiology and immunology, pharmacology and neuroscience, or integrative physiology. Master's degrees are also offered in these areas, as well as in the areas of science education, clinical research and education, biotechnology, forensic genetics, clinical research management, and medical sciences (premedical preparation). All integrated curricula include biochemistry, biomedical cell and molecular biology, biomedical ethics, immunology, microbiology, physiology, pharmacology, and statistics. In addition, specific readings, seminars, and discussions are assigned to each candidate individually through an advisory committee. Successful completion of an oral qualifying exam as well as a written and orally defended grant application is required of all Ph.D. students at the end of their formal course work. Both the thesis and dissertation must be defended orally. Joint degrees (D.O./M.S., D.O./Ph.D., P.A./Ph.D., and M.P.H./Ph.D.) are available.

Research Facilities

UNT Health Science Center has 50,000 square feet of prime research space distributed among the different departments. Offices for the faculty, support staff, and administration occupy 6,000 square feet of additional space. The research facilities include state-of-the-art laboratories in a superb research environment. Individual laboratories are equipped with microcomputers (DEC, Macintosh, and IBM PCs) for online data collection and manipulation. Mainframe facilities are available by phone-modem and Internet interfaces with the system at UNT. Support facilities include a large multianimal vivarium (AALAC approved) and the Gibson D. Lewis Health Science Library. All UNT Health Science Center faculty members and students receive a full complement of library services, including borrowing privileges, use of individual and group study areas, photocopying, document delivery/interlibrary loan, expert instruction in the use of information resources, and access to professionally trained librarians for reference and search assistance.

Financial Aid

Students pursuing either the M.S. or Ph.D. may receive financial support from an academic department on a competitive basis. The current level of support may be up to $20,770 over twelve months. Employment by UNT Health Science Center as a graduate teaching assistant ensures eligibility for in-state tuition and fees. As part of their training, many students are required to act as teaching assistants in teaching laboratories, regardless of their financial support. The Office of Financial Aid administers a variety of federally sponsored programs.

Cost of Study

In 2006–07, tuition and fees for residents of Texas are $79 per semester hour, with a $120 minimum for a regular semester. Tuition for nonresidents is $335 per semester hour but is waived for certain qualified graduate students, with resident tuition charged. These figures do not include board-designated tuition or course fees. (All costs are subject to change.)

Living and Housing Costs

Housing facilities in Fort Worth are moderately priced and range from $400 to $600 for a one-bedroom apartment to $550 to $700 for a two-bedroom apartment. No campus housing is available. General cost-of-living figures are below the national average but are equal to the cost of living within the state of Texas.

Student Group

Total enrollment at UNT Health Science Center is approximately 1,000; biomedical sciences students account for approximately 25 percent of the total enrollment.

Student Outcomes

Biomedical scientists work in industry, academia, and government agencies. Recent graduates have been named to postdoctoral fellowships nationally, including Harvard, Johns Hopkins, and the National Institutes of Health, and are employed as faculty members at universities and research scientists in industry.

Location

UNT Health Science Center is in Fort Worth, a city known for its colorful Western heritage and friendly, casual lifestyle. The campus is near Fort Worth's downtown area and adjacent to its cultural district, which includes the Amon Carter Museum of Western Art, Kimbell Art Museum, Fort Worth Art Museum, Fort Worth Museum of Science and History, and the Cowgirl Hall of Fame and Museum. The area is served by the Dallas–Fort Worth International Airport and Dallas Love Field.

The Health Science Center

UNT Health Science Center, a public institution, is comprised of the Graduate School of Biomedical Sciences, the School of Public Health, and the Texas College of Osteopathic Medicine. The Center is nationally recognized for osteopathic medical education, biomedical sciences education, multidisciplinary biomedical research, public health education, and primary and specialized patient care.

Applying

Admission requirements include a baccalaureate degree in biology or a related field, a competitive GPA in undergraduate work, and a competitive GRE score. International students must have a TOEFL score of 213 (computer-based) or more. Two letters of evaluation, an essay regarding recent scientific advances, and a statement of personal career goals are also required.

Correspondence and Information

Graduate School of Biomedical Sciences
University of North Texas Health Science Center at Fort Worth
3500 Camp Bowie Boulevard
Fort Worth, Texas 76107-2699

Phone: 817-735-2560
      800-511-GRAD (toll-free)
E-mail: gsbs@hsc.unt.edu
Web site: http://www.hsc.unt.edu

University of North Texas Health Science Center at Fort Worth

AREAS OF RESEARCH

Biochemistry and Molecular Biology
Research addresses prominent disease states such as cancer, diabetes, cardiovascular disease, and Alzheimer's. Specific research interests include cancer cell biochemistry, tumor invasion/angiogenesis/metastasis, anticancer drug development, mechanisms of cancer cell chemotherapy, apoptosis and drug resistance, molecular basis of muscle contraction, cell-cell and cell-matrix interactions, human tissue engineering, regulation of apolipoprotein E expression, hyperbaric oxygen and atherosclerosis, caveolin and cholesterol transport, enzymology of plasma lipoprotein metabolism, signal transduction mechanisms of insulin biosynthesis and secretion, molecular mechanisms of aging/antioxidants/Alzheimer's, proteases action in natural killer cells, blood cell differentiation, structure and function of parasitic enzymes.

Biomedical Sciences
The M.S. and Ph.D. degrees in biomedical sciences are designed to give students a broad base of knowledge in those disciplines that flourish in the environment of a health science center. Students are required to pursue specialized research of an interdisciplinary nature. Advanced courses focus on the individual student's particular interests and involve offerings from more than one discipline.

Biotechnology
The biotechnology program trains individuals for careers in industry and research by providing the tools and experience needed for the highly technological positions offered in emerging biotechnology companies and research institutions. All students in biotechnology are required to train in molecular, cellular, physiological, and pharmacological techniques and to complete an internship in a research or industrial laboratory setting that forms the basis of the laboratory internship practicum.

Cell Biology and Genetics
Cell biology research focuses on diseases of the eye such as degenerative retinal diseases, glaucoma, diabetic retinopathy, retinal neovascularization, and cataracts. Other research includes angiogenesis, apoptosis, cell secretory mechanisms, cell signaling, glial cell biology, growth factors and neurotrophins, neuroinflammation, nuclear function, and stem cell research.

Clinical Research and Education–Osteopathic Manipulative Medicine
The Master of Science in clinical research and education–osteopathic manipulative medicine track, is directed toward students who have completed or are completing graduate-level training in a clinical health-care discipline who want to participate in advancing osteopathic medicine and medical principles through teaching and/or research. The degree is designed to build on the student's clinical skills by fostering the development of additional skills in educational methodology and research techniques. While the degree can help students planning a clinical career by helping them to be more sophisticated consumers of the latest research, it is designed to be of particular value to students planning a career in graduate medical education or in academic medicine.

Clinical Research Management
The master's program in Clinical Research Management provides a strong foundation upon which to build a career. The rigorous curriculum focuses on providing students a broad-based view of the biomedical sciences as well as in-depth knowledge of regulatory requirements (code of federal regulations, good clinical practices), ethical issues, and both the medical writing and administrative skills necessary to conduct clinical research. As part of the program, all students complete an internship practicum in clinical studies and use this experience to write the thesis pursuant to receiving the Master of Science degree.

Forensic Genetics
This field of study concentrates on the methods of analysis and procedures used in genetic identity testing of evidentiary materials from human and nonhuman sources, utilizing advanced and state-of-the-art technologies, including microsatellite analysis, mitochondrial DNA, RT-PCR, and SNP technologies. Training provides a firm foundation in population genetics, data analysis, and legal testimony.

Health Psychology
The program in health psychology is designed to train students in careers in either basic or applied research into factors influencing physical and psychological health. The emphasis is broadly defined according to faculty representation with areas of research that includes diabetes, obesity, cardiovascular disease, cancer, neurological/neurodegenerative conditions, neurobehavioral toxicology, HIB, addictions, chronic pain, stress, pediatric health, aging, psychoneuroimmunology, psychophysiological processes, multicultural health practices, and alternative and complementary approaches to health.

Integrative Physiology
Research interests include microcirculation, cardioprotection, myocardial energy metabolism, cardiac electrophysiology and endocrinology, coronary flow and flow regulation, cardiovascular responses to exercise and obesity, mechanisms of blood pressure and blood volume regulation, and sleep physiology.

Microbiology and Immunology
Faculty in microbiology and immunology maintain outstanding research programs with special emphasis on infectious disease, microbiology, cancer, and immunology. Research opportunities for students include regulation of prokaryotic and eukaryotic gene expression, molecular biology of microbial virulence and metabolism, respiratory immunology, molecular immunology, tumor immunology, vaccine development, and cancer biology and metastasis.

Pharmacology and Neuroscience
Research interests include neurodegeneration and aging, molecular pharmacology and toxicology, behavioral pharmacology, substance abuse, vision biology, and signal transduction.

Primary-Care Clinical Research
The M.S. in primary-care clinical research is designed to provide medical students and physicians with clinical research training pertinent to family medicine and other issues involving primary care and osteopathic medicine. Specific research interests include health promotion and disease prevention; behavioral aspects of chronic disease; osteopathic health policy; osteopathic manipulative treatment; consumer health informatics; and clinical trials, cohort studies, case-control studies, and metanalyses as they pertain to any aspects of primary care or osteopathic medicine.

Science Education
The science education program provides advanced skills to individuals who have chosen careers in middle and high school science teaching. Students take integrated biomedical science courses that offer training in disciplines that range from molecular biology to whole organisms, a specialized course in which students design scientific demonstrations/experiences for middle/secondary school classes, and an internship practicum in which they spend a minimum of 10 hours per week assisting in the classroom teaching of middle or high school science.

UNIVERSITY OF NOTRE DAME

Department of Biological Sciences

Programs of Study

The Department of Biological Sciences offers programs leading to the M.S. and Ph.D. degrees, but concentration is on the latter. Both programs require original research and the presentation of a thesis or dissertation. Most students complete the master's program in two years and the doctoral program in four to five years.

Research programs offer training in animal behavior, cellular and molecular biology, comparative physiology and biochemistry, developmental biology, ecology and evolution, genetics and cytogenetics, immunology, microbiology, and neuroscience. Programs of special emphasis, described in more detail in the Special Current Research Programs section, focus on aquatic and terrestrial ecology, evolution and environmental biology, bioinformatics, cellular and molecular biology, cancer biology, comparative physiology, parasitology, and vector biology. Courses offered by the South Bend Center for Medical Education, a division of the Indiana University School of Medicine, are open to graduate students on a limited basis, and certain members of the faculty of the center can serve as research advisers or members of the thesis committee.

Research Facilities

The Galvin Life Science Center has spacious and well-equipped research laboratories. Facilities and equipment include the Freimann Life Science animal center, greenhouses, controlled-environment rooms, automated water chemistry analysis, robotic liquid handler, a microarray and scanner, nucleic acid capillary sequencers, scintillation counters, phosphorimager, and a zebrafish facility. The Department's Center for Tropical Disease Research and Training emphasizes research on vectors and agents of important tropical diseases, including malaria. An electron microscope suite houses a Hitachi H-600 high-resolution microscope, a JEOL scanning electron microscope, and a BioRad MRC1024 Confocal Imaging System. The Department houses and operates the University Flow Cytometry Center, with instruments for both cell-cycle analysis and cell sorting. The Life Science Center houses an excellent specialized teaching and research branch of the campus library. The Hank Family Center for Environmental Studies houses faculty members and students studying the impacts of human activities on the environment.

The Department's Greene-Nieuwland Herbaria contain about 250,000 specimens. The University of Notre Dame Environmental Research Center at Land O'Lakes, Wisconsin, a private 2,830-hectare wooded tract containing more than nineteen lakes and streams as well as laboratories and living quarters, has a wide range of habitats for ecological, limnological, and entomological research. The Center for Environmental Science and Technology contains a large array of analytical equipment for analyzing environmental samples of water, air, or solids. In addition, the Radiation Laboratory, a University institute for high-energy radiation studies, provides facilities for biological research. The University publishes the journal *The American Midland Naturalist*. The University's Office of Information Technology provides full access to PCs, Macintoshes, Silicon Graphics, Sun workstations, and an SGI supercomputer. DNA and protein sequence databases and retrieval software, including licenses for GCG and MacVector, are maintained. The Department has a full complement of data processing equipment and maintains IBM PC and Macintosh local area networks that form part of the campus network system.

Financial Aid

Financial support is available to all qualified graduate students. Teaching or research assistantships and fellowships are available; remuneration included tuition and fees plus a minimum cash stipend of $16,000 for the 2005–06 academic year. The Department has funds for research-related travel and other research expenses. Most students also receive a summer stipend.

Cost of Study

Tuition was $31,000 for the 2005–06 academic year. Part-time and nondegree students paid $1722 per credit hour. Most students received full-tuition scholarships.

Living and Housing Costs

The estimated cost for University Village apartments for the year 2005–06 ranged between $465 and $708 per month, including utilities. The total cost of on-campus housing, including utilities, ranged from $3998 to $4819 per year, depending on the choice of housing complex. There are public cafeterias, and various meal plans are available in the dining halls.

Student Group

Graduate students in the Department of Biological Sciences have a wide range of educational and geographic backgrounds. The 125 students now in residence are graduates of private and public colleges and universities in the United States and other countries. The high faculty-student ratio is maintained through selective enrollment, which allows maximum student-faculty interaction in both classrooms and research laboratories.

Student Outcomes

Graduates typically go on to postdoctoral research appointments followed by academic careers in teaching and research; alternatively, graduates seek research careers with the government or private industry.

Location

Metropolitan South Bend, with a population of nearly 200,000, offers opportunities for an active social and cultural life. The area's resources include a resident symphony, libraries, historical and art museums, a zoo, and many parks and recreational areas. Descriptive brochures are available from the Chamber of Commerce, 401 East Colfax, South Bend, Indiana 46601. The four major hospitals in the area provide excellent medical facilities. In addition to Notre Dame, there are three other institutions of higher learning in South Bend: Bethel College, St. Mary's College, and Indiana University.

The University

Founded in 1842, the University of Notre Dame is a private, independent, coeducational institution located on a 500-hectare campus noted for its beauty. There are approximately 2,000 students in the graduate and professional schools, with a total student enrollment at the University of 10,000. Notre Dame has long had a reputation for intellectual freedom at all levels and in every area of contemporary thought. The campus is exceptionally active, with a rich cultural program.

Applying

Although applications for admission can be considered at any time, applications for financial support should be completed by January 15. In addition to appropriate training in basic biology (a minimum of 20 credit hours), applicants must have taken two years of chemistry, a year of physics, and a year of mathematics, including calculus. Exceptions are made under some circumstances, but all deficiencies must be made up during the first year of residence. Scores on the General Test of the Graduate Record Examinations are required in support of an application. The Subject Test of the Graduate Record Examinations is recommended.

Correspondence and Information

Graduate Admissions Office
University of Notre Dame
Notre Dame, Indiana 46556
Phone: 574-631-7706

Director of Graduate Studies
Department of Biological Sciences
University of Notre Dame
P.O. Box 369
Notre Dame, Indiana 46556-0369
Phone: 574-631-6552
E-mail: biology.biosadm.1@nd.edu
Web site: http://www.biology.nd.edu/

University of Notre Dame

THE FACULTY AND THEIR RESEARCH

John H. Adams, Ph.D., Illinois. Expression and function of molecules associated with how protozoan parasites invade host cells.

Gary E. Belovsky, Ph.D., Harvard. Population ecology; ecological theory; plant-animal interactions.

Nora J. Besansky, Ph.D., Yale. Molecular evolutionary genetics of mosquitoes.

Sunny K. Boyd, Ph.D., Oregon State. Neuroendocrine regulation of vertebrate behavior.

Frank H. Collins, Ph.D., California, Davis. Genetics of arthropod vectors of human pathogens.

Crislyn D'Souza-Schorey, Ph.D., Texas. Cell signaling and GTPases.

John G. Duman, Ph.D., California, San Diego (Scripps). Comparative physiology and biochemistry of subzero-temperature adaptations in poikilothermic animals, especially insects and plants; ice nucleator and antifreeze proteins.

Jeffrey L. Feder, Ph.D., Michigan State. Population, ecological, and evolutionary genetics; speciation in Tephritid fruit flies.

Michael T. Ferdig, Ph.D., Wisconsin–Madison. Integrated genomic approaches to identify genes underlying complex phenotypes in the malaria parasite *Plasmodium falciparum*.

Malcolm J. Fraser Jr., Ph.D., Ohio State. Molecular biology and genetics of viruses; invertebrate cell culture; transposons.

Paul R. Grimstad, Ph.D., Wisconsin. Medical entomology; arbovirus ecology and epidemiology; vector competence; mosquito bionomics, systematics, and zoogeography; arthropod-borne zoonoses.

Kristin M. Hager, Ph.D., Alabama. Cell biology and genetics of Apicomplexan protein secretion.

Ronald A. Hellenthal, Ph.D., Minnesota. Systematics and ecology of aquatic insects; computer applications to biology.

Jessica J. Hellmann, Ph.D., Stanford. Population ecology; conservation biology.

Edward H. Hinchcliffe, Ph.D., Minnesota. Cell-cycle regulation of centrosome reproduction and mitotic spindle formation.

Hope Hollocher, Ph.D., Washington (St. Louis). Population and evolutionary genetics.

David R. Hyde, Ph.D., Penn State. Molecular genetics of the developing *Drosophila* visual system; identification and analysis of molecules involved in phototransduction and signal transduction in the *Drosophila* brain.

Alan L. Johnson, Ph.D., Cornell. Cell and molecular biology of ovarian follicle differentiation and atresia; ovarian cancer.

Charles F. Kulpa, Ph.D., Michigan. Bacterial physiology; microbiology of wastewater treatment; microbial degradation of xenobiotic compounds; applications of mixed cultures; genetic exchange in mixed cultures.

Gary A. Lamberti, Ph.D., Berkeley. Stream and river ecology; plant-herbivore interactions; nutrient cycling; exotic species.

Lei Li, Ph.D., Georgia State. Molecular genetics of retinal degeneration and visual modulation.

David M. Lodge, D.Phil., Oxford. Ecology of aquatic communities; benthic-pelagic links; predation; grazing; invading species.

Mary Ann McDowell, Ph.D., Wisconsin–Madison. Immunology of leishmaniasis: host-parasite interactions.

* Edward E. McKee, Ph.D., Penn State. Mitochondrial biogenesis, bioenergetics and transport.

* Kenneth R. Olson, Ph.D., Michigan State. Fish physiology: gill structure and function, heavy-metal toxicity; comparative physiology; evolution of blood pressure regulation.

Joseph E. O'Tousa, Ph.D., Washington (Seattle). *Drosophila* neurogenetics; molecular biology of photoreceptors.

Jeanne Romero-Severson, Ph.D., Wisconsin–Madison. Plant genetics.

Jeffrey S. Schorey, Ph.D., Texas. Immunology and cell biology of mycobacterium–host cell interactions.

David W. Severson, Ph.D., Wisconsin. Quantitative and population genetics of mosquitoes.

Kristin Shrader-Frechete, Ph.D., Notre Dame. Philosophy of science; ethics; social philosophy.

Jennifer L. Tank, Ph.D., Virginia Tech. Stream ecology and biogeochemistry.

Martin Tenniswood, Ph.D., Queen's at Kingston. Prostate and breast cancer; glycoprotein biogenesis and apoptosis.

Kevin T. Vaughan, Ph.D., Cornell. Dynactin and cytoplasmic dynein-mediated motility.

JoEllen Welsh, Ph.D., Cornell. Breast cancer; mechanism of action of vitamin D_3 and its analogs.

SPECIAL CURRENT RESEARCH PROGRAMS

The special programs described below are independent, but they are also closely interconnected and mutually enhancing.

Cellular and Molecular Biology. The department offers a broad training program in cell and molecular biology, emphasizing genetic and molecular approaches to a variety of research problems, including protein-DNA interactions, signal transduction, vesicular transport, development, and speciation. More than fifteen labs are currently active in this area. In addition, the interdisciplinary Biochemistry, Biophysics and Molecular Biology (BBMB) program links more than thirty laboratories, with interests in the biomolecular sciences in three departments on campus.

Center for Tropical Disease Research and Training. The Center for Tropical Disease Research and Training Laboratory coordinates the efforts of an active group of faculty members and graduate students working with mosquitoes. Training in this area reflects the application of modern biology (cell, developmental, genetic, physiological, biochemical, endocrinological, behavioral, and population) to the traditional fields of medical entomology and parasitology. This multidisciplinary program is augmented by fieldwork in the control of vector-borne disease in Indiana and elsewhere.

Ecology, Evolution, and Environmental Biology. The program is strong in areas of stream, lake, and wetland ecology; terrestrial ecology, emphasizing global change; and ecological and evolutionary genetics. Basic and applied questions are addressed using physiological, genetic, population, community, and ecosystem approaches. Research facilities include the laboratories at the environmental research center (UNDERC) near Land O'Lakes, Wisconsin, and greenhouses in the Hank Family Center for Environmental Studies. Fully equipped laboratories are also available on campus for doing molecular genetic research (PCR amplification, DNA sequencing, isozyme electrophoresis) to augment field studies.

IGERT Program. The Department of Biological Sciences houses a new Integrative Graduate Education and Research Training (IGERT) Program, funded by the National Science Foundation and starting in fall 2006. The program studies global linkages of biology, the environment, and society (GLOBES), with the goal of integrating Notre Dame faculty members and graduate students across the biological and social sciences in a team-based approach to solving human and environmental heath problems. The program will support 20 Ph.D. students, each at a stipend level of $30,000 per year, with complete tuition remission, as well as related research supplies. These special fellowships are only available to U.S. citizens and permanent residents. More information about the GLOBES program and how to apply can be found on the Web at http://globes.nd.edu.

Integrative Cell and Molecular Physiology. Research focuses on the molecular biology, cell physiology, and endocrinology of birds, fish, and amphibians; the physiology and biochemistry of cold-hardiness in insects and plants; and cardiovascular regulation and reproduction in fish and birds. In addition, an interdisciplinary program in neuroscience links multiple laboratories in the department with several other departments on campus.

Parasitology. The research programs focus on understanding parasite biology using biochemical, physiological, immunological, and genetic analysis. Particular areas of interest are specialized biochemical pathways of helminths, immunologic defense mechanisms elicited by intracellular protozoan parasites, molecular interactions of host cell invasion by the Apicomplexa, and field studies of parasitic diseases.

**Faculty member of the South Bend Center for Medical Education.*

UNIVERSITY OF PITTSBURGH

School of Medicine
Interdisciplinary Biomedical Graduate Program

Program of Study	The Interdisciplinary Biomedical Graduate Program offers access to seven Ph.D. degree–granting programs in the School of Medicine. Ph.D. students spend their first year taking courses in the interdisciplinary program, participating in laboratory rotations, and identifying their adviser and area of research. Students then move into one of the following degree-granting programs: biochemistry and molecular genetics, cell biology and molecular physiology, cellular and molecular pathology, immunology, molecular pharmacology, molecular virology and microbiology, or neuroscience. Students continue with more specialized course work in their second year and begin their dissertation research project in the laboratory of choice.
Research Facilities	Research training facilities are located primarily in the Biomedical Science Towers, Scaife Hall, and the Hillman Cancer Center. Laboratories are completely equipped for state-of-the-art studies and include highly automated DNA sequencers; recording spectrophotometers; analytical and preparative centrifuges; gamma radioactivity and liquid scintillation counters; flow cytometers; gas and high-performance liquid chromatographic systems; amino acid analyzers; protein sequencers; mass spectrometers; peptide synthesizers; polymerase chain reaction instrumentation; proteomic instrumentation; individual and centralized computer facilities; electron, confocal, and fluorescence microscopy; and tissue culture facilities.
	There are excellent animal facilities, including a primate laboratory; cold rooms for preparative work; surgical suites; and a pathogen-free facility capable of supporting immunosuppressed and transgenic mice.
	The Maurice and Laura Falk Library of the Health Sciences is the main library of the Medical Center. The library receives more than 2,600 periodicals and has approximately 220,000 volumes. In addition, there are seven specialized libraries in the Medical Center to which students have access.
Financial Aid	All full-time graduate students receive a stipend, tuition remission, and individual health insurance (with an option to purchase additional family coverage) during their graduate training. The stipend is competitive and totals $23,000 for the 2006–07 academic year.
Cost of Study	A tuition waiver is granted to all full-time graduate students.
Living and Housing Costs	Most graduate students find housing in residential areas in the vicinity of the School of Medicine. Apartment costs range from $400 to $500 for an efficiency, $500 to $700 for a one-bedroom apartment, and $650 to $900 for a two-bedroom unit. Many students share larger apartments to minimize housing costs. An excellent public transportation system, free to students, connects the University with all surrounding residential communities.
Student Group	There are approximately 250 graduate students seeking degrees in the seven basic science programs in the School of Medicine and Public Health. They are represented by an active student organization, the Biomedical Graduate Student Association (BGSA), which hosts an annual symposium and lecture. The BGSA hosts many social events throughout the year.
Location	Pittsburgh is a dynamic metropolitan area with many cultural attractions, including the Pittsburgh Symphony, Opera, and Ballet as well as theaters, museums, art galleries, and restaurants. The city's parks and its wooded western Pennsylvania surroundings offer excellent recreational facilities for skiing, boating, mountain biking, rock climbing, white-water rafting, hiking, backpacking, cycling, hunting, and fishing.
The School	University of Pittsburgh School of Medicine Ph.D. students enjoy access to world-class basic research programs and clinical resources. The rigor and comprehensiveness of the School's academic program prepare graduates to meet the challenges of the future and excel in their chosen fields. The first-year shared core curriculum and fast-track approach to lab experience help graduate investigators move from understanding the basics to tackling new challenges in biomedicine to reexamining long-standing questions using recent discoveries and newly developed techniques.
Applying	Requirements for admission include a baccalaureate degree in a natural or physical science or engineering program with a minimum grade point average of 3.0 (on a scale of 4.0), scores from the Graduate Record Examinations (GRE) General Test, and three letters of recommendation. International applicants must take the Test of English as a Foreign Language (TOEFL) and receive a score of 250 or greater on the computer-based test. The online application is available at the Graduate Studies Office Web site.
Correspondence and Information	Interdisciplinary Biomedical Graduate Program Graduate Studies Office School of Medicine 524 Scaife Hall University of Pittsburgh Pittsburgh, Pennsylvania 15261-0001 Phone: 412-648-8957 Fax: 412-648-1077 E-mail: gradstudies@medschool.pitt.edu Web site: http://www.gradbiomed.pitt.edu

University of Pittsburgh

GRADUATE TRAINING PROGRAM DIRECTORS

Interdisciplinary Biomedical Graduate Program: John P. Horn, Ph.D., Professor.

Biochemistry and Molecular Genetics: Thomas Smithgall, Ph.D., Professor.
Cell Biology and Molecular Physiology: Simon Watkins, Ph.D., Associate Professor.
Cellular and Molecular Pathology: Robert Bowser, Ph.D., Associate Professor.
Immunology: Russell Salter, Ph.D., Associate Professor.
Molecular Pharmacology: Donald B. DeFranco, Ph.D., Professor.
Molecular Virology and Microbiology: JoAnne L. Flynn, Ph.D., Associate Professor.
Neuroscience: Codirectors Daniel J. Simons, Ph.D., Professor, and Bita Maghaddam, Ph.D., Professor.
M.D./Ph.D. Program: Clayton Wiley, M.D., Ph.D., Professor and Associate Dean.

UNIVERSITY OF PUERTO RICO, MEDICAL SCIENCES CAMPUS

School of Medicine
Division of Graduate Studies

Programs of Study

The Division of Graduate Studies offers programs leading to the Master of Science (M.S.) and Ph.D. degrees in anatomy, biochemistry, microbiology, pharmacology and toxicology, and physiology. The Master of Science program normally takes two to three years of study, whereas the Ph.D. program may take four to five years. The required course credits for the Ph.D. program are completed during the first four semesters. At the end of the fourth semester or after completion of the required course credits, doctoral students must pass a qualifying examination. A graduate proposal is presented to the Thesis Committee, and completion of the degree is evaluated in the defense of the dissertation. Students participate in a wide array of seminars, workshops, and analogous activities scheduled by the various departments. The graduate faculty members are committed to interdisciplinary collaboration between basic and clinical scientists in order to broaden the exposure of the students. The diversity of the curriculum further permits students research and training in areas such as neurosciences, cardiovascular biology, toxicology, immunology and virology, molecular biology and pathogenesis, exercise physiology, and molecular genetics.

The Division collaborates with the Department of Biology of the Río Piedras campus in sponsoring an intercampus Ph.D. program in biology. The intercampus program allows students to benefit from the scientists, facilities, equipment, and course offerings of the two largest research institutions in the Caribbean.

Research Facilities

The Division's facilities are housed in the Medical Sciences Campus building, with ancillary facilities in the Institute of Neurobiology, the Caribbean Primate Center, the Cancer Center, the Center for the Study of Sexually Transmitted Diseases, the Center for Energy and Environmental Research, the Veterans Administration Hospital, the University Hospital, the University Pediatric Hospital, and affiliated hospitals. These facilities house research and teaching laboratories, faculty offices, lecture rooms, and specialized libraries. A central library serves the general needs of the academic community, and there are linkages with other local and national libraries. Each department has its own laboratories and office space for faculty members and students as well as specialized equipment. A system of core laboratories serves the needs of several departments, providing facilities for tissue culture, electron microscopy, flow cytometry and cell sorting, molecular biology, and state-of-the-art animal facilities, including BL3 areas for nonhuman primates.

Financial Aid

Graduate students can apply for financial assistance through teaching or research assistantships funded by the University, the National Institutes of Health (NIH), the National Science Foundation (NSF), the Ford Foundation, and the Department of Energy, among others.

Cost of Study

Tuition for Puerto Rico residents is $100 per credit hour. Nonresidents from other parts of the United States pay the same tuition that Puerto Rico residents pay at their respective state universities. For international nonresidents, tuition is $3500 per year, plus additional fees. An additional fee of approximately $630 is necessary for all students to cover the construction and laboratory fees as well as medical insurance.

Living and Housing Costs

Housing costs vary widely in the vicinity of the Medical Sciences Campus. Apartments generally rent for between $400 and $600 per month. Prices for food and other articles are similar to those in major United States cities.

Student Group

Total enrollment at the Medical Sciences Campus is approximately 3,200 students. The Division of Graduate Studies has 49 Ph.D. and 23 M.S. students.

Location

The Medical Sciences Campus is located in a suburb of San Juan, Puerto Rico's capital. A great variety of cultural activities can be found in and around this city of more than 1.5 million residents. Large shopping malls, restaurants, condominiums, fine arts centers, and beaches are easily accessible by means of expressways and mass transportation facilities.

The University and The Campus

The University of Puerto Rico is a state-funded, public, coeducational institution of higher learning. The official objectives and guiding principles of the University guarantee equal opportunities to everyone. The institution comprises a system of University colleges, regional colleges, and other units located throughout the island. The Medical Sciences Campus is located on the grounds of the Puerto Rico Medical Center and grants degrees in all the principal fields of health sciences. It houses the Schools of Medicine, Dentistry, Pharmacy, and Public Health and the Division of the College of Allied Health Professions. The Cancer Center and the Center for Sexually Transmitted Diseases are located nearby. The Institute of Neurobiology is located in Old San Juan. The Caribbean Primate Center is located on an offshore island, but part of its colony is kept in a town close to San Juan.

Applying

Applicants should hold a bachelor's degree or its equivalent in biology, chemistry, or physics, with a grade index of at least 3.0 overall and in science subjects on a 4.0 scale. The applicant must also be proficient in English and Spanish and must submit scores on the GRE General Test and the Subject Test in biology, chemistry, or physics. Three letters of recommendation, two from professors in the major field and one from a professor in another department, must also be submitted. The graduate faculty members of the department concerned personally interview the student. The application deadline for admission in January is September 15 and for admission in August is February 15. Requests for more information and application forms should be addressed to the Division of Graduate Studies.

Correspondence and Information

Associate Dean for Biomedical Sciences and Director of Graduate Studies
School of Medicine
University of Puerto Rico
Medical Sciences Campus
P.O. Box 365067
San Juan, Puerto Rico 00936-5067
Phone: 787-758-4639
Fax: 787-767-8693
Web site: http://medweb.rcm.upr.edu/gradstudies/

University of Puerto Rico, Medical Sciences Campus

THE FACULTY AND THEIR RESEARCH

University of Puerto Rico, Medical Sciences Campus, has been abbreviated in the listings below as PR-MSC.

Anatomy
Jennifer Barreto, Assistant Professor; Ph.D., Puerto Rico, Río Piedras, 2001. CNS molecular/cellular changes linked to reproductive health after exposure to androgens during puberty.
Rosa E. Blanco, Professor; Ph.D., Cambridge, 1987. Nerve regeneration and visual system.
Manuel Díaz, Assistant Professor; Ph.D., PR-MSC, 2003. Central pattern generators and control of locomotion in mammals.
Donald C. Dunbar, Professor; Ph.D., Oregon, 1980. Functional morphology of mammals; morphology, biomechanics, and neural control of locomotion and posture.
Juan Carlos Jorge, Assistant Professor; Ph.D., Brandeis, 1997. Behavioral neuroendocrinology; steroid effects in synaptic physiology, neural structure, and behavior in vertebrates.
Robert W. Kensler, Professor; Ph.D., SUNY at Stony Brook, 1978. Macromolecular structure of muscle thick filaments.
Earl Kicliter, Professor; Ph.D., SUNY Upstate Medical Center, 1973. Structure and function of vertebrate visual systems.
Nidza Lugo, Professor; Ph.D., PR-MSC, 1982. Mammalian visual system; expression of neuroactive substances in the circadian visual system.
Mark W. Miller, Professor; Ph.D., Connecticut, 1980. Cellular basis of natural behavior patterns in invertebrates; neuropeptides.
N. L. Pérez-Acevedo, Assistant Professor; Ph.D., Puerto Rico, 2001. Synaptic physiology and cellular basis of emotional memory.
Maria A. Sosa, Associate Professor and Acting Chairman; Ph.D., Florida, 1993. Synaptic physiology and neural basis of aggressive behavior in crustaceans.
Jean E. Turnquist, Professor; Ph.D., Pennsylvania, 1975. Skeletal aging and functional morphology of nonhuman primates.

Biochemistry
Dipak K. Banerjee, Professor; Ph.D., Calcutta, 1976. Cell signaling and angiogenesis; dolichol cycle and cell-cycle dynamics; glycoprotein biochemistry; catecholamine homeostasis.
Carmen L. Cadilla, Professor; Ph.D., Tennessee, Knoxville, 1986. Hormonal regulation of gene expression; genetic diseases affecting Puerto Ricans.
Elsa M. Cora, Professor; Ph.D., PR-MSC, 1984. Molecular and genetic alterations during tumorigenesis.
Sixto García-Castiñeiras, Professor; M.D., Complutense (Madrid), 1967; Ph.D., PR-MSC, 1976. Mechanisms of lens aging and cataract formation; redox active components in aqueous humor; oxidative stress; protein biochemistry.
Braulio D. Jiménez, Professor; Ph.D., Puerto Rico, Mayagüez, 1981. Molecular and environmental toxicology.
Alan Preston, Professor; Ph.D., Purdue, 1971. Nutritional epidemiology.
Carlos Basilio Reyes, Professor; M.D., Chile, 1956. Modulation of eukaryotic transcription by stressor agents.
José R. Rodríguez-Medina, Professor and Chairman; Ph.D., Brandeis, 1986. Function of myosin II in yeast.
José F. Rodríguez-Orengo, Professor; Ph.D., Texas A&M, 1989. Biochemical and pharmacological processes of antiretroviral and anticancer drugs.

Microbiology and Medical Zoology
Edna E. Aquino, Assistant Professor; Ph.D., PR-MSC, 2000. Serpentine receptor trafficking and lipid raft membrane microdomains in glial and lymphoid cells.
Benjamin Bolaños, Associate Professor; Ph.D., Duke, 1983. Pathobiology of *Cryptococcus neoformans* and host response in cryptococcosis.
Ana María Díaz, Associate Professor; D.Sc., Buenos Aires, 1981. Allergy to mites hybridoma and monoclonal antibody technologies.
Ana M. Espino, Assistant Professor; Ph.D., Institute of Tropical Medicine (Havana), 1997. Characterization of T- and B-cell epitopes of multiple antigenic peptides for a *Fasciola/Schistosoma* cross-reactive vaccine.
Wieslaw J. Kozek, Professor; Ph.D., Tulane, 1969. Parasitology: antigenicity and ultrastructure of medically important nematodes, parasite-host communication, and *Wolbachia* symbionts of filariae.
Edmundo N. Kraiselburd, Professor; Ph.D., SUNY at Buffalo, 1972. Design and evaluation of DNA vaccines against human and simian immunodeficiency viruses.
Julio A. Lavergne, Professor; Ph.D., Texas Health Science Center at San Antonio, 1979. Immune functions and pathogenesis of HIV infections; mechanisms of apoptosis in HIV-infected cells.
Idalí Martínez Martínez, Associate Professor; Ph.D., Rutgers, 1995. Development of dengue DNA vaccine.
Loyda M. Meléndez, Professor; Ph.D., Emory, 1990. Immunology of human immunodeficiency virus; HIV tropism and role of placenta in vertical transmission; monocyte immunity and HIV dementia.
Iraida E. Robledo, Associate Professor; Ph.D., Puerto Rico, Río Piedras, 2000. Antimicrobial resistance in gram-positive cocci and in *Helicobacter pylori;* bacteria; genotypic characterization of ESBLs.
Nuri Rodríguez–del Valle, Professor; Ph.D., PR-MSC, 1978. Microbial physiology; biochemical aspects of fungal dimorphism.
Adelfa E. Serrano, Professor; Ph.D., Georgia, 1987. Molecular biology and immunology of parasites, malaria; molecular mechanisms of drug resistance in *Plasmodia;* molecular diagnosis of parasitic infections.
Luis J. Torres-Bauzá, Professor; Ph.D., PR-MSC, 1980. Microbial genetics; plasmid-mediated resistance to antibiotics *(Neisseria gonorrhoeae); Candida albicans* dimorphism.
Guillermo J. Vázquez, Professor and Chairman; M.D., Jefferson Medical, 1974. Therapy of HIV/AIDS and its complications; antimicrobial resistance; *Helicobacter pylori.*

Pharmacology and Toxicology
Sylvette Ayala, Assistant Professor; Ph.D., Texas, 1998. Molecular biology; biochemistry; aging and oxidative stress.
Adriana Báez, Professor; Ph.D., Madrid, 1977. Molecular basis of differentiation; induction by antitopoisomerase drugs; molecular epidemiology of head and neck cancer.
Walmor C. De Mello, Professor and Chairman; M.D., 1955, Ph.D., 1964, Rio de Janeiro. Intercellular communication: variation in junctional permeability.
Emma Fernández-Repollet, Professor; Ph.D., PR-MSC, 1979. Role of T-cells in diabetes mellitus.
Diógenes Herreño-Sáenz, Associate Professor; Ph.D., PR-MSC, 1986. Toxicology, chemical carcinogenesis, biomarkers, and risk assessment.
José G. Ortiz, Professor; Ph.D., Connecticut, 1982. Modulation of neurotransmitter uptake and release in experimental neurological models.
Philip C. Specht, Associate Professor; Ph.D., SUNY Upstate Medical Center, 1972. Computer simulation of pharmacokinetics; contagious diseases.
Susan Corey Specht, Associate Professor; Ph.D., SUNY Upstate Medical Center, 1971. Na pump studies.

Physiology
Jaime Bernstein, Professor; Ph.D., Toronto, 1974. Mechanisms of ion transport across biological membranes.
Jonathan Blagburn, Associate Professor; Ph.D., Thames Polytechnic (London), 1982. Physiology of invertebrate ganglia.
María José Crespo, Associate Professor; Ph.D., PR-MSC, 1993. Vascular alterations in cardiovascular disease.
Nelson Escobales, Professor and Chairman; Ph.D., PR-MSC, 1982. Membrane physiology.
Carlos Jiménez, Associate Professor; Ph.D., New Mexico, 1986. Neurophysiology.
Damien Kuffler, Associate Professor; Ph.D., UCLA, 1975. Nerve regeneration: how axons find their target.
Jorge D. Miranda, Associate Professor; Ph.D., Baylor College of Medicine, 1996. Axonal regeneration in the adult spinal cord.
Guido Santacana, Professor; Ph.D., PR-MSC, 1982. Changes in the physiology and pharmacology of airway smooth muscle that are induced by mechanisms of temperature and drugs of abuse.
Annabell C. Segarra, Professor; Ph.D., NYU, 1988. Neuroendocrinology of the reproductive system; estrogen, opioids, and cocaine sensitization.
Walter I. Silva, Professor; Ph.D., NYU, 1986. Cellular and molecular physiology of vascular and brain cells.
Carlos A. Torres, Assistant Professor; Ph.D., Texas Medical Branch, 1996. Repair of AP sites and aging in *Saccharomyces cerevisiae.*
Conchita Zuazaga, Professor; Ph.D., Minnesota, 1974. Electrical excitability of biological membranes.

UNIVERSITY OF ROCHESTER

School of Medicine and Dentistry
Graduate Education in the Biomedical Sciences Program

Programs of Study

The School of Medicine and Dentistry offers graduate programs leading to the M.A., M.S., M.P.H., and Ph.D. degrees. It also has a number of combined-degree opportunities (M.D./M.S., M.D./M.P.H., M.D./Ph.D, M.B.A./Ph.D.) available. Degrees may be earned in biochemistry, biophysics, epidemiology, genetics, health services research and policy, microbiology and immunology, neurobiology and anatomy, neuroscience, pathology, pharmacology, physiology, and toxicology. There are terminal M.S. degree programs in dental science, marriage and family therapy, microbiology, pharmacology, physiology, public health, and statistics. Other interdisciplinary degree programs may be arranged. First-year students enroll in a core curriculum designed to provide fundamental concepts and knowledge in the biomedical sciences, including biochemistry, cell biology, and molecular biology and genetics. Students admitted into the graduate program have the opportunity to pursue doctoral research with more than 200 faculty members dedicated to excellence in biomedical research. At the end of the first year, students choose their Ph.D. mentor and the Ph.D. degree to be earned.

Interdisciplinary training clusters serve as the route of admission to the Ph.D. program in the biomedical sciences. The arrangement provides students with an exceptionally diverse choice of research areas that encompasses virtually all of modern biomedical sciences. The training clusters in the graduate education program include biochemistry, molecular and cell biology; biophysics and structural biology; cellular and molecular basis of medicine; genetics, genomics, and development; immunology, microbiology, and vaccine biology; neuroscience; and toxicology.

Studies leading to the doctoral degree usually require approximately five years, and those for the master's degree usually two. The combined-degree program permits completion of the requirements for both the M.D. and Ph.D. degrees in approximately seven years. Application to the combined M.D. and Ph.D. program is usually made at the time of application to the medical school but may be made after a student is in residence.

Research Facilities

The School of Medicine and Dentistry has modern, well-equipped laboratories in each department. Many special core facilities, such as the cell-sorting facility, heteroduplex sequencing laboratories, electron microscopes, oligonucleotide synthesizer, protein/peptide sequencers, high-performance liquid chromatography and gas chromatography equipment, fluorescent atomic absorption spectrometers, tunable dye lasers, cell-culture and virus preparation laboratories, extensive computing capability, nuclear magnetic resonance spectrometers, and an X-ray crystallography facility are available. The Medical Center has a large vivarium and a transgenetic animal core facility.

The Aab Institute of Biomedical Sciences is the centerpiece of a ten-year, $400-million strategic plan to expand the Medical Center's research programs in the basic sciences. The Institute, headquartered in a 240,000-square-foot building, is on the Medical Center campus. It enhances the University's strong biomedical research program by creating an organizational structure and professional environment that fosters outstanding interdisciplinary research.

Financial Aid

Most doctoral candidates are supported by training grants, research grants, or University funds. The University has several interdepartmental training grants from the National Institutes of Health, and there are also private sources of financial aid. Most Ph.D. students are awarded support packages that include competitive stipends ($23,000 per year for 2006–07), tuition scholarship, and health-care insurance.

Cost of Study

Full-time registration (16 credit hours per semester) carried a tuition fee of $30,528 for the full year in 2005–06.

Living and Housing Costs

Several types of University housing, ranging from furnished single rooms through town-house units, are available. In addition, there are relatively inexpensive accommodations in private residences within walking distance of the Medical Center. National cost-of-living indexes usually place the Rochester area in the midrange.

Student Group

Approximately 475 graduate students are training in the School of Medicine and Dentistry.

Location

Metropolitan Rochester, situated in upstate New York close to Lake Ontario, has a population of nearly a million. It is the home of Eastman Kodak, Xerox, Bausch & Lomb, and a number of other technically oriented companies. The University and the city complement one another to provide a cultural atmosphere for academic development. Rochester is also close to the Finger Lakes region, which provides opportunities for year-round recreation, with skiing in winter, water sports during other seasons, and hiking throughout the year.

The University

The University of Rochester is an independent, privately endowed institution founded in 1850. The School of Medicine and Dentistry was established in 1925. The Medical Center, comprising the medical school and Strong Memorial Hospital, is entirely under one roof and is located adjacent to the University's River Campus, permitting close interaction among the colleges. Of a total matriculated student population of 8,500, about 2,650 are enrolled in advanced-degree programs. The University's major divisions are the College (Arts and Sciences and Engineering and Applied Sciences), Education and Human Development, Liberal and Applied Sciences, the Eastman School of Music, the Simon School of Business and Management, the School of Medicine and Dentistry, and the School of Nursing.

Applying

Applicants for full-time study must submit completed application materials online by January 1. Paper applications are not accepted. The Graduate Record Examinations General Test is required for admission. The TOEFL is required of international applicants. Application may be made at any time, but most applications are submitted in the late fall and early winter. Admission decisions are generally made in early spring for admission the following September.

Candidates are strongly encouraged to submit applications for admission using the University's online application system. For additional information and the online application, students should visit the Graduate Education Web site, listed in this In-Depth Description.

Correspondence and Information

Offices for Graduate Education
University of Rochester School of Medicine and Dentistry
601 Elmwood Avenue, Box 316A
Rochester, New York 14642

Phone: 585-275-4522
E-mail: gradadm@urmc.rochester.edu
Web site: http://www.urmc.rochester.edu/smd/grad

University of Rochester

PROGRAM DIRECTORS AND GRADUATE PROGRAM COORDINATORS

Ph.D. PROGRAMS

Biochemistry
Director: Mark Dumont, Ph.D.; mark_dumont@urmc.rochester.edu
Coordinator: Rose Burgholzer; rose_burgholzer@urmc.rochester.edu
Telephone: 585-275-3417
Address: 601 Elmwood Avenue
Box 712
Rochester, New York 14642

Biophysics
Director: William Bernhard, Ph.D.; william_bernhard@urmc.rochester.edu
Coordinator: Rose Burgholzer; rose_burgholzer@urmc.rochester.edu
Telephone: 585-275-3730
Address: 601 Elmwood Avenue
Box 712
Rochester, New York 14642

Epidemiology
Director: Susan Fisher, Ph.D.; susan_fisher@urmc.rochester.edu
Coordinator: Pattie Kolomic; pattie_kolomic@urmc.rochester.edu
Telephone: 585-275-7882
Address: 255 Crittenden Boulevard
Helen Wood Hall Box 644
Rochester, New York 14642

Genetics
Director: Dirk Bohmann, Ph.D.; dirk_bohmann@urmc.rochester.edu
Coordinator: Katie Scoville; katie_scoville@urmc.rochester.edu
Telephone: 585-275-1441
Address: 601 Elmwood Avenue
Box 633
Rochester, New York 14642

Health Services Research and Policy
Director: Bruce Friedman, Ph.D.; bruce_friedman@urmc.rochester.edu
Coordinator: Pattie Kolomic; pattie_kolomic@urmc.rochester.edu
Telephone: 585-273-2618
Address: 255 Crittenden Boulevard
Helen Wood Hall Box 644
Rochester, New York 14642

Microbiology and Immunology (M.S. program also)
Director: Dennis McCance, Ph.D.; dennis_mccance@urmc.rochester.edu
Coordinator: Brenda Knorr; brenda_knorr@urmc.rochester.edu
Telephone: 585-275-3402
Address: 601 Elmwood Avenue
Box 672
Rochester, New York 14642

Neuroscience
Director: Robert Freeman, Ph.D.; robert_freeman@urmc.rochester.edu
Coordinator: Barbara Elliott; barbara_elliott@urmc.rochester.edu
Telephone: 585-275-5788
Address: 601 Elmwood Avenue
Box 603
Rochester, New York 14642

Pathology
Director: Robert Mooney, Ph.D.; robert_mooney@urmc.rochester.edu
Coordinator: Taimi Marple; taimi_marple@urmc.rochester.edu

Telephone: 585-273-4580
Address: 601 Elmwood Avenue
Box 626
Rochester, New York 14642

Pharmacology and Physiology (M.S. program also)
Director: David Yule, Ph.D.; david_yule@urmc.rochester.edu
Coordinator: Linda Fullington; linda_fullington@urmc.rochester.edu
Telephone: 585-275-0447
Address: 601 Elmwood Avenue
Box 711
Rochester, New York 14642

Toxicology
Director: Ned Ballatori, Ph.D.; ned_ballatori@urmc.rochester.edu
Coordinator: Muriel Stanley; muriel_stanley@urmc.rochester.edu
Telephone: 585-275-6702
Address: 601 Elmwood Avenue
Box EHSC
Rochester, New York 14642

MASTER'S PROGRAMS

Dental Science
Director: Jim Melvin, Ph.D.; james_melvin@urmc.rochester.edu
Coordinator: Barbara Sperduto; barbara_sperduto@urmc.rochester.edu
Telephone: 585-275-3441
Address: Center for Oral Biology
601 Elmwood Avenue
Box 611
Rochester, New York 14642

Marriage and Family Therapy
Director: Pieter le Roux, D. Litt. et Phil.; pieter_leroux@urmc.rochester.edu
Coordinator: Sherri Seeger, sherri_seeger@urmc.rochester.edu
Telephone: 585-275-2532
Address: 300 Crittenden Boulevard
Box PSYCH M&F
Rochester, New York 14642

Public Health
Director: Sally Trafton, J.D.; sarah_trafton@urmc.rochester.edu
Coordinator: Pattie Kolomic; pattie_kolomic@urmc.rochester.edu
Telephone: 585-275-7882
Address: 255 Crittenden Boulevard
Helen Wood Hall Box 644
Rochester, New York 14642

Statistics
Director: Michael McDermott, Ph.D.; mikem@bst.rochester.edu
Coordinator: Cheryl Bliss Clark; cheryl_blissclark@urmc.rochester.edu
Telephone: 585-275-6696
Address: 601 Elmwood Avenue
Box 630
Rochester, New York 14642

UNIVERSITY OF ROCHESTER

Department of Biology
Programs in Genetics, Evolutionary Biology,
and Molecular, Cellular, and Developmental Biology

Programs of Study	The Department offers programs leading to the Ph.D. degree in a broad spectrum of disciplines, with special emphasis in the areas of genetics; molecular, cellular, and developmental biology; and evolutionary biology.
	The Ph.D. program is intended to educate students to become effective research scientists. The curriculum varies according to the interests and background of the student and usually requires six years for completion. The first two semesters are devoted to course work and research rotations in the laboratories of three different faculty members. Each student is assigned a faculty member who assists in composing a course program designed to help define areas of special interest and intense study. Although the student normally selects most courses from those offered by the Department of Biology, he or she may also take advantage of offerings in other departments, including the School of Medicine and Dentistry. Formal courses are supplemented by a variety of seminars and colloquia. The student usually selects a mentor at the end of the second semester.
	Admission to candidacy for the Ph.D. requires successful completion of an oral examination, which includes defense of a thesis proposal. This exam is normally completed by the end of the second year. Periodic meetings with a thesis advisory committee are required to aid the student in critically evaluating results, assigning priorities, and considering alternative experimental strategies. A limit of seven years is imposed for completion of all requirements for the Ph.D., including 90 credit hours distributed among formal courses, seminars, and research.
	Programs leading to the master's degree in biology, requiring about three semesters for completion, are intended to prepare the recipient for secondary school and undergraduate college teaching or for certain research positions. The degree requires successful completion of 30 credit hours (two semesters of a full course load) and a comprehensive written examination.
Research Facilities	Modern biological research requires a variety of experimental tools, and the facilities available within both the Department of Biology and the University are excellent. Equipment used for contemporary biological research is available, including ultracentrifuges, electrophoresis equipment, PCR machines, oligonucleotide synthesizers, an automatic DNA sequencer, a peptide synthesizer, a peptide sequencer, high-pressure liquid chromatography apparatuses, micromanipulators, scintillation counters, a confocal microscope, electron microscopes, phosphorimagers, graphics workstations, and computerized systems for microscopic image analysis.
	Several mainframe computers can be accessed by students from terminals within the Department, and most personal computers have access to the Internet.
	A renovated greenhouse is attached to Hutchison Hall, home of the Department of Biology. Shops are available for specialized electronics and instrumentation construction. In the adjacent Medical Center, a vivarium houses a variety of research animals and an inbred-mouse colony. The Cell Sorting Facility includes two computer-interfaced cell sorters and a Zeiss Axiomat.
	Library facilities at the University of Rochester are excellent. The biology research collection includes subscriptions and online access to more than 360 journals.
Financial Aid	Graduate students in the Ph.D. program are supported by teaching and research assistantships, assuming satisfactory progress toward the degree. The Department offers a competitive stipend plus individual health insurance and a full-tuition scholarship. In addition, exceptionally well-qualified students are eligible for fellowships that provide an additional $2000 per year.
Cost of Study	The 2005–06 tuition was $954 per credit hour for master's students. Students admitted to the Ph.D. program in the Department receive a full-tuition scholarship.
Living and Housing Costs	The University has housing facilities for married and unmarried graduate students. Monthly rents range from $520 for an unfurnished studio apartment to $770 for a two-bedroom, two-bath furnished unit. Non-University properties are available throughout the area at rates lower than those found in most metropolitan areas.
Student Group	There are currently 42 graduate students and more than 15 postdoctoral fellows and research associates in the Department, drawn from various geographical regions and diverse undergraduate backgrounds.
Location	Rochester is situated in upstate New York, where the Genesee River meets the southern shore of Lake Ontario. The metropolitan area has a population of nearly a million people. The city's commercial base is primarily technological. Major firms include Eastman Kodak Company, Xerox Corporation, and Bausch & Lomb. The area presents cultural opportunities unusual for its size and is especially noted for its musical activities, which center on the University's Eastman School of Music and the Rochester Philharmonic Orchestra. The city and five colleges and universities provide theater, dance, film series, art galleries, and museums. Rochester is close to the Finger Lakes and Adirondack wilderness regions, and there are numerous parks and recreational areas.
The University	The University of Rochester, a private university founded in 1850, maintains a moderate size with an emphasis on graduate education and research. There are approximately 4,435 undergraduates and 2,310 full-time students in advanced degree programs.
Applying	The College of Arts and Sciences requires all applicants to apply online. Prospective students should link to the Admission and Financial Support site on the graduate home page. The Graduate Record Examinations, including an appropriate Subject Test, are required. Students from non-English-speaking countries are required to take the TOEFL. Applications should be submitted by January 1 to guarantee consideration for September enrollment.
Correspondence and Information	Graduate Program Secretary Department of Biology University of Rochester RC Box 270211 Rochester, New York 14627-0211 Phone: 585-275-7991 E-mail: cyly@mail.rochester.edu Web site: http://www.rochester.edu/College/BIO/graduate

University of Rochester

THE FACULTY AND THEIR RESEARCH

Cheeptip Benyajati, Associate Professor; Ph.D., Princeton, 1977. Regulation of gene expression during development of *Drosophila;* transcription, chromatin structure, and mechanisms of developmental chromatin structure remodeling.

Xin Bi, Assistant Professor; Ph.D., Johns Hopkins, 1994. Molecular biology and genetics: epigenetic regulation of eukaryotic gene expression (structural and functional domains of the genome; chromatin boundary elements; gene silencing).

Thomas H. Eickbush, Professor and Chairman; Ph.D., Johns Hopkins, 1979. Molecular biology and evolution: mechanism of integration of retrotransposable elements; control over mobile element activity; regulation of rRNA genes.

James D. Fry, Assistant Professor; Ph.D., Michigan, 1988. Genetics of ecological adaptation in *Drosophila;* evolutionary effects of deleterious mutations; quantitative-genetic theory and methodology.

Richard E. Glor, Assistant Professor; Ph.D., Washington (St. Louis), 2004. Evolutionary ecology: evolution of species diversity during the adaptive radiation of Caribbean *Anolis* lizards.

David S. Goldfarb, Professor; Ph.D., California, Davis, 1983. Molecular cell biology of the nucleus.

Vera Gorbunova, Assistant Professor; Ph.D., Weizmann (Israel), 1999. Mechanisms of aging, DNA repair, and genomic instability.

David C. Hinkle, Professor; Ph.D., Berkeley, 1971. Molecular biology and genetics: enzymatic mechanisms of DNA replication; repair and mutagenesis; yeast enzymes involved in error-prone repair.

John Jaenike, Professor; Ph.D., Princeton, 1975. Ecology and evolution of host-parasite interactions; sex chromosome meiotic drive; *Wolbachia* as male killers and causes of reproductive isolation in *Drosophila;* ecology of mycophagous *Drosophila.*

Heinrich Jasper, Assistant Professor; Ph.D., Heidelberg, 2002. Development, cell biology, and genetics: signal transduction networks governing development, stress responses, metabolism, and aging of *Drosophila.*

Rulang Jiang, Associate Professor; Ph.D., Wesleyan, 1995. Mouse developmental genetics: using mouse models to understand the molecular basis of embryonic development and birth defects.

J. David Lambert, Assistant Professor; Ph.D., Arizona, 2001. Evolution of developmental mechanisms; early patterning in mollusks and related groups; cytoskeletal basis of asymmetric cell divisions; evolution of novel phenotypes.

Rita K. Miller, Assistant Professor; Ph.D., Northwestern, 1993. Cell biology and molecular genetics: cytoskeletal organization of *Saccharomyces cerevisiae;* coordination of cell polarity and positioning of the mitotic spindle; nuclear migration; posttranslational modifications of microtubule-associated proteins.

Joanna B. Olmsted, Professor; Ph.D., Yale, 1971. Cell biology: biochemical, molecular, and developmental studies of cellular regulation of cytoskeleton assembly and organization, particularly microtubule-associated proteins.

H. Allen Orr, Professor; Ph.D., Chicago, 1990. Evolutionary genetics: genetics of speciation in *Drosophila;* genetics and theory of adaptation; population genetics.

Terry Platt, Professor of Biochemistry and Biophysics and of Biology; Ph.D., Harvard, 1972. Molecular mechanisms of gene expression; protein–nucleic acid interactions; transcription termination and mRNA 3' end formation in *Escherichia coli* and yeast.

Daven C. Presgraves, Assistant Professor; Ph.D., Rochester, 2003. Evolutionary genetics: speciation genetics; molecular population genetics; genome evolution.

Justin Ramsey, Assistant Professor; Ph.D., Washington (Seattle), 2003. Evolutionary ecology: mechanisms of adaptation and speciation in flowering plants; polyploidy and chromosome evolution.

Elaine A. Sia, Assistant Professor; Ph.D., Columbia, 1994. Molecular biology and genetics: mutagenesis and repair of the mitochondrial genome of *Saccharomyces cerevisiae.*

John H. Werren, Professor; Ph.D., Utah, 1980. Evolutionary research: microbial-host interactions with emphasis on *Wolbachia* in arthropods, genetics of speciation, and evolution of development; genetics, genomics, and evolution of *Nasonia;* the role of "selfish" or "parasitic" DNA in evolution.

Professors Emeriti

Thomas T. Bannister, Ph.D., Illinois, 1958.
Martin A. Gorovsky, Ph.D., Chicago, 1968.
Stanley M. Hattman, Ph.D., MIT, 1965.
George E. Hoch, Ph.D., Wisconsin–Madison, 1958.
Jerome S. Kaye, Ph.D., Columbia, 1957.
William B. Muchmore, Ph.D., Washington (St. Louis), 1950.
Uzi Nur, Ph.D., Berkeley, 1962.

UNIVERSITY OF SOUTH ALABAMA

College of Medicine
Interdisciplinary Graduate Program in Basic Medical Sciences

Program of Study

The Interdisciplinary Graduate Program in Basic Medical Sciences offers the Doctor of Philosophy degree for students interested in pursuing a career in biomedical research. The program provides excellent training, enabling graduates to pursue academic careers in medical institutions and universities or research positions in government and industry laboratories. Specific training opportunities are available within traditional research programs such as biochemistry and molecular biology, cancer biology, cell biology and neuroscience, microbiology and immunology, molecular and cellular pharmacology, and physiology. In addition, students may elect to direct their education and training within focus groups in the areas of lung biology, cardiovascular pathobiology, gene expression and genetics of disease, integrated biology, mechanisms of signal transduction, microbial genetics, neurotoxicity, sickle-cell pathobiology, and structural biology and drug design.

Students entering the program complete a one-year interdisciplinary core curriculum, which includes Fundamentals of Basic Medical Sciences, Introduction to Research Methods, and laboratory rotations, before beginning their advanced studies.

Research Facilities

More than fifty laboratories, equipped with state-of-the-art instrumentation for research training, as well as the Laboratory of Molecular Biology, a Level-3 research facility, are supported by several multidisciplinary core facilities, including the Flow Cytometry Core, the Biopolymer Core, the Mass Spectrometry Core, the Primate Research Laboratory, and the Transgenic Animal/ES Cell Core. The University of South Alabama (USA) Cancer Research Institute and Center for Lung Biology offer focused interdisciplinary research training.

Financial Aid

The entry stipend for full-time Ph.D. students is $20,000 per year. Loans are available from conventional sources through the University Financial Aid Office.

Cost of Study

All students admitted into the Ph.D. program are granted a full waiver of tuition. There is a University fee of approximately $350 each semester. Single coverage in a group health insurance policy is provided for full-time students.

Living and Housing Costs

The University operates dormitory housing, with costs starting at $1067 per semester. Affordable off-campus housing is available nearby. Cafeteria meal plans range from $980 to $1090 per semester.

Student Group

Enrollment in all graduate programs at the University of South Alabama is approximately 2,700. In addition, there are 263 students enrolled in the M.D. program in the College of Medicine. The Interdisciplinary Graduate Program in Basic Medical Sciences has a current roster of about 40 students and accepts up to 15 new students each year.

Location

Located in southwest Alabama on Mobile Bay, Mobile presents an ideal location for study. White-sand beaches along the Gulf of Mexico at Gulf Shores, Alabama, and Pensacola, Florida, are approximately an hour's drive from the campus and provide varied recreational opportunities, including golfing and fishing. New Orleans, located approximately 100 miles west, offers additional cultural and recreational opportunities. Mobile's population of about 500,000 enjoys a mild climate year-round, with temperatures averaging 70 degrees Fahrenheit. The city is one of the nation's largest seaports and is served by interstate highways and major airlines.

The University

The University of South Alabama, which consists of the Colleges of Allied Health, Arts and Sciences, Education, Engineering, Nursing, and Medicine; the Mitchell College of Business; the Schools of Computer and Information Sciences and Continuing Education and Special Programs; and the Graduate School, was created by an act of the Alabama state legislature in 1963. The University's total enrollment is about 13,500, and the number of students in the health-related professions of nursing, allied health, medicine, and the graduate program in basic medical sciences is approximately 3,700. The total number of faculty members employed by the University is 987, with 223 in the College of Medicine.

The 1,200-acre main campus is located in the western section of Mobile. In addition, the University also owns the USA Springhill and USA Baldwin County campuses, the Dauphin Island Marine Sciences Sea Lab, and the USA Brookley campus on Mobile Bay, which houses the Educational Research and Development Center. The USA Medical Center and USA Children's and Women's Hospital are components of the USA Health System.

Applying

The deadline for submitting applications for the fall is March 31. Matriculation is recommended for summer or fall terms, with fall classes beginning in August. Requirements for consideration include a USA admission application, a Graduate Program supplemental application, certified transcripts of all university work, official scores on the GRE General Test and TOEFL (if applicable), and recommendations from 3 individuals qualified to evaluate the student's academic performance and potential. A baccalaureate degree is required from an accredited college or university.

Correspondence and Information

Director, Basic Medical Sciences Graduate Program
College of Medicine (MSB 3316)
University of South Alabama
Mobile, Alabama 36688-0002
Phone: 251-460-6153
Web site: http://southmed.usouthal.edu/com/bmsphd

University of South Alabama

THE FACULTY

Biochemistry and Molecular Biology

N. N. Aronson Jr., Professor and Chair; Ph.D., Duke, 1966. S. Barik, Professor; Ph.D., Calcutta, 1982. J. D. Funkhouser, Professor; Ph.D., Arkansas, 1975. J. W. Gaubatz, Associate Professor; Ph.D., Texas at Dallas, 1975. R. E. Honkanen, Professor; Ph.D., Georgia, 1986. R. S. Lane, Associate Professor; Ph.D., Michigan, 1969. M. G. Nair, Professor; Ph.D., Florida State, 1969.

Research interests: Biochemistry, molecular genetics, and human diseases of glycoprotein degradation; organic synthesis and biochemical pharmacology of anticancer agents; fetal lung development; gene regulation; DNA damage and repair; viral transcription; biochemistry and cell biology of epithelial cells; mechanisms of signal transduction; role of serine/threonine protein phosphatases in cell division and cancer; parasite biochemistry.

Cancer Biology

G. L. Wilson, Professor; Ph.D., Illinois, 1976. O. Fodstad, Professor; M.D., 1967, Ph.D., 1980, Norway. J. Ju, Assistant Professor, USA Cancer Research Institute; Ph.D., Yale, 1996. R. Samant, Assistant Professor, USA Cancer Research Institute; Ph.D., Indian Institute of Technology, 1999. L. Shevde-Samant, Assistant Professor, USA Cancer Research Institute; Ph.D., Bombay, 1999. Advisory Committee: R. D. Balczon, Associate Professor; Ph.D., Florida State, 1984. R. E. Honkanen, Professor; Ph.D., Georgia, 1986. R. N. Lausch, Professor; Ph.D., Florida, 1966. J. A. Tucker, Professor and Chair of Pathology; M.D., Vanderbilt, 1981.

Research interests: Signal transduction; gene expression; mechanisms of carcinogenesis; drug development; gene therapy.

Cell Biology and Neuroscience

G. L. Wilson, Professor and Chair; Ph.D., Illinois, 1976. M. F. Alexeyev, Assistant Professor; Ph.D., Kiev, 1992. R. D. Balczon, Associate Professor; Ph.D., Florida State, 1984. Y. M. Bhatnagar, Associate Professor; Ph.D., Boston College, 1974. K. M. Chou, Assistant Professor; Ph.D., Vermont, 1997. S. D. Critz, Associate Professor; Ph.D., Texas, 1988. P. A. Fields, Associate Professor; Ph.D., Texas A&M, 1975. A. L. Gard, Professor; Ph.D., Iowa, 1985. S. G. Kayes, Professor; Ph.D., Iowa, 1977. S. P. LeDoux, Professor and Vice Chair; Ph.D., South Alabama, 1986. S. F. Ofori-Acquah, Assistant Professor; Ph.D., London, 1999.

Research interests: Gene expression and cancer; diabetes and other endocrine diseases; sickle-cell disease; neurodegeneration and aging; DNA repair mechanisms; gene therapy.

Lung Biology

T. Stevens, Professor and Director; Ph.D., Colorado State, 1991. A. B. Al-Mehdi, Assistant Professor; M.D., Ph.D., Crimea Medical Institute (Ukraine), 1990. M. F. Alexeyev, Assistant Professor; Ph.D., Kiev, 1992. R. D. Balczon, Associate Professor; Ph.D., Florida State, 1984. S. Barik, Professor; Calcutta, 1982. S. T. Ballard, Professor; Ph.D., North Carolina State, 1989. F. G. Eyal, Professor; M.D., Hebrew (Jerusalem), 1973. B. W. Fouty, Associate Professor; M.D., Washington (Seattle), 1988. J. D. Funkhouser, Professor; Ph.D., Arkansas, 1975. M. N. Gillespie, Professor; Ph.D., Kentucky, 1981. J. Haynes Jr., Professor; M.D., South Alabama, 1980. R. E. Honkanen, Professor; Ph.D., Georgia, 1986. S. G. Kayes, Professor; Ph.D., Iowa, 1977. J. A. C. King, Associate Professor; Ph.D., East Tennessee State, 1987; M.D., South Carolina, 1992. S. P. LeDoux, Professor; Ph.D., South Alabama, 1986. T. M. Lincoln, Professor; Ph.D., Tennessee, 1971. S. F. Ofori-Acquah, Assistant Professor; Ph.D., London, 1999. J. W. Olson, Professor; Ph.D., USC, 1977. J. C. Parker, Professor; Ph.D., Mississippi, 1972. T. C. Rich, Assistant Professor; Ph.D., Vanderbilt, 1996. K. L. Schaphorst, Associate Professor; M.D., Nebraska, 1988. A. E. Taylor, Professor Emeritus; Ph.D., Mississippi, 1964. M. I. Townsley, Professor; Ph.D., California, Davis, 1982. W. W. Wagner Jr., Professor; Ph.D., Colorado State, 1974. R. M. Whitehurst Jr., Associate Professor; M.D., South Alabama, 1987. G. L. Wilson, Professor; Ph.D., Illinois, 1976. S. Wu, Assistant Professor; M.D., 1986, Henan Medical (China), 1986.

Research interests: Endothelia cell biology; pulmonary circulation; acute lung injury; chronic obstructive pulmonary disease (emphysema); cystic fibrosis; environmental exposures to toxins; pulmonary hypertension; lung cancer; pulmonary problems in premature infants; sickle-cell disease.

Microbiology and Immunology

J. H. Coggin, Professor and Chair; Ph.D., Chicago, 1966. J. P. Audia, Assistant Professor; Ph.D., South Alabama, 2002. A. L. Barsoum, Assistant Professor; Ph.D., Kiel (Germany), 1965. J. W. Foster, Professor; Ph.D., Hahnemann, 1978. R. B. Hester, Associate Professor; Ph.D., Mississippi, 1972. R. N. Lausch, Professor; Ph.D., Florida, 1966. J. E. Oakes, Professor; Ph.D., Tennessee, 1975. J. W. Rohrer, Associate Professor; Ph.D., Kansas, 1974. H. H. Winkler, Professor and Vice Chair; Ph.D., Harvard, 1966. D. O. Wood, Professor; Ph.D., Georgia, 1978.

Research interests: Mechanisms of virulence and infection; host response to infectious agents; characterization of oncofetal antigens; immune regulation; gene expression; obligate intracellular bacteria (rickettesiae); select agents; ocular infection and immune responses; oncogenesis; stress responses in enteric bacteria.

Molecular and Cellular Pharmacology

M. N. Gillespie, Professor and Chair; Ph.D., Kentucky, 1981. A. B. Al-Mehdi, Assistant Professor; M.D., Ph.D., Crimea Medical Institute (Ukraine), 1990. J. E. Ayling, Professor; Ph.D., Berkeley, 1966. M. Chinkers, Associate Professor; Ph.D., Vanderbilt, 1981. J. W. Olson, Professor; Ph.D., USC, 1977. T. C. Rich, Assistant Professor; Ph.D., Vanderbilt, 1996. J. G. Scammell, Professor; Ph.D., Florida, 1981. S. W. Schaffer, Professor; Ph.D., Minnesota, 1970. T. Stevens, Professor; Ph.D., Colorado State, 1991. S. J. Strada, Professor; Ph.D., Vanderbilt, 1970. W. W. Wagner Jr., Professor; Ph.D., Colorado State, 1974. R. M. Whitehurst, Associate Professor; M.D., South Alabama, 1987. S. Wu, Assistant Professor; M.D., 1986, Henan Medical (China), 1986.

Research interests: Cellular and molecular pharmacology; second messenger action; mechanisms of signal transduction; molecular basis of drug action; cardiovascular pharmacology and heart failure; molecular endocrinology; toxicology; pulmonary endothelial cell biology; molecular basis for ion channel function; structure-based drug design.

Physiology

T. M. Lincoln, Professor and Chair; Ph.D., Tennessee, 1971. S. T. Ballard, Professor; Ph.D., North Carolina State, 1989. M. V. Cohen, Professor; M.D., Harvard, 1968. J. M. Downey, Professor; Ph.D., Illinois, 1971. J. C. Parker, Professor; Ph.D., Mississippi, 1972. A. E. Taylor, Professor Emeritus; Ph.D., Mississippi, 1964. M. S. Taylor, Assistant Professor; Ph.D., South Alabama, 1999. M. I. Townsley, Professor; Ph.D., California, Davis, 1982. D. S. Weber, Assistant Professor; Ph.D., Medical College of Wisconsin, 1998. J. Yang, Assistant Professor; Ph.D., Ohio State, 1998.

Research interests: Cardiovascular physiology; lung, cardiac microcirculations; neural control of the heart and vasculature; airway solute and water transport; mechanisms of ischemia/reperfusion injury; vascular cell imaging; vascular smooth-muscle physiology; vascular smooth-muscle signal transduction and function; regulation of cell calcium; lipid metabolism; metabolic disorders (obesity and diabetes).

UNIVERSITY OF SOUTH CAROLINA

School of Medicine
Graduate Program in Biomedical Science

Programs of Study	The Graduate Program in Biomedical Science at the University of South Carolina School of Medicine offers a course of study and significant research opportunities leading to the Master of Biomedical Science (M.B.S.) degree and the Ph.D. degree. The goal of the Ph.D. program is to prepare students to become productive biomedical researchers and highly qualified teachers.

The purpose of the M.B.S. program is to provide broad-based interdisciplinary training in biomedical science to individuals who wish to expand or change their educational background and training to fulfill personal, preprofessional, or other career advancement goals. A combined M.D./Ph.D. degree is also offered in which students attend the first two years of medical school before joining a research laboratory. After successfully completing and defending their thesis, M.D./Ph.D. students return to the medical curriculum for their clinical years.

The first-year core curriculum of the M.B.S. and Ph.D. programs provides all students with a broad overview of biomedical science that promotes later flexibility. During the first year, students carry out laboratory rotations leading to the selection of a mentor in whose laboratory they undertake advanced training and dissertation research. Areas of research include molecular oncology, neuroscience, vision science, developmental biology and anatomy, reproductive development, and microbiology and immunology. The research groups are composed of faculty members from several units in the University and School of Medicine. The program sponsors an annual student research symposium and supports students' travel to scientific meetings.

Research Facilities

The School of Medicine is located on the Veterans Administration Hospital campus, approximately 5 miles from the main campus of the University. Research laboratories are equipped with advanced instrumentation and equipment necessary for modern biomedical research. A broad range of shared core facilities on both campuses includes DNA sequencing and oligonucleotide synthesis, gene array manufacture and analysis, electron and confocal fluorescence microscopy, fluorescence-activated cell sorting and analysis, mass spectrometry, and a computer and communications resource center. The School of Medicine also has a state-of-the-art animal resource center. The School library currently contains more than 93,000 volumes and a large number of print and electronic biomedical serial publications. Libraries on the main campus house more than 7 million items. Access to most major biomedical journals is available electronically.

Financial Aid

Graduate research assistantships are available; stipends are $20,000 to $22,000 per year for 2006–07. All Ph.D. students hold graduate research assistantships. The cost of health insurance is also paid.

Cost of Study

In 2004–05, tuition and fees were $3445 per semester for state residents and $7460 for nonresidents. Fees for students holding graduate research assistantships were reduced to $950 per semester.

Living and Housing Costs

The area near the campus offers a variety of living accommodations at rents ranging upward from $400 per month. Living expenses average $13,500 per year for a single student living in private housing.

Student Group

Thirty-six students were enrolled in the Ph.D. program and 10 in the M.B.S. program in the 2004–05 academic year. All were studying full-time, and 36 students received financial support for living expenses.

Student Outcomes

The biomedical science Ph.D. program has had 64 graduates since 1995, with an average time for degree completion of 5.1 years. Graduates have found employment in private biotechnology industry (Roche Bioscience, Merck, Mojave Therapeutics), government research labs (NIH, CDC, FAA), faculty positions and academic research (Harvard, Johns Hopkins, Washington (Seattle), Michigan, Northwestern, Tufts, Virginia, Providence, North Carolina, Baylor), and clinical research settings (Memorial Sloan Kettering, Imperial Cancer Research (London), M. D. Anderson, Cleveland Clinic).

Location

Situated in central South Carolina, Columbia, the state's capital, has a population of more than 384,000. Coastal and mountain recreational areas are within easy driving distance. Columbia has mild winters and warm summers.

The University

The University is located on a 222-acre campus in the heart of Columbia, adjacent to the state capitol. Chartered in 1801, it is one of the older state universities. Of the approximately 20,000 students enrolled at the main campus, more than one third are graduate students. The School of Medicine is located on a satellite campus adjacent to the Veterans Administration Hospital.

Applying

Admission decisions are made on the basis of an overall evaluation of the applicant's preparation and ability to complete advanced study in biomedical science. Applicants must hold or be pursuing a baccalaureate degree. Undergraduate preparation should include one year each of biology, physics, inorganic chemistry, organic chemistry, and mathematics. A grade point average of 3.0 (B) or better is essential both in the sciences and overall. Scores on the General Test of the Graduate Record Examinations (GRE) that fall above the 50th percentile are also required.

Although there is no deadline for applications, students are encouraged to apply early because the assistantships are limited in number and are awarded on a competitive basis. Applications should be submitted to the Graduate School, University of South Carolina, Columbia, South Carolina 29208. Electronic applications are available on the Web at http://www.gradschool.sc.edu. Informal inquiries and preliminary applications may be made on the Web site listed in the Correspondence and Information section.

Correspondence and Information

Dr. Richard Hunt
Director of the Biomedical Graduate Programs
School of Medicine
University of South Carolina
Columbia, South Carolina 29208
Phone: 803-733-3100
Fax: 803-733-3168
E-mail: biomed@med.sc.edu
Web site: http://pathmicro.med.sc.edu/graduate/biomed.htm

University of South Carolina

THE FACULTY AND THEIR RESEARCH

James R. Augustine, Associate Professor; Ph.D., Alabama at Birmingham, 1973. Neuroanatomy of the primate nervous system.

Troy Baudino, Assistant Professor; Ph.D., Saint Louis, 1999. The role of myc in regulating angiogenesis in tumorigenesis and development.

Charles A. Blake, Professor; Ph.D., UCLA, 1972. Neuroendocrinology of reproduction and development.

Thomas K. Borg, Professor; Ph.D., Wisconsin, 1969. Heart development and disease.

Gregory L. Brower, Associate Professor; D.V.M., Texas A&M, 1987; Ph.D., Auburn, 1998. Mechanisms of heart failure secondary to diabetes: the role of the cardiac mast cell.

Philip Buckhaults, Assistant Professor; Ph.D., Georgia, 1996. Genetics of colorectal cancer.

James Buggy, Associate Professor; Ph.D., Pittsburgh, 1974. Neurophysiology; hypothalamic homeostatic regulation.

Wayne Carver, Associate Professor; Ph.D., South Carolina, 1988. Molecular biology of cell-cell and cell-matrix interactions in heart development.

James R. Coleman, Adjunct Professor; Ph.D., UCLA, 1974. Sensory physiology and development.

Kim E. Creek, Professor; Ph.D., Purdue, 1980. Retinoids, growth factors, and control of proliferation of normal and cancer cells.

Craig W. Davis, Associate Professor; Ph.D., Emory, 1977. Molecular mechanisms of neurotransmission in the central nervous system.

Jim R. Fadel, Assistant Professor; Ph.D., Ohio State, 1998. Forebrain dopamine and acetylcholine systems; hypothalamic neuropeptides.

Janet L. Fisher, Assistant Professor; Ph.D., North Carolina at Chapel Hill, 1994. Functional properties of recombinant GABA-A receptors.

Alvin Fox, Professor; Ph.D., Leeds (England), 1976. Bacterial pathogenesis and analytical microbiology; mass spectrometry.

Karen Fox, Research Associate Professor; Ph.D., South Carolina, 1989. Molecular taxonomy of bacteria.

Jason D. Gardner, Assistant Professor; Ph.D., Louisiana Tech, 1997. Mechanisms of female cardioprotection: the role of female hormones.

Edie C. Goldsmith, Assistant Professor; Ph.D., North Carolina at Chapel Hill, 1996. Cardiac development, hypertrophy, and failure.

Richard Goodwin, Assistant Professor; Ph.D., South Carolina, 1998. Developmental biology; molecular evolution; heart development.

Gregory A. Hand, Adjunct Assistant Professor; Ph.D., Texas Southwestern Medical Center at Dallas, 1995. Stress interactions in brain; intervention strategies.

Paul R. Housley, Associate Professor; Ph.D., Michigan, 1981. Molecular mechanisms of drug action and steroid hormone action.

William J. M. Hrushesky, Adjunct Professor; M.D., SUNY at Buffalo, 1973. Cancer chemotherapy and chronobiology.

Margaret Hunt, Associate Professor; Ph.D., Cambridge, 1973. Growth factors and extracellular matrix in human retinal cell transformation.

Richard C. Hunt, Professor; Ph.D., Cambridge, 1973. Antioxidants in the human retina; transcription factors in retinal cell proliferation.

Joseph S. Janicki, Professor; Ph.D., Alabama at Birmingham, 1974. Mechanisms of heart failure, with an emphasis on the myocardial extracellular matrix.

Sandra J. Kelly, Adjunct Professor; Ph.D., McGill, 1985. Neurobiology of social interaction; CNS actions of ethanol.

Holly A. LaVoie, Assistant Professor; Ph.D., Virginia, 1994. Pituitary hormones, growth factors, and genes controlling ovarian steroid production.

Susan M. Lessner, Research Assistant Professor; Ph.D., MIT, 2000. Angiogenesis in atherosclerosis.

Patsy Lill, Associate Professor; Ph.D., University of Health Sciences (Chicago), 1975. Tumor immunology.

Eugene P. Mayer, Associate Professor; Ph.D., Marquette, 1973. Mechanisms of cell-cell interactions; effects of bacterial membrane components on lymphocytes; characterization of cells from atherosclerotic plaque.

Alexander J. McDonald, Professor; Ph.D., West Virginia, 1978. Neurobiology of the limbic system.

Robert McKallip, Assistant Professor; Ph.D., George Washington, 2000. Role of CD44 in the interaction of activated lymphocytes with melanoma; effect of cannabinoid exposure on tumor growth and the antitumor immune response.

Clarke F. Millette, Professor; Ph.D., Rockefeller, 1975. Biochemistry and cell biology of mammalian spermatogenesis.

Stephanie J. Muga, Assistant Professor; Ph.D., Texas at Austin, 1995. Carcinogenesis and chemopreventive agents.

Mitzi Nagarkatti, Professor; Ph.D., Defense R. & D. Establishment, Gwalior (India), 1981. Cancer biology, tumor immunology and immunotherapy; T–cell, endothelial-cell interactions; immunity against toxins in biodefense.

Prakash Nagarkatti, Professor; Ph.D., Jiwaji (India), 1980. Effect of environmental pollutants on the immune system; drug abuse and apoptosis in immune cells; cannabinoids in cancer therapy and autoimmunity.

Chandrashekhar Patel, Research Assistant Professor; Ph.D., Indian Institute of Science, 1987. Molecular mechanisms regulating chronic inflammatory disorders in the vasculature.

Jeffrey R. Patton, Associate Professor; Ph.D., North Carolina, 1984. RNA-protein interaction, mRNA splicing, and RNA modification.

Lance E. Paulman, Assistant Professor; Ph.D., South Carolina, 1998. Extracellular matrix in the developing central nervous system.

Lucia A. Pirisi-Creek, Professor; M.D., Sassari (Italy), 1983. Role of human papillomaviruses in human neoplasia.

Jay D. Potts, Research Assistant Professor; Ph.D., Iowa, 1991. Molecular and biochemical aspects of ocular development and disease.

Robert L. Price, Research Professor; Ph.D., Southern Illinois, 1984. Ultrastructure of heart development and disease.

Ann Ramsdell, Assistant Professor; Ph.D., Medical University of South Carolina, 1996. Developmental biology; molecular embryology.

Lawrence P. Reagan, Assistant Professor; Ph.D., Pennsylvania, 1995. Effects of stress on hippocampal neurons.

Sarah M. Sweitzer, Assistant Professor; Ph.D., Dartmouth, 2001. Pain processing in the developing nervous system.

Brenda J. Tripathi, Professor; Ph.D., London, 1971. Functional cell biology of the eye.

Ramesh Tripathi, Professor; M.D., Agra (India), 1959; Ph.D., London, 1970. Experimental ophthalmology; aqueous humor and growth factors.

Kenneth B. Walsh, Associate Professor; Ph.D., Cincinnati, 1986. Regulation of ion channels by protein kinases and guanine nucleotide-binding proteins.

Michael J. Wargovich, Professor; Ph.D., Texas Tech, 1981. Chemoprevention of colon cancer; herbals and alternative medicines.

L. Britt Wilson, Associate Professor; Ph.D., Texas Southwestern Medical Center at Dallas, 1990. Autonomic regulation of cardiovascular function.

Marlene A. Wilson, Associate Professor; Ph.D., Illinois, 1985. Hormone- and drug-induced alterations in neural systems.

Steven P. Wilson, Professor; Ph.D., Duke, 1976. Posttranscriptional processing of neuropeptide precursors; gene therapy.

Matthew B. Wolf, Professor; Ph.D., UCLA, 1967. Fluid-electrolyte balance.

Michael Yost, Research Assistant Professor; Ph.D., South Carolina, 1999. Bioengineering; artificial myocardium.

UNIVERSITY OF SOUTHERN CALIFORNIA

Ph.D. Programs in Biomedical and Biological Sciences

Programs of Study	Ph.D. Programs in Biomedical and Biological Sciences (PIBBS) is an entryway into fifteen Ph.D. programs in the biomedical and biological sciences at the University of Southern California (USC). The aim of PIBBS is to attract highly motivated, exceptionally qualified students by offering access through a single admissions process to the entire range of biomedical and biological research training programs at USC. PIBBS and its associated Ph.D. programs train students to become independent scientists for careers in biomedical research and related professions in settings such as academia, industry, and government. Areas of research training include biochemistry, biomedical engineering, biophysics, genomic studies, biostatistics, cancer biology, cell biology, computational biology, developmental biology, epidemiology, genetics, immunology, marine biology, molecular biology, microbiology, neurobiology, pathology, pharmacology, pharmaceutics, physiology, and toxicology. Students admitted to PIBBS spend two semesters taking courses from a flexible curriculum. During their initial year, most PIBBS students take courses in biochemistry, molecular biology, cell biology, and physiology or one of several courses on the molecular basis of disease; however, the curriculum for each student is specifically tailored according to his or her interests and background. Also in their first year, students complete three or more laboratory rotations, which are selected by each student from a list of approximately 200 potential faculty mentors. At the end of the second semester, each student chooses a faculty member to serve as his or her thesis research advisor and is enrolled as a second-year student in the Ph.D. program of his or her choice. Subsequently, each student focuses on the completion of course requirements and qualifying examinations for the specific Ph.D. program chosen and develops and completes an original research project that serves as the basis for a doctoral dissertation. Design and conduct of the research project is in close consultation with the faculty mentor and an advisory committee of 2 or more additional faculty members who are chosen by the student. Most students complete the Ph.D. degree in 5 to 5½ years.
Research Facilities	As a top research institution, the University of Southern California has state-of-the-art facilities for all of the disciplines mentioned above. The University library system is a leader in network-accessible catalogs and research journals. The University's many outstanding professional school programs (including medicine, pharmacy, dentistry, nursing, law, engineering, and business) and its close affiliation with more than twenty hospitals provides an exceptionally rich environment for both basic and translationally oriented research and training.
Financial Aid	All students are given research assistantships or fellowships and receive an annual stipend of $24,876, plus an additional $1432 to cover all standard student fees, health insurance, and dental costs.
Cost of Study	PIBBS pays for all standard student fees and the standard University student health insurance policy. All tuition costs are also covered.
Living and Housing Costs	Housing arrangements are the responsibility of each student. Housing costs vary greatly depending on the type and location of the accommodations. The stipend is sufficient to allow comfortable, convenient housing. There are ample housing facilities in the many communities that surround the campuses. A limited number of on-campus units are available.
Student Group	Between 30 and 35 full-time students are admitted to the program each year.
Student Outcomes	Most graduates seek additional research training as postdoctoral research fellows. Students who receive their Ph.D. degrees from USC have been highly successful in securing fellowships at top national and international research universities and institutions. Eventual career destinations have included research, teaching, or research-related careers in academia, industry, and government.
Location	The University's research facilities are located primarily on three campuses within a 5-mile radius of downtown Los Angeles: the University Park (undergraduate) Campus, the Health Sciences Campus, and Children's Hospital of Los Angeles. The Los Angeles metropolitan area is one of the most culturally and geographically diverse and exciting locations in the world. Los Angeles is the center of the world's entertainment industry and is internationally recognized for its diverse mix of cultures and cultural activities. The mild climate is conducive to year-round outdoor activities. Opportunities for everything from surfing to skiing are within an hour's drive of the campuses.
The University and The Program	The University of Southern California is one of the nation's top private research universities. Diversity is the hallmark of the USC student community, with 15,000 undergraduate and 12,000 graduate students from all fifty states of the U.S. and more than 100 other countries. More than 200 faculty members, with active, externally funded programs in the biomedical and biological sciences, are found in several different academic units: the College of Letters, Arts, and Sciences and the professional Schools of Medicine, Pharmacy, Dentistry, and Engineering. PIBBS is an interschool program sponsored by the Provost's Office, which provides students with access to the entire scope of biomedical and biological sciences at USC. PIBBS students can choose from at least fifteen different Ph.D. programs and their participating faculty members for their research-oriented doctoral training experience at USC.
Applying	PIBBS seeks highly motivated American and international students with a bachelor's or master's degree with major emphasis in the natural sciences; outstanding academic records; competitive Graduate Record Examination scores; and prior research experience. Students can apply online and print preapplication forms through the University's Web site. The completed preapplication form, official transcripts from all colleges and universities attended, three letters of recommendation, GRE scores, and (for international students) TOEFL scores should be submitted to the address listed in this description. Initial enrollment is in the fall semester only. All applications should be submitted by the December 15 deadline.
Correspondence and Information	PIBBS Admissions Committee Office of Scientific Affairs Keck School of Medicine University of Southern California 1975 Zonal Avenue, KAM 110 Los Angeles, California 90089-9023 E-mail: pibbs@usc.edu Web site: http://www.usc.edu/pibbs

University of Southern California

THE FACULTY AND THEIR RESEARCH

For information on participating Ph.D. programs and research interests of individual faculty members, students should visit the PIBBS Web site at http://www.usc.edu/pibbs. The participating Ph.D. programs, faculty heads of each program, and telephone numbers and e-mail addresses for obtaining information on specific Ph.D. programs are listed in this description.

Ph.D. Programs in the College of Letters, Arts, and Sciences
Marine Sciences: David Caron, Ph.D. (e-mail: marinebio@usc.edu).
Molecular Biology: Oscar Aparicio, Ph.D. (e-mail: molecule@usc.edu).
Neurosciences: Sarah Bottjer, Ph.D. (e-mail: neurosci@usc.edu).

Ph.D. Programs in the School of Dentistry
Craniofacial Molecular Biology: Charles Shuler, D.M.D., Ph.D. (telephone: 323-442-3174; e-mail: lmatsumo@hsc.usc.edu).

Ph.D. Programs in the School of Engineering
Biomedical Engineering: Bartlett Mel, Ph.D. (telephone: 213-740-7237; e-mail: marubaya@bcf.usc.edu).

Ph.D. Programs in the Keck School of Medicine
Biochemistry and Molecular Biology: Deborah Johnson, Ph.D. (telephone: 323-442-1146; e-mail: annvazqu@hsc.usc.edu).
Biostatistics: Brian Langholz, Ph.D. (telephone: 323-442-1810; e-mail: mtrujill@bsc.usc.edu).
Cell and Neurobiology: Thomas McNeill, Ph.D. (telephone: 323-442-1881; e-mail: cnb@hsc.usc.edu).
Developmental Biology: Cheng-Ming Chuong, M.D., Ph.D. (telephone: 323-442-1296; e-mail: chuong@pathfinder.hsc.usc.edu).
Epidemiology: Anna Wu, Ph.D. (telephone: 323-442-1810; e-mail: mtrujill@hsc.usc.edu).
Molecular Microbiology and Immunology: Stanley Tahara, Ph.D. (telephone: 323-442-2337; e-mail: microbio@usc.edu).
Pathobiology: Cheng-Ming Chuong, M.D., Ph.D. (telephone: 323-442-1168; e-mail: doumak@pathfinder.hsc.usc.edu).
Physiology and Biophysics: Alicia McDonough, Ph.D. (telephone: 323-442-1040; e-mail: physio1@hsc.usc.edu).
Preventive Medicine: Kim Reynolds, Ph.D. (telephone: 626-457-6648; e-mail: barovich@hsc.usc.edu).

Ph.D. Programs in the School of Pharmacy
Molecular Pharmacology and Toxicology: Roger Duncan, Ph.D. (telephone: 323-442-1474; e-mail: pharmgrd@hsc.usc.edu).
Pharmaceutical Sciences: Wei-Chiang Shen, Ph.D. (telephone: 323-442-1474; e-mail: pharmgrd@hsc.usc.edu).

UNIVERSITY OF SOUTHERN MISSISSIPPI

Department of Biological Sciences

Programs of Study	Graduate study in the biological sciences leads to the M.S. and Ph.D. degrees. The Department is characterized by active research programs in cellular and molecular biology, ecology and environmental biology, marine biology, and microbiology as evidenced by extramural funding, scholarship of discovery and application, and graduate student activity. A program of study and research for each student is developed in consultation with a faculty mentor and advisory committee. Students are expected to complete a Ph.D. program in five to six years and a master's program in two or three years.
Research Facilities	Well-equipped laboratories provide facilities for research in molecular, cellular, organismal, and ecological fields. Modern transmission and scanning EM facilities are shared with the School of Polymer Science, located adjacent to Johnson Science Tower. An animal research facility, located in Johnson Science Tower, includes large and small wards, a laboratory, and quarantine/hazard wards. The Department maintains a herbarium, greenhouse, and teaching garden. The Museum of Ichthyology holds one of the largest fish collections in the Southeast region. The Department's wet lab facility includes environmental chambers with programmable light and temperature regimes, four labs equipped with automatic water-quality monitoring and recording equipment, and a modular steam system.
	The Department plays the leadership role in the state of Mississippi's NIH-funded Functional Genomics Network. The University's Molecular and Cellular Visualization Center, which is part of this collaborative biomedical research effort, includes a Zeiss LSM 510 confocal microscope, computer visualization technology, bioinformatics server, and support staff.
	The Department operates a nearby 100-acre natural science park for field research and instruction, which includes lab space, flight aviary, and fields for growing plants. The Ragland Hills Nature Conservancy, two national forests, the Crosby Arboretum, and Black Creek Wilderness Area are within minutes of the University. Mississippi has 369 miles of tidal shoreline protected by a chain of barrier islands lying 9 to 12 miles offshore, several of which are included within the Gulf Islands National Seashore. This coast is part of the largest estuarine region in North America and supports a diverse flora and fauna.
	The Department is affiliated with the Mississippi-Alabama Sea Grant Consortium, the U.S. Naval Oceanographic Laboratories at the Stennis Space Center, and the Waterways Experiment Station in Vicksburg, Mississippi.
Financial Aid	All graduate students in the Department are supported by fellowships or assistantships. Students are eligible for teaching assistantships, which carry stipends of $8750 and $10,500 for M.S. and Ph.D. students, respectively, for the fall and spring semesters. Some teaching assistants continue as TAs during the summer semester at comparable rates, while others receive aid via other mechanisms, such as fellowships and research assistantships. Grant-supported research assistantships carry comparable levels of financial support. A supplement is provided to all graduate assistants to cover the cost of health insurance. Participation in regional, national, and international scientific meetings is encouraged, and the Department provides travel grants.
Cost of Study	Tuition is $1937 per semester. Nonresidents are charged an additional fee of $2439 per semester. All current graduate students are provided graduate assistantships in addition to a tuition waiver.
Living and Housing Costs	The cost of living is average for the Southeast and lower than the national average. Many doctoral students have taken advantage of low housing costs to purchase homes in the area. Students pay $200 to $400 per month for off-campus housing, depending on the number of students sharing the apartment or house. Housing on campus, including some for married students, is also available at costs that range from approximately $300 to $375 per month.
Student Group	In 2004, 76 full-time students were pursuing graduate degrees; 49 percent are women and 25 percent are international. There were 47 master's and 29 Ph.D. candidates.
Student Outcomes	Graduates are enjoying careers in governmental agencies, health facilities, industry, academic and research institutions, and public education. Recent graduates are doing postdoctoral research at North Carolina State University, Ohio State University, University of Colorado, University of North Carolina, and Virginia Polytechnic Institute and State University. Others have gained professional appointments with the Nevada Division of Wildlife, U.S. Army Corp of Engineers, U.S. Department of Agriculture, Mississippi Museum of Natural Sciences, Grand Bay National Estuarine Research Reserve (NERR), the Nature Conservancy, Alabama A&M University, University of South Florida, University of Tennessee, and Xavier University.
Location	The main campus of the University is located in Hattiesburg, a prosperous residential community with an area population of 50,000. Situated at the confluence of the Bowie and Leaf Rivers, Hattiesburg is an educational, medical, commercial, and recreational hub of south Mississippi. The Magnolia Trace, one of the country's longest "rails-to-trails," starts adjacent to the campus and stretches over 40 miles. The DeSoto National Forest, Black Creek Wilderness Area, and the Pascagoula River Management Area are only a few minutes from campus and provide camping, canoeing, and hiking. The Mississippi barrier islands, beaches, and opportunities for marine activities are located an hour's drive south. New Orleans, with its French Quarter, Creole cuisine, jazz musicians, Aquarium of the Americas, and Audubon Zoo, is less than a 2-hour drive from campus, while the Mississippi Delta, home of the Blues, and the Mississippi River are within easy driving distance.
The University and The Department	The University of Southern Mississippi (USM) is a comprehensive university with an enrollment of more than 14,000. USM offers the variety of academic and social opportunities expected of a comprehensive university, while maintaining a sense of community that is most often found in smaller institutions. The University is recognized by the Carnegie Foundation as a Carnegie institution.
	The Department, which is part of the College of Science and Technology, occupies offices and laboratories in the ten-story Johnson Science Tower and the three-story Walker Science Building. Biochemists are also located in the Johnson Science Tower, which facilitates interdisciplinary cooperation. Close ties with faculty members at the University's Gulf Coast Research Laboratory in Ocean Springs, Mississippi, provide additional research opportunities in marine biology.
	The University of Southern Mississippi subscribes to a policy of equal opportunity. Admission to and participation in the University's academic programs are open to all candidates solely on the basis of their qualifications.
Applying	Applicants should have an undergraduate major or its equivalent in the biological sciences. Applicants for admission are required to present transcripts, scores from the Graduate Record Examinations General Test and the TOEFL (international students only), a statement of research interests, and three letters of reference. The target date for fall admission is February 15, while application for spring semester should be received by September 15. Students are encouraged to contact individual faculty members concerning their research programs and to visit the campus, view its facilities, and meet graduate students and faculty members. Application material is available through the University's Graduate School Web site at http://www.usm.edu/graduatestudies. Domestic students may apply online, while international students should submit a hard copy of the application. The application form, transcripts, and test scores should be submitted to the Office of Graduate Admissions, whereas the statement of research interests and letters of reference should be submitted directly to the Department of Biological Sciences.
Correspondence and Information	Dr. Shiao Wang, Coordinator of Graduate Studies Department of Biological Sciences University of Southern Mississippi 118 College Drive, #5018 Hattiesburg, Mississippi 39406-5018 Phone: 601-266-4748 Web site: http://www.usm.edu/biology

University of Southern Mississippi

THE FACULTY AND THEIR RESEARCH

Gary Anderson, Professor; Ph.D., South Carolina, 1974. Physiological ecology; biology of parasitic epicaridean isopods and freshwater anostracans; instructional technology.

David Beckett, Associate Professor; Ph.D., Cincinnati, 1983. Freshwater ecology; ecology of freshwater invertebrates; ecology of large rivers; natural history of bats.

Lawrence Bellipanni, Associate Professor Emeritus; Ed.D., Mississippi State, 1994. Science education.

Patricia Biesiot, Associate Professor; Ph.D., MIT, 1986. Marine invertebrate zoology; biochemical and physiological aspects of marine organisms.

Kenneth Curry, Associate Professor; Ph.D., USC, 1982. Developmental studies of fungi and plants, including problems in plant pathology.

Micheal Davis, Assistant Professor; Ph.D., Auburn, 1999. Forest communities and global climate change; plant community response to stress; plant-animal interactions.

Youping Deng, Assistant Professor; Ph.D., Peking Union Medical College, 1998. Bioinformatics; computational biology; functional genomics; diabetes; cancer.

Robert Diehl, Research Associate; Ph.D. Illinois, 2003. Ecological and behavioral modeling; application of weather surveillance radar to study animal movement.

Mohamed Elasri, Assistant Professor; Ph.D., Oklahoma State, 1998. Pathogenic microbiology; molecular mechanisms of virulence in *Staphylococcus aureus.*

R. D. Ellender, Professor and Assistant Dean for Research and Development, College of Science and Technology; Ph.D., Texas A&M, 1969. Environmental microbiology/virology; bacterial source tracking.

Yanlin Guo, Assistant Professor; Ph.D. Texas at Austin, 1996. Cell biology; cellular signaling transduction.

Rosalina Hairston, Professor; Ph.D., Texas at Austin, 1976. Relationship between teaching practices and methods of assessment; environmental science education.

Jazmir Hernandez, Assistant Professor; Ph.D., Penn State, 1999. Exercise physiology; intracellular signaling pathways.

Sherry Herron, Assistant Professor; Ph.D., Southern Mississippi, 1999. Biology education; curriculum development; professional development of science teachers.

Fred Howell, Professor; Ph.D., Texas Tech, 1972. Entomology; aquatic insect ecology; environmental bioassessment and monitoring.

Brian Kreiser, Assistant Professor; Ph.D., Colorado, 1999. Population genetics; geographic patterns of genetic variation; application of molecular techniques to assess phylogeographic structure.

Aimee Lee, Instructor and Freshman Lab Coordinator; M.S., Southern Mississippi, 1999. Environmental science education.

Cynthia Littlejohn, Instructor and Microbiology Lab Coordinator; M.S., Jacksonville State, 2002. Microbiology; science education.

Bobby Middlebrooks, Professor; Ph.D., Texas Health Science Center at Dallas, 1966. Comparative immunology; development of immunoassay methodology; viral pathogenesis in lower vertebrates.

Frank Moore, Professor and Chair; Ph.D., Clemson, 1978. Animal behavior and behavioral ecology; stopover biology and conservation biology of migratory birds.

Donald Norris, Professor Emeritus; Ph.D., Tulane, 1969. Parasitology; behavior of helminths; pathology of parasitic infections.

George Pessoney, Professor; Ph.D., Texas at Austin, 1968. Algal taxonomy and morphology; relationships between algae and water quality.

Carl Qualls, Assistant Professor; Ph.D., Sydney, 1995. Evolution of viviparity and the determinants of reproductive mode in reptiles; management and conservation of endangered and threatened species.

Jennifer Regan, Instructor; Ph.D., Houston, 1999. Population genetics; heritability and reproductive behavior; sexual selection theory.

Stephen Ross, Professor Emeritus; Ph.D., South Florida, 1974. Ecology and systematics of fishes; control and persistence of fish assemblage structure and function.

Samuel Rosso, Professor Emeritus; Ph.D., Washington (St. Louis), 1963. Anatomy of vascular plants; microtechniques; natural history of rare and endangered plant species.

George Santangelo, Professor; Ph.D., Yale, 1984. Eukaryotic gene regulation; mechanism of transcription initiation; molecular biology of budding yeasts.

Jacob Schaefer, Assistant Professor; Ph.D., Oklahoma, 1999. Fish ecology; competitive interactions; modeling movement and dispersal; community structure; physiological ecology; hybridization.

Raymond Scheetz, Professor Emeritus; Ph.D., Delaware, 1969. Role of cytoskeletal elements in cell structure and function; science education.

Glen Shearer, Professor; Ph.D., Oklahoma, 1983. Microbial physiology; molecular biology of medically important fungi.

Shiao Wang, Associate Professor; Ph.D., LSU, 1986. Molecular biology of aquatic organisms; physiological ecology.

Judith Williams, Assistant Professor; Ph.D., South Carolina, 1992. Physiological ecology; diapause in free-living copepods; taxonomy and distribution of Cladocera.

THE UNIVERSITY OF TENNESSEE HEALTH SCIENCE CENTER

Integrated Program in Biomedical Sciences

Programs of Study

The Integrated Program in Biomedical Sciences (IPBS), a research-oriented interdisciplinary program, involves faculty members from the University of Tennessee Health Science Center (UTHSC) and affiliate or joint faculty members from nearby St. Jude Children's Research Hospital and the Veterans Affairs Medical Center. Unlike traditional, department-based graduate programs, the IPBS provides the Ph.D. or M.D./Ph.D. degree-seeking student a broad-based, cross-disciplinary training that is essential in today's competitive research environment. The IPBS consists of seven tracks: cancer and developmental biology; cell biology and biochemistry; genetics, functional genomics, and proteomics; integrative systems biology; microbial pathogenesis, immunology, and inflammation; molecular therapeutics and cell signaling; and neurosciences. Significantly, faculty members based in several different departments contribute to a single track, thereby enhancing the interdisciplinary training of students. All students apply directly to the IPBS and complete a first-year curriculum composed of courses in cellular and molecular biology, systems biology, and IPBS Seminar. Students select a track and laboratory for their dissertation research after completing four rotations in prospective mentor laboratories during the first year. After their track selection, students are required to take two elective courses in addition to the core curriculum. Students who identify a research area of interest early may opt to take elective courses during their first year. All students must pass an admission-to-candidacy exam before the end of their third year. Completion of the Ph.D. program is anticipated to take four to six years.

Research Facilities

All three participating institutions offer state-of-the-art facilities, and members of the graduate faculty have excellent support from the National Institutes of Health and other recognized funding agencies. The UTHSC campus is home to five Centers of Excellence that contribute laboratory and monetary support to the research enterprise. St. Jude Children's Research Hospital is an internationally recognized treatment and research center dedicated to finding cures for childhood catastrophic diseases. Faculty members from the hospital have particular strengths in the areas of cancer, HIV/AIDS, immunodeficiencies, genetic diseases, and hematologic disorders and include a Nobel laureate and several Howard Hughes Investigators. The Veterans Affairs Hospital in Memphis is one of the largest in the country and houses an active research program.

Financial Aid

IPBS students receive a waiver of tuition and fees for six years. In addition, the College of Graduate Health Sciences offers two years of stipend support to all incoming IPBS students. After this, students receive an annual stipend from their supporting mentor. Currently, stipends are $20,000 per year, and students may also receive cost-of-living increases. Outstanding graduates of U.S. institutions are eligible for Alumni Endowment Scholarships that contribute up to $3000 of additional annual support.

Cost of Study

Tuition and fees in 2005–06 were $8324 for two semesters of study for Tennessee residents. Full tuition and fees for out-of-state students were $20,800. Tuition and fees are waived for six years of residence. Books and supplies are estimated at about $1000 per year. The cost of health insurance is approximately $1000 per year.

Living and Housing Costs

UTHSC provides single-student housing in the Goodman Residence Hall, which offers one bedroom in a four-bedroom, apartment-style facility. On-campus housing for single students cost $4320 for twelve months in 2005–06. Several apartment complexes are located within walking distance of the campus, and a wide variety of housing options can be found throughout the city. For a single student living off campus, annual living expenses are estimated to be about $16,200.

Student Group

About 300 graduate students join approximately 1,800 medical, dental, pharmacy, nursing, and allied health students on the UTHSC campus.

Student Outcomes

Graduates of the IPBS are well trained for jobs in academia, industry, or government. Recent graduates have gone on to postdoctoral positions at institutions such as Harvard, Rockefeller, Stanford, and Washington Universities; the University of California at San Francisco; Jonas Salk Research Facility; and Scripps Research Institute.

Location

Greater Memphis, with 1.2 million residents, is the educational, cultural, and commercial hub of the mid-South. UTHSC and its affiliated institutions are located in the heart of the South's largest medical complex. Memphis is synonymous with music but also offers many other cultural opportunities, including excellent art and science museums, the National Civil Rights Museum, a ballet company, a symphony, repertory theaters, and the Memphis Zoo, one of only four U.S. zoos to exhibit giant pandas. Autozone Park, home to the Memphis Redbirds, a Triple-A professional baseball team, is within walking distance of the campus, and the nearby FedExForum hosts the Memphis Grizzlies of the National Basketball Association. The mild climate invites visitors to historic Beale Street, home of the blues, and to numerous outdoor events, such as Arts in the Parks, the World Championship Barbecue Cooking Contest, and the Memphis in May International Festival, which salutes a different country each year and brings a sampling of its culture to the banks of the Mississippi River.

The University

The University of Tennessee is a system of four campuses throughout the state, the oldest and largest of which was founded in Knoxville in 1794. The campus in Memphis originated in 1911 as a medical college and has grown into one of the largest centers in the country for training health professionals and biomedical scientists.

Applying

The minimum requirements for admission include a baccalaureate degree with a grade point average of 3.0. In addition, official scores on the General Test of the Graduate Record Examinations (GRE) and three letters of recommendation must be submitted. A minimum score of 1000 is required on the verbal and quantitative sections of the GRE. To be competitive, international applicants should have a GRE score above 1200 and must also demonstrate proficiency in English by a TOEFL score above 213 on the computer-based exam. The deadline is March 1 for submitting an application. Prospective students should fill out the online application found at UTHSC's Web site and request that the following information be sent directly (not from the student) to the Office of Enrollment Services: three letters of recommendation, official transcripts and proof of degree, and official GRE and TOEFL (if applicable) scores. International applicants must have their transcripts certified by one of the agencies listed at the school's Web site.

Correspondence and Information

Dr. Patrick Ryan, Director
Integrated Program in Biomedical Sciences
The University of Tennessee Health Science Center
858 Madison Avenue
Memphis, Tennessee 38163
Phone: 901-448-8764
E-mail: ipbs@utmem.edu
Web site: http://www.utmem.edu/grad/

The University of Tennessee Health Science Center

THE FACULTY

Cancer and Developmental Biology
Suzanne Baker, John Cunningham, Mustafa Dabbous, James Downing, Linda Harris, Seema Khurana, Richard Kriwacki, Jill Lahti, Leonard Lothstein, Peter McKinnon, Paul Ney, Guillermo Oliver, Lawrence Pfeffer, Martine Roussel, Robert Scott, Andrzej Slominski, Beatriz Sosa-Pineda, Parker Suttle, Trevor Sweatman, Gerard Zambetti, Xin Zhang.

Cell Biology and Biochemistry
Mary-Ann Bjornsti, John Cox, Mary Dahmer, John Fain, Zheng Fan, Arthur Geller, Linda Hendershot, Polly Hofmann, Rod Hori, James Ihle, Suzanne Jackowski, Lisa Jennings, Howard Jernigan, Leonard Johnson, Abbas Kitabchi, Katsumi Kitagawa, Anjaparavanda Naren, Ken Nishimoto, Tayebeh Pourmotabbed, Gadiparthi Rao, Radhakrishna Rao, Brenda Schulman, Susan Senogles, Charles Sherr, Solomon Solomon, Dario Vignali, Mitchell Watsky, Stephen White, Jie Zheng.

Genetics, Functional Genomics, and Proteomics
Ron Adkins, Cheng Cheng, Terrance Cooper, Yan Cui, Dominic Desiderio, Weikuan Gu, Martha Howe, Tina Izard, Clayton Naeve, David Nelson, Mary Relling, Charles Rock, Rob Williams.

Integrative Systems Biology
Clark Blatteis, Thomas Chiang, Ioannis Dragatsis, Jonathan Jaggar, Charles Leffler, Hiroko Nishimura, David Nutting, Edward Schneider, Brian Sorrentino, Donald Thomason, Lester VanMiddlesworth, Christopher Waters.

Microbial Pathogenesis, Immunology, and Inflammation
Elisabeth Adderson, Lorraine Albritton, Robert Belland, James Bina, David Brand, Gerald Byrne, Mary Ellen Conley, James Dale, Elizabeth Fitzpatrick, James Fleckenstein, Terrance Geiger, David Hasty, Thomas Hatch, George Hilliard, Julia Hurwitz, Malak Kotb, Tony Marion, Jonathan McCullers, Mark Miller, Peter Murray, Allen Portner, Marko Radic, Fabio Re, Edward Rosloniec, Charles Russell, Patrick Ryan, Clare Sample, Jeff Sample, Elaine Tuomanen, Richard Webby, Robert Webster, Michael Whitt.

Molecular Therapeutics and Cell Signaling
Marcello Arsura, Suleiman Bahouth, Lauren Cagen, George Cook, Alex Dopico, Marshall Elam, Ramareddy Guntaka, Kafait Malik, Shannon Matta, Rennolds Ostrom, Edwards Park, Rajendra Raghow, Burt Sharp, Jeffery Steketee, Ted Strom, Steven Tavalin, Gabor Tigyi, Fu-Ming Zhou.

Neurosciences
William Armstrong, John Boughter, Susan Brasser, Joseph Callaway, Angela Cantrell, Edward Chaum, Tom Curran, Alessandra d'Azzo, Andrea Elberger, Matthew Ennis, Bob Foehring, Eldon Geisert, Dan Goldowitz, Kristin Hamre, Detlef Heck, Paul Herron, Ramin Homayouni, Marcia Honig, Monica Jablonski, Dianna Johnson, Hitoshi Kita, Mark LeDoux, James Morgan, Randall Nelson, Thad Nowak, Mel Park, William Pulsinelli, Anton Reiner, Thomas Schikorski, Reese Scroggs, Richard Smeyne, David Smith, Robert Waters, Jian Zuo.

THE UNIVERSITY OF TEXAS AT SAN ANTONIO

Department of Biology

Programs of Study

The Department of Biology offers the Ph.D. degree in biology, with a concentration in either neurobiology or cell and molecular biology, as well as the M.S. degree in biology or biotechnology. Master's students may pursue research in molecular and cellular neuroscience, biochemistry, molecular cell biology, enzymology, membrane biology, molecular genetics, protein and nucleic acid structure, developmental biology, tumor biology, aging, molecular virology, medical microbiology, bioremediation, endocrinology, and parasitology. The M.S. program in biotechnology is specifically designed to enable the graduate to enter the biotechnology industry. Doctoral students must obtain a minimum of 91 semester hours of graduate credit beyond the bachelor's degree and must also complete rotations in at least two laboratories. At the end of the second year, doctoral students take a qualifying examination. Upon passing the examination, the student submits a research proposal under the guidance of the dissertation adviser. In general, four to five years of full-time study are required to complete the doctoral program, while the programs leading to the M.S. degree require two years.

Research Facilities

The Department of Biology possesses state-of-the-art laboratories (both faculty research and core laboratories) for students to pursue graduate research projects. Core equipment available to the department includes scanning and transmission electron microscopy, confocal microscopy, fluorescence-activated cell sorting, gene chip microarray, and phosphoimagery, among many others. In addition to campuswide computing facilities, the department has Sun Microsystems and Silicon Graphics workstations.

Financial Aid

In 2005–06, all doctoral students were supported with $21,000, which was a combination of research and teaching support. In addition to this support, all tuition and fees are paid for 21 credit hours per year. For qualified students, the Minority Biomedical Research Support/Research Initiative for Scientific Enhancement (MBRS/RISE) Program supports the stipend, tuition and fees, and travel to scientific meetings. Teaching assistantships are available to qualified M.S. students.

Cost of Study

In the 2005–06 academic year, tuition and fees for a full-time master's degree student (9 semester hours) were $2463 per semester for Texas residents and $7413 per semester for nonresidents.

Living and Housing Costs

University housing, which includes Chisholm Hall and University Oaks Apartments, is available. New dormitory buildings are currently being built. Off-campus housing includes many apartments adjacent to the University as well as a large number located within a 5-mile drive. The rate for a one-bedroom apartment is approximately $500 per month.

Student Group

In the 2005 fall semester, the University enrolled more than 27,000 students, of whom more than 2,500 were graduate students. The Department of Biology admits doctoral students each academic year. Approximately 100–150 M.S. students are admitted each semester. The student group is comprised of students from Texas, from out of state, and from other countries.

Location

San Antonio, with a population of 1.5 million, is one of the nation's major metropolitan areas. As the home of the Alamo and numerous other missions built by the Franciscans, the city is historically and culturally diverse. The Guadalupe Cultural Arts Center, McNay Art Museum, the San Antonio Museum of Art, and the Witte Museum enrich the city. The performing arts are represented by the San Antonio Symphony, the annual Tejano Music Festival and Tejano Music Awards, and performances by opera and ballet companies. Also notable are Sea World, Six Flags Fiesta Texas, Brackenridge Park, the Botanical Gardens, and the downtown Riverwalk. The San Antonio Zoo has the third-largest collection in North America. A city landmark is the Tower of the Americas, which was built for the 1968 World's Fair. San Antonio is home to the National Basketball Association's Spurs, league champions in 2000 and 2003. Numerous nearby lakes allow almost year-round outdoor activity, and the beaches of the Texas Gulf coast are within a 1½- to 2-hour drive.

The University

The University was founded in 1969 and has since become a comprehensive metropolitan institution. Its research expenditures place it in the top 25 percent of public universities in Texas. The University has entered a new building and recruitment phase with a view to greatly expand the research effort in the biosciences.

Applying

To ensure full consideration, doctoral students interested in the neurobiology program are encouraged to submit their applications for admission along with all supporting documentation by December 15 for acceptance the following fall semester. Doctoral students interested in the cell and molecular program are encouraged to submit their applications for admission along with all supporting documentation by February 1. Information on applying may be obtained from the Office of Graduate Studies. Applications may be done on the Internet at https://apply.embark.com/grad/utsa/32/. The deadlines for the M.S. application are July 1 for the fall semester, November 1 for the spring semester, and May 1 for the summer semester. International M.S. student deadlines are April 1 for the fall semester, September 1 for the spring semester, and March 1 for the summer semester.

Correspondence and Information

For application information:
Office of Graduate Studies
The University of Texas at San Antonio
One UTSA Circle
San Antonio, Texas 78249
Phone: 210-458-4330
Web site: http://www.utsa.edu/graduate/Admission/index.html

For program information:
Chair, Biology Department
The University of Texas at San Antonio
One UTSA Circle
San Antonio, Texas 78249
Phone: 210-458-4463 or 4459
E-mail: roberta.hernandez@utsa.edu
Web site: http://www.bio.utsa.edu

The University of Texas at San Antonio

THE FACULTY AND THEIR RESEARCH

Deborah L. Armstrong, Professor of Neurophysiology; Ph.D., Syracuse, 1982. Hippocampal synaptic modulation.

Bernard P. Arulanandam, Associate Professor of Immunology; Ph.D., Medical College of Ohio, 1999. Cellular immunology; mucosal immunity.

Edwin J. Barea-Rodriguez, Associate Professor of Neurobiology; Ph.D., Southern Illinois, 1992. Neuroplasticity; learning and memory; animal cognition.

James Bower, Professor of Neurocomputation; Ph.D., Wisconsin–Madison, 1981. Relationship between structure and function in neural circuits.

J. Aaron Cassill, Associate Professor of Cell/Molecular Biology; Ph.D., California, San Diego, 1988. Modulation of the *Drosophila* visual signaling cascade; retinal degeneration and receptor structure.

James P. Chambers, Professor of Biochemistry; Ph.D., Texas Health Science Center at San Antonio, 1975. Development of biosensor sensing elements using phage displayed libraries; calcium channel expression; molecular motors.

G. Jilani Chaudry, Assistant Professor of Biotechnology; Ph.D., Texas at Dallas, 1991. Mammalian cell intoxication by anthrax toxin.

Brenda J. Claiborne, Professor of Neurobiology; Ph.D., California, San Diego, 1981. Developmental neurobiology; neuronal circuits; computational modeling.

Gary Cole, Professor of Biotechnology; Ph.D., Waterloo, 1969. Mechanisms of fungal virulence and host immunity.

Brian E. Derrick, Associate Professor of Neurobiology; Ph.D., Berkeley, 1993. Neurobiology of learning and memory; neurocomputation; neuropharmacology of addictive drugs.

Thomas Forsthuber, Professor of Immunology; Ph.D., Tübingen (Germany), 1989. Cellular immunology; T-cell immunity; autoimmune diseases.

Gary O. Gaufo, Assistant Professor of Endocrinology; Ph.D., Berkeley, 1995.

Matthew J. Gdovin, Associate Professor of Evolutionary Biology; Ph.D., Dartmouth, 1995. Control of respiration, development, evolution, neurobiology, and neural networks.

M. Neal Guentzel, Professor of Microbiology; Ph.D., Texas at Austin, 1972. Bioremediation involving the use of microorganisms for removing toxic contaminants.

Luis S. Haro, Associate Professor of Cell/Molecular Biology; Ph.D., California, Santa Cruz, 1985. Growth hormones; hormone receptors; biochemistry; endocrinology; neuroendocrinology; biotechnology; molecular biology; evolution; molecular modeling.

Hans W. Heidner, Associate Professor of Microbiology; Ph.D., California, Davis, 1991. Replication of togaviruses; molecular determinants of viral-host range, virulence, and pathogenesis.

David B. Jaffe, Associate Professor of Neuroscience; Ph.D., Baylor College of Medicine, 1992. Neuronal modulation, synaptic integration, and synaptic plasticity are studied using a combination of electrophysiological, imaging, and computational methods.

Karl Klose, Professor of Microbiology; Ph.D., Berkeley, 1993. Bacterial pathogenesis.

Jose Lopez-Ribot, Associate Professor of Microbiology; M.D., Ph.D., Valencia (Spain), 1991. The opportunistic pathogenic fungus *Candida albicans*.

Richard G. LeBaron, Associate Professor of Cell/Molecular Biology; Ph.D., Alabama at Birmingham, 1988. Extracellular matrix; cell adhesion and migration; vision; cancer.

Martha J. Lundell, Associate Professor of Molecular Genetics; Ph.D., UCLA, 1988. Development and function of serotonin neurons in *Drosophila*.

Andrew O. Martinez, Professor of Genetics; Ph.D., Arizona, 1974. Gene regulation; aging; cancer.

Joe L. Martinez Jr., Professor of Neuroscience and Director; Ph.D., Delaware, 1971. Learning and memory.

John McCarrey, Professor of Molecular and Cellular Biology; Ph.D., California, Davis, 1981. Development, differentiation, and manipulation of mammalian germ cells.

Paul Mueller, Assistant Professor of Biology; Ph.D., Caltech, 1990. Cell cycle regulation; developmental biology.

Carlos Paladini, Assistant Professor of Neurobiology; Ph.D., Rutgers, 1999. Dopamine neuron physiology and addiction.

Clyde F. Phelix, Associate Professor of Anatomy; Ph.D., Missouri, 1988. Functional neuroanatomical research in limbic modulation of the visceral and somatic motor system; clinical correlations as relevant to cardiovascular, Alzheimer's, and Huntington's diseases and to alcohol, nicotine, and cocaine abuse.

Rama Ratnam, Assistant Professor of Cognitive/Systems Neuroscience; Ph.D., Illinois, 1998.

Robert R. Renthal, Professor of Biochemistry; Ph.D., Columbia, 1972. Protein biochemistry; sensory receptors.

Paul H. Rodriguez, Professor of Genetics; Ph.D., Rhode Island, 1970. Insect and mosquito genetics; mutagenesis; host-parasite relationships; vector competence; biochemical and molecular genetics; genetics of disease transmission; human genetics.

David M. Senseman, Associate Professor of Biology; Ph.D., Princeton, 1976. Parallel processing of visual, auditory, and olfactory information within the vertebrate CNS.

Janakirim Seshu, Assistant Professor of Microbiology; Ph.D., Washington State, 1996.

Valerie Sponsel, Associate Professor of Biology; Ph.D., Wales, 1972; D.Sc., Bristol, 1984. Regulation of plant growth and development by plant hormones; characterization of new gibberellin-responsive semidwarf mutants of Arabidopsis.

Garry Sunter, Assistant Professor of Cell/Molecular Biology; Ph.D., Imperial College (London), 1985. Plant gene expression; DNA replication and plant-pathogen interactions.

Judy Teale, Professor of Immunology; Ph.D., Virginia, 1976. Immunoparasitology; neuroimmunology; immune response to *Francisella tularensis*.

Andrew T. C. Tsin, Professor of Biochemistry and Physiology; Ph.D., Alberta, 1979. Neurobiology of the eye; biochemistry and physiology of the visual cycle; cell biology of the retina and the retinal pigment epithelium.

Oscar Van Auken, Professor of Plant Ecology; Ph.D., Utah, 1969. Plant ecology; species interactions; community composition and structure; rare species.

Yufeng Wang, Assistant Professor of Computational Biology; Ph.D., Iowa State, 2001. Bioinformatics.

Matthew J. Wayner, Professor of Neurobiology; Ph.D., Illinois, 1953. Hypothalamic-hippocampal interactions: learning and memory.

Tao Wei, Assistant Professor of Biotechnology; Ph.D., Uppsala (Sweden), 2000. Microbiology; instability of bacterial and yeast genomes during DNA replication and recombination.

Nicole Y. Wicha, Assistant Professor of Cognitive Science; Ph.D., California, San Diego, 2003. Cognitive neuroscience; human brain imaging and cognition.

Charles J. Wilson, Professor of Neurocomputation; Ph.D., Colorado at Boulder, 1979. Computational neuroscience; nonlinear dynamics of neurons and neuronal networks; neurocomputing; reinforcement learning.

Floyd Wormley, Assistant Professor of Microbiology/Immunology; Ph.D., LSU Health Sciences Center, 2001.

THE UNIVERSITY OF TEXAS
GRADUATE SCHOOL OF BIOMEDICAL SCIENCES
AT HOUSTON

UT Health Science Center at Houston
and UT M. D. Anderson Cancer Center

Programs of Study

The Graduate School of Biomedical Sciences (GSBS) offers the M.S. and Ph.D. degrees to prepare students for careers in research, education, and other areas that require advanced training in modern biomedical sciences. An M.D./Ph.D. program is offered by special arrangement with the University of Texas Medical School at Houston.

Training is offered in behavioral science, biochemistry, bioinformatics, biostatistics, cancer biology, cell biology, developmental biology, genetics, genetic counseling, immunology, medical physics, microbiology, molecular biology, pathology, neuroscience, pharmacology, physiology, reproductive biology, structural biology, toxicology, and virology or students may design an individual multidisciplinary or interdisciplinary program of study and research.

The graduate faculty of more than 500 members is drawn from the components of the University of Texas Health Science Center at Houston (Medical School, Dental Branch, School of Public Health, and School of Health Information Sciences), the University of Texas M. D. Anderson Cancer Center at Houston and its Science Park-Research Division at Smithville, and the Texas A&M University Institute of Biosciences and Technology.

The GSBS organization is unique in that students are admitted to the graduate school itself rather than a single department or program. This allows an admitted student to join an established program of study or design an individualized multidisciplinary or interdisciplinary program and provides the flexibility so that this choice can be made before, during, or after the first year of study.

Research Facilities

The GSBS is located in the Texas Medical Center (TMC), one of the world's largest centers for biomedical education and research and patient care. In addition to training at the University of Texas institutions, GSBS students can take courses, attend seminars, and utilize certain facilities at other institutions in the TMC area, including Rice University, the University of Houston, and Baylor College of Medicine. An outstanding library serves the largest student and faculty population of any medical library in the nation, and computing facilities in the TMC institutions are world class.

Financial Aid

All Ph.D. students are supported throughout their studies by research assistantships and traineeships that include stipends ($23,000 in 2005–06), required fees, medical insurance, and tuition. More than $100,000 of supplemental scholarships and fellowships is available each year as well as teaching assistantships and internships for those who want the experience. Travel awards are available for GSBS students to attend scientific meetings. Some stipends are available for students studying for the M.S. degree.

Cost of Study

Tuition for full-time students was $946 per semester for Texas residents and approximately $3430 for nonresidents in 2005–06. Research assistants and trainees are considered state residents for the purpose of tuition assessment. Tuition and fees are provided for all full-time Ph.D. students.

Living and Housing Costs

Safe, convenient, and affordable housing, beginning at $500 per month for a single apartment, is available within a mile of the campus, and University housing is also available. Houston has the second-lowest cost of living among major American cities and the lowest cost of housing of the ten largest metropolitan areas in the country.

Student Group

Currently, 514 students are enrolled, of whom 57 percent are women. The multiethnic student population consists of 198 Texas residents, 144 from other states, and 172 from two dozen other countries.

Location

The GSBS campus is the Texas Medical Center, among the world's largest and most modern facilities for biomedical education and research. It is 2 miles from downtown Houston, adjacent to Rice University and a large, student-friendly area for shopping and dining. Houston, one of the cultural centers of the Southwest, is ethnically diverse and within 1 hour of the Gulf of Mexico.

The School

The GSBS was established in 1963 and has awarded more than 1,500 M.S. and Ph.D. degrees in the biomedical sciences since its inception. Alumni currently hold faculty positions in major universities and undergraduate colleges; pharmaceutical, biotech, and other industries; and government or private institutes. They are involved in research, teaching, scientific administration, and other activities such as patent law and scientific journalism. GSBS provides numerous career development activities in addition to scientific training.

Applying

The preferred application date for U.S. citizens seeking fall admission is January 15. Applications for the fall, spring, or summer terms must be completed two months prior to the anticipated enrollment date. All credentials from U.S. applicants should be forwarded to the Office of the Registrar. Applicants who are not U.S. citizens or permanent residents must submit the International Student Application by December 15. For more information, students should refer to the School's Web site, listed in this In-Depth Description.

Correspondence and Information

For application materials:
Office of the Registrar
The University of Texas Graduate School of Biomedical
  Sciences
P.O. Box 20036
Houston, Texas 77225-0036

Phone: 713-500-3333
E-mail: GSBS.Admissions@uth.tmc.edu
Web site: http://gsbs.uth.tmc.edu

For Graduate School information:
Admissions
The University of Texas Graduate School of Biomedical
  Sciences
P.O. Box 20334
Houston, Texas 77225-0334

Phone: 800-UTH-GSBS (884-4727; toll-free)
E-mail: GSBS.Admissions@uth.tmc.edu

The University of Texas Graduate School of Biomedical Sciences at Houston

THE FACULTY AND AREAS OF RESEARCH

The research interests of the more than 500 GSBS faculty members may be conveniently grouped into the scientific areas below. Specific information about the faculty and opportunities for study in any of these areas may be obtained from the individuals listed or from UT–Houston's GSBS home page at http://gsbs.uth.tmc.edu.

* **Biochemistry and Molecular Biology:** Dr. Michael Blackburn, Department of Biochemistry and Molecular Biology, UT–Houston Medical School, 6431 Fannin, Room 6.104, Houston, Texas 77030 (e-mail: michael.r.blackburn@uth.tmc.edu).

* **Biomathematics and Biostatistics:** Dr. Li Zhang, Department of Biostatistics and Applied Mathematics, M. D. Anderson Cancer Center, 1515 Holcombe Boulevard, Unit 447, Houston, Texas 77030 (e-mail: zhangli@odin.mdacc.tmc.edu).

Biophysics: Dr. Raymond Meyn, Department of Experimental Radiation Oncology, M. D. Anderson Cancer Center, 1515 Holcombe Boulevard, Unit 066, Houston, Texas 77030 (e-mail: rmeyn@mdanderson.org).

* **Cancer Biology:** Dr. Dihua Yu, Department of Surgical Oncology, M. D. Anderson Cancer Center, 1515 Holcombe Boulevard, Unit 107, Houston, Texas 77030 (e-mail: dyu@mdanderson.org).

* **Cell and Regulatory Biology:** Dr. Edgar T. Walters, Department of Integrative Biology and Pharmacology. UT–Houston Medical School, 6431 Fannin, Room 4.116, Houston, Texas 77030 (e-mail: edgar.t.walters@uth.tmc.edu).

* **Genes and Development:** Dr. Michelle Barton, Department of Molecular Genetics, M. D. Anderson Cancer Center, 1515 Holcombe Boulevard, Unit 1000, Houston, Texas 77030 (e-mail: mbarton@mdanderson.org).

† **Genetic Counseling:** Dr. Jacqueline Hecht, Department of Pediatrics, UT–Houston Medical School, 6431 Fannin, Room 3.136, Houston, Texas 77030 (e-mail: jacqueline.t.hecht@uth.tmc.edu).

* **Human and Molecular Genetics:** Dr. Ralf Krahe, Molecular Genetics, M. D. Anderson Cancer Center, 1515 Holcombe Boulevard, Unit 1006, Houston, Texas 77030 (e-mail: rkrahe@mdanderson.org).

* **Immunology:** Dr. Stephanie Watowich, Department of Immunology, M. D. Anderson Cancer Center, 1515 Holcombe Boulevard, Unit 902, Houston, Texas 77030 (e-mail: swatowic@mdanderson.org).

* †**Medical Physics:** Dr. Edward Jackson, Department of Imaging Physics, M. D. Anderson Cancer Center, 1515 Holcombe Boulevard, Unit 56, Houston, Texas 77030 (e-mail: ejackson@mdanderson.org).

* **Microbiology and Molecular Genetics:** Dr. Samuel Kaplan, Department of Microbiology and Molecular Genetics, UT–Houston Medical School, 6431 Fannin, Room MSB 1.200, Houston, Texas 77030 (e-mail: samuel.kaplan@uth.tmc.edu).

* **Molecular Carcinogenesis:** Dr. Ellen R. Richie, M. D. Anderson Cancer Center, Science Park–Research Division, P.O. Box 389, Smithville, Texas 78957 (e-mail: erichie@odin.mdacc.tmc.edu).

* **Molecular Pathology:** Dr. Diane Hickson-Bick, Department of Pathology and Laboratory Medicine, UT–Houston Medical School, 6431 Fannin, Room 2.288, Houston, Texas 77030 (e-mail: diane.l.bick@uth.tmc.edu).

* **Neuroscience:** Dr. M. Neal Waxham, Department of Neurobiology and Anatomy, UT–Houston Medical School, 6431 Fannin, Room 7.254, Houston, Texas 77030 (e-mail: m.n.waxham@uth.tmc.edu).

Pharmacology: Dr. Carmen Dessauer, Department of Integrative Biology and Pharmacology. UT–Houston Medical School, 6431 Fannin, Room 4.220, Houston, Texas 77030 (e-mail: carmen.w.dessauer@uth.tmc.edu).

Physiology: Dr. Edgar T. Walters, Department of Integrative Biology and Pharmacology. UT–Houston Medical School, 6431 Fannin, Room 4.116, Houston, Texas 77030 (e-mail: edgar.t.walters@uth.tmc.edu).

Radiation Biology: Dr. Raymond Meyn, Department of Experimental Radiation Oncology, M. D. Anderson Cancer Center, 1515 Holcombe Boulevard, Unit 066, Houston, Texas 77030 (e-mail: rmeyn@mdanderson.org).

Reproductive Biology: Dr. Lovell A. Jones, Gynecologic Oncology, M. D. Anderson Cancer Center, 1515 Holcombe Boulevard, Unit 304, Houston, Texas 77030 (e-mail: lajones@mdanderson.org).

* **Toxicology:** Dr. David McConkey, Department of Cancer Biology, M. D. Anderson Cancer Center, 1515 Holcombe Boulevard, Unit 173, Houston, Texas 77030 (e-mail: dmcconke@mdanderson.org).

* **Virology and Gene Therapy:** Dr. Jagannadha K. Sastry, Department of Immunology, M. D. Anderson Cancer Center, 1515 Holcombe Boulevard, Unit 901, Houston, Texas 77030 (e-mail: jsastry@mdanderson.org).

** One of the formal doctoral programs of study established by the graduate faculty.*
† Specialized M.S. degree programs.

THE UNIVERSITY OF TEXAS
SOUTHWESTERN MEDICAL CENTER AT DALLAS

Southwestern Graduate School of Biomedical Sciences
Division of Basic Science

Program of Study	The University of Texas (UT) Southwestern Medical Center at Dallas provides opportunities for students to prepare for careers in the biomedical sciences through study and research leading to the Ph.D. degree under the auspices of the Southwestern Graduate School of Biomedical Sciences. The Division of Basic Science comprises eight graduate programs in the areas of biological chemistry, cell regulation, genetics and development, immunology, integrative biology, molecular biophysics, molecular microbiology, and neuroscience. Additional areas of graduate study include cancer biology, chemistry, computational biology, and pharmacology. These graduate programs are interdisciplinary by design, and each reflects an area of research strength of the Graduate School faculty. Many members of the faculty have received recognition for their scientific contributions: 4 have been awarded the Nobel Prize, 17 are members of the National Academy of Sciences, 14 are members of the American Academy of Arts and Sciences, and 11 are members of the Howard Hughes Medical Institute. The most important element shared by the programs is an intense and exciting research experience in an active, productive, and critical scientific environment; this is the essence of graduate education at UT Southwestern. The goal of the Division of Basic Science is to provide both a broad, integrated understanding of contemporary biomedical science and in-depth training in a specific area. Students entering the Division of Basic Science may concentrate their study in any of the areas of graduate study indicated above. Specific information about courses of study and the research interests of the faculty members of individual graduate programs may be obtained by visiting the Web site. Collaboration among program faculty members in the basic science and medical departments provides students with the opportunity to contribute to research efforts in the areas of arthritis, cancer, diabetes, endocrinology, metabolism, neurobiology, nutrition, and vascular and heart disease, among others. During the first year of training, students participate in a core course that provides an integrated approach to the study of biochemistry, biophysics, molecular biology, genetics, biological regulation, cell biology, and organismal biology, appropriate for students with interests in any area of study. Course work is supplemented by a rich schedule of seminars offered on wide-ranging topics. A substantial benefit of this approach is to prepare students for the increasingly interdisciplinary nature of biomedical science. This eight-month course is taught by faculty members from all of the graduate programs. The standard first-year curriculum also includes completion of three 8- to 10-week laboratory rotations and advanced course work. A student may select any of the 242 members of the Division faculty as a preceptor for a research rotation. Formal selection of a specific graduate program is postponed until the end of the first academic year in order to foster flexibility and allow an informed choice of the most appropriate program for advanced study; however, students may associate informally with a program during the first year to the extent desired. During the second year of study, students complete advanced and sharply focused parts of the curriculum, which are organized by the faculty members of the individual graduate programs, and focus on developing a research project. A coordinated design of advanced specialized course work is another characteristic of the Division of Basic Science's integrative approach to graduate education. Advanced courses include those with pan-program appeal, as well as a variety of more specialized courses.
Research Facilities	State-of-the-art equipment and facilities required for modern biomedical research are available. A central computing facility, animal facilities, and a comprehensive library are available. There are core research facilities and services for DNA microarray analysis, molecular and cellular imaging, transgenic animals, flow cytometry, structural biology, protein chemistry, mass spectrometry, rapid biochemical kinetics, analytical ultracentrifugation, DNA sequencing, and antibody production.
Financial Aid	The Division provides every matriculated student with a research assistantship. Currently, students receive a nationally competitive stipend ($23,000 for fall 2006) plus a liberal health insurance allowance. Tuition is paid for all full-time Division of Basic Science students. There are employment opportunities in the medical and business communities for spouses.
Cost of Study	In 2005–06, tuition and fees were $3451 for the full year. All Division of Basic Science graduate students are reimbursed for tuition, health insurance, and student service fees. Student parking is currently $70 per year.
Living and Housing Costs	University student housing (one- and two-bedroom apartments) is available to incoming graduate students. In addition, students find attractive accommodations in the many privately owned apartment complexes, duplexes, and houses nearby that may be rented at reasonable rates. In 2005–06, the cost of an average one-bedroom apartment was approximately $700 per month.
Student Group	In fall 2005, 2,367 students were enrolled at the University of Texas Southwestern Medical Center. Students enrolled in the Graduate School numbered 1,155. Also on the campus were 904 medical students and 308 students in the Southwestern Allied Health Sciences School. A very active Graduate Student Organization is present on campus, and an International Visitors Network exists to assist students from other countries. There are 704 graduate students and 451 postdoctoral fellows currently in training with the 242 faculty members of the Division of Basic Science.
Location	Dallas provides an interesting contrast of cosmopolitan sophistication and easy access to rural and wilderness retreats. The city's many cultural offerings include art galleries, symphony concerts, the ballet, theaters, and opera. There are several museums in Dallas and in nearby Fort Worth. Professional football, baseball, basketball, and hockey can be enjoyed nearby. Dallas has a mild climate that is suitable for many outdoor activities. Area parks, lakes, and rivers provide numerous recreational facilities.
The University and The School	The University includes Southwestern Graduate School of Biomedical Sciences, Southwestern Medical School, and the Southwestern Allied Health Sciences School. A Howard Hughes Medical Institute facility is on campus. Graduate programs were initially established as part of Southwestern Medical School's teaching program in the basic sciences, with the first graduate degree being awarded in 1947. In 1972, establishment of the graduate school under the name Graduate School of Biomedical Sciences was authorized. Consistently among the top ten medical schools in the country as measured by NIH research grant support to basic science faculty members, UT Southwestern provides a rich scholarly environment for Ph.D. training. In 2006, *U.S. News & World Report* ranked UT Southwestern nineteenth in the nation among institutions that grant Ph.D. degrees in the biological sciences. More recently, *Science Watch,* which ranks research institutions in six specific scientific disciplines, ranked UT Southwestern in the top ten in the country in four of the six disciplines.
Applying	Applicants must submit an online application at http://www.utsouthwestern.edu/dbsapp, including a description of prior research experiences and a statement that indicates educational and professional goals. Applicants must send GRE General Test scores, transcripts of undergraduate and graduate work, and a minimum of three letters of recommendation. International students whose native language is not English must also submit TOEFL scores. Most successful applicants have had one or more significant research experiences prior to applying. Students matriculate in the fall of each year; there is no spring or summer matriculation.

Correspondence and Information	Nancy E. Street, Ph.D.
	Associate Dean
	Division of Basic Science
	Southwestern Graduate School of Biomedical Sciences
	5323 Harry Hines Boulevard
	Dallas, Texas 75390-9004
	Phone: 214-648-6708
	Fax: 214-648-2102
	E-mail: nancy.street@utsouthwestern.edu
	    dbsinfo@utsouthwestern.edu
	Web site: http://www.utsouthwestern.edu/gradschool
		Nancy McKinney
		Educational Coordinator
		Division of Basic Science
		Southwestern Graduate School of Biomedical Sciences
		5323 Harry Hines Boulevard
		Dallas, Texas 75390-9042
		Phone: 214-648-8099
		    800 833-6134 (toll-free)
		Fax: 214-648-2978
		E-mail: nancy.mckinney@utsouthwestern.edu
		Web site: http://www.utsouthwestern.edu/gradschool

The University of Texas Southwestern Medical Center at Dallas

OFFICERS AND PROGRAM HEADS

Melanie H. Cobb, Ph.D., Dean, Southwestern Graduate School of Biomedical Sciences. Telephone: 214-648-3365.
Michael G. Roth, Ph.D., Associate Dean, Division of Basic Science. Telephone: 214-648-3276.
Dennis M. McKearin, Ph.D., Associate Dean, Medical Scientist Training Program. Telephone: 214-648-6764.
Nancy E. Street, Ph.D., Associate Dean, Division of Basic Science. Telephone: 214-648-6708.
Stuart E. Ravnik, Ph.D., Assistant Dean, Division of Basic Science. Telephone: 214-648-6791.
Robert Rawson, Ph.D., Assistant Dean, Division of Basic Science. Telephone: 214-648-6663.

Division of Basic Science
Nancy E. McKinney, M.S., Educational Coordinator. Telephone: 214-648-8099.

Biological Chemistry Graduate Program
Meg Phillips, Ph.D., Chair. Telephone: 214-645-6165.

Cell Regulation Graduate Program
Paul Sternweis, Ph.D., Chair. Telephone: 214-645-6129.

Genetics and Development Graduate Program
John M. Abrams, Ph.D., Chair. Telephone: 214-648-2224.

Immunology Graduate Program
Nicolai Van Oers, Ph.D., Chair. Telephone: 214-648-1899.

Integrative Biology Graduate Program
Yi Liu, Ph.D., Chair. Telephone: 214-645-6021.

Molecular Biophysics Graduate Program
Kevin H. Gardner, Ph.D., Chair. Telephone: 214-645-6365.

Molecular Microbiology Graduate Program
Michael Gale, M.D., Chair. Telephone: 214-648-1899.

Neuroscience Graduate Program
Jane Johnson, Ph.D., Chair. Telephone: 214-648-1802.

THE UNIVERSITY OF TEXAS
SOUTHWESTERN MEDICAL CENTER AT DALLAS
Medical Scientist Training Program

Program of Study

The Medical Scientist Training Program (MSTP) at the University of Texas (UT) Southwestern Medical Center is intended to prepare a select group of individuals with interests in basic biomedical science and medicine for careers as physician-scientists. The UT Southwestern MSTP is directed by Nobel laureate Dr. Michael Brown and Dr. Robert Munford and is one of a group of M.D./Ph.D. programs supported by the National Institute of General Medical Sciences of the NIH. The program (http://www.utsouthwestern.edu/mstp) is designed to provide training leading to both the M.D. and Ph.D. degrees through highly integrated tracks that offer the complete breadth of medical school and graduate school experiences. The course of study begins with the first two years of medical school and includes laboratory research rotations during the summer months, followed by full-time laboratory research leading to the Ph.D. degree (usually three to four years), and concludes with the final two years of medical school clinical clerkships. Integrated into these two tracks of study are activities designed specifically for the MSTP student. These include a comprehensive weekly basic science seminar series, a student journal club, an annual MSTP retreat, selected internal medicine grand rounds and post-rounds discussion of the underlying basic science of a clinical problem, and numerous social/professional activities. Integral to the program is a system of continuous counseling and guidance, with special emphasis on selecting a research mentor, and a series of personalized tutoring sessions in the clinic for students returning to medical school after completion of their Ph.D. dissertations.

MSTP fellows earn the Ph.D. degree through one of eight graduate programs in the Division of Basic Science: biological chemistry, cell regulation, genetics and development, immunology, integrative biology, molecular biophysics, molecular microbiology, and neuroscience (http://www8.utsouthwestern.edu/utsw/home/education/dcmb/index.html). These programs are organized around the research interests of more than 200 selected faculty members of the division. Participating faculty members are, in many instances, members of more than one program, which offers students ample flexibility in selecting a graduate program of study.

Research Facilities

UT Southwestern Medical Center is one of the nation's elite biomedical research institutions and contains more than 5 million square feet of space dedicated to teaching and research. The facilities include the Howard Hughes Medical Institute, a number of research centers dedicated to the study of specific human disorders, and fully equipped laboratories for conducting state-of-the-art biomedical research. The clinical facilities at the Medical Center's 60-acre campus include the UT Southwestern Hospitals and Parkland Memorial Hospital, one of the busiest public hospitals in the country, and Children's Medical Center of Dallas. Among the other clinical training facilities are Presbyterian Hospital and the Dallas Veterans Affairs Medical Center.

Financial Aid

All students enrolled in the MSTP at UT Southwestern receive full fellowships throughout their training. The fellowships include full tuition coverage in both medical and graduate school and a stipend of $22,500 per year for all years in the program. Fellowships also include a comprehensive health insurance plan.

Cost of Study

Tuition costs for all students are covered by their fellowships.

Living and Housing Costs

New housing that is dedicated to students and staff is conveniently located adjacent to the Medical Center. Rent is competitive with other housing in the Dallas area. Additional housing is available in nearby neighborhoods or other communities in Dallas and its suburbs. MSTP students benefit from favorable cost-of-living circumstances by renting nearby apartments or purchasing condominiums or houses in the Dallas area.

Student Group

Current fellows come from all areas of the country, with the majority from the East and West Coasts and Texas. On average, the program enrolls 10 to 15 students each year. There are currently 103 students in the MSTP. Approximately 700 other graduate students at the Medical Center are currently pursuing Ph.D. degrees in the basic biomedical sciences, while the medical school enrolls 884 students. The total population of the Medical Center, including the School of Allied Health Sciences, is about 2,300 students.

Student Outcomes

The earliest graduates of the program are now in faculty positions, usually in clinical departments, at the major biomedical research institutions around the country, while the more recent graduates are in residency or fellowship training. Those individuals in faculty positions spend the majority of their time in research activities. In keeping with the goal of training future academic physician-scientists, almost all MSTP graduates go on to clinical residency training programs in such disciplines as internal medicine, pathology, neurology, psychiatry, dermatology, and pediatrics.

Location

Dallas provides an interesting contrast of urban sophistication and easy access to rural retreats. The city's cultural offerings include art galleries, symphony and jazz concerts, ballet, theater, opera, literary events, and outdoor festivals throughout the year. There are four major art museums in Dallas and Fort Worth, attracting world-renowned exhibits. Professional baseball, basketball, football, and hockey also are available in the metropolitan area.

The University

Established in 1943, UT Southwestern is a young institution that continues to experience dramatic growth in its facilities and the number of outstanding faculty members and students. It is home to 4 Nobel laureates, 17 members of the National Academy of Sciences, and 11 investigators of the Howard Hughes Medical Institute. In addition, the UT Southwestern faculty includes 17 members of the Institute of Medicine and a number of primary editors of several of the most widely used medical textbooks in the world. Consistently placing among the elite medical schools in the country, as measured by NIH research grant support to basic science faculty members and quality of clinical training, the institution provides a rich scholarly environment for M.D./Ph.D. training.

Applying

There are no state residency requirements for admission to the MSTP. International students are also welcome and are supported at the same levels as U.S. resident students. Application to the MSTP is initiated by applying to the medical school through the American Medical College Application Service (AMCAS) system. Interested students should call UT Southwestern's toll-free number or visit its Web site, both listed in this Close-Up. MCAT scores are required, but GRE scores are optional. The application deadline is November 1 for admission for the following August.

Correspondence and Information

Dennis M. McKearin, Ph.D.
Associate Dean
Medical Scientist Training Program
The University of Texas Southwestern Medical Center at Dallas
5323 Harry Hines Boulevard
Dallas, Texas 75390-9033
Phone: 800-633-6787 (toll-free)
Fax: 214-648-2814
E-mail: robin.downing@utsouthwestern.edu
Web site: http://www.utsouthwestern.edu/mstp

The University of Texas Southwestern Medical Center at Dallas

THE FACULTY AND THEIR RESEARCH

More than 230 faculty members of the Division of Basic Science participate in the MSTP. Below is a partial list of faculty members who have recently served or are currently serving as mentors to MSTP fellows.

Leon Avery. Genetics of feeding behavior in *Caenorhabditis elegans*.

Michael Bennett. Genetics and biology of bone marrow transplants, with emphasis on natural killer (NK) cells.

James Bibb. Signal transduction pathways underlying neuronal plasticity.

Michael Brown. Mechanisms by which cells coordinate synthesis and disposition of mevalonate; purification, characterization, and cloning cDNAs for the a- and b-subunits of protein farnesyltransferase.

Steve Cannon. How ion channels regulate electrical excitability of cells; how defects in these channels lead to human disease.

Zhijian (James) Chen. Understanding the mechanisms and pathways of ubiquitin signaling, especially in regard to immunity and cancer.

Melanie Cobb. Definition of roles of MAP kinase pathways in glucose-dependent insulin secretion from beta cells.

Johann Deisenhofer. Three-dimensional protein structures by X-ray crystallography.

Michael Gale. Molecular virology of hepatitis C virus.

David Garbers. Molecular basis of fertilization, nuclear reprogramming by the egg, and biology of the male germ stem cell.

J. Victor Garcia-Martinez. Genetics of HIV pathogenesis.

Kevin Gardner. Biophysical, chemical, and biochemical investigations of sensory protein modules.

Dan Garry. Transcriptional regulation of stem cell populations, especially in heart and skeletal muscle lineages.

Alfred Gilman. G proteins; adenylyl cyclase.

Joseph Goldstein. Mechanisms by which cells coordinate synthesis and disposition of mevalonate; purification, characterization, and cloning cDNAs for the a- and b-subunits of protein farnesyltransferase.

Jonathan Graff. Cell signaling mechanisms dictating patterning in the development of vertebrates.

Stephen Hammes. Transcription-independent, steroid-mediated signaling; maturation of mammalian oocytes; ovarian pathology associated with states of androgen excess, such as polycystic ovarian syndrome.

Eric Hansen. Investigation of bacterial virulence mechanisms, using DNA microarray and molecular genetic methods.

Patrick Harran. Synthetic organic chemistry and chemical biology.

Mark Henkemeyer. Eph receptors and Ephrin ligands as signaling molecules controlling cell migration and adhesion.

Joachim Herz. Mechanisms of signaling by lipoprotein receptors and its implications for vascular biology, brain development, and Alzheimer's disease.

Joseph Hill. Molecular signaling processes in cardiac hypertrophy and failure.

Helen Hobbs. Genetic determinants of plasma lipid levels; LDL metabolism; role of ABC transporters in lipid transport.

Kimberly Huber. Cellular mechanisms of synaptic plasticity; metabotropic glutamate receptors and dendritic protein synthesis; how plasticity is altered in mouse models of mental retardation.

Jane Johnson. Transcriptional control of neurogenesis and neuronal specification in the spinal cord and brain.

Nitin Karandikar. Role of CD8+ and CD4+ T cells in the pathogenesis and regulation of immune-mediated disease, especially multiple sclerosis, hepatitis C infection, and graft-versus-host disease.

Thomas Kodadek. Enzymology and regulation of eukaryotic gene expression; chemical biology and proteomics.

Michael Kyba. Factors involved in self-renewal and differentiation of stem cells; mouse models of therapy for degenerative diseases.

Kristen Lynch. mRNA splicing regulation of protein expression and cellular function upon antigen stimulation of T cells.

David Mangelsdorf. Translation of extracellular signals into transcriptional responses.

Dennis McKearin. Use of molecular genetics to study stem cell biology during *Drosophila* oogenesis.

Steven McKnight. Regulation of gene expression: identification and purification of transcription factors, biochemical and biophysical studies of protein assemblies that control gene expression, intracellular signaling pathways relating to cytokine function, regulation of cell proliferation, and terminal cell differentiation.

Robert Munford. Molecular signal information in lipopolysaccharide; developing new methods for preventing severe inflammation with gene transfer technology.

Eric Nestler. Gene expression and addictive behaviors.

Eric Olson. Genetic control of morphogenesis.

Kim Orth. Biochemistry of infectious disease: analysis of how bacterial virulence factors manipulate host signaling machinery during infection.

Luis Parada. Elucidation of regulatory pathways that control the complex process of vertebrate development.

Keith Parker. Development and function of the adrenal glands, ovaries, and testes, focusing on roles of the orphan nuclear receptor steroidogenic factor 1.

Meg Phillips. Structural basis of enzyme function.

Rama Ranganathan. Molecular structure and biological activity.

Joyce Repa. Effects of diet and drugs on nuclear receptor regulation and relationship to atherosclerosis and type 2 diabetes.

Michael Rosen. Understanding structural, biochemical, and cell biological mechanisms that regulate the actin cytoskeleton.

David Russell. Genetics and physiological roles of steroid hormone metabolizing enzymes and bile acid biosynthetic enzymes.

Jerry Shay. Role of telomeres and telomerase in aging and cancer.

Stephen Sprang. Study of signal transduction mechanisms by crystallographic methods and other physical techniques.

Thomas Sudhof. How presynaptic neurons recognize a postsynaptic cell to form a synapse; how synapses are organized via cell-adhesion molecules; how neurotransmitters are released at a synapse.

Michelle Tallquist. Use of mouse molecular genetics to unravel signal transduction pathways during embryonic development.

Ellen Vitetta. Use of immunotoxins to inactivate cancer cells and HIV-infected cells.

Ward Wakeland. Genetic basis of lupus erythematosus.

Xiaodong Wang. Molecular mechanisms of programmed cell death.

Masashi Yanagisawa. Vascular endothelial cells.

Helen Yin. Role of phosphoinositides in membrane transport, cell motility, and signaling; phosphoinositide-regulated pathways as potential targets for therapeutic intervention.

Hongtao Yu. Spindle checkpoint function, especially how the spindle checkpoint signals to inhibit APC in response to spindle damage.

THE UNIVERSITY OF TOLEDO

Department of Biological Sciences

Programs of Study	The Department of Biological Sciences offers the Master of Science (M.S.) and the Ph.D. in biology, as well as a dual degree, the Master of Science and the Master of Education (M.S./M.Ed.). Study leading to a Master of Science usually takes between 2 and 2½ years to complete. The first year involves identifying a professor in the major, taking course work, and beginning the thesis research project. In the second year, the student focuses on completion of the research project. For the dual degree of Master of Science and Master of Education, students must meet requirements for the degree as stated by the College of Education. Students doing their thesis in biology rather than education must fulfill the same thesis-related requirements as other biology M.S. candidates. The doctoral degree in biology in the cell-molecular track is awarded to students who demonstrate proficiency in the discipline and display the potential to make substantial contributions to the field. It is not awarded merely as a result of courses taken or for years spent in studying or research. The quality of work and the resourcefulness of the student must be such that the faculty mentor can expect a continuing effort toward the advancement of knowledge and significant achievement in research and related activities. In general, work for the Ph.D. typically takes five years and 90 semester hours of study beyond the bachelor's degree. A substantial portion of this time is spent in independent research leading to a dissertation. Work done toward a master's degree may apply as part of the student's doctoral program. Depending on their research focus, doctoral students complete an individualized program of course work in advanced cell and molecular biology and biochemistry, with participation in the departmental seminar series. Ph.D. students must pass a written qualifying examination during the first two years of study and an oral comprehensive examination involving a defense of their research proposal after gaining admission to candidacy. Completion of the degree requires an oral defense of the thesis (M.S.) or dissertation (Ph.D.). The Department of Biological Sciences considers experience in teaching to be a vital and significant component of graduate education. Thus, all graduate students are required to complete at least one semester of formal teaching experience prior to graduation.
Research Facilities	The Department of Biological Sciences has excellent research and support facilities. The new $33-million Wolfe Hall building houses state-of-the-art research laboratories for the Departments of Biological Sciences and Chemistry and the College of Pharmacy. Students have access to outstanding instrumentation facilities in the Wolfe Hall-Bowman-Oddy complex, including the Arts and Sciences Instrumentation Center and the new state-of-the-art Center for Molecular Biology, with gene-chip microarray capability and confocal microscopy. These facilities parallel or exceed in quality those of the finest graduate institutions in the country.
Financial Aid	Most full-time graduate students receive some financial support. College fellowships, teaching assistantships, and research assistantships, which include a stipend and a tuition waiver, are available for qualified students on a competitive basis. The out-of-state tuition surcharge normally charged to out-of-state and international students is waived for students whose permanent address is within one of the following Michigan counties: Hillsdale, Lenawee, Macomb, Oakland, Washtenaw, and Wayne. In addition, The University of Toledo offers an out-of-state tuition surcharge waiver to cities and regions that are a part of the Sister Cities Agreement. These regions include Toledo, Spain; Londrina, Brazil; Qinhuangdao, China; Csongrad County, Hungary; Delmenhorst, Germany; Toyohashi, Japan; Tanga, Tanzania; Bekaa Valley, Lebanon; and Poznan, Poland. The University of Toledo Graduate College offers a variety of memorial and minority scholarship awards, including the Ronald E. McNair Postbaccalaureate Achievement Scholarship, the Graduate Minority Assistantship Award, and two full University fellowships.
Cost of Study	The graduate tuition rate for the 2006–07 academic year is $390.05 per semester credit hour for in-state students. For nonresidents, the out-of-state surcharge is $367.15 per semester credit hour. Additional fees are required and include the general fee, technology fee, and mandatory insurance.
Living and Housing Costs	The University of Toledo has a diverse offering of student housing options, including suite-style and traditional residential halls. Housing is offered to graduate students through Residence Life or contracted individually by the student. Affordable, high-quality off-campus apartment-style housing within walking distance of campus is abundant.
Student Group	There are approximately 20,000 students at the University of Toledo. About 4,000 are graduate and professional students. The University has a rich diversity of student organizations. Students join groups that are organized around common cultural, religious, athletic, and educational interests.
Student Outcomes	Master's and doctoral students go on to excellent postdoctoral placements, academic positions, or careers in industry or government. Recent M.S. graduates have landed research positions in industry at companies such as Pfizer, Procter & Gamble, Pharmacia, Abbott Labs, and GlaxoSmithKline. Recent Ph.D. graduates have accepted outstanding postdoctoral placements at institutions such as the NIH, Berkeley, Duke, Johns Hopkins, Harvard, Michigan, Northwestern, and Cold Spring Harbor.
Location	The University of Toledo has several campus sites in the city of Toledo. Most engineering graduate students take classes on the Main campus, which is located in suburban western Toledo. With a population of more than 330,000, Toledo is the fiftieth-largest city in the United States. It is located on the western shores of Lake Erie, within a 2-hour drive of Cleveland and Detroit.
The University	The University of Toledo was founded by Jessup W. Scott in 1872 as a municipal institution and became part of the state of Ohio's system of higher education in 1967. On July 1, 2006, The University of Toledo merged with the Medical University of Ohio becoming one of only seventeen American universities to offer professional and graduate academic programs in medicine, law, pharmacy, nursing, health sciences, engineering, and business.
Applying	Students entering the program are expected to have a strong background in biology with related areas in biochemistry but may be admitted on a provisional basis if they lack such a background. Successful applicants work with nationally and internationally recognized faculty members on research foci in cell/molecular biology. A bachelor's or professional degree earned from a department of approved standing and granted by an accredited college or university is required. All applicants with less than a 2.7 cumulative grade point average on all undergraduate work are required to forward results of the Graduate Record Examinations Aptitude and/or other appropriate qualifying examinations as specified by the department. On the basis of the results of these examinations, the department makes a recommendation to the College of Graduate Studies. In addition to the requirements for regular admission, all students from non-English-speaking countries must achieve satisfactory scores on the Test of English as a Foreign Language and the Graduate Record Examinations. All international students must also demonstrate that they have adequate financial resources for their graduate education if not awarded an assistantship. Students may wish to submit online applications through the College of Graduate Studies.
Correspondence and Information	Department of Biological Sciences Mail Stop 601 University of Toledo 2801 West Bancroft Street Toledo, Ohio 43606-3390 Phone: 419-530-2065 Fax: 419-530-7737 E-mail: doug.leaman@utoledo.edu Web site: http://www.biosciences.utoledo.edu

The University of Toledo

THE FACULTY AND THEIR RESEARCH

Brian Ashburner, Assistant Professor; Ph.D., Loyola Chicago, 1995. Transcription regulation.
Deborah Chadee, Assistant Professor; Ph.D., Manitoba. 1999. Signal transduction.
Charles Creutz, Associate Professor; Ph.D., Pennsylvania, 1971. Biological chemistry; physics; immunology.
Fan Dong, Associate Professor; M.D., Ph.D., Suzhou Medical College, 1988. Cytokine receptor signaling.
Ernest Dubrul, Associate Professor; Ph.D., Washington (St. Louis), 1969. Science education.
Emilio Duran, Assistant Professor; Ph.D., Toledo, 1996. Science education.
John Gray, Associate Professor; Ph.D., Purdue, 1992. Plant molecular biology.
Patricia Komuniecki, Department Chair; Ph.D., Massachusetts, 1977. Developmental regulation in parasitic helminths.
Richard Komuniecki, Distinguished University Professor; Ph.D., Massachusetts, 1976. Biochemistry/molecular biology of parasites.
Douglas Leaman, Associate Professor; Ph.D., Missouri, 1993. Molecular biology of cytokines.
Scott Leisner, Associate Professor; Ph.D., Purdue, 1989. Cell/molecular biology/plant virology.
John Plenefisch, Associate Professor; Ph.D., MIT, 1990. Developmental genetics; cell interactions.
Donald Pribor, Professor; Ph.D., Catholic University, 1964. Creativity and dialogue collaboration.
Anthony Quinn, Associate Professor; Ph.D., Oklahoma Health Sciences Center, 1994. Autoimmune disease/T-cell regulation.
Lirim Shemshedini, Associate Professor; Ph.D., Vermont, 1989. Cell/molecular biology/transcription regulation.
William Taylor, Assistant Professor; Ph.D., Manitoba, 1994. Mechanisms controlling the cell cycle.
Deborah Vestal, Associate Professor; Ph.D., Syracuse, 1988. Molecular/cell biology of cytokines.

THE UNIVERSITY OF TOLEDO

Health Science Campus
Graduate Program in Biomedical Sciences

Programs of Study

The biomedical sciences program at The University of Toledo offers the Ph.D. degree for graduate studies in three interdisciplinary areas: cellular and molecular neurobiology (CMN), molecular and cellular biology (MCB), and molecular basis of disease (MBD). The program normally requires four to five years. The primary goal of the doctoral program is to train students for independent, creative careers in research and/or teaching. Students initially follow an integrated core curriculum that consists of molecular biology, cell and developmental biology, and genetics. This program is augmented by seminars and laboratory experience. Research rotations provide an opportunity for the students to gain exposure to a thriving research community and to identify a major adviser for their thesis research. For more information about the program, students should visit the University's Web site.

The CMN program encompasses areas of graduate study specializing in furthering understanding of the nervous system. Faculty members are associated with a range of departments and divisions, including biochemistry and cancer biology, neurology, neurosciences, neurosurgery, pharmacology, psychiatry, public health, and radiology. Ongoing research at The University of Toledo's Health Science Campus is providing insights into nervous system function, development and plasticity, regeneration and repair following damage, regulation of ion-channel structure and function, the basis of neural disease, and behavior. Specialized courses are offered in cellular and systems neuroscience. Research rotations provide students exposure to a broad range of research questions, experimental techniques, and state-of-the-art equipment. The program is flexible; it allows students to customize their program of study to their research interests and goals. For more information about the program, students should contact Dr. Elizabeth Tietz (phone: 419-383-4182; fax: 419-383-2871; e-mail: etietz@meduohio.edu).

The MCB program prepares students for careers as productive scientists in university-based research laboratories or pharmaceutical/biotechnology companies. The program aims to provide a collegial and supportive environment in which students can interact closely with faculty members to achieve their individual goals. Graduate training in MCB focuses on the development of intellectual and technical skills that are needed to make new discoveries related to fundamental biological processes, including the regulation of gene expression, control of cell growth and metabolism, programmed cell death, differentiation of cells and tissues, and responses of cells to hormones, drugs, and environmental agents. Through a combination of formal courses and supervised research projects in well-equipped laboratories, students are taught to apply a wide range of modern experimental approaches to test hypotheses and acquire new information. The research strategies may cross several disciplines, including genetics and molecular biology, protein chemistry and enzymology, cell biology and tissue culture, microscopic analysis, and the emerging areas of genomics, proteomics, and bioinformatics. Faculty members participating in the MCB program are based in several different departments, and their research interests cover a broad range of topics of medical significance, including cancer biology, gene therapy, hormone and neurotransmitter receptor signaling mechanisms, and the molecular biology and genetics of hypertension and heart disease. Students are encouraged to sample several research opportunities before selecting a faculty mentor for their major research project. For more information about the program, students should contact Dr. Dorothea Sawicki via e-mail at dsawicki@meduohio.edu.

The MBD program provides students with the necessary tools to pursue a career investigating the basis of disease processes. The program encompasses a unique interdisciplinary approach to elucidate the underlying molecular mechanisms of diseases that have profound impact on human health, including cancer; diabetes; infertility; immunological, microbial, bacterial, and genetic diseases; and many others. The program draws on the faculty research strengths in signal transduction, genetics, molecular biology, gene microarrays, genomics, gene knockout and transgenics, proteomics, tissue culture, cell biology, and protein and carbohydrate biochemistry. Students are trained using an integrated approach. Both the M.S. and the Ph.D. degree are offered, individually and in combination with an M.D. degree. For more information, students should contact Dr. Sonia M. Najjar, Director of the Molecular Basis of Disease Ph.D. Program (phone: 419-383-5196; fax: 419-383-2871; e-mail: snajjar@meduohio.edu).

The M.S. degree in biomedical sciences (including physician assistant studies and physical therapy) and in occupational health, Master of Science in Nursing, a doctoral degree in occupational therapy, Master of Public Health, and combined M.D./Ph.D., M.D./M.S., and M.D./M.P.H. degrees are also offered.

Research Facilities

The University's modern laboratories are equipped for state-of-the-art investigations in all areas of biomedical research. Specialized facilities for tissue culture, flow cytometry, imaging, and microscopy (confocal, light, and electron) are available. Core laboratories for microarray and proteomics analysis opened in 2002, and a bioinformatics computer facility opened in 2003. Institutional resources include animal research facilities with P3 biocontainment rooms, a computer center, and an outstanding library that participates in the OhioLINK statewide online library system.

Financial Aid

Most full-time graduate students receive some financial support. College fellowships, teaching assistantships, and research assistantships, which include a stipend and a tuition waiver, are available for qualified students on a competitive basis. The out-of-state tuition surcharge normally charged to out-of-state and international students is waived for students whose permanent address is within one of the following Michigan counties: Hillsdale, Lenawee, Macomb, Oakland, Washtenaw, and Wayne. In addition, the University of Toledo offers an out-of-state tuition surcharge waiver to cities and regions that are a part of the Sister Cities Agreement. These regions include Toledo, Spain; Londrina, Brazil; Qinhuangdao, China; Csongrad County, Hungary; Delmenhorst, Germany; Toyohashi, Japan; Tanga, Tanzania; Bekaa Valley, Lebanon; and Poznan, Poland. The University of Toledo Graduate College offers a variety of memorial and minority scholarship awards, including the Ronald E. McNair Postbaccalaureate Achievement Scholarship, the Graduate Minority Assistantship Award, and two full University fellowships.

Cost of Study

The graduate tuition rate for the 2006–07 academic year is $390.05 per semester credit hour for in-state students. For nonresidents, the out-of-state surcharge is $367.15 per semester credit hour. Additional fees are required and include the general fee, technology fee, and mandatory insurance.

Living and Housing Costs

The University of Toledo has a diverse offering of student housing options, including suite-style and traditional residential halls. Housing is offered to graduate students through Residence Life or contracted individually by the student. Affordable, high-quality off-campus apartment-style housing within walking distance of campus is abundant.

Student Group

There are approximately 20,000 students at the University of Toledo. About 4,000 are graduate and professional students. The University has a rich diversity of student organizations. Students join groups that are organized around common cultural, religious, athletic, and educational interests.

Location

The University of Toledo has several campus sites in the city of Toledo. The Main campus is located in suburban western Toledo and the Health Science Campus is located in southwest Toledo. With a population of more than 330,000, Toledo is the fiftieth-largest city in the United States. It is located on the western shores of Lake Erie, within a 2-hour drive of Cleveland and Detroit.

The University

The University of Toledo was founded by Jessup W. Scott in 1872 as a municipal institution and became part of the state of Ohio's system of higher education in 1967. On July 1, 2006, the University of Toledo merged with the Medical University of Ohio becoming one of only seventeen American universities to offer professional and graduate academic programs in medicine, law, pharmacy, nursing, health sciences, engineering, and business.

Applying

Applicants should have at least a baccalaureate degree by the time of enrollment, with a strong academic background in the biological, chemical, and physical sciences. Application for admission may be completed on the University's Web site and should be submitted along with an application fee of $30, a personal statement, and official transcripts from each institution attended. Scores on the GRE General Test and a minimum of three letters of recommendation are also required for some programs. TOEFL scores are required for international students. For the combined degree programs, students should submit an application to both the College of Graduate Studies and the College of Medicine.

Correspondence and Information

Admissions Office
College of Graduate Studies
The University of Toledo Health Science Campus
3045 Arlington Avenue
Toledo, Ohio 43614-5805

Phone: 419-383-4107
E-mail: graduatestudies@meduohio.edu
Web site: http://hsc.utoledo.edu/grad/index.html

The University of Toledo

THE FACULTY AND THEIR RESEARCH

David Allison, M.D., Michigan, 1967; Ph.D., Chicago, 1976; Surgery (MCB). Aneuploidy and cancer.

Amir Askari, Ph.D., Cornell, 1960; Pharmacology (MCB). Regulation of cardiac NaK-ATPase.

Venkatesha Basrur, Ph.D., Indian Institute of Science (Bangalore), 1998; Medical Microbiology and Immunology (MBD). Bioinformatics, proteomics, and genomics.

Andrew Beavis, Ph.D., Bristol (England), 1978; Pharmacology (MCB). Bioenergetics and transport.

Carol A. Bennett-Clarke, Ph.D., Rochester, 1980; Neurosciences (CMN). Role of serotonin in sensory cortical development.

Michael S. Bisesi, Ph.D., SUNY College of Environmental Science and Forestry, 1987; Public Health (CMN). Neurotoxicology.

Robert M. Blumenthal, Ph.D., Michigan, 1977; Medical Microbiology and Immunology (MBD). Gene regulation and DNA methylation.

Paul H. Brand, Ph.D., Rochester, 1972; Physiology and Cardiovascular Genomics (MBD). Regulation of arterial blood pressure and body fluid volumes.

Donald P. Braun, Ph.D., Illinois Medical Center, 1976; Surgery (MCB/MBD). Cancer immunology; cellular immunology.

Joana Chakraborty, Ph.D., Institute for Nuclear Physics (Calcutta), 1962; Physiology and Cardiovascular Genomics (MCB). Reproductive physiology; HIV/AIDS.

Nicholas L. Chiaia, Ph.D., Kent State, 1985; Neurosciences (CMN). Development and injury-induced reorganization of somatosensory system.

George Cicila, Ph.D., Pennsylvania, 1986; Physiology and Cardiovascular Genomics (MCB/MBD). Genetics of high blood pressure and other cardiovascular traits.

Robert S. Crissman, Ph.D., North Dakota, 1973; Neurosciences (CMN). Ultrastructure and physiology of developing trigeminal system.

Keith A. Crist, Ph.D., California, Davis, 1983; Surgery (MBD). Chemoprevention and cell-cycle regulation.

Howard Dene, Ph.D., Wayne State, 1976; Physiology and Cardiovascular Genomics (MCB). Genetics of hypertension.

Michael J. Dennis, Ph.D., Texas Health Science Center at San Antonio, 1979; Radiology (CMN). Diagnostic imaging methods; functional MRI.

J. David Dignam, Ph.D., Texas, 1977; Biochemistry and Cancer Biology (MCB). Structure-function of aminoacyl-tRNA synthetases; AAV rep proteins.

Han-Fei Ding, M.D., Anhui Medical College (China), 1982, Ph.D., South Alabama, 1993; Biochemistry and Cancer Biology (MCB/MBD). Biochemical dissection of p53-dependent apoptosis.

Linda A. Dokas, Ph.D., Michigan, 1973; Neurology (CMN). Regulation of neuronal signal transduction in aging and Alzheimer's disease.

Lee E. Faber, Ph.D., Indiana, 1970; Physiology and Cardiovascular Genomics (MCB). Endocrinology and gene regulation.

Alexei Fedorov, Ph.D., Russian Academy of Sciences, 1993; Medicine (MCB). Bioinformatics.

David R. Giovannucci, Ph.D., Wayne State, 1993; Neurosciences (CMN/MCB). Regulation of neurotransmitter secretion and calcium signaling.

Donald A. Godfrey, Ph.D., Harvard, 1972; Surgery (CMN). Neurotransmission in auditory and vestibular systems.

L. John Greenfield Jr., M.D., 1988, Ph.D, 1989, Virginia; Neurology (CMN). Epilepsy; electrophysiology of ion channels.

William T. Gunning, Ph.D., Medical College of Ohio, 1991; Biochemistry and Cancer Biology (MBD). Experimental carcinogenesis and chemoprevention in murine models.

Mark H. Hankin, Ph.D., Case Western Reserve, 1985; Neurosciences (CMN). Developmental neurobiology of the retina and visual system.

Marthe J. Howard, Ph.D., California, Irvine, 1984; Neurosciences (CMN/MCB). Role of transcription and growth factors in neural crest development.

Chiung-Yu Hung, Ph.D., Texas at Austin, 1995; Medical Microbiology and Immunology (MCB/MBD). Pathogenesis of fungal infection.

Jerzy Jankun, Ph.D., Poznan (Poland), 1977; Urology and Physiology and Cardiovascular Genomics (MCB). Antitumor targeting of PAI-1 and A-chain ricin conjugates.

Bina Joe, Ph.D., Mysore (India), 1996; Physiology and Cardiovascular Genomics (MCB). Molecular genetics of polygenic diseases.

Boyd M. Koffman, M.D., George Washington, 1990; Ph.D., Maryland, 1990; Neurology (CMN). Neuromuscular diseases.

Thaddeus W. Kurczynski, Ph.D., 1969, M.D., 1970, Case Western Reserve; Pediatrics (CMN). Genetics of neurological disorders.

Eric R. Lafontaine, Ph.D., Calgary, 1997; Medical Microbiology and Immunology (MBD/MCB). Bacterial pathogenesis.

Richard D. Lane, Ph.D., Virginia Commonwealth, 1980; Neurosciences (CMN). Reorganization of sensory pathways following injury.

Abraham Lee, Ph.D., Arizona State, 1991; Physical Therapy (MBD). Skeletal muscle carbohydrate metabolism; glucose transport; diabetes.

Soon Jin Lee, Ph.D., Pennsylvania, 1985; Physiology and Cardiovascular Genomics (MCB). Molecular genetics in hypertension; molecular mechanism of cardiac fibrosis.

Paul F. Lehmann, Ph.D., Cambridge, 1974; Medical Microbiology and Immunology (MBD). Molecular characterization of pathogenic fungi.

Deepak Malhotra, Ph.D., 1984, M.D., 1985, Case Western Reserve; Medicine (MBD/MCB). Nephrology.

Krishna Mallik, M.D., Virginia, 1996; Orthopaedic Surgery (MBD). Orthopedic sports surgery.

William Maltese, Ph.D., Syracuse, 1977; Biochemistry and Cancer Biology (MCB/MBD). Rab proteins and their roles in cellular trafficking and cell signaling.

Maurice Manning, Ph.D., London, 1961; D.Sc., University College, Galway (Ireland), 1974; Biochemistry and Cancer Biology (MCB). Peptide synthesis and design.

Joseph F. Margiotta, Ph.D., SUNY at Stony Brook, 1980; Neurosciences (CMN/MCB). Molecular aspects of ion-channel function and regulation.

Angele V. McGrady, Ph.D., Toledo, 1972; Psychiatry (CMN). Psychophysiology; behavioral medicine; biofeedback.

A. John McSweeny, Ph.D., Northern Illinois, 1975; Psychiatry (CMN). Neuropsychology.

Azedine Medhkour, M.D., Algiers, 1977; Surgery (CMN). Neurosurgery; hormonal and chemical modulation of brain tumors.

Ronald Mellgren, Ph.D., Iowa State, 1976; Pharmacology (MCB). Regulated intracellular proteolysis; cell-cycle control.

Nikoli Modyanov, Ph.D., Shemyakin Institute (Moscow), 1970; Pharmacology (MCB). Molecular basis of membrane transport.

Richard D. Mooney, Ph.D., Minnesota, 1976; Neurosciences (CMN). Synaptic connectivity in the developing and mature mammalian visual system.

Sonia Najjar, Ph.D., Stanford, 1989; Pharmacology (MBD/MCB). Obesity, at the crossroads between cancer and diabetes.

Isabella Novella, Ph.D., Oviedo (Spain), 1990; Medical Microbiology and Immunology (MBD/MCB). Virology; virus evolution.

Jean Overmeyer, Ph.D., Kentucky, 1991; Biochemistry and Cancer Biology (MBD/MCB). Rab escort protein function in membrane targeting.

Ana Marie Oyarce, Ph.D., Georgetown, 1991; Pharmacology (CMN/MCB/MBD). Trafficking of dopamine beta-monooxygenase and dopaminergic regulation of neuropeptides.

Zhixing Kevin Pan, M.D., 1982, Ph.D., 1991, Shanghai Medical; Medical Microbiology and Immunology (MBD/MCB). Molecular mechanisms of host-pathogen interactions.

Thomas Papadimos, M.D., Medical College of Ohio, 1978; M.P.H., Johns Hopkins, 1984; Medicine (MBD). Anesthesiology; critical-care medicine; general preventive medicine.

Sumudra Periyasamy, Ph.D., Medical College of Ohio, 1988; Pharmacology (MCB). Signal transduction and steroid hormone receptors.

Douglas L. Pittman, Ph.D., Iowa, 1996; Physiology and Cardiovascular Genomics (MBD/MCB). Mammalian molecular genetics.

Manohar Ratnam, Ph.D., Indian Institute of Science (Bangalore), 1983; Biochemistry and Cancer Biology (MBD/MCB). Folate receptors in malignancy and hematopoiesis; GPI modification.

Michael Rees, M.D., Michigan, 1991; Urology (MBD/MCB). Xenotransplantation.

Howard Rosenberg, Ph.D., 1975, M.D., 1976, Cornell; Pharmacology (CMN). Neuropharmacology; drugs of abuse.

Randall J. Ruch, Ph.D., Medical College of Ohio, 1988; Biochemistry and Cancer Biology (MBD). Experimental carcinogenesis and cell growth regulation.

Yasser Saad, Ph.D., Medical College of Ohio, 2000; Physiology and Cardiovascular Genomics (MCB). Molecular genetics.

Edwin Sánchez, Ph.D., Michigan, 1983; Pharmacology (MCB). Signal transduction; steroid hormone receptors.

Dorothea Sawicki, Ph.D., Columbia, 1972; Medical Microbiology and Immunology (MCB/MBD). RNA virus transcription.

Stanley Sawicki, Ph.D., Columbia, 1974; Medical Microbiology and Immunology (MCB). RNA virus transcription and persistence.

Joseph Shapiro, M.D., University of Medicine and Dentistry of New Jersey, 1980; Medicine (MCB/MBD). Effects of acid-base and electrolyte disturbances on cardiac energy metabolism, function, and growth.

Ewa Skrzypczak-Jankun, Ph.D., A. Mickiewicz University (Poland), 1976; Urology (MBD). Crystallography: structure analysis of molecules and structure-function relation in proteins and enzymes; targeted drug development.

Cynthia Smas, D.Sc., Harvard, 1994; Biochemistry and Cancer Biology (MBD/MCB). Molecular mechanisms of adipocyte function and differentiation.

Lianhui Tao, M.D., 1982, D.Sc., 1988, Hunan Medical University (China); Pathology (MBD). Pathophysiology and environmental toxicology.

Elizabeth Tietz, Ph.D., Wayne State, 1983; Pharmacology (CMN). Neuropharmacology regulation of structure and function of ionotropic receptors; epilepsy; drugs of abuse.

James P. Trempe, Ph.D., Wright State, 1984; Biochemistry and Cancer Biology (MCB/MBD). Gene-therapy vectors; virus-cell interactions.

Robert Trumbly, Ph.D., California, Davis, 1980; Biochemistry and Cancer Biology (MCB). Transcription repression and signal transduction in yeast.

John W. Turner Jr., Ph.D., Cornell, 1970; Physiology and Cardiovascular Genomics (MBD). Neuroendocrinology and reproductive physiology.

John T. Wall, Ph.D., Virginia, 1979; Neurosciences (CMN). Brain reorganization after injury.

Hardress J. Waller, Ph.D., Washington (Seattle), 1957; Surgery (CMN). Electrophysiology of auditory brainstem.

Xiaodong Wang, Ph.D., Texas Southwestern Medical Center at Dallas, 2001; Pharmacology (MCB). Folding assembly and trafficking of transporters and cell-surface receptors.

M. A. Julia Westerink, M.D., Vrije (Amsterdam), 1980; Medicine (MBD). Vaccine development for infectious diseases.

R. Douglas Wilkerson, Ph.D., Medical University of South Carolina, 1972; Pharmacology (CMN). Neural control of cardiovascular function.

James C. Willey, M.D., Medical College of Ohio, 1978; Medicine (MBD). Effects of toxins on lung function.

Kandace Williams, Ph.D., Dartmouth, 1987; Biochemistry and Cancer Biology (MCB/MBD). DNA damage and repair; cancer.

R. Mark Wooten, Ph.D., Mississippi Medical Center, 1995; Medical Microbiology and Immunology (MCB). Host-pathogen interactions in Lyme disease.

Zijian Xie, Ph.D., Medical College of Ohio, 1990; Pharmacology (MCB). Molecular biology of membrane transporters and signal transduction.

Kam Yeung, Ph.D., South Alabama, 1990; Biochemistry and Cancer Biology (MCB/MBD). Regulation of Pol II transcription; Raf-1 kinase inhibitors regulating cell growth and apoptosis.

UNIVERSITY OF UTAH

Department of Biology

Programs of Study

The Department offers M.S. and Ph.D. degree programs designed to prepare students for careers as research biologists and as university or college faculty members. Research activities in the Department are diverse, ranging from mechanisms of protein folding to the ecology of tropical rain forests. The general areas of cellular and molecular biology, biochemistry, genetics, development, neurobiology, physiology, ecology, and evolution are well represented, as are plant, animal, and microbial systems. The particular research interests of faculty members are listed in the faculty section.

Students may enter the graduate program in biology directly through admission to the Department of Biology or indirectly through admission to the campuswide graduate programs in biological chemistry, molecular biology, and neuroscience. Students seeking a Ph.D. in organismal biology or ecology should apply directly to the Department, as should students seeking a master's degree in any field of biology. Students interested in other areas may apply either to the Department of Biology or to one of the campuswide programs. Prospective students should see the University of Utah's descriptions in *Peterson's Graduate Programs in the Biological Sciences* under the headings of Cell, Molecular, and Structural Biology; Genetics and Developmental Biology; and Microbiological Sciences. Each of these programs has a Web site through which the most up-to-date information can be obtained.

Research Facilities

The Department has all of the necessary equipment and facilities for modern research in biology. In addition to major equipment in individual laboratories, there are Department-wide facilities, including computational, microscopy (confocal, scanning, and transmission), stable isotope mass spectrometry, and animal and plant growth facilities. The Department is a member of campuswide graduate programs in biological chemistry, neuroscience, and molecular biology. For studies of ecology and evolution, the University is situated in proximity to alpine, grassland, and desert ecosystems. Adjacent to campus is the Red Butte Canyon Research Natural Area, a 5,100-acre preserve. The Department of Biology is a member of the Organization for Tropical Studies (OTS).

Financial Aid

Graduate students receive financial support from teaching and research assistantships, training grants, and fellowships. For the 2006–07 year, teaching assistants are scheduled to receive $7500 to $8000 per semester, up to a maximum of $16,000 per year. Support from teaching assistantships may be supplemented by funds from other sources (including research assistantships and fellowships), up to a maximum of $24,000. Provided that progress toward the degree is satisfactory, the Department makes every effort to ensure that Ph.D. students are supported for a total of five years at a minimum level of $15,000 per year. Ph.D. students are also eligible for a total of five years tuition support. At this time, the Department is unable to provide teaching assistantships for M.S. students, though such students may be supported by research assistantships.

Cost of Study

For the few students not receiving tuition waivers, the cost of tuition in 2005–06 for 9 hours of course work was $1761.52 per academic semester for Utah residents and $5470.64 for nonresidents. Costs are subject to increase for 2006–07.

Living and Housing Costs

The cost of dormitory suites for single students in the University residence halls ranges from $5500 to $6000 (including meals) per academic year. Residential hall furnished suites rent from $393 (4 single students) to $429 (2 single students) per month per person. A variety of optional meal plans are also available through University Food Services. Rents for unfurnished University apartments range from $454 to $900 per month, including all utilities. Off-campus housing is available near the University. Costs are subject to increase.

Student Group

The University has an enrollment of approximately 28,900 students, including students drawn from all parts of the United States and many other countries. The Department of Biology has an enrollment of 1,000 undergraduate majors and about 73 graduate students.

Location

Salt Lake City, with a metropolitan population of more than 1 million, is the cultural and economic center of the intermountain region. The Utah Symphony is internationally known. The University Theatre, University Special Events Center, Ballet West, Repertory Dance Theatre, Salt Palace, and Symphony Hall present a variety of outstanding programs. Known as the "Crossroads of the West," Salt Lake City is only a day's drive from a number of national parks, including Yellowstone, Grand Teton, Rocky Mountain, Mesa Verde, Grand Canyon, Zion, Bryce Canyon, and Canyonlands. The University is located at the base of the Wasatch Mountains. Within 15 miles of the campus are peaks that rise above 11,000 feet and provide the finest in year-round recreational opportunities, such as good fishing, hunting, mountain climbing, golfing, and some of the best skiing in the world.

The University

Founded in 1850, the University of Utah is the oldest state university west of the Missouri River. A vigorous institution with a teaching faculty of 1,500 and an adjunct, research, and clinical faculty of 1,800, it is a comprehensive university that includes colleges of medicine, law, education, business, mines and mineral industries, science, humanities, social and behavioral science, health, fine arts, engineering, nursing, pharmacy, and social work in addition to the Graduate School. Among its many assets are an excellent library and an outstanding computer center.

Applying

Applications are invited from students with strong career interests in biology. Final decisions on admission are normally made following personal interviews with prospective applicants, arranged at the Department's expense. Applications should be submitted directly to the Department. The application deadline for fall semester is January 12, 2007.

The University is an Equal Opportunity/Affirmative Action employer and encourages applications from women and members of minority groups and provides reasonable accommodation to the known disabilities of applicants and employees.

Correspondence and Information

Graduate Program Coordinator
Department of Biology
University of Utah
Salt Lake City, Utah 84112-0840
Phone: 801-581-5636
Fax: 801-581-4668
E-mail: shannon.nielsen@bioscience.utah.edu
Web site: http://www.biology.utah.edu

University of Utah

THE FACULTY AND THEIR RESEARCH

Frederick R. Adler, Professor; Ph.D., Cornell, 1991. Evolutionary ecology; mathematical biology; behavioral ecology.
Markus Babst, Assistant Professor; Ph.D., Zurich, 1995. Protein sorting and transport in eukaryotic cells.
Michael Bastiani, Professor; Ph.D., California, Davis, 1981. Developmental neurobiology.
Mary S. Beckerle, Professor; Ph.D., Colorado, 1982. Cell adhesion and signaling; cell motility.
David F. Blair, Professor; Ph.D., Caltech, 1986. Bacterial motility; bioenergetics; membranes.
Lynn A. Bohs, Associate Professor; Ph.D., Harvard, 1986. Plant systematics; economic botany.
David R. Bowling, Assistant Professor; Ph.D., Colorado, 1999. Ecosystem ecology, biogeochemistry.
Dennis M. Bramble, Professor; Ph.D., Berkeley, 1971. Functional and evolutionary vertebrate morphology; biomechanics.
Mario R. Capecchi, Distinguished Professor; Ph.D., Harvard, 1967. Developmental genetics; mouse models of human disease; neural development.
David R. Carrier, Professor; Ph.D., Michigan, 1988. Functional morphology; biomechanics.
Thure E. Cerling, Distinguished Professor; Ph.D., Berkeley, 1977. Paleoecology; stable and cosmogenic isotopes; environmental studies.
Dale H. Clayton, Professor; Ph.D., Chicago, 1989. Host-parasite coevolution; systematics of birds and ectoparasites.
Phyllis D. Coley, Professor; Ph.D., Chicago, 1981. Plant ecology and evolutionary biology; herbivory; tropical communities.
Colin Dale, Assistant Professor; Ph.D., Liverpool (England), 1997. Microbiology and molecular evolution; insects and endosymbionts.
Diane W. Davidson, Professor; Ph.D., Utah, 1976. Community ecology; ant ecology, evolution, behavior.
M. Denise Dearing, Associate Professor; Ph.D., Utah, 1995. Physiological ecology; plant-mammal interactions; disease ecology.
Gary N. Drews, Professor; Ph.D., UCLA, 1989. Molecular genetics of plant development in *Arabidopsis thaliana.*
James R. Ehleringer, Distinguished Professor; Ph.D., Stanford, 1977. Plant ecology.
Donald H. Feener Jr., Professor; Ph.D., Texas at Austin, 1978. Insect behavior, ecology, and evolution; host-parasite interactions.
David L. Gard, Professor; Ph.D., Caltech, 1982. Regulation of microtubule assembly; *Xenopus* development.
Raymond F. Gesteland, Distinguished Professor; Ph.D., Harvard, 1965. Molecular genetics of decoding and human genome project.
David P. Goldenberg, Professor; Ph.D., MIT, 1981. Protein folding, dynamics, and function; NMR spectroscopy.
Kent G. Golic, Professor; Ph.D., Washington (Seattle), 1986. Chromosome organization and function in *Drosophila melanogaster.*
Franz Goller, Associate Professor; Ph.D., Notre Dame, 1991. Behavioral physiology of sound communication.
Martin P. Horvath, Assistant Professor; Ph.D., Chicago, 1994. Macromolecular recognition; X-ray crystallography.
Kelly T. Hughes, Professor; Ph.D., Utah, 1984. Microbial physiology and genetics; gene regulation; flagellum assembly; type III secretion.
Erik M. Jorgensen, Professor; Ph.D., Washington (Seattle), 1989. Genetic analysis of neurotransmission in *C. elegans.*
Darryl L. Kropf, Professor; Ph.D., Colorado Health Sciences Center, 1983. Cell polarity; plant embryogenesis and development.
Thomas A. Kursar, Associate Professor; Ph.D., Chicago, 1982. Physiological ecology of tropical rainforest plants.
Andres Villu Maricq, Professor; Ph.D., Berkeley, 1987; M.D., California, San Francisco, 1990. Neurobiology: synaptic function.
Lawrence M. Okun, Professor; Ph.D., Stanford, 1969. Neurobiology; fluorescent dye design.
Baldomero M. Olivera, Distinguished Professor; Ph.D., Caltech, 1966. Conotoxins; ion channels; molecular neurobiology.
John S. Parkinson, Professor; Ph.D., Caltech, 1969. Molecular biology and genetics of signal transduction.
Wayne K. Potts, Professor; Ph.D., Washington (Seattle), 1986. Host-pathogen coevolution; behavior; immunogenetics and evolutionary genetics.
Martin C. Rechsteiner, Professor; Ph.D., Johns Hopkins, 1969. Ubiquitin, proteasomes, and proteolysis.
Gary J. Rose, Professor; Ph.D., Cornell, 1983. Neuroethology of audition and electroreception.
Jon Seger, Professor; Ph.D., Harvard, 1980. Evolutionary ecology and genetics.
Michael D. Shapiro, Assistant Professor; Ph.D., Harvard, 2001. Vertebrate genetics, development, and evolution.
Janet M. Shaw, Professor; Ph.D., UCLA, 1988. GTPases and mitochondrial membrane fission and fusion.
Leslie E. Sieburth, Associate Professor; Ph.D., Georgia, 1990. Leaf developmental genetics; root-shoot signaling in *Arabidopsis*.
John S. Sperry, Professor; Ph.D., Harvard, 1985. Plant organismal structure and function; water relations.
Neil J. Vickers, Associate Professor; Ph.D., California, Riverside, 1992. Neuroethology of olfaction and odor-mediated behavior.
Stanly B. Williams, Assistant Professor; Ph.D., Cornell, 1997. Microbial biology: bacterial circadian rhythms and signal transduction.
David R. Wolstenholme, Professor; Ph.D., 1961, D.Sc., 1973, Sheffield (England). Animal mitochondrial DNAs: structure and evolution.
Doju Yoshikami, Professor; Ph.D., Cornell, 1970. Cellular neurobiology; ion channels; neurotoxins and nociception.

Adjunct and Research Faculty

Wolfgang Baehr, Adjunct Professor; Ph.D., Heidelberg, 1970. Molecular biology of vertebrate phototransduction.
Roy D. Bloebaum, Adjunct Professor; Ph.D., Western Australia, 1981. Bone biology: structure, function, and adaptation.
Sherwood R. Casjens, Adjunct Professor; Ph.D., Stanford, 1972. Virus assembly; Lyme disease agent genomics.
Colleen G. Farmer, Research Assistant Professor; Ph.D., Brown, 1997. Comparative physiology.
Naomi C. Franklin, Research Professor; Ph.D., Yale, 1954. Genetic regulation of transcription and development in bacteriophage.
Glenn Herrick, Adjunct Professor; Ph.D., Princeton, 1973. Conflict-of-interest-driven evolution; telomeres and transposons of ciliates.
Wai Mun Huang, Adjunct Professor; Ph.D., Johns Hopkins, 1967. Chromosome structure, topoisomerases, protelomerases.
Michael McIntosh, Research Professor; M.D., UCLA, 1983. Pharmacology and therapeutic development of neuroactive peptides.
Alan R. Rogers, Adjunct Professor; Ph.D., New Mexico, 1982. Population genetics; evolutionary ecology.
Eric W. Schmidt, Assistant Professor; Ph.D., California, San Diego, 1999. Marine symbiosis and biosynthesis of natural products.
Irene Terry, Research Associate Professor; Ph.D., North Carolina State, 1983. Plant-herbivore interactions and genetics; insect behavior.

UNIVERSITY OF VERMONT

Biological and Biomedical Sciences

Programs of Study

Graduate studies in the biological and biomedical sciences at the University of Vermont offer comprehensive and specialized training leading to a Ph.D. or M.S. degree in the cellular and molecular mechanisms of physiological function and disease. For more information, students should visit the University's Web site. A combined M.D./Ph.D. option is also available. Particular areas of focus include asthma and pulmonary function, cancer biology, cardiovascular function and disease, cell signaling, environmental pathology, immunobiology, microbial and viral pathogenesis, molecular biophysics, molecular biology and genetics, neuroscience, and structural biology. An internationally recognized faculty from the Departments of Anatomy and Neurobiology, Biochemistry, Microbiology and Molecular Genetics, Molecular Physiology and Biophysics, Pathology, and Pharmacology and the Cell and Molecular Biology Program provide an outstanding interdisciplinary environment in which to pursue a career in the biological and biomedical sciences.

Research Facilities

The College of Medicine at the University of Vermont consists of three major research buildings, including the Health Sciences Research Facility. A new Ambulatory Care Center and Medical Education Building, including a new multimedia library facility, is currently under construction. In addition, the College of Medicine is contiguous with the Fletcher Allen Medical Center and Hospital, providing a seamless interface for clinical training and research opportunities. The latest state-of-the-art research techniques are available to study contemporary problems in the biomedical sciences, including core facilities for atomic force microscopy, cell imaging, confocal microscopy, cryo-electron microscopy, flow cytometry, instrumentation and modeling, transgenic mice, NMR spectroscopy, and X-ray crystallography. Numerous facilities also exist for state-of-the-art work in cell culture, molecular biology, molecular genetics, and protein biochemistry.

The research programs in the biological and biomedical sciences are well supported by a variety of sources, including the National Institutes of Health, the National Science Foundation, the American Cancer Society, the American Heart Association, the Burroughs Wellcome Foundation, the Pew Charitable Trust, the Howard Hughes Foundation, the Department of Energy, and the Environmental Protection Agency.

Financial Aid

Graduate students are supported by a combination of teaching and research fellowships, including four NIH-funded training grants, during all years of study in the program. The stipend level during 2005–06 was about $20,770, depending on the type of fellowship, for a twelve-month appointment. Students holding fellowships also receive full-tuition scholarships.

Cost of Study

Tuition costs are provided for all students for the duration of their enrollment.

Living and Housing Costs

Estimated costs for food, housing, books, supplies, personal expenses, and transportation in the Burlington area are approximately $800 to $1000 per month for a single student. Housing is available in the surrounding community. A limited amount of housing is available in University residence halls for single students and in University apartments for married students.

Student Group

There are approximately 120 highly motivated and well-qualified doctoral students in the biological and biomedical sciences at the University of Vermont, plus an additional 400 students enrolled in the medical degree program. Overall, the student body is geographically, ethnically, and culturally diverse.

Student Outcomes

Graduates from the degree programs in the biological and biomedical sciences at the University of Vermont are well prepared for research and teaching careers in both academic and industrial settings. Many alumni have competed very successfully for positions at the top research and academic institutions in the country and have established themselves as leaders in their fields.

Location

The University of Vermont is located in Burlington on the scenic shores of Lake Champlain. With its proximity to the lake, the Green Mountains in Vermont, and the Adirondack Mountains in New York, Burlington offers an abundance of outdoor recreation opportunities, including boating, fishing, swimming, hiking, and skiing. Chittenden County has a population of approximately 150,000, and Burlington, its Queen City, is the social and cultural hub of Vermont, offering a wide variety of events such as concerts, theater, and festivals. Major cities, including Montreal (100 miles north), Albany (150 miles southwest), Boston (150 miles south), and New York (250 miles southwest), are short drives away.

The University

The University of Vermont, founded in 1791, is a comprehensive university that offers a wide variety of degree programs and research opportunities. Its moderate size promotes an environment in which students can work side-by-side with leading scientists in their field of choice.

Applying

Applicants must demonstrate excellence in their undergraduate performance in the physical and biological sciences. Graduate Record Examinations (GRE) scores on the General Test must be submitted along with transcripts and three letters of recommendation by January 15 to the Graduate College through the Graduate Admissions Office, 333 Waterman Building, University of Vermont, Burlington, Vermont 05405-0160.

Correspondence and Information

Dr. Christopher L. Berger, Director
Biological and Biomedical Sciences Program
123B Health Sciences Research Facility
College of Medicine
University of Vermont
Burlington, Vermont 05405-0075
Phone: 802-656-0832
Fax: 802-656-0747
E-mail: bbs@uvm.edu
Web site: http://www.med.uvm.edu/bbs/

University of Vermont

THE FACULTY AND THEIR RESEARCH

Anatomy and Neurobiology (http://www.uvm.edu/annb)
The Department of Anatomy and Neurobiology faculty members are involved in research programs concerning mechanisms regulating neuronal degeneration, regeneration, and plasticity; development and pattern formation in the autonomic nervous system; organization of somatosensory and autonomic pathways; neurotransmitter and neuropeptide expression and secretion; specific synaptic actions of neuroactive compounds; modification of calcium and other intracellular signaling pathways in excitable cells; and cardiovascular and gastrointestinal functions in normal and diseased states.

Biochemistry (http://biochem.uvm.edu)
The Department of Biochemistry features a broad diversity of research interests, including major emphases in cell biology, blood coagulation, protein-DNA interactions, protein-protein interactions, and structural biology. The department offers a highly interdisciplinary research environment supported by a full spectrum of technologies, including biochemistry and biophysics, cell and molecular biology, genomics and proteomics, macromolecular structure, and molecular genetics.

Microbiology and Molecular Genetics (http://www.uvm.edu/microbiology)
The Department of Microbiology and Molecular Genetics faculty members are asking fundamental questions in eukaryotic and prokaryotic cell and molecular biology. The department applies the methods of microbiology, genetics, biochemistry, bioinformatics, and structural biology to elucidate the molecular mechanisms of the cell. The interdisciplinary nature of these fields means that a prospective graduate student is offered a wide choice of research opportunities. Cross-departmental, interdisciplinary collaborations are facilitated by regular meetings and journal clubs focusing on nucleic acid biochemistry, DNA repair, signal transduction, bacterial pathogenesis, and structural biology and bioinformatics.

Molecular Physiology and Biophysics (http://physioweb.med.uvm.edu)
Within the Department of Molecular Physiology and Biophysics a powerful group of interacting laboratories focuses their research on the molecular basis of cardiovascular and muscle function as well as biological motility. This group's research covers model systems from the whole heart down to the level of a single molecular motor using state-of-the-art methodologies such as X-ray crystallography, molecular biology, and single-molecule biophysics. Areas of emphasis include excitation and intracellular signals, protein structure-function relationships with emphasis on molecular motors, and genetic control of differentiation.

Pathology (http://www.fahc.org/pathology/index.html)
The Department of Pathology offers research training programs in environmental and experimental pathology. Areas of interest include molecular and cellular approaches for studying mechanisms of environmental disease, cancer biology, cardiovascular function and disease, lung biology and disease, DNA damage and repair, cell signaling, control of mitogenesis, and cell death.

Pharmacology (http://pharmweb.med.uvm.edu/gradstud.htm)
Research interests in the Department of Pharmacology include cellular and molecular pharmacology of the vascular system and its neural control, drug treatment of hypertension and stroke, biochemical pharmacology, chemotherapy of cancer, AIDS therapy, membrane receptor and ion channel pharmacology, molecular biological approaches to drug design, molecular neuropharmacology, and signal transduction.

Cell and Molecular Biology Program (http://uvm.edu)
The Cell and Molecular Biology Program is directed toward an understanding of the cell, its components and its architecture, the functional relations of its parts, and the properties that have given it a central position in biological theory. Research programs focus on an array of topics, including cancer biology, cell injury and repair, the cytoskeleton and motility, DNA replication and repair, gene regulation, host-pathogen relationships, immunobiology, membrane function, neurobiology, nucleic acid structure and function, plant biology and development, protein structure and function, signal transduction, and virology.

VANDERBILT UNIVERSITY

Department of Biological Sciences

Programs of Study	The Department of Biological Sciences formed from the coalescence of the Departments of Biology and Molecular Biology at Vanderbilt University. Research activities in the department encompass the study of biology at the molecular, subcellular, cellular, organismal, population, and community levels. Fields of study available to graduate students include ecology and evolutionary biology, neurobiology and behavior, biological clocks, developmental biology, cell biology, molecular genetics, biochemistry, molecular biophysics, and structural biology. Depending upon the specific interests of the student, undergraduate backgrounds emphasizing biological sciences, chemistry, mathematics, or physics course work are recommended. Students interested in obtaining a Ph.D. degree in biological sciences may apply for direct admission in the Biological Sciences Graduate Program, or they may enter through the Interdisciplinary Graduate Program (IGP) in the Biomedical Sciences. During the first year in the program, students engage in course work that reflects their scientific interests and also gain valuable experience in research by completing at least two laboratory rotations. Students typically finalize their selection of a mentor and begin research in the laboratory of their choice by May of their first year.

Research Facilities — The department is housed in a recently completed $100-million, state-of-the-art building with an adjoining molecular biology wing. This building complex is shared with several programs in the Medical Center and is adjacent to the Departments of Chemistry, Physics, and Mathematics. Excellent library and computer facilities are also nearby. The department is well equipped with major instrumentation, which allows students and faculty members to pursue studies requiring electron or fluorescence microscopy, automated DNA sequencing, X-ray diffraction, and nuclear magnetic resonance. Shared resources include a research greenhouse and a 2,000-tank zebrafish facility.

Financial Aid — Awards to students carry stipends and also cover tuition. In addition, at the time of application, applicants may compete for Harold S. Vanderbilt graduate scholarships, which provide an additional $3000–$5000 stipend to students of exceptional accomplishment and high promise.

Cost of Study — Tuition and fees are paid by the program for all students; support is from traineeships, fellowships, and University assistantships.

Living and Housing Costs — Graduate students may choose to live on or off campus. University apartments are available for graduate students and married couples. Information on housing is available from the Office of Housing, Box 1677, Station B.

Student Group — Vanderbilt has about 9,500 students, 4,000 of whom are studying for graduate or professional degrees. The Interdisciplinary Graduate Program admits approximately 75 students and the Department of Biological Sciences directly admits approximately 5 students each year. The department has 50 to 60 students in the Ph.D. program.

Student Outcomes — Graduates have pursued a variety of career options. Many initially undertake postdoctoral studies at Yale, Harvard, Stanford, the University of Chicago, the National Institutes of Health (NIH), and La Jolla Cancer Center, among others. Many of these students have moved into more permanent academic positions at such institutions as Stanford, LSU, NIH, La Jolla Cancer Center, Texas A&M, and Case Western Reserve. Others have pursued industrial positions at companies such as Sterling Research Group; Merck, Sharp, and Dohme; and Hoffmann-LaRoche. A few graduates have pursued careers in biotechnology/patent law by going to law school at such institutions as Berkeley and Columbia.

Location — Nashville is the capital of Tennessee. In this city of about half a million, there are more than a dozen institutions of higher learning that enrich the intellectual interests and social outlook of the community. There is a rich cultural and musical life based on the Tennessee Performing Arts Center (host to plays, musicals, concerts, ballets, operas, and symphony performances), the Grand Ole Opry, and an extensive recording industry for country, rock, and gospel music. Old Hickory and Percy Priest Lakes and surrounding parks offer recreational facilities less than 30 minutes from the University; boating, fishing, hiking, and camping areas abound.

The University — Vanderbilt University is a private, coeducational institution founded in 1873. The University consists of the College of Arts and Science, School of Engineering, Divinity School, School of Nursing, School of Medicine, George Peabody College of Teachers, Graduate School, School of Law, Owen Graduate School of Management, and Blair School of Music. Vanderbilt has a long tradition of academic excellence and a distinguished record in scholarship and research.

Applying — Applications are due by January 15, but those received subsequently are considered if vacancies exist. Students holding degrees in the physical or biological sciences may apply. Students are advised to take the Graduate Record Examinations as early as possible. Detailed information on research programs, curricula, and financial aid is sent upon request. Updated research interests can be found on the department's Web site.

Correspondence and Information

Dr. Douglas G. McMahon
Director of Graduate Studies
Department of Biological Sciences
Vanderbilt University
Nashville, Tennessee 37235-1634
Phone: 615-936-3933
E-mail: bioscidgs@vanderbilt.edu
Web site: http://sitemason.vanderbilt.edu/biosci

Vanderbilt University

THE FACULTY AND THEIR RESEARCH

Research interests of the faculty members range from biomolecular structure determination at atomic resolution to the study of ecosystems and evolutionary biology. Individual projects include protein structure determination by X-ray crystallography, fiber diffraction, or nuclear magnetic resonance; protein chaperone influence on the cytoskeleton; protein transport and sorting; glycoprotein biosynthesis; regulation of gene expression through pre-mRNA packaging and splicing; regulation of DNA replication in mammalian cells and synthesis of telomeres in yeast; calcium homeostasis; thiamine utilization; *Dictyostelium* development; embryonic and neural development in zebrafish; molecular and neural basis of biological clocks; organization of mammalian sensory systems; nervous system development and function in *Drosophila;* olfaction in insect disease vectors; insect neuropeptides and hormones; parasitology and immune response to parasites; evolution in structured populations; evolutionary ecology of species interactions; ecology of speciation; physiological and community ecology of plants.

Patrick Abbot, Assistant Professor; Ph.D., Arizona. Arthropod behavior and speciation; molecular evolutionary genetics of microbes.
Bruce H. Appel, Assistant Professor; Ph.D., Utah. Genetic, molecular, and cellular analysis of neural development in zebrafish.
Kendal Broadie, Professor; Ph.D., Cambridge. Genetic dissection of synapse development; function and plasticity.
Clint E. Carter, Professor; Ph.D., UCLA. Immune system regulation during infection with *Trypanosoma cruzi* (Chagas' disease); genetic analysis of risk factors for disease morbidity.
Kenneth C. Catania, Assistant Professor; Ph.D., California, San Diego. Mammalian sensory systems, with a focus on the organization and development of sensory cortex and its relationship to behavior.
Brandt F. Eichman, Assistant Professor; Ph.D., Oregon State. Structure and function of protein complexes involved in DNA repair and replication; X-ray crystallography.
Ellen Fanning, Stevenson Professor of Molecular Biology; Dr.rer.nat., Cologne (Germany). Molecular elucidation of the mechanisms that control DNA replication in mammalian cells.
Katherine L. Friedman, Assistant Professor; Ph.D., Washington (Seattle). Telomerase function in *Saccharomyces cerevisiae.*
Daniel J. Funk, Assistant Professor; Ph.D., SUNY at Stony Brook. Ecology and evolution of specialized insect herbivores.
Joshua T. Gamse, Assistant Professor; Ph.D., MIT. Development and genetics of left-right asymmetry in the zebrafish brain.
Todd R. Graham, Associate Professor; Ph.D., Saint Louis. Protein transport and sorting in the secretory pathway.
Christopher J. Janetopoulos, Assistant Professor; Ph.D., Texas A&M. Establishment of polarity during cell migration and division.
Carl H. Johnson, Professor; Ph.D., Stanford. Molecular, cellular, and evolutionary analyses of circadian clocks.
Daniel L. Kaplan, Assistant Professor; Ph.D., Yale. Molecular motors that catalyze DNA replication, recombination, and repair.
Andrezej M. Krezel, Assistant Professor; Ph.D., Wisconsin–Madison. The use of multidimensional heteronuclear NMR to investigate protein-protein interactions exemplified by TGF-beta receptors and their ligands.
Wallace M. LeStourgeon, Professor; Ph.D., Texas at Austin. Molecular mechanisms of premessenger RNA processing and the function of RNA chaperonins in normal and cancer cells.
David E. McCauley, Professor; Ph.D., SUNY at Stony Brook. Studies of the influence of population genetic structure on evolution using a combination of theoretical, field, and molecular approaches.
Douglas McMahon, Professor; Ph.D., Virginia. Molecular neurobiology of the visual and circadian systems of the brain.
Tom N. Oeltmann, Associate Professor; Ph.D., Georgia. Biosynthesis, processing, and sorting of lysosomal enzymes in lower eukaryotes, with an emphasis on the human parasite *Trypanosoma cruzi.*
Terry L. Page, Professor; Ph.D., Texas at Austin. Regulation by the nervous system of daily (circadian) rhythms in physiology and behavior.
James G. Patton, Associate Professor; Ph.D., Mayo. Control of gene expression at the RNA level, focusing on the mechanisms of pre-mRNA splicing and the regulation of alternative splicing.
Charles K. Singleton, Professor; Ph.D., Purdue. Regulation of timing and pathway choice during development by sensory histidine kinases; mechanisms of thiamine utilization and thiamine deficiency–induced brain diseases in humans.
Lilianna Solnica-Krezel, Associate Professor; Ph.D., Wisconsin. Genetic and cellular basis of inductive and morphogenetic events that establish the body plan of zebrafish embryos.
Gerald Stubbs, Professor; D.Phil., Oxford. X-ray crystallography; fiber diffraction; filamentous plant virus structure; virus-host interaction.
Donna J. Webb, Assistant Professor; Ph.D., Virginia. Processes that underline cell migration, including the signaling pathways that regulate adhesion and cytoskeletal dynamics in normal and cancer cells and the formation of dendritic spines and synapses in the central nervous system.
Laurence J. Zwiebel, Assistant Professor; Ph.D., Brandeis. Molecular basis of olfaction and its role in host selection of *Anopheline* mosquitoes, which are the major vectors for malaria transmission.

Professors Emeriti
Burton J. Bogitsh, Professor; Ph.D., Virginia. Cytochemistry of helminthic parasites.
William G. Eickmeier, Associate Professor; Ph.D., Wisconsin. Plant ecology, particularly physiological mechanisms of plant adaptation to stressful physical environments such as deserts.
Sidney Fleischer, Professor; Ph.D., Indiana. Cell signaling via calcium as second messenger; heart and muscle physiology; intracellular calcium release channels; gene knockout and transgenic mice.
Hans-Willi Honegger, Professor; Ph.D., Tübingen (Germany). Establishment of the cellular location, mechanism of release, and receptor for the neurohormone bursicon initiation cuticle sclerotization in insects and crustacea.
Oscar Touster, Ph.D., Illinois. Processing of glycoproteins, particularly in regard to Golgi membrane functions, lysosomal diseases, and nerve-cell metabolism.
John H. Venable, Ph.D., Yale. Biophysical investigation of the structure and function of proteins and nucleic acids.
Dean P. Whittier, Ph.D., Harvard. Morphology and development of fern gametophytes, with emphasis on the mycorrhizal gametophytes of the Psilotaceae, Lycopodiaceae, and Ophioglossaceae.
Robley C. Williams Jr., Professor; Ph.D., Rockefeller. Biochemical and biophysical approaches to molecular chaperones and protein folding, emphasizing their involvement in formation of the cytoskeleton.

VILLANOVA UNIVERSITY

Department of Biology

Programs of Study	The Department of Biology offers a program of study that leads to the Master of Science degree in biology (with thesis) or the Master of Arts degree in biology (nonthesis). For the Master of Science degree, a minimum of 30 semester credit hours is required, up to 10 of which may be contributed as research credits. Course work is determined by the student, the research adviser, and the student's advisory committee. Thesis research may be undertaken in the areas of behavioral biology, biogeochemistry, cellular physiology, developmental biology, ecology, endocrinology, evolutionary morphology, genetics, herpetology, molecular biology, neurophysiology, ornithology, plant ecology, plant physiology, protozoology, or virology. For the Master of Arts degree, a minimum of 33 semester hours of course work is required. Certificates of Graduate Study (16 credit hours) and Advanced Graduate Study (24 credit hours) are also available in cell, molecular, and developmental biology and ecology, evolution, and organismal biology.
Research Facilities	The laboratories of the Department of Biology are well equipped for graduate instruction and research in many areas of biology. The Department maintains a geographic information system with ARC/INFO and IDRISI, an animal-care facility, a darkroom, and an electron microscope facility. The equipment includes three automated sequencers; a liquid scintillation spectrometer and a gamma counter for work with radioisotopes; a still-video imaging system for molecular biology documentation; both scanning and transmission electron microscopes; biomaterials testing apparatus; a cabinet X-ray facility; a cell-culture laboratory; chromatographic and electrophoretic instruments; a variety of centrifuges, including a preparative ultracentrifuge; and spectrophotometric, photographic, bioelectronic, and other instruments. Extensive computer facilities are available within the Department; the University Computer Center, located in the same building as the Department, is easily accessible.
Financial Aid	Teaching assistantships carrying a stipend of $11,750 and waiver of tuition were available in 2005–06, as were government-supported Federal Work-Study funds and loans. A limited number of research assistantships are awarded on a competitive basis. Additional summer support is also available to qualified students.
Cost of Study	Tuition in 2005–06 was $600 per semester credit hour.
Living and Housing Costs	Privately owned rooms and apartments are available along the Main Line and in other nearby communities. Costs for rooms start at approximately $100 per week, depending on kitchen privileges. Apartments range between $550 and $900 per month, depending on their size and their location. Students who share apartments may have costs as low as $300 per month.
Student Group	The Department enrolls approximately 50 graduate students a year. Nearly 2,000 students from the United States and abroad are enrolled in various graduate programs at Villanova.
Location	Villanova University is situated in Villanova, Pennsylvania, on U.S. Route 30, 6 miles west of the Philadelphia city line. The University can be reached in half an hour from central Philadelphia by two commuter rail lines; each of these systems has a station at the University. Buses pass the campus, which is one mile east of I-476 and easily reached by car from I-95, I-76 (the Schuylkill), and the Mid-County exit of the Pennsylvania Turnpike.
The University	Villanova University is a Roman Catholic university, sponsored by the religious order of St. Augustine. It benefits from more than 150 years of excellent educational tradition. The University recognizes its responsibility to disseminate knowledge and to seek new knowledge through research and scholarship. It has been the University's tradition to foster close and warm relationships between its students and faculty members. Villanova is an equal opportunity employer. With more than 220 landscaped acres in a beautiful residential section, the Villanova campus is one of the showplaces of the Philadelphia area and is a registered arboretum. The campus has facilities for a variety of athletic activities.
Applying	Graduates of accredited colleges who wish to work for their Master of Science or Master of Arts degree in biology must have completed a minimum of 24 semester hours of undergraduate work in biology and 6 to 8 hours each of calculus, physics, general chemistry, and organic chemistry. The selection of applicants is based on academic record, supporting letters, and Graduate Record Examinations (GRE) scores on the General Test. The GRE Subject Test in biology or biochemistry, cell and molecular biology is strongly recommended but not required.

Correspondence and Information

For information:
Director of the Graduate Program
Department of Biology
Villanova University
Villanova, Pennsylvania 19085
Web site:
   http://www.biology.villanova.edu

For financial aid:
Financial Aid Office
Kennedy Hall
Villanova University
Villanova, Pennsylvania 19085

For admissions and assistantships:
Dean of Graduate Studies
Villanova University
Villanova, Pennsylvania 19085

Villanova University

THE FACULTY AND THEIR RESEARCH

Ronald A. Balsamo, Ph.D., California, Riverside, 1994. Relationship of leaf architecture and biomechanics to drought tolerance; whole plant response to environmental stress.

Anil Bamezai, Ph.D., All India Institute of Medical Sciences, 1987. Regulation of CD4+ T-lymphocyte development in the thymus and responses to protein antigens in the peripheral lymphoid tissues; focus on set of glycosylphosphatidyl-inositol (GPI)–anchored proteins housed in lipid rafts that function as signaling foci on the plasma membrane.

Aaron M. Bauer, Ph.D., Berkeley, 1986. Phylogenetic systematics of squamate reptiles, especially geckos and skinks; evolutionary morphology of reptilian integumentary and musculoskeletal systems; historical biogeography of tropical and Southern Hemisphere reptiles and amphibians; herpetology of southern Africa, islands of the southwestern Pacific, India, and Sri Lanka; eighteenth- and nineteenth-century history of herpetology.

Robert L. Curry, Ph.D., Michigan, 1987. Vertebrate behavioral, population, and molecular ecology; ornithology; conservation biology; ecological, behavioral, and genetic aspects of hybridization in chickadees; ecology of the Florida Scrub-Jay (collaboration at Archbold Biological Station); conservation ecology of insular Mimidae.

Mary Desmond, Ph.D., Colorado at Boulder, 1973. Developmental biology, developmental genetics, and neuroembryology; basic cellular mechanisms underlying normal development of the brain and spinal cord, using chick and mouse embryos as experimental models; findings: brain growth requires occlusion of the spinal cord and intraluminal pressure generated from cerebrospinal fluid (CSF).

Angela J. DiBenedetto, Ph.D., Cornell, 1989. Molecular and cellular biology; genetics; developmental neurobiology, especially programmed cell death.

Norman R. Dollahon, Ph.D., Nebraska–Lincoln, 1971. Parasitology, protozoology, and electron microscopy.

John Friede, Ph.D., Minnesota, 1970. Microbiology; effect of potential therapeutic agents on energy-generating mechanisms; transport and exotoxin production in pathogenic bacteria.

Russell M. Gardner, Ph.D., Indiana, 1975. Endocrinology; pharmacology; mechanisms of hormone action, development of hormone responses; hormonal control of uterine growth and differentiation.

Vikram K. Iyengar, Ph.D., Cornell, 2001. Behavioral ecology; entomology; chemical ecology; sexual selection in insects and other arthropods; how the costs and benefits of mate choice shape mating systems.

Todd Jackman, Ph.D., Berkeley, 1993. Evolutionary genetics of salamanders and lizards using DNA sequence data in combination with other data to provide a robust historical framework for examining evolutionary processes; population studies of wandering salamanders; speciation and biodiversity of the New Caledonian herpetofauna.

Janice E. Knepper, Ph.D., Brown, 1979. Molecular biology; virology; molecular mechanisms of viral oncogenesis.

John Olson, Ph.D., Michigan, 1990. Metabolic and muscle physiology; ecophysiology; functional and structural maturation of the effector tissues for thermogenesis and substrate mobilization in birds and mammals; mechanical performance of muscles during locomotion in both invertebrate and vertebrate species.

Joseph A. Orkwiszewski, Ph.D., Bryn Mawr, 1971. Regulation of plant development; control mechanisms of plant growth and development.

Michael Russell, Ph.D., Berkeley, 1990. Investigations at the intersection of marine ecology, population biology, and fisheries science; sustainable management and conservation of commercially important marine invertebrate fisheries.

Louise A. Russo, Ph.D., Penn State Hershey Medical Center, 1987. Cell biology and physiology; understanding the establishment of the uterine receptive state and fertility; hormone-induced uterine growth and remodeling, using an in vivo model system in the rat; analysis of regulated expression of specific cell-surface adhesion receptor proteins and degradative enzymes.

Philip J. Stephens, Ph.D., Aberdeen (Scotland), 1977. Physiology; use of computer technology and animations in traditional and asynchronous teaching, distance learning, animations, interactive, and student-centered learning.

Jacqueline F. Webb, Ph.D., Boston University, 1988. Fish biology and sensory biology; evolutionary and developmental morphology of fish sensory systems, especially the mechanosensory lateral line system; analysis of sensory specializations in coral reef butterfly fishes; development of the lateral line system in cichlids and zebrafish.

R. Kelman Wieder, Ph.D., West Virginia, 1982. Ecosystem ecology and biogeochemistry, wetland ecology, constructed wetland for mine drainage or stormwater treatment; biotic and abiotic factors influencing carbon cycling and accumulation/release in boreal peatland ecosystems, including continental bogs and fens, in the discontinuous permafrost region of western Canada.

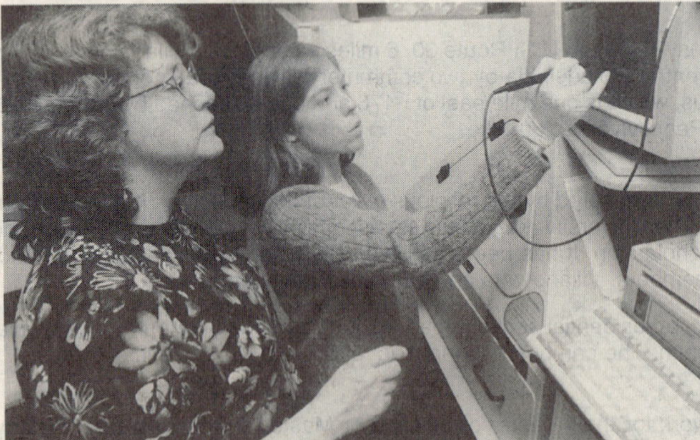

A graduate student and professor analyze the results of DNA amplification by scanning a gel on the image analyzer.

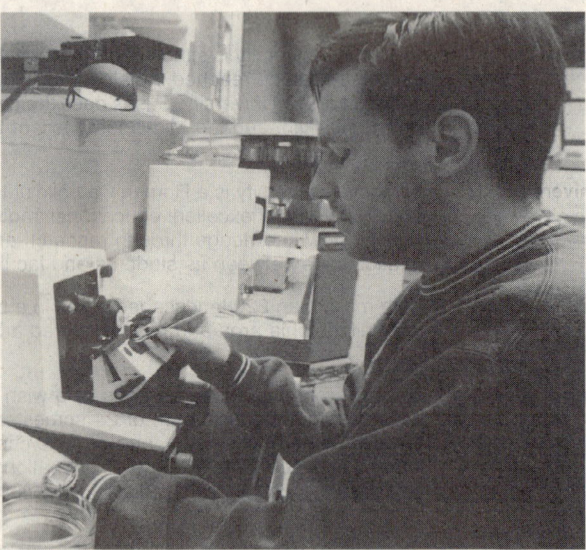

A graduate student prepares histological sections of sensory specialization organs of a coral reef fish.

VIRGINIA COMMONWEALTH UNIVERSITY

Medical College of Virginia Campus
Graduate Programs in the School of Medicine

Programs of Study

The School of Medicine administers graduate programs through a number of departments. Programs at the Ph.D. level include anatomy, biochemistry, biostatistics, epidemiology (through the Department of Epidemiology and Community Health), human genetics, microbiology and immunology, pathology, pharmacology and toxicology, and physiology.

Master's-level training includes programs in anatomy, biochemistry, biostatistics, genetic counseling (through the Department of Human Genetics), human genetics, microbiology and immunology, pharmacology and toxicology, physiology, and public health (as the M.P.H. program through the Department of Epidemiology and Community Health).

Interdisciplinary curriculum tracks in immunology, molecular biology and genetics, neuroscience, and structural biology are available through participating departments. The School administers additional interdisciplinary training programs in anatomy or physiology/physical therapy and combined-degree programs (M.D./Ph.D. and M.D./M.P.H.).

The School also offers the Preprofessional Basic Sciences Certificate Program with six curricula administered by the Departments of Anatomy, Biochemistry, Human Genetics, Microbiology and Immunology, Pharmacology and Toxicology, and Physiology.

Completion of M.S. requirements usually takes two years, and completion of Ph.D. requirements usually takes four to five years. In 2005–06, 35 students completed Ph.D. requirements in the basic health sciences departments and 50 earned master's degrees. For additional information, prospective students can visit the School of Medicine's Web site (listed below).

Research Facilities

Research resources appropriate to current biomedical research, including bioinformatics and proteomics, are available within departments or in resource centers. Library holdings are extensive and accessible electronically.

Financial Aid

Doctoral applicants are eligible for teaching/research assistantships or fellowships. For 2005–06, most stipends ranged from $20,000 to $24,000 per year; costs of tuition and fees are usually provided when an offer of financial support is made.

Cost of Study

Tuition and fees for 2005–06 were $22,970 per year for out-of-state students and $9978 per year for Virginia residents. Tuition and fees are usually covered by assistantships or fellowships.

Living and Housing Costs

Virginia Commonwealth University (VCU) is an urban institution with limited graduate housing on campus. The University is adjacent to residential areas of the city, with a broad spectrum of apartment housing available. Information about both on-campus and off-campus housing can be found on the VCU Web site at the address listed below.

Student Group

All students at the Medical College of Virginia (MCV) campus are pursuing degrees in the health professions: medicine, dentistry, pharmacy, nursing, and allied health. There are graduate programs in nursing and allied health professions as well as in the basic health sciences, pharmacy, and biomedical engineering.

Location

The MCV Campus is located in downtown Richmond. Metropolitan Richmond has a population of 900,000. In addition to being a historical center and the seat of Virginia's state government, Richmond offers many cultural opportunities, including museums, a symphony orchestra, and theaters. The Richmond Coliseum, where sporting events, spectaculars, musical performances, and other forms of entertainment take place, is located at the edge of the campus.

The University and The College

VCU is located in Richmond, Virginia, on two campuses. The MCV Campus is the home of a comprehensive health sciences educational center. VCU is a Carnegie Foundation Doctoral/Research-Extensive university, with particular research strengths in the biomedical sciences. The University has an overall enrollment of 26,000 students, with 2,500 graduate and professional students on the MCV Campus.

Applying

Applications may be filed at any time but preferably by January 6 for admission in the next academic year, which begins in the latter part of August. Late application limits the possibility of consideration for financial support. All applicants must have an earned baccalaureate degree or its equivalent at the time of enrollment. Additional entrance requirements for specific programs are stipulated by the departments.

A completed application should be submitted to the address given below, along with an official transcript of all graduate and undergraduate records mailed from the college or university registrar, recommendations (on forms supplied with the application) from teachers who can evaluate the applicant's ability to carry out graduate study, and a letter from the applicant indicating motivation and educational preparation for graduate study, as well as objectives of graduate study. The GRE General Test is required. International students whose native language is not English must have a TOEFL score of 550 or higher. Acceptance of an applicant is based on the recommendation of the department. International applicants should submit complete applications before January 1 to allow adequate time for credential review.

Correspondence and Information

School of Graduate Studies
Virginia Commonwealth University
P.O. Box 843051
Richmond, Virginia 23284-3051
Phone: 804-828-6916
Web site: http://www.vcu.edu (VCU)
       http://www.medschool.vcu.edu (School of Medicine)

Virginia Commonwealth University

DEPARTMENTS AND AREAS OF RESEARCH

Anatomy and Neurobiology

Dr. John T. Povlishock, Chair (plus 21 faculty members). The department's research interests are primarily in neurobiology and immunobiology and include the study of morphological correlates of traumatic brain injury, neural injury and regeneration, neural plasticity in the visual system, neuroanatomy and neurophysiology of eye movement, oculomotor and tongue nerve-muscle interactions, CNS multisensory integration, neuro-oncology, ontogeny of the immune system, immune dysfunction and malignancy, endothelial cell biology, angiogenesis and wound healing, tissue engineering and gene expression in cardiac myocytes, and neuroendocrine mechanisms of growth hormone. Contact: Dr. Leichnetz (804-828-9512; gleichne@hsc.vcu.edu).

Biochemistry

Dr. Sarah Spiegel, Chair (plus 20 faculty members). Research interests are cellular and molecular signaling, structural biology, eukaryotic molecular biology, lipid and membrane biochemistry, molecular genetics, enzymology, and tumor biology. Emphasis is now on developments in proteomics, genomics, metabolomics, and lipidomics. Contact: Committee for Graduate Studies (804-828-9762; biochemgrad@vcu.edu).

Biostatistics

Dr. W. H. Carter, Chair (plus 12 faculty members). Current research interests are health services research; linear models; response surfaces; design and analysis of drug/chemical combination experiments; robust statistics; clinical trials; biomedical applications; epidemiology, including cancer epidemiology and the epidemiology of chronic disease; multivariate analysis; survival analysis; quantitative bioinformatics; and sequential statistical analysis. Contact: Mr. Russ Boyle (804-828-9824; bisinfo@hsc.vcu.edu).

Epidemiology

Dr. Betsy Turf, Director. This Ph.D. program is administered through Epidemiology and Community Health and builds on training for the Master of Public Health degree. Contact: Dr. Turf (804-828-9785; eturf@vcu.edu).

Human Genetics

Dr. P. O'Connell, Chair (plus 24 faculty members). Areas of research involve human cytogenetics, molecular genetics, genetic epidemiology, and biochemical genetics. Clinical interests of departmental members include linkage studies, studies of human twins and their offspring, hereditary deafness, delineation of new genetic syndromes, diagnosis and treatment of metabolic diseases, genetic counseling, and behavior genetics. Contact: Dr. Shiang (804-828-9632 ext. 124; rshiang@hsc.vcu.edu). Genetic Counseling Program Contact: Ms. Vanner-Nicely (804-828-9632 ext. 135; lvanner@hsc.vcu.edu).

Microbiology and Immunology

Dr. Dennis E. Ohman, Chair (plus 50 faculty members). Research interests in microbiology include mechanisms of microbial pathogenesis; host-parasite relationships; antibiotic resistance; membrane transport; phages; and the physiology, genetics, and molecular biology of bacteria, parasites, viruses, and fungi. Research interests in immunology include antigen presentation, antibody synthesis, immediate hypersensitivity, phospholipid metabolism, and cancer and cellular biology. Contact: Dr. Cabral (804-828-2306; gacabral@hsc.vcu.edu).

Pathology

Dr. David Wilkerson, Chair (plus 29 faculty members). Current research interests include carcinogenesis, tumor cell invasion and metastasis, infectious disease and virology, molecular diagnostics, clinical chemistry, toxicology, development of diagnostics immunoassays, clinical outcomes research, and basic molecular and cellular pathobiology. Contact: Dr. Holt (804-828-4802; seholt@hsc.vcu.edu).

Pharmacology and Toxicology

Dr. Billy R. Martin, Chair (plus 40 faculty members). Research interests include molecular and cellular pharmacology, signal transduction, alcohol and drug abuse, behavioral pharmacology, cancer pharmacology, immunotoxicology, central nervous system pharmacology, cardiovascular pharmacology, traumatic brain injury, and DNA repair mechanisms. Contact: Dr. Sawyer (804-828-7918; ssawyer@hsc.vcu.edu).

Physiology

Dr. Margaret Biber, Chair (plus 60 faculty members). Research emphasizes cardiovascular, gastrointestinal, and sensory (olfaction and taste) function using modern cellular and molecular approaches and interdisciplinary collaboration. Specific areas include cardiovascular: microcirculation, effect of disease on voltage-gated potassium channels, cardiac cell volume and stretch-activated ion channels; gastrointestinal: neural networks, signaling molecules (growth factors, hormones, transmitters, gases), and intracellular signaling cascades involved in development and regulation of motility and secretory activity; and chemical senses: transduction mechanisms, development, plasticity, and regeneration. Contact: Dr. Ford (804-828-9501; ford@hsc.vcu.edu).

Epidemiology and Community Health

Dr. Tilahun Adera, Chair (plus 11 faculty members). The department offers programs of study leading to a doctoral degree in epidemiology and the Master of Public Health degree. Research interests include etiological studies of low back pain, hearing loss, breast cancer, violence prevention program evaluation, health policy analysis, nutrition and health, evaluation of environmental and occupational exposures, and maternal and child. Contact: Ms. Bryant (804-828-9785; kpbryant@vcu.edu).

Genetic Counseling Program

Lauren Vanner-Nicely, Director. This is a two-year master's degree program in the Department of Human Genetics leading to the Master of Genetic Counseling degree. The curriculum combines course work and practical experience in human genetics, counseling and related disciplines. Genetic Counselors are professional members of a medical genetics team, who participate in diagnostic evaluation, counseling, education, administration, and clinical research. Contact: Ms. Vanner-Nicely (804-828-9632 Ext. 135; lvanner@hsc.vcu.edu).

Molecular Biology and Genetics

Dr. Gail Christie, Director. The program coordinates scholarly activity in molecular biology and genetics among participating departments in the School of Medicine. Contact: Dr. Christie (804-828-9093; christie@hsc.vcu.edu).

Immunology

Dr. John Tew, Director. The program coordinates scholarly activity in the field of immunology among participating departments in the School of Medicine. Contact: Dr. Tew (804-828-9715; tew@hsc.vcu.edu).

Neuroscience

Dr. Les Satin, Director. The program coordinates scholarly activity in the neurosciences among participating departments in the School of Medicine. Contact: Dr. Satin (804-828-7823; lsatin@hsc.vcu.edu).

M.D./Ph.D. Program

Dr. Gordon Archer, Director. The M.D./Ph.D. program offers training to students with career objectives as physician scientists. Ph.D. training in any of the basic science programs is initiated after the second year of medical school and allows access to a variety of research opportunities. Contact: Ms. Sorrell (804-828-9711; ssorrell@hsc.vcu.edu).

WESLEYAN UNIVERSITY

Department of Biology

Program of Study

The Department of Biology offers graduate studies in cell and developmental biology, evolutionary and population biology, genetics, neuroscience, and bioinformatics, leading to the degree of Doctor of Philosophy in biology. A Master of Arts degree may be awarded under certain conditions (see below). Although the emphasis is on an intensive research opportunity, leading to a thesis, students are also expected to acquire a broad knowledge of related biological fields through an individual program of courses, seminars, and reading. A weekly series of research seminars is presented by distinguished scientists from other universities and is supplemented by regular seminars presented by Wesleyan graduate students and faculty members. A high faculty-student ratio in the Department ensures close interaction among students, faculty members, and postdoctoral fellows. In addition, the modest size of the program permits the requirements to be tailored to fit the different needs of individual students. All graduate students have the opportunity for instructing undergraduate laboratory sections and classroom teaching with faculty guidance.

The M.A. degree may be awarded only under circumstances in which a student enrolled in the Ph.D. program is unable to continue pursuing the Ph.D. degree for academic or personal reasons; the program requires two years and a minimum of eight semester courses, including graduate seminars and thesis research. The Ph.D. degree normally requires at least four years beyond the baccalaureate; although mainly a research degree, it involves an individualized program of courses adapted to the student's proposed field of interest and his or her background in the biological and physical sciences.

The Department staff includes 14 professors, 1 adjunct faculty member, several postdoctoral fellows, and a number of research associates and assistants. The Department of Biology is closely affiliated with the Department of Molecular Biology and Biochemistry, housed in the same complex.

Research Facilities

The Department is housed in Shanklin and Hall-Atwater Laboratories. Various specialized facilities include animal rooms, a computer-regulated greenhouse, NMR instruments, PCR and microarray equipment, and an imaging facility that includes electron and confocal microscopes. State-of-the-art University computer facilities support teaching and research.

Financial Aid

All students in good standing in the program are provided an annual stipend in the form of a research fellowship or teaching assistantship. All summer stipends are research fellowships, free of teaching duties. The 2005–06 stipend for a twelve-month period was $20,451. To this is added free tuition.

Cost of Study

Since all graduate degree candidates hold fellowships or assistantships and the tuition and fees are waived, the only academic costs to students are books and other educational materials and thesis expenses.

Living and Housing Costs

For a single student, room and board are estimated at $8930 per year, with miscellaneous expenses averaging an additional $1500 to $1800. For married students, the costs would be greater, but a dependency allowance is available for those who qualify. Both University accommodations and privately owned housing are available.

Student Group

The relatively small scale of the programs allows intensive interaction between students, postdoctoral fellows, and faculty members in close-knit laboratory groups. In 2005–06, there were 24 graduate students from colleges in the United States and abroad.

Student Outcomes

Ph.D. recipients over the past five years have made diverse career choices, including postdoctoral positions at Brigham and Women's HHMI, Brown, Columbia, Duke, Harvard, Jackson Laboratory, Johns Hopkins, NIMH, Rockefeller, SUNY at Albany, Sussex University (England), University of Connecticut, University of Massachusetts, University of Michigan, University of Stockholm (Sweden), Czech Academy of Sciences, University of Toronto, Vanderbilt, and Yale and research positions at Bristol-Myers Squibb, Cato Corporation, Merck, and Pfizer, Inc.

Location

Wesleyan University is situated in Middletown, Connecticut, a small city with a population of 45,000 on the west bank of the picturesque Connecticut River, about 19 miles south of Hartford and 25 miles north of New Haven. It lies approximately midway between New York and Boston and about 25 miles from Long Island Sound. The University provides many recreational and cultural opportunities, including many made possible by the University's outstanding graduate program in ethnomusicology. These are supplemented by opportunities available in the surrounding countryside and in the large cities nearby.

The University

Founded in 1831, the University is an independent institution of liberal arts and sciences. Since 1963, Wesleyan has developed selective programs leading to the Ph.D. degree in a number of departments. Current registration includes approximately 2,700 undergraduates and about 250 graduate students. The University has total resources, including endowment, of approximately $350 million.

Applying

No specific courses are required for admission, but successful applicants usually have a strong undergraduate major in biology or in a closely related science. The complete application consists of Graduate Record Examinations scores, transcripts of all previous academic work at the college level, and three letters of recommendation from college instructors familiar with the applicant's record and ability. Personal interviews are encouraged. To qualify for primary consideration, completed applications should be received by January 15. Late applications are considered if vacancies exist.

Correspondence and Information

Graduate Admissions Committee
Department of Biology
Wesleyan University
Middletown, Connecticut 06459-0170

Phone: 860-685-2140
Fax: 860-685-3279
E-mail: biophd@wesleyan.edu
Web site: http://www.wesleyan.edu/bio

Wesleyan University

THE FACULTY AND THEIR RESEARCH

Professors
David Bodznick, Ph.D., Washington (Seattle). Neuroethology; sensory information processing in vertebrate brains.
Barry Chernoff, Ph.D., Michigan. Systematics and biogeography of freshwater fishes of Latin America; morphological evolution; conservation of aquatic ecosystems.
Frederick M. Cohan, Ph.D., Harvard. Population biology of bacteria.
J. James Donady, Ph.D., Iowa. Genetic regulation of differentiation; role of nuclear structure in gene regulation.
Laura Grabel, Ph.D., California, San Diego. Role of cell-matrix interactions during teratocarcinoma stem cell differentiation; identification of lineage-determining genes in early mouse embryogenesis.
Michael Weir, Ph.D., Pennsylvania. Developmental genetic studies of *Drosophila* development; bioinformatics.
Jason S. Wolfe, Ph.D., Berkeley. Cellular reproduction; cell interactions; cell death.

Associate Professors
Ann C. Burke, Ph.D., Harvard. Development and evolution of vertebrates.
Stephen H. Devoto, Ph.D., Rockefeller. The development of zebrafish muscle cell identity.
John Kirn, Ph.D., Cornell. Developmental neurobiology of vocal learning in songbirds.
Janice Naegele, Ph.D., MIT. Developmental cell death and neuronal survival in the mammalian nervous system.
Sonia E. Sultan, Ph.D., Harvard. Evolutionary implications of plant response to the environment.

Assistant Professors
Gloster Aaron, Ph.D., Pennsylvania. Roles of interneurons in the organization of cortical activity.
Michael Singer, Ph.D., Arizona. Ecological causes of herbivore foraging behavior; consequences of carnivore-herbivore-plant interactions for biodiversity.

Associated Faculty in the Department of Molecular Biology and Biochemistry
Mark Flory, Ph.D., Washington (Seattle). Proteomic and cell biological analysis of fission yeast centrosomes and telomeres.
Manju M. Hingorani, Ph.D., Ohio State. Enzymology of DNA mismatch repair and DNA replication in *E. coli* and *S. cerevisiae.*
Scott Holmes, Ph.D., Virginia. Molecular genetics of gene silencing in *Saccharomyces cerevisiae;* chromosome structure and function.
Anthony I. Infante, Ph.D., Pennsylvania. Gene regulation in embryogenesis; molecular mechanisms regulating protein synthesis in sea urchins.
Robert Lane, Ph.D., Caltech. The olfactory system: studies in gene regulation and genome evolution.
Michael McAlear, Ph.D., McGill. Genetic and biochemical analysis of DNA replication and DNA repair in budding yeast.
Ishita Mukerji, Ph.D., Berkeley. Protein and nucleic acid structure; structure-function relationships in protein–nucleic acid interactions using UV resonance Raman spectroscopy.
Donald B. Oliver, Ph.D., Tufts. Analysis of protein export pathways in bacteria: genetics, regulation, and biochemistry; structural and functional dissection of *E. coli* SecA ATPase.

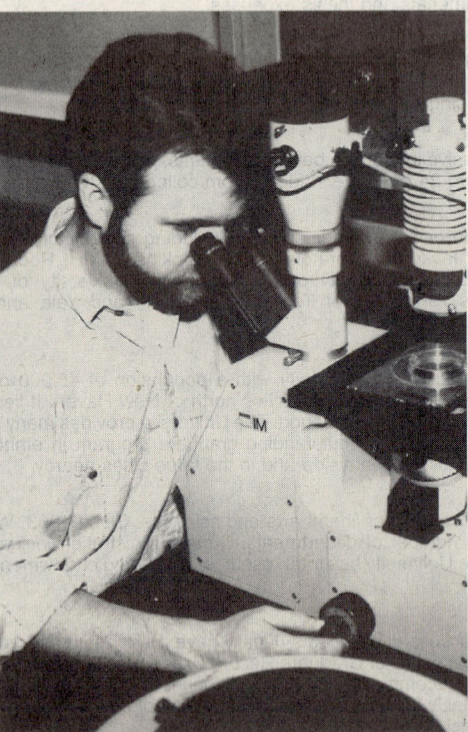

Inspecting newly transfected mouse teratocarcinoma cells.

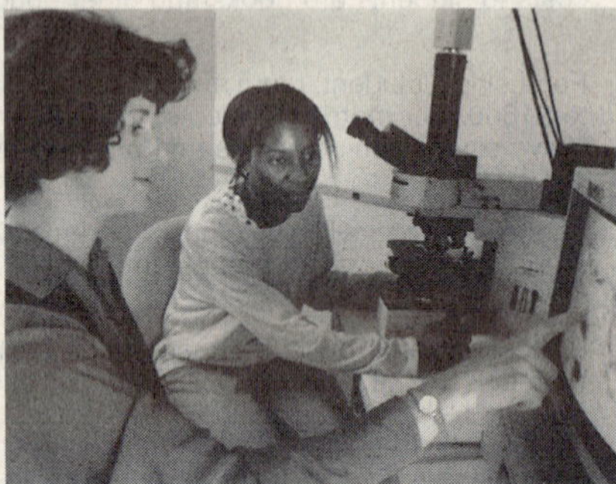

Visualizing neurons of the rat basal ganglia stained with monoclonal antibodies.

Harvesting inbred achenes from field-collected *Polygonum* populations in the research greenhouse.

WEST VIRGINIA UNIVERSITY

Department of Biology

Programs of Study	The Department of Biology offers programs that lead to M.S. and Ph.D. degrees in the focal areas of Ecology and Evolutionary Biology (EEB), Forensic Biology (FB), Genetics and Genome Biology (GGB), or Neurobiology and Endocrinology (NE). The EEB program offers research opportunities in the areas of wetland and forest biogeochemistry, urban ecology, conservation of rare and endangered species, plant systematics, mycorrhizal-plant interactions, global change effects in forests and agriculture, and invasive species biology. Research in the FB program includes methods of individual and species identification, as well as novel applications of molecular techniques for biological evidence. The GGB program offers research opportunities in the areas of gene regulation, chromosome structure, chromatin remodeling, developmental signaling pathways, neuronal patterning, plant genomics and molecular biology, mobile genetic elements, and bioinformatics. In the NE program, research emphases include sensory neuroscience, ethology, neuroethology, invertebrate neurobiology, developmental neurobiology, neural plasticity, toxicology and pharmacology, reproductive physiology, and endocrinology.
Research Facilities	The Department of Biology is housed in the 192,000-square-foot Life Sciences Building on the downtown campus, which contains modern research and teaching facilities. Recombinant DNA core facilities and monoclonal antibody, protein analysis, and image analysis centers are available in the Departments of Biochemistry and Anatomy in the Health Sciences Center and the Mary Babb Randolph Cancer Center. Animal care facilities are located within the building. The biological research, instruction, and seminars available in the School of Medicine and the College of Agriculture provide further opportunities for students in the Department of Biology. Six greenhouses and a growth chamber facility are available for classes and environmental research. The Core Arboretum, West Virginia University Herbarium, Agronomy Farm, Horticulture Farm, Animal Science Farm, West Virginia University Forest, and the Fernow Experimental Forest are available for ecological studies.
	Computer facilities are located throughout the Morgantown campuses, and computing services are provided through the West Virginia Network for Education Telecomputing. The combined libraries of the University contain more than 1 million volumes and 1.5 million microfilms, with a strong collection in the biological sciences.
Financial Aid	The major source of support is the teaching assistantship, which offers stipends that range from $10,600 (M.S.) to between $15,100 and $17,100 (Ph.D.) for nine months, plus tuition waivers and the abatement of some fees. Inexpensive health care is available for graduate students. Several competitive awards of $2000 above the minimum stipend are granted either by the Department or the College of Arts and Sciences. Stipends for research may be available from faculty grant support. Swiger Doctoral Fellowships, valued at $20,000 per annum, plus remission of tuition and some fees, are available on a competitive basis for Ph.D. applicants. William Ford Federal Direct Loans can provide low-interest assistance. Inquiries concerning loans may be directed to the Financial Aid Office.
Cost of Study	Department of Biology graduate teaching and research assistants are granted full tuition waivers and the abatement of some fees.
Living and Housing Costs	University apartments are available for graduate students (see http://www.sa.wvu.edu/housing/index.htm). However, most students find off-campus apartments or houses. Rents typically range from $400 to $600 per month.
Student Group	Graduate students with a variety of backgrounds come to West Virginia University (WVU) from many parts of the United States and the world.
Location	Morgantown has won numerous accolades for its high-quality, small city environment. With a population of 45,000, Morgantown is situated in the beautiful mountainous terrain of the Appalachians. State parks in the vicinity offer opportunities for swimming, mountain biking, kayaking, rafting, waterskiing, fishing, hiking, skiing, and horseback riding. The University and Morgantown provide many cultural activities, including concerts, operas, and repertory theater. Pittsburgh, 70 miles to the north, and the Washington-Baltimore area, 200 miles to the east, are easily accessible by four-lane interstate highways and offer a wealth of cultural and recreational activities. Direct air service from Morgantown to each of these areas provides connections with virtually all major airlines.
	A major research resource in the Appalachian region, Morgantown has the research facilities of four federal agencies: the Department of Health and Human Services (Appalachian Laboratory for Occupational Safety and Health), the U.S. Forest Service (Northeastern Forest Experiment Station), the National Energy Technology Laboratory of the U.S. Department of Energy, and the Soil Conservation Service (West Virginia headquarters).
The University	West Virginia University, a land-grant institution founded in 1867, is the center of graduate and professional education, research, and extension programs in the state. With about 25,000 students and 1,300 full-time faculty members, the University is a comprehensive public institution that offers 171 degree programs from the bachelor's through the doctoral level. There are fourteen undergraduate, graduate, and professional schools and three interdisciplinary programs, two of which relate to the biological sciences. All WVU programs are fully accredited by the North Central Association of Colleges and Schools for ten-year periods, in recognition of the University's status as one of a limited number of mature, doctoral degree–granting institutions in the United States.
Applying	Applicants should request application forms from the address below. Prospective students must demonstrate potential and a strong motivation for graduate studies in addition to a sound undergraduate preparation in the natural sciences. Complete applications include the WVU application and fee, GRE scores, a personal statement (one- to two-page) regarding research experience and interests, three reference letters, and official transcripts. International students must show competence in English by submitting scores on the Test of English as a Foreign Language. For fall admission, applications must be completed by January 1.
Correspondence and Information	Dr. William T. Peterjohn Associate Chair of Graduate Studies Department of Biology West Virginia University P.O. Box 6057 Morgantown, West Virginia 26506-6057 Phone: 304-293-5201 Ext. 31510 E-mail: bpj@wvu.edu Web site: http://www.as.wvu.edu/biology/graduate/programs.htm

West Virginia University

THE FACULTY AND THEIR RESEARCH

C. Barth, Assistant Professor; Ph.D., Heinrich Heine (Germany). Stress physiology: plant defense signaling in response to biotic and abiotic stress.

A. P. Bidwai, Associate Professor; Ph.D., Utah State. Molecular genetic analysis of protein kinase CK2 in *Drosophila*.

C. P. Bishop, Associate Professor; Ph.D., Virginia. Molecular genetics; developmental biology; forensic biology.

J. R. Cumming, Associate Professor and Chair; Ph.D., Cornell. Plant-soil interactions, mycorrhizae, and metal stress; forest ecology in the urban environment.

K. C. Daly, Assistant Professor; Ph.D., Arizona. Behavioral, neurophysiological, and computational analysis of olfactory processing; learning and memory.

S. P. DiFazio, Assistant Professor; Ph.D., Oregon State. Plant genomics; molecular ecology; plant population genetics; biotechnology risk assessment.

S. M. Farris, Assistant Professor; Ph.D., Illinois. Development and evolution of brain and behavior.

J. A. Flores, Associate Professor; Ph.D., George Washington. Mammalian reproductive physiology; follicular development and corpus luteum physiology.

C. M. Foran, Assistant Professor; Ph.D., Cornell. Environmental physiology; neuroendocrine control of reproduction; endocrine disruption.

D. Ford-Werntz, Clinical Associate Professor; Ph.D., Washington–St. Louis. Plant systematics: Portulacaceae; West Virginia flora.

K. Garbutt, Associate Professor; Ph.D., U.C.N., Wales. Population genetics: ecological genetics and population biology of weedy plants.

P. E. Keeting, Associate Professor; Ph.D., University of Medicine and Dentistry of New Jersey. Cellular and molecular biology: molecular mechanisms of hormonal regulation of cellular biology.

J. B. McGraw, Eberly Professor of Biology; Ph.D., Duke. Plant demography; remote sensing; conservation biology.

W. T. Peterjohn, Associate Professor; Ph.D., Duke. Biogeochemistry; global change; nitrogen saturation and cycling; soil microbiology.

D. A. Ray, Assistant Professor; Ph.D., Texas Tech. Molecular evolution; population genetics; forensic genetics; mobile element biology; genomics.

R. B. Thomas, Associate Professor; Ph.D., Clemson. Physiological ecology: plant response to environmental stress, global change, and multiple resource limitations.

M. R. Walbridge, Professor; Ph.D., North Carolina. Wetland ecology; biogeochemistry; urban ecosystems.

K. S. Weiler, Assistant Professor; Ph.D., Princeton. Chromosome structure and epigenetic regulatory mechanisms in *Drosophila*.

J. D. Wells, Associate Professor; Ph.D., Illinois at Chicago. Forensic biology and molecular systematics.

WEST VIRGINIA UNIVERSITY

M.D./Ph.D. Scholars Program

Program of Study	West Virginia University School of Medicine's M.D./Ph.D. Scholars Program prepares students for academic careers that combine the practice and teaching of clinical medicine with laboratory investigation of disease mechanisms. The goal is to train future independent investigators to function as physician-scientists. This dual training program requires at least seven years to complete. Students receive both the M.D. and Ph.D. degrees. As M.D./Ph.D. Scholars, students benefit from individual attention by faculty members, within a research environment that is dynamic, collaborative, and interdisciplinary. Students enter the program in the summer before beginning medical school, with an orientation to the various areas of research and with the selection of a physician-scientist role model. They choose one research laboratory and spend six weeks conducting research before beginning medical school. In Years 1 and 2, trainees take the integrated medical school basic science curriculum and participate in biweekly research forums with all of the M.D./Ph.D. trainees. At these forums, students present their research, learn from other physician-scientist role models, and discuss academic career opportunities. During the summer after Year 1, trainees complete a second research rotation to facilitate their final selection of a faculty research adviser and a specific graduate program. All M.D./Ph.D. Scholars work in state-of-the-art, federally funded research laboratories. After successful completion of Years 1 and 2 of the medical curriculum as well as successful completion of Step 1 of the United States Medical Licensing Examination (USMLE), students enter the research portion of their Ph.D. training, which takes at least three years to complete. Research opportunities are numerous and include cell and molecular biology, integrative physiology, immunology, exercise physiology, cardiovascular sciences, receptor biochemistry, bacterial pathogenesis, lung cell biology, environmental exposures, inflammation, molecular genetics, pharmaceutical and pharmacological sciences, neuroendocrine and reproductive biology, developmental biology, tumor invasion and angiogenesis, neurodegenerative disorders and stroke, functional brain imaging and cognitive behavior, memory and learning, and outcomes- and epidemiology-based studies in public health. Research training is supervised by graduate faculty members from basic science, public health, and clinical departments and who are members of one of seven graduate programs in the biomedical sciences (Biochemistry and Molecular Biology, Cancer Cell Biology, Cellular and Integrative Physiology, Exercise Physiology, Immunology and Microbial Pathogenesis, Neuroscience, and Pharmaceutical and Pharmacological Sciences) or are members of the Graduate Program in Public Health Sciences. After completing their Ph.D. research and defending their doctoral dissertation, students begin clinical rotations in Years 3 and 4 of the M.D. training period on the Health Sciences campus in Morgantown.
Research Facilities	Institutional facilities include a computer-based learning center, a centralized animal facility with a transgenic barrier, and a library housing more than 205,000 volumes and 2,400 journals. Core facilities are available for examining gene expression or genetic variation (Affymetrix platform), image analysis, confocal and electron microscopy and laser capture microdissection, live-cell imaging, mass spectrometry, flow cytometry and high-speed cell sorting, proteomics, recombinant DNA technology, transgenic rodent biology, and functional neuroimaging (fMRI, PET/CT). Clinical facilities include West Virginia University Ruby Memorial Hospital, the Jon Michael Moore Trauma Center, and Chestnut Ridge Psychiatric Hospital. Affiliated research centers include National Institute for Occupational Safety and Health (NIOSH), Center for Advanced Imaging, Blanchette Rockefeller Neurosciences Institute, Sensory Neuroscience Research Center, and Mary Babb Randolph Cancer Center.
Financial Aid	M.D./Ph.D. students receive full financial support during their training, provided they remain in good academic standing and excel in research. Such support includes health insurance, an annual stipend of $22,000, and full tuition coverage during both the M.D. and Ph.D. training periods.
Cost of Study	Students' tuition costs are covered.
Living and Housing Costs	The cost of an efficiency apartment in University-owned housing is approximately $400 per month. A limited number of University apartments are available for married students. Privately owned apartments in Morgantown cost $400 to $600 per month. In general, the cost of living is lower compared to larger cities.
Student Group	There are currently 20 students in the M.D./Ph.D. Scholars Program, and 2 to 3 students are accepted each year. The medical school has more than 100 other graduate students pursuing Ph.D. degrees in the biomedical sciences and approximately 400 medical students. University enrollment is 25,000 students, which includes 6,500 graduate and professional students who come from all parts of the United States and many other countries.
Location	With an appealing balance to life, Morgantown is a vibrant university community of 80,000 residents in northern West Virginia. Located near the Pennsylvania border at the western edge of the Appalachian Mountains, abundant opportunities exist for activities such as world-class white-water rafting and kayaking, hiking and camping, mountain biking, fishing, and skiing. Morgantown has a cosmopolitan atmosphere with a range of activities usually found in much larger cities. It enjoys proximity to major metropolitan centers; Pittsburgh is a 90-minute drive north, and Washington, D.C., is a 3-hour drive east.
The University	West Virginia University (WVU) is a comprehensive, land-grant, Carnegie-designated Doctoral/Research University–Extensive public institution. The University's academic Health Sciences Center includes the Schools of Medicine, Dentistry, Nursing, and Pharmacy, all of which offer graduate degree programs. The research enterprise at WVU Health Sciences Center has been organized around six interdisciplinary research centers in the health-related areas of cancer cell biology, cardiovascular sciences, diabetes and obesity, immunopathology and microbial pathogenesis, neuroscience, and respiratory biology and lung diseases. As a member of the Big East Conference, WVU participates in NCAA Division I sports. WVU also offers a variety of creative arts, theater, and entertainment opportunities.
Applying	Admission requires successful application to the Medical School Admissions Committee through the American Medical College Application Service (AMCAS), followed by a separate application to the M.D./Ph.D. Scholars Program. The M.D./Ph.D. Admissions Committee requires a personal interview and two additional letters of recommendation specifically addressing the student's research potential before making a final decision on acceptance. Prospective students should visit WVU's Web site at http://www.hsc.wvu.edu/som/resoff/gradprograms/MDPhD.asp for more information and an online application.
Correspondence and Information	Fred L. Minnear, Ph.D. Professor and Assistant Dean for Graduate Studies Director, M.D./Ph.D. Scholars Program Office of Research and Graduate Studies West Virginia University School of Medicine P.O. Box 9104 Morgantown, West Virginia 26506 Phone: 304-293-6229 E-mail: fminnear@hsc.wvu.edu Web site: http://www.hsc.wvu.edu/som/resoff/gradprograms/mdphd.asp

West Virginia University

RESEARCH EMPHASIS AREAS

Research training opportunities exist in the laboratories of graduate faculty members with academic appointments in various basic science, public health, and clinical departments in the Schools of Medicine and Pharmacy. These faculty members have graduate appointments in one of eight graduate programs: Biochemistry and Molecular Biology, Cancer Cell Biology, Cellular and Integrative Physiology, Exercise Physiology, Immunology and Microbial Pathogenesis, Neuroscience, Pharmaceutical and Pharmacological Sciences, and Public Health Sciences. Faculty research interests focus on one or more of the following six major interdisciplinary biomedical research areas:

Cancer Cell Biology

The three main areas of research emphasis are cellular signaling, tumor microenvironment, and cancer therapeutics. Cellular signaling focuses on protein and lipid-based signals that influence tumor growth, survival, motility, and invasion. Tumor microenvironment addresses the mechanisms by which tumor cells interact with other cells in the stroma to promote tumor survival, angiogenesis, and inflammation. Therapeutics addresses the mechanisms by which novel cancer therapeutic compounds block tumor cell growth and metastasis, as well as strategies for the translational development and delivery of conventional chemotherapeutics and targeted small-molecule compounds. Public health research focuses on epidemiology, control, and prevention of tobacco use.

Neuroscience

Research emphasis is in sensory neuroscience; cognitive neuroscience and functional brain imaging; behavorial neuroscience, including learning and memory and psychotherapeutic drugs; neurodegeneration and stroke; aging and Alzheimer's disease; neural injury development and disorders of the visual and auditory systems; neuroendocrinology; autonomic control of feeding and respiration; and environmental influences on brain function, such as exposure to solvents and neurologic morbidity. Diseases studied include Alzheimer's and related dementias, anxiety and stress-related disorders, asthma, autism, blindness, depression, epilepsy, obesity and eating disorders, schizophrenia, and stroke and related cerebrovascular events.

Cardiovascular Sciences

Research interests encompass the physiological and pharmacological sciences and incorporate biochemical, immunological, electrophysiological, cellular, and molecular approaches to address contemporary hypotheses. Researchers study the integrative control of cardiovascular, microvascular, endocrine, neural, and respiratory function, with an emphasis on heart, vascular, and lung diseases or processes such as aging, Alzheimer's, angiogenesis, asthma, COPD, coronary artery disease, congestive heart failure, diabetes and obesity, hypertension, and tissue edema. Specific research interests include endothelial permeability, vascular integrity, and tissue edema; hypertension and nitric oxide; diabetes, obesity, and exercise; lipid metabolism and gene expression; regulation of angiogenesis and vascular development; local and neural control of blood flow; and epidemiology and prevention of obesity, diabetes, and other cardiovascular morbidities.

Respiratory Biology and Lung Diseases

Research interests include inhalant-mediated lung epithelial injury; airway inflammation and asthma; alveolar macrophage activation and cytokines; lung endothelial and epithelial permeability; oxidant-antioxidant balance in the lung; cancer and fibrosis: environmental/occupational mediators; and epidemiology and prevention of lung diseases.

Immunopathology and Microbial Pathogenesis

Faculty members and students explore diverse areas of inquiry related to the medical implications of microbes and the human body's response to them. Research areas include the physiology of pathogenic microbes; microbial virulence factors; the biochemistry of inflammatory cytokines; interactions between microbes and their hosts; immune response in bacterial and viral diseases; effects of man-made pesticides and herbicides on the immune system; molecular aspects of cell signaling as it relates to cancer chemotherapy and cell growth control; cytotoxic T cells in transplant rejection and disease pathogenesis; bacterial diseases, including Lyme disease, cystic fibrosis, and streptococcal infections; peptide and DNA vaccines for contraception; and science-based prevention of infectious diseases.

Diabetes and Obesity

Research interests include metabolic dysfunction in diabetes; relationships of obesity to diabetes; effects of exercise on muscle development and metabolism; cellular adaptations of skeletal muscle to overload and disease; genetic influences on response to exercise; neuroendocrine control of food intake; adipose tissue biology and lipid transport; and population-based research, epidemiology, and prevention research as related to diabetes and obesity.

WICHITA STATE UNIVERSITY

Department of Biological Sciences

Program of Study	The Master of Science program offered by the Department of Biological Sciences provides an advanced education with a variety of specializations in the broad areas of avian, cell, endocrine, environmental, molecular, plant, and reproductive biology. All incoming students are assigned to a temporary graduate adviser, and typically by the end of the first year, students choose a permanent graduate adviser and committee. The advisers work with students to develop a program of studies that meets their educational goals.
	All students are required to attend the Biology Seminar course each semester and must give a minimum of two oral presentations. Candidates must complete 30 credit hours of graduate work, including the presentation and oral defense of a thesis based on original research. In addition, all students must demonstrate proficiency in at least one research tool, such as knowledge of a modern foreign language or completion of acceptable course work in statistics or calculus.
Research Facilities	The Department is housed primarily on 2½ floors of Hubbard Hall, which contains modern, well-equipped research laboratories. Core laboratories have been established for protein analysis, bacteriology, bioinformatics, environmental toxicology, and microscopic imaging. Each core facility is staffed by a faculty director. Major instrumentation is available for the sequencing and synthesis of DNA and protein, PCR-dependent procedures, digital photomicroscopy, and carbohydrate analysis, as well as state-of-the-art chromatography and electrophoresis systems. Additional facilities include isolated animal maintenance and procedure rooms, a climate-controlled greenhouse, an environmental growth chamber room, a small herbarium, and vertebrate study collections. Large areas of relatively undisturbed prairie habitats are located nearby, and a biological field station provides excellent opportunities for aquatic and field research. The library contains more than 22,190 titles of books and periodicals that cover all areas of biology.
Financial Aid	A number of graduate teaching assistantships are available to qualified students. Recipients are required to teach 8 to 10 hours per week, primarily in laboratories, receiving a stipend of $800 per month for the academic year. In addition, recipients are eligible for a tuition waiver. Nonresident graduate students who are awarded graduate teaching assistantships are assessed tuition and fees at the Kansas resident's rate. A limited number of graduate research fellowships are awarded annually, but recipients are not eligible for a tuition waiver.
Cost of Study	For 2006–07, tuition fees are $198 per credit hour for resident students and $548 per credit hour for nonresident students. In addition, a nonrefundable registration fee of $17 is required each semester.
Living and Housing Costs	In 2006–07, on-campus housing costs for the academic year range from $3960 to $6476 for single occupancy and from $2860 to $5376 for double occupancy, depending on the room size and meal plan. The University is located in an urban setting, and many convenient off-campus rentals in various price ranges are available.
Student Group	As an urban institution, Wichita State University enrolls approximately 15,000 students. More than 10,000 students are employed either full-time or part-time, and nearly 1,900 are over age 30. A total of approximately 3,100 graduate students are enrolled in programs of the eight schools and colleges of the University. Currently, approximately 25 students are enrolled full-time in the graduate program of the Department of Biological Sciences.
Location	Wichita is the largest city in Kansas, with a metropolitan population of more than 400,000 residents, and is known as the "Air Capital of the World," since it is the home of Beech (now Raytheon), Boeing, Cessna, and Learjet. It is also a regional medical center, home to energy and agricultural industries, and the industrial and educational center of Kansas. Thus, Wichita enjoys a diversified economy that provides outstanding career opportunities in a variety of fields. Wichita offers the cultural and economic advantages of a big city but maintains the friendly atmosphere of a smaller town.
The University and The Department	Wichita State University became part of the Kansas State University System in 1964 after operating as a private institution (Fairmount College) since 1895 and as a municipal university since 1895. In addition to diverse academic programs, the University provides many cultural opportunities for its students through the Ulrich Art Museum, the University Symphony, and the University Theater. As part of the Fairmount College of Liberal Arts and Sciences, the Department of Biological Sciences participates in college educational programs, such as exploring new ways to increase interdepartmental graduate contacts and cooperation, and offers basic and applied research opportunities in various biological fields.
Applying	Completed application forms and two official transcripts of all previous academic work must be submitted to the graduate school by April 1 (for international students) or June 1 (for citizens or permanent residents) for the fall semester. For the spring semester, deadlines are August 1 (for international students) or December 1 (for citizens or permanent residents). Admission as a student in full standing requires the completion of 24 semester hours in biological sciences and 15 semester hours in chemistry, an overall minimum grade point average of 2.75 on a 4.0 scale for the most recent 60 semester hours completed, a grade point average of at least 3.0 on a 4.0 scale for all undergraduate biological science courses, three letters of reference from science faculty members, receipt of satisfactory scores from the General Test and Subject Test in biology of the Graduate Record Examinations (GRE), and satisfactory TOEFL scores from students whose native language is not English. Nonrefundable application fees are $35 for domestic applicants and $50 for international applicants.
Correspondence and Information	Graduate Coordinator Department of Biological Sciences Wichita State University Wichita, Kansas 67260-0026 Phone: 316-978-3111 Fax: 316-978-3772 E-mail: biology@wichita.edu Web site: http://www.webs.wichita.edu/biology/

Wichita State University

THE FACULTY AND THEIR RESEARCH

George Bousfield, Professor; Ph.D., Indiana University, 1981. Reproductive endocrinology: glycoprotein hormones; mechanism of gonadotropic action; carbohydrate biochemistry.

Karen L. Brown, Associate Professor; Ph.D., University of Georgia, 1982. Ecology and population genetics: genetic, behavioral, and ecological interactions involved in the regulation of animal populations; effects of environmental stress on genetic variation of populations.

Donald A. Distler, Associate Professor; Ph.D., University of Kansas, 1966. Aquatic biology and ecology: limnological characteristics of river basins; invertebrate distributions and life histories in river basins.

William J. Hendry III, Professor; Ph.D., Worcester Foundation for Experimental Biology/Clark University, 1982. Cellular and molecular endocrinology: endocrine disruption; estrogenic control of normal and neoplastic uterine morphogenesis; mechanisms of glucocorticoid-regulated tumor cell growth and gene expression.

Li Jia, Research Assistant Professor; Ph.D., University of California at Riverside, 2001. Bioinformatics and computational molecular biology: intracellular signaling network simulation; gene regulatory network modeling; molecular evolution analysis; molecular pattern discovery; comparative genomics; biological database construction.

Wendell W. Leavitt, Professor; Ph.D., University of New Hampshire, 1963. Cell biology and molecular endocrinology: regulation of steroid receptor systems; endocrine signaling mechanisms in pregnancy.

Jeffrey V. May, Associate Professor; Ph.D., University of Rhode Island, 1979. Reproductive endocrinology and cell biology: intraovarian regulation of mammalian folliculogenesis; autocrine/paracrine regulation of ovarian function by polypeptide growth factors.

David J. McDonald, Professor; Ph.D., Kansas State University, 1988. Mammalian genetics, mutagenesis, and molecular genetics: animal disease model production; comparative pathology; comparative genetics.

Christopher M. Rogers, Associate Professor; Ph.D., Indiana University, 1988. Avian biology; cost-benefit analysis of energy storage strategies; population biology: environmental factors affecting nest success and population trajectory; migratory behavior.

F. Leland Russell, Assistant Professor; Ph.D., University of Texas at Austin, 1999. Plant population and community ecology; plant-animal interaction; herbivores' effect on plant population and communities; grassland, savanna, and woodland dynamics.

Mark A. Schneegurt, Associate Professor; Ph.D., Brown University, 1989. Applied and environmental microbiology: microbial ecology; bioremediation; bioprospecting; cyanobacteria.

Bin Shuai, Assistant Professor; Ph.D., University of California at Riverside, 2003. Molecular and cellular mechanisms underlying pollen development and function; signaling pathway mediated by pollen-specific receptor-like kinases during pollen tube growth; transcriptional regulation of pollen-specific gene expression.

Arthur L. Youngman, Assistant Professor; Ph.D., University of Texas at Austin, 1965. Plant physiological ecology and environmental biology: developmental and physiological responses of *Suaeda depressa* and *Atriplex triangularis* to soil salinity; nitrogen allocation patterns among inland halophytes, with particular reference to the role of proline and quaternary ammonium compounds; interspecific and intraspecific interactions among species of inland halophytes.

WILLIAM PATERSON UNIVERSITY
OF NEW JERSEY
School of Science and Health
Department of Biology
Graduate Programs in Biology and Biotechnology

Programs of Study

The Department of Biology offers graduate programs leading to the M.S. degree in biology and M.S. degree in biotechnology. Thesis and nonthesis options are available. Graduate courses are normally scheduled for the late afternoons, evenings, or Saturdays. Classes are small, and students have a close working relationship with the faculty. Full-time students can expect to complete the course of study in about two years, while part-time students may take longer. The degree must be completed within six years of matriculation.

For the biology M.S. degree, students may choose from four areas of concentration, each offering a coherent course of study and research: physiology with an emphasis on neurobiology; limnology and terrestrial ecology; molecular biology with an emphasis on biotechnology; and general biology. The concentration in general biology is designed for the student who wants a broad graduate program. The M.S. program provides courses and research opportunities in the areas of animal behavior, molecular biology, biotechnology, biochemistry, cell biology, ecology, endocrinology, evolution, genetics, immunology, limnology, microbiology, neurobiology, plant physiology, plant tissue culture, reproductive biology, scanning and transmission electron microscopy, and virology. The thesis option requires completion of 30 graduate credits in biology and related fields, including 6 credits of thesis research. There are two nonthesis options, each requiring 30 graduate credits, including 3 credits of graduate independent study or independent reading. All students are required to pass a comprehensive examination.

The biotechnology M.S. degree program prepares students for a variety of opportunities in the area of biotechnology and related fields such as molecular biology, immunology, genetic engineering, and protein biochemistry. Graduates of the program are well grounded in molecular biology, including nucleic acid structure and function, gene sequencing and synthesis; plant and animal cell culture; hybridoma technology; immunological techniques; Southern, Northern, Western, and dot blots; use of radioisotopes; DNA, RNA, and protein isolation, separation, and purification; genetic engineering of plant and animal cells; and other advanced techniques. There are also a variety of electives. The 36-credit program includes two required molecular biology courses and four required laboratory courses in biotechnology. Independent study or a thesis may be elected, giving students the opportunity to develop and complete a research project in biotechnology. All students are required to pass an exit examination.

Research Facilities

The Department of Biology has state-of-the-art equipment necessary for a wide range of research activities. Major facilities and their associated equipment include animal facilities with computerized equipment for data collection and analysis; a neurobiology facility; microscopy facilities, including transmission and scanning electron microscopes; a well-equipped ecology laboratory with both stationary and field equipment; and an automated, climate-controlled greenhouse. The fully equipped biotechnology laboratories contain equipment for the sequencing, synthesis, and PCR amplification of DNA; ultracentrifuges; electrophoresis equipment; computerized spectrophotometers; HPLCs; and a radioisotope laboratory as well as a modern tissue-culture laboratory equipped and used for mammalian, plant, and insect cell culture.

Financial Aid

Several half-time graduate assistantships that include a full waiver of tuition are available. The Monroe Spivak Fellowship is available to outstanding candidates who wish to pursue graduate studies in ecology.

Cost of Study

Tuition and fees are currently $476 per credit for New Jersey residents and $717 per credit for nonresidents.

Living and Housing Costs

There are many apartments in the surrounding area; rents vary considerably depending on the community. Public transportation is available, but an automobile is a convenience because of the suburban location of William Paterson University. Housing is also available in on-campus dormitories.

Student Group

The Department enrolls about 50 graduate students. Many are from outside the area, including about 15 percent who are international students. Many of the students are pursuing their studies on a part-time basis. The numerous pharmaceutical and biotechnological research facilities in northern New Jersey have provided employment for many of the Department's interested graduates.

Location

The University is located in northern New Jersey about 20 miles west of New York City. The area abounds in opportunities for such outdoor activities as swimming, canoeing, bicycling, sailing, hiking, cross-country and downhill skiing, bird-watching, and fishing. Cultural and educational resources are almost boundless, and there are many historic sites in the area.

The University

William Paterson is one of eight college and university campuses in the New Jersey state system. There are now sixteen graduate and thirty-six undergraduate degree programs offered by the University's four schools. The University's 280-acre campus sits on a lovely hilltop graced by woods, cliffs, streams, a waterfall, and two ponds.

Applying

The minimum requirements for admission are a bachelor's degree from an accredited college or university, preferably with a major in biology or a related field. Undergraduate courses should include 16 credits of biology (including genetics) and one year each of chemistry, organic chemistry, physics, and calculus. In addition, biochemistry is required for the biotechnology program. Some deficiencies can be made up during the first year of graduate study. A cumulative scholastic average of no less than 2.75 on a 4.0 scale is expected. Scores on the General Test of the Graduate Record Examinations, as well as two letters of recommendation, must be submitted. An application form and catalog are available online or from the Office of Graduate Services. Prospective students wishing to visit the campus should contact the Graduate Director.

Correspondence and Information

Dr. Robert Chesney, Graduate Director
Graduate Program in Biological Sciences
Department of Biology
Science Hall, Room 434
William Paterson University
300 Pompton Road
Wayne, New Jersey 07470-2103
Phone: 973-720-3455 or 2245
E-mail: chesneyr@wpunj.edu

Office of Graduate Services
Raubinger Hall, Room 102
William Paterson University
300 Pompton Road
Wayne, New Jersey 07470-2103
Phone: 973-720-3578 or 3579

William Paterson University of New Jersey

THE FACULTY AND THEIR RESEARCH

The Department of Biology is strongly committed to teaching and research. Faculty members are listed below under the general area of their research interests. Due to the interdisciplinary nature of much of this research, these divisions are only approximate, and there is much overlap among areas.

Biotechnology/Molecular Biology/Genetics
Robert H. Chesney, Professor and Graduate Director; Ph.D., Virginia, 1974. Molecular biology; bacterial and phage genetics.
Jeffrey Dudycha, Assistant Professor; Ph.D., Michigan State, 1999. Genetics of senescence in *Daphnia*.
E. Eileen Gardner, Associate Professor; Ph.D., Texas, 1980. Protein biochemistry; cytoskeletal proteins.
Claire M. Leonard, Assistant Professor; Ph.D., New York Medical College, 1987. Molecular biology; human metabolic diseases.
Pradeep K. Patnaik, Assistant Professor; Ph.D., Indiana, 1989. Molecular genetics of parasitic protozoa.
David H. Slaymaker, Assistant Professor; Ph.D., California, Riverside, 1999. Plant defense responses; plant molecular biology.
Miryam Z. Wahrman, Professor; Ph.D., Cornell, 1981. Molecular biology of development; bioactive marine compounds.
Carey Waldburger, Assistant Professor; Ph.D., USC, 1992. Bacterial signal transduction.

Ecology/Evolution
Kendall Martin, Assistant Professor; Ph.D., Oregon State, 2001. Microbial ecology.
Michael S. Peek, Assistant Professor; Ph.D., Maryland, 2002. Plant ecology.
Lance S. Risley, Associate Professor; Ph.D., Georgia, 1987. Ecosystem ecology; forest canopies; bats.
Michael J. Sebetich, Professor; Ph.D., Rutgers, 1972. Ecology of lakes and streams.
Stephen G. Vail, Associate Professor; Ph.D., California, Davis, 1990. Population ecology; plant-animal interactions; Lyme disease.

Neuroscience/Biopsychology/Physiology
Robert H. Benno, Professor; Ph.D., Iowa, 1978. Computer-assisted imaging; developmental neurobiology.
Danielle Desroches, Professor; Ph.D., CUNY, Hunter, 1981. Neuroendocrinology; teratogenic agents and development.
Martin E. Hahn, Professor; Ph.D., Miami (Ohio), 1970. Animal behavior; genetics of aggression.
Jaishri Menon, Assistant Professor; Ph.D., Baroda (India), 1978. Environmental physiology, vertebrate morphogenesis.
Emmanuel Onaivi, Assistant Professor; Ph.D., Bradford (England), 1987. Neuropharmacology; cannabinoids and emotional disturbance.

Genetically engineered plants grown in plant biotechnology facilities are transferred to departmental greenhouse.

Students conducting limnological research on nearby reservoir.

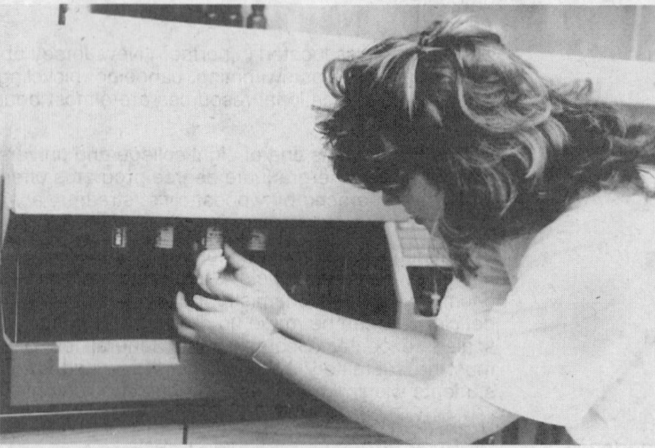

Student working with DNA synthesizer in fully equipped biotechnology laboratory.

WORCESTER POLYTECHNIC INSTITUTE

Department of Biology and Biotechnology

Programs of Study

The Department of Biology and Biotechnology at Worcester Polytechnic Institute (WPI) offers a fulltime research-oriented program leading to either a doctoral or master's degree. Major research strengths in the Department target areas of cell, molecular, and developmental/regenerative biology; ecology/evolution; computational biology; and applied microbial systems. Students may choose one of these areas or explore the possibility of an innovative interdisciplinary area available through the Interdisciplinary Program.

The goals of these degree programs are to give the student a broad background in a particular scientific area, a working knowledge of modern research tools, a strong appreciation for scientific research in theoretical and experimental areas, and a foundation for lifelong learning and experimenting—both individually and as part of a team. The Ph.D. program is designed to provide a basis for careers not only in science but also in diverse areas such as management of technology, technical communication, secondary education, and medicine. The research-based program leading to the M.S. in biotechnology specifically focuses on developing practical skills required for success in the vibrant biotech industry. There is a low student-faculty ratio, ensuring that each student receives individual attention and advising; a unique program of research and course work offers flexibility to address the needs and goals of each student.

The most important requirement of both M.S. and Ph.D. programs is a research thesis or dissertation that applies the basic principles of biology and biotechnology. Every graduate student is required to complete and defend a research thesis. With their broad knowledge of the field of biology and biotechnology and their detailed knowledge and applied research skills in their area of specialization, graduates are well prepared for further graduate education or for employment in academics or industry.

All graduate students are required to successfully complete three of the four Departmental core courses, give annual presentations of their research, and register for the graduate seminar each semester. By the end of the first year, students choose a faculty research mentor and assemble a faculty advisory committee; these faculty members guide the student's research and approve the completed research thesis or dissertation. A meeting with the student's advisory committee is required at least one time per semester to review progress.

For the Ph.D. in Biotechnology Program, a student must complete a total of 90 credit hours of course work and dissertation research. To become a candidate for the Ph.D. in biotechnology, a graduate student must successfully defend before the dissertation committee an original research proposal in an oral qualifying examination, which should be taken after the first year of study. The student must complete a Ph.D. dissertation and successfully defend it before the dissertation committee in a final oral examination. In addition, Ph.D. students must complete a teaching requirement and a cultural studies requirement. In order to graduate, a student should author at least one manuscript submitted for publication in a refereed journal and must present at least one paper at a national or international conference.

For the M.S. degree in biology and biotechnology, students must complete a minimum of 30 credit hours of course and laboratory research work, 6 of which must be thesis-research credits. A completed thesis project is required and must be approved by the thesis advisory committee. An oral defense of the thesis research is required.

One of the most attractive features of the Department is the opportunity for close interaction among graduate students and faculty members. In addition to formal teaching and research programs, the Department offers a seminar series and journal clubs. In addition, students participate and present their work in scientific clubs/meetings in the regional Boston area.

Research Facilities

Students in all of the programs benefit from an excellent faculty and some of the finest facilities anywhere. As a part of WPI's Gateway Park project, a new four-story laboratory will house all life science research. The graduate experience is further enhanced by the availability of several world-class research centers across the campus. These include the Nanotechnology Center, the Bioengineering Institute, the Tissue Engineering Center, and the state-of-the-art Bioprocess Center that includes an IF-75L fermenter computer control and monitoring systems. A sophisticated imaging center containing several inverted microscopes, fluorescent microscopes, micromanipulators, and an atomic force microscope is fully equipped to handle both static- and living-cell microscopy. The Department has a variety of modern equipment for research in cellular, molecular, and developmental biology; biochemistry, computational biology; and neuroscience. A modern greenhouse and several environmentally controlled growth chambers are also available for plant research. In addition, the Department has access to a variety of other modern research and teaching laboratories, sophisticated computing facilities, lecture and seminar rooms, instrument and chemical storage areas, machine shop services, and office space.

Financial Aid

The majority of graduate students are financially supported by either teaching or research assistantships. All students are expected to devote their full time toward research and teaching responsibilities. For the academic year 2006–07, teaching assistantships carry a stipend of $1667 per month, and research assistantship stipends vary between $1667 and $2500 per month. Both assistantships provide remission of tuition for up to 20 credits. Additional assistance may be available for the summer. U.S. citizens with exceptional qualifications are encouraged to apply for the Robert F. Goddard Fellowship. This prestigious award offers full-time tuition for one year (up to 20 credits) and a stipend of $1667 per month for twelve months. Other fellowship opportunities are also available. Information may be found online at: http://www.grad.wpi.edu/Financial/.

Cost of Study

Graduate tuition for the 2006–07 academic year is $997 per credit hour. There are nominal extra charges for the thesis, health insurance, and other fees.

Living and Housing Costs

On-campus graduate student housing is limited to a space-available basis. There is no on-campus housing for married students. Apartments and rooms in private homes near the campus are available at varying costs. For further information and apartment listings, students should visit the Residential Services Office Web site at http://www.wpi.edu/Admin/RSO/Offcampus/.

Student Group

Worcester Polytechnic Institute has a student body of 3,869, of whom 837 are full- or part-time graduate students. Most states and nearly fifty countries are represented. Recent graduates from the Department of Biology and Biotechnology have accepted academic or industrial research positions or gone on to further graduate studies.

Location

WPI is located on an 80-acre campus in a residential section of Worcester. The city, the second-largest in New England, has many colleges and an unusual variety of cultural opportunities. Located three blocks from the campus, the nationally famous Worcester Art Museum contains one of the finest permanent collections in the country and offers many special activities of interest to students. The community also provides outstanding programs in music and theater. The DCU Center offers rock concerts and semiprofessional athletic events. Easily reached for recreation are Boston and Cape Cod to the east and the Berkshires to the west, and good skiing is nearby to the north. Complete athletic and recreational facilities and a program of concerts and special events are available on campus to graduate students.

The Institute

Worcester Polytechnic Institute, founded in 1865, is the third-oldest independent university of engineering and science in the United States. Graduate study has been available at the Institute for more than 100 years. Classes are small and provide for close student-faculty relationships. Graduate students frequently interact in research with undergraduates, participating in WPI's innovative project-based program of education.

Applying

Applicants must submit WPI application forms, official college transcript(s), three letters of recommendation, a $70 application fee (waived for WPI alumni), scores from the GRE General Test, and a Statement of Purpose. International students whose primary language is not English must also submit proof of English language proficiency. WPI accepts either the Test of English as a Foreign Language (TOEFL) or the International English Language Testing System (IELTS). A TOEFL score of at least 550 on the paper-based test (213 on the computer-based test or 79–80 on the Internet-based test) or an IELTS overall band score of 6.5 (no band score below 6.0) is required for admission. To be considered for funding (assistantships and fellowships), complete applications must be on file by January 15 for fall admission, and October 15 for spring admission. Files completed after those deadlines are reviewed on a rolling basis and may not be considered for funding. Some fellowships require an additional application. Prospective students should visit http://www.grad.wpi.edu/Financial/ for more information. Inquiries should be directed to the head of the Department or to the Graduate Studies and Enrollment Office at grad_studies@wpi.edu.

Correspondence and Information

Dr. Eric W. Overström
Department of Biology and Biotechnology
Worcester Polytechnic Institute
Worcester, Massachusetts 01609

Phone: 508-831-6646
Fax: 508-831-5936
E-mail: ewo@wpi.edu
Web site: http://www.wpi.edu/Academics/Depts/Bio/Graduate/

Worcester Polytechnic Institute

THE FACULTY AND THEIR RESEARCH

Eric W. Overström, Professor and Department Head; Ph.D. (developmental biology), Massachusetts Amherst, 1981. Cellular and molecular determinants of oocyte developmental competence; regulation of M-phase exit during meiosis; role of key spindle- and centrosome-associated enabling factors in somatic cell cloning.

David S. Adams, Professor; Ph.D. (molecular biology), Texas at Austin, 1979. Role of neurotrophic factors (NTFs) in neuronal regeneration and their potential use as therapeutics in Alzheimer's and Parkinson's diseases; identification of signal-transduction pathways, transcription-factor activations, and changes in gene expression induced by NTFs.

Joseph C. Bagshaw, Professor; Ph.D. (molecular biology), Tennessee–Oak Ridge, 1969. Eukaryotic gene structure and function; organization and expression of histone and tubulin genes in the brine shrimp *Artemia;* molecular aspects of evolution and systematics; molecular genetics of marine shrimp; applications of recombinant DNA technology in agriculture and aquaculture.

Ted C. Crusberg, Associate Professor; Ph.D. (chemistry), Clark, 1968. Heavy-metal binding and other forms of heavy-metal removal (biomineralization) from industrial wastewaters using microorganisms; development of the technology to demonstrate feasibility of this new heavy metal bioremediation system.

Daniel G. Gibson III, Assistant Professor; Ph.D. (biology), Boston University, 1980. Comparative endocrinology and neurobiology in arthropods; study of growth and molting hormones and neuromuscular transmission in chelicerates, such as the American horseshoe crab (research model).

Lauren M. Mathews, Assistant Professor; Ph.D. (evolutionary biology), Louisiana, 2001. Population genetics and phylogeography of marine and aquatic species; behavioral ecology and social-system evolution.

Samuel M. Politz, Associate Professor; Ph.D. (molecular biology), UCLA, 1978. Genetics, developmental biology, and immunology; genetic control of surface antigen expression in the nematode *C. elegans* studied by molecular genetic and immunological approaches; role of nematode surface in avoiding host immune responses and infections by pathogenic microbes.

Reeta Prusty Rao, Assistant Professor; Ph.D. (biochemistry and molecular biology), Penn State, 1999. Understanding how fungi perceive secondary metabolites as quorum sensing cues to initiate pathogenesis; using the genetic model organism *S. cerevisiae* to identify unique fungal genes as potential targets of antifungal therapy.

Jill Rulfs, Associate Professor; Ph.D. (biochemistry), Tufts, 1982. Development of cell culture model systems, regulation of cellular transdifferentiation; anabolic steroid precursor effects on cell differentiation and tumorigenesis.

Elizabeth Ryder, Associate Professor; Ph.D. (genetics), Harvard, 1993. Signal-transduction pathways in control of neuronal migration, using *C. elegans* as a model system; computational approaches to understanding cell-specific expression of genes in *C. elegans;* simulation of early development in the *C. elegans* embryo using computational techniques.

Pamela J. Weathers, Professor; Ph.D. (botany and plant pathology), Michigan State, 1974. Through biochemical studies, the use of transformed plants, and novel bioreactors, focus is on the study of overproduction of terpenoid drugs such as the antimalarial drug, artemisinin.

Affiliated Faculty and Their Research

Michael A. Buckholt, Adjunct Assistant Professor and Laboratory Instructor; Ph.D. (biomedical sciences), Worcester Polytechnic, 1992. Molecular parasitology of *Microsporidia* and *Cryptosporidium;* parasite-host interactions; waterborne parasite detection and inactivation; development of model systems.

Alex DiIorio, Affiliate Assistant Professor and Director of Bioprocess Center; Ph.D. (biochemical engineering), Worcester Polytechnic, 1991. Biopolymers for heavy metal remediation, coupled fermentation, and product-recovery systems.

JoAnn Whitefleet-Smith, Adjunct Assistant Professor; Ph.D. (biochemistry), Wisconsin–Madison, 1984. Protein structure-function relationship; natural product isolation and characterization; recombinant protein expression.

WRIGHT STATE UNIVERSITY

Biomedical Sciences Ph.D. Program

Program of Study

The Biomedical Sciences (BMS) Ph.D. Program trains students for a career in research. Through an interdisciplinary approach to the biomedical sciences, the program recognizes the interrelatedness of the various traditional disciplines and seeks to provide students with the skills and perspective to investigate biomedical questions that cross defined disciplines. The program of study, while individualized, provides an integrated background in biology, chemistry, and mathematics; a mastery of skills at one or more advanced levels of study; and the competence to do independent research. The necessary breadth of basic knowledge is obtained in an interdisciplinary core, which consists of biochemistry and molecular biology, cell biology, cell physiology and biophysics, neuroscience and immunology, and biostatistics. The core curriculum and laboratory rotations occupy the first year and prepare the student to make a knowledgeable decision regarding specialization in succeeding years. The advanced curriculum is organized into areas of concentration that are interdisciplinary in nature. They provide for the in-depth study of a particular area within the biomedical sciences, such as bioengineering, immunobiology, molecular genetics, biochemistry, neuroscience, cell physiology and biophysics, cell biology, NMR and other forms of medical imaging, pharmacology and toxicology, applied biomedical computation, and chemical and structural biomedical sciences.

The Program also offers a dual M.D./Ph.D. degree in conjunction with the Wright State University School of Medicine. It combines rigorous training in clinical medicine and basic sciences during a seven-year course of study. It is for students with an intense interest in research and practical aspects of medicine.

The Biomedical Sciences Ph.D. Program offers considerable flexibility for advanced training in a variety of contemporary research specializations. The program's faculty members, currently numbering more than 70, are drawn from several departments in the College of Science and Mathematics, the Boonshoft School of Medicine, the College of Engineering and Computer Science, and from affiliate off-campus organizations, including the Armstrong Laboratory at Wright-Patterson Air Force Base, the Dayton Veterans Administration Center, and the Kettering/Scott Magnetic Resonance Laboratory. Contributing departments include Biochemistry and Molecular Biology; Biological Sciences; Biomedical and Human Factors Engineering; Chemistry; Community Health; Computer Science; Mathematics and Statistics; Medicine; Neuroscience, Cell Biology and Physiology; Pharmacology and Toxicology; Physics; and Psychology.

Research Facilities

Wright State University is equipped with the laboratories and equipment needed to carry out modern research in the biomedical sciences. The extensive facilities include the Brehm Laboratory for environmental studies and the Kettering/Scott Magnetic Resonance Laboratory. The Armstrong Laboratory at Wright-Patterson Air Force Base permits the selective use of its specialized facilities by Wright State students and staff. Researchers from Central State University, the University of Dayton, and the Veterans Administration Center are members of the program faculty. Modern, well-equipped, and Association for Assessment and Accreditation of Laboratory Animal Care (AAALAC)-accredited facilities for laboratory animal care are located both on campus and at several off-campus locations. Both the University Library and the Health Sciences Library have extensive holdings related to the biomedical sciences. There is a modern central computer facility and access to the Internet.

Financial Aid

Assistance is awarded to qualified applicants on a competitive basis. In 2006–07, research and teaching assistantships provide a stipend of $22,000 per year and a full tuition fee waiver.

Cost of Study

In 2005–06, tuition per credit hour was $271 for residents of Ohio and $468 for nonresidents. Full-time BMS students enroll for 15 credits per quarter.

Living and Housing Costs

Housing is available on campus for University students, with a special section for graduate students. Off-campus rooms and apartments are available and vary widely in cost. The cost of living in the Dayton area is less than in other metropolitan areas.

Student Group

Fifty-one students are currently enrolled in the program. There are more than 16,000 students on the Wright State campus, including 3,854 graduate students.

Location

Wright State University is located 6 miles northeast of Dayton, 50 miles north of Cincinnati, and 75 miles southwest of Columbus. Recreational and cultural opportunities in Dayton are numerous and include films, restaurants, the Dayton Philharmonic, the Dayton Opera, an art museum, and the Dayton Ballet Company. Three state parks, skiing and boating facilities, and the Kings Island amusement park provide outdoor activities. Nearby Cincinnati is the home of two professional sports teams (the baseball Reds and the football Bengals), the Cincinnati Zoo, the Cincinnati Orchestra, and many fine restaurants.

The University

Wright State University is a state-supported university and is fully accredited by the North Central Association of Colleges and Schools. It is composed of the Colleges of Science and Mathematics, Engineering and Computer Science, Business and Administration, Liberal Arts, and Education and Human Services and the Schools of Nursing, Medicine, and Professional Psychology. The 600-acre campus just outside Dayton includes excellent research, teaching, and recreational facilities, all housed in modern buildings.

Applying

Detailed information about the program, including prerequisites and financial aid, may be obtained by contacting the director of the program. Application materials, including references and GRE scores, must be complete by March 1 to be considered for the class entering the following September.

Correspondence and Information

Dr. Gerald M. Alter, Director
Biomedical Sciences Ph.D. Program
136 Biological Sciences Building
Wright State University
Dayton, Ohio 45435-0001

Phone: 937-775-2504
Fax: 937-775-3485
E-mail: director.bms@wright.edu
Web site: http://www.wright.edu/academics/biomed/

Wright State University

AREAS OF CONCENTRATION

Applied and Predictive Toxicology
The effects of chemicals on health and well being and on our environment range from miraculous to disastrous. Faculty members in this area are actively engaged in research addressing toxicological problems. The latest molecular, biological, chemical detection, and immunocytochemical techniques are used. Faculty member research expertise includes aquatic toxicology, dermal toxicology, ecotoxicology, environmental toxicology, immunotoxicology, risk assessment, toxicogenomics, and stress/toxicant interactions.

Applied Biomedical Computation
Faculty members within this area of concentration conduct research emphasizing modeling of macromolecular structures, modeling of biological processes, and construction, as well as mining, of large databases. Computational methods allow researchers to rationally propose structures of complex molecules and systems; to quantitatively test hypotheses regarding multifaceted molecular, cellular, organismic, and population processes; and to recognize, as well as test, relationships in vast and complex data sets.

Cell Biology and Physiology
Researchers in this area study processes that are fundamental to the understanding, prevention, and eventual treatment of diseases of the cardiovascular system, skin, blood, kidneys, lungs, gastrointestinal tract, and brain. A wide range of biophysical, biochemical, and biological approaches are used in studies of molecular, cell, organ, and whole organism processes. Specific areas of current research are membrane transport related to cell volume and ion regulation, cell differentiation, intracellular sorting and secretion of hormones, comparative aspects of kidney function, cellular growth control, intracellular signaling pathways, membrane channels, transporters and receptors, neural control of respiration, effects of hyperoxia and hyperbaria on neural-cell function, mitochondrial energy production, nuclear transport, brain edema, immunity, and wound healing.

Chemical and Structural Biomedical Sciences
Chemistry plays a pivotal role as the function properties of biologically relevant molecules are encoded in their covalent and noncovalent structures. Scientists in this area conduct research encompassing the entire breadth of the chemical/biological interface. The faculty members are involved in a number of diverse projects involving molecular modeling and design, development of novel polymeric supports and immobilized reagents, combinatorial chemistry/solid phase organic synthesis, high throughput purification and analysis, analytical methods for drugs and biological macromolecules, quantitative structure activity relationships, cheminformatics, protein and enzyme structure and dynamics, structural basis of catalysis, and metabolic regulation.

Epidemiology
Faculty members in this area of concentration examine the relationships of human development and body composition to risk for heart and other disease. Research areas include genetic epidemiology; the development, implementation, and validation of new methods for the study of body composition; new statistical methods and models; and determination of causal relationships involving body composition, adipose tissue distribution, lifestyle, and risk factors for cardiovascular disease.

Immunology
This area provides opportunities to probe fundamental cellular and molecular mechanisms related to immunology and infection. Present faculty member research interests are indoor allergies, basic and clinical immunology, retrovirology (retroviral variation, HIV, endogenous retroviruses), immunotoxicology, viral pathogenicity, vaccine development, immunoparasitology of ectoparasites, microbial ecology, immune modulation, algal toxins, inflammatory and immune effector cell function, and cytokine signaling and apoptosis.

Medical Physics and Engineering
Faculty members in this area of concentration apply engineering methods and principles to problems in medicine and biology. Current efforts in this concentration include the development of medical and surgical instrumentation systems, the design of rehabilitative devices, the interacting of complex systems in data collection and analysis, and the adaptation of computer technology to assist the health-care industry.

Molecular Biology/Biochemistry
Faculty members in these areas investigate the molecular biology and genetics of cellular metabolic processes, signal transduction pathways, and gene expression and development. Major areas of interest include DNA replication, repair and transcription, human molecular genetics, protein/enzyme and polynucleotide structure and function, molecular evolution, mechanisms of oncogenesis, retroviral recombination, and signal transduction mechanisms.

Neuroscience
The neuroscience laboratories associated with the program use many techniques in in vivo and in vitro studies at the molecular, cellular, and systems levels. Current research opportunities are available in the following areas: ion channel, ion transporter, and neurotransmitter receptor expression and localization; development of synaptic connections; hyperbaric physiology; cardiovascular and respiratory control; neuroendocrinology; regulation of ion channel; and receptor function and cell volume.

THE FACULTY

Norma Adragna, Ph.D.	Robert Fyffe, Ph.D.	Daniel Organisciak, Ph.D.
William Albery, Ph.D.	Roger Gilpin, Ph.D.	Oleg Paliy, Ph.D.
Gerald Alter, Ph.D.	Melvyn Goldfinger, Ph.D.	John Paietta, Ph.D.
Francisco Alvarez, Ph.D.	David Goldstein, Ph.D.	Steve Patrick, Ph.D.
Francisco J. Alvarez-Leefmans, Ph.D.	Julian Gomez-Cambronero, Ph.D.	John Pearson, Ph.D.
Larry Arlian, Ph.D.	Keith Grasman, Ph.D.	Chandler Phillips, M.D.
Scott Baird, Ph.D.	Robert Grubbs, Ph.D.	Lawrence Prochaska, Ph.D.
Michael Baumann, Ph.D.	Dan Halm, Ph.D.	Robert Putnam, Ph.D.
Steven Berberich, Ph.D.	Thomas Hangartner, Ph.D.	Michael Raymer, Ph.D.
Jack Bernstein, Ph.D.	Ping He, Ph.D.	Nicholas Reo, Ph.D.
Nancy Bigley, Ph.D.	Richard Henderson, M.D.	David Reynolds, Ph.D.
Thomas Brown, Ph.D.	Michael Hennessy, Ph.D.	Mateen Rizki, Ph.D.
Paula Bubulya, Ph.D.	Barbara Hull, Ph.D.	Munsup Seoh, Ph.D.
Wayne Carmichael, Ph.D.	Madhavi Kadakia, Ph.D.	Roger Siervogel, Ph.D.
William C. Chumlea, Ph.D.	Harry Khamis, Ph.D.	Shumei Sun, Ph.D.
David Cool, Ph.D.	Dan Krane, Ph.D.	Julie Skipper, Ph.D.
Timothy Cope, Ph.D.	Peter Lauf, M.D.	Stephanie Smith, Ph.D.
Adrian Corbett, Ph.D.	Michael Leffak, Ph.D.	Courtney Sulentic, Ph.D.
Jay Dean, Ph.D.	James Lucot, Ph.D.	Thomas Svobodny, Ph.D.
Patrick Dennis, Ph.D.	Mark Mamrack, Ph.D.	Ken Turnbull, Ph.D.
Guozhu Dong, Ph.D.	James McDougal, Ph.D.	Thomas Van't Hof, Ph.D.
Travis Doom, Ph.D.	Mill Miller, Ph.D.	Scott Watamaniuk, Ph.D.
Eric Fossum, Ph.D.	Sidney Miller, M.D.	Michele Wheatly, Ph.D.
Brent Foy, Ph.D.	Mariana Morris, Ph.D.	Dawn Wooley, Ph.D.
John Frazier, Ph.D.	James Olson, Ph.D.	

YALE UNIVERSITY

Combined Program in the Biological and Biomedical Sciences

Programs of Study	The Graduate School of Arts and Sciences and the School of Medicine at Yale University offer a combined interdepartmental graduate program in the Biological and Biomedical Sciences (BBS), which provides students flexible opportunities to study and conduct thesis research with any of the more than 280 biological science faculty members at the School of Medicine and on the University's Science Hill, which are close to one another in the center of New Haven. BBS consists of eight interest-based tracks that serve to organize research and educational activities. The tracks include computational biology and bioinformatics; molecular cell biology, genetics, and development; molecular biophysics and biochemistry; neuroscience; immunology; microbiology; pharmacological sciences and molecular medicine; and physiology and integrative medical biology. More detailed information for each track as well as a listing of the faculty members who are affiliated with each can be found on the BBS Web site at http://www.bbs.yale.edu. Students applying to BBS should identify the track that best represents their interests; applications are processed and reviewed by that track. Affiliation does not, however, limit a student's opportunities to work with faculty members or participate in training activities of the other tracks, but simply establishes smaller groups of students and faculty members with common interests. The course of study is configured individually according to a student's interests, needs, background, and departmental requirements. All educational activities, seminars, retreats, and other functions are closely coordinated to maximize opportunities for interactions among students and faculty members throughout BBS. Students have the opportunity to perform at least three laboratory rotations during the first year with faculty members of their choosing. In addition to allowing students to receive hands-on experience in different laboratories, the rotations assist students in selection of a thesis adviser, typically by the start of the second year. At this time, students (in consultation with their advisers) choose the department from which they will ultimately receive their degrees. Time required for completion of the Ph.D. degree averages 5.5 years. A qualifying examination is administered in the second year.
Research Facilities	Yale is one of the few major research institutions with closely situated University and Medical School campuses, an arrangement that promotes active interchange among students and faculty members throughout the entire community. There are exceptional laboratory facilities for research and training in all aspects of the biological and biomedical sciences. The physical plant has been enhanced by several major construction and renovation projects completed within the past few years. In addition, there are excellent centers for instrumentation, which are accessible to all students. These include the Keck Protein Chemistry Laboratory and DNA Synthesis and Microarray Facility; the Center for Structural Biology, which houses state-of-the-art X-ray diffraction equipment and molecular graphics facility; the Center for Cell Imaging, which includes facilities for electron microscopy, immunocytochemistry, scanning laser confocal microscopy, and computer-generated image analysis; central laboratories for the production of transgenic mice and monoclonal antibodies; and the Magnetic Resonance Center. These facilities provide opportunities for both hands-on and theoretical training to graduate students in all tracks.
Financial Aid	Students accepted for admission normally receive health insurance, full tuition, and a twelve-month stipend ($27,000 in 2006–07). Students are generally expected to serve as teaching fellows for one or two semesters during their time at Yale.
Cost of Study	The cost of study and health care is provided for all students for the duration of their enrollment.
Living and Housing Costs	Single rooms and apartments are available on campus for rents that range from $350 to $800 per month. A wide variety of off-campus housing is also available in all price ranges. These are available in neighborhoods immediately adjacent to Yale as well as in suburban, rural, and shoreline communities adjacent to New Haven.
Student Group	Approximately 70–80 students enter BBS each year, divided among the various tracks. There are currently nearly 450 graduate students in the biological and biomedical sciences at Yale.
Location	Yale University is situated in New Haven, a small (population 130,000) historic New England city located directly on Long Island Sound. New Haven has an active cultural life and is situated only 1½ hours from New York and 2½ hours from Boston, both of which are easily visited by car or by train. Although part of a major metropolitan and cultural center, New Haven is surrounded by picturesque communities and scenic rural areas that offer easily accessible opportunities for outdoor activities and housing.
The University	Yale University was founded in 1701 and has grown to become a large and diverse campus consisting of Yale College, the Graduate School of Arts and Sciences, the School of Medicine, and nine other professional schools. The University has a total enrollment of more than 11,000 students, nearly half of whom are studying at the graduate level.
Applying	Procedures for applying and the online application can be found on the BBS Web site.
Correspondence and Information	Lynn Cooley, Director or John Alvaro, Administrative Director Program in Biological and Biomedical Sciences Yale University P.O. Box 208084 New Haven, Connecticut 06520-8084 Phone: 203-785-3735 Fax: 203-785-3734 E-mail: bbs@yale.edu Web site: http://www.bbs.yale.edu

Yale University

FACULTY HEADS AND GRADUATE RESEARCH PROGRAMS

Computational Biology and Bioinformatics Track

The Computational Biology and Bioinformatics Track is designed to provide an educational and training program for students who wish to participate in a broad interdisciplinary program at the interface of biology, computer science, statistics, and mathematics. In particular, emphasis is placed on large-scale biological research, including genomics, proteomics, and comparative biology. The track combines research and training with the goal of developing competency in three core areas: computational biology and bioinformatics, biological sciences, and informatics (including computer science, statistics, and applied mathematics).

Perry Miller, Director of Graduate Studies. BBS/Computational Biology and Bioinformatics Track, P.O. Box 208103, New Haven, Connecticut 06520-8103. E-mail: dgs.bioinfo@yale.edu; Web site: http://www.bbs.yale.edu.

Immunology Track

Training and research in the Immunology Track focus on the molecular, cellular, and genetic underpinnings of immune system function and development and on host-pathogen interactions. Specific areas of interest include T-cell development, activation and effector functions, the role of cytokines in immunoregulation, intracellular signaling and the control of transcription in lymphocytes, antigen processing and presentation, immunoglobulin and T-cell receptor gene rearrangement, immunological memory, the immunobiology of vascular endothelial cells, and B- and T-cell tolerance. In addition, mechanisms of autoimmunity and immunodeficiency are a major interest. A number of important human diseases are under study, including diabetes, systemic lupus erythematosus, multiple sclerosis, AIDS, and a variety of other infectious diseases.

Al Bothwell, Director of Graduate Studies. Barbara Giamattei, Student Services Officer. BBS/Immunology Track, P.O. Box 208011, New Haven, Connecticut 06520-8011. Telephone: 203-785-3857; fax 203-737-5637; e-mail: barbara.giamattei@yale.edu; Web site: http://www.bbs.yale.edu.

Microbiology Track

The Microbiology Track is an individualized training program utilizing cellular, molecular, and genetic approaches in the study of microorganisms and the effects on their hosts. Students can specialize in various areas, including bacteriology, virology, microbe-host interactions, vector biology, microbial pathogenesis and parasitology, and microbial genetics, ecology, evolution, and physiology. Students in the track will be formally affiliated with one of the degree-granting departments that participate in the Microbiology Program and the BBS: Epidemiology and Public Health and Pathology.

Joann Sweasy, Director of Graduate Studies. Darlene Smith, Graduate Registrar. BBS/Microbiology Track, P.O. Box 360812, New Haven, Connecticut 06536-0812. Telephone: 203-737-2404; fax: 203-737-2630; e-mail: joann.sweasy@yale.edu.

Molecular Biophysics and Biochemistry Track

The Molecular Biophysics and Biochemistry (MB&B) Track at Yale is designed to prepare students for careers as independent investigators in the broad area of molecular and structural biology. The faculty members have diverse interests and are drawn from several departments, including Molecular Biophysics and Biochemistry, Biology, Chemistry, and Genetics. Current areas of research include control of the cell cycle and animal development; structure and function of proteins and nucleic acids; catalytic RNAs and RNA processing; signal transduction in plants and mammals; protein folding; mechanisms of transcription, replication, transposition, and recombination; cell motility; molecular immunology; regulation of metabolism; and the application of advanced computational methods to the study of macromolecules.

Mark Soloman, Director of Graduate Studies. Nessie Stewart, Graduate Registrar. BBS/Molecular Biophysics and Biochemistry Track, P.O. Box 208114, New Haven, Connecticut 06520-8114. Telephone: 203-432-5662; fax: 203-432-5832; e-mail: mbb.grad@yale.edu; Web site: http://www.bbs.yale.edu.

Molecular Cell Biology, Genetics, and Development Track

Research interests of the faculty members in the MCGD Track cover a broad but integrated array of topics, and research projects often combine approaches from several fields. For example, some faculty members use genetics and molecular biology to investigate the cellular basis of developmental processes, while others combine biochemistry and genetics to study basic cellular processes or the molecular basis of disease. The experimental approaches used are similarly rich and broad in scope, involving the use of microscopy and biochemistry as well as classical and molecular genetics to analyze function in yeast, *Drosophila, C. elegans, Arabidopsis,* and mammalian cells and tissues.

Anne Scott, Graduate Registrar. BBS/MCGD Track, P.O. Box 208103, New Haven Connecticut 06520-8103. Telephone: 203-432-3538. E-mail: anne.scott@yale.edu; Web site: http://www.bbs.yale.edu.

Neuroscience Track

The Neuroscience Track offers training and research opportunities in virtually all areas of neurobiology ranging from the mathematical to the molecular to the integrative.

Charles Greer, Co-Director of Graduate Studies. Amy Arnsten, Co-Director of Graduate Studies. Carol Russo, Student Services Officer. BBS/Neuroscience Track, P.O. Box 208074, New Haven, Connecticut 06520-8074. Telephone: 203-785-5932; fax: 203-785-5971; e-mail: bbs.neuro@yale.edu.

Pharmacological Sciences and Molecular Medicine Track

Recent advances in genetics, immunology, cell, and molecular biology have revealed the mechanistic underpinnings of many diseases. The Pharmacology and Molecular Medicine Track offers the opportunity to use tools from these disciplines in an integrative approach to investigate the mechanisms of disease pathogenesis and to apply these findings to diagnosis and treatment. Specific areas of investigation include cancer and viral diseases, atherosclerosis, inflammation and wound healing, neurophysiology, psychopharmacology, disorders of coagulation, and immune and genetic disorders.

Elias Lolis, Co-Director of Graduate Studies. David F. Stern, Co-Director of Graduate Studies. JoAnn Falato, Graduate Registrar. BBS/Pharmacology and Molecular Medicine Track, P.O. Box 208066, New Haven, Connecticut 06520. Telephone: 203-785-6721; fax: 203-785-7467; e-mail: bbs.pharm@yale.edu or bbs.molmed@yale.edu.

Physiology and Integrative Medical Biology Track

The Physiology and Integrative Medical Biology Track offers an interdisciplinary course of study that leads to the Ph.D. degree. This multidimensional track includes activities in cell, systems, and integrative physiology. Research in physiology and integrative medical biology integrates information from genetics, functional genomics, and functional proteomics into whole-animal and human biology, including pathophysiology, pharmacology, translational research, and biomedical engineering. Areas of current interest include ion channels, solute transporters and pumps, membrane biophysics, electrophysiology, epithelial transport, cellular and systems neurobiology, protein trafficking (synapse function, endocytosis and exocytosis, secretion), biosynthesis of membrane proteins, structural biology of membrane proteins, physiological genomics, signal transduction pathways, vascular biology, organ physiology, small-animal physiology, genetic models of human disease, and pathophysiology.

Emile Boulpaep, Director of Graduate Studies. Physiology and Integrative Medical Biology Track, P.O. Box 208026, New Haven, Connecticut 06520-8026. Telephone: 203-737-2215; fax: 203-785-4951. Web site: http://www.bbs.yale.edu.

Section 2
Anatomy

This section contains a directory of institutions offering graduate work in anatomy, followed by in-depth entries submitted by institutions that chose to prepare detailed program descriptions. Additional information about programs listed in the directory but not augmented by an in-depth entry may be obtained by writing directly to the dean of a graduate school or chair of a department at the address given in the directory.

For programs offering related work, see also in this book Biological and Biomedical Sciences; Cell, Molecular, and Structural Biology; Genetics, Developmental Biology, and Reproductive Biology; Neuroscience and Neurobiology; Pathology and Pathobiology; Physiology; and Zoology. In Book 2, see Sociology, Anthropology, and Archaeology; and in Book 6, see Allied Health, Dentistry and Dental Sciences, and Veterinary Medicine and Sciences.

CONTENTS

Program Directory

Announcement

Close-Ups

Anatomy

Albert Einstein College of Medicine, Sue Golding Graduate Division of Medical Sciences, Department of Anatomy and Structural Biology, Bronx, NY 10461. Offers anatomy (PhD); cell and developmental biology (PhD). *Degree requirements:* For doctorate, thesis/dissertation. *Entrance requirements:* For doctorate, GRE General Test. Additional exam requirements/recommendations for international students: Required—TOEFL. Electronic applications accepted. *Faculty research:* Cell motility, cell membranes and membrane-cytoskeletal interactions as applied to processing of pancreatic hormones, mechanisms of secretion.

Auburn University, College of Veterinary Medicine and Graduate School, Graduate Programs in Veterinary Medicine, Auburn University, AL 36849. Offers biomedical sciences (MS, PhD), including anatomy, physiology and pharmacology (MS), biomedical sciences (PhD), clinical sciences (MS), large animal surgery and medicine (MS), pathobiology (MS), radiology (MS), small animal surgery and medicine (MS). Part-time programs available. *Faculty:* 76 full-time (24 women). *Students:* 11 full-time (8 women), 33 part-time (20 women); includes 3 minority (all Hispanic Americans), 14 international. 41 applicants, 37% accepted, 11 enrolled. In 2005, 9 master's, 3 doctorates awarded. *Degree requirements:* For doctorate, thesis/dissertation. *Entrance requirements:* For master's, GRE General Test; for doctorate, GRE General Test, GRE Subject Test. *Application deadline:* For fall admission, 7/7 for domestic students; for spring admission, 11/24 for domestic students. Applications are processed on a rolling basis. *Application fee:* $25 ($50 for international students). Electronic applications accepted. *Financial support:* Research assistantships, teaching assistantships, Federal Work-Study available. Support available to part-time students. Financial award application deadline: 3/15. *Application contact:* Dr. Stephen L. McFarland, Acting Dean of the Graduate School, 334-844-4700.

Barry University, School of Graduate Medical Sciences, Program in Anatomy, Miami Shores, FL 33161-6695. Offers MS. *Students:* 2. *Entrance requirements:* For master's, GRE. *Expenses:* Tuition: Full-time $12,330; part-time $685 per credit. *Unit head:* Dr. Ramjeet Pemsingh, Chair, 305-899-3264, Fax: 305-899-3253, E-mail: rpemsingh@mail.barry.edu. *Application contact:* Marc A. Weiner, Director of Graduate and Medical Sciences Admissions and Marketing, 305-899-3130, Fax: 305-899-3253, E-mail: mweiner@mail.barry.edu.

Boston University, Sargent College of Health and Rehabilitation Sciences, Department of Health Sciences, Boston, MA 02215. Offers applied anatomy and physiology (MS, PhD); nutrition (MS). Part-time programs available. *Faculty:* 11 full-time (9 women), 6 part-time/adjunct (5 women). *Students:* 57 full-time (49 women), 5 part-time (all women); includes 7 minority (all Asian Americans or Pacific Islanders), 8 international. Average age 26. 67 applicants, 40% accepted, 20 enrolled. *Degree requirements:* For master's, thesis or alternative; for doctorate, one foreign language, thesis/dissertation. *Entrance requirements:* For master's, GRE General Test, minimum GPA of 3.0; for doctorate, GRE General Test. Additional exam requirements/recommendations for international students: Required—TOEFL (minimum score 550 paper-based). *Application deadline:* For fall admission, 3/1 for domestic students; for spring admission, 10/1 for domestic students. Applications are processed on a rolling basis. Application fee: $65. Electronic applications accepted. *Expenses:* Tuition: Full-time $31,530; part-time $985 per credit. Required fees: $316; $40 per semester. Tuition and fees vary according to course level and program. *Financial support:* In 2005–06, 20 fellowships with full tuition reimbursements, 7 research assistantships with full tuition reimbursements, 6 teaching assistantships with full tuition reimbursements were awarded; career-related internships or fieldwork, Federal Work-Study, institutionally sponsored loans, scholarships/grants, and tuition waivers (partial) also available. Support available to part-time students. Financial award application deadline: 4/15. *Faculty research:* Muscle metabolism, body acid-base balance, human performance, physical conditioning, diabetes. *Unit head:* Dr. Eileen O'Keefe, Chair, 617-353-7532, E-mail: ebokeefe@bu.edu. *Application contact:* Sharon Sankey, Director, Student Services, 617-353-2713, Fax: 617-353-7500, E-mail: ssankey@bu.edu.

Boston University, School of Medicine, Division of Graduate Medical Sciences, Department of Anatomy and Neurobiology, Boston, MA 02118. Offers MA, PhD, MD/PhD. Part-time programs available. *Faculty:* 13 full-time (3 women), 6 part-time/adjunct (0 women). *Students:* 32 full-time (19 women); includes 7 minority (2 African Americans, 3 Asian Americans or Pacific Islanders, 2 Hispanic Americans), 4 international. Average age 28.Terminal master's awarded for partial completion of doctoral program. *Degree requirements:* For master's and doctorate, thesis/dissertation, qualifying exam. *Entrance requirements:* For master's and doctorate, GRE General Test, GRE Subject Test. Additional exam requirements/recommendations for international students: Required—TOEFL. *Application deadline:* For spring admission, 10/15 priority date for domestic students. Electronic applications accepted. *Expenses:* Tuition: Full-time $31,530; part-time $985 per credit. Required fees: $316; $40 per semester. Tuition and fees vary according to course level and program. *Financial support:* Fellowships with tuition reimbursements, research assistantships with tuition reimbursements, Federal Work-Study, scholarships/grants, and traineeships available. *Faculty research:* Neuroanatomy, development of the nervous system, aging, respiratory system, reproductive system. *Unit head:* Dr. Mark Moss, Chairman, 617-638-4200, Fax: 617-638-4216.

Case Western Reserve University, School of Medicine and School of Graduate Studies, Graduate Programs in Medicine, Department of Anatomy, Cleveland, OH 44106. Offers applied anatomy (MS); biological anthropology (MS, PhD); cellular biology (MS, PhD); developmental biology (PhD); molecular biology (PhD). Part-time programs available. *Faculty:* 20 full-time (5 women), 16 part-time/adjunct (5 women). *Students:* 6 full-time (0 women), 49 part-time (19 women); includes 20 minority (7 African Americans, 12 Asian Americans or Pacific Islanders, 1 Hispanic American). Average age 25. 39 applicants, 92% accepted, 29 enrolled. In 2005, 14 master's awarded. *Median time to degree:* Of those who began their doctoral program in fall 1997, 67% received their degree in 8 years or less. *Degree requirements:* For master's, thesis (for some programs), comprehensive exam; for doctorate, thesis/dissertation. *Entrance requirements:* For master's, GRE General Test; for doctorate, GRE General Test, GRE Subject Test. Additional exam requirements/recommendations for international students: Required—TOEFL. *Application deadline:* For fall admission, 5/1 for domestic students; for spring admission, 8/1 priority date for domestic students. Applications are processed on a rolling basis. Application fee: $50. *Financial support:* In 2005–06, 6 research assistantships with full tuition reimbursements (averaging $23,000 per year) were awarded; fellowships, scholarships/grants also available. *Faculty research:* Hypoxia, cell injury, biochemical aberration occurrences in ischemic tissue, human functional morphology, evolutionary morphology. Total annual research expenditures: $691,974. *Unit head:* Dr. Joseph C. LaManna, Chairman, 216-368-1100, Fax: 216-368-8669, E-mail: jcl4@po.cwru.edu. *Application contact:* Morley Schwebel, Administrator, 216-368-2433, Fax: 216-368-8669, E-mail: mxs86@po.cwru.edu.

Columbia University, College of Physicians and Surgeons and Graduate School of Arts and Sciences, Graduate School of Arts and Sciences at the College of Physicians and Surgeons, Department of Anatomy and Cell Biology, New York, NY 10032. Offers anatomy (M Phil, MA, PhD); anatomy and cell biology (PhD). Only candidates for the PhD are admitted. Terminal master's awarded for partial completion of doctoral program. *Degree requirements:* For doctorate, thesis/dissertation, oral exam. *Entrance requirements:* For master's and doctorate, GRE General Test. Additional exam requirements/recommendations for international students: Required—TOEFL. *Expenses:* Tuition: Full-time $31,448. Tuition and fees vary according to course level, course load, campus/location and program. *Faculty research:* Protein sorting, membrane biophysics, muscle energetics, neuroendocrinology, developmental biology, cytoskeleton, transcription factors.

Cornell University, Graduate School, Graduate Fields of Agriculture and Life Sciences, Field of Zoology, Ithaca, NY 14853-0001. Offers animal cytology (MS, PhD); comparative and functional anatomy (MS, PhD); developmental biology (MS, PhD); ecology (MS, PhD); histology (MS, PhD). *Faculty:* 26 full-time (4 women). *Students:* 4 full-time (3 women), 3 international.

9 applicants, 11% accepted, 1 enrolled. *Degree requirements:* For doctorate, thesis/dissertation, 2 semesters of teaching experience, comprehensive exam. *Entrance requirements:* For doctorate, GRE General Test, GRE Subject Test (biology), 2 letters of recommendation. Additional exam requirements/recommendations for international students: Required—TOEFL (minimum score 550 paper-based; 213 computer-based). *Application deadline:* For fall admission, 2/1 for domestic students. Application fee: $60. Electronic applications accepted. *Financial support:* In 2005–06, 4 students received support, including 1 fellowship with full tuition reimbursement available, 1 research assistantship with full tuition reimbursement available, 2 teaching assistantships with full tuition reimbursements available; institutionally sponsored loans, scholarships/grants, health care benefits, tuition waivers (full and partial), and unspecified assistantships also available. Financial award applicants required to submit FAFSA. *Faculty research:* Organismal biology, functional morphology, biomechanics, comparative vertebrate anatomy, comparative invertebrate anatomy, paleontology. *Unit head:* Director of Graduate Studies, 607-253-3276, Fax: 607-253-3756. *Application contact:* Graduate Field Assistant, 607-253-3276, Fax: 607-253-3756, E-mail: graduate_edcvm@cornell.edu.

Dalhousie University, Faculty of Graduate Studies and Faculty of Medicine, Graduate Programs in Medicine, Department of Anatomy and Neurobiology, Halifax, NS B3H 4R2, Canada. Offers M Sc, PhD. *Degree requirements:* For master's and doctorate, thesis/dissertation. *Entrance requirements:* For master's and doctorate, GRE (recommended), minimum A- average. Additional exam requirements/recommendations for international students: Required—TOEFL. *Faculty research:* Neuroscience histology, cell biology, neuroendocrinology, evolutionary biology.

Duke University, Graduate School, Department of Biological Anthropology and Anatomy, Durham, NC 27710. Offers cellular and molecular biology (PhD); gross anatomy and physical anthropology (PhD), including comparative morphology of human and non-human primates, primate social behavior, vertebrate paleontology; neuroanatomy (PhD). *Faculty:* 6 full-time. *Students:* 14 full-time (9 women); includes 1 minority (African American), 2 international. 39 applicants, 18% accepted, 3 enrolled. In 2005, 3 doctorates awarded. *Degree requirements:* For doctorate, one foreign language, thesis/dissertation. *Entrance requirements:* For doctorate, GRE General Test. Additional exam requirements/recommendations for international students: Required—IELT (preferred) or TOEFL. *Application deadline:* For fall admission, 12/31 for domestic students, 12/31 for international students. Application fee: $75. Electronic applications accepted. *Financial support:* Fellowships, teaching assistantships, Federal Work-Study available. Financial award application deadline: 12/31. *Unit head:* Daniel Schmitt, Director of Graduate Studies, 919-684-5664, Fax: 919-684-8034, E-mail: l.squires@baa.mc.duke.edu.

East Carolina University, Brody School of Medicine, Department of Anatomy and Cell Biology, Greenville, NC 27858-4353. Offers PhD. *Faculty:* 13 full-time (4 women), 1 part-time/adjunct (0 women). *Students:* Average age 29. 13 applicants, 23% accepted. In 2005, 1 degree awarded. *Median time to degree:* Of those who began their doctoral program in fall 1997, 100% received their degree in 8 years or less. *Degree requirements:* For doctorate, thesis/dissertation, comprehensive exam, registration. *Entrance requirements:* For doctorate, GRE General Test. Additional exam requirements/recommendations for international students: Required—TOEFL. *Application deadline:* For fall admission, 6/1 for domestic students. Applications are processed on a rolling basis. Application fee: $50. *Expenses:* Tuition, state resident: full-time $2,516. Tuition, nonresident: full-time $12,832. *Financial support:* In 2005–06, 8 fellowships with full tuition reimbursements (averaging $21,500 per year) were awarded; health care benefits also available. Financial award application deadline: 6/1. *Faculty research:* Kinesin motors during slow matogensis, mitochondria and peroxisomes in obesity, ovarian innervation, tight junction function and regulation. *Unit head:* Dr. Charles Hodson, Interim Chairman, 252-744-2851, Fax: 252-744-2850, E-mail: hodsonc@ccu.edu. *Application contact:* Dr. Ron Dudek, Senior Director of Graduate Studies, 252-744-3284, Fax: 252-744-2850, E-mail: dudekr@ecu.edu.

East Tennessee State University, James H. Quillen College of Medicine, Biomedical Science Graduate Program, Johnson City, TN 37614. Offers anatomy (MS, PhD); biochemistry (MS, PhD); biophysics (MS, PhD); microbiology (MS, PhD); pharmacology (MS, PhD); physiology (MS, PhD). Part-time programs available. *Faculty:* 49 full-time (12 women), 1 (woman) part-time/adjunct. *Students:* 30 full-time (19 women), 5 part-time (4 women); includes 2 minority (1 African American, 1 Asian American or Pacific Islander), 11 international. Average age 31. 78 applicants, 13% accepted, 9 enrolled. In 2005, 1 master's, 4 doctorates awarded. Terminal master's awarded for partial completion of doctoral program. *Degree requirements:* For master's, one foreign language, thesis, comprehensive qualifying exam; for doctorate, 2 foreign languages, thesis/dissertation. *Entrance requirements:* For master's, GRE General Test, minimum GPA of 3.0, bachelor's degree in biological or related science; for doctorate, GRE General Test, GRE Subject Test. Additional exam requirements/recommendations for international students: Required—TOEFL (minimum score 550 paper-based; 213 computer-based). *Application deadline:* For fall admission, 3/15 for domestic students; for spring admission, 3/1 for domestic students. Application fee: $25 ($35 for international students). *Expenses:* Contact institution. *Financial support:* In 2005–06, 7 research assistantships with full tuition reimbursements (averaging $15,000 per year) were awarded; teaching assistantships with full tuition reimbursements, career-related internships or fieldwork, Federal Work-Study, institutionally sponsored loans, scholarships/grants, and tuition waivers (full) also available. Financial award application deadline: 7/1; financial award applicants required to submit FAFSA. Total annual research expenditures: $2.1 million. *Unit head:* Dr. Mitchell E. Robinson, Assistant Dean, Director, 423-439-4658, E-mail: robinson@etsu.edu.

Howard University, Graduate School of Arts and Sciences, Department of Anatomy, Washington, DC 20059-0002. Offers MS, PhD. *Faculty:* 20 full-time (5 women). *Students:* 13 full-time (8 women); 9 African Americans, 4 Hispanic Americans. Average age 25. 15 applicants, 40% accepted, 3 enrolled. In 2005, 1 master's, 1 doctorate awarded. *Median time to degree:* Of those who began their doctoral program in fall 1997, 100% received their degree in 8 years or less. *Degree requirements:* For master's and doctorate, thesis/dissertation, teaching experience, comprehensive exam. *Entrance requirements:* For master's and doctorate, GRE General Test, minimum GPA of 2.5. Additional exam requirements/recommendations for international students: Required—TOEFL. *Application deadline:* For fall admission, 6/1 for domestic students; for spring admission, 11/1 priority date for domestic students. Applications are processed on a rolling basis. Application fee: $45. *Financial support:* In 2005–06, 4 students received support, including 1 fellowship with full tuition reimbursement available (averaging $15,000 per year), 3 teaching assistantships with full tuition reimbursements available (averaging $14,000 per year); research assistantships, career-related internships or fieldwork, institutionally sponsored loans, and scholarships/grants also available. Financial award application deadline: 6/1. *Faculty research:* Neural control of function, mammalian evolution and paleontology, cellular differentiation, cellular and neuronal communication, ovarian cancer. Total annual research expenditures: $200,000. *Unit head:* Dr. James H. Baker, Chairman, 202-806-6555, Fax: 202-267-7055. *Application contact:* Dr. Thomas Heinbockel, Chair of the Graduate Committee and Director of Graduate Studies, 202-806-9773, Fax: 202-265-7055, E-mail: theinbockel@howard.edu.

Indiana University Bloomington, Medical Sciences Program, Bloomington, IN 47405-7000. Offers anatomy and cell biology (MA, PhD); pharmacology (MS, PhD); physiology (MA, PhD). *Students:* 11 full-time (2 women), 9 part-time (5 women); includes 5 minority (1 African American, 4 Asian Americans or Pacific Islanders), 2 international. Average age 29. *Entrance requirements:* For master's, GRE, minimum GPA of 3.0; for doctorate, GRE. Additional exam requirements/recommendations for international students: Required—TOEFL. *Application deadline:* For fall admission, 1/15 for domestic students; for spring admission, 9/1 for domestic students. Application fee: $45 ($55 for international students). *Expenses:* Tuition, state resident: full-time $5,437; part-time $227 per credit hour. Tuition, nonresident: full-time $15,836; part-time $660 per credit hour. Required fees: $821. Tuition and fees vary according to campus/location and program. *Financial support:* Fellowships available. *Unit head:* Dr. John B. Watkins, Assistant Dean/Director, 812-855-0616. *Application contact:* Kimberly Bunch, Director of Graduate Admissions, 812-855-1119, E-mail: kbunch@indiana.edu.

Indiana University–Purdue University Indianapolis, Indiana University School of Medicine, Department of Anatomy and Cell Biology, Indianapolis, IN 46202-2896. Offers MS, PhD, MD/PhD. *Faculty:* 14 full-time (1 woman). *Students:* 1 (woman) full-time, 13 part-time (6 women), 9 international. Average age 31. *Degree requirements:* For master's, thesis or alternative; for doctorate, thesis/dissertation. *Entrance requirements:* For master's and doctorate, GRE General Test. *Application deadline:* For fall admission, 1/15 for domestic students. Application fee: $50 ($60 for international students). *Expenses:* Tuition, state resident: full-time $5,159; part-time $215 per credit hour. Tuition, nonresident: full-time $14,890; part-time $620 per credit hour. Required fees: $614. Tuition and fees vary according to campus/location and program. *Financial support:* In 2005–06, 1 fellowship was awarded; research assistantships, Federal Work-Study, institutionally sponsored loans, tuition waivers (partial), and stipends also available. Financial award application deadline: 2/15. *Faculty research:* Acoustic reflex control, osteoarthritis and bone disease, diabetes, kidney diseases, cellular and molecular neurobiology. *Unit head:* Dr. David B. Burr, Chairman, 317-274-7494, Fax: 317-278-2040, E-mail: dburr@indyvax.iupui.edu. *Application contact:* Dr. James Williams, Graduate Adviser, 317-274-3423, Fax: 317-278-2040, E-mail: williams@anatomy.iupui.edu.

The Johns Hopkins University, School of Medicine, Graduate Programs in Medicine, Center for Functional Anatomy and Evolution, Baltimore, MD 21218-2699. Offers PhD. *Faculty:* 5 full-time (1 woman), 1 (woman) part-time/adjunct. *Students:* 8 full-time (2 women). Average age 25. 16 applicants, 13% accepted, 2 enrolled. In 2005, 3 degrees awarded. *Degree requirements:* For doctorate, thesis/dissertation, oral exams, comprehensive exam. *Entrance requirements:* For doctorate, GRE. Additional exam requirements/recommendations for international students: Required—TOEFL. *Application deadline:* For fall admission, 1/1 for domestic students, 1/1 for international students. Application fee: $60. *Expenses:* Tuition: Full-time $30,960. Tuition and fees vary according to degree level and program. *Financial support:* In 2005–06, 8 teaching assistantships with full tuition reimbursements (averaging $24,600 per year) were awarded; fellowships, career-related internships or fieldwork, institutionally sponsored loans, health care benefits, and tuition waivers (full) also available. *Faculty research:* Vertebrate evolution, functional anatomy, primate evolution, vertebrate paleobiology, vertebrate morphology. *Unit head:* Dr. Kenneth D. Rose, Director, 410-955-7172, Fax: 410-614-9030, E-mail: kdrose@jhmi.edu. *Application contact:* Catherine L. Will, Coordinator, Graduate Student Affairs, 410-614-3385, E-mail: grad_study@som.adm.jhu.edu.

Kansas State University, College of Veterinary Medicine and Graduate School, Graduate Programs in Veterinary Medicine, Department of Anatomy and Physiology, Manhattan, KS 66506. Offers biomedical science (MS); physiology (PhD). *Faculty:* 14 full-time (3 women), 1 part-time/adjunct (0 women). *Students:* 16 full-time (5 women), 7 part-time (4 women); includes 1 minority (Hispanic American), 12 international. Average age 30. 4 applicants, 100% accepted, 4 enrolled. In 2005, 2 degrees awarded. Terminal master's awarded for partial completion of doctoral program. *Degree requirements:* For master's, thesis; for doctorate, one foreign language, thesis/dissertation. *Entrance requirements:* Additional exam requirements/recommendations for international students: Required—TOEFL (minimum score 550 paper-based; 213 computer-based). *Application deadline:* For fall admission, 2/1 for domestic students. Applications are processed on a rolling basis. Application fee: $30 ($55 for international students). *Expenses:* Tuition, state resident: full-time $5,160; part-time $215 per credit hour. Tuition, nonresident: full-time $12,816; part-time $534 per credit hour. Required fees: $564. *Financial support:* In 2005–06, 12 research assistantships (averaging $21,325 per year) were awarded; teaching assistantships with partial tuition reimbursements, Federal Work-Study, institutionally sponsored loans, and scholarships/grants also available. Financial award application deadline: 3/1. *Faculty research:* Cardiovascular and pulmonary, immunophysiology, neuroscience, pharmacology, epithelial. Total annual research expenditures: $4 million. *Unit head:* Frank Blecha, Head, 785-532-2741, Fax: 785-532-4557, E-mail: blecha@vet.ksu.edu. *Application contact:* Chris Ross, Director, 785-532-4507, Fax: 785-432-4557, E-mail: ross@vet.ksu.edu.

Loma Linda University, School of Medicine, Department of Pathology and Human Anatomy, Loma Linda, CA 92350. Offers MS, PhD. Part-time programs available. *Faculty:* 33 full-time (5 women), 26 part-time/adjunct (4 women). *Students:* 7 full-time (6 women), 3 part-time (1 woman); includes 2 African Americans, 1 Asian American or Pacific Islander, 3 Hispanic Americans. Terminal master's awarded for partial completion of doctoral program. *Degree requirements:* For master's, thesis; for doctorate, 2 foreign languages, thesis/dissertation. *Entrance requirements:* For master's and doctorate, GRE General Test. *Application deadline:* Applications are processed on a rolling basis. Application fee: $40. *Financial support:* Tuition waivers (full and partial) available. Support available to part-time students. *Faculty research:* Neuroendocrine system, histochemistry and image analysis, effect of age and diabetes on PNS, electron microscopy, histology. *Unit head:* Dr. Kenneth Wright, Coordinator, 909-824-301.

Louisiana State University Health Sciences Center, School of Graduate Studies in New Orleans, Department of Cell Biology and Anatomy, New Orleans, LA 70112-2223. Offers cell biology and anatomy (MS, PhD), including cell biology, developmental biology, neurobiology and anatomy. *Degree requirements:* For master's and doctorate, thesis/dissertation. *Entrance requirements:* For master's and doctorate, GRE General Test, GRE Subject Test, minimum undergraduate GPA of 3.0. Additional exam requirements/recommendations for international students: Required—TOEFL. *Faculty research:* Visual system organization, neural development, plasticity of sensory systems, information processing through the nervous system, visuomotor integration.

Louisiana State University Health Sciences Center at Shreveport, Department of Cellular Biology and Anatomy, Shreveport, LA 71130-3932. Offers MS, PhD, MD/PhD. *Faculty:* 16 full-time, 7 part-time/adjunct. *Students:* 13 full-time (8 women); includes 1 minority (African American) Average age 32. 19 applicants, 11% accepted. In 2005, 1 degree awarded. Terminal master's awarded for partial completion of doctoral program. *Degree requirements:* For master's and doctorate, thesis/dissertation. *Entrance requirements:* For master's and doctorate, GRE General Test. Additional exam requirements/recommendations for international students: Required—TOEFL. *Application deadline:* For fall admission, 5/1 for domestic students. Applications are processed on a rolling basis. Application fee: $30. *Financial support:* In 2005–06, 11 students received support, including 1 fellowship, 10 research assistantships; institutionally sponsored loans also available. Financial award application deadline: 5/1. *Faculty research:* Alcohol and immunity, neuroscience, olfactory physiology, extracellular matrix, cancer cell biology and gene therapy. *Unit head:* Dr. Leonard L. Seelig, Chairman, 318-675-5312. *Application contact:* Dr. Kathryn A. Hamilton, Graduate Coordinator, 318-675-5312.

See Close-Up on page 163.

Loyola University Chicago, Graduate School, Department of Cell Biology, Neurobiology and Anatomy, Maywood, IL 60153. Offers MS, PhD, MD/PhD. Part-time programs available. *Faculty:* 17 full-time (6 women), 5 part-time/adjunct (1 woman). *Students:* 17 full-time (8 women), 1 part-time; includes 5 minority (1 African American, 3 Asian Americans or Pacific Islanders, 1 Hispanic American), 3 international. Average age 27. 30 applicants, 23% accepted, 7 enrolled. In 2005, 4 master's, 4 doctorates awarded. Terminal master's awarded for partial completion of doctoral program. *Median time to degree:* Of those who began their doctoral program in fall 1997, 100% received their degree in 8 years or less. *Degree requirements:* For master's, thesis/dissertation; for doctorate, thesis/dissertation, comprehensive exam. *Entrance requirements:* For master's, GRE General Test, minimum GPA of 3.0; for doctorate, GRE General Test, GRE Subject Test (biology), minimum GPA of 3.0. Additional exam requirements/recommendations for international students: Required—TOEFL (minimum score 600 paper-based; 250 computer-based). *Application deadline:* For fall admission, 5/1 priority date for domestic students, 5/1 priority date for international students. Applications are processed on a rolling basis. Applica-

tion fee: $0. Electronic applications accepted. *Expenses:* Tuition: Full-time $11,610; part-time $645 per credit. Required fees: $55 per semester. *Financial support:* In 2005–06, 8 fellowships with full tuition reimbursements (averaging $22,000 per year), 5 research assistantships with full tuition reimbursements (averaging $22,000 per year) were awarded; Federal Work-Study and unspecified assistantships also available. Financial award application deadline: 5/1; financial award applicants required to submit FAFSA. *Faculty research:* Brain steroids, immunology, neuroregeneration, cytokines. Total annual research expenditures: $1 million. *Unit head:* Dr. John Clancy, Chair, 708-216-3352. *Application contact:* Dr. Tharkery S. Gray, Graduate Program Director, 708-216-3345.

See Close-Up on page 575.

McGill University, Faculty of Graduate and Postdoctoral Studies, Faculty of Medicine, Department of Anatomy and Cell Biology, Montréal, QC H3A 2T5, Canada. Offers M Sc, PhD. *Degree requirements:* For master's, thesis; for doctorate, one foreign language, thesis/dissertation. *Entrance requirements:* For master's, minimum GPA of 3.2. *Faculty research:* Molecular cell biology, neurobiology, spermatogenesis, developmental biology, aging.

Medical College of Georgia, School of Graduate Studies, Department of Cellular Biology and Anatomy, Augusta, GA 30912-1500. Offers PhD. *Faculty:* 22 full-time (7 women). *Students:* 7 full-time (4 women), 4 part-time (1 woman); includes 2 minority (both African Americans), 2 international. Average age 30. In 2005, 1 degree awarded. *Degree requirements:* For doctorate, thesis/dissertation. *Entrance requirements:* For doctorate, GRE General Test. Additional exam requirements/recommendations for international students: Required—TOEFL. *Application deadline:* For fall admission, 6/30 for domestic students, 4/15 for international students. Applications are processed on a rolling basis. Application fee: $30. *Financial support:* In 2005–06, 2 fellowships with partial tuition reimbursements (averaging $25,500 per year), 5 research assistantships with partial tuition reimbursements (averaging $22,500 per year), 1 teaching assistantship with partial tuition reimbursement (averaging $22,500 per year) were awarded; institutionally sponsored loans and scholarships/grants also available. Support available to part-time students. Financial award application deadline: 5/31; financial award applicants required to submit FAFSA. *Faculty research:* Eye disease, diabetic infections, developmental biology, cell injury and death, stroke and neurotoxicity. *Unit head:* Dr. Sally S. Atherton, Chair, 706-721-3731, Fax: 706-721-6120, E-mail: satherton@mail.mcg.edu. *Application contact:* Dr. Zheng Dong, Program Director, 706-721-2825, Fax: 706-721-6120, E-mail: zdong@mcg.edu.

Medical University of South Carolina, College of Graduate Studies, Program in Molecular and Cellular Biology and Pathobiology, Charleston, SC 29425-0002. Offers cell biology and anatomy (PhD); marine biomedicine (PhD). *Faculty:* 173 full-time (41 women). *Students:* 62 full-time (32 women); includes 8 minority (3 African Americans, 1 American Indian/Alaska Native, 4 Asian Americans or Pacific Islanders), 7 international. Average age 28. 181 applicants, 34% accepted, 43 enrolled. In 2005, 20 doctorates awarded. *Degree requirements:* For doctorate, thesis/dissertation, teaching and research seminar, oral and written exams. *Entrance requirements:* For doctorate, GRE General Test, interview, minimum GPA of 3.2. Additional exam requirements/recommendations for international students: Required—TOEFL (minimum score 600 paper-based; 250 computer-based). *Application deadline:* For fall admission, 1/15 priority date for domestic students, 1/15 priority date for international students. Applications are processed on a rolling basis. Application fee: $0 ($75 for international students). Electronic applications accepted. *Financial support:* In 2005–06, 12 students received support, including fellowships with partial tuition reimbursements available (averaging $21,000 per year); Federal Work-Study and scholarships/grants also available. Financial award application deadline: 3/15; financial award applicants required to submit FAFSA. Total annual research expenditures: $10 million. *Unit head:* Dr. Donald R. Menick, Director, 843-856-5045, Fax: 843-792-6590, E-mail: menickd@musc.edu. *Application contact:* Cheryl Brown, 843-792-4620, Fax: 843-792-4645, E-mail: brownche@musc.edu.

New York Medical College, Graduate School of Basic Medical Sciences, Department of Cell Biology, Valhalla, NY 10595-1691. Offers cell biology and neuroscience (MS, PhD). Part-time and evening/weekend programs available. Terminal master's awarded for partial completion of doctoral program. *Degree requirements:* For master's, thesis/dissertation; for doctorate, thesis/dissertation, comprehensive exam. *Entrance requirements:* For master's and doctorate, GRE General Test. Additional exam requirements/recommendations for international students: Required—TOEFL. *Faculty research:* Mechanisms of growth control in skeletal muscle, cartilage differentiation, cytoskeletal functions, signal transduction pathways, neuronal development and plasticity.

See Close-Up on page 577.

The Ohio State University, College of Medicine and Public Health and Graduate School, Graduate Programs in the Basic Medical Sciences, Department of Biomedical Informatics, Columbus, OH 43210. Offers anatomy (MS, PhD). Terminal master's awarded for partial completion of doctoral program. *Degree requirements:* For doctorate, thesis/dissertation. *Entrance requirements:* For master's and doctorate, GRE General Test, GRE Subject Test. Additional exam requirements/recommendations for international students: Required—TOEFL (minimum score 600 paper-based; 250 computer-based). Electronic applications accepted. *Faculty research:* Cell biology, biomechanical trauma, computer-assisted instruction.

The Ohio State University, College of Veterinary Medicine and Graduate School, Program in Veterinary Medicine, Department of Veterinary Biosciences, Columbus, OH 43210. Offers anatomy and cellular biology (MS, PhD); pathobiology (MS, PhD); pharmacology (MS, PhD); toxicology (MS, PhD); veterinary physiology (MS, PhD). *Faculty research:* Microvasculature, muscle biology, neonatal lung and bone development.

Palmer College of Chiropractic, Division of Graduate Studies, Davenport, IA 52803-5287. Offers anatomy (MS); clinical research (MS). *Faculty:* 136 full-time (44 women). *Students:* 12 full-time (4 women), 1 part-time. In 2005, 3 degrees awarded. *Degree requirements:* For master's, thesis, comprehensive exam, registration. *Entrance requirements:* For master's, GRE General Test, minimum GPA of 2.5. Additional exam requirements/recommendations for international students: Required—TOEFL, TSE. *Application deadline:* For fall admission, 9/1 for domestic students; for spring admission, 5/28 for domestic students. Applications are processed on a rolling basis. Application fee: $50. Electronic applications accepted. *Expenses:* Contact institution. *Financial support:* In 2005–06, 5 students received support, including teaching assistantships with full and partial tuition reimbursements available (averaging $6,269 per year); research assistantships, Federal Work-Study, institutionally sponsored loans, tuition waivers (full), and stipends also available. Support available to part-time students. Financial award application deadline: 4/1; financial award applicants required to submit FAFSA. *Unit head:* Dr. Dennis Marchiori, Administrator of Graduate Studies, 563-884-5307, Fax: 563-884-5505, E-mail: marchiori_d@palmer.edu. *Application contact:* Janet Zink, Graduate Studies Coordinator, 563-884-5307, Fax: 563-884-5226, E-mail: zink_j@palmer.edu.

The Pennsylvania State University Milton S. Hershey Medical Center, Graduate School Programs in the Biomedical Sciences, Program in Anatomy, Hershey, PA 17033-2360. Offers MS, PhD. *Students:* 4 full-time (2 women), 1 international. Average age 27. *Degree requirements:* For master's, thesis or alternative, registration. *Entrance requirements:* For master's, GRE General Test or MCAT. Additional exam requirements/recommendations for international students: Required—TOEFL (minimum score 500 paper-based; 213 computer-based). *Application deadline:* Applications are processed on a rolling basis. Application fee: $45. Electronic applications accepted. *Financial support:* In 2005–06, 4 research assistantships with full tuition reimbursements were awarded; scholarships/grants, health care benefits, and unspecified assistantships also available. Financial award applicants required to submit FAFSA. *Faculty research:* Developmental biology, stem cell, cancer-basic science and clinical application, wound healing, angiogenesis. *Unit head:* Dr. Patricia J. McLaughlin, Program Director, 717-531-6411, Fax: 717-531-5184, E-mail: grad-hmc@psu.edu. *Application contact:* Dee Clarke, Program Assistant, 717-531-8651, Fax: 717-531-8651, E-mail: grad-hmc@psu.edu.

Anatomy

Purdue University, School of Veterinary Medicine and Graduate School, Graduate Programs in Veterinary Medicine, Department of Basic Medical Sciences, West Lafayette, IN 47907. Offers anatomy (MS, PhD); pharmacology (MS, PhD); physiology (MS, PhD). Part-time programs available. *Faculty:* 16 full-time (4 women), 4 part-time/adjunct (1 woman). *Students:* 24 full-time (12 women), 3 part-time (all women); includes 1 minority (Hispanic American), 18 international. Average age 27. 18 applicants, 28% accepted, 4 enrolled. In 2005, 1 master's, 3 doctorates awarded. Terminal master's awarded for partial completion of doctoral program. *Median time to degree:* Of those who began their doctoral program in fall 1997, 33% received their degree in 8 years or less. *Degree requirements:* For master's and doctorate, thesis/dissertation. *Entrance requirements:* For master's and doctorate, GRE General Test. Additional exam requirements/recommendations for international students: Required—TOEFL. *Application deadline:* For fall admission, 12/31 priority date for domestic students, 12/31 priority date for international students. Application fee: $55. Electronic applications accepted. *Financial support:* In 2005–06, 4 fellowships with partial tuition reimbursements (averaging $13,251 per year), 12 research assistantships with partial tuition reimbursements (averaging $15,012 per year), 2 teaching assistantships with partial tuition reimbursements (averaging $17,800 per year) were awarded. Financial award application deadline: 3/1; financial award applicants required to submit FAFSA. *Faculty research:* Development and regeneration, tissue injury and shock, biomedical engineering, ovarian function, bone and cartilage biology, cell and molecular biology. *Unit head:* Dr. Gordon L. Coppoc, Head, 765-494-8592, Fax: 765-494-0781, E-mail: coppoc@purdue.edu. *Application contact:* Dr. Kevin M. Hannon, Chairman, Graduate Committee, 765-494-5949, Fax: 765-494-0781, E-mail: bmsgrad@purdue.edu.

Queen's University at Kingston, School of Graduate Studies and Research, Faculty of Health Sciences, Department of Anatomy and Cell Biology, Kingston, ON K7L 3N6, Canada. Offers M Sc, PhD. Part-time programs available. *Degree requirements:* For master's, thesis; for doctorate, one foreign language, thesis/dissertation, comprehensive exam. *Entrance requirements:* Additional exam requirements/recommendations for international students: Required—TOEFL. Electronic applications accepted. *Faculty research:* Human kinetics, neuroscience, reproductive biology, cardiovascular.

Rosalind Franklin University of Medicine and Science, School of Graduate and Postdoctoral Studies, Department of Cell Biology and Anatomy, Program in Anatomy, North Chicago, IL 60064-3095. Offers MS, PhD, MD/MS, MD/PhD. Part-time programs available. Terminal master's awarded for partial completion of doctoral program. *Degree requirements:* For master's, thesis, qualifying exam; for doctorate, thesis/dissertation, original research project, comprehensive exam. *Entrance requirements:* For master's and doctorate, GRE General Test, minimum GPA of 3.0. Additional exam requirements/recommendations for international students: Required—TOEFL, TWE. *Faculty research:* Molecular neuroscience, cell biology.

Rush University, Graduate College, Division of Anatomy and Cell Biology, Chicago, IL 60612-3832. Offers MS, PhD, MD/MS, MD/PhD. *Faculty:* 7 full-time (1 woman), 4 part-time/adjunct (3 women). *Students:* 6 full-time (2 women); includes 1 minority (Asian American or Pacific Islander), 2 international. Average age 22. 8 applicants, 13% accepted, 1 enrolled. In 2005, 2 master's awarded. Terminal master's awarded for partial completion of doctoral program. *Median time to degree:* Of those who began their doctoral program in fall 1997, 100% received their degree in 8 years or less. *Degree requirements:* For master's, thesis; for doctorate, thesis/dissertation, preliminary exam, dissertation proposal, comprehensive exam. *Entrance requirements:* For master's, GRE General Test, minimum GPA of 3.0, bachelor's degree in biology or chemistry (preferred), interview; for doctorate, GRE General Test, minimum GPA of 3.0, interview. Additional exam requirements/recommendations for international students: Required—TOEFL. *Application deadline:* For spring admission, 4/1 for domestic students, 4/1 for international students. Applications are processed on a rolling basis. Application fee: $25. Electronic applications accepted. *Financial support:* In 2005–06, 2 students received support, including fellowships with full tuition reimbursements available (averaging $21,000 per year), research assistantships with full tuition reimbursements available (averaging $2,500 per year), teaching assistantships with full tuition reimbursements available (averaging $2,500 per year); Federal Work-Study, institutionally sponsored loans, tuition waivers (full), and unspecified assistantships also available. Support available to part-time students. Financial award applicants required to submit FAFSA. *Faculty research:* Incontinence following vaginal distension, knee replacement, biomimetric materials, injured spinal motoneurons, implant fixation. Total annual research expenditures: $1 million. *Unit head:* Dr. Frank Hughes, Director, 312-942-6783, Fax: 312-942-5744, E-mail: frank_f_hughes@rush.edu. *Application contact:* Connie M. Lambert, Department Administrator, 312-563-2563, Fax: 312-563-3552, E-mail: connie_m_lambert@rush.edu.

Saint Louis University, Graduate School, School of Medicine and Graduate School, Center for Anatomical Science and Education, St. Louis, MO 63103-2097. Offers anatomy (MS-R, PhD). *Faculty:* 10 full-time (1 woman). *Students:* 10 full-time (4 women), 2 part-time; includes 2 minority (1 African American, 1 Hispanic American), 2 international. Average age 27. 16 applicants, 81% accepted, 6 enrolled. In 2005, 3 degrees awarded. *Degree requirements:* For master's, thesis, comprehensive exam; for doctorate, thesis/dissertation, departmental qualifying exams, comprehensive exam. *Entrance requirements:* For master's and doctorate, GRE General Test, GRE Subject Test, letters of recommendation, resumé. Additional exam requirements/recommendations for international students: Required—TOEFL (minimum score 550 paper-based; 213 computer-based). *Application deadline:* For fall admission, 7/1 for domestic students, 7/1 for international students; for spring admission, 11/1 for domestic students, 11/1 for international students. Applications are processed on a rolling basis. Application fee: $40. *Expenses:* Tuition: Part-time $760 per credit hour. Required fees: $55 per semester. *Financial support:* In 2005–06, 3 students received support; teaching assistantships with tuition reimbursements available, Federal Work-Study, institutionally sponsored loans, health care benefits, and unspecified assistantships available. Support available to part-time students. Financial award application deadline: 6/1; financial award applicants required to submit FAFSA. *Faculty research:* Neurodegenerative diseases, cerebellar cortical circuitry, neurogenesis. *Unit head:* Dr. Daniel L Tolbert, Director, 314-977-8035, Fax: 314-977-5127, E-mail: tolbertd@slu.edu. *Application contact:* Gary Behrman, Associate Dean of the Graduate School, 314-977-3827, E-mail: behrmang@slu.edu.

State University of New York at Buffalo, Graduate School, School of Medicine and Biomedical Sciences, Graduate Programs in Medicine and Biomedical Sciences, Department of Pathology and Anatomical Sciences, Buffalo, NY 14260. Offers anatomical sciences (MA, PhD); pathology (MA, PhD). *Faculty:* 15 full-time (2 women), 7 part-time/adjunct (2 women). *Students:* 7 full-time (5 women), 1 part-time, 1 international. In 2005, 1 master's, 1 doctorate awarded. *Entrance requirements:* For doctorate, GRE, 3 letters of recommendation. Additional exam requirements/recommendations for international students: Required—TOEFL. *Application deadline:* For fall admission, 2/1 priority date for domestic students, 2/1 priority date for international students. Application fee: $35. *Faculty research:* Immunopathology-immunobiology, experimental hypertension, neuromuscular disease, molecular pathology, cell motility and cytoskeleton. *Unit head:* Dr. Francisco Valazquez, Chairman, 716-829-2846, Fax: 716-829-2086, E-mail: fvelazquez@kaleidahealth.org. *Application contact:* Dr. Peter Nickerson, Director of Graduate Studies.

State University of New York Upstate Medical University, College of Graduate Studies, Department of Anatomy and Cell Biology, Syracuse, NY 13210-2334. Offers MS, PhD, MD/PhD. Terminal master's awarded for partial completion of doctoral program. *Degree requirements:* For master's, thesis/dissertation; for doctorate, thesis/dissertation, comprehensive exam. *Entrance requirements:* For master's, GRE General Test, GRE Subject Test, interview; for doctorate, GRE General Test, interview. Additional exam requirements/recommendations for international students: Required—TOEFL.

Stony Brook University, State University of New York, Health Sciences Center, School of Medicine and Graduate School, Graduate Programs in Medicine, Department of Anatomical Sciences, Stony Brook, NY 11794. Offers PhD. *Faculty:* 9 full-time (4 women), 1 part-time/

adjunct (0 women). *Students:* 8 full-time (2 women). Average age 30. 15 applicants, 27% accepted. In 2005, 1 degree awarded. *Degree requirements:* For doctorate, thesis/dissertation, comprehensive exam. *Entrance requirements:* For doctorate, GRE General Test, GRE Subject Test, BA in life sciences, minimum GPA of 3.0. Additional exam requirements/recommendations for international students: Required—TOEFL. *Application deadline:* For fall admission, 1/15 for domestic students. Application fee: $50. *Expenses:* Tuition, state resident: full-time $6,900; part-time $288 per credit. Tuition, nonresident: full-time $10,920; part-time $455 per credit. Required fees: $704. *Financial support:* In 2005–06, 2 fellowships, 4 research assistantships, 1 teaching assistantship were awarded; Federal Work-Study also available. Financial award application deadline: 3/15. *Faculty research:* Biological membranes, biomechanics of locomotion, systematics and evolutionary history of primates. Total annual research expenditures: $428,949. *Unit head:* Dr. Jack Stern, Chair, 631-444-2350, Fax: 631-444-3947, E-mail: jstern@mail.som.sunnysb.edu. *Application contact:* Dr. Catherine Forster, Graduate Program Director, 631-444-8203, Fax: 631-444-3947, E-mail: cforster@mail.som.sunnysb.edu.

Announcement: Paleontology, functional morphology, and systematics of vertebrates are the foci of the curriculum leading to a PhD in Anatomical Sciences. Students in this program pursue research analyzing vertebrate structure and function in relation to adaptation and evolutionary history, including such questions as the origin of birds and the evolution of human upright walking.

See Close-Up on page 347.

Temple University, Health Sciences Center, School of Medicine and Graduate School, Graduate Programs in Medicine, Department of Anatomy and Cell Biology, Philadelphia, PA 19122-6096. Offers MS, PhD. *Faculty:* 18 full-time (6 women). *Students:* 7 full-time (2 women), 2 part-time (1 woman), 7 international. 10 applicants, 60% accepted, 2 enrolled. *Degree requirements:* For doctorate, thesis/dissertation, research seminars. *Entrance requirements:* For master's and doctorate, GRE General Test, GRE Subject Test, minimum GPA of 3.0. Additional exam requirements/recommendations for international students: Required—TOEFL. *Application deadline:* For fall admission, 1/15 for domestic students, 12/15 for international students; for spring admission, 9/1 for domestic students, 8/1 for international students. Application fee: $50. Electronic applications accepted. *Expenses:* Tuition, state resident: full-time $8,694; part-time $483 per credit. Tuition, nonresident: full-time $12,672; part-time $704 per credit. Required fees: $500; $122 per semester. Tuition and fees vary according to course level, campus/location and program. *Financial support:* Fellowships, Federal Work-Study available. Financial award application deadline: 1/15; financial award applicants required to submit FAFSA. *Faculty research:* Neurobiology, reproductive biology, cardiovascular system, musculoskeletal biology, developmental biology. Total annual research expenditures: $632,373. *Unit head:* Dr. Steven Popoff, Chair, 215-707-3163, Fax: 215-707-2966, E-mail: spopoff@temple.edu.

Texas A&M University, College of Veterinary Medicine, Graduate Programs in Veterinary Medicine, Department of Veterinary Integrative Biosciences, College Station, TX 77843. Offers epidemiology (MS); food safety/toxicology (MS); veterinary anatomy (MS, PhD); veterinary public health (MS). *Faculty:* 13 full-time (3 women), 3 part-time/adjunct (1 woman). *Students:* 26 full-time (19 women), 4 part-time (3 women); includes 3 minority (2 Asian Americans or Pacific Islanders, 1 Hispanic American), 6 international. Average age 30. 15 applicants, 80% accepted, 11 enrolled. In 2005, 3 master's, 1 doctorate awarded. Terminal master's awarded for partial completion of doctoral program. *Degree requirements:* For master's and doctorate, thesis/dissertation, comprehensive exam, registration. *Entrance requirements:* For master's and doctorate, GRE General Test, minimum undergraduate GPA of 3.0. Additional exam requirements/recommendations for international students: Required—TOEFL. *Application deadline:* For fall admission, 7/15 priority date for domestic students, 4/1 priority date for international students; for spring admission, 10/1 priority date for domestic students, 9/15 priority date for international students. Applications are processed on a rolling basis. Application fee: $50 ($75 for international students). Electronic applications accepted. *Expenses:* Tuition, state resident: full-time $4,488; part-time $187 per credit hour. Tuition, nonresident: full-time $11,112; part-time $463 per credit hour. Required fees: $1,974. *Financial support:* In 2005–06, fellowships (averaging $18,000 per year), research assistantships (averaging $15,600 per year), teaching assistantships (averaging $15,600 per year) were awarded; institutionally sponsored loans, unspecified assistantships, and clinical associateships also available. Financial award application deadline: 7/15; financial award applicants required to submit FAFSA. *Faculty research:* Metal toxicology, reproductive biology, genetics of neural development, developmental biology, environmental toxicology. *Unit head:* Dr. Evelyn Tiffany-Castiglioni, Head, 979-845-2828, Fax: 979-847-8981, E-mail: ecastiglioni@cvm.tamu.edu. *Application contact:* Dr. Jane Welsh, Chair, Fax: 979-847-8981, E-mail: jwelsh@cum.tamu.edu.

Université Laval, Faculty of Medicine, Department of Anatomy and Physiology, Québec, QC G1K 7P4, Canada. Offers M Sc, PhD. Terminal master's awarded for partial completion of doctoral program. *Degree requirements:* For master's, thesis (for some programs); for doctorate, thesis/dissertation, comprehensive exam. Electronic applications accepted.

Université Laval, Faculty of Medicine, Post-Professional Programs in Medical Studies, Québec, QC G1K 7P4, Canada. Offers anatomy–pathology (DESS); anesthesia–resuscitation (DESS); cardiology (DESS); care of older people (Diploma); clinical research (DESS); community health (DESS); dermatology (DESS); diagnostic radiology (DESS); emergency medicine (Diploma); family medicine (DESS); general surgery (DESS); geriatrics (DESS); hematology (DESS); internal medicine (DESS); maternal and fetal medicine (Diploma); medical biochemistry (DESS); medical microbiology and infectious diseases (DESS); medical oncology (DESS); nephrology (DESS); neurology (DESS); neurosurgery (DESS); obstetrics and gynecology (DESS); ophthalmology (DESS); orthopedic surgery (DESS); oto-rhino-laryngology (DESS); palliative medicine (Diploma); pediatrics (DESS); plastic surgery (DESS); psychiatry (DESS); pulmonary medicine (DESS); radiology–oncology (DESS); thoracic surgery (DESS); urology (DESS). *Degree requirements:* For other advanced degree, comprehensive exam. *Entrance requirements:* For degree, knowledge of French. Electronic applications accepted.

The University of Arizona, College of Medicine, Graduate Programs in Medicine, Department of Cell Biology and Anatomy, Tucson, AZ 85721. Offers PhD. *Degree requirements:* For doctorate, thesis/dissertation. *Entrance requirements:* For doctorate, GRE General Test. *Faculty research:* Heart development, neural development, cellular toxicology and microcirculation; membrane traffic and cytoskeleton; cell-surface receptors.

University of Arkansas for Medical Sciences, College of Medicine and Graduate School, Graduate Programs in Medicine, Department of Neurobiology and Developmental Sciences, Little Rock, AR 72205-7199. Offers MS, PhD, MD/PhD. *Faculty:* 31 full-time (10 women), 7 part-time/adjunct (4 women). *Students:* 14 full-time, 5 part-time. *Degree requirements:* For master's and doctorate, thesis/dissertation. *Entrance requirements:* For master's, GRE General Test; for doctorate, GRE General Test, GRE Subject Test. Additional exam requirements/recommendations for international students: Required—TOEFL. Application fee: $0. *Financial support:* Research assistantships available. Support available to part-time students. *Unit head:* Dr. Gwen Childs, Chair, 501-686-5180. *Application contact:* Dr. David Davies, Graduate Coordinator, 501-686-5184, E-mail: dldavies@uams.edu.

See Close-Up on page 1031.

The University of British Columbia, Faculty of Medicine and Faculty of Graduate Studies, Graduate Programs in Medicine, Department of Cellular and Physiological Sciences, Division of Anatomy and Cell Biology, Vancouver, BC V6T 1Z3, Canada. Offers M Sc, PhD. *Faculty:* 12 full-time (2 women), 2 part-time/adjunct (0 women). *Students:* 17 full-time (6 women), 2 international. Average age 27. 12 applicants, 17% accepted, 2 enrolled. In 2005, 3 master's, 1 doctorate awarded. *Degree requirements:* For master's, thesis/dissertation, oral defense; for doctorate, thesis/dissertation, oral defense, comprehensive exam. *Entrance requirements:* Additional exam requirements/recommendations for international students: Required—TOEFL (minimum score 550 paper-based; 213 computer-based). *Application deadline:* For fall admission, 4/1 priority date for domestic students, 3/1 priority date for international students. Applications are processed on a rolling basis. Application fee: $90 Canadian dollars ($150 Canadian dollars for international students). Electronic applications accepted. *Financial support:* In 2005–06, 3 fellowships (averaging $17,000 per year), 8 research assistantships with partial tuition reimbursements (averaging $17,000 per year), 6 teaching assistantships (averaging $5,000 per year) were awarded; Federal Work-Study, institutionally sponsored loans, scholarships/grants, traineeships, tuition waivers (full and partial), and unspecified assistantships also available. *Faculty research:* Cell and developmental biology, membrane biophysics, cellular immunology, cancer, fetal alcohol syndrome. Total annual research expenditures: $3 million Canadian dollars. *Application contact:* Dr. John Church, Graduate Adviser, 604-822-2751, Fax: 604-822-2316, E-mail: jchurch@interchange.ubc.ca.

University of California, Irvine, College of Medicine and School of Biological Sciences, Department of Anatomy and Neurobiology, Irvine, CA 92697. Offers biological sciences (MS, PhD). *Degree requirements:* For doctorate, thesis/dissertation. *Entrance requirements:* For master's and doctorate, GRE General Test, GRE Subject Test. Additional exam requirements/recommendations for international students: Required—TOEFL (minimum score 550 paper-based; 213 computer-based), TSE. Electronic applications accepted. *Faculty research:* Neurotransmitter immunocytochemistry, intracellular physiology, molecular neurobiology, forebrain organization and development, structure and function of sensory and motor systems.

University of California, Los Angeles, School of Medicine and Graduate Division, Graduate Programs in Medicine, Department of Neurobiology, Los Angeles, CA 90095. Offers anatomy and cell biology (PhD). *Degree requirements:* For doctorate, thesis/dissertation, oral and written qualifying exams. *Entrance requirements:* For doctorate, GRE General Test, GRE Subject Test, bachelor's degree in physical or biological science. *Faculty research:* Neuroendocrinology, neurophysiology.

University of California, San Francisco, Graduate Division, Biomedical Sciences Graduate Group, Program in Anatomy, San Francisco, CA 94143. Offers PhD. *Degree requirements:* For doctorate, variable foreign language requirement, thesis/dissertation. *Entrance requirements:* For doctorate, GRE General Test. *Faculty research:* Cell biology, neurohistology, mammalian development.

University of Chicago, Division of the Biological Sciences, Department of Darwinian Sciences: Ecological, Integrative and Evolutionary Biology, Department of Organismal Biology and Anatomy, Chicago, IL 60637-1513. Offers functional and evolutionary biology (PhD); organismal biology and anatomy (PhD). *Faculty:* 12 full-time (1 woman), 5 part-time/adjunct (3 women). *Students:* 12 full-time (3 women), 3 international. Average age 28. In 2005, 2 doctorates awarded. *Degree requirements:* For doctorate, thesis/dissertation, registration. *Entrance requirements:* For doctorate, GRE General Test. Additional exam requirements/recommendations for international students: Required—TOEFL. *Application deadline:* For fall admission, 12/28 priority date for domestic students, 12/28 priority date for international students. Application fee: $55. Electronic applications accepted. *Financial support:* In 2005–06, 10 students received support, including fellowships with tuition reimbursements available (averaging $26,301 per year), research assistantships with tuition reimbursements available (averaging $26,301 per year); institutionally sponsored loans, scholarships/grants, traineeships, and health care benefits also available. Financial award applicants required to submit FAFSA. *Faculty research:* Ecological physiology, evolution of fossil reptiles, vertebrate paleontology. *Unit head:* Dr. Neil Shubin, Chairman, 773-702-7472, Fax: 773-702-0037, E-mail: nshubin@uchicago.edu. *Application contact:* Carolyn Johnson, Graduate Administrative Director, 773-702-9474, Fax: 773-702-4699, E-mail: csjohnso@uchicago.edu.

University of Florida, College of Medicine, Department of Anatomy and Cell Biology, Gainesville, FL 32611. Offers PhD. *Faculty:* 13 full-time (4 women), 1 (woman) part-time/adjunct. *Degree requirements:* For doctorate, thesis/dissertation. *Entrance requirements:* For doctorate, GRE General Test, minimum GPA of 3.0. Additional exam requirements/recommendations for international students: Required—TOEFL. *Application deadline:* For fall admission, 2/15 for domestic students. Applications are processed on a rolling basis. Application fee: $30. Electronic applications accepted. *Expenses:* Tuition, state resident: full-time $6,234. Tuition, nonresident: full-time $21,359. Tuition and fees vary according to program. *Financial support:* In 2005–06, research assistantships (averaging $24,047 per year); fellowships *Faculty research:* Structure and function of intracellular organelles, cell adhesion, differentiation. *Unit head:* Dr. Stephen P. Sugrue, Chairman, 352-392-3569, Fax: 352-392-3305. *Application contact:* Dr. Wayne McCormack, Associate Dean of Graduate Education, 352-392-7413, Fax: 352-846-3466, E-mail: mccormac@pathology.ufl.edu.

University of Georgia, College of Veterinary Medicine and Graduate School, Graduate Programs in Veterinary Medicine, Department of Veterinary Anatomy and Radiology, Athens, GA 30602. Offers veterinary anatomy (MS). *Faculty:* 2 full-time (1 woman). *Students:* 1 applicant, 100% accepted, 0 enrolled. *Degree requirements:* For master's, thesis. *Entrance requirements:* For master's, GRE General Test. *Application deadline:* For fall admission, 7/1 for domestic students; for spring admission, 11/15 for domestic students. Application fee: $50. Electronic applications accepted. *Financial support:* Fellowships, research assistantships, teaching assistantships, unspecified assistantships available. *Unit head:* Dr. Royce E. Roberts, Head, Fax: 706-542-0051, E-mail: rroberts@vet.uga.edu. *Application contact:* Dr. Sharon L. Crowell-Davis, Graduate Coordinator, 706-542-8343, Fax: 706-542-0051, E-mail: scrowell@vet.uga.edu.

University of Guelph, Ontario Veterinary College and Graduate Program Services, Graduate Programs in Veterinary Sciences, Department of Biomedical Sciences, Guelph, ON N1G 2W1, Canada. Offers morphology (M Sc, DV Sc, PhD); pharmacology (M Sc, DV Sc, PhD); physiology (M Sc, DV Sc, PhD); toxicology (M Sc, DV Sc, PhD). Part-time programs available. *Faculty:* 25. *Students:* 43 (27 women). In 2005, 6 master's, 2 doctorates awarded. *Median time to degree:* Of those who began their doctoral program in fall 1997, 100% received their degree in 8 years or less. *Degree requirements:* For master's, thesis/dissertation; for doctorate, thesis/dissertation, comprehensive exam. *Entrance requirements:* For master's, honors B Sc, minimum 75% average in last 20 courses; for doctorate, M Sc with thesis from accredited institution. Additional exam requirements/recommendations for international students: Required—TOEFL (minimum score 550 paper-based; 213 computer-based). *Application deadline:* Applications are processed on a rolling basis. Application fee: $75. Electronic applications accepted. *Financial support:* Fellowships, research assistantships, teaching assistantships available. *Faculty research:* Cellular morphology; endocrine, vascular and reproductive physiology; clinical pharmacology; veterinary toxicology; developmental biology. Total annual research expenditures: $3.5 million. *Unit head:* Dr. N. Maclusky, Chair, 519-824-4120 Ext. 54904, Fax: 511-767-1450. *Application contact:* Dr. G. Kirby, Graduate Coordinator, 519-824-4120 Ext. 54948, Fax: 519-767-1450, E-mail: gkirby@uoguelph.ca.

University of Hawaii at Manoa, John A. Burns School of Medicine and Graduate Division, Graduate Programs in Biomedical Sciences, Department of Anatomy, Biochemistry, Physiology and Reproductive Biology, Honolulu, HI 96822. Offers anatomy and reproductive biology (MS); physiology (MS, PhD); reproductive biology (PhD). Part-time programs available. *Students:* 2 applicants, 100% accepted. *Degree requirements:* For doctorate, thesis/dissertation. *Entrance requirements:* For doctorate, GRE General Test, GRE Subject Test. Application fee: $50. *Expenses:* Tuition, state resident: full-time $8,400; part-time $200 per credit hour. Tuition, nonresident: full-time $11,088; part-time $462 per credit hour. Tuition and fees vary according to program. *Financial support:* Fellowships, research assistantships, teaching assistantships available. *Faculty research:* Biology of gametes and fertilization, reproductive endocrinology. *Unit head:* Scott Lozanoff, Chair, 808-956-7131, Fax: 808-956-9481, E-mail: lozanoffs@jabsom.biomed.hawaii.edu.

University of Illinois at Chicago, College of Medicine and Graduate College, Graduate Programs in Medicine, Department of Anatomy and Cell Biology, Chicago, IL 60607-7128.

Offers MS, PhD, MD/PhD. *Degree requirements:* For master's and doctorate, thesis/dissertation. *Entrance requirements:* For master's and doctorate, GRE General Test. Additional exam requirements/recommendations for international students: Required—TOEFL. Electronic applications accepted. *Faculty research:* Neuroanatomy, functional morphology, cytoskeleton, synapses, neural transplants.

The University of Iowa, Roy J. and Lucille A. Carver College of Medicine and Graduate College, Graduate Programs in Medicine, Department of Anatomy and Cell Biology, Iowa City, IA 52242-1316. Offers PhD. *Faculty:* 24 full-time (6 women). *Students:* 16 full-time (5 women); includes 1 minority (Asian American or Pacific Islander), 4 international. Average age 28. 115 applicants, 2% accepted. In 2005, 2 degrees awarded. *Median time to degree:* Of those who began their doctoral program in fall 1997, 100% received their degree in 8 years or less. *Degree requirements:* For doctorate, thesis/dissertation, comprehensive exam, registration. *Entrance requirements:* For doctorate, GRE General Test, minimum GPA of 3.0. Additional exam requirements/recommendations for international students: Required—TOEFL. *Application deadline:* For fall admission, 2/1 priority date for domestic students, 2/1 priority date for international students. Applications are processed on a rolling basis. Application fee: $60 ($85 for international students). Electronic applications accepted. *Expenses:* Tuition, state resident: part-time $1,882 per term. Tuition, nonresident: full-time $17,338; part-time $4,907 per term. Tuition and fees vary according to course load and program. *Financial support:* In 2005–06, 16 students received support, including fellowships with full tuition reimbursements available (averaging $22,000 per year), 14 research assistantships with full tuition reimbursements available (averaging $22,000 per year), teaching assistantships with full tuition reimbursements available (averaging $22,000 per year); institutionally sponsored loans, scholarships/grants, and health care benefits also available. Financial award application deadline: 3/1. *Faculty research:* Biology of differentiation and transformation, developmental and vascular cell biology, neurobiology. Total annual research expenditures: $5.7 million. *Unit head:* Dr. John F. Engelhardt, Professor and Interim Head, 319-335-7744, Fax: 319-335-7198, E-mail: john-engelhardt@uiowa.edu. *Application contact:* Julie A. Stark, Program Assistant, 319-335-7744, Fax: 319-335-7198, E-mail: julie-stark@uiowa.edu.

University of Kansas, Graduate Studies Medical Center, Interdisciplinary Graduate Program in Biomedical Sciences, Department of Anatomy and Cell Biology, Lawrence, KS 66045. Offers MA, PhD, MD/PhD. Part-time programs available. *Faculty:* 1 part-time/adjunct. *Students:* Average age 28. In 2005, 2 master's, 1 doctorate awarded. Terminal master's awarded for partial completion of doctoral program. *Degree requirements:* For master's, comprehensive oral exam, oral defense of thesis; for doctorate, thesis/dissertation, comprehensive exam. *Entrance requirements:* Additional exam requirements/recommendations for international students: Required—TOEFL, TSE. *Application deadline:* For fall admission, 1/15 for domestic students. Applications are processed on a rolling basis. Application fee: $0. Electronic applications accepted. *Expenses:* Tuition, state resident: full-time $4,859. Tuition, nonresident: full-time $12,000. Required fees: $589. Tuition and fees vary according to program. *Financial support:* Fellowships, research assistantships with partial tuition reimbursements, teaching assistantships with full and partial tuition reimbursements, institutionally sponsored loans available. *Faculty research:* Kidney development, immunobiology of pregnancy, vascular formation, developmental neurobiology, cell-matrix interactions. *Unit head:* Dr. Dale R. Abrahamson, Chairman, 913-588-7000, Fax: 913-588-2710, E-mail: dabrahamson@kume.edu. *Application contact:* Dr. Michael Werle, Graduate Adviser, 913-588-7491, Fax: 913-588-2710, E-mail: mwerle@kumc.edu.

University of Kentucky, Graduate School, Graduate School Programs from the College of Medicine, Program in Anatomy and Neurobiology, Lexington, KY 40506-0032. Offers anatomy (PhD). *Faculty:* 32 full-time (8 women). *Students:* 21 full-time (6 women), 3 part-time (all women); includes 4 minority (1 African American, 1 Asian American or Pacific Islander, 2 Hispanic Americans), 8 international. Average age 27. 13 applicants, 92% accepted, 10 enrolled. In 2005, 4 degrees awarded. *Median time to degree:* Of those who began their doctoral program in fall 1997, 83.3% received their degree in 8 years or less. *Degree requirements:* For doctorate, thesis/dissertation, comprehensive exam. *Entrance requirements:* For doctorate, GRE General Test, minimum undergraduate GPA of 3.0. Additional exam requirements/recommendations for international students: Required—TOEFL (minimum score 550 paper-based; 213 computer-based). *Application deadline:* For fall admission, 7/17 priority date for domestic students, 2/1 priority date for international students; for spring admission, 12/13 priority date for domestic students, 6/15 priority date for international students. Applications are processed on a rolling basis. Application fee: $40 ($55 for international students). Electronic applications accepted. *Expenses:* Tuition, state resident: full-time $6,308; part-time $331 per credit hour. Tuition, nonresident: full-time $13,968; part-time $756 per credit hour. Tuition and fees vary according to course load, degree level and program. *Financial support:* In 2005–06, 6 fellowships with full tuition reimbursements (averaging $5,000 per year), 22 research assistantships with full tuition reimbursements (averaging $21,000 per year), 14 teaching assistantships with full tuition reimbursements (averaging $10,500 per year) were awarded; tuition waivers (full) also available. Financial award application deadline: 3/15. *Faculty research:* Neuroendocrinology, developmental neurobiology, neurotrophic substances, neural plasticity and trauma, neurobiology of aging. Total annual research expenditures: $1.4 million. *Unit head:* Dr. Douglas Gould, Director of Graduate Studies, 859-323-5484, Fax: 859-323-5946, E-mail: djgould@uky.edu. *Application contact:* Dr. Brian Jackson, Senior Associate Dean, 859-257-8176, Fax: 859-323-1928.

University of Louisville, School of Medicine, Department of Anatomical Sciences and Neurobiology, Louisville, KY 40292-0001. Offers MS, PhD. *Students:* 26 full-time (9 women), 7 part-time (3 women); includes 1 minority (African American), 17 international. Average age 28. In 2005, 13 master's, 4 doctorates awarded. *Degree requirements:* For master's and doctorate, thesis/dissertation. *Entrance requirements:* For master's and doctorate, GRE General Test. Additional exam requirements/recommendations for international students: Required—TOEFL. *Application deadline:* For fall admission, 1/15 for domestic students; for spring admission, 4/15 for domestic students. Application fee: $50. Electronic applications accepted. *Expenses:* Tuition, state resident: full-time $6,006; part-time $334 per credit hour. Tuition, nonresident: full-time $16,554; part-time $920 per credit hour. Tuition and fees vary according to course load, degree level and program. *Financial support:* Fellowships with full tuition reimbursements, research assistantships with full tuition reimbursements, teaching assistantships with full tuition reimbursements available. *Unit head:* Dr. Fred J. Roisen, Chair, 502-852-5165, Fax: 502-852-6228, E-mail: fjrois01@gwise.louisville.edu. *Application contact:* Dr. Kathleen Klueber, Director of Graduate Studies, 502-852-5172, Fax: 502-852-6228, E-mail: kmklue01@gwise.louisville.edu.

University of Manitoba, Faculty of Medicine and Faculty of Graduate Studies, Graduate Programs in Medicine, Department of Human Anatomy and Cell Science, Winnipeg, MB R3T 2N2, Canada. Offers M Sc, PhD. *Degree requirements:* For master's, thesis; for doctorate, one foreign language, thesis/dissertation.

University of Maryland, Graduate School, Graduate Programs in Medicine, Department of Anatomy and Neurobiology, Baltimore, MD 21201. Offers MS, PhD, MD/PhD. Part-time and evening/weekend programs available. *Faculty:* 16 full-time (2 women). *Students:* 7 full-time (4 women); includes 1 minority (African American), 3 international. Average age 29. 1 applicant, 0% accepted, 0 enrolled. *Degree requirements:* For master's, thesis optional; for doctorate, one foreign language, thesis/dissertation. *Entrance requirements:* For master's, GRE General Test, minimum GPA of 3.0; for doctorate, GRE General Test, GRE Subject Test (recommended), minimum GPA of 3.0. Additional exam requirements/recommendations for international students: Required—TOEFL, TOEFL or IELTS; Recommended—IELT. *Application deadline:* For fall admission, 5/31 for domestic students, 1/15 for international students. Application fee: $50. Electronic applications accepted. *Expenses:* Tuition, state resident: full-time $8,079; part-time $409 per credit hour. Tuition, nonresident: full-time $18,384; part-time $731 per credit hour. Required fees: $695; $10 per credit hour. Tuition and fees vary according to degree level and program. *Financial support:* Research assistantships available.

Anatomy

University of Maryland *(continued)*
Support available to part-time students. Financial award application deadline: 2/15. *Faculty research:* Neural networks, chemical sensory pathways, electrophysiology, developmental neurobiology. *Unit head:* Dr. Michael T. Shipley, Chair, 410-706-3590, Fax: 410-706-2512, E-mail: mshipley@umaryland.edu. *Application contact:* Dr. George Markelonis, Program Director, 410-706-3713, Fax: 410-706-2512, E-mail: gmarkel@umaryland.edu.

University of Maryland, Graduate School, Interdisciplinary Training Program in Muscle Biology, Baltimore, MD 21201.

See Close-Up on page 431.

University of Mississippi Medical Center, School of Graduate Studies in the Health Sciences, Department of Anatomy, Jackson, MS 39216-4505. Offers MS, PhD, MD/PhD. *Faculty:* 14 full-time (4 women), 1 part-time/adjunct (0 women). *Students:* 6 full-time (3 women); includes 2 minority (1 African American, 1 Asian American or Pacific Islander), 4 international. Average age 27. 6 applicants, 83% accepted, 1 enrolled. In 2005, 2 degrees awarded. Terminal master's awarded for partial completion of doctoral program. *Median time to degree:* Of those who began their doctoral program in fall 1997, 100% received their degree in 8 years or less. *Degree requirements:* For master's, thesis; for doctorate, thesis/dissertation, first authored publication, comprehensive exam. *Entrance requirements:* For master's and doctorate, GRE General Test, minimum GPA of 3.0. Additional exam requirements/recommendations for international students: Required—TOEFL. *Application deadline:* For fall admission, 1/31 priority date for domestic students, 1/31 priority date for international students. Application fee: $10. *Financial support:* In 2005–06, research assistantships (averaging $16,559 per year) Financial award application deadline: 3/15. *Faculty research:* Systems neuroscience with emphasis on motor and sensory, cell biology with emphasis on cell-matrix interactions, development of cardiovascular system, biology of glial cells. Total annual research expenditures: $800,000. *Unit head:* , Dr. Duane E. Haines, Chairman, 601-984-1640, Fax: 601-984-1655, E-mail: dhaines@anatomy.umsmed.edu. *Application contact:* Dr. James C. Lynch, Director of Graduate Program, 601-984-1657, Fax: 601-984-1655, E-mail: jclynch@anatomy.umsmed.edu.

University of Nebraska Medical Center, Graduate Studies, Department of Genetics, Cell Biology and Anatomy, Omaha, NE 68198. Offers MS, PhD. Part-time programs available. *Faculty:* 20 full-time (6 women), 1 part-time/adjunct (0 women). *Students:* 12 full-time (4 women); includes 7 minority (1 African American, 5 Asian Americans or Pacific Islanders, 1 Hispanic American), 5 international. Average age 28. 8 applicants, 38% accepted, 3 enrolled. In 2005, 1 degree awarded. Terminal master's awarded for partial completion of doctoral program. *Degree requirements:* For master's and doctorate, thesis/dissertation, comprehensive exam, registration. *Entrance requirements:* For master's and doctorate, GRE General Test. Additional exam requirements/recommendations for international students: Required—TOEFL. *Application deadline:* For fall admission, 3/1 for domestic students. Applications are processed on a rolling basis. Application fee: $45. Electronic applications accepted. *Expenses:* Tuition, area resident: Part-time $200 per hour. Tuition, nonresident: part-time $538 per hour. Required fees: $308; $59 per term. *Financial support:* In 2005–06, 4 students received support, including 11 fellowships with full tuition reimbursements available (averaging $21,000 per year), 11 research assistantships with full tuition reimbursements available (averaging $21,000 per year), teaching assistantships with full tuition reimbursements available (averaging $21,000 per year); institutionally sponsored loans also available. Support available to part-time students. Financial award application deadline: 3/1. *Faculty research:* Hematology, immunology, developmental biology, cardiovascular biology, neuroscience. *Unit head:* Dr. Shantaram S. Joshi, Graduate Committee Chair, 402-559-4165, Fax: 402-559-3400, E-mail: ssjoshi@unmc.edu. *Application contact:* Penni Davis, Graduate Student Support, 402-559-3316, Fax: 402-559-3144, E-mail: pkdavis@unmc.edu.

University of North Dakota, School of Medicine and Graduate School, Graduate Programs in Medicine, Department of Anatomy, Grand Forks, ND 58202. Offers MS, PhD. *Faculty:* 10 full-time (1 woman). *Students:* 1 full-time (0 women), 9 part-time (3 women). 11 applicants, 18% accepted, 2 enrolled. In 2005, 1 degree awarded. *Degree requirements:* For master's, thesis/dissertation, final exam; for doctorate, thesis/dissertation, final exam, comprehensive exam. *Entrance requirements:* For master's and doctorate, GRE General Test, minimum GPA of 3.0. Additional exam requirements/recommendations for international students: Required—TOEFL (minimum score 550 paper-based; 213 computer-based). *Application deadline:* For fall admission, 2/15 priority date for domestic students, 2/15 priority date for international students; for spring admission, 10/15 priority date for domestic students, 10/15 priority date for international students. Applications are processed on a rolling basis. Application fee: $35. Electronic applications accepted. *Financial support:* In 2005–06, 7 teaching assistantships with full tuition reimbursements (averaging $14,135 per year) were awarded; fellowships, research assistantships, Federal Work-Study, institutionally sponsored loans, scholarships/grants, and tuition waivers (full and partial) also available. Support available to part-time students. Financial award applicants required to submit FAFSA. *Faculty research:* Coronary vessel, vasculogenesis, acellular glomerular and retinal microvessel membranes, ependymal cells, cardiac muscle. *Unit head:* Dr. Ken Ruit, Director, 701-777-2570, Fax: 701-777-3527, E-mail: kenruit@medicine.nodak.edu. *Application contact:* Brenda Halle, Admissions Specialist, 701-777-2947, Fax: 701-777-3619, E-mail: brendahalle@mail.und.nodak.edu.

University of North Texas Health Science Center at Fort Worth, Graduate School of Biomedical Sciences, Fort Worth, TX 76107-2699. Offers anatomy and cell biology (MS, PhD); biochemistry and molecular biology (MS, PhD); biomedical sciences (MS, PhD); biotechnology (MS); forensic genetics (MS); integrative physiology (MS, PhD); medical science (MS); microbiology and immunology (MS, PhD); pharmacology (MS, PhD); science education (MS). *Faculty:* 57 full-time (19 women), 2 part-time/adjunct (0 women). *Students:* 177 full-time (98 women), 42 part-time (33 women); includes 61 minority (15 African Americans, 3 American Indian/Alaska Native, 27 Asian Americans or Pacific Islanders, 16 Hispanic Americans), 53 international. Average age 28. 237 applicants, 62% accepted, 91 enrolled. In 2005, 37 master's, 15 doctorates awarded. Terminal master's awarded for partial completion of doctoral program. *Degree requirements:* For master's and doctorate, thesis/dissertation. *Entrance requirements:* For master's and doctorate, GRE General Test. Additional exam requirements/recommendations for international students: Required—TOEFL. *Application deadline:* For fall admission, 5/1 for domestic students. Application fee: $25 ($50 for international students). *Expenses: Contact institution. Financial support:* In 2005–06, 80 research assistantships (averaging $16,000 per year) were awarded; fellowships, teaching assistantships, career-related internships or fieldwork, Federal Work-Study, institutionally sponsored loans, scholarships/grants, and traineeships also available. Support available to part-time students. Financial award application deadline: 4/1; financial award applicants required to submit FAFSA. *Faculty research:* Alzheimer's disease, aging, eye diseases, cancer, cardiovascular disease. Total annual research expenditures: $21 million. *Unit head:* Dr. Thomas Yorio, Dean, 817-735-2560, Fax: 817-735-0243, E-mail: yoriot@hsc.unt.edu. *Application contact:* Carla Lee, Director of Graduate Admissions and Services, 817-735-2560, Fax: 817-735-0243, E-mail: gsbs@hsc.unt.edu.

See Close-Up on page 279.

University of Prince Edward Island, Atlantic Veterinary College, Graduate Program in Veterinary Medicine, Charlottetown, PE C1A 4P3, Canada. Offers anatomy (M Sc, PhD); bacteriology (M Sc, PhD); clinical pharmacology (M Sc, PhD); clinical sciences (M Sc, PhD); epidemiology (M Sc, PhD), including reproduction; fish health (M Sc, PhD); food animal nutrition (M Sc, PhD); immunology (M Sc, PhD); microanatomy (M Sc, PhD); parasitology (M Sc, PhD); pathology (M Sc, PhD); pharmacology (M Sc, PhD); physiology (M Sc, PhD); toxicology (M Sc, PhD); veterinary science (M Vet Sc); virology (M Sc, PhD). Part-time programs available. *Faculty:* 76 full-time (25 women), 49 part-time/adjunct (8 women). *Students:* 54 full-time (32 women), 2 part-time. Average age 30. In 2005, 7 master's, 6 doctorates awarded. *Degree requirements:* For master's and doctorate, thesis/dissertation. *Entrance*

requirements: For master's, DVM, B Sc honors degree, or equivalent; for doctorate, M Sc. *Application deadline:* Applications are processed on a rolling basis. Application fee: $50. *Expenses:* Contact institution. Tuition charges are reported in Canadian dollars. Part-time tuition and fees vary according to course level, degree level, campus/location, program and student level. *Financial support:* In 2005–06, 4 fellowships (averaging $25,000 Canadian dollars per year), 4 research assistantships (averaging $16,500 Canadian dollars per year) were awarded; career-related internships or fieldwork also available. *Faculty research:* Animal health management, infectious diseases, fin fish and shellfish health, basic biomedical sciences, ecosystem health. Total annual research expenditures: $1.2 million Canadian dollars. *Unit head:* Dr. James Bellamy, Associate Dean of Graduate Studies and Research, 902-566-0856, E-mail: bellamy@upei.ca. *Application contact:* Cheryl Gaudet, Registrar's Office, 902-566-0781, Fax: 902-566-0795, E-mail: registrar@upei.ca.

University of Puerto Rico, Medical Sciences Campus, School of Medicine, Division of Graduate Studies, Department of Anatomy, San Juan, PR 00936-5067. Offers MS, PhD. *Faculty:* 13 full-time (6 women). *Students:* 5 full-time (4 women); all minorities (all Hispanic Americans) Average age 25. 1 applicant, 0% accepted, 0 enrolled. In 2005, 1 degree awarded. *Degree requirements:* For master's and doctorate, one foreign language, thesis/dissertation, comprehensive exam, registration. *Entrance requirements:* For master's and doctorate, GRE General Test, GRE Subject Test, interview, minimum GPA of 3.0, 3 letters of recommendation. *Application deadline:* For fall admission, 2/15 for domestic students, 2/15 for international students; for spring admission, 9/15 for domestic students, 9/15 for international students. Application fee: $15. *Expenses:* Tuition, state resident: full-time $3,600; part-time $100 per credit hour. Tuition, nonresident: full-time $4,655. Required fees: $1,734. Tuition and fees vary according to class time, degree level and program. *Financial support:* In 2005–06, 3 research assistantships (averaging $9,354 per year), 3 teaching assistantships (averaging $9,354 per year) were awarded; fellowships, Federal Work-Study, institutionally sponsored loans, and tuition waivers (full) also available. Support available to part-time students. Financial award application deadline: 4/30. *Faculty research:* Neurobiology, primatology, visual system, muscle structure, craniofacial. *Unit head:* Dr. Maria Sosa, Interim Director, 787-758-2525 Ext. 1514, Fax: 787-767-0788, E-mail: masosa@neuro.upr.edu. *Application contact:* Dr. Earl Kicliter, Coordinator, 787-721-4527, Fax: 787-721-1236, E-mail: ekiclite@neuro.upr.edu.

University of Rochester, School of Medicine and Dentistry, Graduate Programs in Medicine and Dentistry, Department of Neurobiology and Anatomy, Program in Neurobiology and Anatomy, Rochester, NY 14627-0250. Offers MS, PhD. *Degree requirements:* For doctorate, thesis/dissertation, qualifying exam. *Entrance requirements:* For master's and doctorate, GRE General Test.

University of Saskatchewan, College of Medicine, Department of Anatomy and Cell Biology, Saskatoon, SK S7N 5A2, Canada. Offers M Sc, PhD. *Faculty:* 17. *Students:* 26, 4 international. *Degree requirements:* For master's and doctorate, thesis/dissertation, registration. *Entrance requirements:* Additional exam requirements/recommendations for international students: Required—TOEFL. *Application deadline:* For fall admission, 7/1 for domestic students. Applications are processed on a rolling basis. Application fee: $50. *Financial support:* Fellowships, research assistantships, teaching assistantships available. Financial award application deadline: 1/31. *Unit head:* Dr. R. Doucette, Head (Acting), 306-966-4075, Fax: 306-966-4298, E-mail: rondouc@duke.usask.ca. *Application contact:* Dr. D. Schreyer, Graduate Chair, 306-966-4298, Fax: 306-966-4298, E-mail: david.shreyer@usask.ca.

University of Saskatchewan, Western College of Veterinary Medicine and College of Graduate Studies and Research, Graduate Programs in Veterinary Medicine, Department of Veterinary Biomedical Sciences, Saskatoon, SK S7N 5A2, Canada. Offers veterinary anatomy (M Sc); veterinary biomedical sciences (M Vet Sc); veterinary physiological sciences (M Sc, PhD). *Faculty:* 15 full-time (5 women). *Students:* 36 full-time (19 women); includes 3 minority (all African Americans) 6 applicants, 33% accepted. In 2005, 5 master's, 1 doctorate awarded. *Degree requirements:* For master's and doctorate, thesis/dissertation. *Faculty research:* Toxicology, animal reproduction, pharmacology, chloride channels, pulmonary pathobiology. *Unit head:* Dr. Barry Blakley, Head, 306-966-7349, Fax: 306-966-7376, E-mail: barry.blakley@usask.ca.

University of South Florida, College of Medicine and College of Graduate Studies, Graduate Programs in Medical Sciences, Tampa, FL 33620-9951. Offers anatomy (PhD); biochemistry and molecular biology (MS, PhD), including biochemistry and molecular biology (PhD); bioinformatics and computational biology (MS); medical microbiology and immunology (PhD); pathology (PhD); pharmacology and therapeutics (PhD), including medical sciences; physiology and biophysics (PhD). *Students:* 108 full-time (53 women), 34 part-time (26 women); includes 31 minority (9 African Americans, 9 Asian Americans or Pacific Islanders, 13 Hispanic Americans), 30 international. 117 applicants, 99% accepted, 116 enrolled. In 2005, 9 master's, 4 doctorates awarded. *Degree requirements:* For doctorate, thesis/dissertation. *Entrance requirements:* For doctorate, GRE General Test, minimum GPA of 3.0. Application fee: $30. *Expenses: Contact institution. Financial support:* Institutionally sponsored loans and scholarships/grants available. Financial award application deadline: 4/1; financial award applicants required to submit FAFSA. *Unit head:* Dr. Joseph J. Krzanowski, Associate Dean for Research and Graduate Affairs, 813-974-4181, Fax: 813-974-4317, E-mail: jkrzanow@com1.med.usf.edu.

The University of Tennessee, Graduate School, College of Agricultural Sciences and Natural Resources, Department of Animal Science, Knoxville, TN 37996. Offers animal anatomy (PhD); breeding (MS, PhD); management (MS, PhD); nutrition (MS, PhD); physiology (MS, PhD). Part-time programs available. *Degree requirements:* For master's and doctorate, thesis/dissertation. *Entrance requirements:* For master's and doctorate, GRE General Test, minimum GPA of 2.7. Additional exam requirements/recommendations for international students: Required—TOEFL. Electronic applications accepted.

The University of Tennessee Health Science Center, College of Graduate Health Sciences, Department of Anatomy and Neurobiology, Memphis, TN 38163-0002. Offers PhD. *Faculty:* 26 full-time (4 women), 1 (woman) part-time/adjunct. *Students:* 13 full-time (8 women); includes 9 minority (8 Asian Americans or Pacific Islanders, 1 Hispanic American). Average age 24. 49 applicants, 12% accepted. In 2005, 3 degrees awarded. *Degree requirements:* For doctorate, thesis/dissertation, oral and written preliminary and comprehensive exams. *Entrance requirements:* For doctorate, GRE General Test, minimum GPA of 3.0. *Application deadline:* For fall admission, 5/15 for domestic students. Application fee: $0. Electronic applications accepted. *Financial support:* Fellowships, research assistantships, teaching assistantships, Federal Work-Study, traineeships, and tuition waivers (full) available. Financial award application deadline: 2/25. *Unit head:* Dr. Stephen T. Kitai, Chairman, 901-448-5957, Fax: 901-448-7193. *Application contact:* Ida W. Mosby, Director, Enrollment Services, 901-448-5560, E-mail: imosby@utmem.edu.

See Close-Up on page 1057.

The University of Toledo, College of Graduate Studies, Department of Anatomy and Neurobiology, Toledo, OH 43606-3390. Offers MS. Part-time programs available. *Degree requirements:* For master's, thesis, qualifying exam. *Entrance requirements:* For master's, GRE General Test, minimum undergraduate GPA of 3.0. *Expenses:* Tuition, state resident: full-time $6,623; part-time $308 per credit hour. Tuition, nonresident: full-time $13,232; part-time $735 per credit hour. *Faculty research:* Organization, development and path of the cardiovascular and nervous systems.

University of Utah, School of Medicine and The Graduate School, Graduate Programs in Medicine, Department of Neurobiology and Anatomy, Salt Lake City, UT 84112-1107. Offers M Phil, MS, PhD. Part-time programs available. Terminal master's awarded for partial completion of doctoral program. *Degree requirements:* For master's and doctorate, one foreign language, thesis/dissertation. *Entrance requirements:* For master's and doctorate, GRE General Test. *Expenses:* Tuition, state resident: full-time $2,932; part-time $369 per credit.

Tuition, nonresident: full-time $10,350; part-time $1,302 per credit. Required fees: $516 per term. Tuition and fees vary according to course load and program. *Faculty research:* Neuroscience, neuroanatomy, developmental neurobiology, neurogenetics.

University of Vermont, College of Medicine and Graduate College, Graduate Programs in Medicine, Department of Anatomy and Neurobiology, Burlington, VT 05405. Offers PhD, MD/PhD. *Students:* 15 (6 women) 4 international. 14 applicants, 43% accepted, 1 enrolled. *Degree requirements:* For doctorate, thesis/dissertation. *Entrance requirements:* For doctorate, GRE General Test. Additional exam requirements/recommendations for international students: Required—TOEFL (minimum score 550 paper-based; 213 computer-based). *Application deadline:* For fall admission, 1/15 for domestic students. Applications are processed on a rolling basis. Application fee: $40. Electronic applications accepted. *Expenses:* Tuition, area resident: Part-time $410 per credit hour. Tuition, nonresident: part-time $1,034 per credit hour. *Financial support:* Teaching assistantships available. Financial award application deadline:3/1. *Faculty research:* Autonomic neurobiology, developmental neurobiology, neurotransmitter expression and release, plasticity and regeneration. *Unit head:* Dr. Rodney L. Parsons, Chairperson, 802-656-2230. *Application contact:* Dr. Rae Nishi, Coordinator, 802-656-2230, E-mail: rae.nishi@uvm.edu.

The University of Western Ontario, Faculty of Graduate Studies, Biosciences Division, Department of Biology, London, ON N6A 5B8, Canada. Offers M Sc, PhD. *Degree requirements:* For master's, thesis/dissertation; for doctorate, thesis/dissertation, comprehensive exam. *Entrance requirements:* For master's, honors degree or equivalent in biological sciences; for doctorate, master's degree. Additional exam requirements/recommendations for international students: Required—TOEFL. *Faculty research:* Cell and molecular biology, developmental biology, neuroscience, immunobiology and cancer.

University of Wisconsin–Madison, School of Veterinary Medicine, Department of Animal Health and Biomedical Sciences, Program in Comparative Biosciences, Madison, WI 53706-1380. Offers anatomy (MS, PhD); biochemistry (MS, PhD); cellular and molecular biology (MS, PhD); environmental toxicology (MS, PhD); neurosciences (MS, PhD); pharmacology (MS, PhD); physiology (MS, PhD). *Degree requirements:* For doctorate, thesis/dissertation.

Virginia Commonwealth University, Graduate School, School of Allied Health Professions, Department of Physical Therapy and Department of Anatomy, Program in Anatomy and Physical Therapy, Richmond, VA 23284-9005. Offers PhD. *Accreditation:* APTA. *Students:* 3 full-time (1 woman); includes 1 minority (African American) 2 applicants, 50% accepted. In 2005, 2 degrees awarded. *Degree requirements:* For doctorate, thesis/dissertation. *Entrance requirements:* For doctorate, GRE General Test. *Application deadline:* For fall admission, 5/1 for domestic students. Application fee: $50. *Expenses:* Tuition, state resident: full-time $6,268; part-time $405 per credit. Tuition, nonresident: full-time $15,904; part-time $940 per credit. Required fees: $751 per semester hour. Tuition and fees vary according to course load and program. *Unit head:* Dr. George R. Leichnetz, Director, Graduate Programs in Anatomy, 804-828-9512, Fax: 804-828-9477, E-mail: grleichn@vcu.edu.

Virginia Commonwealth University, Medical College of Virginia-Professional Programs, School of Medicine and Graduate Programs, School of Medicine Graduate Programs, Department of Anatomy, Richmond, VA 23284-9005. Offers anatomy (MS, PhD, CBHS); anatomy and physical therapy (PhD); neuroscience (MS, PhD). *Faculty:* 31 full-time (6 women). *Students:* 53 full-time (23 women), 3 part-time (all women). 97 applicants, 52% accepted. In 2005, 3 master's, 3 doctorates, 16 other advanced degrees awarded. *Degree requirements:* For master's, thesis; for doctorate, thesis/dissertation, comprehensive oral and written exams. *Entrance requirements:* For master's, DAT, GRE General Test, or MCAT; for doctorate, DAT, GRE General Test, MCAT. *Application deadline:* For fall admission, 2/15 for domestic students. Application fee: $50. *Expenses:* Tuition, state resident: full-time $6,268; part-time $405 per credit. Tuition, nonresident: full-time $15,904; part-time $940 per credit. Required fees: $751 per semester hour. Tuition and fees vary according to course load and program. *Financial support:* Fellowships available. *Unit head:* Dr. John T. Povlishock, Chair, 804-828-9623, Fax: 804-828-9477, E-mail: jtpovlis@vcu.edu. *Application contact:* Dr. George R. Leichnetz, Director, Graduate Programs in Anatomy, 804-828-9512, Fax: 804-828-9477, E-mail: grleichn@vcu.edu.

See Close-Up on page 349.

Wake Forest University, School of Medicine and Graduate School, Graduate Programs in Medicine, Department of Neurobiology and Anatomy, Winston-Salem, NC 27109. Offers PhD. *Degree requirements:* For doctorate, thesis/dissertation. *Entrance requirements:* For doctorate, GRE General Test. Additional exam requirements/recommendations for international students: Required—TOEFL. Electronic applications accepted. *Faculty research:* Sensory neurobiology, reproductive endocrinology, regulatory processes in cell biology.

Wayne State University, School of Medicine and Graduate School, Graduate Programs in Medicine, Department of Anatomy and Cell Biology, Detroit, MI 48202. Offers anatomy (MS, PhD). *Faculty:* 21 full-time (4 women), 1 (woman) part-time/adjunct. *Students:* 11 full-time (8 women), 2 part-time (1 woman); includes 2 minority (both Asian Americans or Pacific Islanders), 7 international. Average age 30. 15 applicants, 33% accepted, 4 enrolled. In 2005, 2 degrees awarded. *Degree requirements:* For doctorate, thesis/dissertation. *Entrance requirements:* For master's and doctorate, GRE General Test, minimum GPA of 3.0. Additional exam requirements/recommendations for international students: Required—TOEFL (minimum score 600 paper-based; 260 computer-based); Recommended—TWE (minimum score 6). *Application deadline:* For fall admission, 7/1 for domestic students, 6/1 for international students. Applications are processed on a rolling basis. Application fee: $30 ($50 for international students). Electronic applications accepted. *Expenses:* Tuition, state resident: part-time $338 per credit hour. Tuition, nonresident: part-time $746 per credit hour. Required fees: $24 per credit hour. Full-time tuition and fees vary according to program. *Financial support:* In 2005–06, 13 research assistantships (averaging $20,246 per year) were awarded; fellowships, teaching assistantships, Federal Work-Study also available. Financial award application deadline: 2/1. *Faculty research:* Inflammation and inflammatory mediators, neuronal plasticity, neural connections and glia, vision and visual neurosciences, cell signaling and receptor interactions. Total annual research expenditures: $4 million. *Unit head:* Dr. Linda Hazlett, Chair, 313-577-1061, Fax: 313-577-3125, E-mail: aa4536@wayne.edu. *Application contact:* Dr. Roberta Pourcho, Graduate Director, 313-577-1002, Fax: 313-577-3125, E-mail: rpourcho@med.wayne.edu.

Wright State University, School of Graduate Studies, College of Science and Mathematics, Department of Anatomy and Physiology, Dayton, OH 45435. Offers anatomy (MS); physiology and biophysics (MS). *Degree requirements:* For master's, thesis optional. *Entrance requirements:* Additional exam requirements/recommendations for international students: Required—TOEFL. *Faculty research:* Reproductive cell biology, neurobiology of pain, neurohistochemistry.

STONY BROOK UNIVERSITY, STATE UNIVERSITY OF NEW YORK

Department of Anatomical Sciences
Ph.D. in Anatomical Sciences

Program of Study

The Department of Anatomical Sciences, within the School of Medicine, offers a multidisciplinary graduate program leading to the Ph.D. degree. Students receive comprehensive training to prepare them for teaching and research in the areas of evolutionary morphology, functional morphology, musculoskeletal biology, and vertebrate paleontology. Graduate students are guided through a program of courses appropriate to their particular needs. In this regard, the Department of Anatomical Sciences interacts with other departments in the School of Medicine as well as those in the Biological Sciences Division and the Anthropology Department.

In addition to providing training in vertebrate paleontology with ready access to appropriate field sites, the program is concerned with the analysis and interpretation of vertebrate structure in relation to adaptation and systematics. Training and research focus on two broad areas: an evolutionary perspective in the analysis of morphology, including the influences of function, structure, and phylogenetic history; and the structural adaptations of bone as load-bearing tissue, including the physiologic mechanisms of osteogenesis and osteolysis. Both the locomotor and the craniodental systems are regions of current interest and investigation within the department. Emphasis is placed on the application of experimental and quantitative techniques to the analysis of the relationship between form and function.

Every student is required to teach medical human gross anatomy at least once before graduation. The Graduate Program in Anatomical Sciences does not accept students whose goal is a master's degree. In exceptional instances, a student already in the program may be awarded an M.S. degree upon completing an approved course of study, including attaining a minimum of 30 graduate credit hours, passing a comprehensive examination, and submitting a master's thesis.

Research Facilities

Studies of skeletal adaptations integrate collaborations with the Department of Biomedical Engineering and the Musculoskeletal Research Laboratory of the Department of Orthopaedics. Questions of systematics are approached at many different levels, ranging from alpha taxonomy to higher-order relationships and utilizing such techniques as quantitative cladistics and numerical taxonomy, as well as more traditional taxonomic methods. Students in the program have the opportunity to master a variety of research methods and analytical strategies, including behavioral ecology; cineradiography; electromyography; finite-element analysis; in vivo bone strain measurement; kinematics and kinetics; principles of paleontological fieldwork; quantitative morphology, including scaling (allometry) and multivariate morphometrics; reflected-light microscopy; scanning electron microscopy and tandem-scanning; and systematic classification techniques. Active paleontological field sites are available in the western U.S., in Madagascar, and in Kenya.

Financial Aid

Because Stony Brook is committed to attracting quality students, the Graduate School provides two competitive fellowships for U.S. citizens and permanent residents. Graduate Council fellowships are for outstanding doctoral candidates studying in any discipline, and the W. Burghardt Turner Fellowships target outstanding African-American, Hispanic-American, and Native American students entering either a doctoral or master's degree program. For doctoral students, both fellowships provide an annual stipend of at least $15,975 for up to five years, as well as a full tuition scholarship. For master's students, the Turner Fellowship provides an annual stipend of $10,000 for up to two years, along with a full tuition scholarship. Health insurance subsidies are also provided within a scale depending on the size of the fellow's dependent family. Departments and degree programs award approximately 900 teaching and graduate assistantships and approximately 600 research assistantships on an annual basis. Full assistantships carry a stipend amount that usually ranges from $11,947 to $18,000, depending on the department.

Cost of Study

In 2006–07, full-time tuition is $3450 per semester for state residents and $5460 per semester for nonresidents. Part-time tuition is $288 per credit hour for residents and $455 per credit hour for nonresidents. Additional charges include an activity fee of $22 and a comprehensive fee of $330 per semester.

Living and Housing Costs

University apartments range in cost from approximately $325 per month to approximately $1400 per month, depending on the size of the unit. Off-campus housing options include furnished rooms to rent and houses and/or apartments to share that can be rented for $350 to $1500 per month.

Location

Stony Brook's campus is approximately 50 miles east of Manhattan on the north shore of Long Island. The cultural offerings of New York City and Suffolk County's countryside and seashore are conveniently located nearby. Cold Spring Harbor Laboratories and Brookhaven National Laboratories are easily accessible from, and have close relationships with, the University.

The University

The University, established in 1957, achieved national stature within a generation. Founded at Oyster Bay, Long Island, the school moved to its present location in 1962. Stony Brook has grown to encompass more than 110 buildings on 1,100 acres. There are more than 1,568 faculty members, and the annual budget is more than $805 million. The Graduate Student Organization oversees the spending of the student activity fee for graduate student campus events. International students find the additional four-week Summer Institute in American Living very helpful. The Intensive English Center offers classes in English as a second language. The Career Development Office assists with career planning and has information on permanent full-time employment. Disabled Student Services has a Resource Center that offers placement testing, tutoring, vocational assessment, and psychological counseling. The Counseling Center provides individual, group, family, and marital counseling and psychotherapy. Day-care services are provided in four on-campus facilities. The Writing Center offers tutoring in all phases of writing.

Applying

Applicants are judged on the basis of distinguished undergraduate records (and graduate records, if applicable), thorough preparation for advanced study and research in the field of interest, candid appraisals from those familiar with the applicant's academic/professional work, potential for graduate study, and a clearly defined statement of purpose and scholarly interest germane to the program. A baccalaureate degree is required, with a minimum overall grade point average of 2.75 and an average grade of B in the major and related courses. Some programs require a higher GPA. Students should submit admission and financial aid applications by February 1 for the fall semester.

Correspondence and Information

Catherine Forster
Department of Anatomical Sciences
Stony Brook University, School of Medicine
Stony Brook, New York 11794-8081
Phone: 631-444-8203
Fax: 631-444-3947
E-mail: catherine.forster@stonybrook.edu
Web site: http://www.uhmc.sunysb.edu/anatomy/gpanat.html

Stony Brook University, State University of New York

THE FACULTY AND THEIR RESEARCH

Peter Brink, Professor (joint appointment) and Chair of Physiology and Biophysics; Ph.D., Illinois, 1976. Intercellular communication: gap junctions; lacrimal gland secretions.

Brigitte Demes, Professor; Ph.D., Bochum (Germany), 1982. Biomechanics and functional morphology; kinematics and kinetics of primate locomotion; influence of body size on locomotion and the structure of the musculoskeletal system; long-bone loading patterns.

Diane Doran, Professor (joint appointment); Ph.D., SUNY at Stony Brook, 1989. Positional behavior of African apes as models for understanding early hominid locomotor behavior; foraging strategy and social organization in western lowland gorillas to assess how ecological factors influence social behavior; niche separation of sympatric gorillas and chimpanzees to ascertain the role of interspecific competition in the evolution of social behavior of closely related species.

John Fleagle, Distinguished Professor; Ph.D., Harvard, 1976. Primate evolution, behavior, and ecology; primate anatomy, growth, and development.

Catherine Forster, Associate Professor; Ph.D., Pennsylvania, 1990. Dinosaur systematics; cladistic and morphometric analyses; Gondwana biogeography; dinosaur origins.

Frederick Grine, Professor (joint appointment) and Chair of Anthropology; Ph.D., Witwatersrand (South Africa), 1984. Early hominid evolution; tooth enamel structure; systematics.

Françoise Jouffroy, Adjunct Professor; Doctorat ès Sciences, Paris IV (Sorbonne), 1963. Evolution of vertebrate locomotion; primate adaptations; experimental analysis of locomotor movements; functional morphology of the musculoskeletal system; immunocytochemistry of myofibers.

William Jungers, Professor; Ph.D., Michigan, 1976. Functional morphology; morphometrics; primate evolution; paleoanthropology.

David Krause, Distinguished Service Professor; Ph.D., Michigan, 1982. Systematics; biogeography; functional morphology; paleoecology; biochronology of Mesozoic and Early Cenozoic mammals; evolutionary and biogeographic history of the Gondwanan vertebrate fauna; biogeographic origins of the extant Malagasy vertebrate fauna.

Susan Larson, Professor; Ph.D., Wisconsin, 1982. Primate and human anatomy; experimental functional morphology and biomechanics; human and primate evolution.

Lawrence Martin, Professor (joint appointment) and Dean of the Graduate School; Ph.D., University College, London, 1983. Evolution of hominoid primates; development, structure, and thickness of dental enamel.

Russell Mittermeier, Adjunct Professor and President of Conservation International; Ph.D., Harvard, 1977. Ecology and behavior of primates; primate conservation.

Maureen O'Leary, Associate Professor; Ph.D., Johns Hopkins, 1997. Evolution and systematics of mammals; vertebrate evolution in sub-Saharan Africa; combined data phylogeny reconstruction; systematic biology; vertebrate paleontology.

Clinton Rubin, Professor (joint appointment) and Chair of Biomedical Engineering; Ph.D., Bristol (England), 1983. Cellular mechanisms responsible for the growth, healing, and homeostasis of bone.

Jack Stern, Distinguished Teaching Professor and Chair; Ph.D., Chicago, 1969. Evolution of postcranial adaptations in primates, with emphasis on the origins of bipedalism; muscle function in nonhuman primates and humans; biomechanics.

Randall Susman, Professor; Ph.D., Chicago, 1976; J.D., Touro, 1988. Fossil record of hominid evolution; ecology and behavior of African apes; experimental studies of ape and human locomotion; comparative morphology.

VIRGINIA COMMONWEALTH UNIVERSITY

Department of Anatomy and Neurobiology

Program of Study

The majority of Ph.D. students are involved in state-of-the-art research in cellular and molecular neuroscience with nationally and internationally recognized faculty members working in neural injury and regeneration, visuomotor sciences, and the cell biology of the nervous system. Interdisciplinary studies are encouraged by affiliations with faculty members in neurosurgery, neurology, neurophysiology, and neuropharmacology. Doctoral students in the Department of Anatomy and Neurobiology have the opportunity to pursue a curriculum in neurobiology with concentrated course work in cellular and molecular neuroscience and developmental neurobiology. Graduates of the Department are prepared for careers in research and teaching in the academic setting. In the second semester of the first year, students take a rotation through one or two prospective laboratories. At the completion of the elective rotations, students are fully engaged in their own research project and take two electives related to their research. Popular electives include cell and molecular biology, biochemistry, mammalian physiology, introduction to neuroscience, immunobiology, neurochemical pharmacology, and tumor biology.

Research Facilities

The Department of Anatomy and Neurobiology maintains several state-of-the-art core facilities. A shared molecular core facility provides technical support and equipment for the analysis and manipulation of RNA and DNA. An imaging core provides access to a Leica TCS-SP2 AOBS confocal microscope equipped with 5 lasers and 3 confocal detectors, Zeiss LSM 510 META NLO multiphoton system with four lasers and two detectors, a Nikon Optiphot-2 microscope outfitted with an MRI-103 driver, a MAC 2000 motorized stage, and a MTI 3CCD video camera interfaced with MicroBrightfield Neurolucida software; Nikon ECLIPSE E800M microscope fitted with a complete filter set interfaced with digital camera; three video-equipped workstations interfaced with the MCID M5+ platform from Imaging Research, Inc., a phosphor imaging and a calcium imaging system, and denistometric analysis of gels and autoradiograms; and ultrastructural analysis. A Jeol JFM-1230 TEM is equipped with a GATAN Ultrascan 4000 digital 4K X 4K CCD camera, Zeiss EM 10CA TEM, and a JEOL JSM-820 scanning electron microscope. An image-analysis suite is equipped with workstations and access to AnalySIS, CAST2 (a stereological toolbox), DigitalMicrograph, IPLab (deconvolution software), Olympus MicroSuite, and Zeiss LSM Image Browser. Technical-support services are available for specimen preparation and sectioning. A digital darkroom is also available.

Financial Aid

All predoctoral students are supported with graduate fellowships or faculty research grants that cover tuition and provide an annual stipend of $20,000. Master's degree and certificate students are not eligible for support.

Cost of Study

Tuition and fees are determined by the University on an annual basis. In 2005–06, tuition and fees were $10,100 for residents and $22,000 for nonresidents.

Living and Housing Costs

Virginia Commonwealth University (VCU) is an urban institution with limited graduate housing on campus. The University is adjacent to residential areas of the city, with a broad spectrum of apartment housing available. Information about both on-campus and off-campus housing can be found on the VCU Web site (http://www.students.vcu.edu).

Student Group

Virginia Commonwealth University has a total enrollment of approximately 22,000 students and is home to one of the largest health science centers in the nation. The medical campus of VCU has more than 2,500 students enrolled in one of five professional schools: the School of Medicine, the School of Dentistry, the School of Pharmacy, the School of Nursing, or the School of Allied Health. Graduate training is provided by the School of Medicine in eight different basic science departments.

Location

VCU is located in Richmond, Virginia. This metropolitan city has central access to Washington, D.C.; the Blue Ridge Mountains; the Chesapeake Bay; and Virginia Beach. Surrounding areas provide a diverse variety of outdoor activities including boating, fishing, and downhill skiing. Richmond and the neighboring areas are home to many historical treasures, including Williamsburg and the Jamestown settlement as well as battlefields and monuments of the Civil and Revolutionary Wars. Richmond also offers cultural opportunities through the art and science museums, a symphony orchestra, and several theaters. Sporting events include professional teams in baseball, ice hockey, soccer, and arena football. NASCAR events are held at Richmond Motor Speedway; horse racing events are held at Colonial Downs, 30 minutes from downtown Richmond.

The University

Shared resources in the University community include access to multiple core facilities, including a transgenic core facility, flow cytometry, biostatistics, virus vector development, monoclonal antibodies, and nucleic acid synthesis and analysis.

Applying

Admission to the graduate programs of the Department of Anatomy and Neurobiology requires a bachelor's degree with curriculum in biology, general chemistry, or organic chemistry. Applicants must take the Graduate Record Examinations (GRE) or MCAT and submit three letters of reference. The Subject Test in biology is not required. Applicants are also required to submit a biographical statement describing their background and reasons for their interest in pursuing graduate studies in anatomy. International applicants whose native language is not English are required to take the Test of English as a Foreign Language (TOEFL). The deadline for applications to the M.S. and Ph.D. programs is May 1 to be considered for admission in the following fall semester. There is no deadline for applying to the postbaccalaureate premedical certificate program.

Correspondence and Information

For application forms:
Admissions
School of Graduate Studies
Box 843051
Virginia Commonwealth University
Richmond, Virginia 23284-3051
Phone: 804-828-6916

For additional program information:
Dr. George R. Leichnetz
Graduate Program Director
Department of Anatomy and Neurobiology
Virginia Commonwealth University Medical Campus
Box 980709
Richmond, Virginia 23298-0709
Web site: http://www.vcu.edu/anatomy/start.html

Virginia Commonwealth University

THE FACULTY AND THEIR RESEARCH

John W. Bigbee, Ph.D. (neurobiology), Stanford, 1982. Examining the role of acetylcholinesterase in neural development, with a specific focus on the morphogenic capacity of this enzyme to support axonal growth; employing cell culture systems, including dorsal root ganglion neurons and neuroblastoma cell lines, to examine the consequences of pharmacological and immunological perturbation of AChE.

Raymond J. Colello, D. Phil., Oxford, 1990. Elucidating the signals that influence oligodendrocyte progenitor cell migration and those that initiate myelination in the central nervous system; employing a host of cell biological techniques to chronicle the events of myelin gene transcription and translation as well as utilizing confocal and electron microscopic techniques to identify the specific time course of axon ensheathment.

Jeff Dupree, Ph.D., Ph.D., Virginia Commonwealth, 1995. Examining how proper communication between neurons and oligodendrocytes, the myelin-forming cell of the central nervous system, is essential for function and survival of these two cell types during development; the less-understood role that neuron-oligodendrocyte interactions play in maintaining the mature nervous system; identifying factors that mediate neuron-oligodendrocyte interactions in the adult nervous system and determining consequences that result when these cell-cell communications are disrupted.

Babette Fuss, Ph.D., Swiss Federal Institute of Technology, 1992. Defining the cell biology of the myelin forming cell of the CNS, the oligodendrocyte, and its vulnerability to immunological damage; characterization of novel oligodendrocyte-specific proteins using molecular, cell biological, and transgenic techniques designed to gain new insight into the mechanisms of myelin sheath formation during normal development and during remyelination under pathological conditions.

Stephen J. Goldberg, Ph.D. (biology), Clark, 1971. Investigating the neurophysiological and mechanical properties of brain stem motor units involved in the tongue and eye movements; examination of motoneuron conduction velocity, input resistance, and rheobase as well as synaptic responses; study of muscle fibers innervated by a single motoneuron using a force transducer, which can measure muscle mechanics in response to intracellular stimulation of the motoneuron.

Stephen A. Gudas, Ph.D. (physical therapy), Virginia Commonwealth, 1976; Ph.D., (anatomy), Virginia Commonwealth, 1988. Licensed physical therapist and practice in the cancer rehabilitation program of the Massey Cancer Center; clinical research involving spinal cord compression, head and neck cancer, and rehabilitation of the metastatic patient.

Jack L. Haar, Ph.D. (anatomy), Ohio State, 1970. Cell migration, research directed toward better understanding the migration of hematopoietic stem cells to the fetal, young adult, and aged thymus in mice; understanding the process of tumor cell migration with the goal of inhibiting tumor metastatic spread.

Caroline Goode Jackson, Ph.D. (anatomy), Virginia Commonwealth, 1973. Teaching of embryology; in the process of converting previously prepared study packages into computer-based units.

Kimberle Jacobs, Ph.D. (neuroscience), Brown, 1994. Neuroscience of cortical elements and circuitry that contribute to plasticity of the nervous system, including alterations in synaptic strengths, cellular differentiation, and neuronal excitability. Laboratory uses of a combination of patch-clamping, field potentials, single cell aRNA amplification, immunohistochemistry, and anatomical techniques to explore how modifications in particular cell types contribute to overall changes in functional circuitry, currently applied to a developmental model of epilepsy.

Richard J. Krieg Jr., Ph.D. (anatomy), UCLA, 1975. Laboratory studies growth and growth hormone (GH) secretion and reproduction and luteinizing hormone (LH) secretion; attempting to find out why growth fails in children with renal insufficiency and why reproductive function is compromised in women with kidney disease; use of blood sampling techniques in freely behaving animals to see what happens to GH and LH secretion in animals with renal insufficiency.

George R. Leichnetz, Ph.D. (anatomy), Ohio State, 1970. Brain connections concerned with the role of the cerebral cortex in the control of eye movement; the development of computer-based instructional packages for the gross and sectional anatomy of the central nervous system; light- and electron-microscopic neurohistology, self-test package for students to evaluate their understanding of neuroanatomy and its clinical correlations.

J. Ross McClung, Ph.D. (anatomy), Texas Medical Branch, 1971. Anatomy of neurons, leading to work in synaptology, degeneration, and regeneration, as well as analyses of the levels of order found in the neuronal organization of the CNS; collaborating with Dr. Stephen Goldberg on the use of intracellular electrophysiological characterization of motoneurons followed by intracellular HRP labeling to generate a complete anatomical analysis of these functionally characterized neurons.

A. Rory McQuiston, Ph.D. (pharmacology), Alberta, 1995. Exploring how activity and information coding in the CNS are regulated by anatomically diverse inhibitory interneurons; study of the neurobiological systems of inhibitory interneurons having crucial roles in controlling CNS network function and behavior (interneuron dysfunction has been associated with CNS disorders such as epilepsy and schizophrenia) utilizing optical (photostimulation and voltage-sensitive dye imaging) and electrophysiological techniques (visualized whole-cell patch clamping in rat brain slices) to determine how drugs of abuse and behaviorally relevant neuromodulators alter the activity of cortical interneurons and ultimately change neuronal network function.

Randall E. Merchant, Ph.D. (anatomy), North Dakota, 1978. Development of immunotherapy as a treatment modality for malignant gliomas, centering efforts on the development and use of tumor cell vaccines to be used alone or in combination with adoptively transferred lymphocytes in an attempt to restore and/or enhance reactivity of the patient's immune system for their tumor; performed both preclinical and Phase I clinical trials of a new adoptive immunotherapy protocol using glioma-sensitized T cells over the past few years, and, more recently, using vaccines that have been transfected with genes encoding for cytokines.

M. Alex Meredith, Ph.D. (anatomy), Virginia Commonwealth, 1981. Examination of the circuitry underlying how the brain processes simultaneous information from different sensory modalities and how the loss or injury to one sensory system might lead to compensatory changes in the others. Laboratory uses of neurophysiological, anatomical, and immunocytochemical techniques to study the architecture and role of inhibition in multisensory circuits of the cerebral cortex.

Alice Pakurar, Ph.D. (anatomy), Michigan, 1968. Computer-aided instructional materials that complement teaching responsibilities; instructor-student verification of the visual discipline of microscopic anatomy using computer applications; headed a team that has produced a CD atlas of digital images taken from glass microscopic loan slides and organized these images into laboratory units.

Linda L. Phillips, Ph.D. (anatomy), Wake Forest, 1980. Laboratory studies the interaction between excessive neuroexcitation and neuronal deafferentation following traumatic brain injury (TBI), looking at how this interaction affects both the ensuing pathology and recovery mechanisms involving synaptic plasticity; rationale for this approach includes its direct relevance to human TBI where excitotoxic insult is often compounded by diffuse deafferentation and poor outcome.

John T. Povlishock, Ph.D. (anatomy), Saint Louis, 1973. Laboratory studies traumatic brain injury in an effort to better understand the initiating mechanisms that lead to traumatically induced vascular and axonal change; employs contemporary structural, functional, behavioral, and cellular/molecular endpoints to evaluate progressive plasmalemmal and cytoskeletal change linked to the activation of cysteine protease and free radical pathways.

Thomas M. Reeves, Ph.D. (developmental biopsychology), Southern Illinois, 1984. Electrophysiology of brain injury and recovery of function; experimental models, including fluid percussion traumatic brain injury (TBI), deafferenting lesions, or a combination of these models; functional measures include intracellular and extracellular electrophysiological parameters, with an emphasis on evoked extracellular field potentials, well suited to the study of brain injury; demonstrations of TBI-induced deficits in long-term potentiation (LTP).

Milton M. Sholley, Ph.D. (anatomy), Temple, 1974. Study of the pathophysiology of the endothelial cells that line blood vessels, using the human umbilical vein endothelial cell (HUVEC) culture system to model endothelial injuries that simulate those that occur during inflammation associated with ischemia/reperfusion injury and sepsis in vivo; investigations during the past few years have involved cellular and molecular mechanisms that alter the endothelium during cytokine-induced activation and oxidative injury.

David G. Simpson, Ph.D. (cell, molecular, and structural biology), Northwestern, 1991. Laboratory uses electrospinning technology to process native proteins into tissue-engineering scaffolds for the fabrication of bioengineered skin and other tissues for use in regenerative medicine. Second project examines how physical perturbations interact with matrix proteins to regulate the expression and activation of matrix metalloproteases, working to define how the compliment of integrin receptors present on the cell surface regulates cellular response to phenotypic signals derived from the extracellular matrix.

Section 3
Biochemistry

This section contains a directory of institutions offering graduate work in biochemistry, followed by in-depth entries submitted by institutions that chose to prepare detailed program descriptions. Additional information about programs listed in the directory but not augmented by an in-depth entry may be obtained by writing directly to the dean of a graduate school or chair of a department at the address given in the directory.

For programs offering related work, see also in this book Biological and Biomedical Sciences; Biophysics; Botany and Plant Biology; Cell, Molecular, and Structural Biology; Genetics, Developmental Biology, and Reproductive Biology; Microbiological Sciences; Neuroscience and Neurobiology; Nutrition; Pathology and Pathobiology; Pharmacology and Toxicology; and Physiology. In Book 4, see Agricultural and Food Sciences, Chemistry, and Physics; in Book 5, see Agricultural Engineering and Bioengineering, Biomedical Engineering and Biotechnology, Chemical Engineering, and Materials Sciences and Engineering; and in Book 6, see Allied Health, Pharmacy and Pharmaceutical Sciences, and Veterinary Medicine and Sciences.

CONTENTS

Biochemistry

Albert Einstein College of Medicine, Sue Golding Graduate Division of Medical Sciences, Department of Biochemistry, Bronx, NY 10461-1602. Offers PhD, MD/PhD. *Degree requirements:* For doctorate, thesis/dissertation. *Entrance requirements:* For doctorate, GRE General Test. Additional exam requirements/recommendations for international students: Required—TOEFL. *Faculty research:* Biochemical mechanisms, enzymology, protein chemistry, bio-organic chemistry, molecular genetics.

See Close-Up on page 377.

American University of Beirut, Graduate Programs, Faculty of Medicine, Beirut, Lebanon. Offers biochemistry (MS); human morphology (MS); microbiology and immunology (MS); neuroscience (MS); pharmacology and therapeutics (MS); physiology (MS). *Degree requirements:* For master's, one foreign language, thesis (for some programs), comprehensive exam, registration. *Entrance requirements:* For master's, GRE, letter of recommendation.

Arizona State University, Division of Graduate Studies, College of Liberal Arts and Sciences, Division of Natural Sciences and Mathematics, Department of Chemistry and Biochemistry, Tempe, AZ 85287. Offers MNS, MS, PhD. *Degree requirements:* For master's, thesis; for doctorate, one foreign language, thesis/dissertation. *Entrance requirements:* For master's and doctorate, GRE. Additional exam requirements/recommendations for international students: Required—TOEFL, TSE.

Baylor College of Medicine, Graduate School of Biomedical Sciences, Department of Biochemistry and Molecular Biology, Houston, TX 77030-3498. Offers PhD, MD/PhD. *Faculty:* 31 full-time (6 women). *Students:* 52 full-time (24 women); includes 9 minority (2 African Americans, 2 Asian Americans or Pacific Islanders, 5 Hispanic Americans), 30 international. Average age 28. 147 applicants, 13% accepted, 9 enrolled. In 2005, 6 doctorates awarded. *Median time to degree:* Of those who began their doctoral program in fall 1997, 89% received their degree in 8 years or less. *Degree requirements:* For doctorate, thesis/dissertation, public defense. *Entrance requirements:* For doctorate, GRE General Test, GRE Subject Test (strongly recommended), minimum GPA of 3.0. Additional exam requirements/recommendations for international students: Required—TOEFL. *Application deadline:* For fall admission, 2/1 for domestic students. Application fee: $30. Electronic applications accepted. *Expenses:* Tuition: Full-time $8,200. Full-time tuition and fees vary according to program. *Financial support:* In 2005–06, 43 students received support, including 18 fellowships (averaging $23,000 per year), 34 research assistantships (averaging $23,000 per year); career-related internships or fieldwork, Federal Work-Study, institutionally sponsored loans, health care benefits, tuition waivers (full), and stipends also available. Financial award applicants required to submit FAFSA. *Faculty research:* Mechanisms of enzyme action, nucleic acid enzymology, and mutagenesis; biochemistry of connective tissue, proteins, and polysaccharides; chemical metabolism of lipids and lipoproteins. *Unit head:* Dr. Adam Kuspa, Director, 713-798-4527. *Application contact:* Charlotte Cherry, Graduate Program Administrator, 713-798-4527, Fax: 713-796-9438, E-mail: ccherry@bcm.tmc.edu.

See Close-Up on page 379.

Baylor College of Medicine, Graduate School of Biomedical Sciences, Interdepartmental Program in Cell and Molecular Biology, Houston, TX 77030-3498. Offers biochemistry (PhD); cell and molecular biology (PhD); genetics (PhD); human genetics (PhD); immunology (PhD); microbiology (PhD); virology (PhD). *Faculty:* 99 full-time (21 women). *Students:* 56 full-time (28 women); includes 20 minority (4 African Americans, 1 American Indian/Alaska Native, 6 Asian Americans or Pacific Islanders, 9 Hispanic Americans), 5 international. Average age 28. 164 applicants, 18% accepted, 12 enrolled. In 2005, 3 doctorates awarded. *Median time to degree:* Of those who began their doctoral program in fall 1997, 63% received their degree in 8 years or less. *Degree requirements:* For doctorate, thesis/dissertation, public defense. *Entrance requirements:* For doctorate, GRE General Test, GRE Subject Test (strongly recommended), minimum GPA of 3.0. Additional exam requirements/recommendations for international students: Required—TOEFL. *Application deadline:* For fall admission, 1/1 for domestic students. Applications are processed on a rolling basis. Application fee: $30. Electronic applications accepted. *Expenses:* Tuition: Full-time $8,200. Full-time tuition and fees vary according to program. *Financial support:* In 2005–06, 52 students received support, including 20 fellowships (averaging $23,000 per year), 36 research assistantships (averaging $23,000 per year); teaching assistantships, Federal Work-Study, institutionally sponsored loans, health care benefits, and tuition waivers (full) also available. Financial award applicants required to submit FAFSA. *Faculty research:* Gene expression and regulation, developmental biology and genetics, signal transduction and membrane biology, aging process, molecular virology. *Unit head:* Dr. Tom Cooper, Director, 713-798-6557. *Application contact:* Lourdes Fernandez, Graduate Program Administrator, 713-798-6557, Fax: 713-798-6325, E-mail: cmbprog@bcm.edu.

See Close-Up on page 545.

Boston College, Graduate School of Arts and Sciences, Department of Biology, Program in Biochemistry, Chestnut Hill, MA 02467-3800. Offers MS, PhD, MBA/MS. Terminal master's awarded for partial completion of doctoral program. *Degree requirements:* For master's and doctorate, thesis/dissertation. *Entrance requirements:* For master's and doctorate, GRE General Test, GRE Subject Test. Additional exam requirements/recommendations for international students: Required—TOEFL (minimum score 550 paper-based; 213 computer-based). *Application deadline:* For fall admission, 1/15 for domestic students. Application fee: $70. Electronic applications accepted. *Financial support:* Fellowships, research assistantships, teaching assistantships, Federal Work-Study and scholarships/grants available. Support available to part-time students. Financial award application deadline: 3/1; financial award applicants required to submit FAFSA.

Boston College, Graduate School of Arts and Sciences, Department of Chemistry, Program in Biochemistry, Chestnut Hill, MA 02467-3800. Offers PhD. *Degree requirements:* For doctorate, 2 foreign languages, thesis/dissertation, comprehensive exam. *Entrance requirements:* For doctorate, GRE General Test, GRE Subject Test. *Application deadline:* For fall admission, 2/1 for domestic students. Application fee: $70. *Financial support:* Fellowships, research assistantships, teaching assistantships, Federal Work-Study and tuition waivers (partial) available. Support available to part-time students. Financial award application deadline: 3/1.

Boston University, Graduate School of Arts and Sciences, Molecular Biology, Cell Biology, and Biochemistry Program (MCBB), Boston, MA 02215. Offers MA, PhD. *Students:* 41 full-time (21 women); includes 3 minority (2 African Americans, 1 Asian American or Pacific Islander), 7 international. Average age 29. 111 applicants, 23% accepted, 8 enrolled. In 2005, 8 master's, 8 doctorates awarded. Terminal master's awarded for partial completion of doctoral program. *Degree requirements:* For master's, one foreign language, thesis (for some programs), registration; for doctorate, one foreign language, thesis/dissertation, comprehensive exam, registration. *Entrance requirements:* For master's and doctorate, GRE General Test, GRE Subject Test. Additional exam requirements/recommendations for international students: Required—TOEFL (minimum score 600 paper-based; 250 computer-based). *Application deadline:* For fall admission, 1/1 for domestic students, 1/1 for international students. Application fee: $60. *Expenses:* Tuition: Full-time $31,530; part-time $985 per credit. Required fees: $316; $40 per semester. Tuition and fees vary according to course level and program. *Financial support:* In 2005–06, 17 students received support, including 1 fellowship (averaging $16,500 per year), 15 research assistantships with full tuition reimbursements available (averaging $16,000 per year), 1 teaching assistantship with full tuition reimbursement available (averaging $16,000 per year); Federal Work-Study, scholarships/grants, and traineeships also available. Financial award application deadline: 1/15; financial award applicants required to submit FAFSA. *Unit head:* Kim McCall, Director, 617-353-2432, Fax: 617-353-6340, E-mail: mccall@bu.edu. *Application contact:* Academic Administrator, 617-353-2432, Fax: 617-353-6340, E-mail: mcbb@bu.edu.

Boston University, School of Medicine, Division of Graduate Medical Sciences, Department of Biochemistry, Boston, MA 02118. Offers MA, PhD, MD/PhD. Part-time programs available. *Faculty:* 22 full-time (6 women), 29 part-time/adjunct (2 women). *Students:* 38 full-time (17 women), 1 part-time; includes 5 minority (4 Asian Americans or Pacific Islanders, 1 Hispanic American), 18 international. Average age 29. Terminal master's awarded for partial completion of doctoral program. *Degree requirements:* For master's, thesis or alternative, qualifying exam; for doctorate, thesis/dissertation, qualifying exam. *Entrance requirements:* For master's and doctorate, GRE General Test, GRE Subject Test. Additional exam requirements/recommendations for international students: Required—TOEFL. *Application deadline:* For fall admission, 1/15 for domestic students; for spring admission, 10/15 priority date for domestic students. Electronic applications accepted. *Expenses:* Tuition: Full-time $31,530; part-time $985 per credit. Required fees: $316; $40 per semester. Tuition and fees vary according to course level and program. *Financial support:* Fellowships, research assistantships, Federal Work-Study, scholarships/grants, and traineeships available. *Faculty research:* Extracellular matrix, gene expression, receptors, growth control. *Application contact:* Dr. Barbara Schreiber, Information Contact, 617-638-5094, Fax: 617-638-5339, E-mail: schreibe@biochem.bumc.bu.edu.

See Close-Up on page 381.

Brandeis University, Graduate School of Arts and Sciences, Programs in Life Sciences, Department of Biochemistry, Waltham, MA 02454-9110. Offers MS, PhD. Part-time programs available. *Faculty:* 11 full-time (12 women). *Students:* 23 full-time (12 women); includes 2 minority (1 Asian American or Pacific Islander, 1 Hispanic American), 7 international. Average age 24. 85 applicants, 14% accepted, 5 enrolled. In 2005, 1 master's, 4 doctorates awarded. *Degree requirements:* For doctorate, thesis/dissertation, area exams. *Entrance requirements:* For doctorate, GRE General Test, resumé, 3 letters of recommendation. Additional exam requirements/recommendations for international students: Required—TOEFL (minimum score 600 paper-based; 250 computer-based). *Application deadline:* For fall admission, 1/15 for domestic students. Applications are processed on a rolling basis. Application fee: $55. Electronic applications accepted. *Financial support:* In 2005–06, 6 fellowships with tuition reimbursements (averaging $26,500 per year), 17 research assistantships with tuition reimbursements (averaging $26,500 per year), 4 teaching assistantships with tuition reimbursements (averaging $6,000 per year) were awarded; career-related internships or fieldwork, scholarships/grants, and tuition waivers (full and partial) also available. Financial award application deadline: 4/15; financial award applicants required to submit CSS PROFILE or FAFSA. *Faculty research:* Enzyme mechanisms, genetics, molecular developmental biology, structural biology, neurobiology. *Unit head:* Dr. Dorothee Kern, Chair, 781-736-2577, Fax: 781-736-3107, E-mail: gelles@brandeis.edu. *Application contact:* Lynn Olsen, Administrative Assistant, 781-736-2300, Fax: 781-736-2349, E-mail: lolsen@brandeis.edu.

Brigham Young University, Graduate Studies, College of Physical and Mathematical Sciences, Department of Chemistry and Biochemistry, Provo, UT 84602-1001. Offers analytical chemistry (MS, PhD); biochemistry (MS, PhD); inorganic chemistry (MS, PhD); organic chemistry (MS, PhD); physical chemistry (MS, PhD). *Faculty:* 35 full-time (2 women). *Students:* 100 full-time (33 women); includes 2 minority (both Asian Americans or Pacific Islanders), 68 international. Average age 29. 85 applicants, 44% accepted, 21 enrolled. In 2005, 9 master's, 9 doctorates awarded. *Median time to degree:* Of those who began their doctoral program in fall 1997, 87% received their degree in 8 years or less. *Degree requirements:* For master's, thesis, registration; for doctorate, thesis/dissertation, degree qualifying exam. *Entrance requirements:* For master's, pass 1 (biochemistry), 4 (chemistry) of 5 area exams, GRE General Test, minimum GPA of 3.0 in last 60 hours; for doctorate, pass 1 (biochemistry) or 4 (chemistry) of 5 area exams, GRE General Test, minimum GPA of 3.0 in last 60 hours. Additional exam requirements/recommendations for international students: Required—TOEFL, TWE. *Application deadline:* For fall admission, 2/1 priority date for domestic students, 2/1 priority date for international students. Applications are processed on a rolling basis. Application fee: $50. Electronic applications accepted. *Financial support:* In 2005–06, 100 students received support, including 12 fellowships with full tuition reimbursements available (averaging $20,500 per year), 47 research assistantships with full tuition reimbursements available (averaging $20,500 per year), 39 teaching assistantships with full tuition reimbursements available (averaging $20,400 per year); institutionally sponsored loans, scholarships/grants, health care benefits, tuition waivers (full), and unspecified assistantships also available. Financial award application deadline: 2/1. *Faculty research:* Separation science, molecular recognition, organic synthesis and biomedical application, biochemistry and molecular biology, molecular spectroscopy. Total annual research expenditures: $3.9 million. *Unit head:* Dr. Paul B. Farnsworth, Chair, 801-422-6502, Fax: 801-422-0153, E-mail: paul_farnsworth@byu.edu. *Application contact:* Dr. David V. Dearden, Graduate Coordinator, 801-422-2355, Fax: 801-422-0153, E-mail: david_dearden@byu.edu.

Brown University, Graduate School, Department of Chemistry, Providence, RI 02912. Offers biochemistry (PhD); chemistry (Sc M, PhD). *Degree requirements:* For master's, thesis; for doctorate, one foreign language, thesis/dissertation, cumulative exam.

Brown University, Graduate School, Division of Biology and Medicine, Program in Molecular Biology, Cell Biology, and Biochemistry, Providence, RI 02912. Offers biochemistry (M Med Sc, Sc M, PhD), including biochemistry (Sc M, PhD), biology (Sc M, PhD), medical science (M Med Sc, PhD); biology (MA); cell biology (M Med Sc, Sc M, PhD), including biochemistry (Sc M, PhD), biology (Sc M, PhD), medical science (M Med Sc, PhD); developmental biology (M Med Sc, Sc M, PhD), including biochemistry (Sc M, PhD), biology (Sc M, PhD), medical science (M Med Sc, PhD); immunology (M Med Sc, Sc M, PhD), including biochemistry (Sc M, PhD), biology (Sc M, PhD), medical science (M Med Sc, PhD); molecular microbiology (M Med Sc, Sc M, PhD), including biochemistry (Sc M, PhD), biology (Sc M, PhD), medical science (M Med Sc, PhD). Part-time programs available. Terminal master's awarded for partial completion of doctoral program. *Degree requirements:* For master's, thesis (for some programs); for doctorate, one foreign language, thesis/dissertation, preliminary exam. *Entrance requirements:* For master's and doctorate, GRE General Test, GRE Subject Test. Additional exam requirements/recommendations for international students: Required—TOEFL. Electronic applications accepted. *Faculty research:* Molecular genetics, gene regulation.

See Close-Up on page 555.

California Institute of Technology, Division of Biology and Division of Chemistry and Chemical Engineering, Biochemistry and Molecular Biophysics Graduate Option, Pasadena, CA 91125-0001. Offers PhD. *Faculty:* 42 full-time (11 women). *Students:* 54 full-time (18 women); includes 14 minority (10 Asian Americans or Pacific Islanders, 4 Hispanic Americans). 119 applicants, 18% accepted, 10 enrolled. In 2005, 7 doctorates awarded. *Degree requirements:* For doctorate, thesis/dissertation, qualifying exam. *Entrance requirements:* For doctorate, GRE General Test. Additional exam requirements/recommendations for international students: Required—TOEFL. *Application deadline:* For fall admission, 1/1 for domestic students. Application fee: $50. Electronic applications accepted. *Financial support:* In 2005–06, fellowships with full tuition reimbursements (averaging $21,195 per year), teaching assistantships with full tuition reimbursements (averaging $4,305 per year) were awarded; research assistantships with full tuition reimbursements, institutionally sponsored loans also available. Financial award application deadline: 1/1. *Unit head:* Dr. Stephen L. Mayo, Executive Officer, 626-395-6408, E-mail: steve@mayo.caltech.edu. *Application contact:* Alison Ross, Graduate Program Coordinator, 626-395-6446, Fax: 626-568-9430, E-mail: biochemop@cco.caltech.edu.

California Institute of Technology, Division of Chemistry and Chemical Engineering, Pasadena, CA 91125-0001. Offers biochemistry and molecular biophysics (MS, PhD); chemical engineering (MS, PhD); chemistry (MS, PhD). *Faculty:* 41 full-time (8 women). *Students:* 345 full-time (121 women). Average age 24. 647 applicants, 27% accepted, 61 enrolled. In 2005, 19

Biochemistry

California Institute of Technology (continued)

master's, 57 doctorates awarded. Terminal master's awarded for partial completion of doctoral program. *Degree requirements:* For master's and doctorate, thesis/dissertation. *Entrance requirements:* Additional exam requirements/recommendations for international students: Required—TOEFL; Recommended—IELT, TWE, TSE. *Application deadline:* For fall admission, 1/1 for domestic students, 1/1 for international students. Application fee: $50. Electronic applications accepted. *Financial support:* In 2005–06, 345 students received support; fellowships, research assistantships, teaching assistantships, Federal Work-Study, institutionally sponsored loans, scholarships/grants, traineeships, health care benefits, and unspecified assistantships available. Financial award application deadline: 1/1. *Faculty research:* Molecular structure and interactions in chemical and biological systems, theoretical studies of fundamental chemical processes, chemical reaction engineering, biochemical engineering, catalysis. *Unit head:* Dr. David A. Tirrell, Chairman, 626-395-3646, Fax: 626-568-8824, E-mail: tirrell@caltech.edu.

California Polytechnic State University, San Luis Obispo, College of Science and Mathematics, Department of Chemistry and Biochemistry, San Luis Obispo, CA 93407. Offers polymers and coating science (MS). Part-time programs available. *Faculty:* 1 full-time (0 women), 2 part-time/adjunct (1 woman). *Students:* 3 full-time (1 woman), 3 part-time (1 woman). 5 applicants, 80% accepted, 4 enrolled. *Degree requirements:* For master's, comprehensive oral exam. *Entrance requirements:* For master's, minimum GPA of 2.5 in last 90 quarter units of course work. Additional exam requirements/recommendations for international students: Required—TOEFL, TWE. *Application deadline:* For fall admission, 6/1 for domestic students, 11/30 for international studentsFor winter admission, 8/1 for domestic students; for spring admission, 12/1 for domestic students. Applications are processed on a rolling basis. Application fee: $55. Electronic applications accepted. *Expenses:* Tuition, nonresident: full-time $226; part-time $226 per unit. Required fees: $4,827; $1,063 per unit. *Financial support:* Career-related internships or fieldwork and Federal Work-Study available. Support available to part-time students. Financial award application deadline: 3/2; financial award applicants required to submit FAFSA. *Unit head:* Dr. Ray Fernando, Graduate Coordinator, 805-756-2395, E-mail: rhfernan@calpoly.edu.

California State University, East Bay, Academic Programs and Graduate Studies, College of Science, Department of Chemistry, Option in Biochemistry, Hayward, CA 94542-3000. Offers biochemistry (MS). *Degree requirements:* For master's, comprehensive exam or thesis. *Entrance requirements:* For master's, minimum GPA of 2.5 in field during previous 2 years of course work. Application fee: $55. *Financial support:* Federal Work-Study and institutionally sponsored loans available. Support available to part-time students. Financial award application deadline: 3/2. *Unit head:* Susan Opp, Graduate Coordinator, 510-885-3475, E-mail: susan.opp@csueastbay.edu. *Application contact:* Deborah Baker, Associate Director, 510-885-3286, Fax: 510-885-4777, E-mail: deborah.baker@csueastbay.edu.

California State University, Fullerton, Graduate Studies, College of Natural Science and Mathematics, Department of Chemistry and Biochemistry, Fullerton, CA 92834-9480. Offers analytical chemistry (MS); biochemistry (MS); geochemistry (MS); inorganic chemistry (MS); organic chemistry (MS); physical chemistry (MS). Part-time programs available. *Students:* 19 full-time (6 women), 20 part-time (9 women); includes 20 minority (1 African American, 13 Asian Americans or Pacific Islanders, 6 Hispanic Americans), 8 international. Average age 28. 36 applicants, 47% accepted, 4 enrolled. In 2005, 11 degrees awarded. *Degree requirements:* For master's, thesis, departmental qualifying exam. *Entrance requirements:* For master's, minimum GPA of 2.5 in last 60 units of course work, major in chemistry or related field. Application fee: $55. *Expenses:* Tuition, area resident: Part-time $2,270 per year. Tuition, state resident: full-time $2,572; part-time $339 per unit. Tuition, nonresident: full-time $339; part-time $339 per unit. International tuition: $339 full-time. *Financial support:* Teaching assistantships, career-related internships or fieldwork, Federal Work-Study, institutionally sponsored loans, and scholarships/grants available. Support available to part-time students. Financial award application deadline: 3/1. *Unit head:* Dr. Maria Linder, Chair, 714-278-3621. *Application contact:* Dr. Gregory Williams, Adviser, 714-278-2170.

California State University, Long Beach, Graduate Studies, College of Natural Sciences and Mathematics, Department of Chemistry, Long Beach, CA 90840. Offers biochemistry (MS); chemistry (MS). Part-time programs available. *Students:* 23 full-time (15 women), 16 part-time (6 women); includes 23 minority (1 African American, 17 Asian Americans or Pacific Islanders, 5 Hispanic Americans). Average age 28. 37 applicants, 49% accepted, 12 enrolled. In 2005, 3 degrees awarded. *Degree requirements:* For master's, thesis, departmental qualifying exam. *Application deadline:* For fall admission, 7/1 for domestic students; for spring admission, 12/1 for domestic students. Applications are processed on a rolling basis. Application fee: $55. Electronic applications accepted. *Expenses:* Tuition, area resident: Full-time $3,102; part-time $1,800 per semester hour. Tuition, state resident: part-time $339 per semester hour. Tuition, nonresident: full-time $339. Required fees: $779. *Financial support:* Research assistantships, teaching assistantships, Federal Work-Study, institutionally sponsored loans, scholarships/grants, and unspecified assistantships available. Financial award application deadline: 3/2. *Faculty research:* Enzymology, organic synthesis, molecular modeling, environmental chemistry, reaction kinetics. *Unit head:* Dr. Douglas D. McAbee, Chair, 562-985-4941, Fax: 562-985-8557, E-mail: dmcabee@csulb.edu. *Application contact:* Dr. Lijuan Li, Graduate Coordinator, 562-985-5068, Fax: 562-985-2315, E-mail: lli@csulb.edu.

California State University, Los Angeles, Graduate Studies, College of Natural and Social Sciences, Department of Chemistry and Biochemistry, Option in Biochemistry, Los Angeles, CA 90032-8530. Offers MS. Part-time and evening/weekend programs available. *Students:* 4 full-time (1 woman), 2 part-time (both women); includes 4 minority (1 African American, 1 Asian American or Pacific Islander, 2 Hispanic Americans). *Degree requirements:* For master's, one foreign language. *Entrance requirements:* Additional exam requirements/recommendations for international students: Required—TOEFL. *Application deadline:* For fall admission, 6/30 for domestic students; for spring admission, 2/1 for domestic students. Applications are processed on a rolling basis. Application fee: $55. *Financial support:* Federal Work-Study available. Support available to part-time students. Financial award application deadline: 3/1. *Faculty research:* Biosynthesis of NAD in bacteria, kinetic and regulatory properties of enzymes, regulation of lipoprotein by dietary cholesterol. *Unit head:* Dr. Jamil Momand, Head, 323-343-2361.

Carnegie Mellon University, Mellon College of Science, Department of Biological Sciences, Pittsburgh, PA 15213-3891. Offers biochemistry (PhD); biophysics (PhD); cell biology (PhD); computational biology (MS, PhD); developmental biology (PhD); genetics (PhD); molecular biology (PhD); neurobiology (PhD). *Degree requirements:* For doctorate, thesis/dissertation, comprehensive exam. *Entrance requirements:* For doctorate, GRE General Test, GRE Subject Test, interview. Electronic applications accepted. *Faculty research:* Genetic structure, function, and regulation; protein structure and function; biological membranes; biological spectroscopy.

See Close-Up on page 93.

Case Western Reserve University, School of Medicine and School of Graduate Studies, Graduate Programs in Medicine, Department of Biochemistry, Cleveland, OH 44106. Offers biochemical research (MS); biochemistry (MS, PhD). Part-time programs available. *Faculty:* 52 full-time (8 women), 3 part-time/adjunct (0 women). *Students:* 64 full-time (33 women); includes 30 minority (all Asian Americans or Pacific Islanders) Average age 24. 154 applicants, 1% accepted. In 2005, 2 master's, 5 doctorates awarded. Terminal master's awarded for partial completion of doctoral program. *Median time to degree:* Of those who began their doctoral program in fall 1997, 100% received their degree in 8 years or less. *Degree requirements:* For master's, thesis (for some programs); for doctorate, thesis/dissertation. *Entrance requirements:* For master's and doctorate, GRE General Test. Additional exam requirements/recommendations for international students: Required—TOEFL. *Application deadline:* For fall admission, 2/1 for domestic students. Applications are processed on a rolling basis. Application fee: $50. Electronic

applications accepted. *Financial support:* In 2005–06, fellowships with full tuition reimbursements (averaging $23,000 per year); research assistantships with full tuition reimbursements *Faculty research:* Regulation of metabolism, regulation of gene expression and protein synthesis, cell biology, molecular biology, structural biology. *Unit head:* Dr. Michael A. Weiss, Chairman, 216-368-5991, Fax: 216-368-3419, E-mail: maw21@po.cwru.edu. *Application contact:* Dr. Hung-Ying L. Kao, Chairman, Graduate Admissions Committee, 216-844-7572, Fax: 216-368-3419, E-mail: hung.kao@case.edu.

Case Western Reserve University, School of Medicine and School of Graduate Studies, Graduate Programs in Medicine, Department of Nutrition, Cleveland, OH 44106. Offers dietetics (MS); nutrition (MS, PhD), including nutrition and biochemistry (PhD); public health nutrition (MS). Part-time programs available. *Faculty:* 19 full-time (11 women), 31 part-time/adjunct (24 women). *Students:* 61 full-time (53 women), 15 part-time (13 women); includes 38 minority (2 African Americans, 31 Asian Americans or Pacific Islanders, 5 Hispanic Americans). Average age 25. 47 applicants, 53% accepted, 21 enrolled. In 2005, 17 master's, 3 doctorates awarded. Terminal master's awarded for partial completion of doctoral program. *Degree requirements:* For master's, thesis (for some programs); for doctorate, thesis/dissertation. *Entrance requirements:* For master's, GRE General Test; for doctorate, GRE General Test, GRE Subject Test. Additional exam requirements/recommendations for international students: Required—TOEFL. *Application deadline:* For fall admission, 3/1 for domestic students; for spring admission, 11/1 for domestic students. Applications are processed on a rolling basis. Application fee: $50. *Financial support:* In 2005–06, 11 students received support, including 3 fellowships with full tuition reimbursements available (averaging $23,000 per year), 9 research assistantships with full tuition reimbursements available (averaging $23,000 per year); career-related internships or fieldwork, Federal Work-Study, and tuition waivers (partial) also available. *Faculty research:* Fatty acid metabolism, application of gene therapy to nutritional problems, dietary intake methodology, nutrition and physical fitness, metabolism during infancy and pregnancy. Total annual research expenditures: $1.5 million. *Unit head:* Dr. Henri Brunengraber, Chairman, 216-368-2440, Fax: 216-368-6644, E-mail: hxb8@case.edu. *Application contact:* Pamela A. Woodruff, Department Assistant III, 216-368-2440, Fax: 216-368-6644, E-mail: paw5@case.edu.

See Close-Up on page 1087.

City College of the City University of New York, Graduate School, College of Liberal Arts and Science, Division of Science, Department of Chemistry, Program in Biochemistry, New York, NY 10031-9198. Offers MA, PhD. *Students:* 7 applicants, 43% accepted, 3 enrolled. In 2005, 2 degrees awarded. Terminal master's awarded for partial completion of doctoral program. *Degree requirements:* For doctorate, one foreign language, thesis/dissertation. *Entrance requirements:* For master's and doctorate, GRE. Additional exam requirements/recommendations for international students: Required—TOEFL (minimum score 500 paper-based; 173 computer-based). *Application deadline:* For fall admission, 5/1 for domestic students; for spring admission, 11/15 for domestic students. Application fee: $125. *Financial support:* Application deadline: 6/1. *Faculty research:* Fatty acid metabolism, lectins, gene structure. *Unit head:* Horst Schulz, Chairman, 212-650-8323.

Clemson University, Graduate School, College of Agriculture, Forestry and Life Sciences, Department of Genetics and Biochemistry, Program in Biochemistry and Molecular Biology, Clemson, SC 29634. Offers MS, PhD. *Students:* 3 full-time (2 women), (all international). 16 applicants, 19% accepted, 3 enrolled. In 2005, 1 master's awarded. *Degree requirements:* For master's, thesis/dissertation; for doctorate, thesis/dissertation, comprehensive exam. *Entrance requirements:* For master's and doctorate, GRE General Test. Additional exam requirements/recommendations for international students: Required—TOEFL. *Application deadline:* For fall admission, 6/1 priority date for domestic students, 4/15 priority date for international students. Applications are processed on a rolling basis. Application fee: $50. *Financial support:* Fellowships, research assistantships, teaching assistantships available. Financial award application deadline: 3/15; financial award applicants required to submit FAFSA. *Faculty research:* Biomembranes, protein structure, molecular biology of plants, APYA and stress response. Total annual research expenditures: $670,000. *Unit head:* Dr. Kerry Smith, Coordinator, 864-656-6935, Fax: 864-656-0435, E-mail: kssmith@clemson.edu.

See Close-Ups on pages 383 and 385.

Colorado State University, Graduate School, College of Natural Sciences, Department of Biochemistry and Molecular Biology, Fort Collins, CO 80523-0015. Offers biochemistry (MS, PhD). Part-time programs available. *Faculty:* 11 full-time (4 women), 2 part-time/adjunct (0 women). *Students:* 17 full-time (8 women), 17 part-time (9 women); includes 4 minority (1 American Indian/Alaska Native, 3 Asian Americans or Pacific Islanders), 10 international. Average age 28. 26 applicants, 42% accepted, 9 enrolled. In 2005, 4 master's, 5 doctorates awarded. Terminal master's awarded for partial completion of doctoral program. *Degree requirements:* For master's, thesis optional; for doctorate, thesis/dissertation. *Entrance requirements:* For master's and doctorate, GRE General Test, minimum GPA of 3.2. Additional exam requirements/recommendations for international students: Required—TOEFL (minimum score 620 paper-based; 260 computer-based). *Application deadline:* For fall admission, 1/7 priority date for domestic students, 1/10 priority date for international students; for spring admission, 11/12 priority date for domestic students, 11/10 priority date for international students. Applications are processed on a rolling basis. Application fee: $50. Electronic applications accepted. *Expenses:* Tuition, state resident: full-time $3,690; part-time $205 per credit. Tuition, nonresident: full-time $14,958; part-time $831 per credit. Required fees: $1,061. *Financial support:* In 2005–06, 32 students received support, including fellowships with full tuition reimbursements available (averaging $21,800 per year), research assistantships with full tuition reimbursements available (averaging $21,800 per year), teaching assistantships with full tuition reimbursements available (averaging $21,800 per year); Federal Work-Study, institutionally sponsored loans, traineeships, and tuition waivers (partial) also available. Financial award application deadline: 2/15; financial award applicants required to submit FAFSA. *Faculty research:* Cellular biology, molecular gene expression, neurobiology, structural biology, yeast genetics. Total annual research expenditures: $4.3 million. *Unit head:* Marvin R. Paule, Interim Chair, 970-491-5566, Fax: 970-491-0494, E-mail: marv.paule@lamar.colostate.edu. *Application contact:* Yvonne Bridgeman, Graduate Program Assistant, 970-491-6841, Fax: 970-491-0494, E-mail: yyvonne.bridgeman@colostate.edu.

Colorado State University-Pueblo, College of Science and Mathematics, Pueblo, CO 81001-4901. Offers applied natural science (MS), including biochemistry, biology, chemistry. Part-time and evening/weekend programs available. *Faculty:* 15 full-time (9 women). *Students:* 14 full-time (9 women), 5 part-time (1 woman); includes 2 minority (both Hispanic Americans) 8 applicants, 100% accepted, 8 enrolled. In 2005, 9 degrees awarded. *Degree requirements:* For master's, thesis (for some programs), internship report (if non-thesis), comprehensive exam (for some programs), registration. *Entrance requirements:* For master's, GRE General Test, minimum GPA of 3.0. Additional exam requirements/recommendations for international students: Required—TOEFL (minimum score 500 paper-based; 173 computer-based). *Application deadline:* For fall admission, 6/15 priority date for domestic students, 6/15 priority date for international students; for spring admission, 10/15 priority date for domestic students, 10/15 priority date for international students. Applications are processed on a rolling basis. Application fee: $35. *Expenses:* Tuition, state resident: full-time $2,177; part-time $121 per credit hour. Tuition, nonresident: full-time $10,157; part-time $564 per credit hour. Required fees: $490; $41 per credit hour. *Financial support:* In 2005–06, 9 students received support, including 1 fellowship (averaging $1,000 per year), 3 teaching assistantships with partial tuition reimbursements available (averaging $9,000 per year); research assistantships, career-related internships or fieldwork, scholarships/grants, and unspecified assistantships also available. Financial award application deadline: 6/1; financial award applicants required to submit FAFSA. *Faculty research:* Fungal cell walls, molecular biology, bioactive materials synthesis, forensic chemistry, atomic force microscopy-surface chemistry. Total annual research expenditures: $400,000. *Unit head:* Dr. Kristina Proctor, Dean, 719-549-2340, Fax: 719-549-2732, E-mail: kristina.

proctor@colostate-pueblo.edu. *Application contact:* Dr. Melvin Druelinger, Director, MSANS Program, 719-549-2325, Fax: 719-549-2071, E-mail: mel.druelinger@colostate-pueblo.edu.

Columbia University, College of Physicians and Surgeons and Graduate School of Arts and Sciences, Graduate School of Arts and Sciences at the College of Physicians and Surgeons, Department of Biochemistry and Molecular Biophysics, New York, NY 10032. Offers biochemistry and molecular biophysics (M Phil, PhD); biophysics (PhD). Only candidates for the PhD are admitted. *Degree requirements:* For doctorate, one foreign language, thesis/dissertation. *Entrance requirements:* For master's and doctorate, GRE General Test. Additional exam requirements/recommendations for international students: Required—TOEFL. *Expenses:* Tuition: Full-time $31,448. Tuition and fees vary according to course level, course load, campus/location and program.

Cornell University, Graduate School, Graduate Fields of Agriculture and Life Sciences, Field of Biochemistry, Molecular and Cell Biology, Ithaca, NY 14853-0001. Offers biochemistry (PhD); biophysics (PhD); cell biology (PhD); molecular and cell biology (PhD); molecular biology (PhD). *Faculty:* 60 full-time (13 women). *Students:* 177 applicants, 25% accepted, 22 enrolled. In 2005, 13 degrees awarded. *Degree requirements:* For doctorate, thesis/dissertation, 2 semesters of teaching experience, comprehensive exam. *Entrance requirements:* For doctorate, GRE General Test, GRE Subject Test (biology, chemistry, physics; or biochemistry, cell and molecular biology), 3 letters of recommendation. Additional exam requirements/recommendations for international students: Required—TOEFL (minimum score 600 paper-based; 250 computer-based). *Application deadline:* For fall admission, 1/5 for domestic students. Application fee: $60. Electronic applications accepted. *Financial support:* In 2005–06, 90 students received support, including 35 fellowships with full tuition reimbursements available, 42 research assistantships with full tuition reimbursements available, 13 teaching assistantships with full tuition reimbursements available; institutionally sponsored loans, scholarships/grants, health care benefits, tuition waivers (full and partial), and unspecified assistantships also available. Financial award applicants required to submit FAFSA. *Faculty research:* Biophysics, structural biology. *Unit head:* Director of Graduate Studies, 607-255-2100, Fax: 607-255-2100. *Application contact:* Graduate Field Assistant, 607-255-2100, Fax: 607-255-2100, E-mail: bmcb@cornell.edu.

Cornell University, Graduate School, Graduate Fields of Arts and Sciences, Field of Chemistry and Chemical Biology, Ithaca, NY 14853-0001. Offers analytical chemistry (PhD); bio-organic chemistry (PhD); biophysical chemistry (PhD); chemical biology (PhD); chemical physics (PhD); inorganic chemistry (PhD); materials chemistry (PhD); organic chemistry (PhD); organometallic chemistry (PhD); physical chemistry (PhD); polymer chemistry (PhD); theoretical chemistry (PhD). *Faculty:* 44 full-time (2 women). *Students:* 190 full-time (73 women); includes 23 minority (4 African Americans, 8 Asian Americans or Pacific Islanders, 11 Hispanic Americans), 65 international. 339 applicants, 35% accepted, 49 enrolled. In 2005, 23 doctorates awarded. *Degree requirements:* For doctorate, thesis/dissertation, comprehensive exam. *Entrance requirements:* For doctorate, GRE General Test, GRE Subject Test (chemistry), 3 letters of recommendation. Additional exam requirements/recommendations for international students: Required—TOEFL (minimum score 600 paper-based; 250 computer-based). *Application deadline:* For fall admission, 1/10 for domestic students. Application fee: $60. Electronic applications accepted. *Financial support:* In 2005–06, 185 students received support, including 33 fellowships with full tuition reimbursements available, 85 research assistantships with full tuition reimbursements available, 67 teaching assistantships with full tuition reimbursements available; institutionally sponsored loans, scholarships/grants, health care benefits, tuition waivers (full and partial), and unspecified assistantships also available. Financial award applicants required to submit FAFSA. *Faculty research:* Analytical, organic, inorganic, physical, materials, chemical biology. *Unit head:* Director of Graduate Studies, 607-255-4139, Fax: 607-255-4137. *Application contact:* Graduate Field Assistant, 607-255-4139, Fax: 607-255-4137, E-mail: chemgrad@cornell.edu.

Cornell University, Joan and Sanford I. Weill Medical College and Graduate School of Medical Sciences, Weill Graduate School of Medical Sciences, Program in Molecular Biology, Cell Biology, Biochemistry and Structural Biology, New York, NY 10021-4896. Offers MS, PhD, MD/PhD. *Faculty:* 84 full-time (22 women). *Students:* 121 full-time (68 women); includes 14 minority (2 African Americans, 7 Asian Americans or Pacific Islanders, 5 Hispanic Americans), 34 international. Average age 22. In 2005, 8 degrees awarded. *Median time to degree:* Of those who began their doctoral program in fall 1997, 100% received their degree in 8 years or less. *Degree requirements:* For doctorate, thesis/dissertation, final exam. *Entrance requirements:* For doctorate, GRE General Test, GRE Subject Test, background in genetics, molecular biology, chemistry, or biochemistry. Additional exam requirements/recommendations for international students: Required—TOEFL. *Application deadline:* For fall admission, 12/15 for domestic students. Application fee: $60. *Expenses:* Tuition: Full-time $32,320. Required fees: $1,025. *Financial support:* Fellowships, stipends available. *Unit head:* Dr. Christopher Lima, Co-Director, 212-639-8205, Fax: 212-717-3047, E-mail: limac@mskcc.org.

Cornell University, Joan and Sanford I. Weill Medical College and Graduate School of Medical Sciences, Weill Graduate School of Medical Sciences, Tri-Institutional Program in Chemical Biology and Computational Biology, New York, NY 10021-4896. Offers PhD. Offered jointly by Cornell University, Weill Graduate School of Medical Sciences, The Rockefeller University and Sloan-Kettering Institute; students must be accepted to Cornell University Graduate Program in Chemistry. *Students:* 17 full-time (9 women); includes 1 minority (Asian American or Pacific Islander), 11 international. *Expenses:* Tuition: Full-time $32,320. Required fees: $1,025. *Unit head:* Timothy Ryan, Director, 212-746-6403.

Dalhousie University, Faculty of Medicine, Department of Biochemistry and Molecular Biology, Halifax, NS B3H 4R2, Canada. Offers M Sc, PhD, MD/PhD. *Degree requirements:* For master's and doctorate, thesis/dissertation, demonstrating/teaching experience, oral defense, seminar, comprehensive exam, registration. *Entrance requirements:* For master's and doctorate, GRE. Additional exam requirements/recommendations for international students: Required—TOEFL. *Faculty research:* Gene expression and cell regulation; lipids, lipoproteins, and membranes; molecular evolution; proteins, molecular cell biology and molecular genetics; comparative genomics, proteomics, and molecular evolution; structure, function, and metabolism of biomolecules.

Dartmouth College, School of Arts and Sciences, Program in Biochemistry, Hanover, NH 03755. Offers PhD, MD/PhD. *Faculty:* 13 full-time (2 women), 2 part-time/adjunct (0 women). *Students:* 42 full-time (21 women); includes 3 minority (2 Asian Americans or Pacific Islanders, 1 Hispanic American), 11 international. Average age 27. 336 applicants, 20% accepted, 32 enrolled. In 2005, 9 doctorates awarded. *Degree requirements:* For doctorate, thesis/dissertation. *Entrance requirements:* For doctorate, GRE General Test, GRE Subject Test. Additional exam requirements/recommendations for international students: Required—TOEFL. *Application deadline:* For fall admission, 1/7 for domestic students. Application fee: $40. Electronic applications accepted. *Expenses:* Tuition: Full-time $31,770. *Financial support:* In 2005–06, 40 students received support, including fellowships with full tuition reimbursements available (averaging $23,000 per year), research assistantships with full tuition reimbursements available (averaging $23,000 per year); Federal Work-Study, scholarships/grants, traineeships, and unspecified assistantships also available. Financial award application deadline: 4/15. *Faculty research:* Gene transcription, hormone regulation, cellular immunology, neurobiology, lipid metabolism. Total annual research expenditures: $9.3 million. *Unit head:* Ta Yuan Chang, Chair, Graduate Committee, 603-650-1622. *Application contact:* Information Contact, 603-650-1612, Fax: 603-650-1006, E-mail: mcb@dartmouth.edu.

See Close-Up on page 387.

DePaul University, College of Liberal Arts and Sciences, Department of Chemistry, Chicago, IL 60604-2287. Offers biochemistry (MS); chemistry (MS); polymer chemistry and coatings technology (MS). Part-time and evening/weekend programs available. *Faculty:* 11 full-time (4 women), 4 part-time/adjunct (1 woman). *Students:* 9 full-time (5 women), 9 part-time (3 women); includes 3 minority (1 American Indian/Alaska Native, 1 Asian American or Pacific Islander, 1 Hispanic American), 3 international. Average age 27. 6 applicants, 100% accepted, 4 enrolled. In 2005, 2 master's awarded. *Degree requirements:* For master's, thesis (for some programs), oral exam for selected programs. *Entrance requirements:* For master's, BS in chemistry or equivalent. Additional exam requirements/recommendations for international students: Required—TOEFL (minimum score 590 paper-based; 243 computer-based). *Application deadline:* For fall admission, 7/15 for domestic students, 5/1 for international studentsFor winter admission, 11/15 for domestic students; for spring admission, 2/15 for domestic students. Applications are processed on a rolling basis. Application fee: $35. Electronic applications accepted. *Financial support:* In 2005–06, 4 students received support, including 6 teaching assistantships with tuition reimbursements available (averaging $8,000 per year) Financial award application deadline: 4/1. *Faculty research:* Polymers, DNA sequencing, computational chemistry, water pollution, diffusion kinetics. Total annual research expenditures: $30,000. *Unit head:* Dr. Wendy S. Wolbach, Chair, 773-325-7420, Fax: 773-325-7421, E-mail: wwolbach@condor.depaul.edu. *Application contact:* Kavitha Chinthada, Director of Graduate Admissions, 773-325-7885, Fax: 773-325-7311, E-mail: kchintha@depaul.edu.

Drexel University, College of Medicine, Biomedical Graduate Programs, Program in Biochemistry, Philadelphia, PA 19104-2875. Offers MS, PhD, MD/PhD. Part-time programs available. Terminal master's awarded for partial completion of doctoral program. *Degree requirements:* For master's, thesis, comprehensive exam; for doctorate, thesis/dissertation, qualifying exam. *Entrance requirements:* For master's, GRE General Test, minimum GPA of 2.75; for doctorate, GRE General Test, minimum GPA of 3.0. Additional exam requirements/recommendations for international students: Required—TOEFL. Electronic applications accepted.

Duke University, Graduate School, Department of Biochemistry, Durham, NC 27710. Offers crystallography of macromolecules (PhD); enzyme mechanisms (PhD); lipid biochemistry (PhD); membrane structure and function (PhD); molecular genetics (PhD); neurochemistry (PhD); nucleic acid structure and function (PhD); protein structure and function (PhD). *Faculty:* 28 full-time. *Students:* 68 full-time (28 women); includes 6 minority (1 African American, 5 Asian Americans or Pacific Islanders), 15 international. 106 applicants, 19% accepted, 5 enrolled. In 2005, 11 doctorates awarded. *Degree requirements:* For doctorate, thesis/dissertation. *Entrance requirements:* For doctorate, GRE General Test, GRE Subject Test (recommended). Additional exam requirements/recommendations for international students: Required—IELT (preferred) or TOEFL. *Application deadline:* For fall admission, 12/31 for domestic students, 12/31 for international students. Electronic applications accepted. *Financial support:* Fellowships, research assistantships, teaching assistantships, Federal Work-Study available. Financial award application deadline: 12/31. *Unit head:* Leonard Spicer, Director of Graduate Studies, 919-681-8770, Fax: 919-684-8885, E-mail: anorfleet@biochem.duke.edu.

Duke University, Graduate School, Program in Biological Chemistry, Durham, NC 27710. Offers PhD, Certificate. Certificate students must be enrolled in a participating PhD program. *Faculty:* 20 full-time. *Students:* 1 (woman) full-time. *Entrance requirements:* For degree, GRE General Test, GRE Subject Test. Additional exam requirements/recommendations for international students: Required—IELT (preferred) or TOEFL. *Application deadline:* For fall admission, 12/31 for domestic students, 12/31 for international students. Application fee: $75. Electronic applications accepted. *Financial support:* Application deadline: 12/31. *Unit head:* Johannes Rudolph, Director of Graduate Studies, 919-668-6188, Fax: 919-684-8346, E-mail: rudolph@biochem.duke.edu.

Duquesne University, Bayer School of Natural and Environmental Sciences, Department of Chemistry and Biochemistry, Pittsburgh, PA 15282-0001. Offers biochemistry (MS, PhD); chemistry (MS, PhD). Part-time programs available. *Faculty:* 17 full-time (5 women). *Students:* 36 full-time (19 women), 5 part-time (2 women); includes 1 minority (Hispanic American), 11 international. Average age 27. 34 applicants, 53% accepted, 8 enrolled. In 2005, 5 master's, 9 doctorates awarded. Terminal master's awarded for partial completion of doctoral program. *Degree requirements:* For master's, thesis (for some programs), comprehensive exam (for some programs), registration; for doctorate, thesis/dissertation, registration. *Entrance requirements:* For master's and doctorate, GRE General Test. Additional exam requirements/recommendations for international students: Required—TOEFL, TSE. *Application deadline:* For fall admission, 2/15 priority date for domestic students, 2/15 priority date for international students; for spring admission, 10/1 priority date for domestic students, 10/1 priority date for international students. Applications are processed on a rolling basis. Application fee: $0. *Expenses:* Contact institution. Tuition and fees vary according to degree level and program. *Financial support:* In 2005–06, 1 fellowship with full and partial tuition reimbursement (averaging $19,100 per year), 8 research assistantships with full tuition reimbursements (averaging $18,850 per year), 26 teaching assistantships with full tuition reimbursements (averaging $18,850 per year) were awarded; institutionally sponsored loans, scholarships/grants, and unspecified assistantships also available. Financial award application deadline: 5/1; financial award applicants required to submit FAFSA. *Faculty research:* Computational physical chemistry, bioinorganic chemistry, analytical chemistry, biophysics, synthetic organic chemistry. Total annual research expenditures: $1.5 million. *Unit head:* Dr. Jeffry Madura, Chair, 412-396-6341, Fax: 412-396-5683, E-mail: madura@duq.edu. *Application contact:* Mary Ann Quinn, Assistant to the Dean Graduate Affairs, 412-396-6339, Fax: 412-396-4881, E-mail: gradinfo@duq.edu.

East Carolina University, Brody School of Medicine, Department of Biochemistry and Molecular Biology, Greenville, NC 27858-4353. Offers PhD. *Faculty:* 10 full-time (1 woman). *Students:* 2 full-time (both women), 3 part-time (all women); includes 1 minority (Asian American or Pacific Islander), 1 international. Average age 25. 8 applicants, 75% accepted. *Median time to degree:* Of those who began their doctoral program in fall 1997, 67% received their degree in 8 years or less. *Degree requirements:* For doctorate, thesis/dissertation, comprehensive exam, registration. *Entrance requirements:* For doctorate, GRE General Test. Additional exam requirements/recommendations for international students: Required—TOEFL. *Application deadline:* For fall admission, 6/1 for domestic students. Applications are processed on a rolling basis. Application fee: $50. *Expenses:* Tuition, state resident: full-time $2,516. Tuition, nonresident: full-time $12,832. *Financial support:* In 2005–06, 7 fellowships with full and partial tuition reimbursements (averaging $21,500 per year) were awarded Financial award application deadline: 6/1. *Faculty research:* Gene regulation, development and differentiation, contractility and motility, macromolecular interactions, cancer. Total annual research expenditures: $856,955. *Unit head:* Dr. Joseph Cory, Chairman, 252-744-2675, Fax: 252-744-3383, E-mail: coryjo@ecu.edu. *Application contact:* Dr. George J. Kasperek, Assistant Dean for Graduate Studies/BSOM, 252-744-3305, Fax: 252-744-0203, E-mail: kasperekg@ecu.edu.

East Tennessee State University, James H. Quillen College of Medicine, Biomedical Science Graduate Program, Johnson City, TN 37614. Offers anatomy (MS, PhD); biochemistry (MS, PhD); biophysics (MS, PhD); microbiology (MS, PhD); pharmacology (MS, PhD); physiology (MS, PhD). Part-time programs available. *Faculty:* 49 full-time (12 women), 1 (woman) part-time/adjunct. *Students:* 30 full-time (19 women), 5 part-time (4 women); includes 2 minority (1 African American, 1 Asian American or Pacific Islander), 11 international. Average age 31. 78 applicants, 13% accepted, 9 enrolled. In 2005, 1 master's, 4 doctorates awarded. Terminal master's awarded for partial completion of doctoral program. *Degree requirements:* For master's, one foreign language, thesis, comprehensive qualifying exam; for doctorate, 2 foreign languages, thesis/dissertation. *Entrance requirements:* For master's, GRE General Test, minimum GPA of 3.0, bachelor's degree in biological or related science; for doctorate, GRE General Test, GRE Subject Test. Additional exam requirements/recommendations for international students: Required—TOEFL (minimum score 550 paper-based; 213 computer-based). *Application deadline:* For fall admission, 3/15 for domestic students; for spring admission, 3/1 for domestic students. Application fee: $25 ($35 for international students). *Expenses:* Contact institution. *Financial support:* In 2005–06, 7 research assistantships with full tuition reimbursements (averaging $15,000 per year) were awarded; teaching assistantships with full tuition reimbursements, career-related internships or fieldwork, Federal Work-Study, institutionally sponsored loans, scholarships/grants, and tuition waivers (full) also available. Financial award application deadline: 7/1; financial award applicants required to submit FAFSA. Total

Biochemistry

East Tennessee State University (continued)
annual research expenditures: $2.1 million. *Unit head:* Dr. Mitchell E. Robinson, Assistant Dean, Director, 423-439-4658, E-mail: robinson@etsu.edu.

Emory University, Graduate School of Arts and Sciences, Division of Biological and Biomedical Sciences, Program in Biochemistry, Cell and Developmental Biology, Atlanta, GA 30322-1100. Offers PhD. *Faculty:* 61 full-time (12 women). *Students:* 56 full-time (42 women); includes 8 minority (5 African Americans, 2 Asian Americans or Pacific Islanders, 1 Hispanic American), 9 international. Average age 27. 96 applicants, 38% accepted, 17 enrolled. In 2005, 6 doctorates awarded. *Median time to degree:* Of those who began their doctoral program in fall 1997, 100% received their degree in 8 years or less. *Degree requirements:* For doctorate, thesis/dissertation, comprehensive exam, registration. *Entrance requirements:* For doctorate, GRE General Test, minimum GPA of 3.0 in science course work. Additional exam requirements/recommendations for international students: Required—TOEFL. *Application deadline:* For fall admission, 1/3 for domestic students, 1/3 for international students. Application fee: $50. Electronic applications accepted. *Expenses:* Tuition: Full-time $14,400. Required fees: $217. *Financial support:* In 2005–06, 26 students received support, including 26 fellowships with full tuition reimbursements available (averaging $23,000 per year); institutionally sponsored loans, health care benefits, and tuition waivers (full) also available. *Faculty research:* Signal transduction, molecular biology, enzymes and cofactors, receptor and ion channel function, membrane biology. *Unit head:* Grace Pavlath, Director, 404-727-3353, Fax: 404-727-2880, E-mail: gpavlat@emory.edu. *Application contact:* 404-727-2545, Fax: 404-727-3322, E-mail: gdbbs@emory.edu.

Florida Atlantic University, Charles E. Schmidt College of Science, Department of Chemistry and Biochemistry, Boca Raton, FL 33431-0991. Offers MS, MST, PhD. Part-time programs available. *Faculty:* 13 full-time (2 women). *Students:* 48 full-time (26 women), 6 part-time (4 women); includes 11 minority (2 African Americans, 2 Asian Americans or Pacific Islanders, 7 Hispanic Americans), 18 international. Average age 31. 24 applicants, 25% accepted, 5 enrolled. In 2005, 8 master's, 2 doctorates awarded. Terminal master's awarded for partial completion of doctoral program. *Degree requirements:* For master's, thesis/dissertation; for doctorate, thesis/dissertation, comprehensive exam. *Entrance requirements:* For master's, GRE General Test, minimum GPA of 3.0; for doctorate, GRE, minimum GPA of 3.0. Additional exam requirements/recommendations for international students: Required—TOEFL. *Application deadline:* For fall admission, 7/1 priority date for domestic students, 2/15 priority date for international students; for spring admission, 4/1 priority date for domestic students, 1/15 priority date for international students. Applications are processed on a rolling basis. Application fee: $30. *Expenses:* Tuition, area resident: Full-time $4,394; part-time $244 per credit. Tuition, state resident: full-time $4,394; part-time $244 per credit. Tuition, nonresident: full-time $16,441; part-time $912 per credit. International tuition: $16,441 full-time. *Financial support:* In 2005–06, 2 research assistantships with full tuition reimbursements (averaging $14,000 per year), 24 teaching assistantships with full tuition reimbursements (averaging $14,000 per year) were awarded; fellowships, Federal Work-Study also available. *Faculty research:* Polymer synthesis and characterization, spectroscopy, geochemistry, environmental chemistry, biomedical chemistry. Total annual research expenditures: $1.2 million. *Unit head:* Dr. Gregg B. Fields, Chair, 561-297-2093, Fax: 561-297-2759, E-mail: fieldsg@fau.edu. *Application contact:* Dr. Salvatore D. Lepore, Professor, 561-297-0330, Fax: 561-297-2759, E-mail: slepore@fau.edu.

Florida State University, Graduate Studies, College of Arts and Sciences, Department of Chemistry and Biochemistry, Specialization in Biochemistry, Tallahassee, FL 32306. Offers MS, PhD. Part-time programs available. *Faculty:* 11 full-time (4 women). *Students:* 47 full-time (17 women); includes 7 minority (2 African Americans, 4 Asian Americans or Pacific Islanders, 1 Hispanic American), 19 international. Average age 25. In 2005, 5 master's, 3 doctorates awarded. Terminal master's awarded for partial completion of doctoral program. *Degree requirements:* For master's and doctorate, thesis/dissertation, cumulative and diagnostic exams. *Entrance requirements:* For master's and doctorate, GRE General Test, minimum B average in undergraduate course work. Additional exam requirements/recommendations for international students: Required—TOEFL (minimum score 515 paper-based; 213 computer-based). *Application deadline:* For fall admission, 4/15 for domestic students, 4/15 for international students. Applications are processed on a rolling basis. Application fee: $30. Electronic applications accepted. *Financial support:* In 2005–06, fellowships with tuition reimbursements (averaging $18,000 per year), 12 research assistantships with tuition reimbursements (averaging $19,000 per year), 15 teaching assistantships with tuition reimbursements (averaging $19,000 per year) were awarded; career-related internships or fieldwork, Federal Work-Study, institutionally sponsored loans, and traineeships also available. Financial award application deadline: 2/15; financial award applicants required to submit FAFSA. *Faculty research:* Metalloenzymes, gene regulation, DNA structure, NMR of synthetic membranes, secondary metabolites. *Application contact:* Dr. Oliver Steinbock, Chair, Graduate Admissions Committee, 888-525-9286, Fax: 850-644-8281, E-mail: gradinfo@chem.fsu.edu.

Florida State University, Graduate Studies, College of Arts and Sciences, Program in Molecular Biophysics, Tallahassee, FL 32306. Offers biochemistry, molecular and cell biology (PhD); computational structural biology (PhD); molecular biophysics (PhD). *Faculty:* 44 full-time (7 women). *Students:* 29 full-time (17 women); includes 1 minority (Hispanic American), 12 international. Average age 30. 14 applicants, 100% accepted, 4 enrolled. In 2005, 3 degrees awarded. *Median time to degree:* Of those who began their doctoral program in fall 1997, 100% received their degree in 8 years or less. *Degree requirements:* For doctorate, thesis/dissertation, comprehensive exam. *Entrance requirements:* For doctorate, GRE General Test. Additional exam requirements/recommendations for international students: Required—TOEFL (minimum score 600 paper-based; 250 computer-based). *Application deadline:* For fall admission, 1/15 for domestic students, 1/15 for international students. Applications are processed on a rolling basis. Application fee: $30. Electronic applications accepted. *Financial support:* In 2005–06, 29 students received support, including 29 research assistantships (averaging $19,500 per year); health care benefits and tuition waivers (partial) also available. Financial award applicants required to submit FAFSA. *Faculty research:* Protein and nucleic acid structure and function, membrane protein structure, computational biophysics, 3-D image reconstruction. Total annual research expenditures: $5.1 million. *Unit head:* Dr. P. Bryant Chase, Director, MOB Graduate Program, 850-644-0056, Fax: 850-644-7244, E-mail: chase@bio.fsu.edu. *Application contact:* Dale E. Leonard, Academic Coordinator, Graduate Programs, 850-644-1012, Fax: 850-644-7244, E-mail: mob@sb.fsu.edu.

Georgetown University, Graduate School of Arts and Sciences, Department of Chemistry, Washington, DC 20057. Offers analytical chemistry (MS, PhD); biochemistry (MS, PhD); chemical physics (MS, PhD); inorganic chemistry (MS, PhD); organic chemistry (MS, PhD); physical chemistry (MS, PhD); theoretical chemistry (MS, PhD). Terminal master's awarded for partial completion of doctoral program. *Degree requirements:* For master's, thesis (for some programs), qualifying exam; for doctorate, thesis/dissertation, comprehensive exam. *Entrance requirements:* For master's and doctorate, GRE General Test. Additional exam requirements/recommendations for international students: Required—TOEFL.

Georgetown University, Graduate School of Arts and Sciences, Programs in Biomedical Sciences, Department of Biochemistry and Molecular Biology, Washington, DC 20057. Offers PhD, MD/PhD. *Degree requirements:* For doctorate, thesis/dissertation, comprehensive exam. *Entrance requirements:* For master's and doctorate, GRE General Test. Additional exam requirements/recommendations for international students: Required—TOEFL.

The George Washington University, Columbian College of Arts and Sciences, Department of Biochemistry and Molecular Biology, Washington, DC 20037. Offers biochemistry (PhD); genomics, proteomics, and bioinformatics (MS). *Degree requirements:* For master's, comprehensive exam; for doctorate, thesis/dissertation, general exam. *Entrance requirements:* For master's, GRE General Test, interview, minimum GPA of 3.0; for doctorate, GRE General Test, minimum

GPA of 3.0. Additional exam requirements/recommendations for international students: Required—TOEFL (minimum score 550 paper-based; 213 computer-based).

See Close-Up on page 389.

The George Washington University, Columbian College of Arts and Sciences, Institute for Biomedical Sciences, Washington, DC 20052. Offers biochemistry and molecular biology (PhD); genetics (MS, PhD); microbiology and immunology (PhD); molecular medicine (PhD); neuroscience (PhD); pharmacology (PhD). Part-time and evening/weekend programs available. *Degree requirements:* For doctorate, thesis/dissertation. *Entrance requirements:* For doctorate, GRE General Test, minimum GPA of 3.0. Additional exam requirements/recommendations for international students: Required—TOEFL (minimum score 600 paper-based; 250 computer-based). Electronic applications accepted.

See Close-Up on page 135.

Georgia Institute of Technology, Graduate Studies and Research, College of Sciences, School of Chemistry and Biochemistry, Atlanta, GA 30332-0001. Offers MS, MS Chem, PhD. Terminal master's awarded for partial completion of doctoral program. *Degree requirements:* For master's, thesis (for some programs); for doctorate, thesis/dissertation. *Entrance requirements:* For master's and doctorate, GRE General Test, GRE Subject Test, minimum GPA of 2.7. Additional exam requirements/recommendations for international students: Required—TOEFL. Electronic applications accepted. *Faculty research:* Inorganic, organic, physical, and analytical chemistry.

Georgia State University, College of Arts and Sciences, Department of Biology, Program in Molecular Genetics and Biochemistry, Atlanta, GA 30303-3083. Offers MS, PhD. *Degree requirements:* For master's, thesis or alternative, exam; for doctorate, thesis/dissertation, exam. *Entrance requirements:* For master's and doctorate, GRE General Test. Additional exam requirements/recommendations for international students: Required—TOEFL. Electronic applications accepted. *Expenses:* Tuition, state resident: full-time $4,368; part-time $182 per semester hour. Tuition, nonresident: full-time $8,732; part-time $728 per semester hour. Required fees: $46 per semester hour.

Graduate School and University Center of the City University of New York, Graduate Studies, Program in Biochemistry, New York, NY 10016-4039. Offers PhD. *Faculty:* 22 full-time (6 women). *Students:* 59 full-time (33 women); includes 7 minority (1 African American, 1 Asian American or Pacific Islander, 5 Hispanic Americans), 41 international. Average age 28. 37 applicants, 46% accepted, 10 enrolled. In 2005, 12 degrees awarded. *Degree requirements:* For doctorate, thesis/dissertation, field experience. *Entrance requirements:* For doctorate, GRE General Test. Additional exam requirements/recommendations for international students: Required—TOEFL. *Application deadline:* For fall admission, 4/15 for domestic students; for spring admission, 11/15 for domestic students. Application fee: $125. Electronic applications accepted. *Financial support:* In 2005–06, 38 students received support, including 37 fellowships, 2 teaching assistantships; research assistantships, career-related internships or fieldwork, Federal Work-Study, institutionally sponsored loans, and tuition waivers (full and partial) also available. Financial award application deadline: 2/1; financial award applicants required to submit FAFSA. *Unit head:* Dr. Lesley Davenport, Executive Officer, 212-817-8086, Fax: 212-817-1503, E-mail: ldvnport@brooklyn.cuny.edu.

Harvard University, Graduate School of Arts and Sciences, Department of Chemistry and Chemical Biology, Cambridge, MA 02138. Offers biochemical chemistry (PhD); inorganic chemistry (PhD); organic chemistry (PhD); physical chemistry (PhD). *Students:* 188 full-time (36 women). 346 applicants, 24% accepted. In 2005, 33 doctorates awarded. *Degree requirements:* For doctorate, thesis/dissertation, cumulative exams. *Entrance requirements:* For doctorate, GRE General Test, GRE Subject Test. Additional exam requirements/recommendations for international students: Required—TOEFL. *Application deadline:* For fall admission, 12/31 for domestic students. Application fee: $60. *Expenses:* Tuition: Full-time $28,752. Full-time tuition and fees vary according to program and student level. *Financial support:* Fellowships, research assistantships, teaching assistantships, career-related internships or fieldwork, Federal Work-Study, and institutionally sponsored loans available. Financial award application deadline: 12/30. *Unit head:* Betsey Cogswell, Administrator, 617-495-5696, Fax: 617-495-5264. *Application contact:* Graduate Admissions Office, 617-496-3208.

Harvard University, Graduate School of Arts and Sciences, Division of Medical Sciences, Boston, MA 02115. Offers biological chemistry and molecular pharmacology (PhD); cell biology (PhD); genetics (PhD); microbiology and molecular genetics (PhD); pathology (PhD), including experimental pathology. *Students:* 433 full-time (210 women). In 2005, 83 doctorates awarded. *Degree requirements:* For doctorate, thesis/dissertation. *Entrance requirements:* For doctorate, GRE General Test, GRE Subject Test. Additional exam requirements/recommendations for international students: Required—TOEFL. Application fee: $60. *Expenses:* Tuition: Full-time $28,752. Full-time tuition and fees vary according to program and student level. *Financial support:* Fellowships, research assistantships, teaching assistantships, institutionally sponsored loans and tuition waivers (full) available. Financial award application deadline: 1/1. *Unit head:* Administrator, 617-432-2029. *Application contact:* Administrator, 617-432-2029.

Howard University, College of Medicine, Department of Biochemistry and Molecular Biology, Washington, DC 20059-0002. Offers biochemistry and molecular biology (PhD); biotechnology (MS). Part-time programs available. *Degree requirements:* For master's, externship; for doctorate, thesis/dissertation, comprehensive exam. *Entrance requirements:* For master's and doctorate, GRE General Test, minimum GPA of 3.0. *Faculty research:* Cellular and molecular biology of olfaction, gene regulation and expression, enzymology, NMR spectroscopy of molecular structure, hormone regulation/metabolism.

See Close-Up on page 391.

Howard University, Graduate School of Arts and Sciences, Department of Chemistry, Washington, DC 20059-0002. Offers analytical chemistry (MS, PhD); atmospheric (MS, PhD); biochemistry (MS, PhD); environmental (MS, PhD); inorganic chemistry (MS, PhD); organic chemistry (MS, PhD); physical chemistry (MS, PhD); polymer chemistry (MS, PhD). Part-time programs available. *Faculty:* 25. *Students:* 36 full-time (14 women); includes 30 minority (all African Americans), 6 international. Average age 28. 17 applicants, 6% accepted. In 2005, 5 degrees awarded. *Degree requirements:* For master's, one foreign language, thesis, teaching experience, comprehensive exam, registration; for doctorate, 2 foreign languages, thesis/dissertation, teaching experience, comprehensive exam, registration. *Entrance requirements:* For master's, GRE General Test, minimum GPA of 2.7; for doctorate, GRE General Test, minimum GPA of 3.0. *Application deadline:* For fall admission, 3/1 for domestic students; for spring admission, 11/1 for domestic students. Applications are processed on a rolling basis. Application fee: $45. *Financial support:* In 2005–06, 1 fellowship (averaging $25,000 per year), 14 research assistantships (averaging $27,000 per year), 15 teaching assistantships (averaging $24,000 per year) were awarded; institutionally sponsored loans, scholarships/grants, and unspecified assistantships also available. Financial award application deadline: 3/1; financial award applicants required to submit FAFSA. *Faculty research:* Stratospheric aerosols, liquid crystals, polymer coatings, terrestrial and extraterrestrial atmospheres, amidogen reaction. Total annual research expenditures: $1.7 million. *Unit head:* Dr. Jesse M. Nicholson, Chairman, 202-806-6900, E-mail: jnicholson@howard.edu. *Application contact:* Dr. Marlene Sherrill, Director, Student Relations and Enrollment Management, 202-806-7469, Fax: 202-462-4053.

Hunter College of the City University of New York, Graduate School, School of Arts and Sciences, Department of Chemistry, Program in Biochemistry, New York, NY 10021-5085. Offers MA. Part-time programs available. *Faculty:* 2 full-time (both women). *Students:* 1 full-time (0 women), 1 (woman) part-time; includes 1 minority (African American) Average age 30. 2 applicants, 100% accepted, 1 enrolled. In 2005, 1 degree awarded. *Degree requirements:* For master's, comprehensive exam or thesis. *Entrance requirements:* For master's, GRE General Test, 1 year of course work in chemistry, quantitative analysis, organic chemistry, physical chemistry, biology, biochemistry lecture and laboratory. Additional exam requirements/recommendations for international students: Required—TOEFL. *Application deadline:* For fall admission, 4/1 for domestic students; for spring admission, 11/1 for domestic students. Application fee: $125. *Expenses:* Tuition, state resident: full-time $6,400; part-time $270 per credit. Tuition, nonresident: part-time $500 per credit. Required fees: $50 per term. Part-time tuition and fees vary according to course load and program. *Financial support:* Teaching assistantships, Federal Work-Study, scholarships/grants, and tuition waivers (partial) available. Support available to part-time students. *Faculty research:* Protein/nucleic acid interactions, physical properties of iron-sulfur proteins, neurotransmitter receptors in *Drosophila*, requirements of DNA synthesis, oncogenes. *Application contact:* William Zlata, Director for Graduate Admissions, 212-772-4482, Fax: 212-650-3336, E-mail: admissions@hunter.cuny.edu.

Illinois Institute of Technology, Graduate College, College of Science and Letters, Department of Biological, Chemical and Physical Sciences, Biology Division, Chicago, IL 60616-3793. Offers biochemistry (MS); biology (MBS, PhD); biotechnology (MS); cell biology (MS); microbiology (MS); molecular biochemistry and biophysics (MS, PhD). Part-time and evening/weekend programs available. Postbaccalaureate distance learning degree programs offered (no on-campus study). Terminal master's awarded for partial completion of doctoral program. *Degree requirements:* For master's, thesis (for some programs), comprehensive exam; for doctorate, thesis/dissertation, comprehensive exam. *Entrance requirements:* For master's and doctorate, GRE General Test, minimum undergraduate GPA of 3.0. Additional exam requirements/recommendations for international students: Required—TOEFL (minimum score 550 paper-based; 213 computer-based). Electronic applications accepted. *Faculty research:* Protein crystallography, small angle x-ray differation of muscle, spectroscopy of multidomain proteins, structure and function of cell cycle proteins, development of anticancer drugs.

Indiana University Bloomington, Graduate School, College of Arts and Sciences, Department of Chemistry, Bloomington, IN 47405-7000. Offers analytical chemistry (PhD); biological chemistry (PhD); chemistry (MAT); inorganic chemistry (PhD); physical chemistry (PhD). PhD offered through the University Graduate School. *Faculty:* 29 full-time (2 women). *Students:* 110 full-time (38 women), 35 part-time (11 women); includes 8 minority (3 African Americans, 3 Asian Americans or Pacific Islanders, 2 Hispanic Americans), 57 international. Average age 26. In 2005, 7 master's, 20 doctorates awarded. Terminal master's awarded for partial completion of doctoral program. *Degree requirements:* For master's and doctorate, thesis/dissertation. *Entrance requirements:* For master's and doctorate, GRE General Test, GRE Subject Test. Additional exam requirements/recommendations for international students: Required—TOEFL. *Application deadline:* For fall admission, 1/15 priority date for domestic students, 12/15 priority date for international students; for spring admission, 9/1 priority date for domestic students, 9/1 priority date for international students. Applications are processed on a rolling basis. Application fee: $50 ($60 for international students). *Expenses:* Tuition, state resident: full-time $5,437; part-time $227 per credit hour. Tuition, nonresident: full-time $15,836; part-time $660 per credit hour. Required fees: $821. Tuition and fees vary according to campus/location and program. *Financial support:* In 2005–06, 23 fellowships with full tuition reimbursements, 57 research assistantships with full tuition reimbursements, 78 teaching assistantships with full tuition reimbursements were awarded; Federal Work-Study and institutionally sponsored loans also available. *Faculty research:* Synthesis of complex natural products, organic reaction mechanisms, organic electrochemistry, transitive-metal chemistry, solid-state and surface chemistry. Total annual research expenditures: $7.7 million. *Unit head:* Dr. David Clemmer, Chairperson, 812-855-2268. *Application contact:* Dr. Jack K. Crandall, Chairperson of Admissions, 812-855-2068, Fax: 812-855-8300, E-mail: chemgrad@indiana.edu.

Indiana University Bloomington, Graduate School, College of Arts and Sciences, Interdepartmental Program in Biochemistry, Bloomington, IN 47405-7000. Offers PhD. Program offered through the University Graduate School. *Students:* 102 full-time (54 women), 76 part-time (40 women); includes 21 minority (8 African Americans, 4 Asian Americans or Pacific Islanders, 9 Hispanic Americans), 34 international. Average age 27. In 2005, 3 degrees awarded. *Degree requirements:* For doctorate, thesis/dissertation. *Entrance requirements:* For doctorate, GRE General Test, GRE Subject Test (biochemistry or chemistry), BA or BE in biochemistry or chemistry. Additional exam requirements/recommendations for international students: Required—TOEFL. *Application deadline:* For fall admission, 1/15 priority date for domestic students, 12/15 priority date for international students; for spring admission, 9/1 priority date for domestic students, 9/1 priority date for international students. Application fee: $45 ($55 for international students). *Expenses:* Tuition, state resident: full-time $5,437; part-time $227 per credit hour. Tuition, nonresident: full-time $15,836; part-time $660 per credit hour. Required fees: $821. Tuition and fees vary according to campus/location and program. *Financial support:* Fellowships with full tuition reimbursements, research assistantships with full tuition reimbursements, teaching assistantships with full tuition reimbursements available. *Faculty research:* Biological membranes, enzymology, bioanalytical chemistry, photosynthesis. *Unit head:* Dr. Carl Bauer, Director, 812-855-1520. *Application contact:* Dr. Jack K. Crandall, Chairperson of Admissions, 812-855-2068, Fax: 812-855-8300, E-mail: chemgrad@indiana.edu.

Indiana University–Purdue University Indianapolis, Indiana University School of Medicine, Department of Biochemistry and Molecular Biology, Indianapolis, IN 46202-2896. Offers PhD, MD/MS, MD/PhD. *Faculty:* 17 full-time (4 women). *Students:* 23 full-time (12 women), 21 part-time (17 women); includes 4 minority (2 African Americans, 1 Asian American or Pacific Islander, 1 Hispanic American), 17 international. Average age 28. In 2005, 5 doctorates awarded. Terminal master's awarded for partial completion of doctoral program. *Degree requirements:* For doctorate, thesis/dissertation. *Entrance requirements:* For doctorate, GRE General Test, GRE Subject Test (recommended), previous course work in organic chemistry. *Application deadline:* For fall admission, 1/15 for domestic students. Applications are processed on a rolling basis. Application fee: $50 ($60 for international students). *Expenses:* Tuition, state resident: full-time $5,159; part-time $215 per credit hour. Tuition, nonresident: full-time $14,890; part-time $620 per credit hour. Required fees: $614. Tuition and fees vary according to campus/location and program. *Financial support:* Fellowships with tuition reimbursements, research assistantships with tuition reimbursements, teaching assistantships, Federal Work-Study, institutionally sponsored loans, scholarships/grants, and tuition waivers (partial) available. Support available to part-time students. Financial award application deadline: 2/1. *Faculty research:* Metabolic regulation, enzymology, peptide and protein chemistry, cell biology, signal transduction. *Unit head:* Dr. Zhong-Yin Zhang, Chairman, 317-274-7151. *Application contact:* Dr. David W. Allmann, Chairperson, Admissions Committee, 317-274-4096, Fax: 317-274-4686, E-mail: dallman@iupui.edu.

See Close-Up on page 393.

Iowa State University of Science and Technology, Graduate College, College of Agriculture and College of Liberal Arts and Sciences, Department of Biochemistry, Biophysics, and Molecular Biology, Ames, IA 50011. Offers biochemistry (MS, PhD); biophysics (MS, PhD); genetics (MS, PhD); molecular, cellular, and developmental biology (MS, PhD); toxicology (MS, PhD). *Faculty:* 24 full-time, 2 part-time/adjunct. *Students:* 97 full-time (42 women), 2 part-time (1 woman); includes 2 minority (1 African American, 1 Asian American or Pacific Islander), 68 international. 28 applicants, 46% accepted, 13 enrolled. In 2005, 9 master's, 8 doctorates awarded. *Degree requirements:* For master's and doctorate, thesis/dissertation. *Entrance requirements:* For master's and doctorate, GRE General Test. Additional exam requirements/recommendations for international students: Required—TOEFL (paper score 550; computer score 213) or IELTS (score 6.5). *Application deadline:* For fall admission, 2/1 priority date for domestic students, 2/1 priority date for international students. Application fee: $30 ($70 for international students). Electronic applications accepted. *Expenses:* Tuition, state resident: full-time $6,410. Tuition, nonresident: full-time $16,422. Tuition and fees vary according to program. *Financial support:* In 2005–06, 91 research assistantships with full tuition reimbursements (averaging $16,290 per year), 2 teaching assistantships with full and partial tuition reimbursements (averaging $14,455 per year) were awarded; scholarships/grants, health care benefits, and unspecified assistantships also available. *Unit head:* Dr. Alan M. Myers, Chair, 515-294-6116, E-mail: biochem@iastate.edu.

See Close-Up on page 395.

The Johns Hopkins University, Bloomberg School of Public Health, Department of Biochemistry and Molecular Biology, Baltimore, MD 21205. Offers biochemistry and molecular biology (MHS, Sc M). *Faculty:* 26 full-time (9 women). *Students:* 69 full-time (44 women), 4 part-time (3 women); includes 17 minority (2 African Americans, 1 American Indian/Alaska Native, 9 Asian Americans or Pacific Islanders, 5 Hispanic Americans), 10 international. Average age 24. 174 applicants, 32% accepted, 39 enrolled. In 2005, 16 master's, 3 doctorates awarded. *Median time to degree:* Of those who began their doctoral program in fall 1997, 69% received their degree in 8 years or less. *Degree requirements:* For master's, thesis, registration; for doctorate, thesis/dissertation, oral and written exams, comprehensive exam, registration. *Entrance requirements:* For master's, MCAT or GRE, 3 letters of recommendation, curriculum vitae; for doctorate, GRE General Test, 3 letters of recommendation, curriculum vitae. Additional exam requirements/recommendations for international students: Required—TOEFL (minimum score 600 paper-based; 250 computer-based). *Application deadline:* For fall admission, 11/10 for domestic students. Applications are processed on a rolling basis. Application fee: $45. Electronic applications accepted. *Expenses:* Tuition: Full-time $30,960. Tuition and fees vary according to degree level and program. *Financial support:* In 2005–06, 7 fellowships with tuition reimbursements (averaging $25,000 per year), 30 research assistantships with tuition reimbursements (averaging $25,000 per year), 4 teaching assistantships (averaging $250 per year) were awarded; Federal Work-Study, institutionally sponsored loans, scholarships/grants, health care benefits, and stipends also available. Financial award application deadline: 3/15. *Faculty research:* DNA replication, repair, structure, carcinogenesis, protein structure, enzyme catalysts, reproductive biology. Total annual research expenditures: $3.7 million. *Unit head:* Dr. Roger McMacken, Chairman, 410-955-3671, Fax: 410-955-2926, E-mail: rmcmacke@jhsph.edu. *Application contact:* Gerry Graziano, Admissions Coordinator, 410-955-3672, Fax: 410-955-2926, E-mail: ggrazian@jhsph.edu.

The Johns Hopkins University, National Institutes of Health Sponsored Programs, Department of Biology, Baltimore, MD 21218-2699. Offers biochemistry (PhD); biophysics (PhD); cell biology (PhD); developmental biology (PhD); genetic biology (PhD); molecular biology (PhD). *Faculty:* 25 full-time (4 women). *Students:* 126 full-time (72 women); includes 36 minority (3 African Americans, 1 American Indian/Alaska Native, 21 Asian Americans or Pacific Islanders, 11 Hispanic Americans), 19 international. 282 applicants, 26% accepted, 36 enrolled. In 2005, 15 degrees awarded. *Median time to degree:* Of those who began their doctoral program in fall 1997, 81.2% received their degree in 8 years or less. *Degree requirements:* For doctorate, thesis/dissertation, comprehensive exam, registration. *Entrance requirements:* For doctorate, GRE General Test. Additional exam requirements/recommendations for international students: Required—TOEFL (minimum score 600 paper-based; 250 computer-based), TWE, TSE. *Application deadline:* For fall admission, 12/15 for domestic students. Application fee: $60. *Expenses:* Tuition: Full-time $30,960. Tuition and fees vary according to degree level and program. *Financial support:* In 2005–06, 24 fellowships (averaging $23,000 per year), 93 research assistantships (averaging $23,000 per year), 22 teaching assistantships (averaging $23,000 per year) were awarded; Federal Work-Study, institutionally sponsored loans, scholarships/grants, traineeships, health care benefits, tuition waivers (partial), and unspecified assistantships also available. Financial award application deadline: 4/15; financial award applicants required to submit FAFSA. *Faculty research:* Protein and nucleic acid biochemistry and biophysical chemistry, molecular biology and development. Total annual research expenditures: $11.2 million. *Unit head:* Dr. Allen Shearn, Chair, 410-516-4693, Fax: 410-516-5213, E-mail: bio_cals@jhu.edu. *Application contact:* Joan Miller, Academic Affairs Manager, 410-516-5502, Fax: 410-516-5213, E-mail: joan@jhu.edu.

See Close-Up on page 155.

The Johns Hopkins University, School of Medicine, Graduate Programs in Medicine, Department of Biological Chemistry, Baltimore, MD 21205. Offers PhD. *Faculty:* 18 full-time (5 women), 1 part-time/adjunct (0 women). *Students:* 38 full-time (15 women), 37 international. 67 applicants, 7% accepted, 3 enrolled. In 2005, 5 degrees awarded. *Median time to degree:* Of those who began their doctoral program in fall 1997, 97% received their degree in 8 years or less. *Degree requirements:* For doctorate, thesis/dissertation. *Entrance requirements:* For doctorate, GRE General Test. Additional exam requirements/recommendations for international students: Required—TOEFL. *Application deadline:* For winter admission, 1/15 for domestic students. Application fee: $75. Electronic applications accepted. *Expenses:* Tuition: Full-time $30,960. Tuition and fees vary according to degree level and program. *Financial support:* Health care benefits and tuition waivers (full) available. Financial award application deadline: 1/1. *Faculty research:* Cell adhesion, genetics, signal transduction and RNA metabolism, enzyme structure and function, gene expression. *Unit head:* Dr. Denise Montell, Professor, 410-614-2016, Fax: 410-614-8375, E-mail: dmontell@jhmi.edu. *Application contact:* Lina Marino, Program Coordinator, 410-614-2976, Fax: 410-614-8375, E-mail: lmarino2@bsjhmi.edu.

The Johns Hopkins University, School of Medicine, Graduate Programs in Medicine, Program in Biochemistry, Cellular and Molecular Biology, Baltimore, MD 21205. Offers PhD. *Faculty:* 91 full-time (25 women), 11 part-time/adjunct (5 women). *Students:* 155 full-time (75 women); includes 19 minority (6 African Americans, 12 Asian Americans or Pacific Islanders, 1 Hispanic American), 51 international. Average age 25. 358 applicants, 18% accepted, 20 enrolled. In 2005, 19 doctorates awarded. *Median time to degree:* Of those who began their doctoral program in fall 1997, 100% received their degree in 8 years or less. *Degree requirements:* For doctorate, thesis/dissertation, comprehensive exam. *Entrance requirements:* For doctorate, GRE General Test. Additional exam requirements/recommendations for international students: Required—TOEFL. *Application deadline:* For winter admission, 1/10 for domestic students. Applications are processed on a rolling basis. Application fee: $90. Electronic applications accepted. *Expenses:* Tuition: Full-time $30,960. Tuition and fees vary according to degree level and program. *Financial support:* In 2005–06, 3 fellowships with partial tuition reimbursements (averaging $32,000 per year), 154 research assistantships with full tuition reimbursements (averaging $24,600 per year) were awarded; traineeships and tuition waivers (full) also available. Financial award application deadline: 12/31. *Faculty research:* Developmental biology, genomics/proteomics, protein targeting, signal transduction, structural biology. *Unit head:* Dr. Craig Montell, Director, 410-955-3466, Fax: 410-614-8842, E-mail: cmontell@jhmi.edu. *Application contact:* Dr. Jeff Corden, Admissions Director, 410-955-3506, Fax: 410-614-8842, E-mail: jcorden@jhmi.edu.

Kansas State University, Graduate School, College of Arts and Sciences, Department of Biochemistry, Manhattan, KS 66506. Offers MS, PhD. Part-time programs available. *Faculty:* 12 full-time (2 women), 1 part-time/adjunct (0 women). *Students:* 24 full-time (9 women), 2 part-time (1 woman); includes 1 minority (Asian American or Pacific Islander), 20 international. 32 applicants, 41% accepted, 8 enrolled. In 2005, 8 degrees awarded. *Degree requirements:* For master's and doctorate, thesis/dissertation. *Entrance requirements:* For master's, GRE General Test, minimum GPA of 3.0 for junior and senior year; for doctorate, GRE General Test, minimum undergraduate GPA of 3.0 or an excellent postgraduate record. Additional exam requirements/recommendations for international students: Required—TOEFL (minimum score 550 paper-based; 213 computer-based). *Application deadline:* For fall admission, 1/10 priority date for domestic students, 1/10 priority date for international students; for spring admission, 10/1 for domestic students, 10/1 for international students. Applications are processed on a rolling basis. Application fee: $30 ($55 for international students). *Expenses:* Tuition, state resident: full-time $5,160; part-time $215 per credit hour. Tuition, nonresident: full-time $12,816; part-time $534 per credit hour. Required fees: $564. *Financial support:* In 2005–06, 26 research assistantships (averaging $14,648 per year), 11 teaching assistantships with partial tuition reimbursements (averaging $7,339 per year) were awarded; fellowships, Federal Work-Study, institutionally sponsored loans, and scholarships/grants also available. Support available to part-time students. Financial award application deadline: 3/1; financial award applicants required to submit FAFSA. *Faculty research:* Protein structure/function, insect biochemistry, cellular signaling cascades, environmental biochemistry, biochemistry of vision. Total annual research expenditures: $2.2 million. *Unit head:* Michael Kanost, Head, 785-532-6964, Fax:

Biochemistry

Kansas State University (continued)

785-532-7278, E-mail: kanost@ksu.edu. *Application contact:* Lawrence C. Davis, Director, 785-532-6121, Fax: 785-532-7278, E-mail: ldavis@ksu.edu.

Kansas State University, Graduate School, College of Arts and Sciences, Department of Chemistry, Manhattan, KS 66506. Offers analytical chemistry (MS); biological chemistry (MS); chemistry (PhD); inorganic chemistry (MS); materials chemistry (MS); organic chemistry (MS); physical chemistry (MS). *Faculty:* 14 full-time (1 woman). *Students:* 65 full-time (21 women), 5 part-time (2 women); includes 1 minority (African American), 48 international. 63 applicants, 67% accepted, 22 enrolled. In 2005, 3 master's, 5 doctorates awarded. Terminal master's awarded for partial completion of doctoral program. *Degree requirements:* For master's and doctorate, thesis/dissertation. *Entrance requirements:* For master's and doctorate, GRE, minimum GPA of 3.0. Additional exam requirements/recommendations for international students: Required—TOEFL (minimum score 550 paper-based; 213 computer-based). *Application deadline:* For fall admission, 2/1 priority date for domestic students, 2/1 priority date for international students; for spring admission, 10/1 for domestic students, 8/1 for international students. Applications are processed on a rolling basis. Application fee: $30 ($55 for international students). *Expenses:* Tuition, state resident: full-time $5,160; part-time $215. Tuition, nonresident: full-time $12,816; part-time $534. International tuition: $12,816 full-time. Required fees: $564. *Financial support:* In 2005–06, 30 research assistantships (averaging $11,956 per year), 26 teaching assistantships with full tuition reimbursements (averaging $11,945 per year) were awarded; fellowships, institutionally sponsored loans and scholarships/grants also available. Support available to part-time students. Financial award application deadline: 3/1; financial award applicants required to submit FAFSA. *Faculty research:* Nanotechnologies, functional materials, bio-organic and bio-physical processes, sensors and separations, synthesis and synthetic methods. Total annual research expenditures: $2.2 million. *Unit head:* Eric Maatta, Head, 785-532-6665, Fax: 785-532-6666, E-mail: eam@ksu.edu. *Application contact:* Christer Aakeröy, Director, 785-532-6096, Fax: 785-532-6666, E-mail: aakeroy@ksu.edu.

Kent State University, College of Arts and Sciences, Department of Chemistry, Kent, OH 44242-0001. Offers analytical chemistry (MS, PhD); biochemistry (MS, PhD); chemistry (MA, MS, PhD); inorganic chemistry (MS, PhD); organic chemistry (MS, PhD); physical chemistry (MS, PhD). Terminal master's awarded for partial completion of doctoral program. *Degree requirements:* For master's and doctorate, thesis/dissertation, comprehensive exam, registration. *Entrance requirements:* For master's and doctorate, placement exam, GRE General Test, GRE Subject Test (recommended), minimum GPA of 2.75. Additional exam requirements/recommendations for international students: Required—TOEFL (minimum score 575 paper-based; 230 computer-based). Electronic applications accepted. *Faculty research:* Biological chemistry, materials chemistry, molecular spectroscopy.

Laurentian University, School of Graduate Studies and Research, Programme in Chemistry and Biochemistry, Sudbury, ON P3E 2C6, Canada. Offers M Sc. Part-time programs available. *Degree requirements:* For master's, thesis or alternative. *Entrance requirements:* For master's, honors degree with minimum second class. *Faculty research:* Cell cycle checkpoints, kinetic modeling, toxicology to metal stress, quantum chemistry, biogeochemistry metal speciation.

Lehigh University, College of Arts and Sciences, Department of Biological Sciences, Bethlehem, PA 18015-3094. Offers biochemistry (PhD); integrative biology (PhD); molecular biology (PhD). Postbaccalaureate distance learning degree programs offered (no on-campus study). *Faculty:* 22 full-time (9 women). *Students:* 31 full-time (21 women), 51 part-time (29 women); includes 2 minority (1 Asian American or Pacific Islander, 1 Hispanic American), 6 international. 65 applicants, 29% accepted. In 2005, 3 doctorates awarded. *Median time to degree:* Of those who began their doctoral program in fall 1997, 100% received their degree in 8 years or less. *Degree requirements:* For doctorate, thesis/dissertation, comprehensive exam. *Entrance requirements:* For doctorate, GRE General Test. Additional exam requirements/recommendations for international students: Required—TOEFL, TSE. *Application deadline:* For fall admission, 1/15 for domestic students. Applications are processed on a rolling basis. Application fee: $50. Electronic applications accepted. *Financial support:* In 2005–06, 30 students received support, including 4 fellowships with tuition reimbursements available (averaging $20,000 per year), 6 research assistantships with tuition reimbursements available (averaging $20,000 per year), 16 teaching assistantships with tuition reimbursements available (averaging $20,000 per year); scholarships/grants, tuition waivers (full and partial), and unspecified assistantships also available. Financial award application deadline: 1/31. *Faculty research:* Gene expression, cytoskeleton and cell structure, cell cycle and growth regulation, neuroscience, animal behavior. *Unit head:* Dr. Neal G. Simon, Chairperson, 610-758-3680, Fax: 610-758-4004. *Application contact:* Dr. Michael R. Kuchka, Graduate Coordinator, 610-758-3687, Fax: 610-758-4004, E-mail: mrk5@lehigh.edu.

Loma Linda University, School of Medicine, Department of Biochemistry/Microbiology, Loma Linda, CA 92350. Offers MS, PhD. Part-time programs available. *Faculty:* 37 full-time (7 women), 5 part-time/adjunct (1 woman). *Students:* 6 full-time (4 women), 6 part-time (3 women); includes 4 Asian Americans or Pacific Islanders, 2 Hispanic Americans, 3 international. *Degree requirements:* For master's, thesis or alternative; for doctorate, thesis/dissertation. *Entrance requirements:* For master's and doctorate, GRE General Test. *Application support:* Tuition waivers (full and partial) available. Support available to part-time students. *Faculty research:* Physical chemistry of macromolecules, biochemistry of endocrine system, biochemical mechanism of bone volume regulation. *Unit head:* Dr. Lawrence C. Sowers, Coordinator, 909-824-4527.

Louisiana State University and Agricultural and Mechanical College, Graduate School, College of Basic Sciences, Department of Biological Sciences, Baton Rouge, LA 70803. Offers biochemistry (MS, PhD); biological science (MS, PhD). Part-time programs available. *Faculty:* 79 full-time (13 women). *Students:* 145 full-time (74 women), 8 part-time (5 women); includes 18 minority (15 African Americans, 2 Asian Americans or Pacific Islanders, 1 Hispanic American), 51 international. Average age 29. 185 applicants, 22% accepted, 145 enrolled. In 2005, 3 master's, 17 doctorates awarded. Terminal master's awarded for partial completion of doctoral program. *Degree requirements:* For doctorate, thesis/dissertation. *Entrance requirements:* For master's and doctorate, GRE General Test, minimum GPA of 3.0. Additional exam requirements/recommendations for international students: Required—TOEFL (minimum score 550 paper-based; 213 computer-based). *Application deadline:* For fall admission, 5/15 for domestic students, 5/15 for international students; for spring admission, 10/15 for domestic students, 10/15 for international students. Applications are processed on a rolling basis. Application fee: $25. Electronic applications accepted. *Financial support:* In 2005–06, 17 fellowships with full and partial tuition reimbursements (averaging $18,054 per year), 48 research assistantships with full and partial tuition reimbursements (averaging $21,414 per year), 74 teaching assistantships with full and partial tuition reimbursements (averaging $18,451 per year) were awarded; Federal Work-Study, institutionally sponsored loans, and unspecified assistantships also available. Support available to part-time students. Financial award applicants required to submit FAFSA. Total annual research expenditures: $8.1 million. *Unit head:* Dr. Terry Bricker, Chair, 225-578-2601, Fax: 225-578-2597, E-mail: btbric@lsu.edu. *Application contact:* Dr. Thomas S. Moore, Associate Chairman, 225-578-1556, Fax: 225-578-7299, E-mail: gradoff@lsu.edu.

See Close-Up on page 161.

Louisiana State University Health Sciences Center at Shreveport, Department of Biochemistry and Molecular Biology, Shreveport, LA 71130-3932. Offers MS, PhD, MD/PhD. *Faculty:* 16 full-time (3 women), 5 part-time/adjunct (0 women). *Students:* 38 full-time (15 women); includes 14 minority (3 African Americans, 11 Asian Americans or Pacific Islanders). Average age 26. 50 applicants, 10% accepted, 5 enrolled. In 2005, 2 master's, 1 doctorate awarded. *Median time to degree:* Of those who began their doctoral program in fall 1997,

100% received their degree in 8 years or less. *Degree requirements:* For master's and doctorate, thesis/dissertation. *Entrance requirements:* For master's and doctorate, GRE General Test. Additional exam requirements/recommendations for international students: Required—TOEFL. *Application deadline:* For fall admission, 4/15 for domestic students. Applications are processed on a rolling basis. Application fee: $30. *Financial support:* In 2005–06, 38 students received support, including research assistantships (averaging $18,000 per year); institutionally sponsored loans also available. *Faculty research:* Metabolite transport, regulation of translation and transcription, prokaryotic molecular genetics, cell matrix biochemistry, yeast molecular genetics, oncogenes. Total annual research expenditures: $160,000. *Unit head:* Dr. Robert E. Rhoads, Head, 318-675-5160, Fax: 318-675-5180. *Application contact:* Dr. Robert L. Smith, Associate Professor and Director of Graduate Studies, 318-675-5168, Fax: 318-675-5180, E-mail: rsmith2@lsuhsc.edu.

See Close-Up on page 163.

Loyola University Chicago, Graduate School, Program in Molecular and Cellular Biochemistry, Chicago, IL 60611-2196. Offers biochemistry (MS, PhD). *Faculty:* 25 full-time (8 women). *Students:* 15 full-time (9 women), 1 (woman) part-time, 5 international. Average age 26. 44 applicants, 11% accepted, 4 enrolled. In 2005, 1 master's, 5 doctorates awarded. *Degree requirements:* For master's, oral and written reports; for doctorate, thesis/dissertation, comprehensive exam. *Entrance requirements:* For master's and doctorate, GRE General Test. *Application deadline:* For fall admission, 5/15 for domestic students. Applications are processed on a rolling basis. Application fee: $0. Electronic applications accepted. *Expenses:* Tuition: Full-time $11,610; part-time $645 per credit. Required fees: $55 per semester. *Financial support:* In 2005–06, 6 students received support, including 6 fellowships with full tuition reimbursements available, 9 research assistantships with full tuition reimbursements available; Federal Work-Study, institutionally sponsored loans, and scholarships/grants also available. Financial award application deadline: 5/15. *Faculty research:* Molecular oncology, molecular neurochemical mechanisms of brain development and alcohol addiction, biochemistry of RNA and protein synthesis and intracellular protein degradation, developmentally regulated genes, neurotransmitters and cell-cell interactions. *Unit head:* Dr. Mary D. Manteuffel, Chief, Division of Molecular and Cellular Biochemistry, 708-216-3370, Fax: 708-216-8523, E-mail: mmanteu@lumc.edu. *Application contact:* Ashyia D. Paul, Administrative Secretary, 708-216-3360, Fax: 708-216-8523, E-mail: apaul@lumc.edu.

See Close-Up on page 397.

Massachusetts Institute of Technology, School of Science, Department of Chemistry, Cambridge, MA 02139-4307. Offers biological chemistry (PhD, Sc D); inorganic chemistry (PhD, Sc D); organic chemistry (PhD, Sc D); physical chemistry (PhD, Sc D). *Faculty:* 30 full-time (6 women). *Students:* 242 full-time (84 women), 3 part-time (2 women); includes 30 minority (5 African Americans, 1 American Indian/Alaska Native, 19 Asian Americans or Pacific Islanders, 5 Hispanic Americans), 77 international. Average age 26. 476 applicants, 27% accepted, 54 enrolled. In 2005, 48 doctorates awarded. *Degree requirements:* For doctorate, thesis/dissertation, comprehensive exam. *Entrance requirements:* For doctorate, GRE General Test. Additional exam requirements/recommendations for international students: Required—TOEFL (minimum score 577 paper-based; 233 computer-based). *Application deadline:* For fall admission, 12/15 for domestic students, 12/15 for international students. Application fee: $70. Electronic applications accepted. *Expenses:* Tuition: Full-time $32,100. Required fees: $200. Part-time tuition and fees vary according to course load. *Financial support:* In 2005–06, 210 students received support, including 36 fellowships with tuition reimbursements available (averaging $29,141 per year), 143 research assistantships with tuition reimbursements available (averaging $24,550 per year), 92 teaching assistantships with tuition reimbursements available (averaging $25,155 per year); career-related internships or fieldwork, Federal Work-Study, institutionally sponsored loans, scholarships/grants, health care benefits, and unspecified assistantships also available. *Faculty research:* Synthetic organic chemistry, enzymatic reaction mechanisms, inorganic and organometallic spectroscopy, high resolution NMR spectroscopy. Total annual research expenditures: $21.5 million. *Unit head:* Prof. Timothy Swager, Department Head, 617-253-1801, E-mail: tswager@mit.edu. *Application contact:* Susan Brighton, Graduate Administrator, 617-253-1845, Fax: 617-258-0241, E-mail: brighton@mit.edu.

Mayo Graduate School, Graduate Programs in Biomedical Sciences, Programs in Biochemistry, Structural Biology, Cell Biology, and Genetics, Rochester, MN 55905. Offers biochemistry and structural biology (PhD); cell biology and genetics (PhD); molecular biology (PhD). *Degree requirements:* For doctorate, oral defense of dissertation, qualifying oral and written exam. *Entrance requirements:* For doctorate, GRE, 1 year of chemistry, biology, calculus, and physics. Additional exam requirements/recommendations for international students: Required—TOEFL. Electronic applications accepted. *Faculty research:* Gene structure and function, membranes and receptors/cytoskeleton, oncogenes and growth factors, protein structure and function, steroid hormonal action.

See Close-Up on page 399.

McGill University, Faculty of Graduate and Postdoctoral Studies, Faculty of Medicine, Department of Biochemistry, Montreal, QC H3G 1Y6, Canada. Offers M Sc, PhD. *Degree requirements:* For master's and doctorate, thesis/dissertation. *Entrance requirements:* For master's, GRE Subject Test, minimum GPA of 3.2; for doctorate, GRE General Test, GRE Subject Test, minimum GPA of 3.2. Additional exam requirements/recommendations for international students: Required—TOEFL. *Faculty research:* Signal transduction, apoptosis, structural biology, membrane biochemistry, molecular genetics.

McMaster University, School of Graduate Studies, Faculty of Science, Department of Biochemistry and Biomedical Sciences, Hamilton, ON L8S 4M2, Canada. Offers M Sc, PhD. Terminal master's awarded for partial completion of doctoral program. *Degree requirements:* For master's, thesis/dissertation; for doctorate, thesis/dissertation, comprehensive exam. *Entrance requirements:* For master's and doctorate, minimum B+ average. Additional exam requirements/recommendations for international students: Required—TOEFL (minimum score 550 paper-based; 213 computer-based). *Faculty research:* Molecular and cell biology, biomolecular structure and function, molecular pharmacology and toxicology.

Medical College of Georgia, School of Graduate Studies, Department of Biochemistry and Molecular Biology, Augusta, GA 30912-1500. Offers PhD. *Faculty:* 10 full-time (1 woman). *Students:* 9 full-time (3 women); includes 1 minority (African American), 5 international. Average age 30. In 2005, 3 degrees awarded. *Degree requirements:* For doctorate, thesis/dissertation. *Entrance requirements:* For doctorate, GRE General Test. Additional exam requirements/recommendations for international students: Required—TOEFL. *Application deadline:* For fall admission, 6/30 for domestic students, 4/15 for international students. Applications are processed on a rolling basis. Application fee: $30. *Financial support:* In 2005–06, 2 students received support, including 8 research assistantships with partial tuition reimbursements available (averaging $22,500 per year); institutionally sponsored loans and scholarships/grants also available. Support available to part-time students. Financial award application deadline: 5/31; financial award applicants required to submit FAFSA. *Faculty research:* Bacterial pathogenesis, eye diseases, vitamins and amino acid transporters, transcriptional control and signal transduction, tumor biology. *Unit head:* Dr. Vadivel Ganapathy, Chair, 706-721-7652, Fax: 706-721-9947, E-mail: vganapat@mail.mcg.edu. *Application contact:* Dr. Darren D. Browning, Program Director, 706-721-9526, Fax: 706-721-6608, E-mail: dbrowning@mail.mcg.edu.

Medical College of Wisconsin, Graduate School of Biomedical Sciences, Department of Biochemistry, Milwaukee, WI 53226-0509. Offers MS, PhD. *Degree requirements:* For doctorate, thesis/dissertation, comprehensive exam, registration. *Entrance requirements:* For doctorate, GRE General Test, GRE Subject Test. Additional exam requirements/recommendations for international students: Required—TOEFL. *Faculty research:* Enzymology, macromolecular structure and synthesis, nucleic acids, molecular and cell biology.

Biochemistry

Medical University of South Carolina, College of Graduate Studies, Program in Biochemistry and Molecular Biology, Charleston, SC 29425-0002. Offers Pharm D, MS, PhD, DMD/PhD, MD/PhD. *Faculty:* 22 full-time (6 women). *Students:* 11 full-time (3 women); includes 1 minority (Hispanic American), 3 international. Average age 28. 181 applicants, 34% accepted, 43 enrolled. In 2005, 2 degrees awarded. *Degree requirements:* For doctorate, thesis/dissertation, teaching and research seminar, oral and written exams. *Entrance requirements:* For doctorate, GRE General Test, interview. Additional exam requirements/recommendations for international students: Required—TOEFL (minimum score 600 paper-based; 250 computer-based). *Application deadline:* For fall admission, 1/15 priority date for domestic students, 1/15 priority date for international students. Applications are processed on a rolling basis. Application fee: $0 ($75 for international students). Electronic applications accepted. *Financial support:* In 2005–06, 4 students received support, including fellowships with partial tuition reimbursements available (averaging $21,000 per year); Federal Work-Study and scholarships/grants also available. Financial award application deadline: 3/15; financial award applicants required to submit FAFSA. *Faculty research:* Lipid biochemistry, DNA replication, nucleic acid/protein interaction, proteases, signal. *Unit head:* Dr. Y. A. Hannun, Chairman, 843-792-9318, Fax: 843-792-6590, E-mail: hannun@musc.edu. *Application contact:* Cheryl Brown, 843-792-4620, Fax: 843-792-4645, E-mail: brownche@musc.edu.

Meharry Medical College, School of Graduate Studies, Department of Biochemistry, Nashville, TN 37208-9989. Offers PhD. *Degree requirements:* For doctorate, thesis/dissertation, comprehensive exam. *Entrance requirements:* For doctorate, GRE. *Faculty research:* Regulation of metabolism, enzymology, signal transduction, physical biochemistry.

Memorial University of Newfoundland, School of Graduate Studies, Department of Biochemistry, St. John's, NL A1C 5S7, Canada. Offers biochemistry (M Sc, PhD); food science (M Sc, PhD). Part-time programs available. *Students:* 26 full-time (14 women), 1 (woman) part-time, 13 international. 19 applicants, 21% accepted, 2 enrolled. In 2005, 4 master's, 1 doctorate awarded. *Degree requirements:* For master's, thesis; for doctorate, thesis/dissertation, oral defense of thesis, comprehensive exam. *Entrance requirements:* For master's, 2nd class degree in related field; for doctorate, M Sc. *Application deadline:* For fall admission, 3/1 for domestic students, 3/1 for international students. For winter admission, 7/1 for domestic students; for spring admission, 11/1 for domestic students. Applications are processed on a rolling basis. Application fee: $40 Canadian dollars. Electronic applications accepted. *Expenses:* Tuition: Part-time $733 per term. Tuition and fees vary according to degree level and program. *Financial support:* Fellowships, research assistantships, teaching assistantships available. *Faculty research:* Toxicology, cell and molecular biology, food engineering, marine biotechnology, lipid biology. Total annual research expenditures: $1.1 million. *Unit head:* Dr. Martin Mulligan, Head, 709-737-8530, E-mail: mulligan@mun.ca. *Application contact:* Dr. Sukhinder Kaur, Graduate Officer, 709-737-8529, Fax: 709-737-2422, E-mail: biochem@mun.ca.

Miami University, Graduate School, College of Arts and Sciences, Department of Chemistry and Biochemistry, Oxford, OH 45056. Offers analytical chemistry (MS, PhD); biochemistry (MS, PhD); chemical education (MS, PhD); chemistry (MS, PhD); inorganic chemistry (MS, PhD); organic chemistry (MS, PhD); physical chemistry (MS, PhD). Part-time programs available. *Degree requirements:* For master's, thesis, final exam; for doctorate, thesis/dissertation, final exams, comprehensive exam. *Entrance requirements:* For master's, minimum undergraduate GPA of 3.0 during previous 2 years or 2.75 overall; for doctorate, minimum undergraduate GPA of 2.75, 3.0 graduate. Additional exam requirements/recommendations for international students: Required—TOEFL (minimum score 550 paper-based; 213 computer-based), TWE(minimum score 4). Electronic applications accepted.

Michigan State University, College of Human Medicine and The Graduate School, Graduate Programs in Human Medicine, East Lansing, MI 48824. Offers biochemistry and molecular biology (MS, PhD); bioethics, humanities, and society (MA); epidemiology (MS, PhD); microbiology (MS); microbiology and molecular genetics (PhD); pharmacology and toxicology (MS, PhD); physiology (MS, PhD). *Students:* 57 full-time (35 women), 22 part-time (17 women); includes 11 minority (6 African Americans, 5 Asian Americans or Pacific Islanders), 32 international. Average age 30. *Entrance requirements:* Additional exam requirements/recommendations for international students: Required—TOEFL (minimum score 550 paper-based; 213 computer-based). *Expenses:* Tuition, state resident: part-time $330 per credit hour. Tuition, nonresident: part-time $685 per credit hour. Tuition and fees vary according to program. *Financial support:* Institutionally sponsored loans available. *Unit head:* Dr. Nigel S. Paneth, Associate Dean, Research and Graduate Studies, 517-432-4789, E-mail: paneth@msu.edu.

Michigan State University, College of Osteopathic Medicine and The Graduate School, Graduate Studies in Osteopathic Medicine, East Lansing, MI 48824. Offers biochemistry and molecular biology (MS, PhD); microbiology (MS); microbiology and molecular genetics (PhD); pharmacology and toxicology (MS, PhD), including pharmacology and toxicology, pharmacology and toxicology-environmental toxicology (PhD); *physiology (MS, PhD). *Students:* 2 full-time (1 woman), 1 international. Average age 27. *Expenses:* Tuition, state resident: part-time $330 per credit hour. Tuition, nonresident: part-time $685 per credit hour. Tuition and fees vary according to program. *Unit head:* Dr. Veronica M. Maher, Associate Dean, Graduate Studies, 517-353-7785, Fax: 517-353-9004, E-mail: maher@msu.edu. *Application contact:* Kathie Schafer, Director of Admissions, 517-353-7740, Fax: 517-355-3296, E-mail: comadm@com.msu.edu.

Michigan State University, The Graduate School, College of Agriculture and Natural Resources, MSU-DOE Plant Research Laboratory, East Lansing, MI 48824. Offers biochemistry and molecular biology (PhD); cellular and molecular biology (PhD); crop and soil sciences (PhD); genetics (PhD); microbiology and molecular genetics (PhD); plant biology (PhD); plant physiology (PhD). Offered jointly with the Department of Energy. *Faculty:* 9 full-time (2 women). *Degree requirements:* For doctorate, thesis/dissertation, laboratory rotation, defense of dissertation, comprehensive exam. *Entrance requirements:* For doctorate, GRE General Test, acceptance into one of the affiliated department programs; 3 letters of recommendation; bachelor's degree or equivalent in life sciences, chemistry, biochemistry, or biophysics; research experience. Application fee: $50. Electronic applications accepted. *Expenses:* Tuition, state resident: part-time $330 per credit hour. Tuition, nonresident: part-time $685 per credit hour. Tuition and fees vary according to program. *Faculty research:* Role of hormones in the regulation of plant development and physiology, molecular mechanisms associated with signal recognition, development and application of genetic methods and materials, protein routing and function. Total annual research expenditures: $7.4 million. *Unit head:* Dr. Kenneth Keegstra, Director, 517-353-2270, Fax: 517-353-9168, E-mail: keegstra@msu.edu. *Application contact:* Janet Taylor, Graduate Program Secretary, 517-353-2270, Fax: 517-353-9168, E-mail: prl@msu.edu.

Michigan State University, The Graduate School, College of Natural Science and Graduate Programs in Human Medicine and Graduate Studies in Osteopathic Medicine, Department of Biochemistry and Molecular Biology, East Lansing, MI 48824. Offers biochemistry and molecular biology (MS, PhD); biochemistry and molecular biology/environmental toxicology (PhD). *Faculty:* 30 full-time (5 women). *Students:* 74 full-time (31 women), 1 part-time; includes 3 minority (1 African American, 1 Asian American or Pacific Islander, 1 Hispanic American), 35 international. Average age 28. 93 applicants, 12% accepted. In 2005, 2 master's, 7 doctorates awarded. *Degree requirements:* For master's, oral defense of thesis, annual review, thesis optional; for doctorate, thesis/dissertation, teaching, annual evaluations, research proposal, comprehensive exam. *Entrance requirements:* For master's, GRE General Test, minimum GPA of 3.2 in last 2 undergraduate years, course work in biochemistry, 3 letters of recommendation from former instructors; for doctorate, GRE General Test, minimum GPA of 3.2; bachelor's or master's degree in chemistry, biochemistry or related science; course work in biochemistry; 3 letters of recommendation from former instructors. Additional exam requirements/recommendations for international students: Required—TOEFL (minimum score 550 paper-based; 213 computer-based), Michigan State University ELT (85), Michigan ELAB (83).

Application deadline: For fall admission, 12/27 for domestic students. Application fee: $50. Electronic applications accepted. *Expenses:* Tuition, state resident: part-time $330 per credit hour. Tuition, nonresident: part-time $685 per credit hour. Tuition and fees vary according to program. *Financial support:* In 2005–06, 18 fellowships with tuition reimbursements (averaging $11,755 per year), 64 research assistantships with tuition reimbursements (averaging $15,738 per year) were awarded; scholarships/grants and unspecified assistantships also available. *Faculty research:* Biochemistry of the cell nucleus; protein structure, function, and design; plant biochemistry; biological modeling; gene expression in development and disease. Total annual research expenditures: $10.7 million. *Unit head:* Dr. S. L. Ferguson-Miller, Chairperson, 517-353-0804, Fax: 517-353-9334, E-mail: fergus20@msu.edu. *Application contact:* Jessica Hudson, Graduate Program Secretary, 517-353-0807, Fax: 517-353-9334, E-mail: jhudson@msu.edu.

See Close-Up on page 401.

Mississippi State University, College of Agriculture and Life Sciences, Department of Biochemistry and Molecular Biology, Mississippi State, MS 39762. Offers biochemistry (MS); molecular biology (PhD). *Faculty:* 8 full-time (1 woman). *Students:* 19 full-time (6 women), 1 (woman) part-time; includes 1 minority (Hispanic American), 13 international. Average age 28. 33 applicants, 42% accepted, 1 enrolled.Terminal master's awarded for partial completion of doctoral program. *Degree requirements:* For master's, thesis (for some programs), comprehensive oral or written exam; for doctorate, thesis/dissertation, comprehensive oral and written exam. *Entrance requirements:* For master's, GRE General Test, minimum GPA of 2.75; for doctorate, GRE. Additional exam requirements/recommendations for international students: Required—TOEFL. *Application deadline:* For fall admission, 7/1 for domestic students; for spring admission, 11/1 for domestic students. Applications are processed on a rolling basis. Application fee: $30. Electronic applications accepted. *Expenses:* Tuition, state resident: full-time $4,312; part-time $240 per hour. Tuition, nonresident: full-time $9,772; part-time $543 per hour. International tuition: $10,102 full-time. Tuition and fees vary according to course load. *Financial support:* Research assistantships with full tuition reimbursements, Federal Work-Study, institutionally sponsored loans, and unspecified assistantships available. Financial award applicants required to submit FAFSA. *Faculty research:* Fish nutrition, plant and animal molecular biology, plant biochemistry, enzymology, lipid metabolism. Total annual research expenditures: $288,720. *Unit head:* Dr. John A. Boyle, Head, 662-325-2640, Fax: 662-325-8664, E-mail: jab@ra.msstate.edu. *Application contact:* Philip G. Bonfanti, Director of Admissions, 662-325-4104, Fax: 662-325-8872, E-mail: admit@msstate.edu.

Montana State University, College of Graduate Studies, College of Letters and Science, Department of Chemistry and Biochemistry, Bozeman, MT 59717. Offers biochemistry (MS, PhD); chemistry (MS, PhD). Part-time programs available. *Faculty:* 16 full-time (4 women), 10 part-time/adjunct (6 women). *Students:* 1 full-time (0 women), 57 part-time (16 women); includes 1 minority (Hispanic American), 9 international. Average age 27. 33 applicants, 79% accepted, 18 enrolled. In 2005, 3 master's, 5 doctorates awarded. *Degree requirements:* For master's, thesis (for some programs), comprehensive exam, registration; for doctorate, thesis/dissertation, comprehensive exam, registration. *Entrance requirements:* For master's and doctorate, GRE General Test. Additional exam requirements/recommendations for international students: Required—TOEFL (minimum score 550 paper-based; 213 computer-based). *Application deadline:* For fall admission, 7/15 priority date for domestic students, 5/15 priority date for international students; for spring admission, 12/1 priority date for domestic students, 10/1 priority date for international students. Applications are processed on a rolling basis. Application fee: $30. Electronic applications accepted. *Expenses:* Tuition, state resident: full-time $4,132. Tuition, nonresident: full-time $1,132. *Financial support:* In 2005–06, 51 students received support, including 27 research assistantships with full tuition reimbursements available (averaging $22,955 per year), 24 teaching assistantships with full tuition reimbursements available (averaging $22,742 per year); unspecified assistantships also available. Financial award application deadline: 3/1; financial award applicants required to submit FAFSA. *Faculty research:* Structure, mechanism, nanotechnology, spectroscopy, synthesis. Total annual research expenditures: $5.4 million. *Unit head:* Dr. David Singel, Interim Department Head, 406-994-3960, Fax: 406-994-5407, E-mail: rchds@chemistry.montana.edu.

Montclair State University, The Graduate School, College of Science and Mathematics, Department of Chemistry and Biochemistry, Montclair, NJ 07043-1624. Offers chemistry (MS), including biochemistry. Part-time and evening/weekend programs available. *Faculty:* 11 full-time (2 women), 1 (woman) part-time/adjunct. *Students:* 8 full-time (5 women), 13 part-time (8 women); includes 8 minority (3 African Americans, 4 Asian Americans or Pacific Islanders, 1 Hispanic American), 4 international. 18 applicants, 72% accepted, 12 enrolled. In 2005, 5 degrees awarded. *Degree requirements:* For master's, comprehensive exam. *Entrance requirements:* For master's, GRE General Test, 24 credits of course work in undergraduate chemistry, 2 letters of recommendation. Additional exam requirements/recommendations for international students: Required—TOEFL (minimum score 83 computer-based). *Application deadline:* Applications are processed on a rolling basis. Application fee: $60. Electronic applications accepted. *Expenses:* Tuition: Full-time $3,001; part-time $409 per credit. Required fees: $56 per credit. Tuition and fees vary according to course load, degree level and program. *Financial support:* In 2005–06, 3 research assistantships with full tuition reimbursements (averaging $7,000 per year) were awarded; Federal Work-Study, scholarships/grants, and unspecified assistantships also available. Support available to part-time students. Financial award application deadline: 3/1; financial award applicants required to submit FAFSA. *Faculty research:* Antimicrobial compounds; marine bacteria. Total annual research expenditures: $5,000. *Unit head:* Dr. Jeff Toney, Chair, 973-655-7121. *Application contact:* Dr. Mark Whitener, Adviser, 973-655-7166, E-mail: whitenerm@mail.montclair.edu.

New Mexico Institute of Mining and Technology, Graduate Studies, Department of Chemistry, Socorro, NM 87801. Offers biochemistry (MS); chemistry (MS); environmental chemistry (PhD); explosives technology and atmospheric chemistry (PhD). Part-time programs available. *Degree requirements:* For master's and doctorate, thesis/dissertation. *Entrance requirements:* For master's, GRE General Test; for doctorate, GRE General Test, GRE Subject Test. Additional exam requirements/recommendations for international students: Required—TOEFL (minimum score 540 paper-based; 207 computer-based). Electronic applications accepted. *Faculty research:* Organic, analytical, environmental, and explosives chemistry.

New Mexico State University, Graduate School, College of Arts and Sciences, Department of Chemistry and Biochemistry, Las Cruces, NM 88003-8001. Offers MS, PhD. Part-time programs available. *Faculty:* 19 full-time (1 woman), 2 part-time/adjunct (0 women). *Students:* 54 full-time (17 women), 8 part-time; includes 13 minority (1 American Indian/Alaska Native, 2 Asian Americans or Pacific Islanders, 10 Hispanic Americans), 29 international. Average age 30. 32 applicants, 34% accepted, 7 enrolled. In 2005, 7 master's, 3 doctorates awarded. *Degree requirements:* For master's and doctorate, thesis/dissertation. *Entrance requirements:* For master's and doctorate, GRE, BS in chemistry or biochemistry, minimum GPA of 3.0. Additional exam requirements/recommendations for international students: Required—TOEFL. *Application deadline:* For fall admission, 7/1 for domestic students; for spring admission, 11/1 for domestic students. Applications are processed on a rolling basis. Application fee: $30 ($50 for international students). *Expenses:* Tuition, state resident: full-time $3,156; part-time $175 per credit. Tuition, nonresident: full-time $12,510; part-time $565 per credit. Required fees: $1,050. *Financial support:* In 2005–06, 1 fellowship, 16 research assistantships, 20 teaching assistantships were awarded; career-related internships or fieldwork and Federal Work-Study also available. Support available to part-time students. Financial award application deadline: 3/1. *Faculty research:* Clays, surfaces, and water structure; electroanalytical and environmental chemistry; organometallic synthesis and organobiomimetics; molecular genetics and enzymology of stress; spectroscopy and reaction kinetics. *Unit head:* Dr. Aravamudan Gopalan, Head, 505-646-5877, Fax: 505-646-2649, E-mail: agopalan@nmsu.edu. *Application contact:* Dr. James Herndon, Associate Professor, Chemistry, 505-646-2738, Fax: 505-646-2649.

New York Medical College, Graduate School of Basic Medical Sciences, Program in Biochemistry and Molecular Biology, Valhalla, NY 10595-1691. Offers MS, PhD, MD/PhD.

Biochemistry

New York Medical College (continued)

Part-time and evening/weekend programs available. Terminal master's awarded for partial completion of doctoral program. *Degree requirements:* For master's, thesis/dissertation; for doctorate, thesis/dissertation, comprehensive exam. *Entrance requirements:* For master's and doctorate, GRE General Test. Additional exam requirements/recommendations for international students: Required—TOEFL. *Faculty research:* Mechanisms of control of blood coagulation, molecular neurobiology, molecular probes for infectious disease, protein-DNA interactions, molecular biology and biochemistry of double-stranded RNA-dependent enzymes.

North Carolina State University, Graduate School, College of Agriculture and Life Sciences, Department of Biochemistry, Raleigh, NC 27695. Offers MS, PhD. Terminal master's awarded for partial completion of doctoral program. *Degree requirements:* For master's, thesis (for some programs); for doctorate, thesis/dissertation. *Entrance requirements:* For master's and doctorate, GRE General Test. Additional exam requirements/recommendations for international students: Required—TOEFL. Electronic applications accepted. *Faculty research:* Regulation of gene expression, structure and function of proteins and nucleic acids, molecular biology, high-field NMR, bioinorganic chemistry.

See Close-Up on page 403.

North Dakota State University, The Graduate School, College of Science and Mathematics, Department of Chemistry, Program in Biochemistry, Fargo, ND 58105. Offers MS, PhD. Part-time programs available. *Faculty:* 5 full-time (0 women). *Students:* 9 full-time (6 women); includes 7 minority (all Asian Americans or Pacific Islanders) Average age 24. 13 applicants, 62% accepted, 6 enrolled. In 2005, 1 degree awarded. *Degree requirements:* For master's and doctorate, thesis/dissertation. *Entrance requirements:* Additional exam requirements/recommendations for international students: Required—TOEFL (minimum score 550 paper-based). *Application deadline:* For fall admission, 4/15 for domestic students. Applications are processed on a rolling basis. Application fee: $45 ($60 for international students). *Financial support:* In 2005–06, 4 research assistantships with full tuition reimbursements (averaging $19,000 per year), 5 teaching assistantships with full tuition reimbursements (averaging $19,000 per year) were awarded; career-related internships or fieldwork, Federal Work-Study, and institutionally sponsored loans also available. Financial award application deadline: 4/15. *Application contact:* Dr. Seth Rasmussen, Chair, Graduate Admissions, 701-231-8747, Fax: 701-231-8831, E-mail: seth.rasmussen@ndsu.nodak.edu.

Northeastern University, College of Arts and Sciences, Department of Chemistry and Chemical Biology, Boston, MA 02115-5096. Offers analytical chemistry (PhD); chemistry (MS, PhD); inorganic chemistry (PhD); organic chemistry (PhD); physical chemistry (PhD). Part-time and evening/weekend programs available. *Faculty:* 11 full-time (2 women), 3 part-time/adjunct (0 women). *Students:* 46 full-time (23 women), 29 part-time (17 women). Average age 28. 113 applicants, 37% accepted. In 2005, 12 master's, 10 doctorates awarded. Terminal master's awarded for partial completion of doctoral program. *Degree requirements:* For master's, thesis (for some programs); for doctorate, thesis/dissertation, qualifying exam in specialty area. *Entrance requirements:* Additional exam requirements/recommendations for international students: Required—TOEFL. *Application deadline:* For fall admission, 4/15 for domestic students. Applications are processed on a rolling basis. Application fee: $50. Electronic applications accepted. *Financial support:* In 2005–06, 27 research assistantships with tuition reimbursements (averaging $16,000 per year) were awarded; fellowships with tuition reimbursements, teaching assistantships with tuition reimbursements, career-related internships or fieldwork, tuition waivers (partial), and unspecified assistantships also available. Financial award application deadline: 4/15; financial award applicants required to submit FAFSA. *Faculty research:* Electron transfer, theoretical chemical physics, analytical biotechnology, mass spectrometry, materials chemistry. Total annual research expenditures: $250,000. *Unit head:* Dr. Graham Jones, Chairman, 617-373-2822, Fax: 617-373-8795, E-mail: chemistry-grad-info@neu.edu. *Application contact:* Dr. Pam Mabrouk, Chair, Graduate Admissions Committee, 617-373-2383, Fax: 617-373-8795, E-mail: chemistry-grad-info@neu.edu.

Northern Arizona University, Graduate College, College of Engineering and Natural Science, Department of Chemistry and Biochemistry, Flagstaff, AZ 86011. Offers chemistry (MS). Part-time programs available. *Degree requirements:* For master's, thesis. *Faculty research:* Biochemistry of exercise, organic and inorganic mechanism studies, inhibition of ice mutation, polymer separation.

Northern Illinois University, Graduate School, College of Liberal Arts and Sciences, Department of Chemistry and Biochemistry, De Kalb, IL 60115-2854. Offers chemistry (MS, PhD). *Faculty:* 16 full-time (1 woman), 3 part-time/adjunct (1 woman). *Students:* 46 full-time (19 women), 8 part-time (5 women); includes 5 minority (2 African Americans, 1 Asian American or Pacific Islander, 2 Hispanic Americans), 21 international. Average age 28. 63 applicants, 46% accepted, 9 enrolled. In 2005, 3 master's, 2 doctorates awarded. Terminal master's awarded for partial completion of doctoral program. *Degree requirements:* For master's, research seminar, thesis optional; for doctorate, one foreign language, thesis/dissertation, candidacy exam, dissertation defense, research seminar. *Entrance requirements:* For master's, GRE General Test, bachelor's degree in mathematics or science, minimum GPA of 2.75; for doctorate, GRE General Test, bachelor's degree in mathematics or science; minimum undergraduate GPA of 2.75, 3.2 graduate. Additional exam requirements/recommendations for international students: Required—TOEFL (minimum score 550 paper-based; 213 computer-based). *Application deadline:* For fall admission, 6/1 for domestic students, 5/1 for international students; for spring admission, 10/1 for domestic students, 10/1 for international students. Applications are processed on a rolling basis. Application fee: $30. Electronic applications accepted. *Expenses:* Tuition, area resident: Full-time $4,565; part-time $19 per credit hour. Tuition, state resident: full-time $4,565; part-time $191 per credit hour. Tuition, nonresident: full-time $9,129; part-time $382 per credit hour. *Financial support:* In 2005–06, 8 research assistantships with full tuition reimbursements, 37 teaching assistantships with full tuition reimbursements were awarded; fellowships with full tuition reimbursements, career-related internships or fieldwork, Federal Work-Study, scholarships/grants, tuition waivers (full), and unspecified assistantships also available. Support available to part-time students. Financial award applicants required to submit FAFSA. *Faculty research:* Viscoelastic properties of polymers, lig and buding tocytochrome coxidases, computational inorganic chemistry, chemistry of organosilanes. *Unit head:* Dr. James Erman, Chair, 815-753-1181, Fax: 815-753-4802, E-mail: jerman@niu.edu. *Application contact:* Dr. Jon Carnahan, Director, Graduate Studies, 815-753-6879, E-mail: carnahan@niu.edu.

Northern Michigan University, College of Graduate Studies, College of Arts and Sciences, Department of Chemistry, Marquette, MI 49855-5301. Offers biochemistry (MS); chemistry (MS). Part-time programs available. *Degree requirements:* For master's, thesis. *Entrance requirements:* For master's, GRE General Test, minimum GPA of 3.0.

Northwestern University, The Graduate School and Judd A. and Marjorie Weinberg College of Arts and Sciences, Interdepartmental Biological Sciences Program (IBiS), Department of Biochemistry, Molecular Biology, and Cell Biology, Evanston, IL 60208. Offers biochemistry (PhD); cell and molecular biology (PhD); molecular biophysics (PhD); structural biology (PhD). Department participates in the Interdepartmental Biological Sciences Program (IBiS). *Entrance requirements:* For doctorate, GRE General Test. Additional exam requirements/recommendations for international students: Required—TOEFL (minimum score 600 paper-based), TSE (minimum score 50). Electronic applications accepted. *Faculty research:* Protein structure, protein-DNA interactions, gene regulation, development of embryos, hormone action and signal transduction.

Northwestern University, The Graduate School, Interdepartmental Degree Programs, Interdepartmental Biological Sciences Program, Evanston, IL 60208.

Northwestern University, Northwestern University Feinberg School of Medicine, Department of Cell and Molecular Biology, Chicago, IL 60611.

Northwestern University, Northwestern University Feinberg School of Medicine, Department of Molecular Pharmacology and Biological Chemistry, Chicago, IL 60611.

Northwestern University, Northwestern University Feinberg School of Medicine and Interdepartmental Degree Programs, Integrated Graduate Programs in the Life Sciences, Chicago, IL 60611. Offers cancer biology (PhD); cell biology (PhD); developmental biology (PhD); evolutionary biology (PhD); immunology and microbial pathogenesis (PhD); molecular biology and genetics (PhD); neurobiology (PhD); pharmacology and toxicology (PhD); structural biology and biochemistry (PhD). *Degree requirements:* For doctorate, thesis/dissertation, written and oral qualifying exams, comprehensive exam. *Entrance requirements:* For doctorate, GRE General Test. Additional exam requirements/recommendations for international students: Required—TOEFL (minimum score 600 paper-based; 250 computer-based). Electronic applications accepted.

See Close-Up on page 189.

OGI School of Science & Engineering at Oregon Health & Science University, Graduate Studies, Department of Environmental and Biomolecular Systems, Program in Biochemistry and Molecular Biology, Beaverton, OR 97006-8921. Offers MS, PhD. *Degree requirements:* For doctorate, dissertation defense. *Entrance requirements:* For master's, GRE General Test, 3 letters of recommendation; for doctorate, GRE General Test, GRE Subject Test, 3 letters of recommendation. Application fee: $65. *Expenses:* Tuition, state resident: full-time $22,760; part-time $625 per credit. Required fees: $350. *Application contact:* Information Contact, 503-748-1070, E-mail: info@bmb.ogi.edu.

The Ohio State University, Graduate School, College of Biological Sciences, Biochemistry Program, Columbus, OH 43210. Offers MS, PhD. Terminal master's awarded for partial completion of doctoral program. *Degree requirements:* For master's and doctorate, thesis/dissertation. *Entrance requirements:* For master's and doctorate, GRE General Test. Additional exam requirements/recommendations for international students: Required—TOEFL (minimum score 620 paper-based; 260 computer-based); Recommended—TSE. Electronic applications accepted. *Faculty research:* Biotechnology, carbohydrates, carcinogenesis, enzymology, intermediary metabolism.

The Ohio State University, Graduate School, College of Biological Sciences, Department of Biochemistry, Columbus, OH 43210. Offers MS. *Degree requirements:* For master's, thesis optional. *Entrance requirements:* For master's, GRE General Test. Additional exam requirements/recommendations for international students: Required—TOEFL (minimum score 600 paper-based; 250 computer-based), TSE. Electronic applications accepted.

Ohio University, Graduate Studies, College of Arts and Sciences, Department of Chemistry and Biochemistry, Athens, OH 45701-2979. Offers MS. *Faculty:* 23 full-time (6 women), 2 part-time/adjunct (0 women). *Students:* 67 full-time (30 women), 7 part-time (2 women); includes 2 minority (1 African American, 1 Asian American or Pacific Islander), 35 international. 47 applicants, 47% accepted, 17 enrolled. In 2005, 6 master's, 9 doctorates awarded. *Degree requirements:* For master's and doctorate, thesis/dissertation, exam. *Entrance requirements:* For master's and doctorate, GRE. Additional exam requirements/recommendations for international students: Required—TOEFL (minimum score 630 paper-based), TWE, TSE. *Application deadline:* For fall admission, 2/15 priority date for domestic students, 2/15 priority date for international students. Applications are processed on a rolling basis. Application fee: $45. *Financial support:* In 2005–06, 34 teaching assistantships with full tuition reimbursements (averaging $18,500 per year); fellowships, research assistantships with full tuition reimbursements, Federal Work-Study, institutionally sponsored loans, and tuition waivers (full and partial) also available. Financial award application deadline: 3/15. *Faculty research:* Materials, RNA, synthesis, science. Total annual research expenditures: $3.5 million. *Unit head:* Dr. Kenneth Brown, Chair, 740-593-1737, Fax: 740-593-0148, E-mail: brownk3@oak.cats.ohiou.edu. *Application contact:* Dr. Mark C. McMills, Graduate Chair, 740-593-1755, Fax: 740-593-0148, E-mail: mcmills@ohio.edu.

Oklahoma State University, College of Agricultural Science and Natural Resources, Department of Biochemistry and Molecular Biology, Stillwater, OK 74078. Offers MS, PhD. *Faculty:* 25 full-time (12 women). *Students:* 11 full-time (5 women), 24 part-time (12 women); includes 7 minority (2 African Americans, 3 American Indian/Alaska Native, 2 Asian Americans or Pacific Islanders), 23 international. Average age 28. 63 applicants, 25% accepted, 4 enrolled. In 2005, 2 master's, 6 doctorates awarded. *Degree requirements:* For master's, thesis, oral exam; for doctorate, one foreign language, thesis/dissertation. *Entrance requirements:* For master's and doctorate, GRE. Additional exam requirements/recommendations for international students: Required—TOEFL. *Application deadline:* For fall admission, 6/1 priority date for domestic students, 3/1 priority date for international students. Applications are processed on a rolling basis. Application fee: $40 ($75 for international students). Electronic applications accepted. *Expenses:* Tuition, state resident: full-time $4,253; part-time $139 per credit hour. Tuition, nonresident: full-time $12,569; part-time $485 per credit hour. Required fees: $43 per credit hour. One-time fee: $20 part-time. Tuition and fees vary according to course load and program. *Financial support:* In 2005–06, 32 research assistantships (averaging $17,690 per year) were awarded; teaching assistantships, career-related internships or fieldwork, Federal Work-Study, scholarships/grants, health care benefits, tuition waivers (partial), and unspecified assistantships also available. Support available to part-time students. Financial award application deadline: 3/1. *Unit head:* Dr. Earl Mitchell, Interim Head, 405-744-6189, Fax: 405-744-7799.

See Close-Up on page 405.

Old Dominion University, College of Sciences, Program in Chemistry, Norfolk, VA 23529. Offers analytical chemistry (MS); biochemistry (MS); clinical chemistry (MS); environmental chemistry (MS); organic chemistry (MS); physical chemistry (MS). Part-time and evening/weekend programs available. *Faculty:* 15 full-time (4 women). *Students:* 5 full-time (3 women), 8 part-time (3 women); includes 8 minority (4 African Americans, 1 American Indian/Alaska Native, 1 Asian American or Pacific Islander, 2 Hispanic Americans). Average age 27. 19 applicants, 74% accepted. In 2005, 1 degree awarded. *Degree requirements:* For master's, thesis, comprehensive exam. *Entrance requirements:* For master's, GRE General Test, minimum GPA of 3.0 in major, 2.5 overall. Additional exam requirements/recommendations for international students: Required—TOEFL. *Application deadline:* For fall admission, 7/1 for domestic students; for spring admission, 11/1 for domestic students. Applications are processed on a rolling basis. Application fee: $30. *Expenses:* Tuition, state resident: part-time $263 per credit hour. Tuition, nonresident: part-time $661 per credit hour. Required fees: $39 per semester. Part-time tuition and fees vary according to campus/location. *Financial support:* In 2005–06, research assistantships with tuition reimbursements (averaging $15,000 per year), teaching assistantships with tuition reimbursements (averaging $15,000 per year) were awarded; fellowships, career-related internships or fieldwork and scholarships/grants also available. Financial award application deadline: 2/15; financial award applicants required to submit FAFSA. *Faculty research:* Organic and trace metal biogeochemistry, materials chemistry, bioanalytical chemistry, computational chemistry. Total annual research expenditures: $854,968. *Unit head:* Dr. John R. Donat, Graduate Program Director, 757-683-4098, Fax: 757-683-4628, E-mail: chemgpd@odu.edu.

Oregon Health & Science University, School of Medicine, Graduate Programs in Medicine, Department of Biochemistry and Molecular Biology, Portland, OR 97239-3098. Offers PhD, MD/PhD. Part-time programs available. *Degree requirements:* For doctorate, thesis/dissertation. *Entrance requirements:* For doctorate, GRE General Test, MCAT (MD/PhD). *Faculty research:* Protein structure and function, enzymology, metabolism, membranes transport.

Oregon State University, Graduate School, College of Science, Department of Biochemistry and Biophysics, Corvallis, OR 97331-6503. Offers MA, MAIS, MS, PhD. *Faculty:* 11 full-time (3 women), 1 part-time/adjunct (0 women). *Students:* 24 full-time (7 women), 1 part-time; includes 1 minority (Asian American or Pacific Islander), 11 international. Average age 31. In 2005, 4 doctorates awarded. *Degree requirements:* For master's, thesis optional; for doctorate, thesis/dissertation, exams. *Entrance requirements:* For master's, GRE General Test, minimum GPA of 3.0; for doctorate, GRE Subject Test, minimum GPA of 3.0. Additional exam requirements/recommendations for international students: Required—TOEFL. *Application deadline:* For fall admission, 4/15 for domestic students. Applications are processed on a rolling basis. Application fee: $50. *Expenses:* Tuition, area resident: Part-time $301 per credit. Tuition, state resident: full-time $8,139; part-time $501 per credit. Tuition, nonresident: full-time $14,376; part-time $532 per credit. Required fees: $1,266. *Financial support:* Research assistantships, teaching assistantships, institutionally sponsored loans available. Support available to part-time students. Financial award application deadline: 2/1. *Faculty research:* DNA and deoxyribonucleotide metabolism, cell growth control, receptors and membranes, protein structure and function. *Unit head:* Dr. Pui Shing Ho, Chair, 541-737-1865, Fax: 541-737-0481. *Application contact:* Dr. W. Curtis Johnson, Chairman, Graduate Committee, 541-737-4143, Fax: 541-737-0481, E-mail: johnsowc@ucs.orst.edu.

The Pennsylvania State University Milton S. Hershey Medical Center, Graduate School Programs in the Biomedical Sciences, Graduate Program in Biochemistry and Molecular Biology, Hershey, PA 17033-2360. Offers MS, PhD, MD/PhD. *Students:* 20 full-time (13 women); includes 2 minority (both Asian Americans or Pacific Islanders), 7 international. Average age 27.Terminal master's awarded for partial completion of doctoral program. *Median time to degree:* Of those who began their doctoral program in fall 1997, 100% received their degree in 8 years or less. *Degree requirements:* For master's, thesis or alternative, registration; for doctorate, thesis/dissertation, oral exam, comprehensive exam, registration. *Entrance requirements:* For master's, GRE General Test; for doctorate, GRE General Test, minimum GPA of 3.0. Additional exam requirements/recommendations for international students: Required—TOEFL (minimum score 550 paper-based; 213 computer-based). *Application deadline:* For fall admission, 2/1 priority date for domestic students, 2/1 priority date for international students. Applications are processed on a rolling basis. Application fee: $45. Electronic applications accepted. *Financial support:* In 2005–06, 3 fellowships with full tuition reimbursements, 17 research assistantships with full tuition reimbursements were awarded; scholarships/grants, health care benefits, and unspecified assistantships also available. Financial award applicants required to submit FAFSA. *Faculty research:* X-ray crystallography of proteins, glycosphingolipid interactions with viruses and toxins, DNA replication and repair, tobacco and environmental carcinogenesis, generegulation. *Unit head:* Dr. Judith S. Bond, Chair, 717-531-8585, Fax: 717-531-7072, E-mail: bchem-grad-hmc@psu.edu. *Application contact:* Ruth Dean, Administrative Assistant, 717-531-8586, Fax: 717-531-7072, E-mail: bchem-grad-hmc@psu.edu.

The Pennsylvania State University University Park Campus, Graduate School, Eberly College of Science, Department of Biochemistry and Molecular Biology, Program in Biochemistry, Microbiology, and Molecular Biology, State College, University Park, PA 16802-1503. Offers MS, PhD. *Degree requirements:* For master's, thesis/dissertation; for doctorate, thesis/dissertation, comprehensive exam. *Entrance requirements:* For master's and doctorate, GRE General Test, GRE Subject Test. Application fee: $45. *Expenses:* Tuition, state resident: full-time $12,518; part-time $522 per credit. Tuition, nonresident: full-time $23,004; part-time $959 per credit. Required fees: $484. Tuition and fees vary according to course load, campus/location and program. *Financial support:* Fellowships, unspecified assistantships available.

See Close-Up on page 407.

The Pennsylvania State University University Park Campus, Graduate School, Intercollege Graduate Programs, Intercollege Graduate Program in Integrative Biosciences, State College, University Park, PA 16802-1503. Offers integrative biosciences (PhD), including biomolecular transport dynamics, cell and developmental biology, cellular and molecular mechanisms of toxicity, chemical biology, ecological and molecular plant physiology, immunobiology, molecular medicine, neuroscience, nutrition science. *Students:* 110 full-time (58 women), 2 part-time (both women); includes 6 minority (3 African Americans, 3 Asian Americans or Pacific Islanders), 61 international. *Entrance requirements:* For master's and doctorate, GRE General Test. Application fee: $45. *Expenses:* Tuition, state resident: full-time $12,518; part-time $522 per credit. Tuition, nonresident: full-time $23,004; part-time $959 per credit. Required fees: $484. Tuition and fees vary according to course load, campus/location and program. *Financial support:* Fellowships available. *Unit head:* Dr. Richard J. Frisque, Co-Director, 814-863-3523, Fax: 814-863-1357, E-mail: rjf6@psu.edu.

See Close-Up on page 197.

Purdue University, College of Pharmacy and Pharmacal Sciences and Graduate School, Graduate Programs in Pharmacy and Pharmacal Sciences, Department of Medicinal Chemistry and Molecular Pharmacology, West Lafayette, IN 47907. Offers analytical medicinal chemistry (PhD); computational and biophysical medicinal chemistry (PhD); medicinal and bioorganic chemistry (PhD); medicinal biochemistry and molecular biology (PhD); molecular pharmacology and toxicology (PhD); natural products and pharmacognosy (PhD); nuclear pharmacy (MS); radiopharmaceutical chemistry and nuclear pharmacy (PhD). *Faculty:* 21 full-time (2 women), 2 part-time/adjunct (0 women). *Students:* 69 full-time (39 women), 2 part-time (1 woman); includes 8 minority (2 African Americans, 2 Asian Americans or Pacific Islanders, 4 Hispanic Americans), 21 international. Average age 26. 125 applicants, 25% accepted, 15 enrolled. In 2005, 2 master's, 11 doctorates awarded. Terminal master's awarded for partial completion of doctoral program. *Degree requirements:* For master's and doctorate, thesis/dissertation. *Entrance requirements:* For master's, GRE General Test, minimum B average; BS in biology, chemistry, or pharmacy; for doctorate, GRE General Test, minimum B average; BS in biology, chemistry, or pharmacology. Additional exam requirements/recommendations for international students: Required—TOEFL. *Application deadline:* Applications are processed on a rolling basis. Application fee: $55. Electronic applications accepted. *Financial support:* Fellowships, research assistantships, teaching assistantships, traineeships available. Support available to part-time students. Financial award applicants required to submit FAFSA. *Faculty research:* Drug design and development, cancer research, drug synthesis and analysis, chemical pharmacology, environmental toxicology. *Unit head:* Dr. R. F. Borch, Graduate Head, 765-494-1403. *Application contact:* Dr. D. E. Bergstrom, Graduate Committee, 765-494-6275, E-mail: bergstrom@pharmacy.purdue.edu.

Purdue University, Graduate School, College of Agriculture, Department of Biochemistry, West Lafayette, IN 47907. Offers MS, PhD. *Faculty:* 15 full-time (3 women), 2 part-time/adjunct (0 women). *Students:* 15 full-time (5 women); includes 1 minority (Hispanic American), 10 international. Average age 27. 80 applicants, 11% accepted, 3 enrolled. In 2005, 1 master's, 4 doctorates awarded. Terminal master's awarded for partial completion of doctoral program. *Degree requirements:* For master's, thesis; for doctorate, thesis/dissertation, preliminary and qualifying exams. *Entrance requirements:* For master's and doctorate, GRE General Test. Additional exam requirements/recommendations for international students: Required—TOEFL. *Application deadline:* For fall admission, 1/15 priority date for domestic students, 1/15 priority date for international students. Applications are processed on a rolling basis. Application fee: $55. Electronic applications accepted. *Financial support:* In 2005–06, 12 fellowships (averaging $17,280 per year), 9 research assistantships (averaging $17,280 per year), 7 teaching assistantships (averaging $17,280 per year) were awarded. Support available to part-time students. Financial award application deadline: 4/15; financial award applicants required to submit FAFSA. *Faculty research:* Molecular biology and post-translational modifications of neuropeptides, membrane transport proteins. *Unit head:* Dr. J. D. Forney, Head, 765-494-1632, Fax: 765-494-7987, E-mail: forney@purdue.edu. *Application contact:* Penny J Pava, Graduate Contact for Admissions, 765-454-1612, E-mail: pavap@purdue.edu.

Purdue University, School of Veterinary Medicine and Graduate School, Graduate Programs in Veterinary Medicine, Department of Veterinary Pathobiology, West Lafayette, IN 47907. Offers biochemistry and molecular biology (MS, PhD); comparative epidemiology (MS, PhD); epidemiology (MS, PhD); immunology (MS, PhD); infectious diseases (MS, PhD); interdisciplinary genetics (PhD); laboratory animal medicine (MS, PhD); microbiology (MS, PhD); molecular

virology (MS, PhD); parasitology (MS, PhD); pathobiology (MS, PhD); public health epidemiology (MS, PhD); toxicology (MS, PhD); veterinary anatomic pathology (MS, PhD); veterinary clinical pathology (MS, PhD); virology (MS, PhD). *Faculty:* 32 full-time (7 women). *Students:* 49 full-time (20 women), 3 part-time (1 woman); includes 2 minority (both African Americans), 31 international. Average age 35. In 2005, 3 master's, 8 doctorates awarded. Terminal master's awarded for partial completion of doctoral program. *Degree requirements:* For master's, thesis (for some programs); for doctorate, thesis/dissertation. *Entrance requirements:* For master's and doctorate, GRE General Test. Additional exam requirements/recommendations for international students: Required—TOEFL (minimum score 575 paper-based), TWE (minimum score 4). *Application deadline:* For fall admission, 8/12 for domestic students, 6/15 for international students; for spring admission, 1/12 for domestic students, 10/15 for international students. Application fee: $55. *Financial support:* Fellowships, research assistantships, teaching assistantships available. Financial award application deadline: 3/1; financial award applicants required to submit FAFSA. *Unit head:* Dr. H. Hogenesch, Head, 765-494-7543.

Queens College of the City University of New York, Division of Graduate Studies, Mathematics and Natural Sciences Division, Department of Chemistry and Biochemistry, Flushing, NY 11367-1597. Offers biochemistry (MA); chemistry (MA). Part-time and evening/weekend programs available. *Faculty:* 14 full-time (4 women). *Students:* 20 applicants, 65% accepted, 9 enrolled. In 2005, 3 degrees awarded. *Degree requirements:* For master's, comprehensive exam. *Entrance requirements:* For master's, GRE, previous course work in calculus and physics, minimum GPA of 3.0. Additional exam requirements/recommendations for international students: Required—TOEFL. *Application deadline:* For fall admission, 4/1 for domestic students; for spring admission, 11/1 for domestic students. Applications are processed on a rolling basis. Application fee: $125. *Expenses:* Tuition, state resident: part-time $270 per credit. Tuition, nonresident: part-time $500 per credit. Required fees: $112 per year. *Financial support:* Career-related internships or fieldwork, Federal Work-Study, institutionally sponsored loans, tuition waivers (partial), and adjunct lectureships available. Support available to part-time students. Financial award application deadline: 4/1; financial award applicants required to submit FAFSA. *Unit head:* Dr. William Hersh, Chairperson, 718-997-4144. *Application contact:* Graduate Adviser, 718-997-4100.

Queen's University at Kingston, School of Graduate Studies and Research, Faculty of Health Sciences, Department of Biochemistry, Kingston, ON K7L 3N6, Canada. Offers M Sc, PhD. Part-time programs available. *Degree requirements:* For master's, thesis/dissertation, research proposal; for doctorate, thesis/dissertation, research proposal, comprehensive exam. *Entrance requirements:* For master's and doctorate, GRE required if undergraduate degree is not from a Canadian University. Additional exam requirements/recommendations for international students: Required—TOEFL (minimum score 580 paper-based; 237 computer-based). Electronic applications accepted. *Faculty research:* Gene expression, protein structure, enzyme activity, signal transduction.

Rensselaer Polytechnic Institute, Graduate School, School of Science, Department of Biology, Troy, NY 12180-3590. Offers biochemistry (MS, PhD); biophysics (MS, PhD); cell biology (MS, PhD); developmental biology (MS, PhD); microbiology (MS, PhD); molecular biology (MS, PhD). Part-time programs available. Terminal master's awarded for partial completion of doctoral program. *Degree requirements:* For master's and doctorate, thesis/dissertation, comprehensive exam, registration. *Entrance requirements:* For master's and doctorate, GRE General Test. Additional exam requirements/recommendations for international students: Required—TOEFL. Electronic applications accepted. *Expenses:* Tuition: Full-time $31,000; part-time $1,320 per credit. Required fees: $1,623. *Faculty research:* Bioinformatics, molecular biology/biochemistry, cell and tissue biology, environment, ecology.

See Close-Up on page 199.

Rensselaer Polytechnic Institute, Graduate School, School of Science, Department of Chemistry and Chemical Biology, Troy, NY 12180-3590. Offers analytical chemistry (MS, PhD); biochemistry (MS, PhD); inorganic chemistry (MS, PhD); organic chemistry (MS, PhD); physical chemistry (MS, PhD); polymer chemistry (MS, PhD). Part-time and evening/weekend programs available. *Faculty:* 19 full-time (3 women). *Students:* 49 full-time (28 women), 2 part-time (1 woman), 33 international. Average age 24. 80 applicants, 23% accepted, 7 enrolled. In 2005, 1 master's, 14 doctorates awarded. Terminal master's awarded for partial completion of doctoral program. *Median time to degree:* Of those who began their doctoral program in fall 1997, 100% received their degree in 8 years or less. *Degree requirements:* For master's, thesis (for some programs), registration; for doctorate, thesis/dissertation, comprehensive exam, registration. *Entrance requirements:* For master's, GRE General Test, GRE Subject Test (strongly recommended); for doctorate, GRE General Test, GRE Subject Test (chemistry or biochemistry strongly recommended). Additional exam requirements/recommendations for international students: Required—TOEFL (minimum score 600 paper-based). *Application deadline:* For fall admission, 2/1 for domestic students; for spring admission, 11/15 for domestic students. Applications are processed on a rolling basis. Application fee: $75. Electronic applications accepted. *Expenses:* Tuition: Full-time $31,000; part-time $1,320 per credit. Required fees: $1,623. *Financial support:* In 2005–06, 49 students received support, including 1 fellowship with full tuition reimbursement available (averaging $30,000 per year), 25 research assistantships with full tuition reimbursements available (averaging $21,500 per year), 30 teaching assistantships with full tuition reimbursements available (averaging $21,500 per year); institutionally sponsored loans and tuition waivers (full and partial) also available. Financial award application deadline: 2/1. *Faculty research:* Synthetic polymer and biopolymer chemistry, physical chemistry of polymeric systems, bioanalytical chemistry, synthetic and computational drug design, protein folding and protein design. Total annual research expenditures: $1.9 million. *Unit head:* Dr. Linda B. McGown, Chair, 518-276-4856, Fax: 518-276-4887, E-mail: mcgowl@rpi.edu. *Application contact:* Beth E. McGraw, Department Admissions Assistant, 518-276-6456, Fax: 518-276-4887, E-mail: mcgrae@rpi.edu.

Rensselaer Polytechnic Institute, Graduate School, School of Science, Program in Biochemistry and Biophysics, Troy, NY 12180-3590. Offers biochemistry (MS); biophysics (MS). Part-time programs available. *Entrance requirements:* For master's, GRE General Test, GRE Subject Test (biology, chemistry or biochemistry); recommendations for international students: Required—TOEFL. Electronic applications accepted. *Expenses:* Tuition: Full-time $31,000; part-time $1,320 per credit. Required fees: $1,623. *Faculty research:* Biopolymers, photosynthesis, cellular bioengineering.

Rice University, Graduate Programs, Wiess School of Natural Sciences, Department of Biochemistry and Cell Biology, Houston, TX 77251-1892. Offers MA, PhD. *Degree requirements:* For master's and doctorate, thesis/dissertation. *Entrance requirements:* For master's and doctorate, GRE. Additional exam requirements/recommendations for international students: Required—TOEFL. Electronic applications accepted. Expenses: Contact institution. *Faculty research:* Steroid metabolism, protein structure NMR, biophysics, cell growth and movement.

Announcement: Graduate fellowships and research assistantships are awarded to all students who do not receive assistance from other sources. Diverse faculty and excellent facilities provide a broad program for a variety of studies in biochemistry, molecular and cell biology, and biophysics.

See Close-Up on page 409.

Rosalind Franklin University of Medicine and Science, School of Graduate and Postdoctoral Studies, Department of Biochemistry and Molecular Biology, Program in Biochemistry, North Chicago, IL 60064-3095. Offers MS, PhD, MD/MS, MD/PhD. Part-time programs available. Terminal master's awarded for partial completion of doctoral program. *Degree requirements:* For master's and doctorate, thesis/dissertation, comprehensive exam. *Entrance requirements:* For master's and doctorate, GRE General Test, minimum GPA of 3.0. Additional exam requirements/recommendations for international students: Required—TOEFL, TWE.

Biochemistry

Rush University, Graduate College, Division of Biochemistry, Chicago, IL 60612-3832. Offers PhD, MD/PhD. *Degree requirements:* For doctorate, thesis/dissertation, preliminary exam. *Entrance requirements:* For doctorate, GRE General Test. Additional exam requirements/recommendations for international students: Required—TOEFL. Electronic applications accepted. *Faculty research:* Biochemistry of extracellular matrix, connective tissue biosynthesis and degradation, molecular biology of connective tissue components, cartilage, arthritis.

Rutgers, The State University of New Jersey, Newark, Graduate School, Program in Chemistry, Newark, NJ 07102. Offers analytical chemistry (MS, PhD); biochemistry (MS, PhD); inorganic chemistry (MS, PhD); organic chemistry (MS, PhD); physical chemistry (MS, PhD). Part-time and evening/weekend programs available. *Faculty:* 22 full-time (5 women). *Students:* 24 full-time (12 women), 28 part-time (10 women); includes 31 minority (2 African Americans, 29 Asian Americans or Pacific Islanders). 69 applicants, 43% accepted, 13 enrolled. In 2005, 7 master's, 9 doctorates awarded. Terminal master's awarded for partial completion of doctoral program. *Degree requirements:* For master's, cumulative exams, thesis optional; for doctorate, thesis/dissertation, exams, research proposal. *Entrance requirements:* For master's and doctorate, GRE General Test, minimum undergraduate B average. Additional exam requirements/recommendations for international students: Required—TOEFL. *Application deadline:* For fall admission, 7/1 for domestic students; for spring admission, 12/1 for domestic students. Applications are processed on a rolling basis. Application fee: $50. Electronic applications accepted. *Expenses:* Tuition, state resident: full-time $10,440; part-time $435 per credit. Tuition, nonresident: full-time $15,520; part-time $637 per credit. *Financial support:* In 2005–06, 35 students received support, including 5 fellowships with partial tuition reimbursements available (averaging $18,000 per year), 6 research assistantships (averaging $16,988 per year), 20 teaching assistantships with partial tuition reimbursements available (averaging $16,988 per year); Federal Work-Study and institutionally sponsored loans also available. Financial award application deadline: 3/1. *Faculty research:* Medicinal chemistry, natural products, isotope effects, biophysics and biorganic approaches to enzyme mechanisms, organic and organometallic synthesis. *Unit head:* Prof. W. Philip Huskey, Chairman and Program Director, 973-353-5741, Fax: 973-353-1264, E-mail: huskey@andromeda.rutgers.edu.

Rutgers, The State University of New Jersey, New Brunswick/Piscataway, Graduate School, Program in Biochemistry, New Brunswick, NJ 08901-1281. Offers biochemistry (MS, PhD); molecular biology (MS, PhD). *Faculty:* 105 full-time. *Students:* 30 full-time (11 women), 1 (woman) part-time; includes 1 minority (Asian American or Pacific Islander), 20 international. Average age 25. 74 applicants, 34% accepted, 14 enrolled. In 2005, 5 doctorates awarded. Terminal master's awarded for partial completion of doctoral program. *Median time to degree:* Of those who began their doctoral program in fall 1997, 86% received their degree in 8 years or less. *Degree requirements:* For master's, thesis, qualifying exam; for doctorate, thesis/dissertation, written qualifying exam. *Entrance requirements:* For master's, GRE General Test, GRE Subject Test; for doctorate, GRE General Test, GRE Subject Test (recommended), minimum GPA of 3.0. Additional exam requirements/recommendations for international students: Required—TOEFL. *Application deadline:* For fall admission, 1/5 for domestic students. Applications are processed on a rolling basis. Application fee: $50. *Expenses:* Tuition, state resident: full-time $10,440; part-time $435 per credit. Tuition, nonresident: full-time $15,520; part-time $647 per credit. Required fees: $129 per credit. Tuition and fees vary according to program. *Financial support:* In 2005–06, 30 students received support, including 11 fellowships with full tuition reimbursements available (averaging $24,000 per year), 14 research assistantships with full tuition reimbursements available (averaging $24,000 per year), 5 teaching assistantships with full tuition reimbursements available (averaging $16,988 per year); Federal Work-Study also available. Support available to part-time students. Financial award application deadline: 1/15; financial award applicants required to submit FAFSA. *Faculty research:* DNA replication and transcription, virus gene expression, tumor biology, structural biochemistry, signal transduction and molecular targeting. *Unit head:* Dr. Abram Gabriel, Acting Director, 732-235-5097, Fax: 732-445-6370, E-mail: gabriel@cabm-rutgers.edu. *Application contact:* Carolyn J. Ambrose, Administrative Assistant, 732-445-3430, Fax: 732-445-6370, E-mail: ambrose@biology.rutgers.edu.

Rutgers, The State University of New Jersey, New Brunswick/Piscataway, Graduate School, Program in Chemistry and Chemical Biology, New Brunswick, NJ 08901-1281. Offers analytical chemistry (MS, PhD); biological chemistry (PhD); chemistry education (MST); inorganic chemistry (MS, PhD); organic chemistry (MS, PhD); physical chemistry (MS, PhD). Part-time and evening/weekend programs available. *Faculty:* 48 full-time. *Students:* 104 full-time (41 women), 22 part-time (4 women); includes 11 minority (2 African Americans, 6 Asian Americans or Pacific Islanders, 3 Hispanic Americans), 51 international. Average age 29. 108 applicants, 51% accepted, 24 enrolled. In 2005, 13 master's, 10 doctorates awarded. Terminal master's awarded for partial completion of doctoral program. *Degree requirements:* For master's, thesis or alternative, exam, comprehensive exam, registration; for doctorate, thesis/dissertation, cumulative exams, 1 year residency, comprehensive exam, registration. *Entrance requirements:* For master's and doctorate, GRE General Test, GRE Subject Test. Additional exam requirements/recommendations for international students: Required—TOEFL. *Application deadline:* For fall admission, 4/15 priority date for domestic students, 4/1 priority date for international students; for spring admission, 12/1 priority date for domestic students, 12/1 priority date for international students. Applications are processed on a rolling basis. Application fee: $50. Electronic applications accepted. *Expenses:* Tuition, state resident: full-time $10,440; part-time $435 per credit. Tuition, nonresident: full-time $15,520; part-time $647 per credit. Required fees: $129 per credit. Tuition and fees vary according to program. *Financial support:* In 2005–06, 104 students received support, including 8 fellowships with full tuition reimbursements available (averaging $22,000 per year), 23 research assistantships with full tuition reimbursements available (averaging $18,347 per year), 59 teaching assistantships with full tuition reimbursements available (averaging $18,347 per year); career-related internships or fieldwork, Federal Work-Study, traineeships, and health care benefits also available. Financial award application deadline: 3/1; financial award applicants required to submit FAFSA. *Faculty research:* Biophysical organic/bioorganic, inorganic/bioinorganic, theoretical, and solid-state/surface chemistry. Total annual research expenditures: $14.5 million. *Unit head:* Dr. Roger Jones, Director and Chair, 732-445-4900, Fax: 732-445-5312, E-mail: jones@rutchem.rutgers.edu. *Application contact:* Dr. Martha A. Cotter, Vice Chair, 732-445-2259, Fax: 732-445-5312, E-mail: cotter@rutchem.rutgers.edu.

Saint Louis University, Graduate School and School of Medicine, Graduate Program in Biomedical Sciences and Graduate School, Department of Biochemistry and Molecular Biology, St. Louis, MO 63103-2097. Offers PhD. *Faculty:* 34 full-time (16 women). *Students:* 18 full-time (11 women); includes 1 minority (Asian American or Pacific Islander), 4 international. Average age 26. 5 applicants, 100% accepted, 3 enrolled. In 2005, 4 degrees awarded. *Degree requirements:* For doctorate, thesis/dissertation, departmental qualifying exams, comprehensive exam. *Entrance requirements:* For doctorate, GRE General Test, letters of recommendation, resumé. Additional exam requirements/recommendations for international students: Required—TOEFL (minimum score 550 paper-based; 213 computer-based). *Application deadline:* For fall admission, 7/1 for domestic students, 7/1 for international students; for spring admission, 11/1 for domestic students, 11/1 for international students. Applications are processed on a rolling basis. Application fee: $40. *Expenses:* Tuition: Full-time $760 per credit hour. Required fees: $55 per semester. *Financial support:* In 2005–06, 10 students received support, including 2 research assistantships (averaging $16,400 per year); fellowships with tuition reimbursements available, teaching assistantships, health care benefits and unspecified assistantships also available. Support available to part-time students. Financial award application deadline: 6/1; financial award applicants required to submit FAFSA. *Faculty research:* Transcription, chromatin modification and regulation of gene expression; structure/function of proteins and enzymes, including x-ray crystallography; inflammatory mediators in pathenogenesis of diabetes and arteriosclerosis; cellular signaling in response to growth factors, opiates and angiogenic mediators; genomics and proteomics of Cryptococcus neoformans. *Unit head:* Dr. William S. Sly, Chairperson, 314-977-9201, Fax: 314-577-8156, E-mail: slyws@slu.edu. *Application contact:* Gary Behrman, Associate Dean of the Graduate School, 314-977-3827, E-mail: behrmang@slu.edu.

San Francisco State University, Division of Graduate Studies, College of Science and Engineering, Department of Chemistry and Biochemistry, San Francisco, CA 94132-1722. Offers chemistry (MS), including biochemistry. Part-time programs available. *Degree requirements:* For master's, thesis, ACS exams in 3 chemical disciplines (including physical chemistry), essay test. *Entrance requirements:* For master's, minimum GPA of 2.5 in last 60 units. Additional exam requirements/recommendations for international students: Required—TOEFL (minimum score 550 paper-based; 213 computer-based). Electronic applications accepted. *Faculty research:* Physical chemistry of macromolecules, physical and synthetic organic chemistry, membrane and enzyme biochemistry, organometallic chemistry.

The Scripps Research Institute, Kellogg School of Science and Technology, Program in Chemistry and Chemical Biology, La Jolla, CA 92037. Offers chemical biology (PhD); chemistry (PhD). *Faculty:* 20 full-time (0 women). *Students:* 104 full-time (29 women). Average age 22. 291 applicants, 23% accepted, 24 enrolled. In 2005, 11 degrees awarded. *Degree requirements:* For doctorate, thesis/dissertation. *Entrance requirements:* For doctorate, GRE General Test, GRE Subject Test. Additional exam requirements/recommendations for international students: Required—TOEFL. *Application deadline:* For fall admission, 1/1 for domestic students, 1/1 for international students. Application fee: $0. *Expenses:* Tuition: Full-time $5,000. *Financial support:* Institutionally sponsored loans and stipends available. *Faculty research:* Synthetic organic chemistry and natural product synthesis, bio-organic chemistry and molecular design, biocatalysis and protein design, chemical biology. *Unit head:* Dr. James R. Williamson, Associate Dean, 858-784-8740. *Application contact:* Marylyn Rinaldi, Administrative Director, 858-784-8469, Fax: 858-784-2802, E-mail: mrinaldi@scripps.edu.

Seton Hall University, College of Arts and Sciences, Department of Chemistry and Biochemistry, South Orange, NJ 07079-2694. Offers analytical chemistry (MS, PhD); biochemistry (MS, PhD); chemistry (MS); inorganic chemistry (MS, PhD); organic chemistry (MS, PhD); physical chemistry (MS, PhD). Part-time and evening/weekend programs available. *Students:* 19 full-time (6 women), 46 part-time (17 women). Average age 34. 33 applicants, 79% accepted, 14 enrolled. In 2005, 11 master's, 7 doctorates awarded. Terminal master's awarded for partial completion of doctoral program. *Degree requirements:* For master's, formal seminar, thesis optional; for doctorate, thesis/dissertation, annual seminars, comprehensive exam. *Entrance requirements:* Additional exam requirements/recommendations for international students: Required—TOEFL. *Application deadline:* For fall admission, 7/1 priority date for domestic students, 7/1 priority date for international students; for spring admission, 11/1 priority date for domestic students, 11/1 priority date for international students. Applications are processed on a rolling basis. Application fee: $50. Electronic applications accepted. *Financial support:* In 2005–06, 1 research assistantship, 19 teaching assistantships were awarded; Federal Work-Study also available. *Faculty research:* DNA metal reactions; chromatography; bioinorganic, biophysical, organometallic, polymer chemistry; heterogeneous catalyst. *Unit head:* Dr. Nicholas Snow, Chair, 973-761-9414, Fax: 973-761-9772, E-mail: snownich@shu.edu. *Application contact:* Dr. Stephen Kelty, Director of Graduate Studies, 973-761-9129, Fax: 973-761-9772, E-mail: keltyste@shu.edu.

Simon Fraser University, Graduate Studies, Department of Molecular Biology and Biochemistry, Burnaby, BC V5A 1S6, Canada. Offers M Sc, PhD. *Degree requirements:* For master's and doctorate, thesis/dissertation. *Entrance requirements:* For master's, minimum GPA of 3.0; for doctorate, minimum GPA of 3.5. Additional exam requirements/recommendations for international students: Required—TWE or IELTS. *Faculty research:* Molecular genetics and development, biochemistry, molecular physiology, genomics, molecular phylogenetics and population genetics, bioinformation.

South Dakota State University, Graduate School, College of Arts and Science and College of Agriculture and Biological Sciences, Department of Chemistry, Brookings, SD 57007. Offers analytical chemistry (MS, PhD); biochemistry (MS, PhD); chemistry (MS, PhD); inorganic chemistry (MS, PhD); organic chemistry (MS, PhD); physical chemistry (MS, PhD). *Degree requirements:* For master's, oral exam; for doctorate, thesis/dissertation, preliminary oral and written exams, research tool. *Entrance requirements:* For master's, bachelor's degree in chemistry or equivalent. Additional exam requirements/recommendations for international students: Required—TOEFL. *Faculty research:* Environmental chemistry, computational chemistry, organic synthesis and photochemistry, novel material development and characterization.

Southern Illinois University Carbondale, Graduate School, College of Science, Department of Chemistry and Biochemistry, Carbondale, IL 62901-4701. Offers MS, PhD. Part-time programs available. *Faculty:* 18 full-time (1 woman), 2 part-time/adjunct (0 women). *Students:* 36 full-time (18 women), 19 part-time (10 women); includes 4 minority (all African Americans), 39 international. Average age 25. 53 applicants, 21% accepted, 6 enrolled. In 2005, 2 master's, 3 doctorates awarded. Terminal master's awarded for partial completion of doctoral program. *Degree requirements:* For master's, one foreign language, thesis; for doctorate, variable foreign language requirement, thesis/dissertation. *Entrance requirements:* For master's, minimum GPA of 2.7; for doctorate, GRE General Test, minimum GPA of 3.25. Additional exam requirements/recommendations for international students: Required—TOEFL. *Application deadline:* Applications are processed on a rolling basis. Application fee: $0. *Financial support:* In 2005–06, 17 research assistantships with full tuition reimbursements, 23 teaching assistantships with full tuition reimbursements were awarded; fellowships with full tuition reimbursements, Federal Work-Study, institutionally sponsored loans, and tuition waivers (full) also available. Support available to part-time students. *Faculty research:* Materials, separations, computational chemistry, synthetics. Total annual research expenditures: $1 million. *Unit head:* Lori Vermeuler, Chair, 618-453-6482, Fax: 618-453-6408. *Application contact:* Steve Scheiner, Chair, Graduate Admissions Committee, 618-453-6476, Fax: 618-453-6408, E-mail: scheiner@chem.siu.edu.

Southern Illinois University Carbondale, Graduate School, College of Science, Program in Molecular Biology, Microbiology, and Biochemistry, Carbondale, IL 62901-4701. Offers MS, PhD. *Faculty:* 16 full-time (2 women). *Students:* 39 full-time (23 women), 37 part-time (19 women); includes 11 minority (9 African Americans, 2 Asian Americans or Pacific Islanders), 40 international. Average age 25. 134 applicants, 16% accepted, 17 enrolled. In 2005, 7 master's, 4 doctorates awarded. *Degree requirements:* For master's and doctorate, thesis/dissertation. *Entrance requirements:* For master's, GRE, minimum GPA of 2.7; for doctorate, GRE, minimum GPA of 3.25. Additional exam requirements/recommendations for international students: Required—TOEFL. *Application deadline:* Applications are processed on a rolling basis. Application fee: $20. *Financial support:* In 2005–06, 40 students received support, including 3 fellowships with full tuition reimbursements available, 24 research assistantships with full tuition reimbursements available, 12 teaching assistantships with full tuition reimbursements available; Federal Work-Study and institutionally sponsored loans also available. Support available to part-time students. Financial award application deadline: 3/1. *Faculty research:* Prokaryotic gene regulation and expression; eukaryotic gene regulation; microbial, phylogenetic, and metabolic diversity; immune responses to tumors, pathogens, and autoantigens; protein folding and structure. *Unit head:* Dr. John Martinko, Director, 618-453-8116, Fax: 618-453-8036, E-mail: martinko.mbmb@science.siu.edu. *Application contact:* Donna Mueller, Office Systems Specialist II, 618-536-2349, Fax: 618-453-8036, E-mail: mueller@micro.siu.edu.

See Close-Up on page 595.

Southern University and Agricultural and Mechanical College, Graduate School, College of Sciences, Department of Chemistry, Baton Rouge, LA 70813. Offers analytical chemistry (MS); biochemistry (MS); environmental sciences (MS); inorganic chemistry (MS); organic chemistry (MS); physical chemistry (MS). *Faculty:* 9 full-time (2 women), 3 part-time/adjunct (2 women). *Students:* 20 full-time (13 women), 8 part-time (5 women); all minorities (20 African Americans, 8 Asian Americans or Pacific Islanders). Average age 23. 30 applicants, 70% accepted, 14 enrolled. In 2005, 3 master's awarded. *Degree requirements:* For master's, thesis. *Entrance requirements:* For master's, GMAT or GRE General Test. Additional exam requirements/recommendations for international students: Required—TOEFL (minimum score 525 paper-based; 193 computer-based). *Application deadline:* For fall admission, 4/15 for domestic students; for spring admission, 11/1 priority date for domestic students. Applications are processed on a rolling basis. Application fee: $5. *Financial support:* In 2005–06, 31 research assistantships (averaging $7,000 per year), 10 teaching assistantships (averaging $7,000 per year) were awarded; scholarships/grants also available. Financial award application deadline: 4/15. *Faculty research:* Synthesis of macrocyclic ligands, latex accelerators, anticancer drugs, biosensors, absorption isotheums, isolation of specific enzymes from plants. Total annual research expenditures: $400,000. *Unit head:* Dr. Ella Kelley, Chair, 225-771-3990, Fax: 225-771-3992.

Stanford University, School of Medicine, Graduate Programs in Medicine, Department of Biochemistry, Stanford, CA 94305-9991. Offers PhD. *Degree requirements:* For doctorate, thesis/dissertation. *Entrance requirements:* For doctorate, GRE General Test, GRE Subject Test (biology or chemistry). Additional exam requirements/recommendations for international students: Required—TOEFL. Electronic applications accepted. *Faculty research:* DNA replication, recombination, and gene regulation; methods of isolating, analyzing, and altering genes and genomes; protein structure, protein folding, and protein processing; protein targeting and transport in the cell; intercellular signaling.

See Close-Up on page 411.

State University of New York at Buffalo, Graduate School, School of Medicine and Biomedical Sciences, Graduate Programs in Medicine and Biomedical Sciences, Department of Biochemistry, Buffalo, NY 14214. Offers MA, PhD. *Faculty:* 17 full-time (5 women). *Students:* 28 full-time (18 women), 1 part-time; includes 1 minority (American Indian/Alaska Native), 15 international. Average age 29. 12 applicants, 33% accepted, 3 enrolled. In 2005, 1 master's, 4 doctorates awarded. Terminal master's awarded for partial completion of doctoral program. *Median time to degree:* Of those who began their doctoral program in fall 1997, 100% received their degree in 8 years or less. *Degree requirements:* For doctorate, thesis/dissertation. *Entrance requirements:* For master's, GRE General Test, GRE Subject Test (biology, chemistry, or biochemistry and molecular biology); for doctorate, GRE General Test, GRE Subject Test (biology, chemistry, or biochemistry and molecular biology), 3 letters of recommendation. Additional exam requirements/recommendations for international students: Required—TOEFL (minimum score 550 paper-based; 213 computer-based). *Application deadline:* For fall admission, 2/1 priority date for domestic students, 2/1 priority date for international students. Applications are processed on a rolling basis. Application fee: $35. Electronic applications accepted. *Financial support:* In 2005–06, 2 fellowships with full tuition reimbursements (averaging $11,850 per year), 23 research assistantships with full tuition reimbursements (averaging $21,000 per year) were awarded; teaching assistantships with full tuition reimbursements, Federal Work-Study, institutionally sponsored loans, scholarships/grants, health care benefits, and unspecified assistantships also available. Financial award application deadline: 2/1; financial award applicants required to submit FAFSA. *Faculty research:* Gene expression, proteins and metalloenzymes, biochemical endocrinology. Total annual research expenditures: $1.8 million. *Unit head:* Dr. Kenneth M. Blumenthal, Chair, 716-829-2727, Fax: 716-829-2725, E-mail: kblumen@buffalo.edu. *Application contact:* Dr. Daniel Kosman, Director of Graduate Studies, 716-829-2842, Fax: 716-829-2725, E-mail: camkos@acsu.buffalo.edu.

State University of New York College of Environmental Science and Forestry, Faculty of Chemistry, Syracuse, NY 13210-2779. Offers biochemistry (MS, PhD); environmental and forest chemistry (MS, PhD); organic chemistry of natural products (MS, PhD); polymer chemistry (MS, PhD). *Faculty:* 14 full-time (1 woman). *Students:* 28 full-time (15 women), 8 part-time (2 women); includes 1 minority (American Indian/Alaska Native), 11 international. Average age 28. 23 applicants, 74% accepted, 2 enrolled. In 2005, 2 master's, 5 doctorates awarded. *Degree requirements:* For master's, thesis/dissertation, registration; for doctorate, thesis/dissertation, comprehensive exam, registration. *Entrance requirements:* For master's and doctorate, GRE General Test, GRE Subject Test, minimum GPA of 3.0. Additional exam requirements/recommendations for international students: Required—TOEFL (minimum score 550 paper-based; 213 computer-based). *Application deadline:* For fall admission, 2/1 priority date for domestic students, 2/1 priority date for international students; for spring admission, 11/1 priority date for domestic students, 11/1 priority date for international students. Applications are processed on a rolling basis. Application fee: $60. Electronic applications accepted. *Expenses:* Tuition, area resident: Full-time $6,900; part-time $288 per credit. Tuition, state resident: full-time $10,920. Tuition, nonresident: full-time $10,920; part-time $455 per credit. International tuition: $10,920 full-time. Required fees: $395; $32 per credit. $20 per term. One-time fee: $145. *Financial support:* In 2005–06, 35 students received support, including 3 fellowships with full tuition reimbursements available (averaging $9,446 per year), 13 research assistantships with full tuition reimbursements available (averaging $12,500 per year), 10 teaching assistantships with full tuition reimbursements available (averaging $11,540 per year); Federal Work-Study, institutionally sponsored loans, scholarships/grants, health care benefits, and unspecified assistantships also available. Financial award application deadline: 6/30; financial award applicants required to submit FAFSA. *Faculty research:* Polymer chemistry, biochemistry. Total annual research expenditures: $1.8 million. *Unit head:* Dr. John P. Hassett, Chair, 315-470-6827, Fax: 315-470-6856, E-mail: jphasset@syr.edu. *Application contact:* Dr. Dudley J. Raynal, Dean, Instruction and Graduate Studies, 315-470-6599, Fax: 315-470-6978, E-mail: esfgrad@esf.edu.

State University of New York Upstate Medical University, College of Graduate Studies, Department of Biochemistry and Molecular Biology, Syracuse, NY 13210-2334. Offers MS, PhD, MD/PhD. Terminal master's awarded for partial completion of doctoral program. *Degree requirements:* For master's, thesis/dissertation; for doctorate, thesis/dissertation, comprehensive exam. *Entrance requirements:* For master's and doctorate, GRE General Test, GRE Subject Test, interview. Additional exam requirements/recommendations for international students: Required—TOEFL. *Faculty research:* Enzymology, membrane structure and functions, developmental biochemistry.

Stevens Institute of Technology, Graduate School, Arthur E. Imperatore School of Sciences and Arts, Department of Chemistry and Chemical Biology, Hoboken, NJ 07030. Offers chemical biology (MS, PhD); chemistry (MS, PhD), including analytical chemistry (MS, PhD, Certificate), chemical biology (MS, PhD, Certificate), organic chemistry, physical chemistry, polymer chemistry (MS, PhD, Certificate); chemistry and chemical biology (Certificate), including analytical chemistry (MS, PhD, Certificate), bioinformatics, biomedical chemistry, chemical biology (MS, PhD, Certificate), chemical physiology, polymer chemistry (MS, PhD, Certificate). Part-time and evening/weekend programs available. *Students:* 17 full-time (8 women), 28 part-time (16 women); includes 12 minority (9 Asian Americans or Pacific Islanders, 3 Hispanic Americans), 14 international. 51 applicants, 75% accepted. Terminal master's awarded for partial completion of doctoral program. *Degree requirements:* For master's, thesis or alternative; for doctorate, one foreign language, thesis/dissertation; for Certificate, project or thesis. *Entrance requirements:* Additional exam requirements/recommendations for international students: Required—TOEFL. *Application deadline:* Applications are processed on a rolling basis. Application fee: $50. Electronic applications accepted. *Expenses:* Tuition: Part-time $920 per credit hour. Tuition and fees vary according to program. *Financial support:* Fellowships, research assistantships, teaching assistantships, Federal Work-Study and institutionally sponsored loans available. Financial award application deadline: 4/1. *Unit head:* Dr. Francis T. Jones, Director, 201-216-5518, Fax: 201-216-8240.

Stony Brook University, State University of New York, Graduate School, College of Arts and Sciences, Department of Biochemistry and Cell Biology, Molecular and Cellular Biology Program, Specialization in Biochemistry and Molecular Biology, Stony Brook, NY 11794. Offers PhD. *Faculty:* 18 full-time (5 women). *Students:* 5 full-time (0 women); includes 1 minority (Asian American or Pacific Islander), 4 international. Average age 31. In 2005, 2 degrees awarded. *Degree requirements:* For doctorate, thesis/dissertation, teaching experience, comprehensive exam. *Entrance requirements:* For doctorate, GRE General Test, GRE Subject Test. Additional exam requirements/recommendations for international students: Required—TOEFL. *Application deadline:* For fall admission, 1/15 for domestic students. Applica-

tion fee: $50. *Expenses:* Tuition, state resident: full-time $6,900; part-time $288 per credit. Tuition, nonresident: full-time $10,920; part-time $455 per credit. Required fees: $704. *Financial support:* Fellowships, research assistantships, teaching assistantships, Federal Work-Study available. *Application contact:* Director, Graduate Program, 631-632-1210, E-mail: mcbprog@life.bio.sunysb.edu.

Announcement: The Molecular and Cellular Biology Program offers a specialization in molecular biology and biochemistry. Students can perform their PhD thesis research with one of nearly 100 outstanding research faculty members.

See Close-Up on page 413.

Stony Brook University, State University of New York, Graduate School, College of Arts and Sciences, Department of Biochemistry and Cell Biology, Program in Biochemistry and Structural Biology, Stony Brook, NY 11794. Offers PhD. *Students:* 31 full-time (17 women); includes 5 minority (1 African American, 3 Asian Americans or Pacific Islanders, 1 Hispanic American), 23 international. Average age 27. 110 applicants, 9% accepted. *Expenses:* Tuition, state resident: full-time $6,900; part-time $288 per credit. Tuition, nonresident: full-time $10,920; part-time $455 per credit. Required fees: $704. *Financial support:* In 2005–06, 22 research assistantships, 6 teaching assistantships were awarded. *Application contact:* Director, Graduate Program, 631-632-8533, Fax: 631-632-9730, E-mail: mcbprog@life.bio.sunysb.edu.

Announcement: The interdepartmental Graduate Program in Biochemistry and Structural Biology stresses biochemical, structural, and computational approaches to solving complex biological problems. With 30 faculty members and proximity to Brookhaven National Laboratory and Cold Spring Harbor Laboratory, the program provides unique opportunities for undertaking world-class research at the interface between biochemistry and structural biology.

See Close-Up on page 415.

Syracuse University, Graduate School, College of Arts and Sciences, Department of Biology and Department of Physics and Department of Chemistry, Program in Structural Biology, Biochemistry and Biophysics, Syracuse, NY 13244. Offers PhD. *Students:* 9 full-time (4 women), 4 international. 44 applicants, 14% accepted, 4 enrolled. *Degree requirements:* For doctorate, thesis/dissertation, exam. *Entrance requirements:* For doctorate, GRE General Test, GRE Subject Test. Additional exam requirements/recommendations for international students: Required—TOEFL. *Application deadline:* For fall admission, 1/10 for domestic students. Applications are processed on a rolling basis. Application fee: $65. Electronic applications accepted. *Financial support:* Fellowships, research assistantships, teaching assistantships, tuition waivers available. *Unit head:* Stewart Loh, Director, 315-464-8731, Fax: 315-443-4070. *Application contact:* Evelyn Lott, Information Contact, 315-443-9154, Fax: 315-443-2012, E-mail: ealott@syr.edu.

Temple University, Health Sciences Center, School of Medicine and Graduate School, Graduate Programs in Medicine, Department of Biochemistry, Philadelphia, PA 19122-6096. Offers MS, PhD. *Faculty:* 13 full-time (2 women). *Students:* 11 full-time (7 women), 6 part-time (3 women); includes 5 minority (1 African American, 3 Asian Americans or Pacific Islanders, 1 Hispanic American), 8 international. 20 applicants, 35% accepted, 5 enrolled. In 2005, 1 master's, 2 doctorates awarded. *Degree requirements:* For master's, thesis, research seminar; for doctorate, GRE General Test, GRE Subject Test, minimum GPA of 3.0. Additional exam requirements/recommendations for international students: Required—TOEFL (minimum score 650 paper-based; 280 computer-based). *Application deadline:* For fall admission, 4/15 priority date for domestic students, 12/15 priority date for international students; for spring admission, 11/15 priority date for domestic students, 8/1 priority date for international students. Applications are processed on a rolling basis. Application fee: $50. Electronic applications accepted. *Expenses:* Tuition, state resident: full-time $8,694; part-time $483 per credit. Tuition, nonresident: full-time $12,672; part-time $704 per credit. Required fees: $500; $122 per semester. Tuition and fees vary according to course level, campus/location and program. *Financial support:* Fellowships, research assistantships, Federal Work-Study and institutionally sponsored loans available. Financial award application deadline: 1/15; financial award applicants required to submit FAFSA. *Faculty research:* Metabolism, enzymology, molecular biology, membranology, biophysics. *Unit head:* Dr. Jimmy Collins, Acting Chair, 215-707-1549, Fax: 215-707-5963, E-mail: jcollins@temple.edu.

See Close-Up on page 417.

Texas A&M University, College of Agriculture and Life Sciences, Department of Biochemistry and Biophysics, College Station, TX 77843. Offers biochemistry (MS, PhD); biophysics (MS). *Faculty:* 16 full-time (3 women), 3 part-time/adjunct (0 women). *Students:* 116 full-time (59 women), 10 part-time (4 women); includes 15 minority (2 African Americans, 6 Asian Americans or Pacific Islanders, 7 Hispanic Americans), 64 international. Average age 27. 133 applicants, 42% accepted, 25 enrolled. In 2005, 3 master's, 13 doctorates awarded. *Entrance requirements:* For master's and doctorate, GRE General Test. Additional exam requirements/recommendations for international students: Required—TOEFL. *Application deadline:* For fall admission, 2/1 priority date for domestic students, 12/1 priority date for international students. Applications are processed on a rolling basis. Application fee: $50 ($75 for international students). Electronic applications accepted. *Expenses:* Tuition, state resident: full-time $4,488; part-time $187 per credit hour. Tuition, nonresident: full-time $11,112; part-time $463 per credit hour. Required fees: $1,974. *Financial support:* In 2005–06, 6 fellowships with tuition reimbursements (averaging $20,000 per year), 70 research assistantships with partial tuition reimbursements (averaging $20,000 per year) were awarded; teaching assistantships with partial tuition reimbursements, institutionally sponsored loans, scholarships/grants, traineeships, and unspecified assistantships also available. Financial award application deadline: 2/1; financial award applicants required to submit FAFSA. *Faculty research:* Enzymology, gene expression, protein structure, plant biochemistry. *Unit head:* Dr. Gregory D. Reinhart, Department Head, 979-845-5032, Fax: 979-845-9274, E-mail: gdr@tamu.edu. *Application contact:* Pat Swigert, Academic Advisor, 979-845-1779, Fax: 979-845-9274, E-mail: pat@tamu.edu.

Texas A&M University System Health Science Center, Graduate School of Biomedical Sciences, Department of Biochemistry and Structural Biology, College Station, TX 77840. Offers PhD. *Degree requirements:* For doctorate, thesis/dissertation. *Entrance requirements:* For doctorate, GRE General Test. *Faculty research:* Immunology, cell and membrane biology, protein biochemistry, molecular genetics, parasitology, vertebrate embryogenesis and microbiology.

Texas State University-San Marcos, Graduate School, College of Science, Department of Chemistry and Biochemistry, Program in Biochemistry, San Marcos, TX 78666. Offers MS. *Students:* 16 full-time (9 women), 7 part-time (2 women); includes 12 minority (1 African American, 2 Asian Americans or Pacific Islanders, 9 Hispanic Americans), 1 international. Average age 26. 9 applicants, 89% accepted, 6 enrolled. In 2005, 2 degrees awarded. *Degree requirements:* For master's, thesis. *Entrance requirements:* For master's, GRE General Test, bachelor's degree in chemistry, minimum GPA of 2.75 in last 60 hours of course work. *Application deadline:* For fall admission, 6/15 priority date for domestic students, 6/1 priority date for international students; for spring admission, 10/15 priority date for domestic students, 10/1 priority date for international students. Applications are processed on a rolling basis. Application fee: $40 ($90 for international students). *Expenses:* Tuition, area resident: Part-time $116 per credit. Tuition, state resident: full-time $3,168; part-time $176 per credit. Tuition, nonresident: full-time $8,136; part-time $452 per credit. Required fees: $1,112; $74 per credit. Full-time tuition and fees vary according to course load. *Financial support:* In 2005–06, 19 students received support; research assistantships, teaching assistantships available. Financial award application deadline: 4/1; financial award applicants required to submit FAFSA.

Texas Tech University Health Sciences Center, Graduate School of Biomedical Sciences, Department of Cell Biology and Biochemistry, Program in Biochemistry and Molecular Genet-

Biochemistry

Texas Tech University Health Sciences Center *(continued)*
ics, Lubbock, TX 79430. Offers MS, PhD, MD/PhD, MS/PhD. *Faculty:* 12 full-time (2 women), 10 part-time/adjunct (5 women). *Students:* 4 full-time (0 women), 3 international. Average age 28. 7 applicants, 0% accepted, 0 enrolled.Terminal master's awarded for partial completion of doctoral program. *Degree requirements:* For master's and doctorate, thesis/dissertation, preliminary, comprehensive, and final exams, comprehensive exam, registration. *Entrance requirements:* For master's and doctorate, GRE General Test, minimum GPA of 3.0. Additional exam requirements/recommendations for international students: Required—TOEFL. *Application deadline:* For fall admission, 5/15 priority date for domestic students, 4/15 priority date for international students; for spring admission, 11/15 for domestic students, 10/15 for international students. Application fee: $45. Electronic applications accepted. *Financial support:* In 2005–06, 1 fellowship with full tuition reimbursement (averaging $20,500 per year), 5 research assistantships (averaging $20,500 per year) were awarded; health care benefits also available. *Faculty research:* Reproductive endocrinology, immunology, developmental biochemistry. Total annual research expenditures: $2.6 million. *Unit head:* Dr. Sandra Whelly, Graduate Director, 806-743-2503, Fax: 806-743-2990, E-mail: sandra.whelly@ttuhsc.edu. *Application contact:* Pam Roddy, Assistant Director, 806-743-2701, Fax: 806-743-2990, E-mail: pam.roddy@ttuhsc.edu.

Thomas Jefferson University, Jefferson College of Graduate Studies, Program in Biochemistry and Molecular Biology, Philadelphia, PA 19107. Offers PhD. *Faculty:* 17 full-time (9 women), 9 part-time/adjunct (1 woman). *Students:* 15 full-time (8 women), 1 (woman) part-time; includes 4 minority (1 African American, 2 Asian Americans or Pacific Islanders, 1 Hispanic American), 1 international. Average age 24. 36 applicants, 17% accepted, 2 enrolled. In 2005, 3 doctorates awarded. *Degree requirements:* For doctorate, thesis/dissertation, preliminary exam, comprehensive exam, registration. *Entrance requirements:* For doctorate, GRE General Test or MCAT, minimum GPA of 3.2. Additional exam requirements/recommendations for international students: Required—TOEFL (minimum score 213 computer-based). *Application deadline:* For fall admission, 3/1 priority date for domestic students, 1/1 priority date for international students. Applications are processed on a rolling basis. Application fee: $50. Electronic applications accepted. *Expenses:* Tuition: Full-time $14,894; part-time $800 per credit. *Financial support:* In 2005–06, 3 students received support; fellowships with full tuition reimbursements available, research assistantships, Federal Work-Study, institutionally sponsored loans, scholarships/grants, traineeships, and stipends available. Financial award application deadline: 5/1; financial award applicants required to submit FAFSA. *Faculty research:* Signal transduction and molecular genetics, translational biochemistry, human mitochondrial genetics, molecular biology of protein-RNA interaction, mammalian mitochondrial biogenesis and function. Total annual research expenditures: $3 million. *Unit head:* Dr. Diane Merry, Program Director, 215-503-4907, Fax: 215-923-9162, E-mail: diane.merry@jefferson.edu. *Application contact:* Jessie F. Pervall, Director of Admissions, 215-503-0155, Fax: 215-503-9920, E-mail: jessie.pervall@jefferson.edu.

See Close-Up on page 419.

Thomas Jefferson University, Jefferson College of Graduate Studies, Program in Biomedical Chemistry, Philadelphia, PA 19107. Offers MS. Part-time and evening/weekend programs available. *Faculty:* 9 full-time (1 woman). *Students:* Average age 26. 32 applicants, 75% accepted, 21 enrolled. In 2005, 8 degrees awarded. *Degree requirements:* For master's, thesis, registration. *Entrance requirements:* For master's, GRE General Test, minimum GPA of 3.0. Additional exam requirements/recommendations for international students: Required—TOEFL (minimum score 213 computer-based). *Application deadline:* For fall admission, 8/1 priority date for domestic students, 3/1 priority date for international students. For winter admission, 12/1 for domestic students; for spring admission, 4/1 for domestic students. Applications are processed on a rolling basis. Application fee: $50. Electronic applications accepted. *Expenses:* Contact institution. *Financial support:* In 2005–06, 12 students received support. Federal Work-Study and institutionally sponsored loans available. Support available to part-time students. Financial award application deadline: 5/1; financial award applicants required to submit FAFSA. *Faculty research:* Genetics, protein chemistry, clinical enzymology, chemical carcinogenesis, autoimmunity. *Unit head:* Dr. Dennis M. Gross, Associate Dean, 215-503-0156, Fax: 215-503-3433, E-mail: dennis.gross@jefferson.edu. *Application contact:* Jessie F. Pervall, Director of Admissions, 215-503-0155, Fax: 215-503-9920, E-mail: jessie.pervall@jefferson.edu.

Tufts University, Sackler School of Graduate Biomedical Sciences, Department of Biochemistry, Boston, MA 02155. Offers PhD. Applications are processed through through integrated studies program. *Faculty:* 28 full-time (5 women). *Students:* 25 full-time (15 women); includes 3 minority (2 Asian Americans or Pacific Islanders, 1 Hispanic American), 11 international. Average age 29. *Degree requirements:* For doctorate, thesis/dissertation. *Financial support:* In 2005–06, 25 students received support, including 25 research assistantships with full tuition reimbursements available (averaging $29,000 per year); scholarships/grants and health care benefits also available. Financial award application deadline: 1/15. *Faculty research:* Enzymes and mechanisms, signal transduction, NMR spectroscopy, DNA biosynthesis, membrane function. *Unit head:* Dr. Larry Feig, Program Director, 617-636-6956, Fax: 617-636-2409, E-mail: larry.feig@tufts.edu. *Application contact:* 617-636-6767, Fax: 617-636-0375, E-mail: sacklerschool@tufts.edu.

Tufts University, Sackler School of Graduate Biomedical Sciences, Integrated Studies Program, Medford, MA 02155. Offers PhD. *Students:* 12 full-time (10 women); includes 2 minority (both Asian Americans or Pacific Islanders), 4 international. Average age 26. 264 applicants, 19% accepted, 12 enrolled. *Entrance requirements:* For doctorate, GRE General Test, 3 letters of reference. Additional exam requirements/recommendations for international students: Required—TOEFL. *Application deadline:* For fall admission, 1/15 priority date for domestic students, 1/15 priority date for international students. Application fee: $65. *Financial support:* In 2005–06, 12 students received support, including 12 research assistantships (averaging $29,000 per year) Financial award application deadline: 1/15.

Tulane University, School of Medicine and Graduate School, Graduate Programs in Medicine, Department of Biochemistry, New Orleans, LA 70118-5669. Offers MS, PhD, MD/PhD. MS and PhD offered through the Graduate School. *Degree requirements:* For master's, thesis; for doctorate, 2 foreign languages, thesis/dissertation. *Entrance requirements:* For master's, GRE General Test, GRE Subject Test, minimum B average in undergraduate course work; for doctorate, GRE General Test, GRE Subject Test. Additional exam requirements/recommendations for international students: Required—TOEFL or TSE. Electronic applications accepted. *Faculty research:* Nucleic acid chemistry, complex carbohydrates biochemistry.

See Close-Up on page 421.

Université de Moncton, Faculty of Science, Department of Chemistry and Biochemistry, Moncton, NB E1A 3E9, Canada. Offers biochemistry (M Sc); chemistry (M Sc). Part-time programs available. *Degree requirements:* For master's, one foreign language, thesis. *Entrance requirements:* For master's, minimum GPA of 3.0. Electronic applications accepted. *Faculty research:* Environmental contaminants, natural products synthesis, nutraceutical, organic catalysis, molecular biology of cancer.

Université de Montréal, Faculty of Medicine and Faculty of Graduate Studies, Graduate Programs in Medicine, Department of Biochemistry, Montréal, QC H3C 3J7, Canada. Offers biochemistry (M Sc, PhD); clinical biochemistry (DEPD). *Faculty:* 36 full-time (7 women). *Students:* 151 full-time (79 women), 4 part-time (1 woman). 102 applicants, 37% accepted, 32 enrolled. In 2005, 26 master's, 13 doctorates awarded. Terminal master's awarded for partial completion of doctoral program. *Degree requirements:* For master's, thesis; for doctorate, thesis/dissertation, general exam. *Entrance requirements:* For master's and doctorate, proficiency in French, knowledge of English; for DEPD, proficiency in French. *Application deadline:* For fall and spring admission, 2/1. For winter admission, 11/1 for domestic students. Application fee: $30. Electronic applications accepted. *Unit head:* Lea Brakier-Gungras, Director, 514-343-

6372, Fax: 514-343-2210, E-mail: lea.brakier.gingras@umontreal.ca. *Application contact:* Luc DesGroseillers, Graduate Chair, 514-343-5802, Fax: 514-343-2210, E-mail: degros@bcm.umontreal.ca.

Université de Montréal, Faculty of Medicine and Faculty of Graduate Studies, Graduate Programs in Medicine, Program in Specialized Studies, Montréal, QC H3C 3J7, Canada. Offers anesthesia (DESS); diagnostic radiology (DESS); family medicine (DESS); medical biochemistry (DESS); medical genetics (DESS); medicine (DESS); microbiology and infectious diseases (DESS); nuclear medicine (DESS); obstetrics and gynecology (DESS); ophthalmology (DESS); pediatrics (DESS); psychiatry (DESS); radiology-oncology (DESS); surgery (DESS). *Faculty:* 159 full-time (37 women), 345 part-time/adjunct (102 women). *Entrance requirements:* For degree, proficiency in French. *Application deadline:* For fall and spring admission, 2/1. For winter admission, 11/1 for domestic students. Application fee: $30. Electronic applications accepted. *Unit head:* Renée Roy, Vice Dean, 514-343-7798.

Université de Sherbrooke, Faculty of Medicine and Health Sciences, Graduate Programs in Medicine, Department of Biochemistry, Sherbrooke, QC J1K 2R1, Canada. Offers M Sc, PhD. *Faculty:* 1 (woman) full-time. *Students:* 18 full-time (11 women), 6 part-time (3 women). Average age 26. 8 applicants, 38% accepted, 2 enrolled. In 2005, 4 master's, 2 doctorates awarded. *Degree requirements:* For master's and doctorate, thesis/dissertation. *Application deadline:* For fall admission, 6/30 for domestic students; for winter admission, 10/31 for domestic students; for spring admission, 2/28 for domestic students. Application fee: $50. Electronic applications accepted. *Unit head:* Dr. Simon Labbé, Director, 819-564-5460, E-mail: simon.labbe@usherbrooke.ca.

Université Laval, Faculty of Medicine, Post-Professional Programs in Medical Studies, Québec, QC G1K 7P4, Canada. Offers anatomy–pathology (DESS); anesthesia–resuscitation (DESS); cardiology (DESS); care of older people (Diploma); clinical research (DESS); community health (DESS); dermatology (DESS); diagnostic radiology (DESS); emergency medicine (Diploma); family medicine (DESS); general surgery (DESS); geriatrics (DESS); hematology (DESS); internal medicine (DESS); maternal and fetal medicine (Diploma); medical biochemistry (DESS); medical microbiology and infectious diseases (DESS); medical oncology (DESS); nephrology (DESS); neurology (DESS); neurosurgery (DESS); obstetrics and gynecology (DESS); ophthalmology (DESS); orthopedic surgery (DESS); oto-rhino-laryngology (DESS); palliative medicine (Diploma); pediatrics (DESS); plastic surgery (DESS); psychiatry (DESS); pulmonary medicine (DESS); radiology–oncology (DESS); thoracic surgery (DESS); urology (DESS). *Degree requirements:* For other advanced degree, comprehensive exam. *Entrance requirements:* For degree, knowledge of French. Electronic applications accepted.

Université Laval, Faculty of Sciences and Engineering, Department of Biochemistry and Microbiology, Programs in Biochemistry, Québec, QC G1K 7P4, Canada. Offers M Sc, PhD. Terminal master's awarded for partial completion of doctoral program. *Degree requirements:* For master's, thesis/dissertation; for doctorate, thesis/dissertation, comprehensive exam. *Entrance requirements:* For master's and doctorate, knowledge of French, comprehension of written English. Electronic applications accepted.

University at Albany, State University of New York, School of Public Health, Department of Biomedical Sciences, Program in Biochemistry, Molecular Biology, and Genetics, Albany, NY 12222-0001. Offers MS, PhD. *Degree requirements:* For master's and doctorate, thesis/dissertation. *Entrance requirements:* For master's and doctorate, GRE General Test, GRE Subject Test. *Application deadline:* For fall admission, 2/15 for domestic students. Application fee: $60. *Financial support:* Application deadline: 2/1. *Unit head:* Dr. James Dias, Chair, Department of Biomedical Sciences, 518-474-2662.

The University of Alabama at Birmingham, Graduate Programs in Joint Health Sciences, Department of Biochemistry and Molecular Genetics, Birmingham, AL 35294. Offers biochemistry (PhD). The department participates in the Cellular and Molecular Biology Graduate Program. *Students:* 47 full-time (17 women); includes 9 minority (4 African Americans, 4 Asian Americans or Pacific Islanders, 1 Hispanic American), 24 international. 43 applicants, 37% accepted. In 2005, 8 degrees awarded. *Degree requirements:* For doctorate, thesis/dissertation. *Entrance requirements:* For doctorate, GRE General Test, interview. *Application deadline:* Applications are processed on a rolling basis. Electronic applications accepted. *Expenses:* Tuition, state resident: part-time $170 per credit hour. Tuition, nonresident: full-time $4,612; part-time $425 per credit hour. International tuition: $10,732 full-time. Required fees: $11 per credit hour. Tuition and fees vary according to course load, degree level and program. *Financial support:* In 2005–06, 8 fellowships were awarded *Unit head:* Dr. Tim M. Townes, Chair, 205-934-5294, E-mail: ttownes@uab.edu. *Application contact:* Information Contact, 205-934-6034, Fax: 205-975-2547.

See Close-Up on page 423.

The University of Alabama at Birmingham, Graduate Programs in Joint Health Sciences, Department of Cell Biology, Graduate Program in Cellular and Molecular Biology, Birmingham, AL 35294.

See Close-Ups on pages 601 and 603.

University of Alaska Fairbanks, College of Natural Sciences and Mathematics, Department of Chemistry and Biochemistry, Fairbanks, AK 99775-7520. Offers biochemistry and molecular biology (MS, PhD); chemistry (MA, MS); environmental chemistry (MS, PhD). Part-time programs available. *Faculty:* 13 full-time (3 women). *Students:* 30 full-time (13 women), 8 part-time (4 women); includes 6 minority (1 African American, 1 American Indian/Alaska Native, 2 Asian Americans or Pacific Islanders, 2 Hispanic Americans), 9 international. Average age 29. 35 applicants, 63% accepted, 9 enrolled. In 2005, 4 master's, 4 doctorates awarded. Terminal master's awarded for partial completion of doctoral program. *Degree requirements:* For master's, thesis, seminar, comprehensive exam, registration; for doctorate, thesis/dissertation, comprehensive exam, registration. *Entrance requirements:* For master's, GRE General Test; for doctorate, GRE General Test, GRE Subject Test (biology or chemistry). Additional exam requirements/recommendations for international students: Required—TOEFL (minimum score 550 paper-based; 213 computer-based). *Application deadline:* For fall admission, 6/1 for domestic students, 3/1 for international students; for spring admission, 12/1 for domestic students, 9/1 for international students. Applications are processed on a rolling basis. Application fee: $50. Electronic applications accepted. *Expenses:* Tuition, state resident: full-time $4,392; part-time $244 per credit. Tuition, nonresident: full-time $8,964; part-time $498 per credit. International tuition: $8,964 full-time. Required fees: $800; $5 per credit. $48 per contact hour. Tuition and fees vary according to course level, course load, campus/location and reciprocity agreements. *Financial support:* In 2005–06, 6 research assistantships with tuition reimbursements (averaging $9,595 per year), 16 teaching assistantships with tuition reimbursements (averaging $9,100 per year) were awarded; fellowships with tuition reimbursements, Federal Work-Study and scholarships/grants also available. Financial award applicants required to submit FAFSA. *Faculty research:* Atmospheric aerosols; plant chemistry; hibernation and neuroprotection; transition metal based drugs for diabetes; liganogated ion channels. *Unit head:* , Dr. Thomas Clausen, Chair, 907-474-5510, Fax: 907-474-5640, E-mail: fychem@uaf.edu.

University of Alberta, Faculty of Medicine and Dentistry and Faculty of Graduate Studies and Research, Graduate Programs in Medicine, Department of Biochemistry, Edmonton, AB T6G 2E1, Canada. Offers M Sc, PhD. Terminal master's awarded for partial completion of doctoral program. *Degree requirements:* For master's and doctorate, thesis/dissertation. *Entrance requirements:* For master's and doctorate, minimum GPA of 7.0 on a 9.0 scale. Tuition and fees charges are reported in Canadian dollars. *Expenses:* Tuition, state resident: part-time $562 Canadian dollars per term. Tuition, nonresident: full-time $3,375 Canadian dollars. Required fees: $573 Canadian dollars; $84 Canadian dollars per term. *Faculty research:* Proteins, lipids, nucleic acids, growth factors, membranes.

Biochemistry

The University of Arizona, College of Medicine, Graduate Programs in Medicine and Graduate College, Department of Biochemistry and Molecular Biophysics, Tucson, AZ 85721. Offers MS, PhD. Terminal master's awarded for partial completion of doctoral program. *Degree requirements:* For master's, thesis optional; for doctorate, thesis/dissertation. *Entrance requirements:* For master's and doctorate, GRE General Test. *Faculty research:* Membrane biochemistry, lipid biochemistry, neurobiochemistry, protein synthesis and degradation, bioenergetics.

University of Arkansas for Medical Sciences, College of Medicine and Graduate School, Graduate Programs in Medicine, Department of Biochemistry and Molecular Biology, Little Rock, AR 72205-7199. Offers MS, PhD, MD/PhD. *Faculty:* 26 full-time (7 women), 7 part-time/adjunct (0 women). *Students:* 16 full-time, 3 part-time. *Degree requirements:* For master's, thesis, comprehensive exam; for doctorate, thesis/dissertation, qualifying exam. *Entrance requirements:* For master's, GRE General Test, bachelor's degree in biology, chemistry, or related field; for doctorate, GRE General Test. Additional exam requirements/recommendations for international students: Required—TOEFL. Application fee: $0. *Financial support:* Research assistantships, unspecified assistantships available. Support available to part-time students. *Faculty research:* Gene regulation, growth factors, oncogenes, metabolic diseases, hormone regulation. *Unit head:* Dr. Alan D. Elbein, Chairman, 501-686-5185. *Application contact:* Dr. Kevin Raney, Graduate Coordinator, 501-686-5244, E-mail: raneykevind@uams.edu.

The University of British Columbia, Faculty of Medicine and Faculty of Graduate Studies, Graduate Programs in Medicine, Department of Biochemistry and Molecular Biology, Vancouver, BC V6T 1Z1, Canada. Offers M Sc, PhD. *Faculty:* 24 full-time (3 women). *Students:* 57 full-time (23 women). Average age 27. 76 applicants, 58% accepted, 11 enrolled. In 2005, 6 master's, 4 doctorates awarded. *Median time to degree:* Of those who began their doctoral program in fall 1997, 100% received their degree in 8 years or less. *Degree requirements:* For master's, thesis/dissertation; for doctorate, thesis/dissertation, comprehensive exam. *Entrance requirements:* For master's, GRE (international students), first class B Sc; for doctorate, GRE (international students), master's or first class honors bachelor's degree in biochemistry. Additional exam requirements/recommendations for international students: Required—TOEFL (minimum score 625 paper-based; 263 computer-based). *Application deadline:* For fall admission, 1/31 priority date for domestic students, 1/31 priority date for international students; for spring admission, 7/1 priority date for domestic students, 7/1 priority date for international students. Applications are processed on a rolling basis. Application fee: $90 ($150 for international students). Electronic applications accepted. *Financial support:* In 2005–06, 23 fellowships (averaging $21,400 per year), 23 research assistantships (averaging $20,300 per year), 15 teaching assistantships (averaging $2,300 per year) were awarded; institutionally sponsored loans, scholarships/grants, traineeships, tuition waivers (partial), and unspecified assistantships also available. *Faculty research:* Membrane biochemistry, protein structure/function, signal transduction, biochemistry. Total annual research expenditures: $12.6 million. *Unit head:* Dr. Christopher A. Proud, Head, 604-822-2792, Fax: 604-822-5227. *Application contact:* Hiltrud M. Vogler, Graduate Secretary, 604-822-5925, Fax: 604-822-5227, E-mail: biograd@interchange.ubc.ca.

University of Calgary, Faculty of Medicine and Faculty of Graduate Studies, Department of Biochemistry and Molecular Biology, Calgary, AB T2N 1N4, Canada. Offers M Sc, PhD. *Faculty:* 51 full-time (8 women), 19 part-time/adjunct (3 women). *Students:* 81 full-time (48 women), 1 part-time. Average age 27. 64 applicants, 17% accepted, 9 enrolled. In 2005, 7 master's, 3 doctorates awarded. *Degree requirements:* For master's, thesis; for doctorate, thesis/dissertation, candidacy exam. *Entrance requirements:* For master's and doctorate, GRE General Test, minimum GPA of 3.2. Additional exam requirements/recommendations for international students: Required—TOEFL. *Application deadline:* For fall admission, 6/15 priority date for domestic students, 5/15 priority date for international students. For winter admission, 10/15 for domestic students; for spring admission, 3/15 for domestic students. Applications are processed on a rolling basis. Application fee: $100 ($130 for international students). Electronic applications accepted. *Financial support:* In 2005–06, 28 fellowships, 19 research assistantships (averaging $4,100 per year), 3 teaching assistantships (averaging $6,110 per year) were awarded; stipends for Ph. D. students is $20,000 and M.Sc. students $18,000 also available. Financial award application deadline: 2/1. *Faculty research:* Molecular and developmental genetics; molecular biology of disease; genomics, proteomics and bioinformatics; ceu signaling and structure. *Unit head:* Dr. Jim McGhee, Graduate Coordinator, Fax: 403-210-8109, E-mail: jmcghee@ucalgary.ca. *Application contact:* Judy E. Gayford, Graduate Program Administrator, 403-220-8306, Fax: 403-210-8109, E-mail: bmbgrad@ucalgary.ca.

University of California, Berkeley, Graduate Division, Group in Comparative Biochemistry, Berkeley, CA 94720-1500. Offers MA, PhD. *Degree requirements:* For doctorate, thesis/dissertation, qualifying exam. *Entrance requirements:* For doctorate, GRE General Test, GRE Subject Test, minimum GPA of 3.0. Additional exam requirements/recommendations for international students: Required—TOEFL.

University of California, Davis, Graduate Studies, Graduate Group in Biochemistry and Molecular Biology, Davis, CA 95616. Offers MS, PhD. *Faculty:* 117 full-time. *Students:* 94 full-time (49 women); includes 28 minority (1 African American, 1 American Indian/Alaska Native, 22 Asian Americans or Pacific Islanders, 4 Hispanic Americans), 27 international. Average age 28. 207 applicants, 29% accepted, 18 enrolled. In 2005, 3 master's, 5 doctorates awarded. Terminal master's awarded for partial completion of doctoral program. *Median time to degree:* Of those who began their doctoral program in fall 1997, 66.7% received their degree in 8 years or less. *Degree requirements:* For master's, thesis (for some programs), comprehensive exam (for some programs); for doctorate, thesis/dissertation. *Entrance requirements:* For master's and doctorate, GRE General Test, GRE Subject Test. Additional exam requirements/recommendations for international students: Required—TOEFL (minimum score 550 paper-based; 213 computer-based). *Application deadline:* For fall admission, 1/15 for domestic students, 1/15 for international students. Application fee: $60. Electronic applications accepted. *Financial support:* In 2005–06, 86 students received support, including 16 fellowships with full and partial tuition reimbursements available (averaging $12,716 per year), 56 research assistantships with full and partial tuition reimbursements available (averaging $15,941 per year), 6 teaching assistantships with partial tuition reimbursements available (averaging $15,082 per year); Federal Work-Study, institutionally sponsored loans, scholarships/grants, tuition waivers (full and partial), and unspecified assistantships also available. Financial award application deadline: 1/15; financial award applicants required to submit FAFSA. *Faculty research:* Gene expression, protein structure, molecular virology, protein synthesis, enzymology, membrane transport and structural biology. *Unit head:* J. Clark Lagarias, Chair, 530-752-1865, E-mail: jclagarias@ucdavis.edu. *Application contact:* Angelina Kuo, Graduate Staff, 530-752-2981, Fax: 530-752-8391, E-mail: abkuo@ucdavis.edu.

University of California, Irvine, College of Medicine and School of Biological Sciences, Department of Biological Chemistry, Irvine, CA 92697. Offers biological sciences (MS, PhD). Students apply through the Graduate Program in Molecular Biology, Genetics, and Biochemistry. *Degree requirements:* For doctorate, thesis/dissertation. *Entrance requirements:* For master's, minimum GPA of 3.0; for doctorate, GRE General Test, GRE Subject Test, minimum GPA of 3.0. Additional exam requirements/recommendations for international students: Required—TOEFL (minimum score 550 paper-based; 213 computer-based), TSE. Electronic applications accepted. *Faculty research:* RNA splicing, mammalian chromosomal organization, membrane-hormone interactions, regulation of protein synthesis, molecular genetics of metabolic processes.

University of California, Irvine, Office of Graduate Studies, School of Biological Sciences, Department of Molecular Biology and Biochemistry, Irvine, CA 92697. Offers biological science (MS); biological sciences (PhD); biotechnology (MS). *Degree requirements:* For doctorate, thesis/dissertation. *Entrance requirements:* For master's, GRE, minimum GPA of 3.0; for doctorate, GRE General Test, GRE Subject Test, minimum GPA of 3.0. Additional exam requirements/recommendations for international students: Required—TOEFL (minimum score

550 paper-based; 213 computer-based), TSE. Electronic applications accepted. *Faculty research:* Structure and synthesis of nucleic acids and proteins, regulation, virology, biochemical genetics, gene organization.

University of California, Irvine, Office of Graduate Studies, School of Biological Sciences and College of Medicine, Graduate Program in Molecular Biology, Genetics, and Biochemistry, Irvine, CA 92697-3915. Offers biological sciences (PhD). *Faculty:* 156 full-time (38 women). *Students:* 53 full-time (30 women). Average age 25. 452 applicants, 26% accepted, 53 enrolled. In 2005, 19 doctorates awarded. *Median time to degree:* Of those who began their doctoral program in fall 1997, 10% received their degree in 8 years or less. *Degree requirements:* For doctorate, thesis/dissertation, teaching assignment, preliminary exam. *Entrance requirements:* For doctorate, GRE General Test, minimum GPA of 3.0, research experience. Additional exam requirements/recommendations for international students: Required—TOEFL, IELT, TSE. *Application deadline:* For fall admission, 1/1 for domestic students, 1/1 for international students. Application fee: $60 ($80 for international students). Electronic applications accepted. *Expenses: Contact institution. Financial support:* In 2005–06, 52 fellowships with full tuition reimbursements (averaging $25,000 per year) were awarded; institutionally sponsored loans, scholarships/grants, tuition waivers (full), and stipends also available. Financial award application deadline: 1/7; financial award applicants required to submit FAFSA. *Faculty research:* Cellular biochemistry; gene structure and expression; protein structure, function, and design; molecular genetics; pathogenesis and inherited disease. *Unit head:* Dr. Peter J. Bryant, Director, 949-824-4714, Fax: 949-824-3571, E-mail: gp-mbgb@uci.edu. *Application contact:* Kimberly McKinney, Administrator, 949-824-8145, Fax: 949-824-1965, E-mail: kamckinn@uci.edu.

University of California, Los Angeles, Graduate Division, College of Letters and Science, Department of Chemistry and Biochemistry, Program in Biochemistry and Molecular Biology, Los Angeles, CA 90095. Offers MS, PhD. MS admission to program only under exceptional circumstances. *Entrance requirements:* For master's, GRE General Test, GRE Subject Test, minimum GPA of 3.0; for doctorate, GRE General Test, GRE Subject Test, minimum undergraduate GPA of 3.0. Electronic applications accepted.

University of California, Los Angeles, School of Medicine and Graduate Division, Graduate Programs in Medicine, Department of Biological Chemistry, Los Angeles, CA 90095. Offers MS, PhD. *Degree requirements:* For master's, comprehensive exam or thesis; for doctorate, thesis/dissertation, oral and written qualifying exams. *Entrance requirements:* For master's and doctorate, GRE General Test.

University of California, Riverside, Graduate Division, Department of Biochemistry, Riverside, CA 92521-0102. Offers biochemistry and molecular biology (MS, PhD). Part-time programs available. *Faculty:* 39 full-time (11 women). *Students:* 50 full-time (22 women); includes 16 minority (1 African American, 13 Asian Americans or Pacific Islanders, 2 Hispanic Americans), 17 international. Average age 27. In 2005, 21 master's, 3 doctorates awarded. Terminal master's awarded for partial completion of doctoral program. *Degree requirements:* For master's, comprehensive exams or thesis; for doctorate, thesis/dissertation, 2 quarters of teaching experience, qualifying exams. *Entrance requirements:* For master's and doctorate, GRE General Test, minimum GPA of 3.2. Additional exam requirements/recommendations for international students: Required—TOEFL (minimum score 550 paper-based; 213 computer-based); Recommended—TSE (minimum score 50). *Application deadline:* For fall admission, 5/1 for domestic students, 2/1 for international students. For winter admission, 9/1 for domestic students; for spring admission, 12/1 for domestic students. Applications are processed on a rolling basis. Application fee: $60 ($75 for international students). Electronic applications accepted. *Expenses:* Tuition, nonresident: full-time $14,694. Required fees: $9,009. Full-time tuition and fees vary according to program. *Financial support:* In 2005–06, 5 fellowships with full tuition reimbursements (averaging $16,000 per year) were awarded; research assistantships, teaching assistantships, career-related internships or fieldwork, Federal Work-Study, institutionally sponsored loans, scholarships/grants, and tuition waivers (full and partial) also available. Financial award application deadline: 2/5; financial award applicants required to submit FAFSA. *Faculty research:* Structural biology and molecular biophysics, signal transduction, plant biochemistry and molecular biology, gene expression and metabolic regulation, molecular toxicology and pathogenesis. *Unit head:* Dr. Jolinda A. Traugh, Chair, 951-827-4239, Fax: 951-827-3590, E-mail: jolinda.traugh@ucr.edu. *Application contact:* Janet Fast, Graduate Program Assistant, 877-480-3487, Fax: 951-827-3590, E-mail: janet.fast@ucr.edu.

University of California, San Diego, Graduate Studies and Research, Department of Chemistry and Biochemistry, La Jolla, CA 92093. Offers chemistry (MS, PhD). *Degree requirements:* For doctorate, thesis/dissertation. *Entrance requirements:* For doctorate, GRE General Test, GRE Subject Test. Electronic applications accepted.

Announcement: Research and training at UCSD are at the cutting edge of the latest fields in chemistry, spanning the areas of biochemistry, biophysics, and analytical, inorganic, organic, physical, and theoretical chemistry. The doctoral program prepares students for research and teaching careers in academia and industry. Course work and research experiences are tailored to meet individual goals. Student support (stipend plus fees and tuition) is provided for all students through teaching and research assistantships, fellowships, and awards. Applications received by January 15 receive the highest priority. For more information, contact 858-534-6870, e-mail: gradinfo@chem.ucsd.edu.

University of California, San Diego, Graduate Studies and Research, Division of Biology, Program in Biochemistry, La Jolla, CA 92093-0348. Offers PhD. Offered in association with the Salk Institute. *Degree requirements:* For doctorate, thesis/dissertation, qualifying exam. Electronic applications accepted.

University of California, San Francisco, Graduate Division and School of Medicine, Department of Biochemistry and Biophysics, Program in Biochemistry and Molecular Biology, San Francisco, CA 94143. Offers PhD, MD/PhD. *Degree requirements:* For doctorate, thesis/dissertation. *Entrance requirements:* For doctorate, GRE General Test, GRE Subject Test. Additional exam requirements/recommendations for international students: Required—TOEFL. Expenses: Contact institution. *Faculty research:* Structural biology, genetics, cell biology, cell physiology, metabolism.

University of California, San Francisco, School of Pharmacy and Graduate Division, Chemistry and Chemical Biology Graduate Program, San Francisco, CA 94143. Offers PhD. *Faculty:* 48 full-time (9 women), 1 part-time/adjunct (0 women). *Students:* 51 full-time (18 women); includes 18 minority (3 African Americans, 11 Asian Americans or Pacific Islanders, 4 Hispanic Americans), 2 international. Average age 27. 136 applicants, 13% accepted. In 2005, 9 degrees awarded. *Median time to degree:* Of those who began their doctoral program in fall 1997, 100% received their degree in 8 years or less. *Degree requirements:* For doctorate, thesis/dissertation. *Entrance requirements:* For doctorate, GRE General Test, GRE Subject Test, minimum GPA of 3.0. Additional exam requirements/recommendations for international students: Required—TOEFL. *Application deadline:* For fall admission, 1/3 for domestic students. Applications are processed on a rolling basis. Application fee: $80. Electronic applications accepted. *Financial support:* In 2005–06, 21 fellowships with partial tuition reimbursements (averaging $16,880 per year), 13 research assistantships with full tuition reimbursements (averaging $25,000 per year), 3 teaching assistantships with partial tuition reimbursements (averaging $7,383 per year) were awarded; institutionally sponsored loans, scholarships/grants, traineeships, and tuition waivers (full) also available. Financial award application deadline: 1/10. *Faculty research:* Biochemistry; macromolecular structure; cellular and molecular pharmacology; physical chemistry and computational biology; synthetic chemistry. *Unit head:* Dr. Charles S. Craik, Director, 415-476-1913, Fax: 415-502-4690. *Application contact:* Christine Olson, Graduate Program Coordinator, 415-476-1914, Fax: 415-514-1546, E-mail: ccb@picasso.ucsf.edu.

University of California, Santa Barbara, Graduate Division, College of Letters and Sciences, Division of Mathematics, Life, and Physical Sciences, Department of Biomolecular Science and Engineering, Santa Barbara, CA 93106. Offers biochemistry and molecular biology (MS,

Biochemistry

University of California, Santa Barbara *(continued)*
PhD); biomolecular science and engineering (PhD); biophysics and bioengineering (MS, PhD). *Faculty:* 39 full-time (4 women). *Students:* 25 full-time (11 women); includes 4 minority (all Asian Americans or Pacific Islanders), 1 international. Average age 25. 70 applicants, 33% accepted, 4 enrolled. In 2005, 5 degrees awarded. Terminal master's awarded for partial completion of doctoral program. *Median time to degree:* Of those who began their doctoral program in fall 1997, 100% received their degree in 8 years or less. *Degree requirements:* For master's, thesis optional; for doctorate, thesis/dissertation, lab rotations, seminars, coursework, TA ships, research units, comprehensive exam, registration. *Entrance requirements:* For master's and doctorate, GRE General Test, GRE Subject Test. Additional exam requirements/recommendations for international students: Required—TOEFL (minimum score 630 paper-based; 267 computer-based). *Application deadline:* For fall admission, 12/15 for domestic students, 12/15 for international students. Application fee: $60. Electronic applications accepted. *Financial support:* In 2005–06, 3 fellowships with full and partial tuition reimbursements (averaging $21,000 per year), 15 research assistantships with full and partial tuition reimbursements (averaging $21,000 per year), 8 teaching assistantships with full and partial tuition reimbursements (averaging $21,000 per year) were awarded; career-related internships or fieldwork, Federal Work-Study, institutionally sponsored loans, scholarships/grants, traineeships, health care benefits, tuition waivers (full and partial), and unspecified assistantships also available. Financial award application deadline: 12/15; financial award applicants required to submit FAFSA. *Faculty research:* Genetics and biochemistry of bacterial gene expression; structure-function relationships in proteins and nucleic acids, protein chemistry; biochemistry and biophysics of marine adhesion. *Unit head:* Prof. Philip A. Pincus, Chair, 805-893-4685, E-mail: fyl@mrl.ucsb.edu. *Application contact:* Krista Grace, Staff Graduate Program Advisor, 805-893-2290, Fax: 805-893-4724, E-mail: grace@lifesci.ucsb.edu.

University of California, Santa Cruz, Division of Graduate Studies, Division of Physical and Biological Sciences, Department of Chemistry and Biochemistry, Santa Cruz, CA 95064. Offers MS, PhD. *Faculty:* 22 full-time (3 women). *Students:* 92 full-time (43 women); includes 27 minority (1 African American, 2 American Indian/Alaska Native, 14 Asian Americans or Pacific Islanders, 10 Hispanic Americans), 10 international. 142 applicants, 41% accepted, 21 enrolled. In 2005, 2 master's, 10 doctorates awarded. *Degree requirements:* For doctorate, one foreign language, thesis/dissertation, qualifying exam. *Entrance requirements:* For master's and doctorate, GRE General Test, GRE Subject Test. *Application deadline:* For fall admission, 1/15 for domestic students. Application fee: $60. *Expenses:* Tuition, nonresident: full-time $14,694. Required fees: $9,437. *Financial support:* Fellowships, research assistantships, teaching assistantships, Federal Work-Study and institutionally sponsored loans available. Financial award application deadline: 1/15. *Faculty research:* Marine chemistry; biochemistry; inorganic, organic, and physical chemistry. *Unit head:* Dr. Joseph Konopelski, Chair, 831-459-2067. *Application contact:* Evie Alloy, Department Assistant, 831-459-2023, E-mail: gradinfo@chemistry.ucsc.edu.

University of Chicago, Division of the Biological Sciences, Department of Molecular Biosciences: Biochemistry, Genetics, Cell and Developmental Biology, Department of Biochemistry and Molecular Biology, Chicago, IL 60637-1513. Offers PhD, MD/PhD. *Faculty:* 31 full-time (8 women). *Students:* 48 full-time (19 women); includes 14 minority (1 African American, 1 American Indian/Alaska Native, 10 Asian Americans or Pacific Islanders, 2 Hispanic Americans), 5 international. Average age 27. In 2005, 7 doctorates awarded. *Degree requirements:* For doctorate, one foreign language, thesis/dissertation, qualifying exam. *Entrance requirements:* For doctorate, GRE General Test, GRE Subject Test. Additional exam requirements/recommendations for international students: Required—TOEFL. *Application deadline:* For fall admission, 12/28 priority date for domestic students, 12/28 priority date for international students. Application fee: $55. Electronic applications accepted. *Financial support:* In 2005–06, 35 students received support, including fellowships with tuition reimbursements available (averaging $26,301 per year), research assistantships with tuition reimbursements available (averaging $26,301 per year); institutionally sponsored loans, scholarships/grants, traineeships, and health care benefits also available. Financial award applicants required to submit FAFSA. *Faculty research:* Molecular biology, gene expression, and DNA-protein interactions; membrane biochemistry, molecular endocrinology, and transmembrane signaling; enzyme mechanisms, physical biochemistry, and structural biology. Total annual research expenditures: $5 million. *Unit head:* Dr. Anthony A. Kossiakoff, Chairman, 773-702-9297, Fax: 773-702-0439, E-mail: koss@cummings.uchicago.edu. *Application contact:* Lisa Alvarez, Graduate Student Administrator, 773-702-9213, Fax: 773-702-0439, E-mail: lalvarez@bsd.uchicago.edu.

University of Cincinnati, Division of Research and Advanced Studies, College of Medicine, Graduate Programs in Biomedical Sciences, Department of Molecular Genetics, Biochemistry and Microbiology, Cincinnati, OH 45267. Offers MS, PhD. Terminal master's awarded for partial completion of doctoral program. *Degree requirements:* For master's, thesis or alternative; for doctorate, thesis/dissertation, qualifying exam. *Entrance requirements:* For master's, GRE General Test; for doctorate, GRE General Test, GRE Subject Test. Additional exam requirements/recommendations for international students: Required—TOEFL (minimum score 590 paper-based; 243 computer-based), TWE. Electronic applications accepted. *Faculty research:* Cancer biology and developmental genetics, gene regulation and chromosome structure, microbiology and pathogenic mechanismis, structural biology, membrane biology and signal transduction.

University of Cincinnati, Division of Research and Advanced Studies, McMicken College of Arts and Sciences, Department of Chemistry, Cincinnati, OH 45221. Offers analytical chemistry (MS, PhD); biochemistry (MS, PhD); inorganic chemistry (MS, PhD); organic chemistry (MS, PhD); physical chemistry (MS, PhD); polymer chemistry (MS, PhD); sensors (PhD). Part-time and evening/weekend programs available. Terminal master's awarded for partial completion of doctoral program. *Degree requirements:* For master's, thesis optional; for doctorate, thesis/dissertation, comprehensive exam, registration. *Entrance requirements:* For master's and doctorate, GRE General Test. Additional exam requirements/recommendations for international students: Required—TOEFL (minimum score 580 paper-based; 237 computer-based); Recommended—TSE(minimum score 50). Electronic applications accepted. *Faculty research:* Biomedical chemistry, laser chemistry, surface science, chemical sensors, synthesis.

University of Colorado at Boulder, Graduate School, College of Arts and Sciences, Department of Chemistry and Biochemistry, Boulder, CO 80309. Offers biochemistry (PhD); chemistry (MS). *Faculty:* 40 full-time (8 women). *Students:* 125 full-time (54 women), 61 part-time (24 women); includes 13 minority (1 African American, 1 American Indian/Alaska Native, 7 Asian Americans or Pacific Islanders, 4 Hispanic Americans), 13 international. Average age 27. 31 applicants, 100% accepted. In 2005, 12 master's, 23 doctorates awarded. *Degree requirements:* For master's, comprehensive exam or thesis; for doctorate, thesis/dissertation, cumulative exam, comprehensive exam. *Entrance requirements:* For master's, GRE General Test, GRE Subject Test, minimum GPA of 2.75; for doctorate, GRE General Test, GRE Subject Test, minimum GPA of 3.0. *Application deadline:* For fall admission, 2/28 priority date for domestic students, 12/1 priority date for international students. Applications are processed on a rolling basis. Application fee: $50 ($60 for international students). *Financial support:* In 2005–06, fellowships with full tuition reimbursements (averaging $6,060 per year), research assistantships with full tuition reimbursements (averaging $14,958 per year), teaching assistantships with full tuition reimbursements (averaging $13,932 per year) were awarded; institutionally sponsored loans, traineeships, and tuition waivers (full) also available. Support available to part-time students. Financial award application deadline: 2/28. *Faculty research:* Biochemistry, atmospheric chemistry, analytical chemistry, biophysical chemistry, chemical physics. Total annual research expenditures: $15.9 million. *Unit head:* Veronica Vaida, Chair, 303-492-6533, Fax: 303-492-5894, E-mail: chemdir@colorado.edu. *Application contact:* Graduate Program Assistant, 303-492-8978, Fax: 303-492-5894, E-mail: chem_gradstudents@colorado.edu.

University of Colorado at Denver and Health Sciences Center, Graduate School, Program in Biomedical Sciences, Department of Biochemistry and Molecular Genetics, Denver, CO

80262. Offers biochemistry (PhD). In 2005, 2 degrees awarded. *Degree requirements:* For doctorate, thesis/dissertation. *Entrance requirements:* For doctorate, GRE General Test. Additional exam requirements/recommendations for international students: Required—TOEFL (minimum score 550 paper-based; 213 computer-based). *Application deadline:* For fall admission, 1/15 for domestic students. Application fee: $50. Electronic applications accepted. *Expenses:* Tuition, state resident: full-time $11,730. Tuition, nonresident: full-time $22,980. Tuition and fees vary according to degree level and program. *Financial support:* Fellowships available. Financial award application deadline: 3/15; financial award applicants required to submit FAFSA. *Faculty research:* DNA damage, cancer and neurodegeneration, molecular mechanisms of pro-mRNA splicing, yeast RNA polymerases, DNA replication. *Unit head:* Dr. Tom Blumenthal, Chairman, 303-724-3203, E-mail: tom.blumenthal@uchsc.edu. *Application contact:* Diane Ross, Administrative Assistant, 303-315-7014, E-mail: diane.ross@uchsc.edu.

University of Connecticut, Graduate School, College of Liberal Arts and Sciences, Department of Molecular and Cell Biology, Field of Biochemistry, Storrs, CT 06269. Offers MS, PhD. *Faculty:* 19 full-time (9 women), 2 part-time (1 woman); includes 4 minority (1 American Indian/Alaska Native, 3 Asian Americans or Pacific Islanders), 3 international. Average age 27. 47 applicants, 21% accepted, 5 enrolled. In 2005, 3 master's, 1 doctorate awarded. Terminal master's awarded for partial completion of doctoral program. *Degree requirements:* For master's, comprehensive exam; for doctorate, thesis/dissertation. *Entrance requirements:* For master's and doctorate, GRE General Test, GRE Subject Test. Additional exam requirements/recommendations for international students: Required—TOEFL (minimum score 550 paper-based; 213 computer-based). *Application deadline:* For fall admission, 2/1 priority date for domestic students, 2/1 priority date for international students; for spring admission, 11/1 for domestic students, 10/1 for international students. Applications are processed on a rolling basis. Application fee: $55. Electronic applications accepted. *Expenses:* Tuition, state resident: part-time $444 per credit hour. Tuition, nonresident: part-time $1,154 per credit hour. Tuition and fees vary according to course load. *Financial support:* In 2005–06, 9 research assistantships with full tuition reimbursements, 7 teaching assistantships with full tuition reimbursements were awarded; fellowships, Federal Work-Study, scholarships/grants, health care benefits, and unspecified assistantships also available. Financial award application deadline: 2/1; financial award applicants required to submit FAFSA. *Application contact:* Anne St. Onje, Graduate Coordinator, 860-486-4314, Fax: 860-486-3943, E-mail: ann.st_onje@uconn.edu.

See Close-Up on page 611.

University of Connecticut Health Center, Graduate School, Programs in Biomedical Sciences, Program in Molecular Biology and Biochemistry, Farmington, CT 06030. Offers PhD, DMD/PhD, MD/PhD. *Degree requirements:* For doctorate, thesis/dissertation, comprehensive exam, registration. *Entrance requirements:* For doctorate, GRE General Test. Additional exam requirements/recommendations for international students: Required—TOEFL (minimum score 600 paper-based; 250 computer-based). Electronic applications accepted.

See Close-Up on page 615.

University of Delaware, College of Arts and Sciences, Department of Chemistry and Biochemistry, Newark, DE 19716. Offers biochemistry (MA, MS, PhD); chemistry (MA, MS, PhD). Part-time programs available. *Faculty:* 32 full-time (5 women), 2 part-time/adjunct (0 women). *Students:* 143 full-time (61 women), 11 part-time (3 women); includes 15 minority (4 African Americans, 8 Asian Americans or Pacific Islanders, 3 Hispanic Americans), 56 international. Average age 27. 147 applicants, 44% accepted, 28 enrolled. In 2005, 7 master's, 14 doctorates awarded. Terminal master's awarded for partial completion of doctoral program. *Degree requirements:* For master's, one foreign language, thesis (for some programs); for doctorate, one foreign language, thesis/dissertation, cumulative exam. *Entrance requirements:* For master's and doctorate, GRE General Test. Additional exam requirements/recommendations for international students: Required—TOEFL (minimum score 600 paper-based; 260 computer-based), TSE(minimum score 50). *Application deadline:* For fall admission, 3/31 priority date for domestic students, 3/31 priority date for international students. Applications are processed on a rolling basis. Application fee: $60. Electronic applications accepted. *Financial support:* In 2005–06, 89 students received support, including 4 fellowships with full tuition reimbursements available (averaging $21,750 per year), 72 research assistantships with full tuition reimbursements available (averaging $21,750 per year), 50 teaching assistantships with full tuition reimbursements available (averaging $21,750 per year) Financial award application deadline: 3/31. *Faculty research:* Protein studies; mechanism of enzymes; synthesis, electronic structure, and bonding of organic, inorganic and organometallic compounds; spectroscopy including single molecule spectroscopy. Total annual research expenditures: $7.8 million. *Unit head:* Dr. Charles G. Riordan, Chairman, 302-831-1247, Fax: 302-831-6335, E-mail: riordan@udel.edu. *Application contact:* Dr. Andrew Teplyakov, Graduate Coordinator, 302-831-1969, Fax: 302-831-6335, E-mail: andrewt@udel.edu.

University of Detroit Mercy, College of Engineering and Science, Department of Chemistry and Biochemistry, Detroit, MI 48219-0900. Offers macromolecular chemistry (MS). Evening/weekend programs available. *Degree requirements:* For master's, thesis. *Entrance requirements:* For master's, GRE General Test, minimum GPA of 3.0. *Faculty research:* Polymer and physical chemistry, industrial aspects of chemistry.

University of Florida, College of Medicine, Department of Biochemistry and Molecular Biology, Gainesville, FL 32611. Offers MS, PhD. *Faculty:* 34. *Students:* 7 applicants. In 2005, 1 degree awarded. *Degree requirements:* For doctorate, thesis/dissertation. *Entrance requirements:* For doctorate, GRE General Test, minimum GPA of 3.0. Additional exam requirements/recommendations for international students: Required—TOEFL. *Application deadline:* For fall admission, 2/15 for domestic students. Applications are processed on a rolling basis. Application fee: $30. Electronic applications accepted. *Expenses:* Tuition, state resident: full-time $6,234. Tuition, nonresident: full-time $21,359. Tuition and fees vary according to program. *Financial support:* In 2005–06, research assistantships with full tuition reimbursements (averaging $24,517 per year); fellowships with full tuition reimbursements, traineeships and unspecified assistantships also available. Financial award application deadline: 3/1. *Faculty research:* Gene expression, metabolic regulation, structural biology, enzyme mechanism, membrane transporters. *Unit head:* Dr. James Flanegan, Director, 352-392-0688, E-mail: flanegan@ufl.edu. *Application contact:* Dr. Wayne McCormack, Program Director, 352-392-7413, Fax: 352-846-3466, E-mail: idp@ufl.edu.

University of Florida, College of Medicine and Graduate School, Interdisciplinary Program in Biomedical Sciences, Concentration in Biochemistry and Molecular Biology, Gainesville, FL 32611. Offers PhD. *Faculty:* 27 full-time (7 women). *Students:* 28 full-time (15 women); includes 6 minority (3 African Americans, 3 Asian Americans or Pacific Islanders). In 2005, 1 degree awarded. *Degree requirements:* For doctorate, thesis/dissertation. *Entrance requirements:* For doctorate, GRE General Test, minimum GPA of 3.0. Additional exam requirements/recommendations for international students: Required—TOEFL. *Application deadline:* For fall admission, 2/5 for domestic students. Applications are processed on a rolling basis. Application fee: $30. Electronic applications accepted. *Expenses:* Tuition, state resident: full-time $6,234. Tuition, nonresident: full-time $21,359. Tuition and fees vary according to program. *Financial support:* In 2005–06, 26 research assistantships with full tuition reimbursements (averaging $24,517 per year) were awarded; fellowships with full tuition reimbursements *Faculty research:* Gene expression, metabolic regulation, structural biology, enzyme mechanism, membrane transporters. *Unit head:* Dr. Michael S. Kilberg, Director, 352-392-2711, E-mail: mkilberg@biochem.med.ufl.edu. *Application contact:* Dr. Wayne McCormack, Associate Dean of Graduate Education, 352-392-7413, Fax: 352-846-3466, E-mail: mccormac@pathology.ufl.edu.

University of Georgia, Graduate School, College of Arts and Sciences, Department of Biochemistry and Molecular Biology, Athens, GA 30602. Offers MS, PhD. *Faculty:* 18 full-time (2 women). *Students:* 45 full-time, 4 part-time; includes 8 minority (5 African Americans, 2 Asian Americans or Pacific Islanders, 1 Hispanic American), 11 international. 94 applicants, 12% accepted, 9 enrolled. In 2005, 4 master's, 7 doctorates awarded. *Degree requirements:* For master's and doctorate, one foreign language, thesis/dissertation. *Entrance requirements:* For master's and doctorate, GRE General Test. Additional exam requirements/recommendations for international students: Required—TOEFL, TSE. *Application deadline:* For fall admission, 1/1 priority date for domestic students, 1/1 priority date for international students. Application fee: $50. Electronic applications accepted. *Financial support:* Fellowships, research assistantships, teaching assistantships, scholarships/grants and unspecified assistantships available. Financial award application deadline: 1/1. *Unit head:* Dr. J. David Puett, Head, 706-542-1676, Fax: 706-542-0182, E-mail: puett@bmb.uga.edu. *Application contact:* Dr. Alan E. Przybyla, Graduate Coordinator, 706-542-1728, Fax: 706-542-1738, E-mail: przybyla@bmb.uga.edu.

University of Guelph, Graduate Program Services, College of Biological Science, Department of Molecular and Cellular Biology, Guelph, ON N1G 2W1, Canada. Offers biochemistry (M Sc, PhD); biophysics (M Sc, PhD); botany (M Sc, PhD); microbiology (M Sc, PhD); molecular biology and genetics (M Sc, PhD). *Faculty:* 43 full-time (6 women). *Students:* 136 full-time (52 women). Average age 26. 56 applicants, 30% accepted, 17 enrolled. In 2005, 4 master's, 2 doctorates awarded. *Degree requirements:* For master's, thesis/dissertation, research proposal; for doctorate, thesis/dissertation, research proposal, comprehensive exam, registration. *Entrance requirements:* Additional exam requirements/recommendations for international students: Required—TOEFL (minimum score 550 paper-based; 213 computer-based), IELT (minimum score 7). *Application deadline:* For fall admission, 7/1 for domestic students, 1/1 for international students. For winter admission, 11/1 for domestic students; for spring admission, 3/1 for domestic students. Applications are processed on a rolling basis. Application fee: $75. Electronic applications accepted. *Financial support:* Fellowships, research assistantships, teaching assistantships available. Support available to part-time students. *Faculty research:* Physiology, structure, genetics, and ecology of microbes; virology and microbial technology. *Unit head:* Dr. Chris Whitfield, Chair, 519-824-4120 Ext. 53361, Fax: 519-827-1802, E-mail: cwhitfie@uoguelph.ca. *Application contact:* Laurie Winn, Graduate Admissions Secretary, 519-824-4320 Ext. 52730, Fax: 519-767-1656, E-mail: lwinn@uoguelph.ca.

University of Guelph, Graduate Program Services, College of Physical and Engineering Science, Guelph-Waterloo Centre for Graduate Work in Chemistry and Biochemistry, Guelph, ON N1G 2W1, Canada. Offers biochemistry (M Sc, PhD); chemistry (M Sc, PhD). Part-time programs available. *Faculty:* 56 full-time (8 women). *Students:* 189 full-time (41 women), 7 part-time (3 women). In 2005, 28 master's, 17 doctorates awarded. *Degree requirements:* For master's and doctorate, thesis/dissertation. *Application deadline:* Applications are processed on a rolling basis. Application fee: $75. *Financial support:* Fellowships, research assistantships, teaching assistantships available. *Faculty research:* Inorganic, analytical, biological, physical/theoretical, polymer, and organic chemistry. *Unit head:* Dr. A. L. Schwan, Director, 519-824-4120 Ext. 53848, Fax: 519-823-8097, E-mail: gwc@uoguelph.ca. *Application contact:* A. Wetmore, Administrative Assistant, 519-824-4120 Ext. 53848, Fax: 519-823-8097, E-mail: gwc@uoguelph.ca.

University of Houston, College of Natural Sciences and Mathematics, Department of Biology and Biochemistry, Houston, TX 77204. Offers biochemistry (MA, MS, PhD); biology (MA, MS, PhD). *Faculty:* 24 full-time (4 women), 2 part-time/adjunct (1 woman). *Students:* 107 full-time (46 women), 6 part-time (1 woman); includes 9 minority (2 African Americans, 1 American Indian/Alaska Native, 4 Asian Americans or Pacific Islanders, 2 Hispanic Americans), 80 international. Average age 28. 37 applicants, 76% accepted, 21 enrolled. In 2005, 8 master's, 11 doctorates awarded. Terminal master's awarded for partial completion of doctoral program. *Degree requirements:* For master's, thesis (for some programs); for doctorate, thesis/dissertation, comprehensive exam. *Entrance requirements:* For master's and doctorate, GRE General Test. Additional exam requirements/recommendations for international students: Required—TSE or SPEAK Test. *Application deadline:* For fall admission, 4/7 for domestic students; for spring admission, 10/2 priority date for domestic students. Applications are processed on a rolling basis. Application fee: $0 ($75 for international students). *Financial support:* In 2005–06, research assistantships with full tuition reimbursements (averaging $14,300 per year), 62 teaching assistantships with full tuition reimbursements (averaging $14,300 per year) were awarded; fellowships with full tuition reimbursements, career-related internships or fieldwork, Federal Work-Study, institutionally sponsored loans, scholarships/grants, health care benefits, and unspecified assistantships also available. Support available to part-time students. Financial award application deadline: 3/10. *Faculty research:* Evolutionary biology, neuroscience, infectious diseases, circadian rhythm, protein structure. *Unit head:* Dr. Stuart Dryer, Chairman, 713-743-2666, Fax: 713-743-2632, E-mail: sdryer@uh.edu. *Application contact:* Elizabeth Bullock, Graduate Adviser and Program Coordinator, 713-743-2633, Fax: 713-743-2899, E-mail: ebullock@uh.edu.

University of Idaho, College of Graduate Studies, College of Agricultural and Life Sciences, Department of Microbiology, Molecular Biology and Biochemistry, Moscow, ID 83844-2282. Offers MS, PhD. *Students:* 30 full-time (15 women), 8 part-time (3 women), 19 international. Average age 30. In 2005, 5 master's, 4 doctorates awarded. *Degree requirements:* For master's, thesis; for doctorate, one foreign language, thesis/dissertation. *Entrance requirements:* For master's, minimum GPA of 2.8; for doctorate, minimum undergraduate GPA of 2.8, 3.0 graduate. *Application deadline:* For fall admission, 8/1 for domestic students; for spring admission, 12/15 for domestic students. Application fee: $55 ($60 for international students). *Expenses:* Tuition, state resident: full-time $4,508. Tuition, nonresident: full-time $8,770; part-time $130 per credit. Required fees: $217 per credit. *Financial support:* Research assistantships, teaching assistantships available. Financial award application deadline: 2/15. *Faculty research:* Environmental fields, food, immunology. *Unit head:* Dr. Patricia Hartzell, Head, 208-885-0572.

University of Illinois at Chicago, College of Medicine and Graduate College, Graduate Programs in Medicine, Department of Biochemistry and Molecular Biology, Chicago, IL 60607-7128. Offers MS, PhD. Terminal master's awarded for partial completion of doctoral program. *Degree requirements:* For master's and doctorate, thesis/dissertation. *Entrance requirements:* For master's and doctorate, GRE General Test. Additional exam requirements/recommendations for international students: Required—TOEFL. Electronic applications accepted. *Faculty research:* Nature of cellular components, control of metabolic processes, regulation of gene expression.

University of Illinois at Urbana–Champaign, Graduate College, College of Liberal Arts and Sciences, School of Chemical Sciences, Champaign, IL 61820. Offers MS, PhD. *Faculty:* 53 full-time (6 women), 2 part-time/adjunct (1 woman). *Students:* 411 full-time (136 women), 4 part-time (1 woman); includes 36 minority (7 African Americans, 25 Asian Americans or Pacific Islanders, 4 Hispanic Americans), 113 international. 629 applicants, 22% accepted, 87 enrolled. In 2005, 34 master's, 47 doctorates awarded. *Degree requirements:* For doctorate, thesis/dissertation. *Entrance requirements:* For master's, minimum GPA of 3.0. *Application deadline:* Applications are processed on a rolling basis. Application fee: $50 ($60 for international students). Electronic applications accepted. *Expenses:* Contact institution. *Financial support:* In 2005–06, 116 fellowships, 251 research assistantships, 267 teaching assistantships were awarded; career-related internships or fieldwork, Federal Work-Study, institutionally sponsored loans, and tuition waivers (full and partial) also available. Financial award application deadline: 2/15. *Unit head:* Thomas B. Rauchfuss, Director, 217-333-5070, Fax: 217-333-3120, E-mail: rauchfuz@uiuc.edu. *Application contact:* Cheryl Kappes, Administrative Aide, 217-333-5070, Fax: 217-333-3120, E-mail: dambache@uiuc.edu.

University of Illinois at Urbana–Champaign, Graduate College, College of Liberal Arts and Sciences, School of Molecular and Cellular Biology, Department of Biochemistry, Champaign, IL 61820. Offers MS, PhD. *Faculty:* 10 full-time (1 woman). *Students:* 100 full-time (41 women); includes 20 minority (3 African Americans, 15 Asian Americans or Pacific Islanders, 2 Hispanic Americans), 38 international. 95 applicants, 20% accepted, 11 enrolled. In 2005, 5

master's, 9 doctorates awarded. *Degree requirements:* For doctorate, one foreign language, thesis/dissertation. *Entrance requirements:* For master's, GRE General Test, GRE Subject Test, minimum GPA of 3.0; for doctorate, GRE General Test, GRE Subject Test. *Application deadline:* Applications are processed on a rolling basis. Application fee: $50 ($60 for international students). Electronic applications accepted. *Financial support:* In 2005–06, 24 fellowships, 74 research assistantships, 63 teaching assistantships were awarded; traineeships and tuition waivers (full and partial) also available. Financial award application deadline: 2/15. *Unit head:* Colin A. Wraight, Interim Head, 217-333-3945, Fax: 217-333-8920, E-mail: cwraight@uiuc.edu. *Application contact:* Louise R. Cox, Assistant to the Head, 217-333-7149, Fax: 217-333-8920, E-mail: l-cox1@uiuc.edu.

See Close-Up on page 425.

The University of Iowa, Roy J. and Lucille A. Carver College of Medicine and Graduate College, Graduate Programs in Medicine, Department of Biochemistry, Iowa City, IA 52242-1316. Offers MS, PhD, MD/PhD. *Faculty:* 17 full-time (4 women), 5 part-time/adjunct (4 women). *Students:* 32 full-time (13 women), 11 international. Average age 26. 191 applicants, 6% accepted, 4 enrolled. In 2005, 4 master's, 2 doctorates awarded. Terminal master's awarded for partial completion of doctoral program. *Median time to degree:* Of those who began their doctoral program in fall 1997, 100% received their degree in 8 years or less. *Degree requirements:* For master's, thesis, registration; for doctorate, thesis/dissertation, research project, comprehensive exam, registration. *Entrance requirements:* For master's, GRE General Test; for doctorate, GRE. Additional exam requirements/recommendations for international students: Required—TOEFL (minimum score 600 paper-based; 250 computer-based). *Application deadline:* For fall admission, 2/1 priority date for domestic students, 2/1 priority date for international students. Applications are processed on a rolling basis. Application fee: $60 ($80 for international students). Electronic applications accepted. *Expenses:* Tuition, state resident: part-time $1,882 per term. Tuition, nonresident: full-time $17,338; part-time $4,907 per term. Tuition and fees vary according to course load and program. *Financial support:* In 2005–06, fellowships with full tuition reimbursements (averaging $22,000 per year), research assistantships with full tuition reimbursements (averaging $22,000 per year), teaching assistantships with full tuition reimbursements (averaging $22,000 per year) were awarded; institutionally sponsored loans, scholarships/grants, traineeships, tuition waivers, and unspecified assistantships also available. *Faculty research:* Regulation of gene expression, protein structure, membrane structure/function, DNA structure and replication. Total annual research expenditures: $5.7 million. *Unit head:* Dr. John E. Donelson, Head, 319-335-7934, Fax: 319-335-9570, E-mail: john-donelson@uiowa.edu. *Application contact:* Admissions Committee, 319-335-7932, Fax: 319-335-9570, E-mail: janis-shields@uiowa.edu.

See Close-Up on page 427.

University of Kansas, Graduate School, College of Liberal Arts and Sciences, Division of Biological Sciences, Department of Molecular Biosciences, Lawrence, KS 66045. Offers biochemistry and biophysics (MA, PhD); microbiology (MA, PhD); molecular, cellular, and developmental biology (MA, PhD). *Faculty:* 44. *Students:* 48 full-time (28 women), 5 part-time (3 women); includes 1 minority (Asian American or Pacific Islander), 26 international. Average age 27. 111 applicants, 19% accepted. In 2005, 5 master's, 5 doctorates awarded. Terminal master's awarded for partial completion of doctoral program. *Degree requirements:* For master's and doctorate, thesis/dissertation, comprehensive exam. *Entrance requirements:* For master's and doctorate, GRE General Test. Additional exam requirements/recommendations for international students: Required—TOEFL, IELT, TOEFL (paper 570; computer 230) or IELT; Recommended—TWE, TSE. *Application deadline:* For fall admission, 1/15 priority date for domestic students, 1/15 priority date for international students. Applications are processed on a rolling basis. Application fee: $55 ($60 for international students). Electronic applications accepted. *Expenses:* Tuition, state resident: full-time $4,859. Tuition, nonresident: full-time $12,000. Required fees: $589. Tuition and fees vary according to program. *Financial support:* Fellowships, research assistantships with tuition reimbursements, teaching assistantships with tuition reimbursements available. Financial award application deadline: 3/1. *Faculty research:* Structure and function of proteins, genetics of organism development, molecular genetics, neurophysiology, molecular virology and pathogenics, developmental biology, cell biology. *Unit head:* Kathy Suprenant, Chair, 785-864-4631, Fax: 785-864-5294, E-mail: ksupre@ku.edu. *Application contact:* John P. Connolly, Information Contact, 785-864-4311, Fax: 785-864-5924, E-mail: jconnolly@ku.edu.

University of Kansas, Graduate Studies Medical Center, Interdisciplinary Graduate Program in Biomedical Sciences, Department of Biochemistry and Molecular Biology, Lawrence, KS 66045. Offers MS, PhD, MD/PhD. Part-time programs available. *Faculty:* 16. *Students:* 2 full-time (0 women), 6 part-time (2 women), includes 1 Hispanic American, 1 International. Average age 29. In 2005, 1 degree awarded. Terminal master's awarded for partial completion of doctoral program. *Median time to degree:* Of those who began their doctoral program in fall 1997, 100% received their degree in 8 years or less. *Degree requirements:* For master's, oral defense of thesis; for doctorate, comprehensive oral and written exam. *Expenses:* Tuition, state resident: full-time $4,859. Tuition, nonresident: full-time $12,000. Required fees: $589. Tuition and fees vary according to program. *Financial support:* Fellowships, research assistantships with partial tuition reimbursements, teaching assistantships with full and partial tuition reimbursements, unspecified assistantships available. *Faculty research:* Regulation of gene expression, molecular genetics of kidney disease, molecular chaperones, glycoproteins, cell cycle control. Total annual research expenditures: $1.9 million. *Unit head:* Dr. Gerald M. Carlson, Chairman, 913-588-6574, Fax: 913-588-7007, E-mail: gcarlson@kumc.edu. *Application contact:* Dr. Glen Andrews, Director of Graduate Studies, 913-588-6935, Fax: 913-588-7440, E-mail: gandrews@kumc.edu.

See Close-Up on page 429.

University of Kentucky, Graduate School, Graduate School Programs from the College of Medicine, Program in Molecular and Cellular Biochemistry, Lexington, KY 40506-0032. Offers biochemistry (PhD). *Faculty:* 26 full-time (4 women), 1 part-time/adjunct (0 women). *Students:* 35 full-time (17 women); includes 1 minority (African American), 11 international. Average age 28. 40 applicants, 90% accepted, 31 enrolled. In 2005, 5 degrees awarded. *Median time to degree:* Of those who began their doctoral program in fall 1997, 84% received their degree in 8 years or less. *Degree requirements:* For doctorate, thesis/dissertation, comprehensive exam. *Entrance requirements:* For doctorate, GRE General Test, minimum undergraduate GPA of 3.0. Additional exam requirements/recommendations for international students: Required—TOEFL (minimum score 550 paper-based; 213 computer-based). *Application deadline:* For fall admission, 7/17 priority date for domestic students, 2/1 priority date for international students; for spring admission, 12/13 priority date for domestic students, 6/15 priority date for international students. Applications are processed on a rolling basis. Application fee: $40 ($55 for international students). Electronic applications accepted. *Expenses:* Tuition, state resident: full-time $6,308; part-time $331 per credit hour. Tuition, nonresident: full-time $13,968; part-time $756 per credit hour. Tuition and fees vary according to course load, degree level and program. *Financial support:* In 2005–06, 4 fellowships with full tuition reimbursements (averaging $10,000 per year), 29 research assistantships with full tuition reimbursements (averaging $21,000 per year), 5 teaching assistantships with full tuition reimbursements (averaging $20,000 per year) were awarded; Federal Work-Study, scholarships/grants, traineeships, health care benefits, tuition waivers (partial), and unspecified assistantships also available. Support available to part-time students. Financial award application deadline: 3/15. *Unit head:* Dr. Kevin Sarge, Director of Graduate Studies, 859-257-5777, Fax: 859-323-1037, E-mail: kdsarge@pop.uky.edu. *Application contact:* Dr. Brian Jackson, Senior Associate Dean, 859-257-8176, Fax: 859-323-1928.

University of Lethbridge, School of Graduate Studies, Lethbridge, AB T1K 3M4, Canada. Offers accounting (MScM); addictions counseling (M Sc); agricultural biotechnology (M Sc); agricultural studies (M Sc, MA); anthropology (MA); archaeology (MA); art (MA); biochemistry (M Sc); biological sciences (M Sc); biomolecular science (PhD); biosystems and biodiversity

Biochemistry

University of Lethbridge *(continued)*
(PhD); Canadian studies (MA); chemistry (M Sc); computer science (M Sc); computer science and geographical information science (M Sc); counseling psychology (M Ed); dramatic arts (MA); earth, space, and physical science (PhD); economics (MA); educational leadership (M Ed); English (MA); environmental science (M Sc); evolution and behavior (PhD); exercise science (M Sc); finance (MScM); French (MA); French/German (MA); French/Spanish (MA); general education (M Ed); general management (MScM); geography (M Sc, MA); German (MA); health sciences (MA); history (MA); human resource management and labour relations (MScM); individualized multidisciplinary (M Sc, MA); information systems (MScM); international management (MScM); kinesiology (M Sc, MA); management (M Sc, MA); marketing (MScM); mathematics (M Sc); music (MA); Native American studies (MA); neuroscience (M Sc, PhD); new media (M Sc); nursing (M Sc); philosophy (M Sc); physics (M Sc); policy and strategy (MScM); political science (MA); psychology (M Sc, MA); religious studies (MA); sociology (MA); theoretical and computational science (PhD); urban and regional studies (MA). Part-time and evening/weekend programs available. *Faculty:* 250. *Students:* 193 full-time, 145 part-time. 35 applicants, 100% accepted, 35 enrolled. In 2005, 40 degrees awarded. *Degree requirements:* For doctorate, thesis/dissertation, comprehensive exam. *Entrance requirements:* For master's, GMAT (M Sc management), bachelor's degree in related field, minimum GPA of 3.0 during previous 20 graded semester courses, 2 years teaching or related experience (M Ed); for doctorate, master's degree, minimum graduate GPA of 3.5. Additional exam requirements/recommendations for international students: Required—TOEFL. Application fee: $60 Canadian dollars. *Expenses:* Tuition, nonresident: part-time $531 per course. Required fees: $83 per year. Tuition and fees vary according to degree level and program. *Financial support:* Fellowships, research assistantships, teaching assistantships, scholarships/grants, health care benefits, and unspecified assistantships available. *Faculty research:* Movement and brain plasticity, gibberellin physiology, photosynthesis, carbon cycling, molecular properties of main-group ring components. *Unit head:* Dr. Shamsul Alam, Dean, 403-329-2121, Fax: 403-329-2097, E-mail: inquiries@uleth.ca. *Application contact:* Kathy Schrage, Administrative Assistant, Office of the Academic Vice President, 403-329-2121, Fax: 403-329-2097, E-mail: inquiries@uleth.ca.

University of Louisville, Graduate School, College of Arts and Sciences, Department of Chemistry, Louisville, KY 40292-0001. Offers analytical chemistry (MS, PhD); biochemistry (MS, PhD); chemical physics (PhD); inorganic chemistry (MS, PhD); organic chemistry (MS, PhD); physical chemistry (MS, PhD). *Students:* 37 full-time (16 women), 16 part-time (7 women); includes 3 minority (2 African Americans, 1 Hispanic American), 24 international. Average age 28. In 2005, 1 master's, 8 doctorates awarded. *Degree requirements:* For master's, thesis/dissertation; for doctorate, thesis/dissertation, comprehensive exam. *Entrance requirements:* For master's and doctorate, GRE General Test. Additional exam requirements/recommendations for international students: Required—TOEFL. *Application deadline:* Applications are processed on a rolling basis. Application fee: $50. *Expenses:* Tuition, state resident: full-time $3,003; part-time $334 per credit hour. Tuition, nonresident: full-time $8,277; part-time $920 per credit hour. Tuition and fees vary according to course load, degree level and program. *Financial support:* In 2005–06, 33 teaching assistantships with tuition reimbursements were awarded; fellowships, research assistantships *Unit head:* Dr. George R. Pack, Chair, 502-852-6798, Fax: 502-852-8149, E-mail: george.pack@louisville.edu.

University of Louisville, School of Medicine, Department of Biochemistry and Molecular Biology, Louisville, KY 40292-0001. Offers MS, PhD, MD/MS, MD/PhD. *Students:* 28 full-time (15 women), 13 part-time (6 women); includes 7 minority (3 African Americans, 3 Asian Americans or Pacific Islanders, 1 Hispanic American), 10 international. Average age 30. In 2005, 2 master's, 3 doctorates awarded. *Degree requirements:* For master's, thesis/dissertation; for doctorate, thesis/dissertation, comprehensive exam. *Entrance requirements:* For master's and doctorate, GRE General Test, minimum GPA of 3.0. Additional exam requirements/recommendations for international students: Required—TOEFL. *Application deadline:* For fall admission, 1/15 for domestic students. Application fee: $50. Electronic applications accepted. *Expenses:* Tuition, state resident: full-time $6,006; part-time $334 per credit hour. Tuition, nonresident: full-time $16,554; part-time $920 per credit hour. Tuition and fees vary according to course load, degree level and program. *Financial support:* Fellowships with tuition reimbursements, research assistantships with tuition reimbursements, teaching assistantships with tuition reimbursements, scholarships/grants, traineeships, tuition waivers (full and partial), and unspecified assistantships available. *Unit head:* Dr. Kenneth S. Ramos, Chair, 502-852-5217, Fax: 502-852-6222, E-mail: ksramos01@louisville.edu. *Application contact:* Dr. Steven R. Ellis, Contact, 502-852-5221, Fax: 502-852-6222, E-mail: srellis@louisville.edu.

University of Maine, Graduate School, College of Natural Sciences, Forestry, and Agriculture, Department of Biochemistry, Molecular Biology, and Microbiology, Orono, ME 04469. Offers biochemistry (MPS, MS); biochemistry and molecular biology (PhD); microbiology (MPS, MS, PhD). *Students:* 40 full-time (23 women), 17 part-time (11 women); includes 2 minority (1 Asian American or Pacific Islander, 1 Hispanic American), 13 international. Average age 31. 42 applicants, 48% accepted, 11 enrolled. In 2005, 5 master's awarded. *Degree requirements:* For doctorate, thesis/dissertation. *Entrance requirements:* For master's and doctorate, GRE General Test. Additional exam requirements/recommendations for international students: Required—TOEFL. *Application deadline:* For fall admission, 2/1 for domestic students. Applications are processed on a rolling basis. Application fee: $50. Electronic applications accepted. *Financial support:* In 2005–06, 5 research assistantships with tuition reimbursements (averaging $18,000 per year), 12 teaching assistantships with tuition reimbursements (averaging $16,000 per year) were awarded; tuition waivers (full and partial) also available. Financial award application deadline: 3/1. *Unit head:* Dr. John Singer, Chair, 207-581-2810, Fax: 207-581-2801. *Application contact:* Scott G. Delcourt, Associate Dean of the Graduate School, 207-581-3219, Fax: 207-581-3232, E-mail: graduate@maine.edu.

University of Manitoba, Faculty of Medicine and Faculty of Graduate Studies, Graduate Programs in Medicine, Department of Biochemistry and Medical Genetics, Winnipeg, MB R3T 2N2, Canada. Offers M Sc, PhD. Terminal master's awarded for partial completion of doctoral program. *Degree requirements:* For master's and doctorate, thesis/dissertation. *Faculty research:* Cancer, gene expression, membrane lipids, metabolic control, genetic diseases.

University of Maryland, Graduate School, Graduate Programs in Medicine, Department of Biochemistry and Molecular Biology, Baltimore, MD 21201. Offers biochemistry (PhD). *Faculty:* 23 full-time (3 women), 2 part-time/adjunct. *Students:* 12 full-time (9 women), 14 part-time (9 women); includes 3 African Americans, 2 Hispanic Americans, 9 international. Average age 28. 25 applicants, 28% accepted, 5 enrolled. In 2005, 7 degrees awarded. *Degree requirements:* For doctorate, thesis/dissertation. *Entrance requirements:* For doctorate, GRE General Test, GRE Subject Test. Additional exam requirements/recommendations for international students: Required—TOEFL, TOEFL or IELTS; Recommended—IELT. *Application deadline:* For fall admission, 7/1 for domestic students, 1/15 for international students. Application fee: $50. Electronic applications accepted. *Expenses:* Tuition, state resident: full-time $8,079; part-time $409 per credit hour. Tuition, nonresident: full-time $18,384; part-time $731 per credit hour. Required fees: $695; $10 per credit hour. Tuition and fees vary according to degree level and program. *Financial support:* Fellowships, research assistantships available. Financial award application deadline: 2/15. *Faculty research:* Membrane transport, hormonal regulation, protein structure, molecular virology. *Unit head:* Dr. Giuseppe Inesi, Chairman, 410-706-3220, Fax: 410-706-8297, E-mail: ginesi@umaryland.edu. *Application contact:* Dr. David Weber, Program Director, 410-706-6304, Fax: 410-706-8297, E-mail: dweber@umaryland.edu.

University of Maryland, Graduate School, Interdisciplinary Training Program in Muscle Biology, Baltimore, MD 21201.

See Close-Up on page 431.

University of Maryland, School of Medicine, Graduate Program in Life Sciences, Program in Biochemistry and Molecular Biology, Baltimore, MD 21201. Offers PhD. *Expenses:* Tuition, state resident: full-time $8,079; part-time $409 per credit hour. Tuition, nonresident: full-time $18,384; part-time $731 per credit hour. Required fees: $695; $10 per credit hour. Tuition and fees vary according to degree level and program.

University of Maryland, Baltimore County, Graduate School, College of Natural and Mathematical Sciences, Department of Chemistry and Biochemistry, Program in Biochemistry, Baltimore, MD 21250. Offers biochemistry (PhD); neuroscience (PhD). *Students:* Average age 25. *Degree requirements:* For doctorate, thesis/dissertation, comprehensive exam. *Entrance requirements:* For doctorate, GRE General Test, minimum GPA of 3.0. Additional exam requirements/recommendations for international students: Required—TOEFL (minimum score 550 paper-based; 213 computer-based). *Application deadline:* For fall admission, 7/1 priority date for domestic students, 1/1 priority date for international students; for spring admission, 12/1 for domestic students. Applications are processed on a rolling basis. Application fee: $45. Electronic applications accepted. *Expenses:* Tuition, state resident: part-time $395 per credit. Tuition, nonresident: part-time $652 per credit. Tuition and fees vary according to course load, program and reciprocity agreements. *Financial support:* Fellowships with full tuition reimbursements, research assistantships with full tuition reimbursements, teaching assistantships with tuition reimbursements, tuition waivers (full) available. *Unit head:* , Dr. Michael F. Summers, Graduate Coordinator, 410-455-2527. *Application contact:* Kathleen Reinecke, Biochemistry Coordinator, 410-706-8417, Fax: 410-706-8297, E-mail: kreineck@umaryland.edu.

See Close-Up on page 433.

University of Maryland, College Park, Graduate Studies, College of Chemical and Life Sciences, Department of Chemistry and Biochemistry, Biochemistry Program, College Park, MD 20742. Offers MS, PhD. Part-time and evening/weekend programs available. *Students:* 25 full-time (15 women), 1 part-time; includes 4 minority (3 Asian Americans or Pacific Islanders, 1 Hispanic American), 11 international. 34 applicants, 12% accepted, 4 enrolled. In 2005, 2 master's, 4 doctorates awarded. Terminal master's awarded for partial completion of doctoral program. *Median time to degree:* Of those who began their doctoral program in fall 1997, 60% received their degree in 8 years or less. *Degree requirements:* For master's, thesis or alternative; for doctorate, thesis/dissertation, 2 seminar presentations, oral exam. *Entrance requirements:* For master's and doctorate, GRE General Test, GRE Subject Test (recommended), minimum GPA of 3.0, 3 letters of recommendation. Additional exam requirements/recommendations for international students: Required—TOEFL. *Application deadline:* For fall admission, 4/1 for domestic students, 2/1 for international students; for spring admission, 10/21 for domestic students, 6/1 for international students. Applications are processed on a rolling basis. Application fee: $60. Electronic applications accepted. *Financial support:* In 2005–06, 1 fellowship (averaging $4,000 per year) was awarded; research assistantships, teaching assistantships with partial tuition reimbursements, Federal Work-Study also available. Support available to part-time students. Financial award applicants required to submit FAFSA. *Faculty research:* Analytical biochemistry, immunochemistry, drug metabolism, biosynthesis of proteins, mass spectrometry. *Application contact:* Dean of Graduate School, 301-405-4190, Fax: 301-314-9305.

University of Massachusetts Amherst, Graduate School, College of Natural Sciences and Mathematics, Department of Biochemistry and Molecular Biology, Amherst, MA 01003. Offers biochemistry (MS, PhD). PhD offered through the Department of Molecular and Cellular Biology. Part-time programs available. *Faculty:* 16 full-time (7 women). *Students:* 3 full-time (2 women), 2 part-time (1 woman); includes 1 minority (Asian American or Pacific Islander) Average age 24. 8 applicants, 100% accepted, 6 enrolled. In 2005, 5 degrees awarded. Terminal master's awarded for partial completion of doctoral program. *Degree requirements:* For master's, thesis or alternative; for doctorate, one foreign language, thesis/dissertation. *Entrance requirements:* Additional exam requirements/recommendations for international students: Required—TOEFL (minimum score 530 paper-based; 197 computer-based). *Application deadline:* For fall admission, 2/1 priority date for domestic students, 2/1 priority date for international students. Applications are processed on a rolling basis. Application fee: $40 ($65 for international students). Electronic applications accepted. *Expenses:* Tuition, state resident: part-time $110 per credit. Tuition, nonresident: part-time $414 per credit. Required fees: $2,824 per term. One-time fee: $250 part-time. Full-time tuition and fees vary according to course load, campus/location, program and reciprocity agreements. *Financial support:* In 2005–06, fellowships with full tuition reimbursements (averaging $8,731 per year), research assistantships with full tuition reimbursements (averaging $10,560 per year), teaching assistantships with full tuition reimbursements (averaging $8,009 per year) were awarded; career-related internships or fieldwork, Federal Work-Study, scholarships/grants, traineeships, and unspecified assistantships also available. Support available to part-time students. Financial award application deadline: 1/15. *Unit head:* Dr. Danny Schnell, Director, 413-545-0353, Fax: 413-545-4490.

University of Massachusetts Amherst, Graduate School, College of Natural Sciences and Mathematics, Program in Molecular and Cellular Biology, Amherst, MA 01003. Offers biological chemistry (PhD); cell and developmental biology (PhD). Part-time programs available. *Faculty:* 1 full-time (0 women). *Students:* 87 full-time (43 women), 1 part-time; includes 9 minority (1 African American, 4 Asian Americans or Pacific Islanders, 4 Hispanic Americans), 41 international. Average age 30. 219 applicants, 23% accepted, 13 enrolled. In 2005, 8 doctorates awarded. *Degree requirements:* For doctorate, one foreign language, thesis/dissertation. *Entrance requirements:* For master's and doctorate, GRE General Test. Additional exam requirements/recommendations for international students: Required—TOEFL (minimum score 530 paper-based; 197 computer-based). *Application deadline:* For fall admission, 1/15 priority date for domestic students, 1/15 priority date for international students. Applications are processed on a rolling basis. Application fee: $40 ($65 for international students). Electronic applications accepted. *Expenses:* Tuition, state resident: part-time $110 per credit. Tuition, nonresident: part-time $414 per credit. Required fees: $2,824 per term. One-time fee: $250 part-time. Full-time tuition and fees vary according to course load, campus/location, program and reciprocity agreements. *Financial support:* In 2005–06, 6 fellowships with full tuition reimbursements (averaging $3,451 per year), 4 teaching assistantships with full tuition reimbursements (averaging $8,696 per year) were awarded; career-related internships or fieldwork, Federal Work-Study, scholarships/grants, traineeships, and unspecified assistantships also available. Support available to part-time students. Financial award application deadline: 1/15. *Unit head:* Dr. Rodney K. Murphey, Head, 413-545-3246, Fax: 413-545-1812.

See Close-Up on page 619.

University of Massachusetts Lowell, Graduate School, College of Arts and Sciences, Department of Biological Sciences, Lowell, MA 01854-2881. Offers biochemistry (PhD); biological sciences (MS); biotechnology (MS). Part-time programs available. *Degree requirements:* For master's and doctorate, thesis/dissertation. *Entrance requirements:* For master's and doctorate, GRE General Test. Electronic applications accepted.

University of Massachusetts Lowell, Graduate School, College of Arts and Sciences, Department of Chemistry, Lowell, MA 01854-2881. Offers biochemistry (PhD); chemistry (MS, PhD); environmental studies (PhD); polymer sciences (MS, PhD). Terminal master's awarded for partial completion of doctoral program. *Degree requirements:* For master's, thesis; for doctorate, 2 foreign languages, thesis/dissertation. *Entrance requirements:* For master's and doctorate, GRE General Test. Electronic applications accepted.

University of Massachusetts Worcester, Graduate School of Biomedical Sciences, Department of Biochemistry and Molecular Pharmacology, Worcester, MA 01655-0115. Offers PhD. *Faculty:* 34 full-time (5 women). *Degree requirements:* For doctorate, thesis/dissertation. *Entrance requirements:* For doctorate, GRE General Test. Additional exam requirements/recommendations for international students: Required—TOEFL (minimum score 600 paper-based; 250 computer-based). *Application deadline:* For fall admission, 12/15 for domestic students, 12/15 for international students. Applications are processed on a rolling basis.

Application fee: $25 ($50 for international students). *Expenses:* Tuition, state resident: full-time $2,640. Tuition, nonresident: full-time $9,856. Required fees: $5,685. *Financial support:* In 2005–06, research assistantships with full tuition reimbursements (averaging $25,235 per year); unspecified assistantships also available. *Faculty research:* Neuropharmacology, control of DNA metabolism, clinical pharmacology and toxicology, chemical biology. *Unit head:* Dr. C. Robert Matthews, Chair, 508-856-2251, Fax: 508-856-2151. *Application contact:* Michael Cole, Director of Admissions and Recruitment, 508-856-4779, Fax: 508-856-3659, E-mail: michael.cole@umassmed.edu.

University of Medicine and Dentistry of New Jersey, Graduate School of Biomedical Sciences, Graduate Programs in Biomedical Sciences–Newark, Department of Biochemistry and Molecular Biology, Newark, NJ 07107. Offers MS, PhD. *Degree requirements:* For master's, thesis; for doctorate, thesis/dissertation, qualifying exam. *Entrance requirements:* For master's and doctorate, GRE General Test. Additional exam requirements/recommendations for international students: Required—TOEFL. Application fee: $40. *Financial support:* Fellowships, research assistantships, Federal Work-Study, institutionally sponsored loans, and tuition waivers (full and partial) available. Financial award application deadline: 5/1. *Unit head:* Dr. Makund Modak, Program Director, 973-972-4053, Fax: 973-972-5594, E-mail: modak@umdnj.edu. *Application contact:* Dr. Henry E. Brezenoff, Acting Dean, 973-972-5333, Fax: 973-972-7148, E-mail: hbrezeno@umdnj.edu.

University of Medicine and Dentistry of New Jersey, Graduate School of Biomedical Sciences, Graduate Programs in Biomedical Sciences–Piscataway, Program in Biochemistry and Molecular Biology, Piscataway, NJ 08854-5635. Offers MS, PhD, MD/PhD. Terminal master's awarded for partial completion of doctoral program. *Degree requirements:* For master's and doctorate, thesis/dissertation, qualifying exam. *Entrance requirements:* For master's and doctorate, GRE General Test. Additional exam requirements/recommendations for international students: Required—TOEFL. *Application deadline:* For fall admission, 1/5 for domestic students. Application fee: $40. *Financial support:* Fellowships, research assistantships, teaching assistantships available. Financial award application deadline: 5/1. *Faculty research:* Signal transduction, regulation of RNA, polymerase II transcribed genes, developmental gene expression. *Unit head:* Dr. Kiran Madura, Director, 732-235-4595, Fax: 732-235-4783, E-mail: maduraki@umdnj.edu.

University of Miami, Graduate School, Miller School of Medicine, Graduate Programs in Medicine, Department of Biochemistry and Molecular Biology, Coral Gables, FL 33124. Offers PhD, MD/PhD. *Faculty:* 29 full-time (6 women). *Students:* 47 full-time (20 women); includes 8 minority (1 African American, 7 Hispanic Americans), 24 international. Average age 29. 84 applicants, 14% accepted, 7 enrolled. In 2005, 6 degrees awarded. *Median time to degree:* Of those who began their doctoral program in fall 1997, 98% received their degree in 8 years or less. *Degree requirements:* For doctorate, thesis/dissertation, proposition exams, comprehensive exam. *Entrance requirements:* For doctorate, GRE General Test. Additional exam requirements/recommendations for international students: Required—TOEFL. *Application deadline:* Applications are processed on a rolling basis. Application fee: $50. Electronic applications accepted. *Financial support:* In 2005–06, 34 fellowships with full tuition reimbursements (averaging $22,000 per year) were awarded; research assistantships, scholarships/grants and tuition waivers (full) also available. *Faculty research:* Macromolecule metabolism, molecular genetics, protein folding and 3-D structure, regulation of gene expression and enzyme function, signal transduction and developmental biology. Total annual research expenditures: $3 million. *Unit head:* Dr. Murray P. Deutscher, Chairman, 305-243-3150, Fax: 305-243-3955, E-mail: mdeutsch@mednet.med.miami.edu. *Application contact:* Dr. Rudolf Werner, Director, 305-243-6998, Fax: 305-243-3955, E-mail: rwerner@molbio.med.miami.edu.

University of Michigan, Horace H. Rackham School of Graduate Studies, Chemical Biology Program, Ann Arbor, MI 48109. Offers PhD. *Faculty:* 40 full-time (5 women). *Students:* 7 full-time (3 women), 1 international. Average age 25. *Entrance requirements:* Additional exam requirements/recommendations for international students: Required—TOEFL (minimum score 600 paper-based; 250 computer-based). *Application deadline:* For fall admission, 1/15 priority date for domestic students, 1/1 priority date for international students. Application fee: $0 ($75 for international students). Electronic applications accepted. *Expenses:* Tuition, state resident: full-time $14,082; part-time $894 per credit hour. Tuition, nonresident: full-time $28,500; part-time $1,675 per credit hour. Required fees: $189; $189 per unit. *Financial support:* In 2005–06, 7 students received support, including fellowships with full tuition reimbursements available (averaging $23,500 per year), research assistantships with full tuition reimbursements available (averaging $23,500 per year) *Faculty research:* Chemical genetics, structural enzymology, signal transduction, biological catalysis, biomolecular structure, function and recognition. *Unit head:* Justine Altman, Program Manager, 734-763-7175, Fax: 734-615-1252, E-mail: justine@umich.edu.

University of Michigan, Medical School and Horace H. Rackham School of Graduate Studies, Program in Biomedical Sciences (PIBS), Department of Biological Chemistry, Ann Arbor, MI 48109. Offers PhD. *Degree requirements:* For doctorate, thesis/dissertation, oral defense of dissertation, preliminary exam. *Entrance requirements:* For doctorate, GRE General Test, GRE Subject Test, minimum GPA of 3.0. Additional exam requirements/recommendations for international students: Required—TOEFL. Electronic applications accepted. *Expenses:* Tuition, state resident: full-time $14,082; part-time $894 per credit hour. Tuition, nonresident: full-time $28,500; part-time $1,675 per credit hour. Required fees: $189; $189 per unit. *Faculty research:* Nucleic acids, gene regulation, mechanistic enzymology, protein biochemistry, signal transduction.

University of Minnesota, Duluth, Medical School, Department of Biochemistry, Molecular Biology and Biophysics, Duluth, MN 55812-2496. Offers MS, PhD. *Faculty:* 8 full-time (1 woman), 2 part-time/adjunct (1 woman). *Students:* 5 full-time (3 women). Average age 26. 5 applicants, 40% accepted, 2 enrolled.Terminal master's awarded for partial completion of doctoral program. *Degree requirements:* For master's and doctorate, thesis/dissertation. *Entrance requirements:* For master's and doctorate, GRE General Test. Additional exam requirements/recommendations for international students: Required—TOEFL. *Application deadline:* For fall admission, 1/2 priority date for domestic students, 1/2 priority date for international students. For winter admission, 3/15 for domestic students. Application fee: $55 ($75 for international students). *Financial support:* In 2005–06, 4 students received support, including fellowships with full tuition reimbursements available (averaging $22,500 per year), research assistantships with full tuition reimbursements available (averaging $18,000 per year); teaching assistantships with tuition reimbursements available (averaging $18,000 per year); Federal Work-Study, institutionally sponsored loans, health care benefits, and unspecified assistantships also available. Financial award application deadline: 4/1. *Faculty research:* Intestinal cancer biology; hepatotoxins and mitochondriopathies; cell cycle regulation in stem cells; neurobiology of brain development, trace metal function and blood-brain barrier; hibernation biology. Total annual research expenditures: $1.5 million. *Unit head:* Dr. Lester R. Drewes, Professor/Head, 218-726-7925, Fax: 218-726-8014, E-mail: ldrewes@d.umn.edu. *Application contact:* Stacy Johnson, Program Associate, 218-726-6354, E-mail: ahcd@d.umn.edu.

University of Minnesota, Twin Cities Campus, Medical School and Graduate School, Graduate Programs in Medicine, Department of Biochemistry, Molecular Biology and Biophysics, Minneapolis, MN 55455-0213. Offers PhD. *Degree requirements:* For doctorate, thesis/dissertation. *Entrance requirements:* For doctorate, GRE General Test. Electronic applications accepted. Expenses: Contact institution. Full-time tuition and fees vary according to class time, course load, program and reciprocity agreements. *Faculty research:* Physical biochemistry, enzymology, physiological chemistry.

University of Mississippi Medical Center, School of Graduate Studies in the Health Sciences, Department of Biochemistry, Jackson, MS 39216-4505. Offers MS, PhD, MD/PhD. *Faculty:* 11 full-time (1 woman). *Students:* 22 full-time (9 women); includes 1 minority (African American), 14 international. Average age 27. 7 applicants, 43% accepted.Terminal master's awarded for partial completion of doctoral program. *Degree requirements:* For master's,

thesis; for doctorate, thesis/dissertation, first authored publication. *Entrance requirements:* For doctorate, GRE General Test, minimum GPA of 3.0. Additional exam requirements/recommendations for international students: Required—TOEFL. *Application deadline:* For fall admission, 3/1 for domestic students. Application fee: $10. *Financial support:* In 2005–06, research assistantships (averaging $16,559 per year) Financial award application deadline: 3/31. *Faculty research:* Structural biology, regulation of gene expression, enzymology of redox reactions, mechanism of anti cancer drugs, function of nuclear substructure. Total annual research expenditures: $1.4 million. *Unit head:* Dr. Mark O. J. Olson, Chairman, 601-984-1500, Fax: 601-984-1501, E-mail: molson@biochem.umsmed.edu. *Application contact:* Dr. Drazen Raucher, Director, 601-984-1500, Fax: 601-984-1501, E-mail: draucher@biochem.umsmed.edu.

See Close-Up on page 435.

University of Missouri–Columbia, School of Medicine and Graduate School, Graduate Programs in Medicine and College of Agriculture, Food and Natural Resources, Department of Biochemistry, Columbia, MO 65211. Offers MS, PhD. *Faculty:* 34 full-time (11 women), 4 part-time/adjunct (3 women). *Students:* 17 full-time (7 women), 21 part-time (9 women); includes 2 minority (1 Asian American, 1 American or Pacific Islander), 18 international. Average age 27.Terminal master's awarded for partial completion of doctoral program. *Degree requirements:* For master's and doctorate, thesis/dissertation. *Entrance requirements:* For master's and doctorate, GRE General Test, minimum GPA of 3.0. *Application deadline:* For fall admission, 2/1 for domestic students. Application fee: $45 ($60 for international students). *Financial support:* Fellowships, research assistantships, teaching assistantships, institutionally sponsored loans available. *Faculty research:* Enzymology, plant biochemistry, molecular biochemistry. *Unit head:* Dr. Steve R. Van Doren, Director of Graduate Studies, 573-882-5113.

University of Missouri–Kansas City, School of Biological Sciences, Program in Molecular Biology and Biochemistry, Kansas City, MO 64110-2499. Offers PhD. Offered through the School of Graduate Studies. *Degree requirements:* For doctorate, thesis/dissertation, comprehensive exam. *Entrance requirements:* For doctorate, GRE General Test, bachelor's degree in chemistry, biology, or a related discipline; minimum GPA of 3.0. Additional exam requirements/recommendations for international students: Required—TOEFL. *Application deadline:* For fall admission, 3/1 for domestic students. Application fee: $35 ($50 for international students). *Expenses:* Tuition, state resident: full-time $4,738; part-time $263 per credit hour. Tuition, nonresident: full-time $12,235; part-time $679 per credit hour. Required fees: $582. Tuition and fees vary according to course load, program and student level. *Financial support:* Research assistantships with full tuition reimbursements, teaching assistantships with full and partial tuition reimbursements, scholarships/grants, tuition waivers (full and partial), and unspecified assistantships available. *Unit head:* Dr. Henry Miziorko, Head, 816-235-2235, E-mail: miziorkoh@umkc.edu. *Application contact:* Laura Batenic, Information Contact, 816-235-2352, Fax: 816-235-5158, E-mail: batenicl@umkc.edu.

See Close-Up on page 629.

University of Missouri–St. Louis, College of Arts and Sciences, Department of Biology, St. Louis, MO 63121. Offers biology (MS, PhD), including animal behavior (MS), biochemistry (MS), biotechnology (MS), conservation biology (MS), development (MS), ecology (MS), environmental studies (PhD), evolution (MS), genetics (MS), molecular/cellular biology (MS), physiology (MS), plant systematics, population biology (MS), tropical biology (MS); biotechnology (Certificate); tropical biology and conservation (Certificate). Part-time programs available. *Faculty:* 50. *Students:* 26 full-time (15 women), 101 part-time (51 women); includes 13 minority (5 African Americans, 6 Asian Americans or Pacific Islanders, 2 Hispanic Americans), 41 international. Average age 32. In 2005, 22 master's, 2 doctorates awarded. *Degree requirements:* For master's, thesis or alternative; for doctorate, one foreign language, thesis/dissertation, 1 semester of teaching experience. *Entrance requirements:* For doctorate, GRE General Test. *Application deadline:* For spring admission, 12/1 priority date for domestic students. Applications are processed on a rolling basis. Application fee: $35 ($40 for international students). Electronic applications accepted. *Expenses:* Tuition, state resident: part-time $263 per credit hour. Tuition, nonresident: part-time $680 per credit hour. Required fees: $53 per credit hour. Tuition and fees vary according to program. *Financial support:* In 2005–06, 11 fellowships with full tuition reimbursements (averaging $30,000 per year), 15 research assistantships with full and partial tuition reimbursements (averaging $16,000 per year), 22 teaching assistantships with full and partial tuition reimbursements (averaging $16,000 per year) were awarded; career-related internships or fieldwork and Federal Work-Study also available. Support available to part-time students. Financial award application deadline: 2/1. *Faculty research:* Molecular biology, microbial genetics. *Unit head:* Zuleyma Tang-Martinez, Director of Graduate Studies, 314-516-6498, Fax: 314-516-6233, E-mail: zuleyma@umsl.edu. *Application contact:* 314-516-5458, Fax: 314-516-5310, E-mail: gradadm@umsl.edu.

University of Missouri–St. Louis, College of Arts and Sciences, Department of Chemistry and Biochemistry, St. Louis, MO 63121. Offers chemistry (MS, PhD), including inorganic chemistry, organic chemistry, physical chemistry. Part-time and evening/weekend programs available. *Faculty:* 16. *Students:* 28 full-time (20 women), 16 part-time (7 women); includes 6 minority (5 African Americans, 1 Hispanic American), 23 international. Average age 32. 2,298 applicants, 74% accepted. In 2005, 9 master's, 4 doctorates awarded. Terminal master's awarded for partial completion of doctoral program. *Degree requirements:* For master's, thesis optional; for doctorate, thesis/dissertation. *Entrance requirements:* For doctorate, GRE General Test, 3 letters of recommendation. Additional exam requirements/recommendations for international students: Required—TOEFL (minimum score 550 paper-based; 213 computer-based). *Application deadline:* For fall admission, 7/1 for domestic students; for spring admission, 12/7 priority date for domestic students. Applications are processed on a rolling basis. Application fee: $35 ($40 for international students). Electronic applications accepted. *Expenses:* Tuition, state resident: part-time $263 per credit hour. Tuition, nonresident: part-time $680 per credit hour. Required fees: $53 per credit hour. Tuition and fees vary according to program. *Financial support:* In 2005–06, 9 research assistantships with full and partial tuition reimbursements (averaging $16,500 per year), 17 teaching assistantships with full and partial tuition reimbursements (averaging $16,500 per year) were awarded; fellowships with full and partial tuition reimbursements *Faculty research:* Metallaborane chemistry, serum transferrin chemistry, natural products chemistry, organic synthesis. *Unit head:* Dr. Cynthia Dupureur, Director of Graduate Studies, 314-516-5311, Fax: 314-516-5342, E-mail: gradchem@umsl.edu. *Application contact:* 314-516-5458, Fax: 314-516-5310, E-mail: gradadm@umsl.edu.

The University of Montana, Graduate School, College of Arts and Sciences, Division of Biological Sciences, Program in Biochemistry and Microbiology, Missoula, MT 59812-0002. Offers biochemistry (MS); integrative microbiology and biochemistry (PhD); microbial ecology (MS, PhD); microbiology (MS). Terminal master's awarded for partial completion of doctoral program. *Degree requirements:* For master's, thesis; for doctorate, variable foreign language requirement, thesis/dissertation. *Entrance requirements:* For master's and doctorate, GRE General Test. *Application deadline:* For fall admission, 2/1 for domestic students. Application fee: $45. *Expenses:* Tuition, state resident: part-time $267 per credit. Tuition, nonresident: part-time $665 per credit. Part-time tuition and fees vary according to course load and degree level. *Financial support:* In 2005–06, research assistantships with tuition reimbursements (averaging $9,400 per year), teaching assistantships with full tuition reimbursements (averaging $9,400 per year) were awarded; Federal Work-Study and tuition waivers (full and partial) also available. Financial award application deadline: 3/1; financial award applicants required to submit FAFSA. *Faculty research:* Ribosome structure, medical microbiology/pathogenesis, microbial ecology/environmental microbiology. *Application contact:* Janean Clark, Graduate Programs Secretary, 406-243-5222, Fax: 406-243-4184, E-mail: jmclark@selway.umt.edu.

University of Nebraska–Lincoln, Graduate College, College of Agricultural Sciences and Natural Resources and College of Arts and Sciences, Department of Biochemistry, Lincoln, NE 68588. Offers MS, PhD. Terminal master's awarded for partial completion of doctoral program. *Degree requirements:* For master's, thesis optional; for doctorate, thesis/dissertation,

Biochemistry

University of Nebraska–Lincoln (continued)
comprehensive exam. *Entrance requirements:* For master's and doctorate, GRE General Test, GRE Subject Test. Additional exam requirements/recommendations for international students: Required—TOEFL (minimum score 550 paper-based; 213 computer-based). Electronic applications accepted. *Faculty research:* Molecular genetics, enzymology, photosynthesis, molecular virology, structural biology.

University of Nebraska Medical Center, Graduate Studies, Department of Biochemistry and Molecular Biology, Omaha, NE 68198. Offers MS, PhD. *Faculty:* 17 full-time (2 women), 28 part-time/adjunct (8 women). *Students:* 31 full-time (15 women); includes 1 minority (Asian American or Pacific Islander), 20 international. Average age 28. 20 applicants, 30% accepted, 5 enrolled. In 2005, 2 doctorates awarded. *Median time to degree:* Of those who began their doctoral program in fall 1997, 75% received their degree in 8 years or less. *Degree requirements:* For master's and doctorate, thesis/dissertation, comprehensive exam. *Entrance requirements:* For master's and doctorate, GRE General Test. Additional exam requirements/recommendations for international students: Required—TOEFL (minimum score 550 paper-based; 215 computer-based). Application fee: $45. Electronic applications accepted. *Expenses:* Tuition, area resident: Part-time $200 per hour. Tuition, nonresident: part-time $538 per hour. Required fees: $308; $59 per term. *Financial support:* In 2005–06, 17 students received support, including 2 fellowships with full tuition reimbursements available (averaging $21,000 per year), 29 research assistantships with full tuition reimbursements available (averaging $21,000 per year); institutionally sponsored loans also available. Support available to part-time students. Financial award application deadline: 2/15. *Faculty research:* Recombinant DNA, cancer biology, diabetes and drug metabolism, biochemical endocrinology. *Unit head:* Dr. Surinder K. Batra, Chairman, Graduate Committee, 402-559-5455, Fax: 402-559-6650, E-mail: biochem@unmc.edu.

University of Nevada, Las Vegas, Graduate College, College of Science, Department of Chemistry, Las Vegas, NV 89154-9900. Offers biochemistry (MS); chemistry (MS); environmental science/chemistry (PhD); radiochemistry (PhD). Part-time programs available. *Faculty:* 19 full-time (3 women), 5 part-time/adjunct (1 woman). *Students:* 17 full-time (6 women), 18 part-time (13 women); includes 6 minority (3 Asian Americans or Pacific Islanders, 3 Hispanic Americans), 11 international. 30 applicants, 70% accepted, 17 enrolled. In 2005, 7 degrees awarded. *Degree requirements:* For master's, thesis. *Entrance requirements:* For master's, GRE General Test, minimum GPA of 3.0 in last 2 years or 2.75 cumulative. Additional exam requirements/recommendations for international students: Required—TOEFL (minimum score 550 paper-based; 213 computer-based). *Application deadline:* For fall admission, 6/15 for domestic students, 5/1 for international students; for spring admission, 11/15 for domestic students, 10/1 for international students. Application fee: $60 ($75 for international students). Electronic applications accepted. *Expenses:* Tuition, state resident: part-time $150 per credit. Tuition, nonresident: part-time $315 per credit. Tuition and fees vary according to course load, program and reciprocity agreements. *Financial support:* In 2005–06, 3 research assistantships with partial tuition reimbursements (averaging $1,000 per year), 9 teaching assistantships with full tuition reimbursements (averaging $10,000 per year) were awarded; career-related internships or fieldwork, Federal Work-Study, institutionally sponsored loans, scholarships/grants, health care benefits, and unspecified assistantships also available. Support available to part-time students. Financial award application deadline: 3/1. *Unit head:* Dr. Spencer Steinberg, Chair, 702-895-3510. *Application contact:* Graduate Coordinator, 702-895-3753, Fax: 702-895-4180, E-mail: gradcollege@unlv.edu.

University of Nevada, Reno, Graduate School, College of Agriculture, Biotechnology and Natural Resources, Interdisciplinary Program in Biochemistry, Reno, NV 89557. Offers MS, PhD. Offered through the College of Liberal Arts, the M. C. Fleischmann College of Agriculture, and the School of Medicine. *Faculty:* 15. *Students:* 10 full-time (7 women), 16 part-time (7 women); includes 5 minority (1 African American, 2 Asian Americans or Pacific Islanders, 2 Hispanic Americans), 6 international. Average age 31. 15 applicants, 60% accepted, 7 enrolled. In 2005, 3 degrees awarded. Terminal master's awarded for partial completion of doctoral program. *Degree requirements:* For master's and doctorate, thesis/dissertation. *Entrance requirements:* For master's, GRE General Test, minimum GPA of 2.75; for doctorate, GRE General Test, minimum GPA of 3.0. Additional exam requirements/recommendations for international students: Required—TOEFL. *Application deadline:* For fall admission, 3/1 for domestic students. Application fee: $60 ($95 for international students). *Expenses:* Tuition, area resident: full-time $2,767; part-time $923 per semester. Tuition, state resident: full-time $5,733; part-time $1,911 per semester. Tuition, nonresident: full-time $12,679; part-time $1,911 per semester. International tuition: $13,878 full-time. Required fees: $404; $202 per term. One-time fee: $90. *Financial support:* In 2005–06, 20 research assistantships were awarded; teaching assistantships, Federal Work-Study and institutionally sponsored loans also available. Financial award application deadline: 2/15. *Faculty research:* Cancer research, insect biochemistry, plant biochemistry, enzymology. *Unit head:* Dr. John Cashman, Graduate Program Director, 775-784-6031.

See Close-Up on page 437.

University of New Hampshire, Graduate School, College of Life Sciences and Agriculture, Department of Biochemistry and Molecular Biology, Durham, NH 03824. Offers MS, PhD. Part-time programs available. *Faculty:* 12 full-time. *Students:* 18 full-time (8 women), 2 part-time; includes 2 minority (both Asian Americans or Pacific Islanders); 6 international. Average age 28. 11 applicants, 27% accepted, 1 enrolled. In 2005, 3 master's, 3 doctorates awarded. Terminal master's awarded for partial completion of doctoral program. *Degree requirements:* For master's, thesis; for doctorate, one foreign language, thesis/dissertation. *Entrance requirements:* For master's and doctorate, GRE General Test. Additional exam requirements/recommendations for international students: Required—TOEFL (minimum score 550 paper-based; 213 computer-based); Recommended—TSE. *Application deadline:* For fall admission, 4/1 priority date for domestic students, 4/1 priority date for international students. Applications are processed on a rolling basis. Application fee: $60. Electronic applications accepted. *Expenses:* Tuition, state resident: full-time $8,010; part-time $445 per credit hour. Tuition, nonresident: full-time $19,730; part-time $810 per credit hour. Required fees: $322 per semester. Tuition and fees vary according to course load and program. *Financial support:* In 2005–06, 1 fellowship, 11 research assistantships, 5 teaching assistantships were awarded; career-related internships or fieldwork, Federal Work-Study, scholarships/grants, and tuition waivers (full and partial) also available. Support available to part-time students. Financial award application deadline: 2/15. *Faculty research:* Developmental biochemistry, biochemistry of natural products, physical biochemistry, biochemical genetics, structure and metabolism of macromolecules. *Application contact:* Stacia Sower, Administrative Assistant, 603-862-2103, E-mail: biochemistry.dept@unh.edu.

University of New Mexico, School of Medicine, Biomedical Sciences Graduate Program, Albuquerque, NM 87131-5196. Offers biochemistry and molecular biology (MS, PhD); cell biology and physiology (MS, PhD); molecular genetics and microbiology (MS, PhD); neuroscience (MS, PhD); pathology (MS, PhD); toxicology (MS, PhD). Part-time programs available. Terminal master's awarded for partial completion of doctoral program. *Degree requirements:* For master's, thesis/dissertation; for doctorate, thesis/dissertation, comprehensive exam. *Entrance requirements:* For master's and doctorate, GRE General Test, minimum undergraduate GPA of 3.0. Additional exam requirements/recommendations for international students: Required—TOEFL. Electronic applications accepted. *Expenses:* Tuition, state resident: full-time $5,676. Tuition, nonresident: full-time $14,974; part-time $238 per credit hour. Required fees: $385 per term. Tuition and fees vary according to course load and program. *Faculty research:* Signal transduction, infectious disease, biology of cancer, structural biology, neuroscience.

The University of North Carolina at Chapel Hill, School of Medicine and Graduate School, Graduate Programs in Medicine, Department of Biochemistry and Biophysics, Chapel Hill, NC 27599. Offers MS, PhD. *Faculty:* 42 full-time (8 women). *Students:* 74 full-time (27 women); includes 2 minority (both African Americans), 17 international. Average age 22. 131 applicants,

21% accepted, 17 enrolled. In 2005, 2 master's, 6 doctorates awarded. Terminal master's awarded for partial completion of doctoral program. *Median time to degree:* Of those who began their doctoral program in fall 1997, 100% received their degree in 8 years or less. *Degree requirements:* For master's and doctorate, thesis/dissertation, comprehensive exam. *Entrance requirements:* For master's and doctorate, GRE General Test, GRE Subject Test (recommended), minimum GPA of 3.0. Additional exam requirements/recommendations for international students: Required—TOEFL. *Application deadline:* For fall admission, 1/1 for domestic students, 1/1 for international students. Application fee: $70. Electronic applications accepted. *Financial support:* In 2005–06, fellowships with full tuition reimbursements (averaging $23,000 per year), research assistantships with full tuition reimbursements (averaging $23,000 per year), teaching assistantships with full tuition reimbursements (averaging $23,000 per year) were awarded. Total annual research expenditures: $6.4 million. *Unit head:* Dr. Henrik Dohlman, Chair, 919-962-8326, Fax: 919-966-2852, E-mail: hendrik-dohlman@med.unc.edu. *Application contact:* Diane M. Harris, Student Services Manager, 919-966-4683, Fax: 919-966-2852, E-mail: diane_harris@med.unc.edu.

The University of North Carolina at Greensboro, Graduate School, College of Arts and Sciences, Department of Chemistry and Biochemistry, Greensboro, NC 27412-5001. Offers biochemistry (MS); chemistry (MS). *Students:* 6 full-time, 19 part-time. *Degree requirements:* For master's, one foreign language, thesis. *Entrance requirements:* For master's, GRE General Test. Additional exam requirements/recommendations for international students: Required—TOEFL. *Application deadline:* For fall admission, 6/15 for domestic students; for spring admission, 3/15 priority date for domestic students. Applications are processed on a rolling basis. *Expenses:* Tuition, state resident: part-time $302 per credit hour. Tuition, nonresident: part-time $1,683 per credit hour. Required fees: $51 per credit hour. Tuition and fees vary according to course load and program. *Financial support:* Fellowships, research assistantships with full tuition reimbursements, teaching assistantships with full tuition reimbursements, career-related internships or fieldwork, scholarships/grants, and traineeships available. *Faculty research:* Synthesis of novel cyclopentadienes, molybdenum hydroxylase-cata ladder polymers, vinyl silicones. *Unit head:* Dr. Jerry Walsh, Interim Head, 336-334-5714, Fax: 336-334-5402, E-mail: jlwalsh@uncg.edu. *Application contact:* Michelle Harkleroad, Director of Graduate Admissions, 336-334-4884, Fax: 336-334-4424, E-mail: mbharkle@uncg.edu.

University of North Dakota, School of Medicine and Graduate School, Graduate Programs in Medicine, Department of Biochemistry, Grand Forks, ND 58202. Offers MS, PhD. *Faculty:* 10 full-time (2 women), 4 part-time/adjunct (0 women). *Students:* 2 full-time (1 woman), 15 part-time (6 women). 21 applicants, 14% accepted, 3 enrolled. In 2005, 1 degree awarded. *Degree requirements:* For master's, thesis/dissertation, final exam; for doctorate, thesis/dissertation, final exam, comprehensive exam. *Entrance requirements:* For master's and doctorate, GRE General Test, minimum GPA of 3.0. Additional exam requirements/recommendations for international students: Required—TOEFL (minimum score 550 paper-based; 213 computer-based). *Application deadline:* For fall admission, 2/15 priority date for domestic students, 2/15 priority date for international students; for spring admission, 10/15 priority date for domestic students, 10/15 priority date for international students. Applications are processed on a rolling basis. Application fee: $35. Electronic applications accepted. *Financial support:* In 2005–06, 7 research assistantships with full tuition reimbursements (averaging $13,997 per year), 5 teaching assistantships with full tuition reimbursements (averaging $13,997 per year) were awarded; fellowships, Federal Work-Study, institutionally sponsored loans, scholarships/grants, and tuition waivers (full and partial) also available. Support available to part-time students. Financial award application deadline: 3/15; financial award applicants required to submit FAFSA. *Faculty research:* Glucose-6-phosphatase, guanine nucleotides, carbohydrate and lipid metabolism, cytoskeletal proteins, chromatin structure. *Unit head:* Dr. Katherine Sukalski, Director, 701-777-4049, Fax: 701-777-3527, E-mail: sukalski@medicine.nodak.edu. *Application contact:* Brenda Halle, Admissions Specialist, 701-777-2947, Fax: 701-777-3619, E-mail: brendahalle@mail.und.edu.

University of North Texas, Robert B. Toulouse School of Graduate Studies, College of Arts and Sciences, Department of Biological Sciences, Division of Biochemistry, Denton, TX 76203. Offers MS, PhD. *Students:* 16 full-time (6 women), 2 part-time (1 woman); includes 8 minority (1 African American, 5 Asian Americans or Pacific Islanders, 2 Hispanic Americans), 4 international. Average age 29. Terminal master's awarded for partial completion of doctoral program. *Degree requirements:* For master's, oral defense of thesis; for doctorate, one foreign language, thesis/dissertation, comprehensive exam. *Entrance requirements:* For master's, GRE General Test, placement exams in 3 areas; for doctorate, GRE General Test, placement exams in 4 areas. *Application deadline:* For fall admission, 7/1 for domestic students; for spring admission, 11/1 for domestic students. Application fee: $50 ($75 for international students). *Expenses:* Tuition, state resident: full-time $3,258; part-time $181 per semester hour. Tuition, nonresident: full-time $8,226; part-time $451 per semester hour. Required fees: $1,219; $68 per semester hour. *Financial support:* Fellowships, research assistantships, teaching assistantships, career-related internships or fieldwork, Federal Work-Study, and institutionally sponsored loans available. Financial award application deadline: 4/1. *Faculty research:* Protein and nucleic acid structure, aging and nutritional biochemical changes, biochemical parasitology, plasma lipoprotein metabolism. *Unit head:* Dr. Robert Pirtle, Head, 940-565-2011, Fax: 940-565-4136, E-mail: pirtle@unt.edu. *Application contact:* Beth Chlapek, Graduate Adviser, 940-565-3627, Fax: 940-565-3821, E-mail: bsloan@unt.edu.

University of North Texas Health Science Center at Fort Worth, Graduate School of Biomedical Sciences, Fort Worth, TX 76107-2699. Offers anatomy and cell biology (MS, PhD); biochemistry and molecular biology (MS, PhD); biomedical sciences (MS, PhD); biotechnology (MS); forensic genetics (MS); integrative physiology (MS, PhD); medical science (MS); microbiology and immunology (MS, PhD); pharmacology (MS, PhD); science education (MS). *Faculty:* 57 full-time (9 women), 2 part-time/adjunct (0 women). *Students:* 177 full-time (98 women), 42 part-time (33 women); includes 61 minority (15 African Americans, 3 American Indian/Alaska Native, 27 Asian Americans or Pacific Islanders, 16 Hispanic Americans), 53 international. Average age 28. 237 applicants, 62% accepted, 91 enrolled. In 2005, 37 master's, 15 doctorates awarded. Terminal master's awarded for partial completion of doctoral program. *Degree requirements:* For master's and doctorate, thesis/dissertation. *Entrance requirements:* For master's and doctorate, GRE General Test. Additional exam requirements/recommendations for international students: Required—TOEFL. *Application deadline:* For fall admission, 5/1 for domestic students. Application fee: $25 ($50 for international students). *Expenses: Contact institution.* *Financial support:* In 2005–06, 80 research assistantships (averaging $16,000 per year) were awarded; fellowships, teaching assistantships, career-related internships or fieldwork, Federal Work-Study, institutionally sponsored loans, scholarships/grants, and traineeships also available. Support available to part-time students. Financial award application deadline: 4/1; financial award applicants required to submit FAFSA. *Faculty research:* Alzheimer's disease, aging, eye diseases, cancer, cardiovascular disease. Total annual research expenditures: $21 million. *Unit head:* Dr. Thomas Yorio, Dean, 817-735-2560, Fax: 817-735-0243, E-mail: yoriot@hsc.unt.edu. *Application contact:* Carla Lee, Director of Graduate Admissions and Services, 817-735-2560, Fax: 817-735-0243, E-mail: gsbs@hsc.unt.edu.

See Close-Up on page 279.

University of Notre Dame, Graduate School, College of Science, Department of Chemistry and Biochemistry, Notre Dame, IN 46556. Offers biochemistry (MS, PhD); inorganic chemistry (MS, PhD); organic chemistry (MS, PhD); physical chemistry (MS, PhD). *Faculty:* 30 full-time (3 women). *Students:* 148 full-time (52 women); includes 5 minority (1 African American, 1 Asian American or Pacific Islander, 3 Hispanic Americans), 65 international. 144 applicants, 38% accepted, 21 enrolled. In 2005, 7 master's, 6 doctorates awarded. Terminal master's awarded for partial completion of doctoral program. *Median time to degree:* Of those who began their doctoral program in fall 1997, 54% received their degree in 8 years or less. *Degree requirements:* For master's, thesis, comprehensive exam; for doctorate, thesis/dissertation, qualifying exam. *Entrance requirements:* For master's and doctorate, GRE General Test, GRE Subject Test (strongly recommended). Additional exam requirements/recommendations for international students: Required—TOEFL. *Application deadline:* For fall admission, 2/1 for domestic students. Applications are processed on a rolling basis. Application fee: $50. Electronic applications accepted. *Financial support:* In 2005–06, 148 students received support, including 19 fellowships with full tuition reimbursements available (averaging $22,000 per year), 57 research assistantships with full tuition reimbursements available (averaging $15,250 per year), 53 teaching assistantships with full tuition reimbursements available (averaging $16,000 per year); tuition waivers (full) also available. Financial award application deadline: 2/1. *Faculty research:* Reaction design and mechanistic studies; reactive intermediates; synthesis, structure and reactivity of organometallic. cluster complexes, and biologically active natural products; bioorganic chemistry; enzymology. Total annual research expenditures: $9.4 million. *Unit head:* Dr. Richard E. Taylor, Director of Graduate Studies, 574-631-7058, Fax: 574-631-6652, E-mail: taylor.61@nd.edu. *Application contact:* Dr. Terrence J. Akai, Director of Graduate Admissions, 574-631-7706, Fax: 574-631-4183, E-mail: gradad@nd.edu.

University of Oklahoma, Graduate College, College of Arts and Sciences, Department of Chemistry and Biochemistry, Norman, OK 73019-0390. Offers MS, PhD. Part-time programs available. *Faculty:* 27 full-time (5 women). *Students:* 114 full-time (45 women), 12 part-time (4 women); includes 4 minority (3 African Americans, 1 Asian American or Pacific Islander), 79 international. 55 applicants, 71% accepted, 23 enrolled. In 2005, 15 master's, 4 doctorates awarded. Terminal master's awarded for partial completion of doctoral program. *Degree requirements:* For master's, thesis optional; for doctorate, thesis/dissertation. *Entrance requirements:* For master's, GRE, BS in chemistry; for doctorate, GRE. Additional exam requirements/recommendations for international students: Required—TOEFL (minimum score 550 paper-based; 213 computer-based). *Application deadline:* For fall admission, 4/1 priority date for domestic students, 4/1 priority date for international students; for spring admission, 9/1 priority date for domestic students, 9/1 priority date for international students. Applications are processed on a rolling basis. Application fee: $40 ($90 for international students). *Expenses:* Tuition, area resident: Full-time $3,029; part-time $126 per credit hour. Tuition, state resident: full-time $3,029; part-time $126 per credit hour. Tuition, nonresident: full-time $10,807; part-time $450 per credit hour. International tuition: $10,807 full-time. Required fees: $1,231; $44 per credit hour. Tuition and fees vary according to course load and program. *Financial support:* In 2005–06, 22 students received support, including 6 fellowships with full tuition reimbursements available (averaging $5,000 per year), 36 research assistantships with partial tuition reimbursements available (averaging $13,115 per year), 74 teaching assistantships with partial tuition reimbursements available (averaging $13,936 per year); career-related internships or fieldwork, Federal Work-Study, scholarships/grants, traineeships, and tuition waivers (partial) also available. Support available to part-time students. Financial award application deadline: 4/1; financial award applicants required to submit FAFSA. *Faculty research:* Genomics, mechanisms of enzyme action bacterial transport processes, protein and nucleic acid structure and function. Total annual research expenditures: $3.1 million. *Unit head:* Dr. Glenn Dryhurst, Chair, 405-325-4811, Fax: 405-325-6111, E-mail: gdryhurst@ou.edu. *Application contact:* Ariene Crawford, Graduate Recruiting Secretary, 405-325-2946, Fax: 405-325-6111, E-mail: admission@chemdept.chem.ou.edu.

University of Oklahoma Health Sciences Center, College of Medicine and Graduate College, Graduate Programs in Medicine, Department of Biochemistry and Molecular Biology, Oklahoma City, OK 73190. Offers biochemistry (MS, PhD); molecular biology (MS, PhD). Part-time programs available. Terminal master's awarded for partial completion of doctoral program. *Degree requirements:* For master's and doctorate, thesis/dissertation. *Entrance requirements:* For master's, GRE General Test, 2 letters of recommendation; for doctorate, GRE General Test, 3 letters of recommendation. Additional exam requirements/recommendations for international students: Required—TOEFL. *Faculty research:* Gene expression, regulation of transcription, enzyme evolution, melanogenesis, signal transduction.

University of Oregon, Graduate School, College of Arts and Sciences, Department of Chemistry, Eugene, OR 97403. Offers biochemistry (MA, MS, PhD); chemistry (MA, MS, PhD). *Faculty:* 34 full-time (7 women), 7 part-time/adjunct (3 women). *Students:* 114 full-time (42 women), 7 part-time (3 women); includes 6 minority (2 African Americans, 6 Asian Americans or Pacific Islanders), 15 international. 18 applicants, 89% accepted. In 2005, 8 master's, 17 doctorates awarded. Terminal master's awarded for partial completion of doctoral program. *Degree requirements:* For doctorate, thesis/dissertation. *Entrance requirements:* For master's and doctorate, GRE General Test, minimum GPA of 3.0. Additional exam requirements/recommendations for international students: Required—TOEFL. *Application deadline:* For fall admission, 1/10 for domestic students. Applications are processed on a rolling basis. Application fee: $50. *Financial support:* In 2005–06, 54 teaching assistantships were awarded; Federal Work-Study and institutionally sponsored loans also available. Financial award application deadline: 4/15. *Faculty research:* Organic chemistry, organometallic chemistry, inorganic chemistry, physical chemistry, materials science, biochemistry, chemical physics, molecular or cell biology. *Unit head:* Tom Dyke, Head, 541-346-4603. *Application contact:* Lynde Ritzow, Graduate Admissions Coordinator, 541-346-4789, E-mail: lynde@oregon.uoregon.edu.

University of Ottawa, Faculty of Graduate and Postdoctoral Studies, Faculty of Medicine, Department of Biochemistry, Microbiology and Immunology, Ottawa, ON K1N 6N5, Canada. Offers biochemistry (M Sc, PhD); microbiology and immunology (M Sc, PhD). *Faculty:* 72 full-time (19 women), 22 part-time/adjunct (8 women). *Students:* 163 full-time, 22 part-time. 35 applicants, 69% accepted, 16 enrolled. In 2005, 16 master's, 17 doctorates awarded. *Degree requirements:* For master's, thesis; for doctorate, thesis/dissertation, seminar, comprehensive exam. *Entrance requirements:* For master's, honors degree or equivalent, minimum B average; for doctorate, master's degree, minimum B+ average. *Application deadline:* For fall admission, 3/1 priority date for domestic students, 2/15 priority date for international students. For winter admission, 11/15 for domestic students; for spring admission, 4/1 for domestic students. Application fee: $75. Electronic applications accepted. *Expenses:* Tuition: Part-time $260 per credit. Tuition and fees vary according to course load and program. *Financial support:* Fellowships, research assistantships with full tuition reimbursements, teaching assistantships with full tuition reimbursements, career-related internships or fieldwork, Federal Work-Study, scholarships/grants, traineeships, tuition waivers (full and partial), and unspecified assistantships available. Financial award application deadline: 2/15. *Faculty research:* General biochemistry, molecular biology, microbiology, host biology, nutrition and metabolism. *Unit head:* Dr. Zemin Yao, Chair, 613-562-5800 Ext. 5459, Fax: 613-562-5440. *Application contact:* Carol Ann Kelly, Academic Assistant, 613-562-5800 Ext. 5424, Fax: 613-562-5440, E-mail: grads@uottawa.ca.

University of Pennsylvania, School of Medicine, Biomedical Graduate Studies, Graduate Group in Biochemistry and Molecular Biophysics, Philadelphia, PA 19104. Offers PhD, MD/PhD, VMD/PhD. *Faculty:* 63. *Students:* 64 full-time (25 women); includes 9 minority (3 African Americans, 4 Asian Americans or Pacific Islanders, 2 Hispanic Americans), 10 international. 101 applicants, 34% accepted, 10 enrolled. In 2005, 9 doctorates awarded. *Degree requirements:* For doctorate, thesis/dissertation. *Entrance requirements:* For doctorate, GRE General Test. Additional exam requirements/recommendations for international students: Required—TOEFL. *Application deadline:* For fall admission, 12/15 priority date for domestic students, 12/1 priority date for international students. Applications are processed on a rolling basis. Application fee: $70. Electronic applications accepted. *Financial support:* In 2005–06, 69 students received support; fellowships, research assistantships, scholarships/grants, traineeships, and unspecified assistantships available. *Faculty research:* Biochemistry of cell differentiation, tissue culture, intermediary metabolism, structure of proteins and nucleic acids, biochemical genetics. *Unit head:* Dr. Joshua Wand, Chairperson, 215-573-7288. *Application contact:* Ruth Keris, Graduate Group Administrator, 215-898-4639, Fax: 215-573-2085, E-mail: keris@mail.med.upenn.edu.

University of Pittsburgh, School of Medicine, Graduate Programs in Medicine, Program in Biochemistry and Molecular Genetics, Pittsburgh, PA 15260. Offers MS, PhD. *Faculty:* 46 full-time (8 women). *Students:* 23 full-time (14 women); includes 1 minority (Hispanic American), 5 international. Average age 28. 415 applicants, 22% accepted, 42 enrolled. In 2005, 5 degrees awarded. *Median time to degree:* Of those who began their doctoral program in fall 1997, 95% received their degree in 8 years or less. *Degree requirements:* For doctorate, thesis/dissertation, comprehensive exam, registration. *Entrance requirements:* For doctorate, GRE General Test, GRE Subject Test, minimum QPA of 3.0. Additional exam requirements/recommendations for international students: Required—TOEFL (minimum score 600 paper-based; 250 computer-based), IELT (minimum score 7). *Application deadline:* For fall admission, 12/15 priority date for domestic students, 12/15 priority date for international students. Application fee: $40. Electronic applications accepted. *Expenses:* Tuition, state resident: full-time $13,194; part-time $537 per credit. Tuition, nonresident: full-time $25,012; part-time $1,026 per credit. Required fees: $700; $164 per term. Tuition and fees vary according to campus/location and program. *Financial support:* In 2005–06, 1 fellowship with full tuition reimbursement (averaging $21,500 per year), 23 research assistantships with full tuition reimbursements (averaging $21,500 per year) were awarded; teaching assistantships with full tuition reimbursements, institutionally sponsored loans, scholarships/grants, traineeships, health care benefits, and unspecified assistantships also available. *Faculty research:* Molecular genetics of cancer, gene expression and signal transduction, genomics and proteomics, human gene therapy, structural dynamics and bioinformatics. *Unit head:* Dr. Thomas Smithgall, Professor, 412-648-9495, Fax: 412-624-1901, E-mail: tsmithga@pitt.edu. *Application contact:* Graduate Studies Administrator, 412-648-8957, Fax: 412-648-1077, E-mail: gradstudies@medschool.pitt.edu.

University of Puerto Rico, Medical Sciences Campus, School of Medicine, Division of Graduate Studies, Department of Biochemistry, San Juan, PR 00936-5067. Offers MS, PhD. *Degree requirements:* For master's, one foreign language, thesis/dissertation; for doctorate, one foreign language, thesis/dissertation, comprehensive exam. *Entrance requirements:* For master's and doctorate, GRE General Test, GRE Subject Test, interview, minimum GPA of 3.0. *Expenses:* Tuition, state resident: full-time $3,600; part-time $100 per credit hour. Tuition, nonresident: full-time $4,655. Required fees: $1,734. Tuition and fees vary according to class time, degree level and program. *Faculty research:* Aging, plasma proteins, vitamins and minerals, protein structure/function, glycosilation of proteins.

University of Regina, Faculty of Graduate Studies and Research, Faculty of Science, Department of Chemistry and Biochemistry, Regina, SK S4S 0A2, Canada. Offers analytical chemistry (M Sc, PhD); biochemistry (M Sc, PhD); inorganic chemistry (M Sc, PhD); organic chemistry (M Sc, PhD); physical chemistry (M Sc, PhD). Part-time programs available. *Faculty:* 10 full-time (2 women), 4 part-time/adjunct (1 woman). *Students:* 14 full-time (6 women), 3 part-time (2 women). 21 applicants, 33% accepted. *Degree requirements:* For master's and doctorate, thesis/dissertation, departmental qualifying exam. *Entrance requirements:* For master's and doctorate, GRE. Additional exam requirements/recommendations for international students: Required—TOEFL (minimum score 580 paper-based; 237 computer-based). *Application deadline:* For fall admission, 1/1 for domestic studentsFor winter admission, 7/1 for domestic students. Applications are processed on a rolling basis. Application fee: $60 ($100 for international students). *Financial support:* In 2005–06, 4 fellowships (averaging $14,886 per year), 1 research assistantship (averaging $12,750 per year), 5 teaching assistantships (averaging $13,501 per year) were awarded; scholarships/grants also available. Financial award application deadline: 6/15. *Faculty research:* Organic synthesis, organic oxidations, ionic liquids theoretical/computational chemistry, protein biochemistry/biophysics, environmental analytical, photophysical/photochemistry. *Unit head:* Dr. Andrew G. Wee, Head, 306-585-4767, Fax: 306-585-4894, E-mail: chem.chair@uregina.ca. *Application contact:* Dr. Allan East, Associate Professor, 306-585-4003, Fax: 306-585-4894, E-mail: allan.east@uregina.ca.

University of Rhode Island, Graduate School, College of the Environment and Life Sciences, Department of Cell and Molecular Biology, Kingston, RI 02881. Offers biochemistry (MS, PhD); microbiology (MS, PhD), including biodegradation (MS), cellular development (MS), electron microscopy and ultrastructure (MS), genetics and molecular biology (MS), immunology (MS), marine and freshwater ecosystems (MS), microbial pathogenesis (MS), microbial physiology (MS), protozoology (MS), virology (MS), water-pollution microbiology (MS); molecular genetics (MS, PhD). In 2005, 2 degrees awarded. *Degree requirements:* For master's and doctorate, thesis/dissertation. *Entrance requirements:* For master's and doctorate, GRE General Test. Additional exam requirements/recommendations for international students: Required—TOEFL. *Expenses:* Tuition, state resident: full-time $5,522; part-time $307 per credit. Tuition, nonresident: full-time $15,992; part-time $888 per credit. Required fees: $1,786; $73 per credit. One-time fee: $80 part-time. *Financial support:* Fellowships, research assistantships, teaching assistantships available. *Unit head:* Dr. Jay Sperry, Chairperson, 401-874-5900.

University of Rochester, School of Medicine and Dentistry, Graduate Programs in Medicine and Dentistry, Department of Biochemistry and Biophysics, Program in Biochemistry, Rochester, NY 14627-0250. Offers MS, PhD. Terminal master's awarded for partial completion of doctoral program. *Degree requirements:* For doctorate, thesis/dissertation, qualifying exam. *Entrance requirements:* For master's and doctorate, GRE General Test.

University of Saskatchewan, College of Medicine, Department of Biochemistry, Saskatoon, SK S7N 5A2, Canada. Offers M Sc, PhD, Diploma. *Faculty:* 24. *Students:* 33. *Degree requirements:* For master's and doctorate, thesis/dissertation, registration. *Entrance requirements:* Additional exam requirements/recommendations for international students: Required—TOEFL. *Application deadline:* For fall admission, 7/1 for domestic students. Applications are processed on a rolling basis. Application fee: $50. *Financial support:* Fellowships, research assistantships, teaching assistantships available. Financial award application deadline: 1/31. *Unit head:* Dr. R. Khandewal, Head, 306-966-4359, Fax: 306-966-4390, E-mail: ramji.khandelwal@usask.ca. *Application contact:* Dr. W. Roesler, Graduate Chair, 306-966-4359, Fax: 306-966-4390, E-mail: bill.roesler@usdsk.ca.

The University of Scranton, Graduate School, Department of Chemistry, Program in Biochemistry, Scranton, PA 18510. Offers MA, MS. Part-time and evening/weekend programs available. *Faculty:* 10 full-time (3 women), 1 part-time/adjunct (0 women). *Students:* 6 full-time (5 women), 12 part-time (6 women); includes 1 minority (Hispanic American) Average age 25. 8 applicants, 100% accepted. In 2005, 10 degrees awarded. *Degree requirements:* For master's, thesis (for some programs), capstone experience, comprehensive exam (for some programs), registration. *Entrance requirements:* For master's, minimum GPA of 2.75. Additional exam requirements/recommendations for international students: Required—TOEFL (minimum score 500 paper-based; 173 computer-based), IELT (minimum score 6). *Application deadline:* Applications are processed on a rolling basis. Application fee: $50. *Expenses:* Tuition: Part-time $647 per credit. Required fees: $25 per term. *Financial support:* In 2005–06, 17 teaching assistantships with full and partial tuition reimbursements (averaging $8,600 per year) were awarded; career-related internships or fieldwork, Federal Work-Study, unspecified assistantships, and teaching fellowships also available. Support available to part-time students. Financial award application deadline: 3/1. *Unit head:* Dr. Christopher A. Baumann, Director, 570-941-6389, Fax: 570-941-7510, E-mail: cab@scranton.edu.

University of South Alabama, College of Medicine and Graduate School, Program in Basic Medical Sciences, Specialization in Biochemistry and Molecular Biology, Mobile, AL 36688-0002. Offers PhD. *Faculty:* 9 full-time (2 women). *Degree requirements:* For doctorate, thesis/dissertation. *Entrance requirements:* For doctorate, GRE General Test or MCAT. *Application deadline:* For fall admission, 4/1 for domestic students. Applications are processed on a rolling basis. Application fee: $25. *Expenses:* Tuition, state resident: full-time $4,008. Tuition, nonresident: full-time $8,016. Required fees: $692. *Financial support:* Fellowships, institutionally sponsored loans available. Financial award application deadline: 4/1. *Faculty research:* Biochemistry of aging, mechanisms of metabolic regulation and oxygen metabolism. *Unit head:* Dr. Nathan N. Aronson, Chair, 251-460-6402. *Application contact:* Lanette Flagge, Academic Advisor, 251-460-6153.

University of South Carolina, The Graduate School, College of Arts and Sciences, Department of Chemistry and Biochemistry, Columbia, SC 29208. Offers IMA, MAT, MS, PhD. IMA and MAT offered in cooperation with the College of Education. Part-time programs available. Terminal master's awarded for partial completion of doctoral program. *Degree requirements:*

Biochemistry

University of South Carolina (continued)
For master's and doctorate, thesis/dissertation, comprehensive exam, registration. *Entrance requirements:* For master's and doctorate, GRE General Test. Additional exam requirements/recommendations for international students: Required—TOEFL. Electronic applications accepted. *Faculty research:* Spectroscopy, crystallography, organic and organometallic synthesis, analytical chemistry, materials.

University of Southern California, Keck School of Medicine and Graduate School, Graduate Programs in Medicine, Department of Biochemistry and Molecular Biology, Los Angeles, CA 90089. Offers MS, PhD. *Faculty:* 44 full-time (10 women). *Students:* 106 full-time (58 women); includes 17 minority (12 Asian Americans or Pacific Islanders, 5 Hispanic Americans), 57 international. Average age 24. 487 applicants, 19% accepted, 44 enrolled. In 2005, 5 master's, 8 doctorates awarded. Terminal master's awarded for partial completion of doctoral program. *Degree requirements:* For master's and doctorate, thesis/dissertation, registration. *Entrance requirements:* For master's and doctorate, GRE General Test, minimum GPA of 3.0. Additional exam requirements/recommendations for international students: Required—TOEFL (minimum score 600 paper-based; 250 computer-based). *Application deadline:* For fall admission, 12/15 priority date for domestic students, 12/5 priority date for international students. Application fee: $65 ($75 for international students). Electronic applications accepted. *Expenses:* Tuition: Full-time $25,416; part-time $1,059 per unit. Required fees: $484; $484 per year. Tuition and fees vary according to course load and program. *Financial support:* In 2005–06, 93 students received support, including 4 fellowships with tuition reimbursements available (averaging $24,876 per year), 87 research assistantships with tuition reimbursements available (averaging $24,876 per year), 2 teaching assistantships with tuition reimbursements available (averaging $24,876 per year); career-related internships or fieldwork, institutionally sponsored loans, scholarships/grants, and tuition waivers (full) also available. Support available to part-time students. Financial award application deadline: 4/15. *Faculty research:* Molecular genetics, gene expression, membrane biochemistry, metabolic regulation, cancer biology. *Unit head:* Dr. Michael R. Stallcup, Chair, 323-442-1145, Fax: 323-442-1224, E-mail: stallcup@usc.edu. *Application contact:* Anne L. Rice, Student Services Coordinator, 323-442-1145, Fax: 323-442-1224, E-mail: annvazqu@usc.edu.

University of Southern Mississippi, Graduate School, College of Science and Technology, Department of Chemistry and Biochemistry, Hattiesburg, MS 39406-0001. Offers analytical chemistry (MS, PhD); biochemistry (MS, PhD); inorganic chemistry (MS, PhD); organic chemistry (MS, PhD); physical chemistry (MS, PhD). *Degree requirements:* For master's and doctorate, thesis/dissertation, comprehensive exam. *Entrance requirements:* For master's, GRE General Test, minimum GPA of 2.75 in last 60 hours; for doctorate, GRE General Test, minimum GPA of 3.5. Additional exam requirements/recommendations for international students: Required—TOEFL. *Faculty research:* Plant biochemistry, photo chemistry, polymer chemistry, x-ray analysis, enzyme chemistry.

University of South Florida, College of Graduate Studies, College of Arts and Sciences, Department of Chemistry, Tampa, FL 33620-9951. Offers analytical chemistry (MS, PhD); biochemistry (MS, PhD); inorganic chemistry (MS, PhD); organic chemistry (MS, PhD); physical chemistry (MS, PhD); polymer chemistry (PhD). Part-time programs available. *Faculty:* 22. *Students:* 102 full-time (45 women), 15 part-time (6 women); includes 13 minority (5 Asian Americans or Pacific Islanders, 8 Hispanic Americans), 60 international. 80 applicants, 71% accepted, 23 enrolled. In 2005, 2 master's awarded. Terminal master's awarded for partial completion of doctoral program. *Degree requirements:* For master's, thesis; for doctorate, 2 foreign languages, thesis/dissertation, colloquium. *Entrance requirements:* For master's and doctorate, GRE General Test, minimum GPA of 3.0 in last 30 hours of chemistry course work, 3 letters of recommendation. Additional exam requirements/recommendations for international students: Required—TOEFL (minimum score 550 paper-based; 213 computer-based), TSE(minimum score 50). *Application deadline:* For fall admission, 5/1 priority date for domestic students, 3/1 priority date for international students; for spring admission, 10/1 priority date for domestic students, 8/1 priority date for international students. Applications are processed on a rolling basis. Application fee: $30. Electronic applications accepted. *Financial support:* In 2005–06, 91 students received support; fellowships with partial tuition reimbursements available, research assistantships with partial tuition reimbursements available, teaching assistantships with partial tuition reimbursements available, career-related internships or fieldwork, institutionally sponsored loans, scholarships/grants, health care benefits, and unspecified assistantships available. Financial award application deadline: 6/30. *Faculty research:* Synthesis, bio-organic chemistry, bioinorganic chemistry, environmental chemistry, NMR. *Unit head:* Dr. Michael Zaworotko, Chairperson, 813-974-4129, Fax: 813-974-3203, E-mail: xtal@usf.edu. *Application contact:* Dr. Brian Space, Graduate Coordinator, 813-974-3397, Fax: 813-974-3203.

University of South Florida, College of Medicine and College of Graduate Studies, Graduate Programs in Medical Sciences, Tampa, FL 33620-9951. Offers anatomy (PhD); biochemistry and molecular biology (MS, PhD), including biochemistry and molecular biology (PhD), bioinformatics and computational biology (MS); medical microbiology and immunology (PhD); pathology (PhD); pharmacology and therapeutics (PhD), including medical sciences; physiology and biophysics (PhD). *Students:* 108 full-time (53 women), 34 part-time (26 women); includes 31 minority (9 African Americans, 9 Asian Americans or Pacific Islanders, 13 Hispanic Americans), 30 international. 117 applicants, 99% accepted, 116 enrolled. In 2005, 9 master's, 4 doctorates awarded. *Degree requirements:* For doctorate, thesis/dissertation. *Entrance requirements:* For doctorate, GRE General Test, minimum GPA of 3.0. Application fee: $30. *Expenses: Contact institution. Financial support:* Institutionally sponsored loans and scholarships/grants available. Financial award application deadline: 4/1; financial award applicants required to submit FAFSA. *Unit head:* Dr. Joseph J. Krzanowski, Associate Dean for Research and Graduate Affairs, 813-974-4181, Fax: 813-974-4317, E-mail: jkrzanow@com1.med.usf.edu.

The University of Tennessee, Graduate School, College of Arts and Sciences, Department of Biochemistry, Cellular and Molecular Biology, Knoxville, TN 37996. Offers MS, PhD. Terminal master's awarded for partial completion of doctoral program. *Degree requirements:* For master's and doctorate, thesis/dissertation. *Entrance requirements:* For master's and doctorate, GRE General Test, minimum GPA of 2.7. Additional exam requirements/recommendations for international students: Required—TOEFL. Electronic applications accepted.

The University of Tennessee Health Science Center, College of Graduate Health Sciences, Department of Molecular Sciences, Memphis, TN 38163-0002. Offers bacterial pathogenesis (PhD); biochemistry (PhD); immunology (PhD); microbiology (PhD); molecular and cell biology–pathology (PhD); molecular biology (PhD); signal transduction (PhD); structural biology (PhD); virology (PhD). *Faculty:* 17 full-time (4 women). *Students:* 25 full-time (11 women); includes 14 minority (1 African American, 13 Asian Americans or Pacific Islanders). Average age 24. 83 applicants, 12% accepted. In 2005, 6 degrees awarded. *Degree requirements:* For doctorate, thesis/dissertation, oral and written preliminary and comprehensive exams. *Entrance requirements:* For doctorate, GRE General Test, GRE Subject Test, minimum GPA of 3.0. Additional exam requirements/recommendations for international students: Required—TOEFL. *Application deadline:* For fall admission, 5/15 for domestic students. Application fee: $0. *Financial support:* In 2005–06, 8 fellowships, 21 research assistantships were awarded; teaching assistantships, traineeships and tuition waivers (full) also available. *Unit head:* Dr. David Hasty, Chairman, 901-448-6150, Fax: 901-448-7360. *Application contact:* Ida W. Mosby, Director, Enrollment Services, 901-448-5560, E-mail: imosby@utmem.edu.

The University of Texas at Austin, Graduate School, College of Natural Sciences, Department of Chemistry and Biochemistry, Program in Biochemistry, Austin, TX 78712-1111. Offers MA, PhD. *Entrance requirements:* For master's and doctorate, GRE General Test.

The University of Texas at Austin, Graduate School, College of Natural Sciences, Institute for Cellular and Molecular Biology, Program in Biochemistry, Austin, TX 78712-1111. Offers PhD.

The University of Texas Health Science Center at Houston, Graduate School of Biomedical Sciences, Program in Biochemistry and Molecular Biology, Houston, TX 77225-0036. Offers MS, PhD, MD/PhD. *Faculty:* 24 full-time (4 women). *Students:* 21 full-time (14 women); includes 1 minority (Hispanic American), 11 international. Average age 24. 14 applicants, 14% accepted, 2 enrolled. In 2005, 5 degrees awarded. Terminal master's awarded for partial completion of doctoral program. *Degree requirements:* For master's and doctorate, thesis/dissertation. *Entrance requirements:* For master's and doctorate, GRE General Test. Additional exam requirements/recommendations for international students: Required—TOEFL, TWE. *Application deadline:* For fall admission, 1/15 for domestic students; for spring admission, 11/1 for domestic students. Applications are processed on a rolling basis. Application fee: $10. Electronic applications accepted. *Financial support:* Fellowships with full tuition reimbursements, research assistantships with full tuition reimbursements, teaching assistantships, institutionally sponsored loans, scholarships/grants, and health care benefits available. Financial award application deadline: 1/15. *Faculty research:* Protein structure and function, regulation of gene expression, cellular signal transduction, regulation of cell growth and differentiation, molecular basis of human disease. *Unit head:* Dr. Michael R Blackburn, Director, 713-500-6087, Fax: 713-500-0652, E-mail: michael.r.blackburn@uth.tmc.edu. *Application contact:* Dr. Victoria P. Knutson, Assistant Dean of Admissions, 713-500-9877, Fax: 713-500-9877, E-mail: victoria.p.knutson@uth.tmc.edu.

The University of Texas Health Science Center at San Antonio, Graduate School of Biomedical Sciences, Department of Biochemistry, San Antonio, TX 78229-3900. Offers MS, PhD. Terminal master's awarded for partial completion of doctoral program. *Degree requirements:* For master's and doctorate, thesis/dissertation. *Entrance requirements:* For master's and doctorate, GRE General Test. *Faculty research:* Protein structure and function, lipid biochemistry, metabolic regulation, immunology, membrane assembly.

The University of Texas Medical Branch, Graduate School of Biomedical Sciences, Program in Biochemistry and Molecular Biology, Galveston, TX 77555. Offers biochemistry (PhD); bioinformatics (PhD); biophysics (PhD); cell biology (PhD); computational biology (PhD); structural biology (PhD). *Students:* 38 full-time (14 women), 1 part-time; includes 5 minority (3 Asian Americans or Pacific Islanders, 2 Hispanic Americans), 18 international. Average age 27. In 2005, 7 degrees awarded. *Degree requirements:* For doctorate, thesis/dissertation. *Entrance requirements:* Additional exam requirements/recommendations for international students: Required—TOEFL (minimum score 550 paper-based; 213 computer-based). *Application deadline:* Applications are processed on a rolling basis. Application fee: $30 ($75 for international students). Electronic applications accepted. *Expenses:* Tuition, state resident: full-time $8,350; part-time $90 per credit hour. Tuition, nonresident: full-time $21,450; part-time $366 per credit hour. Required fees: $1,027; $11 per credit hour. $60 per term. *Financial support:* In 2005–06, fellowships (averaging $23,000 per year), research assistantships (averaging $23,000 per year) were awarded. Financial award applicants required to submit FAFSA. *Unit head:* Dr. Leenian L. Chan, Director, 409-772-2861, Fax: 409-772-9679, E-mail: lchan@utmb.edu. *Application contact:* Debora Botting, Co-ordinator II Special Programs, 409-772-2769, Fax: 409-747-0552, E-mail: dmbottin@utmb.edu.

The University of Texas Southwestern Medical Center at Dallas, Southwestern Graduate School of Biomedical Sciences, Division of Basic Science, Program in Biological Chemistry, Dallas, TX 75390. Offers PhD. *Faculty:* 44 full-time (9 women), 1 part-time/adjunct (0 women). *Students:* 37 full-time (13 women), 1 part-time; includes 1 minority (Hispanic American), 9 international. Average age 28. In 2005, 3 doctorates awarded. *Degree requirements:* For doctorate, thesis/dissertation, qualifying exam. *Entrance requirements:* For doctorate, GRE General Test, minimum GPA of 3.0. Additional exam requirements/recommendations for international students: Required—TOEFL. *Application deadline:* For fall admission, 1/5 for domestic students. Application fee: $0. Electronic applications accepted. *Expenses:* Tuition, state resident: full-time $6,550; part-time $50 per credit hour. Tuition, nonresident: full-time $19,650; part-time $326 per credit hour. Required fees: $42 per credit hour. Tuition and fees vary according to degree level and program. *Financial support:* Fellowships, research assistantships, institutionally sponsored loans available. *Faculty research:* Regulation of gene expression, protein trafficking, molecular neurobiology, protein structure and function, metabolic regulation. *Unit head:* Dr. Margaret A. Phillips, Chair, 214-645-6164, Fax: 214-645-6166, E-mail: margaret.phillips@utsouthwestern.edu. *Application contact:* Dr. Nancy E. Street, Associate Dean, 214-648-6708, Fax: 214-648-2102, E-mail: nancy.street@utsouthwestern.edu.

University of the Sciences in Philadelphia, College of Graduate Studies, Program in Chemistry, Biochemistry and Pharmacognosy, Philadelphia, PA 19104-4495. Offers biochemistry (MS, PhD); chemistry (MS, PhD); medicinal chemistry (MS, PhD); pharmacognosy (MS, PhD). Part-time programs available. *Faculty:* 13 full-time (3 women). *Students:* 8 full-time (4 women), 7 part-time (3 women); includes 2 minority (both Asian Americans or Pacific Islanders), 5 international. Average age 28. In 2005, 3 master's, 4 doctorates awarded. *Degree requirements:* For master's, thesis/dissertation, qualifying exams; for doctorate, thesis/dissertation, qualifying exams, comprehensive exam. *Entrance requirements:* For master's and doctorate, GRE General Test, GRE Subject Test. Additional exam requirements/recommendations for international students: Required—TOEFL, TWE, TSE. *Application deadline:* Applications are processed on a rolling basis. Application fee: $50. *Expenses:* Tuition: Part-time $999 per credit. Tuition and fees vary according to program. *Financial support:* In 2005–06, 19 students received support, including 1 fellowship with full tuition reimbursement available (averaging $19,000 per year), 1 research assistantship with full tuition reimbursement available (averaging $19,000 per year), 12 teaching assistantships with full tuition reimbursements available (averaging $18,500 per year); institutionally sponsored loans, scholarships/grants, and tuition waivers (full) also available. Financial award application deadline: 5/1. *Faculty research:* Organic and medicinal synthesis, mass spectroscopy use in protein analysis, study of analogues of taxol, cholesteryl esters. Total annual research expenditures: $341,700. *Unit head:* Dr. James McKee, Director, 215-596-8847, Fax: 215-596-8543, E-mail: jmckee@usip.edu. *Application contact:* Joyce D'Angelo, Administrative Assistant, 215-596-8937, E-mail: j.dangel@usip.edu.

The University of Toledo, Graduate School, College of Arts and Sciences, Department of Chemistry, Toledo, OH 43606-3390. Offers analytical chemistry (MS, PhD); biological chemistry (MS, PhD); inorganic chemistry (MS, PhD); organic chemistry (MS, PhD); physical chemistry (MS, PhD). Part-time programs available. *Faculty:* 18. *Students:* 80 full-time (35 women), 8 part-time (5 women); includes 3 minority (2 African Americans, 1 Asian American or Pacific Islander), 43 international. Average age 27. 32 applicants, 75% accepted, 19 enrolled. In 2005, 7 master's awarded. *Degree requirements:* For master's and doctorate, thesis/dissertation. *Entrance requirements:* For master's and doctorate, GRE General Test, GRE Subject Test. Additional exam requirements/recommendations for international students: Required—TOEFL. *Application deadline:* For fall admission, 8/1 for domestic students. Applications are processed on a rolling basis. Application fee: $45. Electronic applications accepted. *Expenses:* Tuition, area resident: Full-time $3,312; part-time $308 per credit hour. Tuition, state resident: full-time $3,312. Tuition, nonresident: full-time $6,616; part-time $735 per credit hour. International tuition: $6,616 full-time. *Financial support:* In 2005–06, 1 research assistantship with full tuition reimbursement (averaging $4,000 per year), 55 teaching assistantships with full tuition reimbursements (averaging $13,336 per year) were awarded; fellowships, Federal Work-Study and institutionally sponsored loans also available. Support available to part-time students. Financial award application deadline: 4/1; financial award applicants required to submit FAFSA. *Faculty research:* Enzymology, materials chemistry, crystallography, theoretical chemistry. *Unit head:* Dr. Alan Pinkerton, Chair, 419-530-7902, Fax: 419-530-4033, E-mail: apinker@uoft02.utoledo.edu. *Application contact:* Charlene Morlock-Hansen, 419-530-2100, E-mail: charlene.hanson@utoledo.edu.

The University of Toledo, College of Graduate Studies, College of Pharmacy, Graduate Programs in Pharmacy, Program in Medicinal and Biological Chemistry, Toledo, OH 43606-3390. Offers MS, PhD. *Faculty:* 10 full-time (2 women), 6 part-time/adjunct (1 woman). *Students:* 17 full-time (5 women); includes 1 minority (African American), 9 international. Average age 26: 36 applicants, 22% accepted, 4 enrolled. In 2005, 3 master's, 2 doctorates awarded. Terminal master's awarded for partial completion of doctoral program. *Degree requirements:* For master's and doctorate, thesis/dissertation. *Entrance requirements:* For master's and doctorate, GRE General Test. Additional exam requirements/recommendations for international students: Required—TOEFL (minimum score 550 paper-based; 213 computer-based). *Application deadline:* For fall admission, 2/1 priority date for domestic students, 1/1 priority date for international students. Application fee: $45. Electronic applications accepted. *Expenses:* Tuition, state resident: full-time $6,623; part-time $308 per credit hour. Tuition, nonresident: full-time $13,232; part-time $735 per credit hour. *Financial support:* Research assistantships with full tuition reimbursements, teaching assistantships with full tuition reimbursements available. *Faculty research:* Neuroscience, molecular modeling, immunotoxicology, organic synthesis, peptide biochemistry. *Unit head:* Dr. Marcia McInerney, Chair, 419-530-1981, Fax: 419-530-7946, E-mail: mmciner@utnet.utoledo.edu. *Application contact:* Linda McPherson, Information Contact, 419-530-2902, Fax: 419-530-7946, E-mail: lmcpher@utnet.utoledo.edu.

See Close-Up on page 439.

The University of Toledo, School of Graduate Studies, Department of Biochemistry and Molecular Biology, Toledo, OH 43606-3390. Offers medical sciences (MS), including biochemistry, genetics, molecular biology. Part-time programs available. *Degree requirements:* For master's, thesis, qualifying exam. *Entrance requirements:* For master's, GRE General Test, minimum undergraduate GPA of 3.0. Additional exam requirements/recommendations for international students: Required—TOEFL. *Expenses:* Tuition, state resident: full-time $6,623; part-time $308 per credit hour. Tuition, nonresident: full-time $13,232; part-time $735 per credit hour. *Faculty research:* Gene regulation, protein structure, receptors, protein phosphorylation, peptides.

University of Toronto, School of Graduate Studies, Life Sciences Division, Department of Biochemistry, Toronto, ON M5S 1A1, Canada. Offers M Sc, PhD. *Degree requirements:* For master's, thesis, oral examination of thesis; for doctorate, thesis/dissertation, oral defense of thesis. *Entrance requirements:* For master's, GRE General Test and GRE Subject Test in biochemistry/molecular biology (international applicants only), B Sc in biochemistry or molecular biology, minimum B+ average, letters of reference. Additional exam requirements/recommendations for international students: Required—TOEFL (minimum score 580 paper-based; 237 computer-based), TWE (minimum score 5).

University of Utah, The Graduate School and Graduate Programs in Medicine, Program in Biological Chemistry, Salt Lake City, UT 84112-1107.

See Close-Up on page 441.

University of Utah, School of Medicine and The Graduate School, Graduate Programs in Medicine, Department of Biochemistry, Salt Lake City, UT 84112-1107. Offers MS, PhD. Terminal master's awarded for partial completion of doctoral program. *Degree requirements:* For master's and doctorate, thesis/dissertation. *Entrance requirements:* For doctorate, GRE Subject Test, minimum GPA of 3.0. Electronic applications accepted. *Expenses:* Tuition, state resident: full-time $2,932; part-time $369 per credit. Tuition, nonresident: full-time $10,350; part-time $1,302 per credit. Required fees: $516 per term. Tuition and fees vary according to course load and program. *Faculty research:* Protein structure and function, nucleic acid structure and function, nucleic acid enzymology, RNA modification, protein turnover.

University of Vermont, College of Medicine and Graduate College, Graduate Programs in Medicine, Department of Biochemistry, Burlington, VT 05405. Offers MS, PhD, MD/MS, MD/PhD. *Students:* 15 (4 women) 3 international. 26 applicants, 15% accepted, 4 enrolled. In 2005, 3 doctorates awarded. *Degree requirements:* For master's and doctorate, thesis/dissertation. *Entrance requirements:* For master's and doctorate, GRE General Test, GRE Subject Test. Additional exam requirements/recommendations for international students: Required—TOEFL (minimum score 550 paper-based; 213 computer-based). *Application deadline:* For fall admission, 4/1 for domestic students. Applications are processed on a rolling basis. Application fee: $40. Electronic applications accepted. *Expenses:* Tuition, area resident: Part-time $410 per credit hour. Tuition, nonresident: part-time $1,034 per credit hour. *Financial support:* Fellowships, research assistantships, teaching assistantships, analytical assistantships available. Financial award application deadline: 3/1. *Faculty research:* Endocrinology, protein chemistry, cell-surface signaling. *Unit head:* , Dr. Kenneth G. Mann, Chairperson, 802-656-0355. *Application contact:* Dr. Scott Morrical, Coordinator, 802-656-2220.

See Close-Up on page 443.

University of Victoria, Faculty of Graduate Studies, Faculty of Science, Department of Biochemistry and Microbiology, Victoria, BC V8W 2Y2, Canada. Offers biochemistry (M Sc, PhD); microbiology (M Sc, PhD). *Faculty:* 14 full-time (3 women), 1 part-time/adjunct (0 women). *Students:* Average age 22. 37 applicants, 43% accepted, 5 enrolled. In 2005, 3 master's, 1 doctorate awarded. *Degree requirements:* For master's, thesis, seminar; for doctorate, thesis/dissertation, seminar, candidacy exam. *Entrance requirements:* For master's, GRE General Test (75th percentile); for doctorate, GRE General Test (75th percentile), minimum B+ average, M Sc. Additional exam requirements/recommendations for international students: Required—TOEFL (minimum score 600 paper-based; 250 computer-based). *Application deadline:* For fall admission, 5/31 for domestic students, 12/15 for international students. Applications are processed on a rolling basis. Application fee: $75 ($125 for international students). Electronic applications accepted. Tuition charges are reported in Canadian dollars. *Expenses:* Tuition, nonresident: full-time $4,492 Canadian dollars; part-time $749 Canadian dollars per term. International tuition: $5,346 Canadian dollars full-time. Tuition and fees vary according to course load, campus/location and program. *Financial support:* In 2005–06, 5 fellowships (averaging $14,250 per year), 16 research assistantships (averaging $11,000 per year), 20 teaching assistantships (averaging $2,000 per year) were awarded; career-related internships or fieldwork, institutionally sponsored loans, and awards, graduate teaching fellowships also available. Financial award application deadline: 2/15. *Faculty research:* Molecular pathogenesis, prokaryotic, eukaryotic, macromolecular interactions, microbial surfaces, virology, molecular genetics. Total annual research expenditures: $2.2 million. *Unit head:* Dr. Claire Cupples, Chair, 250-721-7077, Fax: 250-721-8855, E-mail: ccupples@uvic.ca. *Application contact:* Melinda Powell, Graduate Secretary, 250-721-8861, Fax: 250-721-8855, E-mail: biocgsec@uvic.ca.

University of Virginia, School of Medicine, Department of Biochemistry and Molecular Genetics, Charlottesville, VA 22903. Offers biochemistry (PhD). *Students:* 38 full-time (21 women); includes 1 minority (African American), 7 international. Average age 27. In 2005, 7 degrees awarded. *Degree requirements:* For doctorate, thesis/dissertation. *Entrance requirements:* For doctorate, GRE General Test, GRE Subject Test. Additional exam requirements/recommendations for international students: Required—TOEFL. Application fee: $60. Electronic applications accepted. *Expenses:* Tuition, state resident: full-time $7,731. Tuition, nonresident: full-time $18,672. Required fees: $1,479. Full-time tuition and fees vary according to degree level and program. *Financial support:* Fellowships, research assistantships, teaching assistantships available. Financial award applicants required to submit FAFSA. *Unit head:* Joyce L. Hamlin, Chairman, 434-924-1940, Fax: 434-924-1667. *Application contact:* Peter C. Brunjes, Associate Dean for Graduate Programs and Research, 434-924-7184, Fax: 434-924-6737, E-mail: grad-a-s@virginia.edu.

University of Washington, School of Medicine and Graduate School, Graduate Programs in Medicine, Department of Biochemistry, Seattle, WA 98195. Offers PhD. *Degree requirements:* For doctorate, thesis/dissertation. *Entrance requirements:* For doctorate, GRE General Test, GRE Subject Test (biology, chemistry, biochemistry, or cell and molecular biology), minimum GPA of 3.0. Additional exam requirements/recommendations for international students: Required—TOEFL. Electronic applications accepted. *Faculty research:* Blood coagulation, structure and function of enzymes, fertilization events, interaction of plants with bacteria, protein structure.

The University of Western Ontario, Faculty of Graduate Studies, Biosciences Division, Department of Biochemistry, London, ON N6A 5B8, Canada. Offers M Sc, PhD. *Degree requirements:* For master's and doctorate, thesis/dissertation. *Entrance requirements:* For master's, minimum B+ average in last 2 years of undergraduate study; for doctorate, M Sc or an external scholarship winner.

University of West Florida, College of Arts and Sciences: Sciences, Division of Life and Health Sciences, Department of Biology, Pensacola, FL 32514-5750. Offers biological chemistry (MS); biology (MS); biology education (MST); coastal zone studies (MS); environmental biology (MS). *Accreditation:* NCATE. *Faculty:* 16 full-time (5 women), 2 part-time/adjunct (0 women). *Students:* 12 full-time (10 women), 31 part-time (19 women); includes 4 minority (1 African American, 2 Asian Americans or Pacific Islanders, 1 Hispanic American), 2 international. Average age 28. 17 applicants, 71% accepted, 8 enrolled. In 2005, 4 degrees awarded. *Degree requirements:* For master's, thesis. *Entrance requirements:* For master's, GRE General Test. Additional exam requirements/recommendations for international students: Required—TOEFL (minimum score 550 paper-based; 213 computer-based). *Application deadline:* For fall admission, 6/1 for domestic students, 5/15 for international students; for spring admission, 11/1 for domestic students, 10/1 for international students. Applications are processed on a rolling basis. Application fee: $30. *Expenses:* Tuition, state resident: full-time $5,833; part-time $243 per credit hour. Tuition, nonresident: full-time $21,204; part-time $884 per credit hour. Tuition and fees vary according to campus/location. *Financial support:* In 2005–06, 20 students received support, including 5 research assistantships with partial tuition reimbursements available (averaging $5,000 per year), 11 teaching assistantships with partial tuition reimbursements available (averaging $8,000 per year) Financial award application deadline: 4/15; financial award applicants required to submit FAFSA.

University of Windsor, Faculty of Graduate Studies and Research, Faculty of Science, Department of Chemistry and Biochemistry, Windsor, ON N9B 3P4, Canada. Offers M Sc, PhD. Part-time programs available. *Faculty:* 21 full-time (1 woman), 5 part-time/adjunct (1 woman). *Students:* 59 full-time (26 women). 71 applicants, 39% accepted. In 2005, 8 master's, 7 doctorates awarded. *Degree requirements:* For master's, thesis/dissertation; for doctorate, thesis/dissertation, comprehensive exam. *Entrance requirements:* For master's and doctorate, minimum B average. Additional exam requirements/recommendations for international students: Required—TOEFL (minimum score 560 paper-based; 220 computer-based), GRE. *Application deadline:* For fall admission, 7/1 for domestic students. Applications are processed on a rolling basis. Application fee: $55. Electronic applications accepted. *Financial support:* In 2005–06, 54 teaching assistantships (averaging $8,956 per year) were awarded; research assistantships, Federal Work-Study, scholarships/grants, unspecified assistantships, and bursaries also available. Financial award application deadline: 2/15. *Faculty research:* Molecular biology/recombinant DNA techniques (PCR, cloning mutagenesis), No/02 detectors, western immunoblotting and detection, CD/NMR protein/peptide structure determination, confocal/electron microscopes. *Unit head:* Dr. Douglas Stephan, Head, 519-253-3000 Ext. 3537, Fax: 519-973-7098, E-mail: stephan@uwindsor.ca. *Application contact:* Marlene Bezaire, Graduate Secretary, 519-253-3000 Ext. 3520, Fax: 519-971-7098, E-mail: spsgrad@uwindsor.ca.

University of Wisconsin–Madison, Graduate School, College of Agricultural and Life Sciences, Department of Biochemistry, Madison, WI 53706-1380. Offers MS, PhD. *Faculty:* 36 full-time (10 women). *Students:* 111 full-time (43 women), 18 international. 241 applicants, 12% accepted, 12 enrolled. In 2005, 5 master's, 9 doctorates awarded. Terminal master's awarded for partial completion of doctoral program. *Degree requirements:* For master's and doctorate, thesis/dissertation. *Entrance requirements:* For master's and doctorate, GRE General Test, GRE Subject Test (recommended). Additional exam requirements/recommendations for international students: Required—TOEFL. *Application deadline:* For fall admission, 1/1 priority date for domestic students, 1/1 priority date for international students. Applications are processed on a rolling basis. Application fee: $45. Electronic applications accepted. *Financial support:* In 2005–06, fellowships with full tuition reimbursements (averaging $20,000 per year), research assistantships with full tuition reimbursements (averaging $22,500 per year) were awarded; traineeships, health care benefits, and tuition waivers (full) also available. Financial award application deadline: 1/1. *Faculty research:* Molecular structure of vitamins and hormones, enzymology, NMR spectroscopy, protein structure, molecular genetics. Total annual research expenditures: $18 million. *Unit head:* Dr. Elizabeth A. Craig, Chair, 608-262-3040, Fax: 608-262-3453, E-mail: chair@biochem.wisc.edu. *Application contact:* Colleen A. Clary, University Services Program Associate, 608-262-3899, Fax: 608-265-9661, E-mail: cclary@biochem.wisc.edu.

University of Wisconsin–Madison, Medical School and Graduate School, Graduate Programs in Medicine, Department of Biomolecular Chemistry, Madison, WI 53706-1380. Offers MS, PhD. *Faculty:* 24 full-time (6 women). *Students:* 68 full-time (33 women); includes 3 minority (1 Asian American or Pacific Islander, 2 Hispanic Americans), 14 international. Average age 25. 176 applicants, 14% accepted, 7 enrolled. In 2005, 4 doctorates awarded. Terminal master's awarded for partial completion of doctoral program. *Median time to degree:* Of those who began their doctoral program in fall 1997, 100% received their degree in 8 years or less. *Degree requirements:* For master's and doctorate, thesis/dissertation. *Entrance requirements:* For doctorate, GRE. *Application deadline:* For fall admission, 1/1 for domestic students. Application fee: $45. Electronic applications accepted. *Financial support:* In 2005–06, fellowships with full tuition reimbursements (averaging $20,000 per year), research assistantships with full tuition reimbursements (averaging $22,500 per year), teaching assistantships with full tuition reimbursements (averaging $27,640 per year) were awarded; traineeships, health care benefits, and tuition waivers (full) also available. *Faculty research:* Membrane biochemistry, protein folding and translocation, gene expression, signal transduction, cell growth and differentiation. Total annual research expenditures: $3.4 million. *Unit head:* Dr. Robert H. Fillingame, Chair, 608-262-1347, Fax: 608-262-5253, E-mail: rhfillin@wisc.edu. *Application contact:* Elyse Meuer, Student Services Coordinator, 608-262-1347, Fax: 608-262-5253, E-mail: eemeuer@wisc.edu.

See Close-Up on page 445.

University of Wisconsin–Madison, School of Veterinary Medicine, Department of Animal Health and Biomedical Sciences, Program in Comparative Biosciences, Madison, WI 53706-1380. Offers anatomy (MS, PhD); biochemistry (MS, PhD); cellular and molecular biology (MS, PhD); environmental toxicology (MS, PhD); neurosciences (MS, PhD); pharmacology (MS, PhD); physiology (MS, PhD). *Degree requirements:* For doctorate, thesis/dissertation.

Utah State University, School of Graduate Studies, College of Science, Department of Chemistry and Biochemistry, Logan, UT 84322. Offers biochemistry (MS, PhD); chemistry (MS, PhD). Part-time programs available. *Faculty:* 16 full-time (3 women). *Students:* 134 full-time (43 women), 3 part-time (2 women); includes 3 minority (all African Americans), 89 international. Average age 28. 40 applicants, 58% accepted, 14 enrolled. In 2005, 3 master's, 4 doctorates awarded. Terminal master's awarded for partial completion of doctoral program. *Degree requirements:* For master's and doctorate, thesis/dissertation, oral and written exams. *Entrance requirements:* For master's and doctorate, GRE General Test, minimum GPA of 3.0. Additional exam requirements/recommendations for international students: Required—TOEFL. *Application deadline:* For fall admission, 4/15 priority date for domestic students, 4/15 priority date for international students; for spring admission, 10/15 for domestic students, 10/15 for international students. Applications are processed on a rolling basis. Application fee: $50 ($60 for international students). *Financial support:* In 2005–06, 29 research assistantships with partial tuition reimbursements (averaging $17,000 per year), 16 teaching assistantships with partial tuition reimbursements (averaging $17,000 per year) were awarded; fellowships with tuition reimbursements, Federal Work-Study, institutionally sponsored loans, and tuition waivers (partial) also available. Support available to part-time students. Financial award application deadline: 4/15. *Faculty research:* Analytical, inorganic, organic, and physical chemistry; iron in asbestos chemistry and carcinogenicity; dicopper complexes; photothermal spectrometry; metal molecule clusters. Total annual research expenditures: $2.1 million. *Unit head:* Dr. Steve

Biochemistry

Utah State University (continued)

Scheiner, Head, 435-797-7419, Fax: 435-797-3390, E-mail: scheiner@cc.usu.edu. *Application contact:* Dr. Lisa M. Berreau, Admissions Chair, 435-797-1625, Fax: 435-797-3390, E-mail: berreau@cc.usu.edu.

Vanderbilt University, Graduate School and School of Medicine, Department of Biochemistry, Nashville, TN 37240-1001. Offers MS, PhD, MD/PhD. *Faculty:* 47 full-time (7 women), 2 part-time/adjunct (1 woman). *Students:* 41 full-time (25 women); includes 4 minority (3 African Americans, 1 Hispanic American), 2 international. In 2005, 2 master's, 9 doctorates awarded. *Degree requirements:* For master's, thesis; for doctorate, thesis/dissertation, preliminary, qualifying, and final exams. *Entrance requirements:* For master's, GRE General Test; for doctorate, GRE General Test, GRE Subject Test (recommended). *Application deadline:* For fall admission, 1/15 for domestic students, 1/15 for international students. Application fee: $0. Electronic applications accepted. *Expenses:* Tuition: Part-time $1,283 per semester hour. Required fees: $2,202; $1,101 per semester. One-time fee: $30. Tuition and fees vary according to course load, program and student level. *Financial support:* Fellowships with full tuition reimbursements, research assistantships, institutionally sponsored loans, traineeships, and tuition waivers (partial) available. Financial award application deadline: 1/15. *Faculty research:* Protein chemistry, carcinogenesis, metabolism, toxicology, receptors and signaling, DNA recognition and transcription. *Unit head:* Michael R. Waterman, Chair, 615-322-3315, Fax: 615-322-4349, E-mail: michael.waterman@vanderbilt.edu. *Application contact:* Scott W. Hiebert, Director of Graduate Studies, 615-343-3315, Fax: 615-322-4349, E-mail: marlene.jayne@mcmail.vanderbilt.edu.

Vanderbilt University, School of Medicine, Program in Chemical and Physical Biology, Nashville, TN 37240-1001. Offers PhD. *Degree requirements:* For doctorate, thesis/dissertation, dissertation defense, comprehensive exam, registration. *Entrance requirements:* For doctorate, GRE, 3 letters of recommendation, interview by invitation. Additional exam requirements/recommendations for international students: Required—TOEFL. Electronic applications accepted. *Expenses:* Tuition: Part-time $1,283 per semester hour. Required fees: $2,202; $1,101 per semester. One-time fee: $30. Tuition and fees vary according to course load, program and student level. *Faculty research:* Mathematical modeling, enzyme kinetics, structural biology, genomics, proteomics and mass spectrometry.

Virginia Commonwealth University, Medical College of Virginia-Professional Programs, School of Medicine and Graduate Programs, School of Medicine Graduate Programs, Department of Biochemistry, Richmond, VA 23284-9005. Offers biochemistry (MS, PhD, CBHS); bioinformatics (MS); molecular biology and genetics (MS, PhD); neurosciences (PhD). *Faculty:* 26 full-time (6 women). *Students:* 48 full-time (24 women), 9 part-time (8 women); includes 12 minority (2 African Americans, 8 Asian Americans or Pacific Islanders, 2 Hispanic Americans), 8 international. 61 applicants, 75% accepted. In 2005, 1 master's, 4 doctorates, 2 other advanced degrees awarded. *Degree requirements:* For master's, thesis; for doctorate, thesis/dissertation, comprehensive oral and written exams. *Entrance requirements:* For master's and doctorate, DAT, GRE General Test, MCAT. *Application deadline:* For fall admission, 2/15 for domestic students. Application fee: $50. *Expenses:* Tuition, state resident: full-time $3,185; part-time $405 per credit. Tuition, nonresident: full-time $7,952; part-time $940 per credit. Required fees: $751 per semester hour. Tuition and fees vary according to course load and program. *Financial support:* Fellowships, research assistantships available. *Faculty research:* Molecular biology, peptide/protein chemistry, neurochemistry, enzyme mechanisms, macromolecular structure determination. Total annual research expenditures: $3.5 million. *Unit head:* Dr. Sarah Spiegel, Chair, 804-828-9762, Fax: 804-828-1473, E-mail: sspiegel@vcu.edu. *Application contact:* Dr. Keith R. Shelton, Program Director, 804-828-9886, Fax: 804-828-1473, E-mail: krshelto@vcu.edu.

See Close-Up on page 447.

Virginia Polytechnic Institute and State University, Graduate School, College of Agriculture and Life Sciences, Department of Biochemistry, Blacksburg, VA 24061. Offers life sciences (MS, PhD). *Faculty:* 16 full-time (2 women). *Students:* 25 full-time (16 women), 1 (woman) part-time; includes 5 minority (1 African American, 1 American Indian/Alaska Native, 1 Asian American or Pacific Islander, 2 Hispanic Americans), 7 international. Average age 28. 33 applicants, 18% accepted, 4 enrolled. In 2005, 3 master's, 5 doctorates awarded. *Entrance requirements:* For master's and doctorate, GRE General Test. Additional exam requirements/recommendations for international students: Required—TOEFL (minimum score 580 paper-based; 237 computer-based). *Application deadline:* Applications are processed on a rolling basis. Application fee: $45. Electronic applications accepted. *Expenses:* Contact institution. *Financial support:* In 2005–06, 1 fellowship with full tuition reimbursement (averaging $6,000 per year), 12 research assistantships with full tuition reimbursements (averaging $13,823 per year), 10 teaching assistantships with full tuition reimbursements (averaging $13,552 per year) were awarded; career-related internships or fieldwork, Federal Work-Study, scholarships/grants, and unspecified assistantships also available. Financial award application deadline: 2/1. *Faculty research:* Molecular biology, molecular entomology, enzymology, signal transduction, protein structure-function. *Unit head:* Dr. Peter Kennelly, Head, 540-231-6315, Fax: 540-231-9070. *Application contact:* Sheila Early, Professor, 540-231-6315, Fax: 540-231-9070, E-mail: smearly@vt.edu.

See Close-Up on page 449.

Wake Forest University, School of Medicine and Graduate School, Graduate Programs in Medicine, Department of Biochemistry, Winston-Salem, NC 27109. Offers PhD. *Degree requirements:* For doctorate, thesis/dissertation. *Entrance requirements:* For doctorate, GRE General Test. Additional exam requirements/recommendations for international students: Required—TOEFL. Electronic applications accepted. *Faculty research:* Biomembranes, cancer, biophysics.

See Close-Up on page 451.

Washington State University, Graduate School, College of Sciences, Department of Chemistry, Pullman, WA 99164. Offers analytical chemistry (MS, PhD); biological systems (MS, PhD); inorganic chemistry (MS, PhD); organic chemistry (MS, PhD); physical chemistry (MS, PhD). *Faculty:* 33. *Students:* 38 full-time (19 women), 2 part-time (1 woman); includes 3 minority (2 Asian Americans or Pacific Islanders, 1 Hispanic American), 11 international. Average age 29. 55 applicants, 33% accepted, 9 enrolled. In 2005, 5 master's, 7 doctorates awarded. Terminal master's awarded for partial completion of doctoral program. *Degree requirements:* For master's, oral exam, teaching experience, thesis optional; for doctorate, thesis/dissertation, oral exam, written exam, comprehensive exam. *Entrance requirements:* For master's and doctorate, GRE General Test, minimum GPA of 3.0, 3 letters of recommendation. *Application deadline:* For fall admission, 3/1 priority date for domestic students, 3/1 priority date for international students; for spring admission, 10/1 priority date for domestic students, 7/1 priority date for international students. Applications are processed on a rolling basis. Application fee: $35. *Expenses:* Tuition, state resident: full-time $6,295; part-time $336 per credit. Tuition, nonresident: full-time $15,949; part-time $819 per credit. Required fees: $933. Part-time tuition and fees vary according to campus/location and program. *Financial support:* In 2005–06, 38 students received support, including 3 fellowships (averaging $8,977 per year), 12 research assistantships with full and partial tuition reimbursements available (averaging $15,999 per year), 23 teaching assistantships with full and partial tuition reimbursements available (averaging $14,747 per year); career-related internships or fieldwork, Federal Work-Study, institutionally sponsored loans, scholarships/grants, health care benefits, unspecified assistantships, and summer support also available. Financial award application deadline: 4/1; financial award applicants required to submit FAFSA. *Faculty research:* Environmental chemistry, materials chemistry, radio chemistry, bio-organic, computational chemistry. Total annual research expenditures: $3.8 million. *Unit head:* Sue B. Clark, Chair, Admissions Committee, 509-335-8866, Fax: 509-335-8867, E-mail: sclark@mail.wsu.edu. *Application contact:* Admissions Committee.

Washington State University, Graduate School, College of Sciences, School of Molecular Biosciences, Program in Biochemistry and Biophysics, Pullman, WA 99164. Offers MS, PhD. *Faculty:* 23 full-time (5 women), 21 part-time/adjunct (4 women). *Students:* 26 full-time (9 women), 1 (woman) part-time; includes 1 minority (Hispanic American), 12 international. Average age 27. 231 applicants, 18% accepted, 13 enrolled. In 2005, 1 master's, 2 doctorates awarded. Terminal master's awarded for partial completion of doctoral program. *Degree requirements:* For master's, thesis or alternative, oral exam; for doctorate, thesis/dissertation, oral exam, written exam, comprehensive exam, registration. *Entrance requirements:* For master's and doctorate, GRE General Test, minimum GPA of 3.0. Additional exam requirements/recommendations for international students: Required—TOEFL (minimum score 550 paper-based; 213 computer-based). *Application deadline:* For fall admission, 12/15 for domestic students, 12/15 for international students. Application fee: $35. Electronic applications accepted. *Expenses:* Tuition, state resident: full-time $6,295; part-time $336 per credit. Tuition, nonresident: full-time $15,949; part-time $819 per credit. Required fees: $933. Part-time tuition and fees vary according to campus/location and program. *Financial support:* In 2005–06, 5 fellowships with full tuition reimbursements (averaging $18,384 per year), 11 research assistantships with full tuition reimbursements (averaging $18,384 per year), 10 teaching assistantships with full tuition reimbursements (averaging $18,384 per year) were awarded; career-related internships or fieldwork, Federal Work-Study, institutionally sponsored loans, traineeships, and health care benefits also available. Financial award application deadline: 4/1; financial award applicants required to submit FAFSA. *Faculty research:* Gene regulation, signal transduction, protein export, reproductive biology, DNA repair. Total annual research expenditures: $5.8 million. *Application contact:* Kelly G. McGovern, Academic Coordinator, 509-335-4566, Fax: 509-335-1907, E-mail: mcgnerk@wsu.edu.

Washington University in St. Louis, Graduate School of Arts and Sciences, Division of Biology and Biomedical Sciences, Program in Biochemistry, St. Louis, MO 63130-4899. Offers PhD. *Degree requirements:* For doctorate, thesis/dissertation. *Entrance requirements:* For doctorate, GRE General Test, GRE Subject Test. Electronic applications accepted.

Washington University in St. Louis, Graduate School of Arts and Sciences, Division of Biology and Biomedical Sciences, Program in Chemical Biology, St. Louis, MO 63130-4899. Offers PhD. *Degree requirements:* For doctorate, thesis/dissertation. *Entrance requirements:* For doctorate, GRE General Test, GRE Subject Test. Electronic applications accepted.

Wayne State University, School of Medicine and Graduate School, Graduate Programs in Medicine, Department of Biochemistry and Molecular Biology, Detroit, MI 48202. Offers MS, PhD. *Faculty:* 13 full-time (11 women). *Students:* 22 full-time (7 women), 1 part-time; includes 1 minority (African American), 20 international. Average age 28. 39 applicants, 18% accepted, 3 enrolled. In 2005, 2 degrees awarded. Terminal master's awarded for partial completion of doctoral program. *Degree requirements:* For master's, thesis; for doctorate, one foreign language, thesis/dissertation. *Entrance requirements:* For master's and doctorate, GRE General Test, GRE Subject Test. Additional exam requirements/recommendations for international students: Required—TOEFL (minimum score 550 paper-based; 213 computer-based); Recommended—TWE (minimum score 6). *Application deadline:* Applications are processed on a rolling basis. Application fee: $30 ($50 for international students). Electronic applications accepted. *Expenses:* Tuition, state resident: part-time $338 per credit hour. Tuition, nonresident: part-time $746 per credit hour. Required fees: $24 per credit hour. Full-time tuition and fees vary according to program. *Financial support:* In 2005–06, 22 research assistantships (averaging $20,159 per year) were awarded; fellowships, teaching assistantships *Faculty research:* Protein structure, molecular biology, molecular genetics, enzymology, x-ray crystallography. Total annual research expenditures: $3.6 million. *Unit head:* Dr. Barry P. Rosen, Chair, 313-577-1512, Fax: 313-577-2765, E-mail: aa1133@wayne.edu. *Application contact:* Marilyn Doscher, Graduate Director, 313-577-1295, E-mail: mdoscher@med.wayne.edu.

Wesleyan University, Graduate Programs, Department of Chemistry, Middletown, CT 06459-0260. Offers biochemistry (MA, PhD); chemical physics (MA, PhD); inorganic chemistry (MA, PhD); organic chemistry (MA, PhD); physical chemistry (MA, PhD); theoretical chemistry (MA, PhD). *Faculty:* 13 full-time (2 women), 2 part-time/adjunct (1 woman). *Students:* 41 full-time (20 women); includes 1 minority (Asian American or Pacific Islander), 22 international. Average age 26. In 2005, 2 master's, 2 doctorates awarded. Terminal master's awarded for partial completion of doctoral program. *Degree requirements:* For master's and doctorate, one foreign language, thesis/dissertation. *Entrance requirements:* For master's, GRE General Test, GRE Subject Test; for doctorate, GRE Subject Test. *Application deadline:* For fall admission, 3/1 for domestic students. Applications are processed on a rolling basis. Application fee: $0. *Expenses:* Tuition: Full-time $24,732. One-time fee: $20 full-time. *Financial support:* Fellowships, research assistantships, teaching assistantships, institutionally sponsored loans available. *Unit head:* Dr. Phillip Bolton, Chair, 860-685-2668. *Application contact:* Karen Karpa, Information Contact, 860-685-2573, Fax: 860-685-2211, E-mail: kkarpa@wesleyan.edu.

Wesleyan University, Graduate Programs, Department of Molecular Biology and Biochemistry, Middletown, CT 06459-0260. Offers biochemistry (PhD); molecular biology (PhD). *Faculty:* 9 full-time (2 women). *Students:* 23 full-time (18 women), 14 international. Average age 28. In 2005, 2 doctorates awarded. *Degree requirements:* For doctorate, thesis/dissertation. *Entrance requirements:* For doctorate, GRE General Test, GRE Subject Test. *Application deadline:* For fall admission, 3/15 for domestic students. Applications are processed on a rolling basis. Application fee: $0. *Expenses:* Tuition: Full-time $24,732. One-time fee: $20 full-time. *Financial support:* Research assistantships, teaching assistantships, institutionally sponsored loans available. *Faculty research:* Genome organization, regulation of gene expression, molecular biology of development, physical biochemistry. *Unit head:* Ishita Mukerji, Chair, 860-685-2422, E-mail: imukerji@wesleyan.edu. *Application contact:* Debra Alexson, Information Contact, 860-685-7640, E-mail: dalexson@wesleyan.edu.

Announcement: Department of Molecular Biology and Biochemistry provides research opportunities in gene expression, DNA replication, protein secretion, chromosomal proteins, protein–nucleic acid interactions, phosphorus chemistry, NMR and Raman spectroscopy, and computer modeling. Extensive faculty-student interaction. Opportunities to teach. All students receive stipend, tuition remission, good fringe benefits.

See Close-Up on page 667.

West Virginia University, School of Medicine, Graduate Programs at the Health Science Center, Biomedical Sciences Graduate Program, Program in Biochemistry and Molecular Biology, Morgantown, WV 26506. Offers MS, PhD. *Faculty:* 20 full-time (3 women). *Students:* 14 full-time (7 women); includes 8 minority (1 African American, 7 Asian Americans or Pacific Islanders), 7 international. Average age 27. In 2005, 1 master's, 1 doctorate awarded. *Median time to degree:* Of those who began their doctoral program in fall 1997, 98% received their degree in 8 years or less. *Degree requirements:* For doctorate, thesis/dissertation, comprehensive exam. *Entrance requirements:* For doctorate, GRE General Test, minimum GPA of 3.0. Additional exam requirements/recommendations for international students: Required—TOEFL. *Application deadline:* For fall admission, 3/1 for domestic students, 1/15 for international students. Applications are processed on a rolling basis. Application fee: $0. Electronic applications accepted. *Financial support:* In 2005–06, research assistantships with full tuition reimbursements (averaging $20 per year); institutionally sponsored loans, traineeships, and health care benefits also available. *Faculty research:* Regulation of gene expression, cell survival mechanisms, signal transduction, regulation of metabolism, sensory neuroscience. Total annual research expenditures: $1 million. *Unit head:* Dr. Diana S. Beattie, Chair, 304-293-7522, Fax: 304-293-6846, E-mail: diana.beattie@hsc.wvu.edu. *Application contact:* Dr. Lisa Salati, Graduate Program Coordinator, 304-293-7759, Fax: 304-293-6846, E-mail: lisa.salati@hsc.wvu.edu.

See Close-Up on page 453.

Worcester Polytechnic Institute, Graduate Studies and Enrollment, Department of Chemistry and Biochemistry, Worcester, MA 01609-2280. Offers biochemistry (MS); bioscience administration (MS); chemistry (MS, PhD). *Faculty:* 11 full-time (2 women). *Students:* 18 full-time (8 women); includes 1 minority (Asian American or Pacific Islander), 9 international. 20 applicants, 70% accepted, 7 enrolled. In 2005, 2 master's, 2 doctorates awarded. *Degree requirements:* For master's, thesis/dissertation; for doctorate, thesis/dissertation, comprehensive exam. *Entrance requirements:* For master's and doctorate, GRE General Test, 3 letters of recommendation. Additional exam requirements/recommendations for international students: Required—TOEFL (minimum score 550 paper-based; 213 computer-based). *Application deadline:* For fall admission, 1/15 for domestic students; for spring admission, 10/15 priority date for domestic students. Applications are processed on a rolling basis. Application fee: $70. Electronic applications accepted. *Expenses:* Tuition: Part-time $997 per credit hour. *Financial support:* In 2005–06, 15 students received support, including 3 research assistantships with full tuition reimbursements available, 10 teaching assistantships with full and partial tuition reimbursements available; fellowships with full tuition reimbursements available, career-related internships or fieldwork, institutionally sponsored loans, scholarships/grants, health care benefits, and unspecified assistantships also available. Financial award application deadline: 1/15. *Faculty research:* Plant biochemistry, membrane biophysics, photochemistry, organic synthesis, materials synthesis. Total annual research expenditures: $176,018. *Unit head:* Dr. James W Pavlik, Interim Head, 508-831-5371, Fax: 508-831-5933, E-mail: jwpavlik@wpi.edu. *Application contact:* Dr. Jose Arguello, Graduate Coordinator, 508-831-5326, Fax: 508-831-5933, E-mail: arguello@wpi.edu.

Wright State University, School of Graduate Studies, College of Science and Mathematics, Department of Biochemistry and Molecular Biology, Dayton, OH 45435. Offers MS. *Degree requirements:* For master's, thesis. *Entrance requirements:* Additional exam requirements/recommendations for international students: Required—TOEFL. *Faculty research:* Regulation of gene expression, macromolecular structural function, NMR imaging, visual biochemistry.

Yale University, Graduate School of Arts and Sciences, Department of Molecular Biophysics and Biochemistry, New Haven, CT 06520. Offers MS, PhD. *Degree requirements:* For doctorate, thesis/dissertation. *Entrance requirements:* For doctorate, GRE General Test, GRE Subject Test.

Yale University, School of Medicine and Graduate School of Arts and Sciences, Combined Program in Biological and Biomedical Sciences (BBS), Molecular Biophysics and Biochemistry Track, New Haven, CT 06520. Offers PhD, MD/PhD. *Students:* 14 full-time. *Degree requirements:* For doctorate, thesis/dissertation. *Entrance requirements:* For doctorate, GRE General Test. Additional exam requirements/recommendations for international students: Required—TOEFL. *Application deadline:* For fall admission, 12/8 for domestic students, 12/8 for international students. Electronic applications accepted. *Financial support:* Fellowships, research assistantships available. *Unit head:* Dr. Mark Solomon, Director of Graduate Studies, 203-432-5562, Fax: 203-432-9782, E-mail: mbb.grad@yale.edu. *Application contact:* Nessie Stewart, Graduate Registrar, 203-432-5662, Fax: 203-432-6178, E-mail: mbb.grad@yale.edu.

Cross-Discipline Announcements

Massachusetts Institute of Technology, School of Engineering, Biological Engineering Division, Cambridge, MA 02139-4307.

Program provides opportunities for study and research at the interface of biology and engineering leading to specialization in bioengineering and applied biosciences. The areas include understanding how biological systems operate, especially when perturbed by genetic, chemical, or materials interventions or subjected to pathogens or toxins, and designing innovative biology-based technologies in diagnostics, therapeutics, materials, and devices for application to human health and diseases, as well as other societal problems and opportunities.

Massachusetts Institute of Technology, School of Science, Department of Biology, Cambridge, MA 02139-4307.

Graduate work in the Department of Biology at MIT leads to the PhD degree. Research opportunities include many areas of modern biology: biochemistry, biophysics, cellular and developmental biology, immunology, microbiology, and neurobiology. Students come from a wide variety of backgrounds. Previous experience in the biological sciences, although desirable, is not a prerequisite for admission. Formal courses in biochemistry, genetics, and the method and logic of molecular biology are required of all students early in their graduate program. Each student is encouraged to follow the program of study that best meets his or her educational goals. Special emphasis is placed on research leading to the PhD thesis.

Oregon Health & Science University, School of Medicine, Graduate Programs in Medicine, Department of Behavioral Neuroscience, Portland, OR 97239-3098.

The department offers an interdisciplinary graduate program consisting of basic science training in behavioral neuroscience with specialization in such areas as physiological psychology, behavioral and molecular genetics, behavioral pharmacology, neuroendocrinology, and biological bases of addiction. Students with degrees in biological sciences, psychology, or neuroscience are encouraged to apply. Visit http://www.ohsu.edu/behneuro/.

Princeton University, Graduate School, Department of Molecular Biology, Princeton, NJ 08544-1019.

Graduate studies in the Department of Molecular Biology at Princeton emphasize training in research and encourage students to apply molecular, biochemical, structural, genetic, and computational approaches to biological problems. Faculty members and students pursue research in a wide variety of areas of biochemistry and cell biology, biological dynamics and computational biology, biomedicine and societal issues, biophysics and structural biology, developmental biology, genetics and genomics, microbiology, neurobiology, oncology, and virology.

University of Pittsburgh, School of Arts and Sciences, Department of Biological Sciences, Program in Molecular, Cellular, and Developmental Biology, Pittsburgh, PA 15260.

In the department's graduate program in molecular, cellular, and developmental biology, faculty research programs use prokaryotic and eukaryotic experimental systems and molecular and genetic approaches to understand the structure and function of genes and proteins, macromolecular interactions, cell-specific gene regulation, regulation of cell proliferation, and embryogenesis.

ALBERT EINSTEIN COLLEGE OF MEDICINE OF YESHIVA UNIVERSITY

Department of Biochemistry

Program of Study

The Department of Biochemistry offers a challenging and broad-based graduate program of research and course work leading to the Ph.D. degree, or the combined M.D./Ph.D. degree for students previously admitted to medical school. The aim of the graduate program is to train scientists who have a strong foundation in the chemical and physical sciences for critical analysis and solution of biomedical problems on the molecular level. The theme of molecular structure and molecular mechanism is recurrent through the diverse research areas represented by the Biochemistry faculty: regulation of gene expression, receptors for and action of hormones and neurotransmitters, analysis of protein–nucleic acid interactions, chemistry and metabolism of simple and complex lipids and saccharides, mechanism of enzyme action, protein and nucleic acid structure and function, protein engineering, design of antibiotics and chemotherapeutics, and structure-function of transmembrane proteins.

The program of study includes lecture courses, laboratory rotations, journal club presentations, a qualifying examination, and thesis research. The course requirements are one year of graduate-level biochemistry, one semester of physical biochemistry of macromolecules, and at least one semester of molecular biology and genetics. Additional courses are selected according to the student's individual interests. Two laboratory rotations (or their equivalent) are required; these facilitate choice of a thesis adviser by exposing the student to different research programs. The qualifying examination is taken after the first or the second year of study. Once admitted to the Department, each student is assigned an advisory committee of faculty members for consultation in the choice of courses and laboratory rotations.

Research Facilities

The Department houses modern equipment for the preparation and analytical characterization of biological molecules, including instrumentation for macromolecule X-ray crystallography, high-field 300- and 600-MHz multinuclear NMR spectrometers, advanced Raman and IR spectrometers with ultrafast temperature jump capability, and networked computers and molecular modeling workstations. The College manages several beamlines at the National Synchrotron Light Source at Brookhaven National Laboratories for Crystallography and X-ray-induced free-radical footprinting and is a partner in the New York Structural Biology High-Field NMR Center. In addition, the College of Medicine has instrumentation centers that are available to all basic scientists: a DNA chip-technology center; a hybridoma facility; an analytical imaging facility; matrix-assisted laser desorption ionization (MALDI) time-of-flight (TOF) and TOF/TOF mass spectrometers, electrospray ionization (ESI) ion trap and QqTOF mass spectrometers, and a Fourier transform ion cyclotron resonance mass spectrometer; continuous-wave, spin echo EPR and ENDOR spectrometers; an analytical ultracentrifuge and a surface plasmon resonance (Biacore) instrument; DNA and peptide synthesizers and sequencers; a fluorescence-activated cell sorter; a transgenic mouse facility; an animal institute; an animal and human imaging center; and a library with a large subscription to current periodicals, including electronic access.

Financial Aid

All students receive financial support throughout their graduate training. This support includes an annual living stipend, which was $25,000 in the 2005–06 academic year, and full payment of tuition, which was $35,925 in 2005–06. No teaching or assisting is required for receipt of this financial aid.

Cost of Study

Full tuition and an annual stipend are paid for all students admitted to the Department of Biochemistry.

Living and Housing Costs

Apartments are available close to the College of Medicine in the Max and Evelynne Low Residence Hall and the Rhinelander Apartments. The average monthly rent in the 2004–05 academic year was $462 for a studio apartment, $646 for a one-bedroom, and $274 per person in a shared two-bedroom apartment. Some apartments are available in the adjacent community at comparable rates. A cafeteria in Mazer Hall is open to all students for lunch.

Student Group

The roster of the Sue Golding Graduate Division of Medical Sciences, of which the Department of Biochemistry is a member, lists 394 students enrolled in its programs. Among these are 117 M.D./Ph.D. candidates. The graduate students interact informally and through a Graduate Student Council. The council elects student representatives to Sue Golding Graduate Division committees, plans social events, and provides a forum for voicing student concerns. In addition to the graduate students, approximately 750 M.D. candidates are enrolled in the College of Medicine.

Student Outcomes

Ph.D. graduates from the Department of Biochemistry now hold professional positions in universities, industry, and government. These include institute directors and research scientists at NIH, faculty members at leading universities and medical schools, and research scientists at most leading pharmaceutical companies. Many other former students hold positions in scientific administration and biotechnology start-up companies.

Location

The Albert Einstein College of Medicine is situated in the residential Pelham Bay area of the northeastern Bronx. This area borders Westchester County on its northern edge, offering numerous recreational opportunities. Manhattan is readily accessible by public transportation, express buses, and automobile and offers the world's greatest variety of cultural activities. The many research institutions in the New York metropolitan area present opportunities for expanding one's scientific awareness and for participating in research collaborations.

The College

The Albert Einstein College of Medicine is a privately endowed, coeducational, nondenominational constituent college of Yeshiva University. Founded in 1955, the Ph.D. granting departments of the College are among the top 10 percent of U.S. medical schools in NIH research support. The Sue Golding Graduate Division was established in 1957 for advancing research training and study leading to the Ph.D. degree. Postdoctoral research training for U.S. and international scientists is coordinated by the Belfer Institute.

Applying

A course of study leading to a bachelor's degree or its equivalent from a college or university of recognized standing is expected. Undergraduate course work should include organic chemistry, calculus, physics, and biology. Courses in physical chemistry and biochemistry as well as research experience in an independent or honors project or through summer employment are strongly recommended. Admission decisions are made by considering each application individually, weighing grades, GRE scores, reference letters, and the personal interview. International students whose native language is not English must submit scores on the Test of English as a Foreign Language (TOEFL). Candidates are encouraged to submit completed applications by January 1 for September enrollment. Application forms may be obtained from the director of the Sue Golding Graduate Division. Specific questions about the Department of Biochemistry should be addressed to the chairman of the Department.

Correspondence and Information

For Departmental information:
Dr. Vern L. Schramm, Chairman
Department of Biochemistry
Albert Einstein College of Medicine
1300 Morris Park Avenue
Bronx, New York 10461
Phone: 718-430-2813

For applications:
Director
Sue Golding Graduate Division of Medical Sciences
Albert Einstein College of Medicine
1300 Morris Park Avenue
Bronx, New York 10461
Phone: 718-430-2345
E-mail: phd@aecom.yu.edu
Web site: http://www.aecom.yu.edu

Albert Einstein College of Medicine of Yeshiva University

THE FACULTY AND THEIR RESEARCH

Ernesto Abel-Santos, Assistant Professor of Biochemistry; Ph.D., Washington (St. Louis), 1997. Combinatorial approaches to identify essential bacterial genes; protein library screening for drug recognition determinants in bacterial multidrug efflux pumps.

Steven C. Almo, Professor of Biochemistry; Ph.D., Harvard, 1990. Structural cell biology; crystallographic and cell biological studies of the actin cytoskeleton and the immune response.

Ruth Hogue Angeletti, Professor of Developmental Biology and Cancer and of Biochemistry and Director, Laboratory for Macromolecular Analysis and Proteomics; Ph.D., Washington (St. Louis), 1969. Functional proteomics of hepatic neoplasia, cytoskeletal scaffolds, and waterborne parasites; mass spectrometry.

John S. Blanchard, Professor of Biochemistry; Ph.D., Wisconsin, 1979. Kinetic and chemical mechanisms of enzyme-catalyzed reactions; structure-function relationships in bacterial N-acetyltransferases; characterization of biosynthetic pathways in *M. tuberculosis*.

Olga O. Blumenfeld, Professor Emerita of Biochemistry and Curator, Blood Group Antigen Gene Mutation Database; Ph.D., NYU, 1957. Molecular basis for allelic diversification of glycophorin family.

Michael D. Brenowitz, Professor of Biochemistry; Ph.D., Duke, 1982. Regulation of gene expression; thermodynamic and kinetic analysis of cooperative protein-nucleic acid interactions; mechanism of DNA recognition; RNA folding.

Anne R. Bresnick, Associate Professor of Biochemistry; Ph.D., Yeshiva (Einstein), 1991. Cellular biochemistry of the actin cytoskeleton; mechanisms of cell motility, cell division, and tumor metastasis.

Robin W. Briehl, Professor of Physiology and Biophysics and of Biochemistry; M.D., Harvard, 1954. Biophysical chemistry of proteins; allostery; physical chemistry and structure of helical polymers, focusing on sickle cell hemoglobin and its aggregation, gelation, and phase changes and the micromechanics of the fibers it forms.

Robert Callender, Professor of Biochemistry; Ph.D., Harvard, 1969. Structure and dynamics of proteins; advanced optical and vibrational spectroscopic techniques; enzymatic catalysis; dynamics of protein folding; photophysics of vision; drug design.

Mark R. Chance, Professor of Physiology and Biophysics and of Biochemistry; Ph.D., Pennsylvania, 1986. Protein-protein interactions; structural and functional proteomics; mass spectroscopy; rapid reaction kinetics; nucleic acid and protein footprinting; RNA folding; actin cytoskeleton dynamics.

Maureen J. Charron, Professor of Biochemistry; Ph.D., CUNY, Queens, 1987. The mammalian family of glucose transporters (regulation of gene expression, pathophysiology, and immunolocalization); diabetes and obesity pathogenesis; insulin and glucagons receptor action and signal transduction.

David E. Cohen, Associate Professor of Biochemistry Voluntary; M.D./Ph.D., Harvard, 1987. Cholesterol metabolism and biliary lipid secretion; hepatocellular transport of biliary phospholipids; structure and function of lipid transfer proteins.

Sasha Englard, Professor Emeritus of Biochemistry; Ph.D., Western Reserve, 1953. Regulation of carnitine biosynthesis and degradation; mechanism of α-ketoglutarate-linked dioxygenases, mechanism of trimethylamine formation from betaines.

Andras Fiser, Assistant Professor of Biochemistry; Ph.D., Hungarian Academy of Sciences, 1997. Bioinformatics and computational biology; modeling of protein structures and functions; structural genomics and proteomics.

Mark E. Girvin, Associate Professor of Biochemistry; Ph.D., Purdue, 1985. Multinuclear, multidimensional NMR structural studies of proteins, membrane protein structure, and molecular dynamics.

Victor B. Hatcher, Professor of Biochemistry and Medicine; Ph.D., McGill, 1969. Signal transmission in human endothelial cells; role of extracellular matrix and fibroblast growth factors in endothelial cell proliferation; bacterial adhesion and infection of endothelial cells.

David S. Lawrence, Professor of Biochemistry; Ph.D., UCLA, 1982. Signal transduction; enzymology of protein kinases and protein phosphatases; design of cell division inhibitors/anticancer agents.

Thomas S. Leyh, Professor of Biochemistry; Ph.D., Pennsylvania, 1983. Energy coupling in metabolism; physical-chemical basis of catalysis and protein function; GTPase-linked activation of sulfate; enzymology and metabolic consequences of sulfuryl-transfer (the regulation of estrogen activity).

Irving Listowsky, Professor of Biochemistry; Ph.D., Polytechnic of Brooklyn, 1964. Protein structure and molecular biology; protein-ligand interactions; structure and tissue specific regulation of expression of glutathione transferase genes; vis-a-vis chemoprotective and redox mechanisms.

Maynard H. Makman, Professor of Biochemistry and Molecular Pharmacology; M.D./Ph.D., Western Reserve, 1962. Molecular mechanisms of hormone and neurotransmitter action in brain, retina, and neuronal cultures; neuroimmune mechanisms in macrophages, granulocytes, and leukemic cells; receptor and second messenger regulation.

Steven L. Roderick, Professor of Biochemistry; Ph.D., Washington (St. Louis), 1984. Protein structure-function relationships; X-ray crystallography; enzymes of cell wall biosynthesis and antibiotic inactivation.

Vern L. Schramm, Professor and Chairman of Biochemistry; Ph.D., Australian National, 1969. Transition-state structure and mechanism of enzymes of purine and pyrimidine metabolism; design of antibiotics and chemotherapeutic agents; action of bacterial and plant toxins; design of transition state inhibitors.

Steven D. Schwartz, Professor of Physiology and Biophysics and of Biochemistry; Ph.D., Berkeley, 1984. Theoretical studies of biophysical systems; quantum mechanics; statistical mechanics; theory of enzymatic action and drug design.

Sam Seifter, Distinguished Professor Emeritus of Biochemistry; Ph.D., Western Reserve, 1944. Chemistry and biology of collagen and extracellular matrices; structure and mechanism of action of bacterial collagenase; new methods for study of carboxymethylation and isomerization of amino acid residues in proteins.

Howard M. Steinman, Professor of Biochemistry; Ph.D., Yale, 1970. Bacterial pathogenesis; biochemical, genetic, and proteomic approaches to define mechanisms by which environmental stress influences pathogenicity.

Ian M. Willis, Professor of Biochemistry; Ph.D., New South Wales (Australia), 1983. Mechanisms of eukaryotic gene transcription; assembly and function of RNA polymerase III transcription complexes; signaling pathways and mechanisms regulating ribosome and tRNA synthesis.

Zhong-Yin Zhang, Professor of Molecular Pharmacology and of Biochemistry; Ph.D., Purdue, 1990. Protein tyrosine phosphatases and protein kinases; mechanism of action; molecular basis of substrate recognition; structure-based rational inhibitor design.

BAYLOR COLLEGE OF MEDICINE

Department of Biochemistry and Molecular Biology

Program of Study

The Department offers a research-oriented program leading to a Ph.D. degree in biochemistry and molecular biology. A wide diversity of research interests are represented, with special strengths in developmental biology and cellular regulation, developmental biology and molecular embryology, genomics, neurobiochemistry, molecular immunology, macromolecular assembly and regulation, electron cryomicroscopy, X-ray crystallography, membrane and lipid biochemistry, molecular genetics and gene regulation, cell-cycle regulation, neurobiochemistry and signal transduction, protein design and engineering, and structural biophysics.

During the first year, students pursue a flexible curriculum, including basic course work in biochemistry and molecular biology as well as more advanced courses in such areas as genetic engineering and gene regulation, macromolecular design and analysis, enzymology, and biomembranes. Computer science and molecular biophysics courses may be pursued by those planning work in structural biophysics or computational biology. In addition, students conduct research during rotations spent in laboratories they select. These laboratory rotations help in the choice of an adviser for the thesis research that is the heart of the Ph.D. program. Students also participate in a lively series of seminars, research meetings, and journal clubs that help develop their skills in presentation and critical analysis of scientific data.

The research activities of the Department are strengthened by intellectual interactions and collaborations with other departments both at Baylor and at neighboring research institutions located in or near the Texas Medical Center. Students may pursue course work offered by these departments as well, including study at Rice University, the University of Texas–Houston Graduate School of Biomedical Sciences, the University of Texas–Houston Medical School, and the University of Houston.

The Department also offers postdoctoral training in a variety of areas, as well as an M.D./Ph.D. program offered in conjunction with the medical curriculum.

Research Facilities

In addition to spacious and modern laboratories, the Department has state-of-the-art facilities for virtually all major areas of modern biochemical research. These include some of the most advanced equipment available for X-ray diffraction studies, high-resolution electron cryomicroscopy, automated DNA synthesis and sequencing, peptide sequencing, laser-induced time-resolved fluorescence and photochemistry, large-scale culture of microorganisms, microinjection of cells, fluorescence microscopy, and the production of transgenic mice. Library facilities are available both in the Department and across the street at the Jesse Jones Medical Library, which has one of the world's largest collections of biomedical literature. The Keck Center for Computational Biology, operated jointly with Rice University, provides the most advanced computer technology for computational biology.

Financial Aid

Tuition costs are covered by scholarships awarded by the College or by national and local granting agencies. Additional stipends for entering students are currently $23,000 per year, provided through training and research grants. Various institutions in the Texas Medical Center offer many employment opportunities for spouses of students. A special office provides assistance to international students and their families.

Cost of Study

Students pay a one-time matriculation fee of $25, a one-time graduation fee of $140, and a yearly fee for the Learning Resource Center (first year, $150; thereafter, $20). Health insurance is provided by Baylor at no cost to the students. Family coverage is available at an additional cost.

Living and Housing Costs

Houston offers one of the lowest costs of living of major metropolitan areas. Several apartments and condominiums are conveniently located near or within the Texas Medical Center. One-bedroom apartments rent for about $475 monthly. Early reservations are suggested for on-campus accommodations.

Student Group

The Department has 45 graduate students and usually admits new students at the rate of 6 to 8 per year. Approximately 35 postdoctoral fellows and trainees, from laboratories in the United States and abroad, are working in the Department. The graduate student enrollment of the College amounts to about 350; the group comprises both U.S. citizens and international students. No limitations are applied on the basis of sex, age, race, or nationality.

Student Outcomes

The Department offers a research-intensive Ph.D. program that is designed to prepare students for careers in biomedical research, biotechnology, and teaching. In addition to the more traditional opportunities at academic institutions, many attractive positions are now available in the pharmaceutical, biotechnology, and agricultural industries. Some students use their backgrounds in biomedical research as the basis for successful careers in intellectual property law, marketing and sales, and science writing and reporting.

Location

With a population of approximately 2 million, Houston is the largest city in Texas. It is a dynamic scientific and industrial center, offering many cultural and recreational activities. The Wortham Center, the Menil Collection, the Museum of Fine Arts, and the Houston Ballet and Grand Opera are only a few of the area's cultural assets. The numerous universities in Houston and its environs offer inexpensive educational and sports events. The Astrodome and Compaq Center provide air-conditioned comfort for the city's thousands of sports enthusiasts. The city's mild climate and location favor outdoor sports; a number of beaches are only 45 miles from Houston.

The College

Baylor College of Medicine, a nonsectarian private corporation, has a strong commitment to training and research in basic biological sciences. The Albert B. Alkek Graduate School of Biomedical Sciences building houses additional classrooms for the Graduate School, a graduate student lounge, a reading room, and additional research space for the College. Baylor prides itself on the research accomplishments of its faculty, students, and fellows. Strong interactions among departments, between faculty members and students, and between basic and clinical researchers provide a rich environment for training and research.

Applying

Admission requires a bachelor's degree in a scientific discipline, strong undergraduate performance, and high scores on the General Test and one advanced Subject Test of the Graduate Record Examinations (GRE). Students are encouraged to obtain a strong background in chemistry, including organic and physical chemistry. Those interested in structural biophysics should also be adequately prepared in mathematics, physics, and computer science. Students are also encouraged to pursue undergraduate research, and can contact the Department about summer research opportunities available to undergraduates. The application deadline is January 1. Electronic applications may be obtained by visiting the biochemistry and molecular biology Web site (http://www.bcm.tmc.edu/biochem). The $30 application fee is waived for applications received via the Web site.

Correspondence and Information

Director of Graduate Studies
Department of Biochemistry and Molecular Biology
Baylor College of Medicine
One Baylor Plaza, MSC BCM125
Houston, Texas 77030-3498

Phone: 713-798-4527
Fax: 713-796-9438
E-mail: ccherry@bcm.tmc.edu
Web site: http://www.bcm.edu/biochem

Baylor College of Medicine

THE FACULTY AND THEIR RESEARCH

Stanley H. Appel, Professor; M.D., Columbia, 1960. Molecular approaches to neurodegenerative disease: ALS, Parkinson's disease, Alzheimer's disease; autoimmune reactions; mechanisms of neuronal cell death.

M. Zouhair Atassi, Professor; Ph.D., 1960, D.Sc., 1973, Birmingham (England). Design and synthesis of protein active sites; molecular and cellular immunology; neurochemistry and neuroimmunology; structure, function, and autoimmune recognition of ion channels and hormone receptors.

Susan M. Berget, Professor; Ph.D., Minnesota, 1974. Molecular mechanism of RNA processing in higher eukaryotes.

A. Craig Chinault, Professor; Ph.D., MIT, 1976. Eukaryotic DNA replication; replication/transcription relationships.

Subrahmanyam Chirala, Associate Professor; Ph.D., Indian Agricultural Research Institute (New Delhi), 1975. Regulation of the genes involved in fatty acid biosynthesis in yeast and higher eukaryotes.

Wah Chiu, Professor; Ph.D., Berkeley, 1975. Structural studies of macromolecules and macromolecular assemblies by electron cryomicroscopy and crystallography.

Gregor Eichele, Professor; Ph.D., Basel (Switzerland), 1981. Molecular embryology and functional genomics.

Mark L. Entman, Professor; M.D., Duke, 1963. Cellular and molecular basis of cardiovascular inflammation.

Henry F. Epstein, Professor; M.D., Stanford, 1968. Molecular and structural biology of protein kinases and molecular chaperones active in cytoskeletal regulation.

H. F. Gilbert, Professor; Ph.D., Wisconsin, 1975. Mechanism of chaperone-mediated protein folding; chemistry and biology of disulfide formation.

Xiangwei He, Assistant Professor; Ph.D., Baylor College of Medicine, 1997. Chromosome segregation: interaction between spindle and kinetochores.

Adam Kuspa, Professor and Chairman, Department of Biochemistry and Molecular Biology; Ph.D., Stanford, 1989. Cell differentiation and functional genomics of *Dictyostelium*.

Steven J. Ludtke, Assistant Professor; Ph.D., Rice, 1996. Electron cryomicroscopy of biological molecules and macromolecular assemblies.

Jianpeng Ma, Assistant Professor; Ph.D., 1996, Boston University. Energetics and dynamics of protein structures; application of conformational dynamics to the refinement of X-ray structure; ligand design for anticancer compounds.

Joel D. Morrisett, Professor; Ph.D., North Carolina at Chapel Hill, 1968. Atherosclerosis; characterization of atherosclerotic lesions by NMR spectroscopy and imaging; expression of atherosclerosis-associated genes and proteins by microarray and immunohistochemical technologies.

Timothy Palzkill, Associate Professor; Ph.D., Iowa, 1988. Genome-wide protein-protein interaction studies.

Xuewen Pan, Assistant Professor; Ph.D., Duke, 1993. Genetic networking, chemical genomics, and technology development; dissection of genetic network and elucidation of chemical mechanisms of action.

B. V. Venkataram Prasad, Professor; Ph.D., Indian Institute of Science (Bangalore), 1981. Three-dimensional structural analysis of medically important animal viruses; understanding structure-mechanism correlation in cell entry, uncoating, assembly, and neutralization processes.

Jun Qin, Assistant Professor; Ph.D., Rockefeller, 1996. Proteomics analysis of the human DNA damage signaling network.

Florante A. Quiocho, Professor; Ph.D., Yale, 1966. X-ray crystallographic analysis of biological macromolecules; molecular recognition.

Susan M. Rosenberg, Professor; Ph.D., Oregon, 1986. Molecular mechanisms of mutation, DNA repair, and recombination; genetic instability; genome rearrangement; antibiotic resistance; adaptive mutation.

Shelley Sazer, Associate Professor; Ph.D., Stanford, 1988. Cell-cycle control in fission yeast.

Michael F. Schmid, Associate Professor; Ph.D., Washington (Seattle), 1974. High-resolution structural studies of crystals and macromolecular complexes by electron cryomicroscopy and image processing.

Zhou Songyang, Assistant Professor; Ph.D., Tufts, 1995. Molecular mechanisms of signal transduction; genetic and biochemical analyses of cell growth and survival; genome-wide examination of protein-protein interactions and networks; specificity and structural analysis of protein domains.

Francis T. F. Tsai, Assistant Professor; D.Phil., Oxford (England), 1997. Structure-function studies of macromolecular complexes and assemblies involved in transcriptional regulation and protein folding.

Salih J. Wakil, Professor; Ph.D., Washington (Seattle), 1952. Mechanisms and hormonal control of fatty acid metabolism; structure-function relationship of fatty acid synthase and acetyl-CoA carboxylase; characterization of genes coding for fatty acid synthase and acetyl-CoA carboxylase; metabolic consequences of FAS and ACC knockout in transgenic mice.

Theodore G. Wensel, Professor; Ph.D., California, Davis, 1984. Biochemistry and biophysics of membranes; mechanisms of signal transduction in the retina; biochemical applications of lasers and fluorescence; G proteins.

John H. Wilson, Professor and Graduate Program Director; Ph.D., Caltech, 1971. DNA repair and recombination; genomic instability induced by triplet repeats; oligonucleotide-directed genome modification; mouse models for treatment of retinal diseases.

E. Lynn Zechiedrich, Associate Professor; Ph.D., Vanderbilt, 1990. DNA topoisomerases and antibiotic resistance in *E. coli.*

Pumin Zhang, Assistant Professor; Ph.D., Wisconsin–Madison. Control of cell division cycles in development and disease; how cell-cycle checkpoints are activated in response to internal and external stimuli and how these checkpoints prevent cell-cycle progression.

Zheng Zhou, Assistant Professor; Ph.D., Baylor College of Medicine, 1994. Molecular genetics and cell biology; identification and characterization of genes controlling the phagocytosis of apoptotic cells and other substances.

BOSTON UNIVERSITY

School of Medicine
Department of Biochemistry

Programs of Study

The graduate program in the Department of Biochemistry at Boston University School of Medicine accepts qualified applicants into M.A., Ph.D., and M.D./Ph.D. programs. The rigorous programs combine both basic and advanced course work in biochemistry, cell biology, and molecular biology with extensive research training. Research interests of the faculty are numerous and focus on the elucidation of the biochemical basis of life processes, with particular emphasis on human biology and disease. Broad topic areas of interest include gene structure, regulation of gene expression, signal transduction, protein/enzyme structure and function, cell-cycle control, differentiation, and development.

The first year of the Ph.D. program is dedicated to course work and laboratory rotations to acquaint the student with the wide breadth of research options available. At the end of the first year, the student chooses a mentor under whose guidance the research work proceeds. At this time, the candidate must pass a written qualifying examination. Additional course work in the second year of study is followed by an oral qualifying examination. At this time, the student is assigned a committee of faculty members who monitor the student's progress on his or her research project, which culminates in preparation of a thesis documenting the student's independent research work. Students in the M.A. program complete formal classes during the first year of study (and are not required to take qualifying examinations). M.A. students choose a laboratory in which to perform research after one semester of laboratory rotations. M.A. candidates are required to write a thesis describing their work.

Research Facilities

The Department of Biochemistry within Boston University School of Medicine is part of a large medical school/hospital complex. In addition to state-of-the-art laboratories, there are an extensive medical library, animal-care facilities, and computer facilities. Additional services available to members of the Department of Biochemistry include confocal laser scanning microscopy, DNA sequencing, mass spectrometry, and transgenic animal facilities.

Financial Aid

Ph.D. students are eligible for full financial aid, including tuition and fees, health insurance, and a monthly living stipend.

Cost of Study

Tuition for the 2005–06 academic year was $31,530. For most Ph.D. students, tuition is covered by financial aid, as described above.

Living and Housing Costs

Most graduate students live in apartments in Boston, Cambridge, and the surrounding areas. Shared apartment rentals typically range from $500 to $800 per month per person.

Student Group

The average total number of biochemistry graduate students at the medical school is 50. Approximately 190 graduate students and 600 medical students are enrolled at the medical school.

Location

The Boston University School of Medicine is located in the South End of Boston. The broad range of biomedical research that takes place here is augmented by the large research community existing in greater Boston. Faculty members take advantage of opportunities to extend the School's borders and establish collaborative arrangements with colleagues at other institutions.

Boston and the surrounding areas offer a rich selection of cultural and recreational activities, including museums, music, major-league sports, and skiing, to name a few. Close by are Cape Cod, the White Mountains, and the picturesque New England countryside.

The University

Boston University is an independent, coeducational, nonsectarian university. Its academic diversity meets the needs of one of the largest bodies of scholars in the world. Boston University was incorporated in 1869. Today, its sixteen schools and colleges provide students with the advantages of a large contemporary educational complex, while maintaining many traditional priorities. The two main campuses are the Charles River and the Medical Center campuses. The Department of Biochemistry is housed at the Medical Center campus.

Applying

Students who have completed an undergraduate degree, usually with a major in biochemistry, biology, or chemistry, and have taken courses in general biology, general chemistry, organic chemistry, and calculus may apply for either an M.A. or a Ph.D. in biochemistry. Students who have completed an M.A. degree in biochemistry or a closely related field can apply for a post-master's Ph.D. Also eligible for admission are M.D./Ph.D. students. Students are admitted only in September. Application forms may be obtained from the Director of Graduate Studies. Although the deadline for submission is March 31, applications are processed on a rolling basis, and early applications are given priority. Additional materials required to support the application include college transcripts, scores on GRE General Test, three letters of recommendation, and a personal statement. International students are required to take the TOEFL.

Correspondence and Information

Barbara M. Schreiber, Ph.D., Director of Graduate Studies
Department of Biochemistry
715 Albany Street
Boston, Massachusetts 02118
Phone: 617-638-5094
Fax: 617-638-5339
E-mail: schreibe@biochem.bumc.bu.edu
Web site: http://www.bumc.bu.edu/busm/bi

Boston University

THE FACULTY AND THEIR RESEARCH

Carmela Abraham, Ph.D., Professor. Defining the role of amyloid and inflammation in the brain during normal aging and Alzheimer's disease using biochemical, immunochemical, and molecular biology techniques.

Lawreen Connors, Ph.D., Assistant Research Professor. Structural studies of the human proteins, transthyretin (TTR), and immunoglobulin light chains (LC) in the context of amyloid disease pathology.

Barbara E. Corkey, Ph.D., Professor. Metabolic regulation of signal transduction in the fat cell and pancreatic β-cell and its relationship to obesity and insulin secretion.

Catherine E. Costello, Ph.D., Research Professor. Development and application of advanced mass spectrometric methods for structural determinations of biopolymers, with particular emphasis placed on glycobiology and the establishment of structure-activity relationships in immunology, parasitology, nervous system development and function, and amyloid deposition.

Stephen R. Farmer, Ph.D., Professor. Expression of the major cytoskeletal protein (i.e., tubulin and actin) genes in mammalian nonmuscle cells; modulation of these genes by changes in cell configuration in normal and transformed cells.

Richard E. Fine, Ph.D., Professor. Intracellular route of receptor trafficking in differentiated cells, including liver and brain; mechanism of Ca^{2+} mobilization in the brain mediated by hormones that stimulate phosphoinositide turnover.

Judith Ann Foster, Ph.D., Professor. Regulation of elastic fiber gene expression in the development and repair of pulmonary and cardiovascular tissues.

Carl Franzblau, Ph.D., Professor. Studies on the chemistry, biosynthesis, and turnover of the extracellular matrix proteins, including elastin and collagen.

Vladamir Gabai, Ph.D., Assistant Research Professor. Role of molecular chaperone Hsp72 in modulation of signal transduction pathways and survival of normal and transformed cells.

Herbert Kagan, Ph.D., Professor. Mechanisms of action, regulation, and molecular biology of amine oxidases, with emphasis on connective-tissue lysyl oxidase.

Konstantin V. Kandror, Ph.D., Professor. Regulated vesicular traffic in different eukaryotic cells.

Kathrin H. Kirsch, Ph.D., Assistant Professor. Cellular and molecular oncology; signal transduction by oncogenes and tumor suppressor genes.

Wande Li, Ph.D., Associate Research Professor. Investigating novel biological functions of lysyl oxidase, a copper-dependent enzyme for ECM cross-linking and its expression regulated by metal homeostasis.

Tova Meshulam, Ph.D., Associate Research Professor. Plasma membrane architecture: the role of caveolin/caveolae and cholesterol in transport of fatty acids and signaling.

Matthew A. Nugent, Ph.D., Professor. Role of extracellular matrix in controlling growth factor-receptor interactions and cell proliferation.

Peter O'Connor, Ph.D., Assistant Research Professor. Inventing and developing new mass spectrometric tools for use in proteomics and analysis of other biomolecules with improved accuracy, sensitivity, and resolution.

Mikhail P. Panchenko, Ph.D., Assistant Research Professor. Receptor serine/threonine kinases and G-protein–coupled receptors signaling in vascular endothelium and smooth muscle.

Paul F. Pilch, Ph.D., Professor. Membrane trafficking and the cell biology of insulin action.

Peter Polgar, Ph.D., Professor. Investigating the structure-function of the bradykinin receptor by site-directed mutagenesis.

Katya Ravid, Ph.D., Professor. The development and proliferation of blood and vascular cells using transgenic mice and culture systems.

Barbara M. Schreiber, Ph.D., Associate Professor. Atherosclerosis and aortic smooth-muscle cells; the effect of atherogenic lipoproteins and serum amyloid A on smooth-muscle-cell function.

Pritam Sengupta, Ph.D., Associate Research Professor. Collagen gene repression in fibroblasts; collagen gene regulation by DNA methylation and histone modification.

Michael Y. Sherman, Ph.D., Associate Professor. Understanding the molecular mechanisms underlying the central role of heat shock protein Hsp72 in prevention of cell death.

Elizabeth R. Simons, Ph.D., Professor. Role of platelet-endothelium interactions in the progression of Alzheimer's disease; stimulus responses in blood cells and acquisition of such responses as precursor cells mature.

Barbara D. Smith, Ph.D., Professor. Changes in gene expression of connective-tissue components associated with transformation and differentiation.

Gail E. Sonenshein, Ph.D., Professor. Role of NK-κB transcription factors and the c-Myc oncogene in control of apoptosis. Regulation of collagen gene expression and proliferation in aortic smooth-muscle cells.

Phillip J. Stone, Ph.D., Professor. The connective-tissue protein, elastin; its enzymatic destruction and repair in vitro and in vivo, associated with pathologic processes.

Karen Symes, Ph.D., Associate Professor. Molecular basis of cell movements during early embryonic development of *Xenopus laevis*, PDGF signaling, small GTP-binding protein signaling, cytoskeleton reorganization, and cell-matrix interactions.

Linda Taylor, Ph.D., Assistant Research Professor. Investigating the interaction of prostaglandins and cytokines on the expression of cyclooxygenase 1 and 2.

Keith Tornheim, Ph.D., Associate Professor. Role of metabolic oscillations in fuel stimulated insulin secretion from pancreatic β-cells; altered fuel metabolism and nitric oxide production in vascular disease in diabetes.

Paul A. Toselli, M.D., Ph.D., Associate Professor. Cellular responses of cultured arterial smooth-muscle cells to laser injury, using ultrastructural and biochemical methods.

Abdulmaged M. Traish, Ph.D., Professor. Mechanism of steroid hormone action at the molecular level; role of hormones in cancer; role of neurotransmitters in regulation of smooth-muscle tone and the neurological regulation of erectile dysfunction.

Vickery Trinkaus-Randall, Ph.D., Professor. Regulation of proteoglycans in response to growth factors and regulation of calcium signaling and phosphorylation in response to injury.

Zhi-Xiong (Jim) Xiao, Ph.D., Associate Professor. Functions of tumor suppressor proteins in cell cycle, cell proliferation, and genomic stability.

Rina Yamin, Ph.D., Assistant Research Professor. Amyloid precursor protein (APP) processing and the degradation of the amyloid-beta peptide in Alzheimer's disease.

Joseph Zaia, Ph.D., Assistant Research Professor. Investigation into the structure and function of heparan sulfate using mass spectrometry as the primary research tool.

Vassilis I. Zannis, Ph.D., Professor. Transcriptional regulation of the apolipoprotein genes; structure and function of human apoB and apoA-I; genetic variation and posttranslational modification of apolipoproteins.

CLEMSON UNIVERSITY

Master of Science in Biochemistry and Molecular Biology and Genetics

Programs of Study

The Department of Genetics and Biochemistry offers the M.S. in both genetics and in biochemistry and molecular biology. Concentrations within the degrees include animal, plant, and microbial molecular biology and genetics, including bioinformatics and population genetics. The degrees prepare students for careers in research, both academic and industrial, and require students to have the maturity, mental discipline, and work habits to be independent and productive researchers, with a demonstrated enthusiasm for their field.

The first year of both programs consists of course work and laboratory research, with many courses in common between the degrees. Students typically select an adviser before entering the program and begin their laboratory research immediately. An advisory committee, selected by the adviser and the student, determines the student's course work requirements and keeps track of the student's progress in the classroom, laboratory, and final examinations. While course requirements vary for each degree in the second year, breadth and depth of preparation are expected of each candidate, and the remainder of study is focused on research.

Students are kept abreast of recent developments in genetics, biochemistry, and molecular biology through seminars, colloquia, journal clubs, and special courses. Much of their education is received informally through frequent discussions with faculty members and other students, both within and outside the department. Students are expected to present research results at regional and national scientific meetings, publish in recognized journals, and submit grant proposals in collaboration with faculty members.

Research Facilities

The laboratories for the Department of Genetics and Biochemistry are housed in Jordan Hall and the Biosystems Research Complex. The department is well equipped with most of the conventional instrumentation expected for modern genetic and biochemical analysis. Graduate students have access to specialized research laboratories, including a DNA sequencing facility with LiCor and ABI sequencing equipment, a computerized bioinstrumentation laboratory, an image analysis facility, a confocal/electron microscopy suite equipped for transmission and scanning electron microscopy, controlled-environment chambers, a tissue-culture laboratory, fully equipped darkrooms, isotope handling and counting facilities, plant growth chambers, a $7-million greenhouse complex suitable for use with transgenic plants, and a small-animal-care facility with surgical capabilities. Newly established proteomics and microarray facilities complete the range of tools available to graduate researchers. In addition, students can access help from staff members.

Financial Aid

Financial support is available through grant-supported research assistantships. The Graduate Admissions Committee does not accept unsupported or part-time students. Stipends for M.S. students are $17,200 for twelve months with reduced tuition, awarded upon acceptance. Fees for graduate students with assistantships are approximately $1000 per session for up to 13 credit hours. Graduate students are required to enroll in fall, spring, and both summer sessions. Summer session fees are approximately $400 per session.

Cost of Study

For 2006–07, graduate assistants pay a flat fee of $1079 per semester and $348 per summer session.

Living and Housing Costs

On-campus housing is available; for information, students should visit http://www.housing.clemson.edu. The cost of living in Clemson is quite low compared to the national average; students who choose to live off-campus typically spend $300–$400 per month for rent, depending on the location, amenities, roommates, and other factors.

Student Outcomes

M.S. students have diverse career paths, with some opting for further training toward the Ph.D., others accepting research positions in industry and academia, and a few pursuing careers in science teaching.

Location

Clemson is a small, beautiful college town near the Blue Ridge Mountains and Lake Hartwell. The Upstate is one of the country's fastest-growing areas and is an important part of the I-85 corridor, a multistate area along Interstate 85 that runs from metro Atlanta to Richmond, Virginia, and encompasses Charlotte, North Carolina, and North Carolina's Research Triangle. Atlanta and Charlotte are each a 2-hour's drive away. Many financial institutions and other industries have a national or major presence in the Upstate, including Wachovia, Bank of America, BMW, Bon Secours St. Francis Health System, Bosch North America, Bowater, Charter Communications, Ernst and Young, Fluor Corporation, IBM, Microsoft, Michelin of North America, and many others.

The University

Clemson is classified by the Carnegie Foundation as Doctoral/Research University–Extensive, a category comprising less than 4 percent of all universities in America. The University's mission is to fulfill the covenant between its founder and the people of South Carolina to establish a "high seminary of learning" through its responsibilities of teaching, research, and extended public service. The University has identified eight areas of academic emphasis that create collaborations that, in turn, help fulfill the University's mission.

Applying

Applicants may apply on the Web at http://www.grad.clemson.edu/p_apply.html. Applications, along with a $50 nonrefundable fee, should be received no later than five weeks prior to registration. Every required item in support of the application must be on file by that date. Students are advised to contact the department for the deadlines of the program of proposed study.

Correspondence and Information

Both programs:
Lisa Pape
Administrative Graduate Coordinator
100 Jordan Hall
Clemson University
Clemson, South Carolina 29634-0318
Phone: 864-656-6877
        866-247-8358 (toll-free)
Fax: 864-656-6879
E-mail: lpape@clemson.edu
Web site: http://www.clemson.edu/
        genbiochem/

Biochemistry and Molecular Biology:
Kerry Smith
Assistant Professor and Graduate
        Coordinator
100 Jordan Hall
Clemson University
Clemson, South Carolina 29634-0318
Phone: 864-656-6935
Fax: 864-656-0393
E-mail: kssmith@clemson.edu

Genetics:
Jim Morris
Assistant Professor/Graduate
        Coordinator
214 Biosystems Research Complex
Clemson University
Clemson, South Carolina 29634-0318
Phone: 864-656-0293
Fax: 864-656-0393
E-mail: jmorri2@clemson.edu

Clemson University

THE FACULTY AND THEIR RESEARCH

Albert G. Abbott, Professor; Ph.D., Brown. Biological sciences.
Weiguo Cao, Assistant Professor; Ph.D., Idaho. Microbiology.
Chin-Fu Chen, Assistant Professor; Ph.D., SUNY at Stony Brook. Genetics.
Julianne Collins, Adjunct Faculty, GGC; Ph.D., Alabama at Birmingham. Medical genetics.
Barbara DuPont, Adjunct Faculty, GGC; Ph.D., Texas at Austin. Zoology/human genetics.
David Everman, Adjunct Faculty, GGC; M.D., Emory. Medical genetics.
Julia Frugoli, Assistant Professor; Ph.D., Dartmouth. Biological sciences.
Richard H. Hilderman, Department Chair and Head; Ph.D., Missouri. Microbiology.
Harry D. Kurtz Jr., Assistant Professor; Ph.D., Idaho. Bacteriology.
Amy Lawton-Rauh, Assistant Professor; Ph.D., North Carolina State. Genetics.
Hong Luo, Associate Professor; Ph.D., Louvain (Belgium). Molecular biology.
William R. Marcotte Jr., Associate Professor; Ph.D., Virginia. Microbiology.
Brandon D. Moore, Assistant Professor; Ph.D., Washington State. Botany.
James C. Morris, Assistant Professor; Ph.D., Georgia. Cellular biology.
Gary L. Powell, Professor; Ph.D., Purdue. Biological sciences.
Curtis Rogers, Adjunct Faculty, GGC; M.D., Medical University of South Carolina. Genetics/pediatrics.
Charles Schwartz, Adjunct Faculty, GGC; Ph.D., Vanderbilt. Biochemistry.
Kerry S. Smith, Assistant Professor; Ph.D., Pennsylvania. Molecular biology.
Anand Srivastava, Adjunct Faculty, GGC; Ph.D., Banaras Hindu (India). Biochemistry.
Roger Stevenson, Adjunct Faculty, GGC; M.D., Wake Forest. Pediatrics.
Jeffrey P. Tomkins, Assistant Professor; Ph.D., Clemson. Genetics.

CLEMSON UNIVERSITY

Ph.D. in Biochemistry and Molecular Biology and Genetics

Programs of Study	The Department of Genetics and Biochemistry offers the Ph.D. in both genetics and in biochemistry and molecular biology. Concentrations within the degrees include human, animal, plant, and microbial molecular biology and genetics, including bioinformatics and population genetics. The degrees prepare students for careers in research, both academic and industrial, and require students to have the maturity, mental discipline, and work habits to be independent and productive researchers, with a demonstrated enthusiasm for their field. The first year of both programs consists of course work and laboratory rotations, with many courses in common between the degrees. Students typically select an adviser and begin their research after a three-laboratory rotation during the first year. An advisory committee, selected by the adviser and the student, determines the student's course work requirements, keeps track of the student's progress in the classroom and laboratory, and administers the oral examination (usually in the third year of study) and final examinations. While course requirements vary for each degree in the second year, breadth and depth of preparation are expected of each candidate, and the remainder of the four to five years of study is focused on research. Students are kept abreast of recent developments in genetics, biochemistry, and molecular biology through seminars, colloquia, journal clubs, and special courses. Much of their education is received informally through frequent discussions with faculty members and other students, both within and outside the department. Students are expected to present research results at regional and national scientific meetings, publish in recognized journals, and submit grant proposals in collaboration with faculty members.
Research Facilities	The laboratories for the Department of Genetics and Biochemistry are housed in Jordan Hall and the Biosystems Research Complex. The department is well equipped with most of the conventional instrumentation expected for modern genetic and biochemical analysis. Graduate students have access to specialized research laboratories, including a DNA sequencing facility with LiCor and ABI sequencing equipment, a computerized bioinstrumentation laboratory, an image analysis facility, a confocal/electron microscopy suite equipped for transmission and scanning electron microscopy, controlled-environment chambers, a tissue-culture laboratory, fully equipped darkrooms, isotope handling and counting facilities, plant growth chambers, a $7-million greenhouse complex suitable for use with transgenic plants, and a small-animal-care facility with surgical capabilities. Newly established proteomics and microarray facilities complete the range of tools available to graduate researchers. In addition, students can access help from staff members.
Financial Aid	Financial support is available through grant-supported research assistantships, University-supported research and teaching assistantships, graduate fellowships, and grants-in-aid. The Graduate Admissions Committee does not accept unsupported or part-time students. Stipends for Ph.D. graduate assistantships are $19,200 for twelve months with reduced tuition, awarded upon acceptance. Fees for graduate students with assistantships are approximately $1000 per session for up to 13 credit hours. Graduate students are required to enroll in fall, spring, and both summer sessions. Summer session fees are approximately $400 per session.
Cost of Study	Tuition for 2006–07 is $4643 per semester for in-state students and $9255 per semester for nonresidents. Off-campus rates are $535 per hour for in-state students and $918 per hour for nonresidents. Graduate assistants pay a flat fee of $1079 per semester and $348 per summer session. Graduate fellows pay South Carolina resident fees.
Living and Housing Costs	On-campus housing is available; for information, students should visit http://www.housing.clemson.edu. The cost of living in Clemson is quite low compared to the national average; students who choose to live off-campus typically spend $300–$400 per month for rent, depending on the location, amenities, roommates, and other factors.
Student Group	There are approximately 30 Ph.D. students in the programs. About 60 percent are women, almost all are full-time, and approximately 40 percent are international students.
Student Outcomes	Most of the Ph.D. graduates in genetics and biochemistry and molecular biology go on to postdoctoral research in either industry or academics.
Location	Clemson is a small, beautiful college town near the Blue Ridge Mountains and Lake Hartwell. The Upstate is one of the country's fastest-growing areas and is an important part of the I-85 corridor, a multistate area along Interstate 85 that runs from metro Atlanta to Richmond, Virginia, and encompasses Charlotte, North Carolina, and North Carolina's Research Triangle. Atlanta and Charlotte are each a 2-hour's drive away. Many financial institutions and other industries have a national headquarters for a major presence in the Upstate, including Wachovia, Bank of America, BMW, Bon Secours St. Francis Health System, Bosch North America, Bowater, Charter Communications, Ernst and Young, Fluor Corporation, IBM, Microsoft, Michelin of North America, and many others.
The University	Clemson is classified by the Carnegie Foundation as Doctoral/Research University–Extensive, a category comprising less than 4 percent of all universities in America. The University's mission is to fulfill the covenant between its founder and the people of South Carolina to establish a "high seminary of learning" through its responsibilities of teaching, research, and extended public service. The University has identified eight areas of academic emphasis that create collaborations that, in turn, help fulfill the University's mission.
Applying	Applicants may apply on the Web at http://www.grad.clemson.edu/p_apply.html. Applications, along with a $50 nonrefundable fee, should be received no later than five weeks prior to registration. Every required item in support of the application must be on file by that date. Students are advised to contact the department for the deadlines of the program of proposed study.

Correspondence and Information

Both programs:
Lisa Pape
Administrative Graduate Coordinator
100 Jordan Hall
Clemson University
Clemson, South Carolina 29634-0318
Phone: 864-656-6877
     866-247-8358 (toll-free)
Fax: 864-656-6879
E-mail: lpape@clemson.edu
Web site: http://www.clemson.edu/
    genbiochem/

Biochemistry and Molecular Biology:
Kerry Smith
Assistant Professor and Graduate
    Coordinator
100 Jordan Hall
Clemson University
Clemson, South Carolina 29634-0318
Phone: 864-656-6935
Fax: 864-656-0393
E-mail: kssmith@clemson.edu

Genetics:
Jim Morris
Assistant Professor/Graduate
    Coordinator
214 Biosystems Research Complex
Clemson University
Clemson, South Carolina 29634-0318
Phone: 864-656-0293
Fax: 864-656-0393
E-mail: jmorri2@clemson.edu

Clemson University

THE FACULTY AND THEIR RESEARCH

Albert G. Abbott, Professor; Ph.D., Brown. Biological sciences.
Weiguo Cao, Assistant Professor; Ph.D., Idaho. Microbiology.
Chin-Fu Chen, Assistant Professor; Ph.D., SUNY at Stony Brook. Genetics.
Julianne Collins, Adjunct Faculty, GGC; Ph.D., Alabama at Birmingham. Medical genetics.
Barbara DuPont, Adjunct Faculty, GGC; Ph.D., Texas at Austin. Zoology/human genetics.
David Everman, Adjunct Faculty, GGC; M.D., Emory. Medical genetics.
Julia Frugoli, Assistant Professor; Ph.D., Dartmouth. Biological sciences.
Richard H. Hilderman, Department Chair and Head; Ph.D., Missouri. Microbiology.
Harry D. Kurtz Jr., Assistant Professor; Ph.D., Idaho. Bacteriology.
Amy Lawton-Rauh, Assistant Professor; Ph.D., North Carolina State. Genetics.
Hong Luo, Associate Professor; Ph.D., Louvain (Belgium). Molecular biology.
William R. Marcotte Jr., Associate Professor; Ph.D., Virginia. Microbiology.
Brandon D. Moore, Assistant Professor; Ph.D., Washington State. Botany.
James C. Morris, Assistant Professor; Ph.D., Georgia. Cellular biology.
Gary L. Powell, Professor; Ph.D., Purdue. Biological sciences.
Curtis Rogers, Adjunct Faculty, GGC; M.D., Medical University of South Carolina. Genetics/pediatrics.
Charles Schwartz, Adjunct Faculty, GGC; Ph.D., Vanderbilt. Biochemistry.
Kerry S. Smith, Assistant Professor; Ph.D., Pennsylvania. Molecular biology.
Anand Srivastava, Adjunct Faculty, GGC; Ph.D., Banaras Hindu (India). Biochemistry.
Roger Stevenson, Adjunct Faculty, GGC; M.D., Wake Forest. Pediatrics.
Jeffrey P. Tomkins, Assistant Professor; Ph.D., Clemson. Genetics.

DARTMOUTH COLLEGE

Dartmouth Medical School
Graduate Program in Biochemistry

Program of Study

The Program in Biochemistry uses a broad range of biophysical, biochemical, and cell biological approaches to cover many areas of biochemistry and cell biology, including membrane protein function, actin and microtubule cytoskeletal dynamics, enzyme kinetics, protein folding, membrane trafficking, cholesterol metabolism and transport, transcription, RNA processing, mitosis, cell motility, hematopoiesis, and circadian rhythms. This diversity results from an interest in cellular and molecular function in general, rather than in a few specialized fields. The program's faculty members use a similarly broad array of techniques in their research, including molecular structure techniques, such as X-ray crystallography and structural electron microscopy; biophysical techniques, such as steady-state and stopped-flow spectrophotometry and spectrofluorimetry, analytical ultracentrifugation, and calorimetry; sophisticated biochemical techniques to probe enzyme kinetics, vesicle transport, transcription, RNA processing, and prion protein dynamics; live cell light and fluorescence microscopy; electron microscopy; and genetics in yeast and mice.

Admission to the Graduate Program in Biochemistry, and to other closely-related life sciences graduate programs (Genetics, Microbiology and Immunology, and Biology), is through the shared admissions path of the Molecular and Cellular Biology (MCB) Program at Dartmouth, enabling students to choose any laboratory in these departments for their thesis work. During their first year, students complete a comprehensive course in the principles of biochemistry and molecular and cellular biology. At the same time, they are introduced to research through a series of one-term rotations in three of the sixty-two laboratories in the MCB Program. After their first year, students select a laboratory and graduate program, such as the Biochemistry Program, for the remainder of their course work and thesis research, thereby providing a focused intellectual environment in their chosen area of study. By the end of the second year, each student must pass a qualifying examination.

Research Facilities

The Biochemistry Program is located in the modern facilities of Dartmouth Medical School on the campus of Dartmouth College. In addition to state-of-the-art facilities for standard biochemistry and cell biology, specialized instrumentation is available, such as LCQ and MALDI mass spectrometers, an analytical ultracentrifuge, an isothermal titration calorimeter, an X-ray diffraction apparatus, confocal microscopes, electron microscopes, animal genetics facilities, microinjection equipment, DNA and peptide synthesis/sequencing, and fluorescence-activated cell sorters (FACS). The Dana Biomedical Library is fully equipped with reference material for cutting-edge biomedical research, and the Kiewit Computation Center of Dartmouth College provides vigorous support for bioinformatics and computational biology studies.

Financial Aid

Students enrolled in the Biochemistry Program receive a support stipend. In 2004–05, annual stipends for first-year students were $24,000, and they increased to $24,500 after the student passed a qualifying examination. In addition, health insurance is fully covered through the Dartmouth Student Group Health Plan. Stipend awards are continued annually, based on satisfactory fulfillment of program standards and requirements.

Cost of Study

All students in the program receive a scholarship for the full cost of tuition.

Living and Housing Costs

The Biochemistry Program helps graduate students find appropriate housing, either in College facilities or in privately owned accommodations in the Hanover area. Single-student housing (usually off campus) costs $400 to $600 per month. Married students can rent College-owned duplexes for $675 to $850 per month plus heat and utilities.

Student Group

There are currently 40 students in the Biochemistry Program and 110 in the MCB Program, while faculty members in these two groups number 22 and 62, respectively. The program is small enough to permit close faculty-student interaction, yet large enough to allow for a variety of educational experiences. Dartmouth has about 4,500 undergraduates and 1,200 graduate/professional students.

Student Outcomes

Students completing the Biochemistry Program go on to postdoctoral positions in leading academic laboratories around the country. After appropriate postdoctoral training, the majority of students find faculty positions at colleges, universities, or medical schools or work in leading biotechnology labs.

Location

Hanover is a pleasant town of about 6,000 located on the Connecticut River, which forms the border between Vermont and New Hampshire, surrounded by picturesque countryside. Unlike many small towns, Hanover has sophisticated dining, excellent bookstores, varied movies and theater, and a vibrant cultural life. The area provides excellent opportunities for all types of outdoor activities, especially hiking and skiing. Drives to Boston, Montreal, and New York take about 2, 3, and 5 hours, respectively. A growing number of technology-based companies, as well as the Dartmouth-Hitchcock Medical Center, provide many employment opportunities.

The College

Dartmouth was founded in 1769 as a college committed to liberal learning and has a long-standing tradition of close student-faculty ties, a tradition strongly endorsed by the Biochemistry Program. Programs of study leading to advanced professional degrees are offered by the Medical School, the Thayer School of Engineering, the Faculty of Arts and Sciences, and the Tuck School of Business Administration.

The Hopkins Center for the Performing Arts serves as the cultural focus of Dartmouth, incorporating concert halls, theaters, art galleries, art studios, and crafts workshops. A wide range of musical, dance, and theatrical productions by visiting artists are sponsored by the HOP.

Applying

Application forms and more detailed information about the Biochemistry Program may be obtained from the Chair of the Graduate Committee at the address given below. Applications are reviewed beginning on January 2. Applicants should send their application, supporting documentation (transcripts, references, and GRE scores), and application fee in time to be received no later than January 4.

It is the long-standing policy of Dartmouth College to actively support equality of opportunity for all persons, regardless of race or ethnic background. No student will be denied admission or be otherwise discriminated against because of race, color, religion, handicap, gender or sexual preference, or national or ethnic origin.

Correspondence and Information

Chair, Graduate Admissions Committee
Graduate Program in Biochemistry
Dartmouth Medical School
7200 Vail Building, Room 413
Hanover, New Hampshire 03755-3844

Phone: 603-650-1616
Fax: 603-650-1128
E-mail: biochemistry@dartmouth.edu
Web site: http://dms.dartmouth.edu/biochem/

Dartmouth College

THE FACULTY AND THEIR RESEARCH

Medical School

Amy C. Anderson, Assistant Professor of Chemistry; Ph.D., Harvard, 1997. Structure-based drug design; protein crystallography; enzyme mechanisms.

Bradley A. Arrick, Associate Professor of Medicine; Ph.D., Rockefeller, 1983; M.D., Cornell, 1984. Regulation of growth factor gene expression; role of growth factors in tumor cell biology; cancer genetics.

Charles K. Barlowe, Professor of Biochemistry; Ph.D., Texas, 1990. Biochemical and molecular analysis of vesicular transport in the secretory pathway.

Charles Brenner, Associate Professor of Genetics and Biochemistry; Ph.D., Stanford, 1993. Enzyme function and structure in tumor suppression, ataxia-oculomotor apraxia, and NAD+ biosynthesis.

Constance E. Brinckerhoff, Professor of Medicine and Biochemistry; Ph.D., SUNY at Buffalo, 1968. Molecular and cell biology of matrix metalloproteinase gene expression in connective tissue disease and tumor invasion.

Ta-Yuan Chang, Professor and Chair of Biochemistry; Ph.D., North Carolina at Chapel Hill, 1973. Regulation of cholesterol metabolism and trafficking in mammalian cells; molecular biology of integral membrane proteins.

Charles N. Cole, Professor of Biochemistry and Genetics; Ph.D., MIT, 1972. Nucleocytoplasmic transport of messenger RNA and proteins in yeast; micro RNAs in human breast cancer.

Duane A. Compton, Professor of Biochemistry; Ph.D., Texas, 1988. Regulation of cell structure during mitosis.

Jay C. Dunlap, Professor of Genetics and Biochemistry and Chair of Genetics; Ph.D., California, Santa Cruz, 1979. Genetics of circadian rhythms.

Leslie P. Henderson, Professor of Physiology and Biochemistry; Ph.D., Stanford, 1982. Molecular and cellular basis for sexual dimorphism in the vertebrate nervous system.

Henry N. Higgs, Assistant Professor of Biochemistry; Ph.D., Washington (Seattle), 1996. Cytoskeletal and membrane dynamics during cell motility.

F. Jon Kull, Assistant Professor of Chemistry; Ph.D., California, San Francisco, 1996. Protein crystallography; molecular motors; cellular transport mechanisms; enzyme mechanisms.

Gustav E. Lienhard, Professor of Biochemistry; Ph.D., Yale, 1964. Signal transduction and membrane trafficking, with emphasis on signaling from the insulin receptor and the regulation of glucose transport by insulin.

Jennifer J. Loros, Professor of Biochemistry and Genetics; Ph.D., California, Santa Cruz, 1984. Fungal genetics and the molecular analysis of circadian clocks.

Dean R. Madden, Associate Professor of Biochemistry; Ph.D., Harvard, 1992. Molecular mechanics of ion channel structure and function.

Robert A. Maue, Professor of Physiology and Biochemistry; Ph.D., California, San Diego, 1985. Cellular and molecular biology of neuronal ion channels; molecular mechanisms underlying regulation of ion channel expression by neurotrophic factors.

Lawrence C. Meyers, Associate Professor of Biochemistry; Ph.D., Harvard, 1995. Molecular mechanisms of transcriptional regulation.

Oscar A. Scornik, Professor Emeritus of Biochemistry; M.D., 1957, Ph.D., 1961, Buenos Aires. Regulation of protein synthesis and degradation in mammalian cells.

Nancy A. Speck, Professor of Biochemistry; Ph.D., Northwestern, 1983. Hematopoiesis and its dysregulation in leukemia.

Surachai Supattapone, Associate Professor of Biochemistry and Medicine; Ph.D., Johns Hopkins, 1988; D.Phil., Oxford, 1991; M.D., Johns Hopkins, 1992. Study of mechanisms by which prions and other misfolded proteins cause neurodegeneration, using a combination of biochemical, cell culture, and genetic techniques.

Bernard L. Trumpower, Professor of Biochemistry; Ph.D., Saint Louis, 1968. Mitochondrial bioenergetics; function of supernumerary subunits in mitochondrial energy transducing enzyme complexes.

William T. Wickner, Professor of Biochemistry; M.D., Harvard, 1973. Protein secretion, membrane assembly, and organelle inheritance during cell division.

Lee A. Witters, Eugene W. Leonard Professor of Medicine and Biochemistry; M.D., Rochester, 1969. Hormonal and substrate regulation of fatty acid metabolism; protein kinase regulation.

Entrance to Dartmouth Medical School, which houses the Department of Biochemistry.

THE GEORGE WASHINGTON UNIVERSITY

Graduate Program in Biochemistry and Molecular Genetics

Programs of Study

The Program in Biochemistry and Molecular Genetics offers a program leading to the Ph.D. In addition, the Department of Biochemistry offers several M.S. degree programs: M.S. in biochemistry with a thesis option, M.S. with a nonthesis option, and M.S. in genomics and bioinformatics. The Ph.D. and M.S. (thesis option) programs are designed to prepare students for careers in research and in teaching in which the principles and methods of biochemistry are applied to the study of biological and medical problems. The M.S. (nonthesis option) is an alternative to conventional graduate education and is designed for those who work full-time but are interested in advancing their careers. A new M.S. degree program in genomics and bioinformatics is offered that couples a genome-wide approach to medicine with rapidly evolving methods of analysis. The programs endeavor to establish for each student a broad general knowledge of modern biochemistry, molecular biology, and genetics, followed by development of an individual program. Faculty members interact with the clinical departments in such areas as cancer, immunology, heart disease, vascular biology, cerebral resuscitation, blood disorders, inherited diseases, and problems of aging.

Candidates for the Ph.D. and M.S. degrees are required to complete formal course work in general biochemistry, laboratory methodologies, enzymology and proteins, and molecular biology plus additional courses planned with the graduate adviser to provide a strong background in the area of specialization.

The program leading to the Ph.D. degree in biochemistry and molecular genetics requires a total of 72 credit hours. An interdisciplinary core curriculum (23–26 credit hours) during the first year of Ph.D. graduate training covers macromolecular interactions of proteins and nucleic acids, cell biology, and physiology and is taken by all Ph.D. candidates in the biomedical sciences. Laboratory rotations enable students to become more familiar with research techniques, scientific communication, and potential dissertation projects. Up to 24 credits may be satisfied by dissertation research and the remaining credits by specialized electives. Additional requirements include a written and an oral comprehensive examination and a written dissertation on original experimental work in a suitable area of biochemistry. Successful oral defense of the dissertation completes the degree requirements. Admission to the George Washington University's (GW's) Institute for Biomedical Sciences is the mechanism whereby students ultimately become affiliated with the Ph.D. programs in biochemistry.

The program leading to the M.S. degree (thesis) requires a total of 30 credit hours, 6 of which are earned by the preparation of a thesis. The program leading to the M.S. degree (nonthesis) requires a total of 36 credit hours, 6 of which are earned by preparing a review paper. A written comprehensive examination is required for each M.S. program. Completion of the M.S. programs generally requires two years.

Research Facilities

The Biochemistry Department resides in Ross Hall, a modern and fully equipped research and teaching facility that also houses other basic science and clinical departments. A close relationship exists with the off-campus teaching and research faculty at the National Institutes of Health, Children's Hospital, and the Institute for Genomic Research.

The School of Medicine and Health Sciences has excellent modern library facilities, with a special section devoted to audiovisual educational aids.

Financial Aid

For Ph.D. students, fellowships are available. They carry a $23,000-per-year stipend and full tuition remission (support during the second year is dependent on a satisfactory record during the first year). After completion of the core curriculum, stipend support transfers to the laboratories in which students are conducting their thesis research.

Cost of Study

For the 2006–07 academic year, the cost of tuition is $970 per credit, and the cost of student association fees is $1 per semester credit hour.

Living and Housing Costs

The cost of living varies widely according to the type of accommodations and the area in which the student chooses to live. University housing is not generally available for graduate students. Information about off-campus housing may be obtained from the director of housing and residence life.

Student Group

The Institute for Biomedical Sciences admits 10–15 students per year. During the first year, the students participate in the core curriculum; a dissertation adviser and program are chosen at the end of the first academic year. These affiliations are important for the students' subsequent studies and research.

Location

Because the George Washington University is located in the nation's capital, there is much of historical note, and many cultural and educational opportunities lie within easy reach. Of particular interest to students in the health sciences is the University's proximity to the National Institutes of Health, the National Library of Medicine, the Library of Congress, the Department of Agriculture, and the Food and Drug Administration. The medical schools of Georgetown, Howard, and the Uniformed Services University of the Health Sciences are also nearby.

The University

The George Washington University is both a national and an international institution, educating students from fifty states and 150 countries in a climate of academic accomplishment and scholarly integrity. More than 10,000 graduate students, along with almost 7,000 undergraduate students, interact with notable faculty members and myriad cultural, governmental, corporate, and research institutions in the vibrant atmosphere of one of the world's most beautiful and influential cities.

The main GW campus is located five blocks from the White House in a historic area of Washington, D.C., known as Foggy Bottom. In addition to the extensive offerings of the on-campus libraries, students may have access through their research committee to a multitude of information and research resources, including the Library of Congress, the Smithsonian Institution, the National Institutes of Health, and other associations, nonprofit organizations, and institutions.

Ready transportation to the city's resources is available through Metro, Washington's extensive rapid transit system. With a station located on the campus, Metro makes the University easily accessible from many parts of the city and the Virginia and Maryland suburbs. The Children's Research Institute of Children's Hospital and the National Institutes of Health are also near Metro stations.

Applying

Applicants benefit from a strong background in chemistry and the biological sciences, including courses in general, analytical, organic, and physical chemistry; mathematics through integral calculus; and one year of college physics. Deficiencies in these courses may be made up after admission. An overall and advanced grade point average of at least a B and a combined score of at least 1100 on the verbal and quantitative portions of the General Test of the Graduate Record Examinations are recommended minimum admission requirements. Ph.D. applicants are referred to the University's Institute for Biomedical Sciences for application directions. The Columbian College of Arts and Sciences operates on a semester system, with terms beginning in August and January. For admission to M.S. programs, application materials should be received by May 1 for fall admission and October 1 for spring admission. The application deadline for the Ph.D. program is January 4.

Correspondence and Information

For Ph.D. program information:
Graduate Adviser
Graduate Program in Biochemistry and
   Molecular Genetics
Ross Hall, Room 605
The George Washington University
2300 Eye Street, NW
Washington, D.C. 20037
Phone: 202-994-2179
Fax: 202-994-0967
E-mail: gwibs@gwu.edu
Web site: http://www.gwumc.edu/ibs

For M.S. program information:
Phone: 202-994-3539
E-mail: bcmjyv@gwumc.edu
   (Biochemistry)
Web site:
   http://www.gwumc.edu/biochem

E-mail:
   bcmfxk@gwumc.edu (Genomics)
Web site:
   http://www.gwumc.edu/bioinformatics

For applications and catalogs:
Dean
The Office of Graduate Studies
Phillips Hall, Room #107
Columbian College of Arts and Sciences
The George Washington University
Washington, D.C. 20052
Phone: 202-994-6210
Fax: 202-994-6213
E-mail: askccas@gwu.edu
Web site:
   http://columbian.gwu.edu

The George Washington University

THE FACULTY AND THEIR RESEARCH

The George Washington University Medical Center, Ross Hall Laboratories

Mahnaz Badamchian, Associate Research Professor; Ph.D., Utah, 1983. Isolation and characterization of thymosins and cholin O-acetyl-transferase. *FASEB J.* 10(6):2185, 1996; *FASEB J.* 10(6):2142, 1996.

J. Martyn Bailey, Professor; Ph.D., 1952, D.Sc., 1971, Wales. Molecular biology of prostaglandins and corticosteroids; antiviral drug development; molecular approaches to diabetes and atherosclerosis. *Biochem. Soc. Trans.* 29:70, 2001; *J. Am. Chem. Soc.* 121:3810, 1999.

Patricia E. Berg, Associate Professor; Ph.D., IIT, 1973. New homeobox gene's role in breast cancer and leukemia; sickle cell anemia therapy. *Breast Cancer Res.* 5:82–7, 2003; *Mol. Cell. Biol.* 22:2505–14, 2002.

Maria Bottazi, Assistant Research Professor; Ph.D., Florida, 1995. Molecular and cellular immunoepidemiology of parasitic infections; mechanisms of signal transduction of infectious organisms. *Int. J. Parasitol.* 33(11):1245–58, 2003; *Mol. Cell. Biol.* 21:7607–16, 2001.

Bernard Bouscarel, Associate Research Professor; Ph.D., Toulouse III (France), 1985. Physiologic and pathophysiologic roles of signal transduction in liver disease and GI cancer. *Gastroenterology* 117:433–52, 1999; *Am. J. Physiol.* 281:C1396–402, 2001.

Michael Bukrinsky, Professor; Ph.D., Russian Academy of Sciences (Moscow), 1984. HIV-related molecular virology and immunology; design of new anti-HIV therapeutic strategies. *Trends Immunol.* 23:323–5, 2002; *Virology* 302:195–206, 2002.

Susan M. Ceryak, Assistant Research Professor of Pharmacology; Ph.D., George Washington, 1994. Signaling pathways involved in cell-cycle regulation and dysregulation in cancer. *Mol. Cell. Biochem.* 255:139–49, 2004; *J. Biol. Chem.* 278:17885–94, 2003.

Sidong Fu, Assistant Research Professor; M.D., Xi'an (China), 1989; Ph.D., Beijing (China), 1994. Homeobox genes in breast cancer and cancer genetics; bioinformatics. *Leukemia* 14(11):1867–75, 2000; *Gastroenterology* 116(6):1319–29, 1999.

Linda L. Gallo, Professor; Ph.D., George Washington, 1969. Mechanism and regulation of cholesterol absorption; chemistry, molecular biology, and physiology of cholesterol esterase and acyl coenzyme A; cholesterol acyl transferase; atherosclerosis; lipid metabolism in wasting disorders. *Br. J. Can.* 72:1173–9, 1995; *Mediators Inflamm.* 2:51–6, 1993.

Allan L. Goldstein, Professor and Chairman; Ph.D., Rutgers, 1964. Chemical and biological properties of the thymosins; neuroimmunology; immunodeficiency diseases; cancer; AIDS; aging. *J. Invest. Dermatol.* 113:364–8, 1999.

Valerie W. Hu, Associate Professor; Ph.D., Caltech, 1978. Genomics and mechanism(s) of autism spectrum disorders; molecular and cellular responses to chronic radiation; cell cycle; apoptosis; stress-response; genomic and proteomic analyses of radiation effects. *FASEB J.* 17:1470–86, 2003; *FASEB J.* 15:1562–8, 2001.

Fatah Kashanchi, Associate Professor; Ph.D., Kansas, 1991. Molecular pathogenesis of AIDS and ATL; cell-cycle associated events related to host cell and human retroviruses, including HIV-1 and HTLV-1; genomics, microarray, and proteomics of infected and uninfected cells. *J. Biol. Chem.* 277(7):4973–80, 2002; *Virology* 289(2):312–26, 2001.

Andrei M. Komarov, Assistant Research Professor; M.D., 1984, Ph.D., 1988, Moscow. Nitric oxide: biochemistry and detection; biomedical applications of electron paramagnetic resonance spectroscopy. *Methods Enzymol.* 359:66–74, 2002; *Mol. Cell. Biochem.* 234/235:387–92, 2002.

Jay H. Kramer, Associate Research Professor; Ph.D., Lehigh, 1982. Oxidative injury, free-radical detection, and dysfunction during myocardial ischemia/reperfusion; oxidative stress during dietary Mg-deficiency and iron overload; antioxidant therapy. *Exp. Biol. Med.* 228:665–73, 2003; *Mol. Cell. Biochem.* 245:141–8, 2003.

Ajit Kumar, Professor; Ph.D., Chicago, 1968. RNA-protein interaction; transactivation of HIV gene expression. *Virology* 216:411–7, 1996; *Proc. Natl. Acad. Sci. U.S.A.* 86:7828–32, 1989.

Raj Lakshman, Research Professor; Ph.D., India. Coronary heart disease; lipids; metabolic and genetic obesity; hepatoxins; gene regulation and expression; retinoids.

David Leitenberg, Assistant Professor; M.D./Ph.D., Iowa, 1990. Regulation of T-lymphocyte activation and development. *Immunity* 15:729–38, 2001; *Immunity* 10:701–11, 1999.

I. Tong Mak, Associate Research Professor; Ph.D., Wisconsin–Madison, 1982. Free radical biology in cardiovascular cells; antioxidant drug therapy and mechanisms; iron overload and lysosomes; endothelial apoptosis. *J. Pharmacol. Exp. Ther.* 308:85–90, 2004; *Cardiovasc. Toxicol.* 4:109–15, 2004.

Philippe Marmillot, Assistant Research Professor; Ph.D., Compiègne (France), 1990. Alcohol consumption; alcoholic liver disease; cardiovascular disease; small GTP-binding proteins, annexins; protein trafficking. *Metabolism* 48:1184–92, 1999; *Metabolism* 49:508–12, 2000.

Timothy A. McCaffrey, Associate Professor; Ph.D., Purdue, 1985. Genomics; cardiovascular disease; growth factors; molecular biology; microarray; apoptosis; biochemistry. *Cytokine Growth Factor Rev.* 11:103–14, 2000; *J. Clin. Invest.* 105(5):653–62, 2000.

Joseph J. Pinzone, Assistant Professor; M.D., Washington (St. Louis), 1992. Cellular differentiation; endocrinology; metabolism; tumor; translational research. *Breast Cancer Res.* 5(4):R82–7; *Mol. Cell. Biol.* 24:4605–12, 2004.

Marcos Rojkind, Research Professor; M.D., 1960, Ph.D., 1971, Mexico. Fibrogenic mechanisms of alcohol and hepatitis C virus; oxidative stress and scarring of the liver; laminin-binding proteins of hepatocytes and hepatomas. *Am. J. Pathol.* 162:1771–80, 2003; *Cytokine* 20 (1–2):12–20, 2003.

Rita Teresa Roy, Assistant Research Professor; M.D., George Washington, 1994. Bioinformatics; distance learning.

Gary L. Simon, Professor, Vice Chairman (Medicine), and Director, Division of Infectious Diseases; Ph.D., Wisconsin–Madison, 1972; M.D., Maryland, 1975. Infectious disease; AIDS.

Jack Y. Vanderhoek, Professor; Ph.D., MIT, 1966. Regulation of eicosanoid metabolism by natural and pharmacological agents; nuclear eicosanoid binding proteins. *Biochim. Biophys. Acta* 1635:75–82; 1640:69–76, 2003.

Glenn A. Walker, Professor; Ph.D., Michigan State, 1963. Techniques in teaching biochemistry for graduate students and medical students.

William B. Weglicki, Professor; M.D., Maryland, 1962. Neuropeptide-induced cardiovascular inflammation associated with dietary Mg-deficiency; oxidative stress and antioxidants in iron-mediated free-radical injury to ischemic/reperfused heart and endothelial cells. *Magnesium Res.* 16(2):91–7, 2003; *Mol. Cell. Biochem.* 245:141–8, 2003.

Children's Hospital Laboratories

Anamaris M. Colberg-Poley, Professor; Ph.D., Penn State, 1980. Molecular analyses of herpesvirus immediate early gene expression and proteins. *J. Gen. Virol.*, in press; *J. Gen. Virol.* 84:3353–8, 2003.

Eric P. Hoffman, Professor; Ph.D., Johns Hopkins, 1986. Molecular basis of inherited muscle and inherited central nervous system disease. *Am. J. Hum. Genet.* 64:934–8; 65:252–4, 1999.

Stephan Ladisch, Professor; M.D., Pennsylvania, 1973. Tumor cell gangliosides: role in tumor formation, regulation of the immune response, and usefulness as circulating markers of malignancy. *J. Biol. Chem.* 275:34213–23, 2000; *JNCI* 92:912–7, 2000.

Mary C. Rose, Associate Research Professor; Ph.D., Case Western Reserve, 1970. Molecular analyses of mucus in airway diseases; regulation of mucin genes, goblet cell hyperplasia, mucin glycoproteins. *Am. J. Respir. Cell Mol. Biol.* 25:533–537, 2001; *J. Aerosol Med.* 13: 245–261, 2000.

Mendel Tuchman, Professor; M.D., Tel Aviv, 1979. Biochemistry and molecular biology of the urea cycle; therapy of hyperammonemia with genetically engineered bacteria; inborn errors of metabolism. *Pediatr. Res.* 48:842–6, 2000; *Eur. J. Pediatr.* 159(3):S196–8, 2000.

The Institute for Genomic Research

William C. Nierman, Professor; Ph.D., Berkeley, 1979. Genomics-based functional analysis of microbial pathogenesis; bioinformatics. *Nature* 423:81–6, 2003; *FEMS Microbiol. Lett.* 218:223–30, 2002.

Scott N. Peterson, Associate Professor; Ph.D., North Carolina, 1992. Microbial genomics. *Cell* 101:146–8, 2000; *Science* 286:2165–9, 1998.

Adjunct Faculty

G. Marius Clore, Adjunct Professor; M.D., University College, London, 1979; Ph.D., National Institute for Medical Research (London), 1982. Structural biology and nuclear magnetic resonance; signal transduction; protein–nucleic acid interactions.

Nicholas Robert S. Hall, Adjunct Associate Professor; Ph.D., Florida, 1976. Interrelationships between the brain and immune system.

Kenneth C. Ingham, Professor; Ph.D., Colorado, 1970. Structure, function, and interactions of blood proteins; fibronectin, extracellular matrix.

Marilyn M. Lightfoote, Adjunct Assistant Professor; Ph.D., Virginia, 1983. AIDS; infectious diseases; retrovirology.

Leonid Medved, Adjunct Professor; Ph.D., 1980, D.Sci., 1991, Kiev (Ukraine). Structure of complex proteins; molecular mechanisms of blood coagulation and fibrinolysis; fibrinogen and fibrin. *J. Biol. Chem.* 278:37154–9, 2003; *Biochemistry* 42:7709–16, 2003.

Jonathan R. Merril, Adjunct Assistant Professor; M.D., George Washington, 1989. Medical informatics; distance learning.

Terry W. Moody, Professor; Ph.D., Caltech, 1977. CNS neuropeptides; growth factor receptors in cancer cells; chemistry and biology of bombesin.

Marshall W. Nirenberg, Adjunct Professor; Ph.D., Michigan, 1957. Genetic code; regulatory mechanisms in synthesis of macromolecules.

Neil F. Notaroberto, Adjunct Assistant Professor; M.D., George Washington, 1992. Informatics; distance learning; retina/vitreous surgery.

Prem S. Sarin, Adjunct Professor; Ph.D., Delhi, 1959; Ph.D., Cambridge, 1962. Molecular virology; AIDS vaccine development.

Alexander Wlodawer, Adjunct Professor; Ph.D., California, 1974. Techniques for X-ray and neutron crystallography and application to structures of biologically important molecules.

Ray A. Wolf, Adjunct Assistant Professor; Pharm.D., Kentucky, 1979. Bioavailability; pharmacokinetics; biomedical informatics; distance learning.

HOWARD UNIVERSITY

College of Medicine
Department of Biochemistry and Molecular Biology

Programs of Study	The Department of Biochemistry and Molecular Biology offers programs leading to the Master of Science and Doctor of Philosophy degrees. The primary objective of the Ph.D. program is to prepare candidates for research and teaching careers. The first few semesters are devoted primarily to core course work designed to give the student a broad background in the fundamental theories and techniques of biochemistry. During this time, the student gains exposure to methods for the solution of research problems by working in the laboratories of various faculty members. During the remaining part of the programs, students become increasingly involved in laboratory research and in the critical analysis of biochemical literature. Emphasis is placed on giving candidates rigorous standards of scholarship and critical attitudes toward the solution of research problems. A broad range of research interests exists within the Department. An active seminar program conducted by visiting scientists gives students and faculty members opportunities to broaden their outlook on current scientific problems. Candidates for the Ph.D. degree must obtain the equivalent of 72 semester hours of credit with a grade point average of 3.0 or better, pass qualifying examinations, and complete and defend a dissertation. The dissertation must advance knowledge in its research area and must be of publishable quality.

The M.S. (biotechnology) program prepares students for biomedical industrial careers. The curriculum emphasizes development of technical skills, does not require a thesis, and features an industrial externship. Candidates for the M.S. (biotechnology) degree must obtain 49 semester hours of credit with a grade point average of 3.0 (B) or better.

Research Facilities The Department has spacious laboratories equipped with modern instruments that are used by both faculty members and students in such research areas as macromolecules (structure and function), including computationally based approaches to biomolecular structure and behavior; enzyme kinetics and mechanisms of action; hormonal control mechanisms; gene organization and expression; cancer research; drug metabolism; lipid metabolism; and clinical and nutritional biochemistry.

Financial Aid The Department has a limited number of teaching assistantships and fellowships. These are awarded to the best-qualified applicants.

Cost of Study Full-time tuition and fees per year (9 semester hours or more) were $14,075 in 2005–06. There was a matriculation fee of $257.50, which included student activities and on-campus health services and benefits. Other fees included a self-help fee and endowment fee, amounting to a total of $277.50.

Living and Housing Costs For assistance in finding University-approved off-campus housing, graduate students should write directly to the Supervisor of Off-Campus Housing. Dormitory cafeterias with low prices are open to all University students.

Student Group More than 12,000 students are registered in the University. In a recent year, the College of Medicine, located on the main campus, had 398 students enrolled. Of these, 65 were doctoral students.

Location The Department of Biochemistry and Molecular Biology is in the College of Medicine, which is located in the middle of the nation's capital. Many government facilities, such as Walter Reed Hospital, the National Institutes of Health, the Library of Congress, and the National Medical Library, are within a few minutes' drive. The Kennedy Center for the Performing Arts, the Folger Library with its Shakespeare collection, and the many museums, art galleries, and exhibitions give the District of Columbia a distinct cultural advantage over most other cities.

The University In the years since its inception in 1867, Howard University has grown from a single frame building to a 75-acre campus with buildings and equipment valued at more than $70 million. The University, which has more than 1,255 faculty members, is fully accredited and offers a full range of graduate and undergraduate programs. Howard University is a member of the Consortium of Universities of the Washington Metropolitan Area.

Applying Because assistantships and fellowships are limited, all applications and other necessary documents should be submitted before March 1. The application fee is $45. Students must present at least a 3.0 cumulative average in order to be considered for financial aid and should have a strong background in biology, calculus, physics, and organic and physical chemistry. Prospective students can apply online at http://www.gs.howard.edu.

Correspondence and Information

For application forms:
Office of Admissions
Johnson Building
Howard University
Washington, D.C. 20059

Phone: 202-806-2752
Web site: http://www.gs.howard.edu

For additional information:
Director, Graduate Studies Committee
Department of Biochemistry and Molecular Biology
College of Medicine
Howard University
Washington, D.C. 20059-0001

Phone: 202-806-6289
Fax: 202-806-5784
E-mail: biwashington@howard.edu
Web site: http://www.med.howard.edu/

Howard University

THE FACULTY AND THEIR RESEARCH

Cynthia K. Abrams, Associate Professor; Ph.D., Maryland. Nutritional biochemistry; clinical nutrition; fat digestion.
Selenium deficiency in long term total parenteral nutrition. *Nutr. Clin. Pract.* 7:175–8, 1992. With Siram et al.

Carolyn Whitfield Broome, Associate Professor; Ph.D., George Washington. Breast cancer genetics and gene expression in African Americans.
Estrogen receptor/progesterone receptor–negative breast cancers of young African-American women have a higher frequency of methylation of multiple genes than those of Caucasian women. *Clin. Cancer Res.* 10:2052–7, 2004. With Mehrotra et al.
Inherited *BRCA2* mutations in African Americans with breast and/or ovarian cancer: A study of familial and early onset cases. *Hum. Genet.* 113:452–60, 2003 (online). With Kanaan et al.
Breast cancer genetics in African Americans. *Cancer* 97:236–45, 2003. With Olopade et al.

W. Malcolm Byrnes, Assistant Professor; Ph.D., LSU. Enzymes of primary and secondary metabolism in prokaryotes.
Structure-specific DNA-induced conformational changes in *Taq* polymerase revealed by small angle neutron scattering. *J. Biol. Chem.* 279:39146–54, 2004. With Ho et al.
Extrinsic factors potassium chloride and glycerol induce thermostability in recombinant anthranilate synthase from *Archaeoglobus fulgidus*. *Extremophiles* 8(6):455–62, 2004. With Vilker.

Marguerite W. Coomes, Associate Professor; Ph.D., Texas Health Science Center at Dallas. Drug metabolism; epidermal enzymes; protein degradation.
Amino acid metabolism. In *Textbook of Biochemistry with Clinical Correlations*, 6th edition, Thomas M. Devlin, ed. New York: John Wiley & Sons, 2006.
Cytochrome P-450 and cyclosporine metabolism in transplant patients. *Transplantation Proc.* 25:1980–2, 1993. With Toussaint.

Felix Friedberg, Professor; Ph.D., Berkeley. Gene expression.
Parvalbumin isoforms in zebrafish. *Mol. Biol. Rep.* 32:167, 2005.
Calmodulin genes in zebrafish (revisited). *Mol. Biol. Rep.* 32:55, 2005. With Taliaferro.
Characteristic phenomena of living things. *J. Wash. Acad. Sci.* 90:71–9, 2004.

Matthew George Jr., Associate Professor and Chairman; Ph.D., Berkeley. Molecular genetics; molecular evolution; mitochondrial DNA.
Signal transduction of phorbol 12-myristate 13-acetate (PMA)–induced growth inhibition of human monocytic leukemia THP-1 cells is reactive oxygen dependent. *Leukemia Res.* 29:863–79, 2005. With Traore et al.

Arvind K. N. Nandedkar, Professor and Director; Ph.D., Delhi (India). Clinical chemistry and toxicology: clinical and biochemical approaches in pathogenesis of microorganisms; epilepsy, drug metabolism, and forensic sciences.
Newborn tyroxine levels and childhood ADHD. *Clin. Biochem.* 35:131–6, 2002. With Soldin et al.

Richard H. Pointer, Professor; Ph.D., Brown. Biochemical endocrinology; mechanism(s) of hormone action; hormonal regulation of carbohydrate and lipid metabolism.
Pesticide inactivation of peanut glutamate dehydrogenase: Biochemical basis of the enzyme's isomerization. *J. Agric. Food Chem.* 47(8):3345–51, 1999. With Osuji, Braithewaite, and Reyes.

Allen R. Rhoads, Professor; Ph.D., Maryland. Calcium regulation/signal transduction mechanisms and pathways.
Expression of IQ-motif genes in human cells and ASPM domain structure. *J. Ethnicity Disease* 15(S5):88–9, 2005. With Kenguele.

Thomas E. Smith, Professor; Ph.D., George Washington. Mechanisms of action and control of enzymes: structure-function relationships.
Molecular cell biology. In *Textbook of Biochemistry with Clinical Correlations*, 6th edition, Thomas M. Devlin, ed. New York: John Wiley & Sons, 2006.

William M. Southerland, Professor; Ph.D., Duke. Molecular modeling, molecular dynamics, and design of therapeutic agents.
Barnase thermal denaturation via molecular dynamics simulations: Detection of early denaturation sites. *J. Mol. Model. Mol. Graphics* 24:233–43, 2006. With Yin and Bowen.

Eric Walters, Associate Professor; Ph.D., Missouri. Cell and molecular biology of olfaction.
Localization and characterization of glutathione-s-transferase isozymes alpha, mu, and pi within the mouse vomeronasal organ. *Neurosci. Lett.* 375:198–202, 2005. With Green and Weech.
Inhibition of tumor growth and prolonged survival of rats with intracranial gliomas following administration of clotrimazole. *J. Neurosurg.* 103:79–86, 2005. With Khalid, Tokunaga, and Caputy.
Zonal expression and activity of glutathione S-transferase enzymes in mouse olfactory mucosa. *Brain Res.* 995:151–7, 2004. With Whitby-Logan and Weech.
Propylthiouracil alters glutathione-dependent enzyme expression and activity in mouse olfactory mucosa. *Brain Res.* 997:149–56, 2003. With Etienne and Maruniak.
Characterization of the mouse olfactory glutathione S-transferases during the acute phase response. *J. Neurosci. Res.* 73:679–85, 2003. With Weech and Quash.

INDIANA UNIVERSITY AT INDIANAPOLIS

Indiana University Medical Center
School of Medicine
Department of Biochemistry and Molecular Biology

Program of Study

The Department of Biochemistry and Molecular Biology offers a graduate research program leading to the Ph.D. degree awarded by the Indiana University (IU) Graduate School. The program provides rigorous theoretical and experimental preparation for a career in modern biochemistry and molecular biology. The research of the faculty includes signal transduction mechanisms, molecular biology, cellular regulation, genetics, regulation of metabolic processes, and the structure and function of proteins. Extensive research interactions occur with other academic and clinical departments on the Indianapolis campus. Much of the research is of fundamental importance to cancer, diabetes, and alcoholism. Most of the faculty members are associated with one or more of the following major research centers: Indiana Alcohol Research Center, IU Center for Diabetes Research, IU Center for Cancer Research, IU Center for Structural Biology, IU Center for Computational Biology and Bioinformatics, and IU Center for Medical Genomics.

The first year of graduate study is devoted principally to advanced course work. However, the student becomes involved in research from the beginning and usually has identified a thesis research project by the end of the first year. Academic requirements are completed after two years, at which point the student may advance as a doctoral candidate and devote full-time study to an original investigation culminating in the Ph.D. thesis. The academic minor can be selected in a related discipline (microbiology, pharmacology, physiology), but students may draw from a broad spectrum of graduate courses to compose a coherent interdisciplinary minor in cancer biology, diabetes, life science, or physical science. Students accepted into medical school may enroll in a combined M.D./Ph.D. program.

The Department has a varied research program that combines the traditional values and rigor of the biochemical approach with the latest techniques of molecular biology. Research projects range from the most basic studies of the physical properties of biomolecules to the application of molecular genetics in diagnosis. The research environment is enriched by numerous interactions and collaborations with faculty members in other basic science and clinical departments at the Medical Center with the chemistry, physics, and biology faculty on the adjacent Indiana University–Purdue University at Indianapolis (IUPUI) campus and with colleagues at nearby universities, most notably Purdue University and Indiana University Bloomington. These interactions are especially reinforced by the multidisciplinary research centers for cancer, diabetes, and alcoholism.

Research Facilities

The Department of Biochemistry and Molecular Biology occupies 20,000 square feet of laboratory space on the fourth floor of the Medical Science Building. The Department has excellent facilities for modern research in biochemistry and molecular biology. This includes state-of-the-art instrumentation for X-ray crystallography, microarray analysis, mass spectrometry, protein-protein interaction, and automated DNA sequencing. Advanced instrumentation available to students on campus includes fluorescence-activated cell sorters, electron microscopes, and a 500-MHz multinuclear NMR spectrometer. The Proteomics Core Facility and the Center for Medical Genomics provide state-of-the-art facilities for training of investigators and students in these disciplines. Research cores include facilities for the production of knockout and transgenic animals. The adjacent Medical Research Building houses a medical library (holding more than 175,000 volumes, including 2,000 biomedical journals) and a professionally staffed animal research facility. The laboratories of several members of the faculty are located in the nearby Center for Cancer Research and the Biotechnology Research and Training Center. Online electronic literature searching is available through personal computers located in all laboratories.

Financial Aid

Stipends for 2005–06 were $21,600 per year. Tuition fees are provided for Ph.D. students in good academic standing.

Cost of Study

In 2005–06, in-state tuition and fees were $214.95 per credit hour, and out-of-state fees were $620.40 per credit hour. Fee scholarships may be used to defray tuition costs.

Living and Housing Costs

Dormitories and apartments for single or married students are available on campus, along with a variety of dining facilities. Many pleasant residential areas are within a 15-minute drive of the Medical Center. The general cost-of-living index is substantially lower than in most metropolitan areas.

Student Group

Health science programs at the Medical Center enroll more than 4,000 students, including 289 graduate students in nine basic science programs. Total enrollment at the Indianapolis campus is 28,000, including undergraduates and part-time students.

Student Outcomes

Over the last ten years, approximately two thirds of graduates have taken academic positions and one third have taken industrial positions.

Location

Indianapolis, a rapidly developing metropolitan area of a million people, has gained nationwide recognition as an attractive place to live. It provides the advantages of a large city in terms of cultural activity and entertainment, but it is small enough to permit freedom of movement and community identification. It is the home of major pharmaceutical and biochemical companies. There are many recreational facilities, including the largest city park in the country, a velodrome, and a zoo near the campus. The city's cultural activities include programs supported by the Indianapolis Museum of Art and the Indianapolis Symphony Orchestra. Diverse year-round programs of theater, ballet, opera, and popular entertainment are presented at Clowes Hall, Conseco Field House, the Indianapolis Convention Center, and RCA Dome. Indianapolis is the home of the 500-mile auto race and of several professional sports teams, including the Indiana Pacers, Indianapolis Colts, Indianapolis Indians, and Indianapolis Ice.

The University and The Center

Indiana University is a comprehensive statewide university system, with core campuses located at Indianapolis and Bloomington. The Indiana University Medical Center, including the Schools of Medicine, Dentistry, Nursing, and Allied Health Sciences, is located on a 100-acre campus a little more than a mile from the center of Indianapolis. The Medical Center is adjacent to an undergraduate campus, VA and county hospitals, and a conference center and hotel. Students have access to outstanding athletics facilities in the campus sports complex, which provides swimming and track facilities that are among the finest in the nation.

Applying

The principal prerequisite for admission into the program is an undergraduate degree, usually in biology, biochemistry, or chemistry, from an accredited institution in the United States or equivalent training from a university outside of the United States. The General Test of the Graduate Record Examinations is required.

Correspondence and Information

Ronald C. Wek, Ph.D.
Chairperson, Admissions Committee
Department of Biochemistry and Molecular Biology
Indiana University School of Medicine
635 Barnhill Drive, Room 4067
Indianapolis, Indiana 46202-5122

Phone: 317-274-0549
Fax: 317-274-4686
E-mail: rwek@iupui.edu
Web site: http://www.biochemistry.iu.edu

Indiana University at Indianapolis

THE FACULTY AND THEIR RESEARCH

Faculty at Indianapolis

Simon J. Atkinson, Ph.D., Associate Professor. Cellular and molecular biology of the actin cytoskeleton; regulation of cell motility.

Martin Bard, Ph.D., Professor. Yeast genetics and molecular biology of sterol synthesis and regulation; protein-protein interactions of sterol biosynthetic enzymes.

William F. Bosron, Ph.D., Professor. Structure and mechanism of alcohol dehydrogenases and carboxylesterases; regulation of ethanol, retinoid, and drug ester (cocaine, heroin, meperidine) metabolism; role of retinoid metabolism in liver fibrosis.

David W. Crabb, M.D., John B. Hickham Professor and Chair, Department of Medicine. Regulation of expression of aldehyde dehydrogenase; control of ethanol metabolism in transduced cells; signaling by way of the peroxisome proliferator activated receptor (PPAR), sterol response element-binding protein (SREBP), and AMP-dependent protein kinase (AMPK) in liver.

Dring N. Crowell, Ph.D., Professor. Regulation of plant growth and development by isoprenoid compounds and isoprenylated proteins.

Mark A. Deeg, M.D., Ph.D., Associate Professor. Regulation of glycosylphosphatidylinositol metabolism in diabetes and atherosclerosis; ethanol regulation of inflammation and atherosclerosis.

Timothy DeGrado, Ph.D., Professor. Molecular imaging of lipid and carbohydrate metabolism; development of radiopharmaceuticals and tracer kinetic methods for cardiac and oncologic imaging.

Anna A. DePaoli-Roach, Ph.D., Professor. Structure, function, and regulation of protein phosphatases; study of function in knockout and overexpressing transgenic mice; role of glycogen in glucose homeostasis and obesity.

A. Keith Dunker, Ph.D., Professor. Bioinformatics and laboratory experiments to understand functions of intrinsic disorder and thereby to challenge the sequence-to-structure-to-function paradigm.

Kenneth W. Dunn, Ph.D., Associate Professor. Endocytic membrane transport in polarized epithelia; quantitative and 3-dimensional microscopy.

Joseph R. Dynlacht, Ph.D., Associate Professor. Mechanisms of radiation-induced cell killing; heat-shock response; DNA double-strand break repair; nuclear matrix structure and composition; radiation-induced cataractogenesis.

Howard J. Edenberg, Ph.D., Chancellor's Professor. Genomics and bioinformatics; regulation of gene expression; genetics of complex diseases; molecular biology of alcoholism and alcohol metabolism.

Jeffrey S. Elmendorf, Ph.D., Associate Professor. Molecular mechanisms of insulin-stimulated glucose transport; signal transduction; vesicular trafficking; role of the cytoskeleton; microscopic and proteomic approaches.

Rose S. Fife, M.D., Professor. Biochemistry of extracellular matrix proteins; effects of nitric oxide synthase and cyclooxygenase on cell growth.

Shao-Ling Fong, Ph.D., Associate Professor. Gene regulation of visual cycle and visual transduction proteins in human retina.

Millie M. Georgiadis, Ph.D., Associate Professor. X-ray crystallographic and biochemical analysis of protein–nucleic acid interactions involved in retroviral replication, nuclear export, and DNA repair.

Mark G. Goebl, Ph.D., Professor. Regulation of the cell cycle by ubiquitin-dependent protein degradation; global analysis of SCF controlled cellular physiology.

Matthew W. Grow, Ph.D., Assistant Professor. Genetic networks involved in vertebrate cardiovascular development.

Maureen A. Harrington, Ph.D., Professor. Regulation of NF-kB controlled gene expression and development of innate immune response.

Robert A. Harris, Ph.D., Distinguished Professor and Showalter Professor. Roles of mitochondrial protein kinases and phosphatases in glucose homeostasis and brain function; regulation of branched chain amino acid catabolism.

Thomas D. Hurley, Ph.D., Professor. X-ray crystallography of dehydrogenases and glycosyltransferases; enzyme engineering using site-directed mutagenesis; kinetic and thermodynamic analysis of protein-ligand interactions.

Hiremagalur N. Jayaram, Ph.D., Professor. Regulation of nucleic acid metabolism; interaction of NAD metabolism in diabetes; biochemical targeting of therapy; gene therapy; mechanisms of drug action and resistance; developing a drug from the bench to the clinic.

Larry R. Jones, Ph.D., M.D., Professor. Purification, cloning, and expression of membrane protein involved in regulation of cardiac contractility and rhythm.

Reuben Kapur, Ph.D., Assistant Professor. Understanding intracellular signaling mechanisms involved in regulating adhesion, migration, and self-renewal of hematopoietic stem and progenitor cells.

Mark R. Kelley, Ph.D., Professor. DNA repair genes for translational applications; targeting DNA repair genes for cancer treatments.

Suk-Hee Lee, Ph.D., Professor. Stress and DNA damage-induced S-phase arrest and regulation of DNA replication and repair in mammalian system.

William J. McBride, Ph.D., Professor. Neurochemistry; interactions of neurotransmitters; neurochemistry of alcoholism.

Harikrishna Nakshatri, B.V.Sc., Ph.D., Associate Professor and Marian J. Morrison Investigator in Breast Cancer Research. Role of the transcription factor NF-kB in breast cancer; metastatic growth of hormone-independent breast cancers; breast cancer stem cells and chemotherapy resistance.

Byron L. Olson, Ph.D., Professor. Nicotine metabolism; effects of nicotine on cell signaling in fibroblasts.

David A. Potter, M.D., Ph.D., Associate Professor. Novel mechanisms of Hsp90 inhibition in breast cancer; calpain mechanisms of cell differentiation and proliferation.

Lawrence A. Quilliam, Ph.D., Associate Professor. Signal transduction; regulation and biological function of Ras family GTPases; Ras-related proteins and cancer; genetic models of GTPase function.

Stephen K. Randall, Ph.D., Associate Professor. Role of calcium and ion-binding proteins in cold and other abiotic stress responses in plants.

Simon Rhodes, Ph.D., Professor. Transcriptional regulation of endocrine organ development and function.

Peter J. Roach, Ph.D., Chancellor's Professor. Hormonal control of glycogen metabolism; glycogenin; protein phosphorylation and protein kinases; signal transduction in mammals and yeast; protein kinases of the cdk family; genomic and proteomic approaches.

Roger W. Roeske, Ph.D., Professor. Bioorganic chemistry; peptide synthesis; hormone-receptor interaction; self-assembly systems.

C. Max Schmidt, M.D., Ph.D., Assistant Professor. MAPK and cyclooxygenase regulation of hepatocellular and pancreatic carcinomas.

Weinian Shou, Ph.D., Associate Professor. Cellular and molecular biology of vertebrate development; regulations in cardiac development and function.

Jay R. Simon, Ph.D., Professor. Neurochemistry; mechanisms involved in regulation of the dopamine transporter; drug abuse.

David G. Skalnik, Ph.D., Professor. Epigenetic regulation of chromatin structure and gene expression in mammals.

Debbie C. Thurmond, Ph.D., Assistant Professor. Regulation of pancreatic insulin secretion and skeletal muscle/adipose tissue glucose uptake (insulin action) by SNARE proteins and the actin cytoskeleton, in vitro and in vivo.

Mark C. Wagner, Ph.D., Assistant Professor. Cellular and molecular biology of unconventional myosins; role in renal function.

James P. Walsh, M.D., Ph.D., Associate Professor. Molecular biology of phosphoinositide-mediated signal transduction.

Ronald C. Wek, Ph.D., Professor. Cellular stress response pathways and their linkage to disease.

Mervin C. Yoder, M.D., Professor. Regulation of yolk sac organogenesis and emergence of hematopoietic and endothelial lineages, including circulating and resident hematopoietic and endothelial progenitor cells.

Zhong-Yin Zhang, Ph.D., Professor and Chairman. Protein tyrosine phosphatases: structure and function, chemical biology, activity-based proteomics, signaling mechanisms, and roles in diseases.

Faculty at Other Centers for Medical Education

W. Marshall Anderson, Ph.D., Professor (Gary). Sequence analysis of gamma diagnostic phage of *B. anthrasis;* composition analysis of the gamma diagnostic phage receptor complex on *B. cereus;* identification of source of *E. coli* in watershed of northwest Indiana.

David L. Daleke, Ph.D., Associate Professor (Bloomington). Structure and function of biological membranes; identification, isolation, and characterization of phospholipid transporters; role of membrane phospholipid asymmetry in diabetes; synthesis of novel phospholipids.

Dipika Gupta, Ph.D., Assistant Professor (Gary). Innate immune response to bacteria.

Michael W. King, Ph.D., Professor (Terre Haute). Tissue regeneration and early embryogenesis.

Edward E. McKee, Ph.D., Associate Professor (South Bend). Mitochondrial biogenesis and gene expression; mitochondrial nucleotide transport and metabolism; mitochondrial drug toxicity; cardiac metabolism.

Barth H. Ragatz, Ph.D., Professor (Fort Wayne). Influence of adenine nucleotide analogues on platelet functionality.

Kent L. Redman, Ph.D., Associate Professor (Fort Wayne). Protein-RNA interactions critical for ribosome assembly and function; the biology and chemistry of RNA methyltransferases that form 5-methylcytosine in rRNA and tRNA.

Godfrey Tunnicliff, Ph.D., Professor (Evansville). Metabolism of brain GABA; characterization of GABA recognition sites; 4-aminobutyrate transaminase.

Claire E. Walczak, Ph.D., Associate Professor (Bloomington). Biochemistry and cell biology of mitotic spindle assembly and chromosome segregation in *Xenopus* egg extracts and in tissue culture cells.

IOWA STATE UNIVERSITY
OF SCIENCE AND TECHNOLOGY

Department of Biochemistry, Biophysics, and Molecular Biology

Programs of Study

The Department of Biochemistry, Biophysics, and Molecular Biology (BBMB) offers programs leading to M.S. and Ph.D. degrees with majors in the fields of biochemistry; biophysics; genetics; immunobiology; molecular, cellular, and developmental biology; plant physiology; and toxicology. Students enter a rotation program that lasts through the end of the first semester. By the second semester, students have identified their area of research and begun work on their research project with their chosen mentor. A minimum of 30 credits is required for the M.S. degree. For the Ph.D. degree, a minimum of three years of full-time study, at least half of which must be spent in residence, is required. Preliminary examinations in the field of specialization are required for admission to candidacy for the Ph.D. degree. The final examination for M.S. and Ph.D. candidates is an oral defense of the thesis. At some time during graduate work, each student is expected to serve as a teaching assistant. More information about individual faculty members' research activities is available on request or from the departmental Web site, listed below.

Research Facilities

The department is housed in the Molecular Biology Building, a unique structure that integrates 204,000 square feet of research laboratories and classrooms with artwork inspired by genetic engineering and crafted by award-winning national artists. The modular design of the laboratories allows scientists to quickly reconfigure their space as research needs change. State-of-the-art instrumentation facilities for protein chemistry, nucleic acids, flow cytometry, antibody and hybridoma production, and macromolecular structure determination support life sciences research on campus and throughout the state. The W. M. Keck Metabolomics Research Facility is the most recent addition. Supporting instrument services, a fabrication shop, and chemical stores are conveniently located in a nearby building. Molecular Biology Building classrooms are fitted with the latest instructional technology equipment. Students are linked to the University's high-speed computing network through approximately thirty workstations strategically placed throughout the building. For students and faculty members who prefer it, wireless networking is available in all classrooms and other public areas.

Financial Aid

Financial aid is available to graduate students in the form of research assistantships, teaching assistantships, and fellowships. Most Ph.D. students are supported by research assistantships that currently provide $20,000 per year along with a 50 percent tuition credit. Students are eligible for a complete tuition credit on a competitive basis.

Cost of Study

In 2006–07, tuition for graduate students with a 50 percent tuition credit is $1484 per semester ($3793 for twelve-month enrollment). Computer fees ($126 for twelve months), Health Center fees ($291 for twelve months), and Activities and Service fees ($552 for twelve months) are paid by the student. Health insurance for graduate students on assistantships is paid by the Graduate College.

Living and Housing Costs

University student apartments are available at rates of $356 to $529 per month. Rooms in the graduate dormitory rent for $464 for a double room and $533 for a single room per month. Private rooms and apartments are available in the vicinity of the campus.

Student Group

The current enrollment of the University is 26,971 students, including 4,741 graduate students. More than 3,000 graduate students come from all areas of the United States; the remainder are international students representing nearly 100 countries. There are approximately 90 graduate students enrolled in the department; most are preparing for research careers in industry or universities.

Student Outcomes

Ph.D. graduates typically pursue careers in research as directors of academic, industry, or government laboratories. In this course, Ph.D. graduates normally go on to postdoctoral research appointments prior to seeking assistant professor/laboratory director positions. Other career courses include college/university teaching or research-related administration. M.S. graduates normally go on to Ph.D. programs or research associate positions in industry or academic laboratories.

Location

Ames is a city of about 50,000 residents, including the student body of Iowa State University. The city is located in a rural area 30 miles north of Des Moines, the capital of Iowa. An active cultural life is provided in Ames by local musical and theatrical groups and by the many internationally acclaimed artists who perform on campus. Athletic events at the University provide a year-round focus of interest. The University has a scenic golf course adjacent to the campus and offers facilities for numerous indoor and outdoor sports. The climate is typical of the Midwest: warm to hot in the summer, generally quite cold in the winter, and mild in spring and fall.

The University

Iowa State University of Science and Technology was founded in 1858 as Iowa State College, one of the first land-grant institutions in the nation. Besides the Graduate College, the University has eight undergraduate colleges—Agriculture, Business, Design, Education, Engineering, Family and Consumer Sciences, Liberal Arts and Sciences, and Veterinary Medicine. The scenic campus is a point of pride for students and faculty members and offers a fitting environment for the varied activities of those who work and study at the University.

Applying

Applicants must be graduates of an accredited institution and must normally rank in the upper half of his or her class. GRE General Test scores are required; the Subject Test in biochemistry, cell and molecular biology; chemistry; biology; or physics is recommended. Undergraduate preparation should include emphasis in chemistry, physics, mathematics, and biology. Visits to the department are encouraged.

Correspondence and Information

Department of Biochemistry, Biophysics, and Molecular Biology
1210 Molecular Biology Building
Iowa State University of Science and Technology
Ames, Iowa 50011
Phone: 515-294-2231
      800-433-3464 (toll-free from within the United States)
Fax: 515-294-0453
E-mail: biochem@iastate.edu
Web site: http://www.bb.iastate.edu/

Iowa State University of Science and Technology

THE FACULTY AND THEIR RESEARCH

Senior Staff of the Department

Linda Ambrosio, Associate Professor; Ph.D., Princeton, 1985. Molecular genetics and developmental mechanisms in *Drosophila;* signal transduction by *D-raf;* cell determination and differentiation.

Amy H. Andreotti, Associate Professor of Biochemistry; Ph.D., Princeton, 1994. Nuclear magnetic resonance of proteins; macromolecular structure and recognition.

Donald C. Beitz, Distinguished Professor of Biochemistry and Animal Science; Ph.D., Michigan State, 1967. Regulation of plasma cholesterol and fat deposition; etiology of milk fever and ketosis in dairy cattle: vitamin D and meat tenderness; role of diet in meat and milk consumption.

Thomas Bobik, Associate Professor of Biochemistry; Ph.D., Illinois at Urbana-Champaign, 1990. Genetics and biochemistry of vitamin B-12.

Alan DiSpirito, Associate Professor; Ph.D., Ohio State, 1983. Bioenergetics of chemoautotrophic and methanotrophic bacteria.

Herbert J. Fromm, Distinguished Professor of Biochemistry; Ph.D., Loyola of Chicago, 1954. Enzyme chemistry; site-directed mutagenesis; enzyme structure-function relationships.

Jack Girton, Associate Professor; Ph.D., Alberta, 1979. Molecular structure and function of chromatin; molecular genetic regulation of cell determination.

Mark S. Hargrove, Associate Professor of Biochemistry and Biophysics; Ph.D., Rice, 1995. Heme protein structure and function; X-ray crystallography.

Richard B. Honzatko, Professor of Biochemistry and Biophysics; Ph.D., Harvard, 1982. X-ray crystallography of proteins; enzyme structure-function.

Ted W. Huiatt, Associate Professor of Biochemistry and Animal Science; Ph.D., Iowa State, 1979. Biochemical regulation of growth and differentiation of striated muscle during embryonic development; assembly of myofibrillar proteins in embryonic striated muscle.

Robert L. Jernigan, Professor of Biochemistry and Director, Laurence H. Baker Center for Bioinformatics and Biological Statistics; Ph.D., Stanford, 1968. Bioinformatics; genomics; computational biology; interfaces between structure, function, and sequence; motions of proteins and large assemblages; protein and drug design.

Jorgen Johansen, Professor; Ph.D., Copenhagen, 1984. Developmental and molecular neurobiology; molecular basis for neuronal cell recognition and axonal guidance.

Kristen M. Johansen, Professor; Ph.D., Yale, 1989. Regulation of nuclear architecture and chromatin structure during the cell cycle.

Gustavo MacIntosh, Assistant Professor; Ph.D., Buenos Aires, 1997. Genetics and proteomics of secreted ribonucleases in plants; functional characteristics of plant ribonucleases in relation to defense, development, and stress; signal transduction during wounding and stress responses.

Alan M. Myers, Professor of Biochemistry and Chair of the Department; Ph.D., Duke, 1983. Molecular mechanisms of cellular morphogenesis in yeast; molecular mechanisms of starch biosynthesis in maize.

Basil J. Nikolau, Professor of Biochemistry; Ph.D., Massey (New Zealand), 1981. Biochemistry and molecular genetics of plant metabolism; functional genomics; metabolomics; acetyl-CoA and biotin metabolic networks; regulation and structure-function studies of biotin-containing enzymes.

Marit Nilsen-Hamilton, Professor of Biochemistry; Ph.D., Cornell, 1973. Regulation of gene expression by growth factors; role of growth factors and their regulated gene products; nucleic acid, aptamer-based analytical technology; imaging gene expression in vivo.

Reuben J. Peters, Assistant Professor; Ph.D., California, San Francisco, 1998. Enzymatic/metabolic mechanisms and engineering in plant terpenoid natural product biosynthesis.

Richard M. Robson, Professor of Biochemistry, Molecular Biology, and Animal Science; Ph.D., Iowa State, 1969. Biochemistry, molecular biology, structure, and function of muscle contractile and cytoskeletal proteins.

John F. Robyt, Professor of Biochemistry; Ph.D., Iowa State, 1962. Carbohydrate chemistry and enzymology; analytical methods; starch, dextran, and sucrose chemistry and enzymology.

Yeon-Kyun Shin, Professor of Biophysics; Ph.D., Cornell, 1990. SNARE complex assembly and mechanisms of membrane fusion; EPR methods.

Michael Shogren-Knaak, Assistant Professor of Biochemistry; Ph.D., Stanford, 1994. Role and establishment of histone modifications; chromatin structure.

Robert Thornburg, Professor of Biochemistry; Ph.D., South Carolina, 1981. Plant molecular biology; eukaryotic gene regulation and expression; plant response to insect attack.

Affiliates and Collaborators Primarily Associated with This Department

Arun B. Barua, Adjunct Professor and Scientist; Ph.D., Gauhati (India), 1965. Metabolism and function of vitamin A and carotenoids.

Martha G. James, Adjunct Associate Professor of Biochemistry; Ph.D., Iowa State, 1989. Plant molecular biology, molecular mechanisms of starch metabolism.

Terry Meyer, Faculty Collaborator and Research Manager, Plant Biochemistry and Molecular Biology at Pioneer Hi-Bred International; Ph.D., Iowa State, 1987. Purification and characterization of insecticidal proteins; cloning of the related genes and of insect receptor molecules; development of transgenic maize with genes for control of insect pests; plant gene expression.

Guru Rao, Collaborator and Associate Professor of Biochemistry, Research Fellow at Pioneer Hi-Bred International; Ph.D., Mysore (India), 1980. Protein engineering for nutritional enhancement: conformation and stability of an extensively-mutated plant protein.

Louisa B. Tabatabai, Professor of Biochemistry and Research Chemist, National Animal Disease Center, USDA/ARS; Ph.D., Iowa State, 1976. Structure and function of bacterial virulence factors; development of vaccines and diagnostic reagents; protein chemistry; proteomics.

Senior Staff Associated Primarily with Other Departments

W. Allen Miller, Professor of Plant Pathology and Biochemistry; Ph.D., Wisconsin–Madison, 1984. RNA virus replication and gene expression; translation mechanisms; barley yellow dwarf virus genomics.

The skylit atrium of the Molecular Biology Building provides natural light for its forty-eight research laboratories and eight instrumentation centers.

The 204,000-square-foot Molecular Biology Building is home to the Department of Biochemistry, Biophysics, and Molecular Biology.

LOYOLA UNIVERSITY CHICAGO

Stritch School of Medicine
Program in Molecular and Cellular Biochemistry

Programs of Study	Programs of study are offered leading to both the M.S. and Ph.D. degrees in the principal fields of cellular biochemistry, molecular biology, tumor biology, and neurochemistry. The Ph.D. program is designed to train students for a career of research and teaching, recognizing that a strong Ph.D. training program can also lead graduates to many interesting and challenging alternative careers. The master's program was created for students who wish to improve their qualifications for more advanced degrees such as the M.D. or Ph.D., or who are interested in obtaining an M.S. degree to meet other career objectives, such as teaching secondary school, working in industry, or obtaining a law degree.
	Each student works closely with a faculty mentor during the research training. Ph.D. candidates must pass a comprehensive qualifying exam that includes both a written test of critical thinking and problem solving skills as well as the preparation and review of a grant related to the proposed dissertation research. Master's students take a year of concentrated course work, then participate in a one-year focused research project that culminates in a departmental seminar and written thesis report.
	Students in the seven biomedical doctoral programs at Loyola University Chicago must take a common curriculum during the first semester. This includes courses in the areas of molecular biochemistry, cell biology, systems biology, and methods. After the first semester, students in the Graduate Program in Molecular and Cellular Biochemistry need to take additional courses in molecular biology, cellular biochemistry, structure and function of proteins, and biostatistics. Electives include neurochemistry, advanced eukaryotic genetics, molecular oncology, signal transduction, and cellular and molecular neurobiology. Additional elective courses are available.
	Major research is being conducted in the biochemistry of apoptosis and transcription factors, second messenger signaling systems, biochemistry of neurodegeneration (especially Parkinson's disease and AIDS), brain development, neurochemical mechanisms of fetal and adult alcohol toxicity, intracellular protein degradation, molecular biology of cancer and metastasis, cell-membrane biochemistry and cell-cell interactions, heat shock proteins, the biosynthesis and action of hormones, gene regulation, and inflammation.
Research Facilities	Major equipment and facilities include recording spectrophotometers and fluorometers, scintillation counters, preparative ultracentrifuges, high-performance liquid chromatographs, gas-liquid chromatographs, tissue-culture facilities, visible and fluorescent microplate readers, image analysis apparatus, networked microcomputers, and electrophoresis, hybridization, thermocycler, and real-time PCR instruments for molecular biology. Core equipment and facilities include a molecular biological core facility with microarray and proteomics capabilities, a fluorescence-activated cell sorting facility, a transgenic animal facility, an electron microscope and image analysis facility, an animal-care facility, and bioinformatics software. The Medical Center Library is a designated resource for the eight-state Greater Midwest Region Medical Library Network and the national Network of Libraries of Medicine, which link health-science libraries throughout the United States. Collaborative research opportunities exist with four research institutes on campus (Oncology, Cardiovascular, Neuroscience and Aging, and Shock/Trauma) as well as with the adjacent Hines Veterans Affairs Hospital and nearby Argonne National Laboratory.
Financial Aid	Graduate students admitted to the Ph.D. program normally receive a fellowship or research assistantship that provides, in addition to complete remission of tuition and fees, a highly competitive stipend.
Cost of Study	Tuition and fees were approximately $17,500 per year in 2005–06. This amount is waived for supported students.
Living and Housing Costs	The cost of housing near Loyola Medical Center in 2005–06 varied from approximately $600 to $800 per month for a one-bedroom apartment to approximately $800 to $1000 per month for a two-bedroom apartment.
Student Group	There are approximately 140 graduate students and 560 medical students at the Medical Center. The Division of Molecular and Cellular Biochemistry has about 17 graduate students from several states and countries; about half are women. There are also doctoral programs in anatomy, microbiology, molecular biology, neuroscience, pharmacology, and physiology at Loyola University Medical Center.
Student Outcomes	Most immediate graduates obtain postdoctoral positions at nationally ranked institutions, such as Harvard, Stanford, and Northwestern Universities; the University of Chicago; the NIH; Mayo Clinic; and others. Some students pursue traineeships in industry. Scientists graduating within the past fifteen years are pursuing careers as teaching–research faculty members at universities, research scientists in industry, or full-time college teachers. Graduates have begun their own technical companies or have obtained further professional training in medicine or law and are utilizing their scientific training as physicians or lawyers.
Location	Loyola Medical Center is located in the western Chicago suburbs, approximately 13 miles west of downtown Chicago. The art, natural history, and science museums as well as the aquarium, planetarium, and two zoos are within 30 minutes' travel time from the Medical Center. Chicago is also the home of the Chicago Symphony Orchestra, the Lyric Opera, and several prominent theater and dance groups. For sports enthusiasts, there are major-league baseball, basketball, football, hockey, and soccer teams in Chicago. Lakes, parks, and ski resorts are easily accessible to Chicagoans.
The University	Loyola University Chicago enjoys a heritage of more than 120 years as an independent urban university. The undergraduate school is located in Chicago. The Medical Center opened its facilities in suburban Maywood in 1968.
Applying	All applications need to be received online. For complete instructions, prospective students should visit https://app.applyyourself.com/?id=loyolamc. Entering graduate students should have completed courses in differential and integral calculus, organic chemistry, and physics and have a strong background in biology. Academic transcripts, GRE General Test scores, and three letters of recommendation, preferably from science teachers, should be submitted at the time of application. GRE Subject Tests are encouraged but not required. For students whose native language is not English, a score of at least 600 on the Test of English as a Foreign Language (TOEFL) is required for consideration for admission.
	To ensure consideration for financial aid, applicants are encouraged to complete their applications no later than March 15. The academic year begins in the fall semester, which starts approximately August 1.
Correspondence and Information	Dr. William Simmons Graduate Admissions Committee Division of Molecular and Cellular Biochemistry Department of Cell Biology, Neurobiology, and Anatomy Stritch School of Medicine Loyola University Chicago 2160 South First Avenue Phone: 708-216-3362 Fax: 708-216-8523 E-mail: egrzeda@lumc.edu Web site: http://www.meddean.luc.edu/lumen/deptwebs/biochem/

Loyola University Chicago

THE FACULTY AND THEIR RESEARCH

Guan Chen, Associate Professor; Ph.D., Heidelberg, 1990. Study of MAPK (ERK, JNK, and p38) signal transduction pathways.

Divaker Choubey, Associate Professor; Ph.D., Indian Institute of Science, 1985. Molecular mechanisms of the regulation of cell proliferation and apoptosis.

Michael A. Collins, Professor; Ph.D., Purdue, 1968. Brain neurodegeneration and/or neuroprotection mechanisms during alcohol intake and parkinsonism.

Mitchell F. Denning, Associate Professor; Ph.D., Wisconsin–Madison, 1991. Signal transduction pathways that regulate normal skin homeostasis; skin carcinogenesis.

Manuel O. Diaz, Professor; M.D., Uruguay, 1976. Expression of gene promoters and transcription factors in T-cell–senescence; role of telomerase and telomeric proteins in regulation of telomere length.

Andrew Dingwall, Associate Professor, Ph.D., Yeshiva (Einstein), 1989. Study of a highly conserved group of proteins that form a complex whose main function is to regulate gene expression through direct effects on chromatin structure.

Luisa DiPietro, Associate Professor; D.D.S., Ph.D., Illinois. Inflammation and angiogenesis in tissue repair.

Lydia DonCarlos, Professor; Ph.D., Kent State, 1985. Sexual differentiation of the brain; effects of gonadal steroid hormones on neural development and brain responses to injury.

Mary J. Druse-Manteuffel, Professor; Ph.D., North Carolina, 1972. Developmental neurochemistry; development of CNS neurotransmitter systems in the offspring of alcoholic rats; neurotrophic effects of serotonin.

Mary Ann Emanuele, Professor; M.D., Loyola Chicago, 1975. Effects of ethanol on rat hypothalamic, pituitary, and gonadal gene transcription and translation.

Allen Frankfater, Professor; Ph.D., Duke, 1968. Lysosomal proteinases, the regulation of their expression in metastatic tumors, and role in cancer cell invasion; role of the IGF-II receptor in lysosomal enzyme trafficking and tumor growth.

Jeffrey R. Kanofsky, Professor; Ph.D., IIT, 1972; M.D., Rush, 1975. Biochemistry of oxygen radicals and singlet oxygen.

Elizabeth J. Kovacs, Associate Professor; Ph.D., Vermont, 1984. Mechanisms by which age, gender, and alcohol exposure affect inflammation and cell-mediated immunity after injury.

Ruben Mestril, Associate Professor; Ph.D., Miami (Florida), 1986. Role of heat shock protein expression in cardiac and myogenic cells during ischemic stress.

Lucio Miele, M.D., Ph.D., Professor and Director, Breast Cancer Basic Research Program. Understanding Notch signaling in cancer apoptosis; developed several Notch-inhibitory agents, both biological and traditional pharmaceutics, which have remarkable anticancer activity in vitro and in vivo; developing clinical applications for the use of Notch inhibitors.

Gregory A. Mignery, Associate Professor; Ph.D., Texas A&M, 1987. Structural and functional characterization of the inositol 1,4,5-trisphosphate receptor protein family in cerebellar and cardiac cells; regulation and role of proteins that govern intracellular calcium signaling dynamics.

Erika Piedras-Renteria, Assistant Professor; Ph.D., Illinois, 1996. Molecular mechanisms of neuronal calcium function in normal and pathological conditions.

Richard M. Schultz, Professor; Ph.D., Brandeis, 1969. Mechanism of cancer-cell metastasis and the involvement of proteolytic enzymes in the metastatic process; oncogene pathways in induction of metastasis.

George J. Siegel, Professor; M.D., Miami, 1961. Expression and regulation of Na, K-ATPase in developing and aging brain, retina, and kidney; growth-factor effects in the nervous system; neurodegeneration and therapy of Alzheimer's disease.

William H. Simmons, Professor; Ph.D., Illinois at Chicago, 1979. Mechanisms of vasoactive peptide metabolism; purification and characterization of peptidases; aminopeptidase P inhibitors as cardiovascular drugs.

Evan Stubbs, Associate Professor; Ph.D., Missouri, 1987. Immune-mediated mechanisms of peripheral neuropathies.

Pamela Witte, Professor; Ph.D., Texas, 1984. B lymphocytes; stromal cells; hematopoiesis; differentiation; aging.

Nancy Zeleznik-Le, Associate Professor; Ph.D., Duke, 1988. Characterization of the differences in function of the MLL chimeric fusion protein with the wild-type MLL protein in both in vitro and in vivo models of MLL-CBP leukemia.

MAYO GRADUATE SCHOOL

Biochemistry and Structural Biology Program
Cell Biology and Genetics Program

Programs of Study

The Department of Biochemistry and Molecular Biology (BMB) in the Mayo Graduate School offers graduate training leading to the Ph.D. Students select a specialization in either biochemistry and structural biology or cell biology and genetics. Both programs require the completion of 42 Mayo Graduate School credits, including 16 credits from the core curriculum.

The Biochemistry and Structural Biology (BSB) Program integrates researchers from across Mayo's diverse scientific strengths and provides an exceptional educational experience. The program gives broad exposure to the fundamental principles of biochemistry, molecular and cell biology, biophysics, and molecular genetics. It offers intermediate and advanced courses emphasizing physical biochemistry, macromolecular structure analysis, and macromolecular engineering. The curriculum includes both fundamental and applied studies of biochemistry and structural biology. Sixteen dedicated faculty members support the program as teachers and mentors. Some current areas of research are macromolecule structure-function interactions and rational drug design; NMR and other spectroscopies applied to macromolecules; biomathematics, modeling of complex phenomena, chaos theory, and kinetics; biochemistry of ion transport; structural biology of nucleic acids, their complexes, and transcription; molecular endocrinology and biochemistry of gene regulation; protein misfolding; and proteomics technology.

The Cell Biology and Genetics (CBG) Program reflects the educational efforts of more than 30 faculty members representing broad research interests in molecular biology, cell biology, and genetics. Students in the CBG Program receive a comprehensive introduction to key topics in molecular research. Intermediate and advanced courses follow, emphasizing topics related to cell function and architecture, the genetics of model organisms, and the biology of membranes. The curriculum includes both fundamental and applied studies of topics in cells, model organisms, and human biology. Some current areas of research are signal transduction from receptor to nucleus; DNA repair, replication, and checkpoint control; calcium transport, signaling, and calcium-regulated cell functions; the biology and regulation of oncogenes; vesicular transport and organelle assembly; lipid and membrane function; and developmental biology.

Laboratory rotations are an important part of the first year of graduate study. Through lab rotations, students try different types of research, experience different labs and mentors, and choose their area of specialization. All students complete at least three 2- to 3-month lab rotations. The first rotation may be done during the summer before classes start. While Mayo Graduate School places primary emphasis on research training, classroom experience is also an integral component of the Ph.D. program, providing the intellectual foundation necessary for well-rounded scientists. Therefore, each area of specialization offers advanced courses built on the framework set by the core curriculum. As with all Mayo Graduate School programs, students in the BSB and CBG Programs are free to select a qualified research mentor from any Mayo program or department. Although each Mayo investigator has an independent laboratory composed of postdoctoral fellows, graduate students, and technicians, active collaborations with other research groups within the institution are common. This provides for a variety of mentorship opportunities in addition to a student's thesis adviser. Shared core facility resources and cooperation among research groups allow students to follow their research, whatever direction it takes.

Research Facilities

Modern biomedical research requires teamwork and highly sophisticated, expensive technologies. Therefore, collaboration and shared resources are central to Mayo's extensive research activities. An annual budget exceeding $296 million, with more than half coming from NIH and other extramural sources, supports a wide range of research programs. The Department of Biochemistry and Molecular Biology occupies laboratories on three floors (60,000 square feet). In addition to well-equipped individual laboratories, the department operates research core facilities with the most modern instrumentation for automated nucleic acid and peptide synthesis and sequencing, nuclear magnetic resonance, mass spectroscopy, fluorescence-activated cell sorting, and electron microscopy. Each core facility is supervised by research faculty members and staffed by highly skilled technical experts. Students are encouraged to use these resources as their research dictates, from occasionally submitting samples for analysis to full training in the instrumentation. Research is also supported by the departmental computer facility, which provides access to a variety of software used for molecular modeling and data analysis. In addition, Mayo Medical Library contains 413,000 volumes and subscribes to 4,300 medical and scientific journals.

Financial Aid

Mayo Graduate School provides stipends and full-tuition scholarships for all of its Ph.D. students. Therefore, the students have exceptional flexibility in choosing research mentors. Mayo Graduate School offers outstanding student benefits. Students appointed to Ph.D. program receive a stipend ($22,250 per year in 2004–05), full tuition scholarships, flexible vacation time, and paid travel to scientific meetings.

Cost of Study

All costs, except those for books, are provided in addition to the yearly stipend. Students pay no tuition or ancillary fees.

Living and Housing Costs

Living costs in Rochester are comparable to those in cities of similar size within the upper Midwest and generally lower than those in urban areas of the East and West Coasts. A single student can live comfortably on the stipend provided.

Student Group

The Department of Biochemistry and Molecular Biology's graduate program is specifically organized to educate a small, select group of students. An average ratio of 2 students per faculty member provides each student with an unusual opportunity for close collaboration with internationally recognized researchers.

Location

The city of Rochester combines the best of two worlds—the warmth and friendliness of a small town with the bustling commerce, entertainment, and conveniences of a metropolis. Lectures, symphony concerts, art exhibits, and a civic theater contribute to a cosmopolitan atmosphere, unusual for a city its size. The city and its environs provide an extensive, four-season calendar of recreation. Attractions in Minneapolis, a major metropolitan area, and St. Paul, just 80 miles north of Rochester, include professional sports, the nationally acclaimed Tyrone Guthrie Theatre, and the Minnesota Institute of Arts.

The Graduate School

The Mayo Graduate School is a division of Mayo Foundation, which also includes Mayo Clinic, Mayo Medical School, Mayo School of Health Sciences, and affiliated hospitals. Mayo Foundation is accredited by the Higher Learning Commission of the North Central Association of Colleges and Schools.

Mayo Graduate School, with campuses in Jacksonville, Florida; Rochester, Minnesota; and Scottsdale, Arizona, has a highly select student body. The 130 Ph.D. and M.D./Ph.D. candidates have easy access to more than 900 research and clinical faculty members. This favorable student-faculty ratio gives students the opportunity for close collaboration with internationally recognized researchers who serve as thesis advisers. Smaller class sizes also allow the programs to adapt easily to students' individual backgrounds and interests.

Applying

Admission to the graduate programs is competitive. Applicants are required to submit official transcripts from each college and university attended, three letters of recommendation, a summary of scientific interests and career goals, and scores from a recent GRE General Test. Qualified applicants are encouraged to schedule a personal interview. Applications for the Ph.D. program should be submitted by December 15. Mayo Foundation is an affirmative action and equal opportunity educator and employer.

Correspondence and Information

Department of Biochemistry and Molecular Biology
Mayo Graduate School
200 First Street, SW
Rochester, Minnesota 55905
Phone: 507-538-1160
Web site: http://www.mayo.edu/mgs/index.html

Mayo Graduate School

THE FACULTY AND THEIR RESEARCH

Zeljko Bajzer, Ph.D., Associate Professor. Mathematical modeling and data analysis, with special emphasis on nonlinear phenomena.

Mark E. Bolander, M.D., Professor. Roles of TGFβ's and FGF's; molecular and cellular events responsible for bone resorption; evaluation of cellular mechanisms by which musculoskeletal tissues respond to mechanical stress.

Thomas P. Burghardt, Ph.D., Professor. Dynamics and structure of contractile proteins.

Roberto B. Cattaneo, Ph.D., Professor. Virus assembly and cell entry; targeting and cytoreductive therapy; pathology and vaccines.

Junjie Chen, Ph.D., Assistant Professor. Understanding tumor suppressor function of BRCA1/BRCA2; elucidating mammalian DNA damage-signaling pathway and its role in tumorigenesis; mechanisms of genomic instability in mammals.

Fergus J. Couch, Ph.D., Assistant Professor. Characterization of the 17q23 amplicon in breast tumors; identification of mediators of cisplatin and taxol resistance in ovarian tumors; structure-function studies of the BRCA2 breast cancer predisposition gene; environmental risk factors influencing breast cancer risk in BRCA1 and BRCA2 mutation carriers.

Norman Eberhardt, Ph.D., Professor. Regulation of gene expression; amyloidogenesis and cell death; follicular thyroid cancer.

Mark J. Federspiel, Ph.D., Assistant Professor. Retroviral vectors; antiviral strategies; molecular medicine.

Joseph P. Grande, M.D., Ph.D., Professor. Signaling pathways in renal fibrogenesis.

Peter C. Harris, Ph.D., Professor. Molecular genetics; polycystic kidney disease research.

Bruce F. Horazdovsky, Ph.D., Associate Professor. Growth factor receptor trafficking in cancer and neurodegenerative disease.

Grazia Isaya, M.D., Ph.D., Associate Professor. Mitochondrial biogenesis; iron homeostasis, oxidative damage, and Friedreich ataxia.

Ralf G. Janknecht, Ph.D., Assistant Professor. Regulation of gene transcription by ETS proteins; intracellular signaling pathways; cancer.

Robert B. Jenkins, M.D., Ph.D., Professor. Genetics of gliomas; prostate cancer genetics; breast cancer genetics.

William E. Karnes Jr., M.D., Associate Professor. Proliferative and apoptotic signaling pathways involving the EGF receptor, protein kinase C, and beta-catenin.

David J. Katzmann, Ph.D., Assistant Professor. Receptor down-regulation; multivesicular body formation.

Rajiv Kumar, M.D., Professor. Structure and function of vitamin D–dependent calcium-binding proteins; vitamin D receptor; metabolism and mechanism of action of 1,25-dihydroxyvitamin D.

Nicholas F. LaRusso, M.D., Professor. Concepts/technologies of cell and molecular biology in understanding hepatic epithelial function (hepatocytes, cholangiocytes) and when disease alters their normal physiology.

Edward B. Leof, Ph.D., Professor. Pathway(s) downstream of transforming growth factor beta (TGFβ) receptor binding; role of cell division cycle (cdc) genes in *Pneumocystis carinii* infection.

Andrew H. Limper, M.D., Professor. Pathogenesis of *Pneumocystis carinii* pneumonia and other infections of the immune-compromised host.

Slobodan I. Macura, Ph.D., Professor. Determination of 3-D structure of polypeptides and small proteins in solution; techniques for separation of structural and dynamical parameters and structure refinement; NMR study of internal mobility of proteins.

L. James Maher III, Ph.D., Professor. DNA flexibility and bending; artificial control of gene expression.

Daniel J. McCormick, Ph.D., Professor. Structure-function relationships of proteins; interaction of protein ligands to receptors on the cell surface.

Cynthia T. McMurray, Ph.D., Professor. Huntington's disease; schizophrenia; receptor-mediated gene expression; trinucleotide expansion; DNA-protein interactions.

Mark A. McNiven, Ph.D., Professor. Regulation of growth factor internalization by endocytosis; molecular mechanisms of tumor cell migration and metastasis; cytoskeletal organization and alterations during transformation.

Georges Mer, Ph.D., Assistant Professor. 3-D structures of biological macromolecules by NMR spectroscopy.

Laurence J. Miller, M.D., Professor. G-protein–coupled GI hormone receptor structure, function, and regulation.

Debabrata Mukhopadhyay, Ph.D., Professor. Tumor angiogenesis and vascular biology, nanotechnology, and nanoscience.

Whyte G. Owen, Ph.D., Professor. Thrombin specificity in platelet signaling; enzymology of mouse thrombin/platelets; membrane-binding structure of prothrombin fragment 1; marked platelets in the presymptomatic diagnosis of atherosclerosis.

Richard E. Pagano, Ph.D., Professor. Synthesis, transport, and function of lipid molecules in cells.

Robin Patel, M.D., Associate Professor. Microbial biofilms, mechanism, diagnostics, and therapeutics; antimicrobial resistance mechanisms.

Larry R. Pease, Ph.D., Professor. Antigen recognition by T cells; immune potentiation of T-cell immunity/repertoire; antitumor and antivirus responses; immunology of multiple sclerosis; antibodies as receptor ligands.

Joseph F. Poduslo, Ph.D., Professor. Delivery of therapeutic/diagnostic proteins across blood-brain barrier; targeting Alzheimer's plaques for MRI; passive immunization in Alzheimer's; antioxidant enzyme therapy for ALS.

Franklyn G. Prendergast, M.D., Ph.D., Professor. Protein structure and dynamics; biochemistry and bioluminescence.

Marina Ramirez-Alvarado, Ph.D., Assistant Professor. Mechanism of protein folding and misfolding associated with amyloid disease.

John R. Riordan, Ph.D., Professor. ABC proteins; ion channels; cystic fibrosis.

Jeffrey L. Salisbury, Ph.D., Professor. Cell-cycle control; centrosomes; mitotic spindle poles; breast cancer.

Robert D. Simari, M.D., Associate Professor. Identifying novel mechanisms of vascular disease; developing novel therapeutic strategies; test vectors in relevant animal models; designing clinical gene transfer studies.

David I. Smith, Ph.D., Professor. Genomic instability in cancer; role of the common fragile sites in cancer development; genetics of ovarian cancer.

Thomas C. Spelsberg, Ph.D., Professor. Tumor suppressor gene TIEG in TGFβ–Smad action pathway in bone/cancer cells; estrogen action on gene expression in bone/cancer cells: role of coregulators and ER isoforms; calcification of heart valves and roles of statins.

Emanuel E. Strehler, Ph.D., Professor. Calcium signaling and transport; calcium-binding proteins in cell differentiation and tumorigenesis.

Amy Tang, Ph.D., Assistant Professor. Influence of toxins, drugs, and proteolysis on cell signaling pathways; innate immunity and cellular defense.

Donald J. Tindall, Ph.D., Professor. Cellular pathways that control gene expression; mechanism of androgen action; prostate cancer progression.

David O. Toft, Ph.D., Professor. Action of molecular chaperones; protein folding and processing; steroid receptors and hormone action.

Raul A. Urrutia, M.D., Associate Professor. Molecular mechanisms of transcriptional regulation; cell differentiation.

Jan van Deursen, Ph.D., Associate Professor. Molecular genetic basis of cancer and aging; mechanisms that regulate development of cancer and aging-related disorders, by combining in vitro biochemistry approaches with genetic experiments in mice.

Stanimir Vuk-Pavlovic, Ph.D., Professor. Cell-cell interactions evolve and can be manipulated by biological response modifiers, including analogs of the naturally occurring hormone somatostatin.

Cynthia Wetmore, M.D., Ph.D., Assistant Professor. Molecular control of normal and neoplastic proliferation in the nervous system; role of sonic hedgehog signaling in proliferation of neural precursors; repair of DNA damage in the nervous system.

Eric D. Wieben, Ph.D., Professor. Regulation of gene expression; pharmacogenomics; genomics technology.

Xiaolei Xu, Ph.D., Assistant Professor. Zebra fish as a genetic model for cardiovascular diseases.

Charles Y. Young, Ph.D., Professor. Understanding fundamental biology of prostate cancer; prostate cancer detection and prevention; development of vaccines.

Zhiguo Zhang, Ph.D., Assistant Professor. Epigenetic silencing in yeast, chromatin structure, and inheritance.

Mayo Clinic Scottsdale Faculty

Xiu-Bao Chang, Ph.D., Assistant Professor. ATP- and glutathione-dependent anticancer drug transport and selective killing of MRP1-over-expressing cancer cells.

Sandra J. Gendler, Ph.D., Professor. Tumor cell biology; mucins in cancer and cystic fibrosis Immunotherapy.

James J. Lee, Ph.D., Associate Professor. Molecular mechanisms regulating cell determination during mammalian hematopoiesis and embryonic development.

Nancy A. Lee, Ph.D., Assistant Professor. Molecular mechanisms in the recruitment of pro-inflammatory cells to sites of chronic inflammation.

Joseph C. Loftus, Ph.D., Associate Professor. Structural basis of ligand recognition by integrins.

David F. Smith, Ph.D., Professor. Mechanisms and physiological functions of molecular chaperones; steroid hormone receptors; reproductive biology and cancer.

MICHIGAN STATE UNIVERSITY

Department of Biochemistry and Molecular Biology

Programs of Study

Ph.D. and M.S. degree programs are both offered, and postdoctoral study is encouraged and supported. The major objectives of the programs are to help students develop their creative potential and to prepare them for effective careers in research and teaching in the biochemical sciences. The Department of Biochemistry and Molecular Biology is an integral part of three colleges (Natural Science, Human Medicine, and Osteopathic Medicine), thus enabling students to select their dissertation research projects from an exceptionally wide range of topics in biochemistry and molecular biology. Opportunities are also available to graduate students in biochemistry for joint programs with neuroscience, biotechnology, cell biology, human medicine, environmental toxicology, genetics, osteopathic medicine, plant sciences, and various other biological and medical sciences.

Doctoral programs are planned on an individual basis by the student in consultation with his or her major professor and a guidance committee. Course-credit requirements vary according to individual backgrounds and interests. The heart of the program is the performance of substantial original and creative research, which forms the basis of the doctoral dissertation. All requirements for the Ph.D. degree are typically completed within four or five years after the baccalaureate; the intermediate attainment of an M.S. degree is not required.

Research Facilities

The six-floor Biochemistry Building at Michigan State University (MSU) provides one of the finest research and teaching facilities of its kind in the nation. It houses thirty-three superbly equipped research laboratories; a tissue-culture facility; an instrument shop; animal rooms; a facility containing eight mass spectrometers, including instruments for protein characterization and peptide sequencing; and a macromolecular structure facility containing an amino acid analyzer, two protein sequenators, nucleic acid synthesizers, and advanced HPLC and PCR equipment. Also available is automated equipment for all types of electrophoretic and chromatographic separations and for the use of radioactive isotopes. Other items include a laser-excited fluorescence spectrometer with subcellular resolution; a differential scanning calorimeter; digital image analyzers; minisupercomputers for computational chemistry and molecular modeling; automated centrifuges; various UV, IR, FT-IR, visible, and fluorescence spectrometers; and numerous other modern instruments.

Financial Aid

Departmental traineeships, assistantships, and fellowships are available to graduate students, with stipends currently ranging upward from $19,656 for twelve months. The Department also provides a tuition waiver of 6 to 9 credits per semester. In addition, students are encouraged to seek graduate fellowships from foundations and granting agencies such as the National Science Foundation and the National Institutes of Health. Fellowships paying less than Departmental assistantships are supplemented by the Department to an equivalent level.

The University and adjacent institutions and industries offer excellent employment opportunities for students' spouses, and incomes of married couples are often augmented in this way.

Cost of Study

Graduate assistants pay in-state tuition after 9 credits. Students usually register for 7 credits per semester in the first two semesters and for about 6 credits in succeeding semesters.

Living and Housing Costs

Single rooms in Owen Hall, the graduate residence center, rent for $2516 per semester. Food may be obtained from any of several campus cafeterias and local restaurants. Privately owned off-campus rooms and apartments are available, and the University owns and operates more than 2,460 one- and two-bedroom apartments to help meet the housing needs of married students. These rent for $552 and $612 per month, respectively, and include all utilities, essential furniture, and a private telephone. MSU housing rates are the lowest of Michigan universities and the third lowest among Big Ten schools.

Student Group

The on-campus enrollment at Michigan State University in fall 2005 was 44,836, including 9,428 graduate students. The Department of Biochemistry and Molecular Biology has 260 undergraduate majors and 68 graduate students. Fifty-one research associates are currently pursuing postdoctoral studies in the Department.

Location

East Lansing is a residential city adjacent to the Michigan State University campus and close to Lansing, the state capital. Graduate students and their families are offered many opportunities for cultural and social development by the University and by neighboring civic groups. The surrounding area has excellent facilities for camping, hiking, canoeing, sailing, salmon fishing, and skiing.

The University

Michigan State University was founded in 1855 as the pioneer land-grant college. It has since grown to be one of America's largest universities. Its parklike campus, which straddles the meandering Red Cedar River, is an arboretum of more than 5,000 acres containing 7,800 species of trees, shrubs, and vines. Through its fourteen colleges and eighty-nine departments, it offers about 200 programs leading to undergraduate and graduate degrees. Cultural facilities on campus include the Kresge Art Center, the University Museum, the Abrams Planetarium, and the Wharton Center for the Performing Arts.

Extensive recreational facilities maintained by the University include an ice arena, two 18-hole golf courses, forty-eight tennis courts and an indoor tennis facility, and several Olympic-size outdoor and indoor swimming pools.

Applying

Candidates with a bachelor's or master's degree in chemistry, biochemistry, biology, or other physical, biological, medical, or agricultural science are invited to apply. Undergraduate preparation should include courses in general, analytical, physical, and organic chemistry and in physics, general biology, basic biochemistry, and mathematics through calculus. Minor deficiencies in these areas may be rectified by taking appropriate undergraduate courses for collateral credit concurrently with graduate courses. Applicants are expected to take the Graduate Record Examinations, including the General Test (including verbal measure, analytical measure, writing measure, and either the mathematical reasoning measure or the quantitative reasoning measure) and a Subject Test in chemistry, biology, or biochemistry, cell and molecular biology. Students are normally admitted to the fall semester, but requests for entry in other semesters are considered. To be considered for a Departmental assistantship, applicants should submit completed applications for the fall semester as early as possible, preferably by February 1.

Correspondence and Information

Graduate Admissions
Department of Biochemistry and Molecular Biology
Michigan State University
East Lansing, Michigan 48824
Phone: 517-353-0807
Fax: 517-353-9334
E-mail: oesterju@msu.edu
Web site: http://www.bch.msu.edu

Michigan State University

THE FACULTY AND THEIR RESEARCH

D. N. Arnosti, Associate Professor; Ph.D., Berkeley. Eukaryotic gene regulation, transcriptional activation, and repression in *Drosophila*; function of retinoblastoma tumor suppressor genes in development.

C. Benning, Associate Professor; Ph.D., Michigan State. Biosynthesis and function of membrane lipids in photosynthetic organisms; regulation of lipid metabolism in developing oil seeds; molecular and biochemical genetics of *Arabidopsis* and photosynthetic bacteria.

Z. F. Burton, Professor; Ph.D., UCLA. Mechanism of elongation by human RNA polymerase II; regulation by transcription factor IIF and hepatitis delta antigen.

D. DellaPenna, Professor; Ph.D., California, Davis. Plant biochemistry, physiology, and genomics; synthesis and function of carotenoids, vitamin E, and other nutritional components in plants and photosynthetic bacteria.

D. L. DeWitt, Associate Professor; Ph.D., Michigan State. Molecular biology of lipid signaling; nuclear signaling; structural biology of membrane proteins.

J. L. Ekstrom, Assistant Professor; Ph.D., Cornell. Protein crystallography; structure and function of transcriptional regulatory complexes; protein disulfide isomerase; structure and redesign of engineered binding proteins; drug design.

M. Feig, Assistant Professor; Ph.D., Houston. Computational modeling of structure and dynamics of proteins and nucleic acids; development of implicit solvent methodologies; multiresolution modeling of supramolecular assemblies.

S. M. Ferguson-Miller, University Distinguished Professor and Chair; Ph.D., Wisconsin–Madison. Mechanism and regulation of energy transduction in mitochondria and bacteria; kinetics, mechanics, and protein chemistry of electron transfer and proton transfer in cytochrome *c* oxidase.

P. J. Fraker, Professor; Ph.D., Illinois at Urbana-Champaign. Immunology; interrelationships between nutritional status and immune function; modulation of lymphopoiesis by glucocorticoids; study of biochemical events leading to apoptosis or programmed cell death in cells of the immune system, using flow cytometry.

K. A. Gallo, Associate Professor; Ph.D., Harvard. Regulation of protein kinases and kinase partners involved in eukaryotic cell growth.

R. M. Garavito, Professor; Ph.D., Purdue. Structure of membrane proteins involved in membrane biogenesis and function; glycosyltransferases; protein crystallography.

R. P. Hausinger, Professor; Ph.D., Minnesota. Mechanism of enzyme action; structure-function studies of enzymes; incorporation of nickel in bacterial urease; biodegradation of sulfonate compounds; DNA repair; microbial physiology.

R. W. Henry, Assistant Professor; Ph.D., Alberta. Transcriptional regulation in eukaryotes; functions and architecture of basal transcription machinery; molecular determination of RNA polymerase specificity.

R. I. Hollingsworth, Professor; Ph.D., West Indies. Application of spectroscopic methods and computational chemistry to the study of the structure and function of glycoconjugates and other macromolecules; chemical synthesis of saccharides.

C. G. Hoogstraten, Assistant Professor; Ph.D., Wisconsin. Biophysical studies of RNA catalysis and RNA-protein interactions; macromolecular NMR and pulsed EPR methods and applications; mechanisms and regulation of eukaryotic mRNA splicing.

G. Howe, Associate Professor; Ph.D., UCLA. Regulation of plant defense responses; biochemistry of jasmonic acid biosynthesis; plant-insect interactions.

D. B. Jump, Professor; Ph.D., Georgetown. Molecular endocrinology; hormonal/nutrient regulation of gene transcription in liver and white adipose tissue.

J. M. Kaguni, Professor; Ph.D., UCLA. Mechanisms of DNA replication and its regulation in *Escherichia coli;* nucleic acid–protein interactions.

L. S. Kaguni, Professor; Ph.D., UCLA. Enzymology of mitochondrial DNA replication; organization and structure of mitochondrial DNA replication origins; DNA-protein interactions; regulation of DNA synthesis during *Drosophila* development.

K. Keegstra, Professor; Ph.D., Colorado. Biosynthesis of plant cell wall polysaccharides; utilization of genomic techniques to identify and characterize genes required for xyloglucan biosynthesis in *Arabidopsis*.

L. Kroos, Professor and Associate Chair; Ph.D., Stanford. Regulation of gene expression during development; prokaryotic RNA polymerases and regulatory proteins; *Bacillus* sporulation; cell-cell interactions during *Myxococcus* fruiting body formation.

L. A. Kuhn, Professor and Co-Director, Quantitative Biology and Modeling Initiative; Ph.D., Pennsylvania. Developing computational approaches for protein folding, flexibility, docking, and inhibitor design; integrating computational and experimental approaches.

M.-H. Kuo, Assistant Professor; Ph.D., Rochester. Modifications and dynamics of chromatin; transcriptional regulation; proteomic interactions involving posttranslational modifications; tumor suppressor protein p53.

J. J. LaPres, Assistant Professor; Ph.D., Northwestern. Toxicology of transition metals and environmental pollutants; development of genomic and high-throughput screens.

R. Larkin, Assistant Professor; Ph.D., Missouri. Plastid-to-nucleus signaling pathways that coordinate nuclear transcription with plastid development and metabolism.

R. Last, Professor; Ph.D., Carnegie Mellon. *Arabidopsis* functional genomics; regulation of plant metabolic pathways for nutritionally important molecules; plant stress tolerance mechanisms; metabolic engineering of plants.

V. M. Maher, Professor; Ph.D., Wisconsin–Madison. Molecular and cellular mechanisms of mutagenesis and of recombination in human cells; DNA repair; chemical- and radiation-induced mutagenesis and carcinogenesis.

J. J. McCormick, Professor; Ph.D., Catholic University. Molecular mechanisms of malignant transformation in human cells; chemical- and radiation-induced oncogene activation, carcinogenesis, and mutagenesis; DNA repair.

E. McGroarty, Professor; Ph.D., Purdue. Interactions of surface-active compounds with cell membranes; structure, composition, and physical properties of bacterial membranes.

B. Montgomery-Kaguri, Assistant Professor; Ph.D., California, Davis. Molecular basis of organ-specific phytochrome responses in higher plants; light-regulated development in cyanobacteria.

J. Preiss, University Distinguished Professor; Ph.D., Duke. Genetic and allosteric regulation of bacterial glycogen and plant starch biosynthetic enzymes and their expression, structure, and function.

G. E. Reid, Assistant Professor; Ph.D., Melbourne. Bioanalytical mass spectrometry; development and application of novel chemical methods and instrumentation for qualitative and quantitative proteomic analysis.

M. S. Schindler, Professor; Ph.D., Weizmann (Israel). Membrane biochemistry and biophysics; fluorescence imaging; cellular micromanipulation with laser tweezers; multidrug resistance in cancer therapy; cytoskeletal organization and function.

S. J. Triezenberg, Professor; Ph.D., Michigan. Mechanisms of gene regulation in eukaryotes; transcriptional activation; herpes simplex virus; viral and cellular regulatory proteins; transcriptional adaptors and histone acetylation in plants.

K. D. Walker, Assistant Professor; Ph.D., Washington (Seattle). Elucidation of biosynthetic pathways of plant-derived products for use (and potential use) as pharmaceuticals, neutraceuticals, and allelochemicals, utilizing an array of methodologies grounded in bioorganic chemistry, organic synthesis, enzymology, molecular genetics, molecular biology, and natural products isolation and characterization (i.e., metabolomics).

J. L. Wang, Professor; Ph.D., Rockefeller. Carbohydrate recognition; analysis of carbohydrate-binding proteins in the cell nucleus; nuclear processing and export of mRNA.

J. T. Watson, Professor; Ph.D., MIT. Gas chromatography–mass spectrometry; peptide structure and posttranslational modification of proteins by electrospray and matrix-assisted laser desorption mass spectrometry.

W. J. Wedemeyer, Assistant Professor; Ph.D., Cornell. Modeling of proteins de novo and from low-resolution data; structural studies of HIV envelope proteins and the mechanism of HIV cell entry; experimental studies of protein folding; peptide/protein design.

H. Yan, Professor; Ph.D., Ohio State. Protein structure, function, and dynamics; protein engineering; multidimensional multinuclear NMR; 6-hydroxymethyl-7, 8-dihydropteroate pyrophosphokinase; shikimate kinase; cytosine deaminase.

T. R. Zacharewski, Associate Professor; Ph.D., Texas A&M. Toxicogenomics; receptor-mediated toxicology; endocrine disruption; expression profiling; ligand-induced protein structure.

J. G. Zeikus, Professor; Ph.D., Indiana. Microbial biochemistry; thermozyme structure and function; bioelectrochemistry; biotechnology and bioengineering.

NORTH CAROLINA STATE UNIVERSITY

Department of Molecular and Structural Biochemistry

Programs of Study

The Department of Molecular and Structural Biochemistry offers a program of graduate study and research leading to the Ph.D. and M.S. degrees. The graduate program is designed to prepare Ph.D. scientists for careers in research and teaching. Emphasis is focused upon laboratory work where students work closely with faculty members to accomplish their thesis research. Prospective students are expected to have sound training at the undergraduate level in chemistry, physics, mathematics, biology, and molecular biology/genetics, but deficiencies can be remedied with appropriate preparatory course work. Students spend their first year taking the biochemistry graduate core courses and participating in a laboratory rotation program, all leading to selection of a faculty member as Ph.D. research adviser by the end of their first year. In conjunction with their thesis advisory committee, students then design a program of advanced graduate study and research.

The Department of Molecular and Structural Biochemistry offers a variety of areas for thesis research. The Department has established research interest groups that share common areas of research focus in molecular and/or structural biology. Laboratories with common research interests include the structural biology of proteins, the structural biology of nucleic acids/nucleic acid–protein complexes, RNA biology, animal and plant virology, plant molecular biology, and the regulation of gene expression at both the transcriptional and translational levels. Techniques used in these investigations include modern molecular biology techniques; analysis of gene expression including microarray analysis, biochemical fractionation, and analysis of proteins and nucleic acids using affinity, electrophoretic, and HPLC chromatographic techniques; and investigation of macromolecular structure using high-field NMR spectroscopy, X-ray crystallography, circular dichroism, fluorometry, and mass spectrometry.

Research Facilities

Research facilities of the Department of Molecular and Structural Biochemistry are centrally located on campus in Polk Hall, near the University library. The Department is well equipped to carry out research in biochemistry, biophysics, molecular biology, and molecular genetics.

Facilities and equipment are available for culturing animal, plant, and bacterial cells and viruses. Research animals are housed in the University's modern Biological Resources Facility. All laboratories are well equipped with standard instruments for molecular and structural biology research. Specialized equipment in the Department includes 500- and 600-MHz NMR spectrometers (Triangle access to an 800-MHz instrument), X-ray generators and detectors, CD spectrometer, fluorometer, mass spectrometers (magnetic sector and ion trap), isothermal titration calorimeter, and Silicon Graphics workstations for macromolecular modeling. On-campus genomics and proteomics facilities are staffed with full-time personnel. Departmental computer facilities provide direct access to online scientific journals in the University library and also access to the Triangle Universities' Cray supercomputer.

Financial Aid

Financial support in the form of research or teaching assistantships is awarded to all Ph.D. students accepted into the graduate program. The twelve-month stipend for the 2004–05 academic year was $18,000. Several institutional training grant opportunities are available, and accepted students are encouraged to apply for national predoctoral fellowships. The Research Triangle area also offers excellent employment opportunities for graduate student spouses.

Cost of Study

Tuition and fees for 2005–06 were $2428.50 per semester for North Carolina residents and $8482.50 per semester for out-of-state students. Tuition, fees, and health insurance are provided for full-time Ph.D. students in good academic standing.

Living and Housing Costs

University housing facilities are available for both single and married graduate students. Most graduate students prefer off-campus housing, which is readily available. The cost of living in the Triangle area is moderate, and rents typically range from $500 to $800 per month for one- and two-bedroom apartments. Public and university transportation serves many apartment complexes located near the University.

Student Group

There are currently more than 29,000 students enrolled at North Carolina State University (NCSU) from across the United States and many other countries. This total includes approximately 5,600 graduate students. The Department of Molecular and Structural Biochemistry has approximately 300 undergraduate majors and a steady enrollment of between 40 and 45 graduate students, most of whom are enrolled in the Ph.D. program. The Department's research environment is enriched with numerous undergraduate, staff, and postdoctoral investigators.

Location

North Carolina State University is located in Raleigh, the capital of North Carolina, and forms one corner of the Research Triangle in conjunction with Duke University and the University of North Carolina at Chapel Hill. Raleigh has a population of approximately 250,000, and the Triangle area provides numerous cultural activities and opportunities. NCSU is within easy driving distance of the North Carolina coast to the east and the Great Smoky Mountains to the west.

The University and The Department

North Carolina State University, a land-grant institution established in 1887, is one of two Research I Universities in the University of North Carolina System. The Department of Molecular and Structural Biochemistry is jointly administered by the College of Agriculture and Life Sciences and the College of Physical and Mathematical Sciences. Its 15 faculty members are committed to maintaining modern, well-funded, productive research programs and fostering collaborative research efforts with other research groups both on and off campus.

Applying

Candidates with bachelor's degrees in chemistry, biochemistry, biology, or other physical or biological science are encouraged to apply. Students seeking fall admission should submit their applications no later than January 15. The completed application also requires a copy of undergraduate transcripts, a statement of research/professional interests, three letters of recommendation, and scores of the GRE General Test. The Subject Test in biochemistry, molecular and cell biology is not required but can be useful in evaluating applicants when personal interviews are not possible. International students are required to submit TOEFL scores. An online application process is available through the NCSU Graduate School. Interested students are encouraged to make initial contact with the Department via the information request form found at the Department's Web site under the "Apply to the Graduate Program" heading.

Correspondence and Information

Dr. E. Stuart Maxwell, Director of Graduate Programs
Department of Molecular and Structural Biochemistry
North Carolina State University
P.O. Box 7622
Raleigh, North Carolina 27695-7622
Phone: 919-515-2581
Fax: 919-515-2047
E-mail: pamela_francis@ncsu.edu
Web site: http://biochem.ncsu.edu

North Carolina State University

THE FACULTY AND THEIR RESEARCH

Paul F. Agris, Professor; Ph.D., MIT. Contributions of nucleic acid chemistry and structure to function in gene expression with application to biomedical sciences; revision of the rules governing decoding of genomic information.
  The role of modifications in codon recognition and discrimination: tRNA$^{Lys}_{UUU}$. *Nature Struct. Biol.* 11:1186–91, 2004.

Dennis Brown, Professor and Department Head; Ph.D., Pennsylvania. Structure, assembly, and function of virus membranes.
  Sindbis virus produced by a mutation in the capsid protein. *J. Virol.* 307:54–66, 2003.

John Cavanagh, Professor; Ph.D., Cambridge. NMR and mass spectrometry analysis of protein structure; inherent flexibility of proteins involved in bacterial protection; protein structure-function relationships in biology.
  The DNA-binding domain in the *Bacillus subtilis* transition-state regulator AbrB employs significant motion for promiscuous DNA recognition. *J. Mol. Biol.* 305:429–43, 2001.

A. Clay Clark, Assistant Professor; Ph.D., Texas A&M. Protein folding, assembly, and maturation of caspases; mechanisms of molecular chaperones in protein folding; folding of six-helix bundle proteins with Greek key topology.
  Equilibrium and kinetic folding of an alpha-helical Greek key protein domain: Caspase recruitment domain (card) of RICK. *Biochemistry* 42:6310–20, 2003.

Michael B. Goshe, Assistant Professor; Ph.D., Case Western Reserve. Biological mass spectrometry applications and method development for proteomics; ascertaining and quantifying sites of protein modifications; characterizing protein-protein, protein-nucleic acid, and protein-ligand interactions.
  Stable isotope-coded proteomic mass spectrometry. *Curr. Opin. Biotechnol.* 14:101–9, 2003.

Linda Hanley-Bowdoin, Professor; Ph.D., Rockefeller. Plant DNA virus; geminivirus; cell division cycle and DNA replication; plant development; protein-protein interactions; gene expression; cell biology; genomics.
  A conserved motif in the geminivirus replication protein interacts with the plant retinoblastoma homolog RBR. *J. Virol.* 78:4817–26, 2004 (with Arguello-Astorga, et al.).

Charles C. Hardin, Associate Professor; Ph.D., Iowa State. Nucleic acid structural energetics.
  Cloning, Gene Expression and Protein Purification: Experimental Procedures and Process Rationale. Oxford University Press, 2001.

Cynthia L. Hemenway, Professor; Ph.D., Columbia. Molecular plant virology; viral replication; RNA replication; protein-RNA interactions; RNA structure.
  Cis-acting regulatory elements in the potato virus X 3' non-translated region differentially affect minus-strand and plus-strand RNA accumulation. *J. Mol. Biol.* 326:701–20, 2003.

James A. Knopp, Associate Professor and Coordinator of Undergraduate Academic Programs; Ph.D., Illinois. Plant hypersensitive response.
  Involvement of nitric oxide in *Ralstonia solanacearum*–induced hypersensitive reaction in tobacco. *Proc. Sec. Int. Bacterial Wilt Symp.*, 1998.

Carla Mattos, Assistant Professor; Ph.D., MIT. Structural analysis of the complete binding surfaces of small monomeric GTPases.
  Crystal structures of Ral-GppNHp and Ral-GDP reveal two binding sites that are also present in *Ras* and *Rap. Structure* 12:2025–36, 2004 (with Nicely, Kosak, and de Serrano).

E. Stuart Maxwell, Professor and Director of Graduate Programs; Ph.D., Massachusetts. Synthesis, structure, and function of the small nucleolar RNAs (snoRNAs); eukaryotic snoRNP/archaeal sRNP structure and function; RNA-guided nucleotide modification and ribosome biogenesis.
  Evolutionary origins of the nucleotide modification guide RNAs: The primordial ribosome? *Trends Biochem. Sci.* 29:343–50, 2004 (with Brown and Tran).

William L. Miller, Professor; Ph.D., Cornell. Regulation of follicle-stimulating hormone (FSH) by transforming growth factor family members; gonadotropin-releasing hormone and steroids.
  The nature of GnRH induction of FSH. *TEMS* 13:257–62, 2002.

Robert Rose, Assistant Professor; Ph.D., California, San Francisco. Protein crystallography; transcriptional regulation.
  Structural basis of dimerization, coactivator recognition, and MODY3 mutations in HNF-1alpha. *Nat. Struct. Biol.* 7:744–8, 2000.

Edward C. Sisler, Professor; Ph.D., North Carolina State. Basic research on the actions of the plant hormone ethylene.
  1-Substituted cyclopropenes: Effective blocking agents for ethylene action in plants. *Plant Growth Regul.* 40:223–8, 2003.

Paul Wollenzien, Professor; Ph.D., Berkeley. Ribosome structure and function; RNA conformational dynamics; molecular modeling; protein synthesis; allosterism.
  Efficiency and pattern of UV pulse laser induced RNA-RNA crosslinking in the ribosome. *Nucl. Acids Res.* 32:1518–26, 2004.

Associate Faculty

Stephan Franzen, Associate Professor, Biophysical and Biological Chemistry, North Carolina State. RNA structure and enzymatic reaction mechanisms.

Hosni M. Hassan, Professor of Microbiology and Department Head, North Carolina State. Molecular biology and gene regulation of metalloenzymes and antioxidant enzymes; oxidative stress.

Jonathan Horowitz, Ph.D., Associate Professor, College of Veterinary Medicine, Department of Biomedical Sciences, North Carolina State. Role of retinoblastoma (Rb) protein in control of mammalian cell proliferation and differentiation.

James W. Moyer, Professor of Plant Pathology and Department Head, North Carolina State. Biological diversity and elucidation of the molecular determinants of the *Tospovirus* genus of Bunyaviridae.

Ronald E. Sederoff, Professor of Forestry, North Carolina State. Molecular genetics and genetic engineering in forestry.

Adjunct Faculty

Kenneth Korack, Scientific Program Director for Environmental Disease and Medical Program and the Chief of Laboratory for Reproductive Disease Testing at the National Institute of Environmental Health Sciences, Research Triangle Park, North Carolina.

James D. Otvos, Director of Scientific Research at LipoMed, Raleigh, North Carolina.

Hanna Gracz, Research Assistant Professor and Laboratory Supervisor.

David Presutti, Teaching Assistant Professor and Director of Teaching Laboratories, North Carolina State.

Avis Sylvia, Lecturer, North Carolina State.

OKLAHOMA STATE UNIVERSITY

Department of Biochemistry and Molecular Biology

Program of Study

The Department of Biochemistry and Molecular Biology offers a program of graduate study leading to the M.S. and Ph.D. degrees. The training is designed for relevance in the twenty-first century. The requirements for the M.S. degree in biochemistry are 24 semester hours of course work, 6 semester hours of research, an acceptable research thesis, and defense of the thesis in a final oral examination. A nonthesis M.S. degree program requiring 32 semester hours of course work is also an option. The requirements for the Ph.D. degree include approximately 30 to 35 semester hours of course work, determined by the student's advisory committee, and 55 to 60 semester hours of thesis research. Additional requirements are participation in the department's seminar program, satisfactory performance on cumulative and qualifying examinations, one year of a foreign language at the undergraduate level, presentation of an acceptable thesis, and successful defense of the thesis in a final oral examination. Candidates for the Ph.D. degree are not required to obtain the M.S. degree.

Research Facilities

The Department of Biochemistry and Molecular Biology is located in the Jess L. and Miriam B. Stevens wing of the Noble Research Center. It has approximately 41,000 square feet of teaching and research space. The department is well equipped for modern biochemical and biophysical research and cooperates closely with the Departments of Chemistry, Physics, Entomology and Plant Pathology, Botany, Microbiology and Molecular Genetics, and Plant and Soil Science. The University Library, located one building from the Noble Research Center, has excellent holdings in chemistry, biochemistry, and related areas. The library has open stacks and is accessible 360 days a year, and many journals are available online. A computer network links the department to other campus departments, the University's mainframe, national computer networks and databases, and the Internet. Many computers are available to students. A state-of-the-art research facility is equipped with special instruments, including VG Instruments Model ZAB-2SE high-resolution, high-resolution NMR spectrometers with pulse FT capabilities (Unity Innova 400 and 600), and a CMX300 solid-state NMR.

The Department houses a recombinant DNA/protein resource facility that provides access to the latest technologies in molecular biology. The facility contains a variety of specialized instrumentation, including a 3700 automated DNA sequencing system, an ABI 491 (Procise) automated protein/peptide sequencing system, an ABI DE Pro MALDI-TOF mass spectrometry proteomics workstation, a Beckman-Coulter Biomek solution-handling robot, an ABI 7500 RT-PCR for quantitating gene expression, and a variety of small instruments, such as thermocyclers. State-of-the-art computer facilities for DNA and protein sequence analysis and database searching, digital imaging analysis (Bio-Rad phosphorimager, Gel Doc, and densitometer) are also available. OSU Microarray Core Facility functions as a service, training, and education facility and provides state-of-the-art instrumentation for microarray production and analysis, including a GeneMachines OmniGrid microarray fabrication system and a PerkinElmer ScanArray Express laser scanner. The facility also provides several robust computers, a Nanodrop spectrophotometer, and a suite of microarray and bioinformatics software for image and data analysis.

Financial Aid

Financial support is available in the form of predoctoral traineeships, fellowships, research assistantships, and teaching assistantships. In 2005–06, stipends ranged from $17,325 to $19,000 per year for half-time, according to the student's abilities and experience. This stipend includes a health insurance plan. In addition, there are employment opportunities for students' spouses in the University and the Stillwater community.

Cost of Study

Tuition and fees in 2005–06 were $1712 per semester for an average course load. Out-of-state tuition was $4830 per semester. Persons appointed to graduate traineeships, fellowships, and assistantships are exempt from out-of-state fees.

Living and Housing Costs

Living costs in Stillwater are modest to average. Furnished and air-conditioned University apartments for family student housing rented for $520 to $1124 per month in 2005–06. The cost of room in dormitories ranges from $1400 to $2600 per semester.

Student Group

On-campus enrollment at Oklahoma State University numbers about 19,350, with approximately 4,300 students in the Graduate College. Of about 1,600 students who are not residents of Oklahoma, 774 are international students from ninety other countries. The Department of Biochemistry and Molecular Biology enrolls 70 to 80 undergraduate majors and 30 to 40 graduate students. In addition, 10 to 15 research associates and visiting professors are on the biochemistry staff each year.

Student Outcomes

The majority of Ph.D. graduates become engaged in postdoctoral research and then obtain positions in universities, research foundations, government, or industry. Recent graduates have gone to UCLA, Stanford, MIT, and Duke. Two graduates have been presidents of Sigma Chemical Company. M.S. graduates usually obtain laboratory positions in universities, government, or industry.

Location

Stillwater is located in north-central Oklahoma, approximately 75 miles west of Tulsa and 60 miles northeast of Oklahoma City. The present population is about 42,000. Payne County has a population of 62,000. The cultural and recreational facilities of Stillwater make possible a variety of athletic activities, concerts, art and drama presentations, and public lectures throughout the year.

The University

Oklahoma State University was founded in 1890 by Act of the First Territorial Legislature as Oklahoma's land-grant university. The main campus of 840 acres has more than 200 permanent buildings, including a $3.5-million physical education building and a $3-million performing arts center. The faculty consists of 1,185 members.

Applying

Application for admission should be made through the Department of Biochemistry and Molecular Biology. Early application and a background in chemical and biological sciences are desirable. Application forms can be submitted online at http://gradcollege.okstate.edu/apply/default.htm.

Correspondence and Information

Dr. Richard C. Essenberg
Department of Biochemistry and Molecular Biology
Oklahoma State University
246 NRC
Stillwater, Oklahoma 74078-3035

Phone: 405-744-6189
Fax: 405-744-7799
E-mail: graduate-info@biochem.okstate.edu
Web site: http://biochem.okstate.edu

Oklahoma State University

THE FACULTY AND THEIR RESEARCH

Patricia Ayoubi, Assistant Professor; Ph.D., Oklahoma State, 1997. Microarrays; bioinformatics; functional genomics; differential gene expression analysis.

Margaret K. Essenberg, Regents Professor Emeritus; Ph.D., Brandeis, 1971. Infection-induced sesquiterpenes in plants: their role in disease resistance; cellular localization; regulation of their biosynthesis.

Richard C. Essenberg, Professor; Ph.D., Harvard, 1971. Factors required for *Brucella abortus* to survive in macrophages; interactions between ticks and their hosts.

Steven D. Hartson, Assistant Research Scientist; Ph.D., Oklahoma State, 1991. Hsp 90 chaperone machinery; proteomics.

Peter R. Hoyt, Research Associate Professor; Ph.D., University of Texas Medical Center at Galveston, 1988. Control of gene expression in relation to cellular systems biology in carcinogenesis, development, and response to environmental stimuli.

Ramamurthy Mahalingam, Assistant Professor, Ph.D., Clemson, 1998. Oxidative stress signal transduction in plants; functional genomics.

Michael A. Massiah, Assistant Professor; Ph.D., Maryland, 1996. NMR studies of protein structure and function.

Robert L. Matts, Professor; Ph.D., Wisconsin–Madison, 1980. Posttranscriptional control of protein synthesis; regulatory mechanisms involving protein phosphorylation and guanine nucleotide exchange; roles of molecular chaperones in regulation, protein folding, and signal transduction.

Ulrich K. Melcher, Professor; Ph.D., Michigan State, 1970. Role of virus-virus interaction through recombination, complementation, and synergy in plant-virus evolution; tobamoviruses; caulimoviruses; virus detection and identification.

Earl D. Mitchell Jr., Professor and Interim Head; Ph.D., Michigan State, 1966.

Andrew J. Mort, Regents Professor, Michigan State, 1978. Growth of plant cell walls; cell walls in disease and symbiosis; structural determination of glycoproteins and polysaccharides.

Margaret L. Pierce, Associate Research Professional; Ph.D., Michigan State, 1981. Recognition of potential pathogens by plants; sensory transduction and information transfer in plants.

Patricia Rayas-Duarte, Professor; Ph.D., Nebraska, 1988. Cereal chemistry.

José L. Soulages, Associate Professor; Ph.D., Universidad Nacional de la Plata (Argentina), 1989. Structure-function relationships of insect lipoproteins.

Chang-An Yu, Regents Professor; Ph.D., Illinois at Urbana-Champaign, 1969. Bioenergetics; membrane structure-function relationships.

Linda Yu, Professor; Ph.D., Illinois at Urbana-Champaign, 1970. Coenzyme Q; protein interactions; membrane bioenergetics; structure-function of photosynthetic cytochrome bc$_1$ complex.

The Jess L. and Miriam B. Stevens Wing of the Noble Research Center—home of the Department of Biochemistry and Molecular Biology.

The University Library.

The Student Union.

THE PENNSYLVANIA STATE UNIVERSITY

Department of Biochemistry and Molecular Biology
Graduate Study in Biochemistry

Program of Study

The Department of Biochemistry and Molecular Biology administers a broad-ranging graduate program in biochemistry, microbiology, and molecular biology that trains doctoral students for independent research and teaching in the molecular and cellular aspects of biology. While biochemistry is one of the scientific disciplines that may become a specific area of emphasis for an individual Ph.D. student, a common core curriculum provides students with training in all of the areas represented within the graduate program. Beginning and advanced graduate-level courses are taught by research faculty members in their area of specialization. Research is conducted in many areas of modern biomolecular sciences. (The faculty's research interests are described in the faculty section of this In-Depth Description.)

Students complete three laboratory rotations and may then begin a research program while in the first year of study. Personal attention is given to each student by an academic and a research adviser. Advancement to Ph.D. candidacy is decided on the basis of course work, research, and performance on an oral candidacy exam in the second year. The research adviser and a dissertation committee chosen from the program's faculty are the student's major consultants on dissertation research. The dissertation committee also administers the student's comprehensive examination and final Ph.D. dissertation defense. Students may earn the M.S. degree during the course of their Ph.D. work, but normally they bypass the master's degree with the approval of the graduate faculty. The Ph.D. usually requires about five years, including the dissertation research.

Research Facilities

The departmental laboratories are located in three modern buildings in the Eberly College of Science complex, adjacent to the main University library. Students, faculty members, and other research personnel utilize spacious, well-equipped, modern laboratory facilities. Several large general facilities are available to the department for shared use. These include facilities for mass spectrometry and proteomics, DNA microarray and genomics, X-ray crystallography, transgenic mice, nucleic acids, electron microscopy, and hybridoma/monoclonal antibodies, as well as the Center for Quantitative Cell Analysis.

Financial Aid

All admitted students receive financial support in the form of graduate assistantships at approximately $19,885 per year, plus a waiver of tuition, or fellowships from various University or outside sources. Strong applicants qualify for special Eberly College of Science Braddock and Roberts Awards, which provide supplements to their fellowships or assistantships. Furthermore, Penn State University pays most of the cost of health and dental benefits for its graduate students. Students can also elect to enroll in a subsidized family health coverage plan to cover eligible dependents, and supplemental dental and vision plans are also available. The department is well supported by research grants from the NIH, NSF, NASA, DOE, and USDA and other agencies.

Cost of Study

Tuition is provided for all admitted students.

Living and Housing Costs

Housing on campus for married graduate students includes one- and two-bedroom apartments ranging from $545 to $935 per month. Residence hall housing is available without meals for $1715 (double occupancy) or $2225 (single occupancy) per semester. Board only is $1435 to $1815 per semester, depending on the meal plan. Numerous off-campus accommodations are available at varying costs.

Student Group

Enrollment at the University Park Campus is about 41,795, including about 11,000 graduate students. The average number of graduate students in the department is 100, most from out of state. About 25 new students are admitted each year, virtually all seeking the Ph.D. degree. Upon graduating, most students obtain postdoctoral fellowships in prominent national and international laboratories for further specialized research. Graduates currently hold posts in major universities, research institutions, and industry.

Location

Penn State University Park Campus is located in State College, in scenic central Pennsylvania. The population of the area is about 80,000. Residents have the benefit of a wide variety of cultural, educational, and recreational resources. Nearby Interstate 80 and the University Park airport allow for excellent year-round access to the community. Pittsburgh, Philadelphia, New York, Washington, and Toronto are all within 250 miles.

The University and The Department

The 50 faculty members in the graduate program of the Department of Biochemistry and Molecular Biology have extensive research support from outside grants and contracts; such support totals $7 million per year at present. The department is closely associated with the University's Biotechnology Institute, which emphasizes application-oriented research in several areas of biotechnology, including aspects of genetic engineering. The facilities and personnel of the institute, supported jointly by the state of Pennsylvania and industry, are located adjacent to the department's buildings in the Eberly College of Science complex.

A faculty–graduate student ratio of about 1:2 permits excellent personal interactions at all levels. Extensive professional and social activities abound within the department and the University—e.g., numerous formal and informal seminars given by prominent visiting scientists; student and faculty symposia, promoting intradepartmental interactions; and the annual Summer Symposium in Molecular Biology, sponsored by the Recombinant DNA Technology Cooperative Program of Pennsylvania State University and biotechnology-related industries.

Applying

Students come from various educational backgrounds, including undergraduate majors in biochemistry, biology, biophysics, cell biology, chemistry, general science, genetics, microbiology, molecular biology, and physics. Preparation should include courses in biology, organic chemistry, calculus, general physics, and, preferably, physical chemistry. Limited deficiencies may be made up concurrently with graduate studies. Students holding master's degrees are encouraged to apply to the program.

Applications are accepted throughout the year, but most offers of admission and assistantships are made by March for the following summer or fall semester. Scores on the General Test and one Subject Test of the Graduate Record Examinations and, in the case of international students, on the TOEFL are required. These tests should be taken early enough so that the scores are received by January 15, preferably earlier, for consideration for admission in the fall. In addition, the department requires its own application form, undergraduate records, and three letters of recommendation for its review of applicants. Departmental application forms and further information should be requested directly from the address listed in this description.

Correspondence and Information

Chairman, Graduate Admissions
Department of Biochemistry and Molecular Biology
108 Althouse Laboratory
The Pennsylvania State University
University Park, Pennsylvania 16802

Phone: 814-865-5497
E-mail: bmmb@psu.edu
Web site: http://www.bmb.psu.edu

The Pennsylvania State University

THE FACULTY AND THEIR RESEARCH

Robert A. Schlegel, Professor and Head; Ph.D., Harvard, 1971. Mechanisms of transbilayer phospholipid movements and recognition of apoptotic blood cells.

Sarah E. Ades, Assistant Professor; Ph.D., MIT, 1995. Stress-induced signal transduction pathways between the cell envelope and cytoplasm in gram-negative bacteria.

Avery August, Associate Professor; Ph.D., Cornell, 1994. Signal transduction in lymphocytes by Src and Tec family tyrosine kinases. Adaptor proteins in leukemogenesis.

Paul L. Babitzke, Associate Professor; Ph.D., Georgia, 1991. Regulation of gene expression in *Bacillus subtilis;* protein-RNA interactions.

J. Martin Bollinger Jr., Associate Professor; Ph.D., MIT, 1993. Catalytic mechanisms of redox metalloenzymes; mechanisms of assembly of complex metal clusters in enzymes and regulatory proteins.

Squire Booker, Associate Professor; Ph.D., MIT, 1994. Mechanisms of cofactor action in enzymatic reactions.

Jean E. Brenchley, Professor; Ph.D., California, Davis, 1970. Discovery, study, and improvement of useful microbial enzymes; characterization of cold-active enzymes and their genes.

Donald A. Bryant, Professor and Ernest C. Pollard Professor of Biotechnology; Ph.D., UCLA, 1977. Molecular genetics of cyanobacteria and green bacteria; structure and function of photosystem I; light-harvesting antennae; phycobiliproteins; control of gene expression.

Craig E. Cameron, Paul Berg Professor of Biochemistry and Molecular Biology; Ph.D., Case Western Reserve, 1993. Genome replication of positive-strand RNA viruses, with emphasis on hepatitis C virus.

Nina V. Fedoroff, Willaman Professor of Life Sciences; Ph.D., Rockefeller, 1972. Plant transposable elements, transposon tagging, and epigenetic regulation of gene expression.

James G. Ferry, Stanley Person Professor of Biochemistry and Molecular Biology; Ph.D., Illinois, 1974. Enzymology and functional genomics.

Richard L. Frisque, Professor; Ph.D., Wisconsin, 1978. Viral functions involved in cellular transformation, tumorigenicity, and gene expression; role of genetic variation in the pathogenesis of disease; viral latency.

Carol V. Gay, Professor; Ph.D., Penn State, 1972. Cell biology of skeletal tissues; calcium metabolism.

David S. Gilmour, Associate Professor; Ph.D., Cornell, 1984. Transcriptional regulation in *Drosophila*.

John H. Golbeck, Professor; Ph.D., Indiana, 1976. Structure-function relationships in iron-sulfur type photosystems, including photosystem I of cyanobacteria and the reaction center of green sulfur bacteria.

Wendy Hanna-Rose, Assistant Professor; Ph.D., Harvard, 1996. Developmental genetics and morphogenesis in *Caenorhabditis elegans*.

Ross C. Hardison, T. Ming Chu Professor of Biochemistry and Molecular Biology; Ph.D., Iowa, 1977. Molecular basis of gene regulation and evolution in eukaryotes.

Eric T. Harvill, Assistant Professor; Ph.D., UCLA, 1995. Interactions between bacterial virulence factors and host immunity.

Teh-Hui Kao, Professor; Ph.D., Yale, 1980. Molecular studies of RNase-based self-incompatibility and pollen development in plants.

Kenneth Keiler, Assistant Professor; Ph.D., MIT, 1996. Temporal and spatial regulation of protein expression in bacterial development.

Emine Koc, Assistant Professor; Ph.D., New Mexico State, 1997. Investigation of mammalian mitochondrial ribosomes in translation, apoptosis, and disease states of the mitochondria by using various proteomics techniques.

Kouacou Konan, Assistant Professor; Ph.D., Indiana, 1995. Hepatitis C and related viruses: how their replication interferes with host protein and membrane trafficking.

Andrey S. Krasilnikov, Assistant Professor; Ph.D., Russian Academy of Sciences (Moscow), 1995. Structural biology of RNA and RNA-protein complexes; X-ray crystallography.

Maria Krasilnikova, Research Assistant Professor; Ph.D., Moscow Institute of Physics and Technology, 1995. Mechanism of expansion in trinucleotide repeat diseases; transcription stalling at GAA repeats in Friedreich's ataxia.

Carsten Krebs, Assistant Professor; Ph.D., Max Planck Institute for Radiation Chemistry (Mülheim), 1997. Bioinorganic chemistry: spectroscopic and kinetic studies on the mechanisms of iron-containing enzymes.

Zhi-Chun Lai, Associate Professor; Ph.D., Yeshiva (Einstein), 1989. Molecular and developmental genetics of the *Drosophila* visual system.

Arthur M. Lesk, Professor; Ph.D., Princeton, 1966. Studies of how amino acid sequences determine protein structures and protein evolution, using bioinformatics techniques.

Bernhard Lüscher, Associate Professor; Ph.D., Zurich, 1986. Structure, function, and postsynaptic targeting of $GABA_A$ receptors.

Christopher Malone, Assistant Professor; Ph.D., Colorado at Boulder, 1999. Interaction of the nucleus with the cytoplasm.

Andrea M. Mastro, Professor; Ph.D., Penn State, 1971. Cell biology; immunobiology—protein kinases and growth factors; lymphocyte mitogenesis and tumor promoters; immune-endocrine interactions.

Pamela J. Mitchell, Associate Professor; Ph.D., Columbia, 1985. Regulation of gene expression during mouse and *Drosophila* embryogenesis.

Katsuhiko Murakami, Assistant Professor; Ph.D., National Institute of Genetics (Japan), 1997. X-ray crystallographic studies of RNA polymerases.

Anton Nekrutenko, Assistant Professor; Ph.D., Texas Tech, 1999. Comparative genomics; developing new computational tools to provide genomic insights into biological phenomena.

B. Tracy Nixon, Associate Professor; Ph.D., MIT, 1982. Molecular genetics and biology of nitrogen fixation; modulation and prokaryotic sensory transduction.

Randen L. Patterson, Assistant Professor; Ph.D., Maryland, Baltimore, 2000. Systems biology approach toward modeling of calcium channels within neural networks.

Robert Paulson, Associate Professor; Ph.D., California, San Francisco, 1993. Genetic and biochemical analysis of cell signaling during hematopoiesis.

Gary H. Perdew, Professor; Ph.D., Oregon State, 1984. Mechanisms of dioxin toxicity; regulation of Ah receptor activation; Ah receptor-mediated gene regulation.

Jeffrey M. Peters, Associate Professor; Ph.D., California, Davis, 1992. Delineating roles for peroxisome proliferator-activated receptors in homeostasis; toxicology and cancer.

Ronald D. Porter, Associate Professor; Ph.D., Duke, 1976. Recombination mechanism in *Escherichia coli;* single-stranded–DNA-binding proteins.

Kathleen Postle, Professor; Ph.D., Wisconsin–Madison, 1978. Gram-negative bacteria; energy transduction; TonB system; membrane protein complexes; siderophore-mediated iron transport.

B. Franklin Pugh, Professor; Ph.D., Wisconsin–Madison, 1987. Regulation of gene transcription; protein–nucleic acid assembly mechanisms.

Joseph C. Reese, Associate Professor; Ph.D., Illinois at Urbana-Champaign, 1992. Study of cell cycle– and DNA damage–induced transcription in *Saccharomyces cerevisiae*.

Lorraine C. Santy, Research Assistant Professor; Ph.D., Harvard, 1997. Small GTPases; motility and migration; epithelial cell morphology and function.

Stephen Schuster, Associate Professor; Ph.D., Munich, 1991. Genome sequencing and analysis of host-adapted and free-living microbial species.

Song Tan, Associate Professor; Ph.D., Cambridge, 1989. Structural biology of eukaryotic gene regulation; X-ray crystallography of multicomponent protein and protein-DNA complexes.

Michael Teng, Assistant Professor; Ph.D., Chicago, 1993. Pathogenesis of respiratory syncytial virus infection; virus–host cell interactions.

Graham H. Thomas, Associate Professor; Ph.D., Edinburgh (Scotland), 1984. Genetics and molecular and cell biology of the cytoskeleton in *Drosophila* development.

Ming Tien, Professor; Ph.D., Michigan State, 1981. Microbiology, biochemistry, enzymology, and molecular biology of lignin-degrading organisms.

C.-P. David Tu, Professor; Ph.D., Cornell, 1976. Gene expression, structure, and function of glutathione S-transferases.

Yanming Wang, Assistant Professor; Ph.D., Iowa State, 2001. Chromatin, epigenetics, gene expression, histone modifications, cell differentiation, and cancer.

RICE UNIVERSITY

Department of Biochemistry and Cell Biology

Program of Study	The research activities of the Department of Biochemistry and Cell Biology at Rice University span a wide range of interests, organisms, and approaches, providing students with a broad spectrum from which to choose specific areas of study. The research areas encompassed by the departmental laboratories include biochemistry, biological chemistry and metabolic engineering, biophysics and structural biology, cell and developmental biology, microbiology, molecular genetics, neurobiology, plant biology, protein folding, proteomics, and signal transduction.
	Graduate students in biochemistry and cell biology at Rice get the best of both worlds—they are part of a small, elite university that nevertheless has Nobel Prize winners on the faculty and is a member institution of the world's largest medical center, the Texas Medical Center. The Texas Medical Center includes M. D. Anderson Cancer Center, Baylor College of Medicine, and the University of Texas Medical School at Houston, in addition to four nursing schools and fourteen hospitals. Johnson Space Center is 30 miles south of the campus, and the Department has collaborations with NASA, pursuing research related to microgravity conditions on the International Space Station.
	The Department emphasizes a five-year program of graduate study leading to the Ph.D. degree; master's degrees are not routinely offered. The first year of study involves course work, laboratory rotations, and selection of an adviser for thesis work; independent research is initiated by the end of the second semester of residence. The written proposal for the candidacy examination is completed at the end of the second year. The exam is based on the students' research, is similar in form to a grant application, and is defended orally. The remaining three years of the program consist of preparation and defense of the thesis. The Department is well supported, with 24 full-time and 3 emeritus faculty appointments, extensive grant support, laboratories for research and instruction, support staff, scientific equipment for research and teaching, and fellowships for graduate students. There is an additional complement of 52 postdoctoral research scientists and visiting scientists.
Research Facilities	The Department is fully equipped for research in biochemistry, biological chemistry and metabolic engineering, biophysics and structural biology, cell and developmental biology, microbiology, molecular genetics, neurobiology, plant biology, and signal transduction. Specialized equipment is available, including apparatuses for visible, ultraviolet, CD, MCD, GC-MS, fluorescence and infrared spectroscopy; phosphorimaging; densitometry; calorimetry; ESR and high-field NMR spectroscopy; X-ray crystallography; cryo-electron microscopy and 3-D image reconstruction facilities; laser photolysis and rapid-mixing apparatus; equipment for electrophysiology; confocal, video, and electron microscopes; liquid scintillation spectrometry; instruments for oligonucleotide synthesis and amino acid analysis; an analytical ultracentrifuge; and computer graphics systems for molecular modeling. Specialized facilities for animal and plant care, genetic experiments, and growth and harvesting of microorganisms on various scales, as well as a variety of equipment for centrifugation, ultracentrifugation, chromatography, tissue culture, and electrophoresis, are available. Extensive computational facilities are provided.
	The Department occupies approximately 62,000 square feet of research and office facilities. In addition, two laboratories with approximately 5,000 square feet of space are available for instruction.
Financial Aid	All biochemistry and cell biology graduate students receive fully funded research fellowships that include a full tuition waiver of $22,700 and a stipend of $24,000 per year in 2006–07. In addition, students are encouraged to apply for independent predoctoral fellowships from national organizations. There are many specialized training internships within the Department of Biochemistry and Cell Biology: the Houston Area Molecular Biophysics Program (HAMBP), the Keck Center for Interdisciplinary Bioscience Training, the Computational Biology and Medicine Training Program (NLM), Computational and Structural Biology in Biodefense, the Nanobiology Training Program, the Pharmacoinformatics Training Program, the Structural and Computational Biology Training Program, the NSF Integrative Graduate Education and Research Traineeship Program (IGERT), and the NIH Biotechnology Training Program. Participants are selected after their first year of enrollment, but even students not supported by these programs have access to their facilities, training, and other opportunities.
Cost of Study	A full tuition waiver and a stipend are included in the graduate research fellowships, as outlined in the Financial Aid section. Additional fees are $500 per year.
Living and Housing Costs	The University provides authorized housing for graduate students, and there is a wide choice of off-campus housing nearby. The cost of living in Houston is below average for cities of its size.
Student Group	The Department of Biochemistry and Cell Biology enrolled 65 graduate students in 2004–05; all were full-time, and 34 were women.
Location	Rice University has an exceptionally beautiful campus of 300 acres in a pleasant residential area of central Houston. The campus is adjacent to other member institutions of the Texas Medical Center complex, which includes Baylor College of Medicine, M. D. Anderson Cancer Center, the University of Texas Medical School at Houston, and a number of additional private and public hospitals and medical institutions.
The University	Rice University is a private research university with a long history of excellence, especially in the areas of science and engineering. The University enrolls approximately 4,800 students; about 1,900 are studying at the graduate level.
	The plan for growth over the next two decades is carefully designed to capitalize on the unique characteristics of the University, including its small size and considerable intellectual and financial resources. Bioscience is one of the areas targeted for enhancement, and recent building construction and faculty recruiting reflect this priority. The multidisciplinary nature of the program promotes the collaborative spirit that is essential for modern biosciences.
Applying	Applications are online only and may be found on the Department's Web site. All students should complete the Graduate Record Examinations. The TOEFL is required for international students. The application deadline is February 1, but early applications are encouraged.
Correspondence and Information	Dr. George N. Bennett, Chair Department of Biochemistry and Cell Biology, MS 140 Rice University P.O. Box 1892 Houston, Texas 77251-1892 Phone: 713-348-4015 Fax: 713-348-5154 E-mail: bioc@rice.edu Web site: http://www.biochem.rice.edu

Rice University

THE FACULTY AND THEIR RESEARCH

Bonnie Bartel, Professor; Ph.D., MIT, 1990 (e-mail: bartel@rice.edu). Molecular genetic dissection of plant responses to and metabolism of the growth hormone auxin; peroxisome biogenesis; roles of microRNAs. The *Arabidopsis* peroxisomal targeting signal type 2 receptor PEX7 is necessary for peroxisome function and dependent on PEX5. *Mol. Biol. Cell* 16:573–83, 2005 (with Woodward).

Kathleen M. Beckingham, Professor; Ph.D., Cambridge, 1972 (e-mail: kate@rice.edu). Molecular studies of calcium/calmodulin signaling and gravitoxic sensing in *Drosophila melanogaster*. Gravitaxis in *Drosophila melanogaster:* A forward genetic screen. *Genes, Brain and Behavior,* in press (with Armstrong et al.).

George N. Bennett, Professor and Chair; Ph.D., Purdue, 1974 (e-mail: gbennett@rice.edu). Acid stress response; biodegradation; metabolic engineering; recombinant DNA techniques. SpoIIE regulates sporulation but does not directly affect solventogenesis in *Clostridium acetobutylicum* ATCC 824. *J. Bacteriol.* 187:1930–6, 2005 (with Scotcher).

Janet Braam, Professor; Ph.D., Cornell, 1985 (e-mail: braam@rice.edu). Molecular biology and genetics of plant development; perception of and response to environmental stimuli; roles of Ca^{2+} and calmodulin in signal transduction; roles of cell-wall modifying enzymes in development. Genome-wide identification of touch- and darkness-regulated Arabidopsis genes: A focus on calmodulin-like and XTH genes. *New Phytol.* 165:429–44, 2005 (with Lee and Polisensky).

Raymon M. Glantz, Professor; Ph.D., Syracuse, 1967 (e-mail: rmg@rice.edu). Transduction in arthropod photoreceptors; receptors for neurotransmitters; polarization vision; computational properties and dynamics of neuronal networks. Analysis and simulation of gain control and precision in crayfish visual interneurons. *J. Neurophysiol.* 92:2747–61, 2004.

Richard H. Gomer, Professor; Ph.D., Caltech, 1983 (e-mail: richard@rice.edu). Studies of development, cell type choice, density sensing factors, and cell number counting, using molecular biology, cell biology, and protein biochemistry techniques. Exposure of cells to a cell-number counting factor decreases the activity of glucose-6-phosphatase to decrease intracellular glucose levels in *Dictyostelium. Eukaryotic Cell* 4:72–81, 2005 (with Jang).

Michael C. Gustin, Associate Professor; Ph.D., Yale, 1981 (e-mail: gustin@rice.edu). Molecular and genetic analysis of growth control mechanisms in cells; osmosensing MAP kinase cascades; transcriptional regulation by nitric oxide. Inducible defense mechanism against nitric oxide in *Candida albicans. Eukaryotic Cell* 3:715–23, 2004 (with Ullmann et al.).

Jordan Konisky, Professor; Ph.D., Wisconsin–Madison, 1968 (e-mail: konisky@rice.edu). Evolution, biochemistry, and molecular biology of extremophiles; mechanism of gene expression in methanogens and archaea. Structures of thermophilic and mesophilic adenylate kinases from the genus *Methanococcus. J. Mol. Biol.* 330:1087–99, 2004 (with Criswell et al.).

Mary Ellen Lane, Assistant Professor; Ph.D., Columbia, 1994 (e-mail: melane@rice.edu). Expression of zebrafish pitx3 in the primordial of the pituitary, lens, olfactory epithelium, and cranial ganglia is differentially modulated by hedgehog and modal signaling. *Genesis* 44:33–40, 2005 (with Zalinski, Shah, and Jamrich).

Kevin R. MacKenzie, Assistant Professor; Ph.D., Yale, 1996 (e-mail: mev@rice.edu). Structure, function, and stability of membrane proteins. Sequence-specific dimerization of the transmembrane domain of the "BH3-only" protein BNIP3 in membranes and detergent. *J. Biol. Chem.* 278:51950–6, 2003 (with Sulistijo and Jaszewski).

Seiichi P. T. Matsuda, Professor; Ph.D., Harvard, 1994 (e-mail: matsuda@ruf.rice.edu). Recombinant biosynthetic approaches to natural products chemistry; molecular biological approaches to mammalian and plant sterol biosynthesis and terpene cyclization and oxidation; directed evolution to investigate steric control of enzymatic oxidosqualene cyclization. Genome mining to identify new plant triterpenoids. *J. Am. Chem. Soc.* 126:5678–9, 2004 (with Fazio and Xu).

Kathleen Shive Matthews, Stewart Memorial Professor and Dean of Natural Sciences; Ph.D., Berkeley, 1970 (e-mail: ksm@rice.edu). Structure and function of genetic regulatory proteins. Integrated insights from simulation, experiment, and mutational analysis yield new details of LacI function. *Biochemistry,* in press (with Zhan and Swint-Kruse).

James A. McNew, Assistant Professor; Ph.D., Texas Southwestern Medical Center at Dallas, 1994 (e-mail: mcnew@rice.edu). Molecular mechanisms of membrane fusion. Sec1p directly stimulates SNARE-mediated membrane fusion in vitro. *J. Cell Biol.* 167:75–83, 2004 (with Scott et al.).

Edward P. Nikonowicz, Associate Professor; Ph.D., Purdue, 1990 (e-mail: edn@rice.edu). Biochemistry of RNA and RNA-protein interactions; correlation of structure, function, and dynamics; structural studies of base-modified RNAs. Metal ion stabilization of the U-turn of the A37 N6-dimethylallyl modified anticodon stem-loops of *Escherichia coli* tRNAPhe. *Biochem.* 43:55–66, 2004 (with Cabello-Villegas and Tworowska).

John Steven Olson, Ralph and Dorothy Looney Professor of Biochemistry and Cell Biology; Ph.D., Cornell, 1972 (e-mail: olson@rice.edu). Biophysical chemistry; kinetics of ligand binding to heme proteins; oxygen transport by erythrocytes; design of heme-protein-based blood substitute. NO scavenging and the hypertensive effect of Hb-based blood substitutes. *Free Radical Biol. Med.* 36:685–97, 2004 (with Foley et al.).

Graham Palmer, Professor Emeritus; Ph.D., Sheffield, 1962 (e-mail: graham@rice.edu). Mechanisms of oxidative enzymes; EPR spectroscopy as applied to biochemical systems; electron transport; cytochrome oxidase. Proton interactions with hemes and *a3* in bovine heart cytochrome c oxidase. *Biochem.* 44:4562–71, 2005 (with Parul and Fabian).

Ronald J. Parry, Professor; Ph.D., Brandeis, 1968 (e-mail: parry@rice.edu). Studies of the biosynthesis of antibiotics, antitumor agents, and toxins in bacteria, using molecular biology, protein biochemistry, and organic synthesis. Molecular characterization and analysis of the biosynthetic gene cluster for the azoxy antibiotic valanimycin. *Mol. Microbiol.* 46:505–17, 2002 (with Garg, Ma, and Hoyt).

Yousif Shamoo, Associate Professor; Ph.D., Yale, 1988 (e-mail: shamoo@rice.edu). Structural biology of DNA replication and its relation to carcinogenesis; X-ray crystallography/biochemistry of proteins in RNA processing; molecular evolution of thermostability. Structural and thermodynamic analysis of human PCNA with peptides derived from DNA polymerase-delta p66 subunit and flap endonuclease. *Structure* 12:2209–19, 2004 (with Bruning).

Jonathan J. Silberg, Assistant Professor; California, Irvine, 2000 (e-mail: joff@rice.edu). Computation-guided protein design; directed evolution; dissecting molecular interactions; biogenesis of iron-sulfur clusters. On the conservative nature of intragenic recombination. *Proc. Natl. Acad. Sci. U.S.A.* 102:5380–5, 2005 (with Drummond et al.).

Michael Stern, Professor; Ph.D., California, San Francisco, 1985 (e-mail: stern@rice.edu). Molecular and genetic control of ion channel activity and synaptic transmission in the *Drosophila* nervous system. Control of *Drosophila* perineural growth by two interacting neurotransmitter-mediated signaling pathways. *Proc. Natl. Acad. Sci. U.S.A.* 98:10445–50, 2001 (with Yager et al.).

Charles R. Stewart, Professor; Ph.D., Stanford, 1967 (e-mail: crs@rice.edu). Bacteriophage molecular genetics; mechanisms of host takeover during viral infection; mechanisms for regulation of sequential gene action. Roles of genes 44, 50, and 51 in regulating gene expression and host-takeover during infection of *Bacillus subtilis* by bacteriophage SPO1. *J. Bacteriol.* 186:1785–92, 2004 (with Sampath).

Yizhi Jane Tao, Assistant Professor; Ph.D., Purdue, 1999 (e-mail: ytao@rice.edu). Assembly and replication of DNA and RNA viruses; three-dimensional cryo-electron microscopy reconstruction; X-ray crystallography. RNA synthesis in a cage: Structural studies on reovirus polymerase lambda λ3. *Cell* 111:733–45, 2002 (with Farsetta, Nibert, and Harrison).

Daniel Wagner, Assistant Professor; Ph.D., Texas–Houston Health Science Center, 1997 (e-mail: dswagner@rice.edu). Vertebrate development; embryonic morphogenesis; zebrafish genetics. Maternal control of development at the midblastula transition and beyond: Mutants from the zebrafish II. *Dev. Cell* 6:781–90, 2004 (with Dosch et al.).

Pernilla Wittung-Stafshede, Associate Professor; Ph.D., Chalmers (Sweden), 1996 (e-mail: pernilla@rice.edu). Biophysical characterization of protein-folding reactions, with focus on the roles of cofactors and assembly. Role of cofactors in protein folding. *Acc. Chem. Res.* 35:201–8, 2002.

STANFORD UNIVERSITY

Department of Biochemistry

Program of Study	The Department of Biochemistry at Stanford University offers a Ph.D. degree. The program of study is designed to prepare students for careers in research and teaching. The major emphasis is on training in research. Students work closely with a dissertation adviser and members of a research group on novel and important biological problems at the molecular level. Group size averages about 10, with a maximum of about 15 students and postdoctoral fellows. To promote interaction among students, postdoctoral fellows, and faculty members, laboratory rooms are shared by members of various research groups. In addition to offering students access to all of the faculty members, this unique arrangement encourages collaboration between groups and has fostered the development of many new technologies. Predoctoral training begins in the fall of each year.

Courses in biochemistry, biophysics, and molecular biology are taught by the faculty, and advanced courses in specialized areas are also offered. These include the chemistry and biology of proteins, nucleic acids, and membranes; biological regulatory mechanisms; mechanistic aspects of enzyme action; bioinformatics and genomics; and molecular and genetic aspects of cellular and developmental biology. The program of study is created in consultation with the adviser to best fulfill each student's educational goals. Graduate students learn about teaching by assisting in the departmental teaching programs.

All of the bioscience departments and programs at Stanford participate in a flexible admissions program. This unique program offers first-year graduate students immersion into a particular research environment that matches the student's interests in a home program while also offering a choice of laboratory rotations and a research adviser from all of the more than 100 bioscience research laboratories at Stanford. For more information about biosciences and about flexible admissions at Stanford, students should consult the Web site for the bioscience programs at http://med.stanford.edu/biosciences/index.html.

Research Facilities

The department is fully equipped to carry out state-of-the-art research in biochemistry and molecular biology and in genetics and developmental biology. Facilities and equipment are available for work with plant and animal cell cultures, bacteria, bacteriophages, yeast, *Dictyostelium, Drosophila,* and animal viruses. The most advanced technology is available for physical biochemistry (500-MHz NMR, CD spectropolarimeter), stopped-flow analysis, laser light trapping, microarray analysis of whole genomes, and video microscopy. Facilities are also available for peptide and oligonucleotide synthesis, protein and DNA sequencing, and sophisticated computer analysis.

The Department of Biochemistry is located in the Arnold and Mabel Beckman Laboratories for Biochemistry in Stanford's Beckman Center for Molecular and Genetic Medicine. The Department of Biochemistry is joined there by three interrelated academic groups: the Departments of Developmental Biology and of Molecular and Cellular Physiology and the Howard Hughes Unit in Molecular and Genetic Medicine. Housing nearly 40 faculty members and their research teams, the Beckman Center is a focal point for biological research at Stanford.

Financial Aid

Students receive a stipend of approximately $27,500 per year, and tuition costs are paid by the department. Students are strongly urged to apply for a predoctoral fellowship from the National Science Foundation. Students should write to the Fellowship Office, National Research Council, 2101 Constitution Avenue, Washington, D.C. 20418 and apply to other agencies for fellowship and scholarship awards.

Cost of Study

Tuition is fully covered by traineeships and fellowships.

Living and Housing Costs

Most students share houses and apartments in the surrounding area at a cost of about $600 to $800 per student per month. Some students live on campus in subsidized apartment units that rent for approximately $350 per month for an individual and $800 per month for couples. Health insurance is provided.

Student Group

The total enrollment at Stanford University is more than 14,000, including approximately 7,500 graduate students. There are nearly 40 students in the biochemistry Ph.D. program.

Location

Stanford University is located in Palo Alto, a community of 60,000, located 35 miles south of San Francisco. Extensive cultural and recreational opportunities are available at the University and in surrounding areas as well as in San Francisco. To the west of the campus lie the Santa Cruz Mountains and the Pacific Ocean, and to the east, the Sierra Nevada range with its many national parks, including Yosemite. This area is famous for hiking, camping, and skiing. The climate on the peninsula is mild, with an average midday temperature of 60°F in winter and 74°F in summer.

The University

Stanford is a private university, founded in 1885 and ranked among the top few in academic excellence. The University's students are about evenly divided between undergraduate and both graduate and professional school students. The Department of Biochemistry, located in the Stanford Medical Center on campus, is part of the graduate division of the University and a department of the School of Medicine.

Applying

Students interested in pursuing a career in biochemistry, biophysics, and molecular and cell biology are invited to apply. Those applying should have a bachelor's degree and should have had at least a year of course work in each of the following areas: calculus, physics, inorganic chemistry, organic chemistry, physical chemistry, biochemistry/molecular biology, and biological sciences. The department is especially interested in students who have carried out original research in physical or biological sciences during their undergraduate studies.

Students must see to it that all application materials (transcripts, scores, etc.) are received by the department before December 12. All applicants are notified by April 15. The department requires scores on the Graduate Record Examinations General Test. A GRE Subject Test in biology, chemistry, or biochemistry, cell, and molecular biology is recommended but not required. Students must take these exams by the October/November test dates to ensure that their scores are received in time.

Correspondence and Information

The Graduate Admissions Program
Old Union, Room 137
Stanford University
Stanford, California 94305-3005

Phone: 650-723-4291
E-mail: ck.gaa@forsythe.stanford.edu
Web site: http://biochem.stanford.edu

Stanford University

THE FACULTY AND THEIR RESEARCH

Robert L. Baldwin, Professor Emeritus; D.Phil., Oxford, 1954.

Paul Berg, Professor Emeritus; Ph.D., Western Reserve, 1952.

Patrick O. Brown, Professor; Ph.D., 1980, M.D., 1982, Chicago. The development and application of microarray-based methods and computational tools for systematic and quantitative studies of gene expression and biological regulation. Genomic and biochemical approaches are used to investigate the basic mechanisms that regulate the genome's expression program, explore basic questions in human biology, and to develop new ways to diagnose and treat cancer and other diseases.

Douglas L. Brutlag, Professor; Ph.D., Stanford, 1972. Application of information science to DNA/protein sequences and structures in order to understand the flow of information from genome to phenotype. Of particular importance are the problems of predicting structure and function from sequence, simulating protein-ligand docking, and protein-protein and protein-DNA interactions. The lab is also involved in discovering conserved protein motifs and conserved DNA motifs. The DNA motifs studied are involved in transcription factor finding sites, transcription start and termination sites, and splice donor/acceptor sites. Critical methods used include comparative genomics, machine learning, simulation, information theory, and statistics.

Gilbert Chu, Professor; Ph.D., MIT, 1973; M.D., Harvard, 1980. Cells recognize and respond to DNA damage in order to survive and maintain genomic integrity. To understand repair pathways for DNA damaged by ultraviolet radiation (UV) and ionizing radiation (IR), purified proteins, cell extracts, and intact cells are utilized. These pathways include nucleotide excision repair and double-strand break repair, which are important for cell survival, suppressing mutagenesis, and generating immunological diversity. To understand transcriptional responses to DNA damage, the lab has developed statistical methods for analyzing microarray data. The lab uses transcriptional responses to predict cancer risk and risk for toxicity from anticancer treatment.

Ronald W. Davis, Professor; Ph.D., Caltech, 1970. Functional genomics and technology development. As a model system, the lab uses *Saccharomyces cerevisiae* to apply DNA microarrays and other new technology to analyze the whole genome for phenotype of deletions, RNA expression levels, protein levels, metabolic regulation, DNA replication, protein-protein interaction, and drug sensitivity. A major program is the quantitative phenotypic analysis of yeast deletion. The lab has deleted 95 percent of the yeast genes and replaced them with a molecular bar code that allows the lab to analyze a mixture of all deletion simultaneously. This program is a model for functional genomics. A program in genome profiling of drug and toxin interactions has been established. In addition, a new program in ecological genomics is being developed. New technology for very high throughput and low-cost simple nucleotide polymorphism has been developed to be applied to complex genetic problems in model organisms and humans.

James E. Ferrell Jr., Professor; Ph.D., 1984, M.D., 1986, Stanford. Cell signaling and cell cycle regulation. The lab is studying signaling proteins that control mitosis and meiosis. The goal is to determine how the biochemical properties of individual signaling proteins and the organization of these proteins into cascades and loops generates systems-level behaviors such as amplification, adaptation, and oscillations. Experimental work is carried out in *Xenopus* oocytes, eggs, and egg extracts—systems which are well suited to quantitative biochemical analysis. These experimental studies are complemented with computational approaches.

Pehr Harbury, Associate Professor; Ph.D., Harvard, 1994. Structural determinants of protein folding, design, and small-molecule recognition. The molecular mechanisms that confer specific shapes on proteins and determine how proteins recognize small molecules are being studied. The goal is to elucidate predictive principles by which novel structures and catalytic properties can be conferred accurately on designed polypeptides and to achieve the rational design of ligands for proteins of known conformation. The lab relies primarily on three tools: the computational engineering of structures at atomic resolution, which is made possible by the advent of classical molecular mechanics potentials; biophysical characterization of peptide proteins composed from an expanded amino acid alphabet; and the generation and screening of combinatorial libraries.

Daniel Herschlag, Professor; Ph.D., Brandeis, 1988. RNA folding; RNA and protein catalytic mechanisms; global RNA processing. Biochemical and biophysical approaches are taken to understand mechanisms of simple and complex RNA and protein enzymes and RNA folding. These approaches include single molecular fluorescence and force measurements, small-angle X-ray scattering, vibrational spectroscopy, and several other techniques. These approaches complement in-depth kinetic and thermodynamic studies. Cellular RNA processing events are dissected using DNA microarrays to simultaneously follow all yeast RNAs.

David S. Hogness, Professor Emeritus; Ph.D., Caltech, 1952.

Dale Kaiser, Professor; Ph.D., Caltech, 1955. Regulation of multicellular development; how cellular patterns are formed and how cells in an embryo coordinate their activities so that the right cell is in the right place at the right time. Myxobacteria are one of the simplest organisms that exhibit multicellular development with cellular differentiation and spatially localized gene expression. They permit both biochemical and genetic studies of cell-cell interactions necessary for development and differentiation. Two extracellular factors needed at different times for transcription of developmentally regulated genes have been purified. Their signal transduction pathways are being dissected.

Arthur Kornberg, Professor; M.D., Rochester, 1941. Inorganic polyphosphate (poly P) is a linear polymer of many tens or hundreds of orthophosphate (P_i) residues linked by high-energy, phosphoanhydride bonds. Likely a prominent precursor in prebiotic evolution, poly P is now found in volcanic condensates, deep-oceanic steam vents, and in every living thing—bacteria, fungi, protozoa, plants, and mammals. Among poly P functions are kinase donor to glucose, nucleoside diphosphates, and proteins; phosphate reservoir; divalent metal (Ca^{++}, Mg^{++}, Mn^{++}) chelator; and component of a membrane complex in metal ion channels and bacterial transformation. Recent studies have disclosed three additional roles: "alarmone" in response to stresses and deficiencies; adaptations for survival in the stationary phase; and motility, quorum sensing, biofilm formation, and virulence in bacterial pathogens tested. Current studies focus on the biochemical mechanisms responsible for those many functions of poly P and in animal cells and in sporulation.

Mark A. Krasnow, Professor; Ph.D., 1983, M.D., 1985, Chicago. Genetic, genomic, and biochemical analysis of epithelial morphogenesis, using the *Drosophila* tracheal (respiratory) system and mouse lung as models. The goals are to understand the cellular and molecular basis of how epithelial cells migrate and find their targets, how the cells assemble into tubes, how tubes interconnect, and how these processes are regulated by tissue oxygen need.

I. Robert Lehman, Professor Emeritus; Ph.D., Johns Hopkins, 1954. DNA replication in eukaryotes. Research in the lab is aimed at understanding the mechanism by which the genome of herpes simplex virus type 1 (HSV-1) is replicated. The linear HSV-1 genome circularizes following infection and undergoes θ-type replication dependent upon one or more of the three origins. In the second phase, the circular products of θ replication engage in rolling-circle replication promoted by the virus-encoded enzymes to produce concatameric molecules that are then cleaved to unit length and packaged into viral particles. The lab has found that a complex containing the virus-encoded enzymes that it has isolated from HSV-1 infected human cells can promote the rolling-circle phase of HSV-1 DNA replication. Research aims are to purify the complex to homogeneity and determine its components, particularly those contributed by the host cell, and to identify the factors that promote the switch in the mode of DNA replication from θ to rolling circle. There is some indication that a recombinational event may be involved.

Suzanne R. Pfeffer, Professor; Ph.D., California, San Francisco, 1983. Biochemistry of intracellular transport. Research is aimed at the molecular mechanisms of protein targeting to distinct intracellular compartments. Protein transport between endosomes and the Golgi apparatus is studied in a cell-free system to discover proteins that catalyze vesicular transport. The *ras*-like GTPase, rab9, and a rab-specific, nucleotide exchanger is required. Also being investigated is how sixty different Rab GTPases are delivered to distinct compartments in mammalian cells and how Rab GTPases function in prostate and breast cancer and in Niemann-Pick Type C disease.

James A. Spudich, Professor; Ph.D., Stanford, 1968. Biochemical, molecular, genetic, and structural studies of actin, myosin, and associated regulatory proteins from eukaryotic cells. Work focuses on the design and development of in vitro assays for ATP-dependent movement of purified myosin on filaments reconstituted from purified actin. This includes the development of laser traps for single-molecule analyses. Myosin gene cloning and expression of site-directed mutagenized forms, which are analyzed for altered functions, is also carried out. Emphasis is on the molecular basis of energy transduction that leads to myosin movements on actin filaments and also on regulation of actin and myosin interactions and of their assembly states, with particular interest in *Dictyostelium* chemotaxis, cytokinesis, and other forms of cell movement.

Aaron Straight, Assistant Professor; Ph.D., California, San Francisco, 1998. Eukaryotic chromosome structure and function. Cells have devised extraordinary mechanisms to ensure that chromosomes are faithfully replicated and equally segregated during cell division. Study is centered on how chromosome structure contributes to the fidelity of cell duplication. Work focuses on understanding how kinetochores control the equal distribution of chromosomes by interacting with microtubules of the mitotic spindle. Current research involves understanding how kinetochore position is specified by genetic and epigenetic determinants, the biochemical mechanisms of kinetochore assembly, and how kinetochores harness the forces required to segregate chromosomes during anaphase.

Julie A. Theriot, Assistant Professor; Ph.D., California, San Francisco, 1993. Mechanisms of actin-based cell motility; cell biology of host-pathogen interactions. Polymerizing networks of actin filaments are capable of generating significant amounts of force that can be used by eukaryotic cells and their prokaryotic pathogens to change shape or to move, using the nonequilibrium protein polymerization reaction to transduce chemical energy into mechanical energy without the assistance of any molecular motor proteins. Projects focus on understanding the biochemical and biophysical mechanisms of force generation by actin polymerization and the large-scale self-organization and polarization of actin networks, both at the leading edge of motile cells and on the surfaces of bacterial intracellular pathogens that include *Listeria monocytogenes* and *Shigella flexneri*.

STONY BROOK UNIVERSITY, STATE UNIVERSITY OF NEW YORK

with Cold Spring Harbor Laboratory and Brookhaven National Laboratory
Graduate Studies in Biochemistry and Molecular Biology

Program of Study

Biochemistry and Molecular Biology is a graduate specialization within the umbrella Program of Molecular and Cellular Biology. Training stresses biochemical and structural approaches to solve biological problems. Students have the opportunity to perform Ph.D. thesis research in a number of laboratories at three collaborative institutions: the State University of New York at Stony Brook, Cold Spring Harbor Laboratory, or Brookhaven National Laboratory. The Ph.D. is granted by the State University of New York at Stony Brook.

The integrated nature of the program offers students an extraordinary range of experimental science. Students are involved in ongoing research as soon as they arrive on campus. During the first academic year, they experience research in four different laboratories of program faculty members. Following the first academic year, a member of the program faculty is selected as a Ph.D. mentor. Diverse biological systems are available to investigate a variety of biological questions, including cancer, neurobiology, gene expression, infectious disease, protein trafficking, DNA replication, structural biology, signal transduction, cell cycle, apoptosis, virology, development, biological membranes, and immune defense.

The program offers the graduate student a unique opportunity to choose between three academic subspecialties: biochemistry and molecular biology, cellular and developmental biology, or immunology and pathology.

During the first academic year, students attend core courses in biochemistry, molecular genetics, and cell biology. In addition, students participate in a journal club/readings course, in which students learn to critically evaluate original research articles. Subsequent specialized courses and electives are then designed to enhance the student's knowledge in their selected academic subspecialty.

In the second and third semesters, graduate students develop teaching skills by serving as a teaching assistant with the guidance of the faculty instructor of the course. In the third and subsequent years, graduate students present their research progress to other students and faculty members in graduate student seminars. These provide opportunities to gain communication skills and learn about ongoing research of students in other laboratories.

In the third year, students prepare a written Ph.D. thesis proposal in consultation with their research faculty adviser. Following successful defense of the proposal, the student advances to candidacy and the Proposal Committee and the faculty adviser become the student's Ph.D. Thesis Committee. This committee meets at least once a year to support the research efforts of the student.

Research Facilities

Research is supported by state-of-the-art facilities at Stony Brook. General computing services are available to students, including an account for e-mail and Internet access. In addition to state-of-the-art equipment in individualized laboratories, several core facilities support program research. The new Proteomics Center is a core facility at Stony Brook whose services include protein sequencing, peptide synthesis, analytical HPLC, preparative HPLC, MALDI-TOF mass spectrometry, and DNA sequencing. The University recently opened a DNA Microarray Center that offers researchers the ability to simultaneously analyze the expression of thousands of genes in humans, rodents, yeast, plants, bacteria, or *Drosophila*. The Center for Structural Biology has advanced instrumentation for NMR spectroscopy and X-ray crystallography. The University Microscopy Imaging Center assists in research projects requiring advanced light, electronic light, and electron microscopy techniques.

Financial Aid

All students accepted into the Biochemistry and Molecular Biology Graduate Program receive full financial support. Students are provided with a stipend of $25,000 per year for cost of living expenses. In addition, tuition is waived for all graduate students accepted into the program. Health insurance is provided.

Cost of Study

University fees are approximately $750 per year.

Living and Housing Costs

Room rent for a single student living in an on-campus apartment ranges from about $325 to $1500 per month, depending on size and on the number of occupants. Privately owned housing is available close to the University.

Student Group

There are approximately 100 students in the Graduate Program. Many enter the program as recent college graduates, and others have had previous graduate school or research experience. Generally, students have undergraduate majors in biology, biochemistry, or chemistry, but some students come from other disciplines. Diverse cultural and ethical interests are supported by the campus community.

Location

Stony Brook is located on the wooded North Shore of Long Island, New York, an area of picturesque villages, harbors, and beaches. It is about 50 miles east of New York City, with easy access provided by the Long Island Railroad or major parkways so that students can take advantage of New York City's commercial, scientific, and cultural resources.

The University

The University was founded in 1957 in Oyster Bay, Long Island. In 1962, it moved to Stony Brook and now occupies more than 1,000 acres with more than 100 buildings. Stony Brook has exceptional strength in the sciences, mathematics, humanities, fine arts, social sciences, engineering, and health professions for both undergraduate and graduate students. It was classified by the Carnegie Foundation as one of the nation's leading research institutions. The University Hospital and Medical Center supports the Schools of Medicine, Dentistry, Nursing, and Health Technology and Management.

An Indoor Sports Complex promotes Division I athletics and Seawolf fans. The Staller Center for the Arts offers a wide variety of professional performances in music, film, dance, theater, and fine art exhibitions in its five theaters and University Art Gallery.

Applying

Applications to the graduate program should be submitted by January 15 for admission the following September, although applications are accepted after that date. Applicants must have a baccalaureate degree from an accredited college or university and submit school transcripts. The General Test of the Graduate Record Examinations (GRE) is required, and a Subject Test in an appropriate field is optional. Three letters of recommendation are required.

Prospective candidates are invited to visit Stony Brook whenever possible. This affords students an opportunity to meet with program faculty members and students. To apply online, students should visit http://www.grad.sunysb.edu/applying/applying.htm.

Correspondence and Information

Director, Graduate Studies in Biochemistry and Molecular Biology
Life Sciences Rooms 336 and 338
State University of New York at Stony Brook
Stony Brook, New York 11794-5215

Phone: 631-632-8533
Fax: 631-632-9730
E-mail: mcbprog@life.bio.sunysb.edu
Web site: http://life.bio.sunysb.edu/mcb

Stony Brook University, State University of New York

THE FACULTY AND THEIR RESEARCH

Department of Biochemistry and Cell Biology
Paul M. Bingham, Ph.D., 1979. Genetic control of development and gene expression in animals. Deborah Brown, Ph.D., 1987. Structure and function of cholesterol/sphingolipid-rich membrane rafts and caveolae. Vitaly Citovsky, Ph.D., 1987. Nuclear targeting and intercellular communication in plants. Neta Dean, Ph.D., 1988. Protein glycosylation; fungal cell wall biosynthesis; fungal pathogenesis. Dale Deutsch, Ph.D., 1972. Molecular neurobiology of anandamide (the endogenous marijuana) regulation. J. Peter Gergen, Ph.D., 1982. Gene expression and development in *Drosophila*. Robert Haltiwanger, Ph.D., 1986. Role of protein glycosylation in signal transduction; Notch signaling. Bernadette Holdener, Ph.D., 1990. Role of protein folding in WNT signal transduction and development. Nancy Hollingsworth, Ph.D., 1988. Chromosome structure and function during meiosis in yeast. Jen-Chih Hsieh, Ph.D., 1994. Molecular mechanism of Wnt signaling. A. Wali Karzai, Ph.D., 1995. Structure and function of RNA-binding proteins and biochemical studies of the SmpB·SsrA quality control system for protein tagging, directed degradation, and ribosome rescue. William J. Lennarz, Ph.D., 1959. Biosynthesis and function of glycoproteins in cell-cell interactions. Erwin London, Ph.D., 1979. Membrane protein structure/translocation/folding; structure and function of sphingolipid/cholesterol rafts in membranes. Harvard Lyman, Ph.D., 1960. Photocontrol of chloroplast development. Kenneth B. Marcu, Ph.D., 1975. Molecular control of innate and adaptive immunity: NF-kappaB kinases and inflammatory response; antibody gene class switch recombination and somatic hypermutation. Aaron Neiman, Ph.D., 1994. Vesicle trafficking and intracellular signaling in yeast. Nisson Schechter, Ph.D., 1971. Homeobox and filament proteins in neuronal differentiation, growth, and regeneration. Sanford R. Simon, Ph.D., 1967. Extracellular matrix degradation by inflammatory and tumor cell proteases. Steven O. Smith, Ph.D., 1985. Structure and function of membrane proteins. James Staros, Ph.D., 1974. Biochemical and biophysical approaches to signal transduction by ErbB-family receptors. Rolf Sternglanz, Ph.D., 1967. Chromatin structure and function; histone modifying enzymes; gene expression. Gerald H. Thomsen, Ph.D., 1988. Regulation of early vertebrate development by growth factor signals, ubiquitin modification, and T-box family transcription factors.

Department of Chemistry
Carlos Simmerling, Ph.D., 1994. Computational studies of biomolecular structure and dynamics. Peter J. Tonge, Ph.D., 1986. Tuberculosis pathogenesis and drug discovery; enzyme mechanisms and rational inhibitor design; fluorescent proteins.

Department of Molecular Genetics and Microbiology
Jorge L. Benach, Ph.D., 1972. Pathogenesis of spirochetal infections; utilization of host macromolecules. Allen Bruce Futcher, Ph.D., 1983. Cell-cycle control in yeast and on analysis of gene expression networks using microarrays and bioinformatics. Michael Hayman, Ph.D., 1973. Signal transduction pathways in cell growth, differentiation, and cancer. Patrick Hearing, Ph.D., 1984. Adenovirus–host cell interactions; adenovirus assembly and vectors for gene therapy. Eugene Katz, Ph.D., 1969. Genetics/development in cellular slime molds. James Konopka, Ph.D., 1985. Hormone signal transduction/yeast cell morphogenesis. Janet Leatherwood, Ph.D., 1993. Cell-cycle control of DNA replication/microarray analyses of fission yeast. Nancy Reich, Ph.D., 1983. Signal transduction and gene expression in response to cytokines and virus. David D. Thanassi, Ph.D., 1995. Virulence factors of pathogenic bacteria. Eckard Wimmer, Ph.D., 1962. RNA virus genetics, replication, and pathogenicity; cellular virus receptors. Wei-Xing Zong, Ph.D., 1999. Molecular regulation of cell-death pathways.

Department of Neurobiology and Behavior
Simon Halegoua, Ph.D., 1978. Molecular control of the neuronal phenotype. Maurice Kernan, Ph.D., 1990. Differentiation and signal transduction in *Drosophila* mechanosensory cilia and sperm. Joel M. Levine, Ph.D., 1980. Cell-surface molecules of the developing nervous system. Gail Mandel, Ph.D., 1977. Gene expression in the nervous system. David McKinnon, Ph.D., 1987. Molecular physiology of sympathetic neurons and cardiac muscle. Howard Sirotkin, Ph.D., 1996. Genetic and molecular analysis of early vertebrate development. Lonnie P. Wollmuth, Ph.D.,1992. Molecular and biophysical mechanisms of excitatory synaptic transmission.

Department of Oral Biology and Pathology
Soosan Ghazizadeh, Ph.D., 1994. Keratinocyte stem cell lineage and use in cell and gene therapy.

Department of Pathology
Howard B. Fleit, Ph.D., 1980. Leukocyte Fc receptors; macrophage differentiation. Martha Furie, Ph.D., 1980. Interactions among pathogenic bacteria, endothelium, and leukocytes. Berhane Ghebrehiwet, D.V.M./D.Sc., 1974. Biochemistry; function of the complement system. Richard R. Kew, Ph.D., 1986. Leukocyte chemotaxis/inflammation. Ute M. Moll, M.D., 1985. The p53 and Rb/E2F tumor suppressor gene family: function/regulation in normal cells and tumor-associated inactivation. Eric D. Spitzer, M.D./Ph.D., 1985. Molecular biology of *Cryptococcus neoformans*. Gary Zieve, Ph.D., 1977. Assembly/transport of snRNP particles.

Department of Pharmacological Sciences
Miguel Berrios, Ph.D., 1983. Nuclear structure and function; cell biology of DNA damage and repair. Daniel Bogenhagen, M.D., 1977. Mitochondrial DNA; mitochondrial proteomics. Holly Colognato, Ph.D., 2000. Extracellular matrix in the brain; roles during development and during neurodegeneration. Michael A. Frohman, M.D./Ph.D., 1986. Signal transduction; membrane vesicle trafficking; regulated exocytosis; diabetes; phototransduction. Arthur P. Grollman, M.D., 1959. Molecular carcinogenesis and DNA repair. Craig C. Malbon, Ph.D., 1976. Wnt-Frizzled signaling, G-proteins, and development. Masaaki Moriya, Ph.D., 1981. Cellular response to DNA damage. Jeffery Pessin, Ph.D., 1980. Identification of insulin-mediated signal cascades regulation of intracellular GLUT4 vesicle trafficking and biogenesis. Joav Prives, Ph.D., 1968. Cytoskeletal membrane interactions in muscle cells. Styliani Anna Tsirka, Ph.D., 1989. Neuronal microglial interactions in the physiology and pathology of the central nervous system. Orlando D. Schärer, Ph.D., 1996. Chemical biology of mammalian DNA repair.

Department of Physiology and Biophysics
Roger A. Johnson, Ph.D., 1968. Adenine nucleoside 3'-polyphosphates, adenylyl cyclases, and signal transduction. Stuart McLaughlin, Ph.D., 1968. Calcium/phospholipid second messenger system. W. Todd Miller, Ph.D., 1989. Tyrosine phosphorylation and signal transduction. Nicholas Nassar, Ph.D., 1992. Crystallographic and biochemical studies of signal proteins. Suzanne Scarlata, Ph.D., 1984. Structure/oligomerization of membrane proteins. Ilan Spector, Ph.D., 1967. Actin cytoskeleton in normal and cancer cells. Hsien-yu Wang, Ph.D., 1989. Wnt/Frizzled and G-protein signal transduction in development. Thomas White, Ph.D., 1994. Molecular biology and physiology of gap-junction channels.

School of Medicine and Other Departments
Wen-Tien Chen, Ph.D., 1979. Proteases and integrins in cancer invasion, metastasis, and angiogenesis. Yaacov Hod, Ph.D., 1977. Hormonal control of gene expression in prostate cancer cells. Jolyon Jesty, D.Phil., 1972. Mechanisms of thrombogenesis. Richard Lin, M.D., 1988. Signal transduction and cell growth. Erich R. Mackow, Ph.D., 1984. Hantavirus and rotavirus pathogenesis; viral regulations of cell signaling pathways and responses; viral attachment and entry; reverse genetics. Wolfgang Quitschke, Ph.D., 1983. Gene regulation of proteins associated with neurodegenerative diseases. Mario J. Rebecchi, Ph.D., 1984. Phospholipases and signal transduction. Roy T. Steigbigel, M.D., 1966. Immune dysfunction induced by HIV infection. William E. VanNostrand, Ph.D., 1985. Physiologic and pathophysiologic vascular functions of the Alzheimer's disease amyloid beta-protein precursor.

Brookhaven National Laboratory
John Dunn, Ph.D., 1970. Genomics and gene expression. Richard Setlow, Ph.D., 1947. DNA damage and repair; carcinogenesis and mutagenesis in fish. F. William Studier, Ph.D., 1963. Molecular genetics of phage T7; structural genomics.

Cold Spring Harbor Laboratory
Gregory Hannon, Ph.D., 1991. Genetics of growth in mammalian cells and dsRNA-induced gene silencing. Leemor Joshua-Tor, Ph.D., 1991. X-ray crystallography; molecular recognition; nucleic acid regulation; RNAi. Adrian Krainer, Ph.D., 1986. mRNA splicing; gene expression; RNA-protein interaction. Yuri Lazebnik, Ph.D., 1986. Molecular mechanisms of cancer and apoptosis. Scott Lowe, Ph.D., 1994. Modulation of apoptosis; chemosensitivity; senescence by cancer genes. Vivek Mittal, Ph.D.,1994. Id transcription factors; tumor-mediated neovascularization. K. Muthuswamy Senthil, Ph.D., 1995. Understanding cancer initiation using three-dimensional epithelial structures. David L. Spector, Ph.D., 1980. Spatial organization of gene expression. Arne Stenlund, Ph.D., 1984. DNA replication of papillomaviruses. Bruce Stillman, Ph.D., 1979. DNA replication and chromatin assembly in human and yeast cells. William P. Tansey, Ph.D., 1991. Regulation of oncoprotein stability. Nicholas Tonks, Ph.D., 1985. Characterization of protein tyrosine phosphatases. Linda Van Aelst, Ph.D., 1991. Role of Ras in mammalian cell transformation. Michael Wigler, Ph.D., 1978. Genomics and cancer.

STONY BROOK UNIVERSITY, STATE UNIVERSITY OF NEW YORK

Graduate Program in Biochemistry and Structural Biology

Programs of Study

The Graduate Program in Biochemistry and Structural Biology is an interdepartmental graduate training program that stresses biochemical, structural, and computational approaches to solving complex biological problems. Only candidates working toward the Ph.D. degree are admitted. Together with the Graduate Program in Molecular and Cell Biology (description in *Peterson's Graduate Programs in the Biological Sciences,* Section 6), the Graduate Program in Biochemistry and Structural Biology broadly covers graduate research in the biochemical sciences at Stony Brook.

The program includes faculty members from a number of departments and from Brookhaven National Laboratory. During the first year, students take core courses in biochemistry, structural biology, computational biology, and either cell biology or molecular biology, depending on their desired area of specialization. The program is research oriented, and participation in research is begun immediately through at least three 2-month rotation periods in which the beginning student participates in ongoing research projects. Before the end of the first year, the student is expected to begin a thesis research project. In later years, students can enroll in specialty courses, and, as part of the Ph.D. program, all students also obtain teaching experience in undergraduate classroom or laboratory courses for two semesters.

Students take a comprehensive written examination in January of their second year. Admission to candidacy for the Ph.D. occurs after students defend their thesis proposal in an oral examination. After admission to candidacy, the student works virtually full-time on thesis research that culminates in the submission and defense of a written Ph.D. dissertation.

Research Facilities

The biological sciences complex at Stony Brook is formed by three adjacent facilities. The Life Sciences Building is a large, modern facility containing 110 research and teaching laboratories, as well as the offices for the Graduate Program in Biochemistry and Structural Biology. Connected to the Life Sciences Building via two skywalks is the Centers for Molecular Medicine, a state-of-the-art research facility that houses interdepartmental programs in structural biology, infectious diseases, developmental genetics, and neurobiology. Several members of the program are located in the adjacent Health Sciences Center, a modern, high-rise complex that houses a large research facility and the University Hospital.

State-of-the-art facilities are available for biochemistry and structural biology. The Center for Structural Biology has several high-field NMR instruments and facilities for X-ray crystallography. With close ties to the Brookhaven National Laboratory, Stony Brook students can take advantage of the high-energy beam lines for diffraction studies. The center also has a large computational core facility, with several multiprocessor computers for structure refinement and analysis, and numerous molecular graphics workstations. Throughout the program, there is state-of-the-art equipment for protein purification and analysis, including Raman, infrared, fluorescence, and CD spectrophotometers and mass spectrometers. Mass spectroscopy and peptide synthesis are also carried out in the Proteomics Center. The biological sciences complex also has tissue-culture facilities and a centralized *Drosophila* facility. These facilities are supported by a wide range of instrumentation for cell and molecular biology, including transmission and scanning electron microscopes, confocal microscopes, and phosphorimagers.

Financial Aid

Students are supported through appointments as trainees, teaching assistants, or research assistants. For 2006–07, these positions carry stipends of $25,000 a year. Health insurance is provided.

Cost of Study

Tuition is waived for all students. However, all students pay approximately $750 per year in University fees.

Living and Housing Costs

Rent for a single student living in an on-campus apartment ranges from about $325 to $1500 per month, depending on the size of the apartment and on the number of occupants. Privately owned housing is also available close to the University.

Student Group

There are approximately 200 graduate students in the biological sciences at Stony Brook. The Graduate Program in Biochemistry and Structural Biology was established in 2000 to train students with interests in biochemistry and structural biology. Generally, students should have undergraduate majors in biochemistry, chemistry, biology, physics, or related areas, but some students come from other disciplines. The faculty has trained a number of graduate and postdoctoral students who are members of minority groups.

Location

Stony Brook is located about 60 miles east of Manhattan on the wooded North Shore of Long Island, within a few miles of picturesque villages, harbors, and beaches. Major parkways and the Long Island Rail Road provide ready access to the cultural, scientific, and commercial resources of New York City. The campus is 15 miles from Brookhaven National Laboratory.

The University

The State University of New York at Stony Brook was founded in 1957 at Oyster Bay, Long Island. In 1960, it was designated a university center and was given a mandate to develop undergraduate and graduate programs through the Ph.D. in the humanities, the physical and biological sciences, the social sciences, and engineering. In 1962, it moved to a 480-acre campus at Stony Brook; there are now more than ninety buildings on almost 1,000 acres in Stony Brook. The most recent addition is the Centers for Molecular Medicine, which houses the Center for Structural Biology and connects to the Life Sciences Building to form a biosciences complex.

Applying

Applicants should apply by January 15 to be admitted for the following semester. However, late applications are accepted. Applicants must have a baccalaureate degree from an accredited college or university. The suggested background includes mathematics through one year of calculus and at least one year of physics, biology, general chemistry, and organic chemistry; one semester of physical chemistry is also suggested. The General Test of the Graduate Record Examinations is required, and the Subject Test in the appropriate field is recommended. Students who do not meet these requirements may be admitted on condition that they remedy deficiencies after admission. Students who are members of minority groups are encouraged to apply. For online application, students should visit http://www.grad.sunysb.edu/applying/applying.htm.

Correspondence and Information

Director, Biochemistry and Structural Biology Graduate Program
Department of Biochemistry and Cell Biology
Stony Brook University
Mail Code 5215
Stony Brook, New York 11794-5215
Phone: 631-632-8533
Fax: 631-632-9730
E-mail: bsb@csb.sunysb.edu
Web site: http://csb.sunysb.edu/bsb

Stony Brook University, State University of New York

THE FACULTY AND THEIR RESEARCH

Department of Biochemistry and Cell Biology
Deborah Brown, Ph.D., Stanford, 1987. Structure and function of cholesterol/sphingolipid-rich membrane rafts and caveolae.
Dale Deutsch, Ph.D., Purdue, 1972. Molecular neurobiology of anandamide (the endogenous marijuana) regulation.
J. Peter Gergen, Ph.D., Brandeis, 1982. Transcriptional regulation in development; structure and function of Runt domain proteins.
Robert Haltiwanger, Ph.D., Duke, 1986. Role of protein glycosylation in signal transduction; Notch signaling.
Bernadette Holdener, Ph.D., Illinois, 1990. Role of protein folding in WNT signal transduction and development.
Jen-Chih Hsieh, Ph.D., Duke, 1994. Wnt signaling in embryo development and carcinogenesis.
A. Wali Karzai, Ph.D., Johns Hopkins, 1995. Structure-function studies of RNA-protein interactions.
William J. Lennarz, Ph.D., Illinois, 1959. Biosynthesis and function of glycoproteins in cell-cell interactions.
Erwin London, Ph.D., Cornell, 1979. Membrane protein structure/translocation/folding; structure and function of sphingolipid/cholesterol rafts in membranes.
Aaron Neiman, Ph.D., California, San Francisco, 1994. Developmental regulation of the secretory pathway.
Sanford R. Simon, Ph.D., Rockefeller, 1967. Extracellular matrix degradation by inflammatory and tumor cell proteases.
Steven O. Smith, Ph.D., Berkeley, 1985. Structure and function of membrane proteins.
James V. Staros, Ph.D., Yale, 1994. Biochemical and biophysical approaches to signal transduction by Erbβ-family receptors.
Rolf Sternglanz, Ph.D., Harvard, 1967. Chromatin structure and function; histone modifying enzymes; gene expression.
Gerald H. Thomsen, Ph.D., Rockefeller, 1988. Regulation of early vertebrate development by growth factor signals, ubiquitin modification, and T box family transcription factors.

Department of Applied Mathematics and Statistics
David F. Green, Ph.D., MIT, 2002. Computational biology of protein-protein interactions at the structural and systems levels.
John B. Reinitz, Ph.D., 1988, Yale. Systems biology of development and transcription in *Drosophila*.
Robert C. Rizzo, Ph.D., Yale, 2001. Computational biology; drug design.

Department of Chemistry
Daniel P. Raleigh, Ph.D., MIT, 1988. Experimental studies of protein folding and amyloid formation.
Nicole Sampson, Ph.D., Berkeley, 1990. Protein structure-function; mammalian fertilization.
Carlos L. Simmerling, Ph.D., Illinois, 1991. Development of tools for efficient simulation of chemical systems and using them to study the structure and dynamics of molecules involved in biological processes.
Peter J. Tonge, Ph.D., Birmingham, 1986. Tuberculosis pathogenesis and drug discovery; enzyme mechanisms and rational inhibitor design; fluorescent proteins.
Jin Wang, Ph.D., Illinois, 1991. Fundamental mechanism of biomolecular folding and recognition, especially protein folding and protein-protein/protein-DNA interactions.

Department of Pharmacological Sciences
Carlos de los Santos, Ph.D., Buenos Aires, 1987. Solution structures of damaged DNA; structural basis of chemical mutagenesis, lesion recognition, and DNA repair.
Arthur Grollman, M.D., Johns Hopkins, 1959. Molecular carcinogenesis and DNA repair.
Orlando D. Schärer, Ph.D., Harvard, 1996. Chemical biology of mammalian DNA repair.

Department of Physiology and Biophysics
Stuart McLaughlin, Ph.D., British Columbia, 1968. Calcium/phospholipid second messenger system.
W. Todd Miller, Ph.D., Rockefeller, 1989. Tyrosine phosphorylation and signal transduction.
Nicholas Nassar, Ph.D., Joseph Fourier (France), 1992. Crystallographic and biochemical studies of signal proteins.
Suzanne Scarlata, Ph.D., Illinois, 1984. Structure and oligomerization of membrane proteins.

Brookhaven National Laboratory
Carl W. Anderson, Ph.D., Washington (St. Louis), 1970. Cellular responses to DNA damage: role of posttranslational modifications in regulating p53 and DNA-PK.
Dax Fu, Ph.D., Mayo, 1995. Structures and mechanisms of membrane channels and transporters; X-ray crystallography.
Huilin Li, Ph.D., University of Sciences and Technology (Beijing), 1994. Structural biology of macromolecular assemblies and membrane proteins by cryo-electron microscopy.
Chang-Jun Liu, Ph.D., Chinese Academy of Sciences, 1999. Elucidation of plant natural product biosynthesis and lignocellulosic modification.
Lisa Miller, Ph.D., Yeshiva (Einstein), 1995. Study of the chemical makeup of tissue in disease, using high-resolution infrared and X-ray imaging.
Jörg Schwender, Ph.D., Karlsruhe (Germany), 1999. Plant biochemistry; metabolic flux analysis; central carbon metabolism; seed development.
John Shanklin, Ph.D., Wisconsin–Madison, 1988. Structure-function relationships of lipid modification enzymes.
Subramanyan Swaminathan, Ph.D., Madras (India), 1980. Structure-function relationships of bacterial toxins.

TEMPLE UNIVERSITY
of the Commonwealth System of Higher Education

School of Medicine
Department of Biochemistry

Programs of Study

The Department of Biochemistry in the School of Medicine offers a program leading to the Doctor of Philosophy degree in biochemistry. A limited number of students, nominated by a faculty sponsor, may be accepted as candidates for the M.S. degree. The graduate faculty has designed a well-balanced program that can be individually tailored to meet every student's interests and needs and fully prepare him or her for a career in academic or industrial biochemistry or related fields. The breadth of research interests within the department gives students an excellent opportunity for graduate training in many areas of biochemistry and molecular biology, including intermediary metabolism, enzymology, lipid biochemistry, blood chemistry, molecular biology, cancer research, polynucleotide biochemistry, hormone research, membrane biochemistry, and physical biochemistry. Members of the Fels Research Institute and the Specialized Center for Thrombosis Research who hold academic appointments in the Department of Biochemistry also participate in the graduate training program.

In addition to completing the required course work, which provides a solid foundation in biochemistry and molecular biology, the student has the opportunity to take related courses in other graduate departments and in the medical school.

Seminar programs involve students with scientists from other leading universities and research institutes and acquaint each student with the diversity of research in biochemistry and related fields and the techniques used to investigate them. Students also receive practice in evaluating and presenting scientific information.

The Ph.D. program is designed to take five years of study. Joint M.D./Ph.D. and M.D./M.S. programs are open to students accepted by both the School of Medicine and the Graduate School.

Research Facilities

The Temple University Medical Research Building houses the basic medical science departments. Excellent research facilities are available to all members of the Department of Biochemistry to conduct modern biochemical and molecular biological research. The Department of Biochemistry houses the Temple University Post-Genomic Laboratory, which provides state-of-the-art instrumentation and expertise in proteomics, tissue microanalysis, and bioinformatics. Equipment includes preparative and continuous-flow centrifuges, gradient density and continuous-flow gradient ultracentrifuges, high-pressure liquid and gas chromatographs, IR and EPR spectrometers, single- and double-beam recording spectrophotometers, a recording spectrofluorometer, liquid scintillation counters, phosphorimagers, confocal microscopy perfusion apparatus, and immunobiochemical identification equipment. Departmental facilities also include cold preparation and general instrumentation rooms, walk-in constant-temperature rooms, and teaching laboratories. The departmental library and the medical school library are located in the adjacent Kresge Science Building.

Financial Aid

Funds for stipends and tuition support are available through various sources.

Cost of Study

Temple is a state-affiliated university. Tuition for the 2005–06 academic year for full-time graduate study was $483 per semester hour for Pennsylvania residents and $706 per semester hour for out-of-state residents.

Living and Housing Costs

Relatively inexpensive privately owned apartments and rooms are available at locations accessible to public transportation and about 10 to 15 minutes from the medical school.

Student Group

Approximately 20 graduate students are enrolled in the Department of Biochemistry. A low student-faculty ratio fosters individual attention and facilitates a close working relationship between students and faculty members.

Student Outcomes

Recent graduates are pursuing postdoctoral training or holding faculty positions at highly reputable institutions, including the National Institutes of Health, Columbia, Emory, Harvard, Ohio State, Stanford, and Berkeley. Some graduates hold positions in the private sector. Such postgraduate appointments attest to the well-balanced graduate program, which prepares students for careers in academic or industrial biochemistry or related fields.

Location

The Temple University Health Sciences Center is only 20 minutes by subway from the center of one of America's most historic cities. Known as the "birthplace of freedom," Philadelphia houses many historic shrines, such as Independence Hall and the Liberty Bell. Excellent restaurants, numerous museums, major-league sports, and theater exemplify the wide variety of diversions that the city offers. In addition, the world-famous Philadelphia Orchestra, along with an outstanding annual series of jazz, rock, and country music performers, makes the Philadelphia area one of the liveliest music centers in the country. For outdoor recreation, there is beautiful Fairmount Park, whose nearly 4,000 acres provide a setting for boating, swimming, biking, and other summer sports.

The University

There are nearly 2,000 students at the Health Sciences Center, which encompasses the Schools of Medicine, Dentistry, and Pharmacy, the College of Allied Health Professions, and Temple University Hospital. The center is located on Broad Street, five blocks from the North Philadelphia station of Amtrak and about 1½ miles north of the main campus of Temple University.

The Academic Center, located at Broad Street and Montgomery Avenue, is the site of the Colleges of Liberal Arts, Science and Technology, Education, Engineering, Music and Dance, and Tourism and Hospitality Management; the Schools of Business and Management, Social Administration, Law, and Communications and Theater; and the Department of Health Studies.

Applying

Students may apply at any time but usually are admitted to begin their studies in September. Scores on the Graduate Record Examinations are required.

Correspondence and Information

Graduate Admissions Committee
Department of Biochemistry
Temple University Health Sciences Center
3420 North Broad Street
Philadelphia, Pennsylvania 19140
Fax: 215-707-7536
E-mail: kathleen.pope@temple.edu
    mdphd@temple.edu (for the M.D./Ph.D. program)
Web site: http://www.medschool.temple.edu/Biochemistry/index.html

Temple University

THE FACULTY AND THEIR RESEARCH

James P. Burke, Associate Professor; Ph.D., CUNY, Mount Sinai, 1974. Effect of trace metals on membrane structure and function. E-mail: jburke@pcpm.edu

Parkson Lee-Gau Chong, Professor; Ph.D., Illinois, 1982. Structure and function of archaebacterial membranes; biochemical application of fluorescence spectroscopy, regulation of membrane surface acting enzymes; cholesterol superlattice. E-mail: parkson.chong@temple.edu

Jimmy H. Collins, Professor and Acting Chairperson; Ph.D., Texas at Austin, 1977. Structure, function, and regulation of membrane-associated molecular motors. E-mail: jimmy.collins@temple.edu

Danny N. Dhanasekaran, Associate Professor; Ph.D., Indian Institute of Science, 1985. Molecular biology of signal transduction; role of G-protein signaling in cell growth, differentiation, and oncogenesis; structure and function analysis of G-proteins. E-mail: danny.dhanasekaran@temple.edu

Kathleen Giangiacomo, Associate Professor; Ph.D., Pennsylvania, 1989. Structure and function of potassium channel proteins. E-mail: kathleen.giangiacomo@temple.edu

Xavier Graña, Associate Professor; Ph.D., Barcelona (Spain), 1991. Regulation of cell-cycle progression, cell-cycle exit to quiescence and cell-cycle exit to undergo cell differentiation in mammalian cells; role of cyclins, cyclin-dependent kinase (CDKs), growth suppressor genes, and cyclin-dependent kinase inhibitors (CKIs). E-mail: xavier.grana-amat@temple.edu

Charles Grubmeyer, Professor and Chairperson, Admissions Committee; Ph.D., Alberta, 1979. Enzymatic reaction mechanisms; enzymes of histidine and nucleotide biosynthesis. E-mail: charles.grubmeyer@temple.edu

Barbara Hoffman, Professor; Ph.D., Michigan, 1973. Role of oncogenes, tumor suppressor genes, and transcription factors in normal blood cell development, programmed cell death, and leukemia. E-mail: barbara.hoffman@temple.edu

Keith E. Latham, Associate Professor; Ph.D., Virginia, 1988. Molecular and genetic analyses of embryonic mouse development. E-mail: keith.latham@temple.edu

Dan A. Liebermann, Professor; Ph.D., Weizmann (Israel), 1980. Proto-oncogenes and tumor suppressor genes in normal cell development, programmed cell death, and oncogenesis. E-mail: dan.liebermann@temple.edu

Warren E. Masker, Professor; Ph.D., Rochester, 1969. Molecular mechanisms of mutagenesis; mechanisms of DNA replication, recombination, and repair. E-mail: warren.masker@thunder.ocis.temple.edu

Salim Merali, Associate Professor; Ph.D., CUNY Graduate Center, 1993. Proteomics and metabolomics studies; tissue profiling by mass spectroscopy; identification of novel biochemical approaches to the treatment and diagnosis of *Pneumocystis* pneumonia (PCP) and cancer. E-mail: smerali@temple.edu

Elizabeth Moran, Professor; Ph.D., New York Medical College, 1983. Regulation of growth and differentiation in normal and cancer cells; interactions of cell growth regulatory proteins and chromatin-modifying complexes with the transforming proteins of DNA tumor viruses. E-mail: elizabeth.moran@temple.edu

E. Premkumar Reddy, Professor and Director of the Fels Institute for Cancer Research and Molecular Biology; Ph.D., Osmania (India), 1971. Role of oncogenes in cell growth, differentiation, and neoplasia; retrovirology and gene therapy. E-mail: premkumar.reddy@temple.edu

Scott K. Shore, Associate Professor; Ph.D., Drexel, 1984. Mechanisms of oncogene activation and their role in cancer; gene therapy for human leukemia. E-mail: scott.shore@temple.edu

Dianne R. Soprano, Professor and Director, Biochemistry Graduate Program; Ph.D., Rutgers, 1980. Molecular mechanism of action of vitamin A; effects of retinol and retinoic acid in the differentiation of cells; molecular biology of the retinoid binding proteins. E-mail: dianne.soprano@temple.edu

Barbara L. Stitt, Associate Professor; Ph.D., Caltech, 1978. Structure and mechanism of *E. coli* transcription termination factor rho and its role in gene expression. E-mail: barbara.stitt@temple.edu

Robert J. Suhadolnik, Professor; Ph.D., Penn State, 1956. Biosynthesis of nucleosides; RNA and DNA synthesis in bacteria, viruses, and mammalian tissue; role of interferon in development of the 2-5A synthetase/RNase L/PKR antiviral-anticancer state in mammalian cells; analogues of 2',5'-oligo(A); mechanism of action of biological response modifiers (including dsRNAs) as anti-HIV and anticancer agents. E-mail: robert.suhadolnik@temple.edu

Peter N. Walsh, Professor; M.D., Washington (St. Louis), 1961; D.Phil., Oxford, 1972. Blood coagulation protein biochemistry; characterization of structural domains of coagulation proteins; enzymology and kinetics of blood coagulation; receptor-mediated interactions of coagulation proteins with platelets; characterization of intrinsic platelet coagulation proteins and inhibitors. E-mail: peter.walsh@temple.edu

THOMAS JEFFERSON UNIVERSITY

Jefferson College of Graduate Studies
Kimmel Cancer Center
Department of Biochemistry and Molecular Biology
Graduate Program in Biochemistry and Molecular Biology

Programs of Study

The Ph.D. program in biochemistry and molecular biology is offered through the Department of Biochemistry and Molecular Biology at Thomas Jefferson University. The Ph.D. program in biochemistry and molecular biology is designed to provide students with the basis for successful careers as independent scientists and scholars in either the academic or industrial sector. Students entering with a baccalaureate degree take core curriculum courses in biochemistry, molecular biology, cell biology, and genetics. Advanced courses in structural biology, molecular pharmacology, molecular biology, and genetics can be taken in both the first and second years, and the curriculum is flexible to accommodate the individual student's background and interests. All course work is generally completed by the end of the second year. Research rotations are included in the first-year curriculum. Current research interests of the faculty members include structure-function relationships for tRNA, mitochondrial genetics and related diseases, the role of SUMO modification in protein function, MAP kinase pathways in meiotic regulatory pathways, protein misfolding in neurologic disease, nucleotide chemistry in therapeutics and gene therapy, combinatorial drug discovery, molecular regulation of G-protein–coupled receptors, biochemical aspects of viral entry and transport, and many more. In addition, several weekly seminar series, frequent laboratory seminars, and invited lectureships form an integral part of the student's education. Students take a preliminary examination at the end of the second year. Success in this examination allows the student to continue with the thesis program and research. The thesis research is performed under the supervision of a faculty adviser chosen by the student, and the written thesis describing the experimental work and its significance is defended before the degree is attained.

The Graduate Program in Biochemistry and Molecular Biology, along with the Programs in Genetics, Molecular Pharmacology and Structural Biology, and Immunology and Microbial Pathogenesis, make up the Joint Graduate Programs of the Kimmel Cancer Center. Students applying to the Joint Graduate Programs may perform research rotations or work with faculty members in any of these programs. There is also an M.S. program in biomedical chemistry. This 40-credit program may be completed on a part-time or full-time basis over a two- to four-year period.

Research Facilities

Research laboratories are primarily located in the Department of Biochemistry and Molecular Biology, which is housed in the Bluemle Life Sciences Building. In addition to extensive basic equipment and facilities, the program provides access to numerous specialized resources. These include facilities for peptide and oligonucleotide synthesis and sequencing; cell sorting by flow cytometry; a CD spectrometer; state-of-the-art X-ray detectors for macromolecular crystallography, protein purification, and characterization; proteomics and microarray analysis; and biomolecular imaging. Interaction with the faculties of other Departments of the University is encouraged as part of the students' education.

Financial Aid

Financial support is available to full-time Ph.D. students in the form of University fellowships. Most types of aid include a stipend ($24,500 in 2006–07) and remission of tuition plus health insurance benefits for a single person. Also available to students demonstrating financial need are Title IV funds and University loan programs.

Cost of Study

Tuition and fees are $15,640 per year in 2006–07.

Living and Housing Costs

Apartments are available in University residences for graduate students enrolled in the program; early rental application is advised. Reasonable alternative housing near the University in the Philadelphia area can also be found.

Student Group

Each year, approximately 15 students are admitted into the Joint Graduate Programs of the Kimmel Cancer Center. The incoming class size in biochemistry and molecular biology is expected to be 3. At present, the University enrolls more than 2,500 students. The College of Graduate Studies enrolls about 630 students, about half of whom are women. A low ratio of students to faculty members is maintained in order to encourage close interaction.

Location

Thomas Jefferson University is centrally located in Philadelphia within walking distance of many places of cultural interest, including concert halls, theaters, museums, art galleries, and historic sites. There are numerous intercollegiate and professional sports events. Convenient bus and subway lines connect the University with other local universities and colleges and with several outstanding libraries. The proximity of the New Jersey shore and the Pennsylvania mountains offers year-round recreational opportunities, and New York City and Washington, D.C., are each just 2 hours away.

The University

Thomas Jefferson University is an academic health center emphasizing the biological sciences. It evolved from the Jefferson Medical College, which was founded in 1824, and includes the College of Graduate Studies, the College of Health Professions, the University hospital, and various affiliated hospitals and institutions. Jefferson Alumni Hall houses the TJU Fitness and Recreation Facility, study lounges, a swimming pool, a gymnasium, and a handball/squash court.

Applying

Applicants must have at least a bachelor's degree, with a strong educational record in biochemistry, chemistry, and molecular biology. Preference is given to applications received after January 1 in the order in which they are received. Admission is based on GRE scores, undergraduate performance, letters of recommendation, research accomplishments, and interview evaluations. Applications from minority group candidates are encouraged. Application forms can be obtained from the Web site and submitted online.

Correspondence and Information

For applications:
Jessie Pervall
Director of Admissions and Recruitment
Jefferson College of Graduate Studies
Thomas Jefferson University
1020 Locust Street, M-60
Philadelphia, Pennsylvania 19107-6799
Phone: 215-503-4400
Fax: 215-503-3433
E-mail: cgs-info@jefferson.edu
Web site: http://www.jefferson.edu/jcgs

For program information:
Joanne Balitzky
Training Programs Coordinator
Kimmel Cancer Center
Thomas Jefferson University
233 South 10th Street, 910 BLSB
Philadelphia, Pennsylvania 19107-5541
Phone: 215-503-6687
Fax: 215-503-0622
E-mail: joanne.balitzky@jefferson.edu
Web site: http://www.jefferson.edu/jcgs/phd/bmb/

Thomas Jefferson University

THE FACULTY AND THEIR RESEARCH

Arthur Allen, Professor; Ph.D., Temple, 1956. Interrelationships between carbohydrate and lipid metabolism.

Emad S. Alnemri, Professor; Ph.D., Temple, 1991. Molecular mechanisms of programmed cell death (apoptosis); regulation of caspase activation during apoptosis and inflammation; role of mitochondria in cell death and survival.

Carol L. Beck, Assistant Professor; Ph.D., Vanderbilt, 1993. Voltage-gated ion channel physiology and pharmacology; molecular genetic basis of ion channel diseases; skeletal muscle voltage-gated chloride channels; calcium-activated chloride channels.

Jeffrey L. Benovic, Professor and Chairperson; Ph.D., Duke, 1986. Molecular and regulatory properties of G-protein–coupled receptors; role of receptor dysregulation in cancer and cardiovascular and neurological disorders.

George C. Brainard, Professor; Ph.D., Wesleyan, 1973. Photobiology and neuroendocrine and circadian regulation; control of melatonin in humans; effects of light on physiology, behavior, and cancer progression; use of light as countermeasure in long-duration space flight.

David Capuzzi, Professor; M.D., Jefferson Medical, 1964; Ph.D., Johns Hopkins, 1969. Controlled studies to examine the safety and efficacy of new pharmacologic and nutritional therapies.

Mon-Li Chu, Professor; Ph.D., Florida, 1975. Role of extracellular matrix genes in development and disease, with emphasis on cardiovascular, skin, and musculoskeletal systems; molecular basis and therapies of congenital muscular dystrophy.

Edgar Davidson, Research Assistant Professor; Ph.D., Glasgow, 1981. Human mitochondrial genetics; DNA mutations in disease; biochemistry and molecular genetics of respiratory chain enzymes; mitochondrial translation; mitochondrial tRNA synthetases.

John L. Farber, Professor; M.D., California, San Francisco, 1966. Biochemical mechanisms of cell injury in ischemia of liver-cell necrosis; biochemical toxicology of activated oxygen; chemical carcinogenesis.

Sam Gandy, Professor; M.D./Ph.D., Medical University of South Carolina, 1982. Protein phosphorylation enzymes (especially PKC, ERK, PP1, PP2A, ROCKs) and the target phosphoproteins that regulate trafficking and processing of membrane proteins, especially the Alzheimer amyloid precursor protein (APP) and the secretase proteases that cleave APP (alpha, beta gamma secretases); neurotransmitters and hormones linked to the phosphorylation enzymes, especially gonadal hormones and ERK; regulated shedding of ectodomains of cell surface proteins; regulated intramembranous proteolysis and assembly of the gamma secretase.

Barry J. Goldstein, Professor; M.D./Ph.D., Rochester, 1982. Insulin receptor signal transduction; role of protein tyrosine phosphatases in the regulation of insulin action and insulin-resistant disease states; vascular signaling of adipose secretory products, including cytokines and adiponectin.

Gerald B. Grunwald, Professor; Ph.D., Wisconsin–Madison, 1981. Developmental biology and neuroscience; eye development and disease; analysis of the role of cadherin cell adhesion molecules in normal and abnormal embryonic development and in proliferative diseases of the nervous system and the eye; investigation of cellular and molecular regulatory mechanisms of cadherin expression and function.

Noreen J. Hickok, Associate Professor; Ph.D., Brandeis, 1981. Development of surfaces for combating peri-prosthetic infection; bacterial-osteoblastic cell interactions; role of extracellular bone matrix in bone formation, bacterial adhesion, and latent infections.

Shiu-Ying Ho, Research Assistant Professor; Ph.D., Rutgers, 2000. Forward and reverse genetic approaches to studying lipid metabolism using zebrafish as a vertebrate model system; the role of fat-free gene in regulating dietary lipid absorption and metabolism.

Jan B. Hoek, Professor; Ph.D., Amsterdam, 1972. Systems biology of intracellular signal transduction networks; deregulation of cytokine and growth-factor signaling in the liver associated with chronic alcohol consumption; early signaling responses during liver regeneration; bioenergetics and mitochondrial metabolism and its role in intracellular signaling and apoptosis.

Ya-Ming Hou, Professor; Ph.D., Berkeley, 1986. Genetic and biochemical studies of tRNA, including structure and function mechanism of tRNA aminoacylation; maintenance of the tRNA 3' end, maturation and processing, editing and repair, decoding on the ribosome, and development of bacterial pathogenesis in infectious diseases; targeting tRNAs as strategies against metabolic and neurodegenerative disorders.

James B. Jaynes, Associate Professor; Ph.D., Washington (Seattle), 1980. Developmental genetics and molecular biology of processes regulated by homeodomain transcription factors and higher-order chromatin structure.

Sergio Jimenez, Professor; M.D., San Marcos (Peru), 1964. Regulation of collagen gene expression; molecular mechanisms of tissue fibrosis; role of non-Smad pathways on transforming growth factor β (TGFβ) regulation of extracellular matrix production; regulation of chondrocyte gene expression and pathogenesis of osteoarthritis.

Erica Johnson, Assistant Professor; Ph.D., MIT, 1992. The ubiquitin-related protein Smt3/SUMO and its conjugation pathway and function in yeast.

Hideko Kaji, Professor; Ph.D., Purdue, 1958. Molecular mechanisms and function of ribosome recycling factor (RRF) in protein synthesis; biochemical pharmacology of human antiretroviral drugs.

James H. Keen, Professor; Ph.D., Cornell, 1976. Molecular mechanisms coupling signal transduction with receptor-mediated endocytosis and exocytosis; membrane transport studied by biochemical, molecular biological, and morphological approaches.

Michael P. King, Associate Professor; Ph.D., Caltech, 1987. Mammalian mitochondrial biogenesis; molecular genetics of mtDNA mutations in human disease; mitochondrial transformation; posttranscriptional modification of mitochondrial RNAs; motor-neuron development.

Jose Martinez, Professor; M.D., Central Madrid, 1957. Role of fibrin in the differentiation of endothelial cells toward the formation of capillary tubes.

Alexander Mazo, Associate Professor; Ph.D., Russian Academy of Sciences, 1976. Chromatin modifications by epigenetic factors and nuclear hormone receptors.

W. Edward Mercer, Professor; Ph.D., Penn State, 1980. Molecular mechanisms of the p53 tumor-suppressor protein in cell-cycle checkpoint control and apoptotic cell death; molecular mechanisms of cell-cycle–dependent kinases, protein kinase inhibitors, and protein phosphatases in cell-cycle checkpoint control and apoptosis.

Diane E. Merry, Associate Professor and Program Director; Ph.D., Pennsylvania, 1991. Molecular pathogenesis of Kennedy's disease and other polyglutamine expansion diseases; protein misfolding and aggregation in neurodegenerative disease; mouse models of neurologic disease; domain function in trafficking and degradation of the androgen receptor; gene and pharmacological therapies for neurodegenerative diseases.

Ulrich Rodeck, Professor; M.D., Germany, 1981. Regulation of epithelial cell survival by the epidermal growth factor receptor; coordinate regulation of cell-cycle progression and cell survival by growth factor– and adhesion-dependent signal transduction; signal transduction events controlling cell fate in the anchorage-independent state; zebrafish as a new in vivo model system for the genotoxic stress response.

Peter Ronner, Associate Professor; Ph.D., Swiss Federal Institute of Technology, 1978. Control of insulin and glucagon release; regulation of ATP-sensitive K channels and voltage-dependent Ca channels; role of cellular metabolism in signal transduction.

Michael Root, Assistant Professor; M.D./Ph.D., Harvard, 1997. Structure and function of glycoproteins involved in viral and cell-cell membrane fusion; design of viral entry inhibitors and immunogens for vaccine development.

Barbara P. Schick, Associate Professor; Ph.D., Bryn Mawr, 1979. Regulation of gene expression and biosynthesis of proteoglycans in hematopoietic and nonhematopoietic cells during cell development; functional aspects of the intracellular and secreted proteoglycans.

Matthias J. Schnell, Professor; Ph.D., Hohenheim (Stuttgart), 1994. Recombinant rhabdovirus-based vectors as potential live and killed vaccines against HIV-1, other infectious diseases, and cancer; molecular biology and pathogenesis of negative-stranded RNA viruses; molecular mechanism of neurotropism of rabies virus.

Stephanie Schulz, Research Assistant Professor; Ph.D., Vermont, 1987. Molecular characterization of guanylyl cyclase receptors; structure-function analysis, signal transduction mechanisms, and regulation of gene expression.

Charles P. Scott, Assistant Professor; Ph.D., Pennsylvania, 1997. Combinatorial drug discovery; rational drug design; chemical genomics of host-pathogen interactions and cancer.

Sandor Shapiro, Professor; M.D., Harvard, 1957. Structure and function of the human filamins.

Thomas Tulenko, Professor; Ph.D., Boston University, 1972. Role of the cell membrane in the pathobiology of arterial wall cells associated with various peripheral vascular disease states, with particular emphasis on atherosclerosis.

Jouni Uitto, Professor; Ph.D./M.D., Helsinki, 1970. Molecular genetics of the cutaneous basement membrane zone; regulation of collagen and elastin gene expression and the molecular basis of heritable and acquired connective tissue diseases.

Scott A. Waldman, Professor; Ph.D., Thomas Jefferson, 1980; M.D., Stanford, 1987. Molecular mechanisms of signal transduction, with emphasis on receptor-effector coupling and postreceptor signaling mechanisms; molecular mechanisms underlying tissue-specific transcriptional regulation; translation of molecular signaling mechanisms to novel diagnostic and therapeutic approaches to patients with cancer.

Philip B. Wedegaertner, Associate Professor; Ph.D., California, San Diego, 1991. G-protein signal transduction; molecular mechanisms and functions of covalent modifications and regulated subcellular localization.

David A. Wenger, Professor; Ph.D., Temple, 1968. Defects in lysosomal storage disorders, including enzymes and studies at the lipid, protein, RNA, and DNA levels; projects related to the treatment of Krabbe disease by gene therapy with the galactocerebrosidase gene in the mouse, dog, and monkey models.

Eric Wickstrom, Professor; Ph.D., Berkeley, 1972. Sensing, imaging, regulation, and control of oncogene expression in cells and animal models with nucleic acid derivatives; permanent bonding of drugs to implants.

Charlene J. Williams, Professor; Ph.D., Rutgers, 1983. Molecular genetics and biochemistry of heritable osteoarthropathies and inflammatory arthropathies.

John C. Williams, Assistant Professor; Ph.D., Columbia, 1995. Biochemical and biophysical studies of dynein function, regulation, and cargo recognition.

Kevin J. Williams, Associate Professor; M.D., Johns Hopkins, 1980. Cell-surface heparan sulfate proteoglycans as receptors mediating novel endocytic pathways; molecular processes involved in retention or trapping of atherogenic lipoproteins with the vessel wall leading to atherosclerosis; development of a synthetic particle that pulls cholesterol out of arteries and delivers it into the liver for disposal.

Edward Winter, Associate Professor; Ph.D., SUNY at Stony Brook, 1984. Meiotic development; chromosome structure and function; MAP kinase signaling pathways in yeast.

Kyonggeun Yoon, Associate Professor; Ph.D., Berkeley, 1976. New gene therapy technologies for potential treatment of genetic or acquired skin diseases.

Allen R. Zeiger, Professor; Ph.D., Johns Hopkins, 1967. Antibiotics covalently bonded to medical implants; opioid receptors and immune function; opiate metabolism.

TULANE UNIVERSITY

School of Medicine
Department of Biochemistry

Programs of Study

The Department of Biochemistry offers programs of study leading to the Ph.D. and M.S. degrees. Entering students usually have a B.S. degree with solid preparation in biochemistry, molecular and cellular biology, genetics, chemistry, physics, and mathematics. After a core curriculum is completed during the first year, students' programs of study are individually developed to meet their specific interests and may include courses in other basic science disciplines. The goal of the Ph.D. program is to develop highly competent scientists with the ability to conduct independent research in academic or other environments. The Department has undergone major expansion over the last few years, including renovations of the physical facilities, development of modern shared instrumentation facilities for molecular biology, and recruitment of new faculty members.

The Ph.D. Program in Biochemistry provides lecture and tutorial courses, seminars, laboratory rotations, and major research projects carried out under the supervision of faculty research advisers. By the end of the second year of training, each student must demonstrate a comprehensive knowledge of biochemistry and molecular biology by successfully completing a qualifying examination. The doctoral degree is awarded following the formal presentation and defense of the dissertation research. There is no foreign language requirement. Normally four years are required to complete the Ph.D.

Research Facilities

The Department of Biochemistry is housed in the School of Medicine, which is located in a large medical complex. Faculty members have research laboratories that are well equipped for modern biochemistry and molecular biology research. There are also major shared instrumentation facilities and cooperative research efforts between faculty members in many basic science and clinical departments of the University as well as at the Tulane Regional Primate Center.

Financial Aid

Stipends (at least $21,000 per year) and tuition scholarships are provided by the University or the Department.

Cost of Study

Tuition payments are generally covered by tuition scholarships.

Living and Housing Costs

The School of Medicine maintains an apartment complex within walking distance of the campus; the average monthly rent for a shared double room in 2004–05 was $545 per person. Most students live in apartments, many of which are available off campus. The cost of living in New Orleans is relatively low for a large metropolitan area.

Student Group

Approximately 4,900 graduate students are enrolled at Tulane. Of these, about 100 are in the basic medical sciences, and about 15 to 20 are in biochemistry. The Department generally has 6 to 8 postdoctoral fellows, who augment the research training of graduate students in biochemistry. The size of the graduate program allows personalized attention, maximum student-faculty interactions, and tutorials as well as lecture courses.

Location

New Orleans offers the advantages of a major cosmopolitan city and a distinctive mixture of cultural and recreational activities. The ambience of the old French Quarter and Uptown Garden District plus the Cajun culture and two-week-long Mardi Gras festivities are unique in the United States. Furthermore, New Orleans is an attractive place to live because of its housing affordability and ease of commuting from various parts of the city and surrounding suburbs.

The University

Tulane University was founded in 1834. Eleven academic divisions now enroll about 13,215 students. In addition to having one of the finest and most diverse academic curricula in the South, Tulane University and the city of New Orleans offer many cultural programs, including theater, opera, art exhibits, concerts, and outstanding lecture series by distinguished visitors.

Applying

Applicants for admission are expected to have a strong educational background in biochemistry, molecular biology, and chemistry. Admission is based on the student's performance in undergraduate studies, scores on the Graduate Record Examinations, and letters of recommendation. A bachelor's degree with a major in biochemistry, biology, genetics, or chemistry from an accredited college or university is a prerequisite. Adequate TOEFL scores are required of students from countries in which English is not the native language.

Correspondence and Information

Graduate Studies Committee
Department of Biochemistry SL43
Tulane University Health Sciences Center
1430 Tulane Avenue
New Orleans, Louisiana 70112
Phone: 504-988-5291
Fax: 504-988-2739
E-mail: thills@tulane.edu
Web site: http://www.tulane.edu/~biochem/

Tulane University

THE FACULTY AND THEIR RESEARCH

Diane A. Blake, Professor; Ph.D., Illinois at Urbana-Champaign, 1977. Effects of extracellular matrix on cell growth and wound healing; isolation and characterization of antimetal antibodies.

Melanie Ehrlich, Professor; Ph.D., SUNY at Stony Brook, 1970. Gene expression and DNA-binding proteins; DNA rearrangements accompanying oncogenic transformation; DNA methylation and stability of human chromosomes.

David L. Hurley, Associate Professor; Ph.D., Penn State, 1986. Neuroendocrine regulation of gene transcription.

Frank E. Jones, Associate Professor; Ph.D., McMaster, 1995. Mouse and cellular models of signal transduction events contributing to breast cancer.

Jim D. Karam, Professor and Chair; Ph.D., North Carolina at Chapel Hill, 1965. RNA-binding proteins and control of DNA replication.

Samuel J. Landry, Associate Professor; Ph.D., LSU, 1988. Protein folding; protein-protein interactions; antigen processing.

Su-Chen Li, Research Professor; Ph.D., Oklahoma, 1966. Catabolism of glycoconjugates.

Yu-Teh Li, Professor; Ph.D., Oklahoma, 1963. Chemistry and chemical pathology of glycoconjugates.

Arthur J. Lustig, Professor; Ph.D., Chicago, 1981. Role of telomere dynamics in chromosome stability and transcriptional regulation.

James M. Nolan, Research Assistant Professor; Ph.D., Duke, 1988. Ribozyme structure-function and RNA-protein interactions.

Jeremy M. Stark, Assistant Professor; Ph.D., Washington (Seattle), 1998. Mechanisms of homologous DNA repair.

Joseph Vaccaro, Assistant Professor; Ph.D., Johns Hopkins, 1995. Molecular mechanisms of signaling and viral replication.

William C. Wimley, Associate Professor; Ph.D., Virginia, 1990. Folding and structure of proteins in membranes.

SELECTED PUBLICATIONS OF THE FACULTY

Delehanty, J. B., R. M. Jones, T. C. Bishop, and **D. A. Blake.** Identification of important residues in metal-chelate recognition by monoclonal antibodies. *Biochemistry* 42:14173–83, 2003.

Blake, R. C., II, et al. **(D. A. Blake).** Allosteric binding properties of a monoclonal antibody and its Fab fragment. *Biochemistry* 42:497–598, 2003.

Darwish, I. A., and **D. A. Blake.** Development and validation of a sensitive one-step immunoassay for determination of cadmium in human serum. *Anal. Chem.* 74:52–8, 2002.

Ehrlich, M. The controversial denouement of vertebrate DNA methylation research. *Biochemistry* 5:568–75, 2005.

Nishiyama, R., et al. **(M. Ehrlich).** A DNA repeat, NBL2, is hypermethylated in some cancers but hypomethylated in others. *Cancer Biol. Ther.* 4(4):440–8, 2005.

Yang, F., C. Shao, V. Vedanarayanan, and **M. Ehrlich.** Cytogenetic and immuno-FISH analysis of the 4q subtelomeric region, which is associated with facioscapulohumeral muscular dystrophy. *Chromosoma* 2:350–9, 2004.

Clements, M. D., H. L. Bart Jr., and **D. L. Hurley.** Isolation and characterization of two distinct growth hormone cDNAs from the tetraploid smallmouth buffalofish *(Ictiobus bubalus). Gen. Comp. Endocrinol.* 136:411–18, 2004.

Hurley, D. L., et al. Reduced hypothalamic neuropeptide Y expression in GH- and PRL-deficient Ames and Snell dwarf mice. *Endocrinology* 144:4783–9, 2003.

Wojtkiewicz, P. W., C. J. Phelps, and **D. L. Hurley.** Transcript abundance in mouse pituitaries with altered GH expression quantified by RT-PCR implicates transcription factor Zn-16 in gene regulation in vivo. *Endocrine* 18:67–74, 2002.

Vidal, G. A., A. Naresh, L. Marrero, and **F. E. Jones.** Presenilin-dependent γ-secretase processing regulates multiple ERBB4/HER4 activities. *J. Biol. Chem.* 280:19777–83, 2005.

Williams, C. C., et al. **(F. E. Jones).** The ERBB4/HER4 receptor tyrosine kinase regulates gene expression by functioning as a STAT5A nuclear chaperone. *J. Cell Biol.* 167:469–78, 2004.

Long, W., et al. **(F. E. Jones).** Impaired differentiation and lactational failure in *ErbB4*-deficient mammary glands identify ErbB4 as an obligate mediator of Stat5. *Development* 130:5257–68, 2003.

Karam, J. D. Bacteriophages: The viruses for all seasons of molecular biology. *Virol. J.* 2:19, 2005.

Petrov V. M., and **J. D. Karam.** Diversity of structure and function of DNA polymerase (gp43) of T4-related bacteriophages. *Biochemistry* 69(11):1213–18, 2004.

Borjac-Natour, J. M., V. M., Petrov, and **J. D. Karam.** Divergence of the mRNA targets for the Ssb proteins of bacteriophages T4 and RB69. *Virol. J.* 1:4, 2004.

Kleen, T. O., et al. **(S. J. Landry).** Tc1 effector diversity shows dissociated expression of granzyme B and interferon-gamma in HIV infection. *AIDS* 18:383–92, 2004.

Curiel, T. J., et al. **(S. J. Landry).** Peptides identified through phage display direct immunogenic antigen to dendritic cells. *J. Immunol.* 172:7425–31, 2004.

Shewmaker, F., et al. **(S. J. Landry).** A mobile loop order-disorder transition modulates the speed of chaperonin cycling. *Protein Sci.* 13:2139–48, 2004.

Landry, S. J. Structure and energetics of an allele-specific genetic interaction between DnaJ and DnaK: Correlation of nuclear magnetic resonance chemical shift perturbations in the J-domain of Hsp40/DnaJ with binding affinity for the ATPase domain of Hsp70/DnaK. *Biochemistry* 42:4926–36, 2003.

Li, S.-C., Y. Hama, and **Y.-T. Li.** Interaction of GM2 activator protein with glycosphingolipids. *Methods Enzymol.* 363:230–41, 2003.

Mauri, L., et al. **(S.-C. Li** and **Y.-T. Li).** Procedure for the separation of the GM2 ganglioside species with different ceramide structures by flash reversed-phase silica gel liquid chromatography. *J. Chromatogr.* 796:1–10, 2003.

Shimada, Y., **Y.-T. Li,** and **S.-C. Li.** Effect of GM2 activator protein on the enzymatic hydrolysis of phospholipids and sphingomyelin. *J. Lipid Res.* 44:342–8, 2003.

Li, Y.-T., et al. **(S.-C. Li).** Presence of an unusual GM2 derivative, taurine-conjugated GM2, in Tay-Sachs brain. *J. Biol. Chem.* 278:35286–91, 2003.

Joseph, I., D. Jia, and **A. J. Lustig.** Ndj1p-dependent epigenetic resetting of telomere size in yeast meiosis. *Curr. Biol.* 15(3):231–7, 2005.

Williams, B., M. K. Bhattacharyya, and **A. J. Lustig.** Mre11p nuclease activity is dispensable for telomeric rapid deletion. *DNA Repair* 4(9):994–1005, 2005.

Lustig, A. J. Clues to catastrophic telomere loss in mammals from yeast telomere rapid deletion. *Nat. Rev. Genet.* 4:916–23, 2003.

Williams, B., and **A. J. Lustig.** The paradoxical relationship between NHEJ and telomeric fusion. *Mol. Cell.* 11:1125–6, 2003.

Rasmussen, T. A., and **J. M. Nolan.** G350 of *Escherichia coli* RNase P RNA contributes to Mg binding near the active site of the enzyme. *Gene* 10:177–85, 2002.

Sharkady, S. M., and **J. M. Nolan.** Bacterial ribonuclease P holoenzyme crosslinking analysis reveals protein interaction sites on the RNA subunit. *Nucleic Acids Res.* 29:3848–56, 2001.

Stark, J. M., et al. Genetic steps of mammalian homologous repair with distinct mutagenic consequences. *Mol. Cell Biol.* 24(21):9305–16, 2004.

Richardson, C., **J. M. Stark,** M. Ommundsen, and M. Jasin. Rad51 overexpression promotes alternative double-strand break repair pathways and genome instability. *Oncogene* 23(2):546–53, 2004.

Stark, J. M., and M. Jasin. Extensive loss of heterozygosity is suppressed during homologous repair of chromosomal breaks. *Mol. Cell Biol.* 23(2):733–43, 2003.

Ray, A. S., et al. **(J. A. Vaccaro).** Probing the molecular mechanisms of AZT drug resistance mediated by HIV-1 reverse transcriptase using a transient kinetic analysis. *Biochemistry* 42:8831–41, 2003.

Vaccaro, J. A., K. M. Parnell, S. A. Terezakis, and K. S. Anderson. Mechanism of inhibition of the human immunodeficiency virus type 1 reverse transcriptase by d4TTP: An equivalent incorporation efficiency relative to the natural substrate dTTP. *Antimicrob. Agents Chemother.* 44:217–21, 2000.

Jeansonne, D. P., et al. **(J. A. Vaccaro).** A rapid ATP affinity-based purification for the human non-receptor tyrosine kinase c-Src. *Protein Expression Purification,* in press.

You, M., E. Li, **W. C. Wimley,** and K. Hristova. Forster resonance energy transfer in liposomes: Measurements of transmembrane helix dimerization in the native bilayer environment. *Anal. Biochem.* 340(1):154–64, 2005.

Sainz, B., Jr, et al. **(W. C. Wimley).** The aromatic domain of the coronavirus class I viral fusion protein induces membrane permeabilization: Putative role during viral entry. *Biochemistry.* 44(3):947–58, 2005.

Wimley, W. C., and S. H. White. Reversible unfolding of beta-sheets in membranes: A calorimetric study. *J. Mol. Biol.* 342(3):703–11, 2004.

THE UNIVERSITY OF ALABAMA AT BIRMINGHAM

Department of Biochemistry and Molecular Genetics

Program of Study

The Department of Biochemistry and Molecular Genetics, in conjunction with the Departments of Cell Biology, Microbiology, and Neurobiology, offers course work and individual laboratory research leading to the Ph.D. degree. The program is designed to provide high-quality interdisciplinary training in cell and molecular biology to a selected group of predoctoral students, preparing them to become independent investigators in these disciplines. Students are immersed in research at the forefront of scientific endeavor and provided with sufficient guidance and course work to place their research in the proper perspective.

This interdisciplinary program includes faculty members from Anesthesiology, the Arthritis Center, Biochemistry and Molecular Genetics, Biology, the Cancer Center, Cell Biology, Endocrinology, Geographic Medicine, Hematology/Oncology, Infectious Disease, Medicine, Microbiology, Neurobiology, Neurology, Neuropsychiatry, Oral Biology, Pathology, Pediatrics, Pharmacology, Physiology and Biophysics, Psychiatry, and Public Health.

The first-year curriculum emphasizes three areas: acquisition of a working knowledge of contemporary cellular and molecular biology through an intensive, integrated course in biochemistry, genetics, cell biology, virology, immunology, and neurobiology; involvement in a diversity of laboratory research training experiences; and the development of skills in reading, writing, and speaking. Advanced students are engaged primarily in research but also take some advanced courses and tutorials in specialized areas of interest and participate in seminars. Completion of requirements for the Ph.D. usually takes six years. No foreign language is required.

Areas of specialization for dissertation research include fundamental molecular biology; biochemistry of nucleic acids; prokaryotic and eukaryotic molecular and cell biology; molecular virology; viral, microbial, and mammalian genetics; immunogenetics; cellular, developmental, and tumor immunology; immunochemistry; biological macromolecules and membranes; host-parasite relationships; infectious diseases; biochemistry of connective tissues; X-ray crystallography; NMR spectroscopy; molecular biophysics; ion-channel structure and biophysics; synaptic function and plasticity; neuronal development; and glial-neuronal signaling. Students are encouraged to present research results at scientific meetings and publish in scientific journals.

Research Facilities

Faculty members participating in the program have more than 400,000 square feet of laboratory space. In addition to well-equipped labs, a number of core facilities are available. These include automated DNA sequencing, DNA chip analysis, NMR spectroscopy, electron microscopy, protein and nucleic acid synthesis and analysis, mass spectrometry, confocal microscopy, fluorescence-activated cell sorting, large-scale bacterial fermentation, X-ray diffraction, a P3 containment facility, computer facilities, and a hybridoma facility.

Financial Aid

All students admitted to the program receive support through national or state granting agencies in the amount of $23,000 plus student health insurance and full payment of tuition and fees.

Cost of Study

Tuition and fees for the 2004–05 academic year were $5500 for in-state students and $11,000 for out-of-state students. As indicated above, tuition and fees are paid for all students.

Living and Housing Costs

The cost of living in Birmingham is slightly lower than the U.S. average. Housing is readily available in the Medical Center area.

Student Group

The graduate program of the Departments of Biochemistry and Molecular Genetics, Cell Biology, Neurobiology, and Microbiology consists of more than 130 faculty members, 190 full-time graduate students, and about 150 postdoctoral fellows and visiting faculty members. The total enrollment at the University of Alabama at Birmingham is more than 15,000.

Student Outcomes

Graduates typically go on to postdoctoral research appointments, followed by careers in academic research and teaching, research in the biotechnology industry, or other science-related professions.

Location

Birmingham is located in the lovely rolling foothills of the Appalachian Mountain range in central Alabama. A metropolitan area that includes 1 million people, Birmingham is only a few hours' drive from Atlanta, Nashville, New Orleans, and the Gulf Coast. The city has excellent art and historical museums, theaters, libraries, a symphony orchestra, a ballet, a zoo, and botanical gardens. A host of recreational opportunities, including camping, swimming, fishing, hiking, golf, tennis, and boating, are available the year round in numerous local and state parks.

The University

The University of Alabama at Birmingham consists of University College, the Graduate School, and the Medical Center. It is located in the largest population center of the state and is heavily engaged in research, instruction, and service programs. Its dedication to excellence in biomedical research is exemplified by its consistent ranking in the top twenty institutions in receipt of federal research funds.

Applying

Applications to the program are evaluated by the Admissions Committee in consultation with other faculty members. The admission decision is based on scores achieved on the Graduate Record Examinations (a combined score of at least 1200, nominally, on the verbal and quantitative portions of the General Test), undergraduate grade point average (consideration is given to the curriculum completed), letters of evaluation, prior research experience, and a personal interview.

To be accepted into the program, the student should have completed a B.S. degree that includes the following undergraduate course work by the time of entrance: calculus, general and organic chemistry, and at least one introductory course in zoology or biology. Courses in physical chemistry, genetics, cell biology, and related topics are also beneficial to the candidate. Any remedial course work must be completed with a grade of B or better before the end of the first full year of doctoral study.

Applications are strongly encouraged from individuals with prior research experience, an M.S. degree in a related area, or a professional degree such as the M.D., D.M.D., D.V.M., or O.D.

The program anticipates admitting 35 to 40 students each year.

Correspondence and Information

Admissions Committee
Cellular and Molecular Biology Graduate Program
Suite 260, Bevill Biomedical Research Building
The University of Alabama at Birmingham
Birmingham, Alabama 35294-2170
Phone: 800-262-7764 (toll-free)
E-mail: cmb@uab.edu
Web site: http://www.cmb.uab.edu

The University of Alabama at Birmingham

THE FACULTY AND THEIR RESEARCH

Cell Physiology, Adhesion, and Signaling
S. Barnes, Ph.D.: bile acids, polyphenols, and metabolism and their effects on protein expression and function in chronic disease.
Z. Bebok, M.D.: membrane protein biogenesis in epithelial cells (CFTR as model); unfolded protein response.
R. Carter, M.D.: immune and autoimmune lymphocytes; structure and function in vivo.
C. Chang, Ph.D.: signal pathways in frog development.
D. Chaplin, M.D., Ph.D.: control of inflammation and lymphoid organ formation by LT and TNF.
J. Chatham, Ph.D.: cardiomyocyte function and metabolism in diabetes and ischemic heart disease.
J. Collawn, Ph.D.: intracellular protein sorting.
C. Gladson, M.D.: molecular mechanisms promoting tumor-cell proliferation; invasion and angiogenesis in gliomas.
J. Hagood, M.D.: fibroblast signaling in lung remodeling and fibrogenesis.
J. Kearney, Ph.D.: B cells; B-cell development; hybridomas; transgenic and knockout mice; immunoregulation; *B. anthracis.*
N. Kedishvili, Ph.D.: regulation of retinoic acid homeostasis.
F. Lin, Ph.D.: regulation of cell growth by G-protein-coupled receptor signaling.
R. Marchase, Ph.D.: calcium regulation and its impairment in diabetes.
G. Marques, Ph.D.: TGF-ß signaling in nervous system development and function.
M. Miller, Ph.D.: function and evolution of intercellular communication mechanisms.
E. Schwiebert, Ph.D.: extracellular nucleotide signaling and epithelial cell biology and physiology.
L. Schwiebert, Ph.D.: airway inflammation.
E. Sztul, Ph.D.: membrane traffic; protein degradation.
J. Thompson, Ph.D.: molecular mechanisms of angiogenesis.
A. Woods, Ph.D.: *Syndecan proteoglycans* in cell adhesion and matrix assembly.

Gene Regulation and Expression
A. Agarwal, M.D.: regulation of heme oxygenase gene expression in kidney and vascular injury.
L. Bridges, M.D., Ph.D.: genetic influences on treatment response in rheumatoid arthritis.
C. Chen, Ph.D.: mechanism and regulation of mammalian mRNA turnover.
X. Chen, Ph.D.: the p53 tumor-suppressor gene family and transcriptional regulation.
D. Crawford, M.D., Ph.D.: the role of G2/M-specific genes in mitosis and G2 DNA damage checkpoint regulation.
P. Higgins, Ph.D.: genetic and biochemical studies of chromosome dynamics.
C. Klug, Ph.D.: hematopoietic stem-cell biology and acute leukemias.
W.-C. Lin, M.D., Ph.D.: cell-cycle control and DNA damage response.
S. Lobo-Ruppert, Ph.D.: role of transcription factor oncogenes in tumor initiation.
R. Mayne, Ph.D.: collagen and collagen-binding proteins in health and disease.
J. McDonald, M.D.: cellular life and death signals in cancer; AIDS and bone disease.
M. Ruppert, M.D., Ph.D.: role of zinc finger transcription factors in tumor progression.
T. Ryan, Ph.D.: gene regulation; stem cells; mouse models; mutagenesis; cell therapies.
C. Turnbough, Ph.D.: *B. anthracis* spore structure-function/bacterial gene regulation.
H. Wang, Ph.D.: role of histone modification in chromatin function.

Immunology
S. Barnum, Ph.D.: role of complement, acute phase proteins, and adhesion molecules in acute and chronic inflammation in the central nervous system.
T. Benveniste, Ph.D.: immune/nervous system interactions.
O. Branch, Ph.D.: malaria molecular epidemiology and immunology.
P. Bucy, M.D.: T-cell regulation of immune responses in vivo.
R. Davis, M.D.: lymphocyte development and mechanisms of lymphomagenesis.
C. Elson, Ph.D.: chronic intestinal inflammation.
K. Fujihashi, D.D.S., Ph.D.: mucosal immunology; regulation of S-igA antibody responses; mucosal vaccine development.
V. Ghanta, Ph.D.: tumor immunology; CNS and immune system interactions.
Z. Hel, Ph.D.: development and testing of novel HIV/AIDS vaccine strategies.
L. Justement, Ph.D.: analysis of molecular mechanisms regulating lymphocyte biology.
J. Kabarowski, Ph.D.: regulation of inflammation and chronic inflammatory disease by lysophospholipid receptors.
J. Katz, Ph.D., D.D.S.: vaccine delivery systems; inflammation; innate immunity.
J. Kapp, Ph.D.: immune regulation and transplantation.
R. Kimberly, M.D.: autoimmunity, molecular mechanisms, and genetic risk.
W. Koopman, M.D.: pathogenesis of immune disease.
H. Kubagawa, M.D.: molecular genetics and immunopathology of host defense.
J. Mestecky, M.D., Ph.D.: mucosal immunity; vaccines.
M. Nahm, M.D.: adaptive and innate immune responses to vaccines against bacteria.
T. Strong, Ph.D.: identification of tumor antigens and development of cancer vaccines.
A. Szalai, Ph.D.: inflammation innate immunity and the acute phase proteins in health and disease.
L. Timares, Ph.D.: engineering dendritic cells for immunotherapy.
M. Walter, Ph.D.: structure and function of cytokines and their signaling molecules.
C. Weaver, M.D.: T-cell development.
Z. Zhang, Ph.D.: molecular regulation of early B-cell development and antibody repertoire formation.
T. Zhou, M.D.: specific induction of apoptosis in autoimmune and inflammatory cells.

Macromolecular Structure and Function
J. Blalock, Ph.D.: rational drug and vaccine design.
C. Brouillette, Ph.D.: protein structural cooperativity and energetics.
D. Chattopadhyay, Ph.D.: structure-function analysis of proteins.
I. Chesnokov, Ph.D.: DNA replication and cell cycle in eukaryotes.
H. Cheung, Ph.D.: regulatory mechanism in cardiac muscle.
L. DeLucas, Ph.D.: protein crystallography/protein crystal growth.
G. Elgavish, Ph.D.: NMR studies of intact hearts.
S. Frank, M.D.: growth hormone action and GH receptor structure and function.
B. Freeman, Ph.D.: tissue metabolism of reactive inflammatory mediators.
R. Krishna, Ph.D.: structural biology and biomolecular NMR spectroscopy.
J. Murphy-Ullrich, Ph.D.: extracellular matrix control of cell and growth-factor function.

C. Raman, Ph.D.: lymphocyte activation, immune tolerance and autoimmunity tolerance, and autoimmunity.
J. Segrest, M.D., Ph.D.: structural biology of supramolecular assemblies, particularly lipoproteins and membranes.
B. Sha, Ph.D.: structure and function of molecular chaperones.
N. Sthanam, Ph.D.: bacterial surface protein anchoring.

Molecular Genetics and Disease
S. Abulkadir, Ph.D.: molecular genetics of prostate cancer.
P. Atkinson, M.D., Ph.D.: primary immunodeficiency/role of infection in chronic diseases.
D. Bedwell, Ph.D.: translation termination; treatment of genetic diseases.
P. Burrows, Ph.D.: B-lymphocyte development and function.
P. Detloff, Ph.D.: mouse models of human genetic disorders.
K. Dybvig, Ph.D.: pathogenic mechanisms of mycoplasmas.
L. Guay-Woodford, M.D.: characterizing molecular determinants involved in PKD pathogenesis.
K. Jiao, M.D., Ph.D.: TGF-ß/BMP signaling during cardiogenesis.
K. Kirk, Ph.D.: the CFTR chloride channel.
J. Kudlow, M.D.: nucleocytoplasmic O-glycosylation in the control of cell function.
J. Mountz, M.D.: gene therapy; T-cell aging; immunogenetics; T-cell imaging.
H. Schroeder, M.D., Ph.D.: development/function of lymphocyte antigen receptors.
R. Serra, Ph.D.: mechanism of TGF-ß action in developmental and disease processes.
T. Townes, Ph.D.: developmental regulation of gene expression.
T. Unnasch, Ph.D.: molecular epidemiology and ecology of vector-borne diseases.
D. Welch, Ph.D.: molecular basis of tumor progression and metastasis.
B. Yoder, Ph.D.: cilia signaling and dysfunction in development and disease.

Molecular Pathogenesis
D. Balkovetz, M.D., Ph.D.: epithelial cell biology; epithelial cell signal regulation; regulation of paracellular transport across epithelial cell tight junctions.
W. Benjamin, Ph.D.: molecular typing of *Streptococcus pneumoniae.*
D. Briles, Ph.D.: bacterial pathogenesis; virulence; immunity; pneumococcus.
N. Childers, D.D.S., Ph.D.: development of a vaccine for the prevention of dental caries.
M. Cooper, M.D.: developmental immunobiology, with emphasis on B-cell and T-cell differentiation; clinical immunology, with emphasis on immunodeficiency diseases and lymphoid malignancies.
S. Hollingshead, Ph.D.: mechanisms of variation in microbial pathogenesis.
R. Lorenz, M.D., Ph.D.: cellular and molecular immunology of the gastrointestinal tract.
S. Michalek, Ph.D.: mucosal vaccines and host mechanisms involved in inflammation.
R. Morrison, Ph.D.: immunobiology of Chlamydia infection.
C. Morrow, Ph.D.: understanding, at the molecular level, virus–host cell interactions.
M. Niederweis, Ph.D.: the role of porins in outer-membrane permeability and drug resistance of mycobacteria.
D. Pritchard, Ph.D.: roles of glycoconjugates in pathogenesis.
J. Rayner, Ph.D.: cell biology of the malaria parasite *Plasmodium falciparum.*
A. Steyn, Ph.D.: mechanism of mycobacterium tuberculosis virulence.
K. Waites, M.D.: diagnostic microbiology; epidemiology and mechanisms of antimicrobial resistance; mycoplasma and ureaplasma diseases; staphylococcal diseases.
H. Wu, Ph.D.: bacteria-host interaction.
J. Yother, Ph.D.: capsular polysaccharides of *Streptococcus pneumoniae.*

Neurobiology
D. Benos, Ph.D.: molecular basis of operation on ion channels and transporters.
M. Brenner, Ph.D.: molecular studies of astrocytes in health and disease.
L. Dobrunz, Ph.D.: synaptic transmission and plasticity in hippocampus.
J. Hablitz, Ph.D.: cellular mechanisms of neurotransmission.
G. Johnson, Ph.D.: molecular mechanisms of neurodegeneration.
R. Jope, Ph.D.: neuronal signaling mechanisms regulating gene expression and cell death.
R. Lester, Ph.D.: molecular and cellular mechanisms of nicotine addiction.
L. Pozzo-Miller, Ph.D.: neurotrophins on Ca2+ signaling, synapse development, and plasticity.
K. Roth, M.D., Ph.D.: molecular regulation of neuronal cell death.
D. Ruden, Ph.D.: environmental and developmental toxicology.
H. Sontheimer, Ph.D.: the role of neuroglia in brain function and disease.
D. Sweatt, Ph.D.: signal transduction mechanisms in learning and memory.
A. Theibert, Ph.D.: role of phosphoinositides in developmental neurobiology.
S. Wilson, Ph.D.: mouse models of neurodegeneration.
M. Wyss, Ph.D.: control of the autonomic nervous system.
Y. Zhou, Ph.D.: modulation of ion channels; regulation of neuronal excitability and synaptic transmission.

Virology
W. Britt, M.D.: human herpesviruses, molecular virology, and pathogenesis.
L. Chow, Ph.D.: human papillomavirus DNA replication and pathogenesis.
T. T. Dokland, Ph.D.: cryoelectron microscopy and X-ray crystallography of virus assembly processes.
J. Engler, Ph.D.: identifying novel diagnostic peptides and cell-specific ligands.
B. Hahn, M.D.: origin and evolution of primate lentiviruses.
J. Kappes, Ph.D.: HIV, molecular virology, and pathogenesis.
R. Kaslow, M.D.: immunogenetic determinants in AIDS and other diseases.
O. Kutsch, Ph.D.: HIV-1 latency and drug screening.
E. Lefkowitz, Ph.D.: bioinformatics; biodefense; microbial genomics and evolution.
M. Luo, Ph.D.: structure-based approaches to anti-infectious agents.
P. Prevelige, Ph.D.: structural biology of viral assembly and infection.
G. Shaw, M.D., Ph.D.: evolution and persistence of HIV-1.
W. Sullender, M.D.: respiratory syncytial virus; antigenic diversity.
S. Thompson, Ph.D.: translation initiation and replication of RNA viruses.
R. Whitley, M.D.: herpesvirus; varicella zoster virus.
A. Zajac, Ph.D.: antiviral immunity; T-cell responses; immunological memory.

UNIVERSITY OF ILLINOIS AT URBANA–CHAMPAIGN

Department of Biochemistry

Programs of Study

The graduate programs of the Department of Biochemistry are designed to assist students in developing their capabilities for independent thought and research. While programs leading to both the Ph.D. and M.S. degrees are offered, the number of M.S. candidates admitted is quite small. Faculty research areas include the most dynamic areas of current research in biological chemistry and molecular biology: physical approaches to the structure and function of proteins, nucleic acids, and carbohydrates; enzymology; membrane biochemistry and protein-lipid interactions; protein–nucleic acid interactions; molecular biological approaches to gene organization and expression; immunology; microbial physiology; and signal transduction.

The Department of Biochemistry is part of the School of Molecular and Cellular Biology that also includes the Departments of Cell and Developmental Biology, Molecular and Integrative Physiology, Microbiology, and the program in biophysics. Students in the School take a common core curriculum in the first semester while participating in a laboratory rotation program. Students may select faculty research advisers from any of the four departments, thereby obtaining a broad range of training and research opportunities. The program includes training in both bioethics and bioinformatics. In addition to completing formal course work, students participate in a variety of seminars and colloquia. This combination of formal course work and informal interaction with faculty members and other graduate students, when coupled with faculty-guided research, fosters individual development. The University offers a business certification program to provide training for those students who seek industrial/business employment after the Ph.D. The University Biotechnology Center also provides an active and successful job placement office to assist students in locating jobs at the Ph.D. level.

Research Facilities

The Department of Biochemistry is housed primarily in Roger Adams Laboratory, a modern facility that contains all of the equipment necessary for biochemical research. Excellent ancillary facilities include electronics, glassblowing, and machine shops; departmental and online computers; and facilities for large-scale tissue culture work and microorganism growth. Researchers in the Department have access to excellent facilities for magnetic resonance spectroscopy, mass spectrometry, X-ray crystallography, amino acid analysis, protein sequencing, and DNA synthesis. The extensive biology and chemistry libraries are readily accessible to graduate students.

Financial Aid

All graduate students in the Ph.D. program who are making satisfactory progress toward the degree are guaranteed full financial support throughout their tenure. Support is provided through fellowships, traineeships, teaching assistantships, and research assistantships. Stipends for 2006–07 range from $22,660 to $24,000 per year (depending on the mode of support) and provided a full tuition and partial fee waiver.

Cost of Study

All available types of graduate student support provide a full tuition and partial fee waiver. Other fees assessed of all students (including health coverage) are currently $633 per semester.

Living and Housing Costs

Room and board costs in University graduate student residence halls currently range from $3900 to $4900 per academic year. University housing for families is available at a cost of $460 to $580 per month. Many privately owned apartments and houses are available near the campus at reasonable rates ($350 to $600 per month). The cost of living is relatively low.

Student Group

There are 98 graduate students (41 women), 244 undergraduate biochemistry majors (121 women), and approximately 25 postdoctoral fellows enrolled in the Department. Graduates of the program typically enter jobs in industry and research.

Location

The twin cities of Champaign and Urbana offer a wide range of cultural activities sponsored by both the University and community groups. The Krannert Center for the Performing Arts, with its complex of concert halls, theaters, and an art museum, is directly across the street from Roger Adams Laboratory, where the biochemistry department is located. There are excellent facilities for participant sports as well as major intercollegiate athletic programs on campus. Swimming, boating, and hiking areas are readily accessible.

The University

The University of Illinois was chartered in 1867. The Urbana-Champaign campus has long been a major center for research in the physical sciences and engineering as well as in the biological sciences and in agriculture. The University of Illinois also has the third-largest academic library in the United States.

Applying

Applications are welcomed from students with a variety of backgrounds in the biological and physical sciences; undergraduate preparation should include courses in chemistry, biology, physics, and calculus. Completion of the application process before the January 1 deadline is strongly advised. International applicants must mail all materials before December 15. Applicants should have a minimum grade point average of 3.0 (on a 4.0 scale) and are required to take the General Test of the Graduate Record Examination. The Subject Test in biochemistry; cell and molecular biology; chemistry; or biology is strongly recommended for international applicants and for fellowship consideration. International applicants from non-English-speaking countries are required to submit a TOEFL score (minimum score is 230 on the computer-based test or 600 on the paper-based test) and a TSE score (minimum score is 50).

Correspondence and Information

Coordinator of Graduate Studies
Department of Biochemistry
420 Roger Adams Laboratory B4
University of Illinois at Urbana-Champaign
600 South Mathews Avenue
Urbana, Illinois 61801
Phone: 217-333-7149
E-mail: biochadm@life.uiuc.edu
Web site: http://www.life.uiuc.edu/biochem

University of Illinois at Urbana-Champaign

THE FACULTY AND THEIR RESEARCH

Listed below are the faculty members of the Department of Biochemistry and their research interests. The departmental brochure and Web site provide additional details concerning departmental research programs.

Lin-Feng Chen, Assistant Professor of Biochemistry; Ph.D., Kyoto (Japan), 1999. Epigenetic regulation of NF-κB; the role of NF-κB in apoptosis and cancer; cross-talk between NF-κB and other signaling pathways. *Mol. Cell. Biol.* 25:7966, 2005; *Nat. Rev. Mol. Cell Biol.* 5(5):392–401, 2004; *Science* 293:1653, 2001.

Robert M. Clegg, Professor of Physics and Affiliate of Biochemistry; Ph.D., Cornell, 1974. Multifarious molecular structures, conformational changes, and thermodynamic stabilities and functions of nucleic acids and protein–nucleic acid complexes. *J. Mol. Biol.* 351(5):1123–45, 2005; *Biochem. Soc. Trans.* 32(1):41–5, 2004 (with McKinney et al.).

Antony R. Crofts, Professor of Biochemistry and Biophysics; Ph.D., Cambridge, 1965. Structure-function relationships in the membrane proteins of bioenergetic systems; mechanism of the bc complex; photosynthesis in intact plants. *Biochim. Biophys. Acta* 1655:77–92, 2004; *Photosystem II: The Water/Plastoquinone Oxido-Reductase in Photosynthesis,* eds. T. Wydrzynski and K. Satoh. The Netherlands: Kluwer Academic Publications, 2003 (with Petrouleas); *Biochim. Biophys. Acta* 277:4605, 2002 (with Samoilova et al.).

John E. Cronan Jr., Professor of Microbiology and Biochemistry; Ph.D., California, Irvine, 1968. Biosynthesis, regulation, and function of membrane lipids and fatty acid–derived coenzymes. *J. Biol. Chem.* 280:34675–83 (with Thomas); *Chem. Biol.* 12:461–8 (with Choi-Rhee); *J. Bacteriol.* 186:1869–78, 2004 (with Lai).

Robert B. Gennis, Professor of Biochemistry and Chemistry; Ph.D., Columbia, 1971. Structure and function of ion-pumping respiratory complexes; mechanism of cytochrome oxidase; prokaryotic ion channels and transporters. *Biochim. Biophys. Acta* 1655:321, 2004 (with McMahon et al.); *Proc. Natl. Acad. Sci. U.S.A.* 101(29):10544, 2004 (with Zaslavsky et al.); *Biochemistry* 43(38):12322, 2004 (with Barquera et al.).

John A. Gerlt, Professor of Biochemistry, Chemistry, and Biophysics and Gutgsell Chair; Ph.D., Harvard, 1974. Mechanisms of enzyme-catalyzed reactions; designed and directed evolution; functional genomics. *Biochemistry* 44:11722, 2005 (with Vick et al.); *Biochemistry* 44:1807, 2005 (with Yew et al.); *Annu. Rev. Biochem.* 70:209, 2001 (with Babbitt).

Richard I. Gumport, Professor and Associate Dean for Academic Affairs, College of Medicine; Ph.D., Chicago, 1968. Nucleic acid enzymology. *Nucl. Acids Res.* 34:806, 2006 (with Thomas); *Mol. Microbiol.* 50:89, 2003 (with Swalla et al.); *Biol. Chem.* 278:26094, 2003 (with Thomas et al.); *Nucleic Acids Res.* 31:805, 2003 (with Swalla et al.).

Taekjip Ha, Associate Professor of Physics and Biophysics and Affiliate of Biochemistry; Ph.D., Berkeley, 1996. Using physical concepts and experimental techniques to study fundamental questions in molecular biology. *Nature* 437:1321–5, 2005 (with Myong et al.); *Biophys. J.* 89:413–7, 2005 (with Balci et al.); *Proc. Nat. Acad. Sci. USA* 102:5715–20, 2005 (with McKinney et al.).

Paul J. Hergenrother, Assistant Professor of Chemistry and Affiliate of Biochemistry; Ph.D., Texas at Austin, 1999. Synthesis and identification of organic compounds with novel biological activity. *J. Am. Chem. Soc.* 127:8686, 2005; *Chem. Biol.* 12:789, 2005; *J. Am. Chem. Soc.* 127:12434, 2005; *J. Am. Chem. Soc.* 126:15402, 2004; *J. Am. Chem. Soc.* 126:9196, 2004; *J. Am. Chem. Soc.* 125:14672, 2003.

Raven H. Huang, Assistant Professor; Ph.D., Seattle, 1995. Mechanistic and structural studies of enzymes involved in RNA modifications and protein toxins involved in site-specific RNA cleavages; design and engineering of small proteins or organic ligands that are useful against certain lethal bacteria or human diseases. *Biochemistry* 44:15488, 2005 (with Phannachet et al.); *Biochemistry* 44:12057, 2005 (with Elias); *Biochemistry* 44:10494, 2005 (with Lin et al.).

Eric Jakobsson, Professor of Biophysics, Bioengineering, Physiology, and Biochemistry; Ph.D., Dartmouth, 1969. Computational studies of biological membranes: life at the interface. *Genome Biol.* 6(1):R4, 2005 (with Tasneem et al.); *Biochemistry* 51:16320, 2004 (with Malmberg et al.); *Biophys. J.* 5:3312, 2004 (with Pandit et al.).

Neil L. Kelleher, Associate Professor of Chemistry and Affiliate of Biochemistry; Ph.D., Cornell, 1997. Proteomics of pathogenic and extremophilic microorganisms using FT mass spectrometry, posttranslational modifications, covalent catalysis, and enzymology. *Mol. Cell. Proteomics* 4:1002, 2005 (with Roth et al.); *Mass Spectrom. Rev.* 24:126, 2005 (with Meng et al.); *J. Am. Chem. Soc.* 126:13265, 2004 (with McLoughlin et al.).

David M. Kranz, Professor; Ph.D., Illinois at Urbana-Champaign, 1982. Structure, function, and engineering of T-cell receptors and antibodies. *J. Mol. Biol.* 353:308, 2005 (with Buonpane et al.); *J. Mol. Biol.* 346:223, 2005 (with Chlewicki et al.); *Nat. Immunol.* 6:130, 2005; *Gene Ther.* 11:1234, 2004 (with Peng et al.); *Mol. Immunol.* 40:1027, 2004 (with Holler).

Yi Lu, Professor of Chemistry and Affiliate of Biochemistry; Ph.D., UCLA, 1992. Protein and biocatalysis: selection of catalytic DNA/RNA as biosensors in environmental and clinical applications; DNA materials in molecular electronics and photonics. *Curr. Opin. Chem. Biol.* 9:118, 2005; *J. Am. Chem. Soc.* 127:12677, 2005 (with Liu); *Biochemistry* 44:6559, 2005 (with Zhao et al.).

William W. Mantulin, Adjunct Professor of Biochemistry and Biophysics; Ph.D., Northeastern, 1972. Cell biology and membrane dynamics. *J. Biol. Chem.* 279(47):49160, 2004 (with Levi); *Biophys. J.* 87:1260, 2004 (with Gratton et al.); optical imaging of tissues. *J. Biomed. Opt.* 9(1):221, 2004 (with Toronov et al.).

Susan A. Martinis, Associate Professor of Biochemistry; Ph.D., Illinois at Urbana-Champaign, 1990. RNA structure and function; RNA-protein interactions; chemical mechanisms of biological reactions; protein synthesis; tRNA synthetases. *Biochemistry* 44:15437, 2005 (with Zhai); *Biochemistry* 43:155, 2004 (with Mursinna et al.); *Mol. Cell* 11:951, 2003 (with Lincecum et al.).

James H. Morrissey, Professor of Biochemistry (also in the College of Medicine); Ph.D., California, San Diego, 1980. Regulation of the blood clotting system and modulation of membrane-bound serine protease function. *J. Biol. Chem.* 279:39745, 2004; *Biochemistry* 41:3364, 2002 (with Neuenschwander et al.); *J. Biol. Chem.* 724:4962, 1999 (with Hamik et al.).

Satish K. Nair, Assistant Professor; Ph.D., Pennsylvania, 1994. X-ray crystallographic studies of molecular interactions that govern recognition events in biological systems. *J. Immunol.* 175(7):4175, 2005 (with Lasker et al.); *Cell* 112:193, 2003 (with Burley); *Cell* 104:901, 2001 (with Campbell et al.); *Nature* 404:715, 2000 (with Burley).

Chad M. Rienstra, Assistant Professor of Chemistry and Biophysics and Affiliate of Biochemistry; Ph.D. MIT. 1999. Solid-state nuclear magnetic resonance (SSNMR), including the development of new pulse sequence methodology and instrumentation and application to studies of protein structure and dynamics. *J. Magn. Reson.* 173: 40, 2005 (with Stringer et al.).

Mary A. Schuler, Professor of Cell and Developmental Biology, Biochemistry, and Plant Biology; Ph.D., Cornell, 1981. Plant and insect P-450 monoxygenases; plant pre-mRNA splicing; plant gene regulation; molecular modeling of catalytic site. *RNA* 11:128, 2005 (with Thimmapuram et al.); *Arch. Biochem. Biophys.* 424:141, 2004 (with Duan et al.); *Ann. Rev. Plant Biol.* 54:629, 2003 (with Werck-Reichhart).

Bradford S. Schwartz, Professor and Dean, College of Medicine; M.D., Rush, 1977. Mechanisms of protease cascade activation and regulation; interactions of proteases and inhibitors. *J. Biol. Chem.* 277:46852, 2002 (with Barker-Carlson et al.); *J. Biol. Chem.* 274:15278, 1999 (with Espana); *J. Biol. Chem.* 269:8319, 1994; *J. Biol. Chem.* 266:14580, 1991 (with Manchanda).

David J. Shapiro, Professor; Ph.D., Purdue, 1972. Screening for small molecule regulators of steroid hormone receptor action; estrogen action in transcription; mRNA stability, apoptosis, and inflammatory disease. *BioTechniques* 37:813, 2004 (with Wang et al.); *J. Biol. Chem.* 279:5025, 2004 (with Krieg et al.); *Nucleic Acids Res.* 31:5644, 2003 (with Goolsby).

Scott K. Silverman, Assistant Professor of Chemistry and Affiliate of Biochemistry; Ph.D., Caltech, 1997. RNA structure, folding, and catalysis; DNA enzymes; DNA constraints for nanotechnology; allosteric nucleic acid enzymes as sensors. *J. Am. Chem. Soc.* 127:10144, 2005 (with Miduturu); *Angew. Chem. Int. Ed.* 44:5863, 2005 (with Wang); *J. Am. Chem. Soc.* 127:13124, 2005 (with Purtha et al.).

Stephen G. Sligar, Professor of Biochemistry, Chemistry, and Biophysics (also in the College of Medicine); Ph.D., Illinois at Urbana-Champaign, 1975. G-protein coupled receptors; proteomics and bioinformatics and therapeutic delivery. *Science* 304:674, 2004; *Protein Sci.* 12:2476, 2003 (with Bayburt); biomolecular chemistry of oxygenases. *Science* 287:1615, 2000 (with Schlichting et al.).

Maria Spies, Assistant Professor of Biochemistry; Ph.D., Osaka (Japan), 2000. Biochemical mechanisms and function of DNA helicases and DNA motor proteins; mechanistic aspects of protein-nucleic acids and protein-protein interactions; homologous genetic recombination; DNA repair; molecular motors. *J. Biol. Chem.,* in press; *Cell* 114(5):647–54, 2003.

Wilfred A. van der Donk, Professor of Chemistry and Affiliate of Biochemistry; Ph.D., Rice, 1994. Antibiotics biosynthesis; radical chemistry in enzyme catalysis. *Chem. Rev.* 105:633, 2005 (with Chatterjee et al.); *Biochemistry* 44:4765, 2005 (with Woorlyer et al.); *Science* 303:679, 2004 (with Xie et al.).

Colin A. Wraight, Professor and Head of Biochemistry, Professor of Biophysics and Plant Biology; Ph.D., Bristol, 1971. Electron and proton transfer in proteins; structure-function of redox cofactor-protein interactions. *Frontiers Biosci.* 9:309, 2004; *Adv. Protein Chem.* 63:71, 2003 (with Moser et al.); *Biochemistry* 42:4064, 2003 (with Wells et al.); *Biochemistry* 40:12584, 2001 (with Shinkarev et al.).

Emeriti Faculty

Michael Glaser, Professor Emeritus and Associate Head of Biochemistry; Ph.D., California, San Diego, 1971. Formation and function of membrane domains; signal transduction; growth factor and hormone regulation of myelin synthesis. *Biochemistry* 42:42, 2003 (with Wanaski et al.); *Mol. Biol. Cell* 11:2283, 2000 (with Chan et al.).

George W. Ordal, Professor Emeritus (also in the College of Medicine); Ph.D., Stanford, 1971. Investigation of the molecular basis of excitation and adaptation during chemotactic sensory transduction in bacteria. *J. Biol. Chem.* 279:21787, 2004 (with Szurmant et al.); *J. Mol. Biol.* 331:941, 2003 (with Bunn).

Robert L. Switzer, Professor Emeritus (also in the College of Medicine); Ph.D., Berkeley, 1966. Regulation of bacterial gene expression by transcriptional attenuation. *J. Bacteriol.* 187:1773, 2005 (with Chander et al.); *Proc. Natl. Acad. Sci. U.S.A.* 101:10943, 2004 (with Meng et al.).

THE UNIVERSITY OF IOWA

Department of Biochemistry

Programs of Study

The Department of Biochemistry offers a program of study and research leading to the Doctor of Philosophy degree. In addition, qualified students can pursue a course of study leading to both M.D. and Ph.D. degrees. Applicants should have training in general, organic, and physical chemistry; basic biochemistry; mathematics through calculus; and at least one year each of biology and physics. A bachelor's degree from an accredited institution is required.

First-year students gain research experience by rotating through four laboratories while taking courses in biophysical chemistry and molecular biology. At the end of the first year, students choose a research laboratory in which to perform their thesis research. Second-year students complete the formal course work and a comprehensive examination, which certifies them as candidates for the doctoral degree. The final years of graduate residence are devoted to the design, execution, and evaluation of experiments, culminating in the Ph.D. dissertation.

Research Facilities

The Department of Biochemistry is part of the College of Medicine, the College of Liberal Arts and Sciences, and the Graduate College. The Department occupies approximately 36,000 square feet of space in the Bowen Science Building and 12,000 square feet of space in the Medical Education and Biomedical Research Facility, as well as labs in selected other College of Medicine buildings. There are twenty-two research groups in the Department of Biochemistry, each well furnished with state-of-the-art equipment and facilities to support its activities. Major instruments used by the Department include electron microscopes, amino acid analyzers, a protein sequenator, DNA and peptide synthesizers, stopped-flow spectrometers, a spectropolarimeter, a spectrofluorimeter, a mass spectrometer, 500-MHz and 600-MHz NMR spectrometers, and X-ray diffraction area detectors.

Financial Aid

Students received an annual stipend of $22,000 in 2005–06. Tuition and fees are offset by the Department.

Cost of Study

As indicated in the Financial Aid section, tuition is paid for each student who holds an appointment in the program. All research supplies and materials are provided by the Department.

Living and Housing Costs

One- and two-bedroom apartments are available for family housing through the University Apartments Office. Rent is approximately $408 to $545 per month. Applicants should contact the University Apartments Office by January of the year in which they wish housing. Off-campus housing, in attractive residential areas surrounding the University, is available at a wide range of prices.

Student Group

The enrollment at the University of Iowa is approximately 29,000, including 9,000 graduate students. The Department enrolls approximately 35 graduate students. A graduate student organization in the Department facilitates communication, makes recommendations, and generally accepts responsibilities concerning student welfare.

Location

The University of Iowa includes more than 100 buildings on a 1,900-acre campus situated on both banks of the Iowa River. Iowa City and adjacent communities have an aggregate population of 85,000 (1990 census) and are located on Interstates 80 and 380, about 235 miles from Chicago and less than 300 miles from Minneapolis, Omaha, Kansas City, and St. Louis. Nine major airlines serve nearby Cedar Rapids Airport. The University is the major enterprise in the community, and its full and varied educational and cultural program attracts speakers and performers from all over the world.

The University

Founded in 1847, the University was the first U.S. public institution to admit men and women on an equal basis. It is a member in the select Association of American Universities and is prominent in research, medicine, and the arts. The campus is located on the Iowa River, with the Department of Biochemistry located within the rapidly growing health sciences complex on the west side of the river.

Applying

The Department normally admits new graduate students for the fall session. Applicants are selected during the spring semester. They may obtain appropriate forms from the Department of Biochemistry. An official transcript from each undergraduate and graduate institution attended must be submitted with these forms. All applicants should take the Graduate Record Examinations. The General Test is required, and a Subject Test in either biochemistry, cell and molecular biology; chemistry; or biology is strongly recommended. These examinations should be taken in October or December so that scores will be available by January. International applicants must provide TOEFL scores.

Correspondence and Information

Department of Biochemistry
4-403 BSB
The University of Iowa
Iowa City, Iowa 52242-1109
Phone: 319-335-7932
      877-846-8569 (toll-free)
E-mail: biochem@uiowa.edu
Web site: http://www.biochem.uiowa.edu

The University of Iowa

THE FACULTY AND THEIR RESEARCH

Robert E. Cohen, Associate Professor; Ph.D. (biochemistry), Berkeley, 1980. Biochemistry of the ubiquitin system; control of intracellular protein degradation; protein-protein interactions; protein stability and conformational equilibriums.

Kris A. DeMali, Assistant Professor; Ph.D. (pharmacology), Colorado, 1999. Molecular basis of cell adhesion and its regulation by signaling and cytoskeletal proteins.

John E. Donelson, Professor and Head; Ph.D. (biochemistry), Cornell, 1971. Mechanisms of eukaryotic gene expression, with an emphasis on tropical parasites; molecular basis of parasite evasion of the mammalian immune response.

Adrian Elcock, Assistant Professor; Ph.D. (chemistry), Oxford, 1994. Computer simulations of macromolecular interactions.

Ernesto Fuentes, Assistant Professor, Ph.D. (biochemistry), Illinois, 1999. Energetics and dynamics of protein-protein interactions; NMR; signal transduction pathways in cancer.

Alice B. Fulton, Professor; Ph.D. (cell biology), Brown, 1977. Developmental controls in muscle, including cytoskeletal assembly and regulation of differentiation.

Pamela K. Geyer, Professor; Ph.D. (biochemistry), Ohio State, 1983. Long-distance regulation of eukaryotic gene expression; chromosome organization and gene expression; insulators.

Shahram Khademi, Assistant Professor; Ph.D. (biochemistry and biophysics), Texas A&M, 2001. Biophysics of membrane protein structure and function; X-ray crystallography.

Rex Montgomery, Professor; Ph.D. (chemistry), 1946, D.Sc., 1963, Birmingham (England). Biochemistry of glycoproteins; microbial polysaccharide antigens; fermentation by-products; carbohydrate structures.

Kenneth P. Murphy, Adjunct Associate Professor; Ph.D. (physical biochemistry), Colorado, 1990. Thermodynamics of protein folding, protein-protein interactions, and protein-ion interaction; microcalorimetry.

Bryce V. Plapp, Professor; Ph.D. (biochemistry), Berkeley, 1966. Mechanisms of alcohol dehydrogenases; protein chemistry, structure, and function; enzyme engineering by directed mutagenesis; control of alcohol metabolism.

David H. Price, Professor; Ph.D. (biochemistry), Florida State, 1980. Eukaryotic transcriptional regulation; RNA polymerase II elongation control by P-TEFb and TTF2; coupling of RNA processing to transcription; involvement of P-TEFb in HIV replication.

Subramanian Ramaswamy, Associate Professor; Ph.D. (biophysics), Indian Institute of Science, 1992. Structural enzymology of dioxygenases and dehydrogenases; protein engineering; macromolecular crystallography; inflammation.

Peter A. Rubenstein, Professor; Ph.D. (biochemistry and molecular biology), Harvard, 1973. Actin structure-function relationships using site-directed mutagenesis; the effect of actin-binding proteins on F-actin structure; actin site-directed mutagenesis; actin-porin interactions.

Madeline A. Shea, Professor; Ph.D. (biophysics), Johns Hopkins, 1984. Protein chemistry; thermodynamics; cooperative ligand binding and conformational change; calmodulin regulation of ion channels and other target proteins.

Arthur A. Spector, Professor of Biochemistry and Internal Medicine; M.D., Pennsylvania, 1960. Polyunsaturated fatty acids; prostaglandins and lipid biomediators; vascular biology.

Nancy C. Stellwagen, Adjunct Professor; Ph.D. (chemistry), Berkeley, 1967. DNA gel electrophoresis; sequence-dependent DNA structure; transient electric birefringence; capillary electrophoresis.

Joseph A. Walder, Adjunct Professor; M.D./Ph.D. (chemistry), Northwestern, 1979. Development of modified hemoglobins as blood substitutes; oligonucleotide synthesis; development of antisense oligonucleotides; substrate specificity and mechanism of RNase H.

Lori L. Wallrath, Associate Professor; Ph.D. (genetics), Michigan State, 1991. Nuclear organization; chromatin structure and gene expression.

M. Todd Washington, Assistant Professor; Ph.D. (biochemistry), Ohio State, 1998. Mechanisms of enzymes involved in DNA replication and DNA repair, particularly the mechanisms of specialized DNA polymerases involved in translesion DNA synthesis.

Daniel L. Weeks, Professor; Ph.D. (molecular biology), Purdue, 1983. Molecular controls of embryonic development.

Marc S. Wold, Professor; Ph.D. (biochemistry), Johns Hopkins, 1984. Molecular mechanisms of eukaryotic DNA replication and repair; biochemistry of eukaryotic replication proteins; eukaryotic single-strand DNA-binding proteins; regulation of chromosomal replication.

Faculty Participants in Joint Programs

Rama K. Mallampalli, Professor; M.D. (internal medicine and biochemistry), Wisconsin, 1984. Phosphatidylcholine biosynthesis in eukaryotic cells; regulation of CTP:phosphocholine cytidylyltransferase; pulmonary surfactant metabolism.

Emeritus Professors

Arthur Arnone, Professor; Ph.D. (physical chemistry), MIT, 1970. Macromolecular structure determination; hemoglobin crystallography; protein structure and function.

Thomas W. Conway, Professor Emeritus; Ph.D. (chemistry), Texas, 1962. Control of translation; mechanisms of interferon action; interferon-related protein kinases.

Earle Stellwagen, Professor Emeritus; Ph.D. (biochemistry), Berkeley, 1963. Electrochemical properties of model DNAs.

Charles A. Swenson, Professor Emeritus; Ph.D. (physical chemistry), Iowa, 1959. Physical chemistry of proteins; spectroscopy; electric birefringence; microcalorimetry; interactions of muscle proteins in relation to mechanisms of control and to energy transduction in muscle contraction.

UNIVERSITY OF KANSAS MEDICAL CENTER

Department of Biochemistry and Molecular Biology

Programs of Study

The Department of Biochemistry and Molecular Biology (BMB) offers programs leading to both Master of Science and doctoral degrees. The combined M.D./Ph.D. degree is available in collaboration with the University of Kansas School of Medicine. Postdoctoral training is an integral part of the program.

The Interdisciplinary Graduate Program in Biomedical Sciences (IGPBS) constitutes the first year of study. IGPBS is an interdepartmental, interdisciplinary program taught by faculty members from the Departments of Anatomy and Cell Biology; Biochemistry and Molecular Biology; Microbiology, Molecular Genetics, and Immunology; Pathology and Laboratory Medicine; Molecular and Integrative Physiology; and Pharmacology, Toxicology, and Therapeutics, as well as from the Training Program in Environmental Toxicology.

The advanced curriculum includes specialized training in physical biochemistry, protein structure-function, and advanced molecular genetics. The Ph.D. candidacy examination usually is taken at the end of the second year. The Ph.D. degree requires approximately 35 credit hours of courses emphasizing theoretical concepts as well as practical aspects of laboratory work. It is especially important that students develop the ability to apply this knowledge and experience to an independent research problem. The program is flexible in accommodating the individual student's prior educational and ultimate goals. Completion of the requirements for a Ph.D. generally requires four or five calendar years. The M.S. requires about one year of course work and a year of supervised research leading to a thesis. Candidates for combined M.D./Ph.D. degrees enter this program through the graduate and medical schools.

Research Facilities

The Department of Biochemistry and Molecular Biology is primarily located on the fourth floor of Wahl Hall, home to many of the basic science departments. One of the most modern facilities is the Lied Biomedical Research Building, opened in 1995. The Lied building is home to several of the Department's faculty members. The Electron Microscopy Facility is also housed in the Lied building. The building is connected to the Archie Dykes Health Sciences Library and the Animal Research Facility. A new research building is under construction, and most BMB faculty members are scheduled to move to it in winter 2006–07.

Financial Aid

Students admitted into the IGPBS program receive financial assistance through the University's graduate teaching assistantship program. This is a financial package worth $21,700 of direct funds plus a full tuition waiver worth approximately $12,800.

Cost of Study

In 2004–05, Kansas residents paid $178 per credit hour for tuition and $166.64 for fees for more than 6 credits per semester. Out-of-state residents paid $459.10 per credit hour for tuition and $166.64 for fees for more than 6 credits per semester. Additional fees may apply depending upon the major. Tuition and fees are set by the Kansas Board of Regents and are subject to change.

Living and Housing Costs

Campus-owned housing is available on a limited basis. The surrounding neighborhoods offer numerous privately owned off-campus housing opportunities. Rents range from $350 to $750 per month for a one-bedroom apartment. Living expenses are comparable to other large Midwestern cities.

Student Group

Approximately 15 students are involved in the Department of Biochemistry and Molecular Biology programs. Students come from throughout the Midwestern United States and several other countries.

Student Outcomes

The master's degree in biochemistry typically leads students to positions in the advanced technical level in academic research, industry, or government and may also lead to teaching positions at the secondary or junior college level.

Upon completion of the Ph.D. program, most students continue with two or more years of postdoctoral training in a variety of research areas. Doctoral holders in biochemistry may find positions in the private sector or in government, or they may obtain a faculty position at the college or university level.

Location

The University of Kansas Medical Center is located in Kansas City, Kansas, on a major interstate highway system. Located in the heart of the nation's Midwest, Kansas City offers a variety of activities to match virtually any interest. Metropolitan Kansas City encompasses seven counties and fifty municipalities in two states, Kansas and Missouri. The city hosts professional sports teams; a zoo; music, art, and history museums; fine restaurants; shopping centers; and outdoor activities.

The University

The University of Kansas Medical Center offers educational programs through its Schools of Medicine, Nursing, Allied Health, Pharmacy, and Graduate Studies. Students develop many of their clinical skills at the University of Kansas Hospital, where they assist physicians and residents in caring for more than 19,000 patients a year. KU Medical Center's mission calls for improving the quality and availability of health-care professionals, clinical services for patients, and research in the health sciences. KU Medical Center is a stakeholder institution in the Kansas City Area Life Sciences Institute, which was formed in 2000, with the goal of facilitating interaction between existing life sciences programs as well as catalyzing new development.

Applying

Applicants should have a bachelor's degree, with a strong background in mathematics through calculus and biological sciences, and two semesters of organic chemistry. Knowledge of physical chemistry is recommended. Applicants are urged to take the general aptitude and advanced sections of the GRE. Applicants from outside the U.S. require a TOEFL score; a minimum score of 570 is needed for consideration.

To ensure consideration for admission in the fall, applications should be received by January 1 to be considered for recruitment in February and March. Applications for assistantships and fellowships should be received by February 1 preceding the academic year for which assistance is requested. In general, students are accepted for the fall semester before late April.

Correspondence and Information

Department of Biochemistry and Molecular Biology
University of Kansas Medical Center
3901 Rainbow Boulevard
4011 WHE, Mailstop 3030
Kansas City, Kansas 66160-7421
Phone: 913-588-7008
Fax: 913-588-7440
Web site: http://www.kumc.edu/biochemistry

University of Kansas Medical Center

THE FACULTY AND THEIR RESEARCH

Glen K. Andrews, Professor; Ph.D., Baylor College of Medicine, 1978. Understanding mechanisms that regulate the metabolism of essential and toxic metals in mammals.

Wilfred N. Arnold, Professor; Ph.D., Cornell, 1962. Biochemistry, biophysics, and ultrastructure of yeast cell envelopes; selenium in health and disease.

James P. Calvet, Professor; Ph.D., Connecticut, 1975. Gene expression; polycystin signaling; polycystic kidney disease.

Gerald M. Carlson, Professor and Chairperson; Ph.D., Iowa State, 1975. Regulation through protein-protein interactions within the multisubunit phosphorylase kinase complex.

Aron W. Fenton, Assistant Professor; Ph.D., Oklahoma State, 1999. Thermodynamic and biophysical probes of the molecular mechanisms of allosteric enzymes.

Harvey F. Fisher, Professor; Ph.D., Chicago, 1952. Enzyme mechanisms; thermodynamics of biochemical systems.

Mark T. Fisher, Associate Professor; Ph.D., Illinois at Urbana-Champaign, 1987. Protein folding, unfolding, and assembly; molecular chaperones; thermodynamics of ligand and protein-protein interactions.

Jared Grantham, Professor; M.D., Kansas, 1962. Polycystic kidney disease, mechanisms of cyst formation and growth; role of a novel neutral lipid in the progression of autosomal dominant polycystic kidney disease; mechanisms of fluid secretion in epithelium of renal cysts; role adenylyl cyclase signal transduction in cyst growth.

George M. Helmkamp Jr; Professor; Ph.D., Harvard, 1970. Structural characterization and molecular biology of phosphatidylinositol transfer proteins.

Todd Holyoak, Assistant Professor; Ph.D., Notre Dame, 2000. Determining mechanisms of catalysis and substrate selectivity in families of highly selective proteases; X-ray crystallography.

Alexy S. Ladokhin, Assistant Professor; Ph.D., National Academy of Sciences (Ukraine), 1989. Membrane protein folding and functioning; bacterial toxins; antimicrobial peptides; biophysical techniques.

Kenneth Peterson, Professor; Ph.D., Arizona, 1987. Developmental regulation of gene expression; molecular and epigenetic mechanisms controlling transcription.

Allen B. Rawitch, Professor; Vice Chancellor, Academic Affairs; and Dean, Graduate Studies; Ph.D., UCLA, 1967. Structure and function of thyroid proteins as they relate to the biosynthesis of thyroid hormone.

Liskin Swint-Kruse, Assistant Professor; Ph.D., Iowa, 1995. Biophysical and computational exploration of dissimilar allosteric mechanisms in protein homologues, including DNA-binding proteins; protein engineering.

Lynwood Yarbrough, Professor; Ph.D., Purdue, 1971. Protein structure-function relationships, in particular, folding and assembly of multisubunit protein complexes such as tubulin.

Stowers Institute for Medical Research—Affiliate Faculty

Peter Baumann, Assistant Professor and Assistant Investigator, SIMR; Ph.D, University College (London), 1998. Functional analysis of telomeres and their roles in cellular immortality and cancer.

Joan W. Conaway, Professor and Investigator, SIMR; Ph.D., Stanford, 1987. Analysis of the molecular mechanism and regulation of gene transcription.

Ronald C. Conaway, Professor and Investigator, SIMR; Stanford, 1984. Analysis of the molecular mechanism and regulation of gene transcription.

Chunying Du, Assistant Professor and Assistant Investigator, SIMR; Ph.D., Iowa State, 1997. Investigation of apoptosis, or programmed cell death, in mammals toward the goal of understanding how disorders of this process cause human disease.

Jennifer E. Gerton, Assistant Professor and Assistant Investigator, SIMR; Ph.D., Stanford, 1997. Genomic and genetic analysis of chromosome segregation and chromosome dynamics.

Sue L. Jaspersen, Assistant Professor and Assistant Investigator, SIMR; Ph.D., California, San Francisco, 2000. Regulation and function of centrosomes in the budding yeast *Saccharomyces cerevisiae*.

Robert E. Krumlauf, Professor and Scientific Director and Investigator, SIMR; Ph.D., Ohio State, 1979. Analysis of molecular pathways that regulate how the mammalian head, brain, and nervous system are built, using a variety of vertebrate model systems.

UNIVERSITY OF MARYLAND

School of Medicine and School of Engineering
Interdisciplinary Training Program in Muscle Biology

Program of Study

This program provides interdisciplinary training in muscle biology for predoctoral students. The structure, function, and development and plasticity of skeletal, cardiac, and smooth muscle are considered on the molecular, subcellular, cellular, tissue, and organ levels. The program's 26 faculty members come from four basic science departments—Anatomy, Biochemistry and Molecular Biology, Pharmacology, and Physiology—in the School of Medicine at University of Maryland, Baltimore, as well as from the University of Maryland Biotechnology Institute and the Department of Mechanical Engineering in the College of Engineering at the University of Maryland, Baltimore County. Students enroll in the Ph.D. program of one of these departments or in the Program in Molecular and Cell Biology, University of Maryland, Baltimore. Reflecting the diversity of faculty backgrounds, the training offered ranges from the molecular biological determinants of muscle development and molecular aspects of structure and function of muscle proteins through cell biological aspects of muscle cytoskeleton and matrix, biophysical and physiological analysis of individual muscle cell function, and biomechanical properties of whole muscles and muscular organs. The faculty members are nationally and internationally recognized in the areas of calcium control of muscle function and muscle cytoskeleton and matrix. Students receive training in these and in a variety of related areas, including the molecular biology of muscle and the application of molecular biological and digital imaging techniques to basic questions in muscle biology, with emphasis on the use of several complementary techniques to approach each question under investigation. The major didactic aspect of the predoctoral training is an interdisciplinary course on muscle, which is regularly offered by the program faculty members, has been well received by past student groups, and provides in-depth consideration of all aspects of muscle biology.

Research Facilities

State-of-the-art facilities are available for molecular biological, biochemical, cell biological, structural, biophysical, physiological, and developmental studies of muscle cells and their components. Application and development of new digital imaging microscopic techniques, including laser scanning confocal methods and computer analysis of digital images, is a common strength of many of the program laboratories.

Financial Aid

All students admitted into the program are assured financial support in the form of stipends (in 2006–07, $23,000 for first-year students, increasing with experience) and waiver of tuition and fees. Support is provided as stipends from the Graduate School or Medical School or as awards from external funding. Support is awarded competitively, based on undergraduate records, GRE or MCAT scores, and letters of recommendation. Stipend supplements of $2000 to $2500 per year are awarded competitively as Graduate School merit awards or may be available through external support.

Cost of Study

Full tuition and fees, including health insurance costs, are paid for all students.

Living and Housing Costs

Most students live in apartments or renovated Baltimore row houses, which are available within walking distance of the University. Rents are moderate, ranging from $400 per month for a small apartment to $800 per month for a house that can be shared by 3 to 4 students. Within a short commute, suburban Baltimore offers a larger variety of residential neighborhoods where rents are modest.

Student Group

There are more than 4,500 students at the University of Maryland, Baltimore, enrolled in the Schools of Law, Nursing, Medicine, Dentistry, Pharmacy, and Social Work and in the graduate programs of the biomedical sciences departments. Biomedical sciences students interact across departments to discuss their course work and research and to formulate suggestions regarding the program. Several courses have been introduced at the suggestion of the students. Students have fared well in postdoctoral positions, and some of the earliest graduates have gone on to take tenure-track faculty positions at leading research institutions.

Location

The University of Maryland, Baltimore, is situated in downtown Baltimore, just a 10-minute walk from the Inner Harbor, which has the National Aquarium and Maryland Science Center, Oriole Park at Camden Yards, and a fine collection of theaters, museums, concert facilities, and restaurants. The nearby business district and neighborhoods rich in historic row houses are also within walking distance, making public transportation and inexpensive housing very accessible. Students have found Baltimore to be a very hospitable city, with a distinct, attractive local character, typified by the Orioles, crab houses, and Pimlico. The proximity to the NIH, Johns Hopkins University, and the growing biotechnology industry in the Baltimore-Washington area offers many professional and scientific advantages.

The University

The Interdisciplinary Training Program in Muscle Biology is offered through the University of Maryland Graduate School, Baltimore. The Graduate School combines the research and graduate programs at the University of Maryland, Baltimore, with those at the University of Maryland, Baltimore County, to create an outstanding array of graduate programs and degree offerings, spanning the arts and humanities, behavioral and social sciences, computer science, engineering, physical and natural sciences, biological sciences, and human services.

Applying

Applicants are evaluated on the basis of their previous course work and training, scores in national examinations, research experience, and letters of recommendation. All students applying for the program need to submit copies of their undergraduate transcripts and their GRE or MCAT scores. Forms for letters of recommendation are included in the application folder. Application forms can be obtained directly from the program office (see below), from any of the individual departments participating in the program, or from the Graduate School office. Written inquiries should be made to the director. Telephone inquiries are also recommended.

Correspondence and Information

Dr. Martin F. Schneider
Department of Biochemistry and Molecular Biology
School of Medicine
University of Maryland
108 North Greene Street
Baltimore, Maryland 21201
Phone: 410-706-8417
Fax: 410-706-8297
E-mail: kreineck@umaryland.edu
Web site: http://medschool.umaryland.edu/musclebiology

University of Maryland

THE FACULTY AND THEIR RESEARCH

Martin F. Schneider, Professor of Biochemistry and Director; Ph.D., Duke, 1969. Skeletal muscle physiology and cell biophysics: excitation-contraction coupling, membrane voltage sensors, intracellular calcium release, calcium binding and transport, calcium stores, muscle activity-dependent gene expression; cut fiber voltage clamp; multiple parameter optical recording; intramembrane charge movements; confocal imaging of calcium sparks and of transcription factors in living cells.

Robert J. Bloch, Professor of Physiology and Co-Director; Ph.D., Harvard, 1972. Molecular interactions between the cytoskeleton and membrane proteins that help to organize and stabilize the plasma membrane and intracellular membrane systems of excitable cells, and how these interactions are altered in human disease.

Mordecai P. Blaustein, Professor of Physiology; M.D., Washington (St. Louis), 1962. Calcium signaling in vascular smooth muscle (in relation to myogenic tone and hypertension) and in neurons and glia; Na^+ and Ca^{2+} ion transport and storage; mechanisms of local Ca^{2+} control; digital imaging with fluorescent dyes and patch clamp; immunocytochemistry; molecular biology, including studies of transgenic materials

Meredith Bond, Professor and Chair of Physiology; Ph.D., Pennsylvania, 1984. β-adrenergic pathways in hypertrophied and failing hearts; cardiac function; modulatory role of kinases; regulation of phosphorylation of myofibrillar proteins in the heart; role of A-kinase anchoring proteins in kinase targeting; gene expression by oligonucleotide arrays in failing and non-failing hearts; genomic and proteomic analysis of human heart failure.

Jim Shaojun Du, Associate Professor, Center of Marine Biotechnology; Ph.D., Toronto, 1993. Molecular mechanisms of skeletal differentiation; signaling molecules Hedgehog, TGF-beta, and Wnt in positive and negative regulation of muscle cell proliferation and differentiation; transgenic fish as models for muscle developmental studies.

Iain K. G. Farrance, Assistant Professor of Biochemistry; Ph.D., Georgia, 1988. Transcriptional regulation of gene expression in normal and hypertrophic cardiac muscle: cloning and characterization of transcription factors in heart, mechanisms of specific transcriptional responses to extracellular signals, and structure of the DNA-binding domains of muscle transcription factors.

Donald L. Gill, Professor of Biochemistry; Ph.D., London, 1978. Ca^{2+} signaling mechanisms in nonmuscle and neural cells: mechanisms of Ca^{2+} regulation within Ca^{2+} pools, Ca^{2+} pool analysis by single cell imaging, relationship between Ca^{2+} pool function and smooth muscle cell growth; cellular and subcellular imaging; analysis of Ca^{2+} fluxes in cells; isolation and expression of Ca^{2+} transport proteins.

Hugo Gonzalez-Serratos, Professor of Physiology; M.D., Mexico, 1958; Ph.D., London, 1967. Skeletal, muscle, and nerve: muscle excitation-contraction coupling, muscle fatigue, intracellular calcium release regulation, sarcolemma mechanical physiology and biophysical properties, cytosolic homeostatic Ca regulation; force recordings; general microscopy; fluorescence imaging microscopy; cine-recordings; microphotography; electrophysiology.

Michael G. Klein, Associate Professor of Biochemistry; Ph.D., Illinois, 1986. Control of intracellular calcium in muscle cells; excitation-contraction coupling; confocal imaging of localized calcium signals; functional characterization of SERCA isoforms.

W. Jonathan Lederer, Professor of Physiology and Director of the Medical Biotechnology Center, UMBI; Ph.D., 1975, M.D., 1976, Yale. Cellular and molecular control of intracellular calcium in heart; excitation-contraction coupling in heart muscle; cellular and molecular operation of the Na/Ca exchanger, sarcolemmal ion channels, and the SR calcium release channels and calcium sparks; molecular defects that contribute to heart failure and cardiac arrhythmias; confocal imaging and patch clamp investigation of excitable cells.

Stephen B. Liggett, Professor of Medicine; M.D., Miami (Florida), 1982. Signal transduction mechanisms and pharmacology of G-protein coupled receptors; genetics, haplotypic genome structure, and functional genomics of GPCR polymorphisms; genetic association and family studies and pharmacogenetics of heart failure and asthma; transgenic mouse mimicry of human GPCR polymorphisms; mechanistic basis of GPCR crosstalk in cardiac and airway smooth muscle using knockouts and transgenics, with in vivo and ex vivo mouse physiology.

William R. Randall, Associate Professor of Pharmacology; Ph.D., California, Davis, 1977. Molecular mechanisms regulating synapse and myotendinous junction formation: molecular biology of acetylcholinesterase expression, selective gene expression at neuromuscular and myotendinous junction, protein targeting; recombinant DNA techniques; production and analysis of cDNA and genomic libraries; analysis of promoter, enhancer regions of genes and analysis of transcription factors; immunocytochemistry; in situ hybridization and confocal microscopy.

Terry B. Rogers, Professor of Biochemistry; Ph.D., California, Davis; 1977. Regulation of excitation-contraction coupling in cardiac myocytes: regulation of cardiac cell function by angiotensin II, regulation of cardiac sarcoplasmic reticulum function by intracellular signaling mechanisms, role of protein kinase C in the regulation of cardiac cell function, diversity of phospholipase C isoform expression in cardiac cells; enzyme assays; second messenger quantitation and Ca^{2+} fluxes; molecular biology; PCR and antisense oligonucleotide; single cell electrophysiology and Ca indicator.

Matthew Trudeau, Assistant Professor of Physiology; Ph.D., Wisconsin–Madison, 1998. Molecular mechanisms of ion channels: properties of HERG, a voltage-activated potassium channel from heart, and its role of in familial and acquired cardiac syndromes; cyclic nucleotide-activated ion channels from retina and their role in vision loss.

David J. Weber, Professor of Biochemistry; Ph.D., North Carolina at Chapel Hill, 1988. Structure-function studies of S100 calcium-binding proteins and their protein-protein interactions, including those that regulate the tumor suppressor protein, p53; methods include NMR spectroscopy, site-directed mutagenesis, protein chemistry, and molecular biology techniques.

W. Gil Wier, Professor of Physiology; Ph.D., Utah, 1978. Ca^{2+} in excitation-contraction coupling of cardiac muscle: control of Ca^{2+}-induced Ca^{2+} release by Ca^{2+} current; digital imaging video microscopy; microfluorimetry.

Affiliate Faculty

Lisa D. Brown, Assistant Professor of Biochemistry and Molecular Biology; Ph.D., Connecticut, 1998. Muscle plasticity; in vitro analysis of muscle activity–dependent gene expression.

John H. Collins, Professor of Biochemistry; Ph.D., Boston University, 1970. Structural genomics of contractile, regulatory, and transport proteins; retrieval of sequences from databases; sequence alignments; identification of evolutionary, structural, and functional relationships; computer modeling.

Charles D. Eggleton, Associate Professor of Mechanical Engineering; Ph.D., Stanford, 1994. Theoretical analysis and computer simulations of cell membrane deformation in response to external forces, cytoskeletal-membrane interactions, cellular adhesion, mechanotransduction, and oxygen transport to muscle tissue.

Joseph P. Y. Kao, Professor of Physiology; Ph.D., Berkeley, 1985. Design, chemical synthesis, and application of fluorescent indicators and photo-releasable messenger molecules (caged compounds) for probing signaling processes in cellular physiology.

Aikaterini Kontrogianni, Assistant Professor of Physiology; Ph.D., Baylor College of Medicine, 1997. Molecular templates in sarcomerogenesis.

Paul W. Luther, Research Assistant Professor of Physiology; Ph.D., Illinois, 1984. Organization of postsynaptic proteins; rotary replication electron microscopy; digital fluorescence microscopy; confocal microscopy.

Neil C. Porter, Assistant Professor of Neurology; M.D., Johns Hopkins, 1990. Pathophysiology of muscle degeneration in muscular dystrophy; improvement in carrier detection of Duchenne muscular dystrophy; mechanisms of motor recovery following traumatic nerve injury.

Carlota M. Sumbilla, Research Assistant Professor of Biochemistry; Ph.D., Maryland College Park, 1975. Structure-function relationships in ion transport proteins; overexpression of functional mutant chimeric ATPases in cultured mammalian systems, including cardiac cells; hypertrophy and apoptosis in cardiac myocytes; experimental methods: site-specific mutagenesis, adenovirus-mediated recombinant DNA techniques, cell culture, gene transfer, enzyme biochemistry.

Christopher W. Ward, Assistant Professor of Nursing and Biochemistry; Ph.D., Virginia-Maryland Regional College of Veterinary Medicine, 1996. Striated muscle physiology: Ca^{2+} homeostatis, excitation-contraction coupling, local Ca^{2+} signaling/Ca^{2+} sparks, cellular basis of muscle performance and dysfunction; fluorescence imaging microscopy; confocal microscopy; in situ/in vitro force recording.

Liang Zhu, Associate Professor of Mechanical Engineering; Ph.D., CUNY, City College, 1995. Theoretical simulation of Ca^{2+} release events in skeletal muscle; effects of blood perfusion and vasculature on local heat transfer in microcirculation; selective brain cooling in humans.

UNIVERSITY OF MARYLAND, BALTIMORE COUNTY

Department of Chemistry and Biochemistry

Programs of Study

The Department of Chemistry and Biochemistry at University of Maryland, Baltimore County (UMBC), offers programs of graduate study in chemistry and biochemistry leading to the M.S. and the Ph.D. degrees. The department also participates in a biotechnology M.S. program (applied molecular biology) in collaboration with the Department of Biological Sciences, an interdisciplinary Ph.D. program in molecular and cellular biology, an Interface Program that provides cross-disciplinary experience in chemistry and biology, and the Meyerhoff Program in biomedical sciences. The department's graduate programs in biochemistry and in molecular and cellular biology benefit from being part of larger intercampus joint programs in those fields sponsored in conjunction with departments at the medical, dental, and pharmacy schools of the downtown Baltimore campus.

Upon entering the graduate program, both M.S. and Ph.D. students are required to take a set of placement examinations designed to test their proficiency at the senior undergraduate level and to indicate any areas of deficiency. Under the guidance of an advisory committee, they next complete a group of courses, constituting a core curriculum, which has been designed to bring them to a minimum level of proficiency in each of the major areas of chemistry. In addition, they are expected to take specialized courses in their field of interest. To fulfill the course requirements normally requires one to two years, depending upon the student's initial level of proficiency as demonstrated by the placement examinations.

To qualify for the degree, M.S. students must pass a comprehensive examination in their major field. Thesis research must be approved by a thesis committee, which also administers an oral examination based on the thesis research. Completing the course requirements, passing the comprehensive and oral examinations, and gaining approval of the thesis constitute fulfillment of the M.S. degree requirements. Candidates for the M.S. also have the option of substituting additional course work for the thesis. Graduate students in the Ph.D. program are expected to pass comprehensive examinations, prepare and defend an acceptable research proposal, present a dissertation based upon original research, and pass an oral defense.

The principal areas of thesis research for both the M.S. and Ph.D. degrees include analytical chemistry; biochemistry; bioinorganic chemistry; enzymology; mass spectrometry; models for enzymic reactions; organic mechanisms; organic synthesis; protein and nucleic acid chemistry; theoretical, physical, bioanalytical, biophysical, and carbohydrate chemistry; and chemistry at the biology interface.

Research Facilities

Extensive facilities are available for cutting-edge research. The specialized research instrumentation available includes calorimetry, chromatography, stopped-flow and temperature-jump kinetics, nanosecond laser flash photolysis, nuclear magnetic resonance spectroscopy (one 200-, one 300-, one 500-, two 600-, and one 800-MHz instruments), electron spin resonance spectroscopy, circular dichroism, X-ray diffraction, infrared spectroscopy, laser fluorescence spectroscopy, atomic absorption, gas chromatography–mass spectrometry, and Fourier transform/ion cyclotron resonance mass spectrometry, as well as extensive molecular modeling computer facilities. Also located in the department is a Center for Structural Biochemistry, which specializes in the structural analysis of biological molecules (e.g., biopolymers, peptides, glycoproteins). In addition to a laser desorption mass spectrometer, and two 500 MHz and 600 MHz NMRs, the center houses one of the few tandem mass spectrometers located in academic institutions worldwide. The Howard Hughes Medical Investigator Suite houses the second 600- and 800-MHz NMRs, which are used for high-dimensional studies of HIV proteins, metallobiomolecules, and macromolecular interactions. The main University library contains more than 2,500 volumes of chemistry texts and subscribes to 150 chemistry and biochemistry periodicals. The department reference room also provides access to the principal journals.

Financial Aid

The University offers financial packages with twelve-month stipends of $17,000 to $21,000, health insurance, and tuition remission to qualified students. Enhanced stipend fellowships are also available.

Cost of Study

Tuition in 2005–06 was $395 per credit hour for Maryland residents and $652 per credit hour for nonresidents. Fees were about $770 per semester.

Living and Housing Costs

There are a limited number of on-campus dormitory rooms available for graduate students. Most graduate students are housed in apartments or rooming houses in the nearby communities of Arbutus and Catonsville. A single graduate student can expect living and educational expenses of approximately $16,000 to $18,000 a year.

Student Group

In fall 2004, the department's graduate students included 34 men and 43 women. All of the students plan careers in chemistry and biochemistry in teaching or research or in associated regulatory and financial areas. Approximately 80 percent of the students are receiving some form of financial aid.

Location

The University has a scenic location on the periphery of the Baltimore metropolitan area. Downtown Baltimore can be reached in 15 minutes by car, while Washington is an hour away. Both Baltimore and Washington have very extensive cultural facilities, including eight major universities, a number of museums and art galleries of international reputation, two major symphony orchestras, and numerous theaters.

The University

The University of Maryland, Baltimore County, was established in 1966 on a 500-acre campus. The Department of Chemistry and Biochemistry of the University of Maryland Graduate School, Baltimore, is located at the UMBC campus. The University has about 11,700 students drawn primarily from Maryland, although an increasing number have enrolled from other states and countries. The undergraduates are predominantly interested in professional or business careers. Because a high percentage of the students are the first members of their families to attend college, UMBC has a very different atmosphere from that encountered in older institutions and has made a particular effort to attract members of minority groups, who now account for about 36 percent of the undergraduate student body.

Applying

Applications should include an academic transcript, three references, and the results of the General Test of the Graduate Record Examinations.

Correspondence and Information

Graduate Program Director
Department of Chemistry and Biochemistry
University of Maryland, Baltimore County
1000 Hilltop Circle
Baltimore, Maryland 21250
Phone: 866-PhD-UMBC (743-8622; toll-free for chemistry and biochemistry only)
E-mail: chemgrad@umbc.edu
Web site: http://www.umbc.edu/chem
http://www.umbc.edu/biochem

University of Maryland, Baltimore County

THE FACULTY AND THEIR RESEARCH

Bradley R. Arnold, Associate Professor; Ph.D., Utah, 1991; postdoctoral studies at the National Research Council of Canada, Ottawa, and the Center for Photoinduced Charge Transfer, Rochester. Physical chemistry; application of time-resolved polarized spectroscopy.

C. Allen Bush, Professor; Ph.D., Berkeley, 1965; postdoctoral studies at Cornell. Biophysical chemistry: conformation and dynamics of carbohydrates, glycoproteins, glycopeptides, and polysaccharides by NMR spectroscopy, computer modeling, and circular dichroism.

Donald Creighton, Professor; Ph.D., UCLA, 1972; postdoctoral studies at the Institute for Cancer Research. Enzyme mechanisms and protein structure, studies on sulfhydryl proteases and glyoxalase enzymes.

Brian M. Cullum, Assistant Professor; Ph.D., South Carolina; postdoctoral studies at Oak Ridge National Laboratory. Analytical chemistry; development of optical sensors and optical sensing techniques for biomedical and environmental research.

Dan Fabris, Associate Professor; Ph.D., Padua (Italy), 1989; postdoctoral studies at the National Research Council, Area of Research of Padua (Italy), and University of Maryland, Baltimore County. Bioanalytical and biomedical applications of mass spectrometry; nucleic acid adducts; protein-nucleic acid interactions.

James C. Fishbein, Professor; Ph.D., Brandeis, 1985; postdoctoral studies at University of Toronto. Mechanisms of organic reactions in aqueous solutions; generation and study of reactive intermediates, particularly those involved in nitrosamine and nitrosamide carcinogenesis.

Colin W. Garvie, Assistant Professor; Ph.D., Leeds (England), 1997; postdoctoral studies at Howard Hughes Medical Institute, Johns Hopkins University School of Medicine. Structural studies of the redox regulation of circadian clock proteins; structural studies of the MHC II enhanceosome.

Susan K. Gregurick, Assistant Professor; Ph.D., Maryland College Park, 1994; postdoctoral studies at Maryland Biotechnology Institute. Development of evolutionary algorithms for use in intricate chemical problems.

Ramachandra S. Hosmane, Professor; Ph.D., South Florida, 1978; postdoctoral studies at Illinois. Organic synthesis; biomedicinal chemistry, with applications in antiviral and anticancer therapy; and biomedical technology, with applications in artificial blood.

Richard L. Karpel, Professor; Ph.D., Brandeis, 1970; postdoctoral studies at Princeton. Interactions of helix destabilizing proteins with nucleic acids and the involvement of such proteins in various aspects of RNA function, metal ion–nucleic acid interactions.

Lisa A. Kelly, Associate Professor; Ph.D., Bowling Green, 1993; postdoctoral studies at Brookhaven National Laboratory. Photoredox-initiated chemical bond cleavage in biological and model systems.

William R. LaCourse, Professor; Ph.D., Northeastern, 1987; postdoctoral studies at Ames Laboratory (USDOE) and Iowa State. Pulsed electrochemical detection techniques for bioanalytical separations.

Joel F. Liebman, Professor; Ph.D., Princeton, 1970; postdoctoral studies at Cambridge and the National Institute of Standards and Technology. Energetics of organic molecules, especially considerations of strain and aromaticity; gaseous ions; noble gas, fluorine, boron, and silicon chemistry; mathematical and quantum chemistry.

Wuyuan Lu, Affiliate Assistant Professor; Ph.D., Purdue, 1994; postdoctoral studies at the Scripps Research Institute and the University of Chicago. Structural and functional relationships of maternin and d-peptides as receptor antagonists and enzyme inhibitors.

Ralph M. Pollack, Professor; Ph.D., Berkeley, 1968; postdoctoral studies at Northwestern. Enzyme reactions, model systems for enzyme mechanisms, organic reaction mechanisms.

Katherine L. Seley, Professor; Ph.D., Auburn, 1996; postdoctoral studies at Auburn. Discovery, design, and synthesis of nucleoside/nucleotide and heterocyclic enzyme inhibitors for use as medicinal agents with chemotherapeutic emphasis in the areas of anticancer, antiviral, antibiotic, and antiparasitic targets.

Veronika Szalai, Assistant Professor; Ph.D., Yale, 1998; postdoctoral studies at North Carolina at Chapel Hill. Spectroscopic characterization of biomolecular assemblies containing paramagnetic transition metals or spin labels.

Paul J. Smith, Associate Professor; Ph.D., Pittsburgh, 1993; postdoctoral studies at Johns Hopkins. Bioorganic and physical organic chemistry; host-guest chemistry; DNA structure and DNA binding by small molecules.

Michael F. Summers, Professor and Howard Hughes Associate Medical Investigator; Ph.D., Emory, 1984; postdoctoral studies at the Center for Drugs and Biologics, FDA, Bethesda, Maryland. Elucidation of structural, dynamic, and thermodynamic features of metallobiomolecules utilizing advanced two-dimensional and multinuclear NMR methods.

James S. Vincent, Associate Professor Emeritus; Ph.D., Harvard, 1963; postdoctoral studies at Caltech. Infrared and Raman spectroscopy of phospholipid membrane systems, magnetic spectroscopy of transition-metal complexes.

Dale L. Whalen, Professor; Ph.D., Berkeley, 1965; postdoctoral studies at UCLA. Reactions of carcinogenic polycyclic aromatic hydrocarbon epoxides, organic reaction mechanisms.

UNIVERSITY OF MISSISSIPPI MEDICAL CENTER

School of Graduate Studies in the Health Sciences
Department of Biochemistry

Program of Study

The program of study leads to the Ph.D. degree. Students receive a broad training in modern biochemistry, including molecular biology, protein chemistry, and biophysical methods. The curriculum is designed to consist of independent study and research after completion of required course work. The aim is to prepare graduates to compete successfully in today's challenging research environment.

A minimum of 55 credit hours of course work is required, which is accumulated by taking about fifteen courses. These include the general biochemistry course, the advanced courses in biochemistry, and additional courses from other departments within the Medical Center. In order to assist them in choosing a research adviser, first-year students also take a course in laboratory techniques and complete laboratory rotations with several faculty members. Dissertation research begins in the summer of the first year. After the second year, students are normally finished with their course work and complete the written comprehensive exam. By the end of the third year, students take an oral exam that tests their ability to develop and defend research proposals. They are then admitted to candidacy for the Ph.D. degree. Graduation requires satisfactory completion, written presentation, and oral defense of an original research project as well as a manuscript that has been accepted for publication in a peer-reviewed scientific journal. Completion of the program usually takes four to five years.

Topics available for dissertation research include regulation of gene expression, nuclear organization and disease, mechanisms of drug resistance, three-dimensional structural analysis of multiprotein complexes, mechanisms of electron and proton transfer through proteins, biophysical analysis of protein-drug-microtubule interactions, assembly of membrane protein complexes, membrane-cytoskeleton interaction, and microtubule polymerization.

Research Facilities

The Department is extremely well equipped for training and research in cell and molecular biology, biochemistry, and biophysics. Each faculty member has generous laboratory space and both basic and specialized equipment for projects. In addition, there is an abundance of shared Departmental equipment and facilities. Of particular note are imaging capabilities (including several fluorescence microscopes, one of which is equipped with a Photometric Coolsnap FX CCD camera and filterwheel, a transmission electron microscope with cryostage, and Silicon Graphics workstations for image analysis and molecular modeling), a DNA sequencing facility (LiCor 4000L), a DNA microarray facility for genomic expression analysis, and biophysical instrumentation (Optima XL-A analytical ultracentrifuge, stopped-flow spectrophotometer, steady-state fluorimeters, differential scanning calorimeters, laser optical tweezers, a rapid-freeze-quench apparatus, a SpectroGenesis CCD-ICP spectrometer, and a Bruker EPR spectrometer with helium temperature capacity).

Financial Aid

Full-time students receive graduate research assistantships. For the 2006–07 academic year, the annual stipend is $23,100.

Cost of Study

Until advanced to candidacy for the degree, a student pays tuition and fees of approximately $2025 per year. Subsequent years are $400 or less. Students must also have health insurance. A specialized student plan is available at a cost of approximately $1450 per year.

Living and Housing Costs

The overall cost of living in Jackson is lower than the average cost across the United States. Stipends are adequate to cover modest living expenses. There are numerous apartments and small houses available for rent within walking distance of the campus. Campus housing is limited, consisting of a women's residence hall and ninety-five low-cost apartments. Married students or single parents enrolled in Medical Center programs get preference on apartment assignments. There is usually a one- to two-year waiting list for campus apartments. For additional information, applicants should contact the Dormitory or Student Apartment Manager, 2590 North State Street, Jackson, Mississippi 39216.

Student Group

The Department maintains an average of 15 to 20 graduate students, with 3 to 5 new students arriving each year. Entering class size is determined by availability of stipend support. Typically, each class is a blend of American and international students. Several have begun their graduate studies after substantial time of employment in a variety of professions, and many are married.

Student Outcomes

The majority of graduate students move into postdoctoral positions at well-known universities and research institutes. Recent examples of institutions attended include Johns Hopkins University, the National Institutes of Health, Duke University, Washington University, Scripps Institute, Yale University, University of Virginia, and Baylor College of Medicine. Other graduates have chosen industrial research settings, with such companies as Amgen, Ciba-Geigy, and New England Biolabs.

Location

Jackson is the capital of Mississippi, with 400,000 people in the metropolitan area. It is the midpoint between Memphis and New Orleans as well as between Dallas and Atlanta. The city and surrounding area are rich in history as well as at the forefront of developing technologies, particularly telecommunications.

With seven colleges in the immediate vicinity, educational opportunities abound. The area is replete with outstanding cultural activities, including historical and art museums, professional theater, opera and ballet companies, a symphony orchestra, and one of the region's finest planetariums. Jackson and surrounding cities host numerous sports events, such as minor league baseball and hockey games, a professional golf tournament, and national rodeo and equestrian competitions. The nearby 30,000-acre reservoir enhances the boating, camping, fishing, hiking, and other outdoor activities available year-round in the numerous local and state parks.

The University and The Department

The University of Mississippi Medical Center in Jackson is the health sciences campus of the University, which is in Oxford. The University is the only publicly funded Mississippi institution of higher learning that has achieved a chartered Phi Beta Kappa Chapter. In 2001, the Graduate Program in the Medical Sciences became the School of Graduate Studies in the Health Sciences. It joins the Schools of Medicine, Dentistry, Nursing, and Health Related Professions; teaching laboratories for all programs; and the University Hospitals and Clinics. Total enrollment in all programs is approximately 1,800.

The Department of Biochemistry is strongly committed to graduate research and teaching. Its size provides for close scientific and personal interaction between faculty members and students. Most faculty members have achieved national and international recognition through their research programs.

Applying

A strong background in biology and chemistry is a minimum prerequisite for admission. Applicants must have a bachelor's degree with an overall grade point average of 3.0 on a 4.0 scale. The GRE general test is required for all students plus the TOEFL for international students.

Application packets are sent to prospective students only after an initial review of qualifications. A separate financial aid application is not required. Initial inquiries should be made by December 1 for entrance the following August. Formal applications should be completed by February 1 to ensure full consideration by the Admissions Committee.

The University of Mississippi Medical Center offers equal opportunity in education and employment, M/F/D/V.

Correspondence and Information

Dr. Drazen Raucher
Director of Graduate Studies for Biochemistry
University of Mississippi Medical Center
2500 North State Street
Jackson, Mississippi 39216

Phone: 601-984-1500
Fax: 601-984-1501
E-mail: BioChem@biochem.umsmed.edu
Web site: http://biochemistry.umc.edu

University of Mississippi Medical Center

THE FACULTY AND THEIR RESEARCH

David T. Brown, Professor; Ph.D., Mississippi Medical Center, 1988. Organization of eukaryotic chromatin; structure and function of histone variants in vivo.

John J. Correia, Professor; Ph.D., Connecticut, 1981. Physical biochemistry; protein-protein interactions; analytical ultracentrifugation; mechanism and regulation of microtubule assembly/disassembly; molecular mechanism of antimitotic drugs; kinesin, ncd-tubulin, stathmin interactions with tubulin.

Victor L. Davidson, Professor; Ph.D., Texas Tech, 1982. Structure and function of enzymes and electron transport proteins; mechanisms of intermolecular electron transfer; biochemistry of methylotrophic bacteria; biosensors.

Michael D. Hebert, Assistant Professor; Ph.D., Georgia State, 1997. Nuclear organization; biogenesis of splicing machinery; neurodegenerative disorders.

Jonathan P. Hosler, Professor; Ph.D., Michigan, 1983. Structure and function of proton transfer pathways in membrane proteins; mechanisms of metal center assembly by chaperones.

Aimin Liu, Assistant Professor; Ph.D., Stockholm, 1999. Structure and function relationship of a nuclear metalloprotein involved in the NF-κB signaling and transcriptional activities; mechanistic enzymology of tryptophan catabolic pathway and biodegradation of toxic compounds; electron paramagnetic resonance spectroscopy.

Sharon Lobert, Adjunct Professor; Ph.D., Vanderbilt, 1989; RN. Structure-function studies of antimitotic drugs and their interaction with tubulin; protein-protein interactions; drug resistance to anticancer drugs; cell culture quantitation of proteins and mRNA; fluorescence microscopy studies.

Joe Maher, Adjunct Associate Professor; M.D., Texas Health Science Center at Dallas, 1987. Signal transduction and the regulation of gene expression; genetic control of cell fate determination; involvement of these processes in human diseases, including cancer; human cancer genetics.

Drazen Raucher, Assistant Professor; Ph.D., Florida State, 1995. Membrane cytoskeleton adhesion measurements using laser optical tweezers; effects of pharmaceutical reagents on cell physiology; targeted delivery of antitumor drugs and oligonucleotides by genetically engineered thermally responsive polymers.

Donald B. Sittman, Professor and Interim Chairman; Ph.D., North Carolina at Chapel Hill, 1978. Function of H1 histone variants; genomic microarray analysis.

Parminder Vig, Adjunct Professor; Ph.D., Chandigarh (India), 1988. Molecular mechanisms of nerve-cell death in neurodegenerative diseases using transgenic mouse models and cell culture.

Charles L. Woodley, Associate Professor; Ph.D., Nebraska, 1972. Initiation of protein synthesis.

UNIVERSITY OF NEVADA, RENO

Department of Biochemistry and Molecular Biology
Graduate Programs in Biochemistry

Programs of Study

The Department of Biochemistry and Molecular Biology at the University of Nevada, Reno (UNR), offers a challenging and broad-based graduate program of research and course studies leading to the M.S. or Ph.D. in biochemistry. The aim of the graduate program is to train scientists for critical analysis and solution of biochemical problems at the molecular level. The diverse research areas represented by the faculty have the common theme of understanding the structures and roles of macromolecules in complex biological systems. Students benefit from exposure to faculty members appointed in both the College of Agriculture, Biotechnology and Natural Resources (CABNR) and the School of Medicine. They have an opportunity for multidisciplinary interactions with graduate students and faculty members in related departments, including the Departments of Physiology and Anatomy, Microbiology and Immunology, Pharmacology, Chemistry, Biology, and Animal Biotechnology. The academic environment is lively and highly interactive, as represented by a diverse, interdisciplinary seminar program sponsored in conjunction with other related departments. The program of study includes lecture courses, laboratory rotations, journal club presentations, a qualifying written and oral examination, dissertation research, and an opportunity to teach. First-year students take a core curriculum and gain research experience by rotating through student-selected research laboratories. Laboratory rotations facilitate the choice of a dissertation adviser. Doctoral and master's research projects are selected by the student in consultation with a major dissertation adviser and an advisory committee. The requirements for the Ph.D. can generally be completed in four or five years. The program, which is designed to prepare students for careers in research and/or teaching, emphasizes a cooperative, personal working environment between students and members of the faculty.

Faculty interests cover a wide range of disciplines in the biomedical sciences and life sciences. Research interests include environmental and biotic stress and rubber and vitamin biosynthesis in plants; insect peptide and lipid hormones and pheromones and insect lipid metabolism; muscle protein enzymology, structure, and signaling; muscle contraction and excitation-contraction coupling; cell motility; insulin signaling pathways and glucose transport; structure of membrane receptors; membrane-cytoskeletal interactions; oxygen toxicity; mammalian lipid metabolism in cancer; receptor-mediated endocytosis; and computational methods in database mining and macromolecular structure. Each faculty member directs an active research program and is dedicated to training postdoctoral associates and doctoral- and master's-level graduate students as well as undergraduate students. Faculty members are funded by the National Institutes of Heath, National Science Foundation, Department of Energy, and other extramural sources in excess of $5 million per year.

Research Facilities

Research in UNR's Department of Biochemistry and Molecular Biology is supported by state-of-the-art approaches to genomics, proteomics, gene transfer, recombinant techniques, bioinformatics, computational biology, electrophysiology, spectroscopy, single-molecule biophysics, protein analytical biochemistry, mass spectrometry, and X-ray crystallography, among others. Facilities and technical staff members are available for analysis of samples by electron, confocal, two-photon confocal, single-molecule, and atomic force microscopy; flow cytometry; mass spectrometry; and high-throughput DNA sequence and mRNA expression analysis. In addition, research centers for genomics, monoclonal antibody production, construction of viral vectors, calcium imaging, proteomics analysis, bioinformatics and molecular modeling, and transgenic mouse generation and housing are also available.

The UNR libraries serve as the primary center for informational resources and services in support of teaching and research. The libraries' Web-based information delivery system provides access to the libraries' physical collections (more than 1 million books, 5,000 print journals, 12,000 videos and DVDs, and 3.3 million microforms); course reserves, most of which are available online; full-text articles from a growing number of electronic journals and magazines (currently around 15,000); approximately 13,000 electronic books; more than 200 general and specialized databases; and high-quality Internet resources selected and organized for the UNR community.

Financial Aid

Graduate fellowships, assistantships, and research awards are available to students admitted to the Graduate School on a competitive basis. Both fellowships and assistantships carry a stipend and a tuition waiver. Assistantship stipends currently start at $20,000 and vary upward depending upon year of study. Information is available from the Graduate School (http://www.vpr.unr.edu/grad2/) or the Department of Biochemistry and Molecular Biology (http://www.ag.unr.edu/biochemistry/).

Cost of Study

Nevada residents pay registration fees only; nonresidents pay tuition in addition to registration fees. In 2005–06, part-time tuition (1–6 credits) was $149.50 per credit, and full-time tuition (7 credits or more) was $4733.50 per semester. The registration fee was $140 per credit. Students awarded research fellowships or teaching assistantships are entitled to a partial fee waiver of nonresident tuition rates and pay only $27.45 per credit.

Living and Housing Costs

A room in the residence halls ranges from $3990 to $4990 per academic year, depending on location. Meal plans range from $2595 to $3395. A listing of off-campus housing—including rooms, apartments, and houses—is also available.

Student Group

There are approximately 30 students enrolled in the graduate biochemistry programs.

Student Outcomes

Ph.D. graduates from the Department of Biochemistry and Molecular Biology now hold professional positions in universities, industry, and government, including federal science administration. These include faculty members at leading universities and medical schools and research scientists at pharmaceutical companies. Many other former students hold positions in scientific administration and biotechnology start-up companies.

Location

The University of Nevada, Reno, is a center of innovation and energy for the thriving Reno-Sparks metropolitan area. Its 255-acre campus of rolling hills features a blend of ivy-covered buildings, sweeping lawns, and functional, progressive architecture. Reno-Sparks is in an unusually attractive natural setting. It is bounded on the west by the majestic Sierra Nevada range and on the east by a rolling basin and range province. Reno-Sparks benefits from a comfortable climate marked by generally cool and dry weather with more than 300 sunny (or cloudless) days per year. The area is a haven for those who love the four seasons and outdoor activities. Recreational activities are easy to find. Students are within less than an hour's driving distance of the many world-class ski resorts of Lake Tahoe and the historic Western realm of Virginia City.

The University

The University of Nevada, Reno, is a constitutionally established land-grant university founded in 1874. The University served the state of Nevada as its only state-supported institution of higher education for almost seventy-five years. In that historical role, it has emerged as a doctoral-granting university that focuses its resources on doing a select number of things well. A diverse student body strengthens the academic atmosphere for the cultural and intellectual development of the student. By fostering creative and scholarly activity, the University encourages and supports faculty research and the application of that research to state and national problems. UNR is growing rapidly and currently enrolls more than 16,000 students, including 3,000 enrolled in graduate programs. The University houses a School of Medicine with a class of 65 medical students. The Northwest Commission on Colleges and Universities (NWCCU) accredits the University.

Applying

There is a $60 nonrefundable graduate program application fee. All GPA and test score information must be included on the application. Interested students should submit two copies of their official undergraduate and graduate school transcripts directly from the institutions previously attended to the University of Nevada, Reno. Applicants who are applying for a graduate assistantship must include three letters of recommendation.

Correspondence and Information

Dr. John C. Cushman
Graduate Director
Department of Biochemistry and Molecular Biology
University of Nevada, Reno
1664 North Virginia Street, MS 200
Reno, Nevada 89557-0014

Phone: 775-784-6911
Fax: 775-784-1650
E-mail: jcushman@unr.edu
Web site: http://www.ag.unr.edu/biochemistry/

University of Nevada, Reno

THE FACULTY AND THEIR RESEARCH

Josh E. Baker, Assistant Professor; Ph.D., Minnesota, Twin Cities, 1999. Use of optical traps, fluorescence microscopy, and single-molecule imaging techniques to study the molecular basis for cell motility.

Brian W. Beck, Associate Director of Structural Bioinformatics and Molecular Modeling, Nevada Center for Bioinformatics; Ph.D., Washington State, 1997. Molecular modeling; computational biochemistry.

Gary Blomquist, Professor and Department Chair; Ph.D., Montana State, 1973. Insect biochemistry; lipid metabolism; biosynthesis and molecular biology of sex pheromone production; comparative biochemistry.

Grant R. Cramer, Professor; Ph.D., California, Davis, 1985. Salt tolerance of plants; growth and ion transport of plants at the whole-plant and cellular levels; interactions with plant hormones.

Christine Cremo, Professor; Ph.D., Oregon State, 1983. Structure and function of motor proteins in smooth muscle.

John C. Cushman, Professor; Ph.D., Rutgers, 1987. Molecular genetics of Crassulacean acid metabolism; molecular mechanisms of signal transduction and adaptive responses to salinity and drought stress in plants; functional genomics of salinity and drought stress tolerance.

Hanna Damke, Research Assistant Professor; Ph.D., Marburg (Germany), 1992. Role of the signaling GTPase dynamin in coordinating endocytosis with other cellular functions to maintain homeostasis.

Patricia Ellison, Research Assistant Professor; Ph.D., Sheffield, 1981. Regulation of smooth muscle myosins by phosphorylation and dephosphorylation; interaction between myosin and actin, myosin light-chain kinase, and phosphatase.

Kevin Facemyer, Research Assistant Professor; Ph.D., Washington State, 1996. Computational biochemistry of motor proteins.

Martin Gollery, Associate Director of Bioinformatics, Nevada Center for Bioinformatics; B.S., California, San Diego, 1981. Development of novel bioinformatics techniques to characterize nucleic and amino acid data that does not have homologous matches to existing database entries; modeling protein families and domains through the use of hidden Markov models.

Jeffrey Harper, Associate Professor; Ph.D., Washington (St. Louis), 1985. Mineral nutrition; calcium signaling; engineering plants to better tolerate abiotic and biotic stress.

Cynthia Corley Mastick, Associate Professor; Ph.D., Carnegie Mellon, 1990. Cellular mechanisms of insulin actions on maturity-onset or type 2 diabetes; molecular mechanisms of signal transduction and signaling specificity; cellular basis of insulin action and peripheral insulin resistance; regulation of glucose and lipid uptake/metabolism; cell biology of adipocytes and muscle.

Grant Mastick, Assistant Professor; Ph.D., Carnegie Mellon, 1992. Formation of neuronal connections in the embryonic brain; developmental neurobiology; axon guidance by cell-adhesion molecules; mouse neural mutants; gene regulatory networks and transcriptional regulation of gene expression.

Kunio Misono, Associate Professor; Ph.D., Vanderbilt, 1978. Structure of cell membrane receptors and their signal transduction mechanisms.

Ron Mittler, Associate Professor; Ph.D., Rutgers, 1993. Involvement of reactive oxygen intermediates in the response of plants to different environmental stimuli and stress (biotic and abiotic); mechanisms underlying the acclimation of desert plants to their harsh environments.

Haruo Ogawa, Research Assistant Professor; Ph.D., Tokyo Institute of Technology, 1998. Understanding receptor structure and signal transduction mechanisms; crystallization of membrane proteins.

Ronald S. Pardini, Professor and Associate Director, Nevada Agriculture Experiment Station; Ph.D., Illinois, 1965. Nutritional intervention with omega-3 fatty acids in the treatment of cancer; understanding mechanisms of omega-3 fatty acid–induced growth inhibition and enhanced response to cancer therapy.

Yue Qiu, Research Assistant Professor; Ph.D., Kochi (Japan), 1991. Hormone-receptor interactions; signal transduction mechanisms; cardiovascular regulation and diseases.

David Quilici, Manager, Nevada Proteomics Center; Ph.D., Nevada, Reno, 1997. Identifying unknown compounds; quantifying known compounds; elucidating structure and chemical properties of molecules.

Kathleen M. Schegg, Research Biochemist, Nevada Proteomics Center; Ph.D., Nevada, Reno, 1980. Proteomics; 2-D gel separation of proteins; amino acid analysis; protein sequencing; peptide synthesis.

David A. Schooley, Professor; Ph.D., Stanford, 1968. Structural and biosynthetic studies on physiologically active materials, chiefly, insect juvenile hormones and peptide hormones; methods for titer determination of hormones; stereochemistry and its analysis; analytical biochemistry.

David Shintani, Associate Professor; Ph.D., Michigan State, 1996. Plant biochemistry and genome research; metabolic and developmental regulation of plant isoprenoid metabolism; vitamin and cofactor biosynthesis in plants.

Claus Tittiger, Associate Professor; Ph.D., Queen's at Kingston, 1994. Insect molecular biology and genomics; isoprenoid pheromone biosynthesis; juvenile hormone regulation; cytochrome P450s; hydrocarbon and lipid metabolism.

Maria L. Valencik, Assistant Professor; Ph.D., UCLA, 1991. Cardiovascular research; integrins and natriuretic peptides in cardiac myocytes.

David C. Ward, Professor; Ph.D., Rockefeller, 1969. Molecular genetics of human and mouse genomes; cytogenetics and human disease diagnosis.

Lee Weber, Professor; Ph.D., SUNY at Albany, 1975. Structure and expression of human stress protein genes; function of heat-shock proteins in stress resistance.

William H. Welch, Professor; Ph.D., Kansas, 1968. Role of cations in enzyme structure and function; structure-function relationships of biological molecules.

THE UNIVERSITY OF TOLEDO

College of Pharmacy
Department of Medicinal and Biological Chemistry

Programs of Study

The Department of Medicinal and Biological Chemistry (MBC) in the University of Toledo's (UT) College of Pharmacy presents opportunities for graduate studies leading to the M.S. and Ph.D. degrees in medicinal chemistry. The focus of the educational program is on the theory and practice of drug design, with additional advanced-level instruction in the appropriate underlying scientific areas. Students normally focus entirely on course work during the first year of the program. This centers on advanced medicinal and biomedicinal chemistry courses, which are offered in small-group tutorial sessions. Students are expected to concentrate on fundamental concepts from the drug design literature and then formulate their own ideas about possible new directions based on recent advances in the field. Each student is responsible for a number of written and oral presentations. Students generally rotate through two faculty laboratories, where they carry out small research projects. These experiences provide a basis for choosing a laboratory in which to conduct subsequent thesis or dissertation research. Fundamental courses in drug design and the biochemical basis of disease are also offered in the first year. In addition to this core program, students normally matriculate in advanced or basic courses in biochemistry, immunology, neurobiology, and synthetic and physical organic chemistry, some of which are individualized to match the students' specific research interests. Fewer courses are taken in each successive year of the program, with attention to research increases. During the second year, M.S. students normally complete a research project and write a thesis. Ph.D. students participate in a comprehensive examination, which includes both a written and an oral component, leading directly to Ph.D. candidacy. Research leading to the Ph.D. dissertation usually requires two to three years to complete following the initial one to two years of course work.

Research Facilities

Construction of a life science complex on the UT campus was completed in 1997. This complex couples approximately 175,000 square feet of new, high-tech classroom and state-of-the-art laboratory space with a somewhat larger amount of prior space, such that the Departments of Biology and Chemistry, along with the entire College of Pharmacy, are now joined within a contiguous arrangement that is conducive to both formal and casual interactions among the various faculty members and students. From this complementary physical setting, basic and applied research within all aspects of the life sciences is conducted at the undergraduate, graduate, and postdoctoral levels. The MBC Department seminar and conference rooms are convenient meeting places for research discussions, seminars by visiting scientists, and social functions. Academic and research activities are further enhanced by the presence of four well-established UT Centers of Excellence, such as the Center for Drug Design and Development (CD3), all located within the new complex. The CD3 occupies three 1,000-square-foot labs in the life science complex. All of the facilities and equipment essential for new drug discovery and development in the biological, chemical, and pharmaceutical sciences are housed within the MBC, CD3, and adjacent laboratories. Specific areas of clinical research and practice for the CD3's network are largely coordinated through, or are conducted in close conjunction with, the Center for Applied Pharmacology (CAP), which specializes in entertaining clinically oriented Phase I, II, and IV studies. Laboratory space and equipment for conducting experimental analytical work and preclinical trials are available at the University. Computer analyses are provided using mainframe or microcomputers available on campus.

Laboratory space and equipment at The University of Toledo are well-suited for conducting biopharmaceutical-related analytical work. Measurements of drug and metabolites in biological samples can be performed using fluorescence polarization immunoassay (FPIA), gas chromatography (GC), and high-pressure liquid chromatography (HPLC). There is a flow cytometry core for immunology research, and advanced molecular biology work is done in the molecular biology core, which consists of real-time PCR instrumentation and DNA microarray analysis equipment.

Financial Aid

Most full-time graduate students receive some financial support. College fellowships, teaching assistantships, and research assistantships, which include a stipend and a tuition waiver, are available for qualified students on a competitive basis. The out-of-state tuition surcharge normally charged to out-of-state and international students is waived for students whose permanent address is within one of the following Michigan counties: Hillsdale, Lenawee, Macomb, Oakland, Washtenaw, and Wayne. In addition, The University of Toledo offers an out-of-state tuition surcharge waiver to cities and regions that are a part of the Sister Cities Agreement. These regions include Toledo, Spain; Londrina, Brazil; Qinhuangdao, China; Csongrad County, Hungary; Delmenhorst, Germany; Toyohashi, Japan; Tanga, Tanzania; Bekaa Valley, Lebanon; and Poznan, Poland. The University of Toledo Graduate College offers a variety of memorial and minority scholarship awards, including the Ronald E. McNair Postbaccalaureate Achievement Scholarship, the Graduate Minority Assistantship Award, and two full University fellowships.

Cost of Study

The graduate tuition rate for the 2006–07 academic year is $390.05 per semester credit hour for in-state students. For nonresidents, the out-of-state surcharge is $367.15 per semester credit hour. Additional fees are required and include the general fee, technology fee, and mandatory insurance.

Living and Housing Costs

The University of Toledo has a diverse offering of student housing options, including suite-style and traditional residential halls. Housing is offered to graduate students through Residence Life or contracted individually by the student. Affordable, high-quality off-campus apartment-style housing within walking distance of campus is abundant.

Student Group

There are approximately 20,000 students at the University of Toledo. About 4,000 are graduate and professional students. The University has a rich diversity of student organizations. Students join groups that are organized around common cultural, religious, athletic, and educational interests.

Location

The University of Toledo has several campus sites in the city of Toledo. Most engineering graduate students take classes on the Main campus, which is located in suburban western Toledo. With a population of more than 330,000, Toledo is the fiftieth-largest city in the United States. It is located on the western shores of Lake Erie, within a 2-hour drive of Cleveland and Detroit.

The University and The Department

The University of Toledo was founded by Jessup W. Scott in 1872 as a municipal institution and became part of the state of Ohio's system of higher education in 1967. On July 1, 2006, The University of Toledo merged with the Medical University of Ohio becoming one of only seventeen American universities to offer professional and graduate academic programs in medicine, law, pharmacy, nursing, health sciences, engineering, and business.

The Department of Medicinal and Biological Chemistry consists of 17 full-time faculty members with a variety of research interests, ranging from treating autoimmune diseases such as diabetes to the development of novel therapeutics for cancer and AIDS.

Applying

The department's deadline for the fall admission is January 31. Interested students must submit the completed application form, three letters of recommendation, and official transcripts of previous course work at the undergraduate and, if applicable, the graduate levels. In addition, students must submit scores from the GRE. Because both the master's and doctoral programs are selective admission programs, only a limited number of students are accepted each year. Meeting the minimum requirements does not guarantee admission into either program. The Admissions Committee may choose to waive any of the admission requirements when, in its judgment, it is in the best interest of the applicant and the program. Applicants are strongly encouraged to visit the department and meet with a faculty member in their area of interest to discuss applying to the program and their plans for graduate school.

Correspondence and Information

Dr. Katherine Wall
Department of Medicinal and Biological Chemistry
College of Pharmacy Mail Stop 606
University of Toledo
2801 West Bancroft Street
Toledo, Ohio 43606-3390
Phone: 419-530-1943
Fax: 419-530-7946
E-mail: mmciner@utnet.utoledo.edu
Web site: http://www.mbc.pharm.utoledo.edu

The University of Toledo

THE FACULTY AND THEIR RESEARCH

The department has a strong research focus in therapeutic drug design directed to immune and central nervous system targets. Some research efforts are aimed at the development of novel therapeutic agents for Alzheimer's disease. Cancer-related research interests include the synthesis of antimitotic agents, the design of radiosensitizing drugs, and modified monoclonal antibodies as vehicles for the site-directed inactivation of cancer cells. Immunological approaches have focused on selective immunotoxic proteins of venom origin, the role of T cells in the development of diabetes, and peptides as selective immunosuppressive agents, including peptides as selective interleukin receptor antagonists and in the selective blockage of antigen presentation. Molecular biological approaches include the development of techniques for selective blockage of gene expression in eukaryotic cells.

Paul W. Erhardt, Professor and Director, Center for Drug Design and Development; Ph.D., Minnesota, Twin Cities, 1974. Design and synthesis of small-molecule therapeutic agents; medicinal chemistry of anticancer and critical-care drugs.

Max O. Funk, Distinguished University Professor; Ph.D., Duke, 1975. Mechanistic enzymology; design of mechanism-based inhibitors of enzymes; studies on lipoxygenase, its inhibition, and its role in leukotriene biosynthesis; leukotriene and lipoxin antagonism.

Stephen L. Goldman, Professor; Ph.D., Missouri, 1968. *Agrobacterium tumefaciens*–mediated genetic alteration of monocotyledonous plants; productive introduction and expression of genes for therapeutic proteins; mechanisms of gene regulation in development.

Ezdihar A. Hassoun, Professor; Ph.D., Uppsala (Sweden), 1985. Oxidative stress and oxygen free radical biochemistry; in vivo and in vitro studies on the toxicities of environmental pollutants.

Channing L. Hinman, Associate Professor; Ph.D., UCLA, 1977. Site-directed action of immunotoxic proteins of venom origin; destruction of selective lymphocyte subtypes by membrane-lytic toxins; animal model studies of the immunologic course and suppression of autoimmune diseases affecting the nervous system, e.g., experimental autoimmune myasthenia gravis; applications of membrane-active toxin components in cancer treatment.

Richard A. Hudson, Professor; Ph.D., Chicago, 1966. Models in drug design; redox-reactive agents for acetylcholine receptors and other sites of cholinergic action; peptides as immunosuppressive and tolerance-inducing agents; pyrroloquinoline quinone (PQQ) analogues as probes in flavin and PQQ-requiring enzyme systems; novel reverse transcriptase inhibitors affecting retroviral mutation rates.

Jon R. Kirchhoff, Professor; Ph.D., Purdue, 1985. Chemically modified electrodes, microelectrodes, biomimetic electrochemistry, electrochemical, and spectroelectrochemical methods for characterization and analysis; redox properties of metal complexes; photochemistry and photophysics of coordination complexes.

Richard W. Komuniecki, Professor; Ph.D., Massachusetts, 1976. Aerobic/anaerobic transitions in parasitic helminths; pyruvated dehydrogenase complex.

Marcia F. McInerney, Professor; Ph.D., Michigan, 1989. Immunoregulation in and basic studies on the mechanism underlying autoimmune diabetes, selective reduction, and modulation of T cells in autoimmune disease; mechanisms of cell movement into the islet; toll-like receptor responses in diabetes to oral pathogens involved in severe periodontitis; development of peptide delivery systems.

William S. Messer Jr., Professor; Ph.D., Rochester, 1985. Development of new treatments for neurological disorders; synthesis of novel M1 agonists for Alzheimer's disease; synthesis of M4 agonists for schizophrenia; development of M5 antagonists for the treatment of drug abuse.

Peter I. Nagy, Research Professor; Ph.D., Eötvös Loránd (Budapest), 1972. Theoretical studies in drug design based on molecular structure in the gas and solution phases; studies of conformation in hydrogen-bonded systems and structural parameters related to drug design.

Steven M. Peseckis, Associate Professor; Ph.D., MIT, 1983. Design, construction, and deployment of relational databases relevant to drug discovery and design; gene array and informatic approaches to investigating heterotrimeric G-protein–mediated signal transduction pathways; use of small-molecule synthesis and molecular biological methodology to explore functional roles and drug target potential of protein lipophilic modifications, such as myristylation and palmitoylation of heterotrimeric G-protein alpha subunit N-termini.

A. Alan Pinkerton, Professor; Ph.D., Alberta, 1971. Structural characterization of drugs, drug precursors, drug-related metabolites, and drug-receptor complexes, using single-crystal X-ray diffraction and solution and solid-state NMR spectroscopy, as well as combining use of structural information with computer molecular modeling into synthetic strategies for new drugs and toward understanding modes of action of known drugs.

James T. Slama, Professor; Ph.D., Berkeley, 1977. Enzymes as targets in drug design; design, synthesis, and enzymology of mechanism-based enzyme inhibitors; ADP-ribosylation and a novel mechanism of metabolic regulation; ADP-ribosyltransferases as targets for drug design; cytochrome P-450 inhibitors.

L. M. Viranga Tillekeratne, Research Professor; D.Phil., Oxford, 1975. Studies of enzyme inhibitors of natural and synthetic origin for development of potential anti-AIDS therapeutics; isolation and characterization of biologically active natural products.

Hermann von Grafenstein, Associate Professor; M.D., Max Planck Institute for Biochemistry (Munich), 1982; Ph.D., Konstanz (Germany), 1983. Structure of MHC-bound antigenic peptides computational simulation and experimental tests; structure-function relationships of cytokine receptors.

Katherine A. Wall, Professor; Ph.D., Berkeley, 1977. Recognition of antigens by T cells; targeting T cells in drug design; molecular analysis of the autoimmune disease myasthenia gravis using a murine model; design of immunomodulatory drugs.

UNIVERSITY OF UTAH

Program in Biological Chemistry

Program of Study

The Graduate Program in Biological Chemistry is an interdisciplinary program leading to the Ph.D. degree. Participating faculty members are drawn from the Departments of Biology, Biochemistry, Chemistry, Medicinal Chemistry, Oncological Sciences, Pathology, and Pharmaceutics and Pharmaceutical Chemistry. The aim of the program is to train students for independent research at the interface of chemistry and biology. A wide variety of research interests are represented, including biochemistry, biophysics and molecular simulation, chemical biology, medicinal chemistry, and structural biology.

Students admitted to the Program in Biological Chemistry take a common set of first-year courses. Entering students rotate in four laboratories in order to become acquainted with possible research supervisors. A student in good standing at the end of the first academic year chooses a faculty research supervisor for further doctoral study and is automatically admitted to the graduate program of the supervisor's department. Additional requirements for the Ph.D. are set by individual departments, but they are similar throughout the University.

Research Facilities

A series of shared core facilities are available to all participating departments. These include cores for oligonucleotide and peptide synthesis, DNA and protein sequencing, mass spectrometry, NMR, transgenic mice, homologous recombination, and various types of microscopy. Participating departments have up-to-date facilities, including X-ray diffraction instrumentation and biophysical equipment (calorimetry, diverse types of spectrometers, and sedimentation ultracentrifugation). There are two comprehensive science libraries, one on the main campus (near the Departments of Biology and Chemistry) and the other at the Health Sciences Center.

Financial Aid

The Program in Biological Chemistry supports students during the first year without requiring teaching. Support after the first year is from individual departments and research programs through teaching assistantships, research assistantships, and NIH Training Grants. One semester of teaching experience is generally required after the first year. The level of support for 2005–06 was $23,000 plus a $1000 starting allocation for the year. The program also covers the cost of health and dental insurance.

Cost of Study

Tuition and fees are paid by the Program in Biological Chemistry for the first year. After the first year, departments or individual research programs support their students; there is currently no cost to students.

Living and Housing Costs

Housing is available within walking distance of the University in private apartments and in University housing. Rent for a typical one-bedroom unfurnished private apartment is $550 per month both on and off campus. University housing includes all utilities except a telephone.

Student Group

The Program in Biological Chemistry admits 10–15 graduate students per year. The University has nearly 6,100 graduate students, with approximately 300 graduate students in disciplines related to biological chemistry. Students come from all states and many countries.

Location

The University is located in Salt Lake City, 1½ miles east of the city center. The site is at an altitude of 4,500 feet, just at the base of the Wasatch Range, which rises precipitously to the east and south of the University. The main campus, home to the Departments of Biology and Chemistry, is less than 1 mile from the Health Sciences Center, which houses the Departments of Biochemistry, Oncological Sciences, and Human Genetics. The Departments of Medicinal Chemistry, Pharmaceutical Chemistry, and Pharmacology and Toxicology are in adjacent buildings. Travel between the two campuses is a pleasant 15-minute walk, and there are frequent free shuttle buses.

Metropolitan Salt Lake City, with a population of more than 820,000, is the economic and cultural center of the intermountain region of the western United States. The city is well known for classical music (Utah Symphony), dance (Repertory Dance Theater, Ballet West), and sports (Utah Jazz and University teams). Salt Lake City is a popular vacation destination for outdoor enthusiasts because of the nearby mountains, forests, and deserts. Nine national parks are within a day's drive.

The University

Founded in 1850, the University of Utah is the oldest state university west of the Missouri River. The total enrollment is approximately 28,000. There are nearly 1,700 teaching faculty members and 1,800 adjunct, research, and clinical faculty members. The University includes colleges of medicine, law, education, business, mines and mineral industries, science, humanities, social and behavioral sciences, health, fine arts, engineering, nursing, pharmacy, and social work.

Applying

Application materials are available on the program's Web site or by request. The program requires a completed application form, transcripts of grades of all university courses, scores on the General Test portion of the Graduate Record Examinations (verbal, quantitative, and analytical writing), three letters of recommendation solicited by the applicant, and a detailed personal statement of career interest. The University's application fee is paid by the program. Applications are due January 15 for entrance the following August. The Admissions Committee of the program invites promising prospective students for personal interviews; travel expenses for invited students are paid by the program.

Correspondence and Information

Darrell Davis, Director
Interdepartmental Graduate Program in Biological Chemistry
University of Utah
15 North 2030 East
533 EIHG, Room 1400
Salt Lake City, Utah 84112-5330
Phone: 801-581-5207
Fax: 801-585-2465
E-mail: barbara.saffel@genetics.utah.edu
Web site: http://www.bioscience.utah.edu

University of Utah

THE FACULTY AND THEIR RESEARCH

Following the faculty member's name, his or her department is listed in parentheses, followed by the specific research interest.

E. Dale Abel (Biochemistry): cardiac metabolism and gene expression by insulin.
Markus Babst (Biology): protein trafficking.
Kuby Balagurunathan (Medicinal Chemistry): glycobiology, carbohydrate biosynthesis.
Brenda Bass (Biochemistry): dsRNA binding proteins.
Peter Beal (Chemistry): nucleic acid chemical biology.
David Blair (Biology): bacterial motility.
Don Blumenthal (Biochemistry): protein kinases.
Susan Bock (Medicinal Chemistry): protein conformational change and targeting.
Cynthia Burrows (Chemistry): nucleic acid chemistry.
Brad Cairns (Oncological Sciences): chromatin; transcription; genomics.
Dana Carroll (Biochemistry): recombination; gene targeting.
Tom Cheatham (Medicinal Chemistry): biomolecular simulation.
Sheila David (Chemistry): DNA repair biochemistry.
Darrell Davis (Medicinal Chemistry): RNA structure.
Peter Flynn (Chemistry): solution NMR.
Tim Formosa (Biochemistry): DNA replication; chromatin.
David Gard (Biology): microtubules and MAPs.
David Goldenberg (Biology): protein dynamics and function.
Barbara Graves (Oncological Sciences): transcription; protein biochemistry.
Charles Grissom (Chemistry): cancer drug delivery and enzyme mechanisms.
Eric Hegg (Chemistry): metalloenzymes; structure and mechanisms.
James Herron (Pharmaceutics and Pharmaceutical Chemistry): structural immunology; biosensors.
Chris Hill (Biochemistry): protein structure and function.
Martin Horvath (Biology): structural biology; X-ray crystallography.
Wai Mun Huang (Pathology): structure and function of topoisomerases.
Chris Ireland (Medicinal Chemistry): anticancer drug discovery.
David Jones (Oncological Sciences and Medicinal Chemistry): colon cancer development.
Michael Kay (Biochemistry): protein design.
Jindrich Kopecek (Pharmaceutics and Pharmaceutical Chemistry): biorecognition and drug delivery.
Carol Lim (Pharmaceutics and Pharmaceutical Chemistry): controlled protein delivery.
Janet Lindsley (Biochemistry): DNA-based enzymes.
Baldomera Olivera (Biology): molecular neurobiology; conotoxins.
Dale Poulter (Chemistry): mechanistic enzymology.
Steve Prescott (Oncological Sciences): signaling; lipids; inflammation.
Glenn Prestwich (Medicinal Chemistry): ligand signaling and biomaterials.
Jon Rainier (Chemistry): synthetic organic chemistry.
Martin Rechsteiner (Biochemistry): intracellular proteolysis.
Jared Rutter (Biochemistry): metabolic signaling.
Eric Schmidt (Medicinal Chemistry): natural products chemistry/biology.
Scott Summers (Biochemistry): lipid metabolism; diabetes.
Wes Sundquist (Biochemistry): HIV assembly.
Jill Trewhella (Chemistry): biomolecule communication.
Than Truong (Chemistry): computer modeling; enzyme mechanisms.
Greg Voth (Chemistry): theory of biomolecular systems.
Dennis Winge (Biochemistry): metalloproteins; metalloregulation.
Bruce Yu (Pharmaceutics and Pharmaceutical Chemistry): peptide engineering.
Ilya Zharov (Chemistry): organic and biological materials.

UVM

UNIVERSITY OF VERMONT

Department of Biochemistry

Program of Study	The Department of Biochemistry offers a program of focused research training in a variety of current topics in biochemistry and molecular biology. Its sister program in cell biology is listed elsewhere in this volume. Students begin with course work that provides broad exposure to biochemical science, but thesis research may begin as soon as a student enters the program. Training is also provided in the preparation and presentation of NIH-style research grant proposals. Ph.D. and M.S. degrees are awarded.

Advanced topic seminar courses and an excellent weekly seminar series, featuring outstanding visiting scientists, give students the opportunity to acquire a working knowledge of the major topics currently under investigation in biochemical research.

Thesis research is conducted with one of the 17 full-time and adjunct faculty members. Students usually select a faculty adviser during the first year of the program. The Department maintains a high level of extramurally funded research in a variety of areas, including protein biochemistry, enzymology, structural biology, cancer biology, gene expression, protein–nucleic acid interactions, cell-surface events, and intracellular and extracellular signaling.

The program requires a series of three qualifying exams in the first 2½ years, including an oral exam covering general biochemistry, written and oral defense of a proposed thesis project, and preparation and defense of an NIH-style grant proposal. Completion of the Ph.D. program requires four to five years.

Research Facilities

The Department of Biochemistry is located in the Given Medical Research Building. Specialized facilities include a molecular visualization room, stop-flow CD, fluorescence, rapid kinetics, X-ray crystallography center, MALDI-TOF mass spectrometer, chromatography systems, analytical ultracentrifuge, monoclonal antibody production facilities, and a core facility for state-of-the-art synthesis, separation, composition analysis, and molecular modeling of polypeptides. In addition, the Department has access to an atomic force microscope, cryo-electron microscopy (cryo-EM), an FT-IR spectrometer, a cell-sorting facility, and a state-of-the-art imaging facility.

Financial Aid

Financial aid consists of research traineeships, research assistantships, and teaching assistantships, all of which carry a tuition waiver and a living stipend of up to $20,772 for a twelve-month appointment in 2006–07. Financial aid is usually available to all Ph.D. students accepted into the program.

Cost of Study

Tuition is $410 per credit hour for Vermont residents and $1034 per credit hour for nonresidents. Fees range from $158 to $700, depending on credit-hour enrollment.

Living and Housing Costs

The stipends mentioned in the Financial Aid section are sufficient to provide students a modest standard of living. Married student housing is available through the University on a first-come basis; single students generally live off campus, near the University. Students often reduce housing costs by sharing off-campus housing.

Student Group

Current enrollment in the Graduate College at the University of Vermont is more than 1,000. There are 14 Ph.D. students in the Department of Biochemistry.

Student Outcomes

For the past ten years, the Biochemistry Graduate Program has had a completion rate greater than 95 percent. Departmental policy expects Ph.D. degree completion within five years, although in rare instances six years have been required. With few exceptions, all graduates have gone on to academic or industrial research careers subsequent to additional postdoctoral training at leading national and international institutions.

Location

Burlington, situated on the shore of Lake Champlain and nestled between the scenic Green and Adirondack Mountains, is renowned for its full range of outdoor recreational and cultural opportunities. Mozart festivals and the year-round George Bishop Lane Performing Artists Series are among the cultural events sponsored by the University, the city, and neighboring colleges. Nearby small villages provide an opportunity to enjoy traditional Vermont activities such as craft fairs and country auctions.

The University and The Department

The University of Vermont, founded in 1791, has a long tradition of excellence in undergraduate and graduate teaching. Total enrollment is approximately 9,500 students.

The Department of Biochemistry is in the College of Medicine and has 13 full-time faculty members as well as 4 adjunct faculty members whose primary appointments are in other departments. The Department is home to several internationally recognized research programs; the continuing growth of extramural grant support indicates the dynamic nature of its faculty members and students.

Applying

The deadline for receipt of applications is March 1 for financial aid; earlier completion of the application process is recommended. Application materials are available from the Department of Biochemistry and on the Web at http://biochem.uvm.edu/graduate_program.php. Students are encouraged to make preliminary contact with the Department.

Correspondence and Information

Scott W. Morrical, Ph.D., Director of Admissions
Department of Biochemistry
College of Medicine
University of Vermont
89 Beaumont Avenue
Burlington, Vermont 05405-0068

Phone: 802-656-2220
E-mail: info@biochem.uvm.edu
Web site: http://biochem.uvm.edu |

University of Vermont

THE FACULTY AND THEIR RESEARCH

Christopher Berger, Associate Professor of Molecular Physiology/Biophysics and Biochemistry; Ph.D., Minnesota, 1991. Protein structural dynamics during muscle contraction.

Beth Bouchard, Research Assistant Professor of Biochemistry; Ph.D. Vermont, 1996. Regulation of the cellular mechanisms mediating the endocytosis of plasma factor V by megakaryocytes.

Kathleen Brummel-Ziedins, Research Assistant Professor of Biochemistry; Ph.D., Maryland, 1997. Enzymatic events of blood coagulation, leading up to and following clot formation, and how they alter in various disease states.

Saulius Butenas, Research Associate Professor of Biochemistry; Ph.D., Kaunas (Lithuania), 1985. Protein-protein interactions in in vitro models of blood coagulation.

Kenneth R. Cutroneo, Professor of Biochemistry; Ph.D., Rhode Island, 1971. Molecular mechanisms of regulation of procollagen synthesis by glucocorticoids and bleomycin; relation of these effects to glucocorticoid-mediated skin atrophy and bleomycin-induced lung fibrosis.

Stephen J. Everse, Associate Professor of Biochemistry; Ph.D., California, San Diego, 1995. Structure determination of proteins (especially blood coagulation factors) by X-ray crystallographic methods.

Christopher Francklyn, Professor of Biochemistry and Microbiology and Molecular Genetics; Ph.D., California, Santa Barbara, 1988. Protein–nucleic acid recognition; structure and function of RNA-binding proteins, especially aminoacyl tRNA synthetases; RNA structure and function; alternative functions of proteins derived from the translation apparatus.

Robert Hondal, Assistant Professor of Biochemistry; Ph.D., Ohio, 1997. Protein semisynthesis of RNase A and thioredoxin reductase; mechanistic enzymology of thioredoxin reductase and other selenium-containing proteins; synthesis of peptides and peptide analogues to study protein structure-function relationships.

Robert Kelm, Associate Professor of Medicine and Biochemistry; Ph.D., Vermont, 1991. Biochemistry and molecular biology of ssDNA/RNA-binding proteins; Vascular gene regulation.

Kenneth G. Mann, Professor of Biochemistry; Ph.D., Iowa, 1967. Proteins of the blood coagulation process; bone biochemistry.

Anne B. Mason, Research Professor of Biochemistry; Ph.D., Boston University, 1979. Investigation of interaction of the iron-binding protein transferrin with its specific protein receptor: physical/chemical measurements, cellular-binding studies, and use of monoclonal antibodies to transferrin and its receptor.

Scott W. Morrical, Professor of Biochemistry and Molecular Genetics and Director of Admissions; Ph.D., Wisconsin–Madison, 1987. Enzymology of DNA replication, recombination, and repair; physical biochemistry and structural biology of protein-DNA and protein-protein interactions.

Thomas Orfeo, Research Associate Professor of Biochemistry; Ph.D., Vermont, 1987. Blood coagulation studies.

Behnaz Parhami-Seren, Research Associate Professor of Biochemistry; Ph.D., Weizmann Institute, 1984. Production and characterization of monoclonal antibodies to haptens and proteins.

Burton E. Sobel, Amidon Professor of Medicine and Professor of Biochemistry; M.D., Massachusetts, 1962. Modulation of fibrinolytic activity in the blood and on cell surfaces at the levels of transcription, translation, and posttranslational modifications and its relation to the pathophysiology of vascular disease.

Paula B. Tracy, Professor and Interim Chairperson of Biochemistry and Professor of Medicine; Ph.D., Syracuse, 1977. Regulation of the blood coagulation process, mediated through the assembly and function of coagulant enzymatic complexes on the surface of platelets, monocytes, and vascular endothelium and through proteolytic and membrane-associated activities intrinsic to these cell populations.

Russell P. Tracy, Professor of Pathology and Biochemistry; Ph.D., Syracuse, 1978. Development of assays for biomarkers and cellular processes related to cardiovascular diseases and aging; application of these assays in epidemiological studies to understand vascular pathophysiology and develop cardiovascular risk markers.

On the campus of the University of Vermont.

UNIVERSITY OF WISCONSIN–MADISON

Medical School
Department of Biomolecular Chemistry

Program of Study	The Department of Biomolecular Chemistry in the School of Medicine provides a broad graduate program in biological chemistry, molecular biology, and molecular genetics, with an emphasis on the structure and function of proteins and nucleic acids, cellular control mechanisms, and hormone action. The diversity of research fields presents numerous opportunities for graduate study in areas ranging from fundamental biochemical problems to molecular mechanisms of human disease. Flexible, individualized programs of study provide a variety of research and learning experiences for graduate students interested in the biomedical sciences. Students become integrally involved in research activities during their first semester on campus. Before selecting a research adviser, students participate in short-term projects through rotations among different laboratories. After acceptance by an adviser, students participate in the research program of the chosen laboratory and initiate their thesis research studies. Programs of academic course work place a major emphasis on biochemistry but are drawn from the entire university curriculum in order to best suit the needs of the individual student. In addition to lecture and lab courses, the Ph.D. program includes research seminars, tutorials on special topics, and journal clubs. Most students satisfy the course requirements for advancement to Ph.D. candidacy by the end of the second year. The preliminary examination consists of an oral examination based on a research proposal prepared by the student. The final examination consists of an oral defense of the submitted thesis, which represents a finished, original investigation.
Research Facilities	Laboratories are well equipped for research in the fields of biological chemistry and molecular biology. In addition to the Department's own state-of-the-art research facilities, the campus also has several centralized facilities to support biological research (i.e., Biotechnology Center, NMR facility, pilot plant fermentation facilities, Biophysics Instrumentation facility, etc.). Numerous opportunities for interdepartmental study and research are available.
Financial Aid	Funds are available to support students throughout their graduate studies. Types of awards include research assistantships, campus and national fellowships, and traineeships. These awards provide stipends and health insurance to students, enabling them to devote their entire efforts to course work and research.
Cost of Study	All students admitted for graduate study receive complete tuition support in addition to an annual stipend.
Living and Housing Costs	Madison's affordable housing and relatively low cost of living offers an economic incentive not found in most large university settings. According to the Campus Assistance Center, the average city rent per month ranges from approximately $470 for an efficiency to $865 for a two-bedroom apartment. The University also provides graduate student family housing, which is conveniently located at the northwest end of campus.
Student Group	The Department maintains a diverse student population. Research assistantships or fellowships support all of the students currently enrolled in the program. The success of alumni in the last several years is indicative of the quality of the program. Graduates have assumed postdoctoral research positions in major research laboratories and universities.
Location	Madison, with a population around 200,000, stretches between the shores of two large lakes and is bordered by a 1,300-acre arboretum. The rolling hills, numerous lakes, and parks of south-central Wisconsin are readily accessible and provide an ideal environment for outdoor recreation. As the state capital and home of the University, Madison offers a rich cultural life, a well-developed public transportation system, and numerous other civic attractions. Madison ranked number one in *Money* magazine's 1998 "Best Places to Live in America" poll.
The University	Founded in 1849 as a land grant institution, the University of Wisconsin–Madison is one of the top ten research universities in the country, a fact that attracts students from around the world. Currently more than 43,000 students are enrolled, nearly 10,000 of whom are studying for advanced degrees. The Madison campus has 125 departments, which engage more than 2,500 faculty members and offer unusual breadth and expertise in nearly all academic disciplines. The life sciences are particularly well represented, with internationally recognized programs in the biological and medical sciences.
Applying	Application materials are available online at http://www.bmolchem.wisc.edu/apply.html. In order to be considered for a University fellowship and some other types of financial aid, all applications should be completed no later than January 1 for fall admission. In preparation for graduate studies, the Department recommends that all applicants complete course work in biology, general genetics, physics, and organic and physical chemistry. If the applicant is unable to obtain course work in physical chemistry prior to beginning graduate studies, an equivalent course is required as part of the graduate curriculum. Graduate Record Examinations aptitude test scores are required of all applicants and one advanced test score is recommended.
Correspondence and Information	Chair, Graduate Admissions Committee Department of Biomolecular Chemistry 587 Medical Sciences Building University of Wisconsin–Madison 1300 University Avenue Madison, Wisconsin 53706-1532 Phone: 608-261-1492 Fax: 608-262-5253 E-mail: bmolchem@mailplus.wisc.edu Web site: http://www.bmolchem.wisc.edu

University of Wisconsin–Madison

THE FACULTY AND THEIR RESEARCH

Richard A. Anderson, Professor (also Pharmacology); Ph.D., Minnesota, 1982; H. I. Romnes Faculty Fellow, Vilas Associate, Kellett Award. The role of cellular signal transduction in cell migration, assembly of cell-cell contacts, nuclear events, and cancer metastasis.
Type Iγ phosphatidylinositol phosphate kinase targets and regulates focal adhesions. *Nature* 420:89–93, 2002.

James D. Bangs, Professor (also Medical Microbiology and Immunology); Ph.D., Johns Hopkins, 1986. Cell and molecular biology of the parasitic protozoan *Trypanosoma brucei* (African sleeping sickness); protein trafficking and secretion; lysosomal biogenesis; lifecycle differentiation.
GPI-dependent trafficking in bloodstream stage African trypanosomes. *Euk. Cell* 2:76, 2002.

Paul J. Bertics, Professor; Ph.D., Wisconsin–Madison, 1984; Shaw Scholar. Regulation of cellular growth and differentiation; enzymology of protein kinases; biochemical characterization and interaction of growth factor, growth hormone, and purinergic receptors; endotoxin signaling and macrophage activation.
Platelet-derived growth factor–stimulated migration of murine fibroblasts is associated with epidermal growth factor receptor expression and tyrosine phosphorylation. *J. Biol. Chem.* 275:2951, 2000.

Emery H. Bresnick, Professor (also Pharmacology); Ph.D., Michigan, 1989; Leukemia Society of America Scholar; Shaw Scientist. Chromatin structure-function; developmental signaling mechanisms; molecular hematology and vascular biology.
Coregulator-dependent facilitation of chromatin occupancy by GATA-1. *Proc. Natl. Acad. Sci. U.S.A.* 101(4):980–5, 2004. With Pal et al.

David A. Brow, Professor; Ph.D., California, San Diego, 1986; Searle Scholar. Nuclear steps in gene expression; transcription by RNA polymerases II and III; RNA dynamics in pre-mRNA splicing.
Allosteric cascade of spliceosome activation. *Annu. Rev. Genet.* 36:333, 2002.

James E. Dahlberg, Professor; Ph.D., Chicago, 1966; Frederick Sanger Professorship National Academy of Sciences. RNA, processing and transport; nuclear export; microRNA control.
Nuclear export of microRNA precursors. *Science* 303:95–8, 2004.

Richard S. Eisenstein, Associate Professor (also Nutritional Sciences); Ph.D., Wisconsin–Madison, 1984. Metal regulation of gene expression; posttranscriptional control of mammalian iron metabolism; iron sulfur cluster formation and disassembly.
Novel role of phosphorylation in Fe-S cluster stability revealed by phosphomimetic mutations at Ser-138 of iron regulatory protein 1. *Proc. Natl. Acad. Sci. U.S.A.* 95:15235, 1998.

Robert H. Fillingame, Professor; Ph.D., Washington (Seattle), 1973; WARF Mid-Career Faculty Researcher Award. Membrane biochemistry; mechanism of H⁺-transporting, rotary ATP synthase of oxidative phosphorylation.
Structure of Ala24/Asp61 → Asp24/Asn61 substituted subunit *c* of *Escherichia coli* ATP synthase: Implications for the mechanism of proton transport and rotary movement in the F₀ complex. *Biochemistry* 41:5537, 2002.

Katrina T. Forest, Associate Professor (also Bacteriology); Ph.D., Princeton, 1993. The pilus-retraction protein PilT: ultrastructure of the biological assembly.
Functional dissection of a conserved motif within the pilus retraction protein PilT. *J. Bacteriol.* 187(2):611–8, 2005. With Aukema, Kron, and Herdendorf.

Catherine A. Fox, Associate Professor; Ph.D., Wisconsin–Madison, 1992; American Cancer Society Research Scholar, 2002. Replication, structure, and expression of eukaryotic chromosomes; conserved mechanisms linking transcription, cell cycle, and cell differentiation.
The origin recognition complex and Sir4 protein recruit Sir1p to yeast silent chromatin through independent interactions requiring a common Sir1p domain. *Mol. Cell Biol.* 24: 774–86, 2004.

Warren Heideman, Professor (also School of Pharmacy, Department of Pharmacology); Ph.D., Washington (Seattle), 1983. Signal transduction by nutrient molecules; fundamental mechanisms of eukaryotic cell-cycle regulation: developmental biology and toxicology of zebrafish.
ACE2 is required for daughter cell-specific G₁ delay in *Saccharomyces cerevisiae*. *Proc. Natl. Acad. Sci. U.S.A.* 100(18):10275–80, 2003.

Christina M. Hull, Assistant Professor; Ph.D., California, San Francisco, 2000; Burroughs Wellcome Fund Career Award in the Biomolecular Sciences 2003; HHMI/UW Madison Start-Up Award 2004; March of Dimes Basil O'Connor Starter Scholar Award 2005. Molecular biology of the human fungal pathogen *Cryptococcus neoformans*; sexual development; cell identity determination; gene regulation; protein-DNA interactions.
Cell identity and sexual development in *Cryptococcus neoformans* are controlled by the mating type-specific homeodomain protein Sxi1alpha. *Genes Dev.* 16:3046, 2002.

Anna Huttenlocher, Associate Professor (also Pediatrics and Pharmacology); M.D., Harvard, 1988; Shaw Scholar. Adhesion-mediated regulation of cell migration, invasion, and cellular signal transduction.
Calpain regulates neutrophil chemotaxis. *Proc. Natl. Acad. Sci. U.S.A.* 100:4006, 2003.

James L. Keck, Assistant Professor; Ph.D., Berkeley, 1997. Structure-function studies of DNA replication, recombination, and repair proteins.
High-resolution structure of the *E. coli* RecQ helicase catalytic core. *EMBO J.* 22:4910, 2003.

Patricia J. Kiley, Professor; Ph.D., Illinois, 1987; NSF Young Investigator. Transcriptional regulation; signal transduction of oxygen deprivation; Fe-S cluster assembly and regulation.
IscR, an Fe-S cluster containing transcription factor, represses expression of *Escherichia coli* genes encoding Fe-S cluster assembly proteins. *Proc. Natl. Acad. Sci. U.S.A.* 98:14895, 2001.

Robert Landick, Professor (also Bacteriology); Ph.D., Michigan, 1983. RNA polymerase structure-function; mechanism and regulation of RNA chain elongation.
Allosteric control of RNA polymerase by a site that contacts nascent RNA hairpins. *Science* 292:730, 2001.

Deane F. Mosher, Professor (also Medicine and Anatomy); M.D., Harvard, 1968; Robert F. Schilling Professorship. Extracellular matrix; cell adhesion.
Contraction of collagen matrices mediated by α2β1A and αvβ3 integrins. *J. Cell Sci.* 113:2375, 2000.

Arthur S. Polans, Professor (also Ophthalmology and Visual Sciences); Ph.D., Wisconsin–Madison, 1980; Jules and Doris Stein RPB Professor of Ophthalmology. Biochemistry and immunology of ocular tumors and paraneoplastic syndromes; transformation and establishment of the metastatic phenotype; structure and function of calcium-binding proteins; phototransduction.
Expression of the receptor tyrosine kinase Ax1 promotes ocular melanoma cell survival. *Cancer Res.* 64:128–34. 2004.

Michael D. Sheets, Associate Professor; Ph.D., Wisconsin–Madison, 1989; March of Dimes Basil O'Connor Scholar. Regulation of vertebrate development and cell differentiation by molecular signaling pathways.
The FGF pathway is required for the trunk-inducing functions of Spemann's organizer. *Dev. Biol.* 237:295, 2001.

Wilmot B. Valhmu, Assistant Professor (also Orthopedics and Rehabilitation); Ph.D., South Carolina, 1990. Biochemistry, cell biology, and molecular biology of musculoskeletal tissues: regulation of cartilage extracellular matrix gene expression; mechanical signal transduction mechanisms of chondrocytes; molecular etiology of osteoarthritis.
Regulatory activities of the 5' and 3' untranslated regions and promoter of the human aggrecan gene. *J. Biol. Chem.* 273:6196, 1998.

David A. Wassarman, Assistant Professor (also Pharmacology); Ph.D., Yale, 1992. Transcriptional regulation in *Drosophila*.
TAF250 is required for multiple developmental events in *Drosophila*. *Proc. Natl. Acad. Sci. U.S.A.* 97:1154, 2000.

Karen M. Wassarman, Assistant Professor (also Bacteriology); Ph.D., Yale, 1992. Examining the role of small RNAs in bacteria.
6S RNA regulates *E. coli* RNA polymerase activity. *Cell* 101:613–23, 2000.

VIRGINIA COMMONWEALTH UNIVERSITY

School of Medicine
Department of Biochemistry

Programs of Study

The Department of Biochemistry prepares students for successful careers as independent scientists in academia, government, and biotechnology. The Department has research efforts of international stature in several areas, including cellular and molecular signaling, tumor biology, structural biology, eukaryotic molecular biology, lipid and membrane biochemistry, and molecular genetics, using state-of-the-art approaches in enzymology, genomics, proteomics, and lipidomics. The Department's academic Ph.D. and M.D./Ph.D. degree programs are highly flexible and individually tailored to meet the needs of each student. Advanced training with a large and diverse faculty is possible in virtually every aspect of modern biochemistry and cellular and molecular biology. During the first year, students pursue research rotations, take formal course work, and become familiar with current research topics through seminars, discussion groups, and lectures by distinguished scientists. By the end of the first year, students choose a faculty adviser and begin thesis research. Following completion of the research project and defense of the doctoral thesis, graduates are equipped to participate in virtually any area of current biomedical research in the most prestigious laboratories.

The Department also offers an M.S. program with thesis research performed under the guidance of a faculty adviser and a postbaccalaureate certificate program that provides graduate training in biochemistry (two semesters, 27 credits) without the necessity of completing a research project or thesis. Most students complete the certificate program in order to improve their academic qualifications for medical or dental school. However, some students view this program as an introduction to graduate education in biochemistry and a stepping stone to the M.S. or Ph.D. programs.

Research Facilities

The Department of Biochemistry occupies a newly renovated floor of Sanger Hall and has additional facilities in other parts of the Medical Center. The faculty has a high level of state and federal financial support that makes excellent facilities available. Most facets of molecular and cellular biochemical research are provided for within the Department. State-of-the-art instrumentation needed to study virtually any aspect of modern biochemistry and cell and molecular biology is available. The core facilities include synthesis and sequencing of oligonucleotides and peptides; facilities for confocal and electron microscopy; a flow cytometry facility; special services for producing transgenic mice; a computer center; spectrometers for NMR, ESR, and CA; high-resolution mass spectrometers; and X-ray crystallography. Additional information is available at http://www.vcu.edu/biochem/department/core.shtml.

Financial Aid

Students accepted into the Ph.D. program are eligible for fellowships that pay tuition, fees, and a stipend. The minimum stipend for fall 2005 was $21,000. In addition, all students are eligible for support from grants awarded to investigators once a research adviser has been selected. Exceptional students are nominated for special fellowships, which provide stipends of up to $25,000 a year plus tuition and fees. Many University jobs are available for spouses of graduate students, and there are many additional employment opportunities in the area.

Cost of Study

Offers of financial support for Ph.D. students include cost of tuition and fees.

Living and Housing Costs

University-sponsored housing includes dormitories located on campus and a number of apartments. Information is available by contacting the University Housing and Residence Education Office, P.O. Box 842517, Richmond, Virginia 23284-2517; e-mail: rgresham@vcu.edu. Other housing is available in a variety of areas in the city. Information about off-campus housing for married students and others can be found at http://www.students.vcu.edu/commons/services/offcampus/index.html.

Student Group

Virginia Commonwealth University, the largest university in the state, has a total enrollment of about 27,500 students. Approximately 3,500 students receive their health science education and training at the Medical Center. There are about 370 graduate students in the basic health sciences departments. An average of 50 students from all areas of the U.S. and many other countries are enrolled in the Department. A roughly equal number of men and women are advanced-degree candidates. The Department endeavors to maintain a graduate student-faculty ratio of 2:1.

Student Outcomes

Recent graduates have been employed primarily in research and teaching positions at academic, government, or industrial institutions following one or more years of postdoctoral training.

Location

Richmond, the capital of Virginia for more than 200 years and with a historical background that dates to the origins of the U.S., has also become a contemporary urban center. The Medical Center's campus enjoys a location in downtown Richmond adjacent to the state capitol and Shockoe Slip, the original center of commercial Richmond, which now houses a variety of restaurants and shops. Richmond is located along the historic James River and, with a metropolitan population of 600,000, offers a wide variety of cultural opportunities that include museums, art galleries, and concerts. Colonial Williamsburg, Jamestown, and Yorktown are nearby, and Richmond is less than a 2-hour drive from the Chesapeake Bay and Atlantic shore at Virginia Beach, the scenic Blue Ridge Mountains, and the nation's capital, Washington, D.C. The blend of tradition, ongoing development, temperate climate, geographic location, and the affordable cost of living provides an attractive environment in which to live and work.

The University

Founded in 1838, the Medical College of Virginia is a comprehensive academic medical center. In 1968, the Medical College became a component of Virginia Commonwealth University, at present one of the largest institutions of higher education in Virginia. The Department of Biochemistry is a component of the School of Medicine, one of the larger medical schools in the country. The Health Sciences campus also includes the Schools of Dentistry, Pharmacy, Nursing, and Allied Health Professions.

Applying

Applicants must have earned a baccalaureate degree or its equivalent at the time of enrollment, with an undergraduate minimum GPA of 3.0. The General Test of the Graduate Record Examinations (GRE) is required. For combined-degree students, the Medical College Admission Test or Dental Admission Test may be substituted for the GRE. For international applicants, the TOEFL is required. Applicants are judged on the basis of their undergraduate record, GRE scores, and letters of recommendation. Applications may be filed at any time but preferably before January 9 for full consideration for financial support. Students are typically admitted for the fall semester, which begins in the latter part of August. Application forms and information are available at http://www.vcu.edu/graduate/ps/apply_domestic.html. International application forms are available at http://www.vcu.edu/graduate/ps/apply_international.html. Further information about the Department of Biochemistry can be found at http://www.vcu.edu/biochem/.

Correspondence and Information

Graduate School
Moseley House
Virginia Commonwealth University
1001 Grove Avenue
P.O. Box 843051
Richmond, Virginia 23284-3051
E-mail: vcu-grad@vcu.edu
Web site: http://www.vcu.edu/graduate

Virginia Commonwealth University

THE FACULTY AND THEIR RESEARCH

Suzanne E. Barbour, Associate Professor; Ph.D., Johns Hopkins, 1990. Role of PLA2 and lipid mediators in the host response in periodontal disease; regulation of cell phospholipid metabolism by phospholipase A2.

William Barton, Assistant Professor; Ph.D., Cornell, 2001. X-ray crystallography of extracellular receptor-ligand complexes and intracellular signaling pathways; regulation of angiogenesis during wound repair by the angiopoietin-Tie2 receptor-ligand system; regulation of angiogenesis and metastasis via Tie2 activation/deactivation during solid-tumor development; identification of small-molecule inhibitors of the angiopoietin-2–Tie2 receptor-ligand system via computational as well as high-throughput screening methodologies.

Matthew J. Beckman, Assistant Professor; Ph.D., Iowa State, 1993. Diseases of bone; vitamin D and other steroids; peptide calciotropic hormones; cartilage and connective tissue matrix growth factors.

Charles E. Chalfant, Assistant Professor; Ph.D., South Florida, 1997. Ceramide regulates the alternative splicing of Bcl-x; role of ceramide-1-phosphate in prostanoid synthesis.

Jason Chen, Associate Professor; Ph.D., Washington (Seattle), 1995. Biology of vertebrate retinal photoreceptors.

Jan F. Chlebowski, Professor; Ph.D., Case Western Reserve, 1969. Regulation of protein biosynthesis; modification of enzyme structure and function; calorimetric and spectroscopic approaches to macromolecular function.

Sumitra Deb, Professor; Ph.D., Calcutta, 1982. Molecular mechanisms of oncogenesis: role of mutations of tumor suppressor p53; characterization of biological pathways used by p53, mutant p53, and related proteins to regulate cell growth by functional genomics.

Swati Deb, Associate Professor; Ph.D., Calcutta, 1982. Mechanism of cell growth deregulation during oncogenesis; role of MDM2 and other oncogenes in cell growth regulation and deregulation using cell-cycle analysis and proteomics.

Paul Dent, Associate Professor; Ph.D., Dundee (Scotland), 1992. Signaling processes that regulate proliferation of normal liver cells in regenerating liver; mechanisms by which EGFR signaling promotes cell growth after ionizing radiation.

Robert F. Diegelmann, Professor; Ph.D., Georgetown, 1970. Regulation of collagen deposition following tissue injury; interaction of inflammatory cells and components with collagen-producing cells.

Frank Fang, Assistant Professor; Ph.D., Toronto, 1994. Role of lysophospholipids in the pathogenesis of human ovarian, breast, and prostate cancers.

Gregorio Gil, Professor; Ph.D., Barcelona, 1981. Molecular mechanisms involved in cholesterol and bile acid regulation of gene expression.

W. McLean Grogan, Professor; Ph.D., Purdue, 1972. Properties and regulatory mechanisms of cholesterol ester hydrolases; regulation of cholesterol metabolism and homeostasis; membrane structural dynamics; steroid 6b-hydroxylases.

Tomasz Kordula, Associate Professor; Ph.D., Jagiellonian (Poland), 1992. Molecular mechanisms regulating both tissue-specific and cytokine-induced gene transcription.

Sheldon Milstein, Adjunct Professor; Ph.D., USC, 1968. Enzymatic mechanisms in genetic disorders of aromatic amino acid metabolism, including PKU and its variant forms; the sphingolipid metabolites, sphingosine-1-phosphate and ceramide, in nitric oxide production and neuronal development; apoptosis.

Darrell L. Peterson, Professor; Ph.D., Notre Dame, 1970. Structure and function of viral proteins, with an emphasis on the structural proteins and enzymes of hepadnaviruses and hepatitis C virus; development of novel immunogens for use as vaccines against viral, bacterial, or protozoan pathogens.

Paul H. Ratz, Professor; Ph.D., Penn State, 1982. Regulation of contraction of vascular and bladder smooth muscles.

James R. Roesser, Assistant Professor; Ph.D., Virginia, 1986. RNA-protein interactions; alternative RNA splicing in mammalian cells; eukaryotic gene regulation.

Carmen Sato-Bigbee, Associate Professor; Ph.D., Buenos Aires, 1985. Regulation of oligodendrocyte proliferation and differentiation; signal transduction systems; mechanism of gene regulation during brain development and nervous system regeneration.

J. Neel Scarsdale, Assistant Professor; Ph.D., Yale, 1989. Conformational analysis of glycoconjugates by nuclear magnetic resonance methods and molecular mechanics.

Sarah Spiegel, Professor and Chair; Ph.D., Weizman (Israel), 1983. Molecular and cellular mechanisms of the novel lipid mediator, sphingosine-1-phosphate, in regulation of cell growth, motility, prevention of apoptosis, angiogenesis, and neuronal development.

Rik Van Antwerpen, Assistant Professor; Ph.D., Utrecht (Netherlands), 1989. Structure and function of plasma lipoproteins; cholesterol transport by LDL and HDL; roles of lipoproteins in heart disease and stroke; lipoprotein-mediated lipid signaling.

H. Tonie Wright, Professor; Ph.D., California, San Diego, 1968. X-ray crystallography, protein engineering, protein folding; structures of serpins, serine hydroxymethyltransferase; protein deamidation.

Zendra E. Zehner, Professor; Ph.D., Baylor College of Medicine, 1979. Regulation of gene expression during growth, cell cycle, and development; aberrant gene expression in cancer; DNA-protein interactions; mechanisms of transcriptional repression; IFP gene expression in breast cancer; myogenesis; nuclear export and localization of cytoskeletal mRNAs; translational control.

Affiliate Faculty

Donald J. Abraham, Professor and Chair, Department of Medicinal Chemistry; Ph.D., Purdue, 1963. Molecular nature of small-molecule interactions with hemoglobin; computational and molecular modeling; X-ray crystallographic analysis.

John N. Clore, Professor, Department of Internal Medicine/Endocrinology; M.D., Virginia Commonwealth, 1982. Regulation of hepatic glucose production, with particular emphasis on diabetes mellitus; regulation of glucose-6-phosphatase in human liver.

Babette Fuss, Associate Professor, Department of Anatomy and Neurobiology; Ph.D., Swiss Federal Institute of Technology (Zurich), 1992. Autotaxin, lysolipids, and oligodendrocytes.

Martin Graham, Professor, Department of Pediatrics; M.D./Ch.B., Cape Town (South Africa), 1973. Intestinal smooth-muscle biology; collagen and collagenase gene expression; inflammatory cytokines; mast cell proteases; lattice contraction; glucocorticoid receptors; glucocorticoid and c-jun interactions; ascorbic acid.

Steven Grant, Professor, Department of Internal Medicine/Hematology-Oncology; M.D., CUNY, Mount Sinai, 1973. Signal transduction and cell-cycle events involved in the regulation of programmed cell death (apoptosis) in human leukemia cells following exposure to nucleoside analogs and other antineoplastic agents.

Hisashi Harada, Assistant Professor, Department of Internal Medicine/Hematology-Oncology; Ph.D., Osaka (Japan), 1991. Intracellular signaling pathways in cancerous cells.

Fred M. Hawkridge, Professor, Department of Chemistry; Ph.D., Kentucky, 1971. Electron transfer kinetics and mechanisms in biological molecules; oxidase enzymes in electrode-supported bilayers.

W. Michael Holmes, Professor, Department of Microbiology and Immunology; Ph.D., Tennessee, 1974. RNA structure and function; regulation of tRNA gene expression; RNA modification; RNA localization.

Phillip B. Hylemon, Professor, Department of Microbiology and Immunology; Ph.D., Virginia Tech, 1972. Regulation of hepatic cholesterol and bile acid synthesis; genetics and enzymology of intestinal bile acid metabolism.

Donald F. Kirby, Professor, Department of Internal Medicine/Gastroenterology; M.D., George Washington, 1979. Early enteral feedings; enteral access techniques; short bowel syndrome; parenteral nutrition; refeeding syndrome; obesity.

Rakesh Kukreja, Professor, Department of Internal Medicine/Cardiology; Ph.D., Kurukshetra (India), 1982. Myocardial ischemia/reperfusion; preconditioning; gene expression of heat stress proteins; free radicals in myocardial injury; signaling mechanisms in myocardial protection.

David Lanning, Assistant Professor, Department of Surgery; M.D., 1995, Ph.D., 2000, Virginia Commonwealth. Characterization of in vitro and in vivo fetal wound healing models (primarily mouse and rabbit) to elucidate the molecular biology and mechanisms of wound contraction and scar formation.

Jason Rife, Assistant Professor, Department of Medicinal Chemistry; Ph.D., South Florida, 1994. Molecular modeling of RNA-drug interactions.

John J. Ryan, Associate Professor, Department of Biology; Ph.D., Virginia Commonwealth, 1993. Signal transduction by the IL-4 and c-Kit receptors as it relates to proliferation and gene expression.

Robert M. Tombes, Associate Professor, Department of Biology; Ph.D., Washington (Seattle), 1986. Regulation of cell growth and differentiation by calcium, calmodulin (CaM), and CaM-dependent protein kinases; mechanisms of alternative mRNA splicing; protein structure-function relationships; protein targeting.

Kristoffer Valerie, Professor, Department of Radiation Oncology; Ph.D., Royal Institute of Technology (Stockholm), 1986. Molecular biology of DNA repair and HIV-1 gene expression, radiation-induced gene expression, and signal transduction mechanisms; genetic approaches to modulate cellular radioresistance.

Dorne Yager, Associate Professor, Department of Surgery; Ph.D., North Carolina at Chapel Hill, 1985. Regulation of hyaluronan synthase gene expression; regulation of gene expression by the myeloperoxidase/halide system during inflammation.

W. Andrew Yeudall, Associate Professor, Philips Institute of Oral and Craniofacial Molecular Biology; Ph.D., Glasgow (Scotland), 1990. Molecular and cellular biology of head and neck squamous cell carcinoma.

VIRGINIA POLYTECHNIC INSTITUTE AND STATE UNIVERSITY
Department of Biochemistry

Programs of Study

The graduate program in biochemistry at Virginia Tech leads to the degree of Doctor of Philosophy (Ph.D.) and aims to provide students with the necessary technical and critical-thinking skills for a career in the biochemical sciences. The diversity of research programs in the department promotes exposure to a variety of important themes in modern biochemistry. This research aims to further the understanding of many significant biological processes, including the molecular genetics and enzymology of nitrogen fixation; enzyme function and mechanism in archaea, cyanobacteria, and solvent-producing fermentative bacteria; signal transduction in plants and bacteria; regulation of prokaryotic gene expression; RNA methylation; cellular quiescence and resistance to water stress; resistance to oxidative stress; transposition in insects; chorion hardening in the mosquito egg; and hemoglobin catabolism by the malaria parasite. In addition, one member of the department faculty conducts research on the teaching and learning of genetics, genomics, and biotechnology as well as on partnerships among K–12 students, their teachers, and research scientists.

In their first year, graduate students become acquainted with advanced principles of biochemistry through course work. In addition, students perform two laboratory rotations to gain familiarity with departmental research. At the end of the first year, students begin their dissertation research with a faculty adviser. An advisory committee is selected in the second year and meets at least yearly to monitor the student's progress. Candidates for the Ph.D. program are required to take a preliminary examination and to prepare and defend a research proposal. The course of study culminates in a public oral defense of the doctoral thesis project.

Research Facilities

The individual laboratories within the department are well equipped for modern biochemical, molecular biological, and microbiological research. A sampling of available equipment includes a liquid chromatography/tandem mass spectrometry system, an electron paramagnetic resonance spectrometer, a fluorometer, stopped-flow equipment for rapid kinetic analysis, a robotic workstation for high-throughput screening, temperature cyclers for real-time PCR, a fluorescence microscope with digital imaging capabilities, liquid chromatography systems, systems for electronic autoradiography and fluorescent/chemiluminescent imaging, and tissue culture and insect rearing facilities.

Numerous shared on-campus resources enhance the research opportunities available to students. These include the Keck Confocal Microscope Facility, the Keck Transgenic Plant Greenhouse, the Fralin Fermentation and Protein Purification Facility, and the Flow Cytometry Core Facility. Access to NMR spectrometers and X-ray diffractometers is available through the Department of Chemistry. The Core Computation and Laboratory Facilities at the Virginia Bioinformatics Institute provide a range of bioinformatic, DNA-sequencing, and genomic/proteomic services. The VT CAVE is a multiperson, room-sized, high-resolution, 3-D video and audio environment that can be used to visualize a wide variety of biological molecules in 3-D space.

Financial Aid

Assistantships are available to qualified candidates. In 2005–06, assistantship stipends were $19,320 for the calendar year. In addition, tuition waivers are granted to all students receiving assistantships.

Cost of Study

Tuition for the 2005–06 academic year was $6558; however, tuition waivers are granted to all students receiving assistantships. Fees for the 2005–06 academic year were $1539.

Living and Housing Costs

Housing is available in dormitories on campus; more information is available through the Graduate Life Center. Apartments, usually unfurnished, are available off campus and start at about $300 for an unfurnished studio. Assistance can be obtained from the Off-Campus Housing Office. Generally, living costs in the area are somewhat below the national average.

Student Group

The graduate program of the Department of Biochemistry consists of approximately 30 graduate students, nearly all of whom are in the Ph.D. program. More than 300 undergraduates are currently enrolled as biochemistry majors, and many participate in departmental research programs.

Location

Blacksburg is located in scenic southwestern Virginia. With a population of 40,000, Blacksburg is a rural college town that offers the convenience of small-town life along with a range of cultural, artistic, and intellectual opportunities. The campus lies 2,200 feet above sea level on a plain between the Blue Ridge and Allegheny mountains. This historic area is noted for its natural beauty and opportunities for outdoor recreation. The Roanoke-Salem metropolitan area lies 40 miles to the east.

The University

Virginia Tech was founded in 1872 as a land-grant university. Today, with some 27,000 students—including 6,000 graduate students—and 2,600 faculty members and researchers, it is Virginia's largest university and one of the leading research institutions in the nation. The spacious campus has more than 100 buildings spread over 2,600 acres. The cultural life of the community is enhanced by several excellent lecture, concert, and theater series and sports events. Excellent recreational facilities are available to students.

Applying

A background in biology, biochemistry, and chemistry is considered the minimum prerequisite for admission, but students with majors in other fields are also accepted. All application materials for admission and financial assistance should be received by February 1 for the following academic year. The General Test of the Graduate Record Examinations is required, and international students must also submit TOEFL scores. Application to the program is made online at the Graduate School's Web site (http://www.grads.vt.edu).

Correspondence and Information

For departmental information:

Dr. Peter J. Kennelly, Head
Department of Biochemistry (0308)
Virginia Polytechnic Institute and State University
Blacksburg, Virginia 24061
Phone: 540-231-6315
Web site: http://www.biochem.vt.edu

Virginia Polytechnic Institute and State University

THE FACULTY

The members of the faculty who are involved in the development of graduate programs are listed below.

David R. Bevan, Associate Professor; Ph.D., Northwestern. Structure and dynamics of proteins and DNA; molecular modeling; computational biology.

Jiann-Shin Chen, Professor; Ph.D., Purdue. Enzymology of acetone-butanol-isopropanol (solvent) production by clostridia; regulation of solvent-production and nitrogen-fixation genes in clostridia.

Dennis R. Dean, Professor and Director, Fralin Biotechnology Center; Ph.D., Purdue. Enzymology and molecular genetics of metallocluster biosynthesis; biological nitrogen fixation.

Erin Dolan, Assistant Professor; Ph.D., California, San Francisco. Genetic, genomics, and biotechnology education; student-teacher-scientist partnerships.

Glenda Gillaspy, Assistant Professor; Ph.D., Case Western Reserve. Growth control and signal transduction in plants.

Eugene M. Gregory, Associate Professor; Ph.D., North Carolina. Occurrence, metabolic activity, and mechanism of action of the detoxifying enzyme superoxide dismutase in various aerobes and anaerobes and in model systems; structure-function relations in enzymes.

Richard Helm, Associate Professor; Ph.D., Wisconsin–Madison. Systems biology of suspended animation; mechanisms of longevity and aging; analytical biochemistry (metabolites and proteins).

Peter J. Kennelly, Professor and Department Head; Ph.D., Purdue. Control of cellular processes by protein phosphorylation-dephosphorylation; identification of new protein kinases and phosphatases; evolution of cellular control mechanisms.

Michael Klemba, Assistant Professor; Ph.D., Yale. Hemoglobin catabolism by the human malaria parasite *Plasmodium falciparum;* protein trafficking to the *P. falciparum* food vacuole.

Timothy J. Larson, Professor; Ph.D., Texas Health Science Center at Houston. Roles of sulfurtransferases in bacterial sulfur metabolism; genetic regulation of glycerol-3-phosphate utilization and incorporation into membrane phospholipids in *Escherichia coli.*

Jianyong Li, Associate Professor; Ph.D., Wisconsin–Madison. Structure/function relationships of proteins involved in tryptophan oxidation pathway in mosquitoes; biochemical mechanism of mosquito chorion formation and hardening.

William E. Newton, Professor; Ph.D., London (England). Structure-function relationships in native and altered nitrogenases and mechanisms of biological nitrogen fixation.

Malcolm Potts, Professor; Ph.D., Durham (England). Desiccation tolerance, quiescence, aging, and longevity of model systems, including bacteria, yeast, nematode, and mammalian cells (including human); molecular ecology of cyanobacteria.

Thomas O. Sitz, Associate Professor; Ph.D., Virginia Tech. Nucleic acid biochemistry; RNA-RNA interactions; mRNA methylation; methyltransferases.

Zhijian Tu, Assistant Professor; Ph.D., Arizona. Insect molecular biology; genomics, evolution, and applications of transposable elements.

Robert H. White, Associate Professor; Ph.D., Illinois at Urbana-Champaign. Application of stable isotopes and mass spectrometry to biochemistry; biosynthesis and function of coenzymes; archaebacterial biochemistry; prebiotic chemistry.

WAKE FOREST UNIVERSITY

School of Medicine
Department of Biochemistry
Graduate Program of Biochemistry and Molecular Biology

Program of Study

The Department offers the Ph.D. degree in biochemistry and molecular biology through the Graduate School of Wake Forest University. There are 23 full-time faculty members, and the Department is well supported by extramural research funding. Faculty members participate in several interdisciplinary programs, including the Structural and Computational Biophysics Program (http://scbtrack.wfu.edu), the Molecular Genetics Program (http://www.wfubmc.edu/molecular_genetics), the basic science division of the Comprehensive Cancer Center (http://www.wfubmc.edu/cancer), the Molecular Medicine Program (http://www.wfubmc.edu/pcr/study-phdmm.html), and the Neuroscience Program (http://www.wfubmc.edu/neuroscience).

Students in the Ph.D. program take approximately 19 hours of formal course work, including biochemistry I and II, molecular biology, Introduction to Biochemical Research, and Special Topics in Biochemical Literature. Students also participate in at least two individual laboratory research rotations in their first year in order to choose a faculty research adviser. At the end of the second year, after the student has passed the required biochemistry courses, the Ph.D. preliminary examination is completed. In subsequent years, students primarily continue with laboratory research under the direction of their research adviser. Completion of the Ph.D. degree requires the student to generate a body of original research and an oral defense of a written research dissertation. Students are given the opportunity to present their research in weekly departmental seminars as well as at various national and international scientific meetings.

Research Facilities

The Department occupies approximately 35,000 square feet of research space in the Hanes Building and Nutrition Research Center on the Bowman Gray Campus and at the Bowman Gray Technical Center, which is located near the Reynolda Campus. Students and faculty members have access to the combined facilities of the Department, the Molecular Genetics Program, the Genomics Center, and the Comprehensive Cancer Center. The Center for Structural Biology (http://csb.wfu.edu) provides renovated lab, office, and conference space at the Bowman Gray Technical Center and includes state-of-the-art facilities for protein crystallography and mass spectrometry as well as shared facilities for biomolecular resources, analytical ultracentrifugation, biomolecular computing and graphics, and rapid reaction kinetics and spectroscopy. A laboratory automation and single-crystal microspectroscopy facility has recently been established. Facilities on the Bowman Gray Campus provide for NMR (600-MHz), circular dichroism and time-resolved fluorescence, analytical imaging and dynamic light scattering as well as an array of mass spectrometers, including a Q-TOF tandem mass spectrometer. They also feature state-of-the-art facilities for genomics and cDNA microarray analysis.

Financial Aid

All first-year students receive stipend support in the form of Dean's fellowships or graduate fellowships ($20,772 each for 2006–07). Fellowships are granted initially for one full calendar year. In second and subsequent years, students are supported by research and training grants and in exceptional cases by the Camillo Artom Scholarship or by the Sandy Lee Cowgill Memorial Scholarship Fund. Students are encouraged to apply for fellowships awarded nationally on a competitive basis. All entering students receive an IBM ThinkPad computer.

Cost of Study

Tuition costs are met in full by the medical school for students holding institutional fellowships.

Living and Housing Costs

Most students live off campus in houses and apartments, many within walking distance of the Medical Center. Rents range from $350 to $600 per month. Single and married graduate students are eligible for inexpensive University-owned apartments on a space-available basis; detailed information on these units is available from the Wake Forest University Graduate School of Arts and Sciences, Biomedical Sciences, Bowman Gray Campus (http://www.wfubmc.edu/graduate).

Student Group

The University has an enrollment of 6,716 students. In 2005–06, there were 709 full-time students enrolled in the Graduate School, and 30 were full-time students in biochemistry.

Location

Winston-Salem (http://www.ideallianceinc.org/winstonsalem.htm), with a population of 194,000, has an active cultural life; the North Carolina School of the Arts, the SciWorks nature center, and the programs of Wake Forest University are among Winston-Salem's features. The presence of Old Salem, a restored Moravian Congregation village, and sophisticated industrial complexes reflect the rich history of the city. Residents enjoy the abundant recreational opportunities of the Piedmont region.

The University

Wake Forest University (http://www.wfu.edu), founded in 1834, is one of the oldest institutions of higher education in North Carolina. Since 1977, the University has graduated 10 Rhodes Scholars, 2 Mellon Fellows, 6 Luce Scholars, 13 Truman Scholars, and several National Science Foundation fellows. The University is a private, independent, fully coeducational school comprised of two campuses. The Reynolda Campus includes the Wake Forest College; the School of Law; and Babcock Graduate School of Management. The Hawthorne Campus, 3 miles away, includes the Wake Forest University School of Medicine as part of the Medical Center complex. Founded in 1902, the medical school has major clinical research programs in cancer, neurological disorders, and aging.

Applying

The Department invites applications from students with strong undergraduate majors in chemistry, biochemistry, biology, or molecular biology. Admission normally requires 20–25 hours of university-level chemistry; at least one semester of calculus-based physical chemistry is recommended. The GRE General Test, required of all applicants, should be taken no later than December, so that results are available by February. Applicants are encouraged, but not required, to take the GRE Subject Test in chemistry; biochemistry, cell and molecular biology; or biology. Complete application materials, including college transcripts and three letters of recommendation, should be submitted by January 15, 2007. Early application is encouraged. Detailed information on the program in biochemistry is available on request. Wake Forest University is committed to administering all its educational and employment opportunities regardless of race, color, sex, national origin, handicap or disability, or status as a Vietnam era or special disabled veteran.

Correspondence and Information

Graduate Recruiter
Department of Biochemistry
Wake Forest University School of Medicine
Medical Center Boulevard
Winston-Salem, North Carolina 27157-1016

Phone: 336-716-0768
Fax: 336-716-7671
E-mail: biochemrecruit@wfubmc.edu
Web site: http://www.wfubmc.edu/biochem

Wake Forest University

THE FACULTY AND THEIR RESEARCH

Additional information, including lists of current publications, can be found in the *Directory of Graduate Research*, published by the American Chemical Society.

Maryam Ahmed, Instructor; Ph.D., Wake Forest, 1997. Mechanisms of virus suppression of host responses; virus-induced killing of cancer cells.

Donald W. Bowden, Professor; Ph.D., Berkeley, 1978. Genetic analysis of human diseases, using DNA polymorphisms; emphasis on diabetes and complications of diabetes.

Al Claiborne, Professor; Ph.D., Duke, 1979. Structural biology, redox thiology; structures and mechanisms of novel flavoprotein peroxide and disulfide reductases that utilize a single redox-active half-cysteine, such as Cys-SOH or Cys-SSCoA.

Larry W. Daniel, Professor; Ph.D., Tennessee Health Sciences, 1978. Arachidonic acid metabolism; mechanism of action of phorbol diester tumor promoters; regulation of protein kinase C; phospholipase D signaling; redox control of intracellular signaling.

Roy R. Hantgan, Associate Professor; Ph.D., Cornell, 1974. Integrin structure and function; molecular and cellular mechanisms of blood coagulation and fibrinolysis.

Thomas Hollis, Assistant Professor; Ph.D., Texas at Austin, 1997. Structure and function of DNA repair proteins and Fanconi anemia–associated proteins; protein crystallography.

David A. Horita, Assistant Professor; Ph.D., Wisconsin–Madison, 1994. Solution-state NMR studies of proteins and nucleic acids; structural biology of protein-lipid interactions; structural biology of DNA repair enzymes.

Susan M. Hutson, Professor; Ph.D., Wisconsin–Madison, 1976. Structure and function of vitamin B6–dependent enzymes; protein-protein interactions; brain neurotransmitter metabolism; amino acid signaling; diet-gene interactions in animal models.

Mark O. Lively, Professor and Director, Biomolecular Resource Laboratories; Ph.D., Georgia Tech, 1978. Analysis of protein structure and function using mass spectrometry, proteomics, and bioinformatics; proteolytic enzymes; cell biology and biochemistry of laryngopharyngeal reflux and gastroesophageal reflux disease.

W. Todd Lowther, Assistant Professor; Ph.D., Florida, 1994. X-ray crystallographic and biochemical analyses of enzymes involved in methionine sulfoxide and cysteine sulfinic acid reduction and glyoxylate and lipid metabolism.

Douglas S. Lyles, Professor and Chair; Ph.D., Mississippi Medical Center, 1975. Virus structure and assembly at cell membranes; viral suppression of host transcription and translation; virus-induced apoptosis.

Linda C. McPhail, Professor; Ph.D., Wake Forest, 1976. Structure, regulation, and function of mammalian NADPH oxidases; intracellular signaling pathways involving phospholipase D and protein and lipid kinases; lipid-protein and protein-protein interactions.

Charles S. Morrow, Associate Professor; Ph.D., Saint Louis, 1980; M.D., Missouri–Columbia, 1983. Mechanisms of cancer drug resistance and carcinogen detoxification and cellular efflux; regulation of electrophilic lipid signaling.

Derek Parsonage, Assistant Professor; Ph.D., Birmingham (England), 1984. Structure-function studies of FAD-containing enzymes and thiol-dependent antioxidant systems; expression of recombinant proteins.

Fred W. Perrino, Associate Professor; Ph.D., Cincinnati, 1986. Structure and mechanism of enzymes in DNA replication and repair; molecular mechanisms of mutagenesis during DNA replication in human cells.

Leslie B. Poole, Associate Professor; Ph.D., Wake Forest, 1988. Antioxidant enzyme systems; cysteine-dependent peroxidases; flavoprotein structure and function; redox-active protein disulfide and sulfenic acid centers; regulation of redox signaling networks.

Lawrence L. Rudel, Professor in Biochemistry and Pathology; Ph.D., Arkansas Medical Sciences, 1969. Plasma lipoprotein metabolism and regulation of hepatic cholesterol transport and metabolism.

Susan Sergeant, Instructor; Ph.D., Missouri–Columbia, 1987. Regulation of intracellular signal transduction.

Peter B. Smith, Professor; Ph.D., Tennessee, 1973. Expression of monooxygenase enzymes for in vitro toxicology.

Andrew J. Sweatt, Research Assistant Professor; Ph.D., Duke, 1983. Metabolism of amino acids in brain and retina; neuron-glia interactions; immunolocalization, transgenic models.

Michael J. Thomas, Professor; Ph.D., UCLA, 1973. Biological oxidations involving oxy-radicals; biological applications of mass spectroscopy; application of mass spectroscopy to the analysis of complex protein structures.

Suzy V. Torti, Associate Professor; Ph.D., Tufts, 1977. Regulatory mechanisms in cellular iron metabolism; relationships between proteins of iron metabolism and cancer; development of novel antineoplastic therapeutics targeting iron metabolism; chemoprevention.

Alan J. Townsend, Professor; Ph.D., North Carolina at Chapel Hill, 1986. Biochemical mechanisms of resistance to cytotoxic and mutagenic agents; enzymatic defenses against oxidative damage; chemoprevention of cancer by dietary or pharmacologic agents.

WEST VIRGINIA UNIVERSITY

Graduate Program in Biochemistry and Molecular Biology

Programs of Study

The Graduate Program in Biochemistry and Molecular Biology is one of seven graduate programs in West Virginia University (WVU) Schools of Medicine and Pharmacy offering interdisciplinary biomedical research training leading to the Ph.D. or M.D./Ph.D. degree. Researchers in the program utilize the analytical tools of biochemistry, molecular biology, and protein chemistry. Research areas include regulation of gene expression, RNA processing, signal transduction, programmed cell death, carcinogenesis, oxidant-induced cellular stress, ion channel pharmacology, structure-function relationships of proteins, development of the auditory and visual systems, mechanotransduction in the auditory system, and the genetic basis for age-related macular degeneration. Underlying diseases studied include atherosclerosis, blindness, cancer, deafness, and diabetes. The goal of the Graduate Program in Biochemistry and Molecular Biology is to teach critical research skills to study the molecular basis of cellular function. This is accomplished by an emphasis on time spent in the laboratory and an educational approach based on developing critical-thinking skills.

Students benefit from individual attention by faculty members within a research environment that is dynamic, collaborative, and interdisciplinary. In addition to course work and laboratory research, students participate in seminars, journal clubs, and research conferences. They also attend national scientific meetings and obtain valuable speaking and teaching experience. The Ph.D. typically takes five years to complete. During Year 1, students matriculate into a common integrated core curriculum. This integrated first year allows students to build competence in key areas of contemporary science, gain exposure to the various training program options, meet potential dissertation advisers, and network scientifically and socially. In the second semester, students customize their course work by selecting from an array of program-specific electives. At the end of Year 1, students select a research adviser and can select Biochemistry and Molecular Biology as their training program. Year 2 consists of advanced course work, research, teaching, and the candidacy examination. Years 3 to 5 are devoted to dissertation research. The Graduate Program in Biochemistry and Molecular Biology also participates in the combined M.D./Ph.D. Scholars Program. As M.D./Ph.D. Scholars, students take the first two years of the medical curriculum, followed typically by three years of research as required for the Ph.D. degree before returning to the M.D. program.

Research Facilities

Institutional facilities include a computer-based learning center, a centralized animal facility with a transgenic barrier, and a library housing more than 205,000 volumes and 2,400 journals. Core facilities are available for examining gene expression or genetic variation (Affymetrix platform), image analysis, confocal and electron microscopy and laser capture dissection, live-cell imaging, mass spectrometry, flow cytometry and high-speed cell sorting, proteomics, recombinant DNA technology, transgenic rodent biology, and functional neuroimaging (fMRI, PET/CT). Affiliated research centers include the National Institute for Occupational Safety and Health (NIOSH), the Center for Advanced Imaging, Blanchette Rockefeller Neurosciences Institute, Sensory Neuroscience Research Center, and Mary Babb Randolph Cancer Center.

Financial Aid

Ph.D. and M.D./Ph.D. students in the biomedical sciences receive financial support during their training, provided they remain in good academic standing and excel in research. Such support includes full tuition, health insurance, and an annual stipend of $22,000. Combined M.D./Ph.D. students also receive medical tuition waivers.

Cost of Study

Students' tuition costs are covered.

Living and Housing Costs

The cost of an efficiency apartment in University-owned housing is approximately $400 per month. A limited number of University apartments are available for married students. Privately owned apartments in Morgantown cost $400 to $600 per month. In general, the cost of living is lower compared to larger cities.

Student Group

Total University enrollment is approximately 26,000 students, which includes 6,500 graduate and professional students. Graduate students come from all parts of the United States and many other countries.

Location

With an appealing balance to life, Morgantown is a vibrant university community of 80,000 residents in northern West Virginia. Located near the Pennsylvania border at the western edge of the Appalachian Mountains, abundant opportunities exist for activities such as world-class white-water rafting and kayaking, hiking and camping, mountain biking, fishing, and skiing. Morgantown has a cosmopolitan atmosphere with a range of activities usually found in much larger cities. It also enjoys proximity to major metropolitan centers: Pittsburgh is a 90-minute drive north and Washington, D.C., is a 3-hour drive east.

The University

West Virginia University is a comprehensive, land-grant, Carnegie-designated Doctoral/Research University–Extensive public institution. The University's academic Health Sciences Center includes the Schools of Medicine, Dentistry, Nursing, and Pharmacy, all of which offer graduate degree programs. There are seven Ph.D. biomedical research training programs in the Schools of Medicine and Pharmacy: Biochemistry and Molecular Biology, Cancer Cell Biology, Cellular and Integrative Physiology, Exercise Physiology, Immunology and Microbial Pathogenesis, Neuroscience, and Pharmaceutical and Pharmacological Sciences. Graduate faculty members in these programs are from various basic science and clinical departments throughout WVU and are members of interdisciplinary research centers in five health-related areas: cancer cell biology, cardiovascular sciences, immunopathology and microbial pathogenesis, neuroscience, and respiratory biology and lung diseases.

As a member of the Big East Conference, WVU participates in NCAA Division I sports and also offers a wide variety of creative arts, theater, and entertainment opportunities.

Applying

Applicants must have a bachelor's degree and excellent GPA and GRE scores. Three letters of recommendation and a personal statement are required. Students are invited in groups of 10 for a paid, two-day visit/interview from January through March. Prospective students should visit the Web site at http://www.hsc.wvu.edu/som/resoff/gradprograms/PhD.asp for more information and an online application.

Correspondence and Information

Lisa Salati, Ph.D., Graduate Director
Graduate Program in Biochemistry and Molecular Biology
West Virginia University
P.O. Box 9142
Morgantown, West Virginia 26506
Phone: 304-293-2494
E-mail: lsalati@hsc.wvu.edu
Web site: http://www.hsc.wvu.edu/som/resoff/gradprograms/
    PhD.asp

Office of Research and Graduate Education
Health Sciences Center
West Virginia University
P.O. Box 9104
Morgantown, West Virginia 26506
Phone: 304-293-7116
E-mail: cnoel@hsc.wvu.edu

West Virginia University

THE FACULTY AND THEIR RESEARCH

Yehenew Agazie, Assistant Professor; Ph.D., Saskatchewan. Signal transduction by tyrosine kinases and phosphatases.

Janet Cyr, Associate Professor; Ph.D., Texas Southwestern Medical Center at Dallas. Molecular basis of mechanotransduction in hair cells.

Jeff Fedan, Adjunct Professor (NIOSH); Ph.D., Alabama at Birmingham. Mechanisms of airway disorders and occupational asthma, particularly the mechanisms in airway epithelium responsible for hyperreactivity.

Steven Frisch, Professor; Ph.D., Berkeley. Anoikis and its regulation; control of cell adhesion.

Stephen Graber, Associate Professor; Ph.D., Vermont. Molecular mechanisms underlying G-protein–mediated signal transduction.

Michael Gunther, Associate Professor; Ph.D., Colorado State. Free radicals and disease.

Brad Hillgartner, Professor; Ph.D., Michigan State. Regulation of genes involved in fatty acid synthesis; thyroid hormone action.

Qiang Ma, Adjunct Professor; Ph.D., Rutgers. Regulation and biology of the aryl hydrocarbon receptor; mechanism of gene regulation by oxidative chemicals.

Peter Mathers, Associate Professor; Ph.D., Caltech. Molecular control of visual and auditory development.

Michael Miller, Professor; Ph.D., Penn State. Mechanisms of chemotaxis/motility in spirochete bacteria and influence of the endocrine system on the immune system.

Visvanathan Ramamurthy, Assistant Professor; Ph.D., Wesleyan. Biochemical mechanisms behind gene mutations that result in photoreceptor cell death; lipid modifications of proteins in photoreceptors; novel treatment for blinding diseases.

Heimo Riedel, Professor; Ph.D., European Molecular Biology Laboratory (Germany) and Heidelberg. Cancer and diabetes, defects in cellular signaling.

Lisa Salati, Professor; Ph.D., Minnesota. Regulation of gene expression by fatty acids–inhibition of RNA processing.

Andrew Shiemke, Associate Professor; Ph.D., Oregon Graduate Institute of Science and Technology. Biological oxidation of methane and ammonia; metalloproteins and respiratory chains.

Maxim Sokolov, Assistant Professor; Ph.D., Weizmann (Israel). Biochemistry of vision and molecular control of vision related proteins.

Knox Van Dyke, Professor; Ph.D., St. Louis. Oxidative stress; diabetes; neurodegenerative disease; chemiluminescence; chemotherapy resistance; antiviral, antimicrobial, and antimalarial drugs.

William Wonderlin, Associate Professor; Ph.D., Johns Hopkins. Ion channel physiology and pharmacology; physiology of the endoplasmic reticulum.

Sepideh Zareparsi, Assistant Professor; Ph.D., Oregon Health Sciences. Genetic analysis of complex ophthalmic disorders; gene-expression analysis of retinal aging and age-related retinal disorders.

Section 4
Biophysics

This section contains a directory of institutions offering graduate work in biophysics, followed by in-depth entries submitted by institutions that chose to prepare detailed program descriptions. Additional information about programs listed in the directory but not augmented by an in-depth entry may be obtained by writing directly to the dean of a graduate school or chair of a department at the address given in the directory.

For programs offering related work, see also in this book Biochemistry; Biological and Biomedical Sciences; Cell, Molecular, and Structural Biology; Neuroscience and Neurobiology; and Physiology. In Book 4, see Chemistry and Physics; in Book 5, see Agricultural Engineering and Bioengineering and Biomedical Engineering and Biotechnology; and in Book 6, see Allied Health, Optometry and Vision Sciences, and Public Health.

CONTENTS

Biophysics

Albert Einstein College of Medicine, Sue Golding Graduate Division of Medical Sciences, Department of Physiology and Biophysics, Bronx, NY 10461. Offers PhD, MD/PhD. *Degree requirements:* For doctorate, thesis/dissertation. *Entrance requirements:* For doctorate, GRE General Test. Additional exam requirements/recommendations for international students: Required—TOEFL. *Faculty research:* Biophysical and biochemical basis of body function at the subcellular, cellular, organ, and whole-body level.

Baylor College of Medicine, Graduate School of Biomedical Sciences, Department of Molecular Physiology and Biophysics, Houston, TX 77030-3498. Offers PhD, MD/PhD. *Faculty:* 33 full-time (10 women). *Students:* 17 full-time (10 women); includes 2 minority (1 African American, 1 Asian American or Pacific Islander), 10 international. Average age 29. 23 applicants, 35% accepted, 3 enrolled. In 2005, 2 degrees awarded. *Median time to degree:* Of those who began their doctoral program in fall 1997, 100% received their degree in 8 years or less. *Degree requirements:* For doctorate, thesis/dissertation, public defense. *Entrance requirements:* For doctorate, GRE General Test, GRE Subject Test (strongly recommended), minimum GPA of 3.0. Additional exam requirements/recommendations for international students: Required—TOEFL. *Application deadline:* For fall admission, 1/1 for domestic students. Application fee: $30. Electronic applications accepted. *Expenses:* Tuition: Full-time $8,200. Full-time tuition and fees vary according to program. *Financial support:* In 2005–06, 17 students received support, including fellowships (averaging $23,000 per year), research assistantships (averaging $20,000 per year); career-related internships or fieldwork, Federal Work-Study, institutionally sponsored loans, health care benefits, and tuition waivers (full) also available. Financial award applicants required to submit FAFSA. *Faculty research:* Multip-photon imaging, Magnetic Resonance Imaging (MRI), structure and function of ion channels and transport proteins, signal transduction, synaptic plasticity, cell-cycle control, reactive oxygen species, neurodegenerative diseases and cardiac development. *Unit head:* Dr. Robia Pautler, Director, 713-798-5630, Fax: 713-798-3475. *Application contact:* Cherry McGlory, Graduate Program Administrator, 713-798-5109, Fax: 713-798-3475, E-mail: molphys@bcm.tmc.edu.

See Close-Up on page 1241.

Boston University, Graduate School of Arts and Sciences, Program in Cellular Biophysics, Boston, MA 02215. Offers PhD. *Students:* 12 full-time (6 women), 5 international. Average age 26. 2 applicants, 0% accepted. In 2005, 2 degrees awarded. *Degree requirements:* For doctorate, one foreign language, thesis/dissertation, comprehensive exam, registration. *Entrance requirements:* For doctorate, GRE General Test, GRE Subject Test, 3 letters of recommendation. Additional exam requirements/recommendations for international students: Required—TOEFL (minimum score 550 paper-based; 213 computer-based). *Application deadline:* For fall admission, 5/1 for domestic students, 5/1 for international students; for spring admission, 10/15 for domestic students, 10/15 for international students. Application fee: $60. *Expenses:* Tuition: Full-time $31,530; part-time $985 per credit. Required fees: $316; $40 per semester. Tuition and fees vary according to course level and program. *Financial support:* Career-related internships or fieldwork available. Support available to part-time students. Financial award application deadline: 1/15; financial award applicants required to submit FAFSA. *Unit head:* Dr. Melvin Carter Cornwall, Director, 617-638-4256, Fax: 617-638-7228, E-mail: cornwall@bu.edu.

Boston University, School of Medicine, Division of Graduate Medical Sciences, Department of Physiology and Biophysics, Boston, MA 02118. Offers MA, PhD, MD/PhD. Part-time programs available. *Faculty:* 24. *Students:* 26 full-time (12 women); includes 2 minority (1 Asian American or Pacific Islander, 1 Hispanic American), 20 international. Average age 27. Terminal master's awarded for partial completion of doctoral program. *Degree requirements:* For master's and doctorate, thesis/dissertation, qualifying exam. *Entrance requirements:* For master's and doctorate, GRE General Test, GRE Subject Test (strongly recommended). Additional exam requirements/recommendations for international students: Required—TOEFL. *Application deadline:* For fall admission, 1/1 for domestic students; for spring admission, 10/15 priority date for domestic students. Electronic applications accepted. *Expenses:* Tuition: Full-time $31,530; part-time $985 per credit. Required fees: $316; $40 per semester. Tuition and fees vary according to course level and program. *Financial support:* Fellowships with tuition reimbursements, research assistantships with tuition reimbursements, scholarships/grants and traineeships available. *Faculty research:* X-ray scattering, NMR spectroscopy, protein crystallography, structural electron microscopy, molecular modeling. *Unit head:* Dr. Donald M. Small, Chairman, 617-638-4001. *Application contact:* Dr. Christopher Akey, Director of Graduate Studies, 617-638-4042, Fax: 617-638-4041, E-mail: cakey@bu.edu.

See Close-Up on page 467.

Brandeis University, Graduate School of Arts and Sciences, Programs in Life Sciences, Program in Biophysics and Structural Biology, Waltham, MA 02454-9110. Offers MS, PhD. *Faculty:* 23 full-time (6 women). *Students:* 10 full-time (3 women); includes 1 minority (Asian American or Pacific Islander), 2 international. Average age 26. 40 applicants, 20% accepted, 2 enrolled. In 2005, 1 master's, 4 doctorates awarded. *Degree requirements:* For doctorate, one foreign language, thesis/dissertation. *Entrance requirements:* For doctorate, GRE General Test, resumé, 3 letters of recommendation. Additional exam requirements/recommendations for international students: Required—TOEFL (minimum score 600 paper-based; 250 computer-based). *Application deadline:* For fall admission, 1/15 for domestic students. Applications are processed on a rolling basis. Application fee: $55. Electronic applications accepted. *Financial support:* In 2005–06, 2 fellowships with tuition reimbursements (averaging $26,500 per year), 8 research assistantships with tuition reimbursements (averaging $26,500 per year), teaching assistantships with tuition reimbursements (averaging $6,000 per year) were awarded; scholarships/grants and tuition waivers (full and partial) also available. Support available to part-time students. Financial award application deadline: 4/15; financial award applicants required to submit CSS PROFILE or FAFSA. *Faculty research:* Biophysical chemistry, macromolecular structure and function, single molecule biophysics, macromolecules. *Unit head:* Dr. Dorothee Kern, Chair, 781-736-2354, Fax: 781-736-2349, E-mail: dkern@brandeis.edu. *Application contact:* Lynn Olsen, Administrative Assistant, 781-736-2300, Fax: 781-736-2349, E-mail: lolsen@brandeis.edu.

California Institute of Technology, Division of Biology, Program in Cell Biology and Biophysics, Pasadena, CA 91125-0001. Offers PhD. *Degree requirements:* For doctorate, thesis/dissertation, qualifying exam. *Entrance requirements:* For doctorate, GRE General Test. *Application deadline:* For fall admission, 1/1 for domestic students. Application fee: $0. *Financial support:* Application deadline: 1/1. *Application contact:* Elizabeth Ayala, Graduate Program Coordinator, 626-395-4497, Fax: 626-683-3343, E-mail: biograd@cco.caltech.edu.

Carnegie Mellon University, Mellon College of Science, Department of Biological Sciences, Pittsburgh, PA 15213-3891. Offers biochemistry (PhD); biophysics (PhD); cell biology (PhD); computational biology (MS, PhD); developmental biology (PhD); genetics (PhD); molecular biology (PhD); neurobiology (PhD). *Degree requirements:* For doctorate, thesis/dissertation, comprehensive exam. *Entrance requirements:* For doctorate, GRE General Test, GRE Subject Test, interview. Electronic applications accepted. *Faculty research:* Genetic structure, function, and regulation; protein structure and function; biological membranes; biological spectroscopy.

See Close-Up on page 93.

Case Western Reserve University, School of Medicine and School of Graduate Studies, Graduate Programs in Medicine, Department of Physiology and Biophysics, Cleveland, OH 44106. Offers cell physiology (PhD); molecular/cellular biophysics (PhD); physiology and biophysics (PhD); physiology and biotechnology (MS); systems physiology (PhD). *Faculty:* 52. *Students:* 34 full-time (15 women); includes 8 minority (3 African Americans, 5 Asian Americans or Pacific Islanders), 9 international. Average age 27. 59 applicants, 25% accepted, 7 enrolled. In 2005, 5 master's, 3 doctorates awarded. Terminal master's awarded for partial completion of

doctoral program. *Degree requirements:* For master's and doctorate, thesis/dissertation. *Entrance requirements:* For master's, GRE General Test, minimum GPA of 3.28; for doctorate, GRE General Test, minimum GPA of 3.6. Additional exam requirements/recommendations for international students: Required—TOEFL. *Application deadline:* For fall admission, 3/15 for domestic students. Applications are processed on a rolling basis. Application fee: $50. Electronic applications accepted. *Financial support:* In 2005–06, fellowships with tuition reimbursements (averaging $22,000 per year), research assistantships (averaging $22,000 per year) were awarded; scholarships/grants, health care benefits, and tuition waivers (full) also available. *Faculty research:* Cardiovascular physiology, calcium metabolism, epithelial cell biology. Total annual research expenditures: $12.1 million. *Unit head:* Dr. Cathleen Carlin, Interim Chairman and Professor, 216-368-8939, Fax: 216-368-5586, E-mail: cxc39@case.edu. *Application contact:* Jean Davis, Coordinator, Graduate Training Programs, 216-368-2084, Fax: 216-368-5586, E-mail: jxd16@po.cwru.edu.

See Close-Up on page 1245.

Clemson University, Graduate School, College of Engineering and Science, Department of Physics and Astronomy, Program in Physics, Clemson, SC 29634. Offers astronomy and astrophysics (MS, PhD); atmospheric physics (MS, PhD); biophysics (MS, PhD). Part-time programs available. *Students:* 53 full-time (15 women), 2 part-time; includes 2 minority (1 Asian American or Pacific Islander, 1 Hispanic American), 19 international. 46 applicants, 41% accepted, 10 enrolled. In 2005, 5 master's, 4 doctorates awarded. Terminal master's awarded for partial completion of doctoral program. *Degree requirements:* For master's, thesis or alternative; for doctorate, thesis/dissertation. *Entrance requirements:* For master's and doctorate, GRE General Test. Additional exam requirements/recommendations for international students: Required—TOEFL. *Application deadline:* For fall admission, 2/15 for domestic students. Applications are processed on a rolling basis. Application fee: $50. *Financial support:* Fellowships, research assistantships, teaching assistantships available. Financial award application deadline: 6/1; financial award applicants required to submit FAFSA. *Faculty research:* Radiation physics, solid-state physics, nuclear physics, radar and lidar studies of atmosphere. *Unit head:* Dr. Brad Myer, Head, 864-656-5320. *Application contact:* Dr. Miguel Larsen, Coordinator, 864-656-5309, Fax: 864-656-0805, E-mail: mlarsen@clemson.edu.

Columbia University, College of Physicians and Surgeons and Graduate School of Arts and Sciences, Graduate School of Arts and Sciences at the College of Physicians and Surgeons, Department of Biochemistry and Molecular Biophysics, New York, NY 10032. Offers biochemistry and molecular biophysics (M Phil, PhD); biophysics (PhD). Only candidates for the PhD are admitted. *Degree requirements:* For doctorate, one foreign language, thesis/dissertation. *Entrance requirements:* For master's and doctorate, GRE General Test. Additional exam requirements/recommendations for international students: Required—TOEFL. *Expenses:* Tuition: Full-time $31,448. Tuition and fees vary according to course level, course load, campus/location and program.

Columbia University, College of Physicians and Surgeons and Graduate School of Arts and Sciences, Graduate School of Arts and Sciences at the College of Physicians and Surgeons, Department of Physiology and Cellular Biophysics, New York, NY 10032. Offers M Phil, MA, PhD, MD/PhD. Only candidates for the PhD are admitted. Terminal master's awarded for partial completion of doctoral program. *Degree requirements:* For doctorate, thesis/dissertation. *Entrance requirements:* For master's and doctorate, GRE General Test. Additional exam requirements/recommendations for international students: Required—TOEFL. *Expenses:* Tuition: Full-time $31,448. Tuition and fees vary according to course level, course load, campus/location and program. *Faculty research:* Membrane physiology, cellular biology, cardiovascular physiology, neurophysiology.

Columbia University, College of Physicians and Surgeons and Graduate School of Arts and Sciences, Graduate School of Arts and Sciences at the College of Physicians and Surgeons, Integrated Program in Cellular, Molecular and Biophysical Studies, New York, NY 10032. Offers M Phil, MA, PhD, MD/PhD. Only candidates for the PhD are admitted. Terminal master's awarded for partial completion of doctoral program. *Degree requirements:* For doctorate, thesis/dissertation. *Entrance requirements:* For master's, GRE General Test; for doctorate, GRE General Test, GRE Subject Test. Additional exam requirements/recommendations for international students: Required—TOEFL. Expenses: Contact institution. Tuition and fees vary according to course level, course load, campus/location and program. *Faculty research:* Transcription, macromolecular sorting, gene expression during development, cellular interaction.

Cornell University, Graduate School, Graduate Field of Biophysics, Ithaca, NY 14853-0001. Offers PhD. *Faculty:* 27 full-time (5 women). *Students:* 14 full-time (4 women); includes 1 minority (African American), 8 international. 33 applicants, 9% accepted, 2 enrolled. *Degree requirements:* For doctorate, thesis/dissertation, comprehensive exam. *Entrance requirements:* For doctorate, GRE General Test, GRE Subject Test (physics or chemistry preferred), 3 letters of recommendation. Additional exam requirements/recommendations for international students: Required—TOEFL (minimum score 550 paper-based; 213 computer-based). *Application deadline:* For fall admission, 1/15 for domestic students. Application fee: $60. Electronic applications accepted. *Financial support:* In 2005–06, 14 students received support, including 4 fellowships with full tuition reimbursements available, 8 research assistantships with full tuition reimbursements available, 2 teaching assistantships with full tuition reimbursements available; institutionally sponsored loans, scholarships/grants, health care benefits, tuition waivers (full and partial), and unspecified assistantships also available. Financial award applicants required to submit FAFSA. *Faculty research:* Protein structure and function, biomolecular and cellular function, membrane biophysics, signal transduction, computational biology. *Unit head:* Director of Graduate Studies, 607-255-2100, Fax: biophysics@cornell.edu. *Application contact:* Graduate Field Assistant, 610-255-2100, E-mail: biophysics@cornell.edu.

Cornell University, Graduate School, Graduate Fields of Agriculture and Life Sciences, Field of Biochemistry, Molecular and Cell Biology, Ithaca, NY 14853-0001. Offers biochemistry (PhD); biophysics (PhD); cell biology (PhD); molecular and cell biology (PhD); molecular biology (PhD). *Faculty:* 60 full-time (13 women). *Students:* 177 applicants, 25% accepted, 22 enrolled. In 2005, 13 degrees awarded. *Degree requirements:* For doctorate, thesis/dissertation, 2 semesters of teaching experience, comprehensive exam. *Entrance requirements:* For doctorate, GRE General Test, GRE Subject Test (biology, chemistry, physics; or biochemistry, cell and molecular biology), 3 letters of recommendation. Additional exam requirements/recommendations for international students: Required—TOEFL (minimum score 600 paper-based; 250 computer-based). *Application deadline:* For fall admission, 1/5 for domestic students. Application fee: $60. Electronic applications accepted. *Financial support:* In 2005–06, 90 students received support, including 35 fellowships with full tuition reimbursements available, 42 research assistantships with full tuition reimbursements available, 13 teaching assistantships with full tuition reimbursements available; institutionally sponsored loans, scholarships/grants, health care benefits, tuition waivers (full and partial), and unspecified assistantships also available. Financial award applicants required to submit FAFSA. *Faculty research:* Biophysics, structural biology. *Unit head:* Director of Graduate Studies, 607-255-2100, Fax: 607-255-2100. *Application contact:* Graduate Field Assistant, 607-255-2100, Fax: 607-255-2100, E-mail: bmcb@cornell.edu.

Cornell University, Joan and Sanford I. Weill Medical College and Graduate School of Medical Sciences, Weill Graduate School of Medical Sciences, Program in Physiology, Biophysics and Systems Biology, New York, NY 10021-4896. Offers PhD, MD/PhD. *Faculty:* 30 full-time (8 women). *Students:* 39 full-time (19 women); includes 4 minority (1 African American, 3 Asian Americans or Pacific Islanders), 21 international. 32 applicants, 25% accepted, 4 enrolled. In 2005, 1 doctorate awarded. *Degree requirements:* For doctorate, thesis/dissertation, final exam. *Entrance requirements:* For doctorate, GRE General Test, GRE Subject Test, introductory courses in biology, inorganic and organic chemistry, physics, and

Biophysics

mathematics. Additional exam requirements/recommendations for international students: Required—TOEFL. *Application deadline:* For fall admission, 12/15 for domestic students. Application fee: $60. *Expenses:* Tuition: Full-time $32,320. Required fees: $1,025. *Financial support:* In 2005–06, 1 fellowship was awarded; stipends also available. *Unit head:* Doris Herzlinger, Director, 212-746-6377, E-mail: daherzli@med.cornell.edu.

Dalhousie University, Faculty of Graduate Studies and Faculty of Medicine, Graduate Programs in Medicine, Department of Physiology and Biophysics, Halifax, NS B3H 4R2, Canada. Offers M Sc, PhD, MD/PhD. *Degree requirements:* For master's and doctorate, thesis/dissertation. *Entrance requirements:* For master's and doctorate, GRE Subject Test. Additional exam requirements/recommendations for international students: Required—TOEFL. *Faculty research:* Computer modeling, reproductive and endocrine physiology, cardiovascular physiology, neurophysiology, membrane biophysics.

East Carolina University, Graduate School, Thomas Harriot College of Arts and Sciences, Department of Physics, Greenville, NC 27858-4353. Offers applied and biomedical physics (MS); medical physics (MS); physics (PhD). Part-time programs available. *Faculty:* 19 full-time (2 women). *Students:* 25 full-time (8 women), 10 part-time (1 woman); includes 6 minority (4 African Americans, 1 American Indian/Alaska Native, 1 Asian American or Pacific Islander), 7 international. Average age 30. 11 applicants, 36% accepted, 4 enrolled. In 2005, 4 master's, 5 doctorates awarded. *Degree requirements:* For master's, one foreign language, comprehensive exam. *Entrance requirements:* For master's, GRE General Test. Additional exam requirements/recommendations for international students: Required—TOEFL. *Application deadline:* Applications are processed on a rolling basis. Application fee: $50. *Expenses:* Tuition, state resident: full-time $2,516. Tuition, nonresident: full-time $12,832. *Financial support:* Research assistantships with partial tuition reimbursements, teaching assistantships with partial tuition reimbursements, Federal Work-Study available. Support available to part-time students. Financial award application deadline: 6/1. *Unit head:* Dr. John Sutherland, Chair, 252-328-6739, Fax: 252-328-6314, E-mail: sutherlandj@ecu.edu. *Application contact:* Dean of Graduate School, 252-328-6012, Fax: 252-328-6071, E-mail: gradschool@ecu.edu.

East Tennessee State University, James H. Quillen College of Medicine, Biomedical Science Graduate Program, Johnson City, TN 37614. Offers anatomy (MS, PhD); biochemistry (MS, PhD); biophysics (MS, PhD); microbiology (MS, PhD); pharmacology (MS, PhD); physiology (MS, PhD). Part-time programs available. *Faculty:* 49 full-time (12 women), 1 (woman) part-time/adjunct. *Students:* 30 full-time (19 women), 5 part-time (4 women); includes 2 minority (1 African American, 1 Asian American or Pacific Islander), 11 international. Average age 31. 78 applicants, 13% accepted, 9 enrolled. In 2005, 1 master's, 4 doctorates awarded. Terminal master's awarded for partial completion of doctoral program. *Degree requirements:* For master's, one foreign language, thesis, comprehensive qualifying exam; for doctorate, 2 foreign languages, thesis/dissertation. *Entrance requirements:* For master's, GRE General Test, minimum GPA of 3.0, bachelor's degree in biological or related science; for doctorate, GRE General Test, GRE Subject Test. Additional exam requirements/recommendations for international students: Required—TOEFL (minimum score 550 paper-based; 213 computer-based). *Application deadline:* For fall admission, 3/15 for domestic students; for spring admission, 3/1 for domestic students. Application fee: $25 ($35 for international students). *Expenses:* Contact institution. *Financial support:* In 2005–06, 7 research assistantships with full tuition reimbursements (averaging $15,000 per year) were awarded; teaching assistantships with full tuition reimbursements, career-related internships or fieldwork, Federal Work-Study, institutionally sponsored loans, scholarships/grants, and tuition waivers (full) also available. Financial award application deadline: 7/1; financial award applicants required to submit FAFSA. Total annual research expenditures: $2.1 million. *Unit head:* Dr. Mitchell E. Robinson, Assistant Dean, Director, 423-439-4658, E-mail: robinson@etsu.edu.

Emory University, Graduate School of Arts and Sciences, Department of Physics, Atlanta, GA 30322-1100. Offers biophysics (PhD); condensed matter physics (PhD); non-linear physics (PhD); radiological physics (PhD); soft condensed matter physics (PhD); solid-state physics (PhD); statistical physics (PhD). *Faculty:* 19 full-time (2 women). *Students:* 17 full-time (4 women); includes 2 minority (1 African American, 1 Hispanic American), 13 international. Average age 24. 40 applicants, 13% accepted. *Degree requirements:* For doctorate, thesis/dissertation, qualifier proposal (PhD qualification). *Entrance requirements:* For doctorate, GRE General Test, minimum GPA of 3.0. Additional exam requirements/recommendations for international students: Required—TOEFL (minimum score 600 paper-based). *Application deadline:* For fall admission, 1/3 priority date for domestic students, 1/3 priority date for international students. Application fee: $50. Electronic applications accepted. *Expenses:* Tuition: Full-time $14,400. Required fees: $217. *Financial support:* In 2005–06, 17 students received support, including 6 fellowships (averaging $20,000 per year); institutionally sponsored loans, scholarships/grants, health care benefits, and tuition waivers (full) also available. Financial award application deadline: 1/3; financial award applicants required to submit FAFSA. *Faculty research:* Experimental studies of the structure and function of metalloproteins, soft condensed matter, granular materials, biophotonics and fluorescence correlation spectroscopy, single molecule studies of DNA-protein systems. Total annual research expenditures: $1.5 million. *Unit head:* Dr. Raymond DuVarney, Chair, 404-727-4296, Fax: 404-727-0873, E-mail: phsrcd@physics.emory.edu. *Application contact:* Dr. Kurt Warncke, Director of Graduate Studies, 404-727-2975, Fax: 404-727-0873, E-mail: kwarncke@physics.emory.edu.

Georgetown University, Graduate School of Arts and Sciences, Programs in Biomedical Sciences, Department of Physiology and Biophysics, Washington, DC 20057. Offers MS, PhD, MD/PhD. *Degree requirements:* For doctorate, thesis/dissertation. *Entrance requirements:* For master's, GRE General Test, MCAT; for doctorate, GRE General Test. Additional exam requirements/recommendations for international students: Required—TOEFL.

Harvard University, Graduate School of Arts and Sciences, Committee on Biophysics, Cambridge, MA 02138. Offers PhD. *Students:* 9 full-time (0 women). 60 applicants, 25% accepted. In 2005, 8 degrees awarded. *Degree requirements:* For doctorate, thesis/dissertation, exam, qualifying paper. *Entrance requirements:* For doctorate, GRE General Test, GRE Subject Test (recommended). Additional exam requirements/recommendations for international students: Required—TOEFL. *Application deadline:* For fall admission, 12/15 for domestic students. Application fee: $60. *Expenses:* Tuition: Full-time $28,752. Full-time tuition and fees vary according to program and student level. *Financial support:* Fellowships, research assistantships, teaching assistantships, career-related internships or fieldwork, Federal Work-Study, and institutionally sponsored loans available. Financial award application deadline: 12/30. *Faculty research:* Structural molecular biology, cell and membrane biophysics, molecular genetics, physical biochemistry, mathematical biophysics. *Unit head:* Betsey Cogswell, Administrator, 617-495-5497, Fax: 617-495-5264. *Application contact:* Office of Admissions and Financial Aid, 617-495-5315.

See Close-Up on page 469.

Howard University, Graduate School of Arts and Sciences, Department of Physiology and Biophysics, Program in Biophysics, Washington, DC 20059-0002. Offers biophysics (PhD); physiology (PhD). Part-time programs available. *Degree requirements:* For doctorate, thesis/dissertation, comprehensive exam, registration. *Entrance requirements:* For doctorate, GRE General Test, minimum B average in field. *Faculty research:* Cardiovascular physiology, pulmonary physiology, neurophysiology, endocrinology.

Indiana University–Purdue University Indianapolis, Indiana University School of Medicine, Department of Cellular and Integrative Physiology, Indianapolis, IN 46202-2896. Offers MS, PhD, MD/PhD. *Faculty:* 12 full-time (4 women). *Students:* 37 full-time (13 women), 11 part-time (3 women); includes 5 minority (1 African American, 4 Asian Americans or Pacific Islanders), 9 international. Average age 26. Terminal master's awarded for partial completion of doctoral program. *Degree requirements:* For doctorate, thesis/dissertation. *Entrance requirements:* For master's, GRE General Test, GRE Subject Test, previous course work in biology, calculus, physical chemistry, and physics; for doctorate, GRE General Test, GRE Subject Test. *Application*

deadline: For fall admission, 4/15 for domestic students. Applications are processed on a rolling basis. Application fee: $50 ($60 for international students). *Expenses:* Tuition, state resident: full-time $5,159; part-time $215 per credit hour. Tuition, nonresident: full-time $14,890; part-time $620 per credit hour. Required fees: $614. Tuition and fees vary according to campus/location and program. *Financial support:* In 2005–06, 14 students received support; fellowships with partial tuition reimbursements available, research assistantships with partial tuition reimbursements available, Federal Work-Study and institutionally sponsored loans available. *Faculty research:* Cardiovascular physiology, cell growth and development, respiratory biology, cell signaling mechanisms, cytoskeleton function. *Unit head:* Dr. Michael Sturek, Chair, 317-274-7772, Fax: 317-274-3318. *Application contact:* Dr. Fred M. Pavalko, Graduate Director, 317-274-3140, Fax: 317-274-3318, E-mail: fpavalko@iupui.edu.

Indiana University–Purdue University Indianapolis, Indiana University School of Medicine, Program in Medical Biophysics, Indianapolis, IN 46202-2896. Offers MS, PhD. *Students:* 1 (woman) full-time, 1 part-time, 1 international. Average age 30. 5 applicants, 40% accepted. In 2005, 1 degree awarded. Terminal master's awarded for partial completion of doctoral program. *Degree requirements:* For master's and doctorate, thesis/dissertation. *Entrance requirements:* For master's and doctorate, GRE General Test, GRE Subject Test. *Application deadline:* For fall admission, 1/15 for domestic students. Applications are processed on a rolling basis. Application fee: $50 ($60 for international students). *Expenses:* Tuition, state resident: full-time $5,159; part-time $215 per credit hour. Tuition, nonresident: full-time $14,890; part-time $620 per credit hour. Required fees: $614. Tuition and fees vary according to campus/location and program. *Financial support:* Fellowships with full tuition reimbursements, research assistantships with full tuition reimbursements, teaching assistantships with full tuition reimbursements, Federal Work-Study, institutionally sponsored loans, and tuition waivers (partial) available. Financial award application deadline: 4/15. *Faculty research:* Membrane biophysics, protein structure and function, biological magnetic resonance, smooth muscle biophysics, photobiology. Total annual research expenditures: $2 million. *Unit head:* Simon Atkinson, Chair, 317-274-3772. *Application contact:* Dr. Richard Haak, Graduate Adviser, 317-274-7626, Fax: 317-274-4090, E-mail: rhaak@iupui.edu.

Iowa State University of Science and Technology, Graduate College, College of Agriculture and College of Liberal Arts and Sciences, Department of Biochemistry, Biophysics, and Molecular Biology, Ames, IA 50011. Offers biochemistry (MS, PhD); biophysics (MS, PhD); genetics (MS, PhD); molecular, cellular, and developmental biology (MS, PhD); toxicology (MS, PhD). *Faculty:* 24 full-time, 2 part-time/adjunct. *Students:* 97 full-time (42 women), 2 part-time (1 woman); includes 2 minority (1 African American, 1 Asian American or Pacific Islander), 68 international. 28 applicants, 46% accepted, 13 enrolled. In 2005, 9 master's, 8 doctorates awarded. *Degree requirements:* For master's and doctorate, thesis/dissertation. *Entrance requirements:* For master's and doctorate, GRE General Test. Additional exam requirements/recommendations for international students: Required—TOEFL (paper score 550; computer score 213) or IELTS (score 6.5). *Application deadline:* For fall admission, 2/1 priority date for domestic students, 2/1 priority date for international students. Application fee: $30 ($70 for international students). Electronic applications accepted. *Expenses:* Tuition, state resident: full-time $6,410. Tuition, nonresident: full-time $16,422. Tuition and fees vary according to program. *Financial support:* In 2005–06, 91 research assistantships with full tuition reimbursements (averaging $16,290 per year), 2 teaching assistantships with full and partial tuition reimbursements (averaging $14,455 per year) were awarded; scholarships/grants, health care benefits, and unspecified assistantships also available. *Unit head:* Dr. Alan M. Myers, Chair, 515-294-6116, E-mail: biochem@iastate.edu.

See Close-Up on page 395.

The Johns Hopkins University, National Institutes of Health Sponsored Programs, Department of Biology, Baltimore, MD 21218-2699. Offers biochemistry (PhD); biophysics (PhD); cell biology (PhD); developmental biology (PhD); genetic biology (PhD); molecular biology (PhD). *Faculty:* 25 full-time (4 women). *Students:* 126 full-time (72 women); includes 36 minority (3 African Americans, 1 American Indian/Alaska Native, 21 Asian Americans or Pacific Islanders, 11 Hispanic Americans), 19 international. 282 applicants, 26% accepted, 36 enrolled. In 2005, 15 degrees awarded. *Median time to degree:* Of those who began their doctoral program in fall 1997, 81.2% received their degree in 8 years or less. *Degree requirements:* For doctorate, thesis/dissertation, comprehensive exam, registration. *Entrance requirements:* For doctorate, GRE General Test. Additional exam requirements/recommendations for international students: Required—TOEFL (minimum score 600 paper-based; 250 computer-based), TWE, TSE. *Application deadline:* For fall admission, 12/15 for domestic students. Application fee: $60. *Expenses:* Tuition: Full-time $30,960. Tuition and fees vary according to degree level and program. *Financial support:* In 2005–06, 24 fellowships (averaging $23,000 per year), 93 research assistantships (averaging $23,000 per year), 22 teaching assistantships (averaging $23,000 per year) were awarded; Federal Work-Study, institutionally sponsored loans, scholarships/grants, traineeships, health care benefits, tuition waivers (partial), and unspecified assistantships also available. Financial award application deadline: 4/15; financial award applicants required to submit FAFSA. *Faculty research:* Protein and nucleic acid biochemistry and biophysical chemistry, molecular biology and development. Total annual research expenditures: $11.2 million. *Unit head:* Dr. Allen Shearn, Chair, 410-516-4693, Fax: 410-516-5213, E-mail: bio_cals@jhu.edu. *Application contact:* Joan Miller, Academic Affairs Manager, 410-516-5502, Fax: 410-516-5213, E-mail: joan@jhu.edu.

See Close-Up on page 155.

The Johns Hopkins University, School of Medicine, Graduate Programs in Medicine, Department of Biophysics and Biophysical Chemistry, Baltimore, MD 21218-2699. Offers MS, PhD. *Degree requirements:* For doctorate, thesis/dissertation, oral exam, thesis defense. *Entrance requirements:* For doctorate, GRE. Electronic applications accepted. *Expenses:* Tuition: Full-time $30,960. Tuition and fees vary according to degree level and program. *Faculty research:* Protein structure and protein folding, membranes, RNA and RNPs, DNA, structure and mechanism in enzymes and metabolic pathways, biophysical methods.

The Johns Hopkins University, Zanvyl Krieger School of Arts and Sciences, Thomas C. Jenkins Department of Biophysics, Baltimore, MD 21218-2699. Offers MA, PhD. *Faculty:* 7 full-time (2 women), 4 part-time/adjunct (0 women). *Students:* 61 full-time (29 women); includes 6 minority (2 African Americans, 3 Asian Americans or Pacific Islanders, 1 Hispanic American), 10 international. Average age 26. 78 applicants, 33% accepted, 18 enrolled. In 2005, 1 master's, 4 doctorates awarded. *Median time to degree:* Of those who began their doctoral program in fall 1997, 87% received their degree in 8 years or less. *Degree requirements:* For doctorate, thesis/dissertation, comprehensive exam, registration. *Entrance requirements:* For master's and doctorate, GRE General Test. Additional exam requirements/recommendations for international students: Required—TOEFL (minimum score 600 paper-based; 250 computer-based); Recommended—IELT, TWE. *Application deadline:* For fall admission, 1/5 priority date for domestic students, 1/5 priority date for international students. Applications are processed on a rolling basis. Application fee: $60. Electronic applications accepted. *Expenses:* Tuition: Full-time $30,960. Tuition and fees vary according to degree level and program. *Financial support:* In 2005–06, 49 students received support, including 25 fellowships with full tuition reimbursements available (averaging $24,600 per year), 21 research assistantships with full tuition reimbursements available (averaging $24,600 per year), 1 teaching assistantship with tuition reimbursement available (averaging $24,600 per year); Federal Work-Study, institutionally sponsored loans, scholarships/grants, traineeships, and tuition waivers (full) also available. Financial award application deadline: 4/15; financial award applicants required to submit FAFSA. *Faculty research:* Chemistry of RNA, signal transduction/protein energies, mucosal protective mechanisms, chemistry of proteins, modeling/simulation of folding. Total annual research expenditures: $2.9 million. *Unit head:* Dr. George Rose, Chair, 410-516-0151, Fax: 410-516-4118. *Application contact:* Ranice Crosby, Coordinator, Graduate Admissions, 410-516-5197, Fax: 410-516-4118, E-mail: tcjenkin@jhu.edu.

Medical College of Wisconsin, Graduate School of Biomedical Sciences, Department of Biophysics, Milwaukee, WI 53226-0509. Offers PhD. *Degree requirements:* For doctorate,

Biophysics

Medical College of Wisconsin *(continued)*
thesis/dissertation, comprehensive exam, registration. *Entrance requirements:* For doctorate, GRE. Additional exam requirements/recommendations for international students: Required—TOEFL. Electronic applications accepted.

Medical College of Wisconsin, Graduate School of Biomedical Sciences, Program in Biophysics, Milwaukee, WI 53226-0509. Offers PhD, MD/PhD. Part-time programs available. *Degree requirements:* For doctorate, thesis/dissertation, oral exam. *Entrance requirements:* For doctorate, GRE General Test. Additional exam requirements/recommendations for international students: Required—TOEFL. Electronic applications accepted. *Faculty research:* X-ray crystallography, electron spin resonance and membrane structure, protein and membrane dynamics, magnetic resonance imaging, free radical biology.

See Close-Up on page 471.

Mount Sinai School of Medicine of New York University, Graduate School of Biological Sciences, New York, NY 10029-6504. Offers biophysics, structural biology and biomathematics (PhD); community medicine (MPH); genetic counseling (MS); genetics and genomic sciences (PhD); mechanisms of disease and therapy (PhD); microbiology (PhD); molecular, cellular, biochemical and developmental sciences (PhD); neurosciences (PhD). *Students:* 218 full-time (109 women). 4,208 applicants, 7% accepted, 117 enrolled. Terminal master's awarded for partial completion of doctoral program. *Degree requirements:* For master's, registration; for doctorate, thesis/dissertation, registration. *Entrance requirements:* For doctorate, GRE General Test, GRE Subject Test, 3 years of college pre-med course work. Additional exam requirements/recommendations for international students: Required—TOEFL. *Application deadline:* For fall admission, 1/15 for domestic students. Application fee: $75. Electronic applications accepted. *Expenses:* Tuition: Full-time $33,250. Required fees: $1,600. Full-time tuition and fees vary according to degree level, program and reciprocity agreements. *Financial support:* In 2005–06, fellowships with full tuition reimbursements (averaging $26,000 per year), research assistantships with full tuition reimbursements (averaging $26,000 per year) were awarded; Federal Work-Study, institutionally sponsored loans, scholarships/grants, health care benefits, and unspecified assistantships also available. Financial award application deadline: 4/5; financial award applicants required to submit FAFSA. *Faculty research:* Cancer, gene therapy, minimally invasive surgery, cardiac translational research. Total annual research expenditures: $162.2 million. *Unit head:* Dr. Diomedes Logothetis, Dean, 212-241-6546, Fax: 212-241-0651, E-mail: diomedes.logothetis@mssm.edu. *Application contact:* Lily Recanati, Manager, 212-241-3267, Fax: 212-241-0651, E-mail: lily.recanati@mssm.edu.

See Close-Up on page 177.

Northwestern University, The Graduate School and Judd A. and Marjorie Weinberg College of Arts and Sciences, Interdepartmental Biological Sciences Program (IBiS), Department of Biochemistry, Molecular Biology, and Cell Biology, Evanston, IL 60208. Offers biochemistry (PhD); cell and molecular biology (PhD); molecular biophysics (PhD); structural biology (PhD). Department participates in the Interdepartmental Biological Sciences Program (IBiS). *Entrance requirements:* For doctorate, GRE General Test. Additional exam requirements/recommendations for international students: Required—TOEFL (minimum score 600 paper-based), TSE (minimum score 50). Electronic applications accepted. *Faculty research:* Protein structure, protein-DNA interactions, gene regulation, development of embryos, hormone action and signal transduction.

The Ohio State University, Graduate School, College of Biological Sciences, Program in Biophysics, Columbus, OH 43210. Offers MS, PhD. *Degree requirements:* For master's, thesis optional; for doctorate, thesis/dissertation. *Entrance requirements:* For master's and doctorate, GRE General Test. Additional exam requirements/recommendations for international students: Required—TOEFL (minimum score 600 paper-based; 250 computer-based); Recommended—TSE.

Oregon State University, Graduate School, College of Science, Department of Biochemistry and Biophysics, Corvallis, OR 97331-6503. Offers MA, MAIS, MS, PhD. *Faculty:* 3 full-time (3 women), 1 part-time/adjunct (0 women). *Students:* 24 full-time (7 women), 1 part-time; includes 1 minority (Asian American or Pacific Islander), 11 international. Average age 31. In 2005, 4 doctorates awarded. *Degree requirements:* For master's, thesis optional; for doctorate, thesis/dissertation, exams. *Entrance requirements:* For master's, GRE General Test, minimum GPA of 3.0; for doctorate, GRE Subject Test, minimum GPA of 3.0. Additional exam requirements/recommendations for international students: Required—TOEFL. *Application deadline:* For fall admission, 4/15 for domestic students. Applications are processed on a rolling basis. Application fee: $50. *Expenses:* Tuition, area resident: Part-time $301 per credit. Tuition, state resident: full-time $8,139; part-time $501 per credit. Tuition, nonresident: full-time $14,376; part-time $532 per credit. Required fees: $1,266. *Financial support:* Research assistantships, teaching assistantships, institutionally sponsored loans available. Support available to part-time students. Financial award application deadline: 2/1. *Faculty research:* DNA and deoxyribonucleotide metabolism, cell growth control, receptors and membranes, protein structure and function. *Unit head:* Dr. Pui Shing Ho, Chair, 541-737-1865, Fax: 541-737-0481. *Application contact:* Dr. W. Curtis Johnson, Chairman, Graduate Committee, 541-737-4143, Fax: 541-737-0481, E-mail: johnsowc@ucs.orst.edu.

Princeton University, Graduate School, Graduate Program in Molecular Biophysics, Princeton, NJ 08544-1019. Offers PhD. Program administered by the Departments of Molecular Biology, Chemistry, and Physics. *Degree requirements:* For doctorate, thesis/dissertation. *Entrance requirements:* For doctorate, GRE General Test, GRE Subject Test (biology, chemistry, or physics). Additional exam requirements/recommendations for international students: Required—TOEFL (minimum score 600 paper-based; 250 computer-based). Electronic applications accepted.

Purdue University, Graduate School, School of Science, Department of Biological Sciences, West Lafayette, IN 47907. Offers biochemistry (PhD); biophysics (PhD); cell and developmental biology (PhD); ecology, evolutionary and population biology (MS, PhD), including ecology, evolutionary biology, population biology; genetics (MS, PhD); microbiology (MS, PhD); molecular biology (PhD); neurobiology (MS, PhD); plant physiology (PhD). *Faculty:* 47 full-time (9 women), 4 part-time/adjunct (1 woman). *Students:* 97 full-time (53 women), 8 part-time (4 women); includes 13 minority (3 African Americans, 1 American Indian/Alaska Native, 4 Asian Americans or Pacific Islanders, 5 Hispanic Americans), 50 international. Average age 28. 168 applicants, 29% accepted, 23 enrolled. In 2005, 18 master's, 9 doctorates awarded. Terminal master's awarded for partial completion of doctoral program. *Degree requirements:* For master's, thesis (for some programs); for doctorate, thesis/dissertation, seminars, teaching experience. *Entrance requirements:* For master's and doctorate, GRE General Test. Additional exam requirements/recommendations for international students: Required—TOEFL, TSE. *Application deadline:* For fall admission, 2/15 for domestic students, 1/31 for international students. Applications are processed on a rolling basis. Application fee: $55. Electronic applications accepted. *Financial support:* In 2005–06, 15 fellowships, 60 research assistantships, 53 teaching assistantships were awarded. Support available to part-time students. Financial award application deadline: 2/15; financial award applicants required to submit FAFSA. *Unit head:* Dr. Richard J Kuhn, Head, 765-494-4407. *Application contact:* Nancy Konopka, Graduate Studies Office Manager, 765-494-8142, Fax: 765-494-0876, E-mail: njk@bilbo.bio.purdue.edu.

Rensselaer Polytechnic Institute, Graduate School, School of Science, Department of Biology, Troy, NY 12180-3590. Offers biochemistry (MS, PhD); biophysics (MS, PhD); cell biology (MS, PhD); developmental biology (MS, PhD); microbiology (MS, PhD); molecular biology (MS, PhD). Part-time programs available. Terminal master's awarded for partial completion of doctoral program. *Degree requirements:* For master's and doctorate, thesis/dissertation, comprehensive exam, registration. *Entrance requirements:* For master's and doctorate, GRE General Test. Additional exam requirements/recommendations for international students: Required—TOEFL. Electronic applications accepted. *Expenses:* Tuition: Full-time $31,000;

part-time $1,320 per credit. Required fees: $1,623. *Faculty research:* Bioinformatics, molecular biology/biochemistry, cell and tissue biology, environment, ecology.

See Close-Up on page 199.

Rensselaer Polytechnic Institute, Graduate School, School of Science, Program in Biochemistry and Biophysics, Troy, NY 12180-3590. Offers biochemistry (MS); biophysics (MS). Part-time programs available. *Entrance requirements:* For master's, GRE General Test, GRE Subject Test (biology, chemistry or biochemistry). Additional exam requirements/recommendations for international students: Required—TOEFL. Electronic applications accepted. *Expenses:* Tuition: Full-time $31,000; part-time $1,320 per credit. Required fees: $1,623. *Faculty research:* Biopolymers, photosynthesis, cellular bioengineering.

The Scripps Research Institute, Kellogg School of Science and Technology, Program in Biology and Biophysical Sciences, La Jolla, CA 92037. Offers biology (PhD); biophysics (PhD). *Faculty:* 99 full-time (29 women). *Students:* 96 full-time (40 women). 299 applicants, 17% accepted, 12 enrolled. In 2005, 10 degrees awarded. *Degree requirements:* For doctorate, thesis/dissertation. *Entrance requirements:* For doctorate, GRE General Test, GRE Subject Test. Additional exam requirements/recommendations for international students: Required—TOEFL. *Application deadline:* For fall admission, 1/1 for domestic students, 1/1 for international students. Application fee: $0. *Expenses:* Tuition: Full-time $5,000. *Financial support:* Institutionally sponsored loans and stipends available. *Faculty research:* Biocatalysis and enzyme engineering, molecular structure and function, neurosciences, immunology, plant biology. *Unit head:* Dr. Stephen P. Mayfield, Associate Dean, 784-9848. *Application contact:* Marylyn Rinaldi, Administrative Director, 858-784-8469, Fax: 858-784-2802, E-mail: mrinaldi@scripps.edu.

Simon Fraser University, Graduate Studies, Faculty of Science, Department of Physics, Burnaby, BC V5A 1S6, Canada. Offers biophysics (M Sc, PhD); chemical physics (M Sc, PhD); physics (M Sc, PhD). *Degree requirements:* For master's and doctorate, thesis/dissertation. *Entrance requirements:* For master's, minimum GPA of 3.0; for doctorate, minimum GPA of 3.5. Additional exam requirements/recommendations for international students: Required—TOEFL or IELTS. *Faculty research:* Solid-state physics, magnetism, energy research, superconductivity, nuclear physics.

Stanford University, School of Humanities and Sciences, Program in Biophysics, Stanford, CA 94305-9991. Offers PhD. *Degree requirements:* For doctorate, thesis/dissertation, oral exam. *Entrance requirements:* For doctorate, GRE General Test, GRE Subject Test. Additional exam requirements/recommendations for international students: Required—TOEFL. Electronic applications accepted.

State University of New York at Buffalo, Graduate School, Graduate Programs in Cancer Research and Biomedical Sciences at Roswell Park Cancer Institute, Department of Molecular and Cellular Biophysics and Biochemistry at Roswell Park Cancer Institute, Buffalo, NY 14214. Offers PhD. *Faculty:* 17 part-time/adjunct (2 women). *Students:* 15 full-time (7 women), 9 international. Average age 25. 20 applicants, 40% accepted, 3 enrolled. In 2005, 1 degree awarded. *Degree requirements:* For doctorate, thesis/dissertation. *Entrance requirements:* For doctorate, GRE General Test, GRE Subject Test. Additional exam requirements/recommendations for international students: Required—TOEFL, TWE, TSE. *Application deadline:* For fall admission, 2/1 for domestic students. Applications are processed on a rolling basis. Application fee: $35. Electronic applications accepted. *Financial support:* In 2005–06, 4 fellowships with full tuition reimbursements (averaging $21,000 per year), 21 research assistantships with full tuition reimbursements (averaging $21,000 per year) were awarded; Federal Work-Study and institutionally sponsored loans also available. Financial award application deadline: 6/1; financial award applicants required to submit FAFSA. *Faculty research:* MRI research, structural and function of biomolecules, photodynamic therapy, DNA damage and repair, heat-shock proteins and vaccine research. Total annual research expenditures: $2.5 million. *Unit head:* Dr. Latif Kazim, Chairman, 716-845-3055, Fax: 716-845-8899, E-mail: latif.kazim@roswellpark.org. *Application contact:* Craig R. Johnson, Director of Admissions, 716-845-2339, Fax: 716-845-8178, E-mail: craig.johnson@roswellpark.edu.

State University of New York at Buffalo, Graduate School, School of Medicine and Biomedical Sciences, Graduate Programs in Medicine and Biomedical Sciences, Department of Physiology and Biophysics, Buffalo, NY 14260. Offers biophysics (MS, PhD); physiology (PhD). *Faculty:* 26 full-time (5 women), 1 part-time/adjunct (0 women). *Students:* 10 full-time (4 women), 2 part-time (1 woman); includes 2 minority (both Asian Americans or Pacific Islanders), 2 international. Average age 26. 9 applicants, 22% accepted, 1 enrolled. In 2005, 2 master's, 4 doctorates awarded. Terminal master's awarded for partial completion of doctoral program. *Degree requirements:* For master's, thesis, oral exam, project; for doctorate, thesis/dissertation, oral and written qualifying exam or 2 research proposals. *Entrance requirements:* For master's and doctorate, GRE General Test. Additional exam requirements/recommendations for international students: Required—TOEFL. *Application deadline:* For fall admission, 2/1 for domestic students. Applications are processed on a rolling basis. Application fee: $35. Electronic applications accepted. *Financial support:* In 2005–06, fellowships with tuition reimbursements (averaging $21,000 per year), 17 research assistantships with tuition reimbursements (averaging $21,000 per year) were awarded; Federal Work-Study, institutionally sponsored loans, health care benefits, and unspecified assistantships also available. Financial award application deadline: 2/1; financial award applicants required to submit FAFSA. *Faculty research:* Neurosciences, ion channels, cardiac physiology, renal/epithelial transport, cardiopulmonary exercise. Total annual research expenditures: $5.6 million. *Unit head:* Dr. Harold C. Strauss, Chair, 716-829-2738, Fax: 716-829-2344, E-mail: hstrauss@buffalo.edu. *Application contact:* Dr. Malcolm Slaughter, Professor, 716-829-3240, Fax: 716-829-2364, E-mail: mslaught@buffalo.edu.

Stony Brook University, State University of New York, Health Sciences Center, School of Medicine and Graduate School, Graduate Programs in Medicine, Department of Molecular Physiology and Biophysics, Stony Brook, NY 11794. Offers physiology and biophysics (PhD). *Faculty:* 17 full-time (3 women). *Students:* 21 full-time (12 women); includes 2 minority (both Asian Americans or Pacific Islanders), 8 international. Average age 29. 24 applicants, 33% accepted. In 2005, 7 degrees awarded. *Degree requirements:* For doctorate, thesis/dissertation, comprehensive exam. *Entrance requirements:* For doctorate, GRE General Test, GRE Subject Test, BS in related field, minimum GPA of 3.0. Additional exam requirements/recommendations for international students: Required—TOEFL. *Application deadline:* For fall admission, 1/15 for domestic students. Application fee: $50. *Expenses:* Tuition, area resident: Full-time $6,900; part-time $288. Tuition, state resident: full-time $6,900. Tuition, nonresident: full-time $10,920; part-time $455. International tuition: $10,920 full-time. Required fees: $704. *Financial support:* In 2005–06, 13 research assistantships, 5 teaching assistantships were awarded; fellowships, Federal Work-Study also available. Financial award application deadline: 3/15. *Faculty research:* Cellular electrophysiology, membrane permeation and transport, metabolic endocrinology. Total annual research expenditures: $6.7 million. *Unit head:* Dr. Peter Brink, Chair, 631-444-2287, Fax: 631-444-3432. *Application contact:* Dr. Leon C. Moore, Graduate Adviser, 631-444-2287, Fax: 631-444-3432, E-mail: moore@pofvax.pnb.sunysb.edu.

See Close-Ups on pages 473 and 1259.

Syracuse University, Graduate School, College of Arts and Sciences, Department of Biology and Department of Physics and Department of Chemistry, Program in Structural Biology, Biochemistry and Biophysics, Syracuse, NY 13244. Offers PhD. *Students:* 9 full-time (4 women), 4 international. 44 applicants, 14% accepted, 4 enrolled. *Degree requirements:* For doctorate, thesis/dissertation, exam. *Entrance requirements:* For doctorate, GRE General Test, GRE Subject Test. Additional exam requirements/recommendations for international students: Required—TOEFL. *Application deadline:* For fall admission, 1/10 for domestic students. Applications are processed on a rolling basis. Application fee: $65. Electronic applications accepted. *Financial support:* Fellowships, research assistantships, teaching assistantships, tuition waivers available. *Unit head:* Stewart Loh, Director, 315-464-8731, Fax: 315-443-

4070. *Application contact:* Evelyn Lott, Information Contact, 315-443-9154, Fax: 315-443-2012, E-mail: ealott@syr.edu.

Texas A&M University, College of Agriculture and Life Sciences, Department of Biochemistry and Biophysics, College Station, TX 77843. Offers biochemistry (MS, PhD); biophysics (MS). *Faculty:* 16 full-time (3 women), 3 part-time/adjunct (0 women). *Students:* 116 full-time (59 women), 10 part-time (4 women); includes 15 minority (2 African Americans, 6 Asian Americans or Pacific Islanders, 7 Hispanic Americans), 64 international. Average age 27. 133 applicants, 42% accepted, 25 enrolled. In 2005, 3 master's, 13 doctorates awarded. *Entrance requirements:* For master's and doctorate, GRE General Test. Additional exam requirements/recommendations for international students: Required—TOEFL. *Application deadline:* For fall admission, 2/1 priority date for domestic students, 12/1 priority date for international students. Applications are processed on a rolling basis. Application fee: $50 ($75 for international students). Electronic applications accepted. *Expenses:* Tuition, state resident: full-time $4,488; part-time $187 per credit hour. Tuition, nonresident: full-time $11,112; part-time $463 per credit hour. Required fees: $1,974. *Financial support:* In 2005–06, 6 fellowships with tuition reimbursements (averaging $20,000 per year), 70 research assistantships with partial tuition reimbursements (averaging $20,000 per year) were awarded; teaching assistantships with partial tuition reimbursements, institutionally sponsored loans, scholarships/grants, traineeships, and unspecified assistantships also available. Financial award application deadline: 2/1; financial award applicants required to submit FAFSA. *Faculty research:* Enzymology, gene expression, protein structure, plant biochemistry. *Unit head:* Dr. Gregory D. Reinhart, Department Head, 979-845-5032, Fax: 979-845-9274, E-mail: gdr@tamu.edu. *Application contact:* Pat Swigert, Academic Advisor, 979-845-1779, Fax: 979-845-9274, E-mail: pat@tamu.edu.

Université de Montréal, Faculty of Medicine and Faculty of Graduate Studies, Graduate Programs in Medicine, Department of Physiology, Montréal, QC H3C 3J7, Canada. Offers biophysics and molecular physiology (M Sc, PhD); neurological sciences (M Sc, PhD); physiology (M Sc, PhD). *Faculty:* 27 full-time (3 women), 8 part-time/adjunct (1 woman). *Students:* 97 full-time (56 women), 1 (woman) part-time. 23 applicants, 30% accepted, 7 enrolled. In 2005, 8 master's, 1 doctorate awarded. Terminal master's awarded for partial completion of doctoral program. *Degree requirements:* For master's, thesis; for doctorate, thesis/dissertation, general exam. *Entrance requirements:* For master's and doctorate, proficiency in French, knowledge of English. *Application deadline:* For fall and spring admission, 2/1. For winter admission, 11/1 for domestic students. Application fee: $30. Electronic applications accepted. *Financial support:* Fellowships, research assistantships available. *Faculty research:* Cardiovascular, neuropeptides, membrane transport and biophysics, signaling pathways. *Unit head:* Allan Smith, Interim Director, 514-343-6347, Fax: 514-343-7072. *Application contact:* Réjean Couture, Graduate Chairman, 514-343-7060, Fax: 514-343-2111.

Université de Sherbrooke, Faculty of Medicine and Health Sciences, Graduate Programs in Medicine, Department of Physiology and Biophysics, Sherbrooke, QC J1K 2R1, Canada. Offers M Sc, PhD. Part-time programs available. *Faculty:* 8 full-time (2 women). *Students:* 20 full-time (11 women), 11 part-time (6 women). Average age 25. 8 applicants, 38% accepted, 2 enrolled. In 2005, 4 degrees awarded. Terminal master's awarded for partial completion of doctoral program. *Degree requirements:* For master's and doctorate, thesis/dissertation. *Application deadline:* For fall admission, 6/30 for domestic students. For winter admission, 10/31 for domestic students; for spring admission, 2/28 for domestic students. Application fee: $50. Electronic applications accepted. *Financial support:* Fellowships available. *Faculty research:* Electrophysiology, physiopathology of bladder, electrocardiography, ionic channels of VSM, cardiac-sarcoplasmic reticulum. *Unit head:* Dr. Nuria Basora, Director, 819-564-5307, E-mail: nuria.basora@usherbrooke.ca.

Université du Québec à Trois-Rivières, Graduate Programs, Program in Biophysics and Cellular Biology, Trois-Rivières, QC G9A 5H7, Canada. Offers M Sc, PhD. Part-time programs available. *Degree requirements:* For master's and doctorate, thesis/dissertation. *Entrance requirements:* For master's, appropriate bachelor's degree, proficiency in French; for doctorate, appropriate master's degree, proficiency in French.

The University of Alabama at Birmingham, Graduate Programs in Joint Health Sciences, Department of Physiology and Biophysics, Birmingham, AL 35294. Offers basic medical sciences (MSBMS); biophysical sciences (PhD); integrative biomedical sciences (PhD). *Students:* 18 full-time (8 women); includes 6 minority (3 African Americans, 1 American Indian/Alaska Native, 2 Asian Americans or Pacific Islanders), 5 international. 34 applicants, 82% accepted. In 2005, 9 degrees awarded. *Entrance requirements:* For doctorate, GRE General Test, interview, minimum GPA of 3.0. Additional exam requirements/recommendations for international students: Required—TOEFL. *Application deadline:* For fall admission, 3/1 for domestic students. Applications are processed on a rolling basis. Application fee: $35 ($60 for international students). Electronic applications accepted. *Expenses:* Contact institution. Tuition and fees vary according to course load, degree level and program. *Financial support:* Fellowships available. *Faculty research:* Standard physiology (neurological, endocrine, cardiovascular, respiratory, and renal), cell and membrane biology. *Unit head:* Dr. Dale J. Benos, Chair, 205-934-6220, Fax: 205-934-2377. *Application contact:* Director of Graduate Studies, 205-934-3969, Fax: 205-975-9028.

University of Arkansas for Medical Sciences, College of Medicine and Graduate School, Graduate Programs in Medicine, Department of Physiology and Biophysics, Little Rock, AR 72205-7199. Offers MS, PhD, MD/PhD. *Faculty:* 16 full-time (2 women), 23 part-time/adjunct (3 women). *Students:* 27 full-time, 6 part-time. *Degree requirements:* For master's and doctorate, thesis/dissertation. *Entrance requirements:* For master's and doctorate, GRE General Test. Additional exam requirements/recommendations for international students: Required—TOEFL. Application fee: $0. *Financial support:* Research assistantships available. Support available to part-time students. *Unit head:* Dr. Michael L. Jennings, Chairman, 501-686-5123. *Application contact:* Dr. Patricia Wight, Graduate Coordinator, 501-686-5366, E-mail: wightpatricia@uams.edu.

University of California, Berkeley, Graduate Division, Group in Biophysics, Berkeley, CA 94720-1500. Offers PhD. *Degree requirements:* For doctorate, thesis/dissertation, qualifying exam. *Entrance requirements:* For doctorate, GRE General Test, GRE Subject Test, minimum GPA of 3.0.

University of California, Davis, Graduate Studies, Graduate Group in Biophysics, Davis, CA 95616. Offers MS, PhD. *Faculty:* 47 full-time. *Students:* 20 full-time (6 women); includes 3 minority (1 Asian American or Pacific Islander, 2 Hispanic Americans), 6 international. Average age 30. 35 applicants, 34% accepted, 3 enrolled. *Median time to degree:* Of those who began their doctoral program in fall 1997, 60% received their degree in 8 years or less. *Degree requirements:* For doctorate, thesis/dissertation. *Entrance requirements:* For master's and doctorate, GRE General Test, GRE Subject Test. Additional exam requirements/recommendations for international students: Required—TOEFL (minimum score 550 paper-based; 213 computer-based). *Application deadline:* For fall admission, 1/15 priority date for domestic students, 1/15 priority date for international students. Application fee: $60. Electronic applications accepted. *Financial support:* In 2005–06, 17 students received support, including 3 fellowships with full and partial tuition reimbursements available (averaging $15,939 per year), 9 research assistantships with full and partial tuition reimbursements available (averaging $15,563 per year), 2 teaching assistantships with partial tuition reimbursements available (averaging $15,082 per year); Federal Work-Study, institutionally sponsored loans, scholarships/grants, tuition waivers (full and partial), and unspecified assistantships also available. Financial award application deadline: 1/15; financial award applicants required to submit FAFSA. *Faculty research:* Molecular structure, protein structure/function relationships, spectroscopy. *Unit head:* Marjorie Longo, Chair, 530-752-6348, Fax: 530-752-8822, E-mail: mllongo@ucdavis.edu. *Application contact:* Ellen Picht, Administrative Assistant, 530-752-4863, E-mail: empicht@ucdavis.edu.

University of California, Irvine, College of Medicine and School of Biological Sciences, Department of Physiology and Biophysics, Irvine, CA 92697. Offers biological sciences (PhD). Students apply through the Graduate Program in Molecular Biology, Genetics, and Biochemistry. *Degree requirements:* For doctorate, thesis/dissertation. *Entrance requirements:* For doctorate, GRE General Test, GRE Subject Test, minimum GPA of 3.0. Additional exam requirements/recommendations for international students: Required—TOEFL (minimum score 550 paper-based; 213 computer-based), TSE. Electronic applications accepted. *Faculty research:* Membrane physiology, exercise physiology, regulation of hormone biosynthesis and action, endocrinology, ion channels and signal transduction.

University of California, San Diego, Graduate Studies and Research, Department of Physics, La Jolla, CA 92093. Offers biophysics (MS, PhD); physics (MS, PhD); physics/materials physics (MS). *Degree requirements:* For doctorate, thesis/dissertation. *Entrance requirements:* For master's and doctorate, GRE General Test, GRE Subject Test. Additional exam requirements/recommendations for international students: Required—TOEFL. Electronic applications accepted.

University of California, San Francisco, School of Pharmacy and School of Medicine, Graduate Group in Biophysics, San Francisco, CA 94143. Offers PhD. *Faculty:* 48 full-time (8 women), 2 part-time/adjunct (0 women). *Students:* 61 full-time (12 women); includes 9 minority (7 Asian Americans or Pacific Islanders, 2 Hispanic Americans). Average age 25. 60 applicants, 33% accepted, 9 enrolled. In 2005, 6 degrees awarded. *Median time to degree:* Of those who began their doctoral program in fall 1997, 100% received their degree in 8 years or less. *Degree requirements:* For doctorate, thesis/dissertation. *Entrance requirements:* For doctorate, GRE General Test, GRE Subject Test. Additional exam requirements/recommendations for international students: Required—TOEFL. *Application deadline:* For fall admission, 12/5 for domestic students. Application fee: $60. Electronic applications accepted. *Financial support:* In 2005–06, fellowships with full tuition reimbursements (averaging $25,000 per year), research assistantships with full tuition reimbursements (averaging $25,000 per year), teaching assistantships with full tuition reimbursements (averaging $25,000 per year) were awarded; traineeships, health care benefits, tuition waivers (full), unspecified assistantships, and stipends also available. *Faculty research:* Structural and computational biology; proteomic, genomic, and cell biology; chemistry. *Unit head:* Dr. David Agard, Chairman, 415-476-2521. *Application contact:* Julie Ransom, Program Administrator, 415-476-6671, Fax: 415-476-1902, E-mail: ransom@cgl.ucsf.edu.

University of California, Santa Barbara, Graduate Division, College of Letters and Sciences, Division of Mathematics, Life, and Physical Sciences, Department of Biomolecular Science and Engineering, Santa Barbara, CA 93106. Offers biochemistry and molecular biology (MS, PhD); biomolecular science and engineering (PhD); biophysics and bioengineering (MS, PhD). *Faculty:* 39 full-time (4 women). *Students:* 25 full-time (11 women); includes 4 minority (all Asian Americans or Pacific Islanders), 1 international. Average age 25. 70 applicants, 33% accepted, 4 enrolled. In 2005, 5 degrees awarded. Terminal master's awarded for partial completion of doctoral program. *Median time to degree:* Of those who began their doctoral program in fall 1997, 100% received their degree in 8 years or less. *Degree requirements:* For master's, thesis optional; for doctorate, thesis/dissertation, lab rotations, seminars, course-work, TA ships, research units, comprehensive exam, registration. *Entrance requirements:* For master's and doctorate, GRE General Test, GRE Subject Test. Additional exam requirements/recommendations for international students: Required—TOEFL (minimum score 630 paper-based; 267 computer-based). *Application deadline:* For fall admission, 12/15 for domestic students, 12/15 for international students. Application fee: $60. Electronic applications accepted. *Financial support:* In 2005–06, 3 fellowships with full and partial tuition reimbursements (averaging $21,000 per year), 15 research assistantships with full and partial tuition reimbursements (averaging $21,000 per year), 8 teaching assistantships with full and partial tuition reimbursements (averaging $21,000 per year) were awarded; career-related internships or fieldwork, Federal Work-Study, institutionally sponsored loans, scholarships/grants, traineeships, health care benefits, tuition waivers (full and partial), and unspecified assistantships also available. Financial award application deadline: 12/15; financial award applicants required to submit FAFSA. *Faculty research:* Genetics and biochemistry of bacterial gene expression; structure-function relationships in proteins and nucleic acids, protein chemistry; biochemistry and biophysics of marine adhesion. *Unit head:* Prof. Philip A. Pincus, Chair, 805-893-4685, E-mail: fyl@mrl.ucsb.edu. *Application contact:* Krista Grace, Staff Graduate Program Advisor, 805-893-2290, Fax: 805-893-4724, E-mail: grace@lifesci.ucsb.edu.

University of Cincinnati, Division of Research and Advanced Studies, College of Medicine, Graduate Programs in Biomedical Sciences, Department of Pharmacology and Cell Biophysics, Cincinnati, OH 45221. Offers cell biophysics (PhD); pharmacology (PhD). *Degree requirements:* For doctorate, thesis/dissertation, qualifying exam. *Entrance requirements:* For doctorate, GRE General Test. Additional exam requirements/recommendations for international students: Required—TOEFL. Electronic applications accepted. *Faculty research:* Lipoprotein research, enzyme regulation, electrophysiology, gene actuation.

University of Colorado at Denver and Health Sciences Center, Graduate School, Program in Biomedical Sciences, Department of Physiology and Biophysics, Denver, CO 80262. Offers PhD. In 2005, 1 degree awarded. *Degree requirements:* For doctorate, thesis/dissertation, comprehensive exam. *Entrance requirements:* For doctorate, GRE General Test, 4 letters of recommendation. Additional exam requirements/recommendations for international students: Required—TOEFL (minimum score 550 paper-based; 213 computer-based). *Application deadline:* For fall admission, 1/15 for domestic students. Application fee: $50. *Expenses:* Tuition, state resident: full-time $11,730. Tuition, nonresident: full-time $22,980. Tuition and fees vary according to degree level and program. *Financial support:* Fellowships, research assistantships, teaching assistantships, Federal Work-Study and institutionally sponsored loans available. Support available to part-time students. Financial award application deadline: 3/15; financial award applicants required to submit FAFSA. *Faculty research:* Nicotonic receptors, immunity, molecular structure, function and regulation of ion channels, calcium influx in the function of cytotoxic T lymphocytes. *Unit head:* Dr. William Betz, Chair, 303-315-8946, E-mail: bill.betz@uchsc.edu. *Application contact:* Kathy Fernandez, Information Contact, 303-315-7831, E-mail: kathy.fernandez@uchsc.edu.

University of Connecticut, Graduate School, College of Liberal Arts and Sciences, Department of Molecular and Cell Biology, Field of Biophysics and Structural Biology, Storrs, CT 06269. Offers MS, PhD. *Faculty:* 10 full-time (5 women). *Students:* 2 full-time (both women); includes 1 minority (African American), 1 international. Average age 25. 15 applicants, 13% accepted, 2 enrolled. In 2005, 1 degree awarded. Terminal master's awarded for partial completion of doctoral program. *Degree requirements:* For master's, comprehensive exam; for doctorate, thesis/dissertation. *Entrance requirements:* For master's and doctorate, GRE General Test, GRE Subject Test. Additional exam requirements/recommendations for international students: Required—TOEFL (minimum score 550 paper-based; 213 computer-based). *Application deadline:* For fall admission, 2/1 priority date for domestic students, 2/1 priority date for international students; for spring admission, 11/1 for domestic students, 10/1 for international students. Applications are processed on a rolling basis. Application fee: $55. Electronic applications accepted. *Expenses:* Tuition, state resident: part-time $444 per credit hour. Tuition, nonresident: part-time $1,154 per credit hour. Tuition and fees vary according to course load. *Financial support:* In 2005–06, 2 research assistantships with full tuition reimbursements were awarded; fellowships, teaching assistantships with full tuition reimbursements, Federal Work-Study, scholarships/grants, health care benefits, and unspecified assistantships also available. Financial award application deadline: 2/1; financial award applicants required to submit FAFSA. *Application contact:* Anne St. Onje, Graduate Coordinator, 860-486-4314, Fax: 860-486-3943, E-mail: ann.st_onje@uconn.edu.

See Close-Up on page 611.

University of Guelph, Graduate Program Services, Biophysics Interdepartmental Group, Guelph, ON N1G 2W1, Canada. Offers M Sc, PhD. *Faculty:* 17 full-time (2 women). *Students:* 26 full-time (10 women). 13 applicants, 54% accepted, 7 enrolled. In 2005, 2 master's, 2

Biophysics

University of Guelph (continued)

doctorates awarded. *Median time to degree:* Of those who began their doctoral program in fall 1997, 100% received their degree in 8 years or less. *Degree requirements:* For master's, thesis/dissertation; for doctorate, thesis/dissertation, comprehensive exam. *Entrance requirements:* For master's, minimum B average during previous 2 years of course work; for doctorate, minimum B+ average. Additional exam requirements/recommendations for international students: Required—TOEFL (minimum score 550 paper-based; 213 computer-based). *Application deadline:* For fall admission, 8/1 priority date for domestic students, 5/1 priority date for international students. For winter admission, 12/1 for domestic students; for spring admission, 4/1 for domestic students. Applications are processed on a rolling basis. Application fee: $75. Electronic applications accepted. *Financial support:* In 2005–06, research assistantships (averaging $14,400 per year), teaching assistantships (averaging $4,600 per year) were awarded. *Faculty research:* Molecular, cellular, structural, and computational biophysics. *Unit head:* Dr. Frances J. Sharom, Director, 519-824-4120 Ext. 52247, Fax: 519-837-1802, E-mail: fsharom@uoguelph.ca.

University of Guelph, Graduate Program Services, College of Biological Science, Department of Molecular and Cellular Biology, Guelph, ON N1G 2W1, Canada. Offers biochemistry (M Sc, PhD); biophysics (M Sc, PhD); botany (M Sc, PhD); microbiology (M Sc, PhD); molecular biology and genetics (M Sc, PhD). *Faculty:* 43 full-time (6 women). *Students:* 136 full-time (52 women). Average age 26. 56 applicants, 30% accepted, 17 enrolled. In 2005, 4 master's, 2 doctorates awarded. *Degree requirements:* For master's, thesis/dissertation, research proposal; for doctorate, thesis/dissertation, research proposal, comprehensive exam, registration. *Entrance requirements:* Additional exam requirements/recommendations for international students: Required—TOEFL (minimum score 550 paper-based; 213 computer-based), IELT (minimum score 7). *Application deadline:* For fall admission, 7/1 for domestic students, 1/1 for international students. For winter admission, 11/1 for domestic students; for spring admission, 3/1 for domestic students. Applications are processed on a rolling basis. Application fee: $75. Electronic applications accepted. *Financial support:* Fellowships, research assistantships, teaching assistantships available. Support available to part-time students. *Faculty research:* Physiology, structure, genetics, and ecology of microbes; virology and microbial technology. *Unit head:* Dr. Chris Whitfield, Chair, 519-824-4120 Ext. 53361, Fax: 519-827-1802, E-mail: cwhitfie@uoguelph.ca. *Application contact:* Laurie Winn, Graduate Admissions Secretary, 519-824-4320 Ext. 52730, Fax: 519-767-1656, E-mail: lwinn@uoguelph.ca.

University of Illinois at Chicago, College of Medicine and Graduate College, Graduate Programs in Medicine, Department of Physiology and Biophysics, Chicago, IL 60607-7128. Offers MS, PhD. Terminal master's awarded for partial completion of doctoral program. *Degree requirements:* For master's and doctorate, thesis/dissertation. *Entrance requirements:* For master's and doctorate, GRE General Test. Additional exam requirements/recommendations for international students: Required—TOEFL. Electronic applications accepted. *Faculty research:* Neuroscience, endocrinology and reproduction, cell physiology, exercise physiology, NMR.

University of Illinois at Urbana–Champaign, Graduate College, College of Liberal Arts and Sciences, School of Molecular and Cellular Biology, Center for Biophysics and Computational Biology, Champaign, IL 61820. Offers PhD. *Students:* 66 full-time (20 women), 2 part-time (1 woman); includes 9 minority (1 African American, 6 Asian Americans or Pacific Islanders, 2 Hispanic Americans), 41 international. 126 applicants, 10% accepted, 7 enrolled. In 2005, 7 degrees awarded. *Degree requirements:* For doctorate, thesis/dissertation. *Entrance requirements:* For doctorate, minimum GPA of 3.0. Additional exam requirements/recommendations for international students: Required—TOEFL. *Application deadline:* For fall admission, 5/15 for domestic students. Application fee: $50 ($60 for international students). Electronic applications accepted. *Financial support:* In 2005–06, 15 fellowships, 55 research assistantships, 17 teaching assistantships were awarded; scholarships/grants and traineeships also available. Financial award application deadline: 1/15. *Unit head:* Martin Gruebele, Director, 217-333-1630, Fax: 217-244-6615, E-mail: gruebele@uiuc.edu. *Application contact:* Cynthia Dodds, Secretary, 217-333-1630, Fax: 217-244-6615, E-mail: biophysics@life.uiuc.edu.

Announcement: The Center for Biophysics and Computational Biology provides a broad selection of interdisciplinary research training opportunities. Students in the program apply physical, experimental, and computational methods to biological problems at all levels of complexity from molecular to organismal. A detailed list of research projects is available at http://www.life.uiuc.edu/biophysics.

The University of Iowa, Roy J. and Lucille A. Carver College of Medicine and Graduate College, Graduate Programs in Medicine, Department of Physiology and Biophysics, Iowa City, IA 52242-1316. Offers physiology and biophysics (PhD); physiology and biophysiology (MS). *Faculty:* 15 full-time (3 women), 12 part-time/adjunct (2 women). *Students:* 20 full-time (10 women); includes 1 minority (Asian American or Pacific Islander), 7 international. Average age 25. 17 applicants, 24% accepted, 4 enrolled. Terminal master's awarded for partial completion of doctoral program. *Degree requirements:* For master's and doctorate, thesis/dissertation, teaching experience, comprehensive exam. *Entrance requirements:* For master's and doctorate, GRE General Test, minimum GPA of 3.0. *Application deadline:* For fall admission, 4/1 for domestic students, 3/1 for international students; for spring admission, 10/1 for domestic students, 9/1 for international students. Applications are processed on a rolling basis. Application fee: $60 ($80 for international students). *Expenses:* Tuition, state resident: part-time $1,882 per term. Tuition, nonresident: full-time $17,338; part-time $4,907 per term. Tuition and fees vary according to course load and program. *Financial support:* In 2005–06, 1 fellowship with full tuition reimbursement (averaging $22,000 per year), 19 research assistantships with full tuition reimbursements (averaging $22,000 per year) were awarded; traineeships also available. Financial award application deadline: 4/1. *Faculty research:* Cellular and molecular endocrinology, membrane structure and function, cardiac cell electrophysiology, regulation of gene expression, neurophysiology. *Unit head:* Kevin P. Campbell, Interim Head, 319-335-7800, Fax: 319-335-7330, E-mail: kevin-campbell@uiowa.edu. *Application contact:* Dr. Thomas Schmidt, Chairman of Graduate Admissions, 319-335-7847, Fax: 319-335-7330, E-mail: thomas-schmidt@uiowa.edu.

See Close-Up on page 1265.

University of Kansas, Graduate School, College of Liberal Arts and Sciences, Division of Biological Sciences, Department of Molecular Biosciences, Lawrence, KS 66045. Offers biochemistry and biophysics (MA, PhD); microbiology (MA, PhD); molecular, cellular, and developmental biology (MA, PhD). *Faculty:* 44. *Students:* 48 full-time (28 women), 5 part-time (3 women); includes 1 minority (Asian American or Pacific Islander), 26 international. Average age 27. 111 applicants, 19% accepted. In 2005, 5 master's, 5 doctorates awarded. Terminal master's awarded for partial completion of doctoral program. *Degree requirements:* For master's and doctorate, thesis/dissertation, comprehensive exam. *Entrance requirements:* For master's and doctorate, GRE General Test. Additional exam requirements/recommendations for international students: Required—TOEFL, IELT, TOEFL (paper 570; computer 230) or IELT; Recommended—TWE, TSE. *Application deadline:* For fall admission, 1/15 priority date for domestic students, 1/15 priority date for international students. Applications are processed on a rolling basis. Application fee: $55 ($60 for international students). Electronic applications accepted. *Expenses:* Tuition, state resident: full-time $4,859. Tuition, nonresident: full-time $12,000. Required fees: $589. Tuition and fees vary according to program. *Financial support:* Fellowships, research assistantships with tuition reimbursements, teaching assistantships with tuition reimbursements available. Financial award application deadline: 3/1. *Faculty research:* Structure and function of proteins, genetics of organism development, molecular genetics, neurophysiology, molecular virology and pathogenics, developmental biology, cell biology. *Unit head:* Kathy Suprenant, Chair, 785-864-4631, Fax: 785-864-5294, E-mail: ksupre@ku.edu. *Application contact:* John P. Connolly, Information Contact, 785-864-4311, Fax: 785-864-5924, E-mail: jconnolly@ku.edu.

University of Louisville, School of Medicine, Department of Physiology and Biophysics, Louisville, KY 40292-0001. Offers MS, PhD. *Students:* 25 full-time (13 women), 10 part-time (3 women); includes 11 minority (7 African Americans, 3 Asian Americans or Pacific Islanders, 1 Hispanic American), 8 international. Average age 27. In 2005, 9 master's, 3 doctorates awarded. *Degree requirements:* For master's and doctorate, thesis/dissertation. *Entrance requirements:* For master's, GRE General Test, 36 hours of graduate course work; for doctorate, GRE General Test, minimum GPA of 3.0. *Application deadline:* For fall admission, 1/15 for domestic students. Applications are processed on a rolling basis. Application fee: $50. Electronic applications accepted. *Expenses:* Tuition, state resident: full-time $6,006; part-time $334 per credit hour. Tuition, nonresident: full-time $16,554; part-time $920 per credit hour. Tuition and fees vary according to course load, degree level and program. *Financial support:* Fellowships with full tuition reimbursements available. *Unit head:* Dr. Irving G. Joshua, Chair, 502-852-5371, Fax: 502-852-6239, E-mail: igjosh01@gwise.louisville.edu.

University of Miami, Graduate School, Miller School of Medicine, Graduate Programs in Medicine, Department of Physiology and Biophysics, Coral Gables, FL 33124. Offers PhD, MD/PhD. *Faculty:* 14 full-time (2 women). *Students:* 15 full-time (8 women); includes 3 minority (1 African American, 2 Hispanic Americans), 9 international. Average age 27. 36 applicants, 8% accepted, 3 enrolled. In 2005, 6 degrees awarded. *Degree requirements:* For doctorate, thesis/dissertation, qualifying exam. *Entrance requirements:* For doctorate, GRE General Test, minimum GPA of 3.0 in sciences. Additional exam requirements/recommendations for international students: Required—TOEFL. *Application deadline:* Applications are processed on a rolling basis. Application fee: $50. Electronic applications accepted. *Financial support:* In 2005–06, fellowships with full tuition reimbursements (averaging $22,000 per year); research assistantships, career-related internships or fieldwork and institutionally sponsored loans also available. *Faculty research:* Cell and membrane physiology, cell-to-cell communication, molecular neurobiology, neuroimmunology, neural development. *Unit head:* Dr. Karl Magleby, Chairman, 305-243-6821, Fax: 305-243-6898, E-mail: kmagleby@miami.edu. *Application contact:* Dr. David M. Landowne, Director of Graduate Studies, 305-243-6821, Fax: 305-243-5931, E-mail: dl@miami.edu.

University of Michigan, Horace H. Rackham School of Graduate Studies, Interdepartmental Program in Biophysics, Ann Arbor, MI 48109. Offers PhD. *Faculty:* 37 full-time (7 women). *Students:* 19 full-time (7 women); includes 1 minority (African American), 7 international. Average age 22. 30 applicants, 17% accepted, 4 enrolled. In 2005, 2 degrees awarded. *Degree requirements:* For doctorate, thesis/dissertation, oral defense of dissertation, preliminary exam. *Entrance requirements:* For doctorate, GRE General Test, GRE Subject Test. Additional exam requirements/recommendations for international students: Required—TOEFL. *Application deadline:* For fall admission, 1/10 for domestic students, 1/10 for international students. Application fee: $55. Electronic applications accepted. *Expenses:* Tuition, state resident: full-time $14,082; part-time $894 per credit hour. Tuition, nonresident: full-time $28,500; part-time $1,675 per credit hour. Required fees: $189; $189 per unit. *Financial support:* In 2005–06, 14 fellowships with full tuition reimbursements (averaging $20,000 per year), 6 research assistantships with full tuition reimbursements (averaging $20,000 per year), teaching assistantships with full tuition reimbursements (averaging $20,000 per year) were awarded; scholarships/grants, traineeships, health care benefits, and unspecified assistantships also available. Financial award application deadline: 3/15. *Faculty research:* Structural biology, computational biophysics, physical chemistry, cellular biophysics. *Unit head:* Dr. James E. Penner-Hahn, Chair, 734-764-5257, Fax: 734-764-3323, E-mail: jeph@umich.edu. *Application contact:* Teresa A Clayton, Graduate Coordinator, 734-763-6722, Fax: 734-764-3323, E-mail: tereclay@umich.edu.

University of Minnesota, Duluth, Medical School, Department of Biochemistry, Molecular Biology and Biophysics, Duluth, MN 55812-2496. Offers MS, PhD. *Faculty:* 8 full-time (1 woman), 2 part-time/adjunct (1 woman). *Students:* 5 full-time (3 women). Average age 26. 5 applicants, 40% accepted, 2 enrolled. Terminal master's awarded for partial completion of doctoral program. *Degree requirements:* For master's and doctorate, thesis/dissertation. *Entrance requirements:* For master's and doctorate, GRE General Test. Additional exam requirements/recommendations for international students: Required—TOEFL. *Application deadline:* For fall admission, 1/2 priority date for domestic students, 1/2 priority date for international students. For winter admission, 3/15 for domestic students. Application fee: $55 ($75 for international students). *Financial support:* In 2005–06, 4 students received support, including fellowships with full tuition reimbursements available (averaging $22,500 per year), research assistantships with full tuition reimbursements available (averaging $18,000 per year), teaching assistantships with tuition reimbursements available (averaging $18,000 per year); Federal Work-Study, institutionally sponsored loans, health care benefits, and unspecified assistantships also available. Financial award application deadline: 4/1. *Faculty research:* Intestinal cancer biology; hepatotoxins and mitochondriopathies; cell cycle regulation in stem cells; neurobiology of brain development, trace metal function and blood-brain barrier; hibernation biology. Total annual research expenditures: $1.5 million. *Unit head:* Dr. Lester R. Drewes, Professor/Head, 218-726-7925, Fax: 218-726-8014, E-mail: ldrewes@d.umn.edu. *Application contact:* Stacy Johnson, Program Associate, 218-726-6354, E-mail: ahcd@d.umn.edu.

University of Minnesota, Twin Cities Campus, Graduate School, Program in Biophysical Sciences and Medical Physics, Minneapolis, MN 55455-0213. Offers MS, PhD. Part-time programs available. *Degree requirements:* For master's, research paper, oral exam, thesis optional; for doctorate, thesis/dissertation, oral/written preliminary exam, oral final exam. *Expenses:* Tuition, state resident: full-time $8,748; part-time $729 per credit. Tuition, nonresident: full-time $15,848; part-time $1,321 per credit. Full-time tuition and fees vary according to class time, course load, program and reciprocity agreements. *Faculty research:* Theoretical biophysics, radiological physics, cellular and molecular biophysics.

University of Minnesota, Twin Cities Campus, Medical School and Graduate School, Graduate Programs in Medicine, Department of Biochemistry, Molecular Biology and Biophysics, Minneapolis, MN 55455-0213. Offers PhD. *Degree requirements:* For doctorate, thesis/dissertation. *Entrance requirements:* For doctorate, GRE General Test. Electronic applications accepted. *Expenses:* Contact institution. Full-time tuition and fees vary according to class time, course load, program and reciprocity agreements. *Faculty research:* Physical biochemistry, enzymology, physiological chemistry.

University of Mississippi Medical Center, School of Graduate Studies in the Health Sciences, Department of Physiology and Biophysics, Jackson, MS 39216-4505. Offers MS, PhD, MD/PhD. *Faculty:* 24 full-time (8 women). *Students:* 5 full-time (3 women); includes 3 minority (2 African Americans, 1 Asian American or Pacific Islander). Average age 25. 9 applicants, 22% accepted, 2 enrolled. *Degree requirements:* For master's, thesis; for doctorate, thesis/dissertation, first authored publication. *Entrance requirements:* For master's and doctorate, GRE General Test, minimum GPA of 3.0. *Application deadline:* For fall admission, 8/1 for domestic students. Applications are processed on a rolling basis. Application fee: $10. *Financial support:* In 2005–06, 5 students received support, including 5 research assistantships (averaging $16,557 per year) Financial award application deadline: 4/1. *Faculty research:* Cardiovascular, renal, endocrine, and cellular neurophysiology; molecular physiology. Total annual research expenditures: $2.8 million. *Unit head:* Dr. John E. Hall, Chairman, 601-984-1801, Fax: 601-984-1817. *Application contact:* Dr. Thomas E. Lohmeier, Director, 601-984-1820, Fax: 601-984-1817, E-mail: tlohmeier@physiology.umsmed.edu.

University of Missouri–Kansas City, School of Biological Sciences, Program in Cell Biology and Biophysics, Kansas City, MO 64110-2499. Offers PhD offered through the School of Graduate Studies. *Degree requirements:* For doctorate, thesis/dissertation, comprehensive exam. *Entrance requirements:* For doctorate, GRE General Test, bachelor's degree in chemistry, biology or related field; minimum GPA of 3.0. Additional exam requirements/recommendations for international students: Required—TOEFL. *Application deadline:* For fall admission, 3/1 for domestic students. Applications are processed on a rolling basis. Application fee: $35 ($50 for international students). Electronic applications accepted. *Expenses:* Tuition, state resident: full-time $4,738; part-time $263 per credit hour. Tuition, nonresident: full-time $12,235; part-time $679 per credit hour. Required fees: $582. Tuition and fees vary according to course load, program and student level. *Financial support:* Fellowships with full tuition reimbursements, research assistantships with full tuition reimbursements, teaching assistantships with full and partial tuition reimbursements, scholarships/grants, tuition waivers (full and partial), and unspecified assistantships available. *Unit head:* Dr. George Thomas, Head, 816-235-5247, E-mail: thomasgj@umkc.edu. *Application contact:* Laura Batenic, Information Contact, 816-235-2352, Fax: 816-235-5158, E-mail: batenicl@umkc.edu.

See Close-Up on page 627.

University of New Mexico, Graduate School, College of Arts and Sciences, Department of Physics and Astronomy, Albuquerque, NM 87131-2039. Offers biomedical physics (MS, PhD); optical science and engineering (MS); optical sciences and engineering (PhD); physics (MS, PhD). *Faculty:* 35 full-time (4 women), 11 part-time/adjunct (1 woman). *Students:* 89 full-time (16 women), 15 part-time (2 women); includes 4 minority (2 Asian Americans or Pacific Islanders, 2 Hispanic Americans), 50 international. Average age 29. 99 applicants, 37% accepted, 18 enrolled. In 2005, 12 master's, 9 doctorates awarded. *Degree requirements:* For master's, thesis (for some programs), comprehensive exam (for some programs); for doctorate, thesis/dissertation, comprehensive exam. *Entrance requirements:* Additional exam requirements/recommendations for international students: Required—TOEFL (minimum score 610 paper-based; 253 computer-based). *Application deadline:* For fall admission, 2/1 for domestic students; for spring admission, 8/1 for domestic students. Application fee: $40. Electronic applications accepted. *Expenses:* Tuition, nonresident: full-time $3,388; part-time $238 per credit hour. Required fees: $385 per term. Tuition and fees vary according to course load and program. *Financial support:* In 2005–06, 15 students received support, including research assistantships with full tuition reimbursements available (averaging $20,000 per year), teaching assistantships with full tuition reimbursements available (averaging $13,500 per year); fellowships with full tuition reimbursements available, career-related internships or fieldwork, scholarships/grants, health care benefits, and unspecified assistantships also available. Financial award application deadline: 2/1; financial award applicants required to submit FAFSA. *Faculty research:* High-energy and particle physics, optical and laser sciences, condensed matter, nuclear and particle physics, surface physics, biomedical physics, quantum information, subatomic physics, nonlinear science, general relativity. Total annual research expenditures: $6.1 million. *Unit head:* Dr. Bernd Bassalleck, Chair, 505-277-1517, Fax: 505-277-1520, E-mail: bassek@unm.edu. *Application contact:* Mary De Witt, Program Advisement Coordinator, 505-277-1514, Fax: 505-277-1514, E-mail: mdewitt@unm.edu.

The University of North Carolina at Chapel Hill, School of Medicine and Graduate School, Graduate Programs in Medicine, Department of Biochemistry and Biophysics, Chapel Hill, NC 27599. Offers MS, PhD. *Faculty:* 42 full-time (8 women). *Students:* 74 full-time (27 women); includes 2 minority (both African Americans), 17 international. Average age 22. 131 applicants, 21% accepted, 17 enrolled. In 2005, 2 master's, 6 doctorates awarded. Terminal master's awarded for partial completion of doctoral program. *Median time to degree:* Of those who began their doctoral program in fall 1997, 100% received their degree in 8 years or less. *Degree requirements:* For master's and doctorate, thesis/dissertation, comprehensive exam. *Entrance requirements:* For master's and doctorate, GRE General Test, GRE Subject Test (recommended), minimum GPA of 3.0. Additional exam requirements/recommendations for international students: Required—TOEFL. *Application deadline:* For fall admission, 1/1 for domestic students, 1/1 for international students. Application fee: $70. Electronic applications accepted. *Financial support:* In 2005–06, fellowships with full tuition reimbursements (averaging $23,000 per year), research assistantships with full tuition reimbursements (averaging $23,000 per year), teaching assistantships with full tuition reimbursements (averaging $23,000 per year) were awarded. Total annual research expenditures: $6.4 million. *Unit head:* Dr. Henrik Dohlman, Chair, 919-962-8326, Fax: 919-966-2852, E-mail: hendrik-dohlman@med.unc.edu. *Application contact:* Diane M. Harris, Student Services Manager, 919-966-4683, Fax: 919-966-2852, E-mail: diane_harris@med.unc.edu.

University of Rochester, School of Medicine and Dentistry, Graduate Programs in Medicine and Dentistry, Department of Biochemistry and Biophysics, Program in Biophysics, Rochester, NY 14627-0250. Offers MS, PhD. Terminal master's awarded for partial completion of doctoral program. *Degree requirements:* For doctorate, thesis/dissertation, qualifying exam. *Entrance requirements:* For master's and doctorate, GRE General Test.

University of Southern California, Keck School of Medicine and Graduate School, Graduate Programs in Medicine, Department of Physiology and Biophysics, Los Angeles, CA 90089. Offers MS, PhD, MD/PhD. Part-time programs available. *Faculty:* 15 full-time (3 women). *Students:* 17 full-time (11 women); includes 6 minority (1 African American, 5 Asian Americans or Pacific Islanders), 5 international. Average age 29. 3 applicants, 100% accepted, 2 enrolled. In 2005, 1 master's, 1 doctorate awarded. Terminal master's awarded for partial completion of doctoral program. *Median time to degree:* Of those who began their doctoral program in fall 1997, 100% received their degree in 8 years or less. *Degree requirements:* For master's, thesis optional; for doctorate, thesis/dissertation, comprehensive exam, registration. *Entrance requirements:* For master's and doctorate, GRE General Test, minimum GPA of 3.0. *Application deadline:* For fall admission, 2/1 priority date for domestic students, 2/1 priority date for international students; for spring admission, 10/1 priority date for domestic students, 10/1 priority date for international students. Applications are processed on a rolling basis. Application fee: $65 ($75 for international students). Electronic applications accepted. *Expenses:* Tuition: Full-time $25,416; part-time $1,059 per unit. Required fees: $484; $484 per year. Tuition and fees vary according to course load and program. *Financial support:* In 2005–06, 16 students received support, including 1 fellowship with partial tuition reimbursement available (averaging $24,876 per year), 14 research assistantships with full tuition reimbursements available (averaging $24,876 per year); teaching assistantships with full tuition reimbursements available, scholarships/grants and traineeships also available. *Faculty research:* Endocrinology, metabolism, cell transport, molecular biology, mathematical modeling. Total annual research expenditures: $6.5 million. *Unit head:* Dr. Harvey Kaslow, Director, Graduate Studies, 323-442-1244, Fax: 323-442-2283, E-mail: hrkaslow@usc.edu. *Application contact:* Elena Camarena, Graduate Coordinator, 323-442-1039, Fax: 323-442-2283, E-mail: physiol@hsc.usc.edu.

University of South Florida, College of Medicine and College of Graduate Studies, Graduate Programs in Medical Sciences, Tampa, FL 33620-9951. Offers anatomy (PhD); biochemistry and molecular biology (MS, PhD), including biochemistry and molecular biology (PhD); bioinformatics and computational biology (MS); medical microbiology and immunology (PhD); pathology (PhD); pharmacology and therapeutics (PhD), including medical sciences; physiology and biophysics (PhD). *Students:* 108 full-time (53 women), 34 part-time (26 women); includes 31 minority (9 African Americans, 9 Asian Americans or Pacific Islanders, 13 Hispanic Americans), 30 international. 117 applicants, 99% accepted, 116 enrolled. In 2005, 9 master's, 4 doctorates awarded. *Degree requirements:* For doctorate, thesis/dissertation. *Entrance requirements:* For doctorate, GRE General Test, minimum GPA of 3.0. Application fee: $30. *Expenses:* Contact institution. *Financial support:* Institutionally sponsored loans and scholarships/grants available. Financial award application deadline: 4/1; financial award applicants required to submit FAFSA. *Unit head:* Dr. Joseph J. Krzanowski, Associate Dean for Research and Graduate Affairs, 813-974-4181, Fax: 813-974-4317, E-mail: jkrzanow@com1.med.usf.edu.

The University of Tennessee Health Science Center, College of Graduate Health Sciences, Department of Physiology, Memphis, TN 38163-0002. Offers MS, PhD. *Faculty:* 21 full-time (4 women), 2 part-time/adjunct (1 woman). *Students:* 3 full-time (2 women); all minorities (all Asian Americans or Pacific Islanders) Average age 25. 13 applicants, 23% accepted. *Degree requirements:* For master's, thesis, comprehensive exam; for doctorate, thesis/dissertation, oral and written preliminary and comprehensive exams. *Entrance requirements:* For master's, GRE General Test, minimum GPA of 3.0; for doctorate, GRE General Test, GRE Subject Test, minimum GPA of 3.0. Additional exam requirements/recommendations for international students: Required—TOEFL. *Application deadline:* For fall admission, 5/15 for domestic students. Application fee: $0. Electronic applications accepted. *Financial support:* Fellowships, research assistant-

ships, teaching assistantships, institutionally sponsored loans and tuition waivers (full) available. Financial award application deadline: 2/25. *Unit head:* Dr. Leonard R. Johnson, Chairman, 901-448-7088, Fax: 901-448-7126, E-mail: ljohnson@utmem.edu. *Application contact:* Ida W. Mosby, Director, Enrollment Services, 901-448-5560, E-mail: imosby@utmem.edu.

The University of Texas Medical Branch, Graduate School of Biomedical Sciences, Program in Biochemistry and Molecular Biology, Galveston, TX 77555. Offers biochemistry (PhD); bioinformatics (PhD); biophysics (PhD); cell biology (PhD); computational biology (PhD); structural biology (PhD). *Students:* 38 full-time (14 women), 1 part-time; includes 5 minority (3 Asian Americans or Pacific Islanders, 2 Hispanic Americans), 18 international. Average age 27. In 2005, 7 degrees awarded. *Degree requirements:* For doctorate, thesis/dissertation. *Entrance requirements:* Additional exam requirements/recommendations for international students: Required—TOEFL (minimum score 550 paper-based; 213 computer-based). *Application deadline:* Applications are processed on a rolling basis. Application fee: $30 ($75 for international students). Electronic applications accepted. *Expenses:* Tuition, state resident: full-time $8,350; part-time $90 per credit hour. Tuition, nonresident: full-time $21,450; part-time $366 per credit hour. Required fees: $1,027; $11 per credit hour. $60 per term. *Financial support:* In 2005–06, fellowships (averaging $23,000 per year), research assistantships (averaging $23,000 per year) were awarded. Financial award applicants required to submit FAFSA. *Unit head:* Dr. Leenian L. Chan, Director, 409-772-2861, Fax: 409-772-9679, E-mail: lchan@utmb.edu. *Application contact:* Debora Botting, Co-ordinator II Special Programs, 409-772-2769, Fax: 409-747-0552, E-mail: dmbottin@utmb.edu.

University of Toronto, School of Graduate Studies, Life Sciences Division, Department of Medical Biophysics, Toronto, ON M5S 1A1, Canada. Offers M Sc, PhD. *Degree requirements:* For master's and doctorate, thesis/dissertation. *Entrance requirements:* For master's, GRE General Test, GRE Subject Test (for international applicants), resumé, 2 letters of reference; for doctorate, GR General Test, GRE Subject Test (for international applicants), resumé, 2 letters of reference. Additional exam requirements/recommendations for international students: Required—TOEFL (minimum score 620 paper-based; 260 computer-based), TWE (minimum score 5).

University of Vermont, College of Medicine and Graduate College, Graduate Programs in Medicine, Department of Molecular Physiology and Biophysics, Burlington, VT 05405. Offers MS, PhD, MD/MS, MD/PhD. *Students:* 4 (2 women) 2 international. 16 applicants, 19% accepted, 2 enrolled. *Degree requirements:* For master's and doctorate, thesis/dissertation. *Entrance requirements:* For master's and doctorate, GRE General Test. Additional exam requirements/recommendations for international students: Required—TOEFL (minimum score 550 paper-based; 213 computer-based). *Application deadline:* For fall admission, 4/1 for domestic students. Applications are processed on a rolling basis. Application fee: $40. Electronic applications accepted. *Expenses:* Tuition, area resident: Part-time $410 per credit hour. Tuition, nonresident: part-time $1,034 per credit hour. *Financial support:* Fellowships, research assistantships, teaching assistantships available. Financial award application deadline: 3/1. *Unit head:* Dr. D. Warshaw, Chairperson, 802-656-2540. *Application contact:* Dr. C. Berger, Coordinator, 802-656-2540.

University of Virginia, School of Medicine, Department of Molecular Physiology and Biological Physics, Charlottesville, VA 22903. Offers biological and physical sciences (MS); molecular medicine and systems biology (PhD); physiology (PhD). *Students:* 32 full-time (13 women); includes 2 minority (both Asian Americans or Pacific Islanders), 6 international. Average age 28. In 2005, 21 master's, 5 doctorates awarded. *Degree requirements:* For doctorate, thesis/dissertation. *Entrance requirements:* For doctorate, GRE General Test, GRE Subject Test. Additional exam requirements/recommendations for international students: Required—TOEFL. *Application deadline:* Applications are processed on a rolling basis. Application fee: $60. Electronic applications accepted. *Expenses:* Tuition, state resident: full-time $7,731. Tuition, nonresident: full-time $18,672. Required fees: $1,479. Full-time tuition and fees vary according to degree level and program. *Financial support:* Fellowships, research assistantships, teaching assistantships available. Financial award applicants required to submit FAFSA. *Unit head:* Dr. Howard Kutchai, Interim Chair, 434-924-5108, Fax: 434-982-1616. *Application contact:* Peter C. Brunjes, Associate Dean for Graduate Programs and Research, 434-924-7184, Fax: 434-924-6737, E-mail: grad-a-s@virginia.edu.

University of Virginia, School of Medicine, Interdisciplinary Program in Biophysics, Charlottesville, VA 22908. Offers PhD. *Students:* 21 full-time (6 women); includes 1 minority (Hispanic American), 13 international. Average age 28. In 2005, 3 degrees awarded. *Degree requirements:* For doctorate, thesis/dissertation. *Entrance requirements:* For doctorate, GRE General Test. Additional exam requirements/recommendations for international students: Required—TOEFL. *Application deadline:* For fall admission, 7/15 for domestic students, 4/1 for international students. Applications are processed on a rolling basis. Application fee: $60. Electronic applications accepted. *Expenses:* Tuition, state resident: full-time $7,731. Tuition, nonresident: full-time $18,672. Required fees: $1,479. Full-time tuition and fees vary according to degree level and program. *Financial support:* Fellowships with full tuition reimbursements, research assistantships with full tuition reimbursements, teaching assistantships with full tuition reimbursements, tuition waivers (full) available. Financial award application deadline: 12/2; financial award applicants required to submit FAFSA. *Faculty research:* Membrane biophysics, macromolecular structure, thermodynamics, spectroscopy, cell biophysics. *Unit head:* John Bushweller, Director, 434-243-6409, Fax: 434-982-1616, E-mail: jhb4v@virginia.edu. *Application contact:* Pam Mullinex, Program Coordinator, 434-243-7248, Fax: 434-982-1616, E-mail: prm8b@virginia.edu.

University of Washington, School of Medicine and Graduate School, Graduate Programs in Medicine, Department of Physiology and Biophysics, Seattle, WA 98195. Offers PhD. *Degree requirements:* For doctorate, thesis/dissertation. *Entrance requirements:* For doctorate, GRE General Test. Additional exam requirements/recommendations for international students: Required—TOEFL. *Faculty research:* Membrane and cell biophysics, neuroendocrinology, cardiovascular and respiratory physiology, systems neurophysiology and behavior, molecular physiology.

The University of Western Ontario, Faculty of Graduate Studies, Biosciences Division, Department of Medical Biophysics, London, ON N6A 5B8, Canada. Offers M Sc, PhD. *Degree requirements:* For master's and doctorate, thesis/dissertation. *Faculty research:* Haemodynamics and cardiovascular biomechanics, microcirculation, orthopedic biomechanics, radiobiology, medical imaging.

University of Wisconsin–Madison, Graduate School, Program in Biophysics, Madison, WI 53706-1380. Offers PhD. *Degree requirements:* For doctorate, thesis/dissertation, comprehensive exam, registration. *Entrance requirements:* For doctorate, GRE General Test, minimum GPA of 3.0. Additional exam requirements/recommendations for international students: Required—TOEFL (minimum score 600 paper-based). Electronic applications accepted. *Faculty research:* NMR spectroscopy, high-speed automated DNA sequencing, digital imaging microscopy, x-ray crystallography.

Vanderbilt University, Graduate School and School of Medicine, Department of Molecular Physiology and Biophysics, Nashville, TN 37240-1001. Offers MS, PhD, MD/PhD. *Faculty:* 53 full-time (16 women). *Students:* 50 full-time (28 women), 1 (woman) part-time; includes 5 minority (1 African American, 1 American Indian/Alaska Native, 2 Asian Americans or Pacific Islanders, 1 Hispanic American), 10 international. In 2005, 1 master's, 6 doctorates awarded. *Degree requirements:* For doctorate, thesis/dissertation, preliminary, qualifying, and final exams. *Entrance requirements:* For doctorate, GRE General Test, GRE Subject Test (recommended). *Application deadline:* For fall admission, 1/15 for domestic students, 1/15 for international students. Application fee: $0. Electronic applications accepted. *Expenses:* Tuition: Part-time $1,283 per semester hour. Required fees: $2,202; $1,101 per semester. One-time fee: $30. Tuition and fees vary according to course load, program and student level. *Financial support:* Fellowships with full tuition reimbursements, research assistantships with full tuition

Biophysics

Vanderbilt University (continued)

reimbursements, Federal Work-Study, institutionally sponsored loans, traineeships, and tuition waivers (partial) available. Financial award application deadline: 1/15. *Faculty research:* Molecular endocrinology, membrane transport biophysics, metabolic regulation, neurobiology. *Unit head:* Alan D. Cherrington, Chair, 615-322-7000, Fax: 615-343-0490, E-mail: alan.cherrington@vanderbilt.edu. *Application contact:* Hussane Mchaourab, Director of Graduate Studies, 615-322-7000, Fax: 615-343-0490, E-mail: hassane.mchaourab@vanderbilt.edu.

Vanderbilt University, School of Medicine, Program in Chemical and Physical Biology, Nashville, TN 37240-1001. Offers PhD. *Degree requirements:* For doctorate, thesis/dissertation, dissertation defense, comprehensive exam, registration. *Entrance requirements:* For doctorate, GRE, 3 letters of recommendation, interview by invitation. Additional exam requirements/recommendations for international students: Required—TOEFL. Electronic applications accepted. *Expenses:* Tuition: Part-time $1,283 per semester hour. Required fees: $2,202; $1,101 per semester. One-time fee: $30. Tuition and fees vary according to course load, program and student level. *Faculty research:* Mathematical modeling, enzyme kinetics, structural biology, genomics, proteomics and mass spectrometry.

Washington State University, Graduate School, College of Sciences, School of Molecular Biosciences, Program in Biochemistry and Biophysics, Pullman, WA 99164. Offers MS, PhD. *Faculty:* 23 full-time (5 women), 21 part-time/adjunct (4 women). *Students:* 26 full-time (9 women), 1 (woman) part-time; includes 1 minority (Hispanic American), 12 international. Average age 27. 231 applicants, 18% accepted, 13 enrolled. In 2005, 1 master's, 2 doctorates awarded. Terminal master's awarded for partial completion of doctoral program. *Degree requirements:* For master's, thesis or alternative, oral exam; for doctorate, thesis/dissertation, oral exam, written exam, comprehensive exam, registration. *Entrance requirements:* For master's

and doctorate, GRE General Test, minimum GPA of 3.0. Additional exam requirements/recommendations for international students: Required—TOEFL (minimum score 550 paper-based; 213 computer-based). *Application deadline:* For fall admission, 12/15 for domestic students, 12/15 for international students. Application fee: $35. Electronic applications accepted. *Expenses:* Tuition, state resident: full-time $6,295; part-time $336 per credit. Tuition, nonresident: full-time $15,949; part-time $819 per credit. Required fees: $933. Part-time tuition and fees vary according to campus/location and program. *Financial support:* In 2005–06, 5 fellowships with full tuition reimbursements (averaging $18,384 per year), 11 research assistantships with full tuition reimbursements (averaging $18,384 per year), 10 teaching assistantships with full tuition reimbursements (averaging $18,384 per year) were awarded; career-related internships or fieldwork, Federal Work-Study, institutionally sponsored loans, traineeships, and health care benefits also available. Financial award application deadline: 4/1; financial award applicants required to submit FAFSA. *Faculty research:* Gene regulation, signal transduction, protein export, reproductive biology, DNA repair. Total annual research expenditures: $5.8 million. *Application contact:* Kelly G. McGovern, Academic Coordinator, 509-335-4566, Fax: 509-335-1907, E-mail: mcgnerk@wsu.edu.

Wright State University, School of Graduate Studies, College of Science and Mathematics, Department of Anatomy and Physiology, Dayton, OH 45435. Offers anatomy (MS); physiology and biophysics (MS). *Degree requirements:* For master's, thesis optional. *Entrance requirements:* Additional exam requirements/recommendations for international students: Required—TOEFL. *Faculty research:* Reproductive cell biology, neurobiology of pain, neurohistochemistry.

Yale University, Graduate School of Arts and Sciences, Department of Molecular Biophysics and Biochemistry, New Haven, CT 06520. Offers MS, PhD. *Degree requirements:* For doctorate, thesis/dissertation. *Entrance requirements:* For doctorate, GRE General Test, GRE Subject Test.

Molecular Biophysics

Baylor College of Medicine, Graduate School of Biomedical Sciences, Program in Structural and Computational Biology and Molecular Biophysics, Houston, TX 77030-3498. Offers PhD, MD/PhD. *Faculty:* 63 full-time (4 women). *Students:* 40 full-time (7 women); includes 5 minority (3 Asian Americans or Pacific Islanders, 2 Hispanic Americans), 22 international. Average age 27. 73 applicants, 22% accepted, 10 enrolled. In 2005, 4 degrees awarded. *Median time to degree:* Of those who began their doctoral program in fall 1997, 80% received their degree in 8 years or less. *Degree requirements:* For doctorate, thesis/dissertation, public defense. *Entrance requirements:* For doctorate, GRE General Test, GRE Subject Test (strongly recommended), minimum GPA of 3.0. Additional exam requirements/recommendations for international students: Required—TOEFL. *Application deadline:* For fall admission, 2/1 for domestic students. Application fee: $30. Electronic applications accepted. *Expenses:* Tuition: Full-time $8,200. Full-time tuition and fees vary according to program. *Financial support:* In 2005–06, 37 students received support, including 26 fellowships (averaging $23,000 per year), 14 research assistantships (averaging $23,000 per year); career-related internships or fieldwork, Federal Work-Study, institutionally sponsored loans, health care benefits, and tuition waivers (full) also available. Financial award applicants required to submit FAFSA. *Faculty research:* X-ray and electron crystallography, light and electron microscopy, computer image reconstruction, molecular spectroscopy. *Unit head:* Dr. Wah Chiu, Director, 713-798-6985. *Application contact:* Wanda Waguespack, Graduate Program Administrator, 713-798-5197, Fax: 713-798-6325, E-mail: wandaw@bcm.edu.

See Close-Up on page 465.

California Institute of Technology, Division of Biology and Division of Chemistry and Chemical Engineering, Biochemistry and Molecular Biophysics Graduate Option, Pasadena, CA 91125-0001. Offers PhD. *Faculty:* 42 full-time (11 women). *Students:* 54 full-time (18 women); includes 14 minority (10 Asian Americans or Pacific Islanders, 4 Hispanic Americans). 119 applicants, 18% accepted, 10 enrolled. In 2005, 7 doctorates awarded. *Degree requirements:* For doctorate, thesis/dissertation, qualifying exam. *Entrance requirements:* For doctorate, GRE General Test. Additional exam requirements/recommendations for international students: Required—TOEFL. *Application deadline:* For fall admission, 1/1 for domestic students. Application fee: $50. Electronic applications accepted. *Financial support:* In 2005–06, fellowships with full tuition reimbursements (averaging $21,195 per year), teaching assistantships with full tuition reimbursements (averaging $4,305 per year) were awarded; research assistantships with full tuition reimbursements, institutionally sponsored loans also available. Financial award application deadline: 1/1. *Unit head:* Dr. Stephen L. Mayo, Executive Officer, 626-395-6408, E-mail: steve@mayo.caltech.edu. *Application contact:* Alison Ross, Graduate Program Coordinator, 626-395-6446, Fax: 626-568-9430, E-mail: biochemop@cco.caltech.edu.

California Institute of Technology, Division of Chemistry and Chemical Engineering, Pasadena, CA 91125-0001. Offers biochemistry and molecular biophysics (MS, PhD); chemical engineering (MS, PhD); chemistry (MS, PhD). *Faculty:* 41 full-time (8 women). *Students:* 345 full-time (121 women). Average age 24. 647 applicants, 27% accepted, 61 enrolled. In 2005, 19 master's, 57 doctorates awarded. Terminal master's awarded for partial completion of doctoral program. *Degree requirements:* For master's and doctorate, thesis/dissertation. *Entrance requirements:* Additional exam requirements/recommendations for international students: Required—TOEFL; Recommended—IELT, TWE, TSE. *Application deadline:* For fall admission, 1/1 for domestic students, 1/1 for international students. Application fee: $50. Electronic applications accepted. *Financial support:* In 2005–06, 345 students received support; fellowships, research assistantships, teaching assistantships, Federal Work-Study, institutionally sponsored loans, scholarships/grants, traineeships, health care benefits, and unspecified assistantships available. Financial award application deadline: 1/1. *Faculty research:* Molecular structure and interactions in chemical and biological systems, theoretical studies of fundamental chemical processes, chemical reaction engineering, biochemical engineering, catalysis. *Unit head:* Dr. David A. Tirrell, Chairman, 626-395-3646, Fax: 626-568-8824, E-mail: tirrell@caltech.edu.

Duke University, Graduate School, Program in Structural Biology and Biophysics, Durham, NC 27710. Offers Certificate. Students must be enrolled in a participating PhD program. *Faculty:* 22 full-time. *Students:* 2 full-time (1 woman). 14 applicants, 57% accepted, 1 enrolled. *Entrance requirements:* For degree, GRE General Test, GRE Subject Test. Additional exam requirements/recommendations for international students: Required—IELT (preferred) or TOEFL. *Application deadline:* For fall admission, 12/31 for domestic students, 12/31 for international students. Application fee: $75. *Financial support:* Application deadline: 12/31. *Unit head:* Lorena Beese, Director of Graduate Studies, 919-681-5267, Fax: 919-684-8346, E-mail: mbp@biochem.duke.edu.

See Close-Up on page 565.

Florida State University, Graduate Studies, College of Arts and Sciences, Program in Molecular Biophysics, Tallahassee, FL 32306. Offers biochemistry, molecular and cell biology (PhD); computational structural biology (PhD); molecular biophysics (PhD). *Faculty:* 44 full-time (7 women). *Students:* 29 full-time (17 women); includes 1 minority (Hispanic American), 14 international. Average age 30. 14 applicants, 100% accepted, 4 enrolled. In 2005, 3 degrees awarded. *Median time to degree:* Of those who began their doctoral program in fall 1997, 100% received their degree in 8 years or less. *Degree requirements:* For doctorate,

thesis/dissertation, comprehensive exam. *Entrance requirements:* For doctorate, GRE General Test. Additional exam requirements/recommendations for international students: Required—TOEFL (minimum score 600 paper-based; 250 computer-based). *Application deadline:* For fall admission, 1/15 for domestic students, 1/15 for international students. Applications are processed on a rolling basis. Application fee: $30. Electronic applications accepted. *Financial support:* In 2005–06, 29 students received support, including 29 research assistantships (averaging $19,500 per year); health care benefits and tuition waivers (partial) also available. Financial award applicants required to submit FAFSA. *Faculty research:* Protein and nucleic acid structure and function, membrane protein structure, computational biophysics, 3-D image reconstruction. Total annual research expenditures: $5.1 million. *Unit head:* Dr. P. Bryant Chase, Director, MOB Graduate Program, 850-644-0056, Fax: 850-644-7244, E-mail: chase@bio.fsu.edu. *Application contact:* Dale E. Leonard, Academic Coordinator, Graduate Programs, 850-644-1012, Fax: 850-644-7244, E-mail: mob@sb.fsu.edu.

Illinois Institute of Technology, Graduate College, College of Science and Letters, Department of Biological, Chemical and Physical Sciences, Biology Division, Chicago, IL 60616-3793. Offers biochemistry (MS); biology (MBS, PhD); biotechnology (MS); cell biology (MS); microbiology (MS); molecular biochemistry and biophysics (MS, PhD). Part-time and evening/weekend programs available. Postbaccalaureate distance learning degree programs offered (no on-campus study). Terminal master's awarded for partial completion of doctoral program. *Degree requirements:* For master's, thesis (for some programs), comprehensive exam; for doctorate, thesis/dissertation, comprehensive exam. *Entrance requirements:* For master's and doctorate, GRE General Test, minimum undergraduate GPA of 3.0. Additional exam requirements/recommendations for international students: Required—TOEFL (minimum score 550 paper-based; 213 computer-based). Electronic applications accepted. *Faculty research:* Protein crystallography, small angle x-ray diffraction of muscle, spectroscopy of multidomain proteins, structure and function of cell cycle proteins, development of anticancer drugs.

University of Michigan, Horace H. Rackham School of Graduate Studies, Interdepartmental Program in Molecular Biophysics, Ann Arbor, MI 48109.

University of Pennsylvania, School of Medicine, Biomedical Graduate Studies, Graduate Group in Biochemistry and Molecular Biophysics, Philadelphia, PA 19104. Offers PhD, MD/PhD, VMD/PhD. *Faculty:* 63. *Students:* 64 full-time (25 women); includes 9 minority (3 African Americans, 4 Asian Americans or Pacific Islanders, 2 Hispanic Americans), 10 international. 101 applicants, 34% accepted, 10 enrolled. In 2005, 9 doctorates awarded. *Degree requirements:* For doctorate, thesis/dissertation. *Entrance requirements:* For doctorate, GRE General Test. Additional exam requirements/recommendations for international students: Required—TOEFL. *Application deadline:* For fall admission, 12/15 priority date for domestic students, 12/1 priority date for international students. Applications are processed on a rolling basis. Application fee: $70. Electronic applications accepted. *Financial support:* In 2005–06, 69 students received support; fellowships, research assistantships, scholarships/grants, traineeships, and unspecified assistantships available. *Faculty research:* Biochemistry of cell differentiation, tissue culture, intermediary metabolism, structure of proteins and nucleic acids, biochemical genetics. *Unit head:* , Dr. Joshua Wand, Chairperson, 215-573-7288. *Application contact:* Ruth Keris, Graduate Group Administrator, 215-898-4639, Fax: 215-573-2085, E-mail: keris@mail.med.upenn.edu.

University of Pittsburgh, School of Medicine and School of Arts and Sciences, Molecular Biophysics and Structural Biology Graduate Program, Pittsburgh, PA 15260. Offers PhD. *Faculty:* 42 full-time (11 women). *Students:* 11 full-time (4 women), 9 international. Average age 25. 10 applicants, 70% accepted, 5 enrolled. *Degree requirements:* For doctorate, thesis/dissertation, comprehensive exam, registration. *Entrance requirements:* For doctorate, GRE General Test. Additional exam requirements/recommendations for international students: Required—TOEFL (minimum score 600 paper-based; 250 computer-based), IELT (minimum score 7). *Application deadline:* For fall admission, 2/1 for domestic students, 2/1 for international students. Application fee: $0. *Expenses:* Tuition, state resident: full-time $13,194; part-time $537 per credit. Tuition, nonresident: full-time $25,012; part-time $1,026 per credit. Required fees: $700; $164 per term. Tuition and fees vary according to campus/location and program. *Financial support:* In 2005–06, 5 fellowships with tuition reimbursements (averaging $23,552 per year) were awarded; institutionally sponsored loans, scholarships/grants, traineeships, and unspecified assistantships also available. *Faculty research:* Structural biology, protein dynamics and folding, computational biophysics, molecular informatics, membrane biophysics and ion channels, NMR, x-ray crystallography cryaelectron microscopy. *Unit head:* Dr. Angela M Gronenborn, Director. *Application contact:* Jennifer L. Walker, MBSB Program Coordinator, 412-648-8957, Fax: 412-648-1077, E-mail: mbsbinfo@medschool.pitt.edu.

The University of Texas Medical Branch, Graduate School of Biomedical Sciences, Program in Cellular Physiology and Molecular Biophysics, Galveston, TX 77555. Offers MS, PhD. *Students:* 10 full-time (4 women); includes 1 minority (African American), 9 international. Average age 31. In 2005, 3 doctorates awarded. *Degree requirements:* For master's, thesis or alternative; for doctorate, thesis/dissertation. *Entrance requirements:* For master's and doctorate, GRE General Test. Additional exam requirements/recommendations for international students: Required—TOEFL (minimum score 550 paper-based; 213 computer-based). *Application deadline:* Applications are processed on a rolling basis. Application fee: $30 ($75 for international students). Electronic applications accepted. *Expenses:* Tuition, state resident: full-time $8,350; part-time $90 per credit hour. Tuition, nonresident: full-time $21,450; part-time $366 per credit hour. Required fees: $1,027; $11 per credit hour. $60 per term. *Financial support:* In 2005–06, fellowships (averaging $23,000 per year), research assistantships with full tuition reimbursements (averaging $23,000 per year) were awarded. Financial award applicants required to submit FAFSA. *Unit head:* Dr. Malcolm S. Brodwick, Director, 409-772-2973, Fax: 409-772-9382, E-mail: mbrodwic@utmb.edu. *Application contact:* Linda Spurger, Information Contact, 409-772-1942, Fax: 409-772-4687, E-mail: llspurge@utmb.edu.

The University of Texas Southwestern Medical Center at Dallas, Southwestern Graduate School of Biomedical Sciences, Division of Basic Science, Program in Molecular Biophysics, Dallas, TX 75390. Offers PhD. *Faculty:* 34 full-time (6 women). *Students:* 45 full-time (14 women), 3 part-time (2 women); includes 8 minority (1 African American, 5 Asian Americans or Pacific Islanders, 2 Hispanic Americans), 21 international. Average age 29. In 2005, 6 doctorates awarded. *Degree requirements:* For doctorate, thesis/dissertation, qualifying exam. *Entrance requirements:* For doctorate, GRE General Test, minimum GPA of 3.0. Additional exam requirements/recommendations for international students: Required—TOEFL. *Application deadline:* For fall admission, 1/5 for domestic students. Applications are processed on a rolling basis. Application fee: $0. Electronic applications accepted. *Expenses:* Tuition, state resident: full-time $6,550; part-time $50 per credit hour. Tuition, nonresident: full-time $19,650; part-time $326 per credit hour. Required fees: $42 per credit hour. Tuition and fees vary according to degree level and program. *Financial support:* Fellowships, research assistantships, institutionally sponsored loans and traineeships available. *Faculty research:* Optical spectroscopy, x-ray crystallography, protein chemistry, ion channels, contractile and cytoskeletal proteins. *Unit head:* Dr. Kevin Gardner, Chair, 214-645-6365, Fax: 214-645-6353, E-mail: kevin.gardner@utsouthwestern.edu. *Application contact:* Dr. Nancy E. Street, Associate Dean, 214-648-6708, Fax: 214-648-2102, E-mail: nancy.street@utsouthwestern.edu.

Vanderbilt University, Graduate School, Interdepartmental Program in Molecular Biophysics, Nashville, TN 37240-1001.

Virginia Commonwealth University, Medical College of Virginia-Professional Programs, School of Medicine and Graduate Programs, School of Medicine Graduate Programs, Department of Biochemistry, Richmond, VA 23284-9005. Offers biochemistry (MS, PhD, CBHS); bioinformatics (MS); molecular biology and genetics (MS, PhD); neurosciences (PhD). *Faculty:* 26 full-time (6 women). *Students:* 48 full-time (24 women), 9 part-time (8 women); includes 12 minority (2 African Americans, 8 Asian Americans or Pacific Islanders, 2 Hispanic Americans), 8 international. 61 applicants, 75% accepted. In 2005, 1 master's, 4 doctorates, 2 other advanced degrees awarded. *Degree requirements:* For master's, thesis; for doctorate, thesis/dissertation, comprehensive oral and written exams. *Entrance requirements:* For master's and doctorate, DAT, GRE General Test, MCAT. *Application deadline:* For fall admission, 2/15 for domestic students. Application fee: $50. *Expenses:* Tuition, state resident: full-time $3,185; part-time $405 per credit. Tuition, nonresident: full-time $7,952; part-time $940 per credit. Required fees: $751 per semester hour. Tuition and fees vary according to course load and program. *Financial support:* Fellowships, research assistantships available. *Faculty research:* Molecular biology, peptide/protein chemistry, neurochemistry, enzyme mechanisms, macromolecular structure determination. Total annual research expenditures: $3.5 million. *Unit head:* , Dr. Sarah Spiegel, Chair, 804-828-9762, Fax: 804-828-1473, E-mail: sspiegel@vcu.edu. *Application contact:* Dr. Keith R. Shelton, Program Director, 804-828-9886, Fax: 804-828-1473, E-mail: krshelto@vcu.edu.

See Close-Up on page 447.

Washington University in St. Louis, Graduate School of Arts and Sciences, Division of Biology and Biomedical Sciences, Program in Molecular Biophysics, St. Louis, MO 63130-4899. Offers PhD. *Degree requirements:* For doctorate, thesis/dissertation. *Entrance requirements:* For doctorate, GRE General Test, GRE Subject Test. Electronic applications accepted.

Yale University, School of Medicine and Graduate School of Arts and Sciences, Combined Program in Biological and Biomedical Sciences (BBS), Molecular Biophysics and Biochemistry Track, New Haven, CT 06520. Offers PhD, MD/PhD. *Students:* 14 full-time. *Degree requirements:* For doctorate, thesis/dissertation. *Entrance requirements:* For doctorate, GRE General Test. Additional exam requirements/recommendations for international students: Required—TOEFL. *Application deadline:* For fall admission, 12/8 for domestic students, 12/8 for international students. Electronic applications accepted. *Financial support:* Fellowships, research assistantships available. *Unit head:* Dr. Mark Solomon, Director of Graduate Studies, 203-432-5562, Fax: 203-432-9782, E-mail: mbb.grad@yale.edu. *Application contact:* Nessie Stewart, Graduate Registrar, 203-432-5662, Fax: 203-432-6178, E-mail: mbb.grad@yale.edu.

Radiation Biology

Auburn University, College of Veterinary Medicine and Graduate School, Graduate Programs in Veterinary Medicine, Auburn University, AL 36849. Offers biomedical sciences (MS, PhD), including anatomy, physiology and pharmacology (MS), biomedical sciences (PhD), clinical sciences (MS), large animal surgery and medicine (MS), pathobiology (MS), radiology (MS), small animal surgery and medicine (MS). Part-time programs available. *Faculty:* 76 full-time (24 women). *Students:* 11 full-time (8 women), 33 part-time (20 women); includes 3 minority (all Hispanic Americans), 14 international. 41 applicants, 37% accepted, 11 enrolled. In 2005, 9 master's, 3 doctorates awarded. *Degree requirements:* For doctorate, thesis/dissertation. *Entrance requirements:* For master's, GRE General Test; for doctorate, GRE General Test, GRE Subject Test. *Application deadline:* For fall admission, 7/7 for domestic students; for spring admission, 11/24 for domestic students. Applications are processed on a rolling basis. Application fee: $25 ($50 for international students). Electronic applications accepted. *Financial support:* Research assistantships, teaching assistantships, Federal Work-Study available. Support available to part-time students. Financial award application deadline: 3/15. *Application contact:* Dr. Stephen L. McFarland, Acting Dean of the Graduate School, 334-844-4700.

Austin Peay State University, Graduate School, College of Science and Mathematics, Department of Biology, Clarksville, TN 37044-0001. Offers biology (MS); clinical laboratory science (MS); radiologic science (MS). Part-time programs available. *Faculty:* 12 full-time (3 women). *Students:* 4 full-time (2 women), 17 part-time (12 women); includes 2 minority (1 African American, 1 Hispanic American), 4 international. Average age 31. In 2005, 4 degrees awarded. *Degree requirements:* For master's, thesis optional. *Entrance requirements:* For master's, GRE General Test, 3 letters of recommendation. Additional exam requirements/recommendations for international students: Required—TOEFL (minimum score 500 paper-based; 173 computer-based). *Application deadline:* For fall admission, 7/31 for domestic students; for spring admission, 12/17 priority date for domestic students. Applications are processed on a rolling basis. Application fee: $25. *Expenses:* Tuition, state resident: full-time $4,936; part-time $340 per credit hour. Tuition, nonresident: full-time $14,248; part-time $744 per credit hour. Required fees: $957. Part-time tuition and fees vary according to course load. *Financial support:* In 2005–06, research assistantships (averaging $9,250 per year), teaching assistantships with partial tuition reimbursements (averaging $6,900 per year) were awarded; career-related internships or fieldwork, Federal Work-Study, institutionally sponsored loans, scholarships/grants, and unspecified assistantships also available. Support available to part-time students. Financial award application deadline: 4/1. *Faculty research:* Nonpaint source pollution, amphibian biomonitoring, aquatic toxicology, biological indicators of water quality, taxonomy. *Unit head:* Dr. Don Dailey, Interim Chair, 931-221-7781, E-mail: daileyd@apsu.edu.

Colorado State University, College of Veterinary Medicine and Biomedical Sciences, Department of Environmental and Radiological Health Sciences, Fort Collins, CO 80523-0015. Offers environmental health (MS, PhD); radiological health sciences (MS, PhD). *Faculty:* 24 full-time (4 women), 1 part-time/adjunct (0 women). *Students:* 32 full-time (18 women), 44 part-time (23 women); includes 2 American Indian/Alaska Native, 5 Asian Americans or Pacific Islanders, 1 Hispanic American, 9 international. Average age 32. 40 applicants, 60% accepted, 15 enrolled. In 2005, 17 master's, 6 doctorates awarded. Terminal master's awarded for partial completion of doctoral program. *Median time to degree:* Of those who began their doctoral program in fall 1997, 100% received their degree in 8 years or less. *Degree requirements:* For master's, thesis (for some programs), publishable paper; for doctorate, thesis/dissertation, publishable paper, comprehensive exam, registration. *Entrance requirements:* For master's and doctorate, GRE General Test, 1 year of course work in biology lab and chemistry lab, 1 semester of course work in organic chemistry, course work in calculus. Additional exam requirements/recommendations for international students: Required—TOEFL. *Application deadline:* For fall admission, 3/1 for domestic students, 2/1 for international students; for spring admission, 10/1 for domestic students. Application fee: $50. Electronic applications accepted. *Expenses:* Tuition, state resident: full-time $3,690; part-time $205 per credit. Tuition, nonresident: full-time $14,958; part-time $831 per credit. Required fees: $1,061. *Financial support:* In 2005–06, 17 students received support, including 5 fellowships with full and partial tuition reimbursements available (averaging $24,000 per year), 9 research assistantships with full tuition reimbursements available (averaging $19,000 per year), 3 teaching assistantships with partial tuition reimbursements available (averaging $10,000 per year); career-related internships or fieldwork, Federal Work-Study, institutionally sponsored loans, traineeships, and unspecified assistantships also available. Support available to part-time students. Financial award application deadline: 2/1. *Faculty research:* Epidemiology, toxicology, industrial hygiene, occupational health, radiation therapy. Total annual research expenditures: $10 million. *Unit head:* Dr. John D. Zimbrick, Head, 970-491-7038, Fax: 970-491-6023, E-mail: zimbrick@colostate.edu.

Georgetown University, Graduate School of Arts and Sciences, Programs in Biomedical Sciences, Department of Health Physics, Washington, DC 20057. Offers health physics (MS); radiobiology (MS). *Degree requirements:* For master's, thesis. *Entrance requirements:* Additional exam requirements/recommendations for international students: Required—TOEFL.

Université de Sherbrooke, Faculty of Medicine and Health Sciences, Graduate Programs in Medicine, Program in Radiobiology, Sherbrooke, QC J1K 2R1, Canada. Offers M Sc, PhD. *Faculty:* 10 full-time (0 women), 3 part-time/adjunct (1 woman). *Students:* 14 full-time (6 women), 12 part-time (8 women). Average age 25. 4 applicants, 75% accepted, 3 enrolled. In 2005, 5 master's, 5 doctorates awarded. *Degree requirements:* For master's and doctorate, thesis/dissertation. *Application deadline:* For fall admission, 6/30 for domestic students. For winter admission, 10/31 for domestic students; for spring admission, 2/28 for domestic students. Application fee: $50. Electronic applications accepted. *Financial support:* Fellowships available. *Faculty research:* DNA repair, physiochemical actions of radiation, radiopharmacy, phototherapy, imaging. *Unit head:* Dr. Mohamed Bentourkia, Director, 819-820-6868 Ext. 11863, E-mail: mohamed.bentourkia@usherbrooke.ca.

The University of Iowa, Roy J. and Lucille A. Carver College of Medicine and Graduate College, Graduate Programs in Medicine, Program in Free Radical and Radiation Biology, Iowa City, IA 52242-1316. Offers MS, PhD. Part-time programs available. *Faculty:* 6 full-time (0 women); includes 2 minority (1 African American, 1 Asian American or Pacific Islander), 12 international. Average age 30. 13 applicants, 46% accepted, 3 enrolled. In 2005, 1 master's, 2 doctorates awarded. *Degree requirements:* For doctorate, thesis/dissertation. *Entrance requirements:* For master's and doctorate, GRE. *Application deadline:* For fall admission, 5/31 for domestic students; for spring admission, 10/31 for domestic students. Applications are processed on a rolling basis. Application fee: $30 ($50 for international students). *Expenses:* Tuition, state resident: part-time $1,882 per term. Tuition, nonresident: full-time $17,338; part-time $4,907 per term. Tuition and fees vary according to course load and program. *Financial support:* In 2005–06, fellowships with partial tuition reimbursements (averaging $21,500 per year), research assistantships with tuition reimbursements (averaging $21,500 per year) were awarded; traineeships, health care benefits, tuition waivers (partial), and unspecified assistantships also available. *Faculty research:* Radiation injury and cellular repair, cell proliferation kinetics, free radical biology, tumor control, PET imaging, EPR. Total annual research expenditures: $1 million. *Unit head:* Larry W. Oberley, Head, 319-335-8015, Fax: 319-335-8039, E-mail: larry-oberley@uiowa.edu. *Application contact:* Jennifer K. DeWitte, Grant/Program Administrator, 319-335-8164, Fax: 319-335-8039, E-mail: jennifer-dewitte@uiowa.edu.

Announcement: The MS and PhD programs are based on in vivo and in vitro study of biological systems and their responses to ionizing and non-ionizing radiations, hyperthermia, and chemical injury, as well as the role of free radicals and antioxidants in cancer and other disease states. For more information, visit the Web site at http://www.uihealthcare.com/depts/med/radiationoncology/frrb/index.

University of Oklahoma Health Sciences Center, College of Medicine and Graduate College, Graduate Programs in Medicine, Department of Radiological Sciences, Oklahoma City, OK 73190. Offers medical radiation physics (MS, PhD), including diagnostic radiology, nuclear medicine, radiation therapy, ultrasound. Part-time programs available. Terminal master's awarded for partial completion of doctoral program. *Degree requirements:* For master's and doctorate, thesis/dissertation. *Entrance requirements:* For master's, GRE General Test; for doctorate, GRE General Test, 3 letters of recommendation. Additional exam requirements/recommendations for international students: Required—TOEFL. *Faculty research:* Monte Carlo applications in radiation therapy, observer performed studies in diagnostic radiology, error analysis in gated cardiac nuclear medicine studies, nuclear medicine absorbed fraction determinations.

The University of Texas Southwestern Medical Center at Dallas, Southwestern Graduate School of Biomedical Sciences, Division of Clinical Science, Radiological Sciences Program, Dallas, TX 75390. Offers MS, PhD. *Faculty:* 23 full-time (0 women), 2 part-time/adjunct (0 women). *Students:* 12 full-time (4 women); includes 2 minority (1 African American, 1 Asian American or Pacific Islander), 6 international. Average age 34. 27 applicants, 7% accepted, 2 enrolled.Terminal master's awarded for partial completion of doctoral program. *Degree requirements:* For master's, thesis/dissertation; for doctorate, thesis/dissertation, comprehensive exam. *Entrance requirements:* For master's and doctorate, GRE General Test. Additional exam requirements/recommendations for international students: Required—TOEFL. *Application deadline:* For fall admission, 5/15 for domestic students. Application fee: $0. *Expenses:* Tuition, state resident: full-time $6,550; part-time $50 per credit hour. Tuition, nonresident: full-time $19,650; part-time $326 per credit hour. Required fees: $42 per credit hour.

Radiation Biology

The University of Texas Southwestern Medical Center at Dallas (continued)
Tuition and fees vary according to degree level and program. *Financial support:* In 2005–06, 2 research assistantships were awarded; institutionally sponsored loans and scholarships/grants also available. Financial award application deadline: 3/1; financial award applicants required to submit FAFSA. *Faculty research:* Medical physics, nuclear medicine, noninvasive NMR methods, ultrasound, pathophysiological *in vivo* investigation. *Unit head:* Dr. Peter P. Antich, Chair, 214-648-2856, Fax: 214-648-2991, E-mail: peter.antich@utsouthwestern.edu. *Application contact:* Kay Emerson, Program Assistant, 214-648-2503, Fax: 214-648-2991, E-mail: kay.emerson@utsouthwestern.edu.

Cross-Discipline Announcements

Princeton University, Graduate School, Department of Molecular Biology, Princeton, NJ 08544-1019.

Graduate studies in the Department of Molecular Biology at Princeton emphasize training in research and encourage students to apply molecular, biochemical, and genetic approaches to biological problems. Faculty members and students pursue research in a wide variety of areas of biochemistry and cell biology, biological dynamics and computational biology, biomedicine and societal issues, biophysics and structural biology, developmental biology, genetics and genomics, microbiology, neurobiology, oncology, and virology.

University of Minnesota, Twin Cities Campus, Graduate School, Program in Biophysical Sciences and Medical Physics, Minneapolis, MN 55455-0213.

This interdisciplinary program provides research training and experience in biophysics, medical imaging, magnetic resonance imaging and spectroscopy, radiobiology, and radiation therapy physics. Thirty-one faculty members from 11 departments provide diverse opportunities for multidisciplinary research. A strong undergraduate physics background is desirable. Limited financial aid is available.

BAYLOR COLLEGE OF MEDICINE

Structural and Computational Biology and Molecular Biophysics

Program of Study

The Structural and Computational Biology and Molecular Biophysics (SCBMB) program is an interdisciplinary and interdepartmental program that offers a Ph.D. in structural and computational biology and molecular biophysics. This program is designed to train students by employing a strong research emphasis in these areas while also providing a solid background in biochemistry and cell and molecular biology. Faculty research activities cover development of state-of-the-art structural and computational techniques, protein design and engineering, biophysical chemistry of macromolecules, DNA structure and topology, membrane biophysics, and genome informatics. Sixty-nine faculty members in this program, from different departments at Baylor College of Medicine (BCM), Rice University (Rice) (within walking distance), the University of Houston (UH), the University of Texas–Houston Health Science Center (UT–HSC), the University of Texas–M. D. Anderson Cancer Center (MDACC), and the University of Texas Medical Branch in Galveston (UTMB) are involved in teaching and supervising students' research. The program seeks applicants with undergraduate degrees in physical, chemical, mathematical, computational, and the engineering sciences as well as students with traditional backgrounds in biochemistry and molecular biology. The first-year curriculum is tailored to each student's background; undergraduate courses in science, mathematics, and engineering are available at Rice University and the University of Houston. Courses not available at Baylor can be taken free of charge at the other two institutions. Students participate in a minimum of three laboratory rotations to experience different research areas. Seminars of student research and distinguished scientists enhance the educational experience.

Research Facilities

Through participating faculty members, students have access to a number of research centers, each uniquely equipped and staffed. These include the W. M. Keck Center for Interdisciplinary Bioscience Training of the Gulf Coast Consortia, National Center for Macromolecular Imaging, Institute for Molecular Design, Center for High Performance Software Research, Houston Area Computational Science Consortium, Computer and Information Technology Institute, and Texas Learning and Computation Center. These facilities offer students access to state-of-the-art hardware and software for X-ray and electron crystallography, magnetic resonance spectroscopy, genomics, proteomics, computational biophysics, and advanced optical and MRI imaging techniques.

Financial Aid

All students enrolled in the program receive an annual stipend in the amount of $23,000 and paid individual health insurance. Separate offices provide assistance to students with additional need.

Cost of Study

Full-tuition scholarships are supported by Baylor College of Medicine. Students pay a one-time matriculation fee of $25, a graduation fee of $140 (due in the fourth year), and an annual student fee of $150 for the first year and $20 for each following year. International students pay additional yearly visa fees: $75 for an F-1 and $100 for a J-1.

Living and Housing Costs

There are numerous apartment complexes located very close to the Texas Medical Center. Students frequently bike or walk to the campus from these apartments. The cost of living in Houston is less than in most major U.S. cities. The cost for food and recreation is modest, and there are ample opportunities for employment of spouses in the Texas Medical Center.

Student Group

The SCBMB program currently has 38 graduate students and enrolls 6 to 8 students per year. Approximately 400 students are enrolled in the graduate school. Students interact with predoctoral and postdoctoral fellows and with staff members in a variety of research centers. These centers sponsor annual symposia, workshops, seminars, and informal discussion groups.

Location

Houston is a dynamic city, with a population of approximately 4 million people. With its large seaport and modern airport, it is a center of international travel. Mexico City is 90 minutes away by air, and the Gulf of Mexico is only an hour's ride by car. The climate offers very pleasant cool and dry weather from fall through spring. The temperature during winter rarely drops below freezing, and it does not snow. Although summer temperatures are in the 90s with moderately high humidity, comfort is ensured by air-conditioning in all homes and workplaces. Symphony, opera, ballet, live theater, year-round major-league sports, and a large number of diverse ethnic groups and restaurants help to make Houston an entertaining and exciting city in which to live and work.

The College

Baylor College of Medicine was established as an independent, private university committed to excellence in the training and education of scholars and physicians. The College is located in the Texas Medical Center, which is composed of more than 645 acres and includes forty-two independent institutions. The University of Texas Health Science Center, the School of Public Health, and the M. D. Anderson Cancer Center are also on the campus, together with twenty-nine hospitals and research institutes. The Texas Medical Center is one of the most actively growing science centers in the country. The influx of new colleagues and the opportunities created by an atmosphere of expansion provide a stimulating academic environment. More information on Baylor College of Medicine can be found on the College's Web site at http://www.bcm.edu.

Applying

Applicants must be in excellent academic standing and have a bachelor's degree, with extensive course work in biology, chemistry, physics, and mathematics. GRE scores less than three years old at the time of application must be provided. Applications should be accompanied by transcripts from all colleges and universities attended, plus three letters of recommendation and a statement of research interest and career goals. Applications, catalogs, and instructions can be obtained via the Internet through the graduate school's Web site at http://www.bcm.edu/gradschool/?PMID=3514 or the SCBMB Web site. The $30 application fee is waived for electronic submission. The application deadline is January 1 for fall admission. Successful candidates are invited to visit BCM to meet with the participating faculty members and students in order to have a firsthand look at the SCBMB program. Expenses for travel and accommodations during the visit are provided by Baylor College of Medicine. Questions regarding the application process can be directed to the admissions e-mail address at gradappboss@bcm.edu.

Correspondence and Information

Dr. Wah Chiu, Director
Ph.D. Program in Structural and Computational Biology and Molecular Biophysics
Graduate School of Biomedical Sciences, Room N204T
Baylor College of Medicine
One Baylor Plaza, MS: BCM 215
Houston, Texas 77030
Phone: 713-798-5197
Fax: 713-798-6325
E-mail: scb@bcm.edu
Web site: http://scbmb.bcm.tmc.edu

Baylor College of Medicine

THE FACULTY AND THEIR RESEARCH

Wah Chiu, Ph.D.; Director.
Aleksandar Milosavljevic, Ph.D.; Co-Director.
Joel Morrisett, Ph.D.; Co-Director.
Timothy G. Palzkill, Ph.D.; Co-Director.

Aladin M. Boriek, Ph.D.; Medicine and Molecular Physiology and Biophysics at BCM. Respiratory system mechanics; muscle mechanics; mechanics of the cytoskeleton; lung mechanics; computational models of tissue mechanics; mechanotransduction in skeletal, smooth, and cardiac muscles; mechanotransduction of the respiratory pump.

James M. Briggs, Ph.D.; Biology and Biochemistry at UH. Computational biochemistry; computer-aided drug design.

William R. Brinkley, Ph.D.; Molecular and Cellular Biology at BCM. Structure and assembly of the mitotic apparatus; molecular mechanisms for aneuploidy and genomic instability in tumor cells.

William E. Brownell, Ph.D.; Otorhinolaryngology and Communicative Sciences at BCM. Biophysics of hearing: the cellular and molecular biology of outer hair cell electromotility.

Joseph Bryan, Ph.D.; Molecular and Cellular Biology at BCM. Molecular biology of ATP-sensitive potassium channels and regulation of insulin release.

Xiaomin Chen, Ph.D.; Biochemistry and Molecular Biology at MDACC. X-ray crystallography and macromolecular interactions.

Wah Chiu, Ph.D.; Biochemistry and Molecular Biology at BCM. Electron cryomicroscopy and computational biology of macromolecular machines and cells.

Dar-Chone Chow, Ph.D.; Chemistry at UH. Thermodynamic and structural studies of protein interactions.

John W. Clark Jr., Ph.D.; Electrical and Computer Engineering at Rice. Mathematical modeling of biological systems.

Cecilia Clementi, Ph.D.; Chemistry at Rice. Theory of protein folding; protein modeling and simulations; folding/function relationship.

Alan Cox, Ph.D.; Computer Science at Rice. Parallel and distributed computing; computational biology.

John A. Dani, Ph.D.; Neuroscience at BCM. Neuronal changes that underlie learning, memory, addiction, and behavior.

Anne H. Delcour, Ph.D.; Biology and Biochemistry at UH. Molecular properties and structure-function relationships of ion channels and pore-forming proteins; electrophysiology of bacterial ion channels.

Mary E. Dickinson, Ph.D.; Molecular Physiology and Molecular Biophysics at BCM. In vivo optical microscopy; analysis of vascular networks; role of mechanical forces in vertebrate development.

Henry F. Epstein, M.D.; Neuroscience and Cell Biology at UTMB. Structure and mechanisms of myosin assembly and protein kinase action.

Mary K. Estes, Ph.D.; Molecular Virology and Microbiology at BCM. Molecular biology, pathogenesis, and vaccines of gastrointestinal viruses.

George E. Fox, Ph.D.; Biology and Biochemistry at UH. RNA structure, function, and evolution; bioinformatics.

Fabrizio Gabbiani, Ph.D.; Neuroscience at BCM. Biophysics of information processing in the nervous system.

Xiaolian Gao, Ph.D.; Biology and Biochemistry at UH. Interdisciplinary research in fields of chemistry, nucleic acid, peptide, and protein biotechnology; bioinformatics; synthetic biology; structure biology; and chemical biology.

Richard A. Gibbs, Ph.D.; Molecular and Human Genetics at BCM. The Human Genome Project; molecular basis of human genetic diseases; molecular evolution.

Hiram Gilbert, Ph.D.; Biochemistry and Molecular Biology at BCM. Disulfide bond formation during protein folding and in biological regulation; catalysis of protein folding and the mechanisms of folding enzymes and molecular chaperones.

Dan Graur, Ph.D.; Biology and Biochemistry at UH. Molecular evolution; evolutionary bioinformatics.

Rudy Guerra, Ph.D.; Statistics at Rice. Statistical genetics; bioinformatics; statistical applications in cancer and cardiovascular disease.

Susan L. Hamilton, Ph.D.; Molecular Physiology and Biophysics at BCM. Structure and function of proteins involved in excitation-contraction coupling in skeletal and cardiac muscle and the elucidation of their roles in human disease.

Vincent J. Hilser, Ph.D.; Biochemistry and Molecular Biology at ZUTMB. Experimental and theoretical studies of protein structure and dynamics.

Raymond H. Jacobson, Ph.D.; Biochemistry and Molecular Biology at MDACC. Structure and function of basal transcription machinery; enzymatic activities and functional properties of general transcription factor TFIID.

S. Lennart Johnsson, Ph.D.; Computer Science at UH. Computational and information grids; high-performance scientific computation; parallel algorithms; middleware for grids and parallel computers.

Lydia E. Kavraki, Ph.D.; Computer Science and Bioengineering at Rice. Computational structural biology; bioinformatics; drug design; algorithms.

Ching-Hwa Kiang, Ph.D.; Physics and Astronomy at Rice. Single-molecule biophysics; mechanical properties of biomolecules; statistical physics.

Marek Kimmel, Ph.D.; Statistics at Rice. Informatics and statistical modeling of genome dynamics.

Kurt L. Krause, M.D., Ph.D.; Biology and Biochemistry at UH. Structure and function of macromolecules; structural biology of infectious diseases, drug design, enzyme mechanism, nucleases, racemases, and flavoproteins.

Ching C. Lau, M.D., Ph.D.; Pediatrics and Cancer Genomics Program at BCM. Genomics and proteomics of cancer, bioinformatics, and development of targeted therapy.

Suzanne M. Leal, Ph.D.; Molecular and Human Genetics at BCM. Statistical genetics/genetic epidemiology; mapping of complex and Mendelian traits; methodological research.

Olivier Lichtarge, M.D., Ph.D.; Molecular and Human Genetics at BCM. Bioinformatics and structural biology; evolutionary studies of structure-function; protein engineering; drug design; functional genomics.

Steven Ludtke, Ph.D.; Biochemistry and Molecular Biology at BCM. Scientific image processing; electron cryomicroscopy; structural biology; scientific databases.

Jianpeng Ma, Ph.D.; Biochemistry and Molecular Biology at BCM and Bioengineering at Rice. Computational and experimental study of structure-function relationships of biological macromolecules.

Michael A. Mancini, Ph.D.; Molecular and Cellular Biology at BCM. Transcription analyses at the single cell level.

Jonathan Miller, Ph.D.; Biochemistry and Molecular Biology at BCM. Bioinformatics; gene regulation; ncRNA identification.

Aleksandar Milosavljevic, Ph.D.; Molecular and Human Genetics at BCM. Comparative genomics and bioinformatics.

P. Read Montague, Ph.D.; Neuroscience at BCM. Theoretical and computational neuroscience; neuroimaging.

Joel D. Morrisett, Ph.D.; Medicine and Biochemistry and Molecular Biology at BCM. Structure and function of proteins involved in extracellular and intracellular lipid transport; in vivo and ex vivo imaging of atherosclerosis in the arterial wall; cardiovascular genomics and proteomics; NMR; MRI.

Timothy G. Palzkill, Ph.D.; Molecular Virology and Microbiology at BCM. Genetic and biochemical analysis of protein structure and function; functional genomics: genome-wide analysis of protein-ligand interactions.

Steen E. Pedersen, Ph.D.; Molecular Physiology and Biophysics at BCM. Structure and function of ligand-gated ion channels; thermodynamics of linked systems; applications of fluorescence spectroscopy to ion channel function.

Pawel A. Penczek, Ph.D.; Biochemistry and Molecular Biology at UT–HSC. Three-dimensional structures of large macromolecular complexes, using stain, cryo-EM, and computer-image processing techniques.

B. Montgomery Pettitt, Ph.D.; Chemistry at UH. Solution theory and thermodynamics of proteins and nucleic acids.

Paul J. Pfaffinger, Ph.D.; Neuroscience at BCM. Regulation, structure, and function of voltage-gated potassium channels in order to understand the basic mechanisms involved in establishing cellular electrophysiological properties.

James W. Pflugrath, Ph.D.; Rigaku Americas Corporation. Computational and macromolecular crystallography; X-ray diffraction technology development; area detectors; synchrotrons.

Henry J. Pownall, Ph.D.; Medicine at BCM. Genetic, structural, and nutritional basis of disorders in lipid metabolism, type 2 diabetes mellitus, and coronary artery disease; physicochemical methods; energy metabolism; structure-function studies.

B. V. Venkataram Prasad, Ph.D.; Biochemistry and Molecular Biology and Molecular Virology and Microbiology at BCM. Three-dimensional structural studies of gasteroenteric viruses: rotaviruses, calicivirus, and Norwalk virus.

Jun Qin, Ph.D.; Biochemistry and Molecular Biology at BCM. Network analysis proteomics: the human DNA damage signaling network.

Florante A. Quiocho, Ph.D.; Biochemistry and Molecular Biology at BCM. X-ray crystallographic analysis of biological macromolecules; molecular recognition.

Robert Raphael, Ph.D.; Bioengineering at Rice. Electromechanical transduction in cochlear outer hair cells; theoretical modeling of biological systems; membrane mechanics; modulation of cell mechanics by signaling molecules; biophysical factors mediating gene delivery.

Peter Saggau, Ph.D.; Neuroscience at BCM. Cellular mechanisms of learning and memory and of epilepsy; advanced optical imaging techniques.

Michael F. Schmid, Ph.D.; Biochemistry and Molecular Biology at BCM. Structures of macromolecular complexes (cytoskeleton, membrane proteins); cryotomography of nanomachines, organelles, and whole cells; computational methods for electron crystallography and image processing.

David W. Scott, Ph.D.; Statistics at Rice. Theoretical studies of multivariate probability density estimation, computational algorithms, and data exploration using computer visualization.

Gad Shaulsky, Ph.D.; Molecular and Human Genetics at BCM. Developmental genetics in *Dictyostelium*; dedifferentiation; functional genomics; microarray analysis of gene expression; evolution of sociality.

Richard N. Sifers, Ph.D.; Pathology and Molecular and Cellular Biology at BCM. Cellular glycobiology as a post-translational checkpoint and modifier of conformational disease.

Jack W. Smith, M.D., Ph.D.; Health Informatics at UT–HSC. Decision support systems and intelligent tutoring systems in health care.

Danny C. Sorensen, Ph.D.; Computational and Applied Mathematics at Rice. Numerical linear algebra and optimization; parallel computing; large-scale Eigenanalysis; dimension reduction for dynamical systems.

John L. Spudich, Ph.D.; Center for Membrane Biology at UT–HSC. Structure-function of membrane receptors; photosensory transduction mechanisms.

Francis T. F. Tsai, D.Phil.; Biochemistry and Molecular Biology at BCM. Structural biochemistry of molecular machines and macromolecular assemblies; protein crystallography; electron microscopy; functional proteomics.

Motonari Uesugi, Ph.D.; Biochemistry and Molecular Biology at BCM. Chemical biology of gene expression and cell differentiation.

Salih J. Wakil, Ph.D.; Biochemistry and Molecular Biology at BCM. Mechanisms and hormonal control of fatty acid metabolism; structure-function relationship of fatty acid synthase and acetyl-CoA carboxylase; generation of ACC2$^{-/-}$ mice; role of ACC2 in fatty acid oxidation and type 2 diabetes; generation of tissue-specific ACC knockout mice for studying its role in fatty acid synthesis..

George M. Weinstock, Ph.D.; Molecular and Human Genetics at BCM. Genomics, molecular genetics, and bioinformatics applied to study function in microbial pathogens and higher organisms.

Theodore G. Wensel, Ph.D.; Biochemistry and Molecular Biology at BCM. Structural dynamics of signal-transducing membranes; biochemical and bioanalytical applications of time-resolved luminescence.

John H. Wilson, Ph.D.; Biochemistry and Molecular Biology at BCM. Instability of trinucleotide repeats in mammalian cells and mice; knock-in mouse models for retinitis pigmentosa; gene therapy of dominant rhodopsin mutations.

Willy R. Wriggers, Ph.D.; Health Informatics at UT-HSC. 3-D structure and dynamics of subcellular biological machines; fast Fourier-accelerated correlation techniques for structural biology; modeling; electron microscopy and imaging; physics-based deformable models; haptic rendering; molecular graphics.

Samuel M. Wu, Ph.D.; Ophthalmology at BCM. Electrophysiology of the vertebrate retina; membrane biophysics; electrical and pharmacological properties of retinal synapses; retinal circuitry and visual perception.

E. Lynn Zechiedrich, Ph.D.; Molecular Virology and Microbiology at BCM. Integrative approaches to understanding DNA topoisomerases, chemotherapeutics that target topoisomerases, and drug resistance.

Z. Hong Zhou, Ph.D.; Pathology and Laboratory Medicine at UT–HSC. Three-dimensional structures of macromolecular machines and viruses by electron cryomicroscopy, tomography, and bioinformatics.

BOSTON UNIVERSITY

School of Medicine
Department of Physiology and Biophysics

Programs of Study

The Department of Physiology and Biophysics at the Boston University School of Medicine offers degree programs leading to the M.A. or Ph.D. degree. Each program stresses laboratory training and original research performed under the direction of a faculty member, as well as satisfactory performance in graduate course work.

In the first two years, a core curriculum of biochemistry, cell biology, and experimental methods in physiology and biophysics is supplemented by courses in biophysics, physiology, and structural and molecular biology. The course program emphasizes flexibility and individual choice. During the first year, students participate in rotations through the laboratories of 3 to 4 faculty members in an effort to gain firsthand knowledge and experience in a variety of experimental approaches and methodologies. At the end of the first year, students select their degree adviser and begin research within that laboratory. For the Ph.D. program, students take a qualifying exam at the end of the second year. Throughout their training, students take active part in a weekly departmental seminar series and annually report their research progress during a departmental retreat.

Research Facilities

Located on the Boston University School of Medicine campus, the Department of Physiology and Biophysics has faculty laboratories and administrative offices located within the Center for Advanced Biomedical Research and the School of Medicine. The Department has extensive facilities and instrumentation for state-of-the-art biophysical and physiological research, including laboratories for NMR spectroscopy, X-ray diffraction/scattering, and cryocrystallography; electron cryomicroscopy, structural electron microscopy, and 3-D image reconstruction; microspectrophotometry; calorimetry; light and fluorescence microscopy; surface chemistry; and tissue culture. In addition, facilities are available for research and studies on actin-myosin and muscle, atherosclerosis, electrophysiology and ion channels, enzymes, filaments and helices, lipoproteins, membrane biology, membrane proteins and lipid-binding proteins, molecular modeling, protein folding, protein crystallography, signaling, structural biology, and vision. The medical campus offers a centralized medical-scientific library, animal-care facilities, cell and molecular biology facilities, and numerous research laboratories in other basic and clinical departments.

Financial Aid

Full financial aid is provided for all students in the Ph.D. program. Aid is in the form of complete tuition remission and a graduate assistantship providing a competitive monthly stipend. The stipend for the 2005–06 academic year was $27,000.

Cost of Study

Tuition for the 2005–06 academic year was $31,530, plus fees and student health insurance.

Living and Housing Costs

Graduate students typically live in apartments within Boston or the surrounding communities. In shared apartments, rents are typically $500 and more per month.

Student Group

Typically, 6 to 8 graduate students matriculate into the Ph.D. program each year. Approximately 800 graduate students are enrolled at the medical school in various departments and programs.

Location

The Boston University School of Medicine is located within the South End neighborhood of Boston. This neighborhood has convenient access to public transportation and is within walking distance of downtown Boston. The Boston area is home to many major universities and colleges and provides a stimulating and exciting environment for biomedical research and collaboration.

The University and The Department

Incorporated in 1869, Boston University is a private, coeducational, nonsectarian, urban university. Nineteen thousand full-time undergraduates, 2,000 graduate students, and more than 2,400 faculty members contribute to the unique academic atmosphere and research excellence of this world-renowned university. The medical school campus is a short ride on a free and convenient shuttle bus from the Charles River campus, allowing easy access and interaction between the two campuses. Along with the School of Medicine, the medical campus includes the Boston Medical Center and its two hospitals, a teaching facility, the Goldman School of Dental Medicine, the School of Public Health, and private research enterprises. The faculty of the Department of Physiology and Biophysics maintains an internationally recognized research and instructional program and shares a deep commitment to biomedical research training and education.

Applying

The general requirements for applying to the Department of Physiology and Biophysics are detailed in the *Boston University School of Medicine, Division of Graduate Medical Sciences Catalogue.* Standard requirements include scores from the GRE General Test (a GRE Subject Test in an area related to physiology or biophysics is optional but recommended). TOEFL scores are required for applicants from other countries. Application can be made to the Graduate School or directly to the Department.

Correspondence and Information

Dr. C. James McKnight, Chair
Admissions and Student Affairs Committee
Department of Physiology and Biophysics
Boston University School of Medicine
700 Albany Street, W-407
Boston, Massachusetts 02118-2526

Phone: 617-638-4042
Fax: 617-638-4041
E-mail: info@bu.edu
Web site: http://www.bumc.bu.edu/Biophysics

Boston University

THE FACULTY AND THEIR RESEARCH

Christopher W. Akey, Professor; Ph.D., Cornell. Structural electron microscopy and X-ray crystallography are being used to study the function of protein translocation channels and their accessory proteins (ER ribosome-channel complex, type IVb secretion system), apoptosomes, signaling platforms, and histone chaperones.

Karen N. Allen, Associate Professor; Ph.D., Brandeis. The focus is the elucidation of enzyme mechanisms through X-ray crystallography and kinetics. Specific interests include: (1) How do enzyme superfamilies evolve to retain common chemistry while achieving substrate specificity? (2) How do protein-protein interactions of isozymes allow differential cellular functions?

David Atkinson, Professor and Research Professor of Biochemistry; Ph.D., Council for National Academic Awards (England). Structure and function of the plasma lipoproteins and apolipoproteins: the structural and molecular basis of lipid transport and the regulation of lipid metabolism in the body are investigated by X-ray crystallography, electron microscopy, and thermodynamic and spectroscopic methods.

Esther Bullitt, Assistant Professor; Ph.D., Brandeis. Research on the structure and function of macromolecular assemblies, using electron microscopy and computer image processing. Current studies focus on understanding how the structure of adhesion pili facilitates their role in the initiation of infection by pathogenic bacteria.

M. Carter Cornwall, Professor; Ph.D., Utah. Cellular and molecular mechanisms of dark adaptation and visual pigment regeneration in vertebrate retinal rod and cone photoreceptors. Optical methods (fluorescence, microspectrophotometry) as well as electrophysiological techniques (extracellular current measurements) are used at the single-cell level.

Fernando Garcia-Diaz, Associate Professor; Ph.D., Malaga (Spain). Electrophysiology of membrane transport; expression and regulation of ion channels; development of cochlear ganglion neurons.

Maria Gomez, Assistant Professor; M.D., Caldas (Colombia); Ph.D., Boston University. Light signaling in the two main evolutionary lineages of photoreceptors. Current emphasis is on the multiple messengers generated by activation of phospholipase C in microvillar photoreceptors and the role of the cyclic nucleotides in light adaptation in ciliary photoreceptors.

Hwai-Chen Guo, Associate Professor; Ph.D., Cornell. Structure and function of proteins; X-ray crystallographic studies of factors involved in genetic regulation; structure-based design of DNA-binding proteins and enzymes by protein engineering techniques.

Olga Gursky, Associate Professor; Ph.D., Brandeis. Folding, structure, and stability of apolipoproteins, their peptide analogues and their complexes with lipids, analyzed by spectroscopic, calorimetric, and X-ray diffraction methods.

James A. Hamilton, Professor of Medicine and Biochemistry; Ph.D., Indiana Bloomington. Laboratory research focuses on fatty acid (FA) transport, atherosclerosis, and lipid deposition (ectopic fat) in obese and diabetic persons. Structures of FA-binding proteins are determined by multinuclear 2-D NMR and 3-D NMR. Transport of FA into cells is studied by new fluorescence approaches. Atherosclerotic plaques and ectopic fat are studied by NMR and MR imaging (MRI) in animals and humans.

James F. Head, Professor; Ph.D., Birmingham (England). Crystallography is used to characterize protein-ligand and protein-protein interactions to understand how these interactions regulate physiological processes. Recent structures include the calcium-dependent photoprotein aequorin and a repressor protein associated with the regulation of antibiotic resistance in *E. coli*, MarR.

Haya Herscovitz, Assistant Professor; Ph.D., Tel-Aviv (Israel). Regulation of the assembly of apolipoprotein B (apoB) with lipids to form and secrete VLDL: chaperone-assisted folding of apoB. Biochemical, cell, and molecular biological techniques are used.

William Lehman, Professor; Ph.D., Princeton. Structural studies are carried out on the assembly, function, and regulation of actin-containing thin filaments in muscle and nonmuscle cells. Electron microscopy and 3-D image reconstruction are used to analyze thin-filament mechanisms.

Simon Levy, Associate Professor; Ph.D., Boston University. Modulation of excitability of nerve cells by second messengers. Current projects include calcium regulation and detection in nerve cells, role of calcium-induced calcium release in the excitability of neurons, and mechanisms of light transduction in photoreceptors.

C. James McKnight, Associate Professor; Ph.D., Texas Southwestern Medical Center at Dallas. Structure and function studies of peptides and proteins; macromolecular interactions; structure determination by multidimensional heteronuclear NMR; actin-binding proteins; lipoproteins.

Jeffrey R. Moore, Assistant Professor; Ph.D., Vermont. Research interests include molecular biomechanics of contractile proteins, the structure and function of motor proteins, and calcium regulation of muscle contraction. Laser traps and fluorescence microscopy are used to study the relationship between the structure, mechanics, and biochemistry of mechanoenzymes.

Enrico Nasi, Professor; Ph.D., Bryn Mawr. Ion channels mediating the receptor potential in visual cells; regulation of the phototransduction cascade. Specific issues include mechanisms of channel gating and permeation, modulation of photoresponse sensitivity, and kinetics. Main techniques: patch clamp recording, Ca fluorescence measurements, immunocytochemistry.

Judith D. Saide, Associate Professor; Ph.D., Boston University. Assembly of proteins in the myofibrils of *Drosophila* flight muscle; identification and interaction of proteins that are associated with the Z-band. Approaches include monoclonal antibody production, immunoelectron microscopy, gene cloning, and bacterial expression of protein fragments.

Barbara A. Seaton, Professor; Ph.D., MIT. Protein-membrane interactions of annexins, lung surfactant proteins, and blood coagulation proteins are studied using X-ray crystallography and complementary approaches such as mutagenesis and spectroscopy.

G. Graham Shipley, Professor and Professor of Biochemistry; Ph.D., D.Sc., Nottingham (England). Major research interests include membrane structure-function, membrane lipids, membrane receptors, and receptor-ligand interactions. Specific systems being studied using structural biology methods (e.g., protein crystallography, electron microscopy) are LDL receptor–LDL, insulin receptor–insulin, and ganglioside GM1–cholera toxin.

Donald M. Small, Professor and Chairman, Department of Physiology and Biophysics, and Professor of Medicine, Biophysics, and Biochemistry; M.D., UCLA. Interests include the fundamental mechanism by which apolipoprotein B (apoB) is translated and translocated across the endoplasmic reticulum and assembled with lipids to form a nascent lipoprotein particle. Also of interest are the physical properties of lipids, fats, oils, detergents, proteins, and lipid-protein assemblies.

Raphael A. Zoeller, Associate Professor; Ph.D., Texas A&M. Generation of mutant animal cell lines deficient in lipid biosynthesis. The use of somatic cell genetics to define roles of lipids in stroke, myocardial infarct, and neuromuscular diseases and to identify genes important for regulation of lipid metabolism.

HARVARD UNIVERSITY

Biophysics Program

Program of Study

The Committee on Higher Degrees in Biophysics offers a program of study leading to the Ph.D. degree. The committee comprises senior representatives of the Departments of Chemistry and Chemical Biology, Physics, and Molecular and Cellular Biology; the Division of Engineering and Applied Physics; and the Division of Medical Sciences. Students receive sufficient training in physics, biology, and chemistry to enable them to apply the concepts and methods of the physical sciences to the solution of biological problems.

An initial goal of the Biophysics Program is to provide an introduction through courses and seminars to several of the diverse areas of biophysics, such as structural molecular biology, cell and membrane biophysics, neurobiology, molecular genetics, physical biochemistry, and theoretical biophysics. The program is flexible, and special effort has been devoted to minimizing course work and other formal requirements. Students engage in several research rotations during their first two years. The qualifying examination is taken at the end of the second year to determine admission to candidacy. Students undertake thesis research as early as possible in the field and subject of their choice. Opportunities for thesis research are available in a number of special fields. The Ph.D. requires not less than three years devoted to advanced studies, including thesis research and the thesis. The Committee on Higher Degrees in Biophysics anticipates on average it takes five years, with the maximum being six years, to complete this program.

Research Facilities

Many more of the University's modern research facilities are available to the biophysics student because of the interdepartmental nature of the program. Research programs may be pursued in the Departments of Chemistry and Chemical Biology, Molecular and Cellular Biology, Applied Physics, and Engineering Sciences in Cambridge, as well as in the Departments of Biological Chemistry and Molecular Pharmacology, Genetics, Microbiology and Molecular Genetics, Neurobiology, Virology, and Cell Biology in the Harvard Medical School Division of Medical Sciences. Research may also be pursued in the Harvard School of Public Health, the Dana Farber Cancer Institute, Children's Hospital, Massachusetts General Hospital, Beth Israel Hospital, and more than ten other Harvard-affiliated institutions located throughout the cities.

Financial Aid

In 2006–07, all graduate students receive a stipend ($28,008 for twelve months) and full tuition and health fees ($32,882). A semester of teaching is required in the second year. Students are strongly encouraged to apply for fellowships from such sources as the National Science Foundation, the NDSEG, the Hertz Foundation, and the Ford Foundation. Full-time Ph.D. candidates in good academic standing are guaranteed full financial support through their sixth year of study or throughout their academic program if less than six years.

Cost of Study

Tuition and health fees for the 2006–07 academic year are $32,882. After two years in residence, students are eligible for a reduced rate (currently $10,476).

Living and Housing Costs

Accommodations in graduate residence halls are available at rents ranging from $4456 to $7003 per academic year. In addition, there are approximately 1,500 apartments available for graduate students in Harvard-owned buildings. Applications may be obtained from the Harvard University Housing Office, which also maintains a list of available private rooms, houses, and apartments in the vicinity.

Student Group

On average, the program enrolls 50 students annually. Currently, 15 women and 8 international students are enrolled in the program. Biophysics students intermingle in both their research and their social life with graduate students from the many other departments where research in the biophysical sciences is carried out.

Location

The Biophysics Program maintains a dual-campus orientation in the neighboring cities of Cambridge and Boston. Their proximity provides for a wide range of academic, cultural, extracurricular, and recreational opportunities, and the large numbers of theaters, museums, libraries, and universities contribute to enrich the scientific and cultural life of students. Because New England is compact in area, it is easy to reach countryside, mountains, and seacoast for winter and summer sports or just for a change of scenery.

The University

Established in 1636 in the Massachusetts Bay Colony, Harvard has grown to become a complex of many facilities whose educational vitality, social commitment, and level of cultural achievement contribute to make the University a leader in the academic world. Comprising more than 15,000 students and 3,000 faculty members, Harvard appeals to self-directed, resourceful students of diverse beliefs and backgrounds.

Applying

Students must apply by December 8, 2006, to be considered for admission in September 2007. Scores on the General Test of the Graduate Record Examinations are required except in special circumstances. GRE Subject Tests are recommended. Due to the early application deadline, applicants should plan to take the GRE test no later than October to ensure that original scores are received by December 8. Information about Graduate School fellowships and scholarships, admission procedures, and graduate study at Harvard may be obtained by writing to the Admissions Office.

Correspondence and Information

For information on the program:
Harvard Biophysics Program
Harvard Medical School Campus
Building C2, Room 122
240 Longwood Avenue
Boston, Massachusetts 02115
E-mail: biophys@fas.harvard.edu
Web site: http://fas.harvard.edu/~biophys

For application forms for admission and financial aid:
Admissions Office
Graduate School of Arts and Sciences
Holyoke Center
Harvard University
1350 Massachusetts Avenue
Cambridge, Massachusetts 02138
E-mail: admiss@fas.harvard.edu
Web site: http://www.gsas.harvard.edu

Harvard University

THE FACULTY AND THEIR RESEARCH

The following faculty members accept students for degree work in biophysics. Thesis research with other faculty members is possible by arrangement.

John Assad, Ph.D., Associate Professor of Neurobiology. Mechanisms of visual processing in the visual cortex of awake behaving monkeys.
Frederick M. Ausubel, Ph.D., Professor of Genetics. Molecular biology of microbial pathogenesis in plants and animals.
Howard Berg, Ph.D., Herchel Smith Professor of Physics and Professor of Molecular and Cellular Biology. Motile behavior of bacteria.
Stephen C. Blacklow, M.D., Ph.D., Associate Professor of Pathology. Molecular basis for specificity in protein folding and protein-protein interactions.
Martha L. Bulyk, Ph.D., Assistant Professor of Pathology. Computational methods, genomic and proteomic technologies in the study of DNA-protein interactions.
Lewis Cantley, Ph.D., Professor of Cell Biology and Systems Biology. Structural basis for specificity in eukaryotic signal transduction pathways.
James J. Chou, Ph.D., Assistant Professor of Biological Chemistry and Molecular Pharmacology. NMR spectroscopy on membrane-associated proteins and peptides.
George McDonald Church, Ph.D., Professor of Genetics. Human and microbial functional genomics; genotyping; gene expression regulatory network models.
David E. Clapham, M.D., Ph.D., Professor of Pediatrics and Professor of Neurobiology. Intracellular signal transduction.
Jon Clardy, Ph.D., Professor of Biological Chemistry and Molecular Pharmacology. Chemical ecology; biosynthesis; structure-based design.
Jonathan B. Cohen, Ph.D., Professor of Neurobiology. Structure and function of ligand-gated ion channels.
R. John Collier, Ph.D., Maude and Lillian Presley Professor of Microbiology and Molecular Genetics. Structure and activity of bacterial toxins.
David P. Corey, Ph.D., Professor of Neurobiology. Ion channels in neural cell membranes.
Bruce F. Demple, Ph.D., Professor of Toxicology. Cellular responses to oxidative stress; repair of free radical–damaged DNA.
John E. Dowling, Ph.D., Llura and Gordon Gund Professor of Neurosciences. Structure, function, genetics, and development of the retina.
Michael J. Eck, M.D., Ph.D., Associate Professor of Biological Chemistry and Molecular Pharmacology. Structural studies of proteins involved in signal transduction pathways.
Florian Engert, Ph.D., Assistant Professor of Molecular and Cellular Biology. Synaptic plasticity and neuronal networks.
Rachelle Gaudet, Ph.D., Assistant Professor of Molecular and Cellular Biology. Structural studies of the stereochemistry of signaling and transport through biological membranes.
David E. Golan, M.D., Ph.D., Professor of Biological Chemistry and Molecular Pharmacology and of Medicine. Membrane dynamics; membrane structure; cellular adhesion.
Jene A. Golovchenko, Ph.D., Professor of Physics. Experimental condensed-matter physics.
Edward E. Harlow, Ph.D., Professor of Biological Chemistry and Molecular Pharmacology. Tumor suppressor genes; cell-cycle control.
Stephen C. Harrison, Ph.D., Professor of Biological Chemistry and Molecular Pharmacology. Structure of viruses and viral membranes; protein-DNA interactions; structural aspects of signal transduction and membrane traffic; X-ray diffraction.
J. Woodland Hastings, Ph.D., Paul C. Mangelsdorf Professor of Natural Sciences. Molecular mechanisms in bioluminescence and circadian rhythms.
James M. Hogle, Ph.D., Professor of Biological Chemistry and Molecular Pharmacology. Structure and function of viruses and virus-related proteins; X-ray crystallography.
Donald E. Ingber, M.D., Ph.D., Judah Folkman Professor of Vascular Biology. Research in integrin signaling, cytoskeleton, and control of angiogenesis.
David Jeruzalmi, Ph.D., Assistant Professor of Molecular and Cellular Biology. Structural studies of nucleoprotein assemblies.
Tomas Kirchhausen, Ph.D., Professor of Cell Biology. Molecular mechanisms of membrane traffic; X-ray crystallography; chemical genetics.
Nancy Kleckner, Ph.D., Herchel Smith Professor of Molecular Biology. Chromosome metabolism in bacteria and yeast.
Roberto G. Kolter, Ph.D., Professor of Microbiology and Molecular Genetics. DNA protection from oxidative damage; cell-cell communication in biofilms; microbial evolution.
David R. Liu, Ph.D., John L. Loeb Associate Professor of the Natural Sciences. Organic chemistry and chemical biology.
Jun S. Liu, Ph.D., Professor of Statistics. Stochastic processes, probability theory, and statistical inference.
Gavin MacBeath, Ph.D., Assistant Professor of Chemistry and Chemical Biology. Molecular recognition in complex processes: protein trafficking, intercellular communication, and apoptosis.
Tom Maniatis, Ph.D., Thomas H. Lee Professor of Molecular and Cellular Biology. Eukaryotic gene expression.
Markus Meister, Ph.D., Jeff C. Tarr Professor of Molecular and Cellular Biology. Function of neuronal circuits.
Keith W. Miller, Ph.D., Mallinckrodt Professor of Pharmacology, Department of Anesthesia. Molecular mechanisms of regulatory conformation changes and drug action on membrane receptors and channels, using rapid kinetics, time-resolved photolabeling, and spectroscopy (EPR, fluorescence, NMR); characterization of lipid-protein interactions in membrane proteins.
Timothy Mitchison, Ph.D., Hasib Sabbagh Professor of Systems Biology. Cytoskeleton dynamics; mechanism of mitosis and cell locomotion; small-molecule inhibitors.
Venkatesh N. Murthy, Ph.D., Morris Khan Associate Professor of Molecular and Cellular Biology. Mechanisms of synaptic transmission and plasticity.
David Pellman, M.D., Associate Professor of Pediatrics. The mechanics and regulation of mitosis.
Mara Prentiss, Ph.D., Professor of Physics. Exploitation of optical manipulation to measure adhesion properties, including virus cell binding.
Tom A. Rapoport, Ph.D., Professor of Cell Biology. Mechanism of how proteins are transported across the endoplasmic reticulum membrane.
Frederick P. Roth, Ph.D., Assistant Professor of Biological Chemistry and Molecular Pharmacology. Computational molecular biology.
Gary Ruvkun, Ph.D., Professor of Genetics. Genetic control of developmental timing, neurogenesis, and neural function.
Bernardo L. Sabatini, Ph.D., Assistant Professor of Neurobiology. Regulation of synaptic transmission and dendritic function in the mammalian brain.
Aravinthan D. T. Samuel, Ph.D., Assistant Professor of Physics. Topics in biophysics, neurobiology, and animal behavior.
Stuart L. Schreiber, Ph.D., Morris Loeb Professor of Chemistry and Chemical Biology. Forward and reverse chemical genetics: using small molecules to explore biology.
Brian Seed, Ph.D., Professor of Genetics. Genetic analysis of signal transduction in the immune system.
Eugene Shakhnovich, Ph.D., Professor of Chemistry and Chemical Biology. Theory and experiments in protein folding and design; theory of molecular evolution; rational drug design and physical chemistry of protein-ligand interactions; theory of complex systems.
Steven E. Shoelson, M.D., Ph.D., Professor of Medicine. Structural and cellular biology of insulin signal transduction, insulin, resistance, diabetes, and obesity.
Pamela Silver, Ph.D., Professor of Systems Biology. Nucleocytoplasmic transport; RNA-protein interactions; protein methylation; cell-based small-molecule screens.
Timothy A. Springer, Ph.D., Latham Family Professor of Pathology. Molecular biology of immune cell interactions.
Shamil R. Sunyaev, Ph.D., Assistant Professor of Genetics. Population genetic variation and genomic divergence, with a focus on protein coding regions.
Jack W. Szostak, Ph.D., Professor of Genetics. Directed evolution; information content and molecular function; self-replicating systems.
Antoine van Oijen, Ph.D., Assistant Professor of Biological Chemistry and Molecular Pharmacology. Single-molecule novel fluorescence and nanomanipulation studies of protein-protein and protein-nucleic acid interactions.
Gregory L. Verdine, Ph.D., Professor of Chemistry and Chemical Biology. Protein–nucleic acid interactions; transcriptional regulation; X-ray crystallography.
Gerhard Wagner, Ph.D., Elkan Blout Professor of Biological Chemistry and Molecular Pharmacology. Protein and nucleic acid structure, interaction, and mobility; NMR spectroscopy.
John R. Wakeley, Ph.D., Thomas D. Cabot Associate Professor of Organismic and Evolutionary Biology. Theoretical population genetics.
Christopher T. Walsh, Ph.D., Hamilton Kuhn Professor of Biological Chemistry and Molecular Pharmacology. Enzymatic reaction and antibiotic synthesis mechanisms.
Thomas Walz, Ph.D., Assistant Professor of Cell Biology. High-resolution electron microscopy.
George M. Whitesides, Ph.D., Mallinckrodt Professor of Chemistry. Molecular pharmacology; biosurface chemistry; virology.
Xiaoliang Sunney Xie, Ph.D., Professor of Chemistry and Chemical Biology. Single-molecule spectroscopy and dynamics; molecular interaction and chemical dynamics in biological systems.
Gary Yellen, Ph.D., Professor of Neurobiology. Molecular physiology of ion channels: functional motions, drug interactions, and electrophysiological mechanisms.
Xaiowei Zhuang, Ph.D., Assistant Professor of Chemistry and Chemical Biology and of Physics. Single-molecule biophysics.

MEDICAL COLLEGE OF WISCONSIN

Graduate School of Biomedical Sciences
Biophysics Graduate Program

Programs of Study	The Biophysics Graduate Program has two tracks, each leading to the Ph.D. degree: molecular biophysics and magnetic resonance biophysics. The molecular biophysics track encompasses the investigation, detection, and use of free radicals and paramagnetic metal ions in biological systems. Free radicals are involved in many disease processes and yet are also an integral part of cellular communication. Free radicals can also be used to label proteins and to map out protein structure, providing information on protein dynamics and conformational changes that cannot be obtained from crystal structure data. In addition, free-radical labels can be used to probe the dynamics of biological membranes. Paramagnetic metal ions are central to most biological processes and electron transfer systems. A major technique used in the above studies is electron paramagnetic resonance (EPR), and the Department of Biophysics houses one of the few National Centers for EPR-related research. Students with a more physical background may specialize in EPR instrumentation. In the magnetic resonance biophysics track, particular emphasis is placed on magnetic resonance imaging (MRI) and magnetic resonance spectroscopy (MRS). Functional magnetic resonance imaging (fMRI) of the human brain is an active research area (neuroscience, contrast mechanisms, technical development). Other opportunities exist in both tracks, as is evident from the research interests of the faculty.

All students are expected to be comfortable with mathematical approaches to biological problems. Molecular biophysics students must demonstrate general proficiency at an undergraduate level in chemistry, biology, and mathematics prior to advancement to candidacy. Magnetic resonance biophysics students must demonstrate proficiency in neuroscience, physics, and mathematics prior to advancement to candidacy. The biophysics program offers a number of core courses as well as courses in each of the two tracks. All graduate students in biophysics are required to take the biophysics seminar and a journal club. Additional courses may be taken from the basic science and clinical departments throughout the Medical College and also from Marquette University and the University of Wisconsin–Milwaukee. Every effort is made to tailor the formal course work to the particular needs of the student.

Research Facilities
The program is composed of faculty members actively involved in research in clinical and basic science departments. This diversity provides excellent research opportunities in a large number of areas. Most faculty members have external research support to provide the most up-to-date equipment and facilities. Students also have access to the National Biomedical EPR Center, the Allen-Bradley Animal Laboratory, the Animal Resource Center, the Todd Wehr Library, the research laboratories at the Clement J. Zablocki VA Medical Center, and the Magnetic Resonance Imaging Center. A 3-Tesla human MRI scanner and a 9.4-Tesla animal MRI scanner are available for research involving human subjects and animal models.

Financial Aid
Every full-time Ph.D. candidate in good academic standing is guaranteed a full scholarship, including tuition and stipend, throughout the academic program. Candidates are only accepted when financial support is available. This support may come from any of several sources, such as the department or program, the graduate school, or an extramural funding agency. Most students are classified as fellows, trainees, or research assistants and spend no time during their academic programs working as teaching assistants.

Cost of Study
All full-time degree candidates in good standing receive a full scholarship, as explained in the Financial Aid section, for at least five years.

Living and Housing Costs
Living accommodations, for both single and married students, are available nearby and in the regional Milwaukee area.

Student Group
The total enrollment in the Graduate School of the Medical College is about 620. Approximately 800 medical students are enrolled in the College. The Biophysics Graduate Program has 20 students.

Location
The Medical College is located on the suburban grounds of the Milwaukee Regional Medical Center. The center houses hospitals, other medical institutions, and several small companies devoted to the development of new medical devices. Milwaukee offers numerous cultural events in music, art, and theater. The city is rich in old-world tradition and has a diverse ethnic heritage leading to an abundant variety of fine restaurants in the area. Milwaukee County has an excellent park system, with more than 125 parks scattered throughout the city and metropolitan area, and an extensive bicycle trail system. The city is bordered on the east by Lake Michigan and lies within commuting distance of 200 inland lakes, which provide recreation throughout the year.

The Medical College
Medical College of Wisconsin was originally chartered as the Marquette University School of Medicine in 1913 and became a freestanding, private corporation in 1967. The Medical College moved into its quarters on the grounds of the Milwaukee Regional Medical Center in 1978; the College now neighbors the majority of its clinical teaching facilities. The major teaching hospitals of the Medical College—Froedtert Memorial Lutheran Hospital, Veterans Administration Medical Center, Children's Hospital of Wisconsin, and the Milwaukee County Mental Health Complex—constitute the core of the Medical Center of southeastern Wisconsin. The faculty members of the clinical and basic science departments frequently interact and have numerous collaborative research investigations.

Applying
Students with an undergraduate major in mathematics, science, or engineering are invited to apply. Applications are also invited from students interested in a biophysical approach to neuroscience. Students interested in the molecular biophysics track should apply to the Interdisciplinary Program in Biomedical Sciences. Students interested in the MRI track should apply directly to the Department of Biophysics. Application forms and additional information about the program can be obtained by writing to the Biophysics Graduate Program address.

Correspondence and Information
Biophysics Graduate Program
Medical College of Wisconsin
8701 Watertown Plank Road
P.O. Box 26509
Milwaukee, Wisconsin 53226-0509

Phone: 414-456-4000
Fax: 414-456-6512
E-mail: gradschool@mcw.edu
Web site: http://www.biophysics.mcw.edu

Medical College of Wisconsin

THE FACULTY AND THEIR RESEARCH

Molecular Biophysics

William E. Antholine, Associate Professor, Department of Biophysics; Ph.D., Iowa State, 1971. Uptake studies of metal antitumor agents; development of radiosensitizing agents.

Brian Bennett, Associate Professor, Department of Biophysics; Ph.D., Sussex (England), 1994. Transition metal–containing enzymes and electron-transfer proteins: the roles of transition icon catalytic and redox-active centers in life-threatening conditions.

Jimmy B. Feix, Professor, Department of Biophysics; Ph.D., Kentucky, 1981. Electron spin resonance studies of model and biological membranes; membrane interactions of photosensitizing dyes; effects of oxidative stress; preparation and use of liposomes.

Neil Hogg, Associate Professor, Department of Biophysics; Ph.D., Essex (England), 1993. Biological chemistry of nitric oxide and its oxidation products in pathophysiology.

James S. Hyde, Professor, Department of Biophysics; Ph.D., MIT, 1959. Spin label probes of rotational, lateral, and transverse diffusion of synthetic and biological membranes; magnetic resonance imaging; functional neuroimaging.

Balaraman Kalyanaraman, Professor and Chairman, Department of Biophysics; Ph.D., Alabama, 1978. Free-radical metabolites in the biological system; generation of free radicals from photodynamic treatment.

Candice S. Klug, Associate Professor, Department of Biophysics; Ph.D., Medical College of Wisconsin, 1999. Protein structure and functional dynamics studies using site-directed spin labeling electron paramagnetic resonance spectroscopy.

Witold K. Subczynski, Associate Professor, Department of Biophysics; Ph.D., Moscow, 1976. Spin label studies on membrane dynamics and organization (raft-domain formation), spin label oximetry and NO-metry.

Jeannette Vásquez Vivar, Assistant Professor, Department of Biophysics; Ph.D., São Paulo, 1992. Mechanisms regulating superoxide and nitric oxide formation from nitric oxide synthase; tetrahydrobiopterin metabolism; applications of electron paramagnetic resonance spin trapping methodology.

MRI/MRS

Jeffrey R. Binder, Associate Professor, Department of Neurology; M.D., Nebraska, 1986. Functional magnetic resonance imaging; human brain auditory and language systems; fMRI in epilepsy patients.

Edgar A. DeYoe III, Professor, Department of Radiology; Ph.D., Rochester, 1983. Functional magnetic resonance imaging mapping of the human visual cortex.

Andrew S. Greene, Professor, Department of Physiology; Ph.D., Johns Hopkins, 1985. Mathematical modeling and simulation of complex biological systems.

Anthony D. Hudetz, Associate Professor, Department of Anesthesiology; Ph.D., Semmelweis (Hungary), 1985. Cerebral microcirculation control.

James S. Hyde, Professor, Department of Biophysics; Ph.D., MIT, 1959. Spin label probes of rotational, lateral, and transverse diffusion of synthetic and biological membranes; magnetic resonance imaging; functional neuroimaging.

Andrzej Jesmanowicz, Associate Professor, Department of Biophysics; Ph.D., Copernicus (Poland), 1975. MRI physics and image processing.

Shi-Jiang Li, Professor, Department of Biophysics; Ph.D., Ohio, 1985. Magnetic resonance imaging and magnetic resonance spectroscopy studies of tumor hypoxia and oxygenation.

Stephen M. Rao, Professor, Department of Neuropsychology; Ph.D., Wayne State, 1979. Functional magnetic resonance imaging.

Daniel B. Rowe, Assistant Professor, Department of Biophysics; Ph.D., California, Riverside, 1998. Multivariate Bayesian statistics for fMRI activation determination from time courses; psychometric Bayesian factor analysis; functional connectivity.

Kathleen M. Schmainda, Associate Professor, Department of Radiology; Ph.D., Harvard–MIT Division of Health Sciences and Technology, 1993. Quantification of biologically relevant parameters with exogenous, contrast-enhanced MRI; development of MRI methods to diagnose disease, with a focus on diabetes and cancer.

Ming Zhao, Assistant Professor, Department of Biophysics; Ph.D., Cambridge, 2001. Developing noninvasive, target-specific imaging methods for the detection and understanding of biological/pathological events.

Affiliated Faculty

Joseph H. Battocletti, Professor, Department of Neurosurgery; Ph.D., UCLA, 1961. Blood flowmeter development; NMR imaging of human fluids and tissues.

Michael T. Gillin, Associate Professor, Department of Radiation Oncology; Ph.D., California, Davis, 1970. Radiation dosimetry and radiation effects; computer interface systems; beam collimation.

Albert W. Girotti, Professor, Department of Biochemistry; Ph.D., Massachusetts, 1965. Mechanism of action of antineoplastic photosensitizing agents; oxygen radical-induced damage in biological membranes; photodynamic effects on biological membranes.

William R. Hendee, Professor and Senior Associate Dean, Department of Radiology; Ph.D., Texas, 1962. Visual perception and cognition: radiation standards and protection.

Donald R. Jacobson, Assistant Professor, Department of Radiology; Ph.D., Medical College of Wisconsin, 1991. Medical imaging; mammography; computed tomography.

John P. Kampine, Professor and Chairman, Department of Anesthesiology; M.D., 1961, Ph.D., 1965, Marquette. Neural control of the circulation; role of cardiopulmonary afferents in cardiovascular regulation.

Gretchen S. Mandel, Associate Professor, Department of Medicine; Ph.D., Pennsylvania, 1972. X-ray crystallography of biologically and medically important molecules; epitaxial growth in renal calculi.

Neil S. Mandel, Professor, Department of Medicine; Ph.D., Pennsylvania, 1971. Molecular basis of crystal-related arthritic, renal, and pulmonary diseases at the atomic level of resolution; crystallographic structures of biologically and medically important molecules.

Henry M. Miziorko, Professor, Department of Biochemistry; Ph.D., Pennsylvania, 1974. Mechanisms of enzyme regulation and catalysis by application of physical techniques to secure structural and kinetic information.

Charles R. Wilson, Associate Professor, Department of Radiology, and Recruitment Director, Biophysics Graduate Program; Ph.D., Wisconsin–Madison, 1972. Body composition analysis using techniques of X-ray photon-absorptiometry, biomechanics of bone, and radiologic medical imaging.

STONY BROOK UNIVERSITY, STATE UNIVERSITY OF NEW YORK

Biophysics Program

Program of Study

Biophysics is administered by a group of faculty members from a number of departments at Stony Brook, Cold Spring Harbor Laboratory, and Brookhaven National Laboratory. Because of the diverse backgrounds of students and the wide interests of the faculty, the program of course work is flexible and can be structured around several areas, such as structural biology. Most students enter the Department of Physiology and Biophysics, though it is also possible to enter any of the other departments with which Biophysics Program faculty members are affiliated. The program of study leads to the Doctor of Philosophy degree from the department in which the student enrolls.

On arrival at Stony Brook, each student is assigned a Student Advisory Committee of 4 faculty members, who are selected to fit the interests of the student. The Advisory Committee and the student set up the program of course work and rotations that take place during the first two years of the program. Rotations consist of three to six months of work in the laboratory of one of the faculty members. During this period, the student performs experiments and does selective readings in literature related to the research interests of the faculty member.

At the end of the third semester, a student usually advances to candidacy for the Ph.D. degree. Advancement to candidacy entails two requirements: the Advisory Committee must certify that the student has adequate preparation in biology, chemistry, mathematics, and physics; and the student must pass a qualifying examination. After advancement to candidacy, the student concentrates on independent research, performed in the laboratory and under the guidance of his or her thesis adviser.

Research Facilities

All of the faculty members have independent research programs and well-equipped laboratories to carry out these programs. There are major shared facilities for X-ray diffraction, nuclear magnetic resonance, electron microscopy, molecular modeling, and molecular biology at Stony Brook and Cold Spring Harbor Laboratory. In addition, Brookhaven National Laboratory offers unique possibilities employing synchrotron radiation and neutron diffraction.

Financial Aid

Students are normally admitted with a graduate stipend, which currently amounts to $25,000 per year, and a tuition waiver.

Cost of Study

Tuition for full-time study for the 2006–07 academic year is $6900 for New York State residents and $10,920 for out-of-state students. Miscellaneous fees can be as high as $700 per year with health insurance.

Living and Housing Costs

On-campus housing is available for either single or married students. The cost of apartments for married students varies according to size. A single student can share a room for $4368 per year. A furnished room in a private house costs $300 to $400 per month. Students sometimes share a complete house at a cost of $300 to $500 per month per student. Private studio apartments are available at $400 to $550 per month, and one-bedroom apartments rent for $750 to $950 per month. Although the cost of living on Long Island is relatively high, the stipend allows a modest but reasonable lifestyle for a single student. Married students need additional income.

Student Group

There are approximately 5,000 graduate students enrolled at Stony Brook. Of these, about 200 are studying in the area of health sciences and 20 are pursuing research and training in biophysics. This student group provides a cosmopolitan environment for academic and social interactions.

Location

Stony Brook is located on the North Shore of Long Island, in a pleasant environment of historic villages, harbors, beaches, and wooded hills. The Long Island Rail Road, with a railway station on the edge of the campus, provides service to Manhattan, about 50 miles away. A nearby ferry to Connecticut gives access to trains and highways to the north. The Cold Spring Harbor Laboratory and the Brookhaven National Laboratory are both within 20 miles of the campus; some of the program's faculty members have laboratories at these institutions. Despite the nearness to many cities, laboratories, and universities, the Stony Brook area retains its physical beauty and the flavor of its peaceful New England heritage. The Fine Arts Center offers exhibitions, concerts, and theatrical performances. There are abundant athletic facilities on campus; beaches and boating are nearby.

The University

The State University of New York at Stony Brook initially attracted an internationally acclaimed faculty in the physical sciences and mathematics, owing in part to the establishment of the Institute of Theoretical Physics and the proximity of the Brookhaven National Laboratory. The development of biological sciences was stimulated by the opening on campus of the Health Sciences Center, which includes the University Hospital and schools for training students in medicine and dentistry. The proximity of medical scientists, biologists, mathematicians, and physical scientists provides many opportunities for interactions and collaborations. The campus now has about 100 buildings on 1,000 acres, housing more than 1,200 faculty members.

Applying

Applications must be received by February 1 for fall admission. All students must arrange to have their scores on the General Test of the Graduate Record Examinations sent to the department, or the application cannot be evaluated. In addition, international applicants must provide their score on the Test of English as a Foreign Language (TOEFL). Online applications can be completed at https://app.applyyourself.com/?id=sunysb-gs.

Correspondence and Information

Biophysics Graduate Admissions
Department of Physiology and Biophysics
Stony Brook University, State University of New York
Stony Brook, New York 11794-8661
Phone: 631-444-2299
Web site: http://www.pnb.sunysb.edu/

Stony Brook University, State University of New York

THE FACULTY AND THEIR RESEARCH

P. R. Brink, Ph.D., Professor and Chairman, Department of Physiology and Biophysics. Biophysical properties of gap junctions.
C. Clausen, Ph.D., Associate Professor, Department of Physiology and Biophysics. Transport properties of epithelia.
I. Cohen, M.D., Ph.D., Leading Professor, Department of Physiology and Biophysics. Cardiac electrophysiology and biophysics.
J. Dilger, Ph.D., Professor, Departments of Anesthesiology and Physiology and Biophysics. Anesthetic effects on synaptic transmission.
R. T. Mathias, Ph.D., Professor, Department of Physiology and Biophysics. Biophysical and electrical properties of the lens and cardiac muscle.
D. McKinnon, Ph.D., Associate Professor, Department of Neurobiology and Behavior. Molecular biology of ion channels.
S. McLaughlin, Ph.D., Professor, Department of Physiology and Biophysics. Electrostatic properties of membranes and second messengers.
L. Mendell, Ph.D., Professor and Chairman, Department of Neurobiology and Behavior. Physiology and modifiability of synapses in the spinal cord.
W. T. Miller, Ph.D., Professor, Department of Physiology and Biophysics. Protein structure and function; molecular mechanisms of signal transduction.
L. Moore, Ph.D., Professor, Department of Physiology and Biophysics. Biophysical mechanisms of kidney function.
N. Nassar, Ph.D., Assistant Professor, Department of Physiology and Biophysics. X-ray crystallography; signal transduction.
M. Rebecchi, Ph.D., Associate Professor, Department of Physiology and Biophysics. Regulation of phosphoinositide specific phospholipase C.
J. R. Sachs, M.D., Professor, Department of Medicine. Mechanism of cation transport in mammalian cells.
S. F. Scarlata, Ph.D., Professor, Department of Physiology and Biophysics. G-protein signaling and HIV assembly.
S. O. Smith, Ph.D., Professor, Department of Biochemistry and Cell Biology. Structure and function of membrane proteins.
W. Van der Kloot, Ph.D., Distinguished Professor Emeritus, Department of Physiology and Biophysics. Synaptic transmission at the neuromuscular junction.

Affiliated Faculty

C. De los Santos, Ph.D., Associate Professor, Department of Pharmacology. NMR structural studies of nucleic acids and proteins.
M. Eisenberg, Ph.D., Associate Professor, Department of Pharmacological Sciences. Computer-assisted modeling of biomolecules.
A. P. Grollman, M.D., Professor, Department of Pharmacological Science. Chemical carcinogenesis and mutagenesis.
C. J. Jacobsen, Ph.D., Assistant Professor and NSF Presidential Faculty Fellow, Department of Physics. Soft X-ray microscopy of cellular structure and materials structure.
L. Joshua-Tor, Ph.D., Assistant Investigator, Cold Spring Harbor Laboratory. Structural biology; X-ray crystallography; molecular recognition; transcription; proteases.
J. Kirz, Ph.D., Professor, Department of Physics. Microscopy and microanalysis of cellular architecture with soft X-rays.
A. Krainer, Ph.D., Professor, Cold Spring Harbor Lab. Mechanisms and regulation of messenger RNA splicing in human cells.
W. J. Lennarz, Ph.D., Professor and Chairman, Department of Biochemistry. Biosynthesis and function of glycoproteins in development.
E. London, Ph.D., Professor, Department of Biochemistry. Membrane lipid-protein interactions; protein toxin structure and function.
C. Malbon, Ph.D., Professor, Department of Pharmacology. Elucidating the genetic basis of developmental and metabolic diseases.
G. Matthews, Ph.D., Professor, Department of Neurobiology and Behavior. Cellular biophysics of electrical signals in the retina.
H. Metcalf, Ph.D., Professor, Department of Physics. Physical analysis of biological systems.
D. Raleigh, Ph.D., Associate Professor, Department of Chemistry. Experimental studies of protein folding and amyloid formation.
C. T. Rubin, Ph.D., Professor, Department of Orthopaedics. Cellular mechanisms responsible for adaptation in bone.
N. Sampson, Ph.D., Associate Professor, Department of Chemistry. Enzyme mechanisms and protein structure-function relationships.
R. Setlow, Ph.D., Senior Scientist, Department of Biology, Brookhaven National Laboratory. DNA damage and repair.
S. M. Sherman, Ph.D., Professor, Department of Neurobiology and Behavior. Plasticity of the developing visual system.
S. R. Simon, Ph.D., Professor, Departments of Biochemistry and Pathology. Interaction of neutrophil elastase with connective tissue.
G. Stell, Ph.D., Professor, Departments of Chemistry and Mechanical Engineering. Theory of the osmotic properties of ionic solutions and membranes.
P. Tonge, Ph.D., Assistant Professor, Department of Chemistry. Enzyme mechanisms in antitubercular drugs and Alzheimer's disease.
S. Wong, Ph.D., Assistant Professor, Department of Chemistry. Fundamental structure correlations in unique nanostructures.

Section 5
Botany and Plant Biology

This section contains a directory of institutions offering graduate work in botany and plant biology, followed by in-depth entries submitted by institutions that chose to prepare detailed program descriptions. Additional information about programs listed in the directory but not augmented by an in-depth entry may be obtained by writing directly to the dean of a graduate school or chair of a department at the address given in the directory.

For programs offering related work, see also in this book Biochemistry; Biological and Biomedical Sciences; Cell, Molecular, and Structural Biology; Ecology, Environmental Biology, and Evolutionary Biology; Entomology; Genetics, Developmental Biology, and Reproductive Biology; and Microbiological Sciences. In Book 2, see Architecture (Landscape Architecture) and Economics (Agricultural Economics and Agribusiness); in Book 4, see Agricultural and Food Sciences; and in Book 5, see Agricultural Engineering and Bioengineering.

CONTENTS

Program Directories

Announcements

Cross-Discipline Announcement

Close-Ups

Botany

Auburn University, Graduate School, College of Sciences and Mathematics, Department of Biological Sciences, Auburn University, AL 36849. Offers botany (MS, PhD); microbiology (MS, PhD); zoology (MS, PhD). *Faculty:* 28 full-time (5 women). *Students:* 44 full-time (23 women), 49 part-time (27 women); includes 8 minority (5 African Americans, 3 Hispanic Americans), 17 international. 65 applicants, 51% accepted, 22 enrolled. In 2005, 11 master's, 1 doctorate awarded. *Entrance requirements:* For master's and doctorate, GRE General Test. Additional exam requirements/recommendations for international students: Required—TOEFL. *Application deadline:* For fall admission, 7/7 for domestic students; for spring admission, 11/24 for domestic students. Electronic applications accepted. *Financial support:* Research assistantships, teaching assistantships available. *Unit head:* Dr. James M Barbaree, Chair, 334-844-1647, Fax: 334-844-1645. *Application contact:* Dr. Stephen L. McFarland, Acting Dean of the Graduate School, 334-844-4700.

California State University, Chico, Graduate School, College of Natural Sciences, Department of Biological Sciences, Program in Botany, Chico, CA 95929-0515. Offers MS. *Degree requirements:* For master's, thesis, seminar presentation. *Entrance requirements:* For master's, GRE General Test, 2 letters of recommendation. Additional exam requirements/recommendations for international students: Required—TOEFL (minimum score 550 paper-based; 213 computer-based). Electronic applications accepted.

California State University, Fullerton, Graduate Studies, College of Natural Science and Mathematics, Department of Biological Science, Fullerton, CA 92834-9480. Offers biological science (MS); botany (MS); microbiology (MS). Part-time programs available. *Students:* 18 full-time (10 women), 41 part-time (26 women); includes 23 minority (3 African Americans, 9 Asian Americans or Pacific Islanders, 11 Hispanic Americans), 3 international. Average age 27. 61 applicants, 39% accepted, 19 enrolled. In 2005, 10 degrees awarded. *Degree requirements:* For master's, thesis. *Entrance requirements:* For master's, DAT, GRE General Test and GRE Subject Test, or MCAT, minimum GPA of 3.0 in biology. Application fee: $55. *Expenses:* Tuition, nonresident: part-time $339 per unit. *Financial support:* Teaching assistantships, career-related internships or fieldwork, Federal Work-Study, institutionally sponsored loans, and scholarships/grants available. Support available to part-time students. Financial award application deadline: 3/1. *Faculty research:* Glycosidase release and the block to polyspermy in ascidian eggs. *Unit head:* Dr. Robert Koch, Chair, 714-278-3614. *Application contact:* Dr. Michael Horn, Adviser, 714-278-3707.

Claremont Graduate University, Graduate Programs, Program in Botany, Claremont, CA 91711-6160. Offers MS, PhD. Part-time programs available. *Faculty:* 3 full-time (1 woman). *Students:* 11 full-time (6 women); includes 1 minority (Asian American or Pacific Islander), 2 international. Average age 37. In 2005, 1 master's, 1 doctorate awarded. Terminal master's awarded for partial completion of doctoral program. *Degree requirements:* For master's and doctorate, 2 foreign languages, thesis/dissertation. *Entrance requirements:* For master's and doctorate, GRE General Test. *Application deadline:* For fall admission, 2/15 for domestic students. Applications are processed on a rolling basis. Electronic applications accepted. *Expenses:* Tuition: Full-time $27,902; part-time $1,214 per unit. Required fees: $1,600; $800 per term. Tuition and fees vary according to degree level and program. *Financial support:* Fellowships, research assistantships, Federal Work-Study, institutionally sponsored loans, and tuition waivers (full) available. Support available to part-time students. Financial award application deadline: 2/15; financial award applicants required to submit FAFSA. *Unit head:* Mark Porter, Acting Chair, 909-625-8767, Fax: 909-626-3489, E-mail: marc.porter@cgu.edu. *Application contact:* Ann Joslin, Program Coordinator, 909-625-8767 Ext. 251, Fax: 909-626-3489, E-mail: ann.joslin@cgu.edu.

Colorado State University, Graduate School, College of Natural Sciences, Department of Biology, Fort Collins, CO 80523-0015. Offers botany (MS, PhD); zoology (MS, PhD). Part-time programs available. *Faculty:* 22 full-time (6 women). *Students:* 18 full-time (11 women), 21 part-time (11 women); includes 4 minority (2 American Indian/Alaska Native, 1 Asian American or Pacific Islander, 1 Hispanic American), 7 international. Average age 28. 43 applicants, 33% accepted, 9 enrolled. In 2005, 5 master's, 4 doctorates awarded. Terminal master's awarded for partial completion of doctoral program. *Degree requirements:* For master's, thesis (for some programs), comprehensive exam; for doctorate, thesis/dissertation, comprehensive exam. *Entrance requirements:* For master's and doctorate, GRE General Test, minimum GPA of 3.0. Additional exam requirements/recommendations for international students: Required—TOEFL. *Application deadline:* For fall admission, 1/15 priority date for domestic students, 1/15 priority date for international students; for spring admission, 11/1 for domestic students, 11/1 for international students. Applications are processed on a rolling basis. Application fee: $50. Electronic applications accepted. *Expenses:* Tuition, state resident: full-time $3,690; part-time $205 per credit. Tuition, nonresident: full-time $14,958; part-time $831 per credit. Required fees: $1,061. *Financial support:* In 2005–06, 127 students received support, including 6 fellowships with full tuition reimbursements available (averaging $22,500 per year), 27 research assistantships with full tuition reimbursements available (averaging $13,000 per year), 94 teaching assistantships with full tuition reimbursements available (averaging $12,888 per year); career-related internships or fieldwork, Federal Work-Study, institutionally sponsored loans, and traineeships also available. Financial award application deadline: 2/15. *Faculty research:* Aquatic and terrestrial ecology, cell biology and genetics, plant/animal physiology, developmental biology, evolutionary biology. Total annual research expenditures: $3.6 million. *Unit head:* Daniel R. Bush, Chair, 970-491-7011, Fax: 970-491-0649. *Application contact:* Dorothy Ramirez, Graduate Coordinator, 970-491-1923, Fax: 970-491-0649, E-mail: dorothy.ramirez@colostate.edu.

Connecticut College, Graduate School, Department of Botany, New London, CT 06320-4196. Offers MA, MAT. Part-time programs available. *Degree requirements:* For master's, thesis. *Entrance requirements:* For master's, GRE or MAT. *Faculty research:* Tidal marsh ecology, upland vegetation dynamics, plant development, halophyte physiology.

Emporia State University, School of Graduate Studies, College of Liberal Arts and Sciences, Department of Biological Sciences, Emporia, KS 66801-5087. Offers botany (MS); environmental biology (MS); general biology (MS); microbial and cellular biology (MS); zoology (MS). Part-time programs available. *Faculty:* 13 full-time (2 women), 4 part-time/adjunct (2 women). *Students:* 3 full-time (1 woman), 17 part-time (7 women), 3 international. 10 applicants, 90% accepted, 7 enrolled. In 2005, 5 degrees awarded. *Degree requirements:* For master's, comprehensive exam or thesis. *Entrance requirements:* For master's, GRE, appropriate undergraduate degree, interview, letters of reference. Additional exam requirements/recommendations for international students: Required—TOEFL (minimum score 450 paper-based; 133 computer-based). *Application deadline:* For fall admission, 8/15 for domestic students. Applications are processed on a rolling basis. Application fee: $30 ($75 for international students). Electronic applications accepted. *Expenses:* Tuition, state resident: full-time $2,890; part-time $132 per credit. Tuition, nonresident: full-time $9,258; part-time $422 per credit. Required fees: $626; $41 per credit. Tuition and fees vary according to degree level. *Financial support:* In 2005–06, 4 research assistantships with full tuition reimbursements (averaging $6,492 per year), 9 teaching assistantships with full tuition reimbursements (averaging $6,492 per year) were awarded; fellowships, career-related internships or fieldwork, Federal Work-Study, institutionally sponsored loans, health care benefits, and unspecified assistantships also available. Financial award application deadline: 3/15; financial award applicants required to submit FAFSA. *Faculty research:* Fisheries, range, and wildlife management; aquatic, plant, grassland, vertebrate, and invertebrate ecology; mammalian and plant systematics, taxonomy, and evolution; immunology, virology, and molecular biology. *Unit head:* Dr. J. Richard Schrock, Chair, 620-341-5311, Fax: 620-341-5607, E-mail: jschrock@emporia.edu. *Application contact:* Dr. Derek Zelmer, Graduate Coordinator, 620-341-5623, Fax: 620-341-5607, E-mail: dzelmer@emporia.edu.

Illinois State University, Graduate School, College of Arts and Sciences, Department of Biological Sciences, Normal, IL 61790-2200. Offers biological sciences (MS); biology (PhD); biotechnology (MS); botany (PhD); ecology (PhD); genetics (PhD); microbiology (PhD); physiology (PhD); zoology (PhD). Part-time programs available. *Faculty:* 27 full-time (6 women). *Students:* 36 full-time (28 women), 25 part-time (17 women); includes 1 minority (Asian American or Pacific Islander), 23 international. 80 applicants, 21% accepted. In 2005, 14 master's, 5 doctorates awarded. *Degree requirements:* For master's, thesis or alternative; for doctorate, variable foreign language requirement, thesis/dissertation, 2 terms of residency. *Entrance requirements:* For master's, GRE General Test, minimum GPA of 2.6 in last 60 hours of course work; for doctorate, GRE General Test. *Application deadline:* Applications are processed on a rolling basis. Application fee: $30. *Expenses:* Tuition, state resident: full-time $3,060; part-time $170 per credit hour. Tuition, nonresident: full-time $6,390; part-time $355 per credit hour. Required fees: $1,411; $47 per credit hour. *Financial support:* In 2005–06, 22 research assistantships (averaging $13,909 per year), 38 teaching assistantships (averaging $12,617 per year) were awarded; Federal Work-Study, tuition waivers (full), and unspecified assistantships also available. Financial award application deadline: 4/1. *Faculty research:* CRUI: osmoregulation in euryhaline fish: physiology, ecology and molecular biology; the PRISM project: enhancing science and math education; cell structure—functions of Na pump assembly. *Unit head:* Dr. Hou Tak Takucheung, Acting Chairperson, 309-438-3669. *Application contact:* Derek A. McCracken, Graduate Adviser, 309-438-3664.

See Close-Up on page 147.

Kent State University, College of Arts and Sciences, Department of Biological Sciences, Program in Botany, Kent, OH 44242-0001. Offers MS. *Degree requirements:* For master's, thesis. *Entrance requirements:* For master's, GRE General Test, minimum GPA of 3.0. Additional exam requirements/recommendations for international students: Required—TOEFL (minimum score 600 paper-based; 257 computer-based). Electronic applications accepted.

Miami University, Graduate School, College of Arts and Sciences, Department of Botany, Oxford, OH 45056. Offers MA, MS, PhD. Part-time programs available. *Degree requirements:* For master's, thesis, final exam; for doctorate, thesis/dissertation, final exams, comprehensive exam. *Entrance requirements:* For master's, GRE General Test, GRE Subject Test, minimum undergraduate GPA of 3.0 during previous 2 years or 2.75 overall; for doctorate, GRE General Test, GRE Subject Test, minimum undergraduate GPA of 2.75, 3.0 graduate. Additional exam requirements/recommendations for international students: Required—TOEFL (minimum score 550 paper-based; 213 computer-based), TWE (minimum score 4). Electronic applications accepted.

See Close-Up on page 493.

North Carolina State University, Graduate School, College of Agriculture and Life Sciences, Department of Botany, Raleigh, NC 27695. Offers MS, PhD. Part-time programs available. Terminal master's awarded for partial completion of doctoral program. *Degree requirements:* For master's, thesis (for some programs); for doctorate, thesis/dissertation. *Entrance requirements:* For master's and doctorate, GRE. Additional exam requirements/recommendations for international students: Required—TOEFL. Electronic applications accepted. *Faculty research:* Plant molecular and cell biology, aquatic ecology, community ecology, restoration, systematics plant pathogen and environmental interactions.

North Dakota State University, The Graduate School, College of Science and Mathematics, Department of Biological Sciences, Fargo, ND 58105. Offers biological sciences (MS); botany (MS, PhD); cellular and molecular biology (PhD); environmental and conservation sciences (MS, PhD); genomics (MS, PhD); natural resource management (MS, PhD); zoology (MS, PhD). *Faculty:* 20. *Students:* 30 full-time (10 women), 5 part-time (1 woman); includes 1 minority (Asian American or Pacific Islander), 3 international. Average age 24. 14 applicants, 43% accepted. In 2005, 4 master's awarded. *Degree requirements:* For master's and doctorate, thesis/dissertation. *Entrance requirements:* For master's and doctorate, GRE General Test. Additional exam requirements/recommendations for international students: Required—TOEFL. *Application deadline:* For fall admission, 3/15 for domestic students; for spring admission, 10/30 priority date for domestic students. Applications are processed on a rolling basis. Application fee: $45 ($60 for international students). Electronic applications accepted. *Financial support:* In 2005–06, 3 fellowships with full tuition reimbursements (averaging $15,000 per year), 9 research assistantships with full tuition reimbursements (averaging $14,400 per year), 19 teaching assistantships with full tuition reimbursements (averaging $9,550 per year) were awarded; career-related internships or fieldwork, Federal Work-Study, institutionally sponsored loans, scholarships/grants, tuition waivers (full), and unspecified assistantships also available. Support available to part-time students. Financial award application deadline: 4/15; financial award applicants required to submit FAFSA. *Faculty research:* Comparative endocrinology, physiology, behavioral ecology, plant cell biology, aquatic biology. Total annual research expenditures: $675,000. *Unit head:* Dr. William J. Bleier, Chair, 701-231-7087, Fax: 701-231-7149, E-mail: william.bleier@ndsu.nodak.edu.

Nova Scotia Agricultural College, Research and Graduate Studies, Truro, NS B2N 5E3, Canada. Offers agriculture (M Sc), including air quality, animal behavior, animal molecular genetics, animal nutrition, animal technology, aquaculture, botany, crop management, crop physiology, ecology, environmental microbiology, food science, horticulture, nutrient management, pest management, physiology, plant biotechnology, plant pathology, soil chemistry, soil fertility, waste management and composting, water quality. Part-time programs available. *Faculty:* 43 full-time (7 women), 21 part-time/adjunct (1 woman). *Students:* 50 full-time (25 women), 15 part-time (11 women); includes 7 minority (3 African Americans, 3 Asian Americans or Pacific Islanders, 1 Hispanic American). Average age 25. In 2005, 23 degrees awarded. *Degree requirements:* For master's, thesis, registration. *Entrance requirements:* For master's, honors B Sc, minimum GPA of 3.0. Additional exam requirements/recommendations for international students: Required—TOEFL (minimum score 580 paper-based; 237 computer-based), Michigan English Language Assessment Battery, IELT, Can Test, CAEL. *Application deadline:* For fall admission, 6/1 for domestic students, 4/1 for international students. For winter admission, 10/31 for domestic students; for spring admission, 2/28 for domestic students. Applications are processed on a rolling basis. Application fee: $70. *Expenses:* Tuition, state resident: part-time $2,328 per year. Tuition, nonresident: full-time $6,984; part-time $7,968 per year. International tuition: $12,624 full-time. Required fees: $481; $46 per course. Tuition and fees vary according to program and student level. *Financial support:* In 2005–06, 48 students received support, including 7 fellowships (averaging $15,000 per year), 10 research assistantships (averaging $15,000 per year), 15 teaching assistantships (averaging $900 per year); career-related internships or fieldwork, scholarships/grants, and unspecified assistantships also available. *Faculty research:* Bio-product development, organic agriculture, nutrient management, air and water quality, agricultural biotechnology. Total annual research expenditures: $4.7 million. *Unit head:* Jill L. Rogers, Manager, 902-893-6360, Fax: 902-893-3430, E-mail: jrogers@nsac.ca. *Application contact:* Marie Law, Administrative Assistant, 902-893-6502, Fax: 902-893-3430, E-mail: mlaw@nsac.ca.

Oklahoma State University, College of Arts and Sciences, Department of Botany, Stillwater, OK 74078. Offers botany (MS); environmental science (PhD); plant science (PhD). *Faculty:* 11 full-time (4 women). *Students:* 1 (woman) full-time, 7 part-time (3 women), 2 international. Average age 35. 6 applicants, 50% accepted, 2 enrolled. *Degree requirements:* For master's and doctorate, thesis/dissertation. *Entrance requirements:* For master's and doctorate, GRE General Test. Additional exam requirements/recommendations for international students: Required—TOEFL. *Application deadline:* For fall admission, 6/1 priority date for domestic students, 3/1 priority date for international students. Applications are processed on a rolling basis. Application fee: $40 ($75 for international students). Electronic applications accepted. *Expenses:* Tuition, state resident: full-time $4,253; part-time $139 per credit hour. Tuition, nonresident: full-time $12,569; part-time $485 per credit hour. Required fees: $43 per credit hour. One-time fee: $20 part-time. Tuition and fees vary according to course load and program. *Financial support:* In 2005–06, 5 research assistantships (averaging $14,339 per year), 11 teaching assistantships (averaging $16,755 per year) were awarded; career-related internships or fieldwork, Federal Work-Study, scholarships/grants, health care benefits, tuition waivers (partial), and unspecified assistantships also available. Support available to part-time students. Financial award application deadline: 3/1. *Faculty research:* Ethnobotany, developmental genetics of Arabidopsis, biological roles of Plasmodesmata, community ecology and biodiversity, nutrient cycling in grassland ecosystems. *Unit head:* Dr. William J. Henley, Head, 405-744-5559, Fax: 405-744-7074.

Oregon State University, Graduate School, College of Science, Department of Botany and Plant Pathology, Corvallis, OR 97331. Offers ecology (MA, MAIS, MS, PhD); genetics (MA, MAIS, MS, PhD); molecular and cellular biology (MA, MAIS, MS, PhD); mycology (MA, MAIS, MS, PhD); plant pathology (MA, MAIS, MS, PhD); plant physiology (MA, MAIS, MS, PhD); structural botany (MA, MAIS, MS, PhD); systematics (MA, MAIS, MS, PhD). Part-time programs available. *Faculty:* 10 full-time (2 women), 4 part-time/adjunct (1 woman). *Students:* 38 full-time (22 women), 2 part-time (1 woman); includes 4 minority (1 American Indian/Alaska Native, 2 Asian Americans or Pacific Islanders, 1 Hispanic American), 4 international. Average age 30. In 2005, 5 master's, 3 doctorates awarded. *Degree requirements:* For master's, variable foreign language requirement, thesis optional; for doctorate, thesis/dissertation. *Entrance requirements:* For master's and doctorate, GRE General Test, minimum GPA of 3.0 in last 90 hours. Additional exam requirements/recommendations for international students: Required—TOEFL. *Application deadline:* For fall admission, 2/1 for domestic students. Applications are processed on a rolling basis. Application fee: $50. *Expenses:* Tuition, area resident: Part-time $301 per credit. Tuition, state resident: full-time $8,139; part-time $501 per credit. Tuition, nonresident: full-time $14,376; part-time $532 per credit. Required fees: $1,266. *Financial support:* Fellowships, research assistantships, teaching assistantships, career-related internships or fieldwork, Federal Work-Study, institutionally sponsored loans, and scholarships/grants available. Support available to part-time students. Financial award application deadline: 2/1. *Faculty research:* Plant ecology, plant molecular biology, systematic botany, epidemiology, host-pathogen interaction. *Unit head:* Dr. Daniel Arp, Chair, 541-737-1297, Fax: 541-737-3573. *Application contact:* Dr. Donald B. Zobel, Professor, 541-737-3451, Fax: 541-737-3573, E-mail: zobeld@bcc.orst.edu.

Purdue University, Graduate School, College of Agriculture, Department of Botany and Plant Pathology, West Lafayette, IN 47907. Offers MS, PhD. Part-time programs available. *Faculty:* 24 full-time (5 women), 4 part-time/adjunct (1 woman). *Students:* 40 full-time (11 women), 3 part-time, 16 international. Average age 27. 21 applicants, 38% accepted, 6 enrolled. In 2005, 2 master's, 1 doctorate awarded. Terminal master's awarded for partial completion of doctoral program. *Degree requirements:* For master's and doctorate, thesis/dissertation. *Entrance requirements:* For master's and doctorate, GRE. Additional exam requirements/recommendations for international students: Required—TOEFL. *Application deadline:* For fall admission, 3/15 priority date for domestic students, 3/15 priority date for international students; for spring admission, 12/15 for domestic students, 11/15 for international students. Applications are processed on a rolling basis. Application fee: $55. Electronic applications accepted. *Financial support:* In 2005–06, 30 students received support, including 1 fellowship with full tuition reimbursement available (averaging $17,500 per year), 15 research assistantships with full tuition reimbursements available (averaging $17,500 per year), 10 teaching assistantships with full tuition reimbursements available (averaging $17,500 per year); career-related internships or fieldwork also available. Support available to part-time students. Financial award application deadline: 3/30; financial award applicants required to submit FAFSA. *Faculty research:* Biotechnology, plant growth, weed control, crop improvement, plant physiology. *Unit head:* Dr. Peter B Goldsbrough, Head, 765-494-4615, Fax: 765-494-0363. *Application contact:* Julie A Pluimer, Graduate Admissions Office, 765-494-4614, Fax: 765-494-0363, E-mail: jpluimer@purdue.edu.

Texas A&M University, College of Science, Department of Biology, Program in Botany, College Station, TX 77843. Offers MS, PhD. *Students:* Average age 28. *Degree requirements:* For master's, thesis or alternative; for doctorate, thesis/dissertation, comprehensive exam. *Entrance requirements:* For master's and doctorate, GRE General Test. Additional exam requirements/recommendations for international students: Required—TOEFL. Application fee: $50 ($75 for international students). *Expenses:* Tuition, state resident: full-time $4,488; part-time $187 per credit hour. Tuition, nonresident: full-time $11,112; part-time $463 per credit hour. Required fees: $1,974. *Financial support:* Application deadline: 4/1; *Unit head:* Dr. Mark Zoran, Head, 979-845-7755.

University of Alaska Fairbanks, College of Natural Sciences and Mathematics, Department of Biology and Wildlife, Fairbanks, AK 99775-7520. Offers biological sciences (MS, PhD), including biology, botany, zoology; biology (MAT); wildlife biology (MS, PhD). Part-time programs available. *Faculty:* 29 full-time (7 women), 2 part-time/adjunct (1 woman). *Students:* 89 full-time (53 women), 24 part-time (15 women); includes 6 minority (1 African American, 4 Asian Americans or Pacific Islanders, 1 Hispanic American), 11 international. Average age 30. 74 applicants, 53% accepted, 17 enrolled. In 2005, 14 master's, 4 doctorates awarded. Terminal master's awarded for partial completion of doctoral program. *Degree requirements:* For master's and doctorate, thesis/dissertation, comprehensive exam, registration. *Entrance requirements:* For master's and doctorate, GRE General Test, GRE Subject Test. Additional exam requirements/recommendations for international students: Required—TOEFL (minimum score 550 paper-based; 213 computer-based); Recommended—TWE, TSE. *Application deadline:* For fall admission, 6/1 for domestic students, 3/1 for international students; for spring admission, 12/1 for domestic students, 9/1 for international students. Applications are processed on a rolling basis. Application fee: $50. Electronic applications accepted. *Expenses:* Tuition, state resident: full-time $4,392; part-time $244 per credit. Tuition, nonresident: full-time $8,964; part-time $498 per credit. Required fees: $800; $5 per credit. $48 per contact hour. Tuition and fees vary according to course level, course load, campus/location and reciprocity agreements. *Financial support:* In 2005–06, 36 research assistantships with tuition reimbursements (averaging $9,207 per year), 23 teaching assistantships with tuition reimbursements (averaging $5,358 per year) were awarded; fellowships with tuition reimbursements, career-related internships or fieldwork, Federal Work-Study, and scholarships/grants also available. Financial award application deadline: 2/1; financial award applicants required to submit FAFSA. *Faculty research:* Plant-herbivore interactions, plant metabolic defenses, insect manufacture of glycerol, ice nucleators, structure and functions of arctic and subarctic freshwater ecosystems. *Unit head:* Dr. Kent E. Schubegerle, Chair, 907-474-7671, Fax: 907-474-6716, E-mail: fybio@uaf.edu.

The University of British Columbia, Faculty of Graduate Studies, Faculty of Science, Department of Botany, Vancouver, BC V6T 1Z1, Canada. Offers M Sc, PhD. *Faculty:* 33 full-time (10 women), 11 part-time/adjunct (3 women). *Students:* 62 full-time (31 women). 40 applicants, 35% accepted, 12 enrolled. In 2005, 6 master's, 2 doctorates awarded. *Median time to degree:* Of those who began their doctoral program in fall 1997, 100% received their degree in 8 years or less. *Degree requirements:* For master's, thesis/dissertation; for doctorate, thesis/dissertation, comprehensive exam. *Entrance requirements:* Additional exam requirements/recommendations for international students: Required—TOEFL. *Application deadline:* For fall admission, 3/31 for domestic students, 3/31 for international students. For winter admission, 9/30 for domestic students. Applications are processed on a rolling basis. Application fee: $90 Canadian dollars ($150 Canadian dollars for international students). Electronic applications accepted. *Financial support:* In 2005–06, 59 students received support, including fellowships (averaging $17,000 per year), research assistantships (averaging $9,000 per year), teaching assistantships (averaging $10,000 per year); institutionally sponsored loans, scholarships/grants, and tuition waivers (full and partial) also available. *Faculty research:* Plant ecology, evolution and systematics, cell and developmental biology, plant physiology/biochemistry, genetics. Total annual research expenditures: $3.5 million Canadian dollars. *Unit head:* Dr. Fred Sack, Head, 604-822-2133, Fax: 604-822-6089, E-mail: botagrad@interchange.

ubc.ca. *Application contact:* Dr. George Haughn, Chairman, Graduate Admissions, 604-822-2133, Fax: 604-822-6089, E-mail: botagrad@interchange.ubc.ca.

University of California, Riverside, Graduate Division, Department of Botany and Plant Sciences, Riverside, CA 92521-0102. Offers plant biology (MS, PhD); plant biology (plant genetics) (PhD). Part-time programs available. *Faculty:* 39 full-time (12 women). *Students:* 39 full-time (23 women); includes 4 minority (2 Asian Americans or Pacific Islanders, 2 Hispanic Americans), 15 international. Average age 30. In 2005, 1 master's, 10 doctorates awarded. Terminal master's awarded for partial completion of doctoral program. *Degree requirements:* For master's, comprehensive exams or thesis; for doctorate, thesis/dissertation, qualifying exams. *Entrance requirements:* For master's and doctorate, GRE General Test, minimum GPA of 3.2. Additional exam requirements/recommendations for international students: Required—TOEFL (minimum score 550 paper-based; 213 computer-based); Recommended—TSE (minimum score 50). *Application deadline:* For fall admission, 5/1 for domestic students, 2/1 for international students. For winter admission, 2/1 for domestic students; for spring admission, 12/1 for domestic students. Applications are processed on a rolling basis. Application fee: $60 ($75 for international students). Electronic applications accepted. *Expenses:* Tuition, nonresident: full-time $14,694. Full-time tuition and fees vary according to program. *Financial support:* In 2005–06, research assistantships (averaging $14,000 per year), teaching assistantships (averaging $15,000 per year) were awarded; fellowships, career-related internships or fieldwork, Federal Work-Study, institutionally sponsored loans, scholarships/grants, and tuition waivers (full and partial) also available. Financial award application deadline: 2/1; financial award applicants required to submit FAFSA. *Faculty research:* Agricultural plant biology; biochemistry and physiology; cellular, molecular and developmental biology; ecology, evolution, systematics and ethnobotany; genetics, genomics and bioinformatics. *Unit head:* Dr. Jodie S. Holt, Chair. *Application contact:* Carole Carpenter, Graduate Program Assistant, 800-735-0717, Fax: 951-827-5517, E-mail: plantbio@ucr.edu.

University of Connecticut, Graduate School, College of Liberal Arts and Sciences, Department of Ecology and Evolutionary Biology, Field of Botany, Storrs, CT 06269. Offers MS, PhD. *Faculty:* 15 full-time (4 women). *Students:* 6 full-time (4 women), 1 (woman) part-time, 1 international. Average age 30. 7 applicants, 43% accepted, 3 enrolled. In 2005, 2 degrees awarded. Terminal master's awarded for partial completion of doctoral program. *Degree requirements:* For master's, comprehensive exam; for doctorate, thesis/dissertation. *Entrance requirements:* For master's and doctorate, GRE General Test, GRE Subject Test. Additional exam requirements/recommendations for international students: Required—TOEFL (minimum score 550 paper-based; 213 computer-based). *Application deadline:* For fall admission, 2/1 priority date for domestic students, 2/1 priority date for international students; for spring admission, 11/1 for domestic students, 10/1 for international students. Applications are processed on a rolling basis. Application fee: $55. Electronic applications accepted. *Expenses:* Tuition, state resident: part-time $444 per credit hour. Tuition, nonresident: part-time $1,154 per credit hour. Tuition and fees vary according to course load. *Financial support:* In 2005–06, 3 research assistantships, 2 teaching assistantships were awarded; fellowships, Federal Work-Study, scholarships/grants, health care benefits, and unspecified assistantships also available. Financial award application deadline: 2/1; financial award applicants required to submit FAFSA. *Application contact:* Anne St. Onje, Graduate Coordinator, 860-486-4314, Fax: 860-486-3943, E-mail: ann.st_onje@uconn.edu.

University of Florida, Graduate School, College of Liberal Arts and Sciences and College of Agricultural and Life Sciences, Department of Botany, Gainesville, FL 32611. Offers M Ag, MS, MST, PhD. Part-time programs available. *Faculty:* 14 full-time (3 women). *Students:* 71 (28 women); includes 6 minority (1 African American, 2 Asian Americans or Pacific Islanders, 3 Hispanic Americans) 24 international. In 2005, 6 master's, 8 doctorates awarded. *Degree requirements:* For doctorate, thesis/dissertation. *Entrance requirements:* For master's and doctorate, GRE General Test, minimum GPA of 3.0. Additional exam requirements/recommendations for international students: Required—TOEFL (minimum score 550 paper-based; 213 computer-based). *Application deadline:* For fall admission, 6/1 for domestic students. Applications are processed on a rolling basis. Application fee: $30. Electronic applications accepted. *Expenses:* Tuition, state resident: full-time $6,234. Tuition, nonresident: full-time $21,359. Tuition and fees vary according to program. *Financial support:* In 2005–06, 7 research assistantships (averaging $15,800 per year), 9 teaching assistantships (averaging $15,544 per year) were awarded; fellowships, Federal Work-Study and institutionally sponsored loans also available. Support available to part-time students. Financial award application deadline: 2/15. *Faculty research:* Ecology, physiology, systematics, biochemistry, ecological genetics. *Unit head:* Dr. George E. Bowes, Chair, 352-392-1175, Fax: 352-392-3993, E-mail: gbowes@botany.ufl.edu. *Application contact:* Dr. Alice C. Harmon, Coordinator, 352-392-3217, Fax: 352-392-3993, E-mail: harmon@botany.ufl.edu.

University of Guelph, Graduate Program Services, College of Biological Science, Department of Integrative Biology, Guelph, ON N1G 2W1, Canada. Offers botany (M Sc, PhD); zoology (M Sc, PhD). Part-time programs available. *Faculty:* 37 full-time (5 women). *Students:* 100 full-time (53 women), 3 part-time (all women); includes 5 minority (all Asian Americans or Pacific Islanders), 6 international. Average age 23. 23 applicants, 83% accepted, 19 enrolled. In 2005, 26 master's, 7 doctorates awarded. *Median time to degree:* Of those who began their doctoral program in fall 1997, 95% received their degree in 8 years or less. *Degree requirements:* For master's, thesis, research proposal; for doctorate, thesis/dissertation, research proposal, qualifying exam. *Entrance requirements:* For master's, minimum B average during previous 2 years of course work; for doctorate, minimum A- average. Additional exam requirements/recommendations for international students: Required—TOEFL (minimum score 550 paper-based; 213 computer-based), IELT (minimum score 7). *Application deadline:* For fall admission, 7/1 for domestic students. For winter admission, 11/1 for domestic students; for spring admission, 3/1 for domestic students. Applications are processed on a rolling basis. Application fee: $75. Electronic applications accepted. *Financial support:* In 2005–06, 50 students received support, including 48 research assistantships, 78 teaching assistantships (averaging $4,606 per year); fellowships *Faculty research:* Aquatic science, environmental physiology, parasitology, wildlife biology, management. Total annual research expenditures: $3.8 million. *Unit head:* Dr. Moira Ferguson, Chair, 519-824-4120 Ext. 53598, Fax: 519-767-1656, E-mail: mmfergus@uoguelph.ca. *Application contact:* Laurie Winn, Graduate Admissions Secretary, 519-824-4320 Ext. 52730, Fax: 519-767-1656, E-mail: lwinn@uoguelph.ca.

University of Guelph, Graduate Program Services, College of Biological Science, Department of Molecular and Cellular Biology, Guelph, ON N1G 2W1, Canada. Offers biochemistry (M Sc, PhD); biophysics (M Sc, PhD); botany (M Sc, PhD); microbiology (M Sc, PhD); molecular biology and genetics (M Sc, PhD). *Faculty:* 43 full-time (6 women). *Students:* 136 full-time (52 women). Average age 26. 56 applicants, 30% accepted, 17 enrolled. In 2005, 4 master's, 2 doctorates awarded. *Degree requirements:* For master's, thesis/dissertation, research proposal; for doctorate, thesis/dissertation, research proposal, comprehensive exam, registration. *Entrance requirements:* Additional exam requirements/recommendations for international students: Required—TOEFL (minimum score 550 paper-based; 213 computer-based), IELT (minimum score 7). *Application deadline:* For fall admission, 7/1 for domestic students, 1/1 for international students. For winter admission, 11/1 for domestic students; for spring admission, 3/1 for domestic students. Applications are processed on a rolling basis. Application fee: $75. Electronic applications accepted. *Financial support:* Fellowships, research assistantships, teaching assistantships available. Support available to part-time students. *Faculty research:* Physiology, structure, genetics, and ecology of microbes; virology and microbial technology. *Unit head:* Dr. Chris Whitfield, Chair, 519-824-4120 Ext. 53361, Fax: 519-827-1802, E-mail: cwhitfie@uoguelph.ca. *Application contact:* Laurie Winn, Graduate Admissions Secretary, 519-824-4320 Ext. 52730, Fax: 519-767-1656, E-mail: lwinn@uoguelph.ca.

University of Hawaii at Manoa, Graduate Division, Colleges of Arts and Sciences, College of Natural Sciences, Department of Botany, Honolulu, HI 96822. Offers MS, PhD. *Faculty:* 25 full-time (6 women), 4 part-time/adjunct (1 woman). *Students:* 47 full-time (30 women), 19 part-time (12 women); includes 13 minority (1 African American, 12 Asian Americans or Pacific

Botany

University of Hawaii at Manoa (continued)

Islanders), 9 international. Average age 29. 57 applicants, 39% accepted, 9 enrolled. In 2005, 6 master's, 3 doctorates awarded. Terminal master's awarded for partial completion of doctoral program. *Median time to degree:* Of those who began their doctoral program in fall 1997, 33% received their degree in 8 years or less. *Degree requirements:* For master's, thesis (for some programs), presentation; for doctorate, one foreign language, thesis/dissertation, presentation. *Entrance requirements:* For master's and doctorate, GRE General Test, GRE Subject Test (biology). *Application deadline:* For fall admission, 2/1 for domestic students, 1/15 for international students; for spring admission, 9/1 for domestic students, 8/1 for international students. Application fee: $50. *Expenses:* Tuition, state resident: full-time $8,400; part-time $200 per credit hour. Tuition, nonresident: full-time $11,088; part-time $462 per credit hour. Tuition and fees vary according to program. *Financial support:* In 2005–06, 20 research assistantships (averaging $17,146 per year), 20 teaching assistantships (averaging $14,792 per year) were awarded; tuition waivers (full and partial) also available. *Faculty research:* Plant ecology, evolution, systematics, conservation biology, ethnobotany. *Unit head:* Gerald Carr, Chairperson, 808-956-8369, Fax: 808-956-3923. *Application contact:* Clifford Mordew, Graduate Field Chair, 808-956-8369, Fax: 808-956-3923, E-mail: cmordew@hawaii.edu.

University of Kansas, Graduate School, College of Liberal Arts and Sciences, Division of Biological Sciences, Department of Ecology and Evolutionary Biology, Lawrence, KS 66045. Offers botany (MA, PhD); ecology and evolutionary biology (MA, PhD); entomology (MA, PhD); systematics and ecology (MA). Part-time programs available. *Faculty:* 42. *Students:* 67 full-time (35 women), 21 part-time (9 women); includes 4 minority (1 American Indian/Alaska Native, 3 Hispanic Americans), 20 international. Average age 30. 56 applicants, 32% accepted. In 2005, 10 master's, 4 doctorates awarded. Terminal master's awarded for partial completion of doctoral program. *Degree requirements:* For master's and doctorate, thesis/dissertation, comprehensive exam. *Entrance requirements:* For master's and doctorate, GRE General Test, GRE Subject Test (recommended). Additional exam requirements/recommendations for international students: Required—TOEFL (paper 570; computer 230) or IELT. *Application deadline:* For fall admission, 1/10 priority date for domestic students, 1/10 priority date for international students. Applications are processed on a rolling basis. Application fee: $55 ($60 for international students). Electronic applications accepted. *Expenses:* Tuition, state resident: full-time $4,859. Tuition, nonresident: full-time $12,000. Required fees: $589. Tuition and fees vary according to program. *Financial support:* Fellowships with tuition reimbursements, research assistantships with partial tuition reimbursements, teaching assistantships with full and partial tuition reimbursements available. Financial award application deadline: 3/1. *Faculty research:* Ecology, evolutionary biology, genetics. *Unit head:* Craig E. Martin, Chair, 785-864-5887, Fax: 785-864-5860, E-mail: ecophys@ku.edu. *Application contact:* Jeannie Houts, Graduate Coordinator, 785-864-2362, Fax: 785-864-5860, E-mail: jmhouts@ku.edu.

University of Maine, Graduate School, College of Natural Sciences, Forestry, and Agriculture, Department of Biological Sciences, Program in Botany and Plant Pathology, Orono, ME 04469. Offers MS. Part-time programs available. *Students:* 4 full-time (2 women), 1 (woman) part-time, 1 international. Average age 28. 4 applicants, 25% accepted, 1 enrolled. In 2005, 5 master's awarded. *Degree requirements:* For master's, thesis. *Entrance requirements:* For master's, GRE General Test. Additional exam requirements/recommendations for international students: Required—TOEFL. *Application deadline:* For fall admission, 2/1 for domestic students. Applications are processed on a rolling basis. Application fee: $50. Electronic applications accepted. *Financial support:* Fellowships, research assistantships, teaching assistantships, career-related internships or fieldwork, Federal Work-Study, institutionally sponsored loans, and tuition waivers (full) available. Financial award application deadline: 3/1. *Faculty research:* Molecular biology of viral and fungal pathogens, marine ecology, paleoecology and acid systematics and evolution. *Unit head:* Dr. Stelbs Tavanteis, Coordinator, 207-581-2986. *Application contact:* Scott G. Delcourt, Associate Dean of the Graduate School, 207-581-3219, Fax: 207-581-3232, E-mail: graduate@maine.edu.

University of Manitoba, Faculty of Graduate Studies, Faculty of Science, Biological Sciences Division, Department of Botany, Winnipeg, MB R3T 2N2, Canada. Offers M Sc, PhD. *Degree requirements:* For master's, thesis; for doctorate, one foreign language, thesis/dissertation.

University of Missouri–St. Louis, College of Arts and Sciences, Department of Biology, St. Louis, MO 63121. Offers biology (MS, PhD), including animal behavior (MS), biochemistry (MS), biotechnology (MS), conservation biology (MS), development (MS), ecology (MS), environmental studies (PhD), evolution (MS), genetics (MS), molecular/cellular biology (MS), physiology (MS), plant systematics, population biology (MS), tropical biology (MS); biotechnology (Certificate); tropical biology and conservation (Certificate). Part-time programs available. *Faculty:* 50. *Students:* 26 full-time (15 women), 101 part-time (51 women); includes 13 minority (5 African Americans, 6 Asian Americans or Pacific Islanders, 2 Hispanic Americans), 41 international. Average age 32. In 2005, 22 master's, 2 doctorates awarded. *Degree requirements:* For master's, thesis or alternative; for doctorate, one foreign language, thesis/dissertation, 1 semester of teaching experience. *Entrance requirements:* For doctorate, GRE General Test. *Application deadline:* For spring admission, 12/1 priority date for domestic students. Applications are processed on a rolling basis. Application fee: $35 ($40 for international students). Electronic applications accepted. *Expenses:* Tuition, state resident: part-time $263 per credit hour. Tuition, nonresident: part-time $680 per credit hour. Required fees: $53 per credit hour. Tuition and fees vary according to program. *Financial support:* In 2005–06, 11 fellowships with full tuition reimbursements (averaging $30,000 per year), 15 research assistantships with full and partial tuition reimbursements (averaging $16,000 per year), 22 teaching assistantships with full and partial tuition reimbursements (averaging $16,000 per year) were awarded; career-related internships or fieldwork and Federal Work-Study also available. Support available to part-time students. Financial award application deadline: 2/1. *Faculty research:* Molecular biology, microbial genetics. *Unit head:* Zuleyma Tang-Martinez, Director of Graduate Studies, 314-516-6498, Fax: 314-516-6233, E-mail: zuleyma@umsl.edu. *Application contact:* 314-516-5458, Fax: 314-516-5310, E-mail: gradadm@umsl.edu.

The University of North Carolina at Chapel Hill, Graduate School, College of Arts and Sciences, Department of Biology, Program in Botany, Chapel Hill, NC 27599. Offers MA, MS, PhD. Terminal master's awarded for partial completion of doctoral program. *Degree requirements:* For master's, thesis (for some programs), comprehensive exam; for doctorate, thesis/dissertation, comprehensive exam. *Entrance requirements:* For master's and doctorate, GRE General Test, GRE Subject Test. Additional exam requirements/recommendations for international students: Required—TOEFL (minimum score 550 paper-based; 213 computer-based).

Announcement: Candidates for the MS, MA, and PhD biology degrees can conduct research in plant genomics, anatomy, biochemistry, ecology, genetics, molecular botany, morphology, paleobotany, phycology, and systematics. Teaching assistantships, research assistantships, fellowships. Well-equipped laboratories, excellent library, large herbarium, botanical garden.

University of North Dakota, Graduate School, College of Arts and Sciences, Department of Biology, Grand Forks, ND 58202. Offers botany (MS, PhD); ecology (MS, PhD); entomology (MS, PhD); environmental biology (MS, PhD); fisheries/wildlife (MS, PhD); genetics (MS, PhD); zoology (MS, PhD). *Faculty:* 16 full-time (3 women). *Students:* 15 applicants, 13% accepted, 2 enrolled. In 2005, 5 degrees awarded. Terminal master's awarded for partial completion of doctoral program. *Degree requirements:* For master's, thesis/dissertation, final exam; for doctorate, thesis/dissertation, final exam, comprehensive exam. *Entrance requirements:* For master's, GRE General Test, GRE Subject Test, minimum GPA of 3.0; for doctorate, GRE General Test, GRE Subject Test, minimum GPA of 3.5. Additional exam requirements/recommendations for international students: Required—TOEFL (minimum score 550 paper-based; 213 computer-based). *Application deadline:* For fall admission, 10/1 for domestic students, 10/1 for international students. Application fee: $35. Electronic applications accepted.

Financial support: In 2005–06, 8 research assistantships with full tuition reimbursements (averaging $11,375 per year), 13 teaching assistantships with full tuition reimbursements (averaging $10,813 per year) were awarded; fellowships, Federal Work-Study, institutionally sponsored loans, scholarships/grants, and tuition waivers (full and partial) also available. Support available to part-time students. Financial award application deadline: 3/15; financial award applicants required to submit FAFSA. *Faculty research:* Population biology, wildlife ecology, RNA processing, hormonal control of behavior. *Unit head:* Dr. Richard Sweitzel, Graduate Director, 701-777-4676, Fax: 701-777-2623, E-mail: richard_sweitzel@und.nodak.edu.

University of Oklahoma, Graduate College, College of Arts and Sciences, Department of Botany and Microbiology, Program in Botany, Norman, OK 73019-0390. Offers MS, PhD. *Students:* 16 full-time (7 women), 2 part-time (both women); includes 2 minority (1 African American, 1 Hispanic American), 7 international. 5 applicants, 60% accepted, 3 enrolled. In 2005, 4 master's, 1 doctorate awarded. Terminal master's awarded for partial completion of doctoral program. *Degree requirements:* For master's, thesis, oral exam; for doctorate, one foreign language, thesis/dissertation, general exam. *Entrance requirements:* Additional exam requirements/recommendations for international students: Required—TOEFL (minimum score 550 paper-based; 213 computer-based). *Application deadline:* For fall admission, 6/1 for domestic students, 4/1 for international students; for spring admission, 12/1 for domestic students, 9/1 for international students. Applications are processed on a rolling basis. Application fee: $40 ($90 for international students). *Expenses:* Tuition, state resident: full-time $3,029; part-time $126 per credit hour. Tuition, nonresident: full-time $10,807; part-time $450 per credit hour. Required fees: $1,231; $44 per credit hour. Tuition and fees vary according to course load and program. *Financial support:* In 2005–06, 3 students received support, including 5 fellowships with full tuition reimbursements available (averaging $5,000 per year); research assistantships with full tuition reimbursements available, teaching assistantships with full tuition reimbursements available, scholarships/grants and unspecified assistantships also available. Financial award applicants required to submit FAFSA. *Faculty research:* Plant molecular systomastics, plant molecular biology, ecosystem ecology, structural botany, grassland ecology. *Application contact:* Adell Hopper, Staff Assistant, 405-325-4322, Fax: 405-325-7619, E-mail: ahopper@ou.edu.

University of South Florida, College of Graduate Studies, College of Arts and Sciences, Department of Biology, Tampa, FL 33620-9951. Offers biology (PhD); botany (MS); ecology (PhD); microbiology (MS); physiology (PhD); zoology (MS). Part-time programs available. *Faculty:* 21. *Students:* 54 full-time (32 women), 26 part-time (17 women); includes 8 minority (2 African Americans, 1 American Indian/Alaska Native, 1 Asian American or Pacific Islander, 4 Hispanic Americans), 14 international. 76 applicants, 34% accepted, 13 enrolled. In 2005, 3 master's, 3 doctorates awarded. *Degree requirements:* For master's, thesis (for some programs), graduate seminar in biology; for doctorate, 2 foreign languages, thesis/dissertation, essay of research interest, comprehensive exam. *Entrance requirements:* For master's, GRE General Test, minimum undergraduate GPA of 3.0 in last 60 hours of course work; for doctorate, GRE General Test, GRE Subject Test in biology, minimum undergraduate GPA of 3.0 in last 60 hours of course work. Additional exam requirements/recommendations for international students: Required—TOEFL (minimum score 570 paper-based), TSE (minimum score 50). *Application deadline:* For fall admission, 2/1 priority date for domestic students, 3/1 priority date for international students; for spring admission, 10/1 for domestic students, 8/1 for international students. Application fee: $30. Electronic applications accepted. *Financial support:* Fellowships with full tuition reimbursements, research assistantships with full tuition reimbursements, teaching assistantships with full tuition reimbursements, Federal Work-Study and unspecified assistantships available. Financial award application deadline: 6/30. *Unit head:* Sydney Pierce, Chairperson, 813-974-3250, Fax: 813-974-3263. *Application contact:* Christine Smith, Graduate Advisor, 813-974-4747, Fax: 813-974-3263, E-mail: csmith2@chuma1.cas.usf.edu.

University of Toronto, School of Graduate Studies, Life Sciences Division, Department of Botany, Toronto, ON M5S 1A1, Canada. Offers M Sc, PhD. *Degree requirements:* For master's, thesis, thesis defense; for doctorate, thesis/dissertation, thesis defense, oral thesis examination. *Entrance requirements:* For master's, minimum B+ average in final year, B overall, 3 letters of reference. Additional exam requirements/recommendations for international students: Required—TOEFL (minimum score 580 paper-based; 237 computer-based), TWE (minimum score 5).

University of Vermont, Graduate College, College of Agriculture and Life Sciences, Department of Botany and Agricultural Biochemistry, Burlington, VT 05405. Offers field naturalist (MS); plant biology (MS, PhD). *Faculty:* 11 full-time (4 women). *Students:* 14 (12 women) 5 international. 32 applicants, 25% accepted, 4 enrolled. In 2005, 5 master's, 1 doctorate awarded. *Degree requirements:* For master's and doctorate, thesis/dissertation, comprehensive exam. *Entrance requirements:* For master's and doctorate, GRE General Test. Additional exam requirements/recommendations for international students: Required—TOEFL (minimum score 550 paper-based; 213 computer-based). *Application deadline:* For fall admission, 1/1 priority date for domestic students, 1/1 priority date for international students. Applications are processed on a rolling basis. Application fee: $55. Electronic applications accepted. *Expenses:* Tuition, area resident: Part-time $410 per credit hour. Tuition, nonresident: part-time $1,034 per credit hour. *Financial support:* In 2005–06, research assistantships with full tuition reimbursements (averaging $21,800 per year), teaching assistantships with full tuition reimbursements (averaging $21,800 per year) were awarded; fellowships, health care benefits, tuition waivers (full and partial), and stipends also available. Financial award application deadline: 1/1. *Faculty research:* Systematics, biochemistry, ecology and evolution, physiology, development and molecular genetics. Total annual research expenditures: $739,761. *Unit head:* Dr. Thomas Vogelmann, Chairperson, 802-656-2930, Fax: 802-656-0440. *Application contact:* Lillian Reade, Coordinator, 802-656-2930, Fax: 802-656-0440, E-mail: lillian.reade@uvm.edu.

University of Washington, Graduate School, College of Arts and Sciences, Department of Botany, Seattle, WA 98195. Offers MS, PhD. *Degree requirements:* For master's, thesis optional; for doctorate, thesis/dissertation. *Entrance requirements:* For master's and doctorate, GRE General Test, GRE Subject Test, minimum GPA of 3.0. Additional exam requirements/recommendations for international students: Required—TOEFL. Electronic applications accepted. *Faculty research:* Molecular and cellular botany, plant physiology, systematics, plant population biology, ecosystems and environmental studies.

University of Wisconsin–Madison, Graduate School, College of Letters and Science, Department of Botany, Madison, WI 53706-1380. Offers MS, PhD. Part-time programs available. Terminal master's awarded for partial completion of doctoral program. *Degree requirements:* For master's, thesis; for doctorate, one foreign language, thesis/dissertation. *Entrance requirements:* For master's and doctorate, GRE General Test. Electronic applications accepted. *Faculty research:* Taxonomy and systematics; ecology; structural botany; physiological, cellular, and molecular biology.

University of Wisconsin–Oshkosh, The School of Graduate Studies, College of Letters and Science, Department of Biology and Microbiology, Oshkosh, WI 54901. Offers biology (MS), including botany, microbiology, zoology. *Degree requirements:* For master's, thesis, comprehensive exam, registration. *Entrance requirements:* For master's, GRE General Test, minimum GPA of 3.0, BS in biology. Additional exam requirements/recommendations for international students: Required—TOEFL (minimum score 550 paper-based; 213 computer-based). Electronic applications accepted.

University of Wyoming, Graduate School, College of Arts and Sciences, Department of Botany, Laramie, WY 82070. Offers botany (MS, PhD); botany/water resources (MS). Part-time programs available. *Faculty:* 11 full-time (2 women), 5 part-time/adjunct (2 women). *Students:* 19 full-time (11 women), 10 part-time (6 women); includes 1 minority (Hispanic American), 2 international. 21 applicants, 67% accepted. In 2005, 4 master's awarded. Terminal master's awarded for partial completion of doctoral program. *Degree requirements:* For master's and doctorate, thesis/dissertation. *Entrance requirements:* For master's and doctorate, GRE General Test, minimum GPA of 3.0. Additional exam requirements/recommendations for international students: Required—TOEFL. *Application deadline:* For fall admission, 2/1 for domestic students. Application fee: $50. Electronic applications accepted. *Expenses:* Tuition, state resident: full-time $3,720; part-time $155 per credit hour. Tuition, nonresident: full-time $10,704; part-time $446 per credit hour. Required fees: $666; $162 per semester. Tuition and fees vary according to course load and program. *Financial support:* In 2005–06, 2 research assistantships with full tuition reimbursements (averaging $10,062 per year), 11 teaching assistantships with full tuition reimbursements (averaging $10,062 per year) were awarded; scholarships/grants also available. Financial award application deadline: 2/1. *Faculty research:* Ecology, systematics, physiology, mycology, genetics. Total annual research expenditures: $1.7 million. *Unit head:* Gregory K. Brown, Head, 307-766-2380, Fax: 307-766-2851, E-mail: gkbrown@uwyo.edu. *Application contact:* William A. Reiners, Associate Professor, 307-766-2235, Fax: 307-766-2851, E-mail: reiners@uwyo.edu.

Virginia Polytechnic Institute and State University, Graduate School, College of Science, Department of biological sciences, Blacksburg, VA 24061. Offers botany (MS, PhD); ecology and evolutionary biology (MS, PhD); genetics and developmental biology (MS, PhD); microbiology (MS, PhD); zoology (MS, PhD). *Faculty:* 38 full-time (9 women). *Students:* 70 full-time (30 women), 4 part-time (1 woman); includes 6 minority (2 African Americans, 1 American Indian/Alaska Native, 2 Asian Americans or Pacific Islanders, 1 Hispanic American), 12 international. Average age 27. 81 applicants, 22% accepted, 14 enrolled. In 2005, 9 master's, 8 doctorates awarded. *Entrance requirements:* For master's and doctorate, GRE General Test. Additional exam requirements/recommendations for international students: Required—TOEFL (minimum score 550 paper-based; 213 computer-based). *Application deadline:* Applications are processed on a rolling basis. Application fee: $45. Electronic applications accepted. *Expenses:* Tuition, state resident: full-time $6,558; part-time $364 per credit. Tuition, nonresident: full-time $11,296; part-time $628 per credit. Required fees: $1,419; $468 per credit. $234 per term. *Financial support:* In 2005–06, 28 research assistantships with full tuition reimbursements (averaging $16,689 per year), 37 teaching assistantships with full tuition reimburse-

ments (averaging $13,902 per year) were awarded; career-related internships or fieldwork, Federal Work-Study, scholarships/grants, and unspecified assistantships also available. *Faculty research:* Freshwater ecology, cell cycle regulation, behavioral ecology, motor proteins. *Unit head:* Dr. Bob Jones, Chairman, 540-231-9514, Fax: 540-231-9307, E-mail: rhjones@vt.edu. *Application contact:* Sue Rasmussen, Graduate Secretary, 540-231-8929, Fax: 540-231-9307, E-mail: sueras@vt.edu.

Washington State University, Graduate School, College of Sciences, School of Biological Sciences, Department of Botany, Pullman, WA 99164. Offers MS, PhD. *Faculty:* 31. *Students:* 27 full-time (11 women), 1 part-time; includes 1 minority (Asian American or Pacific Islander), 5 international. Average age 30. 30 applicants, 13% accepted, 4 enrolled. In 2005, 2 master's, 3 doctorates awarded. *Degree requirements:* For master's, oral exam, thesis optional; for doctorate, thesis/dissertation, oral exam. *Entrance requirements:* For master's and doctorate, GRE General Test, GRE Subject Test (recommended), minimum GPA of 3.0, 3 letters of recommendation. Additional exam requirements/recommendations for international students: Required—TOEFL. *Application deadline:* For fall admission, 1/15 priority date for domestic students, 1/15 priority date for international students; for spring admission, 9/15 for domestic students, 7/1 for international students. Applications are processed on a rolling basis. Application fee: $35. *Expenses:* Tuition, state resident: full-time $6,295; part-time $336 per credit. Tuition, nonresident: full-time $15,949; part-time $819 per credit. Required fees: $933. Part-time tuition and fees vary according to campus/location and program. *Financial support:* In 2005–06, 3 fellowships (averaging $4,000 per year), 4 research assistantships with full and partial tuition reimbursements (averaging $14,383 per year), 21 teaching assistantships with full and partial tuition reimbursements (averaging $14,481 per year) were awarded; career-related internships or fieldwork, Federal Work-Study, institutionally sponsored loans, and tuition waivers (partial) also available. Financial award application deadline: 4/1; financial award applicants required to submit FAFSA. *Faculty research:* Molecular biology, plant physiology, systematics, ecology-evolution. *Unit head:* Dr. R. Alan Black, Associate Director, 509-335-0179, E-mail: blackra@wsu.edu. *Application contact:* Graduate Secretary, E-mail: sbsgrad@wsu.edu.

Plant Biology

Clemson University, Graduate School, College of Agriculture, Forestry and Life Sciences, Department of Biological Sciences, Program in Plant and Environmental Sciences, Clemson, SC 29634. Offers MS, PhD. *Students:* 35 full-time (11 women), 14 part-time (8 women); includes 2 minority (1 American Indian/Alaska Native, 1 Hispanic American), 12 international. 20 applicants, 45% accepted, 8 enrolled. In 2005, 8 master's, 1 doctorate awarded. *Degree requirements:* For master's, thesis. *Entrance requirements:* For master's, GRE General Test, bachelor's degree in biological science or chemistry. Additional exam requirements/recommendations for international students: Required—TOEFL. *Application deadline:* For fall admission, 6/1 for domestic students. Application fee: $50. *Financial support:* Teaching assistantships available. Financial award application deadline: 3/15; financial award applicants required to submit FAFSA. *Faculty research:* Systematics, aquatic botany, plant ecology, plant-fungus interactions, plant developmental genetics. *Unit head:* Dr. Halina Knapp, Coordinator, 864-656-3523, Fax: 864-656-7594, E-mail: hskrpsk@clemson.edu.

See Close-Ups on pages 489 and 491.

Cornell University, Graduate School, Graduate Fields of Agriculture and Life Sciences, Field of Plant Biology, Ithaca, NY 14853-0001. Offers cytology (MS, PhD); paleobotany (MS, PhD); plant cell biology (MS, PhD); plant ecology (MS, PhD); plant molecular biology (MS, PhD); plant morphology, anatomy and biomechanics (MS, PhD); plant physiology (MS, PhD); systematic botany (MS, PhD). *Faculty:* 69 full-time (15 women). *Students:* 50 full-time (27 women); includes 2 minority (both African Americans), 16 international. 33 applicants, 45% accepted, 10 enrolled. In 2005, 1 master's, 7 doctorates awarded. *Degree requirements:* For doctorate, thesis/dissertation, comprehensive exam. *Entrance requirements:* For doctorate, GRE General Test, GRE Subject Test in biology (recommended), 3 letters of recommendation. Additional exam requirements/recommendations for international students: Required—TOEFL (minimum score 610 paper-based; 253 computer-based). *Application deadline:* For fall admission, 1/15 for domestic students. Application fee: $60. Electronic applications accepted. *Financial support:* In 2005–06, 50 students received support, including 10 fellowships with full tuition reimbursements available, 23 research assistantships with full tuition reimbursements available, 17 teaching assistantships with full tuition reimbursements available; institutionally sponsored loans, scholarships/grants, health care benefits, tuition waivers (full and partial), and unspecified assistantships also available. Financial award applicants required to submit FAFSA. *Faculty research:* Plant cell biology/cytology; plant molecular biology; plant morphology/anatomy/biomechanics; plant physiology, systematic botany, paleobotany; plant ecology, ethnobotany, plant biochemistry, photosynthesis. *Unit head:* Director of Graduate Studies, 607-255-2131. *Application contact:* Graduate Field Assistant, 607-255-2131, E-mail: plbio@cornell.edu.

Florida State University, Graduate Studies, College of Arts and Sciences, Department of Biological Science, Program in Plant Biology, Tallahassee, FL 32306. Offers MS, PhD. *Faculty:* 6 full-time (2 women). *Students:* 15 full-time (10 women); includes 3 minority (2 Asian Americans or Pacific Islanders, 1 Hispanic American), 6 international. *Degree requirements:* For master's and doctorate, thesis/dissertation, teaching experience, seminar presentation, comprehensive exam, registration. *Entrance requirements:* For master's and doctorate, GRE General Test (minimum 1100: U-500, G-500), minimum upper division GPA of 3.0. Additional exam requirements/recommendations for international students: Required—TOEFL (minimum score 600 paper-based; 250 computer-based), IB 100. *Application deadline:* For fall admission, 1/15 for domestic students, 12/1 for international students; for spring admission, 10/15 for domestic students, 9/1 for international students. Application fee: $30. *Financial support:* In 2005–06, fellowships with full tuition reimbursements (averaging $19,000 per year), research assistantships with full tuition reimbursements (averaging $19,000 per year), teaching assistantships with full tuition reimbursements (averaging $17,600 per year) were awarded. Financial award application deadline: 1/15; financial award applicants required to submit FAFSA. *Faculty research:* Photosynthetic mechanisms, cell development, physiology, ecology, cytogenetics. *Application contact:* Judy Bowers, Coordinator, Graduate Affairs, 850-644-3023, Fax: 850-644-9829, E-mail: gradinfo@bio.fsu.edu.

Indiana University Bloomington, Graduate School, College of Arts and Sciences, Department of Biology, Program in Plant Sciences, Bloomington, IN 47405-7000. Offers MA, PhD. PhD offered through the University Graduate School. Part-time programs available. *Faculty:* 14 full-time (4 women). *Students:* 17 full-time (10 women); includes 1 minority (Asian American or Pacific Islander) In 2005, 1 degree awarded. Terminal master's awarded for partial completion of doctoral program. *Degree requirements:* For master's, thesis or alternative; for doctorate, thesis/dissertation. *Entrance requirements:* For master's and doctorate, GRE General Test. Additional exam requirements/recommendations for international students: Required—TOEFL. *Application deadline:* For fall admission, 1/5 for domestic students; for spring admission, 9/1 priority date for domestic students. Applications are processed on a rolling basis. Application fee: $45. Electronic applications accepted. *Expenses:* Tuition, state resident: full-time $5,437; part-time $227 per credit hour. Tuition, nonresident: full-time $15,836; part-time $660 per credit hour. Required fees: $821. Tuition and fees vary according to campus/location and program. *Financial support:* In 2005–06, 17 students received support, including

fellowships with tuition reimbursements available (averaging $18,000 per year), research assistantships with tuition reimbursements available (averaging $18,000 per year), teaching assistantships with tuition reimbursements available (averaging $18,000 per year) Financial award application deadline: 1/15. *Faculty research:* Molecular biology, physiology, systematics, genetics and evolutionary biology. *Unit head:* Dr. Keith Clay, Head, 812-855-8158, E-mail: clay@indiana.edu. *Application contact:* Gretchen Clearwater, Adviser for Graduate Affairs, 812-855-1861, Fax: 812-855-6705, E-mail: biograd@bio.indiana.edu.

Miami University, Graduate School, College of Arts and Sciences, Department of Botany, Oxford, OH 45056. Offers MA, MS, PhD. Part-time programs available. *Degree requirements:* For master's, thesis, final exam; for doctorate, thesis/dissertation, final exams, comprehensive exam. *Entrance requirements:* For master's, GRE General Test, GRE Subject Test, minimum undergraduate GPA of 3.0 during previous 2 years or 2.75 overall; for doctorate, GRE General Test, GRE Subject Test, minimum undergraduate GPA of 2.75, 3.0 graduate. Additional exam requirements/recommendations for international students: Required—TOEFL (minimum score 550 paper-based; 213 computer-based), TWE (minimum score 4). Electronic applications accepted.

See Close-Up on page 493.

Michigan State University, The Graduate School, College of Agriculture and Natural Resources, MSU-DOE Plant Research Laboratory, East Lansing, MI 48824. Offers biochemistry and molecular biology (PhD); cellular and molecular biology (PhD); crop and soil sciences (PhD); genetics (PhD); microbiology and molecular genetics (PhD); plant biology (PhD); plant physiology (PhD). Offered jointly with the Department of Energy. *Faculty:* 9 full-time (2 women). *Degree requirements:* For doctorate, thesis/dissertation, laboratory rotation, defense of dissertation, comprehensive exam. *Entrance requirements:* For doctorate, GRE General Test, acceptance into one of the affiliated department programs; 3 letters of recommendation; bachelor's degree or equivalent in life sciences, chemistry, biochemistry, or biophysics; research experience. Application fee: $50. Electronic applications accepted. *Expenses:* Tuition, state resident: part-time $330 per credit hour. Tuition, nonresident: part-time $685 per credit hour. Tuition and fees vary according to program. *Faculty research:* Role of hormones in the regulation of plant development and physiology, molecular mechanisms associated with signal recognition, development and application of genetic methods and materials, protein routing and function. Total annual research expenditures: $7.4 million. *Unit head:* Dr. Kenneth Keegstra, Director, 517-353-2270, Fax: 517-353-9168, E-mail: keegstra@msu.edu. *Application contact:* Janet Taylor, Graduate Program Secretary, 517-353-2270, Fax: 517-353-9168, E-mail: prl@msu.edu.

Michigan State University, The Graduate School, College of Natural Science and College of Agriculture and Natural Resources, Department of Plant Biology, East Lansing, MI 48824. Offers plant biology (MS, PhD); plant breeding and genetics—plant biology (MS, PhD). *Faculty:* 17 full-time (5 women). *Students:* 33 full-time (17 women), 4 part-time (2 women); includes 3 minority (1 Asian American or Pacific Islander, 2 Hispanic Americans), 7 international. Average age 29. 68 applicants, 15% accepted. In 2005, 1 master's, 4 doctorates awarded. *Degree requirements:* For master's, variable foreign language requirement, teaching, thesis optional; for doctorate, variable foreign language requirement, thesis/dissertation, teaching experience, preliminary exam, oral defense of dissertation. *Entrance requirements:* For master's, GRE General Test, minimum GPA of 3.0; course work in chemistry, mathematics and physics, and biological sciences; 3 letters of recommendation; for doctorate, GRE General Test, master's degree or equivalent, minimum GPA of 3.0, training in the biological sciences, 3 letters of recommendation. Additional exam requirements/recommendations for international students: Required—TOEFL (minimum score 550 paper-based; 213 computer-based), Michigan State University ELT (85), Michigan ELAB (83). *Application deadline:* For fall admission, 12/27 for domestic students. Application fee: $50. Electronic applications accepted. *Expenses:* Tuition, state resident: part-time $330 per credit hour. Tuition, nonresident: part-time $685 per credit hour. Tuition and fees vary according to program. *Financial support:* In 2005–06, 15 fellowships with tuition reimbursements (averaging $8,760 per year), 10 research assistantships with tuition reimbursements (averaging $14,873 per year), 18 teaching assistantships with tuition reimbursements (averaging $14,244 per year) were awarded; scholarships/grants and unspecified assistantships also available. *Faculty research:* Physiological, molecular, and biochemical mechanisms; systematics; inheritance; ecology and geohistory. Total annual research expenditures: $4.2 million. *Unit head:* Dr. Richard E. Triemer, Chairperson, 517-353-9400, Fax: 517-353-1926, E-mail: triemer@msu.edu. *Application contact:* Kasey Crawford, Graduate Secretary, 517-432-4429, Fax: 517-353-1926, E-mail: plntbiol@mu.edu.

New York University, Graduate School of Arts and Science, Department of Biology, New York, NY 10012-1019. Offers biology (PhD); biomedical journalism (MS); cancer and molecular biology (PhD); computational biology (PhD); computers in biological research (MS); developmental genetics (PhD); general biology (MS); immunology and microbiology (PhD); molecular genetics (PhD); neurobiology (PhD); oral biology (MS); plant biology (PhD); recombinant DNA technology (MS). Part-time programs available. *Faculty:* 24 full-time (5 women), 8 part-time/

Plant Biology

New York University (continued)

adjunct. *Students:* 104 full-time (52 women), 41 part-time (23 women); includes 28 minority (2 African Americans, 20 Asian Americans or Pacific Islanders, 6 Hispanic Americans), 47 international. Average age 27. 349 applicants, 56% accepted, 39 enrolled. In 2005, 59 master's, 4 doctorates awarded. Terminal master's awarded for partial completion of doctoral program. *Degree requirements:* For master's, thesis or alternative, qualifying paper; for doctorate, thesis/dissertation, comprehensive exam. *Entrance requirements:* For master's, GRE General Test; for doctorate, GRE General Test, GRE Subject Test. Additional exam requirements/recommendations for international students: Required—TOEFL. *Application deadline:* For fall admission, 1/4 for domestic students. Application fee: $80. *Financial support:* Fellowships with tuition reimbursements, research assistantships with tuition reimbursements, teaching assistantships with tuition reimbursements, career-related internships or fieldwork, Federal Work-Study, institutionally sponsored loans, scholarships/grants, health care benefits, and unspecified assistantships available. Financial award application deadline: 1/4; financial award applicants required to submit FAFSA. *Faculty research:* Genomics, molecular and cell biology, development and molecular genetics, molecular evolution of plants and animals. *Unit head:* Gloria Coruzzi, Chairman, 212-998-8200, Fax: 212-995-4015, E-mail: biology@nyu.edu. *Application contact:* Stephen Small, Director of Graduate Studies, 212-998-8200, Fax: 212-995-4015, E-mail: biology@nyu.edu.

The Ohio State University, Graduate School, College of Biological Sciences, Department of Plant Cellular and Molecular Biology, Columbus, OH 43210. Offers plant biology (MS, PhD). *Degree requirements:* For master's, thesis optional; for doctorate, one foreign language, thesis/dissertation. *Entrance requirements:* For master's and doctorate, GRE General Test. Additional exam requirements/recommendations for international students: Required—TOEFL (minimum score 600 paper-based; 250 computer-based), TSE. Electronic applications accepted. *Faculty research:* Regulatory, environmental, structural, systematic, and evolutionary botany.

Ohio University, Graduate Studies, College of Arts and Sciences, Department of Environmental and Plant Biology, Athens, OH 45701-2979. Offers MS, PhD. Part-time programs available. *Faculty:* 12 full-time (3 women), 1 (woman) part-time/adjunct. *Students:* 32 full-time (16 women); includes 4 minority (2 Asian Americans or Pacific Islanders, 2 Hispanic Americans), 6 international. Average age 25. 29 applicants, 62% accepted, 8 enrolled. In 2005, 6 master's, 3 doctorates awarded. *Median time to degree:* Of those who began their doctoral program in fall 1997, 100% received their degree in 8 years or less. *Degree requirements:* For master's, thesis/dissertation, 2 quarters of teaching experience; for doctorate, thesis/dissertation, 2 quarters of teaching experience, comprehensive exam. *Entrance requirements:* For master's, GRE General Test, minimum GPA of 3.0; for doctorate, GRE General Test, minimum GPA of 3.2. Additional exam requirements/recommendations for international students: Required—TOEFL (minimum score 620 paper-based; 260 computer-based). *Application deadline:* For fall admission, 1/15 priority date for domestic students, 1/15 priority date for international students. Applications are processed on a rolling basis. Application fee: $45. Electronic applications accepted. *Financial support:* In 2005–06, 3 fellowships with full tuition reimbursements (averaging $17,000 per year), 4 research assistantships with full tuition reimbursements, 21 teaching assistantships with tuition reimbursements were awarded; Federal Work-Study, institutionally sponsored loans, and scholarships/grants also available. Financial award application deadline: 1/15. *Faculty research:* Cellular and molecular biology, ecology, paleobotany, systematics, development. Total annual research expenditures: $859,166. *Unit head:* Dr. Gar W. Rothwell, Chair, 740-593-1126, Fax: 740-593-1130, E-mail: rothwell@ohio.edu. *Application contact:* Dr. Morgan L. Vis, Graduate Chair, 740-593-1126, Fax: 740-593-1130, E-mail: vischia@ohio.edu.

Rutgers, The State University of New Jersey, New Brunswick/Piscataway, Graduate School, Program in Plant Biology, New Brunswick, NJ 08901-1281. Offers horticulture (MS, PhD); molecular biology and biochemistry (MS, PhD); pathology (MS, PhD); plant ecology (MS, PhD); plant genetics (PhD); plant physiology (MS, PhD); production and management (MS); structure and plant groups (MS, PhD). Part-time programs available. *Faculty:* 65 full-time, 1 part-time/adjunct (0 women). *Students:* 33 full-time (14 women), 13 part-time (6 women); includes 3 minority (1 African American, 2 Asian Americans or Pacific Islanders), 10 international. Average age 31. 56 applicants, 23% accepted, 10 enrolled. In 2005, 1 master's, 4 doctorates awarded. Terminal master's awarded for partial completion of doctoral program. *Median time to degree:* Of those who began their doctoral program in fall 1997, 90% received their degree in 8 years or less. *Degree requirements:* For master's, thesis or alternative, comprehensive exam; for doctorate, thesis/dissertation, comprehensive exam. *Entrance requirements:* For master's and doctorate, GRE General Test, GRE Subject Test (recommended). Additional exam requirements/recommendations for international students: Required—TOEFL (minimum score 600 paper-based; 250 computer-based). *Application deadline:* For fall admission, 4/1 for domestic students, 4/1 for international students. Application fee: $50. Electronic applications accepted. *Expenses:* Tuition, state resident: full-time $10,440; part-time $435 per credit. Tuition, nonresident: full-time $15,520; part-time $647 per credit. Required fees: $129 per credit. Tuition and fees vary according to program. *Financial support:* In 2005–06, 42 students received support, including 9 fellowships with full tuition reimbursements available (averaging $24,000 per year), 22 research assistantships with full tuition reimbursements available (averaging $16,500 per year), 10 teaching assistantships with full tuition reimbursements available (averaging $16,988 per year) Financial award application deadline: 1/15; financial award applicants required to submit FAFSA. *Faculty research:* Molecular biology and biochemistry of plants, plant development and genomics, plant protection, plant improvement, plant management of horticultural and field crops. Total annual research expenditures: $10 million. *Unit head:* Dr. Thomas Leustek, Director, 732-932-8165 Ext. 326, Fax: 732-932-9377, E-mail: leustek@aesop.rutgers.edu. *Application contact:* Barbara Mulder, Program Associate, 732-932-9375 Ext. 358, Fax: 732-932-9377, E-mail: plantbio@aesop.rutgers.edu.

Southern Illinois University Carbondale, Graduate School, College of Science, Department of Plant Biology, Carbondale, IL 62901-4701. Offers MS, PhD. *Faculty:* 13 full-time (2 women). *Students:* 9 full-time (5 women), 24 part-time (12 women); includes 1 minority (Hispanic American), 10 international. Average age 25. 10 applicants, 80% accepted, 5 enrolled. In 2005, 4 master's, 2 doctorates awarded. *Degree requirements:* For master's, thesis; for doctorate, one foreign language, thesis/dissertation. *Entrance requirements:* For master's, GRE General Test, minimum GPA of 2.7; for doctorate, GRE General Test, minimum GPA of 3.25. Additional exam requirements/recommendations for international students: Required—TOEFL. *Application deadline:* Applications are processed on a rolling basis. Application fee: $20. *Financial support:* In 2005–06, 24 students received support, including 4 fellowships with full tuition reimbursements available, 6 research assistantships with full tuition reimbursements available, 13 teaching assistantships with full tuition reimbursements available; Federal Work-Study, institutionally sponsored loans, and tuition waivers (full) also available. Support available to part-time students. *Faculty research:* Algal toxins, ethnobotany, community and wetland ecology, morphogenesis, systematics and evolution. Total annual research expenditures: $524,140. *Unit head:* Dr. Dale Vitt, Chairperson, 618-453-3210. *Application contact:* Dr. Walter Schmid, Graduate Coordinator, 618-536-2331.

Announcement: The Department of Plant Biology at Southern Illinois University Carbondale offers the degrees of MS and PhD in plant biology. The program prepares students for careers as plant biologists in the food and pharmaceutical industries as well as other positions.

See Close-Up on page 495.

Texas A&M University, College of Agriculture and Life Sciences, Department of Soil and Crop Sciences, Intercollegiate Faculty of Molecular and Environmental Plant Sciences, College Station, TX 77843. Offers MS, PhD. *Students:* Average age 29. *Degree requirements:* For master's and doctorate, thesis/dissertation, seminar. *Entrance requirements:* For master's and doctorate, GRE General Test, letters of reference. Additional exam requirements/recommendations for international students: Required—TOEFL. *Application deadline:* For fall admission, 3/1 for domestic students; for spring admission, 8/1 for domestic students. Applica-

tions are processed on a rolling basis. Application fee: $50 ($75 for international students). Electronic applications accepted. *Expenses:* Tuition, state resident: full-time $4,488; part-time $187 per credit hour. Tuition, nonresident: full-time $11,112; part-time $463 per credit hour. Required fees: $1,974. *Financial support:* In 2005–06, fellowships with tuition reimbursements (averaging $20,000 per year), research assistantships (averaging $17,000 per year), teaching assistantships (averaging $18,200 per year) were awarded. Financial award application deadline: 3/1; financial award applicants required to submit FAFSA. *Faculty research:* Functional genomics, bioremediation, physiological ecology, transformation systems, abiotic stress. *Unit head:* Dr. Marla L. Binzel, Chair, 979-845-8938, Fax: 979-458-0533, E-mail: m-binzel@tamu.edu. *Application contact:* Dr. Jean Gould, Admissions Chair, 979-845-5078, Fax: 979-845-6049, E-mail: gould@tamu.edu.

Université Laval, Faculty of Agricultural and Food Sciences, Program in Plant Biology, Québec, QC G1K 7P4, Canada. Offers M Sc, PhD. Terminal master's awarded for partial completion of doctoral program. *Degree requirements:* For master's, thesis (for some programs); for doctorate, thesis/dissertation, comprehensive exam. *Entrance requirements:* For master's and doctorate, knowledge of French and English. Electronic applications accepted.

University of Alberta, Faculty of Graduate Studies and Research, Department of Biological Sciences, Edmonton, AB T6G 2E1, Canada. Offers environmental biology and ecology (M Sc, PhD); microbiology and biotechnology (M Sc, PhD); molecular biology and genetics (M Sc, PhD); physiology and cell biology (M Sc, PhD); plant biology (M Sc, PhD); systematics and evolution (M Sc, PhD). *Faculty:* 72 full-time (15 women), 15 part-time/adjunct (4 women). *Students:* 238 full-time (117 women), 32 part-time (15 women), 31 international. 206 applicants, 42% accepted. In 2005, 29 master's, 31 doctorates awarded. Terminal master's awarded for partial completion of doctoral program. *Degree requirements:* For master's and doctorate, thesis/dissertation, registration. *Entrance requirements:* Additional exam requirements/recommendations for international students: Required—TOEFL. *Application deadline:* For fall admission, 3/1 for domestic students. Applications are processed on a rolling basis. Application fee: $0. Tuition and fees charges are reported in Canadian dollars. *Expenses:* Tuition, state resident: part-time $562 Canadian dollars per term. Tuition, nonresident: full-time $3,375 Canadian dollars. Required fees: $573 Canadian dollars; $84 Canadian dollars per term. *Financial support:* In 2005–06, 4 research assistantships with partial tuition reimbursements (averaging $12,000 per year), 103 teaching assistantships with partial tuition reimbursements (averaging $12,300 per year) were awarded; career-related internships or fieldwork and scholarships/grants also available. *Unit head:* Laura Frost, Chair, 780-492-1904. *Application contact:* Dr. John P. Chang, Associate Chair for Graduate Studies, 780-492-1257, Fax: 780-492-9457, E-mail: bio.grad.coordinator@ualberta.ca.

University of California, Berkeley, Graduate Division, College of Natural Resources, Department of Plant and Microbial Biology, Berkeley, CA 94720-1500. Offers plant biology (PhD). *Degree requirements:* For doctorate, thesis/dissertation, qualifying exam, seminar presentation. *Entrance requirements:* For doctorate, GRE General Test, minimum GPA of 3.0. *Faculty research:* Development, molecular biology, genetics, microbial biology, mycology.

University of California, Davis, Graduate Studies, Graduate Group in Plant Biology, Davis, CA 95616. Offers MS, PhD. *Faculty:* 94 full-time. *Students:* 68 full-time (33 women); includes 8 minority (6 Asian Americans or Pacific Islanders, 2 Hispanic Americans), 19 international. Average age 29. 88 applicants, 18% accepted, 10 enrolled. In 2005, 4 master's, 10 doctorates awarded. *Median time to degree:* Of those who began their doctoral program in fall 1997, 75% received their degree in 8 years or less. *Degree requirements:* For master's, thesis (for some programs), comprehensive exam (for some programs); for doctorate, thesis/dissertation. *Entrance requirements:* For master's, GRE General Test, GRE Subject Test (biology), minimum GPA of 3.0; for doctorate, GRE General Test, GRE Subject Test (biology). Additional exam requirements/recommendations for international students: Required—TOEFL (minimum score 550 paper-based; 213 computer-based). *Application deadline:* For fall admission, 1/1 for domestic students, 1/1 for international students. Application fee: $60. Electronic applications accepted. *Financial support:* In 2005–06, 60 students received support, including 10 fellowships with full and partial tuition reimbursements available (averaging $17,189 per year), 32 research assistantships with full and partial tuition reimbursements available (averaging $16,116 per year), 11 teaching assistantships with partial tuition reimbursements available (averaging $15,345 per year); Federal Work-Study, institutionally sponsored loans, scholarships/grants, tuition waivers (full and partial), and unspecified assistantships also available. Financial award application deadline: 1/15; financial award applicants required to submit FAFSA. *Faculty research:* Cell and molecular biology, ecology, systematics and evolution, integrative plant and crop physiology, plant development and structure. *Unit head:* John Harada, Graduate Program Chair, 530-752-0673, Fax: 530-752-5410, E-mail: jjharada@ucdavis.edu. *Application contact:* Torri Hollowell, Administrative Assistant, 530-752-7094, Fax: 530-752-5410, E-mail: trhollowell@ucdavis.edu.

University of California, Riverside, Graduate Division, Department of Botany and Plant Sciences, Riverside, CA 92521-0102. Offers plant biology (MS, PhD); plant biology (plant genetics) (PhD). Part-time programs available. *Faculty:* 39 full-time (12 women). *Students:* 39 full-time (23 women); includes 4 minority (2 Asian Americans or Pacific Islanders, 2 Hispanic Americans), 15 international. Average age 30. In 2005, 1 master's, 10 doctorates awarded. Terminal master's awarded for partial completion of doctoral program. *Degree requirements:* For master's, comprehensive exams or thesis; for doctorate, thesis/dissertation, qualifying exams. *Entrance requirements:* For master's and doctorate, GRE General Test, minimum GPA of 3.2. Additional exam requirements/recommendations for international students: Required—TOEFL (minimum score 550 paper-based; 213 computer-based); Recommended—TSE (minimum score 50). *Application deadline:* For fall admission, 5/1 for domestic students, 2/1 for international students. For winter admission, 2/1 for domestic students; for spring admission, 12/1 for domestic students. Applications are processed on a rolling basis. Application fee: $60 ($75 for international students). Electronic applications accepted. *Expenses:* Tuition, nonresident: full-time $14,694. Full-time tuition and fees vary according to program. *Financial support:* In 2005–06, research assistantships (averaging $14,000 per year), teaching assistantships (averaging $15,000 per year) were awarded; fellowships, career-related internships or fieldwork, Federal Work-Study, institutionally sponsored loans, scholarships/grants, and tuition waivers (full and partial) also available. Financial award application deadline: 2/1; financial award applicants required to submit FAFSA. *Faculty research:* Agricultural plant biology; biochemistry and physiology; cellular, molecular and developmental biology; ecology, evolution, systematics and ethnobotany; genetics, genomics and bioinformatics. *Unit head:* Dr. Jodie S. Holt, Chair. *Application contact:* Carole Carpenter, Graduate Program Assistant, 800-735-0717, Fax: 951-827-5517, E-mail: plantbio@ucr.edu.

University of California, San Diego, Graduate Studies and Research, Division of Biology, Program in Plant Systems Biology, La Jolla, CA 92093. Offers PhD.

University of Connecticut, Graduate School, College of Liberal Arts and Sciences, Department of Molecular and Cell Biology, Storrs, CT 06269. Offers applied genomics (MS, PSM); biobehavioral science (PhD); biochemistry (MS, PhD); biophysics and structural biology (MS, PhD); biotechnology (MS); cell and developmental biology (MS, PhD); genetics, genomics, and bioinformatics (MS); including genetics (MS, PhD); genetics, genomics, and bioinformation (PhD), including genetics (MS, PhD); microbial systems analysis (MS, PSM); microbiology (MS, PhD); plant cell and molecular biology (MS, PhD). *Faculty:* 64 full-time (13 women). *Students:* 124 full-time (58 women), 15 part-time (8 women); includes 17 minority (5 African Americans, 1 American Indian/Alaska Native, 9 Asian Americans or Pacific Islanders, 2 Hispanic Americans), 36 international. Average age 27. 285 applicants, 28% accepted, 55 enrolled. In 2005, 23 master's, 10 doctorates awarded. Terminal master's awarded for partial completion of doctoral program. *Degree requirements:* For master's, comprehensive exam; for doctorate, thesis/dissertation. *Entrance requirements:* For master's and doctorate, GRE General Test, GRE Subject Test. Additional exam requirements/recommendations for international students: Required—TOEFL (minimum score 550 paper-based; 213 computer-based). *Application

Plant Biology

deadline: For fall admission, 2/1 priority date for domestic students, 2/1 priority date for international students; for spring admission, 11/1 for domestic students, 10/1 for international students. Applications are processed on a rolling basis. Application fee: $55. Electronic applications accepted. *Expenses:* Tuition, state resident: part-time $444 per credit hour. Tuition, nonresident: part-time $1,154 per credit hour. Tuition and fees vary according to course load. *Financial support:* In 2005–06, 35 research assistantships with full tuition reimbursements, 59 teaching assistantships with full tuition reimbursements were awarded; fellowships, Federal Work-Study, scholarships/grants, health care benefits, and unspecified assistantships also available. Financial award application deadline: 2/1; financial award applicants required to submit FAFSA. *Unit head:* Philip L. Yeagle, Head, 860-486-4329, Fax: 860-486-4331, E-mail: yeagle@uconnvm.uconn.edu. *Application contact:* Anne St. Onje, Graduate Coordinator, 860-486-4314, Fax: 860-486-3943, E-mail: ann.st_onje@uconn.edu.

University of Florida, Graduate School, College of Agricultural and Life Sciences and College of Liberal Arts and Sciences, Program in Plant Molecular and Cellular Biology, Gainesville, FL 32611. Offers MS, PhD. *Faculty:* 34. *Students:* 15 applicants, 53% accepted. *Degree requirements:* For master's and doctorate, thesis/dissertation. *Entrance requirements:* For master's and doctorate, GRE General Test, minimum GPA of 3.0. *Application deadline:* For fall admission, 2/1 for domestic students. Applications are processed on a rolling basis. Application fee: $20. Electronic applications accepted. *Expenses:* Tuition, state resident: full-time $6,234. Tuition, nonresident: full-time $21,359. Tuition and fees vary according to program. *Financial support:* Fellowships, research assistantships, teaching assistantships, unspecified assistantships available. *Faculty research:* Plant pathology, genetics, biochemistry, microbiology. *Unit head:* Dr. Dave G. Clark, Director, 352-392-1831 Ext. 370, Fax: 352-392-3870, E-mail: dgclark@ifas.ufl.edu. *Application contact:* Dr. Dave G. Clark, Director, 352-392-1831 Ext. 370, Fax: 352-392-3870, E-mail: dgclark@ifas.ufl.edu.

University of Georgia, Graduate School, College of Arts and Sciences, Department of Plant Biology, Athens, GA 30602. Offers MS, PhD. *Faculty:* 18 full-time (5 women), 1 (woman) part-time/adjunct. *Students:* 49 full-time, 2 part-time; includes 2 minority (1 Asian American or Pacific Islander, 1 Hispanic American), 21 international. 47 applicants, 40% accepted, 15 enrolled. In 2005, 1 master's, 8 doctorates awarded. *Degree requirements:* For master's, thesis; for doctorate, one foreign language, thesis/dissertation. *Entrance requirements:* For master's and doctorate, GRE General Test. *Application deadline:* For fall admission, 1/1 for domestic students. Application fee: $50. Electronic applications accepted. *Financial support:* Fellowships, research assistantships, teaching assistantships, unspecified assistantships available. *Unit head:* Dr. Russell L. Malmberg, Head, 706-542-1850, Fax: 706-542-1805, E-mail: russell@plantbio.uga.edu. *Application contact:* Dr. Chris Peterson, Graduate Coordinator, 706-542-1809, Fax: 706-542-1805, E-mail: chris@plantbio.uga.edu.

University of Illinois at Chicago, Graduate College, College of Liberal Arts and Sciences, Department of Biological Sciences, Chicago, IL 60607-7128. Offers cell and developmental biology (PhD); ecology and evolution (MS, DA, PhD); genetics and development (PhD); molecular biology (MS, PhD); neurobiology (MS, PhD); plant biology (MS, DA, PhD). *Degree requirements:* For master's, thesis; for doctorate, thesis/dissertation, preliminary exam. *Entrance requirements:* For master's and doctorate, GRE General Test, GRE Subject Test, previous course work in physics, calculus, and organic chemistry; minimum GPA of 2.75. Additional exam requirements/recommendations for international students: Required—TOEFL. Electronic applications accepted.

University of Illinois at Urbana–Champaign, Graduate College, College of Liberal Arts and Sciences, School of Integrative Biology, Department of Plant Biology, Champaign, IL 61820. Offers MS, PhD. *Faculty:* 14 full-time (2 women). *Students:* 18 full-time (8 women), 2 part-time (both women); includes 3 minority (1 American Indian/Alaska Native, 2 Asian Americans or Pacific Islanders), 5 international. 33 applicants, 15% accepted, 5 enrolled. In 2005, 5 master's, 2 doctorates awarded. *Degree requirements:* For doctorate, one foreign language, thesis/dissertation. *Entrance requirements:* For master's, GRE General Test, GRE Subject Test, minimum GPA of 3.0. *Application deadline:* Applications are processed on a rolling basis. Application fee: $50 ($60 for international students). Electronic applications accepted. *Financial support:* In 2005–06, 13 research assistantships, 10 teaching assistantships were awarded; fellowships Financial award application deadline: 2/15. *Unit head:* Evan DeLucia, Head, 217-333-3261, Fax: 217-244-7246, E-mail: delucia@uiuc.edu. *Application contact:* Lisa Boise, Administrative Secretary, 217-333-3261, Fax: 217-244-7246, E-mail: l-boise@uiuc.edu.

University of Illinois at Urbana–Champaign, Graduate College, College of Liberal Arts and Sciences, School of Integrative Biology, Program in Physiological and Molecular Plant Biology, Champaign, IL 61820. Offers PhD. *Faculty:* 14 full-time (2 women). *Students:* 17 full-time (8 women), 2 part-time (1 woman); includes 5 minority (3 Asian Americans or Pacific Islanders, 2 Hispanic Americans), 6 international. 22 applicants, 9% accepted, 0 enrolled. In 2005, 3 degrees awarded. *Application deadline:* For fall admission, 3/15 for domestic students; for spring admission, 9/15 for domestic students. Applications are processed on a rolling basis. Application fee: $50 ($60 for international students). Electronic applications accepted. *Financial support:* In 2005–06, 2 fellowships, 11 research assistantships, 9 teaching assistantships were awarded. Financial award application deadline: 2/15. *Unit head:* Hans Bohnert, Acting Director, 217-265-5475, Fax: 217-244-1224, E-mail: bohnerth@uiuc.edu. *Application contact:* Carol Hall, Secretary, 217-333-8208, Fax: 217-244-1224, E-mail: c-hall@uiuc.edu.

The University of Iowa, Graduate College, College of Liberal Arts and Sciences, Department of Biological Sciences, Iowa City, IA 52242-1316.

See Close-Up on page 257.

The University of Iowa, Graduate College, College of Liberal Arts and Sciences, Department of Biological Sciences, Iowa City, IA 52242-1316.

University of Maine, Graduate School, College of Natural Sciences, Forestry, and Agriculture, Department of Biological Sciences, Program in Plant Science, Orono, ME 04469. Offers PhD. Part-time programs available. *Students:* 9 full-time (4 women), 4 part-time (3 women), 2 international. Average age 35. 5 applicants, 40% accepted, 1 enrolled. *Degree requirements:* For doctorate, thesis/dissertation. *Entrance requirements:* For doctorate, GRE General Test. Additional exam requirements/recommendations for international students: Required—TOEFL. *Application deadline:* For fall admission, 2/1 for domestic students. Applications are processed on a rolling basis. Application fee: $50. Electronic applications accepted. *Financial support:* Fellowships, research assistantships with tuition reimbursements, teaching assistantships with tuition reimbursements, career-related internships or fieldwork, Federal Work-Study, institutionally sponsored loans, and tuition waivers (full) available. Financial award application deadline: 3/1. *Unit head:* , Dr. Stelbs Tavanteis, Coordinator, 207-581-2986. *Application contact:* Scott G. Delcourt, Associate Dean of the Graduate School, 207-581-3219, Fax: 207-581-3232, E-mail: graduate@maine.edu.

University of Maine, Graduate School, College of Natural Sciences, Forestry, and Agriculture, Department of Plant, Soil, and Environmental Sciences, Orono, ME 04469. Offers biological sciences (PhD); ecology and environmental sciences (MS, PhD); forest resources (PhD); horticulture (MS); plant science (PhD); plant, soil, and environmental sciences (MS); resource utilization (MS). *Faculty:* 25. *Students:* 15 full-time (8 women), 8 part-time (6 women), 3 international. Average age 32. 6 applicants, 33% accepted, 1 enrolled. In 2005, 9 master's awarded. *Entrance requirements:* For master's and doctorate, GRE General Test. Additional exam requirements/recommendations for international students: Required—TOEFL. *Application deadline:* Applications are processed on a rolling basis. Application fee: $50. Electronic applications accepted. *Financial support:* In 2005–06, 9 research assistantships with tuition reimbursements (averaging $12,180 per year) were awarded; teaching assistantships, scholarships/grants, tuition waivers (full and partial), and unspecified assistantships also available. *Unit head:* Greg Porter, Chair, 207-581-2943, Fax: 207-581-3207. *Application*

contact: Scott G. Delcourt, Associate Dean of the Graduate School, 207-581-3219, Fax: 207-581-3232, E-mail: graduate@maine.edu.

University of Maryland, College Park, Graduate Studies, College of Chemical and Life Sciences, Department of Cell Biology and Molecular Genetics, Program in Plant Biology, College Park, MD 20742. Offers MS, PhD. Part-time and evening/weekend programs available. *Students:* 1 full-time (0 women). In 2005, 3 degrees awarded. *Application deadline:* For fall admission, 1/11 for domestic students, 1/11 for international students. Applications are processed on a rolling basis. Application fee: $60. Electronic applications accepted. *Financial support:* Fellowships, research assistantships, teaching assistantships available. Financial award applicants required to submit FAFSA. *Faculty research:* Genetics and molecular biology, virology, plant pathology, mycology, nematology. *Application contact:* Dean of Graduate School, 301-405-4190, Fax: 301-314-9305.

University of Massachusetts Amherst, Graduate School, Interdisciplinary Programs, Program in Plant Biology, Amherst, MA 01003. Offers MS, PhD. *Students:* 14 full-time (8 women), 2 part-time (1 woman); includes 3 minority (2 Asian Americans or Pacific Islanders, 1 Hispanic American), 5 international. 35 applicants, 26% accepted, 8 enrolled. In 2005, 1 master's, 1 doctorate awarded. *Degree requirements:* For master's and doctorate, thesis/dissertation. *Entrance requirements:* For master's and doctorate, GRE General Test. Additional exam requirements/recommendations for international students: Required—TOEFL (minimum score 530 paper-based; 197 computer-based). *Application deadline:* For fall admission, 1/2 priority date for domestic students, 1/2 priority date for international students; for spring admission, 10/1 for domestic students, 10/1 for international students. Applications are processed on a rolling basis. Application fee: $40 ($65 for international students). Electronic applications accepted. *Expenses:* Tuition, state resident: part-time $110 per credit. Tuition, nonresident: part-time $414 per credit. Required fees: $2,824 per term. One-time fee: $250 part-time. Full-time tuition and fees vary according to course load, campus/location, program and reciprocity agreements. *Financial support:* In 2005–06, 1 fellowship with full tuition reimbursement (averaging $806 per year), 5 research assistantships with full tuition reimbursements (averaging $8,576 per year) were awarded; teaching assistantships with full tuition reimbursements, career-related internships or fieldwork, Federal Work-Study, scholarships/grants, traineeships, and unspecified assistantships also available. Support available to part-time students. Financial award application deadline: 1/15. *Unit head:* Dr. Elsbeth Walker, Head, 413-577-3217, Fax: 413-545-3243. *Application contact:* Information Contact, 413-577-3217, Fax: 413-545-3243.

See Close-Up on page 497.

University of Minnesota, Twin Cities Campus, Graduate School, College of Biological Sciences, Program in Plant Biological Sciences, Minneapolis, MN 55455-0213. Offers MS, PhD. Part-time programs available. Terminal master's awarded for partial completion of doctoral program. *Degree requirements:* For master's, thesis or alternative, registration; for doctorate, thesis/dissertation, written and oral preliminary exams. *Entrance requirements:* For master's and doctorate, GRE General Test. Additional exam requirements/recommendations for international students: Required—TOEFL. Electronic applications accepted. *Expenses:* Tuition, state resident: full-time $8,748; part-time $729 per credit. Tuition, nonresident: full-time $15,848; part-time $1,321 per credit. Full-time tuition and fees vary according to class time, course load, program and reciprocity agreements. *Faculty research:* Cell and molecular biology; plant physiology; plant structure, diversity, and development; ecology, systematics, evolution and genomics.

University of Missouri–Columbia, Graduate School, College of Agriculture, Food and Natural Resources, Division of Plant Sciences, Department of Plant Pathology and Microbiology, Columbia, MO 65211. Offers MS, PhD. Terminal master's awarded for partial completion of doctoral program. *Degree requirements:* For master's and doctorate, thesis/dissertation. *Entrance requirements:* For master's and doctorate, GRE General Test, minimum GPA of 3.0. *Application deadline:* For fall admission, 3/1 for domestic students. Applications are processed on a rolling basis. Application fee: $45 ($60 for international students). *Financial support:* Research assistantships, teaching assistantships, institutionally sponsored loans available. *Unit head:* Dr. Jeanne Mihail, Director of Graduate Studies, 573-882-0574, E-mail: mihailj@missouri.edu.

University of New Hampshire, Graduate School, College of Life Sciences and Agriculture, Department of Plant Biology, Durham, NH 03824. Offers MS, PhD. Part-time programs available. *Faculty:* 22 full-time. *Students:* 10 full-time (7 women), 11 part-time (9 women), 5 international. Average age 30. 11 applicants, 55% accepted, 2 enrolled. In 2005, 9 master's, 3 doctorates awarded. Terminal master's awarded for partial completion of doctoral program. *Degree requirements:* For master's and doctorate, thesis/dissertation. *Entrance requirements:* For master's and doctorate, GRE General Test, GRE Subject Test. Additional exam requirements/recommendations for international students: Required—TOEFL (minimum score 550 paper-based; 213 computer-based); Recommended—TSE. *Application deadline:* For fall admission, 4/1 priority date for domestic students, 4/1 priority date for international students. Applications are processed on a rolling basis. Application fee: $60. Electronic applications accepted. *Expenses:* Tuition, state resident: full-time $8,010; part-time $445 per credit hour. Tuition, nonresident: full-time $19,730; part-time $810 per credit hour. Required fees: $322 per semester. Tuition and fees vary according to course load and program. *Financial support:* In 2005–06, 1 fellowship, 6 research assistantships, 8 teaching assistantships were awarded; career-related internships or fieldwork, Federal Work-Study, scholarships/grants, and tuition waivers (full and partial) also available. Support available to part-time students. Financial award application deadline: 2/15. *Unit head:* Garrett Crow, Chairperson, 603-862-2865. *Application contact:* Flora Joyal, Administrative Assistant, 603-862-4095, E-mail: flora.joyal@unh.edu.

University of Pennsylvania, School of Arts and Sciences, Program in Plant and Microbial Biology, Philadelphia, PA 19104. Offers PhD. *Degree requirements:* For doctorate, thesis/dissertation. *Entrance requirements:* For doctorate, GRE General Test, GRE Subject Test. Additional exam requirements/recommendations for international students: Required—TOEFL. Electronic applications accepted. *Faculty research:* Plant-cell interaction, plant developmental genetics, cell fine structure, reproductive biology, tropical animal-plant interactions.

The University of Texas at Austin, Graduate School, College of Natural Sciences, School of Biological Sciences, Program in Plant Biology, Austin, TX 78712-1111. Offers MA, PhD. *Entrance requirements:* For master's and doctorate, GRE General Test, minimum GPA of 3.0. Additional exam requirements/recommendations for international students: Required—TOEFL. Electronic applications accepted. *Faculty research:* Systematics, plant molecular biology, psychology, ecology, evolution.

The University of Western Ontario, Faculty of Graduate Studies, Biosciences Division, Department of Plant Sciences, London, ON N6A 5B8, Canada. Offers plant and environmental sciences (M Sc); plant sciences (M Sc, PhD); plant sciences and environmental sciences (PhD); plant sciences and molecular biology (M Sc, PhD). *Degree requirements:* For master's and doctorate, thesis/dissertation. *Entrance requirements:* For doctorate, M Sc or equivalent. *Faculty research:* Ecology systematics, plant biochemistry and physiology, yeast genetics, molecular biology.

Washington University in St. Louis, Graduate School of Arts and Sciences, Division of Biology and Biomedical Sciences, Program in Plant Biology, St. Louis, MO 63130-4899. Offers PhD. *Degree requirements:* For doctorate, thesis/dissertation. *Entrance requirements:* For doctorate, GRE General Test, GRE Subject Test. Electronic applications accepted.

Yale University, Graduate School of Arts and Sciences, Department of Molecular, Cellular, and Developmental Biology, Program in Plant Sciences, New Haven, CT 06520. Offers PhD. *Degree requirements:* For doctorate, thesis/dissertation. *Entrance requirements:* For doctorate, GRE General Test, GRE Subject Test.

See Close-Up on page 499.

Plant Molecular Biology

Cornell University, Graduate School, Graduate Fields of Agriculture and Life Sciences, Field of Plant Biology, Ithaca, NY 14853-0001. Offers cytology (MS, PhD); paleobotany (MS, PhD); plant cell biology (MS, PhD); plant ecology (MS, PhD); plant molecular biology (MS, PhD); plant morphology, anatomy and biomechanics (MS, PhD); plant physiology (MS, PhD); systematic botany (MS, PhD). *Faculty:* 69 full-time (15 women). *Students:* 50 full-time (27 women); includes 2 minority (both African Americans), 16 international. 33 applicants, 45% accepted, 10 enrolled. In 2005, 1 master's, 7 doctorates awarded. *Degree requirements:* For doctorate, thesis/dissertation, comprehensive exam. *Entrance requirements:* For doctorate, GRE General Test, GRE Subject Test in biology (recommended), 3 letters of recommendation. Additional exam requirements/recommendations for international students: Required—TOEFL (minimum score 610 paper-based; 253 computer-based). *Application deadline:* For fall admission, 1/15 for domestic students. Application fee: $60. Electronic applications accepted. *Financial support:* In 2005–06, 50 students received support, including 10 fellowships with full tuition reimbursements available, 23 research assistantships with full tuition reimbursements available, 17 teaching assistantships with full tuition reimbursements available; institutionally sponsored loans, scholarships/grants, health care benefits, tuition waivers (full and partial), and unspecified assistantships also available. Financial award applicants required to submit FAFSA. *Faculty research:* Plant cell biology/cytology; plant molecular biology; plant morphology/anatomy/biomechanics; plant physiology, systematic botany, paleobotany; plant ecology, ethnobotany, plant biochemistry, photosynthesis. *Unit head:* Director of Graduate Studies, 607-255-2131. *Application contact:* Graduate Field Assistant, 607-255-2131, E-mail: plbio@cornell.edu.

Michigan Technological University, Graduate School, School of Forest Resources and Environmental Science, Program in Forest Molecular Genetics and Biotechnology, Houghton, MI 49931-1295. Offers MS, PhD. Part-time programs available. *Faculty:* 23 full-time (4 women), 16 part-time/adjunct (2 women). *Students:* 13 full-time (4 women), 4 part-time (all women), 16 international. Average age 30. 10 applicants, 70% accepted, 3 enrolled. In 2005, 1 master's, 1 doctorate awarded. Terminal master's awarded for partial completion of doctoral program. *Median time to degree:* Of those who began their doctoral program in fall 1997, 100% received their degree in 8 years or less. *Degree requirements:* For master's, thesis (for some programs), registration; for doctorate, thesis/dissertation, comprehensive exam, registration. *Entrance requirements:* For master's, GRE. Additional exam requirements/recommendations for international students: Required—TOEFL (minimum score 550 paper-based; 213 computer-based). *Application deadline:* Applications are processed on a rolling basis. Application fee: $40 ($45 for international students). Electronic applications accepted. *Expenses:* Tuition, nonresident: full-time $11,232; part-time $468 per credit. Required fees: $754; $377 per semester. Full-time tuition and fees vary according to course load, degree level and program. *Financial support:* In 2005–06, 14 students received support, including fellowships with full tuition reimbursements available (averaging $9,542 per year), 14 research assistantships with full tuition reimbursements available (averaging $9,542 per year), teaching assistantships with full tuition reimbursements available (averaging $9,542 per year); career-related internships or fieldwork, Federal Work-Study, scholarships/grants, health care benefits, tuition waivers (partial), unspecified assistantships, and co-op also available. Financial award applicants required to submit FAFSA. *Application contact:* Dr. Chandrashekhar P. Joshi, Associate Professor and Graduate Program Coordinator, 906-487-3480, Fax: 906-487-2915, E-mail: cpjoshi@mtu.edu.

Rutgers, The State University of New Jersey, New Brunswick/Piscataway, Graduate School, Program in Plant Biology, New Brunswick, NJ 08901-1281. Offers horticulture (MS, PhD); molecular biology and biochemistry (MS, PhD); pathology (MS, PhD); plant ecology (MS, PhD); plant genetics (PhD); plant physiology (MS, PhD); production and management (MS); structure and plant groups (MS, PhD). Part-time programs available. *Faculty:* 65 full-time, 1 part-time/adjunct (0 women). *Students:* 33 full-time (14 women), 13 part-time (6 women); includes 1 minority (1 African American, 2 Asian Americans or Pacific Islanders), 10 international. Average age 31. 56 applicants, 23% accepted, 10 enrolled. In 2005, 1 master's, 4 doctorates awarded. Terminal master's awarded for partial completion of doctoral program. *Median time to degree:* Of those who began their doctoral program in fall 1997, 90% received their degree in 8 years or less. *Degree requirements:* For master's, thesis or alternative, comprehensive exam; for doctorate, thesis/dissertation, comprehensive exam. *Entrance requirements:* For master's and doctorate, GRE General Test, GRE Subject Test (recommended). Additional exam requirements/recommendations for international students: Required—TOEFL (minimum score 600 paper-based; 250 computer-based). *Application deadline:* For fall admission, 4/1 for domestic students, 4/1 for international students. Application fee: $50. Electronic applications accepted. *Expenses:* Tuition, state resident: full-time $10,440; part-time $435 per credit. Tuition, nonresident: full-time $15,520; part-time $647 per credit. Required fees: $129 per credit. Tuition and fees vary according to program. *Financial support:* In 2005–06, 42 students received support, including 9 fellowships with full tuition reimbursements available (averaging $24,000 per year), 22 research assistantships with full tuition reimbursements available (averaging $16,500 per year), 10 teaching assistantships with full tuition reimbursements available (averaging $16,988 per year) Financial award application deadline: 1/15; financial award applicants required to submit FAFSA. *Faculty research:* Molecular biology and biochemistry of plants, plant development and genomics, plant protection, plant improvement, plant management of horticultural and field crops. Total annual research expenditures: $10 million. *Unit head:* Dr. Thomas Leustek, Director, 732-932-8165 Ext. 326, Fax: 732-932-9377, E-mail: leustek@aesop.rutgers.edu. *Application contact:* Barbara Mulder, Program Associate, 732-932-9375 Ext. 358, Fax: 732-932-9377, E-mail: plantbio@aesop.rutgers.edu.

University of California, Los Angeles, Graduate Division, College of Letters and Science, Department of Organismic Biology, Ecology and Evolution, Los Angeles, CA 90095. Offers biology (MA, PhD); plant molecular biology (PhD). *Degree requirements:* For master's, comprehensive exam or thesis; for doctorate, thesis/dissertation, oral and written qualifying exams. *Entrance requirements:* For master's, GRE General Test, GRE Subject Test (biology), minimum GPA of 3.0; for doctorate, GRE General Test, GRE Subject Test (biology), minimum undergraduate GPA of 3.0. Electronic applications accepted. *Faculty research:* Molecular, cell, and developmental biology; interactive biology; organisms and populations.

University of California, San Diego, Graduate Studies and Research, Division of Biology, Program in Plant Molecular Biology, La Jolla, CA 92093. Offers PhD. Offered in association with the Salk Institute. *Degree requirements:* For doctorate, thesis/dissertation, qualifying exam. Electronic applications accepted.

University of Connecticut, Graduate School, College of Liberal Arts and Sciences, Department of Molecular and Cell Biology, Storrs, CT 06269. Offers applied genomics (MS, PSM); biobehavioral science (PhD); biochemistry (MS, PhD); biophysics and structural biology (MS, PhD); biotechnology (MS); cell and developmental biology (MS, PhD); genetics, genomics, and bioinformatics (MS), including genetics (MS, PhD); genetics, genomics, and bioinformation (PhD), including genetics (MS, PhD); microbial systems analysis (MS, PSM); microbiology (MS, PhD); plant cell and molecular biology (MS, PhD). *Faculty:* 64 full-time (13 women). *Students:* 124 full-time (58 women), 15 part-time (8 women); includes 17 minority (5 African Americans, 1 American Indian/Alaska Native, 9 Asian Americans or Pacific Islanders, 2 Hispanic Americans), 36 international. Average age 27. 285 applicants, 28% accepted, 55 enrolled. In 2005, 23 master's, 10 doctorates awarded. Terminal master's awarded for partial completion of doctoral program. *Degree requirements:* For master's, comprehensive exam; for doctorate, thesis/dissertation. *Entrance requirements:* For master's and doctorate, GRE General Test, GRE Subject Test. Additional exam requirements/recommendations for international students: Required—TOEFL (minimum score 550 paper-based; 213 computer-based). *Application deadline:* For fall admission, 2/1 priority date for domestic students, 2/1 priority date for international students; for spring admission, 11/1 for domestic students, 10/1 for international students. Applications are processed on a rolling basis. Application fee: $55. Electronic applications accepted. *Expenses:* Tuition, state resident: part-time $444 per credit hour. Tuition, nonresident: part-time $1,154 per credit hour. Tuition and fees vary according to course load. *Financial support:* In 2005–06, 35 research assistantships with full tuition reimbursements, 59 teaching assistantships with full tuition reimbursements were awarded; fellowships, Federal Work-Study, scholarships/grants, health care benefits, and unspecified assistantships also available. Financial award application deadline: 2/1; financial award applicants required to submit FAFSA. *Unit head:* Philip L. Yeagle, Head, 860-486-4329, Fax: 860-486-4331, E-mail: yeagle@uconnvm.uconn.edu. *Application contact:* Anne St. Onje, Graduate Coordinator, 860-486-4314, Fax: 860-486-3943, E-mail: ann.st_onje@uconn.edu.

University of Florida, Graduate School, College of Agricultural and Life Sciences and College of Liberal Arts and Sciences, Program in Plant Molecular and Cellular Biology, Gainesville, FL 32611. Offers MS, PhD. *Faculty:* 34. *Students:* 15 applicants, 53% accepted. *Degree requirements:* For master's and doctorate, thesis/dissertation. *Entrance requirements:* For master's and doctorate, GRE General Test, minimum GPA of 3.0. *Application deadline:* For fall admission, 2/1 for domestic students. Applications are processed on a rolling basis. Application fee: $20. Electronic applications accepted. *Expenses:* Tuition, state resident: full-time $6,234. Tuition, nonresident: full-time $21,359. Tuition and fees vary according to program. *Financial support:* Fellowships, research assistantships, teaching assistantships, unspecified assistantships available. *Faculty research:* Plant pathology, genetics, biochemistry, microbiology. *Unit head:* Dr. Dave G. Clark, Director, 352-392-1831 Ext. 370, Fax: 352-392-3870, E-mail: dgclark@ifas.ufl.edu. *Application contact:* Dr. Dave G. Clark, Director, 352-392-1831 Ext. 370, Fax: 352-392-3870, E-mail: dgclark@ifas.ufl.edu.

University of Massachusetts Amherst, Graduate School, Interdisciplinary Programs, Program in Plant Biology, Amherst, MA 01003. Offers MS, PhD. *Students:* 14 full-time (8 women), 2 part-time (1 woman); includes 3 minority (2 Asian Americans or Pacific Islanders, 1 Hispanic American), 5 international. 35 applicants, 26% accepted, 8 enrolled. In 2005, 1 master's, 1 doctorate awarded. *Degree requirements:* For master's and doctorate, thesis/dissertation. *Entrance requirements:* For master's and doctorate, GRE General Test. Additional exam requirements/recommendations for international students: Required—TOEFL (minimum score 530 paper-based; 197 computer-based). *Application deadline:* For fall admission, 1/2 priority date for domestic students, 1/2 priority date for international students; for spring admission, 10/1 for domestic students, 10/1 for international students. Applications are processed on a rolling basis. Application fee: $40 ($65 for international students). Electronic applications accepted. *Expenses:* Tuition, state resident: part-time $110 per credit. Tuition, nonresident: part-time $414 per credit. Required fees: $2,824 per term. One-time fee: $250 part-time. Full-time tuition and fees vary according to course load, campus/location, program and reciprocity agreements. *Financial support:* In 2005–06, 1 fellowship with full tuition reimbursement (averaging $806 per year), 5 research assistantships with full tuition reimbursements (averaging $8,576 per year) were awarded; teaching assistantships with full tuition reimbursements, career-related internships or fieldwork, Federal Work-Study, scholarships/grants, traineeships, and unspecified assistantships also available. Support available to part-time students. Financial award application deadline: 1/15. *Unit head:* Dr. Elsbeth Walker, Head, 413-577-3217, Fax: 413-545-3243. *Application contact:* Information Contact, 413-577-3217, Fax: 413-545-3243.

See Close-Up on page 497.

Washington State University, Graduate School, College of Agricultural, Human, and Natural Resource Sciences, Program in Molecular Plant Sciences, Pullman, WA 99164. Offers MS, PhD. *Faculty:* 25. *Students:* 8 full-time (2 women), 3 international. Average age 27. 68 applicants, 13% accepted, 6 enrolled. In 2005, 3 degrees awarded. Terminal master's awarded for partial completion of doctoral program. *Median time to degree:* Of those who began their doctoral program in fall 1997, 100% received their degree in 8 years or less. *Degree requirements:* For master's and doctorate, thesis/dissertation, oral exam, written exam. *Entrance requirements:* For master's and doctorate, GRE General Test, minimum GPA of 3.0, 3 letters of recommendation. Additional exam requirements/recommendations for international students: Required—TOEFL. *Application deadline:* For fall admission, 1/1 priority date for domestic students, 1/1 priority date for international students. Applications are processed on a rolling basis. Application fee: $35. *Expenses:* Tuition, state resident: full-time $6,295; part-time $336 per credit. Tuition, nonresident: full-time $15,949; part-time $819 per credit. Required fees: $933. Part-time tuition and fees vary according to campus/location and program. *Financial support:* In 2005–06, 4 fellowships (averaging $4,500 per year), 18 research assistantships with full and partial tuition reimbursements (averaging $14,508 per year), 4 teaching assistantships with full and partial tuition reimbursements (averaging $12,759 per year) were awarded; career-related internships or fieldwork, Federal Work-Study, institutionally sponsored loans, and tuition waivers (partial) also available. Financial award application deadline: 4/1; financial award applicants required to submit FAFSA. *Faculty research:* Cell response to environmental signals, transport of amino acids, regulation of synthesis of defense proteins. Total annual research expenditures: $3.7 million. *Unit head:* Dr. David Kramer, Chair, 509-335-8382. *Application contact:* Paula Marston, Graduate Program Assistant, 509-335-3412, Fax: 509-335-7643, E-mail: plph@wsu.edu.

Plant Pathology

Auburn University, Graduate School, College of Agriculture, Department of Entomology and Plant Pathology, Auburn University, AL 36849. Offers entomology (M Ag, MS, PhD); plant pathology (M Ag, MS, PhD). Part-time programs available. *Faculty:* 21 full-time (6 women). *Students:* 18 full-time (9 women), 6 part-time (3 women), 11 international. 14 applicants, 71% accepted, 6 enrolled. In 2005, 2 master's, 3 doctorates awarded. *Degree requirements:* For master's, thesis (for some programs); for doctorate, one foreign language, thesis/dissertation. *Entrance requirements:* For master's, GRE General Test; for doctorate, GRE General Test, GRE Subject Test, master's degree with thesis. *Application deadline:* For fall admission, 7/7 for

domestic students; for spring admission, 11/24 for domestic students. Applications are processed on a rolling basis. Application fee: $25 ($50 for international students). Electronic applications accepted. *Financial support:* Research assistantships, teaching assistantships, Federal Work-Study available. Support available to part-time students. Financial award application deadline: 3/15. *Faculty research:* Pest management, biological control, systematics, medical entomology. *Unit head:* Dr. Michael L. Williams, Chair, 334-844-5006. *Application contact:* Dr. Stephen L. McFarland, Acting Dean of the Graduate School, 334-844-4700.

Colorado State University, Graduate School, College of Agricultural Sciences, Department of Bioagricultural Sciences and Pest Management, Fort Collins, CO 80523-0015. Offers entomology (MS, PhD); plant pathology and weed science (MS, PhD). *Faculty:* 18 full-time (3 women). *Students:* 21 full-time (12 women), 18 part-time (8 women); includes 3 minority (1 American Indian/Alaska Native, 1 Asian American or Pacific Islander, 1 Hispanic American), 1 international. Average age 34. 16 applicants, 44% accepted, 6 enrolled. In 2005, 5 master's, 4 doctorates awarded. *Degree requirements:* For master's, thesis (for some programs); registration; for doctorate, thesis/dissertation, registration. *Entrance requirements:* For master's and doctorate, GRE General Test, minimum GPA of 3.0. Additional exam requirements/recommendations for international students: Required—TOEFL (minimum score 550 paper-based; 213 computer-based). *Application deadline:* For fall admission, 4/1 priority date for domestic students, 4/1 priority date for international students; for spring admission, 9/1 priority date for domestic students, 9/1 priority date for international students. Applications are processed on a rolling basis. Application fee: $50. Electronic applications accepted. *Expenses:* Tuition, state resident: full-time $3,690; part-time $205 per credit. Tuition, nonresident: full-time $14,958; part-time $831 per credit. Required fees: $1,061. *Financial support:* In 2005–06, 1 student received support, including fellowships (averaging $2,500 per year), research assistantships with full tuition reimbursements available (averaging $17,500 per year), teaching assistantships with full tuition reimbursements available (averaging $12,402 per year); scholarships/grants, traineeships, and unspecified assistantships also available. Financial award application deadline: 4/1; financial award applicants required to submit FAFSA. *Faculty research:* Biological control of post-insect plant pathogens and weeds, integrated pest management, weed ecology and biology, and pests genome's of plants. Total annual research expenditures: $2.2 million. *Unit head:* Thomas O. Holtzer, 970-491-5261, Fax: 970-491-3862, E-mail: tholtzer@lamar.colostate.edu. *Application contact:* Janet Dill, Graduate Program Coordinator, 970-491-0402, Fax: 970-491-3862, E-mail: janet.dill@colostate.edu.

Cornell University, Graduate School, Graduate Fields of Agriculture and Life Sciences, Field of Plant Pathology, Ithaca, NY 14853-0001. Offers ecological and environmental plant pathology (MPS, MS, PhD); epidemiological plant pathology (MPS, MS, PhD); molecular plant pathology (MPS, MS, PhD); mycology (MPS, MS, PhD); plant disease epidemiology (MPS, MS, PhD); plant pathology (MPS, MS, PhD). *Faculty:* 57 full-time (16 women). *Students:* 42 full-time (28 women); includes 3 minority (2 Asian Americans or Pacific Islanders, 1 Hispanic American), 14 international. 21 applicants, 29% accepted, 5 enrolled. In 2005, 1 master's, 6 doctorates awarded. *Degree requirements:* For master's, thesis (MS), project paper (MPS); for doctorate, thesis/dissertation, comprehensive exam. *Entrance requirements:* For master's and doctorate, GRE General Test, GRE Subject Test (biology recommended), 3 letters of recommendation. Additional exam requirements/recommendations for international students: Required—TOEFL (minimum score 550 paper-based; 213 computer-based). *Application deadline:* For fall admission, 1/15 for domestic students. Applications are processed on a rolling basis. Application fee: $60. Electronic applications accepted. *Financial support:* In 2005–06, 39 students received support, including 5 fellowships with full tuition reimbursements available, 28 research assistantships with full tuition reimbursements available, 6 teaching assistantships with full tuition reimbursements available; institutionally sponsored loans, scholarships/grants, health care benefits, tuition waivers (full and partial), and unspecified assistantships also available. Financial award applicants required to submit FAFSA. *Faculty research:* Plant pathology; mycology; molecular plant pathology; plant disease epidemiology, ecological and environmental plant pathology; plant disease epidemiology and simulation modeling. *Unit head:* Director of Graduate Studies, 607-255-3259, Fax: 607-255-4471. *Application contact:* Graduate Field Assistant, 607-255-3259, Fax: 607-255-4471, E-mail: plpathology@cornell.edu.

Iowa State University of Science and Technology, Graduate College, College of Agriculture, Department of Plant Pathology, Ames, IA 50011. Offers MS, PhD. *Faculty:* 16 full-time. *Students:* 27 full-time (17 women), 3 part-time; includes 3 minority (2 African Americans, 1 Hispanic American), 17 international. 9 applicants, 44% accepted, 4 enrolled. In 2005, 4 master's, 4 doctorates awarded. *Degree requirements:* For master's, thesis or alternative; for doctorate, thesis/dissertation. *Entrance requirements:* For master's and doctorate, GRE General Test, resumé. Additional exam requirements/recommendations for international students: Required—TOEFL (paper score 530; computer score 197) or IELTS (score 6.0). *Application deadline:* For fall admission, 1/1 priority date for domestic students, 1/1 priority date for international students. Applications are processed on a rolling basis. Application fee: $30 ($70 for international students). Electronic applications accepted. *Expenses:* Tuition, state resident: full-time $6,410. Tuition, nonresident: full-time $16,422. Tuition and fees vary according to program. *Financial support:* In 2005–06, 23 research assistantships with full and partial tuition reimbursements (averaging $16,060 per year), 2 teaching assistantships with full and partial tuition reimbursements (averaging $14,830 per year) were awarded; fellowships, scholarships/grants, health care benefits, and unspecified assistantships also available. *Unit head:* Dr. Thomas Baum, Chair, 515-294-1741, Fax: 515-294-9420, E-mail: plantpath@iastate.edu.

Kansas State University, Graduate School, College of Agriculture, Department of Plant Pathology, Manhattan, KS 66506. Offers MS, PhD. *Faculty:* 16 full-time (3 women), 4 part-time/adjunct (0 women). *Students:* 20 full-time (8 women), 2 part-time (both women), 14 international. Average age 25. In 2005, 4 master's, 5 doctorates awarded. Terminal master's awarded for partial completion of doctoral program. *Degree requirements:* For master's, thesis, oral exam; for doctorate, thesis/dissertation, preliminary exams. *Entrance requirements:* For master's and doctorate, minimum undergraduate GPA of 3.0. Additional exam requirements/recommendations for international students: Required—TOEFL (minimum score 550 paper-based; 213 computer-based). *Application deadline:* For fall admission, 2/1 for domestic students; for spring admission, 10/1 for domestic students. Applications are processed on a rolling basis. Application fee: $30 ($55 for international students). *Expenses:* Tuition, state resident: full-time $5,160; part-time $215 per credit hour. Tuition, nonresident: full-time $12,816; part-time $534 per credit hour. Required fees: $564. *Financial support:* In 2005–06, 36 research assistantships (averaging $19,828 per year) were awarded; teaching assistantships, Federal Work-Study, institutionally sponsored loans, and scholarships/grants also available. Support available to part-time students. Financial award application deadline: 3/1; financial award applicants required to submit FAFSA. *Faculty research:* Plant molecular biology, microbial genetics, microbiology plant genetics. Total annual research expenditures: $4.5 million. *Unit head:* Dr. Scot Hulbert, 785-532-6176, Fax: 785-532-5692, E-mail: shulbrt@plantpath.ksu.edu. *Application contact:* Dr. Bill Bockus, Director, 785-532-1378, Fax: 785-532-5692, E-mail: bockus@plant-path.ksu.edu.

Louisiana State University and Agricultural and Mechanical College, Graduate School, College of Agriculture, Department of Plant Pathology and Crop Physiology, Baton Rouge, LA 70803. Offers plant health (MS, PhD). *Faculty:* 15 full-time (0 women). *Students:* 9 full-time (5 women), 4 part-time (1 woman); includes 1 American Indian/Alaska Native, 6 international. Average age 31. 3 applicants, 67% accepted, 1 enrolled. Terminal master's awarded for partial completion of doctoral program. *Degree requirements:* For master's and doctorate, thesis/dissertation. *Entrance requirements:* For master's and doctorate, GRE General Test, minimum GPA of 3.0. Additional exam requirements/recommendations for international students: Required—TOEFL (minimum score 550 paper-based; 213 computer-based). *Application deadline:* For fall admission, 1/25 priority date for domestic students, 5/15 priority date for international students. Applications are processed on a rolling basis. Electronic applications accepted. *Financial support:* In 2005–06, 11 students received support, including 9 research assistantships with partial tuition reimbursements available (averaging $15,778 per year); fellowships, teaching assistantships with partial tuition reimbursements available, career-related internships or fieldwork, Federal Work-Study, and tuition waivers (full) also available. Support available to part-time students. Financial award applicants required to submit FAFSA. *Faculty research:* Plant health and protection, weed biology and management, crop physiology and biotechnology. *Unit head:* Dr. Jerry Berggren, Head, 225-765-2876, Fax: 225-578-1415, E-mail: gberggren@agctr.lsu.edu. *Application contact:* Dr. Rodrigo Valverde, Graduate Adviser, 225-578-8303, Fax: 225-578-2526, E-mail: rvalverde@agctr.lsu.edu.

Michigan State University, The Graduate School, College of Agriculture and Natural Resources and College of Natural Science, Department of Plant Pathology, East Lansing, MI 48824. Offers MS, PhD. *Faculty:* 9 full-time (2 women). *Students:* 15 full-time (9 women), 1 (woman) part-time; includes 4 minority (2 African Americans, 2 Asian Americans or Pacific Islanders), 1 international. Average age 26. 3 applicants, 33% accepted. In 2005, 3 master's, 3 doctorates awarded. *Degree requirements:* For master's, teaching experience, final exam, thesis optional; for doctorate, thesis/dissertation, teaching experience, oral exam in defense of dissertation, comprehensive exam. *Entrance requirements:* For master's and doctorate, GRE General Test, minimum GPA of 3.0, prior scientific training, 3 letters of recommendation. Additional exam requirements/recommendations for international students: Required—TOEFL (minimum score 600 paper-based; 250 computer-based). *Application deadline:* For fall admission, 1/15 for domestic students. Application fee: $0. *Expenses:* Tuition, state resident: part-time $330 per credit hour. Tuition, nonresident: part-time $685 per credit hour. Tuition and fees vary according to program. *Financial support:* In 2005–06, 1 fellowship with tuition reimbursement (averaging $15,000 per year), 13 research assistantships with tuition reimbursements (averaging $13,310 per year), 3 teaching assistantships with tuition reimbursements (averaging $13,470 per year) were awarded; institutionally sponsored loans, scholarships/grants, and unspecified assistantships also available. *Faculty research:* Epidemiology, mycology, host-parasite interactions, soil microbiology, molecular biology. Total annual research expenditures: $3.9 million. *Unit head:* Dr. Ray Hammerschmidt, Chairperson, 517-355-8624, Fax: 517-353-1781, E-mail: hammers1@msu.edu. *Application contact:* Linda Colon, Graduate Admissions Secretary, 517-432-4592, Fax: 517-353-1781, E-mail: colon@msu.edu.

Mississippi State University, College of Agriculture and Life Sciences, Department of Entomology and Plant Pathology, Mississippi State, MS 39762. Offers agricultural pest management (MS); entomology (MS, PhD); plant pathology (MS, PhD). *Faculty:* 19 full-time (1 woman). *Students:* 14 full-time (5 women), 7 part-time (4 women); includes 1 minority (Hispanic American), 6 international. Average age 33. 12 applicants, 25% accepted, 2 enrolled. In 2005, 1 master's awarded. *Degree requirements:* For master's and doctorate, thesis/dissertation, comprehensive oral or written exam. *Entrance requirements:* For master's, GRE General Test, minimum GPA of 2.75; for doctorate, GRE General Test. Additional exam requirements/recommendations for international students: Required—TOEFL. *Application deadline:* For fall admission, 7/1 for domestic students; for spring admission, 11/1 for domestic students. Applications are processed on a rolling basis. Application fee: $30. Electronic applications accepted. *Expenses:* Tuition, state resident: full-time $4,312; part-time $240 per hour. Tuition, nonresident: full-time $9,772; part-time $543 per hour. International tuition: $10,102 full-time. Tuition and fees vary according to course load. *Financial support:* Research assistantships, teaching assistantships, Federal Work-Study, institutionally sponsored loans, and unspecified assistantships available. Financial award applicants required to submit FAFSA. *Faculty research:* Ecology and population dynamics, physiology, biochemistry and behavior, systematics. Total annual research expenditures: $584,995. *Unit head:* Dr. Clarence H. Collison, Head, 662-325-2085, Fax: 662-325-8837, E-mail: chc2@ra.msstate.edu. *Application contact:* Philip G. Bonfanti, Director of Admissions, 662-325-4104, Fax: 662-325-8872, E-mail: admit@msstate.edu.

Montana State University, College of Graduate Studies, College of Agriculture, Department of Plant Sciences and Plant Pathology, Bozeman, MT 59717. Offers plant pathology (MS); plant science (MS, PhD). Part-time programs available. *Faculty:* 24 full-time (3 women), 4 part-time/adjunct (3 women). *Students:* 5 full-time (2 women), 18 part-time (7 women), 8 international. Average age 29. 5 applicants, 80% accepted, 4 enrolled. In 2005, 6 master's, 1 doctorate awarded. *Degree requirements:* For master's, thesis (for some programs), comprehensive exam, registration; for doctorate, thesis/dissertation, comprehensive exam, registration. *Entrance requirements:* For master's and doctorate, GRE General Test. Additional exam requirements/recommendations for international students: Required—TOEFL (minimum score 550 paper-based; 213 computer-based). *Application deadline:* For fall admission, 7/15 priority date for domestic students, 5/15 priority date for international students; for spring admission, 12/1 priority date for domestic students, 10/1 priority date for international students. Applications are processed on a rolling basis. Application fee: $30. Electronic applications accepted. *Expenses:* Tuition, state resident: full-time $4,132. Tuition, nonresident: full-time $1,132. *Financial support:* In 2005–06, 25 students received support, including 25 research assistantships with full tuition reimbursements available (averaging $15,000 per year), 10 teaching assistantships with full tuition reimbursements available (averaging $2,000 per year); Federal Work-Study, health care benefits, and unspecified assistantships also available. Financial award application deadline: 3/1; financial award applicants required to submit FAFSA. *Faculty research:* Plant genetics, plant-microbe interactions, plant physiology, plant taxonomy. Total annual research expenditures: $2.2 million. *Unit head:* Dr. John Sherwood, Department Head, 406-994-5153, Fax: 406-994-7600, E-mail: sherwood@montana.edu.

New Mexico State University, Graduate School, College of Agriculture and Home Economics, Department of Entomology, Plant Pathology and Weed Science, Las Cruces, NM 88003-8001. Offers agricultural biology (MS). Part-time programs available. *Faculty:* 12 full-time (5 women), 1 part-time/adjunct (0 women). *Students:* 14 full-time (12 women), 8 part-time (4 women); includes 7 minority (1 African American, 1 American Indian/Alaska Native, 1 Asian American or Pacific Islander, 4 Hispanic Americans), 1 international. Average age 30. 9 applicants, 67% accepted, 6 enrolled. In 2005, 4 degrees awarded. *Degree requirements:* For master's, thesis, comprehensive exam, registration. *Entrance requirements:* For master's, GRE General Test. *Application deadline:* For fall admission, 7/1 for domestic students; for spring admission, 11/1 priority date for domestic students. Applications are processed on a rolling basis. Application fee: $30 ($50 for international students). Electronic applications accepted. *Expenses:* Tuition, state resident: full-time $3,156; part-time $175 per credit. Tuition, nonresident: full-time $12,510; part-time $565 per credit. Required fees: $1,050. *Financial support:* In 2005–06, 1 fellowship, 11 research assistantships with partial tuition reimbursements, 4 teaching assistantships with partial tuition reimbursements were awarded; career-related internships or fieldwork also available. Financial award application deadline: 3/1. *Faculty research:* Integrated pest management, pesticide application and safety, livestock ectoparasite research, biotechnology, nematology. *Unit head:* Dr. Grant Kinzer, Head, 505-646-3225, Fax: 505-646-8087, E-mail: gkinzer@nmsu.edu.

North Carolina State University, Graduate School, College of Agriculture and Life Sciences, Department of Plant Pathology, Raleigh, NC 27695. Offers MS, PhD. Terminal master's awarded for partial completion of doctoral program. *Degree requirements:* For master's, thesis (for some programs); for doctorate, thesis/dissertation. *Entrance requirements:* For master's and doctorate, GRE. Additional exam requirements/recommendations for international students: Required—TOEFL. Electronic applications accepted. *Faculty research:* Microbe-plant interactions, biology of plant pathogens, pathogen evaluation, host-plant resistance, genomics.

North Dakota State University, The Graduate School, College of Agriculture, Food Systems, and Natural Resources, Department of Plant Pathology, Fargo, ND 58105. Offers MS, PhD. Part-time programs available. *Faculty:* 14 full-time (1 woman), 4 part-time/adjunct (0 women). *Students:* 20 full-time (5 women), 4 part-time (1 woman), 21 international. Average age 23. 12 applicants, 58% accepted, 7 enrolled. In 2005, 2 degrees awarded. *Degree requirements:* For master's and doctorate, thesis/dissertation. *Entrance requirements:* Additional exam requirements/recommendations for international students: Required—TOEFL. *Application deadline:* Applications are processed on a rolling basis. Application fee: $45 ($60 for international students). *Financial support:* In 2005–06, 19 research assistantships with full tuition reimbursements were awarded; Federal Work-Study and institutionally sponsored loans also available. Financial award application deadline: 4/15. *Faculty research:* Electron microscopy, disease physiology, molecular biology, genetic resistance, tissue culture. *Unit head:* Dr. Jack Rasmussen, Chair, 701-231-7058, Fax: 701-231-7851, E-mail: jack.rasmussen@ndsu.edu.

Nova Scotia Agricultural College, Research and Graduate Studies, Truro, NS B2N 5E3, Canada. Offers agriculture (M Sc), including air quality, animal behavior, animal molecular genetics, animal nutrition, animal technology, aquaculture, botany, crop management, crop physiology, ecology, environmental microbiology, food science, horticulture, nutrient manage-

Plant Pathology

Nova Scotia Agricultural College (continued)
ment, pest management, physiology, plant biotechnology, plant pathology, soil chemistry, soil fertility, waste management and composting, water quality. Part-time programs available. *Faculty:* 43 full-time (7 women), 21 part-time/adjunct (1 woman). *Students:* 50 full-time (25 women), 15 part-time (11 women); includes 7 minority (3 African Americans, 3 Asian Americans or Pacific Islanders, 1 Hispanic American). Average age 25. In 2005, 23 degrees awarded. *Degree requirements:* For master's, thesis, registration. *Entrance requirements:* For master's, honors B Sc, minimum GPA of 3.0. Additional exam requirements/recommendations for international students: Required—TOEFL (minimum score 580 paper-based; 237 computer-based), Michigan English Language Assessment Battery, IELT, Can Test, CAEL. *Application deadline:* For fall admission, 6/1 for domestic students, 4/1 for international students. For winter admission, 10/31 for domestic students; for spring admission, 2/28 for domestic students. Applications are processed on a rolling basis. Application fee: $70. *Expenses:* Tuition, state resident: part-time $2,328 per year. Tuition, nonresident: full-time $6,984; part-time $7,968 per year. International tuition: $12,624 full-time. Required fees: $481; $46 per course. Tuition and fees vary according to program and student level. *Financial support:* In 2005–06, 48 students received support, including 7 fellowships (averaging $15,000 per year), 10 research assistantships (averaging $15,000 per year), 15 teaching assistantships (averaging $900 per year); career-related internships or fieldwork, scholarships/grants, and unspecified assistantships also available. *Faculty research:* Bio-product development, organic agriculture, nutrient management, air and water quality, agricultural biotechnology. Total annual research expenditures: $4.7 million. *Unit head:* Jill L. Rogers, Manager, 902-893-6360, Fax: 902-893-3430, E-mail: jrogers@nsac.ca. *Application contact:* Marie Law, Administrative Assistant, 902-893-6502, Fax: 902-893-3430, E-mail: mlaw@nsac.ca.

The Ohio State University, Graduate School, College of Food, Agricultural, and Environmental Sciences, Department of Plant Pathology, Columbus, OH 43210. Offers MS, PhD. *Degree requirements:* For master's, thesis optional; for doctorate, thesis/dissertation. *Entrance requirements:* For master's and doctorate, GRE General Test. Additional exam requirements/recommendations for international students: Required—TOEFL (paper 550; computer 213) or IELT (7) or Michigan English Language Assessment Battery (88). Electronic applications accepted.

Oklahoma State University, College of Agricultural Science and Natural Resources, Department of Entomology and Plant Pathology, Program in Plant Pathology, Stillwater, OK 74078. Offers PhD. *Degree requirements:* For doctorate, thesis/dissertation. *Entrance requirements:* For doctorate, GRE. Additional exam requirements/recommendations for international students: Required—TOEFL. *Application deadline:* For fall admission, 6/1 priority date for domestic students, 3/1 priority date for international students. Applications are processed on a rolling basis. Application fee: $40 ($75 for international students). Electronic applications accepted. *Expenses:* Tuition, state resident: full-time $4,253; part-time $139 per credit hour. Tuition, nonresident: full-time $12,569; part-time $485 per credit hour. Required fees: $43 per credit hour. One-time fee: $20 part-time. Tuition and fees vary according to course load and program. *Financial support:* Research assistantships, career-related internships or fieldwork, Federal Work-Study, scholarships/grants, health care benefits, tuition waivers (partial), and unspecified assistantships available. Support available to part-time students. Financial award application deadline: 3/1. *Unit head:* Dr. Jack Dillwith, Coordinator, 405-744-9412.

Oregon State University, Graduate School, College of Science, Department of Botany and Plant Pathology, Corvallis, OR 97331. Offers ecology (MA, MAIS, MS, PhD); genetics (MA, MAIS, MS, PhD); molecular and cellular biology (MA, MAIS, MS, PhD); mycology (MA, MAIS, MS, PhD); plant pathology (MA, MAIS, MS, PhD); plant physiology (MA, MAIS, MS, PhD); structural botany (MA, MAIS, MS, PhD); systematics (MA, MAIS, MS, PhD). Part-time programs available. *Faculty:* 10 full-time (2 women), 4 part-time/adjunct (1 woman). *Students:* 38 full-time (22 women), 2 part-time (1 woman); includes 4 minority (1 American Indian/Alaska Native, 2 Asian Americans or Pacific Islanders, 1 Hispanic American), 4 international. Average age 30. In 2005, 5 master's, 3 doctorates awarded. *Degree requirements:* For master's, variable foreign language requirement, thesis optional; for doctorate, thesis/dissertation. *Entrance requirements:* For master's and doctorate, GRE General Test, minimum GPA of 3.0 in last 90 hours. Additional exam requirements/recommendations for international students: Required—TOEFL. *Application deadline:* For fall admission, 2/1 for domestic students. Applications are processed on a rolling basis. Application fee: $50. *Expenses:* Tuition, area resident: Part-time $301 per credit. Tuition, state resident: full-time $8,139; part-time $501 per credit. Tuition, nonresident: full-time $14,376; part-time $532 per credit. Required fees: $1,266. *Financial support:* Fellowships, research assistantships, teaching assistantships, career-related internships or fieldwork, Federal Work-Study, institutionally sponsored loans, and scholarships/grants available. Support available to part-time students. Financial award application deadline: 2/1. *Faculty research:* Plant ecology, plant molecular biology, systematic botany, epidemiology, host-pathogen interaction. *Unit head:* Dr. Daniel Arp, Chair, 541-737-1297, Fax: 541-737-3573. *Application contact:* Dr. Donald B. Zobel, Professor, 541-737-3451, Fax: 541-737-3573, E-mail: zobeld@bcc.orst.edu.

The Pennsylvania State University University Park Campus, Graduate School, College of Agricultural Sciences, Department of Plant Pathology, State College, University Park, PA 16802-1503. Offers M Agr, MS, PhD. *Students:* 23 full-time (11 women), 1 (woman) part-time; includes 2 minority (both Hispanic Americans), 10 international. *Entrance requirements:* For master's and doctorate, GRE General Test. *Expenses:* Tuition, state resident: full-time $12,518; part-time $522 per credit. Tuition, nonresident: full-time $23,004; part-time $959 per credit. Required fees: $484. Tuition and fees vary according to course load, campus/location and program. *Unit head:* Dr. Barbara J. Christ, Unit Leader, 814-865-7448, E-mail: ebf@psu.edu. *Application contact:* Dr. Gretchen Kuldau, Professor, 814-863-7232, E-mail: kuldau@psu.edu.

Purdue University, Graduate School, College of Agriculture, Department of Botany and Plant Pathology, West Lafayette, IN 47907. Offers MS, PhD. Part-time programs available. *Faculty:* 24 full-time (5 women), 4 part-time/adjunct (1 woman). *Students:* 40 full-time (11 women), 3 part-time, 16 international. Average age 27. 21 applicants, 38% accepted, 6 enrolled. In 2005, 2 master's, 1 doctorate awarded. Terminal master's awarded for partial completion of doctoral program. *Degree requirements:* For master's and doctorate, thesis/dissertation. *Entrance requirements:* For master's and doctorate, GRE. Additional exam requirements/recommendations for international students: Required—TOEFL. *Application deadline:* For fall admission, 3/15 priority date for domestic students, 3/15 priority date for international students; for spring admission, 12/15 for domestic students, 11/15 for international students. Applications are processed on a rolling basis. Application fee: $55. Electronic applications accepted. *Financial support:* In 2005–06, 30 students received support, including 1 fellowship with full tuition reimbursement available (averaging $17,500 per year), 15 research assistantships with full tuition reimbursements available (averaging $17,500 per year), 10 teaching assistantships with full tuition reimbursements available (averaging $17,500 per year); career-related internships or fieldwork also available. Support available to part-time students. Financial award application deadline: 3/30; financial award applicants required to submit FAFSA. *Faculty research:* Biotechnology, plant growth, weed control, crop improvement, plant physiology. *Unit head:* Dr. Peter B Goldsbrough, Head, 765-494-4615, Fax: 765-494-0363. *Application contact:* Julie A Pluimer, Graduate Admissions Office, 765-494-4614, Fax: 765-494-0363, E-mail: jpluimer@purdue.edu.

Rutgers, The State University of New Jersey, New Brunswick/Piscataway, Graduate School, Program in Plant Biology, New Brunswick, NJ 08901-1281. Offers horticulture (MS, PhD); molecular biology and biochemistry (MS, PhD); pathology (MS, PhD); plant ecology (MS, PhD); plant genetics (PhD); plant physiology (MS, PhD); production and management (MS); structure and plant groups (MS, PhD). Part-time programs available. *Faculty:* 65 full-time, 1 part-time/adjunct (0 women). *Students:* 33 full-time (14 women), 13 part-time (6 women); includes 3 minority (1 African American, 2 Asian Americans or Pacific Islanders), 10 international. Average age 31. 56 applicants, 23% accepted, 10 enrolled. In 2005, 1 master's, 4 doctorates awarded. Terminal master's awarded for partial completion of doctoral program. *Median time to degree:* Of those who began their doctoral program in fall 1997, 90% received their degree in 8 years or less. *Degree requirements:* For master's, thesis or alternative, comprehensive exam; for doctorate, thesis/dissertation, comprehensive exam. *Entrance requirements:* For master's and doctorate, GRE General Test, GRE Subject Test (recommended). Additional exam requirements/recommendations for international students: Required—TOEFL (minimum score 600 paper-based; 250 computer-based). *Application deadline:* For fall admission, 4/1 for domestic students, 4/1 for international students. Application fee: $50. Electronic applications accepted. *Expenses:* Tuition, state resident: full-time $10,440; part-time $435 per credit. Tuition, nonresident: full-time $15,520; part-time $647 per credit. Required fees: $129 per credit. Tuition and fees vary according to program. *Financial support:* In 2005–06, 42 students received support, including 9 fellowships with full tuition reimbursements available (averaging $24,000 per year), 22 research assistantships with full tuition reimbursements available (averaging $16,500 per year), 10 teaching assistantships with full tuition reimbursements available (averaging $16,988 per year) Financial award application deadline: 1/15; financial award applicants required to submit FAFSA. *Faculty research:* Molecular biology and biochemistry of plants, plant development and genomics, plant protection, plant improvement, plant management of horticultural and field crops. Total annual research expenditures: $10 million. *Unit head:* Dr. Thomas Leustek, Director, 732-932-8165 Ext. 326, Fax: 732-932-9377, E-mail: leustek@aesop.rutgers.edu. *Application contact:* Barbara Mulder, Program Associate, 732-932-9375 Ext. 358, Fax: 732-932-9377, E-mail: plantbio@aesop.rutgers.edu.

South Dakota State University, Graduate School, College of Agriculture and Biological Sciences, Department of Plant Science, Program in Plant Pathology, Brookings, SD 57007. Offers MS. *Degree requirements:* For master's, thesis, oral exam. *Entrance requirements:* For master's, GRE General Test. Additional exam requirements/recommendations for international students: Required—TOEFL. *Faculty research:* Small grain and row crop pathology, virology, nematology, molecular biology, host/pathogen systems, epidemiology and disease management.

State University of New York College of Environmental Science and Forestry, Faculty of Environmental and Forest Biology, Syracuse, NY 13210-2779. Offers chemical ecology (MPS, MS, PhD); conservation biology (MPS, MS, PhD); ecology (MPS, MS, PhD); entomology (MPS, MS, PhD); environmental interpretation (MPS, MS, PhD); environmental physiology (MPS, MS, PhD); fish and wildlife biology (MPS, MS, PhD); forest pathology and mycology (MPS, MS, PhD); plant science and biotechnology (MPS, MS, PhD). *Faculty:* 27 full-time (4 women), 4 part-time/adjunct (0 women). *Students:* 81 full-time (50 women), 62 part-time (33 women); includes 4 minority (1 Asian American or Pacific Islander, 3 Hispanic Americans), 16 international. Average age 30. 84 applicants, 54% accepted, 17 enrolled. In 2005, 17 master's, 3 doctorates awarded. *Degree requirements:* For master's, thesis (for some programs), registration; for doctorate, thesis/dissertation, comprehensive exam, registration. *Entrance requirements:* For master's and doctorate, GRE General Test, GRE Subject Test, minimum GPA of 3.0. Additional exam requirements/recommendations for international students: Required—TOEFL (minimum score 550 paper-based; 213 computer-based). *Application deadline:* For fall admission, 2/1 priority date for domestic students, 2/1 priority date for international students; for spring admission, 11/1 priority date for domestic students, 11/1 priority date for international students. Applications are processed on a rolling basis. Application fee: $60. *Expenses:* Tuition, area resident: Full-time $6,900; part-time $288 per credit. Tuition, nonresident: full-time $10,920; part-time $455 per credit. Required fees: $395; $32 per credit. $20 per term. One-time fee: $145. *Financial support:* In 2005–06, 86 students received support, including 13 fellowships with full and partial tuition reimbursements available (averaging $9,446 per year), 40 research assistantships with full and partial tuition reimbursements available (averaging $11,000 per year), 32 teaching assistantships with full and partial tuition reimbursements available (averaging $9,446 per year); Federal Work-Study, institutionally sponsored loans, scholarships/grants, health care benefits, and unspecified assistantships also available. Financial award application deadline: 6/30. *Faculty research:* Ecology, fish and wildlife biology and management, plant science, entomology. Total annual research expenditures: $4.1 million. *Unit head:* Dr. Donald J. Leopold, Chair, 315-470-6770, Fax: 315-470-6934, E-mail: dendro@esf.edu. *Application contact:* Dr. Dudley J. Raynal, Dean, Instruction and Graduate Studies, 315-470-6599, Fax: 315-470-6978, E-mail: esfgrad@esf.edu.

Texas A&M University, College of Agriculture and Life Sciences, Department of Plant Pathology and Microbiology, College Station, TX 77843. Offers plant pathology (MS, PhD); plant protection (M Agr). Part-time programs available. Postbaccalaureate distance learning degree programs offered. *Faculty:* 7 full-time (1 woman), 2 part-time/adjunct (0 women). *Students:* 21 full-time (10 women), 4 part-time; includes 2 minority (both Hispanic Americans), 10 international. Average age 31. 17 applicants, 59% accepted, 8 enrolled. In 2005, 3 master's, 4 doctorates awarded. *Degree requirements:* For master's, thesis/dissertation, comprehensive exam (for some programs), registration; for doctorate, thesis/dissertation, comprehensive exam, registration. *Entrance requirements:* For master's and doctorate, GRE General Test, letters of recommendation, BS/BA in biological sciences. *Application deadline:* Applications are processed on a rolling basis. Application fee: $50 ($75 for international students). *Expenses:* Tuition, state resident: full-time $4,488; part-time $187 per credit hour. Tuition, nonresident: full-time $11,112; part-time $463 per credit hour. Required fees: $1,974. *Financial support:* In 2005–06, research assistantships with partial tuition reimbursements (averaging $16,800 per year), teaching assistantships with partial tuition reimbursements (averaging $16,800 per year) were awarded; fellowships, career-related internships or fieldwork, Federal Work-Study, institutionally sponsored loans, and unspecified assistantships also available. Support available to part-time students. Financial award application deadline: 4/1; financial award applicants required to submit FAFSA. *Faculty research:* Plant disease control, population biology of plant pathogens, disease epidemiology, molecular genetics of host/parasite interactions. *Unit head:* Dr. Dennis C. Gross, Professor and Head, 979-845-7311, Fax: 979-845-6483, E-mail: plpm-head@ppserver.tamu.edu. *Application contact:* Dr. Clint Magill, Chair, Graduate Program Committee, 979-845-7311.

The University of Arizona, Graduate College, College of Agriculture and Life Sciences, Division of Plant Pathology, Tucson, AZ 85721. Offers MS, PhD. Part-time programs available. *Degree requirements:* For master's, thesis optional; for doctorate, thesis/dissertation. *Entrance requirements:* For master's and doctorate, GRE (recommended), minimum GPA of 3.0. Additional exam requirements/recommendations for international students: Required—TOEFL. *Faculty research:* Fungal molecular biology, ecology of soil-borne plant pathogens, plant virology, plant bacteriology, plant/pathogen interactions.

University of Arkansas, Graduate School, Dale Bumpers College of Agricultural, Food and Life Sciences, Department of Plant Pathology, Fayetteville, AR 72701-1201. Offers MS. *Students:* 9 full-time (4 women), 2 part-time (both women), 5 international. 5 applicants, 40% accepted. In 2005, 1 degree awarded. *Degree requirements:* For master's, thesis. Application fee: $40 ($50 for international students). *Financial support:* In 2005–06, 9 research assistantships were awarded; fellowships, teaching assistantships, career-related internships or fieldwork and Federal Work-Study also available. Support available to part-time students. Financial award application deadline: 4/1; financial award applicants required to submit FAFSA. *Unit head:* Dr. Sung M. Lim, Chair, 479-575-2446. *Application contact:* David TeBeest, Graduate Coordinator, 479-575-2678.

University of California, Davis, Graduate Studies, Program in Plant Pathology, Davis, CA 95616. Offers MS, PhD. *Faculty:* 41 full-time. *Students:* 33 full-time (21 women); includes 1 minority (Asian American or Pacific Islander), 16 international. Average age 31. 37 applicants, 35% accepted, 5 enrolled. In 2005, 4 master's, 8 doctorates awarded. Terminal master's awarded for partial completion of doctoral program. *Median time to degree:* Of those who began their doctoral program in fall 1997, 71% received their degree in 8 years or less. *Degree requirements:* For master's, thesis (for some programs), comprehensive exam (for some programs); for doctorate, thesis/dissertation. *Entrance requirements:* For master's and doctorate, GRE General Test. Additional exam requirements/recommendations for international students: Required—TOEFL (minimum score 550 paper-based; 213 computer-based). *Application deadline:* For fall admission, 4/1 for domestic students, 4/1 for international students. Application fee: $60. Electronic applications accepted. *Financial support:* In 2005–06, 33 students received support, including fellowships with full and partial tuition reimbursements available (averaging $9,290 per year), research assistantships with full and partial tuition reimbursements available (averaging $9,290 per year); Federal Work-Study, institutionally sponsored loans, scholarships/grants, tuition waivers (full and partial), unspecified assistantships, and graduate student researcher (GSR) or teaching assistantships for consideration of University-funded fellowships is January 15 also available. Financial award application deadline: 1/15; financial award applicants required to submit FAFSA. *Faculty research:* Soil microbiology; diagnosis etiology and control of plant diseases; genomics and molecular biology of plant microbe interactions; biotechnology, ecology of plant pathogens and epidemiology of diseases in agricultural and native ecosystems. *Unit head:* Thomas R. Gordon, Graduate Chair, 530-754-9893, Fax: 530-752-5674, E-mail: trgordon@ucdavis.edu. *Application contact:* Elizabeth M. Jeffery, Administrative Assistant, 530-752-0300, Fax: 530-752-5674, E-mail: emjeffery@ucdavis.edu.

University of California, Riverside, Graduate Division, Department of Plant Pathology, Riverside, CA 92521-0102. Offers MS, PhD. *Faculty:* 14 full-time (2 women). *Students:* 16 full-time (5 women); includes 2 minority (1 Asian American or Pacific Islander, 1 Hispanic American), 18 international. Average age 31. In 2005, 1 master's, 3 doctorates awarded. Terminal master's awarded for partial completion of doctoral program. *Degree requirements:* For master's, comprehensive exams or thesis; for doctorate, thesis/dissertation, qualifying exams. *Entrance requirements:* For master's and doctorate, GRE General Test, minimum GPA of 3.2. Additional exam requirements/recommendations for international students: Required—TOEFL (minimum score 550 paper-based; 213 computer-based); Recommended—TSE (minimum score 50). *Application deadline:* For fall admission, 5/1 for domestic students, 2/1 for international students. For winter admission, 9/1 for domestic students; for spring admission, 12/1 for domestic students. Applications are processed on a rolling basis. Application fee: $60 ($75 for international students). Electronic applications accepted. *Expenses:* Tuition, nonresident: full-time $14,694. Required fees: $9,009. Full-time tuition and fees vary according to program. *Financial support:* In 2005–06, fellowships with tuition reimbursements (averaging $12,000 per year), research assistantships with tuition reimbursements (averaging $12,000 per year), teaching assistantships with tuition reimbursements (averaging $15,000 per year) were awarded; career-related internships or fieldwork, institutionally sponsored loans, scholarships/grants, health care benefits, tuition waivers (full and partial), and unspecified assistantships also available. Financial award application deadline: 3/1; financial award applicants required to submit FAFSA. *Faculty research:* Host-pathogen interactions, biological control and integrated approaches to disease management, fungicide behavior, molecular genetics. *Unit head:* Dr. Michael Allen, Chair, 951-827-5494. *Application contact:* Zina Romero, Graduate Program Assistant, 800-735-0717, Fax: 951-827-5913, E-mail: plantpa@urc.edu.

University of Florida, Graduate School, College of Agricultural and Life Sciences, Department of Plant Pathology, Gainesville, FL 32611. Offers MS, PhD. *Faculty:* 13 full-time (3 women), 1 part-time/adjunct (0 women). *Students:* 67 (32 women); includes 10 minority (5 African Americans, 1 Asian American or Pacific Islander, 4 Hispanic Americans) 17 international. 26 applicants, 58% accepted. In 2005, 3 master's, 10 doctorates awarded. *Degree requirements:* For master's, thesis optional; for doctorate, thesis/dissertation. *Entrance requirements:* For master's and doctorate, GRE General Test, minimum GPA of 3.0. *Application deadline:* For fall admission, 6/1 for domestic students. Applications are processed on a rolling basis. Application fee: $20. Electronic applications accepted. *Expenses:* Tuition, state resident: full-time $6,234. Tuition, nonresident: full-time $21,359. Tuition and fees vary according to program. *Financial support:* In 2005–06, 20 students received support, including 16 research assistantships (averaging $18,286 per year); fellowships, teaching assistantships, career-related internships or fieldwork also available. *Faculty research:* Causes of development of disease in plants, molecular and biochemical aspects of disease, biological control of pathogens and weeds, genetic engineering of resistant plants. *Application contact:* Dr. Jeffrey B. Jones, Coordinator, 352-392-3631 Ext. 346, Fax: 352-392-6532, E-mail: jbj@ifas.ufl.edu.

University of Georgia, Graduate School, College of Agricultural and Environmental Sciences, Department of Plant Pathology, Athens, GA 30602. Offers plant pathology (MS, PhD); plant protection and pest management (MPPPM). *Faculty:* 20 full-time (2 women). *Students:* 13 full-time, 2 part-time; includes 1 minority (African American), 7 international. 11 applicants, 45% accepted, 7 enrolled. In 2005, 4 master's, 3 doctorates awarded. *Degree requirements:* For master's, thesis (MS); for doctorate, one foreign language, thesis/dissertation. *Entrance requirements:* For master's and doctorate, GRE General Test. *Application deadline:* For fall admission, 7/1 for domestic students; for spring admission, 11/15 for domestic students. Application fee: $50. Electronic applications accepted. *Financial support:* Fellowships, research assistantships, teaching assistantships, unspecified assistantships available. *Unit head:* Dr. John L. Sherwood, Head, 706-542-2571, Fax: 706-542-1262, E-mail: sherwood@uga.edu. *Application contact:* Dr. Charles W. Mims, Graduate Coordinator, 706-542-1291, Fax: 706-542-1262, E-mail: cwmims@uga.edu.

University of Guelph, Graduate Program Services, Ontario Agricultural College, Department of Environmental Biology, Guelph, ON N1G 2W1, Canada. Offers entomology (M Sc, PhD); environmental biology and biotechnology (M Sc, PhD); environmental toxicology (M Sc, PhD); plant and forest systems (M Sc, PhD); plant pathology (M Sc, PhD). Part-time programs available. *Faculty:* 20 full-time (2 women), 21 part-time/adjunct (2 women). *Students:* 60 full-time (32 women), 13 part-time (8 women). 37 applicants, 35% accepted, 13 enrolled. In 2005, 12 master's, 7 doctorates awarded. *Median time to degree:* Of those who began their doctoral program in fall 1997, 79.5% received their degree in 8 years or less. *Degree requirements:* For master's, thesis/dissertation, registration; for doctorate, thesis/dissertation, comprehensive exam, registration. *Entrance requirements:* For master's, minimum B- average during previous 2 years of course work; for doctorate, minimum B average. Additional exam requirements/recommendations for international students: Required—TOEFL or IELT. *Application deadline:* Applications are processed on a rolling basis. Application fee: $75. Electronic applications accepted. *Financial support:* In 2005–06, research assistantships (averaging $16,500 per year), teaching assistantships (averaging $4,606 per year) were awarded; fellowships *Faculty research:* Entomology, environmental microbiology and biotechnology, environmental toxicology, forest ecology, plant pathology. Total annual research expenditures: $3 million. *Unit head:* Dr. M. A. Dixon, Chair, 519-824-4120 Ext. 52555, Fax: 519-837-0442, E-mail: mdixon@uoguelph.ca. *Application contact:* Dr. H. Lee, Admissions Coordinator, 519-824-4120 Ext. 53828, Fax: 519-837-0442, E-mail: hlee@uoguelph.ca.

University of Hawaii at Manoa, Graduate Division, College of Tropical Agriculture and Human Resources, Department of Plant and Environmental Protection Sciences, Program in Tropical Plant Pathology, Honolulu, HI 96822. Offers MS, PhD. *Faculty:* 14 full-time (1 woman), 3 part-time/adjunct (0 women). *Students:* 11 full-time (6 women), 3 part-time; includes 2 minority (1 African American, 1 Asian American or Pacific Islander), 8 international. 3 applicants, 100% accepted, 1 enrolled. In 2005, 1 master's awarded. *Median time to degree:* Of those who began their doctoral program in fall 1997, 50% received their degree in 8 years or less. *Application deadline:* For fall admission, 3/1 for domestic students, 3/1 for international students; for spring admission, 9/1 for domestic students, 9/1 for international students. Application fee: $50. *Expenses:* Tuition, state resident: full-time $8,400; part-time $200 per credit hour. Tuition, nonresident: full-time $11,088; part-time $462 per credit hour. Tuition and fees vary according to program. *Unit head:* Dr. Brent Sipes, Graduate Chairperson, 808-956-7076, Fax: 808-956-2832, E-mail: sipes@hawaii.edu. *Application contact:* Dr. Brent Sipes, Graduate Chairperson, 808-956-7076, Fax: 808-956-2832, E-mail: sipes@hawaii.edu.

University of Kentucky, Graduate School, Graduate School Programs in the College of Agriculture, Program in Plant Pathology, Lexington, KY 40506-0032. Offers MS, PhD. *Faculty:* 16 full-time (3 women). *Students:* 18 full-time (12 women); includes 1 minority (African American), 11 international. Average age 30. 47 applicants, 53% accepted, 21 enrolled. In 2005, 1

master's, 3 doctorates awarded. *Median time to degree:* Of those who began their doctoral program in fall 1997, 50% received their degree in 8 years or less. *Degree requirements:* For master's and doctorate, thesis/dissertation, comprehensive exam. *Entrance requirements:* For master's, GRE General Test, minimum undergraduate GPA of 2.5; for doctorate, GRE General Test, minimum graduate GPA of 3.0. Additional exam requirements/recommendations for international students: Required—TOEFL (minimum score 550 paper-based; 213 computer-based). *Application deadline:* For fall admission, 7/17 priority date for domestic students, 2/1 priority date for international students; for spring admission, 12/13 priority date for domestic students, 6/15 priority date for international students. Applications are processed on a rolling basis. Application fee: $40 ($55 for international students). Electronic applications accepted. *Expenses:* Tuition, state resident: full-time $3,159; part-time $331 per credit hour. Tuition, nonresident: full-time $6,984; part-time $756 per credit hour. Tuition and fees vary according to course load, degree level and program. *Financial support:* In 2005–06, 7 fellowships with full tuition reimbursements (averaging $4,500 per year), 11 research assistantships with full tuition reimbursements (averaging $16,500 per year) were awarded; teaching assistantships with full tuition reimbursements, career-related internships or fieldwork, Federal Work-Study, institutionally sponsored loans, scholarships/grants, traineeships, health care benefits, tuition waivers (partial), and unspecified assistantships also available. Support available to part-time students. Financial award application deadline: 3/15. *Faculty research:* Molecular biology of viruses and fungi, biochemistry and physiology of disease resistance, plant transformation, disease ecology, forest pathology. Total annual research expenditures: $1.5 million. *Unit head:* Dr. Lisa Vaillancourt, Director of Graduate Studies, 859-257-7445, Fax: 859-323-1961, E-mail: vaillan@uky.edu. *Application contact:* Dr. Brian Jackson, Senior Associate Dean, 859-257-8176, Fax: 859-323-1928, E-mail: lance.brunner@uky.edu.

University of Maine, Graduate School, College of Natural Sciences, Forestry, and Agriculture, Department of Biological Sciences, Program in Botany and Plant Pathology, Orono, ME 04469. Offers MS. Part-time programs available. *Students:* 4 full-time (2 women), 1 (woman) part-time, 1 international. Average age 28. 4 applicants, 25% accepted, 1 enrolled. In 2005, 5 master's awarded. *Degree requirements:* For master's, thesis. *Entrance requirements:* For master's, GRE General Test. Additional exam requirements/recommendations for international students: Required—TOEFL. *Application deadline:* For fall admission, 2/1 for domestic students. Applications are processed on a rolling basis. Application fee: $50. Electronic applications accepted. *Financial support:* Fellowships, research assistantships, teaching assistantships, career-related internships or fieldwork, Federal Work-Study, institutionally sponsored loans, and tuition waivers (full) available. Financial award application deadline: 3/1. *Faculty research:* Molecular biology of viral and fungal pathogens, marine ecology, paleoecology and acid systematics and evolution. *Unit head:* Dr. Stelbs Tavanteis, Coordinator, 207-581-2986. *Application contact:* Scott G. Delcourt, Associate Dean of the Graduate School, 207-581-3219, Fax: 207-581-3232, E-mail: graduate@maine.edu.

University of Minnesota, Twin Cities Campus, Graduate School, College of Agricultural, Food, and Environmental Sciences, Department of Plant Pathology, Minneapolis, MN 55455-0213. Offers MS, PhD. Part-time programs available. *Faculty:* 16 full-time (5 women), 7 part-time/adjunct (2 women). *Students:* 16 full-time (5 women), 3 part-time (1 woman); includes 11 minority (3 Asian Americans or Pacific Islanders, 8 Hispanic Americans), 5 international. Average age 30. 9 applicants, 22% accepted, 2 enrolled. In 2005, 4 master's, 3 doctorates awarded. Terminal master's awarded for partial completion of doctoral program. *Degree requirements:* For master's, thesis (for some programs); for doctorate, thesis/dissertation. *Entrance requirements:* For master's and doctorate, GRE General Test. Additional exam requirements/recommendations for international students: Required—TOEFL. *Application deadline:* Applications are processed on a rolling basis. Application fee: $55 ($75 for international students). Electronic applications accepted. *Expenses:* Tuition, state resident: full-time $8,748; part-time $729 per credit. Tuition, nonresident: full-time $15,848; part-time $1,321 per credit. Full-time tuition and fees vary according to class time, course load, program and reciprocity agreements. *Financial support:* In 2005–06, 4 students received support, including research assistantships with full tuition reimbursements available (averaging $17,400 per year); teaching assistantships Support available to part-time students. *Faculty research:* Plant disease management, biological control, product deterioration, international agriculture, molecular biology. Total annual research expenditures: $1.8 million. *Unit head:* Dr. Carol A. Ishimara, Head, 612-625-9736, Fax: 612-625-9728. *Application contact:* Dr. Brian J. Steffenson, Professor, 612-625-4735, Fax: 612-625-9728, E-mail: bsteffen@umn.edu.

University of Missouri–Columbia, Graduate School, College of Agriculture, Food and Natural Resources, Division of Plant Sciences, Department of Plant Pathology and Microbiology, Columbia, MO 65211. Offers MS, PhD. Terminal master's awarded for partial completion of doctoral program. *Degree requirements:* For master's and doctorate, thesis/dissertation. *Entrance requirements:* For master's and doctorate, GRE General Test, minimum GPA of 3.0. *Application deadline:* For fall admission, 3/1 for domestic students. Applications are processed on a rolling basis. Application fee: $45 ($60 for international students). *Financial support:* Research assistantships, teaching assistantships, institutionally sponsored loans available. *Unit head:* Dr. Jeanne Mihail, Director of Graduate Studies, 573-882-0574, E-mail: mihailj@missouri.edu.

University of Rhode Island, Graduate School, College of the Environment and Life Sciences, Department of Plant Sciences, Program in Entomology, Kingston, RI 02881. Offers MS, PhD. *Degree requirements:* For master's, thesis/dissertation, professional seminar, comprehensive exam; for doctorate, thesis/dissertation, professional seminar. *Entrance requirements:* For master's and doctorate, GRE General Test. *Application deadline:* For fall admission, 4/15 for domestic students. Applications are processed on a rolling basis. Application fee: $35. *Expenses:* Tuition, state resident: full-time $5,522; part-time $307 per credit. Tuition, nonresident: full-time $15,992; part-time $888 per credit. Required fees: $1,786; $73 per credit. One-time fee: $80 part-time. *Faculty research:* Physiology and fine structure of host-parasite relations, etiology of plant disease, biological control of insects, integrated pest management. *Unit head:* Dr. Richard Casagrande, Chairman, Department of Plant Sciences, 401-874-2924.

The University of Tennessee, Graduate School, College of Agricultural Sciences and Natural Resources, Department of Entomology and Plant Pathology, Knoxville, TN 37996. Offers entomology (MS, PhD); integrated pest management and bioactive natural products (PhD); plant pathology (MS, PhD). Part-time programs available. *Degree requirements:* For master's, thesis, seminar. *Entrance requirements:* For master's, GRE General Test, minimum GPA of 2.7, 3 reference letters, letter of intent; for doctorate, GRE General Test, minimum GPA of 2.7, 3 reference letters, letter of intent, proposed dissertation research. Additional exam requirements/recommendations for international students: Required—TOEFL. Electronic applications accepted.

University of Wisconsin–Madison, Graduate School, College of Agricultural and Life Sciences, Department of Plant Pathology, Madison, WI 53706-1380. Offers MS, PhD. Part-time programs available. *Faculty:* 20 full-time (7 women). *Students:* 29 full-time (16 women), 11 international. Average age 24. 37 applicants, 30% accepted, 4 enrolled. In 2005, 9 master's, 10 doctorates awarded. Terminal master's awarded for partial completion of doctoral program. *Degree requirements:* For master's and doctorate, thesis/dissertation. *Entrance requirements:* For master's and doctorate, GRE. Additional exam requirements/recommendations for international students: Required—TOEFL. *Application deadline:* For fall admission, 1/15 priority date for domestic students, 1/15 priority date for international students. Application fee: $45. Electronic applications accepted. *Financial support:* In 2005–06, 2 fellowships with full tuition reimbursements (averaging $17,772 per year), 15 research assistantships with full tuition reimbursements (averaging $17,772 per year) were awarded; career-related internships or fieldwork, institutionally sponsored loans, and tuition waivers (full) also available. *Faculty research:* Plant disease, plant health, plant-microbe interactions, plant disease management, biological control. Total annual research expenditures: $3 million. *Unit head:* Murray Clayton, Chair, 608-262-1009. *Application contact:* Cathy Davis Gray, Student Services Coordinator, 608-262-9926, Fax: 608-263-2626, E-mail: admissions@plantpath.wisc.edu.

Virginia Polytechnic Institute and State University, Graduate School, College of Agriculture and Life Sciences, Department of Plant Pathology, Physiology and Weed Science, Blacks-

Plant Pathology

Virginia Polytechnic Institute and State University *(continued)*
burg, VA 24061. Offers plant pathology (MS, PhD); plant physiology and weed science (MS, PhD); plant protection (MS). *Faculty:* 13 full-time (3 women), 1 (woman) part-time/adjunct. *Students:* 16 full-time (9 women), 2 part-time (1 woman); includes 1 minority (African American), 9 international. Average age 28. 14 applicants, 29% accepted, 3 enrolled. In 2005, 3 master's, 2 doctorates awarded. *Entrance requirements:* For master's and doctorate, GRE. Additional exam requirements/recommendations for international students: Required—TOEFL (minimum score 575 paper-based; 231 computer-based). *Application deadline:* Applications are processed on a rolling basis. Application fee: $45. *Expenses:* Tuition, state resident: full-time $6,558; part-time $364 per credit. Tuition, nonresident: full-time $11,296; part-time $628 per credit. Required fees: $1,419; $468 per credit. $234 per term. *Financial support:* In 2005–06, 13 research assistantships with full tuition reimbursements (averaging $13,823 per year), 1 teaching assistantship with full tuition reimbursement (averaging $13,552 per year) were awarded; career-related internships or fieldwork, Federal Work-Study, scholarships/grants, and unspecified assistantships also available. Financial award application deadline: 4/1. *Faculty research:* Biotechnology, Dutch elm disease, weed control, plant pathogenic microorganisms, agronomic crop resistance to fungal and viral pathogens. *Unit head:* Dr. Erik Stromberg, Head, 540-231-7871, Fax: 540-231-7477. *Application contact:* Donna Ford, 540-231-6361, Fax: 540-231-7477.

Washington State University, Graduate School, College of Agricultural, Human, and Natural Resource Sciences, Department of Plant Pathology, Pullman, WA 99164. Offers MS, PhD. *Faculty:* 17. *Students:* 24 full-time (14 women), 1 part-time; includes 1 minority (Asian American or Pacific Islander), 12 international. Average age 31. 55 applicants, 13% accepted, 7 enrolled. In 2005, 5 master's, 4 doctorates awarded. Terminal master's awarded for partial completion of doctoral program. *Degree requirements:* For master's and doctorate, thesis/dissertation, oral exam. *Entrance requirements:* For master's, minimum GPA of 3.0, 3 letters of recommendation; for doctorate, GRE, minimum GPA of 3.0, 3 letters of recommendation. Additional exam requirements/recommendations for international students: Required—TOEFL (minimum score 550 paper-based; 213 computer-based). *Application deadline:* For fall admission, 3/1 priority date for domestic students, 3/1 priority date for international students; for spring admission, 7/1 for domestic students, 7/1 for international students. Applications are processed on a rolling basis. Application fee: $35. Electronic applications accepted. *Expenses:* Tuition, state resident: full-time $6,295; part-time $336 per credit. Tuition, nonresident: full-time $15,949; part-time $819 per credit. Required fees: $933. Part-time tuition and fees vary according to campus/location and program. *Financial support:* In 2005–06, 25 students received support, including 1 fellowship (averaging $14,057 per year), 24 research assistantships with full and partial tuition reimbursements available (averaging $12,975 per year); teaching assistantships with full and partial tuition reimbursements available, career-related internships or fieldwork, Federal Work-Study, institutionally sponsored loans, scholarships/grants, and teaching associateships also available. Financial award application deadline: 4/1; financial award applicants required to submit FAFSA. *Faculty research:* Biology of fungi, bacteria, and viruses; diseases of plants; genetics of fungi, bacteria, and viruses. Total annual research expenditures: $2.3 million. *Unit head:* Dr. Timothy D. Murray, Chair, 509-335-9541, Fax: 509-335-9581, E-mail: tim_murray@wsu.edu. *Application contact:* Dana Martin, Coordinator, 509-335-9542, Fax: 509-335-5902, E-mail: plpathstudents@wsu.edu.

West Virginia University, Davis College of Agriculture, Forestry and Consumer Sciences, Division of Plant and Soil Sciences, Morgantown, WV 26506. Offers agronomy (MS); entomology (MS); environmental microbiology (MS); horticulture (MS); plant pathology (MS). *Faculty:* 18 full-time (1 woman), 1 part-time/adjunct (0 women). *Students:* 17 full-time (11 women), 3 part-time (2 women), 2 international. Average age 27. In 2005, 5 degrees awarded. *Degree requirements:* For master's, thesis. *Entrance requirements:* For master's, GRE, minimum GPA of 2.5. Additional exam requirements/recommendations for international students: Required—TOEFL. *Application deadline:* Applications are processed on a rolling basis. Application fee: $45. *Expenses:* Tuition, state resident: full-time $4,582; part-time $258 per credit hour. Tuition, nonresident: full-time $13,820; part-time $741 per credit hour. *Financial support:* In 2005–06, 13 research assistantships with full tuition reimbursements (averaging $9,936 per year), 4 teaching assistantships with full tuition reimbursements (averaging $9,936 per year) were awarded; Federal Work-Study, institutionally sponsored loans, and tuition waivers (full and partial) also available. Financial award application deadline: 2/1; financial award applicants required to submit FAFSA. *Faculty research:* Water quality, reclamation of disturbed land, crop production, pest control, environmental protection. Total annual research expenditures: $1 million. *Unit head:* Dr. Barton S. Baker, Chair and Division Director, 304-293-4817 Ext. 4342, Fax: 304-293-2960, E-mail: barton.baker@mail.wvu.edu.

Plant Physiology

Colorado State University, Graduate School, College of Agricultural Sciences, Department of Horticulture and Landscape Architecture, Fort Collins, CO 80523-0015. Offers floriculture (M Agr, MS, PhD); horticultural food crops (M Agr, MS, PhD); nursery and landscape management (M Agr, MS, PhD); plant genetics (MS, PhD); plant physiology (MS, PhD); turf management (M Agr, MS, PhD). Part-time programs available. *Faculty:* 18 full-time (3 women). *Students:* 11 full-time (5 women), 9 part-time (3 women), 4 international. Average age 35. 13 applicants, 69% accepted, 4 enrolled. In 2005, 1 master's, 2 doctorates awarded. *Median time to degree:* Of those who began their doctoral program in fall 1997, 90% received their degree in 8 years or less. *Degree requirements:* For master's, thesis or alternative; for doctorate, thesis/dissertation. *Entrance requirements:* For master's and doctorate, GRE General Test, minimum GPA of 3.0. Additional exam requirements/recommendations for international students: Required—TOEFL (minimum score 550 paper-based; 213 computer-based). *Application deadline:* For fall and winter admission, 2/1 for domestic students, 1/1 for international students; for spring admission, 9/1 priority date for domestic students, 7/1 priority date for international students. Applications are processed on a rolling basis. Application fee: $50. Electronic applications accepted. *Expenses:* Tuition, state resident: full-time $3,690; part-time $205 per credit. Tuition, nonresident: full-time $14,958; part-time $831 per credit. Required fees: $1,061. *Financial support:* Fellowships with full tuition reimbursements, research assistantships with partial tuition reimbursements, teaching assistantships with partial tuition reimbursements, career-related internships or fieldwork, Federal Work-Study, institutionally sponsored loans, and traineeships available. Financial award application deadline: 8/16. *Faculty research:* Antioxidants in food crops, environmental physiology, water conservation, tissue culture, rhizosphere biology, cancer prevention through dietary intervention. Total annual research expenditures: $600,000. *Unit head:* Dr. Stephen J. Wallner, Head, 970-491-7018, Fax: 970-491-7745. *Application contact:* Gretchen L. DeWeese, Administrative Assistant III, 970-491-7018, Fax: 970-491-7745, E-mail: gretchen.deweese@colostate.edu.

Cornell University, Graduate School, Graduate Fields of Agriculture and Life Sciences, Field of Plant Biology, Ithaca, NY 14853-0001. Offers cytology (MS, PhD); paleobotany (MS, PhD); plant cell biology (MS, PhD); plant ecology (MS, PhD); plant molecular biology (MS, PhD); plant morphology, anatomy and biomechanics (MS, PhD); plant physiology (MS, PhD); systematic botany (MS, PhD). *Faculty:* 69 full-time (15 women). *Students:* 50 full-time (27 women); includes 2 minority (both African Americans), 16 international. 33 applicants, 45% accepted, 10 enrolled. In 2005, 1 master's, 7 doctorates awarded. *Degree requirements:* For doctorate, thesis/dissertation, comprehensive exam. *Entrance requirements:* For doctorate, GRE General Test, GRE Subject Test in biology (recommended), 3 letters of recommendation. Additional exam requirements/recommendations for international students: Required—TOEFL (minimum score 610 paper-based; 253 computer-based). *Application deadline:* For fall admission, 1/15 for domestic students. Application fee: $60. Electronic applications accepted. *Financial support:* In 2005–06, 50 students received support, including 10 fellowships with full tuition reimbursements available, 23 research assistantships with full tuition reimbursements available, 17 teaching assistantships with full tuition reimbursements available; institutionally sponsored loans, scholarships/grants, health care benefits, tuition waivers (full and partial), and unspecified assistantships also available. Financial award applicants required to submit FAFSA. *Faculty research:* Plant cell biology/cytology; plant molecular biology; plant morphology/anatomy/biomechanics; plant physiology, systematic botany, paleobotany; plant ecology, ethnobotany, plant biochemistry, photosynthesis. *Unit head:* Director of Graduate Studies, 607-255-2131. *Application contact:* Graduate Field Assistant, 607-255-2131, E-mail: plbio@cornell.edu.

Iowa State University of Science and Technology, Graduate College, Interdisciplinary Programs, Program in Plant Physiology, Ames, IA 50011. Offers MS, PhD. *Students:* 23 full-time (9 women), 18 international. 19 applicants, 37% accepted, 2 enrolled. In 2005, 2 master's, 4 doctorates awarded. *Degree requirements:* For master's and doctorate, thesis/dissertation. *Entrance requirements:* For master's and doctorate, GRE General Test. Additional exam requirements/recommendations for international students: Required—TOEFL (paper score 540; computer score 207) or IELTS (score 6). *Application deadline:* For fall admission, 1/1 priority date for domestic students, 1/1 priority date for international students; for spring admission, 9/1 for domestic students, 9/1 for international students. Applications are processed on a rolling basis. Application fee: $30 ($70 for international students). Electronic applications accepted. *Expenses:* Tuition, state resident: full-time $6,410. Tuition, nonresident: full-time $16,422. Tuition and fees vary according to program. *Financial support:* In 2005–06, 18 research assistantships with partial tuition reimbursements (averaging $16,593 per year), 2 teaching assistantships with partial tuition reimbursements (averaging $15,000 per year) were awarded; scholarships/grants, health care benefits, and unspecified assistantships also available. *Unit head:* Dr. Robert Thornburg, Supervisory Committee Chair, 515-294-7885, E-mail: ippm@iastate.edu.

Nova Scotia Agricultural College, Research and Graduate Studies, Truro, NS B2N 5E3, Canada. Offers agriculture (M Sc), including air quality, animal behavior, animal molecular genetics, animal nutrition, animal technology, aquaculture, botany, crop management, crop physiology, ecology, environmental microbiology, food science, horticulture, nutrient management, pest management, physiology, plant biotechnology, plant pathology, soil chemistry, soil fertility, waste management and composting, water quality. Part-time programs available. *Faculty:* 43 full-time (7 women), 21 part-time/adjunct (1 woman). *Students:* 50 full-time (25 women), 15 part-time (11 women); includes 7 minority (3 African Americans, 3 Asian Americans or Pacific Islanders, 1 Hispanic American). Average age 25. In 2005, 23 degrees awarded. *Degree requirements:* For master's, thesis, registration. *Entrance requirements:* For master's, honors B Sc, minimum GPA of 3.0. Additional exam requirements/recommendations for international students: Required—TOEFL (minimum score 580 paper-based; 237 computer-based), Michigan English Language Assessment Battery, IELT, Can Test, CAEL. *Application deadline:* For fall admission, 6/1 for domestic students, 4/1 for international students. For winter admission, 10/31 for domestic students; for spring admission, 2/28 for domestic students. Applications are processed on a rolling basis. Application fee: $70. *Expenses:* Tuition, state resident: part-time $2,328 per year. Tuition, nonresident: full-time $6,984; part-time $7,968 per year. International tuition: $12,624 full-time. Required fees: $481; $46 per course. Tuition and fees vary according to program and student level. *Financial support:* In 2005–06, 48 students received support, including 7 fellowships (averaging $15,000 per year), 10 research assistantships (averaging $15,000 per year), 15 teaching assistantships (averaging $900 per year); career-related internships or fieldwork, scholarships/grants, and unspecified assistantships also available. *Faculty research:* Bio-product development, organic agriculture, nutrient management, air and water quality, agricultural biotechnology. Total annual research expenditures: $4.7 million. *Unit head:* Jill L. Rogers, Manager, 902-893-6360, Fax: 902-893-3430, E-mail: jrogers@nsac.ca. *Application contact:* Marie Law, Administrative Assistant, 902-893-6502, Fax: 902-893-3430, E-mail: mlaw@nsac.ca.

Oregon State University, Graduate School, College of Science, Department of Botany and Plant Pathology, Corvallis, OR 97331. Offers ecology (MA, MAIS, MS, PhD); genetics (MA, MAIS, MS, PhD); molecular and cellular biology (MA, MAIS, MS, PhD); mycology (MA, MAIS, MS, PhD); plant pathology (MA, MAIS, MS, PhD); plant physiology (MA, MAIS, MS, PhD); structural botany (MA, MAIS, MS, PhD); systematics (MA, MAIS, MS, PhD). Part-time programs available. *Faculty:* 10 full-time (2 women), 4 part-time/adjunct (1 woman). *Students:* 38 full-time (22 women), 2 part-time (1 woman); includes 4 minority (1 American Indian/Alaska Native, 2 Asian Americans or Pacific Islanders, 1 Hispanic American), 4 international. Average age 30. In 2005, 5 master's, 3 doctorates awarded. *Degree requirements:* For master's, variable foreign language requirement, thesis optional; for doctorate, thesis/dissertation. *Entrance requirements:* For master's and doctorate, GRE General Test, minimum GPA of 3.0 in last 90 hours. Additional exam requirements/recommendations for international students: Required—TOEFL. *Application deadline:* For fall admission, 2/1 for domestic students. Applications are processed on a rolling basis. Application fee: $50. *Expenses:* Tuition, area resident: Part-time $301 per credit. Tuition, state resident: full-time $8,139; part-time $501 per credit. Tuition, nonresident: full-time $14,376; part-time $532 per credit. Required fees: $1,266. *Financial support:* Fellowships, research assistantships, teaching assistantships, career-related internships or fieldwork, Federal Work-Study, institutionally sponsored loans, and scholarships/grants available. Support available to part-time students. Financial award application deadline: 2/1. *Faculty research:* Plant ecology, plant molecular biology, systematic botany, epidemiology, host-pathogen interaction. *Unit head:* Dr. Daniel Arp, Chair, 541-737-1297, Fax: 541-737-3573. *Application contact:* Dr. Donald B. Zobel, Professor, 541-737-3451, Fax: 541-737-3573, E-mail: zobeld@bcc.orst.edu.

Oregon State University, Graduate School, Program in Plant Physiology, Corvallis, OR 97331. Offers MS, PhD. *Students:* 3 full-time (all women), (all international). Average age 31. *Degree requirements:* For master's and doctorate, thesis/dissertation. *Entrance requirements:* For master's, BS in related area; for doctorate, BS or MS in related area, minimum GPA of 3.0 in last 90 hours of course work. Additional exam requirements/recommendations for international students: Required—TOEFL. *Application deadline:* For fall admission, 3/1 for domestic students. Applications are processed on a rolling basis. Application fee: $50. *Expenses:* Tuition, area resident: Part-time $301 per credit. Tuition, state resident: full-time $8,139; part-time $501 per credit. Tuition, nonresident: full-time $14,376; part-time $532 per credit. Required fees: $1,266. *Financial support:* Fellowships, research assistantships, teaching assistantships, career-related internships or fieldwork, Federal Work-Study, and institutionally sponsored loans available. Support available to part-time students. Financial award application deadline: 2/1. *Faculty research:* Nitrogen metabolism, physiological ecology, phloem transport, mineral nutrition, plant hormones. *Unit head:* Dr. Daniel Arp, Director, 541-737-1297, Fax: 541-737-3479. *Application contact:* Dr. Daniel Arp, Chair, 541-737-1297, Fax: 541-737-3573.

The Pennsylvania State University University Park Campus, Graduate School, Intercollege Graduate Programs, Intercollege Graduate Program in Integrative Biosciences, State College, University Park, PA 16802-1503. Offers integrative biosciences (PhD), including biomolecular transport dynamics, cell and developmental biology, cellular and molecular mechanisms of toxicity, chemical biology, ecological and molecular plant physiology, immunobiology, molecular medicine, neuroscience, nutrition science. *Students:* 110 full-time (58 women), 2 part-time (both women); includes 6 minority (3 African Americans, 3 Asian Americans or Pacific Islanders), 61 international. *Entrance requirements:* For master's and doctorate, GRE General Test. Application fee: $45. *Expenses:* Tuition, state resident: full-time $12,518; part-time $522 per credit. Tuition, nonresident: full-time $23,004; part-time $959 per credit. Required fees: $484. Tuition and fees vary according to course load, campus/location and program. *Financial support:* Fellowships available. *Unit head:* Dr. Richard J. Frisque, Co-Director, 814-863-3523, Fax: 814-863-1357, E-mail: rjf6@psu.edu.

See Close-Up on page 197.

The Pennsylvania State University University Park Campus, Graduate School, Intercollege Graduate Programs, Intercollege Graduate Program in Plant Physiology, State College, University Park, PA 16802-1503. Offers MS, PhD. *Students:* 20 full-time (10 women); includes 1 minority (Hispanic American), 14 international. *Entrance requirements:* For master's and doctorate, GRE General Test. Application fee: $45. *Expenses:* Tuition, state resident: full-time $12,518; part-time $522 per credit. Tuition, nonresident: full-time $23,004; part-time $959 per credit. Required fees: $484. Tuition and fees vary according to course load, campus/location and program. *Unit head:* Dr. Teh-hui Kao, Chair, 814-865-2626, Fax: 814-863-9416, E-mail: txk3@psu.edu.

Purdue University, Graduate School, School of Science, Department of Biological Sciences, West Lafayette, IN 47907. Offers biochemistry (PhD); biophysics (PhD); cell and developmental biology (PhD); ecology, evolutionary and population biology (MS, PhD), including ecology, evolutionary biology, population biology; genetics (MS, PhD); microbiology (MS, PhD); molecular biology (PhD); neurobiology (MS, PhD); plant physiology (PhD). *Faculty:* 47 full-time (9 women), 4 part-time/adjunct (1 woman). *Students:* 97 full-time (53 women), 8 part-time (4 women); includes 13 minority (3 African Americans, 1 American Indian/Alaska Native, 4 Asian Americans or Pacific Islanders, 5 Hispanic Americans), 50 international. Average age 28. 168 applicants, 29% accepted, 23 enrolled. In 2005, 18 master's, 9 doctorates awarded. Terminal master's awarded for partial completion of doctoral program. *Degree requirements:* For master's, thesis (for some programs); for doctorate, thesis/dissertation, seminars, teaching experience. *Entrance requirements:* For master's and doctorate, GRE General Test. Additional exam requirements/recommendations for international students: Required—TOEFL, TSE. *Application deadline:* For fall admission, 2/15 for domestic students, 1/31 for international students. Applications are processed on a rolling basis. Application fee: $55. Electronic applications accepted. *Financial support:* In 2005–06, 15 fellowships, 60 research assistantships, 53 teaching assistantships were awarded. Support available to part-time students. Financial award application deadline: 2/15; financial award applicants required to submit FAFSA. *Unit head:* Dr. Richard J Kuhn, Head, 765-494-4407. *Application contact:* Nancy Konopka, Graduate Studies Office Manager, 765-494-8142, Fax: 765-494-0876, E-mail: njk@bilbo.bio.purdue.edu.

Rutgers, The State University of New Jersey, New Brunswick/Piscataway, Graduate School, Program in Plant Biology, New Brunswick, NJ 08901-1281. Offers horticulture (MS, PhD); molecular biology and biochemistry (MS, PhD); pathology (MS, PhD); plant ecology (MS, PhD); plant genetics (PhD); plant physiology (MS, PhD); production and management (MS); structure and plant groups (MS, PhD). Part-time programs available. *Faculty:* 65 full-time, 1 part-time/adjunct (0 women). *Students:* 33 full-time (14 women), 13 part-time (6 women); includes 3 minority (1 African American, 2 Asian Americans or Pacific Islanders), 10 international. Average age 31. 56 applicants, 23% accepted, 10 enrolled. In 2005, 1 master's, 4 doctorates awarded. Terminal master's awarded for partial completion of doctoral program. *Median time to degree:* Of those who began their doctoral program in fall 1997, 90% received their degree in 8 years or less. *Degree requirements:* For master's, thesis or alternative, comprehensive exam; for doctorate, thesis/dissertation, comprehensive exam. *Entrance requirements:* For master's and doctorate, GRE General Test, GRE Subject Test (recommended). Additional exam requirements/recommendations for international students: Required—TOEFL (minimum score 600 paper-based; 250 computer-based). *Application deadline:* For fall admission, 4/1 for domestic students, 4/1 for international students. Application fee: $50. Electronic applications accepted. *Expenses:* Tuition, state resident: full-time $10,440; part-time $435 per credit. Tuition, nonresident: full-time $15,520; part-time $647 per credit. Required fees: $129 per credit. Tuition and fees vary according to program. *Financial support:* In 2005–06, 42 students received support, including 9 fellowships with full tuition reimbursements available (averaging $24,000 per year), 22 research assistantships with full tuition reimbursements available (averaging $16,500 per year), 10 teaching assistantships with full tuition reimbursements available (averaging $16,988 per year) Financial award application deadline: 1/15; financial award applicants required to submit FAFSA. *Faculty research:* Molecular biology and biochemistry of plants, plant development and genomics, plant protection, plant improvement, plant management of horticultural and field crops. Total annual research expenditures: $10 million. *Unit head:* Dr. Thomas Leustek, Director, 732-932-8165 Ext. 326, Fax: 732-932-9377, E-mail: leustek@aesop.rutgers.edu. *Application contact:* Barbara Mulder, Program Associate, 732-932-9375 Ext. 358, Fax: 732-932-9377, E-mail: plantbio@aesop.rutgers.edu.

University of Kentucky, Graduate School, Graduate School Programs in the College of Agriculture, Program in Plant Physiology, Lexington, KY 40506-0032. Offers PhD. *Faculty:* 3 full-time (0 women). *Students:* 17 full-time (11 women), 1 (woman) part-time; includes 1 minority (Asian American or Pacific Islander), 13 international. Average age 32. 27 applicants, 67% accepted, 15 enrolled. In 2005, 4 degrees awarded. *Median time to degree:* Of those who began their doctoral program in fall 1997, 85.7% received their degree in 8 years or less. *Degree requirements:* For doctorate, thesis/dissertation, comprehensive exam. *Entrance requirements:* For doctorate, GRE General Test, minimum graduate GPA of 3.0. Additional exam requirements/recommendations for international students: Required—TOEFL (minimum score 550 paper-based; 213 computer-based). *Application deadline:* For fall admission, 7/17 priority date for domestic students, 2/1 priority date for international students; for spring admission, 12/13 priority date for domestic students, 6/15 priority date for international students. Applications are processed on a rolling basis. Application fee: $40 ($55 for international students). Electronic applications accepted. *Expenses:* Tuition, state resident: full-time $6,308; part-time $331 per credit hour. Tuition, nonresident: full-time $13,968; part-time $756 per credit hour. Tuition and fees vary according to course load, degree level and program. *Financial support:* In 2005–06, 13 students received support, including 13 research assistantships with full tuition reimbursements available (averaging $18,000 per year); fellowships with full tuition reimbursements available, teaching assistantships with full tuition reimbursements available, Federal Work-Study, scholarships/grants, traineeships, health care benefits, tuition waivers (partial), and unspecified assistantships also available. Support available to part-time students. Financial award application deadline: 3/15; financial award applicants required to submit FAFSA. *Faculty research:* Biochemistry and biophysics of photosynthesis, biochemical and molecular basis for resistance of plants to pathogens, plant gene expression, physiological aspects of crop production. *Unit head:* Dr. Arthur G. Hunt, Director of Graduate Studies, 859-257-5020, Fax: 859-257-7125, E-mail: arthur.hunt@uky.edu. *Application contact:* Dr. Brian Jackson, Senior Associate Dean, 859-257-8176, Fax: 859-323-1928.

University of Massachusetts Amherst, Graduate School, Interdisciplinary Programs, Program in Plant Biology, Amherst, MA 01003. Offers MS, PhD. *Students:* 14 full-time (8 women), 2 part-time (1 woman); includes 3 minority (2 Asian Americans or Pacific Islanders, 1 Hispanic American), 5 international. 35 applicants, 26% accepted, 8 enrolled. In 2005, 1 master's, 1 doctorate awarded. *Degree requirements:* For master's and doctorate, thesis/dissertation. *Entrance requirements:* For master's and doctorate, GRE General Test. Additional exam requirements/recommendations for international students: Required—TOEFL (minimum score 530 paper-based; 197 computer-based). *Application deadline:* For fall admission, 1/2 priority date for domestic students, 1/2 priority date for international students; for spring admission, 10/1 for domestic students, 10/1 for international students. Applications are processed on a rolling basis. Application fee: $40 ($65 for international students). Electronic applications accepted. *Expenses:* Tuition, state resident: part-time $110 per credit. Tuition, nonresident: part-time $414 per credit. Required fees: $2,824 per term. One-time fee: $250 part-time. Full-time tuition and fees vary according to course load, campus/location, program and reciprocity agreements. *Financial support:* In 2005–06, 1 fellowship with full tuition reimbursement (averaging $806 per year), 5 research assistantships with full tuition reimbursements (averaging $8,576 per year) were awarded; teaching assistantships with full tuition reimbursements, career-related internships or fieldwork, Federal Work-Study, scholarships/grants, traineeships, and unspecified assistantships also available. Support available to part-time students. Financial award application deadline: 1/15. *Unit head:* Dr. Elsbeth Walker, Head, 413-577-3217, Fax: 413-545-3243. *Application contact:* Information Contact, 413-577-3217, Fax: 413-545-3243.

See Close-Up on page 497.

The University of Tennessee, Graduate School, College of Arts and Sciences, Program in Life Sciences, Knoxville, TN 37996. Offers genome science and technology (MS, PhD); plant physiology and genetics (MS, PhD). *Degree requirements:* For doctorate, one foreign language, thesis/dissertation. *Entrance requirements:* For master's and doctorate, GRE General Test, minimum GPA of 2.7. Additional exam requirements/recommendations for international students: Required—TOEFL. Electronic applications accepted.

Virginia Polytechnic Institute and State University, Graduate School, College of Agriculture and Life Sciences, Department of Plant Pathology, Physiology and Weed Science, Blacksburg, VA 24061. Offers plant pathology (MS, PhD); plant physiology and weed science (MS, PhD); plant protection (MS). *Faculty:* 13 full-time (3 women), 1 (woman) part-time/adjunct. *Students:* 16 full-time (9 women), 2 part-time (1 woman); includes 1 minority (African American), 9 international. Average age 28. 14 applicants, 29% accepted, 3 enrolled. In 2005, 3 master's, 2 doctorates awarded. *Entrance requirements:* For master's and doctorate, GRE. Additional exam requirements/recommendations for international students: Required—TOEFL (minimum score 575 paper-based; 231 computer-based). *Application deadline:* Applications are processed on a rolling basis. Application fee: $45. Electronic applications accepted. *Expenses:* Tuition, state resident: full-time $6,558; part-time $364 per credit. Tuition, nonresident: full-time $11,296; part-time $628 per credit. Required fees: $1,419; $468 per credit. $234 per term. *Financial support:* In 2005–06, 13 research assistantships with full tuition reimbursements (averaging $13,823 per year), 1 teaching assistantship with full tuition reimbursement (averaging $13,552 per year) were awarded; career-related internships or fieldwork, Federal Work-Study, scholarships/grants, and unspecified assistantships also available. Financial award application deadline: 4/1. *Faculty research:* Biotechnology, Dutch elm disease, weed control, plant pathogenic microorganisms, agronomic crop resistance to fungal and viral pathogens. *Unit head:* Dr. Erik Stromberg, Head, 540-231-7871, Fax: 540-231-7477. *Application contact:* Donna Ford, 540-231-6361, Fax: 540-231-7477.

Cross-Discipline Announcement

University of Pittsburgh, School of Arts and Sciences, Department of Biological Sciences, Program in Ecology and Evolution, Pittsburgh, PA 15260.

Department offers master's and doctoral programs in ecology and evolution. Research areas: ecological and population genetics, tropical ecology, community ecology, evolutionary ecology, plant reproductive ecology, metapopulation structure, life history and breeding system evolution, and plant-animal interaction.

CLEMSON UNIVERSITY

M.S. Program in Plant and Environmental Sciences

Program of Study	The Master of Science (M.S.) Program in Plant and Environmental Sciences (PES) at Clemson University is an interdisciplinary program comprising faculty members housed in the Departments of Biological Sciences; Entomology, Soils, and Plant Sciences; and Horticulture. Faculty research emphasizes seven major research areas: plant biotechnology, genetics and breeding; diversity and ecology; physiology; sustainable environment and landscape; soil sciences; pathology and pest management; and crop and turf management. Each student's degree program is tailored to their professional goals and is guided by an adviser and graduate advisory committee. The advisory committee determines the student's specific course work requirements, monitors the student's progress in research and the classroom, and administers the final examinations. It is recommended that the GS-2 Form (a plan of course work) be approved during the first semester of an M.S. program. The admissions requirements and the curriculum ensure that the students gain an outstanding education in their chosen field within plant and environmental sciences. Research projects concentrate on timely issues in basic and applied plant and environmental sciences and show strong connectedness with public interest to sustain, improve, and wisely use biological resources. PES students have won awards for research and teaching in national and international competitions, and they are successful in academia and professional assignments in public and private horticultural and agricultural industries.
Research Facilities	Laboratories of PES faculty members are affiliated with the Departments of Biological Sciences; Horticulture; and Entomology, Soils, and Plant Sciences. Facilities include the Biosystems Research Complex, in which 40,000 square feet of greenhouse provide varied levels of controlled environment for plant-growth experiments; the Electron Microscopy Facility, equipped with transmission, scanning, and confocal microscopes; the Analytical Laboratory, for quantification of pesticide residues, secondary metabolites, and hormones; and the DNA Sequencing Facility. The CU Genomics Institute is a state-of-the-art facility focusing on the structural and functional genomics of agricultural and horticultural crops and their pests. In proximity to the campus are Musser Farm, a 240-acre fruit tree research facility; Simpson Experimental Station, for plant genetics and breeding and crop ecology; and the South Carolina Botanical Garden, a teaching and research facility. Three Research and Education Centers offer instruction and practical experience in crop, turf, and vegetable physiology; pathology; entomology; and production, while enhancing environmental quality. These research centers are located in the Piedmont region and the southern Appalachian Mountains—two of the most biologically diverse areas in North America.
Financial Aid	Research assistantships are awarded on a competitive basis. External funds granted to PES faculty members are the major source of assistantships and research support. Thus, admission preference is given to applicants whose expressed research interest matches that of one of the faculty members. The applicants are encouraged to contact individual faculty members to address common research interests and funding availability. Applicants without sponsors are generally not accepted.
Cost of Study	Tuition for 2006–07 is $4643 per semester for in-state students and $9255 per semester for nonresidents. Off-campus rates are $535 per hour for in-state students and $918 per hour for nonresidents. Graduate assistants pay a flat fee of $1079 per semester and $348 per summer session. Graduate fellows pay South Carolina resident fees.
Living and Housing Costs	On-campus housing is available. For information, students should visit http://www.housing.clemson.edu. The cost of living in Clemson is quite low compared to the national average. Students who choose to live off the campus typically spend $300–$400 per month for rent, depending on location, amenities, roommates, etc.
Student Group	The program has 53 students in the master's and doctoral tracks combined. Thirty-six percent of the students are women, and 25 percent are international students.
Student Outcomes	Some M.S. degree program graduates choose to further their studies at such institutions as California, Davis; Rutgers; and the Medical University of South Carolina. Other PES graduates accept employment in various research institutions, state agencies, and horticultural and agricultural companies.
Location	Clemson is a small, beautiful college town near the Blue Ridge Mountains and Lake Hartwell in upstate South Carolina. The Upstate is one of the country's fastest-growing areas and is the midpoint of the Charlotte-to-Atlanta I-85 corridor, a multistate area along Interstate 85 that runs from metro Atlanta to Richmond, Virginia, and encompasses Charlotte, North Carolina, and North Carolina's Research Triangle. Atlanta and Charlotte are each a 2-hour's drive away. Many financial institutions and other industries have national headquarters for a major presence in the Upstate, including Wachovia, Bank of America, BMW, Bon Secours St. Francis Health System, Bosch North America, Bowater, Charter Communications, Ernst and Young, Fluor Corporation, IBM, Microsoft, Michelin of North America, and many others.
The University	Clemson is classified by the Carnegie Foundation as Doctoral/Research University–Extensive, a category comprising less than 4 percent of all universities in America. The University's mission is to fulfill the covenant between its founder and the people of South Carolina to establish a "high seminary of learning" through its responsibilities of teaching, research, and extended public service. The University has identified eight areas of academic emphasis that create collaborations that, in turn, help fulfill the University's mission.
Applying	Candidates should have a strong undergraduate background in the biological and agricultural sciences as appropriate to their focus area. Undergraduate curricula that may provide this background are botany, biology, or chemistry or one of the agricultural plant and soil environmental sciences, such as agronomy, horticulture, forest resources, and plant pathology. The program also welcomes students with nontraditional backgrounds who may need to complete some relevant undergraduate courses to supplement their graduate program. Applicants may apply on the Web at http://www.grad.clemson.edu/p_apply.html. Applications, with a $50 nonrefundable fee, should be received no later than five weeks prior to registration. Every required item in support of the application must be on file by that date. Students are advised to contact the department for the deadlines of the program of proposed study.
Correspondence and Information	Halina Knap, Ph.D. Program Coordinator Department of Entomology, Soils, and Plant Sciences 272 Poole Agricultural Center Box 340315 Clemson University Clemson, South Carolina 29634-0315 Phone: 864-656-3443 Fax: 864-656-0795 E-mail: hskrpsk@ces.clemson.edu Web site: http://www.clemson.edu/plantenvgrad

Clemson University

THE FACULTY AND THEIR RESEARCH

A. G. Abbott, Professor of Biological Sciences; Ph.D., Brown. Plant molecular genetics.
J. W. Adelberg, Assistant Professor of Horticulture; Ph.D., Clemson. Plant physiology, in vitro propagation.
J. G. Andrae, Assistant Professor of Entomology, Soils, and Plant Sciences; Ph.D., Idaho. Forage establishment, management and grazing systems.
W. V. Baird, Professor of Horticulture; Ph.D., Virginia. Plant molecular genetics.
R. E. Ballard, Professor of Biological Sciences; Ph.D., Iowa. Plant systematics.
W. L. Bauerle, Assistant Professor of Horticulture; Ph.D., Cornell. Tree stress physiology.
D. G. Bielenberg, Assistant Professor of Horticulture; Ph.D., Penn State. Stress physiology and stress molecular mechanisms.
D. W. Bradshaw, Professor of Horticulture; Ph.D., Virginia Tech. Sustainable vegetable production.
J. C. Caldwell, Associate Professor of Horticulture; Ph.D., Arkansas. Arboriculture and urban forestry.
N. D. Camper, Professor of Entomology, Soils, and Plant Sciences; Ph.D., N.C. State. Biology of medicinal plants.
J. D. Culin, Professor of Entomology, Soils, and Plant Sciences; Ph.D., Kentucky. Plant-insect interaction.
S. J. DeWalt, Assistant Professor of Biological Sciences; Ph.D., LSU. Plant population, community ecology.
R. J. Dufault, Professor of Horticulture; Ph.D., Kansas State. Vegetable physiology and cultural management.
L. A. Dyck, Professor Emeritus of Biological Sciences; Ph.D., Washington (St. Louis). Shoreline ecology and restoration.
J. E. Faust, Associate Professor of Horticulture; Ph.D., Michigan State. Floriculture physiology.
B. A. Fortnum, Professor of Entomology, Soils, and Plant Sciences; Ph.D., Clemson. Etiology and epidemiology of solanaceous diseases.
T. C. Hale, Assistant Professor of Horticulture; Ph.D., Texas A&M. Turf management.
M. T. Haque, Professor of Horticulture; Ph.D., Clemson. Landscape ecology and design.
R. L. Hassell, Associate Professor of Horticulture; Ph.D., Ohio State. Vegetable physiology and production.
S. N. Jeffers, Associate Professor of Entomology, Soils, and Plant Sciences; Ph.D., Cornell. Diseases of ornamental crops and trees.
A. P. Keinath, Professor of Entomology, Soils, and Plant Sciences; Ph.D., Michigan. Vegetable pathology.
H. T. Knap, Professor of Entomology, Soils, and Plant Sciences; Ph.D., Academy of Agriculture (Poland). Cytogenetics and plant molecular genetics.
D. R. Layne, Associate Professor of Horticulture; Ph.D., Michigan State. Peach production and physiology.
S. A. Lewis, Professor of Entomology, Soils, and Plant Sciences; Ph.D., Arizona. Plant parasitic nematology, ecology and management.
H. Liu, Associate Professor of Horticulture; Ph.D., Rhode Island. Turfgrass physiology, stress and pest management.
S. B. Martin, Professor of Entomology, Soils, and Plant Sciences; Ph.D., N.C. State. Diseases of turfgrasses.
L. B. McCarty, Professor of Horticulture; Ph.D., Clemson. Turfgrass management, physiology and pest management.
J. D. Mueller, Professor of Entomology, Soils, and Plant Sciences; Ph.D., Illinois. Crop disease control, nematode biology and ecology.
J. K. Norsworthy, Assistant Professor of Entomology, Soils, and Plant Sciences; Ph.D., Arkansas. Weed science and species allelopathy.
V. L. Quisenberry, Professor of Entomology, Soils, and Plant Sciences; Ph.D., Kentucky. Soil physics.
N. C. Rajapakse, Professor of Horticulture; Ph.D., Texas A&M. Photomorphogenesis, postharvest physiology, phytochemicals.
G. L. Reighard, Professor of Horticulture; Ph.D., Michigan State. Fruit tree culture, physiology, and genetics.
M. B. Riley, Professor of Entomology, Soils, and Plant Sciences; Ph.D., Clemson. Plant physiology, pesticide residues, hormones.
J. W. Rushing, Professor of Horticulture; Ph.D., Florida. Postharvest physiology.
G. Schnabel, Assistant Professor of Entomology, Soils, and Plant Sciences; Ph.D., Hohenheim (Germany). Fungal biology and genetics.
S. W. Scott, Professor of Entomology, Soils, and Plant Sciences; Ph.D., Wales (England). Plant virology.
B. M. Shepard, Professor of Entomology, Soils, and Plant Sciences; Ph.D., Texas A&M. Integrated pest management.
E. R. Shipe, Professor of Entomology, Soils, and Plant Sciences; Ph.D., Virginia Tech. Soybean breeding/genetics.
T. P. Spira, Professor of Biological Sciences; Ph.D., Berkeley. Plant ecology, evolution and conservation.
W. C. Stringer, Associate Professor of Entomology, Soils, and Plant Sciences; Ph.D., Virginia Tech. Native grass ecology and management.
J. P. Tomkins, Assistant Professor of Genetics and Biochemistry; Ph.D., Clemson. Crop genomics.
C. Wells, Assistant Professor of Horticulture; Ph.D., Penn State. Root demography, physiology and senescence.
T. Whitwell, Professor of Horticulture; Ph.D., Oklahoma State. Environmental remediation and weed science.
G. Zehnder, Professor of Entomology, Soils, and Plant Sciences; Ph.D., California, Riverside. Sustainable agriculture.

CLEMSON UNIVERSITY

Ph.D. Program in Plant and Environmental Sciences

Program of Study

The Ph.D. Program in Plant and Environmental Sciences (PES) at Clemson University is an interdisciplinary program comprising faculty members housed in the Departments of Biological Sciences; Entomology, Soils, and Plant Sciences; and Horticulture. Faculty research emphasizes seven major research areas: plant biotechnology, genetics and breeding; diversity and ecology; physiology; sustainable environment and landscape; soil sciences; pathology and pest management; and crop and turf management. A dissertation based on original research is required for the Ph.D. degree. There is no semester credit hour requirement; the course work plan is based on student interests and dissertation emphasis, as determined in consultation with the major adviser and graduate committee.

The admissions requirements and the curriculum ensure that the students gain an outstanding education in their chosen field within plant and environmental sciences. Research projects concentrate on timely issues in basic and applied plant and environmental sciences and show strong connectedness with public interest to sustain, improve, and wisely use biological resources. PES students have won awards for research and teaching in national and international competitions, and they are successful in academia and professional assignments in public and private horticultural and agricultural industries.

Research Facilities

Laboratories of PES faculty members are affiliated with the Departments of Biological Sciences; Horticulture; and Entomology, Soils, and Plant Sciences. Facilities include the Biosystems Research Complex, in which 40,000 square feet of greenhouse provide varied levels of controlled environment for plant-growth experiments; the Electron Microscopy Facility, equipped with transmission, scanning, and confocal microscopes; the Analytical Laboratory, for quantification of pesticide residues, secondary metabolites, and hormones; and the DNA Sequencing Facility.

The CU Genomics Institute is a state-of-the-art facility focusing on the structural and functional genomics of agricultural and horticultural crops and their pests. In proximity to the campus are Musser Farm, a 240-acre fruit tree research facility; Simpson Experimental Station, for plant genetics and breeding and crop ecology; and the South Carolina Botanical Garden, a teaching and research facility.

Three Research and Education Centers offer instruction and practical experience in crop, turf, and vegetable physiology; pathology; entomology; and production, while enhancing environmental quality. These research centers are located in the Piedmont region and the southern Appalachian Mountains—two of the most biologically diverse areas in North America.

Financial Aid

Research assistantships are awarded on a competitive basis. External funds granted to PES faculty members are the major source of assistantships and research support. Thus, admission preference is given to applicants whose expressed research interest matches that of one of the faculty members. The applicants are encouraged to contact individual faculty members to address common research interests and funding availability. Applicants without sponsors are generally not accepted.

Cost of Study

Tuition for 2006–07 is $4643 per semester for in-state students and $9255 per semester for nonresidents. Off-campus rates are $535 per hour for in-state students and $918 per hour for nonresidents. Graduate assistants pay a flat fee of $1079 per semester and $348 per summer session. Graduate fellows pay South Carolina resident fees.

Living and Housing Costs

On-campus housing is available. For information, students should visit http://www.housing.clemson.edu. The cost of living in Clemson is quite low compared to the national average. Students who choose to live off the campus typically spend $300–$400 per month for rent, depending on location, amenities, roommates, etc.

Student Group

The program has 53 students in the master's and doctoral tracks combined. Thirty-six percent of the students are women, and 25 percent are international students.

Student Outcomes

Several Ph.D. graduates have attained university or college professorship positions. PES graduates are employed in various research institutions, state agencies, and horticultural and agricultural companies.

Location

Clemson is a small, beautiful college town near the Blue Ridge Mountains and Lake Hartwell in upstate South Carolina. The Upstate is one of the country's fastest-growing areas and is the midpoint of the Charlotte-to-Atlanta I-85 corridor, a multistate area along Interstate 85 that runs from metro Atlanta to Richmond, Virginia, and encompasses Charlotte, North Carolina, and North Carolina's Research Triangle. Atlanta and Charlotte are each a 2-hour's drive away. Many financial institutions and other industries have national headquarters for a major presence in the Upstate, including Wachovia, Bank of America, BMW, Bon Secours St. Francis Health System, Bosch North America, Bowater, Charter Communications, Ernst and Young, Fluor Corporation, IBM, Microsoft, Michelin of North America, and many others.

The University

Clemson is classified by the Carnegie Foundation as Doctoral/Research University–Extensive, a category comprising less than 4 percent of all universities in America. The University's mission is to fulfill the covenant between its founder and the people of South Carolina to establish a "high seminary of learning" through its responsibilities of teaching, research, and extended public service. The University has identified eight areas of academic emphasis that create collaborations that, in turn, help fulfill the University's mission.

Applying

Candidates should have a strong background in the biological and agricultural sciences as appropriate to their focus area. Curricula that may provide this background are botany, biology, or chemistry or one of the agricultural plant and soil environmental sciences, such as agronomy, horticulture, forest resources, and plant pathology. The program also welcomes students with nontraditional backgrounds who may need to complete some relevant undergraduate courses to supplement their graduate program.

Applicants may apply on the Web at http://www.grad.clemson.edu/p_apply.html. Applications, with a $50 nonrefundable fee, should be received no later than five weeks prior to registration. Every required item in support of the application must be on file by that date. Students are advised to contact the department for the deadlines of the program of proposed study.

Correspondence and Information

Halina Knap, Ph.D., Program Coordinator
Department of Entomology, Soils, and Plant Sciences
272 Poole Agricultural Center
Box 340315
Clemson University
Clemson, South Carolina 29634-0315
Phone: 864-656-3443
Fax: 864-656-0795
E-mail: hskrpsk@ces.clemson.edu
Web site: http://www.clemson.edu/plantenvgrad

Clemson University

THE FACULTY AND THEIR RESEARCH

A. G. Abbott, Professor of Biological Sciences; Ph.D., Brown. Plant molecular genetics.

J. W. Adelberg, Assistant Professor of Horticulture; Ph.D., Clemson. Plant physiology, in vitro propagation.

J. G. Andrae, Assistant Professor of Entomology, Soils, and Plant Sciences; Ph.D., Idaho. Forage establishment, management and grazing systems.

W. V. Baird, Professor of Horticulture; Ph.D., Virginia. Plant molecular genetics.

R. E. Ballard, Professor of Biological Sciences; Ph.D., Iowa. Plant systematics.

W. L. Bauerle, Assistant Professor of Horticulture; Ph.D., Cornell. Tree stress physiology.

D. G. Bielenberg, Assistant Professor of Horticulture; Ph.D., Penn State. Stress physiology and stress molecular mechanisms.

D. W. Bradshaw, Professor of Horticulture; Ph.D., Virginia Tech. Sustainable vegetable production.

J. C. Caldwell, Associate Professor of Horticulture; Ph.D., Arkansas. Arboriculture and urban forestry.

N. D. Camper, Professor of Entomology, Soils, and Plant Sciences; Ph.D., N.C. State. Biology of medicinal plants.

J. D. Culin, Professor of Entomology, Soils, and Plant Sciences; Ph.D., Kentucky. Plant-insect interaction.

S. J. DeWalt, Assistant Professor of Biological Sciences; Ph.D., LSU. Plant population, community ecology.

R. J. Dufault, Professor of Horticulture; Ph.D., Kansas State. Vegetable physiology and cultural management.

L. A. Dyck, Professor Emeritus of Biological Sciences; Ph.D., Washington (St. Louis). Shoreline ecology and restoration.

J. E. Faust, Associate Professor of Horticulture; Ph.D., Michigan State. Floriculture physiology.

B. A. Fortnum, Professor of Entomology, Soils, and Plant Sciences; Ph.D., Clemson. Etiology and epidemiology of solanaceous diseases.

T. C. Hale, Assistant Professor of Horticulture; Ph.D., Texas A&M. Turf management.

M. T. Haque, Professor of Horticulture; Ph.D., Clemson. Landscape ecology and design.

R. L. Hassell, Associate Professor of Horticulture; Ph.D., Ohio State. Vegetable physiology and production.

S. N. Jeffers, Associate Professor of Entomology, Soils, and Plant Sciences; Ph.D., Cornell. Diseases of ornamental crops and trees.

A. P. Keinath, Professor of Entomology, Soils, and Plant Sciences; Ph.D., Michigan. Vegetable pathology.

H. T. Knap, Professor of Entomology, Soils, and Plant Sciences; Ph.D., Academy of Agriculture (Poland). Cytogenetics and plant molecular genetics.

D. R. Layne, Associate Professor of Horticulture; Ph.D., Michigan State. Peach production and physiology.

S. A. Lewis, Professor of Entomology, Soils, and Plant Sciences; Ph.D., Arizona. Plant parasitic nematology, ecology and management.

H. Liu, Associate Professor of Horticulture; Ph.D., Rhode Island. Turfgrass physiology, stress and pest management.

S. B. Martin, Professor of Entomology, Soils, and Plant Sciences; Ph.D., N.C. State. Diseases of turfgrasses.

L. B. McCarty, Professor of Horticulture; Ph.D., Clemson. Turfgrass management, physiology and pest management.

J. D. Mueller, Professor of Entomology, Soils, and Plant Sciences; Ph.D., Illinois. Crop disease control, nematode biology and ecology.

J. K. Norsworthy, Assistant Professor of Entomology, Soils, and Plant Sciences. Ph.D., Arkansas. Weed science and species allelopathy.

V. L. Quisenberry, Professor of Entomology, Soils, and Plant Sciences; Ph.D., Kentucky. Soil physics.

N. C. Rajapakse, Professor of Horticulture; Ph.D., Texas A&M. Photomorphogenesis, postharvest physiology, phytochemicals.

G. L. Reighard, Professor of Horticulture; Ph.D., Michigan State. Fruit tree culture, physiology, and genetics.

M. B. Riley, Professor of Entomology, Soils, and Plant Sciences; Ph.D., Clemson. Plant physiology, pesticide residues, hormones.

J. W. Rushing, Professor of Horticulture; Ph.D., Florida. Postharvest physiology.

G. Schnabel, Assistant Professor of Entomology, Soils, and Plant Sciences; Ph.D., Hohenheim (Germany). Fungal biology and genetics.

S. W. Scott, Professor of Entomology, Soils, and Plant Sciences; Ph.D., Wales (England). Plant virology.

B. M. Shepard, Professor of Entomology, Soils, and Plant Sciences; Ph.D., Texas A&M. Integrated pest management.

E. R. Shipe, Professor of Entomology, Soils, and Plant Sciences; Ph.D., Virginia Tech. Soybean breeding/genetics.

T. P. Spira, Professor of Biological Sciences; Ph.D., Berkeley. Plant ecology, evolution and conservation.

W. C. Stringer, Associate Professor of Entomology, Soils, and Plant Sciences; Ph.D., Virginia Tech. Native grass ecology and management.

J. P. Tomkins, Assistant Professor of Genetics and Biochemistry; Ph.D., Clemson. Crop genomics.

C. Wells, Assistant Professor of Horticulture; Ph.D., Penn State. Root demography, physiology and senescence.

T. Whitwell, Professor of Horticulture; Ph.D., Oklahoma State. Environmental remediation and weed science.

G. Zehnder, Professor of Entomology, Soils, and Plant Sciences; Ph.D., California, Riverside. Sustainable agriculture.

MIAMI UNIVERSITY

College of Arts and Science
Department of Botany

Programs of Study

The M.A. and M.S. in botany are two-year degrees with emphasis on completion of either an internship experience (M.A.) or a thesis experience based on original research (M.S.). The M.A. is designed as a technical professional path for individuals pursuing industrial or research support as a career objective. The M.S. is designed for those who aspire to the Ph.D., in most cases. A major professor and graduate advisory committee guide each student through the program. Each master's student is expected to demonstrate a broad knowledge of botany in an oral examination before his or her graduate committee. The student proposes an internship/thesis project to the committee and, after conducting the necessary research and writing, presents and defends the project before the general public and the committee.

The Ph.D. is a three- to five-year research degree with emphasis on the successful completion of an original dissertation. A major professor guides the student through the program, with the assistance of a graduate committee tailored to the student's individual objectives. During the first two years, the student is expected to complete course work, present a dissertation proposal, and pass a preliminary examination. The remainder of a student's program is devoted primarily to independent research, culminating in a dissertation defended before the general public and the committee.

Departmental expertise is available in the areas of developmental anatomy, ecology, evolution, molecular biology, cell biology, bioinformatics, developmental and comparative morphology, mycology, physiology, and systematics. Each student is expected to develop a broad understanding of basic biological principles, in addition to obtaining the specialized skills of the contemporary researcher.

The Department participates in interdepartmental graduate programs in ecology (http://www.muohio.edu/ecology) and molecular biology (http://www.muohio.edu/molbio). The Department conducts advanced seminars and workshops both on and off campus in specific topics that vary from year to year.

Research Facilities

The Electron Microscopy Facility houses two transmission electron microscopes and two scanning electron microscopes, two confocal microscopes, a deconvolution microscope, and a multimode light microscope system. Microanalysis can be performed with energy dispersive X-ray spectroscopy or EBSD systems. In addition to standard sample preparation, the facility is equipped for cryofixation and freeze substitution. Specializing in scientific imaging the EMF has a specialized computer graphics laboratory for digital imaging and analysis, publication, and presentation preparation.

The Center for Bioinformatics and Functional Genomics is an interdepartmental facility for molecular biology research that is located in Pearson Hall. It has several automated genetic analyzers, as well as bioinformatics servers and software, for DNA sequencing and fragment analysis; real-time PCR thermocyclers, a microarray scanner, robotic sample preparation, and analysis software for gene expression studies; and several thermocyclers, spectrophotometers, a DNA fluorometer, a phosphorimager, a robotic station, and a digital gel imaging system for preparation and analysis of molecular biology samples.

The Willard Sherman Turrell Herbarium is Ohio's largest herbarium with more than 620,000 specimens. Active exchange programs are carried on with herbaria throughout the world. The herbarium has an endowment fund to support research in systematic botany. More than 13,000 square feet of glasshouse space and a plant growth chamber facility with fourteen growth chambers provide controlled environmental conditions for experimentation. The Ecology Research Center, located 2 miles from the main campus, consists of six experimental ponds and 168 acres (68 hectares) of land used for field manipulative studies. Additional opportunities for field research include nearby streams and rivers, the lake and forests at Hueston Woods State Park, and University-owned natural areas near the campus.

Numerous microcomputers within the Department and the University mainframe computers provide access to the Internet. The Brill Science Library contains a wide range of research journals and online computerized literature search and interlibrary loan services. Graduate students can access up to $2000 per semester for research expenses by applying to the Department's Academic Challenge Grant Program. The Department funds student travel to academic meetings and has a separate funding program for field-oriented research travel.

Financial Aid

Graduate assistantships or teaching associateships are offered to nearly all students admitted into the Ph.D., M.S., and M.A. programs. Most awardees teach laboratory sections associated with undergraduate courses, but some assist faculty members in research. Stipends for continuous year-round registration in the 2006–07 academic year are $10,377 for master's degree students and $18,923 for Ph.D. degree students (includes a summer scholarship of $1800). Instructional fees (currently $4548.48 per semester) are waived. Supplemental nonrenewable Graduate Academic Achievement Scholarships of $5000 are available from the Graduate School for outstanding applicants on a competitive basis.

Cost of Study

Half of the University general fee (which totals $1532.64 per year) is waived for graduate assistants and associates. Students are able to purchase health insurance at a cost of $782 per year.

Living and Housing Costs

Adequate housing is available in Oxford, as long as arrangements are made by the middle of the summer. Monthly cost of housing ranges from $350 to $500 for single rooms or apartments and from $300 to $500 for shared apartments or houses. Monthly individual living expenses average about $900.

Student Group

There are currently 33 resident graduate students in the program; 14 are men and 19 are women.

Student Outcomes

Many of the Department's Ph.D. recipients go on to faculty positions at colleges and universities, often after a postdoctoral position. Others are employed in private-sector research and development or as researchers or administrators in federal agencies such as NASA, the USDA, and the EPA. Most M.S. recipients continue on to pursue Ph.D.'s; others are employed in the pharmaceutical and biotechnology industries or in other biological or environmental science professions. The employment rate for recent graduates approaches 100 percent.

Location

Oxford is a classic college town (population 20,000) with numerous small shops adjacent to campus. The city is located 35 miles north of Cincinnati and 45 miles southwest of Dayton, offering ready access to the diversity of city life.

The University and The Department

Miami University, a state-assisted university, was established in 1809 and is the second-oldest institution of higher learning in Ohio. The Department of Botany, housed in the College of Arts and Science, has a reputation of academic excellence in both graduate and undergraduate education and research. The Department has many distinguished alumni actively working within the discipline throughout the world.

Applying

Applicants for the master's degrees must have anticipated completion of a bachelor's degree. Applications should be completed by January 1. A complete application includes a completed Departmental preapplication form, a completed graduate school application form, two official copies of transcripts for all undergraduate and graduate work, GRE scores (recommended but not required), three letters of recommendation, and a statement about the area of specialization in botany and the student's career objectives. International students must submit TOEFL scores.

Correspondence and Information

Dr. R. J. Hickey, Graduate Student Advisor
Department of Botany
Miami University
Oxford, Ohio 45056

Phone: 513-529-4200
Fax: 513-529-4243
E-mail: botany@muohio.edu (application material)
Web site: http://www.muohio.edu/botany

Miami University

THE FACULTY AND THEIR RESEARCH

Applicants are strongly encouraged to contact graduate faculty members whose research interests suggest they may be willing to direct that student's graduate research. Applicants are not accepted into a program unless a faculty sponsor has been established. Prospective applicants are invited to learn more about Miami's faculty research interests and research facilities by visiting http://www.muohio.edu/botany; this Web site includes links to the faculty members' Web pages.

Telephone numbers and e-mail access are in parentheses following each faculty member's entry. The area code in all cases is 513.

Susan R. Barnum, Ph.D., Iowa State, 1983. Evolution of cyanobacteria and nitrogen fixation genes; structure, function, and evolution of DNA insertion elements in cyanobacteria, evolution of symbiotic cyanobacteria and plant-cyanobacteria interactions. (529-4254, BarnumSR@muohio.edu)

Richard E. Edelmann, Ph.D., Michigan State, 1993. Light and electron microscopy; cellular and developmental ultrastructure; mycology. (529-5712, EdelmaRE@muohio.edu)

Daniel K. Gladish, Ph.D., California, Davis, 1995. Development of root systems; environmental effects on development, especially in roots; programmed cell death. (785-3244, GladisDK@muohio.edu)

David L. Gorchov, Ph.D., Michigan, 1987. Plant ecology, especially invasive plants; population biology of rare and harvested plants; tropical forest management. (529-4205, GorchoDL@muohio.edu)

R. James Hickey, Ph.D., Connecticut, 1985. Taxonomy and systematics of vascular plants, especially pteridophyte groups (Isoetes, ferns, and Lycopodium); flora of the Bahamas. (529-6000, HickeyRJ@muohio.edu)

Alfredo J. Huerta, Ph.D., California, Riverside, 1987. Plant stress metabolism and environmental physiology. (529-4257, HuertaAJ@muohio.edu)

Carolyn Howes Keiffer, Ph.D., Ohio, 1996. Plant ecology, with emphasis on restoration ecology and phytoremediation; physiological ecology; American chestnut reintroduction. (727-3243, KeiffeCH@muohio.edu)

John Z. Kiss, Ph.D., Rutgers, 1987. Cell biology; plant physiology; gravitropism and phototropism in higher and lower plants; space biology; ultrastructure and cryotechniques in electron microscopy. (529-5428, KissJZ@muohio.edu)

Q. Quinn Li, Ph.D., Kentucky, 1995. Molecular biology, functional genomics, and bioinformatics of plant mRNA polyadenylation; plant biotechnology. (529-4256, liq@muohio.edu, http://www.polyA.org)

Chun Liang, Ph.D., Georgia, 1999. Plant genomics and bioinformatics; biological databases and data mining; transcriptome analysis for novel gene discovery and gene expression in plants. (529-2336, LiangC@muohio.edu)

Roger D. Meicenheimer, Ph.D., Washington State, 1980. Developmental plant anatomy; plant morphogenesis; pattern formation in plants; phyllotaxis. (529-7012, MeicenRD@muohio.edu, http://www.cas.muohio.edu/~meicenrd)

Nicholas P. Money, Ph.D., Exeter (England), 1986. Biomechanics of fungal growth, reproduction, and pathogenesis; biology of indoor molds. (529-2140, MoneyNP@muohio.edu)

Richard C. Moore, Ph.D., Penn State, 1999. Plant evolutionary biology; evolutionary genetics and genomics; duplicate gene evolution; evolution of sexual reproduction systems; evolution of plant development (evo-devo). (529-4278, MooreRC@muohio.edu)

Elisabeth E. Schussler, Ph.D., LSU, 1997. Botany education; undergraduate biology education; development of aerenchyma tissue in wetland plants. (529-4204, SchussE@muohio.edu)

Nancy L. Smith-Huerta, Ph.D., California, Riverside, 1983. Floral and pollen development; pollination biology. (529-4257, SmithHN@muohio.edu)

M. Henry H. Stevens, Ph.D., Pittsburgh, 1999. Community ecology; effects of resource supply and variability on biodiversity and food webs. (529-4206, HStevens@muohio.edu, http://www.cas.muohio.edu/~stevenmh)

Michael A. Vincent, Ph.D., Miami (Ohio), 1991. Plant taxonomy; floristics; Caribbean flora; endangered species; invasive species; herbarium curation. (529-2755, VincenMA@muohio.edu)

Linda E. Watson, Ph.D., Oklahoma, 1989. Plant systematics and evolution; speciation; phylogenetics. (529-4200, WatsonLE@muohio.edu)

Affiliate Faculty

Adolph M. Greenberg, Ph.D., Wayne State, 1978. Institute of Environmental Sciences. Ecological, medical, and applied anthropology; national parks and protected areas; ethnoecology and ethnobotany. (529-5022, GreenbAM@muohio.edu)

Chris Makaroff, Ph.D., Purdue, 1986. Department of Chemistry and Biochemistry. Protein structure-function relationships; chromatin structure. (529-1659, MakaroCA@muohio.edu)

Kimberly E. Medley, Ph.D., Michigan State, 1990. Department of Geography. Environmental and human influences on the distribution and ecology of temperate and tropical vegetation; ethnobotany. (529-1558, MedleyKE@muohio.edu)

Adjunct Faculty

Diana J. Davis, Ph.D., Colorado State, 1995. Department of Chemistry and Physical Science, College of Mount St. Joseph. Biochemistry of fungal growth, reproduction, and pathogenesis. (244-4478, diana_davis@mail.msj.edu)

Charles Kwit, Ph.D., LSU, 2000. Frugivory and seed dispersal; disturbance ecology; plant conservation, restoration, and management. (529-4258, KwitC@muohio.edu)

Vivian Negrón-Ortiz, Ph.D., Miami (Ohio), 1994. Reproductive biology and systematics of vascular plants of the Caribbean; dioecy, rare, and endangered species; Rubiaceae and Cactaceae. (529-4200, NegronV@muohio.edu)

Neal Sullivan, Ph.D., Missouri–Columbia, 2001. Forest ecology and landscape-level modeling of forest processes. (529-3181, neal.sullivan@muohio.edu, http://www.nsforestry.com)

Miami University provides equal opportunity in education and employment.

SOUTHERN ILLINOIS UNIVERSITY CARBONDALE

Department of Plant Biology
Ph.D. in Plant Biology

Programs of Study

The Department of Plant Biology offers the degree of Ph.D. in plant biology. The need for well-trained specialists in certain areas of plant biology has never been greater, and the outlook for the future is very promising. For example, existing demands on all natural resources have resulted in a demand for people trained in the botanical aspects of environmental sciences, agriculture, conservation, biogeography, land reclamation, water resources, and many other fields. The advent of the biotechnology era has brought with it a demand for specialists in plant physiology and biochemistry, especially in the areas of nutrition, metabolism, growth regulators, tissue culture, biomass production, and environmental stress. There is currently an increased demand for specialists in molecular genetics. Since society's greatest need is for a constant supply of wholesome, nutritious food, the biotechnology industry continues to direct its interests and energies toward the production and improvement of food products. Running a close second to the need for food is the need for new and better medicinal drugs and new methods for their mass production. Since plants are the only primary source of food and since most medicinal drugs are obtained from plants, the food and pharmaceutical industries need a continuous supply of well-trained plant biologists.

Course work for the degree consists of a minimum of 20 semester hours at the 400 and 500 levels in the plant biology program or related disciplines but excludes seminar, readings, research, dissertation, and research tool requirements. Students may select a minor area of specialization once the major area has been declared. A course proposal including core courses must be approved by the student's Advisory Committee and the respective departmental chair and be submitted to the Director of Graduate Studies within the first semester. Changes made after the first semester of the student's program must be approved by the majority of the graduate student's Advisory Committee. Students enroll in the plant biology departmental seminar and a seminar in a related discipline each year. Students also take, either prior to or during their program, courses in all of the following four categories: general plant biology, systematics, physiology, and ecology. Courses in plant anatomy and genetics are strongly recommended for students pursuing careers in teaching and/or research. Students must pass a preliminary examination, demonstrate proficiency in a foreign language, and successfully complete a dissertation.

Research Facilities

In addition to modern laboratories maintained by individual faculty members, research facilities in other departments are also available for use as are centrally administered University facilities. An example of the latter is IMAGE (Integrated Microscopy and Graphics Expertise), which supports a wide variety of analytical and quantitative electron microscopic methodologies. Within the Department of Plant Biology, individual faculty members maintain research equipment that permits a wide range of modern analyses in the areas of molecular biology, systematics, cell biology, and ecology. An excellent collection of science journals and books are located on the fifth and sixth floors of Morris Library. In addition to individual laboratories directed by PLB faculty members, the department maintains a centralized laboratory that is used for teaching various courses in molecular biology. This Plant Molecular Biology Laboratory is outfitted to allow ultralow storage of tissue samples, tissue extraction (proteins, DNA, RNA), and gel electrophoresis (isozymes and DNA). Computer labs are readily available in several buildings.

The Southern Illinois University Herbarium currently houses about 250,000 specimens. As might be expected, the herbarium's holdings are especially rich for southern Illinois, which includes the southernmost eighteen counties of the state. However, the collections also come from other areas of the state and country as well as from several different parts of the world, including Australia, Hawaii, China, Israel, Mexico, Belize, Montserrat, and Brazil. The plant families with the largest number of holdings in the herbarium are the grasses, sedges, legumes, and composites. In addition to the mounted plant specimens, the herbarium also houses ethnobotanical collections, the Walter Welch slide collection of approximately 5,500 kodachromes, and a library of books and reprints.

The Plant Biology Greenhouses comprise four interconnected houses, an office, and a head house. Most growing space in main house consists of in-ground beds, whereas the three attached houses (south, middle, and north houses) contain fixed elevated benches. The collection is taxonomically diverse and consists of representatives of all extant land plants (bryophytes, lycophytes, ferns, gymnosperms, and angiosperms). The collection, maintained by Richard Cole and Karen Frailey, is used for both teaching and research. The Phytotron provides climate-controlled growing space for plants that is ideal for conducting physiological experiments. This facility is currently being utilized by Dr. Stephen Ebbs, who is investigating heavy metal accumulation in *Thlaspi* and phytoremediation in *Salix*.

Financial Aid

The department offers aid in the form of research assistantships and fellowships and teaching assistantships as well as a number of fellowships awarded through the graduate school as a whole.

Cost of Study

In-state graduate tuition is $243 per credit hour in 2006–07. Out-of-state tuition is 2.5 times the in-state tuition rate ($607.50 per credit hour). Graduate students with at least a 25 percent appointment as a graduate assistant receive a tuition waiver. Fees vary from $441.62 (1 credit hour) to $987.30 (12 credit hours).

Living and Housing Costs

For married couples, students with families, and single graduate students, the University has 589 efficiency and one-, two-, and three-bedroom apartments that rent for $438 to $505 per month in 2006–07. Residence halls for single graduate students are also available, as are accessible residence hall rooms and apartments for students with disabilities.

Student Group

Total University enrollment exceeds 21,000, including more than 4,000 graduate students. Men and women come from all fifty states and more than 100 other countries. About 53 percent of the graduate students are women, 23 percent are international students, and 13 percent are members of American minority groups.

Location

SIUC is 350 miles south of Chicago and 100 miles southeast of St. Louis. Nestled in rolling hills bordered by the Ohio and Mississippi Rivers and enhanced by a mild climate, the area has state parks, national forests and wildlife refuges, and large lakes for outdoor recreation. Much of the area is a part of the 240,000-acre Shawnee National Forest. Cultural offerings include theater, opera, concerts, art exhibits, and cinema. Educational facilities for the families of students are excellent.

The University

Southern Illinois University Carbondale is a comprehensive public university with a variety of general and professional education programs. The University offers bachelor's and associate degrees, master's and doctoral degrees, the J.D. degree, and the M.D. degree. The University is fully accredited by the North Central Association of Colleges and Schools. The Graduate School has an essential role in the development and coordination of graduate instruction and research programs. The Graduate Council has academic responsibility for determining graduate standards, recommending new graduate programs and research centers, and establishing policies to facilitate the research effort. Southern Illinois University, Carbondale is a state-funded university founded in 1869.

Applying

Applicants to the doctoral degree program must have a plant sciences–related master's degree (or equivalent). Criteria for admission include GPA, GRE scores, letters of recommendation, transcripts, and availability of faculty, space, and facilities. To be admitted into the program a prospective student must have a minimum GPA of 3.25, and at least one PLB faculty member must be willing to serve as their Major Advisor. Co-advisers may be included if the student wishes to work in the Departments of Forestry or Plant, Soil, and General Agriculture. Deadlines for financial aid are February 15 for fall assistantships and November 15 for January graduate assistantships.

Correspondence and Information

Dr. Dale Vitt, Department Chair
Department of Plant Biology
Southern Illinois University Carbondale
Carbondale, Illinois 62901-4317

E-mail: dvitt@plant.siu.edu
Web site: http://www.plant.siu.edu

Southern Illinois University Carbondale

THE FACULTY AND THEIR RESEARCH

Sara G. Baer, Ph.D., Kansas State, 2001. Ecosystem ecology.
Loretta Battaglia, Ph.D., Georgia, 1998. Wetlands ecology; invasive species; community ecology.
John Bozzola, Ph.D., Southern Illinois Carbondale, 1975. Electron microscopy and cytology.
Stephen Ebbs, Ph.D., Cornell, 1997. Plant physiology; phytoremediation; ecotoxicology; mineral nutrition.
David Gibson, Ph.D., Wales, 1984. Plant population and community ecology; grassland ecology; multivariate methods; competition.
Daniel L. Nickrent, Ph.D., Miami (Ohio), 1984. Plant molecular phylogeny and evolution; biology of flowering parasitic plants.
Karen Renzaglia, Ph.D., Southern Illinois Carbondale, 1981. Biodiversity ultrastructure and phylogeny of early land plants.
Sedonia Sipes, Ph.D., Utah State, 2001. Pollination ecology; molecular systematics; plant evolution.
Dale Vitt, Ph.D., Michigan, 1970. Biogeochemistry and paleoecology of peatlands; biosystematics and taxonomy.
Andrew Wood, Ph.D., Purdue, 1994. Stress physiology.

UNIVERSITY OF MASSACHUSETTS AMHERST

Plant Biology Graduate Program

Program of Study	The Plant Biology Graduate Program at the University of Massachusetts Amherst (UMass Amherst) is an interdisciplinary program that trains master's and doctoral students whose research interests are focused on the study of plants. This distinctive program includes faculty members from the Departments of Biochemistry and Molecular Biology, Biology, Chemical Engineering, Food Science, Microbiology, Natural Resources Conservation, and Plant, Soil, and Insect Sciences. In addition, faculty members from Amherst, Mount Holyoke, and Smith Colleges also participate. Many aspects of modern research in the plant sciences are represented, including evolution, ecology, development, genetics, physiology, biochemistry, and cellular and molecular biology. The emphasis is on multidisciplinary research and involves approaches that range from practical to theoretical.

The program stresses research training. Most Ph.D. students complete two 1-semester rotations in the first year, prior to selecting a thesis adviser. Students must complete a core of three graduate-level courses designed to provide basic knowledge in plant biology, as well as advanced courses in specific areas of interest. Students are also expected to participate actively in journal clubs and seminars. To a large extent, specific course requirements are determined on an individual basis.

For Ph.D. candidates, an oral qualifying examination is given at the end of the first academic year upon the student's completion of the basic core courses in plant biology. An oral defense of an original research proposal constitutes the second qualification examination and is usually completed before the end of the second academic year. All students enrolled in the master's program are required to perform independent research and prepare a thesis. There is a final oral defense of the written thesis or Ph.D. dissertation. It is expected that the master's program should be completed within two years and the Ph.D. program within four to five years.

Research Facilities — Research facilities and equipment are housed in several buildings across campus. Major research capabilities include microbial and plant growth facilities, tissue culture facilities, radioisotope laboratories, facilities for bioinformatics, facilities for monoclonal antibody and recombinant DNA work, a DNA sequencing service, a high-field NMR facility, an imaging facility with confocal and electron microscopy, and a mass spectrometry facility. There are seven greenhouse facilities on campus, with a combined total of more than 17,000 square feet of bench, ground, and bed space. Approximately half of the greenhouse space is devoted to research projects, and the other half is used to maintain a general teaching collection representing more than 680 genera and 1,100 species for courses ranging from aquatic vascular plants to plant cell biology. The University of Massachusetts Herbarium is a regional resource with approximately 224,000 mounted vascular plants, algae, and bryophytes as well as a fruit and seed collection providing material for systematic studies and plant identification.

Financial Aid — All students are guaranteed financial support. In 2005–06, students received approximately $19,345 per year. Most students are appointed to research and teaching assistantships, while some are supported by fellowships. Competitive University fellowships are available, and the program benefits from an endowment fund in honor of Dr. Constantine Gilgut and offers one to two full graduate student fellowships per year specifically in plant biology. Students also receive a tuition waiver and health and dental insurance. All Ph.D. students are required to serve as teaching assistants in an instructive capacity for at least two semesters during their study. Doctoral degree students normally receive support for up to five years; master's students are supported for two years.

Cost of Study — For spring 2005, the in-state tuition was $110 per credit. For non-Massachusetts residents, tuition was $414 per credit. As stated above, tuition and curriculum fees are waived for students supported by teaching and research assistantships or fellowships.

Living and Housing Costs — Most graduate students live off campus. Graduate student housing is available in Prince/Crampton House, a graduate complex; the cost was approximately $2000 per semester in 2005–06. Family housing is also available. Off-campus studio and efficiency apartments cost approximately $500 per month; a two-bedroom apartment costs approximately $600–$700 per month. The support level of research and teaching assistantships allows for a modest but comfortable standard of living.

Student Group — Amherst is the main graduate campus of the University of Massachusetts System and has more than 5,400 graduate students in sixty-eight master's programs and fifty doctoral programs, including those that are part-time. There are currently 17 graduate students in the Plant Biology Graduate Program. All receive financial support. Most students are interested in careers in research and/or teaching, and many go on to postdoctoral appointments and to faculty positions at leading institutions in the United States and abroad.

Location — Amherst is situated in the scenic Connecticut River Valley in rural western Massachusetts. The rich academic and cultural environment of the University is enhanced by its close association with the four nearby colleges (Amherst, Hampshire, Mount Holyoke, and Smith) in the Five College Consortium. Northampton, noted for its cultural diversity, is nearby, and Boston and New York City are easily reached by car, train, or bus. The Pioneer Valley is noted for the variety and quality of its music, theater, art, and outdoor recreational activities.

The University and The Program — The University of Massachusetts Amherst is a land-grant research university and is the largest public university in the Northeast, with an enrollment of more than 24,000 students. The Plant Biology Graduate Program, founded in 1996, represents an integration of the diverse interests and expertise of 38 faculty members from UMass Amherst and Amherst, Mount Holyoke, and Smith Colleges, as well as research activities from seven academic areas within the Five College Consortium.

Applying — Applications are due January 7 for September admission and October 1 for January admission. Prospective students should have substantial undergraduate training in biology or plant science and a minimum overall cumulative average of 3.0. Applicants should submit the standard University of Massachusetts Application to the Graduate School along with a brief narrative that addresses work experience and scholarship. Scores from the GRE General Test are required. Students should also include reprints of papers published in journals and/or abstracts presented at meetings.

Correspondence and Information

For information:
Plant Biology Graduate Program
217 Morrill Science Center South
University of Massachusetts Amherst
611 North Pleasant Street
Amherst, Massachusetts 01003-9297

Phone: 413-577-3217
Fax: 413-545-3243
E-mail: pb@bio.umass.edu
Web site: http://www.bio.umass.edu/plantbio

To obtain and submit application materials:
Graduate Admissions Office
530 Goodell Building
University of Massachusetts Amherst
140 Hicks Way
Amherst, Massachusetts 01003-9332

Phone: 413-545-0721 or 0722
E-mail: gradadm@resgs.umass.edu
Web site: http://www.umass.edu/gradschool

University of Massachusetts Amherst

THE FACULTY AND THEIR RESEARCH

Lynn Adler, Assistant Professor of Plant, Soil, and Insect Sciences; Ph.D., California, Davis, 2000. Ecology and evolution of plant–insect interactions.

Peter Alpert, Associate Professor of Biology; Ph.D., Harvard, 1982. Ecology of invasive plants; resource capture and utilization by clonal plants.

Allen V. Barker, Professor of Plant, Soil, and Insect Sciences; Ph.D., Cornell, 1962. Plant nutrition; stress ethylene; compost utilization.

Tobias Baskin, Associate Professor of Biology; Ph.D., Stanford, 1986. Regulation of plant cell and organ expansion.

Robert Bernatzky, Associate Professor of Plant, Soil, and Insect Sciences; Ph.D., New Mexico State, 1985. Molecular markers for plant genome analyses.

Magdalena Bezanilla, Assistant Professor of Biology; Ph.D., Johns Hopkins, 2000. Role of the actin cytoskeleton in the morphogenesis and development of plant cells.

Thomas H. Boyle, Associate Professor of Plant, Soil, and Insect Sciences and Associate Director, Plant Biology Graduate Program; Ph.D., Maryland, 1986. Reproductive biology of crop plants; germplasm conservation; plant breeding.

Maura C. Cannon, Research Associate Professor of Biochemistry; Ph.D., University College (Ireland), 1972. Patterns of gene expression in the development of *Arabidopsis*.

Frank Caruso, Associate Professor of Plant, Soil, and Insect Sciences; Ph.D., Kentucky, 1978. Biological, chemical, and cultural control of cranberry diseases.

Alice Cheung, Professor of Biochemistry and Molecular Biology; Ph.D., Yale, 1982. Biochemistry and molecular analysis of sexual reproduction.

John Marshall Clark, Professor of Veterinary and Animal Sciences; Ph.D., Michigan State, 1981. Pesticide and environmental toxicology.

Daniel R. Cooley, Associate Professor of Plant, Soil, and Insect Sciences; Ph.D., Massachusetts, 1986. Ecology of plant diseases; integrated pest management; sustainable agriculture.

Lyle E. Craker, Professor of Plant, Soil, and Insect Sciences; Ph.D., Minnesota, 1967. Physiology and biochemistry of oils and essences in herbs.

Aaron Ellison, Senior Research Fellow, Harvard Forest, Harvard University; Ph.D., Brown, 1986. Community and forest ecology.

Amy Frary, Assistant Professor of Biological Sciences, Mount Holyoke College; Ph.D., Cornell, 1996. Basic and applied aspects of genome analysis.

Paul J. Godfrey, Associate Professor of Biology; Ph.D., Duke, 1969. Ecology of vegetation of stressed coastal environments.

Susan Han, Associate Professor of Plant, Soil, and Insect Sciences; Ph.D., California, Davis, 1988. Physiology of flowering and postharvest care of floricultural crops.

Robin Harrington, Associate Professor of Forestry; Ph.D., Wisconsin–Madison, 1987. Forest ecology: canopy processes and regeneration of native and invasive species.

Peter K. Hepler, Professor Emeritus of Biology; Ph.D., Wisconsin, 1964. Cell division and pollen tube growth; the role of calcium and the cytoskeleton.

Matthew Kelty, Associate Professor of Forestry; Ph.D., Yale, 1984. Forest ecology, management, and silviculture.

Guy Lanza, Professor of Microbiology; Ph.D., Virginia Tech, 1972. Bioremediation/restoration of damaged ecosystems using phytoremediation.

Frank X. Mangan, Assistant Professor of Plant, Soil, and Insect Sciences; Ph.D., Massachusetts Amherst, 1998. Vegetable production with an emphasis on soil fertility and ethnic crops.

William J. Manning, Professor of Plant, Soil, and Insect Sciences; Ph.D., Delaware, 1968. Assessing air pollution effects on plant growth and reproduction; bioindicators; plant growth in urban environments.

Michael Marcotrigiano, Professor of Biological Sciences and Director of the Botanic Garden at Smith College; Ph.D., Maryland, 1983. Plant development, micropropagation, and breeding.

Jill Miller, Assistant Professor of Biology at Amherst College; Ph.D., Arizona, 2000. Ecology and evolution of plant sexual systems; floral evolution; incompatibility systems; phenotypic plasticity.

Jennifer Normanly, Associate Professor of Biochemistry and Molecular Biology; Ph.D., Caltech, 1989. Auxin biosynthesis and high-throughput metabolic profiling.

Benjamin Normark, Assistant Professor of Plant, Soil, and Insect Sciences; Ph.D., Cornell, 1984. Evolution of alternative genetic systems in insects.

Klaus Nuesslein, Assistant Professor of Microbiology; Ph.D., Michigan State, 1998. Microbial ecology of terrestrial and aquatic environments; relating environmental influences to community structure and function.

Om Parkash (Dhankher), Assistant Professor of Plant, Soil, and Insect Sciences; Ph.D., Durham (England), 1998. Phytoremediation of heavy metals by genetically engineered plants.

William A. Patterson III, Professor of Forestry and Wildlife Management; Ph.D., Minnesota, 1978. Fire management; forest ecology; paleoecology.

Susan Roberts, Assistant Professor of Chemical Engineering; Ph.D., Cornell, 1998. Optimization of secondary metabolite accumulation in plant cell tissue cultures.

Danny Schnell, Professor of Biochemistry and Molecular Biology; California, Davis, 1987. Intracellular protein targeting and chloroplast biogenesis.

Karen B. Searcy, Lecturer in Biology; Ph.D., Massachusetts, 1984. Herbarium curator; taxonomy; plant ecology.

Kalidas Shetty, Professor of Food Science; Ph.D., Idaho, 1989. Plant and environmental biotechnology; plant secondary metabolism; tissue culture.

Justine Vanden Heuvel, Extension Assistant Professor of Plant, Soil, and Insect Sciences; Ph.D., Guelph, 2002. Carbon production and partitioning in cranberries and grapevines.

Elsbeth L. Walker, Associate Professor of Biology and Director, Plant Biology Graduate Program; Ph.D., Rockefeller, 1990. Molecular genetic approaches to studying mechanisms of metal ion homeostasis in plants.

Carolyn Wetzel, Assistant Professor of Biological Sciences at Smith College; Ph.D., Cornell, 1992. Plant chloroplast development and function; plant cell development; plant physiology.

Robert Wick, Professor of Plant, Soil, and Insect Sciences; Ph.D., Virginia Tech, 1981. Bacterial and fungal diseases of vegetable and greenhouse crops; nematode diseases of turf; plant disease diagnostics.

Christopher L. Woodcock, Professor of Biology; Ph.D., University College (London), 1966. Nuclear structure and function, with special emphasis on chromatin architecture and dynamics.

YALE UNIVERSITY

Department of Molecular, Cellular, and Developmental Biology
Graduate Studies in Plant Sciences

Program of Study

The Program in Plant Sciences offers an interdisciplinary course of study leading to the Ph.D. degree. Research topics include the molecular genetics of plant development, the physiology and molecular biology of hormone action, transposable elements, sex determination, plant-microbe interactions, signal transduction, and the evolution of plants. The Program in Plant Sciences is based in the Department of Molecular, Cellular, and Developmental Biology (MCDB), but it involves plant biologists and facilities from all areas of biology, including the Department of Ecology and Evolutionary Biology, the School of Forestry and Environmental Studies, and the Department of Geology (paleobiology), as well as collaborative interactions with the nearby Connecticut Agriculture Experiment Station. Students are encouraged to develop, with the assistance of faculty advisers, a specific program of course work, independent reading, and research that fits their individual backgrounds and interests.

Graduate study in the Program in Plant Sciences is a full-time, four-year doctoral program. The first year of study generally includes three laboratory rotations and course work aimed at establishing a foundation for dissertation research. By the second year, students decide upon a research topic and select a thesis adviser. The specific academic requirements for the Ph.D. are a grade of honors in two 1-semester courses, completion of the research rotations, and satisfactory progress as determined by the faculty of the Department of Molecular, Cellular, and Developmental Biology. During the second year, students are expected to present and defend a dissertation prospectus as well as demonstrate satisfactory performance on an oral examination on background material. Upon submission and defense of an acceptable dissertation, the Ph.D. degree is awarded through MCDB. The most extensive information about research and training opportunities is available on the World Wide Web site listed below.

Research Facilities

The Program in Plant Sciences, as part of MCDB, is housed in the Osborn Memorial Laboratory and the Kline Biology Tower. Special facilities available include a herbarium, plant growth rooms, reach-in growth chambers, garden plots and greenhouses, excellent botanical and paleobiological collections, equipment for molecular and biochemical studies, electron and confocal microscopes, and facilities for tissue culture, monoclonal antibody production, oligonucleotide and oligopeptide synthesis, and microarray analysis. Other facilities include the Marsh Botanical Gardens, the Yale Natural Preserve, the Yale Marine Biology Laboratory, the Yale Forest, and, by arrangement, facilities at the Connecticut Agriculture Experiment Station.

The Kline Science Library, the Ornithology Library at the Peabody Museum, the Kline Geology Library (paleobiology), and the School of Forestry and Environmental Studies Library (forest biology) contain approximately 1 million volumes. All of these libraries are located in or close to MCDB Department buildings.

Financial Aid

Students admitted to the Ph.D. program normally receive full tuition plus a twelve-month stipend of $27,000 for living expenses.

Cost of Study

The cost of study is covered for all students as described above.

Living and Housing Costs

The cost of living is comparable to that in other large northeastern American cities. The University has some dormitory facilities for single students and apartments for married students. Many students live in adjacent off-campus areas. Information on University housing and a list of private off-campus accommodations are available from the Yale Housing Department, 155 Whitney Avenue, Third Floor, New Haven, Connecticut 06520.

Student Group

In 2005–06, there were 83 graduate students in residence, all working toward the Ph.D.; 48 of these students were women. Each entering class contains approximately 40 students, some of whom specialize in plant sciences. In 2005–06, thirteen Ph.D. degrees were awarded to Department students.

Location

Facilities of the Department of Molecular, Cellular, and Developmental Biology are situated on a grassy hill on the Yale College campus, adjacent to botanical gardens and a large residential neighborhood that serves as home to many graduate students and other members of the academic community. New Haven is situated on the Long Island Sound. New York City is 90 minutes and Boston is 3 hours away by train. In addition, major highways, a local airport, and a national bus line facilitate transportation. New Haven's central location makes it possible for many distinguished scholars from the United States and abroad to visit and allows the University to be an integral part of the worldwide community of scholars.

The creative and performing arts are very active in New Haven. More than 175 musical events take place during the year. Theaters include the Yale Repertory Theater, the Long Wharf Theater, and the Shubert Theater.

The University

The Yale University community is a large and diverse one, consisting of about 5,000 undergraduate students, 5,000 graduate and professional students, 1,500 postdoctoral fellows, and 2,000 faculty members.

Within its urban setting, Yale University has unusually extensive facilities for athletic and recreational activities. The Payne Whitney Gymnasium and Ingalls Skating Rink are located on the central campus. Away from the campus, there are recreational facilities at Yale Bowl, an eighteen-hole golf course, the Yale Sailing Center, and the Yale Outdoor Recreation Center.

Applying

Admission to the Program in Plant Sciences is administered by the Molecular Cell Biology, Genetics, and Development Track of the Yale Program in Biological and Biomedical Sciences (see separate listing for more information). Application forms and information about graduate programs and financial aid can be requested from the Molecular Cell Biology, Genetics, and Development Track/BBS at the address below. Students should indicate their interest in the Program in Plant Sciences on their application. Completed forms as well as GRE General and Subject Test scores, official grade transcripts, and three letters of recommendation must be submitted by December 8, 2006.

Correspondence and Information

Anne Scott, Graduate Registrar
Molecular Cell Biology, Genetics, and Development Track/BBS
Yale University
P.O. Box 208103
New Haven, Connecticut 06520-8103
Phone: 203-432-3538
E-mail: anne.scott@yale.edu
Web site: http://www.biology.yale.edu
      http://info.med.yale.edu/bbs

Yale University

THE FACULTY AND THEIR RESEARCH

Department of Molecular, Cellular, and Developmental Biology Faculty Members with Interests in the Plant Sciences
Stephen L. Dellaporta. Sex determination and cell death in plants.
Xing-Wang Deng. Molecular genetic and genomic analysis of light signaling and development mechanism.
Vivian F. Irish. Developmental genetics of flowering in *Arabidopsis;* evolution of floral homeotic genes.
S. P. Dinesh-Kumar. Molecular and genetic analysis of host-pathogen interactions.
Timothy M. Nelson. Cellular differentiation in leaf development; patterning of venation.
Joel L. Rosenbaum. Cell organelle assembly; IFT and flagellar assembly; sensory function of cilia/flagella.

Additional Yale Faculty Members with Interests in the Plant Sciences
Gary Brudvig (Chemistry). Photosynthetic water oxidation; functions of carotenoids in photosynthesis; application of electron paramagnetic resonance
 spectroscopy to the study of metalloproteins.
Michael Donoghue (Ecology and Evolutionary Biology, Peabody Museum). Plant phylogeny, the evolution of plant diversity, and biogeography.
Leo J. Hickey (Geology). Angiosperm origin and paleoecology; leaf architecture of the dicotyledons.
Oswald Schmitz (Ecology and Evolutionary Biology). How diversity and productivity of plant species is influenced by interactions among top predators,
 herbivores, and plants in ecological food webs.
Melinda Smith (Ecology and Evolutionary Biology). How patterns, determinants, and dynamics of plant diversity and species abundance relate to the
 functioning of ecosystems.
Hongyu Zhao (Genetics). Development of computational methods to dissect biological pathways using diverse types of high-throughput genomics and
 proteomics data.
Steven W. Zucker (Computer Science). Computational modeling of leaf venation, growth, and shape.

Additional Department of Molecular, Cellular, and Developmental Biology Faculty Members
Sidney Altman. Role in vivo of an enzyme with a catalytic RNA subunit; mechanism of action of that enzyme.
*Kim Bottomly. Factors regulating the selective induction of Th1 and Th2 effector T cells.
Ronald R. Breaker. The study of riboswitches and engineering new RNA and DNA enzymes by in vitro evolution.
John R. Carlson. Function and development of the *Drosophila* olfactory system.
*Lynn Cooley. Molecular genetics of *Drosophila* oogenesis; control of oocyte growth; ring canals.
Craig M. Crews. Exploration and control of signal transduction pathways using chemical probes.
Paul Forscher. Molecular mechanisms of axon guidance: Cytoskeletal protein dynamics and related signal transduction.
Martín García-Castro. Origin, induction, and potential of neural crest stem cells.
*Mark W. Hochstrasser. Dynamics of cell differentiation; ubiquitin-proteasome system.
Scott A. Holley. Molecular, genetic, and embryological analysis of segmentation in the zebrafish.
Christine Jacobs-Wagner. Cell shape and cell polarity in *Caulobacter crescentus.*
Douglas R. Kankel. Nervous system development and function in *Drosophila melanogaster.*
*Michael Kashgarian. Na, KATPase expression: epithelial cell polarity; heat shock protein functions.
Haig S. Keshishian. Factors governing the formation of synaptic connections during development.
Mark S. Mooseker. Molecular mechanisms of motility in eukaryotic cells.
*Jon S. Morrow. Molecular basis of polarized membrane and cytoskeletal assembly.
L. Nicholas Ornston. Evolution of metabolic pathways; bacterial physiology.
Thomas D. Pollard. Molecular mechanisms of actin-based cellular movements.
Shirleen Roeder. Meiosis in yeast: homolog pairing, chromosome synapsis, and cell-cycle checkpoints.
*Alanna Schepartz. Chemical Biology: protein design and evolution; molecular mechanism of transcriptional accessory factors; proline-rich motifs in cell
 signaling.
*Steven S. Segal. Cellular and molecular signaling pathways that underlie blood flow control in the microcirculation.
Frank Slack. The role of microRNAs in development, aging, and cancer.
Michael Snyder. Genomics and proteomics in yeast and humans; regulatory networks.
Elke Stein. Axon guidance and synapse formation in the mammalian nervous system.
*Hugh S. Taylor. The molecular regulation of reproductive track development and function.
David G. Wells. Cellular and molecular mechanisms regulating synaptic plasticity in mammalian CNS.
Robert J. Wyman. Molecular biology and neurophysiology of gap junctions; genetic control of neural circuit development.
Weimin Zhong. Regulation of neural stem cells and development of the mammalian neocortex.

**Joint appointees with primary appointment in another department.*

Section 6
Cell, Molecular, and Structural Biology

This section contains a directory of institutions offering graduate work in cell, molecular, and structural biology, followed by in-depth entries submitted by institutions that chose to prepare detailed program descriptions. Additional information about programs listed in the directory but not augmented by an in-depth entry may be obtained by writing directly to the dean of a graduate school or chair of a department at the address given in the directory.

For programs offering related work, see also in this book Anatomy; Biochemistry; Biological and Biomedical Sciences; Biophysics; Botany and Plant Biology; Genetics, Developmental Biology, and Reproductive Biology; Microbiological Sciences; Pathology and Pathobiology; Pharmacology and Toxicology; and Physiology. In Book 4, see Chemistry; in Book 5, see Agricultural Engineering and Bioengineering and Biomedical Engineering and Biotechnology; and in Book 6, see Pharmacy and Pharmaceutical Sciences and Veterinary Medicine and Sciences.

CONTENTS

Cancer Biology/Oncology

Brown University, Graduate School, Division of Biology and Medicine, Program in Pathology and Laboratory Medicine, Providence, RI 02912. Offers biology (PhD); cancer biology (PhD); immunology and infection (PhD); medical science (PhD); pathobiology (Sc M); toxicology and environmental pathology (PhD). Terminal master's awarded for partial completion of doctoral program. *Degree requirements:* For doctorate, thesis/dissertation, preliminary exam. *Entrance requirements:* For master's and doctorate, GRE General Test, GRE Subject Test. Additional exam requirements/recommendations for international students: Required—TOEFL. Electronic applications accepted. *Faculty research:* Environmental pathology, carcinogenesis, immunopathology, signal transduction, innate immunity.

Drexel University, College of Medicine, Biomedical Graduate Programs, Program in Radiation Oncology, Philadelphia, PA 19104-2875. Offers radiation (MS); radiation biology (MS); radiation physics (PhD); radiation science (PhD); radiopharmaceutical science (MS, PhD). Part-time programs available. Terminal master's awarded for partial completion of doctoral program. *Degree requirements:* For master's, thesis, comprehensive exam; for doctorate, one foreign language, thesis/dissertation, qualifying exam. *Entrance requirements:* For master's, GRE General Test, minimum GPA of 2.75; for doctorate, GRE General Test, minimum GPA of 3.0. Additional exam requirements/recommendations for international students: Required—TOEFL. Electronic applications accepted. *Faculty research:* Improved cancer therapy by linear accelerators and internal, sealed radiation sources; algorithms for improved superminicomputer-assisted radiation therapy simulation and tumor imaging; molecular and cellular mechanisms of radiation damage.

Duke University, Graduate School, Department of Molecular Cancer Biology, Durham, NC 27710. Offers PhD. *Faculty:* 33 full-time. *Students:* 52 full-time (32 women); includes 3 minority (1 African American, 2 Hispanic Americans), 12 international. 63 applicants, 13% accepted, 4 enrolled. In 2005, 6 doctorates awarded. *Degree requirements:* For doctorate, thesis/dissertation. *Entrance requirements:* For doctorate, GRE General Test, GRE Subject Test (recommended). Additional exam requirements/recommendations for international students: Required—IELT (preferred) or TOEFL. *Application deadline:* For fall admission, 12/31 for domestic students, 12/31 for international students. Application fee: $75. Electronic applications accepted. *Financial support:* Fellowships, research assistantships available. Financial award application deadline: 12/31. *Unit head:* Ann Marie Pendergast, Director of Graduate Studies, 919-681-8086, Fax: 919-681-7767.

Georgetown University, Graduate School of Arts and Sciences and School of Medicine, Tumor Biology Training Program, Washington, DC 20057.

Mayo Graduate School, Graduate Programs in Biomedical Sciences, Program in Tumor Biology, Rochester, MN 55905. Offers PhD. *Degree requirements:* For doctorate, oral defense of dissertation, qualifying oral and written exam. *Entrance requirements:* For doctorate, GRE, 1 year of chemistry, biology, calculus, and physics. Additional exam requirements/recommendations for international students: Required—TOEFL. Electronic applications accepted.

McMaster University, Faculty of Health Sciences and School of Graduate Studies, Program in Medical Sciences, Molecular Biology, Genetics, and Cancer Area, Hamilton, ON L8S 4M2, Canada. Offers M Sc, PhD. *Students:* 11 full-time, 2 part-time. In 2005, 2 master's, 2 doctorates awarded. *Degree requirements:* For master's, thesis/dissertation; for doctorate, thesis/dissertation, comprehensive exam. *Entrance requirements:* For master's, honors B Sc, B+ average in related field; for doctorate, M Sc, minimum B+ average, students with proven research experience and an A average may be admitted with a B Sc degree. Additional exam requirements/recommendations for international students: Required—TOEFL (minimum score 580 paper-based; 237 computer-based). *Application deadline:* For fall admission, 9/30 for domestic students. For winter admission, 3/31 for domestic students. Applications are processed on a rolling basis. Application fee: $85. *Financial support:* Teaching assistantships available. *Unit head:* , Dr. John Waye, Coordinator, 905-525-9140 Ext. 76273. *Application contact:* Dr. Carl Richards, Associate Dean, 905-525-9140 Ext. 22983, Fax: 905-546-1129.

New York University, Graduate School of Arts and Science, Department of Biology, New York, NY 10012-1019. Offers biology (PhD); biomedical journalism (MS); cancer and molecular biology (PhD); computational biology (PhD); computers in biological research (MS); developmental genetics (PhD); general biology (MS); immunology and microbiology (PhD); molecular genetics (PhD); neurobiology (PhD); oral biology (MS); plant biology (PhD); recombinant DNA technology (MS). Part-time programs available. *Faculty:* 24 full-time (5 women), 8 part-time/adjunct. *Students:* 104 full-time (52 women), 41 part-time (23 women); includes 28 minority (2 African Americans, 20 Asian Americans or Pacific Islanders, 6 Hispanic Americans), 47 international. Average age 27. 349 applicants, 56% accepted, 39 enrolled. In 2005, 59 master's, 4 doctorates awarded. Terminal master's awarded for partial completion of doctoral program. *Degree requirements:* For master's, thesis or alternative, qualifying paper; for doctorate, thesis/dissertation, comprehensive exam. *Entrance requirements:* For master's, GRE General Test; for doctorate, GRE General Test, GRE Subject Test. Additional exam requirements/recommendations for international students: Required—TOEFL. *Application deadline:* For fall admission, 1/4 for domestic students. Application fee: $80. *Financial support:* Fellowships with tuition reimbursements, research assistantships with tuition reimbursements, teaching assistantships with tuition reimbursements, career-related internships or fieldwork, Federal Work-Study, institutionally sponsored loans, scholarships/grants, health care benefits, and unspecified assistantships available. Financial award application deadline: 1/4; financial award applicants required to submit FAFSA. *Faculty research:* Genomics, molecular and cell biology, development and molecular genetics, molecular evolution of plants and animals. *Unit head:* Gloria Coruzzi, Chairman, 212-998-8200, Fax: 212-995-4015, E-mail: biology@nyu.edu. *Application contact:* Stephen Small, Director of Graduate Studies, 212-998-8200, Fax: 212-995-4015, E-mail: biology@nyu.edu.

New York University, School of Medicine and Graduate School of Arts and Science, Sackler Institute of Graduate Biomedical Sciences, Graduate Programs in Molecular Oncology and Immunology, New York, NY 10012-1019. Offers immunology (PhD); molecular oncology (PhD). *Degree requirements:* For doctorate, one foreign language, thesis/dissertation, qualifying exam. *Entrance requirements:* For doctorate, GRE General Test, GRE Subject Test. Additional exam requirements/recommendations for international students: Required—TOEFL. Electronic applications accepted.

See Close-Up on page 891.

Northwestern University, Northwestern University Feinberg School of Medicine and Interdepartmental Degree Programs, Integrated Graduate Programs in the Life Sciences, Chicago, IL 60611. Offers cancer biology (PhD); cell biology (PhD); developmental biology (PhD); evolutionary biology (PhD); immunology and microbial pathogenesis (PhD); molecular biology and genetics (PhD); neurobiology (PhD); pharmacology and toxicology (PhD); structural biology and biochemistry (PhD). *Degree requirements:* For doctorate, thesis/dissertation, written and oral qualifying exams, comprehensive exam. *Entrance requirements:* For doctorate, GRE General Test. Additional exam requirements/recommendations for international students: Required—TOEFL (minimum score 600 paper-based; 250 computer-based). Electronic applications accepted.

See Close-Up on page 189.

Stanford University, School of Medicine, Graduate Programs in Medicine, Program in Cancer Biology, Stanford, CA 94305-9991. Offers PhD. *Degree requirements:* For doctorate, thesis/dissertation, qualifying examination. *Entrance requirements:* For doctorate, GRE General Test, GRE Subject Test. Additional exam requirements/recommendations for international students: Required—TOEFL. Electronic applications accepted.

State University of New York at Buffalo, Graduate School, Graduate Programs in Cancer Research and Biomedical Sciences at Roswell Park Cancer Institute, Department of Molecular Pharmacology and Cancer Therapeutics at Roswell Park Cancer Institute, Program in Molecular Pharmacology and Cancer Therapeutics, Buffalo, NY 14263. Offers PhD. *Faculty:* 28 full-time (8 women). *Students:* 25 full-time (16 women); includes 10 minority (2 African Americans, 8 Asian Americans or Pacific Islanders), 3 international. Average age 26. 26 applicants, 35% accepted. In 2005, 7 degrees awarded. *Degree requirements:* For doctorate, thesis/dissertation, departmental qualifying exam, grant proposal. *Entrance requirements:* For doctorate, GRE General Test (recommended). Additional exam requirements/recommendations for international students: Required—TOEFL; Recommended—TWE. *Application deadline:* For fall admission, 2/1 for domestic students. Applications are processed on a rolling basis. Application fee: $35. Electronic applications accepted. *Financial support:* In 2005–06, 25 students received support, including 6 fellowships with full tuition reimbursements available (averaging $20,772 per year), 19 research assistantships with full tuition reimbursements available (averaging $20,772 per year) *Faculty research:* Molecular pharmacology, cancer cell biology, molecular biology, biochemistry, chemotherapy. Total annual research expenditures: $6.5 million. *Unit head:* Dr. Enrico Mihich, Chair, 716-845-8226, Fax: 716-845-8857. *Application contact:* Dr. David W. Goodrich, Director of Graduate Studies, 716-845-4506, Fax: 716-845-8857, E-mail: david.goodrich@roswellpark.org.

See Close-Up on page 1181.

Université de Montréal, Faculty of Medicine and Faculty of Graduate Studies, Graduate Programs in Medicine, Program in Specialized Studies, Montréal, QC H3C 3J7, Canada. Offers anesthesia (DESS); diagnostic radiology (DESS); family medicine (DESS); medical biochemistry (DESS); medical genetics (DESS); medicine (DESS); microbiology and infectious diseases (DESS); nuclear medicine (DESS); obstetrics and gynecology (DESS); ophthalmology (DESS); pediatrics (DESS); psychiatry (DESS); radiology-oncology (DESS); surgery (DESS). *Faculty:* 159 full-time (37 women), 345 part-time/adjunct (102 women). *Entrance requirements:* For degree, proficiency in French. *Application deadline:* For fall and spring admission, 2/1. For winter admission, 11/1 for domestic students. Application fee: $30. Electronic applications accepted. *Unit head:* Renée Roy, Vice Dean, 514-343-7798.

Université Laval, Faculty of Medicine, Post-Professional Programs in Medical Studies, Québec, QC G1K 7P4, Canada. Offers anatomy–pathology (DESS); anesthesia–resuscitation (DESS); cardiology (DESS); care of older people (Diploma); clinical research (DESS); community health (DESS); dermatology (DESS); diagnostic radiology (DESS); emergency medicine (Diploma); family medicine (DESS); general surgery (DESS); geriatrics (DESS); hematology (DESS); internal medicine (DESS); maternal and fetal medicine (Diploma); medical biochemistry (DESS); medical microbiology and infectious diseases (DESS); medical oncology (DESS); nephrology (DESS); neurology (DESS); neurosurgery (DESS); obstetrics and gynecology (DESS); ophthalmology (DESS); orthopedic surgery (DESS); oto-rhino-laryngology (DESS); palliative medicine (Diploma); pediatrics (DESS); plastic surgery (DESS); psychiatry (DESS); pulmonary medicine (DESS); radiology–oncology (DESS); thoracic surgery (DESS); urology (DESS). *Degree requirements:* For other advanced degree, comprehensive exam. *Entrance requirements:* For degree, knowledge of French. Electronic applications accepted.

University of Alberta, Faculty of Medicine and Dentistry and Faculty of Graduate Studies and Research, Graduate Programs in Medicine, Department of Oncology, Edmonton, AB T6G 2E1, Canada. Offers medical sciences (M Sc, PhD). Terminal master's awarded for partial completion of doctoral program. *Degree requirements:* For master's and doctorate, thesis/dissertation. *Entrance requirements:* For master's and doctorate, minimum GPA of 7.0 on a 9.0 scale. Additional exam requirements/recommendations for international students: Required—TOEFL. Electronic applications accepted. Tuition and fees charges are reported in Canadian dollars. *Expenses:* Tuition, state resident: part-time $562 Canadian dollars per term. Tuition, nonresident: full-time $3,375 Canadian dollars. Required fees: $573 Canadian dollars; $84 Canadian dollars per term. *Faculty research:* Molecular oncology, radiation oncology, medical physics, cellular oncology.

The University of Arizona, Graduate College, Graduate Interdisciplinary Programs, Graduate Interdisciplinary Program in Cancer Biology, Tucson, AZ 85721. Offers PhD. *Degree requirements:* For doctorate, thesis/dissertation, comprehensive exam. *Entrance requirements:* For doctorate, GRE General Test, 3 letters of recommendation. Additional exam requirements/recommendations for international students: Required—TOEFL. *Faculty research:* Differential gene expression, DNA-protein cross linking, cell growth regulation steroid, receptor proteins.

University of Calgary, Faculty of Medicine and Faculty of Graduate Studies, Department of Medical Science, Calgary, AB T2N 1N4, Canada. Offers cancer biology (M Sc, PhD); immunology (M Sc, PhD); joint injury and arthritis research (M Sc, PhD); medical education (M Sc, PhD); medical science (M Sc, PhD); mountain medicine and high altitude physiology (M Sc). *Faculty:* 114 full-time (17 women), 5 part-time/adjunct (0 women). *Students:* 121 full-time (69 women), 1 part-time. 68 applicants, 29% accepted, 19 enrolled. In 2005, 19 master's, 7 doctorates awarded. *Median time to degree:* Of those who began their doctoral program in fall 1997, 100% received their degree in 8 years or less. *Degree requirements:* For master's, thesis; for doctorate, thesis/dissertation, candidacy exam. *Entrance requirements:* For master's, minimum undergraduate GPA of 3.2; for doctorate, minimum graduate GPA of 3.2. Additional exam requirements/recommendations for international students: Required—TOEFL (minimum score 600 paper-based; 250 computer-based). *Application deadline:* For fall admission, 6/15 priority date for domestic students, 5/15 priority date for international students. For winter admission, 10/15 for domestic students; for spring admission, 3/15 for domestic students. Applications are processed on a rolling basis. Application fee: $100 ($130 for international students). Electronic applications accepted. *Financial support:* In 2005–06, 30 students received support, including 22 research assistantships, 2 teaching assistantships; scholarships/grants and tuition waivers (partial) also available. *Faculty research:* Cancer biology, immunology, joint injury and arthritis, medical education, population genomics. *Unit head:* Dr. Francine Smith, Graduate Coordinator, 403-220-6852, Fax: 403-210-8109, E-mail: fsmith@ucalgary.ca. *Application contact:* Christine Szefer, Graduate Program Administrator, 403-220-6852, Fax: 403-210-8109, E-mail: cszefer@ucalgary.ca.

University of California, San Diego, Graduate Studies and Research, Division of Biology, Program in Immunology, Virology, and Cancer Biology, La Jolla, CA 92093. Offers PhD. Offered in association with the Salk Institute. *Degree requirements:* For doctorate, thesis/dissertation, qualifying exam. Electronic applications accepted.

University of California, San Diego, School of Medicine and Graduate Studies and Research, Molecular Pathology Program, La Jolla, CA 92093. Offers bioinformatics (PhD); cancer biology/oncology (PhD); cardiovascular sciences and disease (PhD); microbiology (PhD); molecular pathology (PhD); neurological disease (PhD); stem cell and developmental biology (PhD); structural biology/drug design (PhD). *Entrance requirements:* For doctorate, GRE General Test, GRE Subject Test. Additional exam requirements/recommendations for international students: Required—TOEFL. Electronic applications accepted.

See Close-Up on page 1119.

University of Chicago, Division of the Biological Sciences, Biomedical Sciences: Cancer, Immunology, Nutrition, Pathology, and Microbiology, Committee on Cancer Biology, Chicago, IL 60637-1513. Offers PhD. *Faculty:* 52 full-time (14 women). *Students:* 34 full-time (19 women); includes 5 minority (2 African Americans, 3 Asian Americans or Pacific Islanders), 4 international. Average age 27. In 2005, 3 degrees awarded. *Degree requirements:* For doctorate, thesis/dissertation, registration. *Entrance requirements:* For doctorate, GRE General Test. Additional exam requirements/recommendations for international students: Required—TOEFL. *Application deadline:* For fall admission, 12/28 priority date for domestic students, 12/28

Cancer Biology/Oncology

University of Chicago (continued)
priority date for international students. Application fee: $55. Electronic applications accepted. *Financial support:* In 2005–06, 24 students received support, including fellowships with full tuition reimbursements available (averaging $26,301 per year), research assistantships with full tuition reimbursements available (averaging $26,301 per year); institutionally sponsored loans and traineeships also available. *Faculty research:* Cancer genetics, apoptosis, signal transduction, tumor biology, cell cycle regulation. Total annual research expenditures: $58 million. *Unit head:* Dr. Geoffrey Greene, Chair, 773-702-6964, E-mail: ggreene@uchicago.edu. *Application contact:* Tracie DeMack, Graduate Administrative Director, 773-834-3899, Fax: 773-702-4634, E-mail: tdemack@huggins.bsd.uchicago.edu.

University of Colorado at Denver and Health Sciences Center, Graduate School, Program in Biomedical Sciences, Department of Pathology, Denver, CO 80262. Offers cancer biology (PhD). *Degree requirements:* For doctorate, thesis/dissertation, 3 laboratory rotations, comprehensive exam. *Entrance requirements:* For doctorate, GRE General Test, interview, minimum undergraduate GPA of 3.0. Additional exam requirements/recommendations for international students: Required—TOEFL (minimum score 550 paper-based; 213 computer-based). *Application deadline:* For fall admission, 2/1 for domestic students. Application fee: $50. Electronic applications accepted. *Expenses:* Tuition, state resident: full-time $11,730. Tuition, nonresident: full-time $22,980. Tuition and fees vary according to degree level and program. *Financial support:* Fellowships, research assistantships, teaching assistantships, Federal Work-Study and institutionally sponsored loans available. Support available to part-time students. Financial award application deadline: 3/1; financial award applicants required to submit FAFSA. *Faculty research:* Signal transduction by tyrosine kinases, estrogen and progesterone receptors in breast cancer, mechanism of mitochondrial DNA replication in the mammalian cell. *Unit head:* Dr. Robert Evans, Director, 303-315-5436, E-mail: robert.evans@uchsc.edu.

University of Delaware, College of Arts and Sciences, Department of Biological Sciences, Newark, DE 19716. Offers biotechnology (MS); cancer biology (MS, PhD); cell and extracellular matrix biology (MS, PhD); cell and systems physiology (MS, PhD); developmental biology (MS, PhD); ecology and evolution (MS, PhD); microbiology (MS, PhD); molecular biology and genetics (MS, PhD). *Faculty:* 39 full-time (11 women). *Students:* 64 full-time (46 women), 2 part-time; includes 7 minority (4 African Americans, 2 Asian Americans or Pacific Islanders, 1 Hispanic American), 18 international. Average age 26. 113 applicants, 31% accepted, 19 enrolled. In 2005, 5 master's, 4 doctorates awarded. Terminal master's awarded for partial completion of doctoral program. *Median time to degree:* Of those who began their doctoral program in fall 1997, 100% received their degree in 8 years or less. *Degree requirements:* For master's, thesis/dissertation, preliminary exam; for doctorate, thesis/dissertation, preliminary exam, comprehensive exam. *Entrance requirements:* For master's and doctorate, GRE General Test. Additional exam requirements/recommendations for international students: Required—TOEFL (minimum score 600 paper-based; 250 computer-based); Recommended—TWE, TSE. *Application deadline:* For fall admission, 4/15 for domestic students, 1/15 for international students; for spring admission, 10/1 for domestic students. Applications are processed on a rolling basis. Application fee: $60. Electronic applications accepted. *Financial support:* In 2005–06, 26 students received support, including fellowships with full tuition reimbursements available (averaging $19,000 per year), 19 research assistantships with full tuition reimbursements available (averaging $19,000 per year), 26 teaching assistantships with full tuition reimbursements available (averaging $19,000 per year); tuition waivers (partial) also available. Financial award application deadline: 4/15. *Faculty research:* Microorganisms, bone, cancer metastasis, developmental biology, cell biology, DNA. Total annual research expenditures: $8.3 million. *Unit head:* Dr. Daniel D. Carson, Chair, 302-831-6977, Fax: 302-831-2281, E-mail: dcarson@udel.edu. *Application contact:* Dr. Melinda K. Duncan, Graduate Coordinator, 302-831-1841, Fax: 302-831-2281, E-mail: danders@udel.edu.

University of Massachusetts Worcester, Graduate School of Biomedical Sciences, Department of Cancer Biology, Worcester, MA 01655-0115. Offers PhD. *Faculty:* 13 full-time (6 women). *Entrance requirements:* For doctorate, GRE General Test. Additional exam requirements/recommendations for international students: Required—TOEFL (minimum score 600 paper-based; 250 computer-based). *Application deadline:* For fall admission, 12/15 for domestic students, 12/15 for international students. Application fee: $25 ($50 for international students). Electronic applications accepted. *Expenses:* Tuition, state resident: full-time $2,640. Tuition, nonresident: full-time $9,856. Required fees: $5,685. *Financial support:* In 2005–06, research assistantships (averaging $25,235 per year) *Unit head:* Dr. Dario Altieri, Chair, 508-856-5775. *Application contact:* Michael Cole, Director of Admissions and Recruitment, 508-856-4779, Fax: 508-856-3659, E-mail: michael.cole@umassmed.edu.

University of Medicine and Dentistry of New Jersey, Training Program in the Molecular and Immunopathologic Mechanisms of Cancer, Newark, NJ 07107-1709.

University of Miami, Graduate School, Miller School of Medicine, Program in Cancer Biology, Coral Gables, FL 33124. Offers PhD.

See Close-Up on page 623.

University of Nebraska Medical Center, Graduate Studies, Program in Cancer Research, Omaha, NE 68198. Offers MS, PhD. *Faculty:* 33 full-time (8 women). *Students:* 20 full-time (10 women); includes 1 minority (Hispanic American), 1 international. 15 applicants, 60% accepted, 5 enrolled. Terminal master's awarded for partial completion of doctoral program. *Degree requirements:* For master's and doctorate, thesis/dissertation, comprehensive exam, registration. *Entrance requirements:* For doctorate, GRE, 3 letters of reference; course work in chemistry, biology, physics and mathematics. Additional exam requirements/recommendations for international students: Required—TOEFL (minimum score 550 paper-based; 213 computer-based). *Application deadline:* For fall admission, 6/1 for domestic students, 4/1 for international students; for spring admission, 10/1 for domestic students, 8/1 for international students. Applications are processed on a rolling basis. Application fee: $45. Electronic applications accepted. *Expenses:* Tuition, area resident: Part-time $200 per hour. Tuition, nonresident: part-time $538 per hour. Required fees: $308; $59 per term. *Financial support:* In 2005–06, 7 fellowships with tuition reimbursements (averaging $21,000 per year), 13 research assistantships with tuition reimbursements (averaging $21,000 per year) were awarded; health care benefits also available. *Faculty research:* DNA repair, tumor immunology, signal transduction, structural biology, gene expression. Total annual research expenditures: $18.1 million. *Unit head:* , Dr. Joyce Solheim, Graduate Committee Chair, 402-559-4539, Fax: 402-559-8270, E-mail: jsolheim@unmc.edu. *Application contact:* Matthew D. Winfrey, CRGP Coordinator, 402-559-4022, Fax: 402-559-4651, E-mail: winfreym@unmc.edu.

The University of North Carolina at Chapel Hill, Cancer Cell Biology Predoctoral Training Program, Chapel Hill, NC 27599.

University of Pennsylvania, School of Medicine, Biomedical Graduate Studies, Graduate Group in Cell and Molecular Biology, Program in Cell Growth and Cancer, Philadelphia, PA 19104. Offers PhD, MD/PhD, VMD/PhD. *Degree requirements:* For doctorate, thesis/dissertation. *Entrance requirements:* For doctorate, GRE General Test. Additional exam requirements/recommendations for international students: Required—TOEFL. *Application deadline:* For fall admission, 12/15 priority date for domestic students, 12/1 priority date for international students. Applications are processed on a rolling basis. Application fee: $70. Electronic applications accepted. *Financial support:* In 2005–06, 9 research assistantships were awarded; fellowships, scholarships/grants, traineeships, and unspecified assistantships also available. *Unit head:* , Dr. James Alwine, Chair, 215-898-3256. *Application contact:* Emily Brady, Coordinator, 215-895-8935, Fax: 215-573-2104, E-mail: camb@mail.med.upenn.edu.

University of South Florida, College of Graduate Studies, Program in Cancer Biology, Tampa, FL 33620-9951. Offers PhD. *Students:* 19 full-time (13 women); includes 1 minority (Hispanic American), 7 international. 70 applicants, 26% accepted, 4 enrolled. *Entrance requirements:* Additional exam requirements/recommendations for international students: Required—TOEFL (minimum score 550 paper-based; 213 computer-based). *Application deadline:* For fall admission, 3/1 for domestic students. *Financial support:* Research assistantships with full tuition reimbursements, career-related internships or fieldwork, health care benefits, and unspecified assistantships available. Financial award application deadline: 2/1. *Unit head:* Erica Rauscher, Program Administrator, 813-903-6876, E-mail: cancerphd@moffitt.usf.edu.

See Close-Up on page 643.

The University of Texas Health Science Center at Houston, Graduate School of Biomedical Sciences, Program in Cancer Biology, Houston, TX 77225-0036. Offers MS, PhD, MD/PhD. *Faculty:* 74 full-time (18 women). *Students:* 57 full-time (33 women); includes 13 minority (2 African Americans, 1 American Indian/Alaska Native, 3 Asian Americans or Pacific Islanders, 7 Hispanic Americans), 21 international. Average age 24. 100 applicants, 56% accepted, 32 enrolled. In 2005, 10 degrees awarded. Terminal master's awarded for partial completion of doctoral program. *Degree requirements:* For master's and doctorate, thesis/dissertation. *Entrance requirements:* For master's and doctorate, GRE General Test. Additional exam requirements/recommendations for international students: Required—TOEFL, TWE. *Application deadline:* For fall admission, 1/15 for domestic students; for spring admission, 11/1 for domestic students. Applications are processed on a rolling basis. Application fee: $10. Electronic applications accepted. *Financial support:* Fellowships with full tuition reimbursements, research assistantships with full tuition reimbursements, institutionally sponsored loans, scholarships/grants, and health care benefits available. Financial award application deadline: 1/15. *Faculty research:* Mechanisms of cancer initiation/progression/metastasis/angiogenesis/therapeutic resistance, activation of host antitumor defense, molecular biology of oncogenes and tumor suppressor genes, cancer cell signal transduction. *Unit head:* Dr. Dihua Yu, Director, 713-794-3636, Fax: 713-794-4830, E-mail: dyu@mdanderson.org. *Application contact:* Dr. Victoria P. Knutson, Assistant Dean of Admissions, 713-500-9860, Fax: 713-500-9877, E-mail: victoria.p.knutson@uth.tmc.edu.

The University of Texas Health Science Center at Houston, Graduate School of Biomedical Sciences, Program in Molecular Carcinogenesis, Houston, TX 77225-0036. Offers MS, PhD, MD/PhD. *Faculty:* 26 full-time (9 women). *Students:* 16 full-time (11 women); includes 1 minority (Asian American or Pacific Islander), 7 international. Average age 24. 10 applicants, 30% accepted, 2 enrolled. In 2005, 1 master's, 3 doctorates awarded. Terminal master's awarded for partial completion of doctoral program. *Degree requirements:* For master's and doctorate, thesis/dissertation. *Entrance requirements:* For master's and doctorate, GRE General Test. Additional exam requirements/recommendations for international students: Required—TOEFL, TWE. *Application deadline:* For fall admission, 1/15 for domestic students; for spring admission, 11/1 for domestic students. Applications are processed on a rolling basis. Application fee: $10. Electronic applications accepted. *Financial support:* Fellowships with full tuition reimbursements, research assistantships with full tuition reimbursements, teaching assistantships, institutionally sponsored loans, scholarships/grants, and health care benefits available. Financial award application deadline: 1/15. *Faculty research:* Genetics and molecular biology in cancer and development. *Unit head:* Dr. Ellen R. Richie, Director, 512-237-9435, Fax: 512-237-2444, E-mail: erichie@mdanderson.org. *Application contact:* Dr. Victoria P. Knutson, Assistant Dean of Admissions, 713-500-9860, Fax: 713-500-9877, E-mail: victoria.P.knutson@uth.tmc.edu.

University of Utah, School of Medicine and The Graduate School, Graduate Programs in Medicine, Department of Oncological Sciences, Salt Lake City, UT 84112-1107. Offers MS, PhD. Terminal master's awarded for partial completion of doctoral program. *Degree requirements:* For master's, thesis (for some programs); for doctorate, thesis/dissertation. *Entrance requirements:* For master's and doctorate, GRE General Test, GRE Subject Test, minimum GPA of 3.0. Additional exam requirements/recommendations for international students: Required—TOEFL. *Expenses:* Tuition, state resident: $2,932; part-time $369 per credit. Tuition, nonresident: full-time $10,350; part-time $1,302 per credit. Required fees: $516 per term. Tuition and fees vary according to course load and program. *Faculty research:* Molecular basis of cell growth and differences, regulation of gene expression, biochemical mechanics of DNA replication, molecular biology and biochemistry of signal transduction, somatic cell genetics.

University of Wisconsin–Madison, Medical School and Graduate School, Graduate Programs in Medicine, Program in Cancer Biology, Madison, WI 53706-1380. Offers PhD. *Faculty:* 37 full-time (10 women). *Students:* 47 full-time (23 women); includes 6 minority (1 African American, 1 American Indian/Alaska Native, 4 Asian Americans or Pacific Islanders), 11 international. 132 applicants, 16% accepted, 10 enrolled. In 2005, 6 doctorates awarded. *Median time to degree:* Of those who began their doctoral program in fall 1997, 100% received their degree in 8 years or less. *Degree requirements:* For doctorate, thesis/dissertation, comprehensive exam, registration. *Entrance requirements:* For doctorate, GRE General Test, GRE Subject Test. Additional exam requirements/recommendations for international students: Required—TOEFL (minimum score 580 paper-based; 237 computer-based). *Application deadline:* For fall admission, 1/1 priority date for domestic students, 1/1 priority date for international students. Applications are processed on a rolling basis. Application fee: $45. Electronic applications accepted. *Financial support:* In 2005–06, 47 students received support, including 6 fellowships with full tuition reimbursements available (averaging $23,000 per year), 41 research assistantships with full tuition reimbursements available (averaging $23,000 per year); traineeships also available. Financial award application deadline: 1/1. *Faculty research:* Cancer genetics, tumor virology, chemical carcinogenesis, signal transduction, cell cycle. Total annual research expenditures: $13 million. *Unit head:* Dr. Norman R. Drinkwater, Director, 608-262-2177, Fax: 608-262-2824, E-mail: drinkwater@oncology.wisc.edu. *Application contact:* Bette Sheehan, Administrative Program Manager, 608-262-8651, Fax: 608-262-2824, E-mail: bsheehan@oncology.wisc.edu.

Vanderbilt University, Graduate School, Department of Cancer Biology, Nashville, TN 37240-1001. Offers MS, PhD. *Faculty:* 15 full-time (7 women), 1 part-time/adjunct (0 women). *Students:* 51 full-time (30 women), 2 part-time (1 woman); includes 5 minority (3 African Americans, 1 Asian American or Pacific Islander, 1 Hispanic American), 19 international. *Degree requirements:* For doctorate, thesis/dissertation, final and qualifying exams. *Application deadline:* For fall admission, 1/15 for domestic students, 1/15 for international students. Application fee: $0. Electronic applications accepted. *Expenses:* Tuition: Part-time $1,283 per semester hour. Required fees: $2,202; $1,101 per semester. One-time fee: $30. Tuition and fees vary according to course load, program and student level. *Financial support:* Fellowships with full and partial tuition reimbursements, research assistantships with tuition reimbursements available. *Unit head:* Lynn M. Matrisian, Chair, 615-322-0375, Fax: 615-936-2911, E-mail: lynn.matrisian@vanderbilt.edu. *Application contact:* Al Reynolds, Director of Graduate Studies, 615-322-0375, Fax: 615-936-2911, E-mail: albert.b.reynolds@vanderbilt.edu.

Vanderbilt University, Graduate School, Graduate Program in Toxicology and Carcinogenesis, Nashville, TN 37240-1001.

Wake Forest University, School of Medicine and Graduate School, Graduate Programs in Medicine, Department of Cancer Biology, Winston-Salem, NC 27109. Offers PhD. *Degree requirements:* For doctorate, thesis/dissertation. *Entrance requirements:* For doctorate, GRE General Test. Additional exam requirements/recommendations for international students: Required—TOEFL. Electronic applications accepted. *Faculty research:* Cancer research, mechanisms of carcinogenesis, signal transduction and regulation of cell growth.

See Close-Up on page 659.

Wayne State University, School of Medicine and Graduate School, Graduate Programs in Medicine, Department of Radiation Oncology, Detroit, MI 48202. Offers medical physics (PhD); radiological physics (MS). Part-time and evening/weekend programs available. *Faculty:* 14

full-time (0 women). *Students:* 21 full-time (8 women), 8 part-time (4 women); includes 4 minority (2 African Americans, 2 Asian Americans or Pacific Islanders), 10 international. Average age 28. 37 applicants, 49% accepted, 10 enrolled. In 2005, 9 master's awarded. Terminal master's awarded for partial completion of doctoral program. *Degree requirements:* For master's, thesis, essay, exit exam; for doctorate, thesis/dissertation, qualifying exam. *Entrance requirements:* For master's, GRE General Test, BS in physics or related area; for doctorate, GRE General Test, GRE Subject Test, BS in physics or related area. Additional exam requirements/recommendations for international students: Required—TOEFL (minimum score 550 paper-based; 213 computer-based); Recommended—TWE (minimum score 6). *Application deadline:* For fall admission, 1/15 for domestic students, 2/15 for international students. Applications are processed on a rolling basis. Application fee: $30 ($50 for international students). Electronic applications accepted. *Expenses:* Tuition, state resident: part-time $338 per credit hour. Tuition, nonresident: part-time $746 per credit hour. Required fees: $24 per credit hour. Full-time tuition and fees vary according to program. *Financial support:* In 2005–06, 2 research assistantships were awarded; fellowships, teaching assistantships, career-related internships or fieldwork also available. Support available to part-time students. Financial award application deadline: 1/15. *Faculty research:* Radiotherapy physics, hyperthermia, magnetic resonance imaging and spectroscopy, clinical ultrasound, x-ray physics. Total annual research expenditures: $687,450. *Unit head:* , Andrew Turrisi, Chair, 313-966-2774, Fax: 313-745-2314, E-mail: ar9642@wayne.edu. *Application contact:* Michael Joiner, Professor, 313-745-2489, E-mail: joinerm@kci.wayne.edu.

Wayne State University, School of Medicine and Graduate School, Graduate Programs in Medicine, Program in Cancer Biology, Detroit, MI 48202. Offers MS, PhD. *Students:* 31 full-time (23 women), 1 part-time; includes 4 minority (3 African Americans, 1 Asian American or Pacific Islander), 7 international. Average age 26. 74 applicants, 12% accepted, 5 enrolled. In 2005, 1 master's, 5 doctorates awarded. *Degree requirements:* For doctorate, thesis/dissertation. *Entrance requirements:* For doctorate, GRE General Test. Additional exam requirements/recommendations for international students: Required—TOEFL (minimum score 550 paper-based; 213 computer-based); Recommended—TWE (minimum score 6). *Application deadline:* For fall admission, 4/1 priority date for domestic students, 6/1 priority date for international students. Applications are processed on a rolling basis. Application fee: $30 ($50 for international students). Electronic applications accepted. *Expenses:* Tuition, state resident: part-time $338 per credit hour. Tuition, nonresident: part-time $746 per credit hour. Required fees: $24 per credit hour. Full-time tuition and fees vary according to program. *Financial support:* In 2005–06, 3 fellowships, 14 research assistantships were awarded; teaching assistantships *Faculty research:* Cell regulation, molecular biology and genetics, carcinogenesis, immunological modulation, drug discovery. Total annual research expenditures: $18 million. *Unit head:* Dr. Samuel Brooks, Director, 313-577-1065, Fax: 313-577-4112, E-mail: scbrooks@cmb.biosci.wayne.edu.

See Close-Up on page 663.

West Virginia University, Davis College of Agriculture, Forestry and Consumer Sciences, Interdisciplinary Program in Genetics and Developmental Biology, Morgantown, WV 26506. Offers animal breeding (MS, PhD); biochemical and molecular genetics (MS, PhD); cytogenetics (MS, PhD); descriptive embryology (MS, PhD); developmental genetics (MS);

experimental morphogenesis teratology (MS); human genetics (MS, PhD); immunogenetics (MS, PhD); life cycles of animals and plants (MS, PhD); molecular aspects of development (MS, PhD); mutagenesis (MS, PhD); oncology (MS, PhD); plant genetics (MS, PhD); population and quantitative genetics (MS, PhD); regeneration (MS, PhD); teratology (PhD); toxicology (MS, PhD). *Students:* 13 full-time (7 women), 6 part-time (4 women), 11 international. Average age 28. In 2005, 2 master's, 4 doctorates awarded. *Degree requirements:* For master's, thesis/dissertation; for doctorate, thesis/dissertation, comprehensive exam. *Entrance requirements:* For master's, GRE or MCAT, minimum GPA of 2.75. Additional exam requirements/recommendations for international students: Required—TOEFL. Application fee: $45. *Expenses:* Tuition, state resident: full-time $4,582; part-time $258 per credit hour. Tuition, nonresident: full-time $13,820; part-time $741 per credit hour. *Financial support:* In 2005–06, 3 research assistantships with tuition reimbursements (averaging $9,936 per year), 4 teaching assistantships with tuition reimbursements (averaging $9,936 per year) were awarded; fellowships, Federal Work-Study, institutionally sponsored loans, and tuition waivers (full and partial) also available. Financial award application deadline: 2/1; financial award applicants required to submit FAFSA. Total annual research expenditures: $1 million. *Unit head:* Dr. J. Nath, Chairman, 304-293-6023 Ext. 4333, Fax: 304-293-2960, E-mail: joginder.nath@mail.wvu.edu.

West Virginia University, School of Medicine, Graduate Programs at the Health Science Center, Biomedical Sciences Graduate Program, Program in Cancer Cell Biology, Morgantown, WV 26506. Offers PhD, MD/PhD. *Faculty:* 20 full-time (5 women). *Students:* 16 full-time (10 women). Average age 27. *Degree requirements:* For doctorate, thesis/dissertation, comprehensive exam. *Entrance requirements:* For doctorate, GRE General Test, minimum GPA of 3.0. Additional exam requirements/recommendations for international students: Required—TOEFL. *Application deadline:* For fall admission, 3/1 for domestic students, 1/15 for international students. Applications are processed on a rolling basis. Application fee: $0. Electronic applications accepted. *Financial support:* In 2005–06, research assistantships with full tuition reimbursements (averaging $20,000 per year); institutionally sponsored loans, traineeships, health care benefits, and stipends also available. *Faculty research:* Cellular signaling, tumor microenvironment, therapeutics. Total annual research expenditures: $4 million. *Unit head:* Dr. Scott Weed, Graduate Director, Department of Microbiology, Immunology, and Cell Biology, 304-293-3046, Fax: 304-293-4667, E-mail: scott.weed@hsc.wvu.edu. *Application contact:* Claire Noel, Graduate Adviser and Assistant Director, 304-293-7116, Fax: 304-293-7038, E-mail: claire.noel@hsc.wvu.edu.

See Close-Up on page 669.

Yale University, School of Medicine and Graduate School of Arts and Sciences, Combined Program in Biological and Biomedical Sciences (BBS), Pharmacological Sciences and Molecular Medicine Track, New Haven, CT 06520. Offers PhD, MD/PhD. *Students:* 6 full-time. *Degree requirements:* For doctorate, thesis/dissertation. *Entrance requirements:* For doctorate, GRE General Test. Additional exam requirements/recommendations for international students: Required—TOEFL. *Application deadline:* For fall admission, 12/8 for domestic students, 12/8 for international students. Electronic applications accepted. *Financial support:* Fellowships, research assistantships available. *Unit head:* Dr. David F. Stern, Co-Director, 203-785-4832, Fax: 203-785-7467, E-mail: kathleen.fisher@yale.edu. *Application contact:* Kathy Fisher, Registrar, 203-785-4545, E-mail: kathleen.fisher@yale.edu.

Cell Biology

Albany Medical College, Graduate Programs in the Biological Sciences, Center for Cell Biology and Cancer Research, Albany, NY 12208-3479. Offers MS, PhD. Part-time programs available. *Faculty:* 11 full-time (4 women). *Students:* 19 full-time (13 women); includes 4 minority (1 African American, 2 Asian Americans or Pacific Islanders, 1 Hispanic American), 4 international. Average age 24. 12 applicants, 100% accepted, 6 enrolled. In 2005, 2 master's, 4 doctorates awarded. Terminal master's awarded for partial completion of doctoral program. *Degree requirements:* For master's, thesis/dissertation; for doctorate, thesis/dissertation, comprehensive exam. *Entrance requirements:* For master's and doctorate, GRE General Test. Additional exam requirements/recommendations for international students: Required—TOEFL. *Application deadline:* For fall admission, 3/15 for domestic students. Applications are processed on a rolling basis. *Financial support:* In 2005–06, 10 research assistantships (averaging $23,000 per year) were awarded; Federal Work-Study, scholarships/grants, and tuition waivers (full) also available. Financial award applicants required to submit FAFSA. *Faculty research:* Cancer cell biology, angiogenesis, tissue remodeling, signal transduction, cell adhesion. *Unit head:* Dr. C. Michael DiPersio, Graduate Director, 518-262-5916, Fax: 518-262-5669, E-mail: dipersm@mail.amc.edu.

Albert Einstein College of Medicine, Sue Golding Graduate Division of Medical Sciences, Department of Anatomy and Structural Biology, Bronx, NY 10461. Offers anatomy (PhD); cell and developmental biology (PhD). *Degree requirements:* For doctorate, thesis/dissertation. *Entrance requirements:* For doctorate, GRE General Test. Additional exam requirements/recommendations for international students: Required—TOEFL. Electronic applications accepted. *Faculty research:* Cell motility, cell membranes and membrane-cytoskeletal interactions as applied to processing of pancreatic hormones, mechanisms of secretion.

Albert Einstein College of Medicine, Sue Golding Graduate Division of Medical Sciences, Division of Biological Sciences, Department of Cell Biology, Bronx, NY 10461. Offers PhD, MD/PhD. *Degree requirements:* For doctorate, thesis/dissertation. *Entrance requirements:* For doctorate, GRE General Test. Additional exam requirements/recommendations for international students: Required—TOEFL. *Faculty research:* Molecular and genetic basis of gene expression in animal cells; expression of differentiated traits of albumin, hemoglobin, myosin, and immunoglobin.

Arizona State University, Division of Graduate Studies, College of Liberal Arts and Sciences, Department of Biology, Cell and Developmental Biology Group, Tempe, AZ 85287. Offers MS, PhD. Terminal master's awarded for partial completion of doctoral program. *Degree requirements:* For master's, thesis; for doctorate, thesis/dissertation, oral exam. *Entrance requirements:* For master's, GRE General Test; for doctorate, GRE General Test, GRE Subject Test. Additional exam requirements/recommendations for international students: Required—TOEFL (minimum score 600 paper-based; 250 computer-based); Recommended—TSE. *Faculty research:* Cytoskeleton assembly, exocytosis, cyclic nucleotides, membrane fusion, chromosome distribution.

Baylor College of Medicine, Graduate School of Biomedical Sciences, Department of Molecular and Cellular Biology, Houston, TX 77030-3498. Offers PhD, MD/PhD. *Faculty:* 78 full-time (15 women). *Students:* 66 full-time (29 women); includes 9 minority (2 African Americans, 5 Asian Americans or Pacific Islanders, 2 Hispanic Americans), 24 international. Average age 27. 99 applicants, 25% accepted, 11 enrolled. In 2005, 7 degrees awarded. *Median time to degree:* Of those who began their doctoral program in fall 1997, 75% received their degree in 8 years or less. *Degree requirements:* For doctorate, thesis/dissertation, public defense, qualifying exam. *Entrance requirements:* For doctorate, GRE General Test, GRE Subject Test (strongly recommended), minimum GPA of 3.0. Additional exam requirements/recommendations for international students: Required—TOEFL. *Application deadline:* For fall admission, 2/1 for domestic students. Application fee: $30. Electronic applications accepted. *Expenses:* Tuition: Full-time $8,200. Full-time tuition and fees vary according to program. *Financial support:* In 2005–06, 64 students received support, including 16 fellowships (averaging $20,000 per year),

50 research assistantships (averaging $20,000 per year); career-related internships or fieldwork, Federal Work-Study, institutionally sponsored loans, health care benefits, and tuition waivers (full) also available. Financial award applicants required to submit FAFSA. *Faculty research:* Gene regulation, cell structure/function, developmental biology, neurobiology, reproductive endocrinology. *Unit head:* Dr. JoAnne Richards, Director, 713-798-4598. *Application contact:* Caroline Kosnik, Graduate Program Administrator, 713-798-4598, Fax: 713-790-0545, E-mail: ckosnik@bcm.edu.

See Close-Up on page 549.

Baylor College of Medicine, Graduate School of Biomedical Sciences, Interdepartmental Program in Cell and Molecular Biology, Houston, TX 77030-3498. Offers biochemistry (PhD); cell and molecular biology (PhD); genetics (PhD); human genetics (PhD); immunology (PhD); microbiology (PhD); virology (PhD). *Faculty:* 99 full-time (21 women). *Students:* 56 full-time (28 women); includes 20 minority (4 African Americans, 1 American Indian/Alaska Native, 6 Asian Americans or Pacific Islanders, 9 Hispanic Americans), 5 international. Average age 28. 164 applicants, 18% accepted, 12 enrolled. In 2005, 3 doctorates awarded. *Median time to degree:* Of those who began their doctoral program in fall 1997, 63% received their degree in 8 years or less. *Degree requirements:* For doctorate, thesis/dissertation, public defense. *Entrance requirements:* For doctorate, GRE General Test, GRE Subject Test (strongly recommended), minimum GPA of 3.0. Additional exam requirements/recommendations for international students: Required—TOEFL. *Application deadline:* For fall admission, 1/1 for domestic students. Applications are processed on a rolling basis. Application fee: $30. Electronic applications accepted. *Expenses:* Tuition: Full-time $8,200. Full-time tuition and fees vary according to program. *Financial support:* In 2005–06, 52 students received support, including 20 fellowships (averaging $23,000 per year), 36 research assistantships (averaging $23,000 per year); teaching assistantships, Federal Work-Study, institutionally sponsored loans, health care benefits, and tuition waivers (full) also available. Financial award applicants required to submit FAFSA. *Faculty research:* Gene expression and regulation, developmental biology and genetics, signal transduction and membrane biology, aging process, molecular virology. *Unit head:* Dr. Tom Cooper, Director, 713-798-6557. *Application contact:* Lourdes Fernandez, Graduate Program Administrator, 713-798-6557, Fax: 713-798-6325, E-mail: cmbprog@bcm.edu.

See Close-Up on page 545.

Baylor College of Medicine, Graduate School of Biomedical Sciences, Program in Developmental Biology, Houston, TX 77030-3498. Offers PhD, MD/PhD. *Faculty:* 42 full-time (10 women). *Students:* 38 full-time (17 women); includes 7 minority (1 American Indian/Alaska Native, 5 Asian Americans or Pacific Islanders, 1 Hispanic American), 21 international. Average age 27. 67 applicants, 25% accepted, 7 enrolled. In 2005, 2 doctorates awarded. *Median time to degree:* Of those who began their doctoral program in fall 1997, 100% received their degree in 8 years or less. *Degree requirements:* For doctorate, thesis/dissertation, public defense. *Entrance requirements:* For doctorate, GRE General Test, GRE Subject Test (strongly recommended), minimum GPA of 3.0. Additional exam requirements/recommendations for international students: Required—TOEFL. *Application deadline:* For fall admission, 1/1 for domestic students. Application fee: $30. Electronic applications accepted. *Expenses:* Tuition: Full-time $8,200. Full-time tuition and fees vary according to program. *Financial support:* In 2005–06, 37 students received support, including 10 fellowships (averaging $23,000 per year), 28 research assistantships (averaging $23,000 per year); career-related internships or fieldwork, Federal Work-Study, institutionally sponsored loans, health care benefits, tuition waivers (full), and stipends also available. *Faculty research:* Molecular and genetic approaches to study pattern formation in *Dictyostelium, Drosophila, C.elegans,* mouse, *Xenopus,* and zebrafish; cross-species approach. *Unit head:* Dr. Hugo Bellen, Director, 713-798-6410. *Application contact:* Catherine Tasnier, Graduate Program Administrator, 713-798-6410, Fax: 713-798-5386, E-mail: cat@bcm.edu.

See Close-Up on page 757.

Cell Biology

Baylor College of Medicine, Program in Cell and Molecular Biology of Aging, Houston, TX 77030-3498.

See Close-Up on page 547.

Boston University, Graduate School of Arts and Sciences, Molecular Biology, Cell Biology, and Biochemistry Program (MCBB), Boston, MA 02215. Offers MA, PhD. *Students:* 41 full-time (21 women); includes 3 minority (2 African Americans, 1 Asian American or Pacific Islander), 7 international. Average age 29. 111 applicants, 23% accepted, 8 enrolled. In 2005, 8 master's, 8 doctorates awarded. Terminal master's awarded for partial completion of doctoral program. *Degree requirements:* For master's, one foreign language, thesis (for some programs), registration; for doctorate, one foreign language, thesis/dissertation, comprehensive exam, registration. *Entrance requirements:* For master's and doctorate, GRE General Test, GRE Subject Test. Additional exam requirements/recommendations for international students: Required—TOEFL (minimum score 600 paper-based; 250 computer-based). *Application deadline:* For fall admission, 1/1 for domestic students, 1/1 for international students. Application fee: $60. *Expenses:* Tuition: Full-time $31,530; part-time $985 per credit. Required fees: $316; $40 per semester. Tuition and fees vary according to course level and program. *Financial support:* In 2005–06, 17 students received support, including 1 fellowship (averaging $16,500 per year), 15 research assistantships with full tuition reimbursements available (averaging $16,000 per year), 1 teaching assistantship with full tuition reimbursement available (averaging $16,000 per year); Federal Work-Study, scholarships/grants, and traineeships also available. Financial award application deadline: 1/15; financial award applicants required to submit FAFSA. *Unit head:* , Kim McCall, Director, 617-353-2432, Fax: 617-353-6340, E-mail: mccall@bu.edu. *Application contact:* Academic Administrator, 617-353-2432, Fax: 617-353-6340, E-mail: mcbb@bu.edu.

Boston University, School of Medicine, Division of Graduate Medical Sciences, Department of Pathology and Laboratory Medicine, Boston, MA 02118. Offers cell and molecular biology (PhD); experimental pathology (PhD); immunology (PhD). Part-time programs available. *Faculty:* 12 full-time (4 women), 13 part-time/adjunct (2 women). *Students:* 19 full-time (10 women), 1 (woman) part-time; includes 4 minority (2 Asian Americans or Pacific Islanders, 2 Hispanic Americans), 6 international. Average age 29. *Degree requirements:* For doctorate, thesis/dissertation, qualifying exam. *Entrance requirements:* For doctorate, GRE General Test, GRE Subject Test. Additional exam requirements/recommendations for international students: Required—TOEFL. *Application deadline:* For fall admission, 1/15 for domestic students; for spring admission, 10/15 priority date for domestic students. Electronic applications accepted. *Expenses:* Tuition: Full-time $31,530; part-time $985 per credit. Required fees: $316; $40 per semester. Tuition and fees vary according to course level and program. *Financial support:* In 2005–06, 10 fellowships, 3 research assistantships were awarded; Federal Work-Study, scholarships/grants, and traineeships also available. *Faculty research:* Toxicology, carcinogenesis, endocytosis, cytogenetics. *Unit head:* Leonard Gottlieb, Chairman, 617-638-4500, Fax: 617-638-4085. *Application contact:* Dr. Adrianne Rogers, Associate Chairman, 617-638-4500, Fax: 617-638-4085, E-mail: aerogers@bu.edu.

Boston University, School of Medicine, Division of Graduate Medical Sciences, Program in Cell and Molecular Biology, Boston, MA 02118. Offers PhD, MD/PhD. *Degree requirements:* For doctorate, thesis/dissertation. *Entrance requirements:* For doctorate, GRE General Test, GRE Subject Test. Additional exam requirements/recommendations for international students: Required—TOEFL. *Application deadline:* For fall admission, 1/15 for domestic students; for spring admission, 10/15 priority date for domestic students. Electronic applications accepted. *Expenses:* Tuition: Full-time $31,530; part-time $985 per credit. Required fees: $316; $40 per semester. Tuition and fees vary according to course level and program. *Financial support:* Fellowships, research assistantships, Federal Work-Study, scholarships/grants, and traineeships available. *Unit head:* Dr. Vickery Trinkaus Randall, Director, 617-638-6099, Fax: 617-638-5337, E-mail: vickery@bu.edu. *Application contact:* Dr. Mary Jo Murnane, Admissions Director, 617-638-4926, Fax: 617-638-4085, E-mail: mmurnane@bu.edu.

See Close-Up on page 551.

Brandeis University, Graduate School of Arts and Sciences, Programs in Life Sciences, Program in Molecular and Cell Biology, Waltham, MA 02454-9110. Offers genetics (PhD); microbiology (PhD); molecular and cell biology (MS, PhD); molecular biology (PhD); neurobiology (PhD). *Faculty:* 18 full-time (8 women). *Students:* 43 full-time (21 women); includes 2 minority (both Asian Americans or Pacific Islanders), 11 international. Average age 27. 173 applicants, 8% accepted, 6 enrolled. In 2005, 2 master's, 4 doctorates awarded. Terminal master's awarded for partial completion of doctoral program. *Median time to degree:* Of those who began their doctoral program in fall 1997, 100% received their degree in 8 years or less. *Degree requirements:* For master's, research project, thesis optional; for doctorate, thesis/dissertation, teaching assistant experience, comprehensive exam, registration. *Entrance requirements:* For master's and doctorate, GRE General Test, resumé, 3 letters of recommendation. Additional exam requirements/recommendations for international students: Required—TOEFL (minimum score 600 paper-based; 250 computer-based). *Application deadline:* For fall admission, 1/15 for domestic students. Applications are processed on a rolling basis. Application fee: $55. Electronic applications accepted. *Financial support:* In 2005–06, 20 fellowships with full tuition reimbursements (averaging $26,500 per year), 23 research assistantships with full tuition reimbursements (averaging $26,500 per year), 5 teaching assistantships with full tuition reimbursements (averaging $3,000 per year) were awarded; scholarships/grants, traineeships, health care benefits, and tuition waivers (full and partial) also available. Financial award application deadline: 4/15; financial award applicants required to submit CSS PROFILE or FAFSA. *Faculty research:* Regulation of gene expression by transcription factors, molecular neurobiology, immunology, molecular mechanisms of genetic recombination, and cell differentiation. *Unit head:* Dr. Piali Sengupta, Chair, 781-736-2686, Fax: 781-736-3107, E-mail: piali@brandeis.edu. *Application contact:* Marcia Cabral, Information Officer, 781-736-3100, Fax: 781-736-3107, E-mail: cabral@brandeis.edu.

Brown University, Graduate School, Division of Biology and Medicine, Program in Molecular Biology, Cell Biology, and Biochemistry, Providence, RI 02912. Offers biochemistry (M Med Sc, Sc M, PhD), including biochemistry (Sc M, PhD), biology (Sc M, PhD), medical science (M Med Sc, PhD); biology (MA); cell biology (M Med Sc, Sc M, PhD), including biochemistry (Sc M, PhD), biology (Sc M, PhD), medical science (M Med Sc, PhD); developmental biology (M Med Sc, Sc M, PhD), including biochemistry (Sc M, PhD), biology (Sc M, PhD), medical science (M Med Sc, PhD); immunology (M Med Sc, Sc M, PhD), including biochemistry (Sc M, PhD), biology (Sc M, PhD), medical science (M Med Sc, PhD); molecular microbiology (M Med Sc, Sc M, PhD), including biochemistry (Sc M, PhD), biology (Sc M, PhD), medical science (M Med Sc, PhD). Part-time programs available. Terminal master's awarded for partial completion of doctoral program. *Degree requirements:* For master's, thesis (for some programs); for doctorate, one foreign language, thesis/dissertation, preliminary exam. *Entrance requirements:* For master's and doctorate, GRE General Test, GRE Subject Test. Additional exam requirements/recommendations for international students: Required—TOEFL. Electronic applications accepted. *Faculty research:* Molecular genetics, gene regulation.

See Close-Up on page 555.

California Institute of Technology, Division of Biology, Program in Cell Biology and Biophysics, Pasadena, CA 91125-0001. Offers PhD. *Degree requirements:* For doctorate, thesis/dissertation, qualifying exam. *Entrance requirements:* For doctorate, GRE General Test. *Application deadline:* For fall admission, 1/1 for domestic students. Application fee: $0. *Financial support:* Application deadline: 1/1. *Application contact:* Elizabeth Ayala, Graduate Program Coordinator, 626-395-4497, Fax: 626-683-3343, E-mail: biograd@cco.caltech.edu.

Carnegie Mellon University, Mellon College of Science, Department of Biological Sciences, Pittsburgh, PA 15213-3891. Offers biochemistry (PhD); biophysics (PhD); cell biology (PhD); computational biology (MS, PhD); developmental biology (PhD); genetics (PhD); molecular biology (PhD); neurobiology (PhD). *Degree requirements:* For doctorate, thesis/dissertation, comprehensive exam. *Entrance requirements:* For doctorate, GRE General Test, GRE Subject

Test, interview. Electronic applications accepted. *Faculty research:* Genetic structure, function, and regulation; protein structure and function; biological membranes; biological spectroscopy.

See Close-Up on page 93.

Case Western Reserve University, School of Medicine and School of Graduate Studies, Graduate Programs in Medicine, Department of Anatomy, Cleveland, OH 44106. Offers applied anatomy (MS); biological anthropology (MS, PhD); cellular biology (MS, PhD); developmental biology (PhD); molecular biology (PhD). Part-time programs available. *Faculty:* 20 full-time (5 women), 16 part-time/adjunct (5 women). *Students:* 6 full-time (0 women), 49 part-time (19 women); includes 20 minority (7 African Americans, 12 Asian Americans or Pacific Islanders, 1 Hispanic American). Average age 25. 39 applicants, 92% accepted, 29 enrolled. In 2005, 14 master's awarded. *Median time to degree:* Of those who began their doctoral program in fall 1997, 67% received their degree in 8 years or less. *Degree requirements:* For master's, thesis (for some programs), comprehensive exam; for doctorate, thesis/dissertation. *Entrance requirements:* For master's, GRE General Test; for doctorate, GRE General Test, GRE Subject Test. Additional exam requirements/recommendations for international students: Required—TOEFL. *Application deadline:* For fall admission, 5/1 for domestic students; for spring admission, 8/1 priority date for domestic students. Applications are processed on a rolling basis. Application fee: $50. *Financial support:* In 2005–06, 6 research assistantships with full tuition reimbursements (averaging $23,000 per year) were awarded; fellowships, scholarships/grants also available. *Faculty research:* Hypoxia, cell injury, biochemical aberration occurrences in ischemic tissue, human functional morphology, evolutionary morphology. Total annual research expenditures: $691,974. *Unit head:* Dr. Joseph C. LaManna, Chairman, 216-368-1100, Fax: 216-368-8669, E-mail: jcl4@po.cwru.edu. *Application contact:* Morley Schwebel, Administrator, 216-368-2433, Fax: 216-368-8669, E-mail: mxs86@po.cwru.edu.

Case Western Reserve University, School of Medicine and School of Graduate Studies, Graduate Programs in Medicine, Department of Molecular Biology and Microbiology, Cleveland, OH 44106-4960. Offers cellular biology (PhD); microbiology (PhD); molecular biology (PhD); molecular virology (MD/PhD). Students are admitted to an integrated Biomedical Sciences Training Program involving 11 basic science programs at Case Western Reserve University. *Degree requirements:* For doctorate, thesis/dissertation. *Entrance requirements:* For doctorate, GRE General Test, GRE Subject Test. Additional exam requirements/recommendations for international students: Required—TOEFL. *Application deadline:* Applications are processed on a rolling basis. Application fee: $25. Electronic applications accepted. *Financial support:* In 2005–06, 24 students received support, including fellowships with full tuition reimbursements available (averaging $23,000 per year); Federal Work-Study, scholarships/grants, traineeships, and tuition waivers (full) also available. *Faculty research:* Gene expression in eukaryotic and prokaryotic systems; microbial physiology; intracellular transport and signaling; mechanisms of oncogenesis; molecular mechanisms of RNA processing, editing, and catalysis. Total annual research expenditures: $4.8 million. *Unit head:* Dr. Jonathan Karn, Chairman, 216-368-3330, Fax: 216-368-3055, E-mail: jonathan.karn@case.edu. *Application contact:* Dr. Patrick Viollier, Admissions Coordinator, 216-368-3420, Fax: 216-368-3055, E-mail: patrick.viollier@case.edu.

See Close-Up on page 557.

Case Western Reserve University, School of Medicine and School of Graduate Studies, Graduate Programs in Medicine, Program in Cell Biology, Cleveland, OH 44106. Offers PhD. *Faculty:* 40. *Students:* 7; includes 3 Asian Americans or Pacific Islanders. Average age 26. *Degree requirements:* For doctorate, thesis/dissertation. *Entrance requirements:* For doctorate, GRE General Test, GRE Subject Test, previous course work in biochemistry. Additional exam requirements/recommendations for international students: Required—TOEFL. *Application deadline:* For fall admission, 1/1 priority date for domestic students, 1/1 priority date for international students. Applications are processed on a rolling basis. Application fee: $25. Electronic applications accepted. *Financial support:* In 2005–06, fellowships with tuition reimbursements (averaging $23,000 per year); tuition waivers (full) also available. *Faculty research:* Macromolecular transport, membrane traffic, signal transduction, nuclear organization, lipid metabolism. *Unit head:* Dr. Alan Tartakoff, Director, 216-368-5544, Fax: 216-368-5484, E-mail: amt10@case.edu. *Application contact:* Deborah Noureddine, Coordinator, 216-368-3347, Fax: 216-368-0795, E-mail: drn2@po.cwru.edu.

Case Western Reserve University, School of Medicine and School of Graduate Studies, Graduate Programs in Medicine, Programs in Molecular and Cellular Basis of Disease, Cleveland, OH 44106. Offers cell biology (MS, PhD); immunology (MS, PhD); pathology (MS, PhD). *Faculty:* 37 full-time (7 women). *Students:* 52 full-time (30 women); includes 16 minority (2 African Americans, 14 Asian Americans or Pacific Islanders). Average age 29. In 2005, 5 degrees awarded. Terminal master's awarded for partial completion of doctoral program. *Median time to degree:* Of those who began their doctoral program in fall 1997, 100% received their degree in 8 years or less. *Degree requirements:* For master's and doctorate, thesis/dissertation. *Entrance requirements:* For master's and doctorate, GRE General Test, GRE Subject Test. Additional exam requirements/recommendations for international students: Required—TOEFL (minimum score 550 paper-based; 213 computer-based). Application fee: $50. Electronic applications accepted. *Financial support:* In 2005–06, fellowships with full tuition reimbursements (averaging $23,000 per year); scholarships/grants, traineeships, and tuition waivers (full) also available. *Faculty research:* Neurobiology, molecular biology, cancer biology, biomaterials, biocompatibility. Total annual research expenditures: $11.4 million. *Unit head:* John B Lowe, Chairman, 216-368-0890, Fax: 216-368-0494, E-mail: jbl21@case.edu. *Application contact:* Tandalea Harney, Graduate Programs Director, 216-368-0890, Fax: 216-368-0494, E-mail: trh2@case.edu.

The Catholic University of America, School of Arts and Sciences, Department of Biology, Program in Cell and Microbial Biology, Washington, DC 20064. Offers cell biology (MS, PhD); microbiology (MS, PhD). Part-time programs available. *Students:* 3 full-time (1 woman), 8 part-time (6 women); includes 1 Asian American or Pacific Islander, 5 international. Average age 30. 14 applicants, 64% accepted, 3 enrolled. Terminal master's awarded for partial completion of doctoral program. *Degree requirements:* For master's, thesis or alternative, comprehensive exam; for doctorate, thesis/dissertation, comprehensive exam. *Entrance requirements:* For master's and doctorate, GRE General Test, GRE Subject Test, 3 letters of recommendation. Additional exam requirements/recommendations for international students: Required—TOEFL (minimum score 580 paper-based; 237 computer-based). *Application deadline:* For fall admission, 2/1 for domestic students; for spring admission, 11/15 priority date for domestic students. Applications are processed on a rolling basis. Application fee: $55. Electronic applications accepted. *Expenses:* Tuition: Full-time $24,800; part-time $940 per credit. Required fees: $1,090; $285 per term. Part-time tuition and fees vary according to course load and program. *Financial support:* Fellowships, research assistantships, teaching assistantships, career-related internships or fieldwork, scholarships/grants, tuition waivers (full and partial), and unspecified assistantships available. Support available to part-time students. Financial award application deadline: 2/1; financial award applicants required to submit FAFSA. *Faculty research:* Cell differentiation, regulation of cell growth, drug resistance, gene cloning and sequencing, developmental biology and neurobiology. *Unit head:* Dr. Venigalla Rao, Chair, Department of Biology, 202-319-5267, Fax: 202-319-5721, E-mail: rao@cua.edu.

Colorado State University, Graduate School, Program in Cell and Molecular Biology, Fort Collins, CO 80523-1618. Offers MS, PhD. *Students:* Average age 28. In 2005, 4 master's, 2 doctorates awarded. Terminal master's awarded for partial completion of doctoral program. *Degree requirements:* For master's and doctorate, thesis/dissertation. *Entrance requirements:* For master's and doctorate, GRE General Test, minimum GPA of 3.0. Additional exam requirements/recommendations for international students: Required—TOEFL. *Application deadline:* For fall admission, 1/15 for domestic students. Applications are processed on a rolling basis. Application fee: $50. Electronic applications accepted. *Expenses:* Tuition, state resident: full-time $3,690; part-time $205 per credit. Tuition, nonresident: full-time $14,958; part-time $831 per credit. Required fees: $1,061. *Financial support:* In 2005–06, fellowships

Cell Biology

with full tuition reimbursements (averaging $17,000 per year), research assistantships with full tuition reimbursements (averaging $17,000 per year), teaching assistantships with full tuition reimbursements (averaging $17,000 per year) were awarded; Federal Work-Study and institutionally sponsored loans also available. Financial award application deadline: 2/1. *Faculty research:* Regulation of gene expression, cancer biology, plant molecular genetics, reproductive physiology, environmental toxicology. Total annual research expenditures: $15 million. *Unit head:* Dr. Michael H. Fox, Chairman, 970-491-7618, Fax: 970-491-0623, E-mail: cmb@colostate.edu. *Application contact:* Norma Jean Bulera, Administrative Assistant, 970-491-0241, Fax: 970-491-0623, E-mail: cmb@colostate.edu.

See Close-Up on page 559.

Columbia University, College of Physicians and Surgeons, Coordinated Doctoral Program in Basic Sciences, New York, NY 10027.

Columbia University, College of Physicians and Surgeons and Graduate School of Arts and Sciences, Graduate School of Arts and Sciences at the College of Physicians and Surgeons, Department of Anatomy and Cell Biology, New York, NY 10032. Offers anatomy (M Phil, MA, PhD); anatomy and cell biology (PhD). Only candidates for the PhD are admitted. Terminal master's awarded for partial completion of doctoral program. *Degree requirements:* For doctorate, thesis/dissertation, oral exam. *Entrance requirements:* For master's and doctorate, GRE General Test. Additional exam requirements/recommendations for international students: Required—TOEFL. *Expenses:* Tuition: Full-time $31,448. Tuition and fees vary according to course level, course load, campus/location and program. *Faculty research:* Protein sorting, membrane biophysics, muscle energetics, neuroendocrinology, developmental biology, cytoskeleton, transcription factors.

Columbia University, College of Physicians and Surgeons and Graduate School of Arts and Sciences, Graduate School of Arts and Sciences at the College of Physicians and Surgeons, Integrated Program in Cellular, Molecular and Biophysical Studies, New York, NY 10032. Offers M Phil, MA, PhD, MD/PhD. Only candidates for the PhD are admitted. Terminal master's awarded for partial completion of doctoral program. *Degree requirements:* For doctorate, thesis/dissertation. *Entrance requirements:* For master's, GRE General Test; for doctorate, GRE General Test, GRE Subject Test. Additional exam requirements/recommendations for international students: Required—TOEFL. Expenses: Contact institution. Tuition and fees vary according to course level, course load, campus/location and program. *Faculty research:* Transcription, macromolecular sorting, gene expression during development, cellular interaction.

Cornell University, Graduate School, Graduate Fields of Agriculture and Life Sciences, Field of Biochemistry, Molecular and Cell Biology, Ithaca, NY 14853-0001. Offers biochemistry (PhD); biophysics (PhD); cell biology (PhD); molecular and cell biology (PhD); molecular biology (PhD). *Faculty:* 60 full-time (13 women). *Students:* 177 applicants, 25% accepted, 22 enrolled. In 2005, 13 degrees awarded. *Degree requirements:* For doctorate, thesis/dissertation, 2 semesters of teaching experience, comprehensive exam. *Entrance requirements:* For doctorate, GRE General Test, GRE Subject Test (biology, chemistry, physics; or biochemistry, cell and molecular biology), 3 letters of recommendation. Additional exam requirements/recommendations for international students: Required—TOEFL (minimum score 600 paper-based; 250 computer-based). *Application deadline:* For fall admission, 1/5 for domestic students. Application fee: $60. Electronic applications accepted. *Financial support:* In 2005–06, 90 students received support, including 35 fellowships with full tuition reimbursements available, 42 research assistantships with full tuition reimbursements available, 13 teaching assistantships with full tuition reimbursements available; institutionally sponsored loans, scholarships/grants, health care benefits, tuition waivers (full and partial), and unspecified assistantships also available. Financial award applicants required to submit FAFSA. *Faculty research:* Biophysics, structural biology. *Unit head:* , Director of Graduate Studies, 607-255-2100, Fax: 607-255-2100. *Application contact:* Graduate Field Assistant, 607-255-2100, Fax: 607-255-2100, E-mail: bmcb@cornell.edu.

Cornell University, Graduate School, Graduate Fields of Agriculture and Life Sciences, Field of Zoology, Ithaca, NY 14853-0001. Offers animal cytology (MS, PhD); comparative and functional anatomy (MS, PhD); developmental biology (MS, PhD); ecology (MS, PhD); histology (MS, PhD). *Faculty:* 26 full-time (4 women). *Students:* 4 full-time (3 women), 3 international. 9 applicants, 11% accepted, 1 enrolled. *Degree requirements:* For doctorate, thesis/dissertation, 2 semesters of teaching experience, comprehensive exam. *Entrance requirements:* For doctorate, GRE General Test, GRE Subject Test (biology), 2 letters of recommendation. Additional exam requirements/recommendations for international students: Required—TOEFL (minimum score 550 paper-based; 213 computer-based). *Application deadline:* For fall admission, 2/1 for domestic students. Application fee: $60. Electronic applications accepted. *Financial support:* In 2005–06, 4 students received support, including 1 fellowship with full tuition reimbursement available, 1 research assistantship with full tuition reimbursement available, 2 teaching assistantships with full tuition reimbursements available; institutionally sponsored loans, scholarships/grants, health care benefits, tuition waivers (full and partial), and unspecified assistantships also available. Financial award applicants required to submit FAFSA. *Faculty research:* Organismal biology, functional morphology, biomechanics, comparative vertebrate anatomy, comparative invertebrate anatomy, paleontology. *Unit head:* Director of Graduate Studies, 607-253-3276, Fax: 607-253-3756. *Application contact:* Graduate Field Assistant, 607-253-3276, Fax: 607-253-3756, E-mail: graduate_edcvm@cornell.edu.

Cornell University, Joan and Sanford I. Weill Medical College and Graduate School of Medical Sciences, Weill Graduate School of Medical Sciences, Program in Molecular Biology, Cell Biology, Biochemistry and Structural Biology, New York, NY 10021-4896. Offers MS, PhD, MD/PhD. *Faculty:* 84 full-time (22 women). *Students:* 121 full-time (68 women); includes 14 minority (2 African Americans, 7 Asian Americans or Pacific Islanders, 5 Hispanic Americans), 34 international. Average age 22. In 2005, 8 degrees awarded. *Median time to degree:* Of those who began their doctoral program in fall 1997, 100% received their degree in 8 years or less. *Degree requirements:* For doctorate, thesis/dissertation, final exam. *Entrance requirements:* For doctorate, GRE General Test, GRE Subject Test, background in genetics, molecular biology, chemistry, or biochemistry. Additional exam requirements/recommendations for international students: Required—TOEFL. *Application deadline:* For fall admission, 12/15 for domestic students. Application fee: $60. *Expenses:* Tuition: Full-time $32,320. Required fees: $1,025. *Financial support:* Fellowships, stipends available. *Unit head:* Dr. Christopher Lima, Co-Director, 212-639-8205, Fax: 212-717-3047, E-mail: limac@mskcc.org.

Dartmouth College, Graduate Program in Molecular and Cellular Biology, Hanover, NH 03755.

See Close-Up on page 561.

Drexel University, College of Medicine, Biomedical Graduate Programs, Interdisciplinary Program in Molecular and Cell Biology, Philadelphia, PA 19104-2875. Offers MS, PhD, MD/PhD. Terminal master's awarded for partial completion of doctoral program. *Degree requirements:* For master's, thesis, comprehensive exam; for doctorate, thesis/dissertation, qualifying exam. *Entrance requirements:* For master's, GRE General Test, minimum GPA of 2.75; for doctorate, GRE General Test, minimum GPA of 3.0. Additional exam requirements/recommendations for international students: Required—TOEFL. Electronic applications accepted. *Faculty research:* Molecular anatomy, biochemistry, medical biotechnology, molecular pathology, microbiology and immunology.

Duke University, Graduate School, Department of Biological Anthropology and Anatomy, Durham, NC 27710. Offers cellular and molecular biology (PhD); gross anatomy and physical anthropology (PhD), including comparative morphology of human and non-human primates, primate social behavior, vertebrate paleontology; neuroanatomy (PhD). *Faculty:* 6 full-time. *Students:* 14 full-time (9 women); includes 1 minority (African American), 2 international. 39 applicants, 18% accepted, 3 enrolled. In 2005, 3 doctorates awarded. *Degree requirements:* For doctorate, one foreign language, thesis/dissertation. *Entrance requirements:* For doctor-

ate, GRE General Test. Additional exam requirements/recommendations for international students: Required—IELT (preferred) or TOEFL. *Application deadline:* For fall admission, 12/31 for domestic students, 12/31 for international students. Application fee: $75. Electronic applications accepted. *Financial support:* Fellowships, teaching assistantships, Federal Work-Study available. Financial award application deadline: 12/31. *Unit head:* , Daniel Schmitt, Director of Graduate Studies, 919-684-5664, Fax: 919-684-8034, E-mail: l.squires@baa.mc.duke.edu.

Duke University, Graduate School, Department of Cell Biology, Durham, NC 27710. Offers PhD. *Faculty:* 21 full-time. *Students:* 35 full-time (22 women); includes 6 minority (2 African Americans, 4 Asian Americans or Pacific Islanders), 8 international. In 2005, 6 doctorates awarded. *Degree requirements:* For doctorate, thesis/dissertation. *Entrance requirements:* For doctorate, GRE General Test, GRE Subject Test (recommended). Additional exam requirements/recommendations for international students: Required—IELT (preferred) or TOEFL. *Application deadline:* For fall admission, 12/31 for domestic students. Application fee: $75. *Financial support:* Fellowships, research assistantships, teaching assistantships, Federal Work-Study available. Financial award application deadline: 12/31. *Unit head:* Blanche Capel, Director of Graduate Studies, 919-684-6390, Fax: 919-684-3590, E-mail: teresa.jenkins@duke.edu.

Duke University, Graduate School, Program in Cellular and Molecular Biology, Durham, NC 27710. Offers PhD, Certificate. Certificate students must be enrolled in a participating PhD program. *Faculty:* 127 full-time. *Students:* 24 full-time (7 women); includes 3 minority (all Asian Americans or Pacific Islanders), 5 international. 259 applicants, 21% accepted, 23 enrolled. *Entrance requirements:* For degree, GRE General Test, GRE Subject Test (recommended). Additional exam requirements/recommendations for international students: Required—IELT (preferred) or TOEFL. *Application deadline:* For fall admission, 12/31 for domestic students, 12/31 for international students. Application fee: $75. Electronic applications accepted. *Financial support:* Fellowships available. Financial award application deadline: 12/31. *Unit head:* Dr. Chris Counter, Director of Graduate Studies, 919-684-9890, Fax: 919-681-8911, E-mail: cmbtgp@biochem.duke.edu.

See Close-Up on page 563.

East Carolina University, Brody School of Medicine, Department of Anatomy and Cell Biology, Greenville, NC 27858-4353. Offers PhD. *Faculty:* 13 full-time (4 women), 1 part-time/adjunct (0 women). *Students:* Average age 29. 13 applicants, 23% accepted. In 2005, 1 degree awarded. *Median time to degree:* Of those who began their doctoral program in fall 1997, 100% received their degree in 8 years or less. *Degree requirements:* For doctorate, thesis/dissertation, comprehensive exam, registration. *Entrance requirements:* For doctorate, GRE General Test. Additional exam requirements/recommendations for international students: Required—TOEFL. *Application deadline:* For fall admission, 6/1 for domestic students. Applications are processed on a rolling basis. Application fee: $50. *Expenses:* Tuition, state resident: full-time $2,516. Tuition, nonresident: full-time $12,832. *Financial support:* In 2005–06, 8 fellowships with full tuition reimbursements (averaging $21,500 per year) were awarded; health care benefits also available. Financial award application deadline: 6/1. *Faculty research:* Kinesin motors during slow matogensis, mitochondria and peroxisomes in obesity, ovarian innervation, tight junction function and regulation. *Unit head:* Dr. Charles Hodson, Interim Chairman, 252-744-2851, Fax: 252-744-2850, E-mail: hodsonc@ccu.edu. *Application contact:* Dr. Ron Dudek, Senior Director of Graduate Studies, 252-744-3284, Fax: 252-744-2850, E-mail: dudekr@ecu.edu.

Emory University, Graduate School of Arts and Sciences, Division of Biological and Biomedical Sciences, Program in Biochemistry, Cell and Developmental Biology, Atlanta, GA 30322-1100. Offers PhD. *Faculty:* 61 full-time (12 women). *Students:* 56 full-time (42 women); includes 8 minority (5 African Americans, 2 Asian Americans or Pacific Islanders, 1 Hispanic American), 9 international. Average age 27. 96 applicants, 38% accepted, 17 enrolled. In 2005, 6 doctorates awarded. *Median time to degree:* Of those who began their doctoral program in fall 1997, 100% received their degree in 8 years or less. *Degree requirements:* For doctorate, thesis/dissertation, comprehensive exam, registration. *Entrance requirements:* For doctorate, GRE General Test, minimum GPA of 3.0 in science course work. Additional exam requirements/recommendations for international students: Required—TOEFL. *Application deadline:* For fall admission, 1/3 for domestic students, 1/3 for international students. Application fee: $50. Electronic applications accepted. *Expenses:* Tuition: Full-time $14,400. Required fees: $217. *Financial support:* In 2005–06, 26 students received support, including 26 fellowships with full tuition reimbursements available (averaging $23,000 per year); institutionally sponsored loans, health care benefits, and tuition waivers (full) also available. *Faculty research:* Signal transduction, molecular biology, enzymes and cofactors, receptor and ion channel function, membrane biology. *Unit head:* Grace Pavlath, Director, 404-727-3353, Fax: 404-727-2880, E-mail: gpavlat@emory.edu. *Application contact:* 404-727-2545, Fax: 404-727-3322, E-mail: gdbbs@emory.edu.

Emporia State University, School of Graduate Studies, College of Liberal Arts and Sciences, Department of Biological Sciences, Emporia, KS 66801-5087. Offers botany (MS); environmental biology (MS); general biology (MS); microbial and cellular biology (MS); zoology (MS). Part-time programs available. *Faculty:* 13 full-time (2 women), 4 part-time/adjunct (2 women). *Students:* 3 full-time (1 woman), 17 part-time (7 women), 3 international. 10 applicants, 90% accepted, 7 enrolled. In 2005, 5 degrees awarded. *Degree requirements:* For master's, comprehensive exam or thesis. *Entrance requirements:* For master's, GRE, appropriate undergraduate degree, interview, letters of reference. Additional exam requirements/recommendations for international students: Required—TOEFL (minimum score 450 paper-based; 133 computer-based). *Application deadline:* For fall admission, 8/15 for domestic students. Applications are processed on a rolling basis. Application fee: $30 ($75 for international students). Electronic applications accepted. *Expenses:* Tuition, state resident: full-time $2,890; part-time $132 per credit. Tuition, nonresident: full-time $9,258; part-time $422 per credit. Required fees: $626; $41 per credit. Tuition and fees vary according to degree level. *Financial support:* In 2005–06, 4 research assistantships with full tuition reimbursements (averaging $6,492 per year), 9 teaching assistantships with full tuition reimbursements (averaging $6,492 per year) were awarded; fellowships, career-related internships or fieldwork, Federal Work-Study, institutionally sponsored loans, health care benefits, and unspecified assistantships also available. Financial award application deadline: 3/15; financial award applicants required to submit FAFSA. *Faculty research:* Fisheries, range, and wildlife management; aquatic, plant, grassland, vertebrate, and invertebrate ecology; mammalian and plant systematics, taxonomy, and evolution; immunology, virology, and molecular biology. *Unit head:* Dr. J. Richard Schrock, Chair, 620-341-5311, Fax: 620-341-5607, E-mail: jschrock@emporia.edu. *Application contact:* Dr. Derek Zelmer, Graduate Coordinator, 620-341-5623, Fax: 620-341-5607, E-mail: dzelmer@emporia.edu.

Florida Institute of Technology, Graduate Programs, College of Science, Department of Biological Sciences, Program in Cell and Molecular Biology, Melbourne, FL 32901-6975. Offers MS, PhD. Part-time programs available. *Students:* Average age 34. *Application deadline:* Applications are processed on a rolling basis. Electronic applications accepted. *Expenses:* Tuition: Part-time $825 per credit. *Financial support:* Research assistantships with full and partial tuition reimbursements, teaching assistantships with full and partial tuition reimbursements, career-related internships or fieldwork and tuition remissions available. Financial award application deadline: 3/1; financial award applicants required to submit FAFSA. *Faculty research:* Changes in DNA molecule and differential expression of genetic information during aging. *Application contact:* Carolyn P. Farrior, Director of Graduate Admissions, 321-674-7118, Fax: 321-723-9468, E-mail: cfarrior@fit.edu.

See Close-Up on page 567.

Florida State University, Graduate Studies, College of Arts and Sciences, Department of Biological Science, Program in Cell Biology, Tallahassee, FL 32306. Offers MS, PhD. *Faculty:* 12 full-time (2 women). *Students:* 27 full-time (9 women); includes 3 minority (1 African

Cell Biology

Florida State University (continued)
American, 1 Asian American or Pacific Islander, 1 Hispanic American), 11 international. *Degree requirements:* For master's and doctorate, thesis/dissertation, teaching experience seminar presentation, comprehensive exam, registration. *Entrance requirements:* For master's and doctorate, GRE General Test (minimum 1100: U-500, Q-500), minimum upper division GPA of 3.0. Additional exam requirements/recommendations for international students: Required—TOEFL (minimum score 600 paper-based; 250 computer-based), IB-100. *Application deadline:* For fall admission, 1/15 for domestic students, 12/1 for international students; for spring admission, 10/15 for domestic students, 9/1 for international students. Application fee: $30. *Financial support:* In 2005–06, fellowships with full tuition reimbursements (averaging $19,000 per year), research assistantships with full tuition reimbursements (averaging $19,000 per year), teaching assistantships with full tuition reimbursements (averaging $17,600 per year) were awarded. Financial award application deadline: 1/15; financial award applicants required to submit FAFSA. *Faculty research:* Cell-to-cell interactions, nitrogen metabolism, carbon metabolism, cell motility, DNA methylation. *Application contact:* Judy Bowers, Coordinator, Graduate Affairs, 850-644-3023, Fax: 850-644-9829, E-mail: gradinfo@bio.fsu.edu.

George Mason University, College of Science, Fairfax, VA 22030. Offers bioinformatics (MS, PhD); climate dynamics (PhD); computational sciences (MS); computational sciences and informatics (PhD); computational social science (PhD); computational techniques and applications (Certificate); earth systems and geoinformation science (PhD); earth systems science (MS); nanotechnology and nanoscience (Certificate); neuroscience (PhD); physical sciences (PhD); remote sensing and earth image processing (Certificate). Part-time and evening/weekend programs available. *Degree requirements:* For doctorate, thesis/dissertation, comprehensive exam, registration. *Entrance requirements:* For master's and doctorate, GRE General Test, minimum GPA of 3.0 in last 60 hours. Additional exam requirements/recommendations for international students: Required—TOEFL. Electronic applications accepted. *Expenses:* Tuition, area resident: Full-time $5,244; part-time $219 per credit. Tuition, state resident: part-time $651 per credit. Tuition, nonresident: full-time $15,636. Required fees: $1,524; $65 per credit. *Faculty research:* Space sciences and astrophysics, fluid dynamics, materials modeling and simulation, bioinformatics, global changes and statistics.

George Mason University, College of Arts and Sciences, Department of Biology, Program in Biology, Fairfax, VA 22030. Offers bioinformatics (MS); ecology, systematics and evolution (MS); interpretive biology (MS); molecular and microbiology (PhD); molecular, microbial, and cellular biology (MS); organismal biology (MS). Part-time programs available. *Degree requirements:* For master's, thesis or alternative. *Entrance requirements:* For master's, GRE General Test, GRE Subject Test, bachelor's degree in biology or equivalent. Electronic applications accepted. *Expenses:* Tuition, area resident: Full-time $5,244; part-time $219 per credit. Tuition, state resident: part-time $651 per credit. Tuition, nonresident: full-time $15,636. Required fees: $1,524; $65 per credit.

Georgetown University, Graduate School of Arts and Sciences, Programs in Biomedical Sciences, Department of Cell Biology, Washington, DC 20057. Offers PhD, MD/PhD. *Degree requirements:* For doctorate, thesis/dissertation, comprehensive exam. *Entrance requirements:* For doctorate, GRE General Test. Additional exam requirements/recommendations for international students: Required—TOEFL.

Georgia State University, College of Arts and Sciences, Department of Biology, Program in Cell Biology and Physiology, Atlanta, GA 30303-3083. Offers MS, PhD. *Degree requirements:* For master's, thesis or alternative, exam; for doctorate, thesis/dissertation, exam. *Entrance requirements:* For master's and doctorate, GRE General Test. Additional exam requirements/recommendations for international students: Required—TOEFL. *Expenses:* Tuition, state resident: full-time $4,368; part-time $182 per semester hour. Tuition, nonresident: full-time $8,732; part-time $728 per semester hour. Required fees: $46 per semester hour.

Harvard University, Graduate School of Arts and Sciences, Department of Molecular and Cellular Biology, Cambridge, MA 02138. Offers PhD. *Students:* 81 full-time (37 women). 325 applicants, 16% accepted. In 2005, 16 degrees awarded. *Degree requirements:* For doctorate, thesis/dissertation, oral exam. *Entrance requirements:* For doctorate, GRE General Test, GRE Subject Test (recommended). Additional exam requirements/recommendations for international students: Required—TOEFL. *Application deadline:* For fall admission, 12/15 for domestic students. Application fee: $60. *Expenses:* Tuition: Full-time $28,752. Full-time tuition and fees vary according to program and student level. *Financial support:* Fellowships, research assistantships, teaching assistantships, career-related internships or fieldwork, Federal Work-Study, and institutionally sponsored loans available. Financial award application deadline: 12/30. *Unit head:* Betsey Cogswell, Administrator, 617-495-5497, Fax: 617-495-5264. *Application contact:* James Wang, Assistant Director, 617-495-1901.

See Close-Up on page 571.

Harvard University, Graduate School of Arts and Sciences, Division of Medical Sciences, Boston, MA 02115. Offers biological chemistry and molecular pharmacology (PhD); cell biology (PhD); genetics (PhD); microbiology and molecular genetics (PhD); pathology (PhD), including experimental pathology. *Students:* 433 full-time (210 women). In 2005, 83 doctorates awarded. *Degree requirements:* For doctorate, thesis/dissertation. *Entrance requirements:* For doctorate, GRE General Test, GRE Subject Test. Additional exam requirements/recommendations for international students: Required—TOEFL. Application fee: $60. *Expenses:* Tuition: Full-time $28,752. Full-time tuition and fees vary according to program and student level. *Financial support:* Fellowships, research assistantships, teaching assistantships, institutionally sponsored loans and tuition waivers (full) available. Financial award application deadline: 1/1. *Unit head:* Administrator, 617-432-2029. *Application contact:* Administrator, 617-432-2029.

Illinois Institute of Technology, Graduate College, College of Science and Letters, Department of Biological, Chemical and Physical Sciences, Biology Division, Chicago, IL 60616-3793. Offers biochemistry (MS); biology (MBS, PhD); biotechnology (MS); cell biology (MS); microbiology (MS); molecular biochemistry and biophysics (MS, PhD). Part-time and evening/weekend programs available. Postbaccalaureate distance learning degree programs offered (no on-campus study). Terminal master's awarded for partial completion of doctoral program. *Degree requirements:* For master's, thesis (for some programs), comprehensive exam; for doctorate, thesis/dissertation, comprehensive exam. *Entrance requirements:* For master's and doctorate, GRE General Test, minimum undergraduate GPA of 3.0. Additional exam requirements/recommendations for international students: Required—TOEFL (minimum score 550 paper-based; 213 computer-based). Electronic applications accepted. *Faculty research:* Protein crystallography, small angle x-ray diffraction of muscle, spectroscopy of multidomain proteins, structure and function of cell cycle proteins, development of anticancer drugs.

Indiana University Bloomington, Graduate School, College of Arts and Sciences, Department of Biology, Program in Molecular, Cellular, and Developmental Biology, Bloomington, IN 47405-7000. Offers PhD. Offered through the University Graduate School. *Faculty:* 25 full-time (4 women). *Students:* 48 full-time (23 women); includes 6 minority (5 African Americans, 1 Hispanic American), 16 international. In 2005, 9 degrees awarded. *Degree requirements:* For doctorate, thesis/dissertation. *Entrance requirements:* For doctorate, GRE General Test. Additional exam requirements/recommendations for international students: Required—TOEFL. *Application deadline:* For fall admission, 1/15 for domestic students; for spring admission, 9/1 priority date for domestic students. Applications are processed on a rolling basis. Application fee: $45. Electronic applications accepted. *Expenses:* Tuition, state resident: full-time $5,437; part-time $227 per credit hour. Tuition, nonresident: full-time $15,836; part-time $660 per credit hour. Required fees: $821. Tuition and fees vary according to campus/location and program. *Financial support:* In 2005–06, 48 students received support; fellowships with tuition reimbursements available, research assistantships with tuition reimbursements available, teaching assistantships with tuition reimbursements available available. Financial award application deadline: 2/15. *Faculty research:* Developmental genetics, molecular evolution, macromolecular structure and function, cell biology and microbial molecular genetics. *Unit*

head: Dr. Susan Strome, Head, 812-855-5450, Fax: 812-855-6705, E-mail: sstrome@bio.indiana.edu. *Application contact:* Gretchen Clearwater, Adviser for Graduate Affairs, 812-855-1861, Fax: 812-855-6705, E-mail: biograd@bio.indiana.edu.

Indiana University Bloomington, Medical Sciences Program, Bloomington, IN 47405-7000. Offers anatomy and cell biology (MA, PhD); pharmacology (MS, PhD); physiology (MA, PhD). *Students:* 11 full-time (2 women), 9 part-time (5 women); includes 5 minority (1 African American, 4 Asian Americans or Pacific Islanders), 2 international. Average age 29. *Entrance requirements:* For master's, GRE, minimum GPA of 3.0; for doctorate, GRE. Additional exam requirements/recommendations for international students: Required—TOEFL. *Application deadline:* For fall admission, 1/15 for domestic students; for spring admission, 9/1 for domestic students. Application fee: $45 ($55 for international students). *Expenses:* Tuition, state resident: full-time $5,437; part-time $227 per credit hour. Tuition, nonresident: full-time $15,836; part-time $660 per credit hour. Required fees: $821. Tuition and fees vary according to campus/location and program. *Financial support:* Fellowships available. *Unit head:* Dr. John B. Watkins, Assistant Dean/Director, 812-855-0616. *Application contact:* Kimberly Bunch, Director of Graduate Admissions, 812-855-1119, E-mail: kbunch@indiana.edu.

Indiana University–Purdue University Indianapolis, Indiana University School of Medicine, Department of Anatomy and Cell Biology, Indianapolis, IN 46202-2896. Offers MS, PhD, MD/PhD. *Faculty:* 14 full-time (1 woman). *Students:* 1 (woman) full-time, 13 part-time (6 women), 9 international. Average age 31. *Degree requirements:* For master's, thesis or alternative; for doctorate, thesis/dissertation. *Entrance requirements:* For master's and doctorate, GRE General Test. *Application deadline:* For fall admission, 1/15 for domestic students. Application fee: $50 ($60 for international students). *Expenses:* Tuition, state resident: full-time $5,159; part-time $215 per credit hour. Tuition, nonresident: full-time $14,890; part-time $620 per credit hour. Required fees: $614. Tuition and fees vary according to campus/location and program. *Financial support:* In 2005–06, 1 fellowship was awarded; research assistantships, Federal Work-Study, institutionally sponsored loans, tuition waivers (partial), and stipends also available. Financial award application deadline: 2/15. *Faculty research:* Acoustic reflex control, osteoarthritis and bone disease, diabetes, kidney diseases, cellular and molecular neurobiology. *Unit head:* Dr. David B. Burr, Chairman, 317-274-7494, Fax: 317-278-2040, E-mail: dburr@indyvax.iupui.edu. *Application contact:* Dr. James Williams, Graduate Adviser, 317-274-3423, Fax: 317-278-2040, E-mail: williams@anatomy.iupui.edu.

Iowa State University of Science and Technology, Graduate College, College of Agriculture and College of Liberal Arts and Sciences, Department of Biochemistry, Biophysics, and Molecular Biology, Ames, IA 50011. Offers biochemistry (MS, PhD); biophysics (MS, PhD); genetics (MS, PhD); molecular, cellular, and developmental biology (MS, PhD); toxicology (MS, PhD). *Faculty:* 24 full-time, 2 part-time/adjunct. *Students:* 97 full-time (42 women), 2 part-time (1 woman); includes 2 minority (1 African American, 1 Asian American or Pacific Islander), 68 international. 28 applicants, 46% accepted, 13 enrolled. In 2005, 9 master's, 8 doctorates awarded. *Degree requirements:* For master's and doctorate, thesis/dissertation. *Entrance requirements:* For master's and doctorate, GRE General Test. Additional exam requirements/recommendations for international students: Required—TOEFL (paper score 550; computer score 213) or IELTS (score 6.5). *Application deadline:* For fall admission, 2/1 priority date for domestic students, 2/1 priority date for international students. Application fee: $30 ($70 for international students). Electronic applications accepted. *Expenses:* Tuition, state resident: full-time $6,410. Tuition, nonresident: full-time $16,422. Tuition and fees vary according to program. *Financial support:* In 2005–06, 91 research assistantships with full tuition reimbursements (averaging $16,290 per year), 2 teaching assistantships with full and partial tuition reimbursements (averaging $14,455 per year) were awarded; scholarships/grants, health care benefits, and unspecified assistantships also available. *Unit head:* Dr. Alan M. Myers, Chair, 515-294-6116, E-mail: biochem@iastate.edu.

See Close-Up on page 395.

Iowa State University of Science and Technology, Graduate College, College of Liberal Arts and Sciences and College of Agriculture, Department of Genetics, Developmental and Cell Biology, Ames, IA 50011. Offers MS, PhD. *Faculty:* 30 full-time, 3 part-time/adjunct. *Students:* 77 full-time (34 women), 55 international. 3 applicants, 100% accepted, 1 enrolled. In 2005, 6 master's, 9 doctorates awarded. *Degree requirements:* For master's and doctorate, thesis/dissertation. *Entrance requirements:* For master's and doctorate, GRE General Test. Additional exam requirements/recommendations for international students: Required—TOEFL (paper score 570; computer score 230) or IELTS (score 6.5). *Application deadline:* For fall admission, 1/1 priority date for domestic students, 1/1 priority date for international students. Application fee: $30 ($70 for international students). Electronic applications accepted. *Expenses:* Tuition, state resident: full-time $6,410. Tuition, nonresident: full-time $16,422. Tuition and fees vary according to program. *Financial support:* In 2005–06, 65 students received support, including 65 research assistantships with full and partial tuition reimbursements available (averaging $17,301 per year), 7 teaching assistantships with full and partial tuition reimbursements available (averaging $15,042 per year); fellowships with full tuition reimbursements available, scholarships/grants, health care benefits, and unspecified assistantships also available. Financial award application deadline: 2/1. *Faculty research:* Animal behavior, animal models of gene therapy, cell biology, comparative physiology, developmental biology. *Unit head:* Dr. Martin Spalding, Chair, 515-294-1749.

Iowa State University of Science and Technology, Graduate College, Interdisciplinary Programs, Program in Molecular, Cellular, and Developmental Biology, Ames, IA 50011. Offers MS, PhD. *Students:* 38 full-time (19 women), 1 part-time, 28 international. 78 applicants, 19% accepted, 7 enrolled. In 2005, 2 degrees awarded. *Degree requirements:* For master's, thesis or alternative; for doctorate, thesis/dissertation. *Entrance requirements:* For master's and doctorate, GRE General Test, resumé. Additional exam requirements/recommendations for international students: Required—TOEFL (paper score 550; computer score 220) or IELTS (score 6.5). *Application deadline:* For fall admission, 1/15 priority date for domestic students, 1/15 priority date for international students. Application fee: $30 ($70 for international students). Electronic applications accepted. *Expenses:* Tuition, state resident: full-time $6,410. Tuition, nonresident: full-time $16,422. Tuition and fees vary according to program. *Financial support:* In 2005–06, 32 research assistantships with full and partial tuition reimbursements (averaging $22,000 per year), 6 teaching assistantships with full and partial tuition reimbursements were awarded; scholarships/grants, health care benefits, and unspecified assistantships also available. *Unit head:* Dr. W Allen Miller, Supervisory Committee Chair, 515-294-7252, E-mail: idgp@iastate.edu. *Application contact:* Mary Jordison, Information Contact, 515-294-7252, Fax: 515-924-6790, E-mail: idgp@iastate.edu.

The Johns Hopkins University, National Institutes of Health Sponsored Programs, Department of Biology, Baltimore, MD 21218-2699. Offers biochemistry (PhD); biophysics (PhD); cell biology (PhD); developmental biology (PhD); genetic biology (PhD); molecular biology (PhD). *Faculty:* 25 full-time (4 women). *Students:* 126 full-time (72 women); includes 36 minority (3 African Americans, 1 American Indian/Alaska Native, 21 Asian Americans or Pacific Islanders, 11 Hispanic Americans), 19 international. 282 applicants, 26% accepted, 36 enrolled. In 2005, 15 degrees awarded. *Median time to degree:* Of those who began their doctoral program in fall 1997, 81.2% received their degree in 8 years or less. *Degree requirements:* For doctorate, thesis/dissertation, comprehensive exam, registration. *Entrance requirements:* For doctorate, GRE General Test. Additional exam requirements/recommendations for international students: Required—TOEFL (minimum score 600 paper-based; 250 computer-based), TWE, TSE. *Application deadline:* For fall admission, 12/15 for domestic students. Application fee: $60. *Expenses:* Tuition: Full-time $30,960. Tuition and fees vary according to degree level and program. *Financial support:* In 2005–06, 24 fellowships (averaging $23,000 per year), 93 research assistantships (averaging $23,000 per year), 22 teaching assistantships (averaging $23,000 per year) were awarded; Federal Work-Study, institutionally sponsored loans, scholarships/grants, traineeships, health care benefits, tuition waivers (partial), and unspecified assistantships also available. Financial award application deadline: 4/15; financial award applicants required to submit FAFSA. *Faculty research:* Protein and nucleic acid biochemistry and biophysical chemistry, molecular biology and development. Total annual research expenditures: $11.2 million. *Unit head:* Dr. Allen Shearn, Chair, 410-516-4693, Fax: 410-516-5213, E-mail: bio_cals@jhu.edu. *Application contact:* Joan Miller, Academic Affairs Manager, 410-516-5502, Fax: 410-516-5213, E-mail: joan@jhu.edu.

See Close-Up on page 155.

The Johns Hopkins University, School of Medicine, Graduate Programs in Medicine, Graduate Program in Cellular and Molecular Medicine, Baltimore, MD 21218-2699. Offers PhD. *Faculty:* 112 full-time (22 women). *Students:* 103 full-time (61 women); includes 31 minority (10 African Americans, 1 American Indian/Alaska Native, 14 Asian Americans or Pacific Islanders, 6 Hispanic Americans), 22 international. Average age 26. 230 applicants, 20% accepted, 25 enrolled. In 2005, 11 degrees awarded. *Degree requirements:* For doctorate, thesis/dissertation, oral exam. *Entrance requirements:* For doctorate, GRE General Test. Additional exam requirements/recommendations for international students: Required—TOEFL. *Application deadline:* For winter admission, 1/10 for domestic students. Applications are processed on a rolling basis. Application fee: $75. Electronic applications accepted. *Expenses:* Tuition: Full-time $30,960. Tuition and fees vary according to degree level and program. *Financial support:* In 2005–06, 12 fellowships (averaging $24,600 per year) were awarded; tuition waivers (full) also available. *Faculty research:* Cellular and molecular basis of disease. Total annual research expenditures: $100 million. *Unit head:* Dr. Pierre A. Coulombe, Director, 410-614-0410, Fax: 410-614-7294, E-mail: coulombe@jhmi.edu. *Application contact:* Theo M. Karpovich, Admissions Coordinator, 410-614-0391, Fax: 410-614-7294, E-mail: karpovic@jhmi.edu.

See Close-Up on page 573.

The Johns Hopkins University, School of Medicine, Graduate Programs in Medicine, Program in Biochemistry, Cellular and Molecular Biology, Baltimore, MD 21205. Offers PhD. *Faculty:* 91 full-time (25 women), 11 part-time/adjunct (5 women). *Students:* 155 full-time (75 women); includes 19 minority (6 African Americans, 12 Asian Americans or Pacific Islanders, 1 Hispanic American), 51 international. Average age 25. 358 applicants, 18% accepted, 20 enrolled. In 2005, 19 doctorates awarded. *Median time to degree:* Of those who began their doctoral program in fall 1997, 100% received their degree in 8 years or less. *Degree requirements:* For doctorate, thesis/dissertation, comprehensive exam. *Entrance requirements:* For doctorate, GRE General Test. Additional exam requirements/recommendations for international students: Required—TOEFL. *Application deadline:* For winter admission, 1/10 for domestic students. Applications are processed on a rolling basis. Application fee: $90. Electronic applications accepted. *Expenses:* Tuition: Full-time $30,960. Tuition and fees vary according to degree level and program. *Financial support:* In 2005–06, 3 fellowships with partial tuition reimbursements (averaging $32,000 per year), 154 research assistantships with full tuition reimbursements (averaging $24,600 per year) were awarded; traineeships and tuition waivers (full) also available. Financial award application deadline: 12/31. *Faculty research:* Developmental biology, genomics/proteomics, protein targeting, signal transduction, structural biology. *Unit head:* Dr. Craig Montell, Director, 410-955-3466, Fax: 410-614-8842, E-mail: cmontell@jhmi.edu. *Application contact:* Dr. Jeff Corden, Admissions Director, 410-955-3506, Fax: 410-614-8842, E-mail: jcorden@jhmi.edu.

Kent State University, School of Biomedical Sciences, Program in Cellular and Molecular Biology, Kent, OH 44242-0001. Offers MS, PhD. Offered in cooperation with Northeastern Ohio Universities College of Medicine. Terminal master's awarded for partial completion of doctoral program. *Degree requirements:* For master's and doctorate, thesis/dissertation. *Entrance requirements:* For master's and doctorate, GRE General Test. *Faculty research:* Molecular genetics, molecular endocrinology, virology and tumor biology, P450 enzymology and catalysis, membrane structure and function.

Louisiana State University Health Sciences Center, School of Graduate Studies in New Orleans, Department of Cell Biology and Anatomy, New Orleans, LA 70112-2223. Offers cell biology and anatomy (MS, PhD), including cell biology, developmental biology, neurobiology and anatomy. *Degree requirements:* For master's and doctorate, thesis/dissertation. *Entrance requirements:* For master's and doctorate, GRE General Test, GRE Subject Test, minimum undergraduate GPA of 3.0. Additional exam requirements/recommendations for international students: Required—TOEFL. *Faculty research:* Visual system organization, neural development, plasticity of sensory systems, information processing through the nervous system, visuomotor integration.

Louisiana State University Health Sciences Center at Shreveport, Department of Cellular Biology and Anatomy, Shreveport, LA 71130-3932. Offers MS, PhD, MD/PhD. *Faculty:* 16 full-time, 7 part-time/adjunct. *Students:* 13 full-time (8 women); includes 1 minority (African American) Average age 32. 19 applicants, 11% accepted. In 2005, 1 degree awarded. Terminal master's awarded for partial completion of doctoral program. *Degree requirements:* For master's and doctorate, thesis/dissertation. *Entrance requirements:* For master's and doctorate, GRE General Test. Additional exam requirements/recommendations for international students: Required—TOEFL. *Application deadline:* For fall admission, 5/1 for domestic students. Applications are processed on a rolling basis. Application fee: $30. *Financial support:* In 2005–06, 11 students received support, including 1 fellowship, 10 research assistantships; institutionally sponsored loans also available. Financial award application deadline: 5/1. *Faculty research:* Alcohol and immunity, neuroscience, olfactory physiology, extracellular matrix, cancer cell biology and gene therapy. *Unit head:* Dr. Leonard L. Seelig, Chairman, 318-675-5312. *Application contact:* Dr. Kathryn A. Hamilton, Graduate Coordinator, 318-675-5312.

See Close-Up on page 163.

Loyola University Chicago, Graduate School, Department of Cell Biology, Neurobiology and Anatomy, Maywood, IL 60153. Offers MS, PhD, MD/PhD. Part-time programs available. *Faculty:* 17 full-time (6 women), 5 part-time/adjunct (1 woman). *Students:* 17 full-time (8 women), 1 part-time; includes 5 minority (1 African American, 3 Asian Americans or Pacific Islanders, 1 Hispanic American), 3 international. Average age 27. 30 applicants, 23% accepted, 7 enrolled. In 2005, 4 master's, 4 doctorates awarded. Terminal master's awarded for partial completion of doctoral program. *Median time to degree:* Of those who began their doctoral program in fall 1997, 100% received their degree in 8 years or less. *Degree requirements:* For master's, thesis/dissertation; for doctorate, thesis/dissertation, comprehensive exam. *Entrance requirements:* For master's, GRE General Test, minimum GPA of 3.0; for doctorate, GRE General Test, GRE Subject Test (biology), minimum GPA of 3.0. Additional exam requirements/recommendations for international students: Required—TOEFL (minimum score 600 paper-based; 250 computer-based). *Application deadline:* For fall admission, 5/1 priority date for domestic students, 5/1 priority date for international students. Applications are processed on a rolling basis. Application fee: $0. Electronic applications accepted. *Expenses:* Tuition: Full-time $11,610; part-time $645 per credit. Required fees: $55 per semester. *Financial support:* In 2005–06, 8 fellowships with full tuition reimbursements (averaging $22,000 per year), 5 research assistantships with full tuition reimbursements (averaging $22,000 per year) were awarded; Federal Work-Study and unspecified assistantships also available. Financial award application deadline: 5/1; financial award applicants required to submit FAFSA. *Faculty research:* Brain steroids, immunology, neuroregeneration, cytokines. Total annual research expenditures: $1 million. *Unit head:* Dr. John Clancy, Chair, 708-216-3352. *Application contact:* Dr. Tharkery S. Gray, Graduate Program Director, 708-216-3345.

See Close-Up on page 575.

Marquette University, Graduate School, College of Arts and Sciences, Department of Biology, Milwaukee, WI 53201-1881. Offers cell biology (MS, PhD); developmental biology (MS, PhD); ecology (MS, PhD); endocrinology (MS, PhD); evolutionary biology (MS, PhD); genetics (MS, PhD); microbiology (MS, PhD); molecular biology (MS, PhD); muscle and exercise physiology (MS, PhD); neurobiology (MS, PhD); reproductive physiology (MS, PhD). Terminal

master's awarded for partial completion of doctoral program. *Degree requirements:* For master's, thesis, 1 year of teaching experience or equivalent, comprehensive exam; for doctorate, thesis/dissertation, 1 year of teaching experience or equivalent, qualifying exam. *Entrance requirements:* For master's and doctorate, GRE General Test, GRE Subject Test. Additional exam requirements/recommendations for international students: Required—TOEFL. *Faculty research:* Microbial and invertebrate ecology, evolution of gene function, DNA methylation, DNA arrangement.

Mayo Graduate School, Graduate Programs in Biomedical Sciences, Programs in Biochemistry, Structural Biology, Cell Biology, and Genetics, Rochester, MN 55905. Offers biochemistry and structural biology (PhD); cell biology and genetics (PhD); molecular biology (PhD). *Degree requirements:* For doctorate, oral defense of dissertation, qualifying oral and written exam. *Entrance requirements:* For doctorate, GRE, 1 year of chemistry, biology, calculus, and physics. Additional exam requirements/recommendations for international students: Required—TOEFL. Electronic applications accepted. *Faculty research:* Gene structure and function, membranes and receptors/cytoskeleton, oncogenes and growth factors, protein structure and function, steroid hormonal action.

See Close-Up on page 399.

McGill University, Faculty of Graduate and Postdoctoral Studies, Faculty of Medicine, Department of Anatomy and Cell Biology, Montréal, QC H3A 2T5, Canada. Offers M Sc, PhD. *Degree requirements:* For master's, thesis; for doctorate, one foreign language, thesis/dissertation. *Entrance requirements:* For master's, minimum GPA of 3.2. *Faculty research:* Molecular cell biology, neurobiology, spermatogenesis, developmental biology, aging.

McMaster University, Faculty of Health Sciences and School of Graduate Studies, Program in Medical Sciences, Cell Biology and Metabolism Area, Hamilton, ON L8S 4M2, Canada. Offers M Sc, PhD. *Students:* 16 full-time. In 2005, 3 master's, 3 doctorates awarded. *Degree requirements:* For master's, thesis/dissertation; for doctorate, thesis/dissertation, comprehensive exam. *Entrance requirements:* For master's, honors B Sc, B+ average in related field; for doctorate, M Sc, minimum B+ average, students with proven research experience and an A average may be admitted with a B Sc degree. Additional exam requirements/recommendations for international students: Required—TOEFL (minimum score 580 paper-based; 237 computer-based). *Application deadline:* For fall admission, 9/30 for domestic students. For winter admission, 3/1 for domestic students. Applications are processed on a rolling basis. Application fee: $85. *Financial support:* Teaching assistantships available. *Unit head:* Dr. Steve Shaughnessy, Coordinator, 905-527-2299 Ext. 43778. *Application contact:* Dr. Carl Richards, Associate Dean, 905-525-9140 Ext. 22983, Fax: 905-546-1129.

Medical College of Georgia, School of Graduate Studies, Department of Cellular Biology and Anatomy, Augusta, GA 30912-1500. Offers PhD. *Faculty:* 22 full-time (7 women). *Students:* 7 full-time (4 women), 4 part-time (1 woman); includes 2 minority (both African Americans), 2 international. Average age 30. In 2005, 1 degree awarded. *Degree requirements:* For doctorate, thesis/dissertation. *Entrance requirements:* For doctorate, GRE General Test. Additional exam requirements/recommendations for international students: Required—TOEFL. *Application deadline:* For fall admission, 6/30 for domestic students, 4/15 for international students. Applications are processed on a rolling basis. Application fee: $30. *Financial support:* In 2005–06, 2 fellowships with partial tuition reimbursements (averaging $25,500 per year), 5 research assistantships with partial tuition reimbursements (averaging $22,500 per year), 1 teaching assistantship with partial tuition reimbursement (averaging $22,500 per year) were awarded; institutionally sponsored loans and scholarships/grants also available. Support available to part-time students. Financial award application deadline: 5/31; financial award applicants required to submit FAFSA. *Faculty research:* Eye disease, diabetic infections, developmental biology, cell injury and death, stroke and neurotoxicity. *Unit head:* Dr. Sally S. Atherton, Chair, 706-721-3731, Fax: 706-721-6120, E-mail: satherton@mail.mcg.edu. *Application contact:* Dr. Zheng Dong, Program Director, 706-721-2825, Fax: 706-721-6120, E-mail: zdong@mcg.edu.

Medical College of Wisconsin, Graduate School of Biomedical Sciences, Program in Cell and Developmental Biology, Milwaukee, WI 53226-0509. Offers MS, PhD. Terminal master's awarded for partial completion of doctoral program. *Degree requirements:* For master's, thesis/dissertation; for doctorate, thesis/dissertation, comprehensive exam, registration. *Entrance requirements:* For master's and doctorate, GRE General Test. Additional exam requirements/recommendations for international students: Required—TOEFL. *Faculty research:* Neurobiology, development, neuroscience, teratology.

Medical University of South Carolina, College of Graduate Studies, Program in Molecular and Cellular Biology and Pathobiology, Charleston, SC 29425-0002. Offers cell biology and anatomy (PhD); marine biomedicine (PhD). *Faculty:* 173 full-time (41 women). *Students:* 62 full-time (32 women); includes 8 minority (3 African Americans, 1 American Indian/Alaska Native, 4 Asian Americans or Pacific Islanders), 7 international. Average age 28. 181 applicants, 34% accepted, 43 enrolled. In 2005, 20 doctorates awarded. *Degree requirements:* For doctorate, thesis/dissertation, teaching and research seminar, oral and written exams. *Entrance requirements:* For doctorate, GRE General Test, interview, minimum GPA of 3.2. Additional exam requirements/recommendations for international students: Required—TOEFL (minimum score 600 paper-based; 250 computer-based). *Application deadline:* For fall admission, 1/15 priority date for domestic students, 1/15 priority date for international students. Applications are processed on a rolling basis. Application fee: $0 ($75 for international students). Electronic applications accepted. *Financial support:* In 2005–06, 12 students received support, including fellowships with partial tuition reimbursements available (averaging $21,000 per year); Federal Work-Study and scholarships/grants also available. Financial award application deadline: 3/15; financial award applicants required to submit FAFSA. Total annual research expenditures: $10 million. *Unit head:* Dr. Donald R. Menick, Director, 843-856-5045, Fax: 843-792-6590, E-mail: menickd@musc.edu. *Application contact:* Cheryl Brown, 843-792-4620, Fax: 843-792-4645, E-mail: brownche@musc.edu.

Michigan State University, The Graduate School, College of Agriculture and Natural Resources, MSU-DOE Plant Research Laboratory, East Lansing, MI 48824. Offers biochemistry and molecular biology (PhD); cellular and molecular biology (PhD); crop and soil sciences (PhD); genetics (PhD); microbiology and molecular genetics (PhD); plant biology (PhD); plant physiology (PhD). Offered jointly with the Department of Energy. *Faculty:* 9 full-time (2 women). *Degree requirements:* For doctorate, thesis/dissertation, laboratory rotation, defense of dissertation, comprehensive exam. *Entrance requirements:* For doctorate, GRE General Test, acceptance into one of the affiliated department programs; 3 letters of recommendation; bachelor's degree or equivalent in life sciences, chemistry, biochemistry, or biophysics; research experience. Application fee: $50. Electronic applications accepted. *Expenses:* Tuition, state resident: part-time $330 per credit hour. Tuition, nonresident: part-time $685 per credit hour. Tuition and fees vary according to program. *Faculty research:* Role of hormones in the regulation of plant development and physiology, molecular mechanisms associated with signal recognition, development and application of genetic methods and materials, protein routing and function. Total annual research expenditures: $7.4 million. *Unit head:* Dr. Kenneth Keegstra, Director, 517-353-2270, Fax: 517-353-9168, E-mail: keegstra@msu.edu. *Application contact:* Janet Taylor, Graduate Program Secretary, 517-353-2270, Fax: 517-353-9168, E-mail: prl@msu.edu.

Michigan State University, The Graduate School, College of Natural Science, Program in Cell and Molecular Biology, East Lansing, MI 48824. Offers MS, PhD. *Faculty:* 78 full-time (26 women). *Students:* 39 full-time (26 women); includes 1 minority (Hispanic American), 18 international. Average age 28. 137 applicants, 6% accepted. In 2005, 6 master's, 3 doctorates awarded. *Degree requirements:* For master's, thesis or alternative, oral exam; for doctorate, thesis/dissertation, teaching, research, rotation, preliminary exam. *Entrance requirements:* For master's, GRE General Test, minimum GPA of 3.2, bachelor's degree in biological science, 3 letters of recommendation; for doctorate, GRE General Test, minimum GPA of 3.0 in science/math courses, 3 letters of recommendation, background in biology and chemistry, 1

Cell Biology

Michigan State University (continued)
year of physics, mathematics through calculus, inorganic and organic chemistry courses . Additional exam requirements/recommendations for international students: Required—TOEFL (minimum score 550 paper-based; 213 computer-based). *Application deadline:* For fall admission, 12/27 for domestic students. Application fee: $50. Electronic applications accepted. *Expenses:* Tuition, state resident: part-time $330 per credit hour. Tuition, nonresident: part-time $685 per credit hour. Tuition and fees vary according to program. *Financial support:* In 2005–06, 7 fellowships with tuition reimbursements (averaging $13,911 per year), 28 research assistantships with tuition reimbursements (averaging $15,662 per year) were awarded; teaching assistantships with tuition reimbursements, scholarships/grants and unspecified assistantships also available. Financial award application deadline: 1/1. *Faculty research:* Cancer research, immunology and virology, molecular genetics, protein structure and function, signal transduction. *Unit head:* Dr. Susan E. Conrad, Director, 517-353-5161, Fax: 517-432-8813, E-mail: conrad@msu.edu. *Application contact:* Christine Vandeuren, Graduate Secretary, 517-353-8916, Fax: 517-432-8813, E-mail: cmb@msu.edu.

Missouri State University, Graduate College, College of Health and Human Services, Department of Biomedical Sciences, Program in Cell and Molecular Biology, Springfield, MO 65804-0094. Offers MS. *Students:* 7 full-time (3 women), 1 (woman) part-time, 2 international. Average age 27. 25 applicants, 36% accepted, 2 enrolled. In 2005, 2 degrees awarded. *Degree requirements:* For master's, thesis or alternative, oral and written exams. *Entrance requirements:* For master's, GRE General Test, 2 semesters of course work in organic chemistry and physics, 1 semester of course work in calculus, minimum GPA of 3.0 in last 60 hours of course work. Additional exam requirements/recommendations for international students: Required—TOEFL (minimum score 550 paper-based; 213 computer-based), IELT (minimum score 6). *Application deadline:* For fall admission, 7/20 for domestic students; for spring admission, 12/20 priority date for domestic students. Applications are processed on a rolling basis. Application fee: $30. Electronic applications accepted. *Expenses:* Tuition, state resident: full-time $3,402; part-time $189 per credit. Tuition, nonresident: full-time $6,804; part-time $378 per credit. Required fees: $207 per semester. Part-time tuition and fees vary according to course level, course load and program. *Financial support:* In 2005–06, research assistantships with full tuition reimbursements (averaging $6,575 per year), 3 teaching assistantships with full tuition reimbursements (averaging $6,575 per year) were awarded; career-related internships or fieldwork, Federal Work-Study, institutionally sponsored loans, scholarships/grants, and unspecified assistantships also available. Support available to part-time students. Financial award application deadline: 3/31; financial award applicants required to submit FAFSA. *Unit head:* Albert R. Gordon, Graduate Director, 417-836-5730, Fax: 417-836-5588, E-mail: albertgordon@missouristate.edu. *Application contact:* Information Contact, 417-836-5603, Fax: 417-836-5588.

Mount Sinai School of Medicine of New York University, Graduate School of Biological Sciences, New York, NY 10029-6504. Offers biophysics, structural biology and biomathematics (PhD); community medicine (MPH); genetic counseling (MS); genetics and genomic sciences (PhD); mechanisms of disease and therapy (PhD); microbiology (PhD); molecular, cellular, biochemical and developmental sciences (PhD); neurosciences (PhD). *Students:* 218 full-time (109 women). 4,208 applicants, 7% accepted, 117 enrolled.Terminal master's awarded for partial completion of doctoral program. *Degree requirements:* For master's, registration; for doctorate, thesis/dissertation, registration. *Entrance requirements:* For doctorate, GRE General Test, GRE Subject Test, 3 years of college pre-med course work. Additional exam requirements/recommendations for international students: Required—TOEFL. *Application deadline:* For fall admission, 1/15 for domestic students. Application fee: $75. Electronic applications accepted. *Expenses:* Tuition: Full-time $33,250. Required fees: $1,600. Full-time tuition and fees vary according to degree level, program and reciprocity agreements. *Financial support:* In 2005–06, fellowships with full tuition reimbursements (averaging $26,000 per year), research assistantships with full tuition reimbursements (averaging $26,000 per year) were awarded; Federal Work-Study, institutionally sponsored loans, scholarships/grants, health care benefits, and unspecified assistantships also available. Financial award application deadline: 4/5; financial award applicants required to submit FAFSA. *Faculty research:* Cancer, gene therapy, minimally invasive surgery, cardiac translational research. Total annual research expenditures: $162.2 million. *Unit head:* Dr. Diomedes Logothetis, Dean, 212-241-6546, Fax: 212-241-0651, E-mail: diomedes.logothetis@mssm.edu. *Application contact:* Lily Recanati, Manager, 212-241-3267, Fax: 212-241-0651, E-mail: lily.recantati@mssm.edu.

See Close-Up on page 177.

New York Medical College, Graduate School of Basic Medical Sciences, Department of Cell Biology, Valhalla, NY 10595-1691. Offers cell biology and neuroscience (MS, PhD). Part-time and evening/weekend programs available. Terminal master's awarded for partial completion of doctoral program. *Degree requirements:* For master's, thesis/dissertation; for doctorate, thesis/dissertation, comprehensive exam. *Entrance requirements:* For master's and doctorate, GRE General Test. Additional exam requirements/recommendations for international students: Required—TOEFL. *Faculty research:* Mechanisms of growth control in skeletal muscle, cartilage differentiation, cytoskeletal functions, signal transduction pathways, neuronal development and plasticity.

See Close-Up on page 577.

New York University, School of Medicine and Graduate School of Arts and Science, Sackler Institute of Graduate Biomedical Sciences, Program in Cellular and Molecular Biology, New York, NY 10012-1019. Offers PhD, MD/PhD. *Degree requirements:* For doctorate, one foreign language, thesis/dissertation, qualifying exams, comprehensive exam. *Entrance requirements:* For doctorate, GRE General Test. Additional exam requirements/recommendations for international students: Required—TOEFL. *Faculty research:* Membrane and organelle structure and biogenesis, intracellular transport and processing of proteins, cellular recognition and cell adhesion, oncogene structure and function, action of growth factors.

North Carolina State University, College of Veterinary Medicine, Program in Comparative Biomedical Sciences, Raleigh, NC 27695. Offers cell biology and morphology (MS, PhD); epidemiology and population medicine (MS, PhD); immunology (MS, PhD); microbiology and immunology (MS, PhD); pathology (MS, PhD); pharmacology (MS, PhD); specialized veterinary medicine (MS). Part-time programs available. *Degree requirements:* For master's and doctorate, thesis/dissertation. *Entrance requirements:* For master's and doctorate, GRE General Test. Additional exam requirements/recommendations for international students: Required—TOEFL (minimum score 550 paper-based; 213 computer-based). Electronic applications accepted. Expenses: Contact institution. *Faculty research:* Infectious diseases, cell biology, pharmacology and toxicology, genomics, pathology and population medicine.

North Dakota State University, The Graduate School, College of Agriculture, Food Systems, and Natural Resources and College of Science and Mathematics, Cellular and Molecular Biology Program, Fargo, ND 58105. Offers PhD. Offered in cooperation with eleven departments in the university. *Degree requirements:* For doctorate, thesis/dissertation. *Entrance requirements:* For doctorate, GRE. Additional exam requirements/recommendations for international students: Required—TOEFL (minimum score 525 paper-based). Electronic applications accepted. *Faculty research:* Plant and animal cell biology, gene regulation, molecular genetics, plant and animal virology.

North Dakota State University, The Graduate School, College of Science and Mathematics, Department of Biological Sciences, Fargo, ND 58105. Offers biological sciences (MS); botany (MS, PhD); cellular and molecular biology (PhD); environmental and conservation sciences (MS, PhD); genomics (MS, PhD); natural resource management (MS, PhD); zoology (MS, PhD). *Faculty:* 20. *Students:* 30 full-time (10 women), 5 part-time (1 woman); includes 1 minority (Asian American or Pacific Islander), 3 international. Average age 24. 14 applicants, 43% accepted. In 2005, 4 master's awarded. *Degree requirements:* For master's and doctorate, thesis/dissertation. *Entrance requirements:* For master's and doctorate, GRE General

Test. Additional exam requirements/recommendations for international students: Required—TOEFL. *Application deadline:* For fall admission, 3/15 for domestic students; for spring admission, 10/30 priority date for domestic students. Applications are processed on a rolling basis. Application fee: $45 ($60 for international students). Electronic applications accepted. *Financial support:* In 2005–06, 3 fellowships with full tuition reimbursements (averaging $15,000 per year), 9 research assistantships with full tuition reimbursements (averaging $14,400 per year), 19 teaching assistantships with full tuition reimbursements (averaging $9,550 per year) were awarded; career-related internships or fieldwork, Federal Work-Study, institutionally sponsored loans, scholarships/grants, tuition waivers (full), and unspecified assistantships also available. Support available to part-time students. Financial award application deadline: 4/15; financial award applicants required to submit FAFSA. *Faculty research:* Comparative endocrinology, physiology, behavioral ecology, plant cell biology, aquatic biology. Total annual research expenditures: $675,000. *Unit head:* Dr. William J. Bleier, Chair, 701-231-7087, Fax: 701-231-7149, E-mail: william.bleier@ndsu.nodak.edu.

Northwestern University, The Graduate School and Judd A. and Marjorie Weinberg College of Arts and Sciences, Interdepartmental Biological Sciences Program (IBiS), Department of Biochemistry, Molecular Biology, and Cell Biology, Evanston, IL 60208. Offers biochemistry (PhD); cell and molecular biology (PhD); molecular biophysics (PhD); structural biology (PhD). Department participates in the Interdepartmental Biological Sciences Program (IBiS). *Entrance requirements:* For doctorate, GRE General Test. Additional exam requirements/recommendations for international students: Required—TOEFL (minimum score 600 paper-based), TSE (minimum score 50). Electronic applications accepted. *Faculty research:* Protein structure, protein-DNA interactions, gene regulation, development of embryos, hormone action and signal transduction.

Northwestern University, Northwestern University Feinberg School of Medicine, Department of Cell and Molecular Biology, Chicago, IL 60611.

Northwestern University, Northwestern University Feinberg School of Medicine and Interdepartmental Degree Programs, Integrated Graduate Programs in the Life Sciences, Chicago, IL 60611. Offers cancer biology (PhD); cell biology (PhD); developmental biology (PhD); evolutionary biology (PhD); immunology and microbial pathogenesis (PhD); molecular biology and genetics (PhD); neurobiology (PhD); pharmacology and toxicology (PhD); structural biology and biochemistry (PhD). *Degree requirements:* For doctorate, thesis/dissertation, written and oral qualifying exams, comprehensive exam. *Entrance requirements:* For doctorate, GRE General Test. Additional exam requirements/recommendations for international students: Required—TOEFL (minimum score 600 paper-based; 250 computer-based). Electronic applications accepted.

See Close-Up on page 189.

Oakland University, Graduate Study and Lifelong Learning, College of Arts and Sciences, Department of Biological Sciences, Rochester, MI 48309-4401. Offers biological sciences (MA, MS); cellular biology of aging (MS). *Faculty:* 7 full-time (1 woman), 1 (woman) part-time/adjunct. *Students:* 10 full-time (5 women), 15 part-time (10 women), 1 international. Average age 26. 11 applicants, 73% accepted, 9 enrolled. In 2005, 7 degrees awarded. *Degree requirements:* For master's, thesis. *Entrance requirements:* For master's, GRE Subject Test, GRE General Test, minimum GPA of 3.0 for unconditional admission. Additional exam requirements/recommendations for international students: Required—TOEFL (minimum score 550 paper-based; 213 computer-based). *Application deadline:* For fall admission, 7/15 priority date for domestic students, 5/1 priority date for international students. For winter admission, 12/1 for domestic students; for spring admission, 3/15 for domestic students. Applications are processed on a rolling basis. Application fee: $30. Electronic applications accepted. *Expenses:* Contact institution. *Financial support:* Federal Work-Study, institutionally sponsored loans, and tuition waivers (full) available. Financial award application deadline: 3/1; financial award applicants required to submit FAFSA. *Faculty research:* Mechanism producing rhythmic beating in cilia and flagella, biosynthesis in Rhodobacter spaeroides, promotor escape by RNA polyerase II, biochemical characterization of carbofuron hydroxylase. Total annual research expenditures: $433,454. *Unit head:* , Dr. Arik Dvir, Chair, 248-370-3375, Fax: 248-370-4225. *Application contact:* Dr. Keith Berven, Coordinator, 248-370-3581, Fax: 248-370-4225, E-mail: berven@oakland.edu.

The Ohio State University, College of Medicine and Public Health and Graduate School, Graduate Programs in the Basic Medical Sciences, Department of Physiology and Cell Biology, Columbus, OH 43210. Offers physiology (MS). Students need to apply through the Integrated Biomedical Science Graduate Program. *Faculty research:* Neurobiology of cell and muscle, intestinal mucosal control, autonomic neurophysiology, cardiovascular regulation, cell biology.

The Ohio State University, College of Veterinary Medicine and Graduate School, Program in Veterinary Medicine, Department of Veterinary Biosciences, Columbus, OH 43210. Offers anatomy and cellular biology (MS, PhD); pathobiology (MS, PhD); pharmacology (MS, PhD); toxicology (MS, PhD); veterinary physiology (MS, PhD). *Faculty research:* Microvasculature, muscle biology, neonatal lung and bone development.

The Ohio State University, Graduate School, College of Biological Sciences, Department of Molecular Genetics, Columbus, OH 43210. Offers cell and developmental biology (MS, PhD); genetics (MS, PhD); molecular biology (MS, PhD). *Degree requirements:* For master's and doctorate, thesis/dissertation. *Entrance requirements:* For master's and doctorate, GRE General Test, GRE Subject Test. Additional exam requirements/recommendations for international students: Required—TOEFL (minimum score 573 paper-based; 230 computer-based). Electronic applications accepted.

See Close-Up on page 785.

The Ohio State University, Graduate School, College of Biological Sciences, Program in Molecular, Cellular and Developmental Biology, Columbus, OH 43210. Offers MS, PhD. Terminal master's awarded for partial completion of doctoral program. *Degree requirements:* For master's and doctorate, thesis/dissertation. *Entrance requirements:* For master's and doctorate, GRE General Test, GRE Subject Test. Additional exam requirements/recommendations for international students: Required—TOEFL (minimum score 600 paper-based; 250 computer-based). Electronic applications accepted. *Faculty research:* Cancer biology, cell biology, developmental biolgy, DNA replication, gene expression.

Ohio University, Graduate Studies, College of Arts and Sciences, Department of Biological Sciences, Athens, OH 45701-2979. Offers biological sciences (MS, PhD); cell biology and physiology (MS, PhD); ecology and evolutionary biology (MS, PhD); exercise physiology and muscle biology (MS, PhD); microbiology (MS, PhD); neuroscience (MS, PhD). *Faculty:* 51 full-time (17 women), 6 part-time/adjunct (1 woman). *Students:* 88 full-time (40 women), 1 part-time; includes 1 minority (Hispanic American), 41 international. Average age 24. 50 applicants, 24% accepted, 10 enrolled. In 2005, 9 master's, 12 doctorates awarded. *Median time to degree:* Of those who began their doctoral program in fall 1997, 90% received their degree in 8 years or less. *Degree requirements:* For master's, thesis, 1 quarter of teaching experience; for doctorate, thesis/dissertation, 2 quarters of teaching experience, comprehensive exam. *Entrance requirements:* For master's and doctorate, GRE General Test. Additional exam requirements/recommendations for international students: Required—TOEFL (minimum score 620 paper-based; 260 computer-based). *Application deadline:* For fall admission, 1/15 for domestic students, 1/15 for international students. Application fee: $45. Electronic applications accepted. *Financial support:* In 2005–06, 87 students received support, including 2 fellowships with full tuition reimbursements available (averaging $15,000 per year), 10 research assistantships with full tuition reimbursements available (averaging $15,500 per year), 75 teaching assistantships with full tuition reimbursements available (averaging $15,500 per year); Federal Work-Study, institutionally sponsored loans, and tuition waivers (full) also available. Financial award application deadline: 1/15. *Faculty research:* Ecology and evolutionary biology, exercise physiology and muscle biology, neurobiology, cell biology, physiology. Total annual research expenditures: $2.8 million. *Unit head:* Dr. Ralph DiCaprio, Chair, 740-593-2290, Fax: 740-593-0300, E-mail: dicaprir@ohio.edu. *Application contact:* Dr. Donald B. Miles, Graduate Chair, 740-593-2317, Fax: 740-593-0300, E-mail: milesd@ohio.edu.

See Close-Up on page 191.

Ohio University, Graduate Studies, College of Arts and Sciences, Interdisciplinary Program in Molecular and Cellular Biology, Athens, OH 45701-2979. Offers MS, PhD. *Faculty:* 46 full-time (14 women). *Students:* 36 full-time (19 women), 26 international. Average age 27. 64 applicants, 39% accepted, 16 enrolled. In 2005, 1 master's, 4 doctorates awarded. *Median time to degree:* Of those who began their doctoral program in fall 1997, 100% received their degree in 8 years or less. *Degree requirements:* For master's and doctorate, thesis/dissertation, research proposal, teaching experience, comprehensive exam, registration. *Entrance requirements:* For master's and doctorate, GRE General Test. Additional exam requirements/recommendations for international students: Required—TOEFL (minimum score 620 paper-based; 260 computer-based); Recommended—TWE. *Application deadline:* For fall admission, 2/1 for domestic students, 2/1 for international students. Applications are processed on a rolling basis. Application fee: $45. *Financial support:* In 2005–06, 21 students received support, including research assistantships with full tuition reimbursements available (averaging $17,000 per year), 10 teaching assistantships with full tuition reimbursements available (averaging $17,000 per year); traineeships, health care benefits, tuition waivers (full and partial), and unspecified assistantships also available. Financial award application deadline: 2/1. *Faculty research:* Animal biotechnology, plant molecular biology, immunology, cellular genetics, biochemistry of signal transduction, cancer research, membrane transport. Total annual research expenditures: $30,000. *Unit head:* Dr. Martin Tuck, Chair, 740-593-1747, E-mail: tuck@ohio.edu. *Application contact:* Dr. Robert A. Colvin, Graduate Chair, 740-593-0198, Fax: 740-593-0300, E-mail: colvin@ohio.edu.

See Close-Up on page 581.

Oregon Health & Science University, School of Medicine, Graduate Programs in Medicine, Department of Cell and Developmental Biology, Portland, OR 97239-3098. Offers PhD, MD/PhD. *Degree requirements:* For doctorate, thesis/dissertation. *Entrance requirements:* For doctorate, GRE General Test, GRE Subject Test, MCAT. *Faculty research:* Developmental mechanisms, molecular biology of cancer, molecular neurobiology, intracellular signaling, growth factors and development.

Oregon State University, Graduate School, Program in Molecular and Cellular Biology, Corvallis, OR 97331. Offers MS, PhD. *Students:* 31 full-time (12 women), 3 part-time (2 women); includes 5 minority (1 American Indian/Alaska Native, 4 Asian Americans or Pacific Islanders), 9 international. Average age 30. In 2005, 2 master's, 7 doctorates awarded. *Degree requirements:* For doctorate, thesis/dissertation, oral and written qualifying exams. *Entrance requirements:* For doctorate, minimum GPA of 3.0 in last 90 hours. Additional exam requirements/recommendations for international students: Required—TOEFL. *Application deadline:* For fall admission, 2/15 for domestic students. Applications are processed on a rolling basis. Application fee: $50. *Expenses:* Tuition, area resident: Part-time $301 per credit. Tuition, state resident: full-time $8,139; part-time $501 per credit. Tuition, nonresident: full-time $14,376; part-time $532 per credit. Required fees: $1,266. *Financial support:* Fellowships, career-related internships or fieldwork, Federal Work-Study, and institutionally sponsored loans available. Support available to part-time students. *Unit head:* Dr. James C. Carrington, Interim Director, 541-737-3769.

The Pennsylvania State University Milton S. Hershey Medical Center, Graduate School Programs in the Biomedical Sciences, Intercollege Graduate Program in Cell and Molecular Biology, Hershey, PA 17033-2360. Offers MS, PhD, MD/PhD. *Students:* 23 full-time (12 women); includes 1 minority (Asian American or Pacific Islander), 4 international. Average age 27. Terminal master's awarded for partial completion of doctoral program. *Median time to degree:* Of those who began their doctoral program in fall 1997, 100% received their degree in 8 years or less. *Degree requirements:* For master's, thesis or alternative, registration; for doctorate, thesis/dissertation, oral exam, comprehensive exam, registration. *Entrance requirements:* For master's, GRE General Test or MCAT; for doctorate, GRE General Test or MCAT, minimum GPA of 3.0. Additional exam requirements/recommendations for international students: Required—TOEFL (minimum score 500 paper-based; 213 computer-based). Application fee: $45. *Financial support:* In 2005–06, 2 fellowships with full tuition reimbursements, 21 research assistantships with full tuition reimbursements were awarded; scholarships/grants, traineeships, health care benefits, and unspecified assistantships also available. Financial award applicants required to submit FAFSA. *Faculty research:* Membrane structure, function and modulators; cell division, differentiation and gene expression; Metastasis; intracellular events in the immune system. *Unit head:* Dr. Henry Donahue, Program Director, 717-531-1045, Fax: 717-531-4139, E-mail: cmb-grad-hmc@psu.edu. *Application contact:* Sue Myers, Administration Assistant, 717-531-1045, Fax: 717-531-4139, E-mail: cmb-grad-hmc@psu.edu.

The Pennsylvania State University University Park Campus, Graduate School, Eberly College of Science, Department of Biochemistry and Molecular Biology, Program in Cell and Developmental Biology, State College, University Park, PA 16802-1503. Offers MS, PhD. *Degree requirements:* For doctorate, thesis/dissertation, comprehensive exam. *Entrance requirements:* For doctorate, GRE General Test, GRE Subject Test. Application fee: $45. *Expenses:* Tuition, state resident: full-time $12,518; part-time $522 per credit. Tuition, nonresident: full-time $23,004; part-time $959 per credit. Required fees: $484. Tuition and fees vary according to course load, campus/location and program.

See Close-Up on page 583.

The Pennsylvania State University University Park Campus, Graduate School, Intercollege Graduate Programs, Intercollege Graduate Program in Integrative Biosciences, State College, University Park, PA 16802-1503. Offers integrative biosciences (PhD), including biomolecular transport dynamics, cell and developmental biology, cellular and molecular mechanisms of toxicity, chemical biology, ecological and molecular plant physiology, immunobiology, molecular medicine, neuroscience, nutrition science. *Students:* 110 full-time (58 women), 2 part-time (both women); includes 6 minority (3 African Americans, 3 Asian Americans or Pacific Islanders), 61 international. *Entrance requirements:* For master's and doctorate, GRE General Test. Application fee: $45. *Expenses:* Tuition, state resident: full-time $12,518; part-time $522 per credit. Tuition, nonresident: full-time $23,004; part-time $959 per credit. Required fees: $484. Tuition and fees vary according to course load, campus/location and program. *Financial support:* Fellowships available. *Unit head:* Dr. Richard J. Frisque, Co-Director, 814-863-3523, Fax: 814-863-1357, E-mail: rjf6@psu.edu.

See Close-Up on page 197.

Purdue University, Graduate School, School of Science, Department of Biological Sciences, West Lafayette, IN 47907. Offers biochemistry (PhD); biophysics (PhD); cell and developmental biology (PhD); ecology, evolutionary and population biology (PhD), including ecology, evolutionary biology, population biology; genetics (MS, PhD); microbiology (MS, PhD); molecular biology (PhD); neurobiology (MS, PhD); plant physiology (PhD). *Faculty:* 47 full-time (9 women), 4 part-time/adjunct (1 woman). *Students:* 97 full-time (53 women), 8 part-time (4 women); includes 13 minority (3 African Americans, 1 American Indian/Alaska Native, 4 Asian Americans or Pacific Islanders, 5 Hispanic Americans), 50 international. Average age 28. 168 applicants, 29% accepted, 23 enrolled. In 2005, 18 master's, 9 doctorates awarded. Terminal master's awarded for partial completion of doctoral program. *Degree requirements:* For master's, thesis (for some programs); for doctorate, thesis/dissertation, seminars, teaching experience. *Entrance requirements:* For master's and doctorate, GRE General Test. Additional exam requirements/recommendations for international students: Required—TOEFL, TSE. *Application deadline:* For fall admission, 2/15 for domestic students, 1/31 for international students. Applications are processed on a rolling basis. Application fee: $55. Electronic applications accepted. *Financial*

support: In 2005–06, 15 fellowships, 60 research assistantships, 53 teaching assistantships were awarded. Support available to part-time students. Financial award application deadline: 2/15; financial award applicants required to submit FAFSA. *Unit head:* Dr. Richard J Kuhn, Head, 765-494-4407. *Application contact:* Nancy Konopka, Graduate Studies Office Manager, 765-494-8142, Fax: 765-494-0876, E-mail: njk@bilbo.bio.purdue.edu.

Queen's University at Kingston, School of Graduate Studies and Research, Faculty of Health Sciences, Department of Anatomy and Cell Biology, Kingston, ON K7L 3N6, Canada. Offers M Sc, PhD. Part-time programs available. *Degree requirements:* For master's, thesis; for doctorate, one foreign language, thesis/dissertation, comprehensive exam. *Entrance requirements:* Additional exam requirements/recommendations for international students: Required—TOEFL. Electronic applications accepted. *Faculty research:* Human kinetics, neuroscience, reproductive biology, cardiovascular.

Quinnipiac University, School of Health Sciences, Program in Molecular and Cell Biology, Hamden, CT 06518-1940. Offers MS. Part-time and evening/weekend programs available. *Faculty:* 8 full-time (3 women), 8 part-time/adjunct (2 women). *Students:* 3 full-time (1 woman), 25 part-time (19 women); includes 3 minority (1 African American, 1 Asian American or Pacific Islander, 1 Hispanic American), 2 international. Average age 26. 14 applicants, 57% accepted, 3 enrolled. In 2005, 9 degrees awarded. *Degree requirements:* For master's, thesis optional. *Entrance requirements:* For master's, bachelor's degree in biological, medical, or health sciences; minimum GPA of 2.5. Additional exam requirements/recommendations for international students: Required—TOEFL (minimum score 575 paper-based; 233 computer-based). *Application deadline:* For fall admission, 7/30 priority date for domestic students, 5/30 priority date for international students; for spring admission, 12/15 priority date for domestic students, 10/15 priority date for international students. Applications are processed on a rolling basis. Application fee: $45. Electronic applications accepted. *Expenses:* Tuition: Part-time $570 per credit. *Financial support:* Unspecified assistantships available. Support available to part-time students. Financial award application deadline: 4/15; financial award applicants required to submit FAFSA. *Unit head:* Dr. Charlotte Hammond, Director, 203-582-8058, E-mail: charlotte.hammond@quinnipiac.edu. *Application contact:* Louise Howe, Associate Director of Graduate Admissions, 800-462-1944, Fax: 203-582-3443, E-mail: graduate@quinnipiac.edu.

See Close-Up on page 587.

Rensselaer Polytechnic Institute, Graduate School, School of Science, Department of Biology, Troy, NY 12180-3590. Offers biochemistry (MS, PhD); biophysics (MS, PhD); cell biology (MS, PhD); developmental biology (MS, PhD); microbiology (MS, PhD); molecular biology (MS, PhD). Part-time programs available. Terminal master's awarded for partial completion of doctoral program. *Degree requirements:* For master's and doctorate, thesis/dissertation, comprehensive exam, registration. *Entrance requirements:* For master's and doctorate, GRE General Test. Additional exam requirements/recommendations for international students: Required—TOEFL. Electronic applications accepted. *Expenses:* Tuition: Full-time $31,000; part-time $1,320 per credit. Required fees: $1,623. *Faculty research:* Bioinformatics, molecular biology/biochemistry, cell and tissue biology, environment, ecology.

See Close-Up on page 199.

Rice University, Graduate Programs, Wiess School of Natural Sciences, Department of Biochemistry and Cell Biology, Houston, TX 77251-1892. Offers MA, PhD. *Degree requirements:* For master's and doctorate, thesis/dissertation. *Entrance requirements:* For master's and doctorate, GRE. Additional exam requirements/recommendations for international students: Required—TOEFL. Electronic applications accepted. Expenses: Contact institution. *Faculty research:* Steroid metabolism, protein structure NMR, biophysics, cell growth and movement.

See Close-Up on page 409.

Rosalind Franklin University of Medicine and Science, School of Graduate and Postdoctoral Studies, Department of Cell Biology and Anatomy, Program in Cell Biology, North Chicago, IL 60064-3095. Offers MS, PhD, MD/MS, MD/PhD. *Degree requirements:* For master's, thesis, qualifying exam; for doctorate, thesis/dissertation, original research project, comprehensive exam. *Entrance requirements:* For master's and doctorate, GRE General Test, minimum GPA of 3.0. Additional exam requirements/recommendations for international students: Required—TOEFL, TWE.

Rush University, Graduate College, Division of Anatomy and Cell Biology, Chicago, IL 60612-3832. Offers MS, PhD, MD/MS, MD/PhD. *Faculty:* 7 full-time (1 woman), 4 part-time/adjunct (3 women). *Students:* 6 full-time (2 women); includes 1 minority (Asian American or Pacific Islander), 2 International. Average age 22. 8 applicants, 13% accepted, 1 enrolled. In 2005, 2 master's awarded. Terminal master's awarded for partial completion of doctoral program. *Median time to degree:* Of those who began their doctoral program in fall 1997, 100% received their degree in 8 years or less. *Degree requirements:* For master's, thesis; for doctorate, thesis/dissertation, preliminary exam, dissertation proposal, comprehensive exam. *Entrance requirements:* For master's, GRE General Test, minimum GPA of 3.0, bachelor's degree in biology or chemistry (preferred), interview; for doctorate, GRE General Test, minimum GPA of 3.0, interview. Additional exam requirements/recommendations for international students: Required—TOEFL. *Application deadline:* For spring admission, 4/1 for domestic students, 4/1 for international students. Applications are processed on a rolling basis. Application fee: $25. Electronic applications accepted. *Financial support:* In 2005–06, 2 students received support, including fellowships with full tuition reimbursements available (averaging $21,000 per year), research assistantships with full tuition reimbursements available (averaging $2,500 per year); teaching assistantships with full tuition reimbursements available (averaging $2,500 per year); Federal Work-Study, institutionally sponsored loans, tuition waivers (full), and unspecified assistantships also available. Support available to part-time students. Financial award applicants required to submit FAFSA. *Faculty research:* Incontinence following vaginal distension, knee replacement, biomimetric materials, injured spinal motoneurons, implant fixation. Total annual research expenditures: $1 million. *Unit head:* Dr. Frank Hughes, Director, 312-942-6783, Fax: 312-942-5744, E-mail: frank_f_hughes@rush.edu. *Application contact:* Connie M. Lambert, Department Administrator, 312-563-2563, Fax: 312-563-3552, E-mail: connie_m_lambert@rush.edu.

Rutgers, The State University of New Jersey, New Brunswick/Piscataway, Graduate School, Core Curriculum in Molecular and Cell Biology, New Brunswick, NJ 08901-1281. Offers PhD. *Degree requirements:* For doctorate, thesis/dissertation. *Entrance requirements:* For doctorate, GRE General Test, GRE Subject Test. Additional exam requirements/recommendations for international students: Required—TOEFL. *Expenses:* Tuition, state resident: full-time $10,440; part-time $435 per credit. Tuition, nonresident: full-time $15,520; part-time $647 per credit. Required fees: $129 per credit. Tuition and fees vary according to program. *Financial support:* Fellowships available. *Unit head:* Michael Leibowitz, Program Director, 732-235-4795.

Rutgers, The State University of New Jersey, New Brunswick/Piscataway, Graduate School, Program in Cell and Developmental Biology, New Brunswick, NJ 08901-1281. Offers cell biology (MS, PhD); developmental biology (MS, PhD); immunology (MS). Part-time programs available. *Faculty:* 121 full-time. *Students:* 37 full-time (17 women), 9 part-time (8 women); includes 12 minority (2 African Americans, 1 American Indian/Alaska Native, 6 Asian Americans or Pacific Islanders, 3 Hispanic Americans), 20 international. Average age 31. 74 applicants, 36% accepted, 13 enrolled. In 2005, 8 master's, 7 doctorates awarded. Terminal master's awarded for partial completion of doctoral program. *Degree requirements:* For master's, thesis or alternative; for doctorate, thesis/dissertation, written qualifying exam. *Entrance requirements:* For master's, GRE General Test; for doctorate, GRE General Test, GRE Subject Test (recommended), minimum GPA of 3.0. Additional exam requirements/recommendations for international students: Required—TOEFL. *Application deadline:* For fall admission, 1/5 for domestic students. Applications are processed on a rolling basis. Application fee: $50. Electronic applications accepted. *Expenses:* Tuition, state resident: full-time $10,440; part-time $435 per

Cell Biology

Rutgers, The State University of New Jersey, New Brunswick/Piscataway *(continued)*
credit. Tuition, nonresident: full-time $15,520; part-time $647 per credit. Required fees: $129 per credit. Tuition and fees vary according to program. *Financial support:* In 2005–06, 36 students received support, including 16 fellowships with full tuition reimbursements available (averaging $24,000 per year), 14 research assistantships with full tuition reimbursements available (averaging $21,500 per year), 6 teaching assistantships with full tuition reimbursements available (averaging $16,988 per year) Financial award application deadline: 1/15; financial award applicants required to submit FAFSA. *Faculty research:* Signal transduction, developmental biology, cell biology, developmental genetics, developmental neurobiology. *Unit head:* Dr. Richard Padgett, Acting Director, 732-445-0251, Fax: 732-445-6370, E-mail: padgett@waksman.rutgers.edu. *Application contact:* Carolyn J. Ambrose, Administrative Assistant, 732-445-3430, Fax: 732-445-6370, E-mail: ambrose@biology.rutgers.edu.

Saint Joseph College, Graduate Division, Department of Biology, West Hartford, CT 06117-2700. Offers biology (MS), including general biology, molecular and cellular biology; biology/chemistry (MS). MS biology (including general biology; molecular and cellular biology) offered online only. Part-time and evening/weekend programs available. Postbaccalaureate distance learning degree programs offered (no on-campus study). *Faculty:* 4 full-time (2 women), 5 part-time/adjunct (3 women). *Students:* Average age 34. 34 applicants, 100% accepted, 11 enrolled. In 2005, 13 degrees awarded. *Degree requirements:* For master's, thesis or alternative, comprehensive exam. *Entrance requirements:* For master's, 2 letters of recommendation. *Application deadline:* Applications are processed on a rolling basis. Application fee: $50. Electronic applications accepted. *Expenses:* Tuition: Part-time $540 per credit. Required fees: $25 per credit. *Financial support:* Career-related internships or fieldwork, health care benefits, and unspecified assistantships available. Support available to part-time students. Financial award application deadline: 7/15; financial award applicants required to submit FAFSA. *Faculty research:* Neurology, cardiology, immunology, mircobiology. *Unit head:* Dr. Charles Morgan, Chair, 860-231-5335, E-mail: cmorgan@sjc.edu.

San Diego State University, Graduate and Research Affairs, College of Sciences, Department of Biological Sciences, San Diego, CA 92182. Offers biology (MA, MS), including ecology (MS), molecular biology (MS), physiology (MS), systematics/evolution (MS); biostatistics and biometry (PhD); cell and molecular biology (PhD); ecology (MS, PhD); microbiology (MS). *Students:* 73 full-time (36 women), 100 part-time (53 women); includes 33 minority (2 African Americans, 16 Asian Americans or Pacific Islanders, 15 Hispanic Americans), 33 international. Average age 26. 228 applicants, 23% accepted, 32 enrolled. In 2005, 27 master's, 9 doctorates awarded. Terminal master's awarded for partial completion of doctoral program. *Degree requirements:* For master's and doctorate, thesis/dissertation. *Entrance requirements:* For master's, GRE General Test, GRE Subject Test, resumé or curriculum vitae, 2 letters of recommendation. Additional exam requirements/recommendations for international students: Required—TOEFL. *Application deadline:* For fall admission, 5/1 for domestic students; for spring admission, 11/1 for domestic students, 10/1 for international students. Applications are processed on a rolling basis. Application fee: $55. Electronic applications accepted. *Financial support:* In 2005–06, 116 teaching assistantships were awarded; fellowships, research assistantships, career-related internships or fieldwork and unspecified assistantships also available. Financial award applicants required to submit FAFSA. Total annual research expenditures: $9.9 million. *Unit head:* Christopher Glembotski, Chair, 619-594-6767, Fax: 619-594-5676. *Application contact:* Terry Frey, Graduate Coordinator, 619-594-6756, Fax: 619-594-5676, E-mail: gradcoor@sciences.sdsu.edu.

San Diego State University, Graduate and Research Affairs, College of Sciences, Molecular Biology Institute, Program in Cell and Molecular Biology, San Diego, CA 92182. Offers PhD. *Students:* 14 full-time (10 women), 29 part-time (14 women); includes 11 minority (7 Asian Americans or Pacific Islanders, 4 Hispanic Americans), 11 international. 45 applicants, 60% accepted, 10 enrolled. *Degree requirements:* For doctorate, thesis/dissertation, oral comprehensive qualifying exam. *Entrance requirements:* For doctorate, GRE General Test, GRE Subject Test, resumé or curriculum vitae, 3 letters of recommendation. *Application deadline:* For fall admission, 5/1 for domestic students; for spring admission, 11/1 for domestic students, 10/1 for international students. Applications are processed on a rolling basis. Application fee: $55. Electronic applications accepted. *Financial support:* In 2005–06, 5 fellowships were awarded; institutionally sponsored loans and unspecified assistantships also available. Financial award applicants required to submit CSS PROFILE or FAFSA. *Faculty research:* Structure/dynamics of protein kinesis, chromatin structure and DNA methylation membrane biochemistry, secretory protein targeting, molecular biology of cardiac myocytes. *Unit head:* Dr. Sanford Bernstein, Director, 619-594-5504, Fax: 619-594-5676, E-mail: sbernstein@sunstroke.sdsu.edu. *Application contact:* Dr. Sanford Bernstein, Graduate Advisor, 619-594-5504, Fax: 619-594-5676, E-mail: sbernstein@sunstroke.sdsu.edu.

See Close-Up on page 591.

San Francisco State University, Division of Graduate Studies, College of Science and Engineering, Department of Biology, Program in Cell and Molecular Biology, San Francisco, CA 94132-1722. Offers MA. *Entrance requirements:* For master's, minimum GPA of 2.5 in last 60 units.

State University of New York at Buffalo, Graduate School, Graduate Programs in Cancer Research and Biomedical Sciences at Roswell Park Cancer Institute, Department of Cellular and Molecular Biology at Roswell Park Cancer Institute, Buffalo, NY 14263. Offers PhD. *Faculty:* 30 part-time/adjunct (5 women). *Students:* 18 full-time (5 women), 5 part-time (3 women); includes 1 minority (Asian American or Pacific Islander), 9 international. Average age 25. 40 applicants, 33% accepted, 3 enrolled. In 2005, 5 degrees awarded. *Degree requirements:* For doctorate, thesis/dissertation, exam project. *Entrance requirements:* For doctorate, GRE General Test, minimum B average in undergraduate coursework. Additional exam requirements/recommendations for international students: Required—TOEFL, TWE, TSE. *Application deadline:* For fall admission, 2/1 for domestic students. Applications are processed on a rolling basis. Application fee: $35. Electronic applications accepted. *Financial support:* In 2005–06, 4 fellowships with full tuition reimbursements (averaging $21,000 per year), 17 research assistantships with full tuition reimbursements (averaging $21,000 per year) were awarded; unspecified assistantships also available. Financial award application deadline: 2/1; financial award applicants required to submit FAFSA. *Faculty research:* Cancer genetics, chromatin structure and replication, regulation of transcription, human gene mapping, genetic and structural approaches to regulation of gene expression. Total annual research expenditures: $4.5 million. *Unit head:* Dr. John Yates, Chair, 716-845-8964, Fax: 716-845-8449, E-mail: john.yates@roswellpark.org. *Application contact:* Craig R. Johnson, Director of Admissions, 716-845-2339, Fax: 716-845-8178, E-mail: craig.johnson@roswellpark.edu.

State University of New York Downstate Medical Center, School of Graduate Studies, Program in Molecular and Cellular Biology, Brooklyn, NY 11203-2098. Offers PhD, MD/PhD. Affiliation with a particular PhD degree-granting program is deferred to the second year. *Faculty:* 50 full-time (10 women). *Students:* 43 full-time (20 women); includes 4 Asian Americans or Pacific Islanders, 26 international. Average age 30. 42 applicants, 26% accepted, 5 enrolled. In 2005, 4 degrees awarded. *Degree requirements:* For doctorate, thesis/dissertation, comprehensive exam, registration. *Entrance requirements:* For doctorate, GRE General Test. *Application deadline:* For fall admission, 4/1 for domestic students. Applications are processed on a rolling basis. Application fee: $35. *Financial support:* In 2005–06, 43 students received support, including 43 teaching assistantships with tuition reimbursements available (averaging $23,500 per year); fellowships, research assistantships, career-related internships or fieldwork, Federal Work-Study, health care benefits, and tuition waivers (full) also available. *Faculty research:* Mechanism of gene regulation, molecular virology. *Unit head:* Dr. William Cuirico, Chair, 718-270-3755, Fax: 718-270-1308, E-mail: wcuirico@downstate.edu. *Application contact:* Denise Sheares, Admissions Officer, 718-270-2738, Fax: 718-270-3378, E-mail: dsheares@downstate.edu.

State University of New York Upstate Medical University, College of Graduate Studies, Department of Anatomy and Cell Biology, Syracuse, NY 13210-2334. Offers MS, PhD, MD/PhD. Terminal master's awarded for partial completion of doctoral program. *Degree requirements:* For master's, thesis/dissertation; for doctorate, thesis/dissertation, comprehensive exam. *Entrance requirements:* For master's, GRE General Test, GRE Subject Test, interview; for doctorate, GRE General Test, interview. Additional exam requirements/recommendations for international students: Required—TOEFL.

State University of New York Upstate Medical University, College of Graduate Studies, Department of Cell and Developmental Biology, Syracuse, NY 13210-2334. Offers PhD, MD/PhD. *Degree requirements:* For doctorate, thesis/dissertation, comprehensive exam. *Entrance requirements:* For doctorate, GRE General Test, GRE Subject Test. Additional exam requirements/recommendations for international students: Required—TSE.

State University of New York Upstate Medical University, College of Graduate Studies, Program in Cell and Molecular Biology, Syracuse, NY 13210-2334. Offers PhD. *Degree requirements:* For doctorate, thesis/dissertation, comprehensive exam. *Entrance requirements:* For doctorate, GRE General Test, interview. Additional exam requirements/recommendations for international students: Required—TOEFL.

Stony Brook University, State University of New York, Graduate School, College of Arts and Sciences, Department of Biochemistry and Cell Biology, Molecular and Cellular Biology Program, Stony Brook, NY 11794. Offers biochemistry and molecular biology (PhD); biological sciences (MA); cellular and developmental biology (PhD); immunology and pathology (PhD); molecular and cellular biology (PhD). *Students:* 116 full-time (69 women), 2 part-time (1 woman); includes 16 minority (3 African Americans, 10 Asian Americans or Pacific Islanders, 3 Hispanic Americans), 62 international. Average age 30. 342 applicants, 14% accepted. In 2005, 7 degrees awarded. *Degree requirements:* For doctorate, thesis/dissertation, teaching experience, comprehensive exam. *Entrance requirements:* For doctorate, GRE General Test, GRE Subject Test. Additional exam requirements/recommendations for international students: Required—TOEFL. *Application deadline:* For fall admission, 1/15 for international students. Application fee: $50. *Expenses:* Tuition, state resident: full-time $6,900; part-time $288 per credit. Tuition, nonresident: full-time $10,920; part-time $455 per credit. Required fees: $704. *Financial support:* Fellowships, research assistantships, teaching assistantships, Federal Work-Study available. *Application contact:* Information Contact, 631-632-8533, Fax: 631-632-9730.

See Close-Up on page 597.

Temple University, Health Sciences Center, School of Medicine and Graduate School, Graduate Programs in Medicine, Department of Anatomy and Cell Biology, Philadelphia, PA 19122-6096. Offers MS, PhD. *Faculty:* 18 full-time (6 women). *Students:* 7 full-time (2 women), 2 part-time (1 woman), 7 international. 10 applicants, 60% accepted, 2 enrolled. *Degree requirements:* For doctorate, thesis/dissertation, research seminars. *Entrance requirements:* For master's and doctorate, GRE General Test, GRE Subject Test, minimum GPA of 3.0. Additional exam requirements/recommendations for international students: Required—TOEFL. *Application deadline:* For fall admission, 1/15 for domestic students, 12/15 for international students; for spring admission, 9/1 for domestic students, 8/1 for international students. Application fee: $50. Electronic applications accepted. *Expenses:* Tuition, state resident: full-time $8,694; part-time $483 per credit. Tuition, nonresident: full-time $12,672; part-time $704 per credit. Required fees: $500; $122 per semester. Tuition and fees vary according to course level, campus/location and program. *Financial support:* Fellowships, Federal Work-Study available. Financial award application deadline: 1/15; financial award applicants required to submit FAFSA. *Faculty research:* Neurobiology, reproductive biology, cardiovascular system, musculoskeletal biology, developmental biology. Total annual research expenditures: $632,373. *Unit head:* Dr. Steven Popoff, Chair, 215-707-3163, Fax: 215-707-2966, E-mail: spopoff@temple.edu.

Texas A&M University, College of Science, Department of Biology, Program in Molecular and Cell Biology, College Station, TX 77843. Offers PhD. Program composed of members from 4 colleges and 11 departments. *Degree requirements:* For doctorate, thesis/dissertation. *Entrance requirements:* For doctorate, GRE General Test. Additional exam requirements/recommendations for international students: Required—TOEFL. Application fee: $50 ($75 for international students). *Expenses:* Tuition, state resident: full-time $4,488; part-time $187 per credit hour. Tuition, nonresident: full-time $11,112; part-time $463 per credit hour. Required fees: $1,974. *Financial support:* Fellowships available. Financial award application deadline: 4/1; financial award applicants required to submit FAFSA. *Application contact:* Graduate Advisor, 979-826-2465, Fax: 979-845-2891.

Texas A&M University System Health Science Center, Graduate School of Biomedical Sciences, Program in Cell and Molecular Biology, College Station, TX 77840. Offers PhD.

Texas Tech University Health Sciences Center, Graduate School of Biomedical Sciences, Department of Cell Biology and Biochemistry, Program in Cell and Molecular Biology, Lubbock, TX 79430. Offers MS, PhD, MD/PhD, MS/PhD. *Faculty:* 12 full-time (5 women), 3 part-time/adjunct (1 woman). *Students:* 17 full-time (9 women), 1 (woman) part-time; includes 1 minority (Asian American or Pacific Islander), 6 international. Average age 28. 24 applicants, 33% accepted, 6 enrolled. In 2005, 1 master's, 1 doctorate awarded. Terminal master's awarded for partial completion of doctoral program. *Degree requirements:* For master's and doctorate, thesis/dissertation, comprehensive exam, registration. *Entrance requirements:* For master's and doctorate, GRE General Test, minimum GPA of 3.0. Additional exam requirements/recommendations for international students: Required—TOEFL. *Application deadline:* For fall admission, 5/15 priority date for domestic students, 4/15 priority date for international students; for spring admission, 11/15 for domestic students, 10/15 for international students. Application fee: $45. *Financial support:* In 2005–06, fellowships with full tuition reimbursements (averaging $20,500 per year), 11 research assistantships (averaging $20,500 per year) were awarded; health care benefits also available. *Faculty research:* Biochemical endocrinology, neurobiology, molecular biology, reproductive biology, biology of developing systems. Total annual research expenditures: $2.6 million. *Unit head:* Dr. Martine Coué, Associate Professor, 806-743-1558, Fax: 806-743-2990, E-mail: martine.coue@ttuhsc.edu. *Application contact:* Pam Roddy, Assistant Director, 806-743-2701, Fax: 806-743-2990, E-mail: pam.roddy@ttuhsc.edu.

Thomas Jefferson University, Jefferson College of Graduate Studies, Graduate Program in Molecular Cell Biology, Philadelphia, PA 19107. Offers developmental biology and teratology (PhD); pathology and cell biology (PhD). *Faculty:* 79 full-time. *Students:* 22 full-time (12 women), 1 part-time; includes 2 minority (1 African American, 1 Asian American or Pacific Islander), 3 international. 34 applicants, 26% accepted, 3 enrolled. In 2005, 2 degrees awarded. *Degree requirements:* For doctorate, thesis/dissertation. *Entrance requirements:* For doctorate, GRE General Test, minimum GPA of 3.2. Additional exam requirements/recommendations for international students: Required—TOEFL (minimum score 213 computer-based). *Application deadline:* For fall admission, 3/1 priority date for domestic students, 1/1 priority date for international students. Applications are processed on a rolling basis. Application fee: $50. Electronic applications accepted. *Expenses:* Tuition: Full-time $14,894; part-time $800 per credit. *Financial support:* In 2005–06, 2 students received support, including 222 fellowships with full tuition reimbursements available; research assistantships, Federal Work-Study, institutionally sponsored loans, scholarships/grants, and traineeships also available. Support available to part-time students. Financial award application deadline: 5/1; financial award applicants required to submit FAFSA. *Unit head:* Dr. Theodore Taraschi, Program Director, 215-503-5020, Fax: 215-503-0206, E-mail: theodore.taraschi@jefferson.edu. *Application contact:* Jessie F. Pervall, Director of Admissions, 215-503-0155, Fax: 215-503-9920, E-mail: jessie.pervall@jefferson.edu.

Tufts University, Sackler School of Graduate Biomedical Sciences, Integrated Studies Program, Medford, MA 02155. Offers PhD. *Students:* 12 full-time (10 women); includes 2 minority (both

Cell Biology

Asian Americans or Pacific Islanders), 4 international. Average age 26. 264 applicants, 19% accepted, 12 enrolled. *Entrance requirements:* For doctorate, GRE General Test, 3 letters of reference. Additional exam requirements/recommendations for international students: Required—TOEFL. *Application deadline:* For fall admission, 1/15 priority date for domestic students, 1/15 priority date for international students. Application fee: $65. *Financial support:* In 2005–06, 12 students received support, including 12 research assistantships (averaging $29,000 per year) Financial award application deadline: 1/15.

Tufts University, Sackler School of Graduate Biomedical Sciences, Program in Cell, Molecular and Developmental Biology, Boston, MA 02155. Offers PhD. Applications are processed through integrated studies program. *Faculty:* 31 full-time (9 women). *Students:* 28 full-time (17 women), 8 international. Average age 29. In 2005, 3 degrees awarded. *Degree requirements:* For doctorate, thesis/dissertation. *Financial support:* In 2005–06, 28 students received support, including 28 research assistantships with full tuition reimbursements available (averaging $29,000 per year); fellowships, scholarships/grants, health care benefits, and tuition waivers (full) also available. Financial award application deadline: 1/15. *Faculty research:* Reproduction and hormone action, control of gene expression, cell-matrix and cell-cell interactions, growth control and tumorigenesis, cytoskeleton and contractile proteins. *Unit head:* Dr. John Castellot, Program Director, 617-636-0303, Fax: 617-636-0375, E-mail: john.castellot@tufts.edu. *Application contact:* 617-636-6767, Fax: 617-636-0375, E-mail: sackler-school@tufts.edu.

Tulane University, Graduate School, Department of Cell and Molecular Biology, New Orleans, LA 70118-5669. Offers MS, PhD. Terminal master's awarded for partial completion of doctoral program. *Degree requirements:* For doctorate, thesis/dissertation. *Entrance requirements:* For master's, GRE General Test, minimum B average in undergraduate course work; for doctorate, GRE General Test. Additional exam requirements/recommendations for international students: Required—TOEFL; Recommended—TSE. Electronic applications accepted.

Tulane University, School of Medicine and Graduate School, Graduate Programs in Medicine, Interdisciplinary Graduate Program in Molecular and Cellular Biology, New Orleans, LA 70118-5669. Offers PhD, MD/PhD. PhD offered through the Graduate School. *Degree requirements:* For doctorate, thesis/dissertation. *Entrance requirements:* For doctorate, GRE General Test, GRE Subject Test. Additional exam requirements/recommendations for international students: Required—TOEFL; Recommended—TSE. Electronic applications accepted. *Faculty research:* Developmental biology, neuroscience, virology.

Uniformed Services University of the Health Sciences, School of Medicine, Programs in Biomedical Sciences, Graduate Program in Molecular and Cell Biology, Bethesda, MD 20814-4799. Offers PhD. *Faculty:* 17 full-time (6 women), 3 part-time/adjunct (0 women). *Students:* 19 full-time (8 women); includes 7 minority (1 African American, 5 Asian Americans or Pacific Islanders, 1 Hispanic American), 2 international. Average age 26. 27 applicants, 37% accepted, 3 enrolled. *Median time to degree:* Of those who began their doctoral program in fall 1997, 100% received their degree in 8 years or less. *Degree requirements:* For doctorate, thesis/dissertation, qualifying exam, comprehensive exam. *Entrance requirements:* For doctorate, GRE General Test, minimum GPA of 3.0. Additional exam requirements/recommendations for international students: Required—TOEFL. *Application deadline:* For fall admission, 1/15 for domestic students. Applications are processed on a rolling basis. Application fee: $0. *Financial support:* In 2005–06, fellowships with full tuition reimbursements (averaging $23,000 per year); career-related internships or fieldwork and tuition waivers (full) also available. *Faculty research:* Immunology, biochemistry. *Unit head:* Dr. Jeffrey Harmon, Director, 301-295-3248, Fax: 301-295-1996, E-mail: jharmon@usuhs.mil. *Application contact:* Janet M. Anastasi, Graduate Program Coordinator, 301-295-9474, Fax: 301-295-6772, E-mail: janastasi@usuhs.mil.

See Close-Up on page 599.

Université de Montréal, Faculty of Medicine and Faculty of Graduate Studies, Graduate Programs in Medicine, Department of Pathology and Cellular Biology, Montréal, QC H3C 3J7, Canada. Offers M Sc, PhD. *Faculty:* 32 full-time (9 women), 1 part-time/adjunct (0 women). *Students:* 45 full-time (25 women). 26 applicants, 12% accepted, 3 enrolled. In 2005, 9 master's, 6 doctorates awarded. Terminal master's awarded for partial completion of doctoral program. *Degree requirements:* For master's, thesis; for doctorate, thesis/dissertation, general exam. *Entrance requirements:* For master's and doctorate, proficiency in French, knowledge of English. *Application deadline:* For fall and spring admission, 2/1. For winter admission, 11/1 for prospective students. Application fee: $30. Electronic applications accepted. *Financial support:* Tuition waivers (full) available. *Faculty research:* Immunopathology, cardiovascular pathology, oncogenetics, cellular neurocytology, muscular dystrophy. *Unit head:* Vincent F. Castellucci, Interim Director, 514-343-6294, Fax: 514-343-5755. *Application contact:* Roger Lippé, Information Contact, 514-343-5616, Fax: 514-343-5755.

Université de Sherbrooke, Faculty of Medicine and Health Sciences, Graduate Programs in Medicine, Department of Anatomy and Cell Biology, Sherbrooke, QC J1K 2R1, Canada. Offers cell biology (M Sc, PhD). *Faculty:* 10 full-time (2 women). *Students:* 15 full-time (10 women), 18 part-time (10 women). Average age 24. 6 applicants, 0% accepted. In 2005, 3 master's, 1 doctorate awarded. Terminal master's awarded for partial completion of doctoral program. *Degree requirements:* For master's and doctorate, thesis/dissertation. *Application deadline:* For fall admission, 6/30 for domestic students. For winter admission, 10/31 for domestic students; for spring admission, 2/28 for domestic students. Application fee: $50. Electronic applications accepted. *Unit head:* Dr. Nathalie Rivard, Director, 819-820-6868 Ext. 15431, E-mail: nathalie.rivard@usherbrooke.ca.

Université Laval, Faculty of Medicine, Graduate Programs in Medicine, Programs in Cellular and Molecular Biology, Québec, QC G1K 7P4, Canada. Offers M Sc, PhD. Terminal master's awarded for partial completion of doctoral program. *Degree requirements:* For master's, thesis/dissertation; for doctorate, thesis/dissertation, comprehensive exam. *Entrance requirements:* For master's and doctorate, knowledge of French, comprehension of written English. Electronic applications accepted. *Faculty research:* Oral bacterial metabolism, sugar transport.

University at Albany, State University of New York, College of Arts and Sciences, Department of Biological Sciences, Specialization in Molecular, Cellular, Developmental, and Neural Biology, Albany, NY 12222-0001. Offers MS, PhD. *Degree requirements:* For master's, one foreign language; for doctorate, one foreign language, thesis/dissertation. *Entrance requirements:* For master's and doctorate, GRE General Test. Application fee: $60. *Financial support:* Minority assistantships available. *Unit head:* Dr. Albert Millis, Chair, Department of Biological Sciences, 518-442-4300.

University at Albany, State University of New York, School of Public Health, Department of Biomedical Sciences, Program in Cell and Molecular Structure, Albany, NY 12222-0001. Offers MS, PhD. *Degree requirements:* For master's and doctorate, thesis/dissertation. *Entrance requirements:* For master's and doctorate, GRE General Test, GRE Subject Test. Application fee: $60. *Financial support:* Application deadline: 2/1. *Unit head:* Dr. James Dias, Chair, Department of Biomedical Sciences, 518-474-2662.

The University of Alabama at Birmingham, Graduate Programs in Joint Health Sciences, Birmingham, AL 35294. Offers biochemistry and molecular genetics (PhD), including biochemistry; cell biology (PhD), including cell biology, cellular and molecular biology, cellular and molecular physiology, neuroscience; genetics (PhD); microbiology (PhD); neurobiology (PhD); pathology (PhD); pharmacology and toxicology (PhD), including pharmacology, toxicology; physiology and biophysics (MSBMS, PhD), including basic medical sciences (MSBMS), biophysical sciences (PhD), integrative biomedical sciences (PhD). *Students:* 439 full-time (214 women), 6 part-time (2 women); includes 74 minority (36 African Americans, 4 American Indian/Alaska Native, 28 Asian Americans or Pacific Islanders, 6 Hispanic Americans), 164 international. Average age 28. 463 applicants, 37% accepted. In 2005, 9 master's, 53 doctorates awarded. *Entrance requirements:* For master's, GRE; for doctorate, GRE, interview. *Application deadline:* Applications are processed on a rolling basis. Application fee: $35 ($60

for international students). Electronic applications accepted. *Expenses:* Tuition, state resident: part-time $170 per credit hour. Tuition, nonresident: full-time $4,612; part-time $425 per credit hour. International tuition: $10,732 full-time. Required fees: $11 per credit hour. $124 per term. Tuition and fees vary according to course load, degree level and program. *Financial support:* Fellowships, career-related internships or fieldwork available. *Unit head:* Dr. Robert R. Rich, Vice President/Dean, School of Medicine, 205-934-1111, Fax: 205-934-0333, E-mail: rrich@uab.edu.

The University of Alabama at Birmingham, Graduate Programs in Joint Health Sciences, Department of Cell Biology, Graduate Program in Cellular and Molecular Biology, Birmingham, AL 35294.

See Close-Ups on pages 601 and 603.

University of Alberta, Faculty of Graduate Studies and Research, Department of Biological Sciences, Edmonton, AB T6G 2E1, Canada. Offers environmental biology and ecology (M Sc, PhD); microbiology and biotechnology (M Sc, PhD); molecular biology and genetics (M Sc, PhD); physiology and cell biology (M Sc, PhD); plant biology (M Sc, PhD); systematics and evolution (M Sc, PhD). *Faculty:* 72 full-time (15 women), 15 part-time/adjunct (4 women). *Students:* 238 full-time (117 women), 32 part-time (15 women), 31 international. 206 applicants, 42% accepted. In 2005, 29 master's, 31 doctorates awarded. Terminal master's awarded for partial completion of doctoral program. *Degree requirements:* For master's and doctorate, thesis/dissertation, registration. *Entrance requirements:* Additional exam requirements/recommendations for international students: Required—TOEFL. *Application deadline:* For fall admission, 3/1 for domestic students. Applications are processed on a rolling basis. Application fee: $0. Tuition and fees charges are reported in Canadian dollars. *Expenses:* Tuition, state resident: part-time $562 Canadian dollars per term. Tuition, nonresident: full-time $3,375 Canadian dollars. Required fees: $573 Canadian dollars; $84 Canadian dollars per term. *Financial support:* In 2005–06, 4 research assistantships with partial tuition reimbursements (averaging $12,000 per year), 103 teaching assistantships with partial tuition reimbursements (averaging $12,300 per year) were awarded; career-related internships or fieldwork and scholarships/grants also available. *Unit head:* , Laura Frost, Chair, 780-492-1904. *Application contact:* Dr. John P. Chang, Associate Chair for Graduate Studies, 780-492-1257, Fax: 780-492-9457, E-mail: bio.grad.coordinator@ualberta.ca.

University of Alberta, Faculty of Medicine and Dentistry and Faculty of Graduate Studies and Research, Graduate Programs in Medicine, Department of Cell Biology, Edmonton, AB T6G 2E1, Canada. Offers M Sc, PhD. Terminal master's awarded for partial completion of doctoral program. *Degree requirements:* For master's and doctorate, thesis/dissertation. *Entrance requirements:* For master's and doctorate, GRE. Additional exam requirements/recommendations for international students: Required—TOEFL (minimum score 600 paper-based; 250 computer-based). Tuition and fees charges are reported in Canadian dollars. *Expenses:* Tuition, state resident: part-time $562 Canadian dollars per term. Tuition, nonresident: full-time $3,375 Canadian dollars. Required fees: $573 Canadian dollars; $84 Canadian dollars per term. *Faculty research:* Protein targeting, membrane trafficking, signal transduction, cell growth and division, cell-cell interaction and development.

The University of Arizona, College of Medicine, Graduate Programs in Medicine, Department of Cell Biology and Anatomy, Tucson, AZ 85721. Offers PhD. *Degree requirements:* For doctorate, thesis/dissertation. *Entrance requirements:* For doctorate, GRE General Test. *Faculty research:* Heart development, neural development, cellular toxicology and microcirculation; membrane traffic and cytoskeleton; cell-surface receptors.

The University of Arizona, Graduate College, College of Science, Department of Molecular and Cellular Biology, Tucson, AZ 85721. Offers MS, PhD. Terminal master's awarded for partial completion of doctoral program. *Degree requirements:* For master's and doctorate, thesis/dissertation. *Entrance requirements:* For master's, GRE General Test, GRE Subject Test; for doctorate, GRE General Test, GRE Subject Test, undergraduate research experience. Additional exam requirements/recommendations for international students: Required—TOEFL. *Faculty research:* Plant molecular biology, cellular and molecular aspects of development, genetics of bacteria and lower eukaryotes.

University of Arkansas, Graduate School, Interdisciplinary Program in Cell and Molecular Biology, Fayetteville, AR 72701-1201. Offers MS, PhD. *Students:* 38 full-time (23 women), 12 part-time (7 women); includes 6 minority (4 African Americans, 1 American Indian/Alaska Native, 1 Asian American or Pacific Islander), 24 international. 39 applicants, 18% accepted. In 2005, 10 master's, 3 doctorates awarded. *Degree requirements:* For doctorate, thesis/dissertation. Application fee: $40 ($50 for international students). *Financial support:* In 2005–06, 2 fellowships with tuition reimbursements, 10 research assistantships were awarded; teaching assistantships Financial award application deadline: 4/1; financial award applicants required to submit FAFSA. *Unit head:* Dr. John Kirby, Head, 479-575-8623, Fax: 479-575-5908, E-mail: jkirby@uark.edu.

The University of British Columbia, Faculty of Medicine and Faculty of Graduate Studies, Graduate Programs in Medicine, Department of Cellular and Physiological Sciences, Division of Anatomy and Cell Biology, Vancouver, BC V6T 1Z3, Canada. Offers M Sc, PhD. *Faculty:* 12 full-time (2 women), 2 part-time/adjunct (0 women). *Students:* 17 full-time (6 women), 2 international. Average age 27. 12 applicants, 17% accepted, 2 enrolled. In 2005, 3 master's, 1 doctorate awarded. *Degree requirements:* For master's, thesis/dissertation, oral defense; for doctorate, thesis/dissertation, oral defense, comprehensive exam. *Entrance requirements:* Additional exam requirements/recommendations for international students: Required—TOEFL (minimum score 550 paper-based; 213 computer-based). *Application deadline:* For fall admission, 4/1 priority date for domestic students, 3/1 priority date for international students. Applications are processed on a rolling basis. Application fee: $90 Canadian dollars ($150 Canadian dollars for international students). Electronic applications accepted. *Financial support:* In 2005–06, 3 fellowships (averaging $17,000 per year), 8 research assistantships with partial tuition reimbursements (averaging $17,000 per year), 6 teaching assistantships (averaging $5,000 per year) were awarded; Federal Work-Study, institutionally sponsored loans, scholarships/grants, traineeships, tuition waivers (full and partial), and unspecified assistantships also available. *Faculty research:* Cell and developmental biology, membrane biophysics, cellular immunology, cancer, fetal alcohol syndrome. Total annual research expenditures: $3 million Canadian dollars. *Application contact:* Dr. John Church, Graduate Adviser, 604-822-2751, Fax: 604-822-2316, E-mail: jchurch@interchange.ubc.ca.

University of California, Berkeley, Graduate Division, College of Letters and Science, Department of Molecular and Cell Biology, Berkeley, CA 94720-1500. Offers PhD. *Faculty:* 89 full-time (20 women), 3 part-time/adjunct (1 woman). *Students:* 265 full-time (139 women); includes 62 minority (5 African Americans, 2 American Indian/Alaska Native, 45 Asian Americans or Pacific Islanders, 10 Hispanic Americans), 23 international. Average age 23. 582 applicants, 25% accepted, 54 enrolled. In 2005, 30 doctorates awarded. *Median time to degree:* Of those who began their doctoral program in fall 1997, 85% received their degree in 8 years or less. *Degree requirements:* For doctorate, thesis/dissertation, qualifying exam, 2 semesters of teaching, 3 grad seminars, comprehensive exam, registration. *Entrance requirements:* For doctorate, GRE General Test, GRE Subject Test (recommended), minimum GPA of 3.0. Additional exam requirements/recommendations for international students: Required—TOEFL (minimum score 570 paper-based; 230 computer-based). *Application deadline:* For fall admission, 12/15 for domestic students, 12/15 for international students. Applications are processed on a rolling basis. Application fee: $60. Electronic applications accepted. *Financial support:* In 2005–06, 150 research assistantships with full tuition reimbursements (averaging $25,000 per year), 80 teaching assistantships with full tuition reimbursements (averaging $15,000 per year) were awarded; fellowships with full tuition reimbursements, scholarships/grants, traineeships, health care benefits, tuition waivers (full), and unspecified assistantships also available. Financial award application deadline: 12/15; financial award applicants required to submit FAFSA. *Faculty research:* Biochemistry and molecular biology, cell and developmental biology, genet-

Cell Biology

University of California, Berkeley (continued)
ics, immunology, neurobiology, genomics, evo/devo. *Unit head:* Barbara J Meyer, Professor. *Application contact:* Berta Parra, Student Affairs Officer, 510-642-5252, Fax: 510-642-7000, E-mail: bparra@berkeley.edu.

University of California, Davis, Graduate Studies, Graduate Group in Cell and Developmental Biology, Davis, CA 95616. Offers MS, PhD. *Faculty:* 56 full-time. *Students:* 47 full-time (20 women); includes 13 minority (10 Asian Americans or Pacific Islanders, 3 Hispanic Americans), 9 international. Average age 28. 77 applicants, 42% accepted, 11 enrolled. In 2005, 3 master's, 5 doctorates awarded. *Median time to degree:* Of those who began their doctoral program in fall 1997, 83.3% received their degree in 8 years or less. *Degree requirements:* For master's, thesis (for some programs), comprehensive exam (for some programs); for doctorate, thesis/dissertation. *Entrance requirements:* For doctorate, GRE General Test, GRE Subject Test. Additional exam requirements/recommendations for international students: Required—TOEFL (minimum score 550 paper-based; 213 computer-based). *Application deadline:* For fall admission, 1/15 for domestic students, 1/15 for international students. Application fee: $60. Electronic applications accepted. *Financial support:* In 2005–06, 46 students received support, including 8 fellowships with full and partial tuition reimbursements available (averaging $12,769 per year), 27 research assistantships with full and partial tuition reimbursements available (averaging $16,276 per year), 6 teaching assistantships with partial tuition reimbursements available (averaging $15,082 per year); Federal Work-Study, institutionally sponsored loans, scholarships/grants, tuition waivers (full and partial), and unspecified assistantships also available. Financial award application deadline: 1/15; financial award applicants required to submit FAFSA. *Faculty research:* Molecular basis of cell function and development. *Unit head:* Richard Tucker, Chair, 530-752-0238, E-mail: rptucker@ucdavis.edu. *Application contact:* Angelina Kuo, Graduate Staff, 530-752-2981, Fax: 530-752-8391, E-mail: abkuo@ucdavis.edu.

University of California, Irvine, Office of Graduate Studies, School of Biological Sciences, Department of Developmental and Cell Biology, Irvine, CA 92697. Offers biological sciences (MS, PhD). Students apply through the Graduate Program in Molecular Biology, Genetics, and Biochemistry. *Degree requirements:* For doctorate, thesis/dissertation. *Entrance requirements:* For master's and doctorate, GRE General Test, GRE Subject Test, minimum GPA of 3.0. Additional exam requirements/recommendations for international students: Required—TOEFL (minimum score 550 paper-based; 213 computer-based), TSE. Electronic applications accepted. *Faculty research:* Genetics and development, oncogene signaling pathways, gene regulation, tissue regeneration and molecular genetics.

University of California, Los Angeles, Graduate Division, College of Letters and Science and School of Medicine, UCLA ACCESS to Programs in the Molecular and Cellular Life Sciences, Los Angeles, CA 90095. Offers PhD. *Degree requirements:* For doctorate, thesis/dissertation, oral and written qualifying exams. *Entrance requirements:* For doctorate, GRE General Test, minimum undergraduate GPA of 3.0. Electronic applications accepted. *Faculty research:* Molecular, cellular, and developmental biology; immunology; microbiology; integrative biology.

See Close-Up on page 605.

University of California, Los Angeles, School of Medicine and Graduate Division, Graduate Programs in Medicine, Department of Molecular, Cell and Developmental Biology, Los Angeles, CA 90095. Offers MA, PhD. *Degree requirements:* For doctorate, thesis/dissertation, qualifying exams. *Entrance requirements:* For doctorate, GRE General Test, GRE Subject Test. Additional exam requirements/recommendations for international students: Required—TOEFL.

University of California, Los Angeles, School of Medicine and Graduate Division, Graduate Programs in Medicine, Department of Neurobiology, Los Angeles, CA 90095. Offers anatomy and cell biology (PhD). *Degree requirements:* For doctorate, thesis/dissertation, oral and written qualifying exams. *Entrance requirements:* For doctorate, GRE General Test, GRE Subject Test, bachelor's degree in physical or biological science. *Faculty research:* Neuroendocrinology, neurophysiology.

University of California, Riverside, Graduate Division, Program in Cell, Molecular, and Developmental Biology, Riverside, CA 92521-0102. Offers MS, PhD. *Faculty:* 55 full-time (18 women). *Students:* 57 full-time (29 women); includes 20 minority (3 African Americans, 15 Asian Americans or Pacific Islanders, 2 Hispanic Americans), 18 international. Average age 28. In 2005, 3 master's, 1 doctorate awarded. *Median time to degree:* Of those who began their doctoral program in fall 1997, 100% received their degree in 8 years or less. *Degree requirements:* For master's, thesis, oral defense of thesis; for doctorate, thesis/dissertation, oral defense of thesis, qualifying exams, 2 quarters of teaching experience. *Entrance requirements:* For master's and doctorate, GRE General Test, minimum GPA of 3.2. Additional exam requirements/recommendations for international students: Required—TOEFL (minimum score 550 paper-based; 213 computer-based); Recommended—TSE. *Application deadline:* For fall admission, 5/1 for domestic students, 2/1 for international students. For winter admission, 9/1 for domestic students; for spring admission, 12/1 for domestic students. Applications are processed on a rolling basis. Application fee: $60 ($75 for international students). Electronic applications accepted. *Expenses:* Tuition, nonresident: full-time $14,694. Required fees:$9,009. Full-time tuition and fees vary according to program. *Financial support:* In 2005–06, fellowships (averaging $18,000 per year), research assistantships (averaging $14,000 per year), teaching assistantships (averaging $15,000 per year) were awarded. Financial award application deadline: 1/5. *Unit head:* Dr. Anthony Norman, Director, 951-827-4777, E-mail: norman@ucr.edu. *Application contact:* Kathy Redd, Graduate Program Assistant, 800-735-0717, Fax: 951-827-5517, E-mail: cell@ucr.edu.

University of California, San Diego, Graduate Studies and Research, Division of Biology, Program in Cell and Developmental Biology, La Jolla, CA 92093-0348. Offers PhD. Offered in association with the Salk Institute. *Degree requirements:* For doctorate, thesis/dissertation, qualifying exam. Electronic applications accepted.

University of California, San Diego, Graduate Studies and Research, Division of Biology, Program in Molecular and Cellular Biology, La Jolla, CA 92093. Offers PhD. Offered in association with the Salk Institute. *Degree requirements:* For doctorate, thesis/dissertation, qualifying exam. Electronic applications accepted.

University of California, San Diego, School of Medicine and Graduate Studies and Research, Graduate Studies in Biomedical Sciences, Program in Molecular Cell Biology, La Jolla, CA 92093-0685. Offers PhD. *Degree requirements:* For doctorate, thesis/dissertation, qualifying exam. *Entrance requirements:* For doctorate, GRE General Test. Additional exam requirements/recommendations for international students: Required—TOEFL. Electronic applications accepted. *Faculty research:* Molecular and cellular pharmacology, cell and organ physiology.

See Close-Up on page 607.

University of California, San Diego, School of Medicine and Graduate Studies and Research, Graduate Studies in Biomedical Sciences, Regulatory Biology Program, La Jolla, CA 92093. Offers PhD. *Degree requirements:* For doctorate, thesis/dissertation, 2 qualifying exams. *Entrance requirements:* For doctorate, GRE General Test, GRE Subject Test. Additional exam requirements/recommendations for international students: Required—TOEFL. Electronic applications accepted. *Faculty research:* Eukaryotic regulatory and molecular biology, molecular and cellular pharmacology, cell and organ physiology.

University of California, San Francisco, Graduate Division and School of Medicine, Department of Biochemistry and Biophysics, Program in Cell Biology, San Francisco, CA 94143. Offers PhD, MD/PhD. *Degree requirements:* For doctorate, thesis/dissertation. *Entrance*

requirements: For doctorate, GRE General Test, GRE Subject Test. Additional exam requirements/recommendations for international students: Required—TOEFL. Expenses: Contact institution.

University of California, Santa Barbara, Graduate Division, College of Letters and Sciences, Division of Mathematics, Life, and Physical Sciences, Department of Molecular, Cellular, and Developmental Biology, Santa Barbara, CA 93106. Offers MA, PhD, MA/PhD. *Faculty:* 62 full-time (30 women), 1 part-time/adjunct (0 women). *Students:* 63 full-time (30 women); includes 10 minority (1 African American, 7 Asian Americans or Pacific Islanders, 2 Hispanic Americans), 4 international. Average age 26. 99 applicants, 31% accepted, 18 enrolled. Terminal master's awarded for partial completion of doctoral program. *Degree requirements:* For master's, thesis (for some programs), comprehensive exam (for some programs), registration; for doctorate, thesis/dissertation, comprehensive exam, registration. *Entrance requirements:* For master's and doctorate, GRE General Test, GRE Subject Test. Additional exam requirements/recommendations for international students: Required—TOEFL (minimum score 630 paper-based; 213 computer-based). *Application deadline:* For fall admission, 12/15 for domestic students, 12/15 for international students. Application fee: $60. Electronic applications accepted. *Financial support:* In 2005–06, 51 teaching assistantships were awarded; fellowships with full tuition reimbursements, research assistantships with full tuition reimbursements, career-related internships or fieldwork, Federal Work-Study, scholarships/grants, and health care benefits also available. Financial award application deadline: 12/15; financial award applicants required to submit FAFSA. *Faculty research:* Signal transduction, bacteria pathegenesis, virology, stem cell, neurodegenerative disease. *Unit head:* Dr. Dennis Clegg, Chair, 805-893-8490, E-mail: clegg@lifesci.ucsb.edu. *Application contact:* Krista Grace, Staff Graduate Program Advisor, 805-893-2290, Fax: 805-893-4724, E-mail: grace@lifesci.ucsb.edu.

University of California, Santa Cruz, Division of Graduate Studies, Division of Physical and Biological Sciences, Department of Molecular, Cellular, and Developmental Biology, Santa Cruz, CA 95064. Offers MA, PhD. *Students:* 50 full-time (29 women); includes 16 minority (1 African American, 11 Asian Americans or Pacific Islanders, 4 Hispanic Americans), 5 international. In 2005, 1 degree awarded. Application fee: $60. *Expenses:* Tuition, nonresident: full-time $14,694. Required fees: $9,437. *Unit head:* John Tamkun, Chairperson. *Application contact:* Cindy Hodges, Information Contact, 831-459-2632, E-mail: cindy@biology.ucsc.edu.

University of Chicago, Division of the Biological Sciences, Department of Molecular Biosciences: Biochemistry, Genetics, Cell and Developmental Biology, Department of Molecular Genetics and Cell Biology, Chicago, IL 60637-1513. Offers PhD. *Faculty:* 31 full-time (8 women). *Students:* 36 full-time (21 women); includes 5 minority (all Asian Americans or Pacific Islanders), 6 international. Average age 27. In 2005, 7 doctorates awarded. *Degree requirements:* For doctorate, thesis/dissertation, registration. *Entrance requirements:* For doctorate, GRE General Test. Additional exam requirements/recommendations for international students: Required—TOEFL. *Application deadline:* For fall admission, 12/28 priority date for domestic students, 12/28 priority date for international students. Application fee: $55. Electronic applications accepted. *Financial support:* In 2005–06, 36 students received support, including fellowships (averaging $26,301 per year), research assistantships (averaging $26,301 per year); institutionally sponsored loans, scholarships/grants, traineeships, and health care benefits also available. Financial award applicants required to submit FAFSA. *Faculty research:* Gene expression, chromosome structure, animal viruses, plant molecular genetics. Total annual research expenditures: $8 million. *Unit head:* Dr. Laurens Mets, Chairman, 773-702-8917. *Application contact:* Kristine Gaston, Graduate Administrative Director, 773-702-8037, Fax: 773-702-3172, E-mail: kristine@cummings.uchicago.edu.

University of Cincinnati, Division of Research and Advanced Studies, College of Medicine, Graduate Programs in Biomedical Sciences, Graduate Program in Cell and Molecular Biology, Cincinnati, OH 45221. Offers PhD. *Degree requirements:* For doctorate, thesis/dissertation, qualifying exam. *Entrance requirements:* For doctorate, GRE General Test. Additional exam requirements/recommendations for international students: Required—TOEFL, TWE; Recommended—TSE. Electronic applications accepted. *Faculty research:* Cancer cell biology, basic cell and molecular biology, neuroscience, developmental biology, endocrinology.

University of Colorado at Boulder, Graduate School, College of Arts and Sciences, Department of Molecular, Cellular, and Developmental Biology, Boulder, CO 80309. Offers cellular structure and function (MA, PhD); developmental biology (MA, PhD); molecular biology (MA, PhD). *Faculty:* 25 full-time (6 women). *Students:* 42 full-time (15 women), 26 part-time (12 women); includes 5 minority (1 American Indian/Alaska Native, 1 Asian American or Pacific Islander, 3 Hispanic Americans), 11 international. Average age 28. 19 applicants, 100% accepted. In 2005, 3 master's, 8 doctorates awarded. Terminal master's awarded for partial completion of doctoral program. *Degree requirements:* For master's, thesis or alternative, comprehensive exam; for doctorate, thesis/dissertation, comprehensive exam. *Entrance requirements:* For master's, GRE General Test, GRE Subject Test, minimum undergraduate GPA of 2.75; for doctorate, GRE General Test, GRE Subject Test. *Application deadline:* For fall admission, 1/15 for domestic students, 1/15 for international students. Application fee: $50 ($60 for international students). *Financial support:* In 2005–06, fellowships (averaging $7,489 per year), research assistantships (averaging $12,470 per year), teaching assistantships (averaging $13,078 per year) were awarded; tuition waivers (full) also available. Financial award application deadline: 3/1. *Faculty research:* Molecular biology of RNA and DNA, molecular genetics, cell motility and cytoskeleton, cell membranes, developmental genetics. Total annual research expenditures: $14.7 million. *Unit head:* Leslie Leinwand, Chair, 303-492-7606, Fax: 303-492-7744, E-mail: leslie.leinwand@stripe.colorado.edu. *Application contact:* Student Affairs Office, 303-492-7230, Fax: 303-492-7744, E-mail: mcdbgradinfo@beagle.colorado.edu.

University of Colorado at Denver and Health Sciences Center, Graduate School, Program in Biomedical Sciences, Program in Cell and Developmental Biology, Denver, CO 80262. Offers PhD. In 2005, 3 degrees awarded. *Degree requirements:* For doctorate, thesis/dissertation. *Entrance requirements:* For doctorate, GRE, minimum GPA of 3.0, 3 letters of reference. Additional exam requirements/recommendations for international students: Required—TOEFL (minimum score 550 paper-based; 213 computer-based). *Application deadline:* For fall admission, 1/15 for domestic students. Application fee: $50. Electronic applications accepted. *Expenses:* Tuition, state resident: full-time $11,730. Tuition, nonresident: full-time $22,980. Tuition and fees vary according to degree level and program. *Financial support:* Fellowships, research assistantships, teaching assistantships, Federal Work-Study and institutionally sponsored loans available. Support available to part-time students. Financial award application deadline: 3/15; financial award applicants required to submit FAFSA. *Faculty research:* Human disease, stem cell biology, neuroscience, molecular biology. Total annual research expenditures: $4.6 million. *Unit head:* Dr. Karl Pfenninger, Chair, 303-724-3424, E-mail: karl.pfenninger@uchsc.edu. *Application contact:* Carmel Hardberg, Program Assistant, 303-315-7009, Fax: 303-724-3420, E-mail: cdb@uchsc.edu.

University of Connecticut, Graduate School, College of Liberal Arts and Sciences, Department of Molecular and Cell Biology, Field of Cell and Developmental Biology, Storrs, CT 06269. Offers MS, PhD. *Faculty:* 27 full-time (8 women). *Students:* 35 full-time (14 women), 6 part-time (3 women); includes 3 minority (2 Asian Americans or Pacific Islanders, 1 Hispanic American), 14 international. Average age 29. 62 applicants, 26% accepted, 14 enrolled. In 2005, 7 master's, 2 doctorates awarded. *Degree requirements:* For doctorate, thesis/dissertation. *Entrance requirements:* For master's and doctorate, GRE General Test, GRE Subject Test. Additional exam requirements/recommendations for international students: Required—TOEFL (minimum score 550 paper-based; 213 computer-based). *Application deadline:* For fall admission, 2/1 priority date for domestic students, 2/1 priority date for international students; for spring admission, 11/1 for domestic students, 10/1 for international students. Applications are processed on a rolling basis. Application fee: $55. Electronic applications accepted. *Expenses:* Tuition, state resident: part-time $444 per credit hour. Tuition, nonresident: part-time $1,154 per credit hour. Tuition and fees vary according to course load. *Financial support:* In 2005–06, 9 research assistantships with full tuition reimbursements, 17 teaching assistantships with full tuition reimbursements were awarded; fellowships, Federal Work-Study, scholarships/grants, health care benefits, and unspecified assistantships also available. Financial award application deadline: 2/1; financial award applicants required to submit FAFSA. *Application contact:* Anne St. Onje, Graduate Coordinator, 860-486-4314, Fax: 860-486-3943, E-mail: ann.st_onje@uconn.edu.

See Close-Up on page 611.

University of Connecticut Health Center, Graduate School, Programs in Biomedical Sciences, Program in Cell Biology, Farmington, CT 06030. Offers PhD, DMD/PhD, MD/PhD. *Degree requirements:* For doctorate, thesis/dissertation, comprehensive exam, registration. *Entrance requirements:* For doctorate, GRE General Test. Additional exam requirements/ recommendations for international students: Required—TOEFL (minimum score 600 paper-based; 250 computer-based). Electronic applications accepted. *Faculty research:* Membrane structure and function, cell regulation, electrophysiology, gene expression and regulation, intracellular messengers.

See Close-Up on page 613.

University of Delaware, College of Arts and Sciences, Department of Biological Sciences, Newark, DE 19716. Offers biotechnology (MS); cancer biology (MS, PhD); cell and extracellular matrix biology (MS, PhD); cell and systems physiology (MS, PhD); developmental biology (MS, PhD); ecology and evolution (MS, PhD); microbiology (MS, PhD); molecular biology and genetics (MS, PhD). *Faculty:* 39 full-time (11 women). *Students:* 64 full-time (46 women), 2 part-time; includes 7 minority (4 African Americans, 2 Asian Americans or Pacific Islanders, 1 Hispanic American), 18 international. Average age 26. 113 applicants, 31% accepted, 19 enrolled. In 2005, 5 master's, 4 doctorates awarded. Terminal master's awarded for partial completion of doctoral program. *Median time to degree:* Of those who began their doctoral program in fall 1997, 100% received their degree in 8 years or less. *Degree requirements:* For master's, thesis/dissertation, preliminary exam; for doctorate, thesis/dissertation, preliminary exam, comprehensive exam. *Entrance requirements:* For master's and doctorate, GRE General Test. Additional exam requirements/recommendations for international students: Required—TOEFL (minimum score 600 paper-based; 250 computer-based); Recommended—TWE, TSE. *Application deadline:* For fall admission, 4/15 for domestic students, 1/15 for international students; for spring admission, 10/1 for domestic students. Applications are processed on a rolling basis. Application fee: $60. Electronic applications accepted. *Financial support:* In 2005–06, 26 students received support, including fellowships with full tuition reimbursements available (averaging $19,000 per year), 19 research assistantships with full tuition reimbursements available (averaging $19,000 per year), 26 teaching assistantships with full tuition reimbursements available (averaging $19,000 per year); tuition waivers (partial) also available. Financial award application deadline: 4/15. *Faculty research:* Microorganisms, bone, cancer metastasis, developmental biology, cell biology, DNA. Total annual research expenditures: $8.3 million. *Unit head:* Dr. Daniel D. Carson, Chair, 302-831-6977, Fax: 302-831-2281, E-mail: dcarson@udel.edu. *Application contact:* Dr. Melinda K. Duncan, Graduate Coordinator, 302-831-1841, Fax: 302-831-2281, E-mail: danders@udel.edu.

University of Florida, College of Medicine, Department of Anatomy and Cell Biology, Gainesville, FL 32611. Offers PhD. *Faculty:* 13 full-time (4 women), 1 (woman) part-time/adjunct. *Degree requirements:* For doctorate, thesis/dissertation. *Entrance requirements:* For doctorate, GRE General Test, minimum GPA of 3.0. Additional exam requirements/recommendations for international students: Required—TOEFL. *Application deadline:* For fall admission, 2/15 for domestic students. Applications are processed on a rolling basis. Application fee: $30. Electronic applications accepted. *Expenses:* Tuition, state resident: full-time $6,234. Tuition, nonresident: full-time $21,359. Tuition and fees vary according to program. *Financial support:* In 2005–06, research assistantships (averaging $24,047 per year); fellowships *Faculty research:* Structure and function of intracellular organelles, cell adhesion, differentiation. *Unit head:* Dr. Stephen P. Sugrue, Chairman, 352-392-3569, Fax: 352-392-3305. *Application contact:* Dr. Wayne McCormack, Associate Dean of Graduate Education, 352-392-7413, Fax: 352-846-3466, E-mail: mccormac@pathology.ufl.edu.

University of Florida, College of Medicine and Graduate School, Interdisciplinary Program in Biomedical Sciences, Concentration in Molecular Cell Biology, Gainesville, FL 32611. Offers PhD. *Faculty:* 79. *Students:* 27 full-time (14 women); includes 5 minority (1 African American, 3 Asian Americans or Pacific Islanders, 1 Hispanic American). In 2005, 4 degrees awarded. *Degree requirements:* For doctorate, thesis/dissertation. *Entrance requirements:* For doctorate, GRE General Test, minimum GPA of 3.0. Additional exam requirements/recommendations for international students: Required—TOEFL. *Application deadline:* For fall admission, 2/15 for domestic students. Application fee: $30. Electronic applications accepted. *Expenses:* Tuition, state resident: full-time $6,234. Tuition, nonresident: full-time $21,359. Tuition and fees vary according to program. *Financial support:* Fellowships with full tuition reimbursements, research assistantships with full tuition reimbursements, scholarships/grants and unspecified assistantships available. *Unit head:* Dr. Phyllis LuValle, Director, 352-392-6261, Fax: 352-392-3305, E-mail: luvalle@ufl.edu. *Application contact:* Dr. Wayne McCormack, Associate Dean of Graduate Education, 352-392-7413, Fax: 352-846-3466, E-mail: mccormac@pathology.ufl.edu.

University of Florida, Graduate School, College of Agricultural and Life Sciences, Department of Microbiology and Cell Science, Gainesville, FL 32611. Offers biochemistry and molecular biology (MS, PhD); microbiology and cell science (MS, PhD). *Faculty:* 18 full-time (5 women). *Students:* 25 (9 women); includes 4 minority (1 African American, 2 Asian Americans or Pacific Islanders, 1 Hispanic American) 6 international. 31 applicants, 26% accepted. In 2005, 1 master's, 4 doctorates awarded. *Degree requirements:* For doctorate, thesis/dissertation. *Entrance requirements:* For master's and doctorate, GRE General Test, minimum GPA of 3.0. *Application deadline:* For fall admission, 6/1 for domestic students. Applications are processed on a rolling basis. Application fee: $20. Electronic applications accepted. *Expenses:* Tuition, state resident: full-time $6,234. Tuition, nonresident: full-time $21,359. Tuition and fees vary according to program. *Financial support:* In 2005–06, 20 students received support, including 14 research assistantships (averaging $15,696 per year), 5 teaching assistantships (averaging $14,677 per year); fellowships *Faculty research:* Biomass conversion, membrane and cell wall chemistry, plant biochemistry and genetics. *Unit head:* Dr. Eric Triplett, Chair, 352-392-1906, Fax: 352-392-5922, E-mail: ewt@ufl.edu. *Application contact:* Dr. James F. Preston, Coordinator, 352-392-5923, Fax: 352-392-5922, E-mail: jpreston@ufl.edu.

University of Georgia, Graduate School, College of Arts and Sciences, Department of Cellular Biology, Athens, GA 30602. Offers MS, PhD. *Faculty:* 16 full-time (4 women), 1 part-time/adjunct. *Students:* 37 full-time, 2 part-time; includes 4 minority (3 African Americans, 1 Hispanic American), 22 international. 42 applicants, 29% accepted, 7 enrolled. In 2005, 1 master's, 2 doctorates awarded. *Degree requirements:* For master's, thesis; for doctorate, one foreign language, thesis/dissertation. *Entrance requirements:* For master's and doctorate, GRE General Test. *Application deadline:* For fall admission, 7/1 for domestic students; for spring admission, 11/15 for domestic students. Application fee: $50. Electronic applications accepted. *Financial support:* Fellowships, research assistantships, teaching assistantships, unspecified assistantships available. *Unit head:* Dr. Mark A. Farmer, Head, 706-542-3413, Fax: 706-542-4271. *Application contact:* Dr. Marcus Fechheimer, Information Contact, 706-542-3338, Fax: 706-542-4271, E-mail: fechheim@cb.uga.edu.

University of Guelph, Graduate Program Services, College of Biological Science, Department of Molecular and Cellular Biology, Guelph, ON N1G 2W1, Canada. Offers biochemistry (M Sc, PhD); biophysics (M Sc, PhD); botany (M Sc, PhD); microbiology (M Sc, PhD); molecular biology and genetics (M Sc, PhD). *Faculty:* 43 full-time (6 women). *Students:* 136 full-time (52 women). Average age 26. 56 applicants, 30% accepted, 17 enrolled. In 2005, 4 master's, 2 doctorates awarded. *Degree requirements:* For master's, thesis/dissertation, research proposal; for doctorate, thesis/dissertation, research proposal, comprehensive exam, registration. *Entrance requirements:* Additional exam requirements/recommendations for international

students: Required—TOEFL (minimum score 550 paper-based; 213 computer-based), IELT (minimum score 7). *Application deadline:* For fall admission, 7/1 for domestic students, 1/1 for international students. For winter admission, 11/1 for domestic students; for spring admission, 3/1 for domestic students. Applications are processed on a rolling basis. Application fee: $75. Electronic applications accepted. *Financial support:* Fellowships, research assistantships, teaching assistantships available. Support available to part-time students. *Faculty research:* Physiology, structure, genetics, and ecology of microbes; virology and microbial technology. *Unit head:* Dr. Chris Whitfield, Chair, 519-824-4120 Ext. 53361, Fax: 519-827-1802, E-mail: cwhitfie@uoguelph.ca. *Application contact:* Laurie Winn, Graduate Admissions Secretary, 519-824-4320 Ext. 52730, Fax: 519-767-1656, E-mail: lwinn@uoguelph.ca.

University of Illinois at Chicago, College of Medicine and Graduate College, Graduate Programs in Medicine, Department of Anatomy and Cell Biology, Chicago, IL 60607-7128. Offers MS, PhD, MD/PhD. *Degree requirements:* For master's and doctorate, thesis/dissertation. *Entrance requirements:* For master's and doctorate, GRE General Test. Additional exam requirements/recommendations for international students: Required—TOEFL. Electronic applications accepted. *Faculty research:* Neuroanatomy, functional morphology, cytoskeleton, synapses, neural transplants.

University of Illinois at Chicago, Graduate College, College of Liberal Arts and Sciences, Department of Biological Sciences, Chicago, IL 60607-7128. Offers cell and developmental biology (PhD); ecology and evolution (MS, DA, PhD); genetics and development (PhD); molecular biology (MS, PhD); neurobiology (MS, PhD); plant biology (MS, DA, PhD). *Degree requirements:* For master's, thesis; for doctorate, thesis/dissertation, preliminary exam. *Entrance requirements:* For master's and doctorate, GRE General Test, GRE Subject Test, previous course work in physics, calculus, and organic chemistry; minimum GPA of 2.75. Additional exam requirements/recommendations for international students: Required—TOEFL. Electronic applications accepted.

University of Illinois at Urbana–Champaign, Graduate College, College of Liberal Arts and Sciences, Cell and Molecular Biology Training Program, Champaign, IL 61820.

University of Illinois at Urbana–Champaign, Graduate College, College of Liberal Arts and Sciences, School of Molecular and Cellular Biology, Department of Cell and Developmental Biology, Champaign, IL 61820. Offers PhD. *Faculty:* 13 full-time (2 women). *Students:* 75 full-time (38 women), 1 (woman) part-time; includes 12 minority (2 African Americans, 1 American Indian/Alaska Native, 7 Asian Americans or Pacific Islanders, 2 Hispanic Americans), 41 international. 110 applicants, 6% accepted, 7 enrolled. In 2005, 7 degrees awarded. *Degree requirements:* For doctorate, thesis/dissertation. *Application deadline:* Applications are processed on a rolling basis. Application fee: $50 ($60 for international students). Electronic applications accepted. *Financial support:* In 2005–06, 12 fellowships, 59 research assistantships, 39 teaching assistantships were awarded. Financial award application deadline: 2/15. *Unit head:* Dr. Martha U. Gillette, Head, 217-333-6118, Fax: 217-244-1648, E-mail: mgillett@uiuc.edu. *Application contact:* Audra Weinstein, Staff Secretary, 217-333-6118, Fax: 217-244-1648, E-mail: audra@uiuc.edu.

The University of Iowa, Graduate College, College of Liberal Arts and Sciences, Department of Biological Sciences, Iowa City, IA 52242-1316.

See Close-Up on page 257.

The University of Iowa, Roy J. and Lucille A. Carver College of Medicine and Graduate College, Graduate Programs in Medicine, Department of Anatomy and Cell Biology, Iowa City, IA 52242-1316. Offers PhD. *Faculty:* 24 full-time (6 women). *Students:* 16 full-time (9 women); includes 1 minority (Asian American or Pacific Islander), 4 international. Average age 28. 115 applicants, 2% accepted. In 2005, 2 degrees awarded. *Median time to degree:* Of those who began their doctoral program in fall 1997, 100% received their degree in 8 years or less. *Degree requirements:* For doctorate, thesis/dissertation, comprehensive exam, registration. *Entrance requirements:* For doctorate, GRE General Test, minimum GPA of 3.0. Additional exam requirements/recommendations for international students: Required—TOEFL. *Application deadline:* For fall admission, 2/1 priority date for domestic students, 2/1 priority date for international students. Applications are processed on a rolling basis. Application fee: $60 ($85 for international students). Electronic applications accepted. *Expenses:* Tuition, state resident: part-time $1,882 per term. Tuition, nonresident: full-time $17,338; part-time $4,907 per term. Tuition and fees vary according to course load and program. *Financial support:* In 2005–06, 16 students received support, including fellowships with full tuition reimbursements available (averaging $22,000 per year), 14 research assistantships with full tuition reimbursements available (averaging $22,000 per year), teaching assistantships with full tuition reimbursements available (averaging $22,000 per year); institutionally sponsored loans, scholarships/grants, and health care benefits also available. Financial award application deadline: 3/1. *Faculty research:* Biology of differentiation and transformation, developmental and vascular cell biology, neurobiology. Total annual research expenditures: $5.7 million. *Unit head:* , Dr. John F. Engelhardt, Professor and Interim Head, 319-335-7744, Fax: 319-335-7198, E-mail: john-engelhardt@uiowa.edu. *Application contact:* Julie A. Stark, Program Assistant, 319-335-7744, Fax: 319-335-7198, E-mail: julie-stark@uiowa.edu.

University of Kansas, Graduate School, College of Liberal Arts and Sciences, Division of Biological Sciences, Department of Molecular Biosciences, Lawrence, KS 66045. Offers biochemistry and biophysics (MA, PhD); microbiology (MA, PhD); molecular, cellular, and developmental biology (MA, PhD). *Faculty:* 44. *Students:* 48 full-time (28 women), 5 part-time (3 women); includes 1 minority (Asian American or Pacific Islander), 26 international. Average age 27. 111 applicants, 19% accepted. In 2005, 5 master's, 5 doctorates awarded. Terminal master's awarded for partial completion of doctoral program. *Degree requirements:* For master's and doctorate, thesis/dissertation, comprehensive exam. *Entrance requirements:* For master's and doctorate, GRE General Test. Additional exam requirements/recommendations for international students: Required—TOEFL, IELT, TOEFL (paper 570; computer 230) or IELT; Recommended—TWE, TSE. *Application deadline:* For fall admission, 1/15 priority date for domestic students, 1/15 priority date for international students. Applications are processed on a rolling basis. Application fee: $55 ($60 for international students). Electronic applications accepted. *Expenses:* Tuition, state resident: full-time $4,859. Tuition, nonresident: full-time $12,000. Required fees: $589. Tuition and fees vary according to program. *Financial support:* Fellowships, research assistantships with tuition reimbursements, teaching assistantships with tuition reimbursements available. Financial award application deadline: 3/1. *Faculty research:* Structure and function of proteins, genetics of organism development, molecular genetics, neurophysiology, molecular virology and pathogenics, developmental biology, cell biology. *Unit head:* Kathy Suprenant, Chair, 785-864-4631, Fax: 785-864-5294, E-mail: ksupre@ku.edu. *Application contact:* John P. Connolly, Information Contact, 785-864-4311, Fax: 785-864-5924, E-mail: jconnolly@ku.edu.

University of Kansas, Graduate Studies Medical Center, Interdisciplinary Graduate Program in Biomedical Sciences, Department of Anatomy and Cell Biology, Lawrence, KS 66045. Offers MA, PhD, MD/PhD. Part-time programs available. *Faculty:* 1 part-time/adjunct. *Students:* Average age 28. In 2005, 2 master's, 1 doctorate awarded. Terminal master's awarded for partial completion of doctoral program. *Degree requirements:* For master's, comprehensive oral exam, oral defense of thesis; for doctorate, thesis/dissertation, comprehensive exam. *Entrance requirements:* Additional exam requirements/recommendations for international students: Required—TOEFL, TSE. *Application deadline:* For fall admission, 1/15 for domestic students. Applications are processed on a rolling basis. Application fee: $0. Electronic applications accepted. *Expenses:* Tuition, state resident: full-time $4,859. Tuition, nonresident: full-time $12,000. Required fees: $589. Tuition and fees vary according to program. *Financial support:* Fellowships, research assistantships with partial tuition reimbursements, teaching assistantships with full and partial tuition reimbursements, institutionally sponsored loans available. *Faculty research:* Kidney development, immunobiology of pregnancy, vascular formation, developmental neurobiology, cell-matrix interactions. *Unit head:* Dr. Dale R. Abrahamson,

Cell Biology

University of Kansas *(continued)*
Chairman, 913-588-7000, Fax: 913-588-2710, E-mail: dabrahamson@kume.edu. *Application contact:* Dr. Michael Werle, Graduate Adviser, 913-588-7491, Fax: 913-588-2710, E-mail: mwerle@kumc.edu.

University of Maryland, School of Medicine, Graduate Program in Life Sciences, Baltimore, MD 21201. Offers biochemistry (MS, PhD); epidemiology (MS, PhD); gerontology (PhD); microbiology (PhD); molecular and cell biology (MS); molecular medicine (PhD); neuroscience (MS, PhD); pharmacology (MS); physiology (MS); rehabilitation sciences (PhD); toxicology (MS, PhD). *Faculty:* 245 full-time (52 women). *Students:* 268 full-time (165 women), 46 part-time (30 women); includes 43 minority (25 African Americans, 13 Asian Americans or Pacific Islanders, 5 Hispanic Americans), 86 international. 435 applicants, 26% accepted, 61 enrolled. In 2005, 18 master's, 46 doctorates awarded. *Median time to degree:* Of those who began their doctoral program in fall 1997, 99% received their degree in 8 years or less. *Degree requirements:* For master's, registration; for doctorate, thesis/dissertation, lab rotations, comprehensive exam, registration. *Entrance requirements:* For master's and doctorate, GRE or MCAT. Additional exam requirements/recommendations for international students: Required—TOEFL (minimum score 550 paper-based; 213 computer-based). *Application deadline:* For winter admission, 1/15 for domestic students. Applications are processed on a rolling basis. Application fee: $50. Electronic applications accepted. *Expenses:* Tuition, state resident: full-time $8,079; part-time $409 per credit hour. Tuition, nonresident: full-time $18,384; part-time $731 per credit hour. Required fees: $695; $10 per credit hour. Tuition and fees vary according to degree level and program. *Financial support:* In 2005–06, 30 fellowships with full tuition reimbursements (averaging $23,000 per year), 22 research assistantships with full tuition reimbursements (averaging $23,000 per year) were awarded; health care benefits also available. *Faculty research:* Cancer, reproduction, neuroscience, cardiovascular, immunology. *Unit head:* Dr. Margaret Merryl McCarthy, Assistant Dean for Graduate Studies, 410-706-2655, Fax: 410-706-8341, E-mail: mmcarthy@umaryland.edu.

University of Maryland, Baltimore County, Graduate School, College of Natural and Mathematical Sciences, Department of Biological Sciences, Program in Molecular and Cell Biology, Baltimore, MD 21250. Offers PhD. *Faculty:* 28 full-time (8 women), 2 part-time/adjunct (0 women). *Students:* 19 full-time (8 women); includes 8 minority (2 African Americans, 6 Asian Americans or Pacific Islanders). 35 applicants, 49% accepted, 4 enrolled. In 2005, 3 degrees awarded. *Degree requirements:* For doctorate, thesis/dissertation. *Entrance requirements:* For doctorate, GRE General Test, GRE Subject Test, minimum GPA of 3.0. Additional exam requirements/recommendations for international students: Required—TOEFL. *Application deadline:* For fall admission, 2/1 for domestic students, 1/1 for international students. Applications are processed on a rolling basis. Application fee: $50. Electronic applications accepted. *Expenses:* Tuition, state resident: part-time $395 per credit. Tuition, nonresident: part-time $652 per credit. Required fees: $82 per credit. Tuition and fees vary according to course load, program and reciprocity agreements. *Financial support:* In 2005–06, 19 students received support, including fellowships with full tuition reimbursements available (averaging $23,000 per year), research assistantships with full tuition reimbursements available (averaging $21,500 per year), teaching assistantships with full tuition reimbursements available (averaging $20,500 per year) *Unit head:* Dr. Phyllis Robinson, Director, 410-455-3669, Fax: 410-455-3875, E-mail: biograd@umbc.edu.

University of Maryland, College Park, Graduate Studies, College of Chemical and Life Sciences, Department of Cell Biology and Molecular Genetics, Program in Cell Biology and Molecular Genetics, College Park, MD 20742. Offers MS, PhD. *Students:* 78 full-time (48 women), 3 part-time (2 women); includes 6 minority (3 African Americans, 3 Hispanic Americans), 30 international. 166 applicants, 19% accepted, 16 enrolled. In 2005, 5 master's, 5 doctorates awarded. *Degree requirements:* For master's, thesis; for doctorate, thesis/dissertation, exams. *Entrance requirements:* For master's and doctorate, GRE General Test, 3 letters of recommendation, minimum GPA of 3.0. Additional exam requirements/recommendations for international students: Required—TOEFL; Recommended—TSE. *Application deadline:* For fall admission, 1/11 for domestic students, 1/11 for international students. Application fee: $60. *Financial support:* In 2005–06, 17 fellowships (averaging $4,252 per year) were awarded; research assistantships, teaching assistantships Financial award applicants required to submit FAFSA. *Faculty research:* Cytoskeletal activity, membrane biology, cell division, genetics and genomics, virology. *Application contact:* Dean of Graduate School, 301-405-4190, Fax: 301-314-9305.

University of Maryland, College Park, Graduate Studies, College of Chemical and Life Sciences, Department of Cell Biology and Molecular Genetics, Program in Molecular and Cellular Biology, College Park, MD 20742. Offers PhD. Part-time and evening/weekend programs available. *Students:* 36 full-time (20 women), 1 (woman) part-time; includes 5 minority (1 African American, 3 Asian Americans or Pacific Islanders, 1 Hispanic American), 20 international. 119 applicants, 18% accepted, 9 enrolled. In 2005, 8 degrees awarded. *Median time to degree:* Of those who began their doctoral program in fall 1997, 75% received their degree in 8 years or less. *Degree requirements:* For doctorate, thesis/dissertation, exam, public service. *Entrance requirements:* For doctorate, GRE General Test, 3 letters of reference. Additional exam requirements/recommendations for international students: Required—TOEFL; Recommended—TSE. *Application deadline:* For fall admission, 1/7 for domestic students, 1/7 for international students. Applications are processed on a rolling basis. Application fee: $60. Electronic applications accepted. *Financial support:* In 2005–06, 8 fellowships with full tuition reimbursements (averaging $3,013 per year) were awarded Financial award applicants required to submit FAFSA. *Faculty research:* Monoclonal antibody production, oligonucleotide synthesis, macronolular processing, signal transduction, developmental biology. *Unit head:* Dr. Stephen Wolniak, 301-405-5430, Fax: 301-314-9489, E-mail: swolniak@umd.edu. *Application contact:* Dean of Graduate School, 301-405-4190, Fax: 301-314-9305.

University of Massachusetts Amherst, Graduate School, College of Natural Sciences and Mathematics, Program in Molecular and Cellular Biology, Amherst, MA 01003. Offers biological chemistry (PhD); cell and developmental biology (PhD). Part-time programs available. *Faculty:* 1 full-time (0 women). *Students:* 87 full-time (43 women), 1 part-time; includes 9 minority (1 African American, 4 Asian Americans or Pacific Islanders, 4 Hispanic Americans), 41 international. Average age 30. 219 applicants, 23% accepted, 13 enrolled. In 2005, 8 doctorates awarded. *Degree requirements:* For doctorate, one foreign language, thesis/dissertation. *Entrance requirements:* For doctorate, GRE General Test. Additional exam requirements/recommendations for international students: Required—TOEFL (minimum score 530 paper-based; 197 computer-based). *Application deadline:* For fall admission, 1/15 priority date for domestic students, 1/15 priority date for international students. Applications are processed on a rolling basis. Application fee: $40 ($65 for international students). Electronic applications accepted. *Expenses:* Tuition, state resident: part-time $110 per credit. Tuition, nonresident: part-time $414 per credit. Required fees: $2,824 per term. One-time fee: $250 part-time. Full-time tuition and fees vary according to course load, campus/location, program and reciprocity agreements. *Financial support:* In 2005–06, 6 fellowships with full tuition reimbursements (averaging $3,451 per year), 4 teaching assistantships with full tuition reimbursements (averaging $8,696 per year) were awarded; career-related internships or fieldwork, Federal Work-Study, scholarships/grants, traineeships, and unspecified assistantships also available. Support available to part-time students. Financial award application deadline: 1/15. *Unit head:* Dr. Rodney K. Murphey, Head, 413-545-3246, Fax: 413-545-1812.

See Close-Up on page 619.

University of Massachusetts Boston, Office of Graduate Studies and Research, College of Science and Mathematics, Track in Molecular, Cellular and Organismal Biology, Boston, MA 02125-3393. Offers PhD.

University of Massachusetts Worcester, Graduate School of Biomedical Sciences, Department of Cell Biology, Worcester, MA 01655-0115. Offers PhD. *Faculty:* 24 full-time (8 women).

Degree requirements: For doctorate, thesis/dissertation. *Entrance requirements:* For doctorate, GRE General Test. Additional exam requirements/recommendations for international students: Required—TOEFL (minimum score 600 paper-based; 250 computer-based). *Application deadline:* For fall admission, 12/15 for domestic students, 12/15 for international students. Applications are processed on a rolling basis. Application fee: $25 ($50 for international students). *Expenses:* Tuition, state resident: full-time $2,640. Tuition, nonresident: full-time $9,856. Required fees: $5,685. *Financial support:* In 2005–06, research assistantships with full tuition reimbursements (averaging $25,235 per year); unspecified assistantships also available. *Faculty research:* Cellular development and function, growth, development, differentiation, molecular mechanisms. *Unit head:* Dr. Gary Stein, Chair, 508-856-5625. *Application contact:* Michael Cole, Director of Admissions and Recruitment, 508-856-4779, Fax: 508-856-3659, E-mail: michael.cole@umassmed.edu.

See Close-Up on page 621.

University of Medicine and Dentistry of New Jersey, Graduate School of Biomedical Sciences, Graduate Programs in Biomedical Sciences–Newark, Department of Cell Biology and Molecular Medicine, Newark, NJ 07107. Offers PhD. *Degree requirements:* For doctorate, thesis/dissertation, qualifying exam. *Entrance requirements:* For doctorate, GRE General Test. Additional exam requirements/recommendations for international students: Required—TOEFL. *Application deadline:* For fall admission, 2/1 for domestic students. Application fee: $40. *Financial support:* Fellowships, research assistantships, Federal Work-Study, institutionally sponsored loans, and tuition waivers (full and partial) available. Financial award application deadline: 5/1. *Unit head:* Dr. Dorothy Vatner, Program Director, 973-972-1339, Fax: 973-972-7489, E-mail: vatnerdo@umdnj.edu.

University of Medicine and Dentistry of New Jersey, Graduate School of Biomedical Sciences, Graduate Programs in Biomedical Sciences–Stratford, Program in Cell and Molecular Biology, Stratford, NJ 08084-5634. Offers MS, PhD, DO/PhD. *Degree requirements:* For master's, thesis; for doctorate, thesis/dissertation, qualifying exam. *Entrance requirements:* For master's and doctorate, GRE General Test. Additional exam requirements/recommendations for international students: Required—TOEFL. *Application deadline:* For fall admission, 2/1 for domestic students; for spring admission, 10/1 for domestic students. Applications are processed on a rolling basis. Application fee: $40. *Financial support:* Application deadline: 5/1. *Unit head:* Dr. Diane M Worrad, Program Coordinator, 856-566-6282, Fax: 856-566-6232, E-mail: worraddi@umdnj.edu.

University of Miami, Graduate School, Miller School of Medicine, Graduate Programs in Medicine, Department of Cell Biology and Anatomy, Coral Gables, FL 33124. Offers molecular cell and developmental biology (PhD). *Faculty:* 20 full-time (6 women). *Students:* 25 full-time (10 women); includes 4 minority (1 Asian American or Pacific Islander, 3 Hispanic Americans), 13 international. Average age 29. 57 applicants, 7% accepted, 4 enrolled. In 2005, 1 degree awarded. *Degree requirements:* For doctorate, thesis/dissertation. *Entrance requirements:* For doctorate, GRE General Test, GRE Subject Test. Additional exam requirements/recommendations for international students: Required—TOEFL. *Application deadline:* For fall admission, 3/1 for domestic students. Applications are processed on a rolling basis. Application fee: $50. Electronic applications accepted. *Financial support:* In 2005–06, 22 fellowships with tuition reimbursements (averaging $22,000 per year) were awarded *Unit head:* Dr. Robert Warren, Interim Chair, 305-243-6691, Fax: 305-545-7166, E-mail: rwarren@med.miami.edu. *Application contact:* Dr. Richard Rotundo, Professor, 305-243-6691, Fax: 305-545-7166.

University of Michigan, Horace H. Rackham School of Graduate Studies, College of Literature, Science, and the Arts, Department of Molecular, Cellular, and Developmental Biology, Ann Arbor, MI 48109. Offers MS, PhD. Part-time programs available. *Faculty:* 24 full-time (6 women), 1 part-time/adjunct (0 women). *Students:* 72 full-time, 9 part-time; includes 5 minority (2 African Americans, 1 Asian American or Pacific Islander, 2 Hispanic Americans), 49 international. Average age 24. 102 applicants, 29% accepted, 18 enrolled. In 2005, 5 master's, 7 doctorates awarded. Terminal master's awarded for partial completion of doctoral program. *Median time to degree:* Of those who began their doctoral program in fall 1997, 60% received their degree in 8 years or less. *Degree requirements:* For master's, registration; for doctorate, thesis/dissertation, preliminary exam, oral defense, dissertation. *Entrance requirements:* For master's and doctorate, GRE General Test. Additional exam requirements/recommendations for international students: Required—TOEFL (minimum score 560 paper-based; 220 computer-based). *Application deadline:* For fall admission, 1/5 for domestic students, 1/5 for international students. For winter admission, 11/1 for domestic students. Applications are processed on a rolling basis. Application fee: $60 ($75 for international students). Electronic applications accepted. *Expenses:* Tuition, state resident: full-time $14,082; part-time $894 per credit hour. Tuition, nonresident: full-time $28,500; part-time $1,675 per credit hour. Required fees: $189; $189 per unit. *Financial support:* In 2005–06, 69 students received support, including 18 fellowships with full tuition reimbursements available (averaging $14,326 per year), 24 research assistantships with full tuition reimbursements available (averaging $14,326 per year), 21 teaching assistantships with full tuition reimbursements available (averaging $14,326 per year); career-related internships or fieldwork, scholarships/grants, traineeships, health care benefits, and unspecified assistantships also available. *Faculty research:* Biochemistry, cell biology, microbiology, neurobiology and physiology, developmental biology and plant molecular biology. Total annual research expenditures: $5.5 million. *Unit head:* Dr. Richard Hume, Chair, 734-764-7427, Fax: 734-746-0884. *Application contact:* Mary Carr, Graduate Coordinator, 734-615-1635, Fax: 734-764-0884, E-mail: carrmm@umich.edu.

See Close-Up on page 625.

University of Michigan, Medical School and Horace H. Rackham School of Graduate Studies, Program in Biomedical Sciences (PIBS), Department of Cell and Developmental Biology, Ann Arbor, MI 48109. Offers MS, PhD. *Degree requirements:* For doctorate, thesis/dissertation, oral defense of dissertation, preliminary exam. *Entrance requirements:* For doctorate, GRE General Test, GRE Subject Test, master's degree. Additional exam requirements/recommendations for international students: Required—TOEFL; Recommended—TWE, TSE. Electronic applications accepted. *Expenses:* Tuition, state resident: full-time $14,082; part-time $894 per credit hour. Tuition, nonresident: full-time $28,500; part-time $1,675 per credit hour. Required fees: $189; $189 per unit. *Faculty research:* Small stress proteins, cellular stress response, muscle, male reproductive, toxicology, cell cytoskeleton.

University of Michigan, Medical School and Horace H. Rackham School of Graduate Studies, Program in Biomedical Sciences (PIBS), Interdisciplinary Program in Cellular and Molecular Biology, Ann Arbor, MI 48109. Offers PhD. *Faculty:* 113 part-time/adjunct (32 women). *Students:* 66 full-time (22 women); includes 24 minority (7 African Americans, 12 Asian Americans or Pacific Islanders, 5 Hispanic Americans). Average age 26. In 2005, 10 degrees awarded. *Median time to degree:* Of those who began their doctoral program in fall 1997, 100% received their degree in 8 years or less. *Degree requirements:* For doctorate, thesis/dissertation, oral defense of dissertation, preliminary exam, comprehensive exam. *Entrance requirements:* For doctorate, GRE General Test, GRE Subject Test. *Expenses:* Tuition, state resident: full-time $14,082; part-time $894 per credit hour. Tuition, nonresident: full-time $28,500; part-time $1,675 per credit hour. Required fees: $189; $189 per unit. *Financial support:* In 2005–06, 66 students received support, including 23 fellowships with tuition reimbursements available (averaging $23,500 per year), 43 research assistantships with tuition reimbursements available (averaging $23,500 per year); teaching assistantships, scholarships/grants also available. *Faculty research:* Genetics, genomics, gene regulation, models of disease, microbes. Total annual research expenditures: $20 million. *Unit head:* Dr. Jessica Schwartz, Director, 734-764-5428, Fax: 734-647-6232, E-mail: jeschwar@umich.edu. *Application contact:* Linda Wilson, Student Services Associate I, 734-764-5428, Fax: 734-647-6232, E-mail: cmbgrad@umich.edu.

University of Minnesota, Twin Cities Campus, Graduate School, Program in Molecular, Cellular, Developmental Biology and Genetics, Minneapolis, MN 55455-0213. Offers genetic counseling (MS); molecular, cellular, developmental biology and genetics (PhD). Part-time programs available. Terminal master's awarded for partial completion of doctoral program. *Degree requirements:* For master's, thesis optional; for doctorate, thesis/dissertation. *Entrance requirements:* For master's and doctorate, GRE General Test. Additional exam requirements/recommendations for international students: Required—TOEFL. *Expenses:* Tuition, state resident: full-time $8,748; part-time $729 per credit. Tuition, nonresident: full-time $15,848; part-time $1,321 per credit. Full-time tuition and fees vary according to class time, course load, program and reciprocity agreements. *Faculty research:* Membrane receptors and membrane transport, cell interactions, cytoskeleton and cell mobility, regulation of gene expression, plant cell and molecular biology.

University of Missouri–Columbia, Graduate School, College of Arts and Sciences, Division of Biological Sciences, Columbia, MO 65211. Offers cellular, molecular and developmental biology (MA, PhD); evolutionary biology and ecology (MA, PhD); neurobiology and behavior (MA, PhD). *Faculty:* 43 full-time (12 women). *Students:* 48 full-time (26 women), 28 part-time (10 women); includes 7 minority (2 African Americans, 3 Asian Americans or Pacific Islanders, 2 Hispanic Americans), 20 international. In 2005, 3 master's, 8 doctorates awarded. Terminal master's awarded for partial completion of doctoral program. *Degree requirements:* For master's, thesis/dissertation; for doctorate, thesis/dissertation, comprehensive exam. *Entrance requirements:* For master's and doctorate, GRE General Test, minimum GPA of 3.0. *Application deadline:* For fall admission, 1/15 for domestic students. Applications are processed on a rolling basis. Application fee: $45 ($60 for international students). *Financial support:* Fellowships, research assistantships, teaching assistantships, institutionally sponsored loans available. *Unit head:* Dr. Ray Semlitsch, Director of Graduate Studies, 573-884-6396, E-mail: semlitschr@missouri.edu. *Application contact:* Nila Emerich, Application Contact, 800-553-5698.

University of Missouri–Kansas City, School of Biological Sciences, Program in Cell Biology and Biophysics, Kansas City, MO 64110-2499. Offers PhD. PhD offered through the School of Graduate Studies. *Degree requirements:* For doctorate, thesis/dissertation, comprehensive exam. *Entrance requirements:* For doctorate, GRE General Test, bachelor's degree in chemistry, biology or related field; minimum GPA of 3.0. Additional exam requirements/recommendations for international students: Required—TOEFL. *Application deadline:* For fall admission, 3/1 for domestic students. Applications are processed on a rolling basis. Application fee: $35 ($50 for international students). Electronic applications accepted. *Expenses:* Tuition, state resident: full-time $4,738; part-time $263 per credit hour. Tuition, nonresident: full-time $12,235; part-time $679 per credit hour. Required fees: $582. Tuition and fees vary according to course load, program and student level. *Financial support:* Fellowships with full tuition reimbursements, research assistantships with full tuition reimbursements, teaching assistantships with full and partial tuition reimbursements, scholarships/grants, tuition waivers (full and partial), and unspecified assistantships available. *Unit head:* Dr. George Thomas, Head, 816-235-5247, E-mail: thomasgj@umkc.edu. *Application contact:* Laura Batenic, Information Contact, 816-235-2352, Fax: 816-235-5158, E-mail: batenicl@umkc.edu.

See Close-Up on page 627.

University of Missouri–St. Louis, College of Arts and Sciences, Department of Biology, St. Louis, MO 63121. Offers biology (MS, PhD), including animal behavior (MS); biochemistry (MS), biotechnology (MS), conservation biology (MS), development (MS), ecology (MS), environmental studies (PhD), evolution (MS), genetics (MS), molecular/cellular biology (MS), physiology (MS), plant systematics, population biology (MS), tropical biology (MS); biotechnology (Certificate); tropical biology and conservation (Certificate). Part-time programs available. *Faculty:* 50. *Students:* 26 full-time (15 women), 101 part-time (51 women); includes 13 minority (5 African Americans, 6 Asian Americans or Pacific Islanders, 2 Hispanic Americans), 41 international. Average age 32. In 2005, 22 master's, 2 doctorates awarded. *Degree requirements:* For master's, thesis or alternative; for doctorate, one foreign language, thesis/dissertation, 1 semester of teaching experience. *Entrance requirements:* For doctorate, GRE General Test. *Application deadline:* For spring admission, 12/1 priority date for domestic students. Applications are processed on a rolling basis. Application fee: $35 ($40 for international students). Electronic applications accepted. *Expenses:* Tuition, state resident: part-time $263 per credit hour. Tuition, nonresident: part-time $680 per credit hour. Required fees: $53 per credit hour. Tuition and fees vary according to program. *Financial support:* In 2005–06, 11 fellowships with full tuition reimbursements (averaging $30,000 per year), 15 research assistantships with full and partial tuition reimbursements (averaging $16,000 per year), 22 teaching assistantships with full and partial tuition reimbursements (averaging $16,000 per year) were awarded; career-related internships or fieldwork and Federal Work-Study also available. Support available to part-time students. Financial award application deadline: 2/1. *Faculty research:* Molecular biology, microbial genetics. *Unit head:* Zuleyma Tang-Martinez, Director of Graduate Studies, 314-516-6498, Fax: 314-516-6233, E-mail: zuleyma@umsl.edu. *Application contact:* 314-516-5458, Fax: 314-516-5310, E-mail: gradadm@umsl.edu.

University of Nebraska Medical Center, Graduate Studies, Department of Genetics, Cell Biology and Anatomy, Omaha, NE 68198. Offers MS, PhD. Part-time programs available. *Faculty:* 20 full-time (6 women), 1 part-time/adjunct (0 women). *Students:* 12 full-time (4 women); includes 7 minority (1 African American, 5 Asian Americans or Pacific Islanders, 1 Hispanic American), 5 international. Average age 28. 8 applicants, 38% accepted, 3 enrolled. In 2005, 1 degree awarded. Terminal master's awarded for partial completion of doctoral program. *Degree requirements:* For master's and doctorate, thesis/dissertation, comprehensive exam, registration. *Entrance requirements:* For master's and doctorate, GRE General Test. Additional exam requirements/recommendations for international students: Required—TOEFL. *Application deadline:* For fall admission, 3/1 for domestic students. Applications are processed on a rolling basis. Application fee: $45. Electronic applications accepted. *Expenses:* Tuition, area resident: Part-time $200 per hour. Tuition, nonresident: part-time $538 per hour. Required fees: $308; $59 per term. *Financial support:* In 2005–06, 4 students received support, including 11 fellowships with full tuition reimbursements available (averaging $21,000 per year), 11 research assistantships with full tuition reimbursements available (averaging $21,000 per year), teaching assistantships with full tuition reimbursements available (averaging $21,000 per year); institutionally sponsored loans also available. Support available to part-time students. Financial award application deadline: 3/1. *Faculty research:* Hematology, immunology, developmental biology, cardiovascular biology, neuroscience. *Unit head:* Dr. Shantaram S. Joshi, Graduate Committee Chair, 402-559-4165, Fax: 402-559-3400, E-mail: ssjoshi@unmc.edu. *Application contact:* Penni Davis, Graduate Student Support, 402-559-3316, Fax: 402-559-3144, E-mail: pkdavis@unmc.edu.

University of Nevada, Reno, School of Medicine and Graduate School, Graduate Programs in Medicine, Interdisciplinary Program in Cell and Molecular Biology, Reno, NV 89557. Offers MS, PhD. *Students:* 19 full-time (11 women), 21 part-time (11 women); includes 7 minority (1 African American, 3 Asian Americans or Pacific Islanders, 3 Hispanic Americans), 9 international. Average age 30. 12 applicants, 92% accepted, 6 enrolled. In 2005, 5 degrees awarded. *Expenses:* Tuition, area resident: Full-time $2,767; part-time $923 per semester. Tuition, state resident: full-time $5,733; part-time $1,911 per semester. Tuition, nonresident: full-time $12,679; part-time $1,911 per semester. International tuition: $13,878 full-time. Required fees: $404; $202 per term. One-time fee: $90. *Unit head:* Dr. Stephen St. Jeor, Graduate Program Director, 775-784-4123, E-mail: stjeor@unr.edu.

University of New Haven, Graduate School, College of Arts and Sciences, Program in Cellular and Molecular Biology, West Haven, CT 06516-1916. Offers MS. *Degree requirements:* For master's, thesis or alternative.

See Close-Up on page 631.

University of New Mexico, School of Medicine, Biomedical Sciences Graduate Program, Albuquerque, NM 87131-5196. Offers biochemistry and molecular biology (MS, PhD); cell biology and physiology (MS, PhD); molecular genetics and microbiology (MS, PhD); neuroscience (MS, PhD); pathology (MS, PhD); toxicology (MS, PhD). Part-time programs available.

Terminal master's awarded for partial completion of doctoral program. *Degree requirements:* For master's, thesis/dissertation; for doctorate, thesis/dissertation, comprehensive exam. *Entrance requirements:* For master's and doctorate, GRE General Test, minimum undergraduate GPA of 3.0. Additional exam requirements/recommendations for international students: Required—TOEFL. Electronic applications accepted. *Expenses:* Tuition, state resident: full-time $5,676. Tuition, nonresident: full-time $14,974; part-time $238 per credit hour. Required fees: $385 per term. Tuition and fees vary according to course load and program. *Faculty research:* Signal transduction, infectious disease, biology of cancer, structural biology, neuroscience.

The University of North Carolina at Chapel Hill, Graduate School, College of Arts and Sciences, Department of Biology, Program in Cell Motility and Cytoskeleton, Chapel Hill, NC 27599. Offers PhD. *Entrance requirements:* For doctorate, GRE, General Test.

See Close-Up on page 633.

The University of North Carolina at Chapel Hill, National Institutes of Health Sponsored Programs, Program in Cell Motility and Cytoskeleton, Chapel Hill, NC 27599. Offers PhD. *Unit head:* Dr. Steven W. Matson, Chair, 919-962-5945, E-mail: smatson@bio.unc.edu. *Application contact:* Graduate Admissions, E-mail: cpasternak@bio.unc.edu.

See Close-Up on page 633.

The University of North Carolina at Chapel Hill, School of Medicine and Graduate School, Graduate Programs in Medicine, Department of Cell and Developmental Biology, Chapel Hill, NC 27599. Offers PhD. *Faculty:* 22 full-time (6 women), 1 part-time/adjunct (0 women). *Students:* 39 full-time (20 women); includes 3 minority (1 African American, 2 Asian Americans or Pacific Islanders), 9 international. Average age 24. 53 applicants, 13% accepted, 4 enrolled. In 2005, 1 degree awarded. *Degree requirements:* For doctorate, thesis/dissertation, comprehensive exam, registration. *Entrance requirements:* For doctorate, GRE General Test, GRE Subject Test. *Application deadline:* For fall admission, 1/1 priority date for domestic students, 1/1 priority date for international students. Applications are processed on a rolling basis. Application fee: $70. Electronic applications accepted. *Financial support:* In 2005–06, 4 students received support, including 35 research assistantships with full tuition reimbursements available (averaging $22,000 per year), 4 teaching assistantships with full tuition reimbursements available (averaging $22,000 per year); fellowships, tuition waivers (full) and unspecified assistantships also available. Financial award application deadline: 2/1; financial award applicants required to submit FAFSA. *Faculty research:* Cell adhesion, motility and cytoskeleton; molecular analysis of signal transduction; development biology and toxicology; reproductive biology; cell and molecular imaging. Total annual research expenditures: $8 million. *Unit head:* Dr. Vytas A. Bankaitis, Chair, 919-966-3026, Fax: 919-966-1856, E-mail: vytas@med.unc.edu. *Application contact:* Dr. Patrick Brennwald, Director of Graduate Studies, 919-843-4995, E-mail: patrick_brennwald@med.unc.edu.

University of Notre Dame, Graduate School, College of Science, Department of Biological Sciences, Notre Dame, IN 46556. Offers aquatic ecology, evolution and environmental biology (MS, PhD); cellular and molecular biology (MS, PhD); genetics (MS, PhD); physiology (MS, PhD); vector biology and parasitology (MS, PhD). *Faculty:* 34 full-time (8 women), 3 part-time/adjunct (0 women). *Students:* 124 full-time (55 women); includes 11 minority (1 African American, 6 Asian Americans or Pacific Islanders, 4 Hispanic Americans), 39 international. 95 applicants, 34% accepted, 22 enrolled. In 2005, 4 master's, 11 doctorates awarded. Terminal master's awarded for partial completion of doctoral program. *Median time to degree:* Of those who began their doctoral program in fall 1997, 61% received their degree in 8 years or less. *Degree requirements:* For master's and doctorate, thesis/dissertation, comprehensive exam. *Entrance requirements:* For master's and doctorate, GRE General Test. Additional exam requirements/recommendations for international students: Required—TOEFL. *Application deadline:* For fall admission, 2/1 for domestic students; for spring admission, 11/1 for domestic students. Applications are processed on a rolling basis. Application fee: $50. Electronic applications accepted. *Financial support:* In 2005–06, 124 students received support, including 24 fellowships with full tuition reimbursements available (averaging $22,000 per year), 47 research assistantships with full tuition reimbursements available (averaging $15,250 per year), 45 teaching assistantships with full tuition reimbursements available (averaging $16,000 per year); traineeships and tuition waivers (full) also available. Financial award application deadline: 2/1. *Faculty research:* Tropical disease, molecular genetics, neurobiology, evolutionary biology, aquatic biology. Total annual research expenditures: $15.3 million. *Unit head:* Dr. Gary A. Lamberti, Director of Graduate Studies, 574-631-6552, Fax: 574-631-7413, E-mail: biology.biosadm.1@nd.edu. *Application contact:* Dr. Terrence J. Akai, Director of Graduate Admissions, 574-631-7706, Fax: 574-631-4183, E-mail: gradad@nd.edu.

See Close-Up on page 281.

University of Oklahoma Health Sciences Center, College of Medicine and Graduate College, Graduate Programs in Medicine, Department of Cell Biology, Oklahoma City, OK 73190. Offers MS, PhD. *Degree requirements:* For master's and doctorate, thesis/dissertation. *Entrance requirements:* For doctorate, GRE General Test, GRE Subject Test, 3 letters of recommendation. Additional exam requirements/recommendations for international students: Required—TOEFL. *Faculty research:* Neurobiology, reproductive, neuronal plasticity, extracellular matrix, neuroendocrinology.

University of Ottawa, Faculty of Graduate and Postdoctoral Studies, Faculty of Medicine, Department of Cellular and Molecular Medicine, Ottawa, ON K1H 8M5, Canada. Offers M Sc, PhD. *Faculty:* 69 full-time (9 women). *Students:* 155 full-time, 10 part-time. 45 applicants, 82% accepted, 29 enrolled. In 2005, 15 master's, 3 doctorates awarded. *Degree requirements:* For master's, thesis/dissertation, seminar; for doctorate, thesis/dissertation, seminar, comprehensive exam. *Entrance requirements:* For master's, honors degree or equivalent, minimum B average; for doctorate, master's degree, minimum B+ average. *Application deadline:* For fall admission, 8/1 priority date for domestic students, 2/15 priority date for international students. For winter admission, 11/15 for domestic students; for spring admission, 4/1 for domestic students. Application fee: $75. Electronic applications accepted. *Expenses:* Tuition: Part-time $260 per credit. Tuition and fees vary according to course load and program. *Financial support:* Fellowships, research assistantships with full tuition reimbursements, teaching assistantships with full tuition reimbursements, career-related internships or fieldwork, Federal Work-Study, scholarships/grants, traineeships, tuition waivers (full and partial), and unspecified assistantships available. Financial award application deadline: 2/15. *Faculty research:* Physiology, pharmacology, growth and development. *Unit head:* Dr. Bernard Jasmin, Chair, 613-562-5425, Fax: 613-562-5434, E-mail: jasmin@uottawa.ca. *Application contact:* Donna Hooper, Academic Assistant, 613-562-5800 Ext. 8396, Fax: 613-562-5636, E-mail: clmolm@uottawa.ca.

University of Pennsylvania, School of Medicine, Biomedical Graduate Studies, Graduate Group in Cell and Molecular Biology, Philadelphia, PA 19104. Offers cell biology and physiology (PhD); cell growth and cancer (PhD); developmental biology (PhD); gene therapy and vaccines (PhD); genetics and gene regulation (PhD); microbiology, virology, and parasitology (PhD). *Faculty:* 294. *Students:* 290 full-time (155 women); includes 67 minority (9 African Americans, 41 Asian Americans or Pacific Islanders, 17 Hispanic Americans), 24 international. 510 applicants, 21% accepted, 38 enrolled. In 2005, 46 doctorates awarded. *Degree requirements:* For doctorate, thesis/dissertation. *Entrance requirements:* For doctorate, GRE General Test. Additional exam requirements/recommendations for international students: Required—TOEFL. *Application deadline:* For fall admission, 12/15 priority date for domestic students, 12/1 priority date for international students. Applications are processed on a rolling basis. Application fee: $70. Electronic applications accepted. *Financial support:* Fellowships, research assistantships, scholarships/grants, traineeships, and unspecified assistantships available. Financial award application deadline: 1/2. *Unit head:* Dr. Jonathan Raper, Graduate

Cell Biology

University of Pennsylvania (continued)
Group Chair, 215-898-2180, E-mail: raperj@mail.med.upenn.edu. *Application contact:* Meagan Schofer, Coordinator, 215-895-9536, Fax: 215-573-2104, E-mail: mschofer@mailmed.upenn.edu.

University of Pittsburgh, School of Arts and Sciences, Department of Biological Sciences, Program in Molecular, Cellular, and Developmental Biology, Pittsburgh, PA 15260. Offers PhD. *Faculty:* 22 full-time (7 women), 1 part-time/adjunct (0 women). *Students:* 48 full-time (32 women); includes 2 minority (both African Americans), 10 international. Average age 23. 215 applicants, 2% accepted, 4 enrolled. In 2005, 7 degrees awarded. *Median time to degree:* Of those who began their doctoral program in fall 1997, 100% received their degree in 8 years or less. *Degree requirements:* For doctorate, thesis/dissertation, comprehensive exam, registration. *Entrance requirements:* For doctorate, GRE General Test, GRE Subject Test. Additional exam requirements/recommendations for international students: Required—TOEFL (minimum score 550 paper-based; 213 computer-based). *Application deadline:* For fall admission, 1/15 priority date for domestic students, 12/15 priority date for international students. Applications are processed on a rolling basis. Application fee: $0 ($40 for international students). Electronic applications accepted. *Expenses:* Tuition, state resident: full-time $13,194; part-time $537 per credit. Tuition, nonresident: full-time $25,012; part-time $1,026 per credit. Required fees: $700; $164 per term. Tuition and fees vary according to campus/location and program. *Financial support:* In 2005–06, 18 fellowships with full tuition reimbursements, 93 research assistantships with full tuition reimbursements (averaging $20,332 per year), 22 teaching assistantships with full tuition reimbursements (averaging $20,332 per year) were awarded; Federal Work-Study, scholarships/grants, traineeships, health care benefits, and tuition waivers (full) also available. *Faculty research:* Structure and function of genes and proteins; macromolecular interactions; cell-specific gene regulation; regulation of cell proliferation; embryogenesis. *Application contact:* Cathleen M. Barr, Graduate Administrator, 412-624-4268, Fax: 412-624-4759, E-mail: cbarr@pitt.edu.

See Close-Up on page 637.

University of Pittsburgh, School of Medicine, Graduate Programs in Medicine, Program in Cell Biology and Molecular Physiology, Pittsburgh, PA 15260. Offers MS, PhD. *Faculty:* 29 full-time (7 women). *Students:* 20 full-time (8 women); includes 3 minority (2 African Americans, 1 Asian American or Pacific Islander), 1 international. Average age 28. 415 applicants, 22% accepted, 42 enrolled. In 2005, 1 degree awarded. *Median time to degree:* Of those who began their doctoral program in fall 1997, 95% received their degree in 8 years or less. *Degree requirements:* For doctorate, thesis/dissertation, comprehensive exam, registration. *Entrance requirements:* For doctorate, GRE General Test, GRE Subject Test, minimum QPA of 3.0. Additional exam requirements/recommendations for international students: Required—TOEFL (minimum score 600 paper-based; 250 computer-based), IELT (minimum score 7). *Application deadline:* For fall admission, 12/15 priority date for domestic students, 12/15 priority date for international students. Application fee: $40. Electronic applications accepted. *Expenses:* Tuition, state resident: full-time $13,194; part-time $537 per credit. Tuition, nonresident: full-time $25,012; part-time $1,026 per credit. Required fees: $700; $164 per term. Tuition and fees vary according to campus/location and program. *Financial support:* In 2005–06, fellowships with full tuition reimbursements (averaging $21,500 per year), research assistantships with tuition reimbursements (averaging $21,500 per year), teaching assistantships with tuition reimbursements (averaging $21,500 per year) were awarded; institutionally sponsored loans, scholarships/grants, traineeships, health care benefits, and unspecified assistantships also available. *Faculty research:* Genetic disorders of ion channels, regulation of gene expression/development, membrane traffic of proteins and lipids, reproductive biology, signal transduction in diabetes and metabolism. *Unit head:* Dr. Simon C. Watkins, Graduate Program Director, 412-648-3051, Fax: 412-648-8330, E-mail: swatkins@pitt.edu. *Application contact:* 412-648-8957, Fax: 412-648-1077, E-mail: gradstudies@medschool.pitt.edu.

University of Rhode Island, Graduate School, College of the Environment and Life Sciences, Department of Cell and Molecular Biology, Kingston, RI 02881. Offers biochemistry (MS, PhD); microbiology (MS, PhD), including biodegradation (MS), cellular development (MS), electron microscopy and ultrastructure (MS), genetics and molecular biology (MS), immunology (MS), marine and freshwater ecosystems (MS), microbial pathogenesis (MS), microbial physiology (MS), protozoology (MS), virology (MS), water-pollution microbiology (MS); molecular genetics (MS, PhD). In 2005, 2 degrees awarded. *Degree requirements:* For master's and doctorate, thesis/dissertation. *Entrance requirements:* For master's and doctorate, GRE General Test. Additional exam requirements/recommendations for international students: Required—TOEFL. *Expenses:* Tuition, state resident: full-time $5,522; part-time $307 per credit. Tuition, nonresident: full-time $15,992; part-time $888 per credit. Required fees: $1,786; $73 per credit. One-time fee: $80 part-time. *Financial support:* Fellowships, research assistantships, teaching assistantships available. *Unit head:* Dr. Jay Sperry, Chairperson, 401-874-5900.

University of Saskatchewan, College of Medicine, Department of Anatomy and Cell Biology, Saskatoon, SK S7N 5A2, Canada. Offers M Sc, PhD. *Faculty:* 17. *Students:* 26, 4 international. *Degree requirements:* For master's and doctorate, thesis/dissertation, registration. *Entrance requirements:* Additional exam requirements/recommendations for international students: Required—TOEFL. *Application deadline:* For fall admission, 7/1 for domestic students. Applications are processed on a rolling basis. Application fee: $50. *Financial support:* Fellowships, research assistantships, teaching assistantships available. Financial award application deadline: 1/31. *Unit head:* Dr. R. Doucette, Head (Acting), 306-966-4075, Fax: 306-966-4298, E-mail: rondouc@duke.usask.ca. *Application contact:* Dr. D. Schreyer, Graduate Chair, 306-966-4298, Fax: 306-966-4298, E-mail: david.shreyer@usask.ca.

University of South Alabama, College of Medicine and Graduate School, Program in Basic Medical Sciences, Specialization in Cell Biology and Neuroscience, Mobile, AL 36688-0002. Offers PhD. *Faculty:* 13 full-time (2 women). In 2005, 1 degree awarded. *Degree requirements:* For doctorate, thesis/dissertation, oral and written preliminary exams, research proposal, qualifying exam. *Entrance requirements:* For doctorate, GRE General Test, minimum GPA of 3.0. Additional exam requirements/recommendations for international students: Required—TOEFL. *Application deadline:* For fall admission, 4/30 for domestic students. Applications are processed on a rolling basis. Application fee: $25. *Expenses:* Tuition, state resident: full-time $4,008. Tuition, nonresident: full-time $8,016. Required fees: $692. *Financial support:* Fellowships, institutionally sponsored loans available. Financial award application deadline:4/1. *Faculty research:* Cytoskeleton-membrane interactions, neural basis of oral motor behavior, microtubule organizing centers, molecular biology of human chromosomes, mechanisms of synaptic transmissions. *Unit head:* Dr. Glen Wilson, Chair, 251-460-6490. *Application contact:* Lanette Flagge, Academic Advisor, 251-460-6153.

University of South Carolina, The Graduate School, College of Science and Mathematics, Department of Biological Sciences, Graduate Training Program in Molecular, Cellular, and Developmental Biology, Columbia, SC 29208. Offers MS, PhD. *Degree requirements:* For master's and doctorate, one foreign language, thesis/dissertation. *Entrance requirements:* For master's and doctorate, GRE General Test, minimum GPA of 3.0 in science. Electronic applications accepted. *Faculty research:* Marine ecology, population and evolutionary biology, molecular biology and genetics, development.

See Close-Up on page 639.

The University of South Dakota, School of Medicine and Health Sciences and Graduate School, Biomedical Sciences Graduate Program, Cellular and Molecular Biology Group, Vermillion, SD 57069-2390. Offers MA, PhD. *Faculty:* 9 full-time (2 women). *Students:* 9 full-time (7 women), 7 international. Average age 28. 26 applicants, 15% accepted, 2 enrolled. In 2005, 1 master's, 2 doctorates awarded. Terminal master's awarded for partial completion of doctoral program. *Median time to degree:* Of those who began their doctoral program in fall 1997, 90% received their degree in 8 years or less. *Degree requirements:* For master's, thesis/dissertation, registration; for doctorate, thesis/dissertation, comprehensive exam,

registration. *Entrance requirements:* For master's and doctorate, GRE General Test, GRE Subject Test, minimum GPA of 3.0. Additional exam requirements/recommendations for international students: Required—TOEFL (minimum score 550 paper-based; 213 computer-based). *Application deadline:* For fall admission, 4/15 priority date for domestic students, 4/15 priority date for international students. Applications are processed on a rolling basis. Application fee: $35. *Expenses:* Tuition, state resident: part-time $116 per credit hour. Tuition, nonresident: part-time $341 per credit hour. Required fees: $85 per credit hour. Tuition and fees vary according to course load, program and reciprocity agreements. *Financial support:* In 2005–06, 9 students received support, including 8 fellowships with partial tuition reimbursements available (averaging $20,772 per year), 1 research assistantship with partial tuition reimbursement available (averaging $10,386 per year) Financial award application deadline: 4/15; financial award applicants required to submit FAFSA. *Faculty research:* Molecular aspects of protein and DNA, neurochemistry and energy transduction, gene regulation, cellular development. Total annual research expenditures: $438,000.

University of Southern California, Keck School of Medicine and Graduate School, Graduate Programs in Medicine, Department of Cell and Neurobiology, Los Angeles, CA 90089. Offers MS, PhD. *Faculty:* 23 full-time (4 women), 8 part-time/adjunct (2 women). *Students:* 9 full-time (3 women); includes 3 minority (all Asian Americans or Pacific Islanders) Average age 25. 15 applicants, 0% accepted, 0 enrolled.Terminal master's awarded for partial completion of doctoral program. *Degree requirements:* For master's, thesis or alternative; for doctorate, thesis/dissertation. *Entrance requirements:* For master's, GRE General Test, minimum GPA of 3.0; for doctorate, GRE General Test. *Application deadline:* For fall admission, 3/5 priority date for domestic students, 3/5 priority date for international students. Application fee: $65 ($75 for international students). Electronic applications accepted. *Expenses:* Tuition: Full-time $25,416; part-time $1,059 per unit. Required fees: $484; $484 per year. Tuition and fees vary according to course load and program. *Financial support:* In 2005–06, 3 research assistantships (averaging $24,876 per year), teaching assistantships (averaging $9,056 per year) were awarded; fellowships, Federal Work-Study, institutionally sponsored loans, and tuition waivers (partial) also available. Support available to part-time students. *Faculty research:* Neurobiology and development, gene therapy in vision, lacrimal glands, neuroendocrinology, signal transduction mechanisms. *Unit head:* Dr. Mikel Henry Snow, Chair, 323-442-1881, Fax: 323-442-3466. *Application contact:* Darlene Marie Campbell, Project Specialist, 323-442-1881, Fax: 323-442-0466, E-mail: dmc@usc.edu.

The University of Tennessee Health Science Center, College of Graduate Health Sciences, Department of Molecular Sciences, Concentration in Molecular and Cell Biology-Pathology, Memphis, TN 38163-0002. Offers PhD. *Faculty:* 26 full-time (4 women), 1 (woman) part-time/adjunct. *Students:* 3 full-time (1 woman); includes 2 minority (both Asian Americans or Pacific Islanders) Average age 25. 83 applicants, 12% accepted. In 2005, 6 degrees awarded. *Degree requirements:* For doctorate, thesis/dissertation, oral and written preliminary and comprehensive exams. *Entrance requirements:* For doctorate, GRE General Test, GRE Subject Test, minimum GPA of 3.0. Additional exam requirements/recommendations for international students: Required—TOEFL. *Application deadline:* For fall admission, 5/15 for domestic students. Application fee: $0. Electronic applications accepted. *Financial support:* Fellowships, research assistantships, traineeships and tuition waivers (full) available. *Application contact:* Ida W. Mosby, Director, Enrollment Services, 901-448-5560, E-mail: imosby@utmem.edu.

The University of Texas at Austin, Graduate School, College of Natural Sciences, Institute for Cellular and Molecular Biology, Program in Cell and Molecular Biology, Austin, TX 78712-1111. Offers PhD. *Degree requirements:* For doctorate, thesis/dissertation, qualifying exam. *Entrance requirements:* For doctorate, GRE General Test. *Faculty research:* Virology and immunology, RNA, molecular genetics, developmental biology, plant science, biochemistry evolution.

See Close-Up on page 645.

The University of Texas at Dallas, School of Natural Sciences and Mathematics, Program in Biology, Richardson, TX 75083-0688. Offers bioinformatics and computational biology (MS); biotechnology (MS); molecular and cell biology (MS, PhD). Part-time and evening/weekend programs available. *Faculty:* 15 full-time (2 women). *Students:* 64 full-time (40 women), 18 part-time (14 women); includes 18 minority (1 African American, 15 Asian Americans or Pacific Islanders, 2 Hispanic Americans), 46 international. Average age 28. 133 applicants, 71% accepted, 39 enrolled. In 2005, 10 master's, 7 doctorates awarded. *Degree requirements:* For master's, thesis optional; for doctorate, thesis/dissertation, publishable paper. *Entrance requirements:* For master's and doctorate, GRE General Test. Additional exam requirements/recommendations for international students: Required—TOEFL (minimum score 550 paper-based; 213 computer-based). *Application deadline:* For fall admission, 7/15 for domestic students; for spring admission, 11/15 for domestic students. Applications are processed on a rolling basis. Application fee: $50 ($100 for international students). Electronic applications accepted. *Expenses:* Tuition, state resident: full-time $5,450; part-time $303 per credit. Tuition, nonresident: full-time $12,648; part-time $703 per credit. Tuition and fees vary according to program. *Financial support:* In 2005–06, 17 research assistantships with tuition reimbursements (averaging $14,125 per year), 16 teaching assistantships with tuition reimbursements (averaging $11,972 per year) were awarded; fellowships, career-related internships or fieldwork, Federal Work-Study, institutionally sponsored loans, and scholarships/grants also available. Support available to part-time students. Financial award application deadline: 4/30; financial award applicants required to submit FAFSA. *Faculty research:* DNA replication, regulation of gene expression, subcellular organelles, physical chemistry of macromolecules, damage and repair of cellular DNA. Total annual research expenditures: $4.1 million. *Unit head:* Dr. Donald Gray, Head, 972-883-2513, Fax: 972-883-2502, E-mail: dongray@utdallas.edu. *Application contact:* Dr. Ernest Hannig, Graduate Advisor, 972-883-2505, Fax: 972-883-2409, E-mail: hannig@utdallas.edu.

See Close-Up on page 647.

The University of Texas at San Antonio, College of Sciences, Department of Biology, Program in Biology, San Antonio, TX 78249-0617. Offers cell and molecular biology (PhD); neurobiology (PhD). *Degree requirements:* For doctorate, dissertation, comprehensive exam. *Entrance requirements:* For doctorate, GRE General Test, minimum GPA of 3.0. Additional exam requirements/recommendations for international students: Required—TOEFL. Expenses: Contact institution. *Faculty research:* Neurophysiology, neurotoxicology, neural circuit analysis, neuroendocrinology, development of biosensors for the detection of toxins.

The University of Texas Health Science Center at Houston, Graduate School of Biomedical Sciences, Program in Cell and Regulatory Biology, Houston, TX 77225-0036. Offers MS, PhD, MD/PhD. *Faculty:* 34 full-time (8 women). *Students:* 13 full-time (11 women); includes 2 minority (1 Asian American or Pacific Islander, 1 Hispanic American), 5 international. Average age 23. 57 applicants, 25% accepted, 7 enrolled. In 2005, 1 degree awarded. Terminal master's awarded for partial completion of doctoral program. *Degree requirements:* For master's and doctorate, thesis/dissertation. *Entrance requirements:* For master's and doctorate, GRE General Test. Additional exam requirements/recommendations for international students: Required—TOEFL, TWE. *Application deadline:* For fall admission, 1/15 for domestic students; for spring admission, 11/1 for domestic students. Applications are processed on a rolling basis. Application fee: $10. Electronic applications accepted. *Financial support:* Fellowships with full tuition reimbursements, research assistantships with full tuition reimbursements, teaching assistantships, institutionally sponsored loans, scholarships/grants, and health care benefits available. Financial award application deadline: 1/15. *Faculty research:* Cell signaling, mechanisms of action of hormones and neurotransmitters, cancer chemotherapy, molecular biology of regulatory processes, regulation of ion channels and excitability. *Unit head:* Dr. Carmen Dessauer, Director, 713-500-6308, Fax: 713-500-7444, E-mail: carmen.w.dessauer@uth.tmc.edu. *Application contact:* Dr. Victoria P. Knutson, Assistant Dean of Admissions, 713-500-9860, Fax: 713-500-9877, E-mail: victoria.p.knutson@uth.tmc.edu.

The University of Texas Health Science Center at San Antonio, Graduate School of Biomedical Sciences, Department of Cellular and Structural Biology, San Antonio, TX 78229-3900. Offers PhD. *Degree requirements:* For doctorate, thesis/dissertation, oral qualifying exam. *Entrance requirements:* For doctorate, GRE General Test, previous course work in biology, chemistry, physics, and calculus. *Faculty research:* Human/molecular genetics, endocrinology and neurobiology, cell biology, cancer biology, biology of aging.

The University of Texas Medical Branch, Graduate School of Biomedical Sciences, Program in Biochemistry and Molecular Biology, Galveston, TX 77555. Offers biochemistry (PhD); bioinformatics (PhD); biophysics (PhD); cell biology (PhD); computational biology (PhD); structural biology (PhD). *Students:* 38 full-time (14 women), 1 part-time; includes 5 minority (3 Asian Americans or Pacific Islanders, 2 Hispanic Americans), 18 international. Average age 27. In 2005, 7 degrees awarded. *Degree requirements:* For doctorate, thesis/dissertation. *Entrance requirements:* Additional exam requirements/recommendations for international students: Required—TOEFL (minimum score 550 paper-based; 213 computer-based). *Application deadline:* Applications are processed on a rolling basis. Application fee: $30 ($75 for international students). Electronic applications accepted. *Expenses:* Tuition, state resident: full-time $8,350; part-time $90 per credit hour. Tuition, nonresident: full-time $21,450; part-time $366 per credit hour. Required fees: $1,027; $11 per credit hour. $60 per term. *Financial support:* In 2005–06, fellowships (averaging $23,000 per year), research assistantships (averaging $23,000 per year) were awarded. Financial award applicants required to submit FAFSA. *Unit head:* Dr. Leenian L. Chan, Director, 409-772-2861, Fax: 409-772-9679, E-mail: lchan@utmb.edu. *Application contact:* Debora Botting, Co-ordinator II Special Programs, 409-772-2769, Fax: 409-747-0552, E-mail: dmbottin@utmb.edu.

The University of Texas Medical Branch, Graduate School of Biomedical Sciences, Program in Cell Biology, Galveston, TX 77555. Offers PhD. *Students:* 22 full-time (16 women), 2 part-time (1 woman); includes 3 minority (1 Asian American or Pacific Islander, 2 Hispanic Americans), 7 international. Average age 29. In 2005, 5 doctorates awarded. *Degree requirements:* For doctorate, thesis/dissertation. *Entrance requirements:* For doctorate, GRE General Test. Additional exam requirements/recommendations for international students: Required—TOEFL (minimum score 550 paper-based; 213 computer-based). *Application deadline:* Applications are processed on a rolling basis. Application fee: $30 ($75 for international students). Electronic applications accepted. *Expenses:* Tuition, state resident: full-time $8,350; part-time $90 per credit hour. Tuition, nonresident: full-time $21,450; part-time $366 per credit hour. Required fees: $1,027; $11 per credit hour. $60 per term. *Financial support:* In 2005–06, fellowships (averaging $23,000 per year), research assistantships with full tuition reimbursements (averaging $23,000 per year) were awarded. Financial award applicants required to submit FAFSA. *Unit head:* Dr. Golda Ann Leonard, Director, 409-772-2722, Fax: 409-772-5893, E-mail: galeonar@utmb.edu. *Application contact:* Linda Spurger, Information Contact, 409-772-1942, Fax: 409-772-4687, E-mail: llspurge@utmb.edu.

The University of Texas Southwestern Medical Center at Dallas, Southwestern Graduate School of Biomedical Sciences, Division of Basic Science, Program in Cell Regulation, Dallas, TX 75390. Offers PhD. *Faculty:* 53 full-time (8 women). *Students:* 50 full-time (24 women), 3 part-time; includes 9 minority (7 Asian Americans or Pacific Islanders, 2 Hispanic Americans), 18 international. Average age 27. In 2005, 10 doctorates awarded. *Degree requirements:* For doctorate, thesis/dissertation, qualifying exam. *Entrance requirements:* For doctorate, GRE General Test, minimum GPA of 3.0. Additional exam requirements/recommendations for international students: Required—TOEFL. *Application deadline:* For fall admission, 1/5 for domestic students. Applications are processed on a rolling basis. Application fee: $0. Electronic applications accepted. *Expenses:* Tuition, state resident: full-time $6,550; part-time $50 per credit hour. Tuition, nonresident: full-time $19,650; part-time $326 per credit hour. Required fees: $42 per credit hour. Tuition and fees vary according to degree level and program. *Financial support:* Fellowships, research assistantships, institutionally sponsored loans and traineeships available. *Faculty research:* Molecular and cellular approaches to regulatory biology; receptor-effector coupling; membrane structure, function, and assembly. *Unit head:* Dr. Paul Sternweis, Chair, 214-645-6149, Fax: 214-645-6131, E-mail: paul.sternweis@utsouthwestern.edu. *Application contact:* Dr. Nancy E. Street, Associate Dean, 214-648-6708, Fax: 214-648-2102, E-mail: nancy.street@utsouthwestern.edu.

University of the Sciences in Philadelphia, College of Graduate Studies, Program in Cell Biology and Biotechnology, Philadelphia, PA 19104-4495. Offers MS. Part-time and evening/weekend programs available. *Faculty:* 7 full-time (3 women), 1 (woman) part-time/adjunct. *Students:* 5 full-time (all women), 12 part-time (6 women); includes 1 minority (Asian American or Pacific Islander), 4 international. Average age 28. In 2005, 7 degrees awarded. *Degree requirements:* For master's, thesis (for some programs), registration. *Entrance requirements:* For master's, GRE General Test. Additional exam requirements/recommendations for international students: Required—TOEFL, TWE, TSE. *Application deadline:* Applications are processed on a rolling basis. Application fee: $50. *Expenses:* Tuition: Part-time $999 per credit. Tuition and fees vary according to program. *Financial support:* In 2005–06, 6 students received support, including 5 teaching assistantships with full tuition reimbursements available (averaging $18,500 per year); scholarships/grants and tuition waivers (partial) also available. Financial award application deadline: 5/1. *Faculty research:* Invertebrate cell adhesion, plant-microbe interactions, natural product mechanisms, cell signal transduction, gene regulation and organization. Total annual research expenditures: $30,000. *Unit head:* Dr. John R. Porter, Director, 215-596-8917, Fax: 215-596-8710, E-mail: j.porter@usip.edu. *Application contact:* Joyce D'Angelo, Administrative Assistant, 215-596-8937, E-mail: j.dangel@usip.edu.

The University of Toledo, College of Graduate Studies, Program in Molecular and Cellular Biology, Toledo, OH 43606-3390. Offers PhD, MD/PhD. *Degree requirements:* For doctorate, thesis/dissertation, qualifying exam. *Entrance requirements:* For doctorate, GRE General Test, minimum undergraduate GPA of 3.0. *Expenses:* Tuition, state resident: full-time $6,623; part-time $308 per credit hour. Tuition, nonresident: full-time $13,232; part-time $735 per credit hour.

See Close-Up on page 311.

University of Utah, The Graduate School, College of Science and Graduate Programs in Medicine, Program in Molecular Biology, Salt Lake City, UT 84112-1107.

University of Vermont, Graduate College, Cell and Molecular Biology Program, Burlington, VT 05405. Offers MS, PhD. *Students:* 42 (18 women); includes 4 minority (2 Asian Americans or Pacific Islanders, 2 Hispanic Americans) 7 international. 45 applicants, 24% accepted, 7 enrolled. In 2005, 3 doctorates awarded. *Degree requirements:* For master's and doctorate, thesis/dissertation. *Entrance requirements:* For master's and doctorate, GRE General Test. Additional exam requirements/recommendations for international students: Required—TOEFL (minimum score 550 paper-based; 213 computer-based). *Application deadline:* For fall admission, 1/15 for domestic students. Applications are processed on a rolling basis. Application fee: $40. Electronic applications accepted. *Expenses:* Tuition, area resident: Part-time $410 per credit hour. Tuition, nonresident: part-time $1,034 per credit hour. *Financial support:* Fellowships, research assistantships, teaching assistantships, career-related internships or fieldwork available. Financial award application deadline: 3/1. *Unit head:* , Dr. C. Berger, Coordinator, 802-656-9673.

See Close-Up on page 653.

University of Virginia, School of Medicine, Department of Cell Biology, Charlottesville, VA 22903. Offers PhD, MD/PhD. *Students:* 20 full-time (12 women); includes 6 minority (2 African Americans, 2 Asian Americans or Pacific Islanders, 2 Hispanic Americans), 8 international. Average age 28. In 2005, 1 degree awarded. *Degree requirements:* For doctorate, one foreign language, thesis/dissertation. *Entrance requirements:* For doctorate, GRE General Test, GRE Subject Test. Additional exam requirements/recommendations for international students: Required—TOEFL. *Application deadline:* Applications are processed on a rolling basis. Application fee: $60. Electronic applications accepted. *Expenses:* Tuition, state resident: full-time $7,731. Tuition, nonresident: full-time $18,672. Required fees: $1,479. Full-time tuition and fees vary according to degree level and program. *Financial support:* Applicants required to submit FAFSA. *Unit head:* Dr. Barry M. Gumbiner, Chairman, 434-924-2731, Fax: 434-982-3912. *Application contact:* Peter C. Brunjes, Associate Dean for Graduate Programs and Research, 434-924-7184, Fax: 434-924-6737, E-mail: grad-a-s@virginia.edu.

University of Washington, School of Medicine and Graduate School, Graduate Programs in Medicine, Program in Molecular and Cellular Biology, Seattle, WA 98195. Offers PhD. Offered jointly with Fred Hutchinson Cancer Research Center. *Degree requirements:* For doctorate, thesis/dissertation. *Entrance requirements:* For doctorate, GRE General Test, GRE Subject Test. Additional exam requirements/recommendations for international students: Required—TOEFL. Electronic applications accepted.

See Close-Up on page 655.

The University of Western Ontario, Faculty of Graduate Studies, Biosciences Division, Department of Biology, London, ON N6A 5B8, Canada. Offers M Sc, PhD. *Degree requirements:* For master's, thesis/dissertation; for doctorate, thesis/dissertation, comprehensive exam. *Entrance requirements:* For master's, honors degree or equivalent in biological sciences; for doctorate, master's degree. Additional exam requirements/recommendations for international students: Required—TOEFL. *Faculty research:* Cell and molecular biology, developmental biology, neuroscience, immunobiology and cancer.

University of Wisconsin–La Crosse, Office of University Graduate Studies, College of Science and Health, Department of Biology, La Crosse, WI 54601-3742. Offers aquatic sciences (MS); biology (MS); cellular and molecular biology (MS); clinical microbiology (MS); microbiology (MS); nurse anesthesia (MS); physiology (MS). *Accreditation:* AANA/CANAEP. Part-time programs available. *Faculty:* 18 full-time (5 women), 1 part-time/adjunct (0 women). *Students:* 23 full-time (10 women), 52 part-time (25 women); includes 5 minority (1 American Indian/Alaska Native, 3 Asian Americans or Pacific Islanders, 1 Hispanic American), 3 international. Average age 26. 61 applicants, 44% accepted, 23 enrolled. In 2005, 14 degrees awarded. *Degree requirements:* For master's, thesis, comprehensive exam, registration. *Entrance requirements:* For master's, GRE General Test, minimum GPA of 2.85. Additional exam requirements/recommendations for international students: Required—TOEFL (minimum score 550 paper-based; 213 computer-based). *Application deadline:* For fall admission, 3/1 for domestic students. Applications are processed on a rolling basis. Application fee: $45. Electronic applications accepted. *Expenses:* Tuition, state resident: part-time $354 per credit. Tuition, nonresident: part-time $943 per credit. Tuition and fees vary according to course load, program and reciprocity agreements. *Financial support:* In 2005–06, 10 students received support, including 4 research assistantships with partial tuition reimbursements available (averaging $10,000 per year), 10 teaching assistantships with partial tuition reimbursements available (averaging $9,600 per year); career-related internships or fieldwork, Federal Work-Study, health care benefits, unspecified assistantships, and grant-funded positions also available. Support available to part-time students. Financial award application deadline: 3/15; financial award applicants required to submit FAFSA. *Faculty research:* Cell and molecular biology, physiology, environmental sciences, mycology, biomedical general. Total annual research expenditures: $700,000. *Unit head:* Dr. Tom Volk, Program Director, 608-785-6972, Fax: 608-785-6959, E-mail: volk.thom@uwlax.edu. *Application contact:* Kathryn Kiefer, Associate Director of Admissions, 608-785-8939, E-mail: admissions@uwlax.edu.

University of Wisconsin–Madison, Graduate School, Program in Cellular and Molecular Biology, Madison, WI 53706-1596. Offers PhD. *Faculty:* 175 full-time (51 women). *Students:* 143 full-time (81 women); includes 12 minority (1 African American, 2 American Indian/Alaska Native, 5 Asian Americans or Pacific Islanders, 4 Hispanic Americans), 27 international. Average age 25. 436 applicants, 13% accepted, 22 enrolled. In 2005, 15 degrees awarded. *Median time to degree:* Of those who began their doctoral program in fall 1997, 100% received their degree in 8 years or less. *Degree requirements:* For doctorate, thesis/dissertation, comprehensive exam. *Entrance requirements:* For doctorate, GRE General Test, GRE Subject Test (recommended), minimum GPA of 3.0, lab experience. Additional exam requirements/recommendations for international students: Required—TOEFL (minimum score 580 paper-based; 237 computer-based). *Application deadline:* For fall admission, 12/15 for domestic students, 12/15 for international students. Application fee: $45. Electronic applications accepted. *Financial support:* In 2005–06, 143 students received support, including 25 fellowships with full tuition reimbursements available (averaging $22,500 per year), 115 research assistantships with full tuition reimbursements available (averaging $22,500 per year), 3 teaching assistantships with full tuition reimbursements available (averaging $22,500 per year); traineeships, health care benefits, and unspecified assistantships also available. *Faculty research:* Virology, cancer biology, transcriptional mechanisms, plant biology, immunology.

See Close-Up on page 657.

University of Wisconsin–Madison, School of Veterinary Medicine, Department of Animal Health and Biomedical Sciences, Program in Comparative Biosciences, Madison, WI 53706-1380. Offers anatomy (MS, PhD); biochemistry (MS, PhD); cellular and molecular biology (MS, PhD); environmental toxicology (MS, PhD); neurosciences (MS, PhD); pharmacology (MS, PhD); physiology (MS, PhD). *Degree requirements:* For doctorate, thesis/dissertation.

Vanderbilt University, Graduate School and School of Medicine, Department of Cell and Developmental Biology, Nashville, TN 37240-1001. Offers MS, PhD, MD/PhD. *Faculty:* 20 full-time (7 women). *Students:* 63 full-time (33 women); includes 4 minority (2 Asian Americans or Pacific Islanders, 2 Hispanic Americans), 16 international. In 2005, 1 master's, 8 doctorates awarded. *Degree requirements:* For master's, thesis; for doctorate, thesis/dissertation, preliminary, qualifying, and final exams. *Entrance requirements:* For master's, GRE General Test; for doctorate, GRE General Test, GRE Subject Test (recommended). *Application deadline:* For fall admission, 1/15 for domestic students, 1/15 for international students. Application fee: $0. Electronic applications accepted. *Expenses:* Tuition: Part-time $1,283 per semester hour. Required fees: $2,202; $1,101 per semester. One-time fee: $30. Tuition and fees vary according to course load, program and student level. *Financial support:* Fellowships with full and partial tuition reimbursements, research assistantships with full and partial tuition reimbursements, career-related internships or fieldwork, institutionally sponsored loans, traineeships, and tuition waivers (partial) available. Financial award application deadline: 1/15. *Faculty research:* Reproductive biology, neurobiology, cancer biology, molecular biology, developmental biology. *Unit head:* Susan R. Wente, Chair, 615-322-2134, Fax: 615-343-4539. *Application contact:* David I. Greenstein, Director of Graduate Studies, 615-322-2134, Fax: 615-343-4539, E-mail: david.greenstein@vanderbilt.edu.

Washington State University, Graduate School, College of Sciences, School of Molecular Biosciences, Program in Genetics and Cell Biology, Pullman, WA 99164. Offers MS, PhD. *Faculty:* 23 full-time (5 women), 21 part-time/adjunct (4 women). *Students:* 23 full-time (15 women), 1 (woman) part-time; includes 2 minority (1 American Indian/Alaska Native, 1 Asian American or Pacific Islander), 7 international. Average age 26. 231 applicants, 18% accepted, 13 enrolled. In 2005, 3 master's, 4 doctorates awarded. Terminal master's awarded for partial completion of doctoral program. *Degree requirements:* For master's, thesis or alternative, oral exam; for doctorate, thesis/dissertation, oral exam, comprehensive exam, registration. *Entrance requirements:* For master's and doctorate, GRE General Test, minimum GPA of 3.0. Additional exam requirements/recommendations for international students: Required—TOEFL (minimum score 550 paper-based; 213 computer-based). *Application deadline:* For fall admission, 12/15 for domestic students, 12/15 for international students. Application fee: $35. Electronic applications accepted. *Expenses:* Tuition, state resident: full-time $6,295; part-time $336 per credit. Tuition, nonresident: full-time $15,949; part-time $819 per credit. Required fees: $933. Part-time tuition and fees vary according to campus/location and program. *Financial support:* In 2005–06, 1 fellowship with full tuition reimbursement (averaging $18,852 per year), 16 research assistantships with full tuition reimbursements (averaging $18,852 per year), 6 teaching assistantships with full tuition reimbursements (averaging $18,852 per year) were awarded; Federal Work-Study, institutionally sponsored loans, health care benefits, and unspecified assistant-

Cell Biology

Washington State University *(continued)*
ships also available. Financial award application deadline: 4/1; financial award applicants required to submit FAFSA. *Faculty research:* Plant molecular biology, growth factors, cancer induction and DNA repair, gene regulation and genetic engineering. Total annual research expenditures: $5.8 million. *Application contact:* Kelly G. McGovern, Academic Coordinator, 509-335-4566, Fax: 509-335-1907, E-mail: smbgrad@wsu.edu.

Washington University in St. Louis, Graduate School of Arts and Sciences, Division of Biology and Biomedical Sciences, Program in Molecular Cell Biology, St. Louis, MO 63110. Offers PhD. *Degree requirements:* For doctorate, thesis/dissertation. *Entrance requirements:* For doctorate, GRE General Test, GRE Subject Test. Electronic applications accepted.

Wesleyan University, Graduate Programs, Department of Biology, Middletown, CT 06459-0260. Offers cell biology (PhD); comparative physiology (PhD); developmental biology (PhD); genetics (PhD); neurophysiology (PhD); population biology (PhD). *Faculty:* 12 full-time (3 women). *Students:* 29 full-time (15 women), 10 international. Average age 26. 131 applicants. In 2005, 2 doctorates awarded. *Degree requirements:* For doctorate, one foreign language, thesis/dissertation. *Entrance requirements:* For doctorate, GRE Subject Test. *Application deadline:* For fall admission, 2/15 for domestic students. Applications are processed on a rolling basis. Application fee: $0. *Expenses:* Tuition: Full-time $24,732. One-time fee: $20 full-time. *Financial support:* Research assistantships, teaching assistantships, stipends available. *Faculty research:* Microbial population genetics, genetic basis of evolutionary adaptation, genetic regulation of differentiation and pattern formation in *drosophila*. *Unit head:* Dr. Michael Weir, Chairman, 860-685-2402, E-mail: mweir@wesleyan.edu. *Application contact:* Marjorie Fitzgibbons, Information Contact, 860-685-2157, E-mail: mfitzgibbons@wesleyan.edu.

See Close-Up on page 323.

West Virginia University, Eberly College of Arts and Sciences, Department of Biology, Morgantown, WV 26506. Offers cell and molecular biology (MS, PhD); environmental and evolutionary biology (MS, PhD); integrative organismal biology (PhD); integrative organismal biology (MS). *Faculty:* 18 full-time (3 women), 4 part-time/adjunct (all women). *Students:* 29 full-time (20 women), 5 part-time (3 women); includes 1 minority (Asian American or Pacific Islander), 9 international. Average age 26. 50 applicants, 10% accepted. In 2005, 1 master's, 2 doctorates awarded. Terminal master's awarded for partial completion of doctoral program. *Degree requirements:* For master's, thesis, final exam; for doctorate, thesis/dissertation,

preliminary and final exams. *Entrance requirements:* For master's, GRE General Test, GRE Subject Test, minimum GPA of 3.0; for doctorate, GRE General Test, minimum GPA of 3.0. Additional exam requirements/recommendations for international students: Required—TOEFL. *Application deadline:* For fall admission, 4/1 for domestic students; for spring admission, 10/1 for domestic students. Applications are processed on a rolling basis. Application fee: $45. *Expenses:* Tuition, state resident: full-time $4,582; part-time $258 per credit hour. Tuition, nonresident: full-time $13,820; part-time $741 per credit hour. *Financial support:* In 2005–06, 4 research assistantships, 22 teaching assistantships were awarded; Federal Work-Study and institutionally sponsored loans also available. Financial award application deadline: 4/1; financial award applicants required to submit FAFSA. *Faculty research:* Environmental biology, genetic engineering, developmental biology, global change, biodiversity. *Unit head:* Dr. Jonathan Cumming, Chair, 304-293-5201 Ext. 2508, Fax: 304-293-6363, E-mail: jonathan.cumming@mail.wvu.edu. *Application contact:* Dr. William T. Peterjohn, Director of Graduate Studies, 304-293-5201 Ext. 2510, Fax: 304-293-6363, E-mail: william.peterjohn@mail.wvu.edu.

See Close-Up on page 325.

West Virginia University, Interdisciplinary Studies in Molecular and Cellular Biology, Morgantown, WV 26506.

Yale University, Graduate School of Arts and Sciences, Department of Cell Biology, New Haven, CT 06520. Offers PhD. *Degree requirements:* For doctorate, thesis/dissertation. *Entrance requirements:* For doctorate, GRE General Test. Expenses: Contact institution.

Yale University, Graduate School of Arts and Sciences, Department of Molecular, Cellular, and Developmental Biology, Program in Cell Biology, New Haven, CT 06520. Offers PhD. *Degree requirements:* For doctorate, thesis/dissertation. *Entrance requirements:* For doctorate, GRE General Test, GRE Subject Test.

Yale University, School of Medicine and Graduate School of Arts and Sciences, Combined Program in Biological and Biomedical Sciences (BBS), Molecular Cell Biology, Genetics, and Development Track, New Haven, CT 06520. Offers PhD, MD/PhD. *Students:* 31 full-time. *Entrance requirements:* Additional exam requirements/recommendations for international students: Required—TOEFL. *Application deadline:* For fall admission, 12/8 for domestic students, 12/8 for international students. *Unit head:* , Dr. Thomas Pollard, Co-Director, 203-432-3565. *Application contact:* Anne Scott, Graduate Registrar, 203-432-3538, Fax: 203-432-3597, E-mail: anne.scott@yale.edu.

Molecular Biology

Albany Medical College, Graduate Programs in the Biological Sciences, Center for Cell Biology and Cancer Research, Albany, NY 12208-3479. Offers MS, PhD. Part-time programs available. *Faculty:* 11 full-time (4 women). *Students:* 19 full-time (13 women); includes 4 minority (1 African American, 2 Asian Americans or Pacific Islanders, 1 Hispanic American), 4 international. Average age 24. 12 applicants, 100% accepted, 6 enrolled. In 2005, 2 master's, 4 doctorates awarded. Terminal master's awarded for partial completion of doctoral program. *Degree requirements:* For master's, thesis/dissertation; for doctorate, thesis/dissertation, comprehensive exam. *Entrance requirements:* For master's and doctorate, GRE General Test. Additional exam requirements/recommendations for international students: Required—TOEFL. *Application deadline:* For fall admission, 3/15 for domestic students. Applications are processed on a rolling basis. *Financial support:* In 2005–06, 10 research assistantships (averaging $23,000 per year) were awarded; Federal Work-Study, scholarships/grants, and tuition waivers (full) also available. Financial award applicants required to submit FAFSA. *Faculty research:* Cancer cell biology, angiogenesis, tissue remodeling, signal transduction, cell adhesion. *Unit head:* Dr. C. Michael DiPersio, Graduate Director, 518-262-5916, Fax: 518-262-5669, E-mail: dipersm@mail.amc.edu.

Albert Einstein College of Medicine, Sue Golding Graduate Division of Medical Sciences, Division of Biological Sciences, Department of Developmental and Molecular Biology, Bronx, NY 10461. Offers PhD, MD/PhD. *Degree requirements:* For doctorate, thesis/dissertation. *Entrance requirements:* For doctorate, GRE General Test. Additional exam requirements/recommendations for international students: Required—TOEFL. *Faculty research:* DNA, RNA, and protein synthesis in prokaryotes and eukaryotes; chemical and enzymatic alteration of RNA; glycoproteins.

Baylor College of Medicine, Graduate School of Biomedical Sciences, Department of Biochemistry and Molecular Biology, Houston, TX 77030-3498. Offers PhD, MD/PhD. *Faculty:* 31 full-time (6 women). *Students:* 52 full-time (24 women); includes 9 minority (2 African Americans, 2 Asian Americans or Pacific Islanders, 5 Hispanic Americans), 30 international. Average age 28. 147 applicants, 13% accepted, 9 enrolled. In 2005, 6 doctorates awarded. *Median time to degree:* Of those who began their doctoral program in fall 1997, 89% received their degree in 8 years or less. *Degree requirements:* For doctorate, thesis/dissertation, public defense. *Entrance requirements:* For doctorate, GRE General Test, GRE Subject Test (strongly recommended), minimum GPA of 3.0. Additional exam requirements/recommendations for international students: Required—TOEFL. *Application deadline:* For fall admission, 2/1 for domestic students. Application fee: $30. Electronic applications accepted. *Expenses:* Tuition: Full-time $8,200. Full-time tuition and fees vary according to program. *Financial support:* In 2005–06, 43 students received support, including 18 fellowships (averaging $23,000 per year), 34 research assistantships (averaging $23,000 per year); career-related internships or fieldwork, Federal Work-Study, institutionally sponsored loans, health care benefits, tuition waivers (full), and stipends also available. Financial award applicants required to submit FAFSA. *Faculty research:* Mechanisms of enzyme action, nucleic acid enzymology, and mutagenesis; biochemistry of connective tissue, proteins, and polysaccharides; chemical metabolism of lipids and lipoproteins. *Unit head:* Dr. Adam Kuspa, Director, 713-798-4527. *Application contact:* Charlotte Cherry, Graduate Program Administrator, 713-798-4527, Fax: 713-796-9438, E-mail: ccherry@bcm.tmc.edu.

See Close-Up on page 379.

Baylor College of Medicine, Graduate School of Biomedical Sciences, Department of Molecular and Cellular Biology, Houston, TX 77030-3498. Offers MS, PhD. *Faculty:* 78 full-time (15 women). *Students:* 66 full-time (29 women); includes 9 minority (2 African Americans, 5 Asian Americans or Pacific Islanders, 2 Hispanic Americans), 24 international. Average age 27. 99 applicants, 25% accepted, 11 enrolled. In 2005, 7 degrees awarded. *Median time to degree:* Of those who began their doctoral program in fall 1997, 75% received their degree in 8 years or less. *Degree requirements:* For doctorate, thesis/dissertation, public defense, qualifying exam. *Entrance requirements:* For doctorate, GRE General Test, GRE Subject Test (strongly recommended), minimum GPA of 3.0. Additional exam requirements/recommendations for international students: Required—TOEFL. *Application deadline:* For fall admission, 2/1 for domestic students. Application fee: $30. Electronic applications accepted. *Expenses:* Tuition: Full-time $8,200. Full-time tuition and fees vary according to program. *Financial support:* In 2005–06, 64 students received support, including 16 fellowships (averaging $20,000 per year), 50 research assistantships (averaging $20,000 per year); career-related internships or fieldwork, Federal Work-Study, institutionally sponsored loans, health care benefits, and tuition waivers (full) also available. Financial award applicants required to submit FAFSA. *Faculty research:* Gene regulation, cell structure/function, developmental biology, neurobiology, reproductive endocrinology. *Unit head:* Dr. JoAnne Richards, Director, 713-798-4598. *Application contact:*

Caroline Kosnik, Graduate Program Administrator, 713-798-4598, Fax: 713-790-0545, E-mail: ckosnik@bcm.edu.

See Close-Up on page 549.

Baylor College of Medicine, Graduate School of Biomedical Sciences, Interdepartmental Program in Cell and Molecular Biology, Houston, TX 77030-3498. Offers biochemistry (PhD); cell and molecular biology (PhD); genetics (PhD); human genetics (PhD); immunology (PhD); microbiology (PhD); virology (PhD). *Faculty:* 99 full-time (21 women). *Students:* 56 full-time (28 women); includes 20 minority (4 African Americans, 1 American Indian/Alaska Native, 6 Asian Americans or Pacific Islanders, 9 Hispanic Americans), 5 international. Average age 28. 164 applicants, 18% accepted, 12 enrolled. In 2005, 3 doctorates awarded. *Median time to degree:* Of those who began their doctoral program in fall 1997, 63% received their degree in 8 years or less. *Degree requirements:* For doctorate, thesis/dissertation, public defense. *Entrance requirements:* For doctorate, GRE General Test, GRE Subject Test (strongly recommended), minimum GPA of 3.0. Additional exam requirements/recommendations for international students: Required—TOEFL. *Application deadline:* For fall admission, 1/1 for domestic students. Applications are processed on a rolling basis. Application fee: $30. Electronic applications accepted. *Expenses:* Tuition: Full-time $8,200. Full-time tuition and fees vary according to program. *Financial support:* In 2005–06, 52 students received support, including 20 fellowships (averaging $23,000 per year), 36 research assistantships (averaging $23,000 per year); teaching assistantships, Federal Work-Study, institutionally sponsored loans, health care benefits, and tuition waivers (full) also available. Financial award applicants required to submit FAFSA. *Faculty research:* Gene expression and regulation, developmental biology and genetics, signal transduction and membrane biology, aging process, molecular virology. *Unit head:* Dr. Tom Cooper, Director, 713-798-6557. *Application contact:* Lourdes Fernandez, Graduate Program Administrator, 713-798-6557, Fax: 713-798-6325, E-mail: cmbprog@bcm.edu.

See Close-Up on page 545.

Baylor College of Medicine, Graduate School of Biomedical Sciences, Program in Developmental Biology, Houston, TX 77030-3498. Offers PhD, MD/PhD. *Faculty:* 42 full-time (10 women). *Students:* 38 full-time (17 women); includes 7 minority (1 American Indian/Alaska Native, 5 Asian Americans or Pacific Islanders, 1 Hispanic American), 21 international. Average age 27. 67 applicants, 25% accepted, 7 enrolled. In 2005, 2 doctorates awarded. *Median time to degree:* Of those who began their doctoral program in fall 1997, 100% received their degree in 8 years or less. *Degree requirements:* For doctorate, thesis/dissertation, public defense. *Entrance requirements:* For doctorate, GRE General Test, GRE Subject Test (strongly recommended), minimum GPA of 3.0. Additional exam requirements/recommendations for international students: Required—TOEFL. *Application deadline:* For fall admission, 1/1 for domestic students. Application fee: $30. Electronic applications accepted. *Expenses:* Tuition: Full-time $8,200. Full-time tuition and fees vary according to program. *Financial support:* In 2005–06, 37 students received support, including 10 fellowships (averaging $23,000 per year), 28 research assistantships (averaging $23,000 per year); career-related internships or fieldwork, Federal Work-Study, institutionally sponsored loans, health care benefits, tuition waivers (full), and stipends also available. *Faculty research:* Molecular and genetic approaches to study pattern formation in *Dictyostelium, Drosophila, C.elegans,* mouse, *Xenopus,* and zebrafish; cross-species approach. *Unit head:* Dr. Hugo Bellen, Director, 713-798-6410. *Application contact:* Catherine Tasnier, Graduate Program Administrator, 713-798-6410, Fax: 713-798-5386, E-mail: cat@bcm.edu.

See Close-Up on page 757.

Boston University, Graduate School of Arts and Sciences, Molecular Biology, Cell Biology, and Biochemistry Program (MCBB), Boston, MA 02215. Offers MA, PhD. *Students:* 41 full-time (21 women); includes 3 minority (2 African Americans, 1 Asian American or Pacific Islander), 7 international. Average age 29. 111 applicants, 23% accepted, 8 enrolled. In 2005, 8 master's, 8 doctorates awarded. Terminal master's awarded for partial completion of doctoral program. *Degree requirements:* For master's, one foreign language, thesis (for some programs), registration; for doctorate, one foreign language, thesis/dissertation, comprehensive exam, registration. *Entrance requirements:* For master's and doctorate, GRE General Test, GRE Subject Test. Additional exam requirements/recommendations for international students: Required—TOEFL (minimum score 600 paper-based; 250 computer-based). *Application deadline:* For fall admission, 1/1 for domestic students, 1/1 for international students. Application fee: $60. *Expenses:* Tuition: Full-time $31,530; part-time $985 per credit. Required fees: $316; $40 per semester. Tuition and fees vary according to course level and program. *Financial support:* In 2005–06, 17 students received support, including 1 fellowship (averaging $16,500 per year), 15 research assistantships with full tuition reimbursements available (averaging $16,000 per year), 1 teaching assistantship with full tuition reimbursement available (averaging $16,000 per year); Federal Work-Study, scholarships/grants, and traineeships also available. Financial award application deadline: 1/15; financial award applicants required to submit FAFSA. *Unit head:* Kim McCall, Director, 617-353-2432, Fax: 617-353-6340, E-mail: mccall@bu.edu. *Application contact:* Academic Administrator, 617-353-2432, Fax: 617-353-6340, E-mail: mcbb@bu.edu.

Boston University, School of Medicine, Division of Graduate Medical Sciences, Department of Pathology and Laboratory Medicine, Boston, MA 02118. Offers cell and molecular biology (PhD); experimental pathology (PhD); immunology (PhD). Part-time programs available. *Faculty:* 12 full-time (4 women), 13 part-time/adjunct (2 women). *Students:* 19 full-time (10 women), 1 (woman) part-time; includes 4 minority (2 Asian Americans or Pacific Islanders, 2 Hispanic Americans), 6 international. Average age 29. *Degree requirements:* For doctorate, thesis/dissertation, qualifying exam. *Entrance requirements:* For doctorate, GRE General Test, GRE Subject Test. Additional exam requirements/recommendations for international students: Required—TOEFL. *Application deadline:* For fall admission, 1/15 for domestic students; for spring admission, 10/15 priority date for domestic students. Electronic applications accepted. *Expenses:* Tuition: Full-time $31,530; part-time $985 per credit. Required fees: $316; $40 per semester. Tuition and fees vary according to course level and program. *Financial support:* In 2005–06, 10 fellowships, 3 research assistantships were awarded; Federal Work-Study, scholarships/grants, and traineeships also available. *Faculty research:* Toxicology, carcinogenesis, endocytosis, cytogenetics. *Unit head:* Leonard Gottlieb, Chairman, 617-638-4500, Fax: 617-638-4085. *Application contact:* Dr. Adrianne Rogers, Associate Chairman, 617-638-4500, Fax: 617-638-4085, E-mail: aerogers@bu.edu.

Boston University, School of Medicine, Division of Graduate Medical Sciences, Program in Cell and Molecular Biology, Boston, MA 02118. Offers PhD, MD/PhD. *Degree requirements:* For doctorate, thesis/dissertation. *Entrance requirements:* For doctorate, GRE General Test, GRE Subject Test. Additional exam requirements/recommendations for international students: Required—TOEFL. *Application deadline:* For fall admission, 1/15 for domestic students; for spring admission, 10/15 priority date for domestic students. Electronic applications accepted. *Expenses:* Tuition: Full-time $31,530; part-time $985 per credit. Required fees: $316; $40 per semester. Tuition and fees vary according to course level and program. *Financial support:* Fellowships, research assistantships, Federal Work-Study, scholarships/grants, and traineeships available. *Unit head:* Dr. Vickery Trinkaus Randall, Director, 617-638-6099, Fax: 617-638-5337, E-mail: vickery@bu.edu. *Application contact:* Dr. Mary Jo Murnane, Admissions Director, 617-638-4926, Fax: 617-638-4085, E-mail: mmurnane@bu.edu.

See Close-Up on page 551.

Brandeis University, Graduate School of Arts and Sciences, Programs in Life Sciences, Program in Molecular and Cell Biology, Waltham, MA 02454-9110. Offers genetics (PhD); microbiology (PhD); molecular and cell biology (MS, PhD); molecular biology (PhD); neurobiology (PhD). *Faculty:* 18 full-time (8 women). *Students:* 43 full-time (21 women); includes 2 minority (both Asian Americans or Pacific Islanders), 11 international. Average age 27. 173 applicants, 8% accepted, 6 enrolled. In 2005, 2 master's, 4 doctorates awarded. Terminal master's awarded for partial completion of doctoral program. *Median time to degree:* Of those who began their doctoral program in fall 1997, 100% received their degree in 8 years or less. *Degree requirements:* For master's, research project, thesis optional; for doctorate, thesis/dissertation, teaching assistant experience, comprehensive exam, registration. *Entrance requirements:* For master's and doctorate, GRE General Test, resumé, 3 letters of recommendation. Additional exam requirements/recommendations for international students: Required—TOEFL (minimum score 600 paper-based; 250 computer-based). *Application deadline:* For fall admission, 1/15 for domestic students. Applications are processed on a rolling basis. Application fee: $55. Electronic applications accepted. *Financial support:* In 2005–06, 20 fellowships with full tuition reimbursements (averaging $26,500 per year), 23 research assistantships with full tuition reimbursements (averaging $26,500 per year), 5 teaching assistantships with full tuition reimbursements (averaging $3,000 per year) were awarded; scholarships/grants, traineeships, health care benefits, and tuition waivers (full and partial) also available. Financial award application deadline: 4/15; financial award applicants required to submit CSS PROFILE or FAFSA. *Faculty research:* Regulation of gene expression by transcription factors, molecular neurobiology, immunology, molecular mechanisms of genetic recombination, and cell differentiation. *Unit head:* Dr. Piali Sengupta, Chair, 781-736-2686, Fax: 781-736-3107, E-mail: piali@bradeis.edu. *Application contact:* Marcia Cabral, Information Officer, 781-736-3100, Fax: 781-736-3107, E-mail: cabral@brandeis.edu.

Brigham Young University, Graduate Studies, College of Biological and Agricultural Sciences, Department of Microbiology and Molecular Biology, Provo, UT 84602-1001. Offers microbiology (MS, PhD); molecular biology (MS, PhD). *Faculty:* 16 full-time (2 women). *Students:* 19 full-time (15 women); includes 1 minority (Asian American or Pacific Islander), 4 international. Average age 23. 16 applicants, 19% accepted, 2 enrolled. In 2005, 7 master's, 1 doctorate awarded. *Degree requirements:* For master's and doctorate, thesis/dissertation, comprehensive exam. *Entrance requirements:* For master's, GRE General Test, minimum GPA of 3.0 during previous 2 years; for doctorate, GRE General Test, minimum GPA of 3.0. Additional exam requirements/recommendations for international students: Required—TOEFL, IELT. *Application deadline:* For fall admission, 2/1 priority date for domestic students, 2/1 priority date for international students. For winter admission, 6/30 for domestic students. Application fee: $50. Electronic applications accepted. *Financial support:* In 2005–06, 14 students received support, including 7 research assistantships with full tuition reimbursements available (averaging $14,100 per year), 7 teaching assistantships with full tuition reimbursements available (averaging $14,100 per year); institutionally sponsored loans and tuition waivers (partial) also available. Financial award application deadline: 2/1. *Faculty research:* Immunobiology, environmental microbiology, molecular genetics, molecular virology, tumor biology. Total annual research expenditures: $399,094. *Unit head:* Dr. Brent L. Nielsen, Chair, 801-422-1102, Fax: 801-422-0519, E-mail: brent_nielsen@byu.edu. *Application contact:* Dr. Laura Bridgewater, Graduate Coordinator, 801-422-2434, Fax: 801-422-0519, E-mail: laura_bridgewater@byu.edu.

Brown University, Graduate School, Division of Biology and Medicine, Program in Molecular Biology, Cell Biology, and Biochemistry, Providence, RI 02912. Offers biochemistry (M Med Sc, Sc M, PhD), including biochemistry (Sc M, PhD), biology (Sc M, PhD), medical science (M Med Sc, PhD); biology (MA); cell biology (M Med Sc, Sc M, PhD), including biochemistry (Sc M, PhD), biology (Sc M, PhD), medical science (M Med Sc, PhD); developmental biology (M Med Sc, Sc M, PhD), including biochemistry (Sc M, PhD), biology (Sc M, PhD), medical science (M Med Sc, Sc M, PhD); immunology (M Med Sc, Sc M, PhD), including biochemistry (Sc M, PhD), biology (Sc M, PhD), medical science (M Med Sc, Sc M, PhD); molecular microbiology (M Med Sc, Sc M, PhD), including biochemistry (Sc M, PhD), biology (Sc M, PhD), medical science (M Med Sc, PhD). Part-time programs available. Terminal master's awarded for partial completion of doctoral program. *Degree requirements:* For master's, thesis (for some programs); for doctorate, one foreign language, thesis/dissertation, preliminary exam. *Entrance requirements:* For master's and doctorate, GRE General Test, GRE Subject Test. Additional exam requirements/recommendations for international students: Required—TOEFL. Electronic applications accepted. *Faculty research:* Molecular genetics, gene regulation.

See Close-Up on page 555.

California Institute of Technology, Division of Biology, Program in Molecular Biology, Pasadena, CA 91125-0001. Offers PhD. *Degree requirements:* For doctorate, thesis/dissertation, qualifying exam. *Entrance requirements:* For doctorate, GRE General Test. *Application deadline:* For fall admission, 1/1 for domestic students. Application fee: $0. *Financial support:* Application deadline: 1/1. *Application contact:* Elizabeth Ayala, Graduate Program Coordinator, 626-395-4497, Fax: 626-683-3343, E-mail: biograd@cco.caltech.edu.

Carnegie Mellon University, Mellon College of Science, Department of Biological Sciences, Pittsburgh, PA 15213-3891. Offers biochemistry (PhD); biophysics (PhD); cell biology (PhD);

computational biology (MS, PhD); developmental biology (PhD); genetics (PhD); molecular biology (PhD); neurobiology (PhD). *Degree requirements:* For doctorate, thesis/dissertation, comprehensive exam. *Entrance requirements:* For doctorate, GRE General Test, GRE Subject Test, interview. Electronic applications accepted. *Faculty research:* Genetic structure, function, and regulation; protein structure and function; biological membranes; biological spectroscopy.

See Close-Up on page 93.

Case Western Reserve University, Center for RNA Molecular Biology, Cleveland, OH 44106.

Case Western Reserve University, School of Medicine and School of Graduate Studies, Graduate Programs in Medicine, Department of Anatomy, Cleveland, OH 44106. Offers applied anatomy (MS); biological anthropology (MS, PhD); cellular biology (MS, PhD); developmental biology (PhD); molecular biology (PhD). Part-time programs available. *Faculty:* 20 full-time (5 women), 16 part-time/adjunct (5 women). *Students:* 6 full-time (0 women), 49 part-time (19 women); includes 20 minority (7 African Americans, 12 Asian Americans or Pacific Islanders, 1 Hispanic American). Average age 25. 39 applicants, 92% accepted, 29 enrolled. In 2005, 14 master's awarded. *Median time to degree:* Of those who began their doctoral program in fall 1997, 67% received their degree in 8 years or less. *Degree requirements:* For master's, thesis (for some programs), comprehensive exam; for doctorate, thesis/dissertation. *Entrance requirements:* For master's, GRE General Test; for doctorate, GRE General Test, GRE Subject Test. Additional exam requirements/recommendations for international students: Required—TOEFL. *Application deadline:* For fall admission, 5/1 for domestic students; for spring admission, 8/1 priority date for domestic students. Applications are processed on a rolling basis. Application fee: $50. *Financial support:* In 2005–06, 6 research assistantships with full tuition reimbursements (averaging $23,000 per year) were awarded; fellowships, scholarships/grants also available. *Faculty research:* Hypoxia, cell injury, biochemical aberration occurrences in ischemic tissue, human functional morphology, evolutionary morphology. Total annual research expenditures: $691,974. *Unit head:* , Dr. Joseph C. LaManna, Chairman, 216-368-1100, Fax: 216-368-8669, E-mail: jcl4@po.cwru.edu. *Application contact:* Morley Schwebel, Administrator, 216-368-2433, Fax: 216-368-8669, E-mail: mxs86@po.cwru.edu.

Case Western Reserve University, School of Medicine and School of Graduate Studies, Graduate Programs in Medicine, Department of Molecular Biology and Microbiology, Cleveland, OH 44106-4960. Offers cellular biology (PhD); microbiology (PhD); molecular biology (PhD); molecular virology (MD/PhD). Students are admitted to an integrated Biomedical Sciences Training Program involving 11 basic science programs at Case Western Reserve University. *Degree requirements:* For doctorate, thesis/dissertation. *Entrance requirements:* For doctorate, GRE General Test, GRE Subject Test. Additional exam requirements/recommendations for international students: Required—TOEFL. *Application deadline:* Applications are processed on a rolling basis. Application fee: $25. Electronic applications accepted. *Financial support:* In 2005–06, 24 students received support, including fellowships with full tuition reimbursements available (averaging $23,000 per year); Federal Work-Study, scholarships/grants, traineeships, and tuition waivers (full) also available. *Faculty research:* Gene expression in eukaryotic and prokaryotic systems; microbial physiology; intracellular transport and signaling; mechanisms of oncogenesis; molecular mechanisms of RNA processing, editing, and catalysis. Total annual research expenditures: $4.8 million. *Unit head:* Dr. Jonathan Karn, Chairman, 216-368-3420, Fax: 216-368-3055, E-mail: jonathan.karn@case.edu. *Application contact:* Dr. Patrick Viollier, Admissions Coordinator, 216-368-3420, Fax: 216-368-3055, E-mail: patrick.viollier@case.edu.

See Close-Up on page 557.

Central Connecticut State University, School of Graduate Studies, School of Technology, Department of Biomolecular Sciences, New Britain, CT 06050-4010. Offers MA. *Faculty:* 7 full-time (2 women), 4 part-time/adjunct (all women). *Students:* 11 full-time (5 women), 6 part-time (4 women); includes 4 minority (1 African American, 1 Asian American or Pacific Islander, 2 Hispanic Americans), 2 international. Average age 27. 12 applicants, 58% accepted, 5 enrolled. *Entrance requirements:* For master's, minimum GPA of 2.7. Additional exam requirements/recommendations for international students: Required—TOEFL. *Application deadline:* For fall admission, 7/1 for domestic students; for spring admission, 12/1 for domestic students. Applications are processed on a rolling basis. Electronic applications accepted. *Expenses:* Tuition, area resident: Full-time $3,780. Tuition, state resident: full-time $5,670; part-time $362 per credit. Tuition, nonresident: full-time $10,530; part-time $362 per credit. Required fees: $3,064. One-time fee: $62 part-time. Tuition and fees vary according to degree level and program. *Financial support:* In 2005–06, 2 students received support, including 2 research assistantships; career-related internships or fieldwork, Federal Work-Study, scholarships/grants, and unspecified assistantships also available. Support available to part-time students. Financial award application deadline: 3/1; financial award applicants required to submit FAFSA. *Unit head:* Dr. Thomas King, Chair, 860-832-3560.

Clemson University, Graduate School, College of Agriculture, Forestry and Life Sciences, Department of Genetics and Biochemistry, Program in Biochemistry and Molecular Biology, Clemson, SC 29634. Offers MS, PhD. *Students:* 3 full-time (2 women), (all international). 16 applicants, 19% accepted, 3 enrolled. In 2005, 1 master's awarded. *Degree requirements:* For master's, thesis/dissertation; for doctorate, thesis/dissertation, comprehensive exam. *Entrance requirements:* For master's and doctorate, GRE General Test. Additional exam requirements/recommendations for international students: Required—TOEFL. *Application deadline:* For fall admission, 6/1 priority date for domestic students, 4/15 priority date for international students. Applications are processed on a rolling basis. Application fee: $50. *Financial support:* Fellowships, research assistantships, teaching assistantships available. Financial award application deadline: 3/15; financial award applicants required to submit FAFSA. *Faculty research:* Biomembranes, protein structure, molecular biology of plants, APYA and stress response. Total annual research expenditures: $670,000. *Unit head:* Dr. Kerry Smith, Coordinator, 864-656-6935, Fax: 864-656-0435, E-mail: kssmith@clemson.edu.

See Close-Ups on pages 383 and 385.

Colorado State University, Graduate School, College of Natural Sciences, Department of Biochemistry and Molecular Biology, Fort Collins, CO 80523-0015. Offers biochemistry (MS, PhD). Part-time programs available. *Faculty:* 11 full-time (4 women), 2 part-time/adjunct (0 women). *Students:* 17 full-time (8 women), 17 part-time (9 women); includes 4 minority (1 American Indian/Alaska Native, 3 Asian Americans or Pacific Islanders), 10 international. Average age 28. 26 applicants, 42% accepted, 9 enrolled. In 2005, 4 master's, 5 doctorates awarded. Terminal master's awarded for partial completion of doctoral program. *Degree requirements:* For master's, thesis optional; for doctorate, thesis/dissertation. *Entrance requirements:* For master's and doctorate, GRE General Test, minimum GPA of 3.2. Additional exam requirements/recommendations for international students: Required—TOEFL (minimum score 620 paper-based; 260 computer-based). *Application deadline:* For fall admission, 1/7 priority date for domestic students, 1/10 priority date for international students; for spring admission, 11/12 priority date for domestic students, 11/10 priority date for international students. Applications are processed on a rolling basis. Application fee: $50. Electronic applications accepted. *Expenses:* Tuition, state resident: full-time $3,690; part-time $205 per credit. Tuition, nonresident: full-time $14,958; part-time $831 per credit. Required fees: $1,061. *Financial support:* In 2005–06, 32 students received support, including fellowships with full tuition reimbursements available (averaging $21,800 per year), research assistantships with full tuition reimbursements available (averaging $21,800 per year), teaching assistantships with full tuition reimbursements available (averaging $21,800 per year); Federal Work-Study, institutionally sponsored loans, traineeships, and tuition waivers (partial) also available. Financial award application deadline: 2/15; financial award applicants required to submit FAFSA. *Faculty research:* Cellular biology, molecular gene expression, neurobiology, structural biology, yeast genetics. Total annual research expenditures: $4.3 million. *Unit head:* Marvin R. Paule, Interim Chair, 970-491-5566, Fax: 970-491-0494, E-mail: marv.paule@lamar.colostate.edu.

Molecular Biology

Colorado State University (continued)
Application contact: Yvonne Bridgeman, Graduate Program Assistant, 970-491-6841, Fax: 970-491-0494, E-mail: yyvonne.bridgeman@colostate.edu.

Colorado State University, Graduate School, Program in Cell and Molecular Biology, Fort Collins, CO 80523-1618. Offers MS, PhD. *Students:* Average age 28. In 2005, 4 master's, 2 doctorates awarded. Terminal master's awarded for partial completion of doctoral program. *Degree requirements:* For master's and doctorate, thesis/dissertation. *Entrance requirements:* For master's and doctorate, GRE General Test, minimum GPA of 3.0. Additional exam requirements/recommendations for international students: Required—TOEFL. *Application deadline:* For fall admission, 1/15 for domestic students. Applications are processed on a rolling basis. Application fee: $50. Electronic applications accepted. *Expenses:* Tuition, state resident: full-time $3,690; part-time $205 per credit. Tuition, nonresident: full-time $14,958; part-time $831 per credit. Required fees: $1,061. *Financial support:* In 2005–06, fellowships with full tuition reimbursements (averaging $17,000 per year), research assistantships with full tuition reimbursements (averaging $17,000 per year), teaching assistantships with full tuition reimbursements (averaging $17,000 per year) were awarded; Federal Work-Study and institutionally sponsored loans also available. Financial award application deadline: 2/1. *Faculty research:* Regulation of gene expression, cancer biology, plant molecular genetics, reproductive physiology, environmental toxicology. Total annual research expenditures: $15 million. *Unit head:* Dr. Michael H. Fox, Chairman, 970-491-7618, Fax: 970-491-0623, E-mail: cmb@colostate.edu. *Application contact:* Norma Jean Bulera, Administrative Assistant, 970-491-0241, Fax: 970-491-0623, E-mail: cmb@colostate.edu.

See Close-Up on page 559.

Columbia University, College of Physicians and Surgeons, Coordinated Doctoral Program in Basic Sciences, New York, NY 10027.

Columbia University, College of Physicians and Surgeons and Graduate School of Arts and Sciences, Graduate School of Arts and Sciences at the College of Physicians and Surgeons, Integrated Program in Cellular, Molecular and Biophysical Studies, New York, NY 10032. Offers M Phil, MA, PhD, MD/PhD. Only candidates for the PhD are admitted. Terminal master's awarded for partial completion of doctoral program. *Degree requirements:* For doctorate, thesis/dissertation. *Entrance requirements:* For master's, GRE General Test; for doctorate, GRE General Test, GRE Subject Test. Additional exam requirements/recommendations for international students: Required—TOEFL. Expenses: Contact institution. Tuition and fees vary according to course level, course load, campus/location and program. *Faculty research:* Transcription, macromolecular sorting, gene expression during development, cellular interaction.

Cornell University, Graduate School, Graduate Fields of Agriculture and Life Sciences, Field of Biochemistry, Molecular and Cell Biology, Ithaca, NY 14853-0001. Offers biochemistry (PhD); biophysics (PhD); cell biology (PhD); molecular and cell biology (PhD); molecular biology (PhD). *Faculty:* 60 full-time (13 women). *Students:* 177 students, 25% accepted, 22 enrolled. In 2005, 13 degrees awarded. *Degree requirements:* For doctorate, thesis/dissertation, 2 semesters of teaching experience, comprehensive exam. *Entrance requirements:* For doctorate, GRE General Test, GRE Subject Test (biology, chemistry, physics; or biochemistry, cell and molecular biology), 3 letters of recommendation. Additional exam requirements/recommendations for international students: Required—TOEFL (minimum score 600 paper-based; 250 computer-based). *Application deadline:* For fall admission, 1/5 for domestic students. Application fee: $60. Electronic applications accepted. *Financial support:* In 2005–06, 90 students received support, including 35 fellowships with full tuition reimbursements available, 42 research assistantships with full tuition reimbursements available, 13 teaching assistantships with full tuition reimbursements available; institutionally sponsored loans, scholarships/grants, health care benefits, tuition waivers (full and partial), and unspecified assistantships also available. Financial award applicants required to submit FAFSA. *Faculty research:* Biophysics, structural biology. *Unit head:* , Director of Graduate Studies, 607-255-2100, Fax: 607-255-2100. *Application contact:* Graduate Field Assistant, 607-255-2100, Fax: 607-255-2100, E-mail: bmcb@cornell.edu.

Cornell University, Joan and Sanford I. Weill Medical College and Graduate School of Medical Sciences, Weill Graduate School of Medical Sciences, Program in Molecular Biology, Cell Biology, Biochemistry and Structural Biology, New York, NY 10021-4896. Offers MS, PhD, MD/PhD. *Faculty:* 84 full-time (16 women). *Students:* 121 full-time (66 women); includes 14 minority (2 African Americans, 7 Asian Americans or Pacific Islanders, 5 Hispanic Americans), 34 international. Average age 22. In 2005, 8 degrees awarded. *Median time to degree:* Of those who began their doctoral program in fall 1997, 100% received their degree in 8 years or less. *Degree requirements:* For doctorate, thesis/dissertation, final exam. *Entrance requirements:* For doctorate, GRE General Test, GRE Subject Test, background in genetics, molecular biology, chemistry, or biochemistry. Additional exam requirements/recommendations for international students: Required—TOEFL. *Application deadline:* For fall admission, 12/15 for domestic students. Application fee: $60. *Expenses:* Tuition: full-time $32,320. Required fees: $1,025. *Financial support:* Fellowships, stipends available. *Unit head:* Dr. Christopher Lima, Co-Director, 212-639-8205, Fax: 212-717-3047, E-mail: limac@mskcc.org.

Dartmouth College, Graduate Program in Molecular and Cellular Biology, Hanover, NH 03755.

See Close-Up on page 561.

Drexel University, College of Medicine, Biomedical Graduate Programs, Interdisciplinary Program in Molecular and Cell Biology, Philadelphia, PA 19104-2875. Offers MS, PhD, MD/PhD. Terminal master's awarded for partial completion of doctoral program. *Degree requirements:* For master's, thesis, comprehensive exam; for doctorate, thesis/dissertation, qualifying exam. *Entrance requirements:* For master's, GRE General Test, minimum GPA of 2.75; for doctorate, GRE General Test, minimum GPA of 3.0. Additional exam requirements/recommendations for international students: Required—TOEFL. Electronic applications accepted. *Faculty research:* Molecular anatomy, biochemistry, medical biotechnology, molecular pathology, microbiology and immunology.

Duke University, Graduate School, Department of Biological Anthropology and Anatomy, Durham, NC 27710. Offers cellular and molecular biology (PhD); gross anatomy and physical anthropology (PhD), including comparative morphology of human and non-human primates, primate social behavior, vertebrate paleontology; neuroanatomy (PhD). *Faculty:* 6 full-time. *Students:* 14 full-time (9 women); includes 1 minority (African American), 2 international. 39 applicants, 18% accepted, 3 enrolled. In 2005, 3 doctorates awarded. *Degree requirements:* For doctorate, one foreign language, thesis/dissertation. *Entrance requirements:* For doctorate, GRE General Test. Additional exam requirements/recommendations for international students: Required—IELT (preferred) or TOEFL. *Application deadline:* For fall admission, 12/31 for domestic students, 12/31 for international students. Application fee: $75. Electronic applications accepted. *Financial support:* Fellowships, teaching assistantships, Federal Work-Study available. Financial award application deadline: 12/31. *Unit head:* , Daniel Schmitt, Director of Graduate Studies, 919-684-5664, Fax: 919-684-8034, E-mail: l.squires@baa.mc.duke.edu.

Duke University, Graduate School, Program in Cellular and Molecular Biology, Durham, NC 27710. Offers PhD, Certificate. Certificate students must be enrolled in a participating PhD program. *Faculty:* 127 full-time. *Students:* 24 full-time (7 women); includes 3 minority (all Asian Americans or Pacific Islanders), 5 international. 259 applicants, 21% accepted, 23 enrolled. *Entrance requirements:* For doctorate, GRE General Test, GRE Subject Test (recommended). Additional exam requirements/recommendations for international students: Required—IELT (preferred) or TOEFL. *Application deadline:* For fall admission, 12/31 for domestic students, 12/31 for international students. Application fee: $75. Electronic applications accepted. *Financial support:* Fellowships available. Financial award application deadline:

12/31. *Unit head:* Dr. Chris Counter, Director of Graduate Studies, 919-684-9890, Fax: 919-681-8911, E-mail: cmbtgp@biochem.duke.edu.

See Close-Up on page 563.

East Carolina University, Brody School of Medicine, Department of Biochemistry and Molecular Biology, Greenville, NC 27858-4353. Offers PhD. *Faculty:* 10 full-time (1 woman). *Students:* 2 full-time (both women), 3 part-time (all women); includes 1 minority (Asian American or Pacific Islander), 1 international. Average age 25. 8 applicants, 75% accepted. *Median time to degree:* Of those who began their doctoral program in fall 1997, 67% received their degree in 8 years or less. *Degree requirements:* For doctorate, thesis/dissertation, comprehensive exam, registration. *Entrance requirements:* For doctorate, GRE General Test. Additional exam requirements/recommendations for international students: Required—TOEFL. *Application deadline:* For fall admission, 6/1 for domestic students. Applications are processed on a rolling basis. Application fee: $50. *Expenses:* Tuition, state resident: full-time $2,516. Tuition, nonresident: full-time $12,832. *Financial support:* 7 fellowships with full and partial tuition reimbursements (averaging $21,500 per year) were awarded Financial award application deadline: 6/1. *Faculty research:* Gene regulation, development and differentiation, contractility and motility, macromolecular interactions, cancer. Total annual research expenditures: $856,955. *Unit head:* Dr. Joseph Cory, Chairman, 252-744-2675, Fax: 252-744-3383, E-mail: coryjo@ecu.edu. *Application contact:* Dr. George J. Kasperek, Assistant Dean for Graduate Studies/BSOM, 252-744-3305, Fax: 252-744-0203, E-mail: kasperekg@ecu.edu.

East Carolina University, Graduate School, Thomas Harriot College of Arts and Sciences, Department of Biology, Greenville, NC 27858-4353. Offers biology (MS); molecular biology/biotechnology (MS). Part-time programs available. *Faculty:* 38 full-time (10 women). *Students:* 24 full-time (9 women), 40 part-time (15 women); includes 6 minority (4 African Americans, 1 Asian American or Pacific Islander, 1 Hispanic American), 2 international. Average age 27. 24 applicants, 13% accepted, 3 enrolled. In 2005, 28 degrees awarded. *Degree requirements:* For master's, one foreign language, thesis, comprehensive exam. *Entrance requirements:* For master's, GRE General Test, GRE Subject Test. Additional exam requirements/recommendations for international students: Required—TOEFL. *Application deadline:* For fall admission, 6/1 for domestic students; for spring admission, 10/15 for domestic students. Applications are processed on a rolling basis. Application fee: $50. *Expenses:* Tuition, state resident: full-time $2,516. Tuition, nonresident: full-time $12,832. *Financial support:* Fellowships with partial tuition reimbursements, research assistantships with partial tuition reimbursements, teaching assistantships with partial tuition reimbursements, career-related internships or fieldwork, Federal Work-Study, scholarships/grants, and unspecified assistantships available. Support available to part-time students. Financial award application deadline: 6/1. *Faculty research:* Biochemistry, microbiology, cell biology. *Unit head:* Dr. Ronald Newton, Chair, 252-328-6204, Fax: 252-328-4178, E-mail: newtonro@ecu.edu. *Application contact:* Interim Dean of Graduate School, 252-328-6012, Fax: 252-328-6071, E-mail: gradschool@ecu.edu.

See Close-Up on page 121.

Emory University, Graduate School of Arts and Sciences, Division of Biological and Biomedical Sciences, Program in Genetics and Molecular Biology, Atlanta, GA 30322-1100. Offers PhD. *Faculty:* 51 full-time (9 women). *Students:* 56 full-time (40 women); includes 10 minority (8 African Americans, 2 Asian Americans or Pacific Islanders), 6 international. Average age 27. 81 applicants, 31% accepted, 17 enrolled. In 2005, 3 degrees awarded. *Median time to degree:* Of those who began their doctoral program in fall 1997, 100% received their degree in 8 years or less. *Degree requirements:* For doctorate, thesis/dissertation, comprehensive exam, registration. *Entrance requirements:* For doctorate, GRE General Test, minimum GPA of 3.0 in science course work. Additional exam requirements/recommendations for international students: Required—TOEFL. *Application deadline:* For fall admission, 1/3 for domestic students, 1/3 for international students. Application fee: $50. Electronic applications accepted. *Expenses:* Tuition: Full-time $14,400. Required fees: $217. *Financial support:* In 2005–06, 27 students received support, including 27 fellowships with full tuition reimbursements available (averaging $23,000 per year); institutionally sponsored loans, health care benefits, and tuition waivers (full) also available. *Faculty research:* Gene regulation, genetic combination, developmental regulation. *Unit head:* Dr. Jerry Boss, Director, 404-717-5973, Fax: 404-727-1719, E-mail: boss@microbio.emory.edu. *Application contact:* 404-727-2545, Fax: 404-727-3322, E-mail: gdbbs@emory.edu.

Florida Institute of Technology, Graduate Programs, College of Science, Department of Biological Sciences, Program in Cell and Molecular Biology, Melbourne, FL 32901-6975. Offers MS, PhD. Part-time programs available. *Students:* Average age 34. *Application deadline:* Applications are processed on a rolling basis. Electronic applications accepted. *Expenses:* Tuition: Part-time $825 per credit. *Financial support:* Research assistantships with full and partial tuition reimbursements, teaching assistantships with full and partial tuition reimbursements, career-related internships or fieldwork and tuition remissions available. Financial award application deadline: 3/1; financial award applicants required to submit FAFSA. *Faculty research:* Changes in DNA molecule and differential expression of genetic information during aging. *Application contact:* Carolyn P. Farrior, Director of Graduate Admissions, 321-674-7118, Fax: 321-723-9468, E-mail: cfarrior@fit.edu.

See Close-Up on page 567.

Florida State University, Graduate Studies, College of Arts and Sciences, Department of Biological Science, Program in Molecular Biology, Tallahassee, FL 32306. Offers MS, PhD. *Faculty:* 19 full-time (3 women). *Students:* 45 full-time (23 women); includes 3 minority (1 African American, 1 Asian American or Pacific Islander, 1 Hispanic American), 18 international. *Degree requirements:* For master's and doctorate, thesis/dissertation, teaching experience, seminar presentation, comprehensive exam, registration. *Entrance requirements:* For master's and doctorate, GRE General Test (minimum 1100: U-500, G-500), minimum upper division GPA of 3.0. Additional exam requirements/recommendations for international students: Required—TOEFL (minimum score 600 paper-based; 250 computer-based), IB 100. *Application deadline:* For fall admission, 1/15 for domestic students, 12/1 for international students; for spring admission, 10/15 for domestic students, 9/1 for international students. Application fee: $30. *Financial support:* In 2005–06, fellowships with full tuition reimbursements (averaging $19,000 per year), research assistantships with full tuition reimbursements (averaging $19,000 per year), teaching assistantships with full tuition reimbursements (averaging $17,600 per year) were awarded. Financial award application deadline: 1/15; financial award applicants required to submit FAFSA. *Faculty research:* Development, fertilization, photosynthesis, nuclei, chromosomes. *Application contact:* Judy Bowers, Coordinator, Graduate Affairs, 850-644-3023, Fax: 850-644-9829, E-mail: gradinfo@bio.fsu.edu.

Florida State University, Graduate Studies, College of Arts and Sciences, Program in Molecular Biophysics, Tallahassee, FL 32306. Offers biochemistry, molecular and cell biology (PhD); computational structural biology (PhD); molecular biophysics (PhD). *Faculty:* 44 full-time (7 women). *Students:* 29 full-time (17 women); includes 1 minority (Hispanic American), 52 international. Average age 30. 14 applicants, 100% accepted, 4 enrolled. In 2005, 3 degrees awarded. *Median time to degree:* Of those who began their doctoral program in fall 1997, 100% received their degree in 8 years or less. *Degree requirements:* For doctorate, thesis/dissertation, comprehensive exam. *Entrance requirements:* For doctorate, GRE General Test. Additional exam requirements/recommendations for international students: Required—TOEFL (minimum score 600 paper-based; 250 computer-based). *Application deadline:* For fall admission, 1/15 for domestic students, 1/15 for international students. Applications are processed on a rolling basis. Application fee: $30. Electronic applications accepted. *Financial support:* In 2005–06, 29 students received support, including 29 research assistantships (averaging $19,500 per year); health care benefits and tuition waivers (partial) also available. Financial award applicants required to submit FAFSA. *Faculty research:* Protein and nucleic acid structure and function, membrane protein structure, computational biophysics, 3-D image reconstruction. Total annual research expenditures: $5.1 million. *Unit head:* Dr. P. Bryant Chase, Director,

MOB Graduate Program, 850-644-0056, Fax: 850-644-7244, E-mail: chase@bio.fsu.edu. *Application contact:* Dale E. Leonard, Academic Coordinator, Graduate Programs, 850-644-1012, Fax: 850-644-7244, E-mail: mob@sb.fsu.edu.

George Mason University, College of Science, Fairfax, VA 22030. Offers bioinformatics (MS, PhD); climate dynamics (PhD); computational sciences (MS); computational sciences and informatics (PhD); computational social science (PhD); computational techniques and applications (Certificate); earth systems and geoinformation science (PhD); earth systems science (MS); nanotechnology and nanoscience (Certificate); neuroscience (PhD); physical sciences (PhD); remote sensing and earth image processing (Certificate). Part-time and evening/weekend programs available. *Degree requirements:* For doctorate, thesis/dissertation, comprehensive exam, registration. *Entrance requirements:* For master's and doctorate, GRE General Test, minimum GPA of 3.0 in last 60 hours. Additional exam requirements/recommendations for international students: Required—TOEFL. Electronic applications accepted. *Expenses:* Tuition, area resident: Full-time $5,244; part-time $219 per credit. Tuition, state resident: part-time $651 per credit. Tuition, nonresident: full-time $15,636. Required fees: $1,524; $65 per credit. *Faculty research:* Space sciences and astrophysics, fluid dynamics, materials modeling and simulation, bioinformatics, global changes and statistics.

George Mason University, College of Arts and Sciences, Department of Biology, Program in Biology, Fairfax, VA 22030. Offers bioinformatics (MS); ecology, systematics and evolution (MS); interpretive biology (MS); molecular and microbiology (PhD); molecular, microbial, and cellular biology (MS); organismal biology (MS). Part-time programs available. *Degree requirements:* For master's, thesis or alternative. *Entrance requirements:* For master's, GRE General Test, GRE Subject Test, bachelor's degree in biology or equivalent. Electronic applications accepted. *Expenses:* Tuition, area resident: Full-time $5,244; part-time $219 per credit. Tuition, state resident: part-time $651 per credit. Tuition, nonresident: full-time $15,636. Required fees: $1,524; $65 per credit.

Georgetown University, Graduate School of Arts and Sciences, Programs in Biomedical Sciences, Department of Biochemistry and Molecular Biology, Washington, DC 20057. Offers PhD, MD/PhD. *Degree requirements:* For doctorate, thesis/dissertation, comprehensive exam. *Entrance requirements:* For doctorate, GRE General Test. Additional exam requirements/recommendations for international students: Required—TOEFL.

The George Washington University, Columbian College of Arts and Sciences, Department of Biochemistry and Molecular Biology, Washington, DC 20037. Offers biochemistry (PhD); genomics, proteomics, and bioinformatics (MS). *Degree requirements:* For master's, comprehensive exam; for doctorate, thesis/dissertation, general exam. *Entrance requirements:* For master's, GRE General Test, interview, minimum GPA of 3.0; for doctorate, GRE General Test, minimum GPA of 3.0. Additional exam requirements/recommendations for international students: Required—TOEFL (minimum score 550 paper-based; 213 computer-based).

See Close-Up on page 389.

Harvard University, Graduate School of Arts and Sciences, Department of Molecular and Cellular Biology, Cambridge, MA 02138. Offers PhD. *Students:* 81 full-time (37 women). 325 applicants, 16% accepted. In 2005, 16 degrees awarded. *Degree requirements:* For doctorate, thesis/dissertation, oral exam. *Entrance requirements:* For doctorate, GRE General Test, GRE Subject Test (recommended). Additional exam requirements/recommendations for international students: Required—TOEFL. *Application deadline:* For fall admission, 12/15 for domestic students. Application fee: $60. *Expenses:* Tuition: Full-time $28,752. Full-time tuition and fees vary according to program and student level. *Financial support:* Fellowships, research assistantships, teaching assistantships, career-related internships or fieldwork, Federal Work-Study, and institutionally sponsored loans available. Financial award application deadline: 12/30. *Unit head:* Betsey Cogswell, Administrator, 617-495-5497, Fax: 617-495-5264. *Application contact:* James Wang, Assistant Director, 617-495-1901.

See Close-Up on page 571.

Harvard University, Graduate School of Arts and Sciences, Program in Chemical Biology, Cambridge, MA 02138. Offers PhD. *Expenses:* Tuition: Full-time $28,752. Full-time tuition and fees vary according to program and student level. *Unit head:* Melitta King, Director of Graduate Studies, 617-432-0984.

Howard University, College of Medicine, Department of Biochemistry and Molecular Biology, Washington, DC 20059-0002. Offers biochemistry and molecular biology (PhD); biotechnology (MS). Part-time programs available. *Degree requirements:* For master's, externship; for doctorate, thesis/dissertation, comprehensive exam. *Entrance requirements:* For master's and doctorate, GRE General Test, minimum GPA of 3.0. *Faculty research:* Cellular and molecular biology of olfaction, gene regulation and expression, enzymology, NMR spectroscopy of molecular structure, hormone regulation/metabolism.

See Close-Up on page 391.

Illinois Institute of Technology, Graduate College, College of Science and Letters, Department of Biological, Chemical and Physical Sciences, Biology Division, Chicago, IL 60616-3793. Offers biochemistry (MS); biology (MBS, PhD); biotechnology (MS); cell biology (MS); microbiology (MS); molecular biochemistry and biophysics (MS, PhD). Part-time and evening/weekend programs available. Postbaccalaureate distance learning degree programs offered (no on-campus study). Terminal master's awarded for partial completion of doctoral program. *Degree requirements:* For master's, thesis (for some programs), comprehensive exam; for doctorate, thesis/dissertation, comprehensive exam. *Entrance requirements:* For master's and doctorate, GRE General Test, minimum undergraduate GPA of 3.0. Additional exam requirements/recommendations for international students: Required—TOEFL (minimum score 550 paper-based; 213 computer-based). Electronic applications accepted. *Faculty research:* Protein crystallography, small angle x-ray differation of muscle, spectroscopy of multidomain proteins, structure and function of cell cycle proteins, development of anticancer drugs.

Indiana University Bloomington, Graduate School, College of Arts and Sciences, Department of Biology, Program in Molecular, Cellular, and Developmental Biology, Bloomington, IN 47405-7000. Offers PhD. Offered through the University Graduate School. *Faculty:* 25 full-time (4 women). *Students:* 48 full-time (23 women); includes 6 minority (5 African Americans, 1 Hispanic American), 16 international. In 2005, 9 degrees awarded. *Degree requirements:* For doctorate, thesis/dissertation. *Entrance requirements:* For doctorate, GRE General Test. Additional exam requirements/recommendations for international students: Required—TOEFL. *Application deadline:* For fall admission, 1/15 for domestic students; for spring admission, 9/1 priority date for domestic students. Applications are processed on a rolling basis. Application fee: $45. Electronic applications accepted. *Expenses:* Tuition, state resident: full-time $5,437; part-time $227 per credit hour. Tuition, nonresident: full-time $15,836; part-time $660 per credit hour. Required fees: $821. Tuition and fees vary according to campus/location and program. *Financial support:* In 2005–06, 48 students received support; fellowships with tuition reimbursements available, research assistantships with tuition reimbursements available, teaching assistantships with tuition reimbursements available available. Financial award application deadline: 2/15. *Faculty research:* Developmental genetics, molecular evolution, macromolecular structure and function, cell biology and microbial molecular genetics. *Unit head:* Dr. Susan Strome, Head, 812-855-5450, Fax: 812-855-6705, E-mail: sstrome@bio.indiana.edu. *Application contact:* Gretchen Clearwater, Adviser for Graduate Affairs, 812-855-1861, Fax: 812-855-6705, E-mail: biograd@bio.indiana.edu.

Indiana University–Purdue University Indianapolis, Indiana University School of Medicine, Department of Biochemistry and Molecular Biology, Indianapolis, IN 46202-2896. Offers PhD, MD/MS, MD/PhD. *Faculty:* 17 full-time (4 women). *Students:* 23 full-time (12 women), 21 part-time (17 women); includes 4 minority (2 African Americans, 1 Asian American or Pacific Islander, 1 Hispanic American), 17 international. Average age 28. In 2005, 5 doctorates awarded. Terminal master's awarded for partial completion of doctoral program. *Degree*

requirements: For doctorate, thesis/dissertation. *Entrance requirements:* For doctorate, GRE General Test, GRE Subject Test (recommended), previous course work in organic chemistry. *Application deadline:* For fall admission, 1/15 for domestic students. Applications are processed on a rolling basis. Application fee: $50 ($60 for international students). *Expenses:* Tuition, state resident: full-time $5,159; part-time $215 per credit hour. Tuition, nonresident: full-time $14,890; part-time $620 per credit hour. Required fees: $614. Tuition and fees vary according to campus/location and program. *Financial support:* Fellowships with tuition reimbursements, research assistantships with tuition reimbursements, teaching assistantships, Federal Work-Study, institutionally sponsored loans, scholarships/grants, and tuition waivers (partial) available. Support available to part-time students. Financial award application deadline: 2/1. *Faculty research:* Metabolic regulation, enzymology, peptide and protein chemistry, cell biology, signal transduction. *Unit head:* Dr. Zhong-Yin Zhang, Chairman, 317-274-7151. *Application contact:* Dr. David W. Allmann, Chairperson, Admissions Committee, 317-274-4096, Fax: 317-274-4686, E-mail: dallman@iupui.edu.

See Close-Up on page 393.

Iowa State University of Science and Technology, Graduate College, College of Agriculture and College of Liberal Arts and Sciences, Department of Biochemistry, Biophysics, and Molecular Biology, Ames, IA 50011. Offers biochemistry (MS, PhD); biophysics (MS, PhD); genetics (MS, PhD); molecular, cellular, and developmental biology (MS, PhD); toxicology (MS, PhD). *Faculty:* 24 full-time, 2 part-time/adjunct. *Students:* 97 full-time (42 women), 2 part-time (1 woman); includes 2 minority (1 African American, 1 Asian American or Pacific Islander), 68 international. 28 applicants, 46% accepted, 13 enrolled. In 2005, 9 master's, 8 doctorates awarded. *Degree requirements:* For master's and doctorate, thesis/dissertation. *Entrance requirements:* For master's and doctorate, GRE General Test. Additional exam requirements/recommendations for international students: Required—TOEFL (paper score 550; computer score 213) or IELTS (score 6.5). *Application deadline:* For fall admission, 2/1 priority date for domestic students, 2/1 priority date for international students. Application fee: $30 ($70 for international students). Electronic applications accepted. *Expenses:* Tuition, state resident: full-time $6,410. Tuition, nonresident: full-time $16,422. Tuition and fees vary according to program. *Financial support:* In 2005–06, 91 research assistantships with full tuition reimbursements (averaging $16,290 per year), 2 teaching assistantships with full and partial tuition reimbursements (averaging $14,455 per year) were awarded; scholarships/grants, health care benefits, and unspecified assistantships also available. *Unit head:* Dr. Alan M. Myers, Chair, 515-294-6116, E-mail: biochem@iastate.edu.

See Close-Up on page 395.

Iowa State University of Science and Technology, Graduate College, Interdisciplinary Programs, Bioinformatics and Computational Biology Program, Ames, IA 50011-3260. Offers MS, PhD. *Degree requirements:* For doctorate, thesis/dissertation. *Entrance requirements:* For doctorate, GRE General Test. Additional exam requirements/recommendations for international students: Required—TOEFL or IELTS. Electronic applications accepted. *Expenses:* Tuition, state resident: full-time $6,410. Tuition, nonresident: full-time $16,422. Tuition and fees vary according to program. *Faculty research:* Functional and structural genomics, genome evolution, macromolecular structure and function, mathematical biology and biological statistics, metabolic and developmental networks.

Iowa State University of Science and Technology, Graduate College, Interdisciplinary Programs, Program in Molecular, Cellular, and Developmental Biology, Ames, IA 50011. Offers MS, PhD. *Students:* 38 full-time (19 women), 1 part-time, 28 international. 78 applicants, 19% accepted, 7 enrolled. In 2005, 2 degrees awarded. *Degree requirements:* For master's, thesis or alternative; for doctorate, thesis/dissertation. *Entrance requirements:* For master's and doctorate, GRE General Test, resumé. Additional exam requirements/recommendations for international students: Required—TOEFL (paper score 550; computer score 220) or IELTS (score 6.5). *Application deadline:* For fall admission, 1/15 priority date for domestic students, 1/15 priority date for international students. Application fee: $30 ($70 for international students). Electronic applications accepted. *Expenses:* Tuition, state resident: full-time $6,410. Tuition, nonresident: full-time $16,422. Tuition and fees vary according to program. *Financial support:* In 2005–06, 32 research assistantships with full and partial tuition reimbursements (averaging $22,000 per year), 6 teaching assistantships with full and partial tuition reimbursements were awarded; scholarships/grants, health care benefits, and unspecified assistantships also available. *Unit head:* Dr. W Allen Miller, Supervisory Committee Chair, 515-294-7252, E-mail: idgp@iastate.edu. *Application contact:* Mary Jordison, Information Contact, 515-294-7252, Fax: 515-924-6790, E-mail: idgp@iastate.edu.

Announcement: The MCDB interdepartmental graduate program includes 55 faculty members in 12 departments, offering students diverse training in molecular, cellular, and developmental biology of plants and animals, molecular microbiology, signal transduction, biochemistry, and bioinformatics. Iowa State University provides excellent research facilities and broad curriculum choices. Visit http://grad.admin.iastate.edu/mcdb/mcdb.html or e-mail idgp@iastate.edu for more information.

The Johns Hopkins University, Bloomberg School of Public Health, Department of Biochemistry and Molecular Biology, Baltimore, MD 21205. Offers biochemistry and molecular biology (MHS, Sc M). *Students:* 26 full-time (9 women). *Students:* 69 full-time (44 women), 4 part-time (3 women); includes 17 minority (2 African Americans, 1 American Indian/Alaska Native, 9 Asian Americans or Pacific Islanders, 5 Hispanic Americans), 10 international. Average age 24. 174 applicants, 32% accepted, 39 enrolled. In 2005, 16 master's, 3 doctorates awarded. *Median time to degree:* Of those who began their doctoral program in fall 1997, 69% received their degree in 8 years or less. *Degree requirements:* For master's, thesis, registration; for doctorate, thesis/dissertation, oral and written exams, comprehensive exam, registration. *Entrance requirements:* For master's, MCAT or GRE, 3 letters of recommendation, curriculum vitae; for doctorate, GRE General Test, 3 letters of recommendation, curriculum vitae. Additional exam requirements/recommendations for international students: Required—TOEFL (minimum score 600 paper-based; 250 computer-based). *Application deadline:* For fall admission, 11/10 for domestic students. Applications are processed on a rolling basis. Application fee: $45. Electronic applications accepted. *Expenses:* Tuition: Full-time $30,960. Tuition and fees vary according to degree level and program. *Financial support:* In 2005–06, 7 fellowships with tuition reimbursements (averaging $25,000 per year), 30 research assistantships with tuition reimbursements (averaging $25,000 per year), 4 teaching assistantships (averaging $250 per year) were awarded; Federal Work-Study, institutionally sponsored loans, scholarships/grants, health care benefits, and stipends also available. Financial award application deadline: 3/15. *Faculty research:* DNA replication, repair, structure, carcinogenesis, protein structure, enzyme catalysts, reproductive biology. Total annual research expenditures: $3.7 million. *Unit head:* Dr. Roger McMacken, Chairman, 410-955-3671, Fax: 410-955-2926, E-mail: rmcmacke@jhsph.edu. *Application contact:* Gerry Graziano, Admissions Coordinator, 410-955-3672, Fax: 410-955-2926, E-mail: ggrazian@jhsph.edu.

The Johns Hopkins University, National Institutes of Health Sponsored Programs, Department of Biology, Baltimore, MD 21218-2699. Offers biochemistry (PhD); biophysics (PhD); cell biology (PhD); developmental biology (PhD); genetic biology (PhD); molecular biology (PhD). *Faculty:* 25 full-time (4 women). *Students:* 126 full-time (72 women); includes 36 minority (3 African Americans, 1 American Indian/Alaska Native, 21 Asian Americans or Pacific Islanders, 11 Hispanic Americans), 19 international. 282 applicants, 26% accepted, 36 enrolled. In 2005, 15 degrees awarded. *Median time to degree:* Of those who began their doctoral program in fall 1997, 81.2% received their degree in 8 years or less. *Degree requirements:* For doctorate, thesis/dissertation, comprehensive exam, registration. *Entrance requirements:* For doctorate, GRE General Test. Additional exam requirements/recommendations for international students: Required—TOEFL (minimum score 600 paper-based; 250 computer-based), TWE, TSE. *Application deadline:* For fall admission, 12/15 for domestic students. Application fee: $60. *Expenses:* Tuition: Full-time $30,960. Tuition and fees vary according to degree level and program.

Molecular Biology

The Johns Hopkins University (continued)
Financial support: In 2005–06, 24 fellowships (averaging $23,000 per year), 93 research assistantships (averaging $23,000 per year), 22 teaching assistantships (averaging $23,000 per year) were awarded; Federal Work-Study, institutionally sponsored loans, scholarships/grants, traineeships, health care benefits, tuition waivers (partial), and unspecified assistantships also available. Financial award application deadline: 4/15; financial award applicants required to submit FAFSA. *Faculty research:* Protein and nucleic acid biochemistry and biophysical chemistry, molecular biology and development. Total annual research expenditures: $11.2 million. *Unit head:* Dr. Allen Shearn, Chair, 410-516-4693, Fax: 410-516-5213, E-mail: bio_cals@jhu.edu. *Application contact:* Joan Miller, Academic Affairs Manager, 410-516-5502, Fax: 410-516-5213, E-mail: joan@jhu.edu.

See Close-Up on page 155.

The Johns Hopkins University, School of Medicine, Graduate Programs in Medicine, Department of Pharmacology and Molecular Sciences, Baltimore, MD 21205. Offers PhD. *Faculty:* 42 full-time (10 women). *Students:* 56 full-time (29 women); includes 11 minority (4 African Americans, 7 Asian Americans or Pacific Islanders), 17 international. 151 applicants, 13% accepted, 11 enrolled. In 2005, 7 degrees awarded. *Degree requirements:* For doctorate, thesis/dissertation, departmental seminar, comprehensive exam. *Entrance requirements:* For doctorate, GRE General Test. Additional exam requirements/recommendations for international students: Required—TOEFL. *Application deadline:* For fall admission, 1/10 for domestic students. Application fee: $75. Electronic applications accepted. *Expenses:* Tuition: Full-time $30,960. Tuition and fees vary according to degree level and program. *Financial support:* In 2005–06, fellowships (averaging $24,600 per year) *Unit head:* Dr. Philip A. Cole, Chairman, 410-614-0540, Fax: 410-614-7717, E-mail: pcole@jhmi.edu. *Application contact:* Dr. James T. Stivers, Director of Admissions, 410-955-7117, Fax: 410-955-3023, E-mail: jstivers@jhmi.edu.

See Close-Up on page 1169.

The Johns Hopkins University, School of Medicine, Graduate Programs in Medicine, Predoctoral Training Program in Human Genetics and Molecular Biology, Baltimore, MD 21218-2699. Offers PhD, MD/PhD. *Faculty:* 59 full-time (14 women). *Students:* 52 full-time (27 women); includes 6 minority (2 African Americans, 3 Asian Americans or Pacific Islanders, 1 Hispanic American), 18 international. Average age 24. 171 applicants, 8% accepted, 16 enrolled. In 2005, 6 doctorates awarded. *Degree requirements:* For doctorate, thesis/dissertation, comprehensive exam. *Entrance requirements:* For doctorate, GRE General Test, GRE Subject Test. *Application deadline:* For fall admission, 1/15 priority date for domestic students, 1/15 priority date for international students. Application fee: $70. *Expenses:* Tuition: Full-time $30,960. Tuition and fees vary according to degree level and program. *Financial support:* In 2005–06, 12 fellowships with full tuition reimbursements (averaging $24,600 per year) were awarded; teaching assistantships with full tuition reimbursements, health care benefits also available. *Faculty research:* Human, mammalian, and molecular genetics. *Unit head:* Dr. David Valle, Director, 410-955-4260, Fax: 410-955-7397, E-mail: muscelli@jhmi.edu. *Application contact:* Sandy Muscelli, Program Administrator, 410-955-4260, Fax: 410-955-7397, E-mail: muscelli@jhmi.edu.

See Close-Up on page 779.

The Johns Hopkins University, School of Medicine, Graduate Programs in Medicine, Program in Biochemistry, Cellular and Molecular Biology, Baltimore, MD 21205. Offers PhD. *Faculty:* 91 full-time (25 women), 11 part-time/adjunct (5 women). *Students:* 155 full-time (75 women); includes 19 minority (6 African Americans, 12 Asian Americans or Pacific Islanders, 1 Hispanic American), 51 international. Average age 25. 358 applicants, 18% accepted, 20 enrolled. In 2005, 19 doctorates awarded. *Median time to degree:* Of those who began their doctoral program in fall 1997, 100% received their degree in 8 years or less. *Degree requirements:* For doctorate, thesis/dissertation, comprehensive exam. *Entrance requirements:* For doctorate, GRE General Test. Additional exam requirements/recommendations for international students: Required—TOEFL. *Application deadline:* For winter admission, 1/10 for domestic students. Applications are processed on a rolling basis. Application fee: $90. Electronic applications accepted. *Expenses:* Tuition: Full-time $30,960. Tuition and fees vary according to degree level and program. *Financial support:* In 2005–06, 3 fellowships with partial tuition reimbursements (averaging $32,000 per year), 154 research assistantships with full tuition reimbursements (averaging $24,600 per year) were awarded; traineeships and tuition waivers (full) also available. Financial award application deadline: 12/31. *Faculty research:* Developmental biology, genomics/proteomics, protein targeting, signal transduction, structural biology. *Unit head:* Dr. Craig Montell, Director, 410-955-3466, Fax: 410-614-8842, E-mail: cmontell@jhmi.edu. *Application contact:* Dr. Jeff Corden, Admissions Director, 410-955-3506, Fax: 410-614-8842, E-mail: jcorden@jhmi.edu.

Kent State University, School of Biomedical Sciences, Program in Cellular and Molecular Biology, Kent, OH 44242-0001. Offers MS, PhD. Offered in cooperation with Northeastern Ohio Universities College of Medicine. Terminal master's awarded for partial completion of doctoral program. *Degree requirements:* For master's and doctorate, thesis/dissertation. *Entrance requirements:* For master's and doctorate, GRE General Test. *Faculty research:* Molecular genetics, molecular endocrinology, virology and tumor biology, P450 enzymology and catalysis, membrane structure and function.

Lehigh University, College of Arts and Sciences, Department of Biological Sciences, Bethlehem, PA 18015-3094. Offers biochemistry (PhD); integrative biology (PhD); molecular biology (PhD). Postbaccalaureate distance learning degree programs offered (no on-campus study). *Faculty:* 22 full-time (9 women). *Students:* 31 full-time (21 women), 51 part-time (29 women); includes 2 minority (1 Asian American or Pacific Islander, 1 Hispanic American), 6 international. 65 applicants, 29% accepted. In 2005, 3 doctorates awarded. *Median time to degree:* Of those who began their doctoral program in fall 1997, 100% received their degree in 8 years or less. *Degree requirements:* For doctorate, thesis/dissertation, comprehensive exam. *Entrance requirements:* For doctorate, GRE General Test. Additional exam requirements/recommendations for international students: Required—TOEFL, TSE. *Application deadline:* For fall admission, 1/15 for domestic students. Applications are processed on a rolling basis. Application fee: $50. Electronic applications accepted. *Financial support:* In 2005–06, 30 students received support, including 4 fellowships with tuition reimbursements available (averaging $20,000 per year), 6 research assistantships with tuition reimbursements available (averaging $20,000 per year), 16 teaching assistantships with tuition reimbursements available (averaging $20,000 per year); scholarships/grants, tuition waivers (full and partial), and unspecified assistantships also available. Financial award application deadline: 1/31. *Faculty research:* Gene expression, cytoskeleton and cell structure, cell cycle and growth regulation, neuroscience, animal behavior. *Unit head:* , Dr. Neal G. Simon, Chairperson, 610-758-3680, Fax: 610-758-4004. *Application contact:* Dr. Michael R. Kuchka, Graduate Coordinator, 610-758-3687, Fax: 610-758-4004, E-mail: mrk5@lehigh.edu.

Louisiana State University Health Sciences Center at Shreveport, Department of Biochemistry and Molecular Biology, Shreveport, LA 71130-3932. Offers MS, PhD, MD/PhD. *Faculty:* 16 full-time (3 women), 5 part-time/adjunct (0 women). *Students:* 38 full-time (15 women); includes 14 minority (3 African Americans, 11 Asian Americans or Pacific Islanders). Average age 26. 50 applicants, 10% accepted, 5 enrolled. In 2005, 2 master's, 1 doctorate awarded. *Median time to degree:* Of those who began their doctoral program in fall 1997, 100% received their degree in 8 years or less. *Degree requirements:* For master's and doctorate, thesis/dissertation. *Entrance requirements:* For master's and doctorate, GRE General Test. Additional exam requirements/recommendations for international students: Required—TOEFL. *Application deadline:* For fall admission, 4/15 for domestic students. Applications are processed on a rolling basis. Application fee: $30. *Financial support:* In 2005–06, 38 students received support, including research assistantships (averaging $18,000 per year); institutionally sponsored loans also available. *Faculty research:* Metabolite transport, regulation of

translation and transcription, prokaryotic molecular genetics, cell matrix biochemistry, yeast molecular genetics, oncogenes. Total annual research expenditures: $160,000. *Unit head:* Dr. Robert E. Rhoads, Head, 318-675-5160, Fax: 318-675-5180. *Application contact:* Dr. Robert L. Smith, Associate Professor and Director of Graduate Studies, 318-675-5168, Fax: 318-675-5180, E-mail: rsmith2@lsuhsc.edu.

See Close-Up on page 163.

Loyola University Chicago, Graduate School, Program in Molecular Biology, Chicago, IL 60626. Offers PhD, MD/PhD. *Faculty:* 28 full-time (6 women). *Students:* 20 full-time (12 women); includes 1 minority (Hispanic American), 12 international. Average age 28. 26 applicants, 35% accepted, 2 enrolled. In 2005, 4 degrees awarded. *Median time to degree:* Of those who began their doctoral program in fall 1997, 90% received their degree in 8 years or less. *Degree requirements:* For doctorate, thesis/dissertation, comprehensive exam. *Entrance requirements:* For doctorate, GRE General Test, 3 letters of recommendation. Additional exam requirements/recommendations for international students: Required—TOEFL. *Application deadline:* For fall admission, 3/1 for domestic students. Applications are processed on a rolling basis. Application fee: $40. *Expenses:* Tuition: Full-time $11,610; part-time $645 per credit. Required fees: $55 per semester. *Financial support:* In 2005–06, 19 students received support, including fellowships (averaging $22,000 per year); research assistantships, Federal Work-Study and institutionally sponsored loans also available. Financial award application deadline: 2/15. *Faculty research:* Cell cycle regulation, molecular immunology, molecular genetics, molecular oncology, molecular virology. Total annual research expenditures: $3,500. *Unit head:* Dr. Manuel O. Diaz, Director, 708-327-3172, Fax: 708-216-6505, E-mail: mdiaz@luc.edu. *Application contact:* Dr. Alan Wolfe, Graduate Program Director, 708-216-5814, Fax: 708-216-6505, E-mail: awolfe@luc.edu.

Marquette University, Graduate School, College of Arts and Sciences, Department of Biology, Milwaukee, WI 53201-1881. Offers cell biology (MS, PhD); developmental biology (MS, PhD); ecology (MS, PhD); endocrinology (MS, PhD); evolutionary biology (MS, PhD); genetics (MS, PhD); microbiology (MS, PhD); molecular biology (MS, PhD); muscle and exercise physiology (MS, PhD); neurobiology (MS, PhD); reproductive physiology (MS, PhD). Terminal master's awarded for partial completion of doctoral program. *Degree requirements:* For master's, thesis, 1 year of teaching experience or equivalent, comprehensive exam; for doctorate, thesis/dissertation, 1 year of teaching experience or equivalent, qualifying exam. *Entrance requirements:* For master's and doctorate, GRE General Test, GRE Subject Test. Additional exam requirements/recommendations for international students: Required—TOEFL. *Faculty research:* Microbial and invertebrate ecology, evolution of gene function, DNA methylation, DNA arrangement.

Mayo Graduate School, Graduate Programs in Biomedical Sciences, Programs in Biochemistry, Structural Biology, Cell Biology, and Genetics, Rochester, MN 55905. Offers biochemistry and structural biology (PhD); cell biology and genetics (PhD); molecular biology (PhD). *Degree requirements:* For doctorate, oral defense of dissertation, qualifying oral and written exam. *Entrance requirements:* For doctorate, GRE, 1 year of chemistry, biology, calculus, and physics. Additional exam requirements/recommendations for international students: Required—TOEFL. Electronic applications accepted. *Faculty research:* Gene structure and function, membranes and receptors/cytoskeleton, oncogenes and growth factors, protein structure and function, steroid hormonal action.

See Close-Up on page 399.

McMaster University, Faculty of Health Sciences and School of Graduate Studies, Program in Medical Sciences, Molecular Biology, Genetics, and Cancer Area, Hamilton, ON L8S 4M2, Canada. Offers M Sc, PhD. *Students:* 11 full-time, 2 part-time. In 2005, 2 master's, 2 doctorates awarded. *Degree requirements:* For master's, thesis/dissertation; for doctorate, thesis/dissertation, comprehensive exam. *Entrance requirements:* For master's, honors B Sc, B+ average in related field; for doctorate, M Sc, minimum B+ average, students with proven research experience and an A average may be admitted with a B Sc degree. Additional exam requirements/recommendations for international students: Required—TOEFL (minimum score 580 paper-based; 237 computer-based). *Application deadline:* For fall admission, 9/30 for domestic students. For winter admission, 3/31 for domestic students. Applications are processed on a rolling basis. Application fee: $85. *Financial support:* Teaching assistantships available. *Unit head:* Dr. John Waye, Coordinator, 905-525-9140 Ext. 76273. *Application contact:* Dr. Carl Richards, Associate Dean, 905-525-9140 Ext. 22983, Fax: 905-546-1129.

Medical College of Georgia, School of Graduate Studies, Department of Biochemistry and Molecular Biology, Augusta, GA 30912-1500. Offers PhD. *Faculty:* 10 full-time (1 woman). *Students:* 9 full-time (3 women); includes 1 minority (African American), 5 international. Average age 30. In 2005, 3 degrees awarded. *Degree requirements:* For doctorate, thesis/dissertation. *Entrance requirements:* For doctorate, GRE General Test. Additional exam requirements/recommendations for international students: Required—TOEFL. *Application deadline:* For fall admission, 6/30 for domestic students, 4/15 for international students. Applications are processed on a rolling basis. Application fee: $30. *Financial support:* In 2005–06, 2 students received support, including 8 research assistantships with partial tuition reimbursements available (averaging $22,500 per year); institutionally sponsored loans and scholarships/grants also available. Support available to part-time students. Financial award application deadline: 5/31; financial award applicants required to submit FAFSA. *Faculty research:* Bacterial pathogenesis, eye diseases, vitamins and amino acid transporters, transcriptional control and signal transduction, tumor biology. *Unit head:* Dr. Vadivel Ganapathy, Chair, 706-721-7652, Fax: 706-721-9947, E-mail: vganapat@mail.mcg.edu. *Application contact:* Dr. Darren D. Browning, Program Director, 706-721-9526, Fax: 706-721-6608, E-mail: dbrowning@mail.mcg.edu.

Medical University of South Carolina, College of Graduate Studies, Program in Biochemistry and Molecular Biology, Charleston, SC 29425-0002. Offers Pharm D, MS, PhD, DMD/PhD, MD/PhD. *Faculty:* 22 full-time (6 women). *Students:* 11 full-time (3 women); includes 1 minority (Hispanic American), 3 international. Average age 28. 181 applicants, 34% accepted, 43 enrolled. In 2005, 2 degrees awarded. *Degree requirements:* For doctorate, thesis/dissertation, teaching and research seminar, oral and written exams. *Entrance requirements:* For doctorate, GRE General Test, interview. Additional exam requirements/recommendations for international students: Required—TOEFL (minimum score 600 paper-based; 250 computer-based). *Application deadline:* For fall admission, 1/15 priority date for domestic students, 1/15 priority date for international students. Applications are processed on a rolling basis. Application fee: $0 ($75 for international students). Electronic applications accepted. *Financial support:* In 2005–06, 4 students received support, including fellowships with partial tuition reimbursements available (averaging $21,000 per year); Federal Work-Study and scholarships/grants also available. Financial award application deadline: 3/15; financial award applicants required to submit FAFSA. *Faculty research:* Lipid biochemistry, DNA replication, nucleic acid/protein interaction, proteases, signal. *Unit head:* Dr. Y. A. Hannun, Chairman, 843-792-9318, Fax: 843-792-6590, E-mail: hannun@musc.edu. *Application contact:* Cheryl Brown, 843-792-4620, Fax: 843-792-4645, E-mail: brownche@musc.edu.

Medical University of South Carolina, College of Graduate Studies, Program in Molecular and Cellular Biology and Pathobiology, Charleston, SC 29425-0002. Offers cell biology and anatomy (PhD); marine biomedicine (PhD). *Students:* 173 full-time (41 women). *Students:* 62 full-time (32 women); includes 8 minority (3 African Americans, 1 American Indian/Alaska Native, 4 Asian Americans or Pacific Islanders), 7 international. Average age 28. 181 applicants, 34% accepted, 43 enrolled. In 2005, 20 doctorates awarded. *Degree requirements:* For doctorate, thesis/dissertation, teaching and research seminar, oral and written exams. *Entrance requirements:* For doctorate, GRE General Test, interview, minimum GPA of 3.2. Additional exam requirements/recommendations for international students: Required—TOEFL (minimum score 600 paper-based; 250 computer-based). *Application deadline:* For fall admission, 1/15 priority date for domestic students, 1/15 priority date for international students. Applications are processed on a rolling basis. Application fee: $0 ($75 for international students). Electronic applications accepted. *Financial support:* In 2005–06, 12 students received support, including fellowships with partial tuition reimbursements available (averaging $21,000 per year); Federal Work-Study and scholarships/grants also available. Financial award application deadline: 3/15; financial award applicants required to submit FAFSA. Total annual research expenditures: $10 million. *Unit head:* Dr. Donald R. Menick, Director, 843-856-5045, Fax: 843-792-6590, E-mail: menickd@musc.edu. *Application contact:* Cheryl Brown, 843-792-4620, Fax: 843-792-4645, E-mail: brownche@musc.edu.

Michigan State University, The Graduate School, College of Agriculture and Natural Resources, MSU-DOE Plant Research Laboratory, East Lansing, MI 48824. Offers biochemistry and molecular biology (PhD); cellular and molecular biology (PhD); crop and soil sciences (PhD); genetics (PhD); microbiology and molecular genetics (PhD); plant biology (PhD); plant physiology (PhD). Offered jointly with the Department of Energy. *Faculty:* 9 full-time (2 women). *Degree requirements:* For doctorate, thesis/dissertation, laboratory rotation, defense of dissertation, comprehensive exam. *Entrance requirements:* For doctorate, GRE General Test, acceptance into one of the affiliated department programs; 3 letters of recommendation; bachelor's degree or equivalent in life sciences, chemistry, biochemistry, or biophysics; research experience. Application fee: $50. Electronic applications accepted. *Expenses:* Tuition, state resident: part-time $330 per credit hour. Tuition, nonresident: part-time $685 per credit hour. Tuition and fees vary according to program. *Faculty research:* Role of hormones in the regulation of plant development and physiology, molecular mechanisms associated with signal recognition, development and application of genetic methods and materials, protein routing and function. Total annual research expenditures: $7.4 million. *Unit head:* Dr. Kenneth Keegstra, Director, 517-353-2270, Fax: 517-353-9168, E-mail: keegstra@msu.edu. *Application contact:* Janet Taylor, Graduate Program Secretary, 517-353-2270, Fax: 517-353-9168, E-mail: prl@msu.edu.

Michigan State University, The Graduate School, College of Natural Science and Graduate Programs in Human Medicine and Graduate Studies in Osteopathic Medicine, Department of Biochemistry and Molecular Biology, East Lansing, MI 48824. Offers biochemistry and molecular biology (MS, PhD); biochemistry and molecular biology/environmental toxicology (PhD). *Faculty:* 30 full-time (5 women). *Students:* 74 full-time (31 women), 1 part-time; includes 3 minority (1 African American, 1 Asian American or Pacific Islander, 1 Hispanic American), 35 international. Average age 28. 93 applicants, 12% accepted. In 2005, 2 master's, 7 doctorates awarded. *Degree requirements:* For master's, oral defense of thesis, annual review, thesis optional; for doctorate, thesis/dissertation, teaching, annual evaluations, research proposal, comprehensive exam. *Entrance requirements:* For master's, GRE General Test, minimum GPA of 3.2 in last 2 undergraduate years, course work in biochemistry, 3 letters of recommendation from former instructors; for doctorate, GRE General Test, minimum GPA of 3.2; bachelor's or master's degree in chemistry, biochemistry or related science; course work in biochemistry; 3 letters of recommendation from former instructors. Additional exam requirements/recommendations for international students: Required—TOEFL (minimum score 550 paper-based; 213 computer-based), Michigan State University ELT (85), Michigan ELAB (83). *Application deadline:* For fall admission, 12/27 for domestic students. Application fee: $50. Electronic applications accepted. *Expenses:* Tuition, state resident: part-time $330 per credit hour. Tuition, nonresident: part-time $685 per credit hour. Tuition and fees vary according to program. *Financial support:* In 2005–06, 18 fellowships with tuition reimbursements (averaging $11,755 per year), 64 research assistantships with tuition reimbursements (averaging $15,738 per year) were awarded; scholarships/grants and unspecified assistantships also available. *Faculty research:* Biochemistry of the cell nucleus; protein structure, function, and design; plant biochemistry; biological modeling; gene expression in development and disease. Total annual research expenditures: $10.7 million. *Unit head:* Dr. S. L. Ferguson-Miller, Chairperson, 517-353-0804, Fax: 517-353-9334, E-mail: fergus20@msu.edu. *Application contact:* Jessica Hudson, Graduate Program Secretary, 517-353-0807, Fax: 517-353-9334, E-mail: jhudson@msu.edu.

See Close-Up on page 401.

Michigan State University, The Graduate School, College of Natural Science, Program in Cell and Molecular Biology, East Lansing, MI 48824. Offers MS, PhD. *Faculty:* 78 full-time (26 women). *Students:* 39 full-time (26 women); includes 1 minority (Hispanic American), 18 international. Average age 28. 137 applicants, 6% accepted. In 2005, 6 master's, 3 doctorates awarded. *Degree requirements:* For master's, thesis or alternative, oral exam; for doctorate, thesis/dissertation, teaching, research, rotation, preliminary exam. *Entrance requirements:* For master's, GRE General Test, minimum GPA of 3.2, bachelor's degree in biological science, 3 letters of recommendation; for doctorate, GRE General Test, minimum GPA of 3.0 in science/math courses, 3 letters of recommendation, background in biology and chemistry, 1 year of physics, mathematics through calculus, inorganic and organic chemistry courses . Additional exam requirements/recommendations for international students: Required—TOEFL (minimum score 550 paper-based; 213 computer-based). *Application deadline:* For fall admission, 12/27 for domestic students. Application fee: $50. Electronic applications accepted. *Expenses:* Tuition, state resident: part-time $330 per credit hour. Tuition, nonresident: part-time $685 per credit hour. Tuition and fees vary according to program. *Financial support:* In 2005–06, 7 fellowships with tuition reimbursements (averaging $13,911 per year), 28 research assistantships with tuition reimbursements (averaging $15,662 per year) were awarded; teaching assistantships with tuition reimbursements, scholarships/grants and unspecified assistantships also available. Financial award application deadline: 1/1. *Faculty research:* Cancer research, immunology and virology, molecular genetics, protein structure and function, signal transduction. *Unit head:* Dr. Susan E. Conrad, Director, 517-353-5161, Fax: 517-432-8813, E-mail: conrad@msu.edu. *Application contact:* Christine Vandeuren, Graduate Secretary, 517-353-8916, Fax: 517-432-8813, E-mail: cmb@msu.edu.

Mississippi State University, College of Agriculture and Life Sciences, Department of Biochemistry and Molecular Biology, Mississippi State, MS 39762. Offers biochemistry (MS); molecular biology (PhD). *Faculty:* 8 full-time (1 woman). *Students:* 19 full-time (6 women), 1 (woman) part-time; includes 1 minority (Hispanic American), 13 international. Average age 28. 33 applicants, 42% accepted, 1 enrolled.Terminal master's awarded for partial completion of doctoral program. *Degree requirements:* For master's, thesis (for some programs), comprehensive oral or written exam; for doctorate, thesis/dissertation, comprehensive oral and written exam. *Entrance requirements:* For master's, GRE General Test, minimum GPA of 2.75; for doctorate, GRE. Additional exam requirements/recommendations for international students: Required—TOEFL. *Application deadline:* For fall admission, 7/1 for domestic students; for spring admission, 11/1 for domestic students. Applications are processed on a rolling basis. Application fee: $30. Electronic applications accepted. *Expenses:* Tuition, state resident: full-time $4,312; part-time $240 per hour. Tuition, nonresident: full-time $9,772; part-time $543 per hour. International tuition: $10,102 full-time. Tuition and fees vary according to course load. *Financial support:* Research assistantships with full tuition reimbursements, Federal Work-Study, institutionally sponsored loans, and unspecified assistantships available. Financial award applicants required to submit FAFSA. *Faculty research:* Fish nutrition, plant and animal molecular biology, plant biochemistry, enzymology, lipid metabolism. Total annual research expenditures: $288,720. *Unit head:* Dr. John A. Boyle, Head, 662-325-2640, Fax: 662-325-8664, E-mail: jab@ra.msstate.edu. *Application contact:* Philip G. Bonfanti, Director of Admissions, 662-325-4104, Fax: 662-325-8872, E-mail: admit@msstate.edu.

Missouri State University, Graduate College, College of Health and Human Services, Department of Biomedical Sciences, Program in Cell and Molecular Biology, Springfield, MO 65804-0094. Offers MS. *Students:* 7 full-time (3 women), 1 (woman) part-time, 2 international. Average age 27. 25 applicants, 36% accepted, 2 enrolled. In 2005, 2 degrees awarded. *Degree requirements:* For master's, thesis or alternative, oral and written exams. *Entrance*

---

requirements: For master's, GRE General Test, 2 semesters of course work in organic chemistry and physics, 1 semester of course work in calculus, minimum GPA of 3.0 in last 60 hours of course work. Additional exam requirements/recommendations for international students: Required—TOEFL (minimum score 550 paper-based; 213 computer-based), IELT (minimum score 6). *Application deadline:* For fall admission, 7/20 for domestic students; for spring admission, 12/20 priority date for domestic students. Applications are processed on a rolling basis. Application fee: $30. Electronic applications accepted. *Expenses:* Tuition, state resident: full-time $3,402; part-time $189 per credit. Tuition, nonresident: full-time $6,804; part-time $378 per credit. Required fees: $207 per semester. Part-time tuition and fees vary according to course level, course load and program. *Financial support:* In 2005–06, research assistantships with full tuition reimbursements (averaging $6,575 per year), 3 teaching assistantships with full tuition reimbursements (averaging $6,575 per year) were awarded; career-related internships or fieldwork, Federal Work-Study, institutionally sponsored loans, scholarships/grants, and unspecified assistantships also available. Support available to part-time students. Financial award application deadline: 3/31; financial award applicants required to submit FAFSA. *Unit head:* Albert R. Gordon, Graduate Director, 417-836-5730, Fax: 417-836-5588, E-mail: albertgordon@missouristate.edu. *Application contact:* Information Contact, 417-836-5603, Fax: 417-836-5588.

Montana State University, College of Graduate Studies, College of Agriculture, Department of Veterinary Molecular Biology, Bozeman, MT 59717. Offers MS, PhD. Part-time programs available. *Faculty:* 4 full-time (0 women), 1 (woman) part-time/adjunct. *Students:* 2 full-time (both women), 16 part-time (9 women); includes 1 minority (Hispanic American), 3 international. Average age 29. 8 applicants, 25% accepted, 2 enrolled. In 2005, 1 master's, 2 doctorates awarded. *Degree requirements:* For master's, comprehensive exam, registration; for doctorate, thesis/dissertation, comprehensive exam, registration. *Entrance requirements:* For master's and doctorate, GRE General Test. Additional exam requirements/recommendations for international students: Required—TOEFL (minimum score 550 paper-based; 213 computer-based). *Application deadline:* For fall admission, 7/15 priority date for domestic students, 5/15 priority date for international students; for spring admission, 12/1 priority date for domestic students, 10/1 priority date for international students. Applications are processed on a rolling basis. Application fee: $30. Electronic applications accepted. *Expenses:* Tuition, state resident: full-time $4,132. Tuition, nonresident: full-time $11,332. *Financial support:* In 2005–06, 17 students received support, including 15 research assistantships with full tuition reimbursements available (averaging $20,086 per year), 2 teaching assistantships with full tuition reimbursements available (averaging $20,000 per year) Financial award application deadline: 3/1; financial award applicants required to submit FAFSA. *Faculty research:* Infectious diseases of livestock and wildlife, immunology, functional genomics, bacterial pathogenesis, cellular, molecular and developmental biology. Total annual research expenditures: $10 million. *Unit head:* Dr. Mark G. Quinn, Interim Department Head, 406-994-5721, Fax: 406-994-4303, E-mail: mquinn@montana.edu.

Montclair State University, The Graduate School, College of Science and Mathematics, Department of Biology and Molecular Biology, Montclair, NJ 07043-1624. Offers biology (MS), including biology science education, molecular biology; molecular biology (Certificate). Part-time and evening/weekend programs available. *Faculty:* 19 full-time (7 women), 19 part-time/adjunct (9 women). *Students:* 14 full-time (9 women), 61 part-time (36 women); includes 20 minority (10 African Americans, 6 Asian Americans or Pacific Islanders, 4 Hispanic Americans), 3 international. 52 applicants, 62% accepted, 23 enrolled. In 2005, 19 master's, 4 other advanced degrees awarded. *Degree requirements:* For master's, thesis or alternative, comprehensive exam. *Entrance requirements:* For master's, GRE General Test, 24 credits of course work in undergraduate biology, 2 letters of recommendation, teaching certificate (biology sciences education concentration). Additional exam requirements/recommendations for international students: Required—TOEFL (minimum score 83 computer-based). *Application deadline:* Applications are processed on a rolling basis. Application fee: $60. Electronic applications accepted. *Expenses:* Tuition, state resident: part-time $409 per credit. Tuition, nonresident: part-time $604 per credit. Required fees: $56 per credit. Tuition and fees vary according to course load, degree level and program. *Financial support:* In 2005–06, 7 research assistantships with full tuition reimbursements (averaging $7,000 per year) were awarded; Federal Work-Study, scholarships/grants, and unspecified assistantships also available. Support available to part-time students. Financial award application deadline: 3/1; financial award applicants required to submit FAFSA. *Faculty research:* Cells, algea blooms, scallops, NJ bays, Barnegat Bay. Total annual research expenditures: $48,000. *Unit head:* Dr. Scott Kight, Chairperson, 973-655-7047. *Application contact:* Dr. Reginald Halaby, Adviser, 973-655-4397, E-mail: halabyr@mail.montclair.edu.

Mount Sinai School of Medicine of New York University, Graduate School of Biological Sciences, New York, NY 10029-6504. Offers biophysics, structural biology and biomathematics (PhD); community medicine (MPH); genetic counseling (MS); genetics and genomic sciences (PhD); mechanisms of disease and therapy (PhD); microbiology (PhD); molecular, cellular, biochemical and developmental sciences (PhD); neurosciences (PhD). *Students:* 218 full-time (109 women). 4,208 applicants, 7% accepted, 117 enrolled.Terminal master's awarded for partial completion of doctoral program. *Degree requirements:* For master's, registration; for doctorate, thesis/dissertation, registration. *Entrance requirements:* For doctorate, GRE General Test, GRE Subject Test, 3 years of college pre-med course work. Additional exam requirements/recommendations for international students: Required—TOEFL. *Application deadline:* For fall admission, 1/15 for domestic students. Application fee: $75. Electronic applications accepted. *Expenses:* Tuition: Full-time $33,250. Required fees: $1,600. Full-time tuition and fees vary according to degree level, program and reciprocity agreements. *Financial support:* In 2005–06, fellowships with full tuition reimbursements (averaging $26,000 per year), research assistantships with full tuition reimbursements (averaging $26,000 per year) were awarded; Federal Work-Study, institutionally sponsored loans, scholarships/grants, health care benefits, and unspecified assistantships also available. Financial award application deadline: 4/5; financial award applicants required to submit FAFSA. *Faculty research:* Cancer, gene therapy, minimally invasive surgery, cardiac translational research. Total annual research expenditures: $162.2 million. *Unit head:* Dr. Diomedes Logothetis, Dean, 212-241-6546, Fax: 212-241-0651, E-mail: diomedes.logothetis@mssm.edu. *Application contact:* Lily Recanati, Manager, 212-241-3267, Fax: 212-241-0651, E-mail: lily.recantati@mssm.edu.

See Close-Up on page 177.

New Mexico State University, Graduate School, Graduate Program in Molecular Biology, Las Cruces, NM 88003-8001. Offers MS, PhD. *Students:* 28 full-time (15 women), 1 part-time; includes 9 minority (2 African Americans, 1 American Indian/Alaska Native, 6 Hispanic Americans), 12 international. Average age 32. 26 applicants, 46% accepted, 5 enrolled. In 2005, 1 master's, 4 doctorates awarded. *Degree requirements:* For master's, thesis/dissertation, oral seminars; for doctorate, thesis/dissertation, oral seminars, comprehensive exam, registration. *Entrance requirements:* For master's and doctorate, GRE General Test, minimum GPA of 3.3. Additional exam requirements/recommendations for international students: Required—TOEFL. *Application deadline:* For fall admission, 12/15 for domestic students, 12/15 for international students. Applications are processed on a rolling basis. Application fee: $30 ($50 for international students). Electronic applications accepted. *Expenses:* Tuition, state resident: full-time $3,156; part-time $175 per credit. Tuition, nonresident: full-time $12,510; part-time $565 per credit. Required fees: $1,050. *Financial support:* In 2005–06, 2 fellowships, 1 research assistantship, 4 teaching assistantships were awarded; career-related internships or fieldwork and unspecified assistantships also available. Financial award application deadline: 3/1. *Faculty research:* Emerging pathogens, plant-molecular biology and virology, molecular symbiotic interactions, cell and organismal biology, applied and environmental microbiology. *Unit head:* Dr. Peter Lammers, Director, 505-646-3437, Fax: 505-646-5170, E-mail: plammers@nmsu.edu. *Application contact:* Nancy McDow, Program Secretary, 505-646-3437, Fax: 505-646-5170, E-mail: nancyt@nmsu.edu.

New York Medical College, Graduate School of Basic Medical Sciences, Program in Biochemistry and Molecular Biology, Valhalla, NY 10595-1691. Offers MS, PhD, MD/PhD.

Molecular Biology

New York Medical College (continued)
Part-time and evening/weekend programs available. Terminal master's awarded for partial completion of doctoral program. *Degree requirements:* For master's, thesis/dissertation; for doctorate, thesis/dissertation, comprehensive exam. *Entrance requirements:* For master's and doctorate, GRE General Test. Additional exam requirements/recommendations for international students: Required—TOEFL. *Faculty research:* Mechanisms of control of blood coagulation, molecular neurobiology, molecular probes for infectious disease, protein-DNA interactions, molecular biology and biochemistry of double-stranded RNA-dependent enzymes.

New York University, Graduate School of Arts and Science, Department of Biology, New York, NY 10012-1019. Offers biology (PhD); biomedical journalism (MS); cancer and molecular biology (PhD); computational biology (PhD); computers in biological research (MS); developmental genetics (PhD); general biology (MS); immunology and microbiology (PhD); molecular genetics (PhD); neurobiology (PhD); oral biology (MS); plant biology (PhD); recombinant DNA technology (MS). Part-time programs available. *Faculty:* 24 full-time (5 women), 8 part-time/adjunct. *Students:* 104 full-time (52 women), 41 part-time (23 women); includes 28 minority (2 African Americans, 20 Asian Americans or Pacific Islanders, 6 Hispanic Americans), 47 international. Average age 27. 349 applicants, 56% accepted, 39 enrolled. In 2005, 59 master's, 4 doctorates awarded. Terminal master's awarded for partial completion of doctoral program. *Degree requirements:* For master's, thesis or alternative, qualifying paper; for doctorate, thesis/dissertation, comprehensive exam. *Entrance requirements:* For master's, GRE General Test; for doctorate, GRE General Test, GRE Subject Test. Additional exam requirements/recommendations for international students: Required—TOEFL. *Application deadline:* For fall admission, 1/4 for domestic students. Application fee: $80. *Financial support:* Fellowships with tuition reimbursements, research assistantships with tuition reimbursements, teaching assistantships with tuition reimbursements, career-related internships or fieldwork, Federal Work-Study, institutionally sponsored loans, scholarships/grants, health care benefits, and unspecified assistantships available. Financial award application deadline: 1/4; financial award applicants required to submit FAFSA. *Faculty research:* Genomics, molecular and cell biology, development and molecular genetics, molecular evolution of plants and animals. *Unit head:* Gloria Coruzzi, Chairman, 212-998-8200, Fax: 212-995-4015, E-mail: biology@nyu.edu. *Application contact:* Stephen Small, Director of Graduate Studies, 212-998-8200, Fax: 212-995-4015, E-mail: biology@nyu.edu.

New York University, School of Medicine and Graduate School of Arts and Science, Sackler Institute of Graduate Biomedical Sciences, Program in Cellular and Molecular Biology, New York, NY 10012-1019. Offers PhD, MD/PhD. *Degree requirements:* For doctorate, one foreign language, thesis/dissertation, qualifying exams, comprehensive exam. *Entrance requirements:* For doctorate, GRE General Test. Additional exam requirements/recommendations for international students: Required—TOEFL. *Faculty research:* Membrane and organelle structure and biogenesis, intracellular transport and processing of proteins, cellular recognition and cell adhesion, oncogene structure and function, action of growth factors.

North Dakota State University, The Graduate School, College of Agriculture, Food Systems, and Natural Resources and College of Science and Mathematics, Cellular and Molecular Biology Program, Fargo, ND 58105. Offers PhD. Offered in cooperation with eleven departments in the university. *Degree requirements:* For doctorate, thesis/dissertation. *Entrance requirements:* For doctorate, GRE. Additional exam requirements/recommendations for international students: Required—TOEFL (minimum score 525 paper-based). Electronic applications accepted. *Faculty research:* Plant and animal cell biology, gene regulation, molecular genetics, plant and animal virology.

North Dakota State University, The Graduate School, College of Science and Mathematics, Department of Biological Sciences, Fargo, ND 58105. Offers biological sciences (MS); botany (MS, PhD); cellular and molecular biology (PhD); environmental and conservation sciences (MS, PhD); genomics (MS, PhD); natural resource management (MS, PhD); zoology (MS, PhD). *Faculty:* 20. *Students:* 30 full-time (10 women), 5 part-time (1 woman); includes 1 minority (Asian American or Pacific Islander), 3 international. Average age 24. 14 applicants, 43% accepted. In 2005, 4 master's awarded. *Degree requirements:* For master's and doctorate, thesis/dissertation. *Entrance requirements:* For master's and doctorate, GRE General Test. Additional exam requirements/recommendations for international students: Required—TOEFL. *Application deadline:* For fall admission, 3/15 for domestic students; for spring admission, 10/30 priority date for domestic students. Applications are processed on a rolling basis. Application fee: $45 ($60 for international students). Electronic applications accepted. *Financial support:* In 2005–06, 3 fellowships with full tuition reimbursements (averaging $15,000 per year), 9 research assistantships with full tuition reimbursements (averaging $14,400 per year), 19 teaching assistantships with full tuition reimbursements (averaging $9,550 per year) were awarded; career-related internships or fieldwork, Federal Work-Study, institutionally sponsored loans, scholarships/grants, tuition waivers (full), and unspecified assistantships also available. Support available to part-time students. Financial award application deadline: 4/15; financial award applicants required to submit FAFSA. *Faculty research:* Comparative endocrinology, physiology, behavioral ecology, plant cell biology, aquatic biology. Total annual research expenditures: $675,000. *Unit head:* Dr. William J. Bleier, Chair, 701-231-7087, Fax: 701-231-7149, E-mail: william.bleier@ndsu.nodak.edu.

North Dakota State University, The Graduate School, College of Science and Mathematics, Department of Chemistry, Fargo, ND 58105. Offers biochemistry (MS, PhD); chemistry (MS, PhD). *Faculty:* 13 full-time (1 woman), 1 part-time/adjunct (0 women). *Students:* 26 full-time (6 women), 12 international. Average age 24. 33 applicants, 58% accepted, 10 enrolled. In 2005, 3 master's, 3 doctorates awarded. Terminal master's awarded for partial completion of doctoral program. *Degree requirements:* For master's and doctorate, thesis/dissertation. *Entrance requirements:* For master's and doctorate, GRE General Test. Additional exam requirements/recommendations for international students: Required—TOEFL (minimum score 600 paper-based; 247 computer-based), GRE Subject. *Application deadline:* For fall admission, 6/1 for domestic students. Applications are processed on a rolling basis. Application fee: $45 ($60 for international students). Electronic applications accepted. *Financial support:* In 2005–06, 2 fellowships with tuition reimbursements (averaging $19,000 per year), 11 research assistantships with tuition reimbursements (averaging $19,000 per year), 13 teaching assistantships with tuition reimbursements (averaging $19,000 per year) were awarded; Federal Work-Study, institutionally sponsored loans, and scholarships/grants also available. Financial award application deadline: 4/15. *Faculty research:* Analytical, syntheticorganic, inorganic, physical, and theoretical chemistry. Total annual research expenditures: $1.6 million. *Unit head:* Dr. John Hershberger, Chair, 701-231-8225, Fax: 701-231-8831, E-mail: john.hershberger@ndsu.nodak.edu. *Application contact:* Dr. Seth Rasmussen, Chair, Graduate Admissions, 701-231-8747, Fax: 701-231-8831, E-mail: seth.rasmussen@ndsu.nodak.edu.

Northwestern University, The Graduate School and Judd A. and Marjorie Weinberg College of Arts and Sciences, Interdepartmental Biological Sciences Program (IBiS), Department of Biochemistry, Molecular Biology, and Cell Biology, Evanston, IL 60208. Offers biochemistry (PhD); cell and molecular biology (PhD); molecular biophysics (PhD); structural biology (PhD). Department participates in the Interdepartmental Biological Sciences Program (IBiS). *Entrance requirements:* For doctorate, GRE General Test. Additional exam requirements/recommendations for international students: Required—TOEFL (minimum score 600 paper-based), TSE (minimum score 50). Electronic applications accepted. *Faculty research:* Protein structure, protein-DNA interactions, gene regulation, development of embryos, hormone action and signal transduction.

Northwestern University, Northwestern University Feinberg School of Medicine, Department of Cell and Molecular Biology, Chicago, IL 60611.

Northwestern University, Northwestern University Feinberg School of Medicine and Interdepartmental Degree Programs, Integrated Graduate Programs in the Life Sciences, Chicago, IL 60611. Offers cancer biology (PhD); cell biology (PhD); developmental biology

(PhD); evolutionary biology (PhD); immunology and microbial pathogenesis (PhD); molecular biology and genetics (PhD); neurobiology (PhD); pharmacology and toxicology (PhD); structural biology and biochemistry (PhD). *Degree requirements:* For doctorate, thesis/dissertation, written and oral qualifying exams, comprehensive exam. *Entrance requirements:* For doctorate, GRE General Test. Additional exam requirements/recommendations for international students: Required—TOEFL (minimum score 600 paper-based; 250 computer-based). Electronic applications accepted.

See Close-Up on page 189.

OGI School of Science & Engineering at Oregon Health & Science University, Graduate Studies, Department of Environmental and Biomolecular Systems, Program in Biochemistry and Molecular Biology, Beaverton, OR 97006-8921. Offers MS, PhD. *Degree requirements:* For doctorate, dissertation defense. *Entrance requirements:* For master's, GRE General Test, 3 letters of recommendation; for doctorate, GRE General Test, GRE Subject Test, 3 letters of recommendation. Application fee: $65. *Expenses:* Tuition, state resident: full-time $22,760; part-time $625 per credit. Required fees: $350. *Application contact:* Information Contact, 503-748-1070, E-mail: info@bmb.ogi.edu.

The Ohio State University, Graduate School, College of Biological Sciences, Department of Molecular Genetics, Columbus, OH 43210. Offers cell and developmental biology (MS, PhD); genetics (MS, PhD); molecular biology (MS, PhD). *Degree requirements:* For master's and doctorate, thesis/dissertation. *Entrance requirements:* For master's and doctorate, GRE General Test, GRE Subject Test. Additional exam requirements/recommendations for international students: Required—TOEFL (minimum score 573 paper-based; 230 computer-based). Electronic applications accepted.

See Close-Up on page 785.

The Ohio State University, Graduate School, College of Biological Sciences, Program in Molecular, Cellular and Developmental Biology, Columbus, OH 43210. Offers MS, PhD. Terminal master's awarded for partial completion of doctoral program. *Degree requirements:* For master's and doctorate, thesis/dissertation. *Entrance requirements:* For master's and doctorate, GRE General Test, GRE Subject Test. Additional exam requirements/recommendations for international students: Required—TOEFL (minimum score 600 paper-based; 250 computer-based). Electronic applications accepted. *Faculty research:* Cancer biology, cell biology, developmental biolgy, DNA replication, gene expression.

Ohio University, Graduate Studies, College of Arts and Sciences, Interdisciplinary Program in Molecular and Cellular Biology, Athens, OH 45701-2979. Offers MS, PhD. *Faculty:* 46 full-time (14 women). *Students:* 36 full-time (19 women), 26 international. Average age 27. 64 applicants, 39% accepted, 16 enrolled. In 2005, 1 master's, 4 doctorates awarded. *Median time to degree:* Of those who began their doctoral program in fall 1997, 100% received their degree in 8 years or less. *Degree requirements:* For master's and doctorate, thesis/dissertation, research proposal, teaching experience, comprehensive exam, registration. *Entrance requirements:* For master's and doctorate, GRE General Test. Additional exam requirements/recommendations for international students: Required—TOEFL (minimum score 620 paper-based; 260 computer-based); Recommended—TWE. *Application deadline:* For fall admission, 2/1 for domestic students, 2/1 for international students. Applications are processed on a rolling basis. Application fee: $45. *Financial support:* In 2005–06, 21 students received support, including research assistantships with full tuition reimbursements available (averaging $17,000 per year), 10 teaching assistantships with full tuition reimbursements available (averaging $17,000 per year); traineeships, health care benefits, tuition waivers (full and partial), and unspecified assistantships also available. Financial award application deadline: 2/1. *Faculty research:* Animal biotechnology, plant molecular biology, immunology, cellular genetics, biochemistry of signal transduction, cancer research, membrane transport. Total annual research expenditures: $30,000. *Unit head:* Dr. Martin Tuck, Chair, 740-593-1747, E-mail: tuck@ohio.edu. *Application contact:* Dr. Robert A. Colvin, Graduate Chair, 740-593-0198, Fax: 740-593-0300, E-mail: colvin@ohio.edu.

See Close-Up on page 581.

Oklahoma State University, College of Agricultural Science and Natural Resources, Department of Biochemistry and Molecular Biology, Stillwater, OK 74078. Offers MS, PhD. *Faculty:* 25 full-time (12 women). *Students:* 11 full-time (5 women), 24 part-time (12 women); includes 7 minority (2 African Americans, 3 American Indian/Alaska Native, 2 Asian Americans or Pacific Islanders), 23 international. Average age 28. 63 applicants, 25% accepted, 4 enrolled. In 2005, 2 master's, 6 doctorates awarded. *Degree requirements:* For master's, thesis, oral exam; for doctorate, one foreign language, thesis/dissertation. *Entrance requirements:* For master's and doctorate, GRE. Additional exam requirements/recommendations for international students: Required—TOEFL. *Application deadline:* For fall admission, 6/1 priority date for domestic students, 3/1 priority date for international students. Applications are processed on a rolling basis. Application fee: $40 ($75 for international students). Electronic applications accepted. *Expenses:* Tuition, state resident: full-time $4,253; part-time $139 per credit hour. Tuition, nonresident: full-time $12,569; part-time $485 per credit hour. Required fees: $43 per credit hour. One-time fee: $20 part-time. Tuition and fees vary according to course load and program. *Financial support:* In 2005–06, 32 research assistantships (averaging $17,690 per year) were awarded; teaching assistantships, career-related internships or fieldwork, Federal Work-Study, scholarships/grants, health care benefits, tuition waivers (partial), and unspecified assistantships also available. Support available to part-time students. Financial award application deadline: 3/1. *Unit head:* Dr. Earl Mitchell, Interim Head, 405-744-6189, Fax: 405-744-7799.

See Close-Up on page 405.

Oklahoma State University Center for Health Sciences, Program in Forensic Sciences, Tulsa, OK 74107-1898. Offers forensic DNA/molecular biology (MS); forensic pathology (MS); forensic psychology (MS); forensic sciences (MFSA); forensic toxicology (MS). Part-time and evening/weekend programs available. Postbaccalaureate distance learning degree programs offered (minimal on-campus study). *Faculty:* 2 full-time (0 women), 12 part-time/adjunct (4 women). *Students:* 6 full-time (4 women), 21 part-time (17 women); includes 4 minority (all African Americans), 1 international. Average age 36. 47 applicants, 26% accepted, 11 enrolled. In 2005, 7 degrees awarded. *Degree requirements:* For master's, thesis (for some programs). *Entrance requirements:* For master's, MAT (MFSA) or GRE General Test, professional experience (MFSA). Additional exam requirements/recommendations for international students: Required—TOEFL (minimum score 600 paper-based; 250 computer-based), TWE (minimum score 5). *Application deadline:* For fall admission, 3/31 for domestic students, 3/31 for international students. Application fee: $40 ($75 for international students). *Expenses:* Tuition, state resident: full-time $16,045. Tuition, nonresident: full-time $31,265. Tuition and fees vary according to program. *Financial support:* In 2005–06, 1 student received support. Federal Work-Study available. Support available to part-time students. *Faculty research:* DNA typing, DNA polymorphism, identification through DNA, disease transmission, forensic dentistry, neurotoxicity of HIV. Total annual research expenditures: $216,000. *Unit head:* Dr. Robert T. Allen, Director, 918-561-1108, Fax: 918-561-8414. *Application contact:* Cathy Newsome, Coordinator, 918-699-8608, Fax: 918-561-8414, E-mail: newsome@osu-med.com.

Oregon Health & Science University, School of Medicine, Graduate Programs in Medicine, Department of Biochemistry and Molecular Biology, Portland, OR 97239-3098. Offers PhD, MD/PhD. Part-time programs available. *Degree requirements:* For doctorate, thesis/dissertation. *Entrance requirements:* For doctorate, GRE General Test, MCAT (MD/PhD). *Faculty research:* Protein structure and function, enzymology, metabolism, membranes transport.

Oregon State University, Graduate School, Program in Molecular and Cellular Biology, Corvallis, OR 97331. Offers MS, PhD. *Students:* 31 full-time (12 women), 3 part-time (2 women); includes 5 minority (1 American Indian/Alaska Native, 4 Asian Americans or Pacific Islanders), 9 international. Average age 30. In 2005, 2 master's, 7 doctorates awarded.

Degree requirements: For doctorate, thesis/dissertation, oral and written qualifying exams. *Entrance requirements:* For doctorate, minimum GPA of 3.0 in last 90 hours. Additional exam requirements/recommendations for international students: Required—TOEFL. *Application deadline:* For fall admission, 2/15 for domestic students. Applications are processed on a rolling basis. Application fee: $50. *Expenses:* Tuition, area resident: Part-time $301 per credit. Tuition, state resident: full-time $8,139; part-time $501 per credit. Tuition, nonresident: full-time $14,376; part-time $532 per credit. Required fees: $1,266. *Financial support:* Fellowships, career-related internships or fieldwork, Federal Work-Study, and institutionally sponsored loans available. Support available to part-time students. *Unit head:* Dr. James C. Carrington, Interim Director, 541-737-3769.

The Pennsylvania State University Milton S. Hershey Medical Center, Graduate School Programs in the Biomedical Sciences, Graduate Program in Biochemistry and Molecular Biology, Hershey, PA 17033-2360. Offers MS, PhD, MD/PhD. *Students:* 20 full-time (13 women); includes 2 minority (both Asian Americans or Pacific Islanders), 7 international. Average age 27.Terminal master's awarded for partial completion of doctoral program. *Median time to degree:* Of those who began their doctoral program in fall 1997, 100% received their degree in 8 years or less. *Degree requirements:* For master's, thesis or alternative, registration; for doctorate, thesis/dissertation, oral exam, comprehensive exam, registration. *Entrance requirements:* For master's, GRE General Test; for doctorate, GRE General Test, minimum GPA of 3.0. Additional exam requirements/recommendations for international students: Required—TOEFL (minimum score 550 paper-based; 213 computer-based). *Application deadline:* For fall admission, 2/1 priority date for domestic students, 2/1 priority date for international students. Applications are processed on a rolling basis. Application fee: $45. Electronic applications accepted. *Financial support:* In 2005–06, 3 fellowships with full tuition reimbursements, 17 research assistantships with full tuition reimbursements were awarded; scholarships/grants, health care benefits, and unspecified assistantships also available. Financial award applicants required to submit FAFSA. *Faculty research:* X-ray crystallography of proteins, glycosphingolipid interactions with viruses and toxins, DNA replication and repair, tobacco and environmental carcinogenesis, generegulation. *Unit head:* Dr. Judith S. Bond, Chair, 717-531-8585, Fax: 717-531-7072, E-mail: bchem-grad-hmc@psu.edu. *Application contact:* Ruth Dean, Administrative Assistant, 717-531-8586, Fax: 717-531-7072, E-mail: bchem-grad-hmc@psu.edu.

The Pennsylvania State University Milton S. Hershey Medical Center, Graduate School Programs in the Biomedical Sciences, Graduate Program in Microbiology and Immunology, Hershey, PA 17033-2360. Offers genetics (PhD); immunology (MS, PhD); microbiology (MS); microbiology/virology (PhD); molecular biology (PhD). *Students:* 23 full-time (11 women); includes 2 minority (both Asian Americans or Pacific Islanders), 2 international. Average age 27.Terminal master's awarded for partial completion of doctoral program. *Median time to degree:* Of those who began their doctoral program in fall 1997, 100% received their degree in 8 years or less. *Degree requirements:* For master's, thesis or alternative, registration; for doctorate, thesis/dissertation, oral exam, comprehensive exam, registration. *Entrance requirements:* For master's, GRE or MCAT; for doctorate, GRE General Test or MCAT, minimum GPA of 3.0. Additional exam requirements/recommendations for international students: Required—TOEFL. *Application deadline:* Applications are processed on a rolling basis. Application fee: $45. Electronic applications accepted. *Financial support:* In 2005–06, 23 research assistantships with full tuition reimbursements were awarded; fellowships with full tuition reimbursements, scholarships/grants, health care benefits, and unspecified assistantships also available. Financial award applicants required to submit FAFSA. *Faculty research:* Virus replication and assembly, oncogenesis, interactions of viruses with host cells and animal model systems. *Unit head:* Dr. Richard J. Courtney, Chair, 717-531-7659, Fax: 717-531-6522, E-mail: micro-grad-hmc@psu.edu. *Application contact:* Billie Burns, Secretary, 717-531-7659, Fax: 717-531-6522, E-mail: micro-grad-hmc@psu.edu.

The Pennsylvania State University Milton S. Hershey Medical Center, Graduate School Programs in the Biomedical Sciences, Intercollege Graduate Program in Cell and Molecular Biology, Hershey, PA 17033-2360. Offers MS, PhD, MD/PhD. *Students:* 23 full-time (12 women); includes 1 minority (Asian American or Pacific Islander), 4 international. Average age 27.Terminal master's awarded for partial completion of doctoral program. *Median time to degree:* Of those who began their doctoral program in fall 1997, 100% received their degree in 8 years or less. *Degree requirements:* For master's, thesis or alternative, registration; for doctorate, thesis/dissertation, oral exam, comprehensive exam, registration. *Entrance requirements:* For master's, GRE General Test or MCAT; for doctorate, GRE General Test or MCAT, minimum GPA of 3.0. Additional exam requirements/recommendations for international students: Required—TOEFL (minimum score 500 paper-based; 213 computer-based). Application fee: $45. *Financial support:* In 2005–06, 2 fellowships with full tuition reimbursements, 21 research assistantships with full tuition reimbursements were awarded; scholarships/grants, traineeships, health care benefits, and unspecified assistantships also available. Financial award applicants required to submit FAFSA. *Faculty research:* Membrane structure, function and modulators; cell division, differentiation and gene expression; Metastasis; intracellular events in the immune system. *Unit head:* Dr. Henry Donahue, Program Director, 717-531-1045, Fax: 717-531-4139, E-mail: cmb-grad-hmc@psu.edu. *Application contact:* Sue Myers, Administration Assistant, 717-531-1045, Fax: 717-531-4139, E-mail: cmb-grad-hmc@psu.edu.

The Pennsylvania State University University Park Campus, Graduate School, Eberly College of Science, Department of Biochemistry and Molecular Biology, Program in Biochemistry, Microbiology, and Molecular Biology, State College, University Park, PA 16802-1503. Offers MS, PhD. *Degree requirements:* For master's, thesis/dissertation; for doctorate, thesis/dissertation, comprehensive exam. *Entrance requirements:* For master's and doctorate, GRE General Test, GRE Subject Test. Application fee: $45. *Expenses:* Tuition, state resident: full-time $12,518; part-time $522 per credit. Tuition, nonresident: full-time $23,004; part-time $959 per credit. Required fees: $484. Tuition and fees vary according to course load, campus/location and program. *Financial support:* Fellowships, unspecified assistantships available.

See Close-Up on page 407.

Princeton University, Graduate School, Department of Molecular Biology, Princeton, NJ 08544-1019. Offers PhD. *Degree requirements:* For doctorate, thesis/dissertation. *Entrance requirements:* For doctorate, GRE General Test. Additional exam requirements/recommendations for international students: Required—TOEFL (minimum score 600 paper-based; 250 computer-based). Electronic applications accepted. *Faculty research:* Genetics, virology, biochemistry.

See Close-Up on page 585.

Purdue University, College of Pharmacy and Pharmacal Sciences and Graduate School, Graduate Programs in Pharmacy and Pharmacal Sciences, Department of Medicinal Chemistry and Molecular Pharmacology, West Lafayette, IN 47907. Offers analytical medicinal chemistry (PhD); computational and biophysical medicinal chemistry (PhD); medicinal and bioorganic chemistry (PhD); medicinal biochemistry and molecular biology (PhD); molecular pharmacology and toxicology (PhD); natural products and pharmacognosy (PhD); nuclear pharmacy (MS); radiopharmaceutical chemistry and nuclear pharmacy (PhD). *Faculty:* 21 full-time (2 women), 2 part-time/adjunct (0 women). *Students:* 69 full-time (39 women), 2 part-time (1 woman); includes 8 minority (2 African Americans, 2 Asian Americans or Pacific Islanders, 4 Hispanic Americans), 21 international. Average age 26. 125 applicants, 25% accepted, 15 enrolled. In 2005, 2 master's, 11 doctorates awarded. Terminal master's awarded for partial completion of doctoral program. *Degree requirements:* For master's and doctorate, thesis/dissertation. *Entrance requirements:* For master's, GRE General Test, minimum B average; BS in biology, chemistry, or pharmacy; for doctorate, GRE General Test, minimum B average; BS in biology, chemistry, or pharmacology. Additional exam requirements/recommendations for international students: Required—TOEFL. *Application deadline:* Applications are processed on a rolling basis. Application fee: $55. Electronic applications accepted. *Financial support:* Fellowships, research assistantships, teaching assistantships, traineeships available. Support available to part-time students. Financial award applicants required to submit FAFSA. *Faculty research:* Drug design and development, cancer research, drug synthesis and analysis, chemical pharmacology, environmental toxicology. *Unit head:* Dr. R. F. Borch, Graduate Head, 765-494-1403. *Application contact:* Dr. D. E. Bergstrom, Graduate Committee, 765-494-6275, E-mail: bergstrom@pharmacy.purdue.edu.

Purdue University, Graduate School, School of Science, Department of Biological Sciences, West Lafayette, IN 47907. Offers biochemistry (PhD); biophysics (PhD); cell and developmental biology (PhD); ecology, evolutionary and population biology (MS, PhD), including ecology, evolutionary biology, population biology; genetics (MS, PhD); microbiology (MS, PhD); molecular biology (PhD); neurobiology (MS, PhD); plant physiology (PhD). *Faculty:* 47 full-time (9 women), 4 part-time/adjunct (1 woman). *Students:* 97 full-time (53 women), 8 part-time (4 women); includes 13 minority (3 African Americans, 1 American Indian/Alaska Native, 4 Asian Americans or Pacific Islanders, 5 Hispanic Americans), 50 international. Average age 28. 168 applicants, 29% accepted, 23 enrolled. In 2005, 18 master's, 9 doctorates awarded. Terminal master's awarded for partial completion of doctoral program. *Degree requirements:* For master's, thesis (for some programs); for doctorate, thesis/dissertation, seminars, teaching experience. *Entrance requirements:* For master's and doctorate, GRE General Test. Additional exam requirements/recommendations for international students: Required—TOEFL, TSE. *Application deadline:* For fall admission, 2/15 for domestic students, 1/31 for international students. Applications are processed on a rolling basis. Application fee: $55. Electronic applications accepted. *Financial support:* In 2005–06, 15 fellowships, 60 research assistantships, 53 teaching assistantships were awarded. Support available to part-time students. Financial award application deadline: 2/15; financial award applicants required to submit FAFSA. *Unit head:* Dr. Richard J Kuhn, Head, 765-494-4407. *Application contact:* Nancy Konopka, Graduate Studies Office Manager, 765-494-8142, Fax: 765-494-0876, E-mail: njk@bilbo.bio.purdue.edu.

Purdue University, School of Veterinary Medicine and Graduate School, Graduate Programs in Veterinary Medicine, Department of Veterinary Pathobiology, West Lafayette, IN 47907. Offers biochemistry and molecular biology (MS, PhD); comparative epidemiology (MS, PhD); epidemiology (MS, PhD); immunology (MS, PhD); infectious diseases (MS, PhD); interdisciplinary genetics (PhD); laboratory animal medicine (MS, PhD); microbiology (MS, PhD); molecular virology (MS, PhD); parasitology (MS, PhD); pathobiology (MS, PhD); public health epidemiology (MS, PhD); toxicology (MS, PhD); veterinary anatomic pathology (MS, PhD); veterinary clinical pathology (MS, PhD); virology (MS, PhD). *Faculty:* 32 full-time (7 women). *Students:* 49 full-time (20 women), 3 part-time (1 woman); includes 2 minority (both African Americans), 31 international. Average age 35. In 2005, 3 master's, 8 doctorates awarded. Terminal master's awarded for partial completion of doctoral program. *Degree requirements:* For master's, thesis (for some programs); for doctorate, thesis/dissertation. *Entrance requirements:* For master's and doctorate, GRE General Test. Additional exam requirements/recommendations for international students: Required—TOEFL (minimum score 575 paper-based), TWE (minimum score 4). *Application deadline:* For fall admission, 8/12 for domestic students, 6/15 for international students; for spring admission, 1/12 for domestic students, 10/15 for international students. Application fee: $55. *Financial support:* Fellowships, research assistantships, teaching assistantships available. Financial award application deadline: 3/1; financial award applicants required to submit FAFSA. *Unit head:* Dr. H. Hogenesch, Head, 765-494-7543.

Quinnipiac University, School of Health Sciences, Program in Molecular and Cell Biology, Hamden, CT 06518-1940. Offers MS. Part-time and evening/weekend programs available. *Faculty:* 8 full-time (3 women), 8 part-time/adjunct (2 women). *Students:* 3 full-time (1 woman), 25 part-time (19 women); includes 3 minority (1 African American, 1 Asian American or Pacific Islander, 1 Hispanic American), 2 international. Average age 26. 14 applicants, 57% accepted, 3 enrolled. In 2005, 9 degrees awarded. *Degree requirements:* For master's, thesis optional. *Entrance requirements:* For master's, bachelor's degree in biological, medical, or health sciences; minimum GPA of 2.5. Additional exam requirements/recommendations for international students: Required—TOEFL (minimum score 575 paper-based; 233 computer-based). *Application deadline:* For fall admission, 7/30 priority date for domestic students, 5/30 priority date for international students; for spring admission, 12/15 priority date for domestic students, 10/15 priority date for international students. Applications are processed on a rolling basis. Application fee: $45. Electronic applications accepted. *Expenses:* Tuition: Part-time $570 per credit. *Financial support:* Unspecified assistantships available. Support available to part-time students. Financial award application deadline: 4/15; financial award applicants required to submit FAFSA. *Unit head:* Dr. Charlotte Hammond, Director, 203-582-8058, E-mail: charlotte.hammond@quinnipiac.edu. *Application contact:* Louise Howe, Associate Director of Graduate Admissions, 800-462-1944, Fax: 203-582-3443, E-mail: graduate@quinnipiac.edu.

See Close-Up on page 587.

Rensselaer Polytechnic Institute, Graduate School, School of Science, Department of Biology, Troy, NY 12180-3590. Offers biochemistry (MS, PhD); biophysics (MS, PhD); cell biology (MS, PhD); developmental biology (MS, PhD); microbiology (MS, PhD); molecular biology (MS, PhD). Part-time programs available. Terminal master's awarded for partial completion of doctoral program. *Degree requirements:* For master's and doctorate, thesis/dissertation, comprehensive exam, registration. *Entrance requirements:* For master's and doctorate, GRE General Test. Additional exam requirements/recommendations for international students: Required—TOEFL. Electronic applications accepted. *Expenses:* Tuition: Full-time $31,000; part-time $1,320 per credit. Required fees: $1,623. *Faculty research:* Bioinformatics, molecular biology/biochemistry, cell and tissue biology, environment, ecology.

See Close-Up on page 199.

Rutgers, The State University of New Jersey, New Brunswick/Piscataway, Graduate School, Core Curriculum in Molecular and Cell Biology, New Brunswick, NJ 08901-1281. Offers PhD. *Degree requirements:* For doctorate, thesis/dissertation. *Entrance requirements:* For doctorate, GRE General Test, GRE Subject Test. Additional exam requirements/recommendations for international students: Required—TOEFL. *Expenses:* Tuition, state resident: full-time $10,440; part-time $435 per credit. Tuition, nonresident: full-time $15,520; part-time $647 per credit. Required fees: $129 per credit. Tuition and fees vary according to program. *Financial support:* Fellowships available. *Unit head:* Michael Leibowitz, Program Director, 732-235-4795.

Rutgers, The State University of New Jersey, New Brunswick/Piscataway, Graduate School, Program in Biochemistry, New Brunswick, NJ 08901-1281. Offers biochemistry (MS, PhD); molecular biology (MS, PhD). *Faculty:* 105 full-time. *Students:* 30 full-time (11 women), 1 (woman) part-time; includes 1 minority (Asian American or Pacific Islander), 20 international. Average age 25. 74 applicants, 34% accepted, 14 enrolled. In 2005, 5 doctorates awarded. Terminal master's awarded for partial completion of doctoral program. *Median time to degree:* Of those who began their doctoral program in fall 1997, 86% received their degree in 8 years or less. *Degree requirements:* For master's, thesis, qualifying exam; for doctorate, thesis/dissertation, written qualifying exam. *Entrance requirements:* For master's, GRE General Test, GRE Subject Test; for doctorate, GRE General Test, GRE Subject Test (recommended), minimum GPA of 3.0. Additional exam requirements/recommendations for international students: Required—TOEFL. *Application deadline:* For fall admission, 1/5 for domestic students. Applications are processed on a rolling basis. Application fee: $50. *Expenses:* Tuition, state resident: full-time $10,440; part-time $435 per credit. Tuition, nonresident: full-time $15,520; part-time $647 per credit. Required fees: $129 per credit. Tuition and fees vary according to program. *Financial support:* In 2005–06, 30 students received support, including 11 fellowships with full tuition reimbursements available (averaging $24,000 per year), 14 research assistantships with full tuition reimbursements available (averaging $24,000 per year), 5 teaching assistantships with full tuition reimbursements available (averaging $16,988 per year); Federal Work-Study also available. Support available to part-time students. Financial award application deadline: 1/15; financial award applicants required to submit FAFSA. *Faculty research:* DNA replication and transcription, virus gene expression, tumor biology, structural biochemistry, signal transduction and molecular targeting. *Unit head:* Dr. Abram Gabriel, Acting Director, 732-235-5097, Fax: 732-445-6370, E-mail: gabriel@cabm-rutgers.edu. *Application contact:*

Molecular Biology

Rutgers, The State University of New Jersey, New Brunswick/Piscataway *(continued)*
Carolyn J. Ambrose, Administrative Assistant, 732-445-3430, Fax: 732-445-6370, E-mail: ambrose@biology.rutgers.edu.

Rutgers, The State University of New Jersey, New Brunswick/Piscataway, Graduate School, Program in Microbiology and Molecular Genetics, New Brunswick, NJ 08901-1281. Offers applied microbiology (MS, PhD); clinical microbiology (MS, PhD); computational molecular biology (PhD); immunology (MS, PhD); microbial biochemistry (MS, PhD); molecular genetics (MS, PhD); virology (MS, PhD). Part-time programs available. *Faculty:* 132 full-time. *Students:* 57 full-time (31 women), 18 part-time (8 women); includes 16 minority (9 Asian Americans or Pacific Islanders, 7 Hispanic Americans), 20 international. Average age 29. 104 applicants, 32% accepted, 19 enrolled. In 2005, 9 master's, 10 doctorates awarded. Terminal master's awarded for partial completion of doctoral program. *Median time to degree:* Of those who began their doctoral program in fall 1997, 100% received their degree in 8 years or less. *Degree requirements:* For master's, thesis or alternative, comprehensive exam, registration; for doctorate, thesis/dissertation, written qualifying exam, comprehensive exam, registration. *Entrance requirements:* For master's, GRE General Test, minimum GPA of 3.0; for doctorate, GRE General Test, GRE Subject Test (recommended), minimum GPA of 3.0. Additional exam requirements/recommendations for international students: Required—TOEFL. *Application deadline:* For fall admission, 1/5 priority date for domestic students, 11/1 priority date for international students. Applications are processed on a rolling basis. Application fee: $50. Electronic applications accepted. *Expenses:* Tuition, state resident: full-time $10,440; part-time $435 per credit. Tuition, nonresident: full-time $15,520; part-time $647 per credit. Required fees: $129 per credit. Tuition and fees vary according to program. *Financial support:* In 2005–06, 48 students received support, including 14 fellowships with full tuition reimbursements available (averaging $24,000 per year), 25 research assistantships with full tuition reimbursements available (averaging $24,000 per year), 9 teaching assistantships with full tuition reimbursements available (averaging $16,988 per year); Federal Work-Study, institutionally sponsored loans, scholarships/grants, and unspecified assistantships also available. Financial award application deadline: 1/5; financial award applicants required to submit FAFSA. *Faculty research:* Molecular genetics and microbial physiology; virology and pathogenic microbiology; applied, environmental and industrial microbiology; computers in molecular biology. *Unit head:* Dr. Andrew K. Vershon, Director, 732-445-2905, Fax: 732-445-6370, E-mail: vershon@waksman.rutgers.edu. *Application contact:* Diane Murano, Administrative Assistant, 732-445-5086, Fax: 732-445-6370, E-mail: murano@biology.rutgers.edu.

Rutgers, The State University of New Jersey, New Brunswick/Piscataway, Graduate School, Programs in the Molecular Biosciences, New Brunswick, NJ 08901-1281.

See Close-Up on page 589.

Saint Joseph College, Graduate Division, Department of Biology, West Hartford, CT 06117-2700. Offers biology (MS), including general biology, molecular and cellular biology; biology/chemistry (MS). MS biology (including general biology; molecular and cellular biology) offered online only. Part-time and evening/weekend programs available. Postbaccalaureate distance learning degree programs offered (no on-campus study). *Faculty:* 4 full-time (2 women), 5 part-time/adjunct (3 women). *Students:* Average age 34. 34 applicants, 100% accepted, 11 enrolled. In 2005, 13 degrees awarded. *Degree requirements:* For master's, thesis or alternative, comprehensive exam. *Entrance requirements:* For master's, 2 letters of recommendation. *Application deadline:* Applications are processed on a rolling basis. Application fee: $50. Electronic applications accepted. *Expenses:* Tuition: Part-time $540 per credit. Required fees: $25 per credit. *Financial support:* Career-related internships or fieldwork, health care benefits, and unspecified assistantships available. Support available to part-time students. Financial award application deadline: 7/15; financial award applicants required to submit FAFSA. *Faculty research:* Neurology, cardiology, immunology, mircobiology. *Unit head:* , Dr. Charles Morgan, Chair, 860-231-5335, E-mail: cmorgan@sjc.edu.

Saint Louis University, Graduate School and School of Medicine, Graduate Program in Biomedical Sciences and Graduate School, Department of Biochemistry and Molecular Biology, St. Louis, MO 63103-2097. Offers PhD. *Faculty:* 34 full-time (16 women). *Students:* 18 full-time (11 women); includes 1 minority (Asian American or Pacific Islander), 4 international. Average age 26. 5 applicants, 100% accepted, 3 enrolled. In 2005, 4 degrees awarded. *Degree requirements:* For doctorate, thesis/dissertation, departmental qualifying exams, comprehensive exam. *Entrance requirements:* For doctorate, GRE General Test, letters of recommendation, resumé. Additional exam requirements/recommendations for international students: Required—TOEFL (minimum score 550 paper-based; 213 computer-based). *Application deadline:* For fall admission, 7/1 for domestic students, 7/1 for international students; for spring admission, 11/1 for domestic students, 11/1 for international students. Applications are processed on a rolling basis. Application fee: $40. *Expenses:* Tuition: Part-time $760 per credit hour. Required fees: $55 per semester. *Financial support:* In 2005–06, 10 students received support, including 2 research assistantships (averaging $16,400 per year); fellowships with tuition reimbursements available, teaching assistantships, health care benefits and unspecified assistantships available. Support available to part-time students. Financial award application deadline: 6/1; financial award applicants required to submit FAFSA. *Faculty research:* Transcription, chromatin modification and regulation of gene expression; structure/function of proteins and enzymes, including x-ray crystallography; inflammatory mediators in pathogenesis of diabetes and arteriosclerosis; cellular signaling in response to growth factors, opiates and angiogenic mediators; genomics and proteomics of Cryptococcus neoformans. *Unit head:* Dr. William S. Sly, Chairperson, 314-977-9201, Fax: 314-577-8156, E-mail: slyws@slu.edu. *Application contact:* Gary Behrman, Associate Dean of the Graduate School, 314-977-3827, E-mail: behrmang@slu.edu.

San Diego State University, Graduate and Research Affairs, College of Sciences, Department of Biological Sciences, San Diego, CA 92182. Offers biology (MA, MS), including ecology (MS), molecular biology (MS), physiology (MS), systematics/evolution (MS); biostatistics and biometry (PhD); cell and molecular biology (PhD); ecology (MS, PhD); microbiology (MS). *Students:* 73 full-time (36 women), 100 part-time (53 women); includes 33 minority (2 African Americans, 16 Asian Americans or Pacific Islanders, 15 Hispanic Americans), 33 international. Average age 26. 228 applicants, 23% accepted, 32 enrolled. In 2005, 27 master's, 9 doctorates awarded. Terminal master's awarded for partial completion of doctoral program. *Degree requirements:* For master's and doctorate, thesis/dissertation. *Entrance requirements:* For master's, GRE General Test, GRE Subject Test, resumé or curriculum vitae, 2 letters of recommendation. Additional exam requirements/recommendations for international students: Required—TOEFL. *Application deadline:* For fall admission, 5/1 for domestic students; for spring admission, 11/1 for domestic students, 10/1 for international students. Applications are processed on a rolling basis. Application fee: $55. Electronic applications accepted. *Financial support:* In 2005–06, 116 teaching assistantships were awarded; fellowships, research assistantships, career-related internships or fieldwork and unspecified assistantships also available. Financial award applicants required to submit FAFSA. Total annual research expenditures: $9.9 million. *Unit head:* Christopher Glembotski, Chair, 619-594-6767, Fax: 619-594-5676. *Application contact:* Terry Frey, Graduate Coordinator, 619-594-6756, Fax: 619-594-5676, E-mail: gradcoor@sciences.sdsu.edu.

San Diego State University, Graduate and Research Affairs, College of Sciences, Molecular Biology Institute, Program in Cell and Molecular Biology, San Diego, CA 92182. Offers PhD. *Students:* 14 full-time (10 women), 29 part-time (14 women); includes 11 minority (7 Asian Americans or Pacific Islanders, 4 Hispanic Americans), 11 international. 45 applicants, 60% accepted, 10 enrolled. *Degree requirements:* For doctorate, thesis/dissertation, oral comprehensive qualifying exam. *Entrance requirements:* For doctorate, GRE General Test, GRE Subject Test, resumé or curriculum vitae, 3 letters of recommendation. *Application deadline:* For fall admission, 5/1 for domestic students; for spring admission, 11/1 for domestic students, 10/1 for international students. Applications are processed on a rolling basis. Application fee: $55.

Electronic applications accepted. *Financial support:* In 2005–06, 5 fellowships were awarded; institutionally sponsored loans and unspecified assistantships also available. Financial award applicants required to submit CSS PROFILE or FAFSA. *Faculty research:* Structure/dynamics of protein kinesis, chromatin structure and DNA methylation membrane biochemistry, secretory protein targeting, molecular biology of cardiac myocytes. *Unit head:* Dr. Sanford Bernstein, Director, 619-594-5504, Fax: 619-594-5676, E-mail: sbernstein@sunstroke.sdsu.edu. *Application contact:* Dr. Sanford Bernstein, Graduate Advisor, 619-594-5504, Fax: 619-594-5676, E-mail: sbernstein@sunstroke.sdsu.edu.

See Close-Up on page 591.

San Francisco State University, Division of Graduate Studies, College of Science and Engineering, Department of Biology, Program in Cell and Molecular Biology, San Francisco, CA 94132-1722. Offers MA. *Entrance requirements:* For master's, minimum GPA of 2.5 in last 60 units.

San Jose State University, Graduate Studies and Research, College of Science, Department of Biological Sciences, San Jose, CA 95192-0001. Offers biological sciences (MA, MS); molecular biology and microbiology (MS); organismal biology, conservation and ecology (MS); physiology (MS). Part-time programs available. *Students:* 44 full-time (35 women), 33 part-time (24 women); includes 39 minority (33 Asian Americans or Pacific Islanders, 6 Hispanic Americans), 12 international. Average age 31. 140 applicants, 40% accepted, 29 enrolled. In 2005, 39 degrees awarded. *Entrance requirements:* For master's, GRE. *Application deadline:* For fall admission, 6/29 for domestic students; for spring admission, 11/30 for domestic students. Applications are processed on a rolling basis. Application fee: $59. Electronic applications accepted. *Expenses:* Tuition, nonresident: part-time $339 per unit. Required fees: $1,286 per semester. Tuition and fees vary according to course load and degree level. *Financial support:* In 2005–06, 13 teaching assistantships were awarded; Federal Work-Study also available. Financial award applicants required to submit FAFSA. *Faculty research:* Systemic physiology, molecular genetics, SEM studies, toxicology, large mammal ecology. *Unit head:* Dr. Sally Veregge, Chair, 408-924-4900, Fax: 408-924-4840. *Application contact:* Dr. Howard Shellhammer, Graduate Coordinator, 408-924-4897.

Seton Hall University, College of Arts and Sciences, Department of Biological Sciences, South Orange, NJ 07079-2697. Offers biology (MS); microbiology (MS); molecular bioscience (PhD). Part-time and evening/weekend programs available. *Students:* 17 full-time (14 women), 35 part-time (18 women). Average age 29. 33 applicants, 70% accepted, 14 enrolled. In 2005, 8 degrees awarded. *Degree requirements:* For master's, research paper or thesis, seminar. *Application deadline:* For fall admission, 7/1 priority date for domestic students, 7/1 priority date for international students; for spring admission, 11/1 priority date for domestic students, 11/1 priority date for international students. Applications are processed on a rolling basis. Application fee: $50. Electronic applications accepted. *Financial support:* Teaching assistantships, career-related internships or fieldwork and Federal Work-Study available. *Faculty research:* Neurobiology, genetics, immunology, molecular biology, cellular physiology, toxicology, microbiology, bioinformatics. *Unit head:* Dr. Carolyn Bentivegna, Chair, 973-761-9044, Fax: 973-761-9596, E-mail: bentivca@shu.edu. *Application contact:* Dr. Carroll D. Rawn, Director of Graduate Studies, 973-761-9054, Fax: 973-761-9596, E-mail: rawncarr@shu.edu.

See Close-Up on page 593.

Simon Fraser University, Graduate Studies, Department of Molecular Biology and Biochemistry, Burnaby, BC V5A 1S6, Canada. Offers M Sc, PhD. *Degree requirements:* For master's and doctorate, thesis/dissertation. *Entrance requirements:* For master's, minimum GPA of 3.0; for doctorate, minimum GPA of 3.5. Additional exam requirements/recommendations for international students: Required—TWE or IELTS. *Faculty research:* Molecular genetics and development, biochemistry, molecular physiology, genomics, molecular phylogenetics and population genetics, bioinformation.

Southern Illinois University Carbondale, Graduate School, College of Science, Program in Molecular Biology, Microbiology, and Biochemistry, Carbondale, IL 62901-4701. Offers MS, PhD. *Faculty:* 16 full-time (2 women). *Students:* 39 full-time (23 women), 37 part-time (19 women); includes 11 minority (9 African Americans, 2 Asian Americans or Pacific Islanders), 40 international. Average age 25. 134 applicants, 16% accepted, 17 enrolled. In 2005, 7 master's, 4 doctorates awarded. *Degree requirements:* For master's and doctorate, thesis/dissertation. *Entrance requirements:* For master's, GRE, minimum GPA of 2.7; for doctorate, GRE, minimum GPA of 3.25. Additional exam requirements/recommendations for international students: Required—TOEFL. *Application deadline:* Applications are processed on a rolling basis. Application fee: $20. *Financial support:* In 2005–06, 40 students received support, including 3 fellowships with full tuition reimbursements available, 24 research assistantships with full tuition reimbursements available, 12 teaching assistantships with full tuition reimbursements available; Federal Work-Study and institutionally sponsored loans also available. Support available to part-time students. Financial award application deadline: 3/1. *Faculty research:* Prokaryotic gene regulation and expression; eukaryotic gene regulation; microbial, phylogenetic, and metabolic diversity; immune responses to tumors, pathogens, and autoantigens; protein folding and structure. *Unit head:* Dr. John Martinko, Director, 618-453-8116, Fax: 618-453-8036, E-mail: martinko.mbmb@science.siu.edu. *Application contact:* Donna Mueller, Office Systems Specialist II, 618-536-2349, Fax: 618-453-8036, E-mail: mueller@micro.siu.edu.

Announcement: The Department of Medical Microbiology and Immunology in Springfield will be moving into new state-of-the-art office and laboratory space in the Springfield Combined Laboratory Facility II. The building is currently under construction and will be ready for occupancy in 2006. It will add approximately 21,000 square feet of new laboratory and office space to the department. The facilities will include enhanced BL-3-level laboratories and will be adjacent to new facilities for electron microscopy, image analysis, flow cytometry, and confocal microscopy.

See Close-Up on page 595.

State University of New York at Buffalo, Graduate School, Graduate Programs in Cancer Research and Biomedical Sciences at Roswell Park Cancer Institute, Department of Cellular and Molecular Biology at Roswell Park Cancer Institute, Buffalo, NY 14263. Offers PhD. *Faculty:* 30 part-time/adjunct (5 women). *Students:* 18 full-time (5 women), 5 part-time (3 women); includes 1 minority (Asian American or Pacific Islander), 9 international. Average age 25. 40 applicants, 33% accepted, 3 enrolled. In 2005, 5 degrees awarded. *Degree requirements:* For doctorate, thesis/dissertation, exam project. *Entrance requirements:* For doctorate, GRE General Test, minimum B average in undergraduate coursework. Additional exam requirements/recommendations for international students: Required—TOEFL, TWE, TSE. *Application deadline:* For fall admission, 2/1 for domestic students. Applications are processed on a rolling basis. Application fee: $35. Electronic applications accepted. *Financial support:* In 2005–06, 4 fellowships with full tuition reimbursements (averaging $21,000 per year), 17 research assistantships with full tuition reimbursements (averaging $21,000 per year) were awarded; unspecified assistantships also available. Financial award application deadline: 2/1; financial award applicants required to submit FAFSA. *Faculty research:* Cancer genetics, chromatin structure and replication, regulation of transcription, human gene mapping, genetic and structural approaches to regulation of gene expression. Total annual research expenditures: $4.5 million. *Unit head:* Dr. John Yates, Chair, 716-845-8964, Fax: 716-845-8449, E-mail: john.yates@roswellpark.org. *Application contact:* Craig R. Johnson, Director of Admissions, 716-845-2339, Fax: 716-845-8178, E-mail: craig.johnson@roswellpark.edu.

State University of New York Downstate Medical Center, School of Graduate Studies, Program in Molecular and Cellular Biology, Brooklyn, NY 11203-2098. Offers PhD, MD/PhD. Affiliation with a particular PhD degree-granting program is deferred to the second year. *Faculty:* 50 full-time (10 women). *Students:* 43 full-time (20 women); includes 4 Asian Americans or Pacific Islanders, 26 international. Average age 30. 42 applicants, 26% accepted, 5 enrolled. In 2005, 4 degrees awarded. *Degree requirements:* For doctorate, thesis/dissertation, comprehensive exam, registration. *Entrance requirements:* For doctorate, GRE General Test.

Application deadline: For fall admission, 4/1 for domestic students. Applications are processed on a rolling basis. Application fee: $35. *Financial support:* In 2005–06, 43 students received support, including 43 teaching assistantships with tuition reimbursements available (averaging $23,500 per year); fellowships, research assistantships, career-related internships or fieldwork, Federal Work-Study, health care benefits, and tuition waivers (full) also available. *Faculty research:* Mechanism of gene regulation, molecular virology. *Unit head:* Dr. William Cuirico, Chair, 718-270-3755, Fax: 718-270-1308, E-mail: wcuirico@downstate.edu. *Application contact:* Denise Sheares, Admissions Officer, 718-270-2738, Fax: 718-270-3378, E-mail: dsheares@downstate.edu.

State University of New York Upstate Medical University, College of Graduate Studies, Department of Biochemistry and Molecular Biology, Syracuse, NY 13210-2334. Offers MS, PhD, MD/PhD. Terminal master's awarded for partial completion of doctoral program. *Degree requirements:* For master's, thesis/dissertation; for doctorate, thesis/dissertation, comprehensive exam. *Entrance requirements:* For master's and doctorate, GRE General Test, GRE Subject Test, interview. Additional exam requirements/recommendations for international students: Required—TOEFL. *Faculty research:* Enzymology, membrane structure and functions, developmental biochemistry.

State University of New York Upstate Medical University, College of Graduate Studies, Program in Cell and Molecular Biology, Syracuse, NY 13210-2334. Offers PhD. *Degree requirements:* For doctorate, thesis/dissertation, comprehensive exam. *Entrance requirements:* For doctorate, GRE General Test, interview. Additional exam requirements/recommendations for international students: Required—TOEFL.

Stony Brook University, State University of New York, Graduate School, College of Arts and Sciences, Department of Biochemistry and Cell Biology, Molecular and Cellular Biology Program, Stony Brook, NY 11794. Offers biochemistry and molecular biology (PhD); biological sciences (MA); cellular and developmental biology (PhD); immunology and pathology (PhD); molecular and cellular biology (PhD). *Students:* 116 full-time (69 women), 2 part-time (1 woman); includes 16 minority (3 African Americans, 10 Asian Americans or Pacific Islanders, 3 Hispanic Americans), 62 international. Average age 30. 342 applicants, 14% accepted. In 2005, 7 degrees awarded. *Degree requirements:* For doctorate, thesis/dissertation, teaching experience, comprehensive exam. *Entrance requirements:* For doctorate, GRE General Test, GRE Subject Test. Additional exam requirements/recommendations for international students: Required—TOEFL. *Application deadline:* For fall admission, 1/15 for domestic students. Application fee: $50. *Expenses:* Tuition, state resident: full-time $6,900; part-time $288 per credit. Tuition, nonresident: full-time $10,920; part-time $455 per credit. Required fees: $704. *Financial support:* Fellowships, research assistantships, teaching assistantships, Federal Work-Study available. *Application contact:* Information Contact, 631-632-8533, Fax: 631-632-9730.

Announcement: The Molecular and Cellular Biology Graduate Program at Stony Brook University offers students the freedom to select an academic course of study in a wide range of areas within molecular and cellular biology. Students can perform their PhD thesis research with one of nearly 100 outstanding research faculty members.

See Close-Up on page 597.

Temple University, Health Sciences Center, School of Medicine and Graduate School, Graduate Programs in Medicine, Program in Molecular Biology and Genetics, Philadelphia, PA 19122-6096. Offers PhD, MD/PhD. *Faculty:* 7 full-time (2 women). *Students:* 15 full-time (11 women), 29 part-time (12 women). In 2005, 3 degrees awarded. *Degree requirements:* For doctorate, thesis/dissertation, presentation research/literature seminars distinct from area of concentration. *Entrance requirements:* For doctorate, GRE General Test, GRE Subject Test, minimum GPA of 3.0. Additional exam requirements/recommendations for international students: Required—TOEFL (minimum score 620 paper-based; 260 computer-based). *Application deadline:* For fall admission, 1/15 for domestic students, 12/15 for international students. Application fee: $50. Electronic applications accepted. *Expenses:* Tuition, state resident: full-time $8,694; part-time $483 per credit. Tuition, nonresident: full-time $12,672; part-time $704 per credit. Required fees: $500; $122 per semester. Tuition and fees vary according to course level, campus/location and program. *Financial support:* Fellowships, research assistantships, Federal Work-Study, institutionally sponsored loans, and tuition waivers (full) available. Financial award application deadline: 1/15; financial award applicants required to submit FAFSA. *Faculty research:* Molecular genetics of normal and malignant cell growth, regulation of gene expression, DNA repair systems and carcinogenesis, hormone-receptor interactions and signal transduction systems, structural biology. *Unit head:* Dr. Scott Shore, Chair, 215-707-3359, Fax: 215 707 2805, E mail: sks@temple.edu.

Texas A&M University, College of Science, Department of Biology, Program in Molecular and Cell Biology, College Station, TX 77843. Offers PhD. Program composed of members from 4 colleges and 11 departments. *Degree requirements:* For doctorate, thesis/dissertation. *Entrance requirements:* For doctorate, GRE General Test. Additional exam requirements/recommendations for international students: Required—TOEFL. Application fee: $50 ($75 for international students). *Expenses:* Tuition, state resident: full-time $4,488; part-time $187 per credit hour. Tuition, nonresident: full-time $11,112; part-time $463 per credit hour. Required fees: $1,974. *Financial support:* Fellowships available. Financial award application deadline: 4/1; financial award applicants required to submit FAFSA. *Application contact:* Graduate Advisor, 979-826-2465, Fax: 979-845-2891.

Texas A&M University System Health Science Center, Graduate School of Biomedical Sciences, Department of Medical Microbiology and Immunology, College Station, TX 77840. Offers immunology (PhD); microbiology (PhD); molecular biology (PhD); virology (PhD). *Degree requirements:* For doctorate, thesis/dissertation. *Entrance requirements:* For doctorate, GRE General Test, minimum GPA of 3.0. *Faculty research:* Molecular pathogenesis, microbial therapeutics.

Texas A&M University System Health Science Center, Graduate School of Biomedical Sciences, Program in Cell and Molecular Biology, College Station, TX 77840. Offers PhD.

Texas Woman's University, Graduate School, College of Arts and Sciences, Department of Biology, Denton, TX 76201. Offers biology (MS); biology teaching (MS); molecular biology (PhD). Part-time programs available. *Students:* 23 full-time (15 women), 13 part-time (12 women); includes 9 minority (6 African Americans, 1 Asian American or Pacific Islander, 2 Hispanic Americans), 19 international. Average age 31. In 2005, 8 master's, 4 doctorates awarded. Terminal master's awarded for partial completion of doctoral program. *Degree requirements:* For master's, thesis (for some programs), comprehensive exam; for doctorate, thesis/dissertation, residency, comprehensive exam. *Entrance requirements:* For master's and doctorate, GRE General Test, 3 letters of reference. Additional exam requirements/recommendations for international students: Required—TOEFL (minimum score 550 paper-based; 213 computer-based). *Application deadline:* Applications are processed on a rolling basis. Application fee: $30 ($50 for international students). Electronic applications accepted. *Expenses:* Tuition, state resident: full-time $5,868; part-time $163 per credit. Tuition, nonresident: full-time $15,948; part-time $443 per credit. Required fees: $20 per credit. $152 per term. *Financial support:* In 2005–06, 8 research assistantships (averaging $10,706 per year), 1 teaching assistantship (averaging $10,706 per year) were awarded; career-related internships or fieldwork, Federal Work-Study, institutionally sponsored loans, scholarships/grants, traineeships, health care benefits, and unspecified assistantships also available. Support available to part-time students. Financial award application deadline: 3/1; financial award applicants required to submit FAFSA. *Faculty research:* Plants and plant viruses, plasmia DNA in bacteria, transcriptional regulation of RNA, hormone actions, serotonin modulation of female reproductive behavior. *Unit head:* Dr. Sarah McIntire, Chair, 940-898-2351, Fax: 940-898-2382, E-mail: smcintire@mail.twu.edu. *Application contact:* Samuel Wheeler, Coordinator of Graduate Admissions, 940-898-3188, Fax: 940-898-3081, E-mail: wheelersr@twu.edu.

Thomas Jefferson University, Jefferson College of Graduate Studies, Program in Biochemistry and Molecular Biology, Philadelphia, PA 19107. Offers PhD. *Faculty:* 17 full-time (9 women), 9 part-time/adjunct (1 woman). *Students:* 15 full-time (8 women), 1 (woman) part-time; includes 4 minority (1 African American, 2 Asian Americans or Pacific Islanders, 1 Hispanic American), 1 international. Average age 24. 36 applicants, 17% accepted, 2 enrolled. In 2005, 3 doctorates awarded. *Degree requirements:* For doctorate, thesis/dissertation, preliminary exam, comprehensive exam, registration. *Entrance requirements:* For doctorate, GRE General Test or MCAT, minimum GPA of 3.2. Additional exam requirements/recommendations for international students: Required—TOEFL (minimum score 213 computer-based). *Application deadline:* For fall admission, 3/1 priority date for domestic students, 1/1 priority date for international students. Applications are processed on a rolling basis. Application fee: $50. Electronic applications accepted. *Expenses:* Tuition: Full-time $14,894; part-time $800 per credit. *Financial support:* In 2005–06, 3 students received support; fellowships with full tuition reimbursements available, research assistantships, Federal Work-Study, institutionally sponsored loans, scholarships/grants, traineeships, and stipends available. Financial award application deadline: 5/1; financial award applicants required to submit FAFSA. *Faculty research:* Signal transduction and molecular genetics, translational biochemistry, human mitochondrial genetics, molecular biology of protein-RNA interaction, mammalian mitochondrial biogenesis and function. Total annual research expenditures: $3 million. *Unit head:* Dr. Diane Merry, Program Director, 215-503-4907, Fax: 215-923-9162, E-mail: diane.merry@jefferson.edu. *Application contact:* Jessie F. Pervall, Director of Admissions, 215-503-0155, Fax: 215-503-9920, E-mail: jessie.pervall@jefferson.edu.

See Close-Up on page 419.

Tufts University, Sackler School of Graduate Biomedical Sciences, Department of Molecular Biology and Microbiology, Boston, MA 02155. Offers molecular microbiology (PhD), including microbiology, molecular biology, molecular microbiology. *Faculty:* 18 full-time (7 women). *Students:* 48 full-time (33 women); includes 12 minority (1 American Indian/Alaska Native, 5 Asian Americans or Pacific Islanders, 6 Hispanic Americans), 12 international. Average age 28. 100 applicants, 12% accepted, 6 enrolled. In 2005, 5 degrees awarded. *Degree requirements:* For doctorate, thesis/dissertation, graduate seminar, journal club, graduate research, comprehensive exam, registration. *Entrance requirements:* For doctorate, GRE General Test, 3 letters of reference. Additional exam requirements/recommendations for international students: Required—TOEFL. *Application deadline:* For fall admission, 1/15 priority date for domestic students, 1/15 priority date for international students. Applications are processed on a rolling basis. Application fee: $65. Electronic applications accepted. *Financial support:* In 2005–06, 48 students received support, including 48 research assistantships with full tuition reimbursements available (averaging $29,000 per year); scholarships/grants, health care benefits, and tuition waivers (full) also available. Financial award application deadline: 1/15. *Faculty research:* Fundamental problems of molecular biology of prokaryotes, eukaryotes and their viruses. *Unit head:* Dr. Abraham L. Sonenshein, Director, 617-636-6761, Fax: 617-636-0337, E-mail: linc.sonenshein@tufts.edu. *Application contact:* 617-636-6767, Fax: 617-633-0375, E-mail: sackler-school@tufts.edu.

Tufts University, Sackler School of Graduate Biomedical Sciences, Program in Cell, Molecular and Developmental Biology, Boston, MA 02155. Offers PhD. Applications are processed through through integrated studies program. *Faculty:* 31 full-time (9 women). *Students:* 28 full-time (17 women), 8 international. Average age 29. In 2005, 3 degrees awarded. *Degree requirements:* For doctorate, thesis/dissertation. *Financial support:* In 2005–06, 28 students received support, including 28 research assistantships with full tuition reimbursements available (averaging $29,000 per year); fellowships, scholarships/grants, health care benefits, and tuition waivers (full) also available. Financial award application deadline: 1/15. *Faculty research:* Reproduction and hormone action, control of gene expression, cell-matrix and cell-cell interactions, growth control and tumorigenesis, cytoskeleton and contractile proteins. *Unit head:* Dr. John Castellot, Program Director, 617-636-0303, Fax: 617-636-0375, E-mail: john.castellot@tufts.edu. *Application contact:* 617-636-6767, Fax: 617-636-0375, E-mail: sackler-school@tufts.edu.

Tulane University, Graduate School, Department of Cell and Molecular Biology, New Orleans, LA 70118-5669. Offers MS, PhD. Terminal master's awarded for partial completion of doctoral program. *Degree requirements:* For doctorate, thesis/dissertation. *Entrance requirements:* For master's, GRE General Test, minimum B average in undergraduate course work; for doctorate, GRE General Test. Additional exam requirements/recommendations for international students: Required—TOEFL; Recommended—TSE. Electronic applications accepted.

Tulane University, School of Medicine and Graduate School, Graduate Programs in Medicine, Interdisciplinary Graduate Program in Molecular and Cellular Biology, New Orleans, LA 70118-5669. Offers PhD, MD/PhD. PhD offered through the Graduate School. *Degree requirements:* For doctorate, thesis/dissertation. *Entrance requirements:* For doctorate, GRE General Test, GRE Subject Test. Additional exam requirements/recommendations for international students: Required—TOEFL; Recommended—TSE. Electronic applications accepted. *Faculty research:* Developmental biology, neuroscience, virology.

Uniformed Services University of the Health Sciences, School of Medicine, Programs in Biomedical Sciences, Graduate Program in Molecular and Cell Biology, Bethesda, MD 20814-4799. Offers PhD. *Faculty:* 17 full-time (6 women), 3 part-time/adjunct (0 women). *Students:* 19 full-time (8 women); includes 7 minority (1 African American, 5 Asian Americans or Pacific Islanders, 1 Hispanic American), 2 international. Average age 26. 27 applicants, 37% accepted, 3 enrolled. *Median time to degree:* Of those who began their doctoral program in fall 1997, 100% received their degree in 8 years or less. *Degree requirements:* For doctorate, thesis/dissertation, qualifying exam, comprehensive exam. *Entrance requirements:* For doctorate, GRE General Test, minimum GPA of 3.0. Additional exam requirements/recommendations for international students: Required—TOEFL. *Application deadline:* For fall admission, 1/15 for domestic students. Applications are processed on a rolling basis. Application fee: $0. *Financial support:* In 2005–06, fellowships with full tuition reimbursements (averaging $23,000 per year); career-related internships or fieldwork and tuition waivers (full) also available. *Faculty research:* Immunology, biochemistry. *Unit head:* Dr. Jeffrey Harmon, Director, 301-295-3248, Fax: 301-295-1996, E-mail: jharmon@usuhs.mil. *Application contact:* Janet M. Anastasi, Graduate Program Coordinator, 301-295-9474, Fax: 301-295-6772, E-mail: janastasi@usuhs.mil.

See Close-Up on page 599.

Université de Montréal, Faculty of Graduate Studies, Program in Molecular Biology, Montréal, QC H3C 3J7, Canada. Offers M Sc, PhD. *Students:* 131 full-time (72 women), 2 part-time. 76 applicants, 38% accepted, 26 enrolled. In 2005, 28 master's, 6 doctorates awarded. Terminal master's awarded for partial completion of doctoral program. *Degree requirements:* For master's, thesis; for doctorate, thesis/dissertation, general exam. *Entrance requirements:* For master's and doctorate, proficiency in French, knowledge of English. *Application deadline:* For fall and spring admission, 2/1. For winter admission, 11/1 for domestic students. Application fee: $30. Electronic applications accepted. *Faculty research:* Protein interactions, intracellular signaling, development and differentiation, hematopoiesis, stem cells. *Unit head:* Trang Hoang, Director, 514-343-6970. *Application contact:* Vivianne Jodoin, Information Contact, 514-343-6111 Ext. 0916, Fax: 514-343-7383, E-mail: vivianne.jodoin@umontreal.ca.

Université Laval, Faculty of Medicine, Graduate Programs in Medicine, Programs in Cellular and Molecular Biology, Québec, QC G1K 7P4, Canada. Offers M Sc, PhD. Terminal master's awarded for partial completion of doctoral program. *Degree requirements:* For master's, thesis/dissertation; for doctorate, thesis/dissertation, comprehensive exam. *Entrance requirements:* For master's and doctorate, knowledge of French, comprehension of written English. Electronic applications accepted. *Faculty research:* Oral bacterial metabolism, sugar transport.

University at Albany, State University of New York, College of Arts and Sciences, Department of Biological Sciences, Specialization in Molecular, Cellular, Developmental, and Neural Biology, Albany, NY 12222-0001. Offers MS, PhD. *Degree requirements:* For master's, one

Molecular Biology

University at Albany, State University of New York (continued)
foreign language; for doctorate, one foreign language, thesis/dissertation. *Entrance requirements:* For master's and doctorate, GRE General Test. Application fee: $60. *Financial support:* Minority assistantships available. *Unit head:* Dr. Albert Millis, Chair, Department of Biological Sciences, 518-442-4300.

University at Albany, State University of New York, School of Public Health, Department of Biomedical Sciences, Program in Biochemistry, Molecular Biology, and Genetics, Albany, NY 12222-0001. Offers MS, PhD. *Degree requirements:* For master's and doctorate, thesis/dissertation. *Entrance requirements:* For master's and doctorate, GRE General Test, GRE Subject Test. *Application deadline:* For fall admission, 2/15 for domestic students. Application fee: $60. *Financial support:* Application deadline: 2/1. *Unit head:* Dr. James Dias, Chair, Department of Biomedical Sciences, 518-474-2662.

The University of Alabama at Birmingham, Graduate Programs in Joint Health Sciences, Birmingham, AL 35294. Offers biochemistry and molecular genetics (PhD), including biochemistry; cell biology (PhD), including cell biology, cellular and molecular biology, cellular and molecular physiology, neuroscience; genetics (PhD); microbiology (PhD); neurobiology (PhD); pathology (PhD); pharmacology and toxicology (PhD), including pharmacology, toxicology; physiology and biophysics (MSBMS, PhD), including basic medical sciences (MSBMS), biophysical sciences (PhD), integrative biomedical sciences (PhD). *Students:* 439 full-time (214 women), 6 part-time (2 women); includes 74 minority (36 African Americans, 4 American Indian/Alaska Native, 28 Asian Americans or Pacific Islanders, 6 Hispanic Americans), 164 international. Average age 28. 463 applicants, 37% accepted. In 2005, 9 master's, 53 doctorates awarded. *Entrance requirements:* For master's, GRE; for doctorate, GRE, interview. *Application deadline:* Applications are processed on a rolling basis. Application fee: $35 ($60 for international students). Electronic applications accepted. *Expenses:* Tuition, state resident: part-time $170 per credit hour. Tuition, nonresident: full-time $4,612; part-time $425 per credit hour. International tuition: $10,732 full-time. Required fees: $11 per credit hour. $124 per term. Tuition and fees vary according to course load, degree level and program. *Financial support:* Fellowships, career-related internships or fieldwork available. *Unit head:* Dr. Robert R. Rich, Vice President/Dean, School of Medicine, 205-934-1111, Fax: 205-934-0333, E-mail: rrich@uab.edu.

The University of Alabama at Birmingham, Graduate Programs in Joint Health Sciences, Department of Cell Biology, Graduate Program in Cellular and Molecular Biology, Birmingham, AL 35294.

See Close-Ups on pages 601 and 603.

The University of Alabama at Birmingham, Graduate Programs in Joint Health Sciences, Department of Cell Biology, Program in Cell Biology, Birmingham, AL 35294. Offers PhD. *Students:* 87 full-time (45 women), 2 part-time; includes 11 minority (4 African Americans, 6 Asian Americans or Pacific Islanders, 1 Hispanic American), 45 international. *Expenses:* Tuition, state resident: part-time $170 per credit hour. Tuition, nonresident: full-time $4,612; part-time $425 per credit hour. International tuition: $10,732 full-time. Required fees: $11 per credit hour. $124 per term. Tuition and fees vary according to course load, degree level and program. *Application contact:* Information Contact, 205-975-7145, Fax: 205-975-6748.

See Close-Ups on pages 601 and 603.

University of Alberta, Faculty of Graduate Studies and Research, Department of Biological Sciences, Edmonton, AB T6G 2E1, Canada. Offers environmental biology and ecology (M Sc, PhD); microbiology and biotechnology (M Sc, PhD); molecular biology and genetics (M Sc, PhD); physiology and cell biology (M Sc, PhD); plant biology (M Sc, PhD); systematics and evolution (M Sc, PhD). *Faculty:* 72 full-time (15 women), 15 part-time/adjunct (4 women). *Students:* 238 full-time (117 women), 32 part-time (15 women), 31 international. 206 applicants, 42% accepted. In 2005, 29 master's, 31 doctorates awarded. Terminal master's awarded for partial completion of doctoral program. *Degree requirements:* For master's and doctorate, thesis/dissertation, registration. *Entrance requirements:* Additional exam requirements/recommendations for international students: Required—TOEFL. *Application deadline:* For fall admission, 3/1 for domestic students. Applications are processed on a rolling basis. Application fee: $0. Tuition and fees charges are reported in Canadian dollars. *Expenses:* Tuition, state resident: part-time $562 Canadian dollars per term. Tuition, nonresident: full-time $3,375 Canadian dollars. Required fees: $573 Canadian dollars; $84 Canadian dollars per term. *Financial support:* In 2005–06, 4 research assistantships with partial tuition reimbursements (averaging $12,000 per year), 103 teaching assistantships with partial tuition reimbursements (averaging $12,300 per year) were awarded; career-related internships or fieldwork and scholarships/grants also available. *Unit head:* Laura Frost, Chair, 780-492-1904. *Application contact:* Dr. John P. Chang, Associate Chair for Graduate Studies, 780-492-1257, Fax: 780-492-9457, E-mail: bio.grad.coordinator@ualberta.ca.

The University of Arizona, Graduate College, College of Science, Department of Molecular and Cellular Biology, Tucson, AZ 85721. Offers MS, PhD. Terminal master's awarded for partial completion of doctoral program. *Degree requirements:* For master's and doctorate, thesis/dissertation. *Entrance requirements:* For master's, GRE General Test, GRE Subject Test; for doctorate, GRE General Test, GRE Subject Test, undergraduate research experience. Additional exam requirements/recommendations for international students: Required—TOEFL. *Faculty research:* Plant molecular biology, cellular and molecular aspects of development, genetics of bacteria and lower eukaryotes.

University of Arkansas, Graduate School, Interdisciplinary Program in Cell and Molecular Biology, Fayetteville, AR 72701-1201. Offers MS, PhD. *Students:* 38 full-time (23 women), 12 part-time (7 women); includes 6 minority (4 African Americans, 1 American Indian/Alaska Native, 1 Asian American or Pacific Islander), 24 international. 39 applicants, 18% accepted. In 2005, 10 master's, 3 doctorates awarded. *Degree requirements:* For doctorate, thesis/dissertation. Application fee: $40 ($50 for international students). *Financial support:* In 2005–06, 2 fellowships with tuition reimbursements, 10 research assistantships were awarded; teaching assistantships Financial award application deadline: 4/1; financial award applicants required to submit FAFSA. *Unit head:* Dr. John Kirby, Head, 479-575-8623, Fax: 479-575-5908, E-mail: jkirby@uark.edu.

University of Arkansas for Medical Sciences, College of Medicine and Graduate School, Graduate Programs in Medicine, Department of Biochemistry and Molecular Biology, Little Rock, AR 72205-7199. Offers MS, PhD, MD/PhD. *Faculty:* 26 full-time (7 women), 7 part-time/adjunct (0 women). *Students:* 16 full-time, 3 part-time. *Degree requirements:* For master's, thesis, comprehensive exam; for doctorate, thesis/dissertation, qualifying exam. *Entrance requirements:* For master's, GRE General Test, bachelor's degree in biology, chemistry, or related field; for doctorate, GRE General Test. Additional exam requirements/recommendations for international students: Required—TOEFL. Application fee: $0. *Financial support:* Research assistantships, unspecified assistantships available. Support available to part-time students. *Faculty research:* Gene regulation, growth factors, oncogenes, metabolic diseases, hormone regulation. *Unit head:* Dr. Alan D. Elbein, Chairman, 501-686-5185. *Application contact:* Dr. Kevin Raney, Graduate Coordinator, 501-686-5244, E-mail: raneykevind@uams.edu.

The University of British Columbia, Faculty of Medicine and Faculty of Graduate Studies, Graduate Programs in Medicine, Department of Biochemistry and Molecular Biology, Vancouver, BC V6T 1Z1, Canada. Offers M Sc, PhD. *Faculty:* 24 full-time (3 women). *Students:* 57 full-time (23 women). Average age 27. 76 applicants, 58% accepted, 11 enrolled. In 2005, 6 master's, 4 doctorates awarded. *Median time to degree:* Of those who began their doctoral program in fall 1991 100% received their degree in 8 years or less. *Degree requirements:* For master's, thesis/dissertation; for doctorate, thesis/dissertation, comprehensive exam. *Entrance requirements:* For master's, GRE (international students), first class B Sc; for doctorate, GRE (international students), master's or first class honors bachelor's degree in biochemistry. Additional exam requirements/recommendations for international students: Required—TOEFL

(minimum score 625 paper-based; 263 computer-based). *Application deadline:* For fall admission, 1/31 priority date for domestic students, 1/31 priority date for international students; for spring admission, 7/1 priority date for domestic students, 7/1 priority date for international students. Applications are processed on a rolling basis. Application fee: $90 ($150 for international students). Electronic applications accepted. *Financial support:* In 2005–06, 23 fellowships (averaging $21,400 per year), 23 research assistantships (averaging $20,300 per year), 15 teaching assistantships (averaging $2,300 per year) were awarded; institutionally sponsored loans, scholarships/grants, traineeships, tuition waivers (partial), and unspecified assistantships also available. *Faculty research:* Membrane biochemistry, protein structure/function, signal transduction, biochemistry. Total annual research expenditures: $12.6 million. *Unit head:* Dr. Christopher A. Proud, Head, 604-822-2792, Fax: 604-822-5227. *Application contact:* Hiltrud M. Vogler, Graduate Secretary, 604-822-5925, Fax: 604-822-5227, E-mail: biograd@interchange.ubc.ca.

University of Calgary, Faculty of Medicine and Faculty of Graduate Studies, Department of Biochemistry and Molecular Biology, Calgary, AB T2N 1N4, Canada. Offers M Sc, PhD. *Faculty:* 51 full-time (8 women), 19 part-time/adjunct (3 women). *Students:* 81 full-time (48 women), 1 part-time. Average age 27. 64 applicants, 17% accepted, 9 enrolled. In 2005, 7 master's, 3 doctorates awarded. *Degree requirements:* For master's, thesis; for doctorate, thesis/dissertation, candidacy exam. *Entrance requirements:* For master's and doctorate, GRE General Test, minimum GPA of 3.2. Additional exam requirements/recommendations for international students: Required—TOEFL. *Application deadline:* For fall admission, 6/15 priority date for domestic students, 5/15 priority date for international students. For winter admission, 10/15 for domestic students; for spring admission, 3/15 for domestic students. Applications are processed on a rolling basis. Application fee: $100 ($130 for international students). Electronic applications accepted. *Financial support:* In 2005–06, 28 fellowships, 19 research assistantships (averaging $4,100 per year), 3 teaching assistantships (averaging $6,110 per year) were awarded; stipends for Ph. D. students is $20,000 and M.Sc. students $18,000 also available. Financial award application deadline: 2/1. *Faculty research:* Molecular and developmental genetics; molecular biology of disease; genomics, proteomics and bioinformatics; ceu signaling and structure. *Unit head:* Dr. Jim McGhee, Graduate Coordinator, Fax: 403-210-8109, E-mail: jmcghee@ucalgary.ca. *Application contact:* Judy E. Gayford, Graduate Program Administrator, 403-220-8306, Fax: 403-210-8109, E-mail: bmbgrad@ucalgary.ca.

University of California, Berkeley, Graduate Division, College of Letters and Science, Department of Molecular and Cell Biology, Berkeley, CA 94720-1500. Offers PhD. *Faculty:* 89 full-time (20 women), 3 part-time/adjunct (1 woman). *Students:* 265 full-time (139 women); includes 62 minority (5 African Americans, 2 American Indian/Alaska Native, 45 Asian Americans or Pacific Islanders, 10 Hispanic Americans), 23 international. Average age 23. 582 applicants, 25% accepted, 54 enrolled. In 2005, 30 doctorates awarded. *Median time to degree:* Of those who began their doctoral program in fall 1997, 85% received their degree in 8 years or less. *Degree requirements:* For doctorate, thesis/dissertation, qualifying exam, 2 semesters of teaching, 3 grad seminars, comprehensive exam, registration. *Entrance requirements:* For doctorate, GRE General Test, GRE Subject Test (recommended), minimum GPA of 3.0. Additional exam requirements/recommendations for international students: Required—TOEFL (minimum score 570 paper-based; 230 computer-based). *Application deadline:* For fall admission, 12/15 for domestic students, 12/15 for international students. Applications are processed on a rolling basis. Application fee: $60. Electronic applications accepted. *Financial support:* In 2005–06, 150 research assistantships with full tuition reimbursements (averaging $25,000 per year), 80 teaching assistantships with full tuition reimbursements (averaging $15,000 per year) were awarded; fellowships with full tuition reimbursements, scholarships/grants, traineeships, health care benefits, tuition waivers (full), and unspecified assistantships also available. Financial award application deadline: 12/15; financial award applicants required to submit FAFSA. *Faculty research:* Biochemistry and molecular biology, cell and developmental biology, genetics, immunology, neurobiology, genomics, evo/devo. *Unit head:* Barbara J Meyer, Professor. *Application contact:* Berta Parra, Student Affairs Officer, 510-642-5252, Fax: 510-642-7000, E-mail: bparra@berkeley.edu.

University of California, Davis, Graduate Studies, Graduate Group in Biochemistry and Molecular Biology, Davis, CA 95616. Offers MS, PhD. *Faculty:* 117 full-time. *Students:* 94 full-time (49 women); includes 28 minority (1 African American, 1 American Indian/Alaska Native, 22 Asian Americans or Pacific Islanders, 4 Hispanic Americans), 27 international. Average age 28. 207 applicants, 29% accepted, 18 enrolled. In 2005, 3 master's, 5 doctorates awarded. Terminal master's awarded for partial completion of doctoral program. *Median time to degree:* Of those who began their doctoral program in fall 1997, 66.7% received their degree in 8 years or less. *Degree requirements:* For master's, thesis (for some programs), comprehensive exam (for some programs); for doctorate, thesis/dissertation. *Entrance requirements:* For master's and doctorate, GRE General Test, GRE Subject Test. Additional exam requirements/recommendations for international students: Required—TOEFL (minimum score 550 paper-based; 213 computer-based). *Application deadline:* For fall admission, 1/15 for domestic students, 1/15 for international students. Application fee: $60. Electronic applications accepted. *Financial support:* In 2005–06, 86 students received support, including 16 fellowships with full and partial tuition reimbursements available (averaging $12,716 per year), 56 research assistantships with full and partial tuition reimbursements available (averaging $15,941 per year), 6 teaching assistantships with partial tuition reimbursements available (averaging $15,082 per year); Federal Work-Study, institutionally sponsored loans, scholarships/grants, tuition waivers (full and partial), and unspecified assistantships also available. Financial award application deadline: 1/15; financial award applicants required to submit FAFSA. *Faculty research:* Gene expression, protein structure, molecular virology, protein synthesis, enzymology, membrane transport and structural biology. *Unit head:* J. Clark Lagarias, Chair, 530-752-1865, E-mail: jclagarias@ucdavis.edu. *Application contact:* Angelina Kuo, Graduate Staff, 530-752-2981, Fax: 530-752-8391, E-mail: abkuo@ucdavis.edu.

University of California, Irvine, Office of Graduate Studies, School of Biological Sciences, Department of Molecular Biology and Biochemistry, Irvine, CA 92697. Offers biological science (MS); biological sciences (PhD); biotechnology (MS). *Degree requirements:* For doctorate, thesis/dissertation. *Entrance requirements:* For master's, GRE, minimum GPA of 3.0; for doctorate, GRE General Test, GRE Subject Test, minimum GPA of 3.0. Additional exam requirements/recommendations for international students: Required—TOEFL (minimum score 550 paper-based; 213 computer-based), TSE. Electronic applications accepted. *Faculty research:* Structure and synthesis of nucleic acids and proteins, regulation, virology, biochemical genetics, gene organization.

University of California, Irvine, Office of Graduate Studies, School of Biological Sciences and College of Medicine, Graduate Program in Molecular Biology, Genetics, and Biochemistry, Irvine, CA 92697-3915. Offers biological sciences (PhD). *Faculty:* 156 full-time (38 women). *Students:* 53 full-time (30 women). Average age 25. 452 applicants, 26% accepted, 53 enrolled. In 2005, 19 doctorates awarded. *Median time to degree:* Of those who began their doctoral program in fall 1997, 10% received their degree in 8 years or less. *Degree requirements:* For doctorate, thesis/dissertation, teaching assignment, preliminary exam. *Entrance requirements:* For doctorate, GRE General Test, minimum GPA of 3.0, research experience. Additional exam requirements/recommendations for international students: Required—TOEFL, IELT, TSE. *Application deadline:* For fall admission, 1/1 for domestic students, 1/1 for international students. Application fee: $60 ($80 for international students). Electronic applications accepted. *Expenses:* Contact institution. *Financial support:* In 2005–06, 52 fellowships with full tuition reimbursements (averaging $25,000 per year) were awarded; institutionally sponsored loans, scholarships/grants, tuition waivers (full), and stipends also available. Financial award application deadline: 1/7; financial award applicants required to submit FAFSA. *Faculty research:* Cellular biochemistry; gene structure and expression; protein structure, function, and design; molecular genetics; pathogenesis and inherited disease. *Unit head:* Dr. Peter J. Bryant, Director, 949-824-4714, Fax: 949-824-3571, E-mail: gp-mbgb@uci.edu. *Application contact:* Kimberly McKinney, Administrator, 949-824-8145, Fax: 949-824-1965, E-mail: kamckinn@uci.edu.

University of California, Los Angeles, Graduate Division, College of Letters and Science, Department of Chemistry and Biochemistry, Program in Biochemistry and Molecular Biology,

Molecular Biology

Los Angeles, CA 90095. Offers MS, PhD. MS admission to program only under exceptional circumstances. *Entrance requirements:* For master's, GRE General Test, GRE Subject Test, minimum GPA of 3.0; for doctorate, GRE General Test, GRE Subject Test, minimum undergraduate GPA of 3.0. Electronic applications accepted.

University of California, Los Angeles, Graduate Division, College of Letters and Science, Department of Physiological Science, Program in Molecular, Cellular and Integrative Physiology, Los Angeles, CA 90095. Offers PhD.

University of California, Los Angeles, Graduate Division, College of Letters and Science, Program in Molecular Biology, Los Angeles, CA 90095. Offers PhD. *Degree requirements:* For doctorate, thesis/dissertation, oral and written qualifying exams. *Entrance requirements:* For doctorate, GRE General Test. Electronic applications accepted.

University of California, Los Angeles, Graduate Division, College of Letters and Science and School of Medicine, UCLA ACCESS to Programs in the Molecular and Cellular Life Sciences, Los Angeles, CA 90095. Offers PhD. *Degree requirements:* For doctorate, thesis/dissertation, oral and written qualifying exams. *Entrance requirements:* For doctorate, GRE General Test, minimum undergraduate GPA of 3.0. Electronic applications accepted. *Faculty research:* Molecular, cellular, and developmental biology; immunology; microbiology; integrative biology.

See Close-Up on page 605.

University of California, Los Angeles, School of Medicine and Graduate Division, Graduate Programs in Medicine, Department of Molecular, Cell and Developmental Biology, Los Angeles, CA 90095. Offers MA, PhD. *Degree requirements:* For doctorate, thesis/dissertation, qualifying exams. *Entrance requirements:* For doctorate, GRE General Test, GRE Subject Test. Additional exam requirements/recommendations for international students: Required—TOEFL.

University of California, Riverside, Graduate Division, Program in Cell, Molecular, and Developmental Biology, Riverside, CA 92521-0102. Offers MS, PhD. *Faculty:* 55 full-time (18 women). *Students:* 57 full-time (29 women); includes 20 minority (3 African Americans, 15 Asian Americans or Pacific Islanders, 2 Hispanic Americans), 18 international. Average age 28. In 2005, 3 master's, 1 doctorate awarded. *Median time to degree:* Of those who began their doctoral program in fall 1997, 100% received their degree in 8 years or less. *Degree requirements:* For master's, thesis, oral defense of thesis; for doctorate, thesis/dissertation, oral defense of thesis, qualifying exams, 2 quarters of teaching experience. *Entrance requirements:* For master's and doctorate, GRE General Test, minimum GPA of 3.2. Additional exam requirements/recommendations for international students: Required—TOEFL (minimum score 550 paper-based; 213 computer-based); Recommended—TSE. *Application deadline:* For fall admission, 5/1 for domestic students, 2/1 for international students. For winter admission, 9/1 for domestic students; for spring admission, 12/1 for domestic students. Applications are processed on a rolling basis. Application fee: $60 ($75 for international students). Electronic applications accepted. *Expenses:* Tuition, nonresident: full-time $14,694. Required fees:$9,009. Full-time tuition and fees vary according to program. *Financial support:* In 2005–06, fellowships (averaging $18,000 per year), research assistantships (averaging $14,000 per year), teaching assistantships (averaging $15,000 per year) were awarded. Financial award application deadline: 1/5. *Unit head:* Dr. Anthony Norman, Director, 951-827-4777, E-mail: norman@ucr.edu. *Application contact:* Kathy Redd, Graduate Program Assistant, 800-735-0717, Fax: 951-827-5517, E-mail: cell@ucr.edu.

University of California, San Diego, Graduate Studies and Research, Division of Biology, Program in Genetics and Molecular Biology, La Jolla, CA 92093-0348. Offers PhD. Offered in association with the Salk Institute. *Degree requirements:* For doctorate, thesis/dissertation, qualifying exam. Electronic applications accepted.

University of California, San Diego, Graduate Studies and Research, Division of Biology, Program in Molecular and Cellular Biology, La Jolla, CA 92093. Offers PhD. Offered in association with the Salk Institute. *Degree requirements:* For doctorate, thesis/dissertation, qualifying exam. Electronic applications accepted.

University of California, San Diego, School of Medicine and Graduate Studies and Research, Graduate Studies in Biomedical Sciences, Program in Molecular Cell Biology, La Jolla, CA 92093-0685. Offers PhD. *Degree requirements:* For doctorate, thesis/dissertation, qualifying exam. *Entrance requirements:* For doctorate, GRE General Test. Additional exam requirements/recommendations for international students: Required—TOEFL. Electronic applications accepted. *Faculty research:* Molecular and cellular pharmacology, cell and organ physiology.

See Close-Up on page 607.

University of California, San Diego, School of Medicine and Graduate Studies and Research, Graduate Studies in Biomedical Sciences, Regulatory Biology Program, La Jolla, CA 92093. Offers PhD. *Degree requirements:* For doctorate, thesis/dissertation, 2 qualifying exams. *Entrance requirements:* For doctorate, GRE General Test, GRE Subject Test. Additional exam requirements/recommendations for international students: Required—TOEFL. Electronic applications accepted. *Faculty research:* Eukaryotic regulatory and molecular biology, molecular and cellular pharmacology, cell and organ physiology.

University of California, San Francisco, Graduate Division and School of Medicine, Department of Biochemistry and Biophysics, Program in Biochemistry and Molecular Biology, San Francisco, CA 94143. Offers PhD, MD/PhD. *Degree requirements:* For doctorate, thesis/dissertation. *Entrance requirements:* For doctorate, GRE General Test, GRE Subject Test. Additional exam requirements/recommendations for international students: Required—TOEFL. Expenses: Contact institution. *Faculty research:* Structural biology, genetics, cell biology, cell physiology, metabolism.

University of California, Santa Barbara, Graduate Division, College of Letters and Sciences, Division of Mathematics, Life, and Physical Sciences, Department of Biomolecular Science and Engineering, Santa Barbara, CA 93106. Offers biochemistry and molecular biology (MS, PhD); biomolecular science and engineering (PhD); biophysics and bioengineering (MS, PhD). *Faculty:* 39 full-time (4 women). *Students:* 25 full-time (11 women); includes 4 minority (all Asian Americans or Pacific Islanders), 1 international. Average age 25. 70 applicants, 33% accepted, 4 enrolled. In 2005, 5 degrees awarded. Terminal master's awarded for partial completion of doctoral program. *Median time to degree:* Of those who began their doctoral program in fall 1997, 100% received their degree in 8 years or less. *Degree requirements:* For master's, thesis optional; for doctorate, thesis/dissertation, lab rotations, seminars, coursework, TA ships, research units, comprehensive exam, registration. *Entrance requirements:* For master's and doctorate, GRE General Test, GRE Subject Test. Additional exam requirements/recommendations for international students: Required—TOEFL (minimum score 630 paper-based; 267 computer-based). *Application deadline:* For fall admission, 12/15 for domestic students, 12/15 for international students. Application fee: $60. Electronic applications accepted. *Financial support:* In 2005–06, 3 fellowships with full and partial tuition reimbursements (averaging $21,000 per year), 15 research assistantships with full and partial tuition reimbursements (averaging $21,000 per year), 8 teaching assistantships with full and partial tuition reimbursements (averaging $21,000 per year) were awarded; career-related internships or fieldwork, Federal Work-Study, institutionally sponsored loans, scholarships/grants, traineeships, health care benefits, tuition waivers (full and partial), and unspecified assistantships also available. Financial award application deadline: 12/15; financial award applicants required to submit FAFSA. *Faculty research:* Genetics and biochemistry of bacterial gene expression; structure-function relationships in proteins and nucleic acids, protein chemistry; biochemistry and biophysics of marine adhesion. *Unit head:* Prof. Philip A. Pincus, Chair, 805-893-4685, E-mail: fyl@mrl.ucsb.edu. *Application contact:* Krista Grace, Staff Graduate Program Advisor, 805-893-2290, Fax: 805-893-4724, E-mail: grace@lifesci.ucsb.edu.

University of California, Santa Barbara, Graduate Division, College of Letters and Sciences, Division of Mathematics, Life, and Physical Sciences, Department of Molecular, Cellular, and Developmental Biology, Santa Barbara, CA 93106. Offers MA, PhD, MA/PhD. *Faculty:* 62 full-time (30 women), 1 part-time/adjunct (0 women). *Students:* 63 full-time (30 women); includes 10 minority (1 African American, 7 Asian Americans or Pacific Islanders, 2 Hispanic Americans), 4 international. Average age 26. 99 applicants, 31% accepted, 18 enrolled. Terminal master's awarded for partial completion of doctoral program. *Degree requirements:* For master's, thesis (for some programs), comprehensive exam (for some programs), registration; for doctorate, thesis/dissertation, comprehensive exam, registration. *Entrance requirements:* For master's and doctorate, GRE General Test, GRE Subject Test. Additional exam requirements/recommendations for international students: Required—TOEFL (minimum score 630 paper-based; 213 computer-based). *Application deadline:* For fall admission, 12/15 for domestic students, 12/15 for international students. Application fee: $60. Electronic applications accepted. *Financial support:* In 2005–06, 51 teaching assistantships were awarded; fellowships with full tuition reimbursements, research assistantships with full tuition reimbursements, career-related internships or fieldwork, Federal Work-Study, scholarships/grants, and health care benefits also available. Financial award application deadline: 12/15; financial award applicants required to submit FAFSA. *Faculty research:* Signal transduction, bacteria pathegenesis, virology, stem cell, neurodegenerative disease. *Unit head:* Dr. Dennis Clegg, Chair, 805-893-8490, E-mail: clegg@lifesci.ucsb.edu. *Application contact:* Krista Grace, Staff Graduate Program Advisor, 805-893-2290, Fax: 805-893-4724, E-mail: grace@lifesci.ucsb.edu.

University of California, Santa Cruz, Division of Graduate Studies, Division of Physical and Biological Sciences, Department of Molecular, Cellular, and Developmental Biology, Santa Cruz, CA 95064. Offers MA, PhD. *Students:* 50 full-time (29 women); includes 16 minority (1 African American, 11 Asian Americans or Pacific Islanders, 4 Hispanic Americans), 5 international. In 2005, 1 degree awarded. Application fee: $60. *Expenses:* Tuition, nonresident: full-time $14,694. Required fees: $9,437. *Unit head:* John Tamkun, Chairperson. *Application contact:* Cindy Hodges, Information Contact, 831-459-2632, E-mail: cindy@biology.ucsc.edu.

University of Central Florida, Burnett College of Biomedical Sciences, Program in Biomolecular Science, Orlando, FL 32816. Offers PhD. *Faculty:* 13 full-time (5 women). *Students:* 38 full-time (21 women), 1 part-time; includes 5 minority (3 Asian Americans or Pacific Islanders, 2 Hispanic Americans), 16 international. Average age 28. 45 applicants, 60% accepted, 19 enrolled. In 2005, 2 degrees awarded. *Degree requirements:* For doctorate, thesis/dissertation, qualifying exam, candidacy exam. *Entrance requirements:* For doctorate, GRE General Test, letters of recommendation. Additional exam requirements/recommendations for international students: Required—TOEFL. *Application deadline:* For fall admission, 2/1 for domestic students. Application fee: $30. Electronic applications accepted. *Expenses:* Tuition, state resident: full-time $5,788. Tuition, nonresident: full-time $21,927. Required fees: $241 per credit hour. *Financial support:* In 2005–06, 11 fellowships with partial tuition reimbursements (averaging $2,300 per year), 38 research assistantships (averaging $9,500 per year), 11 teaching assistantships with partial tuition reimbursements (averaging $4,700 per year) were awarded. *Application contact:* Dr. Antonis Zervos, Coordinator, 407-737-2583, E-mail: biomoldoc@mail.ucf.edu.

University of Central Florida, Burnett College of Biomedical Sciences, Program in Molecular Biology and Microbiology, Orlando, FL 32816. Offers microbiology (MS); molecular biology (MS). Part-time and evening/weekend programs available. *Faculty:* 21. *Students:* 16 full-time (11 women), 12 part-time (9 women); includes 9 minority (1 African American, 5 Asian Americans or Pacific Islanders, 3 Hispanic Americans), 7 international. Average age 28. 22 applicants, 32% accepted, 6 enrolled. In 2005, 7 degrees awarded. *Degree requirements:* For master's, thesis, comprehensive exam. *Entrance requirements:* For master's, GRE General Test, minimum GPA of 3.0 in last 60 hours. Additional exam requirements/recommendations for international students: Required—TOEFL. *Application deadline:* For fall admission, 3/15 for domestic students; for spring admission, 12/1 for domestic students. Application fee: $30. Electronic applications accepted. *Expenses:* Tuition, state resident: full-time $5,788. Tuition, nonresident: full-time $21,927. Required fees: $241 per credit hour. *Financial support:* In 2005–06, fellowships with partial tuition reimbursements (averaging $2,800 per year), research assistantships with partial tuition reimbursements (averaging $4,700 per year), teaching assistantships with partial tuition reimbursements (averaging $4,400 per year) were awarded; career-related internships or fieldwork, Federal Work-Study, institutionally sponsored loans, tuition waivers (partial), and unspecified assistantships also available. Financial award application deadline: 3/1; financial award applicants required to submit FAFSA. *Application contact:* Dr. Karl X. Chai, Coordinator, 407-823-4846, E-mail: kxchai@mail.ucf.edu.

University of Chicago, Division of the Biological Sciences, Department of Molecular Biosciences: Biochemistry, Genetics, Cell and Developmental Biology, Department of Biochemistry and Molecular Biology, Chicago, IL 60637-1513. Offers PhD, MD/PhD. *Faculty:* 31 full-time (8 women). *Students:* 48 full-time (19 women); includes 14 minority (1 African American, 1 American Indian/Alaska Native, 10 Asian Americans or Pacific Islanders, 2 Hispanic Americans), 5 international. Average age 27. In 2005, 7 doctorates awarded. *Degree requirements:* For doctorate, one foreign language, thesis/dissertation, qualifying exam. *Entrance requirements:* For doctorate, GRE General Test, GRE Subject Test. Additional exam requirements/recommendations for international students: Required—TOEFL. *Application deadline:* For fall admission, 12/28 priority date for domestic students, 12/28 priority date for international students. Application fee: $55. Electronic applications accepted. *Financial support:* In 2005–06, 35 students received support, including fellowships with tuition reimbursements available (averaging $26,301 per year), research assistantships with tuition reimbursements available (averaging $26,301 per year); institutionally sponsored loans, scholarships/grants, traineeships, and health care benefits also available. Financial award applicants required to submit FAFSA. *Faculty research:* Molecular biology, gene expression, and DNA-protein interactions; membrane biochemistry, molecular endocrinology, and transmembrane signaling; enzyme mechanisms, physical biochemistry, and structural biology. Total annual research expenditures: $5 million. *Unit head:* Dr. Anthony A. Kossiakoff, Chairman, 773-702-9297, Fax: 773-702-0439, E-mail: koss@cummings.uchicago.edu. *Application contact:* Lisa Alvarez, Graduate Student Administrator, 773-702-9213, Fax: 773-702-0439, E-mail: lalvarez@bsd.uchicago.edu.

University of Cincinnati, Division of Research and Advanced Studies, College of Medicine, Graduate Programs in Biomedical Sciences, Department of Environmental Health, Programs in Environmental Genetics and Molecular Toxicology, Cincinnati, OH 45221. Offers MS, PhD. *Degree requirements:* For doctorate, thesis/dissertation. *Entrance requirements:* For master's, GRE; a minimum grade-point average (GPA) of 3.00; 3 letters of recommendation. Additional exam requirements/recommendations for international students: Required—TOEFL (minimum score 520 paper-based; 190 computer-based).

University of Cincinnati, Division of Research and Advanced Studies, College of Medicine, Graduate Programs in Biomedical Sciences, Department of Molecular Genetics, Biochemistry and Microbiology, Cincinnati, OH 45267. Offers MS, PhD. Terminal master's awarded for partial completion of doctoral program. *Degree requirements:* For master's, thesis or alternative; for doctorate, thesis/dissertation, qualifying exam. *Entrance requirements:* For master's, GRE General Test; for doctorate, GRE General Test, GRE Subject Test. Additional exam requirements/recommendations for international students: Required—TOEFL (minimum score 590 paper-based; 243 computer-based), TWE. Electronic applications accepted. *Faculty research:* Cancer biology and developmental genetics, gene regulation and chromosome structure, microbiology and pathogenic mechanisms, structural biology, membrane biochemistry and signal transduction.

University of Cincinnati, Division of Research and Advanced Studies, College of Medicine, Graduate Programs in Biomedical Sciences, Department of Pediatrics Developmental Biology, Program in Molecular and Developmental Biology, Cincinnati, OH 45221. Offers PhD. *Degree requirements:* For doctorate, thesis/dissertation, qualifying exam. *Entrance requirements:* For doctorate, GRE General Test, minimum GPA of 3.2. Additional exam requirements/

Molecular Biology

University of Cincinnati (continued)
recommendations for international students: Required—TOEFL (minimum score 520 paper-based; 190 computer-based). Electronic applications accepted. *Faculty research:* Cancer biology, cardiovascular biology, developmental biology, human genetics, gene therapy, genomics and bioinformatics, immunobiology, molecular medicine, neuroscience, pulmonary biology, reproductive biology, stem cell biology.

See Close-Up on page 793.

University of Cincinnati, Division of Research and Advanced Studies, College of Medicine, Graduate Programs in Biomedical Sciences, Graduate Program in Cell and Molecular Biology, Cincinnati, OH 45221. Offers PhD. *Degree requirements:* For doctorate, thesis/dissertation, qualifying exam. *Entrance requirements:* For doctorate, GRE General Test. Additional exam requirements/recommendations for international students: Required—TOEFL, TWE; Recommended—TSE. Electronic applications accepted. *Faculty research:* Cancer cell biology, basic cell and molecular biology, neuroscience, developmental biology, endocrinology.

University of Colorado at Boulder, Graduate School, College of Arts and Sciences, Department of Molecular, Cellular, and Developmental Biology, Boulder, CO 80309. Offers cellular structure and function (MA, PhD); developmental biology (MA, PhD); molecular biology (MA, PhD). *Faculty:* 25 full-time (6 women). *Students:* 42 full-time (15 women), 26 part-time (12 women); includes 5 minority (1 American Indian/Alaska Native, 1 Asian American or Pacific Islander, 3 Hispanic Americans), 11 international. Average age 28. 19 applicants, 100% accepted. In 2005, 3 master's, 8 doctorates awarded. Terminal master's awarded for partial completion of doctoral program. *Degree requirements:* For master's, thesis or alternative, comprehensive exam; for doctorate, thesis/dissertation, comprehensive exam. *Entrance requirements:* For master's, GRE General Test, GRE Subject Test, minimum undergraduate GPA of 2.75; for doctorate, GRE General Test, GRE Subject Test. *Application deadline:* For fall admission, 1/15 for domestic students, 1/15 for international students. Application fee: $50 ($60 for international students). *Financial support:* In 2005–06, fellowships (averaging $7,489 per year), research assistantships (averaging $12,470 per year), teaching assistantships (averaging $13,078 per year) were awarded; tuition waivers (full) also available. Financial award application deadline: 3/1. *Faculty research:* Molecular biology of RNA and DNA, molecular genetics, cell motility and cytoskeleton, cell membranes, developmental genetics. Total annual research expenditures: $14.7 million. *Unit head:* Leslie Leinwand, Chair, 303-492-7606, Fax: 303-492-7744, E-mail: leslie.leinwand@stripe.colorado.edu. *Application contact:* Student Affairs Office, 303-492-7230, Fax: 303-492-7744, E-mail: mcdbgradinfo@beagle.colorado.edu.

University of Colorado at Denver and Health Sciences Center, Graduate School, Program in Biomedical Sciences, Program in Molecular Biology, Denver, CO 80262. Offers PhD. *Faculty:* 45. *Students:* 50 (30 women). In 2005, 6 degrees awarded. *Degree requirements:* For doctorate, thesis/dissertation. *Entrance requirements:* For doctorate, GRE, minimum GPA of 3.0, 4 letters of recommendation. Additional exam requirements/recommendations for international students: Required—TOEFL (minimum score 550 paper-based; 213 computer-based). *Application deadline:* For fall admission, 1/15 for domestic students. Application fee: $50. Electronic applications accepted. *Expenses:* Tuition, state resident: full-time $11,730. Tuition, nonresident: full-time $22,980. Tuition and fees vary according to degree level and program. *Financial support:* Fellowships, research assistantships, teaching assistantships, Federal Work-Study and institutionally sponsored loans available. Support available to part-time students. Financial award application deadline: 3/15; financial award applicants required to submit FAFSA. *Faculty research:* Gene transcription, RNA processing, chromosome dynamics, DNA damage and repair, chromatin assembly. *Unit head:* Dr. James DeGregori, Director, 303-724-3245, E-mail: james.degregori@uchsc.edu. *Application contact:* Jean Sibley, Administrator, 303-724-3245, Fax: 303-724-3247, E-mail: jean.sibley@uchsc.edu.

See Close-Up on page 609.

University of Connecticut, Graduate School, College of Liberal Arts and Sciences, Department of Molecular and Cell Biology, Field of Microbial Systems Analysis, Storrs, CT 06269. Offers MS, PSM. *Faculty:* 7 full-time (1 woman). *Students:* Average age 44. *Expenses:* Tuition, state resident: part-time $444 per credit hour. Tuition, nonresident: part-time $1,154 per credit hour. Tuition and fees vary according to course load. *Application contact:* Anne St. Onje, Graduate Coordinator, 860-486-4314, Fax: 860-486-3943, E-mail: ann.st_onje@uconn.edu.

University of Connecticut Health Center, Graduate School, Programs in Biomedical Sciences, Program in Molecular Biology and Biochemistry, Farmington, CT 06030. Offers PhD, DMD/PhD, MD/PhD. *Degree requirements:* For doctorate, thesis/dissertation, comprehensive exam, registration. *Entrance requirements:* For doctorate, GRE General Test. Additional exam requirements/recommendations for international students: Required—TOEFL (minimum score 600 paper-based; 250 computer-based). Electronic applications accepted.

See Close-Up on page 615.

University of Delaware, College of Arts and Sciences, Department of Biological Sciences, Newark, DE 19716. Offers biotechnology (MS); cancer biology (MS, PhD); cell and extracellular matrix biology (MS, PhD); cell and systems physiology (MS, PhD); developmental biology (MS, PhD); ecology and evolution (MS, PhD); microbiology (MS, PhD); molecular biology and genetics (MS, PhD). *Faculty:* 39 full-time (11 women). *Students:* 64 full-time (46 women), 2 part-time; includes 7 minority (4 African Americans, 2 Asian Americans or Pacific Islanders, 1 Hispanic American), 18 international. Average age 26. 113 applicants, 31% accepted, 19 enrolled. In 2005, 5 master's, 4 doctorates awarded. Terminal master's awarded for partial completion of doctoral program. *Median time to degree:* Of those who began their doctoral program in fall 1997, 100% received their degree in 8 years or less. *Degree requirements:* For master's, thesis/dissertation, preliminary exam; for doctorate, thesis/dissertation, preliminary exam, comprehensive exam. *Entrance requirements:* For master's and doctorate, GRE General Test. Additional exam requirements/recommendations for international students: Required—TOEFL (minimum score 600 paper-based; 250 computer-based); Recommended—TWE, TSE. *Application deadline:* For fall admission, 4/15 for domestic students, 1/15 for international students; for spring admission, 10/1 for domestic students. Applications are processed on a rolling basis. Application fee: $60. Electronic applications accepted. *Financial support:* In 2005–06, 26 students received support, including fellowships with full tuition reimbursements available (averaging $19,000 per year), 19 research assistantships with full tuition reimbursements available (averaging $19,000 per year), 26 teaching assistantships with full tuition reimbursements available (averaging $19,000 per year); tuition waivers (partial) also available. Financial award application deadline: 4/15. *Faculty research:* Microorganisms, bone, cancer metastasis, developmental biology, cell biology, DNA. Total annual research expenditures: $8.3 million. *Unit head:* Dr. Daniel D. Carson, Chair, 302-831-6977, Fax: 302-831-2281, E-mail: dcarson@udel.edu. *Application contact:* Dr. Melinda K. Duncan, Graduate Coordinator, 302-831-1841, Fax: 302-831-2281, E-mail: danders@udel.edu.

University of Florida, College of Medicine, Department of Biochemistry and Molecular Biology, Gainesville, FL 32611. Offers MS, PhD. *Faculty:* 34. *Students:* 7 applicants. In 2005, 1 degree awarded. *Degree requirements:* For doctorate, thesis/dissertation. *Entrance requirements:* For doctorate, GRE General Test, minimum GPA of 3.0. Additional exam requirements/recommendations for international students: Required—TOEFL. *Application deadline:* For fall admission, 2/15 for domestic students. Applications are processed on a rolling basis. Application fee: $30. Electronic applications accepted. *Expenses:* Tuition, state resident: full-time $6,234. Tuition, nonresident: full-time $21,359. Tuition and fees vary according to program. *Financial support:* In 2005–06, research assistantships with full tuition reimbursements (averaging $24,517 per year); fellowships with full tuition reimbursements, traineeships and unspecified assistantships also available. Financial award application deadline: 3/1. *Faculty research:* Gene expression, metabolic regulation, structural biology, enzyme mechanism, membrane transporters. *Unit head:* Dr. James Flanegan, Director, 352-392-0688, E-mail: flanegan@ufl.

edu. *Application contact:* Dr. Wayne McCormack, Program Director, 352-392-7413, Fax: 352-846-3466, E-mail: idp@ufl.edu.

University of Florida, College of Medicine and Graduate School, Interdisciplinary Program in Biomedical Sciences, Concentration in Biochemistry and Molecular Biology, Gainesville, FL 32611. Offers PhD. *Faculty:* 27 full-time (7 women). *Students:* 28 full-time (15 women); includes 6 minority (3 African Americans, 3 Asian Americans or Pacific Islanders). In 2005, 1 degree awarded. *Degree requirements:* For doctorate, thesis/dissertation. *Entrance requirements:* For doctorate, GRE General Test, minimum GPA of 3.0. Additional exam requirements/recommendations for international students: Required—TOEFL. *Application deadline:* For fall admission, 2/5 for domestic students. Applications are processed on a rolling basis. Application fee: $30. Electronic applications accepted. *Expenses:* Tuition, state resident: full-time $6,234. Tuition, nonresident: full-time $21,359. Tuition and fees vary according to program. *Financial support:* In 2005–06, 26 research assistantships with full tuition reimbursements (averaging $24,517 per year) were awarded; fellowships with full tuition reimbursements *Faculty research:* Gene expression, metabolic regulation, structural biology, enzyme mechanism, membrane transporters. *Unit head:* Dr. Michael S. Kilberg, Director, 352-392-2711, E-mail: mkilberg@biochem.med.ufl.edu. *Application contact:* Dr. Wayne McCormack, Associate Dean of Graduate Education, 352-392-7413, Fax: 352-846-3466, E-mail: mccormac@pathology.ufl.edu.

University of Georgia, Graduate School, College of Arts and Sciences, Department of Biochemistry and Molecular Biology, Athens, GA 30602. Offers MS, PhD. *Faculty:* 18 full-time (2 women). *Students:* 45 full-time, 4 part-time; includes 8 minority (5 African Americans, 2 Asian Americans or Pacific Islanders, 1 Hispanic American), 11 international. 94 applicants, 12% accepted, 9 enrolled. In 2005, 4 master's, 7 doctorates awarded. *Degree requirements:* For master's and doctorate, one foreign language, thesis/dissertation. *Entrance requirements:* For master's and doctorate, GRE General Test. Additional exam requirements/recommendations for international students: Required—TOEFL, TSE. *Application deadline:* For fall admission, 1/1 priority date for domestic students, 1/1 priority date for international students. Application fee: $50. Electronic applications accepted. *Financial support:* Fellowships, research assistantships, teaching assistantships, scholarships/grants and unspecified assistantships available. Financial award application deadline: 3/1. *Unit head:* Dr. J. David Puett, Head, 706-542-1676, Fax: 706-542-0182, E-mail: puett@bmb.uga.edu. *Application contact:* Dr. Alan E. Przybyla, Graduate Coordinator, 706-542-1728, Fax: 706-542-1738, E-mail: przybyla@bmb.uga.edu.

University of Guelph, Graduate Program Services, College of Biological Science, Department of Molecular and Cellular Biology, Guelph, ON N1G 2W1, Canada. Offers biochemistry (M Sc, PhD); biophysics (M Sc, PhD); botany (M Sc, PhD); microbiology (M Sc, PhD); molecular biology and genetics (M Sc, PhD). *Faculty:* 43 full-time (6 women). *Students:* 136 full-time (52 women). Average age 26. 56 applicants, 30% accepted, 17 enrolled. In 2005, 4 master's, 2 doctorates awarded. *Degree requirements:* For master's, thesis/dissertation, research proposal; for doctorate, thesis/dissertation, research proposal, comprehensive exam, registration. *Entrance requirements:* Additional exam requirements/recommendations for international students: Required—TOEFL (minimum score 550 paper-based; 213 computer-based), IELT (minimum score 7). *Application deadline:* For fall admission, 7/1 for domestic students, 1/1 for international students. For winter admission, 11/1 for domestic students; for spring admission, 3/1 for domestic students. Applications are processed on a rolling basis. Application fee: $75. Electronic applications accepted. *Financial support:* Fellowships, research assistantships, teaching assistantships available. Support available to part-time students. *Faculty research:* Physiology, structure, genetics, and ecology of microbes; virology and microbial technology. *Unit head:* Dr. Chris Whitfield, Chair, 519-824-4120 Ext. 53361, Fax: 519-827-1802, E-mail: cwhitfie@uoguelph.ca. *Application contact:* Laurie Winn, Graduate Admissions Secretary, 519-824-4320 Ext. 52730, Fax: 519-767-1656, E-mail: lwinn@uoguelph.ca.

University of Hawaii at Manoa, John A. Burns School of Medicine and Graduate Division, Graduate Programs in Biomedical Sciences, Department of Cell and Molecular Biology, Honolulu, HI 96822. Offers MS, PhD. *Faculty:* 61 full-time (21 women), 7 part-time/adjunct (3 women). *Students:* 28 full-time (13 women); 5 part-time (3 women); includes 12 minority (11 Asian Americans or Pacific Islanders, 1 Hispanic American), 6 international. Average age 29. 62 applicants, 19% accepted, 10 enrolled. In 2005, 2 master's, 3 doctorates awarded. Terminal master's awarded for partial completion of doctoral program. *Degree requirements:* For master's, thesis (for some programs); for doctorate, thesis/dissertation. *Entrance requirements:* For master's and doctorate, GRE, minimum GPA of 3.0. *Application deadline:* For fall admission, 2/15 for domestic students, 2/15 for international students. Applications are processed on a rolling basis. Application fee: $50. *Expenses:* Tuition, state resident: full-time $8,400; part-time $200 per credit hour. Tuition, nonresident: full-time $11,088; part-time $462 per credit hour. Tuition and fees vary according to program. *Financial support:* In 2005–06, 20 research assistantships (averaging $16,774 per year), 3 teaching assistantships (averaging $14,382 per year) were awarded; Federal Work-Study and institutionally sponsored loans also available. Financial award application deadline: 2/1. *Unit head:* Dr. David Haymer, 808-956-7862, Fax: 808-956-5506, E-mail: dhaymer@hawaii.edu.

University of Idaho, College of Graduate Studies, College of Agricultural and Life Sciences, Department of Microbiology, Molecular Biology and Biochemistry, Moscow, ID 83844-2282. Offers MS, PhD. *Students:* 30 full-time (15 women), 8 part-time (3 women), 19 international. Average age 30. In 2005, 5 master's, 4 doctorates awarded. *Degree requirements:* For master's, thesis; for doctorate, one foreign language, thesis/dissertation. *Entrance requirements:* For master's, minimum GPA of 2.8; for doctorate, minimum undergraduate GPA of 2.8, 3.0 graduate. *Application deadline:* For fall admission, 8/1 for domestic students; for spring admission, 12/15 for domestic students. Application fee: $55 ($60 for international students). *Expenses:* Tuition, state resident: full-time $4,508. Tuition, nonresident: full-time $8,770; part-time $130 per credit. Required fees: $217 per credit. *Financial support:* Research assistantships, teaching assistantships available. Financial award application deadline: 2/15. *Faculty research:* Environmental fields, food, immunology. *Unit head:* Dr. Patricia Hartzell, Head, 208-885-0572.

University of Illinois at Chicago, College of Medicine and Graduate College, Graduate Programs in Medicine, Department of Biochemistry and Molecular Biology, Chicago, IL 60607-7128. Offers MS, PhD. Terminal master's awarded for partial completion of doctoral program. *Degree requirements:* For master's and doctorate, thesis/dissertation. *Entrance requirements:* For master's and doctorate, GRE General Test. Additional exam requirements/recommendations for international students: Required—TOEFL. Electronic applications accepted. *Faculty research:* Nature of cellular components, control of metabolic processes, regulation of gene expression.

University of Illinois at Chicago, Graduate College, College of Liberal Arts and Sciences, Department of Biological Sciences, Laboratory for Molecular Biology, Chicago, IL 60607-7128. Offers MS, PhD. *Degree requirements:* For master's, thesis; for doctorate, thesis/dissertation, preliminary exam. *Entrance requirements:* For master's and doctorate, GRE General Test, minimum undergraduate GPA of 3.0 during final 2 years. Additional exam requirements/recommendations for international students: Required—TOEFL.

University of Illinois at Urbana–Champaign, Graduate College, College of Liberal Arts and Sciences, Cell and Molecular Biology Training Program, Champaign, IL 61820.

The University of Iowa, Graduate College, Program in Molecular Biology, Iowa City, IA 52242-1316. Offers PhD, MD/PhD. *Students:* 9 full-time (5 women), 23 part-time (9 women); includes 5 minority (3 Asian Americans or Pacific Islanders, 2 Hispanic Americans), 8 international. 40 applicants, 25% accepted, 5 enrolled. In 2005, 9 degrees awarded. *Degree requirements:* For doctorate, thesis/dissertation, comprehensive exam, registration. *Entrance requirements:* For doctorate, GRE General Test, minimum GPA of 3.0. Additional exam requirements/recommendations for international students: Required—TOEFL (minimum score 600 paper-based; 250 computer-based). *Application deadline:* For fall admission, 2/1 for domestic students,

2/1 for international students. Application fee: $60 ($85 for international students). Electronic applications accepted. *Expenses:* Tuition, state resident: part-time $1,882 per term. Tuition, nonresident: full-time $17,338; part-time $4,907 per term. Tuition and fees vary according to course load and program. *Financial support:* In 2005–06, 1 fellowship, 29 research assistantships with partial tuition reimbursements were awarded; teaching assistantships with partial tuition reimbursements, scholarships/grants also available. Financial award applicants required to submit FAFSA. *Faculty research:* Regulation of gene expression, inherited human genetic diseases, signal transduction mechanisms, structural biology and function. *Unit head:* , Minnetta Gardinier, Director, 319-335-7748, E-mail: curt-sigmund@uiowa.edu. *Application contact:* Paulette Scheler, Program Associate, 800-551-7748.

See Close-Up on page 617.

The University of Iowa, Opportunities for Interdisciplinary Research in Biomolecular Science Program (ORB), Iowa City, IA 52242-1316.

University of Kansas, Graduate School, College of Liberal Arts and Sciences, Division of Biological Sciences, Department of Molecular Biosciences, Lawrence, KS 66045. Offers biochemistry and biophysics (MA, PhD); microbiology (MA, PhD); molecular, cellular, and developmental biology (MA, PhD). *Faculty:* 44. *Students:* 48 full-time (28 women), 5 part-time (3 women); includes 1 minority (Asian American or Pacific Islander), 26 international. Average age 27. 111 applicants, 19% accepted. In 2005, 5 master's, 5 doctorates awarded. Terminal master's awarded for partial completion of doctoral program. *Degree requirements:* For master's and doctorate, thesis/dissertation, comprehensive exam. *Entrance requirements:* For master's and doctorate, GRE General Test. Additional exam requirements/recommendations for international students: Required—TOEFL, IELT, TOEFL (paper 570; computer 230) or IELT; Recommended—TWE, TSE. *Application deadline:* For fall admission, 1/15 priority date for domestic students, 1/15 priority date for international students. Applications are processed on a rolling basis. Application fee: $55 ($60 for international students). Electronic applications accepted. *Expenses:* Tuition, state resident: full-time $4,859. Tuition, nonresident: full-time $12,000. Required fees: $589. Tuition and fees vary according to program. *Financial support:* Fellowships, research assistantships with tuition reimbursements, teaching assistantships with tuition reimbursements available. Financial award application deadline: 3/1. *Faculty research:* Structure and function of proteins, genetics of organism development, molecular genetics, neurophysiology, molecular virology and pathogenics, developmental biology, cell biology. *Unit head:* , Kathy Suprenant, Chair, 785-864-4631, Fax: 785-864-5294, E-mail: ksupre@ku.edu. *Application contact:* John P. Connolly, Information Contact, 785-864-4311, Fax: 785-864-5924, E-mail: jconnolly@ku.edu.

University of Kansas, Graduate Studies Medical Center, Interdisciplinary Graduate Program in Biomedical Sciences, Department of Biochemistry and Molecular Biology, Lawrence, KS 66045. Offers MS, PhD, MD/PhD. Part-time programs available. *Faculty:* 16. *Students:* 2 full-time (0 women), 6 part-time (2 women); includes 1 Hispanic American, 1 international. Average age 29. In 2005, 1 degree awarded. Terminal master's awarded for partial completion of doctoral program. *Median time to degree:* Of those who began their doctoral program in fall 1997, 100% received their degree in 8 years or less. *Degree requirements:* For master's, oral defense of thesis; for doctorate, comprehensive oral and written exam. *Expenses:* Tuition, state resident: full-time $4,859. Tuition, nonresident: full-time $12,000. Required fees: $589. Tuition and fees vary according to program. *Financial support:* Fellowships, research assistantships with partial tuition reimbursements, teaching assistantships with full and partial tuition reimbursements, unspecified assistantships available. *Faculty research:* Regulation of gene expression, molecular genetics of kidney disease, molecular chaperones, glycoproteins, cell cycle control. Total annual research expenditures: $1.9 million. *Unit head:* Dr. Gerald M. Carlson, Chairman, 913-588-6574, Fax: 913-588-7007, E-mail: gcarlson@kumc.edu. *Application contact:* Dr. Glen Andrews, Director of Graduate Studies, 913-588-6935, Fax: 913-588-7440, E-mail: gandrews@kumc.edu.

See Close-Up on page 429.

University of Lethbridge, School of Graduate Studies, Lethbridge, AB T1K 3M4, Canada. Offers accounting (MScM); addictions counseling (M Sc); agricultural biotechnology (M Sc); agricultural studies (M Sc, MA); anthropology (MA); archaeology (MA); art (MA); biochemistry (M Sc); biological sciences (M Sc); biomolecular science (PhD); biosystems and biodiversity (PhD); Canadian studies (MA); chemistry (M Sc); computer science (M Sc); computer science and geographical information science (M Sc); counseling psychology (M Ed); dramatic arts (MA); earth, space, and physical science (PhD); economics (MA); educational leadership (M Ed); English (MA); environmental science (M Sc); evolution and behavior (PhD); exercise science (M Sc); finance (MScM); French (MA); French/German (MA); French/Spanish (MA); general education (M Ed); general management (MScM); geography (M Sc, MA); German (MA); health sciences (M Sc, MA); history (MA); human resource management and labour relations (MScM); individualized multidisciplinary (M Sc, MA); information systems (MScM); international management (MScM); kinesiology (M Sc, MA); management (M Sc, MA); marketing (MScM); mathematics (M Sc); music (MA); Native American studies (MA); neuroscience (M Sc, PhD); new media (MA); nursing (M Sc); philosophy (MA); physics (M Sc); policy and strategy (MScM); political science (MA); psychology (M Sc, MA); religious studies (MA); sociology (MA); theoretical and computational science (PhD); urban and regional studies (MA). Part-time and evening/weekend programs available. *Faculty:* 250. *Students:* 193 full-time, 145 part-time. 35 applicants, 100% accepted, 35 enrolled. In 2005, 40 degrees awarded. *Degree requirements:* For doctorate, thesis/dissertation, comprehensive exam. *Entrance requirements:* For master's, GMAT (M Sc management), bachelor's degree in related field, minimum GPA of 3.0 during previous 20 graded semester courses, 2 years teaching or related experience (M Ed); for doctorate, master's degree, minimum graduate GPA of 3.5. Additional exam requirements/recommendations for international students: Required—TOEFL. Application fee: $60 Canadian dollars. *Expenses:* Tuition, nonresident: part-time $531 per course. Required fees: $83 per year. Tuition and fees vary according to degree level and program. *Financial support:* Fellowships, research assistantships, teaching assistantships, scholarships/grants, health care benefits, and unspecified assistantships available. *Faculty research:* Movement and brain plasticity, gibberellin physiology, photosynthesis, carbon cycling, molecular properties of main-group ring components. *Unit head:* Dr. Shamsul Alam, Dean, 403-329-2121, Fax: 403-329-2097, E-mail: inquiries@uleth.ca. *Application contact:* Kathy Schrage, Administrative Assistant, Office of the Academic Vice President, 403-329-2121, Fax: 403-329-2097, E-mail: inquiries@uleth.ca.

University of Louisville, School of Medicine, Department of Biochemistry and Molecular Biology, Louisville, KY 40292-0001. Offers MS, PhD, MD/MS, MD/PhD. *Students:* 28 full-time (15 women), 13 part-time (6 women); includes 7 minority (3 African Americans, 3 Asian Americans or Pacific Islanders, 1 Hispanic American), 10 international. Average age 30. In 2005, 2 master's, 3 doctorates awarded. *Degree requirements:* For master's, thesis/dissertation; for doctorate, thesis/dissertation, comprehensive exam. *Entrance requirements:* For master's and doctorate, GRE General Test, minimum GPA of 3.0. Additional exam requirements/recommendations for international students: Required—TOEFL. *Application deadline:* For fall admission, 1/15 for domestic students. Application fee: $50. Electronic applications accepted. *Expenses:* Tuition, state resident: full-time $6,006; part-time $334 per credit hour. Tuition, nonresident: full-time $16,554; part-time $920 per credit hour. Tuition and fees vary according to course load, degree level and program. *Financial support:* Fellowships with tuition reimbursements, research assistantships with tuition reimbursements, teaching assistantships with tuition reimbursements, scholarships/grants, traineeships, tuition waivers (full and partial), and unspecified assistantships available. *Unit head:* Dr. Kenneth S. Ramos, Chair, 502-852-5217, Fax: 502-852-6222, E-mail: ksramos01@louisville.edu. *Application contact:* Dr. Steven R. Ellis, Contact, 502-852-5221, Fax: 502-852-6222, E-mail: srellis@louisville.edu.

University of Maine, Graduate School, College of Natural Sciences, Forestry, and Agriculture, Department of Biochemistry, Molecular Biology, and Microbiology, Orono, ME 04469. Offers biochemistry (MPS, MS); biochemistry and molecular biology (PhD); microbiology (MPS, MS,

PhD). *Students:* 40 full-time (23 women), 17 part-time (11 women); includes 2 minority (1 Asian American or Pacific Islander, 1 Hispanic American), 13 international. Average age 31. 42 applicants, 48% accepted, 11 enrolled. In 2005, 5 master's awarded. *Degree requirements:* For doctorate, thesis/dissertation. *Entrance requirements:* For master's and doctorate, GRE General Test. Additional exam requirements/recommendations for international students: Required—TOEFL. *Application deadline:* For fall admission, 2/1 for domestic students. Applications are processed on a rolling basis. Application fee: $50. Electronic applications accepted. *Financial support:* In 2005–06, 5 research assistantships with tuition reimbursements (averaging $18,000 per year), 12 teaching assistantships with tuition reimbursements (averaging $16,000 per year) were awarded; tuition waivers (full and partial) also available. Financial award application deadline: 3/1. *Unit head:* Dr. John Singer, Chair, 207-581-2810, Fax: 207-581-2801. *Application contact:* Scott G. Delcourt, Associate Dean of the Graduate School, 207-581-3219, Fax: 207-581-3232, E-mail: graduate@maine.edu.

University of Maryland, Graduate School, Interdisciplinary Training Program in Muscle Biology, Baltimore, MD 21201.

See Close-Up on page 431.

University of Maryland, School of Medicine, Graduate Program in Life Sciences, Program in Biochemistry and Molecular Biology, Baltimore, MD 21201. Offers PhD. *Expenses:* Tuition, state resident: full-time $8,079; part-time $409 per credit hour. Tuition, nonresident: full-time $18,384; part-time $731 per credit hour. Required fees: $695; $10 per credit hour. Tuition and fees vary according to degree level and program.

University of Maryland, Baltimore County, Graduate School, College of Natural and Mathematical Sciences, Department of Biological Sciences, Program in Applied Molecular Biology, Baltimore, MD 21250. Offers MS. *Faculty:* 28 full-time (8 women), 2 part-time/ adjunct (0 women). *Students:* 11 full-time (6 women); includes 2 minority (1 African American, 1 Asian American or Pacific Islander). 30 applicants, 53% accepted, 11 enrolled. In 2005, 9 degrees awarded. *Entrance requirements:* For master's, GRE General Test, GRE Subject Test (recommended), minimum GPA 3.0. Additional exam requirements/recommendations for international students: Required—TOEFL. *Application deadline:* For fall admission, 4/1 priority date for domestic students, 4/1 priority date for international students. Applications are processed on a rolling basis. Application fee: $50. Electronic applications accepted. *Expenses:* Tuition, state resident: part-time $395 per credit. Tuition, nonresident: part-time $652 per credit. Required fees: $82 per credit. Tuition and fees vary according to course load, program and reciprocity agreements. *Financial support:* In 2005–06, 6 students received support, including 6 teaching assistantships with tuition reimbursements available (averaging $12,000 per year); fellowships, research assistantships, tuition waivers (partial) also available. *Faculty research:* Structure-function of RNA, genetics and molecular biology, biological chemistry. *Unit head:* , Dr. Richard E. Wolf, Director, Graduate Program, 410-455-3669, Fax: 410-455-3875, E-mail: biograd@umbc.edu. *Application contact:* Dr. Richard E. Wolf, Director, Graduate Program, 410-455-3669, Fax: 410-455-3875, E-mail: biograd@umbc.edu.

University of Maryland, Baltimore County, Graduate School, College of Natural and Mathematical Sciences, Department of Biological Sciences, Program in Molecular and Cell Biology, Baltimore, MD 21250. Offers PhD. *Faculty:* 28 full-time (8 women), 2 part-time/ adjunct (0 women). *Students:* 19 full-time (8 women); includes 8 minority (2 African Americans, 6 Asian Americans or Pacific Islanders). 35 applicants, 49% accepted, 4 enrolled. In 2005, 3 degrees awarded. *Degree requirements:* For doctorate, thesis/dissertation. *Entrance requirements:* For doctorate, GRE General Test, GRE Subject Test, minimum GPA of 3.0. Additional exam requirements/recommendations for international students: Required—TOEFL. *Application deadline:* For fall admission, 2/1 for domestic students, 1/1 for international students. Applications are processed on a rolling basis. Application fee: $50. Electronic applications accepted. *Expenses:* Tuition, state resident: part-time $395 per credit. Tuition, nonresident: part-time $652 per credit. Required fees: $82 per credit. Tuition and fees vary according to course load, program and reciprocity agreements. *Financial support:* In 2005–06, 19 students received support, including fellowships with full tuition reimbursements available (averaging $23,000 per year), research assistantships with full tuition reimbursements available (averaging $21,500 per year), teaching assistantships with full tuition reimbursements available (averaging $20,500 per year) *Unit head:* Dr. Phyllis Robinson, Director, 410-455-3669, Fax: 410-455-3875, E-mail: biograd@umbc.edu.

University of Maryland, College Park, Graduate Studies, College of Chemical and Life Sciences, Department of Cell Biology and Molecular Genetics, Program in Molecular and Cellular Biology, College Park, MD 20742. Offers PhD. Part-time and evening/weekend programs available. *Students:* 36 full-time (20 women), 1 (woman) part-time; includes 5 minority (1 African American, 3 Asian Americans or Pacific Islanders, 1 Hispanic American), 20 international. 119 applicants, 18% accepted, 9 enrolled. In 2005, 8 degrees awarded. *Median time to degree:* Of those who began their doctoral program in fall 1997, 75% received their degree in 8 years or less. *Degree requirements:* For doctorate, thesis/dissertation, exam, public service. *Entrance requirements:* For doctorate, GRE General Test, 3 letters of reference. Additional exam requirements/recommendations for international students: Required—TOEFL; Recommended—TSE. *Application deadline:* For fall admission, 1/7 for domestic students, 1/7 for international students. Applications are processed on a rolling basis. Application fee: $60. Electronic applications accepted. *Financial support:* In 2005–06, 8 fellowships with full tuition reimbursements (averaging $3,013 per year) were awarded Financial award applicants required to submit FAFSA. *Faculty research:* Monoclonal antibody production, oligonucleotide synthesis, macronular processing, signal transduction, developmental biology. *Unit head:* Dr. Stephen Wolniak, Chair, 301-405-5430, Fax: 301-314-9489, E-mail: swolniak@umd.edu. *Application contact:* Dean of Graduate School, 301-405-4190, Fax: 301-314-9305.

University of Massachusetts Amherst, Graduate School, College of Natural Sciences and Mathematics, Program in Molecular and Cellular Biology, Amherst, MA 01003. Offers biological chemistry (PhD); cell and developmental biology (PhD). Part-time programs available. *Faculty:* 1 full-time (0 women). *Students:* 87 full-time (43 women), 1 part-time; includes 9 minority (1 African American, 4 Asian Americans or Pacific Islanders, 4 Hispanic Americans), 41 international. Average age 30. 219 applicants, 23% accepted, 13 enrolled. In 2005, 8 doctorates awarded. *Degree requirements:* For doctorate, one foreign language, thesis/dissertation. *Entrance requirements:* For doctorate, GRE General Test. Additional exam requirements/recommendations for international students: Required—TOEFL (minimum score 530 paper-based; 197 computer-based). *Application deadline:* For fall admission, 1/15 priority date for domestic students, 1/15 priority date for international students. Applications are processed on a rolling basis. Application fee: $40 ($65 for international students). Electronic applications accepted. *Expenses:* Tuition, state resident: part-time $110 per credit. Tuition, nonresident: part-time $414 per credit. Required fees: $2,824 per term. One-time fee: $250 part-time. Full-time tuition and fees vary according to course load, campus/location, program and reciprocity agreements. *Financial support:* In 2005–06, 6 fellowships with full tuition reimbursements, 22 research assistantships with full tuition reimbursements (averaging $3,451 per year), 4 teaching assistantships with full tuition reimbursements (averaging $8,696 per year) were awarded; career-related internships or fieldwork, Federal Work-Study, scholarships/grants, traineeships, and unspecified assistantships also available. Support available to part-time students. Financial award application deadline: 1/15. *Unit head:* Dr. Rodney K. Murphey, Head, 413-545-3246, Fax: 413-545-1812.

See Close-Up on page 619.

University of Massachusetts Boston, Office of Graduate Studies and Research, College of Science and Mathematics, Track in Molecular, Cellular and Organismal Biology, Boston, MA 02125-3393. Offers PhD.

University of Medicine and Dentistry of New Jersey, Graduate School of Biomedical Sciences, Graduate Programs in Biomedical Sciences–Newark, Department of Biochemistry and Molecular Biology, Newark, NJ 07107. Offers MS, PhD. *Degree requirements:* For master's, thesis; for doctorate, thesis/dissertation, qualifying exam. *Entrance requirements:* For master's

Molecular Biology

University of Medicine and Dentistry of New Jersey *(continued)*
and doctorate, GRE General Test. Additional exam requirements/recommendations for international students: Required—TOEFL. Application fee: $40. *Financial support:* Fellowships, research assistantships, Federal Work-Study, institutionally sponsored loans, and tuition waivers (full and partial) available. Financial award application deadline: 5/1. *Unit head:* Dr. Makund Modak, Program Director, 973-972-4053, Fax: 973-972-5594, E-mail: modak@umdnj. edu. *Application contact:* Dr. Henry E. Brezenoff, Acting Dean, 973-972-5333, Fax: 973-972-7148, E-mail: hbrezeno@umdnj.edu.

University of Medicine and Dentistry of New Jersey, Graduate School of Biomedical Sciences, Graduate Programs in Biomedical Sciences–Piscataway, Program in Biochemistry and Molecular Biology, Piscataway, NJ 08854-5635. Offers MS, PhD, MD/PhD. Terminal master's awarded for partial completion of doctoral program. *Degree requirements:* For master's and doctorate, thesis/dissertation, qualifying exam. *Entrance requirements:* For master's and doctorate, GRE General Test. Additional exam requirements/recommendations for international students: Required—TOEFL. *Application deadline:* For fall admission, 1/5 for domestic students. Application fee: $40. *Financial support:* Fellowships, research assistantships, teaching assistantships available. Financial award application deadline: 5/1. *Faculty research:* Signal transduction, regulation of RNA, polymerase II transcribed genes, developmental gene expression. *Unit head:* Dr. Kiran Madura, Director, 732-235-4595, Fax: 732-235-4783, E-mail: maduraki@umdnj.edu.

University of Medicine and Dentistry of New Jersey, Graduate School of Biomedical Sciences, Graduate Programs in Biomedical Sciences–Stratford, Program in Cell and Molecular Biology, Stratford, NJ 08084-5634. Offers MS, PhD, DO/PhD. *Degree requirements:* For master's, thesis; for doctorate, thesis/dissertation, qualifying exam. *Entrance requirements:* For master's and doctorate, GRE General Test. Additional exam requirements/recommendations for international students: Required—TOEFL. *Application deadline:* For fall admission, 2/1 for domestic students; for spring admission, 10/1 for domestic students. Applications are processed on a rolling basis. Application fee: $40. *Financial support:* Application deadline: 5/1. *Unit head:* Dr. Diane M Worrad, Program Coordinator, 856-566-6282, Fax: 856-566-6232, E-mail: worraddi@umdnj.edu.

University of Medicine and Dentistry of New Jersey, Graduate School of Biomedical Sciences, Programs in the Molecular Biosciences, Piscataway, NJ 08854-5696.

See Close-Up on page 589.

University of Miami, Graduate School, Miller School of Medicine, Graduate Programs in Medicine, Department of Biochemistry and Molecular Biology, Coral Gables, FL 33124. Offers PhD, MD/PhD. *Faculty:* 29 full-time (6 women). *Students:* 47 full-time (20 women); includes 8 minority (1 African American, 7 Hispanic Americans), 24 international. Average age 29. 84 applicants, 14% accepted, 7 enrolled. In 2005, 6 degrees awarded. *Median time to degree:* Of those who began their doctoral program in fall 1997, 98% received their degree in 8 years or less. *Degree requirements:* For doctorate, thesis/dissertation, proposition exams, comprehensive exam. *Entrance requirements:* For doctorate, GRE General Test. Additional exam requirements/recommendations for international students: Required—TOEFL. *Application deadline:* Applications are processed on a rolling basis. Application fee: $50. Electronic applications accepted. *Financial support:* In 2005–06, 34 fellowships with full tuition reimbursements (averaging $22,000 per year) were awarded; research assistantships, scholarships/grants and tuition waivers (full) also available. *Faculty research:* Macromolecule metabolism, molecular genetics, protein folding and 3-D structure, regulation of gene expression and enzyme function, signal transduction and developmental biology. Total annual research expenditures: $3 million. *Unit head:* , Dr. Murray P. Deutscher, Chairman, 305-243-3150, Fax: 305-243-3955, E-mail: mdeutsch@mednet.med.miami.edu. *Application contact:* Dr. Rudolf Werner, Director, 305-243-6998, Fax: 305-243-3955, E-mail: rwerner@molbio.med.miami.edu.

University of Miami, Graduate School, Miller School of Medicine, Graduate Programs in Medicine, Department of Cell Biology and Anatomy, Coral Gables, FL 33124. Offers molecular cell and developmental biology (PhD). *Faculty:* 20 full-time (6 women). *Students:* 25 full-time (10 women); includes 4 minority (1 Asian American or Pacific Islander, 3 Hispanic Americans), 13 international. Average age 29. 57 applicants, 7% accepted, 4 enrolled. In 2005, 1 degree awarded. *Degree requirements:* For doctorate, thesis/dissertation. *Entrance requirements:* For doctorate, GRE General Test, GRE Subject Test. Additional exam requirements/recommendations for international students: Required—TOEFL. *Application deadline:* For fall admission, 3/1 for domestic students. Applications are processed on a rolling basis. Application fee: $50. Electronic applications accepted. *Financial support:* In 2005–06, 22 fellowships with tuition reimbursements (averaging $22,000 per year) were awarded *Unit head:* Dr. Robert Warren, Interim Chair, 305-243-6691, Fax: 305-545-7166, E-mail: rwarren@med.miami.edu. *Application contact:* Dr. Richard Rotundo, Professor, 305-243-6691, Fax: 305-545-7166.

University of Michigan, Horace H. Rackham School of Graduate Studies, College of Literature, Science, and the Arts, Department of Molecular, Cellular, and Developmental Biology, Ann Arbor, MI 48109. Offers MS, PhD. Part-time programs available. *Faculty:* 24 full-time (6 women), 1 part-time/adjunct (0 women). *Students:* 72 full-time, 9 part-time; includes 5 minority (2 African Americans, 1 Asian American or Pacific Islander, 2 Hispanic Americans), 49 international. Average age 24. 102 applicants, 29% accepted, 18 enrolled. In 2005, 5 master's, 7 doctorates awarded. Terminal master's awarded for partial completion of doctoral program. *Median time to degree:* Of those who began their doctoral program in fall 1997, 60% received their degree in 8 years or less. *Degree requirements:* For master's, registration; for doctorate, thesis/dissertation, preliminary exam, oral defense, dissertation. *Entrance requirements:* For master's and doctorate, GRE General Test. Additional exam requirements/recommendations for international students: Required—TOEFL (minimum score 560 paper-based; 220 computer-based). *Application deadline:* For fall admission, 1/5 for domestic students, 1/5 for international students. For winter admission, 11/1 for domestic students. Applications are processed on a rolling basis. Application fee: $60 ($75 for international students). Electronic applications accepted. *Expenses:* Tuition, state resident: full-time $14,082; part-time $894 per credit hour. Tuition, nonresident: full-time $28,500; part-time $1,675 per credit hour. Required fees: $189; $189 per unit. *Financial support:* In 2005–06, 66 students received support, including 18 fellowships with full tuition reimbursements available (averaging $14,326 per year), 24 research assistantships with full tuition reimbursements available (averaging $14,326 per year), 21 teaching assistantships with full tuition reimbursements available (averaging $14,326 per year); career-related internships or fieldwork, scholarships/grants, traineeships, health care benefits, and unspecified assistantships also available. *Faculty research:* Biochemistry, cell biology, microbiology, neurobiology and physiology, developmental biology and plant molecular biology. Total annual research expenditures: $5.5 million. *Unit head:* Dr. Richard Hume, Chair, 734-764-7427, Fax: 734-746-0884. *Application contact:* Mary Carr, Graduate Coordinator, 734-615-1635, Fax: 734-764-0884, E-mail: carrmm@umich.edu.

See Close-Up on page 625.

University of Michigan, Medical School and Horace H. Rackham School of Graduate Studies, Program in Biomedical Sciences (PIBS), Interdisciplinary Program in Cellular and Molecular Biology, Ann Arbor, MI 48109. Offers PhD. *Faculty:* 113 part-time/adjunct (32 women). *Students:* 66 full-time (22 women); includes 24 minority (7 African Americans, 12 Asian Americans or Pacific Islanders, 5 Hispanic Americans). Average age 26. In 2005, 10 degrees awarded. *Median time to degree:* Of those who began their doctoral program in fall 1997, 100% received their degree in 8 years or less. *Degree requirements:* For doctorate, thesis/dissertation, oral defense of dissertation, preliminary exam, comprehensive exam. *Entrance requirements:* For doctorate, GRE General Test, GRE Subject Test. *Expenses:* Tuition, state resident: full-time $14,082; part-time $894 per credit hour. Tuition, nonresident: full-time $28,500; part-time $1,675 per credit hour. Required fees: $189; $189 per unit. *Financial support:* In 2005–06, 66 students received support, including 23 fellowships with tuition reimbursements available (averaging $23,500 per year), 43 research assistantships with tuition reimbursements available (averaging $23,500 per year); teaching assistantships, scholarships/grants also available. *Faculty research:* Genetics, genomics, gene regulation, models of disease, microbes. Total annual research expenditures: $20 million. *Unit head:* Dr. Jessica Schwartz, Director, 734-764-5428, Fax: 734-647-6232, E-mail: jeschwar@umich.edu. *Application contact:* Linda Wilson, Student Services Associate I, 734-764-5428, Fax: 734-647-6232, E-mail: cmbgrad@umich. edu.

University of Minnesota, Duluth, Medical School, Department of Biochemistry, Molecular Biology and Biophysics, Duluth, MN 55812-2496. Offers MS, PhD. *Faculty:* 8 full-time (1 woman), 2 part-time/adjunct (1 woman). Average age 26. 5 applicants, 40% accepted, 2 enrolled.Terminal master's awarded for partial completion of doctoral program. *Degree requirements:* For master's and doctorate, thesis/dissertation. *Entrance requirements:* For master's and doctorate, GRE General Test. Additional exam requirements/recommendations for international students: Required—TOEFL. *Application deadline:* For fall admission, 1/2 priority date for domestic students, 1/2 priority date for international students. For winter admission, 3/15 for domestic students. Application fee: $55 ($75 for international students). *Financial support:* In 2005–06, 4 students received support, including fellowships with full tuition reimbursements available (averaging $22,500 per year), research assistantships with full tuition reimbursements available (averaging $18,000 per year), teaching assistantships with tuition reimbursements available (averaging $18,000 per year); Federal Work-Study, institutionally sponsored loans, health care benefits, and unspecified assistantships also available. Financial award application deadline: 4/1. *Faculty research:* Intestinal cancer biology; hepatotoxins and mitochondriopathies; cell cycle regulation in stem cells; neurobiology of brain development, trace metal function and blood-brain barrier; hibernation biology. Total annual research expenditures: $1.5 million. *Unit head:* Dr. Lester R. Drewes, Professor/Head, 218-726-7925, Fax: 218-726-8014, E-mail: ldrewes@d.umn.edu. *Application contact:* Stacy Johnson, Program Associate, 218-726-6354, E-mail: ahcd@d.umn.edu.

University of Minnesota, Twin Cities Campus, Graduate School, Program in Molecular, Cellular, Developmental Biology and Genetics, Minneapolis, MN 55455-0213. Offers genetic counseling (MS); molecular, cellular, developmental biology and genetics (PhD). Part-time programs available. Terminal master's awarded for partial completion of doctoral program. *Degree requirements:* For master's, thesis optional; for doctorate, thesis/dissertation. *Entrance requirements:* For master's and doctorate, GRE General Test. Additional exam requirements/recommendations for international students: Required—TOEFL. *Expenses:* Tuition, state resident: full-time $8,748; part-time $729 per credit. Tuition, nonresident: full-time $15,848; part-time $1,321 per credit. Full-time tuition and fees vary according to class time, course load, program and reciprocity agreements. *Faculty research:* Membrane receptors and membrane transport, cell interactions, cytoskeleton and cell mobility, regulation of gene expression, plant cell and molecular biology.

University of Minnesota, Twin Cities Campus, Medical School and Graduate School, Graduate Programs in Medicine, Department of Biochemistry, Molecular Biology and Biophysics, Minneapolis, MN 55455-0213. Offers PhD. *Degree requirements:* For doctorate, thesis/dissertation. *Entrance requirements:* For doctorate, GRE General Test. Electronic applications accepted. Expenses: Contact institution. Full-time tuition and fees vary according to class time, course load, program and reciprocity agreements. *Faculty research:* Physical biochemistry, enzymology, physiological chemistry.

University of Missouri–Columbia, Graduate School, College of Arts and Sciences, Division of Biological Sciences, Columbia, MO 65211. Offers cellular, molecular and developmental biology (MA, PhD); evolutionary biology and ecology (MA, PhD); neurobiology and behavior (MA, PhD). *Faculty:* 43 full-time (12 women). *Students:* 48 full-time (26 women), 28 part-time (10 women); includes 7 minority (2 African Americans, 3 Asian Americans or Pacific Islanders, 2 Hispanic Americans), 20 international. In 2005, 3 master's, 8 doctorates awarded. Terminal master's awarded for partial completion of doctoral program. *Degree requirements:* For master's, thesis/dissertation; for doctorate, thesis/dissertation, comprehensive exam. *Entrance requirements:* For master's and doctorate, GRE General Test, minimum GPA of 3.0. *Application deadline:* For fall admission, 1/15 for domestic students. Applications are processed on a rolling basis. Application fee: $45 ($60 for international students). *Financial support:* Fellowships, research assistantships, teaching assistantships, institutionally sponsored loans available. *Unit head:* Dr. Ray Semlitsch, Director of Graduate Studies, 573-884-6396, E-mail: semlitschr@missouri. edu. *Application contact:* Nila Emerich, Application Contact, 800-553-5698.

University of Missouri–Kansas City, School of Biological Sciences, Program in Molecular Biology and Biochemistry, Kansas City, MO 64110-2499. Offers PhD. Offered through the School of Graduate Studies. *Degree requirements:* For doctorate, thesis/dissertation, comprehensive exam. *Entrance requirements:* For doctorate, GRE General Test, bachelor's degree in chemistry, biology, or a related discipline; minimum GPA of 3.0. Additional exam requirements/recommendations for international students: Required—TOEFL. *Application deadline:* For fall admission, 3/1 for domestic students. Application fee: $35 ($50 for international students). *Expenses:* Tuition, state resident: full-time $4,738; part-time $263 per credit hour. Tuition, nonresident: full-time $12,235; part-time $679 per credit hour. Required fees: $582. Tuition and fees vary according to course load, program and student level. *Financial support:* Research assistantships with full tuition reimbursements, teaching assistantships with full and partial tuition reimbursements, scholarships/grants, tuition waivers (full and partial), and unspecified assistantships available. *Unit head:* Dr. Henry Miziorko, Head, 816-235-2235, E-mail: miziorkoh@umkc.edu. *Application contact:* Laura Batenic, Information Contact, 816-235-2352, Fax: 816-235-5158, E-mail: batenicl@umkc.edu.

See Close-Up on page 629.

University of Missouri–St. Louis, College of Arts and Sciences, Department of Biology, St. Louis, MO 63121. Offers biology (MS, PhD), including animal behavior (MS), biochemistry (MS), biotechnology (MS), conservation biology (MS), development (MS), ecology (MS), environmental studies (PhD), evolution (MS), genetics (MS), molecular/cellular biology (MS), physiology (MS), plant systematics, population biology (MS), tropical biology (MS); biotechnology (Certificate); tropical biology and conservation (Certificate). Part-time programs available. *Faculty:* 50. *Students:* 26 full-time (15 women), 101 part-time (51 women); includes 13 minority (5 African Americans, 6 Asian Americans or Pacific Islanders, 2 Hispanic Americans), 41 international. Average age 32. In 2005, 22 master's, 2 doctorates awarded. *Degree requirements:* For master's, thesis or alternative; for doctorate, one foreign language, thesis/dissertation, 1 semester of teaching experience. *Entrance requirements:* For doctorate, GRE General Test. *Application deadline:* For spring admission, 12/1 priority date for domestic students. Applications are processed on a rolling basis. Application fee: $35 ($40 for international students). Electronic applications accepted. *Expenses:* Tuition, state resident: part-time $263 per credit hour. Tuition, nonresident: part-time $680 per credit hour. Required fees: $53 per credit hour. Tuition and fees vary according to program. *Financial support:* In 2005–06, 11 fellowships with full tuition reimbursements (averaging $30,000 per year), 15 research assistantships with full and partial tuition reimbursements (averaging $16,000 per year), 22 teaching assistantships with full and partial tuition reimbursements (averaging $16,000 per year) were awarded; career-related internships or fieldwork and Federal Work-Study also available. Support available to part-time students. Financial award application deadline: 2/1. *Faculty research:* Molecular biology, microbial genetics. *Unit head:* Zuleyma Tang-Martinez, Director of Graduate Studies, 314-516-6498, Fax: 314-516-6233, E-mail: zuleyma@umsl.edu. *Application contact:* 314-516-5458, Fax: 314-516-5310, E-mail: gradadm@umsl.edu.

University of Nebraska Medical Center, Graduate Studies, Department of Biochemistry and Molecular Biology, Omaha, NE 68198. Offers MS, PhD. *Faculty:* 17 full-time (2 women), 28 part-time/adjunct (8 women). *Students:* 31 full-time (15 women); includes 1 minority (Asian American or Pacific Islander), 20 international. Average age 28. 20 applicants, 30% accepted, 5 enrolled. In 2005, 2 doctorates awarded. *Median time to degree:* Of those who began their doctoral program in fall 1997, 75% received their degree in 8 years or less. *Degree requirements:* For master's and doctorate, thesis/dissertation, comprehensive exam. *Entrance requirements:*

For master's and doctorate, GRE General Test. Additional exam requirements/recommendations for international students: Required—TOEFL (minimum score 550 paper-based; 215 computer-based). Application fee: $45. Electronic applications accepted. *Expenses:* Tuition, area resident: Part-time $200 per hour. Tuition, nonresident: part-time $538 per hour. Required fees: $308; $59 per term. *Financial support:* In 2005–06, 17 students received support, including 2 fellowships with full tuition reimbursements available (averaging $21,000 per year), 29 research assistantships with full tuition reimbursements available (averaging $21,000 per year); institutionally sponsored loans also available. Support available to part-time students. Financial award application deadline: 2/15. *Faculty research:* Recombinant DNA, cancer biology, diabetes and drug metabolism, biochemical endocrinology. *Unit head:* Dr. Surinder K. Batra, Chairman, Graduate Committee, 402-559-5455, Fax: 402-559-6650, E-mail: biochem@unmc.edu.

University of Nevada, Reno, School of Medicine and Graduate School, Graduate Programs in Medicine, Interdisciplinary Program in Cell and Molecular Biology, Reno, NV 89557. Offers MS, PhD. *Students:* 19 full-time (11 women), 21 part-time (11 women); includes 7 minority (1 African American, 3 Asian Americans or Pacific Islanders, 3 Hispanic Americans), 9 international. Average age 30. 12 applicants, 92% accepted, 6 enrolled. In 2005, 5 degrees awarded. *Expenses:* Tuition, area resident: Full-time $2,767; part-time $923 per semester. Tuition, state resident: full-time $5,733; part-time $1,911 per semester. Tuition, nonresident: full-time $12,679; part-time $1,911 per semester. International tuition: $13,878 full-time. Required fees: $404; $202 per term. One-time fee: $90. *Unit head:* Dr. Stephen St. Jeor, Graduate Program Director, 775-784-4123, E-mail: stjeor@unr.edu.

University of New Hampshire, Graduate School, College of Life Sciences and Agriculture, Department of Biochemistry and Molecular Biology, Durham, NH 03824. Offers MS, PhD. Part-time programs available. *Faculty:* 12 full-time. *Students:* 18 full-time (8 women), 2 part-time; includes 2 minority (both Asian Americans or Pacific Islanders), 6 international. Average age 28. 11 applicants, 27% accepted, 1 enrolled. In 2005, 3 master's, 3 doctorates awarded. Terminal master's awarded for partial completion of doctoral program. *Degree requirements:* For master's, thesis; for doctorate, one foreign language, thesis/dissertation. *Entrance requirements:* For master's and doctorate, GRE General Test. Additional exam requirements/recommendations for international students: Required—TOEFL (minimum score 550 paper-based; 213 computer-based); Recommended—TSE. *Application deadline:* For fall admission, 4/1 priority date for domestic students, 4/1 priority date for international students. Applications are processed on a rolling basis. Application fee: $60. Electronic applications accepted. *Expenses:* Tuition, state resident: full-time $8,010; part-time $445 per credit hour. Tuition, nonresident: full-time $19,730; part-time $810 per credit hour. Required fees: $322 per semester. Tuition and fees vary according to course load and program. *Financial support:* In 2005–06, 1 fellowship, 11 research assistantships, 5 teaching assistantships were awarded; career-related internships or fieldwork, Federal Work-Study, scholarships/grants, and tuition waivers (full and partial) also available. Support available to part-time students. Financial award application deadline: 2/15. *Faculty research:* Developmental biochemistry, biochemistry of natural products, physical biochemistry, biochemical genetics, structure and metabolism of macromolecules. *Application contact:* Stacia Sower, Administrative Assistant, 603-862-2103, E-mail: biochemistry.dept@unh.edu.

University of New Haven, Graduate School, College of Arts and Sciences, Program in Cellular and Molecular Biology, West Haven, CT 06516-1916. Offers MS. *Degree requirements:* For master's, thesis or alternative.

See Close-Up on page 631.

University of New Mexico, School of Medicine, Biomedical Sciences Graduate Program, Albuquerque, NM 87131-5196. Offers biochemistry and molecular biology (MS, PhD); cell biology and physiology (MS, PhD); molecular genetics and microbiology (MS, PhD); neuroscience (MS, PhD); pathology (MS, PhD); toxicology (MS, PhD). Part-time programs available. Terminal master's awarded for partial completion of doctoral program. *Degree requirements:* For master's, thesis/dissertation; for doctorate, thesis/dissertation, comprehensive exam. *Entrance requirements:* For master's and doctorate, GRE General Test, minimum undergraduate GPA of 3.0. Additional exam requirements/recommendations for international students: Required—TOEFL. Electronic applications accepted. *Expenses:* Tuition, state resident: full-time $5,676. Tuition, nonresident: full-time $14,974; part-time $238 per credit hour. Required fees: $385 per term. Tuition and fees vary according to course load and program. *Faculty research:* Signal transduction, infectious disease, biology of cancer, structural biology, neuroscience.

The University of North Carolina at Chapel Hill, Graduate School, College of Arts and Sciences, Department of Biology, Chapel Hill, NC 27599. Offers botany (MA, MS, PhD); cell biology, development, and physiology (MA, MS, PhD); cell motility and cytoskeleton (PhD); ecology and behavior (MA, MS, PhD); genetics and molecular biology (MA, MS, PhD); morphology, systematics, and evolution (MA, MS, PhD). Terminal master's awarded for partial completion of doctoral program. *Degree requirements:* For master's, thesis (for some programs), comprehensive exam; for doctorate, thesis/dissertation, comprehensive exam. *Entrance requirements:* For master's, GRE General Test, GRE Subject Test, 2 semesters of calculus or statistics, 2 semesters of physics, organic chemistry, 3 semesters of biology; for doctorate, GRE General Test, GRE Subject Test, 2 semesters calculus or statistics, 2 semesters physics, organic chemistry, 3 semesters of biology. Additional exam requirements/recommendations for international students: Required—TOEFL (minimum score 550 paper-based; 213 computer-based). Electronic applications accepted. *Faculty research:* Gene expression, biomechanics, yeast genetics, plant ecology, plant molecular biology.

See Close-Up on page 277.

The University of North Carolina at Chapel Hill, School of Medicine and Graduate School, Graduate Programs in Medicine, Curriculum in Genetics and Molecular Biology, Chapel Hill, NC 27599. Offers MS, PhD. *Faculty:* 84 full-time (30 women), 2 part-time/adjunct (0 women). *Students:* 86 full-time (45 women); includes 16 minority (6 African Americans, 1 American Indian/Alaska Native, 7 Asian Americans or Pacific Islanders, 2 Hispanic Americans), 5 international. 100 applicants, 30% accepted, 18 enrolled. In 2005, 15 doctorates awarded. *Median time to degree:* Of those who began their doctoral program in fall 1997, 97% received their degree in 8 years or less. *Degree requirements:* For doctorate, thesis/dissertation, comprehensive exam, registration. *Entrance requirements:* For doctorate, minimum GPA of 3.0. Additional exam requirements/recommendations for international students: Required—TOEFL. *Application deadline:* For fall admission, 1/1 for domestic students. Applications are processed on a rolling basis. Application fee: $65. Electronic applications accepted. *Financial support:* In 2005–06, 15 fellowships with tuition reimbursements (averaging $22,000 per year), 54 research assistantships with tuition reimbursements (averaging $22,000 per year) were awarded; teaching assistantships with tuition reimbursements, traineeships and tuition waivers (full) also available. *Unit head:* Dr. Robert T. Duronio, Director, 919-966-3548. *Application contact:* Cara F. Marlow, Administrative Assistant, 919-966-2681, Fax: 919-966-0401, E-mail: cara-marlow@med.unc.edu.

University of North Texas, Robert B. Toulouse School of Graduate Studies, College of Arts and Sciences, Department of Biological Sciences, Program in Molecular Biology, Denton, TX 76203. Offers MA, MS, PhD. *Students:* 13 full-time (6 women), 6 part-time (4 women); includes 4 minority (2 Asian Americans or Pacific Islanders, 2 Hispanic Americans), 6 international. Average age 31. *Degree requirements:* For master's, oral defense of thesis; for doctorate, one foreign language, thesis/dissertation, comprehensive exam. *Entrance requirements:* For master's and doctorate, GRE General Test. *Application deadline:* For fall admission, 7/15 for domestic students. Applications are processed on a rolling basis. Application fee: $50 ($75 for international students). *Expenses:* Tuition, state resident: full-time $3,258; part-time $181 per semester hour. Tuition, nonresident: full-time $8,226; part-time $451 per semester hour. Required fees: $1,219; $68 per semester hour. *Financial support:* Research assistantships, teaching assistantships available. *Faculty research:* Adenosine diphosphate ribosylation reactions, mam-

malian gene structure, pyrimidine metabolism, enzymology. *Application contact:* Beth Chlapek, Graduate Adviser, 940-565-3627, Fax: 940-565-3821, E-mail: bsloan@unt.edu.

University of North Texas Health Science Center at Fort Worth, Graduate School of Biomedical Sciences, Fort Worth, TX 76107-2699. Offers anatomy and cell biology (MS, PhD); biochemistry and molecular biology (MS, PhD); biomedical sciences (MS, PhD); biotechnology (MS); forensic genetics (MS); integrative physiology (MS, PhD); medical science (MS); microbiology and immunology (MS, PhD); pharmacology (MS, PhD); science education (MS). *Faculty:* 57 full-time (9 women), 2 part-time/adjunct (0 women). *Students:* 177 full-time (98 women), 42 part-time (33 women); includes 61 minority (15 African Americans, 3 American Indian/Alaska Native, 27 Asian Americans or Pacific Islanders, 16 Hispanic Americans), 53 international. Average age 28. 237 applicants, 62% accepted, 91 enrolled. In 2005, 37 master's, 15 doctorates awarded. Terminal master's awarded for partial completion of doctoral program. *Degree requirements:* For master's and doctorate, thesis/dissertation. *Entrance requirements:* For master's and doctorate, GRE General Test. Additional exam requirements/recommendations for international students: Required—TOEFL. *Application deadline:* For fall admission, 5/1 for domestic students. Application fee: $25 ($50 for international students). *Expenses:* Contact institution. *Financial support:* In 2005–06, 80 research assistantships (averaging $16,000 per year) were awarded; fellowships, teaching assistantships, career-related internships or fieldwork, Federal Work-Study, institutionally sponsored loans, scholarships/grants, and traineeships also available. Support available to part-time students. Financial award application deadline: 4/1; financial award applicants required to submit FAFSA. *Faculty research:* Alzheimer's disease, aging, eye diseases, cancer, cardiovascular disease. Total annual research expenditures: $21 million. *Unit head:* Dr. Thomas Yorio, Dean, 817-735-2560, Fax: 817-735-0243, E-mail: yoriot@hsc.unt.edu. *Application contact:* Carla Lee, Director of Graduate Admissions and Services, 817-735-2560, Fax: 817-735-0243, E-mail: gsbs@hsc.unt.edu.

See Close-Up on page 279.

University of Notre Dame, Graduate School, College of Science, Department of Biological Sciences, Notre Dame, IN 46556. Offers aquatic ecology, evolution and environmental biology (MS, PhD); cellular and molecular biology (MS, PhD); genetics (MS, PhD); physiology (MS, PhD); vector biology and parasitology (MS, PhD). *Faculty:* 34 full-time (8 women), 3 part-time/adjunct (0 women). *Students:* 124 full-time (55 women); includes 11 minority (1 African American, 6 Asian Americans or Pacific Islanders, 4 Hispanic Americans), 39 international. 95 applicants, 34% accepted, 22 enrolled. In 2005, 4 master's, 11 doctorates awarded. Terminal master's awarded for partial completion of doctoral program. *Median time to degree:* Of those who began their doctoral program in fall 1997, 61% received their degree in 8 years or less. *Degree requirements:* For master's and doctorate, thesis/dissertation, comprehensive exam. *Entrance requirements:* For master's and doctorate, GRE General Test. Additional exam requirements/recommendations for international students: Required—TOEFL. *Application deadline:* For fall admission, 2/1 for domestic students; for spring admission, 11/1 for domestic students. Applications are processed on a rolling basis. Application fee: $50. Electronic applications accepted. *Financial support:* In 2005–06, 124 students received support, including 24 fellowships with full tuition reimbursements available (averaging $22,000 per year), 47 research assistantships with full tuition reimbursements available (averaging $15,250 per year), 45 teaching assistantships with full tuition reimbursements available (averaging $16,000 per year); traineeships and tuition waivers (full) also available. Financial award application deadline: 2/1. *Faculty research:* Tropical disease, molecular genetics, neurobiology, evolutionary biology, aquatic biology. Total annual research expenditures: $15.3 million. *Unit head:* Dr. Gary A. Lamberti, Director of Graduate Studies, 574-631-6552, Fax: 574-631-7413, E-mail: biology. biosadm.1@nd.edu. *Application contact:* Dr. Terrence J. Akai, Director of Graduate Admissions, 574-631-7706, Fax: 574-631-4183, E-mail: gradad@nd.edu.

See Close-Up on page 281.

University of Oklahoma Health Sciences Center, College of Medicine and Graduate College, Graduate Programs in Medicine, Department of Biochemistry and Molecular Biology, Oklahoma City, OK 73190. Offers biochemistry (MS, PhD); molecular biology (MS, PhD). Part-time programs available. Terminal master's awarded for partial completion of doctoral program. *Degree requirements:* For master's and doctorate, thesis/dissertation. *Entrance requirements:* For master's, GRE General Test, 2 letters of recommendation; for doctorate, GRE General Test, 3 letters of recommendation. Additional exam requirements/recommendations for international students: Required—TOEFL. *Faculty research:* Gene expression, regulation of transcription, enzyme evolution, melanogenesis, signal transduction.

University of Oregon, Graduate School, College of Arts and Sciences, Department of Biology, Eugene, OR 97403. Offers ecology and evolution (MA, MS, PhD); marine biology (MA, MS, PhD); molecular, cellular and genetic biology (PhD); neuroscience and development (PhD). *Faculty:* 42 full-time (14 women), 5 part-time/adjunct (2 women). *Students:* 92 full-time (43 women), 2 part-time (both women); includes 9 minority (2 African Americans, 1 American Indian/Alaska Native, 4 Asian Americans or Pacific Islanders, 2 Hispanic Americans), 6 international. 18 applicants, 83% accepted. In 2005, 11 master's, 7 doctorates awarded. Terminal master's awarded for partial completion of doctoral program. *Degree requirements:* For master's, thesis (for some programs); for doctorate, thesis/dissertation. *Entrance requirements:* For master's and doctorate, GRE General Test, minimum GPA of 3.2. Additional exam requirements/recommendations for international students: Required—TOEFL. *Application deadline:* For fall admission, 12/15 for domestic students. *Financial support:* In 2005–06, 36 teaching assistantships were awarded; research assistantships, Federal Work-Study, institutionally sponsored loans, and scholarships/grants also available. Financial award application deadline: 2/1. *Faculty research:* Developmental neurobiology; evolution, population biology, and quantitative genetics; regulation of gene expression; biochemistry of marine organisms. *Unit head:* George Sprague, Head, 541-346-6051, Fax: 541-346-6056. *Application contact:* Lynne Romans, Admissions Contact, 541-346-4252, Fax: 541-346-6056, E-mail: lromans@uoregon.edu.

University of Ottawa, Faculty of Graduate and Postdoctoral Studies, Faculty of Medicine, Department of Cellular and Molecular Medicine, Ottawa, ON K1H 8M5, Canada. Offers M Sc, PhD. *Faculty:* 69 full-time (9 women). *Students:* 155 full-time, 10 part-time. 45 applicants, 82% accepted, 29 enrolled. In 2005, 15 master's, 3 doctorates awarded. *Degree requirements:* For master's, thesis/dissertation, seminar; for doctorate, thesis/dissertation, seminar, comprehensive exam. *Entrance requirements:* For master's, honors degree or equivalent, minimum B average; for doctorate, master's degree, minimum B+ average. *Application deadline:* For fall admission, 8/1 priority date for domestic students, 2/15 priority date for international students. For winter admission, 11/15 for domestic students; for spring admission, 4/1 for domestic students. Application fee: $75. Electronic applications accepted. *Expenses:* Tuition: Part-time $260 per credit. Tuition and fees vary according to course load and program. *Financial support:* Fellowships, research assistantships with full tuition reimbursements, teaching assistantships with full tuition reimbursements, career-related internships or fieldwork, Federal Work-Study, scholarships/grants, traineeships, tuition waivers (full and partial), and unspecified assistantships available. Financial award application deadline: 2/15. *Faculty research:* Physiology, pharmacology, growth and development. *Unit head:* Dr. Bernard Jasmin, Chair, 613-562-5425, Fax: 613-562-5434, E-mail: jasmin@uottawa.ca. *Application contact:* Donna Hooper, Academic Assistant, 613-562-5800 Ext. 8396, Fax: 613-562-5636, E-mail: clmolm@uottawa.ca.

University of Pennsylvania, School of Medicine, Biomedical Graduate Studies, Graduate Group in Cell and Molecular Biology, Philadelphia, PA 19104. Offers cell biology and physiology (PhD); cell growth and cancer (PhD); developmental biology (PhD); gene therapy and vaccines (PhD); genetics and gene regulation (PhD); microbiology, virology, and parasitology (PhD). *Faculty:* 294. *Students:* 290 full-time (155 women); includes 67 minority (9 African Americans, 41 Asian Americans or Pacific Islanders, 17 Hispanic Americans), 24 international. 510 applicants, 21% accepted, 38 enrolled. In 2005, 46 doctorates awarded. *Degree requirements:* For doctorate, thesis/dissertation. *Entrance requirements:* For doctorate, GRE

Molecular Biology

University of Pennsylvania (continued)

General Test. Additional exam requirements/recommendations for international students: Required—TOEFL. *Application deadline:* For fall admission, 12/15 priority date for domestic students, 12/1 priority date for international students. Applications are processed on a rolling basis. Application fee: $70. Electronic applications accepted. *Financial support:* Fellowships, research assistantships, scholarships/grants, traineeships, and unspecified assistantships available. Financial award application deadline: 1/2. *Unit head:* Dr. Jonathan Raper, Graduate Group Chair, 215-898-2180, E-mail: raperj@mail.med.upenn.edu. *Application contact:* Meagan Schofer, Coordinator, 215-895-9536, Fax: 215-573-2104, E-mail: mschofer@mailmed.upenn.edu.

University of Pittsburgh, School of Arts and Sciences, Department of Biological Sciences, Program in Molecular, Cellular, and Developmental Biology, Pittsburgh, PA 15260. Offers PhD. *Faculty:* 22 full-time (7 women), 1 part-time/adjunct (0 women). *Students:* 48 full-time (32 women); includes 2 minority (both African Americans), 10 international. Average age 23. 215 applicants, 2% accepted, 4 enrolled. In 2005, 7 degrees awarded. *Median time to degree:* Of those who began their doctoral program in fall 1997, 100% received their degree in 8 years or less. *Degree requirements:* For doctorate, thesis/dissertation, comprehensive exam, registration. *Entrance requirements:* For doctorate, GRE General Test, GRE Subject Test. Additional exam requirements/recommendations for international students: Required—TOEFL (minimum score 550 paper-based; 213 computer-based). *Application deadline:* For fall admission, 1/15 priority date for domestic students, 12/15 priority date for international students. Applications are processed on a rolling basis. Application fee: $0 ($40 for international students). Electronic applications accepted. *Expenses:* Tuition, state resident: full-time $13,194; part-time $537 per credit. Tuition, nonresident: full-time $25,012; part-time $1,026 per credit. Required fees: $700; $164 per term. Tuition and fees vary according to campus/location and program. *Financial support:* In 2005–06, 18 fellowships with full tuition reimbursements, 93 research assistantships with full tuition reimbursements (averaging $20,332 per year), 22 teaching assistantships with full tuition reimbursements (averaging $20,332 per year) were awarded; Federal Work-Study, scholarships/grants, traineeships, health care benefits, and tuition waivers (full) also available. *Faculty research:* Structure and function of genes and proteins; macromolecular interactions; cell-specific gene regulation; regulation of cell proliferation; embryogenesis. *Application contact:* Cathleen M. Barr, Graduate Administrator, 412-624-4268, Fax: 412-624-4759, E-mail: cbarr@pitt.edu.

See Close-Up on page 637.

University of Pittsburgh, School of Medicine and School of Arts and Sciences, Program in Integrative Molecular Biology, Pittsburgh, PA 15260. Offers PhD. *Faculty:* 27 full-time (11 women). *Entrance requirements:* For doctorate, GRE, minimum GPA of 3.7, 3 letters of reference. Additional exam requirements/recommendations for international students: Required—TOEFL (minimum score 650 paper-based; 280 computer-based). *Application deadline:* For fall admission, 12/1 for domestic students, 12/1 for international students. *Expenses:* Tuition, state resident: full-time $13,194; part-time $537 per credit. Tuition, nonresident: full-time $25,012; part-time $1,026 per credit. Required fees: $700; $164 per term. Tuition and fees vary according to campus/location and program. *Application contact:* Jennifer L. Walker, MBSB Program Coordinator, 412-648-8957, Fax: 412-648-1077, E-mail: mbsbinfo@medschool.pitt.edu.

See Close-Up on page 635.

University of Rhode Island, Graduate School, College of the Environment and Life Sciences, Department of Cell and Molecular Biology, Kingston, RI 02881. Offers biochemistry (MS, PhD); microbiology (MS, PhD), including biodegradation (MS), cellular development (MS), electron microscopy and ultrastructure (MS), genetics and molecular biology (MS), immunology (MS), marine and freshwater ecosystems (MS), microbial pathogenesis (MS), microbial physiology (MS), protozoology (MS), virology (MS), water-pollution microbiology (MS); molecular genetics (MS, PhD). In 2005, 2 degrees awarded. *Degree requirements:* For master's and doctorate, thesis/dissertation. *Entrance requirements:* For master's and doctorate, GRE General Test. Additional exam requirements/recommendations for international students: Required—TOEFL. *Expenses:* Tuition, state resident: full-time $5,522; part-time $307 per credit. Tuition, nonresident: full-time $15,992; part-time $888 per credit. Required fees: $1,786; $73 per credit. One-time fee: $80 part-time. *Financial support:* Fellowships, research assistantships, teaching assistantships available. *Unit head:* Dr. Jay Sperry, Chairperson, 401-874-5900.

University of South Alabama, College of Medicine and Graduate School, Program in Basic Medical Sciences, Specialization in Biochemistry and Molecular Biology, Mobile, AL 36688-0002. Offers PhD. *Faculty:* 9 full-time (2 women). *Degree requirements:* For doctorate, thesis/dissertation. *Entrance requirements:* For doctorate, GRE General Test or MCAT. *Application deadline:* For fall admission, 4/1 for domestic students. Applications are processed on a rolling basis. Application fee: $25. *Expenses:* Tuition, state resident: full-time $4,008. Tuition, nonresident: full-time $8,016. Required fees: $692. *Financial support:* Fellowships, institutionally sponsored loans available. Financial award application deadline: 4/1. *Faculty research:* Biochemistry of aging, mechanisms of metabolic regulation and oxygen metabolism. *Unit head:* Dr. Nathan N. Aronson, Chair, 251-460-6402. *Application contact:* Lanette Flagge, Academic Advisor, 251-460-6153.

University of South Carolina, The Graduate School, College of Science and Mathematics, Department of Biological Sciences, Graduate Training Program in Molecular, Cellular, and Developmental Biology, Columbia, SC 29208. Offers MS, PhD. *Degree requirements:* For master's and doctorate, one foreign language, thesis/dissertation. *Entrance requirements:* For master's and doctorate, GRE General Test, minimum GPA of 3.0 in science. Electronic applications accepted. *Faculty research:* Marine ecology, population and evolutionary biology, molecular biology and genetics, development.

See Close-Up on page 639.

The University of South Dakota, School of Medicine and Health Sciences and Graduate School, Biomedical Sciences Graduate Program, Cellular and Molecular Biology Group, Vermillion, SD 57069-2390. Offers MA, PhD. *Faculty:* 9 full-time (7 women), 7 international. Average age 28. 26 applicants, 15% accepted, 2 enrolled. In 2005, 1 master's, 2 doctorates awarded. Terminal master's awarded for partial completion of doctoral program. *Median time to degree:* Of those who began their doctoral program in fall 1997, 90% received their degree in 8 years or less. *Degree requirements:* For master's, thesis/dissertation, registration; for doctorate, thesis/dissertation, comprehensive exam, registration. *Entrance requirements:* For master's and doctorate, GRE General Test, GRE Subject Test, minimum GPA of 3.0. Additional exam requirements/recommendations for international students: Required—TOEFL (minimum score 550 paper-based; 213 computer-based). *Application deadline:* For fall admission, 4/15 priority date for domestic students, 4/15 priority date for international students. Applications are processed on a rolling basis. Application fee: $35. *Expenses:* Tuition, state resident: part-time $116 per credit hour. Tuition, nonresident: part-time $341 per credit hour. Required fees: $85 per credit hour. Tuition and fees vary according to course load, program and reciprocity agreements. *Financial support:* In 2005–06, 9 students received support, including 8 fellowships with partial tuition reimbursements available (averaging $20,772 per year), 1 research assistantship with partial tuition reimbursement available (averaging $10,386 per year) Financial award application deadline: 4/15; financial award applicants required to submit FAFSA. *Faculty research:* Molecular aspects of protein and DNA, neurochemistry and energy transduction, gene regulation, cellular development. Total annual research expenditures: $438,000.

University of Southern California, Graduate School, College of Letters, Arts and Sciences, Department of Biological Sciences, Program in Molecular and Computational Biology, Los Angeles, CA 90089. Offers PhD. *Degree requirements:* For doctorate, thesis/dissertation. *Entrance requirements:* For doctorate, GRE General Test. Additional exam requirements/

recommendations for international students: Required—TOEFL. *Expenses:* Tuition: Full-time $25,416; part-time $1,059 per unit. Required fees: $484; $484 per year. Tuition and fees vary according to course load and program. *Faculty research:* Genomic instabilities, molecular studies in aging, genetic/biochemical studies of DNA repair, molecular/genetic approaches to developmental biology, computation and experimental genomics.

See Close-Up on page 641.

University of Southern California, Keck School of Medicine and Graduate School, Graduate Programs in Medicine, Department of Biochemistry and Molecular Biology, Los Angeles, CA 90089. Offers MS, PhD. *Faculty:* 44 full-time (10 women). *Students:* 106 full-time (58 women); includes 17 minority (12 Asian Americans or Pacific Islanders, 5 Hispanic Americans), 57 international. Average age 24. 487 applicants, 19% accepted, 44 enrolled. In 2005, 5 master's, 8 doctorates awarded. Terminal master's awarded for partial completion of doctoral program. *Degree requirements:* For master's and doctorate, thesis/dissertation, registration. *Entrance requirements:* For master's and doctorate, GRE General Test, minimum GPA of 3.0. Additional exam requirements/recommendations for international students: Required—TOEFL (minimum score 600 paper-based; 250 computer-based). *Application deadline:* For fall admission, 12/15 priority date for domestic students, 12/5 priority date for international students. Application fee: $65 ($75 for international students). Electronic applications accepted. *Expenses:* Tuition: Full-time $25,416; part-time $1,059 per unit. Required fees: $484; $484 per year. Tuition and fees vary according to course load and program. *Financial support:* In 2005–06, 93 students received support, including 4 fellowships with tuition reimbursements available (averaging $24,876 per year), 87 research assistantships with tuition reimbursements available (averaging $24,876 per year), 2 teaching assistantships with tuition reimbursements available (averaging $24,876 per year); career-related internships or fieldwork, institutionally sponsored loans, scholarships/grants, and tuition waivers (full) also available. Support available to part-time students. Financial award application deadline: 4/15. *Faculty research:* Molecular genetics, gene expression, membrane biochemistry, metabolic regulation, cancer biology. *Unit head:* Dr. Michael R. Stallcup, 323-442-1145, Fax: 323-442-1224, E-mail: stallcup@usc.edu. *Application contact:* Anne L. Rice, Student Services Coordinator, 323-442-1145, Fax: 323-442-1224, E-mail: annvazqu@usc.edu.

University of Southern California, Keck School of Medicine and Graduate School, Graduate Programs in Medicine, Department of Preventive Medicine, Division of Biostatistics, Los Angeles, CA 90089. Offers applied biostatistics/epidemiology (MS); biostatistics (MS, PhD); epidemiology (PhD); genetic epidemiology and statistical genetics (PhD); molecular epidemiology (MS, PhD). *Faculty:* 74 full-time (33 women). *Students:* 97 full-time (63 women); includes 25 minority (18 Asian Americans or Pacific Islanders, 7 Hispanic Americans), 48 international. Average age 30. 109 applicants, 70% accepted, 33 enrolled. In 2005, 17 master's, 10 doctorates awarded. Terminal master's awarded for partial completion of doctoral program. *Median time to degree:* Of those who began their doctoral program in fall 1997, 100% received their degree in 8 years or less. *Degree requirements:* For master's and doctorate, thesis/dissertation. *Entrance requirements:* For master's, GRE General Test, GRE Subject Test, minimum GPA of 3.0; for doctorate, GRE General Test, GRE Subject Test, minimum GPA of 3.5. Additional exam requirements/recommendations for international students: Required—TOEFL (minimum score 550 paper-based; 200 computer-based). *Application deadline:* For fall admission, 1/15 for domestic students. Applications are processed on a rolling basis. Application fee: $65 ($75 for international students). Electronic applications accepted. *Expenses:* Tuition: Full-time $25,416; part-time $1,059 per unit. Required fees: $484; $484 per year. Tuition and fees vary according to course load and program. *Financial support:* In 2005–06, 3 fellowships with tuition reimbursements (averaging $24,876 per year), 39 research assistantships with tuition reimbursements (averaging $24,876 per year), 15 teaching assistantships with tuition reimbursements (averaging $24,876 per year) were awarded; career-related internships or fieldwork, Federal Work-Study, institutionally sponsored loans, scholarships/grants, and tuition waivers (partial) also available. Financial award application deadline: 4/1. *Faculty research:* Clinical trials in ophthalmology and cancer research, methods of analysis for epidemiological studies, genetic epidemiology. Total annual research expenditures: $1.3 million. *Unit head:* Dr. Stanley P. Azen, Co-Director, 323-442-1810, Fax: 323-442-2993, E-mail: mtrujill@usc.edu. *Application contact:* Mary L. Trujillo, Student Adviser, 323-442-1810, Fax: 323-442-2993, E-mail: mtrujill@usc.edu.

University of Southern Maine, School of Applied Science, Engineering, and Technology, Program in Applied Immunology and Molecular Biology, Portland, ME 04104-9300. Offers MS. Part-time programs available. *Degree requirements:* For master's, thesis. *Entrance requirements:* For master's, GRE General Test, GRE Subject Test, minimum GPA of 3.0. Additional exam requirements/recommendations for international students: Required—TOEFL. Electronic applications accepted. *Faculty research:* Flow cytometry, cancer, epidemiology, monoclonal antibodies, DNA diagnostics.

University of Southern Mississippi, Graduate School, College of Science and Technology, Department of Biological Sciences, Hattiesburg, MS 39406-0001. Offers environmental biology (MS, PhD); marine biology (MS, PhD); microbiology (MS, PhD); molecular biology (MS, PhD). *Degree requirements:* For master's and doctorate, thesis/dissertation, comprehensive exam. *Entrance requirements:* For master's, GRE General Test, minimum GPA of 3.0; for doctorate, GRE General Test, minimum GPA of 3.5. Additional exam requirements/recommendations for international students: Required—TOEFL.

See Close-Up on page 297.

University of South Florida, College of Medicine and College of Graduate Studies, Graduate Programs in Medical Sciences, Tampa, FL 33620-9951. Offers anatomy (PhD); biochemistry and molecular biology (MS, PhD), including biochemistry and molecular biology (PhD); bioinformatics and computational biology (MS); medical microbiology and immunology (PhD); pathology (PhD); pharmacology and therapeutics (PhD), including medical sciences; physiology and biophysics (PhD). *Students:* 108 full-time (53 women), 34 part-time (26 women); includes 31 minority (9 African Americans, 9 Asian Americans or Pacific Islanders, 13 Hispanic Americans), 30 international. 117 applicants, 99% accepted, 116 enrolled. In 2005, 9 master's, 4 doctorates awarded. *Degree requirements:* For doctorate, thesis/dissertation. *Entrance requirements:* For doctorate, GRE General Test, minimum GPA of 3.0. Application fee: $30. *Expenses: Contact institution. Financial support:* Institutionally sponsored loans and scholarships/grants available. Financial award application deadline: 4/1; financial award applicants required to submit FAFSA. *Unit head:* Dr. Joseph J. Krzanowski, Associate Dean for Research and Graduate Affairs, 813-974-4181, Fax: 813-974-4317, E-mail: jkrzanow@com1.med.usf.edu.

The University of Tennessee Health Science Center, College of Graduate Health Sciences, Department of Molecular Sciences, Concentration in Molecular and Cell Biology-Pathology, Memphis, TN 38163-0002. Offers PhD. *Faculty:* 26 full-time (4 women), 1 (woman) part-time/adjunct. *Students:* 3 full-time (1 woman); includes 2 minority (both Asian Americans or Pacific Islanders) Average age 25. 83 applicants, 12% accepted. In 2005, 6 degrees awarded. *Degree requirements:* For doctorate, thesis/dissertation, oral and written preliminary and comprehensive exams. *Entrance requirements:* For doctorate, GRE General Test, GRE Subject Test, minimum GPA of 3.0. Additional exam requirements/recommendations for international students: Required—TOEFL. *Application deadline:* For fall admission, 5/15 for domestic students. Application fee: $0. Electronic applications accepted. *Financial support:* Fellowships, research assistantships, traineeships and tuition waivers (full) available. *Application contact:* Ida W. Mosby, Director, Enrollment Services, 901-448-5560, E-mail: imosby@utmem.edu.

The University of Texas at Austin, Graduate School, College of Natural Sciences, Institute for Cellular and Molecular Biology, Program in Cell and Molecular Biology, Austin, TX 78712-1111. Offers PhD. *Degree requirements:* For doctorate, thesis/dissertation, qualifying exam. *Entrance requirements:* For doctorate, GRE General Test. *Faculty research:* Virology and immunology, RNA, molecular genetics, developmental biology, plant science, biochemistry evolution.

See Close-Up on page 645.

The University of Texas at Dallas, School of Natural Sciences and Mathematics, Program in Biology, Richardson, TX 75083-0688. Offers bioinformatics and computational biology (MS); biotechnology (MS); molecular and cell biology (MS, PhD). Part-time and evening/weekend programs available. *Faculty:* 15 full-time (2 women). *Students:* 64 full-time (40 women), 18 part-time (14 women); includes 18 minority (1 African American, 15 Asian Americans or Pacific Islanders, 2 Hispanic Americans), 46 international. Average age 28. 133 applicants, 71% accepted, 39 enrolled. In 2005, 10 master's, 7 doctorates awarded. *Degree requirements:* For master's, thesis optional; for doctorate, thesis/dissertation, publishable paper. *Entrance requirements:* For master's and doctorate, GRE General Test. Additional exam requirements/recommendations for international students: Required—TOEFL (minimum score 550 paper-based; 213 computer-based). *Application deadline:* For fall admission, 7/15 for domestic students; for spring admission, 11/15 for domestic students. Applications are processed on a rolling basis. Application fee: $50 ($100 for international students). Electronic applications accepted. *Expenses:* Tuition, state resident: full-time $5,450; part-time $303 per credit. Tuition, nonresident: full-time $12,648; part-time $703 per credit. Tuition and fees vary according to program. *Financial support:* In 2005–06, 17 research assistantships with tuition reimbursements (averaging $14,125 per year), 16 teaching assistantships with tuition reimbursements (averaging $11,972 per year) were awarded; fellowships, career-related internships or fieldwork, Federal Work-Study, institutionally sponsored loans, and scholarships/grants also available. Support available to part-time students. Financial award application deadline: 4/30; financial award applicants required to submit FAFSA. *Faculty research:* DNA replication, regulation of gene expression, subcellular organelles, physical chemistry of macromolecules, damage and repair of cellular DNA. Total annual research expenditures: $4.1 million. *Unit head:* Dr. Donald Gray, Head, 972-883-2513, Fax: 972-883-2502, E-mail: dongray@utdallas.edu. *Application contact:* Dr. Ernest Hannig, Graduate Advisor, 972-883-2505, Fax: 972-883-2409, E-mail: hannig@utdallas.edu.

See Close-Up on page 647.

The University of Texas at San Antonio, College of Sciences, Department of Biology, Program in Biology, San Antonio, TX 78249-0617. Offers cell and molecular biology (PhD); neurobiology (PhD). *Degree requirements:* For doctorate, thesis/dissertation, comprehensive exam. *Entrance requirements:* For doctorate, GRE General Test, minimum GPA of 3.0. Additional exam requirements/recommendations for international students: Required—TOEFL. Expenses: Contact institution. *Faculty research:* Neurophysiology, neurotoxicology, neural circuit analysis, neuroendocrinology, development of biosensors for the detection of toxins.

The University of Texas Health Science Center at Houston, Graduate School of Biomedical Sciences, Program in Biochemistry and Molecular Biology, Houston, TX 77225-0036. Offers MS, PhD, MD/PhD. *Faculty:* 24 full-time (4 women). *Students:* 21 full-time (14 women); includes 1 minority (Hispanic American), 11 international. Average age 24. 14 applicants, 14% accepted, 2 enrolled. In 2005, 5 degrees awarded. Terminal master's awarded for partial completion of doctoral program. *Degree requirements:* For master's and doctorate, thesis/dissertation. *Entrance requirements:* For master's and doctorate, GRE General Test. Additional exam requirements/recommendations for international students: Required—TOEFL, TWE. *Application deadline:* For fall admission, 1/15 for domestic students; for spring admission, 11/1 for domestic students. Applications are processed on a rolling basis. Application fee: $10. Electronic applications accepted. *Financial support:* Fellowships with full tuition reimbursements, research assistantships with full tuition reimbursements, teaching assistantships, institutionally sponsored loans, scholarships/grants, and health care benefits available. Financial award application deadline: 1/15. *Faculty research:* Protein structure and function, regulation of gene expression, cellular signal transduction, regulation of cell growth and differentiation, molecular basis of human disease. *Unit head:* Dr. Michael R Blackburn, Director, 713-500-6087, Fax: 713-500-0652, E-mail: michael.r.blackburn@uth.tmc.edu. *Application contact:* Dr. Victoria P. Knutson, Assistant Dean of Admissions, 713-500-9860, Fax: 713-500-9877, E-mail: victoria.p.knutson@uth.tmc.edu.

The University of Texas Health Science Center at Houston, Graduate School of Biomedical Sciences, Program in Cell and Regulatory Biology, Houston, TX 77225-0036. Offers MS, PhD, MD/PhD. *Faculty:* 34 full-time (8 women). *Students:* 13 full-time (11 women); includes 2 minority (1 Asian American or Pacific Islander, 1 Hispanic American), 5 international. Average age 23. 57 applicants, 25% accepted, 7 enrolled. In 2005, 1 degree awarded. Terminal master's awarded for partial completion of doctoral program. *Degree requirements:* For master's and doctorate, thesis/dissertation. *Entrance requirements:* For master's and doctorate, GRE General Test. Additional exam requirements/recommendations for international students: Required—TOEFL, TWE. *Application deadline:* For fall admission, 1/15 for domestic students; for spring admission, 11/1 for domestic students. Applications are processed on a rolling basis. Application fee: $10. Electronic applications accepted. *Financial support:* Fellowships with full tuition reimbursements, research assistantships with full tuition reimbursements, teaching assistantships, institutionally sponsored loans, scholarships/grants, and health care benefits available. Financial award application deadline: 1/15. *Faculty research:* Cell signaling, mechanisms of action of hormones and neurotransmitters, cancer chemotherapy, molecular biology of regulatory processes, regulation of ion channels and excitability. *Unit head:* Dr. Carmen Dessauer, Director, 713-500-6308, Fax: 713-500-7444, E-mail: carmen.w.dessauer@uth.tmc.edu. *Application contact:* Dr. Victoria P. Knutson, Assistant Dean of Admissions, 713-500-9860, Fax: 713-500-9877, E-mail: victoria.p.knutson@uth.tmc.edu.

The University of Toledo, College of Graduate Studies, Department of Biochemistry and Molecular Biology, Toledo, OH 43606-3390. Offers medical sciences (MS), including biochemistry, genetics, molecular biology. Part-time programs available. *Degree requirements:* For master's, thesis, qualifying exam. *Entrance requirements:* For master's, GRE General Test, minimum undergraduate GPA of 3.0. Additional exam requirements/recommendations for international students: Required—TOEFL. *Expenses:* Tuition, state resident: full-time $6,623; part-time $308 per credit hour. Tuition, nonresident: full-time $13,232; part-time $735 per credit hour. *Faculty research:* Gene regulation, protein structure, receptors, protein phosphorylation, peptides.

The University of Toledo, College of Graduate Studies, Program in Molecular and Cellular Biology, Toledo, OH 43606-3390. Offers PhD, MD/PhD. *Degree requirements:* For doctorate, thesis/dissertation, qualifying exam. *Entrance requirements:* For doctorate, GRE General Test, minimum undergraduate GPA of 3.0. *Expenses:* Tuition, state resident: full-time $6,623; part-time $308 per credit hour. Tuition, nonresident: full-time $13,232; part-time $735 per credit hour.

See Close-Up on page 311.

University of Utah, The Graduate School, College of Science and Graduate Programs in Medicine, Interdisciplinary Studies in Molecular Evolutionary Biology, Salt Lake City, UT 84112-1107.

University of Utah, The Graduate School and Graduate Programs in Medicine, Program in Molecular Biology, Salt Lake City, UT 84132.

See Close-Up on page 651.

University of Vermont, Graduate College, Cell and Molecular Biology Program, Burlington, VT 05405. Offers MS, PhD. *Students:* 42 (18 women); includes 4 minority (2 Asian Americans or Pacific Islanders, 2 Hispanic Americans) 7 international. 45 applicants, 24% accepted, 7 enrolled. In 2005, 3 doctorates awarded. *Degree requirements:* For master's and doctorate, thesis/dissertation. *Entrance requirements:* For master's and doctorate, GRE General Test. Additional exam requirements/recommendations for international students: Required—TOEFL (minimum score 550 paper-based; 213 computer-based). *Application deadline:* For fall admission, 1/15 for domestic students. Applications are processed on a rolling basis. Application fee: $40. Electronic applications accepted. *Expenses:* Tuition, area resident: Part-time $410 per credit hour. Tuition, nonresident: part-time $1,034 per credit hour. *Financial support:* Fellow-

ships, research assistantships, teaching assistantships, career-related internships or fieldwork available. Financial award application deadline: 3/1. *Unit head:* , Dr. C. Berger, Coordinator, 802-656-9673.

See Close-Up on page 653.

University of Washington, School of Medicine, Biomolecular Structure and Design Program, Seattle, WA 98195.

University of Washington, School of Medicine and Graduate School, Graduate Programs in Medicine, Program in Molecular and Cellular Biology, Seattle, WA 98195. Offers PhD. Offered jointly with Fred Hutchinson Cancer Research Center. *Degree requirements:* For doctorate, thesis/dissertation. *Entrance requirements:* For doctorate, GRE General Test, GRE Subject Test. Additional exam requirements/recommendations for international students: Required—TOEFL. Electronic applications accepted.

See Close-Up on page 655.

The University of Western Ontario, Faculty of Graduate Studies, Biosciences Division, Department of Plant Sciences, London, ON N6A 5B8, Canada. Offers plant and environmental sciences (M Sc); plant sciences (M Sc, PhD); plant sciences and environmental sciences (PhD); plant sciences and molecular biology (M Sc, PhD). *Degree requirements:* For master's and doctorate, thesis/dissertation. *Entrance requirements:* For doctorate, M Sc or equivalent. *Faculty research:* Ecology systematics, plant biochemistry and physiology, yeast genetics, molecular biology.

University of Wisconsin–La Crosse, Office of University Graduate Studies, College of Science and Health, Department of Biology, La Crosse, WI 54601-3742. Offers aquatic sciences (MS); biology (MS); cellular and molecular biology (MS); clinical microbiology (MS); microbiology (MS); nurse anesthesia (MS); physiology (MS). *Accreditation:* AANA/CANAEP. Part-time programs available. *Faculty:* 18 full-time (5 women), 1 part-time/adjunct (0 women). *Students:* 23 full-time (10 women), 52 part-time (25 women); includes 5 minority (1 American Indian/Alaska Native, 3 Asian Americans or Pacific Islanders, 1 Hispanic American), 3 international. Average age 26. 61 applicants, 44% accepted, 23 enrolled. In 2005, 14 degrees awarded. *Degree requirements:* For master's, thesis, comprehensive exam, registration. *Entrance requirements:* For master's, GRE General Test, minimum GPA of 2.85. Additional exam requirements/recommendations for international students: Required—TOEFL (minimum score 550 paper-based; 213 computer-based). *Application deadline:* For fall admission, 3/1 for domestic students. Applications are processed on a rolling basis. Application fee: $45. Electronic applications accepted. *Expenses:* Tuition, state resident: part-time $354 per credit. Tuition, nonresident: part-time $943 per credit. Tuition and fees vary according to course load, program and reciprocity agreements. *Financial support:* In 2005–06, 10 students received support, including 4 research assistantships with partial tuition reimbursements available (averaging $10,000 per year), 10 teaching assistantships with partial tuition reimbursements available (averaging $9,600 per year); career-related internships or fieldwork, Federal Work-Study, health care benefits, unspecified assistantships, and grant-funded positions also available. Support available to part-time students. Financial award application deadline: 3/15; financial award applicants required to submit FAFSA. *Faculty research:* Cell and molecular biology, physiology, environmental sciences, mycology, biomedical general. Total annual research expenditures: $700,000. *Unit head:* , Dr. Tom Volk, Program Director, 608-785-6972, Fax: 608-785-6959, E-mail: volk.thom@uwlax.edu. *Application contact:* Kathryn Kiefer, Associate Director of Admissions, 608-785-8939, E-mail: admissions@uwlax.edu.

University of Wisconsin–Madison, Graduate School, Program in Cellular and Molecular Biology, Madison, WI 53706-1596. Offers PhD. *Faculty:* 175 full-time (51 women). *Students:* 143 full-time (81 women); includes 12 minority (1 African American, 2 American Indian/Alaska Native, 5 Asian Americans or Pacific Islanders, 4 Hispanic Americans), 27 international. Average age 25. 436 applicants, 13% accepted, 22 enrolled. In 2005, 15 degrees awarded. *Median time to degree:* Of those who began their doctoral program in fall 1997, 100% received their degree in 8 years or less. *Degree requirements:* For doctorate, thesis/dissertation, comprehensive exam. *Entrance requirements:* For doctorate, GRE General Test, GRE Subject Test (recommended), minimum GPA of 3.0, lab experience. Additional exam requirements/recommendations for international students: Required—TOEFL (minimum score 580 paper-based; 237 computer-based). *Application deadline:* For fall admission, 12/15 for domestic students, 12/15 for international students. Application fee: $45. Electronic applications accepted. *Financial support:* In 2005–06, 143 students received support, including 25 fellowships with full tuition reimbursements available (averaging $22,500 per year), 115 research assistantships with full tuition reimbursements available (averaging $22,500 per year), 3 teaching assistantships with full tuition reimbursements available (averaging $22,500 per year); traineeships, health care benefits, and unspecified assistantships also available. *Faculty research:* Virology, cancer biology, transcriptional mechanisms, plant biology, immunology.

See Close-Up on page 657.

University of Wisconsin–Madison, School of Veterinary Medicine, Department of Animal Health and Biomedical Sciences, Program in Comparative Biosciences, Madison, WI 53706-1380. Offers anatomy (MS, PhD); biochemistry (MS, PhD); cellular and molecular biology (MS, PhD); environmental toxicology (MS, PhD); neurosciences (MS, PhD); pharmacology (MS, PhD); physiology (MS, PhD). *Degree requirements:* For doctorate, thesis/dissertation.

University of Wisconsin–Parkside, College of Arts and Sciences, Program in Applied Molecular Biology, Kenosha, WI 53141-2000. Offers MAMB. Part-time programs available. *Faculty:* 9 full-time (2 women). *Students:* 9 full-time (2 women), 4 part-time (1 woman); includes 3 minority (1 Asian American or Pacific Islander, 2 Hispanic Americans). Average age 25. 4 applicants, 50% accepted, 2 enrolled. In 2005, 3 degrees awarded. *Degree requirements:* For master's, thesis, oral exam. *Entrance requirements:* For master's, GRE General Test, minimum GPA of 3.0; course work in biology, chemistry, math, physics. Additional exam requirements/recommendations for international students: Required—TOEFL (minimum score 550 paper-based; 213 computer-based). *Application deadline:* For fall admission, 7/1 for domestic students. Applications are processed on a rolling basis. Application fee: $45. Electronic applications accepted. *Expenses:* Tuition, state resident: full-time $6,344; part-time $344 per credit. Tuition, nonresident: full-time $16,954; part-time $934 per credit. Tuition and fees vary according to course load and program. *Financial support:* Career-related internships or fieldwork, Federal Work-Study, and unspecified assistantships available. Financial award application deadline: 7/1. *Faculty research:* Gene cloning, genome structure, cell cycle effects on gene expression, molecular biology of plant hormones, laboratory toxin production and resistance. Total annual research expenditures: $200,000. *Unit head:* Dr. David Higgs, Chair, Molecular Biology Programs Committee, 262-595-2786, Fax: 262-595-2056, E-mail: david.higgs@uwp.edu.

University of Wyoming, Graduate School, College of Agriculture, Department of Molecular Biology, Laramie, WY 82070. Offers MS, PhD. *Faculty:* 13 full-time (3 women), 2 part-time/adjunct (0 women). *Students:* 13 full-time (7 women), 8 part-time (3 women); includes 1 minority (Asian American or Pacific Islander), 9 international. Average age 31. 29 applicants, 17% accepted, 5 enrolled. In 2005, 5 master's, 1 doctorate awarded. Terminal master's awarded for partial completion of doctoral program. *Degree requirements:* For master's and doctorate, thesis/dissertation, comprehensive exam. *Entrance requirements:* For master's and doctorate, GRE General Test, GRE Subject Test (recommended), minimum GPA of 3.0. Additional exam requirements/recommendations for international students: Required—TOEFL. *Application deadline:* For fall admission, 2/1 for domestic students. Applications are processed on a rolling basis. Application fee: $50. Electronic applications accepted. *Expenses:* Tuition, state resident: full-time $3,720; part-time $155 per credit hour. Tuition, nonresident: full-time $10,704; part-time $446 per credit hour. Required fees: $666; $162 per semester. Tuition and fees vary according to course load and program. *Financial support:* In 2005–06, 8 students received support, including 8 research assistantships with full tuition reimbursements available

Molecular Biology

University of Wyoming (continued)

(averaging $17,894 per year); institutionally sponsored loans also available. Financial award application deadline: 3/1. *Faculty research:* Protein structure/function, developmental regulation, yeast genetics, bacterial pathogenesis. Total annual research expenditures: $1.9 million. *Unit head:* Dr. Jordanka Zlatanova, Chair, 307-766-3300, Fax: 307-766-5098, E-mail: uwmbio@uwyo.edu. *Application contact:* Dr. Peter Thorsness, Graduate Program Chairperson, 307-766-2038, Fax: 307-766-5098.

Utah State University, School of Graduate Studies, College of Agriculture, Department of Nutrition and Food Sciences, Logan, UT 84322. Offers dietetic administration (MDA); food microbiology and safety (MFMS); nutrition and food sciences (MS, PhD); nutrition science (MS, PhD), including molecular biology. Postbaccalaureate distance learning degree programs offered. *Faculty:* 13 full-time (6 women), 1 (woman) part-time/adjunct. *Students:* 55 full-time (26 women), 13 part-time (12 women); includes 4 minority (all African Americans), 15 international. Average age 27. 14 applicants, 57% accepted, 4 enrolled. In 2005, 8 master's, 3 doctorates awarded. *Degree requirements:* For master's, thesis, BS core competency courses BS core competency courses; for doctorate, thesis/dissertation, teaching experience, inst 7920 and BS core competency courses, comprehensive exam, registration. *Entrance requirements:* For master's, GRE General Test, minimum GPA of 3.0, course work in chemistry; for doctorate, GRE General Test, minimum GPA of 3.2, course work in chemistry, MS or manuscript in referred journal. Additional exam requirements/recommendations for international students: Required—TOEFL (minimum score 550 paper-based). *Application deadline:* For fall admission, 6/15 priority date for domestic students, 6/15 priority date for international students; for spring admission, 10/15 priority date for domestic students, 10/15 priority date for international students. Applications are processed on a rolling basis. Application fee: $50 ($60 for international students). Electronic applications accepted. *Financial support:* In 2005–06, 19 students received support, including fellowships with partial tuition reimbursements available (averaging $14,484 per year), 22 research assistantships with partial tuition reimbursements available (averaging $12,600 per year), 3 teaching assistantships with partial tuition reimbursements available (averaging $6,000 per year); Federal Work-Study, institutionally sponsored loans, scholarships/grants, tuition waivers (full and partial), unspecified assistantships, and fellowships, international students get support from their countries also available. Financial award application deadline: 3/1. *Faculty research:* Mineral balance, meat microbiology and nitrate interactions, milk ultrafiltration, lactic culture, milk coagulation. Total annual research expenditures: $319,280. *Unit head:* Dr. Charles C. Carpenter, Head, 435-797-2126, Fax: 435-797-2379, E-mail: chuck@cc.usu.edu. *Application contact:* Pam Zetterquist, Staff Assistant II, 435-797-4041, Fax: 435-797-2379, E-mail: pzett@cc.usu.edu.

Vanderbilt University, Graduate School and School of Medicine, Department of Molecular Physiology and Biophysics, Nashville, TN 37240-1001. Offers MS, PhD, MD/PhD. *Faculty:* 53 full-time (16 women). *Students:* 50 full-time (28 women), 1 (woman) part-time; includes 5 minority (1 African American, 1 American Indian/Alaska Native, 2 Asian Americans or Pacific Islanders, 1 Hispanic American), 10 international. In 2005, 1 master's, 6 doctorates awarded. *Degree requirements:* For doctorate, thesis/dissertation, preliminary, qualifying, and final exams. *Entrance requirements:* For doctorate, GRE General Test, GRE Subject Test (recommended). *Application deadline:* For fall admission, 1/15 for domestic students, 1/15 for international students. Application fee: $0. Electronic applications accepted. *Expenses:* Tuition: Part-time $1,283 per semester hour. Required fees: $2,202; $1,101 per semester. One-time fee: $30. Tuition and fees vary according to course load, program and student level. *Financial support:* Fellowships with full tuition reimbursements, research assistantships with full tuition reimbursements, Federal Work-Study, institutionally sponsored loans, traineeships, and tuition waivers (partial) available. Financial award application deadline: 1/15. *Faculty research:* Molecular endocrinology, membrane transport biophysics, metabolic regulation, neurobiology. *Unit head:* Alan D. Cherrington, Chair, 615-322-7000, Fax: 615-343-0490, E-mail: alan.cherrington@vanderbilt.edu. *Application contact:* Hussane Mchaourab, Director of Graduate Studies, 615-322-7000, Fax: 615-343-0490, E-mail: hassane.mchaourab@vanderbilt.edu.

Virginia Commonwealth University, Medical College of Virginia-Professional Programs, School of Medicine and Graduate Programs, School of Medicine Graduate Programs, Department of Biochemistry, Richmond, VA 23284-9005. Offers biochemistry (MS, PhD, CBHS); bioinformatics (MS); molecular biology and genetics (MS, PhD); neurosciences (PhD). *Faculty:* 26 full-time (6 women). *Students:* 48 full-time (24 women), 9 part-time (8 women); includes 12 minority (2 African Americans, 8 Asian Americans or Pacific Islanders, 2 Hispanic Americans), 8 international. 61 applicants, 75% accepted. In 2005, 1 master's, 4 doctorates, 2 other advanced degrees awarded. *Degree requirements:* For master's, thesis; for doctorate, thesis/dissertation, comprehensive oral and written exams. *Entrance requirements:* For master's and doctorate, DAT, GRE General Test, MCAT. *Application deadline:* For fall admission, 2/15 for domestic students. Application fee: $50. *Expenses:* Tuition, state resident: full-time $3,185; part-time $405 per credit. Tuition, nonresident: full-time $7,952; part-time $940 per credit. Required fees: $751 per semester hour. Tuition and fees vary according to course load and program. *Financial support:* Fellowships, research assistantships available. *Faculty research:* Molecular biology, peptide/protein chemistry, neurochemistry, enzyme mechanisms, macromolecular structure determination. Total annual research expenditures: $3.5 million. *Unit head:* Dr. Sarah Spiegel, Chair, 804-828-9762, Fax: 804-828-1473, E-mail: sspiegel@vcu.edu. *Application contact:* Dr. Keith R. Shelton, Program Director, 804-828-9886, Fax: 804-828-1473, E-mail: krshelto@vcu.edu.

See Close-Up on page 447.

Virginia Commonwealth University, Medical College of Virginia-Professional Programs, School of Medicine and Graduate Programs, School of Medicine Graduate Programs, Department of Human Genetics, Richmond, VA 23284-9005. Offers genetic counseling (MS); human genetics (PhD, CBHS); molecular biology and genetics (MS, PhD). *Faculty:* 18 full-time (10 women). *Students:* 21 full-time (15 women), 4 part-time (1 woman); includes 2 minority (both Asian Americans or Pacific Islanders), 4 international. 52 applicants, 46% accepted. In 2005, 6 master's, 3 doctorates, 2 other advanced degrees awarded. *Degree requirements:* For master's, thesis; for doctorate, thesis/dissertation, comprehensive oral and written exams. *Entrance requirements:* For master's, DAT, GRE General Test, or MCAT; for doctorate, GRE General Test, DAT, MCAT. *Application deadline:* For fall admission, 2/15 for domestic students. Application fee: $50. *Expenses:* Tuition, state resident: full-time $3,185; part-time $405 per credit. Tuition, nonresident: full-time $7,952; part-time $940 per credit. Required fees: $751 per semester hour. Tuition and fees vary according to course load and program. *Financial support:* Fellowships available. *Faculty research:* Genetic epidemiology, biochemical genetics, quantitative genetics, human cytogenetics, molecular genetics. *Unit head:* Dr. Walter E. Nance, Chair, 804-828-9632, Fax: 804-828-3760, E-mail: wenance@vcu.edu. *Application contact:* Dr. Linda A. Corey, Graduate Program Director, 804-828-8759, Fax: 804-828-3760, E-mail: lacorey@vcu.edu.

Virginia Commonwealth University, Medical College of Virginia-Professional Programs, School of Medicine and Graduate Programs, School of Medicine Graduate Programs, Department of Microbiology and Immunology, Richmond, VA 23284-9005. Offers microbiology and genetics (MS); microbiology and immunology (MS, PhD, CBHS); molecular biology and genetics (PhD). *Faculty:* 33 full-time (8 women). *Students:* 66 full-time (43 women), 9 part-time (6 women); includes 21 minority (12 African Americans, 5 Asian Americans or Pacific Islanders, 4 Hispanic Americans), 11 international. 95 applicants, 48% accepted. In 2005, 25 master's, 7 doctorates, 3 other advanced degrees awarded. *Degree requirements:* For master's, thesis; for doctorate, thesis/dissertation, comprehensive oral and written exams. *Entrance requirements:* For master's, GRE General Test or MCAT; for doctorate, GRE General Test, MCAT. *Application deadline:* For fall admission, 2/15 for domestic students. Application fee: $50. *Expenses:* Tuition, state resident: full-time $6,268; part-time $405 per credit. Tuition, nonresident: full-time $15,904; part-time $940 per credit. Required fees: $751 per semester hour. Tuition and fees vary according to course load and program. *Financial support:* Fellowships, research assistantships, teaching assistantships available. *Faculty research:* Microbial physiology and genetics,

molecular biology, crystallography of biological molecules, antibiotics and chemotherapy, membrane transport. *Unit head:* Dr. Dennis E. Ohman, Chair, 804-828-9728, Fax: 804-828-9946, E-mail: deohman@vcu.edu. *Application contact:* Dr. Guy A. Cabral, Chair, Graduate Program, 804-828-2306, E-mail: gacabral@vcu.edu.

Virginia Commonwealth University, Medical College of Virginia-Professional Programs, School of Medicine and Graduate Programs, School of Medicine Graduate Programs, Department of Pharmacology and Toxicology, Richmond, VA 23284-9005. Offers molecular biology and genetics (PhD); neurosciences (PhD); pharmacology (PhD, CBHS); pharmacology and toxicology (MS). *Faculty:* 36 full-time (9 women). *Students:* 35 full-time (18 women), 11 part-time (all women); includes 7 minority (2 African Americans, 4 Asian Americans or Pacific Islanders, 1 Hispanic American), 5 international. 90 applicants, 33% accepted. In 2005, 6 master's, 2 doctorates awarded. Terminal master's awarded for partial completion of doctoral program. *Degree requirements:* For master's, thesis; for doctorate, thesis/dissertation, comprehensive oral and written exams. *Entrance requirements:* For master's, DAT, GRE General Test or MCAT; for doctorate, GRE General Test, MCAT, DAT. *Application deadline:* For fall admission, 4/1 for domestic students. Application fee: $50. *Expenses:* Tuition, state resident: full-time $6,268; part-time $405 per credit. Tuition, nonresident: full-time $15,904; part-time $940 per credit. Required fees: $751 per semester hour. Tuition and fees vary according to course load and program. *Financial support:* Fellowships, teaching assistantships available. *Faculty research:* Drug abuse, drug metabolism, pharmacodynamics, peptide synthesis, receptor mechanisms. *Unit head:* Dr. Billy R. Martin, Chair, 804-828-8407, Fax: 804-828-2117, E-mail: brmartin@vcu.edu. *Application contact:* Sheryol Cox, Graduate Program Coordinator, 804-828-8400, Fax: 804-828-2117, E-mail: swcox@vcu.edu.

See Close-Up on page 1219.

Wake Forest University, School of Medicine and Graduate School, Graduate Programs in Medicine, Molecular Genetics Program, Winston-Salem, NC 27109. Offers PhD. *Degree requirements:* For doctorate, thesis/dissertation. *Entrance requirements:* For doctorate, GRE General Test. Additional exam requirements/recommendations for international students: Required—TOEFL. Electronic applications accepted. *Faculty research:* Control of gene expression, molecular pathogenesis, protein biosynthesis, cell development, clinical cytogenetics.

See Close-Up on page 805.

Washington State University, Graduate School, College of Sciences, School of Molecular Biosciences, Pullman, WA 99164. Offers biochemistry and biophysics (MS, PhD); genetics and cell biology (MS, PhD); microbiology (MS, PhD). *Faculty:* 25 full-time (5 women), 21 part-time/adjunct (4 women). *Students:* 61 full-time (33 women), 2 part-time (both women); includes 5 minority (1 American Indian/Alaska Native, 1 Asian American or Pacific Islander, 3 Hispanic Americans), 23 international. Average age 26. 231 applicants, 18% accepted, 13 enrolled. In 2005, 5 master's, 7 doctorates awarded. Terminal master's awarded for partial completion of doctoral program. *Degree requirements:* For master's, thesis or alternative, oral exam; for doctorate, thesis/dissertation, oral exam, comprehensive exam, registration. *Entrance requirements:* For master's and doctorate, GRE General Test, minimum GPA 3.0. Additional exam requirements/recommendations for international students: Required—TOEFL (minimum score 550 paper-based; 213 computer-based). *Application deadline:* For fall admission, 12/15 for domestic students, 12/15 for international students. Application fee: $35. Electronic applications accepted. *Expenses:* Tuition, state resident: full-time $6,295; part-time $336 per credit. Tuition, nonresident: full-time $15,949; part-time $819 per credit. Required fees: $933. Part-time tuition and fees vary according to campus/location and program. *Financial support:* In 2005–06, 6 fellowships with full tuition reimbursements (averaging $18,384 per year), 29 research assistantships with full and partial tuition reimbursements (averaging $18,384 per year), 26 teaching assistantships with full and partial tuition reimbursements (averaging $18,384 per year) were awarded; career-related internships or fieldwork, Federal Work-Study, institutionally sponsored loans, traineeships, health care benefits, and unspecified assistantships also available. Financial award application deadline: 4/1; financial award applicants required to submit FAFSA. Total annual research expenditures: $5.8 million. *Unit head:* Dr. John H. Nilson, Director, 509-335-8724, Fax: 509-335-9688, E-mail: jhn@wsu.edu. *Application contact:* Kelly G. McGovern, Academic Coordinator, 509-335-4566, Fax: 509-335-1907, E-mail: smbgrad@wsu.edu.

Washington University in St. Louis, Graduate School of Arts and Sciences, Division of Biology and Biomedical Sciences, Program in Molecular Cell Biology, St. Louis, MO 63110. Offers PhD. *Degree requirements:* For doctorate, thesis/dissertation. *Entrance requirements:* For doctorate, GRE General Test, GRE Subject Test. Electronic applications accepted.

Wayne State University, Graduate School, Program in Molecular Biology and Genetics, Detroit, MI 48202. Offers MS, PhD. *Faculty:* 27. *Students:* 22 full-time (11 women), 12 international. Average age 29. 46 applicants, 4% accepted, 2 enrolled. In 2005, 4 degrees awarded. Terminal master's awarded for partial completion of doctoral program. *Degree requirements:* For master's and doctorate, thesis/dissertation. *Entrance requirements:* For master's and doctorate, GRE General Test. Additional exam requirements/recommendations for international students: Required—TOEFL (minimum score 550 paper-based; 213 computer-based); Recommended—TWE (minimum score 6). *Application deadline:* Applications are processed on a rolling basis. Application fee: $30 ($50 for international students). Electronic applications accepted. *Expenses:* Tuition, state resident: part-time $338 per credit hour. Tuition, nonresident: part-time $746 per credit hour. Required fees: $24 per credit hour. Full-time tuition and fees vary according to program. *Financial support:* In 2005–06, 1 fellowship, 16 research assistantships were awarded; teaching assistantships *Faculty research:* Human gene mapping, genome organization and sequencing, gene regulation, molecular evolution. Total annual research expenditures: $3.3 million. *Unit head:* Dr. David Womble, Director, 313-577-2374, Fax: 313-577-5218, E-mail: aa3330@wayne.edu. *Application contact:* Alexander Gow, Associate Professor, 313-577-9401, E-mail: agow@genetics.wayne.edu.

See Close-Up on page 665.

Wayne State University, School of Medicine and Graduate School, Graduate Programs in Medicine, Department of Biochemistry and Molecular Biology, Detroit, MI 48202. Offers MS, PhD. *Faculty:* 13 full-time (11 women). *Students:* 23 full-time (7 women), 1 part-time; includes 1 minority (African American), 20 international. Average age 28. 39 applicants, 18% accepted, 3 enrolled. In 2005, 2 degrees awarded. Terminal master's awarded for partial completion of doctoral program. *Degree requirements:* For master's, thesis; for doctorate, one foreign language, thesis/dissertation. *Entrance requirements:* For master's and doctorate, GRE General Test, GRE Subject Test. Additional exam requirements/recommendations for international students: Required—TOEFL (minimum score 550 paper-based; 213 computer-based); Recommended—TWE (minimum score 6). *Application deadline:* Applications are processed on a rolling basis. Application fee: $30 ($50 for international students). Electronic applications accepted. *Expenses:* Tuition, state resident: part-time $338 per credit hour. Tuition, nonresident: part-time $746 per credit hour. Required fees: $24 per credit hour. Full-time tuition and fees vary according to program. *Financial support:* In 2005–06, 22 research assistantships (averaging $20,159 per year) were awarded; fellowships, teaching assistantships *Faculty research:* Protein structure, molecular biology, molecular genetics, enzymology, x-ray crystallography. Total annual research expenditures: $3.6 million. *Unit head:* Dr. Barry P. Rosen, 313-577-1512, Fax: 313-577-2765, E-mail: aa1133@wayne.edu. *Application contact:* Marilyn Doscher, Graduate Director, 313-577-1295, E-mail: mdoscher@med.wayne.edu.

Wesleyan University, Graduate Programs, Department of Molecular Biology and Biochemistry, Middletown, CT 06459-0260. Offers biochemistry (PhD); molecular biology (PhD). *Faculty:* 9 full-time (2 women). *Students:* 23 full-time (18 women), 14 international. Average age 28. In 2005, 2 doctorates awarded. *Degree requirements:* For doctorate, thesis/dissertation. *Entrance requirements:* For doctorate, GRE General Test, GRE Subject Test. *Application deadline:* For fall admission, 3/15 for domestic students. Applications are processed on a rolling basis. Application fee: $0. *Expenses:* Tuition: Full-time $24,732. One-time fee: $20 full-time. *Financial* *support:* Research assistantships, teaching assistantships, institutionally sponsored loans available. *Faculty research:* Genome organization, regulation of gene expression, molecular biology of development, physical biochemistry. *Unit head:* Ishita Mukerji, Chair, 860-685-2422, E-mail: imukerji@wesleyan.edu. *Application contact:* Debra Alexson, Information Contact, 860-685-7640, E-mail: dalexson@wesleyan.edu.

See Close-Up on page 667.

West Virginia University, Eberly College of Arts and Sciences, Department of Biology, Morgantown, WV 26506. Offers cell and molecular biology (MS, PhD); environmental and evolutionary biology (MS, PhD); integrative organismal biology (PhD); integrative organismal, biology (MS). *Faculty:* 18 full-time (3 women), 4 part-time/adjunct (all women). *Students:* 29 full-time (20 women), 5 part-time (3 women); includes 1 minority (Asian American or Pacific Islander), 9 international. Average age 26. 50 applicants, 10% accepted. In 2005, 1 master's, 2 doctorates awarded. Terminal master's awarded for partial completion of doctoral program. *Degree requirements:* For master's, thesis, final exam; for doctorate, thesis/dissertation, preliminary and final exams. *Entrance requirements:* For master's, GRE General Test, GRE Subject Test, minimum GPA of 3.0; for doctorate, GRE General Test, minimum GPA of 3.0. Additional exam requirements/recommendations for international students: Required—TOEFL. *Application deadline:* For fall admission, 4/1 for domestic students; for spring admission, 10/1 for domestic students. Applications are processed on a rolling basis. Application fee: $45. *Expenses:* Tuition, state resident: full-time $4,582; part-time $258 per credit hour. Tuition, nonresident: full-time $13,820; part-time $741 per credit hour. *Financial support:* In 2005–06, 4 research assistantships, 22 teaching assistantships were awarded; Federal Work-Study and institutionally sponsored loans also available. Financial award application deadline: 4/1; financial award applicants required to submit FAFSA. *Faculty research:* Environmental biology, genetic engineering, developmental biology, global change, biodiversity. *Unit head:* Dr. Jonathan Cumming, Chair, 304-293-5201 Ext. 2508, E-mail: jonathan.cumming@mail.wvu.edu. *Application contact:* Dr. William T. Peterjohn, Director of Graduate Studies, 304-293-5201 Ext. 2510, Fax: 304-293-6363, E-mail: william.peterjohn@mail.wvu.edu.

See Close-Up on page 325.

West Virginia University, Interdisciplinary Studies in Molecular and Cellular Biology, Morgantown, WV 26506.

West Virginia University, School of Medicine, Graduate Programs at the Health Science Center, Biomedical Sciences Graduate Program, Program in Biochemistry and Molecular Biology, Morgantown, WV 26506. Offers MS, PhD. *Faculty:* 20 full-time (3 women). *Students:* 14 full-time (7 women); includes 8 minority (1 African American, 7 Asian Americans or Pacific Islanders), 7 international. Average age 27. In 2005, 1 master's, 1 doctorate awarded. *Median time to degree:* Of those who began their doctoral program in fall 1997, 98% received their degree in 8 years or less. *Degree requirements:* For doctorate, thesis/dissertation, comprehensive exam. *Entrance requirements:* For doctorate, GRE General Test, minimum GPA of 3.0. Additional exam requirements/recommendations for international students: Required—TOEFL. *Application deadline:* For fall admission, 3/1 for domestic students, 1/15 for international students. Applications are processed on a rolling basis. Application fee: $0. Electronic applications accepted. *Financial support:* In 2005–06, research assistantships with full tuition reimbursements (averaging $20 per year); institutionally sponsored loans, traineeships, and health care benefits also available. *Faculty research:* Regulation of gene expression, cell survival mechanisms, signal transduction, regulation of metabolism, sensory neuroscience. Total annual research expenditures: $1 million. *Unit head:* , Dr. Diana S. Beattie, Chair, 304-293-7522, Fax: 304-293-6846, E-mail: diana.beattie@hsc.wvu.edu. *Application contact:* Dr. Lisa Salati, Graduate Program Coordinator, 304-293-7759, Fax: 304-293-6846, E-mail: lisa.salati@hsc.wvu.edu.

See Close-Up on page 453.

William Paterson University of New Jersey, College of Science and Health, Department of Biology, General Biology Program, Wayne, NJ 07470-8420. Offers general biology (MA); limnology and terrestrial ecology (MA); molecular biology (MA); physiology (MA). Part-time and evening/weekend programs available. *Students:* 2 full-time (1 woman), 13 part-time (9 women); includes 4 Hispanic Americans. In 2005, 4 degrees awarded. *Degree requirements:* For master's, independent study or thesis. *Entrance requirements:* For master's, GRE General Test, minimum GPA of 2.75. *Application deadline:* Applications are processed on a rolling basis. Application fee: $50. Electronic applications accepted. *Expenses:* Tuition, state resident: full-time $476. Tuition, nonresident: full-time $717. *Financial support:* Research assistantships, career-related internships or fieldwork and unspecified assistantships available. Financial award application deadline: 4/1; financial award applicants required to submit FAFSA. *Application contact:* Danielle Liautaud, Assistant Director, 973-720-3579, Fax: 973-720-2035, E-mail: liautaudd@wpunj.edu.

See Close-Up on page 331.

Wright State University, School of Graduate Studies, College of Science and Mathematics, Department of Biochemistry and Molecular Biology, Dayton, OH 45435. Offers MS. *Degree requirements:* For master's, thesis. *Entrance requirements:* Additional exam requirements/recommendations for international students: Required—TOEFL. *Faculty research:* Regulation of gene expression, macromolecular structural function, NMR imaging, visual biochemistry.

Yale University, Graduate School of Arts and Sciences, Department of Molecular, Cellular, and Developmental Biology, Program in Molecular Biology, New Haven, CT 06520. Offers PhD. *Degree requirements:* For doctorate, thesis/dissertation. *Entrance requirements:* For doctorate, GRE General Test, GRE Subject Test.

Yale University, School of Medicine and Graduate School of Arts and Sciences, Combined Program in Biological and Biomedical Sciences (BBS), Molecular Cell Biology, Genetics, and Development Track, New Haven, CT 06520. Offers PhD, MD/PhD. *Students:* 31 full-time. *Entrance requirements:* Additional exam requirements/recommendations for international students: Required—TOEFL. *Application deadline:* For fall admission, 12/8 for domestic students, 12/8 for international students. *Unit head:* Dr. Thomas Pollard, Co-Director, 203-432-3565. *Application contact:* Anne Scott, Graduate Registrar, 203-432-3538, Fax: 203-432-3597, E-mail: anne.scott@yale.edu.

Molecular Medicine

Baylor College of Medicine, Graduate School of Biomedical Sciences, Program in Translational Biology and Molecular Medicine, Houston, TX 77030-3498. Offers PhD. *Faculty:* 118 full-time (33 women). *Students:* 10 full-time (5 women); includes 1 African American, 2 Hispanic Americans, 3 international. 25 applicants, 44% accepted, 9 enrolled. *Degree requirements:* For doctorate, thesis/dissertation, public defense. *Entrance requirements:* For doctorate, GRE, minimum GPA of 3.0. Additional exam requirements/recommendations for international students: Required—TOEFL. *Application deadline:* For fall admission, 2/1 for domestic students. Application fee: $30. Electronic applications accepted. *Expenses:* Tuition: Full-time $8,200. Full-time tuition and fees vary according to program. *Financial support:* In 2005–06, 10 students received support, including 9 fellowships (averaging $23,000 per year), 1 research assistantship (averaging $23,000 per year); career-related internships or fieldwork, Federal Work-Study, health care benefits, and tuition waivers (full) also available. Financial award applicants required to submit FAFSA. *Unit head:* Dr. David Huston, Director, 713-798-1077, Fax: 713-798-3586, E-mail: tbmm@bcm.edu. *Application contact:* Paula Ramirez, Graduate Program Administrator, 713-798-1077, Fax: 713-798-3586, E-mail: pramirez@bcm.edu.

Boston University, School of Medicine, Division of Graduate Medical Sciences, Program in Molecular Medicine, Boston, MA 02215. Offers PhD, MD/PhD. *Degree requirements:* For doctorate, thesis/dissertation, qualifying exam. *Application deadline:* For fall admission, 1/15 for domestic students; for spring admission, 10/15 priority date for domestic students. Electronic applications accepted. *Expenses:* Tuition: Full-time $31,530; part-time $985 per credit. Required fees: $316; $40 per semester. Tuition and fees vary according to course level and program. *Financial support:* Fellowships, research assistantships, Federal Work-Study, scholarships/grants, and traineeships available. *Unit head:* Dr. Joseph Loscalzo, Director, 617-414-1519, Fax: 617-414-1515, E-mail: gpmm@med-med1.bu.edu.

See Close-Up on page 553.

Cleveland State University, College of Graduate Studies, College of Science, Department of Biological, Geological, and Environmental Sciences, Cleveland, OH 44115. Offers biology (MS); environmental science (MS); molecular medicine (PhD); regulatory biology (PhD). Part-time programs available. *Faculty:* 19 full-time (3 women), 34 part-time/adjunct (7 women). *Students:* 54 full-time (34 women), 34 part-time (17 women); includes 8 minority (6 African Americans, 2 Asian Americans or Pacific Islanders), 32 international. Average age 30. 34 applicants, 65% accepted, 18 enrolled. In 2005, 1 master's, 3 doctorates awarded. Terminal master's awarded for partial completion of doctoral program. *Median time to degree:* Of those who began their doctoral program in fall 1997, 100% received their degree in 8 years or less. *Degree requirements:* For master's, thesis (for some programs); for doctorate, thesis/dissertation, comprehensive exam. *Entrance requirements:* For master's and doctorate, GRE General Test, 2 letters of recommendation. Additional exam requirements/recommendations for international students: Required—TOEFL (minimum score 525 paper-based; 197 computer-based); Recommended—TSE. *Application deadline:* For fall admission, 4/1 priority date for domestic students, 4/1 priority date for international students; for spring admission, 12/1 priority date for domestic students. Applications are processed on a rolling basis. Application fee: $30. Electronic applications accepted. *Expenses:* Tuition, state resident: full-time $10,700. Tuition, nonresident: full-time $14,628. Tuition and fees vary according to program. *Financial support:* In 2005–06, 29 students received support, including research assistantships with full and partial tuition reimbursements available (averaging $16,500 per year), teaching assistantships with full and partial tuition reimbursements available (averaging $16,500 per year); institutionally sponsored loans and unspecified assistantships also available. *Faculty research:* Molecular and cell biology, immunology. *Unit head:* Dr. Michael Gates, Chairperson, 216-687-3917, Fax: 216-687-6972, E-mail: m.gates@csuohio.edu. *Application contact:* Dr. Jeffrey Dean, Graduate Program Director, 216-687-2440, Fax: 216-687-6972, E-mail: gpd.bges@csuohio.edu.

Cornell University, Graduate School, Graduate Fields of Comparative Biomedical Sciences, Field of Comparative Biomedical Sciences, Ithaca, NY 14853-0001. Offers cellular and molecular medicine (MS, PhD); developmental and reproductive biology (MS, PhD); infectious diseases (MS, PhD); population medicine and epidemiology (MS); population medicine and epidemiology sciences (PhD); structural and functional biology (MS, PhD). *Faculty:* 135 full-time (38 women). *Students:* 41 full-time (23 women); includes 4 minority (1 African American, 3 Asian Americans or Pacific Islanders), 19 international. 58 applicants, 60% accepted, 33 enrolled. In 2005, 4 degrees awarded. *Degree requirements:* For master's, thesis/dissertation; for doctorate, thesis/dissertation, comprehensive exam. *Entrance requirements:* For master's and doctorate, GRE General Test, 2 letters of recommendation. Additional exam requirements/recommendations for international students: Required—TOEFL (minimum score 550 paper-based; 213 computer-based). *Application deadline:* For fall admission, 12/15 for domestic students. Application fee: $60. Electronic applications accepted. *Financial support:* In 2005–06, 41 students received support, including 13 fellowships with full tuition reimbursements available, 28 research assistantships with full tuition reimbursements available; teaching assistantships with full tuition reimbursements available, institutionally sponsored loans, scholarships/grants, health care benefits, tuition waivers (full and partial), and unspecified assistantships also available. Financial award applicants required to submit FAFSA. *Faculty research:* Receptors and signal transduction, viral and bacterial infectious diseases, tumor metastasis, clinical sciences/nutritional disease, development/neurologic disorders. *Unit head:* Director of Graduate Studies, 607-253-3276, Fax: 607-253-3756. *Application contact:* Graduate Field Assistant, 607-253-3276, Fax: 607-253-3756, E-mail: graduate_edcvm@cornell.edu.

Dartmouth College, Dartmouth Medical School, Program in Experimental and Molecular Medicine, Hanover, NH 03755. Offers experimental and molecular medicine (PhD); pharmcology and toxicology (PhD); psychological and brain sciences (PhD); physiology (PhD). *Degree requirements:* For doctorate, thesis/dissertation, comprehensive exam, registration. *Entrance requirements:* For doctorate, GRE General Test, 3 letters of recommendation. Additional exam requirements/recommendations for international students: Required—TOEFL. Electronic applications accepted. *Expenses:* Tuition: Full-time $31,770.

The Feinstein Institute for Medical Research, Graduate School of Molecular Medicine, Manhasset, NY 11030. Offers PhD. *Faculty:* 30 full-time (11 women). *Students:* 3 full-time (1 woman); includes 1 minority (Hispanic American) Average age 30. 2 applicants, 50% accepted, 1 enrolled. *Degree requirements:* For doctorate, thesis/dissertation, comprehensive exam. *Entrance requirements:* For doctorate, MD. *Application deadline:* Applications are processed on a rolling basis. Application fee: $25. Students receive $50,000 annual fellowship. *Financial support:* In 2005–06, 3 students received support, including 3 fellowships with tuition reimbursements available (averaging $50,000 per year); health care benefits and tuition waivers (full) also available. *Faculty research:* Cardiopulmonary disease, cancer, inflammation, genetics of complex disorders, cytokine biology. *Unit head:* Dr. Barbara M. Steinberg, Chief Scientific Officer, 516-562-1159, Fax: 516-562-1022, E-mail: bsteinbe@lij.edu. *Application contact:* Dr. Annette T. Lee, Associate Dean, 516-562-1108, Fax: 516-562-1153, E-mail: anlee@nshs.edu.

The George Washington University, Columbian College of Arts and Sciences, Institute for Biomedical Sciences, Program in Molecular Medicine, Washington, DC 20052. Offers molecular and cellular oncology (PhD); neurosciences (PhD); pharmacology and physiology (PhD). *Students:* 3 full-time (2 women), 9 part-time (7 women), 3 international. Average age 28. In 2005, 4 degrees awarded. *Degree requirements:* For doctorate, thesis/dissertation, general exams, comprehensive exam. *Entrance requirements:* For doctorate, GRE General Test, interview, minimum GPA of 3.0. Additional exam requirements/recommendations for international students: Required—TOEFL (minimum score 600 paper-based; 250 computer-based). *Application deadline:* For fall admission, 1/2 priority date for domestic students, 1/2 priority date for international students. Applications are processed on a rolling basis. Application fee: $60. Electronic applications accepted. *Financial support:* In 2005–06, 1 student received support, including 1 fellowship with tuition reimbursement available; Federal Work-Study and institutionally sponsored loans also available. Financial award application deadline:

Molecular Medicine

The George Washington University (continued)
2/1. *Unit head:* Dr. Steven R. Patierno, Director, 202-994-3286. *Application contact:* 202-994-2179, Fax: 202-994-0967, E-mail: gwibs@gwu.edu.

See Close-Up on page 569.

The Johns Hopkins University, School of Medicine, Graduate Programs in Medicine, Graduate Program in Cellular and Molecular Medicine, Baltimore, MD 21218-2699. Offers PhD. *Faculty:* 112 full-time (22 women). *Students:* 103 full-time (61 women); includes 31 minority (10 African Americans, 1 American Indian/Alaska Native, 14 Asian Americans or Pacific Islanders, 6 Hispanic Americans), 22 international. Average age 26. 230 applicants, 20% accepted, 25 enrolled. In 2005, 11 degrees awarded. *Degree requirements:* For doctorate, thesis/dissertation, oral exam. *Entrance requirements:* For doctorate, GRE General Test. Additional exam requirements/recommendations for international students: Required—TOEFL. *Application deadline:* For winter admission, 1/10 for domestic students. Applications are processed on a rolling basis. Application fee: $75. Electronic applications accepted. *Expenses:* Tuition: Full-time $30,960. Tuition and fees vary according to degree level and program. *Financial support:* In 2005–06, 12 fellowships (averaging $24,600 per year) were awarded; tuition waivers (full) also available. *Faculty research:* Cellular and molecular basis of disease. Total annual research expenditures: $100 million. *Unit head:* Dr. Pierre A. Coulombe, Director, 410-614-0410, Fax: 410-614-7294, E-mail: coulombe@jhmi.edu. *Application contact:* Theo M. Karpovich, Admissions Coordinator, 410-614-0391, Fax: 410-614-7294, E-mail: karpovic@jhmi.edu.

See Close-Up on page 573.

Medical College of Georgia, School of Graduate Studies, Program in Molecular Medicine, Augusta, GA 30912. Offers PhD. *Faculty:* 34 full-time (6 women). *Students:* 22 full-time (13 women); includes 3 minority (2 African Americans, 1 Hispanic American), 12 international. Average age 29. In 2005, 6 degrees awarded. *Degree requirements:* For doctorate, thesis/dissertation. *Entrance requirements:* For doctorate, GRE General Test. Additional exam requirements/recommendations for international students: Required—TOEFL. *Application deadline:* For fall admission, 6/30 for domestic students. Applications are processed on a rolling basis. Application fee: $30. *Financial support:* In 2005–06, 5 students received support, including 1 fellowship with partial tuition reimbursement (averaging $25,500 per year), 19 research assistantships with partial tuition reimbursements available (averaging $22,500 per year); teaching assistantships, Federal Work-Study, institutionally sponsored loans, and scholarships/grants also available. Support available to part-time students. Financial award application deadline: 5/31; financial award applicants required to submit FAFSA. *Faculty research:* Developmental neurobiology, cancer, regenerative medicine, molecular chaperones molecular immunology. *Unit head:* Dr. Robert Yu, Director of Institute of Molecular Medicine and Genetics, 706-721-0699, Fax: 706-721-7915, E-mail: ryu@mail.mcg.edu. *Application contact:* Dr. Wendy Bollag, Program Director, 706-721-0691, Fax: 706-721-7915, E-mail: wbollag@mail.mcg.edu.

The Pennsylvania State University University Park Campus, Graduate School, Intercollege Graduate Programs, Intercollege Graduate Program in Integrative Biosciences, State College, University Park, PA 16802-1503. Offers integrative biosciences (PhD), including biomolecular transport dynamics, cell and developmental biology, cellular and molecular mechanisms of toxicity, chemical biology, ecological and molecular plant physiology, immunobiology, molecular medicine, neuroscience, nutrition science. *Students:* 110 full-time (58 women), 2 part-time (both women); includes 6 minority (3 African Americans, 3 Asian Americans or Pacific Islanders), 61 international. *Entrance requirements:* For master's and doctorate, GRE General Test. Application fee: $45. *Expenses:* Tuition, state resident: full-time $12,518; part-time $522 per credit. Tuition, nonresident: full-time $23,004; part-time $959 per credit. Required fees: $484. Tuition and fees vary according to course load, campus/location and program. *Financial support:* Fellowships available. *Unit head:* Dr. Richard J. Frisque, Co-Director, 814-863-3523, Fax: 814-863-1357, E-mail: rjf6@psu.edu.

See Close-Up on page 197.

University of Cincinnati, Division of Research and Advanced Studies, College of Medicine, Graduate Programs in Biomedical Sciences, Program in Pathobiology and Molecular Medicine, Cincinnati, OH 45267. Offers pathology (PhD), including anatomic pathology, laboratory medicine, pathobiology and molecular medicine. *Degree requirements:* For doctorate, thesis/dissertation, qualifying exam. *Entrance requirements:* For doctorate, GRE General Test. Additional exam requirements/recommendations for international students: Required—TOEFL (minimum score 620 paper-based; 260 computer-based). Electronic applications accepted. *Faculty research:* Cardiovascular and lipid disorders, digestive and kidney disease, endocrine and metabolic disorders, hematological and oncogenic, immunology and infectious disease.

University of Maryland, School of Medicine, Graduate Program in Life Sciences, Program in Molecular Medicine, Baltimore, MD 21201. Offers PhD. *Expenses:* Tuition, state resident: full-time $8,079; part-time $409 per credit hour. Tuition, nonresident: full-time $18,384; part-time $731 per credit hour. Required fees: $695; $10 per credit hour. Tuition and fees vary according to degree level and program.

University of Medicine and Dentistry of New Jersey, Graduate School of Biomedical Sciences, Graduate Programs in Biomedical Sciences–Newark, Department of Cell Biology and Molecular Medicine, Newark, NJ 07107. Offers PhD. *Degree requirements:* For doctorate, thesis/dissertation, qualifying exam. *Entrance requirements:* For doctorate, GRE General Test. Additional exam requirements/recommendations for international students: Required—TOEFL. *Application deadline:* For fall admission, 2/1 for domestic students. Application fee: $40. *Financial support:* Fellowships, research assistantships, Federal Work-Study, institutionally sponsored loans, and tuition waivers (full and partial) available. Financial award application deadline: 5/1. *Unit head:* Dr. Dorothy Vatner, Program Director, 973-972-1339, Fax: 973-972-7489, E-mail: vatnerdo@umdnj.edu.

University of Rochester, School of Medicine and Dentistry, Program in Pathobiology and Molecular Medicine, Rochester, NY 14627-0250.

The University of Texas Health Science Center at San Antonio, Graduate School of Biomedical Sciences, Program in Molecular Medicine, San Antonio, TX 78229-3900. Offers MS, PhD. *Degree requirements:* For doctorate, oral qualifying exam. *Entrance requirements:* For master's and doctorate, GRE General Test. Additional exam requirements/recommendations for international students: Required—TOEFL. *Faculty research:* DNA repair, tumor suppressor genes, vision in drosophila, *gene expression (nervous system), cell-type specific gene regulation and development.*

See Close-Up on page 649.

University of Virginia, School of Medicine, Department of Molecular Physiology and Biological Physics, Program in Molecular Medicine and Systems Biology, Charlottesville, VA 22903. Offers PhD. *Students:* 1 (woman) full-time. Application fee: $60. *Expenses:* Tuition, state resident: full-time $7,731. Tuition, nonresident: full-time $18,672. Required fees: $1,479. Full-time tuition and fees vary according to degree level and program. *Application contact:* Peter C. Brunjes, Associate Dean for Graduate Programs and Research, 434-924-7184, Fax: 434-924-6737, E-mail: grad-a-s@virginia.edu.

University of Washington, Graduate School, School of Public Health and Community Medicine, Graduate Program in Pathobiology, Seattle, WA 98195. Offers MS, PhD. *Faculty:* 22 full-time (8 women), 11 part-time/adjunct (4 women). *Students:* 47 full-time (31 women); includes 7 minority (1 African American, 4 Asian Americans or Pacific Islanders, 2 Hispanic Americans), 6 international. Average age 47. 55 applicants, 31% accepted, 12 enrolled. In 2005, 1 master's, 2 doctorates awarded. Terminal master's awarded for partial completion of doctoral program. *Median time to degree:* Of those who began their doctoral program in fall 1997, 99% received their degree in 8 years or less. *Degree requirements:* For master's, thesis/dissertation, registration; for doctorate, thesis/dissertation, comprehensive exam, registration. *Entrance requirements:* For master's and doctorate, GRE General Test, minimum GPA of 3.0. Additional exam requirements/recommendations for international students: Required—TOEFL. *Application deadline:* For fall admission, 12/15 for domestic students, 11/15 for international students. Application fee: $50. *Financial support:* In 2005–06, 2 fellowships with tuition reimbursements (averaging $22,560 per year), 18 research assistantships with tuition reimbursements (averaging $22,560 per year) were awarded; career-related internships or fieldwork, institutionally sponsored loans, scholarships/grants, traineeships, tuition waivers (full and partial), and unspecified assistantships also available. Financial award application deadline: 3/1; financial award applicants required to submit FAFSA. *Faculty research:* Pathogenesis of chlamydiae, molecular biology of parasites, signal transduction, antigenic analysis, molecular biology of tumor viruses. *Unit head:* Dr. Andreas Stergachis, Acting Chair, 206-543-8350, Fax: 206-543-3873, E-mail: stergach@u.washington.edu. *Application contact:* Joseph A. Daniels, Manager of Student Services, 206-543-4338, Fax: 206-543-3873, E-mail: pathobio@u.washington.edu.

See Close-Up on page 1129.

Wake Forest University, School of Medicine and Graduate School, Graduate Programs in Medicine, Molecular Genetics Program, Winston-Salem, NC 27109. Offers PhD. *Degree requirements:* For doctorate, thesis/dissertation. Additional exam requirements/recommendations for international students: Required—TOEFL. Electronic applications accepted. *Faculty research:* Control of gene expression, molecular pathogenesis, protein biosynthesis, cell development, clinical cytogenetics.

See Close-Up on page 805.

Wake Forest University, School of Medicine and Graduate School, Graduate Programs in Medicine, Program in Molecular Medicine, Winston-Salem, NC 27109. Offers MS, PhD. *Degree requirements:* For master's and doctorate, thesis/dissertation. *Entrance requirements:* For master's and doctorate, GRE General Test. Additional exam requirements/recommendations for international students: Required—TOEFL. Electronic applications accepted. *Faculty research:* Human biology and disease, scientific basis of medicine, cellular and molecular mechanisms of health and disease.

See Close-Up on page 661.

Yale University, School of Medicine and Graduate School of Arts and Sciences, Combined Program in Biological and Biomedical Sciences (BBS), Pharmacological Sciences and Molecular Medicine Track, New Haven, CT 06520. Offers PhD, MD/PhD. *Students:* 6 full-time. *Degree requirements:* For doctorate, thesis/dissertation. *Entrance requirements:* For doctorate, GRE General Test. Additional exam requirements/recommendations for international students: Required—TOEFL. *Application deadline:* For fall admission, 12/8 for domestic students, 12/8 for international students. Electronic applications accepted. *Financial support:* Fellowships, research assistantships available. *Unit head:* Dr. David F. Stern, Co-Director, 203-785-4832, Fax: 203-785-7467, E-mail: kathleen.fisher@yale.edu. *Application contact:* Kathy Fisher, Registrar, 203-785-4545, E-mail: kathleen.fisher@yale.edu.

Structural Biology

Baylor College of Medicine, Graduate School of Biomedical Sciences, Program in Structural and Computational Biology and Molecular Biophysics, Houston, TX 77030-3498. Offers PhD, MD/PhD. *Faculty:* 63 full-time (4 women). *Students:* 40 full-time (7 women); includes 5 minority (3 Asian Americans or Pacific Islanders, 2 Hispanic Americans), 22 international. Average age 27. 73 applicants, 22% accepted, 10 enrolled. In 2005, 4 degrees awarded. *Median time to degree:* Of those who began their doctoral program in fall 1997, 80% received their degree in 8 years or less. *Degree requirements:* For doctorate, thesis/dissertation, public defense. *Entrance requirements:* For doctorate, GRE General Test, GRE Subject Test (strongly recommended), minimum GPA of 3.0. Additional exam requirements/recommendations for international students: Required—TOEFL. *Application deadline:* For fall admission, 2/1 for domestic students. Application fee: $30. Electronic applications accepted. *Expenses:* Tuition: Full-time $8,200. Full-time tuition and fees vary according to program. *Financial support:* In 2005–06, 37 students received support, including 26 fellowships (averaging $23,000 per year), 14 research assistantships (averaging $23,000 per year); career-related internships or fieldwork, Federal Work-Study, institutionally sponsored loans, health care benefits, and tuition waivers (full) also available. Financial award applicants required to submit FAFSA. *Faculty research:* X-ray and electron crystallography, light and electron microscopy, computer image reconstruction, molecular spectroscopy. *Unit head:* Dr. Wah Chiu, Director, 713-798-6985. *Application contact:* Wanda Waguespack, Graduate Program Administrator, 713-798-5197, Fax: 713-798-6325, E-mail: wandaw@bcm.edu.

See Close-Up on page 465.

Brandeis University, Graduate School of Arts and Sciences, Programs in Life Sciences, Program in Biophysics and Structural Biology, Waltham, MA 02454-9110. Offers MS, PhD. *Faculty:* 23 full-time (6 women). *Students:* 10 full-time (3 women); includes 1 minority (Asian American or Pacific Islander), 2 international. Average age 26. 40 applicants, 20% accepted, 2 enrolled. In 2005, 1 master's, 4 doctorates awarded. *Degree requirements:* For doctorate, one foreign language, thesis/dissertation. *Entrance requirements:* For doctorate, GRE General Test, resumé, 3 letters of recommendation. Additional exam requirements/recommendations for international students: Required—TOEFL (minimum score 600 paper-based; 250 computer-based). *Application deadline:* For fall admission, 1/15 for domestic students. Applications are processed on a rolling basis. Application fee: $55. Electronic applications accepted. *Financial support:* In 2005–06, 2 fellowships with tuition reimbursements (averaging $26,500 per year), 8 research assistantships with tuition reimbursements (averaging $26,500 per year), teaching assistantships with tuition reimbursements (averaging $6,000 per year) were awarded; scholarships/grants and tuition waivers (full and partial) also available. Support available to part-time students. Financial award application deadline: 4/15; financial award applicants required to submit CSS PROFILE or FAFSA. *Faculty research:* Biophysical chemistry, macromolecular structure and function, single molecule biophysics, macromolecules. *Unit head:* Dr. Dorothee Kern, Chair, 781-736-2354, Fax: 781-736-2349, E-mail: dkern@brandeis.edu. *Application contact:* Lynn Olsen, Administrative Assistant, 781-736-2300, Fax: 781-736-2349, E-mail: lolsen@brandeis.edu.

Cornell University, Graduate School, Graduate Fields of Comparative Biomedical Sciences, Field of Comparative Biomedical Sciences, Ithaca, NY 14853-0001. Offers cellular and molecular medicine (MS, PhD); developmental and reproductive biology (MS, PhD); infectious diseases (MS, PhD); population medicine and epidemiology (MS); population medicine and epidemiology sciences (PhD); structural and functional biology (MS, PhD). *Faculty:* 135 full-time (38 women). *Students:* 41 full-time (23 women); includes 4 minority (1 African American, 3 Asian Americans or Pacific Islanders), 19 international. 58 applicants, 60% accepted, 33 enrolled. In 2005, 4 degrees awarded. *Degree requirements:* For master's, thesis/dissertation; for doctorate, thesis/dissertation, comprehensive exam. *Entrance requirements:* For master's and doctorate, GRE General Test, 2 letters of recommendation. Additional exam requirements/recommendations for international students: Required—TOEFL (minimum score 550 paper-based; 213 computer-based). *Application deadline:* For fall admission, 12/15 for domestic students. Application fee: $60. Electronic applications accepted. *Financial support:* In 2005–06, 41 students received support, including 13 fellowships with full tuition reimbursements available, 28 research assistantships with full tuition reimbursements available; teaching assistantships with full tuition reimbursements available, institutionally sponsored loans, scholarships/grants, health care benefits, tuition waivers (full and partial), and unspecified assistantships also available. Financial award applicants required to submit FAFSA. *Faculty research:* Receptors and signal transduction, viral and bacterial infectious diseases, tumor metastasis, clinical sciences/nutritional disease, development/neurologic disorders. *Unit head:* Director of Graduate Studies, 607-253-3276, Fax: 607-253-3756. *Application contact:* Graduate Field Assistant, 607-253-3276, Fax: 607-253-3756, E-mail: graduate_edcvm@cornell.edu.

Cornell University, Joan and Sanford I. Weill Medical College and Graduate School of Medical Sciences, Weill Graduate School of Medical Sciences, Program in Molecular Biology, Cell Biology, Biochemistry and Structural Biology, New York, NY 10021-4896. Offers MS, PhD, MD/PhD. *Faculty:* 84 full-time (22 women). *Students:* 121 full-time (68 women); includes 14 minority (2 African Americans, 7 Asian Americans or Pacific Islanders, 5 Hispanic Americans), 34 international. Average age 22. In 2005, 8 degrees awarded. *Median time to degree:* Of those who began their doctoral program in fall 1997, 100% received their degree in 8 years or less. *Degree requirements:* For doctorate, thesis/dissertation, final exam. *Entrance requirements:* For doctorate, GRE General Test, GRE Subject Test, background in genetics, molecular biology, chemistry, or biochemistry. Additional exam requirements/recommendations for international students: Required—TOEFL. *Application deadline:* For fall admission, 12/15 for domestic students. Application fee: $60. *Expenses:* Tuition: Full-time $32,320. Required fees: $1,025. *Financial support:* Fellowships, stipends available. *Unit head:* Dr. Christopher Lima, Co-Director, 212-639-8205, Fax: 212-717-3047, E-mail: limac@mskcc.org.

Duke University, Graduate School, Program in Structural Biology and Biophysics, Durham, NC 27710. Offers Certificate. Students must be enrolled in a participating PhD program. *Faculty:* 22 full-time. *Students:* 2 full-time (1 woman). 14 applicants, 57% accepted, 1 enrolled. *Entrance requirements:* For degree, GRE General Test, GRE Subject Test. Additional exam requirements/recommendations for international students: Required—IELT (preferred) or TOEFL. *Application deadline:* For fall admission, 12/31 for domestic students, 12/31 for international students. Application fee: $75. *Financial support:* Application deadline: 12/31. *Unit head:* Lorena Beese, Director of Graduate Studies, 919-681-5267, Fax: 919-684-8346, E-mail: mbp@biochem.duke.edu.

See Close-Up on page 565.

Florida State University, Graduate Studies, College of Arts and Sciences, Program in Molecular Biophysics, Tallahassee, FL 32306. Offers biochemistry, molecular and cell biology (PhD); computational structural biology (PhD); molecular biophysics (PhD). *Faculty:* 44 full-time (7 women). *Students:* 29 full-time (17 women); includes 1 minority (Hispanic American), 12 international. Average age 30. 14 applicants, 100% accepted, 4 enrolled. In 2005, 3 degrees awarded. *Median time to degree:* Of those who began their doctoral program in fall 1997, 100% received their degree in 8 years or less. *Degree requirements:* For doctorate, thesis/dissertation, comprehensive exam. *Entrance requirements:* For doctorate, GRE General Test. Additional exam requirements/recommendations for international students: Required—TOEFL (minimum score 600 paper-based; 250 computer-based). *Application deadline:* For fall admission, 1/15 for domestic students, 1/15 for international students. Applications are processed on a rolling basis. Application fee: $30. Electronic applications accepted. *Financial support:* In 2005–06, 29 students received support, including 29 research assistantships (averaging $19,500 per year); health care benefits and tuition waivers (partial) also available. Financial award applicants required to submit FAFSA. *Faculty research:* Protein and nucleic acid structure and function, membrane protein structure, computational biophysics, 3-D image reconstruction. Total annual research expenditures: $5.1 million. *Unit head:* Dr. P. Bryant Chase, Director, MOB Graduate Program, 850-644-0056, Fax: 850-644-7244, E-mail: chase@bio.fsu.edu. *Application contact:* Dale E. Leonard, Academic Coordinator, Graduate Programs. 850-644-1012, Fax: 850-644-7244, E-mail: mob@sb.fsu.edu.

Harvard University, Graduate School of Arts and Sciences, Department of Systems Biology, Cambridge, MA 02138. Offers PhD. *Students:* 9. 100 applicants, 12% accepted, 9 enrolled. *Degree requirements:* For doctorate, thesis/dissertation, lab rotation, qualifying examination. *Entrance requirements:* For doctorate, GRE. Additional exam requirements/recommendations for international students: Required—TOEFL. *Application deadline:* For fall admission, 12/8 priority date for domestic students, 12/8 priority date for international students. Application fee: $90. Electronic applications accepted. *Expenses:* Tuition: Full-time $28,752. Full-time tuition and fees vary according to program and student level. *Financial support:* Institutionally sponsored loans, scholarships/grants, unspecified assistantships, and all students receive a stipend and full tuition reimbursements available. *Unit head:* Judy Finkelstein, Graduate Coordinator, 617-432-5876, Fax: 617-432-5012, E-mail: pamela_silver@dfci.harvard.edu. *Application contact:* Jodi Finkelstein, Coordinator, 617-432-5202, Fax: 617-432-5012, E-mail: jodi_finkelstein@hms.harvard.edu.

See Close-Up on page 675.

Iowa State University of Science and Technology, Graduate College, Interdisciplinary Programs, Bioinformatics and Computational Biology Program, Ames, IA 50011-3260. Offers MS, PhD. *Degree requirements:* For doctorate, thesis/dissertation. *Entrance requirements:* For doctorate, GRE General Test. Additional exam requirements/recommendations for international students: Required—TOEFL or IELTS. Electronic applications accepted. *Expenses:* Tuition, state resident: full-time $6,410. Tuition, nonresident: full-time $16,422. Tuition and fees vary according to program. *Faculty research:* Functional and structural genomics, genome evolution, macromolecular structure and function, mathematical biology and biological statistics, metabolic and developmental networks.

Mayo Graduate School, Graduate Programs in Biomedical Sciences, Programs in Biochemistry, Structural Biology, Cell Biology, and Genetics, Rochester, MN 55905. Offers biochemistry and structural biology (PhD); cell biology and genetics (PhD); molecular biology (PhD). *Degree requirements:* For doctorate, oral defense of dissertation, qualifying oral and written exam. *Entrance requirements:* For doctorate, GRE, 1 year of chemistry, biology, calculus, and physics. Additional exam requirements/recommendations for international students: Required—TOEFL. Electronic applications accepted. *Faculty research:* Gene structure and function, membranes and receptors/cytoskeleton, oncogenes and growth factors, protein structure and function, steroid hormonal action.

See Close-Up on page 399.

Mount Sinai School of Medicine of New York University, Graduate School of Biological Sciences, New York, NY 10029-6504. Offers biophysics, structural biology and biomathematics (PhD); community medicine (MPH); genetic counseling (MS); genetics and genomic sciences (PhD); mechanisms of disease and therapy (PhD); microbiology (PhD); molecular, cellular, biochemical and developmental sciences (PhD); neurosciences (PhD). *Students:* 218 full-time (109 women). 4,208 applicants, 7% accepted, 117 enrolled.Terminal master's awarded for partial completion of doctoral program. *Degree requirements:* For master's, registration; for doctorate, thesis/dissertation, registration. *Entrance requirements:* For doctorate, GRE General

Test, GRE Subject Test, 3 years of college pre-med course work. Additional exam requirements/recommendations for international students: Required—TOEFL. *Application deadline:* For fall admission, 1/15 for domestic students. Application fee: $75. Electronic applications accepted. *Expenses:* Tuition: Full-time $33,250. Required fees: $1,600. Full-time tuition and fees vary according to degree level, program and reciprocity agreements. *Financial support:* In 2005–06, fellowships with full tuition reimbursements (averaging $26,000 per year), research assistantships with full tuition reimbursements (averaging $26,000 per year) were awarded; Federal Work-Study, institutionally sponsored loans, scholarships/grants, health care benefits, and unspecified assistantships also available. Financial award application deadline: 4/5; financial award applicants required to submit FAFSA. *Faculty research:* Cancer, gene therapy, minimally invasive surgery, cardiac translational research. Total annual research expenditures: $162.2 million. *Unit head:* Dr. Diomedes Logothetis, Dean, 212-241-6546, Fax: 212-241-0651, E-mail: diomedes.logothetis@mssm.edu. *Application contact:* Lily Recanati, Manager, 212-241-3267, Fax: 212-241-0651, E-mail: lily.recantati@mssm.edu.

See Close-Up on page 177.

New York University, National Institutes of Health Sponsored Programs, Program in Structural Biology, New York, NY 10012-1019. Offers PhD. *Degree requirements:* For doctorate, thesis/dissertation, qualifying examination. *Entrance requirements:* For doctorate, GRE, GRE Subject Test in biology or chemistry (recommended). Additional exam requirements/recommendations for international students: Required—TOEFL. *Application deadline:* For fall admission, 2/1 for domestic students. Application fee: $60. *Financial support:* Research assistantships with full tuition reimbursements, institutionally sponsored loans and health care benefits available.

See Close-Up on page 579.

New York University, School of Medicine and Graduate School of Arts and Science, Sackler Institute of Graduate Biomedical Sciences, Program in Structural Biology, New York, NY 10012-1019. Offers PhD. *Degree requirements:* For doctorate, thesis/dissertation, qualifying examination. *Entrance requirements:* For doctorate, GRE, recommended GRE Subject Test in Biology or Chemistry. Additional exam requirements/recommendations for international students: Required—TOEFL.

See Close-Up on page 579.

Northwestern University, The Graduate School and Judd A. and Marjorie Weinberg College of Arts and Sciences, Interdepartmental Biological Sciences Program (IBiS), Department of Biochemistry, Molecular Biology, and Cell Biology, Evanston, IL 60208. Offers biochemistry (PhD); cell and molecular biology (PhD); molecular biophysics (PhD); structural biology (PhD). Department participates in the Interdepartmental Biological Sciences Program (IBiS). *Entrance requirements:* For doctorate, GRE General Test. Additional exam requirements/recommendations for international students: Required—TOEFL (minimum score 600 paper-based), TSE (minimum score 50). Electronic applications accepted. *Faculty research:* Protein structure, protein-DNA interactions, gene regulation, development of embryos, hormone action and signal transduction.

Northwestern University, Northwestern University Feinberg School of Medicine and Interdepartmental Degree Programs, Integrated Graduate Programs in the Life Sciences, Chicago, IL 60611. Offers cancer biology (PhD); cell biology (PhD); developmental biology (PhD); evolutionary biology (PhD); immunology and microbial pathogenesis (PhD); molecular biology and genetics (PhD); neurobiology (PhD); pharmacology and toxicology (PhD); structural biology and biochemistry (PhD). *Degree requirements:* For doctorate, thesis/dissertation, written and oral qualifying exams, comprehensive exam. *Entrance requirements:* For doctorate, GRE General Test. Additional exam requirements/recommendations for international students: Required—TOEFL (minimum score 600 paper-based; 250 computer-based). Electronic applications accepted.

See Close-Up on page 189.

Stanford University, School of Medicine, Graduate Programs in Medicine, Department of Structural Biology, Stanford, CA 94305-9991. Offers PhD. *Degree requirements:* For doctorate, thesis/dissertation. *Entrance requirements:* For doctorate, GRE General Test, GRE Subject Test. Additional exam requirements/recommendations for international students: Required—TOEFL. Electronic applications accepted.

State University of New York at Buffalo, Graduate School, School of Medicine and Biomedical Sciences, Graduate Programs in Medicine and Biomedical Sciences, Department of Structural Biology, Buffalo, NY 14260. Offers MS, PhD. *Faculty:* 6 part-time/adjunct (1 woman). *Students:* 6 full-time (2 women), 1 part-time, 1 international. Average age 22. In 2005, 1 degree awarded. *Degree requirements:* For master's, thesis/dissertation; for doctorate, thesis/dissertation, comprehensive exam, registration. *Entrance requirements:* For doctorate, GRE General Test. Additional exam requirements/recommendations for international students: Required—TOEFL (minimum score 600 paper-based; 250 computer-based). *Application deadline:* For fall admission, 2/1 priority date for domestic students, 2/1 priority date for international students. Applications are processed on a rolling basis. Application fee: $35. Electronic applications accepted. *Financial support:* Federal Work-Study, scholarships/grants, traineeships, and unspecified assistantships available. Financial award application deadline: 2/1; financial award applicants required to submit FAFSA. *Unit head:* Dr. George T. DeTitta, Department Chair and Professor, 716-856-9608, Fax: 716-852-6086. *Application contact:* Amy J. Sierocki, Office Manager, 716-898-8585, E-mail: asieracki@hwi.buffalo.edu.

Stony Brook University, State University of New York, Graduate School, College of Arts and Sciences, Department of Biochemistry and Cell Biology, Program in Biochemistry and Structural Biology, Stony Brook, NY 11794. Offers PhD. *Students:* 31 full-time (17 women); includes 5 minority (1 African American, 3 Asian Americans or Pacific Islanders, 1 Hispanic American), 23 international. Average age 27. 110 applicants, 9% accepted. *Expenses:* Tuition, state resident: full-time $6,900; part-time $288 per credit. Tuition, nonresident: full-time $10,920; part-time $455 per credit. Required fees: $704. *Financial support:* In 2005–06, 22 research assistantships, 6 teaching assistantships were awarded. *Application contact:* Director, Graduate Program, 631-632-8533, Fax: 631-632-9730, E-mail: mcbprog@life.bio.sunysb.edu.

See Close-Up on page 415.

Syracuse University, Graduate School, College of Arts and Sciences, Department of Biology and Department of Physics and Department of Chemistry, Program in Structural Biology, Biochemistry and Biophysics, Syracuse, NY 13244. Offers PhD. *Students:* 9 full-time (4 women), 4 international. 44 applicants, 14% accepted, 4 enrolled. *Degree requirements:* For doctorate, thesis/dissertation, exam. *Entrance requirements:* For doctorate, GRE General Test, GRE Subject Test. Additional exam requirements/recommendations for international students: Required—TOEFL. *Application deadline:* For fall admission, 1/10 for domestic students. Applications are processed on a rolling basis. Application fee: $65. Electronic applications accepted. *Financial support:* Fellowships, research assistantships, teaching assistantships, tuition waivers available. *Unit head:* Stewart Loh, Director, 315-464-8731, Fax: 315-443-4070. *Application contact:* Evelyn Lott, Information Contact, 315-443-9154, Fax: 315-443-2012, E-mail: ealott@syr.edu.

Texas A&M University System Health Science Center, Graduate School of Biomedical Sciences, Department of Biochemistry and Structural Biology, College Station, TX 77840. Offers PhD. *Degree requirements:* For doctorate, thesis/dissertation. *Entrance requirements:* For doctorate, GRE General Test. *Faculty research:* Immunology, cell and membrane biology, protein biochemistry, molecular genetics, parasitology, vertebrate embryogenesis and microbiology.

Thomas Jefferson University, Jefferson College of Graduate Studies, Program in Molecular Pharmacology and Structural Biology, Philadelphia, PA 19107. Offers PhD. *Faculty:* 48 full-time. *Students:* 15 full-time (9 women), 1 part-time; includes 3 minority (2 African Americans, 1

Structural Biology

Thomas Jefferson University (continued)
Asian American or Pacific Islander), 3 international. 15 applicants, 20% accepted, 0 enrolled. In 2005, 4 degrees awarded. *Degree requirements:* For doctorate, thesis/dissertation, comprehensive exam, registration. *Entrance requirements:* For doctorate, GRE General Test, minimum GPA of 3.2. Additional exam requirements/recommendations for international students: Required—TOEFL (minimum score 213 computer-based). *Application deadline:* For fall admission, 3/1 priority date for domestic students, 3/1 priority date for international students. Applications are processed on a rolling basis. Application fee: $50. Electronic applications accepted. *Expenses:* Tuition: Full-time $14,894; part-time $800 per credit. *Financial support:* In 2005–06, 1 student received support, including 15 fellowships with full tuition reimbursements available; research assistantships, Federal Work-Study, institutionally sponsored loans, scholarships/grants, and traineeships also available. Support available to part-time students. Financial award application deadline: 5/1; financial award applicants required to submit FAFSA. *Faculty research:* Biochemistry and cell, molecular and structural biology of cell-surface and intracellular receptors, molecular modeling, signal transduction. *Unit head:* Dr. Jeffrey L. Benovic, program Director, 215-503-4607, Fax: 215-923-1098, E-mail: jeff.benovic@jefferson.edu. *Application contact:* Jessie F. Pervall, Director of Admissions, 215-503-0155, Fax: 215-503-9920, E-mail: jessie.pervall@jefferson.edu.

See Close-Up on page 1185.

Tulane University, School of Medicine and Graduate School, Graduate Programs in Medicine, Department of Structural and Cellular Biology, New Orleans, LA 70118-5669. Offers MS, PhD, MD/PhD. MS and PhD offered through the Graduate School. *Degree requirements:* For master's, one foreign language, thesis; for doctorate, 2 foreign languages, thesis/dissertation. *Entrance requirements:* For master's, GRE General Test, minimum B average in undergraduate course work; for doctorate, GRE General Test. Additional exam requirements/recommendations for international students: Required—TOEFL or TSE. Electronic applications accepted. *Faculty research:* Reproductive endocrinology, visual neuroscience, neural response to altered hormones.

University at Albany, State University of New York, School of Public Health, Department of Biomedical Sciences, Program in Cell and Molecular Structure, Albany, NY 12222-0001. Offers MS, PhD. *Degree requirements:* For master's and doctorate, thesis/dissertation. *Entrance requirements:* For master's and doctorate, GRE General Test, GRE Subject Test. Application fee: $60. *Financial support:* Application deadline: 2/1. *Unit head:* Dr. James Dias, Chair, Department of Biomedical Sciences, 518-474-2662.

University of California, San Diego, School of Medicine and Graduate Studies and Research, Molecular Pathology Program, La Jolla, CA 92093. Offers bioinformatics (PhD); cancer biology/oncology (PhD); cardiovascular sciences and disease (PhD); microbiology (PhD); molecular pathology (PhD); neurological disease (PhD); stem cell and developmental biology (PhD); structural biology/drug design (PhD). *Entrance requirements:* For doctorate, GRE General Test, GRE Subject Test. Additional exam requirements/recommendations for international students: Required—TOEFL. Electronic applications accepted.

See Close-Up on page 1119.

University of Connecticut, Graduate School, College of Liberal Arts and Sciences, Department of Molecular and Cell Biology, Field of Biophysics and Structural Biology, Storrs, CT 06269. Offers MS, PhD. *Faculty:* 10 full-time (5 women). *Students:* 2 full-time (both women); includes 1 minority (African American), 1 international. Average age 25. 15 applicants, 13% accepted, 2 enrolled. In 2005, 1 degree awarded. Terminal master's awarded for partial completion of doctoral program. *Degree requirements:* For master's, comprehensive exam; for doctorate, thesis/dissertation. *Entrance requirements:* For master's and doctorate, GRE General Test, GRE Subject Test. Additional exam requirements/recommendations for international students: Required—TOEFL (minimum score 550 paper-based; 213 computer-based). *Application deadline:* For fall admission, 2/1 priority date for domestic students, 2/1 priority date for international students; for spring admission, 11/1 for domestic students, 10/1 for international students. Applications are processed on a rolling basis. Application fee: $55. Electronic applications accepted. *Expenses:* Tuition, state resident: part-time $444 per credit hour. Tuition, nonresident: part-time $1,154 per credit hour. Tuition and fees vary according to course load. *Financial support:* In 2005–06, 2 research assistantships with full tuition reimbursements were awarded; fellowships, teaching assistantships with full tuition reimbursements, Federal Work-Study, scholarships/grants, health care benefits, and unspecified assistantships also available. Financial award application deadline: 2/1; financial award applicants required to submit FAFSA. *Application contact:* Anne St. Onje, Graduate Coordinator, 860-486-4314, Fax: 860-486-3943, E-mail: ann.st_onje@uconn.edu.

See Close-Up on page 611.

University of Pennsylvania, School of Medicine, Biomedical Graduate Studies, Graduate Group in Cell and Molecular Biology, Program in Cell Biology and Physiology, Philadelphia, PA 19104. Offers PhD, MD/PhD, VMD/PhD. *Degree requirements:* For doctorate, thesis/dissertation. *Entrance requirements:* For doctorate, GRE General Test. Additional exam requirements/recommendations for international students: Required—TOEFL. *Application deadline:* For fall admission, 12/15 priority date for domestic students, 12/1 priority date for international students. Applications are processed on a rolling basis. Application fee: $70. Electronic applications accepted. *Financial support:* Fellowships, research assistantships, scholarships/grants, traineeships, and unspecified assistantships available. *Unit head:* Dr. Morris Birnbaum, Head,

215-898-5001. *Application contact:* Emily Brady, Coordinator, 215-895-8935, Fax: 215-573-2104, E-mail: camb@mail.med.upenn.edu.

University of Pittsburgh, School of Medicine and School of Arts and Sciences, Molecular Biophysics and Structural Biology Graduate Program, Pittsburgh, PA 15260. Offers PhD. *Faculty:* 42 full-time (11 women). *Students:* 11 full-time (4 women), 9 international. Average age 25. 10 applicants, 70% accepted, 5 enrolled. *Degree requirements:* For doctorate, thesis/dissertation, comprehensive exam, registration. *Entrance requirements:* For doctorate, GRE General Test. Additional exam requirements/recommendations for international students: Required—TOEFL (minimum score 600 paper-based; 250 computer-based), IELT (minimum score 7). *Application deadline:* For fall admission, 2/1 for domestic students, 2/1 for international students. Application fee: $0. *Expenses:* Tuition, state resident: full-time $13,194; part-time $537 per credit. Tuition, nonresident: full-time $25,012; part-time $1,026 per credit. Required fees: $700; $164 per term. Tuition and fees vary according to campus/location and program. *Financial support:* In 2005–06, 5 fellowships with tuition reimbursements (averaging $23,552 per year) were awarded; institutionally sponsored loans, scholarships/grants, traineeships, and unspecified assistantships also available. *Faculty research:* Structural biology, protein dynamics and folding, computational biophysics, molecular informatics, membrane biophysics and ion channels, NMR, x-ray crystallography cryaelectron microscopy. *Unit head:* Dr. Angela M Gronenbarn, Director. *Application contact:* Jennifer L. Walker, MBSB Program Coordinator, 412-648-8957, Fax: 412-648-1077, E-mail: mbsbinfo@medschool.pitt.edu.

The University of Tennessee Health Science Center, College of Graduate Health Sciences, Department of Molecular Sciences, Memphis, TN 38163-0002. Offers bacterial pathogenesis (PhD); biochemistry (PhD); immunology (PhD); microbiology (PhD); molecular and cell biology-pathology (PhD); molecular biology (PhD); signal transduction (PhD); structural biology (PhD); virology (PhD). *Faculty:* 17 full-time (4 women). *Students:* 25 full-time (11 women); includes 14 minority (1 African American, 13 Asian Americans or Pacific Islanders). Average age 24. 83 applicants, 12% accepted. In 2005, 6 degrees awarded. *Degree requirements:* For doctorate, thesis/dissertation, oral and written preliminary and comprehensive exams. *Entrance requirements:* For doctorate, GRE General Test, GRE Subject Test, minimum GPA of 3.0. Additional exam requirements/recommendations for international students: Required—TOEFL. *Application deadline:* For fall admission, 5/15 for domestic students. Application fee: $0. *Financial support:* In 2005–06, 8 fellowships, 21 research assistantships were awarded; teaching assistantships, traineeships and tuition waivers (full) also available. *Unit head:* Dr. David Hasty, Chairman, 901-448-6150, Fax: 901-448-7360. *Application contact:* Ida W. Mosby, Director, Enrollment Services, 901-448-5560, E-mail: imosby@utmem.edu.

The University of Texas Health Science Center at San Antonio, Graduate School of Biomedical Sciences, Department of Cellular and Structural Biology, San Antonio, TX 78229-3900. Offers PhD. *Degree requirements:* For doctorate, thesis/dissertation, oral qualifying exam. *Entrance requirements:* For doctorate, GRE General Test, previous course work in biology, chemistry, physics, and calculus. *Faculty research:* Human/molecular genetics, endocrinology and neurobiology, cell biology, cancer biology, biology of aging.

The University of Texas Medical Branch, Graduate School of Biomedical Sciences, Program in Biochemistry and Molecular Biology, Galveston, TX 77555. Offers biochemistry (PhD); bioinformatics (PhD); biophysics (PhD); cell biology (PhD); computational biology (PhD); structural biology (PhD). *Students:* 38 full-time (14 women), 1 part-time; includes 5 minority (3 Asian Americans or Pacific Islanders, 2 Hispanic Americans), 18 international. Average age 27. In 2005, 7 degrees awarded. *Degree requirements:* For doctorate, thesis/dissertation. *Entrance requirements:* Additional exam requirements/recommendations for international students: Required—TOEFL (minimum score 550 paper-based; 213 computer-based). *Application deadline:* Applications are processed on a rolling basis. Application fee: $30 ($75 for international students). Electronic applications accepted. *Expenses:* Tuition, state resident: full-time $8,350; part-time $90 per credit hour. Tuition, nonresident: full-time $21,450; part-time $366 per credit hour. Required fees: $1,027; $11 per credit hour. $60 per term. *Financial support:* In 2005–06, fellowships (averaging $23,000 per year), research assistantships (averaging $23,000 per year) were awarded. Financial award applicants required to submit FAFSA. *Unit head:* Dr. Leenian L. Chan, Director, 409-772-2861, Fax: 409-772-9679, E-mail: lchan@utmb.edu. *Application contact:* Debora Botting, Co-ordinator II Special Programs, 409-772-2769, Fax: 409-747-0552, E-mail: dmbottin@utmb.edu.

University of Washington, School of Medicine and Graduate School, Graduate Programs in Medicine, Department of Biological Structure, Seattle, WA 98195. Offers PhD. *Degree requirements:* For doctorate, thesis/dissertation. *Faculty research:* Cellular and developmental biology, experimental immunology and hematology, molecular structure and molecular biology, neurobiology, x-rays.

Yale University, School of Medicine and Graduate School of Arts and Sciences, Combined Program in Biological and Biomedical Sciences (BBS), Molecular Biophysics and Biochemistry Track, New Haven, CT 06520. Offers PhD, MD/PhD. *Students:* 14 full-time. *Degree requirements:* For doctorate, thesis/dissertation. *Entrance requirements:* For doctorate, GRE General Test. Additional exam requirements/recommendations for international students: Required—TOEFL. *Application deadline:* For fall admission, 12/8 for domestic students, 12/8 for international students. Electronic applications accepted. *Financial support:* Fellowships, research assistantships available. *Unit head:* Dr. Mark Solomon, Director of Graduate Studies, 203-432-5562, Fax: 203-432-9782, E-mail: mbb.grad@yale.edu. *Application contact:* Nessie Stewart, Graduate Registrar, 203-432-5662, Fax: 203-432-6178, E-mail: mbb.grad@yale.edu.

Cross-Discipline Announcements

Iowa State University of Science and Technology, Graduate College, Interdisciplinary Programs, Bioinformatics and Computational Biology Program, Ames, IA 50011-3260.

An innovative graduate training program offers the PhD degree in bioinformatics and computational biology. Interdisciplinary research is emphasized. Students with degrees in computational, mathematical, biological, or physical sciences may apply. Assistantships are provided to all students.

Loyola University Chicago, Graduate School, Program in Molecular and Cellular Biochemistry, Chicago, IL 60611-2196.

A modest-size, dynamic faculty offers exciting opportunities for research leading to the MS and PhD in areas that include molecular oncology and molecular neurochemical mechanisms of brain development and alcohol addiction; biochemistry of apoptosis, transcription factors, intracellular protein degradation, cell membranes, neurotransmitters, signal transduction, and cell-cell interactions; and biosynthesis and action of hormones. All students accepted into the PhD program receive very competitive stipend support and tuition remission. A focused program of formal study allows PhD students to initiate research during their first year. The 2-year MS program includes a year of intensive research.

Massachusetts Institute of Technology, School of Engineering, Biological Engineering Division, Cambridge, MA 02139-4307.

Program provides opportunities for study and research at the interface of biology and engineering leading to specialization in bioengineering and applied biosciences. The areas include understanding how biological systems operate, especially when perturbed by genetic, chemical, or materials interventions or subjected to pathogens or toxins, and designing innovative biology-based technologies in diagnostics, therapeutics, materials, and devices for application to human health and diseases, as well as other societal problems and opportunities.

Massachusetts Institute of Technology, School of Science, Department of Biology, Cambridge, MA 02139-4307.

Graduate work in the Department of Biology at MIT leads to the PhD degree. Research opportunities include many areas of modern biology: biochemistry, biophysics, cellular and developmental biology, immunology, microbiology, and neurobiology. Students come from a wide variety of backgrounds. Previous experience in the biological sciences, although desirable, is not a prerequisite for admission. Formal courses in biochemistry, genetics, and the method and logic of molecular biology are required of all students early in their graduate program. Each student is encouraged to follow the program of study that best meets his or her educational goals. Special emphasis is placed on research leading to the PhD thesis.

Oregon Health & Science University, School of Medicine, Graduate Programs in Medicine, Department of Behavioral Neuroscience, Portland, OR 97239-3098.

The department offers an interdisciplinary graduate program consisting of basic science training in behavioral neuroscience with specialization in such areas as physiological psychology, behavioral and molecular genetics, behavioral pharmacology, neuroendocrinology, and biological bases of addiction. Students with degrees in biological sciences, psychology, or neuroscience are encouraged to apply. Visit http://www.ohsu.edu/behneuro/.

Rutgers, The State University of New Jersey, New Brunswick/Piscataway, Graduate School, Program in Chemistry and Chemical Biology, New Brunswick, NJ 08901-1281.

Department of Chemistry and Chemical Biology areas of specialization include bioinformatics; protein engineering; drug and vaccine design; molecular modeling, nucleic acid, peptide, natural product, and biomedical polymer synthesis; theoretical and NMR investigations of protein folding; biomimetics; X-ray crystallographic characterization of human/viral proteins and nucleic acids; and physical and theoretical investigations of protein–nucleic acid and protein-protein interactions.

University of California, Los Angeles, School of Medicine and Graduate Division, Graduate Programs in Medicine, Department of Biomathematics, Los Angeles, CA 90095.

University of California, Los Angeles, Department of Biomathematics offers a graduate program that leads to the PhD degree to train creative, fully independent investigators who can initiate research in mathematical biology/applied mathematics and their chosen biomedical specialty, including genetics, molecular biology, neurosciences, physiology, pharmacology, and immunology.

University of Washington, School of Medicine and Graduate School, Graduate Programs in Medicine, Department of Pharmacology, Seattle, WA 98195.

The department is engaged in research in neuropharmacology, electrical excitability, secondary messenger systems, neural-specific gene expression, and molecular genetic approaches to dissecting the in vivo functions of proteins in the nervous system.

University of Wisconsin–Madison, Medical School and Graduate School, Graduate Programs in Medicine, Molecular and Cellular Pharmacology Program, Madison, WI 53706-1380.

The Molecular and Cellular Pharmacology Program at the University of Wisconsin–Madison has a multidisciplinary focus and leads to a PhD degree. This program is based in the Department of Pharmacology but is designed as an interdisciplinary program and includes 39 faculty members from the Departments of Pharmacology, Biochemistry, Oncology, Biomolecular Chemistry, Genetics, Physiology, Pediatrics, Medicine, Pathology, Anatomy, Anesthesiology, Zoology, and Comparative Bioscience and the School of Pharmacy. The common thread of research in the program is the study of signal transduction mechanisms. The MCP Program offers trainees a variety of biological systems and state-of-the-art experiment opportunities. The limited size of the program ensures strong interactions between students and faculty members.

BAYLOR COLLEGE OF MEDICINE

Interdepartmental Program in Cell and Molecular Biology

Program of Study

In 1988, a small group of faculty members at Baylor College of Medicine set out to design the ideal biomedical graduate program from a student's perspective. What resulted is a multidisciplinary environment that provides the brightest and most ambitious students with the skills needed to navigate multiple scientific disciplines. More than 95 participating faculty members, from eleven different departments, provide students in the Interdepartmental Program in Cell and Molecular Biology (CMB) with a diverse set of choices for their thesis research, leading to the Ph.D. degree. The range of research interests includes molecular mechanisms of inherited diseases, cancer and cell-cycle regulation, biology of aging, human gene therapy, signal transduction and membrane biology, the Human Genome Project, functional genomics, structural and computational biology, gene expression and regulation, developmental biology, molecular virology, and immunology.

A wide range of courses are available during the first year. Both core courses (taught by the Graduate School) and specialty courses (offered by individual departments) allow students to acquire a depth and breadth in a number of different areas, combined with more intensive investigation of topics of particular interest to each student. Course selection is flexible, with multiple choices for each requirement. The course requirements can be easily achieved by the end of the first year. This means that, during the second year, there are no course requirements that stand in the way of the research that the student came to graduate school to do.

An additional unique feature of the CMB program is the first-year Director's Course. In this small seminar course, consisting of 10 to 12 students and taught by four faculty members (one for each term), students develop both practical and intellectual skills as they learn to critically evaluate the primary scientific literature, design and interpret experiments, and give lucid presentations. The intimate format also enables students to get to know the CMB Co-Directors at the beginning of their graduate career, and encourages close working relationships with fellow first-year CMB classmates. The program is supported by a competitive training grant from the National Institute of General Medical Sciences (GM 008231), which is in its seventeenth consecutive year.

Research Facilities

The participating faculty members in the program occupy extensive research space with state-of-the-art equipment and core facilities. In addition to a large number of regular laboratory instruments such as ultracentrifuges, scintillation counters, spectrophotometers, and cell-culture facilities, the faculty members are also in charge of sophisticated equipment for transmission and scanning electron microscopy, gas-phase protein sequencing, mass spectrometry, microarray construction and data analysis, peptide and nucleic acid synthesis, knockout and transgenic mouse facilities, and X-ray diffraction. There is also extensive computing equipment.

Financial Aid

All students in the CMB program receive competitive stipends of $23,000 per year. This stipend is provided for every year of study, and there are no linked teaching requirements. In addition, full health insurance and tuition are completely covered. Top performing students each receive a $500 Claude W. Smith Fellowship Award. Although the stipend is always guaranteed, students are also encouraged to apply for outside funding. The Dean recognizes every student who receives outside funding with an additional $2000 per year supplement to the $23,000 yearly stipend.

Cost of Study

Tuition is supported by Baylor College of Medicine and the training grant. Students pay a one-time matriculation fee of $25, a one-time graduation fee of $140, and an annual student fee of $150 for the first year and $20 for each following year.

Living and Housing Costs

Most students and faculty members live within a few miles of the Medical Center. Housing is not provided for graduate students because a variety of affordable outstanding housing options are readily available. Some CMB students pool their resources to rent a nearby private house, while others choose to rent their own one-bedroom apartments. There are numerous apartment complexes located very close to the Texas Medical Center. Houston's cost of living is also well below that of every major city in the U.S. This means that even graduate students can afford to enjoy the many recreational and cultural opportunities available in Houston.

Student Group

Unlike other interdisciplinary programs, the CMB program accepts only 10 to 12 carefully selected students each year. This means that training, particularly during the first year, involves one-on-one interactions with some of the best scientists in the country in an intimate environment that teaches students how to think like a scientist. In addition, starting in the second year, CMB students present their research once a year in a formal seminar setting. This seminar course, which is attended by CMB students at all stages of their thesis research, helps students develop intellectual vigor as well as learn how to present research in a lucid way. CMB students leave Baylor with seminar skills that rival those of any Ph.D. student in the country.

Student Outcomes

Recently graduated students from these investigators' laboratories have subsequently pursued postdoctoral training in excellent laboratories and high-quality institutions.

Location

Houston is a dynamic city that is both affordable and fun for graduate student life. Symphony concerts, opera, ballet, live theater, year-round major-league sports, and great restaurants are all a part of living in Houston, the fourth-largest city in the United States. From fall through spring, the average temperature ranges between 50°F and 75°F, with the temperatures rarely dropping below freezing, and it does not snow. Average highs during the summer are in the 90s, although it cools down into the 70s each evening, and air conditioning is present in virtually all homes, the Medical Center, indoor sporting events, etc.

The College

Baylor College of Medicine was established as an independent, private university committed to excellence in the training and education of scholars and physicians. The major area of growth for the College continues to be research. The College is located in the Texas Medical Center, which comprises more than 675 acres and includes forty-two independent institutions. The University of Texas Health Science Center, the School of Public Health, and the M. D. Anderson Tumor Institute are also on campus. The Texas Medical Center is one of the most actively growing science centers in the country. The influx of new colleagues and the opportunities created by an atmosphere of expansion provide a stimulating academic environment. More information on Baylor College of Medicine can be found on the College's Web site at http://www.bcm.edu/.

Applying

Applicants are required to have a bachelor's degree or the equivalent in a relevant area of science. Most students who join the CMB program have their undergraduate degree in some aspect of the biological sciences or in chemistry, although students with degrees in other areas, such as engineering, have also joined the program. Applications are due on January 1 and should be accompanied by three letters of recommendation from people who are familiar with the applicant's scholastic qualifications and/or research abilities, as well as a personal statement that describes research experience and career goals. Official GRE scores (not more than three years old) and transcripts from all colleges and universities attended must also be provided. Applications can be made online through the Graduate School Web site at http://www.bcm.edu/gradschool/ or the CMB Web site listed below. There is no fee for online applications. Applicants are invited to visit Baylor to meet with the participating faculty members and students, in order to have a firsthand look at the research and educational opportunities available to students in the CMB program. Expenses for travel and accommodations during the visit are provided by Baylor College of Medicine. Questions regarding the application process can be directed to the e-mail address listed below.

Correspondence and Information

Interdepartmental Program in Cell and Molecular Biology
Graduate School of Biomedical Sciences
Baylor College of Medicine
One Baylor Plaza, MS: BCM215
Houston, Texas 77030

Phone: 713-798-6557
E-mail: cmbprog@bcm.edu
Web site: http://www.bcm.edu/cmb/

Baylor College of Medicine

THE FACULTY AND THEIR RESEARCH

Department of Biochemistry and Molecular Biology
Wah Chiu, Ph.D. Structural and computational biology of macromolecular machines.
Adam Kuspa, Ph.D. Genomics of pathogen recognition and development in *Dictyostelium.*
B. V. Venkataram Prasad, Ph.D. Structural studies of viruses and viral proteins.
Jun Qin, Ph.D. Network analysis proteomics.
Florante A. Quiocho, Ph.D. Protein structure, molecular recognition and function.
Shelley Sazer, Ph.D. Eukaryotic cell-cycle control.
Zhou Songyang, Ph.D. Molecular mechanisms of signal transduction.
Francis T. F. Tsai, D.Phil. Structure and function of macromolecular complexes and molecular machines.
Salih Wakil, Ph.D. Mechanism and regulation of fatty acid metabolism.
Theodore G. Wensel, Ph.D. G-protein signaling in neurons.
John H. Wilson, Ph.D. Instability of trinucleotide repeats; knock-in mouse models for retinitis pigmentosa; gene therapy of dominant rhodopsin mutations.
Zheng Zhou, Ph.D. Molecular genetic studies of clearance of apoptotic cells.

Department of Immunology
Shuhua Han, M.D. Mechanisms of B-cell development and differentiation; autoimmunity; B-cell malignancies.
David M. Spencer, Ph.D. Prostate cancer progression and immunotherapy; gene therapy.
Tse-Hua Tan, Ph.D. Signal transduction by MAP kinases and phosphatases in cancer and immunity.
Jin Wang, Ph.D. Molecular regulation of apoptosis in the immune system.
Biao Zheng, M.D., Ph.D. Somatic genetics and development of immune responses; immunosenescence; autoimmunity.

Department of Medicine
Lawrence Chan, D.Sc. Molecular biology, genetics, and gene therapy of atherosclerosis; lipid disorders and diabetes mellitus.
Jae Woon Lee, Ph.D. Combinatorial transcriptional regulatory code.
Henry Pownall, Ph.D. High-density lipoproteins structure and biogenesis; protein stability; mass spectrometry; proteomics..
Michael D. Schneider, M.D. Molecular genetics of cardiac growth, signal transduction in cardiac muscle, heart repair by cardiogenic stem cells.
Li-Yuan Yu-Lee, Ph.D. Signaling pathways regulating cell division.

Department of Molecular and Cellular Biology
Adam Antebi, Ph.D. Regulation of life history in *C. elegans* by nuclear receptor signaling.
William (B. R.) Brinkley, Ph.D. Factors in the nucleus and mitotic apparatus affecting genomic instability in cancer.
Joseph Bryan, Ph.D. Molecular biology of ATP-sensitive potassium channels and regulation of insulin release.
Eric Chang, Ph.D. Growth and signaling regulation by Ras G-proteins.
Kwang-Wook Choi, Ph.D. Genetic control of patterning and cell polarity in *Drosophila* eye development.
Orla Conneely, Ph.D. Developmental control mechanisms.
Ronald L. Davis, Ph.D. Molecular and cellular biology of learning.
Francisco J. DeMayo, Ph.D. Molecular and developmental biology of the lung and uterus.
Xin-Hua Feng, Ph.D. TGF-ß/SMAD signaling in cell growth control, tissue differentiation and tumorigenesis..
Milan A. Jamrich, Ph.D. Role of homeobox and forkhead genes in vertebrate eye development.
Soo-Kyung Lee, Ph.D. Transcriptional regulatory network in CNS development.
Kathleen A. Mahon, Ph.D. Early development and patterning of the brain and pituitary.
Estela E. Medrano, Ph.D. Molecular biology of aging and transformation in human melanocytes.
David D. Moore, Ph.D. Function of nuclear hormone receptors.
Bert W. O'Malley, M.D. Steroid receptors and coactivators regulate gene expression.
Fred A. Pereira, Ph.D. Signaling pathways regulating development and aging of the hearing and balance systems.
Tony A. Pham, M.D., Ph.D. Molecular biology of neural circuit development and plasticity.
JoAnne S. Richards, Ph.D. Hormonal control of ovarian gene expression.
Dennis R. Roop, Ph.D. Molecular mechanisms regulating normal and abnormal skin development.
Jeffrey M. Rosen, Ph.D. Mammary gland development and breast cancer.
David R. Rowley, Ph.D. Tumor microenvironment in cancer progression.
Ming-Jer Tsai, Ph.D. Transcription factors in development and diseases.
Ming Zhang, Ph.D. Functional study of maspin in mouse development and tumor progression.
Thomas P. Zwaka, M.D., Ph.D. The nature of embryonic stem cell pluripotency

Department of Molecular and Human Genetics
Arthur L. Beaudet, M.D. Role of genomic imprinting in evolution and disease, including Prader-Willi and Angelman syndromes and autism; hepatocyte gene therapy.
Juan Botas, Ph.D. *Drosophila* as a model to investigate neural degeneration; genetic control of normal development.
Si Yi Chen, M.D., Ph.D. Gene therapy and antigen presentation for vaccine development.
Richard A. Gibbs, Ph.D. Genomics; genome sequencing; molecular basis of human genetic diseases.
Xiangwei He, Ph.D. Chromosome segregation: Interaction between spindle and kinetochores.

Christophe Herman, Ph.D. Regulation of cellular processes and quality control by an ATP dependent membrane protease.
Monica Justice, Ph.D. Genetic analysis of mouse development and disease.
Brendan Lee, M.D., Ph.D. Genetic pathways that specify development and homeostasis: translational studies of skeletal and kidney development and therapy for metabolic diseases.
Olivier Lichtarge, M.D., Ph.D. Computational biology and evolutionary predictions of protein-binding surfaces and interactions.
James R. Lupski, M.D., Ph.D. Determination of molecular mechanisms for disease using human genetic approaches to investigate clinical phenotypes.
Michael L. Metzker, Ph.D. Next-generation technology for genome sequencing; novel fluorescence imaging; molecular genetics of diabetics; phylogenetic analysis of HIV-1 transmission between individuals.
David L. Nelson, Ph.D. Genetics of the human X chromosome and DNA instability.
Scott D. Pletcher, Ph.D. The genetics and molecular analysis of aging in *Drosophila.*
Susan Rosenberg, Ph.D. Molecular mechanisms of genome instability, mutation, DNA repair, genome rearrangement, evolution.
Armin Schumacher, Ph.D. Developmental genetics of the mouse.
Gad Shaulsky, Ph.D. Developmental genetics in *Dictyostelium;* intercellular communication during development; functional genomics; microarray analysis of gene expression; evolution of sociality.
Hui Zheng, Ph.D. Molecular genetics of Alzheimer's disease and age-related disorders.

Department of Molecular Physiology and Biophysics
Mary Dickinson, Ph.D. Imaging the role of fluid mechanics in early cardiovascular development.
Jeannette Kunz, Ph.D. Phosphoinositide phosphate kinases in cell polarization and cell motility.
Steen E. Pedersen, Ph.D. Ion channel function and structure.
Pumin Zhang, Ph.D. Cell-cycle regulation in development and disease.

Department of Molecular Virology and Microbiology
Janet S. Butel, Ph.D. Polyomaviruses and pathogenesis of human disease.
Lawrence A. Donehower, Ph.D. Tumor suppressors and mouse cancer and aging models.
Mary K. Estes, Ph.D. Molecular mechanisms regulating virus–intestinal cell interactions and pathogenesis.
Richard A. Hull, Ph.D. Molecular genetics of bacterial virulence.
Ronald Javier, Ph.D. Adenoviruses and viral oncology.
Jason T. Kimata, Ph.D. Retroviral replication and pathogenesis.
Richard E. Lloyd, Ph.D. Translational control in apoptosis and viral infection.
Susan Marriott, Ph.D. Molecular regulation of viral gene expression and cellular transformation.
Timothy G. Palzkill, Ph.D. Protein structure-function; protein-protein interactions.
Robert F. (Frank) Ramig, Ph.D. Genetics, replication, and pathogenesis of *Reoviridae.*
Andrew P. Rice, Ph.D. Viral gene expression.
Betty L. Slagle, Ph.D. Hepatitis B-virus pathogenesis.
Richard E. Sutton, M.D., Ph.D. Lentiviral vectors for gene therapy and the study of HIV.
E. Lynn Zechiedrich, Ph.D. Protein-DNA interactions, genomic instability, and antibiotic resistance.

Department of Neurology
Jeffrey L. Noebels, M.D., Ph.D. Gene control of neuronal excitability.

Department of Neuroscience
Ellen Lumpkin, Ph.D. Molecular basis of mechanotransduction in touch and pain receptors.

Department of Pathology
Thomas A. Cooper, M.D. Alternative splicing regulation in development and disease.
Gretchen J. Darlington, Ph.D. Mouse models of aging.
Scott Goode, Ph.D. Molecular genetics of cell migration and tumor invasion in *Drosophila.*
Graeme Mardon, Ph.D. Molecular genetics of neural development in *Drosophila* and mice.
Richard N. Sifers, Ph.D. Glycobiology; posttranslational disease modifiers; conformational disease.
James Versalovic, M.D., Ph.D. Functional genomics of probiotics; probiotic therapy for oral and intestinal inflammation.

Department of Pediatrics
Margaret A. Goodell, Ph.D. Hematopoietic stem cells: basic biology and gene therapy.
Karen Hirschi, Ph.D. Molecular regulation of vascular development; vascular regeneration and engineering.
Kendal Hirschi, Ph.D. Plant biology related to human nutrition and environmental issues.
Sharon E. Plon, M.D., Ph.D. Control of genomic stability in response to DNA damage; syndromes resulting in an inherited predisposition to cancer.
Huda Y. Zoghbi, M.D. Molecular pathogenesis of neurodegenerative and neurodevelopmental disorders.

Department of Pharmacology
Pui-Kwong Chan, Ph.D. Anti-tumor agents and mechanism.
Anil Jaiswal, Ph.D. Oxidative stress-activated factors and signaling.

BAYLOR COLLEGE OF MEDICINE

Huffington Center on Aging
Program in Cell and Molecular Biology of Aging

Program of Study

The Program in Cell and Molecular Biology of Aging is a subspecialty of the interdisciplinary Cell and Molecular Biology (CMB) program. The goal of this training program is to provide predoctoral and postdoctoral training in the field of aging at the molecular and cellular levels. A training grant award from the National Institute on Aging provides resources for students and program support. Participating faculty members hold appointments in several of the basic science departments, including the Departments of Molecular and Cellular Biology, Molecular and Human Genetics, Pathology, Cardiovascular Sciences, Molecular Virology, and Immunology. The faculty members are actively engaged in various aspects of modern cell and molecular biology research, are well supported by federal grants, and are experienced in graduate education training in their respective departments.

The first year of the graduate program is filled with didactic courses, seminars, and laboratory rotations. During the second year, laboratory work commences, and qualifying examination requirements are fulfilled. The remainder of the graduate education program is oriented toward full-time laboratory work with participation in departmental seminars and student-oriented functions. In addition, students attend the Biology of Aging seminar series and the Journal Club in Aging.

Students seeking a Ph.D. or M.D./Ph.D. degree may enter the program in their first year of graduate school through the Cell and Molecular Biology program or in the second year of graduate school by matriculating in one of the graduate school departments in which aging training mentors hold a primary appointment.

Research Facilities

The participating faculty members in the program occupy extensive research space with modern equipment. In addition to a large number of such regular laboratory instruments as ultracentrifuges, scintillation counters, spectrophotometers, and cell-culture facilities, participating facilities are also in charge of sophisticated equipment for transmission and scanning electron microscopy, gas-phase protein sequencing, peptide and nucleic acid synthesis, microinjection to generate transgenic mice, and X-ray diffraction. There is also extensive computing equipment.

Financial Aid

Predoctoral students enrolled in the program receive initial stipends of $23,000. Baylor College of Medicine also maintains a financial aid office.

Cost of Study

For 2006–07, tuition costs of $8200 as well as the matriculation fee and learning resources center fee are covered by a tuition scholarship. Medical insurance coverage is also provided by the program.

Living and Housing Costs

A limited number of on-campus apartments are available for single students and rent for $450 to $775 per month. There are two dormitories for unmarried women students. Apartments that are in the same price range are also available near the Texas Medical Center, which is located in a pleasant residential section of south-central Houston. Job opportunities for spouses are available at the many institutions of the Texas Medical Center.

Student Group

Approximately 535 students are enrolled in the Graduate School, the Medical School has more than 675 students, and there are more than 500 postdoctoral fellows in the College as a whole.

Location

The College, part of the world's largest medical center, is located in Houston, Texas. Houston, with a population of more than 4.8 million, offers diverse urban amenities, including the symphony, the ballet, live theater, year-round professional and college sports, parks and recreation centers, and many ethnic restaurants. There is very pleasant weather from fall through spring, and the temperature rarely drops below freezing in winter. Gulf Coast beaches are within an hour's drive.

The College

Baylor College of Medicine, the only private, nonsectarian, nonprofit medical school in the greater Southwest, is dedicated to excellence in graduate and medical education. Baylor is committed to major growth in the Graduate School. Today, the College consistently ranks among the country's top 125 medical schools. For 2007, *U.S. News & World Report* has ranked the College tenth overall among the nation's top medical schools for research and among the top 10 percent of American graduate schools. BCM is also listed eleventh among all U.S. medical schools for National Institutes of Health funding and number one for research expenditures in biological science by the National Science Foundation. There is a high degree of interdisciplinary cooperation among the basic science and clinical departments within the College and with other institutions in the Texas Medical Center. A reciprocity agreement allows Baylor graduate students to take courses at Rice University, the University of Texas Health Science Center at Houston, and the University of Houston.

Applying

Applicants must have earned a baccalaureate degree and have a strong background in biology and biochemistry. Candidates for admission must complete the application form for the Graduate School. The application must contain current Graduate Record Examinations (GRE) scores, letters of recommendation, and official undergraduate transcripts. Applications receive three reviews, one by the Admissions Committee of the cell and molecular biology program or the departmental program, a second by a faculty committee of the Cell and Molecular Biology of Aging Training Program, and the third by the Admissions Committee of the Graduate School. The Graduate School application deadline is January 1. The academic year begins July 31. Applicants are responsible for monitoring the progress of their applications. Admission policies at Baylor offer equal opportunity without regard to race, sex, age, religion, or handicap.

Correspondence and Information

Admissions Office
Baylor College of Medicine
One Baylor Plaza
Houston, Texas 77030
Phone: 713-798-4028

For program information, prospective students should contact:
Estela E. Medrano, Ph.D.
Director, Aging Training Program
Huffington Center on Aging
Baylor College of Medicine
One Baylor Plaza, N804
Houston, Texas 77030
Phone: 713-798-1569
E-mail: medrano@bcm.edu
Web site: http://www.hcoa.org

Baylor College of Medicine

THE FACULTY AND THEIR RESEARCH

Adam Antebi, Ph.D., Assistant Professor of Molecular and Cellular Biology. Endocrine regulation of development and aging. Hormonal control of *C. elegans* dauer formation and life span by a Rieske-like oxygenase. *Dev. Cell* 10:473–82, 2006 (with Rottiers). Identification of hormonal ligands for DAF-12 that govern dauer formation and reproduction in *C. elegans*. *Cell* 124:1209–23, 2006 (with Motola). The tick-tock of aging. *Science* 310:1911–3, 2005. *mab-2* encodes RNT-1, a *C. elegans* Runx homologue essential for controlling cell proliferation in a stem cell–like developmental lineage. *Development* 132:5043–54, 2005 (with Nimmo and Woollard). The prepared mind of the worm. *Cell Metab.* 1:157–8, 2005. Identification of DAF-12 binding sites, response elements, and target genes. *Genes Dev.* 18:2529–44, 2004 (with Shostak). A novel coregulator/nuclear receptor complex controls *C. elegans* larval development, fat metabolism, and aging. *Genes Dev.* 18:2120–33, 2004 (with Ludewig). Hormonal signals produced by DAF-9/cytochrome P450 regulate *C. elegans* dauer diapause in response to environmental cues. *Development* 131:1765–76, 2004 (with Gerisch). Tipping the balance toward longevity. *Dev. Cell* 6:315–6, 2004. Long life: A matter of taste (and smell). *Neuron* 41:1–3, 2004. Endocrine regulation of aging by insulin-like signals. *Science* 299:1346–51, 2003 (with Tatar et al.). A hormonal signaling pathway influencing *C. elegans* metabolism, reproductive development, and life span. *Dev. Cell* 1:841–51, 2001 (with Gerisch).

Gretchen J. Darlington, Ph.D., Professor of Pathology, Molecular and Human Genetics, Cell Biology, and Pediatrics. Gene transcription; age-related gene expression; mechanisms of hepatocyte differentiation. C/EBPα is required for proteolytic cleavage of cyclin A by calpain 3 in myeloid precursor cells. *J. Biol. Chem.* 277(37):33848–56, 2002. C/EBPα is required for differentiation of white, but not brown, adipose tissue. *Proc. Natl. Acad. Sci. U.S.A.* 98(22):12532–7, 2001. Molecular mechanisms of liver development and differentiation. *Curr. Opin. Cell Biol.* 11:678–82, 1999. CCAAT/enhancer binding protein α regulates p21 protein and hepatocyte proliferation in newborn mice. *Mol. Cell. Biol.* 17(12):7353–61, 1997. C/EBPα regulates generation of C/EBPβ isoforms through activation of specific proteolytic cleavage. *Mol. Cell. Biol.* 17(12):7353–61, 1997. CCAAT/enhancer binding protein α(C/EBPα) inhibits cell proliferation through the p21 (WAF-1-CIP-1/SDI-1) protein. *Genes Dev.* 10:804–15, 1996. Impaired energy homeostasis in C/EBPα knockout mice. *Science* 269:1108–12, 1995 (with Wang et al.).

Ronald L. Davis, Ph.D., Professor of Molecular and Cellular Biology. Molecular and cellular biology of learning. Olfactory learning. *Neuron* 44:31–48, 2005. Olfactory memory formation in *Drosophila*: From molecular to systems neuroscience. *Ann. Rev. Neurosci.* 28:275–302, 2005. Altered representation of the spatial code for odors after olfactory classical conditioning: Memory trace formation by synaptic recruitment. *Neuron* 42:437-49, 2004 (with Yu et al.). P, a novel system for spatial and temporal control of gene expression in *Drosophila melanogaster*. *Proc. Natl. Acad. Sci. U.S.A.* 98:12602–7, 2001 (with Roman et al.). The role of *Drosophila* mushroom body signaling in olfactory memory. *Science* 10:1126–9, 2001 (with McGuire et al. *Drosophila* fasciclin II is required for formation of odor memories for normal sensitivity to alcohol. *Cell* 105:757–76, 2001 (with Cheng et al.). Learning performance of *Drosophila* after repeated conditioning trials with discrete stimuli. *J. Neurosci.* 20:2944–53, 2000 (with Beck et al.). Integrin-mediated short-term memory in *Drosophila*. *Nature* 391:455–60, 1998 (with Grotewiel et al.).

Lawrence A. Donehower, Ph.D., Professor of Molecular Virology. Role of the tumor suppressor p53 in aging and tumorigenesis. PPM1D dephosphorylates Chk1 and p53 and abrogates cell cycle checkpoints. *Genes Dev.* 19:1162–74, 2005 (with Lu et al.). p53: Guardian AND suppressor of longevity? *Exp. Gerontol.* 40:7–9, 2005. Cooperativity between p53, mdm2, and p19ARF in tumorigenesis. *Oncogene* 22:7831–7, 2003 (with Moore et al.). p53 mutant mice that display early aging-associated phenotypes. *Nature* 415:45–53, 2002 (with Tyner et al.). Does p53 affect organismal aging? *J. Cell. Physiol.* 192:23–33, 2002.

Margaret A. Goodell, Ph.D., Associate Professor of Pediatrics and of Molecular and Human Genetics, Center for Gene Therapy. Hematopoietic stem cells; basic biology and gene therapy. Hematopoietic stem cells do not engraft with absolute efficiencies. *Blood* 107(2):501–7, 2006. Molecular signatures of proliferation and quiescence in the hematopoietic stem cells. *PLoS Biol.*, 2004 (online). Hematopoietic stem cells generate hepatocytes via fusion of myeloid cells. *J. Clin. Invest.* 113:1266–70, 2004. Hematopoietic stem cells generate skeletal muscle. *Nature Med.* 9:1520–7, 2003. Transient RNA interference in hematopoietic progenitors with functional consequences. *Genesis* 36:203–8, 2003. Muscle-derived hematopoietic stem cells are hematopoietic in origin. *Proc. Natl. Acad. Sci. U.S.A.* 99:1341–6, 2002. Regeneration of ischemic cardiac muscle and vascular endothelium by adult stem cells. *J. Clin. Invest.* 107:1395–402, 2001.

Michael Ittmann, M.D., Ph.D., William D. Tigertt Professor of Pathology. Cellular senescence in the pathogenesis of prostate cancer and benign prostatic hyperplasia. Interleukin-8 expression is increased in senescent prostatic epithelial cells and promotes the development of benign prostatic hyperplasia. *Prostate* 60:153–9, 2004. Cellular senescence in the pathogenesis of benign prostatic hyperplasia. *Prostate* 55:30–8, 2003. IL-8 is a paracrine inducer of FGF2, a stromal and epithelial growth factor in benign prostatic hyperplasia. *Am. J. Pathol.* 159:139–47, 2001. IL-1a is a paracrine inducer of FGF-7, a key epithelial growth factor in benign prostatic hyperplasia. *Am. J. Pathol.* 157:249–55, 2000.

Estela E. Medrano, Ph.D., Professor of Molecular and Cellular Biology and of Dermatology. Molecular mechanisms of senescence and transformation. The age of skin cancers. *Sci. Aging Knowledge Environ.* 9:pe13, 2006 (with Desai, Krathen, and Orengo). The low molecular weight cyclin E isoforms augment angiogenesis and metastasis of human melanoma cells in vivo. *Cancer Res.* 65:692-7, 2005 (with Bales et al.). Overexpression of HDAC1 confers resistance to sodium butyrate–mediated apoptosis in melanoma cells through a p53-mediated pathway. *Cancer Res.* 64:7706–10, 2004 (with Bandyopadhyay et al.). SKI regulates Wnt/β-catenin signaling in human melanomas. *Cancer Res.* 63:6626–34 (with Chen et al.). The emerging role of epigenetics in cellular and organismal aging. *Exp. Gerontol.* 38:1299–307 (with Bandyopadhyay). Down-regulation of p300/CBP activates a senescence checkpoint in human melanocytes. *Cancer Res.* 62:6231–9, 2002 (with Bandyopadhyay et al.). Cytoplasmic localization of the oncogenic protein SKI in human cutaneous melanomas in vivo: Functional implications for TGF-beta signaling. *Cancer Res.* 61(22):8074–8, 2001 (with Reed et al.). SKI acts as a co-repressor of Smad2 and Smad3: A mechanism to regulate the response to TGF-beta. *Proc. Natl. Acad. Sci. U.S.A.* 97:5924–9, 2000 (with Xu et al.).

Scott D. Pletcher, Ph.D., Assistant Professor of Molecular and Human Genetics. Genetic mechanisms of aging and lifespan extension via dietary restriction in *Drosophila*. Demography and the short-term benefits of diet manipulation. *Science* 301:1731–3, 2003 (with Mair et al.). Patterns of gene expression throughout adult life in *Drosophila* are dynamic and identify a role for the innate immune response in aging. *Curr. Biol.* 12:712–23, 2002.

Roy G. Smith, Ph.D., Professor of Molecular and Cellular Biology and of Medicine and Director, Huffington Center on Aging. Regulation of pulsatile hormone release; age-related neurodegeneration and frailty; neuroendocrinology; aging and metabolism. Ghrelin amplifies dopamine signaling by crosstalk involving formation of GHS-R/D1R heterodimers. *Mol. Endocrinol.*, in press (with Jiang et al.). Development of growth hormone secretagogues. *Endocr. Rev.* 26(3):346–60, 2006. Cyclin D3 maintains growth inhibitory activity of C/EBPα by stabilizing C/EBPα-cdk2 and C/EBPα-Brm complexes. *Mol. Cell. Biol.* 26:2570–82, 2006 (with Wang et al.). Des-acyl ghrelin induces food intake by a mechanism independent of the growth hormone secretagogue receptor. *Endocrinology* 47(5):2306–14, 2006 (with Toshinai et al.). Ablation of ghrelin gene improves the diabetic but not obese phenotype of *ob/ob* mice. *Cell Metab.* 3:379–86, 2006 (with Sun et al.). Active ghrelin levels and active/total ghrelin ratio in cancer-induced cachexia. *J. Clin. Endocrinol. Metab.* 90(5):2920–6, 2005 (with Garcia et al.). Developments in ghrelin biology and potential clinical relevance. *Trends Endocrinol. Metab.* 16:436–42, 2005. Molecular endocrinology and physiology of the aging central nervous system. *Endocrinol. Rev.* 26:203–50, 2005. Ghrelin stimulation of growth hormone release and appetite is mediated through the growth hormone secretagogue receptor. *Proc. Natl. Acad. Sci. U.S.A.* 101:4679–84, 2004 (with Sun et al.). Agonist-specific coupling of growth hormone secretagogue receptor type 1A to different intracellular signaling systems: Role for adenosine. *Neuroendocrinology* 79:13–25, 2004 (with Carreira et al.). Deletion of ghrelin impairs neither growth nor appetite. *Mol. Cell. Biol.* 23:7973–81, 2003 (with Sun et al.). The distribution and mechanism of action of ghrelin in the CNS demonstrates a novel hypothalamic circuit regulating energy homeostasis. *Neuron* 37:649–61, 2003 (with Cowley et al.). Ligand activation domain of human orphan growth hormone secretagogue receptor (GHS-R) conserved from pufferfish to humans. *Mol. Endocrinol.* 14:160–9, 2000 (with Palyha et al.).

George E. Taffet, M.D., Associate Professor and Chief of Medicine, Geriatrics, and Cardiovascular Sciences. Physiology and biochemistry of aging cardiovascular system. Noninvasive ultrasonic measurement of arterial wall motion in mice. *Am. J. Physiol. Heart Circ. Physiol.* 287:H1426-32, 2004 (with Hartley et al.). The age-associated alterations in late diastolic function in mice are improved by caloric restriction. *J. Gerontol. A Biol. Sci. Med. Sci.* 52(6): B285–90, 1997. Noninvasive indexes of cardiac systolic and diastolic function in hyperthyroid and senescent mouse. *Am. J. Physiol.* 270:H2204–9, 1996. CaATPase content is lower in cardiac sarcoplasmic reticulum isolated from old rats. *Am. J. Physiol.* 264:H1609–14, 1993 (with Tate).

Hui Zheng, Ph.D., Associate Professor of Molecular and Human Genetics. Investigation of Alzheimer's disease using mouse models. Regulation of tyrosinase trafficking and processing by presenilins: Partial loss of function by familial Alzheimer's disease mutation. *Proc. Natl. Acad. Sci. U.S.A.* 103:353–8, 2006 (with Wang et al.). Deletion of presenilin 1 hydrophilic loop sequence leads to impaired gamma-secretase activity and exacerbated amyloid pathology. *J. Neurosci.* 26:3845–54, 2006 (with Deng et al.). Defective neuromuscular synapses in mice lacking amyloid precursor protein (APP) and APP-like protein 2. *J. Neurosci.* 25:1219–25, 2005 (with Wang et al.). Presenilin 1 Alzheimer's disease mutation leads to impaired adult neurogenesis and defective associative learning. *Neuroscience* 126:305–12, 2004 (with Wang et al.). Myeloproliferative disease in mice with reduced presenilin gene dosage: Effect of gamma-secretase blockage. *Biochemistry* 43: 5352–9, 2004 (with Qyang et al.). Presenilins are required for the formation of comma- and S-shaped bodies during nephrogenesis. *Development* 130:5019–29, 2003 (with Wang et al.). Accelerated plaque accumulation, associative learning deficits and up-regulation of α7 nicotinic acetylcholine receptor protein in transgenic mice co-expressing mutant human presenilin 1 and amyloid precursor proteins. *J. Biol. Chem.* 277:22768–80, 2002 (with Dineley et al.). The aspartate-257 of presenilin 1 is indispensable for mouse development and production of β-amyloid peptides through β-catenin independent mechanisms. *Proc. Natl. Acad. Sci. U.S.A.* 99:8760–5, 2002 (with Xia et al.). A Wnt-autonomous β-catenin pool is dependent on presenilin activity for degradation: Implications for β-catenin activation in tumorigenesis. *Cell* 110:751–62, 2002 (with Kang et al.). Loss of presenilin 1 is associated with enhanced β-catenin signaling and skin tumorigenesis. *Proc. Natl. Acad. Sci. U.S.A.* 98:10863–8, 2001 (with Xia et al.).

BAYLOR COLLEGE OF MEDICINE

Department of Molecular and Cellular Biology
Graduate Program

Program of Study	The Department of Molecular and Cellular Biology (MCB) features a diverse, internationally known, and well-funded faculty. The graduate program, which leads to the Ph.D. degree, provides expertise in many areas of gene regulation during cell differentiation and growth. The Department is a participant in two Specialized Programs of Research Excellence (SPORE), one for prostate cancer and one for breast cancer, as well as a program of excellence in gene therapy on cardiovascular disease and a Specialized Cooperative Center Program in Reproductive Research (SCCPRR). Members of the faculty hold appointments in the Center for Cell and Gene Therapy, the Breast Care Center, the Howard Hughes Medical Institute (HHMI), the Huffington Center on Aging, and the Center for Comparative Medicine. Department of Molecular and Cellular Biology faculty members also participate in the Center for Brain and Behavior in association with the Department of Psychiatry, and a Center for Skin Research with the Department of Dermatology. The graduate program is designed to prepare scientists for competitive careers in basic and biomedical research. Entering graduate students spend the first year of study in a sequence of required courses covering the function and organization of the eukaryotic cell, genes (regulation and organization), and the structural interactions of macromolecules. Elective courses include cellular signaling, molecular concepts of learning and memory, developmental biology, molecular aspects of reproductive biology, cancer biology, integrated microscopy, and computational methods in molecular biology. To introduce entering graduate students to the research options available, a series of faculty research talks, reading tutorials, and laboratory rotations are offered. A vigorous program of seminars, workshops, and journal clubs provide students with a current knowledge of specialized topics. Second-year students are admitted to candidacy for the degree upon successful completion of the core curriculum and a qualifying examination. The dissertation is based on original research carried out in the major adviser's laboratory. The progress and quality of each student's research, as well as the general quality of the student's graduate education, are maintained by continuous interactions between the student, the major adviser, and the thesis advisory committee. The MCB Department also participates in a medical school–graduate school training program leading to a combined M.D./Ph.D. degree. The academic year begins in early August and is arranged on a twelve-month basis.
Research Facilities	The Department of Molecular and Cellular Biology occupies extensive, modern facilities that are generously equipped with a full range of instrumentation required for research in cellular, molecular, developmental, and endocrine biology. Special facilities include laboratories for recombinant DNA research, multiple tissue culture facilities, digital fluorescence microscopy, laser scanning confocal microscopy, transmission and scanning electron microscopy, computer facilities for structural/computational biology, and histopathology laboratory. A modern transgenic mouse facility provides state-of-the-art technology to generate transgenic and knockout mutant mouse models that are used by many faculty members and students. The high level of cooperation among the various departments at Baylor College of Medicine and the institutions of the Houston scientific community provide access to additional facilities.
Financial Aid	Students enrolled in the program receive initial stipends of $23,000 per year plus health insurance. Tuition scholarships of $8200 are provided by Baylor College of Medicine to all students accepted into the Department.
Cost of Study	In 2005–06, tuition of $8200 per year was covered by the tuition scholarship, as described above. Students paid a one-time matriculation fee of $25, a one-time graduation fee of $140, and an annual student fee of $150 for the first year and $20 for each of the following years. In 2005–06, health insurance was paid by the Graduate School for first-year students and by the mentor thereafter.
Living and Housing Costs	Privately owned furnished and unfurnished apartments and houses, offered at a wide range of rents, are available nearby. A residence hall adjacent to the College offers moderately priced, fully furnished rooms and apartments. The cost of living in Houston is lower than most major U.S. cities.
Student Group	In 2005–06, the MCB Department had 67 graduate students, including 11 pursuing the M.D./Ph.D. There is a vigorous postdoctoral research training program in the Department, with 100 research fellows and visiting scientists. The Graduate School has 533 students. A low student-faculty ratio favors direct involvement of Baylor graduate students in current clinical and basic science projects. Baylor places no restrictions in its admissions policies on the basis of race, sex, age, religion, country of origin, or handicap.
Student Outcomes	Many graduates are now postdoctoral fellows and professors at prestigious institutions such as Cornell, Duke, Harvard, the Whitehead Institute, MIT, the Salk Institute, the University of California at Berkeley and at San Diego, Vanderbilt, Yale, M. D. Anderson, and the University of North Carolina. Other graduates now hold key positions at biotechnology companies.
Location	Houston is an exciting city with a vibrant scientific community as well as an internationally recognized symphony orchestra, grand opera, ballet, theater companies, and art museums. Recreational opportunities abound, with facilities for a wide range of professional and amateur sports.
The College	Baylor College of Medicine was established as an independent, private university committed to excellence in the training and education of biomedical scholars and physicians. The College is on the 200-acre campus of the Texas Medical Center. The University of Texas Health Science Center, the School of Public Health, the M. D. Anderson Cancer Institute, and Rice University are also part of the Texas Medical Center.
Applying	Applicants should have a bachelor's degree with sound preparation in biology, chemistry, and biochemistry. They must submit GRE scores (the General Test) that are not more than two years old at the time of application. Applications should be accompanied by transcripts from all schools attended, three letters of recommendation, and a statement of research interests and career goals and must be complete in all respects by January 1, with a preferred deadline of December 15. Applications are evaluated and ranked by a faculty-student committee; notification of action regarding admission and financial support is generally made by April 15.
Correspondence and Information	Department of Molecular and Cellular Biology, Room 128A Baylor College of Medicine Houston, Texas 77030 Phone: 713-798-4598 E-mail: mcbgrad@bcm.edu Web site: http://www.bcm.edu/mcb/

Baylor College of Medicine

THE FACULTY AND THEIR RESEARCH

Bert W. O'Malley, Professor and Chairman; M.D., Pittsburgh. Steroid hormone action; gene regulation; transcription factors (coactivators and corepressors); breast cancer; genetics of neurobehavior.

William (Bill) R. Brinkley, Dean, Graduate School of Biomedical Sciences; Ph.D., Iowa State. Mitosis in normal and neoplastic cells; genomic instability; nuclear organization, nucleoskeleton; cell motility; cytoskeleton.

Adam Antebi, Assistant Professor; Ph.D., MIT. Endocrine regulation of *C. elegans* developmental age and aging.

Cassius Bordelon, Associate Professor and Director, Post-Graduate Medical Education Opportunities; Ph.D., Baylor. Anatomical teaching.

Joseph Bryan, Professor; Ph.D., Pennsylvania. Molecular biology of ATP-sensitive potassium channels and regulation of insulin secretion.

Lawrence Chan, Professor; D.Sc., Hong Kong. Cell biology; molecular and gene therapy of diabetes, obesity, and dyslipidemia.

Eric C. Chang, Assistant Professor; Ph.D., SUNY at Buffalo. Molecular and genetic studies of the regulation of signaling and mitotic fidelity.

Kwang-W. Choi, Associate Professor; Ph.D., Princeton. Molecular neurogenetics; pattern formation and cell polarity in *Drosophila* eye development.

Orla M. Conneely, Professor; Ph.D., Ireland. Genetic regulation of vertebrate development.

Austin J. Cooney, Associate Professor; Ph.D., Ireland. Nuclear receptors and their roles in gene regulation in embryonic stem cells.

Ronald L. Davis, Professor; Ph.D., California, Davis. Molecular and cellular biology of learning and memory.

Francisco J. DeMayo, Professor; Ph.D., Michigan State. Molecular developmental biology of the lung; cancer; reproductive biology.

Dean P. Edwards, Professor; Ph.D., Medical College of Georgia. Mechanism of action of steroid receptors in female reproductive tissues and breast cancer.

Xin-Hua Feng, Associate Professor; Ph.D., Maryland College Park. TGF-Beta/SMAD signaling in cell growth control; tissue differentiation and tumorigenesis.

Milan Jamrich, Professor; Ph.D., Heidelberg. Molecular basis of embryonic pattern formation.

Richard Kelley, Associate Professor; Ph.D., Stanford. Role of noncoding RNA in chromatin structure.

Soo-Kyung Lee, Assistant Professor; Ph.D., Chonnam National (Korea). Transcriptional regulation in developing central nervous system.

John P. Lydon, Assistant Professor; Ph.D., National (Ireland). Progesterone receptor in mammary gland development and tumorigenesis.

Kathleen A. Mahon, Associate Professor; Ph.D., Yale. Role of homeobox genes in forebrain and pituitary development in the mouse.

Michael A. Mancini, Associate Professor; Ph.D., Texas Health Science Center at San Antonio. Transcription; nuclear receptors; microscopy.

Shaila K. Mani, Associate Professor; Ph.D., Nagpur (India). Neurotransmitter-hormone interactions in brain and behavior.

Barry M. Markaverich, Associate Professor; Ph.D., Southern Mississippi. Environmental estrogens; endocrine disrupters; cancer.

Daniel Medina, Professor; Ph.D., Berkeley. Mammary tumor biology; preneoplasia; cell-cycle regulation; selenium chemoprevention.

Estela E. Medrano, Professor; Ph.D., Buenos Aires. Molecular mechanisms of melanocyte senescence and transformation.

David D. Moore, Professor; Ph.D., Wisconsin–Madison. Functions of the nuclear hormone receptor superfamily.

Paul A. Overbeek, Professor; Ph.D., Michigan. Transgenic mice; ocular morphogenesis; retinoblastoma models; transposon-mediated insertional mutagenesis.

Fred A. Pereira, Ph.D., Assistant Professor; Ph.D., Manitoba. Signaling pathways regulating development/aging of hearing.

Tony A. Pham, Assistant Professor; M.D., Ph.D., Baylor College of Medicine. Regulation of brain circuit development by genes and experience.

Jun Qin, Associate Professor; Ph.D., Rockefeller. Proteomics analysis of tumor suppressor signaling networks that maintain the human genome.

JoAnne S. Richards, Professor; Ph.D., Brown. Hormonal control of gene expression in the mammalian ovary.

Dennis R. Roop, Professor; Ph.D., Tennessee, Knoxville. Gene regulation and function during normal and abnormal skin development.

Jeffrey M. Rosen, Professor; Ph.D., SUNY at Buffalo. Mammary gland development, stem cells, and breast cancer.

David R. Rowley, Professor; Ph.D., Iowa. Molecular mechanisms of reactive stoma and angiogenesis in the tumor microenvironment.

Tae-Ho Shin, Assistant Professor; Ph.D., Alabama at Birmingham. *C. elegans* embryonic development.

Carolyn L. Smith, Associate Professor; Ph.D., Western Ontario. Estrogen receptor regulation of gene expression.

Roy G. Smith, Professor and Director, Huffington Center on Aging; Ph.D., London. Neuroendocrinology; aging research.

Nancy Smyth Templeton, Assistant Professor; Ph.D., Wesleyan. Nonviral delivery for the treatment of disease.

Ming-Jer Tsai, Professor; Ph.D., California, Davis. Pancreas and neural development; organogenesis steroid hormone action; prostate cancer.

Sophia Y. Tsai, Professor; Ph.D., California, Davis. Nuclear orphan receptor in mouse development and organogenesis.

Nancy L. Weigel, Professor; Ph.D., Johns Hopkins. Steroid receptor phosphorylation; androgen and vitamin D receptors in prostate cancer.

Jiemin Wong, Associate Professor; Ph.D., Vermont. Chromatin remodeling and action of nuclear hormone receptors in chromatin.

Jianming Xu, Associate Professor; Ph.D., Clarkson. Transcriptional coregulators and hormonally promoted breast and prostate cancers.

Ming Zhang, Assistant Professor; Ph.D., SUNY at Albany. Tumor metastasis and angiogenesis; mouse development.

Thomas P. Zwaka, Assistant Professor; M.D., Ph.D., Ulm (Germany). Nature of embryonic stem cell pluripotency.

Joint Faculty Members and Their Research

Arthur Beaudet, Professor (Molecular and Human Genetics); M.D., Yale. Molecular human genetics; epigenetics and imprinting; gene therapy.

Hugo Bellen, Professor (Molecular and Human Genetics and HHMI Investigator); D.V.M., Ph.D., California, Davis. Genetic analysis of neurotransmitter release and neural development in *Drosophila.*

Juan Botas, Associate Professor (Molecular and Human Genetics); Ph.D., Madrid. Neurodegeneration and genetic control of normal development.

Powel Brown, Associate Professor (Breast Center); M.D., Ph.D., NYU. Gene regulation; transcription; breast cancer; cancer prevention; gene therapy.

Wah Chiu, Professor (Biochemistry and Molecular Biology); Ph.D., Berkeley. Electron cryomicroscopy of macromolecular assemblies.

Thomas A. Cooper, Professor (Pathology); M.D., Temple. Regulation of pre-mRNA splicing in development and disease; pathogenesis of myotonic dystrophy.

Gretchen Darlington, Professor (Pathology); Ph.D., Michigan. Tissue-specific gene transcription and differentiation; aging and extended longevity in mice.

Lawrence A. Donehower, Professor; Ph.D., George Washington. Tumor suppressors and mouse cancer and aging models.

Suzanne A. W. Fuqua, Professor; Ph.D., Texas–Houston Health Science Center. Estrogen receptors in breast cancer initiation, progression, and metastasis.

Margaret A. Goldstein, Professor (Medicine); Ph.D., Rice. Image analysis; vision and symbolic language.

Scott Goode, Assistant Professor (Pathology); Ph.D., Chicago. Analysis of cell migrations and tumor cell invasions in *Drosophila.*

Darryl Hadsell, Associate Professor (Pediatrics); Ph.D., Penn State. Growth factor signaling in mammary gland development.

Susan J. Henning, Professor (Pediatrics); Ph.D., Melbourne. Molecular biology of intestinal development; characterization of intestinal stem cells.

Karen K. Hirschi, Associate Professor (Pediatrics); Ph.D., Arizona. Regulation of vascular development; vascular potential of adult stem cells.

Richard Hurwitz, Associate Professor (Pediatrics); M.D., Albany Medical College. Gene therapy of ocular disorders; molecular biology of phototransduction.

Peter J. Koch, Assistant Professor (Dermatology); Ph.D., Heidelberg. Cell adhesion molecules in mammalian development and diseases.

Frank Kretzer, Associate Professor (Ophthalmology); Ph.D., UCLA. Retinal neovascularization.

Dolores J. Lamb, Professor (Urology); Ph.D., Texas–Houston Health Science Center. Molecular endocrinology; male infertility.

Adrian Lee, Associate Professor (Breast Center, Medicine); Ph.D., Surrey (England). Growth factor–hormone interactions in breast cancer.

Jae W. Lee, Associate Professor (Medicine); Ph.D., Texas A&M. Transcriptional regulatory circuits in mammals.

Martin M. Matzuk, Professor (Pathology); M.D., Ph.D., Washington (St. Louis). Transgenic mouse models to study reproduction and oncogenesis.

Franck Mauvais-Jarvis, Assistant Professor (Medicine, Endocrinology); M.D., Ph.D., Paris. Sex steroid receptors and diabetes/obesity.

Steffi Oesterreich, Associate Professor (Breast Center, Medicine); Ph.D., Humboldt (Berlin). Hormone receptor cofactors in breast tumorigenesis.

C. Kent Osborne, Professor (Medicine) and Director (Breast Center); M.D., Missouri. Breast cancer biology; actions of estrogen and antiestrogen.

Pyong Woo Park, Ph.D., Assistant Professor (Medicine); Ph.D., Washington (St. Louis). ECM in microbial pathogenesis and host defense.

Shelley Sazer, Associate Professor (Biochemistry); Ph.D., Stanford. Cell-cycle control and nucleocytoplasmic transport in fission yeast.

Michael D. Schneider, Professor (Medicine); M.D., Pennsylvania. Molecular biology of cardiac growth; adult cardiac progenitor/stem cells.

H. David Shine, Associate Professor (Neurosurgery, Neuroscience); Ph.D., Texas Medical Branch. Molecular biology of nervous system disease.

Richard N. Sifers, Associate Professor (Pathology); Ph.D., Oklahoma. Glycobiology; conformational disease; alpha1-antitrypsin deficiency.

Timothy C. Thompson, Professor (Urology, Radiotherapy); Ph.D., Colorado. Molecular mechanisms of prostate cancer progression; gene therapy.

Li-yuan Yu-Lee, Professor (Medicine); Ph.D., Columbia. Signaling pathways to cell division; 16-kDa prolactin and antiangiogenesis.

Hui Zheng, Associate Professor (Molecular and Human Genetics); Ph.D., Baylor College of Medicine. Molecular genetics of Alzheimer's disease.

BOSTON UNIVERSITY

School of Medicine
Division of Graduate Medical Sciences
Program in Cell and Molecular Biology

Programs of Study

The Division of Graduate Medical Sciences at Boston University School of Medicine accepts qualified applicants into M.A., Ph.D., and M.D./Ph.D. programs. The programs focus on basic scientific and clinical issues related to molecular and cellular biology. The interdisciplinary Program in Cell and Molecular Biology takes advantage of the individual resources of six academic departments in the medical school and one department in the dental school. The six academic departments are anatomy and neurobiology, biochemistry, microbiology, pathology and laboratory medicine, pharmacology and experimental therapeutics, and physiology and biophysics. Students may also take advantage of the Oral Biology Department in the Goldman School of Dental Medicine. The participation of more than 90 faculty members offers a diversity of research opportunities and a wealth of productive interactions.

Research Facilities

The Division of Graduate Medical Sciences is housed within a large medical school–hospital complex, which also houses a scientific-medical library, animal facilities, and a host of diverse research laboratories. Computer facilities are available for students, as is access to the University's mainframe. In addition, state-of-the-art equipment necessary for modern biological research exists in the form of DNA cores, confocal microscopes, and X-ray crystal facilities. The many intensive research projects that are being conducted offer numerous opportunities to students and investigative talents.

Financial Aid

Students in the Ph.D. and M.D./Ph.D. programs are eligible for full financial aid, including tuition and fees, a monthly stipend, and health insurance. Funds are provided from research grants, teaching grants, and departmental budgets. Graduate student stipends are at the rate of $27,000 per year ($28,000 after the student passes the qualifying exam).

Cost of Study

Tuition for the 2006–07 academic year is $33,330. Tuition is covered for most students by financial aid, as described above. Graduate student stipends for the 2006–07 academic year are $27,000 before the qualifying exams and $28,000 after the qualifying exams.

Living and Housing Costs

Most graduate students live in apartments in Boston and the surrounding areas. Shared apartment rentals typically range from $500 to $900 per month.

Student Group

Approximately 5 students enter the Program in Cell and Molecular Biology each year. In addition, students enrolled in other programs also join the program. More than 450 other graduate students and 600 medical students are also enrolled at the Medical Center campus.

Location

The Boston University School of Medicine is located in the South End of Boston and is part of a large medical complex. The broad range of biological research that takes place is augmented by the large research community existing in the greater Boston area. The community includes Harvard University, Tufts University, Massachusetts Institute of Technology, and Brandeis University. Boston is also home to numerous biotechnology companies.

Boston has long been known as "The Hub" and has a rich selection of cultural activities, major-league sports, and restaurants. Close by are Cape Cod, the White Mountains, and the picturesque New England countryside.

The University

Boston University is an independent, coeducational, nonsectarian university. Its academic diversity meets the needs of one of the largest bodies of scholars in the world. A student at Boston University finds a nearly limitless range of educational, social, and civic resources. The University was incorporated in 1869. The University now has sixteen schools and colleges but manages to provide students with the advantages of a large, contemporary educational complex while maintaining many traditional priorities. The two main campuses are the Charles River Campus and the Medical Center.

Applying

Students are ordinarily admitted only in September. Application should be made to the Program in Cell and Molecular Biology at the address given below and should be received early, preferably by mid-February. The GRE General Test is required. It is helpful to take the examination before December 1 because of the delay of about six weeks in receiving examination scores.

Correspondence and Information

Vickery Trinkaus-Randall, Ph.D.
Graduate Program Director
Program in Cell and Molecular Biology
Boston University School of Medicine
715 Albany Street, Room L903
Boston, Massachusetts 02118

Phone: 617-638-5099
Fax: 617-638-5337
E-mail: vickery@bu.edu

Boston University

THE FACULTY AND THEIR RESEARCH

The following is a selection from the 90 faculty members who participate in the Program in Cell and Molecular Biology.

Department of Anatomy and Neurobiology
Julie H. Sandell, Ph.D., Assistant Professor. Development and inherited degeneration in zebra fish and primate retina.

Department of Biochemistry
Peter Brecher, Ph.D., Professor. Effect of hypertension on gene expression in the cardiovascular system.
Stephen Farmer, Ph.D., Professor. Tissue-specific gene expression; role of matrix interactions and cell morphology.
Judith Foster, Ph.D., Professor. Gene regulation.
Matthew Nugent, Ph.D., Associate Professor. Cell proliferation; growth factor–receptor interactions; proteoglycans; extracellular matrix.
Katya Ravid, Ph.D., Associate Professor. Genetic and signaling mechanisms regulating blood cell development; vascular biology.
Barbara Schreiber, Ph.D., Associate Professor. Smooth-muscle cell function/lipids and extracellular matrix.
Gail Sonenshein, Ph.D., Professor. Role of oncogenes in control of cell proliferation, apoptosis, and neoplastic transformation.
Karen Symes, Ph.D., Assistant Professor. Role of cell-cell interactions in embryonic development.
Vickery Trinkaus-Randall, Ph.D., Associate Professor. Growth factor–proteoglycan interactions; extracellular matrix; integrins; wand repair.
Jim Xiao, Ph.D., Assistant Professor. Tumor suppressor genes in cell cycle, proliferation, and differentiation.

Department of Genetics and Genomics
Michael Christman, Ph.D., Professor. Molecular basis of genome instability in cancer.
Landon Moore, Ph.D., Assistant Professor. Chromosome segregation.
Cyrus Vaziri, Ph.D., Assistant Professor. Cell-cycle regulation; DNA damage checkpoints; DNA replication.

Department of Microbiology
Iih-Nan Chou, Ph.D., Professor. Cytoskeletal perturbations and mechanisms of cell injury by carcinogenic metal ions.
R. B. Corley, Ph.D., Professor. B-cell differentiation; immunoglobulin assembly; antibody structure and function.
Sue Fisher, Ph.D., Associate Professor. Molecular genetics and regulation of nitrogen assimilation in *B. subtilis*.
Jack Murphy, Ph.D., Professor. Protein folding; gene regulation.
Ann Marshak Rothstein, Ph.D., Professor. Role of exogenous and endogenous Fas ligand in lymphocyte apoptosis, inflammation, and autoimmunity.
David Seldin, M.D./Ph.D., Assistant Professor. Oncogenes and tumorigenesis as modeled in transgenic mice.
Gregory A. Viglianti, Ph.D., Assistant Professor. Regulation of HIV infectivity and transcriptional regulation of HIV LTR.

Department of Molecular and Cell Biology
Miklos Sahin-Toth, Ph.D., Assistant Professor. Proteases and their inhibitors in the pathogenesis of pancreatitis; crystallography of mutant trypsinogens.

Department of Molecular Medicine
Kenneth Albrecht, Ph.D., Assistant Professor. Mammalian gonadal sex determination.
Victoria Bolotina, Ph.D., Associate Professor. Ion channels and mechanisms of calcium signaling in the cardiovascular system.
Herbert Cohen, Ph.D., Assistant Professor. Molecular basis of renal cancer, renal cystic disease, and renal development.

Department of Pathology and Laboratory Medicine
Kryzysztof Blusztajn, Ph.D., Associate Professor. Acetylcholine synthesis and release; signal transduction by lipid second messengers.
Wellington Cardoso, Ph.D., Associate Professor. Control branching morphogenesis and epithelial cell differentiation in the embryonic lung.
Mary Jo Murnane, Ph.D., Associate Professor. Proteolytic enzymes as tumor markers within a proteolytic cascade in tumorigenesis.

Department of Pharmacology and Experimental Therapeutics
David Farb, Ph.D., Professor. Pharmacology and regulation of gene expression of the GABA/benzodiazepine receptor in the vertebrate central nervous system; studies of the mechanism of action of neuromodulators and elucidation of the structure, function, and cellular dynamics of amino acid receptors in the brain and spinal cord.
Kevin Jarrell, Ph.D., Assistant Professor. Molecular mechanisms of RNA splicing and catalysis.
Shelley Russek, Ph.D., Assistant Professor. Gene expression in neurons/tissue-specific promoters as targets for therapeutic agents.
Bryan Yammamoto, Ph.D., Professor. Neuropharmacology and neurotoxicology of drug abuse; neurodegenerative disorders; psychotherapeutic drugs; studies on the mechanisms and consequences of amphetamine neurotoxicity.

Department of Physiology and Biophysics
Karen Allen, Ph.D., Assistant Professor. Protein structure and function using X-ray crystallography and kinetics.
Hwai-Chen Gou, Ph.D., Assistant Professor. Protein structure and function using X-ray crystallography and molecular biology.
James Head, Ph.D., Professor. Regulatory role of high-affinity intracellular calcium–binding proteins.
Michael Holick, M.D., Ph.D., Professor. Physiology and molecular biology of skin and bone, vitamin D, and peptide hormones.
William Lehman, Ph.D., Professor. Calcium regulatory systems in muscle; image reconstruction of muscle thin filaments.
Hector Lucero, Ph.D., Assistant Professor. Molecular biology of a calcium-binding protein in endoplasmic reticulum.
C. James McKnight, Ph.D., Assistant Professor. Protein structure and function using NMR.
Neil Ruderman, Ph.D., D.Phil., Professor. Insulin action, gene expression, and diacylglycerol protein kinase C in skeletal muscle.
Barbara Seaton, Ph.D., Associate Professor. Structure and function using X-ray crystallography and other biophysical/biochemical techniques.
R. Andrew Zoeller, Ph.D., Assistant Professor. Somatic-cell mutants in phospholipid biosynthesis and transport.

BOSTON UNIVERSITY

Graduate Program in Molecular Medicine

Program of Study

The Graduate Program in Molecular Medicine curriculum consists of one year of basic science courses in the Division of Graduate Medical Sciences. This is followed in the second year by the innovative Molecular Medicine Core Curriculum, which consists of four fall courses: Genetics and Epidemiology of Disease, Cancer Biology, Immunity and Infection, and Organ System Diseases and a single spring course entitled Molecules to Molecular Therapeutics: The Translation of Molecular Observations to Clinical Implementation. After a qualifying examination, candidates carry out dissertation research in the Department of Medicine or affiliated laboratories at the medical school. Degree candidates also participate in a journal club, the Evans Seminars in Molecular Medicine, and the annual Evans Medicine Research Days.

Research Facilities

The Department of Medicine research laboratories are located on the Boston University Medical Center campus primarily in the Evans Biomedical Research Center or in the Center for Advanced Biomedical Research, which also houses core facilities, including the Structural Biology Facility, mass spec, MRI, and the Transgenic Facility. These buildings provide state-of-the-art research space in an open, spacious environment that is fully supported by modern laboratory and computer facilities.

Financial Aid

Students admitted to the program are offered full tuition support. Stipends are provided by the Department of Medicine or by the research laboratories.

Cost of Study

Tuition and fees for the 2005–06 year were approximately $31,500 for a full-time student.

Living and Housing Costs

Five University residence halls are open to graduate students at a cost of $5930 for room and board for two semesters. Many students share houses or apartments in the Boston area at a monthly cost of $500 to $700 per student. The University housing office can assist in rental arrangements.

Student Group

The Graduate Program in Molecular Medicine accepts 3–6 students each year and currently has a total enrollment of 30 students. Some students matriculate directly from bachelor's degree programs; the program also accepts students seeking a combined M.D./Ph.D. degree. A unique component of the student body are students with M.D.s who desire to pursue rigorous scientific training in preparation for a career in academic medicine and research.

Location

The Boston University School of Medicine (BUSM) and its graduate programs in the Division of Medical Sciences, as well as the Boston University Medical Center–affiliated hospitals, are based in Boston's historic South End, an area approximately 1 mile south of the center of Boston in a vibrantly renewed urban community. With three medical schools and many major universities, Boston is a rich and interactive biomedical community and also a center of the biotechnology industry. The Boston University Goldman School of Dentistry and the School of Public Health are on the same South End campus, while the Boston University undergraduate campus and other graduate programs are near the Charles River in the Kenmore Square area of the city. Boston is a cosmopolitan city with a rich academic and intellectual environment and a panoply of cultural, recreational, and sports activities.

The University and The Department

Boston University is a private institution founded in 1839. It is among the top fifteen institutions in the country in NIH-derived research support per faculty member. Faculty members of the Evans Department of Medicine of the Boston University School of Medicine conduct research programs in basic biomedical sciences, translational medicine, and clinical outcomes and epidemiology.

Applying

For an application, interested students should visit the Division of Graduate Medical Sciences Web site at http://www.bumc.bu.edu. An interview is required.

Correspondence and Information

David C. Seldin, M.D., Ph.D.
Program Director
Hematology-Oncology Section
88 East Newton Street, E-113
Boston, Massachusetts 02118
Phone: 617-414-1519
Fax: 617-638-8728
E-mail: dehall@bu.edu
Web site: http://www.bumc.bu.edu/medicine/

John Murphy, Ph.D.
Associate Program Director and Chief,
Section of Molecular Medicine
88 East Newton Street, E-113
Boston, Massachusetts 02118
Phone: 617-414-1519
Fax: 617-638-8728
E-mail: dehall@bu.edu
Web site: http://www.bumc.bu.edu/medicine/ gpmm

Boston University

SECTIONS AND AREAS OF RESEARCH

Cardiology: Wilson S. Colucci, M.D., Chief. The Cardiology Section and affiliated Cardiovascular Research Institute, directed by Dr. Joseph Loscalzo, pursue basic and clinical research in the areas of atherogenesis and thrombosis, ischemia, mechanisms of heart failure and arrhythmogenesis, and apoptosis in the heart. The Cardiology Section is also involved in epidemiological studies of cardiovascular disease through the Framingham Heart Study.

Clinical Epidemiology Research and Training: David T. Felson, M.D., M.P.H., Chief. Research spans epidemiologic and clinical research with a focus on musculoskeletal diseases, their causes, impacts, and treatments. Specific projects include large-scale epidemiologic studies and clinical trials in osteoarthritis, studies of work disability in those with disease, causes of recurrent gout attacks, and evaluating diagnostic testing and outcomes in rheumatic diseases.

Endocrinology, Diabetes & Nutrition: Shalendar Bhasin, M.D., Chief. Major research interests of the Bone Unit include basic and clinical research on vitamin D metabolism, bone growth, and osteoporosis (Michael Holick, Ph.D., M.D.). The Diabetes and Metabolism Research Unit studies mechanisms by which fuels create signals in cells and how these signals are altered by diabetes, exercise, and insulin resistance (Neil Ruderman, M.D.). The Thyroid Research Unit has major interests in the physiology and pathophysiology of thyroid disease, especially the role of iodine and abnormalities in the thyroid hormone receptor (Lewis E. Braverman, M.D., and Joshua D. Safer, M.D.).

Gastroenterology: M. Michael Wolfe, M.D., Chief. Research interests of the section include basic biology of gut peptides and mucins, gallstone formation, and colon carcinogenesis. Clinical research involves new strategies for the prevention of gastroduodenal mucosal damage associated with the use of nonsteroidal anti-inflammatory drugs, screening and surveillance strategies for colorectal cancer, and novel antiviral therapies for hepatitis C.

General Internal Medicine: Jeffrey Samet, M.D., M.A., M.P.H., and Chief. Research activities focus on care of urban populations, with several content areas emphasized and particular methodological strengths localized in collaborative units. The Clinical Addiction Research and Education (CARE) Unit studies the clinical care and health services of alcohol and drug users, HIV-infected patients, and victims of violence. The Health Care Research Unit conducts health services research using Medicate, Veteran's Administration, and other large databases. The Women's Health Unit addresses primary care and preventive medicine issues among women. The Medical Informatics Systems Unit (MISU) utilizes telephone systems to promote improved health outcomes for a variety of chronic medical conditions.

Geriatrics: Rebecca Silliman, M.D., Ph.D., Chief. Research interests of the section include basic research on the mechanisms through which loss of fat tissue and changes in fat tissue function occur in old age and clinical research related to chronic disease management in old age, including diabetes and breast cancer.

Hematology-Oncology: Jack Ansell, M.D., Acting Chief. Research is focused upon innate immunity; lymphocyte biology and apoptosis; cytokines and hematopoiesis; leukemia, lymphoma, and amyloidosis; and breast cancer growth signaling, estrogen regulation, and genetics. Affiliated research programs and their directors include the Immunobiology Unit, T. Rothstein; the Boston Sickle Cell Center, M. Steinberg; the BUSM Cancer Center, D. Faller; and the Amyloidosis Program, M. Skinner. The section also carries out translational cancer research and clinical trials.

Hypertension: Haralambos Gavras, M.D., Chief. Research into the genetics of hypertension, mechanisms of salt-induced hypertension, and the participation of bradykinin in cardiovascular homeostasis is ongoing, as are projects for pharmacologic testing of drugs in experimental animals and human subjects and molecular studies of α2 adrenergic receptors.

Infectious Diseases: Paul Skolnik, M.D., Chief. Research activities are divided principally into four areas: sexually transmitted diseases (STDs); pathogenic fungi, particularly *Cryptococcus neoformans;* basic cell biology and molecular epidemiology of endotoxin and bacterial infections; and clinical studies in acquired immunodeficiency syndrome (AIDS) and in *Mycobacterium avium complex,* an important bacterial pathogen that complicates this disease.

Molecular Medicine: John R. Murphy, Ph.D., Chief. Includes the following units and programs with the Department of Medicine: the Biomolecular Medicine Unit (John Murphy, Ph.D., Unit Head); the Center for the Molecular Stress Response (Stuart Calderwood, M.D., Ph.D., Director); the Molecular Genetics Unit (Vassilis Zannis, Ph.D., Unit Head); the Obesity Research Center (Barbara Corkey, Ph.D., Director); the Genetics program (Lindsay Farrer, Ph.D., Director) and the Whitaker CVI Digital Imaging Microscopy Center (Michael Kirber, M.D., Director). The section of Molecular Medicine represents a diverse set of basic science interests focused on the application of state-of-the-art technologies toward the understanding of the underlying molecular basis of disease. Accordingly, the research focus within the section ranges from basic genetics to the development of novel biotherapies for the targeted elimination of disease-causing cells. Common to each unit, center, or program is the application of molecular biologic and molecular genetic methods towards the long-term goal of bringing new therapies and approaches to the management of refractory human disease.

Preventive Medicine and Epidemiology: R. Curtis Ellison, M.D., Chief. The section is involved primarily in applied epidemiology in the fields of cardiovascular disease, cancer, and problems of aging, including Alzheimer's disease. One of the major sources of research is the Framingham Study. The section also studies ophthalmologic disease, pulmonary disease, cancer, and osteoarthritis and directs the Framingham Children's Study.

Pulmonary: David Center, M.D., Chief. Major research programs are based around the Specialized Center of Research in Pulmonary Fibrosis, the Program in Lung Development, and the Asthma, Allergy and Immunologic Diseases Cooperative Research Center. In addition, active research efforts focus on endothelial biology, alveolar macrophages, genetics and epidemiology of lung cancer, and cardiovascular diseases and sleep disturbances.

Renal: David J. Salant, M.D., Chief. Basic research efforts include immunopathogenesis of glomerular diseases, hemoglobin substitutes and renal function, apoptosis in renal cell injury, the biology of glomerular cell surface proteins, and regulation of genes controlling renal development and tumorigenesis. Clinical research efforts include ESRD treatment outcomes and quality of life evaluation and multicenter studies of novel immunosuppressives in renal transplantation.

Rheumatology: Robert Simms, M.D., Acting Chief. Research focuses on studies of scleroderma, including fibroblast function and processes of inflammation and repair; population-based epidemiological studies of major rheumatic diseases; inflammation and cellular immune function and animal models of rheumatic disease; rheumatoid synovial cell function and integrin-based regulation; and the basic and clinical research activities of the Amyloid Treatment Center, directed by Dr. Martha Skinner.

Vascular Biology: Richard A. Cohen, M.D., Director. Research interests include the mechanisms by which vascular risk factors, including diabetes, hypertension, and hypercholesterolemia, lead to vascular disease. The function of nitric oxide in the setting of elevated levels of oxidants in diseased arteries is a major focus, with projects directed towards understanding their influence on vasodilator mechanisms, cellular signaling pathways, ion channels and transporters, oxidative modification of protein structure, and atherogenesis.

BROWN UNIVERSITY

Graduate School
Division of Biology and Medicine
Graduate Program in Molecular Biology, Cell Biology, and Biochemistry

Program of Study	The Graduate Program in Molecular Biology, Cell Biology, and Biochemistry (MCB) is an interdisciplinary program, with faculty members drawn from the areas of biochemistry; molecular, cell, and developmental biology; neurobiology; pathology; pharmacology; chemistry; and medicine. Students enter the program with strong undergraduate backgrounds in biology, calculus, physics, and chemistry, including organic chemistry and biochemistry. Students rotate in different laboratories in the first year to sample various projects and experimental approaches. Advanced students participate in one year of teaching as an assistant. Five years are generally required to complete the Ph.D. degree. The research interests of the faculty encompass a broad range of investigations at the molecular and cellular levels, using a variety of prokaryotic and eukaryotic cell types. Areas of current investigation include gene expression and targeting, RNA functions, carcinogenesis, developmental genetics, photosynthesis and bioenergetics, cell differentiation, organelle development, pattern formation, cellular and molecular immunology, receptors and signal transduction, and ultrastructural studies. To supplement research activities, the program provides regular opportunities for outside speakers, campus faculty members, and graduate students to give seminars on their current work.
Research Facilities	Graduate student research is conducted in faculty research laboratories, both on campus and at nearby affiliated hospitals and research buildings. In addition to the basic research equipment and facilities within each laboratory, major shared facilities include a computer graphics imaging and microscopic core facility with high-resolution transmission and scanning electron microscopes and laser-scanning confocal microscopes; a professionally staffed animal-care facility fully equipped for animal maintenance, large-animal surgery, and experimentation; a transgenic mouse core facility; a micro array core facility; typhoon digital imagers for radioactive, fluorescent, and chemiluminescent samples; an automated DNA sequencer; a fluorescence-activated cell sorter; a phosphoimager; a facility for histologic tissue examination, including frozen sections; a greenhouse; X-ray crystallographic instrumentation; a 600-MHz NMR spectroscope; a hybridoma laboratory; a 100-liter-capacity fermenter; and a molecular modeling center. The fourteen-story Sciences Library houses approximately 4,000 current periodicals, 530,000 bound volumes, and study space for 300 students. A campuswide broadband communications network provides high-speed data communications on campus, and a very high-speed connection to the Internet backbone is maintained by the University.
Financial Aid	All Ph.D. students in this program are supported by University fellowships, teaching or research assistantships, or traineeships awarded by the National Institutes of Health (NIH) to the program. Stipends for 2006 were $25,000 for twelve months in addition to full remission of tuition and health insurance costs.
Cost of Study	Tuition in 2005–06 was $32,264 for eight courses. Students who have completed twenty-four graduate courses are charged an enrollment fee of $1752 per semester, but no tuition. All Brown students are assessed a health services fee of $526. For more information, students should see the Financial Aid section.
Living and Housing Costs	Apartments in pleasant residential areas nearby are available for about $700 to $1000 per month; rents are often lower for students who share apartments.
Student Group	Approximately 5,400 undergraduates and 1,500 graduate students are enrolled in the University. Students come from all regions of the United States and from more than fifty countries worldwide. About 52 full-time students are working for the Ph.D. degree in this graduate program.
Student Outcomes	Graduates typically accept postdoctoral research appointments, followed by academic careers in teaching and research, or governmental research positions. Brown University also has close ties to many biotechnology and pharmaceutical firms, where some graduates pursue careers in industry.
Location	Brown University is in a Colonial restoration district at the head of Narragansett Bay, within walking distance of downtown Providence, the capital of Rhode Island. The city offers many cultural activities, including concerts, theater, museums, parks, and art galleries, which complement Brown's seminars, colloquia, and social and cultural events. In addition to the University's athletic facilities, many students enjoy ocean sports and the various recreational opportunities available throughout Rhode Island. Boston and New York are easily accessible by car, bus, or train.
The University and The Division	Assembled in 1764 as the seventh college in America and the third in New England, Brown University began offering graduate courses in 1850. The first Ph.D. was awarded in 1889. In 1903, a Graduate Department was created, and in 1927 the Graduate School was established as a formal organization. Currently, all education and research in medicine and the biological sciences are administered by the Division of Biology and Medicine. Faculty members from all elements of the Division participate in one or more graduate programs that offer research degrees.
Applying	Completed applications are due in early January to receive full consideration for financial aid. Applications received later are also considered, but no application for admission to the fall semester can be considered after August 2. Graduate Record Examinations (GRE) scores are required on the General Test and on a Subject Test in biology, in chemistry, or in biochemistry, cell and molecular biology. The latest examination date that allows results to be considered is December. Students whose primary language is not English must also submit scores on the Test of English as a Foreign Language (TOEFL).
	Brown University does not discriminate on the basis of sex, race, color, religion, age, handicap, status as a veteran, national or ethnic origin, or sexual orientation in the administration of its educational policies, admission policies, scholarship and loan programs, or other school-administered programs.
Correspondence and Information	Director, MCB Graduate Program Box G-J364 Brown University Providence, Rhode Island 02912 Phone: 401-863-1661 Fax: 401-863-1348 E-mail: mcbprogram@brown.edu Web site: http://www.brown.edu/Departments/Molecular_Biology/Grad_Program/

Brown University

THE FACULTY AND THEIR RESEARCH

Walter J. Atwood, Ph.D., Associate Professor. Neurovirology; HIV infection of the central nervous system; cellular receptors for viral pathogens.

Elaine Bearer, M.D., Ph.D., Associate Professor. Molecular mechanisms of actin dynamics in cell motility and embryogenesis.

Alex Brodsky, Ph.D., Assistant Professor. Experimental and genomic identification of functional RNA protein interaction sites in vivo.

Laurent Brossay, Ph.D., Assistant Professor. Pathobiology; T-cell receptors.

David Cane, Ph.D., Professor. Biosynthesis of natural products; enzymology and genetics of biosynthetic transformations; stereochemical aspects of biosynthetic processes.

Qian Chen, Ph.D., Professor. Signal transduction in bone growth and regeneration.

Y. Eugene Chin, Ph.D., M.D., Assistant Professor. Identification and characterization of cell growth/death–related genes activated in response to STAT activation.

Robbert Creton, Ph.D., Assistant Research Professor. Mechanisms regulating brain regionalization during early development.

Albert E. Dahlberg, M.D., Ph.D., Professor. Ribosomes and ribosomal RNA structure and function.

Alison DeLong, Ph.D., Assistant Professor. Protein phosphorylation and signal transduction in *Arabidopsis.*

Justin Fallon, Ph.D., Professor. Regulation of synapse formation and plasticity; muscular dystrophies.

William Fairbrother, Ph.D., Assistant Professor. Bioinformatic identification of splicing enhancers.

Richard Freiman, Ph.D., Assistant Professor. Chromatin and transcriptional control; regulation of tissue-specific gene expression in the mouse reproductive system.

Susan A. Gerbi, Ph.D., Professor. RNA-RNA interactions (rRNA, U3 snRNA, 7SL of SRP); chromosome structure and DNA replication; molecular evolution.

Philip A. Gruppuso, Ph.D., Professor. Regulation of fetal hepatic growth.

Stephen Hajduk, Ph.D., Program Director, Global Infectious Diseases. Molecular and biochemical basis of pathogenesis.

Edward Hawrot, Ph.D., Professor. Structure and function of acetylcholine receptors and of protein neurotoxins.

Stephen Helfand, Ph.D., Professor of Biology. Mechanisms of aging.

Douglas C. Hixson, Ph.D., Associate Professor. Hepatocyte cell adhesion molecules; cellular origins of liver cancer; antigenic alterations during hepatocarcinogenesis.

Gerwald Jogl, Ph.D., Assistant Professor. Protein structure; X-ray crystallography.

Mark Johnson, Ph.D., Assistant Professor. Molecular genetic analysis of plant reproductive development; mechanisms of targeted pollen tube growth in *Arabidopsis.*

Arthur Landy, Ph.D., Professor. Mechanism of site-specific recombination; protein-DNA interactions; structure of higher-order protein-DNA complexes.

Jeffrey Laney, Ph.D., Assistant Professor. Ubiquitin-mediated proteolysis; cell differentiation in yeast.

Charles Lawrence, Ph.D., Professor of Applied Mathematics; Director, Center for Computational Molecular Biology. Bioinformatic identification of regulatory elements in DNA.

Diane Lipscombe, Ph.D., Associate Professor. Regulation of voltage-gated Ca channel activity in the nervous system by alternative splicing.

John Marshall, Ph.D., Associate Professor. Glutamate receptor function and regulation; use of green fluorescent protein in protein targeting.

Andrew McArthur, Ph.D., Assistant Professor of MCB, Brown University; Assistant Scientist, Marine Biological Laboratory at Woods Hole. Molecular evolution, microbial diversity, and global infectious disease.

Michael McKeown, Ph.D., Professor. Behavioral genetics in *Drosophila.*

Kimberly Mowry, Ph.D., Associate Professor. mRNA localization during oogenesis; *Xenopus* development.

Eduardo Nillni, Ph.D., Assistant Professor. Protein processing and intracellular trafficking.

Rebecca Page, Ph.D., Assistant Professor. Understanding the molecular basis of protein function using X-ray crystallography.

Wolfgang Peti, Ph.D., Assistant Professor. Structure, dynamics, and interactions of proteins, using NMR spectroscopy.

David Rand, Ph.D., Professor. Molecular evolution; coevolution in mitochondrial genomes; hybrid zones and speciation.

Alan G. Rosmarin, M.D., Associate Professor. Transcription regulation in myeloid cells; molecular basis of hematopoiesis.

Robert Sabatini, Ph.D., Assistant Scientist, Marine Biological Laboratory at Woods Hole. Biological function of a base modification in kinetoplastid DNA.

Arthur Salomon, Ph.D., Assistant Professor. Cellular signaling networks important for understanding and treating disease.

John Sedivy, Ph.D., Professor. Proto-oncogene signaling (c-myc, c-raf); cell-cycle control; molecular mechanisms of cellular senescence; cell culture genetics; gene targeting.

Tricia Serio, Ph.D., Assistant Professor. Protein-only inheritance; yeast prions.

Surendra Sharma, Ph.D., Associate Professor. Cytokines and viruses in B-cell growth.

Jeffery Singer, Ph.D., Assistant Professor. Mammalian cell cycle.

Robert J. Smith, M.D., Professor of Medicine and Director, Division of Endocrinology and the Hallett Center for Diabetes and Endocrinology. Novel proteins which may serve as regulators of signaling or may define new signaling pathways.

Mitchell Sogin, Ph.D., Professor. Evolution of the protists.

Jack Wands, M.D., Professor. Growth regulatory pathways in the liver.

Gary Wessel, Ph.D., Professor. Cell and molecular biology of fertilization; organelle biogenesis.

Kristi Wharton, Ph.D., Associate Professor. Role of growth factors and signal transduction in cell fate determination and differentiation during development; *Drosophila* developmental genetics.

CASE WESTERN RESERVE UNIVERSITY

Department of Molecular Biology and Microbiology

Program of Study

The Department of Molecular Biology and Microbiology at Case Western Reserve University sponsors a program of study leading to the Ph.D. degree and offering training in molecular biology, microbiology, molecular virology, and cell biology. Faculty members' research encompasses a wide variety of experimental systems and addresses fundamental questions at many levels of resolution, from molecular to multicellular. Areas of particular strength include microbial and viral pathogenesis, cell division and organization, signal transduction and oncogenesis, transcriptional regulation of gene expression, and mechanisms and regulation of RNA processing. Further information on the department's faculty members may be found in the Faculty and Their Research section of this description or at the department's Web site, http://www.cwru.edu/med/microbio/mbio.htm. The Department of Molecular Biology and Microbiology offers an interactive and exciting atmosphere for research and learning. The department maintains strong ties to the Case Molecular Virology Program, with which it shares a number of faculty trainers. Other interdepartmental research opportunities include the Case/University Hospitals Center for AIDS Research and the Case/University Hospitals Ireland Cancer Center.

The department's students are admitted through the integrated Biomedical Sciences Training Program (BSTP), which encompasses many related training programs, thereby providing students with expanded opportunities to explore diverse areas of research. The BSTP is an umbrella program that not only admits students seeking a Ph.D. but also oversees their first year of graduate study. Additional information about the BSTP can be obtained from the Coordinator of the Biomedical Sciences Training Program (telephone: 216-368-3347; Web site: http://www.cwru.edu/med/BSTP/index.html).

In general, first-year students in the program divide their time between laboratory research rotations and one intensive course. The first-year core course provides a broad exposure to modern biological principles and a foundation for advanced courses. It consists of an integrated series of lectures, presented by faculty members from several bioscience departments, that includes material drawn from eukaryotic and prokaryotic molecular biology and genetics, developmental biology, biochemistry, biophysics, virology, immunology, and cell biology. For more advanced students, the department offers courses in microbial physiology and genetics, eukaryotic molecular biology and virology, and special topics courses that cover areas of current interest. Students also present seminars in the department and participate in interdepartmental journal clubs. The major emphasis during the second and subsequent years is on thesis research. Because the department's student-faculty ratio is low, students interact closely with their research advisers as they develop expertise in planning, performing, and interpreting experiments. Students in the molecular biology and microbiology program are encouraged to participate in collaborative projects that involve other University departments. Interactions across departmental boundaries are facilitated by an active NIH-sponsored Cell and Molecular Biology Training Grant that supports predoctoral fellows. This grant also provides travel support for presentation of research results at national meetings.

Students seeking the combined Ph.D./M.D. degree may perform their thesis research in this department under the aegis of the Medical Scientist Training Program.

Research Facilities

Departmental facilities include a state-of-the-art tissue culture suite with BSL2+ containment laboratories, ultracentrifuges, phosphorimagers, and real-time PCR instruments as well as fluorescent microscopy equipment, including a DeltaVision deconvolution microscope and associated computer that have the ability to track, at high resolution, changes in fluorescent reporters within living cells. The department has multiple seminar and conference rooms, with digital projectors and document cameras. Equipment shared with nearby departments includes electron and confocal microscopes, microarray spotters and readers, NMR machines, proteomics facilities, and state-of-the-art, multichannel FACS instruments. All laboratories are equipped with gigabyte Ethernet access. The University has campuswide wireless access.

Financial Aid

Ph.D. students are supported mainly as research assistants; in addition, some may be appointed to the NIH-sponsored Cell and Molecular Biology Training Grant as predoctoral fellows. The current stipend level is $23,000 per year.

Cost of Study

Tuition was approximately $20,000 in 2003–04 and is waived for full-time graduate students.

Living and Housing Costs

Rooms for unmarried students are available on campus for approximately $4600 to $5910 per year. Numerous apartments are available within a 2-mile radius of the campus.

Student Group

Approximately 9,600 students are enrolled in the University; 5,050 are enrolled in the graduate and professional schools. About 25 graduate students are in residence in the department. Another 1,000 students attend adjacent institutes of music and art.

Location

Greater Cleveland is a cosmopolitan community of approximately 2 million people. The University is located in University Circle, a 500-acre campus on which more than thirty educational, scientific, medical, cultural, social service, and religious institutions are located, including the world-famous Cleveland Orchestra and the Cleveland Museum of Art. Both downtown Cleveland and Cleveland Hopkins International Airport can be reached by rapid transit from University Circle. In addition, Cleveland offers a wide range of national sports events and entertainment, including theater. Numerous winter and summer recreational facilities are available in the surrounding area. Residential neighborhoods are adjacent to the campus, and a 450-acre farm 20 minutes from the campus is open to students and faculty members.

The University

In 1967, Western Reserve College and its neighboring institution, Case Institute of Technology, joined to form Case Western Reserve University. Case Western Reserve soon became a major center for research and training. The medical school is located on the campus; as a result, there is significant cooperation in both research and teaching between the basic science and clinical departments of the medical school and the science and engineering departments of the University.

Applying

Prerequisites for entrance into the program include one year of organic chemistry; a course in biochemistry is also required but may be taken after matriculation. Students are admitted for the fall term only. Applications should be submitted in late autumn or early winter for an anticipated entrance the following autumn semester, but late applications are accepted occasionally. Application forms may be obtained from the Biomedical Sciences Training Program, but students are encouraged to apply online. Applicants are required to take the Graduate Record Examinations, including one Subject Test.

Admission to the Ph.D. program is determined by the Biomedical Sciences Training Program Admissions Committee.

Correspondence and Information

Biomedical Sciences Training Program
School of Medicine, WG 46
Case Western Reserve University
10900 Euclid Avenue
Cleveland, Ohio 44106-4934
E-mail: drn2@po.cwru.edu
Web site: http://www.cwru.edu/med/microbio/mbio.htm

Case Western Reserve University

THE FACULTY AND THEIR RESEARCH

For more information on the research interests of individual faculty members, students should visit the Department of Molecular Biology and Microbiology's Web site at http://www.cwru.edu/med/microbio/mbio.htm.

Primary Faculty

Erik Andrulis, Assistant Professor; Ph.D., SUNY at Stony Brook, 1998. RNA surveillance mechanisms; exosome structure and function; regulation of transcription elongation; heat shock gene expression.

Susann Brady-Kalnay, Associate Professor; Ph.D., Cincinnati, 1991. Receptor protein tyrosine phosphatases, cell adhesion, and signal transduction; tyrosine phosphatases and the regulation of neuronal pathfinding; tyrosine phosphatases as tumor suppressors in cancer.

Lloyd A. Culp, Professor; Ph.D., MIT, 1969. Tumor biology; adhesion; metastasis as functions of *ras* or *N-myc* oncogenes; molecular mechanisms of tumor cell matrix adhesion; histochemical marker genes to track tumor cell metastasis.

Piet A. J. de Boer, Associate Professor; Ph.D., Amsterdam, 1993. Bacterial molecular genetics; biochemistry and cell biology; focus on bacterial cytokinesis and related cell-cycle events; detailed study, using a variety of techniques, of a number of genes and proteins involved in the selection of the correct site of cell constriction or in the constriction process itself.

Jonathan Karn, Reinberger Professor and Chairman; Ph.D., Rockefeller, 1976. Virology; transcriptional regulation; regulation of human immunodeficiency virus gene expression by Tat.

David McDonald, Assistant Professor; Ph.D., California, San Francisco, 1993. Mechanisms of trafficking of HIV between and within cells of the immune system and dendritic cells, using high-resolution video microscopy.

Catherine Patterson, Assistant Professor; Ph.D., Princeton, 1999. Virus infection of the CNS; molecular virology; viral immunology.

Patrick Viollier, Assistant Professor; Ph.D., Basel (Switzerland), 1999. Genetics of differentiation and pilus development in bacteria.

Jo Ann Wise, Professor; Ph.D., Yale, 1981. Genetic and genomic strategies to understand the recognition and recruitment of premessenger RNAs to the splicing pathway in the fission yeast *Schizosaccharomyces pombe,* a unicellular eukaryote in which splicing signals and factors closely resemble their counterparts in metazoa.

Secondary Faculty

Eric J. Arts, Associate Professor; Ph.D., McGill, 1994. Retrovirology; HIV-1 evolution/fitness within a patient and in the global epidemic; HIV-1 inhibition and drug resistance.

Amiya Banerjee, Cleveland Clinic Foundation; Ph.D., Calcutta, 1964. Molecular basis of pathogenicity for negative-strand RNA viruses, including vesicular stomatitis virus and human parainfluenza virus type 3, with an emphasis on mechanisms of transcription and replication.

Robert Bonomo, Research Services, Veterans Affairs Medical Center; M.D., Case Western Reserve, 1983. Structure function studies of class A beta-lactamases; understanding enzymological factors that permit the successful evolution of beta-lactamases in the clinic; development of immunological tools to study beta-lactamase expression in enteric bacilli.

Mark G. Caprara, Assistant Professor; Ph.D., Temple, 1993. Use of molecular, biochemical, and genetic techniques to define the structural and thermodynamic basis for sequence- and structure-specific recognition of nucleic acids by proteins; emphasis on catalytic group I intron RNAs and protein cofactors required for splicing and DNA-endonucleases that function as mobile genetic elements.

Cheng-Ming Chiang, Associate Professor; Ph.D., Rochester, 1991. Mechanisms of transcriptional regulation in mammalian cells and in human papillomaviruses; role of general cofactors TAFs, Mediator, and PC4 in gene activation and repression with DNA and chromatin templates; transcription complex assembly on human TATA-less promoters.

Michael W. Cho, Assistant Professor, Department of Medicine, Division of Infectious Diseases; Ph.D., Utah. Proteins and enzymes; structural biology.

Pamela B. Davis, Professor; Ph.D./M.D., Duke, 1974. Character of inflammation in the cystic fibrosis lung; nonviral gene therapy for lung disorders; response of the lung to lack of CFTR.

Koh Fujinaga, Assistant Professor; Ph.D., Hokkaido, 1997. Regulation of HIV transcription at the elongation step; use of HIV as a model to elucidate mechanisms that regulate cellular transcription; current emphasis on a cellular cofactor of HIV transactivator Tat protein, the positive transcription elongation factor b (P-TEFb).

Jonatha M. Gott, Associate Professor; Ph.D., SUNY at Albany, 1987. RNA editing in the acellular slime mold *Physarum polycephalum;* emphasis on elucidation of mechanistic aspects of editing using molecular biological, biochemical, and genetic approaches.

Michael Lederman, Professor; M.D., CUNY, Mount Sinai, 1974. Exploring the mechanisms of immune deficiency; strategies to enhance immune function in HIV-1 infection.

Donal Luse, Department of Molecular Genetics, Cleveland Clinic; Ph.D., Johns Hopkins. Mechanism of transcript initiation and elongation by RNA polymerase II.

Michael Maguire, Professor; Ph.D., Indiana, 1973. Multiple roles of Mg2+ and Mn2+ in bacterial metabolism and pathogenesis using primarily *Salmonella enterica* serovar Typhimurium (S. Typhimurium) as a model system; investigations of transport systems for each cation at several levels, from structure to physiology.

Sanford Markowitz, Professor; M.D./Ph.D., Yale, 1980. Molecular biology of colon cancer; functional influence of oncogenes and suppressor genes on transformation, metastasis, and response to therapies.

Philip E. Pellett, Cleveland Clinic Foundation; Ph.D., Chicago. Fundamental understanding of the relationships among the various herpesviruses, the processes by which human herpesviruses replicate and persist in populations on an evolutionary time scale, and the mechanisms by which these viruses cause disease; use of this knowledge to prevent or treat herpesvirus-associated disease.

Robert H. Silverman, Professor, Department of Cancer Biology, Cleveland Clinic Foundation; Ph.D., Iowa State. Antitumor and antiviral mechanisms of interferons; regulation of RNA stability.

COLORADO STATE UNIVERSITY

Graduate Program in Cell and Molecular Biology

Programs of Study

The Graduate Program in Cell and Molecular Biology (CMB) is an interdisciplinary degree-granting program that involves nearly 70 faculty members from ten departments and four colleges who share common interests in cell and molecular biology. The program offers training leading to the M.S. and Ph.D. degrees in cell and molecular biology. The program includes a core of lecture courses; elective courses in laboratory research techniques, ethical conduct of science, and grant writing; a graduate seminar series in which students present their research; and a weekly seminar series that annually includes presentations by Colorado State University (CSU) faculty members and approximately 20 nationally prominent scientists. Core courses are completed during the first year. The M.S. degree is usually completed in two years and the Ph.D. degree within five years.

Current focus areas of research include, but are not limited to, cancer biology, infectious diseases, metabolism, neuroscience, plant biology, regulation of gene expression, reproductive biology, and structural biology. Additional information about individual faculty members is listed in the Faculty and Their Research section.

Research Facilities

Facilities include an electron microscope center (TEM, STEM, SEM, freeze-fracture, X-ray microanalysis) and other research electron microscopes; a flow cytometry and cell-sorting laboratory; an image analysis laboratory, including a confocal laser scanning microscope; fluorescence in situ hybridization (FISH) equipment; an NMR spectroscopy center; and Keck Foundation X-ray diffraction and protein purification facilities. Macromolecular Resources, a University core facility, conducts DNA and protein sequencing, produces polyclonal antibodies, and houses instrumentation for proteomic, genomic, metabolomic, and bioinformatic analyses.

Financial Aid

Graduate fellowships and teaching assistantships are available for newly admitted students in the CMB program on a competitive basis, and tuition is paid for those students on assistantships. Stipends for teaching assistantships are a minimum of $1500 per month. In subsequent years, students are fully funded as research assistants from the grants awarded to their thesis adviser.

Cost of Study

Tuition is paid for students supported by fellowships and graduate teaching or research assistantships. Tuition in 2005–06 for Colorado residents registered full-time was $1845 per semester; tuition for nonresidents was $7479 per semester. Fees were $535 per semester. Tuition and fees are subject to change.

Living and Housing Costs

University housing for single or married graduate students is available in a wide variety of apartments with rents that start at $600 per month for a two-bedroom furnished apartment. Privately owned unfurnished apartments start at $700 per month.

Student Group

The program, which began in 1996, currently has 65 students enrolled. About 60 percent of the students are women. Most students receive full financial support. Colorado State University has an enrollment of approximately 25,000 students, of which approximately 3,800 are graduate students.

Location

Fort Collins, with a population of approximately 135,000, is located at the foot of the Rocky Mountains. Rocky Mountain National Park is nearby, and several ski areas are easily accessible. A wide variety of outdoor activities are readily available in the parks and national forests in the nearby mountains, including skiing, hiking, camping, fishing, white-water rafting, sailing, and mountain climbing. At an elevation of 5,000 feet, Fort Collins has a moderate climate with low humidity and more than 300 days of sunshine annually. Fort Collins has been rated nationally as one of the best cities in which to live. High-technology industry is increasing, although much of the surrounding area is devoted to agriculture and recreation. The Lincoln Center for the Performing Arts, along with the University Center for the Performing Arts, is the home of the cultural activity in Fort Collins, including a resident symphony, ballet companies, and repertory theater. An interstate highway provides rapid access to Denver, which is an hour away. Airline service is provided through the Denver International Airport.

The University

Colorado State University is a land-grant institution that was established in 1870. The main campus, with many new buildings, covers 642 acres near the center of Fort Collins. The 1,730-acre foothills campus has special research facilities for animals, including the Animal Reproduction and Biotechnology Laboratory. The Molecular and Radiological Biosciences building, completed in 1989, provides excellent laboratory and classroom space. The Veterinary Teaching Hospital is located about 1 mile south of the main campus and houses an internationally recognized canine cancer therapy center. There is a state-of-the-art athletics center for the students on campus.

Applying

Applications received by January 15 receive priority for financial support. Students are required to submit scores on the GRE General Test, and a Subject Test score in an area of biology is recommended. Application materials and additional information concerning the graduate program are available from the address listed in this Close-Up.

Correspondence and Information

Dr. Norman P. Curthoys, Director
Graduate Program in Cell and Molecular Biology
Molecular and Radiological Biosciences Building
Campus Delivery 1618
Colorado State University
Fort Collins, Colorado 80523
Phone: 970-491-0241
Fax: 970-491-0623
E-mail: cmb@colostate.edu
Web site: http://cmb.colostate.edu

Colorado State University

THE FACULTY AND THEIR RESEARCH

Ramesh K. Akkina, Professor (Microbiology, Immunology & Pathology); Ph.D., Minnesota, 1982. Virology; AIDS gene therapy; stem cells; viral vectors; RNA therapeutics.

Kenneth G. D. Allen, Professor (Food Science & Human Nutrition); Ph.D., Montana, 1973. Omega-3 fatty acids; gene expression; nutrition and eicosanoids.

Lorinda K. Anderson, Assistant Professor (Biology); Ph.D., Colorado State, 1993. Meiotic recombination in plants and animals; role of recombination-related proteins in homologous synapsis and the control of crossing over; genome structure and evolution.

Susan M. Bailey, Assistant Professor (Environmental & Radiological Health Sciences); Ph.D., New Mexico, 2000. Potential role of dysfunctional (uncapped) telomeres in tumorigenesis, studied with fluorescence in situ hybridization.

James R. Bamburg, Professor (Biochemistry & Molecular Biology); Ph.D., Wisconsin, 1969. Regulation of the cytoskeleton in neuronal growth and pathfinding; signal transduction pathways regulating actin dynamics; abnormalities in actin behavior in neurodegenerative diseases.

B. George Barisas, Professor (Chemistry); Ph.D., Yale, 1971. Biomedical instrumentation; cellular immunology; molecular endocrinology.

Joel S. Bedford, Professor (Environmental & Radiological Health Sciences); Ph.D., Oxford, 1966. Cellular radiation biology; radiation cytogenetics.

Patricia A. Bedinger, Professor (Biology); Ph.D., California, San Francisco, 1982. Pollen development and function at the molecular and cellular level.

Michael E. Bizeau, Assistant Professor (Food Science & Human Nutrition); Ph.D., Arizona State, 1999. Dietary nutrients and exercise impact on animal's physiology and susceptibility to disease.

Daniel Bush, Professor and Chair (Biology); Ph.D., Berkeley, 1984. Signal transduction pathways that regulate the systemic distribution of organic nutrients in plants.

Jonathan O. Carlson, Professor (Microbiology, Immunology & Pathology); Ph.D., Berkeley, 1974. Molecular virology; molecular biology of mosquitoes and arboviruses.

Chaoping Chen, Assistant Professor (Biochemistry & Molecular Biology); Ph.D. Molecular and cell biology of retrovirus assembly and budding.

Debbie C. Crans, Professor (Chemistry); Ph.D., Harvard, 1985. Biological chemistry; vanadium and transition metal chemistry relating to insulin mimetic effects; vanadium compounds with bone-stimulating activities; enzyme mechanisms; phosphorus metabolism.

Norman P. Curthoys, Professor (Biochemistry & Molecular Biology) and Director, Graduate Program in Cell and Molecular Biology; Ph.D., Berkeley, 1970. Effect of acidosis on renal gene expression; mRNA stability; structure of glutaminase.

William S. Dernell, Associate Professor (Clinical Sciences); D.V.M., Illinois, 1985. Preclinical testing of anticancer and antimicrobial chemotherapy agents.

Steven W. Dow, Associate Professor (Microbiology, Immunology & Pathology and Clinical Sciences); Ph.D., Colorado State, 1992. Innate immunity and the lung; bacterial pathogenesis; cancer immunology; vaccines.

Scott Earley, Assistant Professor (Biomedical Sciences); Ph.D., New Mexico. Regulation of vasomotor activity by ion channel expression.

Gregory L. Florant, Professor (Biology); Ph.D., Stanford, 1978. Mammalian physiology; lipid metabolism and energetics.

Michael H. Fox, Professor (Environmental & Radiological Health Sciences); Ph.D., Kansas State, 1977. Flow cytometry; hyperthermia; cell cycle; apoptosis; mutagenesis.

David Frisbie, Assistant Professor (Clinical Sciences–Orthopedic Research); Ph.D., Colorado State, 1999. In vitro and in vivo approaches to diagnostic and therapeutic musculoskeletal disease, with an emphasis on molecular and surgical techniques.

Deborah N. Garrity, Assistant Professor (Biology); Ph.D., Cornell, 1998. Molecular genetic approaches to heart development in zebrafish.

Mercedes Gonzalez-Juarrero (Microbiology, Immunology & Pathology); Ph.D., Madrid, 1990. Study of macrophages and dendritic cells during chronic inflammatory responses.

David W. Grainger, Professor (Chemistry); Ph.D., Utah, 1987. Cell-surface and bacteria-surface adhesion and signaling mechanisms related to biomaterials and biotechnology applications.

Robert J. Handa, Professor (Biomedical Sciences); Ph.D., UCLA, 1983. Molecular mechanisms of estrogen and androgen receptor action in the developing brain.

William H. Hanneman, Assistant Professor (Environmental & Radiological Health Sciences); Ph.D., Texas A&M, 1995. Developmental neurotoxicology, identification and characterization of developmental genes involved in response to hazardous environmental chemicals.

Charles S. Henry, Assistant Professor (Chemistry); Ph.D., Arkansas, 1998. Bioanalytical chemistry; chemical separations and chemical nature of disease.

Douglas N. Ishii, Professor (Biomedical Sciences & Biochemistry and Molecular Biology); Ph.D., Stanford, 1974. Neurobiology of insulin-like growth factors; diabetic neuropathy; blood-brain barrier and brain disorders.

John Kisiday, Assistant Professor (Clinical Sciences–Orthopedic Research); Ph.D., MIT, 2003. Mechanobiology of cartilage regeneration; cartilage tissue engineering.

Paul Kugrens, Professor (Biology); Ph.D., Berkeley, 1971. Algal ultrastructure, development, and molecular phylogeny.

Nora Lapitan, Professor (Soil & Crop Sciences); Ph.D., Kansas State, 1986. Molecular genetics and genomics of plants; molecular genetic techniques to clone genes for economically important traits in wheat and barley.

Susan M. LaRue, Associate Professor (Environmental & Radiological Health Sciences); D.V.M., Georgia, Ph.D., Colorado, 1992. Experimental therapeutics; hyperthermia; tumor physiology; tumor cytogenetics.

Paul J. Laybourn, Associate Professor (Biochemistry & Molecular Biology); Ph.D., California, Davis, 1989. Mechanism of transcription regeneration in a chromatin context.

Marie M. Legare, Assistant Professor (Environmental & Radiological Health Sciences); Ph.D., Texas A&M, 1995. Genetic and molecular approaches to studying neurotoxicology.

Howard L. Liber, Professor (Environmental & Radiological Health Sciences); Ph.D., MIT, 1980. Molecular mechanisms of mutagenesis in human cells.

Karolin Luger, Associate Professor (Biochemistry & Molecular Biology); Ph.D., Basel (Switzerland), 1989. X-ray crystallography of macromolecular assemblies; crystallographic and biochemical analysis of transcription regulation in a chromatin context.

June I. Medford, Associate Professor (Biology); Ph.D., Yale, 1986. Molecular and genetic studies of *Arabidopsis;* in vivo imaging; plant sentinels.

Donald L. Mykles, Professor (Biology); Ph.D., Berkeley, 1979. Regulation of protein turnover; calcium-dependent and ATP/ubiquitin-dependent proteinases; myofibrillar proteins.

Jennifer K. Nyborg, Professor (Biochemistry & Molecular Biology); Ph.D., California, Riverside, 1986. Mechanism of transcriptional deregulation by the human T-cell leukemia virus (HTLV-I) tax protein.

Michael J. Pagliassotti, Professor (Food Science & Human Nutrition); Ph.D., USC, 1988. Nutrient regulation of hepatic glucose metabolism and gene expression.

Marvin R. Paule, Professor and Chair (Biochemistry & Molecular Biology); Ph.D., California, Davis, 1970. Molecular mechanisms of transcription initiation and control in eukaryotic cells; coordinate expression of ribosomal components.

Leonard D. Pearson, Professor (Microbiology, Immunology & Pathology); Ph.D., California, Davis, 1970. Immunology (immunity to lentiviruses, herpes viruses, pox viruses); monoclonal antibodies.

Olve B. Peersen, Assistant Professor (Biochemistry & Molecular Biology); Ph.D., Yale, 1994. Structural and biophysical studies of viral polymerases and picornaviral replication complexes.

Marinus Pilon, Special Assistant Professor (Biology); Ph.D., Utrecht (Netherlands), 1992. Intracellular protein trafficking; metal cofactor transport and metal cofactor assembly of proteins involved in photosynthesis.

Rajinder S. Ranu, Professor (Bioagricultural Sciences & Pest Management); Ph.D., Pennsylvania, 1971. Eukaryotic protein synthesis; plant molecular biology; plant gene expression.

Anireddy S. N. Reddy, Professor (Biology); Ph.D., Jawaharlal Nehru (India), 1984. Signal transduction mechanisms; regulation of gene expression; crop improvement by genetic engineering.

Deborah A. Roess, Professor (Biomedical Sciences); Ph.D., Saint Louis, 1982. Plasma membrane events associated with membrane signaling and cell activation.

Barbara M. Sanborn, Professor and Head (Biomedical Sciences); Ph.D., Boston University, 1968. Signal transduction; molecular mechanisms of hormone action; calcium homeostasis.

Herbert P. Schweizer, Professor (Microbiology, Immunology & Pathology); Ph.D., Konstanz (Germany), 1983. Molecular genetics and biology of pathogenic bacteria; cell-cell communication; multidrug resistance.

George E. Seidel Jr., Professor (Biomedical Sciences); Ph.D., Cornell, 1970. In vitro oocyte maturation, fertilization, metabolism, microsurgery, and cryopreservation of mammalian embryos; genes regulating embryonic development.

Stephen M. Stack, Professor (Biology); Ph.D., Texas, 1969. Structure and behavior of chromosomes in both the mitotic and meiotic cell cycles; the synaptonemal complex; recombination nodules; genetic crossing over.

Erica L. Suchman, Associate Professor (Microbiology, Immunology & Pathology); Ph.D., California, Irvine, 1994. Effects of densonucleosis viruses on mosquitoes for use as possible biocontrol agents.

Michael M. Tamkun, Professor (Biomedical Sciences); Ph.D., Washington (Seattle), 1983. Ion channel molecular biology.

Douglas H. Thamm, Assistant Professor (Clinical Sciences); V.M.D., Pennsylvania, 1995. Signal transduction and its inhibition in comparative cancer models.

Henry J. Thompson, Professor (Horticulture & Landscape Architecture) and Head, Cancer Prevention Laboratory; Ph.D., Rutgers, 1975. Biochemical and molecular approaches to cancer prevention; preclinical models and clinical investigations.

Ronald L. Tjalkens, Assistant Professor (Environmental & Radiological Health Sciences); Ph.D., Michigan, 2001. Molecular neurotoxicology; regulation of nitric synthase in mammalian astroglia; role of astroglia cells in Parkinsonian syndromes.

Stuart A. Tobet, Associate Professor (Biomedical Sciences); Ph.D., MIT, 1985. Development and differentiation of the neuroendocrine brain.

Robert L. Ullrich, Professor (Environmental & Radiological Health Sciences); Ph.D., Rochester, 1975. Carcinogenesis and genetic modifiers of susceptibility.

David M. Vail, Professor (Clinical Sciences); D.V.M., Saskatchewan, 1984. Investigations of novel antineoplastic modalities using companion animal (pet) dogs and cats with spontaneous malignancies as translational models.

Jorge M. Vivanco, Associate Professor (Horticulture & Landscape Architecture); Ph. D., Pennsylvania, 1999. Biochemical, molecular, and metabolic profiling approaches to root exudation processes; biology and biochemistry of ribosome-inactivating proteins in plants.

Michael M. Weil, Associate Professor (Environmental & Radiological Health Sciences); Ph.D., Texas at Austin, 1987. Genetic susceptibility to radiation-induced cancers.

Jeffrey Wilusz, Professor and Head (Microbiology, Immunology & Pathology); Ph.D., Duke, 1985. Mechanisms of regulated posttranscriptional control in mammalian cells.

Stephen J. Withrow, Professor and Director, Animal Cancer Center (Clinical Sciences); D.V.M., Minnesota, 1972. Comparative pet animal models for cancer, with an emphasis on sarcomas and limb-sparing techniques using a multidisciplinary approach.

Raymond S. H. Yang, Professor (Environmental & Radiological Health Sciences); Ph.D., North Carolina State, 1970. Molecular and cellular aspects of chemical carcinogenesis; integration of mathematical modeling and biomedical experimentation.

Mark Zabel, Assistant Professor (Microbiology, Immunology & Pathology); Ph.D., Utah, 2001. Role of complement in peripheral prion pathogenesis of mouse scrapie; mouse models of chronic wasting disease.

DARTMOUTH COLLEGE

Molecular and Cellular Biology Graduate Program

Program of Study

The Molecular and Cellular Biology (MCB) Graduate Program at Dartmouth seeks to train highly qualified students for productive careers in research and teaching through the completion of a Ph.D. degree. MCB is an umbrella program encompassing four graduate programs—biochemistry, biological sciences, genetics, and microbiology and immunology—which allow breadth of research and education experience. Graduate training in the MCB Program is organized around the research expertise of the faculty members involved, who have interests in the following broadly defined areas: biophysics and structural biology, biotechnology, cell biology, circadian biology, computational biology and genomics, developmental biology, evolutionary biology, genetics, immunology, microbiology, molecular pathogenesis and host-microbe interactions, neurobiology, plant molecular biology, regulation of gene expression, and signal transduction and cellular metabolism. The program provides students with a wide range of research opportunities, related course offerings, and broad exposure to research outside of Dartmouth through journal clubs and a number of seminar series featuring internationally recognized speakers. During their first year, all students attend a three-term course in Biochemistry, Cell Biology, Genetics, and Molecular Biology. Additional required courses, as well as electives, are selected from a comprehensive group of advanced graduate courses. Three 1-term rotations in individual faculty members' laboratories train students in research techniques and allow students to select a thesis adviser from among their rotation sponsors. Students choose the laboratories where they perform three research rotations, selecting a thesis adviser at the end of their third rotation.

Research Facilities

Members of the MCB Program are located in two research facilities: in the Cummings, Gilman, Remsen, and Vail buildings at the west and north ends of the undergraduate campus in Hanover, New Hampshire, and in the Borwell and Rubin buildings at the Dartmouth-Hitchcock Medical Center, approximately 2 miles south in Lebanon, New Hampshire. The proximity to other graduate programs and the undergraduate campus makes the Hanover site an attractive and convenient facility. The Borwell and Rubin buildings on the Lebanon campus—northern New England's largest health-care facility—are part of one of the few new medical centers in the U.S. Laboratories at both facilities contain the standard biochemical, immunological, and molecular instrumentation, and dedicated multiuser facilities contain state-of-the-art equipment for light, confocal, and electron microscopy; nucleic acid and protein sequencing; DNA microarraying and scanning; peptide and oligonucleotide synthesis; phosphorimaging; nuclear magnetic resonance; microinjection; and fluorescence-activated cell sorting. The Dana Biomedical (Hanover) and Matthews Fuller Health Sciences (Lebanon) Libraries house extensive collections of books and journals in the biomedical sciences. Dartmouth is known for its outstanding Computational Center, which develops and licenses cutting-edge software, offers both regular and short courses, and maintains a campuswide network that includes all of the laboratories.

Financial Aid

Each student enrolled in the program receives a Dartmouth Fellowship that provides a full-tuition scholarship, a prepaid health-insurance plan, and a student stipend. The stipend for 2006–07 is approximately $23,500 for incoming students and $24,000 for those who have completed their qualifying exams.

Cost of Study

Tuition for 2006–07 is $44,396.

Living and Housing Costs

Dartmouth assists graduate students in arranging for appropriate housing, either in College facilities or in privately owned accommodations in the Hanover area. Single-student housing (usually off campus) costs $580 to $855 per month plus heat and utilities. College-owned as well as private duplex apartments rent for $620 to $780 per month plus heat and utilities.

Student Group

The student population in the MCB Program currently numbers more than 143 full-time students. The MCB Program is large enough to allow for a variety of educational experiences yet permits close faculty-student interaction, which is a hallmark of a Dartmouth education.

Student Outcomes

Students completing the molecular and cellular biology program most often go on to postdoctoral positions in leading academic laboratories around the country. Eventually, the majority find faculty positions at colleges, universities, or medical schools or direct research in industrial (e.g., biotechnology) labs or take positions in business or law.

Location

Hanover lies in the Connecticut River Valley, which forms the border between Vermont and New Hampshire. Hanover is the quintessential New England college town, situated in the heart of the winter skiing and summer lake resort areas of New Hampshire and Vermont. Life in Hanover offers an attractive combination of cultural activities in a beautiful rural setting, free of the inconveniences of modern urban life. In addition to the usual athletic facilities, Dartmouth's Hopkins Center maintains an active year-round program in the arts, including music, drama, and film. Hiking in the White Mountains of New Hampshire and the Green Mountains of Vermont is excellent and is supported by an extensive system of cabins and trails owned and maintained by Dartmouth. In addition, the Appalachian Trail passes through Hanover. The Dartmouth Skiway is 12 miles north of the campus, and major downhill areas are within an hour's drive in both New Hampshire and Vermont. Boston, Montreal, and New York can be reached by car in 2, 3, and 5 hours, respectively.

The College

Dartmouth College offers great opportunities for high-quality graduate education in one of America's most beautiful environments. Dartmouth was founded as a liberal arts college in 1769 but has had students in advanced-degree programs since Dartmouth Medical School opened in 1797. Currently, more than 1,375 students are enrolled in the graduate programs of Dartmouth Arts and Sciences, Dartmouth Medical School, Thayer School of Engineering, and Tuck School of Business Administration. The undergraduate College enrolls approximately 4,000 students.

Applying

Application forms and more detailed information about the MCB Program may be obtained from the Graduate Admissions Committee or from the program's Web site. Application review begins on January 7, when all supporting documentation, including transcripts, references, GRE scores (General Test required), and the application fee, must be received.

It is the long-standing policy of Dartmouth College to actively support equality of opportunity for all persons regardless of race or ethnic background. No student will be denied admission or be otherwise discriminated against because of race, color, sex, religion, handicap, or national or ethnic origin.

Correspondence and Information

Graduate Admissions Committee
Molecular and Cellular Biology Graduate Program
7560 Remsen Building, Room 239
Dartmouth College
Hanover, New Hampshire 03755-3842

Phone: 603-650-1612
Fax: 603-650-1006
E-mail: molecular.and.cellular.biology@dartmouth.edu
Web site: http://www.dartmouth.edu/~mcb

Dartmouth College

THE FACULTY AND THEIR RESEARCH

Yashi Ahmed, M.D., Ph.D. (Developmental Biology, Genetics). The APC/Beta-Catenin signaling pathway in *Drosophila*.

Victor R. Ambros, Ph.D. (Developmental Biology, Genetics). Genetic control of animal development: Developmental timing pathways and the role of small noncoding RNAs.

Bradley A. Arrick, M.D./Ph.D. (Regulation of Gene Expression). Regulation of growth factor gene expression; role of growth factors in tumor cell biology; cancer genetics.

Charles K. Barlowe, Ph.D. (Cell Biology). Biochemical and genetic analysis of vesicular transport in the early secretory pathway.

Edward M. Berger, Ph.D. (Developmental Biology, Regulation of Gene Expression). Genetics and molecular biology: ecdysterone and juvenile hormone action in *Drosophila;* mechanisms of transcriptional regulation.

Brent Berwin, Ph.D. (Cell Biology, Immunology). Immunogenic mechanisms of molecular chaperones.

Sharon E. Bickel, Ph.D. (Cell Biology). Molecular and genetic analysis of chromosome segregation in *Drosophila*.

Charles Brenner, Ph.D. (Biophysics and Structural Biology, Genetics, Signal Transduction and Cellular Metabolism). Enzyme function and structure in cancer, cell cycle, and aging.

Constance E. Brinckerhoff, Ph.D. (Regulation of Gene Expression). Molecular and cell biology of matrix metalloproteinase gene expression in connective tissue disease and tumor invasion.

David J. Bzik, Ph.D. (Microbiology, Molecular Pathogenesis and Host-Microbe Interactions). Molecular basis of pathogenesis in apicomplexa; drug targets in pyrimidine and purine metabolism.

Ta-Yuan Chang, Ph.D. (Signal Transduction and Cellular Metabolism). Cholesterol sensing, trafficking, and esterification with relation to Alzheimer's disease and the Niemann-Pick type C disease.

Ambrose L. Cheung, M.D. (Microbiology, Molecular Pathogenesis and Host-Microbe Interactions). Regulation of virulence determinants, stress-induced response, and in vivo gene expression in *Staphylococcus aureus*.

Charles N. Cole, Ph.D. (Cell Biology, Genetics). Nucleocytoplasmic transport of messenger RNA and proteins in yeast; coordination of RNA processing and RNA export; micro RNAs in breast cancer.

Michael Cole, Ph.D. (Genetics, Regulation of Gene Expression). Molecular basis of cancer; transcription factors; mechanisms of chromosome-mediated transcriptional control; target genes for oncogenic pathways.

Duane A. Compton, Ph.D. (Cell Biology). Regulation of cell structure during mitosis.

José R. Conejo-Garcia (Immunology). Contribution of leukocytes to tumor vascularization and growth; tumor immunotherapy.

Barbara Conradt, Ph.D. (Developmental Biology, Genetics, Regulation of Gene Expression). Regulation of programmed cell death in *C. elegans*.

Patrick J. Dolph, Ph.D. (Signal Transduction and Cellular Metabolism). Genetic and molecular analysis of phototransduction and retinal degeneration in *Drosophila melanogaster*.

Jay C. Dunlap, Ph.D. (Circadian Biology, Genetics, Regulation of Gene Expression). Molecular genetics of eukaryotic regulatory circuits; molecular genetics and mechanism of the circadian biological clock.

Albert J. Erives, Ph.D. (Developmental Biology, Evolutionary Biology). Gene regulation in the evolution and development of metazoan systems.

Patricia Ernst, Ph.D. (Developmental Biology, Genetics, Immunology, Regulation of Gene Expression). Pathways regulating hematopoietic stem-cell development and differentiation.

Michael W. Fanger, Ph.D. (Immunology, Immunotherapy, Molecular Pathogenesis and Host-Microbe Interactions). Fc receptors; tumor-associated antigens and targeted vaccines.

Steven N. Fiering, Ph.D. (Genetics, Regulation of Gene Expression). Chromatin-based regulation of the mammalian genome; transcriptional regulation of the β-globin locus; transgenic mice.

Scott A. Gerber, Ph.D. (Genetics). Quantitative mass spectrometry and proteomics.

Tillman U. Gerngross, Ph.D. (Biotechnology). Engineering of posttranslational processing pathways in yeast, protein production technologies, and protein engineering.

Amy S. Gladfelter, Ph.D. (Cell Biology). Cell biology and comparative genomics in fungi; the evolution of cell cycle.

James D. Gorham, M.D., Ph.D. (Immunology). Immune effects of TGF-β; Th1/Th2 development; liver immunology.

William R. Green, Ph.D. (Immunology, Microbiology, Molecular Pathogenesis and Host-Microbe Interactions). Retroviral pathogenesis and escape from immune responses; cell-mediated immunity to retrovirus-induced leukemia or immunodeficiency; vaccine development, including for smallpox.

Robert H. Gross, Ph.D. (Computational Biology and Genomics, Genetics). Computational molecular biology.

Mary Lou Guerinot, Ph.D. (Genetics, Microbiology, Molecular Pathogenesis and Host-Microbe Interactions, Plant Molecular Biology). Genetics and molecular biology: genetic regulation of the nitrogen-fixing symbiosis between rhizobia and legumes; metal regulation of gene expression in plants and bacteria.

Paul M. Guyre, Ph.D. (Immunology, Microbiology, Molecular Pathogenesis and Host-Microbe Interactions). Mechanisms of steroid hormone and cytokine regulation of immunity, inflammation, and sepsis; functions of scavenger, toll-like, and Fc receptors in human monocytes, macrophages, and dendritic cells.

Leslie P. Henderson, Ph.D. (Neurobiology). Regulation of synaptic transmission in the vertebrate nervous system by anabolic steroids.

Henry N. Higgs, Ph.D. (Cell Biology). Actin and membrane dynamics in lymphocytes.

Deborah Hogan, Ph.D. (Microbiology, Molecular Pathogenesis and Host-Microbe Interactions). Microbial pathogenesis and microbial interactions.

Mark A. Israel, M.D. (Genetics, Signal Transduction and Cellular Metabolism). Regulation of cellular differentiation and proliferation during development and tumorigenesis.

Thomas Jack, Ph.D. (Developmental Biology, Genetics, Plant Molecular Biology). Molecular genetics of flower development in *Arabidopsis*.

Lloyd H. Kasper, M.D. (Immunology, Microbiology, Molecular Pathogenesis and Host-Microbe Interactions). Autoimmunity; experimental inflammatory bowel disease and multiple sclerosis.

F. Jon Kull, Ph.D. (Biophysics and Structural Biology). X-ray crystallography and physical biochemistry of motor proteins and their cytoskeletal partners.

Eric J. Lambie, Ph.D. (Developmental Biology, Genetics). Developmental genetics of gonadal development in *Caenorhabditis elegans*.

Gustav E. Lienhard, Ph.D. (Signal Transduction and Cellular Metabolism). Signal transduction and membrane trafficking in insulin action.

Jennifer J. Loros, Ph.D. (Circadian Biology, Developmental Biology, Genetics). Fungal genetics and the molecular analysis of circadian clocks.

Lee R. Lynd, D.E. (Biotechnology). Biochemical engineering, microbiology, and molecular biology applied to conversion of plant biomass into fuels and chemicals; fundamentals of microbial cellulose utilization; sustainable resource utilization.

Dean R. Madden, Ph.D. (Biophysics and Structural Biology). X-ray crystallographic and biophysical analysis of ion-channel activation, desensitization, and intracellular trafficking.

Robert A. Maue, Ph.D. (Neurobiology). Cellular and molecular biology of ion channels during neuronal development and neurodegenerative disease.

C. Robertson McClung, Ph.D. (Circadian Biology, Developmental Biology, Evolutionary Biology, Genetics, Plant Molecular Biology). Genetics and molecular biology of circadian rhythms in *Arabidopsis*.

Mark A. McPeek, Ph.D. (Developmental Biology, Evolutionary Biology). Molecular genetics of speciation and adaptation.

Jason H. Moore, Ph.D. (Computational Biology, Genetics). Computational genetics; bioinformatics; systems biology; genetics of common human diseases.

Lawrence C. Myers, Ph.D. (Regulation of Gene Expression). Molecular mechanisms of transcriptional regulation in yeast.

Randolph J. Noelle, Ph.D. (Immunology, Microbiology, Molecular Pathogenesis and Host-Microbe Interactions). β-lymphocyte activation; T-helper cell action; CD40 signaling; β-cell memory and plasma development autoimmunity.

George A. O'Toole, Ph.D. (Microbiology, Molecular Pathogenesis and Host-Microbe Interactions). Molecular genetic basis of biofilm formation and antibiotic resistance; signaling in biofilms; role of biofilms in bacterial pathogenesis and biological control.

Kevin J. Peterson, Ph.D. (Developmental Biology, Evolutionary Biology). Origin and early evolution of animals.

Claudio Pikielny, Ph.D. (Cell Biology, Genetics). Cellular and molecular basis of pheromone response in *Drosophila*.

William F. C. Rigby, M.D. (Regulation of Gene Expression). Regulation of CD40 ligand (CD154) and TNF mRNA turnover; novel targets of Von Hippel-Lindau tumor-suppressor gene.

R. Mako Saito, Ph.D. (Developmental Biology, Genetics). Developmental control of cell cycle in *C. elegans*.

George E. Schaller, Ph.D. (Cell Biology, Plant Molecular Biology). Signaling by the plant hormones ethylene and cytokinin.

Charles L. Sentman, Ph.D. (Immunology). Function and regulation of NK cells; role of NK cells during responses to infection and tumors; cancer immunotherapy.

Roger D. Sloboda, Ph.D. (Cell Biology). Microtubule-dependent particle motility using intraflagellar transport in *Chlamydomonas* as a model system.

Elizabeth F. Smith, Ph.D. (Cell Biology). Regulation of eukaryotic flagellar assembly and motility using the biflagellate green alga *Chlamydomonas reinhardtii* as a model system.

Nancy A. Speck, Ph.D. (Genetics, Regulation of Gene Expression). Transcriptional regulation of hematopoiesis and leukemia.

Paula Sundstrom, Ph.D. (Microbiology, Molecular Pathogenesis and Host-Microbe Interactions). Virulence gene regulation and microbe-host interactions in *candidiasis*.

Surachai Supattapone, M.D./Ph.D./D.Phil. (Microbiology, Molecular Pathogenesis and Host-Microbe Interactions). Molecular pathogenesis of prion diseases.

Ronald K. Taylor, Ph.D. (Microbiology, Molecular Pathogenesis and Host-Microbe Interactions, Genetics). Bacterial colonization mechanisms; virulence gene regulation; vaccine and antimicrobial design.

Sergei G. Tevosian, Ph.D. (Developmental Biology, Genetics). Developmental regulators for heart and gonadal development.

Mary Jo Turk, Ph.D. (Immunology). Tumor immunology.

Edward J. Usherwood, Ph.D. (Immunology). Interaction between murine gammaherpesvirus 68 and the immune system.

Michael L. Whitfield, Ph.D. (Computational Biology and Genomics, Genetics). Genome-wide approaches to the study of biological systems.

William T. Wickner, M.D. (Cell Biology, Genetics). Mechanisms of membrane fusion in intracellular traffic.

Lee A. Witters, M.D. (Signal Transduction and Cellular Metabolism). Cellular regulation during metabolic stress and in response to hormonal stimuli.

DUKE UNIVERSITY

Program in Cell and Molecular Biology

Program of Study

The goal of the University Program in Cell and Molecular Biology is to prepare promising young scientists for careers in research and teaching in various areas of cell, molecular, and developmental biology. In recognition of the interdisciplinary nature of these rapidly expanding areas, the program involves the participation of faculty members from the Departments of Biochemistry, Biology, Cell Biology, Immunology, Molecular Genetics and Microbiology, Neurobiology, Pathology, and Pharmacology and Cancer Biology and offers a special combined Ph.D. degree course of study in cell and molecular biology and one of the aforementioned fields. During the first year of study, each student admitted to the Program in Cell and Molecular Biology is expected to do a research rotation among three representative laboratories of program faculty members to gain experience in a broad range of research disciplines. The student also takes courses presenting the fundamental knowledge and recent research developments in various aspects of cell and molecular biology, including the areas of nucleic acids, proteins, membranes, and molecular genetics. In addition, students participate in the program seminar series. At the end of the first year, the student selects a dissertation adviser, a doctoral committee is appointed, and an affiliation with one of the participating departments is established. In all cases, the student is encouraged to plan a training program that takes advantage of the breadth of expertise of the program faculty. Completion of the doctoral program requires approximately 5½ years, of which the last 3½ are devoted to research.

Research Facilities

The laboratories of the 139 faculty members in the eight departments participating in the program constitute an enormous resource, with equipment and expertise in virtually every area of research in cell and molecular biology. The special research emphases of the different departments are described on the reverse of this page.

Financial Aid

A Cell and Molecular Biology Training Grant provides National Research Service Award fellowships to entering students. Applicants to the program are considered for one of these fellowships, which are awarded on the basis of merit. These fellowships provide full tuition and fees and a stipend of $24,000 per year for a period of twenty-two months. After this time, the student continues to receive full support from the dissertation adviser's research funds or from departmental funds. In addition to the training grant fellowships, outstanding students may be nominated for James B. Duke Fellowships, for which there is University-wide competition.

Cost of Study

Tuition and fees are $37,300 for the 2006–07 academic year. This includes tuition, registration fees, a health fee, a transcript fee, a student activities fee, and health insurance.

Living and Housing Costs

A wide variety of housing arrangements are available in Durham and nearby Chapel Hill. Housing is relatively inexpensive and easily accessible compared to major metropolitan areas.

Student Group

Duke University has a total enrollment of approximately 12,000 full-time students. Of this number, more than 5,200 are engaged in graduate or professional studies. There are more than 100 students in the Program in Cell and Molecular Biology, and interaction within this group is actively encouraged.

Student Outcomes

More than 90 percent of Duke CMB students who have completed their Ph.D. in the last ten years remain active in science, and most of the others are active in science-related professions. Forty-two percent are employed by academic or research institutions (including industry), and 50 percent are in postdoctoral training.

Location

The University is located in Durham, a city of 200,000 in a metropolitan area of 800,000. The University of North Carolina at Chapel Hill and the Research Triangle Park are both 12 miles away. A variety of cultural and entertainment activities are provided by touring music, theater, and dance groups of international reputation. In the summer, Duke is the home of the American Dance Festival. The pleasant climate and rural surroundings invite outdoor activity. The ocean and the Appalachian Mountains are 2 to 4 hours to the east and west, respectively, and Washington, D.C., is about 5 hours away by car.

The University

Duke, a private university, dates as a corporate entity from 1924; however, its roots go back more than a century and a half to the Union Institute, founded in 1838. Duke's Graduate School has a faculty of approximately 1,000 members. The campus is located on the edge of Duke Forest, 8,500 acres of rolling, wooded land on the southwestern fringe of Durham. The proximity of all the biological sciences, including the preclinical departments of the medical center, facilitates interdepartmental cooperation and results in a wide range of courses, research opportunities, and seminars. Duke has a reciprocal arrangement with the University of North Carolina at Chapel Hill and North Carolina State University at Raleigh that allows enrollment in courses at these nearby institutions.

Applying

Students apply directly to the Program in Cell and Molecular Biology for admission with the understanding that they establish a departmental affiliation before the beginning of their second year of graduate study. Students have the option of indicating their interest in a specific department or they can reserve that choice until a doctoral adviser is chosen. Applications for admission and fellowship support must be received by December 31, but early applications are welcome. Information about the Program in Cell and Molecular Biology and the various participating departments may be obtained by writing to the director of graduate studies of the program.

Correspondence and Information

Program in Cell and Molecular Biology
Box 3553
Duke University Medical Center
Durham, North Carolina 27710-3553
Phone: 919-684-6559
E-mail: cmbtgp@biochem.duke.edu
Web site: http://cmb.duke.edu

Duke University

THE RESEARCH INTERESTS OF THE FACULTY

Biochemistry
The Department of Biochemistry has 20 faculty members who are members of the Program in Cell and Molecular Biology. Their research interests include the structure-function relationships and biosynthesis of proteins, glycoproteins, nucleic acids, and lipids; enzyme mechanisms; the biogenesis of cell membranes and mitochondria; the regulation of metabolism; the mechanism of hormone action; and the control of gene expression. The graduate program offered to cell and molecular biology trainees by the Department of Biochemistry includes an opportunity for in-depth study in enzymology, immunochemistry, cell differentiation, genetic mechanisms, and X-ray and NMR analysis of molecular structure and in advanced spectroscopic techniques.

Biology
The Department of Biology has 12 faculty members who are members of the Program in Cell and Molecular Biology. Their research interests are mitosis and chromosome motility, mechanisms of cell recognition during embryonic development, cellular mechanisms of pattern formation, translational control of gene expression, fertilization, regulation of the cell cycle, and cloning of developmentally important genes. In plants, major areas of research include studies of plant oxidative processes, characterization of the plant mitochondrial electron transfer chain, biochemistry of photosynthesis and photosynthetic adaptation, targeting of proteins in plant cells, molecular aspects of host-pathogen interactions, and molecular genetics of *Arabidopsis*. The department offers graduate-level courses in advanced cell and developmental biology, genetics, and plant biochemistry.

Cell Biology
The Department of Cell Biology includes the Division of Physiology and has 22 faculty members who belong to the Program in Cell and Molecular Biology. A broad spectrum of research typifies the multidisciplinary characteristic of its graduate program. Major areas of research are muscle biology, cell motility and cytoskeleton, biophysics and physiology of membranes, protein trafficking, high-resolution electron microscopy, membrane receptors and signal transduction, cell-cell interactions, photoreception, metabolic regulation, cell physiology, and marine biology. In addition to applying modern techniques in cell culture, biochemistry, and molecular biology, the department specializes in high-resolution electron microscopy, computer image processing, and advanced video-enhanced light microscopy.

Immunology
The Department of Immunology has 8 faculty members who are members of the Program in Cell and Molecular Biology. Modern research in immunology draws on recent advances in cell and molecular biology, protein chemistry, and virology to determine how the components of the immune system function. Special areas of research interest of the department include immunogenetics, neuroimmunology, tumor immunology, and a wide variety of studies in molecular and cellular immunology.

Molecular Genetics and Microbiology
The Department of Molecular Genetics and Microbiology has 23 faculty members who belong to the Program in Cell and Molecular Biology. Their research highlights the use of genetics and microbial systems to approach diverse issues, including intracellular organization, signal transduction, mechanisms of disease, RNA biology, developmental biology, molecular virology and viral oncology, and pathogenesis of viruses, bacteria, and fungi.

Neurobiology
The Department of Neurobiology has 15 faculty members who belong to the Program in Cell and Molecular Biology. The research interests include a large proportion of sensory neurobiology including vision, gustation, audition, and olfaction. Other interests include neural pathways, neural networks, behavioral neurobiology, neural plasticity, and learning.

Pathology
The Department of Pathology has 15 faculty members who are members of the Program in Cell and Molecular Biology. Their research interests are cardiovascular pathology, neuropathology, immunopathology, cell biology/cytology, biochemical/toxicological/clinical pathology, pulmonary pathology, and tumor biology.

Pharmacology and Cancer Biology
The Department of Pharmacology and Cancer Biology has 21 faculty members who belong to the Program in Cell and Molecular Biology. Their research interests include the mechanism of action of neuropeptides and neurotransmitters; ontogeny of signaling pathways in nervous and cardiovascular tissue; cellular signaling mechanisms; receptor function and cell signaling mechanisms regulating cell growth, proliferation, and death; and the molecular basis of rational drug design.

Nicholas School of the Environment (NSOE)
In addition to the departments, 1 member of the Cell and Molecular Biology Program is also in the NSOE. Research interests include aspects of environmental toxicity on cells, particularly free-radical interactions with respiratory proteins and responses of cells to toxic concentrations of transition metals.

DUKE UNIVERSITY

University Program in Structural Biology and Biophysics

Program of Study

The University Program in Structural Biology and Biophysics (formerly known as Molecular Biophysics) is an interdisciplinary program offering a Certificate in Structural Biology and Biophysics. The Ph.D. is awarded by one of the five participating departments (Biochemistry, Cell Biology, Chemistry, Neurobiology, and Pharmacology and Cancer Biology) from which program faculty members are drawn. Students may apply directly to the program; graduate training is administered jointly by the program and the department in which the student pursues thesis research. Students admitted to the University Program in Structural Biology and Biophysics are free to pursue their Ph.D. research in any program laboratory and will receive their degree from an associated department. The first year of study is devoted to course work and rotations in three laboratories. At the end of the first year, the student chooses a thesis adviser and topic and a department from which to receive the Ph.D. degree. The preliminary examination is taken during the second year.

The program centers on those research endeavors that use physical measurements to study biological macromolecules and their interactions where the details of molecular structure are critical to understanding the biological problem in question. Research problems that fit the paradigm include three-dimensional structure determination by crystallography and NMR; many problems approached by various diffraction and microscopy techniques; molecular modeling and design studies that are tied to direct experimental tests; many spectroscopic studies where construction of a molecular model is vital to planning further experiments; and functional studies in such areas as biochemistry, genetic mechanisms, drug interactions, and membrane systems for which the details of molecular geometry are central to interpreting the experiments.

Research Facilities

Students have access to the research facilities of all the faculty members participating in the program as well as to various research resources. These include magnetic resonance spectroscopy, X-ray crystallography, peptide synthesis, microsequencing, macromolecular graphics, electron microscopy, and computational chemistry. Duke University libraries hold 4 million volumes and subscribe to more than 2,500 periodicals in the biological and medical areas. Resources at the North Carolina Supercomputing Center, the University of North Carolina at Chapel Hill, and North Carolina State University are also available.

Financial Aid

Most students receive a stipend plus tuition and fees. The stipend for 2006–07 is $24,000. These awards permit full-time graduate study for twelve months per year and require satisfactory progress toward completion of the Ph.D. degree for renewal.

Cost of Study

Tuition fees and health insurance for 2006–07 total $37,300. Students are charged a flat tuition rate for six semesters, after which, only registration and health fees are charged. Normally, financial aid pays these as well.

Living and Housing Costs

A wide variety of housing arrangements are available in Durham near the Duke campus. Housing is relatively inexpensive and affordable compared to larger metropolitan areas.

Student Group

There are approximately 5,200 graduate students enrolled at Duke University. The University Program in Structural Biology and Biophysics currently enrolls 25 students and supports 8 students. In addition, program faculty members currently supervise some 40 additional.

Location

Durham is a city of approximately 200,000 people located in a metropolitan area of more than 1.2 million, midway between the Atlantic Ocean and the Appalachian Mountains. The climate is moderate, and the cultural life of the community is well developed. The University of North Carolina at Chapel Hill and North Carolina State University at Raleigh, the state capital, are nearby. Twelve miles away is the well-known Research Triangle Park of North Carolina, where a number of firms have large research establishments. The National Center for Health Statistics and the National Institute of Environmental Health Sciences are also located there. Many of the scientists employed by these organizations live in Durham and add to the scientific and cultural climate generated by the close relations among the universities. The weather allows outdoor recreation all year round.

The University

Duke is a private university that dates as a corporate entity from 1924; however, its roots go back to the Union Institute founded in 1838. The graduate school has a faculty of more than 800. The campus is located on the edge of the Duke Forest, 8,500 acres of rolling, wooded land on the southwestern fringe of Durham. The proximity of all departments in the biological and basic medical sciences facilitates interdepartmental cooperation and results in a wide range of courses, research opportunities, and seminars. Duke has reciprocal arrangements with the University of North Carolina at Chapel Hill and North Carolina State University, which enable students to enroll in courses there.

Applying

Students may apply directly to the University Program in Structural Biology and Biophysics with the understanding that they will establish a departmental affiliation at the end of their first year. Students may also apply to any of the participating departments and indicate their interest in fulfilling the requirements for the Certificate in Structural Biology and Biophysics. Students should apply using the online electronic application available on the graduate school's Web site (http://www.gradschool.duke.edu). All applications must be submitted by December 31.

Correspondence and Information

Professor David Richardson
Director of Graduate Studies
University Program in Structural Biology and Biophysics
Box 3567 DUMC
Duke University
Durham, North Carolina 27710
Phone: 919-681-8825
E-mail: sbb@biochem.duke.edu
Web site: http://sbb.duke.edu

Duke University

THE FACULTY AND THEIR RESEARCH

Lorena S. Beese, Professor (Biochemistry); Ph.D., Brandeis, 1984. Structure and mechanism of proteins and macromolecular assemblies central to DNA replication, DNA repair, and cellular signaling.
Error-prone replication of oxidatively damaged DNA by a high-fidelity DNA polymerase. *Nature* 431(7005):217–21, 2004. With Hsu, Ober, and Carell.

David N. Beratan, Professor (Chemistry); Ph.D., Caltech, 1986. Exploring the molecular mechanisms that enable the function of complex biological machines and molecular materials.
Protein dynamics and electron transfer: Electronic decoherence and non-Condon effects. *Proc. Natl. Acad. Sci. U.S.A.* 102:3552–7, 2005. With Skourtis, Balabin, and Kawatsu.

Glenn Edwards, Professor (Chemistry) and Director of Duke Free Electron Laser Lab; Ph.D., Maryland, 1984.
Oxidation potentials of human eumelanosomes and pheomelanosomes. *Photochem. Photobiol.* 81:145–8, 2005. With Samokhvalov et al.

Harold P. Erickson, Professor (Cell Biology); Ph.D., Johns Hopkins, 1968. Cytoskeleton and extracellular matrix; cell motility and adhesion; bacterial cell division.
Mutants of FtsZ targeting the protofilament interface: Effects on cell division and GTPase activity. *J. Bacteriol.* 187:2727–36, 2005. With Redick, Stricker, and Briscoe.

Katherine Franz, Assistant Professor (Chemistry); Ph.D., MIT, 2000. Bioinorganic chemistry of posttranslational protein modifications and the role of metal ions in neurochemistry.
Fe(III) coordination properties of neuromelanin components: 5,6-dihydroxyindole and 5,6-dihydroxyindole-2-carboxylic acid. *Inorg. Chem.* 45(9):3657–64, 2006. With Charkoudian.

Gordon G. Hammes, Professor (Biochemistry); Ph.D., Wisconsin, 1959. Single-molecule fluorescence studies of enzyme mechanisms and protein dynamics.
Interaction of dihydrofolate reductase with methotrexate: Ensemble and single-molecule kinetics. *Proc. Natl. Acad. Sci.* 99:13481, 2002. With Ravi et al.

Homme Hellinga, Professor (Biochemistry); Ph.D., Cambridge, 1986. Combined theoretical and experimental approaches to protein and drug design; molecular simulation; protein engineering.
Computational design of a biologically active enzyme. *Science* 304(5679):1967–71, 2004. With Dwyer and Looger.

Tao-shih Hsieh, Professor (Biochemistry); Ph.D., Berkeley, 1977. Chromosome structure and function; structure, function, and mechanism of DNA topoisomerase.
D. melanogaster topoisomerase IIIα preferentially relaxes a positively or negatively supercoiled bubble substrate and is essential during development. *J. Biol. Chem.* 280:3564–73, 2005. With Plank, Chu, Pohlhaus, and Wilson-Sali.

Piotr Marszalek, Associate Professor (Mechanical Engineering and Materials Science); Ph.D., Electrotechnical (Warsaw), 1991. Nanomechanics of single molecules and mesoscopic-scale systems.
Reverse engineering of the giant muscle protein titin. *Nature* 418:998–1002, 2002. With Li et al.

Thomas J. McIntosh, Professor (Cell Biology); Ph.D., Carnegie-Mellon, 1973. Membrane structure.
Roles of bilayer material properties in function and distribution of membrane proteins. *Ann. Rev. Biophys. Biomolecular Struct.* 35:177–98, 2006. With Simon.

Terrence G. Oas, Associate Professor (Biochemistry); Ph.D., Oregon, 1986. Protein folding; structure-function relationships of proteins; multidimensional NMR structure determination of macromolecules.
Phosphorylation of RNA polymerase II CTD fragments results in tight binding to the WW domain from the yeast prolyl isomerase Ess1. *Biochemistry* 40:8479–86, 2001. With Myers, Morris, and Greenleaf.

Richard A. Palmer, Professor (Chemistry); Ph.D., Illinois at Urbana-Champaign, 1965. Use of FTIR spectroscopy in the investigation of dynamic processes on the nanosecond-to-millisecond time scale.
Dynamic mechanical analysis and dynamic infrared linear dichroism study of the frequency–dependent viscoelastic behavior of a poly(ester urethane). *Vibrational Spectrosc.*, 2006, doi:10.1016/j.vibspec.2006.04.017. With Wang, Schoonover, and Aubuchon.

David C. Richardson, Professor (Biochemistry); Ph.D., MIT, 1967. Crystallography; analysis and design of protein structure.
The backrub motion: How protein backbone shrugs when a sidechain dances. *Structure* 14:265–74, 2006. With Davis, Arendall III, and Richardson.

Jane S. Richardson, Professor (Biochemistry); M.A., Harvard, 1966. Analysis and design of protein structure.
A test of enhancing model accuracy in high-throughput crystallography. *J. Struct. Funct. Genomics* 6:1–11, 2005. With Arendall III et al.

Johannes J. Rudolph, Assistant Professor (Biochemistry); Ph.D., MIT, 1993. Enzymology of dual-specificity phosphatases and kinases involved in cell-cycle control; protein-protein interactions: surfaces and docking.
Structural mechanism of oxidative regulation of the phosphatase Cdc25B via an intramolecular disulfide bond. *Biochemistry* 44:5307–16, 2005. With Buhrman, Parker, Sohn, and Mattos.

Barbara Ramsay Shaw, Professor (Chemistry); Ph.D., Washington (Seattle), 1973. DNA structure and reactivity.
Convenient synthesis of nucleoside borane diphosphate analogues: Deoxy- and ribonucleoside 5'-P –boranodiphosphates. *J. Org. Chem.* 69(2):7051–7, 2004. With Li.

John Simon, Professor (Chemistry); Ph.D., Harvard. The study of structure and functions of melanins.
Comparison of the structural, chemical, and spectroscopic properties of human black and red hair melanosomes. *Photochem. Photobiol.* 81:135–44, 2005. With Liu et al.

Sidney A. Simon, Professor (Neurobiology); Ph.D., Northwestern, 1973. Mechanisms of sensory transduction.
Orbitofrontal ensemble activity predicts licking and distinguishes among rewards. *J. Neurophysiol.* 95:119–33, 2006. With Gutiérrez, Carmena, and Nicolelis.

Leonard D. Spicer, Professor (Biochemistry); Ph.D., Yale, 1964. NMR studies of molecular structure and function.
Multidimensional NMR spectroscopy for protein characterization and assignment inside cells. *J. Am. Chem. Soc.* 127:10848–9, 2005. With Reardon.

Eric J. Toone, Professor (Chemistry); Ph.D., Toronto, 1988. Structure and energetics of protein-carbohydrate complexes.
A small molecule inhibitor of isoprenylcysteine carboxylmethyl transferase with antitumor activity in cancer cells. *Proc. Natl. Acad. Sci. U.S.A.*, 102(12):4336–41, 2005. With Winter et al.

Antonius VanDongen, Associate Professor (Pharmacology and Cancer Biology); Ph.D., Leiden (Netherlands), 1988. Structure/function relationships of ion channels.
K-channel gating by an affinity-switching selectivity filter. *Proc. Natl. Acad. Sci. U.S.A.* 101:3248–52, 2004.

Weitao Yang, Professor (Chemistry); Ph.D., North Carolina at Chapel Hill, 1986. Quantum mechanical simulations of biological systems and nanostructures.
Contact atomic structure and electron transport through molecules. *J. Chem. Phys.* 122:14502/1–7, 2005. With Ke and Baranger.

John D. York, Associate Professor (Pharmacology and Cancer Biology); Ph.D., Washington (St. Louis), 1993. Inositol lipids and cell signaling.
Regulation of nuclear processes by inositol polyphosphates. *Biochim. Biophys. Acta* 1761(5–6);552–9, 2006.

Stefan Zauscher, Assistant Professor (Mechanical Engineering and Materials Science); Ph.D., Wisconsin–Madison, 1991. Atomic force microscopy, colloidal probe microscopy, and single-molecule force.
The fabrication of stimulus-responsive nanopatterned polymer brushes by scanning probe lithography. *Nano Lett.* 4:373–6, 2004. With Kaholek et al.

Pei Zhou, Assistant Professor (Biochemistry); Ph.D., Harvard, 1998. Regulation of cellular processes by protein-protein interaction.
Solution structure of the Set2-Rpb1 interacting domain of human Set2 and its interaction with the hyperphosphorylated C-terminal domain of Rpb1. *Proc Natl Acad Sci U.S.A.* 102:17636–41, 2005. With Li et al.

FLORIDA INSTITUTE OF TECHNOLOGY

College of Science and Liberal Arts
Department of Biological Sciences
Graduate Program in Cell and Molecular Biology

Program of Study

The Department of Biological Sciences offers a program of graduate study leading to the Doctor of Philosophy degree in cell and molecular biology. The principal areas of study and research are molecular and cellular biology, biochemistry, genetics, neurobiology, endocrinology, reproductive biology, developmental biology, immunology, and microbiology. These programs emphasize the preparation of scientists for research careers in academic or industrial settings.

Each student's program is designed independently and is based on his or her background and needs. The diverse interests of the participating faculty members and the excellent student–faculty member ratio ensure a broad spectrum of research opportunities and a great deal of personal attention.

Comprehensive exams are usually taken in the second year, following the completion of formal course work. Students are encouraged to choose a faculty sponsor and begin their independent research work as early as possible.

Research Facilities

The Department of Biology occupies approximately 45,000 square feet of laboratory and office space. Available facilities support numerous investigative techniques, including gene cloning, ultrastructural analysis, tissue culture, protein isolation and characterization, and various membrane and physiological preparations. Specialized laboratories are available for studies requiring recombinant DNA technology, radioimmunoassay, scintillation spectroscopy, and ultracentrifugation.

Financial Aid

Graduate teaching and research assistantships are available to qualified students. For 2006–07, stipends range from $7600 to $12,500 for nine months. Computer-based information on scholarships, loan funds, and other student assistance may be obtained from the Financial Aid Office. A limited number of assistantships providing tuition remission only or stipend only are also available.

Cost of Study

The 2006–07 tuition is $900 per semester credit hour for all graduate students. As noted above, however, tuition is remitted for some graduate assistants.

Living and Housing Costs

Room and board on campus cost approximately $3000 per semester in 2006–07. On-campus housing (dormitories and apartments) is available for full-time single and married graduate students, but priority for dormitory rooms is given to undergraduate students. Many apartment complexes and rental houses are available near the campus.

Student Group

The department currently has an enrollment of 65 graduate students from colleges throughout the United States. Approximately one half of the graduate students are women, and approximately one fourth are married. Most graduate students receive financial support.

Student Outcomes

Graduates of the Department of Biological Sciences are employed by such facilities as the Florida Department of Environmental Protection; St. John's Water Management District; Brevard County; Sea World; Walt Disney World–The Living Seas; Dynamac; Bionetics; Nantucket Marine Lab; ETT Environmental, Inc.; Brigham and Women's Hospital; Dartmouth Medical School; University of Miami Medical School; Pfizer Foundation; Sri International; Autec; Kistler-Morse; Great Lakes Environmental Center; DuPont Pharmaceuticals; and the South Atlantic Fishery Management.

Location

Florida Tech's main campus is located in Melbourne, a residential community on Florida's Space Coast. Melbourne is the key city in south Brevard County, which also encompasses nine other smaller communities on the mainland and beachside. The Kennedy Space Center and Disney World are within a 90-minute drive of the Institute. The area's economy is a well-balanced mix of electronics, aviation, light manufacturing, opticals, communications, agriculture, and tourism.

The Institute

Florida Tech was founded in 1958 and has developed rapidly into a university that provides both undergraduate and graduate education in the sciences and engineering for selected students from throughout the United States and many other countries. In addition to cell and molecular biology, Florida Tech offers graduate programs in marine biology, conservation biology, applied mathematics, chemical engineering, chemistry, civil engineering, computer science, electrical engineering, environmental engineering, mechanical engineering, ocean engineering, oceanography, operations research, physics, science education, space sciences, and systems engineering.

Applying

Further information and application forms for admission to the Graduate School may be obtained from the Graduate Admissions Office. Applicants must take the Graduate Record Examinations and arrange to have the scores sent to the Graduate Admissions Office. Separate application for financial aid must be made on forms available from the department or the Graduate School and must be submitted to the department by March 1.

Correspondence and Information

Graduate Admissions Office
Florida Institute of Technology
150 West University Boulevard
Melbourne, Florida 32901

Phone: 321-674-8027
      800-944-4348 (toll-free)
Fax: 321-723-9468
E-mail: grad-admissions@fit.edu
Web site: http://www.fit.edu/grad

Dr. G. N. Wells, Head
Department of Biological Sciences
Florida Institute of Technology
150 West University Boulevard
Melbourne, Florida 32901

Phone: 321-674-8034
E-mail: gwells@fit.edu
Web site: http://www.bio.fit.edu

Florida Institute of Technology

THE FACULTY AND THEIR RESEARCH

David J. Carroll, Associate Professor; Ph.D., Connecticut, 1996. Molecular control of signal transduction at fertilization, with focus on egg activation during fertilization and how this activation relieves the cell-cycle block to initiate development.
Calcium-mediated inactivation of MAP kinase pathway in sea urchin eggs at fertilization. *Dev. Biol.* 236:244–57, 2001. With Jaffe et al.

Michael S. Grace, Associate Professor; Ph.D., Emory, 1991. Molecular control of photoreceptors in the retina and nonretinal photoreceptors of the brain, pineal, and parietal organ.
Timing of reproductive immigration in salamanders: Roles of environmental cues and endogenous biological clocks. *Herpetological Rev.* 34:17–20, 2003.

Julia E. Grimwade, Associate Professor; Ph.D., SUNY at Buffalo, 1987. Control of DNA replication; cell-cycle control; DNA-protein interactions.
IHF redistributes bound initiator protein, DnaA on supercoiled *Escherichia coli. Mol. Microbiol.* 35:835–44, 2000. With Ryan and Leonard.

Charles E. Helmstetter, Professor Emeritus; Ph.D., Chicago, 1961. Identification of macromolecular components involved in the regulation of cellular growth and the cell division cycle of bacteria, yeast, and mammalian cells; investigation of bacteria, including the determinants of the timing of initiation of replication from the *E. coli* chromosomal origin, *ori*C, and the mechanism by which chromosomes are partitioned into daughter cells at division; similar studies with eukaryotic cell systems, employing a new technique for analysis of the cell cycle and cell aging.
Synchrony in human, mouse and bacterial cultures: A comparison. *Cell Cycle* 2:42–5, 2003. With Thornton et al.

Alan C. Leonard, Professor; Ph.D., SUNY at Buffalo, 1979. Use of molecular genetic techniques to study the biology of microbial growth control; DNA-protein interaction and temporal gene expression during the bacterial cell division cycle; molecular regulation of plasmid and minichromosome replication in *E. coli.*
IHF and HU stimulate assembly of pre-replication complexes at *Escherichia coli ori*C by two different mechanisms. *Mol. Microbiol.* 4:113–24, 2002. With Ryan et al.

Charles D. A. Polson, Associate Professor; Ph.D., Florida Tech, 1979. Development and application of biotechnology techniques in undergraduate education, especially in the areas of cloning, synthesis, and sequencing; electrophoretic separation of large DNA molecules by pulse-field electrophoresis.

Russell C. Weigel, Associate Professor; Ph.D., Maryland, 1970. Plant physiology, plant tissue culture; basic and applied studies on the formation of organized tissues from cell cultures; preservation of rare species using tissue culture techniques.
In vitro propagation of *Conradina etonia. Fla. Scientist* 65:201–7, 2002. With Peterson.

THE GEORGE WASHINGTON UNIVERSITY

Graduate Program in Molecular Medicine

Program of Study

The Ph.D. program in molecular medicine encompasses the fields of molecular and cellular oncology, neuroscience, and pharmacology and physiology. The program was developed in recognition of the overlapping nature of these three fields and the growing emphasis on interdisciplinary approaches that span the boundaries of the conventional disciplines. Admission is through the interdepartmental Institute for Biomedical Sciences (IBS). In the first year, students learn about common themes in biomedical sciences: nucleic acids, transcription, translation, proteins, cell biology, signal transduction, and others. In the second semester, the molecular medicine core course focuses on the physiological basis of disease, leading students to select a specific concentration by the end of their first year. Specific research strengths are in molecular and cellular oncology (signaling pathways and functional genomics and proteomics, anticancer drugs and stem cells, SNP, and gene array); neuroscience (site-directed mutagenesis and structure/function of proteins, behavioral and cognitive response associated to neuropathology); and pharmacology (mechanism of drug action, normal and pathological signaling responses of cells and organs upon pharmacological drug actions).

The program leading to a Ph.D. degree is designed to permit flexibility. After completing the first-year core curriculum, students select one of the three possible concentrations (molecular and cellular oncology, neuroscience, or pharmacology and physiology). Then, in consultation with the concentration advisers, each student designs a plan of study for the doctoral program that meets his or her individual interests and goals. During the first year, students also have laboratory rotations that provide opportunities to work in several research laboratories before they choose a dissertation research adviser. A research advisory committee is then formed to evaluate the student's technical and intellectual development. The committee consists of the student's research adviser and at least 2 other faculty members who can best evaluate the research. The student prepares and presents a semiannual research progress report that is reviewed by the committee. Students are actively involved in a weekly seminar series that gives participants added opportunities for experience in oral presentations. Each student's academic and research performance is reviewed by the entire program faculty. Toward the end of the second year, students take a comprehensive examination covering didactic material as well as evaluation of laboratory data and knowledge on current scientific literature. After successful completion of this qualifying examination and completion of 48 credit hours of course work, students advance to candidacy. During the following year they prepare and defend a formal dissertation proposal. Upon completion of 72 credit hours and the dissertation, the student has a final examination, which consists of an oral defense of the dissertation research. For updated information about the programs, prospective students should visit the Program's Web site (http://www.gwumc.edu/ibs/).

Research Facilities

The program has recognized the need for highly trained scientists and comprises integrated faculty members and scientists from The George Washington University School of Medicine and Health Sciences, the Columbian College of Arts and Sciences, the National Institutes of Health, the National Cancer Institute, Children's National Medical Center, and The Institute for Genomic Research (TIGR). Therefore, the program encourages interdisciplinary research endeavors among basic science faculty members and clinical faculty members and among basic science faculty members from multiple University departments both in the Medical Center and on the main campus.

Financial Aid

Institute Fellowships are available on a competitive basis. They carry a $23,000-per-year stipend and 24-credit-hour tuition for the first two years (support during the second year is dependent on a satisfactory record during the first year). Subsequently, students are supported by extramural fellowships, scholarships, or research grants to the laboratory in which they are doing their thesis research project. Student tuition and fees are provided in full.

The George Washington University offers one Presidential Merit Fellowship in the Biomedical Sciences that is available on a competitive basis. It carries a $24,000-per-year stipend and 24-credit-hour tuition for two years.

Cost of Study

Tuition for 2006–07 is $970 per credit hour, and the cost of student association fees is $1 per credit hour.

Living and Housing Costs

University housing is not generally available to graduate students. Information on off-campus housing is available through the Office of Campus Life, which hosts several apartment-hunting weekends during the summer. The cost of living in the Washington area is comparable to that of other major metropolitan areas.

Student Group

The Institute for Biomedical Sciences admits 10–15 students per year.

Location

The University benefits from the abundant cultural, historical, and educational offerings of metropolitan Washington, D.C. The University is close to the National Institutes of Health, the National Library of Medicine, the Smithsonian Institution, the Food and Drug Administration, and the Environmental Protection Agency.

The University

The George Washington University, chartered by Congress in 1821, is private and nonsectarian. It holds regional accreditation from the Middle States Association of Colleges and Schools and has received professional recognition for specific programs. Diversified offerings at all levels of the University associate it with the people and activities of many organizations that are exclusive to the Washington area. The campus, located four blocks west of the White House, is a mixture of large modern buildings, traditional town houses, and other classroom and dormitory buildings, a reflection of the varied character of the surrounding area. Within the campus are located all the major facilities of the University, including the University hospital and medical school complex. A safe, clean, and modern Metro-rail system connects the Medical Center with urban Washington D.C., and suburban Virginia and Maryland, including the NIH.

Applying

Ph.D. graduate students are admitted to the Institute for Biomedical Sciences after a review of their qualifications and an interview. The Admissions Committee seeks students with broad interests and enthusiasm for further in-depth study in the biomedical sciences. It is important that applicants have course work that prepares them for a rigorous core curriculum in modern molecular and cellular biology. Normally, a minimum of a B average is required (3.0 on a 4.0 scale). All applicants must submit GRE General Test scores, three letters of recommendation, transcripts from all institutions attended, and a statement of purpose. International applicants should note that a minimum TOEFL score of 600 on the paper-based test (250 on the computer-based test) is required. Applications are accepted for the fall semester only. The deadline for application materials is January 4.

Correspondence and Information

For program information:

Graduate Program in Molecular Medicine
Ross Hall, Room 605
The George Washington University
2300 Eye Street, NW
Washington, D.C. 20037

Phone: 202-994-2179
Fax: 202-994-0967
E-mail: gwibs@gwu.edu
Web site: http://www.gwumc.edu/ibs/

For application forms and catalogs:

Office of Graduate Studies
Columbian College of Arts and Sciences
Phillips Hall, Room 107
The George Washington University
Washington, D.C. 20052

Phone: 202-994-6210
Fax: 202-994-6213
E-mail: askccas@gwu.edu
Web site: http://columbian.gwu.edu

The George Washington University

THE FACULTY AND THEIR RESEARCH

Department of Anatomy and Cell Biology

Anne Chiaramello, Associate Professor; Ph.D., California, San Diego, 1990. Roles of the basic helix-loop-helix transcription factors during vertebrate neurogenesis; gene expression that controls neuronal differentiation and plasticity.

Robert G. Hawley, Professor; Ph.D., Toronto, 1984. Molecular mechanisms in leukemic hematopoiesis.

Janette M. Krum, Associate Professor; Ph.D., George Washington, 1987. Blood and brain development; astroglial/neuronal metabolic interactions; neuronal transplantation.

Sally A. Moody, Professor; Ph.D., Florida, 1981. Molecular and cellular determination of neuronal phenotypes; regulation of neurotransmitters in the developing retina.

Kenna O. Peusner, Professor; Ph.D., Harvard, 1974. Role of synaptic transmission in the development of central vestibular neural circuit.

Jeffrey M. Rosenstein, Professor of Anatomy and Neurosurgery; Ph.D., Penn State, 1976. Mechanisms of vascular and barrier changes in experimental brain tumors.

Mary Ann Stepp, Associate Professor; Ph.D., Boston University, 1986. Cell-cell and cell-substrate interaction; integrins; wound healing; re-epithelialization.

Xi Zhan, Assistant Professor; Ph.D., Columbia, 1988. Fibroblast growth factor; signal transduction from cell surface to cytoskeleton.

Department of Biochemistry and Molecular Biology

Patricia E. Berg, Associate Professor; Ph.D., IIT, 1973. Role of BP1 and other homeobox genes in the progression of breast and prostate cancer.

Bernard Bouscarel, Associate Research Professor; Ph.D., D.Sc., Toulouse III (France), 1985. Role of signal transduction in hepatocellular disorders and cancer.

Sidney W. Fu, Assistant Research Professor; Ph.D., Peking Union Medical College, 1994; M.D., Xi'an Medical (People's Republic of China), 1989. Gene expression and regulation in mammalian cells and transgenic models; human genome structure and cancer genetics; bioinformatics.

Valerie Hu, Professor; Ph.D., Caltech, 1978. Protein biochemistry and Alzheimer's.

Fatah Kashanchi, Associate Professor; Ph.D., Kansas, 1991. Molecular pathogenesis of AIDS and ATL; cell-cycle associated events related to host-cell and human retroviruses, including HIV-1 and HTLV-1; genomics, microarray, and proteomics of infected and uninfected cells.

Ajit Kumar, Professor; Ph.D., Chicago, 1968. Viral carcinogenesis and gene transactivation.

Glenn T. Merlino, Assistant Professor; Ph.D., Michigan, 1980. Growth factors and cell transformation.

Marcos Rojkind, Research Professor; M.D., National of Mexico, 1962; Ph.D., Mexico Polytechnic, 1970. Extracellular matrix in health and disease; molecular mechanisms involved in liver fibrosis; cell-matrix interactions, with special emphasis in laminin-binding proteins of hepatocytes and hepatomas.

Yan An Su, Associate Professor; M.D., Lanzhou Medical College (People's Republic of China), 1982; Ph.D., Michigan, 1992. Genomics of tumor suppressor genes, tumorigenesis, and human cancer.

William B. Weglicki, Professor; M.D., Maryland, 1962. Neuropeptides and cardiovascular inflammation; novel antioxidants; prooxidant stress in Mg deficiency and Fe overload.

Department of Immunology

Edward C. DeFabo, Research Professor; Ph.D., George Washington, 1974. UV light carcinogenesis and cellular immunity.

Achsah D. Keegan, Assistant Professor; Ph.D., Johns Hopkins, 1989. Signal transduction by interleukin-4; regulation of lymphocyte homeostasis by cytokines.

Frances P. Noonan, Professor; Ph.D., Queensland (Australia), 1977. UV light carcinogenesis and cellular immunity.

Yufang Shi, Associate Professor; Ph.D., Alberta (Canada), 1992. Role of CD 28 in apoptosis and septic shock.

Department of Medicine

James D. Ahlgren, Professor; M.D., Georgetown, 1977. Oncology; chemotherapy of solid tumors: P53 and other genetic mutations in colorectal carcinoma; pharmacology of 5FU modulators.

Joao Ascensao, Professor; M.D., Lisbon, 1972. Hematopoietic stem cell transplantation; immunotherapy post-transplant; biology of human natural killer cells.

Michael Bell, Assistant Professor; M.D., SUNY Health Science Center at Brooklyn, 1987. Effects of inflammation of developing brain, with particular emphasis on glial cell development.

Louis A. DePalma, Professor; M.D., Naples (Italy), 1982. Neonatal immunobiology; oncology pathology.

Robert S. Siegal, Associate Professor; M.D., George Washington, 1977. Development of tumor markers; breast cancer screening and diagnosis; growth factors and receptor blockers in breast cancer therapy.

David Reiss, Professor; M.D., Harvard, 1962. Effects of environment and genetics on adolescent development; study of family interactions.

Department of Pathology

Anamaris M. Colber-Poley, Associate Professor; Ph.D., Penn State, 1980. Role of viral genes in pathogenesis; effects of viral proteins on cellular physiology.

Nancy L. DiFronzo, Assistant Professor; Ph.D., Georgia, 1988. Molecular mechanisms of tumorigenesis by nonacute retroviruses.

Christian C. Haudenschild, Research Professor; M.D., Basel (Switzerland), 1968. Vascular pathobiology.

S. Percy Ivy, Assistant Professor; M.D., Tulane, 1980. Molecular mechanisms of drug resistance in pediatric malignancies.

Stephan Ladisch, Professor; M.D., Pennsylvania, 1973. Ganglioside shedding by tumor cells and role in abrogating tumor immunity.

Patricia S. Latham, Associate Professor; M.D., USC, 1972. Gene regulation and cytokine response of tumoricidal monocytes.

Jan M. Orenstein, Professor; Ph.D., 1969, M.D., 1971, SUNY Downstate Medical Center. Ultrastructural clinicopathological correlation of malignancies; pathology of HIV disease.

Gregory H. Reaman, Professor; M.D., Loyola Chicago, 1973. Immunophenotypic and molecular genotypic characterization of childhood acute lymphoblastic leukemia.

Arnold M. Schwartz, Professor; Ph.D., MIT, 1973; M.D., Miami (Florida), 1978. Immunohistochemical and molecular markers of tumor origin and type; molecular diagnosis.

Department of Pharmacology

Susan Ceryak, Associate Research Professor; Ph.D., George Washington, 1994. Signaling pathways involved in cell cycle regulation and dysregulation in cancer.

Vincent A. Chiappinelli, Ralph E. Loewy Professor and Chair of Pharmacology; Ph.D., Connecticut, 1977. Patch clamp electrophysiology; functional and pharmacological studies of nicotinic receptors and presynaptic nerve terminals.

Clare M. Fraser, Professor; Ph.D., SUNY at Buffalo, 1981. Microbial genome sequencing and analysis; evolution of gene families and species.

Vittorio Gallo, Professor; Ph.D., Rome (Italy), 1979. Glial cell development; transcriptional regulation of neural gene expression during development.

Gordon Hager, Professor; Ph.D., Washington (Seattle) 1970; Nuclear receptors, chromatin, and transcription; organization of the mammalian nucleus; subcellular trafficking.

Tim G. Hales, Associate Professor; Ph.D., Dundee (Scotland), 1990. Molecular neuropharmacology and functional genomics; using electrophysiology and imaging to study ion channel/receptor function.

Eric P. Hoffman, Professor; M.D., Johns Hopkins, 1987. Molecular basis of inherited muscle and CNS disease utilizing DNA gene chip technology.

S. Percy Ivy, Assistant Professor; M.D., Tulane, 1980. Molecular mechanisms of drug resistance in pediatric malignancies; mdl gene expression in leukemia.

Achsah D. Keegan, Assistant Professor; Ph.D., Johns Hopkins, 1989. Signal transduction by interleukin-4; regulation of lymphocyte homeostasis by cytokines.

Norman H. Lee, Associate Professor; Ph.D., Maryland, 1990. Studying mRNA regulation and global patterns of gene expression with DNA microarrays.

H. George Mandel, Professor; Ph.D., Yale, 1949. Cancer drug metabolism chemotherapy; molecular carcinogenesis.

David Mendelowitz, Professor; Ph.D. Washington (Seattle), 1989. Neuropharmacology of cardiorespiratory pathways in the brainstem; modulation by anesthetics and nicotinic receptors.

Steven R. Patierno, Professor; Ph.D., Texas, 1985. Molecular and cellular oncology; DNA damage and repair; cell death and survival signaling; invasion, angiogenesis and metastasis; molecular therapeutics.

David C. Perry, Professor; Ph.D. California, San Francisco, 1981. Nicotinic receptor subtypes; localization and regulation of neurotransmitter receptors.

Eva M. Sorenson, Research Associate Professor; Ph.D., St. Louis, 1990. Localization of nicotinic receptors in neuronal circuitry; nicotinic receptor regulation of neuronal activity.

Margaret Sutherland, Assistant Professor; Ph.D., Open University (UK), 1993. Transgenics; glutamate transporter; electrophysiology; embryonic stem cells; amyotrophic lateral sclerosis; epilepsy; neuronal cell death; apoptosis; glioma cancer biology.

Linda L. Werling, Professor and Director of the IBS; Ph.D., Duke, 1983. Signaling in regulation of transmitter release; PCP and nicotine in brain function; neurochemistry of schizophrenia.

HARVARD UNIVERSITY

Department of Molecular and Cellular Biology

Programs of Study	The Department of Molecular and Cellular Biology (MCB) is the home of an interdisciplinary group of world-class scientists and laboratories. Its mission to advance biological research beyond traditional boundaries is supported by innovative research centers and state-of-the-art resources located on an academic campus enriched by museums, libraries, symposia, and events. It is this interdisciplinary and collaborative culture—motivated by a passion for scientific discovery—that makes MCB an exciting place to study the unsolved questions in biology. MCB trains its graduate students to be the next generation of life scientists: creative, independent, and productive researchers working in academe, medicine, industry, law, business, or nonprofit sector. The Engineering and Physical Biology Training Program (EPB) is offered in partnership with the Department of Physics and the Division of Engineering and Applied Sciences. EPB trains a new generation of scientists to view living systems through the lens of physics and engineering. The Genetics and Genomics Training Program (GGTP), offered in partnership with the Department of Organismic and Evolutionary Biology, prepares young scientists for a new generation of genetic research from subcellular signaling mechanisms to organisms in populations. The Molecular, Cellular, and Chemical Biology Training Program (MCCB) is offered in partnership with the Department of Chemistry and Chemical Biology. MCCB prepares students to solve scientific problems through both chemical and biological approaches.
Research Facilities	MCB is affiliated with the Bauer Center for Genomics Research, the Center for Brain Science, and the Harvard Stem Cell Institute. In addition, MCB students have full access to HILS labs and resources.
Financial Aid	MCB requires all applicants to complete the Graduate School of Arts and Sciences (GSAS) Statement of Financial Resources and strongly encourages all applicants to seek graduate fellowships (such as NSF, NIH, A*STAR, NSERC) as part of their professional development. That said, an offer of admission includes MCB's guarantee of full support—tuition, health insurance, fees, and stipend—for five years. Additional funds for equipment, supplies, and conference travel are also available. The stipend rate is set at \$28,008 for 2006–07 and is adjusted annually for the cost of living in Boston.
Cost of Study	Tuition for 2005–06 was \$31,280 for the year.
Living and Housing Costs	The annual cost of living is estimated at a minimum of \$18,270 for 2006–07. Incoming Ph.D. students are guaranteed on-campus housing in GSAS residence halls, starting at \$3965 for the academic year. Meal plans are available at an additional cost of \$1922 for the academic year. Most students move off-campus or into affiliated housing after the first year.
Student Group	MCB matriculates between 15 and 20 Ph.D. candidates each year, with an overall enrollment of 90 to 100 graduate students from more than twenty countries. The term of study is five to six years. The department takes pride in its small, adaptive programs that train a highly talented and diverse group of young scientists.
Student Outcomes	Traditionally, MCB graduates choose an academic career path into postdoctoral fellowships and faculty positions at leading universities or institutes. However, an increasing number enter industry, law, business, journalism, or the non-profit sector.
Location	MCB is located in Cambridge, Massachusetts, on the historic campus of Harvard University, right next to legendary Harvard Square. MCB is up the road from MIT and across the river from Boston.
The University	Harvard University was established in 1636. It is the oldest institution of higher education in America. Prospective students can take a virtual tour at http://www.harvard.edu/about/.
Applying	This Ph.D. program (0900) is part of the Harvard Integrated Life Sciences Program (HILS). Students seeking admission to the MCB Ph.D. program apply online to the Harvard University Graduate School of Arts and Sciences.
	MCB welcomes applications from all students who meet the academic requirements of GSAS, including international students. Applications from students who identify themselves as African American; Mexican American; Native American; Native Pacific Islander (from Hawaii, Samoa, Guam, or Micronesia); or Puerto Rican are strongly encouraged. Undergraduate study in chemistry, biology, biochemistry, physics, engineering, or mathematics completing a baccalaureate degree is required.
	To apply online for admission to EPB, students should select the subject code "0950—Engineering and Physical Biology." To apply for admission to GGTP or MCCB, students should select the subject code "0900—Unspecified." Students are asked to select one of these two training tracks on acceptance.
	Applicants are required to take the GRE General Test by November 2006 in order to report scores on the application. The GRE Subject Test in a science or math discipline is strongly recommended. All applicants for whom English is a non-native language and who do not hold a degree from an institution at which English is the language of instruction are required to take the TOEFL by November 2006 in order to report scores on the application. Applications without scores are considered to be incomplete. All official ETS score reports should be directed to Harvard University Graduate School of Arts and Sciences (code 3451).
	Complete application forms and supporting materials (including letters of recommendation, scores, and transcripts) are due to GSAS by December 8, 2006, for fall 2007 admissions. Late or incomplete applications are be reviewed by the MCB Admissions Committee.
	All finalists are invited to attend one of two Recruitment Events hosted by MCB in mid-February 2006. During this three-day event, recruits meet with faculty and students, learn about the training programs, tour the department and campus, and sightsee Cambridge and Boston. Underrepresented minority students who receive an offer of admission are also be invited to attend a special informational visit hosted by GSAS in early-April to learn about minority student life at Harvard University.
Correspondence and Information	For more information: Department of Molecular and Cellular Biology Harvard University 7 Divinity Avenue Cambridge, Massachusetts 02138 Phone: 617-495-4107 Web site: http://www.mcb.harvard.edu For applications: Office of Admissions and Financial Aid Graduate School of Arts and Sciences Byerly Hall, 2nd floor Harvard University 8 Garden Street Cambridge, Massachusetts 02138 Web site: http://www.gsas.harvard.edu/admissions

Harvard University

THE FACULTY

Howard Berg, Herchel Smith Professor of Physics and Professor of Molecular and Cellular Biology. Motile behavior of bacteria.
Daniel Branton, Higgins Professor of Biology, Emeritus. Molecular probing with nanopore technology.
John Dowling, Llura and Gordon Gund Professor of Neurosciences. Structure, function, development, and genetics of the vertebrate retina.
Victoria D' Souza, Assistant Professor of Molecular and Cellular Biology. Structural biology of retrovirus replication.
Catherine Dulac, Professor of Molecular and Cellular Biology and Howard Hughes Medical Institute Investigator. Molecular and developmental biology of olfactory and pheromone sensing.
Kevin Eggan, Assistant Professor of Molecular and Cellular Biology. Cloning by nuclear transplantation; epigenetic reprogramming; and human ES cell-based models of disease.
Florian Engert, Associate Professor of Molecular and Cellular Biology. Synaptic plasticity in tadpole and zebrafish.
Raymond Erikson, John F. Drum American Cancer Society Professor of Cellular and Developmental Biology. Reversible phosphorylation in cell proliferation.
Nicole Francis, Assistant Professor of Molecular and Cellular Biology. Biochemistry of epigenetic inheritance by polycomb group proteins.
Rachelle Gaudet, Associate Professor of Molecular and Cellular Biology. Structural studies of ion channels and transporters; X-ray crystallography.
William Gelbart, Professor of Molecular and Cellular Biology. Developmental genetics; genomics; bioinformatics.
Guido Guidotti, Higgins Professor of Biochemistry. Structure and function of membrane proteins.
J. Woodland Hastings, Paul C. Mangelsdorf Professor of Natural Sciences. Biochemistry of bioluminescence; mechanism of the circadian cellular biological clock.
Craig P. Hunter, Professor of Molecular and Cellular Biology. *C. elegans* genetics and genomics and intercellular RNA transport.
David Jeruzalmi, Associate Professor of Molecular and Cellular Biology. Structure and function of the nucleo-protein complexes that are utilized to replicate chromosomal DNA.
Nancy Kleckner, Herchel Smith Professor of Molecular Biology. Basic chromosomal processes in yeast and bacteria.
Samuel Kunes, Professor of Molecular and Cellular Biology. Neural development, function, and behavior.
Jeff Lichtman, Professor of Molecular and Cellular Biology. Synaptic structure and competition.
Richard Losick, Harvard College Professor, Maria Moors Cabot Professor of Biology, and Howard Hughes Medical Institute Professor. Gene regulation and development in microorganisms.
Tom Maniatis, Thomas H. Lee Professor of Molecular and Cellular Biology. Mechanisms of gene regulation.
Andrew McMahon, Frank B. Baird, Jr. Professor of Science. Role of cell signaling in the development of the vertebrate body plan.
Markus Meister, Jeff C. Tarr Professor of Molecular and Cellular Biology. Function of neuronal circuits.
Douglas Melton, Thomas Dudley Cabot Professor in the Natural Sciences, and Howard Hughes Medical Institute Investigator. Developmental biology of the pancreas with the long term aim of making insulin producing beta cells for the treatment of diabetes.
Matthew Meselson, Thomas Dudley Cabot Professor of the Natural Sciences. Molecular genetics and evolution.
Matthew Michael, Associate Professor of Molecular and Cellular Biology. Genomic instability.
Andrew Murray, Herchel Smith Professor of Molecular Genetics, Chair of Molecular and Cellular Biology, and Director of the Bauer Center for Genomics Research. Mitosis, meiosis, experimental evolution, and signal transduction.
Venkatesh Murthy, Morris Kahn Associate Professor of Molecular and Cellular Biology. Neuronal cell biology; synaptic transmission and plasticity.
Axel Nohturfft, Associate Professor of Molecular and Cellular Biology. Regulation of membrane lipid homeostasis.
Erin O'Shea, Professor of Molecular and Cellular Biology, Co-Director of the Bauer Center for Genomics Research, and Howard Hughes Medical Institute Investigator. Systems levels and molecular analysis of signaling pathways; transcriptional regulation; proteomics.
Joshua Sanes, Professor of Molecular and Cellular Biology and Director of the Center for Brain Science. Synapse formation.
Alexander Schier, Professor of Molecular and Cellular Biology. Developmental genetics and neurobiology.
Jack Strominger, Higgins Professor of Biochemistry. Molecular basis of immune recognition.

Associate Members

Stuart L. Schreiber, Morris Loeb Professor of Chemical Biology and Howard Hughes Medical Institute Investigator. Chemical genetics and chemical genomics.
Gregory Verdine, Harvard College Professor and Erving Professor of Chemistry. Structural biology; chemical biology; structure and function of DNA-and RNA-processing enzymes; small-molecule RNA interference.

Affiliate Members

David R. Nelson, Arthur K. Soloman Professor of Biophysics and Professor of Physics and Applied Physics. Force-induced denaturation of DNA; sequence heterogeneity and the dynamics of motor proteins; population growth and mutation in disordered media.
Hidde Ploegh, Mallinckrodt Professor of Immunopathology, and Professor of Pathology. Cell biology and biochemistry of proteolysis and antigen presentation.

THE JOHNS HOPKINS UNIVERSITY

School of Medicine
Graduate Program in Cellular and Molecular Medicine

Program of Study	The Graduate Program in Cellular and Molecular Medicine offers graduate training in the broad area of modern biology with an emphasis on human disease. The faculty consists of more than 100 members from both basic science and clinical departments. The program's objective is to train outstanding doctoral students for careers in research and teaching. In addition to rigorous training in scientific research, graduates will have a thorough knowledge of human biology and human disease to prepare them to capitalize upon advances in biological research to advance the knowledge of human disease. During the first year, each student conducts research in at least three different laboratories. These laboratory rotations, each three months in duration, are undertaken concurrently with core courses. The core courses include introduction to the human body, molecular biology and genomics, fundamentals of genetics, cell structure and dynamics, metabolic pathways, immunology, basic principles of disease, and a seminar series, Topics in Cellular and Molecular Medicine. Additional elective courses are available on many topics including immunology, bioorganic chemistry, pharmacology, and neuroscience. At the end of the first year, students select a research adviser and begin original research leading to their doctoral dissertation. An oral examination, administered by the Graduate Board of the University, must be completed by the end of the second year of study. Thereafter the students write a doctoral dissertation based on their original research and present their work at a public seminar. Close interaction between faculty members and students is also encouraged by a large number of journal clubs and seminar programs, organized both at the departmental level and by special scientific interest. Graduates of this program are expected to produce evidence of creative scientific research in cellular and molecular medicine in the form of original research papers published in peer-reviewed scientific journals.
Research Facilities	Students work in the well-equipped laboratories of the faculty that are located throughout the medical school campus. These individual research programs are supported by several shared facilities including microscopy, molecular biology, and protein chemistry. Students have direct access to the School of Medicine's Welch Medical Library, which houses one of the country's best medical and research collections.
Financial Aid	Students admitted to the program are supported financially by a combination of federal and private funding that provides a stipend and tuition for the first year. After the first year, the students' research advisers are then responsible for the expenses. The stipend level is $25,200 and increases annually.
Cost of Study	Tuition for 2006–07 is $32,200 and the matriculation fee is $640; both are covered by the program, as are health and dental insurance.
Living and Housing Costs	The cost of living in Baltimore is comparable to that in other large northeastern American cities. A limited number of rooms for single students are available on campus starting at $290 per month. Most graduate students share apartments or houses that are reasonably priced and located near Homewood, the undergraduate campus of the University. There is a free shuttle bus service between the Homewood campus and the medical school; it runs hourly and is only a 15-minute ride.
Student Group	Now in its thirteenth year, the program enrollment totals 117. It is anticipated that students will take five or six years to complete the program. Student selection is based on academic achievement, aptitude exhibited by scores on the Graduate Board Examinations, supportive letters of recommendation, accomplishments in research and evidence of creativity and maturity. The total number of graduate students in all programs at the School of Medicine is approximately 600.
Location	Baltimore, the largest city in Maryland, is the center of a metropolitan area of 600,000 people. The city is located on the Chesapeake Bay, and its picturesque Inner Harbor, as the focus of one of the nation's most successful urban development programs, is the center of restaurant, shopping, and business districts. It is also the site of a science center, the National Aquarium, and Oriole Park at Camden Yards. The Appalachian Trail passes through Maryland, 1 hour west of the campus. The city has a symphony orchestra, opera and ballet companies, several repertory theaters, more than a dozen museums, and excellent bookstores. The cultural opportunities in New York, Philadelphia, and Washington, D.C., are easily reached by car or train. Baltimore supports professional teams in baseball, soccer, hockey, and football. At the medical school, graduate students are offered membership to the Cooley Athletic Center, a full recreation facility with a pool.
The University	Johns Hopkins University was founded in 1876 and was the first university in the United States to stress graduate education and original research—an emphasis that has been maintained ever since. The School of Medicine is one of the world's foremost centers for biomedical research. The vitality and diversity of its research, ranging from the study of molecular interactions to the study of human societies, provide an exceptionally stimulating environment for graduate students.
Applying	The application deadline is December 31 for matriculation in September. Prerequisites include college-level courses in biology, general and inorganic chemistry, physics, and calculus. Applicants should submit results of the General Test of the Graduate Record Examinations (code 5316), transcripts, three letters of recommendation (preferably from research supervisors), and their own statement of purpose. An undergraduate degree is required before matriculation.
Correspondence and Information	Admissions Coordinator Graduate Program in Cellular and Molecular Medicine School of Medicine The Johns Hopkins University 1830 East Monument Street, Suite 2-103 Baltimore, Maryland 21205 Phone: 410-614-3640 E-mail: ckgraham@jhmi.edu Web site: http://www.hopkinsmedicine.org/graduateprograms/CMM

The Johns Hopkins University

THE FACULTY AND THEIR RESEARCH

Steering Committee: Martin Abeloff, Pierre Coulombe (Program Director), John Griffin, Diane Hayward, Ann Hubbard, John Isaacs, Dan Lane, Don Price, Antony Rosen, Sol Snyder, and Myron Weisfeldt.

Cancer Research: Rhoda Alani: genetic mechanisms of skin cancer development. Richard F. Ambinder: herpesviruses in the pathogenesis of human malignancies; therapeutic treatment strategies. Robert Arceci: chromatin structure and epigenetic change during development to cancer cell survival. Stephen B. Baylin: molecular determinants of human tumor progression. Zaver Bhujwalla: functional and molecular imaging of cancer. Robert A. Casero Jr.: molecular pharmacology; polyamine metabolism; antineoplastic drug development. Curtis Civin: stem cell biology, hematopoietic transplantation, and leukemia research. Chi V. Dang: molecular and cellular biology of oncogenes. Nancy Davidson: biology and treatment of human breast cancer. James Eshleman: mutator phenotypes in colon cancer; novel therapeutics. Andrew P. Feinberg: epigenetics of cancer and common disease; methylation and chromatin mechanisms. Alan D. Friedman: myeloid leukemogenesis; transcriptional control of hematopoietic differentiation. Edward Gabrielson: functional genomics and molecular genetics of breast and lung cancer. Robert Getzenberg: hypothesis-driven proteomics to develop and characterize novel cancer biomarkers. Carol Greider: telomerase biochemistry and telomere length regulation. James Herman: epigenetic change of DNA methylation in cancer. David L. Huso: molecular and cellular basis for modulating genetic disease. John T. Isaacs: programmed (apoptotic) cell death as therapy for cancer. William B. Isaacs: molecular biology and genetics of prostate cancer. Elizabeth Jaffee: development of vaccines for treatment of cancer. Steven T. Leach: regulation of exocrine pancreatic differentiation. William Nelson: the molecular pathogenesis of prostate cancer. John Nicholas: molecular biology of oncogenic herpesviruses. Charles Rudin: cellular responses to genotoxic stress and apoptotic pathways contributing to cancer. Saul J. Sharkis: cellular and molecular biology of hematopoietic stem cell. David Sidransky: molecular progression and early detection of cancer. Donald Small: molecular biology of leukemia; molecularly targeted therapeutics; hematopoiesis; signal transduction. Victor Velculescu: molecular genetic approaches to cancer. Bert Vogelstein: molecular genetics of human cancer and its translational implications.

Cardiopulmonary and Vascular Biology: Subroto Chatterjee: cell adhesion molecules; inflammation and atherosclerosis. Harry C. Dietz: molecular etiology of structural heart disease. Gregory G. Germino: identification and characterization of genes involved in hereditary renal diseases. William B. Guggino: genetic disorders associated with defective fluid and electrolyte movement. Joshua Hare: diseases of the myocardium including myocardial infarction. Landon King: regulation of aquaporin expression in the respiratory tract. Ronald Li: human embryonic stem cell and ion channel engineering. Charles Lowenstein: oxygen radicals and inflammation in the cardiovascular system. Eduardo Marban: molecular physiology of ion channels. Brian O'Rourke: mitochondrial ion channels, bioenergetics, and cardiac excitation-contraction coupling. Gordon F. Tomaselli: structure-function of ion channels. David T. Yue: molecular physiology of ionic channels. Pamela L. Zeitlin: lung development; cystic fibrosis; gene therapy.

Cell Biology and Genetics of Human Diseases: Fred Bunz: genetics of human cell-cycle regulation and differentiation. Shukti Chakravarti: extracellular matrix functions in normal and diseased connective tissues. Pierre A. Coulombe (Program Director): cellular and molecular biology of wound healing in skin. Linzhao Cheng: human stem cell biology, engineering, and transplantation. Garry R. Cutting: inherited diseases of ion transport. Mark Donowitz: signal transduction in epithelial biology. Peter Espenshade: regulation of cholesterol homeostasis; therapeutics for heart disease. Neal S. Fedarko: extracellular matrix and cellular metabolism. Shannon Fisher: developmental genetics. John D. Gearhart: mammalian developmental genetics. Stephen J. Gould: analysis of peroxisome assembly in yeast and humans. Ann L. Hubbard: vesicle traffic and polarity in mammalian cells. Ethylin W. Jabs: craniofacial development and disorders; chromosome structure and function. Ken Kinzler: cancer genetics and its application to the clinic. M. Daniel Lane: adipogenic stem cell commitment-/differentiation-regulated gene expression. Se-Jin Lee: growth and differentiation factors in mammalian development. Susan Michaelis: yeast protein trafficking and human disease; ER quality control; endocytosis; ABC transporters. Peter L. Pedersen: cell energetics and its application to molecular medicine. Rajini Rao: identification and molecular mechanisms of novel ion transporters in yeast and humans. Roger H. Reeves: disruption of development by gene dosage imbalance in Down syndrome. Ronald Rodriguez: urologic application of gene therapy. Lewis Romer: cell-matrix adhesion; tyrosine kinases and integrin signaling; endothelial cell injury. Gregg L. Semenza: transcriptional regulation of gene expression.

Immunology, Virology, and Infectious Disease: William R. Bishai: bacterial gene regulation; pathogenesis; tuberculosis; latent infection; sigma factors. Lieping Chen: costimulatory molecules; lymphocyte activation and deactivation; immunotherapy; transplant rejection. Janice E. Clements: molecular biology and pathogenesis of HIV and SIV. Brendan Cormack: host-pathogen interaction and molecular mechanisms in candidiasis. Stephen Desiderio: development and activation of the immune system. J. Stephen Dumler: molecular, cellular, and immunopathogenesis of tick-borne infectious agents. Paul T. Englund: biochemistry and molecular biology of parasitic protozoa. Susan H. Eshleman: genetic diversity of HIV; HIV drug resistance; HIV mother-to-child transmission. Steve N. Georas: molecular regulation of gene expression in the immune system. Diane E. Griffin: molecular pathogenesis and immunology of viral infections. J. Marie Hardwick: molecular mechanisms of programmed cell death. Gary S. Hayward: molecular biology and genetics of herpesvirus disease processes. S. Diane Hayward: Epstein-Barr virus; latency and B-cell immortalization. Shau-Ku Huang: molecular and genetic mechanisms of asthma and allergic diseases. Hyam I. Levitsky: immunologic mechanisms of vaccine-induced antitumor immunity. Drew M. Pardoll: T-cell development and receptor function; tumor immunology. Paula M. Pitha: role of cytokines and chemokines in HIV pathogenicity and HIV-associated cancer. Hamid Rabb: role of T and B cells in transplantation injury. Noel Rose: basic mechanisms of autoimmunity and autoimmune disease. Antony Rosen: mechanisms of autoimmunity, with particular emphasis on the roles of apoptosis. Jonathan Schneck: vaccine development, interaction of T cells, and major histocompatibility molecules. Robert Siliciano: viral dynamics in HIV infection; viral reservoirs; drug resistance. Mark J. Soloski: immune response to intracellular bacteria; innate immunity. M. Christine Zink: animal models of HIV neurological and pulmonary disease.

Neurobiology: Jay M. Baraban: molecular mechanisms of neuronal plasticity. Arthur L. Burnett: neurophysiology of the pelvis and lower genitourinary tract. Michael Caterina: molecular mechanisms of thermosensation and nociception. Ted M. Dawson: molecular mechanisms of neuronal cell death and neurodegenerative diseases. Valina L. Dawson: cell death and cell survival signaling pathways in models of stroke, ALS, and Parkinson's disease. Nicholas Gaiano: regulation of mammalian neural progenitor stem cells. Richard L. Huganir: molecular mechanisms in the regulation of synaptic function. David C. Johns: role of ion channels in diseases of excitability. Kwang Sik Kim: brain endothelial cell biology and pathogenesis of central nervous system infections. Russell L. Margolis: genetic etiology and pathophysiology of neurological and psychiatric disorders. Guo-li Ming: molecular mechanisms regulating nerve growth and guidance. Timothy Moran: biological basis of food intake, energy balance, and obesity. Donald L. Price: mechanisms of neurodegenerative diseases and experimental therapeutics in models. Randall R. Reed: molecular genetics of olfaction; neuronal differentiation. Christopher A. Ross: genes with triplet repeats; Huntington's disease. Jeffrey D. Rothstein: excitoxicity; glutamate transport; apoptosis; neurodegeneration. Akira Sawa: molecular pathogenesis of mental illness, including schizophrenia, bipolar disorder, and dementia. Solomon H. Snyder: molecular aspects of neurotransmission. Shan Sockanathan: cell-fate specification in the developing nervous system. Hongjun Song: regulation mechanisms of adult neural stem cells. Paul Watkins: defects in fatty-acid metabolism in human genetic neurodegenerative disorders. Philip Wong: molecular mechanisms of Alzheimer's disease and motor neuron disease. Paul F. Worley: gene programs in activity-dependent neural plasticity.

LOYOLA UNIVERSITY CHICAGO

Stritch School of Medicine
Department of Cell Biology, Neurobiology and Anatomy

Programs of Study	The Department of Cell Biology, Neurobiology and Anatomy at Loyola University Chicago offers programs leading to the M.S. and Ph.D. degrees. The Ph.D. degree is essential for students who intend to pursue an academic career in the biomedical sciences. The M.S. program is designed to improve the credentials of students whose goal is to seek an M.D. or Ph.D. or a biomedical sciences position in private industry. An M.D./Ph.D. program is also available to those accepted into the M.D. program at Loyola. In the first semester, all new students take courses in the core curriculum with all other graduate students in the biomedical sciences at the Medical Center. Students continue to take courses that examine the relationship between structure and function, including basic immunology, gross anatomy, histology, and neuroscience.

In all programs, primary emphasis is placed on preparing students as well-trained biomedical scientists. Research training is available in two general areas, cell biology and neurobiology, in the laboratories of most research-intensive faculty members at the Medical Center. Research areas in cell biology include growth factors and gene expression, vascular biology and angiogenesis, transplantation and cellular immunology, cellular and molecular regulation of lymphocyte development, tumor cell biology, alcohol effects on immune function after injury, and estrogen effects on immune responses. Active research in neurobiology includes neural development, plasticity, steroid regulation of neural injury, neurotoxic and protective effects of alcohol, models of Parkinson's disease, neurotransmitters and stress, development of somatosensory and reproductive pathways, neuropeptide localization and gene expression, steroid and reproductive hormone neuroendocrinology, and neuroimmunology.

The Ph.D. degree requires 48 semester hours and normally takes four to five years to complete. During the first year, students devote most of their time to research rotations and to scheduled course work on the structure and function of the human body. During the second year, students continue with additional course work and begin a research program. During the subsequent years, students concentrate on completing their dissertation project and may take more specialized courses appropriate for their individual programs.

The M.S. program requires 24 semester hours and normally takes two years to complete. The first year of study is similar to that of the doctoral program. During the second year, students take additional courses as appropriate but usually concentrate on a research project for their thesis.

The purpose of the programs is to graduate scientists of the highest quality who have obtained both a broad background in their subdisciplines and a depth of knowledge and understanding in the specialty of their choice. Fulfillment of this purpose, especially in the Ph.D. and M.S. programs, ensures that graduates are well prepared for a stimulating and successful career in biomedical sciences in an academic environment or the private sector.

Research Facilities	A wide range of facilities are available for studies requiring state-of-the-art techniques. These include core facilities for nucleic acid synthesis and the necessary equipment for standard cellular, molecular, and neural biological research. The electron microscopy core facility houses modern scanning and transmission electron microscopes and image analysis facilities. The building houses a modern health sciences library; other extensive library facilities are available in the Chicago area. An on-campus academic computing center is available for use. Graduate students have access to computer systems located in the Department.
Financial Aid	The University provides basic science fellowships for Ph.D. candidates. Minority fellowships also are awarded by the University to qualified applicants. Fellowships provide an annual stipend of approximately $22,000, plus full tuition remission. Additional support is available through extramural research funding that has been awarded to individual faculty members.
Cost of Study	Tuition in 2005–06 was $685 per semester hour credit. In addition, there was a health fee of $115 per semester, a student activity fee of $32 per semester, and a health and fitness center membership fee of $119 per semester. Students who cannot provide proof of health insurance are required to obtain the University's student health insurance policy for $1287 per year. The standard course load is 12 credit hours per semester. Tuition costs are subject to change each year.
Living and Housing Costs	All students reside in apartments or houses located in nearby suburbs or Chicago. A student housing office is located on campus for help in finding a place to live. The Medical Center also offers an online housing site at http://www.meddean.luc.edu/templates/ssom/studres/aptgazette/search.cfm. One- and two-bedroom apartments in the area range between $600 and $1200 per month and usually include some utilities, so sharing an apartment with another student is very popular.
Student Group	Loyola University Chicago is a private institution with a student enrollment of about 15,000. There are approximately 90 graduate students at the Medical Center campus, and 12 students are currently enrolled in the Ph.D. program in the Department. Most graduates proceed to an academic career in research and teaching after pursuing postdoctoral training.
Student Outcomes	Recent graduates of the Ph.D. program have obtained postdoctoral training positions for two to five years. Subsequently, most graduates have found full-time faculty positions at universities and medical schools throughout the United States. Some have taken investigative or supervisory positions in industry. Students who obtain master's degrees do so to improve their current job status or to become more competitive for acceptance to medical school.
Location	Loyola University is an urban institution with three main campuses. The Stritch School of Medicine is located about 12 miles from downtown Chicago in pleasant surroundings in the suburb of Maywood. An expressway provides ready access to downtown Chicago, which offers a full range of cultural opportunities. The Brookfield Zoo and the historic Frank Lloyd Wright architecture in Oak Park are located near the campus. Many lakes, parks, and ski resorts are located within a few hours' driving distance.
The School and The Department	The Stritch School of Medicine is more than 70 years old and has developed a fine reputation for the training of physicians, teachers, and researchers. The move to a new campus in 1969 has been followed by unprecedented growth. Graduates of the Department of Cell Biology, Neurobiology and Anatomy and other basic science departments hold prominent positions in academic life across the nation. The Department, particularly its research and other scholarly activities, has expanded during recent years.
Applying	The deadline for the receipt of applications is May 1, but early application is advised. Students contemplating graduate work in the Department are encouraged to acquire a broad experience in the physical and biological sciences as well as in mathematics and statistics. The cumulative GPA for both the undergraduate and graduate degrees should be at least 3.0. The Graduate Record Examinations must be taken before students are accepted into the program, and above-average scores are expected. The Subject Test in biology is optional.

The application process is no longer paper based but totally electronic; it is part of the ApplyYourself Application Network. Applicants receive a PIN and password so that they can return to work on their application over several sessions. The information is transmitted through a secured server and is kept confidential until the application is submitted. The application is available for review by the Department's admissions staff only after submission. Students can apply online at https://app.applyyourself.com/?id=loyolamc.

Correspondence and Information	Graduate Program Director Department of Cell Biology, Neurobiology and Anatomy Stritch School of Medicine Loyola University Chicago 2160 South First Avenue Phone: 708-216-3353 Fax: 708-216-3913 E-mail: grad-medcntr@lumc.edu Web site: http://www.luc.edu/biomed

Loyola University Chicago

THE FACULTY AND THEIR RESEARCH

Anthony J. Castro, Professor, Neuroscience Course Co-Director; Ph.D., Florida, 1970. Neuronal plasticity after brain or spinal cord damage in the newborn; neuronal transplants in the repair of brain damage.

Lee M. Cera, Assistant Professor of Pathology; Assistant Dean, Comparative Studies; and Director, Comparative Medicine Facility; D.V.M., Illinois, 1975; Ph.D., Chicago, 1992. New and emerging infectious/zoonotic diseases.

John Clancy Jr., Professor and Chair; Ph.D., Iowa, 1970. Cellular immunology; role of cytokines and heat shock proteins in transplantation and tumor immunology; neuroimmunology.

Michael A. Collins, Professor, Division of Molecular and Cellular Biochemistry; Ph.D., Purdue 1968. Neurotoxic and neuroprotective mechanisms of ethanol; environmental neurotoxic factors in Parkinson's disease.

Lydia L. DonCarlos, Professor; Ph.D., Kent State, 1985. Development and plasticity of steroid receptor-containing neural circuits; sexual differentiation of the brain; neural substrates underlying reproductive function; neuroprotective effects of gonadal steroid hormones.

Allen Frankfater, Professor, Division of Molecular and Cellular Biochemistry; Ph.D., Duke, 1968. Lysosomal proteinases, regulation of their expression in metastatic tumors, and role in cancer cell invasion; role of the IGF-II receptor in lysosomal enzyme trafficking and tumor growth.

Thackery S. Gray, Professor and Cell Biology, Neurobiology and Anatomy Graduate Program Director; Ph.D., Ohio State, 1979. Neuroanatomical, neurochemical, and behavioral studies of limbic brain structures that mediate anxiety, depression, and mood-related disorders.

Howard P. Greisler, Professor of Surgery and Cell Biology and of Neurobiology and Anatomy; M.D., Penn State, 1975. Vascular biology and tissue engineering; growth-factor regulation of endothelial cell and smooth-muscle cell proliferation in vascular surgery; restenosis and angiogenesis.

Kathryn J. Jones, Professor; Ph.D., Illinois at Chicago, 1983. Molecular mechanisms of neuronal survival and regulation following injury; therapeutic use of steroid hormones in neuronal regeneration; immune regulation of neuronal injury and repair.

Gwendolyn L. Kartje, Associate Professor; Ph.D., 1984, M.D., 1988, Loyola of Chicago. Neuronal plasticity and functional recovery after ischemic brain damage; role of neurite inhibitory factors in stroke recovery.

Elizabeth J. Kovacs, Professor; Ph.D., Vermont, 1984. Control of cytokine gene expression in macrophages; effects of estrogen on inflammatory and immunoregulatory responses; effects of gender, aging, and alcohol consumption on immune function following injury.

Phong T. Le, Associate Professor; Ph.D., Ohio State, 1985. Biology of human thymic epithelial (TEC) in T-cell development; mechanism of TEC-mediated apoptosis of immature T cells; molecular mechanism of thymic involution.

Mary Druse Manteuffel, Professor and Director, Division of Molecular and Cellular Biochemistry; Ph.D., North Carolina at Chapel Hill, 1972. Effects of in utero ethanol exposure on the developing serotonin system; mechanisms of damage and therapeutic intervention.

John A. McNulty, Professor, Vice Chair, Structure of Human Body Course Director, and Core Imaging Facility Director; Ph.D., USC, 1976. Neuroendocrinology of the pineal gland; neuroimmunomodulation; photoneuroendocrine transduction and secretion; cell ultrastructure; computerized image analysis; computer-aided instruction.

Edward J. Neafsey, Professor and Neuroscience Graduate Program Director; Ph.D., UCLA, 1976. Neurotoxicity of ethanol, HIV gp120, amyloid-beta, and beta-carbolines; emotional and autonomic changes in PTSD.

Richard M. Schultz, Professor, Division of Molecular and Cellular Biochemistry; Ph.D., Brandeis, 1969. Oncogene pathway regulation of the transformed and metastatic cellular phenotypes; protease mechanisms in metastasis; protein and enzyme structure and function.

William Simmons, Professor, Division of Molecular and Cellular Biochemistry; Ph.D., Illinois at Chicago, 1979. Metabolism of peptide hormones; reduction of blood pressure and heart attack damage by inhibitors of a bradykinin-degrading enzyme.

Frederick H. Wezeman, Professor of Orthopedic Surgery and Cell Biology, Neurobiology and Anatomy and Associate Dean, Graduate School; Ph.D., Illinois at Chicago, 1969. Alcohol-induced bone disease; molecular mechanisms of osteogenesis; skeletal repair.

Fletcher A. White, Assistant Professor; Ph.D., Medical College of Ohio, 1995. Mechanisms of neuropathic pain; neuroinflammation; development of somatosensory system.

Pamela L. Witte, Professor; Ph.D., Texas, 1984. Cellular and molecular regulation of lymphocyte differentiation; characterization of hemopoietic and lymphopoietic microenvironments; aging.

Representative Publications

Castro, A. J., M. Meyer, A. Dahl Jorgensen, and J. Zimmer. Axonal outgrowth from fetal porcine neocortical tissue xenografted into newborn rats. *Cell Transplantation* 12:733–41, 2003.

Wright, M. H., **L. M. Cera,** N. A. Sarich, and J. A. Lednicky. Reverse transcription–polymerase chain reaction detection and nucleic acid sequence confirmation of reovirus infection in laboratory mice with discordant serologic indirect immunofluorescence assay and enzyme-linked immunosorbent assay results. *Comp. Med.* 54(4):410–7, 2004.

Goral, J., H. L. Mathews, S. G. Nadler, and **J. Clancy Jr.** Reduced levels of Hsp70 result in a therapeutic effect of 15-deoxyspergualin on acute graft-versus-host disease in (DA x LEW)F1 rats. *Immunobiology* 202:254–66, 2000.

DonCarlos, L. L., et al. Androgen receptor immunoreactivity in forebrain axons and dendrites in the rat. *Endocrinology* 144(8):3632–8, 2003.

Szpaderska, A. M., et al. **(A. Frankfater).** Sp1 regulates cathepsin B transcription and invasiveness in murine B16 melanoma cells. *Anticancer Res.* 24: 3887–92, 2004.

Van de Kar, L. D., et al. **(T. S. Gray).** 5-HT2A receptors stimulate ACTH, corticosterone, oxytocin, renin, and prolactin release and activate hypothalamic CRF and oxytocin expressing cells. *J. Neurosci.* 21:3572–9, 2001.

Brewster, L. P., et al. **(H. P. Greisler).** Heparin-independent mitogenicity in an endothelial and smooth muscle cell chimeric growth factor (S130K-HBGAM). *Am. J. Surg.* 188(5):575–9, 2004.

Jones, K. J. Olfactory ensheathing cells: Therapeutic potential for spinal cord regeneration. *New Anat.* 271B:39, 2003.

Markus, T. M., et al. **(G. L. Kartje).** Recovery and brain reorganization after stroke in adult and aged rats. *Ann. Neurol.* 58(6):950–3, 2005.

Goral, J., et al. **E. J. Kovacs.** In vivo ethanol exposure down-regulates TLR2-, TLR4-, and TLR9-mediated macrophage inflammatory response by limiting p38 and ERK1/2 activation. *J. Immunol.* 174:456–63, 2005.

Ortman, C. L., et al. **(P. T. Le).** Molecular characterization of the involuted thymus: Aberrations in expression of transcription regulators in thymocyte and epithelial components. *Int. J. Immunol.* 14:813–22, 2002.

Manteuffel, M. J. Druse, N. F. Tajuddin, R. A. Gillespie, and **P. T. Le.** Signaling pathways involved with serotonin1A agonist-mediated neuroprotection against ethanol-induced apoptosis of fetal rhombencephalic neurons. *Dev. Brain Res.* 159:18–28, 2005.

Tsai, S.-Y., et al. **(J. A. McNulty).** Microglia play a role in mediating the effects of cytokines on the structure and function of the rat pineal gland. *Cell Tissue Res.* 303:423–31, 2001.

Belmadani, A., **E. J. Neafsey,** and **M. A. Collins.** HIV-1 gp120 and ethanol co-exposure in rat brain slice cultures: Curtailment of gp120-induced neurotoxic mediators and neurotoxicity by moderate but not high ethanol concentrations. *J. Neurovirol.* 9:45–54, 2003.

Janulis, M., et al. **(R. M. Schultz).** Role of the mitogen-activated protein kinases and c-Jun/AP-1 *trans*-activating activity in the regulation of protease mRNAs and the malignant phenotype in NIH 3T3 fibroblasts. *J. Biol. Chem.* 274:801–13, 1999.

Ersahin, C., A. M. Szpaderska, A. T. Orawski, and **W. H. Simmons.** Aminopeptidase P isozyme expression in human tissues and peripheral blood mononuclear cell fractions. *Arch. Biochem. Biophys.* 435:303–10, 2005.

Callaci, J., et al. **(F. H. Wezeman).** Binge alcohol treatment increases vertebral bone loss following ovariectomy: Compensation by intermittent parathyroid hormone. *Alcohol. Clin. Exp. Res.,* in press.

White, F. A., S. K. Bhangoo, and R. J. Miller. Chemokines: Integrators of pain and inflammation. *Nat. Rev. Drug Discovery* 4(10):834–44, 2005.

Pifer, J., R. P. Stephan, D. A. Lill-Elghanian, and **P. L. Witte.** Role of stromal cells and their products in protecting B-lineage precursors from dexamethasone-induced apoptosis. *Mech. Ageing Dev.* 124:207–18, 2003.

NEW YORK MEDICAL COLLEGE

Graduate Programs in Cell Biology

Programs of Study

Ph.D. and M.S. degree programs, with an emphasis on cell biology and neuroscience, are offered within an active and expanding department at New York Medical College. The graduate programs prepare students for careers in research and teaching in both academia and industry. Students become involved in state-of-the-art research during the first year through research rotations in faculty laboratories; work-in-progress seminars presented by students, postdoctoral fellows, and faculty members; and journal clubs, where current scientific literature is discussed. In addition, formal courses are taken in biochemistry, cell biology, developmental biology, neuroscience, molecular biology, and immunology.

Following completion of the interdisciplinary core curriculum (usually after 1½ years), Ph.D. students take a qualifying examination in selected areas of cell biology and/or neuroscience. By the beginning of the second year, they select a research sponsor and present a thesis proposal as soon as possible. Upon completion of a written dissertation, the thesis research is presented in a departmental seminar and defended before a thesis committee.

Research Facilities

State-of-the-art research facilities are available in virtually all areas of cell biology, molecular biology, and neuroscience, including recombinant DNA, protein sequencing and mass spectroscopy, electron microscopy, tissue culture, fluorescent digital image analysis, confocal and 2-photon microscopy, electrophysiology and phosphorimaging, subcellular fractionation, and biochemistry. There are also excellent institutional facilities, including a central animal complex, a computer center, and instrumentation and photography units. The medical sciences library contains more than 148,700 volumes, most major journals, and a computer-based retrieval system.

Financial Aid

Ph.D. students are supported through a comprehensive program of financial aid coming from institutional sources as well as various granting agencies. Most students are awarded attractive stipends for the calendar year, which include full tuition waivers.

Cost of Study

In fall 2006, tuition is $645 per credit.

Living and Housing Costs

Apartments are available on campus. Students should contact the Director of Housing, Office of Student Affairs, Administration Building, Valhalla, New York 10595 (telephone: 914-594-4832), to make housing arrangements. Apartments are also available off campus in Valhalla, Elmsford, Hawthorne, and other nearby towns.

Student Group

In 2005–06, there were 57 Ph.D. and 106 M.S. students enrolled in the Graduate School of Basic Medical Sciences (GSBMS) and 4 Ph.D. candidates and 1 M.S. candidate in cell biology.

Location

New York Medical College is located on an attractive 135-acre campus in suburban Westchester, close to the cultural and educational resources of New York City.

The College

New York Medical College, one of the largest medical schools in the country, was established in 1860 under the leadership of William Cullen Bryant. The Graduate School was established in 1968. The College moved to the modern Westchester campus from Manhattan in the early 1970s.

Applying

Students are admitted to the Ph.D. degree program through the Graduate School of Basic Medical Sciences Interdisciplinary Ph.D. Program. Admission into the Ph.D. program is only in the fall semester. Application instructions, forms, and requirements can be found online at the New York Medical College Web site. The GRE General Test and a Subject Test (usually biology, chemistry, or biochemistry) and letters of recommendation are required as part of the application. International applicants must also take the Test of English as a Foreign Language (TOEFL). Application for the M.S. program is also made through the GSBMS office, and students are accepted directly into the Department of Cell Biology and Anatomy. Students are usually admitted in the fall semester, although they are permitted to enter the M.S. program at other times during the year. Application instructions and forms can be obtained from the GSBMS office or Web site.

Correspondence and Information

Victor A. Fried, Ph.D.
Graduate Program Director
Department of Cell Biology and Anatomy
New York Medical College
Valhalla, New York 10595

Phone: 914-594-4025
E-mail: victor_fried@nymc.edu
Web site: http://www.nymc.edu

New York Medical College

THE FACULTY AND THEIR RESEARCH

Praveen Ballabh, Associate Professor; M.D., Institute of Medical Sciences, BHU Varanasi (India), 1984. Pathogenesis of germinal matrix hemorrhage; neuroprotection in a rabbit model of germinal matrix hemorrhage.

Anna B. Drakontides, Professor Emerita; Ph.D., Cornell, 1971. Pathogenesis of early and late changes at the neuromuscular junction and skeletal muscle induced by chemical irritants.

Joseph D. Etlinger, Professor and Chairman; Ph.D., Chicago, 1974. Skeletal muscle growth and atrophy, molecular mechanisms, and selectivity of intracellular proteolysis in erythroid and muscle cells; role of proteasomes and ubiquitin; spinal cord injury.

Reza Farahani, Assistant Professor; Ph.D., Ottawa, 1997. Effect of hypoxia on growth and development; longitudinal study of myelin disorder.

Victor A. Fried, Professor and Graduate Program Director; Ph.D., Oregon, 1970. Role of ubiquitin in cytoskeletal and cell-surface receptor function, posttranslational modifications, and protein sequencing.

Frances Hannan, Assistant Professor; Ph.D., Melbourne, 1991. *Drosophila* model for neurofibromatosis type 1; function of NF1 and TSC in *Drosophila* learning and memory role of NF2 in *Drosophila* hearing.

Jian Kang, Associate Professor; M.D., China, 1982; Ph.D., Florida, 1993. Interplay between glial cells and neurons; roles of astrocytic glutamate release in epileptic seizures; mechanism underlying glutamate-stimulated astrocytic glutamate release; roles of astrocytic glutamate release in synaptic plasticity; kainite receptor-mediated modulation of GABAergic synapses; activity-dependent potentiation of intrinsic neuronal excitability.

Mithun Kumarasiri, Assistant Professor; Ph.D., Columbia, 1986. Skeletal muscle growth and differentiation in response to endocrine factors and activity in tissue culture.

Kenneth M. Lerea, Associate Professor; Ph.D., Rochester, 1984. Mechanisms of signal transduction; role of protein seryl-threonyl and tyrosyl kinases and phosphatases in integrin functions and platelet activation.

Koko Murakami, Research Assistant Professor; Ph.D., Tokyo, 1973. Biochemical and cellular mechanisms for the function of ubiquitin system in skeletal neuronal and auditory development. .

Stuart A. Newman, Professor; Ph.D., Chicago, 1970. Physical and molecular mechanisms of development and evolution; pattern formation in the vertebrate limb; evolution of adipocyte differentiation.

Matthew A. Pravetz, Associate Professor; Ph.D., New York Medical College, 1988. Neural cytoskeleton; tropomyosin-binding proteins in brain; urogenital and gastrointestinal development.

Renato Rozental, Associate Professor; M.D./Ph.D., Rio de Janeiro, 1991. Developmental neurobiology and diseases mediated by cell-cell communication via gap-junction channels; impact of hypoxic-ischemic insults in the developing brain.

Pravin B. Sehgal, Professor; M.D., Bombay, 1973; Ph.D., Rockefeller, 1977. Interleukin-6; interferons; gene expression; signal transduction; STAT transcription factors; pulmonary hypertension; Golgi trafficking; endosomal signaling.

Sansar C. Sharma, Professor; Ph.D., Edinburgh, 1967. Development and regeneration in the visual system and spinal cord; glaucoma and cascade of cell death.

Harry T. Smith, Research Assistant Professor; Ph.D., Arkansas, 1980. Regulatory functions of the ubiquitin system; cytoskeleton structure and function.

Alan D. Springer, Professor; Ph.D., CUNY, 1973. Virtual engineering simulations of the mechanisms underlying retinal and foveal development; causes of retinal detachments in premature infants; causes of strabismus.

Patric K. Stanton, Professor; Ph.D., Uniformed Services University of the Health Sciences, 1985. Cellular mechanisms of learning and memory; synaptic plasticity; long-term potentiation and depression of synaptic strength; optical imaging of vesicular transmitter release, calcium and dendritic motility; cerebral ischemia and neuronal death.

Gerardo Suarez, Research Associate Professor; M.D., Chile, 1960. Protein fructation and oxidative stress; crystallin and collagen self-assembly; molecular mechanisms of diabetic complications and generation of Alzheimer's β-amyloid protein.

Richard J. Zeman, Associate Professor; Ph.D., NYU, 1979. Role of B_2-adrenoceptors in musculoskeletal growth; mechanisms of spinal cord injury; regulation of intracellular calcium.

NEW YORK UNIVERSITY / NATIONAL INSTITUTES OF HEALTH

Graduate Program in Structural Biology

Program of Study	Structural biology has experienced unprecedented growth over the last decade with individual molecular structures leaving their marks on virtually every field of biomedical research. Major advances in the technologies have fueled this growth for X-ray crystallography, electron microscopy, magnetic resonance imaging, mass spectrometry, and computational methods for structure prediction and design. With this in mind, The National Institutes of Health (NIH) and New York University's (NYU) Sackler Institute have combined their resources in establishing a new doctoral training program in structural biology. This program trains students to study the structural basis of biology at both the molecular and cellular levels, using the latest technologies and methodologies. The curriculum includes a broad base of course work in cell and molecular biology together with specialty classes in concepts and techniques of structural biology. Students begin the program with a laboratory rotation at NIH the summer before classes begin. During their first academic year, students take classes and perform research rotations with faculty members at NYU, followed by a second summer lab rotation at NIH. Near the end of this second summer, students choose a dissertation mentor at NIH or establish a collaborative dissertation that allows them to work with both an NIH and an NYU research mentor over the course of their training. Dissertation advisory committees are composed of both NIH and NYU faculty members. An alternative program is available for students wishing to pursue research solely at NYU and, in both cases, students receive their Ph.D. degree from New York University.
Research Facilities	The New York University Medical Center is one of the largest teaching and research complexes in the country. Research laboratories are located in the Basic Medical Science Building; at Tisch, Bellevue, and VA Hospitals; at the Skirball Institute of Biomedical Sciences; and at the newly completed Smilow Research Center. The participating faculty members utilize state-of-the-art research laboratories and have access to centralized core facilities, including a modern protein chemistry laboratory (with HPLC, FPLC, low-speed and high-speed centrifuges, etc.); nucleic acid and protein sequencing and synthesizing instrumentation; microscopic facilities for laser confocal, transmission, and scanning electron microscopy; X-ray crystallography equipment; MRI, NMR, and mass spectrometry facilities; and an image-processing center with a full range of computers (ranging from individual personal computers to a 256-node IBM supercomputer called Max). As the federal government's primary agency for the support and conduct of biomedical research, the NIH's many Institutes and Centers employ approximately 1,200 tenured or tenure-track faculty members and more than 3,700 postdoctoral scientists with medical, dental, or graduate degrees. The NIH intramural research program, located on a 300-acre campus in Bethesda, Maryland, 10 miles from downtown Washington, D.C., has been the scene of many exciting scientific advances. Four Nobel Laureates made their prize-winning discoveries in NIH laboratories on topics that range from breaking the genetic code to discoveries concerning the chemical transmitters in nerve terminals, including the mechanisms for their storage, release, and deactivation.
Financial Aid	Students accepted into the program are fully supported throughout their time spent in the program (usually five years) with stipend, tuition, and fee waivers and medical insurance funding. Low-interest housing loans of $1500 per year are also available for students during their studies at NYU, as are loans for the purchase of personal computers and annual travel funds to attend scientific meetings.
Cost of Study	As described in the Financial Aid section, financial aid awards usually meet the cost of study.
Living and Housing Costs	University housing in dormitory facilities and apartment complexes within the immediate vicinity of the NYU Medical Center is guaranteed for all incoming graduate students. Depending on the type of accommodation, rents range from $630 per month for a dormitory room to $710 to $840 per month for apartments. The services of the Medical Center's housing office are also available to students who have special needs or wish to live off campus. The cost of living in the Washington, D.C., area is comparable to other major cities in the United States. Near NIH, there are various houses, apartments, and rooms in private residences available for rent. For information on available housing near NIH, applicants should visit the Web sites of local newspapers or contact the Graduate Partnerships Program office at the NIH Web site at http://gpp.nih.gov.
Student Group	Students selected to participate in the NYU-NIH program are actually part of a larger group of graduate students in the NYU Program in Structural Biology. Students are drawn from a pool of highly qualified national and international applicants. About 16 percent of the students are from underrepresented minority groups, 50 percent are women, and 30 percent are international students. There are more than 400 graduate students at NIH from more than thirty-five universities. While at NIH, graduate students enjoy services and activities sponsored by the Graduate Partnerships Program, similar to a university campus, which build a strong graduate student community.
Location	The New York University Medical Center, where the Sackler Institute is located, is in the heart of midtown Manhattan, on First Avenue between 23rd and 34th Streets. The location provides easy access to the unparalleled myriad of cultural, social, and educational opportunities of New York City with its world-class museums, libraries, theater, dance, music, and sporting events. The National Institutes of Health is in a northwest suburban area of the metropolitan region, approximately 10 miles from the cultural life of Washington, D.C. The Metro transit system provides easy access to the museums, theater, and other events within the Washington metropolitan area.
The University and The National Institutes of Health	New York University, which was founded in 1831, is the largest private university in the country, with an enrollment of more than 50,000 students. The University includes thirteen colleges and schools at five major centers in Manhattan. The main campus is located near Washington Square in the historic Greenwich Village area of the city. University-sponsored bus service connects the main campus to the Medical Center. Scientists at the National Institutes of Health have played a major role in the development of the understanding of genetics from the breaking of the genetic code in the 1960s to the first successful use of gene therapy to cure a genetic disease in the 1990s. Many graduate students through the years have enjoyed the richness of the unparalleled facilities and research opportunities at NIH through arrangements between the NIH and local universities.
Applying	To apply, prospective students must be U.S. citizens or noncitizen nationals of the United States or must have been lawfully admitted for permanent residence (i.e., they must possess a valid Alien Registration Receipt Card or some other verification of such status). They should hold a bachelor's degree from an accredited university. The student's academic record, GRE scores, and prior research should reflect strong academic ability and evidence of superior promise. Applicants must supply letters of recommendation and a personal statement describing their goals for graduate study, research interests and experiences, and reasons for applying for graduate study. An entering class typically includes a variety of different undergraduate majors in the biological, chemical, or physical sciences. Information on submitting an application to the NYU-NIH program may be found at the NIH Web site. Applications to NIH-NYU Sackler Institute partnership program should be completed and received by the posted University deadline. Early submissions are encouraged.

Correspondence and Information

Graduate Partnerships Program
Building 2, Room 2E06
National Institutes of Health–DHHS
2 Center Drive
Bethesda, Maryland 20892-0234

Phone: 301-594-9605
Fax: 301-594-9606
E-mail: gpp@nih.gov
Web site: http://gpp.nih.gov

NIH Partnership Director
Dr. Nico Tjandra
NHLBI/NIH

Phone: 301-402-3029
E-mail: nico@helix.nih.gov

NYU Partnership Director
Dr. David L. Stokes
Skirball Institute
NYU School of Medicine
540 First Avenue
New York, New York 10016

Phone: 212-263-1580
Fax: 646-219-0300
E-mail: stokes@nyu.edu
Web site: http://saturn.med.nyu.edu/
education/sb/sackler

New York University / National Institutes of Health

THE FACULTY

Students do their dissertation research in the laboratories of both NIH and NYU researchers who have a strong interest in structural biology, including such areas as X-ray crystallography, electron microscopy, magnetic resonance imaging, mass spectrometry, and computational methods. A full description of the program can be found at http://saturn.med.nyu.edu/programs/sb/Sackler/. Currently, there are more than 40 faculty members in the program. For details about specific NIH and NYU researchers, respectively, students should visit the Web sites http://gpp.nih.gov/ or http://saturn.med.nyu.edu/programs/sb/Sackler/Faculty.html. Listed below is a representative sample of structural biologists at NYU and NIH, but there are many more investigators at NIH who are also conducting structural biology–related research.

X-RAY CRYSTALLOGRAPHY

New York University
Stevan Hubbard. Receptor tyrosine kinases.
Moosa Mohammadi. Signaling by fibroblast growth factors.
Da-Neng Wang. Transmembrane nutrient transporters.
Rui-Ming Xu. DNA replication and gene expression.

National Institutes of Health
Susan Buchanan. Membrane protein structure.
Fred Dyda. Mechanisms of catalytic regulation.
David Garboczi. X-ray diffraction and immunology.
Traci Hall. Cell signaling and RNA-protein interactions.
James Hurley. Signal transduction and lipid signaling.
Peter Kwong. HIV vaccines.
Wei Yang. DNA recombination and repair.

ELECTRON IMAGING OF MACROMOLECULAR COMPLEXES AND MEMBRANE PROTEINS

New York University
Xiangpeng Kong. Epithelial membrane structure.
David Stokes. ATP-dependent ion transport and cell-cell junctions.

National Institutes of Health
Jenny Hinshaw. Cryo-EM of membrane trafficking.
Thomas Reese. Light and electron microscopy of transporting proteins in axons.
Alasdair Steven. Cryo-EM of virus replication.
Sriram Subramaniam. Cryo-EM of macromolecular assemblies.

NUCLEAR MAGNETIC RESONANCE SPECTROSCOPY AND IMAGING

New York University
Oded Gonen. Functional MRI on human brains.
Joseph Helpern. Functional MRI on human brains.
Alexej Jerchow. Ion binding to DNA and proteins; methods development.
Daniel Turnbull. MRI and ultrasound on transgenic mice.

National Institutes of Health
Adriaan Bax. Protein structure determination and development of new methods.

Andrew Byrd. Gene expression.
Nico Tjandra. Application of nuclear magnetic resonance in biological systems.
Robert Tycko. Solid-state NMR of prion proteins.

COMPUTATIONAL METHODS

New York University
Stuart Brown. Bioinformatics.
Timothy Cardozo. Development of computational tools targeted at visualizing and engineering 3-D structural features of molecules.
Tamar Schlick. Protein structure prediction and modeling.
Yingkai Zhang. Protein catalysis and recognition.

National Institutes of Health
Stephen Bryant. Computational biology; protein structure prediction.
Tom Darden. Molecular modeling.
Chris Michejda. Drug design.

PROTEOMICS

New York University
Thomas Neubert. Mass spectrometry and posttranslational modification.

National Institutes of Health
Sanford Markey. Proteomics of neurodegenerative processes.

OTHER METHODS

New York University
Joel Belasco. Interactions between RNA and regulatory proteins.
Nicholas Cowan. Protein folding in the eukaryotic cytosol.
Neville Kallenbach. Protein engineering and protein folding.
Nadrian Seeman. DNA nanotechnology.

National Institutes of Health
Debbie Hinton. Mechanisms of transcription initiation and activation.
Adrian Parsegian. Intermolecular force methods.
Alfred Yergey. Mass spectrometric studies of posttranslational modifications.

The Dale and Betty Bumpers Vaccine Research Center (VRC), one of more than thirty major research buildings on the NIH campus, was established to facilitate research in vaccine development.

The National Human Genome Research Institute led the Human Genome Project for the National Institutes of Health, which culminated in the completion of the full human genome sequence in April 2003.

OHIO UNIVERSITY

College of Arts and Sciences
Interdisciplinary Master's/Doctoral Program
in Molecular and Cellular Biology

Program of Study	The Interdisciplinary Master's/Doctoral Program in Molecular and Cellular Biology offers the M.S. and Ph.D. degrees in a wide range of areas of molecular and cellular biology through the cooperative efforts of four departments at Ohio University. Program faculty members from the Departments of Biological Sciences, Biomedical Sciences, Chemistry and Biochemistry, and Environmental and Plant Biology may be selected as advisers for thesis/dissertation research. (Research areas are outlined on the reverse of this page.) Students complete a core curriculum of molecular and cellular biology courses. In consultation with their master's/doctoral advisory committee, students may tailor the remaining course work to their specific interests by choosing among a broad range of offerings in the four departments. Lab rotations are encouraged for beginning students. By the end of the first/second year of study, M.S./Ph.D. students will have developed their dissertation research proposals and have had the proposals approved by their committees. Graduates are prepared to pursue careers in academic instruction and research or in industrial research and development.
Research Facilities	Facilities available include several preparative ultracentrifuges, high-speed centrifuges, gamma and liquid scintillation counters, environmentally controlled rooms and chambers, microbial and tissue culture rooms, laminar flow hoods, research microscopes, inverted microscopes, electron microscopes, a confocal microscope, protein purification and electrophoresis equipment, microcomputers, and a high-speed parallel-processing computer cluster. Faculty laboratories are equipped appropriately for specialized research activities of the faculty members. Facilities for mass spectrometry, NMR, and computer-assisted DNA, RNA, and protein sequence analysis as well as housing for research animals and plants are available to the program participants. Library holdings include 2.5 million volumes and full online access to most scientific journals. Ohio University participates in Ohio Link, providing access to an additional 40 million volumes and full online access to most scientific journals.
Financial Aid	Program students are supported by teaching or research associateships or by fellowships. In 2005–06, program associateships provided an annual stipend of $18,336 for Ph.D. students and $17,300 for M.S. students and a waiver of tuition and fees. A number of employment opportunities are available for students' spouses.
Cost of Study	In 2005–06, tuition and fees for residents of Ohio were $2977 per quarter and insurance was $1080 per year. For nonresidents, tuition and fees were $5641 per quarter and insurance was $1612 per year. Except for general and recreational fees ($516 per quarter), tuition and fees were waived for students awarded associateships. Students with graduate positions, such as graduate assistant or teaching assistant, were given a $185 credit per quarter to cover part of their general fee.
Living and Housing Costs	Unmarried students may choose to live in graduate dormitory housing; the cost per quarter for room and board ranged from approximately $1580 (without air conditioning) to $1617 (with air conditioning) in 2005–06. Meal plans range from $1033 to $1630 per quarter, although meal plans are not a requirement for graduate students. Married students may choose to live in the University Apartments; monthly rents varied from $535 to $740. Houses and apartments in the community are also available.
Student Group	There are 19,704 students enrolled on the main campus of Ohio University. The graduate enrollment of Ohio University is about 2,635. At present, 34 students are enrolled in the Interdisciplinary Master's/Doctoral Program in Molecular and Cellular Biology.
Location	The city of Athens, with a population of about 21,340 , is the home of Ohio University and is located about 75 miles southeast of Columbus. The community and the University offer a broad range of cultural activities, including art, dance, music, and theater. The rural setting offers many outdoor recreational activities, and four attractive state parks are within a short drive of Athens.
The University	Founded in 1804, Ohio University is the oldest institution of higher education in the old Northwest Territory. It is accredited by the North Central Association of Colleges and Schools. Historically, Ohio University has emphasized excellence in instruction and service, but, over the past three decades, a major emphasis on excellence in research has brought the University recognition for its programs of research and graduate study. The Interdisciplinary Master's/Doctoral Program in Molecular and Cellular Biology has been recognized as a Center of Excellence in Ohio by the Ohio Board of Regents. Ohio University operates on the quarter system, with a summer quarter in addition to the three academic-year quarters.
Applying	Students entering the program must also have official graduate status in one of the three participating departments (Biological Sciences, Chemistry and Biochemistry, or Environmental and Plant Biology). Generally, it is recommended that students enter the program in the fall quarter. Applications should be received by February 1 for entrance in the following fall. Application forms may be obtained from the address below or from the Office of Graduate Studies, 206 McKee House, Ohio University, Athens, Ohio 45701. The forms may also be obtained online at the MCB Web site listed below. Students are required to take the GRE General Test. In the case of international students, the Test of English as a Foreign Language (TOEFL) is also required.
Correspondence and Information	Dr. Xiaozhuo Chen Chair, MCB Graduate Studies 109 Konneker Research Laboratories Ohio University Athens, Ohio 45701-2979 Phone: 740-593-9699 Fax: 740-593-4795 E-mail: chenx@mail.biotech.ohiou.edu Web site: http://www.ohio.edu/mcb

Ohio University

THE FACULTY AND THEIR RESEARCH

Department of Biological Sciences

Mary Chamberlin, Professor of Physiology; Ph.D., British Columbia. Control of oxidative phosphorylation during insect development and programmed cell death.

Robert A. Colvin, Associate Professor of Neurobiology; Ph.D., Rutgers. Mechanism of heavy-metal transport in neurons, especially zinc; molecular biology of zinc in neurons and zinc homeostasis; role of zinc in neural degeneration and Alzheimer's disease.

Lisa Crockett, Associate Professor of Physiology; Ph.D., Maine. Comparative physiology and biochemistry; metabolic and membrane adaptations to extreme and varying physical environments.

Janet S. Duerr, Assistant Professor of Cell Biology; Ph.D., Princeton. Genetics and cell biology of protein and neurotransmitter trafficking in neurons; modulation of behavior by neurotransmitters in *C. elegans.*

Donald Holzschu, Associate Professor of Molecular Biology; Ph.D., California, Davis. Mechanisms by which retroviruses induce tumors and contribute to tumor regression in walleye fish.

Daewoo Lee, Assistant Professor of Neurobiology; Ph.D., California, Riverside. Cellular and molecular mechanisms of central synaptic transmission and plasticity; neuromodulation in *Drosophila;* degeneration of dopaminergic neurons in Parkinson's disease.

Robert F. Rakowski, Professor, Department of Biological Sciences; Ph.D., Rochester. Mechanism of ion translocation by membrane proteins; transport protein structure and function; membrane biophysics.

Tomohiko Sugiyama, Assistant Professor, Department of Biological Sciences; Ph.D., Osaka (Japan). Mechanisms of meiotic recombination and DNA double-strand break repair.

Soichi Tanda, Associate Professor of Genetics and Developmental Biology; Ph.D., Hokkaido (Japan). Gene regulation and signal transduction during *Drosophila* development; *Drosophila* hematopoiesis; biological functions of pre-mRNA secondary structures.

Matthew M. White, Associate Professor, Department of Biological Sciences; Ph.D., Virginia Tech. Population genetics of freshwater fishes; fisheries genetics; conservation genetics; biochemical systematics.

Department of Biomedical Sciences

Mark Berryman, Assistant Professor of Microanatomy; Ph.D., Virginia. Role of the actin cytoskeleton in epithelial cell morphogenesis; regulation of membrane-cytoskeletal interactions; characterization of a new family of human chloride channel proteins.

Bonita J. Biegalke, Associate Professor of Microbiology; Ph.D., Washington (Seattle). Molecular biology of herpesviruses; regulation of gene expression; mechanisms of pathogenicity.

Jack Blazyk, Professor of Biochemistry; Ph.D., Brown. Development of novel anti-infective agents; molecular mechanism of antimicrobial peptides.

Xiaozhuo Chen, Associate Professor of Biomedical Science; Ph.D., Ohio. Molecular and cellular biology of metabolic diseases; transgenic animal models for human diseases and identification of biological functions of genes; biology of cancer and apoptosis.

Peter W. Coschigano, Associate Professor of Microbiology; Ph.D., MIT. Molecular and genetic analysis of anaerobic biodegradation of toxic compounds.

Karen T. Coschigano, Assistant Professor of Molecular and Cellular Biology; Ph.D., Brandeis. Molecular basis of growth hormone action and diabetes, with emphasis on the kidney

Kenneth J. Goodrum, Associate Professor of Immunology; Ph.D., Texas at Austin. Role of nitric oxide in innate immune responses to microbial infections.

Mario J. Grijalva, Assistant Professor of Immunology; Ph.D., Ohio. Molecular diagnosis, immunology, and epidemiology of tropical diseases.

Frank M. Horodyski, Professor of Molecular Biology; Ph.D., California, San Diego. Molecular biology of insect neuropeptides and their roles in development and metamorphosis.

Sharon R. Inman, Assistant Professor of Physiology; Ph.D., Louisville. Ischemia/reperfusion injury in renal transplantation.

Calvin B. L. James, Associate Professor of Virology; Ph.D., Howard. Molecular biology of adenoviruses: study of those aspects of signal transduction pathways relevant to both viral and cellular gene regulation.

Peter Johnson, Professor of Biochemistry and Biomedical Sciences; Ph.D., Birmingham (England). Free radical damage in cells and tissues, with focal areas on excitotoxic damage in neurons and lymphocytes, ethanol-induced damage in erythrocytes, and prevention of ischemia-reperfusion injury in lung and heart.

Leonard Kohn, J. O. Watson D. O. Endowed Research Chair and Distinguished Senior Research Scientist, Department of Biomedical Sciences; M.D., Columbia.

John J. Kopchick, Professor of Molecular and Cellular Biology; Ph.D., Texas. Molecular mechanisms of growth hormone action; cloning and expression of growth, diabetes, and obesity-related genes.

Yang V. Li, Assistant Professor of Neuroscience; M.B., Ph.D., Southern Illinois. Cellular and molecular mechanisms of neural transmissions, plasticity, and intracellular signaling with emphasis on role of zinc.

Felicia V. Nowak, Associate Professor of Molecular Neuroendocrinology; Ph.D., Wisconsin–Madison; M.D., Washington (St. Louis). Neuropeptides in mammalian brain development, function, and aging; effects of gender and hormones on neuropeptide gene expression; molecular markers of diabetic injury in the kidney.

Edwin C. Rowland, Associate Professor of Immunoparasitology; Ph.D., Wake Forest. Immunobiology of *Trypanosoma cruzi* infection in mice.

Department of Chemical Engineering

Tingyue Gu, Associate Professor of Chemical Engineering; Ph.D., Purdue. Preparative and large-scale purification of recombinant proteins; dynamics of liquid chromatography; fermentation; microbially induced corrosion.

Department of Chemistry and Biochemistry

Kenneth L. Brown, Professor of Chemistry; Ph.D., Pennsylvania. Chemistry and enzymology of vitamin B_{12}.

Susan Evans, Assistant Professor of Chemistry and Biochemistry; Ph.D., Texas. Molecular and genetic pathways of cancer and embryonic development.

Jennifer V. Hines, Associate Professor of Chemistry and Biochemistry; Ph.D., Michigan. Investigations into the medicinal chemistry and structural biology of RNA leading to the design and development of novel medicinal agents.

Marcia J. Kieliszewski, Associate Professor of Biochemistry; Ph.D., Michigan State. Structure, function, and posttranslational modifications of plant hydroxyproline-rich glycoproteins; glycoprotein design.

Mark McMills, Associate Professor of Chemistry; Ph.D., Michigan State. Synthesis of bioactive natural products; development of novel synthetic methodology.

Malinski Tadeusz, Marvin and Ann Dilley White Professor of Biomedical Science; Ph.D., Poznan (Poland). Nanomedicine, biomedical nanosensors; nitric oxide signaling and nitro-oxidative stress in Parkinson's, Alzheimer's, failing heart, ischemia/reperfusion, aging, atherosclerosis, diabetes, obesity, and wound healing.

Shiyong Wu, Associate Professor of Chemistry and Biochemistry, Ph.D., Nebraska. Translation regulation and endoplasmic reticulum (ER)–stress signal transduction.

Department of Environmental and Plant Biology

Harvey E. Ballard Jr., Associate Professor of Environmental and Plant Biology; Ph.D., Wisconsin–Madison. Molecular systematics and evolution of violets *(Violaceae).*

Ahmed Faik, Assistant Professor, Department of Environmental and Plant Biology; Ph.D., Joseph Fourier (France). Functional genomics of plant cell-wall polysaccharide metabolism; seed development.

Allan M. Showalter, Professor of Plant Molecular Biology; Ph.D., Rutgers. Molecular biology and biochemistry of hydroxyproline-rich glycoproteins in plant cell walls; structure and function of arabinogalactan proteins; molecular adaptations of halophytes to saline environments.

Morgan L. Vis, Associate Professor of Aquatic Botany; Ph.D., Memorial of Newfoundland. Systematics and phylogeography of freshwater red algae.

Sarah Wyatt, Assistant Professor of Plant Developmental Biology; Ph.D., Purdue. The use of molecular and genetic tools to study plant responses to environmental stimuli.

THE PENNSYLVANIA STATE UNIVERSITY

Department of Biochemistry and Molecular Biology
Graduate Study in Cell and Developmental Biology

Program of Study

The Department of Biochemistry and Molecular Biology administers a broad-ranging graduate program that trains students for independent research and teaching in the molecular and cellular aspects of biology. Cell and developmental biology is one of the emphasized scientific disciplines in which students may pursue Ph.D. training. Advanced courses are taught by research faculty members in this area of specialization and are supplemented with courses in biochemistry and molecular biology to provide broad graduate training. Research is conducted in many aspects of cell and developmental biology as well as in other areas of modern molecular biosciences. (The faculty's research interests are described in the faculty section of this In-Depth Description.)

Students complete three laboratory rotations and may then begin a research program while in the first year of study. Personal attention is given to each student by an academic and a research adviser. Advancement to Ph.D. candidacy is decided on the basis of course work, research, and performance on an oral candidacy exam in the second year. The research adviser and a dissertation committee chosen from the program's faculty are the student's major consultants on dissertation research. The dissertation committee also administers the student's comprehensive examination and final Ph.D. dissertation defense. Students may earn the M.S. degree during the course of their Ph.D. work, but normally they bypass the master's degree with the approval of the graduate faculty. The Ph.D. usually requires about five years, including the dissertation research.

Research Facilities

The departmental laboratories are located in three modern buildings in the Eberly College of Science complex, adjacent to the main University library. Students, faculty members, and other research personnel utilize spacious, well-equipped, modern laboratory facilities. Several large general facilities are available in the department for shared use. These include facilities for mass spectrometry and proteomics, DNA microarray and genomics, X-ray crystallography, transgenic mice, nucleic acids, electron microscopy, and hybridoma/monoclonal antibodies, as well as the Center for Quantitative Cell Analysis.

Financial Aid

All admitted students receive financial support in the form of graduate assistantships at approximately $19,885 per year, plus a waiver of tuition, or fellowships from various University or outside sources. Strong applicants qualify for special Eberly College of Science Braddock and Roberts Awards, which provide supplements to their fellowships or assistantships. Furthermore, Penn State University pays most of the cost of health and dental benefits for its graduate students. Students can also elect to enroll in a subsidized family health coverage plan to cover eligible dependents, and supplemental dental and vision plans are also available. The department is well supported by research grants from the NIH, NSF, NASA, DOE, and USDA and other agencies.

Cost of Study

Tuition is provided for all admitted students.

Living and Housing Costs

Housing on campus for married graduate students includes one- and two-bedroom apartments ranging from $545 to $935 per month. Residence hall housing is available without meals for $1715 (double occupancy) or $2225 (single occupancy) per semester. Board only is $1435 to $1815 per semester, depending on the meal plan. Numerous off-campus accommodations are available at varying costs.

Student Group

Enrollment at the University Park Campus is about 41,795, including about 11,000 graduate students. The average number of graduate students in the department is 100, most from out of state. About 25 new students are admitted each year, virtually all seeking the Ph.D. degree. Upon graduating, most students obtain postdoctoral fellowships in prominent national and international laboratories for further specialized research. Graduates currently hold posts in major universities, research institutions, and industry.

Location

Penn State University Park Campus is located in State College, in scenic central Pennsylvania. The population of the area is about 80,000. Residents have the benefit of a wide variety of cultural, educational, and recreational resources. Nearby Interstate 80 and the University Park airport allow for excellent year-round access to the community. Pittsburgh, Philadelphia, New York, Washington, and Toronto are all within 250 miles.

The University and The Department

The 50 faculty members in the graduate program of the Department of Biochemistry and Molecular Biology have extensive research support from outside grants and contracts; such support totals $7 million per year at present. The department is closely associated with the University's Biotechnology Institute, which emphasizes application-oriented research in several areas of biotechnology, including aspects of genetic engineering. The facilities and personnel of the institute, supported jointly by the state of Pennsylvania and industry, are located adjacent to the department's buildings in the Eberly College of Science complex.

A faculty–graduate student ratio of about 1:2 permits excellent personal interactions at all levels. Extensive professional and social activities abound within the department and the University—e.g., numerous formal and informal seminars given by prominent visiting scientists; student and faculty symposia, promoting intradepartmental interactions; and the annual Summer Symposium in Molecular Biology, sponsored by the Recombinant DNA Technology Cooperative Program of Pennsylvania State University and biotechnology-related industries.

Applying

Students come from various educational backgrounds, including undergraduate majors in biochemistry, biology, biophysics, cell biology, chemistry, general science, genetics, microbiology, molecular biology, and physics. Preparation should include courses in biology, organic chemistry, calculus, general physics, and, preferably, physical chemistry. Limited deficiencies may be made up concurrently with graduate studies. Students holding master's degrees are encouraged to apply to the program.

Applications are accepted throughout the year, but most offers of admission and assistantships are made by March for the following summer or fall semester. Scores on the General Test and one Subject Test of the Graduate Record Examinations and, in the case of international students, on the TOEFL are required. These tests should be taken early enough so that the scores are received by January 15, preferably earlier, for consideration for admission in the fall. In addition, the department requires its own application form, undergraduate records, and three letters of recommendation for its review of applicants. Departmental application forms and further information should be requested directly from the address listed in this description.

Correspondence and Information

Chairman, Graduate Admissions
Department of Biochemistry and Molecular Biology
108 Althouse Laboratory
The Pennsylvania State University
Phone: 814-865-5497
E-mail: bmmb@psu.edu
Web site: http://www.bmb.psu.edu

The Pennsylvania State University

THE FACULTY AND THEIR RESEARCH

Robert A. Schlegel, Professor and Head; Ph.D., Harvard, 1971. Mechanisms of transbilayer phospholipid movements and recognition of apoptotic blood cells.

Sarah E. Ades, Assistant Professor; Ph.D., MIT, 1995. Stress-induced signal transduction pathways between the cell envelope and cytoplasm in gram-negative bacteria.

Avery August, Associate Professor; Ph.D., Cornell, 1994. Signal transduction in lymphocytes by Src and Tec family tyrosine kinases. Adaptor proteins in leukemogenesis.

Paul L. Babitzke, Associate Professor; Ph.D., Georgia, 1991. Regulation of gene expression in *Bacillus subtilis;* protein-RNA interactions.

J. Martin Bollinger Jr., Associate Professor; Ph.D., MIT, 1993. Catalytic mechanisms of redox metalloenzymes; mechanisms of assembly of complex metal clusters in enzymes and regulatory proteins.

Squire Booker, Associate Professor; Ph.D., MIT, 1994. Mechanisms of cofactor action in enzymatic reactions.

Jean E. Brenchley, Professor; Ph.D., California, Davis, 1970. Discovery, study, and improvement of useful microbial enzymes; characterization of cold-active enzymes and their genes.

Donald A. Bryant, Professor and Ernest C. Pollard Professor of Biotechnology; Ph.D., UCLA, 1977. Molecular genetics of cyanobacteria and green bacteria; structure and function of photosystem I; light-harvesting antennae; phycobiliproteins; control of gene expression.

Craig E. Cameron, Paul Berg Professor of Biochemistry and Molecular Biology; Ph.D., Case Western Reserve, 1993. Genome replication of positive-strand RNA viruses, with emphasis on hepatitis C virus.

Nina V. Fedoroff, Willaman Professor of Life Sciences; Ph.D., Rockefeller, 1972. Plant transposable elements, transposon tagging, and epigenetic regulation of gene expression.

James G. Ferry, Stanley Person Professor of Biochemistry and Molecular Biology; Ph.D., Illinois, 1974. Enzymology and functional genomics.

Richard L. Frisque, Professor; Ph.D., Wisconsin, 1978. Viral functions involved in cellular transformation, tumorigenicity, and gene expression; role of genetic variation in the pathogenesis of disease; viral latency.

Carol V. Gay, Professor; Ph.D., Penn State, 1972. Cell biology of skeletal tissues; calcium metabolism.

David S. Gilmour, Associate Professor; Ph.D., Cornell, 1984. Transcriptional regulation in *Drosophila.*

John H. Golbeck, Professor; Ph.D., Indiana, 1976. Structure-function relationships in iron-sulfur type photosystems, including photosystem I of cyanobacteria and the reaction center of green sulfur bacteria.

Wendy Hanna-Rose, Assistant Professor; Ph.D., Harvard, 1996. Developmental genetics and morphogenesis in *Caenorhabditis elegans.*

Ross C. Hardison, T. Ming Chu Professor of Biochemistry and Molecular Biology; Ph.D., Iowa, 1977. Molecular basis of gene regulation and evolution in eukaryotes.

Eric T. Harvill, Assistant Professor; Ph.D., UCLA, 1995. Interactions between bacterial virulence factors and host immunity.

Teh-Hui Kao, Professor; Ph.D., Yale, 1980. Molecular studies of RNase-based self-incompatibility and pollen development in plants.

Kenneth Keiler, Assistant Professor; Ph.D., MIT, 1996. Temporal and spatial regulation of protein expression in bacterial development.

Emine Koc, Assistant Professor; Ph.D., New Mexico State, 1997. Investigation of mammalian mitochondrial ribosomes in translation, apoptosis, and disease states of the mitochondria by using various proteomics techniques.

Kouacou Konan, Assistant Professor; Ph.D., Indiana, 1995. Hepatitis C and related viruses: how their replication interferes with host protein and membrane trafficking.

Andrey S. Krasilnikov, Assistant Professor; Ph.D., Russian Academy of Sciences (Moscow), 1995. Structural biology of RNA and RNA-protein complexes; X-ray crystallography.

Maria Krasilnikova, Research Assistant Professor; Ph.D., Moscow Institute of Physics and Technology, 1995. Mechanism of expansion in trinucleotide repeat diseases; transcription stalling at GAA repeats in Friedreich's ataxia.

Carsten Krebs, Assistant Professor; Ph.D., Max Planck Institute for Radiation Chemistry (Mülheim), 1997. Bioinorganic chemistry: spectroscopic and kinetic studies on the mechanisms of iron-containing enzymes.

Zhi-Chun Lai, Associate Professor; Ph.D., Yeshiva (Einstein), 1989. Molecular and developmental genetics of the *Drosophila* visual system.

Arthur M. Lesk, Professor; Ph.D., Princeton, 1966. Studies of how amino acid sequences determine protein structures and protein evolution, using bioinformatics techniques.

Bernhard Lüscher, Associate Professor; Ph.D., Zurich, 1986. Structure, function, and postsynaptic targeting of GABA$_A$ receptors.

Christopher Malone, Assistant Professor; Ph.D., Colorado at Boulder, 1999. Interaction of the nucleus with the cytoplasm.

Andrea M. Mastro, Professor; Ph.D., Penn State, 1971. Cell biology; immunobiology–protein kinases and growth factors; lymphocyte mitogenesis and tumor promoters; immune-endocrine interactions.

Pamela J. Mitchell, Associate Professor; Ph.D., Columbia, 1985. Regulation of gene expression during mouse and *Drosophila* embryogenesis.

Katsuhiko Murakami, Assistant Professor; Ph.D., National Institute of Genetics (Japan), 1997. X-ray crystallographic studies of RNA polymerases.

Anton Nekrutenko, Assistant Professor; Ph.D., Texas Tech, 1999. Comparative genomics; developing new computational tools to provide genomic insights into biological phenomena.

B. Tracy Nixon, Associate Professor; Ph.D., MIT, 1982. Molecular genetics and biology of nitrogen fixation; modulation and prokaryotic sensory transduction.

Randen L. Patterson, Assistant Professor; Ph.D., Maryland, Baltimore, 2000. Systems biology approach toward modeling of calcium channels within neural networks.

Robert Paulson, Associate Professor; Ph.D., California, San Francisco, 1993. Genetic and biochemical analysis of cell signaling during hematopoiesis.

Gary H. Perdew, Professor; Ph.D., Oregon State, 1984. Mechanisms of dioxin toxicity; regulation of Ah receptor activation; Ah receptor-mediated gene regulation.

Jeffrey M. Peters, Associate Professor; Ph.D., California, Davis, 1992. Delineating roles for peroxisome proliferator-activated receptors in homeostasis; toxicology and cancer.

Ronald D. Porter, Associate Professor; Ph.D., Duke, 1976. Recombination mechanism in *Escherichia coli;* single-stranded–DNA-binding proteins.

Kathleen Postle, Professor; Ph.D., Wisconsin–Madison, 1978. Gram-negative bacteria; energy transduction; TonB system; membrane protein complexes; siderophore-mediated iron transport.

B. Franklin Pugh, Professor; Ph.D., Wisconsin–Madison, 1987. Regulation of gene transcription; protein–nucleic acid assembly mechanisms.

Joseph C. Reese, Associate Professor; Ph.D., Illinois at Urbana-Champaign, 1992. Study of cell cycle– and DNA damage–induced transcription in *Saccharomyces cerevisiae.*

Lorraine C. Santy, Research Assistant Professor; Ph.D., Harvard, 1997. Small GTPases; motility and migration; epithelial cell morphology and function.

Stephen Schuster, Associate Professor; Ph.D., Munich, 1991. Genome sequencing and analysis of host-adapted and free-living microbial species.

Song Tan, Associate Professor; Ph.D., Cambridge, 1989. Structural biology of eukaryotic gene regulation; X-ray crystallography of multicomponent protein and protein-DNA complexes.

Michael Teng, Assistant Professor; Ph.D., Chicago, 1993. Pathogenesis of respiratory syncytial virus infection; virus–host cell interactions.

Graham H. Thomas, Associate Professor; Ph.D., Edinburgh (Scotland), 1984. Genetics and molecular and cell biology of the cytoskeleton in *Drosophila* development.

Ming Tien, Professor; Ph.D., Michigan State, 1981. Microbiology, biochemistry, enzymology, and molecular biology of lignin-degrading organisms.

C.-P. David Tu, Professor; Ph.D., Cornell, 1976. Gene expression, structure, and function of glutathione S-transferases.

Yanming Wang, Assistant Professor; Ph.D., Iowa State, 2001. Chromatin, epigenetics, gene expression, histone modifications, cell differentiation, and cancer.

PRINCETON UNIVERSITY

Department of Molecular Biology

Program of Study

The Graduate Program in Molecular Biology fosters the intellectual development of modern biologists. The program welcomes students from a variety of educational backgrounds and offers an educational program that goes well beyond traditional biology. The Department of Molecular Biology at Princeton is a tightly knit, cohesive group of scientists that includes undergraduate and graduate students, postdoctoral fellows, and faculty members with diverse but overlapping interests. Graduate students have a wide choice of advisers with a broad spectrum of interdisciplinary interests and research objectives.

The graduate program offers each entering student the opportunity, with the help of faculty advisers, to design the intellectual program that best meets his or her unique scientific interests. Each student chooses a series of research rotations with faculty members in molecular biology and associated departments (chemistry, computer science, ecology and evolutionary biology, engineering, and physics. Entering students, with the aid of the graduate committee, select core and elective courses from a large number of offerings in a variety of departments and disciplines. This combination of a cohesive department, one-on-one advising, and individualized programs of course work and research provides an ideal environment for graduate students to flourish as independent scientists. Areas of concentration include biochemistry and cell biology, biological dynamics and computational biology, biomedicine and societal issues, biophysics and structural biology, developmental biology, genetics and genomics, microbiology, neurobiology and neuroscience, and oncology and virology.

The general examination is normally administered at the end of the first semester of the second year of study when a student has met all formal course requirements. The exam is a 3-hour oral examination administered by 3 faculty members of the graduate program, none of whom may be the student's thesis adviser. The exam consists of two parts. The first part, the thesis proposal, probes depth of knowledge in the chosen research field and examines the ability of the student to justify and defend the goals, significance, and experimental logic and methods of the proposed plan. The second part of the exam tests the overall skill of the student in the field of molecular biology. The Ph.D. degree is awarded after a public oral examination in which the student defends a written thesis describing original experimental work; attainment of this degree normally requires 5½ years. Degree candidates assist in teaching two-term courses as part of their training.

Research Facilities

All areas of the department in Lewis Thomas Laboratory, Guyot Hall, Moffett Laboratory, George LaVie Schultz Laboratory, Hoyt Laboratory, and the Icahn Labs are fully equipped for contemporary research and include tissue-culture facilities, controlled-temperature rooms, animal quarters, and instrument rooms. In addition, the department houses several special facilities, which are maintained and run by specially trained technicians and are available for use by all department members. These include facilities for the production of transgenic mice and monoclonal antibodies. A flow cytometry facility allows analysis of individual cells. The Syn/Seq facility performs nucleic acid and peptide sequencing and synthesis as well as protein analysis by mass spectrometry. The imaging facility contains several confocal microscopes, a deconvolution microscope, and a transmission electron microscope. Microarray and computation facilities located in the Icahn Building are available for the entire department. Members of the department also share the use of specialized equipment, such as X-ray crystallography facilities and electronics and machine shops. The department has extensive networked computing facilities, connecting hundreds of individual laboratory computers, a departmental computer facility with numerous personal computers, the department's new multiprocessor Linux cluster, and the University's Redhat Linux cluster, which is named Hats.

Reference books, periodicals, and texts that are pertinent to molecular biology are available in the biology and chemistry libraries. Computer terminals in every lab, the molecular biology computer cluster, and the library provide online access to many extensive bibliographic databases and online journals. Many databases and journals are also available through the University computer network.

Financial Aid

Students who have been admitted to the program are encouraged to apply for fellowship support through such funding agencies as the National Science Foundation, the Howard Hughes Medical Institute, the Ford Foundation, or other sponsoring agencies that fund graduate work in the life sciences. Foreign nationals should apply for support from their home governments, employers, or the appropriate international foundations. Students enrolled in the program receive support from a combination of sources, such as individual fellowships, traineeships, research grants, corporate and foundation funds, or University and departmental funds.

An interdepartmental Training Program in Biological Dynamics provides an educational program and support for students with strong undergraduate preparation in physical science and mathematics and lesser preparation in biology.

Cost of Study

The tuition is $34,000 for the 2006–07 academic year.

Living and Housing Costs

In 2006–07, rooms at the New Graduate College cost from $4569 to $5891 for the academic year of thirty-five weeks. Several meal plans are available and are priced at $2969 to $4526. University apartments for married students rent for $664 to $1236 per month. Accommodations are also available in the surrounding community.

Student Group

At Princeton today, a student body of 4,700 men and women undergraduates and 2,000 graduate students works closely with a faculty of about 750. The students are chosen from all disciplines; they come from all over the United States and from forty to fifty other countries.

Location

A lively and attractive academic community of about 25,000, situated at the edge of a piedmont, the town of Princeton lies midway between the metropolitan centers of New York and Philadelphia. Residents can take advantage of an unusually broad array of intellectual activities and recreational opportunities. The University's McCarter Theatre has an excellent repertory company in residence and also presents both classic and contemporary films. In addition, there are concerts by outstanding soloists and professional concert groups. Regular performances are also given by Westminster Choir College, the Princeton Chamber Orchestra, Theatre Intime, and other amateur and student organizations. Local cinemas present both American and international films. Fields, streams, open countryside, and woods are all within walking distance. Buses to New York City leave each half hour from the campus. A shuttle runs from the campus to the main line of Amtrak.

The University

Princeton University was founded in 1746 as the College of New Jersey. The college was moved to Princeton in 1756. At its 150th anniversary in 1896, the trustees changed the name to Princeton University. Princeton's Graduate School was organized in 1901 and has since won international recognition in mathematics, the natural sciences, philosophy, and the humanities. The undergraduate student body included women as regular students for the first time in 1969. Currently, about one third of all graduate students are women.

Applying

December 31 is the deadline for receipt of applications, transcripts, and letters of recommendation from applicants who wish to enter the following September. December 1 is the application deadline for applicants outside North America. Admission information and online application forms are available on the World Wide Web at http://gradschool.princeton.edu/admission/.

Correspondence and Information

Office of Admission
Graduate School
Box 270
Princeton University
Princeton, New Jersey 08544

Laura Gallagher-Katz
Graduate Studies
Department of Molecular Biology
Lewis Thomas Laboratory
Princeton University
Princeton, New Jersey 08544-1014
Phone: 609-258-2803
E-mail: lgallagh@princeton.edu
Web site: http://www.molbio2.princeton.edu

Princeton University

THE FACULTY AND THEIR RESEARCH

Bonnie L. Bassler, Ph.D., Johns Hopkins. Intercellular communication and gene regulation in marine bacteria.
Michael J. Berry, Ph.D., Berkeley. Neural computation in the retina.
David Botstein, Ph.D., Michigan. Mapping and sequencing the genomes of yeast, bacteria and human as well as the genetics of cancer.
James R. Broach, Ph.D., Berkeley. Gene expression in the yeast *Saccharomyces cerevisiae*.
Carlos Brody, Ph.D., Caltech. Computational neuroscience.
Rebecca Burdine, Ph.D., Yale. Left-right patterning in the vertebrate embryo.
Hilary Coller, Ph.D., MIT. Molecular characterization of quiescence in human cells.
Edward C. Cox, Ph.D., Pennsylvania. Developmental genetics and the control of mutation rates.
Jonathan Eggenschwiler, Ph.D., Columbia. Genetic analysis of mouse neural development.
Lynn W. Enquist, Ph.D., Medical College of Virginia. Neurovirology.
Jane Flint, Ph.D., London. Regulation of viral and cellular gene expression in adenovirus-infected cells.
Jacques R. Fresco, Ph.D., NYU. Mutagenesis and structure-function relations in nucleic acids.
Elizabeth Gavis, Ph.D., Stanford. RNA localization and translational regulation during development in *Drosophila*.
Zemer Gitai, Ph.D., California, San Francisco. Bacterial cell biology: fundamentals of cytoskeletal dynamics; polarity and mitosis.
John Hopfield, Ph.D., Cornell. Computational neurobiology/biophysics.
Fred Hughson, Ph.D., Stanford. Protein structure, function, and folding.
Yibin Kang, Ph.D., Duke. Molecular mechanisms of cancer metastasis.
Ihor R. Lemischka, Ph.D., MIT. Analysis of hematopoietic differentiation, using retroviruses as markers.
Manuel Llinas, Ph.D., Berkeley. Molecular mechanisms of the malaria parasite *Plasmodium falciparum*.
Coleen Murphy, Ph.D., Stanford. Molecular mechanisms of aging.
Austin Newton, Ph.D., Berkeley. Molecular and genetic analysis of cell differentiation.
Mark D. Rose, Ph.D., MIT. Mechanism of nuclear fusion in *Saccharomyces cerevisiae*.
Hays S. Rye, Ph.D., Berkeley. Biochemistry of protein folding in cells.
Paul Schedl, Ph.D., Stanford. Control of gene expression and early development in *Drosophila melanogaster*.
Gertrud Schupbach, Ph.D., Zurich. Signal transduction and pattern formation during oogenesis of *Drosophila melanogaster*.
Jean E. Schwarzbauer, Ph.D., Wisconsin. Functions of extracellular-matrix proteins, particularly fibronectin, in vivo, using cell and molecular biological
    techniques.
Thomas E. Shenk, Ph.D., Rutgers. Tumor virus gene expression and oncogenesis.
Yigong Shi, Ph.D., Johns Hopkins. Structural biology of tumor suppressors involved in cell-cycle regulation and signal transduction.
Thomas J. Silhavy, Ph.D., Harvard. Molecular mechanisms of protein export; transmembrane regulation; genetic analysis using gene fusions.
Lee M. Silver, Ph.D., Harvard. Organization, evolution, and expression of mouse and human gene families involved in germ-cell differentiation.
Malcolm S. Steinberg, Ph.D., Minnesota. Adhesive cell recognition and morphogenesis.
Jeffry Stock, Ph.D., Johns Hopkins. Membrane receptors and sensory processing.
David W. Tank, Ph.D., Cornell. Measurement and analysis of neural circuit dynamics.
Saeed Tavazoie, Ph.D., Harvard. Whole genome approaches to understanding.
Samuel Sheng-Hung Wang, Ph.D., Stanford. Dynamics and learning in neural circuits.
Eric Wieschaus, Ph.D., Yale. Oogenesis and early embryogenesis in *Drosophila*.
Ned Wingreen, Ph.D., Cornell. Intracellular networks in bacteria.
Virginia Zakian, Ph.D., Yale. Molecular genetics: DNA replication and chromosome structure in yeast; structure and replication of telomeres.

ASSOCIATED FACULTY MEMBERS FROM OTHER DEPARTMENTS

The following faculty members from the Departments of Chemical Engineering, Chemistry, Computer Science, Ecology and Evolutionary Biology, Electrical Engineering, and Physics are associated with the Department of Molecular Biology. They participate regularly in its teaching and research activities and can sponsor departmental graduate students.

Chemical Engineering
Jeffrey Carbeck, Ph.D., MIT. New tools for the thermodynamic and kinetic characterizations of molecular recognition in biology.
Stas Shvartsman, Ph.D., Princeton. Dynamics of living tissues.
David Wood, Ph.D., Caltech. New technologies from self-modifying proteins.

Chemistry
Jannette Carey, Ph.D., Illinois. Protein and nucleic acid structure, function, and interactions.
Michael Hecht, Ph.D., MIT. Protein structure and the design of novel proteins.
Joshua Rabinowitz, Ph.D., Stanford. Toward a holistic understanding of cellular metabolism.
Clarence Schutt, Ph.D., Harvard. High-resolution structure of actin.

Computer Science
Mona Singh, Ph.D., MIT. Computational molecular biology, as well as its interface with machine learning and algorithms.
Olga Troyanskaya, Ph.D., Stanford. Bioinformatics and genomics: analysis of high-throughput biological data, biological functions and networks, and
    Saccharomyces cerevisiae genomics.

Ecology and Evolutionary Biology
Leonid Kruglyak, Ph.D., Berkeley. Genetic basis of phenotypic variation.
Laura Landweber, Ph.D., Harvard. Molecular evolution of "scrambled genes," the genetic code, and RNA; RNA editing; in vitro evolution; DNA computers.
David L. Stern, Ph.D., Princeton. Evolutionary developmental biology.

Electrical Engineering
Ron Weiss, Ph.D., MIT. Cellular computation, communications, and signal processing.

Physics
William Bialek, Ph.D., Berkeley. Coding, computation, and learning in the nervous system.

QUINNIPIAC UNIVERSITY

Molecular and Cell Biology Program

Programs of Study

The Master of Science in Molecular and Cell Biology Program at Quinnipiac University provides specialized training in laboratory research in the rapidly growing biomedical field. It is only since the development of recombinant DNA technology that the new and exciting field of biotechnology has emerged, and the need for specialists in this area has increased nationally. Biotechnology applies molecular and cell biology to the detection and treatment of human disease and is now a major division of pharmaceutical and research laboratories.

Quinnipiac's Molecular and Cell Biology Program provides the student with specialized training in laboratory research as well as the broad education necessary to function in this new and rapidly growing field. All students must complete a core curriculum of courses, a number of electives, and either a thesis or a comprehensive examination. Quinnipiac's core curriculum provides comprehensive training in advanced biochemistry, molecular genetics, cell biology, and laboratory methods. Electives cover industrial microbiology, virology, immunology, oncology, and molecular pathology. Students who choose the thesis option (37 to 39 credits) conduct original laboratory research either on campus or in an industrial or hospital setting, working under the supervision of a faculty member. Students in the nonthesis track (40 to 45 credits) take additional elective courses to gain competency in a specific area of molecular and cell biology, and they must demonstrate on a comprehensive exam that they have mastered both broad and specific knowledge expected of someone holding a master's degree.

This program prepares students to secure technical positions in biomedical research laboratories (primarily pharmaceutical or biotechnology companies and hospital-based research laboratories) or enter Ph.D. programs. For students already employed by a biotechnology laboratory or pharmaceutical company, the curriculum significantly enhances their knowledge of the state-of-the-art techniques used in the growing field of molecular and cell biology.

Research Facilities

Research and laboratory facilities are located primarily in Tator Hall, which houses a state-of-the-art training laboratory for students in the Molecular and Cell Biology Program. The laboratory contains all the equipment that is used in professional research laboratories, including ultracentrifuges, spectrophotometers, a Kodak imaging station, electron microscopes, and quantitative PCR, HPLC, and immunoflourescence imaging equipment. Students in the Molecular and Cell Biology Program also spend much of their time in Echlin Health Sciences Center. Echlin Center houses classrooms designed for clinical practice in the health sciences and extensive computer and robotics equipment as well as lecture halls and seminar rooms. All of the classrooms are wired, so that the faculty members can utilize the Internet, specialized computer programs, and the library databases and resources in their classes.

The University also provides an impressive collection of health science resources. The newly renovated and expanded Arnold Bernhard Library is one of the most technologically advanced centers for electronic information and learning resources anywhere in the country. The library houses hundreds of health science, medical, and science journals as well as a national citation database of scholarly articles in the health sciences. Students also have access to scores of online databases and resources, including Basic Biosis, *Biochemical Journal,* Biological Sciences from Cambridge Scientific Abstracts, CINAHL, FirstSearch, Harrison's Online Medical Clinic, Health Reference Center, *Journal of Chemical Education, Journal of Chemical Investigation,* Medline, National Science Digital Library, *New England Journal of Medicine,* and *Science Direct.* Off-campus e-mail users have full access to the University's computer network.

Financial Aid

Several avenues are available to help both full- and part-time students fund their education. Students may be eligible for Federal Stafford Student Loans. Graduate assistantships are available on a limited basis to both full- and part-time students.

Cost of Study

Tuition in 2005–06 was $570 per credit hour. In addition, part-time students paid a $30-per-credit student fee, and full-time students paid $275 per semester in student fees.

Living and Housing Costs

On-campus housing is available during the summer. Privately owned housing is available near the campus. For more information concerning off-campus housing, students should contact the Office of Residential Life or visit the University Web site.

Student Group

Approximately 30 full- and part-time students are enrolled in this program. Many are employed in laboratory settings at such facilities as pharmacy and medical schools and research laboratories. There are other students who are not employed in the field who want a career change, and there are some students who are pursuing their master's degrees as a foundation for pursuing a doctorate.

Location

The University is located on a beautiful campus in Hamden, Connecticut, a suburb of New Haven. It is approximately 30 minutes from Hartford, 90 minutes from New York City, and 2 hours from Boston.

The University

Quinnipiac University is nationally recognized as one of the leading centers for higher learning and health services education in the Northeast and is consistently ranked among the best master's-level universities in the north in *U.S. News & World Report's* Guide to America's Best Colleges. All programs have integrated computer technology into academic and campus life, and Quinnipiac has been recognized in *Yahoo! Internet Life* for its achievements in technology.

The University enrolls about 5,000 undergraduate and 1,800 graduate students and offers a full range of undergraduate and graduate programs through the School of Health Sciences, the School of Communications, the School of Business, the College of Liberal Arts, and the School of Law.

Applying

Anyone who holds a baccalaureate degree in the biological, medical, or health sciences is eligible to apply for admission. A detailed autobiography of personal, professional, and educational achievements as well as two letters of reference must be submitted with the application forms. Applications are accepted on a rolling basis for admission in the spring, fall, or summer. The program is selective, but there is no entering class limit.

Correspondence and Information

Office of Graduate Admissions
Quinnipiac University
275 Mount Carmel Avenue
Hamden, Connecticut 06518
Phone: 203-582-8672
       800-462-1944 (toll-free)
Fax: 203-582-3443
E-mail: graduate@quinnipiac.edu
Web site: http://www.quinnipiac.edu

Quinnipiac University

THE FACULTY AND THEIR RESEARCH

Full-time Faculty
Thomas Brady, Ph.D., Connecticut. Clinical pathology.
Deborah Clark, Ph.D., Cornell. Biochemistry.
Charlotte Hammond, Ph.D., Connecticut Health Center. Molecular biology.
Kenneth Kaloustian, Ph.D., New Hampshire. Physiology and endocrinology.
Edward O'Connor, Ph.D., Albany Medical College. Neuropharmacology.
Dennis Opheim, Ph.D., Minnesota. Molecular genetics.
Gene Wong, Ph.D., Alberta. Developmental biology.

Adjunct Faculty
Kenneth Carley, Dr.P.H., Alabama. Epidemiology.
Ron Dulac, M.S., Quinnipiac.
Janet Emanuel, Ph.D., Yale. Human genetics.
John Howe, M.D., Yale.
Edward McDonough, M.D., New York Medical College. Forensic pathology.
Lena Prisco, Ph.D., Connecticut. Molecular biology.
Thomas Tinghitella, Ph.D., Notre Dame. Microbiology.

THESIS TITLES FROM MASTER'S PROGRAM IN MOLECULAR AND CELL BIOLOGY
The Effect of Non-Thermal Therapeutic Ultrasound on T-Cell Proliferation and Gene Activation
Allosteric Ribozymes that Discriminate Between DNA and RNA Oligonucleotide Effectors
Expression of Tissue Factor and Proteinase-Activated Receptor 2 in the MCF-7 Cell Line and Their Interaction in Transmembrane Signaling
Allele Specific PCR Detection of UGT1A1 Mutations Causing Toxicity in Patients Taking Irinotecan
Regulation of Hox Genes in the Female Reproductive Tract
The Use of Real-Time RT-PCR to Assess Cytochrome P450 Expression
Development of a Synthetic Peptide Research Immunoassay to Detect Antibodies against HIV-1 Subtype O
In Vitro Evaluation of Transgene Expression After Iterative Retrovial Transduction

RUTGERS, THE STATE UNIVERSITY OF NEW JERSEY, NEW BRUNSWICK/ PISCATAWAY/UNIVERSITY OF MEDICINE AND DENTISTRY OF NEW JERSEY

Programs in Molecular Biosciences

Program of Study

Molecular Biosciences is an umbrella group of graduate programs in the life sciences division at Rutgers University and the University of Medicine and Dentistry of New Jersey–Graduate School of Biomedical Sciences. This program affords the utmost in flexibility and research opportunities by integrating more than 270 faculty members in nine degree-granting graduate programs at the two Universities. The Rutgers/UMDNJ Joint Graduate Programs in Biochemistry, Cell and Developmental Biology, Cellular and Molecular Pharmacology, Microbiology and Molecular Genetics, and Physiology and Integrative Biology and the UMDNJ-GSBS Graduate Programs in Biochemistry and Molecular Biology; Cellular and Molecular Pharmacology; Molecular Genetics, Microbiology, and Immunology; and Physiology and Integrative Biology use the molecular biosciences umbrella program as a mechanism to admit and train graduate students through their first year of studies.

All students take a series of core courses in molecular biology, molecular genetics, biochemistry, microbiology, and molecular aspects of cell and developmental biology. During their first year, incoming graduate students also do eight-week research rotations in three or four different laboratories to broaden their perspective and help them select a thesis adviser. A written qualifying examination, based on current research articles, must be passed and a proposition paper must be written and defended orally. At that time, the student chooses a specific graduate program, dependent on research interest, and follows specialized curriculum tracks as specified by that graduate program. The defining requirement for the Ph.D. degree is the completion and successful defense of a thesis based on original research.

Research Facilities

The extensive research facilities of both Universities provide opportunities for studying virtually all aspects of molecular, cell, and developmental biology. Research facilities of participating faculty members are located in approximately fifteen academic and clinical departments and research institutes at Rutgers and the Robert Wood Johnson Medical School–UMDNJ. Research institutes and centers include the Waksman Institute, the Cancer Institute of New Jersey, Center for Advanced Biotechnology and Medicine, Center for Agricultural Molecular Biotechnology, Center for Alcohol Studies, Environmental and Occupational Health Sciences Institute, and the Human Genetics Institute. A major computer facility, the Center for Computer and Information Services, provides powerful computing capabilities for advanced research. Microcomputers are available to students in most laboratories. The library system has holdings of more than 3 million bound volumes and 1 million documents; major scientific holdings are located at the Library of Science and Medicine and departmental and research institute libraries.

Financial Aid

All incoming Ph.D. graduate students are provided with financial support, including a stipend of $24,000 and tuition remission. NIH-sponsored training grants in biotechnology and in molecular biophysics support some students, and University Excellence Fellowships are available for candidates with outstanding credentials.

Cost of Study

In 2005–06, the tuition was $10,440 per year for New Jersey residents and $15,520.80 for out-of-state students. Graduate student fees were $753.50 per semester.

Living and Housing Costs

Furnished on-campus apartments are available for about $7075 per person (double occupancy) per calendar year. Residence halls for men and women are available. Family housing and meal plans at dining halls are also available. Information on private housing in the New Brunswick/Piscataway area is available through the Department of Housing (telephone: 732-445-2215; Web site: http://www.housing.rutgers.edu).

Student Group

The total graduate enrollment in the two Universities in New Brunswick/Piscataway is approximately 4,600 students, who participate in sixty-three programs in the two schools. The programs in molecular biosciences expect to admit 45 to 50 students in the fall of each year.

Location

The Graduate Programs in Molecular Biosciences are based at New Brunswick, with the research faculty and facilities centered on the Busch Campus, which serves as the location of the Medical School (UMDNJ–Robert Wood Johnson Medical School), the science facilities of Rutgers University, the Center for Advanced Biotechnology and Medicine (CABM), the Waksman Institute for Microbiology, the Cancer Institute of New Jersey, and the Environmental and Occupational Health Sciences Institute. Additional facilities are located on Cook Campus in New Brunswick. The Busch Campus is located in a suburban setting and is conveniently close to New York City and Philadelphia (about an hour away by train, bus, or car). Nearby recreational facilities include the Jersey Shore and the Pocono Mountains.

The Universities

Chartered as Queens College in 1766, Rutgers was the eighth institution of higher education founded in Colonial America. The name was changed to Rutgers College in 1825, and graduate instruction began at Rutgers in 1876. The University of Medicine and Dentistry of New Jersey was founded in 1970, and its Graduate School of Biomedical Sciences was organized in 1985. The Robert Wood Johnson Medical School (formerly Rutgers Medical School) is the medical school unit in New Brunswick/Piscataway. Students should go to http://lifesci.rutgers.edu/history for more details.

Applying

A bachelor's degree or its equivalent with at least a 3.0 average in academic work is required for admission. The most appropriate preparation is an undergraduate concentration in biology, biochemistry, or chemistry. Admission is based on academic work, letters of recommendation, and scores on the GRE General Test and Subject Test in biology, chemistry, or biochemistry, cell and molecular biology. TOEFL scores are required of students from countries where English is not the native language. Applications received by January 5 receive first consideration. The application fee is $50.

Rutgers University and UMDNJ are equal opportunity/affirmative action employers. Both recruit minority and women candidates, consider all applicants, and, once students are accepted, ensure that they will not be discriminated against in any area.

Correspondence and Information

For applications and program information:

Joint Graduate Programs in Molecular Biosciences
Nelson Biological Laboratories
Room A202
Piscataway, New Jersey 08854

Phone: 732-445-5086 or 3430
E-mail: gradoffice@biology.rutgers.edu
Web site: http://lifesci.rutgers.edu/~molbiosci/

Rutgers, New Brunswick / Piscataway / University of Medicine and Dentistry of New Jersey

THE FACULTY AND THEIR RESEARCH

Cory Abate-Shen. Homeodomain proteins; transcriptional regulation.

Juan P. Advis. Hypothalamic regulation of luteinizing hormone; reproduction physiology.

Alan Antoine. Metabolism of microorganisms.

Edward Arnold. HIV; AIDS; drugs; vaccines; crystallography; structural biology.

Gail Arnold. Development of chimeric human rhinoviruses as vaccines.

David Axelrod. Cellular and molecular oncology.

Bruce Babiarz. Mammalian developmental genetics.

Debarata Banerjee. Experimental therapeutics; drug resistance; molecular imaging.

Tamar Barkay. Microbial ecology of interactions of microbes with toxic materials.

Jean Baum. NMR; protein folding.

Helen Berman. Nucleic acid and protein structure.

Elisabetta Bini. Regulation of gene expression in response to heavy metal exposure.

Leonard Borack. Biochemical genetics of *Drosophila*.

Mark Brenneman. DNA repair and recombination.

Kenneth Breslauer. Drug-DNA interactions.

Gary Brewer. Posttranscriptional control of gene expression in disease.

Steven Brill. Eukaryotic DNA replication.

Barbara Brodsky. Structural analysis of collagen.

Marco Brotto. E-C coupling and muscle plasticity.

Linda Brzustowicz. Human genetic disorders.

Salvatore Caradonna. Human DNA repair.

George Carman. Regulation of phospholipids metabolism; signaling in yeast.

Chavela Carr. Mechanism and regulation of membrane fusion.

Rocco Carsia. Effect of stress on adrenocortical function.

Patrizia Casaccia-Bonnefil. Cell-cycle regulation; apoptotic signaling in neurodevelopment and disease.

Kiran Chada. Mammalian developmental biology.

Theodore Chase. Microbial and plant enzymology.

J. Don Chen. Gene regulation; molecular genetics of leukemia.

Kuang Yu Chen. Molecular basis of cellular growth regulation as it relates to cell differentiation; transformation; aging.

Suzie Chen. Cellular growth control.

Xuemei Chen. Molecular genetic analysis of flower development in *Arabidopsis*.

Khew-Voon Chin. Drug resistance.

Chi-hua Chiu. Evolutionary and developmental genetics.

Wendie Cohick. Cell biology (growth factors, signal transduction, mammary gland).

Allan Conney. Drug metabolism; chemical carcinogenesis.

Paul Copeland. Regulation of gene expression at the translational level.

Lori Covey. Molecular immunology of B- and T-cell interactions.

Dan Cowen. Understanding the downstream actions of antidepressants that selectively increase serotonin.

Kiron Das. Pathogenesis of inflammatory bowel disease.

Robin Davis. Sensory signaling and ontogeny, as related to molecular mechanisms.

David Denhardt. Gene regulation; nitric oxide and cancer.

Emanuel DiCicco-Bloom. Regulation of developmental and adult neurogenesis; cell-cycle mechanisms.

Hugo Dooner. Plant molecular genetics: transposons.

Joseph Dougherty. Retroviral replication and gene transfer.

Cheryl Dreyfus. Role of environmental factors on brain neuron ontogeny and glial function.

Monica Driscoll. Molecular genetics of neurodegeneration.

Richard Ebright. Protein-DNA interaction.

Isaac Edery. Biological clocks.

M. David Egger. Neurophysiology.

Douglas Eveleigh. Industrial microbiology.

Julie Fagan. Function of proteolytic enzymes.

Huizhou Fan. Cell signaling under physiological and pathological conditions.

Martin Farach-Colton. Computational biology.

Bonnie Firestein. Targeting of neuronal proteins.

Joseph Fondell. Regulation of gene expression by nuclear hormone receptors.

Dunne Fong. Molecular cell biology of telomerase; proteinase and proteinase inhibitors.

David Foran. Machine vision systems.

Ramsey Foty. Biophysical and molecular aspects of tumorigenesis and malignant invasion.

Abram Gabriel. Eukaryotic retrotransposition.

Marc Gartenberg. Nuclear organization of DNA.

Celine Gelinas. Function of Rel oncoproteins.

Donald Gerecke. Collagen; lung; liver; heart; fibrosis; hypertension.

F. Joseph Germino. Protein-protein interactions; cell-cycle control.

Marion Gordon. Collagen expression during corneal development; tumor cell metastasis.

David Gorski. Homeobox genes in tumor biology; angiogenesis inhibition in tumor therapy.

Alice Gottlieb. Psoriasis; clinical pharmacology/research; immunotherapy/immunodermatology.

Barth Grant. Endocytosis and recycling; protein/membrane transport; molecular genetics/genomics.

Martin Grumet. Molecular mechanisms of cell adhesion in the nervous system.

Samuel Gunderson. RNA-protein interactions.

Raymond Habas. Wnt signaling.

Max Häggblom. Environmental microbiology.

Beatrice Haimovich. Cell-surface interactions.

William Hait. Calmodulin-mediated signal transduction and drug resistance.

Michael Hampsey. Eukaryotic gene expression; yeast genetics.

Ronald Hart. Neuroimmunology of central nervous system injury.

Charles Hewitt. Posttranslational mechanisms in uracil-DNA repair.

Jody Hey. Evolutionary origins of species and adaptations using DNA sequences.

Sarah Hitchcock-DeGregori. Study of contractile proteins.

Shu-Chan Hsu. Molecular mechanisms of vesicle trafficking underlying neuronal function.

Masayori Inouye. Cellular adaptation to stress.

Sumiko Inouye. Molecular biology of *Myxobacteria*.

Kenneth Irvine. Cell signaling; pattern formation; growth control; developmental glycobiology.

Stephen Isied. Modification of electron transfer proteins.

Estela Jacinta. Signal transduction; growth control; actin cytoskeleton.

Shengkan Jin. Tumor suppressor genes; programmed cell death; tumorigenesis.

William Johnson. Human genetics; nervous system traits and diseases.

Frank Jordan. Bioorganic and biophysical chemistry.

Dennis Joslyn. Structure and function of eukaryotic chromosomes.

Peter Kahn. Protein biophysics; dioxins.

Barton Kamen. Pediatric oncology; floate metabolism; anti-folates; pharmacology.

Stanley Katz. Agricultural microbiology.

Joseph Kedem. Myocardial function.

Lee Kerkhof. Marine microbiology and molecular ecology; microbial population dynamics.

Randall Kerstetter. Molecular genetics of leaf development; dorsal-ventral polarity in *Arabidopsis*.

Mike Kiledjian. Posttranscriptional regulation of gene expression.

Terri Goss Kinzy. Regulation of gene expression; mechanisms of protein synthesis.

Joachim Kohn. Artificial biopolymers.

Ah-Ng Tony Kong. Pharmacogenomics and toxicogenomics of cancer chemopreventive compounds; MAP kinases.

Mary Konsolaki. Alzheimer's disease.

Marilyn Kozak. Mechanism of translation in eukaryotic cells.

Sunita Gupta Kramer. Cell migration and signaling; extracellular matrix; muscle and heart development.

Jerome Kukor. Biochemistry and genetics of microbial biodegradation processes.

Casimir Kulikowski. Machine learning in molecular biology.

Eric Lam. Regulation of gene expression in higher plants.

Jerome Langer. Protein recognition and cytokine biology.

Debra Laskin. Immunology; macrophages; nitric oxide; inflammation.

Jeffrey Laskin. Carcinogenesis and differentiation in culture.

Edmund Lattime. Basic immunologic mechanisms involved in host-tumor interactions.

Michael Lawton. Plant defense and signal transduction.

Hsin-Yi Lee. Developmental biology.

Michael Leibowitz. Molecular genetics of *Pneumocystis carinii*.

John Lenard. Viral entry and assembly.

Thomas Leustek. Sulfur metabolism in higher plants.

Ronald Levy. Computer modeling of protein structure.

Honghua Li. Genome-scale understanding genetic basis of breast cancer.

Alice Liu. Function of cAMP-dependent protein kinase.

Fang Liu. Signal transduction and gene regulation.

Leroy Liu. DNA topoisomerases.

Peter Lobel. Targeting lysosomes in mammalian cells.

Stuart Lutzker. P53, apoptosis, and DNA repair.

Jianjie Ma. Function of membrane proteins.

Kiran Madura. Proteolytic pathways in yeast.

Pal Maliga. Molecular genetics of plastids.

Paul Manowitz. Biochemistry and genetics of mental illness.

Charles Martin. Regulation of membrane assembly.

Michael Matise. Molecular genetic control of vertebrate CNS development.

Tara Matise. Computational genetics.

Fumio Matsumura. Control of cell division.

Karl Matthews. Biotechnology to address virulence and survival of foodborne pathogens.

Michael McCormack. Human genetics.

Terry McGuire. Behavioral genetics and neurogenetics.

Kenneth McKeever. Comparative exercise and cardiovascular physiology.

Kim McKim. Regulation of meiotic recombination.

Randall McKinnon. Molecular neurobiology.

Sally Meiners. Extracellular matrix and control of neural growth and regeneration.

Gary Merrill. Coronary circulation and ventricular rhythmicity.

Joachim Messing. Gene structure in higher plants.

James Millonig. Development of vertebrate CNS.

Prabhas Moghe. Tissue engineering and cell-biomaterial interactions.

Gaetano Montelione. Solution structures of proteins and DNA.

Thomas Montville. Food and fermentation microbiology.

William Moyle. Hormone-receptor interactions.

Robert Nagele. Early embryonic development.

Joseph Naus. Scan statistic probabilities of clustering.

Robert Niederman. Photosynthetic membranes.

Richard Nowakowski. Development of the vertebrate cerebral cortex.

Wilma Olson. Biopolymer conformation and properties.

Tim Otto. Neurobiology of learning and memory.

Richard Padgett. Signal transduction in *Drosophila* and *C. elegans*.

Zui Pan. Calcium signaling in apoptosis and muscle functions.

Jerome Parness. Regulation of intracellular calcium pools.

Nicola Partridge. Hormonal regulation of gene expression.

Smita Patel. Enzymatic mechanisms of DNA and RNA helicases and their role in replication and transcription.

Garth Patterson. Signal transduction in *C. elegans*.

Henrik Pedersen. Biophotonics in bioprocess technology.

Stuart Peltz. Posttranscriptional control mechanisms.

Sidney Pestka. Interferons and cytokines.

Catherine Phillips. Regulation of RNA3 end formation/polyadenylation.

George Pieczenik. Theory of genotypic selection.

Daniel Pilch. Development of topoisomerase I–directed anticancer drugs.

John Pintar. Insulin-like growth factors.

Vince Pirrotta. Chromatin structure and dynamics; polycomb-silencing mechanisms.

Mark Plummer. Kinetic studies of neuronal calcium channels.

Larissa Pohorecky. Pharmacology of alcohol and of stress.

Ronald Poretz. Molecular basis of cell recognition.

Jamshid Rabii. Dynamics of brain-pituitary dynamics.

Arnold Rabson. Molecular biology of human retroviruses.

Cordelia Rauskolb. Regulation and mechanism of segmentation during *Drosophila* development.

Danny Reinberg. Regulation of gene expression.

Michael Reiss. Role of transforming growth factor–β (TGFβ) in human cancer; cancer therapy.

Susan Rittling. Role of osteopontin in tumorigenesis and bone metabolism.

Yacov Ron. Cellular mechanisms of autoimmunity.

Christopher Rongo. Formation and plasticity of synapses.

Martin Rosenberg. Viral gene expression.

Monica Roth. Molecular biology of retroviruses.

Charles Roth. Molecular bioengineering; gene-based technologies; inflammatory diseases and therapies.

Hong Ruan. Adipose tissue–secreted cytokines; adipocyte gene transcription.

Eric Rubin. Clinical and molecular approaches to inhibition of DNA topoisomerases.

Loren Runnels. Signal transduction; role of channel-kinase TRPM7 in cell adhesion.

Alexey Ryazanov. Cell-cycle regulation of protein synthesis.

Amrik Sahota. Kidney diseases; Alzheimer's disease; transplant tolerance; molecular diagnostics.

Dipak Sarkar. Influences of stress and drug abuse on neuroendocrine and neuroimmune systems.

Donald Schaffner. Predictive food microbiology; quantitative microbial risk assessment.

Kathleen Scott. Mammalogy; vertebrate paleontology.

David Seiden. Skeletal and cardiac muscle.

Nagarajan Selvamurugan. TGFβ signaling; breast cancer metastasis.

Federico Sesti. Physiology, structure, and function of potassium channels.

Konstantin Severinov. Transcription mechanisms in microorganisms.

Daniel Shain. Molecular and cellular development of annelids.

Changshun Shao. Genetic, epigenetic , and environmental factors genesis of somatic mutation.

Aaron Shatkin. Viral cytopathogenesis.

Michael Shen. Mammalian embryology.

Yufang Shi. Regulation of immune responses; apoptosis in lymphocytes; tumor immunology; neuroimmunology.

Gleb Shumyatsky. Mechanisms of learning and memory.

Leonard Sigal. Immunopathogenesis of Lyme disease.

Lee Simon. Protein degradation in *Escherichia coli*.

Andrew Singson. Molecular mechanisms of sperm-egg interations in *C. elegans*.

William Sofer. Genetic algorithms and genetic programming.

Patricia Sonsalla. Neurotoxicology.

Martha C. Soto. *C. elegans* embryos to investigate polarized cell divisions and migrations during development.

Stanley Stein. Protein purification and sequencing.

Ruth Steward. Pattern formation in *Drosophila*.

Ann Stock. Structure-function analysis of signal transduction proteins.

Victor Stollar. Virus-host cell interactions.

Roger Strair. Retroviral resistance to antiviral agents.

Vasily Studitsky. Transcription regulation.

Nanjoo Suh. Mechanistic study of cancer.

Mikhail Tchikindas. Microbiology; food safety; bacteriocins.

Thresia Thomas. Breast cancer therapeutics.

T. J. Thomas. Regulation of ornithine decarboxylase in autoimmune diseases.

Moti Lai Tiku. Immunobiology of autoimmune diseases.

Jay Tischfield. Medical and human genetics: mouse models; somatic mutation; complex human disorders.

Robert Trelstad. Patterning processes in development.

Chih-Cheng Tsai. Transcriptional regulation; nuclear receptor corepressors; *Drosophila*.

Nilgun Tumer. Viral replication in transgenic plants.

Lynn Vales. Regulation of gene expression.

Theodorus Van Es. Carbohydrate chemistry.

Andrew Vershon. Transcriptional regulation in yeast.

Costantino Vetriani. Deep-sea microbiology; extremophiles; molecular ecology; adaptations to extreme environments.

William Wadsworth. Guidance of cell migrations in *C. elegans*.

Nancy Walworth. Cell-cycle progression in yeast.

Yuh-Hwa Wang. Triplet-repeat diseases.

William Ward. Application of bioluminescence.

Harvey Weiss. Nitric oxide and cyclic GMP as a "brake" on myocardial function.

William Welsh. Drug discovery; computer-aided molecular modeling; bioinformatics; cheminformatics.

Guy Werlen. Signaling networks and mechanisms that control life and death of developing T lymphocytes.

Eileen White. Programmed cell death.

Julie Williams. Circadian and homeostatic mechanisms of sleep.

Frank Wilson. Functions of motility-related proteins.

Lori White. Molecular mechanisms of xenobiotic-induced pathologies.

Donald Winkelmann. Macromolecular assembly of myosin actin.

Nancy Woychik. Eukaryotic gene regulation; RNA polymerase II.

Mengqing Xiang. Molecular bases of sensorineural development.

Gutian Xiao. Signal transduction pathways in development and function of immunity.

Chung Yang. Nitrosamines; carcinogenesis; cytochrome P-450.

Guofeng You. Molecular and cellular pharmacology; drug/toxin elimination; transporters.

Lily Young. Microbial metabolism of pollutants.

Wise Young. Brain and spinal cord injury.

Peter Yurchenco. Structure and function of basement membranes.

James Zheng. Formation of neuronal circuitry.

X. F. Steven Zheng. Growth control; signal transduction; cancer; chemical genetics and genomics; drug discovery.

Renping Zhou. Brain development.

Barbara Zilinskas. Plant molecular biology.

Gerben Zylstra. Molecular analysis of microbial catabolic pathways.

SAN DIEGO STATE UNIVERSITY

Joint Ph.D. Program in Biology
Emphasis in Cell and Molecular Biology

Programs of Study

In cooperation with the University of California, San Diego (UCSD), San Diego State University (SDSU) offers a doctoral degree program in biology. This joint program involves research training under the supervision of participating SDSU faculty, whose members are drawn from the Departments of Biology and Chemistry or the Molecular Biology Institute and whose interests cover a variety of areas in the molecular life sciences. After admission the student must spend at least one academic year in full-time residence at UCSD and the remainder at SDSU.

There are no formal course requirements except for a one-year graduate course that includes genetics, cell biology, and molecular biology, which is often taken during the student's first year of residence in the program. In consultation with the student, an advising committee, however, determines which additional courses may be needed (from offerings at SDSU or UCSD). During their first year, students usually carry out research project rotations through at least three biology or chemistry laboratories at SDSU, which are selected in consultation with their committee. Following this, students spend the major portion of their time on dissertation research with a selected SDSU faculty research mentor. A limited amount of work as a graduate teaching assistant is also required from every student. Seminars given by visiting scientists, participation in journal clubs, and close student-faculty contacts also serve to widen the student's perspective and grasp of the current frontiers in the molecular life sciences.

By the end of the second year, a Joint Ph.D. Qualifying Committee conducts an oral comprehensive qualifying examination focusing on two areas related to the student's major research interest and presentation of a dissertation proposal. Satisfactory completion and oral defense of a dissertation consisting of original and significant research fulfill the requirements for the degree.

A master's degree program in molecular biology is also offered. This program is described in the separate Announcement under SDSU programs listed in the Biological and Biomedical Sciences section of *Peterson's Graduate Programs in the Biological Sciences.*

Research Facilities

The recently renovated laboratories of the Department of Biology and the newly built chemistry laboratories are fully equipped for modern research in cell and molecular biology and biochemistry. Major items include an automated DNA sequencing and synthesis facility, phosphorimagers, electron microscopes, image processing facilities, NMR and mass spectrometers, spectrofluorometers, and other modern analytical and separation instrumentation; controlled-temperature and animal rooms; research support shops; and multiuser confocal and deconvolution microscopes. SDSU is also the only participant from the California State University (CSU) system to be a member of the regional Cray Supercomputer Consortium. The new SDSU BioScience Center contains research space to house the Center for Microbial Sciences and the SDSU Heart Institute.

Financial Aid

Annual stipends (currently ranging from $20,000 to $23,000 per year) and funds for research supplies are provided for all joint Ph.D. students. A full health benefit package is also provided.

Cost of Study

The full cost of tuition is provided for all students in the joint Ph.D. program.

Living and Housing Costs

There are seven residence halls/complexes at San Diego State University, including the Villa Alvarado apartments. Graduate students also have the option of living in the Piedra del Sol apartments, a sixty-six-unit complex that offers entirely independent apartment living. In addition, numerous apartments and shared single-unit houses are available in the nearby community.

Student Group

At present, 42 students are enrolled in the Joint Ph.D. Program in Biology. The unusually high faculty-to-Ph.D.-student ratio fosters close contact and high-quality training for incoming students. Approximately 65 graduate students are also enrolled in the molecular biology M.S. program under the direction of Molecular Biology Institute faculty members. In addition, more than 100 graduate students are enrolled in other degree programs in biology and chemistry.

Location

SDSU is located 12 miles east of downtown San Diego and the Pacific Ocean. San Diego is host to one of the major biomedical research communities in the United States, with close interactions among UCSD, Salk Institute, Scripps Institute of Oceanography, Scripps Research Institute, Burnham Institute, numerous biotechnology companies, and SDSU. San Diego enjoys a reputation for easy living and has a highly desirable climate. Major cultural attractions include a variety of music and dance clubs; outdoor theme parks, including the world-famous San Diego Zoo; theater; opera; popular and classical music concerts; and a rich art scene. Spectator-sports enthusiasts can enjoy both college and professional sports teams. The nearby ocean, mountains, and deserts allow for an unusually wide variety of year-round outdoor activities.

The University

More than 30,000 students and 1,850 faculty members make SDSU one of the largest of twenty-three institutions within the CSU system, which is itself the largest public university system in the United States. SDSU is also unique within the CSU system in offering a number of joint Ph.D. programs.

Applying

Application for admission should be sent directly to the joint doctoral program coordinator at San Diego State University. A complete application to this program requires the following information to be provided by January 3: the appropriate application form, a statement of purpose, three letters of recommendation, a set of official transcripts of academic work already completed, and results of the Graduate Record Examinations.

Correspondence and Information

Coordinator, Joint Doctoral Program in Cell and Molecular Biology
Department of Biology
San Diego State University
San Diego, California 92182-4614
E-mail: biojtdoc@sunstroke.sdsu.edu
Web site: http://www.bio.sdsu.edu/molbio.html

San Diego State University

THE FACULTY AND THEIR RESEARCH

César Arenas-Mena, Assistant Professor of Biology; Ph.D., Barcelona, 1995. Genomic regulatory networks and evolution of developmental processes in marine invertebrate embryos. (e-mail: arenas@sciences.sdsu.edu)

Sanford I. Bernstein, Professor of Biology; Ph.D., Wesleyan, 1979. Molecular analysis of gene expression during *Drosophila* muscle development; molecular and ultrastructural defects of *Drosophila* muscle mutants; function of muscle protein isoforms; mechanism of alternative RNA splicing. (e-mail: sbernstein@sunstroke.sdsu.edu)

Richard L. Bizzoco, Professor of Biology; Ph.D., Indiana, 1972. Mechanism of cell fusion in the unicellular biflagella algae *Chlamydomonas:* characterization of the components that provide the molecular basis for cell recognition and cell fusion, using immunological, biochemical, and structural approaches with wild-type and mutant strains. (e-mail: rbizzoco@sunstroke.sdsu.edu)

Roger A. Davis, Professor of Biology; Ph.D., Washington State, 1971. Secretory protein targeting and degradation; regulation of expression of liver-specific genes; gene therapy for metabolic disease. (e-mail: rdavis@sunstroke.sdsu.edu)

Terrence G. Frey, Professor of Biology; Ph.D., UCLA, 1975. Structure of cellular organelles, biological macromolecules, membranes, and membrane proteins; bioenergetics: electron microscopy; image processing. (e-mail: tfrey@sunstroke.sdsu.edu)

Christopher Glembotski, Professor of Biology; Ph.D., UCLA, 1979. Cell and molecular biology of cardiac myocytes; hormonal and developmental regulation of cardiac-specific gene expression; mechanisms of cardiac natriuretic peptide gene expression; signal transduction pathways responsible for regulating cell growth. (e-mail: cglembotski@sunstroke.sdsu.edu)

Greg L. Harris, Professor of Biology; Ph.D., North Carolina at Chapel Hill, 1981. Analysis of neuromuscular and olfactory system function in *Drosophila.* (e-mail: gharris@sunstroke.sdsu.edu)

Tom Huxford, Assistant Professor of Chemistry and Biochemistry; Ph.D., California, San Diego, 2001. Structural biology of proteins and protein complexes involved in signaling to NF-kappaB. (e-mail: thuxford@sciences.sdsu.edu)

Scott Kelley, Assistant Professor of Biology; Ph.D., Colorado, 1998. Phylogenetic approaches to RNA structure prediction, DNA and protein motif pattern recognition, and genome sequence analysis; molecular systematic studies of microbial communities. (e-mail: skelley@sciences.sdsu.edu)

David A. Lipson, Assistant Professor of Biology; Ph.D., Colorado, 1998. Soil microbial ecology; plant-microbe interactions; biogeochemistry; linking microbial diversity to ecosystem processes. (e-mail: dlipson@sciences.sdsu.edu)

John Love, Assistant Professor of Chemistry and Biochemistry; Ph.D., California, San Diego, 1998. Protein engineering: driving novel protein-protein associations by computational and experimental design. (e-mail: jlove@sciences.sdsu.edu)

Stanley Maloy, Professor of Biology; Ph.D., California, Irvine, 1981. Genetic, molecular, biochemical, and genomic approaches to salmonella biology and evolution of pathogenesis. (e-mail: smaloy@sciences.sdsu.edu)

Kathleen McGuire, Professor of Biology; Ph.D., Texas Health Science Center at Dallas, 1985. Role of the HTLV-I proteins *Tax* and *Rex* in transformation of human T lymphocytes. (e-mail: kmcguire@sunstroke.sdsu.edu)

Paul J. Paolini, Professor of Biology; Ph.D., California, Davis, 1968. Signal transduction in cardiac myocytes and skeletal fibers; muscle cell mechanics; video-enhanced and fluorescence light microscopy, optical diffractometry, digital image analysis, and X-ray microscopy; computer methods in biological research. (e-mail: ppaolini@sunstroke.sdsu.edu)

Jacques Perrault, Professor of Biology; Ph.D., California, San Diego, 1972. Molecular biology of eukaryotic RNA viruses; regulation of vesicular stomatitis and measles virus genome replication and transcription; defective interfering virus particles and virus persistence; RNA viruses as gene vectors. (e-mail: jperrault@sunstroke.sdsu.edu)

Forest Rohwer, Assistant Professor of Biology; Ph.D., San Diego State/California, San Diego, 1997. Genomic analysis of marine phage; opportunistic infections and coral disease; diversity of coral-associated bacteria. (e-mail: forest@sunstroke.sdsu.edu)

Roger A. Sabbadini, Professor of Biology; Ph.D., California, Davis, 1974. Apoptosis in striated muscle cells; mechanism of action of TNFα in cardiac cells; role of sphingolipid second messengers in the control of muscle excitation-contraction coupling and apoptosis; structure and function of membrane-bound cation-translocating ATPases, calcium release channels, and protein kinases in cardiac and skeletal muscle. (e-mail: rsabbadini@sunstroke.sdsu.edu)

Thomas R. Scott, Professor of Psychology; Ph.D., Duke, 1970. Neurophysiology of feeding and the neural mechanisms of taste; single neuron electrophysiology; behavior. (e-mail: trscott@sciences.sdsu.edu)

Anca M. Segall, Professor of Biology; Ph.D., Utah, 1987. The mechanism of site-specific recombination; structure-function analysis of recombination proteins. (e-mail: asegall@sunstroke.sdsu.edu)

William E. Stumph, Professor of Chemistry and Biochemistry; Ph.D., Caltech, 1979. Regulation of transcription in eukaryotic cells; characterization of genes encoding the small nuclear RNAs U1, U2, and U4; characterization of transcription factors; sequence-specific protein-DNA interactions and assembly of the transcriptional complex. (e-mail: wstumph@sciences.sdsu.edu)

Mark A. Sussman, Professor of Biology; Ph.D., USC, 1989. Structural and molecular basis of heart failure; cell signaling, structural remodeling, cell survival, and cytoskeletal regulation in heart disease. (e-mail: sussman@sunstroke.sdsu.edu)

Constantine D. Tsoukas, Professor of Biology; Ph.D., California, San Francisco, 1975. Signal transduction via the T-cell receptor/CD3 molecular complex; emphasis on human T cells and thymocytes; receptors for Epstein-Barr virus (EBV) on cells of the T lineage. (e-mail: ctsoukas@sunstroke.sdsu.edu)

Peter Van Der Geer, Assistant Professor of Biochemistry; Ph.D., Amsterdam, 1987. Molecular, biological, and biochemical analysis of signal transduction by protein tyrosine kinases. (e-mail: pvanderg@sciences.sdsu.edu)

Elizabeth Waters, Assistant Professor of Biology; Ph.D., Washington (St. Louis), 1993. Plant evolution; origin of land plants; molecular evolution. (e-mail: ewaters@sciences.sdsu.edu)

Robert Zeller, Assistant Professor of Biology; Ph.D., Caltech, 1995. The developmental biology of ascidians; evolution of developmental gene regulatory networks in primitive chordates. (e-mail: rzeller@sunstroke.sdsu.edu)

SETON HALL UNIVERSITY

Ph.D. in Molecular Bioscience and M.S. in Biology and Microbiology

Programs of Study

The Biology Graduate Programs in the College of Arts and Sciences at Seton Hall University prepare students for basic research and teaching applicable for careers in academic institutions or the pharmaceutical and biotechnology industry. The master's and doctoral degrees can be completed on a full- or part-time basis. Many of the part-time students are employed by local pharmaceutical and biotechnology companies and interested in career advancement or changing careers. Graduate classes are offered in the evening and on Saturday in order to accommodate work schedules. The hands-on, laboratory-based courses are a unique aspect of the program and allow graduate-level students to develop new technical skills.

The Ph.D. in Molecular Bioscience program emphasizes the application of molecular and cellular biotechnology in studies on living systems and provides students with a strong foundation in research and teaching. The doctoral program consists of two phases: foundation course work and dissertation research. Students must pass a comprehensive or qualifying examination after completion of the 18 credits of required courses and present and defend an oral and written doctoral dissertation. Students must complete a total of 72 credits, including 49 required credits in course work (21 credits in required courses and 28 credits in dissertation and seminar courses) and 23 credits of electives. The required courses provide students with a strong foundation in subject content and training in research techniques. Electives provide breadth to the students' training in the various subdisciplines of molecular bioscience.

The M.S. in biology has three programs of study: Plan A, with research thesis; Plan B, with library thesis; and Plan C, with library thesis and business minor. Plan A is recommended for students who intend to continue their studies at the Ph.D. level or pursue a career in research. Students must complete 31 credits of course work, thesis research, and seminar. Plan B is primarily a course work–oriented degree program. Students must complete 33 credits of course work and seminar and 1 credit of Selected Topics in conjunction with the library thesis. Plan C, with its component of business administration courses, is for students interested in acquiring knowledge about the technical aspects as well as the business aspects of the biological and pharmaceutical industries. For this plan of study, students must complete 18 credits of course work and seminar, 15 credits in graduate business courses, and 1 credit of Selected Topics in conjunction with the library thesis.

Students may also attain the M.S. in biology with a neuroscience track: Plan F, with research thesis, and Plan G, with library thesis. For Plan F, students must complete 15 credits of selected course work for neuroscience plus 16 credits in electives, research thesis, and seminar. Plan G requires 15 credits of selected course work for neuroscience plus 18 credits in electives and seminar and 1 credit of Selected Topics in Neuroscience in conjunction with their library thesis.

The M.S. in Microbiology program has two programs of study: Plan D is recommended for students who intend to continue their studies in microbiology or biotechnology at the doctoral level or pursue a career in research. Plan E is primarily a course work–oriented degree program and includes a library thesis. For Plan D, students must complete 13 credits in elective course work, research thesis, and seminar plus 15 credits from select microbiology courses and 3 credits of molecular biology, metabolic pathways, or general biochemistry. For Plan E, students must complete 11 credits in elective course work, library thesis, and seminar plus 20 credits in microbiology courses and 3 credits from molecular biology, metabolic pathways, or general biochemistry.

All M.S. students submit a written thesis and make an oral presentation of their thesis in Graduate Seminar. In addition, the research thesis must be defended before a selected committee.

Research Facilities

The faculty members within the Department of Biology maintain active research programs that are sponsored by federal funding, including NIH, NSF, and EPA. In addition, collaborative research programs with local biotechnology and academic institutions further provide opportunities for research training.

The Department of Biology has both modern research labs and teaching labs of various sizes, ranging from 600–800 square feet. Graduate students are trained in the use of state-of-the-art methodologies and carry out cutting-edge research projects. For example, the acquisition of the Cytofluor 4000 plate reader, the STORM 860 imager, and the fluorescent/phase contrast microscope enable students to expand their research projects into the area of fluorescence-based techniques. A top line Olymus FV1000 confocal microscopy system funded through the National Science Foundation is available for both research and teaching.

The Walsh Library provides publications in molecular bioscience, including those from the EBSCO database, and a subscription to Science Direct. Clinical journals related to molecular bioscience are available in the library of the University of Medicine and Dentistry of New Jersey–New Jersey Medical School (UMDNJ-NJMS), located in nearby Newark and are accessible to faculty members and students.

Financial Aid

Teaching assistantships and research assistantships, which include tuition benefits and stipend, are available on a competitive basis. To obtain an application form, students should call 973-761-9044 or send an e-mail to artsci@shu.edu.

Other financial aid options include loans, payment plans, and campus jobs. For more information, prospective students should visit http://www.shu.edu/fawebrsrc.html.

Cost of Study

In 2005–06, tuition was $743 per credit. Full-time students paid $305 in University and technology fees; part-time students paid $185.

Living and Housing Costs

Housing and living costs in South Orange and surrounding towns are comparable to most suburban cities, with studio and one-bedroom apartments renting for $650 to $900 per month. Off-campus listings are available through the Department of Housing and Residence Life.

Location

Seton Hall University is located in South Orange, New Jersey, close to New York City via direct train. The University's proximity to New York City allows students to take advantage of all the city has to offer, while still living in a suburban area. The headquarters and laboratories for many of the major pharmaceutical and biotechnology companies are located in proximity to Seton Hall, thus providing students with opportunities to interact and collaborate with industry researchers while pursuing their graduate degrees.

The University and The Department

Seton Hall University is New Jersey's only Catholic university. The University's diverse academic program is characterized by a strong teaching faculty and a wide range of academic choices. The Department of Biology is part of the College of Arts and Sciences, home to more than 4,000 students and faculty members. The Biology M.S. programs generally enroll 60 to 70 students, while the Ph.D. in Molecular Bioscience program enrolls a combination of 15 to 18 full- and part-time students. Seton Hall has approximately 4,500 graduate students overall. At Seton Hall, students find people who are willing to listen, offer support, and help them get the most out of their education.

Applying

The M.S. and Ph.D. programs follow the general University requirements for admission to graduate studies. In addition, the Department of Biology requires a B.S. in a biological science or related science (a minimum 3.0 GPA), a minimum of 24 credits in biology with laboratory; 16 credits in chemistry, including organic chemistry with laboratory; 8 credits of physics with laboratory; and 8 credits in mathematics beyond pre-calculus. The submission of GRE scores (at least the 50th percentile) is required for the Ph.D. program unless the applicant has completed an M.S. in biology or a related science. For the M.S. programs, GREs are required only if the applicant's GPA is below 3.0 or if the applicant received a B.S. degree outside the United States. International students must provide proper immigration documentation and have a minimum TOEFL score of 550. Students should submit three letters of recommendation, a resume, and a personal statement that describes career goals as well as their scientific background, including previous laboratory training from course work or work experience.

Applications for the doctoral program are reviewed beginning February 1 and are completed once the seats are filled. Individuals interested in applying for a teaching or research assistantship should apply by February 1. All graduate applications for the fall semester are due July 1. Applications are also accepted for the spring semester (deadline: November 1) and for the summer semester (deadline: May 1).

Correspondence and Information

Graduate Program in Biology
College of Arts and Sciences
Fahy Hall, Room 130
Seton Hall University
400 South Orange Avenue
South Orange, New Jersey 07042

Phone: 973-761-9430
Fax: 973-761-9453
E-mail: artsci@shu.edu
Web site: http://artsci.shu.edu/graduateprograms/biology.htm

Seton Hall University

THE FACULTY AND THEIR RESEARCH

Ghaysuddin Ahmad, Ph.D., Immunologist. Immunology of cytoskeleton proteins; study of the effects of heavy metals (air pollutants) on cytoskeleton alterations in animal and cell cultures; study of immunoglobulin Fc receptors on white blood cells and platelets. Course: Immunology, Immunology Lab, Cellular Immunology.

Carolyn S. Bentivegna, Ph.D., Environmental Toxicologist. Development of biological assays: use of gene expression and levels of carbohydrates in *Chironomus riparius* (midge fly larva) as an indicator of exposure to and biological effects of aquatic pollutants; study of genetic diversity; use of molecular biological tools to study genetic relationships between laboratory and wetland populations of chironomids. Course: Fundamentals of Toxicology.

Allan D. Blake, Ph.D., Molecular Neurobiologist/Cell Biologist/Molecular Pharmacologist. Actions of neuropeptides in controlling inflammation and intercellular signaling; structure-function of somatostatin receptor proteins; functional role of somatostatin receptor subtypes in intracellular signaling; modifying intracellular signaling pathways using RNAi. Course: Fundamentals of Neuroscience, Introduction to Pharmacology, Cell Culture Techniques, Signal Transduction.

Sulie L. Chang, Ph.D., Biochemist/Cell Biologist/Neuroimmuologist. Neuroimmune pharmacology of substance of abuse; intracellular convergence of mu opioid receptor dependent pathways and cytokine receptor dependent pathways; how substances of abuse affect vascular endothelial cell barriers; studies of leukocyte-endothelial interaction in drug addicts; brain activities in health and disease; opiate-HIV-1 interaction on inflammation in the CNS. Course: Readings in Molecular Bioscience, Methods in Neuroscience, Cancer Biology, Signal Transduction.

Marian Glenn, Ph.D., Microbial Physiologist. Conservation biology; microbial physiology and microbial ecology, focusing on the microbial community as a bioindicator of ecosystem change; responses of microbial communities to pollution. Course: Microbial Ecology, Microbial Physiology.

Jane Ko, Ph.D., Molecular Biologist/Cell Pharmacologist. Regulation of the G-protein coupled receptor gene; mechanisms underlying the cell/tissue-specific gene expression; study of cross-talk between the signal transduction pathway and the transcriptional regulation; Course: Molecular Biology, Recombinant DNA Technology Laboratory, Microbial Genetics, Methods in Neuroscience.

Eliot Krause, Ph.D., Geneticist. Population genetics/cytogenetics; population genetic studies: control of population size between Tribolium species and within Tribolium species, using various genetic strains and factors that affect competition between genetic strains of Tribolium; genetics and cytogenetics: importance of fragile sites and involvement of sister chromatid exchange with expression of gene products.

Roberta L. Moldow, Ph.D., Neuroendocrinologist. Factors influencing the regulation of the circadian rhythmicity of the hypothalamic pituitary adrenal axis; study of the stress response, including posttraumatic stress disorder; addictive drugs and neuroendocrinology; Gulf War syndrome. Course: Vertebrate Endocrinology.

Anne Pumfery, Ph.D., Virologist/Microbiologist. Role of viral and cellular cyclins and cyclin-dependent kinases in Kaposi's sarcoma–associated herpesvirus/human herpesvirus 8 gene expression and DNA replication; regulation of viral gene expression by chromatin and histone modification. Course: Virology, Cell Culture.

Carroll Rawn, Ph.D. Mycology; plant pathology; metabolism; growth, reproduction, and physiology of fungi, especially plant pathogenic fungi; production of mycotoxins and effects of mycotoxins on fungi. Course: Mycology, Metabolic Pathways in Living Systems.

Heping Zhou, Ph.D., Molecular Biologist. How morphine-endotoxin interaction affects neuroinflammation; HIV-induced neuroinflammation during development; microarray technology; tumorigenesis. Course: Cancer Biology, Bioinformatics, Recombinant DNA Technology Laboratory.

Adjunct Faculty

Vincent DeBari, Ph.D. Course: Biostatistics.
Andrew Giovanni, Ph.D. Course: Reading in Molecular Bioscience; Special Topics in Biological Research (Ph.D. lab rotation).
Wen-Zhe Ho, Ph.D. Course: Special Topics in Biological Research (Ph.D. lab rotation).

SOUTHERN ILLINOIS UNIVERSITY CARBONDALE

Molecular Biology, Microbiology, and Biochemistry Program

Program of Study

The Molecular Biology, Microbiology, and Biochemistry Program offers graduate training leading to the Ph.D. degree. The program is jointly sponsored by the Departments of Microbiology (College of Science at Carbondale), Biochemistry and Molecular Biology (School of Medicine at Carbondale), and Medical Microbiology, Immunology and Cell Biology (School of Medicine at Springfield). Research opportunities are available with faculty members in any of these departments. Students are expected to discuss their research interests with prospective faculty members and begin a project sometime during the first year of graduate study. Students electing to pursue research training with a Springfield faculty member may fulfill course requirements at Carbondale or Springfield. Research activities, including the training of graduate students, represent a major effort of all three departments.

The formal core course requirements for the Ph.D. degree can be met by taking a set of core courses, including classes on biochemistry, bacterial genetics, and research methods and a 1-credit-hour seminar. Ph.D. students must pass a preliminary examination before being advanced to candidacy and then must earn at least 24 dissertation credit hours and prepare and defend a dissertation.

Research Facilities

The departments have modern, well-equipped research laboratories and supporting facilities for P3 physical containment, electron microscopy, X-ray crystallography, NMR spectroscopy, and flow cytometry. The University library at Carbondale is ranked among the top 100 research libraries in the United States. It contains more than 2 million volumes and maintains an extensive collection of research journals in all areas of biological and molecular science. There is also an extensive biological science and medical library at Springfield.

Financial Aid

Graduate teaching and research assistantships that include a stipend and tuition waiver are available to qualified students. Stipend amounts are maintained at a competitive level; for example, incoming students received $1342 per month in 2004–05. Special fellowships are also available. Nearly all active students are supported by teaching or research assistantships or other forms of support. Student loans and a number of scholarships for international students are available. Graduate students' spouses may receive help in finding employment through the Student Work and Financial Assistance Program.

Cost of Study

In-state graduate tuition is $243 per credit hour in 2006–07. Out-of-state tuition is 2.5 times the in-state tuition rate ($607.50 per credit hour). Graduate students with at least a 25 percent appointment as a graduate assistant receive a tuition waiver. Fees vary from $441.62 (1 credit hour) to $987.30 (12 credit hours).

Living and Housing Costs

For married couples, students with families, and single graduate students, the University has 589 efficiency and one-, two-, and three-bedroom apartments that rent for $438 to $505 per month in 2006–07. Residence halls for single graduate students are also available, as are accessible residence hall rooms and apartments for students with disabilities.

Student Group

Enrollment in the combined Carbondale-Springfield graduate program is approximately 80 students.

Student Outcomes

Students with graduate degrees in molecular biology, microbiology, and biochemistry are in demand in several occupational areas. Many graduates are employed by pharmaceutical or biotechnology firms. Ph.D. graduates can expect to be employed by pharmaceutical, chemical, biochemical, or biotechnology firms or continue for advanced academic training as postdoctoral fellows. Many Southern Illinois University Carbondale (SIUC) Ph.D. graduates are successful faculty members at colleges and universities throughout the United States.

Location

SIUC is 350 miles south of Chicago and 100 miles southeast of St. Louis. Nestled in rolling hills bordered by the Ohio and Mississippi Rivers and enhanced by a mild climate, the area has state parks, national forests and wildlife refuges, and large lakes for outdoor recreation. Much of the area is a part of the 240,000-acre Shawnee National Forest. Cultural offerings include theater, opera, concerts, art exhibits, and cinema. Educational facilities for the families of students are excellent.

The University

Southern Illinois University Carbondale is a comprehensive public university with a variety of general and professional education programs. The University offers bachelor's and associate degrees, master's and doctoral degrees, the J.D. degree, and the M.D. degree. The University is fully accredited by the North Central Association of Colleges and Schools. The Graduate School has an essential role in the development and coordination of graduate instruction and research programs. The Graduate Council has academic responsibility for determining graduate standards, recommending new graduate programs and research centers, and establishing policies to facilitate the research effort. Southern Illinois University Carbondale is a state-funded university founded in 1869.

Applying

Applications may be submitted throughout the academic year for admission at the beginning of either the fall or the spring semester; most students matriculate in the fall semester. Applications should be completed six months before expected enrollment.

Correspondence and Information

Ramesh Gupta, Ph.D., Director
Molecular Biology, Microbiology, and Biochemistry Program
MC 4413
Southern Illinois University Carbondale
Carbondale, Illinois 62901-4413
Phone: 618-453-1543
Fax: 618-453-1657
E-mail: mbmb@siumed.edu
Web site: http://www.siu.edu/~mbmb/

Southern Illinois University Carbondale

THE FACULTY AND THEIR RESEARCH

Department of Microbiology, Carbondale
The research goals of the faculty members of the Department of Microbiology are basic, emphasizing projects that promise to enhance fundamental knowledge of biological systems at the molecular level. Major areas of research include aspects of the molecular biology, biochemistry, and genetics of *Escherichia coli;* the physiology of phototrophic bacteria; the molecular biology of T-cell receptors in mice; the molecular biology of soybean pathogens; the molecular biology of mutations in *E. coli;* bioremediation of environmental pollutants; and the biology of ferric iron–reducing and related bacteria.

Laurie A. Achenbach, Professor; Ph.D., Illinois at Urbana-Champaign. Molecular genetics of metabolic pathways involved in the bioremediation of environmental contaminants; bacterial diversity and evolution; molecular anaerobic microbiology.
David P. Clark, Professor; Ph.D., Bristol (England). Genetics and regulation of anaerobic growth in *Escherichia coli.*
Douglas Fix, Associate Professor; Ph.D., Indiana. Molecular mechanisms of mutagenesis in *Escherichia coli.*
John D. Haddock, Associate Professor; Ph.D., Virginia Tech. Physiology and biochemistry of aerobic and anaerobic bacteria that degrade organic pollutants and naturally occurring aromatic compounds.
Michael T. Madigan, Professor; Ph.D., Wisconsin–Madison. General microbiology; bacterial diversity; phototrophic bacteria; microbiology of extreme environments; nitrogen fixation.
John Martinko, Associate Professor and Chair of Microbiology; Ph.D., SUNY at Buffalo. Immunology; biochemistry, molecular biology, and evolution of mouse major histocompatibility complex molecules; mouse T-cell repertoires; molecular evolution of ABO glycosyltransferases in primates.
Jack Parker, Professor and Dean of the College of Science; Ph.D., Purdue. Molecular genetics; use of in vivo and in vitro genetics and biochemical techniques to investigate regulation, function, and evolution of the translational apparatus in *Escherichia coli.*

Department of Medical Microbiology, Immunology and Cell Biology, Springfield
Areas of specialization in the Department of Medical Microbiology, Immunology and Cell Biology include molecular and cellular biology, molecular genetics, immunology of reproduction and type 1 diabetes, membrane biology, cancer biology, microbial immunology, and molecular virology. The research goals of the department are both basic and applied, adding to the scientific knowledge base as well as improving methods of diagnosis and therapy. Close working relationships are maintained with members of the clinical faculty to create a scientist-physician team approach in relevant areas of medical research.

Peter T. Borgia, Professor; Ph.D., Illinois at Urbana-Champaign. Cloning and characterization of genes for chitin synthesis in *Aspergillus.*
Gregory J. Brewer, Professor; Ph.D., California, San Diego. Alzheimer's disease; neuron development and adhesion; neurobiology of synaptogenesis; 2-D and 3-D neuronal networks.
Deliang Cao, Assistant Professor; Ph.D., Institute of Molecular Biology (Hong Kong). Investigation of metabolic pathways and molecular mechanisms of antitumor activity of cytotoxic agents using gene transfer, RNA interference, and gene knockout technologies.
David F. Carpenter, Associate Professor; Ph.D., New Hampshire. Diagnostics for emerging infectious diseases of public health significance.
Subhas Chakraborty, Professor and Associate Director of Basic Science, SIU Cancer Institute; Ph.D., Manitoba. Molecular mechanisms underlying malignant progression; chemoprevention of cancer; cancer diagnostic and prognostic markers; cancer therapeutics.
Morris D. Cooper, Professor and Chair of Medical Microbiology, Immunology and Cell Biology; Ph.D., Georgia. Mucosal immune responses of the human fallopian tube to *Neisseria gonorrhoeae* and *Chlamydia trachomatis* infections; topical microbicide activity against sexually transmitted disease pathogens.
Nancy Khardori, Professor; Ph.D., M.D., India. Microbial adherence and biofilms: Study of the microbial adherence to prosthetic devices; factors facilitating and inhibiting adherence to devices.
Mary McAsey, Assistant Professor; Ph.D, Arizona. Mechanisms of action of steroids in the brain, induction of apoptosis in gynecologic cancers, hormones in diseases of pregnancy.
Yin-Yuan Mo, Assistant Professor; Ph.D., Washington State. Tumor drug resistance and tumor cell biology.
Walter L. Myers, Professor Emeritus; Ph.D., Wisconsin–Madison. Molecular immunology of host antitumor responses; effects of B7-1 gene transfection on tumor immunogenicity.
Daotai Nie, Assistant Professor; Ph.D., South Carolina. Molecular and cellular biology of cancer; tumor radiotherapy; tumor metastasis.
Mary Pauza, Assistant Professor; Ph.D., Minnesota. Autoimmune diabetes; immunopathology and gene therapy; use of transgenic and NOD models.
Sophia Ran, Assistant Professor; Ph.D., Weizmann (Israel). Tumor angiogenesis and lymphangiogenesis; breast cancer metastasis.
Donald S. Torry, Associate Professor; Ph.D., Southern Illinois. Human reproductive biology; cellular biology of angiogenic growth factors and immune cytokines during pregnancy; molecular biology of placental gene expression.
Kounsouke Watabe, Professor; Ph.D., Kyoto (Tokyo). Molecular biology of tumor metastasis; regulation of gene expression of tumor metastasis genes.

Department of Biochemistry and Molecular Biology, Carbondale
The research goals of the faculty members of the Department of Biochemistry and Molecular Biology emphasize projects that focus on various aspects of biological systems at the molecular level. Major areas of research include protein structure, protein bioengineering, eukaryotic gene regulation, DNA repair, NMR studies of proteins and nucleic acids, cell membrane ion channels, transcription regulation and RNA processing in Archaea, the biochemistry of iron metabolism in bacteria, and the genetics of biochemistry and biosynthetic enzymes in yeast.

Blaine B. Bartholomew, Professor; Ph.D., California, Davis. Eukaryotic gene regulation; mechanisms of promoter-selective transcription; structure-function of yeast RNA polymerases; DNA-protein complexes.
Sukesh R. Bhaumik, Assistant Professor; Ph.D., Bombay (India). Regulation of eukaryotic gene expression; transcription-coupled ubiquitination and DNA repair; NMR structural studies of proteins and nucleic acids.
Ramesh Gupta, Associate Professor and Director of the Molecular Biology, Microbiology, and Chemistry Program; Ph.D., Illinois at Urbana-Champaign. Archaea; transcription and RNA processing in extreme halophiles and hyperthermophiles.
Peter M. D. Hardwicke, Professor; Ph.D., King's College. Structure and function in the membrane calcium pump protein; relationship of a membrane-associated proteolipid to BRCA 1 breast cancer gene product.
Jodi I. Huggenvik, Associate Professor; Ph.D., Washington State. Molecular biology of mammalian gene expression and structure-function analysis of tumor suppressor genes.
David A. Lightfoot, Professor; Ph.D., Leeds (England). Molecular agronomist.
Eric C. Niederhoffer, Associate Professor; Ph.D., Texas A&M. Metal ion metabolism and stress responses in microorganisms; virulence factors and swarmer cell differentiation of the urinary tract pathogen *Proteus.*
Joseph C. Schmit, Associate Professor and Chair of Biochemistry and Molecular Biology; Ph.D., Purdue. Developmental biochemistry and genetics of enzymatic activity; control of amino acid metabolism.

STONY BROOK UNIVERSITY, STATE UNIVERSITY OF NEW YORK

with Cold Spring Harbor Laboratory and Brookhaven National Laboratory
Molecular and Cellular Biology Graduate Program

Program of Study	The Molecular and Cellular Biology Graduate Program strives to provide students with the knowledge to become skilled experimentalists, effective communicators, and creative thinkers. Students have the opportunity to perform Ph.D. thesis research in a number of laboratories at three collaborative institutions: The State University of New York at Stony Brook, Cold Spring Harbor Laboratory, or Brookhaven National Laboratory. The Ph.D. in molecular and cellular biology is granted by the State University of New York at Stony Brook.
	The Molecular and Cellular Biology Graduate Program offers the student an extraordinary range of experimental science. Students are involved in ongoing research as soon as they arrive on campus. During the first academic year, they experience research in four different laboratories of program faculty members. Following the first academic year, a member of the program faculty is selected as a Ph.D. mentor. Diverse biological systems are available to investigate a variety of biological questions, including cancer neurobiology, gene expression, infectious disease, protein trafficking, DNA replication, structural biology, protein folding, enzymology, signal transduction, cell cycle, apoptosis, virology, development, biological membranes, and immune defense.
	The Molecular and Cellular Biology Graduate Program offers the graduate student a unique opportunity to choose between three academic specializations: molecular biology and biochemistry, cellular and developmental biology, or immunology and pathology. During the first academic year, students attend core courses in biochemistry, molecular genetics, and cell biology. In addition, students participate in a journal club/readings course in which students learn to critically evaluate original research articles. Subsequent courses and electives are then designed to enhance the student's knowledge in their selected academic specialization.
	In the second and third semesters, graduate students develop teaching skills by serving as a teaching assistant with the guidance of the faculty instructor of the course. In the third and subsequent years, graduate students present their research progress to other students and faculty members in graduate student seminars. These provide opportunities to gain communication skills and learn about ongoing research of students in other laboratories.
	In the third year, students prepare a written Ph.D. thesis proposal in consultation with their research faculty adviser. Following successful defense of the proposal, the student advances to candidacy, and the Proposal Committee and the faculty adviser become the student's Ph.D. Thesis Committee. This committee meets at least once a year to support the research efforts of the student.
Research Facilities	Research is supported by state-of-the-art-facilities at Stony Brook. General computing services are available to students, including an account for e-mail and Internet access. In addition to state-of-the-art equipment in individualized laboratories, several core facilities support program research. The new Proteomics Center is a core facility at Stony Brook whose services include protein sequencing, peptide synthesis, analytical HPLC, preparative HPLC, MALDI-TOF mass spectrometry, and DNA sequencing. The University recently opened a DNA Microarray Center that offers researchers the ability to simultaneously analyze the expression of thousands of genes in humans, rodents, yeast plants, bacteria, or *Drosophila*. The Center for Structural Biology has advanced instrumentation for NMR spectroscopy and X-ray crystallography. The University Microscopy Imaging Center assists in research projects requiring advanced light, electronic light, and electron microscopy techniques. There is also a Cell Culture and Hybridoma Facility.
Financial Aid	All students accepted into the Molecular and Cellular Biology Graduate Program receive full financial support. Students are provided with a stipend of $25,000 per year for cost of living expenses. In addition, tuition is waived for all graduate students accepted into the program. Health insurance is provided.
Cost of Study	University fees are approximately $750 per year.
Living and Housing Costs	Room rent for a single student living in an on-campus apartment ranges from about $325 to $1500 per month, depending on size and on the number of occupants. Privately owned housing is available close to the University.
Student Group	There are approximately 100 graduate students in the Molecular and Cellular Biology Program. Many enter the program as recent college graduates, and others have had previous graduate school or research experience. Generally, students have undergraduate majors in biology, biochemistry, or chemistry, but some students come from other disciplines. Diverse cultural and official interests are supported by the campus community.
Location	Stony Brook is located on the wooded North Shore of Long Island, New York, an area of picturesque villages, harbors, and beaches. It is about 50 miles east of New York City, with easy access provided by the Long Island Railroad or major parkways so that students can take advantage of New York City's commercial, scientific, and cultural resources.
The University	The University was founded in 1957 in Oyster Bay, Long Island. In 1962, it moved to Stony Brook and now occupies more than 1,000 acres with more than 100 buildings. Stony Brook has exceptional strength in the sciences, mathematics, humanities, fine arts, social sciences, engineering, and health professions for both undergraduate and graduate students. It was classified by the Carnegie Foundation as one of the nation's leading research institutions. The University Hospital and Medical Center supports the Schools of Medicine, Dentistry, Nursing, and Health Technology and Management.
	An Indoor Sports Complex promotes Division I athletics and Seawolves fans, and facilities are available for student use. The Staller Center for the Arts offers a wide variety of professional performances in music, film, dance, theater, and fine art exhibitions in its five theaters and 5,000-square-foot University Art Gallery.
Applying	Applications to the Molecular and Cellular Biology Graduate Program should be submitted by January 15 for admission the following September, although applications are accepted after that date. Applicants must have a baccalaureate degree from an accredited college or university and submit school transcripts. The General Test of the Graduate Record Examinations (GRE) is required, and a Subject Test in an appropriate field is optional. Three letters of recommendation are required.
	Prospective candidates are invited to visit Stony Brook whenever possible. This affords students an opportunity to meet with program faculty members and students. To apply online, students should visit http://www.grad.sunysb.edu/applying/applying.htm.
Correspondence and Information	Director, Molecular and Cellular Biology Graduate Program Stony Brook University, State University of New York Life Sciences Rooms 336 and 338 Stony Brook, New York 11794-5215 Phone: 631-632-8533 Fax: 631-632-9730 E-mail: mcbprog@life.bio.sunysb.edu Web site: http://life.bio.sunysb.edu/mcb

Stony Brook University, State University of New York

THE FACULTY AND THEIR RESEARCH

Department of Biochemistry and Cell Biology
Paul M. Bingham, Ph.D., 1979. Genetic control of development and gene expression in animals. Deborah Brown, Ph.D., 1987. Structure and function of cholesterol/sphingolipid-rich membrane rafts and caveolae. Vitaly Citovsky, Ph.D., 1987. Nuclear targeting and intercellular communication in plants. Neta Dean, Ph.D., 1988. Protein glycosylation; fungal cell wall biosynthesis; fungal pathogenesis. Dale Deutsch, Ph.D., 1972. Molecular neurobiology of anandamide (the endogenous marijuana) regulation. J. Peter Gergen, Ph.D., 1982. Gene expression and development in *Drosophila*. Robert Haltiwanger, Ph.D., 1986. Role of protein glycosylation in signal transduction; Notch signaling. Bernadette Holdener, Ph.D., 1990. Role of protein folding in WNT signal transduction and development. Nancy Hollingsworth, Ph.D., 1988. Chromosome structure and function during meiosis in yeast. Jen-Chih Hsieh, Ph.D., 1994. Molecular mechanism of Wnt signaling. A. Wali Karzai, Ph.D., 1995. Structure and function of RNA-binding proteins and biochemical studies of the SmpB-SsrA quality control system for protein tagging, directed degradation, and ribosome rescue. William J. Lennarz, Ph.D., 1959. Biosynthesis and function of glycoproteins in cell-cell interactions. Erwin London, Ph.D., 1979. Membrane protein structure/translocation/folding; structure and function of sphingolipid/cholesterol rafts in membranes. Harvard Lyman, Ph.D., 1960. Photocontrol of chloroplast development. Kenneth B. Marcu, Ph.D., 1975. Molecular control of innate and adaptive immunity: NF-kappaB kinases and inflammatory response; antibody gene class switch recombination and somatic hypermutation. Aaron Neiman, Ph.D., 1994. Vesicle trafficking and intracellular signaling in yeast. Nisson Schechter, Ph.D., 1971. Homeobox and filament proteins in neuronal differentiation, growth, and regeneration. Sanford R. Simon, Ph.D., 1967. Extracellular matrix degradation by inflammatory and tumor cell proteases. Steven O. Smith, Ph.D., 1985. Structure and function of membrane proteins. James Staros, Ph.D., 1974. Biochemical and biophysical approaches to signal transduction by ErbB-family receptors. Rolf Sternglanz, Ph.D., 1967. Chromatin structure and function; histone modifying enzymes; gene expression. Gerald H. Thomsen, Ph.D., 1988. Regulation of early vertebrate development by growth factor signals, ubiquitin modification, and T-box family transcription factors.

Department of Chemistry
Carlos Simmerling, Ph.D., 1994. Computational studies of biomolecular structure and dynamics. Peter J. Tonge, Ph.D., 1986. Tuberculosis pathogenesis and drug discovery; enzyme mechanisms and rational inhibitor design; fluorescent proteins.

Department of Molecular Genetics and Microbiology
Jorge L. Benach, Ph.D., 1972. Pathogenesis of spirochetal infections; utilization of host macromolecules. Allen Bruce Futcher, Ph.D., 1983. Cell-cycle control in yeast and on analysis of gene expression networks using microarrays and bioinformatics. Michael Hayman, Ph.D., 1973. Signal transduction pathways in cell growth, differentiation, and cancer. Patrick Hearing, Ph.D., 1984. Adenovirus–host cell interactions; adenovirus assembly and vectors for gene therapy. Eugene Katz, Ph.D., 1969. Genetics/development in cellular slime molds. James Konopka, Ph.D., 1985. Hormone signal transduction/yeast cell morphogenesis. Janet Leatherwood, Ph.D., 1993. Cell-cycle control of DNA replication/microarray analyses of fission yeast. Nancy Reich, Ph.D., 1983. Signal transduction and gene expression in response to cytokines and virus. David D. Thanassi, Ph.D., 1995. Virulence factors of pathogenic bacteria. Eckard Wimmer, Ph.D., 1962. RNA virus genetics, replication, and pathogenicity; cellular virus receptors. Wei-Xing Zong, Ph.D., 1999. Molecular regulation of cell-death pathways.

Department of Neurobiology and Behavior
Simon Halegoua, Ph.D., 1978. Molecular control of the neuronal phenotype. Maurice Kernan, Ph.D., 1990. Differentiation and signal transduction in *Drosophila* mechanosensory cilia and sperm. Joel M. Levine, Ph.D., 1980. Cell-surface molecules of the developing nervous system. Gail Mandel, Ph.D., 1977. Gene expression in the nervous system. David McKinnon, Ph.D., 1987. Molecular physiology of sympathetic neurons and cardiac muscle. Howard Sirotkin, Ph.D., 1996. Genetic and molecular analysis of early vertebrate development. Lonnie P. Wollmuth, Ph.D.,1992. Molecular and biophysical mechanisms of excitatory synaptic transmission.

Department of Oral Biology and Pathology
Soosan Ghazizadeh, Ph.D., 1994. Keratinocyte stem cell lineage and use in cell and gene therapy.

Department of Pathology
Howard B. Fleit, Ph.D., 1980. Leukocyte Fc receptors; macrophage differentiation. Martha Furie, Ph.D., 1980. Interactions among pathogenic bacteria, endothelium, and leukocytes. Berhane Ghebrehiwet, D.V.M./D.Sc., 1974. Biochemistry; function of the complement system. Richard R. Kew, Ph.D., 1986. Leukocyte chemotaxis/inflammation. Ute M. Moll, M.D., 1985. The p53 and Rb/E2F tumor suppressor gene family: function/regulation in normal cells and tumor-associated inactivation. Eric D. Spitzer, M.D./Ph.D., 1985. Molecular biology of *Cryptococcus neoformans*. Gary Zieve, Ph.D., 1977. Assembly/transport of snRNP particles.

Department of Pharmacological Sciences
Miguel Berrios, Ph.D., 1983. Nuclear structure and function; cell biology of DNA damage and repair. Daniel Bogenhagen, M.D., 1977. Mitochondrial DNA; mitochondrial proteomics. Holly Colognato, Ph.D., 2000. Extracellular matrix in the brain; roles during development and during neurodegeneration. Michael A. Frohman, M.D./Ph.D., 1986. Signal transduction; membrane vesicle trafficking; regulated exocytosis; diabetes; phototransduction. Arthur P. Grollman, M.D., 1959. Molecular carcinogenesis and DNA repair. Craig C. Malbon, Ph.D., 1976. Wnt-Frizzled signaling, G-proteins, and development. Masaaki Moriya, Ph.D., 1981. Cellular response to DNA damage. Jeffery Pessin, Ph.D., 1980. Identification of insulin-mediated signal cascades regulation of intracellular GLUT4 vesicle trafficking and biogenesis. Joav Prives, Ph.D., 1968. Cytoskeletal membrane interactions in muscle cells. Styliani Anna Tsirka, Ph.D., 1989. Neuronal microglial interactions in the physiology and pathology of the central nervous system. Orlando D. Schärer, Ph.D., 1996. Chemical biology of mammalian DNA repair.

Department of Physiology and Biophysics
Roger A. Johnson, Ph.D., 1968. Adenine nucleoside 3'-polyphosphates, adenylyl cyclases, and signal transduction. Stuart McLaughlin, Ph.D., 1968. Calcium/phospholipid second messenger system. W. Todd Miller, Ph.D., 1989. Tyrosine phosphorylation and signal transduction. Nicholas Nassar, Ph.D., 1992. Crystallographic and biochemical studies of signal proteins. Suzanne Scarlata, Ph.D., 1984. Structure/oligomerization of membrane proteins. Ilan Spector, Ph.D., 1967. Actin cytoskeleton in normal and cancer cells. Hsien-yu Wang, Ph.D., 1989. Wnt/Frizzled and G-protein signal transduction in development. Thomas White, Ph.D., 1994. Molecular biology and physiology of gap-junction channels.

School of Medicine and Other Departments
Wen-Tien Chen, Ph.D., 1979. Proteases and integrins in cancer invasion, metastasis, and angiogenesis. Yaacov Hod, Ph.D., 1977. Hormonal control of gene expression in prostate cancer cells. Jolyon Jesty, D.Phil., 1972. Mechanisms of thrombogenesis. Richard Lin, M.D., 1988. Signal transduction and cell growth. Erich R. Mackow, Ph.D., 1984. Hantavirus and rotavirus pathogenesis; viral regulations of cell signaling pathways and responses; viral attachment and entry; reverse genetics. Wolfgang Quitschke, Ph.D., 1983. Gene regulation of proteins associated with neurodegenerative diseases. Mario J. Rebecchi, Ph.D., 1984. Phospholipases and signal transduction. Roy T. Steigbigel, M.D., 1966. Immune dysfunction induced by HIV infection. William E. VanNostrand, Ph.D., 1985. Physiologic and pathophysiologic vascular functions of the Alzheimer's disease amyloid beta-protein precursor.

Brookhaven National Laboratory
John Dunn, Ph.D., 1970. Genomics and gene expression. Richard Setlow, Ph.D., 1947. DNA damage and repair; carcinogenesis and mutagenesis in fish. F. William Studier, Ph.D., 1963. Molecular genetics of phage T7; structural genomics.

Cold Spring Harbor Laboratory
Gregory Hannon, Ph.D., 1991. Genetics of growth in mammalian cells and dsRNA-induced gene silencing. Leemor Joshua-Tor, Ph.D., 1991. X-ray crystallography; molecular recognition; nucleic acid regulation; RNAi. Adrian Krainer, Ph.D., 1986. mRNA splicing; gene expression; RNA-protein interaction. Yuri Lazebnik, Ph.D., 1986. Molecular mechanisms of cancer and apoptosis. Scott Lowe, Ph.D., 1994. Modulation of apoptosis; chemosensitivity; senescence by cancer genes. Vivek Mittal, Ph.D.,1994. Id transcription factors; tumor-mediated neovascularization. K. Muthuswamy Senthil, Ph.D., 1995. Understanding cancer initiation using three-dimensional epithelial structures. David L. Spector, Ph.D., 1980. Spatial organization of gene expression. Arne Stenlund, Ph.D., 1984. DNA replication of papillomaviruses. Bruce Stillman, Ph.D., 1979. DNA replication and chromatin assembly in human and yeast cells. William P. Tansey, Ph.D., 1991. Regulation of oncoprotein stability. Nicholas Tonks, Ph.D., 1985. Characterization of protein tyrosine phosphatases. Linda Van Aelst, Ph.D., 1991. Role of Ras in mammalian cell transformation. Michael Wigler, Ph.D., 1978. Genomics and cancer.

UNIFORMED SERVICES UNIVERSITY OF THE HEALTH SCIENCES

F. Edward Hébert School of Medicine
Graduate Program in Molecular and Cell Biology

Program of Study

The program of study is designed for full-time students who wish to obtain a Ph.D. degree in the area of molecular and cell biology. This interdepartmental graduate program, which includes faculty members from both basic and clinical departments, offers research expertise in a wide range of areas, including bacteriology, immunology, genetics, biochemistry, regulation of gene expression, and cancer biology. The program includes core courses in molecular and cell biology that provide necessary knowledge for modern biomedical research, as well as advanced electives in areas of faculty expertise. The first-year curriculum includes courses in biochemistry, genetics, immunology, cell biology, and experimental methodology. During the first summer, students participate in laboratory rotations in two laboratories of their choice, leading to the choice of a mentor for their doctoral research. The second-year curriculum offers advanced elective courses in a variety of disciplines, including virology, biochemistry, immunology, molecular endocrinology, and cell biology, and marks the transition from classwork to original laboratory research. Throughout their graduate experience, students participate in journal clubs designed to foster interaction across disciplines and to develop the critical skills needed for data presentation and analysis. A year-round seminar series brings renowned scientists to the Uniformed Services University of the Health Sciences (USUHS) to share their results and to meet with students and faculty members. Students may also take advantage of seminars hosted by other programs and departments as well as those presented at the National Institutes of Health. Completion of the research project and preparation and successful defense of a written dissertation leads to the degree of Doctor of Philosophy.

Research Facilities

The University possesses outstanding facilities for research in molecular and cell biology. Well-equipped laboratories and extramurally funded faculty members provide an outstanding environment in which to pursue state-of-the-art research. Shared equipment in a modern biomedical instrumentation core facility includes oligonucleotide and peptide synthesizers and sequencers; a variety of imaging equipment, including laser confocal and electron microscopes; fluorescent-activated cell sorters; and an ACAS workstation. A recently added proteomics facility contains both MADLI-TOF and ESI tandem mass spectrometers. All offices, laboratories, and the Learning Resource Center are equipped with high-speed Internet connectivity and have access to an extensive online journal collection.

Financial Aid

Stipends are available for civilian applicants. Awards of stipends are competitive and may be renewed. For the 2006–07 academic year, stipends for entering students begin at $24,000. Outstanding students may be nominated for the Dean's Special Fellowship, which supports a stipend of $29,000.

Cost of Study

Graduate students are not required to pay tuition. Civilian graduate students do not incur any obligation to the United States government for service after completion of their graduate training programs. Active-duty military personnel incur an obligation for additional military service by Department of Defense regulations that govern sponsored graduate education. Students are required to maintain health insurance.

Living and Housing Costs

The University does not have housing for graduate students. However, there is an abundant supply of rental housing in the area. Living costs in the greater Washington, D.C., area are comparable to those of other East Coast metropolitan areas.

Student Group

The first graduate students in the interdisciplinary Graduate Program in Molecular and Cell Biology at USUHS were admitted in 1995. Over the last decade, the Graduate Program in Molecular and Cell Biology has grown significantly; 17 students are currently enrolled. Thirteen Ph.D. degrees in molecular and cell biology have been awarded over the past seven years.

Location

Metropolitan Washington has a population of about 2.7 million residents in the District of Columbia and the surrounding areas of Maryland and Virginia. The region is a center of education and research and is home to five major universities, four medical schools, the National Library of Medicine and the National Institutes of Health (next to the USUHS campus), Walter Reed Army Medical Center, the Armed Forces Institute of Pathology, the Library of Congress, the Smithsonian Institution, the National Bureau of Standards, and many other private and government research centers. The cultural advantages of the area are many and include the theater, a major symphony orchestra, major-league sports, and world-famous museums. The Metro subway system has a station near campus and provides a convenient connection from the University to the museums and cultural attractions of downtown Washington. For outdoor activities, the Blue Ridge Mountains, the Chesapeake Bay, and the Atlantic coast beaches are all within a few hours' drive.

The University

USUHS is located just outside Washington, D.C., in Bethesda, Maryland. The campus is situated on an attractive, wooded site at the National Naval Medical Center and is close to several major federal health research facilities. Through various affiliation agreements, these institutes provide additional resources to enhance the educational experience of graduate students at USUHS.

Applying

Both civilians and military personnel are eligible to apply for graduate study at USUHS. Before matriculation, each applicant must complete a baccalaureate degree that includes college-level courses in mathematics, biology, physics, inorganic chemistry, and organic chemistry. Advanced courses in biology, chemistry, or related fields, such as biochemistry, cell biology, physical chemistry, microbiology, molecular biology, immunology, and genetics, are desirable but not essential. Each applicant must submit a USUHS graduate training application form, complete academic transcripts of postsecondary education, GRE scores (in addition to the aptitude sections, one advanced examination is recommended), three letters of recommendation from individuals familiar with the academic achievements or research experience of the applicant, and a personal statement expressing the applicant's career objectives. Active-duty military personnel must obtain the approval and sponsorship of their parent military department in addition to acceptance from USUHS. USUHS subscribes fully to the policy of equal educational opportunity and selects students on a competitive basis without regard to race, sex, creed, or national origin. Application forms may be obtained from the Web site at http://cim.usuhs.mil/geo/application.htm. Completed applications must be received before January 15 for matriculation in August.

Correspondence and Information

Associate Dean for Graduate Education
Uniformed Services University of the Health Sciences
4301 Jones Bridge Road
Bethesda, Maryland 20814
Phone: 301-295-3913
    800-772-1747 (toll-free)
Web site: http://www.usuhs.mil/mcb/

For application and information about the molecular and cell biology program:

Dr. Jeffrey Harmon, Director
Graduate Program in Molecular and Cell Biology
Uniformed Services University of the Health Sciences
4301 Jones Bridge Road
Bethesda, Maryland 20814
Phone: 301-295-3642
Fax: 301-295-1996
E-mail: nfinley@usuhs.mil
Web site: http://www.usuhs.mil/mcb/

Uniformed Services University of the Health Sciences

THE FACULTY AND THEIR RESEARCH

Mark R. Adelman, Associate Professor; Ph.D., Chicago, 1969. Roles of the cytoskeleton in the amoeboflagellate transformation of *Physarum polycephalum* and in other systems; confocal microscopy in biomedical research. (http://bicmra.usuhs.mil/)

Regina C. Armstrong, Professor; Ph.D., North Carolina, 1987. Molecular and cellular mechanisms of glial cell development and regeneration. (http://www.usuhs.mil/nes/armstronglab.htm)

Jorge Blanco, Adjunct Assistant Professor; Ph.D., Buenos Aires, 1991. Molecular mechanisms of pathogenesis of respiratory viruses. *J. Infect. Dis.* 185(12): 1780–5, 2002.

Christopher C. Broder, Associate Professor; Ph.D., Florida, 1989. Interactions between enveloped viruses and their cellular receptors: HIV, Hendra virus, and Nipah virus. *J. Virol.* 79:3358–69, 2005. *Blood* 103:1586–94, 2004.

Rolf Bünger, Professor; M.D./Ph.D., Munich, 1979. Cellular, molecular, and metabolic mechanisms of heart and brain circulation and resuscitation at various levels of organization: intact animal/isolated perfused organs/subcellular compartments of cytosol and mitochondria. *J. Mol. Cell Biochem.* 216:37–46, 2001. *Am. J. Physiol.* 281:H854–64, 2001.

Daniel J. Carucci, Director, Malaria Program; M.D., Virginia, 1984; Ph.D., London School of Hygiene and Tropical Medicine, 1995. Navy malaria vaccine research, focusing on genomic and DNA-based vaccine approaches. *Parasitol. Today* 16(10):434–8, 2000.

Mary Lou Cutler, Associate Professor; Ph.D., Hahnemann, 1980. Role of molecules that suppress transformation by the Ras oncogene; Ras signal transduction and human carcinogenesis. *J. Cell. Physiol.* 201:244–58, 2004. *J. Neurooncol.* 60:201–11, 2002.

Peter D'Arpa, Assistant Professor; Ph.D., George Washington, 1986. DNA topoisomerases as antitumor drug targets; functional genomics of topoisomerases-interacting proteins. *J. Biol. Chem.* 277:40020–6, 2002.

Thomas N. Darling, Assistant Professor; M.D., Ph.D., Duke, 1990. Molecular and genetic analysis of disseminated neoplastic cells in lymphangioleiomyomatosis. *Proc. Natl. Acad. Sci. U.S.A.* 101:17462–7, 2004. *J. Am. Acad. Dermatol.* 51:S9–11, 2004.

Saibal Dey, Assistant Professor; Ph.D., Uniformed Services Health Sciences, 1999. Allosteric modulation of the human multidrug transporter MDR1, which alters bioavailability of chemotherapeutic drugs and confers multidrug resistance in cancer cells; biricodar. *Curr. Opin. Invest. Drugs* 3(5):818–23, 2002.

Teresa M. Dunn, Professor; Ph.D., Brandeis, 1984. Sphingolipid synthesis and function. *J. Biol. Chem.* 277:11481–8, 2002; 277:10194–200, 2002. *Mol. Cell Biol.* 21:109–25, 2001. *Methods Enzymol.* 312:317–30, 2000.

Gabriela S. Dveksler, Associate Professor; Ph.D., Uniformed Services Health Sciences, 1991. Murine CD9 is the receptor for pregnancy-specific glycoprotein 17. *J. Leukocyte Biol.*, in press. *Curr. Drug Targets Inflamm. Allergy,* in press.

Chou Zen Giam, Professor; Ph.D., Connecticut Health Center, 1983. Molecular biology and pathogenesis of human T-lymphotropic virus type I, Kaposi's sarcoma herpes virus, and hepatitis C virus; cell-cycle controls, transcriptional regulation, signal transduction, cell transformation, and RNA-interference/posttranscriptional gene silencing. *Proc. Natl. Acad. Sci. U.S.A.* 102:63–8, 2005. *Mol. Cell. Biol.* 23(15):5269–81, 2003.

David A. Grahame, Associate Professor; Ph.D., Ohio State, 1984. Metalloenzyme structure and function in Archaea. *J. Biol. Chem.* 278:6101–10, 2003. *Biochemistry* 40:13068–78, 2001.

Philip M. Grimley, Professor; M.D., Albany Medical College, 1961. Mechanisms of trophic factor signaling and regulation of cell survival/apoptosis, including the role of cell-cycle checkpoints.

Jeffrey M. Harmon, Professor and Director, Molecular and Cell Biology; Ph.D., Rochester, 1976. Mechanism(s) by which steroid hormones regulate gene expression and the role of steroid hormones in the development and treatment of malignant tumors. *Cancer Res.* 60:2056–62, 2000.

Susan R. Haynes, Assistant Professor; Ph.D., Rockefeller, 1982. RNA-binding proteins and posttranscriptional regulation in the control of male and female fertility. *Development* 127:1715–25, 2000.

Ann E. Jerse, Associate Professor; Ph.D., Maryland, Baltimore, 1991. Adaptation of *Neisseria gonorrhoeae* to the female lower genital tract. *Infect. Immun.* 70:2549–58, 2002; 179:911–20, 1999.

Guangyong Ji, Assistant Professor; Ph.D., Illinois at Chicago, 1992. Molecular mechanisms of staphylococcal pathogenesis. *J. Biol. Chem.* 277:34736, 2002 (with Zhang et al.).

Sharon L. Juliano, Professor; Ph.D., Pennsylvania, 1982. Mechanisms of development and plasticity in the cerebral cortex, with particular emphasis on the migration of neurons into the cortical plate and factors maintaining the function and morphology of radial glia and Cajal-Retzius cells.

Robert Lechleider, Assistant Professor; M.D., Illinois at Chicago, 1990. Mechanisms of transforming growth factor beta signal transduction. *Genes Dev.* 12:1587–92, 1998.

Stanley Lipkowitz, Associate Professor; Ph.D., 1983; M.D., 1984, Cornell. Regulation of proliferation and apoptosis in normal and malignant epithelial cells. *Oncogene* 23:7104–15, 2004.

Radha K. Maheshwari, Professor; Ph.D., Kanpur (India), 1974. Protective effect of a polyherbal preparation, Brahma rasayana against tumor growth and lung metastasis in rat prostate model system. *J. Exp. Ther. Oncol.* 4:203–12, 2004. *BioMetals* 16:359–68, 2003. *Shock* 19:150–6, 2003.

Anthony T. Maurelli, Professor; Ph.D., Alabama at Birmingham, 1983. Molecular genetics and regulation of virulence gene expression in *Shigella;* molecular biology and pathogenesis of *Chlamydia. J. Bacteriol.* 185:1218–28, 2003 (with McCoy); 184:4409–19, 2002 (with Kane).

Joseph T. McCabe, Professor and Vice Chair; Ph.D., CUNY Graduate Center, 1983. Progesterone differentially regulates pro- and anti-apoptotic gene expression in cerebral cortex following traumatic brain injury (TBI) in rats. *J. Neurotrauma,* in press (with Yao et al.).

Eleanor S. Metcalf, Professor; Ph.D., Pennsylvania, 1976. The role of intestinal epithelial cells and CD8+CTLs in immune responses to *Salmonella. Nature Med.* 6:215–18, 2000 (with Lo and Soloski). *Infect. Immun.* 66:2310–18, 1998.

Alison D. O'Brien, Professor and Chair; Ph.D., Ohio State, 1976. The role of *E. coli Shiga* toxins in the pathogenesis of hemorrhagic colitis and the hemolytic uremic syndrome; analysis of the mode of action of the Rho-modifying cytotoxic necrotizing factor and its role in the pathogenesis of *E. coli*–mediated urinary tract infections; identification of spore-surface antigens of *Bacillus anthracis* as potential vaccine candidates. *J. Biol. Chem.* 279:33751–8, 2004 (with Sinclair).

Harvey Pollard, Professor and Chair; Ph.D., 1969, M.D., 1973, Chicago. Molecular biology of secretory processes. *Proc. Natl. Acad. Sci. U.S.A.* 98: 4575–80, 2001.

Ignacio Provencio, Assistant Professor; Ph.D., Virginia, 1996. Nonvisual photoreception in mammals. *Nature* 415:493, 2002. *Science* 298:2213–6, 2002.

Gerald Quinnan, Professor; M.D., Saint Louis, 1973. Understanding the significance of mutations in the envelope gene of HIV in determining the resistance of virus to neutralization, as well as neutralizing antibody responses to mutant envelopes expressed in vivo using a Venezuelan equine encephalitis replicon system.

Paul D. Rick, Professor; Ph.D., Minnesota, 1971. Identification and biosynthesis of cyclic enterobacterial common antigen in *Escherichia coli. J. Bacteriol.,* in press; 183:6509–16, 2001.

Brian C. Schaefer, Assistant Professor; Ph.D., Harvard, 1995. Biology of lymphocyte activation, particularly antigen receptor-mediated activation of the transcription factor NF-kappaB; imaging approaches to understand signal transduction mechanisms. *Proc. Natl. Acad. Sci. U.S.A.* 101:1004–9, 2004.

Lisa M. Schwartz, Research Assistant Professor; Ph.D., Uniformed Services Health Sciences, 1986. Utilization of proteomic approaches to decipher signaling pathways involved in cardioprotection. *Circulation* 110(3):106, 2004. *J. Mol. Cell. Cardiol.* 35:A47, 2003.

Yoshitatsu Sei, Research Assistant Professor; M.D., 1983, Ph.D., 1988, Kurume (Japan). Mutation of ion channels and human genetic disease. *J. Immunol.* 167:4887–94, 2001.

Michael Shamblott, Assistant Professor; Ph.D., Johns Hopkins, 2001. Human stem cells and regenerative medicine; development of new tools for bioinformatic research. *Proc. Natl. Acad. Sci. U.S.A.* 98:113–8, 2001.

Ishaiahu Shechter, Professor and Chair; Ph.D., UCLA, 1969. Regulation of cholesterogenesis both in hepatic and nonhepatic cells/tissues, as well as cholesterol homeostasis. *J. Lipid Res.* 37:1406–21, 1996 (with Welch).

Clifford M. Snapper, Professor; M.D., Albany Medical College, 1981. In vivo regulation of protein- and polysaccharide-specific humoral immunity to extracellular bacteria. *Infect. Immun.* 73(7):4427–31, 2005.

Shiv Srivastava, Professor and Co-Director, Center for Prostate Disease Research, Department of Surgery; Ph.D., Indian Institute of Technology (New Delhi), 1980. Molecular genetics of human cancer, with current focus on prostate cancer. *Clin. Chem.* 51:102–12, 2005. *Oncogene* 23:605–11, 2004. *Cancer Res.* 63:4299–304, 2003.

Aviva Symes, Associate Professor; Ph.D., University College, London, 1990. Mechanism of cytokine action in the nervous system. *J. Biol. Chem.* 276: 19966–73, 2001.

Daniel R. TerBush, Assistant Professor; Ph.D., Michigan, 1991. Mechanisms of secretory vesicle targeting and fusion in *Saccharomyces cerevisiae.*

George Tsokos, Professor; M.D., 1975, Ph.D., 1977, Athens. Immune cell signaling aberrations in patients with lupus; mechanisms that regulate the transcription of the CR2 gene. *J. Immunol.* 159:5492–501, 1997.

Robert W. Williams, Associate Professor; Ph.D., Washington State, 1980. Vibrational spectroscopy of peptides and proteins; quantum mechanical calculations on peptide force fields. *J. Mol. Struct. (Theochem.)* 685:101–7, 2004.

T. John Wu, Assistant Professor; Ph.D., Texas A&M, 1991. Molecular and cellular regulation of neuroendocrinology and behavior. *Endocrinology* 146: 280–6, 2005. *Am. J. Obstet. Gynecol.* 192:586–91, 2005. *Brain Res.* 862: 238–41, 2000.

Xin Xiang, Associate Professor; Ph.D., University of Medicine and Dentistry of New Jersey, 1991. In vivo regulation of a microtubule motor, cytoplasmic dynein. *Mol. Biol. Cell* 14:1479–88, 2003.

THE UNIVERSITY OF ALABAMA AT BIRMINGHAM

Department of Cell Biology

Program of Study	The Department of Cell Biology, in conjunction with the Departments of Biochemistry and Molecular Genetics, Microbiology, and Neurobiology, offers course work and individual laboratory research leading to the Ph.D. degree. The program is designed to provide high-quality interdisciplinary training in cell and molecular biology to a selected group of predoctoral students, preparing them to become independent investigators in these disciplines. Students are immersed in research at the forefront of scientific endeavor and provided with sufficient guidance and course work to place their research in the proper perspective.

This interdisciplinary program includes faculty members from Anesthesiology, the Arthritis Center, Biochemistry and Molecular Genetics, Biology, the Cancer Center, Cell Biology, Endocrinology, Geographic Medicine, Hematology/Oncology, Infectious Disease, Medicine, Microbiology, Neurobiology, Neurology, Neuropsychiatry, Oral Biology, Pathology, Pediatrics, Pharmacology, Physiology and Biophysics, Psychiatry, and Public Health.

The first-year curriculum emphasizes three areas: acquisition of a working knowledge of contemporary cellular and molecular biology through an intensive, integrated course in biochemistry, genetics, cell biology, virology, immunology, and neurobiology; involvement in a diversity of laboratory research training experiences; and the development of skills in reading, writing, and speaking. Advanced students are engaged primarily in research but also take some advanced courses and tutorials in specialized areas of interest and participate in seminars. Completion of requirements for the Ph.D. usually takes six years. No foreign language is required.

Areas of specialization for dissertation research include fundamental molecular biology; biochemistry of nucleic acids; prokaryotic and eukaryotic molecular and cell biology; molecular virology; viral, microbial, and mammalian genetics; immunogenetics; cellular, developmental, and tumor immunology; immunochemistry; biological macromolecules and membranes; host-parasite relationships; infectious diseases; biochemistry of connective tissues; X-ray crystallography; NMR spectroscopy; molecular biophysics; ion-channel structure and biophysics; synaptic function and plasticity; neuronal development; and glial-neuronal signaling. Students are encouraged to present research results at scientific meetings and publish in scientific journals. |

Research Facilities	Faculty members participating in the program have more than 400,000 square feet of laboratory space. In addition to well-equipped labs, a number of core facilities are available. These include automated DNA sequencing, DNA chip analysis, NMR spectroscopy, electron microscopy, protein and nucleic acid synthesis and analysis, mass spectrometry, confocal microscopy, fluorescence-activated cell sorting, large-scale bacterial fermentation, X-ray diffraction, a P3 containment facility, computer facilities, and a hybridoma facility.
Financial Aid	All students admitted to the program receive support through national or state granting agencies in the amount of $23,000 plus student health insurance and full payment of tuition and fees.
Cost of Study	Tuition and fees for the 2004–05 academic year were $5500 for in-state students and $11,000 for out-of-state students. As indicated above, tuition and fees are paid for all students.
Living and Housing Costs	The cost of living in Birmingham is slightly lower than the U.S. average. Housing is readily available in the Medical Center area.
Student Group	The graduate program of the Departments of Biochemistry and Molecular Genetics, Cell Biology, Neurobiology, and Microbiology consists of more than 130 faculty members, 190 full-time graduate students, and about 150 postdoctoral fellows and visiting faculty members. The total enrollment at the University of Alabama at Birmingham is more than 15,000.
Student Outcomes	Graduates typically go on to postdoctoral research appointments, followed by careers in academic research and teaching, research in the biotechnology industry, or other science-related professions.
Location	Birmingham is located in the lovely rolling foothills of the Appalachian Mountain range in central Alabama. A metropolitan area that includes 1 million people, Birmingham is only a few hours' drive from Atlanta, Nashville, New Orleans, and the Gulf Coast. The city has excellent art and historical museums, theaters, libraries, a symphony orchestra, a ballet, a zoo, and botanical gardens. A host of recreational opportunities, including camping, swimming, fishing, hiking, golf, tennis, and boating, are available the year round in numerous local and state parks.
The University	The University of Alabama at Birmingham consists of University College, the Graduate School, and the Medical Center. It is located in the largest population center of the state and is heavily engaged in research, instruction, and service programs. Its dedication to excellence in biomedical research is exemplified by its consistent ranking in the top twenty institutions in receipt of federal research funds.
Applying	Applications to the program are evaluated by the Admissions Committee in consultation with other faculty members. The admission decision is based on scores achieved on the Graduate Record Examinations (a combined score of 1200, nominally, on the verbal and quantitative portions of the General Test), undergraduate grade point average (consideration is given to the curriculum completed), letters of evaluation, prior research experience, and a personal interview.

To be accepted into the program, the student should have completed a B.S. degree that includes the following undergraduate course work by the time of entrance: calculus, general and organic chemistry, and at least one introductory course in zoology or biology. Courses in physical chemistry, genetics, cell biology, and related topics are also beneficial to the candidate. Any remedial course work must be completed with a grade of B or better before the end of the first full year of doctoral study.

Applications are strongly encouraged from individuals with prior research experience, an M.S. degree in a related area, or a professional degree such as the M.D., D.M.D., D.V.M., or O.D.

The program anticipates admitting 35 to 40 students each year. |
| **Correspondence and Information** | Admissions Committee
Cellular and Molecular Biology Graduate Program
Suite 260, Bevill Biomedical Research Building
The University of Alabama at Birmingham
Birmingham, Alabama 35294-2170
Phone: 800-262-7764 (toll-free)
E-mail: cmb@uab.edu
Web site: http://www.cmb.uab.edu |

The University of Alabama at Birmingham

THE FACULTY AND THEIR RESEARCH

Cell Physiology, Adhesion, and Signaling
S. Barnes, Ph.D.: bile acids, polyphenols, and metabolism and their effects on protein expression and function in chronic disease.
Z. Bebok, M.D.: membrane protein biogenesis in epithelial cells (CFTR as model); unfolded protein response.
R. Carter, M.D.: immune and autoimmune lymphocytes; structure and function in vivo.
C. Chang, Ph.D.: signal pathways in frog development.
D. Chaplin, M.D., Ph.D.: control of inflammation and lymphoid organ formation by LT and TNF.
J. Chatham, Ph.D.: cardiomyocyte function and metabolism in diabetes and ischemic heart disease.
J. Collawn, Ph.D.: intracellular protein sorting.
C. Gladson, M.D.: molecular mechanisms promoting tumor-cell proliferation; invasion and angiogenesis in gliomas.
J. Hagood, M.D.: fibroblast signaling in lung remodeling and fibrogenesis.
J. Kearney, Ph.D.: B cells; B-cell development; hybridomas; transgenic and knockout mice; immunoregulation; *B. anthracis.*
N. Kedishvili, Ph.D.: regulation of retinoic acid homeostasis.
F. Lin, M.D., Ph.D.: regulation of cell growth by G-protein-coupled receptor signaling.
R. Marchase, Ph.D.: calcium regulation and its impairment in diabetes.
G. Marques, Ph.D.: TGF-ß signaling in nervous system development and function.
M. Miller, Ph.D.: function and evolution of intercellular communication mechanisms.
E. Schwiebert, Ph.D.: extracellular nucleotide signaling and epithelial cell biology and physiology.
L. Schwiebert, Ph.D.: airway inflammation.
E. Sztul, Ph.D.: membrane traffic; protein degradation.
J. Thompson, Ph.D.: molecular mechanisms of angiogenesis.
A. Woods, Ph.D.: *Syndecan proteoglycans* in cell adhesion and matrix assembly.

Gene Regulation and Expression
A. Agarwal, M.D.: regulation of heme oxygenase gene expression in kidney and vascular injury.
L. Bridges, M.D., Ph.D.: genetic influences on treatment response in rheumatoid arthritis.
C. Chen, Ph.D.: mechanism and regulation of mammalian mRNA turnover.
X. Chen, Ph.D.: the p53 tumor-suppressor gene family and transcriptional regulation.
D. Crawford, M.D., Ph.D.: the role of G2/M-specific genes in mitosis and G2 DNA damage checkpoint regulation.
P. Higgins, Ph.D.: genetic and biochemical studies of chromosome dynamics.
C. Klug, Ph.D.: hematopoietic stem-cell biology and acute leukemias.
W.-C. Lin, M.D., Ph.D.: cell-cycle control and DNA damage response.
S. Lobo-Ruppert, Ph.D.: role of transcription factor oncogenes in tumor initiation.
R. Mayne, Ph.D.: collagen and collagen-binding proteins in health and disease.
J. McDonald, M.D.: cellular life and death signals in cancer; AIDS and bone disease.
M. Ruppert, Ph.D.: role of zinc finger transcription factors in tumor progression.
T. Ryan, Ph.D.: gene regulation; stem cells; mouse models; mutagenesis; cell therapies.
C. Turnbough, Ph.D.: *B. anthracis* spore structure-function/bacterial gene regulation.
H. Wang, Ph.D.: role of histone modification in chromatin function.

Immunology
S. Barnum, Ph.D.: role of complement, acute phase proteins, and adhesion molecules in acute and chronic inflammation in the central nervous system.
T. Benveniste, Ph.D.: immune/nervous system interactions.
O. Branch, Ph.D.: malaria molecular epidemiology and immunology.
P. Bucy, M.D., Ph.D.: T-cell regulation of immune responses in vivo.
R. Davis, M.D.: lymphocyte development and mechanisms of lymphomagenesis.
C. Elson, M.D.: chronic intestinal inflammation.
K. Fujihashi, D.D.S., Ph.D.: mucosal immunology; regulation of S-igA antibody responses; mucosal vaccine development.
V. Ghanta, Ph.D.: tumor immunology; CNS and immune system interactions.
Z. Hel, Ph.D.: development and testing of novel HIV/AIDS vaccine strategies.
L. Justement, Ph.D.: analysis of molecular mechanisms regulating lymphocyte biology.
J. Kabarowski, Ph.D.: regulation of inflammation and chronic inflammatory disease by lysophospholipid receptors.
J. Kapp, Ph.D.: immune regulation and transplantation.
J. Katz, Ph.D., D.D.S.: vaccine delivery systems; inflammation; innate immunity.
R. Kimberly, M.D.: autoimmunity, molecular mechanisms, and genetic risk.
W. Koopman, M.D.: pathogenesis of immune disease.
H. Kubagawa, M.D.: molecular genetics and immunopathology of host defense.
J. Mestecky, M.D., Ph.D.: mucosal immunity; vaccines.
M. Nahm, M.D.: adaptive and innate immune responses to vaccines against bacteria.
T. Strong, Ph.D.: identification of tumor antigens and development of cancer vaccines.
A. Szalai, Ph.D.: inflammation innate immunity and the acute phase proteins in health and disease.
L. Timares, Ph.D.: engineering dendritic cells for immunotherapy.
M. Walter, Ph.D.: structure and function of cytokines and their signaling molecules.
C. Weaver, M.D.: T-cell development.
Z. Zhang, Ph.D.: molecular regulation of early B-cell development and antibody repertoire formation.
T. Zhou, M.D.: specific induction of apoptosis in autoimmune and inflammatory cells.

Macromolecular Structure and Function
J. Blalock, Ph.D.: rational drug and vaccine design.
C. Brouillette, Ph.D.: protein structural cooperativity and energetics.
D. Chattopadhyay, Ph.D.: structure-function analysis of proteins.
I. Chesnokov, Ph.D.: DNA replication and cell cycle in eukaryotes.
H. Cheung, Ph.D.: regulatory mechanism in cardiac muscle.
L. DeLucas, Ph.D.: protein crystallography/protein crystal growth.
G. Elgavish, Ph.D.: NMR studies of intact hearts.
S. Frank, M.D.: growth hormone action and GH receptor structure and function.
B. Freeman, Ph.D.: tissue metabolism of reactive inflammatory mediators.
R. Krishna, Ph.D.: structural biology and biomolecular NMR spectroscopy.
J. Murphy-Ullrich, Ph.D.: extracellular matrix control of cell and growth-factor function.

C. Raman, Ph.D.: lymphocyte activation, immune tolerance and autoimmunity tolerance, and autoimmunity.
J. Segrest, M.D., Ph.D.: structural biology of supramolecular assemblies, particularly lipoproteins and membranes.
B. Sha, Ph.D.: structure and function of molecular chaperones.
N. Sthanam, Ph.D.: bacterial surface protein anchoring.

Molecular Genetics and Disease
S. Abulkadir, Ph.D.: molecular genetics of prostate cancer.
P. Atkinson, M.D., Ph.D.: primary immunodeficiency/role of infection in chronic diseases.
D. Bedwell, Ph.D.: translation termination; treatment of genetic diseases.
P. Burrows, Ph.D.: B-lymphocyte development and function.
P. Detloff, Ph.D.: mouse models of human genetic disorders.
K. Dybvig, Ph.D.: pathogenic mechanisms of mycoplasmas.
L. Guay-Woodford, M.D.: characterizing molecular determinants involved in PKD pathogenesis.
K. Jiao, M.D., Ph.D.: TGF-ß/BMP signaling during cardiogenesis.
K. Kirk, Ph.D.: the CFTR chloride channel.
J. Kudlow, M.D.: nucleocytoplasmic O-glycosylation in the control of cell function.
J. Mountz, M.D., Ph.D.: gene therapy; T-cell aging; immunogenetics; T-cell imaging.
H. Schroeder, M.D., Ph.D.: development/function of lymphocyte antigen receptors.
R. Serra, Ph.D.: mechanism of TGF-ß action in developmental and disease processes.
T. Townes, Ph.D.: developmental regulation of gene expression.
D. Welch, Ph.D.: molecular basis of tumor progression and metastasis.
T. Unnasch, Ph.D.: molecular epidemiology and ecology of vector-borne diseases.
B. Yoder, Ph.D.: cilia signaling and dysfunction in development and disease.

Molecular Pathogenesis
D. Balkovetz, M.D., Ph.D.: epithelial cell biology; epithelial cell signal regulation; regulation of paracellular transport across epithelial cell tight junctions.
W. Benjamin, Ph.D.: molecular typing of *Streptococcus pneumoniae.*
D. Briles, Ph.D.: bacterial pathogenesis; virulence; immunity; pneumococcus.
N. Childers, D.D.S., Ph.D.: development of a vaccine for the prevention of dental caries.
M. Cooper, M.D.: developmental immunobiology, with emphasis on B-cell and T-cell differentiation; clinical immunology, with emphasis on immunodeficiency diseases and lymphoid malignancies.
S. Hollingshead, Ph.D.: mechanisms of variation in microbial pathogenesis.
R. Lorenz, M.D., Ph.D.: cellular and molecular immunology of the gastrointestinal tract.
S. Michalek, Ph.D.: mucosal vaccines and host mechanisms involved in inflammation.
R. Morrison, Ph.D.: immunobiology of Chlamydia infection.
C. Morrow, Ph.D.: understanding, at the molecular level, virus–host cell interactions.
M. Niederweis, Ph.D.: the role of porins in outer-membrane permeability and drug resistance of mycobacteria.
D. Pritchard, Ph.D.: roles of glycoconjugates in pathogenesis.
J. Rayner, Ph.D.: cell biology of the malaria parasite *Plasmodium falciparum.*
A. Steyn, Ph.D.: mechanism of mycobacterium tuberculosis virulence.
K. Waites, M.D.: diagnostic microbiology; epidemiology and mechanisms of antimicrobial resistance; mycoplasma and ureaplasma diseases; staphylococcal diseases.
H. Wu, Ph.D.: bacteria-host interaction.
J. Yother, Ph.D.: capsular polysaccharides of *Streptococcus pneumoniae.*

Neurobiology
D. Benos, Ph.D.: molecular basis of operation on ion channels and transporters.
M. Brenner, Ph.D.: molecular studies of astrocytes in health and disease.
L. Dobrunz, Ph.D.: synaptic transmission and plasticity in hippocampus.
J. Hablitz, Ph.D.: cellular mechanisms of neurotransmission.
G. Johnson, Ph.D.: molecular mechanisms of neurodegeneration.
R. Jope, Ph.D.: neuronal signaling mechanisms regulating gene expression and cell death.
R. Lester, Ph.D.: molecular and cellular mechanisms of nicotine addiction.
L. Pozzo-Miller, Ph.D.: neurotrophins on Ca2+ signaling, synapse development, and plasticity.
K. Roth, M.D., Ph.D.: molecular regulation of neuronal cell death.
D. Ruden, Ph.D.: environmental and developmental toxicology.
H. Sontheimer, Ph.D.: the role of neuroglia in brain function and disease.
D. Sweatt, Ph.D.: signal transduction mechanisms in learning and memory.
A. Theibert, Ph.D.: role of phosphoinositides in developmental neurobiology.
S. Wilson, Ph.D.: mouse models of neurodegeneration.
M. Wyss, Ph.D.: control of the autonomic nervous system.
Y. Zhou, Ph.D.: modulation of ion channels; regulation of neuronal excitability and synaptic transmission.

Virology
W. Britt, M.D.: human herpesviruses, molecular virology, and pathogenesis.
L. Chow, Ph.D.: human papillomavirus DNA replication and pathogenesis.
T. T. Dokland, Ph.D.: cryoelectron microscopy and X-ray crystallography of virus assembly processes.
J. Engler, Ph.D.: identifying novel diagnostic peptides and cell-specific ligands.
B. Hahn, M.D.: origin and evolution of primate lentiviruses.
J. Kappes, Ph.D.: HIV, molecular virology, and pathogenesis.
R. Kaslow, M.D.: immunogenetic determinants in AIDS and other diseases.
O. Kutsch, Ph.D.: HIV-1 latency and drug screening.
E. Lefkowitz, Ph.D.: bioinformatics; biodefense; microbial genomics and evolution.
M. Luo, Ph.D.: structure-based approaches to anti-infectious agents.
P. Prevelige, Ph.D.: structural biology of viral assembly and infection.
G. Shaw, M.D., Ph.D.: evolution and persistence of HIV-1.
W. Sullender, M.D.: respiratory syncytial virus; antigenic diversity.
S. Thompson, Ph.D.: translation initiation and replication of RNA viruses.
R. Whitley, M.D.: herpesvirus; varicella zoster virus.
A. Zajac, Ph.D.: antiviral immunity; T-cell responses; immunological memory.

THE UNIVERSITY OF ALABAMA AT BIRMINGHAM

Cellular and Molecular Biology Graduate Program

Program of Study

The Cellular and Molecular Biology Graduate Program, administered by the Departments of Biochemistry and Molecular Genetics, Cell Biology, Microbiology, and Neurobiology, offers course work and individual laboratory research leading to the Ph.D. degree. The program is designed to provide high-quality interdisciplinary training in cell and molecular biology to a selected group of predoctoral students, preparing them to become independent investigators in these disciplines. Students are immersed in research at the forefront of scientific endeavor and provided with sufficient guidance and course work to place their research in the proper perspective.

This interdisciplinary program includes faculty members from Anesthesiology, the Arthritis Center, Biochemistry and Molecular Genetics, Biology, the Cancer Center, Cell Biology, Endocrinology, Geographic Medicine, Hematology/Oncology, Infectious Disease, Medicine, Microbiology, Neurobiology, Neurology, Neuropsychiatry, Oral Biology, Pathology, Pediatrics, Pharmacology, Physiology and Biophysics, Psychiatry, and Public Health.

The first-year curriculum emphasizes three areas: acquisition of a working knowledge of contemporary cellular and molecular biology through an intensive, integrated course in biochemistry, genetics, cell biology, virology, immunology, and neurobiology; involvement in a diversity of laboratory research training experiences; and the development of skills in reading, writing, and speaking. Advanced students are engaged primarily in research but also take some advanced courses and tutorials in specialized areas of interest and participate in seminars. Completion of requirements for the Ph.D. usually takes six years. No foreign language is required.

Areas of specialization for dissertation research include fundamental molecular biology; biochemistry of nucleic acids; prokaryotic and eukaryotic molecular and cell biology; molecular virology; viral, microbial, and mammalian genetics; immunogenetics; cellular, developmental, and tumor immunology; immunochemistry; biological macromolecules and membranes; host-parasite relationships; infectious diseases; biochemistry of connective tissues; X-ray crystallography; NMR spectroscopy; molecular biophysics; ion-channel structure and biophysics; synaptic function and plasticity; neuronal development; and glial-neuronal signaling. Students are encouraged to present research results at scientific meetings and publish in scientific journals.

Research Facilities

Faculty members participating in the program have more than 400,000 square feet of laboratory space. In addition to well-equipped labs, a number of core facilities are available. These include automated DNA sequencing, DNA chip analysis, NMR spectroscopy, electron microscopy, protein and nucleic acid synthesis and analysis, mass spectrometry, confocal microscopy, fluorescence-activated cell sorting, large-scale bacterial fermentation, X-ray diffraction, a P3 containment facility, computer facilities, and a hybridoma facility.

Financial Aid

All students admitted to the program receive support through national or state granting agencies in the amount of $23,000 plus student health insurance and full payment of tuition and fees.

Cost of Study

Tuition and fees for the 2004–05 academic year were $5500 for in-state students and $11,000 for out-of-state students. As indicated above, tuition and fees are paid for all students.

Living and Housing Costs

The cost of living in Birmingham is slightly lower than the U.S. average. Housing is readily available in the Medical Center area.

Student Group

The graduate program of the Departments of Biochemistry and Molecular Genetics, Cell Biology, Neurobiology, and Microbiology consists of more than 130 faculty members, 190 full-time graduate students, and about 150 postdoctoral fellows and visiting faculty members. The total enrollment at the University of Alabama at Birmingham is more than 15,000.

Student Outcomes

Graduates typically go on to postdoctoral research appointments, followed by careers in academic research and teaching, research in the biotechnology industry, or other science-related professions.

Location

Birmingham is located in the lovely rolling foothills of the Appalachian Mountain range in central Alabama. A metropolitan area that includes 1 million people, Birmingham is only a few hours' drive from Atlanta, Nashville, New Orleans, and the Gulf Coast. The city has excellent art and historical museums, theaters, libraries, a symphony orchestra, a ballet, a zoo, and botanical gardens. A host of recreational opportunities, including camping, swimming, fishing, hiking, golf, tennis, and boating, are available the year round in numerous local and state parks.

The University

The University of Alabama at Birmingham consists of University College, the Graduate School, and the Medical Center. It is located in the largest population center of the state and is heavily engaged in research, instruction, and service programs. Its dedication to excellence in biomedical research is exemplified by its consistent ranking in the top twenty institutions in receipt of federal research funds.

Applying

Applications to the program are evaluated by the Admissions Committee in consultation with other faculty members. The admission decision is based on scores achieved on the Graduate Record Examinations (a combined score of at least 1200, nominally, on the verbal and quantitative portions of the General Test), undergraduate grade point average (consideration is given to the curriculum completed), letters of evaluation, prior research experience, and a personal interview.

To be accepted into the program, the student should have completed a B.S. degree that includes the following undergraduate course work by the time of entrance: calculus, general and organic chemistry, and at least one introductory course in zoology or biology. Courses in physical chemistry, genetics, cell biology, and related topics are also beneficial to the candidate. Any remedial course work must be completed with a grade of B or better before the end of the first full year of doctoral study.

Applications are strongly encouraged from individuals with prior research experience, an M.S. degree in a related area, or a professional degree such as the M.D., D.M.D., D.V.M., or O.D.

The program anticipates admitting 35 to 40 students each year.

Correspondence and Information

Admissions Committee
Cellular and Molecular Biology Graduate Program
Suite 260, Bevill Biomedical Research Building
The University of Alabama at Birmingham
Birmingham, Alabama 35294-2170

Phone: 800-262-7764 (toll-free)
E-mail: cmb@uab.edu
Web site: http://www.cmb.uab.edu

The University of Alabama at Birmingham

THE FACULTY AND THEIR RESEARCH

Cell Physiology, Adhesion, and Signaling
S. Barnes, Ph.D.: bile acids, polyphenols, and metabolism and their effects on protein expression and function in chronic disease.
Z. Bebok, M.D.: membrane protein biogenesis in epithelial cells (CFTR as model); unfolded protein response.
R. Carter, M.D.: immune and autoimmune lymphocytes; structure and function in vivo.
C. Chang, Ph.D.: signal pathways in frog development.
D. Chaplin, M.D., Ph.D.: control of inflammation and lymphoid organ formation by LT and TNF.
J. Chatham, Ph.D.: cardiomyocyte function and metabolism in diabetes and ischemic heart disease.
J. Collawn, Ph.D.: intracellular protein sorting.
C. Gladson, M.D.: molecular mechanisms promoting tumor-cell proliferation; invasion and angiogenesis in gliomas.
J. Hagood, M.D.: fibroblast signaling in lung remodeling and fibrogenesis.
J. Kearney, Ph.D.: B cells; B-cell development; hybridomas; transgenic and knockout mice; immunoregulation; *B. anthracis.*
N. Kedishvili, Ph.D.: regulation of retinoic acid homeostasis.
F. Lin, M.D., Ph.D.: regulation of cell growth by G-protein-coupled receptor signaling.
R. Marchase, Ph.D.: calcium regulation and its impairment in diabetes.
G. Marques, Ph.D.: TGF-ß signaling in nervous system development and function.
M. Miller, Ph.D.: function and evolution of intercellular communication mechanisms.
E. Schwiebert, Ph.D.: extracellular nucleotide signaling and epithelial cell biology and physiology.
L. Schwiebert, Ph.D.: airway inflammation.
E. Sztul, Ph.D.: membrane traffic; protein degradation.
J. Thompson, Ph.D.: molecular mechanisms of angiogenesis.
A. Woods, Ph.D.: *Syndecan proteoglycans* in cell adhesion and matrix assembly.

Gene Regulation and Expression
A. Agarwal, M.D.: regulation of heme oxygenase gene expression in kidney and vascular injury.
L. Bridges, M.D., Ph.D.: genetic influences on treatment response in rheumatoid arthritis.
C. Chen, Ph.D.: mechanism and regulation of mammalian mRNA turnover.
X. Chen, Ph.D.: the p53 tumor-suppressor gene family and transcriptional regulation.
D. Crawford, M.D., Ph.D.: the role of G2/M-specific genes in mitosis and G2 DNA damage checkpoint regulation.
P. Higgins, Ph.D.: genetic and biochemical studies of chromosome dynamics.
C. Klug, Ph.D.: hematopoietic stem-cell biology and acute leukemias.
W.-C. Lin, M.D., Ph.D.: cell-cycle control and DNA damage response.
S. Lobo-Ruppert, Ph.D.: role of transcription factor oncogenes in tumor initiation.
R. Mayne, Ph.D.: collagen and collagen-binding proteins in health and disease.
J. McDonald, M.D.: cellular life and death signals in cancer; AIDS and bone disease.
M. Ruppert, M.D., Ph.D.: role of zinc finger transcription factors in tumor progression.
T. Ryan, Ph.D.: gene regulation; stem cells; mouse models; mutagenesis; cell therapies.
C. Turnbough, Ph.D.: *B. anthracis* spore structure-function/bacterial gene regulation.
H. Wang, Ph.D.: role of histone modification in chromatin function.

Immunology
S. Barnum, Ph.D.: role of complement, acute phase proteins, and adhesion molecules in acute and chronic inflammation in the central nervous system.
T. Benveniste, Ph.D.: immune/nervous system interactions.
O. Branch, Ph.D.: malaria molecular epidemiology and immunology.
P. Bucy, M.D., Ph.D.: T-cell regulation of immune responses in vivo.
R. Davis, M.D.: lymphocyte development and mechanisms of lymphomagenesis.
C. Elson, M.D.: chronic intestinal inflammation.
K. Fujihashi, D.D.S., Ph.D.: mucosal immunology; regulation of S-igA antibody responses; mucosal vaccine development.
V. Ghanta, Ph.D.: tumor immunology; CNS and immune system interactions.
Z. Hel, Ph.D.: development and testing of novel HIV/AIDS vaccine strategies.
L. Justement, Ph.D.: analysis of molecular mechanisms regulating lymphocyte biology.
J. Kabarowski, Ph.D.: regulation of inflammation and chronic inflammatory disease by lysophospholipid receptors.
J. Kapp, Ph.D.: immune regulation and transplantation.
J. Katz, Ph.D., D.D.S.: vaccine delivery systems; inflammation; innate immunity.
R. Kimberly, M.D.: autoimmunity, molecular mechanisms, and genetic risk.
W. Koopman, M.D.: pathogenesis of immune disease.
H. Kubagawa, M.D.: molecular genetics and immunopathology of host defense.
J. Mestecky, M.D., Ph.D.: mucosal immunity; vaccines.
M. Nahm, M.D.: adaptive and innate immune responses to vaccines against bacteria.
T. Strong, Ph.D.: identification of tumor antigens and development of cancer vaccines.
A. Szalai, Ph.D.: inflammation innate immunity and the acute phase proteins in health and disease.
L. Timares, Ph.D.: engineering dendritic cells for immunotherapy.
M. Walter, Ph.D.: structure and function of cytokines and their signaling molecules.
C. Weaver, M.D.: T-cell development.
Z. Zhang, Ph.D.: molecular regulation of early B-cell development and antibody repertoire formation.
T. Zhou, M.D.: specific induction of apoptosis in autoimmune and inflammatory cells.

Macromolecular Structure and Function
J. Blalock, Ph.D.: rational drug and vaccine design.
C. Brouillette, Ph.D.: protein structural cooperativity and energetics.
D. Chattopadhyay, Ph.D.: structure-function analysis of proteins.
I. Chesnokov, Ph.D.: DNA replication and cell cycle in eukaryotes.
H. Cheung, Ph.D.: regulatory mechanism in cardiac muscle.
L. DeLucas, Ph.D.: protein crystallography/protein crystal growth.
G. Elgavish, Ph.D.: NMR studies of intact hearts.
S. Frank, M.D.: growth hormone action and GH receptor structure and function.
B. Freeman, Ph.D.: tissue metabolism of reactive inflammatory mediators.
R. Krishna, Ph.D.: structural biology and biomolecular NMR spectroscopy.
J. Murphy-Ullrich, Ph.D.: extracellular matrix control of cell and growth-factor function.

C. Raman, Ph.D.: lymphocyte activation, immune tolerance and autoimmunity tolerance, and autoimmunity.
J. Segrest, M.D., Ph.D.: structural biology of supramolecular assemblies, particularly lipoproteins and membranes.
B. Sha, Ph.D.: structure and function of molecular chaperones.
N. Sthanam, Ph.D.: bacterial surface protein anchoring.

Molecular Genetics and Disease
S. Abulkadir, Ph.D.: molecular genetics of prostate cancer.
P. Atkinson, M.D., Ph.D.: primary immunodeficiency/role of infection in chronic diseases.
D. Bedwell, Ph.D.: translation termination; treatment of genetic diseases.
P. Burrows, Ph.D.: B-lymphocyte development and function.
P. Detloff, Ph.D.: mouse models of human genetic disorders.
K. Dybvig, Ph.D.: pathogenic mechanisms of mycoplasmas.
L. Guay-Woodford, M.D.: characterizing molecular determinants involved in PKD pathogenesis.
K. Jiao, M.D., Ph.D.: TGF-ß/BMP signaling during cardiogenesis.
K. Kirk, Ph.D.: the CFTR chloride channel.
J. Kudlow, M.D.: nucleocytoplasmic O-glycosylation in the control of cell function.
J. Mountz, M.D., Ph.D.: gene therapy; T-cell aging; immunogenetics; T-cell imaging.
H. Schroeder, M.D., Ph.D.: development/function of lymphocyte antigen receptors.
R. Serra, Ph.D.: mechanism of TGF-ß action in developmental and disease processes.
T. Townes, Ph.D.: developmental regulation of gene expression.
D. Welch, Ph.D.: molecular basis of tumor progression and metastasis.
T. Unnasch, Ph.D.: molecular epidemiology and ecology of vector-borne diseases.
B. Yoder, Ph.D.: cilia signaling and dysfunction in development and disease.

Molecular Pathogenesis
D. Balkovetz, M.D., Ph.D.: epithelial cell biology; epithelial cell signal regulation; regulation of paracellular transport across epithelial cell tight junctions.
W. Benjamin, Ph.D.: molecular typing of *Streptococcus pneumoniae.*
D. Briles, Ph.D.: bacterial pathogenesis; virulence; immunity; pneumococcus.
N. Childers, D.D.S., Ph.D.: development of a vaccine for the prevention of dental caries.
M. Cooper, M.D.: developmental immunobiology, with emphasis on B-cell and T-cell differentiation; clinical immunology, with emphasis on immunodeficiency diseases and lymphoid malignancies.
S. Hollingshead, Ph.D.: mechanisms of variation in microbial pathogenesis.
R. Lorenz, M.D., Ph.D.: cellular and molecular immunology of the gastrointestinal tract.
S. Michalek, Ph.D.: mucosal vaccines and host mechanisms involved in inflammation.
R. Morrison, Ph.D.: immunobiology of Chlamydia infection.
C. Morrow, Ph.D.: understanding, at the molecular level, virus–host cell interactions.
M. Niederweis, Ph.D.: the role of porins in outer-membrane permeability and drug resistance of mycobacteria.
D. Pritchard, Ph.D.: roles of glycoconjugates in pathogenesis.
J. Rayner, Ph.D.: cell biology of the malaria parasite *Plasmodium falciparum.*
A. Steyn, Ph.D.: mechanism of mycobacterium tuberculosis virulence.
K. Waites, M.D.: diagnostic microbiology; epidemiology and mechanisms of antimicrobial resistance; mycoplasma and ureaplasma diseases; staphylococcal diseases.
H. Wu, Ph.D.: bacteria-host interaction.
J. Yother, Ph.D.: capsular polysaccharides of *Streptococcus pneumoniae.*

Neurobiology
D. Benos, Ph.D.: molecular basis of operation on ion channels and transporters.
M. Brenner, Ph.D.: molecular studies of astrocytes in health and disease.
L. Dobrunz, Ph.D.: synaptic transmission and plasticity in hippocampus.
J. Hablitz, Ph.D.: cellular mechanisms of neurotransmission.
G. Johnson, Ph.D.: molecular mechanisms of neurodegeneration.
R. Jope, Ph.D.: neuronal signaling mechanisms regulating gene expression and cell death.
R. Lester, Ph.D.: molecular and cellular mechanisms of nicotine addiction.
L. Pozzo-Miller, Ph.D.: neurotrophins on Ca2+ signaling, synapse development, and plasticity.
K. Roth, M.D., Ph.D.: molecular regulation of neuronal cell death.
D. Ruden, Ph.D.: environmental and developmental toxicology.
H. Sontheimer, Ph.D.: the role of neuroglia in brain function and disease.
D. Sweatt, Ph.D.: signal transduction mechanisms in learning and memory.
A. Theibert, Ph.D.: role of phosphoinositides in developmental neurobiology.
S. Wilson, Ph.D.: mouse models of neurodegeneration.
M. Wyss, Ph.D.: control of the autonomic nervous system.
Y. Zhou, Ph.D.: modulation of ion channels; regulation of neuronal excitability and synaptic transmission.

Virology
W. Britt, M.D.: human herpesviruses, molecular virology, and pathogenesis.
L. Chow, Ph.D.: human papillomavirus DNA replication and pathogenesis.
T. T. Dokland, Ph.D.: cryoelectron microscopy and X-ray crystallography of virus assembly processes.
J. Engler, Ph.D.: identifying novel diagnostic peptides and cell-specific ligands.
B. Hahn, M.D.: origin and evolution of primate lentiviruses.
J. Kappes, Ph.D.: HIV, molecular virology, and pathogenesis.
R. Kaslow, M.D.: immunogenetic determinants in AIDS and other diseases.
O. Kutsch, Ph.D.: HIV-1 latency and drug screening.
E. Lefkowitz, Ph.D.: bioinformatics; biodefense; microbial genomics and evolution.
M. Luo, Ph.D.: structure-based approaches to anti-infectious agents.
P. Prevelige, Ph.D.: structural biology of viral assembly and infection.
G. Shaw, M.D., Ph.D.: evolution and persistence of HIV-1.
W. Sullender, M.D.: respiratory syncytial virus; antigenic diversity.
S. Thompson, Ph.D.: translation initiation and replication of RNA viruses.
R. Whitley, M.D.: herpesvirus; varicella zoster virus.
A. Zajac, Ph.D.: antiviral immunity; T-cell responses; immunological memory.

UNIVERSITY OF CALIFORNIA, LOS ANGELES

UCLA ACCESS
Programs in Molecular, Cellular and Integrative Life Sciences

Programs of Study

UCLA ACCESS represents a program of maximal choice in research opportunities in the molecular and cellular life sciences on the UCLA campus. It provides incoming graduate students with the opportunity to choose from 261 faculty members, comprising twenty affinity groups, for studies leading to the Ph.D. degree. The first-year program for ACCESS students consists of three 1-quarter laboratory rotations together with a highly flexible interdisciplinary curriculum. At the end of this period, students choose their thesis adviser, thereby becoming a member of the corresponding department or interdepartmental program (IDP).

In addition to the first-year curriculum, requirements include training in teaching for two quarters (in years two and three, respectively), as well as fulfilling the examination requirements of the thesis adviser's department or IDP. These generally include an oral qualifying examination, a midstream seminar, and submission of a dissertation based on the results of original research. Most students complete their research and receive their Ph.D. degree near the end of the fifth year.

Admission to UCLA ACCESS is the mechanism whereby students can ultimately become affiliated with the following departments: Biochemistry and Molecular Biology; Microbiology, Immunology, and Molecular Genetics; Molecular Biology IDP; Molecular, Cell, and Developmental Biology; and Molecular, Cellular, and Integrative Physiology within the College of Letters and Science as well as Biological Chemistry, Cellular and Molecular Pathology, Human Genetics, Molecular and Medical Pharmacology, Molecular Toxicology, Neurobiology, and Oral Biology within the School of Medicine.

Research Facilities

Laboratories of UCLA ACCESS faculty members are conveniently clustered within a Life Science/Health Science complex on the southern end of the UCLA campus. Facilities within the College of Letters and Science as well as the School of Medicine provide students with state-of-the-art laboratories, equipment, core services, and computer and library resources.

Financial Aid

Student support is guaranteed for five years and is derived from a variety of University sources, including individual fellowships, training grants, teaching assistantships, and research assistantships. The 2005–06 annual stipend (four quarters, or twelve months) was $25,000. Student registration fees and nonresident tuition (where applicable) are also paid in full. Students who receive funding from outside agencies (minimum amount to be determined) are awarded an annual bonus for the duration of the fellowship.

Cost of Study

In addition to the stipend and fee awards outlined above, funds to support research supplies and scholarly travel are also available to graduate students in the programs.

Living and Housing Costs

New University-owned town homes, in the heart of Westwood Village, are now available for rental. In addition, convenient University-owned apartments are available for both single and married students. Numerous non-University-owned rentals are available adjacent to the campus in Westwood as well as in local outside communities, such as Santa Monica, West Los Angeles, Mar Vista, and Beverly Hills. For additional information, students should contact the Housing Assignment Office at 310-825-4271.

Student Group

The diverse student population of UCLA is made up of 24,000 undergraduates and 12,000 graduate and professional students. UCLA ACCESS admits approximately 60 students per year. During the first year, students associate with small faculty and student affinity groups until a thesis adviser and department are chosen.

Location

UCLA is located at the foot of the Santa Monica Mountains, 5 miles from the Pacific Ocean, in one of the most attractive and affluent suburban neighborhoods in Los Angeles. Within a short driving distance are local ski resorts, the high desert and numerous bike trails, beaches, and other recreational amenities. Lively interactions exist between UCLA and other outstanding universities and research institutes in southern California, including Caltech; Salk and Scripps institutes; the UC campuses at San Diego, Irvine, Riverside, and Santa Barbara; and the University of Southern California.

The University

Known for its academic excellence, UCLA ranks among the top ten research universities, and many of UCLA's programs are rated among the best in the nation. UCLA is ranked in the top six nationwide for extramural research funding. The UCLA library is ranked third among all research libraries in the nation and contains more than 5 million volumes.

Applying

Applications are accepted for the fall quarter only. The deadline is December 15, and applicants are encouraged to apply early. Students are also urged to apply for the National Science Foundation and Ford Foundation fellowships, if eligible, by the deadline in November. The application comprises official transcripts, scores on the GRE General Test (the Subject Test is optional), three letters of recommendation, and a statement of purpose. International students whose native language is not English must submit results of the TOEFL. Applications are available online and can be completed at UCLA ACCESS's Web site.

Correspondence and Information

UCLA ACCESS
Programs in Molecular, Cellular and Integrative Life Sciences
172 Boyer Hall
University of California, Los Angeles
P.O. Box 951570
Los Angeles, California 90095-1570
Web site: http://www.uclaaccess.ucla.edu

University of California, Los Angeles

FACULTY RESEARCH

ACCESS offers research leading to the Ph.D. through the following affinity-group areas. The numbers in parentheses indicate the number of faculty members who are currently participating. For further information, students should consult the Web site at http://www.uclaaccess.ucla.edu.

Biochemistry and Molecular Biology (15)
Bioinformatics (2)
Cell Biology (6)
Cell Physiology and Biophysics (10)
Developmental Biology (18)
Gene Regulation (18)
Genetics and Genomics (12)
Immunology (18)
Microbial Physiology and Pathogenesis (9)
Molecular and Medical Pharmacology (18)
Molecular Basis of Disease (20)
Molecular, Cellular, and Integrative Physiology (21)
Molecular Evolution and Computational Biology (3)
Molecular Parasitology (6)
Neurobiology (24)
Plant Molecular Biology (4)
Structural Biology and Proteomics (7)
Tumor Cell Biology and Signal Transduction (13)
Virology and Gene Therapy (11)

UNIVERSITY OF CALIFORNIA, SAN DIEGO

School of Medicine
Graduate Studies in Biomedical Sciences
Molecular Cell Biology

Program of Study

The School of Medicine of the University of California, San Diego (UCSD), offers broad opportunities for advanced study in molecular cell biology leading to a Ph.D. degree. The program emphasizes areas of high current interest in cell biology and their application to the understanding of the molecular and cellular basis of various diseases. Areas of investigation that are particularly well developed include cell-cycle control, mitosis, RNA splicing, signaling during membrane trafficking, molecular motors in vesicular trafficking, transcriptional regulation, cancer cell biology, glycobiology, and immunobiology. These topics are investigated in a wide variety of organisms, from yeast to mammals. Many investigators are taking advantage of powerful genetic approaches using yeast, flies, and mice to study normal cell processes as well as to generate disease models by knocking out specific genes. The program offers abundant opportunities to investigate these problems with investigators who are internationally recognized and working at the cutting edge of their discipline. Many of these faculty members are associated with the Department of Cellular and Molecular Medicine and Medicine, the Howard Hughes Medical Institute, the Ludwig Cancer Research Institute, the Cancer Center, the Genetics Program, and the Glycobiology Research and Training Center, as well as other departments of the School of Medicine.

Ordinarily, five years of study are required to complete the program, which requires three rotations in different laboratories and core courses in the first year, after which the student joins the laboratory of a thesis adviser and begins thesis research.

Research Facilities

The faculty members of the program are concentrated in several buildings, including the George Palade Laboratories for Cellular and Molecular Medicine Building, Cellular and Molecular Medicine East Building, the Basic Science Building, the Leichtag Building, the Cancer Center, and the Stein Research Center. Individual laboratories are well equipped to carry out modern biomedical research. Many common facilities are also available to the faculty, including modern facilities for genomics; proteomics; mass spectrometry; bioinformatics; confocal, deconvolution, and electron microscopy; cell imaging; DNA sequencing; cell sorting; and mouse genetics.

Financial Aid

All students receive full financial support (tuition, fees, health insurance, and a stipend) throughout their Ph.D. training. The current stipend level is $26,000 for twelve months. Because most students admitted are competitive for National Science Foundation and other fellowships, everyone is encouraged to apply for such awards by early November of the senior year.

Cost of Study

All graduate students receive full financial support.

Living and Housing Costs

In 2005–06, limited on-campus housing was available from $550 per month for a single to $1250 per month for a family. A wide range of off-campus housing is available in San Diego and neighboring communities that are within a 15-minute drive of the campus and the ocean. The off-campus housing office maintains up-to-date information on rental listings.

Student Group

In 2005–06, there were 3,629 graduate and 19,763 undergraduate students at UCSD. Students in the basic biomedical sciences program were selected from outstanding colleges throughout the nation and abroad and include recent college graduates as well as some having substantial postcollege research experience. Currently, 160 students are enrolled in the program. Graduates routinely obtain postdoctoral positions in excellent laboratories and move on to challenging careers in academia, research institutes, and industry.

Location

The 1,300-acre UCSD campus lies atop high bluffs in La Jolla, on the northern edge of San Diego, overlooking the Pacific Ocean. The intellectual activities on the campus are enriched by the proximity of the Salk Institute, the Scripps Research Institute, and a burgeoning group of companies engaged in biotechnological research and development. San Diego offers a year-round mild Mediterranean climate, miles of shoreline, and ready access to mountains, deserts, and Mexico. Cultural pleasures include the Old Globe Theatre and La Jolla Playhouse, the San Diego Symphony and Opera, and a wide variety of fine art galleries.

The University

In the forty-six years of its existence, UCSD has already become one of the premier research institutions in the United States. UCSD's ability to attract outstanding faculty members and comprehensive research support together with its strong graduate programs places it among the top ten American universities. The scientific environment is full of excitement with the recent additions of the Department of Cellular and Molecular Medicine, the Howard Hughes Medical Institute, the Ludwig Institute for Cancer Research, and the multimillion-dollar supercomputer center. Several new research buildings are currently under construction on the School of Medicine campus and on the main campus. The faculty includes 4 Nobel laureates and 60 members of the National Academy of Sciences.

Applying

Applications are encouraged from outstanding students who have majored in any laboratory science or in mathematics. Desirable features include a minimum GPA of 3.25 and GRE scores above the 80th percentile on the verbal, quantitative, and analytical sections of the General Test (required) and on the Subject Test in biology; chemistry; or biochemistry, cell and molecular biology. Undergraduate prerequisites include organic chemistry, biochemistry, physical chemistry, calculus, and biology; some exposure to cell or molecular biology and mammalian physiology is strongly advised. Prior research experience is desirable. The preapplication is required before the application is considered. Applications, considered for the fall quarter only, are due December 1; earlier submission is advantageous.

Correspondence and Information

Graduate Studies in Biomedical Sciences
School of Medicine, 0685
University of California, San Diego
9500 Gilman Drive
La Jolla, California 92093-0685
Phone: 858-534-3982
Web site: http://biomedsci.ucsd.edu/

University of California, San Diego

THE FACULTY AND THEIR RESEARCH

Joan Heller Brown, Professor; Ph.D., Yeshiva (Einstein). G-protein receptors; phospholipases; gene expression.

Laurence L. Brunton, Professor; Ph.D., Virginia. Cyclic nucleotide metabolism; regulatory actions of protein kinases.

Dennis A. Carson, Professor; M.D., Columbia. Immunology of autoimmune, immunodeficiency, and neoplastic diseases.

Webster K. Cavenee, Professor; Ph.D., Kansas. Human cancer genetics; tumor suppression; oncogene function.

Kenneth R. Chien, Professor; M.D., Ph.D., Temple. Control of cardiac gene expression during growth and development; molecular characterization of cardiac Na$^+$ channels.

Don Cleveland, Professor; Ph.D., Princeton. Microtubule motors and chromosome movement.

Edward A. Dennis, Professor; Ph.D., Harvard. Phospholipase regulation and mechanism; prostaglandin generation.

Arshad Desai, Assistant Professor; Ph.D., California, San Francisco. Mechanisms that maintain genomic integrity during cell division.

Wolfgang H. Dillmann, Professor; M.D., Munich (Germany). Thyroid hormone effects on sarcoplasmic reticulum Ca^{2+}; ATPase function of heat shock proteins in the heart.

Jack Dixon, Professor; Ph.D., California, Santa Barbara. Cellular signal transduction mechanisms and bacterial pathogenesis.

Steven Dowdy, Professor; Ph.D., California, Irvine. G-cell-cycle progression.

Jeffrey D. Esko, Professor; Ph.D., Wisconsin. Glycobiology with emphasis on chemical and genetic approaches for understanding the biology and assembly of proteoglycans and glycoproteins.

Scott D. Emr, Professor; Ph.D., Harvard. Cell biology of intracellular protein sorting and organelle biogenesis.

Ronald M. Evans, Professor; Ph.D., UCLA. Molecular genetics of steroid, thyroid, and retinoid receptors.

Marilyn G. Farquhar, Professor; Ph.D., Berkeley. Signaling in intracellular membrane traffic and protein sorting; cellular and molecular basis of renal function and disease.

James R. Feramisco, Professor; Ph.D., California, Davis. Role of protogenes and oncogenes in cell growth and differentiation.

Theodore Friedmann, Professor; M.D., Pennsylvania. Development of model systems for human gene therapy; efficient vector-mediated transfer and expression of genes in mammalian cells; large-scale genome characterization.

Xiang-Dong Fu, Professor; Ph.D., Case Western Reserve. Mechanisms of pre–messenger RNA processing.

Mark Ginsberg, Professor; M.D., SUNY Downstate Medical Center. Molecular cell biology; molecular pharmacology.

Christopher K. Glass, Professor; M.D., Ph.D., California, San Diego. Transcriptional control of macrophage development.

Lawrence S. B. Goldstein, Professor; Ph.D., Washington (Seattle). Intracellular modality; cell division; cytoskeleton; neurobiology.

Bruce Hamilton, Professor; Ph.D., Caltech. Genetics and functional genomics.

Paul A. Insel, Professor; M.D., Michigan. Signal transduction by catecholamines, ATP, and G proteins.

Martin F. Kagnoff, Professor; M.D., Harvard. Immunoglobulin isotype regulation; HLA class II D-region genes.

Michael Karin, Professor; Ph.D., UCLA. Regulation of gene transcription; cell type specific gene expression; control of cell differentiation; signal transduction from membrane to nucleus.

Thomas J. Kipps, Professor; M.D., Ph.D., Harvard. Human cell physiology, signal transduction, and gene expression.

Richard D. Kolodner, Professor; Ph.D., California, Irvine. Genetics and biochemistry of DNA repair and recombination.

Hyam L. Leffert, Professor; M.D., Yeshiva (Einstein). Growth processes of hepatocytes; role of ionic fluxes and gene expression; gene therapy.

Jamey Marth, Professor; Ph.D., Washington (Seattle). Molecular and developmental biology of vertebrate oligosaccharides.

Pamela Mellon, Professor; Ph.D., Berkeley. Developmental and hormonal regulation of gene expression.

Alexandra Newton, Professor; Ph.D., Stanford. Signal transduction and protein kinase C.

Sanjay Nigam, Professor; M.D., Pennsylvania. Identifying key growth factors and genes.

Daniel T. O'Connor, Professor; M.D., California, Davis. Exocytosis of catecholamine storage vesicle proteins (chromagranins) and their role in hypertension.

Karen Oegema, Assistant Professor; Ph.D., California, San Francisco. Kinetochore assembly, centromere maintenance, centrosome maturation, and cytokinesis in *C. elegans*.

Jerrold M. Olefsky, Professor; M.D., Illinois. Molecular and cellular mechanisms of insulin action; pathophysiology of diabetes mellitus.

George E. Palade, Professor Emeritus; M.D., Bucharest (Romania). Regulation of membrane traffic in eukaryotic cells; transport of macromolecules across the vascular endothelium.

Bing Ren, Assistant Professor; Ph.D., Harvard. Gene regulatory networks in mammalian cells control cellular proliferation and differentiation.

Michael G. Rosenfeld, Professor; M.D., Rochester. Development of neuroendocrine cell phenotypes; hormonal regulation of gene expression; alternate RNA splicing regulation; molecular biology of neuroendocrine gene expression; growth factor receptors.

Sanford Shattil, Professor; M.D., Illinois. Integrin adhesion receptors in blood cell development and function.

Deborah Spector, Professor; Ph.D., MIT. Molecular cellular biology.

Daniel Steinberg, Professor Emeritus; M.D., Wayne State; Ph.D., Harvard. Metabolism of lipids and lipoproteins, with emphasis on regulatory mechanisms; role of macrophages in atherogenesis.

Palmer Taylor, Professor; Ph.D., Wisconsin. Molecular basis for the specificity of drug-receptor interactions; regulation of enzymes and receptors involved in cholinergic neurotransmission.

Susan Taylor, Professor; Ph.D., Wisconsin–Madison. Biochemistry: structure and function of protein kinases.

Roger Y. Tsien, Professor; Ph.D., Cambridge. Cell biology of intracellular signaling; molecular engineering.

Robert H. Tukey, Professor; Ph.D., Iowa. Use of recombinant DNA to characterize rabbit and human cytochromes P-450.

Wylie W. Vale, Professor; Ph.D., Baylor College of Medicine. Isolation and characterization of hypothalamic-releasing factors; physiology and pharmacology of regulatory peptides and synthesis of potent analogues.

Ajit P. Varki, Professor; M.D., Christian Medical College, Vellore (India). Glycobiology: biochemistry and molecular biology of the oligosaccharides in development and cancer.

Jean Wang, Professor; Ph.D., Berkeley. Molecular cellular biology.

Nicholas J. G. Webster, Professor; Ph.D., Stanford. Transcriptional regulation of insulin receptor gene.

Virgil L. Woods, Professor; M.D., California, San Francisco. Structure and function of adhesion-mediating cell-surface receptors: the integrins.

Tony Wynshaw-Boris, Professor; M.D., Ph.D., Case Western Reserve. Genetic and biochemical pathways.

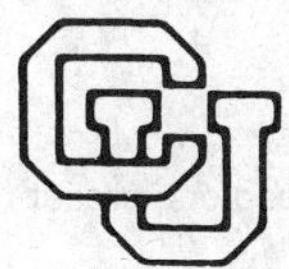

UNIVERSITY OF COLORADO AT DENVER AND HEALTH SCIENCES CENTER

Program in Molecular Biology

Program of Study

The University of Colorado Health Sciences Center has formed an interdisciplinary program that crosses classic department lines to provide an integrative approach to meet the training and research challenges of the rapidly expanding field of molecular biology. The program faculty includes members of the Departments of Biochemistry, Cell and Developmental Biology, Genetics, Medicine, Microbiology, Pathology, Pharmacology, and Physiology in the Graduate School and in the School of Medicine, enabling students to receive more diverse training than is typically available. Students take a limited number of core courses, thus having ample time to pursue individualized programs. Core courses include a first-year core curriculum with course work in biochemistry, molecular biology, genetics, and cell biology. Elective courses cover a wide variety of advanced topics. Students also participate in journal clubs, research seminars, and a broad selection of courses and other programs within the graduate and medical schools.

During their first year, students typically take courses and spend one quarter in each of three different laboratories to familiarize themselves with laboratories in which they may choose to do their thesis research. A general written qualifying examination is taken early in the first summer; students may then begin their selected laboratory research programs. At the end of the second year, students take an oral comprehensive examination in which a research proposal is defended. After successful completion of this examination, students are advanced to candidacy for the Ph.D. degree. Thesis research is generally completed within the next two to three years. Throughout their course of study, students are given the opportunity to develop their analytical and presentation skills through various seminar series, the annual off-campus retreat, and presentations at research-group and national meetings.

Research Facilities

The University of Colorado Health Sciences Center is fully equipped with specialized instrumentation and core facilities required to conduct contemporary biomedical investigations. Core facilities include biostatistics, cytogenetics, DNA sequencing, fermentation, flow cytometry, immunology, transgenics, protein microsequencing, oligonucleotide synthesis, microarray screening, radiological sciences, tissue culture/monoclonal antibody production, and tissue procurement. Shared and available instrumentation includes NMR, X-ray crystallography, fluorescence-activated cell sorter, luminometer, phosphor imager, densitometer with dedicated computer, and an analytical ultracentrifuge. The library offers more than 1,400 journals and maintains a collection of more than 234,000 volumes. The library computer facility contains several IBM and Macintosh personal computers for student use; MEDLINE searches are provided free of charge to students.

Financial Aid

Students are fully supported in their graduate work through the molecular biology program. In 1999, the program was awarded a highly competitive NIH predoctoral training grant, which was renewed in 2004, providing funding to support and train students for years to come. Tuition, health insurance, and a generous stipend are provided to those selected to join the molecular biology program. Stipends for 2006–07 are $22,500 per year.

Cost of Study

The cost of study is met by the support described in the Financial Aid section above.

Living and Housing Costs

Rents in the surrounding community generally range from $600 to $700 per month for a one-bedroom unit. Commuting costs can be kept to a minimum, since shops and housing are within bicycling and walking distance.

Student Group

The 60–70 students currently working within the laboratories participating in the molecular biology program come from diverse academic backgrounds and have degrees in chemistry, biochemistry, physics, engineering, microbiology, and related disciplines. Most students enter the program directly after their undergraduate training, but some have had significant outside work experience or have earned master's degrees. Approximately half of the students are women, and there are several international students. Most students are preparing for research careers in academia or industry.

Student Outcomes

Forty-six students have graduated from the program since its inception in 1987. All have gone on to competitive, high-quality postdoctoral positions at universities, at the NIH, and in industry.

Location

The University is situated in a pleasant residential area and benefits from Denver's ideal geographical setting. The adjacent Rocky Mountains offer a spectacular array of outdoor recreational opportunities, including skiing, camping, hiking, hunting, and a variety of water sports. Because of the protection afforded by the mountains, the climate is relatively mild and dry; the city averages more than 300 days of sunshine per year. Denver offers all of the advantages of any major metropolitan center, including resident theater groups, a symphony orchestra, a variety of other cultural activities, excellent restaurants, and major-league sports.

The Center

The University of Colorado Health Sciences Center contains the health sciences schools and departments of the University of Colorado System. These include the Schools of Medicine, Dentistry, and Pharmacy and the health sciences branch of the Graduate School. The Center provides the opportunity for significant interdisciplinary interaction in an active academic setting.

Applying

Students will be more favorably viewed for admission if they have a combined GRE General Test score of 1200 or higher and an undergraduate GPA of 3.5 or higher. Students seeking admission should have taken organic chemistry (2 semesters, plus 1 semester of laboratory), physical chemistry, general physics, and college-level mathematics through calculus. Students lacking specific courses may apply and be admitted, but the outstanding courses must be rectified during the first year. Completed applications should be received by December 15 to ensure consideration for full financial support.

Correspondence and Information

Molecular Biology Program Admissions
University of Colorado at Denver and Health Sciences Center
Mail Stop 8122
P.O. Box 6511
Aurora, Colorado 80045
Phone: 303-724-3245
Fax: 303-724-3247
E-mail: jean.sibley@uchsc.edu
Web site: http://www.uchsc.edu/molbio

University of Colorado at Denver and Health Sciences Center

THE FACULTY AND THEIR RESEARCH

Steven M. Anderson, Professor of Pathology; Ph.D., Rockefeller, 1981. Signal transduction and oncology; signal transduction by cytokine receptors; breast cancer.

Bruce W. Banfield, Assistant Professor of Microbiology; Ph.D., British Columbia, 1994. Molecular basis of herpesvirus pathways.

David J. Barton, Associate Professor of Microbiology; Ph.D., Medical College of Ohio, 1989. Enterovirus RNA replication.

K. Ulrich (Ulli) Bayer, Assistant Professor of Pharmacology; Ph.D., Hamburg, 1996. Molecular memory and synaptic plasticity; cellular signal transduction and neuronal function.

David L. Bentley, Professor of Biochemistry; Ph.D., Cambridge, 1982. Transcriptional activation and mRNA processing.

Thomas Blumenthal, Professor and Chair of Biochemistry and Molecular Genetics; Ph.D., Johns Hopkins, 1970. Molecular biology; trans-splicing and complex transcription units in *C. elegans*.

Andrew P. Bradford, Assistant Professor of Obstetrics and Gynecology; Ph.D., Newcastle upon Tyne, 1986. Hormone/growth factor signaling and the regulation of gene transcription.

Steven G. Britt, Associate Professor of Cell and Developmental Biology; M.D., Texas Medical Branch, 1986. Visual system development and function using molecular-genetic approaches in *Drosophila*.

Carlos Enrique Catalano, Associate Professor, School of Pharmacy; Ph.D., California, San Francisco, 1987. Mechanisms of viral assembly.

Mair E. A. Churchill, Associate Professor of Pharmacology; Ph.D., Johns Hopkins, 1988. Structure and function of chromosomal proteins; mutagenesis, biophysical and structural characterization of protein–nucleic acid complexes and multiprotein complexes.

Richard E. Davis, Associate Professor of Pediatrics; Ph.D., Massachusetts Amherst, 1982. Spliced leader RNA trans-splicing and its role in post-transcriptional gene expression; structure/function of mRNA cap-binding or interacting proteins.

James DeGregori, Associate Professor of Biochemistry and Molecular Genetics and Director, Molecular Biology Program; Ph.D., MIT, 1993. E2F transcription factor in thymocyte cell death.

Thomas C. Evans, Assistant Professor of Cell and Developmental Biology; Ph.D., Wisconsin, 1988. The control of embryonic polarity.

Heide Ford, Assistant Professor of Obstetrics and Gynecology; Ph.D., Rochester, 1995. The SIX family of homeobox genes and their role in development, cell-cycle control, and tumorigenesis.

Robert Garcea, Professor of Pediatrics; M.D., California, San Francisco, 1974. Structural biology of viruses; human polyomaviruses and their relation to disease; papovavirus assembly.

Arthur Gutierrez-Hartmann, Professor of Medicine, Biochemistry, and Molecular Genetics; M.D., Texas Health Science Center at Dallas, 1975. Molecular neuroendocrinology; mechanisms of tissue-specific and hormonally regulated gene transcription.

James R. Hagman, Associate Professor of Immunology; Ph.D., Washington (Seattle), 1989. Regulation of B cell–specific gene expression.

Kathryn V. Holmes, Professor of Microbiology; Ph.D., Rockefeller, 1968. Molecular virology of RNA viruses; virus receptors and their roles in virus pathogenesis and evolution.

Randall K. Holmes, Professor and Chair of Microbiology; M.D./Ph.D., NYU, 1968. Microbiology and infectious diseases; molecular characterization of bacterial toxins—gene regulation in pathogenic bacteria.

Joan E. Hooper, Associate Professor of Cell and Developmental Biology; Ph.D., California, San Francisco, 1983. Molecular genetics of pattern formation–*Drosophila* segmentation.

Kathryn B. Horwitz, Professor of Medicine and Pathology; Ph.D., Texas Health Science Center at Dallas, 1975. Structure and transcriptional functions of estrogen and progesterone receptors; breast cancer, hormone resistance, and steroid antagonists.

Kathryn Howell, Associate Professor of Cell and Developmental Biology and of Medicine; Ph.D., Rutgers, 1971. Cell-free systems used to identify assay molecules controlling membrane traffic.

Mingxia Huang, Assistant Professor of Biochemistry; Ph.D., Columbia, 1995. Function and regulation of the DNA damage response.

David Jones, Associate Professor of Pharmacology; Ph.D., Cambridge, 1990. The molecular mechanisms that underlie the effects of alcohols in the human brain.

Jeffrey S. Kieft, Assistant Professor of Biochemistry and Molecular Genetics; Ph.D., Berkeley, 1997. Control of translation by RNA structure; RNA-protein interacting.

Leslie Krushel, Assistant Professor of Pharmacology; Ph.D., Toronto, 1990. Mechanisms and regulation of protein synthesis in the nervous system.

Tatiana Kutateladze, Assistant Professor of Pharmacology; Ph.D., Moscow State, 1988. Molecular mechanisms of protein-phospholipid interactions; structure and function of proteins implicated in cancer by NMR spectroscopy; structure-based drug design.

Robert Low, Associate Professor of Pathology; Ph.D., 1975, M.D., 1977, Chicago. Mechanism and control of eukaryotic DNA replication.

James L. Maller, Professor of Pharmacology; Ph.D., Berkeley, 1974. Biochemistry of cell division and early development; second messengers and protein phosphorylation; insulin action; oncogenes.

Sandra Martin, Associate Professor of Cell and Developmental Biology; Ph.D., Berkeley, 1982. Mechanism of LINE-1 retrotransposition; differential gene expression during hibernation.

Charles McHenry, Professor of Biochemistry and Molecular Genetics; Ph.D., California, San Francisco/Santa Barbara, 1974. Enzymology, mechanism, regulation of DNA replication; structure of multienzyme complexes; HIV replication.

James McManaman, Assistant Professor of Physiology; Ph.D., Colorado at Boulder, 1978. Lactatin, lipid secretion.

Paul Megee, Assistant Professor of Biochemistry; Ph.D., Virginia, 1995. Molecular basis of sister chromatid cohesion and chromosome segregation.

Lee A. Niswander, Professor of Pediatrics; Ph.D., Case Western Reserve, 1990. Genetic and cellular control of vertebrate embryonic development.

Steven Nordeen, Associate Professor of Pathology; Ph.D., Rochester, 1977. Molecular mechanism of hormone action; regulation of oncogene transcription; interaction of transcription factors with DNA.

Rytis Prekeris, Assistant Professor of Cell and Developmental Biology; Ph.D., East Carolina, 1997. The role of endocytic membrane and protein traffic in receptor recycling.

Mary E. Reyland, Associate Professor of Craniofacial Biology; Ph.D., Virginia Commonwealth, 1983. Role of protein kinase C isoform–specific signal transduction in indication of programmed cell death.

Jerome Schaack, Associate Professor of Microbiology and Immunology; Ph.D., Yale, 1983. Nuclear matrix association and regulation of transcription; mechanisms of transcriptional activation by adenovirus E1A.

Robert A. Sclafani, Professor of Biochemistry and Molecular Genetics; Ph.D., Columbia, 1981. Eukaryotic chromosomal DNA replication and cell-cycle control in human cancer cells; genetic and molecular analysis of yeast cells.

Lori Sussel, Associate Professor of Biochemistry and Molecular Genetics; Ph.D., Columbia, 1993. Transcriptional regulation of embryonic pancreatic development using mouse gene knockout techniques.

Jessica K. Tyler, Assistant Professor of Biochemistry; Ph.D., Glasgow (Scotland), 1994. Chromatin assembly, structure, and function.

Linda van Dyk, Assistant Professor of Microbiology; Ph.D., Texas Southwestern Medical Center at Dallas, 1994. Control of growth regulation and differentiation of host cells in viral infection and persistence.

Michael Vasil, Professor of Microbiology; Ph.D., Texas Health Science Center at Dallas, 1975. Structure-function studies of bacterial toxins and virulence factors; molecular mechanisms of regulation of virulence factor gene expression, with emphasis on *Pseudomonas aeruginosa*.

Trevor Williams, Associate Professor of Craniofacial Biology; Ph.D., London, 1986. Transcriptional regulation of mouse embryonic development and the role of transcription factors in mammary gland development and breast cancer.

Rui Zhao, Assistant Professor of Biochemistry and Molecular Genetics; Ph.D., Purdue, 1996. Molecular mechanism of pre-mRNA splicing.

UNIVERSITY OF CONNECTICUT

Department of Molecular and Cell Biology

Programs of Study	The faculty members in the Department of Molecular and Cell Biology at the University of Connecticut participate in five interrelated fields of study—biochemistry, biophysics/structural biology, cell biology, genetics, and microbiology—which each lead to the M.S. and Ph.D. degrees. The faculty members maintain federally and privately funded research programs, generally with research groups of moderate size, assuring direct faculty guidance for graduate students. Individual faculty members also participate in the interdisciplinary area of applied genetics. Within each field of study, a variety of specific research themes exists.

The field of biochemistry furnishes students with an extensive education in the structure and function of biological macromolecules, particularly in the areas of protein/peptide structure and macromolecular interactions. Biophysics/structural biology provides an exciting program that focuses on the structure and interactions of proteins, both soluble and membrane-bound, and uses state-of-the-art instrumentation. Cell biology emphasizes the interaction of cells with the extracellular environment and the homeostatic mechanisms that maintain fundamental cell functions, such as motility, signal transduction, and membrane integrity. Genetics offers research training with an emphasis on the structure and regulation of heritable macromolecules, and on the practical application of genetic investigations. Microbiology trains students in many areas of modern microbial biology, including the physiology, genetics, molecular biology, and ecology of prokaryotes; biotechnology; bioremediation; and pathogenesis. Offering these diverse fields of study within the Department results in excellent opportunities for interdisciplinary research experience.

Research Facilities

The principal research facilities are housed in the newly constructed Biological Sciences/Physics Building, the Torrey Life Sciences Building, and Beach Hall. Exceptional facilities within the Department include a center for analytical ultracentrifugation, a cell-culture facility, a laboratory for flow cytometry and confocal microscopy, facilities for macromolecular characterization (including peptide and DNA sequencing and synthesis), and a protein crystallographic laboratory. An extensive collection of journals is available in the University of Connecticut Library and via electronic journal subscription. The University has an excellent computer center, and Internet access is available in all buildings used by the Department. The University has been nationally recognized for its campuswide network.

Financial Aid

Financial support is available to graduate students in several forms. For 2006–07, nine-month teaching assistantships pay $18,270 for beginning graduate students, $19,226 for those with an M.S. or the equivalent, and $21,371 for Ph.D. candidates who have passed the General Examination. All students with an assistantship also receive the option to purchase highly subsidized health benefits and a tuition waiver. University fellowships and research assistantships are also offered, and a summer stipend is available for most doctoral students.

Cost of Study

Graduate tuition for 2006–07 is $7992 for legal residents of Connecticut and $20,772 for out-of-state students. Fees of about $1500 are also assessed. Tuition is prorated for students registering for fewer than 9 credits per semester. Tuition, but not the general University fee, is waived for graduate assistants.

Living and Housing Costs

University-operated and privately owned apartments are available on and near the campus at moderate rents. Houses and apartments for rent can also be found in surrounding communities.

Student Group

Approximately 16,000 undergraduates and 8,000 graduate students are enrolled on the main campus at Storrs. About 80 percent of the undergraduate students and 33 percent of the graduate students are from Connecticut. The rest of the student body represents numerous other states and more than 100 different countries. The Department of Molecular and Cell Biology has a total of approximately 100 graduate students in six fields of study.

Location

The University is located in a scenic countryside setting of small villages, streams, and rolling hills. There is easy access by automobile and bus to major urban and cultural centers, such as Hartford, New Haven, Boston, and New York, and to other educational institutions, such as Yale, Harvard, and MIT. Bradley International Airport is 40 minutes from campus by car.

Recreational opportunities in the area include fishing, skiing, sailing, hiking, ice-skating, and kayaking. The community's cultural life includes film series, plays, concerts, public lectures, and art exhibits. A small shopping center is within walking distance of the campus, and several large centers are nearby.

The University

The University of Connecticut is a land-grant institution, so designated in 1893. The lineage of the University reaches back to 1881, when the Storrs Agricultural School was founded. The transition to university status, achieved in 1939, has been followed by substantial growth of the University, including an ongoing $2.3-billion building and renovation effort.

Applying

A completed application form, transcripts of all previous college or university work, scores on the Graduate Record Examinations, a personal statement of research interests, and three letters of reference are required before consideration of the application can begin. For the academic year 2007–08 the deadline for applicants to the doctoral program is January 15, 2007, and for applicants to the master's program the deadline is May 1, 2007. The University does not discriminate in admissions on the basis of race, sex, age, or national origin.

Correspondence and Information

Anne St. Onge
Department of Molecular and Cell Biology, U-3042
University of Connecticut
Storrs, Connecticut 06269-3042

Phone: 860-486-4314
E-mail: Anne.St_Onge@uconn.edu
Web site: http://www.mcb.uconn.edu

University of Connecticut

THE FACULTY AND THEIR RESEARCH

Arlene D. Albert, Professor; Ph.D. Biochemistry of vision; signal transduction in rod outer segments; lipid organization in photoreceptors; rhodopsin structure; protein-lipid interactions.

Andrei T. Alexandrescu, Assistant Professor; Ph.D. High-resolution solution NMR investigations of protein structure, folding, dynamics, and association; conserved physical properties of OB-fold proteins; HIV regulatory/accessory proteins.

David R. Benson, Professor; Ph.D. Microbial physiology, biochemistry, and molecular genetics of symbiotic nitrogen fixation; steroidal transformations in foods.

Robert Birge, Professor (Chemistry, joint MCB appointment); Ph.D. Molecular mechanisms of light absorption.

Emory H. Braswell, Professor Emeritus and Head, National Analytical Ultracentrifugation Facility; Ph.D. Biophysical chemistry; interactions of macromolecules.

Peter Burkhard, Associate Professor; Ph.D. Structure-based rational design of self-assembling small proteins; biophysical properties and medical applications of these nanoparticles.

Thomas T. Chen, Professor and Director, Biotechnology Center; Ph.D. Structure, evolution, regulation, and molecular actions of growth hormone and insulin-like growth factor genes; regulation of foreign genes in transgenic fish; development of model transgenic fish.

James Cole, Associate Professor; Ph.D. Biophysical and biochemical analysis of enzyme regulation; structure and function of protein kinase PKR; mechanism of HIV entry; structure and function of retroviral integrases; analytical ultracentrifugation.

Daniel J. Gage, Assistant Professor; Ph.D. Molecular genetics of plant-microbe interactions; bacterial physiology; regulation of bacterial gene expression in response to extracellular signals.

Steven J. Geary, Professor (Pathobiology, joint appointment in MCB); Ph.D. Molecular biology of mycoplasmas.

Charles Giardina, Associate Professor; Ph.D. Studies of eukaryotic RNA polymerase II transcription regulation; mechanisms governing transcription of stress-response genes; the nature of RNA polymerase II interactions at the promoter.

Walter Godchaux III, Professor-in-Residence; Ph.D. Mechanisms of gliding motility in bacteria; metabolism of biological sulfates.

J. Peter Gogarten, Professor; Ph.D. Membrane transport; function and evolution of vacuolar type H$^+$-ATPases; the role of cell-wall-bound enzymes in the long-distance transport in plants.

David J. Goldhamer, Associate Professor; Ph.D. Regulation of cell fates in mammalian development; transcriptional control and function of skeletal muscle regulatory genes; muscle stem cell function and plasticity; mechanism of heterotopic bone formation in human disease.

Jörg Graf, Assistant Professor; Ph.D. Molecular genetics of bacteria-animal interactions; identification of bacterial genes required for host symbiosis; pathogenesis, evolution of virulence factors.

Lawrence E. Hightower, Professor and Associate Department Head; Ph.D. Molecular and cellular responses of animal cell to environmental stress; heat shock proteins, growth factors, and cellular stress responses; analysis of regulatory signals, and functions of induced proteins.

Judith A. Kelly, Professor Emerita; Ph.D. Enzyme structure and function, using X-ray crystallography, kinetic studies, and interactive computer graphics.

Debra A. Kendall, Board of Trustees Distinguished Professor; Ph.D. Protein export in prokaryotes; analysis of signal sequence function using molecular biology and physical methods.

David A. Knecht, Professor; Ph.D. Actin cytoskeleton dynamics; small G proteins and signal transduction; phagocytosis; endocytosis; cell motility and chemotaxis.

James R. Knox, Professor Emeritus; Ph.D. Protein structure determination by X-ray crystallography; drug target enzymes of bacterial cell wall synthesis.

Hans Laufer, Professor Emeritus; Ph.D. Developmental and cell biology; hormonal control and signal transduction during development.

Edward R. Leadbetter, Professor and Director of Undergraduate Studies; Ph.D. Microbial ecology, physiology, and biochemistry; gliding motility in prokaryotes; sulfonate biotransformations.

Juliet Lee, Associate Professor; Ph.D. The regulation of cell movement; mechanochemical signal transduction; the role of intracellular calcium; cellular force production and its relationship to adhesion formation and cytoskeletal dynamics.

Michael A. Lynes, Professor and Director of Graduate Studies; Ph.D. Genetic and biochemical control of the immune response; membrane structure and function during development; mechanisms of autoimmune dysfunction.

Philip I. Marcus, Board of Trustees Distinguished Professor; Ph.D. Animal virus-cell interactions; viral interference; the interferon system; regulation of interferon induction-suppression by viruses; cell killing by viruses; development of the avian interferon system; inducible ds RNase.

Craig E. Nelson, Assistant Professor; Ph.D. Molecular biological, genetic, genomic, and computational analysis of the evolution of developmental processes and organismal complexity.

Kenneth M. Noll, Associate Professor; Ph.D. Biochemistry and molecular biology of thermophilic bacteria and archaebacteria; physiology of extremely thermophilic anaerobes; evolution of genome organization in prokaryotes.

Michael J. O'Neill, Associate Professor; Ph.D. Molecular genetics of vertebrate development; molecular mechanisms of genomic imprinting; evolution of genomic imprinting; genetics of imprinting and behavior.

Rachel J. Waugh O'Neill, Associate Professor; Ph.D. Genetics of speciation; mammalian chromosome evolution; genome evolution and remodeling; transposable elements and retroelements; hybridogenesis and clonal inheritance in vertebrates; epigenetics.

Wolf-Dieter D. Reiter, Associate Professor; Ph.D. Molecular genetics of cell-wall synthesis in *Arabidopsis;* structure and function of plant cell walls; developmental regulation of cell-wall deposition; biochemistry of plant carbohydrates.

Margaret J. Sekellick, Professor-in-Residence; Ph.D. Animal virology and the interferon system; cloning and genomic organization of avian interferons; role of interferon system and growth-factor expression in embryonic development.

Linda D. Strausbaugh, Professor; Ph.D. Structure, function, and evolution of multigene families; molecular genetics of insects.

Carolyn M. Teschke; Associate Professor; Ph.D. Biochemical, biophysical, and genetic analyses of protein folding; interaction of folding intermediates with molecular chaperones; assembly of viral capsids.

Robert T. Vinopal, Associate Professor; Ph.D. Bacterial physiology and genetics.

Philip L. Yeagle, Professor and Department Head; Ph.D. Structure and function of biological membranes; high-resolution structure determination of a G-protein receptor, rhodopsin; cholesterol in mammalian cell biology.

David A. Yphantis, Professor Emeritus; Ph.D. Characterization and interactions of macromolecules.

Ping Zhang, Associate Professor; Ph.D. *Drosophila* chromosome structure and function; P element insertional mutagenesis; unusual transcriptional regulation of heterochromatin; Y chromosome genes required for spermatogenesis.

Adam Zweifach, Associate Professor, Ph.D. Lymphocyte physiology and cell biology; role of intracellular calcium dynamics in lymphocyte function; molecular mechanisms of exocytosis; signaling in the immune system.

Associated and Adjunct Faculty

Sydney P. Craig III, Adjunct Professor; Ph.D.

Hedley C. Freake, Professor (Nutritional Sciences); Ph.D. Regulation of fatty-acid synthesis; zinc metabolism.

Carll Ladd, Adjunct Assistant Professor; Ph.D. Connecticut State Police Forensic Science Laboratory.

Thomas Laue, Adjunct Professor; Ph.D. University of New Hampshire.

Alexandros Makriyannis, Professor (Pharmaceutical Sciences); Ph.D. Drug design; membrane biophysics; NMR spectroscopy.

Mary M. McGrane, Associate Professor (Nutritional Sciences); Ph.D. Hormonal regulation of gene expression.

Theodore Rasmussen, Assistant Professor (Animal Sciences, Center for Regenerative Biology); Ph.D.

Daniel Rosenberg, Professor (Medicine/UCONN Health Center); Ph.D. Chemically induced cancer and the genetics of susceptibility.

Lawrence Silbart, Associate Professor (Animal Science); Ph.D. Mucosal immunity and vaccine development.

Xiuchun (Cindy) Tian, Assistant Professor (Animal Sciences, Center for Regenerative Biology); Ph.D.

Pieter Visscher, Professor (Marine Sciences); Ph.D. Applied environmental microbiology.

Susanne Beck von Bodman, Associate Professor (Plant Science); Ph.D. Host-pathogen interactions; bacterial virulence factors.

Xiangzhong (Jerry) Yang, Professor and Director, Center for Regenerative Biology (Animal Sciences); Ph.D.

Steven A. Zinn, Associate Professor (Animal Science); Ph.D. Genetics and physiology of growth factor expression.

UNIVERSITY OF CONNECTICUT HEALTH CENTER

Graduate Program in Cell Biology

Program of Study	The program offers training leading to a Ph.D. in biomedical sciences and includes faculty members from the Department of Cell Biology as well as eight other Health Center departments. Faculty members' research spans a broad range of interests in the areas of eukaryotic cell biology and related clinical aspects. The program is particularly strong in the following areas of research: angiogenesis, cancer biology, gene expression, molecular medicine, reproductive biology, signal transduction, vascular biology, optical methods, proteomics, and computer modeling of complex biological systems. The curriculum for the first year is tailored to the individual student and can include core courses in the basic biomedical sciences that have been specially formulated to acquaint the student with the principles and practice of modern biomedical research as well as more specialized, analytical courses. In consultation with their advisory committee, students work out a supplementary program of advanced courses, laboratory experiences, and independent study designed to prepare them for general examinations near the end of their second year. Thesis research begins in the second or third year, and research and thesis writing normally occupy the third and fourth years.
Research Facilities	The program is situated in the modern Health Center in Farmington. This complex provides excellent physical facilities for research in both basic and clinical sciences, a computer center, and the Lyman Maynard Stowe Library. The program provides research facilities and guidance for graduate and postdoctoral work in cell biology—particularly membrane and surface function, membrane protein synthesis and turnover, cytoskeleton structure and function, stimulus-response coupling, gene expression and regulation, vascular biology, fertilization, bone biology, molecular medicine, early development, signal transduction, angiogenesis, computer modeling, and tumor biology. Facilities for training in cell culture, electron microscopy, electrophysiology, fluorescence spectroscopy, molecular biology, molecular modeling, fluorescence imaging, and intravital microscopy are available.
Financial Aid	Support for doctoral students engaged in full-time degree programs at the Health Center is provided on a competitive basis. Graduate research assistantships for 2006–07 provide a stipend of $26,000 per year, which includes a waiver of tuition/University fees for the fall and spring semesters and a student health insurance plan. While financial aid is offered competitively, the Health Center makes every possible effort to address the financial needs of all students during their period of training.
Cost of Study	For 2006–07, tuition was $3996 per semester ($7992 per year) for full-time students who are Connecticut residents and $10,368 per semester ($20,772 per year) for full-time out-of-state residents. General University fees are added to the cost of tuition for students who do not receive a tuition waiver. These costs are usually met by traineeships or research assistantships for doctoral students.
Living and Housing Costs	There is a wide range of affordable housing options in the greater Hartford area within easy commuting distance of the campus, including an extensive complex that is adjacent to the Health Center. Costs range from $600 to $800 per month for a one-bedroom unit; 2 or more students sharing an apartment usually pay less. University housing is not available at the Health Center.
Student Group	Currently, 20 students are pursuing doctoral studies in the program. The total number of Ph.D. students at the Health Center is approximately 150, while the medical and dental schools combined currently enroll 130 students per class.
Location	The Health Center is located in the historic town of Farmington, Connecticut. Set in the beautiful New England countryside on a hill overlooking the Farmington Valley, it is close to ski areas, hiking trails, and facilities for boating, fishing, and swimming. Connecticut's capital city of Hartford, 7 miles east of Farmington, is the center of an urban region of approximately 800,000 people. The beaches of the Long Island Sound are about 50 minutes away to the south, and the beautiful Berkshires are a short drive to the northwest. New York City and Boston can be reached within 2½ hours by car. Hartford is the home of the acclaimed Hartford Stage Company, TheatreWorks, the Hartford Symphony and Chamber orchestras, two ballet companies, an opera company, the Wadsworth Atheneum (the oldest public art museum in the nation), the Mark Twain house, the Hartford Civic Center, and many other interesting cultural and recreational facilities. The area is also home to several branches of the University of Connecticut, Trinity College, and the University of Hartford, which includes the Hartt School of Music. Bradley International Airport (about 20 minutes from campus) serves the Hartford/Springfield area with frequent airline connections to major cities in this country and abroad. Frequent bus and rail service is also available from Hartford.
The Health Center	The 200-acre Health Center campus at Farmington houses a division of the University of Connecticut Graduate School, as well as the School of Medicine and Dental Medicine. The campus also includes the John Dempsey Hospital, associated clinics, and extensive medical research facilities, all in a centralized facility with more than 1 million square feet of floor space. The Health Center's newest research addition, the Academic Research Building, was opened in 1999. This impressive eleven-story structure provides 170,000 square feet of state-of-the-art laboratory space. The faculty at the center includes more than 260 full-time members. The institution has a strong commitment to graduate study within an environment that promotes social and intellectual interaction among the various educational programs. Graduate students are represented on various administrative committees concerned with curricular affairs, and the Graduate Student Organization (GSO) represents graduate students' needs and concerns to the faculty and administration, in addition to fostering social contact among graduate students in the Health Center.
Applying	Applications for admission should be submitted on standard forms obtained from the Graduate Admissions Office at the UConn Health Center or on the Web site. The application should be filed together with transcripts, three letters of recommendation, a personal statement, and recent results from the General Test of the Graduate Record Examinations. International students must take the Test of English as a Foreign Language (TOEFL) to satisfy Graduate School requirements. The deadline for completed applications and receipt of all supplemental materials is December 15. In accordance with the laws of the state of Connecticut and of the United States, the University of Connecticut Health Center does not discriminate against any person in its educational and employment activities on the grounds of race, color, creed, national origin, sex, age, or physical disability.
Correspondence and Information	Dr. Linda Shapiro Director, Cell Biology Graduate Program MC 3501 University of Connecticut Health Center Farmington, Connecticut 06030-3501 Phone: 860-679-4373 E-mail: lshapiro@neuron.uchc.edu Web site: http://grad.uchc.edu

University of Connecticut Health Center

THE FACULTY AND THEIR RESEARCH

Andrew Arnold, Professor and Director, Center for Molecular Medicine; M.D., Harvard, 1978. Structure and function of the cyclin D1 oncogene and cell-cycle regulator; molecular genetics and biology of endocrine tumors; inherited endocrine neoplastic diseases.

Rashmi Bansal, Associate Professor of Neuroscience; Ph.D., Central Drug Research Institute, 1976. Developmental, cellular, and molecular biology of oligodendrocytes (OLs), the cells that synthesize myelin membrane in the central nervous system.

Gordon G. Carmichael, Professor of Microbiology; Ph.D., Harvard, 1975. Regulation of gene expression in eukaryotes.

Joan M. Caron, Assistant Professor of Cell Biology; Ph.D., Connecticut, 1982. Biochemistry and cell biology of microtubules; palmitoylation of tubulin and cell function; functional role of palmitoylation of signaling proteins.

Kevin P. Claffey, Associate Professor of Cell Biology and Center for Vascular Biology; Ph.D., Boston University, 1989. Angiogenesis in human cancer progression and metastasis; vascular endothelial growth factor (VEGF) expression; hypoxia-mediated gene regulation.

Robert B. Clark, Associate Professor of Medicine, Division of Rheumatic Diseases; M.D., Stanford, 1975. Basic T-lymphocyte biology, especially as it relates to autoimmune diseases, such as multiple sclerosis and rheumatoid arthritis; molecular biology and structure of the T-cell antigen receptor; T-cell function; T-cell activation.

Ann Cowan, Assistant Professor of Biochemistry and Deputy Director of the Center for Biomedical Imaging Technology; Ph.D., Colorado, 1984. Mammalian sperm development.

Kimberly Dodge-Kafka, Assistant Professor of Cell Biology, Center for Cardiology and Cardiovascular Research; Ph.D., Texas–Houston Health Science Center, 1999. Molecular mechanism of signaling pathways in the heart.

David I. Dorsky, Assistant Professor of Medicine; M.D./Ph.D., Harvard, 1982. The structure and function of herpesvirus DNA polymerases and their roles in viral DNA replication.

Alan Fein, Professor of Cell Biology; Ph.D., Johns Hopkins, 1973. Molecular basis of visual excitation and adaptation; signal transduction and calcium homeostasis in platelets.

Maurice B. Feinstein, Professor of Pharmacology; Ph.D., SUNY Downstate Medical Center, 1960. Function and regulation of blood platelets and their role in hemostasis and thrombosis; arachidonic acid biochemistry in platelets.

Guo-Hua Fong, Assistant Professor of Cell Biology and Center for Vascular Biology; Ph.D., Illinois, 1988. Cardiovascular biology.

Henry M. Furneaux, Assistant Professor of Biochemistry; Ph.D., Aberdeen (Scotland), 1978. Identification of proteins that regulate the stability of mRNA.

Brenton R. Graveley, Assistant Professor, Department of Genetics and Developmental Biology; Ph.D., Vermont, 1996. Regulation of alternative splicing in the mammalian nervous system and mechanisms of alternative splicing.

David Han, Associate Professor of Cell Biology and Center for Vascular Biology; Ph.D., George Washington, 1994. Proteomic analysis of complex protein mixtures.

Marc Hansen, Professor of Medicine; Ph.D., Cincinnati, 1986. Analysis of genes involved in the development of the bone tumor osteosarcoma.

Timothy Hla, Professor of Cell Biology and Director, Center for Vascular Biology; Ph.D., George Washington, 1988. Gene expression in endothelial cells as it relates to angiogenesis; G-protein–coupled receptor signaling; biology of cyclooxygenase-2.

Marja Hurley, Professor of Medicine; M.D., Connecticut Health Center, 1972. Molecular mechanisms by which members of the fibroblast growth factor (FGFs) and fibroblast growth factor receptor (FGFR) families (produced by osteoblasts, osteoclasts, and stromal cells) regulate bone development, remodeling, and disorders of bone: Fgf2 knockout and Fgf2 transgenic mice are utilized in loss and gain of function experiments to elucidate the role of FGF-2 in disorders of bone, including osteoporosis.

Laurinda A. Jaffe, Professor of Cell Biology; Ph.D., UCLA, 1977. Physiology of fertilization, in particular the mechanisms by which membrane potential regulates sperm-egg fusion; transduction mechanisms coupling sperm-egg interaction to egg exocytosis; opening of ion channels in the egg membrane.

Stephen M. King, Associate Professor of Biochemistry; Ph.D., London, 1982. Cell biology; biochemistry and function of molecular motors; dynein structure and function.

Dennis E. Koppel, Professor of Biochemistry; Ph.D., Columbia, 1973. Application of biophysical techniques to membrane dynamics; mechanisms by which specialized cell-surface domains are produced and maintained.

Bruce Liang, Professor of Cardiopulmonary Medicine; M.D., Harvard, 1982. Signal transduction; cardiac and vascular cell biology; receptors; G proteins; transgenic mice.

Leslie M. Loew, Professor of Cell Biology and Director, Center for Cell Analysis and Modeling; Ph.D., Cornell, 1974. Spectroscopic methods for measuring spatial and temporal variations in membrane potential; electric field effects on cell membranes; membrane pores induced by toxins and antibiotics.

Nilanjana Maulik, Associate Professor of Surgery; Ph.D., Calcutta, 1990. Molecular and cellular signaling during myocardial ischemia and reperfusion.

Lisa Mehlman, Assistant Professor of Cell Biology; Ph.D., Kent State, 1996. Cell signaling events that regulate oocyte maturation and fertilization; maintenance of oocyte meiotic arrest by G-protein receptors; hormonal regulation of oocyte maturation.

Antoine Menoret, Assistant Professor of Cell Biology; Ph.D., Nantes (France), 1994. Immunological role of heat-shock proteins (HSP) has recently been related to their ability to chaperone immunogenic peptides and to specifically immunize against tumor and virus-infected cells.

Flavia O'Rourke, Assistant Professor of Pharmacology; Ph.D., Connecticut, 1976. Signal transduction in human platelets, with specific interest in the inositol phosphate signaling pathway and its regulation.

Joel Pachter, Professor of Pharmacology; Ph.D., NYU, 1983. Elucidating the mechanisms by which leukocytes and pathogens invade the central nervous system.

John J. Peluso, Professor of Cell Biology and Obstetrics and Gynecology; Ph.D., West Virginia, 1974. Cell and molecular mechanisms involving the regulating ovarian cell mitosis and apoptosis; cell-cell interaction as a regulator of ovarian cell function; identification and characterization of a putative membrane receptor for progesterone.

Steven Pfeiffer, Professor of Microbiology; Ph.D., Washington (St. Louis), 1967. Central nervous system myelination and remyelination: regulation of myelin gene expression during development of progenitor cells into mature oligodendrocytes; cell transplantation in culture and in vivo.

Carol L. Pilbeam, Associate Professor of Medicine; M.D./Ph.D., Yale, 1982. Regulation and function of prostaglandins in bone; transcriptional regulation of cyclooxygenase-2; role of cytokines and estrogen in bone physiology and osteoporosis.

Vladimir Rodionov, Assistant Professor of Cell Biology; Ph.D., Moscow, 1980. Dynamics of cytoskeleton; self-organization of microtubule arrays; regulation of the activity of microtubule motors.

Daniel Rosenberg, Professor of Medicine; Ph.D., Michigan. Molecular genetics of colorectal cancer; signaling pathways in the development of tumors; toxicogenomics.

David W. Rowe, Professor of Pediatrics; M.D., Vermont, 1969. Hormonal and genetic regulation of Type I collagen synthesis in bone, using molecular biological techniques.

John B. Schenkman, Professor of Pharmacology; Ph.D., SUNY Upstate Medical Center, 1964. The cytochrome P450 monooxygenase system; homeostatic control of the hepatic microsomal enzymes.

Ramadan I. Sha'afi, Professor of Cell Biology; Ph.D., Illinois at Urbana-Champaign, 1965. Molecular mechanism of neutrophil activation; identification and sequence of the biophysical and biochemical changes that are involved in the excitation-response cascade of cell activation.

Linda Shapiro, Associate Professor of Cell Biology and Center for Vascular Biology; Ph.D., Michigan, 1984. Regulation and function of CD 13/aminopeptidase N in angiogenic vasculature and early myeloid cells; control of tumor and myocardial angiogenesis by peptidases; inflammatory regulation of angiogenesis.

Mark R. Terasaki, Assistant Professor of Cell Biology; Ph.D., Berkeley, 1983. Structure and function of the endoplasmic reticulum; confocal microscopy.

Jennifer Tirnauer, Assistant Professor of Medicine, Center for Molecular Medicine; M.D., Maryland, 1989. Role of the microtubule cytoskeleton in cancer biology; molecular mechanisms of microtubule regulation.

Dudley Watkins, Professor of Anatomy; M.D., 1966, Ph.D., 1967, Western Reserve. Clarifying the exact biochemical sequence of events that occur on the beta-adrenergic stimulation of amylase secretion from the parotid gland.

James Watras, Associate Professor of Medicine; Ph.D., Washington State, 1979. The mechanisms by which the sarcoplasmic reticulum regulates intracellular calcium concentration in vascular smooth muscle.

Bruce A. White, Professor of Cell Biology; Ph.D., Berkeley, 1980. Regulation of prolactin gene expression by Ca and calmodulin in rat pituitary tumor cells; examination of nuclear DNA-binding proteins, nuclear calmodulin-binding proteins, and nuclear Ca-calmodulin-dependent protein kinase activity.

Charles Wolgemuth, Assistant Professor of Cell Biology; Ph.D., Arizona. Using physics to understand biological systems; morphology; propulsion; growth and fluid dynamics.

Catherine H.-y. Wu, Associate Professor of Medicine; Ph.D., CUNY, Brooklyn, 1976. Mechanisms of procollagen propeptide feedback inhibition of collagen synthesis; pretranslational control.

Dianqing Wu, Associate Professor of Genetics and Developmental Biology; Ph.D., Clarkson, 1991. Molecular basis of signal transduction.

George Y. Wu, Professor of Medicine; M.D./Ph.D., Yeshiva (Einstein), 1976. Receptor-mediated endocytosis of glycoproteins; drug delivery by endocytic targeting; targeted gene delivery and expression.

Lixia Yue, Assistant Professor of Cell Biology and Center for Cardiovascular Research; Ph.D., McGill, 1999. TRP channels and Ca^{2+} signaling mechanisms in cardiac remodeling.

UNIVERSITY OF CONNECTICUT HEALTH CENTER

Graduate Program in Molecular Biology and Biochemistry

Program of Study	The Graduate Program in Molecular Biology and Biochemistry uniquely bridges modern molecular biology, microbiology, biochemistry, cell biology, and structural biology, leading to a Ph.D. in the biomedical sciences. The goals of the graduate program are to provide rigorous research training in an environment dedicated to advancing excellence in teaching and research. Whether graduates enter academic research, the biotechnology industry, liberal arts college teaching, patent law, or other disciplines, they bring to that career a solid base of knowledge, an ability to learn independently and think independently, and an enduring desire to use their full range of professional skills and experience in creative ways. Graduates are expected to have demonstrated a high degree of competence in research, as judged by publications in first-rank journals, and to have developed essential skills in identifying important research problems, planning research projects and scientific writing. In addition, students are expected to have incorporated ethical principles of scientific conduct into their professional attitudes and activities and to be sensitive to such issues throughout their careers. The success of this training approach is indicated by the high percentage of students who have developed successful independent careers in biomedical research. The current program offers an unparalleled opportunity to study a wide variety of biological problems at the biochemical, molecular, cellular, and structural levels. The interests of the faculty are summarized below.
Research Facilities	In addition to the general facilities of the Health Center (see page describing programs in the Biological and Biomedical Sciences), the program offers complete physical research facilities. There is research equipment, as well as expertise, for all areas of genetic, biochemical, molecular, cellular, and biophysical investigation. The department houses the UConn Health Center NMR Structural Biology Facility (http://structbio.uchc.edu), which includes a 400-MHz NMR spectrometer and cryoprobe-equipped 500- and 600-MHz NMR spectrometers, as well as a circular dichroism spectropolarimeter, isothermal titration calorimeter, and multi-angle laser light scattering facilities. An 800-MHz NMR spectrometer and X-ray crystallography facilities are planned. The department also houses the UConn Health Center Structural Biology Computational Facility, which includes a bank of Mac and Linux desktop computers connected to ultrafast servers with the latest structural biology software. Facilities are also available for electron and confocal laser scanning microscopy, low-light-level imaging microscopy (in the state-of-the-art Center for Cell Analysis and Modeling), protein purification and sequencing, cell culture, monoclonal antibody production, DNA oligonucleotide and peptide synthesis and sequencing, and gene silencing using RNAI.
Financial Aid	Support for doctoral students engaged in full-time degree programs at the Health Center is provided on a competitive basis. Graduate research assistantships for 2006–07 provide a stipend of $26,000 per year, which includes a waiver of tuition/University fees for the fall and spring semesters and a student health insurance plan. While financial aid is offered competitively, the Health Center makes every possible effort to address the financial needs of all students during their period of training.
Cost of Study	For 2006–07, tuition is $3996 per semester ($7992 per year) for full-time students who are Connecticut residents and $10,368 per semester ($20,772 per year) for full-time out-of-state residents. General University fees are added to the cost of tuition for students who do not receive a tuition waiver. These costs are usually met by traineeships or research assistantships for doctoral students.
Living and Housing Costs	There is a wide range of affordable housing options in the greater Hartford area within easy commuting distance of the campus, including an extensive complex that is adjacent to the Health Center. Costs range from $600 to $800 per month for a one-bedroom unit; 2 or more students sharing an apartment usually pay less. University housing is not available at the Health Center.
Student Group	There are approximately 30 graduate students in the molecular biology and biochemistry program. There are approximately 150 graduate students in Ph.D. programs on the Health Center campus, and the total enrollment is about 1,000.
Location	The Health Center is located in the historic town of Farmington, Connecticut. Set in the beautiful New England countryside on a hill overlooking the Farmington Valley, it is close to ski areas, hiking trails, and facilities for boating, fishing, and swimming. Connecticut's capital city of Hartford, 7 miles east of Farmington, is the center of an urban region of approximately 800,000 people. The beaches of the Long Island Sound are about 50 minutes away to the south, and the beautiful Berkshires are a short drive to the northwest. New York City and Boston can be reached within 2½ hours by car. Hartford is the home of the acclaimed Hartford Stage Company, TheatreWorks, the Hartford Symphony and Chamber orchestras, two ballet companies, an opera company, the Wadsworth Atheneum (the oldest public art museum in the nation), the Mark Twain house, the Hartford Civic Center, and many other interesting cultural and recreational facilities. The area is also home to several branches of the University of Connecticut, Trinity College, and the University of Hartford, which includes the Hartt School of Music. Bradley International Airport (about 20 minutes from campus) serves the Hartford/Springfield area with frequent airline connections to major cities in this country and abroad. Frequent bus and rail service is also available from Hartford.
The Health Center	The 200-acre Health Center campus at Farmington houses a division of the University of Connecticut Graduate School, as well as the School of Medicine and Dental Medicine. The campus also includes the John Dempsey Hospital, associated clinics, and extensive medical research facilities, all in a centralized facility with more than 1 million square feet of floor space. The Health Center's newest research addition, the Academic Research Building, was opened in 1999. This impressive eleven-story structure provides 170,000 square feet of state-of-the-art laboratory space. The faculty at the center includes more than 260 full-time members. The institution has a strong commitment to graduate study within an environment that promotes social and intellectual interaction among the various educational programs. Graduate students are represented on various administrative committees concerned with curricular affairs, and the Graduate Student Organization (GSO) represents graduate students' needs and concerns to the faculty and administration, in addition to fostering social contact among graduate students in the Health Center.
Applying	Applications should be submitted on standard forms obtained from the Graduate Admissions Office at the UConn Health Center or the Web site. The application should be filed together with transcripts, three letters of recommendation, a personal statement, and recent results from the General Test of the Graduate Record Examinations. International students must take the Test of English as a Foreign Language (TOEFL) to satisfy Graduate School requirements. The deadline for completed applications and receipt of all supplemental materials is December 15. In accordance with the laws of the state of Connecticut and of the United States, the University of Connecticut Health Center does not discriminate against any person in its educational and employment activities on the grounds of race, color, creed, national origin, sex, age, or physical disability.
Correspondence and Information	Dr. Henry Furneaux Program Director for Molecular Biology and Biochemistry University of Connecticut Health Center Farmington, Connecticut 06030-3305 Phone: 860-679-2374 Fax: 860-679-1862 E-mail: furneaux@nso.uchc.edu Web site: http://grad.uchc.edu

University of Connecticut Health Center

THE FACULTY AND THEIR RESEARCH

Choukri Ben Mamoun, Associate Professor; Ph.D., Paris XI (South). Biology of the malaria parasite *Plasmodium falciparum,* including genomic analysis using cDNA microarrays.

Gordon G. Carmichael, Professor; Ph.D., Harvard. Regulation of viral gene expression and function.

John H. Carson, Professor; Ph.D., MIT. RNA transport in cells of the nervous system.

Ann Cowan, Associate Professor; Ph.D., Colorado at Boulder. Plasma membrane proteins in sperm.

Asis Das, Professor; Ph.D., Calcutta. Gene control in bacterial adaptive response.

Betty Eipper, Professor; Ph.D., Harvard. Biosynthesis and secretion of peptides by neurons and endocrine cells.

Shlomo Eisenberg, Professor; Ph.D., McGill. Biochemistry of DNA replication in yeast.

Henry M. Furneaux, Associate Professor; Ph.D., Aberdeen (Scotland). Regulation of gene expression by microRNAs.

Michael Gryk, Assistant Professor; Ph.D., Stanford. Three-dimensional structure and function of proteins involved in DNA repair.

Arthur Gunzl, Associate Professor; Ph.D., Tübingen (Germany). Transcription and antigenic variation in the mammalian parasite *Trypanosoma brucei.*

Timothy Hla, Professor and Director, Center for Vascular Biology; Ph.D., George Washington. Gene expression in endothelial cells as it relates to angiogenesis; G-protein–coupled receptor signaling; biology of cyclooxygenase-2.

Jeffrey Hoch, Associate Professor; Ph.D., Harvard. Biophysical chemistry of proteins.

Glenn F. King, Professor; Ph.D., Sydney. Bacterial cell division and control of vector-borne disease.

Stephen M. King, Associate Professor; Ph.D., University College, London. Structure and function of microtubule-based molecular motor proteins.

Lawrence A. Klobutcher, Professor; Ph.D., Yale. DNA rearrangement, programmed translational frameshifting, and phagocytosis in ciliated protozoa.

Dennis E. Koppel, Professor; Ph.D., Columbia. Biophysical studies of membrane dynamics.

Mark Maciejewski, Assistant Professor; Ph.D., Ohio State. Enzymes of DNA replication, repair, and recombination.

Mary Jane Osborn, Professor, Department of Microbiology; Ph.D., Washington (Seattle). Biogenesis of the outer membrane of *Salmonella.*

Juris Ozols, Professor; Ph.D., Washington (Seattle). Isolation and structure of membranous proteins.

Steven E. Pfeiffer, Professor; Ph.D., Washington (St. Louis). Central nervous system myelin membrane biogenesis, axon-myelin signaling, and multiple sclerosis; proteomics; function of lipid rafts.

Lawrence I. Rothfield, Professor; Ph.D., NYU. Membrane biology and biochemistry; bacterial cell division.

Martin R. Schiller, Assistant Professor; Ph.D., Utah State. Structure-function of RhoGEFs; neurotrophin signaling.

Peter Setlow, Professor; Ph.D., Brandeis. Biochemistry of bacterial spore germination.

Hung Ton-That, Assistant Professor; Ph.D., UCLA. Pilus assembly and sortase-mediated anchoring of surface proteins in gram-positive bacteria.

Sandra K. Weller, Professor and Department Head; Ph.D., Wisconsin. Mechanisms of DNA replication and DNA encapsidation in herpes simplex virus; virus-host interactions.

Stephen K. Wikel, Professor; Ph.D., Saskatchewan. Molecular and cellular immunology of blood-feeding arthropod-host-pathogen interactions.

THE UNIVERSITY OF IOWA

Interdisciplinary Graduate Program in Molecular Biology

Program of Study	The Interdisciplinary Graduate Program in Molecular Biology is administered through the Graduate College at the University of Iowa and offers the Ph.D. degree in molecular biology. This program brings together faculty members from the Departments of Anatomy and Cell Biology, Biochemistry, Biological Sciences, Chemistry, Dermatology, Internal Medicine, Microbiology, Neurology, Ophthalmology, Orthodontics, Pathology, Pediatrics, Pharmacology, Physiology and Biophysics, and Radiation Oncology. The diversity of the research faculty permits molecular biology students to identify research areas in molecular and cellular biology that match their own particular interests, and it increases student access to faculty laboratories irrespective of classical departmental boundaries. Students entering the program complete three 8- to 10-week laboratory rotations and are encouraged to identify a thesis laboratory by the end of their second semester. Didactic course work is completed during the first two years. Required course work includes a fundamental course entitled Principles of Molecular and Cell Biology, followed by one course each in molecular and cellular biology. Depending on their interests and research area, students also broaden their experience by taking one core course in an area of chemistry/biochemistry or pharmacology/physiology. Two additional elective courses are taken to tailor each student's studies to his or her research area. This curriculum provides a focused exposure within the fields of molecular and cellular biology, as well as cross-training opportunities to enhance each student's professional development. Following successful completion of written and oral comprehensive examinations between the second and third years of study, students focus primarily on their thesis research project. Students actively participate in seminars, workshops, and journal clubs throughout their training in the molecular biology program. The average time for completion of the Ph.D. degree is approximately five years. The goal of the program is to provide students with instruction and research training to make them highly competitive as scientists in an academic, industrial, or government setting.
Research Facilities	Many students and faculty members of the Graduate Program in Molecular Biology are housed in the newly renovated Bowen Science Building. Other research laboratories are located in the Biology Building, the Eckstein Medical Research Building, Medical Laboratories, and the newly opened Medical Education and Research Facility. The University of Iowa combines modern laboratories and core facilities to provide state-of-the-art instrumentation and equipment for biomolecular research. The core research facilities, generously supported by the College of Medicine, include the Central Microscopy Research Facility, DNA Sequencing Facility, Flow Cytometry Facility, Gene Targeting Facility, Gene Transfer Vector Core, Molecular Analysis Facility, Nuclear Magnetic Resonance Facility, Tissue Culture/Hybridoma Facility, and Transgenic Animal Facility. These core laboratories, Information Technology Services, and the Hardin Library for the Health Sciences are located in buildings that are in proximity to Bowen Science Building, with easy access for faculty members and students. Equipment includes confocal and epi-illumination fluorescence microscopes, scanning and transmission electron microscopes, gas and liquid high-performance chromatographs, scintillation/gamma counters, spectrophotometers, ultracentrifuges, and electron paramagnetic resonance spectrometers.
Financial Aid	All students admitted into the program receive an annual stipend, which was $22,000 for the 2005–06 academic year. The program also pays tuition costs for each student. This support is provided through the Graduate College, institutional training grants, or other research grants. Stipend support is renewed based on satisfactory progress toward the Ph.D. degree.
Cost of Study	As indicated above, tuition is paid for each student in the program. All research supplies and materials are provided by the program and/or faculty advisers.
Living and Housing Costs	The cost of living in Iowa City is low compared to that of most major health science or research institutions. Most students find affordable housing in residential areas, often within walking distance of the University. One- and two-bedroom apartments, available as family housing, rent for $400 to $545 per month. Applicants should contact Family Housing for these units as soon as their plans are known. Travel on the Cambus system, a modern bus system serving the University area, is free.
Student Group	Enrollment at Iowa is nearly 30,000, with approximately one third in graduate and professional programs. The program currently has 35 students (15 women, 20 men), including 26 domestic, 3 permanent resident, and 6 international students. Students have varied academic backgrounds, but all have a B.S. or B.A. degree in the biological, biochemical, or physical sciences. Most graduates opt for postdoctoral positions in leading research labs, and all move on to challenging research careers in academia, industry, or government.
Location	Iowa City, a university community of 65,000 residents, offers full and varied cultural, social, and recreational opportunities. The city is served by Interstate 80 and is within easy driving distance of Chicago, Kansas City, St. Louis, and Minneapolis–St. Paul. Four major airlines serve the nearby Cedar Rapids airport, which is a 25-minute drive from the campus.
The University and The Program	Programs in the health sciences are offered by the Graduate College and the Colleges of Medicine, Dentistry, Pharmacy, Public Health, and Nursing. The presence of these colleges on a single campus encourages interdepartmental and intercollegiate cooperation in research and other educational activities. The University is recognized as a major institution for graduate studies in many areas. The College of Medicine ranks eleventh among the nation's public medical schools in NIH grants and grant dollars per faculty member.
Applying	Prospective students must apply before February 1 for admission in the summer or fall. Applications after this deadline are reviewed based on space available in the program. Acceptance is based on academic record, a personal statement of interest, letters of recommendation, and an interview. The GRE General Test is required. The TOEFL is also required for international students. All participants must have a bachelor's degree from an accredited university. Laboratory research experience is a positive factor in gaining admission. Applications from underrepresented racial and ethnic groups are encouraged.
Correspondence and Information	Molecular Biology Program Office 1190 Medical Laboratories The University of Iowa Iowa City, Iowa 52242 Phone: 319-335-7748 800-551-6787 (toll-free) Fax: 319-335-7656 E-mail: molecular-biology@uiowa.edu Web site: http://molbio.grad.uiowa.edu

The University of Iowa

THE FACULTY AND THEIR RESEARCH

Lee-Ann Allen, Associate Professor; Internal Medicine, Ph.D., Wisconsin–Madison. Biology of macrophages and neutrophils and phagocyte interactions with pathogenic bacteria including *Helicobacter pylori* and *Francisella tularensis.*

Michael Apicella, Professor and Head of Microbiology; M.D., SUNY Downstate Medical Center. Pathogenesis of human infection by pathogenic *Neisseria* and *Haemophilus.*

Nikolai Artemyev, Professor of Physiology and Biophysics; Ph.D., Russian Academy of Sciences. Molecular mechanisms of sensory signal transduction; signaling via heterotrimeric G proteins; photoreceptors and visual transduction.

Jackie Bickenbach, Associate Professor of Anatomy and Cell Biology; Ph.D., Iowa. Epithelial stem cells and their use in tissue engineering; their potential use as cell therapy agents or a gene delivery system.

Gail Bishop, Professor of Microbiology; Ph.D., Michigan. Molecular mechanisms of signal transduction in B lymphocytes.

Kevin Campbell, Professor and Interim Head of Physiology and Biophysics; Ph.D., Rochester. Biochemistry of muscular dystrophy; molecular studies of Ca^{2+} channels and dystrophin-glycoprotein complex.

Steven Clegg, Professor of Microbiology; Ph.D., Dundee (Scotland). Molecular biology of bacterial adherence factors.

Michael Cohen, Professor and Head of Pathology; M.D., Albany Medical College. Experimental biology of prostate cancer.

Beverly Davidson, Professor of Internal Medicine; Ph.D., Michigan. Gene transfer and gene therapy in disorders involving the central nervous system.

Frederick Domann, Associate Professor of Radiation Oncology; Ph.D., Wisconsin–Madison. Mechanisms of altered gene expression in cancer.

John Donelson, Professor and Head of Biochemistry; Ph.D., Cornell. Mechanisms of gene expression in protozoan parasites.

Martine Dunnwald, Assistant Professor of Dermatology; Ph.D., Laval. Skin and hair follicle stem cells.

John Engelhardt, Professor and Interim Head of Anatomy and Cell Biology; Ph.D., Johns Hopkins. Lung molecular and cellular biology as it relates to the pathogenesis and treatment of cystic fibrosis.

Sarah England, Associate Professor of Physiology and Biophysics; Ph.D., Medical College of Wisconsin. Role of potassium channels in regulating smooth-muscle excitability.

Michael Feiss, Professor of Microbiology; Ph.D., Washington (Seattle). Virus DNA packaging.

Rory Fisher, Professor of Pharmacology; Ph.D., Iowa State. Molecular biology and signal transduction of G-protein–coupled receptors and RGS proteins.

Sonya Franklin, Associate Professor of Chemistry; Ph.D., Berkeley. De novo protein design of artificial endonucleases, transcription factors, and related metalloproteins.

Minnetta Gardinier, Associate Professor of Pharmacology; Ph.D., LSU Medical Center. Oligodendroglial cell biology and CNS myelination.

Pamela Geyer, Professor of Biochemistry; Ph.D., Ohio State. Mechanisms involved in long-distance regulation of gene expression.

Prabhat Goswami, Assistant Professor of Radiation Oncology; Ph.D., Gauhati (India). Intracellular redox and the mammalian cell cycle.

Steven Green, Associate Professor of Biological Sciences; Ph.D., Caltech. Neurotrophic stimuli and the relevant signal transduction pathways.

Gregory Hageman, Professor of Ophthalmology; Ph.D., USC. Retinal cell and molecular cell biology.

Raymond Hohl, Professor of Internal Medicine; M.D., Ph.D., Rush. Role of isoprene compounds in cell proliferation and differentiation.

Mary Horne, Assistant Professor of Pharmacology; Ph.D., Munich. Function of the atypical cyclin G family in cell-cycle arrest, differentiation, and programmed cell death.

Mei-Yu Hsu, Assistant Professor; M.D. Kaohsiung Medical College (Taiwan); Ph.D., Pennsylvania. Morphogens in human melanoma.

Aloysius Klingelhutz, Assistant Professor of Microbiology; Ph.D., Wisconsin–Madison. Molecular mechanisms of immortalization and transformation by human papillomavirus.

Michael Knudson, Assistant Professor of Pathology; M.D., Ph.D. Iowa. Understanding the regulation of programmed cell death.

Amnon Kohen, Assistant Professor of Chemistry; D.Sc., Israel Institute of Technology. Enzymes: mechanistic studies of biological catalysis in DNA biosynthesis and nitrogen fixation.

John Koland, Associate Professor of Pharmacology; Ph.D., Illinois at Urbana-Champaign. Growth factor receptor signal transduction and cellular transformation.

David Kusner, Associate Professor of Internal Medicine; M.D., Harvard; Ph.D., Case Western Reserve. Signal transduction mechanisms that regulate the activation and antimicrobial properties of macrophages, dendritic cells, and neutrophils.

Tomomi Kuwana, Assistant Professor of Pathology; M.D., Niigata (Japan); Ph.D., Cambridge. Mitochondrial regulation of apoptosis.

Gloria Lee, Associate Professor of Internal Medicine; Ph.D., Harvard. Structure and function of the neuronal cytoskeleton and its roles in signal transduction and neurodegenerative diseases.

Steven Lentz, Professor of Internal Medicine; M.D., Ph.D., Washington (St. Louis). Endothelial function in vascular diseases and tissue injury.

Andrew Lidral, Associate Professor of Orthodontics; D.D.S., North Carolina; Ph.D., Iowa. Genetics of craniofacial anomalies.

Jim Jung-Ching Lin, Professor of Biological Sciences; Ph.D., Connecticut. Structure and function of protein components of the cytoskeleton; developmental genetics/ eukaryotic gene expression.

John Logsdon, Assistant Professor of Biological Sciences; Ph.D., Indiana. Molecular genetic aspects of evolution.

Wendy Maury, Assistant Professor of Microbiology; Ph.D., Virginia. Regulation of retroviral pathogenesis.

Anton McCaffrey, Assistant Professor of Internal Medicine; Ph.D., Stanford. The recognition of viruses by host cells and the development of RNA interference bases for antiviral therapeutics.

Jeffrey Meier, Associate Professor of Internal Medicine; M.D., Iowa. Viral pathogenesis at the molecular level.

W. Scott Moye-Rowley, Professor of Physiology and Biophysics; Ph.D., Purdue. Transcriptional control of gene expression; multidrug and oxidative stress resistance in yeast.

Jeffrey Murray, Professor of Pediatrics; M.D., Tufts. Identification and characterization of genes involved in inherited human disorders.

William Nauseef, Professor of Internal Medicine; M.D., SUNY Upstate Medical Center. Molecular and cell biology of human neutrophils.

Henry Paulson, Associate Professor of Neurology; M.D., Ph.D., Yale. Molecular mechanisms of neurodegenerative diseases, with the aim of developing therapeutics.

David Price, Professor of Biochemistry; Ph.D., Florida State. Control of elongation phase of transcription by RNA polymerase II.

Dawn Quelle, Assistant Professor of Pharmacology; Ph.D., Penn State. Cell-cycle control and molecular mechanisms of tumorigenesis.

Subramanian Ramaswamy, Associate Professor of Biochemistry; Ph.D., Indian Institute of Science (Bangalore). Protein crystallography, structural enzymology of naphthalene dioxygenase and alcohol dehydrogenase.

Michael Rebagliati, Assistant Professor of Anatomy and Cell Biology; Ph.D., Harvard. Roles of cyclops and other nodal-related growth factors in neurogenesis, dorsal-ventral polarity, and left-right asymmetry in the zebrafish.

Richard Roller, Associate Professor of Microbiology; Ph.D., Harvard. Molecular biology of HSV infection.

Peter Rubenstein, Professor of Biochemistry; Ph.D., Harvard. Actin site-directed mutagenesis and actin structure-function relationships.

Andrew Russo, Professor of Physiology and Biophysics; Ph.D., Berkeley. Molecular control of neuronal gene expression.

David Sheff, Assistant Professor of Pharmacology; M.D./Ph.D., Iowa. Polarized sorting in endosomes.

Curt Sigmund, Professor of Internal Medicine; Ph.D., SUNY at Buffalo. Gene regulation in cardiovascular homeostasis; transgenic animal models of human cardiovascular disease.

Mark Stamnes, Associate Professor of Physiology and Biophysics; Ph.D., California, San Diego. Protein sorting and transport.

Jack Stapleton, Professor of Internal Medicine; M.D., Kansas. Basic and translational research on GB virus C and hepatitis C virus, and their reactions with HIV.

Mark Stinski, Professor of Microbiology; Ph.D., Michigan State. Regulation of human cytomegalovirus gene expression.

C. Martin Stoltzfus, Professor of Microbiology; Ph.D., Wisconsin–Madison. Regulation of retrovirus RNA splicing, transport, and translation.

Edwin Stone, Professor of Ophthalmology; M.D., Ph.D., Baylor College of Medicine. Molecular genetics of inherited eye diseases.

Stefan Strack, Assistant Professor of Pharmacology; Ph.D., SUNY at Albany. Neuronal signal transduction; role of protein phosphatase 2A in brain function.

Christie Thomas, Associate Professor of Internal Medicine; M.D., Christian Medical College (India). Organization and transcriptional regulation of the epithelial sodium channel genes.

Lubomir Turek, Professor of Pathology; M.D., Karlova (Prague). Human papillomaviruses and cancer.

Lori Wallrath, Associate Professor of Biochemistry; Ph.D., Michigan State. Chromatin structure and gene expression.

Todd Washington, Assistant Professor of Biochemistry; Ph.D., Ohio State. Mechanisms of eukaryotic DNA polymerases that carry out error-prone or error-free replication of damaged DNA.

Daniel Weeks, Professor of Biochemistry; Ph.D., Purdue. Molecular control of embryology.

David Weiss, Associate Professor of Microbiology; Ph.D., Berkeley. Cell division and protein localization in bacteria.

Michael Welsh, Professor of Internal Medicine; M.D., Iowa. Biology of cystic fibrosis; gene therapy; molecular biology of ion channels.

Mary Wilson, Professor of Internal Medicine; M.D., Rochester. Molecular mechanisms of host-parasite interactions in leishmaniasis.

Marc Wold, Professor of Biochemistry; Ph.D., Johns Hopkins. Molecular mechanisms of eukaryotic chromosomal DNA replication; eukaryotic single-strand DNA-binding proteins.

Charles Yeaman, Assistant Professor of Anatomy and Cell Biology; Ph.D., Wisconsin–Madison. Biogenesis and maintenance of epithelial cell polarity.

Joseph Zabner, Professor of Internal Medicine; M.D., Venezuela. Gene transfer to human airway epithelia, in particular, to develop gene therapy for cystic fibrosis.

UNIVERSITY OF MASSACHUSETTS AMHERST

Molecular and Cellular Biology Graduate Program
Cell and Developmental Biology Track

Program of Study

The Molecular and Cellular Biology (MCB) Graduate Program at University of Massachusetts Amherst is an interdisciplinary Ph.D. program with areas of concentration in biological chemistry and molecular biophysics, molecular biology, and cell and developmental biology. Participating faculty members are drawn from eight University departments; from Amherst, Smith, and Mount Holyoke Colleges; and from the Baystate Medical Center in Springfield.

The biological chemistry and molecular biophysics concentration fuses the cultures of biology and chemistry, bringing the synthetic, mechanistic, and analytical powers of chemistry to bear on new areas of biology. Faculty members in this area are particularly strong in fields such as the mechanisms of signal transduction, protein folding, RNA structure and function, protein–nucleic acid interaction, structural biology, and bioorganic and bioinorganic chemistry. Students in this area are eligible for participation in the Chemistry-Biology Interface Training Program.

The molecular biology concentration emphasizes the molecular basis of gene expression and function in viruses, bacteria, fungi, plants, and animals. Interests of participating faculty members include gene organization and regulation, the molecular mechanisms of gene expression, microbial genetics and physiology, functional genomics, molecular genetics of plant development, and the molecular biology of disease, with a focus on the mechanisms of susceptibility and mutation in human disease.

The cell and developmental biology concentration focuses on the genetic control of single cells, as well as their assembly during development and disruption during disease. Faculty members in this area study the structure and function of basic cellular processes, such as cell division, protein trafficking, nuclear architecture, early development and reproduction, cell death and degeneration, and neural development. Model systems include *Drosophila*, zebrafish, *Arabidopsis*, and mouse. Students in this area have the opportunity to couple basic and biomedical approaches through a collaborative research program with the Baystate Medical Center.

All Ph.D. candidates rotate through two research laboratories in the first year and take core courses in biochemistry, molecular biology, and cell biology. A written qualifying exam, the first part of the comprehensive examination, is taken at the end of the first year. An oral defense of an original research proposal, completed before the end of the second year, constitutes the second part of the comprehensive examination. The program hosts a weekly seminar series attended by faculty members and students. After the first year, students participate in informal seminars and journal clubs and take appropriate advanced elective courses. There is also a final oral defense of the written dissertation. Completion of the requirements for the Ph.D. degree takes about five years.

Research Facilities

Major research capabilities include an imaging facility with confocal and electron microscopy, a DNA sequencing service, and facilities for high-field NMR, mass spectrometry, and X-ray diffraction. A variety of specialized equipment is available in the adjoining Departments of Physics, Computer Science, and Polymer Science and Engineering, as well as in the physical and life science departments at the other four institutions in the Five Colleges, Inc., consortium, which includes Amherst, Mount Holyoke, Hampshire, and Smith Colleges, in addition to the University of Massachusetts.

Financial Aid

Financial aid is available to all accepted Ph.D. candidates. Students are eligible for teaching and research assistantships that include tuition, curriculum fee, health insurance, dental coverage, and an annual stipend. Once a dissertation lab is selected, students are supported by research assistantships or fellowships. All students are required to serve as teaching assistants, generally during their first year. MCB students with a strong chemistry background are eligible to participate in the NIH-funded Chemistry-Biology Interface Training Program.

Cost of Study

In 2005–06, resident tuition and fees at the University of Massachusetts were $1320 and $3458.50, respectively. Nonresident tuition and fees were $4968.50 and $4034.50, respectively. Massachusetts residence requirements are usually met after one year's residence and employment. Tuition and some fees are waived for students supported by teaching and research assistantships.

Living and Housing Costs

Most students in the MCB program live off campus. Studio and efficiency apartments cost approximately $550 per month, while a two-bedroom apartment costs approximately $700 to $1000 per month. Graduate student housing is also available in Prince House, a graduate dormitory, at $2437.50 per semester.

Student Group

Amherst is the main graduate campus of the University of Massachusetts system, with 2,747 graduate students enrolled in forty-seven doctoral programs. In fall 2005, there were 76 doctoral candidates in the MCB program, of whom 49 percent were women, 51 percent were men, and 54 percent were international students. All received financial support. Most graduates of the program go on to postdoctoral appointments at leading universities and research institutions or to responsible positions in industry, both in the United States and abroad.

Student Outcomes

Nine Ph.D. degrees were awarded in 2005. Graduates have gone on to postdoctoral research at institutions such as Cold Spring Harbor Laboratory, Dana-Farber Cancer Institute, Harvard, MIT, National Cancer Institute, Stanford, and Yale as well as universities in the United Kingdom, Korea, Taiwan, and the Netherlands. Others have chosen to take positions in industry, including Burroughs-Wellcome, Johnson & Johnson, and Eli Lilly.

Location

Amherst lies in the Connecticut River valley in western Massachusetts. Northampton, noted for its cultural diversity, is nearby, and Boston and New York City are easily reached by car, bus, or train. The Pioneer Valley is well-known for the variety and quality of its music, theater, art, and outdoor recreational activities.

The University and The Program

The Amherst campus is the flagship campus of the University of Massachusetts system, with numerous award-winning faculty members and a rich array of academic and extracurricular programs. It is ranked by the NRC in the top fifty research-doctoral institutions both for its reputational rating and for the number of citations and awards its programs receive. The Molecular and Cellular Biology Graduate Program, inaugurated in 1983 to integrate faculty members, academic activities, and research efforts of a number of well-established departmental programs is among the highly rated group of programs.

Applying

Applications are due on January 15. A late application is accepted if places remain at the time of its receipt. Applicants should have had one year of physics with lab and mathematics through calculus, or applicants should be prepared to take the requisite courses after entering the program. Prior laboratory or research experience is highly desirable. Prospective students should visit the Web site at http://www.bio.umass.edu/mcb for further information. Applications may be obtained and submitted online at http://www.umass.edu/gradschool.

Correspondence and Information

For additional information, students should contact:

Director
MCB Graduate Program
435 Morrill I North
University of Massachusetts
Amherst, Massachusetts 01003
Phone: 413-545-3246
Fax: 413-545-1812
E-mail: mcb@mcb.umass.edu
Web site: http://www.bio.umass.edu/mcb

To submit application materials by mail:

Graduate Admissions Office
530 Goodell
University of Massachusetts
Amherst, Massachusetts 01003
Phone: 413-545-0721

University of Massachusetts Amherst

THE FACULTY AND THEIR RESEARCH

Dominique R. Alfandari, Ph.D., Paris VI (Curie). ADAM metalloprotease function during embryonic development.

Juan Anguita, Ph.D., Leon (Spain). Proinflammatory signals in response to infection.

Kathleen F. Arcaro, Ph.D., Rutgers. Environmental toxicants and the disruption of endocrine function.

Richard B. Arenas (Baystate), M.D., University of Medicine and Dentistry of New Jersey–Robert Wood Johnson Medical School/Rutgers. APC gene replacement in colorectal cancer.

Sarah J. Bacon (Mount Holyoke College), Ph.D., Chicago. Reproductive biology: maternal-fetal histocompatibility.

Cynthia L. Baldwin, Ph.D., Cornell. Cellular immunity: intracellular microbes and γ/δ T cells.

Michael J. Barresi (Smith College), Ph.D., Wesleyan. How glial cells help wire the nervous system in the embryonic zebrafish brain.

Tobias I. Baskin, Ph.D., Stanford. Regulation of plant morphogenesis during growth and development.

Magdalena Bezanilla, Ph.D., Johns Hopkins. Molecular mechanisms behind plant cell growth.

David Bickar (Smith College), Ph.D., Duke. Systems for electron transport and oxygen utilization.

Anthony C. Bishop (Amherst College), Ph.D., Princeton. Bioorganic chemistry and chemical biology.

Eric L. Bittman, Ph.D., Berkeley. Neural basis of circadian rhythms, reproductive cycles, and seasonal breeding.

Samuel J. Black, Ph.D., Edinburgh (Scotland). Molecular basis of innate immunity against parasitic protozoa.

Jeffrey L. Blanchard, Ph.D., Georgia. Microbial cellular and community networks; systems biology; bioinformatics.

Jeffrey D. Blaustein, Ph.D., Massachusetts. Neuroendocrine regulation of behavior.

John P. Burand, Ph.D., Washington State. Molecular biology of insect baculoviruses.

Maura C. Cannon, Ph.D., University College (Ireland). Polyhydroxyalkanoic acids in bacteria and plants; plant embryogenesis.

Alice Y. Cheung, Ph.D., Yale. Molecular and biochemical studies on sexual plant reproduction.

J. Marshall Clark, Ph.D., Michigan State. Molecular action of pesticides and cellular basis for resistance.

Elizabeth A. Connor, Ph.D., Vermont. Development and regeneration of the neuromuscular junction.

Susan Cumberledge, Ph.D., California, Santa Barbara. Signal transduction in development and tumorigenesis.

Sean M. Decatur (Mount Holyoke College), Ph.D., Stanford. Biochemistry of nitrosylhemes and vibrational studies of helical peptides.

Geert J. De Vries, Ph.D., Amsterdam. Cellular basis of hormone effects on brain function and behavior.

Gerald B. Downes, Ph.D., Washington (St. Louis). Development and function of spinal cord networks.

Rachel D. Fink (Mount Holyoke College), Ph.D., Duke. Cell migration and rearrangement at gastrulation.

Rafael A. Fissore, Ph.D., Buenos Aires. Mammalian gametogenesis and cell signaling.

Katherine V. Fite, Ph.D., Brown. Comparative neurobiology of visual systems.

Molly Fitzgerald-Hayes, Ph.D., Connecticut Health Center. Chromosome segregation in yeast.

Neil S. Forbes, Ph.D., Berkeley. Engineering tumor-targeted therapies.

Nancy G. Forger, Ph.D., Berkeley. Development of neural sex differences; hormones and neutrophic factors.

Maurille J. Fournier, Ph.D., Dartmouth. Function of yeast snRNAs; genetic engineering of protein-based polymeric materials.

Scott C. Garman, Ph.D., Harvard. Structural biology of glycoproteins in human disease.

Lila M. Gierasch, Ph.D., Harvard. Biophysical approaches to protein folding and localization.

Richard A. Goldsby (Amherst College), Ph.D., Berkeley. Idiotype/anti-idiotype network interactions.

David J. Gross, Ph.D., Illinois. Signal transduction in individual cells.

Adam C. Hall (Smith College), Ph.D., Imperial College (London). Acute effects of volatile anesthetics using electrophysiological techniques.

David E. Hansen (Amherst College), Ph.D., Harvard. Antibody catalysis; rational drug design.

Jeanne A. Hardy, Ph.D., Berkeley. Design of allosteric pathways in caspases and phosphatases; X-ray crystallography.

Daniel N. Hebert, Ph.D., Massachusetts Medical Center. Protein folding, quality control, and degradation of membrane glycoproteins.

Alejandro P. Heuck, Ph.D., Buenos Aires. Pore-forming toxins and translocation of virulence factors in bacterial pathogenesis.

Abbie M. Jensen, Ph.D., Wisconsin–Madison. Developmental neurobiology.

D. Joseph Jerry, Ph.D., Penn State. Tumor suppressor genes and the cellular basis for susceptibility to breast cancer.

Igor A. Kaltashov, Ph.D., Moscow. Bioanalytical applications of mass spectrometry.

Rolf O. Karlstrom, Ph.D., Utah. Developmental neurobiology; axon guidance and forebrain patterning.

Young-Cheul Kim, Ph.D., Tennessee. Nutrient regulation of adipocyte differentiation and metabolism in obesity and diabetes; phytochemicals and gene expression.

Michele M. Klingbeil, Ph.D., Toledo. Parasitology; replication and repair of mitochondrial DNA (kinetoplast DNA network) in African trypanosomes.

Michael J. Knapp, Ph.D., Berkeley. Enzymology of biological O_2 sensing.

Joseph G. Kunkel, Ph.D., Case Western Reserve. Genetic and biochemical analysis of metamorphosis.

George M. Langford, Ph.D., ITT. Actin-dependent organelle motors and the cytoskeleton.

Wei-Lih Lee, Ph.D., Johns Hopkins. Mechanism of dynein-mediated nuclear migration and spindle positioning.

Susan B. Leschine, Ph.D., Pittsburgh. Microbial physiology, ecology, and molecular phylogeny.

Michael J. Maroney, Ph.D., Washington (Seattle). Bioinorganic chemistry of metalloproteins.

Craig T. Martin, Ph.D., Caltech. Enzymology of protein-DNA interactions.

Lynne McLandsborough, Ph.D., Minnesota. Involvement of bacterial surface molecules in attachment to food and food-processing surfaces.

Jerrold S. Meyer, Ph.D., Brown. Endocrine and pharmacological regulation of brain development.

Murugappan Muthukumar, Ph.D., Chicago. Pattern recognition by macromolecules.

John R. Nambu, Ph.D., Stanford. Nervous system development.

Leonard C. Norkin, Ph.D., Columbia. Polyomaviral persistent infections.

Jennifer Normanly, Ph.D., Caltech. Auxin biosynthesis and signal transduction in *Arabidopsis thaliana.*

Barbara A. Osborne, Ph.D., Stanford. Evolution of immunoglobulin gene families.

Yeonhwa Park, Ph.D., Wisconsin–Madison. Conjugated linoleic acid (CLA) research and biologically active compounds from natural or dietary sources.

Om Parkash, Ph.D., Durham (England). Plant molecular genetics.

Sandra L. Petersen, Ph.D., Oregon State. Transsynaptic regulation of LHRH biosynthesis and release by ovarian hormones.

Randall W. Phillis, Ph.D., Indiana. Genetics and biology of transposable elements in *Drosophila.*

Dominic L. Poccia (Amherst College), Ph.D., Harvard. Structure and activity of male germ line nuclei.

Pablo Pomposiello, Ph.D., Michigan. Global gene regulation response to oxidative damage.

David I. Ratner (Amherst College), Ph.D., Harvard. Differentiation expression in *Dictyostelium.*

Margaret A. Riley, Ph.D., Harvard. Microbial molecular evolution and ecology.

Vincent M. Rotello, Ph.D., Yale. Synthetic models of biomolecular activity.

Steven J. Sandler, Ph.D., Berkeley. Molecular mechanisms of DNA replication and homologous recombination.

Sallie W. Smith Schneider, Ph.D., Massachusetts Amherst. Signaling during mammary gland development and carcinogenesis.

Danny J. Schnell, Ph.D., California, Davis. Protein import into chloroplasts.

Lawrence M. Schwartz, Ph.D., Washington (Seattle). Molecular mechanisms controlling developmentally programmed cell death.

Stylianos P. Scordilis (Smith College), Ph.D., SUNY at Albany. Biochemistry and development of contractile proteins.

Dennis G. Searcy, Ph.D., UCLA. Cellular evolution and physiology.

Rong Shao (Baystate), Ph.D., Philadelphia College of Pharmacy. Molecular mechanisms in breast cancer development.

Kalidas Shetty, Ph.D., Idaho. Role of antioxidant metabolites/enzymes and genes during plant cellular differentiation.

Janice C. Telfer, Ph.D., Harvard. Role of RUNX family transcription factors in immune system development.

S. Thai Thayumanavan, Ph.D., Illinois. Design and synthesis of biomimetic organic molecules and polymers; applications of custom-designed molecules in biology and material science.

Karsten Theis, Ph.D., Berlin. DNA repair and molecular motors.

Lynmarie K. Thompson, Ph.D., MIT. Biophysical studies of membrane proteins: mechanisms of receptors and ion channels.

Pablo E. Visconti, Ph.D., Buenos Aires. Signal transduction pathways during sperm capacitation.

Patricia Wadsworth, Ph.D., Dartmouth. Dynamics of mitotic and interphase microtubules.

Elsbeth L. Walker, Ph.D., Rockefeller. Plant genetics and epigenetics.

Robert M. Weis, Ph.D., Stanford. Physical chemistry of membranes and mechanisms of signal transduction.

Christine A. White-Ziegler (Smith College), Ph.D., Utah. Environmental regulation of gene expression in *Escherichia coli.*

Steven A. Williams (Smith College), Ph.D., California, Davis. Genome structure in filarial parasites.

Patrick L. Williamson (Amherst College), Ph.D., Harvard. Plasma membrane organization in eukaryotes.

Craig T. Woodard (Mount Holyoke College), Ph.D., Yale. Steroid hormonal regulation of development in *Drosophila.*

Christopher L. F. Woodcock, Ph.D., London. Structure-function relationships in the eukaryotic nucleus.

Hen-ming Wu, Ph.D., Yale. Molecular biology of plant development.

Robert A. Zimmermann, Ph.D., MIT. Structure and function of ribosomes.

R. Thomas Zoeller, Ph.D., Oregon State. Molecular mechanisms of signal integration in neuroendocrine systems.

UNIVERSITY OF MASSACHUSETTS WORCESTER

Medical School
Department of Cell Biology
Graduate Study in Molecular Cell Biology

Program of Study

The Department of Cell Biology offers a program of study and research in molecularly oriented cell biology leading to a Ph.D. degree in cell biology or in biomedical sciences. The research interests of the faculty members focus on the use of cellular and molecular approaches to address structure-function relationships associated with cell growth, development, cell signaling, gene expression, and cell motility. The department teaches graduate courses in cell biology, maintains an active basic research program, and strongly supports graduate participation in this research. The faculty has expertise in many molecular and cell-biological techniques, including three-dimensional image analysis, immunofluorescence and confocal microscopy, freeze-fracture and rapid-freeze specimen preparation, electron microscopy, protein chemistry, recombinant DNA, site-directed mutagenesis, construction of transgenic animal models, in situ hybridization, transfection of chimeric genes into cultured cells, chromatin structure analysis, gene mapping, isolation and characterization of transcription factors, messenger RNA isolation and cloning, and microinjection and micromanipulation of cells. Cell, molecular, and developmental biologists in the Cancer Center and the Shriver Center participate in the graduate program and add a further dimension to the expertise available in the department.

Incoming students should have good preparation in biological, chemical, and physical principles. Formal course requirements for the Ph.D. include an integrated core graduate-level course covering biochemistry, molecular biology, and cell biology. In the first year, students are required to undertake two to three laboratory rotations, which can be in any department of the Medical School, prior to selecting a laboratory for thesis research. In addition to completing formal course work and research, students are expected to attend departmental and Medical School seminars and to participate in the general intellectual life of the school. Admission to candidacy occurs following successful completion of a qualifying exam administered during the second year. A faculty member is appointed for each student to provide guidance in selecting courses and laboratory rotations and to evaluate student performance. The Ph.D. degree generally requires five to six years to obtain and is awarded following completion and successful defense of thesis research. The training received through this program is intended to provide doctoral graduates with the academic background, experimental tools, and analytical skills necessary to conduct independent research and teach in a university or medical school environment or to pursue a biotechnology career in industry.

Research Facilities

The Medical School is generously endowed with laboratory, office, and classroom space and a large research library. A 360,000-square-foot research building opened in 2001. The department occupies 35,000 square feet of state-of-the-art research space. Facilities and equipment are the most modern available for biomedical research and include core facilities for peptide synthesis; protein and nucleic acid sequencing; confocal, scanning, transmission, and transmission/cryoelectron microscopy; tissue culture; fluorescence cell sorting; gene profiling using microarrays; digital image analysis; transgenic animals; gene therapy; and recombinant DNA. A large and well-maintained experimental animal-care facility, a computer center, and modern electronics and machine shops are available.

Financial Aid

In 2003–04, beginning full-time students were offered a graduate assistantship at an annual stipend of $23,000. Research assistantships paying $23,000 were available to students starting thesis research. Tuition is waived for both graduate and research assistants; health insurance is provided. Application for assistantships is automatically made on the application for admission.

Cost of Study

Tuition waivers and stipends are available for all incoming students.

Living and Housing Costs

Apartments and rooms are available in the Worcester area and nearby communities. The large number of students in the area makes sharing of apartments feasible at costs that can be borne by students receiving financial aid.

Student Group

Currently, 250 graduate students from throughout the United States and many other countries are enrolled. Other graduate students are working at the Medical School through cooperative programs with the Worcester Consortium and the University of Massachusetts at Amherst. The department currently has 25 graduate students.

Location

Worcester is the second-largest city in New England and is 40 miles west of Boston. Worcester and the surrounding towns offer a variety of living conditions and a wide range of educational and cultural activities. Located nearby are numerous other educational institutions, including Clark University, Holy Cross College, and Worcester Polytechnic Institute.

The University and The Medical School

The University of Massachusetts has five campuses: the main campus at Amherst, about 70 miles west of Worcester; campuses in Boston, Lowell, and Dartmouth; and the Medical School in Worcester. The Medical School opened in 1970; there are now 100 students in each class. The Medical School joins regional institutions and the University of Massachusetts Amherst in the training of other health science professionals. The Graduate School of Nursing opened in 1985.

Applying

The department accepts graduate students through the Graduate School of Biomedical Sciences of the University of Massachusetts at Worcester. Applicants should have a bachelor's degree in a physical, chemical, or biological science. Specific course requirements include a year each of calculus, organic chemistry, physics, and biology. Applicants must take the General Test of the Graduate Record Examinations. Three letters of recommendation and an application fee are required. Applications should be received as soon as possible, preferably by January 2.

The University of Massachusetts at Worcester is an Equal Opportunity/Affirmative Action employer. The Graduate School of Biomedical Sciences considers all applicants, encourages applications from minority and women candidates, and, once students are accepted, ensures that they will not be discriminated against in any area.

The graduate catalog and application forms may be obtained from the address listed in this In-Depth Description. Potential applicants may wish to contact the Graduate Director, Dr. Anthony Imbalzano (e-mail: anthony.imbalzano@umassmed.edu), for additional information about the department's program.

Correspondence and Information

Dean, Graduate School of Biomedical Sciences
University of Massachusetts Worcester
Medical School
Worcester, Massachusetts 01655-0116
Phone: 508-856-4135
Fax: 508-856-3659
E-mail: gsbs@umassmed.edu
Web site: http://www.umassmed.edu/gsbs/

University of Massachusetts Worcester

THE GRADUATE FACULTY AND THEIR RESEARCH

Athena Andreadis, Associate Professor; Ph.D., MIT, 1984. Regulation and functional consequences of alternative splicing in brain. *J. Biol. Chem.* 280:14230–9, 2005. *Gene* 359:63–72, 2005.

Neil Aronin, Professor; M.D., Pennsylvania, 1974. RNAi therapy in neurodegenerative diseases; mechanisms of trinucleotide repeat diseases. *J. Neurosci.* 24:269–81, 2004. *Cell* 115:199–208, 2003. *Science* 277:1990–3, 1997.

Zheng-Zheng Bao, Assistant Professor; Ph.D., Illinois at Urbana-Champaign, 1993. Mechanisms of vertebrate development. *J. Neurosci.* 25:3432, 2005. *Development* 131:1553, 2004. *Development* 130:1037, 2003.

Susan Billings-Gagliardi, Professor; Ph.D., Harvard, 1971. Evidence that CNS hypomyelination does not cause death of jimpy-msd mutant mice. *Dev. Neurosci.* 21:473–82, 1999 (with Wolf and Nunnari).

Roger W. Craig, Professor; Ph.D., London, 1975. Molecular mechanism of muscle contraction; cryoelectron microscopy and image processing of macromolecular assemblies. *Nature* 436:1195, 2005. *J. Mol. Biol.* 346:761, 2005. *J. Mol. Biol.* 342:1223, 2004.

Stephen J. Doxsey, Professor; Ph.D., Yale, 1987. Role of centrosomes in spindle assembly, chromosome segregation, cytokinesis, cell-cycle progression, checkpoint control, autoimmunity, and cancer. *Cell* 23:75, 2005. *J. Cell Biol.* 166:637, 2004.

Andrew H. Fischer, Associate Professor; M.D., Brown, 1984. Cancer diagnosis; oncogene-induced alteration of nuclear envelope structure and function. *Lab. Invest.* 81:501–7, 2001. *Histol. Histopathol.* 16:1–14, 2001. *Am. J. Pathol.* 153:1443–50, 1998.

Harvey M. Florman, Professor; Ph.D., Pennsylvania, 1981. Mechanisms of fertilization; sperm ion channel structure and function. *Nature Cell Biol.* 3:499–502, 2001. *Mol. Biol. Cell* 11:1571–84, 2000.

Mark I. Furman, Associate Professor and Associate Director, Center for Platelet Function Studies; M.D., Emory, 1987. Circulating monocyte-platelet aggregates area more sensitive marker of platelet activation than platelet surface P-selectin: Studies in baboons, human coronary intervention, and human myocardial infarction. *Circulation* 104:1533–7, 2001. Dissociation of glycoprotein IIb/IIa antagonists from platelets does not result in fibrinogen binding or platelet aggregation. *Circulation* 104:1374–9, 2001.

Timothy R. Henion, Assistant Professor; Ph.D., Medical College of Pennsylvania, 1996. Molecular control of glycan expression and function during mammalian nervous system development. *J. Neurosci.* 25:1894, 2005. *J. Biol. Chem.* 276:30261, 2001.

Anthony N. Imbalzano, Associate Professor; Ph.D., Harvard, 1991. Effects of chromatin structure on gene expression, cellular differentiation, and oncogenesis. *Mol. Cell. Biol.* 21:3997, 2005. *Mol. Cell. Biol.* 24:4651, 2004. *Oncogene* 23:3462, 2004. *Nature Genet.* 27:187, 2001.

Y. Tony Ip, Associate Professor; Ph.D., Iowa, 1989. Molecular genetics of innate immune response and development in *Drosophila melanogaster. Development* 128:4757–67, 2001. *Mol. Cell. Biol.* 20:5987–95, 2000. *Genes Dev.* 13:792–7, 1999.

Stephen N. Jones, Associate Professor; Ph.D., Vanderbilt, 1988. Analysis of signal transduction and cancer using genetically modified mice. *Mol. Cell. Biol.* 24:976, 2004. *Oncogene* 23:304, 2004. *Cancer Cell* 4:349, 2003.

Stephen Lambert, Assistant Professor; Ph.D., Witwatersrand, 1992. Membrane-cytoskeletal interactions in axonal growth and differentiation. *J. Cell. Biol.* 162:489, 2003. *J. Biol. Chem.* 278:27333, 2003. *J. Cell. Biol.* 1295:304, 1998.

Jeanne B. Lawrence, Professor; Ph.D., Brown, 1982. Human genetics; X inactivation; structural arrangement of gene transcription RNA splicing. Ectopic human XIST gene can induce chromosome inactivation in postdifferentiation cells. Clustering of multiple specific genes and gene-rich R-bands around SC-35 domains: Evidence for local euchromatic neighborhoods. *Mol. Biol. Cell* 15:197, 2004. *Semin. Cell Dev. Biol.* 14:369, 2003. *J. Cell. Biol.* 162:981, 2003. *Hum. Mol. Genet.* 11:315, 2002. *J. Cell. Biol.* 150:417, 2000.

Mary M. Lee, Professor; M.D., SUNY at Buffalo, 1983. Hormonal regulation of Leydig cell development and differentiation; disorders of sexual differentiation. *Biol. Reprod.* 70:600–7, 2004. *Mol. Cell. Endocrinol.* 211:91–8, 2003.

Jane B. Lian, Professor; Ph.D., Boston University, 1972. Transcriptional control of osteogenesis and bone metastasis. *J. Biol. Chem.* 280:33132–40, 2005 (with Gaur). *J. Biol. Chem.* 280:15872–79, 2005 (with Lengner). *Mol. Cell. Biol.* 25:8581–91, 2005 (with Pratap). *Mol. Cell. Biol.* 24:9248–61, 2004 (with Hassan).

Elizabeth J. Luna, Professor; Ph.D., Stanford, 1977. Membrane skeletons in motile cells, e.g., neutrophils, tumors, and muscle. *J. Cell. Sci.* 117:5043–57, 2004 (with Gangopadhyay et al.). *J. Cell Sci.* 116:2261, 2003. *J. Biol. Chem.* 278:46094–106, 2003.

Jodi K. Maranchie, Assistant Professor; M.D., Northwestern, 1993. Role of redox signaling via Nox 4 in renal tumorigenesis. *Cancer Res.* 65(20):9191, 2005. *Cancer Cell* 1(3):247–55, 2002.

Zdenka Matijasevic, Assistant Professor; Ph.D., Zagreb (Croatia), 1982. Regulation of cell growth and cellular response to stress. *J. Neurosurg.* 102:98, 2005. *Carcinogenesis* 2:589, 2003.

Peter J. McCaffery, Associate Professor; Ph.D., Otago (New Zealand), 1987. Role of vitamin A in the developing and adult brain. *Proc. Natl. Acad. Sci. U.S.A.* 101:5111–16, 2004. *J. Neurosci.* 23:7610–20, 2003.

Craig C. Mello, Professor; Ph.D., Harvard, 1990. Molecular genetic analysis of *C. elegans* embryogenesis. *Nature* 431:338, 2004. *Cell* 111:991, 2002. *Cell* 106:23, 2001.

Jeffrey A. Nickerson, Assistant Professor; Ph.D., Michigan State, 1986. Nuclear structure; architectural organization of RNA metabolism. *Mol. Cell* 10:1235–46, 2002.

Paul R. Odgren, Assistant Professor; Ph.D., Massachusetts Medical Center, 1996. Development and function of skeletal and cartilage cells; genetic models of disorders of skeletal formation and remodeling. *Proc. Natl. Acad. Sci. U.S.A.* 99:14303–8, 2002.

Gerald A. Schwarting, Professor; Ph.D., Munich, 1974. Axon guidance and cell migration mechanisms during olfactory development. *Eur. J. Neurosci.* 19:1800–10, 2004.

Greenfield Sluder, Professor; Ph.D., Pennsylvania, 1976. Centrosome duplication and its regulation in the higher animal cell. In *Centrosomes in Development and Disease,* ed. E. A. Nigg, chap. 9, pp. 167–89. Weinheim: Wiley-VCH, 2004. *J. Cell. Biol.* 165:609, 2004. *Curr. Opin. Cell. Biol.* 16:49, 2004.

Gary S. Stein, Professor and Chair; Ph.D., Vermont, 1969. Control of proliferation, differentiation, and tumorigenesis. *Proc. Natl. Acad. Sci. USA* 102:7174–9, 2005. *Mol. Cell. Biol.* 25:6140–53, 2005. *Proc. Natl. Acad. Sci. USA* 102(5):1454–9, 2005.

Janet L. Stein, Professor; Ph.D., Princeton, 1975. Regulation of gene expression during proliferation and differentiation. *Mol. Cell. Biol.* 23:8110–23, 2003. *Mol. Cell. Biol.* 23:1460–9, 2003.

Janet L. Stein, Professor; Ph.D., Princeton, 1975. Regulation of gene expression during proliferation and differentiation. *Mol. Cell. Biol.* 23:8110–23, 2003. *Mol. Cell. Biol.* 23:1460–9, 2003.

Hiroomi Tada, Assistant Professor; M.D./Ph.D., Thomas Jefferson, 1994. Viral pathogenesis of human colon cancer; novel treatments for metastatic gastrointestinal cancers. *J. Clin. Invest.* 108:83, 2001. *Mol. Ther.* 4:29, 2001.

Andre J. van Wijnen, Associate Professor; Ph.D., Massachusetts Medical Center, 1991. Cell-cycle control of gene expression; transcriptional control of skeletal development and bone formation; subnuclear targeting of proteins. *Cancer Res.* 63:5357, 2003.

George B. Witman, Professor; Ph.D., Yale, 1972. Assembly and function of cilia and flagella; microtubule motors; role of cilia in disease; flagellar proteome. *J. Cell. Biol.* 170:103–13, 2005. *Mol. Biol. Cell* 15:4382–94, 2004. *Curr. Opin. Cell. Biol.* 15:105, 2003. *Nat. Rev. Mol. Cell. Biol.* 3:813–25, 2002.

Merrill K. Wolf, Professor; M.D., Western Reserve, 1956. Quaking shiverer double-mutant mice survive for at least 100 days with no CNS myelin. *Dev. Neurosci.* 21:483–90, 1999 (with Billings-Gagliardi and Nunnari).

John L. Woodhead, Assistant Professor; Ph.D., Massey (New Zealand), 1979. Structural biology of the cardiovascular system; smooth-muscle myosin; blood clotting proteins. *Nature* 436:1195–9, 2005. *J. Mol. Biol.* 342:1223–36, 2004.

Robert B. Zurier, Professor; M.D., Texas Southwestern Medical Center at Dallas, 1962. Prospects for cannobinoids as anti-inflammatory agents. *J. Cell. Biochem.* 88:462-66, 2003.

UNIVERSITY OF MIAMI

Sheila and David Fuente Graduate Program in Cancer Biology

Program of Study

The Sheila and David Fuente Graduate Program in Cancer Biology focuses on the biology of cancer and the development of novel diagnostic and therapeutic approaches. This interdisciplinary program involves faculty members from multiple programs and schools at the University of Miami (UM) and incorporates concepts and state-of-the—art techniques from molecular biology, biochemistry, genetics, genomics, proteomics, structural biology, cell biology, and biostatistics.

An important goal of the program is to provide students with a strong background in basic biomedical research coupled with clinical aspects of cancer, i.e., diagnosis, prognosis, and therapeutic intervention. To achieve this goal the program utilizes a two-tier mentoring system. During their first year of study, students select a research mentor and a physician mentor. Through this dual mentorship students conduct their doctoral research and obtain clinical knowledge in their area of study. The program aims to instill in students the ability to design multidisciplinary research programs in which scientific research is driven by unmet clinical needs.

The curriculum includes core courses in tumor biology, clinical oncology, scientific reasoning and logic in clinical cancer biology, and research methods, followed by electives in cancer epidemiology, cellular and molecular biology, immunology, pharmacology, and microbiology. During the first year, students attend a series of lectures in which physician mentors discuss clinical aspects of cancer detection and treatment. Rotations through faculty laboratories provide students with hands-on experience in various research areas. The rotations also provide the background necessary to select an area of dissertation research and a research mentor, who will also be the student's dissertation adviser. During the second year of study, students formulate a dissertation proposal and take a qualifying exam. Their subsequent work is guided by an individually tailored dissertation committee, including the research and physician mentors.

The University of Miami School of Medicine also offers a course of study leading to both the M.D. and Ph.D. degrees that can be completed in approximately seven years. Applicants interested in the combined degree program should contact the M.D./Ph.D. Program office.

Research Facilities

Academic core facilities and individual faculty research laboratories are located throughout the medical campus. The majority of the research and clinical faculty have offices on the main campus and belong to various basic science and clinical departments and programs, including Biochemistry and Molecular Biology, Cell Biology and Anatomy, Epidemiology and Public Health, Microbiology and Immunology, Molecular and Cellular Pharmacology, Physiology and Biophysics, and Neuroscience.

The Sylvester Comprehensive Cancer Center at the Leonard M. Miller School of Medicine serves as the nucleus for cancer-related research, diagnosis, and treatment at the University of Miami. As the only university-based cancer center serving the five counties of South Florida, it handles more than 1,300 inpatient admissions, performs 3,000 surgical procedures, and treats 3,000 new cancer patients annually. In addition, more than 150 physicians and scientists affiliated with the center are engaged in clinical research and receive more than $30 million annually in research grants.

The Louis Calder Memorial Library holds more than 200,000 volumes and receives more than 1,800 print journals and over 5,000 electronic journals. It houses medical textbooks and monographs, directories and other reference books, and more than 3,000 videotapes, audiotapes, videodisks, and other nonprint material, as well as core collections in anesthesiology, dermatology, otolaryngology, pediatrics, and psychiatry. Additional resources are available in the University libraries on the Coral Gables campus.

Financial Aid

Full stipends of $22,000 per annum are awarded to all accepted students.

Cost of Study

All students are awarded full-tuition fellowships.

Living and Housing Costs

There is no on campus housing available for graduate students.

Student Group

The student body includes individuals from all parts of the United States, as well as from other countries.

Student Outcomes

Graduates of the program have been exposed to ongoing efforts in clinical research programs, including the development and implementation of diagnostic, prognostic, and therapeutic applications. This prepares them to pursue independent research and teaching careers in the field of cancer biology.

Location

The University of Miami is located in Miami-Dade County, where the subtropical environment makes for great outdoor activity year-round, both on and off the thousands of miles of beach. The area's ethnic and cultural diversity invites a wide variety of shopping, dining, music, and theater, as well as festivals and fairs such as Carnaval Miami, a ten-day festival celebrating Latin American culture.

The University

The University of Miami was chartered in 1925 by a group of citizens who felt an institution of higher learning was needed for the development of their young and growing community. Today, this privately supported, nonsectarian institution enrolls more than 15,200 students from around the world in approximately 340 areas of study. In 2004, *U.S. News & World Report*'s Best Hospitals issue listed eight UM specialties among the nation's best, including the Bascom Palmer Eye Institute as the number one eye hospital.

Applying

To be considered for admission, applicants must submit an application for admission, official transcripts showing a GPA of 3.0 or higher and an undergraduate degree in one of the biological or natural sciences, GRE scores, two letters of recommendation, a personal statement, a resume, and a $50 application fee. Admission is based on an applicant's academic record, recommendations from their mentors, and an in-person interview. The application deadline is January 10.

Correspondence and Information

Jeffrey Milner, Graduate Program Coordinator
Sylvester Comprehensive Cancer Center
Miller School of Medicine
University of Miami
P.O. Box 016960
Miami, Florida 33101

Phone: 305-243-8533
Fax: 305-243-5555
E-mail: jmilner@med.miami.edu
Web site: http://www.biomed.miami.edu/cab

University of Miami

THE FACULTY AND THEIR RESEARCH

Michael H. Antoni, Professor of Psychology and Psychiatry; Ph.D., Miami, 1986. Psychooncology; psychoneuroimmunology.

Glen Barber, Associate Professor of Microbiology and Immunology; Ph.D., London, 1988. Viral oncology; viruses as therapeutic agents.

Lisa L. Baumbach-Reardon, Associate Research Professor of Pediatrics, Neurology, and Biochemistry and Molecular Biology; Ph.D., Florida, 1986. Genetic basis and molecular pathophysiology of breast cancer.

Larry Boise, Associate Professor of Microbiology and Immunology; Ph.D., Virginia Commonwealth, 1991. Cell death; multiple myeloma; proteasome inhibitor.

Karoline Briegel, Assistant Professor of Biochemistry and Molecular Biology; Ph.D., Vienna, 1995. Transcription; mammary development and breast cancer; mouse models of mammogenesis and breast cancer.

Kerry Burnstein, Professor of Molecular and Cellular Pharmacology; Ph.D., North Carolina at Chapel Hill, 1986. Signaling mechanisms; prostate cancer cell cycle; steroid hormone responsiveness; vitamin D.

Kermit Carraway, Professor of Cell Biology and Anatomy; Ph.D., Illinois, 1966. Mammary epithelial and cancer cells; ligand for receptor tyrosine kinase ErbB2/HER2/Neu; glycoprotein complex, Muc4.

Gennaro D'Urso, Assistant Professor of Molecular and Cellular Pharmacology; Ph.D., Washington, 1990. Molecular mechanisms; DNA polymerase replication initiation; mitosis; Pol epsilon.

Ahmjad Farooq, Assistant Professor of Biochemistry and Molecular Biology; Ph.D., London, 1988. Protein structure; function and mechanism of signal transduction.

Edward Harhaj, Assistant Professor of Microbiology and Immunology; Ph.D., Penn State, 1999. Retrovirus; T-cell leukemia (ATL); HTLV-I–associated myelopathy/tropical spastic paraparesis.

T. K. Harris, Assistant Professor of Biochemistry and Molecular Biology; Ph.D., Mississippi Medical Center, 1994. NMR and kinetic studies of enzyme signaling mechanisms; tumor-suppressor gene.

David Helfman, Professor of Cell Biology and Anatomy; Ph.D., Emory, 1981. Oncogenic *ras*; cytoskeleton; signal transduction; apoptosis.

Jennifer Hu, Professor of Epidemiology and Public Health; M.P.H., Ph.D, University of Medicine and Dentistry of New Jersey, 1988. Molecular epidemiology; DNA damage/repair; human cancer risk assessment and prevention.

Roland Jurecic, Associate Professor of Microbiology and Immunology; Ph.D., Zagreb (Croatia), 1991. Stem cells; pluripotent cells; stem cell transplantation; gene therapy; hematopoietic stem cells.

Erin Kobetz, Assistant Professor of Epidemiology and Public Health; Ph.D., North Carolina at Chapel Hill, 2004. Racial and socioeconomic disparities in cancer outcomes; epidemiology and public health intervention.

Leonidas Koniaris, Associate Professor of Surgical Oncology and of Cell Biology and Anatomy; M.D., Johns Hopkins. Growth regulation in the liver; mechanisms of cancer-associated wasting.

Ted Lampidis, Professor of Cell Biology and Anatomy; Ph.D., Miami, 1974. Chemotherapeutic agents; p-glycoprotein (P-gp); mediated multiple drug resistance; hypoxia; mitochondrial.

Robert Levy, Professor of Microbiology and Immunology; Ph.D., Ohio State, 1975. T cells in bone marrow transplantation and cancer; graft-versus-host disease in models of allogeneic bone-marrow transplantation.

Jie Li, Assistant Professor of Dermatology and Cutaneous Surgery and of Cell Biology and Anatomy; Ph.D., Cincinnati, 1995. Angiogenesis; tumor biology; skin disease; cutaneous biology.

Balakrishna L. Lokeshwar, Associate Professor of Urology and of Radiation Oncology; Ph.D., Indian Institute of Science, 1984. Prostate cancer; diagnostic markers; mechanism of growth.

Vinata Lokeshwar, Associate Professor of Urology and of Cell Biology and Anatomy; Ph.D., Saint Louis, 1989. Diagnostic markers; mechanism of growth.

Diana Lopez, Professor of Microbiology and Immunology; Ph.D., Miami, 1970. Breast cancer; mammary tumors; tumor immunology; T lymphocytes; natural killer cells; immunotherapeutic.

Izidore Lossos, Associate Professor of Clinical Medicine and of Molecular and Cellular Pharmacology; M.D., Hebrew University Hadassah (Israel), 1987. Immunology; lymphoma; signal transduction.

Thomas Malek, Vice Chair of Microbiology and Immunology; Ph.D., Illinois, 1977. Cytokine receptor regulation of T-lymphocyte development, activation, and memory; T-regulatory cells in suppression of autoimmunity.

Enrique A. Mesri, Associate Professor of Microbiology and Immunology; Ph.D., Buenos Aires (Argentina), 1990. Kaposi's sarcoma; angiogenesis; vGPCR angiogenesis; tumorigenesis; VEGF.

Carlos Moraes, Professor of Neurology and of Cell Biology and Anatomy; Ph.D., Columbia, 1993. Human mitochondrial DNA.

Zafar Nawaz, Associate Professor of Biochemistry and Molecular Medicine; Ph.D., North Texas, 1992. Mechanisms of steroid hormone receptor; estrogen receptor (ER) regulation in breast cancer and androgen receptor (AR) regulation in prostate cancer.

Eckhard R. Podack, Professor and Chair, Microbiology and Immunology; Ph.D., Göttingen (Germany), 1972. Immunotherapy for non–small-cell lung cancer (NSCLC), T cells; transgenic expression; heat-shock proteins.

Joseph Rosenblatt, Professor of Medicine; M. D., UCLA, 1990. Hematologic malignancies; development of novel approaches to breast cancer and solid tumors; gene therapy and immunotherapy of cancer; human retroviruses; immune therapies for cancer and human gene therapy; human T-cell leukemia virus type 1.

Michael Schmale, Professor, Division of Marine Biology and Fisheries; Ph.D., Miami, 1985. Marine animal models; cancer; molecular biology; virology.

Sean Scully, Professor of Orthopaedics and Rehabilitation, of Cell Biology and Anatomy, and of Cellular and Molecular Pharmacology; M.D., Ph.D., Rochester, 1986. Ewing's sarcoma; chondrosarcoma; TGF-beta1 and ECM signals; extracellular matrix; metastasis; signaling; cell-signal transduction.

Rakesh Singal, Associate Professor of Medicine and of Molecular and Cellular Pharmacology; M.B.B.S., 1984, M.D., 1987, Maulana Azad Medical College (India); MRCP. Prostate cancer; epigenetics; transcriptional regulation; DNA methylation; chromatin structure; biomarkers.

Joyce Slingerland, Professor of Medicine and of Biochemistry and Molecular Biology; M.D., 1983, Ph.D., 1992, Toronto; FRCP(C). Breast cancer; molecular mechanisms; molecular genetics; epidemiology; cell cycle; estrogen receptors; FRCP.

Keith Webster, Associate Professor of Molecular and Cellular Pharmacology; Ph.D., York (England). Therapeutic angiogenesis; molecular mechanisms of hypoxia/ischemia-regulated gene expression; pathology of vascular disease; cell death.

Teresa Zimmers, Assistant Professor of Medicine and Co-Director of Research, Division of Surgical Oncology; M.D., Johns Hopkins, 2001. Cancer cachexia; muscle wasting; transcription regulation; interleukin-6; STAT3; myostatin; insulin.

Physician Mentors and Specialties

Eli Avisar, Assistant Professor of Surgery and Co-Leader, Breast Site Disease Group; M.D. Cancer surgery, including breast cancer, melanoma and sarcoma, esophageal, stomach, liver, and pancreatic tumors.

Peter Cassileth, Professor of Medicine; M.D. Lymphoma; leukemia; bone marrow transplantation; breast cancer.

Lynn Feun, Professor of Medicine; M.D. Neuro-oncology and specialization in melanoma, gliomas, hepatomas, regional perfusion therapy.

Elizabeth Franzmann, Assistant Professor of Otolaryngology; M.D. All aspects of head and neck surgery.

Stefan Gluck, Professor of Medicine and Clinical Director, Braman Family Breast Cancer Institute; M.D. Breast cancer; head and neck cancer; hematopoietic stem cell transplantation; phase-I studies with pharmacokinetics; therapeutic drug monitoring.

Mark Goodman, Assistant Professor of Medicine; M.D. Hematologic malignancies; solid tumors; bone marrow transplant.

W. Jarrard Goodwin, Professor of Otolaryngology and Director, UM/Sylvester; M.D. Head and neck cancer surgery.

William Harrington, Professor of Medicine; M.D. Hematology; hematological malignancies; HIV- and human T-cell leukemia virus type I (HTLV-I)-related lymphoma.

Merce Jorda, Assistant Professor of Pathology; M.D., Ph.D. Cytopathology, molecular pathology, immunocytochemistry, pediatric pathology, cytogenetics.

Leonidas Koniaris, Associate Professor of Surgery, Associate Professor of Cell Biology and Anatomy, and Alan Livingstone Chair in Surgical Oncology; M.D. Growth regulation in the liver; mechanisms of cancer associated wasting; hepatobiliary diseases; pancreatic disease; gastrointestinal malignancies; breast cancer.

Izidore Lossos, Associate Professor of Clinical Medicine, Division of Hematology/Oncology and Associate Professor of Molecular and Cellular Pharmacology; M.D. Immunology; lymphoma; signal transduction; Hodgkin's disease; chronic lymphocytic leukemia (CLL).

Joseph Lucci, Clinical Professor (pending rank) and Director, Division of Gynecologic Oncology; M.D. Cervial cancer; cancer immunology; human papillomavirus; tumor vaccines.

Caio Rocha-Lima, Associate Professor of Clinical Medicine; M.D. GI oncology; thoracic oncology; phase I/drug development.

Joseph D. Rosenblatt, Professor of Medicine, Division of Hematology-Oncology, and UM/Sylvester Associate Director for Clinical and Translational Research; M.D. Hematologic malignancies; development of novel approaches to breast cancer and solid tumors; gene therapy and immunotherapy of cancer; human retroviruses; immune therapies for cancer and human gene therapy; human T-cell leukemia virus type 1.

Niramol Savaraj, Adjunct Professor of Medicine; M.D. Brain tumors; lung, colon, and head and neck cancers.

Sean P. Scully, Professor of Orthopaedics and Rehabilitation, Professor of Cell Biology and Anatomy, and Professor of Cellular and Molecular Pharmacology; M.D., Ph.D. Ewing's sarcoma; chondrosarcoma; TGF-beta1and ECM signals; extra-cellular matrix, metastasis; signaling; cell signal transduction.

Rakesh Singal, Associate Professor of Medicine, Division of Hematology-Oncology and Associate Professor of Molecular and Cellular Pharmacology; M.D. Prostate cancer; epigenetics; transcriptional regulation; DNA methylation; chromatin structure; biomarkers.

Joyce Slingerland, Professor of Medicine and Director, Braman Family Breast Cancer Institute, UM/Sylvester; M.D., Ph.D.; FRCP(C). Breast cancer; molecular mechanisms; molecular genetics; epidemiology; cell cycle; estrogen receptors.

UNIVERSITY OF MICHIGAN

Department of Molecular, Cellular and Developmental Biology

Programs of Study

The goal of the Ph.D. program in the Department of Molecular, Cellular and Developmental Biology (MCDB) at the University of Michigan is to train scientists who perform cutting-edge research in a diverse set of biological questions. The MCDB faculty is broadly interested in how organisms, cells, molecules, and genomes function, develop, and evolve. The faculty shares technical approaches, such as recombinant DNA, genetics, biochemistry, and specialized imaging. Due to the extensive pursuit of research in the life sciences at the University of Michigan, students have available to them an unusually rich program of courses and seminars and a wide range of technical expertise to complement their training within MCDB.

A master's degree program in molecular, cellular, and developmental biology is also offered. This program is flexible so that students can meet a variety of professional objectives, such as teaching, assisting in laboratory research, or technological training. The master's degree is not a prerequisite for admission to the doctoral program, nor is admission to the master's program intended as probationary admission to the doctoral program.

Research Facilities

The Kraus Natural Science Building is equipped for experimental work in cell and molecular biology, genetics, neurobiology, physiology, and related fields. This building has undergone a complete renovation that makes its laboratories some of the best-equipped in the nation.

Financial Aid

The Department fully supports Ph.D. students during their graduate work through a combination of fellowships, graduate research assistantships, and graduate teaching assistantships. Financial support includes stipends, tuition and fees, and health benefits. The Department does not support master's students. Domestic students may contact the Financial Aid Office via its Web site (http://www.ofa.umich.edu).

Cost of Study

During 2006–07, full-time tuition at the University of Michigan is listed tentatively at $7496 per semester for residents and $15,069 per semester for nonresidents of the state of Michigan.

Living and Housing Costs

Accommodations for students are available both on and off campus. Rates for University housing vary, depending upon the type chosen. The cost of off-campus rooms and apartments varies according to the type chosen and its proximity to the campus. Rents are, however, comparable to those commonly found in other university communities.

Student Group

The MCDB Department currently has 67 Ph.D. and 13 M.S. students; of the total, 49 are women and 46 are international students.

Location

The University and the Ann Arbor community support an abundance of cultural activities that appeal to many interests. Hundreds of musical and theatrical performances are offered each year, and film aficionados enjoy an immense range of viewing fare. Artists, painters, potters, photographers, sculptors, and weavers all flourish in Ann Arbor, a community that hosts one of the country's leading art fairs every year. Equidistant from Chicago and Toronto and only 50 miles from Detroit, the University of Michigan derives many advantages from its geographical location. Access to Detroit's renowned cultural centers and its rich ethnic heritage greatly enhances student life.

The University

The Ann Arbor campus currently enrolls more than 35,000 students. The hub of this campus is the Central Campus area, which is connected by a free bus system to the North, Medical, and Athletic Campuses. The University offers an almost unlimited choice of fields of study in undergraduate, graduate, and professional education. It is also one of the nation's major research institutions. Its large number of institutes, laboratories, collections, museums, centers, and libraries provide excellent facilities for research in all fields of academic and scientific endeavor.

Applying

New Ph.D. students are admitted only in the fall term. Applications received by January 5 are given priority. Offers of admission are generally made in the second half of March.

M.S. students are admitted in the fall and winter terms. Applications for the fall term must be received by April 1; for the winter term, by November 1.

Correspondence and Information

Graduate Coordinator
Department of Molecular, Cellular and Developmental Biology
University of Michigan
830 North University Avenue
Ann Arbor, Michigan 48109-1048
Phone: 734-615-1635
Web site: http://www.mcdb.lsa.umich.edu

University of Michigan

THE FACULTY AND THEIR RESEARCH

Julian Adams, Professor; Ph.D., California, Davis, 1969. Population genetics; microbial evolution.
Robert A. Bender, Professor; Ph.D., MIT, 1976. Regulation of gene expression.
Richard I. Hume, Professor; Ph.D., Stanford, 1980. Cellular, molecular, and developmental neurobiology.
Daniel Klionsky, Professor; Ph.D., Stanford, 1986. Cell biology.
John Y. Kuwada, Professor; Ph.D., Stanford, 1980. Molecular genetic analysis of vertebrate development.
Eran Pichersky, Professor; Ph.D., California, Davis, 1984. Plant molecular biology and evolution.
Pamela Raymond, Professor; Ph.D., Michigan, 1976. Developmental biology.
John W. Schiefelbein Jr., Professor; Ph.D., Wisconsin–Madison, 1987. Molecular and developmental genetics.
Charles F. Yocum, Professor; Ph.D., Indiana, 1971. Plant biochemistry; photosynthesis.
Jim Bardwell, Associate Professor; Ph.D., Wisconsin–Madison, 1987. Catalysis of protein folding.
Kenneth Cadigan, Associate Professor; Ph.D., Dartmouth, 1989. Genetic analysis of *Drosophila* development; signal transduction.
Amy Chang, Associate Professor; Ph.D., Yale, 1986. Cell biology.
Steven Clark, Associate Professor; Ph.D., Chicago, 1991. Plant development; molecular genetics; signal transduction.
Robert Denver, Associate Professor; Ph.D., Berkeley, 1989. Developmental neuroendocrinology.
Cunming Duan, Associate Professor; Ph.D., Tokyo, 1991. Molecular animal physiology.
Jianming Li, Associate Professor; Ph.D., Virginia, 1995. Plant molecular physiology.
Janine Maddock, Associate Professor; Ph.D., Carnegie Mellon, 1991. Microbial development.
Laura J. Olsen, Associate Professor; Ph.D., Wisconsin–Madison, 1989. Plant cell and molecular biology.
Mohammed Akaaboune, Assistant Professor; Ph.D., Paris VI (Curie), 1996. Neurobiology.
Matt Chapman, Assistant Professor; Ph.D., Indiana, 1998. Molecular physiology.
Gyorgyi Csankovszki, Assistant Professor; Ph.D., MIT, 2001. Cell biology.
Jonathan Demb, Assistant Professor, Ph.D., Stanford, 1997. Developmental neurobiology.
Ursula Jakob, Assistant Professor; Ph.D., Regensburg (Germany), 1995. Protein biochemistry; molecular cell biology.
Anuj Kumar, Assistant Professor; Ph.D., Wright State, 1997. Yeast genomics.
Tzvi Tzfira, Assistant Professor; Ph.D., Hebrew (Israel), 1999. Plant biology.
Yanghuzng Wang, Assistant Professor; Ph.D., Heidelberg, 1999. Developmental neurobiology.
Gisela Wilson, Assistant Professor; Ph.D., Wisconsin–Madison, 1990. Animal physiology.

UNIVERSITY OF MISSOURI–KANSAS CITY

School of Biological Sciences
Program in Cell Biology and Biophysics

Program of Study	The graduate program in cell biology and biophysics at the University of Missouri–Kansas City (UMKC) leads to the Ph.D. degree. The program functions within the interdisciplinary Ph.D. framework of the University and is associated with the M.S. program in cell and molecular biology. The graduate program is designed to prepare students for research-oriented careers in academe, government, or the private sector. An original independent research project under the supervision of a faculty adviser is the core of these programs. Programs of study provide a background of course work tailored to the interests of each student. Opportunity for research experience begins immediately as a component of the first-year curriculum, with each student being assigned short research projects. By the end of the first academic year, the student is also expected to have acquired a general understanding of the basis of molecular and cellular biology. At that time, the student selects a faculty research adviser and makes further course selections. To qualify for doctoral degree candidacy, students take a written comprehensive examination and prepare and defend an original research proposal. The culmination of the graduate degree programs is the preparation and oral defense of a research dissertation, typically five years after entry into the program. Areas of research interest are included in the list of participating faculty below. Extensive possibilities for collaboration exist with the School's program in molecular biology and biochemistry and with regional research associates. Opportunities for postdoctoral research are abundant.
Research Facilities	Research facilities for cell, molecular and structural biology, and biochemistry are primarily located in the Biological Sciences and Chemistry buildings. Modern research is conducted in laboratories assigned to individual faculty members and in specialized central facilities. Sophisticated instrumentation in these facilities includes automated DNA and protein synthesizers and sequencers, mass spectrometers, macromolecular X-ray, low-intensity electron microscope and 600-MHz NMR imaging facilities, molecular graphics equipment, and Fourier-transform infrared and EPR spectrometers. Raman and UV-resonance Raman spectrometers, differential scanning and titration microcalorimeters, analytical ultracentrifuge, HPLCs, amino acid and carbohydrate analyzers, low-intensity fluorescence imaging and confocal microscopes, and a large assortment of scanning spectrophotometers, ELISA readers, gel scanners, centrifuges, and related instrumentation associated with modern biochemical research are available. Students also enjoy the use of Linda Hall Library, one of the country's premier private science libraries; central animal facilities; and a fully integrated computer network with on-site and off-site access to national and international databases and the Internet.
Financial Aid	All fully admitted U.S. citizen and resident doctoral students receive financial support as teaching or research assistants. Support is provided for up to five years for students who are progressing satisfactorily. For the 2004–05 year, stipends were $19,000–$20,000 (depending on qualifications). Other forms of financial aid may be available through the Student Financial Aid Office. The metropolitan area offers many career and educational opportunities for spouses and family.
Cost of Study	In 2004–05, in-state tuition was about $4000 per year, while out-of-state fees were approximately $12,000 per year. Full-time graduate students, as a general rule, receive basic tuition support.
Living and Housing Costs	A wide variety of off-campus housing is available in every price range. The overall cost of living in Kansas City is low compared with metropolitan areas in other parts of the country.
Student Group	The cell biology and biophysics graduate program has a very active graduate student organization. UMKC has approximately 10,000 students, of whom about half are graduate and professional students. The School of Biological Sciences currently has about 60 graduate students and 30 postdoctoral fellows as well as more than 200 undergraduate majors. Ten new doctoral students are admitted each year.
Student Outcomes	The majority of doctoral graduates transfer to nationally known research institutions, typically as postdoctoral associates, or undertake advanced professional training. A short transitional postdoctoral research period within the School is not uncommon.
Location	Kansas City, "The Heart of America," is the center of a metropolitan area with a population of more than 1 million. The University is adjacent to the elegant Country Club Plaza, the city's entertainment and shopping center. Major-league sports, historical and art museums, and many musical, theatrical, and cultural events as well as an extensive parks system provide entertainment throughout the year. A relaxed, Midwestern lifestyle is also an advantage of the setting, which, with its many fountains, boulevards, unusually clean air, and more days of sunshine than in most large U.S. cities, provides an enjoyable quality of life.
The University and The School	UMKC is part of the four-campus University of Missouri System. UMKC is the only comprehensive research university in western Missouri. It has a strong life science mission. The School of Biological Sciences was established in 1985 to develop strong research and graduate programs in modern life sciences. The School has been cited by the Board of Curators of the University of Missouri System as an area of eminence for its programs in molecular biology and biochemistry and in cell biology and biophysics. Program improvement funds have facilitated the hiring of many research-oriented faculty members and the creation of excellent research facilities. An innovative interdisciplinary doctoral program has also been initiated, creating a stimulating environment that offers outstanding research opportunities to graduate students.
Applying	The deadline for applications from U.S. applicants is July 1, but applications received before March 1 have priority for financial support. The deadline for international applications is February 15. A bachelor's degree in biology, chemistry, physics, or a related discipline with a minimum 3.0 grade point average is required for full admission. The General Test of the Graduate Record Examinations is also required. The TOEFL is required for international applicants whose native language is not English.
Correspondence and Information	Graduate Advisor School of Biological Sciences University of Missouri–Kansas City Kansas City, Missouri 64110-2499 Phone: 816-235-2352 Fax: 816-235-5158 E-mail: sbs-grad@umkc.edu Web site: http://sbs.umkc.edu/graduate/

University of Missouri–Kansas City

THE FACULTY AND THEIR RESEARCH

Professors

Lawrence A. Dreyfus, Ph.D., Kansas. Molecular biology; bacterial toxin structure-function.
Alfred F. Esser, Ph.D., Frankfurt (Germany). Membrane biochemistry; complement proteins; cell aging.
William T. Morgan, Ph.D., California, Santa Barbara. Protein structure; site-directed mutagenesis; protein mechanisms of action.
George J. Thomas Jr., Ph.D., MIT. Protein–nucleic acid recognition; virus structure and assembly; Raman spectroscopy of biological molecules.

Associate Professors

Karen J. Bame, Ph.D., UCLA. Metabolism of heparan sulfate proteoglycans.
Antony A. Cooper, Ph.D., McGill. Intracellular protein trafficking; yeast genetics and molecular biology.
Edward P. Gogol, Ph.D., Yale. Structure of macromolecular assemblies; cryoelectron microscopy.
Jeffrey P. Gorski, Ph.D., Wisconsin–Madison. Matrix phosphoproteins structure and function.
Thomas M. Menees, Ph.D., Yale. Replication of retroviral elements and transposons; yeast molecular genetics.
Anthony J. Persechini, Ph.D., Carnegie Mellon. Calcium-calmodulin signaling pathways; intracellular interactions.
Lynda S. Plamann, Ph.D., Iowa. Cell-cell communication during fruiting body formation and sporulation in the soil bacterium *Myxococcus xanthus*.
G. Sullivan Read, Ph.D., Penn State. RNA turnover control; gene regulation; herpes virus.
Ann Smith, Ph.D., London. Receptor-mediated endocytosis; protein-receptor interactions; intercellular heme transport.
Jakob H. Waterborg, Ph.D., Nijmegen (Netherlands). Plant histones; chromatin conformation and gene expression.
Marilyn D. Yoder, Ph.D., California, Riverside. X-ray crystallography; protein structure.

Assistant Professors

Leonard L. Dobens Jr., Ph.D., Dartmouth. Pattern formation; cell-cell signaling.
Michael B. Ferrari, Ph.D., Texas at Austin. Regulation of skeletal muscle development.
Saul M. Honigberg, Ph.D., Yale. Signal transduction; cell-cycle control and cell differentiation.
Stephen J. King, Ph.D., Colorado at Boulder. Protein interactions during intracellular motility.
John H. Laity, Ph.D., Cornell. Molecular recognition; NMR spectroscopy; protein biophysical chemistry.
Michael O'Connor, Ph.D., Ireland. Structure and function of the bacterial ribosome, ribosomal subunits, and the translational reading frame.
Michael D. Plamann, Ph.D., Iowa. Microtubule-associated motors; organelle movement; growth polarity; cytoskeleton.
Jeffrey L. Price, Ph.D., Johns Hopkins. *Drosophila* genes involved in chronobiology and circadian rhythms.
Gerald J. Wyckoff, Ph.D., Chicago. Bioinformatics and study of molecular evolution through large-scale comparative genomics in sexual selection.
Xiao-Qiang Yu, Ph.D., Kansas State. Insect molecular biology and biochemistry of immune responses, pattern recognition proteins and protein-protein/ligand interactions.

Regional Associates

James Coffman, Ph.D., Duke. Regulation of cell proliferation and differentiation during sea urchin development.
Mark Fisher, Ph.D., Illinois. Chaperonin-assisted protein folding and oligomer assembly.
George Helmkamp Jr., Ph.D., Harvard. Transport of phospholipids; phosphatidylinositol transfer proteins.
Lynwood Yarbrough, Ph.D., Purdue. Microtubule structure; cardiac gene expression in stress and aging.

UNIVERSITY OF MISSOURI–KANSAS CITY

School of Biological Sciences
Program in Molecular Biology and Biochemistry

Programs of Study

The graduate program in molecular biology and biochemistry at the University of Missouri–Kansas City (UMKC) leads to the Ph.D. degree. The program functions within the interdisciplinary Ph.D. framework of the University and is associated with the M.S. program in cell and molecular biology. The graduate program is designed to prepare students for research-oriented careers in academe, government, or the private sector. An original independent research project under the supervision of a faculty adviser is the core of these programs.

Programs of study provide a background of course work tailored to the interests of each student. Opportunity for research experience begins immediately as a component of the first-year curriculum, with each student being assigned short research projects. By the end of the first academic year, the student is also expected to have acquired a general understanding of the basis of molecular and cellular biology. At that time, the student selects a faculty research adviser and makes further course selections. To qualify for doctoral degree candidacy, students take a written comprehensive examination and prepare and defend an original research proposal. The culmination of the graduate degree programs is the preparation and oral defense of a research dissertation, typically five years after entry into the program.

Areas of research interest are included in the list of participating faculty members below. Extensive possibilities for collaboration exist with the School's program in cell biology and biophysics and with regional research associates. Opportunities for postdoctoral research are abundant.

Research Facilities

Research facilities for cell, molecular, and structural biology and biochemistry are located primarily in the Biological Sciences and Chemistry buildings. Modern research is done in laboratories assigned to individual faculty members or in specialized central facilities. Sophisticated instrumentation in these facilities includes automated DNA and protein synthesizers and sequencers, mass spectrometers, macromolecular X-ray, low-intensity electron microscope and 600-MHz NMR imaging facilities, molecular graphics equipment, and Fourier-transform infrared and EPR spectrometers. Raman and UV-resonance Raman spectrometers, differential scanning and titration microcalorimeters, analytical ultracentrifuge, HPLCs, amino-acid and carbohydrate analyzers, low-intensity fluorescence imaging and confocal microscopes, and a large assortment of scanning spectrophotometers, ELISA readers, gel scanners, centrifuges, and related instrumentation associated with modern biochemical research are also available. Students enjoy the use of Linda Hall Library, one of the country's premier private science libraries; central animal facilities; and a fully integrated computer network with on-site and off-site access to national and international databases and the Internet.

Financial Aid

All fully admitted U.S. citizen and resident doctoral students receive financial support as teaching or research assistants. Support is provided up to five years for students who are progressing satisfactorily. For the 2004–05 year, stipends were $19,000 to $20,000 (depending on qualifications). Other forms of financial aid may be available through the Student Financial Aid Office. The metropolitan area offers many career and educational opportunities for spouses and families.

Cost of Study

In 2004–05, in-state tuition was about $4000 per year, while out-of-state fees were approximately $12,000 per year. Full-time graduate students, as a general rule, receive basic tuition support.

Living and Housing Costs

A wide variety of off-campus housing is available in every price range. The overall cost of living in Kansas City is low compared with metropolitan areas in other parts of the country.

Student Group

The molecular biology and biochemistry graduate program has a very active graduate student organization. UMKC has approximately 10,000 students, of whom about half are graduate and professional students. The School of Biological Sciences currently has about 60 graduate students and 30 postdoctoral fellows as well as more than 200 undergraduate majors. Ten new doctoral students are admitted each year.

Student Outcomes

The majority of doctoral graduates transfer to nationally known research institutions, typically as postdoctoral associates, or undertake advanced professional training. A short transitional postdoctoral research period within the School is not uncommon.

Location

Kansas City, "The Heart of America," is the center of a metropolitan area with a population of more than 1 million. The University is adjacent to the elegant Country Club Plaza, the city's entertainment and shopping center. Major-league sports, historic and art museums, and many musical, theatrical, and cultural events as well as an extensive parks system provide entertainment throughout the year. A relaxed, Midwestern lifestyle is also an advantage of the setting, which, with its many fountains, boulevards, unusually clean air, and more days of sunshine than in most large U.S. cities, provides an enjoyable quality of life.

The University and The School

UMKC is part of the four-campus University of Missouri system, and it is the only comprehensive research university in western Missouri. It has a strong life science mission. The School of Biological Sciences was established in 1985 to develop strong research and graduate programs in the modern life sciences. The School has been cited by the Board of Curators of the University of Missouri System as an area of eminence for its programs in molecular biology and biochemistry and in cell biology and biophysics. Program improvement funds have facilitated the hiring of many research-oriented faculty members and the creation of excellent research facilities. An innovative interdisciplinary doctoral program has also been initiated, creating a stimulating environment that offers outstanding research opportunities to graduate students.

Applying

The deadline for applications from U.S. applicants is July 1, but applications received before March 1 have priority for financial support. The deadline for international applications is February 15. A bachelor's degree with a minimum 3.0 grade point average in biology, chemistry, physics, or a related discipline is required for full admission. The Graduate Record Examinations General Test is required. The TOEFL is required for international applicants whose native language is not English.

Correspondence and Information

Graduate Advisor
School of Biological Sciences
University of Missouri–Kansas City
Kansas City, Missouri 64110-2499

Phone: 816-235-2352
Fax: 816-235-5158
E-mail: sbs-grad@umkc.edu
Web site: http://sbs.umkc.edu/graduate/

University of Missouri–Kansas City

THE FACULTY AND THEIR RESEARCH

Professors
Lawrence A. Dreyfus, Ph.D., Kansas. Molecular biology; bacterial toxin structure-function.
Alfred F. Esser, Ph.D., Frankfurt (Germany). Membrane biochemistry; complement proteins; cell aging.
William T. Morgan, Ph.D., California, Santa Barbara. Protein structure; site-directed mutagenesis; protein mechanisms of action.
George J. Thomas Jr., Ph.D., MIT. Protein–nucleic acid recognition; virus structure and assembly; Raman spectroscopy of biological molecules.

Associate Professors
Karen J. Bame, Ph.D., UCLA. Metabolism of heparan sulfate proteoglycans.
Antony A. Cooper, Ph.D., McGill. Intracellular protein trafficking; yeast genetics and molecular biology.
Edward P. Gogol, Ph.D., Yale. Structure of macromolecular assemblies; cryoelectron microscopy.
Jeffrey P. Gorski, Ph.D., Wisconsin–Madison. Matrix phosphoproteins structure and function.
Thomas M. Menees, Ph.D., Yale. Replication of retroviral elements and transposons; yeast molecular genetics.
Anthony J. Persechini, Ph.D., Carnegie Mellon. Calcium-calmodulin signaling pathways; intracellular interactions.
Lynda S. Plamann, Ph.D., Iowa. Cell-cell communication during fruiting body formation and sporulation in the soil bacterium *Myxococcus xanthus*.
G. Sullivan Read, Ph.D., Penn State. RNA turnover control; gene regulation; herpes virus.
Ann Smith, Ph.D., London. Receptor-mediated endocytosis; protein-receptor interactions; intercellular heme transport.
Jakob H. Waterborg, Ph.D., Nijmegen (Netherlands). Plant histones; chromatin conformation and gene expression.
Marilyn D. Yoder, Ph.D., California, Riverside. X-ray crystallography; protein structure.

Assistant Professors
Leonard L. Dobens Jr., Ph.D., Dartmouth. Pattern formation; cell-cell signaling.
Michael B. Ferrari, Ph.D., Texas at Austin. Regulation of skeletal muscle development.
Saul M. Honigberg, Ph.D., Yale. Signal transduction; cell-cycle control and cell differentiation.
Stephen J. King, Ph.D., Colorado at Boulder. Protein interactions during intracellular motility.
John H. Laity, Ph.D., Cornell. Molecular recognition; NMR spectroscopy; protein biophysical chemistry.
Michael O'Connor, Ph.D., Ireland. Structure and function of the bacterial ribosome; ribosomal subunits; and the translational reading frame.
Michael D. Plamann, Ph.D., Iowa. Microtubule-associated motors; organelle movement; growth polarity; cytoskeleton.
Jeffrey L. Price, Ph.D., Johns Hopkins. *Drosophila* genes involved in chronobiology and circadian rhythms.
Gerald J. Wyckoff, Ph.D., Chicago. Bioinformatics and study of molecular evolution through large-scale comparative genomics in sexual selection.
Xiao-Qiang Yu, Ph.D., Kansas State. Insect molecular biology and biochemistry of immune responses, pattern recognition proteins, and protein-protein/ligand interactions.

UNIVERSITY OF NEW HAVEN

Program in Cellular and Molecular Biology

Program of Study	The Master of Science (M.S.) Program in Cellular and Molecular Biology at the University of New Haven (UNH) is intended for those individuals interested in the rapidly expanding fields of genomics, proteomics, and bioinformatics, as well as pharmacological research. The level of experience required for an individual to contribute in these fields is not satisfied by an undergraduate degree, thus individuals with advanced training are in demand.

This program's strong emphasis on cell and molecular-biology techniques provides students with the preparation necessary to meet this need for advanced training. The central curriculum consists of courses in cell and molecular biology, genomics, proteomics, bioinformatics, biochemistry, and pharmacology. These courses help develop the student's ability to function as an independent scientist by stressing both the conceptual and technical aspects of each subject.

A minimum of 38 credit hours of graduate work must be completed to earn the degree. This includes eight required courses and at least four elective courses. The research requirement may be satisfied by completion of a research project or an internship for those who are employed or interested in learning about academic or industrial research environments, or the requirement may be met by a thesis for students who intend to pursue a doctoral degree.

Students who elect to write a thesis must take Thesis I and II for 6 credits and complete a minimum of 41 credit hours of graduate work. Thesis preparation and submission must comply with the Graduate School policy on theses as well as all specific department requirements. Successful defense of the thesis must be completed at least three weeks prior to the date of commencement.

The program at UNH is on an accelerated schedule with three full trimesters a year. Part-time students typically complete the degree in three years and full-time students in about sixteen months.

Research Facilities

The Department of Biology and Environmental Sciences has several research and teaching laboratories, including a cell-culture facility and graduate research and computer laboratories. The Marvin K. Peterson Library contains more than 500,000 volumes, subscribes to hundreds of journals, and is a U.S. government documents depository library. Computer workstations are available in the library to access online subscription databases, CDs, and DVDs. In addition, there are more than a dozen computer labs for student use and teaching on campus. Students who complete a research project may do so either at the University or at a laboratory that has been approved by the adviser.

Financial Aid

About 75 percent of all University students receive some form of financial aid. Qualified graduate students are entitled to receive up to $18,500 per year in subsidized and unsubsidized federal student loans. Teaching, research, or administrative assistantships are available to full-time students. Compensation includes $7 per hour as well as a 50 percent tuition reduction; students typically work 15–20 hours per week.

Cost of Study

Tuition for both full-time and part-time students is $520 per credit hour or $1560 per 3-credit course. There is also a technology fee of $20 and a Graduate Student Council fee of $15 each term. Laboratory fees range from $25 to $350.

Living and Housing Costs

On-campus housing for graduate students is currently not available. However, off-campus housing is available in the area at a cost of $575–$1000 per month for a one-bedroom apartment or $775–$1200 for a two-bedroom unit. The Office of Residential Life maintains a listing of vacancies submitted by landlords from the surrounding areas.

Student Group

Students in the program attend on both a full-time and part-time basis. Many of the part-time students take advantage of the "evening only" or weekend schedule of classes, which allows them to continue their full-time employment in the New Haven area's growing biotechnical and pharmaceutical industries.

Student Outcomes

Graduates of the program typically pursue further study in a related doctoral program or find employment in an academic or industrial research environment in fields such as biotechnology and pharmacological research.

Location

The UNH campus is in West Haven, Connecticut, which is contiguous to New Haven. New Haven has numerous art museums, a deepwater harbor and beaches, fine restaurants, parks and walking trails, and three Tony award–winning regional theaters. New Haven is served by a local airport and major railroads and is located at the junction of Interstates 91 and 95. The University of New Haven is within easy driving distance of New York, Boston, Cape Cod, and the ski areas of New England.

The University

The University of New Haven was founded on the Yale University campus in 1920 and became New Haven College in 1926. Today, it includes five undergraduate schools and a Graduate School, with a combined population of 4,500 students. Its programs prepare students to advance in their careers and meet the ever-changing demands of their respective fields. The University offers thirty master's degrees and thirty certificates as well as associate and bachelor's degrees.

Applying

Candidates must submit a completed application form, transcripts from all colleges and universities previously attended, two letters of recommendation, a resume, and a $50 application fee. GRE scores are recommended but not required. Candidates are expected to have a bachelor's degree in biology, chemistry, or a related discipline. Undergraduate course work should include general biology, advanced biology electives, general chemistry, and organic chemistry. It is recommended that applicants have taken courses in introductory statistics, calculus, molecular biology, and biochemistry. Applications may be submitted at any time, but full-time admission is granted for the fall trimester only.

Correspondence and Information

Dr. Eva Sapi, Coordinator
Graduate Programs in Cellular and Molecular Biology
University of New Haven
300 Boston Post Road
West Haven, Connecticut 06516

Phone: 203-479-4552
    800-DIAL-UNH Ext. 4552 (toll-free)
E-mail: esapi@newhaven.edu
Web site: http://www.evasapi.com
    http://www.unh-cmb.org (CMB Web site)
    http://www.unh-lyme.org (Lyme research Web site)

Dr. Pamela Sommers
Director of Graduate Admissions
University of New Haven
300 Boston Post Road
West Haven, Connecticut 06516

Phone: 203-932-7448
Fax: 203-932-7137
E-mail: psommers@newhaven.edu
Web site: http://admissions.newhaven.edu/grad

University of New Haven

THE FACULTY AND THEIR RESEARCH

Michael J. Rossi, Associate Professor and Chair; Ph.D. (biology), Kentucky. Cellular biology and biochemistry; genetic profile of differentiation pathways in smooth muscle cells.

Eva Sapi, Assistant Professor and Coordinator; Ph.D. (genetics), Eötvös Loránd (Budapest). Novel pathogens in Lyme and associated diseases; molecular mechanism of tumorigenicity; novel therapeutic approaches in ovarian cancer.

Ayesha Alvero, Associate Research Fellow; M.D., Yale.

Chiquito Crasto, Associate Research Scientist; Ph.D. (chemistry), Yale. Modeling and dynamic simulations of the olfactory receptors and odor molecules; genomics/proteomics study of olfactory receptors.

Krati Khanna-Gupta, Research Scientist; Ph.D. (biochemistry), Yale. Role played by C/EBPe and other members of the CCAAT/enhancer binding protein (C/EBP) family of transcription factors; downstream targets of C/EBPe; defects associated with the C/EBPe pathway in patients with the rare secondary granule deficiency (SGD) disease.

Joan B. Levy, Director of Cancer Research, Bayer Pharmaceutical; Ph.D. (biochemistry), Vermont. Contribution of tyrosine kinases and tyrosine phosphatases to malignant transformation; role of specific tyrosine kinases and signaling pathways in development of osteoporosis; identifying novel chemotherapeutics for treatment of cancer.

Anthony Melillo, Systems Programmer, Yale; M.S. (computer science), UNH. Integration of custom research databases with public genomic data repositories; storage and retrieval of high-throughput genotyping data.

Gil G. Mor, Associate Professor, Director of Reproductive Immunology Unit, and Director of Translational Research in Gynecologic Oncology; M.D., Hebrew University (Israel); Ph.D. (immunoendocrinology), Weizmann (Israel). Regulation of apoptosis in normal tissue remodeling and cancer of the female reproductive tract; sex hormones and the immune system; immunology of gynecologic tumors; immunology of implantation.

Pauline M. Schwartz, Associate Professor; Ph.D. (medicinal chemistry), Michigan. Pharmacology and medicinal chemistry; chemotherapeutic agents of human papilloma virus infections and cervical cancer; undergraduate research on kinetic models for chemical reactions that are analogous to a mathematical paradox.

David E. E. Sloane, Professor, Departments of English and Education; Ph.D., Duke. Dr. Sloane teaches Power Writing and Speaking for the UNH Graduate School and for business and industry.

UNIVERSITY OF NORTH CAROLINA AT CHAPEL HILL / NATIONAL INSTITUTES OF HEALTH

Graduate Program in Cell Motility and the Cytoskeleton

Program of Study	The movement of cells and organelles within cells is an essential feature of life. The partnership of the intramural research laboratories of the National Institutes of Health (NIH) and the University of North Carolina at Chapel Hill (UNC) in the area of cell motility and the cytoskeleton creates the novel opportunity to receive training with two major groups of faculty members with distinguished research records in understanding the basis of cell motility. An additional benefit of this program is the access to state-of-the-art imaging equipment available on both the NIH and UNC campuses. Students in this joint NIH/UNC program take their course work at the University of North Carolina at Chapel Hill and receive their Ph.D. degree from UNC. Courses at UNC are available in broad areas of cell and molecular biology, biochemistry, and neurobiology. Students take laboratory rotations with faculty members on both campuses and choose mentors based on their laboratory rotations. Joint projects between NIH and UNC faculty members are also featured, so students do their research under the codirection of 2 mentors. Advanced students participate in journal clubs and seminars to enhance their knowledge of the field and are supported to attend national and international meetings. The training program is open to students admitted to the departments represented by any of the faculty members in the program.
Research Facilities	The NIH is the world's premier biomedical research institution. As the federal government's primary agency for biomedical research, the NIH employs nearly 1,200 tenured or tenure-track investigators and more than 3,700 postdoctoral scientists with medical, dental, or graduate degrees. The NIH intramural research program, located on a 300-acre campus in Bethesda, Maryland, 10 miles from downtown Washington, D.C., has been the scene of many exciting scientific advances. It houses more than thirty research buildings comprising twenty-seven Institutes/Centers in a broad spectrum of biomedical and related scientific research. Four Nobel laureates made their prize-winning discoveries in NIH laboratories, and more than 100 received training at NIH. Basic research in the biomedical sciences at the NIH is complemented by an active clinical research program at the unique 250-bed research hospital and laboratory complex, the Warren Grant Magnuson Clinical Center. The NIH campus is also home to the National Library of Medicine, the world's largest medical library. The facilities at UNC Chapel Hill are complementary to those at NIH. Investigators at UNC Chapel Hill have state-of-the-art laboratories equipped for all of the modern techniques used to study cell and organelle motility. Further information is available at the UNC Web site (http://www.bio.unc.edu/graduate/MCDB/), which includes many of the investigators in this program, or through the Web sites of the individual investigators.
Financial Aid	UNC/NIH graduate students are supported jointly in a variety of ways. Sources may include fellowships, training grant support, research assistantships, and/or teaching assistantships. The level of support is highly competitive with other leading institutions and always includes a stipend, health benefits, and tuition support. Stipends for first-year students in this program are $24,800 and increase yearly based on the student's performance. Support is guaranteed for the duration of satisfactory graduate education.
Cost of Study	Students entering the program are completely supported by UNC/NIH mechanisms.
Living and Housing Costs	Apartments, houses, and rooms in private residences are available for rent near the UNC and NIH campuses. For information on available housing, prospective students should visit the Web sites of the University of North Carolina Chapel Hill (http://housing.unc.edu/grad/index.html), the Graduate Partnerships Program (GPP) (http://gpp.nih.gov), or the Web sites of local newspapers or contact the GPP.
Student Group	Students coming to NIH/UNC join thousands of other graduate students who are doing their research in UNC and NIH laboratories. While at UNC/NIH, graduate students enjoy services and activities sponsored by UNC and the Graduate Partnerships Program that ensure student success and create a strong graduate student community. A comfortable and stimulating academic setting is created for students in Chapel Hill at the University and at NIH by a new, centrally located Graduate Student Lounge, bookstore, and auditoriums for research seminars. Student enrolled in this partnership program join a larger group of UNC graduate students studying the biological sciences. While at the NIH, students join over 400 other graduate students from more than 100 universities who are doing their research in NIH laboratories. The Graduate Partnerships Program Office at the NIH sponsors graduate student services and activities similar to those at the University to ensure student success and create a strong graduate student community.
Location	Chapel Hill, a city of about 60,000 residents, is in the center of the state, between the Atlantic Ocean and the Appalachian Mountains. The climate is moderate, and the cultural life of the University and the community is excellent. The well-known Research Triangle Park, where many firms have large scientific research facilities and the USPHS has its Environmental Health Sciences Research Center, is nearby. The small-city atmosphere and proximity to other universities, the Research Triangle, and the state capital create a stimulating scientific and cultural climate. The 300-acre campus of NIH in Bethesda, Maryland, is close to Washington, D.C., affording a spectacular cultural and community environment.
The University and The National Institutes of Health	The NIH and UNC have long histories of training scientists and physicians. Thousands of scientists have completed postdoctoral training in NIH laboratories, and many NIH-trained scientists have received international recognition for their work. The University of North Carolina at Chapel Hill was the first state university to admit students. It was chartered in 1789 and formally opened in 1795. The University's first building, Old East, is a national landmark. The University has a planetarium, an art museum, free concerts and movies, faculty and student art shows, University forums, and many other outstanding cultural activities. Excellent sports facilities are available to students. The University is committed to the principle of equal opportunity, and it does not discriminate on the basis of race, sex, color, national origin, religion, or handicap in its relationship with students, employees, or applicants for admission or employment. The UNC/NIH collaborative environment is rich in scientific exchange, and it provides opportunities for a broad biomedical research experience.
Applying	To apply, prospective students must be U.S. citizens or noncitizen nationals of the United States or must have been lawfully admitted for permanent residence (i.e., they must possess a valid Alien Registration Receipt Card or some other verification of such status). An entering class typically includes a variety of undergraduate majors in the biological, chemical, quantitative, or physical sciences. Information on submitting an application to the NIH/UNC Program in Cell Motility and the Cytoskeleton is available on the Web at http://gradschool.unc.edu/domesticapplicant.html. Information about the Graduate Partnerships Program or joining an NIH laboratory for dissertation research can be found on the Web at http://gpp.nih.gov.
Correspondence and Information	Graduate Partnerships Program Building 2, Room 2E06 National Institutes of Health–DHHS 2 Center Drive Bethesda, Maryland 20892-0234 Phone: 301-594-9605 Fax: 301-594-9606 E-mail: gpp@nih.gov Web site: http://gpp.nih.gov	Dr. Herbert Geller Partnership Director, NIH E-mail: gellerh@nhlbi.nih.gov Dr. Kerry S. Bloom Partnership Director, UNC Chapel Hill E-mail: kerry_bloom@unc.edu

University of North Carolina at Chapel Hill/National Institutes of Health

THE FACULTY

NIH Investigators
Robert S. Adelstein, NHLBI. Function of cytoplasmic myosins: transgenic approaches.
Matthew Daniels, NHLBI. Mechanisms controlling differentiation of nerve and muscle cells at the synapse.
Julie Donaldson, NHLBI. Membrane and organelle trafficking in cellular function.
Thomas Friedman, NIDCD. Identification and characterization of dominant and recessive mutations that cause hereditary hearing impairment.
Herbert M. Geller, NHLBI. Extracellular matrix and cellular motility.
John Hammer, NHLBI. Regulation of cell motility.
Robert Horowits, NIAMS. Myofibril assembly, maintenance, and force generation.
Bechara Kachar, NIDCD. Mechanosensory transduction in auditory and vestibular sensory organs.
Edward Korn, NHLBI. Myosin motors: molecular mechanisms and in vivo functions.
Jennifer Lippincott-Schwartz, NICHD. Global principles underlying secretory membrane trafficking, sorting, and compartmentalization within eukaryotic cells.
Harish Pant, NINDS. Mechanisms of topographic regulation of neuronal cytoskeletal proteins.
Evelyn Ralston. Subcellular architecture of skeletal muscle.
James Sellers, NHLBI. Molecular mechanisms of intracellular motility.
Kuan Wang, NIAMS. Structure and function of muscle proteins.

University of North Carolina at Chapel Hill Professors
Eva Anton, Cell and Molecular Physiology. Molecular analysis of neuronal migration and layer formation in cerebral cortex.
Victoria Bautch, Biology. Blood vessel formation in cancer and development.
Kerry Bloom, Biology. Chromosome and spindle dynamics.
Keith Burridge, Cell and Developmental Biology. Interactions of cells with extracellular matrix.
Richard E. Cheney, Cell and Molecular Physiology. Motor proteins, cytoskeleton, and cell motility.
Robert P. (Bob) Goldstein. Generation of cell diversity in development.
Kenneth A. Jacobson, Cell and Developmental Biology. Membrane proteins and lipids in cell motility.
Carol A. Otey, Cell and Molecular Physiology. Mechanisms of cell motility and adhesion.
Mark Peifer, Biology. Adhesion, signal transduction, and motility.
Steve Rogers, Cell Biology. Mitosis, cytoskeletal dynamics, and cell signaling.
Edward D. (Ted) Salmon. Microtubule assembly, microtubule motors, and the mechanism of chromosome movement.

The Dale and Betty Bumpers Vaccine Research Center (VRC), one of more than thirty major research buildings on the NIH campus, was established to facilitate research in vaccine development.

The National Human Genome Research Institute led the Human Genome Project for the National Institutes of Health, which culminated in the completion of the full human genome sequence in April 2003.

UNIVERSITY OF PITTSBURGH

Program in Integrative Molecular Biology

Program of Study

The Program in Integrative Molecular Biology (PIMB) is an innovative new Ph.D. training program that is designed for well-prepared students with a focused and developed interest in the structure and function of molecules that comprise living systems. The program draws on faculty members from the University of Pittsburgh School of Medicine and from the Department of Biological Sciences in the School of Arts and Sciences. This is an accelerated program that provides the opportunity for students to complete their degrees in approximately four years. Students typically enter the program in the summer session and, after performing three rotations, identify an adviser and area of research. Areas of research focus include genomics, proteomics, and gene function and cellular and developmental dynamics. Required course work is completed during the first year. At the end of the first year, students take a comprehensive examination that includes the submission of a research proposal to national fellowship programs. Students receive career mentoring during the third and fourth years to ensure a seamless transition to the postdoctoral level. Training in the PIMB prepares students for positions as postdoctoral fellows and ultimately for leadership positions in academia and industry.

Research Facilities

Research training facilities are located in the Biomedical Science Towers, Crawford Hall, the Hillman Cancer Center, and Scaife Hall. Laboratories are completely equipped for state-of-the-art studies in molecular biology and include automated DNA sequencers; recording spectrophotometers; analytical and preparative ultracentrifuges; gamma radioactivity and liquid scintillation counters; flow cytometers; gas and high-performance liquid chromatographic systems; amino acid analyzers; protein sequencers; mass spectrometers; peptide synthesizers; PCR instrumentation; proteomic instrumentation; individual and centralized computer facilities, including the Pittsburgh supercomputer facility; electron, confocal, and fluorescence microscopy; tissue culture facilities; gene and protein array analysis; high-throughput screening facilities; X-ray crystallography and cryoelectron microscopy instrumentation; and NMR facilities.

Animal facilities include primate laboratories, surgical suites, and a pathogen-free facility for work with immunosuppressed and transgenic mice.

Library facilities include the Langley Library and the Maurice and Laura Falk Library of the Health Sciences. Combined, these libraries house more than 3,000 periodicals and approximately 300,000 volumes. Online access is provided to hundreds of scientific journals.

Financial Aid

All full-time students receive a stipend, educational enrichment fund, computing and network service, and individual health insurance (with an option to purchase additional family coverage) during their graduate training. The current stipend for graduate students is $23,000.

Cost of Study

A tuition waver is granted to all full-time students.

Living and Housing Costs

Most graduate students find housing in residential areas in proximity to the University of Pittsburgh. Costs range from $400 to $550 for efficiency apartments and from $500 to $900 for two-bedroom apartments. Many students share larger apartments to minimize housing costs. Students are provided free access to a public transportation service that connects the University with all surrounding residential communities.

Student Group

The University of Pittsburgh School of Medicine and Department of Biological Sciences have approximately 300 full-time students. Students are encouraged to participate in the Biomedical Graduate Student Association, which hosts an annual symposium and lecture, as well as several social events throughout the year.

Location

Pittsburgh is a dynamic metropolitan area with a diverse array of cultural attractions, including museums, a world-renowned orchestra, and regional ballet, opera, and theater arts. In addition, the city has a lively club scene, numerous indoor and outdoor recreational activities, and several national sports franchises that play in state-of-the-art facilities.

The University and The Program

The University of Pittsburgh is a member of the Association of American Universities (AAU) and is one of the top recipients of National Institutes of Health (NIH) research dollars (seventh in the nation for 2005). It is home to the University of Pittsburgh Medical Center Health System, an internationally renowned medical center, and home to one of the leading organ transplant centers in the world. Ph.D. students in the program receive rigorous and comprehensive training in molecular biology to prepare them for careers as future leaders in the fields of biological and biomedical sciences.

Applying

Requirements for admission include a baccalaureate degree from a natural or physical science or engineering program, a minimum grade point average of 3.7 (on a scale of 4), combined average GRE scores (quantitative and verbal sections) greater than the 80th percentile, scores from a GRE Subject Test, and three letters of recommendation. Applicants who are citizens of countries where English is not the official language (and the Province of Quebec in Canada) are required to submit evidence of English language proficiency by submitting the official results of the Test of English as a Foreign Language (TOEFL) or the International English Language Testing System (IELTS). A minimum TOEFL score of 650 (paper), 280 (computer), or 114 (iBT) or an IELTS score of at least 7.5 is required for admission to the program.

Correspondence and Information

Program in Integrative Molecular Biology
Graduate Studies Office
524 Scaife Hall
University of Pittsburgh
Pittsburgh, Pennsylvania 15261-0001
Phone: 412-648-8957
Fax: 412-648-1077
E-mail: pimbinfo@medschool.pitt.edu
Web site: http://www.pimb.pitt.edu

University of Pittsburgh

THE FACULTY AND THEIR RESEARCH

Susan Amara, Professor; Ph.D., California, San Diego, 1983. Molecular and cellular biology of neurotransmitter transporters.
Gerard Apodaca, Associate Professor; Ph.D., California, San Francisco, 1989. Membrane traffic in polarized epithelial cells.
Karen Arndt, Associate Professor, Ph.D., Berkeley, 1988. Mechanisms of eukaryotic transcriptional regulation.
Ivet Bahar, Professor; Ph.D., Istanbul Technical, 1986. Modeling and simulations of biomolecular structure and dynamics.
Jeffrey Brodsky, Associate Professor; Ph.D., Harvard, 1990. Molecular chaperones and protein quality control.
Gerard Campbell, Assistant Professor; Ph.D., Leicester, 1987. Molecular genetics of development in *Drosophila*.
Richard Chaillet, Associate Professor; M.D./Ph.D., Yale, 1984. Epigenetic inheritance and genomic imprinting.
Yuan Chang, Professor; M.D., Utah, 1987. Viral oncogenesis and new pathogen discovery.
Deborah Chapman, Assistant Professor; Ph.D., Columbia, 1993. Formation and patterning of paraxial mesoderm.
Donald deFranco, Professor; Ph.D., Yale, 1981. Trafficking and function of glucocorticoid receptors.
Susan Gilbert, Associate Professor; Ph.D., Dartmouth, 1986. Mechanisms of microtubule-dependent ATPases.
Paula Grabowski, Professor; Ph.D., Colorado, 1983. Alternative pre-mRNA splicing.
Graham Hatfull, Professor; Ph.D., Edinburgh, 1981. Molecular genetics of mycobacteria and site-specific recombination.
Roger Hendrix, Professor; Ph.D., Harvard, 1970. Virion structure and assembly and genomics and evolution.
Jeff Hildebrand, Assistant Professor; Ph.D., Virginia, 1995. Control of cellular and embryonic morphology.
Neil Hukriede, Assistant Professor; Ph.D., Rochester, 1997. Origin and regulation of kidney progenitor cells.
Judith Klein-Seetharaman, Assistant Professor; Ph.D., MIT, 2000. Folding and function of membrane proteins.
Jeffrey Lawrence, Associate Professor; Ph.D., Washington (St. Louis), 1991. Evolutionary genetics.
Patrick Moore, Professor; M.D., Utah, 1985. Tumor virology of oncogenesis and KSHV.
James Pipas, Professor; Ph.D., Florida State, 1975. Molecular biology of tumor viruses and tumorigenicity.
William Saunders, Associate Professor; Ph.D., Johns Hopkins, 1990. Chromosomal segregation defects in cancer cells.
Tom Smithgall, Professor; Ph.D., Pennsylvania, 1986. Protein tyrosine kinase signaling in development and disease.
Linton Traub, Associate Professor; Ph.D., Weizmann (Israel), 1992. Molecular mechanisms governing receptor-mediated endocytosis.
Ora Weisz, Associate Professor; Ph.D., Johns Hopkins, 1990. Regulation of apical membrane traffic in renal epithelial cells.
Richard Wood, Professor; Ph.D., Berkeley, 1981. Cellular responses to DNA damage in human cells.

UNIVERSITY OF PITTSBURGH

Department of Biological Sciences
Program in Molecular, Cellular, and Developmental Biology

Programs of Study	Programs of study lead to the Ph.D. degree in molecular, cellular, and developmental biology (MCDB) and ecology and evolution (EE). Study leading to the M.S. degree is also offered in the EE program. The organization of these programs allows students to obtain intensive training within one field of study and makes individual diversification possible. In all three areas, the goal is to train independent scientists in research and teaching. In the MCDB program, the first year includes an intensive core course composed of a number of modules covering current research problems, with an emphasis on critical thinking and breadth. In the past year, these module topics were molecular technology, DNA transactions, transcription, RNA transactions, protein biosynthesis, virology, molecular immunology, physical methods, protein structure and function, cell structure and function, cell proliferation, development, and molecular evolution. The core courses are taught by a large number of research faculty members. Modular courses are presented by other graduate programs within this department, providing a flexible system in which students can supplement their education with modules from other programs. Advanced topics courses are also offered. In the first year, the students explore the scientific opportunities available in the department by attending seminars, including the departmental retreat, and by research rotations in three laboratories. Once an area of specialization is chosen, the major effort is on the student's dissertation research. A variety of informal research meetings and seminars given by local and visiting scientists are held in the department and in several other local departments. Students and faculty members interact with neighboring scientists at the University of Pittsburgh School of Medicine, Pittsburgh Cancer Institute, Pittsburgh Genetics Institute, Graduate School of Public Health, Carnegie Mellon University, and the Carnegie Museum of Natural History, providing additional breadth to the program. Several advanced courses are also offered by the Department of Biological Sciences as well as by other departments. The student's progress is followed by a faculty adviser in the first year and by the thesis adviser and committee thereafter. Near the end of the second year, Ph.D. students are required to pass a comprehensive examination that emphasizes the development of an original research proposal.
Research Facilities	The Department has a newly constructed building, Life Sciences Annex, which provides research and research support facilities for the Departments of Biological Sciences and Neuroscience. The Department is well equipped with modern instrumentation and numerous shared facilities, including microarray printers and readers, ultracentrifuges, electron and confocal microscopes with computer imaging and 3-D reconstruction, mass spectrometers, nucleic acid synthesizers, X-ray diffractometers, minicomputers and microcomputers, an isothermal titration calorimeter, controlled-environment rooms, tissue and virus culture facilities, an animal facility, protein and DNA sequencing facilities, greenhouses, a mechanics shop, and ecology field centers. Excellent libraries are close at hand, including the Carnegie Mellon, Falk (medical school), Hillman, and Langley libraries and that of the Department of Chemistry.
Financial Aid	Full financial aid is available to first-year graduate students. Stipends in 2005–06 were set at $22,386 for twelve months and carried substantial tuition reductions as well. The stipend for 2006–07 is anticipated to be somewhat higher. A limited number of Mellon fellowships are available on the basis of a University-wide competition.
Cost of Study	All students receiving stipends, as described above, are exempt from tuition. In 2005–06, tuition was $12,776 per term for nonresidents and $6967 per term for residents of Pennsylvania. Cost of tuition may be higher in 2006–07.
Living and Housing Costs	Off-campus housing is available near the campus and in surrounding communities. Rents vary depending on the type of accommodations selected. Many opportunities to share expenses are also available. Additional information may be obtained from the University Housing Office (412-624-7116).
Student Group	Forty-two of the department's 55 graduate students are currently in the MCDB program. All of them are working toward the Ph.D. Equal numbers of men and women are recruited each year; their average age is 23. Most students have done undergraduate work in biology, chemistry, physics, or biochemistry at an institution in the United States. Every year a few students come from abroad. A research career in industry, academic institutions, or the civil service is the long-term goal of most students.
Location	Pittsburgh, with a population of 334,563, is part of a greater metropolitan area of 2.3 million people. The University occupies more than 120 acres of Oakland, the cultural hub of the city. Pittsburgh is of sufficient size to support a wide variety of cultural and athletics events, including performances of the Pittsburgh Symphony Orchestra, the Pittsburgh Ballet Theatre, the Pittsburgh Opera, the O'Reilly Theater, the Three Rivers Arts Festival, and the International Poetry Forum, as well as home games of the Pittsburgh Pirates, Steelers, and Penguins. The city also has an extensive system of large parks. Schenley Park, adjacent to the University and to Carnegie Mellon, is the site of the renowned Phipps Conservatory and Botanical Gardens, and Highland Park contains the Pittsburgh Zoo and PPG Aquarium. North of the downtown area are the Allegheny Observatory, the Carnegie Science Center, Andy Warhol Museum, and the National Aviary. Close to the campus are the Carnegie Museums of Natural History and Art and the Carnegie Library. Golf, fishing, camping, hiking, skiing, ice-skating, and boating are available in the area.
The University	The University of Pittsburgh, founded eleven years after the signing of the Declaration of Independence, is a nonsectarian, coeducational institution. As part of the Commonwealth System of Higher Education, it receives financial support from the state. At the end of 2003, the University of Pittsburgh was ranked eighth nationally among all universities in NIH-funded support. The University has 17,181 undergraduate and 9,550 graduate students and 4,145 faculty members. The library system maintains excellent collections totaling more than 5.6 million volumes. Intercollegiate athletics events are frequent, and student tickets are available to many of these events. The Petersen Events Center is a new athletic facility, open to all students, that houses exercise facilities and an Olympic-size swimming pool. The Petersen Events Center is also home to basketball games and concerts.
Applying	Candidates should have an undergraduate degree in biology, chemistry, physics, or mathematics. Candidates should submit the application form, undergraduate transcript, and at least three letters of recommendation. Graduate Record Examinations scores are sent electronically from Educational Testing Services to the University database and downloaded by the Department. Applications for the U.S. and Canada for the fall term are available on the department's Web site and should be submitted by January 15 (December 15 for international students). Early applications are considered first.
Correspondence and Information	Cathy Barr Recruitment and Admissions Committee Department of Biological Sciences A-234 Langley Hall University of Pittsburgh Pittsburgh, Pennsylvania 15260 Phone: 412-624-4268 Fax: 412-624-4759 E-mail: biophd@pitt.edu Web site: http://www.pitt.edu/~biology/

University of Pittsburgh

THE FACULTY AND THEIR RESEARCH

MOLECULAR, CELLULAR, AND DEVELOPMENTAL BIOLOGY PROGRAM

Karen M. Arndt, Associate Professor; Ph.D., Berkeley, 1988. Mechanisms of transcriptional initiation and elongation.

Jeffrey L. Brodsky, Associate Professor; Ph.D., Harvard, 1990. Molecular chaperones.

Gerard L. Campbell, Assistant Professor; Ph.D., Leicester (England), 1987. Developmental genetics in *Drosophila*.

Deborah L. Chapman, Assistant Professor; Ph.D., Columbia, 1993. Molecular control of mouse paraxial mesoderm formation.

Robert L. Duda, Research Assistant Professor; Ph.D., UCLA, 1983. Genetics, structure, and assembly of bacteriophages.

Susan P. Gilbert, Associate Professor; Ph.D., Dartmouth, 1986. Mechanistic analysis of microtuble-based motors kinesin and Ncd.

Paula J. Grabowski, Professor; Ph.D., Colorado, 1983. Regulated splicing of messenger RNA precursors.

Graham F. Hatfull, Professor and Chair; Ph.D., Edinburgh, 1981. Site-specific recombination; molecular genetics of mycobacteria.

John Hempel, Research Associate Professor; Ph.D., Rutgers, 1981. Protein structure-function and evolution; aldehyde dehydrogenases.

Roger W. Hendrix, Professor; Ph.D., Harvard, 1970. Bacteriophage assembly and evolution.

Jeffrey D. Hildebrand, Assistant Professor; Ph.D., Virginia, 1995. Cell biology and genetics of mouse development.

Lewis A. Jacobson, Associate Professor; Ph.D., Illinois at Urbana-Champaign, 1967. Neuromuscular signal transduction and protein degradation.

Linda Jen-Jacobson, Professor; Ph.D., Illinois at Urbana-Champaign, 1967. Molecular mechanisms of sequence-specific DNA-protein interactions; problems in molecular recognition.

Kirill I. Kiselyov, Assistant Professor; Ph.D., Institute of Cytology, St. Petersburg (Russia). Structure and function of calcium channels in mammalian cells.

Jeffrey G. Lawrence, Associate Professor; Ph.D., Washington (St. Louis), 1991. Molecular evolution of bacterial genomes.

Joseph A. Martens, Assistant Professor; Ph.D., Western Ontario, 1990. Transcription of intergenic DNA and its role in regulating gene expression.

Valerie Oke, Assistant Professor; Ph.D., Harvard, 1994. Bacterial differentiation during a plant-microbe interaction.

Craig L. Peebles, Professor; Ph.D., Chicago, 1978. Group II intron self-splicing; yeast pre-tRNA splicing; bacterial RNA processing.

James M. Pipas, Professor; Ph.D., Florida State, 1975. Molecular genetics of tumorigenesis; molecular biology of DNA viruses.

John M. Rosenberg, Professor; Ph.D., MIT, 1973. DNA-protein interactions; molecular mechanisms of gene regulation.

M. Teresa Sáenz-Robles, Research Assistant Professor; Ph.D., Madrid (Spain), 1986. Molecular genetics of tumorigenesis; molecular and developmental biology.

William S. Saunders, Associate Professor; Ph.D., Johns Hopkins, 1990. Chromosomal segregation defects in oral cancer cells.

Anthony Schwacha, Assistant Professor; Ph.D., Harvard, 1996. Eukaryotic DNA replication and nuclear structure.

Beth Stronach, Assistant Professor; Ph.D., Utah, 1997. Signal transduction and morphogenesis.

Richard D. Wood, Adjunct Professor; Ph.D., Berkeley, 1981. Cellular responses to DNA damage in human cells.

Recent Representative Faculty Publications

Sheldon, K. E., D. M. Mauger, and **(K. M. Arndt).** A requirement for the *Saccharomyces cerevisiae* Paf1 complex in snoRNA 3' end formation. *Mol. Cell* 20:225–36, 2005.

Kruse, K. B., **J. L. Brodsky,** and A. A. McCracken. Characterization of an ERAD gene as VPS30/ATG6 reveals two alternative and functionally distinct protein quality control pathways: One for soluble A1PiZ and another for aggregates of A1PiZ. *Mol. Biol. Cell* 17:203–12, 2006.

Campbell, G., and A. Tomlinson. Transducing the Dpp morphogen gradient in the *Drosophila* wing: Regulation of Dpp targets by brinker. *Cell* 96:553–62, 1999.

White, P. H., D. R. Farkas, E. E. McFadden, and **D. L. Chapman.** Defective somite patterning in mouse embryos with reduced levels of Tbx6. *Development* 130:1681–90, 2003.

Wikoff, W. R., et al. **(R. L. Duda).** Topologically linked protein rings in the bacteriophage HK97 capsid. *Science* 289:2129–33, 2000.

Sproul, L. R., et al. **(S. P. Gilbert).** Cik1 targets the minus-end kinesin depolymerase Kar3 to the microtubule plus-ends. *Curr. Biol.* 15:1420–7, 2005.

Han, K., et al. **(P. J. Grabowski).** A combinatorial code for splicing silencing: UAGG and GGGG motifs. *PloS Biol.* 3:e158, 2005.

Ojha, A., et al. **(G. F. Hatfull).** GroEL1: A dedicated chaperone involved in mycolic acid biosynthesis during biofilm formation in mycobacteria. *Cell* 123:861–73, 2005.

Hempel, J., et al. Aldehyde dehydrogenase: Maintaining critical active site geometry at motif 8 in the class 3 enzyme. *Eur. J. Biochem.* 268:722–6, 2001.

Morgan, G. J., et al. **(R. W. Hendrix).** Bacteriophage Mu genome sequence: Analysis and comparison with Mu-like prophages in *Haemophilus, Neisseria and Deinococcus. J. Mol. Biol.* 317:337–59, 2002.

Hildebrand, J. D. Shroom regulates epithelial cell shape via the apical positioning of an actomyosin network. *J. Cell Sci.* 118:5191–203, 2005.

Szewczyk, N. J., and **L. A. Jacobson.** Signal-transduction networks and the regulation of protein degradation in muscle. *Int. J. Biochem. Cell Biol.* 37:1997–2011, 2005.

Sapienza, P. J., et al. **(L. Jen-Jacobson).** Thermodynamic and kinetic basis for relaxed DNA sequence specificity in "promiscuous" mutant EcoRI endonucleases. *J. Mol. Biol.* 348:307–24, 2005.

Kiselyov, K., D. M. Shin, and S. Muallem. Signaling specificity in GPCR-dependent Ca2+ signaling. *Cell Signal* 15:243–53, 2003.

Lawrence, J. G. Shared strategies in gene organization among prokaryotes and eukaryotes. *Cell* 100:407–13, 2002.

Martens, J. A., L. Laprade, and F. Winston. Intergenic transcription is required to repress the *Saccharomyces cerevisiae SER3* gene. *Nature* 429:571–4, 2004.

Oke, V., et al. Identification of the heat shock sigma factor RpoH and a second RpoH-like protein in *Sinorhizobium meliloti. Microbiology* 147:2399–408, 2001.

Podar, M., et al. **(C. L. Peebles).** Domain 5 binds near a highly conserved dinucleotide in the joiner linking domains 2 and 3 of a group II intron. *RNA* 4:151–66, 1998.

Sullivan, C. S., P. Cantalupo, and **J. M. Pipas.** The molecular chaperone activity of SV40 large T antigen is required to disrupt Rb/E2F-family complexes by an ATP-dependent mechanism. *Mol. Cell. Biol.* 20:6233–43, 2000.

Grigorescu, A., et al. **(J. M. Rosenberg).** The integration of recognition and cleavage: X-ray structures of EcoRI endonuclease. In *Restriction Endonucleases,* pp. 137–77, ed. A. Pingoud. Springer Verlag, 2004.

Rempel, R. E., et al. **(M. T. Sáenz-Robles** and **J. M. Pipas).** Loss of E2F4 activity leads to abnormal development of multiple cellular lineages. *Mol. Cell* 6:293–306, 2000.

Quintyne, N. J., et al. **(W. S. Saunders).** Spindle multipolarity is prevented by centrosomal clustering. *Science* 307:127–9, 2005.

Schwacha, A., and S. P. Bell. Interactions between two catalytically distinct MCM subgroups are essential for coordinated ATP hydrolysis and DNA replication. *Mol. Cell* 8:1093–104, 2001.

Stronach, B., and N. Perrimon. Activation of the JNK pathway during dorsal closure in *Drosophila* requires the mixed lineage kinase, slipper. *Genes Dev.* 16:377–87, 2002.

Seki, M., et al. **(R. D. Wood).** High efficiency bypass of DNA damage by human DNA polymerase Q. *EMBO J.* 23:4484–94, 2004.

UNIVERSITY OF SOUTH CAROLINA

Department of Biological Sciences
Graduate Training Program in Molecular, Cellular,
and Developmental Biology

Program of Study

The University of South Carolina offers both M.S. and Ph.D. degrees in biological sciences with an emphasis in molecular, cellular, and developmental biology. The first year is primarily devoted to course work emphasizing biochemistry, cell biology, developmental biology, genetics, and molecular biology. After this, students concentrate on a selected area of biology, conducting original research under the guidance of a faculty member. Requirements for the master's program include one semester of teaching, an oral examination, and a written thesis. Requirements for the doctoral program include a comprehensive exam taken after the completion of course work and a final defense of the thesis.

Research Facilities

Modern, well-equipped laboratories are located in the Coker Life Sciences Building. Also located in Coker Life Sciences is the Electron Microscopy Center, with transmission and scanning electron microscopes and a confocal/multiphoton microscope. Services such as the production of polyclonal antibodies, oligonucleotide synthesis, and DNA sequencing are available within the department. Students have access to extensive computer and graphics facilities, greenhouses, environmental chambers, and animal maintenance facilities. Additional resources include histology, flow cytometry, and imaging facilities and an NSF-supported Deermouse Genetic Stock Center.

Financial Aid

The department provides full support in the form of teaching or research assistantships for all graduate students in good standing. The current stipend is $20,000 for twelve months. Additional funds are available for outstanding applicants, and special fellowships are available for applicants who are members of minority groups.

Cost of Study

In 2006–07, all full-time resident graduate students not holding assistantships pay $4144 per semester for tuition and fees; nonresidents pay $8958. For students holding assistantships, the department is currently paying $2466 in tuition. The University reserves the right to change fees without notice.

Living and Housing Costs

Some housing is available on the campus for single students and those with families. In 2006–07, costs range from $500 to $800 per month. Most students take advantage of the large variety of private living accommodations that are available in the vicinity of the University.

Student Group

About 30 graduate students are currently enrolled in the molecular, cellular, and developmental track in the Department of Biological Sciences. The total student body averages between 65 and 70 students, which includes students in the integrative biology: ecology, evolution, and organismal biology track. All areas of the United States and several other countries are represented. The University of South Carolina as a whole enrolls about 25,700 students, one third of whom are doing graduate work.

Location

Situated in central South Carolina where the Saluda and Broad Rivers merge to form the Congaree River, Columbia is one of the fastest-growing metropolitan areas in the Southeast. With a current population between 500,000 and 600,000, Columbia, the state capital, provides a pleasant balance among urban, suburban, and rural lifestyles. Lake, ocean, and mountain areas are within easy driving distance, and the mild climate of this Sun Belt state encourages a wide variety of year-round recreational activities. The University and community offer numerous historic landmarks and diverse cultural events.

The University

The University of South Carolina is located on a 220-acre campus adjacent to the state capitol, in the center of Columbia. Founded in 1801 as South Carolina College, it has a rich tradition as one of the oldest state universities. The University grants bachelor's degrees in sixty-three majors, and the Graduate School offers both M.S. and Ph.D. degrees in many fields. The School of Law awards the J.D. degree, and the nearby School of Medicine awards the M.D.

Applying

Scores on the General Test of the Graduate Record Examinations, academic transcripts, at least two letters of reference, and a statement of purpose indicating the applicant's specific research interests are also required. Completed applications should be submitted via the Graduate School Web site (http://www.gradschool.sc.edu/) before February 15 to ensure consideration for teaching and research assistantships, although applications are considered after this date. Information concerning research opportunities in specific fields may be obtained by writing to the appropriate faculty members, who are listed on the faculty page.

Correspondence and Information

Director of Graduate Studies
Department of Biological Sciences
University of South Carolina
Columbia, South Carolina 29208

Phone: 803-777-2755
Web site: http://www.biol.sc.edu/

University of South Carolina

THE FACULTY AND THEIR RESEARCH

Prospective students are strongly encouraged to correspond with faculty members who have compatible research interests.

Franklin G. Berger, Professor; Ph.D., Purdue, 1974. Mammalian molecular genetics; mechanism of resistance to anticancer agents; mouse models of colorectal cancer.

Franklyn F. Bolander, Associate Professor; M.D./Ph.D., Duke, 1977. Molecular endocrinology; mammary gland biology; hormone receptor regulation; second messengers.

Lewis H. Bowman, Associate Professor; Ph.D., Virginia, 1979. Gene expression during muscle differentiation; regulation of mRNA stability; plant molecular biology.

Erin L. Connolly, Associate Professor; Ph.D., California, Davis, 1997. Metal transport in plants; iron-deficiency responses; gene regulation; phytoremediation.

Michael J. Dewey, Associate Professor and Director, Peromyscus Genetic Stock Center; Ph.D., Pennsylvania, 1973. Mouse genetics; transgenic mice; mosaic mice; models for studying development, physiology, and habitat adaptation.

Dan A. Dixon, Assistant Professor; Ph.D., Northwestern, 1994. Colon carcinogenesis; inflammation; posttranscriptional gene regulation.

Bert Ely, Professor; Ph.D., Johns Hopkins, 1973. Molecular and population genetics; mitochondrial and Y chromosome inheritance in humans and fish.

Michael R. Felder, Professor; Ph.D., California, Davis, 1970. Mammalian molecular genetics; genes of alcohol metabolism.

Robert Friedman, Assistant Professor; Ph.D., South Carolina, 2002. Bioinformatics; molecular evolution.

Austin L. Hughes, Professor; Ph.D., Indiana, 1984. Molecular evolution and bioinformatics; evolution of the vertebrate immune system; host-parasite coevolution.

Loren W. Knapp, Research Associate Professor and Assistant Dean for Undergraduate Studies; Ph.D., Chicago, 1980. Molecular biology of biomineralization; cytoskeletal structure and function.

Beth A. Krizek, Associate Professor; Ph.D., Johns Hopkins, 1993. *Arabidopsis* flower development; floral patterning; organ size control; gene regulation; plant transcription factors.

Robert P. Lawther, Associate Professor; Ph.D., Pittsburgh, 1974. Analysis of mechanisms regulating gene expression.

David E. Lincoln, Professor; Ph.D., California, Santa Cruz, 1978. Plant-herbivore interactions; global change; biodiversity; chemical ecology; convergent evolution.

Richard Long, Assistant Professor; Ph.D., California, San Diego (Scripps), 2001. Microbial ecology with focus on how bacteria interact with each other and with their environment.

Charles R. Lovell, Professor and Chair; Ph.D., Purdue, 1984. Microbial ecology; molecular ecology; physiology and biochemistry of anaerobic bacteria.

Laszlo Marton, Professor; Ph.D., Szeged Medical (Hungary), 1976. Plant molecular genetics; plant tissue culture; genetic engineering and transgene expression in plants; phytoremediation; wetland plants.

Lydia Matesic, Assistant Professor; Ph.D., Johns Hopkins, 2000. Role of ubiquitination in human disease.

Rekha C. Patel, Associate Professor; Ph.D., Indian Institute of Science, 1987. Protein kinase R; translational regulation; apoptosis; regulation of vascular smooth muscle proliferation; regulation of cell cycle.

Gail J. Pruss, Research Associate Professor; Ph.D., Berkeley, 1977. Molecular biology, molecular genetics; DNA topology.

Joseph M. Quattro, Associate Professor; Ph.D., Rutgers, 1991. Conservation biology; population genetics of marine and freshwater fishes; molecular evolution.

David Reisman, Associate Professor; Ph.D., Wisconsin–Madison, 1986. Molecular biology of oncogenic transformation; regulation of expression of the p53 tumor suppressor gene.

Roger H. Sawyer, Professor and Associate Dean for Natural Sciences; Ph.D., Massachusetts, 1970. Molecular, cellular, and developmental biology of the scales and feathers of reptiles and birds.

Richard M. Showman, Associate Professor; Ph.D., Washington (Seattle), 1979. Developmental biology; parasitology; gene regulation during early development; mitochondrial biogenesis; tRNA transport; apoptosis.

Deanna Smith, Assistant Professor; Ph.D., Stanford, 1994. Cytoskeletal organization and cell motility in health and disease.

Johannes W. Stratmann, Associate Professor; Ph.D., Regensburg (Germany), 1994. Plant stress signal transduction; MAP kinases; UV-B signaling.

Vicki Bowman Vance, Professor; Ph.D., Washington (St. Louis), 1983. Plant molecular genetics; plant molecular virology; gene silencing.

Richard Vogt, Associate Professor and Graduate Director; Ph.D., Washington (Seattle), 1984. Neurobiology of odor detection in insects and vertebrates (e.g., moths, *Drosophila*, mosquitoes, sea turtles, zebrafish); molecular, developmental, and evolutionary biology.

Alan S. Waldman, Professor; Ph.D., Johns Hopkins, 1985. Mammalian molecular genetics; mechanisms of homologous recombination and DNA repair in mammalian cells; genetic rearrangements in cancer cells; gene targeting.

Barbara Criscuolo Waldman, Research Associate Professor; Ph.D., Johns Hopkins, 1985. Glycobiology; membrane proteins and glycoproteins; regulation of glycosylation in mammalian cells; nucleotide sugar transport systems in the Golgi; poly (ADP-ribosylation); genetic recombination.

UNIVERSITY OF SOUTHERN CALIFORNIA

Graduate Program in Molecular and Computational Biology

Program of Study

The Graduate Program in Molecular and Computational Biology is a research-based program leading to the Ph.D. degree. It provides a comprehensive framework of training and experience for careers in research and teaching. The program focuses on molecular mechanisms of DNA processing and on mechanisms of regulation at the subcellular, cellular, and multicellular levels. The relationship between molecular structure and biological function is emphasized.

Areas for advanced study include eukaryotic and prokaryotic molecular genetics, control of gene expression, DNA-protein interactions, mechanisms of recombination, developmental genetics, evolutionary molecular biology, molecular neurobiology, and the biology of aging. The molecular biology program also offers opportunities for participation in a strong interdisciplinary program of research in computational molecular biology. Areas of study in computational biology include molecular sequence analysis, proteomics, population and statistical genetics, molecular evolution, directed evolution, and genotyping technologies. These subdisciplines are correlated and interwoven during the first year in courses that deal analytically with the broad areas of molecular organization and function, eukaryotic and prokaryotic molecular genetics, and techniques in biochemistry and cell biology. First-year students gain initial research experience through a program of laboratory rotations. Following the successful completion of the core courses, each degree candidate selects an area of research concentration and, in the second year, completes qualifying examinations. During the second, third, and fourth years, students pursue independent research and participate in elective seminar courses in the Department of Biological Sciences and other departments throughout the University.

Other fields of study offered in the Department of Biological Sciences include marine environmental biology and neuroscience.

Research Facilities

The research and graduate training laboratories are located in the Ahmanson Center for Biological Research and adjacent buildings. Facilities are well equipped for all types of experimental work in molecular biology, cell biology, and biochemistry, including work with animal cell cultures, bacteria and bacteriophages, yeast, *Drosophila*, and animal viruses. Related research programs in neurobiology, gerontology, and marine biology are conducted nearby and associated through the Department of Biological Sciences.

A sophisticated computer network with software for molecular biology, along with a molecular graphics facility, is available. The well-equipped Center for Electron Microscopy and Microanalysis is in an adjacent building.

Financial Aid

Most graduate students receive appointments as teaching assistants, with stipends of $17,160 per academic year (2005–06) plus full tuition remission; these may be renewed on an annual basis. Most students are also eligible for summer support as research or teaching assistants; the total yearly stipend for this support ranges from $21,000 to $24,100. Highly qualified students may be eligible for Predoctoral Merit Fellowships or for supplementary fellowships from the Division of Natural Sciences and Mathematics.

Cost of Study

Tuition in 2005–06 for graduate work was $1059 per semester unit. Full-time students can carry up to 12 units per semester. Teaching and research assistants receive full tuition remission.

Living and Housing Costs

University apartments are available for married couples, and apartments are easily found elsewhere at moderate rents ($600 to $800 per month). Living costs are comparable to those in other large urban areas.

Student Group

The total on-campus enrollment at the University of Southern California is about 30,000 full- and part-time students. Approximately 14,000 of these students are enrolled in graduate or professional programs.

Location

The University is centrally located in the Los Angeles area, where students can take advantage of many cultural opportunities. In addition to the University of Southern California School of Medicine, whose facilities are available to students and faculty members, there are a number of major universities nearby that cooperate in seminar and library programs. The recreational opportunities provided by the University are supplemented by the outdoor activities for which southern California is famous. The Pacific Ocean and local ski areas are readily available.

The University

The University of Southern California is the oldest and largest independent coeducational university in the West. The main campus now consists of 150 acres and 190 buildings. A major development plan is in progress. The faculty-student ratio is being increased each year, and student participation in the achievement of University goals is encouraged.

Applying

Admission to the graduate program requires an undergraduate degree in one of the natural sciences, with a demonstration of competence and achievement in these studies and an undergraduate grade point average of 3.0 or better. Applicants must take the Graduate Record Examinations; the General Test is mandatory and the Subject Test in either chemistry or biology is recommended. Applications for admission in the fall semester are due the preceding January 15. Application forms and information can be obtained by contacting the Graduate Program in Molecular and Computational Biology.

Correspondence and Information

Graduate Program in Molecular and Computational Biology
Molecular and Computational Biology, Room 201B
University of Southern California
University Park, MC-2910
Los Angeles, California 90089-2910
Phone: 213-821-1088
E-mail: molecule@usc.edu
Web site: http://www.usc.edu/dept/LAS/biosci/mcb

University of Southern California

THE FACULTY AND THEIR RESEARCH

Leonard M. Adleman, Distinguished Henry Salvatori Professor of Computer Science; Ph.D., Berkeley, 1976. Computational aspects of molecular biology; molecular computing; computational complexity; number theory; cryptography. *Science* 266:1021–4, 1994. In *Lecture Notes in Mathematics*, p. 1512, Springer-Verlag, 1992. *Inventiones Mathematicae* 79:409–16, 1985.

Oscar M. Aparicio, Assistant Professor; Ph.D., Chicago, 1993. Regulation of eukaryotic DNA replication, chromosome dynamics, chromatin structure, protein-DNA interactions, cell cycle, and checkpoints. *Science* 294:2357–60, 2001. *Proc. Natl. Acad. Sci. U.S.A.* 96:9130–5, 1999. *Cell* 91:59–69, 1997.

Michelle Arbeitman, Assistant Professor; Ph.D., Stanford, 1998. Using the model system *Drosophila* melanogaster to understand how the potential for reproductive behaviors are built into the nervous system. *Science* 297(5590):2270–5, 2002.

Norman Arnheim, Professor; Ph.D., Berkeley, 1965. Human molecular genetics; single-cell PCR analysis; recombination hot spots; germline mutations. *Proc. Natl. Acad. Sci. U.S.A.* 100:8834–8, 2003; 99:14952–6, 2002. *Am. J. Hum. Genet.* 73:5–16, 2003. *Hum. Mol. Genet.* 12:1021–8, 2003. *Nucleic Acids Res.* 31(3):974–80, 2003.

Don B. Arnold, Assistant Professor; Ph.D., Johns Hopkins, 1993. Neurobiology; subcellular transport and targeting of ion channels and associated proteins in neurons. *Nature Neurosci.* 6:243–50, 2003. *Neuron* 20(4):709–26, 1998; 23:149–57, 1997. *Proc. Natl. Acad. Sci. U.S.A.* 94:8842–7, 1997; 91:9970–74, 1994.

Amy Barrios, Assistant Professor; Ph.D., MIT, 2000. Emphasis on exploring the chemical basis for the effects exerted by small molecules, including metal ions in complex biological systems. *Bioorg. Med. Chem. Lett.* 14:511–16, 2004. *Inorg. Chem.* 40:1250–5, 2001.

Samantha Butler, Assistant Professor; Ph.D., Princeton, 1996. Axonal guidance; growth cones; BMPs; development. *Neuron* 38(3): 389–401, 2003. *Neuron* 24(1):127–41, 1999.

Ting Chen, Assistant Professor; Ph.D., SUNY at Stony Brook, 1997. Algorithmic design in molecular biology; DNA and protein sequence analysis; microarray and mass spectrometry data analysis; functional annotation of genes and proteins. *Genome Res.* 12:1540–8, 2002. *J. Comput. Biol.* 8(6):571–83, 2001; 8(3):325–37, 2001.

Xiaojiang Chen, Associate Professor; Ph.D., California, Davis, 1992. Cancer cell biology; DNA replication, molecular machines; virology; immunology; structural biology; X-ray crystallography. *Cell* 119:47–60, 2004. *Genes Dev.* 18:2039–45, 2004.

Kelvin J. A. Davies, Joint Professor and Associate Dean for Research of the Ethel Percy Andrus Gerontology Center; Ph.D., Berkeley, 1981; D.Sc., Moscow, 1993. Free radicals and oxidative stress; antioxidants; protein damage and degradation; apoptosis; adaptation to stress; regulation of gene expression; aging; senescence. *J. Biol. Chem.* 278:311–8, 2003. *Nature Cell Biol.* 4:674–80, 2002. *J. Biol. Chem.* 276:38787–94, 24129–36, 2001.

Caleb E. Finch, Joint Professor; Ph.D., Rockefeller, 1969. Neurobiology and gerontology; brain aging, Alzheimer's, and neuroendocrinology. *Science* 299:1342–6, 2003. *Exp. Neurol.* 182:135–41, 2003. *Endocrinology* 143:636–46, 2002. *Trends Neurosci.* 24:219–34, 2001.

Steven E. Finkel, Assistant Professor; Ph.D., UCLA, 1994. Mechanisms of long-term survival and evolution of bacteria; generation of microbial diversity; genome evolution; DNA-protein interactions. *J. Bacteriol.* 185:7044–52, 2003. *Proc. Natl. Acad. Sci. U.S.A.* 99:8737–41, 2002. *J. Bacteriol.* 183:6288–93, 2001. *Proc. Natl. Acad. Sci. U.S.A.* 96:4023–7, 1999. *Nature* 400:83–5, 1999.

Susan L. Forsburg, Associate Professor; Ph.D., MIT, 1989. MCM proteins; helicase; DNA replication; recombination; chromatin structure; chromosome segregation; genome integrity; meiosis. *Microbiol Mol. Biol. Rev.* 68:109–31, 2004. *Nat. Cell Biol.* 5 1111–16, 2003.

Myron F. Goodman, Professor; Ph.D., Johns Hopkins, 1968. Molecular biology; biochemical and genetic studies of mutagenesis; DNA replication fidelity and SOS error-prone repair; characterization of normal and aberrant DNA polymerases; biochemical basis of somatic hypermutation. *Nature* 424:103–7, 2003. *Proc. Natl. Acad. Sci. U.S.A.* 100:4102–7, 2003. *J. Biol. Chem.* 278:44361–8, 2003. *Proc. Natl. Acad. Sci. U.S.A.* 99:11061–6, 2002; 99:8737–41, 2002.

Steven D. Goodman, Joint Associate Professor; Ph.D., Johns Hopkins, 1988. Prokaryotic molecular biology; mechanisms of protein-DNA interactions, including gene regulation, recombination, and DNA repair. *Infect. Immun.* 71:1972–9, 2003. *J. Bacteriol.* 184:3442–9, 2002. *J. Biol. Chem.* 274:37004–11, 1999. *Nature* 341:251–7, 1989.

Lei Li, Assistant Professor; Ph.D., Berkeley, 1998. Mathematical and statistical modeling of DNA sequencing and genotyping; computational biology; time series analysis and signal processing. *Statistica Sinica* 12:179–202, 2002. *Ann. Statistics* 28:1279–301, 2000. *Electrophoresis* 20:1433–42, 1999. *J. Time Ser.* 17:65–84, 1996.

Michael R. Lieber, Professor; Ph.D., 1981, M.D., 1983, Chicago. Human DNA repair in the immune system, aging, and cancer. *Nature Immunol.* 4:442–51, 2003. *Curr. Biol.* 12:397–402, 2002. *Cell* 109:807–9, 2002; 108:781–94, 2002.

Emily R. Liman, Assistant Professor; Ph.D., Harvard, 1992. Ion channels; pheromone transduction; taste transduction; molecular evolution. *Proc. Natl. Acad. Sci. U.S.A.* 100:15160–5, 2003; 100:3328–32, 2003. *J. Physiol.* 548:777–87, 2003. *Nature Neurosci.* 6:243–50, 2003.

Valter Longo, Assistant Professor; Ph.D., UCLA, 1997 Aging; Alzheimer's disease; oxidative damage; genetics; signal transduction; SOD.

Magnus Nordborg, Assistant Professor; Ph.D., Stanford, 1994. Population genetics and molecular evolution; ecological genetics of *Arabidopsis*. *Trends Genet.* 18:83–90, 2002. *Nature Genet.* 30:190–3, 2002. In *Handbook of Statistical Genetics*, pp. 179–212, 2001. *Genetics* 154:923–9, 2000. *Am. J. Hum. Genet.* 63:1237–40, 1998.

John A. Petruska, Professor; Ph.D., Chicago, 1962. Biochemistry of inheritance and evolution; DNA replication, methylation, and mutagenesis; triplet repeat expansions and neurological diseases. *J. Biol. Chem.* 273:5204–10, 1998; 270:746–50, 1995. *Nucleic Acids Res.* 24:1992–8, 1996.

Peter Z. Qin, Assistant Professor; Ph.D., Columbia, 1999. Role of RNA in mechanical motion; biophysical and biochemical studies of RNA structure and function. *Biochemistry* 40:6929–36, 2001. *J. Mol. Biol.* 291:15–27, 1999. *Biochemistry* 36:4718–30, 1997.

Michael W. Quick, Associate Professor; Ph.D., Emory, 1992. Drug abuse; neurotransmitter transporters; nicotinic acetylcholine receptors; protein-protein interactions; subcellular trafficking. *Proc. Natl. Acad. Sci. U.S.A.* 99:5686–91, 2002. *J. Neurosci.* RC192:1–6, 2002. *J. Biol. Chem.* 276:42932–7, 2001.

Fengzhu Sun, Associate Professor; Ph.D., USC, 1994. Statistical genetics; linkage and associate studies; mathematical modeling of PCR-related biotechnologies; directed evolution; protein-protein interactions. *Nucleic Acids Res.* 31:974–80, 2003. *Proc. Natl. Acad. Sci. U.S.A.* 99:7335–9, 2002. *Ann. Hum. Genet.* 65:207–19, 2001. *Am. J. Epidemiol.* 150:97–104, 1999. *Nucleic Acids Res.* 23:3034–40, 1995.

Simon Tavaré, Professor; Ph.D., Sheffield (England), 1979. Statistics and probability in molecular biology, population and human genetics, molecular evolution, and bioinformatics. *Nature* 416:726–9, 2002. *Trends Genet.* 18:83–90, 2002. *Proc. Natl. Acad. Sci. U.S.A.* 98:10839–44, 2001; 97:1236–41, 2000.

John Tower, Associate Professor; Ph.D., Johns Hopkins, 1988. Molecular biology of aging in *Drosophila; Drosophila* chorion gene amplification. *Genetics* 161:661–72, 2002; 158:1167–76, 2001. *Genome Biol.* 3:0021.1–10, 2002. *Genes Dev.* 15:134–46, 2001. *Mech. Aging Dev.* 118:1–14, 2000.

Jeffrey D. Wall, Assistant Professor; Ph.D., Chicago, 2000. Population genetics, molecular evolution, and human genetics. *Nature Rev. Genet.* 4:587–97, 2003. *Genetics* 162:203–16, 2002. *Mol. Biol. Evolution* 17:156–63, 2000.

Michael S. Waterman, Professor; Ph.D., Michigan State, 1969. DNA sequence comparisons; RNA secondary structure prediction; sequence pattern identification; identification of common molecular subsequences; analysis of genomic mapping strategies. *Am. J. Hum. Genet.* 73:63–73, 2003. *Genome Res.* 13:1916–22, 2003. *J. Comput. Biol.* 10:803–20, 2003.

Xianghong Zhou, Assistant Professor; Ph.D., Swiss Federal Institute of Technology (ETH Zurich), 2000. Computational functional genomics; cellular regulatory and metabolic network; microarray gene expression analysis; comparative genomics; evolution of the biological systems. *Physiol. Genomics* 13:69–78, 2003. *Proc. Natl. Acad. Sci. U.S.A.* 99(20):12783–8, 2002.

UNIVERSITY OF SOUTH FLORIDA

H. Lee Moffitt Cancer Center and Research Institute
Cancer Biology Ph.D. Program

Program of Study

The Ph.D. Program in Cancer Biology is a collaborative effort between the University of South Florida and H. Lee Moffitt Cancer Center designed to prepare young scientists to meet the future challenges of cancer research. The program was founded in 2000 and embraces the concept that curing cancer is based on two challenges: unraveling the molecular and biological basis for tumor development and devising new detection and treatment approaches based on those discoveries. To meet these challenges, the program provides an integrated curriculum that incorporates training in multiple disciplines encompassing immunology, cancer genetics, cell and molecular biology, signal transduction, drug discovery, functional genomics, proteomics, bioinformatics, chemistry, and translational cancer therapies.

The Cancer Biology Ph.D. Program draws participating faculty members from multiple university departments and colleges. The majority of the faculty is housed within the Moffitt Cancer Center as part of the Department of Interdisciplinary Oncology and includes both basic science and clinical research investigators. The program is strengthened by faculty members from the College of Medicine, College of Liberal Arts and Sciences, and the College of Engineering. Members encompass five broad areas of interest: immunology, molecular oncology, drug discovery, cancer control, and clinical investigations. Each area shares the common goal of understanding, preventing, and curing cancer.

Students begin research during the first semester, typically rotating through two to three laboratories. Formal course work is usually completed by the end of the second academic year. Qualifying examinations consist of a written exam and the development and defense of a unique research proposal. Most students complete the qualifying examinations before the beginning of the third academic year. Past graduates have earned their Ph.D. degrees within four to five years of entering the program.

Research Facilities

Academic core facilities and individual faculty research laboratories are located in two crosswalk-connected structures: the Moffitt Research Center and the new Vincent A. Stabile Research Tower. The H. Lee Moffitt Cancer Center Hospital is adjacent to these research buildings, facilitating the Center's goal to rapidly move laboratory discoveries into the clinical setting. In addition to well-equipped faculty laboratories, a full range of core facilities, including high-throughput drug screening and molecular modeling, experimental pathology, biostatistics and bioinformatics, analytic microscopy, flow cytometry, molecular biology, molecular imaging, proteomics, and mouse modeling are provided for the faculty members and students. In addition, the Moffitt Cancer Center is home to the National Functional Genomics Center, which is an integrated program focusing on gene expression analysis, genetics, proteomics, and translational research to develop new technologies in molecular medicine.

Financial Aid

All students in good standing receive full financial support for tuition, fees, and health insurance. For the academic year 2006–07, a stipend of $20,000 is awarded entering students to cover living expenses. The stipend increases to $20,720 when the student enters the second year. After passing qualifying exams (written and oral), third-year students and beyond receive an additional $1000, bringing the total stipend to $21,720 for the remainder of the student's time in the program.

Cost of Study

The cost of tuition and fees are typically covered by the financial support described above.

Living and Housing Costs

USF campus housing for single and married students is available on a first-come, first-served basis. The majority of the students in the program live off campus, and numerous apartment complexes ring the USF campus, many of which are serviced cost free by the University's bus transportation system. Discounts and special rental packages are available from a number of these apartment complexes. Rent for a one-bedroom/one-bath apartment can start as low as $450 a month.

Student Group

The USF campuses had total enrollment of 42,950 for fall 2004: 32,442 undergraduate and 7,366 graduate students (with 2,112 international students from 116 countries). The Cancer Biology Ph.D. Program currently has 21 students from across the United States, the UK, Spain, Puerto Rico, Canada, Bulgaria, and India.

Location

USF and H. Lee Moffitt Cancer Center are part of a burgeoning geographic region situated off the Gulf of Mexico referred to as the Tampa Bay area. Tampa Bay boasts a booming employment rate, with 60 percent of Florida's high-tech industries located along the Interstate 4 corridor. A climate that is consistently sunny and pleasant encourages year-round festivals and community activities. Broadway shows and traveling art exhibits stop in Tampa on a regular basis. *Hispanic* magazine rates Tampa the "Top City in the U.S. for Hispanics." Tampa is also one of the most broadband-wired cities in America, ranking second in a recent AOL poll.

The University and The Center

USF is a top-tier, metropolitan research university distinguished by excellent research and graduate education. Strong interdisciplinary programs put USF on the leading edge of research in a number of areas. The doctoral program in Cancer Biology blends USF research capacity with the strengths of the nationally recognized H. Lee Moffitt Cancer Center and Research Institute, the only National Cancer Institute Comprehensive Cancer Center in the state of Florida.

Applying

The application deadline for fall admission is February 1 for domestic applicants and January 2 for international applicants. Applications must be received on or before the deadline to be considered for the fall semester. Applications must be sent directly to the Cancer Biology Program office. This is important to ensure the application is reviewed as quickly as possible. Applicants are required to have at least a bachelor's degree or its equivalent from a college or university of recognized standing. All applicants should possess a strong background in the biological, chemical, and physical sciences. Evaluation for admission is based on previous academic record, letters of recommendation, scores on the GRE and TOEFL (if applicable), and acquired research experience.

Correspondence and Information

Cancer Biology Ph.D. Program
H. Lee Moffitt Cancer Center & Research Institute
12902 Magnolia Drive, MRC-4East
Tampa, Florida 33612

Phone: 813-745-6876
Fax: 813-745-7264
E-mail: cancerphd@moffitt.usf.edu
Web site: http://cancerbio.hsc.usf.edu

University of South Florida

THE FACULTY AND THEIR RESEARCH

Deepak Agrawal, Ph.D.; Assistant Professor, Department of Interdisciplinary Oncology. Cell growth regulation; protein-protein interactions; bioengineered substrates; computational biology; bioinformatics.

Mark G. Alexandrow, Ph.D.; Assistant Professor, Interdisciplinary Oncology Program, and Member-in-Residence, H. Lee Moffitt Cancer Center. Understanding the mechanisms by which growth factor signals, or inhibitory TGF-beta signals and the Ras-Rb pathway, regulate the assembly and function of pre-(DNA) Replication Complexes (preRCs) in late G1 phase, and how the DNA replication machinery and preRCs utilize chromatin remodeling complexes to gain access to the DNA substrate during late G1 and S-phase.

Scott J. Antonia, M.D., Ph.D.; Associate Professor, Department of Interdisciplinary Oncology. Developing novel immunotherapeutic strategies for the treatment of cancer patients.

Wenlong Bai, Ph.D.; Associate Professor, Department of Pathology. Role of steroid and vitamin D receptors in the development and treatment of human cancers.

Darrin Beaupre, M.D., Ph.D.; Assistant Professor, Department of Interdisciplinary Oncology. Hematopoietic tumors, most particularly in the disease multiple myeloma.

Amer A. Beg, Ph.D.; Professor, Department of Interdisciplinary Oncology. Molecular mechanisms involved in dendritic cell and T lymphocyte activation; regulation of innate and adaptive immune response genes by NF-B transcription factors.

Kapil N. Bhalla, M.D.; Professor, Department of Interdisciplinary Oncology. Molecular mechanisms that set the threshold and regulate the extrinsic (death receptor) and intrinsic (mitochondrial) pathways of apoptosis (programmed cell death).

George Blanck, Ph.D.; Professor, Department of Biochemistry and Molecular Biology. Understanding the interferon cell-signaling pathway in normal and tumor cells.

Esteban Celis, M.D., Ph.D.; Professor, Department of Interdisciplinary Oncology. Recognition and destruction of tumor cells by T lymphocytes. Development of immune-based therapies for cancer.

Srikumar Chellappan, Ph.D.; Professor, Department of Interdisciplinary Oncology. Mechanisms by which extracellular signals regulate the cell-cycle machinery and how a loss of this regulation leads to oncogenesis.

Jiandong Chen, Ph.D.; Associate Professor, Department of Interdisciplinary Oncology. How p53 is normally regulated and how it is inactivated in tumors without undergoing mutations; identify compounds to block MDM2 function and activate p53 in tumor cells.

William Douglas Cress Jr., Ph.D.; Associate Professor, Department of Interdisciplinary Oncology. Understanding the role of members of the E2F and Bcl-2 families in cell growth and survival, with particular interest in the role of these factors in lung cancer chemotherapy.

William S. Dalton, M.D., Ph.D.; Director of the H. Lee Moffitt Cancer Center. Drug resistance in cancer cells: mechanism of multidrug resistance due to transport genes, role of the Fas/Fas ligand in drug response, and cell adhesion molecules and their role in drug resistance.

Julie Y. Djeu, Ph.D.; Professor, Department of Interdisciplinary Oncology. Molecular mechanisms by which NK cells kill tumor cells.

Gloria Ferreira, Ph.D.; Professor, Department of Biochemistry and Molecular Biology. Biochemistry and molecular regulation of the biosynthesis of heme, the major component of proteins such as hemoglobin, myoglobin, and cytochromes.

Dmitry I. Gabrilovich, M.D., Ph.D.; Professor, Department of Interdisciplinary Oncology. Mechanisms of tumor-associated immunosuppression; development of new effective cancer vaccines.

Anna R. Giuliano, Ph.D.; Professor, Department of Interdisciplinary Oncology. Human papillomavirus (HPV)–related carcinogenesis; establish a cohort to assess the natural history of HPV in men and evaluate the utility of a novel biomarker of cervical cancer risk to be utilized in conjunction with HPV testing.

Dmitry B. Goldgof, Ph.D.; Professor, Department of Computer Science and Engineering. Motion and deformation analysis; computer vision; image processing and its biomedical applications; bioinformatics; pattern recognition.

Thomas M. Guadagno, Ph.D.; Assistant Professor, Department of Interdisciplinary Oncology. Biochemical controls that regulate cell division: determining the signaling pathways that regulate the mitotic spindle apparatus.

Lawrence O. Hall, Ph.D.; Professor, Department of Computer Science and Engineering. Hybrid reasoning systems; machine learning; data mining; pattern recognition; integrating AI into medical image processing.

Jonathan A. Harton, Ph.D.; Assistant Professor, Department of Medical Microbiology and Immunology. Structure/function relationships in the molecular control of immune responses.

John R. Hassell, Ph.D.; Professor, Department of Biochemistry and Molecular Biology. Chondrocyte differentiation and bone growth; wound healing and fibrosis.

Eric Haura, M.D.; Assistant Professor Department of Interdisciplinary Oncology. Oncogenic signal transduction pathways involved in the pathogenesis of lung cancer: understanding the role of tyrosine kinase pathways and STAT signaling in lung cancer.

Lori Hazlehurst, Ph.D.; Assistant Professor, Department of Interdisciplinary Oncology. Delineating mechanisms of de-novo drug resistance associated with anthracyclines and anthracenediones.

Richard Heller, Ph.D.; Professor, Department of Surgery. Development of therapies to facilitate the treatment of human diseases not amenable to standard therapies; development of drug and gene delivery systems.

Kyung Woon Jung, Ph.D.; Associate Professor, Department of Chemistry. Syntheses of biologically important molecules, such as anticancer drugs, antiviral agents, and biomolecular analogues; development of novel synthetic methodologies toward the designed syntheses.

William Kerr, Ph.D.; Associate Professor, Department of Interdisciplinary Oncology. Biology of hematopoietic stem cells (HSC) and identifying genes that are important in HSC function and transplantation.

Johnathan M. Lancaster, M.D.; Assistant Professor, Department of Interdisciplinary Oncology. Molecular genetic etiology of ovarian cancer development, progression, and response to therapy.

Ji-Hyun Lee, Ph.D.; Assistant Professor, Department of Interdisciplinary Oncology. Developing methods for repeated measurement/longitudinal data; tools for testing goodness-of-fit of models for correlated binary data.

Gary Litman, Ph.D.; Professor, Department of Pediatrics. Understanding both the mechanisms that diversify the adaptive immune response and the basis for variation in novel immune-like receptors that affect innate immunity.

Noreen Luetteke, Ph.D.; Assistant Professor, Department of Interdisciplinary Oncology. Regulation of epithelial function and transformation by polypeptide growth factors and their tyrosine kinase receptors.

Shyam S. Mohapatra, Ph.D.; Professor, Department of Internal Medicine. Prophylactic and therapeutic modulation of upper- and lower-airway diseases using gene expression therapy.

Alvaro N. A. Monteiro, Ph.D.; Associate Professor, Department of Interdisciplinary Oncology. Role of genes involved in the development of breast and ovarian cancer: function of the tumor suppressor gene BRCA1.

Dave Morgan, Ph.D.; Professor, Department of Pharmacology and Therapeutics. Fundamental changes in aging brain and age-related neurological disorders; role of astrocytes and microglial cells in the brain's reaction to injury.

Pamela N. Munster, M.D.; Assistant Professor, Department of Interdisciplinary Oncology. Development of novel targeted therapies involving histone deacetylase inhibitors (HDACi) for the treatment of breast cancer and their integration into current treatment strategies.

Tuya Pal, M.D.; Assistant Professor, Department of Interdisciplinary Oncology. Ovarian cancer and mismatch repair deficiency; genetic and hormonal risk factors for breast cancer in African American women.

Jong Park, Ph.D.; Assistant Professor, Department of Interdisciplinary Oncology. Differential cancer risk among ethnic groups using gene profiling and genotyping analysis.

W. J. Pledger, Ph.D.; Professor, Department of Interdisciplinary Oncology. Regulatory mechanisms controlling cellular proliferation, differentiation, and the mechanisms involved in tumor development

Richard S. Pollenz, Ph.D.; Associate Professor, Department of Biology. Response of organisms to environmental stress (chemical contaminants, hydrocarbons, low oxygen) at the molecular level.

Huntington Potter, Ph.D.; Professor, Department of Biochemistry and Molecular Biology. Mechanisms of chromosome mis-segration leading to aneuploid cells and the role in Alzheimer's disease.

Gary W. Reuther, Ph.D.; Assistant Professor, Department of Interdisciplinary Oncology. Identifying novel oncogenes that contribute to the formation and progression of cancer.

Saïd M. Sebti, Ph.D.; Professor, Department of Interdisciplinary Oncology. Mechanisms by which normal cells become malignant, with a major focus on growth factor signal transduction pathways and novel cancer drugs.

Ed Seto, Ph.D.; Professor, Department of Interdisciplinary Oncology. Regulation of cellular and viral gene expression related to cancer.

Larry P. Solomonson, Ph.D.; Professor and Chair, Department of Biochemistry and Molecular Biology. Elucidation of molecular determinants and mechanisms important in regulating production of the biological effector molecule nitric oxide using endothelial cells as the model system.

Eduardo M. Sotomayor, M.D.; Associate Professor, Department of Interdisciplinary Oncology. Cellular and molecular mechanism(s) involved in tolerance induction; design strategies aimed to prevent and/or revert this unresponsive state and therefore enhance the efficacy of the current generation of immunotherapeutic strategies.

Daniel M. Sullivan, M.D.; Professor, Department of Interdisciplinary Oncology. How cancer cells become resistant to the antitumor drugs used to treat human malignancies.

Melvyn S. Tockman, M.D., Ph.D.; Professor, Department of Interdisciplinary Oncology. Translational research into discovery of markers of neoplastic transformation and confirmation of these markers in specimens from human populations.

Hong-Gang Wang, Ph.D.; Associate Professor, Department of Interdisciplinary Oncology. Mechanisms that control programmed cell death and cell-cycle checkpoints in the context of oncogenesis; development of small molecules that directly target the core components of the cell death machinery for drug discovery.

Sheng Wei, M.D.; Associate Professor, Department of Interdisciplinary Oncology. Elucidation of the signal pathway for activation of human neutrophil function by cytokines and bacterial products.

Kenneth L. Wright, Ph.D.; Associate Professor, Department of Interdisciplinary Oncology. Transcriptional and epigenetic mechanisms controlling genes involved in antigen presentation and immune function.

Jie Wu, Ph.D.; Associate Professor, Department of Interdisciplinary Oncology. Reveal the signal transduction mechanisms that control cell proliferation, differentiation, survival, and motility and explore key signaling molecules as targets for cancer therapy.

Tim Yeatman, Ph.D.; Professor, Department of Interdisciplinary Oncology. Characterization of human c-Src in the development and progression of human colon cancer; developing gene expression profiles to improve molecular staging of cancer.

Xue-Zhong Yu, M.D.; Assistant Professor, Department of Interdisciplinary Oncology. T-cell response to transplantation antigen and tumor; how signals through antigen receptors and co-stimulatory receptors modulate T-cell activation or tolerance.

Mike Zaworotko, Ph.D.; Professor and Chair, Department of Chemistry. Crystal engineering; nanotechnology; supramolecular chemistry; X-ray crystallography.

THE UNIVERSITY OF TEXAS AT AUSTIN

Graduate Program in Cell and Molecular Biology

Programs of Study

The Graduate Program in Cell and Molecular Biology is designed for those who anticipate careers as independent research investigators, possibly combined with teaching or other activities. It is primarily for those seeking a Ph.D. Research is supported in a number of areas, reflecting the diverse interests of more than 100 affiliated faculty members. The multidisciplinary program cuts across departmental lines, with faculty members drawn primarily from the Departments of Chemistry and Biochemistry, Molecular Cell and Development Biology, and Molecular Genetics and Microbiology, as well as additional members from Engineering, Integrative Biology, Neurobiology, Nutritional Sciences, Pharmacy, and Physics.

The doctoral degree requires the student to accomplish creative, independent research and to document the research in a scholarly dissertation. In preparation, the student must acquire a strong foundation in biochemistry, molecular genetics, and cell biology and a working knowledge of the area of biology in which research is to be conducted. There are minimal course requirements for the Ph.D., although students are expected to take the Cell and Molecular Biology Graduate Core Courses and one laboratory course, as well as related courses that provide both depth and breadth. At the end of their first year, students select a specialization track (cell and developmental biology, molecular genetics, neurobiology, structural biology, biochemical technology and drug discovery, or bioinformatics, biotechnology, and biosensors) based on their primary research interests. To be admitted to candidacy for the degree, the student must present an acceptable list of courses, must formulate a feasible research program, and must pass a qualifying examination.

Research Facilities

The biology departments are housed in several adjacent buildings near the center of the Austin campus, including the Cell and Molecular Biology Building that opened in 1997. Laboratories are modern, with state-of-the-art equipment. The Graduate Program in Cell and Molecular Biology is supported by the Institute for Cellular and Molecular Biology, which is located in the same building. Support facilities include staffed core facilities for nucleic acid sequencing and synthesis, peptide sequencing and synthesis, flow cytometry, electron and confocal microscopy, molecular modeling, graphics and computing, and a mouse genetic engineering facility. The University of Texas (UT) at Austin library is the fifth-largest academic library in the United States, with nearly 7 million volumes.

Financial Aid

In the first-year appointment as a graduate research assistant, a stipend of $24,000 per year (2005–06 academic year) as well as health insurance coverage and remission of tuition and fees can be expected. Continuing graduate students are usually appointed as graduate research assistants (paid from the research grants of their faculty adviser) or have appointments as teaching assistants.

It is recommended that qualified applicants to the Graduate Program in Cell and Molecular Biology apply for funding from extramural sources, such as NSF Graduate Research Fellowships and NRSA graduate fellowships. Applicants with a specific research interest in computational phylogenetics, neurosciences, or optical imaging should also apply for support to UT Austin's NSF Integrative Graduate Research and Education Training (IGERT) programs or NIH training grants in these areas.

Cost of Study

Tuition and fees for graduate students during the 2005–06 academic year were approximately $4500 (resident tuition for nine months). Research assistants and University fellowship recipients qualify for resident tuition.

Living and Housing Costs

The estimated cost of living for students residing off campus for 2005–06 (nine-month academic year) was $12,900 to $17,000 (tuition and fees, books, room and board, transportation, and personal/miscellaneous expenses). The cost of off-campus housing in Austin ranges from $400 to $800 per month. Furnished and unfurnished University apartments for married graduate students rent for $340 to $500 per month.

Student Group

There are more than 50,000 students at the University of Texas at Austin, including more than 11,000 graduate students in ninety-five fields of study. About 800 doctoral degrees and more than 2,500 master's degrees are awarded each year. Approximately 200 graduate students are engaged in research on some facet of cell and molecular biology through various degree programs.

Location

Austin, the state capital, has a high proportion of professional workers and relatively little industry. It is a clean, attractive city located in the Hill Country of central Texas. With a population of about 600,000 and a metropolitan population of 1 million, Austin provides a wide variety of cultural activities. A series of lakes on the Colorado River stretches 100 kilometers to the northwest and offers excellent facilities for boating, fishing, and swimming. The climate is temperate, with winter temperatures that seldom fall below freezing and summer temperatures that reach into the 90s and, occasionally, higher.

The University

The University of Texas at Austin was established in 1883 and is the academic flagship of the UT System's fifteen component institutions. It is a state-supported institution known nationally for the quality of its academic programs, research, and public service.

Applying

The priority deadline is December 15. Applications received after that date are considered, but it may not be feasible to complete the review of late applications in time for invitations to recruiting weekends. Applications include scores from the GRE General Test, transcripts, three letters of recommendation, and a statement of purpose. A faculty committee reviews all applications and recommends admission and offers of financial assistance beginning in the middle of February. For additional information, prospective students should write to the Graduate Recruitment Coordinator.

Correspondence and Information

Graduate Recruitment Coordinator
Graduate Program in Cell and Molecular Biology
1 University Station A4810
The University of Texas at Austin
Austin, Texas 78712-0160
Phone: 512-471-0934
Fax: 512-471-2149
E-mail: grad.program@icmb.utexas.edu
Web site: http://www.icmb.utexas.edu/cmb

The University of Texas at Austin

THE FACULTY AND THEIR RESEARCH

Creed Abell, Ph.D., Wisconsin. Molecular biology of proteins in the nervous system.

Seema Agarwala, Ph.D., Stony Brook, SUNY. Cellular and molecular mechanisms of neuronal development.

Orly Alter, Ph.D., Stanford. Genomic signal processing and systems biology.

Lauren Ancel-Meyers, Ph.D., Stanford. Theoretical and experimental evolutionary biology; microbial evolution and ecology; RNA secondary structure evolution; theoretical epidemiology.

Eric Anslyn, Ph.D., Caltech. Bioorganic chemistry.

Dean Appling, Ph.D., Vanderbilt. Organization, regulation, and engineering of metabolic pathways.

Karen Artzt, Ph.D., Cornell. Mouse molecular developmental genetics.

Nigel Atkinson, Ph.D., Penn State. Regulation of ion channel gene expression.

Susan Bergeson, Ph.D., Oregon Health Sciences. Behavioral and molecular analysis of alcohol withdrawal syndrome.

George Bittner, Ph.D., Stanford. Cellular, molecular, and developmental neurobiology.

Henry Bose, Ph.D., Indiana. Transformation of lymphoid cells by the *v-rel* oncogene.

Shawn Bratton, Ph.D., Texas at Austin. Basic mechanisms of apoptosis.

Malcolm Brown, Ph.D., Texas at Austin. Cellulose biosynthesis; atomic and molecular resolution with transmission electron microscopy and electronic paper.

Karen Browning, Ph.D., Illinois at Urbana-Champaign. Initiation of protein synthesis.

James Bull, Ph.D., Utah. Evolutionary genetics.

Clarence S. M. Chan, Ph.D., Cornell. Yeast molecular and cell biology; genetic and molecular analysis of chromosome segregation and cellular morphogenesis in yeast.

Jeff Chen, Ph.D., Texas A&M. Polyploidy; epigenetics; genomics; evolution.

David Crews, Ph.D., Rutgers. Brain mechanisms controlling reproduction and behavior; signal transduction and temperature-dependent sex determination; origins of phenotypic and neural plasticity; endocrine disruption.

Maria Croyle, Ph.D., Michigan. Virus-mediated gene delivery.

Kevin Dalby, Ph.D., Cambridge. Mechanisms of catalysis and regulation of protein kinases.

Arturo De Lozanne, Ph.D., Stanford. Molecular analysis of cytokinesis in *Dictyostelium.*

Jaquelin Dudley, Ph.D., Baylor College of Medicine. Transcriptional regulation in eukaryotes; mechanisms of carcinogenesis and pathogenesis by retroviruses.

Charles Earhart Jr., Ph.D., Purdue. Bacterial physiology: iron assimilation; assembly and modification of cell envelopes.

Andy Ellington, Ph.D., Harvard. Therapeutic and diagnostic applications of aptamers; protein and metabolic engineering.

Walter Fast, Ph.D., Northwestern. Enzyme mechanisms; protein engineering; inhibitor design.

Janice Fischer, Ph.D., Harvard. Molecular genetics of *Drosophila* development.

Ernst-Ludwig Florin, Ph.D., Munich Technical. Physics.

George Georgiou, Ph.D., Cornell. Molecular biology and biotechnology; protein synthesis.

Nace Golding, Ph.D., Wisconsin–Madison. Dendritic integration and synaptic plasticity in central neurons.

Andrea Gore, Ph.D., Wisconsin–Madison. Cellular and molecular mechanisms by which the brain controls reproductive development and aging.

Ellen Gottlieb, Ph.D., Yale. Eukaryotic gene expression, with emphasis on mRNA sorting, transport, and localization.

David Graham Ph.D., Illinois at Urbana-Champaign. Biochemical evolution of biosynthetic pathways.

Jeffrey Gross, Ph.D., Duke. Developmental biology of the visual system, using zebrafish.

Robin Gutell, Ph.D., California, Santa Cruz. Computational biology/bioinformatics: RNA structure and folding, molecular evolution.

Marvin Hackert, Ph.D., Iowa State. Protein crystallography; regulation of polyamines; protein chemistry.

Adron Harris, Ph.D., North Carolina at Chapel Hill. Molecular mechanisms of alcohol action and drug addiction.

Rasika Harshey, Ph.D., Indian Institute of Science. DNA transposition; signal transduction in microbial development.

Graeme Henkleman, Ph.D., Washington (Seattle). Theoretical investigations of enzyme reactions, using ab initio calculations.

David Herrin, Ph.D., South Florida. Ribozymes; molecular genetics of chloroplasts/circadian rhythms.

David Hillis, Ph.D., Kansas. Molecular evolution.

David Hoffman, Ph.D., Duke. NMR spectroscopy; RNA-protein interactions.

Jon Huibregtse, Ph.D., Michigan. Ubiquitin proteolysis system and ubiquitin-like modifiers.

Enamul Huq, Ph.D., Purdue. Light signal transduction in plants.

Brent Iverson, Ph.D., Caltech. Bioorganic chemistry; antibody engineering technologies; enzyme engineering technologies; large synthetic molecules that fold, assemble, and bind DNA.

Vishwanath Iyer, Ph.D., Harvard. Genome-wide transcriptional regulatory networks; targets of oncogenic transcription factors; functional genomics; bioinformatics.

Robert Jansen, Ph.D., Ohio State. Chloroplast genome evolution, molecular systematics, and evolution.

Makkuni Jayaram, Ph.D., Indian Institute of Science. Site-specific recombination; DNA rearrangements in yeast; minichromosome partitioning in yeast; mechanism of persistence of selfish genomes.

Arlen Johnson, Ph.D., Harvard. Yeast molecular genetics and biochemistry; RNA turnover and translational control.

Kenneth Johnson, Ph.D., Wisconsin. DNA polymerase mechanisms; HIV reverse transcriptase; kinesin microtubule-dependent motor ATPase pathway; transient-state kinetic analysis.

Daniel Johnston, Ph.D., Duke. Synaptic integration and synaptic plasticity in hippocampus; cellular mechanisms of epileptogenesis.

Christopher Jolly, Ph.D., Texas A&M. Effect of diet and aging on T-cell signal transduction.

Thomas Juenger, Ph.D., Chicago. Plant evolutionary ecology and genetics in natural populations.

Sean Kerwin, Ph.D., Berkeley. Bioorganic chemistry; molecular recognition and rational drug design.

G. Barrie Kitto, Ph.D., Brandeis. Evolution of protein structure; applied biochemistry.

Kimberly Kline, Ph.D., Texas at Austin. Role of vitamin E in regulation of tumor cell growth.

Robert Krug, Ph.D., Rockefeller. Molecular biology of human influenza viruses, including viral replication and gene expression, viral-host interactions, and the role of protein modifications in the interferon response.

Alan Lambowitz, Ph.D., Yale. Self-splicing introns; intron mobility and evolution; retroelements and reverse transcriptases; functional genomics; gene therapy.

Michelle Lane, Ph.D., Rutgers. Role of retinoids in gastrointestinal cancer and development.

Ben Liu, Ph.D., Columbia. Bioorganic chemistry; bacterial cell wall formation; biosynthesis of antibiotics; lipid metabolism; posttranslational modification of nuclear proteins.

Alan Lloyd, Ph.D., Stanford. Molecular genetics and evolution of development of *Arabidopsis.*

Paul MacDonald, Ph.D., Vanderbilt. Localization and spatially regulated translation of mRNAs in *Drosophila* development.

Lara Mahal, Ph.D., Berkeley. Creating new tools for glycobiology.

Dmitrii Makarov, Ph.D., Institute of Chemical Physics (Moscow). Kinetics of RNA folding; mechanical properties of single protein molecules.

Edward M. Marcotte, Ph.D., Texas at Austin. Protein function and interactions; bioinformatics; proteomics; biological networks; evolution of genomes, proteomes, and protein interactions.

Stephen Martin, Ph.D., Princeton. Bioorganic chemistry; enzyme mechanism and specificity; molecular recognition.

Mona Mehdy, Ph.D., California, San Diego. Gene regulation in plants induced by pathogens and environmental stress.

Richard Meyer, Ph.D., Pennsylvania. Transfer of broad host-range plasmids by conjugation.

John Mihic, Ph.D., Toronto. Molecular sites of action of alcohols and anesthetics on neurotransmitter-activated ion channels; mechanisms of ion channel activation and desensitization.

Edward Mills, Ph.D., Purdue. Thermoregulation, metabolism, and aging.

Ian Molineux, Ph.D., Oxford. Biochemical mechanisms of DNA translocation; host-parasite molecular and evolutionary genetics.

Hitoshi Morikawa, M.D., Ph.D., Kyoto (Japan). Neurophysiology of the brain reward circuit and drug addiction.

Richard Morrisett, Ph.D., Alabama at Birmingham. Amino acid–derived neurotransmitter systems and synaptic transmission.

Theresa O'Halloran, Ph.D., North Carolina at Chapel Hill. Molecular genetics of membrane traffic and cytoskeleton.

Tanya Paull, Ph.D., UCLA. DNA repair and genomic stability.

Shelley Payne, Ph.D., Texas Health Science Center at Dallas. Genetic and molecular basis of bacterial pathogenicity.

Martin Poenie, Ph.D., Stanford. Ionic controls of cell cycle and development.

George Pollak, Ph.D., Maryland Medical School. Neurophysiology of the auditory system in echolocating bats.

John Richburg, Ph.D., Rutgers. Mechanism(s) of toxicant-induced germ cell death.

Austen Riggs, Ph.D., Harvard. Molecular biology and biochemistry of proteins.

Mendell Rimer, Ph.D., Maryland at Baltimore. Molecular mechanisms of synapse formation and plasticity.

Jon Robertus, Ph.D., California, San Diego. Protein crystallography; genetic engineering; structure-based drug design.

Stan Roux, Ph.D., Yale. Mechanisms of stimulus-response coupling in plant growth and development.

Krishendu Roy, Ph.D., Johns Hopkins. Polymer-based DNA delivery; DNA vaccines.

Rick Russell, Ph.D., Johns Hopkins. RNA-protein enzymes: formation and function.

Bob Sanders, Ph.D., Penn State. Immunogenetics.

Christine Schmidt, Ph.D., Illinois at Urbana-Champaign. Biomaterials and neural engineering.

Marty Shankland, Ph.D., Berkeley. Developmental pattern formation; body plan evolution.

Jason Shear, Ph.D., Stanford. Bioanalytical technique development, to characterize fundamental chemical properties of individual neurons.

John Sisson, Ph.D., Stanford. Cell and molecular biology of early *Drosophila* development; molecular mechanisms controlling animal cell cytokinesis.

David Stein, Ph.D., Stanford. Pattern formation in the *Drosophila* embryo.

Scott Stevens, Ph.D., North Carolina. Eukaryotic ribonucleoprotein structure and function.

Laura Suggs, Ph.D., Rice. Cardiovascular tissue engineering; extracellular matrix analogs; adult progenitor cells; vasculogenesis.

Paul Szaniszlo, Ph.D., North Carolina. General/medical mycology; fungal biochemistry; molecular biology.

Wesley Thompson, Ph.D., Berkeley. Developmental neurobiology.

Ming Tian, Ph.D., Harvard. Mechanisms of antibody diversification in B cells.

Philip Tucker, Ph.D., MIT. Molecular immunology; developmental gene regulation.

James Walker, Ph.D., Texas at Austin. DNA polymerization; initiation of replication; gene expression; cell division in bacteria.

John Wallingford, Ph.D., Texas at Austin. Molecular and cellular basis of embryonic morphogenesis.

Tandy Warnow, Ph.D., Berkeley. Computational and statistical aspects of phylogenetic reconstruction.

Christian Whitman, Ph.D., California, San Francisco. Enzyme mechanisms; stereochemistry and inhibition.

Whitney Yin, Ph.D., North Carolina at Chapel Hill. Structural and mechanistic properties of enzymes and other proteins involved with RNA transcription and DNA replication.

Harold Zakon, Ph.D., Cornell. Development, evolution, and neurophysiology of electrosensory systems.

Muhammad Zaman, Ph.D., Chicago. Cellular engineering; tumor cell interactions with extracellular matrix.

Bing Zhang, Ph.D., Cornell. Genetic dissection of synaptic vesicle recycling and synaptic plasticity.

Zhiwen Zhang, Ph.D., Texas at Austin. Chemical biology; bioorganic chemistry; bioengineering; glycobiology.

THE UNIVERSITY OF TEXAS AT DALLAS

Graduate Programs in Molecular and Cell Biology

Programs of Study

The mission of the Department of Molecular and Cell Biology at The University of Texas at Dallas (UTD) is to maintain a nationally competitive research program, to train students to become outstanding scientists, and to provide an exceptional education that prepares students for careers and continued education in the life sciences, health, and medicine.

The Ph.D. in Molecular and Cell Biology Program helps students develop a critical and analytical understanding of current developments in the field. Candidates are required to complete 90 semester credit hours (SCH), including 24 in core courses and 9 in graded general electives. The remaining credit hours are devoted to special electives, research, and dissertation. At the end of the second year, students take a qualifying examination, which consists of a research proposal on the subject of the student's dissertation and an oral defense of that proposal. Prior to a final oral defense, students are required to publish at least one manuscript based on their dissertation research. All requirements must be completed within ten years.

The Master of Science (M.S.) in Molecular and Cell Biology Program is designed for students who wish to learn the methodology of research and the fundamentals of problem solving in molecular and cell biology. Students are required to complete 36 SCH, including 17 in core courses and 6 in graded general electives. The remaining credit hours are typically devoted to research and thesis. A nonthesis option is available. All requirements must be completed within six years.

The M.S. in Biotechnology Program, which is administered by the Department of Molecular and Cell Biology and an interdisciplinary advisory Committee on Biotechnology, requires 36 hours of courses, including 12 hours in core courses. Students may also elect to prepare and defend a thesis, which may require more than 36 hours. The program prepares students for careers in biotechnology and assists currently employed professionals in enhancing their career opportunities. Biotechnology captures the exciting possibilities made possible by the decoding of the human genome and by advances in bioanalytical instrumentation. The M.S. in Biotechnology Program is designed so that students may enter with a wide range of prior disciplinary backgrounds.

The M.S. in Bioinformatics and Computational Biology (BCBM) Program is offered jointly by the Departments of Mathematical Sciences (MMS) and Molecular and Cell Biology (MCB). This program combines course work from the disciplines of biology, computer science, and mathematical sciences. Faculty members from both MMS and MCB participate in this program, with the Department of Mathematical Sciences serving as the administrative unit. Both departments advise students. For more information, students should contact the Department of Mathematical Sciences.

Research Facilities

Major items of equipment used by the faculty and available for graduate student research include a Leica TCS SP2 AOBS confocal microscope system, ThermoFinnigan LCQDECA XP ion trap mass spectrometer, complete Spectra-Physics femtosecond laser system, Becton Dickson fluorescence activated cell sorter, Veeco MultiMode SPM atomic force microscope, Perkin Elmer DNA chip reader, Molecular Dynamics PhosphoImagers, BioRad real-time polymerase chain reaction instruments, Beckman scintillation counters, Beckman Optima ultracentrifuges, Jasco J-715 spectropolarimeter, and mammalian cell culture facilities. The Eugene McDermott Library maintains extensive subscriptions to electronic journals.

The areas of faculty research include studies of eukaryotic and prokaryotic gene expression and regulation, genetic recombination, insect toxin receptors, bionanotechnology, the membrane skeleton's role in blood and brain disorders, mitochondrial biogenesis, molecular control of neuronal apoptosis, sickle cell disease, signal molecules in bacteria-plant symbiosis, and DNA-protein complexes. Departmental faculty members are active in the Institute of Biomedical Sciences and Technology, which emphasizes cutting-edge interdisciplinary projects that seek to enhance human health and quality of life by integrating the multidisciplinary expertise of scientists, mathematicians, engineers, and computer and social scientists. In addition, the UTD Sickle Cell Disease Research Center aims to find better therapeutics for sickle cell disease and ultimately a cure by focusing at the molecular, cellular, and whole organism level. The Director and staff are also committed to training promising biomedical researchers. The Center at UTD maintains joint research efforts with the UT Southwestern Comprehensive Sickle Cell Program.

Financial Aid

Outstanding Ph.D. degree applicants receive a teaching assistantship covering full tuition and fees and a stipend of $11,986 for the academic year. For master's degree students, a number of financial aid options are available, including Federal Perkins Loans, subsidized and unsubsidized Federal Stafford Students Loans, and the Texas Public Educational Grant.

Cost of Study

In 2006–07, graduate tuition is $675 for the first SCH for residents and $950 for nonresidents. Each additional SCH costs about $325 for residents and about $600 for nonresidents. Students also pay $75–$155 in parking fees.

Living and Housing Costs

Students living on campus pay $450 to $620 per month for a one-bedroom apartment or $860 to $935 per month for a two-bedroom unit. Meal plans are not available, but students can use designated cards to purchase meals at one of several dining areas on campus. Off-campus apartments are available, ranging from $600 to $1400 per month, depending on size and location.

Student Group

There are more than 5,000 graduate students and 9,400 undergraduate students at UTD, representing most states in the U.S. and many other countries. In the biology graduate programs, there are 94 students; 54 are women, and 57 are international students.

Student Outcomes

Graduates of the program are prepared for academic careers in colleges and universities, including medical and dental schools, and for careers in industrial, hospital, public-health, environmental, and governmental laboratories and organizations.

Location

The University is located on the north side of Richardson, a suburb about 17 miles north of downtown Dallas. The Dallas metropolitan area offers a wide range of cultural, social, sports, and recreational activities.

The University

The University of Texas at Dallas offers educational and research programs of the highest quality that address the multidimensional needs of a society driven by the development, diffusion, understanding, and management of advanced technology. Founded in 1969 as an outgrowth of the Southwest Center for Advanced Studies, UTD provides a unique learning environment to its more than 14,000 students, with seven schools, over thirty research centers and institutes, an array of interdisciplinary degree programs, and a world-class faculty that includes two Nobel laureates.

Applying

For full participation in the Graduate Programs in Molecular and Cell Biology, the student should have a good background in calculus, general physics, organic chemistry, biochemistry, and general biology, including genetics. Entering students not having this background should take additional course work in their first year or in the summer immediately preceding entry.

Applicants must submit an application for admission, official transcripts from all colleges/universities attended, official GRE scores of 1000 or higher, three letters of recommendation, a brief narrative of academic interests and long-range research interests, and a $50 application fee ($150 for international students). International students must also submit a TOEFL score of at least 550 on the paper-based test or 213 on the computer-based test. The deadline to apply is July 1 for fall admission or November 1 for spring admission. Additional application information, including links for the application and reference letter forms, can be found at http://www.utdallas.edu/whyutd/prospective/graduate.html.

Students seeking further information or advisement should go online to the graduate catalog at http://www.utdallas.edu/student/catalog/gradcurrent/ or contact the Department of Molecular and Cell Biology.

Correspondence and Information

<table>
<tr>
<td>

Dr. Ernest M. Hannig, Co-Chair of the Graduate Education
   Committee
Department of Molecular and Cell Biology
University of Texas at Dallas
P.O. Box 830688, FO 31
Richardson, Texas 75083-0688

Phone: 972-883-2505
E-mail: hannig@utdallas.edu
Web site: http://www.utdallas.edu/biology/

</td>
<td>

Dr. Lawrence J. Reitzer, Co-Chair of the Graduate Education
   Committee
Department of Molecular and Cell Biology
University of Texas at Dallas
P.O. Box 830688, FO 31
Richardson, Texas 75083-0688

Phone: 972-883-2502
E-mail: reitzer@utdallas.edu
Web site: http://www.utdallas.edu/biology/

</td>
</tr>
</table>

The University of Texas at Dallas

THE FACULTY AND THEIR RESEARCH

Gail A. M. Breen, Associate Professor; Ph.D. (neuroscience), UCLA. Biogenesis of the mammalian mitochondrial oxidative phosphorylation system; the regulation of mitochondrial gene expression; analysis of the mitochondrial proteome.

Lee A. Bulla, Professor; Ph.D. (microbiology and biochemistry), Oregon State. Invertebrate and microbial molecular biology, with particular focus on biochemical and biophysical characterization of insecticidal toxin receptors in insects.

John G. Burr, Associate Professor; Ph.D. (molecular biology), Berkeley. Eukaryotic cell-growth regulation and oncogensis; oncogenic transformation of cells by Rous sarcoma virus.

Vincent P. Cirillo, Senior Lecturer; Ph.D. (zoology), UCLA. Mechanism of sugar transport; wild type and mutant yeast.

Jeff L. DeJong, Associate Professor; Ph.D. (biochemistry), Penn State. Factors and mechanisms responsible for the transcription of eukaryotic genes.

Santosh D'Mello, Professor; Ph.D. (biology), Pittsburgh. Understanding how apoptosis (programmed cell death) is regulated in neurons of the mammalian brain.

Rockford K. Draper, Professor; Ph.D. (biological chemistry), UCLA. Molecular pathogenesis of protein toxins, such as cholera toxin, membranes trafficking in eukaryotic cells, and bionanotechnology.

Juan González, Professor; Ph.D. (microbiology and molecular genetics), UCLA. Role of exopolysaccharides in the nodulation of legumes by rhizobia and the molecular genetics of plant-microbe interactions.

Steven Goodman, C. L. and Amelia A. Lundell Professor of Life Sciences, Professor of Molecular and Cell Biology, and Director, Institute of Biomedical Sciences and Technology. Ph.D. (biochemistry), St. Louis. A macromolecular structure on the cytoplasmic surface of eukaryotic membranes called the spectrin membrane skeleton; sickle-cell disease.

Donald M. Gray, Professor and Department Head; Ph.D. (molecular biophysics), Yale. Structures of polynucleotides and DNA-protein complexes studied by circular dichroism spectroscopy.

Ernest M. Hannig, Associate Professor; Ph.D. (molecular genetics and microbiology), Rutgers. Protein-protein interactions; genetic and biochemical analysis of translation initiation factors; protein synthesis and its regulation in eukaryotes.

Stephen D. Levene, Associate Professor; Ph.D. (chemistry), Yale. Protein-DNA interactions in site-specific recombination and the structure and dynamics of nucleic acids in solution.

Robert C. Marsh, Senior Lecturer; Ph.D. (molecular biology), Vanderbilt. Subcellular structure; identification and characterization of nuclear matrix proteins, cell-surface lectins, and the protein cross-linking enzyme transglutaminase.

Dennis L. Miller, Associate Professor and Associate Department Head; Ph.D. (biochemistry), Iowa. Structure and organization of mitochondrial DNA, mitochondrial gene expression, RNA editing, and mitochondrial biogenesis; extent and mechanism of RNA editing as a step in the mitochondrial gene expression of *Physarum polycephalum*.

John H. Moltz, Senior Lecturer; Ph.D. (physiology), Texas Health Science Center at Dallas. Human anatomy and physiology curriculum for preprofessional health-care students.

Betty Pace, Professor; M.D., Medical College of Wisconsin. Director, Sickle Cell Disease Research Center. Design of novel treatments for sickle-cell disease; molecular mechanisms involved in fetal hemoglobin synthesis; signal transduction pathways that mediate gamma gene reactivation.

Lawrence J. Reitzer, Professor; Ph.D. (molecular and cell biology), Washington (St. Louis). Regulation of gene expression and metabolism in *Escherichia coli* and pathogenic bacteria; pathways of the catabolism of polyamines; function of transaminases.

Scott A. Rippel, Senior Lecturer; Ph.D. (molecular and cell biology), Texas at Dallas. Biochemistry laboratory techniques and coordination of undergraduate laboratory facilities.

Ilya Sapozhnikov, Senior Lecturer; Ph.D., Academy of Medical Sciences (USSR); M.D., Leningrad Medical School (Russia). Epidemiology, prevention, and treatment of arterial hypertension.

Stephen Spiro, Associate Professor; Ph.D. (molecular biology), Sheffield (UK). Regulation of bacterial gene expression by environmental signals; consequences of gene regulation for physiological adaptation to stress.

Tianbing Xia, Assistant Professor; Ph.D. (biophysical chemistry), Rochester. Understanding the principles of molecular recognition, biomolecular structures, folding and dynamics, and correlations between structures, energetics, dynamics, and functions of important biomolecules; development of a technique using ultrafast laser spectroscopy to capture molecular movements so they can be viewed and analyzed.

THE UNIVERSITY OF TEXAS
HEALTH SCIENCE CENTER AT SAN ANTONIO

Graduate School of Biomedical Sciences
Program in Molecular Medicine

Programs of Study

The Program in Molecular Medicine offers a research-oriented, interdisciplinary course of study leading to the M.S. and Ph.D. degrees. The faculty is composed of both basic and clinical scientists drawn from the Departments of Biochemistry, Molecular Medicine, Cellular and Structural Biology, Medicine, Surgery, Pathology, and Physiology. The objective of the program is to train future scholars in the use of molecular biological approaches for the investigation of fundamental biomedical questions associated with the diagnosis and treatment of human diseases. Upon completion of the program, students are prepared for careers as independent investigators and teachers in cellular and molecular medicine.

During the first year, students attend core courses in advanced molecular biology, molecular medicine, and laboratory techniques. At the same time, they are introduced to research through a series of rotations in the laboratories of individual faculty members. At the end of the first year, students must pass an oral comprehensive examination covering material presented in the first-year classes. Following successful completion of the comprehensive examination, each student selects a faculty adviser and begins doctoral research. During their third year in the program, students must pass the qualifying evaluation, which consists of a written dissertation proposal followed by defense of the proposal in an oral examination. Completion of course work, the comprehensive and qualifying examinations, and doctoral research should take four to five years.

The research interests of the faculty cover many areas of molecular and cell biology, including the molecular genetic basis of cancer and tumorigenesis, mechanisms of cancer metastasis, animal models of disease, transcriptional regulation, development of anticancer drugs, control of mammalian development, bone cell biology in health and disease, mouse genetics, molecular biological basis of aging, DNA repair, genetic recombination, eukaryotic cell-cycle regulation, protein structure, protein degradation, and signal transduction.

Research Facilities

The laboratories of the molecular medicine program faculty members are located at the University of Texas Institute of Biotechnology and the Institute for Drug Development in the Texas Research Park as well as at the main campus of the Health Science Center. Also available are state-of-the-art facilities for cellular and molecular biological research and biochemistry as well as specialized instrumentation required for electron, fluorescence, confocal, and atomic force microscopy; the generation of transgenic and chimeric mice; biomolecular interaction studies; biopolymer synthesis; peptide and nucleic acid sequencing; and protein purification.

Financial Aid

All students accepted into the Ph.D. program receive graduate assistantships with stipends of $21,500 for 2006–07, with no residency requirements. Special Institute of Biotechnology predoctoral awards are competitively available for domestic students in addition to the stipend.

Cost of Study

The 2006–07 tuition is $96 per semester hour for Texas residents and $326 per semester hour for nonresidents. Nonresident students receiving assistantships are eligible for resident tuition rates.

Living and Housing Costs

Affordable student housing is available within a 5-minute walk of the Institute of Biotechnology. Monthly rent averages about $400 for a one-bedroom apartment, and additional housing is available within a short driving distance of the Institute. Housing costs in San Antonio are generally much lower than in most major U.S. cities.

Student Group

The graduate program has a current enrollment of 41, of whom 22 are women. It is anticipated that 10 to 15 students are to be admitted each year. The enrollment is small enough to permit close association with faculty members yet large enough to allow for a variety of educational experiences.

Student Outcomes

Since the program was established in 1993, 42 students have completed Ph.D. training and 16 have completed work for their master's degree.

Location

San Antonio is a dynamic city of more than 1 million people. Located in the subtropical climate of south-central Texas, San Antonio provides excellent opportunities for all types of outdoor activities year-round. The Gulf Coast and Mexico are within 150 miles of San Antonio. The city is surrounded by many rivers and lakes as well as the scenic hill country. San Antonio offers outstanding cultural opportunities, including museums, theaters, a symphony orchestra, the ballet, historic Spanish missions, Sea World, Fiesta Texas theme park, the downtown Riverwalk, and an excellent zoo. The Spurs, of the National Basketball Association, play in the new SBC Center.

The Health Science Center

The primary mission of the Health Science Center is to provide instruction and research experience for scientists, physicians, dentists, nurses, and other health professionals. The current student enrollment of more than 2,700 includes 800 medical students, 400 dental students, 1,000 health professional students, 300 graduate students, and 150 postdoctoral fellows. The faculty numbers more than 1,000 members.

Applying

Applications and supporting materials should be submitted by February 15 for fall admission; however, applications received after February 15 still receive consideration. Applicants must have a bachelor's degree or the equivalent, satisfactory scores on the Graduate Record Examinations, a good academic record, and letters of recommendation from individuals in academic or professional positions.

Correspondence and Information

Chair, Committee on Admissions
Department of Molecular Medicine
Institute of Biotechnology
The University of Texas Health Science Center at San Antonio
15355 Lambda Drive
San Antonio, Texas 78245-3207
Phone: 210-567-7200
E-mail: molecularmedicine@uthscsa.edu
Web site: http://www.molecularmedicine.uthscsa.edu

The University of Texas Health Science Center at San Antonio

THE FACULTY AND THEIR RESEARCH

Hanna E. Abboud, Professor; M.D., Alexandria (Egypt). Role of cytokines in renal disease.

David H. Boldt, Professor; M.D., Tufts. Lymphocyte activation, proliferation, and differentiation; hematologic malignancies.

Thomas G. Boyer, Associate Professor; Ph.D., SUNY at Buffalo. Role of oncoproteins and tumor suppressor proteins in transcription regulation and DNA repair.

Bandana Chatterjee, Professor; Ph.D., Nebraska. Nuclear receptors and metabolic regulation; androgen receptor in prostate cancer and gene regulation.

Barbara Christy, Associate Professor; Ph.D., Johns Hopkins. Function and interactions of proteins regulating cell growth, differentiation, and tumor formation.

Robert A. Clark, Professor; M.D., Columbia. Neutrophil signal transduction and activation of oxygen-dependent microbicidal and cytotoxic systems.

Maria Gaczynska, Associate Professor; Ph.D., Lodz (Poland). Biochemistry and biophysics of macromolecular biological assemblies, with special emphasis on proteolytic systems.

Shou-Jiang Gao, Associate Professor; Ph.D., Bordeaux (France). Kaposi's sarcoma–associated herpesvirus (KHSV) epidemiology and pathogenesis.

David Haile, Associate Professor; M.D., Johns Hopkins. Regulation of genes of iron metabolism by inflammation.

P. John Hart, Associate Professor; Ph.D., Texas at Austin. Metalloprotein structure, action, and redesign; role of copper-zinc superoxide dismutase in Lou Gehrig's disease; structural biology of metal trafficking; blue copper proteins; protein crystallography.

E. Paul Hasty, Associate Professor; D.V.M., Texas A&M. Genetics of chromosomal metabolism, with emphasis on cancer and aging.

Andrew P. Hinck, Associate Professor; Ph.D., Wisconsin–Madison. Solution NMR spectroscopy of proteins and nucleic acids; transforming growth factor B and its interaction with the ligand-binding domain of the TGF-B type I and type II receptors; protein-RNA interactions.

Robert J. Klebe, Professor; Ph.D., Yale. Biochemical mechanism of cell adhesion and cellular morphogenesis.

Jeffrey Kreisberg, Professor; Ph.D., Maryland. Pathogenesis of diabetic glomerulopathy and molecular biology of mesangial cells; signal transduction molecules in cancer.

Sang Eun Lee, Assistant Professor; Ph.D., Brown. Molecular genetics of DNA damage response.

Mary Pat Moyer, Professor; Ph.D., Texas at Austin. Gastrointestinal cell biology, oncology, and toxicology.

Hai Rao, Assistant Professor; Ph.D., SUNY at Stony Brook. Functions of the ubiquitin system.

Arlan Richardson, Professor; Ph.D., Oklahoma. Role of oxidative stress and DNA damage in aging and age-related pathologies.

Z. Dave Sharp, Associate Professor; Ph.D., Arkansas Medical Sciences. Gene regulation in development and aging.

Rajeshwar Rao Tekmal, Professor; Ph.D., Kurukshetra (India). Breast cancer and gynecological malignancies; tumor biology; steroid hormone and growth factor action and regulation.

Manjeri Venkatachalam, Professor; M.B., Calcutta. Cell survival signaling through the PI3 kinase/PDK1/Akt pathway; molecular mechanisms of necrotic and apoptotic cell death by hypoxia/reoxygenation.

P. Renee Yew, Associate Professor; Ph.D., UCLA. Cell-cycle regulation and DNA replication in vertebrates; protein degradation and its role in cancer.

UNIVERSITY OF UTAH

Program in Molecular Biology

Programs of Study	The Program in Molecular Biology has a number of faculty members drawn from the Departments of Biochemistry, Biology, Chemistry, Human Genetics, Medicinal Chemistry, Neurobiology and Anatomy, Oncological Sciences, and Pathology. Research interests include genetics, developmental biology, neurobiology, biochemistry/structural biology, cell biology, cancer biology, gene expression, and microbiology/immunology. Students admitted by the Program in Molecular Biology take a common set of first-year courses. Each student chooses four different research laboratories. In order to become acquainted with possible research supervisors. A student in good standing at the end of the first academic year chooses a faculty research supervisor for further doctoral study and is automatically admitted to the graduate program of the supervisor's department. Additional requirements for the Ph.D. are set by individual departments, but they are similar throughout the University.
Research Facilities	The participating departments all have up-to-date facilities. Shared research equipment includes electron microscopes, X-ray crystallographic diffraction instrumentation, NMR and mass spectrometers, oligonucleotide synthesizers, and protein sequencing equipment. For more information, students should visit http://www.cores.utah.edu. There are centralized facilities for production of monoclonal antibodies and for cell microinjection. There is an extensive network linking most microcomputers and several large mainframe computers. There are two comprehensive science libraries, one on the main campus (near the Departments of Biology and Chemistry) and the other at the Medical Center.
Financial Aid	The Program in Molecular Biology supports students during the first year without requiring teaching. Support after the first year is from individual departments and research programs through teaching assistantships, research assistantships, and NIH training grants. One semester of teaching experience is generally required after the first year, regardless of the source of support. The level of support for 2005–06 was $23,000, plus a $1000 starting allocation for the year. The program also covers the cost of health and dental insurance.
Cost of Study	Tuition and fees are paid by the Program in Molecular Biology for the first year and after that by departments or by individual research programs; currently there is no cost to students.
Living and Housing Costs	Housing is available within walking distance of the University in private apartments and in University housing. Rent for a typical one-bedroom unfurnished private apartment is $550 per month both on and off campus. University housing includes all utilities except a telephone.
Student Group	The Program in Molecular Biology admits 30–40 graduate students per year. The University as a whole has about 6,100 graduate students, with approximately 300 graduate students in disciplines related to the program. Students come from all states and many countries.
Location	The University is located in Salt Lake City, 1½ miles east of the city center, in a quiet residential neighborhood. The site is at an altitude of 4,500 feet just at the base of the Wasatch Range, which rises precipitously to the east and south of the University. The main campus, home to the Department of Biology, is less than 1 mile from the Medical Center, which houses the Departments of Biochemistry, Oncological Sciences, Human Genetics, Neurobiology, and Pathology. Travel between the two campuses is a pleasant 15-minute walk; there are also frequent free shuttle buses. Metropolitan Salt Lake City, with a population of about 820,000, is the economic and cultural center of the intermountain region of the western United States. The city is well-known for classical music (Utah Symphony), dance (Repertory Dance Theater, Ballet West), and sports (Utah Jazz and University teams). Salt Lake City is a popular vacation destination for outdoor enthusiasts because of the nearby mountains, forests, and deserts, with opportunities for skiing, climbing, hiking, fishing, and camping close at hand. Nine national parks are within a day's drive.
The University	Founded in 1850, the University of Utah is the oldest state university west of the Missouri. The total enrollment is approximately 28,000. There are about 1,700 teaching faculty members and 1,800 adjunct, research, and clinical faculty members. The University includes colleges of medicine, law, education, business, mines and mineral industries, science, humanities, social and behavioral sciences, health, fine arts, engineering, nursing, pharmacy, and social work.
Applying	Application materials are available on the program's Web site, listed in this description, or by request. The program requires a completed application form, transcripts of grades of all University courses, a report on the General Test portion of the Graduate Record Examinations (verbal, quantitative, and analytical writing), three letters of recommendation solicited by the applicant, and a detailed personal statement of career interests. The University's application fee is paid by the program. Applications are due January 15 for entrance the following August. The program's Admissions Committee invites promising prospective students for personal interviews; travel expenses for invited students are paid by the program.
Correspondence and Information	Anthea Letsou, Director Interdepartmental Graduate Program in Molecular Biology 533 EIHG Room 1400 University of Utah 15 North 2030 East Salt Lake City, Utah 84112 Phone: 801-581-5207 Fax: 801-585-2465 E-mail: tami.brunson@genetics.utah.edu Web site: http://www.bioscience.utah.edu

University of Utah

FACULTY INTEREST GROUPS

Biochemistry/Structural Biology
E. Dale Abel. Molecular mechanisms in diabetes.
Brenda Bass. dsRNA binding proteins.
David Blair. Bacterial motility.
Don Blumenthal. Protein kinases.
Darrell Davis. RNA structure.
Tim Formosa. DNA replication; chromatin.
David Goldenberg. Protein dynamics and function.
Charles Grissom. Cancer drug delivery and enzymology.
Chris Hill. Protein structure and function.
Martin Horvath. Structural biology; X-ray crystallography.
Michael Kay. Protein design.
Janet Lindsley. DNA-based enzymes.
John Phillips. Heme biosynthesis.
Glenn Prestwich. Lipid signaling and biomaterials.
Martin Rechsteiner. Intracellular proteolysis.
Jared Rutter. Metabolic signaling.
Wes Sundquist. Retrovirus biochemistry.
Stan Williams. Prokaryotic circadian rhythms.
Dennis Winge. Metalloproteins; metalloregulation.

Cancer Biology
Doug Grossman. Apoptosis and skin cancer.
David Jones. Colon cancer development.
Stephen Lessnick. Pediatric cancer development.
Nadeem Moghal. Growth factors; development; cancer.
Steve Prescott. Signaling; lipids; inflammation.
Wolfram Samlowski. Cancer immunotherapy.
Nikolaus Trede. Zebrafish immunology cancer.
David Virshup. Signal transduction.

Cell Biology
Rick Ash. Membrane transport.
Markus Babst. Protein trafficking.
Mary Beckerle. Cell adhesion; cytoskeleton.
David Gard. Microtubules and MAPs.
Don McClain. Molecular biology of diabetes.
Shannon Odelberg. Regeneration and cellular plasticity.
Janet Shaw. Mechanisms of mitochondrial dynamics.
Katharine Ullman. RNA export; cell division.

Developmental Biology
Mario Capecchi. Molecular genetics.
Richard Dorsky. Developmental neurobiology.
Sabine Fuhrmann. Regulation of early eye development.
David Grunwald. Developmental genetics in zebrafish.
Gabrielle Kardon. Musculoskeletal development.
Darryl Kropf. Plant development.
Anthea Letsou. Signaling and developmental biology.
Edward Levine. Developmental neurobiology.
Dean Li. Pathogenesis of vascular disease.
Susan Mango. Development and physiology of organs.
Susanne Mansour. Fgf signaling; ear development.
Anne Moon. Models human congenital malformations.
Charles Murtaugh. Pancreas development.
Tatjana Piotrowski. Sensory organ development in zebrafish.
Yukio Saijoh. Mouse embryogenesis.

Gary Schoenwolf. Pattering during embryogenesis.
Leslie Sieburth. Plant developmental genetics.
Carl Thummel. Development genetics; transcription.
Monica Vetter. Neural development.
Qiang Wu. Gene expression; bioinformatics.
Joseph Yost. Cancer genetics; developmental biology.

Gene Expression
John Atkins. Reprogrammed genetic decoding.
Don Ayer. Transcription; nuclear oncogenes.
Wolfgang Baehr. Mammalian phototransduction.
Brad Cairns. Chromatin; transcription; genomics.
Barbara Graves. Transcription; protein biochemistry.
Betty Leibold. Iron regulation of gene expression.
David Stillman. Gene regulation.

Genetics
Art Brothman. Molecular mechanisms of prostate cancer.
Dana Carroll. Recombination; gene targeting.
Christiane Fauron. Plant mitochondrial genomes.
Kevin Flanigan. Molecular neurogenetics.
Ray Gesteland. Proteosome and ribosome decoding.
Kent Golic. Chromosome structure-function.
Sandra Hasstedt. Genetic epidemiology.
Lynn Jorde. Human population genetics.
Jerry Kaplan. Iron metabolism; intracellular organelle.
Rajendra Kumar-Singh. Gene therapy.
Jean-Marc Lalouel. Human genetics.
Mark Leppert. Human genetics.
Wayne Potts. Immunogenetics.
Shige Sakonju. Regulation of gene expression.
Kang Zhang. Genetics of human vision.

Microbiology/Immunology
Sherwood Casjens. Bacteriophage molecular genetics.
Ray Daynes. Cellular and mucosal immunology.
Robert Fujinami. Viral pathogenesis/autoimmunity.
Lorise Gahring. Neuroimmune interactions.
Wai Mun Huang. Nucleic acid enzymology.
Diane McVey-Ward. Receptor-mediated endocytosis.
Matthew Mulvey. Bacterial pathogenesis.
Vicente Planelles. HIV-1 pathogenesis.
Gerald Spangrude. Stem cell biology.
Janis Weis. Lyme arthritis.
John Weis. Genetics of immune response.

Neurobiology
Chi-Bin Chien. Zebrafish axon guidance.
Maureen Condic. Sensory neural development.
Erik Jorgensen. Genetic analysis/synaptic transmission.
Andres Villu Maricq. Synaptic function.
Baldomera Olivera. Molecular neurobiology; conotoxins.
Scott Rogers. Neurobiology.
Gary Rose. Neurobiology and behavior.
A. Sanchez Alvarado. Molecular basis of regeneration.
Aloisia Schmid. Pathology of neurodegeneration.

UNIVERSITY OF VERMONT

uvm

Cell and Molecular Biology Program

Program of Study

The Graduate College offers an interdepartmental program leading to the degrees of Master of Science and Doctor of Philosophy in the field of cell and molecular biology. The program, which spans the Colleges of Arts and Sciences, Agriculture and Life Sciences, and Medicine, is composed of faculty members drawn from the Departments of Agricultural Biochemistry, Anatomy and Neurobiology, Animal and Food Science, Biochemistry, Biology, Biomedical Technology, Botany, Medicine, Microbiology and Molecular Genetics, Molecular Physiology and Biophysics, Obstetrics and Gynecology, Pathology, Pediatrics, Pharmacology, Physics, and Surgery. The Cell and Molecular Biology Program is directed toward an understanding of the cell, its components and architecture, the functional relations of its parts, and the properties that have given it a central position in biological theory. The goal of the multidisciplinary program is to bring together the University of Vermont's diverse facilities and students and faculty members whose concerns are the activities of the cell and how these activities are integrated within the organism.

Students typically work with a particular faculty member in one of the participating departments and a 4-member studies committee to initiate a thesis project and arrange a course program. The extensive research facilities of all of these departments are available to all graduate students in the program.

Research Facilities

Students have access to well-equipped research facilities for both biochemical and biophysical disciplines. Cell imaging and structural biology facilities are also available on campus. This program operates in conjunction with departments affiliated with the Vermont State Agricultural Experiment Station, the Vermont Cancer Center, and the Fletcher Allen Health Care Center. The combined collections of the libraries on campus, including the main UVM library, Bailey-Howe Library, Dana Medical Library, and the individual physical science and medical libraries, total 1,003,000 volumes.

Financial Aid

Financial aid consists of research assistantships, provided by the program, departments, or individual research grants, and teaching assistantships within the program or from individual departments. Assistantships for 2005–06 generally provided the equivalent of full tuition plus a living stipend of $20,772. Some traineeships and work-study or loan opportunities are also available.

Cost of Study

Tuition is $394 per credit hour for Vermont residents and $985 per credit hour for nonresidents. Fees range up to $574, depending upon credit-hour enrollment.

Living and Housing Costs

The University has housing for married graduate students on a space-available basis. Single graduate students generally live in off-campus rooms or apartments in the vicinity of the University. The optimal time for apartment hunting is May 15 through August 1.

Student Group

Current enrollment in the Graduate College is approximately 1,200; about 350 of these students are pursuing the doctorate. Students come from most states and many countries. There are currently 36 degree candidates in the Cell and Molecular Biology Program.

Student Outcomes

Graduates have been extremely successful in obtaining highly competitive postdoctoral positions at leading academic, governmental, and industrial laboratories throughout the world. Several recent graduates have advanced beyond the postdoctoral level and are part of university faculty or at the senior investigator level at research laboratories at the NIH or in industry.

Location

Burlington, situated on the east shore of Lake Champlain, serves as the metropolitan and cultural hub of Vermont. Lake Champlain, together with the nearby Green Mountains, provides recreational opportunities ranging from the largest ski resort to the best freshwater sailing in the Northeast. Burlington is located 200 miles northwest of Boston; 100 miles south of Montreal, Canada; and 300 miles north of New York City.

The University

Founded in 1791, the University of Vermont is a state-supported institution. It is an amalgam of a traditional New England liberal arts college, a land-grant agricultural college with its associated experiment station, and the state medical college with its affiliated hospitals. This combination of functions provides for diverse orientations among faculty members and students. The total enrollment is approximately 10,000.

Applying

The deadline for receipt of completed applications (and supporting materials) for admission in the fall semester is January 15. Applications must be supported by two official transcripts from each college or university attended, three letters of recommendation, and average GRE General Test scores at or above the 60th percentile. Undergraduate requirements include courses in biology, organic chemistry, calculus, and physics, although minor deficiencies can be made up after matriculation. An interview is recommended. Interested students may write to the address in the Correspondence and Information section for further information concerning the program.

Correspondence and Information

Cell and Molecular Biology Program
C242 Given
University of Vermont
Burlington, Vermont 05405
Phone: 802-656-9673
E-mail: cmb@uvm.edu

University of Vermont

THE FACULTY AND THEIR RESEARCH

Richard Albertini. Human mutagenicity studies.

Christopher Berger. Structural dynamics of molecular motors.

Jeffrey Bond. Structural biology of proteins and DNA.

Marcus Bosenberg. Determination of the cellular and molecular changes responsible for melanoma metastasis.

Jonathan Boyson. iNKT cells; innate immunity; the maternal-fetal interface.

Ralph Budd. Death receptor regulation of apoptosis in the immune system.

John Burke. Ribozymes and RNA structure.

Kelvin Chu. Kinetic crystallography.

Kenneth Cutroneo. Molecular mechanisms of wound healing.

Terrence Delaney. Pathogen-induced disease resistance in plants.

Rona Delay. Neuroscience/olfaction.

Wolfgang Dostmann. cGMP/PKG signaling in vascular smooth muscle.

Sylvie Doublié. X-ray crystallography.

Richard Dutton. Immune cell responses to pathogens and tumors.

Stephen Everse. X-ray crystallography of blood coagulation proteins and transferrins.

Barry Finette. Mutagenic mechanisms in pediatric diseases.

Christopher Francklyn. Structure and function of nucleic acid–binding proteins.

Naomi Fukagawa. Nutritional biochemistry; aging and oxidative stress.

Jeanne Harris. Developmental genetics of legume root nodules.

Nicholas Heintz. Cell-cycle control.

Russ Hovey. Mammary gland molecular and cellular endocrinology.

Alan Howe. Regulation of cell migration and division by extracellular matrix.

Sally Huber. Coxsackievirus-induced autoimmune heart disease and atherosclerosis.

Christopher Huston. *Entamoeba histolytica* phagocytosis of apoptotic host cells.

Charlie Irvin. Asthma pathogenesis.

Yvonne Janssen-Heininger. Transcriptional regulators of lung responses to injury and inflammation.

Douglas Johnson. Signaling pathways regulating yeast morphogenesis and virulence.

Abrar Khan. Antigen presentation and function of CD52.

David Krag. Development of targeted therapeutics.

Steven Lidofsky. Mechanisms of cell volume regulation in liver.

Karen Lounsbury. Vascular smooth muscle signaling: tumor angiogenesis and atherosclerosis.

Mariana Matrajt. Microbial pathogenesis.

Dwight Matthews. Analytical and clinical applications of mass spectrometry.

Scott Morrical. DNA repair: structure and function of protein-DNA complexes.

Brooke Mossman. Cell signaling in lung responses to environmental pathogens.

David Pederson. Regulation of transcription and replication.

Matthew Poynter. Airway epithelium in innate and adaptive immune responses.

Matthew Rand. Notch receptor pathway.

Mercedes Rincon. Signaling and gene regulation in thymus development, activation, differentiation, and death of CD4+ and CD8+ T cells.

Mark Rould. Crystallographic analysis of protein-ligand interactions.

Dinender Singla. Embryonic stem cells derived cardiac differentiation and heart regeneration.

Benjamin Suratt. Mechanisms of neutrophil regulation in homeostasis and inflammation.

Doug Taatjes. Cell imaging and disease.

Cory Teuscher. Genetics of susceptibility to infectious and autoimmune disease.

Markus Thali. HIV assembly and exit.

Mary Tierney. Plant growth and development.

Paula Tracy. Biochemistry of platelet function.

Albert van der Vliet. Redox regulation of airway inflammation; chemistry and biology.

Judy Van Houten. Chemosensory signal transduction in mouse olfactory neurons and paramecium.

Jim Vigoreaux. Biology of insect flight muscle and vertebrate cardiac muscle.

Susan Wallace. DNA damage and repair.

Gary Ward. Cell biology of host-parasite interaction.

David Warshaw. Myosin molecular motors.

Daniel Weiss. Primary focus on cystic fibrosis; asthma and emphysema.

Laurie Whittaker. Airway inflammation in asthma and cystic fibrosis.

David Yandell. Genetic mutations linked to cancer.

Feng-Qi Zhao. Oct transcription factors in mammary gland development, lactation, and breast cancer.

Lake Champlain and the Adirondacks.

Billings Student Center.

Fountain on the campus green.

UNIVERSITY OF WASHINGTON/ FRED HUTCHINSON CANCER RESEARCH CENTER

Molecular and Cellular Biology Program

Program of Study

The University of Washington and the Fred Hutchinson Cancer Research Center offer a program of graduate studies in molecular and cellular biology leading to the Ph.D. degree. More than 200 faculty members participate in the program and are located on the University of Washington campus in the Departments of Biochemistry, Bioengineering, Biological Structure, Biology, Genome Sciences, Immunology, Microbiology, Pathobiology, Pathology, Pharmacology, and Physiology and Biophysics, and on the Day campus at the Hutchinson Center, primarily in the Division of Basic Sciences and the Division of Human Biology. Recently, the Institute for Systems Biology (ISB), a nonprofit research institute headed by Dr. Leroy Hood, has joined the Molecular and Cellular Biology Program.

The goals of the program are to give the student a sound background in molecular and cellular biology and to provide access to the research expertise of all faculty members and laboratories working in this area. These goals are accomplished through the basic elements of the program, which include three quarters of core conjoint courses, a two-quarter literature review course, one quarter of grant writing, three or more quarter-long lab rotations, advanced elective courses in molecular and cellular biology, and a series of informal workshops and seminars on topics in diverse areas of molecular biology and cellular biology. Emphasis is placed on critical evaluation of the literature, exposure to current research methods, and creative thinking through independent research. Students are expected to begin active research in their first year through their lab rotations and to choose a permanent thesis adviser at the end of their first year.

Research Facilities

The program uses the research facilities of the individual departments and the Hutchinson Center. The School of Medicine is housed in the Health Sciences Center. The University Hospital and the College of Arts and Sciences are located in adjoining or nearby buildings. The Hutchinson Center's Day campus is a 15-minute shuttle ride from the University. The laboratories of participating faculty members are well equipped with the latest in research equipment and are funded by external support. The Institute for Systems Biology is located within easy commuting distance. Some of the other facilities available are two Howard Hughes Medical Institute research units, the Lucille P. Markey Genetic Medicine Center, the computer and information systems center, animal quarters, shared major instrument facilities, oligonucleotide and peptide synthesis facilities, a marine biology station at Friday Harbor in the San Juan Islands, and an extensive Health Sciences Library.

Financial Aid

The program offers a salary of approximately $25,010 for twelve months. Students with satisfactory academic progress can anticipate funding that includes tuition and health insurance for the duration of their studies.

Cost of Study

Tuition, salary, and medical, dental, and vision benefits are funded for the duration of the program for students in good standing.

Living and Housing Costs

The University has a wide variety of housing available for single and married students as well as families. Students should call the University Housing Office at 206-543-4059 for further information. Private accommodations may be found within easy walking or bicycling distance.

Student Group

At the University of Washington, approximately 2,500 full-time faculty members serve a student population of 35,000 that is drawn from all over the United States and many other countries.

Up to 25 new students are admitted to the program each year. There is a total of approximately 500 graduate students in the biological sciences at the University of Washington.

Student Outcomes

The Molecular and Cellular Biology Program received degree-granting status in 1994. The majority of students upon earning Ph.D. degrees secure postdoctoral research positions.

Location

All around Seattle, there is an abundance of opportunity for outdoor recreation. Unsurpassed sailing, hiking, mountain climbing, skiing, and camping are all a short distance away. Because of the saltwater expanse of the Puget Sound and the mountains both to the east and to the west, Seattle enjoys a moderate climate, with precipitation averaging 32 inches per year, mostly during the winter and early spring. The city's downtown area offers many cultural and educational advantages, including theater, symphony, films, and opera, while the waterfront is home to a large marketplace, galleries, and fresh seafood restaurants. The University itself sponsors many public lectures, concerts, exhibits, film festivals, and theatrical performances.

The University

The University of Washington is located in a residential section of Seattle near the downtown area. It is bordered by two lakes and is one of the largest and most scenic institutions of higher education in the country. The University is a research-intensive institution, regularly ranking first overall among public universities in externally funded research programs. It is recognized for graduate instruction of high quality, offering more than ninety graduate and professional programs that enroll more than 7,300 graduate students on campus. The Hutchinson Center's research laboratories are located by Lake Union near downtown Seattle.

Applying

Applicants must have completed a baccalaureate or advanced degree by the time of matriculation; degrees emphasizing biology, physical or natural sciences, and mathematics are preferred. It is advisable to take the GRE (the General Test and one Subject Test) no later than October so that scores can be recorded before the deadline (code 0206 on the GRE registration form). New students enter the graduate program in the autumn quarter. The deadline for completion of applications is currently December 15 of the academic year preceding entrance. Students may apply via the online application available at the MCB Program Web site.

Correspondence and Information

Graduate Program Specialist
Molecular and Cellular Biology Program, Box 357275
University of Washington
Seattle, Washington 98195-7275

Phone: 206-685-3155
Fax: 206-685-8174
E-mail: mcb@u.washington.edu
Web site: http://www.mcb-seattle.edu

University of Washington/Fred Hutchinson Cancer Research Center

THE FACULTY AND THEIR RESEARCH

UNIVERSITY OF WASHINGTON

Signal Transduction
Karol Bomsztyk, Dan Bowen-Pope, Rose Ann Cattolico, Jeffrey Chamberlain, Luca Comai, Lucio Costa, Michael Cunningham, Earl Davie, Jean Feagin, Stanley Fields, Sharona Gordon, Ted Gross, Sen-Itiroh Hakomori, Kenneth Kaushansky, Stanley McKnight, Randall Moon, David Morris, Charles Murry, Krzysztof Palczewski, Henk Roelink, Michael Rosenfeld, Carol Sibley, Nephi Stella, Ken Stuart, Ted White, Zhengui Xia, Wenqing Xu, Elton Young.

Molecular Genetics
Arnold Bendich, Peter Byers, Stanley Froehner, Milton Gordon, Benjamin Hall, Marshall Horwitz, Mary-Claire King, Charles Laird, Albert LaSpada, Åke Lernmark, Mary Lidstrom, Ray Monnat, James Mullins, Maynard Olson, Junko Oshima, Leo Pallanck, Ram Samudrala, Wesley Van Voorhis, Christophe Verlinde, Edith Wang, Sam Waterston, Alan Weiner.

Microbiology
Lee Ann Campbell, Carleen Collins, Richard Darveau, Nancy Freitag, David Goodlett, Peter Greenberg, Nancy Haigwood, Amanda Jones, Michael Lagunoff, John Leigh, Colin Manoil, Samuel Miller, Steve Moseley, Lalita Ramakrishnan, Marilyn Roberts, Evgeni Sokurenko, James Staley, Beth Traxler.

DNA Replication/Repair and Recombination
James Champoux, Brian Kennedy, Lawrence Loeb, George Martin.

Developmental Biology
Aimée Bakken, Celeste Berg, Michael Bevan, Mark Bothwell, Robert Braun, Beverly Dale-Crunk, Christine Disteche, Andrew Farr, Elaine Faustman, Nelson Fausto, Pamela Fink, Cecilia Giachelli, Joan Goverman, Stephen Hauschka, Merrill Hille, Raj Kapur, David Kimelman, Dina Mandoli, Richard Palmiter, David Parichy, David Raible, Thomas Reh, Lynn Riddiford, Tim Rose, Hannele Ruohola-Baker, Gerold Schubiger, Stephen Schwartz, Billie Swalla, James Thomas, Keiko Torii, Elizabeth Van Volkenburgh, Barbara Wakimoto, Norman Wolf.

Virology
Lawrence Corey, Michael Katze, Jaisri Lingappa, David Russell, Leonidas Stamatatos, Jeff Vieira.

Immunology
Alan Aderem, Mark Bix, Daniel Campbell, Patrick Concannon, Brad Cookson, Patrick Duffy, Keith Elkon, Ferric Fang, Yansong Gu, Murali-Krishna Kaja, Nancy Maizels, David Rawlings, Andrew Scharenberg, Kelly Smith, Amy Weinmann, Chris Wilson, Steven Ziegler.

Systems Biology
John Aitchison, Chris Amemiya, Nitin Baliga, Leroy Hood, Ilya Shmulevich.

Cell Cycle
Karin Bornfeldt, Trisha Davis, Marilyn Parsons, Peter Rabinovitch, Alexander Rudensky, Edith Wang, Zipora Yablonka-Reuveni.

Cell Motility/Cytoskeleton/Biomechanics
John Clark, Mark Cooper, Linda Wordeman.

Neurobiology
Sandra Bajjalieh, Joe Beavo, Albert Berger, Steven Carlson, William Catterall, Jeffrey Chamberlain, Charles Chavkin, Peter Detwiler, Katherine Graubard, Bertil Hille, James Hurley, William Moody, Neil Nathanson, John Neumaier, Robert Steiner, Daniel Storm, Jane Sullivan, Bruce Tempel, James Truman, A. O. Dennis Willows, William Zagotta.

Extracellular Matrix
James Bassingthwaighte, Paul Bornstein, Daniel Luchtel, Thomas Wight.

Molecular Structure
David Baker, Clement Furlong, Michael Gelb, Wim Hol, Rachel Klevit, Patrick Stayton, Ronald Stenkamp, Gabriel Varani.

Biochemistry
Susan Brockerhoff, Alexey Merz, David Sherman.

Gerontology
Stephen Plymate.

Pharmacology
Ning Zheng.

FRED HUTCHINSON CANCER RESEARCH CENTER

Signal Transduction
Jonathan Cooper, William Grady, Paul Lampe, Suzanne Rutherford.

Molecular Genetics
Antonio Bedalov, Sue Biggins, Steven Collins, Robert Eisenman, Daniel Gottschling, Norman Greenberg, Mark Groudine, Steven Henikoff, David Hockenbery, Christopher Kemp, Harmit Malik, Dusty Miller, Paul Neiman, Peter Nelson, Katie Peichel, Brian Reid, Mark Roth, Gerald Smith, Barbara Trask, Toshiyasu Taniguchi, Muneesh Tewari, Toshio Tsukiyama, Marc Van Gilst, Meng-Chao Yao, Helmut Zarbl.

Developmental Biology
Cecilia Moens, Susan Parkhurst, James Priess, Larry R. Rohrschneider, Philippe Soriano.

Virology
Laura Beretta, Michael Emerman, Denise Galloway, Adam Geballe, Keith Jerome, Maxine Linial, Julie Overbaugh.

Cell Cycle
Linda Breeden, Bruce Clurman, Bruce Edgar, Matthew Fero, Peggy Porter, James Roberts.

Cell Adhesion
William Carter, Valeri Vasioukhin.

Neurobiology
Linda Buck, James Olson, Stephen Tapscott, Edus Warren.

Molecular Structure
Adrian Ferré-D'Amaré, Steven Hahn, Julian Simon, Barry Stoddard, Roland Strong.

Infectious Diseases
M. Juliana McElrath, Nina Salama.

Immunology
J. Lee Nelson.

Stem Cell
Hans-Peter Kiem.

UNIVERSITY OF WISCONSIN–MADISON

Program in Cellular and Molecular Biology

Program of Study

Graduate study in cellular and molecular biology (CMB) at the University of Wisconsin–Madison is a multidisciplinary program leading to the Ph.D. degree; there is no formal master's degree program. The faculty consists of more than 170 participating members drawn from more than forty different departments. This allows students to select a thesis advisor and a thesis research topic from a wide variety of fields. All students form and utilize a thesis committee. Each student's committee members serve as advisers regarding course work and research progress. In addition, the committee evaluates the student's two major exams: the oral preliminary exam, taken the second year, and the final thesis defense, taken in approximately the fifth year of the program. The program is designed to encourage each student to develop an independent and creative approach to science. Students are required to obtain 10 credits in the CMB core curriculum, which consists of both cellular and molecular biology course work. In addition, students may take courses related to their specific area of expertise, ranging from pharmacology to zoology. Beyond formal course work, seminars and journal clubs are an integral part of the program. The combination of course work and seminar experience allows students to obtain a solid foundation in cellular and molecular biology that is also tailored to the professional objectives of each student. Students are represented on all standing committees in the program and play an active role in academic affairs and program development.

Research Facilities

The laboratory in which the student does research depends upon the departmental affiliation of the Thesis Advisor. All laboratories are well equipped and have excellent supporting facilities. There is also a Biotechnology Center and an Integrated Microscopy Resource on campus. A tradition of collaboration and cooperation among the many departments and laboratories at Wisconsin makes available to the students all the modern equipment, techniques, and expertise necessary for successful research in cellular and molecular biology.

Financial Aid

Initially, all students accepted into the Ph.D. degree program receive financial support from a Graduate School Fellowship, an interdepartmental training grant, or a research assistantship. Students are strongly encouraged to apply for a National Science Foundation scholarship at the time of application to Graduate School or during their first semester on campus. Once admitted into the program, all students, including international students, are guaranteed a gross stipend, which is $22,750 in 2006–07.

Cost of Study

Tuition is remitted for all students. Research assistants are required to pay segregated fees, which average $600 per year for predissertator students and $250 per year for dissertators.

Living and Housing Costs

A variety of housing options can be found both on campus and in the surrounding area of Madison. Rents for apartments, ranging from efficiencies to three bedrooms, vary from $400 to $2400 per month. The University maintains single- and double-room graduate dormitories during the academic year. University housing is available for married students, and some one-bedroom units are available for single students.

Student Group

The present total enrollment at UW–Madison is approximately 41,000, including almost 9,000 graduate students. The steady state enrollment in the cellular and molecular biology program is 150 students. An average of 20 new students are admitted to the program each year. Students are drawn from all regions of the United States and about a dozen other countries.

Location

The city of Madison has a population of approximately 208,000. It is the capital of the state and is situated within a superb setting surrounded by five lakes. The population is cosmopolitan and enjoys both city and University-promoted events. Abundant lakes and parks offer an array of outdoor activities for all seasons. The campus has well-maintained facilities for indoor sports.

The University

Founded in 1849, the University of Wisconsin has consistently placed among the top ten universities in the country. It is characterized by a dynamic spirit of inquiry and an atmosphere of academic freedom. The University has unusual breadth in its offerings and has pioneered in placing its facilities and staff at the disposal of the state and the nation.

Applying

Completed applications must be received by December 15. All applicants are required to take the Graduate Record Examinations (GRE). The GRE Subject Test in biology, chemistry, or biochemistry and molecular biology is recommended but not required. International applicants are required to take the Test of English as Foreign Language (TOEFL). Admission to the program is based on demonstrated ability and interest in mathematics, physical sciences, chemistry, and biology. Prior laboratory research experience and three electronic letters of recommendation are required.

Correspondence and Information

Student Services Coordinator
Program in Cellular and Molecular Biology
413 Bock Labs
1525 Linden Drive
University of Wisconsin–Madison
Madison, Wisconsin 53706
Phone: 608-262-3203
E-mail: cmb@bocklabs.wisc.edu
Web site: http://www.cmb.wisc.edu

University of Wisconsin–Madison

THE FACULTY AND THEIR RESEARCH

Cancer Biology
Nihal Ahmad: mechanism of cancer development; novel approaches for cancer management. Elaine T. Alarid: molecular mechanisms of steroid hormone action. Caroline Alexander: mammary tumor cell biology. Paul Bertics: regulation of cell proliferation and immune responses. William F. Dove: molecular cell biology; organismal biology; genetics of the intestinal epithelium and its neoplasms. Norman R. Drinkwater: genetics and molecular biology of carcinogenesis. William R. Engels: genetics and molecular biology of *Drosophila* transposable elements and recombination. Michael K. Fritsch: models for studying chromatin remodeling and transcriptional activation. Michael N. Gould: breast cancer modifiers in the non–protein coding genome. Paul M. Harari: molecular modulation of cellular growth factor receptors in cancer therapy. Michael Hoffmann: chemical genetics approaches to anticancer drug discovery. Paul F. Lambert: molecular genetics of papillomaviruses. Janet E. Mertz: Epstein-Barr virus, estrogen-related receptors, and cancer. Shigeki Miyamoto: regulatory mechanisms of Rel/NF-kappaB transcription factors. Jeff Ross: breast cancer induced by mRNA-binding protein. Eric P. Sandgren: genetics of solid-tissue growth disorders and neoplasia. Linda A. Schuler: Prolactin-related hormones; Signal transduction, receptor trafficking, and role in breast cancer. David C. Schwartz: structural variation in mammalian genomes. Bill Sugden: molecular biology of transforming functions of Epstein-Barr virus. Randal Tibbetts: DNA damage responses; genetic instability; ataxia-telangiectasia and PI3 kinase-related kinases. Ajit K. Verma: signal transduction pathways to oncogenesis. Wei Xu: epigenetic control of estrogen receptor (ER) regulated transcription.

Cell Adhesion & Cytoskeleton
Kurt J. Amann: cytoskeletal control of cell shape and motility. Richard Anderson: molecular and cellular signaling and cancer. William Bement: signal transduction, cell division, and cytoskeleton. James M. Ervasti: dystrophin-glycoprotein complex structure and function. Daniel S. Greenspan: eukaryotic genes in development and genetic diseases. Anna Huttenlocher: cellular and molecular mechanisms that regulate cell migration and chemotaxis in cancer and inflammation. Patricia J. Keely: spatial and mechanical signaling determinants of breast epithelial cell differentiation within extracellular matrices. Laura L. Kiessling: multivalent binding events, biomolecular recognition, and signal transduction. Deane F. Mosher: biochemistry of cell adhesion and the extracellular matrix. Chris J. Murphy: impact of topographic cueing on cell behaviors. Donna Pesciotta Peters: role of extracellular matrix and integrin signaling in human eye. George Phillips: establishing connections between structure, dynamics, and the functions of proteins using X-ray crystallographic, biophysical, and computational methods. Alan C. Rapraeger: cell adhesion and signaling; fibroblast growth factors. Nader Sheibani: cell adhesion and signaling in vascular cells. Jeffery W. Walker: G-protein-coupled receptor signaling in normal and diseased heart. John G. White: generation of cell diversity in *Caenorhabditis elegans*. Christiane Wiese: regulation of mitotic spindle assembly; microtubule nucleation.

Developmental Biology
Lynn Allen-Hoffman: cell biology of adult epidermal stem cells, tissue-specific adaptive responses, therapeutic genetic and tissue engineering. David Andes: drug resistance development in fungal pathogens using *Candida albicans* growth in biofilms as a model system. Craig Atwood: hormonal regulation of aging and disease. Seth S. Blair: development and signaling in *Drosophila*. Robert D. Blank: genetic and structural basis of bone strength. Grace Boekhoff-Falk: limb and neural development. Wade A. Bushman: signaling pathways in prostate development and cancer. Sean B. Carroll: understanding genetic regulatory mechanisms governing animal development and evolution. Karen M. Downs: formation of the murine allantois during mouse gastrulation. Ian D. Duncan: phenotype/genotype of the myelin mutants. John F. Fallon: developmental biology, pattern formation, limb development. Barry Ganetzky: genetic and molecular analysis of neuronal function, development, and survival in *Drosophila*. Thaddeus G. Golos: maternal immune recognition of pregnancy; molecular and cellular biology of placental development. Timothy M. Gomez: regulation of axon guidance and cytoskeletal dynamics. Anne E. Griep: regulation of mammalian development by tumor suppressors. Yevgenya Grinblat: neural development in zebrafish: pattern formation and morphogenesis. Mary C. Halloran: mechanisms of axon guidance and neural crest cell migration during zebrafish neural development. Jeff Hardin: morphogenesis and pattern formation during early development. Judith E. Kimble: molecular genetics of animal development. Allen Laughon: developmental genetics of *Drosophila*. Youngsook Lee: molecular pathways regulating cardiac development/diseases; stem cell research on cardiomyogenesis. Gary E. Lyons: axotrophin, a stem cell gene, has important functions in neurons and T cells. Margaret McFall-Ngai: dynamics of beneficial animal-microbe interactions. Albee Messing: astrocytes and disorders of intermediate filaments. Jon S. Odorico: embryonic stem cells to generate insulin-producing cells for the treatment of diabetes and to study pancreatic islet development. Francisco Pelegri: genetic control of zebrafish embryogenesis. Michael D. Sheets: molecular mechanisms regulating early vertebrate development. Susan M. Smith: molecular regulation of vertebrate embryogenesis: retinoid receptor function, ethanol and signal transduction. Xin Sun: molecular genetics of vertebrate organogenesis. James A. Thomson: human ES cells and the biology of pluripotency. Su-Chun Zhang: neural pathway of human embryonic stem cells.

Immunology
Zsuzsa Fabry: neuroimmunology; autoimmunity and inflammation in the central nervous system (CNS); immunology of the blood-brain barrier; T-cell activation in the CNS; experimental models for Multiple Sclerosis. Jenny Gumperz: biology of innate lymphocytes, particularly CD1d-restricted NKT cells. Colleen E. Hayes: regulation of lymphocyte development and autoimmune disease. Laura J. Knoll: pathogenesis and developmental regulation in the parasite toxoplasma. Miroslav Malkovsky: nonpeptidic anticancer and antiviral vaccines. John M. Mansfield: cellular and molecular immunology; molecular biology of infectious diseases. David O'Connor: pathogenetics of SIV and HIV infections. Donna M. Paulnock: role of macrophages in infectious diseases; innate immune responses during infection with African trypanosomes. Richard A. Proctor: cellular and molecular basis of bacterial toxin infections. Erik Ranheim: role of the Wnt/Frizzled signaling pathway and related molecules in B-cell development and lymphomagenesis. Matyas Sandor: T cells in chronic infectious diseases, granulomatous immune response. Christine M. Seroogy: role of GRAIL (RNF128, a novel E3 ubiquitin ligase) in CD25+ T regulatory cell biological function and dominant tolerance. Paul M. Sondel: clinical immunotherapy of cancer and tumor immunology. Gary A. Splitter: molecular mechanisms of host defense and pathogen survival strategies. M. Suresh: T cell responses to viral infections. David I. Watkins: immune responses to the AIDS virus and vaccine development. Jyoti J. Watters: signal transduction in microglia by extracellular ATP and estrogens. Jon P. Woods: fungal pathogenesis, gene regulation, and host interactions.

Membrane Biology & Protein Trafficking
Alan Attie: molecular genetics of diabetes and insulin resistance; cell biology of lipoprotein assembly; cholesterol trafficking. Jay Bangs: molecular, biochemical, and cellular parasitology. Maureen M. Barr: *C. elegans* sensation and disease models. Edwin R. Chapman: molecular mechanism of exocytosis. Nansi Jo Colley: molecular basis of protein trafficking in the photoreceptors of *Drosophila*. Elizabeth A. Craig: Function of molecular chaperones in protein folding and translocation across membranes. Cynthia Czajkowski: structure and function of the GABAA receptor. David Eide: molecular mechanisms of mineral nutrient homeostasis. Guy E. Groblewski: cellular mechanisms of epithelial cell secretion. Meyer B. Jackson: electrophysiology and imaging techniques are used to explore the basic mechanisms of neuronal signaling. Timothy J. Kamp: cardiac ion channels function and regulation; basic mechanisms of heart failure and arrhythmias; stem cells and cardiac regeneration. Ching Kung: ion channels and sensory transduction in microbes. Sally Leong: molecular genetics of plant-pathogen interactions; metal ion homeostasis. Thomas F. J. Martin: mechanisms of hormone action; regulation of hormone/neurotransmitter secretion. Luigi Puglielli: lipid-mediated signaling during aging and Alzheimer's disease. Gail A. Robertson: molecular mechanisms of ion channel disease. Antony O. W. Stretton: nematode neurobiology.

Molecular and Genome Biology of Microbes
Richard H. Burgess: biochemistry of RNA polymerase/transcription. Silvia Cavagnero: in vivo and in vitro protein folding. Michael M. Cox: molecular biology and enzymology of recombinational DNA repair; function and regulation of RecA protein; mechanisms of genome reconstitution in the radiation-resistant bacterium *Deinococcus radiodurans*. Timothy Donohue: genomic and molecular analysis of metabolic pathways and regulatory circuits. Katrina T. Forest: structural biology of bacterial pathogenesis. Catherine A. Fox: mechanisms linking chromosomal replication and heritable chromatin structures. Richard L. Gourse: prokaryotic transcription; regulation of ribosome synthesis; control of gene expression. Charles W. Kaspar: microbial stress-protection proteins; microbial extremophiles; enterohemorrhagic *Escherichia coli*. James L. Keck: DNA replication, recombination, and repair. Patricia J. Kiley: molecular genetics and biochemistry of oxygen-regulated gene expression and transcription activation. Robert Landick: control of transcription elongation; RNA polymerase structure/function. William S. Reznikoff: regulation of transcription and transposition in *E. coli*.

Plant Biology
Richard Amasino: seasonal control of flowering. Jean-Michel Ane: symbiotic plant-microbe associations (nodulation and mycorrhization). Sebastian Bednarek: membrane trafficking; organelle biogenesis; cytokinesis; cell expansion. Andrew F. Bent: molecular basis of plant disease resistance. Donna E. Fernandez: plant reproductive biology. Dennis Allen Halterman: molecular aspects of disease resistance in potato. Jiming Jiang: cloning and molecular characterization of plant centromeres; developing plant artificial chromosomes. Patrick J. Krysan: MAP kinase signal transduction in *Arabidopsis*. Patrick H. Masson: molecular genetics of root growth behavior in response to mechanical stimuli. Marisa Otegui: endosomal trafficking in plants. Sara E. Patterson: regulation of cell separation and adhesion in *Arabidopsis*. Edgar P. Spalding: ion channels and photosensory transduction in plants. Michael R. Sussman: genome technologies applied to the plasma membrane of eukaryotes: signal transduction and bioenergetics. Richard D. Vierstra: protein degradation in plants; functions of the ubiquitin-dependent proteolytic pathway; plant photomorphogenesis and phytochrome.

RNA
Philip Anderson: molecular mechanisms of gene expression. David A. Brow: DNA transcription and RNA splicing in yeast. Michael R. Culbertson: posttranscriptional control of gene expression in yeast. James E. Dahlberg: RNA processing; nuclear transport of RNA and proteins. Richard S. Eisenstein: regulation of mRNA translation and stability; intracellular trafficking of iron; iron sulfur proteins; iron homeostasis. Scott Kennedy: genetic analysis of siRNAs, microRNAs, and RNAi. James S. Malter: posttranscriptional gene regulation; mRNA binding proteins. Ronald T. Raines: chemical biology; protein design and engineering; protein cytotoxins. Lloyd M. Smith: development and application of novel bioanalytical methods; new instrumentation and chemistries for biological mass spectrometry and biologically modified surfaces. Karen Wassarman: small RNA regulation of gene expression. Marvin P. Wickens: RNA and gene control; role of RNA regulation in development and the nervous system. Jerry Chi-Ping Yin: cellular/molecular mechanisms of memory formation and psychiatric dysfunction.

Transcriptional Mechanisms
Aseem Ansari: understanding and controlling gene expression. Emery H. Bresnick: regulation of hematopoiesis; vascular biology; chromatin structure/function. Dongsheng Cai: linkage between inflammatory pathways and metabolism. Marcin Filutowicz: regulation of DNA replication and transcription and the development of new antimicrobial approaches. Audrey Gasch: genomic analysis of fungal stress responses. Jeffrey A. Johnson: signal transduction, neurotoxicity, and transcriptional control of neuroprotective genes. Thomas Record: protein-DNA interactions in transcription initiation. Anath Shalev: molecular biology of pancreatic islets: transcription, translation, apoptosis. John Svaren: transcriptional regulation of peripheral nerve myelination. David A. Wassarman: transcriptional regulation in *Drosophila*. Bernard Weisblum: inducible antibiotic resistance and drug discovery.

Virology
Paul Ahlquist: molecular mechanisms of virus replication, gene expression and host interactions. Judd Aiken: prion diseases, mitochondria in aging. Curtis R. Brandt: pathogenesis of herpes simplex virus; virulence genes in herpetic eye disease and herpes encephalitis; antivirals; interactions between cytokines and herpes viruses, and gene delivery, gene therapy. Teresa Compton: virus-cell interactions; virally-activated, receptor-mediated, signal transduction; cellular integrins as virus receptors; cell biology of virus entry; toll-like receptors; innate immune pathways. Paul D. Friesen: DNA viruses; programmed cell death (apoptosis); transcriptional regulation. James E. Gern: viral respiratory infections, wheezing illnesses, asthma, and developmental immunology. Robert F. Kalejta: manipulation of the cell cycle by human cylomegalovirus. Yoshihiro Kawaoka: molecular pathogenesis of influenza and Ebola viruses. Daniel D. Loeb: mechanism of replication of hepadnaviruses. Christopher W. Olsen: molecular epidemiology and pathogenesis of influenza viruses. Ann C. Palmenberg: molecular biology of RNA picornaviruses. Stacey Schultz-Cherry: influenza and astrovirus pathogenesis. Robert T. Striker: biochemistry and genetics of *Flaviviridae* replication. John Yin: molecular virology; systems biology; biotechnology.

Miscellaneous
Heidi Goodrich-Blair: molecular mechanisms of nematode-bacterium symbiosis. Tomas A. Prolla: DNA repair mechanisms; mouse models of tumorigenesis. Rodney A. Welch: molecular biology and genetics of bacterial virulence factors and protein secretion.

WAKE FOREST UNIVERSITY

School of Medicine
The Bowman Gray Campus
Department of Cancer Biology

Program of Study	The interdisciplinary program in cancer biology offers the degree of Doctor of Philosophy (Ph.D.) through the Graduate School of Wake Forest University. The primary foci are to give students the multidisciplinary scientific background and familiarity with clinical issues related to human cancer necessary to become contributing career cancer biologists. The research foci of the program are the research interests of the department faculty members. There is a broad range of research activities in the department, most of which falls into two subdisciplines of cancer research—mechanisms of DNA damage, the cellular defense against DNA damage and carcinogenesis; and mechanisms of cell growth and death, some of which involves translational research on novel therapeutic interventions. Both of these subdisciplines are high-priority research topics nationwide and are likely to continue to be so as long as the cancer problem remains. Therefore, graduates of the program have a solid research base with which to build cancer biology careers. During the first year, students participate in core courses in cancer biology, including courses in carcinogens and DNA damage, molecular pathogenesis of cancer, cell biology of breast and prostate cancer, and molecular targets of cancer therapy. In addition, students participate in a core course in clinical issues of cancer biology. Students perform research rotations during their first year in 2 to 3 laboratories of cancer biology faculty members. At the end of the first year, students choose a faculty thesis adviser. Students are admitted to candidacy upon completion of courses and an oral defense of a written research proposal, usually at the end of the second year. Students are expected to complete their degree in four to five years.
Research Facilities	Resources supporting research in the Cancer Biology Program are centrally located on the fourth and fifth floors of the Hanes Research Building. In addition, each program faculty member has well-equipped laboratory space tailored to his or her specific research needs. All faculty members are also members of the Wake Forest University Comprehensive Cancer Center. Membership in the Comprehensive Cancer Center allows access to many core facilities. Computer facilities give students and faculty members access to DNA and protein sequence databases and software for analysis of sequence data. A molecular graphics workstation is available for structural modeling of proteins and nucleic acids. There are many core laboratory facilities that provide services that include automated DNA sequencing and genotyping, oligonucleotide synthesis, peptide sequencing and synthesis, electron microscopy, nuclear magnetic resonance, cytogenetics, fluorescence-activated cell sorting, and tissue culture. The Carpenter Library of the medical center maintains an extensive collection of basic science and clinical journals and is equipped with computer-assisted information retrieval systems.
Financial Aid	The Wake Forest University Graduate School provides financial support in the form of graduate fellowships, which totaled $45,547 in the 2004–05 academic year. The fellowship includes stipend support ($20,772 in 2004–05) and a tuition scholarship ($24,775 in 2004–05). After the first year, students receive full stipend support from their thesis adviser's research funds.
Cost of Study	Tuition costs of $23,610 in 2003–04 were met in full by Wake Forest University School of Medicine for all students holding fellowships.
Living and Housing Costs	Most students live in apartments or rental houses within walking distance of the medical center. Rent ranges from $350 to $450 per month for one-bedroom apartments and from $450 to $550 per month for two-bedroom apartments. There are University-owned efficiency apartments for single and married graduate students on a space-available basis.
Student Group	In the fall of 2004, there were 684 full-time graduate students, with 234 full-time graduate students enrolled on the medical school campus. The total graduate and undergraduate student enrollment at Wake Forest University is 6,444.
Student Outcomes	The Cancer Biology Program graduated its first students in 2002. In addition, the faculty members of the program have broad graduate training experience in other programs. Graduates of these laboratories typically obtain excellent postdoctoral positions at universities, biotechnology research laboratories, or the National Institutes of Health. Following their postdoctoral training, many continue with academic careers as professors and independent biomedical researchers. Others secure research and staff positions in private industry or government laboratories. A few graduates now hold leadership positions in both academia and industry.
Location	The Wake Forest University School of Medicine is located in a residential area of Winston-Salem on the Bowman Gray campus of Wake Forest University, 3 miles from the main campus. The city is in the northwestern region of North Carolina in the foothills of the Blue Ridge Mountains. A population of 150,000 enjoys a mild climate, with early springs and late autumns allowing outdoor activities during most of the year. Cultural attractions include the Southeast Center for Contemporary Art, the Reynolda House Museum of American Art, and performances presented by the nationally recognized North Carolina School of the Arts, the Winston-Salem Symphony, and the Little Theatre. The Sci Works Center and Old Salem area are educational and historic attractions. The city is home to minor-league baseball and hockey teams and has excellent recreational facilities.
The University and The Program	Wake Forest is a private university that was founded in 1834. The Wake Forest University School of Medicine, founded in 1902, has major clinical research programs in cancer, neurological disorders, hypertension, aging, and atherosclerosis. The research facilities at the medical center complex have expanded dramatically over the past several years. Six floors have been added atop the Hanes Research Building, and construction of a new eleven-story building for the Center on Nutrition and Chronic Disease Research was recently completed. The establishment of the Cancer Biology Program exemplifies the commitment of this institution to remain in the forefront of scientific research. New faculty members have been, and continue to be, recruited in an ongoing effort to strengthen the School's capabilities in cancer biology research.
Applying	The program invites applications from students who have completed undergraduate degrees. Completed applications should be received by January 15 for admission in the fall semester. Early application is encouraged. Applications should include scores from the Graduate Record Examinations. Also required are official undergraduate transcripts, three letters of reference, and a statement of personal interests. Application forms and additional information about the Cancer Biology Program may be obtained upon request.
Correspondence and Information	Pam Pitts Department of Cancer Biology Wake Forest University Health Sciences Medical Center Boulevard Winston-Salem, North Carolina 27157 Phone: 336-716-2693 Fax: 336-716-0255 E-mail: ppitts@wfubmc.edu Web site: http://www.wfubmc.edu/canbio

Wake Forest University

THE FACULTY AND THEIR RESEARCH

Steven A. Akman, Professor of Internal Medicine-Hematology/Oncology and Cancer Biology and Program Director; M.D., Yeshiva (Einstein), 1975. Site-specific mutagenicity of alkyl-DNA adducts; unusual DNA structures and genomic instability.

Rebecca W. Alexander, Assistant Professor of Chemistry and Cancer Biology; Ph.D., Pennsylvania, 1996. Structure-function studies of aminoacyl-tRNA synthetases and translational factors; in vitro selection of substrate mimics and inhibitors.

Isabelle M. Berquin, Assistant Professor of Pathology and Cancer Biology; Ph.D., Wayne State, 1996. Molecular pathogenesis of breast cancer; regulation of gene expression; signal transduction; proteases in cancer.

Ulrich Bierbach, Assistant Professor of Chemistry and Associate in Cancer Biology; Ph.D., Oldenburg (Germany), 1992. Metallopharmaceuticals; DNA-targeted cancer chemotherapy; small-molecule hybrid agents in gene regulation; medicinal applications of metal chelating agents.

Bernard A. Brown II, Assistant Professor of Chemistry, Associate in Cancer Biology, and Associate of WFU Center for Structural Biology; Ph.D., North Carolina State, 1997. Biochemistry, biophysics, and structural biology of protein-nucleic acid interactions; small nucleolar RNAs and RNP complexes; RNA modifications; ribonucleotide reductases; noncanonical nucleic acid structures.

Yong Q. Chen, Associate Professor of Cancer Biology; Ph.D., Free University of Brussels, 1989. Signaling mechanism in tumor-suppressor gene DCC-induced apoptosis; comprehensive gene expression profiling of prostate; lipid signaling in prostate.

Scott D. Cramer, Associate Professor of Cancer Biology, Associate in Urology, and Program Co-Director; Ph.D., California, Santa Cruz, 1992. Endocrine and paracrine signaling in the prostate; vitamin D and prostate cancer; regulation of prostate-specific antigen expression.

Zheng Cui, Associate Professor of Pathology and Cancer Biology; Ph.D., Massachusetts, 1988. Mechanisms of host resistance to cancer; lipidomics.

Waldemar Debinski, Professor of Neurosurgery, Radiation Oncology, and Cancer Biology and Director, Brain Tumor Center of Excellence; M.D., Warsaw, 1981; Ph.D., McGill, 1989. Molecular targeting for diagnosis, imaging, and treatment of cancer.

Karin Drotschmann, Assistant Professor of Cancer Biology; Ph.D., Göttingen (Germany), 1997. Defects in mismatch repair proteins; effects on cancer.

William H. Gmeiner, Professor of Cancer Biology and Adjunct Professor in Chemistry; Ph.D., Utah, 1989. Structure, function, and delivery of activated forms of nucleoside antimetabolites.

Suzanne M. Hess, Assistant Professor of Radiation/Oncology and Cancer Biology; Ph.D., Penn State, 1993. Radiation sensitizing agents; novel therapeutic approaches in the treatment for glioblastomas; complementary/alternative approaches to cancer treatment.

Dora Il'yasova, Assistant Professor of Cancer Biology; Ph.D., North Carolina at Chapel Hill, 2001. Epidemiological research on modifiable risk factors in cancer.

S. Bruce King, Associate Professor of Chemistry and Associate in Cancer Biology; Ph.D., Cornell, 1993. Design and synthesis of new compounds capable of interaction with nitric oxide synthase; design, synthesis, and evaluation of new nitric oxide and nitroxyl delivery agents; total synthesis of biologically active natural products; reactions of hydroxyurea with hemeproteins.

Costas Koumenis, Assistant Professor of Radiation Oncology and Cancer Biology; Ph.D., Houston, 1994. Regulation of gene expression and function by ionizing radiation and tumor hypoxia; hypoxia-induced apoptosis; gene therapy–based radiosensitization of malignant gliomes; natural products as radiosensitizers.

Steven J. Kridel, Assistant Professor of Cancer Biology; Ph.D., California, Irvine, 1997. Regulation of fatty acid synthase activity in prostate cancer; biochemical analysis of matrix metalloproteinases.

Gregory L. Kucera, Assistant Professor of Internal Medicine-Hematology/Oncology; Ph.D., Wake Forest, 1987. Effects of novel anticancer drugs on phospholipid metabolism and alterations in signal transduction pathways within neoplasms.

George Kulik, Assistant Professor of Cancer Biology; Ph.D., Kiev, 1991. Anti-apoptotic mechanisms in prostate cancer; novel signal transduction pathway that protect from apoptosis; in vivo imaging of prostate cancer metastases.

Mark Steven Miller, Professor of Cancer Biology and Physiology and Pharmacology; Ph.D., Columbia, 1983. Role of drug metabolic enzymes in determining individual susceptibility to cancer induction by environmental toxicants; molecular pathogenesis of tumor formation.

Charles S. Morrow, Associate Professor of Biochemistry and Cancer Biology; M.D., Missouri–Columbia, 1983; Ph.D., Saint Louis, 1980. Synergy between drug efflux pumps and conjugating enzymes in cancer drug and carcinogen resistance; regulation of genes associated with xenobiotic detoxification.

David A. Ornelles, Associate Professor of Microbiology and Immunology and Cancer Biology; Ph.D., MIT, 1987. The role of the E1B and E4 oncoproteins in the cell cycle–restricted adenovirus replication; virus-mediated control of mRNA transport; oncolytic human adenoviruses.

G. L. Prasad, Associate Professor of General Surgery and Cancer Biology; Ph.D., Indian Institute of Science, 1985. Cytoskeletal regulation of growth and differentiation; modulation of microfilament organization and tumor suppression by tropomyosins; control of apoptosis during breast cancer progression; signal transduction; development of novel chemotherapeutics and biomarkers.

Mike E. C. Robbins, Professor of Radiation Oncology, Neurosurgery, and Cancer Biology; Ph.D., London, 1980. Pathogenesis of radiation-induced late normal tissue injury; role of chronic oxidative stress in radiation-induced brain and kidney injury; role of renin-angiotensin system and PPARs in radiation-induced brain late normal tissue injury; role of antioxidant enzymes in the response of normal and malignant brain cells to oxidative stress.

Gary G. Schwartz, Associate Professor of Cancer Biology and Public Health Sciences; M.P.H., 1988, Ph.D., 1993, North Carolina at Chapel Hill; Ph.D., SUNY Downstate Medical Center, 1985. Role of the vitamin D endocrine system in the etiology, treatment, and prevention of prostate cancer.

Vijayasaradhi Setaluri, Associate Professor of Dermatology and Cancer Biology; Ph.D., Osmania (India), 1985. Intracellular protein sorting and processing of melanoma antigens; cell and molecular biology of melanoma and melanocyte differentiation.

Frank M. Torti, Professor and Chairman of Cancer Biology, Professor of Internal Medicine–Hematology/Oncology Urology, and Director, Comprehensive Cancer Center; M.D., M.P.H., Harvard, 1974. Natural history and therapeutic interventions in patients with genitourinary malignancies; molecular mechanisms of regulation of iron homeostasis in cancer and inflammation.

Suzy V. Torti, Associate Professor of Biochemistry; Ph.D., Tufts, 1977. Regulation of proteins of iron metabolism in normal and transformed cells; development of novel antineoplastic therapeutics targeting iron metabolism.

Alan J. Townsend, Professor of Biochemistry and Cancer Biology; Ph.D., North Carolina at Chapel Hill, 1986. Biochemical and molecular mechanisms of resistance to cytotoxic and mutagenic agents; enzymes of glutathione metabolism; chemoprevention of cancer; oxidative stress and antioxidant defenses.

James P. Vaughn, Assistant Professor of Cancer Biology; Ph.D., Virginia, 1992. Therapeutic strategies against breast cancer; studies of antisense DNA and antibodies targeted against the erbB2 receptor; molecular cloning; development of gene chips and expression arrays.

Mark E. Welker, Professor of Chemistry and Associate in Cancer Biology; Ph.D., Florida State, 1985. Synthesis and evaluation of anticarcinogenic enzyme inducers, particularly glutathione S-transferase; quinone oxoreductase.

John Wilkinson IV, Assistant Professor of Cancer Biology; Ph.D., Boston University, 1994. Development of transgenic models to explore ferritin biology; the chemopreventative antioxidant response.

Jianfeng Xu, Professor of Public Health Sciences and Cancer Biology; M.D., Shanghai Medical, 1984; Dr.Ph., Johns Hopkins, 1997. Mapping prostate cancer genes using genetic linkage and association studies.

REQUIRED AREAS OF STUDY WITHIN THE GRADUATE TRAINING PROGRAM OF THE DEPARTMENT OF CANCER BIOLOGY

Advanced Topics in Cancer Biology. New and important aspects of research in cancer biology, with an emphasis on current literature.

Cancer Cell Biology. Advanced cell biology of cell growth differentiation and apoptosis.

Carcinogens, DNA Damage and Repair. Identification and reaction mechanisms of environmental carcinogens; DNA damage and mutagenesis by endogenous and exogenous mechanisms; nucleotide excision repair, base excision repair, and mismatch repair.

Molecular Pathogenesis of Cancer. Discovery, biochemistry, and function of oncogenes and tumor suppressor genes, signal transduction pathways, and regulators of the cell cycle.

Molecular Targets of Cancer Therapy. Various treatment modalities for cancer, emphasizing the cellular and molecular targets for therapeutic agents. Includes current concepts in chemotherapy, immunological approaches to cancer treatment, and developing gene therapy applications.

Research in Cancer Biology. Opportunities for investigation in a variety of the facets of cancer biology.

Topics in Cancer. Weekly lecture series; comprehensive overview of clinical oncology.

WAKE FOREST UNIVERSITY

School of Medicine
Ph.D. Graduate Program in Molecular Medicine

Program of Study	The Graduate Program in Molecular Medicine is a research-based program leading to the Ph.D. degree. This program addresses the critical need for training talented research scientists whose careers lie at the interface of basic and clinical investigative medicine. Two unique aspects of this training program set it apart from more traditional graduate programs, such as biochemistry or physiology, and indicate that a different type of student should enter the program. The first is that a concerted effort is made throughout the program to integrate clinical medicine into basic science training. The goal of this approach is to educate the student about the basic pathobiology of a disease and to emphasize the important issues and questions about a disease that are not understood. The second aspect is that a strong emphasis is placed on understanding basic mechanisms that underlie the conditions that lead to human diseases rather than limiting study to a specific cellular process. In this manner, it is hoped that the student will begin to see similarities in the basic mechanisms of different disease processes and can move easily between disease-focused research programs. Thus, the first-year student gains initial research experience through a program of laboratory rotations while taking core basic science courses. During the second year, the student focuses on human health, with core lectures on the molecular basis of human diseases and clinical rotations through the affiliated major teaching hospital. The student pursues independent research in the second through fifth years. The Graduate Program in Molecular Medicine provides a comprehensive framework for training and experience in translational research focused on human genetics and gene expression or human cell biology, and it concentrates on different medical fields, such as aging, arthritis, cardiovascular biology, cancer, diabetes, infection and inflammation, neurobiology, pharmacology, and tissue regeneration. The Graduate Program in Molecular Medicine at Wake Forest University is one of a few such programs in the country funded through a National Institutes of Health (NIH) training grant.
Research Facilities	The research facilities are located on the campus of Wake Forest University School of Medicine. Facilities are extensive and well equipped for experiments in inherited and functional genomics; molecular biology; cell biology; neurobiology; lipids and biochemistry; microbiology and immunology; animal manipulation, including transgenic facilities; and large-scale clinical epidemiology. Core facilities for electron and confocal microscopy, lipid analysis, proteomics, DNA analysis/synthesis, X-ray crystallography, high-field NMR spectroscopy, single nucleotide polymorphisms, and imaging equipment (e.g., PET and MRI) are located within the medical school. Some faculty members are located in research facilities on the main campus of Wake Forest. Research interests and publications of faculty members can be found at the program's Web site. Sophisticated computer facilities are available and all students have access to high-speed Internet communications. All students are given laptop computers upon matriculation.
Financial Aid	Most entering graduate students receive full tuition remission scholarships and a stipend, which is projected at $20,772 per year. In addition, the University pays half of the health-care benefits (approximately $1800 per year).
Cost of Study	Tuition for graduate work for the entering 2007 class is projected at $28,600 per year, and students carry between 11 and 13 credit hours. Students generally receive full tuition remission.
Living and Housing Costs	Reasonable, low-cost housing is located nearby, and students may apply for a Medical Foundation apartment near the medical school campus. Living costs are typically lower than those in larger cities, and an apartment averages around $600 per month.
Student Group	Enrollment in all schools of Wake Forest University exceeds 6,000 students and is split between the main campus of Wake Forest and the Medical School campuses.
Location	Wake Forest University is located in Winston-Salem, North Carolina. This is a geographically beautiful area, with the Blue Ridge Mountains about 1 hour west of the University and the Atlantic Ocean beaches 4 hours to the east. Major airports are located within a half-hour's drive of the city. Music, art, and theater are rich within the city. Outdoor sports include winter skiing, hiking, and camping. Professional sports in nearby cities include basketball, hockey, and soccer. Graduate students receive free tickets to University sporting events, including basketball.
The University	Wake Forest University was founded in 1834, and the School of Medicine was founded in 1902. The University now includes six different schools with an emphasis on excellence in academics and research, and a culturally rich and diverse student body. The school is well-endowed with resources and a high faculty-student ratio. The Ph.D. Graduate Program in Molecular Medicine is the fastest-growing graduate program in the University and has a strong sense of community among its students.
Applying	Admission requires completion of the undergraduate degree. Generally, excellent to superior grades (B or better) and strong letters of recommendation are needed. In addition, scores from the Graduate Record Examinations General Test are required, and the Subject Test is recommended. Applicants from non-English-speaking countries must also have satisfactory scores on the Test of English as a Foreign Language. Applications for the fall semester are due the preceding February 1. Personal interviews are usually required; they occur in March. Application information can be obtained by contacting the Graduate Program in Molecular Medicine or from the Web site at http://www1.wfubmc.edu/PTCR/MolMed/PhD+Program.
Correspondence and Information	Kevin High, M.D., and Linda McPhail, Ph.D. c/o Sherrie E. Brazier Graduate Program in Molecular Medicine Wake Forest University School of Medicine Medical Center Boulevard Winston-Salem, North Carolina 27157 Phone: 336-713-4259 E-mail: sbrazier@wfubmc.edu Web site: http://www1.wfubmc.edu/PTCR/MolMed/PhD+Program

Wake Forest University

THE FACULTY

Students should visit the Web site for a complete listing of faculty research and publications.

Steven Akman, M.D., Department of Hematology and Oncology–Cancer Biology.
Martha Alexander-Miller, Ph.D., Department of Microbiology and Immunology.
Anthony Atala, M.D., Department of Surgery–Urology/Institute for Regenerative Medicine.
Eugene Bleecker, M.D., Department of Internal Medicine–Pulmonary/Center for Human Genomics.
Donald W. Bowden, Ph.D., Departments of Biochemistry and Internal Medicine–Endocrinology and Metabolism.
K. Bridget Brosnihan, Ph.D., Department of Surgical Sciences/Hypertension Center.
David Busija, Ph.D., Department of Physiology and Pharmacology/Neuroscience Program.
Jeff Carr, M.D., Department of Radiologic Sciences–Radiology.
Mark Chappell, Ph.D., Departments of Physiology and Pharmacology and Surgical Sciences/Hypertension Center.
Che-ping Cheng, M.D., Ph.D., Department of Internal Medicine–Cardiology.
Ski Chilton, Ph.D., Department of Physiology and Pharmacology.
George Christ, Ph.D., Institute for Regenerative Medicine.
John Robin Crouse, M.D., Departments of Internal Medicine–Endocrinology and Metabolism and Public Health Sciences–Epidemiology.
Zheng Cui, M.D., Ph.D., Department of Biochemistry–Cancer Biology.
Larry Daniel, Ph.D., Department of Biochemistry.
Waldemar Debinski, M.D., Ph.D., Department of Neurobiology–Brain Tumor Center of Excellence.
Purnima Dubey, Ph.D., Department of Pathology.
Thomas Dubose, M.D., Department of Internal Medicine
Delrae Eckman, Ph.D., Department of Pediatrics.
James Eisenach, M.D., Department of Anesthesiology–Obstetrics.
Randolph Geary, M.D., Departments of Surgical Sciences and Comparative Medicine.
Roy Hantgan, Ph.D., Department of Biochemistry.
Gregory Hawkins, Ph.D., Department of Pulmonary and Critical Care/Center for Human Genomics.
David Herrington, M.D., Department of Internal Medicine–Cardiology.
Kevin P. High, M.D., Department of Internal Medicine–Infectious Diseases.
Duncan Hite, M.D., Department of Internal Medicine–Pulmonary.
Timothy D. Howard, Ph.D., Center for Human Genomics.
Gregory Hundley, M.D., Department of Internal Medicine–Cardiology.
Susan Hutson, Ph.D., Department of Biochemistry.
Kazushi Inoue, M.D., Ph.D., Department of Pathology.
Tamison Jewett, M.D., Department of Pediatrics–Medical Genetics.
Danny Kim-Shapiro, Ph.D., Department of Physics.
Bruce King, Ph.D., Department of Chemistry.
Gregory Kucera, Ph.D., Department of Hematology and Oncology–Cancer Biology.
Richard Loeser, M.D., Department of Internal Medicine–Molecular Medicine.
Charles E. McCall, M.D., Department of Internal Medicine–Molecular Medicine.
Maria McGee, M.D., Department of Internal Medicine–Rheumatology.
Linda McPhail, Ph.D., Department of Biochemistry.
Deborah Meyers, Ph.D., Department of Pediatrics–Medical Genetics/Center for Human Genomics.
Mark Miller, Ph.D., Department of Cancer Biology.
Nilamadhab Mishra, M.D., Department of Internal Medicine–Rheumatology and Immunology.
Peter Morris, M.D., Department of Internal Medicine–Pulmonary and Critical Care.
Charles Morrow, M.D., Ph.D., Department of Biochemistry.
John Parks, Ph.D., Department of Pathology.
Raymond Penn, Ph.D., Department of Internal Medicine–Pulmonary and Critical Care.
Mark Pettenati, Ph.D., Department of Pediatrics–Medical Genetics.
G. L. Prasad, Ph.D., Department of General Surgical Sciences.
Stephen Rich, Ph.D., Department of Public Health Sciences–Epidemiology.
Bruce Rubin, M.D., M.Eng., Department of Pediatrics–Allergy and Immunology.
Michele Sale, Ph.D., Department of Internal Medicine/Center for Human Genomics.
David Sane, M.D., Department of Internal Medicine–Cardiology.
Gary Schwartz, Ph.D., M.P.H., Department of Cancer Biology.
Michael Seeds, Ph.D., Department of Internal Medicine–Molecular Medicine and Pulmonary and Critical Care.
Robert Sherertz, M.D., Department of Internal Medicine–Infectious Diseases.
James Smith, Ph.D., Department of Physiology and Pharmacology.
Thomas L. Smith, Ph.D., Department of Surgery–Orthopedics.
Ed Swords, Ph.D., Department of Microbiology and Immunology.
Joseph Tobin, M.D., Department of Anesthesiology.
Frank Torti, Ph.D., Department of Internal Medicine/Comprehensive Cancer Center.
Suzy Torti, Ph.D., Department of Biochemistry.
Alan Townsend, Ph.D., Department of Biochemistry.
James P. Vaughn, Ph.D., Department of Neurobiology and Anatomy.
Mary Lou Voytko, Ph.D., Department of Neurobiology and Anatomy.
Reidar Wallin, Ph.D., Department of Internal Medicine–Rheumatology.
Richard B. Weinberg, M.D., Department of Internal Medicine–Gastroenterology and Biochemistry.
Jianfeng Xu, Dr.P.H., M.D., Department of Internal Medicine/Center for Human Genomics.
Barbara Yoza, Ph.D., Department of Surgical Sciences.

Recent Graduates and Current Postdoctoral Positions

Monica Basehore, Ph.D., American Board of Medical Genetics (ABMG), Clinical Molecular Genetics Fellow; Baylor College of Medicine, Houston, Texas.
Abbie Connoy Beltz, Ph.D., Clinical Research Director; Carolinas Medical Center, Charlotte, North Carolina.
Jing Li Jing, Ph.D., Postdoctoral Fellow.

WAYNE STATE UNIVERSITY

School of Medicine
Graduate Program in Cancer Biology

Program of Study

This is a multidisciplinary graduate program in cancer biology at the Barbara Ann Karmanos Cancer Institute (BAKCI) of Wayne State University (WSU). The Ph.D. degree is awarded in cancer biology. The program involves faculty members from basic and medical science departments, and their research is are well supported with national competitive funding. Faculty members are actively engaged in various aspects of clinical and basic research involving modern cell and molecular biology as it relates to cutting-edge cancer biology, such as control of cell proliferation and apoptosis, mechanisms of invasion and metastasis, and cell interactions. The research interests of the faculty members in the program can be divided into six general areas of investigation: drug discovery and development, breast cancer, proteases and metastasis, molecular oncology and human genetics, cancer control and prevention, and chemical carcinogenesis and toxicology. Cancer biology is a multidisciplinary field of study that requires training in many disciplines, including biochemistry, molecular biology, pharmacology, cellular biology, pathology, physiology, and oncology. The first year of the program is devoted primarily to courses, seminars, and research opportunity explorations. At the end of the first year, the student chooses a graduate adviser and develops a plan for the completion of the dissertation. During the second year, the student completes most of the course material, submits an outline of the dissertation, and begins the predissertation research. Subsequent years are devoted primarily to performing doctoral dissertation research.

Research Facilities

The faculty members participating in the program have at their disposal more than 100,000 square feet of modern laboratory space. In addition to a large number of typical laboratory instruments, there are a number of specialized laboratories and equipment, including those for X-ray crystallography, spectroscopy (NMR, MS, EPR), flow cytometry, image analysis, and DNA and protein synthesis and sequencing. Excellent libraries and computer facilities are also available.

Financial Aid

Fellowships are available on a competitive basis for first-year graduate students. For 2006–07, fellowships provide full tuition and medical and dental insurance, as well as stipend support of at least $19,500 a year. Students awarded fellowships can expect to receive full financial support until they complete the program, provided that they meet the standards of the program. In addition, GPCB students have been awarded NCI training grant and individual grant fellowships.

Cost of Study

Students awarded fellowships have their tuition and fees (which are approximately $8800 per year for in-state students and $18,300 per year for nonresidents in 2006–07) paid by the fellowships.

Living and Housing Costs

WSU has a variety of furnished and unfurnished apartments available for graduate students. The Detroit metropolitan area offers many apartments in a variety of prices. The University Housing Office is available to assist students in locating housing. The cost of food and recreation in Detroit is modest. There are ample opportunities for the employment of spouses at the Detroit Medical Center and at WSU.

Student Group

The GPCB has about 30 students enrolled in the program. Although the cancer biology program is relatively new, having started in 1989, the participating faculty members are experienced in the education of graduate students. More than 10,000 graduate students are enrolled at WSU; virtually all states and many countries are represented.

Student Outcomes

The majority of students receiving a Ph.D. take postdoctoral training in excellent laboratories at institutions of high quality and then obtain employment as university faculty members or as investigators at research institutes or in pharmaceutical companies. Graduates have obtained postdoctoral positions at institutions such as Harvard, Yale, Johns Hopkins, Duke, University of Michigan, Parke-Davis, Scripps Research Institute, and the National Institutes of Health.

Location

WSU and the School of Medicine are located in the University Cultural Center. The Cultural Center contains the Detroit Institute of Arts, the Detroit Public Library, theaters, schools of art and music, the African-American Museum, and the Detroit Historical Museum. Also in the area are Orchestra Hall, Joe Louis Arena, Comerica Park, and Ford Field. The Detroit riverfront, lakes, and parks are within driving distance.

The University

WSU is nationally recognized as a major urban university. BAKCI is, as a National Cancer Institute–designated Comprehensive Cancer Center, an excellent research institution. WSU is a state-supported institution with nine major schools and colleges.

Applying

Applicants must hold a bachelor's degree from an accredited school or present evidence that the degree requirements will be completed before enrollment in the graduate program. The overall grade point average should be at least 3.0 (on a 4.0 scale). Applicants are expected to have had one full year each of general chemistry, organic chemistry, biology, calculus, and physics (exceptions can be made). The GRE General Test is required. International applicants are required to take the TOEFL or an equivalent exam. Three letters of evaluation and a brief description of the applicant's background, goals, and objectives should be submitted by March 1. A description of research experiences is recommended.

Correspondence and Information

Dr. Robert J. Pauley, Director
Graduate Program in Cancer Biology
School of Medicine
Wayne State University
550 East Canfield Avenue, Room 329
Detroit, Michigan 48202
Phone: 313-577-1065
E-mail: rpauley@med.wayne.edu
    ad3340@wayne.edu (to send applications)
Web site: http://www.med.wayne.edu/cancer/

Wayne State University

THE FACULTY AND THEIR RESEARCH

Ayad M. Al-Katib, Professor of Medicine; M.D., Mosul Medical College (Iraq), 1974. Biology and experimental therapeutics of human lymphoid tumors.
Marc D. Basson, Professor of Surgery; M.D., Michigan, 1981; Ph.D., Yale, 1992. Cell-matrix interactions; physical force effects; cell motility; intracellular signaling in intestinal epithelial cells.
George S. Brush, Associate Professor of Pathology, Karmanos Cancer Institute; Ph.D., Johns Hopkins, 1991. DNA damage and DNA replication checkpoints.
Ben D.-M. Chen, Professor of Medicine; Ph.D., Vanderbilt, 1977. Regulation of macrophage production and differentiation by hematopoietic growth factor; growth-factor receptor; signal transduction.
Michael L. Cher, Professor of Urology and Pathology, Karmanos Cancer Institute; M.D., Washington (St. Louis), 1986. Biology of prostate cancer bone metastasis.
Ping Dou, Professor of Pathology; Ph.D., Rutgers, 1988. Chemoprevention and molecular targeting.
James F. Eliason, Associate Professor of Medicine and Oncology; Ph.D., Chicago, 1978. Mechanisms of drug resistance and prediction of patient response to therapy.
Stephen P. Ethier, Professor of Pathology; Associate Center Director, Basic Research; and Deputy Director, Karmanos Cancer Institute; Ph.D., Tennessee–Oak Ridge, 1982. Breast cancer genetics, cell biology, and cell signaling.
David R. Evans, Professor of Biochemistry; Ph.D., Wayne State, 1968. Structure and control mechanisms of enzymes that regulate mammalian pyrimidine biosynthesis.
Richard B. Everson, Professor of Internal Medicine and Pathology; M.D., Rochester, 1972; M.P.H., North Carolina, 1985. Molecular epidemiology of breast and prostate cancer.
Joseph A. Fontana, Professor of Medicine; Ph.D., Johns Hopkins, 1969; M.D., Pennsylvania, 1975. Retinoids and their signaling pathways.
Rafael Fridman, Professor of Pathology; Ph.D., Jerusalem, 1986. Role of tumor proteases in tumor cell invasion.
Craig N. Giroux, Associate Professor of Institute of Environmental Health; Ph.D., MIT, 1979. Molecular biology of germ line differentiation; genome stability; developmental genetics; mechanisms of mutation and tumor prevention.
Miriam L. Greenberg, Professor of Biological Sciences; Ph.D., Yeshiva (Einstein), 1980. Regulation of membrane biogenesis; genetic control of phospholipid biosynthesis; inositol phosphate metabolism in yeast.
Ahmad R. Heydari, Associate Professor of Nutrition and Food Science; Ph.D., Illinois State, 1990. Nutrient-gene interactions in aging and neoplasia; nutrients and DNA damage and repair.
Kenneth V. Honn, Professor of Radiation Oncology and Pathology; Ph.D., Wayne State, 1977. Cancer biology: role of kinases, eicosanoids, and integrin receptors in tumor invasion/metastasis.
Michael C. Joiner, Professor of Radiation Oncology; Ph.D., London, 1980. Mechanisms underlying variation in response to ionizing radiation.
June Kan-Mitchell, Associate Professor of Immunology; Ph.D., Yale, 1977. Mimicking epitope-based vaccines for HIV and cancers.
David H. Kessel, Professor of Pharmacology and of Medicine; Ph.D., Michigan, 1959. Photosensitization of neoplastic cells; photobiology; mechanisms of drug resistance.
Hyeong-Reh C. Kim, Professor of Pathology; Ph.D., Northwestern, 1989. Growth factor signaling and regulation of apoptosis.
Thomas A. Kocarek, Associate Professor, Institute of Chemical Toxicology; Ph.D., Ohio State, 1988. Regulation of cytochrome P450 gene expression.
Omer Kucuk, Professor of Internal Medicine; M.D., Hacettepe (Turkey), 1975. Effects of micronutrients in the prevention of cancer.
Joshua Liao, Assistant Professor of Pathology, Karmanos Cancer Institute; Ph.D., Karolinska (Stockholm), 1996. Mechanisms of mammary gland carcinogenesis.
Adhip N. Majumdar, Professor of Internal Medicine; Ph.D., London, 1968. Aging and carcinogenesis of the gastrointestinal tract, in particular, the role of EGF-receptor family in regulating growth and transformation.
Larry H. Matherly, Professor of Pharmacology and Associate Member, Karmanos Cancer Institute; Ph.D., Penn State, 1981. Cancer chemotherapy: mechanisms of action of antitumor agents; mechanisms of drug resistance.
Raymond R. Mattingly, Associate Professor of Pharmacology; Ph.D., Virginia, 1993. Signal transduction through Ras and heterotrimeric GTP-binding proteins.
Fred R. Miller, Professor, Karmanos Cancer Institute; Ph.D., Wisconsin–Madison, 1976. Progression of preneoplastic breast disease; stromal-epithelial interactions; mechanisms of metastasis.
Ramzi Mohammad, Professor of Hematology and Oncology; Ph.D., Utah State, 1987. Developmental therapeutic program.
Raymond F. Novak, Professor of Pharmacology; Ph.D., Case Western Reserve, 1973. Solvent and carcinogen effects on regulation of cytochrome P-450 and glutathione S-transferase gene expression.
Robert J. Pauley, Professor; Ph.D., Marquette, 1975. Molecular oncology: murine mammary cancer; *Mtv* (mammary tumor virus) and oncogenes; human breast cancer.
Stuart Ratner, Assistant Professor, Karmanos Cancer Institute; Ph.D., Massachusetts, 1980. Tumor immunotherapy; lymphocyte traffic.
Avraham Raz, Professor of Radiation Oncology and Pathology and Director, Cancer Metastasis Program, Karmanos Cancer Institute; Ph.D., Weizmann (Israel), 1978. Tumor metastasis: role of adhesion molecule in tumor spread.
Kaladhar B. Reddy, Associate Professor of Pathology; Ph.D., Osmania (India), 1984. Role of oncogenes, growth factor receptors, and signal transduction in tumor progression and metastasis.
John J. Reiners Jr., Professor of Pharmacology and Associate Professor, Environmental Health Science; Ph.D., Purdue, 1977. Mechanisms of chemical-induced carcinogenesis, signal transduction, and immunomodulation.
James H. Rigby, Professor of Chemistry; Ph.D., Wisconsin–Madison, 1977. Total synthesis and structure-activity studies on tumor-promoting diterpenes and alkaloids; total synthesis of antitumor natural products.
Arun K. Rishi, Associate Professor of Internal Medicine; Ph.D., London, 1987. Retinoid-dependent and -independent cell-cycle and apoptosis regulatory pathways.
Louis J. Romano, Professor of Chemistry; Ph.D., Rutgers, 1976. Chemical carcinogenesis; mutagenesis; replication of damaged DNA.
John C. Ruckdeschel, Professor of Medicine and Oncology and President and CEO, Karmanos Cancer Institute; M.D., Albany Medical College, 1971. Lung cancer and therapy.
Melissa Runge-Morris, Associate Professor; Institute of Chemical Toxicology; M.D., Michigan, 1979. Molecular regulation of the sulfotransferase multigene family.
Fazlul H. Sarkar, Professor of Pathology; Ph.D., Banaras Hindu (India), 1978. Molecular biology of human adenocarcinoma; gene expression and regulation, activation and inactivation, and mutation; tumor angiogenesis; invasion and metastasis.
Eva M. Schmelz, Assistant Professor of Food and Nutrition Science; Ph.D., Giessen (Germany), 1994. Sphingolipid signaling in normal and cancer cells.
Malathy Shekhar, Associate Professor of Pathology; Ph.D., Indian Institute of Science, 1985. Breast cancer.
Shijie Sheng, Associate Professor of Pathology; Ph.D., Florida, 1993. Tumor invasion and metastasis/proteolysis.
Anthony Shields, Professor of Internal Medicine; M.D., Harvard, 1979; Ph.D., MIT, 1979. Positron emission tomography.
Debra F. Skafar, Associate Professor of Physiology; Ph.D., Vanderbilt, 1983. Estrogen receptor signaling mechanisms.
Bonnie F. Sloane, Professor and Chair of Pharmacology; Ph.D., Rutgers, 1976. Cancer biology: role of cysteine proteinases and their inhibitors in malignant progression.
Michael A. Tainsky, Professor of Pathology and Member, Karmanos Cancer Institute; Ph.D., Cornell, 1977. Molecular oncology and genetics.
Guri Tzivion, Associate Professor of Pathology and Member, Karmanos Cancer Institute; Ph.D., Hebrew (Israel), 1998. Regulation of signaling pathways that mediate cell growth and cell death.
Anil Wali, Assistant Professor of Surgery and Pathology; Ph.D., Postgraduate Institute of Medical Education and Research (India), 1990. Genomic and proteomic analysis of lung cancer and mesothelioma.
Wei-Zen Wei, Professor of Immunology, Karmanos Cancer Institute; Ph.D., Brown, 1978. Host immunity in mammary tumorigenesis; modulation of mammary-tumor progression.
Gen Sheng Wu, Assistant Professor of Cancer Biology and Immunology; Ph.D., Chinese Academy of Medical Sciences (Beijing), 1992. Tumor suppressor genes and chemosensitivity.
Hai-Young Wu, Associate Professor of Pharmacology; Ph.D., CUNY Graduate Center, 1985. DNA topology, DNA conformation, gene expression regulation.
Youming Xie, Assistant Professor of Pathology, Karmanos Cancer Institute; Ph.D., Texas, 1996. The ubiquitin-proteasome system.
Fayth Yoshimura, Associate Professor of Immunology and Microbiology; Ph.D., Yale, 1972. Pathogenesis of oncogenic murine retroviruses.

WAYNE STATE UNIVERSITY

School of Medicine
Division of Research and Graduate Studies
Center for Molecular Medicine and Genetics

Program of Study

Students are admitted to a program that leads to a Ph.D. degree in molecular biology and genetics. The program emphasizes research training in eukaryotic molecular and cellular biology with applications to molecular medicine and genetics. Students are prepared for research careers in either academic or industrial settings. Ph.D. students spend the first year taking the core curriculum and doing laboratory rotations to sample the research environment in laboratories of potential interest to them. The core curriculum provides students with a solid foundation in the areas of molecular and cellular biology and genetics. By the end of the first year, students begin thesis research, and during the second year they complete additional course work that is tailored to each student's background and area of research. Faculty laboratories provide opportunities for research in many areas of molecular biology, medicine, and genetics that are relevant to the understanding, diagnosis, treatment, and prevention of human diseases. Multidisciplinary approaches include the use of model organisms for the study of developmental and reproductive biology, cancer, neurobiology, and other areas. Students participate in weekly research presentations and attend a seminar series sponsored by the Center that covers the breadth of research areas in current molecular biology, molecular medicine, and genetics. The Center also offers graduate programs that lead to M.S. degrees in genetic counseling and applied genomic technology.

Further details about the graduate programs, the Center for Molecular Medicine and Genetics (CMMG), and the research interests of the faculty members may be found online at the Center's Web site.

Research Facilities

Newly renovated, state-of-the-art laboratories are located at the medical school. Additional laboratories at the medical school campus are located in the Karmanos Cancer Institute, the Mott Center for Human Growth and Development, the Elliman and Lande buildings, and the Detroit Veterans Administration Hospital. Core facilities for automated DNA sequencing, liquid robotics handling, yeast two-hybrid protein interaction hunts, bioinformatics, high-throughput robotics, DNA microarray and other genomics technologies, fluorescence microscopy, and tissue culture are supported by the Center. Fully computerized libraries are located adjacent to the medical school and within the hospitals of the Detroit Medical Center.

Financial Aid

Graduate research assistantships are available through the Center and provide tuition, an annual salary of about $19,500, and health/dental insurance. No separate application is required for financial aid; however, the student should indicate whether he or she wishes to be considered for a position if financial aid is not available.

Cost of Study

Tuition and fees were about $8300 for state residents ($17,275 for nonresidents) for the twelve-month 2005–06 year. Tuition is paid in full for candidates receiving graduate research assistantships.

Living and Housing Costs

A variety of living accommodations are available both on campus and in the surrounding communities. The cost of living in the metropolitan Detroit area is low.

Student Group

The program in molecular biology and genetics currently has 19 doctoral students (including two M.D./Ph.D. students) and 11 master's students, and it expects to expand with the addition of new faculty members and resources. Of the current doctoral students, 7 (37 percent) are studying under externally competitive fellowship awards.

Student Outcomes

Since the University's graduate training program began in 1987, Ph.D. graduates have taken postdoctoral positions at the National Institutes of Health (NIH) (5); Rockefeller (2); Yale (2); Baylor; Stanford; Caltech; Emory; Vanderbilt; the Universities of Michigan, Washington, and Massachusetts; M. D. Anderson Cancer Center; Karmanos Cancer Institute (3); and the MRC Human Genetics Unit, Edinburgh. Three have completed medical school. CMMG graduates have established successful careers in academia and industry.

Location

Wayne State University is located in the University Cultural Center, adjacent to several museums, including the Detroit Institute of Arts, which has a weekly foreign film series; the Detroit Science Center; the Detroit Historical Museum; and the Museum of African American History. Top entertainment in all areas of the performing arts, as well as four professional sports teams, is offered in metropolitan Detroit. In addition, there is easy access to Ontario, the Great Lakes, and a variety of outdoor activities.

The University

Wayne State University is a major public university with more than fifty doctoral programs in the arts, sciences, engineering, business, and law. With more than 34,000 students, it is one of the largest U.S. urban universities. The School of Medicine is the fourth-largest medical school in the country.

Applying

Applicants are expected to have a B.S. or M.S. with a GPA greater than 3.0, a strong background in basic science, and upper-percentile scores on the GRE General Test. International students must show proficiency in English with a minimum TOEFL score of 600. To submit an application online, students should visit the Web site at http://www.med.wayne.edu/gradprog/onlineapplication.htm. Additional documentation required to complete the online application includes original transcripts from all colleges and universities attended, GRE scores, three letters of recommendation, and the TOEFL score (for international students); this should be sent to the Graduate Officer at the address provided in the Correspondence and Information section of this description. Applicants are encouraged to visit the campus and may be invited for an interview.

Correspondence and Information

Graduate Officer
Center for Molecular Medicine and Genetics
Wayne State University School of Medicine
540 East Canfield Avenue, Room 3127
Detroit, Michigan 48201
Phone: 313-577-5323
Fax: 313-577-5218
E-mail: apply@genetics.wayne.edu
Web site: http://www.genetics.wayne.edu/

Wayne State University

THE FACULTY AND THEIR RESEARCH

Leon Carlock, Associate Professor (also with Anatomy and Cell Biology); Ph.D., Purdue, 1981. Molecular neurobiology; Huntington's disease; neuron-specific gene expression; demyelinating disease.

Gerald Feldman, Professor; M.D./Ph.D., Richmond, 1984. Use of molecular technologies in the diagnosis of genetic diseases, clinical genetics, and dysmorphology; educational programs in medical genetics residency training.

Russell L. Finley, Associate Professor (also with Biochemistry and Molecular Biology); Ph.D., SUNY at Syracuse, 1990. Regulatory networks that control cell proliferation; development of yeast technology for genome characterization; cell-cycle regulation during development; protein interaction networks.

James Y. Garbern, Associate Professor (also with Neurology); Ph.D., 1979, M.D., 1981, Baylor. Neurogenetics; molecular biology of myelin disorders; molecular pathogenesis of glial tumors; homeobox genes.

Morris Goodman, Professor (also with Anatomy and Cell Biology); Ph.D., Wisconsin, 1951. Molecular evolution, with emphasis on globin genes and primate phylogeny gene.

Alexander Gow, Associate Professor (also with Pediatrics and Neurology); Ph.D., Queensland (Australia), 1990. Role of the unfolded protein response in neurodegenerative diseases; molecular characterization of the regulation of axoglial junction assembly in CNS myelin; molecular characterization of the claudin family of integral membrane tight-junction proteins during development in brain, testes, and inner ear using transgenic and homologous recombination in embryonic stem cells.

Anne E. Greb, Assistant Professor; M.S., Wisconsin, 1986. Educational programs in human genetics and genetics counseling; multicultural issues related to genetic counseling and the provision of clinical genetics services.

Lawrence I. Grossman, Professor and Director (also with Internal Medicine); Ph.D., Yeshiva (Einstein), 1971. Molecular genetics and evolution of the electron transport chain; cytochrome c oxidase; mitochondria and mitochondrial diseases.

Henry H. Q. Heng, Associate Professor (also with Karmanos Cancer Institute and Pathology); Ph.D., Toronto, 1994. Molecular cytogenetics; high-order structure of chromosomes; genome structure and function; genetic and physical mapping.

Maik Hüttemann, Assistant Professor; Ph.D., Marburg (Germany), 1999. Function of cytochrome c oxidase isoforms, oxygen sensing, and role of mitochondria in cancer.

John Kamholz, Professor (also with Neurology); M.D., 1980, Ph.D., 1984, Pennsylvania. Regulation of myelination; molecular pathophysiology of demyelinating disease.

Gregory Kapatos, Professor (also with Psychiatry and Behavioral Neurosciences); Ph.D., Pittsburgh, 1978. Cellular and molecular biology of monoamine neurotransmitter secreting neurons.

Stephen A. Krawetz, Charlotte B. Failing Professor (also with Obstetrics and Gynecology and the Institute for Scientific Computing); Ph.D., Toronto, 1983. Gene therapy; control of development and differentiation; expression of connective tissue genes and genes controlling spermatogenesis; computer-assisted sequence analysis.

S. Helena Kuivaniemi, Professor (also with Surgery) and Associate Director; M.D., 1984, Ph.D., 1985, Oulu (Finland). Molecular biology and genetics of aortic aneurysms.

Markku Kurkinen, Professor (also with Pathology); Ph.D., Helsinki, 1979. Extracellular matrix; metalloproteinases; gene regulation; development.

Wayne Lancaster, Professor (also with Obstetrics and Gynecology); Ph.D., Wayne State, 1973. Papillomaviruses: molecular biology, evolution, and role in human carcinogenesis; cancer cell genome instability; ovarian carcinogenesis.

Li Li, Associate Professor (also with Internal Medicine); Ph.D., Texas, 1991. Gene regulation during cardiovascular and hematopoietic development.

Jeffrey A. Loeb, Associate Professor (also with Neurology); Ph.D., 1987, M.D., 1989, Chicago. Developmental neuroscience; molecular mechanisms in synapse formation; neuroregulins and neurotrophins in synaptic development; functional genomic study of the pathogenesis of epilepsy.

Laura S. Martin, Associate Professor (also with Pediatrics); M.D., Florida, 1981. Maternal-fetal-placenta unit and its relationship in inborn errors in metabolism; inborn errors of metabolism; etiology of mental retardation; educational programs for genetic counseling and medical genetics training programs; consultant for children/adults with their families for inherited/genetic disorders.

Orlando J. Miller, Professor Emeritus (also with Obstetrics and Gynecology); M.D., Yale, 1950. Human/mammalian genetics and cytogenetics; chromosome structure and function; genome organization.

Jodi Parrish, Research Scientist; Ph.D., Wisconsin–Madison, 1998. Protein interaction networks; bacterial physiology.

Michael E. Shy, Professor (also with Neurology); M.D., SUNY at Albany, 1979. Molecular biology of Schwann cell axonal interactions; gene therapy in the peripheral nervous system.

Michael Tainsky, Professor (also with Karmanos Cancer Institute and Pathology); Ph.D., Cornell, 1977. Cancer genetic studies of familial cancers to determine the inherited cancer genes, analyze the mechanism of their action, and factors affecting genetic penetrance of the cancer phenotype.

Angela Trepanier, Assistant Professor; M.S., Minnesota, 1994. Educational programs in human genetics and genetic counseling; cancer genetics.

Gerard Tromp, Associate Professor; Ph.D., Rutgers, 1989. Genetic epidemiology and linkage analysis; molecular biology and genetics of intracranial aneurysms; the Blau syndrome; bioinformatics and genomics.

Eva Monica Uddin, Research Scientist; Ph.D., NYU, 2003. Molecular evolution and comparative genomics; molecular evolution in reproduction; mitochondria and aerobic energy metabolism.

Derek Wildman, Assistant Professor; Ph.D., NYU, 2000. Molecular evolution and comparative genomics; molecular evolution in reproduction; mitochondria and aerobic energy metabolism.

David D. Womble, Associate Professor (also with the Institute for Scientific Computing); Ph.D., Wisconsin, 1976. Control of DNA replication and segregation; regulation of the cell division cycle; bioinformatics.

WESLEYAN UNIVERSITY

Department of Molecular Biology and Biochemistry

Program of Study

The graduate program of the Department of Molecular Biology and Biochemistry is designed to lead to the degree of Doctor of Philosophy. A Master of Arts degree is awarded only under special circumstances. The program's primary emphasis is on an intensive research experience culminating in a thesis and at least one publication. The program of study includes a series of graduate courses covering major areas of molecular biology, biochemistry, and biophysics; practicums designed to introduce first-year students to the research interests of faculty members; journal clubs in which current research is discussed informally; and several seminar series in which distinguished outside speakers present their research. A low student-faculty ratio allows close contact between each student and the faculty. The programs of study are individually designed based on each student's previous training and experience. All graduate students have the opportunity to do some undergraduate teaching under faculty supervision. Funds are also available to enable students to attend outside laboratories at Woods Hole, Massachusetts; Cold Spring Harbor, New York; and Bar Harbor, Maine.

The Department staff consists of 8 professors with postdoctoral fellows and research support personnel. In addition, there are a number of associated faculty members in the Departments of Biology and Chemistry with overlapping research interests. The Department is also part of the Molecular Biophysics Program, which is supported by a training grant from the National Institutes of Health. More information is available on the Department's Web site.

Research Facilities

The Department is housed in Shanklin and Hall-Atwater Laboratories. Research facilities include three electron microscopes, an automated DNA/RNA synthesizer, a phosphoimager, a large-scale fermentor, an ELISA reader, a UV-photo system for DNA gels, high-pressure liquid chromatography systems, two nuclear magnetic resonance spectrometers (400 MHz and 500 MHz), a UV resonance Raman spectrometer, preparative and analytical ultracentrifuges, scintillation counters, a fluorometer and optical spectroscopy equipment, an electron paramagnetic resonance spectrometer, animal facilities, a greenhouse, and an advanced center for sequence analysis and visualization of biological macromolecules.

Financial Aid

All graduate students are awarded a teaching or research assistantship when they are accepted into the program. Teaching assistants spend approximately 10 hours per week in course instruction. The twelve-month stipend for 2006–07 is approximately $21,269, to which is added tuition remission and a waiver for health-care and insurance fees. An increase of approximately 3 percent is given for each year of additional study.

Cost of Study

The only financial costs are for books and other educational materials associated with the thesis and graduation.

Living and Housing Costs

For a single student, housing costs are approximately $7000 per year; miscellaneous expenses add about $1800. For married students, the costs are higher, but there is a dependency allowance for spouse and children. The allowance varies depending upon the size of the family and employment of the spouse. Both University and privately owned accommodations are available.

Student Group

Because the Department is relatively small, there is close interaction among students, faculty members, postdoctoral fellows, and technicians. There are approximately 23 graduate students in the Department of Molecular Biology and Biochemistry, a similar number in the Department of Biology, and approximately 30 in the Department of Chemistry.

Student Outcomes

The students come from many different undergraduate institutions and have an excellent opportunity to further their careers in teaching and research positions at universities or in industry. Recent graduates have positions at institutions such as Johns Hopkins, Columbia, Yale, and the University of California and in industrial firms such as Pfizer and Merck.

Location

Wesleyan University is situated in Middletown, Connecticut, a city with a population of 45,000 on the west bank of the picturesque Connecticut River. Middletown lies about 19 miles south of Hartford and 25 miles north of New Haven, nearly midway between New York and Boston and about 25 miles from Long Island Sound. The University provides many cultural and recreational opportunities, which are supplemented by those available in the surrounding countryside and in the large cities nearby.

The University

Founded in 1831, Wesleyan University is an independent institution of liberal arts and sciences, coeducational since 1969. Since 1963, Wesleyan has developed selective programs leading to the Ph.D. degree in a number of departments. Present registration includes approximately 2,600 undergraduates and about 250 students in various master's or doctoral programs. The University's endowment is valued at approximately $500 million. A $15-million Science Center, including a central Science Library, was completed in 1975.

Applying

No specific courses are required for admission, but successful applicants usually have a strong undergraduate major in biology, chemistry, or biochemistry. To be complete, the application must include scores on the General Test of the Graduate Record Examinations, transcripts of all previous academic work at the college level, and three letters of recommendation from college instructors familiar with the applicant's record and ability. Applications should be submitted before March 15 in order to receive adequate consideration; notification of the admission decision is sent by April 15. Personal interviews are recommended.

Correspondence and Information

Admissions Committee
Department of Molecular Biology and Biochemistry
Hall-Atwater/Shanklin Laboratories
Wesleyan University
Middletown, Connecticut 06459-0175

E-mail: mbbgrad@wesleyan.edu
Web site: http://www.wesleyan.edu/mbb/

Wesleyan University

THE FACULTY AND THEIR RESEARCH

Professors
Anthony A. Infante, Ph.D., Pennsylvania. Control of gene expression in a developing multicellular organism, the sea urchin.
Donald B. Oliver, Ph.D., Tufts. Protein targeting in *E. coli* and *B. subtilis;* molecular genetics of Lyme disease.

Associate Professors
Manju M. Hingorani, Ph.D., Ohio State. Enzymology of DNA replication and repair in *E. coli.* and *S. cerevisiae.*
Scott Holmes, Ph.D., Virginia. Control of gene expression by chromatin structure in the yeast *S. cerevisiae.*
Michael McAlear, Ph.D., McGill. Molecular biology and genetics of cell-cycle control in the yeast *S. cerevisiae;* execution of DNA replication and DNA repair.
Ishita Mukerji, Ph.D., Berkeley. Protein and nucleic acid structure; structure-function relationships in protein–nucleic acid interactions using UV resonance Raman spectroscopy.

Assistant Professors
Mark Flory, Ph.D., Washington. Proteomic and cell biological analysis of fission yeast centrosomes and telomeres.
Robert P. Lane, Ph.D., CIT. The olfactory system and new frontiers in genome research.

Associated Faculty in the Department of Biology
Steven H. Devoto, Ph.D., Rockefeller. Cell biology and molecular genetics of muscle development in zebrafish.
Laura Grabel, Ph.D., California, San Diego. Role of cell and matrix interactions during tetratocarcinoma stem cell differentiation.
Michael Weir, Ph.D., Pennsylvania. Genetic analysis of pattern formation in early embryos of *Drosophila.*
Jason Wolfe, Ph.D., Berkeley. Molecular mechanisms of cell-cell communication.

Associated Faculty in the Department of Chemistry
David Beveridge, Ph.D., Cincinnati. Computer simulation of liquids; role of water in biological structures and processes.
Philip Bolton, Ph.D., California, San Diego. Studies of damaged DNA, DNA repair, and quadruplex DNAs.
Rex Pratt, Ph.D., Melbourne. Enzyme mechanisms; bioorganic chemistry; protein chemistry; β-lactamases and β-lactam antibodies.
Irina Russu, Ph.D., Pittsburgh. Allosteric mechanisms in multimeric proteins; molecular mechanisms of DNA/RNA recognition by drugs and proteins studied by NMR spectroscopy.
T. David Westmoreland, Ph.D., North Carolina at Chapel Hill. Electronic structure and mechanism in metalloenzyme active sites; EPR spectroscopy.

WEST VIRGINIA UNIVERSITY

Graduate Program in Cancer Cell Biology

Programs of Study	The Graduate Program in Cancer Cell Biology is one of seven graduate programs in West Virginia University (WVU) Schools of Medicine and Pharmacy offering interdisciplinary biomedical research training leading to the Ph.D. or M.D./Ph.D. degree. Research is focused on the molecular basis of cancer etiology, progression, and translational applications. The three main areas of emphasis are cellular signaling, tumor microenvironment, and cancer therapeutics. Cellular signaling focuses on protein and lipid-based signals that influence tumor cell growth, survival, motility, and invasion. Tumor microenvironment addresses the mechanisms by which tumor cells interact with other cells in the stroma to promote tumor survival, angiogenesis, and inflammation. Therapeutics addresses the mechanisms by which novel cancer therapeutic compounds block tumor cell growth and metastasis, as well as strategies for the translational development and delivery of conventional chemotherapeutics and targeted small molecule compounds.

Students benefit from individual attention by faculty members within a research environment that is dynamic, collaborative, and interdisciplinary. In addition to course work and laboratory research, students participate in seminars, journal clubs, and research conferences. Graduate trainees also attend national scientific meetings and obtain valuable speaking and teaching experience. The Ph.D. typically takes five years to complete. During Year 1, all new graduate students matriculate into a common integrated core curriculum. This integrated first year allows students to build competence in key areas of contemporary science, gain exposure to the various training program options, meet potential dissertation advisers, and network scientifically and socially. In the second semester, students customize their course work by selecting from an array of program-specific electives. At the end of Year 1, students select a research adviser and can select Cancer Cell Biology as their training program. Year 2 consists of advanced course work, research, teaching, and the candidacy examination. Years 3 to 5 are devoted to dissertation research. The Graduate Program in Cancer Cell Biology also participates in the combined M.D./Ph.D. Scholars Program. M.D./Ph.D. Scholars take the first two years of the medical curriculum, followed typically by three years of research as required for the Ph.D. degree before returning to the M.D. program.

Research Facilities Institutional facilities include a computer-based learning center, a centralized animal facility with a transgenic barrier, and a library housing more than 205,000 volumes and 2,400 journals. Core facilities are available for examining gene expression or genetic variation (Affymetrix platform), image analysis, confocal and electron microscopy and laser-capture microdissection, live-cell imaging, mass spectrometry, flow cytometry and high-speed cell sorting, proteomics, recombinant DNA technology, transgenic rodent biology, and functional neuroimaging (fMRI, PET/CT). Affiliated research centers include the National Institute for Occupational Safety and Health (NIOSH), the Center for Advanced Imaging, Blanchette Rockefeller Neurosciences Institute, Sensory Neuroscience Research Center, and Mary Babb Randolph Cancer Center.

Financial Aid Ph.D. and M.D./Ph.D. students in the biomedical sciences receive financial support during their training, provided they remain in good academic standing and excel in research. Such support includes full tuition, health insurance, and an annual stipend of $22,000. Combined M.D./Ph.D. students also receive medical tuition waivers.

Cost of Study Students' tuition costs are covered.

Living and Housing Costs The cost of an efficiency apartment in University-owned housing is approximately $400 per month. A limited number of University apartments are available for married students. Privately owned apartments in Morgantown cost $400 to $600 per month. In general, the cost of living is lower compared to larger cities.

Student Group Total University enrollment is approximately 26,000 students, which includes 6,500 graduate and professional students. Graduate students come from all parts of the U.S. and many other countries.

Location Morgantown is a vibrant university community, with an appealing balance to life, of 80,000 residents in northern West Virginia. Located near the Pennsylvania border at the western edge of the Appalachian Mountains, abundant opportunities exist for activities such as world-class white-water rafting and kayaking, hiking and camping, mountain biking, fishing, and skiing. Morgantown has a cosmopolitan atmosphere with a range of activities usually found in much larger cities. It also enjoys proximity to major metropolitan centers: Pittsburgh is a 90-minute drive north, and Washington, D.C., is a 3-hour drive east.

The University West Virginia University is a comprehensive, land-grant, Carnegie-designated Doctoral/Research University–Extensive public institution. The University's academic Health Sciences Center includes the Schools of Medicine, Dentistry, Nursing, and Pharmacy, all of which offer graduate degree programs. There are seven Ph.D. biomedical research training programs in the Schools of Medicine and Pharmacy that benefit from the common, undifferentiated first-year: Biochemistry and Molecular Biology, Cancer Cell Biology, Cellular and Integrative Physiology, Exercise Physiology, Immunology and Microbial Pathogenesis, Neuroscience, and Pharmaceutical and Pharmacological Sciences. Graduate faculty members in these programs are from various basic science and clinical departments throughout WVU and are members of interdisciplinary research centers in five health-related areas: cancer cell biology, cardiovascular sciences, immunopathology and microbial pathogenesis, neuroscience, and respiratory biology and lung diseases.

As a member of the Big East Conference, WVU participates in NCAA Division I sports. WVU also offers a wide variety of creative arts, theater, and entertainment opportunities.

Applying Applicants must have a bachelor's degree and excellent GPA and GRE scores. Three letters of recommendation and a personal statement are required. Students are invited in groups of 10 for a paid, two-day visit/interview in January through March. Interested students should visit http://www.hsc.wvu.edu/som/resoff/gradprograms/PhD.asp for more information and an online application.

Correspondence and Information

Scott A. Weed, Ph.D., Graduate Director
Graduate Program in Cancer Cell Biology
West Virginia University
P.O. Box 9300
Morgantown, West Virginia 26506
Phone: 304-293-3016
E-mail: sweed@hsc.wvu.edu
Web site: http://www.hsc.wvu.edu/som/resoff/gradprograms.
  PhD.asp

Office of Research and Graduate Education
Health Sciences Center
West Virginia University
P.O. Box 9104
Morgantown, West Virginia 26506
Phone: 304-293-7116
E-mail: cnoel@hsc.wvu.edu

West Virginia University

THE FACULTY AND THEIR RESEARCH

Yehenew Agazie, Assistant Professor; Ph.D., Saskatchewan. Signal transduction by tyrosine kinases and phosphatases.

Daniel Flynn, Professor; Ph.D., North Carolina State. Signaling networks that regulate cellular motility and invasion in breast cancer.

Steven Frisch, Professor; Ph.D., Berkeley. Anoikis and the role of cell adhesion in cellular survival and mechanism of tumor suppression by E1a.

Peter Gannett, Professor; Ph.D., Wisconsin. Chemical carcinogenesis; computational biology; nanotechnology.

Laura Gibson, Associate Professor; Ph.D., West Virginia. Effects of dose-intensive chemotherapy on the bone marrow microenvironment and the subsequent effect on hematopoietic recovery.

Lan Guo, Research Assistant Professor; Ph.D., West Virginia. Bioinformatics and computer software applications that integrate information exchange.

Bing-Hua Jiang, Assistant Professor; Ph.D., Mississippi. PI 3-kinase and angiogenesis.

Jun Liu, Assistant Professor; D. Phil., Oxford; M.D., China Medical. Caveolin-1, endothelial cell motility, and angiogenesis.

Jia Luo, Associate Professor; Ph.D., Iowa. Effects of ethanol on EGF receptor signaling in breast cancer.

William Petros, Associate Professor; Pharm.D., Philadelphia College of Pharmacy and Science. Pharmacokinetics; pharmacodynamics; pharmacogenomics and oncology.

Yong Qian, Adjunct Assistant Professor; Ph.D., West Virginia. Heavy metals, signal transduction, cytoskeleton, and cell motility.

Heimo Riedel, Professor; Ph.D., European Molecular Biology Laboratory (Heidelberg). Adaptor proteins in mitogenesis and cancer.

Christian Stehlik, Research Assistant Professor; Ph.D., Vienna. Death domain folds in cancer; inflammation and immunity.

Grazyna Szklarz, Associate Professor; Ph.D., Clarkson. Cytochrome P450 and molecular modeling.

Linda Vona-Davis, Adjunct Associate Professor; Ph.D., West Virginia. Leptin, obesity, and cancer.

Weixin Wang, Assistant Professor; Ph.D., Shanghai Institute for Biochemistry. DNA damage, apoptosis, and oncology.

Scott Weed, Associate Professor; Ph.D., Yale. Regulation of the actin cytoskeleton in cell motility and tumor cell invasion.

Robert Wysolmerski, Professor; Ph.D., St. Louis. Endothelial cell biology; myosin II-based contraction; cellular tension; vasculogenesis.

Jing Jie Yu, Research Assistant Professor; M.D., Beijing Medical. Mutagenesis of EGF receptor in lung cancer.

Section 7
Computational and Systems Biology

This section contains a directory of institutions offering graduate work in computational and systems biology, followed by in-depth entries submitted by institutions that chose to prepare detailed program descriptions. Additional information about programs listed in the directory but not augmented by an in-depth entry may be obtained by writing directly to the dean of a graduate school or chair of a department at the address given in the directory.

CONTENTS

Program Directories

Computational Biology

Arizona State University, Division of Graduate Studies, College of Liberal Arts and Sciences, Division of Natural Sciences and Mathematics, Program in Computational Biosciences, Tempe, AZ 85287. Offers MS, PSM.

Baylor College of Medicine, Graduate School of Biomedical Sciences, Program in Structural and Computational Biology and Molecular Biophysics, Houston, TX 77030-3498. Offers PhD, MD/PhD. *Faculty:* 63 full-time (4 women). *Students:* 40 full-time (7 women); includes 5 minority (3 Asian Americans or Pacific Islanders, 2 Hispanic Americans), 22 international. Average age 27. 73 applicants, 22% accepted, 10 enrolled. In 2005, 4 degrees awarded. *Median time to degree:* Of those who began their doctoral program in fall 1997, 80% received their degree in 8 years or less. *Degree requirements:* For doctorate, thesis/dissertation, public defense. *Entrance requirements:* For doctorate, GRE General Test, GRE Subject Test (strongly recommended), minimum GPA of 3.0. Additional exam requirements/recommendations for international students: Required—TOEFL. *Application deadline:* For fall admission, 2/1 for domestic students. Application fee: $30. Electronic applications accepted. *Expenses:* Tuition: Full-time $8,200. Full-time tuition and fees vary according to program. *Financial support:* In 2005–06, 37 students received support, including 26 fellowships (averaging $23,000 per year), 14 research assistantships (averaging $23,000 per year); career-related internships or fieldwork, Federal Work-Study, institutionally sponsored loans, health care benefits, and tuition waivers (full) also available. Financial award applicants required to submit FAFSA. *Faculty research:* X-ray and electron crystallography, light and electron microscopy, computer image reconstruction, molecular spectroscopy. *Unit head:* Dr. Wah Chiu, Director, 713-798-6985. *Application contact:* Wanda Waguespack, Graduate Program Administrator, 713-798-5197, Fax: 713-798-6325, E-mail: wandaw@bcm.edu.

See Close-Up on page 465.

Carnegie Mellon University, Mellon College of Science, Department of Biological Sciences, Program in Computational Biology, Pittsburgh, PA 15213-3891. Offers MS. *Entrance requirements:* For master's, GRE General Test, GRE Subject Test, interview.

Claremont Graduate University, Graduate Programs, School of Mathematical Sciences, Claremont, CA 91711-6160. Offers computational and systems biology (PhD); computational science (PhD); engineering mathematics (PhD); operations research and statistics (MA, MS); physical applied mathematics (MA, MS); pure mathematics (MA, MS); scientific computing (MA, MS); systems and control theory (MA, MS). Part-time programs available. *Faculty:* 4 full-time (0 women), 6 part-time/adjunct (2 women). *Students:* 54 full-time (13 women), 13 part-time (1 woman); includes 20 minority (1 African American, 14 Asian Americans or Pacific Islanders, 5 Hispanic Americans), 16 international. Average age 38. In 2005, 5 master's, 5 doctorates awarded. Terminal master's awarded for partial completion of doctoral program. *Degree requirements:* For doctorate, 2 foreign languages, thesis/dissertation. *Entrance requirements:* For master's and doctorate, GRE General Test. *Application deadline:* For fall admission, 2/15 for domestic students. Applications are processed on a rolling basis. Electronic applications accepted. *Expenses:* Tuition: Full-time $27,902; part-time $1,214 per term. Required fees: $1,600; $800 per term. Tuition and fees vary according to degree level and program. *Financial support:* Fellowships, research assistantships, career-related internships or fieldwork, Federal Work-Study, institutionally sponsored loans, and tuition waivers (full and partial) available. Support available to part-time students. Financial award application deadline: 2/15; financial award applicants required to submit FAFSA. *Unit head:* John Angus, Dean, 909-621-8080, Fax: 909-607-8261, E-mail: john.angus@cgu.edu. *Application contact:* Susan Townzen, Program Coordinator, 909-621-8080, Fax: 909-607-8261, E-mail: susan.n.townzen@cgu.edu.

Cornell University, Joan and Sanford I. Weill Medical College and Graduate School of Medical Sciences, Weill Graduate School of Medical Sciences, Tri-Institutional Program in Chemical Biology and Computational Biology, New York, NY 10021-4896. Offers PhD. Offered jointly by Cornell University, Weill Graduate School of Medical Sciences, The Rockefeller University and Sloan-Kettering Institute; students must be accepted to Cornell University Graduate Program in Chemistry. *Students:* 17 full-time (9 women); includes 1 minority (Asian American or Pacific Islander), 11 international. *Expenses:* Tuition: Full-time $32,320. Required fees: $1,025. *Unit head:* Timothy Ryan, Director, 212-746-6403.

Florida State University, Graduate Studies, College of Arts and Sciences, Program in Molecular Biophysics, Tallahassee, FL 32306. Offers biochemistry, molecular and cell biology (PhD); computational structural biology (PhD); molecular biophysics (PhD). *Faculty:* 44 full-time (7 women). *Students:* 29 full-time (17 women); includes 1 minority (Hispanic American), 12 international. Average age 30. 14 applicants, 100% accepted, 4 enrolled. In 2005, 3 degrees awarded. *Median time to degree:* Of those who began their doctoral program in fall 1997, 100% received their degree in 8 years or less. *Degree requirements:* For doctorate, thesis/dissertation, comprehensive exam. *Entrance requirements:* For doctorate, GRE General Test. Additional exam requirements/recommendations for international students: Required—TOEFL (minimum score 600 paper-based; 250 computer-based). *Application deadline:* For fall admission, 1/15 for domestic students, 1/15 for international students. Applications are processed on a rolling basis. Application fee: $30. Electronic applications accepted. *Financial support:* In 2005–06, 29 students received support, including 29 research assistantships (averaging $19,500 per year); health care benefits and tuition waivers (partial) also available. Financial award applicants required to submit FAFSA. *Faculty research:* Protein and nucleic acid structure and function, membrane protein structure, computational biophysics, 3-D image reconstruction. Total annual research expenditures: $5.1 million. *Unit head:* Dr. P. Bryant Chase, Director, MOB Graduate Program, 850-644-0056, Fax: 850-644-7244, E-mail: chase@bio.fsu.edu. *Application contact:* Dale E. Leonard, Academic Coordinator, Graduate Programs, 850-644-1012, Fax: 850-644-7244, E-mail: mob@sb.fsu.edu.

Iowa State University of Science and Technology, Graduate College, Interdisciplinary Programs, Bioinformatics and Computational Biology Program, Ames, IA 50011-3260. Offers MS, PhD. *Degree requirements:* For doctorate, thesis/dissertation. *Entrance requirements:* For doctorate, GRE General Test. Additional exam requirements/recommendations for international students: Required—TOEFL or IELTS. Electronic applications accepted. *Expenses:* Tuition, state resident: full-time $6,410. Tuition, nonresident: full-time $16,422. Tuition and fees vary according to program. *Faculty research:* Functional and structural genomics, genome evolution, macromolecular structure and function, mathematical biology and biological statistics, metabolic and developmental networks.

Massachusetts Institute of Technology, School of Engineering and School of Science, Program in Computational and Systems Biology, Cambridge, MA 02139-4307. Offers PhD. *Students:* 12 full-time (5 women); includes 2 minority (1 African American, 1 Asian American or Pacific Islander), 5 international. Average age 24. 170 applicants, 6% accepted, 8 enrolled. *Degree requirements:* For doctorate, thesis/dissertation, comprehensive exam. *Entrance requirements:* For doctorate, GRE General Test. Additional exam requirements/recommendations for international students: Required—TOEFL (minimum score 577 paper-based; 233 computer-based). *Application deadline:* For fall admission, 1/1 for domestic students, 1/1 for international students. Application fee: $70. Electronic applications accepted. *Expenses:* Tuition: Full-time $32,100. Required fees: $200. Part-time tuition and fees vary according to course load. *Financial support:* In 2005–06, 5 students received support, including 12 fellowships (averaging $24,310 per year); Federal Work-Study, institutionally sponsored loans, scholarships/grants, health care benefits, and unspecified assistantships also available. *Unit head:* Prof. Paul Matsudaira, Director, 617-258-5188. *Application contact:* Darlene Ray, 617-253-3874, Fax: 617-324-4870, E-mail: csbphd@mit.edu.

New Jersey Institute of Technology, Office of Graduate Studies, College of Computing Science, Program in Computational Biology, Newark, NJ 07102. Offers MS. Part-time and evening/weekend programs available. *Students:* 16 full-time (7 women), 14 part-time (2 women); includes 7 minority (1 African American, 5 Asian Americans or Pacific Islanders, 1 Hispanic American), 13 international. Average age 30. 33 applicants, 61% accepted, 6 enrolled. In 2005, 29 degrees awarded. *Entrance requirements:* For master's, GRE General Test. Additional exam requirements/recommendations for international students: Required—TOEFL (minimum score 550 paper-based; 213 computer-based). *Application deadline:* For fall admission, 6/5 for domestic students; for spring admission, 10/15 for domestic students. Applications are processed on a rolling basis. Application fee: $60. Electronic applications accepted. *Expenses:* Tuition, state resident: full-time $9,620; part-time $520 per credit. Tuition, nonresident: full-time $13,542; part-time $715 per credit. Required fees: $78; $54 per credit. $78 per year. Tuition and fees vary according to course load. *Financial support:* Fellowships with full and partial tuition reimbursements, research assistantships with full and partial tuition reimbursements, teaching assistantships with full and partial tuition reimbursements, career-related internships or fieldwork, Federal Work-Study, institutionally sponsored loans, and unspecified assistantships available. Financial award application deadline: 3/15. *Faculty research:* Technological, computational, and mathematical aspects of biology and bioengineering. *Unit head:* Dr. Michael L. Recce, Director, 973-596-5535, E-mail: michael.l.recce@njit.edu. *Application contact:* Kathryn Kelly, Director of Admissions, 973-596-3300, Fax: 973-596-3461, E-mail: admissions@njit.edu.

New York University, Graduate School of Arts and Science, Department of Biology, Program in Computational Biology, New York, NY 10012-1019. Offers PhD. *Students:* 5 full-time (4 women), 7 part-time (1 woman), 1 international. Average age 29. 36 applicants, 31% accepted, 6 enrolled. *Entrance requirements:* For doctorate, GRE. Additional exam requirements/recommendations for international students: Required—TOEFL. Application fee: $80. *Financial support:* Fellowships, research assistantships, teaching assistantships, Federal Work-Study, scholarships/grants, health care benefits, and unspecified assistantships available. *Unit head:* Tamar Schlick, Director, 212-998-3596, Fax: 212-995-3590, E-mail: fas.computational.biology@nyu.edu. *Application contact:* Prof. Carol Leong, Program Administrator, 212-998-3590, Fax: 212-995-3590.

See Close-Up on page 677.

New York University, School of Medicine and Graduate School of Arts and Science, Sackler Institute of Graduate Biomedical Sciences, New York, NY 10012-1019. Offers cellular and molecular biology (PhD); computational biology (PhD); developmental genetics (PhD); medical and molecular parasitology (PhD); microbiology (PhD); molecular oncology and immunology (PhD), including immunology, molecular oncology; neuroscience and physiology (PhD), including neuroscience, physiology; pharmacology (PhD), including molecular pharmacology and signal transduction; structural biology (PhD). *Faculty:* 150 full-time (35 women). *Students:* 240 full-time (124 women), 6 part-time (4 women); includes 54 minority (13 African Americans, 29 Asian Americans or Pacific Islanders, 12 Hispanic Americans), 84 international. Average age 28. 569 applicants, 8% accepted, 46 enrolled. In 2005, 38 degrees awarded. *Degree requirements:* For doctorate, one foreign language, thesis/dissertation, qualifying exam, comprehensive exam. *Entrance requirements:* For doctorate, GRE General Test. Additional exam requirements/recommendations for international students: Required—TOEFL. *Application deadline:* For fall admission, 1/4 for domestic students. Applications are processed on a rolling basis. Application fee: $80. *Expenses: Contact institution. Financial support:* In 2005–06, fellowships with tuition reimbursements (averaging $26,000 per year), research assistantships with tuition reimbursements (averaging $26,000 per year), teaching assistantships with tuition reimbursements (averaging $25,000 per year) were awarded; career-related internships or fieldwork, Federal Work-Study, institutionally sponsored loans, scholarships/grants, health care benefits, tuition waivers (full and partial), and unspecified assistantships also available. Financial award application deadline: 2/1; financial award applicants required to submit FAFSA. *Unit head:* Dr. Joel D. Oppenheim, Senior Associate Dean for Graduate Studies, 212-263-8001, Fax: 212-263-7600. *Application contact:* Lizabeth Greene, Administrative Manager, 212-263-5648, Fax: 212-263-7600, E-mail: sackler-info@med.nyu.edu.

See Close-Up on page 181.

Northwestern University, McCormick School of Engineering and Applied Science, Program in Computational Biology and Bioinformatics, Evanston, IL 60208. Offers MS. Part-time programs available. *Faculty:* 40 full-time (5 women). *Degree requirements:* For master's, thesis, registration. *Entrance requirements:* For master's, GRE General Test, 2 letters of reference. Additional exam requirements/recommendations for international students: Required—TOEFL (minimum score 600 paper-based; 250 computer-based); Recommended—TSE. *Application deadline:* For fall admission, 3/1 priority date for domestic students, 3/1 priority date for international students. Applications are processed on a rolling basis. Application fee: $60 ($75 for international students). Electronic applications accepted. *Faculty research:* Mathematical models of protein signaling, high throughput DNA sequencing, macromolecule interactions, chemoinformatics, genome DNA sequence evolution. *Unit head:* Dr. Ming Yang Kao, Director, 847-563-0426, Fax: 847-491-5258, E-mail: kao@cs.northwestern.edu. *Application contact:* Dr. Dawn M. Graunke, Assistant Program Director, 847-467-1972, Fax: 847-491-5258, E-mail: d-graunke@cs.northwestern.edu.

Rutgers, The State University of New Jersey, Newark, Graduate School, Program in Computational Biology, Newark, NJ 07102. Offers MS. In 2005, 1 degree awarded. *Entrance requirements:* For master's, GRE, minimum undergraduate B average. Additional exam requirements/recommendations for international students: Required—TOEFL. *Application deadline:* For fall admission, 6/1 for domestic students, 6/1 for international students; for spring admission, 12/1 for domestic students, 12/1 for international students. Application fee: $50. *Expenses:* Tuition, state resident: full-time $10,440; part-time $435 per credit. Tuition, nonresident: full-time $15,520; part-time $637 per credit.

University of Idaho, College of Graduate Studies, College of Science, Department of Biological Sciences, Program in Bioinformatics and Computational Biology, Moscow, ID 83844-2282. Offers MS, PhD. *Students:* 3 full-time (0 women), 1 international. *Entrance requirements:* For master's, GRE, minimum GPA of 2.8. *Application deadline:* For fall admission, 8/1 for domestic students; for spring admission, 12/15 for domestic students. Application fee: $55 ($60 for international students). *Expenses:* Tuition, nonresident: full-time $8,770; part-time $130 per credit. Required fees: $4,508; $217 per credit. *Financial support:* Application deadline: 2/15. *Unit head:* Larry J. Forney, Chair, Department of Biological Sciences, 208-885-6280.

University of Illinois at Urbana–Champaign, Graduate College, College of Liberal Arts and Sciences, School of Molecular and Cellular Biology, Center for Biophysics and Computational Biology, Champaign, IL 61820. Offers PhD. *Students:* 66 full-time (20 women), 2 part-time (1 woman); includes 9 minority (1 African American, 6 Asian Americans or Pacific Islanders, 2 Hispanic Americans), 41 international. 126 applicants, 10% accepted, 7 enrolled. In 2005, 7 degrees awarded. *Degree requirements:* For doctorate, thesis/dissertation. *Entrance requirements:* For doctorate, minimum GPA of 3.0. Additional exam requirements/recommendations for international students: Required—TOEFL. *Application deadline:* For fall admission, 5/15 for domestic students. Application fee: $50 ($60 for international students). Electronic applications accepted. *Financial support:* In 2005–06, 15 fellowships, 55 research assistantships, 17 teaching assistantships were awarded; scholarships/grants and traineeships also available. Financial award application deadline: 1/15. *Unit head:* Martin Gruebele, Director, 217-333-1630, Fax: 217-244-6615, E-mail: gruebele@uiuc.edu. *Application contact:* Cynthia Dodds, Secretary, 217-333-1630, Fax: 217-244-6615, E-mail: biophysics@life.uiuc.edu.

University of Pennsylvania, School of Medicine, Biomedical Graduate Studies, Graduate Group in Genomics and Computational Biology, Philadelphia, PA 19104. Offers PhD, MD/PhD, VMD/PhD. *Faculty:* 51. *Students:* 26 full-time (5 women); includes 6 minority (2 African Americans, 4 Asian Americans or Pacific Islanders), 6 international. 47 applicants, 30% accepted, 8 enrolled. *Degree requirements:* For doctorate, thesis/dissertation optional. *Entrance* *requirements:* For doctorate, GRE. Additional exam requirements/recommendations for international students: Required—TOEFL. *Application deadline:* For fall admission, 12/15 priority date for domestic students, 12/1 priority date for international students. Applications are processed on a rolling basis. Application fee: $70. Electronic applications accepted. *Financial support:* In 2005–06, 19 students received support; fellowships, research assistantships, scholarships/grants, traineeships, and unspecified assistantships available. *Unit head:* Dr. Warren Ewens, Chairperson, 215-898-7109. *Application contact:* Sean Dalton, Graduate Coordinator, 215-746-2807, E-mail: sdalton@mail.med.upenn.edu.

University of Pittsburgh, School of Medicine, Joint Program in Computational Biology, Pittsburgh, PA 15260. Offers PhD. *Faculty:* 60 full-time (11 women). *Students:* 6 full-time (0 women), (all international). Average age 27. 35 applicants, 26% accepted, 6 enrolled. *Degree requirements:* For doctorate, thesis/dissertation, comprehensive exam, registration. *Entrance requirements:* For doctorate, GRE General Test, GRE Subject Test. Additional exam requirements/recommendations for international students: Required—TOEFL (minimum score 600 paper-based; 250 computer-based), GRE, AGRE. *Application deadline:* For fall admission, 1/15 priority date for domestic students, 1/15 priority date for international students. Electronic applications accepted. *Expenses:* Tuition, state resident: full-time $13,194; part-time $537 per credit. Tuition, nonresident: full-time $25,012; part-time $1,026 per credit. Required fees: $700; $164 per term. Tuition and fees vary according to campus/location and program. *Financial support:* In 2005–06, 6 students received support, including 6 fellowships with full tuition reimbursements available (averaging $21,500 per year) *Faculty research:* Computational biology, bioinformatics, genomics, bioimage informatics, systems biology. *Unit head:* Dr. Ivet Bahar, Chair, Department of Computational Biology, 412-648-3333, Fax: 412-648-3163, E-mail: bahar@ccbb.pitt.edu. *Application contact:* Dr. Judy Wieber, Research and Education Administrator, Department of Computational Biology, 412-648-8646, Fax: 412-648-3163, E-mail: jwieber@ccbb.pitt.edu.

University of Rochester, School of Medicine and Dentistry, Graduate Programs in Medicine and Dentistry, Department of Biostatistics and Computational Biology, Rochester, NY 14627-0250. Offers medical statistics (MS); statistics (MA, PhD). Terminal master's awarded for partial completion of doctoral program. *Degree requirements:* For doctorate, thesis/dissertation, qualifying exam. *Entrance requirements:* For master's and doctorate, GRE General Test. Additional exam requirements/recommendations for international students: Required—TOEFL.

University of Southern California, Graduate School, College of Letters, Arts and Sciences, Department of Biological Sciences, Program in Molecular and Computational Biology, Los Angeles, CA 90089. Offers PhD. *Degree requirements:* For doctorate, thesis/dissertation. *Entrance requirements:* For doctorate, GRE General Test. Additional exam requirements/recommendations for international students: Required—TOEFL. *Expenses:* Tuition: Full-time $25,416; part-time $1,059 per unit. Required fees: $484; $484 per year. Tuition and fees vary according to course load and program. *Faculty research:* Genomic instabilities, molecular studies in aging, genetic/biochemical studies of DNA repair, molecular/genetic approaches to developmental biology, computation and experimental genomics.

See Close-Up on page 641.

The University of Texas Medical Branch, Graduate School of Biomedical Sciences, Program in Biochemistry and Molecular Biology, Galveston, TX 77555. Offers biochemistry (PhD); bioinformatics (PhD); biophysics (PhD); cell biology (PhD); computational biology (PhD); structural biology (PhD). *Students:* 38 full-time (14 women), 1 part-time; includes 5 minority (3 Asian Americans or Pacific Islanders, 2 Hispanic Americans), 18 international. Average age 27. In 2005, 7 degrees awarded. *Degree requirements:* For doctorate, thesis/dissertation. *Entrance requirements:* Additional exam requirements/recommendations for international students: Required—TOEFL (minimum score 550 paper-based; 213 computer-based). *Application deadline:* Applications are processed on a rolling basis. Application fee: $30 ($75 for international students). Electronic applications accepted. *Expenses:* Tuition, state resident: full-time $8,350; part-time $90 per credit hour. Tuition, nonresident: full-time $21,450; part-time $366 per credit hour. Required fees: $1,027; $11 per credit hour. $60 per term. *Financial support:* In 2005–06, fellowships (averaging $23,000 per year), research assistantships (averaging $23,000 per year) were awarded. Financial award applicants required to submit FAFSA. *Unit head:* Dr. Leenian L. Chan, Director, 409-772-2861, Fax: 409-772-9679, E-mail: lchan@utmb.edu. *Application contact:* Debora Botting, Co-ordinator II Special Programs, 409-772-2769, Fax: 409-747-0552, E-mail: dmbottin@utmb.edu.

Virginia Polytechnic Institute and State University, Graduate School, Intercollege, Program in Genetics, Bioinformatics and Computational Biology, Blacksburg, VA 24061. Offers PhD. *Students:* 27 full-time (14 women), 4 part-time; includes 4 minority (1 African American, 2 Asian Americans or Pacific Islanders, 1 Hispanic American), 20 international. Average age 29. 61 applicants, 20% accepted, 8 enrolled. *Entrance requirements:* For doctorate, GRE. Additional exam requirements/recommendations for international students: Required—TOEFL (minimum score 550 paper-based; 213 computer-based). *Application deadline:* Applications are processed on a rolling basis. Application fee: $45. Electronic applications accepted. *Expenses:* Tuition, state resident: full-time $6,558; part-time $364 per credit. Tuition, nonresident: full-time $11,296; part-time $628 per credit. Required fees: $1,419; $468 per credit. $234 per term. *Financial support:* Fellowships with full tuition reimbursements, research assistantships with full tuition reimbursements, teaching assistantships with full tuition reimbursements, career-related internships or fieldwork, Federal Work-Study, scholarships/grants, and unspecified assistantships available. *Unit head:* Dr. David Bevan, Chair, 540-231-5040, Fax: 540-231-3010, E-mail: drbevan@vt.edu. *Application contact:* Dennie Munson, Information Contact, 540-231-1928, Fax: 540-231-3010, E-mail: dennie@vt.edu.

Washington University in St. Louis, Graduate School of Arts and Sciences, Division of Biology and Biomedical Sciences, Program in Computational Biology, St. Louis, MO 63130-4899. Offers PhD. *Degree requirements:* For doctorate, thesis/dissertation. Electronic applications accepted.

Yale University, School of Medicine and Graduate School of Arts and Sciences, Combined Program in Biological and Biomedical Sciences (BBS), Computational Biology and Bioinformatics Track, New Haven, CT 06520. Offers PhD, MD/PhD. *Students:* 5 full-time. *Entrance requirements:* Additional exam requirements/recommendations for international students: Required—TOEFL. *Application deadline:* For fall admission, 12/8 for domestic students, 12/8 for international students. *Unit head:* Dr. Perry Miller, Director of Graduate Studies, 203-432-8189, E-mail: dgs.bioinfo@yale.edu.

Systems Biology

Cornell University, Joan and Sanford I. Weill Medical College and Graduate School of Medical Sciences, Weill Graduate School of Medical Sciences, Program in Physiology, Biophysics and Systems Biology, New York, NY 10021-4896. Offers PhD, MD/PhD. *Faculty:* 30 full-time (8 women). *Students:* 39 full-time (19 women); includes 4 minority (1 African American, 3 Asian Americans or Pacific Islanders), 21 international. 32 applicants, 25% accepted, 4 enrolled. In 2005, 1 doctorate awarded. *Degree requirements:* For doctorate, thesis/dissertation, final exam. *Entrance requirements:* For doctorate, GRE General Test, GRE Subject Test, introductory courses in biology, inorganic and organic chemistry, physics, and mathematics. Additional exam requirements/recommendations for international students: Required—TOEFL. *Application deadline:* For fall admission, 12/15 for domestic students. Application fee: $60. *Expenses:* Tuition: Full-time $32,320. Required fees: $1,025. *Financial support:* In 2005–06, 1 fellowship was awarded; stipends also available. *Unit head:* Doris Herzlinger, Director, 212-746-6377, E-mail: daherzli@med.cornell.edu.

Harvard University, Graduate School of Arts and Sciences, Department of Systems Biology, Cambridge, MA 02138. Offers PhD. *Students:* 9. 100 applicants, 12% accepted, 9 enrolled. *Degree requirements:* For doctorate, thesis/dissertation, lab rotation, qualifying examination. *Entrance requirements:* For doctorate, GRE. Additional exam requirements/recommendations for international students: Required—TOEFL. *Application deadline:* For fall admission, 12/8 priority date for domestic students, 12/8 priority date for international students. Application fee: $90. Electronic applications accepted. *Expenses:* Tuition: Full-time $28,752. Full-time tuition and fees vary according to program and student level. *Financial support:* Institutionally sponsored loans, scholarships/grants, unspecified assistantships, and all students receive a stipend and full tuition reimbursements available. *Unit head:* Judy Finkelstein, Graduate Coordinator, 617-432-5876, Fax: 617-432-5012, E-mail: pamela_silver@dfci.harvard.edu. *Application contact:* Jodi Finkelstein, Coordinator, 617-432-5202, Fax: 617-432-5012, E-mail: jodi_finkelstein@hms.harvard.edu.

See Close-Up on page 675.

Massachusetts Institute of Technology, School of Engineering and School of Science, Program in Computational and Systems Biology, Cambridge, MA 02139-4307. Offers PhD.

Students: 12 full-time (5 women); includes 2 minority (1 African American, 1 Asian American or Pacific Islander), 5 international. Average age 24. 170 applicants, 6% accepted, 8 enrolled. *Degree requirements:* For doctorate, thesis/dissertation, comprehensive exam. *Entrance requirements:* For doctorate, GRE General Test. Additional exam requirements/recommendations for international students: Required—TOEFL (minimum score 577 paper-based; 233 computer-based). *Application deadline:* For fall admission, 1/1 for domestic students, 1/1 for international students. Application fee: $70. Electronic applications accepted. *Expenses:* Tuition: Full-time $32,100. Required fees: $200. Part-time tuition and fees vary according to course load. *Financial support:* In 2005–06, 5 students received support, including 12 fellowships (averaging $24,310 per year); Federal Work-Study, institutionally sponsored loans, scholarships/grants, health care benefits, and unspecified assistantships also available. *Unit head:* Prof. Paul Matsudaira, Director, 617-258-5188. *Application contact:* Darlene Ray, 617-253-3874, Fax: 617-324-4870, E-mail: csbphd@mit.edu.

University of California, San Diego, Graduate Studies and Research, Division of Biology, Program in Plant Systems Biology, La Jolla, CA 92093. Offers PhD.

Announcement: The Plant Systems Biology PhD program is an interdisciplinary venture between UCSD, Salk Institute, and Scripps Research Institute. Students with different backgrounds will be trained at the interface of systems modeling, computational genomics, and plant sciences and will be positioned at the frontier of systems biology. Visit http://www.biology.ucsd.edu/psbigert/index.htm for more information.

University of Virginia, School of Medicine, Department of Molecular Physiology and Biological Physics, Program in Molecular Medicine and Systems Biology, Charlottesville, VA 22903. Offers PhD. *Students:* 1 (woman) full-time. Application fee: $60. *Expenses:* Tuition, state resident: full-time $7,731. Tuition, nonresident: full-time $18,672. Required fees: $1,479. Full-time tuition and fees vary according to degree level and program. *Application contact:* Peter C. Brunjes, Associate Dean for Graduate Programs and Research, 434-924-7184, Fax: 434-924-6737, E-mail: grad-a-s@virginia.edu.

HARVARD UNIVERSITY

Ph.D. Program in Systems Biology

Program of Study

The Ph.D. program in systems biology aims to recruit students from a variety of different backgrounds, including all areas of biology, physics, chemistry, computer science, engineering, and mathematics, who will work together to forge a new approach to biology that combines theoretical and experimental approaches. The program explains how the higher-level properties of complex systems materialize from the interactions among their parts. Because both the field and the program are new, this program requires unusual levels of independence and creativity from its participants.

Students meet participating faculty members to hear about their research, both in formal lectures and in informal settings. After the first two years, students may choose a single faculty member as their adviser or may elect to initiate a collaboration between two or more labs. The research topic chosen may be entirely theoretical, entirely experimental, or anything in between. Subject to program approval, students can choose an adviser from outside of the Harvard community, including the eleven affiliated hospitals.

Each student's program of graduate study is planned in consultation with faculty advisers. The degree program, made up of three parts (course work, lab rotations, and independent research), is designed to be completed in a maximum of six years. Other program activities include a seminar program, a retreat, a weekly discussion session known as Theory Lunch, and a range of informal and semiformal opportunities to discuss results, practice presentation skills, and interact with the faculty.

Second-year students must teach one term as part of their academic requirements. By the spring of the second year, students are required to submit a brief discussion of the research they wish to undertake and form a Dissertation Advisory Committee (DAC). At the first meeting of the DAC, the research proposal is reviewed and the student is required to defend it. Independent research begins once the qualifying examination is successfully completed. The DAC, in consultation with the student's faculty adviser(s), periodically evaluates the progress of the student's dissertation research and determines at what point the student is ready to defend his or her dissertation.

Research Facilities

The Nikon Imaging Center at Harvard Medical School (NIC@HMS) is a cell biology department and systems biology department core facility for light microscopy developed in partnership with Nikon Instruments, Inc., and Micro Video Instruments, Inc. The NIC@HMS maintains laser scanning confocals, spinning-disk confocals, a total internal fluorescence system, and other basic fluorescence and transmitted-light microscopes. All users are fully trained to use the equipment; no prior imaging experience is necessary.

Financial Aid

All students accepted into the program are awarded full support, including a stipend, full tuition, and health fees.

Cost of Study

For 2005–06, tuition and health fees were $31,280. As described in the Financial Aid section, these fees, along with a stipend, are provided for enrolled students.

Living and Housing Costs

Accommodations in graduate residence halls are available. In addition, there are approximately 1,500 apartments available for graduate students in Harvard-owned buildings. Applications may be obtained from the Housing Office, which also maintains a list of available private rooms, houses, and apartments in the vicinity. The 2005–06 on-campus housing rates ranged from $4679 to $7353.

Student Group

The Ph.D. Program in Systems Biology is a new program. It enrolled its first class of students in fall 2005.

Location

Harvard is located in Cambridge, Massachusetts, on the east coast of the United States. With a population of about 95,800, Cambridge is the state's seventh-largest city. With a rich history and long-established neighborhoods with strong ethnic roots and traditions, Cambridge is also the birthplace of the state's high-technology industry. The presence of both Harvard and the Massachusetts Institute of Technology has encouraged a wide variety of technical, research, and professional firms to locate in the city. Harvard benefits greatly from the vitality and culture of Cambridge and the Boston area, which helps attract talented faculty members and students.

The University and The Program

Harvard University, which celebrated its 350th anniversary in 1986, is the oldest institution of higher learning in the United States. Founded sixteen years after the arrival of the Pilgrims at Plymouth, the University has grown from 9 students with a single master to an enrollment of more than 18,000 degree candidates, including undergraduates and students in ten graduate and professional schools.

The goal of the Ph.D. Program in Systems Biology is to prepare investigators with diverse backgrounds for independent research careers in which combined theoretical and experimental approaches are used to address biological problems.

Applying

Students should submit the completed application form, the application fee, transcripts from all colleges and universities attended, three letters of recommendation (at least one should be from a faculty member at the last school attended), and scores on the General Test of the Graduate Record Examinations (GRE). Applicants are strongly urged to submit a resume or curriculum vitae. Admission is for the fall term only. Other application requirements can be obtained by contacting the department.

Correspondence and Information

All application materials should be mailed to:

Office of Admissions
The Graduate School of Arts and Sciences
Harvard University
P.O. Box 9129
Cambridge, Massachusetts 02238-9129

Jodi Finkelstein, Coordinator of the Systems Biology Program
Harvard Medical School
200 Longwood Avenue, Alpert 536
Boston, Massachusetts 02115

Phone: 617-432-5202
Fax: 617-432-5012
E-mail: jodi_finkelstein@hms.harvard.edu
Web site: http://sysbio.med.harvard.edu/phd/index.html

Harvard University

THE FACULTY AND THEIR RESEARCH

Michael Brenner, Gordon McKay Professor of Applied Mathematics and Applied Physics. Quantitative modeling of complex phenomena in science and engineering.

Lewis Cantley, Professor of Medicine and Systems Biology. Biochemical pathways that regulate normal mammalian cell growth and the defects that cause cell transformation.

George Church, Professor of Genetics and Director, Center for Computational Genetics. Synthetic biology design of 3-D, multicell, and new translational codes; stem-cell, aging, and cancer epigenetics; ecosystem models; personal genomics.

Daniel Fisher, Professor of Physics and Applied Physics. Condensed-matter theory; geophysics; biology.

Walter Fontana, Professor of Systems Biology. Formalization and emergence of functional organization; genotype-phenotype mappings; distributed molecular control.

Jeremy Gunawardena, Senior Lecturer on Systems Biology. Theoretical and experimental approaches to in silico systems biology.

Marc Kirschner, Professor of Systems Biology and Chair, Department of Systems Biology. Regulation of the cell cycle; role of the cytoskeleton in cell morphogenesis; mechanisms of establishing the basic vertebrate body plan.

Roy Kishony, Lecturer on Systems Biology. Combining theoretical and experimental approaches to understand how biological function emerges in complex genetic and chemical networks; using population genetics approaches to understand the interplay between biological design and the evolutionary process.

Galit Lahav, Assistant Professor of Systems Biology. Studying the dynamics of conserved network motifs in diverse signaling systems in humans by stimulating the proteins of interest and accurately monitoring their expression level and localization in individual living cells.

Gavin MacBeath, Assistant Professor of Chemistry and Chemical Biology. Systems-level investigation of protein-protein interactions in intracellular signaling networks, using protein microarrays; emphasis on receptor tyrosine kinase–mediated signaling and presynaptic and postsynaptic signaling.

Lakshminarayanan Mahadevan, Gordon McKay Professor of Applied Mathematics and Mechanics. The applications of mathematics to understand the mechanical behavior of matter in all its forms, with a particular emphasis on soft materials and biological systems.

Timothy Mitchison, Hasib Sabbagh Professor of Systems Biology and Deputy Chairman, Department of Systems Biology. Cytoskeleton dynamics, in particular, the mechanism of mitosis and the mechanism of cell motility dependent on actin polymerization.

Vamsi Mootha, Assistant Professor of Systems Biology. Biochemical adaptation at the level of the mitochondrion, assessed through physiology, functional genomics (microarrays, proteomics), and computation; integration of genome-scale data sets to discover gene networks underlying rare and common human metabolic diseases.

Andrew Murray, Professor of Molecular and Cellular Biology and Director, Bauer Center for Genomics Research. Mitosis; meiosis; experimental evolution; signal transduction.

Radhika Nagpal, Assistant Professor of Computer Science and Instructor in Systems Biology. Developing programming paradigms for robust collective behavior, inspired by biology; understanding robust collective behavior in biological systems.

Martin Nowak, Professor of Mathematics and Biology. Theoretical biology; somatic evolution of cancer.

Erin O'Shea, Professor of Molecular and Cellular Biology and Co-Director, Bauer Center for Genomic Research. Systems-level and molecular analysis of signaling pathways; transcriptional regulation; developing methods for expressing and assaying the entire complement of proteins derived from an organism.

Kevin Kit Parker, Assistant Professor of Biomedical Engineering. Cellular mechanotransduction in the heart.

Johan Paulsson, Lecturer on Systems Biology. Mathematical theory for noise in intracellular networks and development of new experimental techniques for counting molecules in single cells; combining theory and experiments in the study of such areas as stochastic gene expression, homeostatic control, near-critical metabolism, and intracellular selfishness.

Tom Rapoport, Professor of Cell Biology. Getting proteins across membranes.

Brian Seed, Professor of Genetics. Using automated and parallel research methods to accelerate the rate of discovery of signaling pathways in mammals.

Jagesh Shah, Assistant Professor of Systems Biology. Scaling molecular events into cell behavior; using molecular techniques and modern biophysical tools in piecing together quantitative models of endogenous and synthetic cellular networks.

Pamela Silver, Professor of Systems Biology and Director, Ph.D. Program in Systems Biology. Systems analysis of genomes, RNA, and nuclear organization; cell-based screens; synthetic biology.

Antoine Van Oijen, Assistant Professor of Biological Chemistry and Molecular Pharmacology. Single-molecule studies of complex multiprotein machineries; DNA replication; viral fusion.

Xiaoliang Sunney Xie, Professor of Chemistry and Chemical Biology. Single-molecule spectroscopy and dynamics; molecular interaction and chemical dynamics in biological systems

Xiaowei Zhuang, Assistant Professor of Chemistry and Chemical Biology and of Physics. Study of complex biological processes at the single-molecule (or single-working-unit) level; development of new imaging techniques.

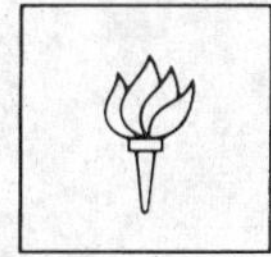

NEW YORK UNIVERSITY

Department of Computational Biology

Program of Study	Computational Biology (COB) is a multidisciplinary program that involves six different divisions and departments at New York University (NYU) and Mount Sinai School of Medicine. Participating departments and divisions are: NYU Graduate School of Arts and Science (Biology, Chemistry, Mathematics, and Neural Science), NYU School of Medicine/The Sackler Institute of Graduate Biomedical Sciences, and Mount Sinai School of Medicine (MSSM) (Biomedical Sciences Program). The goal of the COB program, funded by the National Science Foundation's Integrative Graduate Education and Research Traineeship Program (IGERT) initiative and recently approved by New York State, is to train a new generation of scientists in the fundamentals and applications of computational methods to biological problems, including macromolecular structure and function, genomics, and physiological systems (cells and organs). The NYU/Sinai COB program is designed to educate students from mathematics and physical science backgrounds, as well as students from the biology/chemistry fields, for productive research in biology and medicine using a variety of modern, multidisciplinary quantitative techniques.

Computational Biology (COB) is a multidisciplinary program that involves six different divisions and departments at New York University (NYU) and Mount Sinai School of Medicine. Participating departments and divisions are: NYU Graduate School of Arts and Science (Biology, Chemistry, Mathematics, and Neural Science), NYU School of Medicine/The Sackler Institute of Graduate Biomedical Sciences, and Mount Sinai School of Medicine (MSSM) (Biomedical Sciences Program). The goal of the COB program, funded by the National Science Foundation's Integrative Graduate Education and Research Traineeship Program (IGERT) initiative and recently approved by New York State, is to train a new generation of scientists in the fundamentals and applications of computational methods to biological problems, including macromolecular structure and function, genomics, and physiological systems (cells and organs). The NYU/Sinai COB program is designed to educate students from mathematics and physical science backgrounds, as well as students from the biology/chemistry fields, for productive research in biology and medicine using a variety of modern, multidisciplinary quantitative techniques.

To realize this broad training in biological and computational areas and provide trainees broad scientific perspectives and work experiences, the COB Ph.D. program includes primary and secondary faculty mentorship for thesis research; interdisciplinary training through flexible and background-tailored tracks in molecular and cell biology (biology and biochemistry), molecular modeling and dynamics (chemistry), scientific computing (computer science and mathematics), and computational biology (Interdisciplinary), trainee-led seminars, and ethics/research conduct courses, while ensuring competitive time to degree (five years); summer internships in industry, academia, government, or international laboratories; and learning environments and activities that promote interdisciplinary interactions and broader collaborations within and outside NYU Arts and Science/NYU School of Medicine-Sackler/MSSM, including trainee-led COB seminars, COB retreats, and shared facilities.

Students in the program are required to complete 72 credits, including eight core courses in molecular and cell biology, molecular modeling and dynamics, scientific computing, and computational biology; one noncredit course in research and ethical conduct; an interdisciplinary student seminar and research seminar courses (every term for the first two years); and four elective courses in biology, chemistry, neural science, biomedicine, and mathematical/computational biology. During the first year, students participate in one or more lab experiences, culminating in selection of the primary and secondary mentors; in the second year, they participate in a public seminar presentation in the fall semester and take the COB Ph.D. qualifying exam (written and oral) at the end of the year. Students are also encouraged to take part in summer internships, typically in the third year. The Ph.D. dissertation defense follows at the end of the program.

Research Facilities

The library of the Courant Institute of Math Sciences is focused on research-level material in mathematics, computer science, and related fields, including fluid mechanics, image processing, and robotics. The collection contains more than 60,000 volumes and receives 220 current journal subscriptions. The Frederick L. Ehrman Medical Library holds more than 180,000 monographs in print and electronic format, subscribes to more than 7,000 journals, and has access to 146 databases. The Elmer Holmes Bobst Library is the flagship of a nine-library, 4.5 million–volume system that provides students and faculty members with access to the world's scholarship and serves as a center for the University community's intellectual life. The library houses more than 3.3 million volumes, 20,000 journals, and over 3.5 million microforms and provides access to thousands of electronic resources both on-site and around the world via the Internet. At Mount Sinai School of Medicine, the Levy Library's print collection totals approximately 165,000 volumes, containing over 36,000 monographs and more than 1,200 current journal subscriptions in the fields of biomedicine and health sciences. The library maintains an extensive online collection of more than 12,000 periodical titles (including nonmedical), approximately 500 health sciences textbooks, and more than 60 databases.

Financial Aid

Fellowships provided are a combination package from the National Science Foundation, participating divisions, and mentor research funds. The National Science Foundation's IGERT fellowships provide up to two years of full support, including a student stipend of $30,000 (for 2006–07) per twelve months, full tuition and fees, and NYU health insurance. Stipend amounts for the other years are based on the departmental base. National Science Foundation's stipends are restricted to U.S. citizens and permanent residents. There are very limited stipends from NYU and Sinai for nondomestic students.

Cost of Study

Normally, all students accepted to the COB program are supported by either assistantships or fellowships that cover all tuition and health insurance fees. Under exceptional circumstances, part-time and self-supported students are admitted to some divisions.

Living and Housing Costs

Graduates can find housing in one of the residence halls. NYU housing fees range from $7820 to $17,820 for space in a studio, one-bedroom, or two-bedroom unit for 1 or 2 persons per academic year. Married student housing is available for $25,860 per academic year at the Stuyvesant Town complex. Meal plans range from $1700 to $3750 per academic year for six to twenty-four meals per week, or up to 225 meals per semester. Off-campus apartments are available for $1000 to more than $2000 per month, depending on size, amenities, and location. At Sinai, the Jane B. Aron Residence Hall, across the street from the Medical Center, provides subsidized housing for students. Amenities include a 24-hour doorman, health fitness room, entertainment lounge with big-screen TV, piano, Ping-Pong table and other items, and laundry facilities. Other Sinai-owned apartments are available in the surrounding area. All are within walking distance of the Medical Center and the 96th street subway stop.

Student Group

The program attracts students from a wide range of disciplines, including biology, chemistry, computer science, mathematics, neural science, and medicine.

Student Outcomes

Graduates of the program learn to successfully model biological systems and receive the grounding in computational techniques to apply to specific biological problems. Students act as catalysts for novel interdisciplinary collaborations and acquire expertise in cutting-edge research areas, thus preparing them uniquely for research and education careers in academia, industry, and government.

Location

The NYU campus, where most classes are held, is located in the heart of New York City's Greenwich Village, a historic neighborhood that has attracted generations of writers, musicians, artists, and intellectuals. New York City has a wide array of cultural and recreational opportunities, from nightlife in the East Village to the Metropolitan Museum of Art and from theater on and off Broadway to relaxing in Central Park. NYU Medical Center is located on 30th Street and First Avenue in midtown Manhattan. The Medical Center consists of two hospitals, Tisch Hospital and Rusk Institute of Rehabilitation Medicine, and New York University School of Medicine. Mount Sinai School of Medicine occupies a four-block area in the historic Carnegie Hill neighborhood of New York. Restaurants, shops, and public transportation are conveniently located nearby.

The University

New York University was founded in 1831 by former Secretary of the Treasury Alfred Gallatin as a center of higher learning that would be open to all, regardless of national origin, religious beliefs, or social background. Today, the University enrolls more than 39,000 students from all fifty states and 133 countries in its fourteen schools and colleges, as well as branch campus programs in Westchester and Rockland Counties. NYU is one of only sixty institutions in the distinguished Association of American Universities. The Mount Sinai Hospital was founded in 1852 as the Jews' Hospital in the City of New York, but it was another century before a school of medicine of created at Mount Sinai. The belief grew that the best way to achieve this new kind of medical teaching was to create at Mount Sinai, a new kind of medical institution, a university of the health sciences. As a result, Mount Sinai School of Medicine was created and opened in 1968.

Applying

Prospective students must submit an application for admission, official GRE scores, transcripts of all undergraduate or prior graduate work, a personal statement, three letters of recommendation, and a $90 application fee ($80 for online applications). In addition, to the application for admission, applicants must submit the answers to the eight COB questions listed at the Computational Biology Web site. Application deadlines usually begin December 15 but vary according to the chosen home department. Applications should be submitted online, with supplemental materials (GRE scores, COB supplemental questions, transcripts, and recommendation letters) mailed to the proposed COB home institution. Student should refer to the participating divisions for more detailed application instructions. Students may also visit the computational biology Web site (http://monod.biomath.nyu.edu/igert/) for more detailed instructions.

Correspondence and Information

Dr. Tamar Schlick
IGERT COB Program Director
Courant Institute of Mathematical Sciences
New York University
251 Mercer Street
New York, New York 10012

Phone: 212-998-3116
E-mail: schlick@nyu.edu
Web site: http://monod.biomath.nyu.edu/igert/

Lala Seidensticker
IGERT COB Program Secretary
New York University
31 Washington Place-1021
New York, New York 10003

Phone: 212-998-3598
E-mail: lala@biomath.nyu.edu

New York University

THE FACULTY AND THEIR RESEARCH

NYU/Courant Institute of Mathematical Sciences Faculty
Leslie Greengard, M.D./Ph.D., Yale, 1987. Scientific computing; fast algorithms; potential theory.
David McLaughlin, Ph.D., Indiana, 1971. Nonlinear waves, mathematical physics.
Bhubaneswar Mishra, Ph.D., Carnegie-Mellon, 1985. Computational and systems biology; mathematical and theoretical computer science.
Charles S. Peskin, Ph.D., Yeshiva, 1972. Physiology; fluid dynamics; numerical methods.
Tamar Schlick, Ph.D., NYU, 1987. Mathematical biology; numerical analysis; computational chemistry.
Daniel Tranchina, Ph.D., Rockefeller, 1981. Visual neuroscience; computational neuroscience.
Olof B. Widlund, Ph.D., Uppsala (Sweden), 1966. Numerical analysis; partial differential equations; parallel computing.
Margaret H. Wright, Ph.D., Stanford, 1976. Computational mathematics; optimization; linear.

NYU/Department of Chemistry Faculty
Hin Hark Gan, Ph.D., McGill, 1989. RNA structures and genomics; graph theory.
Tamar Schlick, Ph.D., NYU, 1987. Mathematical biology; numerical analysis; computational chemistry.
Mark Tuckerman, Ph.D., Columbia, 1993. Molecular dynamics; statistical mechanics.
Michael Ward, Ph.D., Princeton, 1981. Crystal engineering, molecular crystals, electrochemistry.
Yingkai Zhang, Ph.D., Duke, 2000. Computational biochemistry biophysics

NYU/Center for Neural Science Faculty
David Heeger, Ph.D., Pennsylvania, 1987. Functional imaging of the human brain (fMRI); computational neuroscience; vision; attention.
Bijan Paseran, Ph.D., Caltech, 2002. Neuronal dynamics and decision making.
Alex D. Reyes, Ph.D., Washington, 1990. Functional interactions of neurons in a network.
John Rinzel, Ph.D., NYU, 1973. Biophysics; neural computations.
Eero Simoncelli, Ph.D., MIT, 1993. Computational modeling of visual processing.
Robert M. Shapley, Ph.D., Rockefeller, 1970. Neuronal basis of visual perception.

NYU/Department of Biology Faculty
Richard Bonneau, Ph.D., Washington, 2001. Modeling global regulatory circuits; art protein folding; distilling functional information from improved genome-wide de novo predictions.
Suse Broyde, Ph.D., Polytechnic of Brooklyn, 1963. Carcinogen-modified DNAs; molecular structure.
Fabio Piano, Ph.D., NYU, 1995. *C. elegans* functional genomics.

NYU Medical School/Sackler Institute of Graduate Biomedical Sciences Faculty
Stuart M. Brown, Ph.D., Cornell, 1992. Bioinformatics in medicine; molecular biology.
Timothy Cardozo, M.D., Ph.D., NYU, 2002. Computational chemistry; protein engineering; rational drug design; bioinformatics.
Brian Dynlacht, Ph.D., Berkeley, 1992. Bioinformatics in medicine; molecular biology.
Yuval Kluger, Ph.D., Tel Aviv, 1992. Computational genomics and proteomics, applications of mathematical physics to biology.
J. Brandon Laflen, Ph.D., Purdue, 2003. Mathematical modeling of human auditory perception.

Mount Sinai School of Medicine Faculty
Aneel K. Aggarwal, Ph.D., King's College (London), 1984. Biophysics; DNA-binding proteins.
Mihaly Mezei, Ph.D., Etövös Loránd (Budapest), 1972. Biophysics; molecular modeling.
Roman Osman, Ph.D., Tel Aviv, 1974. Molecular mechanisms of DNA damaged repair and structure function of peptide receptors.
Roberto Sanchez, Ph.D., Rockefeller, 2000. Bioinformatics and computational structural biology.
Ming-Ming Zhou, Ph.D., Purdue, 1993. Molecular mechanisms of healthy and diseased cells.

Section 8
Ecology, Environmental Biology, and Evolutionary Biology

This section contains a directory of institutions offering graduate work in ecology, environmental biology, and evolutionary biology, followed by in-depth entries submitted by institutions that chose to prepare detailed program descriptions. Additional information about programs listed in the directory but not augmented by an in-depth entry may be obtained by writing directly to the dean of a graduate school or chair of a department at the address given in the directory.

For programs offering related work, see also in this book Biological and Biomedical Sciences; Botany and Plant Biology; Entomology; Genetics, Developmental Biology, and Reproductive Biology; Microbiological Sciences; Pharmacology and Toxicology; and Zoology. In Book 2, see Sociology, Anthropology, and Archaeology; in Book 4, see Agricultural and Food Sciences, Geosciences, Marine Sciences and Oceanography, and Mathematical Sciences; in Book 5, see Civil and Environmental Engineering, Management of Engineering and Technology, and Ocean Engineering; and in Book 6, see Public Health.

CONTENTS

Program Directories

Announcements

Cross-Discipline Announcements

Close-Ups

See also:

Conservation Biology

Arizona State University, Division of Graduate Studies, College of Liberal Arts and Sciences, Department of Biology, Program in Conservation, Tempe, AZ 85287. Offers MS, PhD. Terminal master's awarded for partial completion of doctoral program. *Degree requirements:* For master's, thesis; for doctorate, thesis/dissertation, oral exam. *Entrance requirements:* For master's and doctorate, GRE General Test, GRE Subject Test. Additional exam requirements/recommendations for international students: Required—TOEFL (minimum score 600 paper-based; 250 computer-based); Recommended—TSE.

Central Michigan University, College of Graduate Studies, College of Science and Technology, Department of Biology, Mount Pleasant, MI 48859. Offers biology (MS); conservation biology (MS). *Faculty:* 25 full-time (7 women). *Students:* 19 full-time (10 women), 31 part-time (13 women). Average age 27. In 2005, 18 degrees awarded. *Degree requirements:* For master's, thesis or alternative, registration. *Entrance requirements:* For master's, bachelor's degree in biology, minimum GPA of 3.0. *Application deadline:* Applications are processed on a rolling basis. Application fee: $35 ($45 for international students). *Expenses:* Tuition, area resident: Part-time $325 per credit hour. Tuition, state resident: part-time $603 per credit hour. Tuition and fees vary according to degree level and reciprocity agreements. *Financial support:* In 2005–06, 2 fellowships with tuition reimbursements, 11 research assistantships with tuition reimbursements, 26 teaching assistantships with tuition reimbursements were awarded; career-related internships or fieldwork and Federal Work-Study also available. Financial award application deadline: 3/7. *Faculty research:* Vertebrates, morphology and taxonomy of aquatic plants, molecular biology and genetics, microbials and invertebrate ecology. *Unit head:* Dr. Claudia B. Douglass, Chairperson, 989-774-3227, Fax: 989-774-3462, E-mail: doug1cb@cmich.edu. *Application contact:* Dr. A. Scott McNaught, Graduate Program Coordinator, 989-774-1335, Fax: 989-774-3462, E-mail: scott.mcnaught@cmich.edu.

Columbia University, Graduate School of Arts and Sciences, Division of Natural Sciences, Department of Ecology and Evolutionary Biology, New York, NY 10027. Offers conservation biology (Certificate); ecology and evolutionary biology (PhD); environmental policy (Certificate). *Faculty:* 7 full-time (1 woman), 43 part-time/adjunct (15 women). *Students:* 9 full-time (7 women); includes 1 minority (African American), 2 international. Average age 28. 45 applicants, 18% accepted. *Degree requirements:* For doctorate, one foreign language, thesis/dissertation, teaching experience. *Entrance requirements:* For doctorate, GRE General Test, previous course work in biology. Additional exam requirements/recommendations for international students: Required—TOEFL. Application fee: $75. Electronic applications accepted. *Expenses:* Tuition: Full-time $31,448. Tuition and fees vary according to course level, course load, campus/location and program. *Financial support:* In 2005–06, 4 students received support; fellowships, career-related internships or fieldwork and institutionally sponsored loans available. Financial award application deadline: 1/5. *Faculty research:* Tropical ecology, ethnobotany, global change, systematics. Total annual research expenditures: $300,000. *Unit head:* Shahid Naeem, Chair, 212-854-7337, E-mail: sn212@columbia.edu.

See Close-Up on page 699.

Columbia University, Graduate School of Arts and Sciences, Program in Conservation Biology, New York, NY 10027. Offers MA. *Degree requirements:* For master's, thesis. Application fee: $75. *Expenses:* Tuition: Full-time $31,448. Tuition and fees vary according to course level, course load, campus/location and program. *Application contact:* Robert Furno, Assistant Dean for Admissions, 212-854-4738, Fax: 212-854-2863, E-mail: ref8@columbia.edu.

See Close-Up on page 699.

Frostburg State University, Graduate School, College of Liberal Arts and Sciences, Department of Biology, Program in Applied Ecology and Conservation Biology, Frostburg, MD 21532-1099. Offers MS. *Faculty:* 8. *Students:* 11 full-time (9 women), 10 part-time (5 women). Average age 27. 15 applicants, 33% accepted, 4 enrolled. In 2005, 7 degrees awarded. *Degree requirements:* For master's, thesis. *Entrance requirements:* For master's, GRE General Test, resumé. *Application deadline:* For fall admission, 7/15 for domestic students. Applications are processed on a rolling basis. Application fee: $30. Electronic applications accepted. *Expenses:* Tuition, state resident: full-time $5,292; part-time $294 per credit hour. Tuition, nonresident: full-time $6,066; part-time $337 per credit hour. Required fees: $67; $67 per credit hour. $9 per term. One-time fee: $30 full-time. *Financial support:* In 2005–06, 8 research assistantships with full tuition reimbursements (averaging $5,000 per year) were awarded; career-related internships or fieldwork and Federal Work-Study also available. Financial award application deadline: 4/1; financial award applicants required to submit FAFSA. *Faculty research:* Forest ecology, microbiology of man-made wetlands, invertebrate zoology and entomology, wildlife and carnivore ecology, aquatic pollution ecology. *Unit head:* Dr. R. Scott Fritz, Coordinator, 301-687-4166. *Application contact:* Patricia C. Spiker, Director, Graduate Services, 301-687-7053, Fax: 301-687-4597, E-mail: pspiker@frostburg.edu.

North Dakota State University, The Graduate School, College of Agriculture, Food Systems, and Natural Resources, Department of Entomology, Fargo, ND 58105. Offers entomology (MS, PhD); environment and conservation science (MS, PhD); natural resource management (MS, PhD). Part-time programs available. *Faculty:* 8 full-time (3 women), 6 part-time/adjunct (0 women). *Students:* 18 full-time (6 women), 4 part-time (2 women); includes 9 minority (2 African Americans, 7 Asian Americans or Pacific Islanders). Average age 34. 5 applicants, 20% accepted, 1 enrolled. In 2005, 1 degree awarded. *Median time to degree:* Of those who began their doctoral program in fall 1997, 100% received their degree in 8 years or less. *Degree requirements:* For master's, thesis/dissertation; for doctorate, thesis/dissertation, comprehensive exam. *Entrance requirements:* For master's and doctorate, minimum GPA of 3.0. Additional exam requirements/recommendations for international students: Required—TOEFL. *Application deadline:* Applications are processed on a rolling basis. Application fee: $45 ($60 for international students). Electronic applications accepted. *Financial support:* In 2005–06, 17 students received support, including 19 research assistantships with full tuition reimbursements available (averaging $13,800 per year); Federal Work-Study, institutionally sponsored loans, and unspecified assistantships also available. Financial award application deadline: 4/15. *Faculty research:* Insect systematics, conservation biology, integrated pest management, insect behavior, insect biology. *Unit head:* Dr. Gary J. Brewer, Chair, 701-231-7908, Fax: 701-231-8557, E-mail: gary.brewer@ndsu.nodak.edu.

North Dakota State University, The Graduate School, College of Science and Mathematics, Department of Biological Sciences, Fargo, ND 58105. Offers biological sciences (MS); botany (MS, PhD); cellular and molecular biology (PhD); environmental and conservation sciences (MS, PhD); genomics (MS, PhD); natural resource management (MS, PhD); zoology (MS, PhD). *Faculty:* 20. *Students:* 30 full-time (10 women), 5 part-time (1 woman); includes 1 minority (Asian American or Pacific Islander), 3 international. Average age 24. 14 applicants, 43% accepted. In 2005, 4 master's awarded. *Degree requirements:* For master's and doctorate, thesis/dissertation. *Entrance requirements:* For master's and doctorate, GRE General Test. Additional exam requirements/recommendations for international students: Required—TOEFL. *Application deadline:* For fall admission, 3/15 for domestic students; for spring admission, 10/30 priority date for domestic students. Applications are processed on a rolling basis. Application fee: $45 ($60 for international students). Electronic applications accepted. *Financial support:* In 2005–06, 3 fellowships with full tuition reimbursements (averaging $15,000 per year), 9 research assistantships with full tuition reimbursements (averaging $14,400 per year), 19 teaching assistantships with full tuition reimbursements (averaging $9,550 per year) were awarded; career-related internships or fieldwork, Federal Work-Study, institutionally sponsored loans, scholarships/grants, tuition waivers (full), and unspecified assistantships also available. Support available to part-time students. Financial award application deadline: 4/15; financial award applicants required to submit FAFSA. *Faculty research:* Comparative endocrinology, physiology, behavioral ecology, plant cell biology, aquatic biology. Total annual research

expenditures: $675,000. *Unit head:* Dr. William J. Bleier, Chair, 701-231-7087, Fax: 701-231-7149, E-mail: william.bleier@ndsu.nodak.edu.

Oklahoma State University, College of Arts and Sciences, Department of Zoology, Interdisciplinary Program in Conservation Science, Stillwater, OK 74078. Offers MS, PhD. *Degree requirements:* For master's and doctorate, thesis/dissertation. *Entrance requirements:* For master's and doctorate, GRE General Test, GRE Subject Test. Additional exam requirements/recommendations for international students: Required—TOEFL. *Application deadline:* For fall admission, 2/15 for domestic students. Applications are processed on a rolling basis. Application fee: $40 ($75 for international students). Electronic applications accepted. *Expenses:* Tuition, state resident: full-time $4,253; part-time $139 per credit hour. Tuition, nonresident: full-time $12,569; part-time $485 per credit hour. Required fees: $43 per credit hour. One-time fee: $20 part-time. Tuition and fees vary according to course load and program. *Financial support:* Research assistantships, teaching assistantships, career-related internships or fieldwork, Federal Work-Study, scholarships/grants, health care benefits, tuition waivers (partial), and unspecified assistantships available. Support available to part-time students. Financial award application deadline: 3/1.

San Francisco State University, Division of Graduate Studies, College of Science and Engineering, Department of Biology, Program in Conservation Biology, San Francisco, CA 94132-1722. Offers MA. *Entrance requirements:* For master's, minimum GPA of 2.5 in last 60 units.

State University of New York College of Environmental Science and Forestry, Faculty of Environmental and Forest Biology, Syracuse, NY 13210-2779. Offers chemical ecology (MPS, MS, PhD); conservation biology (MPS, MS, PhD); ecology (MPS, MS, PhD); entomology (MPS, MS, PhD); environmental interpretation (MPS, MS, PhD); environmental physiology (MPS, MS, PhD); fish and wildlife biology (MPS, MS, PhD); forest pathology and mycology (MPS, MS, PhD); plant science and biotechnology (MPS, MS, PhD). *Faculty:* 27 full-time (4 women), 4 part-time/adjunct (0 women). *Students:* 81 full-time (50 women), 62 part-time (33 women); includes 4 minority (1 Asian American or Pacific Islander, 3 Hispanic Americans), 16 international. Average age 30. 84 applicants, 54% accepted, 17 enrolled. In 2005, 17 master's, 3 doctorates awarded. *Degree requirements:* For master's, thesis (for some programs), registration; for doctorate, thesis/dissertation, comprehensive exam, registration. *Entrance requirements:* For master's and doctorate, GRE General Test, GRE Subject Test, minimum GPA of 3.0. Additional exam requirements/recommendations for international students: Required—TOEFL (minimum score 550 paper-based; 213 computer-based). *Application deadline:* For fall admission, 2/1 priority date for domestic students, 2/1 priority date for international students; for spring admission, 11/1 priority date for domestic students, 11/1 priority date for international students. Applications are processed on a rolling basis. Application fee: $60. *Expenses:* Tuition, area resident: Full-time $6,900; part-time $288 per credit. Tuition, nonresident: full-time $10,920; part-time $455 per credit. Required fees: $395; $32 per credit. $20 per term. One-time fee: $145. *Financial support:* In 2005–06, 86 students received support, including 13 fellowships with full and partial tuition reimbursements available (averaging $9,446 per year), 40 research assistantships with full and partial tuition reimbursements available (averaging $11,000 per year), 32 teaching assistantships with full and partial tuition reimbursements available (averaging $9,446 per year); Federal Work-Study, institutionally sponsored loans, scholarships/grants, health care benefits, and unspecified assistantships also available. Financial award application deadline: 6/30. *Faculty research:* Ecology, fish and wildlife biology and management, plant science, entomology. Total annual research expenditures: $4.1 million. *Unit head:* Dr. Donald J. Leopold, Chair, 315-470-6770, Fax: 315-470-6934, E-mail: dendro@esf.edu. *Application contact:* Dr. Dudley J. Raynal, Dean, Instruction and Graduate Studies, 315-470-6599, Fax: 315-470-6978, E-mail: esfgrad@esf.edu.

Texas State University-San Marcos, Graduate School, College of Science, Department of Biology, Program in Population and Conservation Biology, San Marcos, TX 78666. Offers MS. *Degree requirements:* For master's, thesis. *Entrance requirements:* For master's, GRE (score of 1000 preferred), minimum GPA of 2.75 in last 60 hours of undergraduate course work. Application fee: $40 ($90 for international students). *Expenses:* Tuition, area resident: Part-time $116 per credit. Tuition, state resident: full-time $3,168; part-time $176 per credit. Tuition, nonresident: full-time $8,136; part-time $452 per credit. Required fees: $1,112; $74 per credit. Full-time tuition and fees vary according to course load. *Unit head:* Dr. Jim Ott, Graduate Advisor, 512-245-2321, E-mail: jmott@txstate.edu.

Tropical Agriculture Research and Higher Education Center, Graduate School, Turrialba, Costa Rica. Offers agroforestry (PhD); ecological agiculture (PhD); ecological agriculture (MS); environmental socioeconomics (MS, PhD); forest management and conservation (MS); forest sciences (PhD); tropical agroforestry (MS); watershed management (MS, PhD). *Entrance requirements:* For master's, GRE, letters of recommendation; for doctorate, GRE, curriculum vitae, letters of recommendation. Additional exam requirements/recommendations for international students: Required—TOEFL (minimum score 550 paper-based; 213 computer-based). Electronic applications accepted. *Faculty research:* Biodiversity in fragmented landscapes, ecosystem management, integrated pest management, environmental livestock production, biotechnology carbon balances in diverse land uses.

University at Albany, State University of New York, College of Arts and Sciences, Department of Biological Sciences, Program in Biodiversity, Conservation, and Policy, Albany, NY 12222-0001. Offers MS. *Degree requirements:* For master's, one foreign language. *Entrance requirements:* For master's, GRE General Test. Application fee: $60. *Faculty research:* Aquatic ecology, plant community ecology, biodiversity and public policy, restoration ecology, costal and estuarine science. *Unit head:* Gary Kleppel, Program Director, 518-442-4338.

University of Alberta, Faculty of Graduate Studies and Research, Department of Renewable Resources, Edmonton, AB T6G 2E1, Canada. Offers agroforestry (M Ag, M Sc, MF); conservation biology (M Sc, PhD); forest biology and management (M Sc, PhD); land reclamation and remediation (M Sc, PhD); protected areas and wildlands management (M Sc, PhD); soil science (M Ag, M Sc, PhD); water and land resources (M Ag, M Sc, PhD); wildlife ecology and management (M Sc, PhD). Part-time programs available. *Faculty:* 26 full-time (4 women), 22 part-time/adjunct (3 women). *Students:* 63 full-time (33 women), 50 part-time (20 women), 14 international. 122 applicants, 24% accepted, 22 enrolled. In 2005, 16 master's, 8 doctorates awarded. *Median time to degree:* Of those who began their doctoral program in fall 1997, 100% received their degree in 8 years or less. *Degree requirements:* For master's, thesis (for some programs); for doctorate, thesis/dissertation, comprehensive exam. *Entrance requirements:* For master's, minimum 2 years of relevant professional experiences, minimum GPA of 3.0; for doctorate, minimum GPA of 3.0. Additional exam requirements/recommendations for international students: Required—TOEFL (minimum score 550 paper-based; 213 computer-based). *Application deadline:* For fall admission, 7/1 priority date for domestic students, 6/1 priority date for international students. Applications are processed on a rolling basis. Application fee: $0. Electronic applications accepted. Tuition and fees charges are reported in Canadian dollars. *Expenses:* Tuition, state resident: part-time $562 Canadian dollars per term. Tuition, nonresident: full-time $3,375 Canadian dollars. Required fees: $573 Canadian dollars; $84 Canadian dollars per term. *Financial support:* In 2005–06, 63 students received support, including 21 research assistantships with partial tuition reimbursements available (averaging $2,800 per year), 28 teaching assistantships with partial tuition reimbursements available (averaging $1,900 per year); scholarships/grants and unspecified assistantships also available. *Faculty research:* Natural and managed landscapes. Total annual research expenditures: $6.1 million. *Unit head:* Dr. John R. Spence, Chair, 780-492-2820, Fax: 780-492-4323, E-mail: john.spence@ualberta.ca. *Application contact:* Sandy Nakashima, Graduate Program Secretary, 780-492-2820, Fax: 780-492-4323, E-mail: sandy.nakashima@ualberta.ca.

University of Central Florida, College of Sciences, Department of Biology, Orlando, FL 32816. Offers biology (MS); conservation biology (PhD, Certificate). Part-time and evening/

weekend programs available. *Faculty:* 19 full-time (6 women), 6 part-time/adjunct (3 women). *Students:* 43 full-time (34 women), 45 part-time (27 women); includes 6 minority (1 Asian American or Pacific Islander, 5 Hispanic Americans), 4 international. Average age 29. 86 applicants, 59% accepted, 35 enrolled. In 2005, 19 degrees awarded. *Degree requirements:* For master's, thesis or alternative, biology field exam, comprehensive exam. *Entrance requirements:* For master's, GRE General Test, minimum GPA of 3.0 in last 60 hours. Additional exam requirements/recommendations for international students: Required—TOEFL. *Application deadline:* For fall admission, 3/1 for domestic students; for spring admission, 10/15 for domestic students. Application fee: $30. Electronic applications accepted. *Expenses:* Tuition, state resident: full-time $5,788. Tuition, nonresident: full-time $21,927. Required fees: $241 per credit hour. *Financial support:* In 2005–06, 18 fellowships with partial tuition reimbursements (averaging $1,903 per year), 30 research assistantships with partial tuition reimbursements (averaging $4,600 per year), 26 teaching assistantships with partial tuition reimbursements (averaging $10,000 per year) were awarded; career-related internships or fieldwork, Federal Work-Study, institutionally sponsored loans, tuition waivers (partial), and unspecified assistantships also available. Financial award application deadline: 3/1; financial award applicants required to submit FAFSA. *Unit head:* Dr. David Borst, Chair, 407-823-2976, Fax: 407-823-5769, E-mail: dborst@mail.ucf.edu. *Application contact:* Dr. John F. Weishampel, Coordinator, 407-823-2141, Fax: 407-823-5769, E-mail: jweisham@mail.ucf.edu.

University of Hawaii at Manoa, Graduate Division, Specialization in Ecology, Evolution and Conservation Biology, Honolulu, HI 96822. Offers MS, PhD. Program is interdisciplinary; degree is in specific discipline with the specialization in Ecology, Evolution and Conservation Biology. *Degree requirements:* For doctorate, thesis/dissertation. Application fee: $50. *Expenses:* Tuition, state resident: full-time $8,400; part-time $200 per credit hour. Tuition, nonresident: full-time $11,088; part-time $462 per credit hour. Tuition and fees vary according to program. *Financial support:* In 2005–06, 15 students received support, including 3 fellowships, 5 research assistantships, 2 teaching assistantships; career-related internships or fieldwork and tuition waivers (full) also available. *Faculty research:* Agronomy and soil science, zoology, entomology, genetics and molecular biology, botanical sciences. *Unit head:* Dr. Robert Kinzie, Chair, 808-956-4602, Fax: 808-956-9608, E-mail: kinzie@hawaii.edu.

University of Maryland, College Park, Graduate Studies, College of Chemical and Life Sciences, Department of Biology, Program in Sustainable Development and Conservation Biology, College Park, MD 20742. Offers MS. Part-time and evening/weekend programs available. *Students:* 29 full-time (19 women), 10 part-time (9 women); includes 5 minority (1 American Indian/Alaska Native, 1 Asian American or Pacific Islander, 3 Hispanic Americans), 11 international. 63 applicants, 41% accepted, 15 enrolled. In 2005, 15 degrees awarded. *Degree requirements:* For master's, internship, scholarly paper. *Entrance requirements:* For master's, GRE General Test, minimum GPA of 3.0, 3 letters of recommendation. *Application deadline:* For fall admission, 2/15 for domestic students, 2/1 for international students; for spring admission, 11/15 for domestic students, 6/1 for international students. Applications are processed on a rolling basis. Application fee: $60. Electronic applications accepted. *Financial support:* In 2005–06, 4 fellowships (averaging $3,465 per year) were awarded; research assistantships, teaching assistantships Financial award application deadline: 2/1; financial award applicants required to submit FAFSA. *Faculty research:* Biodiversity, global change, conservation. *Unit head:* Dr. David W. Inouye, Director, 301-405-6946, Fax: 301-314-9358, E-mail: inouye@umd.edu. *Application contact:* Dean of Graduate School, 301-405-4190, Fax: 301-314-9305.

University of Minnesota, Twin Cities Campus, Graduate School, College of Natural Resources, Department of Fisheries, Wildlife, and Conservation Biology, Minneapolis, MN 55455-0213. Offers conservation biology (MS, PhD); wildlife conservation (MS, PhD). Terminal master's awarded for partial completion of doctoral program. *Degree requirements:* For master's, thesis optional; for doctorate, thesis/dissertation, comprehensive exam, registration. *Entrance requirements:* For master's and doctorate, GRE. Additional exam requirements/recommendations for international students: Required—TOEFL (minimum score 550 paper-based; 213 computer-based). *Expenses:* Tuition, state resident: full-time $8,748; part-time $729 per credit. Tuition, nonresident: full-time $15,848; part-time $1,321 per credit. Full-time tuition and fees vary according to class time, course load, program and reciprocity agreements. *Faculty research:* Management, ecology, physiology, genetics, and computer modeling of fish and wildlife.

University of Missouri–St. Louis, College of Arts and Sciences, Department of Biology, St. Louis, MO 63121. Offers biology (MS, PhD), including animal behavior (MS), biochemistry

(MS), biotechnology (MS), conservation biology (MS), development (MS), ecology (MS), environmental studies (PhD), evolution (MS), genetics (MS), molecular/cellular biology (MS), physiology (MS), plant systematics, population biology (MS), tropical biology (MS); biotechnology (Certificate); tropical biology and conservation (Certificate). Part-time programs available. *Faculty:* 50. *Students:* 26 full-time (15 women), 101 part-time (51 women); includes 13 minority (5 African Americans, 6 Asian Americans or Pacific Islanders, 2 Hispanic Americans), 41 international. Average age 32. In 2005, 22 master's, 2 doctorates awarded. *Degree requirements:* For master's, thesis or alternative; for doctorate, one foreign language, thesis/dissertation, 1 semester of teaching experience. *Entrance requirements:* For doctorate, GRE General Test. *Application deadline:* For spring admission, 12/1 priority date for domestic students. Applications are processed on a rolling basis. Application fee: $35 ($40 for international students). Electronic applications accepted. *Expenses:* Tuition, state resident: part-time $263 per credit hour. Tuition, nonresident: part-time $680 per credit hour. Required fees: $53 per credit hour. Tuition and fees vary according to program. *Financial support:* In 2005–06, 11 fellowships with full tuition reimbursements (averaging $30,000 per year), 15 research assistantships with full and partial tuition reimbursements (averaging $16,000 per year), 22 teaching assistantships with full and partial tuition reimbursements (averaging $16,000 per year) were awarded; career-related internships or fieldwork and Federal Work-Study also available. Support available to part-time students. Financial award application deadline: 2/1. *Faculty research:* Molecular biology, microbial genetics. *Unit head:* Zuleyma Tang-Martinez, Director of Graduate Studies, 314-516-6498, Fax: 314-516-6233, E-mail: zuleyma@umsl.edu. *Application contact:* 314-516-5458, Fax: 314-516-5310, E-mail: gradadm@umsl.edu.

University of Nevada, Reno, Graduate School, College of Science, Interdisciplinary Program in Ecology, Evolution, and Conservation Biology, Reno, NV 89557. Offers PhD. Offered through the College of Arts and Science, the M. C. Fleischmann College of Agriculture, and the Desert Research Institute. *Faculty:* 5. *Students:* 11 full-time (9 women), 24 part-time (12 women); includes 3 minority (1 Asian American or Pacific Islander, 2 Hispanic Americans), 2 international. Average age 33. 6 applicants, 67% accepted, 3 enrolled. In 2005, 4 degrees awarded. *Degree requirements:* For doctorate, thesis/dissertation. *Entrance requirements:* For doctorate, GRE General Test, GRE Subject Test, minimum GPA of 3.0. Additional exam requirements/recommendations for international students: Required—TOEFL. *Application deadline:* For fall admission, 2/15 for domestic students. Application fee: $60 ($95 for international students). *Expenses:* Tuition, area resident: Full-time $2,767; part-time $923 per semester. Tuition, state resident: full-time $5,733; part-time $1,911 per semester. Tuition, nonresident: full-time $12,679; part-time $1,911 per semester. International tuition: $13,878 full-time. Required fees: $404; $202 per term. One-time fee: $90. *Financial support:* In 2005–06, 1 research assistantship was awarded; teaching assistantships Financial award application deadline: 3/1. *Faculty research:* Population biology, behavioral ecology, plant response to climate change, conservation of endangered species, restoration of natural ecosystems. *Unit head:* Dr. Peter Brussard, Director, 775-784-1360.

University of Wisconsin–Madison, Graduate School, Gaylord Nelson Institute for Environmental Studies, Conservation Biology and Sustainable Development Program, Madison, WI 53706-1380. Offers MS. Part-time programs available. *Faculty:* 2 full-time (1 woman), 60 part-time/adjunct (15 women). *Students:* 26 full-time, 4 part-time; includes 2 minority (1 African American, 1 Asian American or Pacific Islander), 3 international. Average age 29. 76 applicants, 42% accepted, 11 enrolled. In 2005, 13 degrees awarded. *Degree requirements:* For master's, thesis or alternative, exit seminar. *Entrance requirements:* For master's, GRE General Test. Additional exam requirements/recommendations for international students: Required—TOEFL (minimum score 550 paper-based; 213 computer-based). *Application deadline:* For fall admission, 2/1 domestic students, 2/1 for international students; for spring admission, 10/15 for domestic students, 10/15 for international students. Application fee: $45. Electronic applications accepted. *Financial support:* In 2005–06, 2 fellowships with full tuition reimbursements (averaging $14,400 per year), 1 research assistantship with full tuition reimbursement (averaging $14,250 per year), 9 teaching assistantships with full tuition reimbursements (averaging $11,260 per year) were awarded; career-related internships or fieldwork, Federal Work-Study, scholarships/grants, health care benefits, unspecified assistantships, and project assistantships also available. Financial award application deadline: 1/2. *Faculty research:* Ornithology, forestry, sociology, rural sociology, plant ecology. *Unit head:* Stanley A. Temple, Chair, 608-263-6827, Fax: 608-262-2273, E-mail: satemple@wisc.edu. *Application contact:* James E. Miller, Associate Student Services Coordinator, 608-263-4373, Fax: 608-262-2273, E-mail: jemiller@wisc.edu.

Ecology

Arizona State University, Division of Graduate Studies, College of Liberal Arts and Sciences, Department of Biology, Program in Ecology, Tempe, AZ 85287. Offers MS, PhD. Terminal master's awarded for partial completion of doctoral program. *Degree requirements:* For master's, thesis; for doctorate, thesis/dissertation, oral exam. *Entrance requirements:* For master's, GRE General Test; for doctorate, GRE General Test, GRE Subject Test. Additional exam requirements/recommendations for international students: Required—TOEFL (minimum score 600 paper-based; 250 computer-based); Recommended—TSE.

Brown University, Graduate School, Division of Biology and Medicine, Program in Ecology and Evolutionary Biology, Providence, RI 02912. Offers PhD. *Degree requirements:* For doctorate, thesis/dissertation, preliminary exam. *Entrance requirements:* For doctorate, GRE General Test, GRE Subject Test. Additional exam requirements/recommendations for international students: Required—TOEFL. Electronic applications accepted. *Faculty research:* Marine ecology, behavioral ecology, population genetics, evolutionary morphology, plant ecology.

California State University, Fresno, Division of Graduate Studies, College of Agricultural Sciences and Technology, Department of Viticulture and Ecology, Fresno, CA 93740-8027. Offers MS. Part-time and evening/weekend programs available. *Degree requirements:* For master's, thesis (for some programs), comprehensive exam (for some programs). *Entrance requirements:* For master's, GRE General Test, minimum GPA of 2.5. Additional exam requirements/recommendations for international students: Required—TOEFL. Electronic applications accepted. *Faculty research:* Ethel carbonate formation, clinical an dphysiological characterization, grape and wine quality.

Clemson University, Graduate School, College of Agriculture, Forestry and Life Sciences, Department of Biological Sciences, Program in Biological Sciences, Clemson, SC 29634. Offers MS, PhD. *Students:* 21 full-time (14 women), 4 part-time (3 women); includes 1 minority (Hispanic American), 3 international. 12 applicants, 42% accepted, 5 enrolled. In 2005, 1 master's, 1 doctorate awarded. *Degree requirements:* For master's, thesis optional; for doctorate, thesis/dissertation, comprehensive exam. *Entrance requirements:* For master's and doctorate, GRE General Test. Additional exam requirements/recommendations for international students: Required—TOEFL. *Application deadline:* For fall admission, 6/1 for domestic students, 4/15 for international students. Application fee: $50. *Financial support:* Research assistantships, teaching assistantships available. Financial award application deadline: 3/15; financial award applicants required to submit FAFSA. *Application contact:* Information Contact, 864-656-3587, Fax: 864-656-0435.

See Close-Ups on pages 99 and 101.

Colorado State University, Graduate School, Program in Ecology, Fort Collins, CO 80523-0015. Offers MS, PhD. *Students:* Average age 31. *Degree requirements:* For master's, thesis,

oral or written exam; for doctorate, thesis/dissertation, oral and written exams. *Entrance requirements:* For master's and doctorate, GRE General Test, minimum GPA of 3.0. *Application deadline:* For fall admission, 2/1 for domestic students; for spring admission, 9/1 for domestic students. Applications are processed on a rolling basis. Application fee: $50. Electronic applications accepted. *Expenses:* Tuition, state resident: full-time $3,690; part-time $205 per credit. Tuition, nonresident: full-time $14,958; part-time $831 per credit. Required fees: $1,061. *Financial support:* Fellowships, research assistantships, teaching assistantships, career-related internships or fieldwork, institutionally sponsored loans, and traineeships available. *Faculty research:* Plant and animal ecology at organismal, population, community, and ecosystem levels. *Unit head:* Dr. Dan Binkley, Director, 970-491-4373, Fax: 970-491-2796. *Application contact:* Sally Dunphy, Program Assistant, 970-491-4373, Fax: 970-491-2796, E-mail: ecology@picea.cnr.colostate.edu.

Columbia University, Graduate School of Arts and Sciences, Division of Natural Sciences, Department of Ecology and Evolutionary Biology, New York, NY 10027. Offers conservation biology (Certificate); ecology and evolutionary biology (PhD); environmental policy (Certificate). *Faculty:* 7 full-time (1 woman), 43 part-time/adjunct (15 women). *Students:* 9 full-time (7 women); includes 1 minority (African American), 2 international. Average age 28. 45 applicants, 18% accepted. *Degree requirements:* For doctorate, one foreign language, thesis/dissertation, teaching experience. *Entrance requirements:* For doctorate, GRE General Test, previous course work in biology. Additional exam requirements/recommendations for international students: Required—TOEFL. Application fee: $75. Electronic applications accepted. *Expenses:* Tuition: Full-time $31,448. Tuition and fees vary according to course level, course load, campus/location and program. *Financial support:* In 2005–06, 4 students received support; fellowships, career-related internships or fieldwork and institutionally sponsored loans available. Financial award application deadline: 1/5. *Faculty research:* Tropical ecology, ethnobotany, global change, systematics. Total annual research expenditures: $300,000. *Unit head:* Shahid Naeem, Chair, 212-854-7337, E-mail: sn212@columbia.edu.

See Close-Up on page 699.

Cornell University, Graduate School, Graduate Fields of Agriculture and Life Sciences, Field of Ecology and Evolutionary Biology, Ithaca, NY 14853-0001. Offers ecology (PhD), including animal ecology, applied ecology, biogeochemistry, community and ecosystem ecology, limnology, oceanography, physiological ecology, plant ecology, population ecology, theoretical ecology, vertebrate zoology; evolutionary biology (PhD), including ecological genetics, paleobiology, population biology, systematics. *Faculty:* 62 full-time (11 women). *Students:* 61 full-time (33 women); includes 6 minority (1 African American, 3 Asian Americans or Pacific Islanders, 2 Hispanic Americans), 10 international. 107 applicants, 15% accepted, 12 enrolled. In 2005, 7

Ecology

Cornell University *(continued)*
doctorates awarded. *Degree requirements:* For doctorate, thesis/dissertation, 2 semesters of teaching experience, comprehensive exam. *Entrance requirements:* For doctorate, GRE General Test, GRE Subject Test (biology), 2 letters of recommendation. Additional exam requirements/recommendations for international students: Required—TOEFL (minimum score 550 paper-based; 213 computer-based). *Application deadline:* For fall admission, 12/15 for domestic students. Application fee: $60. Electronic applications accepted. *Financial support:* In 2005–06, 59 students received support, including 25 fellowships with full tuition reimbursements available, 10 research assistantships with full tuition reimbursements available, 24 teaching assistantships with full tuition reimbursements available; institutionally sponsored loans, scholarships/grants, health care benefits, tuition waivers (full and partial), and unspecified assistantships also available. Financial award applicants required to submit FAFSA. *Faculty research:* Population and organismal biology, population and evolutionary genetics, systematics and macroevolution, biochemistry, conservation biology. *Unit head:* Director of Graduate Studies, 607-254-4230. *Application contact:* Graduate Field Assistant, 607-254-4230, E-mail: eeb_grad_req@cornell.edu.

Cornell University, Graduate School, Graduate Fields of Agriculture and Life Sciences, Field of Zoology, Ithaca, NY 14853-0001. Offers animal cytology (MS, PhD); comparative and functional anatomy (MS, PhD); developmental biology (MS, PhD); ecology (MS, PhD); histology (MS, PhD). *Faculty:* 26 full-time (4 women), 3 international. *Students:* 4 full-time (3 women), 3 international. 9 applicants, 11% accepted, 1 enrolled. *Degree requirements:* For doctorate, thesis/dissertation, 2 semesters of teaching experience, comprehensive exam. *Entrance requirements:* For doctorate, GRE General Test, GRE Subject Test (biology), 2 letters of recommendation. Additional exam requirements/recommendations for international students: Required—TOEFL (minimum score 550 paper-based; 213 computer-based). *Application deadline:* For fall admission, 2/1 for domestic students. Application fee: $60. Electronic applications accepted. *Financial support:* In 2005–06, 4 students received support, including 1 fellowship with full tuition reimbursement available, 1 research assistantship with full tuition reimbursement available, 2 teaching assistantships with full tuition reimbursements available; institutionally sponsored loans, scholarships/grants, health care benefits, tuition waivers (full and partial), and unspecified assistantships also available. Financial award applicants required to submit FAFSA. *Faculty research:* Organismal biology, functional morphology, biomechanics, comparative vertebrate anatomy, comparative invertebrate anatomy, paleontology. *Unit head:* Director of Graduate Studies, 607-253-3276, Fax: 607-253-3756. *Application contact:* Graduate Field Assistant, 607-253-3276, Fax: 607-253-3756, E-mail: graduate_edcvm@cornell.edu.

Duke University, Graduate School, Department of Ecology, Durham, NC 27708-0342. Offers PhD, Certificate. *Faculty:* 29 full-time. *Students:* 43 full-time (19 women); includes 6 minority (1 African American, 4 Asian Americans or Pacific Islanders, 1 Hispanic American), 6 international. 80 applicants, 11% accepted, 5 enrolled. In 2005, 3 doctorates awarded. *Entrance requirements:* For doctorate, GRE General Test. Additional exam requirements/recommendations for international students: Required—IELT (preferred) or TOEFL. *Application deadline:* For fall admission, 12/31 for domestic students, 12/31 for international students. Application fee: $75. Electronic applications accepted. *Unit head:* Dan Richter, Director of Graduate Studies, 919-613-8031, Fax: 919-660-7425, E-mail: nmm@duke.edu.

Duke University, Graduate School, Department of Environment, Durham, NC 27708. Offers natural resource economics/policy (AM, PhD); natural resource science/ecology (AM, PhD); natural resource systems science (AM, PhD). Part-time programs available. *Faculty:* 36 full-time. *Students:* 71 full-time (44 women); includes 4 minority (3 African Americans, 1 Hispanic American), 26 international. 135 applicants, 16% accepted, 14 enrolled. In 2005, 3 master's, 9 doctorates awarded. *Degree requirements:* For doctorate, variable foreign language requirement, thesis/dissertation. *Entrance requirements:* For master's and doctorate, GRE General Test. Additional exam requirements/recommendations for international students: Required—IELT (preferred) or TOEFL. *Application deadline:* For fall admission, 12/31 for domestic students, 12/31 for international students. Application fee: $75. Electronic applications accepted. *Financial support:* Fellowships, research assistantships, teaching assistantships, Federal Work-Study available. Financial award application deadline: 12/31. *Unit head:* Kenneth Knoerr, Director of Graduate Studies, 919-613-8030, Fax: 919-684-8741, E-mail: nmm@duke.edu.

Duke University, Nicholas School of the Environment and Earth Sciences, Durham, NC 27708-0328. Offers coastal environmental management (MEM); environmental economics and policy (MEM); environmental health and security (MEM); forest resource management (MF); global environmental change (MEM); resource ecology (MEM); water and air resources (MEM). *Accreditation:* SAF (one or more programs are accredited). Part-time programs available. *Degree requirements:* For master's, thesis, registration. *Entrance requirements:* For master's, GRE General Test, previous course work in biology or ecology, calculus, statistics, and microeconomics; computer familiarity with word processing and data analysis. Additional exam requirements/recommendations for international students: Required—TOEFL (minimum score 550 paper-based; 213 computer-based). Electronic applications accepted. Expenses: Contact institution. *Faculty research:* Ecosystem management, conservation ecology, earth systems, risk assessment.

Eastern Kentucky University, The Graduate School, College of Arts and Sciences, Department of Biological Sciences, Richmond, KY 40475-3102. Offers biological sciences (MS); ecology (MS). Part-time programs available. *Degree requirements:* For master's, thesis. *Entrance requirements:* For master's, GRE General Test, minimum GPA of 2.5. *Faculty research:* Systematics, ecology, and biodiversity; animal behavior; protein structure and molecular genetics; biomonitoring and aquatic toxicology; pathogenesis of microbes and parasites.

Emory University, Graduate School of Arts and Sciences, Division of Biological and Biomedical Sciences, Program in Population Biology, Ecology and Evolution, Atlanta, GA 30322-1100. Offers PhD. *Faculty:* 28 full-time (5 women). *Students:* 22 full-time (15 women); includes 3 minority (2 Asian Americans or Pacific Islanders, 1 Hispanic American), 3 international. Average age 27. 18 applicants, 28% accepted, 4 enrolled. In 2005, 1 doctorate awarded. *Degree requirements:* For doctorate, thesis/dissertation, comprehensive exam, registration. *Entrance requirements:* For doctorate, GRE General Test, minimum GPA of 3.0 in science course work. Additional exam requirements/recommendations for international students: Required—TOEFL. *Application deadline:* For fall admission, 1/3 for domestic students, 1/3 for international students. Application fee: $50. Electronic applications accepted. *Expenses:* Tuition: Full-time $14,400. Required fees: $217. *Financial support:* In 2005–06, 11 students received support, including 11 fellowships with full tuition reimbursements available (averaging $23,000 per year); institutionally sponsored loans, health care benefits, and tuition waivers (full) also available. *Faculty research:* Evolution of microbes, infectious disease, the immune system, genetic disease in humans, evolution of behavior. *Unit head:* Dr. Leslie Real, Director, 404-727-4099, Fax: 404-727-2880, E-mail: lreal@biology.emory.edu. *Application contact:* 404-727-2545, Fax: 404-727-3322, E-mail: gdbbs@emory.edu.

Florida Institute of Technology, Graduate Programs, College of Science, Department of Biological Sciences, Program in Ecology, Melbourne, FL 32901-6975. Offers MS. Part-time programs available. *Students:* Average age 26. *Degree requirements:* For master's, thesis. *Entrance requirements:* For master's, GRE General Test, minimum GPA of 3.0. *Application deadline:* Applications are processed on a rolling basis. Electronic applications accepted. *Expenses:* Tuition: Part-time $825 per credit. *Financial support:* In 2005–06, research assistantships with full and partial tuition reimbursements (averaging $5,844 per year), teaching assistantships (averaging $4,267 per year) were awarded; career-related internships or fieldwork and tuition remissions also available. Financial award application deadline: 3/1; financial award applicants required to submit FAFSA. *Faculty research:* Endangered or threatened avian and mammalian species, hydroacoustics and feeding preference of the West Indian manatee, habitat preference of the Florida scrub jay. *Application contact:* Carolyn P. Farrior, Director of Graduate Admissions, 321-674-7118, Fax: 321-723-9468, E-mail: cfarrior@fit.edu.

See Close-Up on page 813.

Florida State University, Graduate Studies, College of Arts and Sciences, Department of Biological Science, Program in Ecology, Tallahassee, FL 32306. Offers MS, PhD. *Faculty:* 20 full-time (3 women). *Students:* 41 full-time (14 women); includes 4 minority (3 Asian Americans or Pacific Islanders, 1 Hispanic American), 6 international. *Degree requirements:* For master's and doctorate, thesis/dissertation, teaching experience seminar presentation, comprehensive exam, registration. *Entrance requirements:* For master's and doctorate, GRE General Test (minimum 1100: U-500, Q-500), minimum upper division GPA of 3.0. Additional exam requirements/recommendations for international students: Required—TOEFL (minimum score 600 paper-based; 250 computer-based), IB 100. *Application deadline:* For fall admission, 1/15 for domestic students, 12/1 for international students; for spring admission, 10/15 for domestic students, 9/1 for international students. Application fee: $30. *Financial support:* In 2005–06, fellowships with full tuition reimbursements (averaging $19,000 per year), research assistantships with full tuition reimbursements (averaging $19,000 per year), teaching assistantships with full tuition reimbursements (averaging $17,600 per year) were awarded. Financial award application deadline: 1/15; financial award applicants required to submit FAFSA. *Faculty research:* Community ecology, biogeography, functional morphology, adaptation groundwater-bearing environments, ecophysiology. *Application contact:* Judy Bowers, Coordinator, Graduate Affairs, 850-644-3023, Fax: 850-644-9829, E-mail: gradinfo@bio.fsu.edu.

Frostburg State University, Graduate School, College of Liberal Arts and Sciences, Department of Biology, Program in Applied Ecology and Conservation Biology, Frostburg, MD 21532-1099. Offers MS. *Faculty:* 8. *Students:* 11 full-time (9 women), 10 part-time (5 women). Average age 27. 15 applicants, 33% accepted, 4 enrolled. In 2005, 7 degrees awarded. *Degree requirements:* For master's, thesis. *Entrance requirements:* For master's, GRE General Test, resumé. *Application deadline:* For fall admission, 7/15 for domestic students. Applications are processed on a rolling basis. Application fee: $30. Electronic applications accepted. *Expenses:* Tuition, state resident: full-time $5,292; part-time $294 per credit hour. Tuition, nonresident: full-time $6,066; part-time $337 per credit hour. Required fees: $67; $67 per credit hour. $9 per term. One-time fee: $30 full-time. *Financial support:* In 2005–06, 8 research assistantships with full tuition reimbursements (averaging $5,000 per year) were awarded; career-related internships or fieldwork and Federal Work-Study also available. Financial award application deadline: 4/1; financial award applicants required to submit FAFSA. *Faculty research:* Forest ecology, microbiology of man-made wetlands, invertebrate zoology and entomology, wildlife and carnivore ecology, aquatic pollution ecology. *Unit head:* Dr. R. Scott Fritz, Coordinator, 301-687-4166. *Application contact:* Patricia C. Spiker, Director, Graduate Services, 301-687-7053, Fax: 301-687-4597, E-mail: pspiker@frostburg.edu.

George Mason University, College of Arts and Sciences, Department of Biology, Program in Biology, Fairfax, VA 22030. Offers bioinformatics (MS); ecology, systematics and evolution (MS); interpretive biology (MS); molecular and microbiology (PhD); molecular, microbial, and cellular biology (MS); organismal biology (MS). Part-time programs available. *Degree requirements:* For master's, thesis or alternative. *Entrance requirements:* For master's, GRE General Test, GRE Subject Test, bachelor's degree in biology or equivalent. Electronic applications accepted. *Expenses:* Tuition, area resident: Full-time $5,244; part-time $219 per credit. Tuition, state resident: part-time $651 per credit. Tuition, nonresident: full-time $15,636. Required fees: $1,524; $65 per credit.

Illinois State University, Graduate School, College of Arts and Sciences, Department of Biological Sciences, Normal, IL 61790-2200. Offers biological sciences (MS); biology (PhD); biotechnology (MS); botany (PhD); ecology (PhD); genetics (PhD); microbiology (PhD); physiology (PhD); zoology (PhD). Part-time programs available. *Faculty:* 27 full-time (6 women). *Students:* 36 full-time (28 women), 25 part-time (17 women); includes 1 minority (Asian American or Pacific Islander), 23 international. 80 applicants, 21% accepted. In 2005, 14 master's, 5 doctorates awarded. *Degree requirements:* For master's, thesis or alternative; for doctorate, variable foreign language requirement, thesis/dissertation, 2 terms of residency. *Entrance requirements:* For master's, GRE General Test, minimum GPA of 2.6 in last 60 hours of course work; for doctorate, GRE General Test. *Application deadline:* Applications are processed on a rolling basis. Application fee: $30. *Expenses:* Tuition, state resident: full-time $3,060; part-time $170 per credit hour. Tuition, nonresident: full-time $6,390; part-time $355 per credit hour. Required fees: $1,411; $47 per credit hour. *Financial support:* In 2005–06, 22 research assistantships (averaging $13,909 per year), 38 teaching assistantships (averaging $12,617 per year) were awarded; Federal Work-Study, tuition waivers (full), and unspecified assistantships also available. Financial award application deadline: 4/1. *Faculty research:* CRUI: osmoregulation in euryhaline fish: physiology, ecology and molecular biology; the PRISM project: enhancing science and math education; cell structure—functions of Na pump assembly. *Unit head:* Dr. Hou Tak Takucheung, Acting Chairperson, 309-438-3669. *Application contact:* Derek A. McCracken, Graduate Adviser, 309-438-3664.

See Close-Up on page 147.

Indiana State University, School of Graduate Studies, College of Arts and Sciences, Department of Life Sciences, Terre Haute, IN 47809-1401. Offers ecology (PhD); life sciences (MS); microbiology (PhD); physiology (PhD); science education (MS); sports medicine (PhD). *Faculty:* 24 full-time (8 women), 12 part-time/adjunct (2 women). *Students:* 51 full-time (24 women), 24 part-time (10 women); includes 3 minority (all Asian Americans or Pacific Islanders), 11 international. Average age 26. 45 applicants, 73% accepted, 29 enrolled. In 2005, 9 master's, 2 doctorates awarded. *Degree requirements:* For master's, thesis (for some programs); for doctorate, thesis/dissertation, comprehensive exam. *Entrance requirements:* For master's and doctorate, GRE General Test. *Application deadline:* For fall admission, 7/1 for domestic students; for spring admission, 11/1 priority date for domestic students. Applications are processed on a rolling basis. Application fee: $35. Electronic applications accepted. *Expenses:* Tuition, state resident: full-time $6,288; part-time $262 per credit hour. Tuition, nonresident: full-time $12,504; part-time $521 per credit hour. *Financial support:* In 2005–06, 26 teaching assistantships with partial tuition reimbursements (averaging $8,005 per year) were awarded; research assistantships with partial tuition reimbursements, Federal Work-Study, institutionally sponsored loans, and tuition waivers (partial) also available. Financial award application deadline: 3/1; financial award applicants required to submit FAFSA. *Unit head:* Dr. Swapan Ghosh, Interim Chairperson, 812-237-2400.

See Close-Up on page 149.

Indiana University Bloomington, Graduate School, College of Arts and Sciences, Department of Biology, Program in Evolution, Ecology, and Behavior, Bloomington, IN 47405-7000. Offers MA, PhD. PhD offered through the University Graduate School. Part-time programs available. *Faculty:* 21 full-time (7 women); includes 6 minority (1 African American, 1 Asian American or Pacific Islander, 4 Hispanic Americans), 7 international. In 2005, 6 degrees awarded. Terminal master's awarded for partial completion of doctoral program. *Degree requirements:* For master's, thesis or alternative; for doctorate, thesis/dissertation. *Entrance requirements:* For master's and doctorate, GRE General Test. Additional exam requirements/recommendations for international students: Required—TOEFL. *Application deadline:* For fall admission, 1/5 for domestic students; for spring admission, 9/1 for domestic students. Applications are processed on a rolling basis. Application fee: $45. Electronic applications accepted. *Expenses:* Tuition, state resident: full-time $5,437; part-time $227 per credit hour. Tuition, nonresident: full-time $15,836; part-time $660 per credit hour. Required fees: $821. Tuition and fees vary according to campus/location and program. *Financial support:* In 2005–06, 72 students received support, including fellowships with tuition reimbursements available (averaging $20,000 per year), research assistantships with tuition reimbursements available (averaging $18,000 per year), teaching assistantships with tuition reimbursements available (averaging $18,000 per year); scholarships/grants and tuition waivers (full) also available. Financial award application deadline: 1/15. *Faculty research:* Ecosystem of community, plant and animal population biology, avian sociobiology, fish ethology, evolutionary genetics. *Unit head:* Mike Wade, Director, 812-856-4680, E-mail: mjwade@indiana.edu. *Application contact:* Gretchen Clearwater, Adviser for Graduate Affairs, 812-855-1861, Fax: 812-855-6705, E-mail: biograd@bio.indiana.edu.

Iowa State University of Science and Technology, Graduate College, College of Liberal Arts and Sciences, Department of Ecology, Evolution, and Organismal Biology, Ames, IA 50011. Offers MS, PhD. *Faculty:* 28 full-time, 3 part-time/adjunct. *Students:* 38 full-time (23 women), 6 part-time (2 women); includes 2 minority (1 African American, 1 American Indian/Alaska Native), 9 international. 11 applicants, 100% accepted, 7 enrolled. In 2005, 8 master's, 3 doctorates awarded. *Degree requirements:* For master's, thesis or alternative; for doctorate, thesis/dissertation. *Entrance requirements:* For master's and doctorate, GRE General Test. Additional exam requirements/recommendations for international students: Required—TOEFL (paper score 530; computer score 197) or IELTS. *Application deadline:* For fall admission, 1/1 priority date for domestic students, 1/1 priority date for international students; for spring admission, 9/1 priority date for domestic students, 9/1 priority date for international students. Application fee: $30 ($70 for international students). Electronic applications accepted. *Expenses:* Tuition, state resident: full-time $6,410. Tuition, nonresident: full-time $16,422. Tuition and fees vary according to program. *Financial support:* In 2005–06, 21 research assistantships with partial tuition reimbursements (averaging $15,968 per year), 7 teaching assistantships with partial tuition reimbursements (averaging $15,224 per year) were awarded; fellowships, scholarships/grants, health care benefits, and unspecified assistantships also available. *Faculty research:* Aquatic and wetland ecology, cytology, ecology, physiology and molecular biology, systematics and evolution. *Unit head:* Dr. Jonathan Wendel, Chair, 515-294-7172.

Iowa State University of Science and Technology, Graduate College, Interdisciplinary Programs, Program in Ecology and Evolutionary Biology, Ames, IA 50011. Offers MS, PhD. *Students:* 45 full-time (27 women), 4 part-time (2 women); includes 1 minority (American Indian/Alaska Native), 8 international. 24 applicants, 0% accepted, 0 enrolled. In 2005, 7 master's, 2 doctorates awarded. *Degree requirements:* For master's, thesis or alternative; for doctorate, thesis/dissertation. *Entrance requirements:* For master's and doctorate, GRE General Test, application to cooperating department. Additional exam requirements/recommendations for international students: Required—TOEFL or IELTS. *Application deadline:* For fall admission, 1/15 priority date for domestic students, 1/15 priority date for international students. Application fee: $30 ($70 for international students). Electronic applications accepted. *Expenses:* Tuition, state resident: full-time $6,410. Tuition, nonresident: full-time $16,422. Tuition and fees vary according to program. *Financial support:* In 2005–06, 33 research assistantships with full and partial tuition reimbursements (averaging $16,188 per year), 8 teaching assistantships with full and partial tuition reimbursements (averaging $15,239 per year) were awarded; scholarships/grants, health care benefits, and unspecified assistantships also available. *Faculty research:* Landscape ecology, aquatic and method ecology, physiological ecology, population genetics and evolution, systematics. *Unit head:* Dr. Carol Vleck, Supervisory Committee Chair, 515-294-6518, E-mail: eeboffice@iastate.edu. *Application contact:* Charles Sauer, Information Contact, 515-294-6518, E-mail: eeboffice@lastate.edu.

Kent State University, College of Arts and Sciences, Department of Biological Sciences, Program in Ecology, Kent, OH 44242-0001. Offers MS, PhD. *Degree requirements:* For master's and doctorate, thesis/dissertation. *Entrance requirements:* For master's, GRE General Test, minimum GPA of 3.0; for doctorate, GRE General Test, minimum GPA of 3.25. Additional exam requirements/recommendations for international students: Required—TOEFL (minimum score 600 paper-based; 287 computer-based). Electronic applications accepted.

Lesley University, Graduate School of Arts and Social Sciences, Cambridge, MA 02138-2790. Offers clinical mental health counseling (MA), including expressive therapies counseling, holistic counseling, school and community counseling; counseling psychology (MA, CAGS), including professional counseling (MA), school counseling (MA); creative arts in learning (CAGS); creative writing (MFA); ecological teaching and learning (MS); environmental education (MS); expressive therapies (MA, PhD, CAGS), including art (MA), dance (MA), expressive therapies, music (MA); independent studies (CAGS); independent study (MA); intercultural relations (MA, CAGS); interdisciplinary studies (MA); visual arts (MFA). Part-time and evening/weekend programs available. Postbaccalaureate distance learning degree programs offered (minimal on-campus study). *Faculty:* 39 full-time (30 women), 139 part-time/adjunct (95 women). *Students:* 592 full-time (539 women), 1,892 part-time (1,728 women); includes 197 minority (147 African Americans, 4 American Indian/Alaska Native, 14 Asian Americans or Pacific Islanders, 32 Hispanic Americans), 57 international. Average age 36. 967 applicants, 76% accepted, 525 enrolled. In 2005, 808 master's, 5 other advanced degrees awarded. *Degree requirements:* For master's, internship, practicum, thesis (expressive therapies); for doctorate and CAGS, thesis/dissertation, arts apprenticeship, field placement; for CAGS, thesis, internship (counseling psychology, expressive therapies). *Entrance requirements:* For master's, MAT (counseling psychology), interview; for doctorate, GRE or MAT; for CAGS, interview, master's degree. Additional exam requirements/recommendations for international students: Required—TOEFL (minimum score 550 paper-based; 213 computer-based). *Application deadline:* Applications are processed on a rolling basis. Application fee: $50. *Expenses:* Tuition: Part-time $695 per credit. Tuition and fees vary according to degree level, campus/location and program. *Financial support:* In 2005–06, 38 students received support, including research assistantships (averaging $3,400 per year), teaching assistantships (averaging $3,400 per year); career-related internships or fieldwork, Federal Work-Study, scholarships/grants, and unspecified assistantships also available. Support available to part-time students. Financial award application deadline: 4/15; financial award applicants required to submit FAFSA. *Faculty research:* Psychotherapy and culture; psychotherapy and psychological trauma; women's issues in art, teaching and psychotherapy; community based art, psycho-spiritual inquiry. *Unit head:* Dr. Julia Halevy, Dean, 617-349-8317, Fax: 617-349-8366, E-mail: jhalevy@lesley.edu. *Application contact:* Dr. Judy Conley, Director of Graduate Admissions, 617-349-8267, Fax: 614-349-8313, E-mail: jconley@lesley.edu.

Marquette University, Graduate School, College of Arts and Sciences, Department of Biology, Milwaukee, WI 53201-1881. Offers cell biology (MS, PhD); developmental biology (MS, PhD); ecology (MS, PhD); endocrinology (MS, PhD); evolutionary biology (MS, PhD); genetics (MS, PhD); microbiology (MS, PhD); molecular biology (MS, PhD); muscle and exercise physiology (MS, PhD); neurobiology (MS, PhD); reproductive physiology (MS, PhD). Terminal master's awarded for partial completion of doctoral program. *Degree requirements:* For master's, thesis, 1 year of teaching experience or equivalent, comprehensive exam; for doctorate, thesis/dissertation, 1 year of teaching experience or equivalent, qualifying exam. *Entrance requirements:* For master's and doctorate, GRE General Test, GRE Subject Test. Additional exam requirements/recommendations for international students: Required—TOEFL. *Faculty research:* Microbial and invertebrate ecology, evolution of gene function, DNA methylation, DNA arrangement.

Michigan State University, The Graduate School, College of Natural Science, Interdepartmental Program in Ecology, Evolutionary Biology and Behavior, East Lansing, MI 48824.

See Close-Up on page 701.

Michigan Technological University, Graduate School, School of Forest Resources and Environmental Science, Program in Applied Ecology, Houghton, MI 49931-1295. Offers MS. Part-time programs available. *Faculty:* 23 full-time (4 women), 16 part-time/adjunct (2 women). *Students:* 2 full-time. Average age 30. 3 applicants, 0% accepted, 0 enrolled. In 2005, 2 degrees awarded. *Degree requirements:* For master's, thesis (for some programs), registration. *Entrance requirements:* For master's, GRE. Additional exam requirements/recommendations for international students: Required—TOEFL (minimum score 550 paper-based; 213 computer-based). *Application deadline:* Applications are processed on a rolling basis. Application fee: $40 ($45 for international students). Electronic applications accepted. *Expenses:* Tuition, nonresident: full-time $11,232; part-time $468 per credit. Required fees: $754; $377 per semester. Full-time tuition and fees vary according to course load, degree level and program. *Financial support:* In 2005–06, 2 students received support, including fellowships with full tuition reimbursements available (averaging $9,542 per year), 2 research assistantships with full tuition reimbursements available (averaging $9,542 per year), teaching assistantships with full tuition reimbursements available (averaging $9,542 per year); career-related internships or fieldwork, Federal

Work-Study, scholarships/grants, health care benefits, tuition waivers (partial), unspecified assistantships, and co-op also available. Financial award applicants required to submit FAFSA. *Application contact:* Dr. Chandrashekhar P. Joshi, Associate Professor and Graduate Program Coordinator, 906-487-3480, Fax: 906-487-2915, E-mail: cpjoshi@mtu.edu.

Minnesota State University Mankato, College of Graduate Studies, College of Science, Engineering and Technology, Department of Biological Sciences, Mankato, MN 56001. Offers biology (MS); biology education (MS); environmental science (MS), including ecology, economic and political systems, human ecosystems, physical science, technology. Part-time programs available. *Students:* 14 full-time (7 women), 17 part-time (10 women). Average age 31. In 2005, 7 degrees awarded. *Degree requirements:* For master's, one foreign language, thesis or alternative, comprehensive exam. *Entrance requirements:* For master's, minimum GPA of 3.0 during previous 2 years of course work. Additional exam requirements/recommendations for international students: Required—TOEFL. *Application deadline:* For fall admission, 7/1 for domestic students; for spring admission, 11/1 for domestic students. Applications are processed on a rolling basis. Application fee: $40. Electronic applications accepted. *Expenses:* Tuition, state resident: part-time $243 per credit. Tuition, nonresident: part-time $400 per credit. Required fees: $30 per credit. *Financial support:* Fellowships, research assistantships with full tuition reimbursements, teaching assistantships with full tuition reimbursements, career-related internships or fieldwork, Federal Work-Study, institutionally sponsored loans, and unspecified assistantships available. Support available to part-time students. Financial award application deadline: 3/15; financial award applicants required to submit FAFSA. *Faculty research:* Limnology, enzyme analysis, membrane engineering, converters. *Unit head:* Dr. Gregg Marg, Chairperson, 507-389-2786. *Application contact:* 507-389-2321, E-mail: grad@mnsu.edu.

Montana State University, College of Graduate Studies, College of Letters and Science, Department of Ecology, Bozeman, MT 59717. Offers biological sciences (MS, PhD); fish and wildlife biology (PhD); fish and wildlife management (MS); land rehabilitation (intercollege) (MS). Part-time programs available. *Faculty:* 15 full-time (4 women). *Students:* 4 full-time (3 women), 61 part-time (23 women); includes 1 minority (American Indian/Alaska Native), 1 international. Average age 31. 8 applicants, 75% accepted, 6 enrolled. In 2005, 9 master's, 4 doctorates awarded. *Degree requirements:* For master's, thesis (for some programs), comprehensive exam, registration; for doctorate, thesis/dissertation, comprehensive exam, registration. *Entrance requirements:* For master's and doctorate, GRE General Test. Additional exam requirements/recommendations for international students: Required—TOEFL (minimum score 550 paper-based; 213 computer-based). *Application deadline:* For fall admission, 7/15 priority date for domestic students, 5/15 priority date for international students; for spring admission, 12/1 priority date for domestic students, 10/1 priority date for international students. Applications are processed on a rolling basis. Application fee: $30. Electronic applications accepted. *Expenses:* Tuition, state resident: full-time $4,132. Tuition, nonresident: full-time $1,132. *Financial support:* In 2005–06, 3 fellowships with full tuition reimbursements (averaging $21,300 per year), 45 research assistantships with full and partial tuition reimbursements (averaging $10,830 per year), 22 teaching assistantships with full and partial tuition reimbursements (averaging $10,268 per year) were awarded; career-related internships or fieldwork, Federal Work-Study, scholarships/grants, health care benefits, tuition waivers, and unspecified assistantships also available. Financial award application deadline: 3/1; financial award applicants required to submit FAFSA. *Faculty research:* Population dynamics, landscape ecology, evolution and genetics, ecological modeling, aquatic ecosystems. Total annual research expenditures: $2.1 million. *Unit head:* Dr. David Roberts, Department Head, 406-994-4548, Fax: 406-994-3190, E-mail: droberts@montana.edu.

North Carolina State University, Graduate School, College of Physical and Mathematical Sciences, Department of Statistics, Program in Biomathematics, Raleigh, NC 27695. Offers biomathematics (M Biomath, MS, PhD); ecology (PhD). Part-time programs available. Terminal master's awarded for partial completion of doctoral program. *Degree requirements:* For master's, thesis (for some programs); for doctorate, thesis/dissertation. *Entrance requirements:* For master's and doctorate, GRE General Test. Additional exam requirements/recommendations for international students: Required—TOEFL. Electronic applications accepted. *Faculty research:* Theory and methods of biological modeling, theoretical biology (genetics, ecology, neurobiology), applied biology (wildlife).

North Dakota State University, The Graduate School, Interdisciplinary Program in Environmental and Conservation Sciences, Fargo, ND 58105. Offers environmental and conservation sciences (PhD); environmental science (MS). *Faculty:* 59. *Degree requirements:* For master's, thesis, comprehensive exam. *Entrance requirements:* Additional exam requirements/recommendations for international students: Required—TOEFL (minimum score 550 paper-based). *Unit head:* Dr. Wei Lin, Director, 701-231-7244, Fax: 701-231-7149, E-mail: wei.lin@ndsu.edu. *Application contact:* Ruth Ann Faulkner, Administrative Assistant, 701-231-6727, E-mail: ruthann.faulkner@ndsu.edu.

Northern Arizona University, Graduate College, College of Engineering and Natural Science, Department of Environmental Sciences and Policy, Flagstaff, AZ 86011. Offers conservation ecology (Certificate); environmental sciences and policy (MS). *Accreditation:* NCA. *Degree requirements:* For master's, thesis optional. *Entrance requirements:* For master's, GRE General Test.

Nova Scotia Agricultural College, Research and Graduate Studies, Truro, NS B2N 5E3, Canada. Offers agriculture (M Sc), including air quality, animal behavior, animal molecular genetics, animal nutrition, animal technology, aquaculture, botany, crop management, crop physiology, ecology, environmental microbiology, food science, horticulture, nutrient management, pest management, physiology, plant biotechnology, plant pathology, soil chemistry, soil fertility, waste management and composting, water quality. Part-time programs available. *Faculty:* 43 full-time (7 women), 21 part-time/adjunct (1 woman). *Students:* 50 full-time (25 women), 15 part-time (11 women); includes 7 minority (3 African Americans, 3 Asian Americans or Pacific Islanders, 1 Hispanic American). Average age 25. In 2005, 23 degrees awarded. *Degree requirements:* For master's, thesis, registration. *Entrance requirements:* For master's, honors B Sc, minimum GPA of 3.0. Additional exam requirements/recommendations for international students: Required—TOEFL (minimum score 580 paper-based; 237 computer-based), Michigan English Language Assessment Battery, IELT, Can Test, CAEL. *Application deadline:* For fall admission, 6/1 for domestic students, 4/1 for international students. For winter admission, 10/31 for domestic students; for spring admission, 2/28 for domestic students. Applications are processed on a rolling basis. Application fee: $70. *Expenses:* Tuition, state resident: part-time $2,328 per year. Tuition, nonresident: full-time $6,984; part-time $7,968 per year. International tuition: $12,624 full-time. Required fees: $481; $46 per course. Tuition and fees vary according to program and student level. *Financial support:* In 2005–06, 48 students received support, including 7 fellowships (averaging $15,000 per year), 10 research assistantships (averaging $15,000 per year), 15 teaching assistantships (averaging $900 per year); career-related internships or fieldwork, scholarships/grants, and unspecified assistantships also available. *Faculty research:* Bio-product development, organic agriculture, nutrient management, air and water quality, agricultural biotechnology. Total annual research expenditures: $4.7 million. *Unit head:* Jill L. Rogers, Manager, 902-893-6360, Fax: 902-893-3430, E-mail: jrogers@nsac.ca. *Application contact:* Marie Law, Administrative Assistant, 902-893-6502, Fax: 902-893-3430, E-mail: mlaw@nsac.ca.

The Ohio State University, Graduate School, College of Biological Sciences, Department of Evolution, Ecology, and Organismal Biology, Columbus, OH 43210. Offers MS, PhD. *Degree requirements:* For master's, thesis optional; for doctorate, thesis/dissertation. *Entrance requirements:* For master's and doctorate, GRE General Test. Additional exam requirements/recommendations for international students: Required—TOEFL (minimum score 600 paper-based; 250 computer-based), TSE (minimum score 60). Electronic applications accepted.

Ohio University, Graduate Studies, College of Arts and Sciences, Department of Biological Sciences, Athens, OH 45701-2979. Offers biological sciences (MS, PhD); cell biology and physiology (MS, PhD); ecology and evolutionary biology (MS, PhD); exercise physiology and

Ecology

Ohio University (continued)

muscle biology (MS, PhD); microbiology (MS, PhD); neuroscience (MS, PhD). *Faculty:* 51 full-time (17 women), 6 part-time/adjunct (1 woman). *Students:* 88 full-time (40 women), 1 part-time; includes 1 minority (Hispanic American), 41 international. Average age 24. 50 applicants, 24% accepted, 10 enrolled. In 2005, 9 master's, 12 doctorates awarded. *Median time to degree:* Of those who began their doctoral program in fall 1997, 90% received their degree in 8 years or less. *Degree requirements:* For master's, thesis, 1 quarter of teaching experience; for doctorate, thesis/dissertation, 2 quarters of teaching experience, comprehensive exam. *Entrance requirements:* For master's and doctorate, GRE General Test. Additional exam requirements/recommendations for international students: Required—TOEFL (minimum score 620 paper-based; 260 computer-based). *Application deadline:* For fall admission, 1/15 for domestic students, 1/15 for international students. Application fee: $45. Electronic applications accepted. *Financial support:* In 2005–06, 87 students received support, including 2 fellowships with full tuition reimbursements available (averaging $15,000 per year), 10 research assistantships with full tuition reimbursements available (averaging $15,500 per year), 75 teaching assistantships with full tuition reimbursements available (averaging $15,500 per year); Federal Work-Study, institutionally sponsored loans, and tuition waivers (full) also available. Financial award application deadline: 1/15. *Faculty research:* Ecology and evolutionary biology, exercise physiology and muscle biology, neurobiology, cell biology, physiology. Total annual research expenditures: $2.8 million. *Unit head:* Dr. Ralph DiCaprio, Chair, 740-593-2290, Fax: 740-593-0300, E-mail: dicaprir@ohio.edu. *Application contact:* Dr. Donald B. Miles, Graduate Chair, 740-593-2317, Fax: 740-593-0300, E-mail: milesd@ohio.edu.

See Close-Up on page 191.

Old Dominion University, College of Sciences, Program in Ecological Sciences, Norfolk, VA 23529. Offers PhD. *Faculty:* 14 full-time (2 women), 41 part-time/adjunct (7 women). *Students:* 4 full-time (2 women), 17 part-time (6 women); includes 2 minority (1 African American, 1 Asian American or Pacific Islander), 3 international. Average age 30. 8 applicants, 63% accepted, 5 enrolled. *Median time to degree:* Of those who began their doctoral program in fall 1997, 100% received their degree in 8 years or less. *Degree requirements:* For doctorate, one foreign language, thesis/dissertation, comprehensive exam. *Entrance requirements:* For doctorate, GRE General Test, 3 letters of recommendation. Additional exam requirements/recommendations for international students: Required—TOEFL (minimum score 550 paper-based). *Application deadline:* For fall admission, 2/1 priority date for domestic students, 2/1 priority date for international students. For winter admission, 6/1 for domestic students; for spring admission, 10/1 for domestic students. Applications are processed on a rolling basis. Application fee: $40. Electronic applications accepted. *Expenses:* Tuition, state resident: part-time $263 per credit hour. Tuition, nonresident: part-time $661 per credit hour. Required fees: $39 per semester. Part-time tuition and fees vary according to campus/location. *Financial support:* In 2005–06, 2 fellowships with full tuition reimbursements (averaging $15,000 per year), 4 research assistantships with full tuition reimbursements (averaging $15,750 per year), 12 teaching assistantships with full tuition reimbursements (averaging $10,833 per year) were awarded; scholarships/grants also available. Financial award application deadline: 2/15; financial award applicants required to submit FAFSA. *Faculty research:* Spiny lobsters in Florida, pollution in marine environment, wetlands and barrier islands, experimental community ecology, evolutionary morphology. Total annual research expenditures: $2 million. *Unit head:* Dr. John Holsinger, Graduate Program Director, 757-683-3606, Fax: 757-683-5283, E-mail: ecolgpd@odu.edu.

The Pennsylvania State University University Park Campus, Graduate School, Intercollege Graduate Programs, Intercollege Graduate Program in Ecology, State College, University Park, PA 16802-1503. Offers MS, PhD. *Students:* 31 full-time (15 women), 7 part-time (3 women), 4 international. *Entrance requirements:* For master's and doctorate, GRE General Test, GRE Subject Test (biology). Application fee: $45. *Expenses:* Tuition, state resident: full-time $12,518; part-time $522 per credit. Tuition, nonresident: full-time $23,004; part-time $959 per credit. Required fees: $484. Tuition and fees vary according to course load, campus/location and program. *Unit head:* David A. Mortensen, Chair, 814-865-1906, Fax: 814-865-9451, E-mail: dmortensen@psu.edu. *Application contact:* David A. Mortensen, Chair, 814-865-1906, Fax: 814-865-9451, E-mail: dmortensen@psu.edu.

Prescott College, Graduate Programs, Program in Environmental Studies, Prescott, AZ 86301. Offers agroecology (MA); ecopsychology (MA); environmental education (MA); environmental studies (MA); sustainability (MA). MA in environmental education offered jointly with Teton Science School. Part-time programs available. Postbaccalaureate distance learning degree programs offered (minimal on-campus study). *Faculty:* 1 full-time (0 women), 32 part-time/adjunct (10 women). *Students:* 16 full-time (9 women), 19 part-time (10 women); includes 2 minority (both Hispanic Americans) Average age 35. In 2005, 12 degrees awarded. *Degree requirements:* For master's, thesis, fieldwork or internship, practicum. *Entrance requirements:* For master's, 2 letters of recommendation, resumé. *Application deadline:* For fall admission, 5/1 for domestic students; for spring admission, 11/1 priority date for domestic students. Applications are processed on a rolling basis. Application fee: $40. Electronic applications accepted. *Expenses:* Tuition: Full-time $12,408; part-time $517 per credit. One-time fee: $103 full-time. *Financial support:* Career-related internships or fieldwork and Federal Work-Study available. Financial award applicants required to submit FAFSA. *Unit head:* Dr. Paul Sneed, Head, 928-350-3204. *Application contact:* Kerstin Alicki, Admissions Counselor, 877-350-2100 Ext. 2102, Fax: 928-776-5242, E-mail: admissions@prescott.edu.

Princeton University, Graduate School, Department of Ecology and Evolutionary Biology, Princeton, NJ 08544-1019. Offers biology (PhD); neuroscience (PhD). *Degree requirements:* For doctorate, thesis/dissertation. *Entrance requirements:* For doctorate, GRE General Test, GRE Subject Test. Additional exam requirements/recommendations for international students: Required—TOEFL (minimum score 600 paper-based; 250 computer-based). Electronic applications accepted.

Announcement: Graduate study in the department leads to the PhD degree. Special areas of focus include evolutionary ecology; ecological and evolutionary genetics; behavioral ecology; theoretical ecology; population and community ecology; physiology; epidemiology of infectious diseases; global interactions among the biosphere, atmosphere, oceans, and climate; and conservation biology. The interests and research of the faculty range widely over these areas, so that incoming students are able to select their advisers from among several professors working in the chosen discipline. Graduate students also have excellent opportunities for combining several areas for innovative interdisciplinary work. The program is designed to develop both the breadth and depth of understanding that will enable students to respond to future advances in the field.

See Close-Up on page 705.

Purdue University, Graduate School, School of Science, Department of Biological Sciences, West Lafayette, IN 47907. Offers biochemistry (PhD); biophysics (PhD); cell and developmental biology (PhD); ecology, evolutionary and population biology (MS, PhD), including ecology, evolutionary biology, population biology; genetics (MS, PhD); microbiology (MS, PhD); molecular biology (PhD); neurobiology (MS, PhD); plant physiology (PhD). *Faculty:* 47 full-time (9 women), 4 part-time/adjunct (1 woman). *Students:* 97 full-time (53 women), 8 part-time (4 women); includes 13 minority (3 African Americans, 1 American Indian/Alaska Native, 4 Asian Americans or Pacific Islanders, 5 Hispanic Americans), 50 international. Average age 28. 168 applicants, 29% accepted, 23 enrolled. In 2005, 18 master's, 9 doctorates awarded. Terminal master's awarded for partial completion of doctoral program. *Degree requirements:* For master's, thesis (for some programs); for doctorate, thesis/dissertation, seminars, teaching experience. *Entrance requirements:* For master's and doctorate, GRE General Test. Additional exam requirements/recommendations for international students: Required—TOEFL, TSE. *Application deadline:* For fall admission, 2/15 for domestic students, 1/31 for international students. Applications are processed on a rolling basis. Application fee: $55. Electronic applications accepted. *Financial support:* In 2005–06, 15 fellowships, 60 research assistantships, 53 teaching assistantships

were awarded. Support available to part-time students. Financial award application deadline: 2/15; financial award applicants required to submit FAFSA. *Unit head:* Dr. Richard J Kuhn, Head, 765-494-4407. *Application contact:* Nancy Konopka, Graduate Studies Office Manager, 765-494-8142, Fax: 765-494-0876, E-mail: njk@bilbo.bio.purdue.edu.

Rice University, Graduate Programs, Wiess School of Natural Sciences, Department of Ecology and Evolutionary Biology, Houston, TX 77251-1892. Offers MA, PhD. Terminal master's awarded for partial completion of doctoral program. *Degree requirements:* For master's, thesis (for some programs), comprehensive exam (for some programs), registration; for doctorate, thesis/dissertation, comprehensive exam, registration. *Entrance requirements:* For master's and doctorate, GRE General Test, GRE Subject Test. Additional exam requirements/recommendations for international students: Required—TOEFL. Electronic applications accepted. *Faculty research:* Trace gas emissions, wetlands, biology, community ecology of forests and grasslands, conservation biology specialization.

Rutgers, The State University of New Jersey, New Brunswick/Piscataway, Graduate School, Program in Ecology and Evolution, New Brunswick, NJ 08901-1281. Offers MS, PhD. Part-time programs available. *Faculty:* 72 full-time, 4 part-time/adjunct. *Students:* 77 full-time (43 women), 19 part-time (4 women); includes 6 minority (1 African American, 1 Asian American or Pacific Islander, 4 Hispanic Americans), 11 international. Average age 31. 71 applicants, 41% accepted, 20 enrolled. In 2005, 5 master's, 7 doctorates awarded. Terminal master's awarded for partial completion of doctoral program. *Degree requirements:* For master's, comprehensive exam; for doctorate, thesis/dissertation, comprehensive exam. *Entrance requirements:* For master's and doctorate, GRE General Test, minimum GPA of 3.0. Additional exam requirements/recommendations for international students: Required—TOEFL (minimum score 550 paper-based; 213 computer-based). *Application deadline:* For fall admission, 4/1 for domestic students, 4/1 for international students. Application fee: $50. Electronic applications accepted. *Expenses:* Tuition, state resident: full-time $10,440; part-time $435 per credit. Tuition, nonresident: full-time $15,520; part-time $647 per credit. Required fees: $129 per credit. Tuition and fees vary according to program. *Financial support:* In 2005–06, 4 fellowships with full tuition reimbursements (averaging $25,000 per year), research assistantships with full tuition reimbursements (averaging $15,400 per year), 22 teaching assistantships with full tuition reimbursements (averaging $16,988 per year) were awarded; Federal Work-Study also available. Financial award application deadline: 1/15; financial award applicants required to submit FAFSA. *Faculty research:* Population and community ecology, population genetics, evolutionary biology, conservation biology, ecosystem ecology. *Unit head:* Dr. Peter J. Morin, Director, Graduate Program, 732-932-3214, Fax: 732-932-8764, E-mail: pjmorin@rci.rutgers.edu. *Application contact:* Marsha Morin, Secretary, 732-932-3213, Fax: 732-932-8764, E-mail: mmorin@aesop.rutgers.edu.

Rutgers, The State University of New Jersey, New Brunswick/Piscataway, Graduate School, Program in Plant Biology, New Brunswick, NJ 08901-1281. Offers horticulture (MS, PhD); molecular biology and biochemistry (MS, PhD); pathology (MS, PhD); plant ecology (MS, PhD); plant genetics (PhD); plant physiology (MS, PhD); production and management (MS); structure and plant groups (MS, PhD). Part-time programs available. *Faculty:* 65 full-time, 1 part-time/adjunct (0 women). *Students:* 33 full-time (14 women), 13 part-time (6 women); includes 3 minority (1 African American, 2 Asian Americans or Pacific Islanders), 10 international. Average age 31. 56 applicants, 23% accepted, 10 enrolled. In 2005, 1 master's, 4 doctorates awarded. Terminal master's awarded for partial completion of doctoral program. *Median time to degree:* Of those who began their doctoral program in fall 1997, 90% received their degree in 8 years or less. *Degree requirements:* For master's, thesis or alternative, comprehensive exam; for doctorate, thesis/dissertation, comprehensive exam. *Entrance requirements:* For master's and doctorate, GRE General Test, GRE Subject Test (recommended). Additional exam requirements/recommendations for international students: Required—TOEFL (minimum score 600 paper-based; 250 computer-based). *Application deadline:* For fall admission, 4/1 for domestic students, 4/1 for international students. Application fee: $50. Electronic applications accepted. *Expenses:* Tuition, state resident: full-time $10,440; part-time $435 per credit. Tuition, nonresident: full-time $15,520; part-time $647 per credit. Required fees: $129 per credit. Tuition and fees vary according to program. *Financial support:* In 2005–06, 42 students received support, including 9 fellowships with full tuition reimbursements available (averaging $24,000 per year), 22 research assistantships with full tuition reimbursements available (averaging $16,500 per year), 10 teaching assistantships with full tuition reimbursements available (averaging $16,988 per year) Financial award application deadline: 1/15; financial award applicants required to submit FAFSA. *Faculty research:* Molecular biology and biochemistry of plants, plant development and genomics, plant protection, plant improvement, plant management of horticultural and field crops. Total annual research expenditures: $10 million. *Unit head:* Dr. Thomas Leustek, Director, 732-932-8165 Ext. 326, Fax: 732-932-9377, E-mail: leustek@aesop.rutgers.edu. *Application contact:* Barbara Mulder, Program Associate, 732-932-9375 Ext. 358, Fax: 732-932-9377, E-mail: plantbio@aesop.rutgers.edu.

San Diego State University, Graduate and Research Affairs, College of Sciences, Department of Biological Sciences, Program in Ecology, San Diego, CA 92182. Offers MS, PhD. *Students:* 25 full-time (16 women), 27 part-time (14 women); includes 11 minority (1 African American, 6 Asian Americans or Pacific Islanders, 4 Hispanic Americans), 5 international. Average age 30. 71 applicants, 25% accepted, 13 enrolled. In 2005, 8 master's, 2 doctorates awarded. *Degree requirements:* For master's and doctorate, thesis/dissertation. *Entrance requirements:* For master's, GRE General Test, resumé or curriculum vitae, 2 letters of recommendation; for doctorate, GRE General Test, GRE Subject Test, resumé or curriculum vitae, 3 letters of recommendation. *Application deadline:* For fall admission, 5/1 for domestic students, 5/1 for international students; for spring admission, 11/1 for domestic students, 10/1 for international students. Applications are processed on a rolling basis. Application fee: $55. Electronic applications accepted. *Financial support:* Research assistantships, teaching assistantships, career-related internships or fieldwork and unspecified assistantships available. *Faculty research:* Conservation and restoration ecology, coastal and marine ecology, global change and ecosystem ecology. Total annual research expenditures: $4 million. *Unit head:* Walter Oechel, Graduate Advisor, 619-594-4818, E-mail: oechel@sunstroke.sdsu.edu. *Application contact:* Walter Oechel, Graduate Advisor, 619-594-4818, E-mail: oechel@sunstroke.sdsu.edu.

See Close-Up on page 707.

San Francisco State University, Division of Graduate Studies, College of Science and Engineering, Department of Biology, Program in Ecology and Systematic Biology, San Francisco, CA 94132-1722. Offers MA. *Entrance requirements:* For master's, minimum GPA of 2.5 in last 60 units.

State University of New York College of Environmental Science and Forestry, Faculty of Environmental and Forest Biology, Syracuse, NY 13210-2779. Offers chemical ecology (MPS, MS, PhD); conservation biology (MPS, MS, PhD); ecology (MPS, MS, PhD); entomology (MPS, MS, PhD); environmental interpretation (MPS, MS, PhD); environmental physiology (MPS, MS, PhD); fish and wildlife biology (MPS, MS, PhD); forest pathology and mycology (MPS, MS, PhD); plant science and biotechnology (MPS, MS, PhD). *Faculty:* 27 full-time (4 women), 4 part-time/adjunct (0 women). *Students:* 81 full-time (50 women), 62 part-time (33 women); includes 4 minority (1 Asian American or Pacific Islander, 3 Hispanic Americans), 16 international. Average age 30. 84 applicants, 54% accepted, 17 enrolled. In 2005, 17 master's, 3 doctorates awarded. *Degree requirements:* For master's, thesis (for some programs), registration; for doctorate, thesis/dissertation, comprehensive exam, registration. *Entrance requirements:* For master's and doctorate, GRE General Test, GRE Subject Test, minimum GPA of 3.0. Additional exam requirements/recommendations for international students: Required—TOEFL (minimum score 550 paper-based; 213 computer-based). *Application deadline:* For fall admission, 2/1 priority date for domestic students, 2/1 priority date for international students; for spring admission, 11/1 priority date for domestic students, 11/1 priority date for international students. Applications are processed on a rolling basis. Application fee: $60. *Expenses:*

Ecology

Tuition, area resident: Full-time $6,900; part-time $288 per credit. Tuition, nonresident: full-time $10,920; part-time $455 per credit. Required fees: $395; $32 per credit. $20 per term. One-time fee: $145. *Financial support:* In 2005–06, 86 students received support, including 13 fellowships with full and partial tuition reimbursements available (averaging $9,446 per year), 40 research assistantships with full and partial tuition reimbursements available (averaging $11,000 per year), 32 teaching assistantships with full and partial tuition reimbursements available (averaging $9,446 per year); Federal Work-Study, institutionally sponsored loans, scholarships/grants, health care benefits, and unspecified assistantships also available. Financial award application deadline: 6/30. *Faculty research:* Ecology, fish and wildlife biology and management, plant science, entomology. Total annual research expenditures: $4.1 million. *Unit head:* Dr. Donald J. Leopold, Chair, 315-470-6770, Fax: 315-470-6934, E-mail: dendro@esf.edu. *Application contact:* Dr. Dudley J. Raynal, Dean, Instruction and Graduate Studies, 315-470-6599, Fax: 315-470-6978, E-mail: esfgrad@esf.edu.

Stony Brook University, State University of New York, Graduate School, College of Arts and Sciences, Department of Ecology and Evolution, Stony Brook, NY 11794. Offers PhD. *Faculty:* 17 full-time (4 women), 1 part-time/adjunct (0 women). *Students:* 49 full-time (20 women), 1 (woman) part-time; includes 7 minority (3 Asian Americans or Pacific Islanders, 4 Hispanic Americans), 8 international. Average age 28. 84 applicants, 36% accepted. In 2005, 4 degrees awarded. *Degree requirements:* For doctorate, one foreign language, thesis/dissertation, teaching experience, comprehensive exam. *Entrance requirements:* For doctorate, GRE General Test, GRE Subject Test. Additional exam requirements/recommendations for international students: Required—TOEFL. *Application deadline:* For fall admission, 1/15 for domestic students. Application fee: $50. *Expenses:* Tuition, state resident: full-time $6,900; part-time $288 per credit. Tuition, nonresident: full-time $10,920; part-time $455 per credit. Required fees: $704. *Financial support:* In 2005–06, 8 fellowships, 13 research assistantships, 23 teaching assistantships were awarded; Federal Work-Study also available. *Faculty research:* Theoretical and experimental population genetics, numerical taxonomy, biostatistics, population and community ecology, plant ecology. Total annual research expenditures: $1.8 million. *Unit head:* Dr. Charles H. Janson, Chair, 631-632-8600. *Application contact:* Dr. Dan Dykhuizen, Director, 631-246-8604, E-mail: dandyk@life.bio.sunysb.edu.

Announcement: The Ecology and Evolution Program at Stony Brook combines cutting-edge research facilities, access to extraordinary field sites, and a world-class faculty. Research opportunities are available across a range of disciplines, and alumni are qualified for a range of positions in academia, industry, government, and nongovernmental organizations.

See Close-Up on page 709.

Texas Christian University, College of Science and Engineering, Department of Biology, Program in Environmental Sciences, Fort Worth, TX 76129-0002. Offers earth sciences (MS); ecology (MS). Part-time and evening/weekend programs available. *Degree requirements:* For master's, thesis optional. *Entrance requirements:* For master's, GRE General Test, GRE Subject Test, 1 year course work in biology and chemistry; 1 semester course work in calculus, government, and physical geology. Additional exam requirements/recommendations for international students: Required—TOEFL. *Application deadline:* For fall admission, 3/1 for domestic students; for spring admission, 12/1 for domestic students. Applications are processed on a rolling basis. Application fee: $0. *Expenses:* Tuition: Part-time $740 per credit hour. *Financial support:* Unspecified assistantships available. Financial award application deadline: 3/1. *Unit head:* Dr. Mike Slattery, Director, 817-257-7506. *Application contact:* Dr. Bonnie Melhart, Associate Dean, College of Science and Engineering, E-mail: b.melhart@tcu.edu.

University at Albany, State University of New York, College of Arts and Sciences, Department of Biological Sciences, Specialization in Ecology, Evolution, and Behavior, Albany, NY 12222-0001. Offers MS, PhD. *Degree requirements:* For master's, one foreign language; for doctorate, one foreign language, thesis/dissertation. *Entrance requirements:* For master's and doctorate, GRE General Test. Application fee: $60. *Financial support:* Minority assistantships available. *Unit head:* Dr. Albert Millis, Chair, Department of Biological Sciences, 518-442-4300.

University of Alberta, Faculty of Graduate Studies and Research, Department of Biological Sciences, Edmonton, AB T6G 2E1, Canada. Offers environmental biology and ecology (M Sc, PhD); microbiology and biotechnology (M Sc, PhD); molecular biology and genetics (M Sc, PhD); physiology and cell biology (M Sc, PhD); plant biology (M Sc, PhD); systematics and evolution (M Sc, PhD). *Faculty:* 72 full-time (15 women), 15 part-time/adjunct (4 women). *Students:* 238 full-time (117 women), 32 part-time (15 women), 31 international. 206 applicants, 42% accepted. In 2005, 29 master's, 31 doctorates awarded. Terminal master's awarded for partial completion of doctoral program. *Degree requirements:* For master's and doctorate, thesis/dissertation, registration. *Entrance requirements:* Additional exam requirements/recommendations for international students: Required—TOEFL. *Application deadline:* For fall admission, 3/1 for domestic students. Applications are processed on a rolling basis. Application fee: $0. Tuition and fees charges are reported in Canadian dollars. *Expenses:* Tuition, state resident: part-time $562 Canadian dollars per term. Tuition, nonresident: full-time $3,375 Canadian dollars. Required fees: $573 Canadian dollars; $84 Canadian dollars per term. *Financial support:* In 2005–06, 4 research assistantships with partial tuition reimbursements (averaging $12,000 per year), 103 teaching assistantships with partial tuition reimbursements (averaging $12,300 per year) were awarded; career-related internships or fieldwork and scholarships/grants also available. *Unit head:* Laura Frost, Chair, 780-492-1904. *Application contact:* Dr. John P. Chang, Associate Chair for Graduate Studies, 780-492-1257, Fax: 780-492-9457, E-mail: bio.grad.coordinator@ualberta.ca.

The University of Arizona, Graduate College, College of Science, Department of Ecology and Evolutionary Biology, Tucson, AZ 85721. Offers MS, PhD. *Degree requirements:* For doctorate, one foreign language, thesis/dissertation, comprehensive exam. *Entrance requirements:* For master's and doctorate, GRE General Test, GRE Subject Test. Additional exam requirements/recommendations for international students: Required—TOEFL. *Faculty research:* Biological diversity, evolutionary history, evolutionary mechanisms, community structure.

University of California, Davis, Graduate Studies, Graduate Group in Ecology, Davis, CA 95616. Offers MS, PhD. *Faculty:* 116 full-time. *Students:* 171 full-time (93 women); includes 18 minority (3 African Americans, 2 American Indian/Alaska Native, 5 Asian Americans or Pacific Islanders, 8 Hispanic Americans), 17 international. Average age 31. 228 applicants, 28% accepted, 32 enrolled. In 2005, 4 master's, 28 doctorates awarded. *Median time to degree:* Of those who began their doctoral program in fall 1997, 61.5% received their degree in 8 years or less. *Degree requirements:* For master's, thesis (for some programs), comprehensive exam (for some programs); for doctorate, thesis/dissertation. *Entrance requirements:* For master's and doctorate, GRE General Test. Additional exam requirements/recommendations for international students: Required—TOEFL (minimum score 550 paper-based; 213 computer-based). *Application deadline:* For fall admission, 12/1 for domestic students, 12/1 for international students. Application fee: $60. Electronic applications accepted. *Financial support:* In 2005–06, 153 students received support, including 72 fellowships with full and partial tuition reimbursements available (averaging $15,944 per year), 44 research assistantships with full and partial tuition reimbursements available (averaging $12,180 per year), 23 teaching assistantships with partial tuition reimbursements available (averaging $15,082 per year); Federal Work-Study, institutionally sponsored loans, scholarships/grants, tuition waivers (full and partial), and unspecified assistantships also available. Financial award application deadline: 1/15; financial award applicants required to submit FAFSA. *Faculty research:* Agricultural conservation, physiological restoration, environmental policy, ecotoxicology. *Unit head:* Mark W. Schwartz, Chair, 530-752-0671, E-mail: mwschwartz@ucdavis.edu. *Application contact:* Silvia Hillyer, Graduate Coordinator, 530-752-6752, Fax: 530-752-3350, E-mail: schillyer@ucdavis.edu.

University of California, Irvine, Office of Graduate Studies, School of Biological Sciences, Department of Ecology and Evolutionary Biology, Irvine, CA 92697. Offers biological sciences (MS, PhD). *Degree requirements:* For master's and doctorate, thesis/dissertation. *Entrance*

requirements: For master's and doctorate, GRE General Test, GRE Subject Test, minimum GPA of 3.0. Additional exam requirements/recommendations for international students: Required—TOEFL (minimum score 550 paper-based; 213 computer-based). Electronic applications accepted. *Faculty research:* Ecological energetics, quantitative genetics, life history evolution, plant-herbivore and plant-pollinator interactions, molecular evolution.

University of California, San Diego, Graduate Studies and Research, Division of Biology, Program in Ecology, Behavior, and Evolution, La Jolla, CA 92093. Offers PhD. *Degree requirements:* For doctorate, thesis/dissertation, qualifying exam. Electronic applications accepted.

University of California, Santa Barbara, Graduate Division, College of Letters and Sciences, Division of Mathematics, Life, and Physical Sciences, Department of Ecology, Evolution, and Marine Biology, Santa Barbara, CA 93106. Offers MA, PhD. *Faculty:* 36 full-time (7 women), 6 part-time/adjunct (1 woman). *Students:* 61 full-time (29 women); includes 10 minority (1 African American, 6 Asian Americans or Pacific Islanders, 3 Hispanic Americans), 3 international. Average age 29. 150 applicants, 9% accepted, 4 enrolled. In 2005, 3 master's, 6 doctorates awarded. Terminal master's awarded for partial completion of doctoral program. *Median time to degree:* Of those who began their doctoral program in fall 1997, 90% received their degree in 8 years or less. *Degree requirements:* For master's, thesis (for some programs), comprehensive exam (for some programs); registration; for doctorate, thesis/dissertation, comprehensive exam, registration. *Entrance requirements:* For master's and doctorate, GRE General Test. Additional exam requirements/recommendations for international students: Required—TOEFL (minimum score 550 paper-based; 213 computer-based). *Application deadline:* For fall admission, 12/15 for domestic students, 12/15 for international students. Application fee: $60. Electronic applications accepted. *Financial support:* In 2005–06, 30 students received support, including 44 fellowships with full and partial tuition reimbursements available, 90 teaching assistantships with partial tuition reimbursements available (averaging $5,025 per year); research assistantships with full and partial tuition reimbursements available, career-related internships or fieldwork, Federal Work-Study, institutionally sponsored loans, scholarships/grants, traineeships, health care benefits, tuition waivers (full and partial), and unspecified assistantships also available. Financial award application deadline: 12/15; financial award applicants required to submit FAFSA. *Faculty research:* Ecology, population genetics, stream ecology, evolution, marine biology. *Unit head:* Alice Alldredge, Chair, 805-893-2415, Fax: 805-893-4724. *Application contact:* Jennifer Chater, Staff Graduate Advisor, 805-893-3023, Fax: 805-893-4724, E-mail: eemb-info@lifesci.ucsb.edu.

University of California, Santa Cruz, Division of Graduate Studies, Division of Physical and Biological Sciences, Department of Ecology and Evolutionary Biology, Santa Cruz, CA 95064. Offers MA, PhD. *Faculty:* 18 full-time (5 women). *Students:* 62 full-time (42 women), 1 (woman) part-time; includes 6 minority (1 American Indian/Alaska Native, 1 Asian American or Pacific Islander, 4 Hispanic Americans), 1 international. 123 applicants, 9% accepted, 8 enrolled. In 2005, 2 master's, 5 doctorates awarded. *Entrance requirements:* Additional exam requirements/recommendations for international students: Required—TOEFL; Recommended—IELT. *Application deadline:* For fall admission, 12/15 for domestic students. Application fee: $60. Electronic applications accepted. *Expenses:* Tuition, nonresident: full-time $14,694. Required fees: $9,437. *Unit head:* Peter Raimondi, Chairperson. *Application contact:* Susan Thuringer, Information Contact, 831-459-4715, E-mail: susan@biology.ucsc.edu.

University of Chicago, Division of the Biological Sciences, Department of Darwinian Sciences: Ecological, Integrative and Evolutionary Biology, Department of Ecology and Evolution, Chicago, IL 60637-1513. Offers PhD. *Faculty:* 14 full-time (2 women). *Students:* 27 full-time (12 women), 8 international. Average age 27. In 2005, 5 doctorates awarded. *Degree requirements:* For doctorate, thesis/dissertation, registration. *Entrance requirements:* For doctorate, GRE General Test. Additional exam requirements/recommendations for international students: Required—TOEFL. *Application deadline:* For fall admission, 12/28 priority date for domestic students, 12/28 priority date for international students. Application fee: $55. Electronic applications accepted. *Financial support:* In 2005–06, 21 students received support, including fellowships with tuition reimbursements available (averaging $26,301 per year), research assistantships with tuition reimbursements available (averaging $26,301 per year); institutionally sponsored loans, scholarships/grants, traineeships, and health care benefits also available. Financial award applicants required to submit FAFSA. *Faculty research:* Population genetics, molecular evolution, behavior. *Unit head:* Dr. Chung-I Wu, Chairman, 773-702-2565, Fax: 773-702-9740, E-mail: ciwu@uchicago.edu. *Application contact:* Carolyn Johnson, Graduate Administrative Director, 773-702-9474, Fax: 773-702-4699, E-mail: csjohnso@uchicago.edu.

University of Colorado at Boulder, Graduate School, College of Arts and Sciences, Department of Ecology and Evolutionary Biology, Boulder, CO 80309. Offers animal behavior (MA); biology (MA, PhD); environmental biology (MA, PhD); evolutionary biology (MA, PhD); neurobiology (MA); population biology (MA); population genetics (PhD). *Faculty:* 26 full-time (5 women). *Students:* 44 full-time (24 women), 22 part-time (14 women); includes 16 minority (1 American Indian/Alaska Native, 3 Asian Americans or Pacific Islanders, 12 Hispanic Americans), 2 international. Average age 30. 20 applicants, 90% accepted. In 2005, 7 master's, 7 doctorates awarded. Terminal master's awarded for partial completion of doctoral program. *Degree requirements:* For master's, thesis or alternative, comprehensive exam; for doctorate, thesis/dissertation, comprehensive exam. *Entrance requirements:* For master's, GRE General Test, GRE Subject Test, minimum undergraduate GPA of 3.0; for doctorate, GRE General Test, GRE Subject Test. *Application deadline:* For fall admission, 1/2 priority date for domestic students, 12/1 priority date for international students. Application fee: $50 ($60 for international students). *Financial support:* In 2005–06, fellowships (averaging $7,844 per year), research assistantships (averaging $15,663 per year), teaching assistantships (averaging $13,829 per year) were awarded; Federal Work-Study, institutionally sponsored loans, and tuition waivers (full) also available. Financial award application deadline: 3/1. *Faculty research:* Behavior, ecology, genetics, morphology, endocrinology. Total annual research expenditures: $5.2 million. *Unit head:* Jeffry Mitton, Chair, 303-492-0505, Fax: 303-492-8699, E-mail: mitton@colorado.edu. *Application contact:* Jill Skarstadt, Graduate Coordinator, 303-492-7654, Fax: 303-492-8699, E-mail: skarstad@colorado.edu.

University of Connecticut, Graduate School, College of Liberal Arts and Sciences, Department of Ecology and Evolutionary Biology, Field of Ecology, Storrs, CT 06269. Offers MS, PhD. *Faculty:* 29 full-time (8 women). *Students:* 31 full-time (18 women), 4 part-time (2 women), 6 international. Average age 33. 44 applicants, 20% accepted, 9 enrolled. In 2005, 1 master's, 7 doctorates awarded. Terminal master's awarded for partial completion of doctoral program. *Degree requirements:* For master's, comprehensive exam; for doctorate, thesis/dissertation. *Entrance requirements:* For master's and doctorate, GRE General Test, GRE Subject Test. Additional exam requirements/recommendations for international students: Required—TOEFL (minimum score 550 paper-based; 213 computer-based). *Application deadline:* For fall admission, 2/1 priority date for domestic students, 2/1 priority date for international students; for spring admission, 11/1 for domestic students, 10/1 for international students. Applications are processed on a rolling basis. Application fee: $55. Electronic applications accepted. *Expenses:* Tuition, state resident: part-time $444 per credit hour. Tuition, nonresident: part-time $1,154 per credit hour. Tuition and fees vary according to course load. *Financial support:* In 2005–06, 12 research assistantships with full tuition reimbursements, 16 teaching assistantships with full tuition reimbursements were awarded; fellowships, Federal Work-Study, scholarships/grants, health care benefits, and unspecified assistantships also available. Financial award application deadline: 2/1; financial award applicants required to submit FAFSA. *Application contact:* Anne St. Onje, Graduate Coordinator, 860-486-4314, Fax: 860-486-3943, E-mail: ann.st_onje@uconn.edu.

University of Connecticut, Graduate School, College of Liberal Arts and Sciences, Department of Psychology, Field of Psychology, Storrs, CT 06269. Offers behavioral neuroscience (PhD); biopsychology (PhD); clinical psychology (MA, PhD); cognition and instruction (PhD); developmental psychology (MA, PhD); ecological psychology (PhD); experimental psychology

Ecology

University of Connecticut *(continued)*
(PhD); general psychology (MA, PhD); industrial/organizational psychology (PhD); language and cognition (PhD); neuroscience (PhD); social psychology (MA, PhD). *Accreditation:* APA (one or more programs are accredited). *Faculty:* 51 full-time (23 women). *Students:* 134 full-time (90 women), 21 part-time (10 women); includes 13 minority (7 African Americans, 2 Asian Americans or Pacific Islanders, 4 Hispanic Americans), 21 international. Average age 28. 510 applicants, 10% accepted, 49 enrolled. In 2005, 22 master's, 15 doctorates awarded. Terminal master's awarded for partial completion of doctoral program. *Degree requirements:* For master's, comprehensive exam; for doctorate, thesis/dissertation. *Entrance requirements:* For master's and doctorate, GRE General Test, GRE Subject Test. Additional exam requirements/recommendations for international students: Required—TOEFL (minimum score 550 paper-based; 213 computer-based). *Application deadline:* For fall admission, 2/1 priority date for domestic students, 2/1 priority date for international students; for spring admission, 11/1 for domestic students, 10/1 for international students. Applications are processed on a rolling basis. Application fee: $55. Electronic applications accepted. *Expenses:* Tuition, state resident: part-time $444 per credit hour. Tuition, nonresident: part-time $1,154 per credit hour. Tuition and fees vary according to course load. *Financial support:* In 2005–06, 67 research assistantships with full tuition reimbursements, 58 teaching assistantships with full tuition reimbursements were awarded; fellowships, career-related internships or fieldwork, Federal Work-Study, scholarships/grants, health care benefits, and unspecified assistantships also available. Financial award application deadline: 2/1; financial award applicants required to submit FAFSA. *Application contact:* Deborah Doucette, Administrative Assistant, 860-486-2057, Fax: 860-486-2760, E-mail: futuregr@psych.psy.uconn.edu.

University of Delaware, College of Agriculture and Natural Resources, Department of Entomology and Wildlife Ecology, Newark, DE 19716. Offers entomology and applied ecology (MS, PhD), including avian ecology, evolution and taxonomy, insect biological control, insect ecology and behavior (MS), insect genetics, pest management, plant-insect interactions, wildlife ecology and management. Part-time programs available. *Faculty:* 9 full-time (1 woman), 4 part-time/adjunct (1 woman). *Students:* 26 full-time (13 women), 6 part-time (1 woman); includes 1 minority (African American) Average age 27. 27 applicants, 44% accepted, 11 enrolled. In 2005, 4 master's, 1 doctorate awarded. *Degree requirements:* For master's, thesis, oral exam, seminar, comprehensive exam; for doctorate, thesis/dissertation, qualifying exam, seminar, comprehensive exam. *Entrance requirements:* For master's, GRE General Test, minimum GPA of 3.0 in field, 2.8 overall; for doctorate, GRE General Test, GRE Subject Test (biology), minimum GPA of 3.0 in field, 2.8 overall. Additional exam requirements/recommendations for international students: Required—TOEFL. *Application deadline:* For fall admission, 2/1 for domestic students; for spring admission, 11/1 for domestic students. Applications are processed on a rolling basis. Application fee: $60. Electronic applications accepted. *Financial support:* In 2005–06, 18 students received support, including 2 fellowships with full tuition reimbursements available (averaging $13,500 per year), 7 research assistantships with full tuition reimbursements available (averaging $16,260 per year), 5 teaching assistantships with full tuition reimbursements available (averaging $12,195 per year); career-related internships or fieldwork, institutionally sponsored loans, scholarships/grants, and tuition waivers (full) also available. Financial award application deadline: 3/1. *Faculty research:* Genetics and resistance, biological control, chemically mediated behavioral ecology, ecology and evolution of plant-insect interactions, ecology of wildlife conservation and management. Total annual research expenditures: $811,092. *Unit head:* Dr. Douglas W. Tallamy, Chair, 302-831-1304, Fax: 302-831-8889, E-mail: dtallamy@udel.edu. *Application contact:* Dr. Charles E. Mason, Graduate Coordinator, 302-831-8888, Fax: 302-831-8889, E-mail: mason@udel.edu.

University of Delaware, College of Arts and Sciences, Department of Biological Sciences, Newark, DE 19716. Offers biotechnology (MS); cancer biology (MS, PhD); cell and extra-cellular matrix biology (MS, PhD); cell and systems physiology (MS, PhD); developmental biology (MS, PhD); ecology and evolution (MS, PhD); microbiology (MS, PhD); molecular biology and genetics (MS, PhD). *Faculty:* 39 full-time (11 women). *Students:* 64 full-time (46 women), 2 part-time; includes 7 minority (4 African Americans, 2 Asian Americans or Pacific Islanders, 1 Hispanic American), 18 international. Average age 26. 113 applicants, 31% accepted, 19 enrolled. In 2005, 5 master's, 4 doctorates awarded. Terminal master's awarded for partial completion of doctoral program. *Median time to degree:* Of those who began their doctoral program in fall 1997, 100% received their degree in 8 years or less. *Degree requirements:* For master's, thesis/dissertation, preliminary exam; for doctorate, thesis/dissertation, preliminary exam, comprehensive exam. *Entrance requirements:* For master's and doctorate, GRE General Test. Additional exam requirements/recommendations for international students: Required—TOEFL (minimum score 600 paper-based; 250 computer-based); Recommended—TWE, TSE. *Application deadline:* For fall admission, 4/15 for domestic students, 1/15 for international students; for spring admission, 10/1 for domestic students. Applications are processed on a rolling basis. Application fee: $60. Electronic applications accepted. *Financial support:* In 2005–06, 26 students received support, including fellowships with full tuition reimbursements available (averaging $19,000 per year), 19 research assistantships with full tuition reimbursements available (averaging $19,000 per year), 26 teaching assistantships with full tuition reimbursements available (averaging $19,000 per year); tuition waivers (partial) also available. Financial award application deadline: 4/15. *Faculty research:* Microorganisms, bone, cancer metastasis, developmental biology, cell biology, DNA. Total annual research expenditures: $8.3 million. *Unit head:* Dr. Daniel D. Carson, Chair, 302-831-6977, Fax: 302-831-2281, E-mail: dcarson@udel.edu. *Application contact:* Dr. Melinda K. Duncan, Graduate Coordinator, 302-831-1841, Fax: 302-831-2281, E-mail: danders@udel.edu.

University of Florida, Graduate School, College of Agricultural and Life Sciences, Department of Wildlife Ecology and Conservation, Gainesville, FL 32611. Offers MS, PhD. *Faculty:* 14 full-time (3 women), 1 part-time/adjunct (0 women). *Students:* 65 (28 women); includes 5 minority (1 American Indian/Alaska Native, 1 Asian American or Pacific Islander, 3 Hispanic Americans) 11 international. 65 applicants, 29% accepted. In 2005, 12 master's, 4 doctorates awarded. *Degree requirements:* For master's, thesis optional; for doctorate, thesis/dissertation. *Entrance requirements:* For master's and doctorate, GRE General Test, minimum GPA of 3.3. *Application deadline:* For fall admission, 6/1 for domestic students; for spring admission, 12/1 for domestic students. Applications are processed on a rolling basis. Application fee: $20. Electronic applications accepted. *Expenses:* Tuition, state resident: full-time $6,234. Tuition, nonresident: full-time $21,359. Tuition and fees vary according to program. *Financial support:* In 2005–06, 46 students received support, including 17 research assistantships (averaging $13,811 per year), 11 teaching assistantships (averaging $14,036 per year); fellowships, institutionally sponsored loans also available. *Faculty research:* Wildlife biology and management, tropical ecology and conservation, conservation biology, landscape ecology and restoration, conservation education. *Unit head:* George Tanner, Interim Chair, 352-846-0570. *Application contact:* Dr. Wiley Kitchens, Coordinator, 352-846-0536, Fax: 352-846-0841, E-mail: kitchensw@wec.ufl.edu.

University of Florida, Graduate School, School of Natural Resources and Environment, Gainesville, FL 32611. Offers interdisciplinary ecology (MS, PhD). *Faculty:* 1 full-time (0 women). *Students:* 119 (63 women); includes 7 minority (1 African American, 3 Asian American or Pacific Islanders, 3 Hispanic Americans) 23 international. In 2005, 18 master's, 4 doctorates awarded. *Degree requirements:* For master's, thesis optional; for doctorate, thesis/dissertation. *Entrance requirements:* For master's and doctorate, GRE General Test, minimum GPA of 3.0. Additional exam requirements/recommendations for international students: Required—TOEFL (minimum score 550 paper-based; 213 computer-based). *Application deadline:* For fall admission, 2/11 for domestic students. Applications are processed on a rolling basis. Application fee: $30. Electronic applications accepted. *Expenses:* Tuition, state resident: full-time $6,234. Tuition, nonresident: full-time $21,359. Tuition and fees vary according to program. *Financial support:* In 2005–06, 9 teaching assistantships (averaging $15,592 per year) were awarded; fellowships, research assistantships *Unit head:* Dr. Stephen R. Humphrey, Director, 352-392-9230, Fax: 352-392-9748, E-mail: humphrey@ufl.edu. *Applica-*

tion contact: Meisha Wade, Coordinator of Academic Programs, 352-392-9230, Fax: 352-392-9748, E-mail: mwade@ufl.edu.

University of Georgia, Graduate School, College of Environment and Design, Institute in Ecology, Athens, GA 30602. Offers conservation ecology and sustainable development (MS); ecology (MS, PhD). *Faculty:* 19 full-time (5 women). *Students:* 82 full-time, 21 part-time; includes 4 minority (2 African Americans, 1 Asian American or Pacific Islander, 1 Hispanic American), 9 international. 134 applicants, 17% accepted, 12 enrolled. In 2005, 14 master's, 10 doctorates awarded. *Degree requirements:* For master's, thesis; for doctorate, one foreign language, thesis/dissertation. *Entrance requirements:* For master's and doctorate, GRE General Test. *Application deadline:* For fall admission, 7/1 for domestic students; for spring admission, 11/15 for domestic students. Application fee: $50. Electronic applications accepted. *Financial support:* Fellowships, research assistantships, teaching assistantships, unspecified assistantships available. *Unit head:* Dr. Alan Covich, Head, 706-542-6018, Fax: 706-542-4819. *Application contact:* Dr. William K. Fitt, Graduate Coordinator, 706-542-3328, Fax: 706-542-6040, E-mail: ecoginfo@uga.edu.

University of Guelph, Graduate Program Services, College of Biological Science, Department of Integrative Biology, Guelph, ON N1G 2W1, Canada. Offers botany (M Sc, PhD); zoology (M Sc, PhD). Part-time programs available. *Faculty:* 37 full-time (5 women). *Students:* 100 full-time (53 women), 3 part-time (all women); includes 5 minority (all Asian Americans or Pacific Islanders), 6 international. Average age 23. 23 applicants, 83% accepted, 19 enrolled. In 2005, 26 master's, 7 doctorates awarded. *Median time to degree:* Of those who began their doctoral program in fall 1997, 95% received their degree in 8 years or less. *Degree requirements:* For master's, thesis, research proposal; for doctorate, thesis/dissertation, research proposal, qualifying exam. *Entrance requirements:* For master's, minimum B average during previous 2 years of course work; for doctorate, minimum A- average. Additional exam requirements/recommendations for international students: Required—TOEFL (minimum score 550 paper-based; 213 computer-based), IELT (minimum score 7). *Application deadline:* For fall admission, 7/1 for domestic students. For winter admission, 11/1 for domestic students; for spring admission, 3/1 for domestic students. Applications are processed on a rolling basis. Application fee: $75. Electronic applications accepted. *Financial support:* In 2005–06, 50 students received support, including 48 research assistantships, 78 teaching assistantships (averaging $4,606 per year); fellowships *Faculty research:* Aquatic science, environmental physiology, parasitology, wildlife biology, management. Total annual research expenditures: $3.8 million. *Unit head:* Dr. Moira Ferguson, Chair, 519-824-4120 Ext. 53598, Fax: 519-767-1656, E-mail: mmfergus@uoguelph.ca. *Application contact:* Laurie Winn, Graduate Admissions Secretary, 519-824-4320 Ext. 52730, Fax: 519-767-1656, E-mail: lwinn@uoguelph.ca.

University of Hawaii at Manoa, Graduate Division, Specialization in Ecology, Evolution and Conservation Biology, Honolulu, HI 96822. Offers MS, PhD. Program is interdisciplinary; degree is in specific discipline with the specialization in Ecology, Evolution and Conservation Biology. *Degree requirements:* For doctorate, thesis/dissertation. Application fee: $50. *Expenses:* Tuition, state resident: full-time $8,400; part-time $200 per credit hour. Tuition, nonresident: full-time $11,088; part-time $462 per credit hour. Tuition and fees vary according to program. *Financial support:* In 2005–06, 15 students received support, including 3 fellowships, 5 research assistantships, 2 teaching assistantships; career-related internships or fieldwork and tuition waivers (full) also available. *Faculty research:* Agronomy and soil science, zoology, entomology, genetics and molecular biology, botanical sciences. *Unit head:* Dr. Robert Kinzie, Chair, 808-956-4602, Fax: 808-956-9608, E-mail: kinzie@hawaii.edu.

University of Illinois at Chicago, Graduate College, College of Liberal Arts and Sciences, Department of Biological Sciences, Chicago, IL 60607-7128. Offers cell and developmental biology (PhD); ecology and evolution (MS, DA, PhD); genetics and development (PhD); molecular biology (MS, PhD); neurobiology (MS, PhD); plant biology (MS, DA, PhD). *Degree requirements:* For master's, thesis; for doctorate, thesis/dissertation, preliminary exam. *Entrance requirements:* For master's and doctorate, GRE General Test, GRE Subject Test, previous course work in physics, calculus, and organic chemistry; minimum GPA of 2.75. Additional exam requirements/recommendations for international students: Required—TOEFL. Electronic applications accepted.

University of Illinois at Urbana–Champaign, Graduate College, College of Liberal Arts and Sciences, School of Integrative Biology, Program in Ecology and Evolutionary Biology, Champaign, IL 61820. Offers MS, PhD. *Students:* 29 full-time (12 women), 7 part-time (4 women); includes 3 minority (1 African American, 2 Asian Americans or Pacific Islanders), 6 international. 54 applicants, 4% accepted, 2 enrolled. In 2005, 2 master's, 1 doctorate awarded. Application fee: $50 ($60 for international students). *Financial support:* In 2005–06, 6 fellowships, 9 research assistantships, 17 teaching assistantships were awarded. *Unit head:* Jeff Brawn, Director, 217-244-5937, Fax: 217-244-1224, E-mail: jbrawn@uiuc.edu. *Application contact:* Carol Hall, Secretary, 217-333-8208, Fax: 217-244-1224, E-mail: cahall@uiuc.edu.

University of Kansas, Graduate School, College of Liberal Arts and Sciences, Division of Biological Sciences, Department of Ecology and Evolutionary Biology, Lawrence, KS 66045. Offers botany (MA, PhD); ecology and evolutionary biology (MA, PhD); entomology (MA, PhD); systematics and ecology (MA). Part-time programs available. *Faculty:* 42. *Students:* 67 full-time (35 women), 21 part-time (9 women); includes 4 minority (1 American Indian/Alaska Native, 3 Hispanic Americans), 20 international. Average age 30. 56 applicants, 32% accepted. In 2005, 10 master's, 4 doctorates awarded. Terminal master's awarded for partial completion of doctoral program. *Degree requirements:* For master's and doctorate, thesis/dissertation, comprehensive exam. *Entrance requirements:* For master's and doctorate, GRE General Test, GRE Subject Test (recommended). Additional exam requirements/recommendations for international students: Required—TOEFL (paper 570; computer 230) or IELT. *Application deadline:* For fall admission, 1/10 priority date for domestic students, 1/10 priority date for international students. Applications are processed on a rolling basis. Application fee: $55 ($60 for international students). Electronic applications accepted. *Expenses:* Tuition, state resident: full-time $4,859. Tuition, nonresident: full-time $12,000. Required fees: $589. Tuition and fees vary according to program. *Financial support:* Fellowships with tuition reimbursements, research assistantships with partial tuition reimbursements, teaching assistantships with full and partial tuition reimbursements available. Financial award application deadline: 3/1. *Faculty research:* Ecology, evolutionary biology, genetics. *Unit head:* Craig E. Martin, Chair, 785-864-5887, Fax: 785-864-5860, E-mail: ecophys@ku.edu. *Application contact:* Jeannie Houts, Graduate Coordinator, 785-864-2362, Fax: 785-864-5860, E-mail: jmhouts@ku.edu.

University of Maine, Graduate School, College of Natural Sciences, Forestry, and Agriculture, Department of Biological Sciences, Program in Ecology and Environmental Sciences, Orono, ME 04469. Offers MS, PhD. Part-time programs available. *Students:* 26 full-time (17 women), 21 part-time (14 women), 2 international. Average age 30. 56 applicants, 20% accepted, 8 enrolled. In 2005, 11 master's, 4 doctorates awarded. *Degree requirements:* For doctorate, thesis/dissertation. *Entrance requirements:* For master's and doctorate, GRE General Test. Additional exam requirements/recommendations for international students: Required—TOEFL. *Application deadline:* For fall admission, 2/1 for domestic students. Applications are processed on a rolling basis. Application fee: $50. Electronic applications accepted. *Financial support:* Fellowships, research assistantships with tuition reimbursements, teaching assistantships with tuition reimbursements, career-related internships or fieldwork, Federal Work-Study, institutionally sponsored loans, and tuition waivers (full) available. Financial award application deadline: 3/1. *Unit head:* Dr. William Glanz, Coordinator, 207-581-2545. *Application contact:* Scott G. Delcourt, Associate Dean of the Graduate School, 207-581-3219, Fax: 207-581-3232, E-mail: graduate@maine.edu.

University of Maine, Graduate School, College of Natural Sciences, Forestry, and Agriculture, Department of Plant, Soil, and Environmental Sciences, Orono, ME 04469. Offers biological sciences (PhD); ecology and environmental sciences (MS, PhD); forest resources (PhD); horticulture (MS); plant science (PhD); plant, soil, and environmental sciences (MS); resource utilization (MS). *Faculty:* 25. *Students:* 15 full-time (8 women), 8 part-time (6 women), 3 international. Average age 32. 6 applicants, 33% accepted, 1 enrolled. In 2005, 9 master's awarded. *Entrance requirements:* For master's and doctorate, GRE General Test. Additional exam requirements/recommendations for international students: Required—TOEFL. *Application deadline:* Applications are processed on a rolling basis. Application fee: $50. Electronic applications accepted. *Financial support:* In 2005–06, 9 research assistantships with tuition reimbursements (averaging $12,180 per year) were awarded; teaching assistantships, scholarships/grants, tuition waivers (full and partial), and unspecified assistantships also available. *Unit head:* Greg Porter, Chair, 207-581-2943, Fax: 207-581-3207. *Application contact:* Scott G. Delcourt, Associate Dean of the Graduate School, 207-581-3219, Fax: 207-581-3232, E-mail: graduate@maine.edu.

University of Maryland, College Park, Graduate Studies, College of Chemical and Life Sciences, Department of Biology, Behavior, Ecology, Evolution, and Systematics Program, College Park, MD 20742. Offers MS, PhD. *Students:* 38 full-time (26 women), 3 part-time (all women); includes 2 minority (both Asian Americans or Pacific Islanders), 3 international. 64 applicants, 9% accepted, 3 enrolled. *Degree requirements:* For master's, thesis, oral defense, seminar; for doctorate, thesis/dissertation, exam, 4 seminars. *Entrance requirements:* For master's and doctorate, GRE General Test, GRE Subject Test in biology, 3 letters of recommendation. Additional exam requirements/recommendations for international students: Required—TOEFL, TSE. *Application deadline:* For fall admission, 1/1 for domestic students, 1/1 for international students. Applications are processed on a rolling basis. Application fee: $60. Electronic applications accepted. *Financial support:* In 2005–06, 9 fellowships with tuition reimbursements (averaging $12,467 per year) were awarded; research assistantships with tuition reimbursements, teaching assistantships with tuition reimbursements, Federal Work-Study and scholarships/grants also available. Support available to part-time students. Financial award applicants required to submit FAFSA. *Faculty research:* Animal behavior, biostatistics, ecology, evolution, neurothology. *Unit head:* Dr. Michele Dudash, Director, 301-405-4552, Fax: 301-314-9358, E-mail: gw10@umail.umd.edu. *Application contact:* Dean of Graduate School, 301-405-4190, Fax: 301-314-9305.

University of Miami, Graduate School, College of Arts and Sciences, Program in Tropical Biology, Ecology, and Behavior, Coral Gables, FL 33124. Offers MS, PhD. Terminal master's awarded for partial completion of doctoral program. *Degree requirements:* For master's, thesis optional; for doctorate, thesis/dissertation, oral and written qualifying exam. *Entrance requirements:* For master's and doctorate, GRE General Test, GRE Subject Test. Additional exam requirements/recommendations for international students: Required—TOEFL. *Application deadline:* For fall admission, 2/1 for domestic students. Applications are processed on a rolling basis. Application fee: $50. *Financial support:* Fellowships, research assistantships, teaching assistantships, career-related internships or fieldwork and institutionally sponsored loans available. Financial award application deadline: 2/15. *Faculty research:* Behavioral ecology, plant-animal and plant-environmental interactions and coevolution, biogeography, conservation biology, genetics. *Unit head:* Dr. Leonel O. Sternberg, Director of Graduate Studies, 305-284-5116, Fax: 305-284-3039, E-mail: leo@bio.miami.edu.

University of Michigan, Horace H. Rackham School of Graduate Studies, College of Literature, Science, and the Arts, Department of Ecology and Evolutionary Biology, Ann Arbor, MI 48109. Offers MS, PhD. Part-time programs available. Terminal master's awarded for partial completion of doctoral program. *Degree requirements:* For master's, thesis (for some programs), registration; for doctorate, thesis/dissertation, two semesters of teaching, comprehensive exam, registration. *Entrance requirements:* For master's and doctorate, GRE. Additional exam requirements/recommendations for international students: Required—TOEFL (minimum score 560 paper-based; 220 computer-based). Electronic applications accepted. *Expenses:* Tuition, state resident: full-time $14,082; part-time $894 per credit hour. Tuition, nonresident: full-time $28,500; part-time $1,675 per credit hour. Required fees: $189; $189 per unit. *Faculty research:* Community ecology, molecular evolution, theoretical ecology, systematics, evolutionary genetics.

University of Minnesota, Twin Cities Campus, Graduate School, College of Agricultural, Food, and Environmental Sciences, Microbial Ecology Program, Minneapolis, MN 55455-0213. Offers MS, PhD. *Faculty:* 10 full-time (2 women). *Students:* 2 full-time (both women), 1 (woman) part-time, 2 international. Average age 30. In 2005, 1 master's, 1 doctorate awarded. Terminal master's awarded for partial completion of doctoral program. *Degree requirements:* For master's and doctorate, thesis/dissertation. *Application deadline:* Applications are processed on a rolling basis. Application fee: $0. *Expenses:* Tuition, state resident: full-time $8,748; part-time $729 per credit. Tuition, nonresident: full-time $15,848; part-time $1,321 per credit. Full-time tuition and fees vary according to class time, course load, program and reciprocity agreements. *Faculty research:* Arthropod-microbe interactions, plant-microbe interactions, resource competition in plants, systematics and ecology of mushrooms, physiology and ecology of oral bacteria, biodegradation. *Unit head:* Dr. Michael J. Sadowsky, Director of Graduate Studies, 612-624-6774, Fax: 612-625-1700, E-mail: sadowsky@umn.edu.

University of Minnesota, Twin Cities Campus, Graduate School, College of Biological Sciences, Department of Ecology, Evolution, and Behavior, Minneapolis, MN 55455-0213. Offers ecology, evolution, and behavior (MS, PhD). Terminal master's awarded for partial completion of doctoral program. *Degree requirements:* For master's, thesis or projects; for doctorate, thesis/dissertation, comprehensive exam. *Entrance requirements:* For master's and doctorate, GRE General Test, minimum GPA of 3.0. Additional exam requirements/recommendations for international students: Required—TOEFL (minimum score 550 paper-based; 213 computer-based), MELAB. Electronic applications accepted. *Expenses:* Tuition, state resident: full-time $8,748; part-time $729 per credit. Tuition, nonresident: full-time $15,848; part-time $1,321 per credit. Full-time tuition and fees vary according to class time, course load, program and reciprocity agreements. *Faculty research:* Behavioral ecology, community ecology, community genetics, ecosystem and global change, evolution and systematics.

University of Missouri–Columbia, Graduate School, College of Arts and Sciences, Division of Biological Sciences, Columbia, MO 65211. Offers cellular, molecular and developmental biology (MA, PhD); evolutionary biology and ecology (MA, PhD); neurobiology and behavior (MA, PhD). *Faculty:* 43 full-time (12 women). *Students:* 48 full-time (26 women), 28 part-time (10 women); includes 7 minority (2 African Americans, 3 Asian Americans or Pacific Islanders, 2 Hispanic Americans), 20 international. In 2005, 3 master's, 8 doctorates awarded. Terminal master's awarded for partial completion of doctoral program. *Degree requirements:* For master's, thesis/dissertation; for doctorate, thesis/dissertation, comprehensive exam. *Entrance requirements:* For master's and doctorate, GRE General Test, minimum GPA of 3.0. *Application deadline:* For fall admission, 1/15 for domestic students. Applications are processed on a rolling basis. Application fee: $45 ($60 for international students). *Financial support:* Fellowships, research assistantships, teaching assistantships, institutionally sponsored loans available. *Unit head:* Dr. Ray Semlitsch, Director of Graduate Studies, 573-884-6396, E-mail: semlitschr@missouri.edu. *Application contact:* Nila Emerich, Application Contact, 800-553-5698.

University of Missouri–St. Louis, College of Arts and Sciences, Department of Biology, St. Louis, MO 63121. Offers biology (MS, PhD), including animal behavior (MS); biochemistry (MS), biotechnology (MS), conservation biology (MS), development (MS), ecology (MS), environmental studies (PhD), evolution (MS), genetics (MS), molecular/cellular biology (MS), physiology (MS), plant systematics, population biology (MS), tropical biology (MS); biotechnology (Certificate); tropical biology and conservation (Certificate). Part-time programs available. *Faculty:* 50. *Students:* 26 full-time (15 women), 101 part-time (51 women); includes 13 minority (5 African Americans, 6 Asian Americans or Pacific Islanders, 2 Hispanic Americans), 41 international. Average age 32. In 2005, 22 master's, 2 doctorates awarded. *Degree requirements:* For master's, thesis or alternative; for doctorate, one foreign language, thesis/dissertation, 1 semester of teaching experience. *Entrance requirements:* For doctorate, GRE General Test. *Application deadline:* For spring admission, 12/1 priority date for domestic students. Applications are processed on a rolling basis. Application fee: $35 ($40 for international students). Electronic applications accepted. *Expenses:* Tuition, state resident: part-

time $263 per credit hour. Tuition, nonresident: part-time $680 per credit hour. Required fees: $53 per credit hour. Tuition and fees vary according to program. *Financial support:* In 2005–06, 11 fellowships with full tuition reimbursements (averaging $30,000 per year), 15 research assistantships with full and partial tuition reimbursements (averaging $16,000 per year), 22 teaching assistantships with full and partial tuition reimbursements (averaging $16,000 per year) were awarded; career-related internships or fieldwork and Federal Work-Study also available. Support available to part-time students. Financial award application deadline: 2/1. *Faculty research:* Molecular biology, microbial genetics. *Unit head:* Zuleyma @umsl.edu, Director of Graduate Studies, 314-516-6498, Fax: 314-516-6233, E-mail: zuleyma@umsl.edu. *Application contact:* 314-516-5458, Fax: 314-516-5310, E-mail: gradadm@umsl.edu.

The University of Montana, Graduate School, College of Arts and Sciences, Division of Biological Sciences, Program in Organismal Biology and Ecology, Missoula, MT 59812-0002. Offers MS, PhD. Terminal master's awarded for partial completion of doctoral program. *Degree requirements:* For master's, one foreign language, thesis; for doctorate, 2 foreign languages, thesis/dissertation. *Entrance requirements:* For master's and doctorate, GRE General Test. *Application deadline:* For fall admission, 2/1 for domestic students. Application fee: $45. *Expenses:* Tuition, state resident: part-time $267 per credit. Tuition, nonresident: part-time $665 per credit. Part-time tuition and fees vary according to course load and degree level. *Financial support:* In 2005–06, research assistantships with full tuition reimbursements (averaging $9,400 per year), teaching assistantships with full tuition reimbursements (averaging $9,400 per year) were awarded; Federal Work-Study also available. Financial award application deadline: 3/1; financial award applicants required to submit FAFSA. *Faculty research:* Conservation biology, ecology and behavior, evolutionary genetics, avian biology. *Application contact:* Janean Clark, Graduate Programs Secretary, 406-243-5222, Fax: 406-243-4184, E-mail: jmclark@selway.umt.edu.

University of Nevada, Reno, Graduate School, College of Science, Interdisciplinary Program in Ecology, Evolution, and Conservation Biology, Reno, NV 89557. Offers PhD. Offered through the College of Arts and Science, the M. C. Fleischmann College of Agriculture, and the Desert Research Institute. *Faculty:* 5. *Students:* 11 full-time (9 women), 24 part-time (12 women); includes 3 minority (1 Asian American or Pacific Islander, 2 Hispanic Americans), 2 international. Average age 33. 6 applicants, 67% accepted, 3 enrolled. In 2005, 4 degrees awarded. *Degree requirements:* For doctorate, thesis/dissertation. *Entrance requirements:* For doctorate, GRE General Test, GRE Subject Test, minimum GPA of 3.0. Additional exam requirements/recommendations for international students: Required—TOEFL. *Application deadline:* For fall admission, 2/15 for domestic students. Application fee: $60 ($95 for international students). *Expenses:* Tuition, area resident: full-time $2,767; part-time $923 per semester. Tuition, state resident: full-time $5,733; part-time $1,911 per semester. Tuition, nonresident: full-time $12,679; part-time $1,911 per semester. International tuition: $13,878 full-time. Required fees: $404; $202 per term. One-time fee: $90. *Financial support:* In 2005–06, 1 research assistantship was awarded; teaching assistantships Financial award application deadline: 3/1. *Faculty research:* Population biology, behavioral ecology, plant response to climate change, conservation of endangered species, restoration of natural ecosystems. *Unit head:* Dr. Peter Brussard, Director, 775-784-1360.

The University of North Carolina at Chapel Hill, Graduate School, College of Arts and Sciences, Curriculum in Ecology, Chapel Hill, NC 27599. Offers MA, MS, PhD. *Degree requirements:* For master's, thesis (for some programs), oral defense of thesis, comprehensive exam, registration; for doctorate, thesis/dissertation, oral exams, oral defense of dissertation, comprehensive exam, registration. *Entrance requirements:* For master's and doctorate, GRE General Test. Additional exam requirements/recommendations for international students: Required—TOEFL (minimum score 550 paper-based; 213 computer-based). Electronic applications accepted. *Faculty research:* Community and population ecology and ecosystems, human ecology, landscape ecology, conservation ecology, marine ecology.

The University of North Carolina at Chapel Hill, Graduate School, College of Arts and Sciences, Department of Biology, Chapel Hill, NC 27599. Offers botany (MA, MS, PhD); cell biology, development, and physiology (MA, MS, PhD); cell motility and cytoskeleton (PhD); ecology and behavior (MA, MS, PhD); genetics and molecular biology (MA, MS, PhD); morphology, systematics, and evolution (MA, MS, PhD). Terminal master's awarded for partial completion of doctoral program. *Degree requirements:* For master's, thesis (for some programs), comprehensive exam; for doctorate, thesis/dissertation, comprehensive exam. *Entrance requirements:* For master's, GRE General Test, GRE Subject Test, 2 semesters of calculus or statistics, 2 semesters of physics, organic chemistry, 3 semesters of biology; for doctorate, GRE General Test, GRE Subject Test, 2 semesters calculus or statistics, 2 semesters physics, organic chemistry, 3 semesters of biology. Additional exam requirements/recommendations for international students: Required—TOEFL (minimum score 550 paper-based; 213 computer-based). Electronic applications accepted. *Faculty research:* Gene expression, biomechanics, yeast genetics, plant ecology, plant molecular biology.

See Close-Up on page 277.

University of North Dakota, Graduate School, College of Arts and Sciences, Department of Biology, Grand Forks, ND 58202. Offers botany (MS, PhD); ecology (MS, PhD); entomology (MS, PhD); environmental biology (MS, PhD); fisheries/wildlife (MS, PhD); genetics (MS, PhD); zoology (MS, PhD). *Faculty:* 16 full-time (3 women). *Students:* 15 applicants, 13% accepted, 2 enrolled. In 2005, 5 degrees awarded. Terminal master's awarded for partial completion of doctoral program. *Degree requirements:* For master's, thesis/dissertation, final exam; for doctorate, thesis/dissertation, final exam, comprehensive exam. *Entrance requirements:* For master's, GRE General Test, GRE Subject Test, minimum GPA of 3.0; for doctorate, GRE General Test, GRE Subject Test, minimum GPA of 3.5. Additional exam requirements/recommendations for international students: Required—TOEFL (minimum score 550 paper-based; 213 computer-based). *Application deadline:* For fall admission, 10/1 for domestic students, 10/1 for international students. Application fee: $35. Electronic applications accepted. *Financial support:* In 2005–06, 8 research assistantships with full tuition reimbursements (averaging $11,375 per year), 13 teaching assistantships with full tuition reimbursements (averaging $10,813 per year) were awarded; fellowships, Federal Work-Study, institutionally sponsored loans, scholarships/grants, and tuition waivers (full and partial) also available. Support available to part-time students. Financial award application deadline: 3/15; financial award applicants required to submit FAFSA. *Faculty research:* Population biology, wildlife ecology, RNA processing, hormonal control of behavior. *Unit head:* Dr. Richard Sweitzel, Graduate Director, 701-777-4676, Fax: 701-777-2623, E-mail: richard_sweitzel@und.nodak.edu.

University of Notre Dame, Graduate School, College of Science, Department of Biological Sciences, Notre Dame, IN 46556. Offers aquatic ecology, evolution and environmental biology (MS, PhD); cellular and molecular biology (MS, PhD); genetics (MS, PhD); physiology (MS, PhD); vector biology and parasitology (MS, PhD). *Faculty:* 34 full-time (8 women), 3 part-time/adjunct (0 women). *Students:* 124 full-time (55 women); includes 11 minority (1 African American, 6 Asian Americans or Pacific Islanders, 4 Hispanic Americans), 39 international. 95 applicants, 34% accepted, 22 enrolled. In 2005, 4 master's, 11 doctorates awarded. Terminal master's awarded for partial completion of doctoral program. *Median time to degree:* Of those who began their doctoral program in fall 1997, 61% received their degree in 8 years or less. *Degree requirements:* For master's and doctorate, thesis/dissertation, comprehensive exam. *Entrance requirements:* For master's and doctorate, GRE General Test. Additional exam requirements/recommendations for international students: Required—TOEFL. *Application deadline:* For fall admission, 2/1 for domestic students; for spring admission, 11/1 for domestic students. Applications are processed on a rolling basis. Application fee: $50. Electronic applications accepted. *Financial support:* In 2005–06, 124 students received support, including 24 fellowships with full tuition reimbursements available (averaging $22,000 per year), 47 research assistantships with full tuition reimbursements available (averaging $15,250 per year), 45 teaching assistantships with full tuition reimbursements available (averaging $16,000 per

Ecology

University of Notre Dame (continued)
year); traineeships and tuition waivers (full) also available. Financial award application deadline: 2/1. *Faculty research:* Tropical disease, molecular genetics, neurobiology, evolutionary biology, aquatic biology. Total annual research expenditures: $15.3 million. *Unit head:* Dr. Gary A. Lamberti, Director of Graduate Studies, 574-631-6552, Fax: 574-631-7413, E-mail: biology. biosadm.1@nd.edu. *Application contact:* Dr. Terrence J. Akai, Director of Graduate Admissions, 574-631-7706, Fax: 574-631-4183, E-mail: gradad@nd.edu.

See Close-Up on page 281.

University of Oregon, Graduate School, College of Arts and Sciences, Department of Biology, Eugene, OR 97403. Offers ecology and evolution (MA, MS, PhD); marine biology (MA, MS, PhD); molecular, cellular and genetic biology (PhD); neuroscience and development (PhD). *Faculty:* 42 full-time (14 women), 5 part-time/adjunct (2 women). *Students:* 92 full-time (43 women), 2 part-time (both women); includes 9 minority (2 African Americans, 1 American Indian/Alaska Native, 4 Asian Americans or Pacific Islanders, 2 Hispanic Americans), 6 international. 18 applicants, 83% accepted. In 2005, 11 master's, 7 doctorates awarded. Terminal master's awarded for partial completion of doctoral program. *Degree requirements:* For master's, thesis (for some programs); for doctorate, thesis/dissertation. *Entrance requirements:* For master's and doctorate, GRE General Test, minimum GPA of 3.2. Additional exam requirements/recommendations for international students: Required—TOEFL. *Application deadline:* For fall admission, 12/15 for domestic students. *Financial support:* In 2005–06, 36 teaching assistantships were awarded; research assistantships, Federal Work-Study, institutionally sponsored loans, and scholarships/grants also available. Financial award application deadline: 2/1. *Faculty research:* Developmental neurobiology; evolution, population biology, and quantitative genetics; regulation of gene expression; biochemistry of marine organisms. *Unit head:* George Sprague, Head, 541-346-6051, Fax: 541-346-6056. *Application contact:* Lynne Romans, Admissions Contact, 541-346-4252, Fax: 541-346-6056, E-mail: lromans@uoregon.edu.

University of Pennsylvania, School of Arts and Sciences, Program in Ecology, Evolution and Biodiversity, Philadelphia, PA 19104. Offers PhD. *Degree requirements:* For doctorate, thesis/dissertation. *Entrance requirements:* For doctorate, GRE General Test, GRE Subject Test. Additional exam requirements/recommendations for international students: Required—TOEFL. Electronic applications accepted.

University of Pittsburgh, School of Arts and Sciences, Department of Biological Sciences, Program in Ecology and Evolution, Pittsburgh, PA 15260. Offers MS, PhD. *Faculty:* 7 full-time (2 women), 3 part-time/adjunct (1 woman). *Students:* 13 full-time (4 women); includes 2 minority (both Hispanic Americans) Average age 23. 34 applicants, 0% accepted, 0 enrolled. In 2005, 1 master's, 2 doctorates awarded. *Median time to degree:* Of those who began their doctoral program in fall 1997, 100% received their degree in 8 years or less. *Degree requirements:* For master's and doctorate, thesis/dissertation, comprehensive exam, registration. *Entrance requirements:* For master's and doctorate, GRE General Test, GRE Subject Test. Additional exam requirements/recommendations for international students: Required—TOEFL (minimum score 550 paper-based; 213 computer-based). *Application deadline:* For fall admission, 1/15 priority date for domestic students, 12/15 priority date for international students. Applications are processed on a rolling basis. Application fee: $0 ($40 for international students). Electronic applications accepted. *Expenses:* Tuition, state resident: full-time $13,194; part-time $537 per credit. Tuition, nonresident: full-time $25,012; part-time $1,026 per credit. Required fees: $700; $164 per term. Tuition and fees vary according to campus/location and program. *Financial support:* In 2005–06, 18 fellowships with full tuition reimbursements (averaging $2,032 per year), 4 research assistantships with full tuition reimbursements (averaging $20,332 per year) were awarded; Federal Work-Study, scholarships/grants, traineeships, health care benefits, and tuition waivers (full) also available. *Faculty research:* Ecological and population genetics, tropical and community ecology, evolutionary ecology, phylogeny of birds, evolution of dispersal and dormancy. *Application contact:* Cathleen M. Barr, Graduate Administrator, 412-624-4268, Fax: 412-624-4759, E-mail: cbarr@pitt.edu.

See Close-Up on page 711.

University of St. Michael's College, Faculty of Theology, Toronto, ON M5S 1J4, Canada. Offers catholic leadership (MA); eastern Christian studies (Certificate, Diploma); religious education (Diploma); theological studies (Diploma); theology (M Div, MA, MRE, MTS, D Min, PhD, Th D); theology and ecology (Certificate); theology and Jewish studies (MA). *Accreditation:* ATS (one or more programs are accredited). Part-time programs available. *Faculty:* 9 full-time (3 women), 15 part-time/adjunct (7 women). *Students:* 106 full-time (37 women), 102 part-time (59 women); includes 24 minority (7 African Americans, 16 Asian Americans or Pacific Islanders, 1 Hispanic American), 24 international. Average age 40. 49 applicants, 84% accepted, 32 enrolled. In 2005, 4 first professional degrees, 11 master's, 8 doctorates, 15 other advanced degrees awarded. *Degree requirements:* For M Div and other advanced degree, thesis optional; for master's, thesis (for some programs), 1 foreign language (MA), 2 foreign languages (Th M); for doctorate, 3 foreign languages, thesis/dissertation, comprehensive exam, registration. *Entrance requirements:* For M Div and other advanced degree, minimum GPA of 2.7; for master's, M Div or BA, course work in an ancient or modern language, minimum GPA of 3.3; for doctorate, MA in theology, Th M, or M Div with thesis, minimum GPA of 3.7. Additional exam requirements/recommendations for international students: Required—TOEFL (minimum score 600 paper-based; 250 computer-based). *Application deadline:* For fall admission, 1/15 for domestic students, 1/15 for international students. Applications are processed on a rolling basis. Application fee: $25 Canadian dollars. Electronic applications accepted. *Expenses:* Tuition: Part-time $1,838 per semester. Required fees: $573 per semester. *Financial support:* In 2005–06, 49 students received support, including fellowships with partial tuition reimbursements available (averaging $2,500 per year), research assistantships with partial tuition reimbursements available (averaging $2,500 per year), 17 teaching assistantships with partial tuition reimbursements available (averaging $2,400 per year); scholarships/grants, tuition waivers (partial), and bursaries also available. Financial award application deadline: 3/1. *Faculty research:* Patristics, eastern Christianity, ecology and theology, ecumenism, Jewish Christian studies. *Unit head:* Dr. Anne Anderson, CSJ, Dean, 416-926-7265, Fax: 416-926-7294, E-mail: anne.anderson@utoronto.ca. *Application contact:* Student Services Officer, 416-926-7140, Fax: 416-926-7294, E-mail: usmetheology.registrar@utoronto.ca.

University of South Carolina, The Graduate School, College of Science and Mathematics, Department of Biological Sciences, Graduate Training Program in Ecology, Evolution, and Organismal Biology, Columbia, SC 29208. Offers MS, PhD. *Degree requirements:* For master's, one foreign language; for doctorate, one foreign language, thesis/dissertation. *Entrance requirements:* For master's and doctorate, GRE General Test, minimum GPA of 3.0 in science. Electronic applications accepted.

See Close-Up on page 713.

University of South Florida, College of Graduate Studies, College of Arts and Sciences, Department of Biology, Tampa, FL 33620-9951. Offers biology (PhD); botany (MS); ecology (PhD); microbiology (MS); physiology (MS); zoology (MS). Part-time programs available. *Faculty:* 21. *Students:* 54 full-time (32 women), 26 part-time (17 women); includes 8 minority (2 African Americans, 1 American Indian/Alaska Native, 1 Asian American or Pacific Islander, 4 Hispanic Americans), 14 international. 76 applicants, 34% accepted, 13 enrolled. In 2005, 3 master's, 3 doctorates awarded. *Degree requirements:* For master's, thesis (for some programs), graduate seminar in biology; for doctorate, 2 foreign languages, thesis/dissertation, essay of research interest, comprehensive exam. *Entrance requirements:* For master's, GRE General Test, minimum undergraduate GPA of 3.0 in last 60 hours of course work; for doctorate, GRE General Test, GRE Subject Test in biology, minimum undergraduate GPA of 3.0 in last 60 hours of course work. Additional exam requirements/recommendations for international students: Required—TOEFL (minimum score 570 paper-based), TSE (minimum score 50). *Application*

deadline: For fall admission, 2/1 priority date for domestic students, 3/1 priority date for international students; for spring admission, 10/1 for domestic students, 8/1 for international students. Application fee: $30. Electronic applications accepted. *Financial support:* Fellowships with full tuition reimbursements, research assistantships with full tuition reimbursements, teaching assistantships with full tuition reimbursements, Federal Work-Study and unspecified assistantships available. Financial award application deadline: 6/30. *Unit head:* Sydney Pierce, Chairperson, 813-974-3250, Fax: 813-974-3263. *Application contact:* Christine Smith, Graduate Advisor, 813-974-4747, Fax: 813-974-3263, E-mail: csmith2@chuma1.cas.usf.edu.

The University of Tennessee, Graduate School, College of Arts and Sciences, Department of Ecology and Evolutionary Biology, Knoxville, TN 37996. Offers behavior (MS, PhD); ecology (MS, PhD); evolutionary biology (MS, PhD). Part-time programs available. *Degree requirements:* For master's and doctorate, thesis/dissertation. *Entrance requirements:* For master's and doctorate, GRE General Test, minimum GPA of 2.7. Additional exam requirements/recommendations for international students: Required—TOEFL. Electronic applications accepted.

The University of Tennessee, Graduate School, College of Arts and Sciences, Department of Mathematics, Knoxville, TN 37996. Offers applied mathematics (MS); mathematical ecology (PhD); mathematics (M Math, MS, PhD). Part-time programs available. *Degree requirements:* For master's, thesis or alternative; for doctorate, one foreign language, thesis/dissertation. *Entrance requirements:* For master's and doctorate, minimum GPA of 2.7. Additional exam requirements/recommendations for international students: Required—TOEFL. Electronic applications accepted.

The University of Texas at Austin, Graduate School, College of Natural Sciences, School of Biological Sciences, Program in Ecology, Evolution and Behavior, Austin, TX 78712-1111. Offers PhD. *Entrance requirements:* For doctorate, GRE General Test. Additional exam requirements/recommendations for international students: Required—TOEFL. Electronic applications accepted.

The University of Toledo, Graduate School, College of Arts and Sciences, Department of Earth, Ecological and Environmental Sciences, Toledo, OH 43606-3390. Offers biology (ecology track) (MS, PhD); geology (MS), including earth surface processes, general geology. Part-time programs available. *Faculty:* 28. *Students:* 8 full-time (6 women), 2 part-time (both women). Average age 31. 6 applicants, 83% accepted, 3 enrolled. In 2005, 2 degrees awarded. *Degree requirements:* For master's, thesis. *Entrance requirements:* For master's, GRE General Test. Additional exam requirements/recommendations for international students: Required—TOEFL. *Application deadline:* For fall admission, 8/1 for domestic students. Applications are processed on a rolling basis. Application fee: $45. Electronic applications accepted. *Expenses:* Tuition, area resident: Part-time $308 per credit hour. Tuition, state resident: full-time $3,312. Tuition, nonresident: full-time $6,616; part-time $735 per credit hour. *Financial support:* In 2005–06, 2 research assistantships (averaging $14,000 per year), 30 teaching assistantships (averaging $11,724 per year) were awarded; Federal Work-Study, institutionally sponsored loans, and tuition waivers (full) also available. Support available to part-time students. Financial award application deadline: 4/1; financial award applicants required to submit FAFSA. *Faculty research:* Environmental geochemistry, geophysics, petrology and mineralogy, paleontology, geohydrology. *Unit head:* Dr. Michael Phillips, Chair, 419-530-4572, Fax: 419-530-4421, E-mail: michael.phillips@utoledo.edu. *Application contact:* Johan Gottgens, 419-530-8451, E-mail: john.gottgens@utoledo.edu.

See Close-Up on page 715.

University of Utah, The Graduate School, College of Science, Department of Biology, Salt Lake City, UT 84112-1107. Offers biology (M Phil); ecology and evolutionary biology (MS, PhD); genetics (MS, PhD); molecular biology (PhD). Part-time programs available. *Faculty:* 42 full-time (8 women), 1 part-time/adjunct (0 women). *Students:* 55 full-time (30 women), 15 part-time (2 women); includes 4 minority (3 Asian Americans or Pacific Islanders, 1 Hispanic American), 19 international. Average age 29. 26 applicants, 54% accepted, 11 enrolled. In 2005, 3 master's, 6 doctorates awarded. Terminal master's awarded for partial completion of doctoral program. *Median time to degree:* Of those who began their doctoral program in fall 1997, 20% received their degree in 8 years or less. *Degree requirements:* For master's and doctorate, thesis/dissertation. *Entrance requirements:* For master's and doctorate, GRE General Test, minimum GPA of 3.0. Additional exam requirements/recommendations for international students: Required—TOEFL (minimum score 500 paper-based; 173 computer-based). *Application deadline:* For fall admission, 1/13 for domestic students, 1/13 for international students. Application fee: $45 ($65 for international students). *Expenses:* Tuition, state resident: full-time $2,932; part-time $369 per credit. Tuition, nonresident: full-time $10,350; part-time $1,302 per credit. Required fees: $516 per term. Tuition and fees vary according to course load and program. *Financial support:* In 2005–06, 2 fellowships with full tuition reimbursements (averaging $30,000 per year), 33 research assistantships with full tuition reimbursements (averaging $22,000 per year), 33 teaching assistantships with full tuition reimbursements (averaging $15,000 per year) were awarded; career-related internships or fieldwork, scholarships/grants, traineeships, and health care benefits also available. Financial award application deadline: 2/15; financial award applicants required to submit FAFSA. *Faculty research:* Behavioral ecology, cellular neurobiology, DNA replication, ecological genetics, herpetology. Total annual research expenditures: $11.4 million. *Unit head:* David R. Wolstenholme, Chair, 801-581-6517, Fax: 801-581-4668, E-mail: wolstenholme@biology.utah.edu. *Application contact:* Shannon Nielsen, Administrative Program Coordinator, 801-581-5636, Fax: 801-581-4668, E-mail: shannon.nielsen@bioscience.utah.edu.

See Close-Up on page 313.

University of Wisconsin–Madison, Graduate School, College of Agricultural and Life Sciences, Department of Wildlife Ecology, Madison, WI 53706-1380. Offers MS, PhD. *Faculty:* 8 full-time (1 woman). *Students:* 16 full-time (6 women), 1 international. 27 applicants, 11% accepted, 3 enrolled. In 2005, 2 master's, 1 doctorate awarded. *Degree requirements:* For master's and doctorate, GRE General Test. *Application deadline:* For fall admission, 1/15 for domestic students; for spring admission, 10/15 priority date for domestic students. Application fee: $90. Electronic applications accepted. *Financial support:* In 2005–06, 1 fellowship with tuition reimbursement (averaging $17,772 per year), 12 research assistantships with tuition reimbursements (averaging $17,772 per year), 3 teaching assistantships with tuition reimbursements (averaging $22,575 per year) were awarded; career-related internships or fieldwork, Federal Work-Study, and institutionally sponsored loans also available. *Faculty research:* Agroecology, ecosystem management, physiological ecology, waterfowl ecology, endangered species. Total annual research expenditures: $1.5 million. *Unit head:* Scott R. Craven, Chair, 608-263-6325, Fax: 608-262-6099, E-mail: srcraven@facstaff.wisc.edu.

Utah State University, School of Graduate Studies, College of Natural Resources, Department of Aquatic, Watershed, and Earth Resources, Logan, UT 84322. Offers ecology (MS, PhD); fisheries biology (MS, PhD); watershed science (MS, PhD). *Faculty:* 12 full-time (3 women), 2 part-time/adjunct (1 woman). *Students:* 118 full-time (42 women), 16 part-time (2 women); includes 4 minority (all Hispanic Americans), 14 international. Average age 31. 31 applicants, 71% accepted, 21 enrolled. In 2005, 8 master's, 4 doctorates awarded. *Degree requirements:* For master's, thesis (for some programs); for doctorate, thesis/dissertation. *Entrance requirements:* For master's and doctorate, GRE General Test, minimum GPA of 3.2. Additional exam requirements/recommendations for international students: Required—TOEFL. *Application deadline:* For fall admission, 2/15 for domestic students; for spring admission, 10/15 for domestic students. Applications are processed on a rolling basis. Application fee: $50 ($60 for international students). Electronic applications accepted. *Financial support:* In 2005–06, 3 fellowships with partial tuition reimbursements (averaging $20,000 per year), 34 research assistantships with partial tuition reimbursements (averaging $14,864 per year) were awarded; teaching assistantships with partial tuition reimbursements, career-related internships or

Environmental Biology

fieldwork, Federal Work-Study, and institutionally sponsored loans also available. Support available to part-time students. Financial award application deadline: 2/15. *Faculty research:* Behavior, population ecology, habitat, conservation biology, restoration, aquatic ecology, fisheries management, fluvial geomorphology, remote sensing, conservation biology. Total annual research expenditures: $4.7 million. *Unit head:* Chris Luecke, Head, 435-797-2463, Fax: 435-797-1871, E-mail: awerinfo@cc.usu.edu. *Application contact:* Brian Bailey, Staff Assistant, 435-797-2459, Fax: 435-797-1871, E-mail: awerinfo@cc.usu.edu.

Utah State University, School of Graduate Studies, College of Natural Resources, Department of Environment and Society, Logan, UT 84322. Offers bioregional planning (MS); geography (MA, MS); human dimensions of ecosystem science and management (MS, PhD); recreation resource management (MS, PhD). *Faculty:* 16 full-time (3 women), 6 part-time/adjunct (2 women). *Students:* 53 full-time (15 women), 53 part-time (28 women), 6 international. Average age 32. 18 applicants, 67% accepted, 10 enrolled. In 2005, 11 degrees awarded. *Degree requirements:* For master's, thesis (for some programs), comprehensive exam. *Entrance requirements:* For master's and doctorate, GRE General Test, minimum GPA of 3.0. Additional exam requirements/recommendations for international students: Required—TOEFL. *Application deadline:* For fall admission, 6/15 for domestic students; for spring admission, 10/15 for domestic students. Applications are processed on a rolling basis. Application fee: $50 ($60 for international students). Electronic applications accepted. *Financial support:* In 2005–06, 14 students received support, including 21 research assistantships with partial tuition reimbursements available (averaging $11,000 per year), 5 teaching assistantships with partial tuition reimbursements available (averaging $10,000 per year); fellowships with partial tuition reimbursements available, career-related internships or fieldwork, Federal Work-Study, tuition waivers (full and partial), and unspecified assistantships also available. Financial award applicants required to submit FAFSA. *Faculty research:* Geographic information systems/geographic and environmental education, bioregional planning, natural resource and environmental policy, outdoor recreation and tourism, natural resource and environmental management. Total annual research expenditures: $1.4 million. *Unit head:* Dr. Terry L. Sharik, Head, 435-797-3270, Fax: 435-797-4048, E-mail: tlsharik@cc.usu.edu. *Application contact:* Dr. Richard E. Toth, Information Contact, 435-797-1790, Fax: 435-797-4048, E-mail: envs.info@cnr.usu.edu.

Utah State University, School of Graduate Studies, College of Natural Resources, Department of Forest, Range, and Wildlife Sciences, Logan, UT 84322. Offers ecology (MS, PhD); forestry (MS, PhD); range science (MS, PhD); wildlife biology (MS, PhD). Part-time programs available. *Faculty:* 22 full-time (4 women), 17 part-time/adjunct (3 women). *Students:* 220 full-time (90 women), 52 part-time (17 women); includes 6 minority (all Asian Americans or Pacific Islanders), 49 international. Average age 24. 56 applicants, 61% accepted, 28 enrolled. In 2005, 11 master's, 4 doctorates awarded. *Degree requirements:* For master's, thesis/dissertation; for doctorate, thesis/dissertation, comprehensive exam. *Entrance requirements:* For master's and doctorate, GRE General Test, minimum GPA of 3.0. Additional exam requirements/recommendations for international students: Required—TOEFL. *Application deadline:* For fall admission, 6/15 for domestic students; for spring admission, 10/15 for domestic students. Applications are processed on a rolling basis. Application fee: $50 ($60 for international students). *Financial support:* In 2005–06, 14 research assistantships with partial tuition reimbursements (averaging $13,600 per year), 6 teaching assistantships (averaging $6,000 per year) were awarded; fellowships, career-related internships or fieldwork, Federal Work-Study, and institutionally sponsored loans also available. *Faculty research:* Range plant ecophysiology, plant community ecology, ruminant nutrition, population ecology. Total annual research expenditures: $3.5 million. *Unit head:* Dr. Johan W. duToit, Head, 435-797-2837, Fax: 435-797-3796, E-mail: ggriffeth@cnr.usu.edu. *Application contact:* Gaye Griffeth, Staff Assistant, 435-797-2503, Fax: 435-797-3796, E-mail: ggriffeth@cnr.usu.edu.

Utah State University, School of Graduate Studies, College of Science, Department of Biology, Logan, UT 84322. Offers biology (MS, PhD); ecology (MS, PhD). Part-time programs available. *Faculty:* 39 full-time (7 women). *Students:* 139 full-time (73 women), 16 part-time (9 women); includes 4 minority (2 African Americans, 1 Asian American or Pacific Islander, 1 Hispanic American), 42 international. Average age 26. 44 applicants, 61% accepted, 21 enrolled. In 2005, 8 master's, 3 doctorates awarded. *Degree requirements:* For master's and doctorate, thesis/dissertation. *Entrance requirements:* For master's and doctorate, GRE General Test, minimum GPA of 3.0. Additional exam requirements/recommendations for inter-

national students: Required—TOEFL (minimum score 575 paper-based). *Application deadline:* For fall admission, 6/15 for domestic students; for spring admission, 10/15 for domestic students. Applications are processed on a rolling basis. Application fee: $50 ($60 for international students). *Financial support:* In 2005–06, 3 fellowships with partial tuition reimbursements (averaging $20,000 per year), research assistantships with partial tuition reimbursements (averaging $16,000 per year), 44 teaching assistantships with partial tuition reimbursements (averaging $11,275 per year) were awarded; career-related internships or fieldwork, Federal Work-Study, and institutionally sponsored loans also available. Support available to part-time students. Financial award application deadline: 2/15. *Faculty research:* Plant, insect, microbial, and animal biology. *Unit head:* Dr. Jon Y. Takemoto, Head, 435-797-1909, E-mail: jon@biology.usu.edu. *Application contact:* Nancy Kay Harrison, Coordinator of Graduate Studies, 435-797-1770, Fax: 435-797-1575, E-mail: nancykay@biology.usu.edu.

Virginia Polytechnic Institute and State University, Graduate School, College of Science, Department of biological Sciences, Blacksburg, VA 24061. Offers botany (MS, PhD); ecology and evolutionary biology (MS, PhD); genetics and developmental biology (MS, PhD); microbiology (MS, PhD); zoology (MS, PhD). *Faculty:* 38 full-time (9 women). *Students:* 70 full-time (30 women), 4 part-time (1 woman); includes 6 minority (2 African Americans, 1 American Indian/Alaska Native, 2 Asian Americans or Pacific Islanders, 1 Hispanic American), 12 international. Average age 27. 81 applicants, 22% accepted, 14 enrolled. In 2005, 9 master's, 8 doctorates awarded. *Entrance requirements:* For master's and doctorate, GRE General Test. Additional exam requirements/recommendations for international students: Required—TOEFL (minimum score 550 paper-based; 213 computer-based). *Application deadline:* Applications are processed on a rolling basis. Application fee: $45. Electronic applications accepted. *Expenses:* Tuition, state resident: full-time $6,558; part-time $364 per credit. Tuition, nonresident: full-time $11,296; part-time $628 per credit. Required fees: $1,419; $468 per credit. $234 per term. *Financial support:* In 2005–06, 28 research assistantships with full tuition reimbursements (averaging $16,689 per year), 37 teaching assistantships with full tuition reimbursements (averaging $13,902 per year) were awarded; career-related internships or fieldwork, Federal Work-Study, scholarships/grants, and unspecified assistantships also available. *Faculty research:* Freshwater ecology, cell cycle regulation, behavioral ecology, motor proteins. *Unit head:* Dr. Bob Jones, Chairman, 540-231-9514, Fax: 540-231-9307, E-mail: rhjones@vt.edu. *Application contact:* Sue Rasmussen, Graduate Secretary, 540-231-8929, Fax: 540-231-9307, E-mail: sueras@vt.edu.

Washington University in St. Louis, Graduate School of Arts and Sciences, Division of Biology and Biomedical Sciences, Program in Evolution, Ecology and Population Biology, St. Louis, MO 63130-4899. Offers ecology (PhD); environmental biology (PhD); evolutionary biology (PhD); genetics (PhD). *Degree requirements:* For doctorate, thesis/dissertation. *Entrance requirements:* For doctorate, GRE General Test, GRE Subject Test. Electronic applications accepted.

William Paterson University of New Jersey, College of Science and Health, Department of Biology, General Biology Program, Wayne, NJ 07470-8420. Offers general biology (MA); limnology and terrestrial ecology (MA); molecular biology (MA); physiology (MA). Part-time and evening/weekend programs available. *Students:* 2 full-time (1 woman), 13 part-time (9 women); includes 4 Hispanic Americans. In 2005, 4 degrees awarded. *Degree requirements:* For master's, independent study or thesis. *Entrance requirements:* For master's, GRE General Test, minimum GPA of 2.75. *Application deadline:* Applications are processed on a rolling basis. Application fee: $50. Electronic applications accepted. *Expenses:* Tuition, state resident: full-time $476. Tuition, nonresident: full-time $717. *Financial support:* Research assistantships, career-related internships or fieldwork and unspecified assistantships available. Financial award application deadline: 4/1; financial award applicants required to submit FAFSA. *Application contact:* Danielle Liautaud, Assistant Director, 973-720-3579, Fax: 973-720-2035, E-mail: liautaudd@wpunj.edu.

See Close-Up on page 331.

Yale University, Graduate School of Arts and Sciences, Department of Ecology and Evolutionary Biology, New Haven, CT 06520. Offers PhD. *Entrance requirements:* For doctorate, GRE General Test, GRE Subject Test (biology).

See Close-Up on page 717.

Environmental Biology

Antioch New England Graduate School, Graduate School, Department of Environmental Studies, Program in Environmental Studies, Keene, NH 03431-3552. Offers conservation biology (MS); environmental advocacy (MS); environmental education (MS); teacher certification in biology (7th-12th grade) (MS); teacher certification in general science (5th-9th grade) (MS). *Degree requirements:* For master's, practicum. *Entrance requirements:* For master's, previous undergraduate course work in biology, chemistry, mathematics (environmental biology). Additional exam requirements/recommendations for international students: Required—TOEFL (minimum score 550 paper-based; 213 computer-based). Electronic applications accepted. Expenses: Contact institution. *Faculty research:* Sustainability, natural resources inventory.

Baylor University, Graduate School, College of Arts and Sciences, Department of Biology, Waco, TX 76798. Offers biology (MA, MS, PhD); environmental biology (MS); limnology (MSL). Part-time programs available. *Faculty:* 13 full-time (3 women). *Students:* 21 full-time (6 women), 16 part-time (8 women); includes 2 minority (1 African American, 1 Hispanic American), 7 international. In 2005, 4 master's, 2 doctorates awarded. *Degree requirements:* For master's, thesis (for some programs); for doctorate, thesis/dissertation. *Entrance requirements:* For master's and doctorate, GRE General Test. *Application deadline:* For fall admission, 1/31 for domestic students. Applications are processed on a rolling basis. Application fee: $25. *Financial support:* Teaching assistantships, career-related internships or fieldwork, Federal Work-Study, institutionally sponsored loans, and tuition waivers (full and partial) available. Support available to part-time students. Financial award application deadline: 2/28. *Faculty research:* Terrestrial ecology, aquatic ecology, genetics. *Unit head:* Dr. Ken Wilkins, Graduate Program Director, 254-710-2911, Fax: 254-710-2969, E-mail: ken_wilkins@baylor.edu. *Application contact:* Sandy Tighe, Administrative Assistant, 254-710-2911, Fax: 254-710-2969, E-mail: sandy_tighe@baylor.edu.

Emporia State University, School of Graduate Studies, College of Liberal Arts and Sciences, Department of Biological Sciences, Emporia, KS 66801-5087. Offers botany (MS); environmental biology (MS); general biology (MS); microbial and cellular biology (MS); zoology (MS). Part-time programs available. *Faculty:* 13 full-time (2 women), 4 part-time/adjunct (2 women). *Students:* 3 full-time (1 woman), 17 part-time (7 women), 3 international. 10 applicants, 90% accepted, 7 enrolled. In 2005, 5 degrees awarded. *Degree requirements:* For master's, comprehensive exam or thesis. *Entrance requirements:* For master's, GRE, appropriate undergraduate degree, interview, letters of reference. Additional exam requirements/recommendations for international students: Required—TOEFL (minimum score 450 paper-based; 133 computer-based). *Application deadline:* For fall admission, 8/15 for domestic students. Applications are processed on a rolling basis. Application fee: $30 ($75 for international students). Electronic applications accepted. *Expenses:* Tuition, state resident: full-time $2,890; part-time $132 per credit. Tuition, nonresident: full-time $9,258; part-time $422 per credit. Required fees: $626; $41 per credit. Tuition and fees vary according to degree level.

Financial support: In 2005–06, 4 research assistantships with full tuition reimbursements (averaging $6,492 per year), 9 teaching assistantships with full tuition reimbursements (averaging $6,492 per year) were awarded; fellowships, career-related internships or fieldwork, Federal Work-Study, institutionally sponsored loans, health care benefits, and unspecified assistantships also available. Financial award application deadline: 3/15; financial award applicants required to submit FAFSA. *Faculty research:* Fisheries, range, and wildlife management; aquatic, plant, grassland, vertebrate, and invertebrate ecology; mammalian and plant systematics, taxonomy, and evolution; immunology, virology, and molecular biology. *Unit head:* Dr. J. Richard Schrock, Chair, 620-341-5311, Fax: 620-341-5607, E-mail: jschrock@emporia.edu. *Application contact:* Dr. Derek Zelmer, Graduate Coordinator, 620-341-5623, Fax: 620-341-5607, E-mail: dzelmer@emporia.edu.

Georgia State University, College of Arts and Sciences, Department of Biology, Program in Applied and Environmental Microbiology, Atlanta, GA 30303-3083. Offers MS, PhD. *Degree requirements:* For master's, thesis or alternative, exam; for doctorate, thesis/dissertation, exam. *Entrance requirements:* For master's and doctorate, GRE General Test. Additional exam requirements/recommendations for international students: Required—TOEFL. Electronic applications accepted. *Expenses:* Tuition, state resident: full-time $4,368; part-time $182 per semester hour. Tuition, nonresident: full-time $8,732; part-time $728 per semester hour. Required fees: $46 per semester hour.

Governors State University, College of Arts and Sciences, Program in Environmental Biology, University Park, IL 60466-0975. Offers MS. Part-time and evening/weekend programs available. *Students:* 5 full-time, 17 part-time. Average age 33. *Degree requirements:* For master's, thesis or alternative. *Application deadline:* For fall admission, 7/15 for domestic students; for spring admission, 11/10 for domestic students. Applications are processed on a rolling basis. Application fee: $25. *Expenses:* Tuition, state resident: full-time $3,768; part-time $157 per semester hour. Tuition, nonresident: full-time $11,304; part-time $471 per semester hour. Required fees: $480; $240 per semester. *Financial support:* Research assistantships, career-related internships or fieldwork, Federal Work-Study, institutionally sponsored loans, and scholarships/grants available. Support available to part-time students. Financial award application deadline: 5/1. *Faculty research:* Animal physiology, cell biology, animal behavior, plant physiology, plant populations. *Unit head:* Dr. Sandra A Mayfield, Interim Dean, College of Arts and Sciences, 708-534-4101.

Hood College, Graduate School, Program in Environmental Biology, Frederick, MD 21701-8575. Offers MS. Part-time and evening/weekend programs available. *Faculty:* 3 full-time (0 women), 4 part-time/adjunct (0 women). *Students:* Average age 33. In 2005, 9 degrees awarded. *Degree requirements:* For master's, thesis. *Entrance requirements:* For master's, minimum GPA of 2.5, 1 year of undergraduate biology and chemistry, 1 semester of mathematics. *Application deadline:* Applications are processed on a rolling basis. Application fee: $35.

Environmental Biology

Hood College (continued)

Expenses: Tuition: Full-time $6,300; part-time $350 per credit. Required fees: $40. *Financial support:* Applicants required to submit FAFSA. *Unit head:* Dr. Drew Ferrier, Director, 301-696-3649, Fax: 301-694-3597, E-mail: dferrier@hood.edu. *Application contact:* Lori White Drega, Associate Dean of Graduate School, 301-696-3811, Fax: 301-696-3597, E-mail: gofurther@hood.edu.

Massachusetts Institute of Technology, School of Engineering, Department of Civil and Environmental Engineering, Cambridge, MA 02139-4307. Offers biological oceanography (PhD, Sc D); chemical oceanography (PhD, Sc D); civil and environmental engineering (M Eng, SM, PhD, Sc D, CE); civil and environmental systems (PhD, Sc D); civil engineering (PhD, Sc D); coastal engineering (PhD, Sc D); construction engineering and management (PhD, Sc D); environmental biology (PhD, Sc D); environmental chemistry (PhD, Sc D); environmental engineering (PhD, Sc D); environmental fluid mechanics (PhD, Sc D); geotechnical and geoenvironmental engineering (PhD, Sc D); hydrology (PhD, Sc D); information technology (PhD, Sc D); oceanographic engineering (PhD, Sc D); structures and materials (PhD, Sc D); transportation (PhD, Sc D). *Faculty:* 34 full-time (3 women). *Students:* 179 full-time (66 women), 1 part-time; includes 15 minority (1 African American, 9 Asian Americans or Pacific Islanders, 5 Hispanic Americans), 116 international. Average age 26. 374 applicants, 38% accepted, 71 enrolled. In 2005, 76 master's, 27 doctorates, 1 other advanced degree awarded. *Degree requirements:* For master's and CE, thesis/dissertation; for doctorate, thesis/dissertation, comprehensive exam. *Entrance requirements:* For master's and doctorate, GRE General Test. Additional exam requirements/recommendations for international students: Required—TOEFL (minimum score 577 paper-based; 233 computer-based). *Application deadline:* For fall admission, 1/2 for domestic students, 1/2 for international students. Application fee: $70. Electronic applications accepted. *Expenses:* Tuition: Full-time $32,100. Required fees: $200. Part-time tuition and fees vary according to course load. *Financial support:* In 2005–06, 134 students received support, including 38 fellowships with tuition reimbursements available (averaging $23,116 per year), 86 research assistantships with tuition reimbursements available (averaging $22,765 per year), 13 teaching assistantships with tuition reimbursements available (averaging $19,753 per year); career-related internships or fieldwork, Federal Work-Study, institutionally sponsored loans, scholarships/grants, health care benefits, and unspecified assistantships also available. Total annual research expenditures: $12.4 million. *Unit head:* Prof. Patrick Jaillet, Department Head, 617-452-3379, Fax: 617-258-6775, E-mail: cee-admissions@mit.edu. *Application contact:* Graduate Admissions, 617-253-7119, Fax: 617-258-6775, E-mail: cee-admissions@mit.edu.

Montana State University, College of Graduate Studies, College of Letters and Science, Department of Ecology, Bozeman, MT 59717. Offers biological science (MS, PhD); fish and wildlife biology (PhD); fish and wildlife management (MS); land rehabilitation (intercollege) (MS). Part-time programs available. *Faculty:* 15 full-time (4 women). *Students:* 4 full-time (3 women), 61 part-time (23 women); includes 1 minority (American Indian/Alaska Native), 1 international. Average age 31. 8 applicants, 75% accepted, 6 enrolled. In 2005, 9 master's, 4 doctorates awarded. *Degree requirements:* For master's, thesis (for some programs), comprehensive exam, registration; for doctorate, thesis/dissertation, comprehensive exam, registration. *Entrance requirements:* For master's and doctorate, GRE General Test. Additional exam requirements/recommendations for international students: Required—TOEFL (minimum score 550 paper-based; 213 computer-based). *Application deadline:* For fall admission, 7/15 priority date for domestic students, 5/15 priority date for international students; for spring admission, 12/1 priority date for domestic students, 10/1 priority date for international students. Applications are processed on a rolling basis. Application fee: $30. Electronic applications accepted. *Expenses:* Tuition, state resident: full-time $4,132. Tuition, nonresident: full-time $1,132. *Financial support:* In 2005–06, 3 fellowships with full tuition reimbursements (averaging $21,300 per year), 45 research assistantships with full and partial tuition reimbursements (averaging $10,830 per year), 22 teaching assistantships with full and partial tuition reimbursements (averaging $10,268 per year) were awarded; career-related internships or fieldwork, Federal Work-Study, scholarships/grants, health care benefits, tuition waivers, and unspecified assistantships also available. Financial award application deadline: 3/1; financial award applicants required to submit FAFSA. *Faculty research:* Population dynamics, landscape ecology, evolution and genetics, ecological modeling, aquatic ecosystems. Total annual research expenditures: $2.1 million. *Unit head:* Dr. David Roberts, Department Head, 406-994-4548, Fax: 406-994-3190, E-mail: droberts@montana.edu.

Morgan State University, School of Graduate Studies, School of Computer, Mathematical, and Natural Sciences, Department of Biology, Program in Bio-Environmental Science, Baltimore, MD 21251. Offers PhD. *Students:* 11 (4 women); includes 5 minority (all African Americans) 6 international. *Degree requirements:* For doctorate, thesis/dissertation, oral defense of dissertation, comprehensive exam. *Entrance requirements:* For doctorate, GRE General Test, GRE Subject Test (biology, chemistry, or related science), bachelor's or master's degree in biology, chemistry, physics or related field; minimum GPA of 3.0. Additional exam requirements/recommendations for international students: Required—TOEFL (minimum score 550 paper-based; 213 computer-based). *Application deadline:* For fall admission, 2/1 for domestic students; for spring admission, 10/1 priority date for domestic students. Applications are processed on a rolling basis. Application fee: $0. *Expenses:* Tuition, state resident: part-time $272 per credit. Tuition, nonresident: part-time $478 per credit. Required fees: $58 per credit. *Unit head:* Dr. Arthur Williams, Chairperson—Department of Biology, 443-885-3070. *Application contact:* Dr. James E. Waller, Admissions and Program Officer, 443-885-3185, Fax: 443-885-8226, E-mail: jwaller@moac.morgan.edu.

See Close-Up on page 703.

Nicholls State University, Graduate Studies, College of Arts and Sciences, Department of Biological Sciences, Thibodaux, LA 70310. Offers marine and environmental biology (MS). Part-time programs available. *Faculty:* 10 full-time (3 women), 1 part-time/adjunct (0 women). *Students:* 21 full-time (5 women), 3 part-time (1 woman); includes 2 minority (1 African American, 1 Asian American or Pacific Islander), 1 international. 21 applicants, 71% accepted, 13 enrolled. In 2005, 6 master's awarded. *Degree requirements:* For master's, thesis, comprehensive exam, registration. *Entrance requirements:* For master's, GRE. Additional exam requirements/recommendations for international students: Required—TOEFL (minimum score 600 paper-based). *Application deadline:* For fall admission, 6/17 for domestic students; for spring admission, 11/15 priority date for domestic students. Application fee: $25. *Expenses:* Tuition, state resident: full-time $3,240. Tuition, nonresident: full-time $8,868. *Financial support:* In 2005–06, 6 students received support, including research assistantships (averaging $12,000 per year), teaching assistantships with tuition reimbursements available (averaging $8,000 per year) *Faculty research:* Biovemediation, ecology, public health, biotechnology, physiology. Total annual research expenditures: $1.2 million. *Unit head:* Dr. Marilyn B. Kilgen, Head, 985-448-4710, E-mail: biol-mbk@nicholls.edu. *Application contact:* Dr. Raj Boopathy, Graduate Coordinator, 985-448-4716, E-mail: biol-rrb@nicholls.edu.

Nova Scotia Agricultural College, Research and Graduate Studies, Truro, NS B2N 5E3, Canada. Offers agriculture (M Sc), including air quality, animal behavior, animal molecular genetics, animal nutrition, animal technology, aquaculture, botany, crop management, crop physiology, ecology, environmental microbiology, food science, horticulture, nutrient management, pest management, physiology, plant biotechnology, plant pathology, soil chemistry, soil fertility, waste management and composting, water quality. Part-time programs available. *Faculty:* 43 full-time (7 women), 21 part-time/adjunct (1 woman). *Students:* 50 full-time (25 women), 15 part-time (11 women); includes 7 minority (3 African Americans, 3 Asian Americans or Pacific Islanders, 1 Hispanic American). Average age 25. In 2005, 23 degrees awarded. *Degree requirements:* For master's, thesis, registration. *Entrance requirements:* For master's, honors of B Sc, minimum GPA of 3.0. Additional exam requirements/recommendations for international students: Required—TOEFL (minimum score 580 paper-based; 237 computer-based), Michigan English Language Assessment Battery, IELT, Can Test, CAEL. *Application deadline:* For fall admission, 6/1 for domestic students, 4/1 for international students. For

winter admission, 10/31 for domestic students; for spring admission, 2/28 for domestic students. Applications are processed on a rolling basis. Application fee: $70. *Expenses:* Tuition, state resident: part-time $2,328 per year. Tuition, nonresident: full-time $6,984; part-time $7,968 per year. International tuition: $12,624 full-time. Required fees: $481; $46 per course. Tuition and fees vary according to program and student level. *Financial support:* In 2005–06, 48 students received support, including 7 fellowships (averaging $15,000 per year), 10 research assistantships (averaging $15,000 per year), 15 teaching assistantships (averaging $900 per year); career-related internships or fieldwork, scholarships/grants, and unspecified assistantships also available. *Faculty research:* Bio-product development, organic agriculture, nutrient management, air and water quality, agricultural biotechnology. Total annual research expenditures: $4.7 million. *Unit head:* Jill L. Rogers, Manager, 902-893-6360, Fax: 902-893-3430, E-mail: jrogers@nsac.ca. *Application contact:* Marie Law, Administrative Assistant, 902-893-6502, Fax: 902-893-3430, E-mail: mlaw@nsac.ca.

Ohio University, Graduate Studies, College of Arts and Sciences, Department of Environmental and Plant Biology, Athens, OH 45701-2979. Offers MS, PhD. Part-time programs available. *Faculty:* 12 full-time (3 women), 1 (woman) part-time/adjunct. *Students:* 32 full-time (16 women); includes 4 minority (2 Asian Americans or Pacific Islanders, 2 Hispanic Americans), 6 international. Average age 25. 29 applicants, 62% accepted, 8 enrolled. In 2005, 6 master's, 3 doctorates awarded. *Median time to degree:* Of those who began their doctoral program in fall 1997, 100% received their degree in 8 years or less. *Degree requirements:* For master's, thesis/dissertation, 2 quarters of teaching experience; for doctorate, thesis/dissertation, 2 quarters of teaching experience, comprehensive exam. *Entrance requirements:* For master's, GRE General Test, minimum GPA of 3.0; for doctorate, GRE General Test, minimum GPA of 3.2. Additional exam requirements/recommendations for international students: Required—TOEFL (minimum score 620 paper-based; 260 computer-based). *Application deadline:* For fall admission, 1/15 priority date for domestic students, 1/15 priority date for international students. Applications are processed on a rolling basis. Application fee: $45. Electronic applications accepted. *Financial support:* In 2005–06, 3 fellowships with full tuition reimbursements (averaging $17,000 per year), 4 research assistantships with full tuition reimbursements, 21 teaching assistantships with tuition reimbursements were awarded; Federal Work-Study, institutionally sponsored loans, and scholarships/grants also available. Financial award application deadline: 1/15. *Faculty research:* Cellular and molecular biology, ecology, plant biology, systematics, development. Total annual research expenditures: $859,166. *Unit head:* Dr. Gar W. Rothwell, Chair, 740-593-1126, Fax: 740-593-1130, E-mail: rothwell@ohio.edu. *Application contact:* Dr. Morgan L. Vis, Graduate Chair, 740-593-1126, Fax: 740-593-1130, E-mail: vischia@ohio.edu.

Rutgers, The State University of New Jersey, New Brunswick/Piscataway, Graduate School, Program in Environmental Sciences, New Brunswick, NJ 08901-1281. Offers air resources (MS, PhD); aquatic biology (MS, PhD); aquatic chemistry (MS, PhD); atmospheric science (MS, PhD); chemistry and physics of aerosol and hydrosol systems (MS, PhD); environmental chemistry (MS, PhD); environmental microbiology (MS, PhD); environmental toxicology (PhD); exposure assessment (PhD); fate and effects of pollutants (MS, PhD); pollution prevention and control (MS, PhD); water and wastewater treatment (MS, PhD); water resources (MS, PhD). *Faculty:* 81 full-time, 7 part-time/adjunct. *Students:* 49 full-time (27 women), 48 part-time (19 women); includes 10 minority (3 African Americans, 6 Asian Americans or Pacific Islanders, 1 Hispanic American), 24 international. Average age 32. 79 applicants, 41% accepted, 15 enrolled. In 2005, 8 master's, 10 doctorates awarded. Terminal master's awarded for partial completion of doctoral program. *Degree requirements:* For master's, thesis or alternative, oral final exam, comprehensive exam; for doctorate, thesis/dissertation, thesis defense, qualifying exam, comprehensive exam. *Entrance requirements:* For master's and doctorate, GRE General Test. Additional exam requirements/recommendations for international students: Required—TOEFL. *Application deadline:* For fall admission, 3/1 for domestic students; for spring admission, 11/1 for domestic students. Applications are processed on a rolling basis. Application fee: $50. Electronic applications accepted. *Expenses:* Tuition, state resident: full-time $10,440; part-time $435 per credit. Tuition, nonresident: full-time $15,520; part-time $647 per credit. Required fees: $129 per credit. Tuition and fees vary according to program. *Financial support:* In 2005–06, 10 fellowships with full tuition reimbursements (averaging $21,887 per year), 34 research assistantships with full tuition reimbursements (averaging $19,367 per year), 3 teaching assistantships with full tuition reimbursements (averaging $17,583 per year) were awarded; career-related internships or fieldwork and Federal Work-Study also available. Financial award application deadline: 1/15; financial award applicants required to submit FAFSA. *Faculty research:* Atmospheric sciences; biological waste treatment; contaminant fate and transport; exposure assessment; air, soil and water quality. Total annual research expenditures: $5.7 million. *Unit head:* John Reinfelder, Director, 732-932-8013, Fax: 732-932-8644, E-mail: reinfelder@envsci.rutgers.edu. *Application contact:* Dr. Paul J. Lioy, Graduate Admissions Committee, 732-932-0150, Fax: 732-445-0116, E-mail: plioy@eohsi.rutgers.edu.

Sonoma State University, School of Science and Technology, Department of Biology, Rohnert Park, CA 94928-3609. Offers environmental biology (MA); general biology (MA). Part-time programs available. *Faculty:* 8 full-time (2 women), 7 part-time/adjunct (4 women). *Students:* 12 full-time (8 women), 1 international. Average age 33. 14 applicants, 36% accepted, 3 enrolled. In 2005, 2 degrees awarded. *Degree requirements:* For master's, thesis or alternative, oral exam. *Entrance requirements:* For master's, GRE General Test, GRE Subject Test, minimum GPA of 3.0. *Application deadline:* For fall admission, 11/30 for domestic students. Applications are processed on a rolling basis. Application fee: $55. *Expenses:* Tuition, nonresident: part-time $339 per unit. Required fees: $2,099 per semester. *Financial support:* In 2005–06, 2 research assistantships, 9 teaching assistantships were awarded; career-related internships or fieldwork and Federal Work-Study also available. Financial award application deadline: 3/2. *Faculty research:* Molecular biology, genetics, riparian and wetland ecology, plant ecology, feeding mechanisms of invertebrates, ichthyology, microbiology. Total annual research expenditures: $32,000. *Unit head:* Dr. James Christmann, Chair, 707-664-2189, E-mail: james.christmann@sonoma.edu. *Application contact:* John Hopkirk, Graduate Adviser, 707-664-2180.

State University of New York College of Environmental Science and Forestry, Faculty of Environmental and Forest Biology, Syracuse, NY 13210-2779. Offers chemical ecology (MPS, MS, PhD); conservation biology (MPS, MS, PhD); ecology (MPS, MS, PhD); entomology (MPS, MS, PhD); environmental interpretation (MPS, MS, PhD); environmental physiology (MPS, MS, PhD); fish and wildlife biology (MPS, MS, PhD); forest pathology and mycology (MPS, MS, PhD); plant science and biotechnology (MPS, MS, PhD). *Faculty:* 27 full-time (4 women), 4 part-time/adjunct (0 women). *Students:* 81 full-time (50 women), 62 part-time (33 women); includes 4 minority (1 Asian American or Pacific Islander, 3 Hispanic Americans), 16 international. Average age 30. 84 applicants, 54% accepted, 17 enrolled. In 2005, 17 master's, 3 doctorates awarded. *Degree requirements:* For master's, thesis (for some programs), registration; for doctorate, thesis/dissertation, comprehensive exam. *Entrance requirements:* For master's and doctorate, GRE General Test, GRE Subject Test, minimum GPA of 3.0. Additional exam requirements/recommendations for international students: Required—TOEFL (minimum score 550 paper-based; 213 computer-based). *Application deadline:* For fall admission, 2/1 priority date for domestic students, 2/1 priority date for international students; for spring admission, 11/1 priority date for domestic students, 11/1 priority date for international students. Applications are processed on a rolling basis. Application fee: $60. *Expenses:* Tuition, area resident: Tuition: Full-time $6,900; part-time $288 per credit. Tuition, nonresident: full-time $10,920; part-time $455 per credit. Required fees: $395; $32 per credit. $20 per term. One-time fee: $145. *Financial support:* In 2005–06, 86 students received support, including 13 fellowships with full and partial tuition reimbursements available (averaging $9,446 per year), 40 research assistantships with full and partial tuition reimbursements available (averaging $11,000 per year), 32 teaching assistantships with full and partial tuition reimbursements available (averaging $9,446 per year); Federal Work-Study, institutionally sponsored loans, scholarships/grants, health care benefits, and unspecified assistantships also available. Financial award application deadline: 6/30. *Faculty research:* Ecology, fish and wildlife biology and

Environmental Biology

management, plant science, entomology. Total annual research expenditures: $4.1 million. *Unit head:* Dr. Donald J. Leopold, Chair, 315-470-6770, Fax: 315-470-6934, E-mail: dendro@esf.edu. *Application contact:* Dr. Dudley J. Raynal, Dean, Instruction and Graduate Studies, 315-470-6599, Fax: 315-470-6978, E-mail: esfgrad@esf.edu.

Tennessee Technological University, Graduate School, College of Arts and Sciences, Department of Biology, Cookeville, TN 38505. Offers environmental biology (MS); fish, game, and wildlife management (MS). Part-time programs available. *Faculty:* 22 full-time (2 women). *Students:* 22 full-time (8 women), 17 part-time (7 women); includes 3 minority (1 African American, 2 Asian Americans or Pacific Islanders). Average age 25. 14 applicants, 50% accepted, 7 enrolled. In 2005, 9 degrees awarded. *Degree requirements:* For master's, thesis. *Entrance requirements:* For master's, GRE General Test. Additional exam requirements/recommendations for international students: Required—TOEFL. *Application deadline:* For fall admission, 3/1 for domestic students; for spring admission, 8/1 for domestic students. Application fee: $25 ($30 for international students). *Expenses:* Tuition, state resident: full-time $8,421; part-time $307 per hour. Tuition, nonresident: full-time $22,389; part-time $711 per hour. *Financial support:* In 2005–06, 22 research assistantships (averaging $9,000 per year), 9 teaching assistantships (averaging $7,500 per year) were awarded. Financial award application deadline: 4/1. *Faculty research:* Aquatics, environmental studies. *Unit head:* Dr. Daniel Combs, Interim Chairperson, 931-372-3134, Fax: 931-372-6257, E-mail: dcombs@tntech.edu. *Application contact:* Dr. Francis O. Otuonye, Associate Vice President for Research and Graduate Studies, 931-372-3233, Fax: 931-372-3497, E-mail: fotuonye@tntech.edu.

University of Alberta, Faculty of Graduate Studies and Research, Department of Biological Sciences, Edmonton, AB T6G 2E1, Canada. Offers environmental biology and ecology (M Sc, PhD); microbiology and biotechnology (M Sc, PhD); molecular biology and genetics (M Sc, PhD); physiology and cell biology (M Sc, PhD); plant biology (M Sc, PhD); systematics and evolution (M Sc, PhD). *Faculty:* 72 full-time (15 women), 15 part-time/adjunct (4 women). *Students:* 238 full-time (117 women), 32 part-time (15 women), 31 international. 206 applicants, 42% accepted. In 2005, 29 master's, 31 doctorates awarded. Terminal master's awarded for partial completion of doctoral program. *Degree requirements:* For master's and doctorate, thesis/dissertation, registration. *Entrance requirements:* Additional exam requirements/recommendations for international students: Required—TOEFL. *Application deadline:* For fall admission, 3/1 for domestic students. Applications are processed on a rolling basis. Application fee: $0. Tuition and fees charges are reported in Canadian dollars. *Expenses:* Tuition, state resident: part-time $562 Canadian dollars per term. Tuition, nonresident: full-time $3,375 Canadian dollars. Required fees: $573 Canadian dollars; $84 Canadian dollars per term. *Financial support:* In 2005–06, 4 research assistantships with partial tuition reimbursements (averaging $12,000 per year), 103 teaching assistantships with partial tuition reimbursements (averaging $12,300 per year) were awarded; career-related internships or fieldwork and scholarships/grants also available. *Unit head:* Laura Frost, 780-492-1904. *Application contact:* Dr. John P. Chang, Associate Chair for Graduate Studies, 780-492-1257, Fax: 780-492-9457, E-mail: bio.grad.coordinator@ualberta.ca.

University of California, Santa Cruz, Division of Graduate Studies, Division of Physical and Biological Sciences, Environmental Toxicology Department, Santa Cruz, CA 95064. Offers MS, PhD. *Faculty:* 5 full-time (2 women). *Students:* 11 full-time (6 women); includes 1 minority (Asian American or Pacific Islander), 3 international. 18 applicants, 33% accepted, 4 enrolled. In 2005, 3 master's, 3 doctorates awarded. Application fee: $60. *Expenses:* Tuition: full-time $14,694. Required fees: $9,437. *Unit head:* Russ Flegal, Chair, 831-459-4719. *Application contact:* Sissy Madden, Information Contact, 831-459-4719, E-mail: madden@etox.ucsc.edu.

University of Guelph, Graduate Program Services, Ontario Agricultural College, Department of Environmental Biology, Guelph, ON N1G 2W1, Canada. Offers entomology (M Sc, PhD); environmental biology and biotechnology (M Sc, PhD); environmental toxicology (M Sc, PhD); plant and forest systems (M Sc, PhD); plant pathology (M Sc, PhD). Part-time programs available. *Faculty:* 20 full-time (2 women), 21 part-time/adjunct (2 women). *Students:* 60 full-time (32 women), 13 part-time (8 women). 37 applicants, 35% accepted, 13 enrolled. In 2005, 12 master's, 7 doctorates awarded. *Median time to degree:* Of those who began their doctoral program in fall 1997, 79.5% received their degree in 8 years or less. *Degree requirements:* For master's, thesis/dissertation, registration; for doctorate, thesis/dissertation, comprehensive exam, registration. *Entrance requirements:* For master's, minimum B- average during previous 2 years of course work; for doctorate, minimum B average. Additional exam requirements/recommendations for international students: Required—TOEFL or IELT. *Application deadline:* Applications are processed on a rolling basis. Application fee: $75. Electronic applications accepted. *Financial support:* In 2005–06, research assistantships (averaging $16,500 per year), teaching assistantships (averaging $4,606 per year) were awarded; fellowships *Faculty research:* Entomology, environmental microbiology and biotechnology, environmental toxicology, forest ecology, plant pathology. Total annual research expenditures: $3 million. *Unit head:* Dr. M. A. Dixon, Chair, 519-824-4120 Ext. 52555, Fax: 519-837-0442, E-mail: mdixon@uoguelph.ca. *Application contact:* Dr. H. Lee, Admissions Coordinator, 519-824-4120 Ext. 53828, Fax: 519-837-0442, E-mail: hlee@uoguelph.ca.

University of Louisiana at Lafayette, Graduate School, College of Sciences, Department of Biology, Lafayette, LA 70504. Offers biology (MS); environmental and evolutionary biology (PhD). *Faculty:* 20 full-time (4 women), 2 part-time/adjunct (0 women). *Students:* 58 full-time (34 women), 12 part-time (10 women); includes 3 minority (2 African Americans, 1 Asian American or Pacific Islander), 23 international. Average age 29. 40 applicants, 38% accepted, 8 enrolled. In 2005, 7 master's, 3 doctorates awarded. Terminal master's awarded for partial completion of doctoral program. *Degree requirements:* For master's, thesis, registration; for doctorate, 2 foreign languages, thesis/dissertation, comprehensive exam, registration. *Entrance requirements:* For master's, GRE General Test, minimum GPA of 2.75; for doctorate, GRE General Test, GRE Subject Test, minimum GPA of 3.0. Additional exam requirements/recommendations for international students: Required—TOEFL (minimum score 550 paper-based; 213 computer-based). *Application deadline:* For fall admission, 5/15 for domestic students, 5/15 for international students; for spring admission, 10/1 for domestic students, 10/1 for international students. Applications are processed on a rolling basis. Application fee: $20 ($30 for international students). Electronic applications accepted. *Expenses:* Tuition, state resident: full-time $3,330; part-time $93 per credit hour. Tuition, nonresident: full-time $9,510; part-time $350 per credit hour. International tuition: $9,646 full-time. *Financial support:* In 2005–06, 17 fellowships with full tuition reimbursements (averaging $14,289 per year), 9 research assistantships with full tuition reimbursements (averaging $8,024 per year), 17 teaching assistantships with full tuition reimbursements (averaging $12,044 per year) were awarded; Federal Work-Study, institutionally sponsored loans, and unspecified assistantships also available. Financial award application deadline: 5/1. *Faculty research:* Structure and ultrastructure, system biology, ecology, processes, environmental physiology. *Unit head:* Dr. Darryl L. Felder, Head, 337-482-6748, Fax: 337-482-5834, E-mail: dlf4517@louisana.edu. *Application contact:* Dr. Paul Leberg, Graduate Coordinator, 337-482-6750, Fax: 337-482-5834, E-mail: leberg@louisiana.edu.

University of Louisville, Graduate School, College of Arts and Sciences, Department of Biology, Program in Environmental Biology, Louisville, KY 40292-0001. Offers PhD. *Students:* 1 (woman) full-time. Average age 33. In 2005, 1 degree awarded. *Degree requirements:* For doctorate, thesis/dissertation. *Entrance requirements:* For doctorate, GRE General Test. *Application deadline:* Applications are processed on a rolling basis. Application fee: $50. *Expenses:* Tuition, state resident: full-time $6,006; part-time $334 per credit hour. Tuition, nonresident: full-time $16,554; part-time $920 per credit hour. Tuition and fees vary according to course load, degree level and program. *Application contact:* Dr. Joseph M. Steffen, Director of Graduate Studies, 502-852-6771, Fax: 502-852-0725, E-mail: joe.steffen@louisville.edu.

See Close-Up on page 261.

University of Massachusetts Amherst, Graduate School, College of Natural Resources and the Environment, Department of Natural Resources Conservation, Program in Wildlife and Fisheries Conservation, Amherst, MA 01003. Offers MS, PhD. Part-time programs available. *Students:* 26 full-time (15 women), 27 part-time (8 women), 8 international. Average age 32. 35 applicants, 37% accepted, 10 enrolled. In 2005, 5 master's, 5 doctorates awarded. Terminal master's awarded for partial completion of doctoral program. *Degree requirements:* For master's, thesis optional; for doctorate, variable foreign language requirement, thesis/dissertation. *Entrance requirements:* For master's and doctorate, GRE General Test. Additional exam requirements/recommendations for international students: Required—TOEFL (minimum score 530 paper-based; 197 computer-based). *Application deadline:* For fall admission, 2/1 priority date for domestic students, 2/1 for international students; for spring admission, 10/1 for domestic students, 10/1 for international students. Applications are processed on a rolling basis. Application fee: $40 ($65 for international students). Electronic applications accepted. *Expenses:* Tuition, state resident: part-time $110 per credit. Tuition, nonresident: part-time $414 per credit. Required fees: $2,824 per term. One-time fee: $250 part-time. Full-time tuition and fees vary according to course load, campus/location, program and reciprocity agreements. *Financial support:* Fellowships with full tuition reimbursements, research assistantships with full tuition reimbursements, teaching assistantships with full tuition reimbursements, career-related internships or fieldwork, Federal Work-Study, scholarships/grants, traineeships, and unspecified assistantships available. Support available to part-time students. Financial award application deadline: 2/1. *Unit head:* Dr. Kevin McGarigal, Director, 413-545-2666, Fax: 413-545-4358.

University of Massachusetts Boston, Office of Graduate Studies and Research, College of Science and Mathematics, Department of Environmental, Coastal and Ocean Sciences, Program in Environmental Biology, Boston, MA 02125-3393. Offers PhD. Part-time and evening/weekend programs available. *Degree requirements:* For doctorate, thesis/dissertation, oral exams, comprehensive exam. *Entrance requirements:* For doctorate, GRE General Test, minimum GPA of 2.75. *Faculty research:* Polychoets biology, predator and prey relationships, population and evolutionary biology, neurobiology, biodiversity.

University of Missouri–Rolla, Graduate School, College of Arts and Sciences, Program in Applied and Environmental Biology, Rolla, MO 65409-0910. Offers MS. *Entrance requirements:* For master's, GRE. Additional exam requirements/recommendations for international students: Required—TOEFL.

University of North Dakota, Graduate School, College of Arts and Sciences, Department of Biology, Grand Forks, ND 58202. Offers botany (MS, PhD); ecology (MS, PhD); entomology (MS, PhD); environmental biology (MS, PhD); fisheries/wildlife (MS, PhD); genetics (MS, PhD); zoology (MS, PhD). *Faculty:* 16 full-time (3 women). *Students:* 15 applicants, 13% accepted, 2 enrolled. In 2005, 5 degrees awarded. Terminal master's awarded for partial completion of doctoral program. *Degree requirements:* For master's, thesis/dissertation, final exam; for doctorate, thesis/dissertation, final exam, comprehensive exam. *Entrance requirements:* For master's, GRE General Test, GRE Subject Test, minimum GPA of 3.0; for doctorate, GRE General Test, GRE Subject Test, minimum GPA of 3.5. Additional exam requirements/recommendations for international students: Required—TOEFL (minimum score 550 paper-based; 213 computer-based). *Application deadline:* For fall admission, 10/1 for domestic students, 10/1 for international students. Application fee: $35. Electronic applications accepted. *Financial support:* In 2005–06, 8 research assistantships with full tuition reimbursements (averaging $11,375 per year), 13 teaching assistantships with full tuition reimbursements (averaging $10,813 per year) were awarded; fellowships, Federal Work-Study, institutionally sponsored loans, scholarships/grants, and tuition waivers (full and partial) also available. Support available to part-time students. Financial award application deadline: 3/15; financial award applicants required to submit FAFSA. *Faculty research:* Population biology, wildlife ecology, RNA processing, hormonal control of behavior. *Unit head:* Dr. Richard Sweitzel, Graduate Director, 701-777-4676, Fax: 701-777-2623, E-mail: richard_sweitzel@und.nodak.edu.

University of Southern Mississippi, Graduate School, College of Science and Technology, Department of Biological Sciences, Hattiesburg, MS 39406-0001. Offers environmental biology (MS, PhD); marine biology (MS, PhD); microbiology (MS, PhD); molecular biology (MS, PhD). *Degree requirements:* For master's and doctorate, thesis/dissertation, comprehensive exam. *Entrance requirements:* For master's, GRE General Test, minimum GPA of 3.0; for doctorate, GRE General Test, minimum GPA of 3.5. Additional exam requirements/recommendations for international students: Required—TOEFL.

See Close-Up on page 297.

University of West Florida, College of Arts and Sciences: Sciences, Division of Life and Health Sciences, Department of Biology, Pensacola, FL 32514-5750. Offers biological chemistry (MS); biology (MS); biology education (MST); coastal zone studies (MS); environmental biology (MS). *Accreditation:* NCATE. *Faculty:* 16 full-time (5 women), 2 part-time/adjunct (0 women). *Students:* 12 full-time (10 women), 31 part-time (19 women); includes 4 minority (1 African American, 2 Asian Americans or Pacific Islanders, 1 Hispanic American), 2 international. Average age 28. 17 applicants, 71% accepted, 8 enrolled. In 2005, 4 degrees awarded. *Degree requirements:* For master's, thesis. *Entrance requirements:* For master's, GRE General Test. Additional exam requirements/recommendations for international students: Required—TOEFL (minimum score 550 paper-based; 213 computer-based). *Application deadline:* For fall admission, 6/1 for domestic students, 5/15 for international students; for spring admission, 11/1 for domestic students, 10/1 for international students. Applications are processed on a rolling basis. Application fee: $30. *Expenses:* Tuition, state resident: full-time $5,833; part-time $243 per credit hour. Tuition, nonresident: full-time $21,204; part-time $884 per credit hour. Tuition and fees vary according to campus/location. *Financial support:* In 2005–06, 20 students received support, including 5 research assistantships with partial tuition reimbursements available (averaging $5,000 per year), 11 teaching assistantships with partial tuition reimbursements available (averaging $8,000 per year) Financial award application deadline: 4/15; financial award applicants required to submit FAFSA.

University of Wisconsin–Madison, Graduate School, College of Agricultural and Life Sciences, Molecular and Environmental Toxicology Center, Madison, WI 53706-1380. Offers MS, PhD. *Students:* 38 full-time (25 women); includes 6 minority (1 African American, 2 American Indian/Alaska Native, 3 Asian Americans or Pacific Islanders), 6 international. Average age 28. In 2005, 1 master's, 4 doctorates awarded. *Degree requirements:* For doctorate, thesis/dissertation. *Entrance requirements:* For master's and doctorate, bachelor's degree in science-related field. Additional exam requirements/recommendations for international students: Required—TOEFL. *Application deadline:* For fall admission, 12/15 priority date for domestic students, 12/15 priority date for international students. Application fee: $45. *Financial support:* In 2005–06, 6 research assistantships with tuition reimbursements (averaging $21,500 per year) were awarded; fellowships with tuition reimbursements, traineeships, health care benefits, and unspecified assistantships also available. *Faculty research:* Toxicology cancer, genetics, cell cycle, xenobotic metabolism. *Unit head:* Dr. Jeffrey A. Johnson, Director, 608-263-4580, Fax: 608-262-5245, E-mail: uwetox@wisc.edu. *Application contact:* Eileen M. Stevens, Program Administrator, 608-263-4580, Fax: 608-262-5245, E-mail: emstevens@wisc.edu.

Washington University in St. Louis, Graduate School of Arts and Sciences, Division of Biology and Biomedical Sciences, Program in Evolution, Ecology and Population Biology, St. Louis, MO 63130-4899. Offers ecology (PhD); environmental biology (PhD); evolutionary biology (PhD); genetics (PhD). *Degree requirements:* For doctorate, thesis/dissertation. *Entrance requirements:* For doctorate, GRE General Test, GRE Subject Test. Electronic applications accepted.

West Virginia University, Davis College of Agriculture, Forestry and Consumer Sciences, Division of Plant and Soil Sciences, Morgantown, WV 26506. Offers agronomy (MS); entomology (MS); environmental microbiology (MS); horticulture (MS); plant pathology (MS). *Faculty:*

Environmental Biology

West Virginia University *(continued)*
18 full-time (1 woman), 1 part-time/adjunct (0 women). *Students:* 17 full-time (11 women), 3 part-time (2 women), 2 international. Average age 27. In 2005, 5 degrees awarded. *Degree requirements:* For master's, thesis. *Entrance requirements:* For master's, GRE, minimum GPA of 2.5. Additional exam requirements/recommendations for international students: Required—TOEFL. *Application deadline:* Applications are processed on a rolling basis. Application fee: $45. *Expenses:* Tuition, state resident: full-time $4,582; part-time $258 per credit hour. Tuition, nonresident: full-time $13,820; part-time $741 per credit hour. *Financial support:* In 2005–06, 13 research assistantships with full tuition reimbursements (averaging $9,936 per year), 4 teaching assistantships with full tuition reimbursements (averaging $9,936 per year) were awarded; Federal Work-Study, institutionally sponsored loans, and tuition waivers (full and partial) also available. Financial award application deadline: 2/1; financial award applicants required to submit FAFSA. *Faculty research:* Water quality, reclamation of disturbed land, crop production, pest control, environmental protection. Total annual research expenditures: $1 million. *Unit head:* Dr. Barton S. Baker, Chair and Division Director, 304-293-4817 Ext. 4342, Fax: 304-293-2960, E-mail: barton.baker@mail.wvu.edu.

West Virginia University, Eberly College of Arts and Sciences, Department of Biology, Morgantown, WV 26506. Offers cell and molecular biology (MS, PhD); environmental and evolutionary biology (MS, PhD); integrative organismal biology (PhD); integrative organismal,

biology (MS). *Faculty:* 18 full-time (3 women), 4 part-time/adjunct (all women). *Students:* 29 full-time (20 women), 5 part-time (3 women); includes 1 minority (Asian American or Pacific Islander), 9 international. Average age 26. 50 applicants, 10% accepted. In 2005, 1 master's, 2 doctorates awarded. Terminal master's awarded for partial completion of doctoral program. *Degree requirements:* For master's, thesis, final exam; for doctorate, thesis/dissertation, preliminary and final exams. *Entrance requirements:* For master's, GRE General Test, GRE Subject Test, minimum GPA of 3.0; for doctorate, GRE General Test, minimum GPA of 3.0. Additional exam requirements/recommendations for international students: Required—TOEFL. *Application deadline:* For fall admission, 4/1 for domestic students; for spring admission, 10/1 for domestic students. Applications are processed on a rolling basis. Application fee: $45. *Expenses:* Tuition, state resident: full-time $4,582; part-time $258 per credit hour. Tuition, nonresident: full-time $13,820; part-time $741 per credit hour. *Financial support:* In 2005–06, 4 research assistantships, 22 teaching assistantships were awarded; Federal Work-Study and institutionally sponsored loans available. Financial award application deadline: 4/1; financial award applicants required to submit FAFSA. *Faculty research:* Environmental biology, genetic engineering, developmental biology, global change, biodiversity. *Unit head:* Dr. Jonathan Cumming, Chair, 304-293-5201 Ext. 2508, Fax: 304-293-6363, E-mail: jonathan.cumming@mail.wvu.edu. *Application contact:* Dr. William T. Peterjohn, Director of Graduate Studies, 304-293-5201 Ext. 2510, Fax: 304-293-6363, E-mail: william.peterjohn@mail.wvu.edu.

See Close-Up on page 325.

Evolutionary Biology

Arizona State University, Division of Graduate Studies, College of Liberal Arts and Sciences, Department of Biology, Program in Evolution, Tempe, AZ 85287. Offers MS, PhD. Terminal master's awarded for partial completion of doctoral program. *Degree requirements:* For master's, thesis; for doctorate, thesis/dissertation, oral exam. *Entrance requirements:* For master's and doctorate, GRE General Test, GRE Subject Test. Additional exam requirements/recommendations for international students: Required—TOEFL (minimum score 600 paper-based; 250 computer-based); Recommended—TSE.

Brown University, Graduate School, Division of Biology and Medicine, Program in Ecology and Evolutionary Biology, Providence, RI 02912. Offers PhD. *Degree requirements:* For doctorate, thesis/dissertation, preliminary exam. *Entrance requirements:* For doctorate, GRE General Test, GRE Subject Test. Additional exam requirements/recommendations for international students: Required—TOEFL. Electronic applications accepted. *Faculty research:* Marine ecology, behavioral ecology, population genetics, evolutionary morphology, plant ecology.

Clemson University, Graduate School, College of Agriculture, Forestry and Life Sciences, Department of Biological Sciences, Program in Biological Sciences, Clemson, SC 29634. Offers MS, PhD. *Students:* 21 full-time (14 women), 4 part-time (3 women); includes 1 minority (Hispanic American), 3 international. 12 applicants, 42% accepted, 5 enrolled. In 2005, 1 master's, 1 doctorate awarded. *Degree requirements:* For master's, thesis optional; for doctorate, thesis/dissertation, comprehensive exam. *Entrance requirements:* For master's and doctorate, GRE General Test. Additional exam requirements/recommendations for international students: Required—TOEFL. *Application deadline:* For fall admission, 6/1 for domestic students, 4/15 for international students. Application fee: $50. *Financial support:* Research assistantships, teaching assistantships available. Financial award application deadline: 3/15; financial award applicants required to submit FAFSA. *Application contact:* Information Contact, 864-656-3587, Fax: 864-656-0435.

See Close-Ups on pages 99 and 101.

Columbia University, Graduate School of Arts and Sciences, Division of Natural Sciences, Department of Ecology and Evolutionary Biology, New York, NY 10027. Offers conservation biology (Certificate); ecology and evolutionary biology (PhD); environmental policy (Certificate). *Faculty:* 7 full-time (1 woman), 43 part-time/adjunct (15 women). *Students:* 9 full-time (7 women); includes 1 minority (African American), 2 international. Average age 28. 45 applicants, 18% accepted. *Degree requirements:* For doctorate, one foreign language, thesis/dissertation, teaching experience. *Entrance requirements:* For doctorate, GRE General Test, previous course work in biology. Additional exam requirements/recommendations for international students: Required—TOEFL. Application fee: $75. Electronic applications accepted. *Expenses:* Tuition: Full-time $31,448. Tuition and fees vary according to course level, course load, campus/location and program. *Financial support:* In 2005–06, 4 students received support; fellowships, career-related internships or fieldwork and institutionally sponsored loans available. Financial award application deadline: 1/5. *Faculty research:* Tropical ecology, ethnobotany, global change, systematics. Total annual research expenditures: $300,000. *Unit head:* Shahid Naeem, Chair, 212-854-7337, E-mail: sn212@columbia.edu.

See Close-Up on page 699.

Cornell University, Graduate School, Graduate Fields of Agriculture and Life Sciences, Field of Ecology and Evolutionary Biology, Ithaca, NY 14853-0001. Offers ecology (PhD), including animal ecology, applied ecology, biogeochemistry, community and ecosystem ecology, limnology, oceanography, physiological ecology, plant ecology, population ecology, theoretical ecology, vertebrate zoology; evolutionary biology (PhD), including ecological genetics, paleobiology, population biology, systematics. *Faculty:* 62 full-time (11 women). *Students:* 61 full-time (33 women); includes 6 minority (1 African American, 3 Asian Americans or Pacific Islanders, 2 Hispanic Americans), 10 international. 107 applicants, 15% accepted, 12 enrolled. In 2005, 7 doctorates awarded. *Degree requirements:* For doctorate, thesis/dissertation, 2 semesters of teaching experience, comprehensive exam. *Entrance requirements:* For doctorate, GRE General Test, GRE Subject Test (biology), 2 letters of recommendation. Additional exam requirements/recommendations for international students: Required—TOEFL (minimum score 550 paper-based; 213 computer-based). *Application deadline:* For fall admission, 12/15 for domestic students. Application fee: $60. Electronic applications accepted. *Financial support:* In 2005–06, 59 students received support, including 25 fellowships with full tuition reimbursements available, 10 research assistantships with full tuition reimbursements available, 24 teaching assistantships with full tuition reimbursements available; institutionally sponsored loans, scholarships/grants, health care benefits, tuition waivers (full and partial), and unspecified assistantships also available. Financial award applicants required to submit FAFSA. *Faculty research:* Population and organismal biology, population and evolutionary genetics, systematics and macroevolution, biochemistry, conservation biology. *Unit head:* Director of Graduate Studies, 607-254-4230. *Application contact:* Graduate Field Assistant, 607-254-4230, E-mail: eeb_grad_req@cornell.edu.

Emory University, Graduate School of Arts and Sciences, Division of Biological and Biomedical Sciences, Program in Population Biology, Ecology and Evolution, Atlanta, GA 30322-1100. Offers PhD. *Faculty:* 28 full-time (5 women). *Students:* 22 full-time (15 women); includes 3 minority (2 Asian Americans or Pacific Islanders, 1 Hispanic American), 3 international. Average age 27. 18 applicants, 28% accepted, 4 enrolled. In 2005, 1 doctorate awarded. *Degree requirements:* For doctorate, thesis/dissertation, comprehensive exam, registration. *Entrance requirements:* For doctorate, GRE General Test, minimum GPA of 3.0 in science course work. Additional exam requirements/recommendations for international students: Required—TOEFL. *Application deadline:* For fall admission, 1/3 for domestic students, 1/3 for international students. Application fee: $50. Electronic applications accepted. *Expenses:* Tuition: Full-time $14,400.

Required fees: $217. *Financial support:* In 2005–06, 11 students received support, including 11 fellowships with full tuition reimbursements available (averaging $23,000 per year); institutionally sponsored loans, health care benefits, and tuition waivers (full) also available. *Faculty research:* Evolution of microbes, infectious disease, the immune system, genetic disease in humans, evolution of behavior. *Unit head:* Dr. Leslie Real, Director, 404-727-4099, Fax: 404-727-2880, E-mail: lreal@biology.emory.edu. *Application contact:* 404-727-2545, Fax: 404-727-3322, E-mail: gdbbs@emory.edu.

Florida State University, Graduate Studies, College of Arts and Sciences, Department of Biological Science, Program in Evolutionary Biology, Tallahassee, FL 32306. Offers MS, PhD. *Faculty:* 20 full-time (3 women). *Students:* 41 full-time (14 women); includes 4 minority (3 Asian Americans or Pacific Islanders, 1 Hispanic American), 6 international. *Degree requirements:* For master's and doctorate, thesis/dissertation, teaching experience seminar presentation, comprehensive exam, registration. *Entrance requirements:* For master's and doctorate, GRE General Test (minimum 1100: U-500, Q-500), minimum upper division GPA of 3.0. Additional exam requirements/recommendations for international students: Required—TOEFL (minimum score 600 paper-based; 250 computer-based), IB 100. *Application deadline:* For fall admission, 1/15 for domestic students, 12/1 for international students; for spring admission, 10/15 for domestic students, 9/1 for international students. Application fee: $30. *Financial support:* In 2005–06, fellowships with full tuition reimbursements (averaging $19,000 per year), research assistantships with full tuition reimbursements (averaging $19,000 per year), teaching assistantships with full tuition reimbursements (averaging $17,600 per year) were awarded. Financial award application deadline: 1/15; financial award applicants required to submit FAFSA. *Faculty research:* Population biology, community ecology, genetics, evolution of protein structure. *Application contact:* Judy Bowers, Coordinator, Graduate Affairs, 850-644-3023, Fax: 850-644-9829, E-mail: gradinfo@bio.fsu.edu.

George Mason University, College of Science, Fairfax, VA 22030. Offers bioinformatics (MS, PhD); climate dynamics (PhD); computational sciences (MS); computational sciences and informatics (PhD); computational social science (PhD); computational techniques and applications (Certificate); earth systems and geoinformation science (PhD); earth systems science (MS); nanotechnology and nanoscience (Certificate); neuroscience (PhD); physical sciences (PhD); remote sensing and earth image processing (Certificate). Part-time and evening/weekend programs available. *Degree requirements:* For doctorate, thesis/dissertation, comprehensive exam, registration. *Entrance requirements:* For master's and doctorate, GRE General Test, minimum GPA of 3.0 in last 60 hours. Additional exam requirements/recommendations for international students: Required—TOEFL. Electronic applications accepted. *Expenses:* Tuition, area resident: Full-time $5,244; part-time $219 per credit. Tuition, state resident: part-time $651 per credit. Tuition, nonresident: full-time $15,636. Required fees: $1,524; $65 per credit. *Faculty research:* Space sciences and astrophysics, fluid dynamics, materials modeling and simulation, bioinformatics, global changes and statistics.

George Mason University, College of Arts and Sciences, Department of Biology, Program in Biology, Fairfax, VA 22030. Offers bioinformatics (MS); ecology, systematics and evolution (MS); interpretive biology (MS); molecular and microbiology (PhD); molecular, microbial, and cellular biology (MS); organismal biology (MS). Part-time programs available. *Degree requirements:* For master's, thesis or alternative. *Entrance requirements:* For master's, GRE General Test, GRE Subject Test, bachelor's degree in biology or equivalent. Electronic applications accepted. *Expenses:* Tuition: area resident: Full-time $5,244; part-time $219 per credit. Tuition, state resident: part-time $651 per credit. Tuition, nonresident: full-time $15,636. Required fees: $1,524; $65 per credit.

Harvard University, Graduate School of Arts and Sciences, Department of Organismic and Evolutionary Biology, Cambridge, MA 02138. Offers biology (PhD). *Students:* 57 full-time (22 women). 98 applicants, 13% accepted. In 2005, 11 doctorates awarded. *Degree requirements:* For doctorate, 2 foreign languages. *Entrance requirements:* For doctorate, GRE General Test, GRE Subject Test (recommended), 7 courses in biology, chemistry, physics, mathematics, computer science, or geology. Additional exam requirements/recommendations for international students: Required—TOEFL. *Application deadline:* For fall admission, 12/15 for domestic students. Application fee: $60. *Expenses:* Tuition: Full-time $28,752. Full-time tuition and fees vary according to program and student level. *Financial support:* Fellowships, research assistantships, teaching assistantships, career-related internships or fieldwork, Federal Work-Study, and institutionally sponsored loans available. Financial award application deadline: 12/30. *Unit head:* Betsey Cogswell, Administrator, 617-495-5497, Fax: 617-495-5264. *Application contact:* Departmental Office, 617-495-2305.

Announcement: Doctoral studies in evolution, behavior, development and physiology, comparative structure and function, ecology and population biology, genetics and molecular evolution, paleobiology, systematics, and biodiversity. Research facilities include modern laboratory facilities, historically valuable botanical and zoological collections, greenhouses and animal-care facilities, a large-animal research laboratory, a 700-acre conservation preserve nearby, and scanning and electron microscopes. Other Harvard resources available include the Arnold Arboretum living collection, the Harvard forestland and laboratory complex, facilities for activities ranging from DNA and protein sequencing to electronic instrument production, computing facilities and Internet connectivity, and machine shops. Applications due December 15. Additional information is available on the Web at http://www.oeb.harvard.edu.

Indiana University Bloomington, Graduate School, College of Arts and Sciences, Department of Biology, Program in Evolution, Ecology, and Behavior, Bloomington, IN 47405-7000. Offers MA, PhD. PhD offered through the University Graduate School. Part-time programs available. *Faculty:* 21 full-time (7 women). *Students:* 72 full-time (41 women); includes 6 minority (1

Evolutionary Biology

African American, 1 Asian American or Pacific Islander, 4 Hispanic Americans), 7 international. In 2005, 6 degrees awarded. Terminal master's awarded for partial completion of doctoral program. *Degree requirements:* For master's, thesis or alternative; for doctorate, thesis/dissertation. *Entrance requirements:* For master's and doctorate, GRE General Test. Additional exam requirements/recommendations for international students: Required—TOEFL. *Application deadline:* For fall admission, 1/5 for domestic students; for spring admission, 9/1 for domestic students. Applications are processed on a rolling basis. Application fee: $45. Electronic applications accepted. *Expenses:* Tuition, state resident: full-time $5,437; part-time $227 per credit hour. Tuition, nonresident: full-time $15,836; part-time $660 per credit hour. Required fees: $821. Tuition and fees vary according to campus/location and program. *Financial support:* In 2005–06, 72 students received support, including fellowships with tuition reimbursements available (averaging $20,000 per year), research assistantships with tuition reimbursements available (averaging $18,000 per year), teaching assistantships with tuition reimbursements available (averaging $18,000 per year); scholarships/grants and tuition waivers (full) also available. Financial award application deadline: 1/15. *Faculty research:* Ecosystem of community, plant and animal population biology, avian sociobiology, fish ethology, evolutionary genetics. *Unit head:* Mike Wade, Director, 812-856-4680, E-mail: mjwade@indiana.edu. *Application contact:* Gretchen Clearwater, Adviser for Graduate Affairs, 812-855-1861, Fax: 812-855-6705, E-mail: biograd@bio.indiana.edu.

Iowa State University of Science and Technology, Graduate College, College of Liberal Arts and Sciences, Department of Ecology, Evolution, and Organismal Biology, Ames, IA 50011. Offers MS, PhD. *Faculty:* 28 full-time, 3 part-time/adjunct. *Students:* 38 full-time (23 women), 6 part-time (2 women); includes 2 minority (1 African American, 1 American Indian/Alaska Native), 9 international. 11 applicants, 100% accepted, 7 enrolled. In 2005, 8 master's, 3 doctorates awarded. *Degree requirements:* For master's, thesis or alternative; for doctorate, thesis/dissertation. *Entrance requirements:* For master's and doctorate, GRE General Test. Additional exam requirements/recommendations for international students: Required—TOEFL (paper score 530; computer score 197) or IELTS. *Application deadline:* For fall admission, 1/1 priority date for domestic students, 1/1 priority date for international students; for spring admission, 9/1 priority date for domestic students, 9/1 priority date for international students. Application fee: $30 ($70 for international students). Electronic applications accepted. *Expenses:* Tuition, state resident: full-time $6,410. Tuition, nonresident: full-time $16,422. Tuition and fees vary according to program. *Financial support:* In 2005–06, 21 research assistantships with partial tuition reimbursements (averaging $15,968 per year), 7 teaching assistantships with partial tuition reimbursements (averaging $15,224 per year) were awarded; fellowships, scholarships/grants, health care benefits, and unspecified assistantships also available. *Faculty research:* Aquatic and wetland ecology, cytology, ecology, physiology and molecular biology, systematics and evolution. *Unit head:* Dr. Jonathan Wendel, Chair, 515-294-7172.

Iowa State University of Science and Technology, Graduate College, Interdisciplinary Programs, Program in Ecology and Evolutionary Biology, Ames, IA 50011. Offers MS, PhD. *Students:* 45 full-time (27 women), 4 part-time (2 women); includes 1 minority (American Indian/Alaska Native), 8 international. 24 applicants, 0% accepted, 0 enrolled. In 2005, 7 master's, 2 doctorates awarded. *Degree requirements:* For master's, thesis or alternative; for doctorate, thesis/dissertation. *Entrance requirements:* For master's and doctorate, GRE General Test, application to cooperating department. Additional exam requirements/recommendations for international students: Required—TOEFL or IELTS. *Application deadline:* For fall admission, 1/15 priority date for domestic students, 1/15 priority date for international students. Application fee: $30 ($70 for international students). Electronic applications accepted. *Expenses:* Tuition, state resident: full-time $6,410. Tuition, nonresident: full-time $16,422. Tuition and fees vary according to program. *Financial support:* In 2005–06, 33 research assistantships with full and partial tuition reimbursements (averaging $16,188 per year), 8 teaching assistantships with full and partial tuition reimbursements (averaging $15,239 per year) were awarded; scholarships/grants, health care benefits, and unspecified assistantships also available. *Faculty research:* Landscape ecology, aquatic and method ecology, physiological ecology, population genetics and evolution, systematics. *Unit head:* Dr. Carol Vleck, Supervisory Committee Chair, 515-294-6518, E-mail: eeboffice@iastate.edu. *Application contact:* Charles Sauer, Information Contact, 515-294-6518, E-mail: eeboffice@lastate.edu.

The Johns Hopkins University, School of Medicine, Graduate Programs in Medicine, Center for Functional Anatomy and Evolution, Baltimore, MD 21218-2699. Offers PhD. *Faculty:* 5 full-time (1 woman), 1 (woman) part-time/adjunct. *Students:* 8 full-time (2 women). Average age 25. 16 applicants, 13% accepted, 2 enrolled. In 2005, 3 degrees awarded. *Degree requirements:* For doctorate, thesis/dissertation, oral exams, comprehensive exam. *Entrance requirements:* For doctorate, GRE. Additional exam requirements/recommendations for international students: Required—TOEFL. *Application deadline:* For fall admission, 1/1 for domestic students, 1/1 for international students. Application fee: $60. *Expenses:* Tuition: Full-time $30,960. Tuition and fees vary according to degree level and program. *Financial support:* In 2005–06, 8 teaching assistantships with full tuition reimbursements (averaging $24,600 per year) were awarded; fellowships, career-related internships or fieldwork, institutionally sponsored loans, health care benefits, and tuition waivers (full) also available. *Faculty research:* Vertebrate evolution, functional anatomy, primate evolution, vertebrate paleobiology, vertebrate morphology. *Unit head:* Dr. Kenneth D. Rose, Director, 410-955-7172, Fax: 410-614-9030, E-mail: kdrose@jhmi.edu. *Application contact:* Catherine L. Will, Coordinator, Graduate Student Affairs, 410-614-3385, E-mail: grad_study@som.adm.jhu.edu.

Marquette University, Graduate School, College of Arts and Sciences, Department of Biology, Milwaukee, WI 53201-1881. Offers cell biology (MS, PhD); developmental biology (MS, PhD); ecology (MS, PhD); endocrinology (MS, PhD); evolutionary biology (MS, PhD); genetics (MS, PhD); microbiology (MS, PhD); molecular biology (MS, PhD); muscle and exercise physiology (MS, PhD); neurobiology (MS, PhD); reproductive physiology (MS, PhD). Terminal master's awarded for partial completion of doctoral program. *Degree requirements:* For master's, thesis, 1 year of teaching experience or equivalent, comprehensive exam; for doctorate, thesis/dissertation, 1 year of teaching experience or equivalent, qualifying exam. *Entrance requirements:* For master's and doctorate, GRE General Test, GRE Subject Test. Additional exam requirements/recommendations for international students: Required—TOEFL. *Faculty research:* Microbial and invertebrate ecology, evolution of gene function, DNA methylation, DNA arrangement.

Michigan State University, The Graduate School, College of Natural Science, Interdepartmental Program in Ecology, Evolutionary Biology and Behavior, East Lansing, MI 48824.

See Close-Up on page 701.

Northwestern University, Northwestern University Feinberg School of Medicine and Interdepartmental Degree Programs, Integrated Graduate Programs in the Life Sciences, Chicago, IL 60611. Offers cancer biology (PhD); cell biology (PhD); developmental biology (PhD); evolutionary biology (PhD); immunology and microbial pathogenesis (PhD); molecular biology and genetics (PhD); neurobiology (PhD); pharmacology and toxicology (PhD); structural biology and biochemistry (PhD). *Degree requirements:* For doctorate, thesis/dissertation, written and oral qualifying exams, comprehensive exam. *Entrance requirements:* For doctorate, GRE General Test. Additional exam requirements/recommendations for international students: Required—TOEFL (minimum score 600 paper-based; 250 computer-based). Electronic applications accepted.

See Close-Up on page 189.

The Ohio State University, Graduate School, College of Biological Sciences, Department of Evolution, Ecology, and Organismal Biology, Columbus, OH 43210. Offers MS, PhD. *Degree requirements:* For master's, thesis optional; for doctorate, thesis/dissertation. *Entrance requirements:* For master's and doctorate, GRE General Test. Additional exam requirements/recommendations for international students: Required—TOEFL (minimum score 600 paper-based; 250 computer-based), TSE (minimum score 60). Electronic applications accepted.

Ohio University, Graduate Studies, College of Arts and Sciences, Department of Biological Sciences, Athens, OH 45701-2979. Offers biological sciences (MS, PhD); cell biology and physiology (MS, PhD); ecology and evolutionary biology (MS, PhD); exercise physiology and muscle biology (MS, PhD); microbiology (MS, PhD); neuroscience (MS, PhD). *Faculty:* 51 full-time (17 women), 6 part-time/adjunct (1 woman). *Students:* 88 full-time (40 women), 1 part-time; includes 1 minority (Hispanic American), 41 international. Average age 24. 50 applicants, 24% accepted, 10 enrolled. In 2005, 9 master's, 12 doctorates awarded. *Median time to degree:* Of those who began their doctoral program in fall 1997, 90% received their degree in 8 years or less. *Degree requirements:* For master's, thesis, 1 quarter of teaching experience; for doctorate, thesis/dissertation, 2 quarters of teaching experience, comprehensive exam. *Entrance requirements:* For master's and doctorate, GRE General Test. Additional exam requirements/recommendations for international students: Required—TOEFL (minimum score 620 paper-based; 260 computer-based). *Application deadline:* For fall admission, 1/15 for domestic students, 1/15 for international students. Application fee: $45. Electronic applications accepted. *Financial support:* In 2005–06, 87 students received support, including 2 fellowships with full tuition reimbursements available (averaging $15,000 per year), 10 research assistantships with full tuition reimbursements available (averaging $15,500 per year), 75 teaching assistantships with full tuition reimbursements available (averaging $15,500 per year); Federal Work-Study, institutionally sponsored loans, and tuition waivers (full) also available. Financial award application deadline: 1/15. *Faculty research:* Ecology and evolutionary biology, exercise physiology and muscle biology, neurobiology, cell biology, physiology. Total annual research expenditures: $2.8 million. *Unit head:* , Dr. Ralph DiCaprio, Chair, 740-593-2290, Fax: 740-593-0300, E-mail: dicaprir@ohio.edu. *Application contact:* Dr. Donald B. Miles, Graduate Chair, 740-593-2317, Fax: 740-593-0300, E-mail: milesd@ohio.edu.

See Close-Up on page 191.

The Pennsylvania State University University Park Campus, Graduate School, Eberly College of Science, Department of Biology, State College, University Park, PA 16802-1503. Offers biology (MS, PhD); molecular evolutionary biology (MS, PhD). *Students:* 51 full-time (24 women), 1 part-time; includes 5 minority (2 African Americans, 2 Asian Americans or Pacific Islanders, 1 Hispanic American), 30 international. *Entrance requirements:* For master's and doctorate, GRE General Test. Application fee: $45. *Expenses:* Tuition, state resident: full-time $12,518; part-time $522 per credit. Tuition, nonresident: full-time $23,004; part-time $959 per credit. Required fees: $484. Tuition and fees vary according to course load, campus/location and program. *Financial support:* Fellowships, research assistantships, teaching assistantships available. *Unit head:* Dr. Douglas R. Cavener, Head, 814-865-5497, Fax: 814-865-9131, E-mail: drc9@psu.edu. *Application contact:* Dr. Douglas R. Cavener, Head, 814-865-5497, Fax: 814-865-9131, E-mail: drc9@psu.edu.

See Close-Up on page 193.

Princeton University, Graduate School, Department of Ecology and Evolutionary Biology, Princeton, NJ 08544-1019. Offers biology (PhD); neuroscience (PhD). *Degree requirements:* For doctorate, thesis/dissertation. *Entrance requirements:* For doctorate, GRE General Test, GRE Subject Test. Additional exam requirements/recommendations for international students: Required—TOEFL (minimum score 600 paper-based; 250 computer-based). Electronic applications accepted.

See Close-Up on page 705.

Purdue University, Graduate School, School of Science, Department of Biological Sciences, West Lafayette, IN 47907. Offers biochemistry (PhD); biophysics (PhD); cell and developmental biology (PhD); ecology, evolutionary and population biology (MS, PhD), including ecology, evolutionary biology, population biology; genetics (MS, PhD); microbiology (MS, PhD); molecular biology (PhD); neurobiology (MS, PhD); plant physiology (PhD). *Faculty:* 47 full-time (9 women), 4 part-time/adjunct (1 woman). *Students:* 97 full-time (53 women), 8 part-time (4 women); includes 13 minority (3 African Americans, 1 American Indian/Alaska Native, 4 Asian Americans or Pacific Islanders, 5 Hispanic Americans), 50 international. Average age 28. 168 applicants, 29% accepted, 23 enrolled. In 2005, 18 master's, 9 doctorates awarded. Terminal master's awarded for partial completion of doctoral program. *Degree requirements:* For master's, thesis (for some programs); for doctorate, thesis/dissertation, seminars, teaching experience. *Entrance requirements:* For master's and doctorate, GRE General Test. Additional exam requirements/recommendations for international students: Required—TOEFL, TSE. *Application deadline:* For fall admission, 2/15 for domestic students, 1/31 for international students. Applications are processed on a rolling basis. Application fee: $55. Electronic applications accepted. *Financial support:* In 2005–06, 15 fellowships, 60 research assistantships, 53 teaching assistantships were awarded. Support available to part-time students. Financial award application deadline: 2/15; financial award applicants required to submit FAFSA. *Unit head:* Dr. Richard J Kuhn, Head, 765-494-4407. *Application contact:* Nancy Konopka, Graduate Studies Office Manager, 765-494-8142, Fax: 765-494-0876, E-mail: njk@bilbo.bio.purdue.edu.

Rice University, Graduate Programs, Wiess School of Natural Sciences, Department of Ecology and Evolutionary Biology, Houston, TX 77251-1892. Offers MA, PhD. Terminal master's awarded for partial completion of doctoral program. *Degree requirements:* For master's, thesis (for some programs), comprehensive exam (for some programs), registration; for doctorate, thesis/dissertation, comprehensive exam, registration. *Entrance requirements:* For master's and doctorate, GRE General Test, GRE Subject Test. Additional exam requirements/recommendations for international students: Required—TOEFL. Electronic applications accepted. *Faculty research:* Trace gas emissions, wetlands, biology, community ecology of forests and grasslands, conservation biology specialization.

Rutgers, The State University of New Jersey, New Brunswick/Piscataway, Graduate School, Program in Ecology and Evolution, New Brunswick, NJ 08901-1281. Offers MS, PhD. Part-time programs available. *Faculty:* 72 full-time, 4 part-time/adjunct. *Students:* 77 full-time (43 women), 19 part-time (4 women); includes 6 minority (1 African American, 1 Asian American or Pacific Islander, 4 Hispanic Americans), 11 international. Average age 31. 71 applicants, 41% accepted, 20 enrolled. In 2005, 5 master's, 7 doctorates awarded. Terminal master's awarded for partial completion of doctoral program. *Degree requirements:* For master's, comprehensive exam; for doctorate, thesis/dissertation, comprehensive exam. *Entrance requirements:* For master's and doctorate, GRE General Test, minimum GPA of 3.0. Additional exam requirements/recommendations for international students: Required—TOEFL (minimum score 550 paper-based; 213 computer-based). *Application deadline:* For fall admission, 4/1 for domestic students, 4/1 for international students. Application fee: $50. Electronic applications accepted. *Expenses:* Tuition, state resident: full-time $10,440; part-time $435 per credit. Tuition, nonresident: full-time $15,520; part-time $647 per credit. Required fees: $129 per credit. Tuition and fees vary according to program. *Financial support:* In 2005–06, 4 fellowships with full tuition reimbursements (averaging $25,000 per year), research assistantships with full tuition reimbursements (averaging $15,400 per year), 22 teaching assistantships with full tuition reimbursements (averaging $16,988 per year) were awarded; Federal Work-Study also available. Financial award application deadline: 1/15; financial award applicants required to submit FAFSA. *Faculty research:* Population and community ecology, population genetics, evolutionary biology, conservation biology, ecosystem ecology. *Unit head:* Dr. Peter J. Morin, Director, Graduate Program, 732-932-3214, Fax: 732-932-8764, E-mail: pjmorin@rci.rutgers.edu. *Application contact:* Marsha Morin, Secretary, 732-932-3213, Fax: 732-932-8764, E-mail: mmorin@aesop.rutgers.edu.

Stony Brook University, State University of New York, Graduate School, College of Arts and Sciences, Department of Ecology and Evolution, Stony Brook, NY 11794. Offers PhD. *Faculty:* 17 full-time (4 women), 1 part-time/adjunct (1 woman). *Students:* 49 full-time (20 women), 1 (woman) part-time; includes 7 minority (3 Asian Americans or Pacific Islanders, 4 Hispanic Americans), 8 international. Average age 28. 84 applicants, 36% accepted. In 2005, 4 degrees awarded. *Degree requirements:* For doctorate, one foreign language, thesis/dissertation, teaching experience, comprehensive exam. *Entrance requirements:* For doctor-

Evolutionary Biology

Stony Brook University, State University of New York *(continued)*
ate, GRE General Test, GRE Subject Test. Additional exam requirements/recommendations for international students: Required—TOEFL. *Application deadline:* For fall admission, 1/15 for domestic students. Application fee: $50. *Expenses:* Tuition, state resident: full-time $6,900; part-time $288 per credit. Tuition, nonresident: full-time $10,920; part-time $455 per credit. Required fees: $704. *Financial support:* In 2005–06, 8 fellowships, 13 research assistantships, 23 teaching assistantships were awarded; Federal Work-Study also available. *Faculty research:* Theoretical and experimental population genetics, numerical taxonomy, biostatistics, population and community ecology, plant ecology. Total annual research expenditures: $1.8 million. *Unit head:* Dr. Charles H. Janson, Chair, 631-632-8600. *Application contact:* Dr. Dan Dykhuizen, Director, 631-246-8604, E-mail: dandyk@life.bio.sunysb.edu.

See Close-Up on page 709.

University at Albany, State University of New York, College of Arts and Sciences, Department of Biological Sciences, Specialization in Ecology, Evolution, and Behavior, Albany, NY 12222-0001. Offers MS, PhD. *Degree requirements:* For master's, one foreign language; for doctorate, one foreign language, thesis/dissertation. *Entrance requirements:* For master's and doctorate, GRE General Test. Application fee: $60. *Financial support:* Minority assistantships available. *Unit head:* Dr. Albert Millis, Chair, Department of Biological Sciences, 518-442-4300.

University of Alberta, Faculty of Graduate Studies and Research, Department of Biological Sciences, Edmonton, AB T6G 2E1, Canada. Offers environmental biology and ecology (M Sc, PhD); microbiology and biotechnology (M Sc, PhD); molecular biology and genetics (M Sc, PhD); physiology and cell biology (M Sc, PhD); plant biology (M Sc, PhD); systematics and evolution (M Sc, PhD). *Faculty:* 72 full-time (15 women), 15 part-time/adjunct (4 women). *Students:* 238 full-time (117 women), 32 part-time (15 women), 31 international. 206 applicants, 42% accepted. In 2005, 29 master's, 31 doctorates awarded. Terminal master's awarded for partial completion of doctoral program. *Degree requirements:* For master's and doctorate, thesis/dissertation, registration. *Entrance requirements:* Additional exam requirements/recommendations for international students: Required—TOEFL. *Application deadline:* For fall admission, 3/1 for domestic students. Applications are processed on a rolling basis. Application fee: $0. Tuition and fees charges are reported in Canadian dollars. *Expenses:* Tuition, state resident: part-time $562 Canadian dollars per term. Tuition, nonresident: full-time $3,375 Canadian dollars. Required fees: $573 Canadian dollars; $84 Canadian dollars per term. *Financial support:* In 2005–06, 4 research assistantships with partial tuition reimbursements (averaging $12,300 per year), 103 teaching assistantships with partial tuition reimbursements (averaging $12,300 per year) were awarded; career-related internships or fieldwork and scholarships/grants also available. *Unit head:* Laura Frost, Chair, 780-492-1904. *Application contact:* Dr. John P. Chang, Associate Chair for Graduate Studies, 780-492-1257, Fax: 780-492-9457, E-mail: bio.grad.coordinator@ualberta.ca.

The University of Arizona, Graduate College, College of Science, Department of Ecology and Evolutionary Biology, Tucson, AZ 85721. Offers MS, PhD. *Degree requirements:* For doctorate, one foreign language, thesis/dissertation, comprehensive exam. *Entrance requirements:* For master's and doctorate, GRE General Test, GRE Subject Test. Additional exam requirements/recommendations for international students: Required—TOEFL. *Faculty research:* Biological diversity, evolutionary history, evolutionary mechanisms, community structure.

University of California, Davis, Graduate Studies, Graduate Group in Population Biology, Davis, CA 95616. Offers PhD. *Faculty:* 37 full-time. *Students:* 34 full-time (18 women); includes 5 minority (1 African American, 4 Asian Americans or Pacific Islanders), 2 international. Average age 28. 70 applicants, 19% accepted, 7 enrolled. In 2005, 3 doctorates awarded. *Median time to degree:* Of those who began their doctoral program in fall 1997, 100% received their degree in 8 years or less. *Degree requirements:* For doctorate, thesis/dissertation. *Entrance requirements:* For doctorate, GRE General Test, GRE Subject Test. Additional exam requirements/recommendations for international students: Required—TOEFL (minimum score 550 paper-based; 213 computer-based). *Application deadline:* For fall admission, 1/2 for domestic students, 1/2 for international students. Application fee: $60. Electronic applications accepted. *Financial support:* In 2005–06, 34 students received support, including 18 fellowships with full and partial tuition reimbursements available (averaging $17,527 per year), 4 research assistantships with full and partial tuition reimbursements available (averaging $15,376 per year), 10 teaching assistantships with partial tuition reimbursements available (averaging $15,191 per year); Federal Work-Study, scholarships/grants, tuition waivers (full and partial), and unspecified assistantships also available. Financial award application deadline: 1/15; financial award applicants required to submit FAFSA. *Faculty research:* Population ecology, population genetics, systematics, evolution, community ecology. *Unit head:* Jay A. Rosenheim, Chair, 530-752-4395, Fax: 530-752-1449, E-mail: jarosenheim@ucdavis.edu. *Application contact:* Stephanie Macey-Gallow, Administrative Assistant, 530-752-1274, E-mail: gradcoordinator@ucdavis.edu.

University of California, Irvine, Office of Graduate Studies, School of Biological Sciences, Department of Ecology and Evolutionary Biology, Irvine, CA 92697. Offers biological sciences (MS, PhD). *Degree requirements:* For master's and doctorate, thesis/dissertation. *Entrance requirements:* For master's and doctorate, GRE General Test, GRE Subject Test, minimum GPA of 3.0. Additional exam requirements/recommendations for international students: Required—TOEFL (minimum score 550 paper-based; 213 computer-based). Electronic applications accepted. *Faculty research:* Ecological energetics, quantitative genetics, life history evolution, plant-herbivore and plant-pollinator interactions, molecular evolution.

University of California, Riverside, Graduate Division, Department of Biology, Riverside, CA 92521-0102. Offers biology (MS, PhD); evolution, ecology and organismal biology (MS, PhD). Department also affiliated with following interdepartmental graduate programs: Cell, Molecular, and Developmental Biology; Evolution and Ecology; Genetics.. *Faculty:* 22 full-time (5 women). *Students:* 49 full-time (25 women), 1 (woman) part-time; includes 3 minority (1 Asian American or Pacific Islander, 2 Hispanic Americans), 6 international. Average age 30. In 2005, 6 master's, 3 doctorates awarded. Terminal master's awarded for partial completion of doctoral program. *Degree requirements:* For master's, oral defense of thesis; for doctorate, thesis/dissertation, 3 quarters of teaching experience, qualifying exams. *Entrance requirements:* For master's and doctorate, GRE General Test, minimum GPA of 3.2. Additional exam requirements/recommendations for international students: Required—TOEFL (minimum score 550 paper-based; 213 computer-based); Recommended—TSE (minimum score 50). *Application deadline:* For fall admission, 5/1 for domestic students, 2/1 for international students. For winter admission, 9/1 for domestic students; for spring admission, 12/1 for domestic students. Applications are processed on a rolling basis. Application fee: $60 ($75 for international students). Electronic applications accepted. *Expenses:* Tuition, nonresident: full-time $14,694. Required fees: $9,009. Full-time tuition and fees vary according to program. *Financial support:* In 2005–06, research assistantships (averaging $14,000 per year), teaching assistantships with tuition reimbursements (averaging $15,000 per year) were awarded; fellowships, career-related internships or fieldwork, Federal Work-Study, institutionally sponsored loans, and tuition waivers (full and partial) also available. Financial award application deadline: 1/5; financial award applicants required to submit FAFSA. *Faculty research:* Molecular genetics, neurophysiology, evolutionary biology, physiology and organismal biology, signal transduction. *Unit head:* Dr. Richard Cardullo, Chair, 951-827-5901, Fax: 951-827-4286. *Application contact:* Zina Romero, Graduate Program Assistant, 800-735-0717, Fax: 951-827-5913, E-mail: biograd@ucr.edu.

University of California, San Diego, Graduate Studies and Research, Division of Biology, Program in Ecology, Behavior, and Evolution, La Jolla, CA 92093. Offers PhD. *Degree requirements:* For doctorate, thesis/dissertation, qualifying exam. Electronic applications accepted.

Announcement: Research in the Ecology, Behavior & Evolution Section is focused on investigating ecological and evolutionary processes operating at the level of populations, species, and communities. Major topics include understanding the distribution and abundance of organisms, behavioral adaptations to environmental and social challenges, the patterns of evolutionary change, and the application of all of these areas to the conservation of biodiversity. Visit http://biology.ucsd.edu/biosections/ebe.html for more information.

University of California, Santa Barbara, Graduate Division, College of Letters and Sciences, Division of Mathematics, Life, and Physical Sciences, Department of Ecology, Evolution, and Marine Biology, Santa Barbara, CA 93106. Offers MA, PhD. *Faculty:* 36 full-time (7 women), 6 part-time/adjunct (1 woman). *Students:* 61 full-time (29 women); includes 10 minority (1 African American, 6 Asian Americans or Pacific Islanders, 3 Hispanic Americans), 3 international. Average age 29. 150 applicants, 9% accepted, 4 enrolled. In 2005, 3 master's, 6 doctorates awarded. Terminal master's awarded for partial completion of doctoral program. *Median time to degree:* Of those who began their doctoral program in fall 1997, 90% received their degree in 8 years or less. *Degree requirements:* For master's, thesis (for some programs), comprehensive exam (for some programs), registration; for doctorate, thesis/dissertation, comprehensive exam, registration. *Entrance requirements:* For master's and doctorate, GRE General Test. Additional exam requirements/recommendations for international students: Required—TOEFL (minimum score 550 paper-based; 213 computer-based). *Application deadline:* For fall admission, 12/15 for domestic students, 12/15 for international students. Application fee: $60. Electronic applications accepted. *Financial support:* In 2005–06, 30 students received support, including 44 fellowships with full and partial tuition reimbursements available, 90 teaching assistantships with partial tuition reimbursements available (averaging $5,025 per year); research assistantships with full and partial tuition reimbursements available, career-related internships or fieldwork, Federal Work-Study, institutionally sponsored loans, scholarships/grants, traineeships, health care benefits, tuition waivers (full and partial), and unspecified assistantships also available. Financial award application deadline: 12/15; financial award applicants required to submit FAFSA. *Faculty research:* Ecology, population genetics, stream ecology, evolution, marine biology. *Unit head:* Dr. Alice Alldredge, Chair, 805-893-2415, Fax: 805-893-4724. *Application contact:* Jennifer Chater, Staff Graduate Advisor, 805-893-3023, Fax: 805-893-4724, E-mail: eemb-info@lifesci.ucsb.edu.

University of California, Santa Cruz, Division of Graduate Studies, Division of Physical and Biological Sciences, Department of Ecology and Evolutionary Biology, Santa Cruz, CA 95064. Offers MA, PhD. *Faculty:* 18 full-time (5 women). *Students:* 62 full-time (42 women), 1 (woman) part-time; includes 6 minority (1 American Indian/Alaska Native, 1 Asian American or Pacific Islander, 4 Hispanic Americans), 1 international. 123 applicants, 9% accepted, 8 enrolled. In 2005, 2 master's, 5 doctorates awarded. *Entrance requirements:* Additional exam requirements/recommendations for international students: Required—TOEFL; Recommended—IELT. *Application deadline:* For fall admission, 12/15 for domestic students. Application fee: $60. Electronic applications accepted. *Expenses:* Tuition, nonresident: full-time $14,694. Required fees: $9,437. *Unit head:* Peter Raimondi, Chairperson. *Application contact:* Susan Thuringer, Information Contact, 831-459-4715, E-mail: susan@biology.ucsc.edu.

University of Chicago, Division of the Biological Sciences, Department of Darwinian Sciences: Ecological, Integrative and Evolutionary Biology, Committee on Evolutionary Biology, Chicago, IL 60637-1513. Offers functional and evolutionary biology (PhD). *Faculty:* 49 full-time (7 women). *Students:* 28 full-time (14 women); includes 4 minority (1 African American, 1 Asian American or Pacific Islander, 2 Hispanic Americans), 5 international. Average age 28. In 2005, 5 doctorates awarded. *Degree requirements:* For doctorate, thesis/dissertation, registration. *Entrance requirements:* For doctorate, GRE General Test. Additional exam requirements/recommendations for international students: Required—TOEFL. *Application deadline:* For fall admission, 12/28 priority date for domestic students, 12/28 priority date for international students. Application fee: $55. Electronic applications accepted. *Financial support:* In 2005–06, 20 students received support, including fellowships with tuition reimbursements available (averaging $26,301 per year); institutionally sponsored loans, scholarships/grants, traineeships, and health care benefits also available. *Faculty research:* Systematics and evolutionary theory, genetics, functional morphology and physiology, behavior, ecology and biogeography. *Unit head:* Dr. David Jablonski, Chairman, 773-702-8163, Fax: 773-702-4699, E-mail: djablons@uchicago.edu. *Application contact:* Carolyn Johnson, Graduate Administrative Director, 773-702-9474, Fax: 773-702-4699, E-mail: csjohnso@uchicago.edu.

University of Colorado at Boulder, Graduate School, College of Arts and Sciences, Department of Ecology and Evolutionary Biology, Boulder, CO 80309. Offers animal behavior (MA); biology (MA, PhD); environmental biology (MA, PhD); evolutionary biology (MA, PhD); neurobiology (MA); population biology (MA); population genetics (PhD). *Faculty:* 26 full-time (5 women). *Students:* 44 full-time (24 women), 22 part-time (14 women); includes 16 minority (1 American Indian/Alaska Native, 3 Asian Americans or Pacific Islanders, 12 Hispanic Americans), 2 international. Average age 30. 20 applicants, 90% accepted. In 2005, 7 master's, 7 doctorates awarded. Terminal master's awarded for partial completion of doctoral program. *Degree requirements:* For master's, thesis or alternative, comprehensive exam; for doctorate, thesis/dissertation, comprehensive exam. *Entrance requirements:* For master's, GRE General Test, GRE Subject Test, minimum undergraduate GPA of 3.0; for doctorate, GRE General Test, GRE Subject Test. *Application deadline:* For fall admission, 1/2 priority date for domestic students, 12/1 priority date for international students. Application fee: $60 ($60 for international students). *Financial support:* In 2005–06, fellowships (averaging $7,844 per year), research assistantships (averaging $15,663 per year), teaching assistantships (averaging $13,829 per year) were awarded; Federal Work-Study, institutionally sponsored loans, and tuition waivers (full) also available. Financial award application deadline: 3/1. *Faculty research:* Behavior, ecology, genetics, morphology, endocrinology. Total annual research expenditures: $5.2 million. *Unit head:* Jeffry Mitton, Chair, 303-492-0505, Fax: 303-492-8699, E-mail: mitton@colorado.edu. *Application contact:* Jill Skarstadt, Graduate Coordinator, 303-492-7654, Fax: 303-492-8699, E-mail: skarstad@colorado.edu.

University of Delaware, College of Arts and Sciences, Department of Biological Sciences, Newark, DE 19716. Offers biotechnology (MS); cancer biology (MS, PhD); cell and extracellular matrix biology (MS, PhD); cell and systems physiology (MS, PhD); developmental biology (MS, PhD); ecology and evolution (MS, PhD); microbiology (MS, PhD); molecular biology and genetics (MS, PhD). *Faculty:* 39 full-time (11 women). *Students:* 64 full-time (46 women), 2 part-time; includes 7 minority (4 African Americans, 2 Asian Americans or Pacific Islanders, 1 Hispanic American), 18 international. Average age 26. 113 applicants, 31% accepted, 19 enrolled. In 2005, 5 master's, 4 doctorates awarded. Terminal master's awarded for partial completion of doctoral program. *Median time to degree:* Of those who began their doctoral program in fall 1997, 100% received their degree in 8 years or less. *Degree requirements:* For master's, thesis/dissertation, preliminary exam; for doctorate, thesis/dissertation, preliminary exam, comprehensive exam. *Entrance requirements:* For master's and doctorate, GRE General Test. Additional exam requirements/recommendations for international students: Required—TOEFL (minimum score 600 paper-based; 250 computer-based); Recommended—TWE, TSE. *Application deadline:* For fall admission, 4/15 for domestic students, 1/15 for international students; for spring admission, 10/1 for domestic students. Applications are processed on a rolling basis. Application fee: $60. Electronic applications accepted. *Financial support:* In 2005–06, 26 students received support, including fellowships with full tuition reimbursements available (averaging $19,000 per year), 19 research assistantships with full tuition reimbursements available (averaging $19,000 per year), 26 teaching assistantships with full tuition reimbursements available (averaging $19,000 per year); tuition waivers (partial) also available. Financial award application deadline: 4/15. *Faculty research:* Microorganisms, bone, cancer metastasis, developmental biology, cell biology, DNA. Total annual research expenditures: $8.3 million. *Unit head:* Dr. Daniel D. Carson, Chair, 302-831-6977, Fax: 302-831-2281, E-mail: dcarson@udel.edu. *Application contact:* Dr. Melinda K. Duncan, Graduate Coordinator, 302-831-1841, Fax: 302-831-2281, E-mail: danders@udel.edu.

Evolutionary Biology

University of Guelph, Graduate Program Services, College of Biological Science, Department of Integrative Biology, Guelph, ON N1G 2W1, Canada. Offers botany (M Sc, PhD); zoology (M Sc, PhD). Part-time programs available. *Faculty:* 37 full-time (5 women). *Students:* 100 full-time (53 women), 3 part-time (all women); includes 5 minority (all Asian Americans or Pacific Islanders), 6 international. Average age 23. 23 applicants, 83% accepted, 19 enrolled. In 2005, 26 master's, 7 doctorates awarded. *Median time to degree:* Of those who began their doctoral program in fall 1997, 95% received their degree in 8 years or less. *Degree requirements:* For master's, thesis, research proposal; for doctorate, thesis/dissertation, research proposal, qualifying exam. *Entrance requirements:* For master's, minimum B average during previous 2 years of course work; for doctorate, minimum A- average. Additional exam requirements/recommendations for international students: Required—TOEFL (minimum score 550 paper-based; 213 computer-based), IELT (minimum score 7). *Application deadline:* For fall admission, 7/1 for domestic students. For winter admission, 11/1 for domestic students; for spring admission, 3/1 for domestic students. Applications are processed on a rolling basis. Application fee: $75. Electronic applications accepted. *Financial support:* In 2005–06, 50 students received support, including 48 research assistantships, 78 teaching assistantships (averaging $4,606 per year); fellowships *Faculty research:* Aquatic science, environmental physiology, parasitology, wildlife biology, management. Total annual research expenditures: $3.8 million. *Unit head:* Dr. Moira Ferguson, Chair, 519-824-4120 Ext. 53598, Fax: 519-767-1656, E-mail: mmfergus@uoguelph.ca. *Application contact:* Laurie Winn, Graduate Admissions Secretary, 519-824-4320 Ext. 52730, Fax: 519-767-1656, E-mail: lwinn@uoguelph.ca.

University of Hawaii at Manoa, Graduate Division, Specialization in Ecology, Evolution and Conservation Biology, Honolulu, HI 96822. Offers MS, PhD. Program is interdisciplinary; degree is in specific discipline with the specialization in Ecology, Evolution and Conservation Biology. *Degree requirements:* For doctorate, thesis/dissertation. Application fee: $50. *Expenses:* Tuition, state resident: full-time $8,400; part-time $200 per credit hour. Tuition, nonresident: full-time $11,088; part-time $462 per credit hour. Tuition and fees vary according to program. *Financial support:* In 2005–06, 15 students received support, including 3 fellowships, 5 research assistantships, 2 teaching assistantships; career-related internships or fieldwork and tuition waivers (full) also available. *Faculty research:* Agronomy and soil science, zoology, entomology, genetics and molecular biology, botanical sciences. *Unit head:* Dr. Robert Kinzie, Chair, 808-956-4602, Fax: 808-956-9608, E-mail: kinzie@hawaii.edu.

University of Illinois at Chicago, Graduate College, College of Liberal Arts and Sciences, Department of Biological Sciences, Chicago, IL 60607-7128. Offers cell and developmental biology (PhD); ecology and evolution (MS, DA, PhD); genetics and development (PhD); molecular biology (MS, PhD); neurobiology (MS, PhD); plant biology (MS, DA, PhD). *Degree requirements:* For master's, thesis; for doctorate, thesis/dissertation, preliminary exam. *Entrance requirements:* For master's and doctorate, GRE General Test, GRE Subject Test, previous course work in physics, calculus, and organic chemistry; minimum GPA of 2.75. Additional exam requirements/recommendations for international students: Required—TOEFL. Electronic applications accepted.

University of Illinois at Urbana–Champaign, Graduate College, College of Liberal Arts and Sciences, School of Integrative Biology, Program in Ecology and Evolutionary Biology, Champaign, IL 61820. Offers MS, PhD. *Students:* 29 full-time (12 women), 7 part-time (4 women); includes 3 minority (1 African American, 2 Asian Americans or Pacific Islanders), 6 international. 54 applicants, 4% accepted, 2 enrolled. In 2005, 2 master's, 1 doctorate awarded. Application fee: $50 ($60 for international students). *Financial support:* In 2005–06, 6 fellowships, 19 research assistantships, 17 teaching assistantships were awarded. *Unit head:* Jeff Brawn, Director, 217-244-5937, Fax: 217-244-1224, E-mail: jbrawn@uiuc.edu. *Application contact:* Carol Hall, Secretary, 217-333-8208, Fax: 217-244-1224, E-mail: cahall@uiuc.edu.

The University of Iowa, Graduate College, College of Liberal Arts and Sciences, Department of Biological Sciences, Iowa City, IA 52242-1316.

See Close-Up on page 257.

University of Kansas, Graduate School, College of Liberal Arts and Sciences, Division of Biological Sciences, Department of Ecology and Evolutionary Biology, Lawrence, KS 66045. Offers botany (MA, PhD); ecology and evolutionary biology (MA, PhD); entomology (MA, PhD); systematics and ecology (MA). Part-time programs available. *Faculty:* 42. *Students:* 67 full-time (35 women), 21 part-time (9 women); includes 4 minority (1 American Indian/Alaska Native, 3 Hispanic Americans), 20 international. Average age 30. 56 applicants, 32% accepted. In 2005, 10 master's, 4 doctorates awarded. Terminal master's awarded for partial completion of doctoral program. *Degree requirements:* For master's and doctorate, thesis/dissertation, comprehensive exam. *Entrance requirements:* For master's and doctorate, GRE General Test, GRE Subject Test (recommended). Additional exam requirements/recommendations for international students: Required—TOEFL (paper 570; computer 230) or IELT. *Application deadline:* For fall admission, 1/10 priority date for domestic students, 1/10 priority date for international students. Applications are processed on a rolling basis. Application fee: $55 ($60 for international students). Electronic applications accepted. *Expenses:* Tuition, state resident: full-time $4,859. Tuition, nonresident: full-time $12,000. Required fees: $589. Tuition and fees vary according to program. *Financial support:* Fellowships with tuition reimbursements, research assistantships with partial tuition reimbursements, teaching assistantships with full and partial tuition reimbursements available. Financial award application deadline: 3/1. *Faculty research:* Ecology, evolutionary biology, genetics. *Unit head:* , Craig E. Martin, Chair, 785-864-5887, Fax: 785-864-5860, E-mail: ecophys@ku.edu. *Application contact:* Jeannie Houts, Graduate Coordinator, 785-864-2362, Fax: 785-864-5860, E-mail: jmhouts@ku.edu.

University of Louisiana at Lafayette, Graduate School, College of Sciences, Department of Biology, Lafayette, LA 70504. Offers biology (MS); environmental and evolutionary biology (PhD). *Faculty:* 20 full-time (4 women), 2 part-time/adjunct (0 women). *Students:* 58 full-time (34 women), 12 part-time (10 women); includes 3 minority (2 African Americans, 1 Asian American or Pacific Islander), 23 international. Average age 29. 40 applicants, 38% accepted, 8 enrolled. In 2005, 7 master's, 3 doctorates awarded. Terminal master's awarded for partial completion of doctoral program. *Degree requirements:* For master's, thesis, registration; for doctorate, 2 foreign languages, thesis/dissertation, comprehensive exam, registration. *Entrance requirements:* For master's, GRE General Test, minimum GPA of 2.75; for doctorate, GRE General Test, GRE Subject Test, minimum GPA of 3.0. Additional exam requirements/recommendations for international students: Required—TOEFL (minimum score 550 paper-based; 213 computer-based). *Application deadline:* For fall admission, 5/15 for domestic students, 5/15 for international students; for spring admission, 10/1 for domestic students, 10/1 for international students. Applications are processed on a rolling basis. Application fee: $20 ($30 for international students). Electronic applications accepted. *Expenses:* Tuition, state resident: full-time $3,330; part-time $93 per credit hour. Tuition, nonresident: full-time $9,510; part-time $350 per credit hour. International tuition: $9,646 full-time. *Financial support:* In 2005–06, 17 fellowships with full tuition reimbursements (averaging $14,289 per year), 9 research assistantships with full tuition reimbursements (averaging $8,024 per year), 17 teaching assistantships with full tuition reimbursements (averaging $12,044 per year) were awarded; Federal Work-Study, institutionally sponsored loans, and unspecified assistantships also available. Financial award application deadline: 5/1. *Faculty research:* Structure and ultrastructure, system biology, ecology, processes, environmental physiology. *Unit head:* Dr. Darryl L. Felder, Head, 337-482-6748, Fax: 337-482-5834, E-mail: dlf4517@louisana.edu. *Application contact:* Dr. Paul Leberg, Graduate Coordinator, 337-482-6750, Fax: 337-482-5834, E-mail: leberg@louisiana.edu.

University of Maryland, College Park, Graduate Studies, College of Chemical and Life Sciences, Department of Biology, Behavior, Ecology, Evolution, and Systematics Program, College Park, MD 20742. Offers MS, PhD. *Students:* 38 full-time (26 women), 3 part-time (all women); includes 2 minority (both Asian Americans or Pacific Islanders), 3 international. 64 applicants, 9% accepted, 3 enrolled. *Degree requirements:* For master's, thesis, oral defense, seminar; for doctorate, thesis/dissertation, exam, 4 seminars. *Entrance requirements:* For master's and doctorate, GRE General Test, GRE Subject Test in biology, 3 letters of recommendation. Additional exam requirements/recommendations for international students: Required—TOEFL, TSE. *Application deadline:* For fall admission, 1/1 for domestic students, 1/1 for international students. Applications are processed on a rolling basis. Application fee: $60. Electronic applications accepted. *Financial support:* In 2005–06, 9 fellowships with tuition reimbursements (averaging $12,467 per year) were awarded; research assistantships with tuition reimbursements, teaching assistantships with tuition reimbursements, Federal Work-Study and scholarships/grants also available. Support available to part-time students. Financial award applicants required to submit FAFSA. *Faculty research:* Animal behavior, biostatistics, ecology, evolution, neurothology. *Unit head:* Dr. Michele Dudash, Director, 301-405-4552, Fax: 301-314-9358, E-mail: gw10@umail.umd.edu. *Application contact:* Dean of Graduate School, 301-405-4190, Fax: 301-314-9305.

University of Massachusetts Amherst, Graduate School, Interdisciplinary Programs, Program in Organismic and Evolutionary Biology, Amherst, MA 01003. Offers MS, PhD. Part-time programs available. *Students:* 32 full-time (21 women), 3 part-time (2 women); includes 3 minority (all Hispanic Americans), 4 international. Average age 30. 55 applicants, 36% accepted, 11 enrolled. In 2005, 6 master's, 6 doctorates awarded. Terminal master's awarded for partial completion of doctoral program. *Degree requirements:* For master's, thesis or alternative; for doctorate, 2 foreign languages, thesis/dissertation. *Entrance requirements:* For master's and doctorate, GRE General Test, GRE Subject Test, 3 letters of recommendation. Additional exam requirements/recommendations for international students: Required—TOEFL (minimum score 530 paper-based; 197 computer-based). *Application deadline:* For fall admission, 12/1 priority date for domestic students, 12/1 priority date for international students. Applications are processed on a rolling basis. Application fee: $40 ($65 for international students). *Expenses:* Tuition, state resident: part-time $110 per credit. Tuition, nonresident: part-time $414 per credit. Required fees: $2,824 per term. One-time fee: $250 part-time. Full-time tuition and fees vary according to course load, campus/location, program and reciprocity agreements. *Financial support:* In 2005–06, 1 fellowship with full tuition reimbursement (averaging $3,855 per year), 2 research assistantships with full tuition reimbursements (averaging $1,847 per year) were awarded; teaching assistantships with full tuition reimbursements, career-related internships or fieldwork, Federal Work-Study, scholarships/grants, traineeships, and unspecified assistantships also available. Support available to part-time students. Financial award application deadline: 1/15. *Unit head:* Dr. Margaret Riley, Director, 413-545-0928, Fax: 413-545-3243.

University of Miami, Graduate School, College of Arts and Sciences, Department of Biology, Coral Gables, FL 33124. Offers biology (MS, PhD); genetics and evolution (MS, PhD). *Faculty:* 21 full-time (3 women), 10 part-time/adjunct (3 women). *Students:* 49 full-time (26 women); includes 2 minority (both Asian Americans or Pacific Islanders), 16 international. Average age 30. 39 applicants, 36% accepted, 10 enrolled. In 2005, 6 degrees awarded. Terminal master's awarded for partial completion of doctoral program. *Median time to degree:* Of those who began their doctoral program in fall 1997, 100% received their degree in 8 years or less. *Degree requirements:* For master's, thesis (for some programs), comprehensive exam (for some programs); for doctorate, thesis/dissertation, oral and written qualifying exam. *Entrance requirements:* For master's and doctorate, GRE General Test, 3 letters of recommendation, research papers. Additional exam requirements/recommendations for international students: Required—TOEFL (minimum score 550 paper-based; 213 computer-based), TSE. *Application deadline:* For fall admission, 1/1 for domestic students, 1/1 for international students. Application fee: $50. Electronic applications accepted. *Financial support:* In 2005–06, 49 students received support, including 9 fellowships with full tuition reimbursements available (averaging $18,000 per year), 15 research assistantships with full tuition reimbursements available (averaging $16,500 per year), 20 teaching assistantships with full tuition reimbursements available (averaging $16,500 per year); career-related internships or fieldwork, Federal Work-Study, institutionally sponsored loans, scholarships/grants, health care benefits, and unspecified assistantships also available. Financial award application deadline: 3/1; financial award applicants required to submit FAFSA. *Faculty research:* Neuroscience to ethology; plants, vertebrates and mycorrhizae; phylogenies, life histories and species interactions; molecular biology, gene expression and populations; cells, auditory neurons and vetebrate locomotion. Total annual research expenditures: $946,278. *Unit head:* Dr. Leonel O. Sternberg, Director of Graduate Studies, 305-284-6436, Fax: 305-284-3039, E-mail: leo@bio.miami.edu. *Application contact:* Beth E. Goad, Graduate Coordinator, 305-284-5116, Fax: 305-284-3039, E-mail: bgoad@bio.miami.edu.

See Close-Up on page 269.

University of Michigan, Horace H. Rackham School of Graduate Studies, College of Literature, Science, and the Arts, Department of Ecology and Evolutionary Biology, Ann Arbor, MI 48100. Offers MS, PhD. Part-time programs available. Terminal master's awarded for partial completion of doctoral program. *Degree requirements:* For master's, thesis (for some programs), registration; for doctorate, thesis/dissertation, two semesters of teaching, comprehensive exam, registration. *Entrance requirements:* For master's and doctorate, GRE. Additional exam requirements/recommendations for international students: Required—TOEFL (minimum score 560 paper-based; 220 computer-based). Electronic applications accepted. *Expenses:* Tuition, state resident: full-time $14,082; part-time $894 per credit hour. Tuition, nonresident: full-time $28,500; part-time $1,675 per credit hour. Required fees: $189; $189 per unit. *Faculty research:* Community ecology, molecular evolution, theoretical ecology, systematics, evolutionary genetics.

University of Minnesota, Twin Cities Campus, Graduate School, College of Biological Sciences, Department of Ecology, Evolution, and Behavior, Minneapolis, MN 55455-0213. Offers ecology, evolution, and behavior (MS, PhD). Terminal master's awarded for partial completion of doctoral program. *Degree requirements:* For master's, thesis or projects; for doctorate, thesis/dissertation, comprehensive exam. *Entrance requirements:* For master's and doctorate, GRE General Test, minimum GPA of 3.0. Additional exam requirements/recommendations for international students: Required—TOEFL (minimum score 550 paper-based; 213 computer-based), MELAB. Electronic applications accepted. *Expenses:* Tuition, state resident: full-time $8,748; part-time $729 per credit. Tuition, nonresident: full-time $15,848; part-time $1,321 per credit. Full-time tuition and fees vary according to class time, course load, program and reciprocity agreements. *Faculty research:* Behavioral ecology, community ecology, community genetics, ecosystem and global change; evolution and systematics.

University of Missouri–Columbia, Graduate School, College of Arts and Sciences, Division of Biological Sciences, Columbia, MO 65211. Offers cellular, molecular and developmental biology (MA, PhD); evolutionary biology and ecology (MA, PhD); neurobiology and behavior (MA, PhD). *Faculty:* 43 full-time (12 women). *Students:* 48 full-time (26 women), 28 part-time (10 women); includes 7 minority (2 African Americans, 3 Asian Americans or Pacific Islanders, 2 Hispanic Americans), 20 international. In 2005, 3 master's, 8 doctorates awarded. Terminal master's awarded for partial completion of doctoral program. *Degree requirements:* For master's, thesis/dissertation; for doctorate, thesis/dissertation, comprehensive exam. *Entrance requirements:* For master's and doctorate, GRE General Test, minimum GPA of 3.0. *Application deadline:* For fall admission, 1/15 for domestic students. Applications are processed on a rolling basis. Application fee: $45 ($60 for international students). *Financial support:* Fellowships, research assistantships, teaching assistantships, institutionally sponsored loans available. *Unit head:* Dr. Ray Semlitsch, Director of Graduate Studies, 573-884-6396, E-mail: semlitschr@missouri.edu. *Application contact:* Nila Emerich, Application Contact, 800-553-5698.

University of Missouri–St. Louis, College of Arts and Sciences, Department of Biology, St. Louis, MO 63121. Offers biology (MS, PhD), including animal behavior (MS), biochemistry (MS), biotechnology (MS), conservation biology (MS), development (MS), ecology (MS), environmental studies (PhD), evolution (MS), genetics (MS), molecular/cellular biology (MS), physiology (MS), plant systematics (MS), population biology (MS), tropical biology (MS); biotechnology (Certificate); tropical biology and conservation (Certificate). Part-time programs available. *Faculty:* 50. *Students:* 26 full-time (15 women), 101 part-time (51 women); includes 13

Evolutionary Biology

University of Missouri–St. Louis (continued)
minority (5 African Americans, 6 Asian Americans or Pacific Islanders, 2 Hispanic Americans), 41 international. Average age 32. In 2005, 22 master's, 2 doctorates awarded. *Degree requirements:* For master's, thesis or alternative; for doctorate, one foreign language, thesis/dissertation, 1 semester of teaching experience. *Entrance requirements:* For doctorate, GRE General Test. *Application deadline:* For spring admission, 12/1 priority date for domestic students. Applications are processed on a rolling basis. Application fee: $35 ($40 for international students). Electronic applications accepted. *Expenses:* Tuition, state resident: part-time $263 per credit hour. Tuition, nonresident: part-time $680 per credit hour. Required fees: $53 per credit hour. Tuition and fees vary according to program. *Financial support:* In 2005–06, 11 fellowships with full tuition reimbursements (averaging $30,000 per year), 15 research assistantships with full and partial tuition reimbursements (averaging $16,000 per year), 22 teaching assistantships with full and partial tuition reimbursements (averaging $16,000 per year) were awarded; career-related internships or fieldwork and Federal Work-Study also available. Support available to part-time students. Financial award application deadline: 2/1. *Faculty research:* Molecular biology, microbial genetics. *Unit head:* Zuleyma Tang-Martinez, Director of Graduate Studies, 314-516-6498, Fax: 314-516-6233, E-mail: zuleyma@umsl.edu. *Application contact:* 314-516-5458, Fax: 314-516-5310, E-mail: gradadm@umsl.edu.

University of Nevada, Reno, Graduate School, College of Science, Interdisciplinary Program in Ecology, Evolution, and Conservation Biology, Reno, NV 89557. Offers PhD. Offered through the College of Arts and Science, the M. C. Fleischmann College of Agriculture, and the Desert Research Institute. *Faculty:* 5. *Students:* 11 full-time (9 women), 24 part-time (12 women); includes 1 minority (1 Asian American or Pacific Islander, 2 Hispanic Americans), 2 international. Average age 33. 6 applicants, 67% accepted, 3 enrolled. In 2005, 4 degrees awarded. *Degree requirements:* For doctorate, thesis/dissertation. *Entrance requirements:* For doctorate, GRE General Test, GRE Subject Test, minimum GPA of 3.0. *Additional exam requirements/recommendations for international students:* Required—TOEFL. *Application deadline:* For fall admission, 2/15 for domestic students. Application fee: 560 ($95 for international students). *Expenses:* Tuition, area resident: Full-time $2,767; part-time $923 per semester. Tuition, state resident: full-time $5,733; part-time $1,911 per semester. Tuition, nonresident: full-time $12,679; part-time $1,911 per semester. International tuition: $13,878 full-time. Required fees: $404; $202 per term. One-time fee: $90. *Financial support:* In 2005–06, 1 research assistantship was awarded; teaching assistantships Financial award application deadline: 3/1. *Faculty research:* Population biology, behavioral ecology, plant response to climate change, conservation of endangered species, restoration of natural ecosystems. *Unit head:* Dr. Peter Brussard, Director, 775-784-1360.

The University of North Carolina at Chapel Hill, Graduate School, College of Arts and Sciences, Department of Biology, Chapel Hill, NC 27599. Offers botany (MA, MS, PhD); cell biology, development, and physiology (MA, MS, PhD); cell motility and cytoskeleton (PhD); ecology and behavior (MA, MS, PhD); genetics and molecular biology (MA, MS, PhD); morphology, systematics, and evolution (MA, MS, PhD). Terminal master's awarded for partial completion of doctoral program. *Degree requirements:* For master's, thesis (for some programs), comprehensive exam; for doctorate, thesis/dissertation, comprehensive exam. *Entrance requirements:* For master's, GRE General Test, GRE Subject Test, 2 semesters of calculus or statistics, 2 semesters of physics, organic chemistry, 3 semesters of biology; for doctorate, GRE General Test, GRE Subject Test, 2 semesters calculus or statistics, 2 semesters physics, organic chemistry, 3 semesters of biology. Additional exam requirements/recommendations for international students: Required—TOEFL (minimum score 550 paper-based; 213 computer-based). Electronic applications accepted. *Faculty research:* Gene expression, biomechanics, yeast genetics, plant ecology, plant molecular biology.

See Close-Up on page 277.

University of Notre Dame, Graduate School, College of Science, Department of Biological Sciences, Notre Dame, IN 46556. Offers aquatic ecology, evolution and environmental biology (MS, PhD); cellular and molecular biology (MS, PhD); genetics (MS, PhD); physiology (MS, PhD); vector biology and parasitology (MS, PhD). *Faculty:* 34 full-time (8 women), 3 part-time/adjunct (0 women). *Students:* 124 full-time (55 women); includes 11 minority (1 African American, 6 Asian Americans or Pacific Islanders, 4 Hispanic Americans), 39 international. 95 applicants, 34% accepted, 22 enrolled. In 2005, 4 master's, 11 doctorates awarded. Terminal master's awarded for partial completion of doctoral program. *Median time to degree:* Of those who began their doctoral program in fall 1997, 61% received their degree in 8 years or less. *Degree requirements:* For master's and doctorate, thesis/dissertation, comprehensive exam. *Entrance requirements:* For master's and doctorate, GRE General Test. Additional exam requirements/recommendations for international students: Required—TOEFL. *Application deadline:* For fall admission, 2/1 for domestic students; for spring admission, 11/1 for domestic students. Applications are processed on a rolling basis. Application fee: $50. Electronic applications accepted. *Financial support:* In 2005–06, 124 students received support, including 24 fellowships with full tuition reimbursements available (averaging $22,000 per year), 47 research assistantships with full tuition reimbursements available (averaging $15,250 per year), 45 teaching assistantships with full tuition reimbursements available (averaging $16,000 per year); traineeships and tuition waivers (full) also available. Financial award application deadline: 2/1. *Faculty research:* Tropical disease, molecular genetics, neurobiology, evolutionary biology, aquatic biology. Total annual research expenditures: $15.3 million. *Unit head:* Dr. Gary A. Lamberti, Director of Graduate Studies, 574-631-6552, Fax: 574-631-7413, E-mail: biosadm.1@nd.edu. *Application contact:* Dr. Terrence J. Akai, Director of Graduate Admissions, 574-631-7706, Fax: 574-631-4183, E-mail: gradad@nd.edu.

See Close-Up on page 281.

University of Oregon, Graduate School, College of Arts and Sciences, Department of Biology, Eugene, OR 97403. Offers ecology and evolution (MA, MS, PhD); marine biology (MA, MS, PhD); molecular, cellular and genetic biology (PhD); neuroscience and development (PhD). *Faculty:* 42 full-time (14 women), 5 part-time/adjunct (2 women). *Students:* 92 full-time (43 women), 2 part-time (both women); includes 9 minority (2 African Americans, 1 American Indian/Alaska Native, 4 Asian Americans or Pacific Islanders, 2 Hispanic Americans), 6 international. 18 applicants, 83% accepted. In 2005, 11 master's, 7 doctorates awarded. Terminal master's awarded for partial completion of doctoral program. *Degree requirements:* For master's, thesis (for some programs); for doctorate, thesis/dissertation. *Entrance requirements:* For master's and doctorate, GRE General Test, minimum GPA of 3.2. Additional exam requirements/recommendations for international students: Required—TOEFL. *Application deadline:* For fall admission, 12/15 for domestic students. *Financial support:* In 2005–06, 36 teaching assistantships were awarded; research assistantships, Federal Work-Study, institutionally sponsored loans, and scholarships/grants also available. Financial award application deadline: 2/1. *Faculty research:* Developmental neurobiology; evolution, population biology, and quantitative genetics; regulation of gene expression; biochemistry of marine organisms. *Unit head:* George Sprague, Head, 541-346-6051, Fax: 541-346-6056. *Application contact:* Lynne Romans, Admissions Contact, 541-346-4252, Fax: 541-346-6056, E-mail: lromans@uoregon.edu.

University of Pennsylvania, School of Arts and Sciences, Program in Ecology, Evolution and Biodiversity, Philadelphia, PA 19104. Offers PhD. *Degree requirements:* For doctorate, thesis/dissertation. *Entrance requirements:* For doctorate, GRE General Test, GRE Subject Test. Additional exam requirements/recommendations for international students: Required—TOEFL. Electronic applications accepted.

University of Pittsburgh, School of Arts and Sciences, Department of Biological Sciences, Program in Ecology and Evolution, Pittsburgh, PA 15260. Offers MS, PhD. *Faculty:* 7 full-time (2 women), 3 part-time/adjunct (1 woman). *Students:* 13 full-time (4 women); includes 2 minority (both Hispanic Americans) Average age 23. 34 applicants, 0% accepted, 0 enrolled. In 2005, 1 master's, 2 doctorates awarded. *Median time to degree:* Of those who began their

doctoral program in fall 1997, 100% received their degree in 8 years or less. *Degree requirements:* For master's and doctorate, thesis/dissertation, comprehensive exam, registration. *Entrance requirements:* For master's and doctorate, GRE General Test, GRE Subject Test. Additional exam requirements/recommendations for international students: Required—TOEFL (minimum score 550 paper-based; 213 computer-based). *Application deadline:* For fall admission, 1/15 priority date for domestic students, 12/15 for international students. Applications are processed on a rolling basis. Application fee: $0 ($40 for international students). Electronic applications accepted. *Expenses:* Tuition, state resident: full-time $13,194; part-time $537 per credit. Tuition, nonresident: full-time $25,012; part-time $1,026 per credit. Required fees: $700; $164 per term. Tuition and fees vary according to campus/location and program. *Financial support:* In 2005–06, 18 fellowships with full tuition reimbursements, 4 research assistantships with full tuition reimbursements (averaging $2,032 per year), 12 teaching assistantships with full tuition reimbursements (averaging $20,332 per year) were awarded; Federal Work-Study, scholarships/grants, traineeships, health care benefits, and tuition waivers (full) also available. *Faculty research:* Ecological and population genetics, tropical and community ecology, evolutionary ecology, phylogeny of birds, evolution of dispersal and dormancy. *Application contact:* Cathleen M. Barr, Graduate Administrator, 412-624-4268, Fax: 412-624-4759, E-mail: cbarr@pitt.edu.

See Close-Up on page 711.

University of South Carolina, The Graduate School of Science and Mathematics, Department of Biological Sciences, Graduate Training Program in Ecology, Evolution, and Organismal Biology, Columbia, SC 29208. Offers MS, PhD. *Degree requirements:* For master's, one foreign language; for doctorate, one foreign language, thesis/dissertation. *Entrance requirements:* For master's and doctorate, GRE General Test, minimum GPA of 3.0 in science. Electronic applications accepted.

See Close-Up on page 713.

The University of Tennessee, Graduate School, College of Arts and Sciences, Department of Ecology and Evolutionary Biology, Knoxville, TN 37996. Offers behavior (MS, PhD); ecology (MS, PhD); evolutionary biology (MS, PhD). Part-time programs available. *Degree requirements:* For master's and doctorate, thesis/dissertation. *Entrance requirements:* For master's and doctorate, GRE General Test, minimum GPA of 2.7. Additional exam requirements/recommendations for international students: Required—TOEFL. Electronic applications accepted.

The University of Texas at Austin, Graduate School, College of Natural Sciences, School of Biological Sciences, Program in Ecology, Evolution and Behavior, Austin, TX 78712-1111. Offers PhD. *Entrance requirements:* For doctorate, GRE General Test. Additional exam requirements/recommendations for international students: Required—TOEFL. Electronic applications accepted.

University of Utah, The Graduate School, College of Science, Department of Biology, Salt Lake City, UT 84112-1107. Offers biology (M Phil); ecology and evolutionary biology (MS, PhD); genetics (MS, PhD); molecular biology (PhD). Part-time programs available. *Faculty:* 42 full-time (8 women), 1 part-time/adjunct (1 woman). *Students:* 55 full-time (30 women), 15 part-time (2 women); includes 4 minority (3 Asian Americans or Pacific Islanders, 1 Hispanic American), 19 international. Average age 29. 26 applicants, 54% accepted, 11 enrolled. In 2005, 3 master's, 6 doctorates awarded. Terminal master's awarded for partial completion of doctoral program. *Median time to degree:* Of those who began their doctoral program in fall 1997, 20% received their degree in 8 years or less. *Degree requirements:* For master's and doctorate, thesis/dissertation. *Entrance requirements:* For master's and doctorate, GRE General Test, minimum GPA of 3.0. Additional exam requirements/recommendations for international students: Required—TOEFL (minimum score 500 paper-based; 173 computer-based). *Application deadline:* For fall admission, 1/13 for domestic students, 1/13 for international students. Application fee: $45 ($65 for international students). *Expenses:* Tuition, state resident: full-time $2,932; part-time $369 per credit. Tuition, nonresident: full-time $10,350; part-time $1,302 per credit. Required fees: $516 per term. Tuition and fees vary according to course load and program. *Financial support:* In 2005–06, 2 fellowships with full tuition reimbursements (averaging $30,000 per year), 33 research assistantships with full tuition reimbursements (averaging $22,000 per year), 33 teaching assistantships with full tuition reimbursements (averaging $15,000 per year) were awarded; career-related internships or fieldwork, scholarships/grants, traineeships, and health care benefits also available. Financial award application deadline: 2/15; financial award applicants required to submit FAFSA. *Faculty research:* Behavioral ecology, cellular neurobiology, DNA replication, ecological genetics, herpetology. Total annual research expenditures: $11.4 million. *Unit head:* David R. Wolstenholme, Chair, 801-581-6517, Fax: 801-581-4668, E-mail: wolstenholme@biology.utah.edu. *Application contact:* Shannon Nielsen, Administrative Program Coordinator, 801-581-5636, Fax: 801-581-4668, E-mail: shannon.nielsen@bioscience.utah.edu.

See Close-Up on page 313.

University of Utah, The Graduate School, College of Science and Graduate Programs in Medicine, Interdisciplinary Studies in Molecular Evolutionary Biology, Salt Lake City, UT 84112-1107.

Virginia Polytechnic Institute and State University, Graduate School, College of Science, Department of biological Sciences, Blacksburg, VA 24061. Offers botany (MS, PhD); ecology and evolutionary biology (MS, PhD); genetics and developmental biology (MS, PhD); microbiology (MS, PhD); zoology (MS, PhD). *Faculty:* 38 full-time (9 women). *Students:* 70 full-time (30 women), 4 part-time (1 woman); includes 6 minority (2 African Americans, 1 American Indian/Alaska Native, 2 Asian Americans or Pacific Islanders, 1 Hispanic American), 12 international. Average age 27. 81 applicants, 22% accepted, 14 enrolled. In 2005, 9 master's, 8 doctorates awarded. *Entrance requirements:* For master's and doctorate, GRE General Test. Additional exam requirements/recommendations for international students: Required—TOEFL (minimum score 550 paper-based; 213 computer-based). *Application deadline:* Applications are processed on a rolling basis. Application fee: $45. Electronic applications accepted. *Expenses:* Tuition, state resident: full-time $6,558; part-time $364 per credit. Tuition, nonresident: full-time $11,296; part-time $628 per credit. Required fees: $1,419; $468 per credit. $234 per term. *Financial support:* In 2005–06, 28 research assistantships with full tuition reimbursements (averaging $16,689 per year), 37 teaching assistantships with full tuition reimbursements (averaging $13,902 per year) were awarded; career-related internships or fieldwork, Federal Work-Study, scholarships/grants, and unspecified assistantships also available. *Faculty research:* Freshwater ecology, cell cycle regulation, behavioral ecology, motor proteins. *Unit head:* , Dr. Bob Jones, Chairman, 540-231-9514, Fax: 540-231-9307, E-mail: rhjones@vt.edu. *Application contact:* Sue Rasmussen, Graduate Secretary, 540-231-8929, Fax: 540-231-9307, E-mail: sueras@vt.edu.

Washington University in St. Louis, Graduate School of Arts and Sciences, Division of Biology and Biomedical Sciences, Program in Evolution, Ecology and Population Biology, St. Louis, MO 63130-4899. Offers ecology (PhD); environmental biology (PhD); evolutionary biology (PhD); genetics (PhD). *Degree requirements:* For doctorate, thesis/dissertation. *Entrance requirements:* For doctorate, GRE General Test, GRE Subject Test. Electronic applications accepted.

Wesleyan University, Graduate Programs, Department of Biology, Middletown, CT 06459-0260. Offers cell biology (PhD); comparative physiology (PhD); developmental biology (PhD); genetics (PhD); neurophysiology (PhD); population biology (PhD). *Faculty:* 12 full-time (3 women). *Students:* 29 full-time (15 women), 10 international. Average age 26. 131 applicants. In 2005, 2 doctorates awarded. *Degree requirements:* For doctorate, one foreign language, thesis/dissertation. *Entrance requirements:* For doctorate, GRE General Test, GRE Subject Test. *Application deadline:* For fall admission, 2/15 for domestic students. Applications are processed on a rolling basis. Application fee: $0. *Expenses:* Tuition: Full-time $24,732. One-time fee: $20 full-time. *Financial support:* Research assistantships, teaching assistantships, stipends available.

Faculty research: Microbial population genetics, genetic basis of evolutionary adaptation, genetic regulation of differentiation and pattern formation in *drosophila. Unit head:* Dr. Michael Weir, Chairman, 860-685-2402, E-mail: mweir@wesleyan.edu. *Application contact:* Marjorie Fitzgibbons, Information Contact, 860-685-2157, E-mail: mfitzgibbons@wesleyan.edu.

See Close-Up on page 323.

West Virginia University, Eberly College of Arts and Sciences, Department of Biology, Morgantown, WV 26506. Offers cell and molecular biology (MS, PhD); environmental and evolutionary biology (MS, PhD); integrative organismal biology (PhD); integrative organismal, biology (MS). *Faculty:* 18 full-time (3 women), 4 part-time/adjunct (all women). *Students:* 29 full-time (20 women), 5 part-time (3 women); includes 1 minority (Asian American or Pacific Islander), 9 international. Average age 26. 50 applicants, 10% accepted. In 2005, 1 master's, 2 doctorates awarded. Terminal master's awarded for partial completion of doctoral program. *Degree requirements:* For master's, thesis, final exam; for doctorate, thesis/dissertation, preliminary and final exams. *Entrance requirements:* For master's, GRE General Test, GRE Subject Test, minimum GPA of 3.0; for doctorate, GRE General Test, minimum GPA of 3.0. Additional exam requirements/recommendations for international students: Required—TOEFL.

Application deadline: For fall admission, 4/1 for domestic students; for spring admission, 10/1 for domestic students. Applications are processed on a rolling basis. Application fee: $45. *Expenses:* Tuition, state resident: full-time $4,582; part-time $258 per credit hour. Tuition, nonresident: full-time $13,820; part-time $741 per credit hour. *Financial support:* In 2005–06, 4 research assistantships, 22 teaching assistantships were awarded; Federal Work-Study and institutionally sponsored loans also available. Financial award application deadline: 4/1; financial award applicants required to submit FAFSA. *Faculty research:* Environmental biology, genetic engineering, developmental biology, global change, biodiversity. *Unit head:* Dr. Jonathan Cumming, Chair, 304-293-5201 Ext. 2508, Fax: 304-293-6363, E-mail: jonathan.cumming@mail.wvu.edu. *Application contact:* Dr. William T. Peterjohn, Director of Graduate Studies, 304-293-5201 Ext. 2510, Fax: 304-293-6363, E-mail: william.peterjohn@mail.wvu.edu.

See Close-Up on page 325.

Yale University, Graduate School of Arts and Sciences, Department of Ecology and Evolutionary Biology, New Haven, CT 06520. Offers PhD. *Entrance requirements:* For doctorate, GRE General Test, GRE Subject Test (biology).

See Close-Up on page 717.

Cross-Discipline Announcements

The Evergreen State College, Graduate Programs, Program in Environmental Studies, Olympia, WA 98505.

The Graduate Program in Environmental Studies at The Evergreen State College combines the environmental sciences with political economy and public policy. The College is nationally recognized for innovative, interdisciplinary teaching; collaborative learning; narrative evaluations; small classes; and projects that bridge theory and practice. Students learn analytical and practical tools needed by environmental professionals.

The University of Alabama at Birmingham, School of Natural Sciences and Mathematics, Department of Biology, Birmingham, AL 35294.

MS and PhD research emphases include comparative and cellular physiology, reproduction and development, microbial ecology and physiology, and aquatic ecology, with particular focus on marine organisms. Facilities at Dauphin Island Sea Laboratory are available for studies of the marine environment.

COLUMBIA UNIVERSITY

Department of Ecology, Evolution, and Environmental Biology

Programs of Study	The Department of Ecology, Evolution, and Environmental Biology (E3B) was established in 2001 in response to the recognition that these fields have their own set of intellectual foci, theoretical foundations, scales of analysis, and methodologies. Its mission is to educate a new generation of scientists and practitioners in the theory and methods of these disciplines. It emphasizes a multidisciplinary perspective on the Earth's declining biodiversity, integrating insights from relevant fields in biology and the social sciences. Faculty members are based at the University or at partner institutions, including the American Museum of Natural History, the New York Botanical Garden, the Wildlife Conservation Society, and Wildlife Trust.

The M.A. in conservation biology integrates the biological sciences with a basic foundation in environmental policy and economics. The program offers four different tracks that can be followed, depending upon student interest. All students are required to complete 48 credits, including two semesters in conservation biology and environmental policy, four semesters of the Research (CERC) Seminar, three electives each in conservation science and environmental policy, and two more electives in either area. Students in the thesis-based option must complete an additional 3 credits with summer field research.

Doctoral candidates may elect to pursue a degree in ecology and evolutionary biology or evolutionary primatology. Unique to the Ph.D. program in ecology and evolutionary biology is course work leading to proficiency in environmental policy and a separate Environmental Policy Certificate. All candidates are required to complete 6 units of full-time residency, three core courses, three to six advanced courses, and two internships; demonstrate proficiency in a foreign language, as needed for their specific fieldwork locations; serve as teaching assistants for two to four semesters; complete two advanced examinations; perform an in-depth review of the scholarly literature that is most relevant to the proposed dissertation research; and orally defend their dissertation. Upon advancing to candidacy, students are expected to submit their proposals to granting agencies for outside funding.

Research Facilities

The Biological Sciences Library houses more than 52,000 volumes, subscribes to about 335 serials, and has digital subscriptions to hundreds of other serials. The print collection is particularly strong in the areas of molecular biology, biochemistry, cell biology, genetics, and neurobiology. The areas of population and evolutionary biology and plant physiology are also collected at a research level. The Center for Environmental Research and Conservation (CERC) conducts research to find long-term solutions to combat the loss of biological diversity and natural resource depletion while meeting the needs of a growing worldwide human population. Black Rock Forest comprises nearly 3,800 acres of land on the west bank of the Hudson River and is dedicated to scientific research and conservation of the ecosystems that once covered the region. The Center for Economy, Environment & Society supports research in environmental economics by affiliated faculty members, develops and supports teaching programs in environmental economics, and serves as a resource for students and scholars with interests in the intersection of economics and environmental sciences.

Financial Aid

Ph.D. fellowships for up to five years are available through the Faculty Fellows program. Students are expected to find their own funding for their dissertation research year (for research and stipend expenses). There are no fellowships for the master's program. However, applicants are strongly encouraged to apply for outside sources of funding, such as the EPA Science to Achieve Results Fellowship Program and the NSF Graduate Fellowship Program).

Cost of Study

In the 2006–07 academic year, full-time tuition is $16,196 per semester, or $32,392 per academic year. Health insurance, health services, and student fees are $2285 for the entire academic year. A part-time option is available for the M.A. program. Full-time second-year master's students register for Extended Residence, which costs half as much as regular full-time tuition.

Living and Housing Costs

In 2006–07, students living on campus can expect to spend approximately $16,060 during the academic year for room and board, $3190 for personal expenses, and $1400 for books and supplies. Students living off campus can expect to spend $850 to $1200 per month for a one-bedroom apartment or $1100 to $1600 per month for a two-bedroom apartment. These are approximate costs and can vary, depending on individual needs.

Student Group

There are currently 34 doctoral candidates and 37 master's students in the program. They come from a variety of backgrounds, but the majority hold undergraduate degrees in the biological sciences and related disciplines.

Student Outcomes

Students develop the skills to conduct ecological, behavioral, systematic, molecular, and other evolutionary biological research as well as to formulate and implement environmental policy. Graduates may pursue academic careers as researchers and teachers or enter the job market directly as scientific researchers, teachers, or administrators in government agencies or in national or international conservation, environmental, and multilateral aid organizations dedicated to the conservation of biodiversity.

Location

The campus is located in New York City's Morningside Heights neighborhood. Residents are a short walk from Riverside Park, Central Park, St. John the Divine Cathedral, and other landmarks, or they can take a subway to world-famous sites like the Empire State Building or Times Square, neighborhoods such as Greenwich Village or Little Italy, or one of New York's many museums, theaters, or restaurants.

The University

Columbia University, the fifth-oldest university in the United States, was founded in 1754 as King's College; the first class, led by Samuel Johnson, had 8 students. Today, the University is one of the most competitive in the nation, enrolling more than 23,000 students in three undergraduate schools, thirteen graduate and professional schools, and a school of continuing education. It is also a premier research institute, where faculty members engage in groundbreaking research in medicine, science, the arts, and the humanities.

Applying

Prospective students must submit an application for admission, a statement of academic purpose, official transcripts, three letters of recommendation, a statistical form, a fellowship statement, and a $90 application fee ($75 if applying online). Applicants should have an undergraduate degree in one of the natural sciences, with course work in calculus, physics, chemistry, statistics, genetics, ecology, and organismal biology. The GRE General Test is required, and the biology Subject Test is strongly recommended. The application deadline for early decision admission to the M.A. program is December 15. The regular decision application deadline is April 1. The application deadline for the Ph.D. program is January 3.

Correspondence and Information

Eleanor Sterling, Director of Graduate Studies
Department of Ecology, Evolution, and Environmental Biology
Columbia University
1200 Amsterdam Avenue
New York, New York 10027
Phone: 212-854-9987
Fax: 212-854-8188
E-mail: es443@columbia.edu
   e3b@columbia.edu (departmental e-mail)
Web site: http://www.columbia.edu/cu/e3b/

James Danoff-Burg, M.A. Program Advisor
Department of Ecology, Evolution, and Environmental Biology
Columbia University
1200 Amsterdam Avenue
New York, New York 10027
Phone: 212-854-0149
Fax: 212-854-8188
E-mail: jd363@columbia.edu
   e3b@columbia.edu (departmental e-mail)
Web site: http://www.columbia.edu/cu/e3b/

Columbia University

THE FACULTY AND THEIR RESEARCH

For a full listing of all E3B faculty members who are approved advisers and are part of the Department's consortium partnership, students should consult the departmental Web site.

Walter J. Bock, Professor of Biological Science; Ph.D., Harvard, 1959. Theoretical aspects of evolution and systematics; feeding apparatus of birds; origin of avian flight.

Hilary S. Callahan, Assistant Professor; Ph.D. (botany), Wisconsin–Madison, 1996. Ability of a genotype to express different phenotypes in response to different environments; phenotypic plasticity of flowering time in the model organism *Arabidopsis thaliana.*

Joel Cohen, Professor; Ph.D. (population sciences and tropical public health), Harvard, 1973. Role of the food web, body sizes, and species abundances in describing ecological communities; new inequalities arising in information theory and operations research; frequency-domain analysis of nonlinear stochastic population models.

Steven A. Cohen, Director, Graduate Program in Earth Systems; Ph.D. (political science), Buffalo, SUNY, 1979. Organizational management; workforce planning; quality management and management innovation.

Marina Cords, Professor; Ph.D. (zoology), Berkeley, 1984. Understanding the mating system of blue monkeys in the Kakamega Forest, western Kenya; social behavior and behavioral ecology of primates; proximate and ultimate explanations of social systems.

James Danoff-Burg, Associate Research Scientist and M.A. Program Advisor; Ph.D., Kansas, 1995. Urbanization and road impacts upon necrophage biodiversity in upstate New York; artificial reefs as a conduit for the spread of invasive species in Punta Cana, Dominican Republic; association between invasive ants and human activity in New York and Hawaii.

James Gibbs, Adjunct Professor; Ph.D. (forestry and environmental studies), Yale, 1995. Biological monitoring, population biology, conservation genetics, and landscape ecology; improving conservation biology education, particularly in tropical, developing countries.

John Glendinning, Associate Professor; Ph.D., Florida, 1989. Chemosensory mechanisms that control the feeding behavior of insects and mammals; contribution of taste and viscerosensory response mechanisms to this coping process.

Kevin Griffin, Associate Professor; Ph.D. (earth and environmental science), Duke, 1994. Plant respiration; global carbon cycle; forest ecology.

Paul E. Hertz, Professor; Ph.D. (biology), Harvard, 1977. Evolution and interaction of behavioral and physiological traits that compensate for geographic and seasonal shifts in operative temperatures; extent and effectiveness of temperature regulation and its effect on resource partitioning in West Indian anoles; role of light intensity in microhabitat selection.

Ralph L. Holloway, Professor; Ph.D. (anthropology), Berkeley, 1964. Brain endocasts of fossil hominids; ape brain endocast morphology and variation; biostereometric analysis of the brain; cerebral asymmetry, lateralization, and cognition.

Darcy Kelley, Professor; Ph.D. (biological sciences), Rockefeller, 1975. Neural systems and behavior; hormonal effects; sexual differentiation.

Don J. Melnick, Professor; Ph.D. (physical anthropology), Yale, 1981. Genetic consequences of habitat fragmentation in vertebrates; genetic indicators for setting conservation priorities.

Brian Morton, Assistant Professor; Ph.D. (genetics), California, Riverside, 1993. How selective constraints on codon usage interact with other structural features; how rate heterogeneity among sites is affected by variation in context or the composition of nucleotides flanking those sites.

Shahid Naeem, Professor and Chair; Ph.D. (zoology), Berkeley, 1989. How extrinsic factors interact with plant biodiversity to regulate the spread of invasive plant species in old fields; how mathematical models developed for reliability engineering can be used for understanding the reliability of ecosystems.

Paul E. Olsen, Arthur D. Stroke Memorial Professor of Geological Sciences; Ph.D. (biology), Yale, 1984. Evolution of continental ecosystems; pattern, causes, and effects of climate change on geological time scales and mass extinctions; effects of evolutionary innovations on global biogeochemical cycles.

Matthew Palmer, Lecturer; Ph.D. (plant diversity), Rutgers, 2005. Community ecology; plant conservation biology; local controls on plant diversity.

Miguel Pinedo-Vasquez, Lecturer and Associate Research Scientist; Ph.D. (forestry and environmental studies), Yale, 1995. Patterns and effects of small-holder management of tropical ecosystems and landscapes.

Jeanne Poindexter, Assistant Professor; Ph.D. (bacteriology), Berkeley, 1963. Ecophysiology of oligotrophic bacteria; effects of nutrient fluxes on physiologic, morphologic, and behavioral properties of bacteria.

Robert E. Pollack, Professor; Ph.D. (biology), Brandeis, 1966. The future of medical research in the U.S.; approaches of medicine to aging; social and political consequences of medical genetics and DNA-based medicine.

William Schuster, Executive Director, Black Rock Forest Consortium; Ph.D. (biology), Colorado, 1989. Forest structure, composition, ecological processes, and how these factors and overall forest health change over timescales; consequences of various human activities for forest ecosystems.

Jill Shapiro, Lecturer; Ph.D. (anthropology), Columbia, 1995. Analysis of interpopulational cranial variation in the orangutan, as compared with that present in the African apes.

Eleanor Sterling, Director, Center for Biodiversity Conservation, American Museum of Natural History; Ph.D. (physical anthropology and forestry), Yale, 1993. Optimal techniques for mammals surveys to studies of the distribution of patterns of biodiversity in tropical regions of the world.

Maria Uriarte, Assistant Professor; Ph.D. (ecology and evolutionary biology), Cornell, 2002. Role that neighborhood interactions play in the assembly and composition of natural plant communities.

Paige West, Assistant Professor; Ph.D. (cultural anthropology), Rutgers. Hunting practices of rural peoples and the population ecology of prey species in Papua New Guinea; commodity ecumene for coffee as the first step in understanding commodity flows from rural Papua New Guinea to urban areas.

MICHIGAN STATE UNIVERSITY

Interdepartmental Graduate Program
in Ecology, Evolutionary Biology and Behavior

Programs of Study

The Interdepartmental Graduate Program in Ecology, Evolutionary Biology and Behavior (EEBB) is a multidisciplinary program of study and research that leads to a dual major in ecology, evolutionary biology, and behavior for the Ph.D. degree or to a specialization in those areas for the M.S. degree in one of the traditional disciplines offered by the participating departments: the Departments of Anthropology, Computer Science and Engineering, Crop and Soil Sciences, Entomology, Fisheries and Wildlife, Forestry, Geography, Geological Sciences, Horticulture, Microbiology, Philosophy, Plant Biology, Plant Pathology, Psychology, Statistics and Probability, and Zoology. The program is built on the premise that in the next several decades, the major research breakthroughs in the fields of ecology, evolution, and behavior will be made by experts trained with a synthetic view of ecology, evolution, and behavior. The EEBB program allows students great flexibility in choosing areas of study. For example, students may study the molecular aspects of evolution, population and community ecology of animals and plants, or the dynamics of natural and managed ecosystems. Each student's course of study is planned with the individual's particular interests, capabilities, and professional goals in mind. The 100 ecology, evolutionary biology, and behavior faculty members represent a wide variety of scientific disciplines. A dual-major Ph.D. or M.S. certification in EEBB is awarded upon completion of the requirements of both the student's affiliated department and the ecology, evolutionary biology, and behavior program. The course work, thesis, and final oral examination can usually be completed in two years for the M.S. degree and five years for the Ph.D. degree program.

Research Facilities

The program provides students with access to research opportunities that range from animal behavior to experimental ecosystem study. One of the strengths of the program is the ease with which students can work in several laboratories.

Laboratories are well equipped with microcomputers, and excellent specialized computing facilities are available in several departments and in the University's Computer Laboratory. Museum collections are extensive at Michigan State University (MSU), particularly for vertebrates, insects, and plants. Collections are housed in the University Museum, the Herbarium, and the Entomology Museum. Perhaps the most remarkable resource at Michigan State, though, is the extensive acreage available for field studies, both on campus and at satellite sites. The campus itself comprises many acres of old fields, woodlots, and farmland that have been extensively utilized for research by the program's students. The Kellogg Biological Station (KBS) provides some of the best-equipped facilities for field research in the country. The combination of excellent laboratories, experimental ponds, experimental forests, and other aquatic and terrestrial sites has placed KBS at the forefront of ecological research.

Financial Aid

A departmental teaching and research assistantship is the most common form of support. Assistantships pay, on average, $1654 per month. A student on an assistantship is assessed in-state tuition. A variety of other competitive assistantships and fellowships are available, including University Distinguished Fellowships, College of Natural Science Recruiting Fellowships, and special research assistantships supported by individual research grants.

Cost of Study

In 2005–06, fees and tuition per semester for an average graduate course load (6 credits) were estimated at $2498 for Michigan residents and $4631 for out-of-state students. Book and supply costs vary, depending on the course of study, but average $1200 per year.

Living and Housing Costs

The charge for accommodations in a single room in Owen Graduate Center for 2005–06 was $2648 per semester; this cost included a $280 meal allowance. University Apartments lease one-bedroom units from $541 to $581 per month. Health services are available through University facilities.

Student Group

Enrollment at MSU for fall 2005 included 34,951 undergraduates, 6,758 graduate students, and 1,492 professional students. Of these, 3,367 were international students who came from 126 different countries; 83 percent were Michigan residents and 10 percent were students from other parts of the United States.

Student Outcomes

Recent graduates of the EEBB program have obtained faculty positions at Dartmouth College, Louisiana State University, College of William and Mary, University of California at Davis, Pennsylvania State University, Augustana College, Ohio State University, Yale University, University of New Hampshire, and Cornell University.

Location

Located on a large peninsula bounded in part by three of the five Great Lakes, East Lansing provides easy access to idyllic scenery and a variety of outdoor activities. The adjacent city of Lansing is the state capital and a manufacturing center. Local grade and high schools are among the best in the nation, and their proximity to MSU encourages numerous cultural and educational opportunities apart from the University functions.

The University

MSU, founded in 1855 as a pioneer land-grant institution, has expanded to offer 200 undergraduate programs and graduate work in seventy-five departments and schools of fourteen colleges. The East Lansing campus covers 2,100 acres, and an additional 3,163 acres are devoted to experimental farms. Located on the campus are the Biotechnology Research Center; a rapidly growing library system that currently holds more than 3 million volumes in the Main Library and fifteen specialized branches; Abrams Planetarium, with a 252-seat sky theater; the MSU Museum, housing special exhibits as well as documented research collections; the Kresge Art Museum; the Wharton Center for Performing Arts, a cultural facility with a 2,500-seat Great Hall and the Festival Stage; a 600-seat theater; and the Center for International Programs, which serves as a center for international activity. Athletics and recreational facilities include the Spartan Stadium, Munn Ice Arena, the Breslin Events Center, Forest Acres golf courses, and the IM Sports facilities, which provide sport gyms, two swimming pools, and several modern weight rooms.

Applying

Students who wish to join the ecology, evolutionary biology, and behavior program must also be admitted to one of the participating degree-granting departments (Anthropology, Computer Science and Engineering, Crop and Soil Sciences, Entomology, Fisheries and Wildlife, Forestry, Geography, Geological Sciences, Horticulture, Microbiology, Philosophy, Plant Biology, Plant Pathology, Psychology, Statistics and Probability, and Zoology). Prospective students should request application materials from and apply directly to one of these departments. They should then notify the Graduate Program in Ecology, Evolutionary Biology and Behavior of their interest, indicating the department to which they applied and the degree they intend to pursue.

Correspondence and Information

Graduate Program in Ecology, Evolutionary Biology and Behavior
103 Giltner Hall
Michigan State University
East Lansing, Michigan 48824-1101
Phone: 517-432-1359
E-mail: eebb@msu.edu
Web site: http://www.msu.edu/~eebb

Michigan State University

THE FACULTY AND THEIR RESEARCH

Gerard C. Adams, Associate Professor; Ph.D., California, Davis, 1981. Plant pathology.

Robert L. Anstey, Professor; Ph.D., Indiana, 1970. Invertebrate paleobiology.

James W. Atkinson, Professor; Ph.D., Emory, 1969. Behavior; developmental and evolutionary biology of mollusks.

Ted R. Batterson, Professor; Ph.D., Michigan State, 1980. Limnology; tropical pond aquaculture.

James R. Bence, Associate Professor; Ph.D., California, Santa Barbara, 1985. Quantitative fishery ecology.

George W. Bird, Professor; Ph.D., Cornell, 1966. Nematology.

Mary T. Bremigan, Associate Professor; Ph.D., Ohio State, 1997. Aquatic ecology and fisheries management.

Catherine M. Bristow, Associate Professor; Ph.D., Princeton, 1982. Insect ecology, mutualisms, and diet choice.

Thomas M. Burton, Professor; Ph.D., Cornell, 1973. Aquatic ecology.

Joseph L. Chartkoff, Professor; Ph.D., UCLA, 1974. Prehistoric archaeology; population ecology; ecosystem evolution.

Jeffrey K. Conner, Professor; Ph.D., Cornell, 1987. Evolutionary ecology; ecological genetics.

Thomas G. Coon, Professor; Ph.D., California, Davis, 1982. Stream fishery ecology.

Frank B. Dazzo, Professor; Ph.D., Florida, 1975. Microbial ecology; rhizobium-legume symbiosis.

Fred C. Dyer, Professor; Ph.D., Princeton, 1984. Ethology and behavioral ecology.

Diane Ebert-May, Professor; Ph.D., Colorado, 1976. Science education; plant ecology.

Heather L. Eisthen, Associate Professor; Ph.D., Indiana, 1992. Neuroethology; evolutionary and developmental neurobiology.

Bryan K. Epperson, Professor; Ph.D., California, Davis, 1983. Population genetics and molecular and forest genetics.

Frank W. Ewers, Professor; Ph.D., Berkeley, 1982. Physiological plant ecology.

Stuart H. Gage, Professor; Ph.D., Michigan State, 1974. Insect population dynamics and human dimensions.

Thomas Getty, Professor; Ph.D., Michigan, 1980. Behavioral ecology of vertebrates.

John P. Giesy, Professor; Ph.D., Michigan State, 1974. Environmental toxicology.

Frederick Gifford, Professor; Ph.D., Pittsburgh, 1984. Philosophy of biology; ethical and social issues concerning science.

Michael D. Gottfried, Associate Professor; Ph.D., Kansas, 1991. Vertebrate paleontology.

Katherine L. Gross, Professor; Ph.D., Michigan State, 1980. Plant population and community ecology.

Robert A. Haack, Adjunct Professor; Ph.D., Florida, 1984. Plant-insect interactions; invasion ecology; conservation biology.

Stephen K. Hamilton, Associate Professor; Ph.D., California, Santa Barbara, 1994. Biogeochemistry and aquatic ecology.

James F. Hancock, Professor; Ph.D., California, Davis, 1977. Physiological and evolutionary genetics of plants.

Daniel Hayes, Associate Professor; Ph.D., Michigan State, 1990. Fish habitat and population interactions.

Richard W. Hill, Professor and Associate Chair, Zoology; Ph.D., Michigan, 1970. Environmental physiology; marine sulfur metabolism.

Kay E. Holekamp, Professor; Ph.D., Berkeley, 1983. Behavioral ecology; ethology and behavioral endocrinology.

Zachary Huang, Associate Professor; Ph.D., Guelph, 1988. Behavioral physiology; insect sociobiology.

Rufus Isaacs, Associate Professor; Ph.D., Imperial College (London), 1994. Entomology.

Andrew M. Jarosz, Associate Professor; Ph.D., Purdue, 1984. Evolutionary ecology of plant-pathogen interactions.

Michael L. Jones, Professor; Ph.D., British Columbia, 1986. Fish population dynamics and management.

Daniel E. Keathley, Professor and Chairperson, Forestry; Ph.D., Ohio State, 1981. Tissue culture and molecular genetics.

Christopher A. Klausmeier, Assistant Professor; Ph.D., Minnesota, 2000. Theoretical and aquatic ecology.

Karen L. Klomparens, Professor; Ph.D., Michigan State, 1977. Plant host-parasite relations; electron microscopy.

Richard K. Kobe, Associate Professor; Ph.D., Connecticut, 1995. Forest ecology.

Douglas A. Landis, Professor; Ph.D., North Carolina State, 1987. Landscape ecology and biological control.

Richard E. Lenski, Hannah Distinguished Professor of Microbial Ecology and Director, Ecology, Evolutionary Biology and Behavior Program; Ph.D., North Carolina, 1983. Ecology; genetics; evolutionary biology.

Weiming Li, Associate Professor; Ph.D., Minnesota, 1994. Fish chemoreception and chemical ecology.

Catherine Lindell, Assistant Professor; Ph.D., Harvard, 1994. Behavioral and landscape ecology.

Elena G. Litchman, Assistant Professor; Ph.D., Minnesota, 1997. Aquatic ecology, community ecology, and global change.

Jianguo (Jack) Liu, University Distinguished Professor and Rachel Carson Chair; Ph.D., Georgia, 1992. Systems modeling and simulation; biodiversity and ecosystem management.

Barbara L. Lundrigan, Associate Professor; Ph.D., Michigan, 1988. Phylogenetics; evolution of behavior and morphology.

David W. MacFarlane, Assistant Professor; Ph.D., Rutgers, 2001. Forest measurements and modeling.

Carolyn Malmstrom, Assistant Professor; Ph.D., Stanford, 1997. Vegetation and ecosystem dynamics.

Brian A. Maurer, Associate Professor; Ph.D., Arizona, 1984. Macroecology; biogeography; quantitative population and community ecology.

Andrew G. McAdam, Assistant Professor; Ph.D., Alberta, 2003. Field ecology; quantitative genetics; evolutionary biology.

Deborah G. McCullough, Associate Professor; Ph.D., Minnesota, 1990. Forest entomology; insect ecology.

Haddish Melakeberhan, Associate Professor; Ph.D., Simon Fraser, 1986. Ecophysiology of plant-nematode interactions.

Richard W. Merritt, Professor; Ph.D., Berkeley, 1974. Aquatic entomology and ecology.

Kelly F. Millenbah, Associate Professor; Ph.D., Michigan State, 1997. Restoration ecology and endangered species management.

James R. Miller, Professor; Ph.D., Penn State, 1975. Behavioral and physiological aspects of insect ecology.

Gary G. Mittelbach, Professor; Ph.D., Michigan State, 1980. Community ecology and aquatics.

Peter G. Murphy, Professor; Ph.D., North Carolina at Chapel Hill, 1970. Plant ecology.

Patrick M. Muzzall, Professor; Ph.D., New Hampshire, 1978. Parasitology.

Antonio A. Nunez, Professor and Associate Chair, Psychology; Ph.D., Florida State, 1977. Biological rhythms; hormone and reproductive behaviors.

Nathaniel Ostrom, Professor; Ph.D., Newfoundland, 1992. Biogeochemistry.

Peggy H. Ostrom, Professor; Ph.D., Newfoundland, 1990. Biogeochemical cycling.

Scott Peacor, Assistant Professor; Ph.D., Michigan, 1992, 2000. Aquatic population and community ecology, modeling.

Robert T. Pennock, Associate Professor; Ph.D., Pittsburgh, 1991. Philosophy of biology; history and philosophy of science; science and values.

L. Alan Prather, Associate Professor and Director, MSU Herbarium; Ph.D., Texas at Austin, 1995. Plant systematics and evolution of floral morphology.

G. Philip Robertson, Professor; Ph.D., Indiana, 1980. Ecosystem and agricultural ecology.

Joan B. Rose, Professor; Ph.D., Arizona, 1985. Environmental microbiology.

David E. Rothstein, Assistant Professor; Ph.D., Michigan, 1999. Ecosystem ecology.

Tao Sang, Associate Professor; Ph.D., Ohio State, 1995. Plant systematics and molecular evolution.

Orlando Sarnelle, Associate Professor; Ph.D., California, Santa Barbara, 1992. Biological limnology; ecology.

Norman J. Sauer, Professor; Ph.D., Michigan State, 1974. Physical and forensic anthropology; human skeletal biology.

Douglas W. Schemske, Hannah Distinguished Professor; Ph.D., Illinois, 1977. Ecological genetics of adaptation and speciation in plants.

Thomas Schmidt, Professor; Ph.D., Ohio State, 1985. Microbial ecology and evolution.

J. Mark Scriber, Professor; Ph.D., Cornell, 1975. Evolutionary ecology of insect-plant interactions.

Kim T. Scribner, Associate Professor; Ph.D., Georgia, 1992. Population genetics, population ecology, and conservation biology.

Alexander W. Shingleton, Assistant Professor; Ph.D., Cambridge, 2001. Ecological and evolutionary developmental biology.

Shin-Han Shiu, Assistant Professor; Ph.D., Wisconsin–Madison, 2001. Evolution of genomes.

Laura Smale, Associate Professor; Ph.D., Berkeley, 1987. Animal behavior and its physiological substrates.

James J. Smith, Associate Professor; Ph.D., Michigan State, 1985. Molecular systematics and evolution.

Richard J. Snider, Professor; Ph.D., Michigan State, 1972. Invertebrate zoology.

Patricia Soranno, Associate Professor; Ph.D., Wisconsin–Madison, 1995. Limnology and ecosystem ecology.

R. Jan Stevenson, Professor; Ph.D., Michigan, 1981. Algal ecology; aquatic ecology; environmental science.

Ralph E. Taggart, Professor; Ph.D., Michigan State, 1971. Botany; paleobotany; palynology.

William W. Taylor, Professor and Chairperson, Fisheries and Wildlife; Ph.D., Arizona State, 1978. Fish population and community dynamics.

Frank W. Telewski, Professor; Ph.D., Wake Forest, 1983. Developmental tree ecophysiology.

James M. Tiedje, University Distinguished Professor; Ph.D., Cornell, 1968. Microbial ecology.

Richard E. Triemer, Professor and Chair, Plant Biology; Ph.D., North Carolina at Chapel Hill, 1975. Algal cell biology, phylogeny, and systematics.

Jean I. Tsao, Assistant Professor; Ph.D., Chicago, 2001. Ecology and evolution of disease organisms.

Merritt R. Turetsky, Assistant Professor; Ph.D., Alberta, 2002. Ecosystem ecology and management; biogeochemistry; plant-soil feedbacks.

Eileen R. Van Tassell, Professor; Ph.D., Catholic University, 1966. Insect systematics and behavior.

Juli Wade, Associate Professor; Ph.D., Texas at Austin, 1992. Sex differences in brain and behavior.

C. Michael Wagner, Assistant Professor; Ph.D., Georgia, 2004. Behavioral and community ecology of fishes.

Edward D. Walker, Professor; Ph.D., Massachusetts, 1984. Medical entomology.

Michael B. Walters, Associate Professor; Ph.D., Minnesota, 1994. Functional ecology of plants; forest community ecology.

Patrick J. Webber, Professor; Ph.D., Queens at Kingston, 1971. Community and landscape ecology.

Mark E. Whalon, Professor; Ph.D., Penn State, 1979. Agricultural pest management.

Thomas S. Whittam, Hannah Distinguished Professor; Ph.D., Arizona, 1981. Evolution of pathogenic microbes.

Scott Winterstein, Professor; Ph.D., New Mexico State, 1985. Wildlife population dynamics.

MORGAN STATE UNIVERSITY

Department of Biology
Ph.D. Program in Bioenvironmental Science

Program of Study

The Ph.D. Program in Bioenvironmental Science at Morgan State University is a didactic and research-driven program with participating faculty members from the Departments of Biology, Chemistry, and Computer Science. A unique strength of the program is the integration of biomedical and quantitative science expertise with classical environmental science. This curriculum allows the student to appreciate environmental impact from the level of the molecule and cell to the level of the whole organism and ecosystem. The program resides within the Department of Biology in the School of Computer, Mathematical and Natural Sciences, and utilizes an integrated interdisciplinary approach that is designed to offer flexibility in areas of specialization and training to meet the changing bioenvironmental needs of the nation and global community in the twenty-first century. The goal of the programs is to produce highly skilled scientists who will contribute knowledge derived from basic and applied research to address the multifaceted concerns of the bioenvironmental science community. The Ph.D. in Bioenvironmental Science offers research opportunities and instruction in six general areas of concentration: environmental toxicology, environmental health sciences, environmental chemistry, environmental biotechnology, environmental sciences, and environmental bioinformatics. The expected timeline for the completion of the Ph.D. degree is as follows: in the first two years, students take general (core) course work and concentration-specific electives; students perform research rotations in faculty laboratories and select a dissertation adviser by the end of the summer of the second year. A comprehensive examination (written and oral) is to be completed after the second year, followed by the presentation of the dissertation proposal. Successful completion of the comprehensive exam and dissertation proposal advance the student to Ph.D. candidacy. The next two to four years until degree completion are dedicated to dissertation research; concentration-specific and elective course work may still be completed. Students have a maximum of seven academic years to complete the degree.

Research Facilities

The School of Computer, Mathematical and Natural Sciences provides state-of-the-art equipment and technical research support in the following facilities: Histology/Cytology Core Laboratory, Cell and Molecular Core Laboratory, Environmental Toxicology Core Laboratory, and the Center for Biosensor Research, which are located in the Science Complex and Richard N. Dixon Science Research Center inaugurated in 2004. Additional research laboratories available to bioenvironmental students are located at the Estuarine Research Center in St. Leonard, Maryland.

Financial Aid

As financial support for the program, competitive stipends are awarded to eligible doctoral candidates along with a tuition waiver.

Cost of Study

Tuition for graduate students is $272 per credit hour for residents and $478 per credit hour for nonresidents. The cost of books and supplies is, on average, $500 per year.

Living and Housing Costs

Residence hall accommodations are available to graduate students, with a room cost of $2345 plus $1320 for the meal plan. Although the University assumes no responsibility for off-campus housing, students are assisted in finding satisfactory accommodations. Students should contact the Director of Housing, Residence Life (telephone: 443-885-3217).

Location

Morgan State University is located in Baltimore, a city rich in education, history, and culture. The seafood cuisine, artistic and natural exhibits, multicultural entertainment, friendly persona, and affordable cost of living help to make life in Baltimore an enriching experience. The geographic location of Baltimore provides easy access to other major cities, such as Philadelphia, New York, and Washington, D.C.

The University

Morgan State University was founded in 1867 as the Centenary Biblical Institute and was renamed Morgan College in 1890. The state of Maryland purchased the institution, making it a public campus, in 1939 and designated it as a university in 1975. In 1988, Maryland reorganized its higher education structure and designated the campus as Maryland's Public Urban University. The School of Graduate Studies commenced in 1963 and initially provided graduate instruction for public school teachers. The curriculum has greatly expanded, and today, the University has established thirteen Ph.D., several professional, and numerous master's degree programs. In 2005, the Carnegie Foundation reclassified the University as a Doctoral, Professions Dominant institution of higher learning.

Applying

Students should obtain application materials by contacting the School of Graduate Studies or the program coordinator. Prospective students should have a bachelor's degree and a cumulative and science GPA of at least 3.0 from an accredited institution and submit satisfactory scores for the Graduate Record Examinations (GRE), three letters of recommendation, official transcripts of undergraduate course work, and, when appropriate, graduate course work. International students whose undergraduate training was not conducted in English are required to submit TOEFL scores.

Correspondence and Information

Dr. Maurice C. Taylor, Dean
School of Graduate Studies and Research
Holmes Hall, Room 206
Morgan State University
1700 East Cold Spring Lane
Baltimore, Maryland 21251
Phone: 443-885-3185
E-mail: mctaylor@moac.morgan.edu

Dr. Christine F. Hohmann
Program Coordinator
Bioenvironmental Science Ph.D. Program
Department of Biology
Morgan State University
1700 East Cold Spring Lane
Baltimore, Maryland 21251
Phone: 443-885-4002
E-mail: chohmann@jewel.morgan.edu

Morgan State University

FACULTY AND THEIR RESEARCH

The faculty members who are available for research guidance demonstrate diverse research interests.

Lisa A. Brown, Ph.D. Muscle plasticity; activity-dependent changes in gene expression in adult skeletal muscle fibers.
Frank Denaro, Ph.D. Neuropathology and neuroscience techniques, including histological and electron microscopy.
Chunlei Fan, Ph.D. Regional studies in sustainable management of coastal and marine habitats.
Cleo Hughes Darden, Ph.D. Molecular biology of neuronal and hormonal regulatory mechanism in stress and hypertension.
Yousef Hijji, Ph.D. Organosilican chemistry and organic microwave synthesis.
Dwayne Hill, Ph.D. Developmental immunotoxicology.
Christine Hohmann, Ph.D. Developmental neurobiology; cellular and organismal environmental influences on brain and cognitive development in mouse
   models.
Mohammad Hokmabadi, Ph.D. Structure of water and aqueous solutions.
Maurice Iwunze, Ph.D. Molecular spectroscopy; electrochemistry/electrocatalysis.
Casonya Johnson, Ph.D. Molecular regulation of developmental process.
Alvin P. Kennedy, Ph.D. Microwave chemistry in microgravity.
Michael Koban, Ph.D. Molecular physiology of stress.
Santosh Mandal, Ph.D. Organometallic chemistry.
Gabrielle L. McLemore, Ph.D. Pharmacology of cannabinoid dependence.
Jochen Mueller, Ph.D. Environmental microbiology.
Richard Ochillo, Ph.D. Pharmacodynamics of autonomic and cardiovascular drugs.
Saroj Pramanik, Ph.D. Translational control of gene expression during cellular differentiation.
Kenneth P. Samuel, Ph.D. Molecular biology of HIV.
James Wachira, Ph.D. Molecular biology of hypertension.
Arthur L. Williams, Ph.D. Molecular biology of gene regulation and biotechnology.
Richard J. Williams, Ph.D. Laser spectroscopy in the areas of near-infrared fluorescence.
Jonathan Wilson, Ph.D. Environmental and aquatic toxicology.

PRINCETON UNIVERSITY

Department of Ecology and Evolutionary Biology

Programs of Study

Princeton University has a very cooperative and congenial group of graduate students and faculty members whose research combines at least two of the fields of ecology, evolution, and behavior. Work is directed toward practically and conceptually important issues in these fields and typically combines theory with observations and experiments, both outdoors and in the laboratory.

During the first two years, students must take the Department's six core courses. These include Fundamental Concepts in Ecology, Evolution, and Behavior I and II; population biology; and tropical ecology, which is an intensive three-week field course in a tropical locality, most recently Kenya. Advanced-topics courses and specialty courses are taken when appropriate. Specialized courses and seminars are developed in response to the particular needs of each cohort of students. New students are expected to begin a research project as soon as possible and work closely with the faculty to find a suitable topic.

By the end of the second year of study, each predoctoral student must stand for the 3-hour General Examination, which includes oral defense of a written analysis of the student's thesis topic results to date and of his or her proposal for thesis research. The student must also demonstrate a general knowledge in at least two of the fields of ecology, evolution, and behavior and of the course work undertaken thus far. After successful completion of the examination, yearly formal reports are required. The course of study and research culminates in a final public oral examination in which the student defends a written thesis describing original work, be it from the field, the laboratory, or theoretical. The Ph.D. degree is awarded after successful defense of the thesis; attainment of the degree normally requires four or five years. Degree candidates assist in teaching for at least two terms as part of their training.

Research Facilities

Offices, labs, and community equipment are located in Eno and Guyot Halls on campus. Spartan facilities are available at the University's 99-acre Stony Ford Ecological Research Preserve, 5 miles from campus. Cordial working relations are maintained with museums, zoos, and other universities, from New York City to Washington, D.C. Routine arrangements are made for fieldwork in Central and South America, East Africa, southern New Jersey, and elsewhere.

An excellent biology library is located in nearby Fine Hall. It has 52,000 volumes and 600 of the principal biological periodicals as well as standard reference works and monographs.

Financial Aid

Entering graduate students who are U.S. citizens are requested to apply for national fellowships (providing a stipend and tuition); information about these fellowships is supplied upon request. Graduate students who do not receive such fellowships, as well as noncitizens, are eligible for fellowships, traineeships, and research and teaching assistantships made available through the Department. Students awarded fellowships, traineeships, and assistantships are also awarded full tuition grants for the academic year. International students are encouraged to apply for fellowships from their native countries. The Department has, in the past, found it possible to support all doctoral candidates for at least four years.

Cost of Study

The tuition was $32,450 for the 2005–06 academic year.

Living and Housing Costs

In 2004–05, rooms at the Graduate College cost from $2900 to $5300 for the academic year. Meal plans ranged in price from $2800 to $4300 for the academic year. University apartments rented for $650 (one bedroom) to $1500 (four bedrooms) per month. Accommodations are also available in surrounding communities.

Student Group

At Princeton, a carefully chosen student body of 4,678 undergraduates and 1,999 graduate students works closely with a faculty of 1,140. Students come from all parts of the United States and from forty to fifty other countries. The Department accepts graduate students with previous specializations not only in biology or one of its branches but also in chemistry, physics, mathematics, anthropology, and engineering.

Location

Located at the edge of a piedmont, Princeton, a lively and attractive academic community of about 26,000, lies midway between the metropolitan centers of New York and Philadelphia. The town offers an unusually broad array of intellectual and recreational activities. The University's McCarter Theatre, winner of a Tony Award for regional theaters, also regularly presents nationally known popular and classical musicians and other performers. Regular performances are given by Westminster Choir College, the Princeton Chamber Orchestra, Theatre Intime, and other amateur and student organizations. Local cinemas present both American and international films, and good restaurants cater to a variety of tastes. Fields, streams, open countryside, and virgin woods are all within walking and bicycling distance. Lake Carnegie, the Millstone River, and the Delaware-Raritan Canal provide opportunities for sailing, canoeing, rowing, and fishing.

The University

Princeton University was founded in 1746 as the College of New Jersey. The college was moved to Princeton in 1756. At its 150th anniversary in 1896, the trustees changed the name to Princeton University. Princeton's long and distinguished scientific tradition began in 1795 with the establishment of the first professorship of chemistry in an American college. The Graduate School was organized in 1901 and has since won international recognition in mathematics, the natural sciences, philosophy, and the humanities.

Applying

December 1 is the deadline for receiving applications, transcripts, scores from the GRE General Test and Subject Test in biology, and letters of recommendation from applicants outside of North America who wish to enter the following September. December 31 is the deadline for applicants within North America. It is to the applicant's benefit to make certain all items arrive at the Graduate School Office of Admission on time.

Application forms for admission and the Graduate School *Announcement* are available from the Graduate School Office of Admission. Detailed information about faculty research is available at http://www.eeb.princeton.edu.

Correspondence and Information

Office of Graduate Studies
Department of Ecology and Evolutionary Biology
Princeton University
Princeton, New Jersey 08544-1003
Phone: 609-258-3977
E-mail: lolly@princeton.edu
Web site: http://www.eeb.princeton.edu

Office of Graduate Admission
Graduate School
Princeton University
P.O. Box 270
Princeton, New Jersey 08544-0270
Web site: http://gso.princeton.edu

Princeton University

THE FACULTY AND THEIR RESEARCH

J. Altmann, Ph.D. Behavioral ecology; effects of the environment, behavior, and social structures on intraspecific differences; life history; field and captive studies of mammals; behavioral aspects of inbreeding and of conservation.

A. P. Dobson, Ph.D. Conservation biology and disease control; population dynamics and coevolution of parasites and their hosts; population dynamics and life-history strategies of birds, primates, and elephants.

J. L. Gould, Ph.D. Animal behavior; communication, navigation, behavioral ecology, sexual selection, and learning.

P. R. Grant, Ph.D. Ecology, evolution, and behavior; speciation, competition, and the direct study of natural selection in populations.

M. Hau, Ph.D. Timing of breeding in birds; behavioral endocrinology; ecophysiology; physiological basis of life-history strategies; circadian rhythms and melatonin; field and captive work with birds in the tropics and temperate zones.

L. O. Hedin, Ph.D. Complex ecological systems; climate change; eutrophication; atmospheric deposition; nutrients and chemical elements influencing ecosystems and their relation to evolution.

H. S. Horn, Ph.D. Adaptive patterns of ecology and social behavior in birds and butterflies; forest succession; dynamics of landscape and of productivity; adaptive patterns of morphology, spatial distribution, and dispersal in trees; local conservation.

L. Kruglyak, Ph.D. Population genetics, intraspecific genetic variation, genetic basis of complex phenotypic variation, regulatory variation, and genetics of gene expression in model systems; signatures of natural selection.

L. Landweber, Ph.D. Molecular evolution of genes and genomes; genetic code origins; early pathways of RNA evolution; RNA editing; gene scrambling; in vitro evolution; DNA computers.

S. A. Levin, Ph.D. Dynamics of populations and communities; spatial heterogeneity and problems of scale; evolutionary ecology; theoretical and mathematical ecology; biodiversity and ecosystem processes.

S. W. Pacala, Ph.D., Director of Graduate Studies. Population biology and community ecology of plants; theoretical and mathematical ecology; global interactions among the biosphere, atmosphere, and hydrosphere.

D. I. Rubenstein, Ph.D., Chair. Behavioral ecology: effect of environmental variation and individual differences on social structure and behavioral relationships within a population; fieldwork with horses, zebras, fish, spiders, and insects; conservation biology.

D. L. Stern, Ph.D. Evolutionary developmental biology of insects; evolutionary genetics of *Drosophila;* molecular causes of morphological variation; developmental biology and genetics of aphid polyphenisms; diapause and endosymbionts.

M. Wikelski, Ph.D. Physiological ecology; physiological characteristics that underlie life-history traits of vertebrates; tropical ecology; tropical conservation biology; Galapagos marine iguanas; evolution of body size; costs of mate choice; shrinkage in body size.

Lecturers

S. Altmann, Ph.D. Behavioral ecology, especially of wild primates; natural history of food and foraging; mathematical models of social behavior.

B. R. Grant, Ph.D. Evolutionary ecology, natural and sexual selection; evolutionary consequences of hybridization, genetic and cultural transmission of traits in the speciation process of Darwin's finches.

Associated Faculty

B. B. Ward, Ph.D. Biological oceanography; marine microbiology; biochemistry of nitrogen in various environments, including oceans, sediments, and Antarctic lakes.

D. S. Wilcove, Ph.D. Conservation biology; conservation of endangered species; policy issues relating to the protection of biodiversity and endangered species; use of economic and regulatory incentives to promote habitat restoration; ecology and conservation of migratory animals.

Graduate student surveying tropical savannah in Kenya during the annual graduate field course.

Studying the evolution of nucleic acid sequences in Professor Laura Landweber's lab.

SAN DIEGO STATE UNIVERSITY

Joint Doctoral Program in Ecology

Programs of Study

San Diego State University (SDSU) offers a doctoral degree program in ecology in cooperation with the Graduate Group in Ecology at the University of California, Davis (UCD). The research interests of the participating faculty members cover a wide range of topics and represent the interdisciplinary nature of ecology. Strong research and graduate programs exist in coastal and marine ecology, physiological ecology, population ecology, restoration and conservation ecology, ecosystem ecology, and global change ecology. There is a growing interest and strength in policy relevance, sustainability, and science education. After formal admission to the program, the student must spend at least three quarters in full-time residence at UCD and two semesters at SDSU.

Upon admission to the program, an advising committee is established to work with the student to develop a course of study, which involves course work at both SDSU and UCD. Prior to taking the qualifying examination, students complete the course of study, including the three quarters at UCD, and develop a firm understanding of ecological principles and research methods. Following successful completion of the qualifying examination (usually at the end of the second year or beginning of the third year), the major remaining requirement for the doctoral degree is satisfactory completion of a dissertation consisting of original and significant research carried out under the guidance of an SDSU faculty major professor and a UCD faculty mentor. This degree program is designed to be completed within five years.

A master's degree program in ecology is also offered. This program is described in the separate description of SDSU programs in biological sciences in this volume.

Research Facilities

The ecology program is well equipped with state-of-the-art facilities and assets, such as a growth chamber facility; computer facilities with instrumentation for photosynthesis respiration, water potential, leaf area index, reflectance, and soil respiration; a microbial ecology research lab; terrestrial ecosystem instrumentation including eddy covariance; boats; and a University-owned environmental aircraft with GIS and remote sensing capabilities. The University operates a world-class field global change facility with elevated CO_2 FACE and CO_2 LT capability and eddy covariance.

Ecology program faculty members operate major field programs in distant locations, including arctic Alaska; Russia; La Paz, Mexico; and the Salton Sea. There are four field stations: the Santa Margarita Ecological Reserve, Sky Oaks Biological Field Station, Fortuna Mountain, and the Pacific Estuarine Research Laboratory. The ecology program also hosts the Biological Resources Division of the USGS.

The ecology program has close interactions with the University of San Diego; University of California, San Diego; Scripps Institute of Oceanography; the California State University, San Marcos and Los Angeles; and many other universities in California, Europe, the Pacific Rim, and elsewhere, nationally and internationally.

Financial Aid

Annual stipends begin at approximately $15,715 for the first year and include a full health benefit package. Travel grants and modest funds for research supplies are available upon request. Additional stipend and research support is provided by faculty research grants and student-obtained fellowships. All doctoral students are encouraged to apply for funding from federal agencies such as NSF, NASA, DOE, EPA, and USDA and from state agencies, as appropriate. Application to NSF for predoctoral research support is strongly encouraged, and past success rates have been high. Students may be supported by other programs, including SDSU's NSF GK-12 PISCES project.

Cost of Study

Currently, the full cost of tuition is provided to all students in the joint doctoral program. Students are not expected to be self-supporting.

Living and Housing Costs

There are seven residence halls/complexes at San Diego State University, including the Villa Alvarado apartments. Graduate students also have the option of living in the Piedra del Sol apartments, a sixty-six-unit complex that offers entirely independent apartment living. In addition, numerous apartments and shared single-unit houses are available in the nearby community. Support levels meet graduate living standards for the region.

Student Group

At present, approximately 24 students are enrolled in the Joint Doctoral Program in Ecology. There is a student graduate association in ecology and representation in meetings and faculty search committees. A high faculty–doctoral student ratio fosters close contact and high-quality training for doctoral students. Approximately 38 graduate students are enrolled in the ecology M.S. program. Ecology graduate students interact closely with graduate students in other departments, including geography, geology, public health, statistics, and biology.

Location

The University is located 12 miles east of the Pacific Ocean and is convenient to all of San Diego, which enjoys a reputation for easy living and a highly desirable climate. Major cultural attractions include the San Diego Globe Theater and the San Diego Symphony Orchestra. The nearby ocean, mountains, and deserts allow an unusually wide variety of ecosystem research sites and year-round outdoor activities.

The University

More than 30,000 students and 1,850 faculty members make SDSU the largest of twenty-two institutions within the California State University (CSU) system, which is itself the largest public university system in the United States. SDSU is also unique within the CSU system in offering a number of joint doctoral programs and maintaining world-class research programs in ecology and a number of other areas.

Applying

Application for admission must be made simultaneously to both SDSU and UCD. A complete application requires the appropriate application form, three letters of recommendation, transcripts of academic work already completed, and scores from the Gradate Record Examinations (GRE), including the Subject Test in advanced biology.

Correspondence and Information

Joint Doctoral Program in Ecology
Department of Biology
San Diego State University
San Diego, California 92182-4614
Web site: http://www.bio.sdsu.edu/ecologyjd/ecology.
docprogram.html (program description)
http://www.bio.sdsu.edu/ecologyjd/ecology.docapp.
html (application instructions)

Sandra Talley, Coordinator
Phone: 619-594-6919
E-mail: stalley@sciences.sdsu.edu
Walter Oechel, Ph.D., Program Coordinator
E-mail: oechel@sunstroke.sdsu.edu

San Diego State University

THE FACULTY AND THEIR RESEARCH

Todd W. Anderson, Assistant Professor of Biology; Ph.D., California, Santa Barbara, 1993. Marine ecology: recruitment processes, population ecology of temperate and tropical reef fishes, influence of habitat structure on demographic processes. (E-mail: todda@sunstroke.sdsu.edu)

Andrew J. Bohonak, Assistant Professor of Biology; Ph.D., Cornell, 1998. Freshwater invertebrates; population genetics; population ecology; estimation of gene flow and dispersal; invasion genetics; computational biology; systematics. (E-mail: bohonak@sciences.sdsu.edu)

Douglas H. Deutschman, Assistant Professor of Biology; Ph.D., Cornell, 1996. Population and community dynamics; plant community structure; population modeling; scaling from individuals to ecosystems; application of statistics and computer models to ecosystem management. (E-mail: ddeutschman@sunstroke.sdsu.edu)

Matthew Edwards, Assistant Professor of Biology; Ph.D., California, Santa Cruz, 2001. Marine ecology: scale-dependent regulation of subtidal kelp forests, recruitment processes in marine macroalgae (with emphasis on the role of dormant stages), impacts of large disturbances (El Niño–Southern Oscillation). (E-mail: edwards@sciences.sdsu.edu)

Janet Franklin, Professor of Biology and Adjunct Professor of Geography; Ph.D., California, Santa Barbara, 1988. Landscape ecology; plant ecology; biogeography; biophysical remote sensing; digital terrain analysis; geographic information science. (E-mail: janet@typhoon.sdsu.edu)

Brian T. Hentschel, Assistant Professor of Biology; Ph.D., Washington (Seattle), 1995. Marine invertebrates; benthic ecology; complex life cycles; ontogenetic niche shifts; larval ecology; deposit feeding; organism-flow interactions. (E-mail: hentsche@sciences.sdsu.edu)

Kevin A. Hovel, Assistant Professor of Biology; Ph.D., William and Mary, 1999. Marine invertebrate conservation ecology: relationship between habitat structure and faunal survival and abundance; effects of habitat fragmentation on fauna in marine systems; crustacean ecology; larval dispersal and recruitment; sea grass community ecology; kelp forest ecology. (E-mail: hovel@sciences.sdsu.edu)

Rebecca Lewison, Assistant Professor of Biology; Ph.D., California, Davis, 2002. Vertebrate conservation; population ecology; behavioral ecology; estimating the impact of anthropogenic disturbances; resource and land use. (E-mail: rlewison@sunstroke.sdsu.edu)

David A. Lipson, Assistant Professor of Biology; Ph.D., Colorado, 1998. Soil microbial ecology; plant-microbe interactions; biogeochemistry; linking microbial diversity to ecosystem processes. (E-mail: dlipson@sciences.sdsu.edu)

Walter C. Oechel, Professor of Biology; Ph.D., California, Riverside, 1970. Plant physiological ecology; ecosystem ecology; effects of global change and elevated carbon dioxide on native ecosystems, especially tundra and Mediterranean-type ecosystems (E-mail: oechel@sunstroke.sdsu.edu)

Tod W. Reeder, Assistant Professor of Biology; Ph.D., Texas at Austin, 1993. Molecular ecology and evolution of amphibians and reptiles, particularly those of the southwestern U.S. and Mexico. (E-mail: treeder@sunstroke.sdsu.edu)

Helen Regan, Assistant Professor of Biology; Ph.D., New England, 1999. Characterization and treatment of epistemic and linguistic uncertainty in ecology and conservation biology, including applying Monte Carlo techniques, dependency bounds convolution, interval analysis, and fuzzy set theory to a variety of problems. (E-mail: hregan@sciences.sdsu.edu)

Kathy S. Williams, Associate Professor of Biology; Ph.D., Stanford, 1981. Insect-plant interactions; effects of food quality on insect population dynamics; insects as indicators of biodiversity and habitat restoration.

STONY BROOK UNIVERSITY, STATE UNIVERSITY OF NEW YORK

Department of Ecology and Evolution
Programs in Applied Ecology and in Ecology and Evolution

Programs of Study

Founded in 1969, the Department of Ecology and Evolution at Stony Brook is one of the oldest departments of its kind in the country. Throughout its history, the department has been recognized as one of the finest in the fields of evolutionary biology and ecology. The study of graduate programs by the National Research Council ranked Stony Brook in the top ten for ecology and evolution programs nationally. Particular areas of strength include population genetics, evolutionary ecology, molecular evolution and phylogenetics, plant ecology, plant-animal interactions, marine ecology, primate evolution and behavior, biometry, and evolutionary theory. Since its inception, the department has offered doctoral degrees in ecology and evolution. The doctoral program combines course work and research, with an emphasis on the latter. Since 1991, the faculty of this program has offered a program leading to a master's degree in biological sciences with a concentration in applied ecology. It was created to meet the increased need for professional training in environmental sciences for positions available in federal, state, county, and township environmental offices as well as in environmental departments of large industrial companies and small environmental consulting firms. Students enrolled in the program need to complete 30 graduate credits and must complete a master's paper to graduate in accordance with the Graduate School requirements. This can be achieved in two to four semesters. For every student in both the doctoral and master's programs, the plan of study is individually tailored to match the student's personal goals within the course offerings and other training opportunities. Courses offered by the Department of Ecology and Evolution provide training in ecology, mathematical methods, statistics, and computer programming as applied to ecological problems.

Research Facilities

The Department of Ecology and Evolution has access to secure areas within Brookhaven National Laboratories for field studies in a preserve of mixed woodlands, ponds, and river habitats. Within the Life Sciences Laboratory building, students in ecology and evolution have access to modern laboratory and computer facilities. These include a new molecular lab with facilities for DNA sequencing, microsatellite analysis, real-time PCR; a *Drosophila* facility; a state-of-the-art laboratory with a diverse array of instrumentation for research in functional, evolutionary, and ecosystem ecology; an extensive greenhouse facility; and a marine ecology laboratory. The department has powerful central computing and graphic capabilities and participates in the campus' NASA Center for Excellence in Remote Sensing. It has statistical software, desktop publishing, and advanced computer power. Offices and laboratories are connected by a local area network that provides access to the Internet, including Internet II.

Financial Aid

The University offers several scholarship programs to graduate students. These fellowships are awarded on the basis of merit. Because Stony Brook is committed to attracting high-quality students, the Graduate School provides two competitive fellowships for U.S. citizens and permanent residents. Graduate Council Fellowships are awarded to exceptional entering doctoral students studying in any discipline. W. Burghardt Turner Fellowships target outstanding African-American, Hispanic-American, and Native American students entering either a doctoral or master's degree program. Both provide five years of stipend support with a teaching assistant requirement one semester per year. For doctoral students, both fellowships provide an annual stipend of at least $15,830 for up to five years, as well as a full tuition scholarship. Health insurance subsidies are also provided within a scale depending on the size of the fellow's dependent family. The program supports graduate students in good standing through teaching assistantships through at least their fifth year of study. Research assistantships are available through the research programs of many individual faculty members. Research assistantships on newly funded grants provide a stipend of at least $25,000. Although numbers vary, about 40 to 50 percent of graduate students beyond their first year are typically supported in this way at any one time. Full teaching assistantships carry a stipend amount that usually ranges from $12,000 to $20,000. In the past, the department has also been able to provide summer support, either by Federal Work-Study Program funds or research assistantships

Cost of Study

As noted above, tuition fees are only waived for graduate students in the Ph.D. program. Optional enrollment in a comprehensive health and dental insurance plan costs $5 per month. In 2005–06, full-time tuition was $3450 per semester for state residents and $5460 per semester for nonresidents. Part-time tuition was $288 per credit hour for residents and $455 per credit hour for nonresidents. Additional charges included an activity fee of $22 and a comprehensive fee of $322.50 per semester.

Living and Housing Costs

University apartments ranged in cost from approximately $350 per month to $1180 per month, depending on the size of the unit. Off-campus housing options include furnished rooms to rent and houses and/or apartments to share that can be rented for $350 to $700 per month.

Student Outcomes

Graduates of the Ph.D. program at Stony Brook have gone on to a variety of successful careers, including faculty positions at numerous prominent research institutions in the U.S. and abroad as well as positions with conservation organizations, small colleges, consulting firms, and various governmental agencies.

Location

Stony Brook is approximately 50 miles east of Manhattan on the north shore of Long Island. The cultural offerings of New York City and Suffolk County's countryside and seashore are located nearby. Cold Spring Harbor Laboratories and Brookhaven National Laboratories are easily accessible from, and have close relationships with, the University.

The University

The University, established in 1957, achieved international stature within a generation. It now ranks among the top 2 percent in the world. Founded at Oyster Bay, Long Island, the school moved to its present location in 1962. Stony Brook has grown to encompass more than 123 buildings on 1,100 acres. There are more than 1,890 faculty members, and the annual budget is more than $1.3 billion. The Graduate Student Organization oversees spending for graduate student campus events. Support services and offices include the four-week Summer Institute in American Living and the Intensive English Center for international students, the Career Development Office, Disabled Student Services, Counseling Center, day-care services, and the Writing Center.

Applying

Applicants are judged on the basis of distinguished undergraduate records (and graduate records, if applicable), thorough preparation for advanced study and research in the field of interest, candid appraisals from those familiar with the applicant's academic/professional work, potential for graduate study, and a clearly defined statement of purpose and scholarly interest germane to the program. The master's and Ph.D. programs do not have any absolute requirements for admission. Current graduate students represent a wide variety of academic backgrounds, nationalities, and research interests. Admission to both the master's and doctoral programs is quite competitive, and applicants are advised to make alternative plans in the event that they do not receive an offer of admission. Students should submit admission and financial aid applications by January 15 for Ph.D. and May 1 for master's degrees. Earlier submissions are requested if possible.

Correspondence and Information

Lev Ginzburg (Master's Program) or Geeta Bharanthan (Ph.D. Program)
Department of Ecology and Evolution
Stony Brook University, State University of New York
Stony Brook, New York 11794-5245

Phone: 631-632-8604
Fax: 631-632-7626
E-mail: geeta@life.bio.sunysb.edu
    risk@life.bio.sunysb.edu
Web site: http://life.bio.sunysb.edu/ee/

Stony Brook University, State University of New York

THE FACULTY AND THEIR RESEARCH

The faculty includes 2 members of the National Academy of Sciences, 4 members of the American Academy of Arts and Sciences, 5 past presidents of the Society for the Study of Evolution and the American Society of Naturalists, and 4 past editors of these societies' journals. Several textbooks written by department faculty members are the standards in their field: Sokal and Rohlf's *Biometry*, Futuyma's *Evolutionary Biology*, and Levinton's *Marine Biology*. In addition, several faculty members have written books that have helped shape modern ecology and evolution, including George Williams' classic *Adaptation and Natural Selection*, Slobodkin's *Growth and Regulation of Animal Populations*, and Sokal and Sneath's *Numerical Taxonomy*. Stony Brook has been the home of the *Quarterly Review of Biology* since the mid-1960s.

Robert Armstrong, Associate Professor of Marine Sciences; Ph.D, Minnesota, 1975. Mathematical modeling in marine ecology and biogeochemistry.
Edwin H. Battley, Associate Professor Emeritus; Ph.D., Stanford, 1956. Microbial ecology and energetics.
Michael Bell, Professor; Ph.D., UCLA, 1976. Evolutionary biology of fishes.
Paul Bingham, Associate Professor of Biochemistry; Ph.D., Harvard, 1979. Gene function.
Ivan Chase, Associate Professor of Sociology; Ph.D., Harvard, 1972. Behavioral ecology.
David Conover, Professor of Marine Sciences; Ph.D., Massachusetts, 1981. Sex determination and life history evolution of fishes; fishery biology.
Walter Eanes, Professor; Ph.D., SUNY at Stony Brook, 1976. Population genetics and molecular evolution.
John Fleagle, Distinguished Professor of Anatomy; Ph.D., Harvard, 1976. Primate behavior and functional morphology.
Catherine A. Forster, Associate Professor of Anatomy; Ph.D., Pennsylvania, 1990. Dinosaur phylogeny; bird phylogeny.
Douglas Futuyma, Distinguished Professor; Ph.D., Michigan, 1969. Evolutionary biology; insect ecology.
R. Geeta, Associate Professor; Ph.D., Arizona, 1993. Systematic patterns of variation in development.
Lev Ginzburg, Professor; Ph.D., Agrophysical Institute (St. Petersburg), 1970. Theoretical ecology; applied ecology.
Catherine Graham, Assistant Professor; Ph.D., Missouri–St. Louis, 2000. Landscape and behavioral ecology.
Jessica Gurevitch, Professor; Ph.D., Arizona, 1982. Plant ecology; physiological ecology.
George Hechtel, Associate Professor; Ph.D., Yale, 1962. Invertebrate zoology; systematics of sponges.
Charles Janson, Professor and Chairman; Ph.D., Washington (Seattle), 1985. Behavioral ecology; primate ecology; seed dispersal.
Manuel Lerdau, Professor; Ph.D., Stanford, 1994. Plant ecology and physiology; global change.
Jeffrey Levinton, Distinguished Professor; Ph.D., Yale, 1971. Marine benthic ecology; macroevolution.
Glenn R. Lopez, Professor of Marine Sciences; Ph.D., SUNY at Stony Brook, 1976. Marine benthic ecology.
Dianna Padilla, Associate Professor; Ph.D., Alberta, 1987. Marine functional ecology; invasion biology.
Massimo Pigliucci, Associate Professor; Ph.D., Connecticut, 1994. Population and quantitative genetics; evolutionary ecology.
F. James Rohlf, Distinguished Professor; Ph.D., Kansas, 1962. Morphometrics; numerical taxonomy; computer applications in systematics and ecology.
Lawrence Slobodkin, Professor Emeritus; Ph.D., Yale, 1951. Ecology.
Robert R. Sokal, Distinguished Professor; Ph.D., Chicago, 1952. Theory of spatial analysis; theory of systematics; physical anthropology.
Randall Susman, Professor of Anatomical Sciences; Ph.D., Chicago, 1976. Primate biology; functional morphology.
John True, Assistant Professor; Ph.D., Duke, 1995. Evolutionary and developmental genetics of color patterning in *Drosophila*.
John J. Wiens, Assistant Professor; Ph.D., Texas at Austin, 1995. Systematics; biology of reptiles and amphibians.
George C. Williams, Professor Emeritus; Ph.D., UCLA, 1955. Evolutionary biology.
Patricia Wright, Professor of Anthropology; Ph.D., CUNY Graduate Center, 1985. Conservation; primate ecology.

UNIVERSITY OF PITTSBURGH

Department of Biological Sciences
Program in Ecology and Evolution

Programs of Study

The Department of Biological Sciences offers a program of study leading to the Ph.D. degree in ecology and evolution (EE). Study leading to the M.S. degree is also offered in the EE program. The organization of the program allows students to specialize in a range of ecological and evolutionary fields of study. The goal is to train independent scientists in research and teaching.

In the EE program, the first two years include intensive core courses in ecology and evolution in alternate years. The core courses are team taught by the EE faculty. Once an area of specialization is chosen, the major effort is on the student's dissertation research. A variety of informal research meetings and seminars given by local and visiting scientists are held in the Department and in several other local departments. Students and faculty members interact with neighboring scientists at Duquesne University, Pittsburgh Genetics Institute, Graduate School of Public Health, Carnegie Mellon University, Carlow College, and the Carnegie Museum of Natural History, thereby providing additional breadth to the program.

The student's progress is followed by a faculty thesis adviser and committee. A comprehensive examination must be passed during the first year for M.S. students. Ph.D. students must pass a preliminary examination during the first year and a comprehensive examination no later than the end of the third year. The format of the test varies depending on the program. In addition to the Departmental Seminar Series, the EE group holds a weekly ecology-evolution seminar together with researchers from local universities and several weekly journal clubs.

The Department of Biological Sciences also offers a program of study leading to the Ph.D. degree in molecular, cellular, and developmental biology (MCDB). The organization of the program allows students to obtain intensive training within one field of study and makes individual diversification possible. In all areas, the goal is to train independent scientists in research and teaching.

Research Facilities

Facilities useful for members of the ecology and evolution group include a modern greenhouse complex, growth chambers for conducting controlled environment experiments, a molecular ecology laboratory, GIS equipment, a microscopy and imaging facility, and an animal-care facility. In addition, the Department is a member of the Organization of Tropical Studies and the Three Rivers Consortium for the Environment. Field research is conducted at the Pymatuning Laboratory of Ecology, the University's field station, and the Powder Mill Nature Reserve of the Carnegie Museum. Excellent libraries are close at hand, including the Carnegie Museum, Falk (medical school), Hillman, and Langley libraries and that of the Department of Chemistry.

Financial Aid

Full financial aid is available to first-year graduate students. Stipends in 2005–06 were set at $22,386 for twelve months and carried substantial tuition reductions as well. The 2006–07 stipend is anticipated to be somewhat higher. A limited number of Mellon fellowships are available on the basis of a University-wide competition.

Cost of Study

All students receiving stipends are exempt from tuition. In 2005–06, tuition was $12,776 per term for nonresidents and $6867 per term for residents of Pennsylvania. Cost of tuition may be higher in 2006–07.

Living and Housing Costs

Off-campus housing is available near the campus and in surrounding communities. Rents vary, depending on the type of accommodations selected. Many opportunities to share expenses are also available. Additional information may be obtained from the University Housing Office by calling 412-624-7116.

Student Group

Thirteen of the Department's 55 graduate students are currently in the EE program. Equal numbers of men and women are recruited each year; their average age is 23. A fair proportion of the entering graduate class is married, and most students have done undergraduate work in biology at an institution in the United States. A research and teaching career in industry, academic institutions, or the civil service is the long-term goal of most students.

Location

Pittsburgh, with a population of 334,563, is part of a greater metropolitan area of 2.3 million people. The University occupies more than 120 acres of Oakland, the cultural hub of the city. Pittsburgh is of sufficient size to support a wide variety of cultural and athletics events, including performances of the Pittsburgh Symphony Orchestra, the Pittsburgh Ballet Theatre, the Pittsburgh Opera, the O'Reilly Theater, the Three Rivers Arts Festival, and the International Poetry Forum as well as home games of the Pittsburgh Pirates, Steelers, and Penguins. The city also has an extensive system of large parks. Schenley Park, adjacent to the University and to Carnegie Mellon, is the site of the renowned Phipps Conservatory and Botanical Gardens, and Highland Park contains the Pittsburgh Zoo and PPG Aquarium. North of the downtown area are the Allegheny Observatory, the Carnegie Science Center, Andy Warhol Museum, and the National Aviary. Close to the campus are the Carnegie Museums of Natural History and Art and the Carnegie Library. Golf, fishing, camping, hiking, skiing, ice-skating, and boating are available in the area.

The University

The University of Pittsburgh, founded eleven years after the signing of the Declaration of Independence, is a nonsectarian, coeducational institution. As part of the Commonwealth System of Higher Education, it receives financial support from the state. At the end of 2003, the University of Pittsburgh was ranked eighth nationally among all universities in NIH-funded support. The University has 17,181 undergraduate and 9,550 graduate students and 4,145 faculty members. The library system maintains excellent collections totaling more than 5.6 million volumes. Intercollegiate athletics events are frequent, and student tickets are available to many of these events. The Petersen Events Center is a new athletic facility that is the venue for basketball games and concerts; it also houses exercise facilities and an Olympic-size swimming pool, which are open to all students.

Applying

The size and interactive nature of the graduate program fosters close working relationships between students and their mentors; this interaction is best served when students and their mentors begin their association before students have been admitted to the program. This process ensures that students have found the best possible mentor for their graduate career. Therefore, it is highly recommended that they establish a dialogue with a potential graduate adviser in ecology or evolution during the application process. Candidates should have an undergraduate degree in biology (other degrees such as chemistry, physics, or mathematics are also considered). Candidates should submit the application form, undergraduate transcript, and at least three letters of recommendation. Graduate Record Examination scores are sent electronically from Educational Testing Services to the University database and are downloaded by the Department. Applications for the U.S. and Canada are available on the Department's Web site and should be submitted by January 15 (December 15 for international applicants) for the fall term. During the application process, prospective students are encouraged to contact faculty members whose labs they are interested in joining. Early applications are considered first.

Correspondence and Information

Cathy Barr
Recruitment and Admissions Committee
Department of Biological Sciences
A-234 Langley Hall
University of Pittsburgh
Pittsburgh, Pennsylvania 15260

Phone: 412-624-4268
Fax: 412-624-4759
E-mail: biophd@pitt.edu
Web site: http://www.pitt.edu/~biology/

University of Pittsburgh

THE PROGRAM FACULTY AND THEIR RESEARCH

Ecology and Evolution Program

Tia-Lynn Ashman, Associate Professor; Ph.D., California, Davis, 1991. Plant evolutionary ecology; pollination; herbivory; plant gender.
Iain M. Campbell, Associate Professor; Ph.D., Glasgow, 1965. Commercial microbiology; gas chromatography–mass spectrometry.
Walter P. Carson, Associate Professor; Ph.D., Cornell, 1993. Plant population and community ecology; herbivory.
William P. Coffman, Associate Professor; Ph.D., Pittsburgh, 1967. Ecology of aquatic insects.
Susan Kalisz, Professor; Ph.D., Chicago, 1985. Evolutionary ecology of plants; molecular genetics of floral development.
Jeffrey G. Lawrence, Associate Professor; Ph.D., Washington, 1991. Molecular evolution of bacterial genomes.
Zhe-Xi Luo, Adjunct Associate Professor; Ph.D., Berkeley, 1989. Evolutionary morphology and phylogeny of mammals.
Rick A. Relyea, Assistant Professor; Ph.D., Michigan, 1998. Ecology and evolution of aquatic communities.
Stephen J. Tonsor, Associate Professor; Ph.D., Chicago, 1983. Evolutionary and ecological genetics.
M. Brian Traw, Assistant Professor; Ph.D., Cornell, 2001. Plant evolutionary ecology.

Recent Representative Faculty Publications

Ashman, T.-L., et al. Scent of a male: The role of floral volatiles in pollination of a gender dimorphic plant. *Ecology* 86:2099–105, 2005.
Carson, W. P., and R. B. Root. Herbivory and plant species coexistence: Community regulation by an outbreaking phytophagous insect. *Ecol. Monogr.* 70(1):73–99, 2000.
Coffman, W. P., and C. L. de la Rosa. Taxonomic composition and temporal organization of tropical and temperate species assemblages of lotic *Chironomidae. J. Kans. Ent. Soc.* 71(4):388–406, 1999.
Goodwille, C., **S. Kalisz,** and C. Eckert. The evolutionary enigma of mixed mating systems in plants: Occurrence, theoretical explanations, and empirical evidence. *Ann. Rev. Ecology Evolution Systematics* 36:47–79, 2005.
Wildeschute, H., D. M. Wolfe, A. Tamewitz, and **J. G. Lawrence.** Protozoan predation, diversifying selection and the evolution of antigenic diversity in *Salmonella. Proc. Natl. Acad. Sci. U.S.A.* 101:10644–9, 2004.
Luo, Z.-X., R. C. Cifelli, and Z. Kielan-Jaworowska. Dual evolution of tribosphenic mammals. *Nature* 409:53–7, 2001.
Relyea, R. A. The impact of insecticides and herbicides on the biodiversity and productivity of aquatic communities. *Ecology Appl.* 15:618–27, 2005.
Tonsor, S. J., C. Alonso-Blanco, and M. Koornneef. Gene function beyond the single trait: Natural variation, gene effects, and evolutionary ecology in *Arabidopsis thaliana. Plant Cell Environ.*, in press.
Morris, W. F., **M. B. Traw,** and J. Bergelson. On testing for a tradeoff between constitutive and induced resistance. *Oikos* 112:102–10, 2005.

UNIVERSITY OF SOUTH CAROLINA

Department of Biological Sciences
Graduate Training Program in Integrative Biology:
Ecology, Evolution, and Organismal Biology

Programs of Study

The University of South Carolina offers both M.S. and Ph.D. degrees in biology, with tracks in Integrative Biology: Ecology, Evolution, and Organismal Biology (IB) and molecular, cellular, and developmental biology (MCDB). Students interested in the MCDB program should consult that listing, which is located in Section 6 of *Peterson's Graduate Programs in the Biological Sciences.*

The IB faculty represents a broad range of research interests and approaches, but always within the context of the organism. Collaborations within the department are common and establish research foci that are flexible to changing times and priorities. Collaborations outside the department have established dynamic research programs with the Marine Science Program (Web site: http://inlet.geol.sc.edu/marine-science/), the School of the Environment (Web site: http://www.sc.edu/environment/), and the School of Public Health (Web site: http://www.sph.sc.edu/). This research atmosphere is enhanced by weekly in-house journal clubs and a program of seminars by scientists invited from other institutions. Students have the opportunity to establish focused research programs within the interests of a single faculty mentor or create novel and semi-independent programs that integrate the strengths of multiple faculty mentors.

The IB track has research concentrations in the following areas: animal behavior, biodiversity, bioinformatics, population genetics, conservation, ecology (ecosystems, community and population dynamics, fish, microbial, plant, and wetlands), ecological physiology, environmental policy, evolution (including genetics and molecular), insect biology, insect-plant interactions, invertebrate zoology, marine biology, neurophysiology, and systematics.

Research Facilities

Students pursuing research at the organismal level have access to outstanding coastal and inland field locations, greenhouse facilities, and laboratories equipped with environmental chambers, fresh and salt water mesocosms, and an extensive seawater system. A full array of instrumentation and equipment is available, including field vehicles, boats, and extensive computer facilities.

Students pursuing research at the cellular and molecular levels have ready access to all requisite equipment, including facilities for chemical purification, tissue culture, confocal and electron microscopy, and other basic equipment. Within the department, the Institute for Biological Research and Biotechnology provides general services, including production of polyclonal antibodies, DNA sequencing, and oligonucleotide synthesis.

Marine-oriented graduate students may use the laboratories, boat fleet, and protected habitats of the Belle W. Baruch Institute for Marine Biology and Coastal Research near Georgetown, South Carolina (Web site: http://inlet.geol.sc.edu/). The field station preserve (17,500 acres) encompasses North Inlet Estuary, with outstanding examples of dunes, salt marsh, maritime forest, mudflats, sandflats, and estuarine habitats. Students also have access to the aquaculture facilities of the South Carolina Department of Wildlife and Marine Resources in Bluffton and the extensive aquaculture facilities of the Baruch Institute at the Wedge Plantation near Georgetown.

Within easy access of campus are a variety of terrestrial research sites, such as the Congaree Swamp National Monument (14,000 acres of river-bottom hardwood forest), the Sandhills Wildlife Refuge, and the U.S. Department of Energy National Research Park, with the Savannah River Ecology Laboratory (250,000 acres with riverine, lake, upland forest, swamp, and stream communities). The Department of Biological Sciences has a large experimental deer mouse colony and the largest active herbarium in South Carolina.

Financial Aid

The department's general policy is to provide sufficient support to ensure that students can fully afford to focus their attention and energies on their graduate studies and research. Financial support is available to all students in the form of teaching and research assistantships; the current stipend is $20,000 for twelve months. Special fellowships are available for outstanding applicants as well as for applicants who are members of underrepresented minority groups.

Cost of Study

In 2006–07, full-time resident graduate students not holding assistantships pay $4144 per semester for tuition and fees; nonresidents pay $8958. Tuition is $2466 per semester for graduate students who hold assistantships. The department is currently paying $2466 for students holding assistantships. The University reserves the right to change fees without notice.

Living and Housing Costs

Some housing is available on the campus for single students and those with families. In 2006–07, costs range from $500 to $800 per month. Most students take advantage of the large variety of private living accommodations that are available in the vicinity of the University; off-campus housing is available at similar rates.

Student Group

About 60 graduate students are currently enrolled in the Department of Biological Sciences. A number of the department's students come from the Southeast, but all areas of the United States and several other countries are represented. The University of South Carolina as a whole enrolls about 25,700 students, a third of whom are doing graduate work.

Location

Situated in central South Carolina where the Saluda and Broad Rivers merge to form the Congaree River, Columbia is one of the fastest-growing metropolitan areas in the Southeast. According to the 2000 census, the population of the greater Columbia area was 508,798, and projections indicate that the population will reach 626,000 by 2010. As the state capital, Columbia provides a pleasant balance between urban, suburban, and rural lifestyles. Lake, ocean, and mountain areas are within easy driving distance, and the mild climate of this Sun Belt state encourages a wide variety of year-round recreational activities. The University and community offer numerous historic landmarks and diverse cultural events.

The University

The University of South Carolina is located on a 220-acre campus adjacent to the state capitol, in the center of Columbia. Founded in 1801 as South Carolina College, it has a rich tradition as one of the oldest public universities in the United States. The University grants bachelor's degrees in sixty-three majors, and the Graduate School offers both M.S. and Ph.D. degrees in many fields. The School of Law awards the J.D. degree, and the School of Medicine awards the M.D.

Applying

Applicants to the Department of Biological Sciences must hold or be pursuing a baccalaureate degree with satisfactory scholastic standing and must submit scores on the General Test of the Graduate Record Examinations, all academic transcripts, and at least two letters of reference. Completed applications should be submitted via the Graduate School Web site (http://www.biol.sc.edu/gradstudies/appl.html) before February 15 to ensure consideration for teaching and research assistantships.

Correspondence and Information

Director of Graduate Studies
Department of Biological Sciences
University of South Carolina
Columbia, South Carolina 29208

Phone: 803-777-2755
Web site: http://www.biol.sc.edu/

University of South Carolina

THE FACULTY AND THEIR RESEARCH

Prospective students are strongly encouraged to correspond with faculty members who have compatible research interests.

Ronald Benner, Professor; Ph.D., Georgia, 1984. Biogeochemistry; microbial ecology; carbon, nitrogen, and phosphorus cycles in aquatic environments; microbial utilization of organic matter.

Franklin G. Berger, Professor; Ph.D., Purdue, 1974. Mammalian molecular genetics; mechanism of resistance to anticancer agents; mouse models of colorectal cancer.

Erin L. Connolly, Associate Professor; Ph.D., California, Davis, 1997. Metal transport in plants; iron-deficiency responses; gene regulation; phytoremediation.

John Mark Dean, Distinguished Professor Emeritus; Ph.D., Purdue, 1962. Age and growth of fishes; fisheries ecology; fishery management.

Patricia J. DeCoursey, Distinguished Professor Emerita; Ph.D., Wisconsin, 1959. Neural, behavioral, and ecological aspects of circadian rhythms in mammals.

Dan A. Dixon, Assistant Professor; Ph.D., Northwestern, 1994. Colon carcinogenesis; inflammation; posttranscriptional gene regulation.

Charles F. Duggins, Research Associate Professor; Ph.D., Florida State, 1980. Molecular systematics of fishes.

Bert Ely, Professor; Ph.D., Johns Hopkins, 1973. Molecular and population genetics of fish.

Robert J. Feller, Professor; Ph.D., Washington (Seattle), 1977. Biological oceanography; benthic ecology; predator-prey interactions; ecological applications of immunoassays; science education; invertebrate feeding dynamics.

Madilyn Fletcher, Professor; Ph.D., Wales (Bangor), 1975. Microbial and estuarine ecology; microbial adhesion to surfaces; physiology of attached bacteria; microbial community structure.

Robert Friedman, Assistant Professor; Ph.D., South Carolina, 2002. Bioinformatics; molecular evolution.

Brian Helmuth, Associate Professor; Ph.D., Washington (Seattle), 1997. Physical biology of marine invertebrates; biomechanics; intertidal and subtidal physiological ecology; coral reef ecology.

Kevin Higgins, Assistant Professor; Ph.D., California, Davis, 1997. Ecological and evolutionary dynamics; metapopulation biology; ecological chaos.

Thomas J. Hilbish, Professor; Ph.D., SUNY at Stony Brook, 1984. Ecological genetics; physiological ecology of marine invertebrates.

Austin L. Hughes, Professor; Ph.D., Indiana, 1984. Molecular evolution and bioinformatics; evolution of the vertebrate immune system; host-parasite coevolution.

David E. Lincoln, Professor; Ph.D., California, Santa Cruz, 1978. Plant-herbivore interactions; global change; biodiversity; chemical ecology; convergent evolution.

Richard Long, Assistant Professor; Ph.D., California, San Diego (Scripps), 2001. Microbial ecology with a focus on how bacteria interact with each other and with their environment.

Charles R. Lovell, Professor and Chair; Ph.D., Purdue, 1984. Microbial ecology; molecular ecology; physiology and biochemistry of anaerobic bacteria.

Laszlo Marton, Professor; Ph.D., Szeged (Hungary), 1976. Plant molecular genetics; plant tissue culture; gene engineering and transgene expression in plants; phytoremediation.

Lydia Matesic, Assistant Professor; Ph.D., Johns Hopkins, 2000. Role of ubiquitination in human disease.

James T. Morris, Professor; Ph.D., Yale, 1979. Wetlands ecology; biogeochemistry; plant physiological ecology; ecosystems ecology.

Timothy A. Mousseau, Professor and Associate Dean for Research; Ph.D., McGill, 1988. Evolution and genetic basis of like histories, maternal effects, phenotypic plasticity, and behavior; measurement of adaptive genetic variation in the wild; entomology; the impact of radioactive contaminants resulting from Chernobyl NPP on evolution processes.

John B. Nelson, Herbarium Curator; Ph.D., Florida State, 1982. Vascular plant taxonomy; rare and threatened species in South Carolina; plant biogeography.

James L. Pinckney, Associate Professor; Ph.D., South Carolina, 1992. Marine ecology; ecosystem processes; microalgal ecophysiology; biological oceanography; biometry.

Joseph M. Quattro, Associate Professor; Ph.D., Rutgers, 1991. Conservation biology; population genetics of marine and freshwater fishes; molecular evolution.

Robert A. Raguso, Assistant Professor; Ph.D., Michigan, 1995. Chemically mediated plant-animal interactions; biosynthesis, physiology, and evolution of fragrance; insect olfaction, behavior, and pollination ecology.

Tammi L. Richardson, Assistant Professor; Ph.D., Dalhousie, 1996. Phytoplankton physiology and ecology; harmful algal blooms; carbon and nitrogen cycling in marine food webs.

Roger H. Sawyer, Professor and Senior Associate Dean for Natural Sciences; Ph.D., Massachusetts, 1970. Molecular, cellular, and developmental biology of the scales and feathers of reptiles and birds.

Richard M. Showman, Associate Professor; Ph.D., Washington (Seattle), 1979. Developmental biology; gene regulation during early development; mitochondrial biogenesis.

Deanna Smith, Assistant Professor; Ph.D., Stanford, 1994. Cytoskeletal organization and cell motility in health and disease.

Stephen E. Stancyk, Professor; Ph.D., Florida, 1974. Marine invertebrate ecology, reproduction, and life histories.

Johannes W. Stratmann, Associate Professor; Ph.D., Regensburg (Germany), 1994. Plant stress signal transduction; MAP kinases; UV-B signaling.

Briana Timmerman, Undergraduate Director; M.S., Washington (Seattle), 1996. Fungal ecology; science education curriculum design.

Richard Vogt, Associate Professor and Graduate Director; Ph.D., Washington (Seattle), 1984. Neurobiology of odor detection in insects and vertebrates (e.g., moths, *Drosophila,* mosquitoes, sea turtles, zebrafish); molecular, developmental, and evolutionary biology.

David S. Wethey, Professor; Ph.D., Michigan, 1979. Population and community ecology; biogeography; rocky intertidal ecology; biophysical ecology; ecological modeling.

Sarah A. Woodin, Professor; Ph.D., Washington (Seattle), 1972. Marine benthic ecology; recruitment processes; functional morphology of polychaetes; chemical ecology.

Duane C. Yoch, Professor Emeritus; Ph.D., Penn State, 1968. Microbial ecology; microbial physiology and biochemistry of salt marsh and estuarine sulfur cycle processes.

Richard G. Zingmark, Professor Emeritus; Ph.D., California, Santa Barbara, 1969. Physiological ecology of estuarine and marine algae; harmful algal blooms; mariculture; sustainable development of tropical coastal ecosystems.

Adjunct Faculty

Paul Bradley, Hydrologist, U.S. Geological Survey, Water Resources Division; Ph.D., South Carolina, 1991.

Justin Congdon, Emeritus Senior Research Scientist, Savannah River Ecology Lab, University of Georgia; Ph.D., Arizona State, 1977.

Travis Glenn, Associate Research Scientist, Savannah River Ecology Laboratory, University of Georgia; Ph.D., Maryland, 1997.

Steven J. Harper, Assistant Research Scientist, Savannah River Ecology Laboratory; Ph.D., Illinois, 1996.

Patricia G. Lincoln, Professor of Biology, Coker College; Ph.D., California, Santa Cruz, 1980.

Alan H. Shoemaker, Curator of Mammals, Riverbanks Zoological Park and Botanical Garden; M.S., South Carolina, 1972.

THE UNIVERSITY OF TOLEDO

Department of Earth, Ecological, and Environmental Sciences

Programs of Study

The University of Toledo (UT) Department of Earth, Ecological, and Environmental Sciences (EEES) offers graduate degrees in geology at the master's level and in biology at the master's and doctoral levels. Students entering the programs are expected to have an adequate background in the natural sciences and mathematics, but may be admitted on a provisional basis if they lack such a background. Successful applicants work with nationally and internationally recognized faculty members on research foci in both basic and applied fields of ecology and geology. Active research groups are currently working in coastal sedimentology and stratigraphy, ecological modeling, environmental geochemistry, environmental microbiology, fisheries genetics, glaciology, hydrogeology, land-water interactions, plant biotechnology, population and community ecology, Quaternary geology, and wetlands ecology. The department emphasizes the interdisciplinary nature of earth, ecological, and environmental sciences. Students, depending on their interest, are encouraged to take courses in related fields including environmental engineering, biochemistry, hydrology, remote sensing, environmental law, and others.

All entering graduate students are assigned to a faculty adviser. Routinely, however, beginning graduate students have already corresponded with EEES faculty members and selected an adviser based on mutual research interests before they start their first semester. Students in the M.S. programs must complete a minimum of 30 semester hours of graduate course work approved by their advisory committee, pass a comprehensive oral qualifying examination prior to completion of the first year in the program (ecology students only), and defend a research thesis consisting of a written report on original, independent research. A nonthesis option is also available for ecology students.

The doctoral degree in biology is awarded to students who have demonstrated mastery in the field of biology and a distinct and superior ability to make substantial contributions to the field. It is not awarded merely as a result of courses taken or for years spent in studying or research. The quality of work and the resourcefulness of the student must be such that the faculty can expect a continuing effort toward the advancement of knowledge and significant achievement in research and related activities. In general, work for the Ph.D. takes at least four years and 90 semester hours of study beyond the bachelor's degree. A substantial portion of this time is spent in independent research leading to a dissertation. Work done toward a master's degree may apply as part of the student's doctoral program.

Doctoral students complete an individualized program of study in advanced ecology, biostatistics, and a selection of courses in a range of ecological subdisciplines and related fields. The curriculum also includes a series of seminars. Ph.D. students must pass a written qualifying examination during the first two years of study and an oral comprehensive examination involving a defense of their research proposal after gaining admission to candidacy.

The department considers experience in teaching to be a vital component of graduate education. Therefore, all graduate students are required to complete at least one semester of formal teaching experience before graduation. This improves the presentation skills of master's students going on to careers in industry and government. Advanced Ph.D. candidates become junior colleagues in the teaching and research mission of the Department, which enhances the professional application of their degree. Moreover, graduates with Ph.D. degrees have accepted excellent positions as postdoctoral researchers or faculty members at academic institutions or leading research corporations.

Research Facilities

Facilities on the main campus include more than 100,000 square feet of research and teaching space in Bowman-Oddy Laboratories, which also houses the College's advanced Instrumentation Center, and a state-of-the-art Plant Science Research Facility (http://www.eeescience.utoledo.edu/facilities/plantcent.htm). Off campus, the Lake Erie Center, the $7-million lakeshore interdisciplinary environmental research facility (http://www.lakeerie.utoledo.edu), and the 47-acre arboretum (http://www.eeescience.utoledo.edu/facilities/stranahan.htm) allows the University to offer research facilities that parallel or exceed in quality those of the finest graduate institutions in the country. Carlson Library's holdings exceed 1.6 million volumes, 1.4 million microforms, 150,000 maps, and 5,000 periodicals. The library is a federal depository for government documents and is a charter member of the statewide cooperative program OhioLINK.

Financial Aid

Most full-time graduate students receive some financial support. College fellowships, teaching assistantships, and research assistantships, which include a stipend and a tuition waiver, are available for qualified students on a competitive basis. The out-of-state tuition surcharge normally charged to out-of-state and international students is waived for students whose permanent address is within one of the following Michigan counties: Hillsdale, Lenawee, Macomb, Oakland, Washtenaw, and Wayne. In addition, The University of Toledo offers an out-of-state tuition surcharge waiver to cities and regions that are a part of the Sister Cities Agreement. These regions include Toledo, Spain; Londrina, Brazil; Qinhuangdao, China; Csongrad County, Hungary; Delmenhorst, Germany; Toyohashi, Japan; Tanga, Tanzania; Bekaa Valley, Lebanon; and Poznan, Poland. The University of Toledo Graduate College offers a variety of memorial and minority scholarship awards, including the Ronald E. McNair Postbaccalaureate Achievement Scholarship, the Graduate Minority Assistantship Award, and two full University fellowships.

Cost of Study

The graduate tuition rate for the 2006–07 academic year is $390.05 per semester credit hour for in-state students. For nonresidents, the out-of-state surcharge is $367.15 per semester credit hour. Additional fees are required and include the general fee, technology fee, and mandatory insurance.

Living and Housing Costs

The University of Toledo has a diverse offering of student housing options, including suite-style and traditional residential halls. Housing is offered to graduate students through Residence Life or contracted individually by the student. Affordable, high-quality off-campus apartment-style housing within walking distance of campus is abundant.

Student Group

There are approximately 20,000 students at the University of Toledo. About 4,000 are graduate and professional students. The University has a rich diversity of student organizations. Students join groups that are organized around common cultural, religious, athletic, and educational interests.

Location

The University of Toledo has several campus sites in the city of Toledo. Most engineering graduate students take classes on the Main campus, which is located in suburban western Toledo. With a population of more than 330,000, Toledo is the fiftieth-largest city in the United States. It is located on the western shores of Lake Erie, within a 2-hour drive of Cleveland and Detroit.

The University

The University of Toledo was founded by Jessup W. Scott in 1872 as a municipal institution and became part of the state of Ohio's system of higher education in 1967. On July 1, 2006, The University of Toledo merged with the Medical University of Ohio becoming one of only seventeen American universities to offer professional and graduate academic programs in medicine, law, pharmacy, nursing, health sciences, engineering, and business.

Applying

Applications may be obtained from the Graduate School or online at http://www.utoledo.edu/grad-school/ and may be submitted directly to the Graduate Admissions Committee online or at the address listed in the Correspondence and Information section. A complete application includes the application form, three letters of recommendation, a copy of all previous undergraduate or graduate transcripts, and scores from the Graduate Record Examinations (GRE). International students whose native language is not English are also required to submit scores from the TOEFL examination. Applications should be completed by March 1 for fall admission and merit fellowship consideration; later applications are considered pending the availability of assistantship funds.

Correspondence and Information

Dr. Johan F. Gottgens (Graduate Coordinator, Ecology) or Dr. James A. Harrell (Graduate Coordinator, Geology)
Department of Earth, Ecological, and Environmental Sciences
Mail Stop 604, University of Toledo
2801 West Bancroft Street
Toledo, Ohio 43606-3390
Phone: 419-530-2009
Fax: 419-530-4421
E-mail: eees@utnet.utoledo.edu
Web site: http://www.eeescience.utoledo.edu

The University of Toledo

THE FACULTY AND THEIR RESEARCH

Jonathon M. Bossenbroek, Assistant Professor of Ecology. Wetlands and landscape ecology; aquatic invasive species.
V. Max Brown, Associate Professor of Geology. Igneous and metamorphic petrology; computer applications in geology.
Mark J. Camp, Associate Professor of Geology. Invertebrate paleontology and paleoecology; Quaternary studies; geology of national and state parks; history of geology.
Jiquan Chen, Associate Professor of Ecology. Landscape and community ecology; ecosystem analysis; ecological modeling.
Daryl F. Dwyer, Associate Professor of Ecology. Environmental microbiology; bioremediation; phytoremediation; microbial physiology; genetics.
Timothy G. Fisher, Associate Professor of Geology. Geomorphology; glaciology; Quaternary reconstructions.
Stephen L. Goldman, Professor of Biology. Plant genetics; genetic engineering in plants; phytoremediation.
Johan F. Gottgens, Associate Professor of Ecology and Ecology Graduate Coordinator. Aquatic ecology; pulse stability; paleolimnological approaches to restoration; the cycling and accumulation of mercury in aquatic environments.
James A. Harrell, Professor of Geology and Geology Graduate Coordinator. Sedimentary geology; archaeological geology of ancient Egypt.
Scott A. Heckathorn, Assistant Professor of Ecology. Plant ecological physiology; biochemistry; heat-shock proteins.
David E. Krantz, Associate Professor of Geology. Coastal and marine geology; oceanography; Quaternary geology.
James M. Martin-Hayden, Associate Professor of Geology. Hydrogeology; hydrogeological field methods; numerical groundwater modeling.
Christine M. Mayer, Assistant Professor of Ecology. Aquatic ecology; invertebrate-fish predator-prey interactions; organism-habitat modification; introduced species.
Daryl L. Moorhead, Associate Professor of Ecology. Mathematical modeling of ecological systems; extreme environments; global change.
Michael W. Phillips, Professor of Geology and Department Chair. Mineralogy; crystallography; crystal chemistry.
Alison L. Spongberg, Associate Professor of Geology. Environmental geochemistry of sediments and water; fate and transport of hazardous contaminants.
Carol A. Stepien, Professor of Ecology and Director of the Lake Erie Center. Fishery genetics in the Great Lakes; evolutionary patterns and population genetics of nonindigenous species invasions in the Great Lakes; systematics of dreissenid mussels and fishes, including the Percidae, Gobiidae, Blennioidei, and relatives.
Donald J. Stierman, Associate Professor of Geology. Applied environmental geophysics; geologic hazards mitigation; earthquake seismology; geological, geophysical, and environmental investigations in Central America.
Elliot J. Tramer, Professor of Ecology. Ornithology and population dynamics of vertebrates and higher plants.
W. Von Sigler, Assistant Professor of Ecology. Molecular ecology; microbial population structure and function in differing environments.
Michael N. Weintraub, Assistant Professor of Ecology. Soil ecology; biogeochemistry; ecosystem ecology.

Affiliated Faculty and Researchers
Thomas Bridgeman: aquatic ecology.
Kim Brosofske: plant ecology.
Thomas R. Crow: forest ecosystems.
Kevin P. Czajkowski: environmental remote sensing.
Stuart L. Dean, Professor Emeritus: structural geology.
Jonathan M. Frantz: horticulture.
Lou Glatzer, Professor Emeritus: environmental microbiology.
Michelle T. Grigore: conservation ecology.
Harry M. Jol: ground-penetrating radar in geomorphology.
Charles R. Krause: plant pathology.
Patrick Lawrence: environmental planning; remote sensing.
James C. Locke: plant pathology.
Parani Madasamy: plant genetics.
Steven McNulty: forest ecology.
Asko Noormets: plant ecophysiology.
Malcolm P. North: forest ecology.
Yude Pan: forest ecology.
Edward Roseman: fish ecology.
Sairam Rudrabhatla: plant genetics.
Rex Strange: molecular evolution and ecology of fishes.
Ge Sun: forest ecology.
Shulu Zhang: plant genetics.
Daolan Zheng: landscape ecology.

YALE UNIVERSITY

Department of Ecology and Evolutionary Biology
Ph.D. Program

Program of Study

The Department of Ecology and Evolutionary Biology (EEB) at Yale University offers training programs leading to the Ph.D. in organismal biology, ecology, and evolutionary biology, including molecular evolution, phylogeny, molecular population genetics, conservation genetics, experimental evolution, developmental evolution, and evolutionary theory and ecology. Applicants should have training in one of the following fields: biology, mathematics, chemistry, physics, statistics, and/or geology. Candidates are selected, regardless of their major, based on overall preparation for a career in research in EEB. Some, planning for careers in applied fields, may have prepared with courses in public policy, economics, and agriculture.

Each entering student, in consultation with a faculty committee, develops a specific program of courses, seminars, laboratory research, and independent reading tailored to the student's interests and background. First-year students carry out at least two research rotations in the first three semesters. Students have the option of a rotation over their first summer. Students must participate in a program of ethics of research and authorship, weekly EEB seminars, and symposia of faculty and graduate student research. In addition, during their first two years of study, students must take three additional graduate-level courses (numbering 500 and above). As required by the Graduate School and the Department, students must achieve an honors grade (A) in two graduate courses. EEB requires an average of at least High Pass in course work during the first two years. All students are required to teach two courses during their first four terms of study.

In the third term of study, each student takes qualifying examinations in ecology and evolutionary biology. By the end of the third term, each student organizes a formal preprospectus consultative meeting with his or her advisory committee to discuss the planned dissertation research. By the end of the fourth term, students present and defend their planned dissertation research at a prospectus meeting. A successful prospectus meeting and completion of course requirements result in candidacy for the Ph.D. Upon a successful defense and submission of the dissertation to the Graduate School, the Ph.D. is awarded.

Research Facilities

Research groups of primary EEB faculty members are on Science Hill, in Osborn Memorial Laboratories (OML) and in the Class of 1954 Environmental Studies Center (ESC; situated between OML and the Peabody Museum). Vivian Irish, a joint faculty member from the Department of Molecular, Cellular, and Developmental Biology, has an office and a lab in OML; Oswald Schmitz and David Skelly, joint faculty members from the School of Forestry and Environmental Studies, are in Greeley Labs; J. Rimas Vaisnys, a joint faculty member from the Department of Electrical Engineering, is in Becton Laboratory; Jacques Gauthier, a joint faculty member from the Department of Geology and Geophysics, is also in ESC; and Gene Likens is the Director of the Institute of Ecosystem Studies at the Mary Flagler Cary Arboretum in Millbrook, New York. State-of-the-art facilities are available for research in various aspects of experimental biology, including a greenhouse facility.

Special facilities include the Yale Institute for Biospheric Studies (YIBS), the YIBS Center for Earth Observation, the YIBS Ecosave Molecular Systematics and Conservation Genetics Laboratory, Peabody Museum of Natural History, Peabody Museum Field Station on Long Island Sound, the Marsh Botanical Garden and greenhouses, the Yale Natural Preserves, the Yale forests, molecular biology facilities, the Statistical Laboratory, and Structural Analysis/Electron Microscopy facilities. There are several science libraries, including the Kline Science Library (biological sciences), the Peabody Museum (ornithological and entomology), the Kline Geology Library (paleobiology), the School of Forestry and Environmental Studies (forest biology), and the Medical Library (biological sciences), representing a total collection of approximately 1 million volumes.

Financial Aid

Students admitted to the Ph.D. program receive five years of financial support (stipend and tuition), which is contingent upon continuing satisfactory progress toward the degree. Sources of financial support include national and University fellowships, National Research Service Awards of the NIH, teaching fellowships, and research assistantships.

Cost of Study

In 2005–06, tuition was $28,000 per year and the stipend was $24,000.

Living and Housing Costs

The cost of living is less than that in other large northeastern American cities. Yale has dormitory facilities for single students and apartments for married students. Many students live in adjacent off-campus areas. Information on University housing and a list of private off-campus accommodations are available from the Yale Housing Department.

Student Group

In 2005–06, there were 30 Ph.D. students in residence.

Location

EEB academic and business offices are located in Osborn Memorial Laboratories on Science Hill, adjacent to the Peabody Museum, the Class of 1954 Environmental Science Center, and the Kline Biology Tower. New Haven, situated on Long Island Sound, is at the junction of I-91, I-95, and the Merritt Parkway. It is roughly midway between New York City and Boston. Bradley International Airport and Tweed New Haven Airport provide direct service to all major national and international cities. Amtrak, MetroNorth, and Greyhound Bus Service operate from Union Station. The creative and performing arts are very active in New Haven. Theaters include the Palace, the Shubert, Yale Rep, and Long Wharf. For two weeks in June each year, the City of New Haven, in partnership with Yale, stages the International Festival of Arts and Ideas, bringing visitors and performers from around the world.

The University

The Yale University community is a large and diverse one, consisting of about 5,300 undergraduate students, 5,500 graduate and professional students, 1,500 postdoctoral fellows, and 2,000 faculty members. Within its urban setting, Yale has extensive facilities for athletic and recreational activities. The Payne Whitney Gymnasium and Ingalls Skating Rink are located on the central campus. Other recreational facilities away from the campus are Yale Bowl, an eighteen-hole golf course, the Yale Sailing Center, and the Yale Outdoor Recreation Center.

Applying

Online applications for the doctoral program in EEB can be found at http://www.yale.edu/graduateschool/admissions/ and are available in late summer. Completed forms, including GRE General Test scores, are due to the Yale Graduate Admissions Office by the beginning of January for matriculation in September. Written decisions concerning admission are sent to each applicant during the month of March.

Correspondence and Information

Director of Graduate Studies
Department of Ecology and Evolutionary Biology
Yale University
P.O. Box 208106
New Haven, Connecticut 06520-8106

Phone: 203-432-3837
Fax: 203-432-2374
E-mail: maureen.cunningham@yale.edu
Web site: http://www.yale.edu/eeb

Yale University

THE FACULTY AND THEIR RESEARCH

Suzanne H. Alonzo. Behavioral and evolutionary ecology; theoretical and empirical research on evolution and ecology of reproductive strategies, conflict between the sexes, predator-prey interactions, and links between population dynamics and life history patterns.

Leo W. Buss. Theoretical research on ontological foundations of evolutionary theory; empirical research on Placozoa and on hydroid allorecognition.

Michael J. Donoghue, Director of the Peabody Museum of Natural History. Plant evolution; phylogenetic biology.

*Jacques Gauthier. Vertebrate paleontology; systematics; lizard evolution.

*Vivian F. Irish. Evolution of floral development; *Arabidopsis* developmental genetics.

Antonia Monteiro. Understanding evolution and development of butterfly wing patterns; combining tools from ethology, population genetics, phylogenetics, and developmental biology to dissect the molecular basis of intraspecific and interspecific variation in developmental mechanisms generating color patterns.

Thomas Near. Evolutionary biology of fishes, retracing how species are related to one another, primarily using DNA sequence data to reconstruct the evolutionary relationships of species represented through branching diagrams or phylogenies.

David M. Post. Aquatic ecology; food-web structure and dynamics; stable isotopes.

Jeffrey R. Powell. Evolutionary genetics; molecular evolution conservation biology.

Richard O. Prum. Evolutionary ornithology, including phylogenetics, behavior, feathers, structural color, evolution and development, sexual selection, and historical biogeography.

*Oswald J. Schmitz. Dynamics and structure of terrestrial food webs; global climate change and ecosystem function; scaling from behavior to ecosystems.

*David E. Skelly. Aquatic ecology; spatial ecology; ecology of disease.

Melinda D. Smith. Biodiversity-ecosystem function; effects of species loss on ecosystems; factors influencing invasion by exotic plant species; grassland ecology; impacts of global change on community and ecosystem processes; scale-dependence of ecological patterns and processes.

Stephen C. Stearns. Life history evolution; evolution of sex; evolutionary genetics; population biology; functional genomics.

Paul E. Turner. Experimental evolution in microbes, especially bacterial and animal viruses; host-parasite interactions; evolution of sex.

*J. Rimas Vaisnys. Identification of dynamics of diverse biological systems.

Gunter Wagner. Population genetics theory; evolution of development.

**Joint appointee with primary appointment in another department.*

Section 9
Entomology

This section contains a directory of institutions offering graduate work in entomology, followed by in-depth entries submitted by institutions that chose to prepare detailed program descriptions. Additional information about programs listed in the directory but not augmented by an in-depth entry may be obtained by writing directly to the dean of a graduate school or chair of a department at the address given in the directory.

For programs offering related work, see also in this book Biochemistry; Biological and Biomedical Sciences; Botany and Plant Biology; Ecology, Environmental Biology, and Evolutionary Biology; Genetics, Developmental Biology, and Reproductive Biology; Microbiological Sciences; Physiology; and Zoology. In Book 2, see Economics (Agricultural Economics and Agribusiness); in Book 4, see Agricultural and Food Sciences and Environmental Sciences and Management; and in Book 5, see Agricultural Engineering and Bioengineering.

CONTENTS

Entomology

Auburn University, Graduate School, College of Agriculture, Department of Entomology and Plant Pathology, Auburn University, AL 36849. Offers entomology (M Ag, MS, PhD); plant pathology (M Ag, MS, PhD). Part-time programs available. *Faculty:* 21 full-time (6 women). *Students:* 18 full-time (9 women), 6 part-time (3 women), 11 international. 14 applicants, 71% accepted, 6 enrolled. In 2005, 2 master's, 3 doctorates awarded. *Degree requirements:* For master's, thesis (for some programs); for doctorate, one foreign language, thesis/dissertation. *Entrance requirements:* For master's, GRE General Test; for doctorate, GRE General Test, GRE Subject Test, master's degree with thesis. *Application deadline:* For fall admission, 7/7 for domestic students; for spring admission, 11/24 for domestic students. Applications are processed on a rolling basis. Application fee: $25 ($50 for international students). Electronic applications accepted. *Financial support:* Research assistantships, teaching assistantships, Federal Work-Study available. Support available to part-time students. Financial award application deadline: 3/15. *Faculty research:* Pest management, biological control, systematics, medical entomology. *Unit head:* Dr. Michael L. Williams, Chair, 334-844-5006. *Application contact:* Dr. Stephen L. McFarland, Acting Dean of the Graduate School, 334-844-4700.

Clemson University, Graduate School, College of Agriculture, Forestry and Life Sciences, Department of Entomology, Clemson, SC 29634. Offers MS, PhD. *Students:* 20 full-time (6 women), 4 part-time (1 woman), 8 international. Average age 31. 6 applicants, 50% accepted, 3 enrolled. In 2005, 3 master's, 3 doctorates awarded. *Degree requirements:* For master's, thesis; for doctorate, one foreign language, thesis/dissertation. *Entrance requirements:* For master's and doctorate, GRE General Test, minimum GPA of 3.0. Additional exam requirements/recommendations for international students: Required—TOEFL. *Application deadline:* For fall admission, 7/1 priority date for domestic students, 4/15 priority date for international students. Applications are processed on a rolling basis. Application fee: $50. *Financial support:* Fellowships, research assistantships, institutionally sponsored loans and unspecified assistantships available. Financial award applicants required to submit FAFSA. *Faculty research:* Aquatic arthropod diversity, crop insect management, medical and veterinary entomology, genetics and biotechnology, urban entomology. *Unit head:* Dr. Joseph Culin, Chair, 864-656-5041, Fax: 864-656-5065, E-mail: jculin@clemson.edu. *Application contact:* John Morse, Coordinator, 864-656-5049, Fax: 864-656-5065, E-mail: jmorse@clemson.edu.

See Close-Ups on pages 725 and 727.

Colorado State University, Graduate School, College of Agricultural Sciences, Department of Bioagricultural Sciences and Pest Management, Fort Collins, CO 80523-0015. Offers entomology (MS, PhD); plant pathology and weed science (MS, PhD). *Faculty:* 18 full-time (3 women). *Students:* 21 full-time (12 women), 18 part-time (8 women); includes 3 minority (1 American Indian/Alaska Native, 1 Asian American or Pacific Islander, 1 Hispanic American), 1 international. Average age 34. 16 applicants, 44% accepted, 6 enrolled. In 2005, 5 master's, 4 doctorates awarded. *Degree requirements:* For master's, thesis (for some programs), registration; for doctorate, thesis/dissertation, registration. *Entrance requirements:* For master's and doctorate, GRE General Test, minimum GPA of 3.0. Additional exam requirements/recommendations for international students: Required—TOEFL (minimum score 550 paper-based; 213 computer-based). *Application deadline:* For fall admission, 4/1 priority date for domestic students, 4/1 priority date for international students; for spring admission, 9/1 priority date for domestic students, 9/1 priority date for international students. Applications are processed on a rolling basis. Application fee: $50. Electronic applications accepted. *Expenses:* Tuition, state resident: full-time $3,690; part-time $205 per credit. Tuition, nonresident: full-time $14,958; part-time $831 per credit. Required fees: $1,061. *Financial support:* In 2005–06, 1 student received support, including fellowships (averaging $2,500 per year), research assistantships with full tuition reimbursements available (averaging $17,500 per year), teaching assistantships with full tuition reimbursements available (averaging $12,402 per year); scholarships/grants, traineeships, and unspecified assistantships also available. Financial award application deadline: 4/1; financial award applicants required to submit FAFSA. *Faculty research:* Biological control of post-insect plant pathogens and weeds, integrated pest management, weed ecology and biology, and pests genome's of plants. Total annual research expenditures: $2.2 million. *Unit head:* Thomas O. Holtzer, Head, 970-491-5261, Fax: 970-491-3862, E-mail: tholtzer@lamar.colostate.edu. *Application contact:* Janet Dill, Graduate Program Coordinator, 970-491-0402, Fax: 970-491-3862, E-mail: janet@colostate.edu.

Cornell University, Graduate School, Graduate Fields of Agriculture and Life Sciences, Field of Entomology, Ithaca, NY 14853-0001. Offers acarology (MS, PhD); apiculture (MS, PhD); applied entomology (MS, PhD); aquatic entomology (MS, PhD); biological control (MS, PhD); insect behavior (MS, PhD); insect biochemistry (MS, PhD); insect ecology (MS, PhD); insect genetics (MS, PhD); insect morphology (MS, PhD); insect pathology (MS, PhD); insect physiology (MS, PhD); insect systematics (MS, PhD); insect toxicology and insecticide chemistry (MS, PhD); integrated pest management (MS, PhD); medical and veterinary entomology (MS, PhD). *Faculty:* 55 full-time (8 women). *Students:* 32 full-time (14 women); includes 3 minority (1 American Indian/Alaska Native, 2 Asian Americans or Pacific Islanders), 13 international. 36 applicants, 28% accepted, 9 enrolled. In 2005, 4 master's, 5 doctorates awarded. *Degree requirements:* For master's, thesis/dissertation; for doctorate, thesis/dissertation, comprehensive exam. *Entrance requirements:* For master's and doctorate, GRE General Test, GRE Subject Test (biology), 3 letters of recommendation. Additional exam requirements/recommendations for international students: Required—TOEFL (minimum score 550 paper-based; 213 computer-based). *Application deadline:* For fall admission, 12/1 for domestic students. Application fee: $60. Electronic applications accepted. *Financial support:* In 2005–06, 31 students received support, including 10 fellowships with full tuition reimbursements available, 7 research assistantships with full tuition reimbursements available, 14 teaching assistantships with full tuition reimbursements available; institutionally sponsored loans, scholarships/grants, health care benefits, tuition waivers (full and partial), and unspecified assistantships also available. Financial award applicants required to submit FAFSA. *Faculty research:* Systematics and biodiversity, integrated pest management, pathology and biological control, toxicology and physiology, ecology and behavior. *Unit head:* Director of Graduate Studies, 607-255-6198, Fax: 607-255-0939. *Application contact:* Graduate Field Assistant, 607-255-6198, Fax: 607-255-0939, E-mail: fieldofent2@cornell.edu.

Florida Agricultural and Mechanical University, Division of Graduate Studies, Research, and Continuing Education, College of Engineering Science, Technology, and Agriculture, Division of Agricultural Sciences, Tallahassee, FL 32307-3200. Offers agribusiness (MS); animal science (MS); engineering technology (MS); entomology (MS); food science (MS); international programs (MS); plant science (MS). *Degree requirements:* For master's, thesis. *Entrance requirements:* For master's, GRE General Test, minimum GPA of 3.0. Additional exam requirements/recommendations for international students: Required—TOEFL (minimum score 500 paper-based).

Iowa State University of Science and Technology, Graduate College, College of Agriculture, Department of Entomology, Ames, IA 50011. Offers MS, PhD. *Faculty:* 14 full-time, 4 part-time/adjunct. *Students:* 23 full-time (13 women), 4 part-time (2 women); includes 1 minority (Hispanic American), 4 international. 11 applicants, 36% accepted, 4 enrolled. In 2005, 7 master's, 3 doctorates awarded. *Degree requirements:* For master's and doctorate, thesis/dissertation. *Entrance requirements:* For master's and doctorate, GRE General Test, GRE Subject Test (biology). Additional exam requirements/recommendations for international students: Required—TOEFL (paper score 550; computer score 213) or IELTS (score 6.5). *Application deadline:* For fall admission, 1/1 priority date for domestic students, 1/1 priority date for international students; for spring admission, 9/1 priority date for domestic students, 9/1 priority date for international students. Application fee: $30 ($70 for international students). Electronic applications accepted. *Expenses:* Tuition, state resident: full-time $6,410. Tuition, nonresident: full-time $16,422. Tuition and fees vary according to program. *Financial support:* In 2005–06, 23 research assistantships with full and partial tuition reimbursements (averaging $15,332 per

year) were awarded; fellowships, teaching assistantships with full and partial tuition reimbursements, scholarships/grants, health care benefits, and unspecified assistantships also available. *Unit head:* Dr. Jon J. Tollefson, Chair, 515-294-7400, Fax: 515-294-2125, E-mail: entomology@iastate.edu. *Application contact:* Dr. Joel Coats, Director of Graduate Education, 515-294-7400, E-mail: entomology@iastate.edu.

Kansas State University, Graduate School, College of Agriculture, Department of Entomology, Manhattan, KS 66506. Offers MS, PhD. *Faculty:* 19 full-time (1 woman), 12 part-time/adjunct (1 woman). *Students:* 29 full-time (14 women), 16 part-time (8 women); includes 3 minority (2 American Indian/Alaska Native, 1 Asian American or Pacific Islander), 23 international. Average age 26. 11 applicants, 64% accepted, 4 enrolled. In 2005, 4 master's, 3 doctorates awarded. *Degree requirements:* For master's, thesis, oral exam; for doctorate, thesis/dissertation, written and oral exams. *Application deadline:* For fall admission, 2/1 for domestic students; for spring admission, 10/1 for domestic students. Applications are processed on a rolling basis. Application fee: $30 ($55 for international students). *Expenses:* Tuition, state resident: full-time $5,160; part-time $215 per credit hour. Tuition, nonresident: full-time $12,816; part-time $534 per credit hour. Required fees: $564. *Financial support:* In 2005–06, 27 research assistantships (averaging $16,360 per year), 1 teaching assistantship with partial tuition reimbursement (averaging $15,217 per year) were awarded; career-related internships or fieldwork, Federal Work-Study, institutionally sponsored loans, scholarships/grants, and tuition waivers (partial) also available. Support available to part-time students. Financial award application deadline: 3/1; financial award applicants required to submit FAFSA. *Faculty research:* Molecular genetics, biologically-based pest management, host plant resistance, ecological genomics, stored product entomology. Total annual research expenditures: $1.8 million. *Unit head:* James Nechols, Head, 785-532-6154, Fax: 785-532-6232. *Application contact:* David Margolies, Director, 785-532-6154, Fax: 785-532-6232.

Louisiana State University and Agricultural and Mechanical College, Graduate School, College of Agriculture, Department of Entomology, Baton Rouge, LA 70803. Offers MS, PhD. *Faculty:* 21 full-time (2 women). *Students:* 19 full-time (9 women), 9 part-time (2 women); includes 1 Asian American or Pacific Islander, 2 Hispanic Americans, 9 international. Average age 31. 11 applicants, 55% accepted, 5 enrolled. In 2005, 5 master's, 3 doctorates awarded. *Degree requirements:* For master's and doctorate, thesis/dissertation. *Entrance requirements:* For master's and doctorate, GRE General Test, minimum GPA of 3.0. Additional exam requirements/recommendations for international students: Required—TOEFL (minimum score 550 paper-based; 213 computer-based). *Application deadline:* For fall admission, 1/25 priority date for domestic students, 5/15 priority date for international students. Applications are processed on a rolling basis. Application fee: $25. Electronic applications accepted. *Financial support:* In 2005–06, 23 students received support, including 18 research assistantships with partial tuition reimbursements available (averaging $15,913 per year); fellowships, teaching assistantships with partial tuition reimbursements available, Federal Work-Study, institutionally sponsored loans, scholarships/grants, and unspecified assistantships also available. Support available to part-time students. Financial award applicants required to submit FAFSA. *Faculty research:* Integrated pest management, ecology and biology, parasitoids and pathogens, host-pest resistance, molecular biology. Total annual research expenditures: $5,929. *Unit head:* Dr. Timothy Schowalter, Head, 225-578-1634, Fax: 225-578-1643.

McGill University, Faculty of Graduate and Postdoctoral Studies, Faculty of Agricultural and Environmental Sciences, Department of Natural Resource Sciences, Montréal, QC H3A 2T5, Canada. Offers agrometeorology (M Sc, PhD); entomology (M Sc, PhD); forest science (M Sc, PhD); microbiology (M Sc, PhD); neotropical environment (M Sc, PhD); soil science (M Sc, PhD); wildlife biology (M Sc, PhD). *Degree requirements:* For master's and doctorate, thesis/dissertation, registration. *Entrance requirements:* For master's, minimum GPA of 3.0 or 3.2 in the last 2 years of university study. Additional exam requirements/recommendations for international students: Required—TOEFL (minimum score 550 paper-based; 213 computer-based), IELT (minimum score 7). Electronic applications accepted. *Faculty research:* Toxicology, reproductive physiology, parasites, wildlife management, genetics.

Michigan State University, The Graduate School, College of Agriculture and Natural Resources and College of Natural Science, Department of Entomology, East Lansing, MI 48824. Offers entomology (MS, PhD); entomology-environmental toxicology (PhD); integrated pest management (MS). *Faculty:* 21 full-time (5 women). *Students:* 24 full-time (12 women), 3 part-time; includes 3 minority (1 African American, 2 Hispanic Americans), 4 international. Average age 30. 22 applicants, 23% accepted. In 2005, 4 master's, 2 doctorates awarded. *Degree requirements:* For master's, thesis or alternative, oral final exam; for doctorate, thesis/dissertation, oral final exam in defense of dissertation, participation in teaching program, comprehensive exam. *Entrance requirements:* For master's, GRE General Test, minimum GPA of 3.0 in last 2 undergraduate years, prior training in physical/biological sciences; for doctorate, GRE General Test, MS with thesis in entomology or related field. Additional exam requirements/recommendations for international students: Required—TOEFL (minimum score 550 paper-based; 213 computer-based), Michigan State University ELT (85), Michigan ELAB (83). *Application deadline:* For fall admission, 12/27 for domestic students. Application fee: $50. Electronic applications accepted. *Expenses:* Tuition, state resident: part-time $330 per credit hour. Tuition, nonresident: part-time $685 per credit hour. Tuition and fees vary according to program. *Financial support:* In 2005–06, 12 fellowships with tuition reimbursements (averaging $2,733 per year), 18 research assistantships with tuition reimbursements (averaging $14,266 per year), 7 teaching assistantships with tuition reimbursements (averaging $12,021 per year) were awarded; scholarships/grants and unspecified assistantships also available. *Faculty research:* Agroaquatic and forest ecology, insect physiology, agricultural entomology, integrated pest management, environmental and analytical toxicology. Total annual research expenditures: $6.8 million. *Unit head:* Dr. Richard Merritt, Chairperson, 517-355-4665, Fax: 517-353-4354, E-mail: merrittr@msu.edu. *Application contact:* Jill Kolp, Graduate Secretary, 517-355-4665, Fax: 517-353-4354, E-mail: kolpj@msu.edu.

Mississippi State University, College of Agriculture and Life Sciences, Department of Entomology and Plant Pathology, Mississippi State, MS 39762. Offers agricultural pest management (MS); entomology (MS, PhD); plant pathology (MS, PhD). *Faculty:* 19 full-time (1 woman). *Students:* 14 full-time (5 women), 7 part-time (4 women); includes 1 minority (Hispanic American), 6 international. Average age 33. 12 applicants, 25% accepted, 2 enrolled. In 2005, 1 master's awarded. *Degree requirements:* For master's and doctorate, thesis/dissertation, comprehensive oral or written exam. *Entrance requirements:* For master's, GRE General Test, minimum GPA of 2.75; for doctorate, GRE General Test. Additional exam requirements/recommendations for international students: Required—TOEFL. *Application deadline:* For fall admission, 7/1 for domestic students; for spring admission, 11/1 for domestic students. Applications are processed on a rolling basis. Application fee: $30. Electronic applications accepted. *Expenses:* Tuition, state resident: full-time $4,312; part-time $240 per hour. Tuition, nonresident: full-time $9,772; part-time $543 per hour. International tuition: $10,102 full-time. Tuition and fees vary according to course load. *Financial support:* Research assistantships, teaching assistantships, Federal Work-Study, institutionally sponsored loans, and unspecified assistantships available. Financial award applicants required to submit FAFSA. *Faculty research:* Ecology and population dynamics, physiology, biochemistry and behavior, systematics. Total annual research expenditures: $584,995. *Unit head:* Dr. Clarence H. Collison, Head, 662-325-2085, Fax: 662-325-8837, E-mail: chc2@ra.msstate.edu. *Application contact:* Philip G. Bonfanti, Director of Admissions, 662-325-4104, Fax: 662-325-8872, E-mail: admit@msstate.edu.

New Mexico State University, Graduate School, College of Agriculture and Home Economics, Department of Entomology, Plant Pathology and Weed Science, Las Cruces, NM 88003-8001. Offers agricultural biology (MS). Part-time programs available. *Faculty:* 12 full-time (5 women), 1 part-time/adjunct (0 women). *Students:* 14 full-time (12 women), 8 part-time (4 women); includes 7 minority (1 African American, 1 American Indian/Alaska Native, 1 Asian

Entomology

American or Pacific Islander, 4 Hispanic Americans), 1 international. Average age 30. 9 applicants, 67% accepted, 6 enrolled. In 2005, 4 degrees awarded. *Degree requirements:* For master's, thesis, comprehensive exam, registration. *Entrance requirements:* For master's, GRE General Test. *Application deadline:* For fall admission, 7/1 for domestic students; for spring admission, 11/1 priority date for domestic students. Applications are processed on a rolling basis. Application fee: $30 ($50 for international students). Electronic applications accepted. *Expenses:* Tuition, state resident: full-time $3,156; part-time $175 per credit. Tuition, nonresident: full-time $12,510; part-time $565 per credit. Required fees: $1,050. *Financial support:* In 2005–06, 1 fellowship, 11 research assistantships with partial tuition reimbursements, 4 teaching assistantships with partial tuition reimbursements were awarded; career-related internships or fieldwork also available. Financial award application deadline:3/1. *Faculty research:* Integrated pest management, pesticide application and safety, livestock ectoparasite research, biotechnology, nematology. *Unit head:* Dr. Grant Kinzer, Head, 505-646-3225, Fax: 505-646-8087, E-mail: gkinzer@nmsu.edu.

North Carolina State University, Graduate School, College of Agriculture and Life Sciences, Department of Entomology, Raleigh, NC 27695. Offers MS, PhD. Terminal master's awarded for partial completion of doctoral program. *Degree requirements:* For master's, thesis (for some programs); for doctorate, thesis/dissertation. *Entrance requirements:* For master's and doctorate, GRE General Test. Electronic applications accepted. *Faculty research:* Physiology, biocontrol, ecology, forest entomology, apiculture.

North Dakota State University, The Graduate School, College of Agriculture, Food Systems, and Natural Resources, Department of Entomology, Fargo, ND 58105. Offers entomology (MS, PhD); environment and conservation science (MS, PhD); natural resource management (MS, PhD). Part-time programs available. *Faculty:* 8 full-time (3 women), 6 part-time/adjunct (0 women). *Students:* 18 full-time (6 women), 4 part-time (2 women); includes 9 minority (2 African Americans, 7 Asian Americans or Pacific Islanders). Average age 34. 5 applicants, 20% accepted, 1 enrolled. In 2005, 1 degree awarded. *Median time to degree:* Of those who began their doctoral program in fall 1997, 100% received their degree in 8 years or less. *Degree requirements:* For master's, thesis/dissertation; for doctorate, thesis/dissertation, comprehensive exam. *Entrance requirements:* For master's and doctorate, minimum GPA of 3.0. Additional exam requirements/recommendations for international students: Required—TOEFL. *Application deadline:* Applications are processed on a rolling basis. Application fee: $45 ($60 for international students). Electronic applications accepted. *Financial support:* In 2005–06, 17 students received support, including 19 research assistantships with full tuition reimbursements available (averaging $13,800 per year); Federal Work-Study, institutionally sponsored loans, and unspecified assistantships also available. Financial award application deadline: 4/15. *Faculty research:* Insect systematics, conservation biology, integrated pest management, insect behavior, insect biology. *Unit head:* Dr. Gary J. Brewer, Chair, 701-231-7908, Fax: 701-231-8557, E-mail: gary.brewer@ndsu.nodak.edu.

The Ohio State University, Graduate School, College of Biological Sciences, Department of Entomology, Columbus, OH 43210. Offers MS, PhD. *Degree requirements:* For master's, variable foreign language requirement, thesis optional; for doctorate, variable foreign language requirement, thesis/dissertation. *Entrance requirements:* For master's and doctorate, GRE General Test. Additional exam requirements/recommendations for international students: Required—TOEFL (minimum score 600 paper-based; 250 computer-based), TSE. Electronic applications accepted. *Faculty research:* Acarology, insect systematics, soil ecology, integrated pest management, chemical ecology.

Oklahoma State University, College of Agricultural Science and Natural Resources, Department of Entomology and Plant Pathology, Program in Entomology, Stillwater, OK 74078. Offers PhD. *Degree requirements:* For doctorate, thesis/dissertation. *Entrance requirements:* For doctorate, GRE. Additional exam requirements/recommendations for international students: Required—TOEFL. *Application deadline:* For fall admission, 6/1 priority date for domestic students, 3/1 priority date for international students. Applications are processed on a rolling basis. Application fee: $40 ($75 for international students). Electronic applications accepted. *Expenses:* Tuition, state resident: full-time $4,253; part-time $139 per credit hour. Tuition, nonresident: full-time $12,569; part-time $485 per credit hour. Required fees: $43 per credit hour. One-time fee: $20 part-time. Tuition and fees vary according to course load and program. *Financial support:* Research assistantships, teaching assistantships, career-related internships or fieldwork, Federal Work-Study, scholarships/grants, health care benefits, tuition waivers (partial), and unspecified assistantships available. Support available to part-time students. Financial award application deadline: 3/1. *Unit head:* Dr. Jack Dillwith, Coordinator, 405-744-9412.

The Pennsylvania State University University Park Campus, Graduate School, College of Agricultural Sciences, Department of Entomology, State College, University Park, PA 16802-1503. Offers M Agr, MS, PhD. *Students:* 22 full-time (10 women), 2 part-time; includes 3 minority (all Hispanic Americans), 5 international. In 2005, 2 degrees awarded. *Entrance requirements:* For master's and doctorate, GRE General Test. *Expenses:* Tuition, state resident: full-time $12,518; part-time $522 per credit. Tuition, nonresident: full-time $23,004; part-time $959 per credit. Required fees: $484. Tuition and fees vary according to course load, campus/location and program. *Unit head:* Dr. Gary W. Felton, Head, 814-863-7789, Fax: 814-865-3048, E-mail: gwf10@psu.edu. *Application contact:* Dr. Gary W. Felton, Head, 814-863-7789, Fax: 814-865-3048, E-mail: gwf10@psu.edu.

Purdue University, Graduate School, College of Agriculture, Department of Entomology, West Lafayette, IN 47907. Offers MS, PhD. Part-time programs available. *Faculty:* 21 full-time (3 women), 6 part-time/adjunct (2 women). *Students:* 37 full-time (19 women); includes 4 minority (1 African American, 1 Asian American or Pacific Islander, 2 Hispanic Americans), 13 international. Average age 28. 26 applicants, 27% accepted, 6 enrolled. In 2005, 5 master's, 3 doctorates awarded. *Degree requirements:* For master's, thesis (for some programs), seminar; for doctorate, thesis/dissertation, seminar. *Entrance requirements:* For master's and doctorate, GRE. Additional exam requirements/recommendations for international students: Required—TOEFL. *Application deadline:* For fall admission, 7/1 priority date for domestic students, 4/1 priority date for international students; for spring admission, 11/1 for domestic students, 10/1 for international students. Applications are processed on a rolling basis. Application fee: $55. Electronic applications accepted. *Financial support:* In 2005–06, 1 fellowship with tuition reimbursement, 20 research assistantships with tuition reimbursements (averaging $14,000 per year), 3 teaching assistantships with tuition reimbursements (averaging $14,000 per year) were awarded; career-related internships or fieldwork also available. Support available to part-time students. Financial award applicants required to submit FAFSA. *Faculty research:* Insect biochemistry, nematology, aquatic diptera, behavioral ecology, insect physiology. *Unit head:* Dr. J. S. Yaninek, Head, 765-494-4553, Fax: 765-494-0535, E-mail: steve_yaninek@purdue.edu. *Application contact:* Amanda L Pendleton, Graduate Admissions Office, 765-494-9061, Fax: 765-494-0535, E-mail: apendle@purdue.edu.

Rutgers, The State University of New Jersey, New Brunswick/Piscataway, Graduate School, Program in Entomology, New Brunswick, NJ 08901-1281. Offers MS, PhD. *Faculty:* 14 full-time (1 woman). *Students:* 9 full-time (4 women), 6 part-time (1 woman); includes 2 minority (1 African American, 1 Asian American or Pacific Islander), 3 international. Average age 32. 11 applicants, 36% accepted, 2 enrolled. In 2005, 2 master's, 1 doctorate awarded. *Degree requirements:* For master's, thesis or alternative; for doctorate, thesis/dissertation. *Entrance requirements:* For master's and doctorate, GRE General Test, GRE Subject Test (recommended). Additional exam requirements/recommendations for international students: Required—TOEFL. *Application deadline:* Applications are processed on a rolling basis. Application fee: $50. Electronic applications accepted. *Expenses:* Tuition, state resident: full-time $10,440; part-time $435 per credit. Tuition, nonresident: full-time $15,520; part-time $647 per credit. Required fees: $129 per credit. Tuition and fees vary according to program. *Financial support:* In 2005–06, 9 students received support, including 2 fellowships with full tuition reimbursements available (averaging $17,000 per year), 5 research assistantships with full

tuition reimbursements available (averaging $14,500 per year), 1 teaching assistantship with full tuition reimbursement available (averaging $16,988 per year) Financial award application deadline: 1/15; financial award applicants required to submit FAFSA. *Faculty research:* Insect toxicology, biolorial control, pathology, IPM and ecology, insect systematics. Total annual research expenditures: $535,595. *Unit head:* Dr. James H. Lashomb, Director, 732-932-9774, Fax: 732-932-7229, E-mail: lashomb@rci.rutgers.edu. *Application contact:* Nancy L. Lyon, Administrative Assistant, 732-932-9774, Fax: 732-952-7229, E-mail: lyon@aesop.rutgers.edu.

Simon Fraser University, Graduate Studies, Faculty of Science, Department of Biological Sciences, Burnaby, BC V5A 1S6, Canada. Offers biological sciences (M Sc, PhD); environmental toxicology (Diploma), including food and drug toxicology, industrial toxicology; pest management (MPM). *Degree requirements:* For masters, doctorate, and Diploma, thesis/dissertation. *Entrance requirements:* For master's and Diploma, minimum GPA of 3.0; for doctorate, minimum GPA of 3.5. Additional exam requirements/recommendations for international students: Required—TOEFL or IELTS. Electronic applications accepted. *Faculty research:* Molecular biology, marine biology, ecology, wildlife biology, endocrinology.

South Dakota State University, Graduate School, College of Agriculture and Biological Sciences, Department of Plant Science, Program in Entomology, Brookings, SD 57007. Offers MS. *Degree requirements:* For master's, thesis, oral exam. *Entrance requirements:* For master's, GRE General Test. Additional exam requirements/recommendations for international students: Required—TOEFL. *Faculty research:* Integrated pest management, biological control, behavioral ecology, biodiversity, systematics.

State University of New York College of Environmental Science and Forestry, Faculty of Environmental and Forest Biology, Syracuse, NY 13210-2779. Offers chemical ecology (MPS, MS, PhD); conservation biology (MPS, MS, PhD); ecology (MPS, MS, PhD); entomology (MPS, MS, PhD); environmental interpretation (MPS, MS, PhD); environmental physiology (MPS, MS, PhD); fish and wildlife biology (MPS, MS, PhD); forest pathology and mycology (MPS, MS, PhD); plant science and biotechnology (MPS, MS, PhD). *Faculty:* 27 full-time (4 women), 4 part-time/adjunct (0 women). *Students:* 81 full-time (50 women), 62 part-time (33 women); includes 4 minority (1 Asian American or Pacific Islander, 3 Hispanic Americans), 16 international. Average age 30. 84 applicants, 54% accepted, 17 enrolled. In 2005, 17 master's, 3 doctorates awarded. *Degree requirements:* For master's, thesis (for some programs), registration; for doctorate, thesis/dissertation, comprehensive exam, registration. *Entrance requirements:* For master's and doctorate, GRE General Test, GRE Subject Test, minimum GPA of 3.0. Additional exam requirements/recommendations for international students: Required—TOEFL (minimum score 550 paper-based; 213 computer-based). *Application deadline:* For fall admission, 2/1 priority date for domestic students, 2/1 priority date for international students; for spring admission, 11/1 priority date for domestic students, 11/1 priority date for international students. Applications are processed on a rolling basis. Application fee: $60. *Expenses:* Tuition, area resident: Full-time $6,900; part-time $288 per credit. Tuition, nonresident: full-time $10,920; part-time $455 per credit. Required fees: $395; $32 per credit. $20 per term. One-time fee: $145. *Financial support:* In 2005–06, 86 students received support, including 13 fellowships with full and partial tuition reimbursements available (averaging $9,446 per year), 40 research assistantships with full and partial tuition reimbursements available (averaging $11,000 per year), 32 teaching assistantships with full and partial tuition reimbursements available (averaging $9,446 per year); Federal Work-Study, institutionally sponsored loans, scholarships/grants, health care benefits, and unspecified assistantships also available. Financial award application deadline: 6/30. *Faculty research:* Ecology, fish and wildlife biology and management, plant science, entomology. Total annual research expenditures: $4.1 million. *Unit head:* Dr. Donald J. Leopold, Chair, 315-470-6770, Fax: 315-470-6934, E-mail: dendro@esf.edu. *Application contact:* Dr. Dudley J. Raynal, Dean, Instruction and Graduate Studies, 315-470-6599, Fax: 315-470-6978, E-mail: esfgrad@esf.edu.

Texas A&M University, College of Agriculture and Life Sciences, Department of Entomology, College Station, TX 77843. Offers M Agr, MS, PhD. *Faculty:* 12 full-time (2 women), 7 part-time/adjunct (0 women). *Students:* 47 full-time (23 women), 10 part-time (6 women); includes 6 minority (1 African American, 1 American Indian/Alaska Native, 4 Hispanic Americans), 14 international. Average age 34. 20 applicants, 85% accepted, 11 enrolled. In 2005, 7 master's, 4 doctorates awarded. *Degree requirements:* For master's, thesis (for some programs), comprehensive exam; for doctorate, thesis/dissertation, comprehensive exam. *Entrance requirements:* For master's and doctorate, GRE General Test. Additional exam requirements/recommendations for international students: Required—TOEFL. *Application deadline:* For fall admission, 2/1 for domestic students; for spring admission, 10/1 for domestic students. Applications are processed on a rolling basis. Application fee: $50 ($75 for international students). Electronic applications accepted. *Expenses:* Tuition, state resident: full-time $4,488; part-time $187 per credit hour. Tuition, nonresident: full-time $11,112; part-time $463 per credit hour. Required fees: $1,974. *Financial support:* In 2005–06, research assistantships with partial tuition reimbursements (averaging $16,500 per year), teaching assistantships with partial tuition reimbursements (averaging $16,500 per year) were awarded; fellowships, Federal Work-Study also available. Financial award application deadline: 3/1; financial award applicants required to submit FAFSA. *Faculty research:* Biology, biological control, integrated pest management, systematics, host plant resistance. *Unit head:* Dr. Kevin Heinz, Head, 979-845-2516, Fax: 979-845-6305. *Application contact:* Jim Woolley, Advisor, 979-845-9349, Fax: 979-845-9938, E-mail: jimwoolley@tamu.edu.

Texas Tech University, Graduate School, College of Agricultural Sciences and Natural Resources, Department of Plant and Soil Science, Lubbock, TX 79409. Offers agronomy (PhD); crop science (MS); entomology (MS); horticulture (MS); soil science (MS). Part-time programs available. *Faculty:* 12 full-time (2 women), 4 part-time/adjunct (0 women). *Students:* 29 full-time (11 women), 23 part-time (9 women), 7 international. Average age 31. 36 applicants, 56% accepted, 14 enrolled. In 2005, 16 master's, 2 doctorates awarded. *Degree requirements:* For doctorate, thesis/dissertation. *Entrance requirements:* For master's and doctorate, GRE General Test. Additional exam requirements/recommendations for international students: Required—TOEFL (minimum score 550 paper-based; 213 computer-based). *Application deadline:* Applications are processed on a rolling basis. Application fee: $50 ($60 for international students). Electronic applications accepted. *Expenses:* Tuition, state resident: full-time $4,296. Tuition, nonresident: full-time $10,920. Required fees: $1,992. Tuition and fees vary according to program. *Financial support:* In 2005–06, 21 students received support, including 21 research assistantships with partial tuition reimbursements available (averaging $12,726 per year), 2 teaching assistantships with partial tuition reimbursements available (averaging $12,478 per year); Federal Work-Study and institutionally sponsored loans also available. Support available to part-time students. Financial award application deadline: 4/15; financial award applicants required to submit FAFSA. *Faculty research:* Molecular and cellular biology of plant stress, physiology/genetics of crop production in semiarid conditions, agricultural bioterrorism, improvement of native plants. Total annual research expenditures: $2.6 million. *Unit head:* Dr. Dick L. Auld, Chair, 806-742-2837, Fax: 806-742-0775, E-mail: dick.auld@ttu.edu. *Application contact:* Dr. Richard E. Zartman, Graduate Adviser, 806-742-2837, Fax: 806-742-0775, E-mail: richard.zartman@ttu.edu.

The University of Arizona, Graduate College, College of Agriculture and Life Sciences, Department of Entomology, Tucson, AZ 85721. Offers MS, PhD. Part-time programs available. *Degree requirements:* For master's and doctorate, thesis/dissertation. *Entrance requirements:* For master's and doctorate, GRE General Test, GRE Subject Test. Additional exam requirements/recommendations for international students: Required—TOEFL (minimum score 550 paper-based; 213 computer-based). *Faculty research:* Toxicology and physiology, plant/insect relations, vector biology, insect pest management, chemical ecology.

The University of Arizona, Graduate College, Graduate Interdisciplinary Programs, Graduate Interdisciplinary Program in Insect Science, Tucson, AZ 85721. Offers PhD. *Degree requirements:* For doctorate, thesis/dissertation. *Entrance requirements:* For doctorate, GRE

Entomology

The University of Arizona (continued)

General Test. Additional exam requirements/recommendations for international students: Required—TOEFL.

University of Arkansas, Graduate School, Dale Bumpers College of Agricultural, Food and Life Sciences, Department of Entomology, Fayetteville, AR 72701-1201. Offers MS, PhD. *Faculty:* 14 full-time (0 women). *Students:* 10 full-time (4 women), 3 part-time (1 woman), 1 international. 9 applicants, 33% accepted. In 2005, 6 master's, 2 doctorates awarded. *Degree requirements:* For master's, thesis; for doctorate, one foreign language, thesis/dissertation. *Entrance requirements:* For master's, GRE, minimum GPA of 3.0; for doctorate, GRE, minimum GPA of 3.25. Application fee: $40 ($50 for international students). *Financial support:* In 2005–06, 10 research assistantships, 1 teaching assistantship were awarded; fellowships with tuition reimbursements, career-related internships or fieldwork and Federal Work-Study also available. Support available to part-time students. Financial award application deadline: 4/1; financial award applicants required to submit FAFSA. *Faculty research:* Integrated pest management, insect virology, insect taxonomy. *Unit head:* Dr. Robert Wiedenmann, Chair, 479-575-6628. *Application contact:* Janet Funk, Graduate Coordinator, 479-575-6628, E-mail: jfunk@uark.edu.

University of California, Davis, Graduate Studies, Graduate Group in Integrated Pest Management, Davis, CA 95616. Offers MS. *Faculty:* 35 full-time. *Students:* 5 full-time (2 women); includes 1 minority (Hispanic American), 1 international. Average age 28. 4 applicants, 50% accepted, 2 enrolled. In 2005, 3 degrees awarded. *Degree requirements:* For master's, thesis (for some programs), comprehensive exam (for some programs). *Entrance requirements:* For master's, GRE General Test, GRE Subject Test (biology), minimum GPA of 3.0. Additional exam requirements/recommendations for international students: Required—TOEFL (minimum score 550 paper-based; 213 computer-based). *Application deadline:* For fall admission, 1/15 for domestic students, 1/15 for international students. Application fee: $60. Electronic applications accepted. *Financial support:* In 2005–06, 5 students received support, including 2 research assistantships with full and partial tuition reimbursements available (averaging $12,211 per year); fellowships with full and partial tuition reimbursements available, teaching assistantships with partial tuition reimbursements available, career-related internships or fieldwork, Federal Work-Study, institutionally sponsored loans, scholarships/grants, and tuition waivers (full and partial) also available. Financial award application deadline: 1/15; financial award applicants required to submit FAFSA. *Unit head:* Howard Ferris, Chair, 530-752-8432, E-mail: hferris@ucdavis.edu. *Application contact:* Christy Hansen, Administrative Assistant, 530-752-0475, E-mail: clhansen@ucdavis.edu.

University of California, Davis, Graduate Studies, Program in Entomology, Davis, CA 95616. Offers MS, PhD. *Faculty:* 28 full-time. *Students:* 34 full-time (24 women); includes 8 minority (1 African American, 5 Asian Americans or Pacific Islanders, 2 Hispanic Americans), 2 international. Average age 29. 35 applicants, 43% accepted, 7 enrolled. In 2005, 1 master's, 1 doctorate awarded. Terminal master's awarded for partial completion of doctoral program. *Median time to degree:* Of those who began their doctoral program in fall 1997, 80% received their degree in 8 years or less. *Degree requirements:* For master's, thesis (for some programs), comprehensive exam (for some programs); for doctorate, thesis/dissertation. *Entrance requirements:* For master's and doctorate, GRE General Test, GRE Subject Test (biology). Additional exam requirements/recommendations for international students: Required—TOEFL (minimum score 550 paper-based; 213 computer-based). *Application deadline:* For fall admission, 11/15 for domestic students, 11/15 for international students. Application fee: $60. Electronic applications accepted. *Financial support:* In 2005–06, 33 students received support, including 11 fellowships with full and partial tuition reimbursements available (averaging $12,828 per year), 13 research assistantships with full and partial tuition reimbursements available (averaging $12,398 per year), 7 teaching assistantships with partial tuition reimbursements available (averaging $12,522 per year); Federal Work-Study, institutionally sponsored loans, scholarships/grants, and tuition waivers (full and partial) also available. Financial award application deadline: 1/15; financial award applicants required to submit FAFSA. *Faculty research:* Bee biology, biological control, systematics, medical/veterinary entomology, pest management. *Unit head:* Robert Washino, Graduate Chair, 530-752-5652, E-mail: rkwashino@ucdavis.edu. *Application contact:* Christy Hansen, Administrative Assistant, 530-752-0475, E-mail: clhansen@ucdavis.edu.

University of California, Riverside, Graduate Division, Department of Entomology, Riverside, CA 92521-0102. Offers MS, PhD. Part-time programs available. *Faculty:* 23 full-time (1 woman). *Students:* 48 full-time (20 women); includes 4 minority (2 Asian Americans or Pacific Islanders, 2 Hispanic Americans), 9 international. Average age 29. In 2005, 4 master's, 4 doctorates awarded. Terminal master's awarded for partial completion of doctoral program. *Degree requirements:* For master's, thesis; for doctorate, thesis/dissertation, qualifying exams. *Entrance requirements:* For master's and doctorate, GRE General Test, minimum GPA of 3.2. Additional exam requirements/recommendations for international students: Required—TOEFL (minimum score 550 paper-based; 213 computer-based); Recommended—TSE (minimum score 50). *Application deadline:* For fall admission, 5/1 for domestic students, 2/1 for international students. For winter admission, 9/1 for domestic students; for spring admission, 12/1 for domestic students. Applications are processed on a rolling basis. Application fee: $60 ($75 for international students). Electronic applications accepted. *Expenses:* Tuition, nonresident: full-time $14,694. Required fees: $9,009. Full-time tuition and fees vary according to program. *Financial support:* Fellowships, research assistantships, teaching assistantships, career-related internships or fieldwork, Federal Work-Study, institutionally sponsored loans, and tuition waivers (full and partial) available. Financial award application deadline: 12/31; financial award applicants required to submit FAFSA. *Faculty research:* Agricultural, urban, medical, and veterinary entomology; biological control; chemical ecology; insect pathogens; novel toxicants. *Unit head:* Dr. Ring Carde, Chair, 951-827-5831, Fax: 951-827-3086. *Application contact:* Deidra Kornfeld, Graduate Program Assistant, 800-735-0717, Fax: 951-827-5517, E-mail: insects@ucr.edu.

University of Connecticut, Graduate School, College of Liberal Arts and Sciences, Department of Ecology and Evolutionary Biology, Field of Entomology, Storrs, CT 06269. Offers MS, PhD. *Faculty:* 9 full-time (2 women). *Students:* 3 full-time (2 women), 1 part-time; includes 1 minority (Hispanic American), 1 international. Average age 30. 2 applicants, 0% accepted. Terminal master's awarded for partial completion of doctoral program. *Degree requirements:* For master's, comprehensive exam; for doctorate, thesis/dissertation. *Entrance requirements:* For master's and doctorate, GRE General Test, GRE Subject Test. Additional exam requirements/recommendations for international students: Required—TOEFL (minimum score 550 paper-based; 213 computer-based). *Application deadline:* For fall admission, 2/1 priority date for domestic students, 2/1 priority date for international students; for spring admission, 11/1 for domestic students, 10/1 for international students. Applications are processed on a rolling basis. Application fee: $55. Electronic applications accepted. *Expenses:* Tuition, state resident: part-time $444 per credit hour. Tuition, nonresident: part-time $1,154 per credit hour. Tuition and fees vary according to course load. *Financial support:* In 2005–06, 3 teaching assistantships were awarded; fellowships, research assistantships, Federal Work-Study, scholarships/grants, health care benefits, and unspecified assistantships also available. Financial award application deadline: 2/1; financial award applicants required to submit FAFSA. *Application contact:* Anne St. Onje, Graduate Coordinator, 860-486-4314, Fax: 860-486-3943, E-mail: ann.st_onje@uconn.edu.

University of Delaware, College of Agriculture and Natural Resources, Department of Entomology and Wildlife Ecology, Newark, DE 19716. Offers entomology and applied ecology (MS, PhD), including avian ecology, evolution and taxonomy, insect biological control, insect ecology and behavior (MS), insect genetics, pest management, plant-insect interactions, wildlife ecology and management. Part-time programs available. *Faculty:* 9 full-time (1 woman), 4 part-time/adjunct (1 woman). *Students:* 26 full-time (13 women), 6 part-time (1 woman); includes 1 minority (African American) Average age 27. 27 applicants, 44% accepted, 11

enrolled. In 2005, 4 master's, 1 doctorate awarded. *Degree requirements:* For master's, thesis, oral exam, seminar, comprehensive exam; for doctorate, thesis/dissertation, qualifying exam, seminar, comprehensive exam. *Entrance requirements:* For master's, GRE General Test, minimum GPA of 3.0 in field, 2.8 overall; for doctorate, GRE General Test, GRE Subject Test (biology), minimum GPA of 3.0 in field, 2.8 overall. Additional exam requirements/recommendations for international students: Required—TOEFL. *Application deadline:* For fall admission, 2/1 for domestic students; for spring admission, 11/1 for domestic students. Applications are processed on a rolling basis. Application fee: $60. Electronic applications accepted. *Financial support:* In 2005–06, 18 students received support, including 2 fellowships with full tuition reimbursements available (averaging $13,500 per year), 7 research assistantships with full tuition reimbursements available (averaging $16,260 per year), 5 teaching assistantships with full tuition reimbursements available (averaging $12,195 per year); career-related internships or fieldwork, institutionally sponsored loans, scholarships/grants, and tuition waivers (full) also available. Financial award application deadline: 3/1. *Faculty research:* Genetics and resistance, biological control, chemically mediated behavioral ecology, ecology and evolution of plant-insect interactions, ecology of wildlife conservation and management. Total annual research expenditures: $811,092. *Unit head:* Dr. Douglas W. Tallamy, Chair, 302-831-1304, Fax: 302-831-8889, E-mail: dtallamy@udel.edu. *Application contact:* Dr. Charles E. Mason, Graduate Coordinator, 302-831-8888, Fax: 302-831-8889, E-mail: mason@udel.edu.

University of Florida, Graduate School, College of Agricultural and Life Sciences, Department of Entomology and Nematology, Gainesville, FL 32611. Offers MS, PhD. *Faculty:* 31 full-time (8 women). *Students:* 94 (38 women); includes 8 minority (2 African Americans, 1 American Indian/Alaska Native, 5 Hispanic Americans) 21 international. 35 applicants, 57% accepted. In 2005, 13 master's, 4 doctorates awarded. Terminal master's awarded for partial completion of doctoral program. *Degree requirements:* For master's, thesis optional; for doctorate, thesis/dissertation. *Entrance requirements:* For master's and doctorate, GRE General Test, GRE Subject Test (biology), minimum GPA of 3.0. *Application deadline:* For fall admission, 6/1 for domestic students. Applications are processed on a rolling basis. Application fee: $20. Electronic applications accepted. *Expenses:* Tuition, state resident: full-time $6,234. Tuition, nonresident: full-time $21,359. Tuition and fees vary according to program. *Financial support:* In 2005–06, 3 fellowships (averaging $12,373 per year), 47 research assistantships (averaging $15,065 per year), 3 teaching assistantships (averaging $15,441 per year) were awarded; career-related internships or fieldwork also available. *Faculty research:* Medical, veterinary, and urban entomology; genetics; biology and management; biocontrol; insect ecology. *Unit head:* Dr. John L. Capinera, Chair, 352-392-1901 Ext. 111, Fax: 352-392-0190, E-mail: jlcap@ifas.ufl.edu. *Application contact:* Dr. Don Hall, Coordinator, 352-392-1901 Ext. 118, Fax: 352-392-0190, E-mail: gradc@ifas.ufl.edu.

University of Georgia, Graduate School, College of Agricultural and Environmental Sciences, Department of Entomology, Athens, GA 30602. Offers entomology (MS, PhD); plant protection and pest management (MPPPM). *Faculty:* 23 full-time (2 women). *Students:* 38 full-time, 1 part-time; includes 1 minority (African American), 16 international. 16 applicants, 44% accepted, 4 enrolled. In 2005, 7 master's, 6 doctorates awarded. *Degree requirements:* For master's, thesis (MS); for doctorate, one foreign language, thesis/dissertation. *Entrance requirements:* For master's and doctorate, GRE General Test. *Application deadline:* For fall admission, 7/1 for domestic students; for spring admission, 11/15 for domestic students. Application fee: $50. Electronic applications accepted. *Financial support:* Unspecified assistantships available. *Faculty research:* Apiculture, acarology, aquatic and soil biology, ecology, systematics. *Unit head:* Dr. Raymond Noblet, Head, 706-542-2816, Fax: 706-542-2279, E-mail: rnoblet@bugs.ent.uga.edu.

University of Guelph, Graduate Program Services, Ontario Agricultural College, Department of Environmental Biology, Guelph, ON N1G 2W1, Canada. Offers entomology (M Sc, PhD); environmental biology and biotechnology (M Sc, PhD); environmental toxicology (M Sc, PhD); plant and forest systems (M Sc, PhD); plant pathology (M Sc, PhD). Part-time programs available. *Faculty:* 20 full-time (1 woman), 21 part-time/adjunct (2 women). *Students:* 60 full-time (32 women), 13 part-time (8 women). 37 applicants, 35% accepted, 13 enrolled. In 2005, 12 master's, 7 doctorates awarded. *Median time to degree:* Of those who began their doctoral program in fall 1997, 79.5% received their degree in 8 years or less. *Degree requirements:* For master's, thesis/dissertation, registration; for doctorate, thesis/dissertation, comprehensive exam, registration. *Entrance requirements:* For master's, minimum B- average during previous 2 years of course work; for doctorate, minimum B average. Additional exam requirements/recommendations for international students: Required—TOEFL or IELT. *Application deadline:* Applications are processed on a rolling basis. Application fee: $75. Electronic applications accepted. *Financial support:* In 2005–06, research assistantships (averaging $16,500 per year), teaching assistantships (averaging $4,606 per year) were awarded; fellowships *Faculty research:* Entomology, environmental microbiology and biotechnology, environmental toxicology, forest ecology, plant pathology. Total annual research expenditures: $3 million. *Unit head:* Dr. M. A. Dixon, Chair, 519-824-4120 Ext. 52555, Fax: 519-837-0442, E-mail: mdixon@uoguelph.ca. *Application contact:* Dr. H. Lee, Admissions Coordinator, 519-824-4120 Ext. 53828, Fax: 519-837-0442, E-mail: hlee@uoguelph.ca.

University of Hawaii at Manoa, Graduate Division, College of Tropical Agriculture and Human Resources, Department of Plant and Environmental Protection Sciences, Program of Entomology, Honolulu, HI 96822. Offers MS, PhD. Part-time programs available. *Faculty:* 13 full-time (1 woman), 10 part-time/adjunct (0 women). *Students:* 14 full-time (8 women), 4 part-time (2 women); includes 5 minority (all Asian Americans or Pacific Islanders) 6 international. Average age 35. 13 applicants, 31% accepted, 4 enrolled. In 2005, 2 degrees awarded. *Median time to degree:* Of those who began their doctoral program in fall 1997, 100% received their degree in 8 years or less. *Degree requirements:* For master's and doctorate, thesis/dissertation. *Entrance requirements:* For master's and doctorate, GRE General Test, GRE Subject Test (biology). *Application deadline:* For fall admission, 3/1 for domestic students, 3/1 for international students; for spring admission, 10/1 for domestic students, 10/1 for international students. Application fee: $50. *Expenses:* Tuition, state resident: full-time $8,400; part-time $200 per credit hour. Tuition, nonresident: full-time $11,088; part-time $462 per credit hour. Tuition and fees vary according to program. *Financial support:* Tuition waivers (full) available. *Faculty research:* Integrated pest management, biological control, urban entomology, medical/forensic entomology resistance. *Unit head:* Dr. Stephen Saul, Graduate Chair, 808-956-7076, Fax: 808-956-2428, E-mail: saul@hawaii.edu. *Application contact:* Ronald Mau, 808-956-7076, Fax: 808-956-5888.

University of Idaho, College of Graduate Studies, College of Agricultural and Life Sciences, Department of Plant, Soil, and Entomological Sciences, Program in Entomology, Moscow, ID 83844-2282. Offers MS, PhD. *Students:* 11 full-time (4 women), 7 part-time (3 women); includes 2 minority (both Hispanic Americans), 5 international. In 2005, 1 degree awarded. *Degree requirements:* For master's, thesis (for some programs); for doctorate, one foreign language, thesis/dissertation. *Entrance requirements:* For master's and doctorate, GRE General Test, minimum GPA of 3.0. *Application deadline:* For fall admission, 8/1 for domestic students; for spring admission, 12/15 for domestic students. Application fee: $55 ($60 for international students). *Expenses:* Tuition, state resident: full-time $4,508. Tuition, nonresident: full-time $8,770; part-time $130 per credit. Required fees: $217 per credit. *Financial support:* Application deadline: 2/15. *Unit head:* Dr. Sanford Eigenbrode, Chair, 208-885-2972.

University of Illinois at Urbana–Champaign, Graduate College, College of Liberal Arts and Sciences, School of Integrative Biology, Department of Entomology, Champaign, IL 61820. Offers entomology (PhD); insect pest management (MS). *Faculty:* 19 full-time (3 women). *Students:* 31 full-time (17 women), 2 part-time; includes 2 minority (1 African American, 1 Asian American or Pacific Islander) 8 international. 26 applicants, 31% accepted, 7 enrolled. In 2005, 1 master's, 4 doctorates awarded. Terminal master's awarded for partial completion of doctoral program. *Degree requirements:* For master's, thesis; for doctorate, one foreign language, thesis/dissertation. *Entrance requirements:* For master's and doctorate, GRE General Test, GRE Subject Test, minimum GPA of 3.0. Additional exam requirements/recommendations for

Entomology

international students: Required—TOEFL. *Application deadline:* For fall and spring admission, 4/15. Applications are processed on a rolling basis. Application fee: $50 ($60 for international students). Electronic applications accepted. *Financial support:* In 2005–06, 22 research assistantships, 14 teaching assistantships were awarded; fellowships, career-related internships or fieldwork, Federal Work-Study, and institutionally sponsored loans also available. Financial award application deadline: 2/15. *Unit head:* Dr. May R. Berenbaum, Head, 217-333-2910, Fax: 217-244-3499, E-mail: maybe@uiuc.edu. *Application contact:* Jackie Bowdry, Administrative Secretary, 217-244-2888, Fax: 217-244-3499, E-mail: jsbowdry@uiuc.edu.

See Close-Up on page 729.

University of Kansas, Graduate School, College of Liberal Arts and Sciences, Division of Biological Sciences, Department of Ecology and Evolutionary Biology, Lawrence, KS 66045. Offers botany (MA, PhD); ecology and evolutionary biology (MA, PhD); entomology (MA, PhD); systematics and ecology (MA). Part-time programs available. *Faculty:* 42. *Students:* 67 full-time (35 women), 21 part-time (9 women); includes 4 minority (1 American Indian/Alaska Native, 3 Hispanic Americans), 20 international. Average age 30. 56 applicants, 32% accepted. In 2005, 10 master's, 4 doctorates awarded. Terminal master's awarded for partial completion of doctoral program. *Degree requirements:* For master's and doctorate, thesis/dissertation, comprehensive exam. *Entrance requirements:* For master's and doctorate, GRE General Test, GRE Subject Test (recommended). Additional exam requirements/recommendations for international students: Required—TOEFL (paper 570; computer 230) or IELT. *Application deadline:* For fall admission, 1/10 priority date for domestic students, 1/10 priority date for international students. Applications are processed on a rolling basis. Application fee: $55 ($60 for international students). Electronic applications accepted. *Expenses:* Tuition, state resident: full-time $4,859. Tuition, nonresident: full-time $12,000. Required fees: $589. Tuition and fees vary according to program. *Financial support:* Fellowships with tuition reimbursements, research assistantships with partial tuition reimbursements, teaching assistantships with full and partial tuition reimbursements available. Financial award application deadline: 3/1. *Faculty research:* Ecology, evolutionary biology, genetics. *Unit head:* Craig E. Martin, Chair, 785-864-5887, Fax: 785-864-5860, E-mail: ecophys@ku.edu. *Application contact:* Jeannie Houts, Graduate Coordinator, 785-864-2362, Fax: 785-864-5860, E-mail: jmhouts@ku.edu.

University of Kentucky, Graduate School, Graduate School Programs in the College of Agriculture, Program in Entomology, Lexington, KY 40506-0032. Offers MS, PhD. *Faculty:* 17 full-time (2 women). *Students:* 31 full-time (15 women), 3 part-time (2 women); includes 4 minority (2 African Americans, 2 Hispanic Americans), 11 international. Average age 29. 39 applicants, 56% accepted, 18 enrolled. In 2005, 3 master's, 2 doctorates awarded. *Median time to degree:* Of those who began their doctoral program in fall 1997, 72% received their degree in 8 years or less. *Degree requirements:* For master's, thesis optional; for doctorate, thesis/dissertation, comprehensive exam. *Entrance requirements:* For master's, GRE General Test, minimum undergraduate GPA of 2.5; for doctorate, GRE General Test, minimum graduate GPA of 3.0. Additional exam requirements/recommendations for international students: Required—TOEFL (minimum score 550 paper-based; 213 computer-based). *Application deadline:* For fall admission, 7/17 priority date for domestic students, 2/1 priority date for international students; for spring admission, 12/13 priority date for domestic students, 6/15 priority date for international students. Applications are processed on a rolling basis. Application fee: $40 ($55 for international students). Electronic applications accepted. *Expenses:* Tuition, state resident: full-time $6,308; part-time $331 per credit hour. Tuition, nonresident: full-time $13,968; part-time $756 per credit hour. Tuition and fees vary according to course load, degree level and program. *Financial support:* In 2005–06, 3 fellowships with full tuition reimbursements (averaging $2,208 per year), 25 research assistantships with full tuition reimbursements (averaging $16,000 per year) were awarded; teaching assistantships, career-related internships or fieldwork, Federal Work-Study, institutionally sponsored loans, scholarships/grants, traineeships, health care benefits, and unspecified assistantships also available. Support available to part-time students. Financial award application deadline: 3/15. *Faculty research:* Applied entomology, behavior, insect biology and ecology, biological control, insect physiology and molecular biology. Total annual research expenditures: $1.5 million. *Unit head:* Dr. Kenneth Yeargan, Chair, 859-257-7450, Fax: 859-323-1120, E-mail: kyeargan@uky.edu. *Application contact:* Dr. Brian Jackson, Senior Associate Dean, 859-257-8176, Fax: 859-323-1928.

University of Maine, Graduate School, College of Natural Sciences, Forestry, and Agriculture, Department of Biological Sciences, Program in Entomology, Orono, ME 04469. Offers MS. Part-time programs available. *Students:* 1 (woman) full-time. In 2005, 2 master's awarded. *Entrance requirements:* For master's, GRE General Test. Additional exam requirements/recommendations for international students: Required—TOEFL. *Application deadline:* For fall admission, 2/1 for domestic students. Applications are processed on a rolling basis. Application fee: $50. Electronic applications accepted. *Financial support:* Fellowships, research assistantships with tuition reimbursements, teaching assistantships with tuition reimbursements, career-related internships or fieldwork, Federal Work-Study, institutionally sponsored loans, and tuition waivers (full) available. Financial award application deadline: 3/1. *Unit head:* Dr. Stelbs Tavanteis, Coordinator, 207-581-2986. *Application contact:* Scott G. Delcourt, Associate Dean of the Graduate School, 207-581-3219, Fax: 207-581-3232, E-mail: graduate@maine.edu.

University of Manitoba, Faculty of Graduate Studies, Faculty of Agriculture, Department of Entomology, Winnipeg, MB R3T 2N2, Canada. Offers M Sc, PhD. *Degree requirements:* For master's, thesis; for doctorate, one foreign language, thesis/dissertation.

University of Maryland, College Park, Graduate Studies, College of Chemical and Life Sciences, Department of Entomology, College Park, MD 20742. Offers MS, PhD. Part-time and evening/weekend programs available. *Faculty:* 34 full-time (12 women), 3 part-time/adjunct (2 women). *Students:* 23 full-time (11 women), 3 part-time (2 women); includes 4 minority (1 African American, 1 Asian American or Pacific Islander, 2 Hispanic Americans), 4 international. 17 applicants, 6% accepted, 1 enrolled. In 2005, 4 master's, 7 doctorates awarded. Terminal master's awarded for partial completion of doctoral program. *Degree requirements:* For master's, thesis; for doctorate, thesis/dissertation, oral qualifying exam. *Entrance requirements:* For master's and doctorate, GRE General Test, minimum GPA of 3.0, 3 letters of recommendation. *Application deadline:* For fall admission, 1/7 for domestic students, 1/7 for international students; for spring admission, 11/1 for domestic students, 6/1 for international students. Applications are processed on a rolling basis. Application fee: $60. Electronic applications accepted. *Financial support:* In 2005–06, 5 fellowships with full tuition reimbursements (averaging $9,941 per year), 3 research assistantships (averaging $20,020 per year), 33 teaching assistantships with tuition reimbursements (averaging $18,499 per year) were awarded; career-related internships or fieldwork and Federal Work-Study also available. Support available to part-time students. Financial award applicants required to submit FAFSA. *Faculty research:* Pest management, biosystematics, physiology and morphology, toxicology. Total annual research expenditures: $1.5 million. *Unit head:* Dr. Charles Mitter, Chair, 301-405-3912, Fax: 301-314-9290, E-mail: cmitter@umd.edu. *Application contact:* Dean of Graduate School, 301-405-4190, Fax: 301-314-9305.

University of Massachusetts Amherst, Graduate School, College of Natural Resources and the Environment, Department of Entomology, Amherst, MA 01003. Offers MS, PhD. Part-time programs available. *Faculty:* 16 full-time (6 women). *Students:* 3 full-time (1 woman), 1 international. Average age 30. 5 applicants, 0% accepted. In 2005, 4 master's, 4 doctorates awarded. Terminal master's awarded for partial completion of doctoral program. *Degree requirements:* For master's, thesis or alternative; for doctorate, thesis/dissertation. *Entrance requirements:* For master's and doctorate, GRE General Test, GRE Subject Test (biology). Additional exam requirements/recommendations for international students: Required—TOEFL (minimum score 530 paper-based; 197 computer-based). *Application deadline:* For fall admission, 2/1 priority date for domestic students, 2/1 priority date for international students; for spring admission, 10/1 for domestic students, 10/1 for international students. Applications are processed on a rolling basis. Application fee: $40 ($65 for international students). Electronic

applications accepted. *Expenses:* Tuition, state resident: part-time $110 per credit. Tuition, nonresident: part-time $414 per credit. Required fees: $2,824 per term. One-time fee: $250 part-time. Full-time tuition and fees vary according to course load, campus/location, program and reciprocity agreements. *Financial support:* Fellowships with full tuition reimbursements, research assistantships with full tuition reimbursements, teaching assistantships with full tuition reimbursements, career-related internships or fieldwork, Federal Work-Study, scholarships/grants, traineeships, and unspecified assistantships available. Support available to part-time students. Financial award application deadline: 2/1. *Unit head:* Dr. Petrus Veneman, Chair, 413-545-1059, Fax: 413-545-2115, E-mail: veneman@pssci.umass.edu.

University of Minnesota, Twin Cities Campus, Graduate School, College of Agricultural, Food, and Environmental Sciences, Department of Entomology, Minneapolis, MN 55455-0213. Offers MS, PhD. Part-time programs available. *Faculty:* 19 full-time (6 women), 5 part-time/adjunct (1 woman). *Students:* 22 full-time (14 women), 12 part-time (6 women); includes 4 minority (2 African Americans, 1 Asian American or Pacific Islander, 1 Hispanic American), 8 international. Average age 32. 14 applicants, 43% accepted, 5 enrolled. In 2005, 6 master's, 7 doctorates awarded. *Median time to degree:* Of those who began their doctoral program in fall 1997, 100% received their degree in 8 years or less. *Degree requirements:* For master's and doctorate, thesis/dissertation, comprehensive exam. *Entrance requirements:* For master's, GRE, minimum undergraduate GPA of 3.0; for doctorate, GRE, minimum undergraduate GPA of 3.0, minimum graduate GPA of 3.5. Additional exam requirements/recommendations for international students: Required—TOEFL. *Application deadline:* For fall admission, 6/15 for domestic students, 6/15 for international students; for spring admission, 10/15 for domestic students, 10/15 for international students. Applications are processed on a rolling basis. Application fee: $55 ($75 for international students). Electronic applications accepted. *Expenses:* Tuition, state resident: full-time $8,748; part-time $729 per credit. Tuition, nonresident: full-time $15,848; part-time $1,321 per credit. Full-time tuition and fees vary according to class time, course load, program and reciprocity agreements. *Financial support:* In 2005–06, 4 fellowships with full tuition reimbursements, 16 research assistantships with full tuition reimbursements, 1 teaching assistantship with full tuition reimbursement were awarded. *Faculty research:* Behavior, ecology, molecular genetics, physiology, systematics and taxonomy. *Unit head:* Mark E. Ascerno, Head, 612-624-3278, Fax: 612-625-5299, E-mail: mascerno@tc.umn.edu. *Application contact:* Marla S. Spivak, Director of Graduate Studies, 612-624-4798, Fax: 612-625-5299, E-mail: spiva001@umn.edu.

University of Missouri–Columbia, Graduate School, College of Agriculture, Food and Natural Resources, Division of Plant Sciences, Department of Entomology, Columbia, MO 65211. Offers MS, PhD. *Degree requirements:* For doctorate, thesis/dissertation. *Application deadline:* Applications are processed on a rolling basis. *Financial support:* Research assistantships, teaching assistantships, institutionally sponsored loans available.

University of Nebraska–Lincoln, Graduate College, College of Agricultural Sciences and Natural Resources, Department of Entomology, Lincoln, NE 68588. Offers MS, PhD. Post-baccalaureate distance learning degree programs offered (no on-campus study). *Degree requirements:* For master's, thesis optional; for doctorate, thesis/dissertation, comprehensive exam. *Entrance requirements:* For master's and doctorate, GRE General Test. Additional exam requirements/recommendations for international students: Required—TOEFL (minimum score 550 paper-based; 213 computer-based). Electronic applications accepted. *Faculty research:* Ecology and behavior, insect-plant interactions, integrated pest management, genetics, urban entomology.

University of North Dakota, Graduate School, College of Arts and Sciences, Department of Biology, Grand Forks, ND 58202. Offers botany (MS, PhD); ecology (MS, PhD); entomology (MS, PhD); environmental biology (MS, PhD); fisheries/wildlife (MS, PhD); genetics (MS, PhD); zoology (MS, PhD). *Faculty:* 16 full-time (3 women). *Students:* 15 applicants, 13% accepted, 2 enrolled. In 2005, 5 degrees awarded. Terminal master's awarded for partial completion of doctoral program. *Degree requirements:* For master's, thesis/dissertation, final exam; for doctorate, thesis/dissertation, final exam, comprehensive exam. *Entrance requirements:* For master's, GRE General Test, GRE Subject Test, minimum GPA of 3.0; for doctorate, GRE General Test, GRE Subject Test, minimum GPA of 3.5. Additional exam requirements/recommendations for international students: Required—TOEFL (minimum score 550 paper-based; 213 computer-based). *Application deadline:* For fall admission, 10/1 for domestic students, 10/1 for international students. Application fee: $35. Electronic applications accepted. *Financial support:* In 2005–06, 8 research assistantships with full tuition reimbursements (averaging $11,375 per year), 13 teaching assistantships with full tuition reimbursements (averaging $10,813 per year) were awarded; fellowships, Federal Work-Study, institutionally sponsored loans, scholarships/grants, and tuition waivers (full and partial) also available. Support available to part-time students. Financial award application deadline: 3/15; financial award applicants required to submit FAFSA. *Faculty research:* Population biology, wildlife ecology, RNA processing, hormonal control of behavior. *Unit head:* Dr. Richard Sweitzel, Graduate Director, 701-777-4676, Fax: 701-777-2623, E-mail: richard_sweitzel@und.nodak.edu.

University of Rhode Island, Graduate School, College of the Environment and Life Sciences, Department of Plant Sciences, Program in Entomology, Kingston, RI 02881. Offers MS, PhD. *Degree requirements:* For master's, thesis/dissertation, professional seminar, comprehensive exam; for doctorate, thesis/dissertation, professional seminar. *Entrance requirements:* For master's and doctorate, GRE General Test. *Application deadline:* For fall admission, 4/15 for domestic students. Applications are processed on a rolling basis. Application fee: $35. *Expenses:* Tuition, state resident: full-time $5,522; part-time $307 per credit. Tuition, nonresident: full-time $15,992; part-time $888 per credit. Required fees: $1,786; $73 per credit. One-time fee: $80 part-time. *Faculty research:* Physiology and fine structure of host-parasite relations, etiology of plant disease, biological control of insects, integrated pest management. *Unit head:* Dr. Richard Casagrande, Chairman, Department of Plant Sciences, 401-874-2924.

The University of Tennessee, Graduate School, College of Agricultural Sciences and Natural Resources, Department of Entomology and Plant Pathology, Knoxville, TN 37996. Offers entomology (MS, PhD); integrated pest management and bioactive natural products (PhD); plant pathology (MS, PhD). Part-time programs available. *Degree requirements:* For master's, thesis, seminar. *Entrance requirements:* For master's, GRE General Test, minimum GPA of 2.7, 3 reference letters, letter of intent; for doctorate, GRE General Test, minimum GPA of 2.7, 3 reference letters, letter of intent, proposed dissertation research. Additional exam requirements/recommendations for international students: Required—TOEFL. Electronic applications accepted.

Announcement: The PhD degree program offers concentrations in entomology, plant pathology, integrated pest management, and bioactive natural products as part of the multidepartment Plants, Soils, and Insects (PSI) program. The MS degree program is designed to provide students with the principles and background for a career in the life sciences. Graduates have proven to be competitive in obtaining positions at academic institutions, in public service or the private sector, and in continuing on to PhD programs. Visit http://eppserver.ag.utk.edu/default.html.

University of Wisconsin–Madison, Graduate School, College of Agricultural and Life Sciences, Department of Entomology, Madison, WI 53706-1380. Offers MS, PhD. *Faculty:* 14 full-time (3 women). *Students:* 30 full-time (13 women), 7 part-time (4 women); includes 1 minority (Hispanic American), 7 international. Average age 30. 30 applicants, 17% accepted, 5 enrolled. In 2005, 6 master's, 3 doctorates awarded. *Degree requirements:* For master's and doctorate, thesis/dissertation. *Entrance requirements:* For master's and doctorate, GRE General Test, minimum GPA of 3.0. Additional exam requirements/recommendations for international students: Required—TOEFL (minimum score 237 computer-based). *Application deadline:* Applications are processed on a rolling basis. Application fee: $45. Electronic applications accepted. *Financial support:* In 2005–06, 2 fellowships with full tuition reimbursements (averaging $19,200 per year), 22 research assistantships with full tuition reimbursements

Entomology

University of Wisconsin–Madison (continued)

(averaging $18,120 per year), 4 teaching assistantships were awarded; institutionally sponsored loans and scholarships/grants also available. Support available to part-time students. *Faculty research:* Ecology, biocontrol, molecular. Total annual research expenditures: $2.2 million. *Unit head:* Walter G. Goodman, Chair, 608-262-1696, Fax: 608-262-3322. *Application contact:* James W. Butts, Graduate Coordinator, 608-262-0625, Fax: 608-262-3322, E-mail: butts@enromology.wisc.edu.

University of Wyoming, Graduate School, College of Agriculture, Department of Renewable Resources, Program in Entomology, Laramie, WY 82070. Offers MS, PhD. *Faculty:* 5 full-time (0 women). *Students:* 2 full-time (both women), 4 part-time (3 women), 2 international. In 2005, 1 master's awarded. *Degree requirements:* For master's and doctorate, thesis/dissertation. *Entrance requirements:* For master's and doctorate, GRE General Test, minimum GPA of 3.0. Additional exam requirements/recommendations for international students: Required—TOEFL. *Application deadline:* For fall admission, 6/1 for domestic students; for spring admission, 12/1 priority date for domestic students. Applications are processed on a rolling basis. Application fee: $50. Electronic applications accepted. *Expenses:* Tuition, state resident: full-time $3,720; part-time $155 per credit hour. Tuition, nonresident: full-time $10,704; part-time $446 per credit hour. Required fees: $666; $162 per semester. Tuition and fees vary according to course load and program. *Financial support:* Research assistantships with full tuition reimbursements available. Financial award application deadline: 3/1. *Faculty research:* Insect pest management, taxonomy, biocontrol of weeds, forest insects, insects affecting humans and animals. *Application contact:* Kimm Mann-Malody, Office Associate, 307-766-2263, Fax: 307-766-6403, E-mail: kimmmann@uwyo.edu.

Virginia Polytechnic Institute and State University, Graduate School, College of Agriculture and Life Sciences, Department of Entomology, Blacksburg, VA 24061. Offers MS, PhD. *Faculty:* 15 full-time (2 women). *Students:* 31 full-time (12 women), 5 part-time (4 women); includes 2 minority (1 American Indian/Alaska Native, 1 Hispanic American), 8 international. Average age 29. 14 applicants, 71% accepted, 10 enrolled. In 2005, 4 master's, 3 doctorates awarded. *Entrance requirements:* For master's and doctorate, GRE. Additional exam requirements/recommendations for international students: Required—TOEFL (minimum score 550 paper-based; 213 computer-based). *Application deadline:* Applications are processed on a rolling basis. Application fee: $45. Electronic applications accepted. *Expenses:* Tuition, state resident: full-time $6,558; part-time $364 per credit. Tuition, nonresident: full-time $11,296; part-time $628 per credit. Required fees: $1,419; $468 per credit. $234 per term. *Financial support:* In 2005–06, 13 research assistantships with full tuition reimbursements (averaging $13,823 per year), 7 teaching assistantships with full tuition reimbursements (averaging $13,552 per year) were awarded; career-related internships or fieldwork, Federal Work-Study, scholarships/grants, and unspecified assistantships also available. Financial award application deadline: 4/1. *Faculty research:* Physiology, ecology, biocontrol, genetics, taxonomy. *Unit head:* Dr. Loke Kok, Head, 540-231-6341, Fax: 540-231-9131, E-mail: ltkok@vt.edu. *Application contact:* Kathy Shelor, Information Contact, 540-231-4010, Fax: 540-231-9131, E-mail: kshelor@vt.edu.

Washington State University, Graduate School, College of Agricultural, Human, and Natural Resource Sciences, Department of Entomology, Pullman, WA 99164. Offers MS, PhD. *Faculty:* 17. *Students:* 18 full-time (6 women), 3 part-time (1 woman); includes 1 minority (Hispanic American), 3 international. Average age 31. 22 applicants, 41% accepted, 6 enrolled. In 2005, 2 master's, 3 doctorates awarded. Terminal master's awarded for partial completion of doctoral program. *Median time to degree:* Of those who began their doctoral program in fall 1997, 100% received their degree in 8 years or less. *Degree requirements:* For master's, oral exam, thesis optional; for doctorate, thesis/dissertation, oral exam, written exam. *Entrance requirements:* For master's, GRE General Test, GRE Subject Test in advanced biology (recommended), minimum GPA of 3.0, 3 letters of recommendation; for doctorate, GRE General Test, minimum GPA of 3.0, 3 letters of recommendation. Additional exam requirements/recommendations for international students: Required—TOEFL (minimum score 550 paper-based; 213 computer-based). *Application deadline:* For fall admission, 3/1 priority date for domestic students, 3/1 priority date for international students; for spring admission, 8/1 priority date for domestic students, 7/1 priority date for international students. Applications are processed on a rolling basis. Application fee: $35. Electronic applications accepted. *Expenses:* Tuition, state resident: full-time $6,295; part-time $336 per credit. Tuition, nonresident: full-time $15,949; part-time $819 per credit. Required fees: $933. Part-time tuition and fees vary according to campus/location and program. *Financial support:* In 2005–06, 2 fellowships (averaging $5,306 per year), 17 research assistantships with full and partial tuition reimbursements (averaging $14,472 per year), 1 teaching assistantship with full and partial tuition reimbursement (averaging $11,637 per year) were awarded; career-related internships or fieldwork, Federal Work-Study, institutionally sponsored loans, tuition waivers (partial), unspecified assistantships, and teaching associateships also available. Financial award application deadline: 4/1; financial award applicants required to submit FAFSA. *Faculty research:* Apiculture, biological control of arthropods, integrated pest management, ecology, physiology and systematics of insects. Total annual research expenditures: $2.9 million. *Unit head:* Dr. Richard S. Zack, Chair, 509-335-5505, Fax: 509-335-1009. *Application contact:* Doris Lohrey-Birch, Academic Secretary, 509-335-5422, Fax: 509-335-1009, E-mail: entom@wsu.edu.

West Virginia University, Davis College of Agriculture, Forestry and Consumer Sciences, Division of Plant and Soil Sciences, Morgantown, WV 26506. Offers agronomy (MS); entomology (MS); environmental microbiology (MS); horticulture (MS); plant pathology (MS). *Faculty:* 18 full-time (1 woman), 1 part-time/adjunct (0 women). *Students:* 17 full-time (11 women), 3 part-time (2 women), 2 international. Average age 27. In 2005, 5 degrees awarded. *Degree requirements:* For master's, thesis. *Entrance requirements:* For master's, GRE, minimum GPA of 2.5. Additional exam requirements/recommendations for international students: Required—TOEFL. *Application deadline:* Applications are processed on a rolling basis. Application fee: $45. *Expenses:* Tuition, state resident: full-time $4,582; part-time $258 per credit hour. Tuition, nonresident: full-time $13,820; part-time $741 per credit hour. *Financial support:* In 2005–06, 13 research assistantships with full tuition reimbursements (averaging $9,936 per year), 4 teaching assistantships with full tuition reimbursements (averaging $9,936 per year) were awarded; Federal Work-Study, institutionally sponsored loans, and tuition waivers (full and partial) also available. Financial award application deadline: 2/1; financial award applicants required to submit FAFSA. *Faculty research:* Water quality, reclamation of disturbed land, crop production, pest control, environmental protection. Total annual research expenditures: $1 million. *Unit head:* Dr. Barton S. Baker, Chair and Division Director, 304-293-4817 Ext. 4342, Fax: 304-293-2960, E-mail: barton.baker@mail.wvu.edu.

CLEMSON UNIVERSITY

Master of Science in Entomology

Program of Study	Providing leadership in environmental entomology, Clemson University offers a Master of Science (M.S.) degree in entomology in four primary research emphasis areas: applied agricultural entomology, arthropod biodiversity, insect molecular physiology and pathology, and urban entomology. Applied agricultural entomology advances environmentally and economically sound management strategies for destructive and beneficial insects in cropping systems. Arthropod biodiversity investigates the biology, complexity, and relationships of arthropod species worldwide and their usefulness in assessing environmental conditions. Insect molecular physiology and pathology elucidates basic molecular and cellular properties of insects as both beneficial and pest organisms and as models for fundamental biological phenomena. Urban entomology develops environmentally sound management practices based on the basic biology, ecology, and behavior of insects and other animals associated with ornamental plants, turf, gardens, homes, and the workplace.

Students in the M.S. degree program generally require two to four years to complete the degree requirements.

Research in one of the four emphasis areas is supervised by a major adviser and an advisory committee selected by the student. The major adviser works especially closely with the graduate researcher in the design, conduct, and analysis of the research.

Professional entomologists find exciting employment opportunities in such areas as university teaching and research; the Cooperative Extension Service; pest control; pest-management consulting; agricultural businesses and other industry; the Armed Forces medical corps; state and federal government, including public health agencies; quarantine and regulatory agencies; and environmental monitoring and research.

Research Facilities

On the Clemson Campus, Long Hall houses laboratories equipped for research in molecular biology, systematics, pathology, fire ant biology, and plant resistance. Long Hall also houses the Clemson University Arthropod Collection and Museum, a weather station, and a photographic darkroom. The Cherry Farm complex includes a self-contained urban entomology laboratory and insectary, rearing rooms, a butterfly house, an apiculture workshop, greenhouses, a 30-acre field research area, and an apiary. The Calhoun Field Laboratory is a field site for long-term research in sustainable farming practices.

Off campus, in other parts of South Carolina, the Coastal Research and Education Center (REC, near Charleston), the Edisto REC (near Blackville), the Pee Dee REC (near Florence), the Musser Farm (near Clemson), and the Simpson Farm (near Pendleton) provide field-research opportunities to study fire ants and insects of vegetables, field crops, turf, and stone fruits. The Archbold Tropical Research and Education Consortium, on the island of Dominica, West Indies, provides opportunities for research in a tropical setting.

Financial Aid

Most graduate students in the Entomology Graduate Program are supported by graduate research assistantships, paid by grants from nonuniversity sources to conduct research on topics of concern to the grant providers. Stipends range from about $14,000 to $22,000 per year for half-time appointments. Besides fellowships and loans offered by the University and the College, the Entomology Graduate Program also offers J. A. Berly and Dr. and Mrs. J. T. Creighton Endowed Memorial Fellowships as incentives for exceptional incoming graduate students. In addition, the E. W. King and W. C. Nettles Memorial Grants help pay costs of graduate research.

Cost of Study

Tuition for 2006–07 is $4643 per semester for in-state students and $9255 per semester for nonresidents. Off-campus rates are $535 per hour for in-state students and $918 per hour for nonresidents. Graduate assistants pay a flat fee of $1079 per semester and $348 per summer session. Graduate fellows pay South Carolina resident fees.

Living and Housing Costs

On-campus housing is available. For information, students should visit http://www.housing.clemson.edu. The cost of living in Clemson is quite low compared to the national average. Students who choose to live off the campus typically spend $300–$400 per month for rent, depending on location, amenities, roommates, etc.

Student Group

The program has approximately 10 students. Forty percent are women, 90 percent attend on a full-time basis, and 10 percent are international students.

Location

Clemson is a small, beautiful college town near the Blue Ridge Mountains and Lake Hartwell in upstate South Carolina. The Upstate is one of the country's fastest-growing areas and is the midpoint of the Charlotte-to-Atlanta I-85 corridor, a multistate area along Interstate 85 that runs from metro Atlanta to Richmond, Virginia, and encompasses Charlotte, North Carolina, and North Carolina's Research Triangle. Atlanta and Charlotte are each a 2-hour's drive away. Many financial institutions and other industries have national offices or a major presence in the Upstate, including Wachovia, Bank of America, BMW, Bon Secours St. Francis Health System, Bosch North America, Bowater, Charter Communications, Ernst and Young, Fluor Corporation, IBM, Microsoft, Michelin of North America, and many others.

The University

Clemson is classified by the Carnegie Foundation as Doctoral/Research University–Extensive, a category comprising less than 4 percent of all universities in America. The University's mission is to fulfill the covenant between its founder and the people of South Carolina to establish a "high seminary of learning" through its responsibilities of teaching, research, and extended public service. The University has identified eight areas of academic emphasis that create collaborations that, in turn, help fulfill the University's mission.

Applying

Academic ability, a capacity for hard work, and a strong desire to learn more about the natural world and to solve human problems are prerequisites for acceptance into the Entomology Graduate Program. Applicants may apply on the Web at http://www.grad.clemson.edu/p_apply.html. Applications with a $50 nonrefundable fee should be received no later than five weeks prior to registration. Every required item in support of the application must be on file by that date. Students are advised to contact the department for the deadlines of the program of proposed study.

Correspondence and Information

Dr. John C. Morse
Entomology Graduate Program Coordinator
Department of Entomology, Soils, and Plant Sciences
Long Hall
Clemson University
Box 340315
Clemson, South Carolina 29634-0315

Phone: 864-656-5049
Fax: 864-656-0274
E-mail: jmorse@clemson.edu
Web site: http://entweb.clemson.edu/

Clemson University

THE FACULTY AND THEIR RESEARCH

Peter H. Adler, Professor; Ph.D., Penn State. Entomology.
Robert G. Bellinger, Associate Professor; Ph.D., Virginia Tech. Entomology.
Eric P. Benson, Professor; Ph.D., Clemson. Entomology.
James J. Camberato, Professor; Ph.D., North Carolina State. Soil science.
Nyal D. Camper, Professor; Ph.D., North Carolina State. Plant physiology and biochemistry.
Gerald R. Carner, Professor; Ph.D., Auburn. Entomology.
Jay W. Chapin, Professor; Ph.D., Clemson. Entomology.
Joseph D. Culin, Professor and Department Chair; Ph.D., Kentucky. Entomology.
Bruce A. Fortnum, Professor; Ph.D., Clemson. Plant pathology.
James R. Frederick, Professor; Ph.D., Illinois. Agronomy.
William M. Hood, Professor; Ph.D., Georgia. Entomology.
Steven Nye Jeffers, Associate Professor; Ph.D., Cornell. Plant pathology.
Albert W. Johnson, Professor; Ph.D., Auburn. Entomology.
Michael A. Jones, Associate Professor; Ph.D., North Carolina State. Crop science.
Anthony P. Keinath, Professor; Ph.D., Cornell. Plant pathology.
Halina T. Knap, Professor; Ph.D., Academy of Agriculture (Poland). Agronomy.
Stephen A. Lewis, Professor; Ph.D., Arizona. Plant pathology.
Robert M. Lippert, Associate Professor; Ph.D., California, Davis. Soil science.
Donald G. Manley, Professor; Ph.D., Arizona. Entomology.
Samuel B. Martin Jr., Professor; Ph.D., North Carolina State. Plant pathology.
Gloria S. McCutcheon, Professor; Ph.D., Georgia. Entomology.
John C. Morse, Professor; Ph.D., Georgia. Entomology.
John D. Mueller, Professor; Ph.D., Illinois. Plant pathology.
Jason K. Norsworthy, Assistant Professor; Ph.D., Arkansas. Agronomy.
Virgil L. Quisenberry, Professor; Ph.D., Kentucky. Soil science.
Melissa B. Riley, Professor; Ph.D., Clemson. Plant physiology.
Guido Schnabel, Assistant Professor; Ph.D., Hohenheim (Stuttgart). Plant pathology.
Simon W. Scott, Professor; Ph.D., Wales. Agricultural botany.
Emerson R. Shipe, Professor; Ph.D., Virginia Tech. Agronomy.
Bill R. Smith, Professor; Ph.D., North Carolina State. Soil science.
John P. Smith, Associate Professor; Ph.D., Clemson. Entomology.
William C. Stringer, Associate Professor; Ph.D., Virginia Tech. Agronomy.
Michael Jack Sullivan, Professor; Ph.D., North Carolina State. Entomology.
Matthew W. Turnbull, Assistant Professor; Ph.D., Kentucky. Entomology.
Samuel G. Turnipseed, Professor; Ph.D., North Carolina State. Entomology and plant pathology.
Geoffrey W. Zehnder, Professor; Ph.D., California, Riverside. Entomology.
Patricia A. Zungoli, Professor; Ph.D., Virginia Tech. Entomology.

CLEMSON UNIVERSITY

Ph.D. in Entomology

Program of Study	Providing leadership in environmental entomology, Clemson University offers a Ph.D. degree in entomology in four primary research emphasis areas: applied agricultural entomology, arthropod biodiversity, insect molecular physiology and pathology, and urban entomology. Applied agricultural entomology advances environmentally and economically sound management strategies for destructive and beneficial insects in cropping systems. Arthropod biodiversity investigates the biology, complexity, and relationships of arthropod species worldwide and their usefulness in assessing environmental conditions. Insect molecular physiology and pathology elucidates basic molecular and cellular properties of insects as both beneficial and pest organisms and as models for fundamental biological phenomena. Urban entomology develops environmentally sound management practices based on the basic biology, ecology, and behavior of insects and other animals associated with ornamental plants, turf, gardens, homes, and the workplace.

Students in the Ph.D. degree program, following an M.S. degree program, generally require three to five years to complete the degree requirements, initially focusing on course work and later on research.

Research in one of the four emphasis areas is supervised by a major adviser and an advisory committee selected by the student. The major adviser works especially closely with the graduate researcher in the design, conduct, and analysis of the research.

Professional entomologists find exciting employment opportunities in such areas as university teaching and research; the Cooperative Extension Service; pest control; pest-management consulting; agricultural businesses and other industry; the Armed Forces medical corps; state and federal government, including public health agencies; quarantine and regulatory agencies; and environmental monitoring and research.

Research Facilities

On the Clemson Campus, Long Hall houses laboratories equipped for research in molecular biology, systematics, pathology, fire ant biology, and plant resistance. Long Hall also houses the Clemson University Arthropod Collection and Museum, a weather station, and a photographic darkroom. The Cherry Farm complex includes a self-contained urban entomology laboratory and insectary, rearing rooms, a butterfly house, an apiculture workshop, greenhouses, a 30-acre field research area, and an apiary. The Calhoun Field Laboratory is a field site for long-term research in sustainable farming practices.

Off campus, in other parts of South Carolina, the Coastal Research and Education Center (REC, near Charleston), the Edisto REC (near Blackville), the Pee Dee REC (near Florence), the Musser Farm (near Clemson), and the Simpson Farm (near Pendleton) provide field-research opportunities to study fire ants and insects of vegetables, field crops, turf, and stone fruits. The Archbold Tropical Research and Education Consortium, on the island of Dominica, West Indies, provides opportunities for research in a tropical setting.

Financial Aid

Most graduate students in the Entomology Graduate Program are supported by graduate research assistantships, paid by grants from nonuniversity sources to conduct research on topics of concern to the grant providers. Stipends range from about $14,000 to $22,000 per year for half-time appointments. Besides fellowships and loans offered by the University and the College, the Entomology Graduate Program also offers J. A. Berly and Dr. and Mrs. J. T. Creighton Endowed Memorial Fellowships as incentives for exceptional incoming graduate students. In addition, the E. W. King and W. C. Nettles Memorial Grants help pay costs of graduate research.

Cost of Study

Tuition for 2006–07 is $4643 per semester for in-state students and $9255 per semester for nonresidents. Off-campus rates are $535 per hour for in-state students and $918 per hour for nonresidents. Graduate assistants pay a flat fee of $1079 per semester and $348 per summer session. Graduate fellows pay South Carolina resident fees.

Living and Housing Costs

On-campus housing is available. For information, students should visit http://www.housing.clemson.edu. The cost of living in Clemson is quite low compared to the national average. Students who choose to live off the campus typically spend $300–$400 per month for rent, depending on location, amenities, roommates, etc.

Student Group

There are approximately 12 Ph.D. students in the program. Seventeen percent are women, 83 percent attend on a full-time basis, and 50 percent are international students.

Location

Clemson is a small, beautiful college town near the Blue Ridge Mountains and Lake Hartwell in upstate South Carolina. The Upstate is one of the country's fastest-growing areas and is the midpoint of the Charlotte-to-Atlanta I-85 corridor, a multistate area along Interstate 85 that runs from metro Atlanta to Richmond, Virginia, and encompasses Charlotte, North Carolina, and North Carolina's Research Triangle. Atlanta and Charlotte are each a 2-hour's drive away. Many financial institutions and other industries have national offices or a major presence in the Upstate, including Wachovia, Bank of America, BMW, Bon Secours St. Francis Health System, Bosch North America, Bowater, Charter Communications, Ernst and Young, Fluor Corporation, IBM, Microsoft, Michelin of North America, and many others.

The University

Clemson is classified by the Carnegie Foundation as Doctoral/Research University–Extensive, a category comprising less than 4 percent of all universities in America. The University's mission is to fulfill the covenant between its founder and the people of South Carolina to establish a "high seminary of learning" through its responsibilities of teaching, research, and extended public service. The University has identified eight areas of academic emphasis that create collaborations that, in turn, help fulfill the University's mission.

Applying

Academic ability, a capacity for hard work, and a strong desire to learn more about the natural world and to solve human problems are prerequisites for acceptance into the Entomology Graduate Program. Applicants may apply on the Web at http://www.grad.clemson.edu/p_apply.html. Applications with a $50 nonrefundable fee should be received no later than five weeks prior to registration. Every required item in support of the application must be on file by that date. Students are advised to contact the department for the deadlines of the program of proposed study.

Correspondence and Information

Dr. John C. Morse
Entomology Graduate Program Coordinator
Department of Entomology, Soils, and Plant Sciences
Long Hall
Clemson University
Box 340315
Clemson, South Carolina 29634-0315
Phone: 864-656-5049
Fax: 864-656-0274
E-mail: jmorse@clemson.edu
Web site: http://entweb.clemson.edu/

Clemson University

THE FACULTY AND THEIR RESEARCH

Peter H. Adler, Professor; Ph.D., Penn State. Entomology.
Robert G. Bellinger, Associate Professor; Ph.D., Virginia Tech. Entomology.
Eric P. Benson, Professor; Ph.D., Clemson. Entomology.
James J. Camberato, Professor; Ph.D., North Carolina State. Soil science.
Nyal D. Camper, Professor; Ph.D., North Carolina State. Plant physiology and biochemistry.
Gerald R. Carner, Professor; Ph.D., Auburn. Entomology.
Jay W. Chapin, Professor; Ph.D., Clemson. Entomology.
Joseph D. Culin, Professor and Department Chair; Ph.D., Kentucky. Entomology.
Bruce A. Fortnum, Professor; Ph.D., Clemson. Plant pathology.
James R. Frederick, Professor; Ph.D., Illinois. Agronomy.
William M. Hood, Professor; Ph.D., Georgia. Entomology.
Steven Nye Jeffers, Associate Professor; Ph.D., Cornell. Plant pathology.
Albert W. Johnson, Professor; Ph.D., Auburn. Entomology.
Michael A. Jones, Associate Professor; Ph.D., North Carolina State. Crop science.
Anthony P. Keinath, Professor; Ph.D., Cornell. Plant pathology.
Halina T. Knap, Professor; Ph.D., Academy of Agriculture (Poland). Agronomy.
Stephen A. Lewis, Professor; Ph.D., Arizona. Plant pathology.
Robert M. Lippert, Associate Professor; Ph.D., California, Davis. Soil science.
Donald G. Manley, Professor; Ph.D., Arizona. Entomology.
Samuel B. Martin Jr., Professor; Ph.D., North Carolina State. Plant pathology.
Gloria S. McCutcheon, Professor; Ph.D., Georgia. Entomology.
John C. Morse, Professor; Ph.D., Georgia. Entomology.
John D. Mueller, Professor; Ph.D., Illinois. Plant pathology.
Jason K. Norsworthy, Assistant Professor; Ph.D., Arkansas. Agronomy.
Virgil L. Quisenberry, Professor; Ph.D., Kentucky. Soil science.
Melissa B. Riley, Professor; Ph.D., Clemson. Plant physiology.
Guido Schnabel, Assistant Professor; Ph.D., Hohenheim (Stuttgart). Plant pathology.
Simon W. Scott, Professor; Ph.D., Wales. Agricultural botany.
Emerson R. Shipe, Professor; Ph.D., Virginia Tech. Agronomy.
Bill R. Smith, Professor; Ph.D., North Carolina State. Soil science.
John P. Smith, Associate Professor; Ph.D., Clemson. Entomology.
William C. Stringer, Associate Professor; Ph.D., Virginia Tech. Agronomy.
Michael Jack Sullivan, Professor; Ph.D., North Carolina State. Entomology.
Matthew W. Turnbull, Assistant Professor; Ph.D., Kentucky. Entomology.
Samuel G. Turnipseed, Professor; Ph.D., North Carolina State. Entomology and plant pathology.
Geoffrey W. Zehnder, Professor; Ph.D., California, Riverside. Entomology.
Patricia A. Zungoli, Professor; Ph.D., Virginia Tech. Entomology.

UNIVERSITY OF ILLINOIS AT URBANA–CHAMPAIGN
Department of Entomology

Programs of Study

The Department offers graduate programs leading to the Master of Science (M.S.) and Doctor of Philosophy (Ph.D.) degrees in entomology. Basic research in insect biology is emphasized within the Department, and major research strengths include plant-insect interactions, evolutionary biology and systematics, neural mechanisms of behavior, development, and sociogenomics. Interactions with Departmental affiliates from the Illinois Natural History Survey (INHS), the College of Veterinary Medicine, and the College of Agricultural, Consumer, and Environmental Sciences provide graduate students with additional research opportunities in agricultural and medical entomology. The Department of Entomology also cooperates in major intercollegiate programs, including the Program in Ecology and Evolutionary Biology (PEEB), the Medical Scholars Program, and the Neuroscience Program.

Both the M.S. and Ph.D. in entomology are designed to provide a combination of intensive research experience and broad training in the core disciplines comprised by entomology. All Ph.D. students take five core courses (systematics, ecology, physiology, genomic analysis, and integrated pest management) as well as courses in statistics and three advanced seminars in topics of their own choosing. M.S. students take their choice of four core choices and complete a research-based thesis. A master's degree (from Illinois or another institution) is a prerequisite for the Ph.D., and doctoral students are also required to take a preliminary examination. The Department is especially interested in attracting students interested in interdisciplinary research to its graduate programs. Most M.S. students complete their work in two to three years. Ph.D. studies typically require another three to four years to complete.

Research Facilities

The Department occupies 20,000 square feet in Morrill Hall, a building completed in 1966. Departmental facilities available to students include laboratories and equipment for most modern biological research, such as a dedicated facility for honey bee research. The University maintains numerous interdepartmental research facilities. The University library has more than 7 million volumes, the fourth-largest holding among university libraries in the world.

The Illinois Natural History Survey has more than 40 acres of agricultural plots available for research. The State of Illinois insect collection, containing more than 5.5 million specimens and more than 3,300 type specimens, is maintained at INHS for use in systematic studies and identification services and as a depository for voucher specimens.

The campus includes large research farms. The University maintains a number of natural areas for research and educational purposes.

Financial Aid

The Department has teaching assistantships available for student support. These are awarded on the basis of academic preparation, interest, and availability. Individual faculty members often provide research assistantship support from their own research grants. Competitive fellowships are available to incoming students who demonstrate outstanding ability. Average assistantship stipends (for nine-month half-time appointments) range from $16,000 to $18,000. International students who do not have a degree from a U.S. institution may qualify as teaching assistants with a satisfactory score in the Test of Spoken English (ETS) or by means of an on-campus SPEAK test, which must be taken in August prior to the start of the academic year.

Cost of Study

All students currently receive some form of financial support. A waiver of tuition and most fees accompanies graduate assistantships.

Living and Housing Costs

The cost of living close to campus in Champaign-Urbana is extremely reasonable. Some on-campus housing for graduate students with families is available, but most graduate students select off-campus housing, which is available at a wide range of prices. Average rent for a one-bedroom apartment ranges from $375 to $500 a month. An award-winning system of public transportation and bicycle trails reduce the need to have a car.

Student Group

The campus enrolls 26,000 undergraduates and 10,000 graduate and professional students. There are currently 37 graduate students in the Department of Entomology. Nineteen are men, 18 are women, and 8 are international students. The majority (17) are Ph.D. candidates. Successful applicants typically have a strong background in biology and undergraduate research experience. Graduate-level introductory courses are available for those without a strong background in entomology.

Student Outcomes

Graduates are employed primarily in academic, extension, museum, and state or federal agency settings. Most students complete the Ph.D. program. Many of these students receive postdoctoral training at Illinois or other institutions.

Location

The campus is located in East Central Illinois in the twin cities of Champaign and Urbana, and is easily accessible by interstate highways, bus, Amtrak, and air. Despite its location in the heart of highly productive agricultural lands, the many resources on the campus devoted to science and technology have given rise to the nickname of "silicon prairie."

The University and The Department

The University of Illinois at Urbana-Champaign is a major research center. Chartered in 1867 as a land-grant university, the campus is now home to more than 150 graduate and professional programs. The current Department was formed in 1909. Entomology is part of the School of Integrative Biology in the College of Liberal Arts and Sciences. Its graduate program is the only program on campus awarding the M.S. and Ph.D. degrees in entomology. All faculty members are active in both research and teaching.

Applying

Applicants must have a baccalaureate degree and meet Graduate College requirements. Applicants must submit GRE General Test scores (advanced test not required), transcripts, a statement describing a rationale for graduate study, and three letters of recommendation. Students should have a background in biology and basic science. Research experience is recommended. TOEFL scores are required for students for whom English is a second language. Applicants should contact the Department and individual faculty members during the application process. Funds are available to defray costs of visiting the campus. Applicants are encouraged to submit their application online. Materials for fall admission should be received by the Department by February 1. Students should inquire concerning admission at other times.

Correspondence and Information

Professor May R. Berenbaum, Head
Department of Entomology
320 Morrill Hall
University of Illinois at Urbana-Champaign
505 South Goodwin Avenue
Urbana, Illinois 61801
Phone: 217-333-2910
Fax: 217-244-3499
E-mail: maybe@uiuc.edu
Web site: http://www.life.uiuc.edu/entomology/

University of Illinois at Urbana–Champaign

THE FACULTY AND THEIR RESEARCH

May R. Berenbaum, Swanlund Professor and Department Head; Ph.D. Cornell, 1980. Plant-insect interactions; biochemical and molecular adaptations of insects to dietary toxins; insect-plant coevolution.

Stewart H. Berlocher, Professor; Ph.D., Texas, 1976. Speciation and host-race formation; insect/plant coevolution; insect systematics.

Samuel N. Beshers, Research Associate/Visiting Lecturer; Ph.D., Boston University, 1993. Sociobiology; ants.

Sydney A. Cameron, Assistant Professor; Ph.D., Kansas, 1985. Hymenopteran systematics; bumblebees; evolution of sociality.

Akira Chiba, Departmental Affiliate; Ph.D., SUNY at Albany, 1986. Molecular basis of neural development in *Drosophila melanogaster*.

Raymond A. Cloyd, Departmental Affiliate; Ph.D., Purdue, 1999. Insect-plant interactions; effects of plants on natural enemy foraging behavior.

Fred Delcomyn, Professor and Director, School of Integrative Biology; Ph.D., Oregon, 1969. Neural mechanisms that control locomotion; robotics.

Christopher H. Dietrich. Departmental Affiliate; Ph.D., North Carolina State, 1990. Systematics and evolution of Membracoidea; biodiversity and conservation.

Catherine E. Eastman, Departmental Affiliate; Ph.D., Louisiana State, 1976. Management of insect pests of vegetable crops.

Bettina M. Francis, Associate Professor; Ph.D., Michigan, 1971. Toxicology of pesticides; development.

Michael E. Gray, Departmental Affiliate; Ph.D., Iowa State, 1986. Population dynamics of western and northern corn rootworms.

Lawrence M. Hanks, Associate Professor; Ph.D., Maryland, 1991. Distribution and abundance of phytophagous insects in ornamental landscapes.

Kevin P. Johnson, Departmental Affiliate; Ph.D., Minnesota, 1997. Insect systematics; biogeography; coevolution of birds and their ectoparasitic lice.

Eli Levine, Departmental Affiliate; Ph.D., Ohio State, 1977. Insect pests of corn and soybeans.

Brenda Molano-Flores, Departmental Affiliate; Ph.D., Iowa, 1997. Prairie restoration; plant-insect interactions; reproductive biology of plants; population biology; conservation biology.

James B. Nardi, Departmental Affiliate; Ph.D., Harvard, 1975. Pattern formation during development; insect olfaction.

Robert J. Novak, Departmental Affiliate; Ph.D., Illinois, 1976. Medical entomology; evolutionary relationships of mosquito-pathogen interactions.

David W. Onstad, Departmental Affiliate; Ph.D., Cornell, 1985. Applied insect ecology.

Hugh M. Robertson, Professor; Ph.D., Witwatersrand (Johannesburg), 1982. Transposable elements in insects; insect olfaction and genomics.

Gene E. Robinson, Professor; Ph.D., Cornell, 1986. Behavioral biology of the honey bee; sociogenomics; insect molecular biology.

Daniel W. Schneider, Departmental Affiliate; Ph.D., Wisconsin, 1990. Community ecology of aquatic invertebrates.

Leellen F. Solter, Departmental Affiliate; Ph.D., Illinois, 1996. Host-parasite-pathogen relationships; classical biological control of insect pests; insect pathology.

Kevin L. Steffey, Departmental Affiliate; Ph.D., Iowa State, 1979. Management of insects in field crops.

Andrew Suarez, Assistant Professor; Ph.D., California, San Diego, 2000. Ant ecology and behavior.

Steven J. Taylor, Departmental Affiliate; Ph.D., Southern Illinois Carbondale, 1996. Life history and biology of waterstriders and related taxa (Heteroptera: Gerromorpha).

David J. Voegtlin, Departmental Affiliate; Ph.D., Berkeley, 1976. Biology and systematics of aphids.

Richard A. Weinzierl, Departmental Affiliate; Ph.D., Oregon State, 1984. Management of insect pests of vegetables, stored grain, and livestock.

James B. Whitfield, Associate Professor; Ph.D., Berkeley. Hymenopteran systematics; host-parasite interactions.

Arthur R. Zangerl, Departmental Affiliate; Ph.D., Illinois, 1982. Chemically mediated insect-plant interactions.

Section 10
Genetics, Developmental Biology, and Reproductive Biology

This section contains a directory of institutions offering graduate work in genetics, developmental biology, and reproductive biology, followed by in-depth entries submitted by institutions that chose to prepare detailed program descriptions. Additional information about programs listed in the directory but not augmented by an in-depth entry may be obtained by writing directly to the dean of a graduate school or chair of a department at the address given in the directory.

For programs offering related work, see also all other sections of this book. In Book 4, see Agricultural and Food Sciences, Chemistry, and Environmental Sciences and Management; in Book 5, see Agricultural Engineering and Bioengineering and Biomedical Engineering and Biotechnology; and in Book 6, see Veterinary Medicine and Sciences.

CONTENTS

Developmental Biology

Albert Einstein College of Medicine, Sue Golding Graduate Division of Medical Sciences, Department of Anatomy and Structural Biology, Bronx, NY 10461. Offers anatomy (PhD); cell and developmental biology (PhD). *Degree requirements:* For doctorate, thesis/dissertation. *Entrance requirements:* For doctorate, GRE General Test. Additional exam requirements/recommendations for international students: Required—TOEFL. Electronic applications accepted. *Faculty research:* Cell motility, cell membranes and membrane-cytoskeletal interactions as applied to processing of pancreatic hormones, mechanisms of secretion.

Albert Einstein College of Medicine, Sue Golding Graduate Division of Medical Sciences, Division of Biological Sciences, Department of Developmental and Molecular Biology, Bronx, NY 10461. Offers PhD, MD/PhD. *Degree requirements:* For doctorate, thesis/dissertation. *Entrance requirements:* For doctorate, GRE General Test. Additional exam requirements/recommendations for international students: Required—TOEFL. *Faculty research:* DNA, RNA, and protein synthesis in prokaryotes and eukaryotes; chemical and enzymatic alteration of RNA; glycoproteins.

Arizona State University, Division of Graduate Studies, College of Liberal Arts and Sciences, Department of Biology, Cell and Developmental Biology Group, Tempe, AZ 85287. Offers MS, PhD. Terminal master's awarded for partial completion of doctoral program. *Degree requirements:* For master's, thesis; for doctorate, thesis/dissertation, oral exam. *Entrance requirements:* For master's, GRE General Test; for doctorate, GRE General Test, GRE Subject Test. Additional exam requirements/recommendations for international students: Required—TOEFL (minimum score 600 paper-based; 250 computer-based); Recommended—TSE. *Faculty research:* Cytoskeleton assembly, exocytosis, cyclic nucleotides, membrane fusion, chromosome distribution.

Baylor College of Medicine, Graduate School of Biomedical Sciences, Program in Developmental Biology, Houston, TX 77030-3498. Offers PhD, MD/PhD. *Faculty:* 42 full-time (10 women). *Students:* 38 full-time (17 women); includes 7 minority (1 American Indian/Alaska Native, 5 Asian Americans or Pacific Islanders, 1 Hispanic American), 21 international. Average age 27. 67 applicants, 25% accepted, 7 enrolled. In 2005, 2 doctorates awarded. *Median time to degree:* Of those who began their doctoral program in fall 1997, 100% received their degree in 8 years or less. *Degree requirements:* For doctorate, thesis/dissertation, public defense. *Entrance requirements:* For doctorate, GRE General Test, GRE Subject Test (strongly recommended), minimum GPA of 3.0. Additional exam requirements/recommendations for international students: Required—TOEFL. *Application deadline:* For fall admission, 1/1 for domestic students. Application fee: $30. Electronic applications accepted. *Expenses:* Tuition: Full-time $8,200. Full-time tuition and fees vary according to program. *Financial support:* In 2005–06, 37 students received support, including 10 fellowships (averaging $23,000 per year), 28 research assistantships (averaging $23,000 per year); career-related internships or fieldwork, Federal Work-Study, institutionally sponsored loans, health care benefits, tuition waivers (full), and stipends also available. *Faculty research:* Molecular and genetic approaches to study pattern formation in *Dictyostelium, Drosophila, C.elegans,* mouse, *Xenopus,* and zebrafish; cross-species approach. *Unit head:* Dr. Hugo Bellen, Director, 713-798-6410. *Application contact:* Catherine Tasnier, Graduate Program Administrator, 713-798-6410, Fax: 713-798-5386, E-mail: cat@bcm.edu.

See Close-Up on page 757.

Brigham Young University, Graduate Studies, College of Biological and Agricultural Sciences, Department of Physiology and Developmental Biology, Provo, UT 84602-1001. Offers neuroscience (MS, PhD); physiology and developmental biology (MS, PhD). Part-time programs available. *Faculty:* 17 full-time (0 women). *Students:* 25 full-time (11 women); includes 1 minority (Asian American or Pacific Islander), 1 international. Average age 26. 26 applicants, 38% accepted, 9 enrolled. In 2005, 5 master's, 2 doctorates awarded. Terminal master's awarded for partial completion of doctoral program. *Degree requirements:* For master's and doctorate, thesis/dissertation. *Entrance requirements:* For master's, GRE General Test, minimum GPA of 3.0 during previous 2 years; for doctorate, GRE General Test, minimum GPA of 3.0 overall. Additional exam requirements/recommendations for international students: Required—TOEFL. *Application deadline:* For fall admission, 2/1 priority date for domestic students, 2/1 priority date for international students. For winter admission, 9/10 for domestic students. Application fee: $50. Electronic applications accepted. *Financial support:* In 2005–06, 26 students received support, including 13 research assistantships with full tuition reimbursements available (averaging $15,500 per year), 13 teaching assistantships with partial tuition reimbursements available (averaging $14,900 per year); fellowships with partial tuition reimbursements available, career-related internships or fieldwork, institutionally sponsored loans, scholarships/grants, tuition waivers (full and partial), unspecified assistantships, and tuition awards also available. Financial award application deadline: 2/1. *Faculty research:* Sex differentiation of the brain, exercise physiology, developmental biology, membrane biophysics, neuroscience. Total annual research expenditures: $892,386. *Unit head:* Dr. James P. Porter, Chair, 801-422-9160, Fax: 801-422-0700, E-mail: james_porter@byu.edu. *Application contact:* Dr. Dixon J. Woodbury, Graduate Coordinator, 801-422-7562, Fax: 801-422-0700, E-mail: dixon_woodbury@byu.edu.

See Close-Up on page 1243.

Brown University, Graduate School, Division of Biology and Medicine, Program in Molecular Biology, Cell Biology, and Biochemistry, Providence, RI 02912. Offers biochemistry (M Med Sc, Sc M, PhD), including biochemistry (Sc M, PhD), biology (Sc M, PhD), medical science (M Med Sc, PhD); biology (MA); cell biology (M Med Sc, Sc M, PhD), including biochemistry (Sc M, PhD), biology (Sc M, PhD), medical science (M Med Sc, PhD); developmental biology (M Med Sc, Sc M, PhD), including biochemistry (Sc M, PhD), biology (Sc M, PhD), medical science (M Med Sc, PhD); immunology (M Med Sc, Sc M, PhD), including biochemistry (Sc M, PhD), biology (Sc M, PhD), medical science (M Med Sc, PhD); molecular microbiology (M Med Sc, Sc M, PhD), including biochemistry (Sc M, PhD), biology (Sc M, PhD), medical science (M Med Sc, PhD). Part-time programs available. Terminal master's awarded for partial completion of doctoral program. *Degree requirements:* For master's, thesis (for some programs); for doctorate, one foreign language, thesis/dissertation, preliminary exam. *Entrance requirements:* For master's and doctorate, GRE General Test, GRE Subject Test. Additional exam requirements/recommendations for international students: Required—TOEFL. Electronic applications accepted. *Faculty research:* Molecular genetics, gene regulation.

See Close-Up on page 555.

California Institute of Technology, Division of Biology, Program in Developmental Biology, Pasadena, CA 91125-0001. Offers PhD. *Degree requirements:* For doctorate, thesis/dissertation, qualifying exam. *Entrance requirements:* For doctorate, GRE General Test. *Application deadline:* For fall admission, 1/1 for domestic students. Application fee: $0. *Financial support:* Application deadline: 1/1. *Application contact:* Elizabeth Ayala, Graduate Program Coordinator, 626-395-4497, Fax: 626-683-3343, E-mail: biograd@cco.caltech.edu.

Carnegie Mellon University, Mellon College of Science, Department of Biological Sciences, Pittsburgh, PA 15213-3891. Offers biochemistry (PhD); biophysics (PhD); cell biology (PhD); computational biology (MS, PhD); developmental biology (PhD); genetics (PhD); molecular biology (PhD); neurobiology (PhD). *Degree requirements:* For doctorate, thesis/dissertation, comprehensive exam. *Entrance requirements:* For doctorate, GRE General Test, GRE Subject Test, interview. Electronic applications accepted. *Faculty research:* Genetic structure, function, and regulation; protein structure and function; biological membranes; biological spectroscopy.

See Close-Up on page 93.

Case Western Reserve University, School of Medicine and School of Graduate Studies, Graduate Programs in Medicine, Department of Anatomy, Cleveland, OH 44106. Offers applied anatomy (MS); biological anthropology (MS, PhD); cellular biology (MS, PhD); developmental biology (PhD); molecular biology (PhD). Part-time programs available. *Faculty:* 20 full-time (5 women), 16 part-time/adjunct (5 women). *Students:* 6 full-time (0 women), 49 part-time (19 women); includes 20 minority (7 African Americans, 12 Asian Americans or Pacific Islanders, 1 Hispanic American). Average age 25. 39 applicants, 92% accepted, 29 enrolled. In 2005, 14 master's awarded. *Median time to degree:* Of those who began their doctoral program in fall 1997, 67% received their degree in 8 years or less. *Degree requirements:* For master's, thesis (for some programs), comprehensive exam; for doctorate, thesis/dissertation. *Entrance requirements:* For master's, GRE General Test; for doctorate, GRE General Test, GRE Subject Test. Additional exam requirements/recommendations for international students: Required—TOEFL. *Application deadline:* For fall admission, 5/1 for domestic students; for spring admission, 8/1 priority date for domestic students. Applications are processed on a rolling basis. Application fee: $50. *Financial support:* In 2005–06, 6 research assistantships with full tuition reimbursements (averaging $23,000 per year) were awarded; fellowships, scholarships/grants also available. *Faculty research:* Hypoxia, cell injury, biochemical aberration occurrences in ischemic tissue, human functional morphology, evolutionary morphology. Total annual research expenditures: $691,974. *Unit head:* Dr. Joseph C. LaManna, Chairman, 216-368-1100, Fax: 216-368-8669, E-mail: jcl4@po.cwru.edu. *Application contact:* Morley Schwebel, Administrator, 216-368-2433, Fax: 216-368-8669, E-mail: mxs86@po.cwru.edu.

Columbia University, College of Physicians and Surgeons and Graduate School of Arts and Sciences, Graduate School of Arts and Sciences at the College of Physicians and Surgeons, Department of Genetics and Development, New York, NY 10032. Offers genetics (M Phil, MA, PhD). Only candidates for the PhD are admitted. Terminal master's awarded for partial completion of doctoral program. *Degree requirements:* For doctorate, thesis/dissertation. *Entrance requirements:* For master's and doctorate, GRE General Test. Additional exam requirements/recommendations for international students: Required—TOEFL. *Expenses:* Tuition: Full-time $31,448. Tuition and fees vary according to course level, course load, campus/location and program. *Faculty research:* Mammalian cell differentiation and meiosis, developmental genetics, yeast and human genetics, chromosome structure, molecular and cellular biology.

Cornell University, Graduate School, Graduate Fields of Agriculture and Life Sciences, Field of Genetics and Development, Ithaca, NY 14853-0001. Offers developmental biology (PhD); genetics (PhD). *Faculty:* 50 full-time (11 women). *Students:* 56 full-time (32 women); includes 10 minority (3 African Americans, 1 American Indian/Alaska Native, 3 Asian Americans or Pacific Islanders, 3 Hispanic Americans), 16 international. 104 applicants, 17% accepted, 9 enrolled. In 2005, 7 doctorates awarded. *Degree requirements:* For doctorate, thesis/dissertation, 2 semesters of teaching experience, comprehensive exam. *Entrance requirements:* For doctorate, GRE General Test, GRE Subject Test in biology or biochemistry (recommended), 2 letters of recommendation. Additional exam requirements/recommendations for international students: Required—TOEFL (minimum score 550 paper-based; 213 computer-based). *Application deadline:* For fall admission, 1/5 for domestic students. Application fee: $60. Electronic applications accepted. *Financial support:* In 2005–06, 56 students received support, including 15 fellowships with full tuition reimbursements available, 30 research assistantships with full tuition reimbursements available, 11 teaching assistantships with full tuition reimbursements available; institutionally sponsored loans, scholarships/grants, health care benefits, tuition waivers (full and partial), and unspecified assistantships also available. Financial award applicants required to submit FAFSA. *Faculty research:* Molecular and general genetics, developmental biology and developmental genetics, evolution and population genetics, plant genetics, microbial genetics. *Unit head:* Director of Graduate Studies, 607-254-2100. *Application contact:* Graduate Field Assistant, 607-254-2100, E-mail: gendev@cornell.edu.

Cornell University, Graduate School, Graduate Fields of Agriculture and Life Sciences, Field of Zoology, Ithaca, NY 14853-0001. Offers animal cytology (MS, PhD); comparative and functional anatomy (MS, PhD); developmental biology (MS, PhD); ecology (MS, PhD); histology (MS, PhD). *Faculty:* 26 full-time (4 women). *Students:* 4 full-time (3 women), 3 international. 9 applicants, 11% accepted, 1 enrolled. *Degree requirements:* For doctorate, thesis/dissertation, 2 semesters of teaching experience, comprehensive exam. *Entrance requirements:* For doctorate, GRE General Test, GRE Subject Test (biology), 2 letters of recommendation. Additional exam requirements/recommendations for international students: Required—TOEFL (minimum score 550 paper-based; 213 computer-based). *Application deadline:* For fall admission, 2/1 for domestic students. Application fee: $60. Electronic applications accepted. *Financial support:* In 2005–06, 4 students received support, including 1 fellowship with full tuition reimbursement available, 1 research assistantship with full tuition reimbursement available, 2 teaching assistantships with full tuition reimbursements available; institutionally sponsored loans, scholarships/grants, health care benefits, tuition waivers (full and partial), and unspecified assistantships also available. Financial award applicants required to submit FAFSA. *Faculty research:* Organismal biology, functional morphology, biomechanics, comparative vertebrate anatomy, comparative invertebrate anatomy, paleontology. *Unit head:* Director of Graduate Studies, 607-253-3276, Fax: 607-253-3756. *Application contact:* Graduate Field Assistant, 607-253-3276, Fax: 607-253-3756, E-mail: graduate_edcvm@cornell.edu.

Cornell University, Graduate School, Graduate Fields of Comparative Biomedical Sciences, Field of Comparative Biomedical Sciences, Ithaca, NY 14853-0001. Offers cellular and molecular medicine (MS, PhD); developmental and reproductive biology (MS, PhD); infectious diseases (MS, PhD); population medicine and epidemiology (MS); population medicine and epidemiology sciences (PhD); structural and functional biology (MS, PhD). *Faculty:* 135 full-time (38 women). *Students:* 41 full-time (23 women); includes 4 minority (1 African American, 3 Asian Americans or Pacific Islanders), 19 international. 58 applicants, 60% accepted, 33 enrolled. In 2005, 4 degrees awarded. *Degree requirements:* For master's, thesis/dissertation; for doctorate, thesis/dissertation, comprehensive exam. *Entrance requirements:* For master's and doctorate, GRE General Test, 2 letters of recommendation. Additional exam requirements/recommendations for international students: Required—TOEFL (minimum score 550 paper-based; 213 computer-based). *Application deadline:* For fall admission, 12/15 for domestic students. Application fee: $60. Electronic applications accepted. *Financial support:* In 2005–06, 41 students received support, including 13 fellowships with full tuition reimbursements available, 28 research assistantships with full tuition reimbursements available; teaching assistantships with full tuition reimbursements available, institutionally sponsored loans, scholarships/grants, health care benefits, tuition waivers (full and partial), and unspecified assistantships also available. Financial award applicants required to submit FAFSA. *Faculty research:* Receptors and signal transduction, viral and bacterial infectious diseases, tumor metastasis, clinical sciences/nutritional disease, development/neurologic disorders. *Unit head:* Director of Graduate Studies, 607-253-3276, Fax: 607-253-3756. *Application contact:* Graduate Field Assistant, 607-253-3276, Fax: 607-253-3756, E-mail: graduate_edcvm@cornell.edu.

Duke University, Graduate School, Program in Developmental Biology, Durham, NC 27710. Offers PhD, Certificate. *Faculty:* 48 full-time. *Students:* 7 full-time (4 women); includes 2 minority (both Asian Americans or Pacific Islanders), 1 international. 41 applicants, 34% accepted, 5 enrolled. *Entrance requirements:* For doctorate, GRE General Test, GRE Subject Test (recommended); for Certificate, GRE General Test, GRE Subject Test. Additional exam requirements/recommendations for international students: Required—IELT (preferred) or TOEFL. *Application deadline:* For fall admission, 12/31 for domestic students, 12/31 for international students. Application fee: $75. *Unit head:* John Klingensmith, Head, 919-684-9402, Fax: 919-681-7767, E-mail: devbio@biochem.duke.edu.

See Close-Up on page 767.

Emory University, Graduate School of Arts and Sciences, Division of Biological and Biomedical Sciences, Program in Biochemistry, Cell and Developmental Biology, Atlanta, GA 30322-1100. Offers PhD. *Faculty:* 61 full-time (12 women). *Students:* 56 full-time (42 women); includes 8 minority (5 African Americans, 2 Asian Americans or Pacific Islanders, 1 Hispanic

Developmental Biology

Emory University (continued)

American), 9 international. Average age 27. 96 applicants, 38% accepted, 17 enrolled. In 2005, 6 doctorates awarded. *Median time to degree:* Of those who began their doctoral program in fall 1997, 100% received their degree in 8 years or less. *Degree requirements:* For doctorate, thesis/dissertation, comprehensive exam, registration. *Entrance requirements:* For doctorate, GRE General Test, minimum GPA of 3.0 in science course work. Additional exam requirements/recommendations for international students: Required—TOEFL. *Application deadline:* For fall admission, 1/3 for domestic students, 1/3 for international students. Application fee: $50. Electronic applications accepted. *Expenses:* Tuition: Full-time $14,400. Required fees: $217. *Financial support:* In 2005–06, 26 students received support, including 26 fellowships with full tuition reimbursements available (averaging $23,000 per year); institutionally sponsored loans, health care benefits, and tuition waivers (full) also available. *Faculty research:* Signal transduction, molecular biology, enzymes and cofactors, receptor and ion channel function, membrane biology. *Unit head:* Grace Pavlath, Director, 404-727-3353, Fax: 404-727-2880, E-mail: gpavlat@emory.edu. *Application contact:* 404-727-2545, Fax: 404-727-3322, E-mail: gdbbs@emory.edu.

Florida State University, Graduate Studies, College of Arts and Sciences, Department of Biological Science, Program in Developmental Biology, Tallahassee, FL 32306. Offers MS, PhD. *Faculty:* 9 full-time (1 woman). *Students:* 24 full-time (14 women); includes 2 minority (1 Asian American or Pacific Islander, 1 Hispanic American), 10 international. *Degree requirements:* For master's and doctorate, thesis/dissertation, teaching experience seminar presentation, comprehensive exam, registration. *Entrance requirements:* For master's and doctorate, GRE General Test (minimum 1100: U-500, Q-500), minimum upper division GPA of 3.0. Additional exam requirements/recommendations for international students: Required—TOEFL (minimum score 600 paper-based; 250 computer-based), IB-100. *Application deadline:* For fall admission, 1/15 for domestic students; for spring admission, 10/15 for domestic students, 9/1 for international students. Application fee: $30. *Financial support:* In 2005–06, fellowships with full tuition reimbursements (averaging $19,000 per year), research assistantships with full tuition reimbursements (averaging $19,000 per year), teaching assistantships with full tuition reimbursements (averaging $18,000 per year) were awarded. Financial award application deadline: 1/15; financial award applicants required to submit FAFSA. *Faculty research:* Cell fusion, fertilization of nucleic acids, meiotic mechanisms, mitosis, DNA repair. *Application contact:* Judy Bowers, Coordinator, Graduate Affairs, 850-644-3023, Fax: 850-644-9829, E-mail: gradinfo@bio.fsu.edu.

Indiana University Bloomington, Graduate School, College of Arts and Sciences, Department of Biology, Program in Molecular, Cellular, and Developmental Biology, Bloomington, IN 47405-7000. Offers PhD. Offered through the University Graduate School. *Faculty:* 25 full-time (4 women). *Students:* 48 full-time (23 women); includes 6 minority (5 African Americans, 1 Hispanic American), 16 international. In 2005, 9 degrees awarded. *Degree requirements:* For doctorate, thesis/dissertation. *Entrance requirements:* For doctorate, GRE General Test. Additional exam requirements/recommendations for international students: Required—TOEFL. *Application deadline:* For fall admission, 1/15 for domestic students; for spring admission, 9/1 priority date for domestic students. Applications are processed on a rolling basis. Application fee: $45. Electronic applications accepted. *Expenses:* Tuition, state resident: full-time $5,437; part-time $227 per credit hour. Tuition, nonresident: full-time $15,836; part-time $660 per credit hour. Required fees: $821. Tuition and fees vary according to campus/location and program. *Financial support:* In 2005–06, 48 students received support; fellowships with tuition reimbursements available, research assistantships with tuition reimbursements available, teaching assistantships with tuition reimbursements available available. Financial award application deadline: 2/15. *Faculty research:* Developmental genetics, molecular evolution, macromolecular structure and function, cell biology and microbial molecular genetics. *Unit head:* Dr. Susan Strome, Head, 812-855-5450, Fax: 812-855-6705, E-mail: sstrome@bio.indiana.edu. *Application contact:* Gretchen Clearwater, Adviser for Graduate Affairs, 812-855-1861, Fax: 812-855-6705, E-mail: biograd@bio.indiana.edu.

Iowa State University of Science and Technology, Graduate College, College of Liberal Arts and Sciences and College of Agriculture, Department of Genetics, Development and Cell Biology, Ames, IA 50011. Offers MS, PhD. *Faculty:* 30 full-time, 3 part-time/adjunct. *Students:* 77 full-time (34 women), 55 international. 3 applicants, 100% accepted, 1 enrolled. In 2005, 6 master's, 9 doctorates awarded. *Degree requirements:* For master's and doctorate, thesis/dissertation. *Entrance requirements:* For master's and doctorate, GRE General Test. Additional exam requirements/recommendations for international students: Required—TOEFL (paper score 570; computer score 230) or IELTS (score 6.5). *Application deadline:* For fall admission, 1/1 priority date for domestic students, 1/1 priority date for international students. Application fee: $30 ($70 for international students). Electronic applications accepted. *Expenses:* Tuition, state resident: full-time $6,410. Tuition, nonresident: full-time $16,422. Tuition and fees vary according to program. *Financial support:* In 2005–06, 65 students received support, including 65 research assistantships with full and partial tuition reimbursements available (averaging $17,301 per year), 7 teaching assistantships with full and partial tuition reimbursements available (averaging $15,042 per year); fellowships with full tuition reimbursements available, scholarships/grants, health care benefits, and unspecified assistantships also available. Financial award application deadline: 2/1. *Faculty research:* Animal behavior, animal models of gene therapy, cell biology, comparative physiology, developmental biology. *Unit head:* Dr. Martin Spalding, Chair, 515-294-1749.

Iowa State University of Science and Technology, Graduate College, Interdisciplinary Programs, Program in Molecular, Cellular, and Developmental Biology, Ames, IA 50011. Offers MS, PhD. *Students:* 38 full-time (19 women), 1 part-time, 28 international. 78 applicants, 19% accepted, 7 enrolled. In 2005, 2 degrees awarded. *Degree requirements:* For master's, thesis or alternative; for doctorate, thesis/dissertation. *Entrance requirements:* For master's and doctorate, GRE General Test, resumé. Additional exam requirements/recommendations for international students: Required—TOEFL (paper score 550; computer score 220) or IELTS (score 6.5). *Application deadline:* For fall admission, 1/15 priority date for domestic students, 1/15 priority date for international students. Application fee: $30 ($70 for international students). Electronic applications accepted. *Expenses:* Tuition, state resident: full-time $6,410. Tuition, nonresident: full-time $16,422. Tuition and fees vary according to program. *Financial support:* In 2005–06, 32 research assistantships with full and partial tuition reimbursements (averaging $22,000 per year), 6 teaching assistantships with full and partial tuition reimbursements were awarded; scholarships/grants, health care benefits, and unspecified assistantships also available. *Unit head:* Dr. W Allen Miller, Supervisory Committee Chair, 515-294-7252, E-mail: idgp@iastate.edu. *Application contact:* Mary Jordison, Information Contact, 515-294-7252, Fax: 515-924-6790, E-mail: idgp@iastate.edu.

The Johns Hopkins University, National Institutes of Health Sponsored Programs, Department of Biology, Baltimore, MD 21218-2699. Offers biochemistry (PhD); biophysics (PhD); cell biology (PhD); developmental biology (PhD); genetic biology (PhD); molecular biology (PhD). *Faculty:* 25 full-time (4 women). *Students:* 126 full-time (72 women); includes 36 minority (3 African Americans, 1 American Indian/Alaska Native, 21 Asian Americans or Pacific Islanders, 11 Hispanic Americans), 19 international. 282 applicants, 26% accepted, 36 enrolled. In 2005, 15 degrees awarded. *Median time to degree:* Of those who began their doctoral program in fall 1997, 81.2% received their degree in 8 years or less. *Degree requirements:* For doctorate, thesis/dissertation, comprehensive exam, registration. *Entrance requirements:* For doctorate, GRE General Test. Additional exam requirements/recommendations for international students: Required—TOEFL (minimum score 600 paper-based; 250 computer-based), TWE, TSE. *Application deadline:* For fall admission, 12/15 for domestic students. Application fee: $60. *Expenses:* Tuition: Full-time $30,960. Tuition and fees vary according to degree level and program. *Financial support:* In 2005–06, 24 fellowships (averaging $23,000 per year), 93 research assistantships (averaging $23,000 per year), 22 teaching assistantships (averaging $23,000 per year) were awarded; Federal Work-Study, institutionally sponsored loans, scholarships/grants, traineeships, health care benefits, tuition waivers (partial), and unspecified assistant-

ships also available. Financial award application deadline: 4/15; financial award applicants required to submit FAFSA. *Faculty research:* Protein and nucleic acid biochemistry and biophysical chemistry, molecular biology and development. Total annual research expenditures: $11.2 million. *Unit head:* Dr. Allen Shearn, Chair, 410-516-4693, Fax: 410-516-5213, E-mail: bio_cals@jhu.edu. *Application contact:* Joan Miller, Academic Affairs Manager, 410-516-5502, Fax: 410-516-5213, E-mail: joan@jhu.edu.

See Close-Up on page 155.

The Johns Hopkins University, School of Medicine, Program in Biochemistry, Cellular and Molecular Biology, Baltimore, MD 21218-2699.

Louisiana State University Health Sciences Center, School of Graduate Studies in New Orleans, Department of Cell Biology and Anatomy, New Orleans, LA 70112-2223. Offers cell biology and anatomy (MS, PhD), including cell biology, developmental biology, neurobiology and anatomy. *Degree requirements:* For master's and doctorate, thesis/dissertation. *Entrance requirements:* For master's and doctorate, GRE General Test, GRE Subject Test, minimum undergraduate GPA of 3.0. Additional exam requirements/recommendations for international students: Required—TOEFL. *Faculty research:* Visual system organization, neural development, plasticity of sensory systems, information processing through the nervous system, visuomotor integration.

Marquette University, Graduate School, College of Arts and Sciences, Department of Biology, Milwaukee, WI 53201-1881. Offers cell biology (MS, PhD); developmental biology (MS, PhD); ecology (MS, PhD); endocrinology (MS, PhD); evolutionary biology (MS, PhD); genetics (MS, PhD); microbiology (MS, PhD); molecular biology (MS, PhD); muscle and exercise physiology (MS, PhD); neurobiology (MS, PhD); reproductive physiology (MS, PhD). Terminal master's awarded for partial completion of doctoral program. *Degree requirements:* For master's, thesis, 1 year of teaching experience or equivalent, comprehensive exam; for doctorate, thesis/dissertation, 1 year of teaching experience or equivalent, qualifying exam. *Entrance requirements:* For master's and doctorate, GRE General Test, GRE Subject Test. Additional exam requirements/recommendations for international students: Required—TOEFL. *Faculty research:* Microbial and invertebrate ecology, evolution of gene function, DNA methylation, DNA arrangement.

Medical College of Wisconsin, Graduate School of Biomedical Sciences, Program in Cell and Developmental Biology, Milwaukee, WI 53226-0509. Offers MS, PhD. Terminal master's awarded for partial completion of doctoral program. *Degree requirements:* For master's, thesis/dissertation; for doctorate, thesis/dissertation, comprehensive exam, registration. *Entrance requirements:* For master's and doctorate, GRE General Test. Additional exam requirements/recommendations for international students: Required—TOEFL. *Faculty research:* Neurobiology, development, neuroscience, teratology.

New York University, Graduate School of Arts and Science, Department of Biology, New York, NY 10012-1019. Offers biology (PhD); biomedical journalism (MS); cancer and molecular biology (PhD); computational biology (PhD); computers in biological research (MS); developmental genetics (PhD); general biology (MS); immunology and microbiology (PhD); molecular genetics (PhD); neurobiology (PhD); oral biology (MS); plant biology (PhD); recombinant DNA technology (MS). Part-time programs available. *Faculty:* 24 full-time (5 women), 8 part-time/adjunct. *Students:* 104 full-time (52 women), 41 part-time (23 women); includes 28 minority (2 African Americans, 20 Asian Americans or Pacific Islanders, 6 Hispanic Americans), 47 international. Average age 27. 349 applicants, 56% accepted, 39 enrolled. In 2005, 59 master's, 4 doctorates awarded. Terminal master's awarded for partial completion of doctoral program. *Degree requirements:* For master's, thesis or alternative, qualifying paper; for doctorate, thesis/dissertation, comprehensive exam. *Entrance requirements:* For master's, GRE General Test; for doctorate, GRE General Test, GRE Subject Test. Additional exam requirements/recommendations for international students: Required—TOEFL. *Application deadline:* For fall admission, 1/4 for domestic students. Application fee: $80. *Financial support:* Fellowships with tuition reimbursements, research assistantships with tuition reimbursements, teaching assistantships with tuition reimbursements, career-related internships or fieldwork, Federal Work-Study, institutionally sponsored loans, scholarships/grants, health care benefits, and unspecified assistantships available. Financial award application deadline: 1/4; financial award applicants required to submit FAFSA. *Faculty research:* Genomics, molecular and cell biology, development and molecular genetics, molecular evolution of plants and animals. *Unit head:* , Gloria Coruzzi, Chairman, 212-998-8200, Fax: 212-995-4015, E-mail: biodgp@nyu.edu. *Application contact:* Stephen Small, Director of Graduate Studies, 212-998-8200, Fax: 212-995-4015, E-mail: biology@nyu.edu.

New York University, School of Medicine and Graduate School of Arts and Science, Sackler Institute of Graduate Biomedical Sciences, New York, NY 10012-1019. Offers cellular and molecular biology (PhD); computational biology (PhD); developmental genetics (PhD); medical and molecular parasitology (PhD); microbiology (PhD); molecular oncology and immunology (PhD), including immunology, molecular oncology; neuroscience and physiology (PhD), including neuroscience, physiology; pharmacology (PhD), including molecular pharmacology and signal transduction; structural biology (PhD). *Faculty:* 150 full-time (35 women). *Students:* 240 full-time (124 women), 6 part-time (4 women); includes 54 minority (13 African Americans, 29 Asian Americans or Pacific Islanders, 12 Hispanic Americans), 84 international. Average age 28. 569 applicants, 8% accepted, 46 enrolled. In 2005, 38 degrees awarded. *Degree requirements:* For doctorate, one foreign language, thesis/dissertation, qualifying exam, comprehensive exam. *Entrance requirements:* For doctorate, GRE General Test. Additional exam requirements/recommendations for international students: Required—TOEFL. *Application deadline:* For fall admission, 1/4 for domestic students. Applications are processed on a rolling basis. Application fee: $80. *Expenses:* Contact institution. *Financial support:* In 2005–06, fellowships with tuition reimbursements (averaging $26,000 per year), research assistantships with tuition reimbursements (averaging $26,000 per year), teaching assistantships with tuition reimbursements (averaging $25,000 per year) were awarded; career-related internships or fieldwork, Federal Work-Study, institutionally sponsored loans, scholarships/grants, health care benefits, tuition waivers (full and partial), and unspecified assistantships also available. Financial award application deadline: 2/1; financial award applicants required to submit FAFSA. *Unit head:* Dr. Joel D. Oppenheim, Senior Associate Dean for Graduate Studies, 212-263-8001, Fax: 212-263-7600. *Application contact:* Lizabeth Greene, Administrative Manager, 212-263-5648, Fax: 212-263-7600, E-mail: sackler-info@med.nyu.edu.

See Close-Up on page 181.

Northwestern University, The Graduate School and Judd A. and Marjorie Weinberg College of Arts and Sciences, Interdepartmental Biological Sciences Program (IBiS), Evanston, IL 60208. Offers biochemistry, molecular biology, and cell biology (PhD), including biochemistry, cell and molecular biology, molecular biophysics, structural biology; biotechnology (PhD); cell and molecular biology (PhD); developmental biology and genetics (PhD); hormone action and signal transduction (PhD); neuroscience (PhD); structural biology, biochemistry, and biophysics (PhD). Participants in the Interdepartmental Biological Sciences Program include the Departments of Biochemistry, Molecular Biology, and Cell Biology; Chemistry; Neurobiology and Physiology; Chemical Engineering; Civil Engineering; and Evanston Hospital. *Entrance requirements:* For doctorate, GRE General Test. Additional exam requirements/recommendations for international students: Required—TOEFL (minimum score 600 paper-based), TSE(minimum score 50). Electronic applications accepted. *Faculty research:* Developmental genetics, gene regulation, DNA-protein interactions, biological clocks, bioremediation.

See Close-Up on page 187.

Northwestern University, Northwestern University Feinberg School of Medicine and Interdepartmental Degree Programs, Integrated Graduate Programs in the Life Sciences, Chicago, IL 60611. Offers cancer biology (PhD); cell biology (PhD); developmental biology (PhD); evolutionary biology (PhD); immunology and microbial pathogenesis (PhD); molecular

Peterson's Graduate Programs in the Biological Sciences 2007

Developmental Biology

biology and genetics (PhD); neurobiology (PhD); pharmacology and toxicology (PhD); structural biology and biochemistry (PhD). *Degree requirements:* For doctorate, thesis/dissertation, written and oral qualifying exams, comprehensive exam. *Entrance requirements:* For doctorate, GRE General Test. Additional exam requirements/recommendations for international students: Required—TOEFL (minimum score 600 paper-based; 250 computer-based). Electronic applications accepted.

See Close-Up on page 189.

The Ohio State University, Graduate School, College of Biological Sciences, Department of Molecular Genetics, Columbus, OH 43210. Offers cell and developmental biology (MS, PhD); genetics (MS, PhD); molecular biology (MS, PhD). *Degree requirements:* For master's and doctorate, thesis/dissertation. *Entrance requirements:* For master's and doctorate, GRE General Test, GRE Subject Test. Additional exam requirements/recommendations for international students: Required—TOEFL (minimum score 573 paper-based; 230 computer-based). Electronic applications accepted.

See Close-Up on page 785.

The Ohio State University, Graduate School, College of Biological Sciences, Program in Molecular, Cellular and Developmental Biology, Columbus, OH 43210. Offers MS, PhD. Terminal master's awarded for partial completion of doctoral program. *Degree requirements:* For master's and doctorate, thesis/dissertation. *Entrance requirements:* For master's and doctorate, GRE General Test, GRE Subject Test. Additional exam requirements/recommendations for international students: Required—TOEFL (minimum score 600 paper-based; 250 computer-based). Electronic applications accepted. *Faculty research:* Cancer biology, cell biology, developmental biolgy, DNA replication, gene expression.

Oregon Health & Science University, School of Medicine, Graduate Programs in Medicine, Department of Cell and Developmental Biology, Portland, OR 97239-3098. Offers PhD, MD/PhD. *Degree requirements:* For doctorate, thesis/dissertation. *Entrance requirements:* For doctorate, GRE General Test, GRE Subject Test, MCAT. *Faculty research:* Developmental mechanisms, molecular biology of cancer, molecular neurobiology, intracellular signaling, growth factors and development.

The Pennsylvania State University University Park Campus, Graduate School, Eberly College of Science, Department of Biochemistry and Molecular Biology, Program in Cell and Developmental Biology, State College, University Park, PA 16802-1503. Offers MS, PhD. *Degree requirements:* For doctorate, thesis/dissertation, comprehensive exam. *Entrance requirements:* For doctorate, GRE General Test, GRE Subject Test. Application fee: $45. *Expenses:* Tuition, state resident: full-time $12,518; part-time $522 per credit. Tuition, nonresident: full-time $23,004; part-time $959 per credit. Required fees: $484. Tuition and fees vary according to course load, campus/location and program.

See Close-Up on page 583.

The Pennsylvania State University University Park Campus, Graduate School, Intercollege Graduate Programs, Intercollege Graduate Program in Integrative Biosciences, State College, University Park, PA 16802-1503. Offers integrative biosciences (PhD), including biomolecular transport dynamics, cell and developmental biology, cellular and molecular mechanisms of toxicity, chemical biology, ecological and molecular plant physiology, immunobiology, molecular medicine, neuroscience, nutrition science. *Students:* 110 full-time (58 women), 2 part-time (both women); includes 6 minority (3 African Americans, 3 Asian Americans or Pacific Islanders), 61 international. *Entrance requirements:* For master's and doctorate, GRE General Test. Application fee: $45. *Expenses:* Tuition, state resident: full-time $12,518; part-time $522 per credit. Tuition, nonresident: full-time $23,004; part-time $959 per credit. Required fees: $484. Tuition and fees vary according to course load, campus/location and program. *Financial support:* Fellowships available. *Unit head:* Dr. Richard J. Frisque, Co-Director, 814-863-3523, Fax: 814-863-1357, E-mail: rjf6@psu.edu.

See Close-Up on page 197.

Purdue University, Graduate School, School of Science, Department of Biological Sciences, West Lafayette, IN 47907. Offers biochemistry (PhD); biophysics (PhD); cell and developmental biology (PhD); ecology, evolutionary and population biology (MS, PhD), including ecology, evolutionary biology, population biology; genetics (MS, PhD); microbiology (MS, PhD); molecular biology (MS, PhD); neurobiology (MS, PhD); plant physiology (PhD). *Faculty:* 47 full-time (9 women), 4 part-time/adjunct (1 woman). *Students:* 97 full-time (53 women), 8 part-time (4 women); includes 13 minority (3 African Americans, 1 American Indian/Alaska Native, 4 Asian Americans or Pacific Islanders, 5 Hispanic Americans), 50 international. Average age 28. 168 applicants, 29% accepted, 23 enrolled. In 2005, 18 master's, 9 doctorates awarded. Terminal master's awarded for partial completion of doctoral program. *Degree requirements:* For master's, thesis (for some programs); for doctorate, thesis/dissertation, seminars, teaching experience. *Entrance requirements:* For master's and doctorate, GRE General Test. Additional exam requirements/recommendations for international students: Required—TOEFL, TSE. *Application deadline:* For fall admission, 2/15 for domestic students, 1/31 for international students. Applications are processed on a rolling basis. Application fee: $55. Electronic applications accepted. *Financial support:* In 2005–06, 15 fellowships, 60 research assistantships, 53 teaching assistantships were awarded. Support available to part-time students. Financial award application deadline: 2/15; financial award applicants required to submit FAFSA. *Unit head:* Dr. Richard J Kuhn, Head, 765-494-4407. *Application contact:* Nancy Konopka, Graduate Studies Office Manager, 765-494-8142, Fax: 765-494-0876, E-mail: njk@bilbo.bio.purdue.edu.

Rensselaer Polytechnic Institute, Graduate School, School of Science, Department of Biology, Troy, NY 12180-3590. Offers biochemistry (MS, PhD); biophysics (MS, PhD); cell biology (MS, PhD); developmental biology (MS, PhD); microbiology (MS, PhD); molecular biology (MS, PhD). Part-time programs available. Terminal master's awarded for partial completion of doctoral program. *Degree requirements:* For master's and doctorate, thesis/dissertation, comprehensive exam, registration. *Entrance requirements:* For master's and doctorate, GRE General Test. Additional exam requirements/recommendations for international students: Required—TOEFL. Electronic applications accepted. *Expenses:* Tuition: Full-time $31,000; part-time $1,320 per credit. Required fees: $1,623. *Faculty research:* Bioinformatics, molecular biology/biochemistry, cell and tissue biology, environment, ecology.

See Close-Up on page 199.

Rutgers, The State University of New Jersey, New Brunswick/Piscataway, Graduate School, Program in Cell and Developmental Biology, New Brunswick, NJ 08901-1281. Offers cell biology (MS, PhD); developmental biology (MS, PhD); immunology (MS). Part-time programs available. *Faculty:* 121 full-time. *Students:* 37 full-time (17 women), 9 part-time (8 women); includes 12 minority (2 African Americans, 1 American Indian/Alaska Native, 6 Asian Americans or Pacific Islanders, 3 Hispanic Americans), 20 international. Average age 31. 74 applicants, 36% accepted, 13 enrolled. In 2005, 8 master's, 7 doctorates awarded. Terminal master's awarded for partial completion of doctoral program. *Degree requirements:* For master's, thesis or alternative; for doctorate, thesis/dissertation, written qualifying exam. *Entrance requirements:* For master's, GRE General Test; for doctorate, GRE General Test, GRE Subject Test (recommended), minimum GPA of 3.0. Additional exam requirements/recommendations for international students: Required—TOEFL. *Application deadline:* For fall admission, 1/5 for domestic students. Applications are processed on a rolling basis. Application fee: $50. Electronic applications accepted. *Expenses:* Tuition, state resident: full-time $10,440; part-time $435 per credit. Tuition, nonresident: full-time $15,520; part-time $647 per credit. Required fees: $129 per credit. Tuition and fees vary according to program. *Financial support:* In 2005–06, 36 students received support, including 16 fellowships with full tuition reimbursements available (averaging $24,000 per year), 14 research assistantships with full tuition reimbursements available (averaging $21,500 per year), 6 teaching assistantships with full tuition reimbursements available (averaging $16,988 per year) Financial award application deadline: 1/15; financial award

applicants required to submit FAFSA. *Faculty research:* Signal transduction, developmental biology, cell biology, developmental genetics, developmental neurobiology. *Unit head:* Dr. Richard Padgett, Acting Director, 732-445-0251, Fax: 732-445-6370, E-mail: padgett@waksman.rutgers.edu. *Application contact:* Carolyn J. Ambrose, Administrative Assistant, 732-445-3430, Fax: 732-445-6370, E-mail: ambrose@biology.rutgers.edu.

Stanford University, School of Medicine, Graduate Programs in Medicine, Department of Developmental Biology, Stanford, CA 94305-9991. Offers PhD. *Degree requirements:* For doctorate, thesis/dissertation, qualifying examination. *Entrance requirements:* For doctorate, GRE General Test, GRE Subject Test. Additional exam requirements/recommendations for international students: Required—TOEFL. Electronic applications accepted. *Faculty research:* Mammalian embryology, developmental genetics with particular emphasis on microbial systems, *Dictyostelium, Drosophila,* the nematode, and the mouse.

State University of New York Upstate Medical University, College of Graduate Studies, Department of Cell and Developmental Biology, Syracuse, NY 13210-2334. Offers PhD, MD/PhD. *Degree requirements:* For doctorate, thesis/dissertation, comprehensive exam. *Entrance requirements:* For doctorate, GRE General Test, GRE Subject Test. Additional exam requirements/recommendations for international students: Required—TSE.

Stony Brook University, State University of New York, Graduate School, College of Arts and Sciences, Department of Biochemistry and Cell Biology, Stony Brook, NY 11794. Offers biochemistry and structural biology (PhD); molecular and cellular biology (MA, PhD), including biochemistry and molecular biology (PhD), biological sciences (MA), cellular and developmental biology (PhD), immunology and pathology (PhD), molecular and cellular biology (PhD). *Faculty:* 24 full-time (5 women). *Students:* 147 full-time (86 women), 2 part-time (1 woman); includes 21 minority (4 African Americans, 13 Asian Americans or Pacific Islanders, 4 Hispanic Americans), 85 international. Average age 28. 452 applicants, 13% accepted. In 2005, 10 master's, 12 doctorates awarded. *Degree requirements:* For doctorate, thesis/dissertation, teaching experience, comprehensive exam. *Entrance requirements:* For doctorate, GRE General Test, GRE Subject Test. Additional exam requirements/recommendations for international students: Required—TOEFL. *Application deadline:* For fall admission, 1/15 for domestic students. Application fee: $50. *Expenses:* Tuition, state resident: full-time $6,900; part-time $288 per credit. Tuition, nonresident: full-time $10,920; part-time $455 per credit. Required fees: $704. *Financial support:* In 2005–06, 5 fellowships, 81 research assistantships, 21 teaching assistantships were awarded; Federal Work-Study also available. *Faculty research:* Genome organization and replication, cell surface dynamics, enzyme structure and mechanism, developmental and regulatory biology. Total annual research expenditures: $9.4 million. *Unit head:* Dr. William J. Lennarz, Chair, 631-632-8560. *Application contact:* Director, Graduate Program, 631-632-8533, Fax: 631-632-9730, E-mail: mcbprog@life.bio.sunysb.edu.

Stony Brook University, State University of New York, Graduate School, College of Arts and Sciences, Department of Biochemistry and Cell Biology, Molecular and Cellular Biology Program, Specialization in Cellular and Developmental Biology, Stony Brook, NY 11794. Offers PhD. *Students:* 8 full-time (6 women), 5 international. Average age 30. In 2005, 2 degrees awarded. *Degree requirements:* For doctorate, one foreign language, thesis/dissertation, teaching experience, comprehensive exam. *Entrance requirements:* For doctorate, GRE General Test, GRE Subject Test. Additional exam requirements/recommendations for international students: Required—TOEFL. *Application deadline:* For fall admission, 1/15 for domestic students. Application fee: $50. *Expenses:* Tuition, state resident: full-time $6,900; part-time $288 per credit. Tuition, nonresident: full-time $10,920; part-time $455 per credit. Required fees: $704. *Financial support:* Fellowships, research assistantships, teaching assistantships available. *Unit head:* Director, Graduate Program, 631-632-8533, Fax: 631-632-9730, E-mail: mcbprog@life.bio.sunysb.edu. *Application contact:* Director, Graduate Program, 631-632-8533, Fax: 631-632-9730, E-mail: mcbprog@life.bio.sunysb.edu.

See Close-Up on page 787.

Thomas Jefferson University, Jefferson College of Graduate Studies, Graduate Program in Molecular Cell Biology, Philadelphia, PA 19107. Offers developmental biology and teratology (PhD); pathology and cell biology (PhD). *Faculty:* 79 full-time. *Students:* 22 full-time (12 women), 1 part-time; includes 2 minority (1 African American, 1 Asian American or Pacific Islander), 3 international. 34 applicants, 26% accepted, 3 enrolled. In 2005, 2 degrees awarded. *Degree requirements:* For doctorate, thesis/dissertation. *Entrance requirements:* For doctorate, GRE General Test, minimum GPA of 3.2. Additional exam requirements/recommendations for international students: Required—TOEFL (minimum score 213 computer-based). *Application deadline:* For fall admission, 3/1 priority date for domestic students, 1/1 priority date for international students. Applications are processed on a rolling basis. Application fee: $50. Electronic applications accepted. *Expenses:* Tuition: Full-time $14,894; part-time $800 per credit. *Financial support:* In 2005–06, 2 students received support, including 222 fellowships with full tuition reimbursements available; research assistantships, Federal Work-Study, institutionally sponsored loans, scholarships/grants, and traineeships also available. Support available to part-time students. Financial award application deadline: 5/1; financial award applicants required to submit FAFSA. *Unit head:* Dr. Theodore Taraschi, Program Director, 215-503-5020, Fax: 215-503-0206, E-mail: theodore.taraschi@jefferson.edu. *Application contact:* Jessie F. Pervall, Director of Admissions, 215-503-0155, Fax: 215-503-9920, E-mail: jessie.pervall@jefferson.edu.

Thomas Jefferson University, Jefferson College of Graduate Studies, Program in Developmental Biology and Teratology, Philadelphia, PA 19107. Offers MS. Part-time and evening/weekend programs available. *Students:* 7 applicants, 100% accepted, 4 enrolled. In 2005, 5 degrees awarded. *Entrance requirements:* For master's, GRE General Test, minimum GPA of 3.0. Additional exam requirements/recommendations for international students: Required—TOEFL (minimum score 213 computer-based). *Application deadline:* For fall admission, 8/1 priority date for domestic students, 3/1 priority date for international students. For winter admission, 12/1 for domestic students; for spring admission, 4/1 for domestic students. Applications are processed on a rolling basis. Application fee: $50. Electronic applications accepted. *Expenses:* Tuition: Full-time $14,894; part-time $800 per credit. *Financial support:* In 2005–06, 2 students received support. Applicants required to submit FAFSA. *Unit head:* Dr. Gerald B. Grunwald, Associate Dean, 215-503-4191, Fax: 215-923-3808, E-mail: gerald.grunwald@jefferson.edu. *Application contact:* Jessie F. Pervall, Director of Admissions, 215-503-0155, Fax: 215-503-9920, E-mail: jessie.pervall@jefferson.edu.

Tufts University, Sackler School of Graduate Biomedical Sciences, Program in Cell, Molecular and Developmental Biology, Boston, MA 02155. Offers PhD. Applications are processed through through integrated studies program. *Faculty:* 31 full-time (9 women). *Students:* 28 full-time (17 women), 8 international. Average age 29. In 2005, 3 degrees awarded. *Degree requirements:* For doctorate, thesis/dissertation. *Financial support:* In 2005–06, 28 students received support, including 28 research assistantships with full tuition reimbursements available (averaging $29,000 per year); fellowships, scholarships/grants, health care benefits, and tuition waivers (full) also available. Financial award application deadline: 1/15. *Faculty research:* Reproduction and hormone action, control of gene expression, cell-matrix and cell-cell interactions, growth control and tumorigenesis, cytoskeleton and contractile proteins. *Unit head:* Dr. John Castellot, Program Director, 617-636-0303, Fax: 617-636-0375, E-mail: john.castellot@tufts.edu. *Application contact:* 617-636-6767, Fax: 617-636-0375, E-mail: sackler-school@tufts.edu.

University at Albany, State University of New York, College of Arts and Sciences, Department of Biological Sciences, Specialization in Molecular, Cellular, Developmental, and Neural Biology, Albany, NY 12222-0001. Offers MS, PhD. *Degree requirements:* For master's, one foreign language; for doctorate, one foreign language, thesis/dissertation. *Entrance requirements:* For master's and doctorate, GRE General Test. Application fee: $60. *Financial support:* Minority assistantships available. *Unit head:* Dr. Albert Millis, Chair, Department of Biological Sciences, 518-442-4300.

Developmental Biology

University of California, Davis, Graduate Studies, Graduate Group in Cell and Developmental Biology, Davis, CA 95616. Offers MS, PhD. *Faculty:* 56 full-time. *Students:* 47 full-time (20 women); includes 13 minority (10 Asian Americans or Pacific Islanders, 3 Hispanic Americans), 9 international. Average age 28. 77 applicants, 42% accepted, 11 enrolled. In 2005, 3 master's, 5 doctorates awarded. *Median time to degree:* Of those who began their doctoral program in fall 1997, 83.3% received their degree in 8 years or less. *Degree requirements:* For master's, thesis (for some programs), comprehensive exam (for some programs); for doctorate, thesis/dissertation. *Entrance requirements:* For doctorate, GRE General Test, GRE Subject Test. Additional exam requirements/recommendations for international students: Required—TOEFL (minimum score 550 paper-based; 213 computer-based). *Application deadline:* For fall admission, 1/15 for domestic students, 1/15 for international students. Application fee: $60. Electronic applications accepted. *Financial support:* In 2005–06, 46 students received support, including 8 fellowships with full and partial tuition reimbursements available (averaging $12,769 per year), 27 research assistantships with full and partial tuition reimbursements available (averaging $16,276 per year), 6 teaching assistantships with partial tuition reimbursements available (averaging $15,082 per year); Federal Work-Study, institutionally sponsored loans, scholarships/grants, tuition waivers (full and partial), and unspecified assistantships also available. Financial award application deadline: 1/15; financial award applicants required to submit FAFSA. *Faculty research:* Molecular basis of cell function and development. *Unit head:* Richard Tucker, Chair, 530-752-0238, E-mail: rptucker@ucdavis.edu. *Application contact:* Angelina Kuo, Graduate Staff, 530-752-2981, Fax: 530-752-8391, E-mail: abkuo@ucdavis.edu.

University of California, Irvine, Office of Graduate Studies, School of Biological Sciences, Department of Developmental and Cell Biology, Irvine, CA 92697. Offers biological sciences (MS, PhD). Students apply through the Graduate Program in Molecular Biology, Genetics, and Biochemistry. *Degree requirements:* For doctorate, thesis/dissertation. *Entrance requirements:* For master's and doctorate, GRE General Test, GRE Subject Test, minimum GPA of 3.0. Additional exam requirements/recommendations for international students: Required—TOEFL (minimum score 550 paper-based; 213 computer-based), TSE. Electronic applications accepted. *Faculty research:* Genetics and development, oncogene signaling pathways, gene regulation, tissue regeneration and molecular genetics.

University of California, Los Angeles, School of Medicine and Graduate Division, Graduate Programs in Medicine, Department of Molecular, Cell and Developmental Biology, Los Angeles, CA 90095. Offers MA, PhD. *Degree requirements:* For doctorate, thesis/dissertation, qualifying exams. *Entrance requirements:* For doctorate, GRE General Test, GRE Subject Test. Additional exam requirements/recommendations for international students: Required—TOEFL.

University of California, Riverside, Graduate Division, Program in Cell, Molecular, and Developmental Biology, Riverside, CA 92521-0102. Offers MS, PhD. *Faculty:* 55 full-time (18 women). *Students:* 57 full-time (29 women); includes 20 minority (3 African Americans, 15 Asian Americans or Pacific Islanders, 2 Hispanic Americans), 18 international. Average age 28. In 2005, 3 master's, 1 doctorate awarded. *Median time to degree:* Of those who began their doctoral program in fall 1997, 100% received their degree in 8 years or less. *Degree requirements:* For master's, thesis, oral defense of thesis; for doctorate, thesis/dissertation, oral defense of thesis, qualifying exams, 2 quarters of teaching experience. *Entrance requirements:* For master's and doctorate, GRE General Test, minimum GPA of 3.2. Additional exam requirements/recommendations for international students: Required—TOEFL (minimum score 550 paper-based; 213 computer-based); Recommended—TSE. *Application deadline:* For fall admission, 5/1 for domestic students, 2/1 for international students. For winter admission, 9/1 for domestic students; for spring admission, 12/1 for domestic students. Applications are processed on a rolling basis. Application fee: $60 ($75 for international students). Electronic applications accepted. *Expenses:* Tuition, nonresident: full-time $14,694. Required fees:$9,009. Full-time tuition and fees vary according to program. *Financial support:* In 2005–06, fellowships (averaging $18,000 per year), research assistantships (averaging $14,000 per year), teaching assistantships (averaging $15,000 per year) were awarded. Financial award application deadline: 1/5. *Unit head:* Dr. Anthony Norman, Director, 951-827-4777, E-mail: norman@ucr.edu. *Application contact:* Kathy Redd, Graduate Program Assistant, 800-735-0717, Fax: 951-827-5517, E-mail: cell@ucr.edu.

University of California, San Diego, Graduate Studies and Research, Division of Biology, Program in Cell and Developmental Biology, La Jolla, CA 92093-0348. Offers PhD. Offered in association with the Salk Institute. *Degree requirements:* For doctorate, thesis/dissertation, qualifying exam. Electronic applications accepted.

University of California, San Diego, School of Medicine and Graduate Studies and Research, Molecular Pathology Program, La Jolla, CA 92093. Offers bioinformatics (PhD); cancer biology/oncology (PhD); cardiovascular sciences and disease (PhD); microbiology (PhD); molecular pathology (PhD); neurological disease (PhD); stem cell and developmental biology (PhD); structural biology/drug design (PhD). *Entrance requirements:* For doctorate, GRE General Test, GRE Subject Test. Additional exam requirements/recommendations for international students: Required—TOEFL. Electronic applications accepted.

See Close-Up on page 1119.

University of California, San Francisco, Graduate Division and School of Medicine, Department of Biochemistry and Biophysics, San Francisco, CA 94143. Offers biochemistry and molecular biology (PhD); cell biology (PhD); developmental biology (PhD); genetics (PhD). *Degree requirements:* For doctorate, thesis/dissertation. *Entrance requirements:* For doctorate, GRE General Test, GRE Subject Test. Additional exam requirements/recommendations for international students: Required—TOEFL. Expenses: Contact institution.

University of California, Santa Barbara, Graduate Division, College of Letters and Sciences, Division of Mathematics, Life, and Physical Sciences, Department of Molecular, Cellular, and Developmental Biology, Santa Barbara, CA 93106. Offers MA, PhD, MA/PhD. *Faculty:* 62 full-time (30 women), 1 part-time/adjunct (0 women). *Students:* 63 full-time (30 women); includes 10 minority (1 African American, 7 Asian Americans or Pacific Islanders, 2 Hispanic Americans), 4 international. Average age 26. 99 applicants, 31% accepted, 18 enrolled. Terminal master's awarded for partial completion of doctoral program. *Degree requirements:* For master's, thesis (for some programs), comprehensive exam (for some programs), registration; for doctorate, thesis/dissertation, comprehensive exam, registration. *Entrance requirements:* For master's and doctorate, GRE General Test, GRE Subject Test. Additional exam requirements/recommendations for international students: Required—TOEFL (minimum score 630 paper-based; 213 computer-based). *Application deadline:* For fall admission, 12/15 for domestic students, 12/15 for international students. Application fee: $60. Electronic applications accepted. *Financial support:* In 2005–06, 51 teaching assistantships were awarded; fellowships with full tuition reimbursements, research assistantships with full tuition reimbursements, career-related internships or fieldwork, Federal Work-Study, scholarships/grants, and health care benefits also available. Financial award application deadline: 12/15; financial award applicants required to submit FAFSA. *Faculty research:* Signal transduction, bacteria pathegenesis, virology, stem cell, neurodegenerative disease. *Unit head:* Dr. Dennis Clegg, Chair, 805-893-8490, E-mail: clegg@lifesci.ucsb.edu. *Application contact:* Krista Grace, Staff Graduate Program Advisor, 805-893-2290, Fax: 805-893-4724, E-mail: grace@lifesci.ucsb.edu.

University of Chicago, Division of the Biological Sciences, Department of Molecular Biosciences: Biochemistry, Genetics, Cell and Developmental Biology, Committee on Developmental Biology, Chicago, IL 60637-1513. Offers cellular differentiation (PhD); developmental endocrinology (PhD); developmental genetics (PhD); developmental neurobiology (PhD); gene expression (PhD). *Faculty:* 23 full-time (11 women), 11 part-time/adjunct (0 women). *Students:* 17 full-time (11 women); includes 6 minority (3 African Americans, 2 Asian Americans or Pacific Islanders, 1 Hispanic American), 2 international. Average age 29. In 2005, 3 degrees awarded. *Degree requirements:* For doctorate, thesis/dissertation, registration. *Entrance requirements:* For doctorate, GRE General Test. Additional exam requirements/

recommendations for international students: Required—TOEFL. *Application deadline:* For fall admission, 12/28 priority date for domestic students, 12/28 priority date for international students. Application fee: $55. Electronic applications accepted. *Financial support:* In 2005–06, fellowships with tuition reimbursements (averaging $26,301 per year), research assistantships with tuition reimbursements (averaging $26,301 per year) were awarded; institutionally sponsored loans, scholarships/grants, traineeships, and health care benefits also available. Financial award applicants required to submit FAFSA. *Faculty research:* Epidermal differentiation, neural lineages, pattern formation. *Unit head:* Dr. Victoria Prince, Chair, 773-834-2100, E-mail: vprince@uchicago.edu. *Application contact:* Kristine Gaston, Graduate Administrative Director, 773-702-8037, Fax: 773-702-3172, E-mail: kristine@cummings.uchicago.edu.

University of Cincinnati, Division of Research and Advanced Studies, College of Medicine, Graduate Programs in Biomedical Sciences, Department of Pediatrics Developmental Biology, Program in Molecular and Developmental Biology, Cincinnati, OH 45221. Offers PhD. *Degree requirements:* For doctorate, thesis/dissertation, qualifying exam. *Entrance requirements:* For doctorate, GRE General Test, minimum GPA of 3.2. Additional exam requirements/recommendations for international students: Required—TOEFL (minimum score 520 paper-based; 190 computer-based). Electronic applications accepted. *Faculty research:* Cancer biology, cardiovascular biology, developmental biology, human genetics, gene therapy, genomics and bioinformatics, immunobiology, molecular medicine, neuroscience, pulmonary biology, reproductive biology, stem cell biology.

See Close-Up on page 793.

University of Colorado at Boulder, Graduate School, College of Arts and Sciences, Department of Molecular, Cellular, and Developmental Biology, Boulder, CO 80309. Offers cellular structure and function (MA, PhD); developmental biology (MA, PhD); molecular biology (MA, PhD). *Faculty:* 25 full-time (6 women). *Students:* 42 full-time (15 women), 26 part-time (12 women); includes 5 minority (1 American Indian/Alaska Native, 1 Asian American or Pacific Islander, 3 Hispanic Americans), 11 international. Average age 28. 19 applicants, 100% accepted. In 2005, 3 master's, 8 doctorates awarded. Terminal master's awarded for partial completion of doctoral program. *Degree requirements:* For master's, thesis or alternative, comprehensive exam; for doctorate, thesis/dissertation, comprehensive exam. *Entrance requirements:* For master's, GRE General Test, GRE Subject Test, minimum undergraduate GPA of 2.75; for doctorate, GRE General Test, GRE Subject Test. *Application deadline:* For fall admission, 1/15 for domestic students, 1/15 for international students. Application fee: $50 ($60 for international students). *Financial support:* In 2005–06, fellowships (averaging $7,489 per year), research assistantships (averaging $12,470 per year), teaching assistantships (averaging $13,078 per year) were awarded; tuition waivers (full) also available. Financial award application deadline: 3/1. *Faculty research:* Molecular biology of RNA and DNA, molecular genetics, cell motility and cytoskeleton, cell membranes, developmental genetics. Total annual research expenditures: $14.7 million. *Unit head:* Leslie Leinwand, Chair, 303-492-7606, Fax: 303-492-7744, E-mail: leslie.leinwand@stripe.colorado.edu. *Application contact:* Student Affairs Office, 303-492-7230, Fax: 303-492-7744, E-mail: mcdbgradinfo@beagle.colorado.edu.

University of Colorado at Denver and Health Sciences Center, Graduate School, Program in Biomedical Sciences, Program in Cell and Developmental Biology, Denver, CO 80262. Offers PhD. In 2005, 3 degrees awarded. *Degree requirements:* For doctorate, thesis/dissertation. *Entrance requirements:* For doctorate, GRE, minimum GPA of 3.0, 3 letters of reference. Additional exam requirements/recommendations for international students: Required—TOEFL (minimum score 550 paper-based; 213 computer-based). *Application deadline:* For fall admission, 1/15 for domestic students. Application fee: $50. Electronic applications accepted. *Expenses:* Tuition, state resident: full-time $11,730. Tuition, nonresident: full-time $22,980. Tuition and fees vary according to degree level and program. *Financial support:* Fellowships, research assistantships, teaching assistantships, Federal Work-Study and institutionally sponsored loans available. Support available to part-time students. Financial award application deadline: 3/15; financial award applicants required to submit FAFSA. *Faculty research:* Human disease, stem cell biology, neuroscience, molecular biology. Total annual research expenditures: $4.6 million. *Unit head:* Dr. Karl Pfenninger, Chair, 303-724-3424, E-mail: karl.pfenninger@uchsc.edu. *Application contact:* Carmel Hardberg, Program Assistant, 303-315-7009, Fax: 303-724-3420, E-mail: cdb@uchsc.edu.

University of Connecticut, Graduate School, College of Liberal Arts and Sciences, Department of Molecular and Cell Biology, Storrs, CT 06269. Offers applied genomics (MS, PSM); biobehavioral science (PhD); biochemistry (MS, PhD); biophysics and structural biology (MS, PhD); biotechnology (MS); cell and developmental biology (MS, PhD); genetics, genomics, and bioinformatics (MS), including genetics (MS, PhD); genetics, genomics, and bioinformation (PhD), including genetics (MS, PhD); microbial systems analysis (MS, PSM); microbiology (MS, PhD); plant cell and molecular biology (MS, PhD). *Faculty:* 64 full-time (13 women). *Students:* 124 full-time (58 women), 15 part-time (8 women); includes 17 minority (5 African Americans, 1 American Indian/Alaska Native, 9 Asian Americans or Pacific Islanders, 2 Hispanic Americans), 36 international. Average age 27. 285 applicants, 28% accepted, 55 enrolled. In 2005, 23 master's, 10 doctorates awarded. Terminal master's awarded for partial completion of doctoral program. *Degree requirements:* For master's, comprehensive exam; for doctorate, thesis/dissertation. *Entrance requirements:* For master's and doctorate, GRE General Test, GRE Subject Test. Additional exam requirements/recommendations for international students: Required—TOEFL (minimum score 550 paper-based; 213 computer-based). *Application deadline:* For fall admission, 2/1 priority date for domestic students, 2/1 priority date for international students; for spring admission, 11/1 for domestic students, 10/1 for international students. Applications are processed on a rolling basis. Application fee: $55. Electronic applications accepted. *Expenses:* Tuition, state resident: part-time $444 per credit hour. Tuition, nonresident: part-time $1,154 per credit hour. Tuition and fees vary according to course load. *Financial support:* In 2005–06, 35 research assistantships with full tuition reimbursements, 59 teaching assistantships with full tuition reimbursements were awarded; fellowships, Federal Work-Study, scholarships/grants, health care benefits, and unspecified assistantships also available. Financial award application deadline: 2/1; financial award applicants required to submit FAFSA. *Unit head:* Philip L. Yeagle, Head, 860-486-4329, Fax: 860-486-4331, E-mail: yeagle@uconnvm.uconn.edu. *Application contact:* Anne St. Onje, Graduate Coordinator, 860-486-4314, Fax: 860-486-3943, E-mail: ann.st_onje@uconn.edu.

University of Connecticut Health Center, Graduate School, Programs in Biomedical Sciences, Program in Genetics and Developmental Biology, Farmington, CT 06030. Offers PhD, DMD/PhD, MD/PhD. *Degree requirements:* For doctorate, thesis/dissertation, comprehensive exam, registration. *Entrance requirements:* For doctorate, GRE General Test, GRE Subject Test. Additional exam requirements/recommendations for international students: Required—TOEFL (minimum score 600 paper-based; 250 computer-based). Electronic applications accepted. *Faculty research:* Limb development/bone formation, genetics and human disease, apoptosis and development, signal transduction in development.

See Close-Up on page 795.

University of Delaware, College of Arts and Sciences, Department of Biological Sciences, Newark, DE 19716. Offers biotechnology (MS); cancer biology (MS, PhD); cell and extracellular matrix biology (MS, PhD); cell and systems physiology (MS, PhD); developmental biology (MS, PhD); ecology and evolution (MS, PhD); microbiology (MS, PhD); molecular biology and genetics (MS, PhD). *Faculty:* 39 full-time (11 women). *Students:* 64 full-time (46 women), 2 part-time; includes 7 minority (4 African Americans, 2 Asian Americans or Pacific Islanders, 1 Hispanic American), 18 international. Average age 26. 113 applicants, 31% accepted, 19 enrolled. In 2005, 5 master's, 4 doctorates awarded. Terminal master's awarded for partial completion of doctoral program. *Median time to degree:* Of those who began their doctoral program in fall 1997, 100% received their degree in 8 years or less. *Degree requirements:* For master's, thesis/dissertation, preliminary exam; for doctorate, thesis/dissertation, preliminary exam, comprehensive exam. *Entrance requirements:* For master's and doctorate, GRE General Test. Additional exam requirements/recommendations for inter-

736 *www.petersons.com* *Peterson's Graduate Programs in the Biological Sciences 2007*

Developmental Biology

national students: Required—TOEFL (minimum score 600 paper-based; 250 computer-based); Recommended—TWE, TSE. *Application deadline:* For fall admission, 4/15 for domestic students, 1/15 for international students; for spring admission, 10/1 for domestic students. Applications are processed on a rolling basis. Application fee: $60. Electronic applications accepted. *Financial support:* In 2005–06, 26 students received support, including fellowships with full tuition reimbursements available (averaging $19,000 per year), 19 research assistantships with full tuition reimbursements available (averaging $19,000 per year), 26 teaching assistantships with full tuition reimbursements available (averaging $19,000 per year); tuition waivers (partial) also available. Financial award application deadline: 4/15. *Faculty research:* Microorganisms, bone, cancer metastasis, developmental biology, cell biology, DNA. Total annual research expenditures: $8.3 million. *Unit head:* Dr. Daniel D. Carson, Chair, 302-831-6977, Fax: 302-831-2281, E-mail: dcarson@udel.edu. *Application contact:* Dr. Melinda K. Duncan, Graduate Coordinator, 302-831-1841, Fax: 302-831-2281, E-mail: danders@udel.edu.

University of Illinois at Chicago, Graduate College, College of Liberal Arts and Sciences, Department of Biological Sciences, Chicago, IL 60607-7128. Offers cell and developmental biology (PhD); ecology and evolution (MS, DA, PhD); genetics and development (PhD); molecular biology (MS, PhD); neurobiology (MS, PhD); plant biology (MS, DA, PhD). *Degree requirements:* For master's, thesis; for doctorate, thesis/dissertation, preliminary exam. *Entrance requirements:* For master's and doctorate, GRE General Test, GRE Subject Test, previous course work in physics, calculus, and organic chemistry; minimum GPA of 2.75. Additional exam requirements/recommendations for international students: Required—TOEFL. Electronic applications accepted.

University of Illinois at Urbana–Champaign, Graduate College, College of Liberal Arts and Sciences, School of Molecular and Cellular Biology, Department of Cell and Developmental Biology, Champaign, IL 61820. Offers PhD. *Faculty:* 13 full-time (2 women). *Students:* 75 full-time (38 women), 1 (woman) part-time; includes 12 minority (2 African Americans, 1 American Indian/Alaska Native, 7 Asian Americans or Pacific Islanders, 2 Hispanic Americans), 41 international. 110 applicants, 6% accepted, 7 enrolled. In 2005, 7 degrees awarded. *Degree requirements:* For doctorate, thesis/dissertation. *Application deadline:* Applications are processed on a rolling basis. Application fee: $50 ($60 for international students). Electronic applications accepted. *Financial support:* In 2005–06, 12 fellowships, 59 research assistantships, 39 teaching assistantships were awarded. Financial award application deadline: 2/15. *Unit head:* Dr. Martha U. Gillette, Head, 217-333-6118, Fax: 217-244-1648, E-mail: mgillett@uiuc.edu. *Application contact:* Audra Weinstein, Staff Secretary, 217-333-6118, Fax: 217-244-1648, E-mail: audra@uiuc.edu.

University of Kansas, Graduate School, College of Liberal Arts and Sciences, Division of Biological Sciences, Department of Molecular Biosciences, Lawrence, KS 66045. Offers biochemistry and biophysics (MA, PhD); microbiology (MA, PhD); molecular, cellular, and developmental biology (MA, PhD). *Faculty:* 44. *Students:* 48 full-time (28 women), 5 part-time (3 women); includes 1 minority (Asian American or Pacific Islander), 26 international. Average age 27. 111 applicants, 19% accepted. In 2005, 5 master's, 5 doctorates awarded. Terminal master's awarded for partial completion of doctoral program. *Degree requirements:* For master's and doctorate, thesis/dissertation, comprehensive exam. *Entrance requirements:* For master's and doctorate, GRE General Test. Additional exam requirements/recommendations for international students: Required—TOEFL, IELT, TOEFL (paper 570; computer 230) or IELT; Recommended—TWE, TSE. *Application deadline:* For fall admission, 1/15 priority date for domestic students, 1/15 priority date for international students. Applications are processed on a rolling basis. Application fee: $55 ($60 for international students). Electronic applications accepted. *Expenses:* Tuition, state resident: full-time $4,859. Tuition, nonresident: full-time $12,000. Required fees: $589. Tuition and fees vary according to program. *Financial support:* Fellowships, research assistantships with tuition reimbursements, teaching assistantships with tuition reimbursements available. Financial award application deadline: 3/1. *Faculty research:* Structure and function of proteins, genetics of organism development, molecular genetics, neurophysiology, molecular virology and pathogenics, developmental biology, cell biology. *Unit head:* Kathy Suprenant, Chair, 785-864-4631, Fax: 785-864-5294, E-mail: ksupre@ku.edu. *Application contact:* John P. Connolly, Information Contact, 785-864-4311, Fax: 785-864-5924, E-mail: jconnolly@ku.edu.

University of Massachusetts Amherst, Graduate School, College of Natural Sciences and Mathematics, Program in Molecular and Cellular Biology, Amherst, MA 01003. Offers biological chemistry (PhD); cell and developmental biology (PhD). Part-time programs available. *Faculty:* 1 full-time (0 women). *Students:* 87 full-time (43 women), 1 part-time; includes 9 minority (1 African American, 4 Asian Americans or Pacific Islanders, 4 Hispanic Americans), 41 international. Average age 30. 219 applicants, 23% accepted, 13 enrolled. In 2005, 8 doctorates awarded. *Degree requirements:* For doctorate, one foreign language, thesis/dissertation. *Entrance requirements:* For doctorate, GRE General Test. Additional exam requirements/recommendations for international students: Required—TOEFL (minimum score 530 paper-based; 197 computer-based). *Application deadline:* For fall admission, 1/15 priority date for domestic students, 1/15 priority date for international students. Applications are processed on a rolling basis. Application fee: $40 ($65 for international students). Electronic applications accepted. *Expenses:* Tuition, state resident: part-time $110 per credit. Tuition, nonresident: part-time $414 per credit. Required fees: $2,824 per term. One-time fee: $250 part-time. Full-time tuition and fees vary according to course load, campus/location, program and reciprocity agreements. *Financial support:* In 2005–06, 6 fellowships with full tuition reimbursements, 22 research assistantships with full tuition reimbursements (averaging $3,451 per year), 4 teaching assistantships with full tuition reimbursements (averaging $8,696 per year) were awarded; career-related internships or fieldwork, Federal Work-Study, scholarships/grants, traineeships, and unspecified assistantships also available. Support available to part-time students. Financial award application deadline: 1/15. *Unit head:* Dr. Rodney K. Murphey, Head, 413-545-3246, Fax: 413-545-1812.

See Close-Up on page 619.

University of Miami, Graduate School, Miller School of Medicine, Graduate Programs in Medicine, Department of Cell Biology and Anatomy, Coral Gables, FL 33124. Offers molecular cell and developmental biology (PhD). *Faculty:* 20 full-time (10 women). *Students:* 58 full-time (10 women); includes 4 minority (1 Asian American or Pacific Islander, 3 Hispanic Americans), 13 international. Average age 29. 57 applicants, 7% accepted, 4 enrolled. In 2005, 1 degree awarded. *Degree requirements:* For doctorate, thesis/dissertation. *Entrance requirements:* For doctorate, GRE General Test, GRE Subject Test. Additional exam requirements/recommendations for international students: Required—TOEFL. *Application deadline:* For fall admission, 3/1 for domestic students. Applications are processed on a rolling basis. Application fee: $50. Electronic applications accepted. *Financial support:* In 2005–06, 22 fellowships with tuition reimbursements (averaging $22,000 per year) were awarded. *Unit head:* Dr. Robert Warren, Interim Chair, 305-243-6691, Fax: 305-545-7166, E-mail: rwarren@med.miami.edu. *Application contact:* Dr. Richard Rotundo, Professor, 305-243-6691, Fax: 305-545-7166.

University of Michigan, Horace H. Rackham School of Graduate Studies, College of Literature, Science, and the Arts, Department of Molecular, Cellular, and Developmental Biology, Ann Arbor, MI 48109. Offers MS, PhD. Part-time programs available. *Faculty:* 24 full-time (6 women), 1 part-time/adjunct (0 women). *Students:* 72 full-time, 9 part-time; includes 5 minority (2 African Americans, 1 Asian American or Pacific Islander, 2 Hispanic Americans), 49 international. Average age 24. 102 applicants, 29% accepted, 18 enrolled. In 2005, 5 master's, 7 doctorates awarded. Terminal master's awarded for partial completion of doctoral program. *Median time to degree:* Of those who began their doctoral program in fall 1997, 60% received their degree in 8 years or less. *Degree requirements:* For master's, registration; for doctorate, thesis/dissertation, preliminary exam, oral defense, dissertation. *Entrance requirements:* For master's and doctorate, GRE General Test. Additional exam requirements/recommendations for international students: Required—TOEFL (minimum score 560 paper-based; 220 computer-based). *Application deadline:* For fall admission, 1/5 for domestic students, 1/5 for international

students. For winter admission, 11/1 for domestic students. Applications are processed on a rolling basis. Application fee: $60 ($75 for international students). Electronic applications accepted. *Expenses:* Tuition, state resident: full-time $14,082; part-time $894 per credit hour. Tuition, nonresident: full-time $28,500; part-time $1,675 per credit hour. Required fees: $189; $189 per unit. *Financial support:* In 2005–06, 66 students received support, including 18 fellowships with full tuition reimbursements available (averaging $14,326 per year), 24 research assistantships with full tuition reimbursements available (averaging $14,326 per year), 21 teaching assistantships with full tuition reimbursements available (averaging $14,326 per year); career-related internships or fieldwork, scholarships/grants, traineeships, health care benefits, and unspecified assistantships also available. *Faculty research:* Biochemistry, cell biology, microbiology, neurobiology and physiology, developmental biology and plant molecular biology. Total annual research expenditures: $5.5 million. *Unit head:* Dr. Richard Hume, Chair, 734-764-7427, Fax: 734-746-0884. *Application contact:* Mary Carr, Graduate Coordinator, 734-615-1635, Fax: 734-764-0884, E-mail: carrmm@umich.edu.

See Close-Up on page 625.

University of Minnesota, Twin Cities Campus, Graduate School, Program in Molecular, Cellular, Developmental Biology and Genetics, Minneapolis, MN 55455-0213. Offers genetic counseling (MS); molecular, cellular, developmental biology and genetics (PhD). Part-time programs available. Terminal master's awarded for partial completion of doctoral program. *Degree requirements:* For master's, thesis optional; for doctorate, thesis/dissertation. *Entrance requirements:* For master's and doctorate, GRE General Test. Additional exam requirements/recommendations for international students: Required—TOEFL. *Expenses:* Tuition, state resident: full-time $8,748; part-time $729 per credit. Tuition, nonresident: full-time $15,848; part-time $1,321 per credit. Full-time tuition and fees vary according to class time, course load, program and reciprocity agreements. *Faculty research:* Membrane receptors and membrane transport, cell interactions, cytoskeleton and cell mobility, regulation of gene expression, plant cell and molecular biology.

University of Missouri–St. Louis, College of Arts and Sciences, Department of Biology, St. Louis, MO 63121. Offers biology (MS, PhD), including animal behavior (MS), biochemistry (MS), biotechnology (MS), conservation biology (MS), development (MS), ecology (MS), environmental studies (PhD), evolution (MS), genetics (MS), molecular/cellular biology (MS), physiology (MS), plant systematics, population biology (MS), tropical biology (MS); biotechnology (Certificate); tropical biology and conservation (Certificate). Part-time programs available. *Faculty:* 50. *Students:* 26 full-time (15 women), 101 part-time (51 women); includes 13 minority (5 African Americans, 6 Asian Americans or Pacific Islanders, 2 Hispanic Americans), 41 international. Average age 32. In 2005, 22 master's, 2 doctorates awarded. *Degree requirements:* For master's, thesis or alternative; for doctorate, one foreign language, thesis/dissertation, 1 semester of teaching experience. *Entrance requirements:* For doctorate, GRE General Test. *Application deadline:* For spring admission, 12/1 priority date for domestic students. Applications are processed on a rolling basis. Application fee: $35 ($40 for international students). Electronic applications accepted. *Expenses:* Tuition, state resident: part-time $263 per credit hour. Tuition, nonresident: part-time $680 per credit hour. Required fees: $53 per credit hour. Tuition and fees vary according to program. *Financial support:* In 2005–06, 11 fellowships with full tuition reimbursements (averaging $30,000 per year), 15 research assistantships with full and partial tuition reimbursements (averaging $16,000 per year), 22 teaching assistantships with full and partial tuition reimbursements (averaging $16,000 per year) were awarded; career-related internships or fieldwork and Federal Work-Study also available. Support available to part-time students. Financial award application deadline: 2/1. *Faculty research:* Molecular biology, microbial genetics. *Unit head:* Zuleyma Tang-Martinez, Director of Graduate Studies, 314-516-6498, Fax: 314-516-6233, E-mail: zuleyma@umsl.edu. *Application contact:* 314-516-5458, Fax: 314-516-5310, E-mail: gradadm@umsl.edu.

The University of North Carolina at Chapel Hill, Graduate School, College of Arts and Sciences, Department of Biology, Chapel Hill, NC 27599. Offers botany (MA, MS, PhD); cell biology, development, and physiology (MA, MS, PhD); cell motility and cytoskeleton (PhD); ecology and behavior (MA, MS, PhD); genetics and molecular biology (MA, MS, PhD); morphology, systematics, and evolution (MA, MS, PhD). Terminal master's awarded for partial completion of doctoral program. *Degree requirements:* For master's, thesis (for some programs), comprehensive exam; for doctorate, thesis/dissertation, comprehensive exam. *Entrance requirements:* For master's, GRE General Test, GRE Subject Test, 2 semesters of calculus or statistics, 2 semesters of physics, organic chemistry, 3 semesters of biology; for doctorate, GRE General Test, GRE Subject Test, 2 semesters calculus or statistics, 2 semesters physics, organic chemistry, 3 semesters of biology. Additional exam requirements/recommendations for international students: Required—TOEFL (minimum score 550 paper-based; 213 computer-based). Electronic applications accepted. *Faculty research:* Gene expression, biomechanics, yeast genetics, plant ecology, plant molecular biology.

See Close-Up on page 277.

The University of North Carolina at Chapel Hill, School of Medicine and Graduate School, Graduate Programs in Medicine, Department of Cell and Developmental Biology, Chapel Hill, NC 27599. Offers PhD. *Faculty:* 22 full-time (6 women), 1 part-time/adjunct (0 women). *Students:* 39 full-time (20 women); includes 3 minority (1 African American, 2 Asian Americans or Pacific Islanders), 9 international. Average age 24. 53 applicants, 13% accepted, 4 enrolled. In 2005, 1 degree awarded. *Degree requirements:* For doctorate, thesis/dissertation, comprehensive exam, registration. *Entrance requirements:* For doctorate, GRE General Test, GRE Subject Test. *Application deadline:* For fall admission, 1/1 priority date for domestic students, 1/1 priority date for international students. Applications are processed on a rolling basis. Application fee: $70. Electronic applications accepted. *Financial support:* In 2005–06, 4 students received support, including 35 research assistantships with full tuition reimbursements available (averaging $22,000 per year), 4 teaching assistantships with full tuition reimbursements available (averaging $22,000 per year); fellowships, tuition waivers (full) and unspecified assistantships also available. Financial award application deadline: 2/1; financial award applicants required to submit FAFSA. *Faculty research:* Cell adhesion, motility and cytoskeleton; molecular analysis of signal transduction; development biology and toxicology; reproductive biology; cell and molecular imaging. Total annual research expenditures: $8 million. *Unit head:* Dr. Vytas A. Bankaitis, Chair, 919-966-3026, Fax: 919-966-1856, E-mail: vytas@med.unc.edu. *Application contact:* Dr. Patrick Brennwald, Director of Graduate Studies, 919-843-4995, E-mail: patrick_brennwald@med.unc.edu.

University of Pennsylvania, School of Medicine, Biomedical Graduate Studies, Graduate Group in Cell and Molecular Biology, Program in Developmental Biology, Philadelphia, PA 19104. Offers PhD, MD/PhD, VMD/PhD. *Degree requirements:* For doctorate, thesis/dissertation. *Entrance requirements:* For doctorate, GRE General Test. Additional exam requirements/recommendations for international students: Required—TOEFL. *Application deadline:* For fall admission, 12/15 priority date for domestic students, 12/1 priority date for international students. Applications are processed on a rolling basis. Application fee: $70. Electronic applications accepted. *Financial support:* Fellowships, research assistantships, scholarships/grants, traineeships, and unspecified assistantships available. *Unit head:* Dr. Daniel Kessler, Chair, 215-898-2180. *Application contact:* Meagan Schofer, Coordinator, 215-895-9536, Fax: 215-573-2104, E-mail: mschofer@mailmed.upenn.edu.

University of Pittsburgh, School of Arts and Sciences, Department of Biological Sciences, Program in Molecular, Cellular, and Developmental Biology, Pittsburgh, PA 15260. Offers PhD. *Faculty:* 22 full-time (7 women), 1 part-time/adjunct (0 women). *Students:* 48 full-time (32 women); includes 2 minority (both African Americans), 10 international. Average age 23. 215 applicants, 2% accepted, 4 enrolled. In 2005, 7 degrees awarded. *Median time to degree:* Of those who began their doctoral program in fall 1997, 100% received their degree in 8 years or less. *Degree requirements:* For doctorate, thesis/dissertation, comprehensive exam, registration. *Entrance requirements:* For doctorate, GRE General Test, GRE Subject Test. Additional exam requirements/recommendations for international students: Required—TOEFL (minimum score

Developmental Biology

University of Pittsburgh (continued)
550 paper-based; 213 computer-based). *Application deadline:* For fall admission, 1/15 priority date for domestic students, 12/15 priority date for international students. Applications are processed on a rolling basis. Application fee: $0 ($40 for international students). Electronic applications accepted. *Expenses:* Tuition, state resident: full-time $13,194; part-time $537 per credit. Tuition, nonresident: full-time $25,012; part-time $1,026 per credit. Required fees: $700; $164 per term. Tuition and fees vary according to campus/location and program. *Financial support:* In 2005–06, 18 fellowships with full tuition reimbursements, 93 research assistantships with full tuition reimbursements (averaging $20,332 per year), 22 teaching assistantships with full tuition reimbursements (averaging $20,332 per year) were awarded; Federal Work-Study, scholarships/grants, traineeships, health care benefits, and tuition waivers (full) also available. *Faculty research:* Structure and function of genes and proteins; macromolecular interactions; cell-specific gene regulation; regulation of cell proliferation; embryogenesis. *Application contact:* Cathleen M. Barr, Graduate Administrator, 412-624-4268, Fax: 412-624-4759, E-mail: cbarr@pitt.edu.

Announcement: In the department's graduate program in molecular, cellular, and developmental biology, faculty research programs use prokaryotic and eukaryotic experimental systems and molecular and genetic approaches to understand the structure and function of genes and proteins, macromolecular interactions, cell-specific gene regulation, regulation of cell proliferation, and embryogenesis.

See Close-Up on page 637.

University of South Carolina, The Graduate School, College of Science and Mathematics, Department of Biological Sciences, Graduate Training Program in Molecular, Cellular, and Developmental Biology, Columbia, SC 29208. Offers MS, PhD. *Degree requirements:* For master's and doctorate, one foreign language, thesis/dissertation. *Entrance requirements:* For master's and doctorate, GRE General Test, minimum GPA of 3.0 in science. Electronic applications accepted. *Faculty research:* Marine ecology, population and evolutionary biology, molecular biology and genetics, development.

See Close-Up on page 639.

The University of Texas at Austin, Graduate School, College of Natural Sciences, Institute for Cellular and Molecular Biology, Program in Genetics and Developmental Biology, Austin, TX 78712-1111. Offers PhD.

The University of Texas Health Science Center at Houston, Graduate School of Biomedical Sciences, Program in Genes and Development, Houston, TX 77030. Offers MS, PhD, MD/PhD. *Faculty:* 38 full-time (11 women). *Students:* 68 full-time (35 women); includes 9 minority (3 Asian Americans or Pacific Islanders, 6 Hispanic Americans), 43 international. Average age 24. 36 applicants, 31% accepted, 3 enrolled. In 2005, 2 master's, 9 doctorates awarded. Terminal master's awarded for partial completion of doctoral program. *Degree requirements:* For master's and doctorate, thesis/dissertation. *Entrance requirements:* For master's and doctorate, GRE General Test. Additional exam requirements/recommendations for international students: Required—TOEFL, TWE. *Application deadline:* For fall admission, 1/15 for domestic students; for spring admission, 11/1 for domestic students. Applications are processed on a rolling basis. Application fee: $10. Electronic applications accepted. *Financial support:* Fellowships with full tuition reimbursements, research assistantships with full tuition reimbursements, teaching assistantships, institutionally sponsored loans, scholarships/grants, and health care benefits available. Financial award application deadline: 1/15. *Faculty research:* Regulation of gene expression; developmental genetics; cell growth, proliferation and death; tissue differentiation and organogenesis; chronmastin. *Unit head:* Dr. Michelle C Barton, Director, 713-834-6268, Fax: 713-834-6273, E-mail: mbarton@mdanderson.org. *Application contact:* Dr. Victoria P. Knutson, Assistant Dean of Admissions, 713-500-9860, Fax: 713-500-9877, E-mail: victoria.p.knutson@uth.tmc.edu.

The University of Texas Southwestern Medical Center at Dallas, Southwestern Graduate School of Biomedical Sciences, Division of Basic Science, Program in Genetics and Development, Dallas, TX 75390. Offers PhD. *Faculty:* 85 full-time (16 women), 2 part-time/adjunct (0 women). *Students:* 80 full-time (40 women), 2 part-time; includes 14 minority (6 Asian Americans or Pacific Islanders, 8 Hispanic Americans), 41 international. Average age 28. In 2005, 16 doctorates awarded. *Degree requirements:* For doctorate, thesis/dissertation, qualifying exam. *Entrance requirements:* For doctorate, GRE General Test, minimum GPA of 3.0. Additional exam requirements/recommendations for international students: Required—TOEFL. *Application deadline:* For fall admission, 1/5 for domestic students. Application fee: $0. Electronic applications accepted. *Expenses:* Tuition, state resident: full-time $6,550; part-time $50 per credit hour. Tuition, nonresident: full-time $19,650; part-time $326 per credit hour. Required fees: $42 per credit hour. Tuition and fees vary according to degree level and program. *Financial support:* Fellowships, research assistantships, institutionally sponsored loans available. *Faculty research:* Human molecular genetics, chromosome structure, gene regulation, molecular biology, gene expression. *Unit head:* Dr. John Abrams, Chair, 214-648-9226, Fax:

214-648-9226, E-mail: john.abrams@utsouthwestern.edu. *Application contact:* Dr. Nancy E. Street, Associate Dean, 214-648-6708, Fax: 214-648-2102, E-mail: nancy.street@utsouthwestern.edu.

Virginia Polytechnic Institute and State University, Graduate School, College of Science, Department of biological Sciences, Blacksburg, VA 24061. Offers botany (MS, PhD); ecology and evolutionary biology (MS, PhD); genetics and developmental biology (MS, PhD); microbiology (MS, PhD); zoology (MS, PhD). *Faculty:* 38 full-time (9 women). *Students:* 70 full-time (30 women), 4 part-time (1 woman); includes 6 minority (2 African Americans, 1 American Indian/Alaska Native, 2 Asian Americans or Pacific Islanders, 1 Hispanic American), 12 international. Average age 27. 81 applicants, 22% accepted, 14 enrolled. In 2005, 9 master's, 8 doctorates awarded. *Entrance requirements:* For master's and doctorate, GRE General Test. Additional exam requirements/recommendations for international students: Required—TOEFL (minimum score 550 paper-based; 213 computer-based). *Application deadline:* Applications are processed on a rolling basis. Application fee: $45. Electronic applications accepted. *Expenses:* Tuition, state resident: full-time $6,558; part-time $364 per credit. Tuition, nonresident: full-time $11,296; part-time $628 per credit. Required fees: $1,419; $468 per credit. $234 per term. *Financial support:* In 2005–06, 28 research assistantships with full tuition reimbursements (averaging $16,689 per year), 37 teaching assistantships with full tuition reimbursements (averaging $13,902 per year) were awarded; career-related internships or fieldwork, Federal Work-Study, scholarships/grants, and unspecified assistantships also available. *Faculty research:* Freshwater ecology, cell cycle regulation, behavioral ecology, motor proteins. *Unit head:* Dr. Bob Jones, Chairman, 540-231-9514, Fax: 540-231-9307, E-mail: rhjones@vt.edu. *Application contact:* Sue Rasmussen, Graduate Secretary, 540-231-8929, Fax: 540-231-9307, E-mail: sueras@vt.edu.

Washington University in St. Louis, Graduate School of Arts and Sciences, Division of Biology and Biomedical Sciences, Program in Developmental Biology, St. Louis, MO 63130-4899. Offers PhD. *Degree requirements:* For doctorate, thesis/dissertation. *Entrance requirements:* For doctorate, GRE General Test, GRE Subject Test. Electronic applications accepted.

Wesleyan University, Graduate Programs, Department of Biology, Middletown, CT 06459-0260. Offers cell biology (PhD); comparative physiology (PhD); developmental biology (PhD); genetics (PhD); neurophysiology (PhD); population biology (PhD). *Faculty:* 12 full-time (3 women). *Students:* 29 full-time (15 women), 10 international. Average age 26. 131 applicants. In 2005, 2 doctorates awarded. *Degree requirements:* For doctorate, one foreign language, thesis/dissertation. *Entrance requirements:* For doctorate, GRE Subject Test. *Application deadline:* For fall admission, 2/15 for domestic students. Applications are processed on a rolling basis. Application fee: $0. *Expenses:* Tuition: Full-time $24,732. One-time fee: $20 full-time. *Financial support:* Research assistantships, teaching assistantships, stipends available. *Faculty research:* Microbial population genetics, genetic basis of evolutionary adaptation, genetic regulation of differentiation and pattern formation in *drosophila*. *Unit head:* Dr. Michael Weir, Chairman, 860-685-2402, E-mail: mweir@wesleyan.edu. *Application contact:* Marjorie Fitzgibbons, Information Contact, 860-685-2157, E-mail: mfitzgibbons@wesleyan.edu.

See Close-Up on page 323.

West Virginia University, Davis College of Agriculture, Forestry and Consumer Sciences, Interdisciplinary Program in Genetics and Developmental Biology, Morgantown, WV 26506. Offers animal breeding (MS, PhD); biochemical and molecular genetics (MS, PhD); cytogenetics (MS, PhD); descriptive embryology (MS, PhD); developmental genetics (MS); experimental morphogenesis teratology (MS); human genetics (MS, PhD); immunogenetics (MS, PhD); life cycles of animals and plants (MS, PhD); molecular aspects of development (MS, PhD); mutagenesis (MS, PhD); oncology (MS, PhD); plant genetics (MS, PhD); population and quantitative genetics (MS, PhD); regeneration (MS, PhD); teratology (MS, PhD); toxicology (MS, PhD). *Students:* 13 full-time (7 women), 6 part-time (4 women), 11 international. Average age 28. In 2005, 2 master's, 4 doctorates awarded. *Degree requirements:* For master's, thesis/dissertation; for doctorate, thesis/dissertation, comprehensive exam. *Entrance requirements:* For master's, GRE or MCAT, minimum GPA of 2.75. Additional exam requirements/recommendations for international students: Required—TOEFL. Application fee: $45. *Expenses:* Tuition, state resident: full-time $4,582; part-time $258 per credit hour. Tuition, nonresident: full-time $13,820; part-time $741 per credit hour. *Financial support:* In 2005–06, 3 research assistantships with tuition reimbursements (averaging $9,936 per year), 4 teaching assistantships with tuition reimbursements (averaging $9,936 per year) were awarded; fellowships, Federal Work-Study, institutionally sponsored loans, and tuition waivers (full and partial) also available. Financial award application deadline: 2/1; financial award applicants required to submit FAFSA. Total annual research expenditures: $1 million. *Unit head:* Dr. J. Nath, Chairman, 304-293-6023 Ext. 4333, Fax: 304-293-2960, E-mail: joginder.nath@mail.wvu.edu.

Yale University, Graduate School of Arts and Sciences, Department of Molecular, Cellular, and Developmental Biology, Program in Developmental Biology, New Haven, CT 06520. Offers PhD. *Degree requirements:* For doctorate, thesis/dissertation. *Entrance requirements:* For doctorate, GRE General Test, GRE Subject Test.

Genetics

Arizona State University, Division of Graduate Studies, College of Liberal Arts and Sciences, Department of Biology, Program in Genetics, Tempe, AZ 85287. Offers MS, PhD. Terminal master's awarded for partial completion of doctoral program. *Degree requirements:* For master's, thesis; for doctorate, thesis/dissertation, oral exam. *Entrance requirements:* For master's and doctorate, GRE General Test, GRE Subject Test. Additional exam requirements/recommendations for international students: Required—TOEFL (minimum score 600 paper-based; 250 computer-based); Recommended—TSE. *Faculty research:* Molecular, developmental, and behavioral genetics; cytogenetics; population studies.

Baylor College of Medicine, Graduate School of Biomedical Sciences, Department of Molecular and Human Genetics, Houston, TX 77030-3498. Offers PhD, MD/PhD. *Faculty:* 52 full-time (8 women). *Students:* 81 full-time (53 women); includes 10 minority (1 African American, 6 Asian Americans or Pacific Islanders, 3 Hispanic Americans), 39 international. Average age 27. 83 applicants, 23% accepted, 11 enrolled. In 2005, 11 doctorates awarded. *Median time to degree:* Of those who began their doctoral program in fall 1997, 70% received their degree in 8 years or less. *Degree requirements:* For doctorate, thesis/dissertation, public defense. *Entrance requirements:* For doctorate, GRE General Test, GRE Subject Test (strongly recommended), minimum GPA of 3.0. Additional exam requirements/recommendations for international students: Required—TOEFL. *Application deadline:* For fall admission, 2/1 for domestic students. Application fee: $30. Electronic applications accepted. *Expenses:* Tuition: Full-time $8,200. Full-time tuition and fees vary according to program. *Financial support:* In 2005–06, 80 students received support, including 23 fellowships (averaging $20,000 per year), 58 research assistantships (averaging $20,000 per year); career-related internships or fieldwork, Federal Work-Study, institutionally sponsored loans, health care benefits, and tuition waivers (full) also available. Financial award applicants required to submit FAFSA. *Faculty research:* Cytogenetics, biochemical genetics, somatic cell genetics, gene therapy. *Unit head:* Dr. Gad Shaulsky, Director, 713-798-5056. *Application contact:* Judi Coleman, Graduate Program Administrator, 713-798-5056, Fax: 713-798-8597, E-mail: genetics-gradprm@bcm.tmc.edu.

See Close-Up on page 759.

Baylor College of Medicine, Graduate School of Biomedical Sciences, Interdepartmental Program in Cell and Molecular Biology, Houston, TX 77030-3498. Offers biochemistry (PhD); cell and molecular biology (PhD); genetics (PhD); human genetics (PhD); immunology (PhD); microbiology (PhD); virology (PhD). *Faculty:* 99 full-time (21 women). *Students:* 56 full-time (28 women); includes 20 minority (4 African Americans, 1 American Indian/Alaska Native, 6 Asian Americans or Pacific Islanders, 9 Hispanic Americans), 5 international. Average age 28. 164 applicants, 18% accepted, 12 enrolled. In 2005, 3 doctorates awarded. *Median time to degree:* Of those who began their doctoral program in fall 1997, 63% received their degree in 8 years or less. *Degree requirements:* For doctorate, thesis/dissertation, public defense. *Entrance requirements:* For doctorate, GRE General Test, GRE Subject Test (strongly recommended), minimum GPA of 3.0. Additional exam requirements/recommendations for international students: Required—TOEFL. *Application deadline:* For fall admission, 1/1 for domestic students. Applications are processed on a rolling basis. Application fee: $30. Electronic applications accepted. *Expenses:* Tuition: Full-time $8,200. Full-time tuition and fees vary according to program. *Financial support:* In 2005–06, 52 students received support, including 20 fellowships (averaging $23,000 per year), 36 research assistantships (averaging $23,000 per year); teaching assistantships, Federal Work-Study, institutionally sponsored loans, health care benefits, and tuition waivers (full) also available. Financial award applicants required to submit FAFSA. *Faculty research:* Gene expression and regulation, developmental biology and genetics, signal transduction and membrane biology, aging process, molecular virology. *Unit head:* Dr. Tom Cooper, Director, 713-798-6557. *Application contact:* Lourdes Fernandez, Graduate Program Administrator, 713-798-6557, Fax: 713-798-6325, E-mail: cmbprog@bcm.edu.

See Close-Up on page 545.

Baylor College of Medicine, Graduate School of Biomedical Sciences, Program in Developmental Biology, Houston, TX 77030-3498. Offers PhD, MD/PhD. *Faculty:* 42 full-time (10 women). *Students:* 38 full-time (17 women); includes 7 minority (1 American Indian/Alaska Native, 5 Asian Americans or Pacific Islanders, 1 Hispanic American), 21 international. Average age 27. 67 applicants, 25% accepted, 7 enrolled. In 2005, 2 doctorates awarded.

Genetics

Median time to degree: Of those who began their doctoral program in fall 1997, 100% received their degree in 8 years or less. *Degree requirements:* For doctorate, thesis/dissertation, public defense. *Entrance requirements:* For doctorate, GRE General Test, GRE Subject Test (strongly recommended), minimum GPA of 3.0. Additional exam requirements/recommendations for international students: Required—TOEFL. *Application deadline:* For fall admission, 1/1 for domestic students. Application fee: $30. Electronic applications accepted. *Expenses:* Tuition: Full-time $8,200. Full-time tuition and fees vary according to program. *Financial support:* In 2005–06, 37 students received support, including 10 fellowships (averaging $23,000 per year), 28 research assistantships (averaging $23,000 per year); career-related internships or fieldwork, Federal Work-Study, institutionally sponsored loans, health care benefits, tuition waivers (full), and stipends also available. *Faculty research:* Molecular and genetic approaches to study pattern formation in *Dictyostelium, Drosophila, C.elegans,* mouse, *Xenopus,* and zebrafish; cross-species approach. *Unit head:* Dr. Hugo Bellen, Director, 713-798-6410. *Application contact:* Catherine Tasnier, Graduate Program Administrator, 713-798-6410, Fax: 713-798-5386, E-mail: cat@bcm.edu.

See Close-Up on page 757.

Boston University, School of Medicine, Department of Genetics and Genomics, Boston, MA 02215. Offers MA, PhD. *Degree requirements:* For master's and doctorate, thesis/dissertation. *Entrance requirements:* For master's and doctorate, GRE, letters of recommendation. Additional exam requirements/recommendations for international students: Required—TOEFL. *Application deadline:* For fall admission, 1/1 for domestic students. Electronic applications accepted. *Expenses:* Tuition: Full-time $31,530; part-time $985 per credit. Required fees: $316; $40 per semester. Tuition and fees vary according to course level and program. *Unit head:* Dr. Shoumita Dasgupta, Assistant Professor and Director of Graduate Studies, 617-414-1580, E-mail: dasgupta@bu.edu.

Brandeis University, Graduate School of Arts and Sciences, Programs in Life Sciences, Program in Molecular and Cell Biology, Waltham, MA 02454-9110. Offers genetics (PhD); microbiology (PhD); molecular and cell biology (MS, PhD); molecular biology (PhD); neurobiology (PhD). *Faculty:* 18 full-time (8 women). *Students:* 43 full-time (21 women); includes 2 minority (both Asian Americans or Pacific Islanders), 11 international. Average age 27. 173 applicants, 8% accepted, 6 enrolled. In 2005, 2 master's, 4 doctorates awarded. Terminal master's awarded for partial completion of doctoral program. *Median time to degree:* Of those who began their doctoral program in fall 1997, 100% received their degree in 8 years or less. *Degree requirements:* For master's, research project, thesis optional; for doctorate, thesis/dissertation, teaching assistant experience, comprehensive exam, registration. *Entrance requirements:* For master's and doctorate, GRE General Test, resumé, 3 letters of recommendation. Additional exam requirements/recommendations for international students: Required—TOEFL (minimum score 600 paper-based; 250 computer-based). *Application deadline:* For fall admission, 1/15 for domestic students. Applications are processed on a rolling basis. Application fee: $55. Electronic applications accepted. *Financial support:* In 2005–06, 20 fellowships with full tuition reimbursements (averaging $26,500 per year), 23 research assistantships with full tuition reimbursements (averaging $26,500 per year), 5 teaching assistantships with full tuition reimbursements (averaging $3,000 per year) were awarded; scholarships/grants, traineeships, health care benefits, and tuition waivers (full and partial) also available. Financial award application deadline: 4/15; financial award applicants required to submit CSS PROFILE or FAFSA. *Faculty research:* Regulation of gene expression by transcription factors, molecular neurobiology, immunology, molecular mechanisms of genetic recombination, and cell differentiation. *Unit head:* Dr. Piali Sengupta, Chair, 781-736-2686, Fax: 781-736-3107, E-mail: piali@bradeis.edu. *Application contact:* Marcia Cabral, Information Officer, 781-736-3100, Fax: 781-736-3107, E-mail: cabral@brandeis.edu.

California Institute of Technology, Division of Biology, Program in Genetics, Pasadena, CA 91125-0001. Offers PhD. *Degree requirements:* For doctorate, thesis/dissertation, qualifying exam. *Entrance requirements:* For doctorate, GRE General Test. *Application deadline:* For fall admission, 1/1 for domestic students. Application fee: $0. *Financial support:* Application deadline: 1/1. *Application contact:* Elizabeth Ayala, Graduate Program Coordinator, 626-395-4497, Fax: 626-683-3343, E-mail: biograd@cco.caltech.edu.

Carnegie Mellon University, Mellon College of Science, Department of Biological Sciences, Pittsburgh, PA 15213-3891. Offers biochemistry (PhD); biophysics (PhD); cell biology (PhD); computational biology (MS, PhD); developmental biology (PhD); genetics (PhD); molecular biology (PhD); neurobiology (PhD). *Degree requirements:* For doctorate, thesis/dissertation, comprehensive exam. *Entrance requirements:* For doctorate, GRE General Test, GRE Subject Test, interview. Electronic applications accepted. *Faculty research:* Genetic structure, function, and regulation; protein structure and function; biological membranes; biological spectroscopy.

See Close-Up on page 93.

Case Western Reserve University, School of Medicine and School of Graduate Studies, Graduate Programs in Medicine, Department of Genetics, Program in Human, Molecular, and Developmental Genetics and Genomics, Cleveland, OH 44106. Offers PhD, MD/PhD. *Students:* Average age 28. *Degree requirements:* For doctorate, thesis/dissertation. *Entrance requirements:* For doctorate, GRE General Test, GRE Subject Test. Additional exam requirements/recommendations for international students: Required—TOEFL. *Application deadline:* For fall admission, 2/28 for domestic students. Applications are processed on a rolling basis. Application fee: $25. *Financial support:* Fellowships, research assistantships available. Financial award application deadline: 2/28. *Faculty research:* Regulation of gene expression, molecular control of development, genomics. Total annual research expenditures: $8 million. *Application contact:* Malana C. Bey, Department Assistant, 216-368-3431, Fax: 216-368-1257, E-mail: mcb19@po.cwru.edu.

See Close-Up on page 761.

Clemson University, Graduate School, College of Agriculture, Forestry and Life Sciences, Department of Genetics and Biochemistry, Program in Genetics, Clemson, SC 29634. Offers MS, PhD. Offered in cooperation with the Department of Animal and Veterinary Sciences. *Students:* 25 full-time (15 women), 1 part-time; includes 1 minority (African American), 8 international. 24 applicants, 29% accepted, 7 enrolled. In 2005, 3 master's, 1 doctorate awarded. *Degree requirements:* For master's and doctorate, thesis/dissertation. *Entrance requirements:* For master's and doctorate, GRE General Test, minimum GPA of 3.2. Additional exam requirements/recommendations for international students: Required—TOEFL. *Application deadline:* For fall admission, 6/1 for domestic students, 4/15 for international students. Applications are processed on a rolling basis. Application fee: $50. *Financial support:* Fellowships, research assistantships, teaching assistantships available. Financial award application deadline: 3/15; financial award applicants required to submit FAFSA. *Faculty research:* Animal, plant, microbial, molecular, and biometrical genetics. *Unit head:* Dr. Julia Frugoli, Coordinator, 864-656-1859, Fax: 864-656-0435, E-mail: jfrugol@clemson.edu.

See Close-Up on page 763.

Colorado State University, Graduate School, College of Agricultural Sciences, Department of Soil and Crop Sciences, Fort Collins, CO 80523-0015. Offers crop science (MS, PhD); plant genetics (MS, PhD); soil science (MS, PhD). Part-time programs available. *Faculty:* 18 full-time (4 women), 1 part-time/adjunct (0 women). *Students:* 18 full-time (6 women), 15 part-time (4 women); includes 2 minority (both Hispanic Americans), 5 international. Average age 33. 16 applicants, 50% accepted, 7 enrolled. In 2005, 6 master's, 1 doctorate awarded. *Median time to degree:* Of those who began their doctoral program in fall 1997, 85% received their degree in 8 years or less. *Degree requirements:* For master's, thesis (for some programs), comprehensive exam, registration; for doctorate, thesis/dissertation, preliminary exam, comprehensive exam, registration. *Entrance requirements:* For master's, minimum GPA of 3.0, appropriate bachelor's degree; for doctorate, minimum GPA of 3.0, appropriate master's degree. Additional exam requirements/recommendations for international students: Required—

TOEFL. *Application deadline:* For fall admission, 2/1 priority date for domestic students, 2/1 priority date for international students; for spring admission, 8/1 priority date for domestic students, 8/1 priority date for international students. Applications are processed on a rolling basis. Application fee: $50. Electronic applications accepted. *Expenses:* Tuition, state resident: full-time $3,690; part-time $205 per credit. Tuition, nonresident: full-time $14,958; part-time $831 per credit. Required fees: $1,061. *Financial support:* In 2005–06, 16 students received support, including 1 fellowship with partial tuition reimbursement available (averaging $15,600 per year), 14 research assistantships with partial tuition reimbursements available (averaging $15,600 per year), 1 teaching assistantship with partial tuition reimbursement available (averaging $15,600 per year); career-related internships or fieldwork and traineeships also available. *Faculty research:* Water quality, soil fertility, soil/plant ecosystems, plant breeding and genetics, information systems/technology. Total annual research expenditures: $3.5 million. *Unit head:* Dr. Gary A. Peterson, Head, 970-491-6501, Fax: 970-491-0564, E-mail: gary.peterson@colostate.edu. *Application contact:* Dr. Pat F. Byrne, Graduate Studies Coordinator, 970-491-6985, Fax: 970-491-0564, E-mail: pbyrne@lamar.colostate.edu.

Columbia University, College of Physicians and Surgeons and Graduate School of Arts and Sciences, Graduate School of Arts and Sciences at the College of Physicians and Surgeons, Department of Genetics and Development, New York, NY 10032. Offers genetics (M Phil, MA, PhD). Only candidates for the PhD are admitted. Terminal master's awarded for partial completion of doctoral program. *Degree requirements:* For doctorate, thesis/dissertation. *Entrance requirements:* For master's and doctorate, GRE General Test. Additional exam requirements/recommendations for international students: Required—TOEFL. *Expenses:* Tuition: Full-time $31,448. Tuition and fees vary according to course level, course load, campus/location and program. *Faculty research:* Mammalian cell differentiation and meiosis, developmental genetics, yeast and human genetics, chromosome structure, molecular and cellular biology.

Cornell University, Graduate School, Graduate Fields of Agriculture and Life Sciences, Field of Genetics and Development, Ithaca, NY 14853-0001. Offers developmental biology (PhD); genetics (PhD). *Faculty:* 50 full-time (11 women). *Students:* 56 full-time (32 women); includes 10 minority (3 African Americans, 1 American Indian/Alaska Native, 3 Asian Americans or Pacific Islanders, 3 Hispanic Americans), 16 international. 104 applicants, 17% accepted, 9 enrolled. In 2005, 7 doctorates awarded. *Degree requirements:* For doctorate, thesis/dissertation, 2 semesters of teaching experience, comprehensive exam. *Entrance requirements:* For doctorate, GRE General Test, GRE Subject Test in biology or biochemistry (recommended), 2 letters of recommendation. Additional exam requirements/recommendations for international students: Required—TOEFL (minimum score 550 paper-based; 213 computer-based). *Application deadline:* For fall admission, 1/5 for domestic students. Application fee: $60. Electronic applications accepted. *Financial support:* In 2005–06, 56 students received support, including 15 fellowships with full tuition reimbursements available, 30 research assistantships with full tuition reimbursements available, 11 teaching assistantships with full tuition reimbursements available; institutionally sponsored loans, scholarships/grants, health care benefits, tuition waivers (full and partial), and unspecified assistantships also available. Financial award applicants required to submit FAFSA. *Faculty research:* Molecular and general genetics, developmental biology and developmental genetics, evolution and population genetics, plant genetics, microbial genetics. *Unit head:* Director of Graduate Studies, 607-254-2100. *Application contact:* Graduate Field Assistant, 607-254-2100, E-mail: gendev@cornell.edu.

Dartmouth College, School of Arts and Sciences, Department of Genetics, Hanover, NH 03755. Offers PhD. Affiliated with the Graduate Program in Molecular and Cellular Biology. *Faculty:* 12 full-time (4 women). *Students:* 20 full-time (10 women), 8 international. Average age 28. 336 applicants, 20% accepted, 32 enrolled. In 2005, 1 degree awarded. *Degree requirements:* For doctorate, thesis/dissertation, teaching experience. *Entrance requirements:* For doctorate, GRE General Test, GRE Subject Test. Additional exam requirements/recommendations for international students: Required—TOEFL. *Application deadline:* For fall admission, 1/7 for domestic students, 1/7 for international students. Application fee: $40. Electronic applications accepted. *Expenses:* Tuition: Full-time $31,770. *Financial support:* In 2005–06, 23 students received support, including fellowships with full tuition reimbursements available (averaging $23,000 per year), research assistantships with full tuition reimbursements available (averaging $23,000 per year); Federal Work-Study, institutionally sponsored loans, scholarships/grants, traineeships, tuition waivers (full and partial), and unspecified assistantships also available. *Faculty research:* Growth factor, regulations. Total annual research expenditures: $6.2 million. *Unit head:* Dr. Jay C. Dunlap, Chair, 603-650-1907, E-mail: genetics.department@dartmouth.edu. *Application contact:* Terri Eastman, Manager, 603-650-1907, Fax: 603-650-1188, E-mail: genetics.department@dartmouth.edu.

See Close-Up on page 765.

Drexel University, College of Medicine, Biomedical Graduate Programs, Interdisciplinary Program in Molecular and Human Genetics, Philadelphia, PA 19104-2875. Offers MS, PhD, MD/PhD. *Degree requirements:* For master's, comprehensive exam; for doctorate, thesis/dissertation, qualifying exam. *Entrance requirements:* For master's and doctorate, GRE General Test. Additional exam requirements/recommendations for international students: Required—TOEFL.

Duke University, Graduate School, Department of Biochemistry, Durham, NC 27710. Offers crystallography of macromolecules (PhD); enzyme mechanisms (PhD); lipid biochemistry (PhD); membrane structure and function (PhD); molecular genetics (PhD); neurochemistry (PhD); nucleic acid structure and function (PhD); protein structure and function (PhD). *Faculty:* 28 full-time. *Students:* 68 full-time (28 women); includes 6 minority (1 African American, 5 Asian Americans or Pacific Islanders), 15 international. 106 applicants, 19% accepted, 5 enrolled. In 2005, 11 doctorates awarded. *Degree requirements:* For doctorate, thesis/dissertation. *Entrance requirements:* For doctorate, GRE General Test, GRE Subject Test (recommended). Additional exam requirements/recommendations for international students: Required—IELT (preferred) or TOEFL. *Application deadline:* For fall admission, 12/31 for domestic students, 12/31 for international students. Electronic applications accepted. *Financial support:* Fellowships, research assistantships, teaching assistantships, Federal Work-Study available. Financial award application deadline: 12/31. *Unit head:* Leonard Spicer, Director of Graduate Studies, 919-681-8770, Fax: 919-684-8885, E-mail: anorfleet@biochem.duke.edu.

Duke University, Graduate School, Program in Genetics and Genomics, Durham, NC 27710. Offers PhD. *Faculty:* 70 full-time. *Students:* 77 full-time (49 women); includes 11 minority (3 African Americans, 3 Asian Americans or Pacific Islanders, 5 Hispanic Americans), 13 international. 101 applicants, 23% accepted, 9 enrolled. In 2005, 7 doctorates awarded. *Degree requirements:* For doctorate, variable foreign language requirement, thesis/dissertation. *Entrance requirements:* For doctorate, GRE General Test, GRE Subject Test (recommended). Additional exam requirements/recommendations for international students: Required—IELT (preferred) or TOEFL. *Application deadline:* For fall admission, 12/31 for domestic students, 12/31 for international students. Application fee: $75. *Financial support:* Fellowships available. Financial award application deadline: 12/31. *Unit head:* Dr. Douglas Marchuk, Director of Graduate Studies, 919-684-3290, Fax: 919-684-0917, E-mail: genetics@biochem.duke.edu.

See Close-Up on page 769.

Emory University, Graduate School of Arts and Sciences, Division of Biological and Biomedical Sciences, Program in Genetics and Molecular Biology, Atlanta, GA 30322-1100. Offers PhD. *Faculty:* 51 full-time (9 women). *Students:* 56 full-time (40 women); includes 10 minority (8 African Americans, 2 Asian Americans or Pacific Islanders), 6 international. Average age 27. 81 applicants, 31% accepted, 17 enrolled. In 2005, 3 degrees awarded. *Median time to degree:* Of those who began their doctoral program in fall 1997, 100% received their degree in 8 years or less. *Degree requirements:* For doctorate, thesis/dissertation, comprehensive exam, registration. *Entrance requirements:* For doctorate, GRE General Test, minimum GPA of 3.0 in science course work. Additional exam requirements/recommendations for international students: Required—TOEFL. *Application deadline:* For fall admission, 1/3 for domestic students, 1/3 for

Genetics

Emory University (continued)

international students. Application fee: $50. Electronic applications accepted. *Expenses:* Tuition: Full-time $14,400. Required fees: $217. *Financial support:* In 2005–06, 27 students received support, including 27 fellowships with full tuition reimbursements available (averaging $23,000 per year); institutionally sponsored loans, health care benefits, and tuition waivers (full) also available. *Faculty research:* Gene regulation, genetic combination, developmental regulation. *Unit head:* Dr. Jerry Boss, Director, 404-717-5973, Fax: 404-727-1719, E-mail: boss@microbio.emory.edu. *Application contact:* 404-727-2545, Fax: 404-727-3322, E-mail: gdbbs@emory.edu.

Florida State University, Graduate Studies, College of Arts and Sciences, Department of Biological Science, Program in Genetics, Tallahassee, FL 32306. Offers MS, PhD. *Faculty:* 21 full-time (5 women). *Students:* 55 full-time (28 women); includes 5 minority (4 Asian Americans or Pacific Islanders, 1 Hispanic American), 14 international. *Degree requirements:* For master's and doctorate, thesis/dissertation, teaching experience, seminar presentation, comprehensive exam, registration. *Entrance requirements:* For master's and doctorate, GRE General Test (minimum 1100: U-500, G-500), minimum upper division GPA of 3.0. Additional exam requirements/recommendations for international students: Required—TOEFL (minimum score 600 paper-based; 250 computer-based), IB 100. *Application deadline:* For fall admission, 1/15 for domestic students, 12/1 for international students; for spring admission, 10/15 for domestic students, 9/1 for international students. Application fee: $30. *Financial support:* In 2005–06, fellowships with full tuition reimbursements (averaging $19,000 per year), research assistantships with full tuition reimbursements (averaging $19,000 per year), teaching assistantships with full tuition reimbursements (averaging $17,600 per year) were awarded. Financial award application deadline: 1/15; financial award applicants required to submit FAFSA. *Faculty research:* Population genetics, fertilization, eukaryotic gene expression, biotechnology, molecular biology. *Application contact:* Judy Bowers, Coordinator, Graduate Affairs, 850-644-3023, Fax: 850-644-9829, E-mail: gradinfo@bio.fsu.edu.

The George Washington University, Columbian College of Arts and Sciences, Institute for Biomedical Sciences, Program in Genetics, Washington, DC 20052. Offers MS, PhD. Part-time and evening/weekend programs available. Terminal master's awarded for partial completion of doctoral program. *Degree requirements:* For master's, thesis, comprehensive exam; for doctorate, thesis/dissertation, general exam. *Entrance requirements:* For master's and doctorate, GRE General Test, interview, minimum GPA of 3.0. Additional exam requirements/recommendations for international students: Required—TOEFL (minimum score 600 paper-based; 250 computer-based). Electronic applications accepted.

Harvard University, Graduate School of Arts and Sciences, Division of Medical Sciences, Boston, MA 02115. Offers biological chemistry and molecular pharmacology (PhD); cell biology (PhD); genetics (PhD); microbiology and molecular genetics (PhD); pathology (PhD), including experimental pathology. *Students:* 433 full-time (210 women). In 2005, 83 doctorates awarded. *Degree requirements:* For doctorate, thesis/dissertation. *Entrance requirements:* For doctorate, GRE General Test, GRE Subject Test. Additional exam requirements/recommendations for international students: Required—TOEFL. Application fee: $60. *Expenses:* Tuition: Full-time $28,752. Full-time tuition and fees vary according to program and student level. *Financial support:* Fellowships, research assistantships, teaching assistantships, institutionally sponsored loans and tuition waivers (full) available. Financial award application deadline: 1/1. *Unit head:* Administrator, 617-432-2029. *Application contact:* Administrator, 617-432-2029.

Harvard University, School of Public Health, Department of Genetics and Complex Diseases, Boston, MA 02115-6096. Offers PhD. *Degree requirements:* For doctorate, thesis/dissertation, qualifying exam. *Entrance requirements:* For doctorate, GRE. Additional exam requirements/recommendations for international students: Required—TOEFL (minimum score 560 paper-based; 220 computer-based); Recommended—IELT (minimum score 7). *Faculty research:* Toxicology, radiation biology.

Howard University, Graduate School of Arts and Sciences, Department of Genetics and Human Genetics, Washington, DC 20059. Offers MS, PhD. Part-time programs available. *Faculty:* 13. *Students:* 23; includes 16 minority (15 African Americans, 1 Asian American or Pacific Islander), 6 international. In 2005, 11 master's, 2 doctorates awarded. *Median time to degree:* Of those who began their doctoral program in fall 1997, 100% received their degree in 8 years or less. *Degree requirements:* For master's, thesis, comprehensive exam; for doctorate, one foreign language, thesis/dissertation, comprehensive exam. *Entrance requirements:* For master's, GRE General Test, minimum GPA of 3.0; course work in biology, chemistry, physics, mathematics, and genetics; for doctorate, GRE General Test, master's degree in genetics or related field, minimum GPA of 3.2. Additional exam requirements/recommendations for international students: Required—TOEFL, GRE required. *Application deadline:* For fall admission, 4/1 for domestic students; for spring admission, 11/1 for domestic students. Applications are processed on a rolling basis. Application fee: $45. Electronic applications accepted. *Financial support:* Fellowships with full tuition reimbursements, research assistantships, teaching assistantships, Federal Work-Study, institutionally sponsored loans, scholarships/grants, tuition waivers (full), and unspecified assistantships available. Financial award applicants required to submit FAFSA. *Unit head:* Dr. Robert F. Murray, Chairman, 202-806-6340. *Application contact:* Dr. Verle E. Headings, Director of Graduate Studies, 202-806-6381.

Hunter College of the City University of New York, Hunter Center for Gene Structure and Function, New York, NY 10021-5085.

Illinois State University, Graduate School, College of Arts and Sciences, Department of Biological Sciences, Normal, IL 61790-2200. Offers biological sciences (MS); biology (PhD); biotechnology (MS); botany (PhD); ecology (PhD); genetics (PhD); microbiology (PhD); physiology (PhD); zoology (PhD). Part-time programs available. *Faculty:* 27 full-time (6 women). *Students:* 36 full-time (28 women), 25 part-time (17 women); includes 1 minority (Asian American or Pacific Islander), 23 international. 80 applicants, 21% accepted. In 2005, 14 master's, 5 doctorates awarded. *Degree requirements:* For master's, thesis or alternative; for doctorate, variable foreign language requirement, thesis/dissertation, 2 terms of residency. *Entrance requirements:* For master's, GRE General Test, minimum GPA of 2.6 in last 60 hours of course work; for doctorate, GRE General Test. *Application deadline:* Applications are processed on a rolling basis. Application fee: $30. *Expenses:* Tuition, state resident: full-time $3,060; part-time $170 per credit hour. Tuition, nonresident: full-time $6,390; part-time $355 per credit hour. Required fees: $1,411; $47 per credit hour. *Financial support:* In 2005–06, 22 research assistantships (averaging $13,909 per year), 38 teaching assistantships (averaging $12,617 per year) were awarded; Federal Work-Study, tuition waivers (full), and unspecified assistantships also available. Financial award application deadline: 4/1. *Faculty research:* CRUI: osmoregulation in euryhaline fish: physiology, ecology and molecular biology; the PRISM project: enhancing science and math education; cell structure—functions of Na pump assembly. *Unit head:* , Dr. Hou Tak Takucheung, Acting Chairperson, 309-438-3669. *Application contact:* Derek A. McCracken, Graduate Adviser, 309-438-3664.

See Close-Up on page 147.

Indiana University Bloomington, Graduate School, College of Arts and Sciences, Department of Biology, Program in Genetics, Bloomington, IN 47405-7000. Offers PhD. Offered through the University Graduate School. *Faculty:* 23 full-time (4 women). *Students:* 16 full-time (6 women); includes 2 minority (1 Asian American or Pacific Islander, 1 Hispanic American). In 2005, 1 degree awarded. *Degree requirements:* For doctorate, thesis/dissertation. *Entrance requirements:* For doctorate, GRE General Test. Additional exam requirements/recommendations for international students: Required—TOEFL. *Application deadline:* For fall admission, 1/5 for domestic students; for spring admission, 9/1 for domestic students. Applications are processed on a rolling basis. Application fee: $45. Electronic applications accepted. *Expenses:* Tuition, state resident: full-time $5,437; part-time $227 per credit hour. Tuition, nonresident: full-time $15,836; part-time $660 per credit hour. Required fees: $821. Tuition and fees vary according to campus/location and program. *Financial support:* In 2005–06, 16

students received support, including fellowships with tuition reimbursements available (averaging $17,000 per year), research assistantships with tuition reimbursements available (averaging $17,000 per year), teaching assistantships with tuition reimbursements available (averaging $17,000 per year); scholarships/grants also available. Financial award application deadline: 1/15. *Faculty research:* Transmission genetics of *Drosophila* and maize, population genetics, viral hybrid DNA techniques, cytogenetics, molecular genetics. *Unit head:* Dr. Susan Strome, Head, 812-855-5450, Fax: 812-855-6705, E-mail: sstrome@bio.indiana.edu. *Application contact:* Gretchen Clearwater, Adviser for Graduate Affairs, 812-855-1861, Fax: 812-855-6705, E-mail: biograd@bio.indiana.edu.

Iowa State University of Science and Technology, Graduate College, College of Liberal Arts and Sciences and College of Agriculture, Department of Genetics, Developmental and Cell Biology, Ames, IA 50011. Offers MS, PhD. *Faculty:* 30 full-time, 3 part-time/adjunct. *Students:* 77 full-time (34 women), 55 international. 3 applicants, 100% accepted, 1 enrolled. In 2005, 6 master's, 9 doctorates awarded. *Degree requirements:* For master's and doctorate, thesis/dissertation. *Entrance requirements:* For master's and doctorate, GRE General Test. Additional exam requirements/recommendations for international students: Required—TOEFL (paper score 570; computer score 230) or IELTS (score 6.5). *Application deadline:* For fall admission, 1/1 priority date for domestic students, 1/1 priority date for international students. Application fee: $30 ($70 for international students). Electronic applications accepted. *Expenses:* Tuition, state resident: full-time $6,410. Tuition, nonresident: full-time $16,422. Tuition and fees vary according to program. *Financial support:* In 2005–06, 65 students received support, including 65 research assistantships with full and partial tuition reimbursements available (averaging $17,301 per year), 7 teaching assistantships with full and partial tuition reimbursements available (averaging $15,042 per year); fellowships with full tuition reimbursements available, scholarships/grants, health care benefits, and unspecified assistantships also available. Financial award application deadline: 2/1. *Faculty research:* Animal behavior, animal models of gene therapy, cell biology, comparative physiology, developmental biology. *Unit head:* Dr. Martin Spalding, Chair, 515-294-1749.

Iowa State University of Science and Technology, Graduate College, Interdisciplinary Programs, Bioinformatics and Computational Biology Program, Ames, IA 50011-3260. Offers MS, PhD. *Degree requirements:* For doctorate, thesis/dissertation. *Entrance requirements:* For doctorate, GRE General Test. Additional exam requirements/recommendations for international students: Required—TOEFL or IELTS. Electronic applications accepted. *Expenses:* Tuition, state resident: full-time $6,410. Tuition, nonresident: full-time $16,422. Tuition and fees vary according to program. *Faculty research:* Functional and structural genomics, genome evolution, macromolecular structure and function, mathematical biology and biological statistics, metabolic and developmental networks.

Iowa State University of Science and Technology, Graduate College, Interdisciplinary Programs, Program in Genetics, Ames, IA 50011. Offers MS, PhD. *Students:* 88 full-time (54 women), 2 part-time (1 woman); includes 6 minority (3 African Americans, 1 Asian American or Pacific Islander, 2 Hispanic Americans), 50 international. 68 applicants, 34% accepted, 10 enrolled. In 2005, 9 master's, 8 doctorates awarded. Terminal master's awarded for partial completion of doctoral program. *Degree requirements:* For master's and doctorate, thesis/dissertation. *Entrance requirements:* For master's and doctorate, GRE General Test. Additional exam requirements/recommendations for international students: Required—TOEFL (paper score 530; computer score 197) or IELTS (score 6.0). *Application deadline:* For fall admission, 1/31 priority date for domestic students, 1/31 priority date for international students. Applications are processed on a rolling basis. Application fee: $30 ($70 for international students). *Expenses:* Tuition, state resident: full-time $6,410. Tuition, nonresident: full-time $16,422. Tuition and fees vary according to program. *Financial support:* In 2005–06, 82 research assistantships with full and partial tuition reimbursements (averaging $16,551 per year), 2 teaching assistantships with full and partial tuition reimbursements (averaging $15,220 per year) were awarded; fellowships, scholarships/grants, health care benefits, and unspecified assistantships also available. *Unit head:* Dr. Patrick Schnable, Supervisory Committee Chair, 515-294-7697, Fax: 515-294-6669, E-mail: genetics@iastate.edu. *Application contact:* Linda Wild, Program Coordinator, 800-499-1972, Fax: 515-294-6669, E-mail: genetics@iastate.edu.

See Close-Up on page 777.

The Johns Hopkins University, Bloomberg School of Public Health, Department of Epidemiology, Baltimore, MD 21205. Offers cancer epidemiology (MHS, Sc M, PhD, Sc D); cardiovascular disease epidemiology (MHS, Sc M, PhD, Sc D); clinical epidemiology (MHS, Sc M, PhD, Sc D); clinical trials (PhD, Sc D); epidemiology (Dr PH, Sc D); epidemiology (general) (MHS, Sc M, PhD); epidemiology of aging (MHS, Sc M, PhD, Sc D); human genetics/genetic epidemiology (MHS, Sc M, PhD, Sc D); infectious disease epidemiology (MHS, Sc M, PhD, Sc D); occupational/environmental epidemiology (MHS, Sc M, PhD, Sc D). *Faculty:* 84 full-time (45 women), 81 part-time/adjunct (33 women). *Students:* 148 full-time (111 women), 52 part-time (37 women); includes 48 minority (14 African Americans, 30 Asian Americans or Pacific Islanders, 4 Hispanic Americans), 36 international. Average age 31. 293 applicants, 37% accepted, 56 enrolled. In 2005, 27 master's, 22 doctorates awarded. *Median time to degree:* Of those who began their doctoral program in fall 1997, 72% received their degree in 8 years or less. *Degree requirements:* For master's, thesis, 1 year full-time residency, comprehensive exam, registration; for doctorate, thesis/dissertation, 1 year full-time residency, oral and written exams, teaching requirement, comprehensive exam, registration. *Entrance requirements:* For master's, GRE General Test, 3 letters of recommendation, curriculum vitae; for doctorate, GRE General Test or MCAT, minimum 1 year work experience, 3 letters of recommendation, curriculum vitae. Additional exam requirements/recommendations for international students: Required—TOEFL (minimum score 600 paper-based; 250 computer-based). *Application deadline:* For fall admission, 12/1 for domestic students. Applications are processed on a rolling basis. Application fee: $45. Electronic applications accepted. *Expenses:* Tuition: Full-time $30,960. Tuition and fees vary according to degree level and program. *Financial support:* In 2005–06, 212 students received support, including 2 fellowships (averaging $28,859 per year); Federal Work-Study, institutionally sponsored loans, scholarships/grants, traineeships, tuition waivers (partial), and stipends also available. Support available to part-time students. Financial award application deadline: 3/15; financial award applicants required to submit FAFSA. *Faculty research:* Cancer and congenital malformations, nutritional epidemiology, AIDS, tuberculosis, cardiovascular disease, risk assessment. Total annual research expenditures: $47.8 million. *Unit head:* Dr. Jonathan M. Samet, Chairman, 410-955-3286, Fax: 410-955-0863, E-mail: jsamet@jhsph.edu. *Application contact:* Frances S. Burman, Senior Academic Coordinator, 410-955-3926, Fax: 410-955-0863, E-mail: fburman@jhsph.edu.

The Johns Hopkins University, National Institutes of Health Sponsored Programs, Department of Biology, Baltimore, MD 21218-2699. Offers biochemistry (PhD); biophysics (PhD); cell biology (PhD); developmental biology (PhD); genetic biology (PhD); molecular biology (PhD). *Faculty:* 25 full-time (4 women). *Students:* 126 full-time (72 women); includes 36 minority (3 African Americans, 1 American Indian/Alaska Native, 21 Asian Americans or Pacific Islanders, 11 Hispanic Americans), 19 international. 282 applicants, 26% accepted, 36 enrolled. In 2005, 15 degrees awarded. *Median time to degree:* Of those who began their doctoral program in fall 1997, 81.2% received their degree in 8 years or less. *Degree requirements:* For doctorate, thesis/dissertation, comprehensive exam, registration. *Entrance requirements:* For doctorate, GRE General Test. Additional exam requirements/recommendations for international students: Required—TOEFL (minimum score 600 paper-based; 250 computer-based), TWE, TSE. *Application deadline:* For fall admission, 12/15 for domestic students. Application fee: $60. *Expenses:* Tuition: Full-time $30,960. Tuition and fees vary according to degree level and program. *Financial support:* In 2005–06, 24 fellowships (averaging $23,000 per year), 93 research assistantships (averaging $23,000 per year), 22 teaching assistantships (averaging $23,000 per year) were awarded; Federal Work-Study, institutionally sponsored loans, scholarships/grants, traineeships, health care benefits, tuition waivers (partial), and unspecified assistantships also available. Financial award application deadline: 4/15; financial award applicants required to submit FAFSA. *Faculty research:* Protein and nucleic acid biochemistry and

Genetics

biophysical chemistry, molecular biology and development. Total annual research expenditures: $11.2 million. *Unit head:* Dr. Allen Shearn, Chair, 410-516-4693, Fax: 410-516-5213, E-mail: bio_cals@jhu.edu. *Application contact:* Joan Miller, Academic Affairs Manager, 410-516-5502, Fax: 410-516-5213, E-mail: joan@jhu.edu.

See Close-Up on page 155.

The Johns Hopkins University, School of Medicine, Program in Biochemistry, Cellular and Molecular Biology, Baltimore, MD 21218-2699.

Kansas State University, Graduate School, Program in Genetics, Manhattan, KS 66506. Offers MS, PhD. *Students:* 12 full-time (5 women), 3 part-time (all women), (all international). Average age 25. In 2005, 3 master's awarded. Terminal master's awarded for partial completion of doctoral program. *Degree requirements:* For master's, thesis, oral exam; for doctorate, one foreign language, thesis/dissertation, preliminary exams, residency. *Entrance requirements:* For master's, GRE General Test, bachelor's degree in science, minimum GPA of 3.0; for doctorate, GRE General Test, minimum GPA of 3.5 in master's studies. Additional exam requirements/recommendations for international students: Required—TOEFL. *Application deadline:* For fall admission, 2/1 priority date for domestic students, 2/1 priority date for international students; for spring admission, 8/1 priority date for domestic students, 8/1 priority date for international students. Applications are processed on a rolling basis. Application fee: $30 ($55 for international students). *Expenses:* Tuition, state resident: full-time $5,160; part-time $215 per credit hour. Tuition, nonresident: full-time $12,816; part-time $534 per credit hour. Required fees: $564. *Financial support:* Research assistantships, teaching assistantships, institutionally sponsored loans, scholarships/grants, and tuition waivers (partial) available. Support available to part-time students. Financial award application deadline: 3/1; financial award applicants required to submit FAFSA. *Faculty research:* Genetics of susceptibility/resistance to animal diseases, insect genetics, genetics of disease resistance in plants, genetics of plant pathogenic fungi, genetics of pathogenicity and virulence plant or animal pathogenic bacteria. *Unit head:* , Barbara Valent, Director, 785-532-2336, Fax: 785-532-6094, E-mail: bvalent@plantpath.ksu.edu. *Application contact:* Merla Brookman, Office Specialist, 785-532-1330, Fax: 785-532-6094, E-mail: genetics@ksu.edu.

Marquette University, Graduate School, College of Arts and Sciences, Department of Biology, Milwaukee, WI 53201-1881. Offers cell biology (MS, PhD); developmental biology (MS, PhD); ecology (MS, PhD); endocrinology (MS, PhD); evolutionary biology (MS, PhD); genetics (MS, PhD); microbiology (MS, PhD); molecular biology (MS, PhD); muscle and exercise physiology (MS, PhD); neurobiology (MS, PhD); reproductive physiology (MS, PhD). Terminal master's awarded for partial completion of doctoral program. *Degree requirements:* For master's, thesis, 1 year of teaching experience or equivalent, comprehensive exam; for doctorate, thesis/dissertation, 1 year of teaching experience or equivalent, qualifying exam. *Entrance requirements:* For master's and doctorate, GRE General Test, GRE Subject Test. Additional exam requirements/recommendations for international students: Required—TOEFL. *Faculty research:* Microbial and invertebrate ecology, evolution of gene function, DNA methylation, DNA arrangement.

Mayo Graduate School, Graduate Programs in Biomedical Sciences, Program in Virology and Gene Therapy, Rochester, MN 55905. Offers PhD.

See Close-Up on page 883.

Mayo Graduate School, Graduate Programs in Biomedical Sciences, Programs in Biochemistry, Structural Biology, Cell Biology, and Genetics, Rochester, MN 55905. Offers biochemistry and structural biology (PhD); cell biology and genetics (PhD); molecular biology (PhD). *Degree requirements:* For doctorate, oral defense of dissertation, qualifying oral and written exam. *Entrance requirements:* For doctorate, GRE, 1 year of chemistry, biology, calculus, and physics. Additional exam requirements/recommendations for international students: Required—TOEFL. Electronic applications accepted. *Faculty research:* Gene structure and function, membranes and receptors/cytoskeleton, oncogenes and growth factors, protein structure and function, steroid hormonal action.

See Close-Up on page 399.

McMaster University, Faculty of Health Sciences and School of Graduate Studies, Program in Medical Sciences, Molecular Biology, Genetics, and Cancer Area, Hamilton, ON L8S 4M2, Canada. Offers M Sc, PhD. *Students:* 11 full-time, 2 part-time. In 2005, 2 master's, 2 doctorates awarded. *Degree requirements:* For master's, thesis/dissertation; for doctorate, thesis/dissertation, comprehensive exam. *Entrance requirements:* For master's, honors B Sc, B+ average in related field; for doctorate, M Sc, minimum B+ average, students with proven research experience and an A average may be admitted with a B Sc degree. Additional exam requirements/recommendations for international students: Required—TOEFL (minimum score 580 paper-based; 237 computer-based). *Application deadline:* For fall admission, 9/30 for domestic students. For winter admission, 3/31 for domestic students. Applications are processed on a rolling basis. Application fee: $85. *Financial support:* Teaching assistantships available. *Unit head:* Dr. John Waye, Coordinator, 905-525-9140 Ext. 76273. *Application contact:* Dr. Carl Richards, Associate Dean, 905-525-9140 Ext. 22983, Fax: 905-546-1129.

Michigan State University, College of Veterinary Medicine and The Graduate School, Graduate Program in Veterinary Medicine and College of Natural Science and Graduate Programs in Human Medicine, Department of Microbiology and Molecular Genetics, East Lansing, MI 48824. Offers industrial microbiology (MS); microbiology (MS, PhD), including environmental toxicology (MS); microbiology and molecular genetics (MS, PhD); microbiology–environmental toxicology (PhD). *Faculty:* 34 full-time (10 women). *Students:* 47 full-time (28 women), 2 part-time (1 woman); includes 3 minority (2 African Americans, 1 Hispanic American), 21 international. Average age 26. 104 applicants, 13% accepted. In 2005, 3 master's, 9 doctorates awarded. *Degree requirements:* For master's, thesis (for some programs); for doctorate, thesis/dissertation, comprehensive exam. *Entrance requirements:* For master's and doctorate, GRE General Test, minimum GPA of 3.0, 3 letters of recommendation. Additional exam requirements/recommendations for international students: Required—TOEFL (minimum score 550 paper-based; 213 computer-based), Michigan State University ELT (85), Michigan ELAB (83). *Application deadline:* For fall admission, 8/27 for domestic students. Application fee: $50. Electronic applications accepted. *Expenses:* Tuition, state resident: part-time $330 per credit hour. Tuition, nonresident: part-time $685 per credit hour. Tuition and fees vary according to program. *Financial support:* In 2005–06, 12 fellowships with tuition reimbursements (averaging $6,638 per year), 29 research assistantships with tuition reimbursements (averaging $15,427 per year), 3 teaching assistantships with tuition reimbursements (averaging $14,823 per year) were awarded; scholarships/grants, traineeships, health care benefits, and unspecified assistantships also available. *Faculty research:* Microbial physiology, ecology and evolution: molecular pathogenesis and infectious diseases; genomics and genetics; cancer, differentiation, immunology and cell biology. Total annual research expenditures: $5.2 million. *Unit head:* Dr. Walter Esselman, Chairperson, 517-355-6463 Ext. 1510, Fax: 517-353-8957, E-mail: mmgchair@msu.edu. *Application contact:* Suzanne Peacock, Graduate Program Coordinator, 517-432-2288, Fax: 517-353-8957, E-mail: micgrad@msu.edu.

See Close-Up on page 887.

Michigan State University, The Graduate School, College of Agriculture and Natural Resources, MSU-DOE Plant Research Laboratory, East Lansing, MI 48824. Offers biochemistry and molecular biology (PhD); cellular and molecular biology (PhD); crop and soil sciences (PhD); genetics (PhD); microbiology and molecular genetics (PhD); plant biology (PhD); plant physiology (PhD). Offered jointly with the Department of Energy. *Faculty:* 9 full-time (2 women). *Degree requirements:* For doctorate, thesis/dissertation, laboratory rotation, defense of dissertation, comprehensive exam. *Entrance requirements:* For doctorate, GRE General Test, acceptance into one of the affiliated department programs; 3 letters of recommendation; bachelor's degree or equivalent in life sciences, chemistry, biochemistry, or biophysics; research experience. Application fee: $50. Electronic applications accepted. *Expenses:* Tuition,

state resident: part-time $330 per credit hour. Tuition, nonresident: part-time $685 per credit hour. Tuition and fees vary according to program. *Faculty research:* Role of hormones in the regulation of plant development and physiology, molecular mechanisms associated with signal recognition, development and application of genetic methods and materials, protein routing and function. Total annual research expenditures: $7.4 million. *Unit head:* Dr. Kenneth Keegstra, Director, 517-353-2270, Fax: 517-353-9168, E-mail: keegstra@msu.edu. *Application contact:* Janet Taylor, Graduate Program Secretary, 517-353-2270, Fax: 517-353-9168, E-mail: prl@msu.edu.

Michigan State University, The Graduate School, College of Natural Science, Program in Genetics, East Lansing, MI 48824. Offers MS, PhD. *Faculty:* 98 full-time (30 women). *Students:* 44 full-time (21 women), 26 international. Average age 28. 25 applicants, 28% accepted. In 2005, 5 degrees awarded. *Degree requirements:* For master's, thesis or alternative, final oral exam, teaching experience; for doctorate, thesis/dissertation, oral defense of dissertation proposal, oral exam in defense of dissertation, teaching experience, comprehensive exam. *Entrance requirements:* For master's, GRE General Test, minimum GPA of 3.0, 3 letters of recommendation; for doctorate, GRE General Test, minimum GPA of 3.3, background in the biological/physical sciences, 3 letters of recommendation. Additional exam requirements/recommendations for international students: Required—TOEFL (minimum score 600 paper-based; 250 computer-based), speak/ITAOI test. *Application deadline:* For fall admission, 12/27 for domestic students. Application fee: $50. Electronic applications accepted. *Expenses:* Tuition, state resident: part-time $330 per credit hour. Tuition, nonresident: part-time $685 per credit hour. Tuition and fees vary according to program. *Financial support:* In 2005–06, 7 fellowships with tuition reimbursements (averaging $14,567 per year), 30 research assistantships with tuition reimbursements (averaging $15,477 per year), 1 teaching assistantship with tuition reimbursement (averaging $14,175 per year) were awarded; scholarships/grants and unspecified assistantships also available. *Faculty research:* Genetics of plant biochemistry and plant-microbe interactions, genetics of human and animal diseases, developmental and behavioral genetics, genomics and proteomics, gene expression and regulation. *Unit head:* Dr. Barbara B. Sears, Director, 517-353-9845, Fax: 517-355-0112, E-mail: sears@msu.edu. *Application contact:* Jeannine Lee, Graduate Secretary, 517-353-9845, Fax: 517-355-0112, E-mail: genetics@msu.edu.

Mount Sinai School of Medicine of New York University, Graduate School of Biological Sciences, New York, NY 10029-6504. Offers biophysics, structural biology and biomathematics (PhD); community medicine (MPH); genetic counseling (MS); genetics and genomic sciences (PhD); mechanisms of disease and therapy (PhD); microbiology (PhD); molecular, cellular, biochemical and developmental sciences (PhD); neurosciences (PhD). *Students:* 218 full-time (109 women). 4,208 applicants, 7% accepted, 117 enrolled. Terminal master's awarded for partial completion of doctoral program. *Degree requirements:* For master's, registration; for doctorate, thesis/dissertation, registration. *Entrance requirements:* For doctorate, GRE General Test, GRE Subject Test, 3 years of college pre-med course work. Additional exam requirements/recommendations for international students: Required—TOEFL. *Application deadline:* For fall admission, 1/15 for domestic students. Application fee: $75. Electronic applications accepted. *Expenses:* Tuition: Full-time $33,250. Required fees: $1,600. Full-time tuition and fees vary according to degree level, program and reciprocity agreements. *Financial support:* In 2005–06, fellowships with full tuition reimbursements (averaging $26,000 per year), research assistantships with full tuition reimbursements (averaging $26,000 per year) were awarded; Federal Work-Study, institutionally sponsored loans, scholarships/grants, health care benefits, and unspecified assistantships also available. Financial award application deadline: 4/5; financial award applicants required to submit FAFSA. *Faculty research:* Cancer, gene therapy, minimally invasive surgery, cardiac translational research. Total annual research expenditures: $162.2 million. *Unit head:* Dr. Diomedes Logothetis, Dean, 212-241-6546, Fax: 212-241-0651, E-mail: diomedes.logothetis@mssm.edu. *Application contact:* Lily Recanati, Manager, 212-241-3267, Fax: 212-241-0651, E-mail: lily.recantati@mssm.edu.

See Close-Up on page 177.

New York University, Graduate School of Arts and Science, Department of Biology, New York, NY 10012-1019. Offers biology (PhD); biomedical journalism (MS); cancer and molecular biology (PhD); computational biology (PhD); computers in biological research (MS); developmental genetics (PhD); general biology (MS); immunology and microbiology (PhD); molecular genetics (PhD); neurobiology (PhD); oral biology (MS); plant biology (PhD); recombinant DNA technology (MS). Part-time programs available. *Faculty:* 24 full-time (5 women), 8 part-time/adjunct. *Students:* 104 full-time (52 women), 41 part-time (23 women); includes 28 minority (2 African Americans, 20 Asian Americans or Pacific Islanders, 6 Hispanic Americans), 47 international. Average age 27. 349 applicants, 56% accepted, 39 enrolled. In 2005, 59 master's, 4 doctorates awarded. Terminal master's awarded for partial completion of doctoral program. *Degree requirements:* For master's, thesis or alternative, qualifying paper; for doctorate, thesis/dissertation, comprehensive exam. *Entrance requirements:* For master's, GRE General Test; for doctorate, GRE General Test, GRE Subject Test. Additional exam requirements/recommendations for international students: Required—TOEFL. *Application deadline:* For fall admission, 1/4 for domestic students. Application fee: $80. *Financial support:* Fellowships with tuition reimbursements, research assistantships with tuition reimbursements, teaching assistantships with tuition reimbursements, career-related internships or fieldwork, Federal Work-Study, institutionally sponsored loans, scholarships/grants, health care benefits, and unspecified assistantships available. Financial award application deadline: 1/4; financial award applicants required to submit FAFSA. *Faculty research:* Genomics, molecular and cell biology, development and molecular genetics, molecular evolution of plants and animals. *Unit head:* Dr. Gloria Coruzzi, Chairman, 212-998-8200, Fax: 212-995-4015, E-mail: biology@nyu.edu. *Application contact:* Stephen Small, Director of Graduate Studies, 212-998-8200, Fax: 212-995-4015, E-mail: biology@nyu.edu.

North Carolina State University, Graduate School, College of Agriculture and Life Sciences, Department of Genetics, Raleigh, NC 27695. Offers MG, MS, PhD. Terminal master's awarded for partial completion of doctoral program. *Degree requirements:* For master's, thesis (for some programs); for doctorate, thesis/dissertation. *Entrance requirements:* For master's and doctorate, GRE General Test, minimum GPA of 3.0. Electronic applications accepted. *Faculty research:* Population and quantitative genetics, plant molecular genetics, developmental genetics.

Northwestern University, The Graduate School and Judd A. and Marjorie Weinberg College of Arts and Sciences, Interdepartmental Biological Sciences Program (IBiS), Evanston, IL 60208. Offers biochemistry, molecular biology, and cell biology (PhD), including biochemistry, cell and molecular biology, molecular biophysics, structural biology; biotechnology (PhD); cell and molecular biology (PhD); developmental biology and genetics (PhD); hormone action and signal transduction (PhD); neuroscience (PhD); structural biology, biochemistry, and biophysics (PhD). Participants in the Interdepartmental Biological Sciences Program include the Departments of Biochemistry, Molecular Biology, and Cell Biology; Chemistry; Neurobiology and Physiology; Chemical Engineering; Civil Engineering; and Evanston Hospital. *Entrance requirements:* For doctorate, GRE General Test. Additional exam requirements/recommendations for international students: Required—TOEFL (minimum score 600 paper-based), TSE(minimum score 50). Electronic applications accepted. *Faculty research:* Developmental genetics, gene regulation, DNA-protein interactions, biological clocks, bioremediation.

See Close-Up on page 187.

Northwestern University, Northwestern University Feinberg School of Medicine, Department of Microbiology-Immunology, Chicago, IL 60611-3008.

Northwestern University, Northwestern University Feinberg School of Medicine and Interdepartmental Degree Programs, Integrated Graduate Programs in the Life Sciences, Chicago, IL 60611. Offers cancer biology (PhD); cell biology (PhD); developmental biology (PhD); evolutionary biology (PhD); immunology and microbial pathogenesis (PhD); molecular

Genetics

Northwestern University (continued)
biology and genetics (PhD); neurobiology (PhD); pharmacology and toxicology (PhD); structural biology and biochemistry (PhD). *Degree requirements:* For doctorate, thesis/dissertation, written and oral qualifying exams, comprehensive exam. *Entrance requirements:* For doctorate, GRE General Test. Additional exam requirements/recommendations for international students: Required—TOEFL (minimum score 600 paper-based; 250 computer-based). Electronic applications accepted.

See Close-Up on page 189.

The Ohio State University, College of Medicine and Public Health and Graduate School, Graduate Programs in the Basic Medical Sciences, Integrated Biomedical Science Graduate Program, Columbus, OH 43210. Offers immunology (MS, PhD); medical genetics (MS, PhD); molecular virology (MS, PhD); pharmacology (MS, PhD). *Degree requirements:* For doctorate, thesis/dissertation. *Entrance requirements:* For master's, GRE General Test; for doctorate, GRE. Additional exam requirements/recommendations for international students: Required—TOEFL (minimum score 600 paper-based; 250 computer-based), TSE. Electronic applications accepted.

The Ohio State University, Graduate School, College of Biological Sciences, Department of Molecular Genetics, Columbus, OH 43210. Offers cell and developmental biology (MS, PhD); genetics (MS, PhD); molecular biology (MS, PhD). *Degree requirements:* For master's and doctorate, thesis/dissertation. *Entrance requirements:* For master's and doctorate, GRE General Test, GRE Subject Test. Additional exam requirements/recommendations for international students: Required—TOEFL (minimum score 573 paper-based; 230 computer-based). Electronic applications accepted.

See Close-Up on page 785.

Oregon Health & Science University, School of Medicine, Graduate Programs in Medicine, Department of Molecular and Medical Genetics, Portland, OR 97239-3098. Offers PhD. *Degree requirements:* For doctorate, thesis/dissertation. *Entrance requirements:* For doctorate, GRE General Test. Additional exam requirements/recommendations for international students: Required—TOEFL. *Faculty research:* Molecular studies of metabolic diseases, gene therapy, control of mycogenesis, regulation of gene expression, DNA replication and repair.

Oregon State University, Graduate School, College of Agricultural Sciences, Program in Genetics, Corvallis, OR 97331. Offers MA, MAIS, MS, PhD. Part-time programs available. *Students:* 12 full-time (5 women), 3 part-time (1 woman), 6 international. Average age 30. In 2005, 3 master's, 1 doctorate awarded. Terminal master's awarded for partial completion of doctoral program. *Degree requirements:* For master's, variable foreign language requirement, thesis or alternative; for doctorate, thesis/dissertation. *Entrance requirements:* For master's and doctorate, GRE General Test, minimum GPA of 3.0 in last 90 hours. Additional exam requirements/recommendations for international students: Required—TOEFL. *Application deadline:* For fall admission, 3/1 for domestic students. Applications are processed on a rolling basis. Application fee: $50. *Expenses:* Tuition, area resident: full-time $301 per credit. Tuition, state resident: full-time $8,139; part-time $501 per credit. Tuition, nonresident: full-time $14,376; part-time $532 per credit. Required fees: $1,266. *Financial support:* Fellowships, research assistantships, teaching assistantships, Federal Work-Study and institutionally sponsored loans available. Financial award application deadline: 2/1. *Faculty research:* Molecular genetics, cytogenetics, population and quantitative genetics, microbial genetics, plant genetics. *Unit head:* , Dr. Lloyd W. Ream, Director, 541-737-3799, Fax: 541-737-3132, E-mail: reamw@bcc.orst.edu.

The Pennsylvania State University Milton S. Hershey Medical Center, Graduate School Programs in the Biomedical Sciences, Graduate Program in Microbiology and Immunology, Hershey, PA 17033-2360. Offers genetics (PhD); immunology (MS, PhD); microbiology (MS); microbiology/virology (PhD); molecular biology (PhD). *Students:* 23 full-time (11 women); includes 2 minority (both Asian Americans or Pacific Islanders), 2 international. Average age 27.Terminal master's awarded for partial completion of doctoral program. *Median time to degree:* Of those who began their doctoral program in fall 1997, 100% received their degree in 8 years or less. *Degree requirements:* For master's, thesis or alternative, registration; for doctorate, thesis/dissertation, oral exam, comprehensive exam, registration. *Entrance requirements:* For master's, GRE or MCAT; for doctorate, GRE General Test or MCAT, minimum GPA of 3.0. Additional exam requirements/recommendations for international students: Required—TOEFL. *Application deadline:* Applications are processed on a rolling basis. Application fee: $45. Electronic applications accepted. *Financial support:* In 2005–06, 23 research assistantships with full tuition reimbursements were awarded; fellowships with full tuition reimbursements, scholarships/grants, health care benefits, and unspecified assistantships also available. Financial award applicants required to submit FAFSA. *Faculty research:* Virus replication and assembly, oncogenesis, interactions of viruses with host cells and animal model systems. *Unit head:* Dr. Richard J. Courtney, Chair, 717-531-7659, Fax: 717-531-6522, E-mail: micro-grad-hmc@psu.edu. *Application contact:* Billie Burns, Secretary, 717-531-7659, Fax: 717-531-6522, E-mail: micro-grad-hmc@psu.edu.

The Pennsylvania State University Milton S. Hershey Medical Center, Graduate School Programs in the Biomedical Sciences, The Huck Institutes of the Life Sciences, Intercollege Graduate Program in Genetics, Hershey, PA 17033-2360. Offers MS, PhD, MD/PhD. *Students:* 31 full-time (20 women); includes 2 minority (1 Asian American or Pacific Islander, 1 Hispanic American), 14 international. Average age 27.Terminal master's awarded for partial completion of doctoral program. *Degree requirements:* For master's, thesis or alternative; for doctorate, thesis/dissertation, oral exam, comprehensive exam, registration. *Entrance requirements:* For master's, GRE General Test; for doctorate, GRE General Test, minimum GPA of 3.0. Additional exam requirements/recommendations for international students: Required—TOEFL (minimum score 500 paper-based; 213 computer-based). *Application deadline:* Applications are processed on a rolling basis. Application fee: $45. Electronic applications accepted. *Financial support:* In 2005–06, 6 fellowships with full tuition reimbursements, 25 research assistantships with full tuition reimbursements were awarded; scholarships/grants, health care benefits, and unspecified assistantships also available. Financial award applicants required to submit FAFSA. *Faculty research:* Genome structure/stability, gene expression, cellular sorting of macromolecules, signal transduction, stem cell differentiation. *Unit head:* Dr. Anita K. Hopper, Co-Chair, 717-531-8982, E-mail: grad-hmc@psu.edu. *Application contact:* Kathy Shuey, Administrative Assistant, 717-531-8982, E-mail: grad-hmc@psu.edu.

The Pennsylvania State University University Park Campus, Graduate School, Intercollege Graduate Programs, Intercollege Graduate Program in Genetics, State College, University Park, PA 16802-1503. Offers MS, PhD. *Students:* 35 full-time (23 women), 2 part-time (1 woman); includes 1 minority (African American), 28 international. *Entrance requirements:* For master's and doctorate, GRE General Test. Application fee: $45. *Expenses:* Tuition, state resident: full-time $12,518; part-time $522 per credit. Tuition, nonresident: full-time $23,004; part-time $959 per credit. Required fees: $484. Tuition and fees vary according to course load, campus/location and program. *Unit head:* Dr. Richard Ordway, Chair, 814-863-5693, Fax: 814-865-9131, E-mail: rordway@psu.edu.

Purdue University, Graduate School, School of Science, Department of Biological Sciences, West Lafayette, IN 47907. Offers biochemistry (PhD); biophysics (PhD); cell and developmental biology (PhD); ecology, evolutionary and population biology (MS, PhD), including ecology, evolutionary biology, population biology; genetics (MS, PhD); microbiology (MS, PhD); molecular biology (PhD); neurobiology (MS, PhD); plant physiology (PhD). *Faculty:* 47 full-time (9 women), 4 part-time/adjunct (1 woman). *Students:* 97 full-time (53 women), 8 part-time (4 women); includes 13 minority (3 African Americans, 1 American Indian/Alaska Native, 4 Asian Americans or Pacific Islanders, 5 Hispanic Americans), 50 international. Average age 28. 168 applicants, 29% accepted, 23 enrolled. In 2005, 18 master's, 9 doctorates awarded. Terminal master's awarded for partial completion of doctoral program. *Degree requirements:* For master's, thesis

(for some programs); for doctorate, thesis/dissertation, seminars, teaching experience. *Entrance requirements:* For master's and doctorate, GRE General Test. Additional exam requirements/recommendations for international students: Required—TOEFL, TSE. *Application deadline:* For fall admission, 2/15 for domestic students, 1/31 for international students. Applications are processed on a rolling basis. Application fee: $55. Electronic applications accepted. *Financial support:* In 2005–06, 15 fellowships, 60 research assistantships, 53 teaching assistantships were awarded. Support available to part-time students. Financial award application deadline: 2/15; financial award applicants required to submit FAFSA. *Unit head:* Dr. Richard J Kuhn, Head, 765-494-4407. *Application contact:* Nancy Konopka, Graduate Studies Office Manager, 765-494-8142, Fax: 765-494-0876, E-mail: njk@bilbo.bio.purdue.edu.

Purdue University, School of Veterinary Medicine and Graduate School, Graduate Programs in Veterinary Medicine, Department of Veterinary Pathobiology, West Lafayette, IN 47907. Offers biochemistry and molecular biology (MS, PhD); comparative epidemiology (MS, PhD); epidemiology (MS, PhD); immunology (MS, PhD); infectious diseases (MS, PhD); interdisciplinary genetics (MS, PhD); laboratory animal medicine (MS, PhD); microbiology (MS, PhD); molecular virology (MS, PhD); parasitology (MS, PhD); pathobiology (MS, PhD); public health epidemiology (MS, PhD); toxicology (MS, PhD); veterinary anatomic pathology (MS, PhD); veterinary clinical pathology (MS, PhD); virology (MS, PhD). *Faculty:* 32 full-time (7 women). *Students:* 49 full-time (20 women), 3 part-time (1 woman); includes 2 minority (both African Americans), 31 international. Average age 35. In 2005, 3 master's, 8 doctorates awarded. Terminal master's awarded for partial completion of doctoral program. *Degree requirements:* For master's, thesis (for some programs); for doctorate, thesis/dissertation. *Entrance requirements:* For master's and doctorate, GRE General Test. Additional exam requirements/recommendations for international students: Required—TOEFL (minimum score 575 paper-based), TWE (minimum score 4). *Application deadline:* For fall admission, 8/12 for domestic students, 6/15 for international students; for spring admission, 1/12 for domestic students, 10/15 for international students. Application fee: $55. *Financial support:* Fellowships, research assistantships, teaching assistantships available. Financial award application deadline: 3/1; financial award applicants required to submit FAFSA. *Unit head:* Dr. H. Hogenesch, Head, 765-494-7543.

Rutgers, The State University of New Jersey, New Brunswick/Piscataway, Graduate School, Program in Microbiology and Molecular Genetics, New Brunswick, NJ 08901-1281. Offers applied microbiology (PhD); clinical microbiology (MS, PhD); computational molecular biology (PhD); immunology (MS, PhD); microbial biochemistry (MS, PhD); molecular genetics (MS, PhD); virology (MS, PhD). Part-time programs available. *Faculty:* 132 full-time. *Students:* 57 full-time (31 women), 18 part-time (8 women); includes 16 minority (9 Asian Americans or Pacific Islanders, 7 Hispanic Americans), 20 international. Average age 29. 104 applicants, 32% accepted, 19 enrolled. In 2005, 9 master's, 10 doctorates awarded. Terminal master's awarded for partial completion of doctoral program. *Median time to degree:* Of those who began their doctoral program in fall 1997, 100% received their degree in 8 years or less. *Degree requirements:* For master's, thesis or alternative, comprehensive exam, registration; for doctorate, thesis/dissertation, written qualifying exam, comprehensive exam, registration. *Entrance requirements:* For master's, GRE General Test, minimum GPA of 3.0; for doctorate, GRE General Test, GRE Subject Test (recommended), minimum GPA of 3.0. Additional exam requirements/recommendations for international students: Required—TOEFL. *Application deadline:* For fall admission, 1/5 priority date for domestic students, 11/1 priority date for international students. Applications are processed on a rolling basis. Application fee: $50. Electronic applications accepted. *Expenses:* Tuition, state resident: full-time $10,440; part-time $435 per credit. Tuition, nonresident: full-time $15,520; part-time $647 per credit. Required fees: $129 per credit. Tuition and fees vary according to program. *Financial support:* In 2005–06, 48 students received support, including 14 fellowships with full tuition reimbursements available (averaging $24,000 per year), 25 research assistantships with full tuition reimbursements available (averaging $24,000 per year), 9 teaching assistantships with full tuition reimbursements available (averaging $16,988 per year); Federal Work-Study, institutionally sponsored loans, scholarships/grants, and unspecified assistantships also available. Financial award application deadline: 1/5; financial award applicants required to submit FAFSA. *Faculty research:* Molecular genetics and microbial physiology; virology and pathogenic microbiology; applied, environmental and industrial microbiology; computers in molecular biology. *Unit head:* Dr. Andrew K. Vershon, Director, 732-445-2905, Fax: 732-445-6370, E-mail: vershon@waksman.rutgers.edu. *Application contact:* Diane Murano, Administrative Assistant, 732-445-5086, Fax: 732-445-6370, E-mail: murano@rutgers.edu.

Rutgers, The State University of New Jersey, New Brunswick/Piscataway, Graduate School, Program in Plant Biology, New Brunswick, NJ 08901-1281. Offers horticulture (MS, PhD); molecular biology and biochemistry (MS, PhD); pathology (MS, PhD); plant ecology (MS, PhD); plant genetics (PhD); plant physiology (MS, PhD); production and management (MS); structure and plant groups (MS, PhD). Part-time programs available. *Faculty:* 65 full-time, 1 part-time/adjunct (0 women). *Students:* 33 full-time (14 women), 13 part-time (6 women); includes 3 minority (1 African American, 2 Asian Americans or Pacific Islanders), 10 international. Average age 31. 56 applicants, 23% accepted, 10 enrolled. In 2005, 1 master's, 4 doctorates awarded. Terminal master's awarded for partial completion of doctoral program. *Median time to degree:* Of those who began their doctoral program in fall 1997, 90% received their degree in 8 years or less. *Degree requirements:* For master's, thesis or alternative, comprehensive exam; for doctorate, thesis/dissertation, comprehensive exam. *Entrance requirements:* For master's and doctorate, GRE General Test, GRE Subject Test (recommended). Additional exam requirements/recommendations for international students: Required—TOEFL (minimum score 600 paper-based; 250 computer-based). *Application deadline:* For fall admission, 4/1 for domestic students, 4/1 for international students. Application fee: $50. Electronic applications accepted. *Expenses:* Tuition, state resident: full-time $10,440; part-time $435 per credit. Tuition, nonresident: full-time $15,520; part-time $647 per credit. Required fees: $129 per credit. Tuition and fees vary according to program. *Financial support:* In 2005–06, 42 students received support, including 9 fellowships with full tuition reimbursements available (averaging $24,000 per year), 22 research assistantships with full tuition reimbursements available (averaging $16,500 per year), 10 teaching assistantships with full tuition reimbursements available (averaging $16,988 per year) Financial award application deadline: 1/15; financial award applicants required to submit FAFSA. *Faculty research:* Molecular biology and biochemistry of plants, plant development and genomics, plant protection, plant improvement, plant management of horticultural and field crops. Total annual research expenditures: $10 million. *Unit head:* Dr. Thomas Leustek, Director, 732-932-8165 Ext. 326, Fax: 732-932-9377, E-mail: leustek@aesop.rutgers.edu. *Application contact:* Barbara Mulder, Program Associate, 732-932-9375 Ext. 358, Fax: 732-932-9377, E-mail: plantbio@aesop.rutgers.edu.

Stanford University, School of Medicine, Graduate Programs in Medicine, Department of Genetics, Stanford, CA 94305-9991. Offers PhD. *Degree requirements:* For doctorate, thesis/dissertation, qualifying examination. *Entrance requirements:* For doctorate, GRE General Test, GRE Subject Test. Additional exam requirements/recommendations for international students: Required—TOEFL. Electronic applications accepted. *Faculty research:* Molecular biology of DNA replication in human cells, analysis of existing and search for new DNA polymorphisms in humans, molecular genetics of prokaryotic and eukaryotic genetic elements, proteins in DNA replication.

Stony Brook University, State University of New York, Graduate School, College of Arts and Sciences, Program in Genetics, Stony Brook, NY 11794. Offers PhD. *Faculty:* 80. *Students:* 61 full-time (36 women); includes 7 minority (2 African Americans, 3 Asian Americans or Pacific Islanders, 2 Hispanic Americans), 29 international. Average age 27. 85 applicants, 7% accepted. In 2005, 8 degrees awarded. *Degree requirements:* For doctorate, thesis/dissertation, teaching experience, comprehensive exam. *Entrance requirements:* For doctorate, GRE General Test, GRE Subject Test. Additional exam requirements/recommendations for international students: Required—TOEFL. *Application deadline:* For fall admission, 1/15 for domestic students. Application fee: $50. *Expenses:* Tuition, state resident: full-time $6,900; part-time $288 per credit. Tuition, nonresident: full-time $10,920; part-time $455 per credit. Required fees: $704. *Financial support:* In 2005–06, 11 fellowships, 49 research assistant-

Genetics

ships, 8 teaching assistantships were awarded; Federal Work-Study also available. *Faculty research:* Gene structure, gene regulation.

Announcement: This interinstitutional program, involving Stony Brook University, Cold Spring Harbor Laboratory, and Brookhaven National Laboratory, offers a highly interactive and nurturing environment for training in all areas of genetics. Specific areas of strength include bioinformatics and structural biology; microbiology; molecular, cellular, and developmental biology; neurobiology; and behavior and evolution.

See Close-Up on page 789.

Temple University, Health Sciences Center, School of Medicine and Graduate School, Graduate Programs in Medicine, Program in Molecular Biology and Genetics, Philadelphia, PA 19122-6096. Offers PhD, MD/PhD. *Faculty:* 7 full-time (2 women). *Students:* 15 full-time (11 women), 29 part-time (12 women). In 2005, 3 degrees awarded. *Degree requirements:* For doctorate, thesis/dissertation, presentation research/literature seminars distinct from area of concentration. *Entrance requirements:* For doctorate, GRE General Test, GRE Subject Test, minimum GPA of 3.0. Additional exam requirements/recommendations for international students: Required—TOEFL (minimum score 620 paper-based; 260 computer-based). *Application deadline:* For fall admission, 1/15 for domestic students, 12/15 for international students. Application fee: $50. Electronic applications accepted. *Expenses:* Tuition, state resident: full-time $8,694; part-time $483 per credit. Tuition, nonresident: full-time $12,672; part-time $704 per credit. Required fees: $500; $122 per semester. Tuition and fees vary according to course level, campus/location and program. *Financial support:* Fellowships, research assistantships, Federal Work-Study, institutionally sponsored loans, and tuition waivers (full) available. Financial award application deadline: 1/15; financial award applicants required to submit FAFSA. *Faculty research:* Molecular genetics of normal and malignant cell growth, regulation of gene expression, DNA repair systems and carcinogenesis, hormone-receptor interactions and signal transduction systems, structural biology. *Unit head:* Dr. Scott Shore, Chair, 215-707-3359, Fax: 215-707-2805, E-mail: sks@temple.edu.

Texas A&M University, College of Agriculture and Life Sciences, Intercollegiate Faculty of Genetics, College Station, TX 77843. Offers MS, PhD. Program composed of members from 4 colleges and 14 departments. *Students:* Average age 29. *Degree requirements:* For master's, thesis optional. *Entrance requirements:* For master's and doctorate, GRE General Test. Additional exam requirements/recommendations for international students: Required—TOEFL. *Application deadline:* For fall admission, 4/15 for domestic students. Applications are processed on a rolling basis. Application fee: $50 ($75 for international students). Electronic applications accepted. *Expenses:* Tuition, state resident: full-time $4,488; part-time $187 per credit hour. Tuition, nonresident: full-time $11,112; part-time $463 per credit hour. Required fees: $1,974. *Financial support:* Fellowships, research assistantships, teaching assistantships available. Financial award application deadline: 4/15; financial award applicants required to submit FAFSA. *Faculty research:* Biochemical genetics, cytogenetics, developmental genetics, immunogenetics, molecular genetics. *Unit head:* Dr. Linda Guarino, Chair, 979-845-1013, Fax: 979-862-9274, E-mail: lguarino@tamu.edu. *Application contact:* Linda Fisher, Administrative Assistant, 979-845-6848, Fax: 979-845-9274.

Texas A&M University, College of Veterinary Medicine, Graduate Programs in Veterinary Medicine, Department of Veterinary Pathobiology, College Station, TX 77843. Offers genetics (MS, PhD); veterinary microbiology (MS, PhD); veterinary parasitology (MS); veterinary pathology (MS, PhD). Part-time programs available. Postbaccalaureate distance learning degree programs offered. *Faculty:* 16 full-time (3 women), 4 part-time/adjunct (1 woman). *Students:* 45 full-time (28 women), 20 part-time (13 women); includes 7 minority (2 African Americans, 5 Hispanic Americans), 19 international. Average age 33. 25 applicants, 76% accepted, 19 enrolled. In 2005, 5 master's, 5 doctorates awarded. Terminal master's awarded for partial completion of doctoral program. *Degree requirements:* For master's and doctorate, thesis/dissertation, seminars. *Entrance requirements:* For master's and doctorate, GRE General Test, minimum GPA of 3.0 in last 60 hours. Additional exam requirements/recommendations for international students: Required—TOEFL. *Application deadline:* For fall admission, 3/1 for domestic students; for spring admission, 8/1 priority date for domestic students. Applications are processed on a rolling basis. Application fee: $50 ($75 for international students). Electronic applications accepted. *Expenses:* Tuition, state resident: full-time $4,488; part-time $187 per credit hour. Tuition, nonresident: full-time $11,112; part-time $463 per credit hour. Required fees: $1,974. *Financial support:* In 2005–06, fellowships with partial tuition reimbursements (averaging $16,000 per year), research assistantships with partial tuition reimbursements (averaging $15,400 per year), teaching assistantships with partial tuition reimbursements (averaging $16,000 per year) were awarded; Federal Work-Study, institutionally sponsored loans, scholarships/grants, traineeships, health care benefits, and unspecified assistantships also available. Support available to part-time students. Financial award applicants required to submit FAFSA. *Faculty research:* Infectious and noninfectious diseases of animals and birds, animal genetics, molecular biology, immunology, virology. *Unit head:* Dr. Ann B. Kier, Head, 979-845-5941, Fax: 979-845-9231, E-mail: akier@cvm.tamu.edu. *Application contact:* Dr. G. G. Wagner, Graduate Advisor, 979-845-2851, Fax: 979-862-1147, E-mail: gwagner@cvm.tamu.edu.

Thomas Jefferson University, Jefferson College of Graduate Studies, Program in Genetics, Philadelphia, PA 19107. Offers PhD. *Faculty:* 36 full-time (6 women). *Students:* 21 full-time (14 women), 1 (woman) part-time; includes 3 minority (2 African Americans, 1 Asian American or Pacific Islander), 1 international. 21 applicants, 29% accepted, 4 enrolled. In 2005, 2 doctorates awarded. *Degree requirements:* For doctorate, thesis/dissertation, comprehensive exam, registration. *Entrance requirements:* For doctorate, GRE General Test, minimum GPA of 3.2. Additional exam requirements/recommendations for international students: Required—TOEFL (minimum score 213 computer-based). *Application deadline:* For fall admission, 3/1 priority date for domestic students, 1/1 priority date for international students. Applications are processed on a rolling basis. Application fee: $50. Electronic applications accepted. *Expenses:* Tuition: Full-time $14,894; part-time $800 per credit. *Financial support:* In 2005–06, 2 students received support, including 21 fellowships with full tuition reimbursements available; research assistantships, Federal Work-Study, institutionally sponsored loans, scholarships/grants, and traineeships also available. Support available to part-time students. Financial award application deadline: 5/1; financial award applicants required to submit FAFSA. *Faculty research:* Functional genomics, cancer susceptibility, cell cycle, regulation oncogenes and tumor suppressor genes, genetics of neoplastic disease. Total annual research expenditures: $17 million. *Unit head:* Dr. Linda Siracusa, Program Director, 215-503-4536, E-mail: linda.siracusa@jefferson.edu. *Application contact:* Jessie F. Pervall, Director of Admissions, 215-503-0155, Fax: 215-503-9920, E-mail: jessie.pervall@jefferson.edu.

See Close-Up on page 791.

Tufts University, Sackler School of Graduate Biomedical Sciences, Graduate Program in Genetics, Boston, MA 02111. Offers PhD. *Faculty:* 43 full-time (15 women). *Students:* 28 full-time (18 women); includes 4 minority (2 Asian Americans or Pacific Islanders, 2 Hispanic Americans), 6 international. Average age 26. 99 applicants, 13% accepted, 5 enrolled. In 2005, 6 degrees awarded. *Degree requirements:* For doctorate, thesis/dissertation, qualifying exam. *Entrance requirements:* For doctorate, GRE General Test, 3 letters of reference. Additional exam requirements/recommendations for international students: Required—TOEFL. *Application deadline:* For fall admission, 1/15 priority date for domestic students, 1/15 priority date for international students. Applications are processed on a rolling basis. Application fee: $65. Electronic applications accepted. *Financial support:* In 2005–06, 28 students received support, including 28 research assistantships with full tuition reimbursements available (averaging $29,000 per year); scholarships/grants and health care benefits also available. Financial award application deadline: 1/15. *Faculty research:* Cancer, human and developmental genetics. *Unit head:* Dr. Ananda Roy, Program Director, 617-636-6715, E-mail: ananda.roy@tufts.edu. *Application contact:* 617-636-6767, Fax: 617-636-0375.

Université de Montréal, Faculty of Medicine and Faculty of Graduate Studies, Graduate Programs in Medicine, Program in Specialized Studies, Montréal, QC H3C 3J7, Canada. Offers anesthesia (DESS); diagnostic radiology (DESS); family medicine (DESS); medical biochemistry (DESS); medical genetics (DESS); medicine (DESS); microbiology and infectious diseases (DESS); nuclear medicine (DESS); obstetrics and gynecology (DESS); ophthalmology (DESS); pediatrics (DESS); psychiatry (DESS); radiology-oncology (DESS); surgery (DESS). *Faculty:* 159 full-time (37 women), 345 part-time/adjunct (102 women). *Entrance requirements:* For degree, proficiency in French. *Application deadline:* For fall and spring admission, 2/1. For winter admission, 11/1 for domestic students. Application fee: $30. Electronic applications accepted. *Unit head:* Renée Roy, Vice Dean, 514-343-7798.

Université du Québec à Chicoutimi, Graduate Programs, Program in Experimental Medicine, Chicoutimi, QC G7H 2B1, Canada. Offers genetics (M Sc). *Degree requirements:* For master's, thesis. *Entrance requirements:* For master's, appropriate bachelor's degree, proficiency in French.

University at Albany, State University of New York, School of Public Health, Department of Biomedical Sciences, Program in Biochemistry, Molecular Biology, and Genetics, Albany, NY 12222-0001. Offers MS, PhD. *Degree requirements:* For master's and doctorate, thesis/dissertation. *Entrance requirements:* For master's and doctorate, GRE General Test, GRE Subject Test. *Application deadline:* For fall admission, 2/15 for domestic students. Application fee: $60. *Financial support:* Application deadline: 2/1. *Unit head:* Dr. James Dias, Chair, Department of Biomedical Sciences, 518-474-2662.

The University of Alabama at Birmingham, Graduate Programs in Joint Health Sciences, Department of Genetics, Birmingham, AL 35294. Offers PhD. *Students:* 14 full-time (12 women); includes 2 minority (1 African American, 1 Asian American or Pacific Islander), 2 international. 17 applicants, 71% accepted. *Degree requirements:* For doctorate, thesis/dissertation. *Entrance requirements:* For doctorate, GRE, interview. *Application deadline:* Applications are processed on a rolling basis. Application fee: $35 ($60 for international students). Electronic applications accepted. *Expenses:* Tuition, state resident: part-time $170 per credit hour. Tuition, nonresident: full-time $4,612; part-time $425 per credit hour. International tuition: $10,732 full-time. Required fees: $11 per credit hour. $124 per term. Tuition and fees vary according to course load, degree level and program. *Financial support:* In 2005–06, 2 fellowships were awarded *Faculty research:* Clinical cytogenetics, cancer cytogenetics, prenatal diagnosis. *Unit head:* Dr. Bruce R. Korf, Chair, 205-934-9411.

University of Alberta, Faculty of Graduate Studies and Research, Department of Biological Sciences, Edmonton, AB T6G 2E1, Canada. Offers environmental biology and ecology (M Sc, PhD); microbiology and biotechnology (M Sc, PhD); molecular biology and genetics (M Sc, PhD); physiology and cell biology (M Sc, PhD); plant biology (M Sc, PhD); systematics and evolution (M Sc, PhD). *Faculty:* 72 full-time (15 women), 15 part-time/adjunct (4 women). *Students:* 238 full-time (117 women), 32 part-time (15 women), 31 international. 206 applicants, 42% accepted. In 2005, 29 master's, 31 doctorates awarded. Terminal master's awarded for partial completion of doctoral program. *Degree requirements:* For master's and doctorate, thesis/dissertation, registration. *Entrance requirements:* Additional exam requirements/recommendations for international students: Required—TOEFL. *Application deadline:* For fall admission, 3/1 for domestic students. Applications are processed on a rolling basis. Application fee: $0. Tuition and fees charges are reported in Canadian dollars. *Expenses:* Tuition, state resident: part-time $562 Canadian dollars per term. Tuition, nonresident: full-time $3,375 Canadian dollars. Required fees: $573 Canadian dollars; $84 Canadian dollars per term. *Financial support:* In 2005–06, 4 research assistantships with partial tuition reimbursements (averaging $12,000 per year), 103 teaching assistantships with partial tuition reimbursements (averaging $12,300 per year) were awarded; career-related internships or fieldwork and scholarships/grants also available. *Unit head:* Laura Frost, Chair, 780-492-1904. *Application contact:* Dr. John P. Chang, Associate Chair for Graduate Studies, 780-492-1257, Fax: 780-492-9457, E-mail: bio.grad.coordinator@ualberta.ca.

University of Alberta, Faculty of Medicine and Dentistry and Faculty of Graduate Studies and Research, Graduate Programs in Medicine, Department of Medical Genetics, Edmonton, AB T6G 2E1, Canada. Offers M Sc, PhD. *Degree requirements:* For master's, thesis/dissertation; for doctorate, thesis/dissertation, comprehensive exam. *Entrance requirements:* For master's and doctorate, minimum GPA of 7.5 on a 9.0 scale. Tuition and fees charges are reported in Canadian dollars. *Expenses:* Tuition, state resident: part-time $562 Canadian dollars per term. Tuition, nonresident: full-time $3,375 Canadian dollars. Required fees: $573 Canadian dollars; $84 Canadian dollars per term. *Faculty research:* Clinical and molecular cytogenetics, ocular genetics, Prader-Willi syndrome, genomic instability, developmental genetics.

The University of Arizona, Graduate College, Graduate Interdisciplinary Programs, Graduate Interdisciplinary Program in Genetics, Tucson, AZ 85721. Offers MS, PhD. *Degree requirements:* For master's, thesis; for doctorate, one foreign language; thesis/dissertation. *Entrance requirements:* For master's, GRE General Test, minimum GPA of 3.5; for doctorate, GRE General Test, master's degree or equivalent. Additional exam requirements/recommendations for international students: Required—TOEFL. *Faculty research:* Cancer research; DNA repair; plant and animal cytogenetics; molecular, population, and ecological genetics.

The University of British Columbia, Faculty of Graduate Studies, Advisory Committee on Genetics, Vancouver, BC V6T 1Z1, Canada. Offers M Sc, PhD. *Faculty:* 67 part-time/adjunct (20 women). *Students:* 93 full-time (45 women); includes 24 minority (1 African American, 22 Asian Americans or Pacific Islanders, 1 Hispanic American). Average age 24. 100 applicants, 30% accepted, 22 enrolled. In 2005, 6 master's, 4 doctorates awarded. *Median time to degree:* Of those who began their doctoral program in fall 1997, 90% received their degree in 8 years or less. *Degree requirements:* For master's, thesis, thesis defense, comprehensive exam, registration; for doctorate, thesis/dissertation, qualifying exam, oral and written comprehensive exams, comprehensive exam, registration. *Entrance requirements:* Additional exam requirements/recommendations for international students: Required—TOEFL. *Application deadline:* For fall admission, 4/30 for domestic students. Applications are processed on a rolling basis. Application fee: $90 ($128 for international students). *Financial support:* In 2005–06, 5 students received support, including research assistantships (averaging $16,500 per year); fellowships, teaching assistantships Financial award application deadline: 1/15. *Faculty research:* Prokaryote and eukaryote genetics. *Unit head:* Dr. H. W. Brock, Director, 604-822-8764, Fax: 604-822-9865, E-mail: genetics@interchange.ubc.ca. *Application contact:* Monica Deutsch, Graduate Admissions Secretary, 604-822-8764, Fax: 604-822-9865, E-mail: genetics@interchange.ubc.ca.

The University of British Columbia, Faculty of Medicine and Faculty of Graduate Studies, Graduate Programs in Medicine, Department of Medical Genetics, Vancouver, BC V6T 1Z3, Canada. Offers M Sc, PhD. *Faculty:* 36 full-time (19 women). *Students:* 35 full-time (21 women). 35 applicants, 54% accepted, 14 enrolled. In 2005, 4 master's, 7 doctorates awarded. *Degree requirements:* For master's and doctorate, thesis/dissertation. *Entrance requirements:* For master's, first class or upper second class B Sc with strong genetics course work; for doctorate, first class B Sc with strong genetics course work or M Sc with strong genetics course work. Additional exam requirements/recommendations for international students: Required—TOEFL. *Application deadline:* For fall admission, 3/1 for domestic students, 2/1 for international students; for spring admission, 9/1 for domestic students, 8/1 for international students. Applications are processed on a rolling basis. Application fee: $90 Canadian dollars ($150 Canadian dollars for international students). Electronic applications accepted. *Financial support:* In 2005–06, 24 students received support, including fellowships (averaging $16,000 per year), research assistantships (averaging $16,000 per year), 4 teaching assistantships (averaging $4,529 per year) *Faculty research:* Human molecular genetics, developmental genetics, clinical genetics, immunogenetics, hematopoiesis. *Unit head:* Dr. Robert McMaster, Head, 604-875-3493, Fax: 604-875-3490, E-mail: robm@interchange.ubc.ca. *Application contact:* Cheryl Bishop, Graduate Secretary, 604-822-5312, Fax: 604-822-5348, E-mail: medgen@interchange.ubc.ca.

Genetics

University of California, Davis, Graduate Studies, Graduate Group in Genetics, Davis, CA 95616. Offers MS, PhD. *Faculty:* 97 full-time. *Students:* 79 full-time (52 women); includes 7 minority (all Asian Americans or Pacific Islanders), 24 international. Average age 29. 73 applicants, 68% accepted, 14 enrolled. In 2005, 5 master's, 6 doctorates awarded. Terminal master's awarded for partial completion of doctoral program. *Median time to degree:* Of those who began their doctoral program in fall 1997, 57.1% received their degree in 8 years or less. *Degree requirements:* For master's, thesis (for some programs), comprehensive exam (for some programs); for doctorate, thesis/dissertation. *Entrance requirements:* For master's and doctorate, GRE General Test, GRE Subject Test. Additional exam requirements/recommendations for international students: Required—TOEFL (minimum score 550 paper-based; 213 computer-based). *Application deadline:* For fall admission, 1/15 for domestic students, 1/15 for international students. Applications are processed on a rolling basis. Application fee: $60. Electronic applications accepted. *Financial support:* In 2005–06, 74 students received support, including 26 fellowships with full and partial tuition reimbursements available (averaging $14,601 per year), 27 research assistantships with full and partial tuition reimbursements available (averaging $15,474 per year), 11 teaching assistantships with partial tuition reimbursements available (averaging $15,768 per year); Federal Work-Study, institutionally sponsored loans, scholarships/grants, tuition waivers (full and partial), and unspecified assistantships also available. Financial award application deadline: 1/15; financial award applicants required to submit FAFSA. *Faculty research:* Molecular, quantitative, and developmental genetics; cytogenetics; plant breeding. *Unit head:* James Murray, Graduate Chair, 530-752-3179, E-mail: jdmurray@ucdavis.edu. *Application contact:* Ellen Picht, Administrative Assistant, 530-752-4863, E-mail: empicht@ucdavis.edu.

University of California, Irvine, Office of Graduate Studies, School of Biological Sciences and College of Medicine, Graduate Program in Molecular Biology, Genetics, and Biochemistry, Irvine, CA 92697-3915. Offers biological sciences (PhD). *Faculty:* 156 full-time (38 women). *Students:* 53 full-time (30 women). Average age 25. 452 applicants, 26% accepted, 53 enrolled. In 2005, 19 doctorates awarded. *Median time to degree:* Of those who began their doctoral program in fall 1997, 10% received their degree in 8 years or less. *Degree requirements:* For doctorate, thesis/dissertation, teaching assignment, preliminary exam. *Entrance requirements:* For doctorate, GRE General Test, minimum GPA of 3.0, research experience. Additional exam requirements/recommendations for international students: Required—TOEFL, IELT, TSE. *Application deadline:* For fall admission, 1/1 for domestic students, 1/1 for international students. Application fee: $60 ($80 for international students). Electronic applications accepted. *Expenses: Contact institution. Financial support:* In 2005–06, 52 fellowships with full tuition reimbursements (averaging $25,000 per year) were awarded; institutionally sponsored loans, scholarships/grants, tuition waivers (full), and stipends also available. Financial award application deadline: 1/7; financial award applicants required to submit FAFSA. *Faculty research:* Cellular biochemistry; gene structure and expression; protein structure, function, and design; molecular genetics; pathogenesis and inherited disease. *Unit head:* Dr. Peter J. Bryant, Director, 949-824-4714, Fax: 949-824-3571, E-mail: gp-mbgb@uci.edu. *Application contact:* Kimberly McKinney, Administrator, 949-824-8145, Fax: 949-824-1965, E-mail: kamckinn@uci.edu.

University of California, Riverside, Graduate Division, Graduate Program in Genetics, Genomics, and Bioinformatics, Riverside, CA 92521-0102. Offers genomics and bioinformatics (PhD); molecular genetics (PhD); population and evolutionary genetics (PhD). *Students:* 19 full-time (7 women); includes 3 minority (2 Asian Americans or Pacific Islanders, 1 Hispanic American), 7 international. Average age 30. In 2005, 1 degree awarded. *Degree requirements:* For doctorate, thesis/dissertation, qualifying exams, teaching experience. *Entrance requirements:* For doctorate, GRE General Test, minimum GPA of 3.2. Additional exam requirements/recommendations for international students: Required—TOEFL (minimum score 550 paper-based; 213 computer-based), Recommended—TSE (minimum score 50). *Application deadline:* For fall admission, 5/1 for domestic students, 2/1 for international students. For winter admission, 9/1 for domestic students; for spring admission, 12/1 for domestic students. Applications are processed on a rolling basis. Application fee: $60 ($75 for international students). Electronic applications accepted. *Expenses:* Tuition, nonresident: full-time $14,694. Full-time tuition and fees vary according to program. *Financial support:* In 2005–06, research assistantships (averaging $14,000 per year), teaching assistantships (averaging $15,000 per year) were awarded; fellowships, career-related internships or fieldwork, Federal Work-Study, institutionally sponsored loans, and tuition waivers (full and partial) also available. Financial award application deadline: 2/1; financial award applicants required to submit FAFSA. *Unit head:* Dr. Katherine Borkovich, Director. *Application contact:* Carole Carpenter, Graduate Program Assistant, 800-735-0717, Fax: 951-827-5517, E-mail: genetics@ucr.edu.

University of California, San Diego, Graduate Studies and Research, Division of Biology, Program in Genetics and Molecular Biology, La Jolla, CA 92093-0348. Offers PhD. Offered in association with the Salk Institute. *Degree requirements:* For doctorate, thesis/dissertation, qualifying exam. Electronic applications accepted.

University of California, San Francisco, Graduate Division and School of Medicine, Department of Biochemistry and Biophysics, Program in Genetics, San Francisco, CA 94143. Offers PhD, MD/PhD. *Degree requirements:* For doctorate, thesis/dissertation. *Entrance requirements:* For doctorate, GRE General Test, GRE Subject Test. Additional exam requirements/recommendations for international students: Required—TOEFL. Expenses: Contact institution. *Faculty research:* Gene expression; chromosome structure and mechanics; medical, somatic cell, and radiation genetics.

University of Chicago, Division of the Biological Sciences, Department of Molecular Biosciences: Biochemistry, Genetics, Cell and Developmental Biology, Committee on Genetics, Chicago, IL 60637-1513. Offers PhD. *Faculty:* 65 full-time (24 women). *Students:* 28 full-time (17 women); includes 4 minority (2 African Americans, 1 Asian American or Pacific Islander, 1 Hispanic American), 5 international. Average age 27. In 2005, 5 doctorates awarded. *Degree requirements:* For doctorate, thesis/dissertation, registration. *Entrance requirements:* For doctorate, GRE General Test, minimum GPA of 3.0. Additional exam requirements/recommendations for international students: Required—TOEFL. *Application deadline:* For fall admission, 12/28 priority date for domestic students, 12/28 priority date for international students. Application fee: $55. Electronic applications accepted. *Financial support:* In 2005–06, 23 students received support, including fellowships with tuition reimbursements available (averaging $26,301 per year), research assistantships with tuition reimbursements available (averaging $26,301 per year); institutionally sponsored loans, scholarships/grants, traineeships, and health care benefits also available. Financial award applicants required to submit FAFSA. *Faculty research:* Molecular genetics, developmental genetics, population genetics, human genetics. *Unit head:* Dr. Douglas Bishop, Chair, 773-702-9211, Fax: 773-702-8093, E-mail: committee-on-genetics@uchicago.edu. *Application contact:* Sue Levison, Administrator, 773-702-2464, Fax: 773-702-3172, E-mail: committee-on-genetics@uchicago.edu.

University of Colorado at Boulder, Graduate School, College of Arts and Sciences, Department of Ecology and Evolutionary Biology, Boulder, CO 80309. Offers animal behavior (MA); biology (MA, PhD); environmental biology (MA, PhD); evolutionary biology (MA, PhD); neurobiology (MA); population biology (MA); population genetics (PhD). *Faculty:* 26 full-time (5 women). *Students:* 44 full-time (24 women), 22 part-time (14 women); includes 16 minority (1 American Indian/Alaska Native, 3 Asian Americans or Pacific Islanders, 12 Hispanic Americans), 2 international. Average age 30. 20 applicants, 90% accepted. In 2005, 7 master's, 7 doctorates awarded. Terminal master's awarded for partial completion of doctoral program. *Degree requirements:* For master's, thesis or alternative, comprehensive exam; for doctorate, thesis/dissertation, comprehensive exam. *Entrance requirements:* For master's, GRE General Test, GRE Subject Test, minimum undergraduate GPA of 3.0; for doctorate, GRE General Test, GRE Subject Test. *Application deadline:* For fall admission, 1/2 priority date for domestic students, 12/1 priority date for international students. Application fee: $50 ($60 for international students). *Financial support:* In 2005–06, fellowships (averaging $7,844 per year), research assistantships (averaging $15,663 per year), teaching assistantships (averaging $13,829 per year) were awarded; Federal Work-Study, institutionally sponsored loans, and tuition waivers (full)

also available. Financial award application deadline: 3/1. *Faculty research:* Behavior, ecology, genetics, morphology, endocrinology. Total annual research expenditures: $5.2 million. *Unit head:* Jeffry Mitton, Chair, 303-492-0505, Fax: 303-492-8699, E-mail: mitton@colorado.edu. *Application contact:* Jill Skarstadt, Graduate Coordinator, 303-492-7654, Fax: 303-492-8699, E-mail: skarstad@colorado.edu.

University of Colorado at Denver and Health Sciences Center, Graduate School, Program in Biomedical Sciences, Program in Human Medical Genetics, Denver, CO 80262. Offers PhD. In 2005, 2 degrees awarded. *Degree requirements:* For doctorate, thesis/dissertation, 3 laboratory rotations. *Entrance requirements:* For doctorate, GRE General Test, minimum GPA of 3.0, 4 letters of recommendation. Additional exam requirements/recommendations for international students: Required—TOEFL (minimum score 550 paper-based; 213 computer-based). *Application deadline:* For fall admission, 1/1 for domestic students. Application fee: $50. *Expenses:* Tuition, state resident: full-time $11,730. Tuition, nonresident: full-time $22,980. Tuition and fees vary according to degree level and program. *Financial support:* Fellowships, research assistantships, teaching assistantships, Federal Work-Study and institutionally sponsored loans available. Support available to part-time students. Financial award application deadline: 3/15; financial award applicants required to submit FAFSA. *Faculty research:* Genetics of colon cancer, cancer cytogenetics, tumor suppressor genes and cancer, molecular basis of inherited human disease, neurodevelopmental genetics. *Unit head:* Dr. Richard A. Spritz, Director, 303-724-3107, E-mail: richard.spritz@uchsc.edu. *Application contact:* MJ Stewart, Administrator, 303-315-7739, Fax: 303-315-0407, E-mail: mj.stewart@uchsc.edu.

University of Connecticut, Graduate School, College of Liberal Arts and Sciences, Department of Molecular and Cell Biology, Field of Genetics, Genomics, and Bioinformatics, Storrs, CT 06269. Offers genetics (MS, PhD). *Students:* 33 full-time (11 women), 3 part-time (2 women); includes 4 minority (2 African Americans, 2 Asian Americans or Pacific Islanders), 7 international. Average age 27. 93 applicants, 25% accepted, 14 enrolled. In 2005, 1 master's, 1 doctorate awarded. Terminal master's awarded for partial completion of doctoral program. *Entrance requirements:* For master's, comprehensive exam; for doctorate, thesis/dissertation. *Entrance requirements:* For master's and doctorate, GRE General Test, GRE Subject Test. Additional exam requirements/recommendations for international students: Required—TOEFL (minimum score 550 paper-based; 213 computer-based). *Application deadline:* For fall admission, 2/1 priority date for domestic students, 2/1 priority date for international students; for spring admission, 11/1 for domestic students, 10/1 for international students. Applications are processed on a rolling basis. Application fee: $55. Electronic applications accepted. *Expenses:* Tuition, state resident: part-time $444 per credit hour. Tuition, nonresident: part-time $1,154 per credit hour. Tuition and fees vary according to course load. *Financial support:* In 2005–06, 7 research assistantships with full tuition reimbursements, 22 teaching assistantships with full tuition reimbursements were awarded; fellowships, Federal Work-Study, scholarships/grants, health care benefits, and unspecified assistantships also available. Financial award application deadline: 2/1; financial award applicants required to submit FAFSA. *Application contact:* Anne St. Onje, Graduate Coordinator, 860-486-4314, Fax: 860-486-3943, E-mail: ann.st_onje@uconn.edu.

See Close-Up on page 611.

University of Connecticut Health Center, Graduate School, Programs in Biomedical Sciences, Program in Genetics and Developmental Biology, Farmington, CT 06030. Offers PhD, DMD/PhD, MD/PhD. *Degree requirements:* For doctorate, thesis/dissertation, comprehensive exam, registration. *Entrance requirements:* For doctorate, GRE General Test, GRE Subject Test. Additional exam requirements/recommendations for international students: Required—TOEFL (minimum score 600 paper-based; 250 computer-based). Electronic applications accepted. *Faculty research:* Limb development/bone formation, genetics and human disease, apoptosis and development, signal transduction in development.

See Close-Up on page 795.

University of Connecticut Health Center, Graduate School, Programs in Biomedical Sciences, Program in Molecular Biology and Biochemistry, Farmington, CT 06030. Offers PhD, DMD/PhD, MD/PhD. *Degree requirements:* For doctorate, thesis/dissertation, comprehensive exam, registration. *Entrance requirements:* For doctorate, GRE General Test. Additional exam requirements/recommendations for international students: Required—TOEFL (minimum score 600 paper-based; 250 computer-based). Electronic applications accepted.

See Close-Up on page 615.

University of Delaware, College of Arts and Sciences, Department of Biological Sciences, Newark, DE 19716. Offers biotechnology (MS); cancer biology (MS, PhD); cell and extracellular matrix biology (MS, PhD); cell and systems physiology (MS, PhD); developmental biology (MS, PhD); ecology and evolution (MS, PhD); microbiology (MS, PhD); molecular biology and genetics (MS, PhD). *Faculty:* 39 full-time (11 women). *Students:* 64 full-time (46 women), 2 part-time; includes 7 minority (4 African Americans, 2 Asian Americans or Pacific Islanders, 1 Hispanic American), 18 international. Average age 26. 113 applicants, 31% accepted, 19 enrolled. In 2005, 5 master's, 4 doctorates awarded. Terminal master's awarded for partial completion of doctoral program. *Median time to degree:* Of those who began their doctoral program in fall 1997, 100% received their degree in 8 years or less. *Degree requirements:* For master's, thesis/dissertation, preliminary exam; for doctorate, thesis/dissertation, preliminary exam, comprehensive exam. *Entrance requirements:* For master's and doctorate, GRE General Test. Additional exam requirements/recommendations for international students: Required—TOEFL (minimum score 600 paper-based; 250 computer-based); Recommended—TWE, TSE. *Application deadline:* For fall admission, 4/15 for domestic students, 1/15 for international students; for spring admission, 10/1 for domestic students. Applications are processed on a rolling basis. Application fee: $60. Electronic applications accepted. *Financial support:* In 2005–06, 26 students received support, including fellowships with full tuition reimbursements available (averaging $19,000 per year), 19 research assistantships with full tuition reimbursements available (averaging $19,000 per year), 26 teaching assistantships with full tuition reimbursements available (averaging $19,000 per year); tuition waivers (partial) also available. Financial award application deadline: 4/15. *Faculty research:* Microorganisms, bone cancer metastasis, developmental biology, cell biology, DNA. Total annual research expenditures: $8.3 million. *Unit head:* Dr. Daniel D. Carson, Chair, 302-831-6977, Fax: 302-831-2281, E-mail: carson@udel.edu. *Application contact:* Dr. Melinda K. Duncan, Graduate Coordinator, 302-831-1841, Fax: 302-831-2281, E-mail: danders@udel.edu.

University of Florida, College of Medicine and Graduate School, Interdisciplinary Program in Biomedical Sciences, Concentration in Genetics, Gainesville, FL 32611. Offers PhD. *Faculty:* 1 full-time (0 women). *Students:* 28 full-time (17 women); includes 4 minority (2 African Americans, 1 Asian American or Pacific Islander, 1 Hispanic American). In 2005, 12 degrees awarded. *Degree requirements:* For doctorate, thesis/dissertation. *Entrance requirements:* For doctorate, GRE General Test, minimum GPA of 3.0. Additional exam requirements/recommendations for international students: Required—TOEFL. *Application deadline:* For fall admission, 2/15 for domestic students. Application fee: $30. Electronic applications accepted. *Expenses:* Tuition, state resident: full-time $6,234. Tuition, nonresident: full-time $21,359. Tuition and fees vary according to program. *Financial support:* Fellowships with full tuition reimbursements, research assistantships with full tuition reimbursements, traineeships and unspecified assistantships available. *Unit head:* Dr. Henry V. Baker, Director, 352-392-0680, E-mail: baker@medmicro.med.ufl.edu. *Application contact:* Dr. Wayne McCormack, Associate Dean of Graduate Education, 352-392-7413, Fax: 352-846-3466, E-mail: mccormac@pathology.ufl.edu.

University of Georgia, Graduate School, College of Arts and Sciences, Department of Genetics, Athens, GA 30602. Offers MS, PhD. *Faculty:* 16 full-time (4 women). *Students:* 48 full-time, 2 part-time; includes 2 minority (1 African American, 1 Hispanic American), 17 international. 102 applicants, 19% accepted, 6 enrolled. In 2005, 9 degrees awarded. Terminal master's awarded for partial completion of doctoral program. *Degree requirements:* For master's,

Genetics

thesis/dissertation; for doctorate, thesis/dissertation, comprehensive exam. *Entrance requirements:* For master's and doctorate, GRE General Test. Additional exam requirements/recommendations for international students: Required—TOEFL. *Application deadline:* For fall admission, 1/1 priority date for domestic students, 1/1 priority date for international students; for spring admission, 11/15 for domestic students. Application fee: $50. Electronic applications accepted. *Financial support:* In 2005–06, fellowships with full tuition reimbursements (averaging $19,000 per year), research assistantships with full tuition reimbursements (averaging $19,000 per year), teaching assistantships with full tuition reimbursements (averaging $19,000 per year) were awarded; scholarships/grants and unspecified assistantships also available. *Unit head:* Dr. Robert Ivarie, Head, 706-542-1424, Fax: 706-542-3910, E-mail: ivarie@uga.edu. *Application contact:* Dr. Daniel Promislow, Graduate Coordinator, 706-542-8000, Fax: 706-542-3910, E-mail: promislo@uga.edu.

University of Hawaii at Manoa, John A. Burns School of Medicine and Graduate Division, Graduate Programs in Biomedical Sciences, Department of Cell and Molecular Biology, Honolulu, HI 96822. Offers MS, PhD. *Faculty:* 61 full-time (21 women), 7 part-time/adjunct (3 women). *Students:* 28 full-time (13 women), 5 part-time (3 women); includes 12 minority (11 Asian Americans or Pacific Islanders, 1 Hispanic American), 6 international. Average age 29. 62 applicants, 19% accepted, 10 enrolled. In 2005, 2 master's, 3 doctorates awarded. Terminal master's awarded for partial completion of doctoral program. *Degree requirements:* For master's, thesis (for some programs); for doctorate, thesis/dissertation. *Entrance requirements:* For master's and doctorate, GRE, minimum GPA of 3.0. *Application deadline:* For fall admission, 2/15 for domestic students, 2/15 for international students. Applications are processed on a rolling basis. Application fee: $50. *Expenses:* Tuition, state resident: full-time $8,400; part-time $200 per credit hour. Tuition, nonresident: full-time $11,088; part-time $462 per credit hour. Tuition and fees vary according to program. *Financial support:* In 2005–06, 20 research assistantships (averaging $16,774 per year), 3 teaching assistantships (averaging $14,382 per year) were awarded; Federal Work-Study and institutionally sponsored loans also available. Financial award application deadline: 2/1. *Unit head:* Dr. David Haymer, 808-956-7862, Fax: 808-956-5506, E-mail: dhaymer@hawaii.edu.

University of Illinois at Chicago, College of Medicine and Graduate College, Graduate Programs in Medicine, Chicago, IL 60607-7128. Offers anatomy and cell biology (MS, PhD); biochemistry and molecular biology (MS, PhD); genetics (PhD), including molecular genetics; health professions education (MHPE); microbiology and immunology (PhD); pharmacology (PhD); physiology and biophysics (MS, PhD); surgery (MS). Part-time programs available. Terminal master's awarded for partial completion of doctoral program. *Degree requirements:* For master's and doctorate, thesis/dissertation. *Entrance requirements:* For master's and doctorate, GRE General Test. Expenses: Contact institution.

University of Illinois at Chicago, Graduate College, College of Liberal Arts and Sciences, Department of Biological Sciences, Chicago, IL 60607-7128. Offers cell and developmental biology (PhD); ecology and evolution (MS, DA, PhD); genetics and development (PhD); molecular biology (MS, PhD); neurobiology (MS, PhD); plant biology (MS, DA, PhD). *Degree requirements:* For master's, thesis; for doctorate, thesis/dissertation, preliminary exam. *Entrance requirements:* For master's and doctorate, GRE General Test, GRE Subject Test, previous course work in physics, calculus, and organic chemistry; minimum GPA of 2.75. Additional exam requirements/recommendations for international students: Required—TOEFL. Electronic applications accepted.

The University of Iowa, Graduate College, College of Liberal Arts and Sciences, Department of Biological Sciences, Iowa City, IA 52242-1316.

See Close-Up on page 257.

The University of Iowa, Graduate College, College of Public Health, Program in Public Health Genetics, Iowa City, IA 52242-1316. Offers statistical genetics (PhD, Certificate). *Faculty:* 5 full-time, 1 part-time/adjunct (0 women). *Students:* 7 full-time (4 women), 2 part-time (1 woman); includes 1 minority (Asian American or Pacific Islander), 6 international. 12 applicants, 33% accepted, 2 enrolled. In 2005, 1 degree awarded. *Degree requirements:* For doctorate, thesis/dissertation, comprehensive exam, registration. *Entrance requirements:* For doctorate, GRE General Test, minimum GPA of 3.0. Additional exam requirements/recommendations for international students: Required—TOEFL (minimum score 600 paper-based; 250 computer-based). *Application deadline:* Applications are processed on a rolling basis. Application fee: $60 ($85 for international students). Electronic applications accepted. *Financial support:* In 2005–06, 1 fellowship, 5 research assistantships with partial tuition reimbursements were awarded; teaching assistantships with partial tuition reimbursements Financial award applicants required to submit FAFSA. *Unit head:* Dr. Veronica J. Vieland, Head, 319-353-4782, Fax: 319-353-3038.

The University of Iowa, Graduate College, Program in Genetics, Iowa City, IA 52242-1316. Offers PhD, MD/PhD. *Students:* 12 full-time (7 women), 25 part-time (11 women); includes 1 minority (Hispanic American), 11 international. 14 applicants, 29% accepted, 3 enrolled. In 2005, 7 degrees awarded. *Degree requirements:* For doctorate, thesis/dissertation, comprehensive exam, registration. *Entrance requirements:* For doctorate, GRE General Test, minimum GPA of 3.0. Additional exam requirements/recommendations for international students: Required—TOEFL (minimum score 550 paper-based; 213 computer-based). *Application deadline:* For fall admission, 1/1 priority date for domestic students, 1/1 priority date for international students. Applications are processed on a rolling basis. Application fee: $60 ($85 for international students). Electronic applications accepted. *Expenses: Contact institution.* Tuition and fees vary according to course load and program. *Financial support:* In 2005–06, 2 fellowships, 31 research assistantships with partial tuition reimbursements, 1 teaching assistantship with partial tuition reimbursement were awarded; traineeships also available. Financial award applicants required to submit FAFSA. *Faculty research:* Developmental genetics, eukaryotic gene expression, human genetics, molecular and biochemical genetics, evolutionary genetics. *Unit head:* Dr. Debashish Bhattacharya, Director, 319-335-3603, E-mail: phd@genetics.uiowa.edu. *Application contact:* Anita Kafer, Director, 319-335-9968, E-mail: phd@genetics.uiowa.edu.

Announcement: The interdisciplinary PhD program in genetics at the University of Iowa offers broad training in genetics. The program is designed to permit students to tailor their curricula to their own research interests and career goals, while providing them with the opportunity to obtain a broad intellectual foundation in many areas of genetics. For further information, e-mail gen-neuro@uiowa.edu or visit the Web site (faculty research descriptions) at http://genetics.grad.uiowa.edu.

The University of Iowa, Roy J. and Lucille A. Carver College of Medicine and Graduate College, Graduate Programs in Medicine, Department of Microbiology, Iowa City, IA 52242-1316. Offers general microbiology and microbial physiology (MS, PhD); immunology (MS, PhD); microbial genetics (MS, PhD); pathogenic bacteriology (MS, PhD); virology (MS, PhD). *Faculty:* 23 full-time (5 women), 1 part-time/adjunct (2 women). *Students:* 54 full-time (25 women); includes 4 minority (3 Asian Americans or Pacific Islanders, 1 Hispanic American), 8 international. 93 applicants, 12% accepted, 4 enrolled. In 2005, 7 degrees awarded. *Median time to degree:* Of those who began their doctoral program in fall 1997, 80% received their degree in 8 years or less. *Degree requirements:* For master's, thesis/dissertation; for doctorate, thesis/dissertation, comprehensive exam. *Entrance requirements:* For master's and doctorate, GRE General Test. Additional exam requirements/recommendations for international students: Required—TOEFL. *Application deadline:* For fall admission, 2/1 for domestic students, 2/1 for international students. Application fee: $50 ($75 for international students). Electronic applications accepted. *Expenses:* Tuition, state resident: part-time $1,882 per term. Tuition, nonresident: full-time $17,338; part-time $4,907 per term. Tuition and fees vary according to course load and program. *Financial support:* In 2005–06, 63 research assistantships with full tuition reimbursements (averaging $22,000 per year) were awarded; institutionally sponsored loans, scholarships/grants, traineeships, and health care benefits also available. *Faculty research:* Biocatalysis and blue jeans, gene regulation, processing and transport of HIV,

retroviral pathogenesis, biodegradation. Total annual research expenditures: $9 million. *Unit head:* Dr. Michael A. Apicella, Head, 319-335-7810, E-mail: grad-micro-info@uiowa.edu.

See Close-Up on page 923.

University of Miami, Graduate School, College of Arts and Sciences, Department of Biology, Coral Gables, FL 33124. Offers biology (MS, PhD); genetics and evolution (MS, PhD). *Faculty:* 21 full-time (3 women), 10 part-time/adjunct (3 women). *Students:* 49 full-time (26 women); includes 2 minority (both Asian Americans or Pacific Islanders), 16 international. Average age 30. 39 applicants, 36% accepted, 10 enrolled. In 2005, 6 degrees awarded. Terminal master's awarded for partial completion of doctoral program. *Median time to degree:* Of those who began their doctoral program in fall 1997, 100% received their degree in 8 years or less. *Degree requirements:* For master's, thesis (for some programs), comprehensive exam (for some programs); for doctorate, thesis/dissertation, oral and written qualifying exam. *Entrance requirements:* For master's and doctorate, GRE General Test, 3 letters of recommendation, research papers. Additional exam requirements/recommendations for international students: Required—TOEFL (minimum score 550 paper-based; 213 computer-based), TSE. *Application deadline:* For fall admission, 1/1 for domestic students, 1/1 for international students. Application fee: $50. Electronic applications accepted. *Financial support:* In 2005–06, 49 students received support, including 9 fellowships with full tuition reimbursements available (averaging $18,000 per year), 15 research assistantships with full tuition reimbursements available (averaging $16,500 per year), 20 teaching assistantships with full tuition reimbursements available (averaging $16,500 per year); career-related internships or fieldwork, Federal Work-Study, institutionally sponsored loans, scholarships/grants, health care benefits, and unspecified assistantships also available. Financial award application deadline: 3/1; financial award applicants required to submit FAFSA. *Faculty research:* Neuroscience to ethology; plants, vertebrates and mycorrhizae; phylogenies, life histories and species interactions; molecular biology, gene expression and populations; cells, auditory neurons and vetebrate locomotion. Total annual research expenditures: $946,278. *Unit head:* Dr. Leonel O. Sternberg, Director of Graduate Studies, 305-284-6436, Fax: 305-284-3039, E-mail: leo@bio.miami.edu. *Application contact:* Beth E. Goad, Graduate Coordinator, 305-284-5116, Fax: 305-284-3039, E-mail: bgoad@bio.miami.edu.

See Close-Up on page 269.

University of Michigan, Horace H. Rackham School of Graduate Studies, Interdisciplinary Program in Genetics, Ann Arbor, MI 48109.

University of Minnesota, Twin Cities Campus, Graduate School, Program in Molecular, Cellular, Developmental Biology and Genetics, Minneapolis, MN 55455-0213. Offers genetic counseling (MS); molecular, cellular, developmental biology and genetics (PhD). Part-time programs available. Terminal master's awarded for partial completion of doctoral program. *Degree requirements:* For master's, thesis optional; for doctorate, thesis/dissertation. *Entrance requirements:* For master's and doctorate, GRE General Test. Additional exam requirements/recommendations for international students: Required—TOEFL. *Expenses:* Tuition, state resident: full-time $8,748; part-time $729 per credit. Tuition, nonresident: full-time $15,848; part-time $1,321 per credit. Full-time tuition and fees vary according to class time, course load, program and reciprocity agreements. *Faculty research:* Membrane receptors and membrane transport, cell interactions, cytoskeleton and cell mobility, regulation of gene expression, plant cell and molecular biology.

University of Missouri–Columbia, Graduate School, Genetics Area Program, Columbia, MO 65211. Offers PhD. *Students:* 8 full-time (4 women), 6 part-time (3 women), 9 international. In 2005, 6 doctorates awarded. *Degree requirements:* For doctorate, thesis/dissertation, comprehensive exam. *Entrance requirements:* For doctorate, GRE General Test, minimum GPA of 3.0. *Application deadline:* For fall admission, 2/1 for domestic students. Applications are processed on a rolling basis. Application fee: $45 ($60 for international students). *Financial support:* Fellowships, research assistantships, teaching assistantships, institutionally sponsored loans available. *Unit head:* , Dr. John F. Cannon, Director of Graduate Studies, 573-852-2780, E-mail: cannonj@missouri.edu.

University of Missouri–St. Louis, College of Arts and Sciences, Department of Biology, St. Louis, MO 63121. Offers biology (MS, PhD), including animal behavior (MS), biochemistry (MS), biotechnology (MS), conservation biology (MS), development (MS), ecology (MS), environmental studies (PhD), evolution (MS), genetics (MS), molecular/cellular biology (MS), physiology (MS), plant systematics, population biology (MS), tropical biology (MS); biotechnology (Certificate); tropical biology and conservation (Certificate). Part-time programs available. *Faculty:* 50. *Students:* 26 full-time (15 women), 101 part-time (51 women); includes 13 minority (5 African Americans, 6 Asian Americans or Pacific Islanders, 2 Hispanic Americans), 41 international. Average age 32. In 2005, 22 master's, 2 doctorates awarded. *Degree requirements:* For master's, thesis or alternative; for doctorate, one foreign language, thesis/dissertation, 1 semester of teaching experience. *Entrance requirements:* For doctorate, GRE General Test. *Application deadline:* For spring admission, 12/1 priority date for domestic students. Applications are processed on a rolling basis. Application fee: $35 ($40 for international students). Electronic applications accepted. *Expenses:* Tuition, state resident: part-time $263 per credit hour. Tuition, nonresident: part-time $680 per credit hour. Required fees: $53 per credit hour. Tuition and fees vary according to program. *Financial support:* In 2005–06, 11 fellowships with full tuition reimbursements (averaging $30,000 per year), 15 research assistantships with full and partial tuition reimbursements (averaging $16,000 per year), 22 teaching assistantships with full and partial tuition reimbursements (averaging $16,000 per year) were awarded; career-related internships or fieldwork and Federal Work-Study also available. Support available to part-time students. Financial award application deadline: 2/1. *Faculty research:* Molecular biology, microbial genetics. *Unit head:* Dr. Zuleyma Tang-Martinez, Director of Graduate Studies, 314-516-6498, Fax: 314-516-6233, E-mail: zuleyma@umsl.edu. *Application contact:* 314-516-5458, Fax: 314-516-5310, E-mail: gradadm@umsl.edu.

University of New Hampshire, Graduate School, College of Life Sciences and Agriculture, Program in Genetics, Durham, NH 03824. Offers MS, PhD. Part-time programs available. *Faculty:* 12 full-time. *Students:* 10 full-time (8 women), 4 part-time (2 women), 2 international. Average age 28. 5 applicants, 60% accepted, 3 enrolled. In 2005, 1 doctorate awarded. *Degree requirements:* For master's and doctorate, thesis/dissertation. *Entrance requirements:* For master's and doctorate, GRE General Test, GRE Subject Test. Additional exam requirements/recommendations for international students: Required—TOEFL (minimum score 550 paper-based; 213 computer-based); Recommended—TSE. *Application deadline:* For fall admission, 4/1 priority date for domestic students, 4/1 priority date for international students. For winter admission, 12/1 for domestic students. Applications are processed on a rolling basis. Application fee: $60. Electronic applications accepted. *Expenses:* Tuition, state resident: full-time $8,010; part-time $445 per credit hour. Tuition, nonresident: full-time $19,730; part-time $810 per credit hour. Required fees: $322 per semester. Tuition and fees vary according to course load and program. *Financial support:* In 2005–06, 2 fellowships, 4 research assistantships, 5 teaching assistantships were awarded; career-related internships or fieldwork, Federal Work-Study, and scholarships/grants also available. Support available to part-time students. Financial award application deadline: 2/15. *Unit head:* Dr. Thomas Davis, Chair, 603-862-3217. *Application contact:* Brenda Lauze, Administrative Assistant, 603-862-2250, E-mail: genetics.dept@unh.edu.

University of New Mexico, School of Medicine, Biomedical Sciences Graduate Program, Albuquerque, NM 87131-5196. Offers biochemistry and molecular biology (MS, PhD); cell biology and physiology (MS, PhD); molecular genetics and microbiology (MS, PhD); neuroscience (MS, PhD); pathology (MS, PhD); toxicology (MS, PhD). Part-time programs available. Terminal master's awarded for partial completion of doctoral program. *Degree requirements:* For master's, thesis/dissertation; for doctorate, thesis/dissertation, comprehensive exam. *Entrance requirements:* For master's and doctorate, GRE General Test, minimum undergraduate GPA of 3.0. Additional exam requirements/recommendations for international students:

Genetics

University of New Mexico (continued)
Required—TOEFL. Electronic applications accepted. *Expenses:* Tuition, state resident: full-time $5,676. Tuition, nonresident: full-time $14,974; part-time $238 per credit hour. Required fees: $385 per term. Tuition and fees vary according to course load and program. *Faculty research:* Signal transduction, infectious disease, biology of cancer, structural biology, neuroscience.

The University of North Carolina at Chapel Hill, Graduate School, College of Arts and Sciences, Department of Biology, Chapel Hill, NC 27599. Offers botany (MA, MS, PhD); cell biology, development, and physiology (MA, MS, PhD); cell motility and cytoskeleton (PhD); ecology and behavior (MA, MS, PhD); genetics and molecular biology (MA, MS, PhD); morphology, systematics, and evolution (MA, MS, PhD). Terminal master's awarded for partial completion of doctoral program. *Degree requirements:* For master's, thesis (for some programs), comprehensive exam; for doctorate, thesis/dissertation, comprehensive exam. *Entrance requirements:* For master's, GRE General Test, GRE Subject Test, 2 semesters of calculus or statistics, 2 semesters of physics, organic chemistry, 3 semesters of biology; for doctorate, GRE General Test, GRE Subject Test, 2 semesters calculus or statistics, 2 semesters physics, organic chemistry, 3 semesters of biology. Additional exam requirements/recommendations for international students: Required—TOEFL (minimum score 550 paper-based; 213 computer-based). Electronic applications accepted. *Faculty research:* Gene expression, biomechanics, yeast genetics, plant ecology, plant molecular biology.

See Close-Up on page 277.

The University of North Carolina at Chapel Hill, School of Medicine and Graduate School, Graduate Programs in Medicine, Curriculum in Genetics and Molecular Biology, Chapel Hill, NC 27599. Offers MS, PhD. *Faculty:* 84 full-time (30 women), 2 part-time/adjunct (0 women). *Students:* 86 full-time (45 women); includes 16 minority (6 African Americans, 1 American Indian/Alaska Native, 7 Asian Americans or Pacific Islanders, 2 Hispanic Americans), 5 international. 100 applicants, 30% accepted, 18 enrolled. In 2005, 15 doctorates awarded. *Median time to degree:* Of those who began their doctoral program in fall 1997, 97% received their degree in 8 years or less. *Degree requirements:* For doctorate, thesis/dissertation, comprehensive exam, registration. *Entrance requirements:* For doctorate, minimum GPA of 3.0. Additional exam requirements/recommendations for international students: Required—TOEFL. *Application deadline:* For fall admission, 1/1 for domestic students. Applications are processed on a rolling basis. Application fee: $65. Electronic applications accepted. *Financial support:* In 2005–06, 15 fellowships with tuition reimbursements (averaging $22,000 per year), 54 research assistantships with tuition reimbursements (averaging $22,000 per year) were awarded; teaching assistantships with tuition reimbursements, traineeships and tuition waivers (full) also available. *Unit head:* Dr. Robert T. Duronio, Director, 919-966-3548. *Application contact:* Cara F. Marlow, Administrative Assistant, 919-966-2681, Fax: 919-966-0401, E-mail: cara-marlow@med.unc.edu.

University of North Dakota, Graduate School, College of Arts and Sciences, Department of Biology, Grand Forks, ND 58202. Offers botany (MS, PhD); ecology (MS, PhD); entomology (MS, PhD); environmental biology (MS, PhD); fisheries/wildlife (MS, PhD); genetics (MS, PhD); zoology (MS, PhD). *Faculty:* 16 full-time (3 women). *Students:* 15 applicants, 13% accepted, 2 enrolled. In 2005, 5 degrees awarded. Terminal master's awarded for partial completion of doctoral program. *Degree requirements:* For master's, thesis/dissertation, final exam; for doctorate, thesis/dissertation, final exam, comprehensive exam. *Entrance requirements:* For master's, GRE General Test, GRE Subject Test, minimum GPA of 3.0; for doctorate, GRE General Test, GRE Subject Test, minimum GPA of 3.5. Additional exam requirements/recommendations for international students: Required—TOEFL (minimum score 550 paper-based; 213 computer-based). *Application deadline:* For fall admission, 10/1 for domestic students, 10/1 for international students. Application fee: $35. Electronic applications accepted. *Financial support:* In 2005–06, 8 research assistantships with full tuition reimbursements (averaging $11,375 per year), 13 teaching assistantships with full tuition reimbursements (averaging $10,813 per year) were awarded; fellowships, Federal Work-Study, institutionally sponsored loans, scholarships/grants, and tuition waivers (full and partial) also available. Support available to part-time students. Financial award application deadline: 3/15; financial award applicants required to submit FAFSA. *Faculty research:* Population biology, wildlife ecology, RNA processing, hormonal control of behavior. *Unit head:* Dr. Richard Sweitzel, Graduate Director, 701-777-4676, Fax: 701-777-2623, E-mail: richard_sweitzel@und.nodak.edu.

University of North Texas Health Science Center at Fort Worth, Graduate School of Biomedical Sciences, Fort Worth, TX 76107-2699. Offers anatomy and cell biology (MS, PhD); biochemistry and molecular biology (MS, PhD); biomedical sciences (MS, PhD); biotechnology (MS); forensic genetics (MS); integrative physiology (MS, PhD); medical science (MS); microbiology and immunology (MS, PhD); pharmacology (MS, PhD); science education (MS). *Faculty:* 57 full-time (9 women), 2 part-time/adjunct (0 women). *Students:* 177 full-time (98 women), 42 part-time (33 women); includes 61 minority (15 African Americans, 3 American Indian/Alaska Native, 27 Asian Americans or Pacific Islanders, 16 Hispanic Americans), 53 international. Average age 28. 237 applicants, 62% accepted, 91 enrolled. In 2005, 37 master's, 15 doctorates awarded. Terminal master's awarded for partial completion of doctoral program. *Degree requirements:* For master's and doctorate, thesis/dissertation. *Entrance requirements:* For master's and doctorate, GRE General Test. Additional exam requirements/recommendations for international students: Required—TOEFL. *Application deadline:* For fall admission, 5/1 for domestic students. Application fee: $25 ($50 for international students). *Expenses:* Contact institution. *Financial support:* In 2005–06, 80 research assistantships (averaging $16,000 per year) were awarded; fellowships, teaching assistantships, career-related internships or fieldwork, Federal Work-Study, institutionally sponsored loans, scholarships/grants, and traineeships also available. Support available to part-time students. Financial award application deadline: 4/1; financial award applicants required to submit FAFSA. *Faculty research:* Alzheimer's disease, aging, eye diseases, cancer, cardiovascular disease. Total annual research expenditures: $21 million. *Unit head:* Dr. Thomas Yorio, Dean, 817-735-2560, Fax: 817-735-0243, E-mail: yoriot@hsc.unt.edu. *Application contact:* Carla Lee, Director of Graduate Admissions and Services, 817-735-2560, Fax: 817-735-0243, E-mail: gsbs@hsc.unt.edu.

See Close-Up on page 279.

University of Notre Dame, Graduate School, College of Science, Department of Biological Sciences, Notre Dame, IN 46556. Offers aquatic ecology, evolution and environmental biology (MS, PhD); cellular and molecular biology (MS, PhD); genetics (MS, PhD); physiology (MS, PhD); vector biology and parasitology (MS, PhD). *Faculty:* 34 full-time (8 women), 3 part-time/adjunct (0 women). *Students:* 124 full-time (55 women); includes 11 minority (1 African American, 6 Asian Americans or Pacific Islanders, 4 Hispanic Americans), 39 international. 95 applicants, 34% accepted, 22 enrolled. In 2005, 4 master's, 11 doctorates awarded. Terminal master's awarded for partial completion of doctoral program. *Median time to degree:* Of those who began their doctoral program in fall 1997, 61% received their degree in 8 years or less. *Degree requirements:* For master's and doctorate, thesis/dissertation, comprehensive exam. *Entrance requirements:* For master's and doctorate, GRE General Test. Additional exam requirements/recommendations for international students: Required—TOEFL. *Application deadline:* For fall admission, 2/1 for domestic students; for spring admission, 11/1 for domestic students. Applications are processed on a rolling basis. Application fee: $50. Electronic applications accepted. *Financial support:* In 2005–06, 124 students received support, including 24 fellowships with full tuition reimbursements available (averaging $22,000 per year), 47 research assistantships with full tuition reimbursements available (averaging $15,250 per year), 45 teaching assistantships with full tuition reimbursements available (averaging $16,000 per year); traineeships and tuition waivers (full) also available. Financial award application deadline: 2/1. *Faculty research:* Tropical disease, molecular genetics, neurobiology, evolutionary biology, aquatic biology. Total annual research expenditures: $15.3 million. *Unit head:* Dr. Gary A. Lamberti, Director of Graduate Studies, 574-631-6552, Fax: 574-631-7413, E-mail: biology.

biosadm.1@nd.edu. *Application contact:* Dr. Terrence J. Akai, Director of Graduate Admissions, 574-631-7706, Fax: 574-631-4183, E-mail: gradad@nd.edu.

See Close-Up on page 281.

University of Oregon, Graduate School, College of Arts and Sciences, Department of Biology, Eugene, OR 97403. Offers ecology and evolution (MA, MS, PhD); marine biology (MA, MS, PhD); molecular, cellular and genetic biology (PhD); neuroscience and development (PhD). *Faculty:* 42 full-time (14 women), 5 part-time/adjunct (2 women). *Students:* 92 full-time (43 women), 2 part-time (both women); includes 9 minority (2 African Americans, 1 American Indian/Alaska Native, 4 Asian Americans or Pacific Islanders, 2 Hispanic Americans), 6 international. 18 applicants, 83% accepted. In 2005, 11 master's, 7 doctorates awarded. Terminal master's awarded for partial completion of doctoral program. *Degree requirements:* For master's, thesis (for some programs); for doctorate, thesis/dissertation. *Entrance requirements:* For master's and doctorate, GRE General Test, minimum GPA of 3.2. Additional exam requirements/recommendations for international students: Required—TOEFL. *Application deadline:* For fall admission, 12/15 for domestic students. *Financial support:* In 2005–06, 36 teaching assistantships were awarded; research assistantships, Federal Work-Study, institutionally sponsored loans, and scholarships/grants also available. Financial award application deadline: 2/1. *Faculty research:* Developmental neurobiology; evolution, population biology, and quantitative genetics; regulation of gene expression; biochemistry of marine organisms. *Unit head:* George Sprague, Head, 541-346-6051, Fax: 541-346-6056. *Application contact:* Lynne Romans, Admissions Contact, 541-346-4252, Fax: 541-346-6056, E-mail: lromans@uoregon.edu.

University of Pennsylvania, School of Medicine, Biomedical Graduate Studies, Graduate Group in Cell and Molecular Biology, Program in Gene Therapy and Vaccines, Philadelphia, PA 19104. Offers PhD, MD/PhD, VMD/PhD. *Degree requirements:* For doctorate, thesis/dissertation. *Entrance requirements:* For doctorate, GRE General Test. Additional exam requirements/recommendations for international students: Required—TOEFL. *Application deadline:* For fall admission, 12/15 priority date for domestic students, 12/1 priority date for international students. Applications are processed on a rolling basis. Application fee: $70. Electronic applications accepted. *Financial support:* Fellowships, research assistantships, scholarships/grants, traineeships, and unspecified assistantships available. *Unit head:* Dr. David M. Weiner, Chair, 215-349-8365. *Application contact:* Emily D. Geib, Coordinator, 215-898-3918, Fax: 215-573-2104, E-mail: eddz@mail.med.upenn.edu.

University of Pennsylvania, School of Medicine, Biomedical Graduate Studies, Graduate Group in Cell and Molecular Biology, Program in Genetics and Gene Regulation, Philadelphia, PA 19104. Offers PhD, MD/PhD, VMD/PhD. *Degree requirements:* For doctorate, thesis/dissertation. *Entrance requirements:* For doctorate, GRE General Test. Additional exam requirements/recommendations for international students: Required—TOEFL. *Application deadline:* For fall admission, 12/15 priority date for domestic students, 12/1 priority date for international students. Applications are processed on a rolling basis. Application fee: $70. Electronic applications accepted. *Financial support:* Fellowships, research assistantships, scholarships/grants, traineeships, and unspecified assistantships available. *Unit head:* Dr. Thomas Kadesch, Chair, 215-898-1047. *Application contact:* Meagan Schofer, Coordinator, 215-895-9536, Fax: 215-573-2104, E-mail: mschofer@mailmed.upenn.edu.

University of Rochester, School of Medicine and Dentistry, Graduate Programs in Medicine and Dentistry, Department of Biomedical Genetics, Rochester, NY 14627-0250. Offers MS, PhD. *Degree requirements:* For doctorate, thesis/dissertation, qualifying exam. *Entrance requirements:* For doctorate, GRE General Test.

University of Southern California, Keck School of Medicine and Graduate School, Graduate Programs in Medicine, Department of Preventive Medicine, Division of Biostatistics, Los Angeles, CA 90089. Offers applied biostatistics/epidemiology (MS); biostatistics (MS, PhD); epidemiology (PhD); genetic epidemiology and statistical genetics (PhD); molecular epidemiology (MS, PhD). *Faculty:* 74 full-time (33 women). *Students:* 97 full-time (63 women); includes 25 minority (18 Asian Americans or Pacific Islanders, 7 Hispanic Americans), 48 international. Average age 30. 109 applicants, 70% accepted, 33 enrolled. In 2005, 17 master's, 10 doctorates awarded. Terminal master's awarded for partial completion of doctoral program. *Median time to degree:* Of those who began their doctoral program in fall 1997, 100% received their degree in 8 years or less. *Degree requirements:* For master's and doctorate, thesis/dissertation. *Entrance requirements:* For master's, GRE General Test, GRE Subject Test, minimum GPA of 3.0; for doctorate, GRE General Test, GRE Subject Test, minimum GPA of 3.5. Additional exam requirements/recommendations for international students: Required—TOEFL (minimum score 550 paper-based; 200 computer-based). *Application deadline:* For fall admission, 1/15 for domestic students. Applications are processed on a rolling basis. Application fee: $65 ($75 for international students). Electronic applications accepted. *Expenses:* Tuition: Full-time $25,416; part-time $1,059 per unit. Required fees: $484; $484 per year. Tuition and fees vary according to course load and program. *Financial support:* In 2005–06, 3 fellowships with tuition reimbursements (averaging $24,876 per year), 39 research assistantships with tuition reimbursements (averaging $24,876 per year), 15 teaching assistantships with tuition reimbursements (averaging $24,876 per year) were awarded; career-related internships or fieldwork, Federal Work-Study, institutionally sponsored loans, scholarships/grants, and tuition waivers (partial) also available. Financial award application deadline: 4/1. *Faculty research:* Clinical trials in ophthalmology and cancer research, methods of analysis for epidemiological studies, genetic epidemiology. Total annual research expenditures: $1.3 million. *Unit head:* Dr. Stanley P. Azen, Co-Director, 323-442-1810, Fax: 323-442-2993, E-mail: mtruxill@usc.edu. *Application contact:* Mary L. Trujillo, Student Adviser, 323-442-1810, Fax: 323-442-2993, E-mail: mtruxill@usc.edu.

The University of Tennessee, Graduate School, College of Arts and Sciences, Program in Life Sciences, Knoxville, TN 37996. Offers genome science and technology (MS, PhD); plant physiology and genetics (MS, PhD). *Degree requirements:* For doctorate, one foreign language, thesis/dissertation. *Entrance requirements:* For master's and doctorate, GRE General Test, minimum GPA of 2.7. Additional exam requirements/recommendations for international students: Required—TOEFL. Electronic applications accepted.

The University of Texas at Austin, Graduate School, College of Natural Sciences, Institute for Cellular and Molecular Biology, Program in Genetics and Developmental Biology, Austin, TX 78712-1111. Offers PhD.

The University of Texas Health Science Center at Houston, Graduate School of Biomedical Sciences, Program in Genes and Development, Houston, TX 77030. Offers MS, PhD, MD/PhD. *Faculty:* 38 full-time (11 women). *Students:* 68 full-time (35 women); includes 9 minority (3 Asian Americans or Pacific Islanders, 6 Hispanic Americans), 43 international. Average age 24. 36 applicants, 31% accepted, 3 enrolled. In 2005, 2 master's, 9 doctorates awarded. Terminal master's awarded for partial completion of doctoral program. *Degree requirements:* For master's and doctorate, thesis/dissertation. *Entrance requirements:* For master's and doctorate, GRE General Test. Additional exam requirements/recommendations for international students: Required—TOEFL, TWE. *Application deadline:* For fall admission, 1/15 for domestic students; for spring admission, 11/1 for domestic students. Applications are processed on a rolling basis. Application fee: $10. Electronic applications accepted. *Financial support:* Fellowships with full tuition reimbursements, research assistantships with full tuition reimbursements, teaching assistantships, institutionally sponsored loans, scholarships/grants, and health care benefits available. Financial award application deadline: 1/15. *Faculty research:* Regulation of gene expression; developmental genetics; cell growth, proliferation and death; tissue differentiation and organogenesis; chrommastin. *Unit head:* Dr. Michelle C Barton, Director, 713-834-6268, Fax: 713-834-6273, E-mail: mbarton@mdanderson.org. *Application contact:* Dr. Victoria P. Knutson, Assistant Dean of Admissions, 713-500-9860, Fax: 713-500-9877, E-mail: victoria.p.knutson@uth.tmc.edu.

Genetics

The University of Texas Medical Branch, Graduate School of Biomedical Sciences, Program in Biochemistry and Molecular Biology, Galveston, TX 77555. Offers biochemistry (PhD); bioinformatics (PhD); biophysics (PhD); cell biology (PhD); computational biology (PhD); structural biology (PhD). *Students:* 38 full-time (14 women), 1 part-time; includes 5 minority (3 Asian Americans or Pacific Islanders, 2 Hispanic Americans), 18 international. Average age 27. In 2005, 7 degrees awarded. *Degree requirements:* For doctorate, thesis/dissertation. *Entrance requirements:* Additional exam requirements/recommendations for international students: Required—TOEFL (minimum score 550 paper-based; 213 computer-based). *Application deadline:* Applications are processed on a rolling basis. Application fee: $30 ($75 for international students). Electronic applications accepted. *Expenses:* Tuition, state resident: full-time $8,350; part-time $90 per credit hour. Tuition, nonresident: full-time $21,450; part-time $366 per credit hour. Required fees: $1,027; $11 per credit hour. $60 per term. *Financial support:* In 2005–06, fellowships (averaging $23,000 per year), research assistantships (averaging $23,000 per year) were awarded. Financial award applicants required to submit FAFSA. *Unit head:* Dr. Leenian L. Chan, Director, 409-772-2861, Fax: 409-772-9679, E-mail: lchan@utmb.edu. *Application contact:* Debora Botting, Co-ordinator II Special Programs, 409-772-2769, Fax: 409-747-0552, E-mail: dmbottin@utmb.edu.

The University of Texas Southwestern Medical Center at Dallas, Southwestern Graduate School of Biomedical Sciences, Division of Basic Science, Program in Genetics and Development, Dallas, TX 75390. Offers PhD. *Faculty:* 85 full-time (16 women), 2 part-time/adjunct (0 women). *Students:* 80 full-time (40 women), 2 part-time; includes 14 minority (6 Asian Americans or Pacific Islanders, 8 Hispanic Americans), 41 international. Average age 28. In 2005, 16 doctorates awarded. *Degree requirements:* For doctorate, thesis/dissertation, qualifying exam. *Entrance requirements:* For doctorate, GRE General Test, minimum GPA of 3.0. Additional exam requirements/recommendations for international students: Required—TOEFL. *Application deadline:* For fall admission, 1/5 for domestic students. Application fee: $0. Electronic applications accepted. *Expenses:* Tuition, state resident: full-time $6,550; part-time $50 per credit hour. Tuition, nonresident: full-time $19,650; part-time $326 per credit hour. Required fees: $42 per credit hour. Tuition and fees vary according to degree level and program. *Financial support:* Fellowships, research assistantships, institutionally sponsored loans available. *Faculty research:* Human molecular genetics, chromosome structure, gene regulation, molecular biology, gene expression. *Unit head:* Dr. John Abrams, Chair, 214-648-9226, Fax: 214-648-9226, E-mail: john.abrams@utsouthwestern.edu. *Application contact:* Dr. Nancy E. Street, Associate Dean, 214-648-6708, Fax: 214-648-2102, E-mail: nancy.street@utsouthwestern.edu.

The University of Toledo, College of Graduate Studies, Department of Biochemistry and Molecular Biology, Toledo, OH 43606-3390. Offers medical sciences (MS), including biochemistry, genetics, molecular biology. Part-time programs available. *Degree requirements:* For master's, thesis, qualifying exam. *Entrance requirements:* For master's, GRE General Test, minimum undergraduate GPA of 3.0. Additional exam requirements/recommendations for international students: Required—TOEFL. *Expenses:* Tuition, state resident: full-time $6,623; part-time $308 per credit hour. Tuition, nonresident: full-time $13,232; part-time $735 per credit hour. *Faculty research:* Gene regulation, protein structure, receptors, protein phosphorylation, peptides.

University of Toronto, School of Graduate Studies, Life Sciences Division, Department of Molecular and Medical Genetics, Toronto, ON M5S 1A1, Canada. Offers genetic counseling (M Sc); molecular and medical genetics (M Sc, PhD). *Degree requirements:* For master's and doctorate, thesis/dissertation. *Entrance requirements:* For master's, B Sc or equivalent, minimum B+ average; for doctorate, M Sc or equivalent, minimum B+ average. Additional exam requirements/recommendations for international students: Required—TOEFL, IELTS (7), MLAB (85) or COPE (4). *Faculty research:* Structural biology, developmental genetics, molecular medicine, genetic counseling.

University of Utah, The Graduate School, College of Science, Department of Biology, Salt Lake City, UT 84112-1107. Offers biology (M Phil); ecology and evolutionary biology (MS, PhD); genetics (MS, PhD); molecular biology (PhD). Part-time programs available. *Faculty:* 42 full-time (8 women), 1 part-time/adjunct (0 women). *Students:* 55 full-time (30 women), 15 part-time (2 women); includes 4 minority (3 Asian Americans or Pacific Islanders, 1 Hispanic American), 19 international. Average age 29. 26 applicants, 54% accepted, 11 enrolled. In 2005, 3 master's, 6 doctorates awarded. Terminal master's awarded for partial completion of doctoral program. *Median time to degree:* Of those who began their doctoral program in fall 1997, 20% received their degree in 8 years or less. *Degree requirements:* For master's and doctorate, thesis/dissertation. *Entrance requirements:* For master's and doctorate, GRE General Test, minimum GPA of 3.0. Additional exam requirements/recommendations for international students: Required—TOEFL (minimum score 500 paper-based; 173 computer-based). *Application deadline:* For fall admission, 1/13 for domestic students, 1/13 for international students. Application fee: $45 ($65 for international students). *Expenses:* Tuition, state resident: full-time $2,932; part-time $369 per credit. Tuition, nonresident: full-time $10,350; part-time $1,302 per credit. Required fees: $516 per term. Tuition and fees vary according to course load and program. *Financial support:* In 2005–06, 2 fellowships with full tuition reimbursements (averaging $30,000 per year), 33 research assistantships with full tuition reimbursements (averaging $22,000 per year), 33 teaching assistantships with full tuition reimbursements (averaging $15,000 per year) were awarded; career-related internships or fieldwork, scholarships/grants, traineeships, and health care benefits also available. Financial award application deadline: 2/15; financial award applicants required to submit FAFSA. *Faculty research:* Behavioral ecology, cellular neurobiology, DNA replication, ecological genetics, herpetology. Total annual research expenditures: $11.4 million. *Unit head:* David R. Wolstenholme, Chair, 801-581-6517, Fax: 801-581-4668, E-mail: wolstenholme@biology.utah.edu. *Application contact:* Shannon Nielsen, Administrative Program Coordinator, 801-581-5636, Fax: 801-581-4668, E-mail: shannon.nielsen@bioscience.utah.edu.

See Close-Up on page 313.

University of Utah, The Graduate School, College of Science and Graduate Programs in Medicine, Program in Molecular Biology, Salt Lake City, UT 84112-1107.

University of Washington, Graduate School, School of Public Health and Community Medicine, Department of Epidemiology, Institute for Public Health Genetics, Seattle, WA 98195. Offers genetic epidemiology (MS); public health genetics (MPH, PhD). Part-time programs available. *Faculty:* 17 full-time (11 women). *Students:* 22 full-time (18 women), 7 part-time (4 women); includes 10 minority (4 African Americans, 5 Asian Americans or Pacific Islanders, 1 Hispanic American), 1 international. Average age 30. 26 applicants, 42% accepted, 5 enrolled. In 2005, 10 degrees awarded. Terminal master's awarded for partial completion of doctoral program. *Degree requirements:* For master's, thesis, practicum (MPH); for doctorate, thesis/dissertation, comprehensive exam, registration. *Entrance requirements:* For master's and doctorate, GRE General Test, experience in health sciences (preferred), minimum GPA of 3.0. Additional exam requirements/recommendations for international students: Required—TOEFL (minimum score 583 paper-based; 237 computer-based). *Application deadline:* For fall admission, 1/15 for domestic students, 11/1 for international students. Application fee: $50. *Financial support:* In 2005–06, 1 student received support, including 1 research assistantship with tuition reimbursement available (averaging $11,943 per year) Financial award application deadline: 1/15. *Faculty research:* Genetic epidemiology; ethical, legal, social issues of genetics; ecogenetics; health policy. *Unit head:* , Dr. Melissa A. Austin, Director, 206-543-0709, Fax: 206-685-9651, E-mail: maustin@u.washington.edu. *Application contact:* Yuko Mera, Student Services Advisor, 206-616-9286, Fax: 206-685-9651, E-mail: phgen@u.washington.edu.

University of Wisconsin–Madison, Graduate School, College of Agricultural and Life Sciences and Graduate Programs in Medicine, Departments of Genetics and Medical Genetics, Program in Genetics, Madison, WI 53706-1380. Offers PhD. *Degree requirements:* For doctorate, thesis/dissertation. Application fee: $45. *Unit head:* Carter L. Denniston, Chair, Departments of Genetics and Medical Genetics, 608-262-1069, Fax: 608-262-2976.

See Close-Up on page 803.

University of Wisconsin–Madison, Graduate School, College of Agricultural and Life Sciences and Graduate Programs in Medicine, Departments of Genetics and Medical Genetics, Program in Medical Genetics, Madison, WI 53706-1380. Offers MS. Application fee: $38. *Unit head:* Carter L. Denniston, Chair, Departments of Genetics and Medical Genetics, 608-262-1069, Fax: 608-262-2976.

See Close-Up on page 803.

University of Wisconsin–Madison, Medical School and Graduate School, Graduate Programs in Medicine, Madison, WI 53706-1380. Offers biomolecular chemistry (MS, PhD); cancer biology (PhD); genetics and medical genetics (MS, PhD), including genetics (PhD), medical genetics (MS); medical physics (MS, PhD), including health physics (MS), medical physics; microbiology (PhD); molecular and cellular pharmacology (PhD); pathology and laboratory medicine (PhD); physiology (PhD); population health (MPH, MS, PhD). Part-time programs available. Postbaccalaureate distance learning degree programs offered (minimal on-campus study). Terminal master's awarded for partial completion of doctoral program. Application fee: $45. Electronic applications accepted. *Expenses: Contact institution. Financial support:* Fellowships with full tuition reimbursements, research assistantships with full tuition reimbursements, teaching assistantships with full tuition reimbursements, scholarships/grants, traineeships, and tuition waivers (full) available. *Unit head:* Dr. Paul M. DeLuca, Associate Dean of Research and Graduate Studies, 608-265-0524, Fax: 608-265-0522, E-mail: pmdeluca@facstaff.wisc.edu.

Virginia Commonwealth University, Medical College of Virginia-Professional Programs, School of Medicine and Graduate Programs, School of Medicine Graduate Programs, Department of Biochemistry, Richmond, VA 23284-9005. Offers biochemistry (MS, PhD, CBHS); bioinformatics (MS); molecular biology and genetics (MS, PhD); neurosciences (PhD). *Faculty:* 26 full-time (6 women). *Students:* 48 full-time (24 women), 9 part-time (8 women); includes 12 minority (2 African Americans, 8 Asian Americans or Pacific Islanders, 2 Hispanic Americans), 8 international. 61 applicants, 75% accepted. In 2005, 1 master's, 4 doctorates, 2 other advanced degrees awarded. *Degree requirements:* For master's, thesis; for doctorate, thesis/dissertation, comprehensive oral and written exams. *Entrance requirements:* For master's and doctorate, DAT, GRE General Test, MCAT. *Application deadline:* For fall admission, 2/15 for domestic students. Application fee: $50. *Expenses:* Tuition, state resident: full-time $3,185; part-time $405 per credit. Tuition, nonresident: full-time $7,952; part-time $940 per credit. Required fees: $751 per semester hour. Tuition and fees vary according to course load and program. *Financial support:* Fellowships, research assistantships available. *Faculty research:* Molecular biology, peptide/protein chemistry, neurochemistry, enzyme mechanisms, macromolecular structure determination. Total annual research expenditures: $3.5 million. *Unit head:* Dr. Sarah Spiegel, Chair, 804-828-9762, Fax: 804-828-1473, E-mail: sspiegel@vcu.edu. *Application contact:* Dr. Keith R. Shelton, Program Director, 804-828-9886, Fax: 804-828-1473, E-mail: krshelto@vcu.edu.

See Close-Up on page 447.

Virginia Commonwealth University, Medical College of Virginia-Professional Programs, School of Medicine and Graduate Programs, School of Medicine Graduate Programs, Department of Human Genetics, Richmond, VA 23284-9005. Offers genetic counseling (MS); human genetics (PhD, CBHS); molecular biology and genetics (MS, PhD). *Faculty:* 18 full-time (10 women). *Students:* 17 full-time (15 women), 4 part-time (1 woman); includes 2 minority (both Asian Americans or Pacific Islanders), 4 international. 52 applicants, 46% accepted. In 2005, 6 master's, 3 doctorates, 2 other advanced degrees awarded. *Degree requirements:* For master's, thesis; for doctorate, thesis/dissertation, comprehensive oral and written exams. *Entrance requirements:* For master's, DAT, GRE General Test, or MCAT; for doctorate, GRE General Test, DAT, MCAT. *Application deadline:* For fall admission, 2/15 for domestic students. Application fee: $50. *Expenses:* Tuition, state resident: full-time $3,185; part-time $405 per credit. Tuition, nonresident: full-time $7,952; part-time $940 per credit. Required fees: $751 per semester hour. Tuition and fees vary according to course load and program. *Financial support:* Fellowships available. *Faculty research:* Genetic epidemiology, biochemical genetics, quantitative genetics, human cytogenetics, molecular genetics. *Unit head:* Dr. Walter E. Nance, Chair, 804-828-9632, Fax: 804-828-3760, E-mail: wenance@vcu.edu. *Application contact:* Dr. Linda A. Corey, Graduate Program Director, 804-828-8759, Fax: 804-828-3760, E-mail: lacorey@vcu.edu.

Virginia Commonwealth University, Medical College of Virginia-Professional Programs, School of Medicine and Graduate Programs, School of Medicine Graduate Programs, Department of Microbiology and Immunology, Richmond, VA 23284-9005. Offers microbiology and genetics (MS); microbiology and immunology (MS, PhD, CBHS); molecular biology and genetics (PhD). *Faculty:* 33 full-time (8 women). *Students:* 66 full-time (43 women), 9 part-time (6 women); includes 21 minority (12 African Americans, 5 Asian Americans or Pacific Islanders, 4 Hispanic Americans), 11 international. 95 applicants, 48% accepted. In 2005, 25 master's, 7 doctorates, 3 other advanced degrees awarded. *Degree requirements:* For master's, thesis; for doctorate, thesis/dissertation, comprehensive oral and written exams. *Entrance requirements:* For master's, GRE General Test or MCAT; for doctorate, GRE General Test, MCAT. *Application deadline:* For fall admission, 2/15 for domestic students. Application fee: $50. *Expenses:* Tuition, state resident: full-time $6,268; part-time $405 per credit. Tuition, nonresident: full-time $15,904; part-time $940 per credit. Required fees: $751 per semester hour. Tuition and fees vary according to course load and program. *Financial support:* Fellowships, research assistantships, teaching assistantships available. *Faculty research:* Microbial physiology and genetics, molecular biology, crystallography of biological molecules, antibiotics and chemotherapy, membrane transport. *Unit head:* Dr. Dennis E. Ohman, Chair, 804-828-9728, Fax: 804-828-9946, E-mail: deohman@vcu.edu. *Application contact:* Dr. Guy A. Cabral, Chair, Graduate Program, 804-828-2306, E-mail: gacabral@vcu.edu.

Virginia Commonwealth University, Medical College of Virginia-Professional Programs, School of Medicine and Graduate Programs, School of Medicine Graduate Programs, Department of Pharmacology and Toxicology, Richmond, VA 23284-9005. Offers molecular biology and genetics (PhD); neurosciences (PhD); pharmacology (PhD, CBHS); pharmacology and toxicology (MS). *Faculty:* 36 full-time (9 women). *Students:* 35 full-time (18 women), 11 part-time (all women); includes 7 minority (2 African Americans, 4 Asian Americans or Pacific Islanders, 1 Hispanic American), 5 international. 90 applicants, 33% accepted. In 2005, 6 master's, 2 doctorates awarded. Terminal master's awarded for partial completion of doctoral program. *Degree requirements:* For master's, thesis; for doctorate, thesis/dissertation, comprehensive oral and written exams. *Entrance requirements:* For master's, DAT, GRE General Test or MCAT; for doctorate, GRE General Test, MCAT, DAT. *Application deadline:* For fall admission, 4/1 for domestic students. Application fee: $50. *Expenses:* Tuition, state resident: full-time $6,268; part-time $405 per credit. Tuition, nonresident: full-time $15,904; part-time $940 per credit. Required fees: $751 per semester hour. Tuition and fees vary according to course load and program. *Financial support:* Fellowships, teaching assistantships available. *Faculty research:* Drug abuse, drug metabolism, pharmacodynamics, peptide synthesis, receptor mechanisms. *Unit head:* Dr. Billy R. Martin, Chair, 804-828-8407, Fax: 804-828-2117, E-mail: brmartin@vcu.edu. *Application contact:* Sheryol Cox, Graduate Program Coordinator, 804-828-8400, Fax: 804-828-2117, E-mail: swcox@vcu.edu.

See Close-Up on page 1219.

Virginia Polytechnic Institute and State University, Graduate School, College of Science, Department of biological Sciences, Blacksburg, VA 24061. Offers botany (MS, PhD); ecology and evolutionary biology (MS, PhD); genetics and developmental biology (MS, PhD); microbiology (MS, PhD); zoology (MS, PhD). *Faculty:* 38 full-time (9 women). *Students:* 70 full-time (30 women), 4 part-time (1 woman); includes 6 minority (2 African Americans, 1 American Indian/Alaska Native, 2 Asian Americans or Pacific Islanders, 1 Hispanic American), 12 international. Average age 27. 81 applicants, 22% accepted, 14 enrolled. In 2005, 9 master's, 8 doctorates awarded. *Entrance requirements:* For master's and doctorate, GRE General Test.

Genetics

Virginia Polytechnic Institute and State University *(continued)*
Additional exam requirements/recommendations for international students: Required—TOEFL (minimum score 550 paper-based; 213 computer-based). *Application deadline:* Applications are processed on a rolling basis. *Application fee:* $45. Electronic applications accepted. *Expenses:* Tuition, state resident: full-time $6,558; part-time $364 per credit. Tuition, nonresident: full-time $11,296; part-time $628 per credit. Required fees: $1,419; $468 per credit. $234 per term. *Financial support:* In 2005–06, 28 research assistantships with full tuition reimbursements (averaging $16,689 per year), 37 teaching assistantships with full tuition reimbursements (averaging $13,902 per year) were awarded; career-related internships or fieldwork, Federal Work-Study, scholarships/grants, and unspecified assistantships also available. *Faculty research:* Freshwater ecology, cell cycle regulation, behavioral ecology, motor proteins. *Unit head:* Dr. Bob Jones, Chairman, 540-231-9514, Fax: 540-231-9307, E-mail: rhjones@vt.edu. *Application contact:* Sue Rasmussen, Graduate Secretary, 540-231-8929, Fax: 540-231-9307, E-mail: sueras@vt.edu.

Virginia Polytechnic Institute and State University, Graduate School, Intercollege, Program in Genetics, Bioinformatics and Computational Biology, Blacksburg, VA 24061. Offers PhD. *Students:* 27 full-time (14 women), 4 part-time; includes 4 minority (1 African American, 2 Asian Americans or Pacific Islanders, 1 Hispanic American), 20 international. Average age 29. 61 applicants, 20% accepted, 8 enrolled. *Entrance requirements:* For doctorate, GRE. Additional exam requirements/recommendations for international students: Required—TOEFL (minimum score 550 paper-based; 213 computer-based). *Application deadline:* Applications are processed on a rolling basis. *Application fee:* $45. Electronic applications accepted. *Expenses:* Tuition, state resident: full-time $6,558; part-time $364 per credit. Tuition, nonresident: full-time $11,296; part-time $628 per credit. Required fees: $1,419; $468 per credit. $234 per term. *Financial support:* Fellowships with full tuition reimbursements, research assistantships with full tuition reimbursements, teaching assistantships with full tuition reimbursements, career-related internships or fieldwork, Federal Work-Study, scholarships/grants, and unspecified assistantships available. *Unit head:* Dr. David Bevan, Chair, 540-231-5040, Fax: 540-231-3010, E-mail: drbevan@vt.edu. *Application contact:* Dennie Munson, Information Contact, 540-231-1928, Fax: 540-231-3010, E-mail: dennie@vt.edu.

Washington State University, Graduate School, College of Sciences, School of Molecular Biosciences, Program in Genetics and Cell Biology, Pullman, WA 99164. Offers MS, PhD. *Faculty:* 23 full-time (5 women), 21 part-time/adjunct (4 women). *Students:* 23 full-time (15 women), 1 (woman) part-time; includes 2 minority (1 American Indian/Alaska Native, 1 Asian American or Pacific Islander), 7 international. Average age 26. 231 applicants, 18% accepted, 13 enrolled. In 2005, 3 master's, 4 doctorates awarded. Terminal master's awarded for partial completion of doctoral program. *Degree requirements:* For master's, thesis or alternative, oral exam; for doctorate, thesis/dissertation, oral exam, comprehensive exam, registration. *Entrance requirements:* For master's and doctorate, GRE General Test, minimum GPA of 3.0. Additional exam requirements/recommendations for international students: Required—TOEFL (minimum score 550 paper-based; 213 computer-based). *Application deadline:* For fall admission, 12/15 for domestic students, 12/15 for international students. *Application fee:* $35. Electronic applications accepted. *Expenses:* Tuition, state resident: full-time $6,295; part-time $336 per credit. Tuition, nonresident: full-time $15,949; part-time $819 per credit. Required fees: $933. Part-time tuition and fees vary according to campus/location and program. *Financial support:* In 2005–06, 1 fellowship with full tuition reimbursement (averaging $18,852 per year), 16 research assistantships with full tuition reimbursements (averaging $18,852 per year), 6 teaching assistantships with full tuition reimbursements (averaging $18,852 per year) were awarded; Federal Work-Study, institutionally sponsored loans, health care benefits, and unspecified assistantships also available. Financial award application deadline: 4/1; financial award applicants required to submit FAFSA. *Faculty research:* Plant molecular biology, growth factors, cancer induction and DNA repair, gene regulation and genetic engineering. Total annual research expenditures: $5.8 million. *Application contact:* Kelly G. McGovern, Academic Coordinator, 509-335-4566, Fax: 509-335-1907, E-mail: smbgrad@wsu.edu.

Washington University in St. Louis, Graduate School of Arts and Sciences, Division of Biology and Biomedical Sciences, Program in Evolution, Ecology and Population Biology, St. Louis, MO 63130-4899. Offers ecology (PhD); environmental biology (PhD); evolutionary biology (PhD); genetics (PhD). *Degree requirements:* For doctorate, thesis/dissertation. *Entrance requirements:* For doctorate, GRE General Test, GRE Subject Test. Electronic applications accepted.

Washington University in St. Louis, School of Medicine, Program in Genetic Epidemiology, St. Louis, MO 63130-4899. Offers MS, Certificate. *Faculty:* 17 full-time (4 women). *Students:* 13 full-time (6 women), 4 part-time (all women); includes 4 minority (2 African Americans, 1 Asian American or Pacific Islander, 1 Hispanic American), 2 international. Average age 30. 15 applicants, 60% accepted, 3 enrolled. In 2005, 9 degrees awarded. *Degree requirements:* For master's, research paper. *Entrance requirements:* For master's, computer programming proficiency, proficiency in statistics and biology/genetics. Additional exam requirements/recommendations for international students: Required—TOEFL. *Application deadline:* For fall admission, 4/1 for domestic students, 4/1 for international students. *Application fee:* $50. *Expenses:* Contact institution. *Financial support:* In 2005–06, 6 students received support, including 6 research assistantships with partial tuition reimbursements available (averaging $14,000 per year); fellowships with partial tuition reimbursements available, Federal Work-Study, institutionally sponsored loans, tuition waivers (full and partial), and unspecified assistantships also available. Financial award application deadline: 4/1; financial award applicants required to submit FAFSA. *Faculty research:* Biostatistics, clinical trials, cardiovascular diseases, genetics, genetic epidemiology. Total annual research expenditures: $4,300. *Unit head:* Dr. Dabeeru C. Rao, Professor, Director of Biostatistics, 314-362-3608, Fax: 314-362-2693, E-mail: rao@wubios.wustl.edu. *Application contact:* June C. Mueller, Program Administrator, 314-362-1052, Fax: 314-362-2693, E-mail: june@wubios.wustl.edu.

Wayne State University, Graduate School, Program in Molecular Biology and Genetics, Detroit, MI 48202. Offers MS, PhD. *Faculty:* 27. *Students:* 22 full-time (11 women), 12 international. Average age 29. 46 applicants, 4% accepted, 2 enrolled. In 2005, 4 degrees awarded. Terminal master's awarded for partial completion of doctoral program. *Degree requirements:* For master's and doctorate, thesis/dissertation. *Entrance requirements:* For master's and doctorate, GRE General Test. Additional exam requirements/recommendations for international students: Required—TOEFL (minimum score 550 paper-based; 213 computer-based); Recommended—TWE (minimum score 6). *Application deadline:* Applications are processed on a rolling basis. *Application fee:* $30 ($50 for international students). Electronic applications accepted. *Expenses:* Tuition, state resident: part-time $338 per credit hour. Tuition, nonresident: part-time $746 per credit hour. Required fees: $24 per credit hour. Full-time tuition and fees vary according to program. *Financial support:* In 2005–06, 1 fellowship, 16 research assistantships were awarded; teaching assistantships *Faculty research:* Human gene mapping, genome organization and sequencing, gene regulation, molecular evolution. Total annual research expenditures: $3.3 million. *Unit head:* Dr. David Womble, Director, 313-577-5218, Fax: 313-577-5218, E-mail: aa3330@wayne.edu. *Application contact:* Alexander Gow, Associate Professor, 313-577-9401, E-mail: agow@genetics.wayne.edu.

Announcement: The Center for Molecular Medicine and Genetics offers a challenging, research-intensive graduate program leading to the PhD degree in molecular biology and genetics. Students study and conduct research in state-of-the-art laboratories located in the 4th-largest medical school in the country. The program emphasizes research training in eukaryotic molecular and cellular biology with applications to molecular medicine and genetics. Multidisciplinary approaches are taken, including the use of model organisms for the study of developmental and reproductive biology, cancer, neurobiology, and other areas. The center also offers graduate programs in genetic counseling and applied genomic technologies.

See Close-Up on page 665.

Wesleyan University, Graduate Programs, Department of Biology, Middletown, CT 06459-0260. Offers cell biology (PhD); comparative physiology (PhD); developmental biology (PhD); genetics (PhD); neurophysiology (PhD); population biology (PhD). *Faculty:* 12 full-time (3 women). *Students:* 29 full-time (15 women), 10 international. Average age 26. 131 applicants. In 2005, 2 doctorates awarded. *Degree requirements:* For doctorate, one foreign language, thesis/dissertation. *Entrance requirements:* For doctorate, GRE Subject Test. *Application deadline:* For fall admission, 2/15 for domestic students. Applications are processed on a rolling basis. *Application fee:* $0. *Expenses:* Tuition: Full-time $24,732. One-time fee: $20 full-time. *Financial support:* Research assistantships, teaching assistantships, stipends available. *Faculty research:* Microbial population genetics, genetic basis of evolutionary adaptation, genetic regulation of differentiation and pattern formation in *drosophila*. *Unit head:* Dr. Michael Weir, Chairman, 860-685-2402, E-mail: mweir@wesleyan.edu. *Application contact:* Marjorie Fitzgibbons, Information Contact, 860-685-2157, E-mail: mfitzgibbons@wesleyan.edu.

See Close-Up on page 323.

West Virginia University, Davis College of Agriculture, Forestry and Consumer Sciences, Interdisciplinary Program in Genetics and Developmental Biology, Morgantown, WV 26506. Offers animal breeding (MS, PhD); biochemical and molecular genetics (MS, PhD); cytogenetics (MS, PhD); descriptive embryology (MS, PhD); developmental genetics (MS); experimental morphogenesis teratology (MS); human genetics (MS, PhD); immunogenetics (MS, PhD); life cycles of animals and plants (MS, PhD); molecular aspects of development (MS, PhD); mutagenesis (MS, PhD); oncology (MS, PhD); plant genetics (MS, PhD); population and quantitative genetics (MS, PhD); regeneration (MS, PhD); teratology (PhD); toxicology (MS, PhD). *Students:* 13 full-time (7 women), 6 part-time (4 women), 11 international. Average age 28. In 2005, 2 master's, 4 doctorates awarded. *Degree requirements:* For master's, thesis/dissertation; for doctorate, thesis/dissertation, comprehensive exam. *Entrance requirements:* For master's, GRE or MCAT, minimum GPA of 2.75. Additional exam requirements/recommendations for international students: Required—TOEFL. *Application fee:* $45. *Expenses:* Tuition, state resident: full-time $4,582; part-time $258 per credit hour. Tuition, nonresident: full-time $13,820; part-time $741 per credit hour. *Financial support:* In 2005–06, 3 research assistantships with tuition reimbursements (averaging $9,936 per year), 4 teaching assistantships with tuition reimbursements (averaging $9,936 per year) were awarded; fellowships, Federal Work-Study, institutionally sponsored loans, and tuition waivers (full and partial) also available. Financial award application deadline: 2/1; financial award applicants required to submit FAFSA. Total annual research expenditures: $1 million. *Unit head:* Dr. J. Nath, Chairman, 304-293-6023 Ext. 4333, Fax: 304-293-2960, E-mail: joginder.nath@mail.wvu.edu.

Announcement: Interdepartmental program in genetics and developmental biology offers MS and PhD degrees. Areas include biochemical and molecular genetics, cytogenetics, mutagenesis, quantitative genetics, human genetics, plant genetics, experimental morphogenesis, and oncology. Write to Dr. J. Nath, Program Chairman, PO Box 6108, Morgantown, WV 26506-6108. joginder.nath@mail.wvu.edu

Yale University, Graduate School of Arts and Sciences, Department of Genetics, New Haven, CT 06520. Offers PhD, MD/PhD. *Degree requirements:* For doctorate, thesis/dissertation. *Entrance requirements:* For doctorate, GRE General Test, GRE Subject Test.

Yale University, Graduate School of Arts and Sciences, Department of Molecular, Cellular, and Developmental Biology, Program in Genetics, New Haven, CT 06520. Offers PhD. *Degree requirements:* For doctorate, thesis/dissertation. *Entrance requirements:* For doctorate, GRE General Test, GRE Subject Test.

Yale University, School of Medicine and Graduate School of Arts and Sciences, Combined Program in Biological and Biomedical Sciences (BBS), Molecular Cell Biology, Genetics, and Development Track, New Haven, CT 06520. Offers PhD, MD/PhD. *Students:* 31 full-time. *Entrance requirements:* Additional exam requirements/recommendations for international students: Required—TOEFL. *Application deadline:* For fall admission, 12/8 for domestic students, 12/8 for international students. *Unit head:* Dr. Thomas Pollard, Co-Director, 203-432-3565. *Application contact:* Anne Scott, Graduate Registrar, 203-432-3538, Fax: 203-432-3597, E-mail: anne.scott@yale.edu.

Genomic Sciences

Boston University, School of Medicine, Department of Genetics and Genomics, Boston, MA 02215. Offers MA, PhD. *Degree requirements:* For master's and doctorate, thesis/dissertation. *Entrance requirements:* For master's and doctorate, GRE, letters of recommendation. Additional exam requirements/recommendations for international students: Required—TOEFL. *Application deadline:* For fall admission, 1/1 for domestic students. Electronic applications accepted. *Expenses:* Tuition: Full-time $31,530; part-time $985 per credit. Required fees: $316; $40 per semester. Tuition and fees vary according to course level and program. *Unit head:* Dr. Shoumita Dasgupta, Assistant Professor and Director of Graduate Studies, 617-414-1580, E-mail: dasgupta@bu.edu.

Case Western Reserve University, School of Medicine and School of Graduate Studies, Graduate Programs in Medicine, Department of Genetics, Program in Human, Molecular, and Developmental Genetics and Genomics, Cleveland, OH 44106. Offers PhD, MD/PhD. *Students:* Average age 28. *Degree requirements:* For doctorate, thesis/dissertation. *Entrance requirements:* For doctorate, GRE General Test, GRE Subject Test. Additional exam requirements/recommendations for international students: Required—TOEFL. *Application deadline:* For fall admission, 2/28 for domestic students. Applications are processed on a rolling basis. *Application fee:* $25. *Financial support:* Fellowships, research assistantships available. Financial award application deadline: 2/28. *Faculty research:* Regulation of gene expression, molecular control of development, genomics. Total annual research expenditures: $8 million. *Application contact:* Malana C. Bey, Department Assistant, 216-368-3431, Fax: 216-368-1257, E-mail: mcb19@po.cwru.edu.

See Close-Up on page 761.

Concordia University, School of Graduate Studies, Faculty of Arts and Science, Department of Biology, Montréal, QC H3G 1M8, Canada. Offers biology (M Sc, PhD); biotechnology and genomics (Diploma). *Students:* 68 full-time (39 women). In 2005, 10 master's, 1 doctorate, 4 other advanced degrees awarded. *Degree requirements:* For master's, thesis; for doctorate, thesis/dissertation, pedagogical training. *Entrance requirements:* For master's, honors degree in biology; for doctorate, M Sc in life science. *Application deadline:* For fall admission, 1/15 for domestic students. For winter admission, 8/31 for domestic students. *Application fee:* $50.

Expenses: Tuition, state resident: full-time $834; part-time $334 per term. Tuition, nonresident: full-time $2,200; part-time $880 per term. Required fees: $680 per term. Tuition and fees vary according to degree level and program. *Faculty research:* Cell biology, animal physiology, ecology, microbiology/molecular biology, plant physiology/biochemistry and biotechnology. *Unit head:* Dr. Luc Varin, Chair, 514-848-2424 Ext. 3390, Fax: 514-848-2881. *Application contact:* Dr. Paul Widden, Director, 514-848-2424 Ext. 3413, Fax: 514-848-2881.

The George Washington University, Columbian College of Arts and Sciences, Department of Biochemistry and Molecular Biology, Program in Genomics, Proteomics, and Bioinformatics, Washington, DC 20052. Offers MS. Part-time programs available. *Entrance requirements:* For master's, GRE General Test, minimum GPA of 3.0. Additional exam requirements/recommendations for international students: Required—TOEFL (minimum score 550 paper-based; 213 computer-based). Electronic applications accepted.

See Close-Up on page 773.

Harvard University, Graduate School of Arts and Sciences, Department of Systems Biology, Cambridge, MA 02138. Offers PhD. *Students:* 9. 100 applicants, 12% accepted, 9 enrolled. *Degree requirements:* For doctorate, thesis/dissertation, lab rotation, qualifying examination. *Entrance requirements:* For doctorate, GRE. Additional exam requirements/recommendations for international students: Required—TOEFL. *Application deadline:* For fall admission, 12/8 priority date for domestic students, 12/8 priority date for international students. Application fee: $90. Electronic applications accepted. *Expenses:* Tuition: Full-time $28,752. Full-time tuition and fees vary according to program and student level. *Financial support:* Institutionally sponsored loans, scholarships/grants, unspecified assistantships, and all students receive a stipend and full tuition reimbursements available. *Unit head:* Judy Finkelstein, Graduate Coordinator, 617-432-5876, Fax: 617-432-5012, E-mail: pamela_silver@dfci.harvard.edu. *Application contact:* Jodi Finkelstein, Coordinator, 617-432-5202, Fax: 617-432-5012, E-mail: jodi_finkelstein@hms.harvard.edu.

See Close-Up on page 675.

Mount Sinai School of Medicine of New York University, Graduate School of Biological Sciences, New York, NY 10029-6504. Offers biophysics, structural biology and biomathematics (PhD); community medicine (MPH); genetic counseling (MS); genetics and genomic sciences (PhD); mechanisms of disease and therapy (PhD); microbiology (PhD); molecular, cellular, biochemical and developmental sciences (PhD); neurosciences (PhD). *Students:* 218 full-time (109 women). 4,208 applicants, 7% accepted, 117 enrolled. Terminal master's awarded for partial completion of doctoral program. *Degree requirements:* For master's, registration; for doctorate, thesis/dissertation, registration. *Entrance requirements:* For doctorate, GRE General Test, GRE Subject Test, 3 years of college pre-med course work. Additional exam requirements/recommendations for international students: Required—TOEFL. *Application deadline:* For fall admission, 1/15 for domestic students. Application fee: $75. Electronic applications accepted. *Expenses:* Tuition: Full-time $33,250. Required fees: $1,600. Full-time tuition and fees vary according to degree level, program and reciprocity agreements. *Financial support:* In 2005–06, fellowships with full tuition reimbursements (averaging $26,000 per year), research assistantships with full tuition reimbursements (averaging $26,000 per year) were awarded; Federal Work-Study, institutionally sponsored loans, scholarships/grants, health care benefits, and unspecified assistantships also available. Financial award application deadline: 4/5; financial award applicants required to submit FAFSA. *Faculty research:* Cancer, gene therapy, minimally invasive surgery, cardiac translational research. Total annual research expenditures: $162.2 million. *Unit head:* Dr. Diomedes Logothetis, Dean, 212-241-6546, Fax: 212-241-0651, E-mail: diomedes.logothetis@mssm.edu. *Application contact:* Lily Recanati, Manager, 212-241-3267, Fax: 212-241-0651, E-mail: lily.recantati@mssm.edu.

See Close-Up on page 177.

North Carolina State University, Graduate School, College of Agriculture and Life Sciences and College of Engineering and College of Natural Resources, Program in Functional Genomics, Raleigh, NC 27695. Offers MFG, MS, PhD. *Degree requirements:* For master's, thesis (for some programs); for doctorate, thesis/dissertation. *Entrance requirements:* For master's and doctorate, GRE, minimum B average. Additional exam requirements/recommendations for international students: Required—TOEFL. Electronic applications accepted. *Faculty research:* Genome structure, genome expression, molecular evolution, nucleic acid structure/function, proteomics.

See Close-Ups on pages 783 and 781.

North Dakota State University, The Graduate School, College of Agriculture, Food Systems, and Natural Resources, Interdisciplinary Program in Genomics, Fargo, ND 58105. Offers genomics and bioinformatics (MS, PhD). Part-time programs available. *Faculty:* 21 full-time (9 women). *Students:* 12 full-time (7 women), 1 part-time; includes 7 Asian Americans or Pacific Islanders. 2 applicants, 100% accepted, 2 enrolled. *Degree requirements:* For master's, thesis/dissertation; for doctorate, thesis/dissertation, comprehensive exam. *Entrance requirements:* For master's and doctorate, minimum GPA of 3.0. Additional exam requirements/recommendations for international students: Required—TOEFL. *Application deadline:* Applications are processed on a rolling basis. Application fee: $45 ($60 for international students). Electronic applications accepted. *Financial support:* In 2005–06, 11 research assistantships with full tuition reimbursements (averaging $15,000 per year) were awarded; unspecified assistantships also available. *Faculty research:* Genome evolution, genome mapping, genome expression, bioinformatics, date mining. Total annual research expenditures: $300,000. *Unit head:* Dr. Phillip E. McClean, Director, 701-231-8443, Fax: 701-231-8474.

North Dakota State University, The Graduate School, College of Science and Mathematics, Department of Biological Sciences, Fargo, ND 58105. Offers biological sciences (MS); botany (MS, PhD); cellular and molecular biology (PhD); environmental and conservation sciences (MS, PhD); genomics (MS, PhD); natural resource management (MS, PhD); zoology (MS, PhD). *Faculty:* 20. *Students:* 30 full-time (10 women), 5 part-time (1 woman); includes 1 minority (Asian American or Pacific Islander), 3 international. Average age 24. 14 applicants, 43% accepted. In 2005, 4 master's awarded. *Degree requirements:* For master's and doctorate, thesis/dissertation. *Entrance requirements:* For master's and doctorate, GRE General Test. Additional exam requirements/recommendations for international students: Required—TOEFL. *Application deadline:* For fall admission, 3/15 for domestic students; for spring admission, 10/30 priority date for domestic students. Applications are processed on a rolling basis. Application fee: $45 ($60 for international students). Electronic applications accepted. *Financial support:* In 2005–06, 3 fellowships with full tuition reimbursements (averaging $15,000 per year), 9 research assistantships with full tuition reimbursements (averaging $14,400 per year), 19 teaching assistantships with full tuition reimbursements (averaging $9,550 per year) were awarded; career-related internships or fieldwork, Federal Work-Study, institutionally sponsored loans, scholarships/grants, tuition waivers (full), and unspecified assistantships also available. Support available to part-time students. Financial award application deadline: 4/15; financial award applicants required to submit FAFSA. *Faculty research:* Comparative endocrinology, physiology, behavioral ecology, plant cell biology, aquatic biology. Total annual research expenditures: $675,000. *Unit head:* Dr. William J. Bleier, Chair, 701-231-7087, Fax: 701-231-7149, E-mail: william.bleier@ndsu.nodak.edu.

University of California, Riverside, Graduate Division, Graduate Program in Genetics, Genomics, and Bioinformatics, Riverside, CA 92521-0102. Offers genomics and bioinformatics (PhD); molecular genetics (PhD); population and evolutionary genetics (PhD). *Students:* 19 full-time (7 women); includes 3 minority (2 Asian Americans or Pacific Islanders, 1 Hispanic American), 7 international. Average age 30. In 2005, 1 degree awarded. *Degree requirements:* For doctorate, thesis/dissertation, qualifying exams, teaching experience. *Entrance requirements:* For doctorate, GRE General Test, minimum GPA of 3.2. Additional exam requirements/recommendations for international students: Required—TOEFL (minimum score 550 paper-based; 213 computer-based); Recommended—TSE (minimum score 50). *Application deadline:*

For fall admission, 5/1 for domestic students, 2/1 for international students. For winter admission, 9/1 for domestic students; for spring admission, 12/1 for domestic students. Applications are processed on a rolling basis. Application fee: $60 ($75 for international students). Electronic applications accepted. *Expenses:* Tuition, nonresident: full-time $14,694. Full-time tuition and fees vary according to program. *Financial support:* In 2005–06, research assistantships (averaging $14,000 per year), teaching assistantships (averaging $15,000 per year) were awarded; fellowships, career-related internships or fieldwork, Federal Work-Study, institutionally sponsored loans, and tuition waivers (full and partial) also available. Financial award application deadline: 2/1; financial award applicants required to submit FAFSA. *Unit head:* Dr. Katherine Borkovich, Director. *Application contact:* Carole Carpenter, Graduate Program Assistant, 800-735-0717, Fax: 951-827-5517, E-mail: genetics@ucr.edu.

University of California, San Francisco, School of Pharmacy and Graduate Division, Pharmaceutical Sciences and Pharmacogenomics Graduate Group, San Francisco, CA 94143. Offers PhD. *Faculty:* 49 full-time (13 women). *Students:* 48 full-time (24 women). Average age 27. 100 applicants, 14% accepted. In 2005, 7 degrees awarded. *Degree requirements:* For doctorate, thesis/dissertation. *Entrance requirements:* For doctorate, GRE General Test, minimum GPA of 3.0. Additional exam requirements/recommendations for international students: Required—TOEFL. *Application deadline:* For fall admission, 1/15 for domestic students. Application fee: $60. *Financial support:* In 2005–06, 4 fellowships with full tuition reimbursements (averaging $25,000 per year), 29 research assistantships with full tuition reimbursements (averaging $23,000 per year), 8 teaching assistantships with full tuition reimbursements (averaging $23,000 per year) were awarded; career-related internships or fieldwork, institutionally sponsored loans, scholarships/grants, traineeships, tuition waivers (full), and unspecified assistantships also available. *Faculty research:* Drug development, drug delivery, molecular pharmacology. *Unit head:* Francis C. Szoka, Program Director, 415-476-3895, Fax: 415-476-0688, E-mail: szoka@cgl.ucsf.edu. *Application contact:* Debbie Acoba, Administrator, 415-476-1947, Fax: 415-476-4929, E-mail: pspg@itsa.ucsf.edu.

University of Cincinnati, Division of Research and Advanced Studies, College of Medicine, Graduate Programs in Biomedical Sciences, Department of Environmental Health, Programs in Environmental Genetics and Molecular Toxicology, Cincinnati, OH 45221. Offers MS, PhD. *Degree requirements:* For doctorate, thesis/dissertation. *Entrance requirements:* For master's, GRE; a minimum grade-point average (GPA) of 3.00; 3 letters of recommendation. Additional exam requirements/recommendations for international students: Required—TOEFL (minimum score 520 paper-based; 190 computer-based).

University of Connecticut, Graduate School, College of Liberal Arts and Sciences, Department of Molecular and Cell Biology, Field of Applied Genomics, Storrs, CT 06269. Offers MS, PSM. *Faculty:* 8 full-time (2 women). *Students:* 8 full-time (5 women), 1 part-time, 1 international. Average age 26. 4 applicants, 100% accepted, 2 enrolled. In 2005, 4 degrees awarded. *Degree requirements:* For master's, comprehensive exam. *Entrance requirements:* For master's, GRE General Test, GRE Subject Test. Additional exam requirements/recommendations for international students: Required—TOEFL (minimum score 550 paper-based; 213 computer-based). *Application deadline:* For fall admission, 2/1 priority date for domestic students, 2/1 priority date for international students; for spring admission, 11/1 for domestic students, 10/1 for international students. Applications are processed on a rolling basis. Application fee: $55. Electronic applications accepted. *Expenses:* Tuition, state resident: part-time $444 per credit hour. Tuition, nonresident: part-time $1,154 per credit hour. Tuition and fees vary according to course load. *Financial support:* In 2005–06, 1 research assistantship with full tuition reimbursement, 2 teaching assistantships with full tuition reimbursements were awarded; fellowships, Federal Work-Study, scholarships/grants, health care benefits, and unspecified assistantships also available. Financial award application deadline: 2/1; financial award applicants required to submit FAFSA. *Application contact:* Anne St. Onje, Graduate Coordinator, 860-486-4314, Fax: 860-486-3943, E-mail: ann.st_onje@uconn.edu.

University of Florida, College of Medicine, Department of Physiology and Functional Genomics, Gainesville, FL 32611. Offers PhD. *Faculty:* 15 full-time (4 women). *Students:* 31 (18 women); includes 3 minority (all Asian Americans or Pacific Islanders) 8 international. In 2005, 10 degrees awarded. *Degree requirements:* For doctorate, thesis/dissertation. *Entrance requirements:* For doctorate, GRE General Test, minimum GPA of 3.0. Additional exam requirements/recommendations for international students: Required—TOEFL. *Application deadline:* For fall admission, 2/1 for domestic students. Application fee: $30. Electronic applications accepted. *Expenses:* Tuition, state resident: full-time $6,234. Tuition, nonresident: full-time $21,359. Tuition and fees vary according to program. *Financial support:* In 2005–06, 13 research assistantships with full tuition reimbursements (averaging $23,321 per year) were awarded; fellowships with full tuition reimbursements, institutionally sponsored loans and traineeships also available. *Faculty research:* Cell and general endocrinology, neuroendocrinology, neurophysiology, respiration, membrane transport and ion channels. *Unit head:* Dr. Charles Wood, Chairman, 352-392-4488, Fax: 352-846-0270, E-mail: cwood@phys.med.ufl.edu. *Application contact:* Dr. Wayne McCormack, Associate Dean of Graduate Education, 352-392-7413, Fax: 352-846-3466, E-mail: mccormac@pathology.ufl.edu.

University of Pennsylvania, School of Medicine, Biomedical Graduate Studies, Graduate Group in Genomics and Computational Biology, Philadelphia, PA 19104. Offers PhD, MD/PhD, VMD/PhD. *Faculty:* 51. *Students:* 26 full-time (5 women); includes 6 minority (2 African Americans, 4 Asian Americans or Pacific Islanders), 6 international. 47 applicants, 30% accepted, 8 enrolled. *Degree requirements:* For doctorate, thesis/dissertation optional. *Entrance requirements:* For doctorate, GRE. Additional exam requirements/recommendations for international students: Required—TOEFL. *Application deadline:* For fall admission, 12/15 priority date for domestic students, 12/1 priority date for international students. Applications are processed on a rolling basis. Application fee: $70. Electronic applications accepted. *Financial support:* In 2005–06, 19 students received support; fellowships, research assistantships, scholarships/grants, traineeships, and unspecified assistantships available. *Unit head:* Dr. Warren Ewens, Chairperson, 215-898-7109. *Application contact:* Sean Dalton, Graduate Coordinator, 215-746-2807, E-mail: sdalton@mail.med.upenn.edu.

The University of Tennessee, Graduate School, College of Arts and Sciences, Program in Life Sciences, Knoxville, TN 37996. Offers genome science and technology (MS, PhD); plant physiology and genetics (MS, PhD). *Degree requirements:* For doctorate, one foreign language, thesis/dissertation. *Entrance requirements:* For master's and doctorate, GRE General Test, minimum GPA of 2.7. Additional exam requirements/recommendations for international students: Required—TOEFL. Electronic applications accepted.

The University of Tennessee–Oak Ridge National Laboratory Graduate School of Genome Science and Technology, Graduate Program, Oak Ridge, TN 37830-8026. Offers life sciences (MS, PhD). *Faculty:* 1 (woman) full-time, 87 part-time/adjunct (19 women). *Students:* 43 full-time (20 women), 2 part-time (1 woman); includes 4 minority (2 African Americans, 2 Asian Americans or Pacific Islanders), 21 international. Average age 30. 49 applicants, 33% accepted, 9 enrolled. In 2005, 3 master's, 3 doctorates awarded. *Median time to degree:* Of those who began their doctoral program in fall 1997, 98% received their degree in 8 years or less. *Degree requirements:* For master's, thesis/dissertation, registration; for doctorate, thesis/dissertation, comprehensive exam, registration. *Entrance requirements:* For master's and doctorate, GRE General Test. Additional exam requirements/recommendations for international students: Required—TOEFL (minimum score 550 paper-based; 213 computer-based). *Application deadline:* For fall admission, 1/15 priority date for domestic students, 2/1 priority date for international students. Applications are processed on a rolling basis. Application fee: $35. Electronic applications accepted. *Expenses:* Tuition, state resident: part-time $296 per hour. Tuition, nonresident: part-time $895 per hour. *Financial support:* In 2005–06, 25 students received support, including 9 research assistantships with full tuition reimbursements available (averaging $18,000 per year); fellowships, institutionally sponsored loans, health care benefits, tuition waivers (full), and unspecified assistantships also available. Financial award application deadline: 3/31. *Faculty research:* Genetics/genomics, structural biology/proteomics, computational

Genomic Sciences

The University of Tennessee–Oak Ridge National Laboratory Graduate School of Genome Science and Technology (continued)
biology/bioinformatics, bioanalytical technologies. *Unit head:* Dr. Cynthia B Peterson, Director, 865-974-4083, Fax: 965-974-0361, E-mail: cbpeters@utk.edu. *Application contact:* Kay Gardner, Program Resource Specialist, 865-574-1227, Fax: 865-574-4812, E-mail: gardnerk@utk.edu.

See Close-Up on page 799.

University of Washington, School of Medicine, Department of Genome Sciences, Seattle, WA 98195. Offers PhD. *Degree requirements:* For doctorate, thesis/dissertation, general exam. *Entrance requirements:* For doctorate, GRE General Test, minimum GPA of 3.0. Additional exam requirements/recommendations for international students: Required—TOEFL. Electronic applications accepted. *Faculty research:* Model organism genetics, human and medical genetics, genomics and proteomics, computational biology.

Wake Forest University, School of Medicine and Graduate School, Graduate Programs in Medicine, Molecular Genetics Program, Winston-Salem, NC 27109. Offers PhD. *Degree requirements:* For doctorate, thesis/dissertation. *Entrance requirements:* For doctorate, GRE General Test. Additional exam requirements/recommendations for international students: Required—TOEFL. Electronic applications accepted. *Faculty research:* Control of gene expression, molecular pathogenesis, protein biosynthesis, cell development, clinical cytogenetics.

See Close-Up on page 805.

Yale University, School of Medicine and Graduate School of Arts and Sciences, Combined Program in Biological and Biomedical Sciences (BBS), Computational Biology and Bioinformatics Track, New Haven, CT 06520. Offers PhD, MD/PhD. *Students:* 5 full-time. *Entrance requirements:* Additional exam requirements/recommendations for international students: Required—TOEFL. *Application deadline:* For fall admission, 12/8 for domestic students, 12/8 for international students. *Unit head:* Dr. Perry Miller, Director of Graduate Studies, 203-432-8189, E-mail: dgs.bioinfo@yale.edu.

Human Genetics

Baylor College of Medicine, Graduate School of Biomedical Sciences, Department of Molecular and Human Genetics, Houston, TX 77030-3498. Offers PhD, MD/PhD. *Faculty:* 52 full-time (8 women). *Students:* 81 full-time (53 women); includes 10 minority (1 African American, 6 Asian Americans or Pacific Islanders, 3 Hispanic Americans), 39 international. Average age 27. 83 applicants, 23% accepted, 11 enrolled. In 2005, 11 doctorates awarded. *Median time to degree:* Of those who began their doctoral program in fall 1997, 70% received their degree in 8 years or less. *Degree requirements:* For doctorate, thesis/dissertation, public defense. *Entrance requirements:* For doctorate, GRE General Test, GRE Subject Test (strongly recommended), minimum GPA of 3.0. Additional exam requirements/recommendations for international students: Required—TOEFL. *Application deadline:* For fall admission, 2/1 for domestic students. Application fee: $30. Electronic applications accepted. *Expenses:* Tuition: Full-time $8,200. Full-time tuition and fees vary according to program. *Financial support:* In 2005–06, 80 students received support, including 23 fellowships (averaging $20,000 per year), 58 research assistantships (averaging $20,000 per year); career-related internships or fieldwork, Federal Work-Study, institutionally sponsored loans, health care benefits, and tuition waivers (full) also available. Financial award applicants required to submit FAFSA. *Faculty research:* Cytogenetics, biochemical genetics, somatic cell genetics, gene therapy. *Unit head:* Dr. Gad Shaulsky, Director, 713-798-5056. *Application contact:* Judi Coleman, Graduate Program Administrator, 713-798-5056, Fax: 713-798-8597, E-mail: genetics-gradprm@bcm.tmc.edu.

See Close-Up on page 759.

Baylor College of Medicine, Graduate School of Biomedical Sciences, Interdepartmental Program in Cell and Molecular Biology, Houston, TX 77030-3498. Offers biochemistry (PhD); cell and molecular biology (PhD); genetics (PhD); human genetics (PhD); immunology (PhD); microbiology (PhD); virology (PhD). *Faculty:* 99 full-time (21 women). *Students:* 56 full-time (28 women); includes 20 minority (4 African Americans, 1 American Indian/Alaska Native, 6 Asian Americans or Pacific Islanders, 9 Hispanic Americans), 5 international. Average age 28. 164 applicants, 18% accepted, 12 enrolled. In 2005, 3 doctorates awarded. *Median time to degree:* Of those who began their doctoral program in fall 1997, 63% received their degree in 8 years or less. *Degree requirements:* For doctorate, thesis/dissertation, public defense. *Entrance requirements:* For doctorate, GRE General Test, GRE Subject Test (strongly recommended), minimum GPA of 3.0. Additional exam requirements/recommendations for international students: Required—TOEFL. *Application deadline:* For fall admission, 1/1 for domestic students. Applications are processed on a rolling basis. Application fee: $30. Electronic applications accepted. *Expenses:* Tuition: Full-time $8,200. Full-time tuition and fees vary according to program. *Financial support:* In 2005–06, 52 students received support, including 20 fellowships (averaging $23,000 per year), 36 research assistantships (averaging $23,000 per year); teaching assistantships, Federal Work-Study, institutionally sponsored loans, health care benefits, and tuition waivers (full) also available. Financial award applicants required to submit FAFSA. *Faculty research:* Gene expression and regulation, developmental biology and genetics, signal transduction and membrane biology, aging process, molecular virology. *Unit head:* Dr. Tom Cooper, Director, 713-798-6557. *Application contact:* Lourdes Fernandez, Graduate Program Administrator, 713-798-6557, Fax: 713-798-6325, E-mail: cmbprog@bcm.edu.

See Close-Up on page 545.

Case Western Reserve University, School of Medicine and School of Graduate Studies, Graduate Programs in Medicine, Department of Genetics, Program in Human, Molecular, and Developmental Genetics and Genomics, Cleveland, OH 44106. Offers PhD, MD/PhD. *Students:* Average age 28. *Degree requirements:* For doctorate, thesis/dissertation. *Entrance requirements:* For doctorate, GRE General Test, GRE Subject Test. Additional exam requirements/recommendations for international students: Required—TOEFL. *Application deadline:* For fall admission, 2/28 for domestic students. Applications are processed on a rolling basis. Application fee: $25. *Financial support:* Fellowships, research assistantships available. Financial award application deadline: 2/28. *Faculty research:* Regulation of gene expression, molecular control of development, genomics. Total annual research expenditures: $8 million. *Application contact:* Malana C. Bey, Department Assistant, 216-368-3431, Fax: 216-368-1257, E-mail: mcb19@po.cwru.edu.

See Close-Up on page 761.

Drexel University, College of Medicine, Biomedical Graduate Programs, Interdisciplinary Program in Molecular and Human Genetics, Philadelphia, PA 19104-2875. Offers MS, PhD, MD/PhD. *Degree requirements:* For master's, comprehensive exam; for doctorate, thesis/dissertation, qualifying exam. *Entrance requirements:* For master's and doctorate, GRE General Test. Additional exam requirements/recommendations for international students: Required—TOEFL.

Howard University, Graduate School of Arts and Sciences, Department of Genetics and Human Genetics, Washington, DC 20059. Offers MS, PhD. Part-time programs available. *Faculty:* 13. *Students:* 23; includes 16 minority (15 African Americans, 1 Asian American or Pacific Islander), 6 international. In 2005, 11 master's, 2 doctorates awarded. *Median time to degree:* Of those who began their doctoral program in fall 1997, 100% received their degree in 8 years or less. *Degree requirements:* For master's, thesis, comprehensive exam; for doctorate, one foreign language, thesis/dissertation, comprehensive exam. *Entrance requirements:* For master's, GRE General Test, minimum GPA of 3.0; course work in biology, chemistry, physics, mathematics, and genetics; for doctorate, GRE General Test, master's degree in genetics or related field, minimum GPA of 3.2. Additional exam requirements/recommendations for international students: Required—TOEFL, GRE required. *Application deadline:* For fall admission, 4/1 for domestic students; for spring admission, 11/1 for domestic students. Applications are processed on a rolling basis. Application fee: $45. Electronic applications accepted. *Financial support:* Fellowships with full tuition reimbursements, research assistantships, teaching assistantships, Federal Work-Study, institutionally sponsored loans, scholarships/grants, tuition waivers (full), and unspecified assistantships available. Financial award applicants required to submit FAFSA. *Unit head:* Dr. Robert F. Murray, Chairman, 202-806-6340. *Application contact:* Dr. Verle E. Headings, Director of Graduate Studies, 202-806-6381.

The Johns Hopkins University, School of Medicine, Graduate Programs in Medicine, Predoctoral Training Program in Human Genetics and Molecular Biology, Baltimore, MD 21218-

2699. Offers PhD, MD/PhD. *Faculty:* 59 full-time (14 women). *Students:* 52 full-time (27 women); includes 6 minority (2 African Americans, 3 Asian Americans or Pacific Islanders, 1 Hispanic American), 18 international. Average age 24. 171 applicants, 18% accepted, 16 enrolled. In 2005, 6 doctorates awarded. *Degree requirements:* For doctorate, thesis/dissertation, comprehensive exam. *Entrance requirements:* For doctorate, GRE General Test, GRE Subject Test. *Application deadline:* For fall admission, 1/15 priority date for domestic students, 1/15 priority date for international students. Application fee: $70. *Expenses:* Tuition: Full-time $30,960. Tuition and fees vary according to degree level and program. *Financial support:* In 2005–06, 12 fellowships with full tuition reimbursements (averaging $24,600 per year) were awarded; teaching assistantships with full tuition reimbursements, health care benefits also available. *Faculty research:* Human, mammalian, and molecular genetics. *Unit head:* Dr. David Valle, Director, 410-955-4260, Fax: 410-955-7397, E-mail: muscelli@jhmi.edu. *Application contact:* Sandy Muscelli, Program Administrator, 410-955-4260, Fax: 410-955-7397, E-mail: muscelli@jhmi.edu.

See Close-Up on page 779.

Louisiana State University Health Sciences Center, School of Graduate Studies in New Orleans, Department of Human Genetics, New Orleans, LA 70112-2223. Offers MS, PhD, MD/PhD. Part-time programs available. Terminal master's awarded for partial completion of doctoral program. *Degree requirements:* For master's and doctorate, thesis/dissertation, comprehensive exam. *Entrance requirements:* For master's and doctorate, GRE General Test. Additional exam requirements/recommendations for international students: Required—TOEFL. *Faculty research:* Genetic epidemiology, segregation and linkage analysis, gene mapping.

McGill University, Faculty of Graduate and Postdoctoral Studies, Faculty of Medicine, Department of Human Genetics, Montréal, QC H3A 2T5, Canada. Offers genetic counseling (M Sc); human genetics (M Sc, PhD). *Degree requirements:* For master's, thesis (human genetics); for doctorate, thesis/dissertation. *Entrance requirements:* For master's and doctorate, minimum GPA of 3.0. Additional exam requirements/recommendations for international students: Required—TOEFL. *Faculty research:* Epidemiological genetics, behavioral genetics, clinical genetics, cytogenetics, developmental genetics.

Memorial University of Newfoundland, Faculty of Medicine and School of Graduate Studies, Graduate Programs in Medicine, Program in Human Genetics, St. John's, NL A1C 5S7, Canada. Offers M Sc, PhD, MD/PhD. *Degree requirements:* For master's, thesis; for doctorate, thesis/dissertation, oral defense of thesis, comprehensive exam. *Entrance requirements:* For master's, MD or B Sc; for doctorate, MD or M Sc. Additional exam requirements/recommendations for international students: Required—TOEFL. Students attend 2 or 3 semesters per year. *Expenses:* Tuition, nonresident: full-time $11,466; part-time $733 per semester. International tuition: $1,906 full-time. Tuition and fees vary according to degree level and program. *Faculty research:* Cancer genetics, gene mapping, medical genetics, birth defects, population genetics.

Sarah Lawrence College, Graduate Studies, Program in Genetic Counseling, Bronxville, NY 10708-5999. Offers human genetics (MS). *Faculty:* 5 full-time, 23 part-time/adjunct. *Students:* 36 full-time (32 women), 6 part-time (all women); includes 7 minority (5 Asian Americans or Pacific Islanders, 2 Hispanic Americans), 11 international. Average age 28. 110 applicants, 41% accepted, 21 enrolled. In 2005, 15 degrees awarded. *Degree requirements:* For master's, thesis, fieldwork. *Entrance requirements:* For master's, previous course work in biology, chemistry, developmental biology, genetics, probability and statistics. *Application deadline:* For fall admission, 4/1 for domestic students. Application fee: $60. *Expenses:* Contact institution. Tuition and fees vary according to course load, program and student level. *Financial support:* In 2005–06, 32 fellowships (averaging $2,782 per year) were awarded; career-related internships or fieldwork, Federal Work-Study, scholarships/grants, and unspecified assistantships also available. Support available to part-time students. Financial award application deadline: 3/1. *Unit head:* , Caroline Lieber, Director, 914-395-2371. *Application contact:* Susan Guma, Dean of Graduate Studies, 914-395-2373, E-mail: sguma@mail.slc.edu.

Tulane University, School of Medicine and Graduate School, Graduate Programs in Medicine, Program in Human Genetics/Genetic Counseling, New Orleans, LA 70118-5669. Offers human genetics (MS, PhD). MS and PhD offered through the Graduate School. *Degree requirements:* For master's and doctorate, thesis/dissertation. *Entrance requirements:* For master's, GRE, MCAT; for doctorate, GRE General Test. Additional exam requirements/recommendations for international students: Required—TOEFL or TSE. Electronic applications accepted. *Faculty research:* Inborn errors of metabolism, DNA methylation, gene therapy.

University of California, Los Angeles, School of Medicine and Graduate Division, Graduate Programs in Medicine, Department of Human Genetics, Los Angeles, CA 90095. Offers MS, PhD. *Entrance requirements:* For master's and doctorate, GRE General Test.

University of Chicago, Division of the Biological Sciences, Department of Molecular Biosciences: Biochemistry, Genetics, Cell and Developmental Biology, Department of Human Genetics, Chicago, IL 60637-1513. Offers PhD. *Faculty:* 19 full-time (11 women). *Students:* 16 full-time (9 women); includes 2 minority (both Asian Americans or Pacific Islanders), 5 international. Average age 27. In 2005, 3 degrees awarded. *Degree requirements:* For doctorate, thesis/dissertation, registration. *Entrance requirements:* For doctorate, GRE General Test. *Application deadline:* For fall admission, 12/28 priority date for domestic students, 12/28 priority date for international students. Application fee: $55. Electronic applications accepted. *Financial support:* In 2005–06, fellowships with tuition reimbursements (averaging $26,301 per year), research assistantships with tuition reimbursements (averaging $26,301 per year) were awarded; institutionally sponsored loans, scholarships/grants, traineeships, and health care benefits also available. Financial award applicants required to submit FAFSA. *Unit head:* Dr. T. Conrad Gilliam, Chairman, 773-834-0525. *Application contact:* Amanda Hughes, Graduate Administrative Director, 773-834-8073, Fax: 773-702-3172, E-mail: ahughes@bsd.uchicago.edu.

University of Manitoba, Faculty of Medicine and Faculty of Graduate Studies, Graduate Programs in Medicine, Department of Biochemistry and Medical Genetics, Winnipeg, MB R3T 2N2, Canada. Offers M Sc, PhD. Terminal master's awarded for partial completion of doctoral program. *Degree requirements:* For master's and doctorate, thesis/dissertation. *Faculty research:* Cancer, gene expression, membrane lipids, metabolic control, genetic diseases.

University of Maryland, Graduate School, Graduate Programs in Medicine, Program in Human Genetics, Baltimore, MD 21201. Offers MS, PhD, MD/PhD. Part-time programs available. *Faculty:* 3 full-time (2 women). *Students:* 9 full-time (7 women); includes 1 minority (African American), 4 international. Average age 26. 11 applicants, 55% accepted, 3 enrolled. In 2005, 2 doctorates awarded. *Degree requirements:* For master's, thesis or alternative; for doctorate, thesis/dissertation, qualifying exam. *Entrance requirements:* For master's and doctorate, GRE General Test, minimum GPA of 3.0. Additional exam requirements/recommendations for international students: Required—TOEFL, TOEFL or IELTS; Recommended—IELT. *Application deadline:* For fall admission, 7/1 for domestic students, 1/15 for international students. Application fee: $50. Electronic applications accepted. *Expenses:* Tuition, state resident: full-time $8,079; part-time $409 per credit hour. Tuition, nonresident: full-time $18,384; part-time $731 per credit hour. Required fees: $695; $10 per credit hour. Tuition and fees vary according to degree level and program. *Financial support:* Fellowships, research assistantships, teaching assistantships, career-related internships or fieldwork available. Support available to part-time students. Financial award application deadline: 2/15. *Faculty research:* Cytogenetics, biochemical genetics, metabolic diseases, population genetics, family data analysis. *Unit head:* Dr. Miriam Blitzer, Interim Chair, 410-706-3480, Fax: 410-706-6105, E-mail: mimi@umaryland.edu.

University of Michigan, Medical School and Horace H. Rackham School of Graduate Studies, Program in Biomedical Sciences (PIBS), Department of Human Genetics, Ann Arbor, MI 48109. Offers MS, PhD. Part-time programs available. *Faculty:* 28 full-time (9 women). *Students:* 34 full-time (26 women); includes 3 minority (2 Asian Americans or Pacific Islanders, 1 Hispanic American), 2 international. Average age 27. 85 applicants, 41% accepted, 16 enrolled. In 2005, 10 master's, 6 doctorates awarded. Terminal master's awarded for partial completion of doctoral program. *Degree requirements:* For doctorate, oral defense of dissertation, oral preliminary exam. *Entrance requirements:* For master's, GRE General Test, 3 letters of recommendation; for doctorate, GRE General Test, GRE Subject Test (biology, biochemistry), 3 letters of recommendation. Additional exam requirements/recommendations for international students: Required—TOEFL. *Application deadline:* For fall admission, 12/31 for domestic students, 11/1 for international students. Application fee: $60 ($75 for international students). Electronic applications accepted. *Expenses:* Tuition, state resident: full-time $14,082; part-time $894 per credit hour. Tuition, nonresident: full-time $28,500; part-time $1,675 per credit hour. Required fees: $189; $189 per unit. *Financial support:* In 2005–06, 11 fellowships with full tuition reimbursements (averaging $20,772 per year), 13 research assistantships with full tuition reimbursements (averaging $23,500 per year), 1 teaching assistantship with full tuition reimbursement (averaging $14,326 per year) were awarded; scholarships/grants, traineeships, and unspecified assistantships also available. *Faculty research:* Molecular, cellular, and population genetics. *Unit head:* Dr. Sally A. Camper, Chair, 734-764-5491, Fax: 734-763-3784, E-mail: scamper@umich.edu. *Application contact:* Janet Miller, Student Services Associate II, 734-764-5490, Fax: 734-763-3784, E-mail: janmil@umich.edu.

University of Pittsburgh, Graduate School of Public Health, Department of Human Genetics, Pittsburgh, PA 15260. Offers genetic counseling (MS); human genetics (MS, PhD); public health genetics (MPH). *Faculty:* 11 full-time (7 women), 9 part-time/adjunct (5 women). *Students:* 28 full-time (12 women), 12 part-time (5 women); includes 5 minority (2 African Americans, 3 Asian Americans or Pacific Islanders), 19 international. Average age 30. 33 applicants, 76% accepted, 11 enrolled. In 2005, 3 master's, 3 doctorates awarded. Terminal master's awarded for partial completion of doctoral program. *Degree requirements:* For master's, thesis/dissertation, comprehensive exam (for some programs); for doctorate, thesis/dissertation, comprehensive exam. *Entrance requirements:* For master's, GRE General Test, previous course work in biochemistry, calculus, and genetics; for doctorate, GRE General Test. Additional exam requirements/recommendations for international students: Required—TOEFL (minimum score 550 paper-based; 213 computer-based). *Application deadline:* Applications are processed on a rolling basis. Application fee: $50 ($60 for international students). Electronic applications accepted. *Expenses:* Tuition, state resident: full-time $13,194; part-time $537 per credit. Tuition, nonresident: full-time $25,012; part-time $1,026 per credit. Required fees: $700; $164 per term. Tuition and fees vary according to campus/location and program. *Financial support:* In 2005–06, 25 students received support, including 1 fellowship (averaging $21,150 per year), 24 research assistantships with full tuition reimbursements available (averaging $20,333 per year); scholarships/grants and tuition waivers (partial) also available. Financial award applicants required to submit FAFSA. *Faculty research:* Genetic mechanisms related to the transition from normal to disease states, how genes and the environment interact to affect the distribution of health and disease in human populations. Total annual research expenditures: $6.2 million. *Unit head:* Dr. Robert E. Ferrell, Chairman, 412-624-3018, Fax: 412-624-3020, E-mail: rferrell@hgen.pitt.edu. *Application contact:* Jeanette Norbut, Administrative Secretary, 412-624-9951, Fax: 412-624-3020, E-mail: jnorbut@hgen.pitt.edu.

University of Pittsburgh, School of Medicine, Graduate Programs in Medicine, Program in Human Genetics, Pittsburgh, PA 15260. Offers PhD. *Faculty:* 34 full-time (11 women). *Students:* 4 full-time (0 women), 3 international. Average age 28. 415 applicants, 22% accepted, 42 enrolled. *Degree requirements:* For doctorate, thesis/dissertation, comprehensive exam, registration. *Entrance requirements:* For doctorate, GRE General Test, GRE Subject Test, minimum QPA of 3.0. Additional exam requirements/recommendations for international students: Required—TOEFL (minimum score 600 paper-based; 250 computer-based), IELT (minimum score 7). *Application deadline:* For fall admission, 12/15 priority date for domestic students, 12/15 priority date for international students. Application fee: $40. Electronic applications accepted. *Expenses:* Tuition, state resident: full-time $13,194; part-time $537 per credit. Tuition, nonresident: full-time $25,012; part-time $1,026 per credit. Required fees: $700; $164 per

term. Tuition and fees vary according to campus/location and program. *Financial support:* In 2005–06, 4 research assistantships with full tuition reimbursements (averaging $21,500 per year) were awarded; fellowships with full tuition reimbursements, teaching assistantships with full tuition reimbursements, institutionally sponsored loans, scholarships/grants, traineeships, tuition waivers (full), and unspecified assistantships also available. *Unit head:* Dr. Eleanor Feingold, Director, 412-383-8599, Fax: 412-624-3020, E-mail: feingold@pitt.edu. *Application contact:* Graduate Studies Administrator, 412-648-8957, Fax: 412-648-1077, E-mail: gradstudies@medschool.pitt.edu.

The University of Texas Health Science Center at Houston, Graduate School of Biomedical Sciences, Program in Human and Molecular Genetics, Houston, TX 77030-3900. Offers MS, PhD, MD/PhD. *Faculty:* 42 full-time (13 women). *Students:* 16 full-time (11 women); includes 5 minority (all Asian Americans or Pacific Islanders), 2 international. Average age 25. 36 applicants, 50% accepted, 6 enrolled. In 2005, 1 master's, 1 doctorate awarded. Terminal master's awarded for partial completion of doctoral program. *Degree requirements:* For master's and doctorate, thesis/dissertation. *Entrance requirements:* For master's and doctorate, GRE General Test. Additional exam requirements/recommendations for international students: Required—TOEFL, TWE. *Application deadline:* For fall admission, 1/15 for domestic students; for spring admission, 11/1 for domestic students. Applications are processed on a rolling basis. Application fee: $10. Electronic applications accepted. *Financial support:* Fellowships with full tuition reimbursements, research assistantships with full tuition reimbursements, teaching assistantships, institutionally sponsored loans, scholarships/grants, and health care benefits available. Financial award application deadline: 1/15. *Faculty research:* Cancer genetics, genetics and epidemiology of complex diseases, bioinformatics and computational genomics, medical genetics, animal models of human disease. *Unit head:* Dr. Ralf Krahe, Director, 713-834-6345, Fax: 713-834-6319, E-mail: rkrahe@mdanderson.org. *Application contact:* Dr. Victoria P. Knutson, Assistant Dean of Admissions, 713-500-9860, Fax: 713-500-9877, E-mail: victoria.p.knutson@uth.tmc.edu.

University of Utah, School of Medicine and The Graduate School, Graduate Programs in Medicine, Department of Human Genetics, Salt Lake City, UT 84112-1107. Offers MS, PhD. *Expenses:* Tuition, state resident: full-time $2,932; part-time $369 per credit. Tuition, nonresident: full-time $10,350; part-time $1,302 per credit. Required fees: $516 per term. Tuition and fees vary according to course load and program.

Virginia Commonwealth University, Medical College of Virginia-Professional Programs, School of Medicine and Graduate Programs, School of Medicine Graduate Programs, Department of Human Genetics, Richmond, VA 23284-9005. Offers genetic counseling (MS); human genetics (PhD, CBHS); molecular biology and genetics (MS, PhD). *Faculty:* 18 full-time (10 women). *Students:* 21 full-time (15 women), 4 part-time (1 woman); includes 2 minority (both Asian Americans or Pacific Islanders), 4 international. 52 applicants, 46% accepted. In 2005, 6 master's, 3 doctorates, 2 other advanced degrees awarded. *Degree requirements:* For master's, thesis; for doctorate, thesis/dissertation, comprehensive oral and written exams. *Entrance requirements:* For master's, DAT, GRE General Test, or MCAT; for doctorate, GRE General Test, DAT, MCAT. *Application deadline:* For fall admission, 2/15 for domestic students. Application fee: $50. *Expenses:* Tuition, state resident: full-time $3,185; part-time $405 per credit. Tuition, nonresident: full-time $7,952; part-time $940 per credit. Required fees: $751 per semester hour. Tuition and fees vary according to course load and program. *Financial support:* Fellowships available. *Faculty research:* Genetic epidemiology, biochemical genetics, quantitative genetics, human cytogenetics, molecular genetics. *Unit head:* Dr. Walter E. Nance, Chair, 804-828-9632, Fax: 804-828-3760, E-mail: wenance@vcu.edu. *Application contact:* Dr. Linda A. Corey, Graduate Program Director, 804-828-8759, Fax: 804-828-3760, E-mail: lacorey@vcu.edu.

Wake Forest University, School of Medicine and Graduate School, Graduate Programs in Medicine, Molecular Genetics Program, Winston-Salem, NC 27109. Offers PhD. *Degree requirements:* For doctorate, thesis/dissertation. *Entrance requirements:* For doctorate, GRE General Test. Additional exam requirements/recommendations for international students: Required—TOEFL. Electronic applications accepted. *Faculty research:* Control of gene expression, molecular pathogenesis, protein biosynthesis, cell development, clinical cytogenetics.

See Close-Up on page 805.

West Virginia University, Davis College of Agriculture, Forestry and Consumer Sciences, Interdisciplinary Program in Genetics and Developmental Biology, Morgantown, WV 26506. Offers animal brooding (MS, PhD); biochemical and molecular genetics (MS, PhD); cytogenetics (MS, PhD); descriptive embryology (MS, PhD); developmental genetics (MS); experimental morphogenesis teratology (MS); human genetics (MS, PhD); immunogenetics (MS, PhD); life cycles of animals and plants (MS, PhD); molecular aspects of development (MS, PhD); mutagenesis (MS, PhD); oncology (MS, PhD); plant genetics (MS, PhD); population and quantitative genetics (MS, PhD); regeneration (MS, PhD); teratology (PhD); toxicology (MS, PhD). *Students:* 13 full-time (7 women), 6 part-time (4 women), 11 international. Average age 28. In 2005, 2 master's, 4 doctorates awarded. *Degree requirements:* For master's, thesis/dissertation; for doctorate, thesis/dissertation, comprehensive exam. *Entrance requirements:* For master's, GRE or MCAT, minimum GPA of 2.75. Additional exam requirements/recommendations for international students: Required—TOEFL. Application fee: $45. *Expenses:* Tuition, state resident: full-time $4,582; part-time $258 per credit hour. Tuition, nonresident: full-time $13,820; part-time $741 per credit hour. *Financial support:* In 2005–06, 3 research assistantships with tuition reimbursements (averaging $9,936 per year), 4 teaching assistantships with tuition reimbursements (averaging $9,936 per year) were awarded; fellowships, Federal Work-Study, institutionally sponsored loans, and tuition waivers (full and partial) also available. Financial award application deadline: 2/1; financial award applicants required to submit FAFSA. Total annual research expenditures: $1 million. *Unit head:* Dr. J. Nath, Chairman, 304-293-6023 Ext. 4333, Fax: 304-293-2960, E-mail: joginder.nath@mail.wvu.edu.

Molecular Genetics

Albert Einstein College of Medicine, Sue Golding Graduate Division of Medical Sciences, Division of Biological Sciences, Department of Molecular Genetics, Bronx, NY 10461. Offers PhD, MD/PhD. *Degree requirements:* For doctorate, thesis/dissertation. *Entrance requirements:* For doctorate, GRE General Test. Additional exam requirements/recommendations for international students: Required—TOEFL. *Faculty research:* Neurologic genetics in *Drosophila*, biochemical genetics of yeast, developmental genetics in the mouse.

Duke University, Graduate School, Department of Molecular Genetics and Microbiology, Durham, NC 27710. Offers PhD. *Faculty:* 22 full-time. *Students:* 36 full-time (22 women); includes 4 minority (2 African Americans, 2 Asian Americans or Pacific Islanders), 10 international. Average age 22. 77 applicants, 26% accepted, 9 enrolled. In 2005, 1 degree awarded. *Degree requirements:* For doctorate, thesis/dissertation. *Entrance requirements:* For doctorate, GRE General Test, GRE Subject Test (recommended). Additional exam requirements/recommendations for international students: Required—IELT (preferred) or TOEFL. *Application deadline:* For fall admission, 12/31 for domestic students, 12/31 for international students. Application fee: $75. Electronic applications accepted. *Financial support:* In 2005–06, fellowships with full tuition reimbursements (averaging $19,350 per year), research assistantships

with full tuition reimbursements (averaging $19,350 per year) were awarded; Federal Work-Study also available. Financial award application deadline: 12/31. *Unit head:* Dr. Hubert Amrein, Director of Graduate Studies, 919-681-1518, Fax: 919-684-2790, E-mail: dgsmicro@mc.duke.edu.

See Close-Up on page 771.

Emory University, Graduate School of Arts and Sciences, Division of Biological and Biomedical Sciences, Program in Microbiology and Molecular Genetics, Atlanta, GA 30322-1100. Offers PhD. *Faculty:* 25 full-time (4 women). *Students:* 25 full-time (12 women); includes 6 minority (2 African Americans, 2 Asian Americans or Pacific Islanders, 2 Hispanic Americans), 3 international. Average age 27. 52 applicants, 17% accepted, 4 enrolled. In 2005, 5 doctorates awarded. *Median time to degree:* Of those who began their doctoral program in fall 1997, 100% received their degree in 8 years or less. *Degree requirements:* For doctorate, thesis/dissertation, comprehensive exam, registration. *Entrance requirements:* For doctorate, GRE General Test, minimum GPA of 3.0 in science course work. Additional exam requirements/recommendations for international students: Required—TOEFL. *Application deadline:* For fall

Molecular Genetics

Emory University (continued)

admission, 1/3 for domestic students, 1/3 for international students. Application fee: $50. Electronic applications accepted. *Expenses:* Tuition: Full-time $14,400. Required fees: $217. *Financial support:* In 2005–06, 11 students received support, including 11 fellowships with full tuition reimbursements available (averaging $23,000 per year); scholarships/grants, health care benefits, and tuition waivers (full) also available. *Faculty research:* Bacterial genetics and physiology, microbial development, molecular biology of viruses and bacterial pathogens, DNA recombination. *Unit head:* Dr. Bill Shafer, Director, 404-728-7688, Fax: 404-329-2210, E-mail: wshafer@emory.edu. *Application contact:* 404-727-2545, Fax: 404-727-3322, E-mail: gdbbs@emory.edu.

Georgia State University, College of Arts and Sciences, Department of Biology, Program in Molecular Genetics and Biochemistry, Atlanta, GA 30303-3083. Offers MS, PhD. *Degree requirements:* For master's, thesis or alternative, exam; for doctorate, thesis/dissertation, exam. *Entrance requirements:* For master's and doctorate, GRE General Test. Additional exam requirements/recommendations for international students: Required—TOEFL. Electronic applications accepted. *Expenses:* Tuition, state resident: full-time $4,368; part-time $182 per semester hour. Tuition, nonresident: full-time $8,732; part-time $728 per semester hour. Required fees: $46 per semester hour.

Harvard University, Graduate School of Arts and Sciences, Division of Medical Sciences, Boston, MA 02115. Offers biological chemistry and molecular pharmacology (PhD); cell biology (PhD); genetics (PhD); microbiology and molecular genetics (PhD); pathology (PhD), including experimental pathology. *Students:* 433 full-time (210 women). In 2005, 83 doctorates awarded. *Degree requirements:* For doctorate, thesis/dissertation. *Entrance requirements:* For doctorate, GRE General Test, GRE Subject Test. Additional exam requirements/recommendations for international students: Required—TOEFL. Application fee: $60. *Expenses:* Tuition: Full-time $28,752. Full-time tuition and fees vary according to program and student level. *Financial support:* Fellowships, research assistantships, teaching assistantships, institutionally sponsored loans and tuition waivers (full) available. Financial award application deadline: 1/1. *Unit head:* Administrator, 617-432-2029. *Application contact:* Administrator, 617-432-2029.

Indiana University–Purdue University Indianapolis, Indiana University School of Medicine, Department of Medical and Molecular Genetics, Indianapolis, IN 46202-2896. Offers genetic counseling (MS); medical and molecular genetics (MS, PhD). Part-time programs available. *Faculty:* 8 full-time (2 women). *Students:* 8 full-time (5 women), 19 part-time (13 women); includes 1 minority (Asian American or Pacific Islander), 4 international. Average age 29. In 2005, 8 master's, 2 doctorates awarded. Terminal master's awarded for partial completion of doctoral program. *Degree requirements:* For master's, thesis optional; for doctorate, thesis/dissertation, research ethics. *Entrance requirements:* For master's and doctorate, GRE General Test, minimum GPA of 3.2. *Application deadline:* For fall admission, 1/15 for domestic students. Application fee: $50 ($60 for international students). *Expenses:* Tuition, state resident: full-time $5,159; part-time $215 per credit hour. Tuition, nonresident: full-time $14,890; part-time $620 per credit hour. Required fees: $614. Tuition and fees vary according to campus/location and program. *Financial support:* In 2005–06, 11 students received support; fellowships with tuition reimbursements available, research assistantships with tuition reimbursements available, Federal Work-Study and institutionally sponsored loans available. Support available to part-time students. Financial award application deadline: 1/15. *Faculty research:* Telomeres, human gene mapping, chromosomes and malignancy, clinical genetics. Total annual research expenditures: $2.1 million. *Unit head:* Dr. Joe Christian, Chairman, 317-274-2241. *Application contact:* Kathleen Wilhelm, Admissions Secretary, 317-274-2241, Fax: 317-274-2387, E-mail: medgen@iupui.edu.

See Close-Up on page 775.

Medical College of Wisconsin, Graduate School of Biomedical Sciences, Department of Microbiology and Molecular Genetics, Milwaukee, WI 53226-0509. Offers MS, PhD, MD/MS, MD/PhD. *Degree requirements:* For doctorate, thesis/dissertation, comprehensive exam, registration. *Entrance requirements:* For doctorate, GRE General Test. Additional exam requirements/recommendations for international students: Required—TOEFL. *Faculty research:* Virology, immunology, bacterial toxins, regulation of gene expression.

Michigan State University, College of Osteopathic Medicine and The Graduate School, Graduate Studies in Osteopathic Medicine, East Lansing, MI 48824. Offers biochemistry and molecular biology (MS, PhD); microbiology (MS); microbiology and molecular genetics (MS, PhD); pharmacology and toxicology (MS, PhD), including pharmacology and toxicology, pharmacology and toxicology-environmental toxicology (PhD); physiology (MS, PhD). *Students:* 2 full-time (1 woman), 1 international. Average age 27. *Expenses:* Tuition, state resident: part-time $330 per credit hour. Tuition, nonresident: part-time $685 per credit hour. Tuition and fees vary according to program. *Unit head:* Dr. Veronica M. Maher, Associate Dean, Graduate Studies, 517-353-7785, Fax: 517-353-9004, E-mail: maher@msu.edu. *Application contact:* Kathie Schafer, Director of Admissions, 517-353-7740, Fax: 517-355-3296, E-mail: comadm@com.msu.edu.

New York University, Graduate School of Arts and Science, Department of Biology, New York, NY 10012-1019. Offers biology (PhD); biomedical journalism (MS); cancer and molecular biology (PhD); computational biology (PhD); computers in biological research (MS); developmental genetics (PhD); general biology (MS); immunology and microbiology (PhD); molecular genetics (PhD); neurobiology (PhD); oral biology (MS); plant biology (PhD); recombinant DNA technology (MS). Part-time programs available. *Faculty:* 24 full-time (5 women), 8 part-time/adjunct. *Students:* 104 full-time (52 women), 41 part-time (23 women); includes 28 minority (2 African Americans, 20 Asian Americans or Pacific Islanders, 6 Hispanic Americans), 47 international. Average age 27. 349 applicants, 56% accepted, 39 enrolled. In 2005, 59 master's, 4 doctorates awarded. Terminal master's awarded for partial completion of doctoral program. *Degree requirements:* For master's, thesis or alternative, qualifying paper; for doctorate, thesis/dissertation, comprehensive exam. *Entrance requirements:* For master's, GRE General Test; for doctorate, GRE General Test, GRE Subject Test. Additional exam requirements/recommendations for international students: Required—TOEFL. *Application deadline:* For fall admission, 1/4 for domestic students. Application fee: $80. *Financial support:* Fellowships with tuition reimbursements, research assistantships with tuition reimbursements, teaching assistantships with tuition reimbursements, career-related internships or fieldwork, Federal Work-Study, institutionally sponsored loans, scholarships/grants, health care benefits, and unspecified assistantships available. Financial award application deadline: 1/4; financial award applicants required to submit FAFSA. *Faculty research:* Genomics, molecular and cell biology, development and molecular genetics, molecular evolution of plants and animals. *Unit head:* Gloria Coruzzi, Chairman, 212-998-8200, Fax: 212-995-4015, E-mail: biology@nyu.edu. *Application contact:* Stephen Small, Director of Graduate Studies, 212-998-8200, Fax: 212-995-4015, E-mail: biology@nyu.edu.

The Ohio State University, Graduate School, College of Biological Sciences, Department of Molecular Genetics, Columbus, OH 43210. Offers cell and developmental biology (MS, PhD); genetics (MS, PhD); molecular biology (MS, PhD). *Degree requirements:* For master's and doctorate, thesis/dissertation. *Entrance requirements:* For master's and doctorate, GRE General Test, GRE Subject Test. Additional exam requirements/recommendations for international students: Required—TOEFL (minimum score 573 paper-based; 230 computer-based). Electronic applications accepted.

See Close-Up on page 785.

Oklahoma State University, College of Arts and Sciences, Department of Microbiology and Molecular Genetics, Stillwater, OK 74078. Offers MS, PhD. *Faculty:* 14 full-time (0 women). *Students:* 10 full-time (2 women), 15 part-time (11 women); includes 2 minority (1 African American, 1 American Indian/Alaska Native), 15 international. Average age 28. 55 applicants, 15% accepted, 4 enrolled. In 2005, 2 degrees awarded. *Degree requirements:* For master's,

thesis/dissertation; for doctorate, thesis/dissertation, comprehensive exam. *Entrance requirements:* For master's and doctorate, GRE General Test. Additional exam requirements/recommendations for international students: Required—TOEFL. *Application deadline:* For fall admission, 6/1 for domestic students, 3/1 for international students. Applications are processed on a rolling basis. Application fee: $40 ($75 for international students). Electronic applications accepted. *Expenses:* Tuition, state resident: full-time $4,253; part-time $139 per credit hour. Tuition, nonresident: full-time $12,569; part-time $485 per credit hour. Required fees: $43 per credit hour. One-time fee: $20 part-time. Tuition and fees vary according to course load and program. *Financial support:* In 2005–06, 9 research assistantships (averaging $14,198 per year), 13 teaching assistantships (averaging $14,042 per year) were awarded; scholarships/grants, health care benefits, tuition waivers (full), and unspecified assistantships also available. Financial award application deadline: 3/1. *Faculty research:* Bioinformatics, genomics-genetics, virology, environmental microbiology, development-molecular mechanisms. *Unit head:* Dr. Robert V. Miller, Head, 405-744-7180.

Rutgers, The State University of New Jersey, New Brunswick/Piscataway, Graduate School, Program in Microbiology and Molecular Genetics, New Brunswick, NJ 08901-1281. Offers applied microbiology (MS, PhD); clinical microbiology (MS, PhD); computational molecular biology (PhD); immunology (MS, PhD); microbial biochemistry (MS, PhD); molecular genetics (MS, PhD); virology (MS, PhD). Part-time programs available. *Faculty:* 132 full-time. *Students:* 57 full-time (31 women), 18 part-time (8 women); includes 16 minority (9 Asian Americans or Pacific Islanders, 7 Hispanic Americans), 20 international. Average age 29. 104 applicants, 32% accepted, 19 enrolled. In 2005, 9 master's, 10 doctorates awarded. Terminal master's awarded for partial completion of doctoral program. *Median time to degree:* Of those who began their doctoral program in fall 1997, 100% received their degree in 8 years or less. *Degree requirements:* For master's, thesis or alternative, comprehensive exam, registration; for doctorate, thesis/dissertation, written qualifying exam, comprehensive exam, registration. *Entrance requirements:* For master's, GRE General Test; for doctorate, GRE General Test, GRE Subject Test (recommended), minimum GPA of 3.0. Additional exam requirements/recommendations for international students: Required—TOEFL. *Application deadline:* For fall admission, 1/5 priority date for domestic students, 11/1 priority date for international students. Applications are processed on a rolling basis. Application fee: $50. Electronic applications accepted. *Expenses:* Tuition, state resident: full-time $10,440; part-time $435 per credit. Tuition, nonresident: full-time $15,520; part-time $647 per credit. Required fees: $129 per credit. Tuition and fees vary according to program. *Financial support:* In 2005–06, 48 students received support, including 14 fellowships with full tuition reimbursements available (averaging $24,000 per year), 25 research assistantships with full tuition reimbursements available (averaging $24,000 per year), 9 teaching assistantships with full tuition reimbursements available (averaging $16,988 per year); Federal Work-Study, institutionally sponsored loans, scholarships/grants, and unspecified assistantships also available. Financial award application deadline: 1/5; financial award applicants required to submit FAFSA. *Faculty research:* Molecular genetics and microbial physiology; virology and pathogenic microbiology; applied, environmental and industrial microbiology; computers in molecular biology. *Unit head:* Dr. Andrew K. Vershon, Director, 732-445-3995, Fax: 732-445-6370, E-mail: vershon@waksman.rutgers.edu. *Application contact:* Diane Murano, Administrative Assistant, 732-445-5086, Fax: 732-445-6370, E-mail: murano@biology.rutgers.edu.

Stony Brook University, State University of New York, Health Sciences Center, School of Medicine and Graduate School, Graduate Programs in Medicine, Department of Molecular Genetics and Microbiology, Stony Brook, NY 11794. Offers molecular microbiology (PhD). *Faculty:* 18 full-time (6 women). *Students:* 33 full-time (22 women); includes 5 minority (2 African Americans, 1 Asian American or Pacific Islander, 2 Hispanic Americans), 7 international. Average age 27. 44 applicants, 16% accepted. In 2005, 7 degrees awarded. *Degree requirements:* For doctorate, thesis/dissertation, comprehensive exam. *Entrance requirements:* For doctorate, GRE General Test, GRE Subject Test. Additional exam requirements/recommendations for international students: Required—TOEFL. *Application deadline:* For fall admission, 1/15 for domestic students. Application fee: $50. *Expenses:* Tuition, state resident: full-time $6,900; part-time $288 per credit. Tuition, nonresident: full-time $10,920; part-time $455 per credit. Required fees: $704. *Financial support:* In 2005–06, 12 fellowships, 29 research assistantships, 2 teaching assistantships were awarded; Federal Work-Study also available. Financial award application deadline: 3/15. *Faculty research:* Adenovirus molecular genetics, molecular biology of tumors, virus SV40, mechanism of tumor infection by SAV virus. Total annual research expenditures: $8.2 million. *Unit head:* Dr. Dafna Bar-Sagi, Chair, 631-632-8812, Fax: 631-632-9797. *Application contact:* Dr. Patrick Hearing, Director, 631-632-8813, Fax: 631-632-9797, E-mail: hearing@asterix.bio.sunysb.edu.

Announcement: Department of Molecular Genetics and Microbiology offers studies in viral and bacterial pathogenesis; signal transduction, cell growth/development, oncogenetics, and cancer; molecular genetics and virology; and structure/biochemistry of macromolecules. The program has 45 interdisciplinary faculty members and 38 graduate students. All students receive tuition scholarships, a salary of $24,000, and health insurance.

See Close-Up on page 895.

Texas Tech University Health Sciences Center, Graduate School of Biomedical Sciences, Department of Cell Biology and Biochemistry, Program in Biochemistry and Molecular Genetics, Lubbock, TX 79430. Offers MS, PhD, MD/PhD, MS/PhD. *Faculty:* 12 full-time (2 women), 10 part-time/adjunct (5 women). *Students:* 4 full-time (0 women), 3 international. Average age 28. 7 applicants, 0% accepted, 0 enrolled. Terminal master's awarded for partial completion of doctoral program. *Degree requirements:* For master's and doctorate, thesis/dissertation, preliminary, comprehensive, and final exams, comprehensive exam, registration. *Entrance requirements:* For master's and doctorate, GRE General Test, minimum GPA of 3.0. Additional exam requirements/recommendations for international students: Required—TOEFL. *Application deadline:* For fall admission, 5/15 priority date for domestic students, 4/15 priority date for international students; for spring admission, 11/15 for domestic students, 10/15 for international students. Application fee: $45. Electronic applications accepted. *Financial support:* In 2005–06, 1 fellowship with full tuition reimbursement (averaging $20,500 per year), 5 research assistantships (averaging $20,500 per year) were awarded; health care benefits also available. *Faculty research:* Reproductive endocrinology, immunology, developmental biochemistry. Total annual research expenditures: $2.6 million. *Unit head:* Dr. Sandra Whelly, Graduate Director, 806-743-2503, Fax: 806-743-2990, E-mail: sandra.whelly@ttuhsc.edu. *Application contact:* Pam Roddy, Assistant Director, 806-743-2701, Fax: 806-743-2990, E-mail: pam.roddy@ttuhsc.edu.

The University of Alabama at Birmingham, Graduate Programs in Joint Health Sciences, Department of Cell Biology, Graduate Program in Cellular and Molecular Biology, Birmingham, AL 35294.

See Close-Ups on pages 601 and 603.

University of California, Irvine, College of Medicine and School of Biological Sciences, Department of Microbiology and Molecular Genetics, Irvine, CA 92697. Offers biological sciences (MS, PhD). Students apply through the Graduate Program in Molecular Biology, Genetics, and Biochemistry. *Degree requirements:* For doctorate, thesis/dissertation. *Entrance requirements:* For doctorate, GRE General Test, GRE Subject Test, minimum GPA of 3.0. Additional exam requirements/recommendations for international students: Required—TOEFL (minimum score 550 paper-based; 213 computer-based), TSE. Electronic applications accepted. *Faculty research:* Molecular biology and genetics of viruses, bacteria, and yeast; immune response; molecular biology of cultured animal cells; genetic basis of cancer; genetics and physiology of infectious agents.

University of California, Los Angeles, School of Medicine and Graduate Division, Graduate Programs in Medicine, Department of Microbiology, Immunology and Molecular Genetics, Los Angeles, CA 90095. Offers MS, PhD. *Degree requirements:* For doctorate, thesis/dissertation, oral and written qualifying exams. *Entrance requirements:* For doctorate, GRE General Test, GRE Subject Test. Additional exam requirements/recommendations for international students: Required—TOEFL.

University of California, Riverside, Graduate Division, Graduate Program in Genetics, Genomics, and Bioinformatics, Riverside, CA 92521-0102. Offers genomics and bioinformatics (PhD); molecular genetics (PhD); population and evolutionary genetics (PhD). *Students:* 19 full-time (7 women); includes 3 minority (2 Asian Americans or Pacific Islanders, 1 Hispanic American), 7 international. Average age 30. In 2005, 1 degree awarded. *Degree requirements:* For doctorate, thesis/dissertation, qualifying exams, teaching experience. *Entrance requirements:* For doctorate, GRE General Test, minimum GPA of 3.2. Additional exam requirements/ recommendations for international students: Required—TOEFL (minimum score 550 paper-based; 213 computer-based); Recommended—TSE (minimum score 50). *Application deadline:* For fall admission, 5/1 for domestic students, 2/1 for international students. For winter admission, 9/1 for domestic students; for spring admission, 12/1 for domestic students. Applications are processed on a rolling basis. Application fee: $60 ($75 for international students). Electronic applications accepted. *Expenses:* Tuition, nonresident: full-time $14,694. Full-time tuition and fees vary according to program. *Financial support:* In 2005–06, research assistantships (averaging $14,000 per year), teaching assistantships (averaging $15,000 per year) were awarded; fellowships, career-related internships or fieldwork, Federal Work-Study, institutionally sponsored loans, and tuition waivers (full and partial) also available. Financial award application deadline: 2/1; financial award applicants required to submit FAFSA. *Unit head:* , Dr. Katherine Borkovich, Director. *Application contact:* Carole Carpenter, Graduate Program Assistant, 800-735-0717, Fax: 951-827-5517, E-mail: genetics@ucr.edu.

University of Chicago, Division of the Biological Sciences, Department of Molecular Biosciences: Biochemistry, Genetics, Cell and Developmental Biology, Department of Molecular Genetics and Cell Biology, Chicago, IL 60637-1513. Offers PhD. *Faculty:* 31 full-time (8 women). *Students:* 36 full-time (21 women); includes 5 minority (all Asian Americans or Pacific Islanders), 6 international. Average age 27. In 2005, 7 doctorates awarded. *Degree requirements:* For doctorate, thesis/dissertation, registration. *Entrance requirements:* For doctorate, GRE General Test. Additional exam requirements/recommendations for international students: Required—TOEFL. *Application deadline:* For fall admission, 12/28 priority date for domestic students, 12/28 priority date for international students. Application fee: $55. Electronic applications accepted. *Financial support:* In 2005–06, 36 students received support, including fellowships (averaging $26,301 per year), research assistantships (averaging $26,301 per year); institutionally sponsored loans, scholarships/grants, traineeships, and health care benefits also available. Financial award applicants required to submit FAFSA. *Faculty research:* Gene expression, chromosome structure, animal viruses, plant molecular genetics. Total annual research expenditures: $8 million. *Unit head:* Dr. Laurens Mets, Chairman, 773-702-8917. *Application contact:* Kristine Gaston, Graduate Administrative Director, 773-702-8037, Fax: 773-702-3172, E-mail: kristine@cummings.uchicago.edu.

University of Cincinnati, Division of Research and Advanced Studies, College of Medicine, Graduate Programs in Biomedical Sciences, Department of Molecular Genetics, Biochemistry and Microbiology, Cincinnati, OH 45267. Offers MS, PhD. Terminal master's awarded for partial completion of doctoral program. *Degree requirements:* For master's, thesis or alternative; for doctorate, thesis/dissertation, qualifying exam. *Entrance requirements:* For master's, GRE General Test; for doctorate, GRE General Test, GRE Subject Test. Additional exam requirements/recommendations for international students: Required—TOEFL (minimum score 590 paper-based; 243 computer-based), TWE. Electronic applications accepted. *Faculty research:* Cancer biology and developmental genetics, gene regulation and chromosome structure, microbiology and pathogenic mechanisms, structural biology, membrane biochemistry and signal transduction.

University of Florida, College of Medicine, Department of Molecular Genetics and Microbiology, Gainesville, FL 32611. Offers MS, PhD. *Faculty:* 26 full-time (7 women). *Students:* 145 (76 women); includes 20 minority (5 African Americans, 9 Asian Americans or Pacific Islanders, 6 Hispanic Americans) 34 international. In 2005, 28 degrees awarded. Terminal master's awarded for partial completion of doctoral program. *Degree requirements:* For master's and doctorate, thesis/dissertation. *Entrance requirements:* For master's and doctorate, GRE General Test, minimum GPA of 3.0. Additional exam requirements/recommendations for international students: Required—TOEFL. *Application deadline:* For fall admission, 2/15 for domestic students. Applications are processed on a rolling basis. Application fee: $30. Electronic applications accepted. *Expenses:* Tuition, state resident: full-time $6,234. Tuition, nonresident: full-time $21,359. Tuition and fees vary according to program. *Financial support:* In 2005–06, 38 research assistantships with full tuition reimbursements (averaging $24,464 per year) were awarded; fellowships with full tuition reimbursements, institutionally sponsored loans, traineeships, and unspecified assistantships also available. *Unit head:* Dr. Henry Baker, Chairman, 352-392-0680, E-mail: hvbaker@ufl.edu. *Application contact:* Dr. Wayne McCormack, Associate Dean of Graduate Education, 352-392-7413, Fax: 352-846-3466, E-mail: mccormac@pathology.ufl.edu.

University of Guelph, Graduate Program Services, College of Biological Science, Department of Molecular and Cellular Biology, Guelph, ON N1G 2W1, Canada. Offers biochemistry (M Sc, PhD); biophysics (M Sc, PhD); botany (M Sc, PhD); microbiology (M Sc, PhD); molecular biology and genetics (M Sc, PhD). *Faculty:* 43 full-time (6 women). *Students:* 136 full-time (52 women). Average age 26. 56 applicants, 30% accepted, 17 enrolled. In 2005, 4 master's, 2 doctorates awarded. *Degree requirements:* For master's, thesis/dissertation, research proposal; for doctorate, thesis/dissertation, research proposal, comprehensive exam, registration. *Entrance requirements:* Additional exam requirements/recommendations for international students: Required—TOEFL (minimum score 550 paper-based; 213 computer-based), IELT (minimum score 7). *Application deadline:* For fall admission, 7/1 for domestic students, 1/1 for international students. For winter admission, 11/1 for domestic students; for spring admission, 3/1 for domestic students. Applications are processed on a rolling basis. Application fee: $75. Electronic applications accepted. *Financial support:* Fellowships, research assistantships, teaching assistantships available. Support available to part-time students. *Faculty research:* Physiology, structure, genetics, and ecology of microbes; virology and microbial technology. *Unit head:* Dr. Chris Whitfield, Chair, 519-824-4120 Ext. 53361, Fax: 519-827-1802, E-mail: cwhitfie@uoguelph.ca. *Application contact:* Laurie Winn, Graduate Admissions Secretary, 519-824-4320 Ext. 52730, Fax: 519-767-1656, E-mail: lwinn@uoguelph.ca.

University of Illinois at Chicago, College of Medicine and Graduate College, Graduate Programs in Medicine, Department of Molecular Genetics, Program in Molecular Genetics, Chicago, IL 60607-7128. Offers PhD. *Degree requirements:* For doctorate, thesis/dissertation. *Entrance requirements:* For doctorate, GRE General Test. Additional exam requirements/ recommendations for international students: Required—TOEFL.

University of Kansas, Graduate Studies Medical Center, Interdisciplinary Graduate Program in Biomedical Sciences, Department of Microbiology, Molecular Genetics and Immunology, Lawrence, KS 66045. Offers PhD, MD/PhD. *Faculty:* 8. *Students:* 2 full-time (both women), 5 part-time (3 women), 1 international. Average age 27. In 2005, 2 degrees awarded. *Degree requirements:* For doctorate, thesis/dissertation, comprehensive exam. *Entrance requirements:* For doctorate, GRE General Test. *Expenses:* Tuition, state resident: full-time $4,859. Tuition, nonresident: full-time $12,000. Required fees: $589. Tuition and fees vary according to program. *Financial support:* Fellowships with tuition reimbursements, research assistantships with partial tuition reimbursements, teaching assistantships with full and partial tuition reimbursements available. *Faculty research:* Bacterial and viral pathogenesis, molecular virology, microbial molecular biology and genetics, immunochemistry. *Unit head:* Dr. Opendra Narayan, Chairman, 913-588-7010, Fax: 913-588-7295, E-mail: bnarayan@kume.edu. *Application contact:* Dr. Joe Lutkenhaus, Director of Graduate Studies, 913-588-7054, Fax: 913-588-7295, E-mail: jlutkenh@kumc.edu.

University of Maryland, College Park, Graduate Studies, College of Chemical and Life Sciences, Department of Cell Biology and Molecular Genetics, Program in Cell Biology and Molecular Genetics, College Park, MD 20742. Offers MS, PhD. *Students:* 78 full-time (48 women), 3 part-time (2 women); includes 6 minority (3 African Americans, 3 Hispanic Americans), 30 international. 166 applicants, 19% accepted, 16 enrolled. In 2005, 5 master's, 5 doctorates awarded. *Degree requirements:* For master's, thesis; for doctorate, thesis/dissertation, exams. *Entrance requirements:* For master's and doctorate, GRE General Test, 3 letters of recommendation, minimum GPA of 3.0. Additional exam requirements/recommendations for international students: Required—TOEFL; Recommended—TSE. *Application deadline:* For fall admission, 1/11 for domestic students, 1/11 for international students. Application fee: $60. *Financial support:* In 2005–06, 17 fellowships (averaging $4,252 per year) were awarded; research assistantships, teaching assistantships Financial award applicants required to submit FAFSA. *Faculty research:* Cytoskeletal activity, membrane biology, cell division, genetics and genomics, virology. *Application contact:* Dean of Graduate School, 301-405-4190, Fax: 301-314-9305.

University of Massachusetts Worcester, Graduate School of Biomedical Sciences, Department of Molecular Genetics and Microbiology, Worcester, MA 01655-0115. Offers PhD. *Faculty:* 21 full-time (3 women). *Degree requirements:* For doctorate, thesis/dissertation. *Entrance requirements:* For doctorate, GRE General Test. Additional exam requirements/recommendations for international students: Required—TOEFL (minimum score 600 paper-based; 250 computer-based). *Application deadline:* For fall admission, 12/15 for domestic students, 12/15 for international students. Applications are processed on a rolling basis. Application fee: $25 ($50 for international students). *Expenses:* Tuition, state resident: full-time $2,640. Tuition, nonresident: full-time $9,856. Required fees: $5,685. *Financial support:* In 2005–06, research assistantships with full tuition reimbursements (averaging $25,235 per year); unspecified assistantships also available. *Faculty research:* Gene structure, regulation of gene expression, pathogenesis. *Unit head:* Dr. Allan Jacobson, Chair, 508-856-2442. *Application contact:* Michael Cole, Director of Admissions and Recruitment, 508-856-4779, Fax: 508-856-3659, E-mail: michael.cole@umassmed.edu.

See Close-Up on page 797.

University of Medicine and Dentistry of New Jersey, Graduate School of Biomedical Sciences, Graduate Programs in Biomedical Sciences–Newark, Department of Microbiology and Molecular Genetics, Newark, NJ 07107. Offers PhD. *Degree requirements:* For doctorate, thesis/dissertation, qualifying exam. *Entrance requirements:* For doctorate, GRE General Test. Additional exam requirements/recommendations for international students: Required—TOEFL. *Application deadline:* For fall admission, 2/1 for domestic students. Application fee: $40. *Financial support:* Fellowships, research assistantships, Federal Work-Study, institutionally sponsored loans, and tuition waivers (full and partial) available. Financial award application deadline: 5/1. *Faculty research:* Molecular genetics of yeast, mutagenesis and carcinogenesis of DNA, bacterial protein synthesis, mammalian cell genetics, adenovirus gene expression. *Unit head:* Dr. Vivian Bellofatto, Program Director, 973-972-4483 Ext. 4406, Fax: 973-972-8981.

See Close-Up on page 935.

University of Medicine and Dentistry of New Jersey, Graduate School of Biomedical Sciences, Graduate Programs in Biomedical Sciences–Piscataway, Program in Molecular Genetics, Microbiology and Immunology, Piscataway, NJ 08854-5635. Offers MS, PhD, MD/PhD. Terminal master's awarded for partial completion of doctoral program. *Degree requirements:* For master's and doctorate, thesis/dissertation, qualifying exam. *Entrance requirements:* For master's and doctorate, GRE General Test. Additional exam requirements/recommendations for international students: Required—TOEFL. *Application deadline:* For fall admission, 1/5 for domestic students. Applications are processed on a rolling basis. Application fee: $40. *Financial support:* Fellowships, research assistantships, teaching assistantships available. Financial award application deadline: 5/1. *Faculty research:* Interferon, receptors, retrovirus evolution, Arbo virus/host cell interactions. *Unit head:* Dr. Joseph P. Dougherty, Director, 732-235-4588, Fax: 732-235-5223, E-mail: doughejp@umdnj.edu.

University of Pittsburgh, School of Medicine, Graduate Programs in Medicine, Program in Biochemistry and Molecular Genetics, Pittsburgh, PA 15260. Offers MS, PhD. *Faculty:* 46 full-time (8 women). *Students:* 23 full-time (14 women); includes 1 minority (Hispanic American), 5 international. Average age 28. 415 applicants, 22% accepted, 42 enrolled. In 2005, 5 degrees awarded. *Median time to degree:* Of those who began their doctoral program in fall 1997, 95% received their degree in 8 years or less. *Degree requirements:* For doctorate, thesis/dissertation, comprehensive exam, registration. *Entrance requirements:* For doctorate, GRE General Test, GRE Subject Test, minimum QPA of 3.0. Additional exam requirements/recommendations for international students: Required—TOEFL (minimum score 600 paper-based; 250 computer-based), IELT (minimum score 7). *Application deadline:* For fall admission, 12/15 priority date for domestic students, 12/15 priority date for international students. Application fee: $40. Electronic applications accepted. *Expenses:* Tuition, state resident: full-time $13,194; part-time $537 per credit. Tuition, nonresident: full-time $25,012; part-time $1,026 per credit. Required fees: $700; $164 per term. Tuition and fees vary according to campus/location and program. *Financial support:* In 2005–06, 1 fellowship with full tuition reimbursement (averaging $21,500 per year), 23 research assistantships with full tuition reimbursements (averaging $21,500 per year) were awarded; teaching assistantships with full tuition reimbursements, institutionally sponsored loans, scholarships/grants, traineeships, health care benefits, and unspecified assistantships also available. *Faculty research:* Molecular genetics of cancer, gene expression and signal transduction, genomics and proteomics, human gene therapy, structural dynamics and bioinformatics. *Unit head:* Dr. Thomas Smithgall, Professor, 412-648-9495, Fax: 412-624-1901, E-mail: tsmithga@pitt.edu. *Application contact:* Graduate Studies Administrator, 412-648-8957, Fax: 412-648-1077, E-mail: gradstudies@medschool.pitt.edu.

University of Rhode Island, Graduate School, College of the Environment and Life Sciences, Department of Cell and Molecular Biology, Kingston, RI 02881. Offers biochemistry (MS, PhD); microbiology (MS, PhD), including biodegradation (MS), cellular development (MS), electron microscopy and ultrastructure (MS), genetics and molecular biology (MS), immunology (MS), marine and freshwater ecosystems (MS), microbial pathogenesis (MS), microbial physiology (MS), protozoology (MS), virology (MS), water-pollution microbiology (MS); molecular genetics (MS, PhD). In 2005, 2 degrees awarded. *Degree requirements:* For master's and doctorate, thesis/dissertation. *Entrance requirements:* For master's and doctorate, GRE General Test. Additional exam requirements/recommendations for international students: Required—TOEFL. *Expenses:* Tuition, state resident: full-time $5,522; part-time $307 per credit. Tuition, nonresident: full-time $15,992; part-time $888 per credit. Required fees: $1,786; $73 per credit. One-time fee: $80 part-time. *Financial support:* Fellowships, research assistantships, teaching assistantships available. *Unit head:* Dr. Jay Sperry, Chairperson, 401-874-5900.

The University of Texas Health Science Center at Houston, Graduate School of Biomedical Sciences, Program in Human and Molecular Genetics, Houston, TX 77030-3900. Offers MS, PhD, MD/PhD. *Faculty:* 42 full-time (13 women). *Students:* 16 full-time (11 women); includes 5 minority (all Asian Americans or Pacific Islanders), 2 international. Average age 25. 36 applicants, 50% accepted, 6 enrolled. In 2005, 1 master's, 1 doctorate awarded. Terminal master's awarded for partial completion of doctoral program. *Degree requirements:* For master's and doctorate, thesis/dissertation. *Entrance requirements:* For master's and doctorate, GRE General Test. Additional exam requirements/recommendations for international students: Required—TOEFL, TWE. *Application deadline:* For fall admission, 1/15 for domestic students; for spring admission, 11/1 for domestic students. Applications are processed on a rolling basis. Application fee: $10. Electronic applications accepted. *Financial support:* Fellowships with full tuition reimbursements, research assistantships with full tuition reimbursements, teaching assistantships, institutionally sponsored loans, scholarships/grants, and health care benefits available. Financial award application deadline: 1/15. *Faculty research:* Cancer genetics, genetics and epidemiology of complex diseases, bioinformatics and computational genomics, medical genet-

Molecular Genetics

The University of Texas Health Science Center at Houston (continued)
ics, animal models of human disease. *Unit head:* Dr. Ralf Krahe, Director, 713-834-6345, Fax: 713-834-6319, E-mail: rkrahe@mdanderson.org. *Application contact:* Dr. Victoria P. Knutson, Assistant Dean of Admissions, 713-500-9860, Fax: 713-500-9877, E-mail: victoria.p.knutson@uth.tmc.edu.

The University of Texas Health Science Center at Houston, Graduate School of Biomedical Sciences, Program in Microbiology and Molecular Genetics, Houston, TX 77225-0036. Offers MS, PhD, MD/PhD. *Faculty:* 18 full-time (5 women). *Students:* 28 full-time (16 women); includes 7 minority (1 African American, 1 Asian American or Pacific Islander, 5 Hispanic Americans), 4 international. Average age 26. 32 applicants, 63% accepted, 8 enrolled. In 2005, 1 degree awarded. Terminal master's awarded for partial completion of doctoral program. *Degree requirements:* For master's and doctorate, thesis/dissertation. *Entrance requirements:* For master's and doctorate, GRE General Test. Additional exam requirements/recommendations for international students: Required—TOEFL, TWE. *Application deadline:* For fall admission, 1/15 for domestic students; for spring admission, 11/1 for domestic students. Applications are processed on a rolling basis. Application fee: $10. Electronic applications accepted. *Financial support:* Fellowships with full tuition reimbursements, research assistantships with full tuition reimbursements, teaching assistantships, institutionally sponsored loans, scholarships/grants, and health care benefits available. Financial award application deadline: 1/15. *Faculty research:* Microbial genomics, microbial diversity, gene regulation, molecular pathogenesis, sensory transduction. *Unit head:* Dr. Samuel Kaplan, Director, 713-500-5502, Fax: 713-500-5499, E-mail: samuel.kaplan@uth.tmc.edu. *Application contact:* Dr. Victoria P. Knutson, Assistant Dean of Admissions, 713-500-9860, Fax: 713-500-9877, E-mail: victoria.p.knutson@uth.tmc.edu.

University of Vermont, College of Medicine and Graduate College, Graduate Programs in Medicine, Department of Microbiology and Molecular Genetics, Burlington, VT 05405. Offers MS, PhD, MD/MS, MD/PhD. *Faculty:* 18 full-time (5 women). *Students:* 13 (6 women); includes 1 minority (Hispanic American) 13 international. 56 applicants, 32% accepted, 5 enrolled. In 2005, 4 doctorates awarded. *Degree requirements:* For master's and doctorate, thesis/dissertation. *Entrance requirements:* For master's and doctorate, GRE General Test. Additional exam requirements/recommendations for international students: Required—TOEFL (minimum score 550 paper-based; 213 computer-based). *Application deadline:* For fall admis-

sion, 2/1 for domestic students. Applications are processed on a rolling basis. Application fee: $40. Electronic applications accepted. *Expenses:* Tuition, area resident: Part-time $410 per credit hour. Tuition, nonresident: part-time $1,034 per credit hour. *Financial support:* Fellowships, research assistantships, teaching assistantships available. Financial award application deadline: 3/1. *Unit head:* Dr. Susan S. Wallace, Chairperson, 802-656-2164. *Application contact:* Dr. S. Doublie, Coordinator, 802-656-2164.

See Close-Up on page 801.

University of Virginia, School of Medicine, Department of Biochemistry and Molecular Genetics, Charlottesville, VA 22903. Offers biochemistry (PhD). *Students:* 38 full-time (21 women); includes 1 minority (African American), 7 international. Average age 27. In 2005, 7 degrees awarded. *Degree requirements:* For doctorate, thesis/dissertation. *Entrance requirements:* For doctorate, GRE General Test, GRE Subject Test. Additional exam requirements/recommendations for international students: Required—TOEFL. Application fee: $60. Electronic applications accepted. *Expenses:* Tuition, state resident: full-time $7,731. Tuition, nonresident: full-time $18,672. Required fees: $1,479. Full-time tuition and fees vary according to degree level and program. *Financial support:* Fellowships, research assistantships, teaching assistantships available. Financial award applicants required to submit FAFSA. *Unit head:* Joyce L. Hamlin, Chairman, 434-924-1940, Fax: 434-924-1667. *Application contact:* Peter C. Brunjes, Associate Dean for Graduate Programs and Research, 434-924-7184, Fax: 434-924-6737, E-mail: grad-a-s@virginia.edu.

Wake Forest University, School of Medicine and Graduate School, Graduate Programs in Medicine, Molecular Genetics Program, Winston-Salem, NC 27109. Offers PhD. *Degree requirements:* For doctorate, thesis/dissertation. *Entrance requirements:* For doctorate, GRE General Test. Additional exam requirements/recommendations for international students: Required—TOEFL. Electronic applications accepted. *Faculty research:* Control of gene expression, molecular pathogenesis, protein biosynthesis, cell development, clinical cytogenetics.

See Close-Up on page 805.

Washington University in St. Louis, Graduate School of Arts and Sciences, Division of Biology and Biomedical Sciences, Program in Molecular Genetics, St. Louis, MO 63130-4899. Offers PhD. *Degree requirements:* For doctorate, thesis/dissertation. *Entrance requirements:* For doctorate, GRE General Test, GRE Subject Test. Electronic applications accepted.

Reproductive Biology

Cornell University, Graduate School, Graduate Fields of Comparative Biomedical Sciences, Field of Comparative Biomedical Sciences, Ithaca, NY 14853-0001. Offers cellular and molecular medicine (MS, PhD); developmental and reproductive biology (MS, PhD); infectious diseases (MS, PhD); population medicine and epidemiology (MS); population medicine and epidemiology sciences (PhD); structural and functional biology (MS, PhD). *Faculty:* 135 full-time (38 women). *Students:* 41 full-time (23 women); includes 4 minority (1 African American, 3 Asian Americans or Pacific Islanders), 19 international. 58 applicants, 60% accepted, 33 enrolled. In 2005, 4 degrees awarded. *Degree requirements:* For master's, thesis/dissertation; for doctorate, thesis/dissertation, comprehensive exam. *Entrance requirements:* For master's and doctorate, GRE General Test, 2 letters of recommendation. Additional exam requirements/recommendations for international students: Required—TOEFL (minimum score 550 paper-based; 213 computer-based). *Application deadline:* For fall admission, 12/15 for domestic students. Application fee: $60. Electronic applications accepted. *Financial support:* In 2005–06, 41 students received support, including 13 fellowships with full tuition reimbursements available, 28 research assistantships with full tuition reimbursements available; teaching assistantships with full tuition reimbursements available, institutionally sponsored loans, scholarships/grants, health care benefits, tuition waivers (full and partial), and unspecified assistantships also available. Financial award applicants required to submit FAFSA. *Faculty research:* Receptors and signal transduction, viral and bacterial infectious diseases, tumor metastasis, clinical sciences/nutritional disease, development/neurologic disorders. *Unit head:* Director of Graduate Studies, 607-253-3276, Fax: 607-253-3756. *Application contact:* Graduate Field Assistant, 607-253-3276, Fax: 607-253-3756, E-mail: graduate_edcvm@cornell.edu.

Eastern Virginia Medical School, Master's Program in Biomedical Sciences (Clinical Embryology and Andrology), Norfolk, VA 23501-1980. Offers MS. *Faculty:* 16. *Students:* 37 full-time (28 women); includes 13 minority (3 African Americans, 7 Asian Americans or Pacific Islanders, 3 Hispanic Americans). 46 applicants, 50% accepted, 23 enrolled. In 2005, 20 degrees awarded. *Application deadline:* For winter admission, 2/15 for domestic students. Applications are processed on a rolling basis. Application fee: $50. *Expenses:* Contact institution. *Financial support:* In 2005–06, 10 students received support. *Unit head:* Dr. Jacob Mayer, Director, 757-446-5049, Fax: 757-446-5905. *Application contact:* Nancy Garcia, Administrator, 757-446-8935, Fax: 757-446-5905, E-mail: garcianw@evms.edu.

New York University, School of Medicine and Graduate School of Arts and Science, Sackler Institute of Graduate Biomedical Sciences, Department of Pharmacology, New York, NY 10012-1019. Offers molecular pharmacology and signal transduction (PhD). *Degree requirements:* For doctorate, one foreign language, thesis/dissertation, qualifying exam, comprehensive exam. *Entrance requirements:* For doctorate, GRE General Test. Additional exam requirements/recommendations for international students: Required—TOEFL. *Faculty research:* Pharmacology and neurobiology, neuropeptides, receptor biochemistry, cytoskeleton, endocrinology.

Northwestern University, The Graduate School and Judd A. and Marjorie Weinberg College of Arts and Sciences, Interdepartmental Biological Sciences Program (IBiS), Evanston, IL 60208. Offers biochemistry, molecular biology, and cell biology (PhD), including biochemistry, cell and molecular biology, molecular biophysics, structural biology; biotechnology (PhD); cell and molecular biology (PhD); developmental biology and genetics (PhD); hormone action and signal transduction (PhD); neuroscience (PhD); structural biology, biochemistry, and biophysics (PhD). Participants in the Interdepartmental Biological Sciences Program include the Departments of Biochemistry, Molecular Biology, and Cell Biology; Chemistry; Neurobiology and Physiology; Chemical Engineering; Civil Engineering; and Evanston Hospital. *Entrance requirements:* For doctorate, GRE General Test. Additional exam requirements/recommendations for international students: Required—TOEFL (minimum score 600 paper-based), TSE(minimum score 50). Electronic applications accepted. *Faculty research:* Developmental genetics, gene regulation, DNA-protein interactions, biological clocks, bioremediation.

See Close-Up on page 187.

The University of British Columbia, Faculty of Medicine and Faculty of Graduate Studies, Graduate Programs in Medicine, Department of Obstetrics and Gynecology, Program in Reproductive and Developmental Sciences, Vancouver, BC V6T 1Z1, Canada. Offers M Sc, PhD. Terminal master's awarded for partial completion of doctoral program. *Degree requirements:* For master's and doctorate, thesis/dissertation. *Entrance requirements:* Additional exam requirements/recommendations for international students: Required—TOEFL (paper score 550; computer score 213) or IELTS (paper score 7). Electronic applications accepted. *Faculty research:* Reproductive and placental endocrinology; immunology of reproductive, fertilization, and embryonic development; perinatal metabolism; neonatal development.

University of Hawaii at Manoa, John A. Burns School of Medicine and Graduate Division, Graduate Programs in Biomedical Sciences, Department of Anatomy, Biochemistry, Physiol-

ogy and Reproductive Biology, Honolulu, HI 96822. Offers anatomy and reproductive biology (MS); physiology (MS, PhD); reproductive biology (PhD). Part-time programs available. *Students:* 2 applicants, 100% accepted. *Degree requirements:* For doctorate, thesis/dissertation. *Entrance requirements:* For doctorate, GRE General Test, GRE Subject Test. Application fee: $50. *Expenses:* Tuition, state resident: full-time $8,400; part-time $200 per credit hour. Tuition, nonresident: full-time $11,088; part-time $462 per credit hour. Tuition and fees vary according to program. *Financial support:* Fellowships, research assistantships, teaching assistantships available. *Faculty research:* Biology of gametes and fertilization, reproductive endocrinology. *Unit head:* Scott Lozanoff, Chair, 808-956-7131, Fax: 808-956-9481, E-mail: lozanoffs@jabsom.biomed.hawaii.edu.

University of Saskatchewan, College of Medicine, Department of Obstetrics, Gynecology and Reproductive Services, Saskatoon, SK S7N 5A2, Canada. Offers M Sc, PhD. *Faculty:* 8. *Students:* 7. *Degree requirements:* For master's and doctorate, thesis/dissertation, registration. *Entrance requirements:* Additional exam requirements/recommendations for international students: Required—TOEFL. *Application deadline:* For fall admission, 7/1 for domestic students. Applications are processed on a rolling basis. Application fee: $50. *Financial support:* Fellowships, research assistantships, teaching assistantships available. Financial award application deadline: 1/31. *Unit head:* Dr. O. Olatunbosun, Head, 306-966-8693, Fax: 306-966-8796. *Application contact:* Dr. R. Pierson, Graduate Chair, 306-966-4458, E-mail: pierson@erato.usask.ca.

University of Wyoming, Graduate School, College of Agriculture, Department of Animal Sciences, Program in Reproductive Biology, Laramie, WY 82070. Offers MS, PhD. *Faculty:* 11 full-time (0 women), 2 part-time/adjunct (0 women). *Students:* 3 full-time (2 women), 1 (woman) part-time. Average age 29. 4 applicants, 25% accepted. In 2005, 1 master's, 2 doctorates awarded. *Degree requirements:* For master's and doctorate, thesis/dissertation. *Entrance requirements:* For master's, GRE General Test, minimum GPA of 3.0; for doctorate, GRE General Test, minimum GPA of 3.0 or MS degree. Additional exam requirements/recommendations for international students: Required—TOEFL. *Application deadline:* For fall admission, 2/1 priority date for domestic students, 2/1 priority date for international students; for spring admission, 9/1 priority date for domestic students, 9/1 priority date for international students. Applications are processed on a rolling basis. Application fee: $50. *Expenses:* Tuition, state resident: full-time $3,720; part-time $155 per credit hour. Tuition, nonresident: full-time $10,704; part-time $446 per credit hour. Required fees: $666; $162 per semester. Tuition and fees vary according to course load and program. *Financial support:* In 2005–06, 4 students received support, including research assistantships with tuition reimbursements available (averaging $12,000 per year); career-related internships or fieldwork, Federal Work-Study, institutionally sponsored loans, scholarships/grants, and unspecified assistantships also available. Financial award application deadline: 3/1. *Faculty research:* Fetal programming, chemical suppression, ovaria function, genetics. *Unit head:* Dr. William Murdoch, Professor, 307-766-3293, Fax: 307-766-2355, E-mail: wmurdoch@uwyo.edu. *Application contact:* Jamie L. Lejambre, Office Assistant, Senior, 307-766-2224, Fax: 307-766-2355, E-mail: animascience@uwyo.edu.

West Virginia University, Davis College of Agriculture, Forestry and Consumer Sciences, Interdisciplinary Program in Genetics and Developmental Biology, Morgantown, WV 26506. Offers animal breeding (MS, PhD); biochemical and molecular genetics (MS, PhD); cytogenetics (MS, PhD); descriptive embryology (MS, PhD); developmental genetics (MS); experimental morphogenesis teratology (MS); human genetics (MS, PhD); immunogenetics (MS, PhD); life cycles of animals and plants (MS, PhD); molecular aspects of development (MS, PhD); mutagenesis (MS, PhD); oncology (MS, PhD); plant genetics (MS, PhD); population and quantitative genetics (MS, PhD); regeneration (MS, PhD); teratology (MS, PhD); toxicology (MS, PhD). *Students:* 13 full-time (7 women), 6 part-time (4 women), 11 international. Average age 28. In 2005, 2 master's, 4 doctorates awarded. *Degree requirements:* For master's, thesis/dissertation; for doctorate, thesis/dissertation, comprehensive exam. *Entrance requirements:* For master's, GRE or MCAT, minimum GPA of 2.75. Additional exam requirements/recommendations for international students: Required—TOEFL. Application fee: $45. *Expenses:* Tuition, state resident: full-time $4,582; part-time $258 per credit hour. Tuition, nonresident: full-time $13,820; part-time $741 per credit hour. *Financial support:* In 2005–06, 3 research assistantships with tuition reimbursements (averaging $9,936 per year), 4 teaching assistantships with tuition reimbursements (averaging $9,936 per year) were awarded; fellowships, Federal Work-Study, institutionally sponsored loans, and tuition waivers (full and partial) also available. Financial award application deadline: 2/1; financial award applicants required to submit FAFSA. Total annual research expenditures: $1 million. *Unit head:* Dr. J. Nath, Chairman, 304-293-6023 Ext. 4333, Fax: 304-293-2960, E-mail: joginder.nath@mail.wvu.edu.

Cross-Discipline Announcements

Massachusetts Institute of Technology, School of Science, Department of Biology, Cambridge, MA 02139-4307.

Graduate work in the Department of Biology at MIT leads to the PhD degree. Research opportunities include many areas of modern biology: biochemistry, biophysics, cellular and developmental biology, immunology, microbiology, and neurobiology. Students come from a wide variety of backgrounds. Previous experience in the biological sciences, although desirable, is not a prerequisite for admission. Formal courses in biochemistry, genetics, and the method and logic of molecular biology are required of all students early in their graduate program. Each student is encouraged to follow the program of study that best meets his or her educational goals. Special emphasis is placed on research leading to the PhD thesis.

Princeton University, Graduate School, Department of Molecular Biology, Princeton, NJ 08544-1019.

Graduate studies in the Department of Molecular Biology at Princeton emphasize training in research and encourage students to apply molecular, biochemical, structural, genetic, and computational approaches to biological problems. Faculty members and students pursue research in a wide variety of areas of biochemistry and cell biology, biological dynamics and computational biology, biomedicine and societal issues, biophysics and structural biology, developmental biology, genetics and genomics, microbiology, neurobiology, oncology, and virology.

University of California, Los Angeles, School of Medicine and Graduate Division, Graduate Programs in Medicine, Department of Biomathematics, Los Angeles, CA 90095.

University of California, Los Angeles, Department of Biomathematics offers a graduate program that leads to the PhD degree to train creative, fully independent investigators who can initiate research in mathematical biology/applied mathematics and their chosen biomedical specialty, including genetics, molecular biology, neurosciences, physiology, pharmacology, and immunology.

BAYLOR COLLEGE OF MEDICINE

Graduate Program in Developmental Biology

Program of Study

The Graduate Program in Developmental Biology (DB Program) awards a Ph.D. degree and is designed to prepare students both intellectually and technologically to pursue a successful career in biological and/or biomedical research. The program also cooperates in the Medical Scientist Training Program, which leads to a combined M.D./Ph.D. degree.

The DB Program provides a wide spectrum of exciting research possibilities and a broad cross-disciplinary training. In order to understand how a single cell develops into a complex organism, the program laboratories use molecular biology, cell biology, biochemistry, imaging, physiology, genetics, and genomics. Studies of organisms as diverse as social molds, worms, flies, frogs, chickens, fish, mice, and humans are conducted using a wide variety of approaches, instruments, and techniques of modern biological research. Members of the DB Program study basic biological mechanisms of direct and fundamental relevance to human development, disease, and stem cell therapy. This allows students to unravel the principles and mechanisms that guide embryonic development, the differentiation of adult cell types, regeneration, and aging. The major research interests are neurobiology; cancer biology; cell death; aging; neurodegenerative and other human diseases; stem cell biology; gene therapy; reproductive development; oogenesis; muscle, heart, blood, kidney, bone, skin, limb, and eye development; cell lineage specification; X chromosome dosage compensation; and plant differentiation.

During their first year, students take core courses in biochemistry, cell biology, and molecular and classical genetics as well as several courses and seminars in developmental biology. They also sample several areas of research by doing rotations in the program's laboratories. Before the end of the first year, students take a qualifying exam and select a laboratory in which they carry out their dissertation research. Subsequently, students meet every six months with their thesis committee to evaluate the research accomplished and redefine goals necessary to complete the thesis project. In the final year, students defend their theses in a public seminar. Study for the Ph.D. degree generally requires five years of graduate work, most of which is spent on the dissertation research. The program is supported by a competitive NIH training grant, the March of Dimes, and the College.

Research Facilities

DB Program faculty members are well funded and drawn from eleven departments and three institutions, Baylor College of Medicine, the University of Texas M. D. Anderson Cancer Center, and Rice University, all within easy walking distance of the Texas Medical Center. They occupy extensive research space with state-of-the-art instrumentation and computing equipment. Cooperative and collaborative interactions among program laboratories and institutions enable students to take full advantage of the facilities of the Texas Medical Center.

Financial Aid

Students enrolled in the program receive a stipend of $23,000 per year, plus health insurance at no extra cost. Tuition scholarships are awarded to all students admitted to the program. Separate offices provide assistance to international students and to students with financial hardships.

Cost of Study

As outlined in the Financial Aid section, tuition is fully covered by the program and the College.

Living and Housing Costs

Numerous options are available for on-campus and off-campus housing, including apartments and houses offered at a wide range of rents. The cost of living in Houston is below that of most large U.S. cities, and there are ample opportunities for employment of spouses in the Texas Medical Center.

Student Group

The DB Program currently has 42 full-time graduate students, including 21 women. The program is committed to excellence and favors a low student-faculty ratio. In addition to the laboratories in the program, students have contact with students, postdoctoral fellows, and faculty members in other programs and departments throughout the school and the medical center. The Baylor graduate school has approximately 500 students, the medical school about 670 students.

Student Outcomes

The Career Resource Center of the graduate school provides career information and counseling for all Baylor graduate and postdoctoral students in biomedical sciences. DB students typically graduate with an excellent to outstanding publication record and go on to successful careers. The average number of publications per graduate student is above 4.5, with an average of more than 2.5 first-author papers. The average impact factor per graduate student publication is more than 10. The DB graduates have subsequently pursued postdoctoral training in excellent laboratories and high-quality institutions.

Location

Houston is a dynamic city with an exciting cultural and metropolitan center. Ballet, opera, symphony, theater, and art museums are excellent and accessible to the general population. In addition, there are more than a thousand bars and restaurants, which are moderately priced. Recreation opportunities abound, with facilities for a wide range of professional and amateur sports. The climate offers very pleasant weather from fall through spring and permits participation in a wide variety of outdoor activities. Gulf Coast beaches are a short drive from the city.

The College

Baylor College of Medicine is an independent, private institution dedicated to training in basic and medical sciences. It is located in the heart of the Texas Medical Center, one of the largest medical centers in the world. It has promoted the development of interdisciplinary, interdepartmental, and interinstitutional programs and has consistently identified and encouraged outstanding investigators. Considered as one of the top research institutions in the nation, the College continues to develop programs and services that meet new needs and trends, making higher education one of the most exciting and rewarding of human experiences.

Applying

Applicants must have a bachelor's degree, preferably with course work in biology and biochemistry. GRE General Test scores less than three years old at the time of application must be provided. Applications should be accompanied by transcripts, three letters of recommendation, and a statement of research interest and career goals; they must be complete by January 1, with a preferred deadline of December 15. Successful candidates are invited to meet with the participating faculty members and students in order to have a firsthand look at the DB Program. Expenses for travel and accommodations during the visit are provided. Admission policies at Baylor offer equal opportunity to all, without regard to race, sex, age, religion, country of origin, or handicap.

Correspondence and Information

Hugo J. Bellen, Director
Graduate Program in Developmental Biology
Baylor College of Medicine, BCM 225
One Baylor Plaza
Houston, Texas 77030
Phone: 713-798-7696
E-mail: cat@bcm.tmc.edu
Web site: http://www.bcm.edu/db/

Baylor College of Medicine

THE FACULTY AND THEIR RESEARCH

Adam Antebi, Assistant Professor of Molecular and Cellular Biology, Huffington Center on Aging; Ph.D., MIT, 1991. Endocrine regulation of *C. elegans* developmental age and aging.

Richard R. Behringer, Professor of Molecular Genetics, University of Texas M. D. Anderson Cancer Center; Ph.D., Columbia (South Carolina), 1986. Molecular genetics of female reproductive tract development.

Hugo J. Bellen, Professor of Molecular and Human Genetics, Molecular and Cellular Biology, and Neuroscience; Director, Program in Developmental Biology; Investigator, Howard Hughes Medical Institute; D.V.M./Ph.D., California, Davis, 1986. Nervous system development and neurotransmitter release in *Drosophila.*

John W. Belmont, Professor of Molecular and Human Genetics and Immunology; M.D./Ph.D., Baylor College of Medicine, 1981. Hematopoietic and immune development and cardiovascular genetics.

Andreas Bergmann, Assistant Professor of Biochemistry and Molecular Biology, University of Texas M. D. Anderson Cancer Center; Ph.D., Max Planck Institute for Developmental Biology (Germany), 1996. Genetic control of programmed cell death (apoptosis) in *Drosophila.*

Juan Botas, Associate Professor of Molecular and Human Genetics; Ph.D., Madrid (Spain), 1986. Comparative analysis of normal development and *Drosophila* models of neurodegenerative diseases.

Janet Braam, Professor of Biochemistry and Cell Biology, Rice University; Ph.D., Cornell, 1985. Molecular and developmental responses of plants to environmental stresses: roles of calmodulin-related proteins and cell wall modifying enzymes.

Rui Chen, Assistant Professor of Molecular and Human Genetics, Human Genome Sequencing Center; Ph.D., Baylor College of Medicine, 1999. System biology; genetics network controlling retinal development in *Drosophila.*

Kwang-Wook Choi, Associate Professor of Molecular and Cellular Biology; Ph.D., Princeton, 1988. Eye development and cell polarity in *Drosophila.*

Gabriella D'Arcangelo, Assistant Professor of Pediatrics and Neuroscience; Ph.D., SUNY at Stony Brook, 1993. Development of the vertebrate nervous system and pediatric neurological diseases.

Ronald L. Davis, Professor of Molecular and Cellular Biology, Neuroscience, and Molecular and Human Genetics; Ph.D., California, Davis, 1979. Molecular and cellular biology of learning and memory.

Francesco J. DeMayo, Professor of Molecular and Cellular Biology; Ph.D., Michigan, 1983. Molecular and developmental biology of the lung and uterus; cancer; reproductive biology.

Mary E. Dickinson, Assistant Professor of Molecular Physiology and Biophysics; Ph.D., Columbia, 1996. Vascular remodeling and heart morphogenesis in early vertebrate embryos.

Scott Goode, Assistant Professor of Pathology, Molecular and Cellular Biology, and Molecular and Human Genetics; Ph.D., Chicago, 1993. Cell migrations and tumor cell invasion in *Drosophila.*

Margaret A. Goodell, Associate Professor of Cell and Gene Therapy, Pediatrics, Molecular and Human Genetics, and Immunology; Ph.D., Cambridge, 1991. Adult and embryonic stem cell biology.

Georg Halder, Associate Professor of Biochemistry and Molecular Biology, University of Texas M. D. Anderson Cancer Center; Ph.D., Basel (Switzerland), 1996. Tumor suppressor genes and organ size control; *Drosophila* genetics.

Karen K. Hirschi, Associate Professor of Pediatrics, Molecular and Cellular Biology, and Cell and Gene Therapy; Ph.D., Arizona, 1990. Vascular development and vascular progenitors in adult tissues.

Milan Jamrich, Professor of Molecular and Human Genetics and Molecular and Cellular Biology; Ph.D., Heidelberg (Germany), 1978. Molecular basis of embryonic pattern formation.

Randy L. Johnson, Associate Professor of Biochemistry and Molecular Biology, University of Texas M. D. Anderson Cancer Center; Ph.D., Columbia, 1991. Mouse developmental genetics.

Monica J. Justice, Associate Professor of Molecular and Human Genetics; Ph.D., Kansas, 1987. Hematovascular development; leukemia; mouse models of human disease.

Richard L. Kelley, Associate Professor of Molecular and Cellular Biology and Molecular and Human Genetics; Ph.D., Stanford, 1984. Role of noncoding RNA in chromatin structure.

Adam Kuspa, Professor of Biochemistry and Molecular Biology and Molecular and Human Genetics; Ph.D., Stanford, 1989. Genomic studies of cell signaling and development in *Dictyostelium.*

Mary Ellen Lane, Assistant Professor of Biochemistry and Cell Biology, Rice University; Ph.D., Columbia, 1994. Molecular genetics of embryonic neural development in zebrafish.

Brendan Lee, Associate Professor of Molecular and Human Genetics and Investigator, Howard Hughes Medical Institute; M.D./Ph.D., SUNY Health Science Center at Brooklyn, 1993. Human and mouse developmental genetics; cartilage and skeletal development.

Soo-Kyung Lee, Assistant Professor of Molecular and Cellular Biology and Molecular and Human Genetics; Ph.D., Chonnam National (Korea), 2001. Transcriptional regulation in CNS development.

Olivier Lichtarge, Associate Professor of Molecular and Human Genetics; M.D./Ph.D., Stanford, 1990. Evolutionary studies of sequence, structure, and function in biological macromolecules; bioinformatics.

Hui-Chen Lu, Assistant Professor of Pediatrics–Neurology; Ph.D., Baylor College of Medicine, 1997. Molecular mechanisms of cortical development.

Kathleen A. Mahon, Associate Professor of Molecular and Cellular Biology; Ph.D., Yale, 1984. Early development of the forebrain and pituitary in mice.

Graeme Mardon, Professor of Pathology, Molecular and Human Genetics, and Neuroscience; Ph.D., MIT, 1990. Neural cell fate determination, development, and degeneration in *Drosophila* and vertebrates.

Martin M. Matzuk, Professor of Pathology, Molecular and Human Genetics, and Molecular and Cellular Biology; M.D./Ph.D., Washington (St. Louis), 1989. Mammalian reproduction, oncogenesis, and development.

David D. Moore, Professor of Molecular and Cellular Biology; Ph.D., Wisconsin–Madison, 1979. Functions of the nuclear hormone receptor superfamily.

Paul A. Overbeek, Professor of Molecular and Cellular Biology, Molecular and Human Genetics, and Neuroscience; Ph.D., Michigan, 1980. Transgenic mice; ocular development; transposon-mediated insertional mutagenesis in mice.

Dennis R. Roop, Professor of Molecular and Cellular Biology; Ph.D., Tennessee, Knoxville, 1977. Gene regulation and function during skin development.

Jeffrey M. Rosen, Professor of Molecular and Cellular Biology; Ph.D., SUNY at Buffalo, 1971. Mammary gland development, stem cells, and breast cancer.

Michael D. Schneider, Professor of Medicine and Molecular and Cellular Biology; M.D., Pennsylvania, 1976. Molecular biology of cardiac muscle cell growth; adult cardiac progenitor/stem cells.

Armin Schumacher, Assistant Professor of Molecular and Human Genetics; M.D., Rhenish-Westphalian Technical (Aachen), 1991. Developmental genetics of the mouse.

Gad Shaulsky, Associate Professor of Molecular and Human Genetics; Ph.D., Weizmann (Israel), 1991. Developmental genetics in *Dictyostelium;* functional genomics; molecular basis of social behavior.

Tae Ho Shin, Assistant Professor of Molecular and Cellular Biology; Ph.D., Alabama at Birmingham, 1994. Embryonic development in *C. elegans.*

Ming-Jer Tsai, Professor of Molecular and Cellular Biology; Ph.D., California, Davis, 1971. Pancreas and neural development; organogenesis steroid hormone action; prostate cancer.

Sophia Tsai, Professor of Molecular and Cellular Biology; Ph.D., California, Davis, 1969. Nuclear orphan receptor in mouse development and organogenesis.

Hui Zheng, Associate Professor of Molecular and Human Genetics, Molecular and Cellular Biology, and Neuroscience, Huffington Center on Aging; Ph.D., Baylor College of Medicine, 1990. Molecular genetics of Alzheimer's disease.

Zheng Zhou, Assistant Professor of Biochemistry; Ph.D., Baylor College of Medicine, 1994. Clearance of apoptotic cells in *C. elegans.*

Huda Y. Zoghbi, Professor of Pediatrics, Molecular and Human Genetics, and Neuroscience and Investigator, Howard Hughes Medical Institute; M.D., Meharry Medical College, 1979. Pathogenesis of polyglutamine neurodegenerative diseases and Rett syndrome; genes essential for neurodevelopment.

Thomas P. Zwaka, Assistant Professor of Cell and Gene Therapy and Molecular and Cellular Biology; M.D./Ph.D., Ulm (Germany), 2000. The nature of embryonic stem cell pluripotency.

BAYLOR COLLEGE OF MEDICINE

Department of Molecular and Human Genetics

Program of Study

The Department of Molecular and Human Genetics offers an NIGMS-supported training program leading to the Ph.D. degree. The Department also cooperates in the Medical Scientist Training Program that leads to a combined M.D./Ph.D. degree.

The research interests of the Department span a broad range, including the principles of DNA replication and repair, DNA recombination, cell division, aging, cancer, development, learning, memory, and social behavior. A variety of model organisms are used, from *E. coli* through yeast and *Dictyostelium* to flies and mice, and there is a strong research program in bioinformatics and genomics as well. Studies in model organisms are tightly integrated with studies on the genetic basis of the human condition. The research program addresses a variety of genetic diseases, and the unique environment of a large medical center allows students to obtain experience in aspects of both basic and clinical research.

The graduate program in molecular and human genetics provides outstanding educational opportunities for students who wish to pursue a career in the broad field of genetics. Students in the program obtain rigorous training in modern biology, with an emphasis on genetics. They participate in cutting-edge research on a variety of topics and publish their work in the some of the best peer-reviewed journals in the world. Students are expected to devote full-time to this course of study. In the first year, they concentrate on course work and laboratory rotations designed to provide a firm basis in fundamental genetic concepts and cover recent developments in molecular biology. The program design permits students to move quickly into their thesis research project. The second year and subsequent years are devoted to independent thesis research and elective course work. Seminar programs, literature review meetings, and research presentations enable students to learn about faculty research programs and the current status of various fields of study in the larger scientific community. An annual weekend-long departmental research retreat provides an excellent opportunity for students and faculty members to share information in an informal setting.

The Department of Molecular and Human Genetics interacts closely with other departments at Baylor and with other institutions in the Texas Medical Center. Course credit reciprocity among Baylor, Rice University, Texas A&M University, the University of Houston, and the University of Texas Health Science Center expands the scholastic horizon of Baylor students.

Research Facilities

Faculty members in the Department occupy modern research laboratories furnished with state-of-the-art equipment. The arrangement of laboratories consists of centralized core facilities in which instrumentation serves several investigators. Cooperative and collaborative interactions among Department faculty members and between the Department of Molecular and Human Genetics and other departments and organizations enable students to take full advantage of the facilities at the Texas Medical Center.

Financial Aid

A stipend and fringe-benefits package is awarded to all graduate students enrolled at Baylor College of Medicine. The stipend is $23,000 per year. Health insurance is provided at no cost. Tuition scholarships are awarded to students who are admitted to the program. Following admission to candidacy, students receive a $1000–travel grant from the Department to initiate participation in national meetings.

Cost of Study

Tuition costs are covered in full by tuition scholarships or training grants.

Living and Housing Costs

Most students and faculty members live within a few miles of the Texas Medical Center. Numerous housing options are available. The cost of living in Houston is lower than most major cities. Job opportunities for spouses are available in the many institutions of the Texas Medical Center.

Student Group

Students in the Department of Molecular and Human Genetics may seek either the Ph.D. degree or the combined M.D./Ph.D. degree. In the past three years, about 15 students per year on average have joined the program. There are approximately 500 students enrolled In the Graduate School; the Medical School has about 670 students. Admission policies at Baylor offer equal opportunity to all, without regard to race, sex, age, religion, country of origin, or handicap.

Location

Baylor College of Medicine is located within the Texas Medical Center, a large and vigorous professional community that also includes the University of Texas Health Science Center, eight teaching hospitals, and the M. D. Anderson Cancer Center. Rice University is nearby. Houston is also the home of Houston Baptist University and the University of Houston. The fourth-largest city in the nation, Houston is an exciting cultural and metropolitan center. Ballet, opera, symphony, and theater are excellent and accessible to the general population. Many fine museums and parks enhance city life. Professional and amateur sports are very popular. The climate permits participation in a wide variety of outdoor activities, and Gulf Coast beaches are a short drive from the city.

The College

Baylor College of Medicine is an independent, private institution dedicated to training in basic and medical sciences. It has promoted the development of interdisciplinary research programs and has consistently identified and encouraged outstanding investigators and medical practitioners. The climate of interaction benefits the graduate student body by expanding its learning opportunities.

Applying

Applicants must have earned a bachelor's degree and have a strong background in biology and biochemistry. Candidates for admission must complete the application form of the Graduate School. The application should contain Graduate Record Examinations scores, including an advanced Subject Test (less than three years old), three letters of recommendation, and official undergraduate transcripts. Applications receive two reviews, one by a faculty committee of the Department and a second by the Admissions Committee of the Graduate School. The application deadline for admission is January 1. Applications can be obtained through the Graduate School Web site in the Correspondence and Information section. The $30 application fee is waived for online applications.

Correspondence and Information

For an application:
Admissions Office
Graduate School
Baylor College of Medicine
Houston, Texas 77030
Phone: 713-798-4060
Web site: http://www.bcm.edu/gradschool/

For information:
Director of Graduate Studies
Department of Molecular and Human Genetics
Baylor College of Medicine
One Baylor Plaza
Houston, Texas 77030
Phone: 713-798-5056
E-mail: genetics-gradprgm@bcm.tmc.edu
Web site: http://www.bcm.tmc.edu/molgen/

Baylor College of Medicine

THE FACULTY AND THEIR RESEARCH

Arthur L. Beaudet, Professor and Chairman; M.D., Yale, 1967. Role of genomic imprinting in evolution and disease, including Prader-Willi and Angelman syndromes and autism; hepatocyte gene therapy.

Antonio Baldini, Professor; M.D., Rome, 1983. Mouse models of congenital heart disease; genetic dissection of DiGeorge syndrome; gene targeting.

Hugo J. Bellen, Professor; D.V.M./Ph.D., California, Davis, 1986. Genetic and molecular analysis of neurotransmitter release and neural development in *Drosophila.*

John W. Belmont, Professor; M.D./Ph.D., Baylor, 1981. Immunogenetics; congenital heart defects.

Colin Bishop, Professor; Ph.D., London Hospital Medical College, 1979. Genetic basis of germ cell development and sex determination in mammals.

Juan Botas, Associate Professor; Ph.D., Madrid, 1986. Molecular mechanisms of pathogenesis in neurodegenerative diseases; comparative molecular genetics of development in *Drosophila* and vertebrates.

Malcolm K. Brenner, Professor; M.D./Ph.D., Cambridge, 1981. Use of gene therapy to improve responses to cancer.

Chester W. Brown, Assistant Professor; M.D./Ph.D., Cincinnati, 1993. Roles of TGF-beta superfamily in body composition, growth, reproduction, and embryogenesis.

Wei-Wen Cai, Assistant Professor; Ph.D., NYU, 1996. Genomic technologies; genetic basis of developmental disabilities; cancer genetics.

Rui Chen, Assistant Professor; Ph.D., Baylor, 1999. System biology; genetic network controlling retinal development in *Drosophila.*

Si-Yi Chen, Associate Professor; M.D./Ph.D., Shanghai, 1983; Ph.D., Beijing, 1988. Development of new technologies for gene therapy and immunotherapy against cancer and infectious disease.

A. Craig Chinault, Professor; Ph.D., MIT, 1976. Eukaryotic DNA replication; research and clinical applications of comparative genomic hybridization (CGH) analyses.

William J. Craigen, Associate Professor; M.D./Ph.D., Baylor, 1988. Regulation of cellular energy metabolism; mouse models of metabolic diseases.

Gretchen J. Darlington, Professor; Ph.D., Michigan, 1970. Molecular mechanisms determining tissue-specific gene expression; gene transcription; molecular basis of cellular and organismal aging.

Ronald L. Davis, Professor; Ph.D., California, Davis, 1979. Molecular and cellular biology of learning and memory.

Richard A. Gibbs, Professor; Ph.D., Melbourne (Australia), 1985. Human genome analysis; molecular analysis of genetic and infectious disease.

Scott Goode, Assistant Professor; Ph.D., Chicago, 1991. Epithelial morphogenesis, cell migrations, and tumor cell invasion of *Drosophila.*

Margaret A. Goodell, Associate Professor; Ph.D., Cambridge, 1991. Murine and human hematopoietic stem cells: regulation, development, and gene therapy.

Philip Hastings, Professor; Ph.D., Cambridge, 1965. Molecular mechanisms of gene amplification, genome instability, and recombination in *E coli.*

Xiangwei He, Assistant Professor; Ph.D., Baylor, 1997. Molecular mechanisms of mitotic chromosome segregation.

Christophe Herman, Assistant Professor; Ph.D., Bruxelles (Belgium), 1996. Proteins degradation; stress response; membrane quality control mechanism.

Kendal D. Hirschi, Associate Professor; Ph.D., Arizona, 1993. Nutrient acquisition in plants.

Grzegorz Ira, Assistant Professor, Ph.D., Copernicus, 1999. Essential genes involved in DNA recombination.

Milan Jamrich, Associate Professor; Ph.D., Heidelberg, 1978. Pattern formation in vertebrate embryos; ocular development; gene therapy.

Monica J. Justice, Associate Professor; Ph.D., Kansas State, 1987. Using mouse mutagenesis to analyze gene function and establish models of human disease.

Richard L. Kelley, Associate Professor; Ph.D., Stanford, 1984. Noncoding RNAs and chromatin structure.

Adam Kuspa, Professor; Ph.D., Stanford, 1989. Genomic studies of cell signaling and development in *Dictyostelium.*

Suzanne M. Leal, Associate Professor; Ph.D., Columbia, 1994. Statistical genetics and genetic epidemiology; the genetics of nonsyndromic hearing loss.

Brendan Lee, Associate Professor; M.D./Ph.D., SUNY Health Science Center at Brooklyn, 1993. Molecular determinants of cartilage and skeletal development and associated human genetic condition; adenoviral hepatocyte gene therapy in human urea cycle disorders.

Soo-Kyung Lee, Assistant Professor; Ph.D., Chonnam National (South Korea), 2001. Transcriptional regulatory network in CNS development.

Olivier Lichtarge, Associate Professor; M.D./Ph.D., Stanford, 1990. Evolutionary studies of sequence, structure, and function in biological macromolecules; protein-ligand interactions; bioinformatics.

James R. Lupski, Professor and Vice Chairman; M.D./Ph.D., NYU, 1985. DNA fingerprinting of bacteria; molecular genetics of Charcot-Marie-Tooth disease and inherited neuropathies; inherited eye diseases; molecular mechanisms for human DNA rearrangements; genomic disorders.

Graeme Mardon, Professor; Ph.D., MIT, 1990. Neural cell fate determination, development, and degeneration in *Drosophila* and vertebrates.

Martin M. Matzuk, Professor; M.D./Ph.D., Washington (St. Louis), 1989. Mammalian reproduction and development.

John D. McPherson, Associate Professor; Ph.D., Queen's at Kingston, 1989. Large-scale mapping; sequencing and analysis; full-length cDNA; resequencing technologies; application of genomics/proteomics to human disease loci.

Michael L. Metzker, Assistant Professor; Ph.D., Baylor, 1996. Next-generation technology for genome sequencing; novel fluorescence imaging; molecular genetics of diabetes; phylogenetic analysis of HIV-1 transmission between individuals.

Aleksandar Milosavljevic, Associate Professor; Ph.D., California, Santa Cruz, 1990. Bioinformatics and comparative genomics.

David D. Moore, Professor; Ph.D., Wisconsin–Madison. Functions of the nuclear hormone receptor superfamily.

David L. Nelson, Professor; Ph.D., MIT, 1984. Human genome mapping and disease gene isolation; fragile-X syndrome, incontinentia pigmenti, and cancer genetics.

Jeffrey Noebels, Professor; M.D./Ph.D., Stanford, 1977; M.D., Yale, 1981. Gene control of neuronal excitability within the developing mammalian CNS.

William E. O'Brien, Professor; Ph.D., Georgia, 1971. Inborn errors of metabolism.

Paul A. Overbeek, Professor; Ph.D., Michigan, 1980. Gene regulation in transgenic mice; ocular development; growth factors; insertional mutagenesis.

Richard Paylor, Associate Professor; Ph.D., Colorado, 1991. Genetic basis of complex behavioral traits in mice.

Leif E. Peterson, Associate Professor; Ph.D., Texas, 1993. Genome-scale modeling of microarray-based gene networks; numerical methods.

Scott D. Pletcher, Assistant Professor; Ph.D., Minnesota, 1998. The genetics and molecular analysis of aging in *Drosophila.*

Sharon Plon, Associate Professor; M.D./Ph.D., Harvard, 1987. Human cell-cycle checkpoint genes; control of genomic stability; cancer genetics.

Susan M. Rosenberg, Professor; Ph.D., Oregon, 1986. Molecular mechanisms of genome instability, mutation, DNA repair, and recombination in *E. coli*; cancer; antibiotic resistance.

Christian Rosenmund, Associate Professor; Ph.D., Oregon Health Sciences, 1993. Molecular mechanisms of synaptic transmission and plasticity at central mammalian synapses.

Armin Schumacher, Assistant Professor; M.D., Aachen (Germany), 1991. Developmental genetics of the mouse.

Gad Shaulsky, Associate Professor; Ph.D., Weizmann (Israel), 1991. Developmental genetics in *Dictyostelium;* functional genomics; the molecular basis of social behavior.

Jeffrey A. Towbin, Professor; M.D., Cincinnati, 1982. Study of familial dilated cardiomyopathy (FDCM), hypertrophic cardiomyopathy (HCM), long QT syndrome (LQTS), Brugada syndrome, viral myocarditis, and inherited congenital heart disease.

Ignatia Van den Veyver, Associate Professor; M.D., Antwerp (Belgium), 1986. Role of epigenetics and genomic imprinting in development and disease; X-linked developmental disorders.

George Weinstock, Professor; Ph.D., MIT, 1977. Genomics of microbes to mammals; infectious disease genetics; bioinformatics.

John H. Wilson, Professor; Ph.D., Caltech, 1971. Instability of trinucleotide repeats; knock-in mouse models for retinitis pigmentosa; gene therapy of dominant rhodopsin mutations.

Hui Zheng, Associate Professor; Ph.D., Baylor, 1990. Molecular genetics of Alzheimer's disease.

Huda Zoghbi, Professor; M.D., Meharry Medical College, 1979. Molecular basis of degenerative and developmental neurologic disorders; nervous system development.

CASE WESTERN RESERVE UNIVERSITY

Department of Genetics
Center for Human Genetics
Program in Human, Molecular, and Developmental Genetics and Genomics

Programs of Study

The Department of Genetics at Case Western Reserve University offers a diversified and highly interactive graduate training program emphasizing molecular and genetic approaches to the study of fundamental problems in biology and medicine. Research interests in the department include the genetic basis of human disease, the molecular biology and genetics of embryonic development, sex determination, and recombination in *Drosophila, C. elegans,* the mouse, and humans; chromosome structure and function; human and mouse genomics; and regulation of gene expression. During the first year, students participate in an integrated interdepartmental course designed to provide an introduction to current problems in modern genetics and developmental, cellular, and molecular biology. Students are then encouraged to pursue a program of research and study that meets their goals and interests. Advanced courses are offered in specialized areas and include Eukaryotic Genetics, Human Genetics, Genetics of Complex Traits, Medical Genetics, Structural Analysis of Complex Genomes, Developmental Genetics, Mammalian Cytogenetics, Chromosome Structure and Function, Principles of Genetic Epidemiology, Yeast Genetics and Cell Biology, and Research in Genetics. Students are encouraged to participate in ongoing journal clubs and research seminars. A program of departmental and interdepartmental seminars by outstanding visiting scientists provides regular exposure to a broad range of current research in genetics.

The graduate program trains students for academic and research careers in the biological and medical sciences. Programs leading to the Ph.D. or combined M.D./Ph.D. are offered by the department. The Department of Genetics accepts direct applications to the program by those who have significant prior research experience in genetics and are committed to training in human genetics, genomics, molecular genetics, and/or developmental genetics. Alternatively, the department also participates in the integrated Biomedical Sciences Training Program. This program coordinates graduate admissions and the first year of graduate training in all basic science graduate programs within the School of Medicine. It allows entering students to choose among more than seventy-five laboratories in participating departments in the School of Medicine. Students may explore different areas through laboratory rotations during the first year. For more information on this integrated program, students should contact the Coordinator of the Biomedical Sciences Training Program, School of Medicine, Case Western Reserve University, WG46, 10900 Euclid Avenue, Cleveland, Ohio 44106.

The Center for Human Genetics is part of the Department of Genetics and is composed of both research and clinical laboratories involved in human and clinical genetics. The diagnostic laboratories and genetic counseling services are jointly administered by the Department of Genetics and University Hospitals of Cleveland. The center supports research programs focusing on chromosome structure and behavior, human genome mapping, the molecular basis of inherited disease, and the genetic dissection of complex disease.

The Developmental Biology Center is also associated with the Department of Genetics. Established in 1961, it was the first university-based center of its kind. Its activities include organizing a program of weekly seminars, journal clubs, and an annual scientific retreat.

Further information on the department and its programs can be obtained from the Web site listed in the Correspondence and Information section.

Research Facilities

The Department of Genetics' state-of-the-art laboratories contain instrumentation essential for all aspects of modern research in molecular genetics. These include microinjection facilities for producing transgenic mice, a *Drosophila* genetics facility, confocal and fluorescence microscopy equipment for gene mapping, and a flow cytometry facility. Other common facilities include instrumentation for automated DNA synthesis, peptide synthesis, protein sequencing, DNA sequencing, and fluorescent genotyping. The Health Sciences Library is housed in the same complex as the department, as are the closely allied Departments of Biochemistry, Molecular Biology and Microbiology, Physiology and Biophysics, Neurosciences, and Pharmacology.

Financial Aid

National Research Service Awards, NIH training grants, and departmental funds cover tuition and fees and, in 2003–04, provided a research stipend of $21,000 per year for all students. No teaching is required. Eligible students are strongly encouraged to apply for competitive national scholarships (e.g., NSF fellowships) during their first year.

Cost of Study

In 2003–04, tuition and fees were $18,946 per year, which is covered for students by the Department of Genetics.

Living and Housing Costs

Most graduate students live off campus in one of the pleasant residential neighborhoods within walking or biking distance of the University. There is a variety of very reasonably priced housing available in these areas. University housing is also available for unmarried graduate students. For the academic year, rooms are $3700, and optional board plans average about $1960.

Student Group

The University's total enrollment is 9,800. Of this number, 1,800 are in the Graduate School and 3,200 in the professional schools. Approximately 50 graduate students are currently enrolled in the Department of Genetics.

Student Outcomes

Of the approximately 56 students who have recently received their Ph.D.'s in the laboratory of the faculty of the Genetics Training Program, 37 percent now hold independent faculty positions, 39 percent are currently in postdoctoral research positions (most in highly competitive laboratories), 12 percent are research employees in industry or in government laboratories, and the remaining 12 percent are in medical training.

Location

Case Western Reserve University is located about 4 miles east of downtown Cleveland in University Circle, which is one of the largest cultural and educational centers in the nation. More than thirty educational, scientific, medical, cultural, social service, and religious institutions, including the world-famous Cleveland Museum of Art and Severance Hall, home of the Cleveland Orchestra, are located in the 500-acre University Circle community. Metropolitan Cleveland is a cosmopolitan community of 2 million people.

The University

Case Western Reserve University is a private nondenominational institution. It was established in 1967 by the joining of Western Reserve University (founded in 1826) and Case Institute of Technology (founded in 1880), which had occupied adjoining campuses since 1883.

Applying

Applications should be marked "Genetics" and should clearly indicate whether the student is seeking admission directly into the Department of Genetics or into the Biomedical Sciences Training Program. Applicants must submit GRE scores, three letters of recommendation, and an official college transcript. Admission requirements are a sound basic training in biology, chemistry, calculus, and physics; courses in biochemistry, genetics, molecular biology, cell biology, and probability and statistics are strongly recommended. Applications for admission should be sent to the address below, preferably before February 28; late applications may also be considered. Students are encouraged to contact individual faculty members directly for a more detailed description of their research programs.

Correspondence and Information

Graduate Admissions
Department of Genetics
Case Western Reserve University
2109 Adelbert Road
Cleveland, Ohio 44106-4955

Phone: 216-368-5847
E-mail: gradadmit@po.cwru.edu
Web site: http://genetics.case.edu/

Case Western Reserve University

THE FACULTY AND THEIR RESEARCH

Research Trainers in Genetics

Mark Adams, Associate Professor; Ph.D., Michigan. Application of genomic-scale technologies; analysis of mouse models of complex disease.

Brian Bai, Assistant Professor; Ph.D., NYU. Control of mammalian neuronal progenitors.

Ronald Conlon, Associate Professor; Ph.D., Texas at Austin. Embryonic patterning and morphogenesis in the mouse; gene targeting; genetic manipulation of development.

Peter J. Harte, Associate Professor; Ph.D., Yale. Regulation of homeotic gene expression during *Drosophila* development; mechanism of action of transcriptional activators and repressors; chromatin structure.

Terry Hassold, Professor; Ph.D., Michigan State. Mechanisms underlying production of human chromosomal abnormalities during meiosis.

Patricia Hunt, Professor; Ph.D., Hawaii. Mammalian gametogenesis; role of sex chromosomes in mammalian development.

Bruce Lamb, Assistant Professor; Ph.D., Pennsylvania. Genetics of Alzheimer's disease; YAC transgenic mice as models to study human disease genes.

Greg Matera, Associate Professor; Ph.D., California, Davis. Nuclear organization and gene expression; subcellular localization of RNA processing machinery; molecular cytogenetics; digital imaging microscopy.

Anne Matthews, Associate Professor; Ph.D., Colorado. Genetic counseling; ethical issues regarding genetic technologies and counseling.

Shawn McCandless, Assistant Professor; M.D., Temple. Treatment of chronic genetically determined disorders: inborn errors of metabolism (fatty oxidation defects and organic acidemias); Prader-Willi syndrome.

Kathleen Molyneaux, Assistant Professor; Ph.D., Cornell. Germ cell development and migration.

Joe Nadeau, Professor; Ph.D., Boston University. Mouse genetics; comparative genome mapping; complex genetic traits.

Kurt Runge, Associate Professor; Ph.D., MIT. Telomere biology; transcriptional silencing; yeast aging.

Helen K. Salz, Professor; Ph.D., California, Davis. Genetic control of RNA splicing; *Drosophila* sex determination; cell fate determination.

Stuart Schwartz, Professor; Ph.D., Indiana Bloomington. Molecular cytogenetics; tumor cytogenetics; etiology of chromosome breakage and aneuploidy.

George Stark, Professor; Ph.D., Columbia. Gene amplification; interferon.

Matthew Warman, Associate Professor; M.D., Cornell. Osteochondrodysplasias; molecular genetic mapping and analysis of inherited disease; inherited metabolic diseases.

Georgia Wiesner, Associate Professor; M.D., Minnesota. Familial cancers; cancer genetics.

Bryan Williams, Professor; Ph.D., Otago (New Zealand). Isolation and characterization of tumor suppressor genes; cytokine signal transduction.

Research Trainers in Other Departments

Mitchell Drumm, Associate Professor, Department of Pediatrics; Ph.D., Michigan. Biology and genetics of cystic fibrosis.

Karl Herrup, Professor, Alzheimer Center; Ph.D., Stanford. Alzheimer's disease; pattern formation; gene regulation; cerebellum; transgenic mice; neuroanatomy.

Phil Morgan, Associate Professor, Department of Anesthesiology; M.D., Colorado. Molecular genetics of the mechanism of volatile anesthetic action in the nematode *Caenorhabditis elegans*.

Margaret Sedensky, Associate Professor, Department of Anesthesiology; M.D., Colorado. Molecular genetics of the mechanism of volatile anesthetic action in the nematode *Caenorhabditis elegans*.

Peter Zimmerman, Assistant Professor, Department of Geographic Medicine; Ph.D., Case Western Reserve. Molecular genetics and infectious disease.

Additional Clinical Faculty in Genetics

Arthur B. Zinn, Associate Professor; M.D., Ph.D., Case Western Reserve. Biochemical genetics: inherited disorders of energy metabolism.

CLEMSON UNIVERSITY

Programs in Genetics and Biochemistry

Programs of Study	The Department of Genetics and Biochemistry offers the M.S. and Ph.D. in both genetics and biochemistry and molecular biology. Concentrations within the degrees include human (Ph.D. only), animal, plant, and microbial molecular biology and genetics. The degrees prepare students for careers in both academic and industrial research. The first year of both programs consists of course work and laboratory rotations (Ph.D. only), with many courses common to both degrees. Ph.D. students typically select an adviser and begin their research after a three-laboratory rotation during the first year. M.S. students choose their adviser before entering the program and begin laboratory research and course work immediately. An advisory committee, selected by the adviser and the student, determines the student's course work requirements, keeps track of the student's progress in the classroom and laboratory, and administers the oral examination (usually the third year of the Ph.D.) and final examinations. While course requirements vary for each degree in the second year, breadth and depth of preparation are expected of each candidate, and the remainder of the two to three years for an M.S. degree or four to five years for a Ph.D. degree is focused on research. Students are kept abreast of recent developments in genetics, biochemistry, and molecular biology through seminars, colloquia, journal clubs, and special courses. Much of their education is received informally through frequent discussions with faculty members and other students, both within and outside the department. Students are expected to present research results at regional and national scientific meetings, publish in recognized journals, and submit grant proposals in collaboration with faculty members.
Research Facilities	The laboratories for the Department of Genetics and Biochemistry are housed in Jordan Hall and the Biosystems Research Complex. The Department is well equipped with most of the conventional instrumentation expected for modern genetic and biochemical analysis. Graduate students have access to specialized research laboratories, including a DNA-sequencing facility with LiCor and ABI-sequencing equipment, a computerized bioinstrumentation laboratory, an image analysis facility, a confocal/electron microscopy suite equipped for transmission and scanning electron microscopy, controlled-environment chambers, a tissue-culture laboratory, fully equipped darkrooms, isotope handling and counting facilities, plant growth chambers, a $7-million greenhouse complex suitable for use with transgenic plants, and a small-animal care facility with surgical capabilities. Newly established proteomics and microarray facilities and their staffs complete the range of resources available to graduate researchers.
Financial Aid	Financial support is available through grant-supported research assistantships, University-supported research and teaching assistantships, graduate fellowships, and grants-in-aid. The Graduate Admissions Committee does not accept unsupported or part-time students. Stipends for Ph.D. graduate assistantships are $19,200 for twelve months, with reduced tuition, and $17,200 for M.S. students for twelve months, with reduced tuition, awarded upon acceptance.
Cost of Study	For 2006–07, graduate assistants pay a flat fee of $1079 per semester and $348 per summer session. Graduate fellows pay South Carolina resident fees.
Living and Housing Costs	On-campus housing is available; for information, students should visit http://www.housing.clemson.edu. The cost of living in Clemson is quite low compared to the national average; students who choose to live off-campus typically spend $300–$400 per month for rent, depending on the location, amenities, roommates, and other factors.
Student Group	The M.S. and Ph.D. degree programs require students to have the maturity, mental discipline, and work habits to be independent and productive researchers, with a demonstrated enthusiasm for their fields.
Location	Clemson is a small, beautiful college town near the Blue Ridge Mountains and Lake Hartwell. The Upstate is one of the country's fastest-growing areas and is an important part of the I-85 corridor, a multistate area along Interstate 85 that runs from metro Atlanta to Richmond, Virginia, and encompasses Charlotte, North Carolina, and North Carolina's Research Triangle. Atlanta and Charlotte are each a 2-hour's drive away. Many financial institutions and other industries have national headquarters for a major presence in the Upstate, including Wachovia, Bank of America, BMW, Bon Secours St. Francis Health System, Bosch North America, Bowater, Charter Communications, Ernst and Young, Fluor Corporation, IBM, Microsoft, Michelin of North America, and many others.
The University	Clemson is classified by the Carnegie Foundation as Doctoral/Research University–Extensive, a category comprising less than 4 percent of all universities in America. The University's mission is to fulfill the covenant between its founder and the people of South Carolina to establish a "high seminary of learning" through its responsibilities of teaching, research, and extended public service. The University has identified eight areas of academic emphasis that create collaborations that, in turn, help fulfill the University's mission.
Applying	This program has unique admission procedures and forms. Students should not apply using the standard Clemson graduate application. For more information, prospective students should visit http://www.clemson.edu/genbiochem/ for the proper procedures, paperwork, and deadlines.

Correspondence and Information	**Both programs:**	**Biochemistry and Molecular Biology:**	**Genetics:**

	Lisa Pape
	Administrative Graduate Coordinator
	100 Jordan Hall
	Clemson University
	Clemson, South Carolina 29634-0318
	Phone: 864-656-6877
	     866-247-8358 (toll-free)
	Fax: 864-656-6879
	E-mail: lpape@clemson.edu
	Web site: http://www.clemson.edu/
	    genbiochem/
		Kerry Smith
		Assistant Professor/Graduate
		   Coordinator
		100 Jordan Hall
		Clemson University
		Clemson, South Carolina 29634-0318
		Phone: 864-656-6935
		Fax: 864-656-0393
		E-mail: kssmith@clemson.edu
		Web site: http://www.clemson.edu/
		    genbiochem/
			Jim Morris
			Assistant Professor/Graduate
			   Coordinator
			214 Biosystems Research Complex
			Clemson University
			Clemson, South Carolina 29634-0318
			Phone: 864-656-0293
			E-mail: jmorri2@clemson.edu
			Web site: http://www.clemson.edu/
			    genbiochem/

Clemson University

THE FACULTY AND THEIR RESEARCH

Albert G. Abbott, Professor; Ph.D., Brown. Biological sciences.
Weiguo Cao, Assistant Professor; Ph.D., Idaho. Microbiology.
Chin-Fu Chen, Assistant Professor; Ph.D., SUNY at Stony Brook. Genetics.
Julianne Collins, Adjunct Faculty, GGC; Ph.D., Alabama at Birmingham. Medical genetics.
Barbara DuPont, Adjunct Faculty, GGC; Ph.D., Texas at Austin. Zoology/human genetics.
David Everman, Adjunct Faculty, GGC; M.D., Emory. Medical genetics.
Julia Frugoli, Assistant Professor; Ph.D., Dartmouth. Biological sciences.
Richard H. Hilderman, Department Chair and Head; Ph.D., Missouri. Microbiology.
Harry D. Kurtz Jr., Assistant Professor; Ph.D., Idaho. Bacteriology.
Amy Lawton-Rauh, Assistant Professor; Ph.D., North Carolina State. Genetics.
Hong Luo, Associate Professor; Ph.D., Louvain (Belgium). Molecular biology.
William R. Marcotte Jr., Associate Professor; Ph.D., Virginia. Microbiology.
Brandon D. Moore, Assistant Professor; Ph.D., Washington State. Botany.
James C. Morris, Assistant Professor; Ph.D., Georgia. Cellular biology.
Gary L. Powell, Professor; Ph.D., Purdue. Biological sciences.
Curtis Rogers, Adjunct Faculty, GGC; M.D., Medical University of South Carolina. Genetics/pediatrics.
Charles Schwartz, Adjunct Faculty, GGC; Ph.D., Vanderbilt. Biochemistry.
Kerry S. Smith, Assistant Professor; Ph.D., Pennsylvania. Molecular biology.
Anand Srivastava, Adjunct Faculty, GGC; Ph.D., Banaras Hindu (India). Biochemistry.
Roger Stevenson, Adjunct Faculty, GGC; M.D., Wake Forest. Pediatrics.
Jeffrey P. Tomkins, Assistant Professor; Ph.D., Clemson. Genetics.

DARTMOUTH COLLEGE

Dartmouth Medical School
Graduate Program in Genetics

Program of Study	The Graduate Program in Genetics is a subgroup of the Molecular and Cellular Biology (MCB) graduate program at Dartmouth College. It is designed to provide advanced study in genetics leading to the Ph.D. degree. MCB is an interdepartmental program that includes faculty members from the Departments of Genetics, Biochemistry, Biological Sciences, Microbiology and Immunology, and Medicine at Dartmouth Medical School. The program offers diverse opportunities for incoming students and provides them with the tools needed to become independent investigators and teachers. All first-year students attend a comprehensive course in the principles of genetics, biochemistry, and molecular and cellular biology, as well as several advanced courses in specific areas of genetics and associated subjects. Three separate research rotations are taken the first year by each student, culminating with the selection of the laboratory in which to pursue a degree.
Research Facilities	Dartmouth Medical School has nucleated a distinguished Department of Genetics, poised to exploit and contribute to the explosive growth in information and its application to solving problems in biological systems. Located in the Medical School complex, the newly renovated department houses state-of-the-art laboratories and equipment and is physically linked to the Dana Biomedical Library and the Gilman Biomedical Center.
Financial Aid	All students enrolled in the MCB graduate program receive a yearly stipend plus health insurance coverage and a tuition waiver. For current stipend information, prospective students should visit the Web site at http://www.dartmouth.edu/~mcb/.
Cost of Study	All students in the program receive a scholarship for the full cost of tuition.
Living and Housing Costs	Dartmouth College assists graduate students in arranging appropriate housing in College facilities or in privately owned accommodations in the Hanover area. Apartment rents for off-campus housing vary widely, but accommodations can often be found at rates similar to those for College housing.
Student Group	Currently more than 120 students are enrolled in the MCB program, about a third in the labs of genetics program members. This allows the unique opportunity for genetics students to interface with a number of genetics-related disciplines while pursuing their degrees. All students in the MCB program participate in student seminar programs and journal clubs.
Student Outcomes	The majority of students continue on to postdoctoral training and then into academic or industrial careers.
Location	Dartmouth College is located in Hanover, New Hampshire. Dartmouth is a coeducational institution with an undergraduate student body of approximately 4,200 and a professional graduate school enrollment of about 1,250. Although small, Hanover boasts a variety of cultural, entertainment, and recreational opportunities as well as a special environment for outdoor activities. Boston, Montreal, and New York are easily accessible by car.
The College and The School	Dartmouth College, a member of the Ivy League, is a private four-year coeducational college with graduate schools of business, engineering, and medicine and eighteen graduate programs in the arts and sciences. The Hopkins Center for the Performing Arts offers concerts and drama programs and sponsors the film society. All students and faculty members have access to numerous arts and crafts workshops and membership in several choral and instrumental groups. Also available to graduate students are the extensive services of the Dartmouth Outing Club and gym facilities. Dartmouth Medical School, founded in 1797, is located on the campus of Dartmouth College in Hanover. It combines modern facilities with tutorial instruction in the basic medical sciences. The Dana Medical Library and the Matthews Fuller Health Science Library house a complete collection of journals in the biomedical sciences.
Applying	Applicants seeking more detailed information about the Graduate Program in Genetics should either contact the Chair of the Graduate Admissions Committee at the address listed in this In-Depth Description or log on to http://www.dartmouth.edu/~mcb/ for an application and general program outline. Complete applications should be received no later than January 7. Dartmouth College and Medical School actively support equal opportunity for all persons regardless of race, ethnic or national background, religion, handicap, color, gender, or sexual preference.
Correspondence and Information	Chair, Graduate Admissions Committee Molecular and Cellular Biology Graduate Program—Genetics 7560 Vail, Room 214A Dartmouth Medical School Hanover, New Hampshire 03755 Phone: 603-650-1612 Fax: 603-650-1318 E-mail: molecular.and.cellular.biology@dartmouth.edu Web site: http://www.dartmouth.edu/~genetics

Dartmouth College

THE FACULTY AND THEIR RESEARCH

Yashi Ahmed, Assistant Professor of Genetics; Ph.D., 1992; M.D., 1993. A *Drosophila* model for signaling by the APC tumor suppressor and beta-catenin oncogene.

Victor Ambros, Professor of Genetics; Ph.D., 1979. Small regulatory RNAs and the control of developmental timing in animals.

Brad Arrick, Associate Professor of Medicine; Ph.D., 1983; M.D., 1984. Growth factors and cancer genetics.

Sharon Bickel, Assistant Professor of Biology; Ph.D., 1991. Regulation of chromosome segregation in *Drosophila.*

Charles Brenner, Associate Professor of Genetics and Biochemistry; Ph.D., 1993. Enzymes involved in tumor suppression, neurological function, and vitamin biosynthesis. (Norris Cotton Cancer Center)

Charles Cole, Professor of Biochemistry and Genetics; Ph.D., 1972. Genetics and molecular biology of mRNA export; micro RNAs and breast cancer.

Michael Cole, Professor of Pharmacology and Toxicology of Genetics; Ph.D., 1978. Molecular basis of cancer; transcription factors; mechanisms of chromosome-mediated transcriptional control; target genes for oncogenic pathways. (Norris Cotton Cancer Center)

Barbara Conradt, Assistant Professor of Genetics; Ph.D., 1994. Regulation of programmed cell death in *C. elegans.*

Patrick Dolph, Associate Professor of Biological Sciences; Ph.D., 1989. Regulation of signal transduction in *Drosophila.*

Jay C. Dunlap, Professor of Genetics and Biochemistry and Chair, Genetics; Ph.D., 1979. Genetics of circadian rhythms.

Patricia Ernst, Assistant Professor of Genetics; Ph.D., 1996. Chromatin regulatory proteins in the development of the hematopoietic and immune systems; genetic pathways involved in leukemogenesis.

Steven N. Fiering, Associate Professor of Microbiology and Immunology and of Genetics; Ph.D., 1990. DNA elements in regulation of mammalian genes during development.

Robert Gross, Associate Professor of Biological Sciences; Ph.D., 1974. Computational molecular biology.

Mary Lou Guerinot, Professor of Biological Sciences; Ph.D., 1979. Genetics and molecular genetics of metal uptake.

Mark A. Israel, Professor of Pediatrics and Genetics and Director, Norris Cotton Cancer Center; M.D., 1973. Regulation of cellular differentiation and proliferation during development and tumorigenesis.

Thomas Jack, Associate Professor of Biological Sciences; Ph.D., 1990. Genetics and molecular genetics of flower development.

Eric Lambie, Associate Professor of Biological Sciences; Ph.D., 1987. Genetics of development in *C. elegans.*

Jennifer J. Loros, Professor of Biochemistry and Genetics; Ph.D., 1984. Fungal genetics; molecular genetics of biological clocks.

Rob McClung, Professor of Biological Sciences; Ph.D., 1986. Genetics and molecular genetics of clocks in plants.

Thuluvancheri K. Mohandas, Professor of Pathology and Genetics; Ph.D., 1972. Human genetics; human cytogenetics; molecular cytogenetics; X-chromosome inactivation.

Jason H. Moore, Associate Professor of Genetics; Ph.D., 1999. Computational genetics; bioinformatics; genetics of common human diseases. (Norris Cotton Cancer Center)

Claudio Pikielny, Assistant Professor of Genetics; Ph.D., 1986. Olfaction and behavioral neurogenetics in animals.

R. Mako Saito, Assistant Professor of Genetics; Ph.D., 1996. Development control of cell cycle in *C. elegans.*

Nancy A. Speck, Professor of Biochemistry; Ph.D., 1983. Hematopoiesis and leukemia.

Ronald K. Taylor, Professor of Microbiology and Immunology; Ph.D., 1984. Molecular mechanisms of bacterial pathogenesis.

Sergei Tevosian, Assistant Professor of Genetics; Ph.D., 1997. Early developmental regulators required for heart and gonadal development in mammals.

Michael Whitfield, Assistant Professor of Genetics; Ph.D., 1999. Regulatory mechanisms governing gene expression and cell biological aspects of cell division and scleroderma.

William Wickner, Professor of Biochemistry; M.D., 1973. Biochemistry and genetics of organelle inheritance.

DUKE UNIVERSITY

Developmental Biology Program

Programs of Study

The curriculum for the interdisciplinary Developmental Biology Program (DBP) at Duke University provides a core of knowledge in developmental biology, while allowing students the flexibility to explore individual interests in related subjects. During the first year, DBP students study developmental biology, an overview of developmental strategies and mechanisms, and a variety of graduate minicourses designed to provide a common level of knowledge that serves as the basis for more advanced courses and dissertation research. Many students take the required advanced graduate-level course in either cell biology or genetics at this time.

In the spring semester of the first and second years, students participate in the Developmental Biology Colloquium course, which is integrated with the spring Developmental Biology Seminar series, where students meet weekly with faculty members and seminar speakers. During the first year, students also participate in three laboratory rotations that introduce students to different scientific approaches, laboratory environments, and possible Ph.D. mentors before selecting their Ph.D. thesis project. By the end of the first year, students usually declare a department in which to earn their Ph.D.

During the first two years, each student is expected to serve as a teaching assistant for one semester, teaching undergraduate courses in developmental biology and the related areas of genetics and cell biology.

Research Facilities

University libraries include more than 5 million volumes, 17.7 million manuscripts, 1.2 million public documents, and thousands of computer files as well as tens of thousands of films, videos, and audio recordings. The Biological Sciences Building houses systematics, population genetics and evolution, animal physiology, and functional morphology; the Duke Phytotron contains plant ecology; and the recently constructed Levine Science Research Center is home to developmental, cellular, and molecular biology. Special facilities include animal rooms, greenhouses, darkrooms, refrigerated and controlled-environment laboratories, scanning and transmission electron microscopes, a Van de Graaf accelerator, X-ray machines, radiation and radioisotope equipment, and a computerized morphometrics laboratory. The Highlands Biological Station is a year-round biological field station located on a high plateau in the southern Appalachian Mountains of southwestern North Carolina. Its principal mission is to promote research and education in biodiversity studies, with special emphasis on the diverse flora and fauna of the region.

Financial Aid

All applications are considered for financial support. NIH Research Service Awards, University fellowships, and teaching and research assistantships provide full support to all students admitted to the program. Financial aid packages for a twelve-month year cover tuition, fees, health insurance, and a stipend. The stipend for 2006–07 is $24,000.

Cost of Study

Tuition and fees are $37,300 for the 2006–07 academic year. This includes tuition, registration fees, a health fee, a transcript fee, a student activity fee, and health insurance.

Living and Housing Costs

A wide variety of housing arrangements are available in Durham and nearby Chapel Hill. Housing is relatively inexpensive and easily accessible compared to major metropolitan areas.

Student Group

In 2005–06, the program had 33 students, 15 of whom were women and 4 of whom were international students. They entered the program with an average 3.5 undergraduate GPA and an average combined GRE score of 1280.

Student Outcomes

The majority of graduates of the program become university faculty members, join research universities, and go on to doctoral universities and liberal arts colleges. Some also go into nonacademic endeavors.

Location

The campus is located in Durham, the fourth-largest city in North Carolina, located between the mountains and the coastal plain. Durham is home to Research Triangle Park, a private, nonprofit park for research-related organizations. One quarter of the workers in Durham are employed in health care.

The University and The Program

In 1924, James Buchanan Duke formalized his family's historic pattern of philanthropy with the establishment of the Duke Endowment. With endowment stipulations to build a university around the local Trinity College, which had been founded as Union Institute by local Methodist and Quaker communities in 1838, the administration was able to rebuild the old campus and create a new one, opening its first graduate school in 1926. Today, it is the home of nine individual schools and nearly 13,000 students and features a world-renowned medical center and the champion Blue Devils basketball team.

The interdisciplinary Developmental Biology Program was created when developmental biologists of the Departments of Biology, Cell Biology, and Genetics began an effort to emphasize developmental biology as an interdepartmental strength. Seven departments are now involved in the program, including the Departments of Biochemistry, Biology, Cell Biology, Immunology, Molecular Genetics and Microbiology, Neurobiology, and Pharmacology and Cancer Biology. The program has more than 40 faculty members, who are well distributed among the various departments, ensuring easy access to different research areas.

Applying

Admission into the Developmental Biology Program requires submitting a completed electronic application, a $75 application fee, three letters of evaluation, copies of an official transcript, and GRE scores. Completed materials should be mailed to the Graduate School Enrollment Services Office, Duke University, 127 Allen Building, Box 90065, Durham, North Carolina 27708-0065.

Correspondence and Information

Director of Graduate Studies
Developmental Biology Program
Duke University Medical Center
DUMC 3565
Durham, North Carolina 27710
Phone: 919-684-6629
E-mail: devbio@biochem.duke.edu
Web site: http://www.devbio.duke.edu

Duke University

THE FACULTY AND THEIR RESEARCH

Hubert Amrein, Associate Professor of Molecular Genetics and Microbiology. Genetic control and molecular mechanisms of dosage compensation in *Drosophila.*

Amy Bejsovec, Associate Professor of Biology. Cell-fate specification and cell-to-cell communication during embryonic development.

Gerard Blobe, Assistant Professor of Pharmacology and Cancer Biology. Role of the TGF-β signaling pathway in cancer and vascular biology.

Blanche Capel, Professor of Cell Biology. Sex determination and mammalian development.

Christopher Counter, Associate Professor of Pharmacology and Cancer Biology. Elucidation of telomere function in cancer.

Clifford Cunningham, Associate Professor of Biology. Evolution and biogeography of marine invertebrates.

Xinnian Dong, Professor of Biology. Response of plants to pathogens; molecular and genetic investigation of the signal transduction pathways activating plant defense genes.

Michael Ehlers, Associate Professor of Neurobiology. Regulation of NMDA-type glutamate receptors by protein phosphorylation and identification of cellular proteins involved in glutamate receptor trafficking.

Sharyn Endow, Professor of Cell Biology. Microtubule motor proteins involved in spindle function; chromosome movement in meiosis and mitosis.

Arno Greenleaf, Professor of Biochemistry. Eukaryotic transcription enzymology and molecular biology; RNA polymerase II, transcription factor, and CTD kinase.

Joseph Heitman, Professor of Molecular Genetics and Microbiology. Genetic analysis of the action of immunosuppressive compounds.

Brigid Hogan, Professor of Cell Biology. Cellular interactions regulating proliferation and differentiation during mammalian organogenesis and development of primordial germ cells.

Erich Jarvis, Assistant Professor of Neurobiology. Molecular pathways of perception and production of learned vocalizations.

Russel Kaufman, Associate Professor of Biochemistry. Regulation of multigene families and developmental hematopoiesis.

Daniel Kiehart, Professor of Biology. Cell motility; muscle contraction; mitosis.

Margaret Kirby, Professor of Pediatrics and of Cell Biology. Neural crest and heart development.

John Klingensmith, Associate Professor of Cell Biology. Mechanisms that establish and pattern the body axes and organ precursors of the mammalian embryo.

Sally Kornbluth, Professor of Pharmacology and Cancer Biology. Examination of cell-cycle regulation using in vitro reconstitution of cell-cycle processes.

Michael Krangel, Professor of Immunology. Rearrangement and expression of T-cell receptor genes; chemotactic cytokines.

Daniel Lew, Associate Professor of Pharmacology and Cancer Biology. Cell-cycle control; control of cell polarity.

Haifan Lin, Associate Professor of Cell Biology. Mechanism of stem-cell division and germline development.

Elwood Linney, Professor of Molecular Genetics and Microbiology. Molecular approaches to retinoic acid and estrogen signaling during zebrafish embryonic development; 3-D imaging of gene expression; environmental biosensing using embryos.

David McClay, Professor of Biology. Problems in evolution and development, speciations genetics, molecular evolution, and population genetics.

Donald McDonnell, Professor of Pharmacology and Cancer Biology. Development and application of novel molecular approaches for the discovery of tissue specific modulators of steroid hormone receptors.

Erik Meyers, Assistant Professor of Cell Biology. Mechanisms of patterning and development during organogenesis of the mammalian embryo; tissue specific genetic manipulation in mice.

Richard Mooney, Associate Professor of Neurobiology. Sensitive period regulation in the brains of songbirds during vocal development.

H. Frederick Nijhout, Professor of Biology. Regulatory processes in development; evolution of developmental mechanisms; Modeling disease and regeneration in zebrafish.

Ken Poss, Assistant Professor of Cell Biology, Modeling disease and regeneration in zebrafish.

Tannishtha Reya, Assistant Professor of Pharmacology and Cancer Biology. Signaling pathways that regulate stem-cell fate.

V. Louise Roth, Associate Professor of Biology. Evolution of body size and shape in mammals.

Theodore Slotkin, Professor of Pharmacology and Cancer Biology. Effects of endocrine status, nutrition, and drugs on nervous system development; neonatal cardiovascular mechanisms; neurobiology of Alzheimer's disease and depression in the elderly.

Kathleen Smith, Professor of Biology. Cranial development; function and evolution in vertebrates.

Tai-ping Sun, Associate Professor of Biology. Molecular genetics of plant growth hormones; environmental and developmental regulation of gibberellin biosynthesis; the mechanisms by which gibberellins affect plant development.

Fan Wang, Assistant Professor of Cell Biology. Neural circuit development in mouse somatic sensory system.

Xiao-Fan Wang, Professor of Pharmacology and Cancer Biology. Structure and function of TGF-β receptors; growth regulation by TGF-β.

Robin Wharton, Professor of Molecular Genetics and Microbiology. Molecular mechanism of pattern formation in the *Drosophila* embryo.

Greg Wray, Professor of Biology. Development and interaction among regulatory genes in echinoderms.

Tso-Pang Yao, Associate Professor of Pharmacology and Cancer Biology. Transcriptional coactivator p300 and CBP using knockout animal and various biochemical approaches.

Yuan Zhuang, Associate Professor of Immunology. Molecular mechanisms by which HLH genes function to determine cell type during development.

DUKE UNIVERSITY

Graduate Program in Genetics and Genomics

Program of Study	The Graduate Program in Genetics and Genomics is an interdepartmental program that spans several basic science and clinical departments and bridges the Medical Center and the College of Arts and Sciences. This program is designed to provide training in genetics that leads to the Ph.D. degree. There are more than 75 faculty members in the Program in Genetics and Genomics. Faculty research is especially strong in such diverse areas as microbial and viral genetics, plant genetics, developmental genetics, human genetics, and population biology and evolutionary genetics. Students are admitted directly to the program and spend their first year taking courses and doing two or three rotations in different laboratories to gain experience and explore the many opportunities for research at Duke before selecting a faculty adviser. Once the student selects an adviser, he/she then has the option of receiving the Ph.D. through the adviser's department or through the University Program in Genetics and Genomics. Students admitted through the program may do their thesis research in the laboratory of any member of the program.

Research Facilities Students in the program have access to all the experimental facilities of the member departments as well as the facilities of other departments when needed. The University Library, the Medical Center Library, and smaller working libraries in the various departments contain over 3 million volumes and subscribe to more than 2,500 periodicals in the biological and medical areas. A phytotron and a primate facility are both on the main campus, and a marine laboratory is located in Beaufort, North Carolina.

Financial Aid All applications are considered for financial support. NIH Research Service Awards, University fellowships, and teaching and research assistantships provide full support to all students admitted to the program. Financial aid packages for a twelve-month year cover tuition, fees, health insurance, and a stipend. The stipend for 2006–07 is $24,000.

Cost of Study Tuition and fees for the academic year 2006–07 total $37,300. This includes tuition, registration fees, a health fee, a transcript fee, a student activity fee, and health insurance.

Living and Housing Costs A wide variety of housing arrangements are available in Durham and nearby Chapel Hill. Housing is relatively inexpensive and easily accessible compared to major metropolitan areas.

Student Group Duke University has a total enrollment of approximately 13,000 full-time students. More than 5,200 are pursuing graduate or professional degrees. There are 84 graduate students in the University Program in Genetics and Genomics.

Student Outcomes The majority of students who receive the Ph.D. continue their training at the postdoctoral level, with the eventual goal of a career in college teaching or academic research. Graduates also find employment in the pharmaceutical and biotech industries and with government and private research organizations.

Location Durham is a city of some 155,000 people located in a metropolitan area of more than 1 million, midway between the Atlantic Ocean and the Appalachian Mountains. The climate is moderate, and the cultural life of the community is well developed. The University of North Carolina at Chapel Hill and North Carolina State University at Raleigh, the state capital, are nearby. Twelve miles away is the well-known Research Triangle Park of North Carolina, where a number of firms have large research establishments. The National Center for Health Statistics and the National Institute of Environmental Health Sciences are also located there. Many of the scientists employed by these organizations live in Durham and add to the scientific and cultural climate generated by the close relations between the universities. The weather is such that outdoor recreation is possible all year round.

The University Duke is a private university that dates as a corporate entity from 1924; however, its roots go back almost 160 years to the Union Institute, founded in 1838. Its graduate school has a faculty of more than 800. The campus is located on the edge of the Duke Forest, 7,700 acres of rolling, wooded land on the southwestern fringe of Durham. The proximity of all departments of the biological sciences, including the preclinical departments of the Medical Center, facilitates interdepartmental cooperation and results in a wide variety of courses, research interests, and seminars. Duke has a reciprocal arrangement with the University of North Carolina at Chapel Hill and North Carolina State University at Raleigh that allows enrollment in courses at these nearby institutions.

Applying Students apply directly to the graduate school and should indicate "Genetics and Genomics Program" as their proposed department. Students may also apply to any of the participating departments, indicating their interest in training through the Graduate Program in Genetics. The online electronic application can be found on the graduate school's Web site (http://www.gradschool.duke.edu). All applications must be postmarked by December 31.

Correspondence and Information

University Program in Genetics and Genomics
Box 3565
Duke University Medical Center
Durham, North Carolina 27710
Phone: 919-684-6629
E-mail: genetics@biochem.duke.edu
Web site: http://upgg.duke.edu/

Dean of the Graduate School
Duke University
Durham, North Carolina 27708
Phone: 919-684-3913

Duke University

THE FACULTY AND THEIR RESEARCH

Alejandro Aballay, Assistant Professor of Molecular Genetics and Microbiology; Ph.D. Models of bacterial pathogenesis.

J. Andrew Alspaugh, Assistant Professor of Medicine; M.D. The mechanisms of microbial pathogenesis using the tools of fungal genetics.

Hubert Amrein, Associate Professor of Genetics; Ph.D. Molecular genetics of sex-specific gene regulation in *Drosophila*.

Michael D. Been, Professor of Biochemistry; Ph.D. RNA catalysis; RNA structure and function.

Amy Bejsovec, Associate Professor of Biology; Ph.D. *Drosophila* genetics for analysis of gene function.

Philip N. Benfey, Professor and Chair of Biology; Ph.D. Molecular-genetic and genomics approaches to identify and characterize the genes that regulate formation of the root in the plant model system, *Arabidopsis thaliana*.

Blanche Capel, Professor of Cell Biology; Ph.D. Mammalian development: molecular mechanisms governing sex determination and organogenesis of the testis.

Maria Cardenas-Carona, Associate Research Professor of Molecular Genetics and Microbiology; Ph.D. The Tor signaling cascade, which senses nutrients and regulates gene expression, translation, and ribosome biogenesis.

Jen-Tsan Ashley Chi, Assistant Professor of Molecular Genetics and Microbiology; M.D./Ph.D. Genomic approaches to analyze biological systems and disease.

Christopher M. Counter, Associate Professor of Pharmacology and Cancer Biology; Ph.D. Molecular mechanisms underlying evolution of normal cells into human cancers; activation of telomerase during tumorigenesis.

Gary M. Cox, Assistant Professor of Medicine; M.D. Molecular pathogenesis of fungal infections.

Bryan R. Cullen, Professor of Genetics; Ph.D. Regulation of human retroviral gene expression.

Clifford W. Cunningham, Associate Professor of Biology; Ph.D. Phylogeny and biogeography of marine invertebrates.

Fred S. Dietrich, Assistant Professor of Genetics; Ph.D. Application of genome-scale sequencing for the comparative analysis of genomes.

Xinnian Dong, Professor of Biology; Ph.D. Mechanisms of plant defense against microbial pathogens using *Arabidopsis thaliana* as a model system.

Sharyn A. Endow, Professor of Cell Biology; Ph.D. Molecular genetics and cell biology of *Drosophila*; genetics, molecular genetics, cell biology, and cytogenetics of meiotic and mitotic chromosome segregation.

Paulo Ferreira, Associate Professor of Molecular Genetics and Microbiology, Ph.D. Molecular, cellular, and genetic basis of retinal function and pathogenesis of inherited retinopathies and allied diseases.

Mariano A. Garcia-Blanco, Professor of Genetics; M.D./Ph.D. Regulation of gene expression in eukaryotes.

David Goldstein, Professor of Molecular Genetics and Microbiology; Ph.D. How genetic diversity contributes to disease susceptibility and variability in response to drugs.

Arno L. Greenleaf, Professor of Biochemistry; Ph.D. Molecular genetics of eukaryotic transcription.

Simon Gregory, Assistant Research Professor of Medicine; Ph.D. Identification of the complex genetic factors that give rise to the development of cardiovascular disease and the detection of genes involved in multiple sclerosis.

Steve Haase, Assistant Professor of Biology; Ph.D. How (Cdks) coordinate duplication and segregation events during the cell cycle.

Michael Hauser, Associate Research Professor of Medicine in the Center for Human Genetics; Ph.D. Performing gene expression analysis to identify susceptibility genes for Parkinson's disease and primary open angle glaucoma.

Joseph Heitman, Professor of Genetics and Pharmacology; Ph.D., M.D. Molecular mechanisms of signal transduction, immunosuppression, and DNA recognition and repair.

Michael S. Hershfield, Professor of Medicine and Assistant Professor of Biochemistry; M.D. Genetics, pathophysiology, and treatment of immunodeficiency diseases caused by inborn errors of purine metabolism; other inherited metabolic diseases.

Brigid L. M. Hogan, Professor of Cell Biology; Ph.D.; FRS. Mammalian development and organogenesis; primordial germ cells and stem cells.

Tao-Shih Hsieh, Professor of Biochemistry; Ph.D. Chromosome structure and function.

Randy L. Jirtle, Professor of Radiation Oncology; Ph.D. Genomic imprinting in carcinogenesis.

Jack D. Keene, Professor of Microbiology; Ph.D. Molecular genetics; RNA metabolism, RNA-binding proteins, autoimmunity, in vitro genetic selection, and combinatorial technologies.

Daniel P. Kiehart, Professor of Biology; Ph.D. Genetics of *Drosophila*; genetics, molecular genetics, and protein chemistry of molecular motors; the cytoskeleton and cell shape changes in cytokinesis and morphogenesis.

John Klingensmith, Associate Professor of Cell Biology; Ph.D. Molecular genetics and cell biology of induction and patterning of the mammalian nervous system.

James Koh, Assistant Research Professor of Surgery; Ph.D. Tumor-suppressor gene function as it relates to the regulation of cell-cycle progression in normal and in neoplastic cells.

Sally Kornbluth, Professor of Pharmacology and Cancer Biology; Ph.D. Cell-cycle regulation; regulation of apoptosis.

Kenneth N. Kreuzer, Professor of Biochemistry; Ph.D. Molecular genetics of bacteriophage replication and recombination; DNA topoisomerases.

Meta J. Kuehn, Assistant Professor of Biochemistry; Ph.D. Protein secretory mechanisms; molecular mechanisms of bacterial pathogenesis; membrane vesicle formation and function.

Daniel J. Lew, Professor of Pharmacology and Cancer Biology; Ph.D. Cell-cycle control and morphogenesis in yeast.

Haifan Lin, Professor of Cell Biology; Ph.D. Genetic, cell biological, and molecular analyses of stem cell division using *Drosophila* germline as a model.

Elwood A. Linney, Professor of Microbiology; Ph.D. Molecular approaches toward the study of gene regulation in mouse embryos and teratocarcinoma cells.

Francois Lutzoni, Associate Professor of Biology; Ph.D. Complexities of lichen's symbiotic relationship.

Paul Magwene, Assistant Professor of Biology; Ph.D. Evolutionary genomics and bioinformatics.

Douglas A. Marchuk, Professor of Genetics; Ph.D. Molecular genetics of human disease.

M. Louise Markert, Associate Professor of Immunology and Pediatrics; Ph.D., M.D. Molecular biology of human inherited immunodeficiency diseases.

Hiroaki Matsunami, Assistant Professor of Genetics; Ph.D. Molecular genetics of sensory perception.

John H. McCusker, Professor of Microbiology; Ph.D. Genetic and molecular analysis of fungal pathogenesis.

Anthony R. Means, Professor of Pharmacology and Cancer Biology; Ph.D. Signal transduction cascades mediated by calcium and calmodulin in cell growth and differentiation.

Thomas Mitchell-Olds, Professor of Biology; Ph.D. Evolutionary and ecological functional genomics.

Paul L. Modrich, James B. Duke Professor of Biochemistry; Ph.D. Molecular mechanisms of protein-DNA interactions; enzymology of mismatch repair and gene conversion.

Joseph R. Nevins, Professor of Genetics; Ph.D. Regulation of eukaryotic gene expression.

R. Bruce Nicklas, Professor of Biology and Cell Biology; Ph.D. Chromosome biology; chromosome movement in mitosis.

H. Frederik Nijhout, Professor of Biology; Ph.D. Endocrinology of insects; pattern formation in development; development and evolution.

Mohammed A. Noor, Associate Professor of Biology; Ph.D. Speciation and evolutionary genetics.

Zhen-Ming Pei, Assistant Professor of Biology; Ph.D. Cell and molecular biology of signal transduction in *Arabidopsis*.

John R. Perfect, Professor of Microbiology; M.D. Genetic and molecular pathogenesis of cryptococcal infections; use of *C. neoformans* as a pathogenic model system.

Margaret A. Pericak-Vance, James B. Duke Professor of Medicine; Ph.D. Genetic analysis of human disease; genetic mapping and identification of inherited disorders.

Thomas D. Petes, Professor of Molecular Genetics and Microbiology; Ph.D. Mechanisms of altering the genetic structure of yeast.

David J. Pickup, Associate Professor of Microbiology; Ph.D. Viral pathogenesis; molecular genetics of pox viruses.

Ken Poss, Assistant Professor of Cell Biology; Ph.D. Mechanisms of heart and fin regeneration in zebrafish; zebrafish models of human disease.

Christian H. Raetz, Professor of Biochemistry; Ph.D./M.D. Membrane biochemistry; molecular genetics of lipid metabolism in animal cells; structure, biosynthesis, and function of bacterial endotoxins.

Mark D. Rausher, Professor of Biology; Ph.D. Ecological and population genetics.

Laura Rusche, Associate Professor of Biochemistry; Ph.D. Formation and function of silenced chromatin.

Frederick H. Schachat, Associate Professor of Cell Biology; Ph.D. Biochemical genetics of contractile protein expression, coordination function, and assembly.

William Scott, Associate Research Professor in Medicine; Ph.D. Genetic epidemiology of infectious diseases and complex diseases associated with aging.

Barbara Ramsay Shaw, Professor of Chemistry; Ph.D. Molecular genetics; mechanisms of mutagenesis, DNA protein interactions, and antisense and gene therapy.

David Sherwood, Assistant Professor of Biology; Ph.D. Genetic analysis of development in *C. elegans*.

Marcy C. Speer, Associate Research Professor of Medicine; Ph.D. Genetic epidemiology and linkage analysis in human disease.

Deborah A. Steege, Professor of Biochemistry; Ph.D. mRNA processing and stability; translational control.

Bruce A. Sullenger, Professor of Experimental Surgery; Ph.D. RNA-based molecular therapeutics.

Beth Sullivan, Assistant Professor of Molecular Genetics and Microbiology; Ph.D. Chromosome inheritance and chromatin biology in *Drosophila* and humans.

Tai-ping Sun, Associate Professor of Biology; Ph.D. Molecular genetics of plant hormone gibberellin.

Dennis Thiele, Professor of Pharmacology and Cancer Biology; Ph.D. Stress-responsive gene expression and diseases of protein malfolding; role of copper and iron in signaling, growth, and development.

Daniel Tracey, Assistant Professor of Cell Biology; Ph.D. Molecular neuroscience of *Drosophila*.

Marcy K. Uyenoyama, Professor of Biology; Ph.D. Mathematical population genetics.

Raphael Valdivia, Assistant Professor of Molecular Genetics and Microbiology; Ph.D. Understanding how intracellular bacterial pathogens exploit the host's normal cellular processes to survive and replicate.

Jeffrey M. Vance, Professor of Medicine and Genetics; Ph.D., M.D. Human genetics and molecular genetics of neurologic diseases.

Rytas J. Vilgalys, Professor of Biology; Ph.D. Molecular evolution and genetics of speciation in fungi.

Fan Wang, Assistant Professor of Cell Biology; Ph.D. Molecular genetic analysis of the development and neuronal connectivity of the somatic sensory system.

Robert Wechsler-Reya, Assistant Professor of Pharmacology and Cancer Biology; Ph.D. Sonic hedgehog signaling in neural development and tumorigenesis.

Robin P. Wharton, Associate Professor of Genetics; Ph.D. Molecular mechanisms governing embryonic pattern formation.

Huntington Willard, Professor and Director, Institute for Genome Sciences & Policy and Vice Chancellor for Genome Sciences; Ph.D. Aspects of the molecular structure and function of human chromosomes and the human genome.

John H. Willis, Associate Professor of Biology; Ph.D. Evolution of reproductive isolation, breeding systems, inbreeding depression, and floral traits in natural plant populations.

Gregory Wray, Associate Professor of Biology; Ph.D. Evolution of developmental mechanisms and gene regulatory systems.

John D. York, Associate Professor of Pharmacology and Cancer Biology; Ph.D. Molecular mechanisms of inositol signal transduction; molecular targets of lithium in manic-depressive disease; protein structure and evolution.

Sally J. York, Assistant Professor of Medicine; M.D., Ph.D. The role of DNA repair in the response and resistance to chemotherapy.

Lingchong You, Assistant Professor of Biomedical Engineering; Ph.D. Research interests include drug and gene delivery, transport, angiogenesis, tumor pathophysiology.

Yuan Zhuang, Associate Professor of Immunology; Ph.D. Molecular mechanisms by which helix-loop-helix (HLH) genes function to determine cell type during development.

DUKE UNIVERSITY

Graduate Program in Molecular Genetics and Microbiology

Program of Study

The Department of Molecular Genetics and Microbiology offers the degree of Doctor of Philosophy (Ph.D.) through the Graduate School. Research and education investigate a wide variety of fundamental problems in genetics and microbiology, with particular focus of interest in microbial pathogenesis, RNA biology, virology, and experimental genetics. Students spend their first year taking courses and doing rotations in different laboratories to explore research opportunities before selecting a faculty adviser. Subsequently, students take an oral preliminary exam and then pursue research full time. As part of its commitment to training young researchers, the program encourages participation in a variety of extra-classroom activities, including departmental research meetings, a departmental retreat, and discussions of topical scientific literature.

Research Facilities

The department is housed in modern facilities in the Jones, CARL, and CIEMAS buildings on the Medical Center campus, adjacent to the other basic science departments. The Center for Genome Technology is part of the department, and it contains modern high-throughput DNA sequencing and microarray technologies. Members of the department have ready access to shared facilities, including those run by the Cancer Center and the Division of Laboratory Animal Resources.

Financial Aid

All applicants for admission are automatically considered for University and departmental fellowships that include a stipend, which was estimated at $23,000 for 2005–06, plus health insurance, fees, and tuition.

Cost of Study

Tuition and fees for academic year 2005–06 were $34,204. This includes registration, health, and summer session fees. The cost of studies is paid through funds from the graduate program.

Living and Housing Costs

A wide variety of housing is available in Durham and nearby Chapel Hill. The atmosphere is suburban, so housing and living costs are generally much lower than those in major metropolitan areas.

Student Group

Duke University has a total enrollment of approximately 12,000 full-time students, 5,200 of whom are pursuing graduate or professional degrees. There are currently 73 students in the Department of Molecular Genetics and Microbiology.

Student Outcomes

Most students continue training at the postdoctoral level, with eventual careers in college teaching and academic or industrial research. In addition to research universities throughout the country, popular destinations include pharmaceutical and biotech industries and government research organizations.

Location

Approximately 155,000 people live in Durham, which enjoys proximity to Chapel Hill and Raleigh (homes of the University of North Carolina and North Carolina State University, respectively). The lifestyle is suburban, but there is ready access to an eclectic mixture of activities ranging from the music clubs in Chapel Hill to sacred music in the Duke Chapel. Research Triangle Park, midway between Durham and Raleigh, draws scientists to a number of large research firms and the National Institute of Environmental Health Sciences.

The University

The diversity at Duke University comprises 8,500 acres of pine forest, Gothic quadrangles, and a prominent Medical Center as well as modern laboratory facilities. The Department of Molecular Genetics and Microbiology was created in 2002 in recognition of the opportunity to pursue graduate education and faculty recruitment in the discipline.

Applying

Students can apply online or request a paper application from the Graduate School at the Web address listed. General Graduate Record Examinations (GRE) scores are required, and Subject Test scores encouraged. Applications must be postmarked by December 31. Interviews are arranged in January. Matriculation is allowed only in the fall semester.

Correspondence and Information

Dr. Hubert Amrein
Director of Graduate Studies
Box 3509
Duke University Medical School
Durham, North Carolina 27710
E-mail: hoa1@duke.edu
Web site: http://www.gradschool.duke.edu
   http://mgm.duke.edu

Duke University

THE FACULTY AND THEIR RESEARCH

Thomas D. Petes, Ph.D., Minnie Geller Professor for Research in Genetics and Chair. Genetic regulation of genomic stability; meiotic recombination hotspots; telomere length regulation.

Alejandro Aballay, Ph.D., Assistant Professor. Genetic analysis of innate immunity; bacterial pathogenesis and host-pathogen interactions; type III secretory systems.

Hubert Amrein, Ph.D., Associate Professor. Genetic analysis of sexual behavior, taste, and pheromone perception in *Drosophila*.

Maria Cardenas-Corona, Ph.D., Research Associate Professor. Nutrient sensing and mechanisms of signaling by the targets of rapamycin: the TOR kinase homologs.

Jen-Tsan Ashley Chi, M.D., Ph.D., Assistant Professor. Genomic approaches to analyze biological systems and disease.

Bryan R. Cullen, Ph.D., James B. Duke Professor. Intrinsic immunity to retroviruses mediated by APOBEC3 proteins; role and mechanism of action of human and viral microRNAs.

Fred Dietrich, Ph.D., Assistant Professor. Comparative analysis of fungal genomes, using *S. cerevisiae, A. gossypii,* and *C. neoformans.*

Mariano A. Garcia-Blanco, M.D., Ph.D., Professor. Role of RNA-protein interactions in the connection of transcription and splicing, the regulation of splicing of nuclear precursor mRNAs (pre-mRNAs), and the expression of flaviviral genes.

David B. Goldstein, Ph.D., Professor. Contribution of genetic diversity to disease susceptibility and variability in response to drugs.

Matthias Gromeier, M.D., Assistant Professor. Molecular basis of enteroviral neuropathogenesis; engineering picornaviruses for the treatment of cancer.

Joseph Heitman, M.D., Ph.D., James B. Duke Professor. Signal transduction cascades as therapeutic targets and in the control of differentiation and virulence in yeast and pathogenic fungi; role of sexual reproduction in the evolution and virulence of microbial pathogens.

Jack D. Keene, Ph.D., James B. Duke Professor. Global coordination of posttranscriptional RNA operons and regulons by RNA-binding proteins and microRNAs.

Elwood A. Linney, Ph.D., Professor. Studies of the basic development of the zebrafish embryonic nervous system, using fluorescent, transgenic reporter embryos; Agilent 22k microarrays; analysis of microRNA expression; studies of the vulnerability of the nervous system to environmental compounds and pharmaceuticals.

Douglas Marchuk, Ph.D., Professor. Molecular genetics of human disease.

Hiroaki Matsunami, Ph.D., Assistant Professor. Molecular biology of olfaction and taste in mammals.

John H. McCusker, Ph.D., Associate Professor. *S. cerevisiae* genetics, fungal pathogenesis, and quantitative genetics using *S. cerevisiae* as a model.

Thomas G. Mitchell, Ph.D., Associate Professor. Medical mycology; models of host-fungal dynamics; mechanisms of resistance (e.g., phagocytes) and determinants of pathogenicity (e.g., fungal capsules, antigens, chemotaxins).

Joseph R. Nevins, Ph.D., James B. Duke Professor. Regulation of eukaryotic gene expression.

David J. Pickup, Ph.D., Associate Professor. Molecular mechanisms of viral pathogenesis; viral interference with innate and adaptive immune responses; development of viral vaccines.

Sue Jinks-Robertson, Ph.D., Professor. Eukaryotic genome stability; recombination and mutagenesis in yeast.

Beth Sullivan, Ph.D., Assistant Professor. Chromosome and chromatin biology; epigenomic organization of eukaryotic centromere regions.

Raphael Valdivia, Ph.D., Assistant Professor. Intracellular membrane trafficking and bacterial pathogenesis.

Robin P. Wharton, Ph.D., Professor. Mechanisms of pattern formation and size regulation during development of *Drosophila;* RNA-protein interactions and translational control.

Huntington F. Willard, Ph.D., Nanaline H. Duke Professor. Organization and function of the human genome; epigenetics; human artificial chromosomes; X chromosome inactivation.

THE GEORGE WASHINGTON UNIVERSITY

School of Medicine and Health Sciences
and Columbian College of Arts and Sciences
Graduate Program in Genomics, Proteomics, and Bioinformatics

Program of Study

The Departments of Biochemistry and Molecular Biology, Microbiology and Tropical Medicine, and Computer Sciences at the George Washington University (GWU) and the Institute for Genomic Research (TIGR), together with the Children's National Medical Center (CNMC), have established an exciting new degree program in the area of functional genomics, which constitutes a basic understanding of genomics, proteomics, and bioinformatics. The program is devoted exclusively to important emerging genome-wide approaches to medicine and biology. The endeavor of the program is for each student to establish a broad knowledge in genomics, proteomics, and bioinformatics and to develop into highly qualified individuals with training in these fields.

The joint M.S. degree program offers two overlapping tracks: biological science and computer science. Students choosing the biological science track of the program take a minimum of seven courses in biological sciences related to their specialization in genomics, proteomics, and bioinformatics and one computer science course to acquire general understanding of the principles of organization, searching, and analysis of large volumes of data.

Students choosing the computer science track of the program take six courses in computer science focusing on specialization in high-performance information processing and six courses in fundamental biology to familiarize themselves with the subjects of genomics and proteomics.

The program leading to the M.S. degree (thesis) requires a total of 32 credit hours, 6 of which are earned by the preparation of a thesis, either at George Washington University Medical Center (GWUMC), at Children's National Medical Center, or at TIGR. The thesis program requires 6 credit hours of hands-on practicum at either of the institutions. The resulting degree has a thesis and/or accepted manuscript in a peer-reviewed journal. The program leading to the M.S. degree (nonthesis) requires a total of 35 credit hours, 6 of which are earned by additional hands-on practicum at either of the institutions. Completion of each program generally requires two years. Additional information concerning the program can be found at http://www.gwumc.edu/bioinformatics.

Research Facilities

The program is housed in Ross and Tompkins Halls. Both facilities contain fully equipped research and teaching facilities, including state-of-the-art classrooms, computing labs, networking labs, and servers. A close relationship exists with the off-campus teaching and research faculty members at the Institute for Genomics Research, the National Institutes of Health, Children's Hospital, and American Red Cross. The School of Medicine and Health Sciences has excellent modern library facilities, with a special section devoted to audiovisual educational aids.

Financial Aid

Financial aid for master's students is not available at this time. Assistantships may become available in the future.

Cost of Study

For the 2005–06 academic year, the cost of tuition was $876 and fees were $34.50 per semester credit hour. Annual increases of 10 percent are anticipated.

Living and Housing Costs

The cost of living varies widely according to the type of accommodations and the area in which the student chooses to live. Information about off-campus housing may be obtained from the Director of Housing and Residence Life.

Student Group

The University has a heterogeneous student body, which includes representatives from all parts of the United States and 100 other countries. The age of students varies widely. There are 15–20 students in each admitting class.

Location

Because the George Washington University is located in the nation's capital, there is much of a historical note, and many cultural and educational opportunities lie within easy reach. Of particular interest to students in the health sciences is the University's proximity to the National Institutes of Health, the National Library of Medicine, the Library of Congress, the Department of Agriculture, and the Food and Drug Administration. The Institute for Genomics Research, Celera, and Human Genome Sciences are also nearby.

The University

George Washington University was chartered by congress in 1821. It is a private, nonsectarian institution located just a few blocks from the White House. A center for faculty and student activities houses a bookstore, travel agency, music listening room, computer room, TV room, study lounge, locker room, vending machines, cafeteria, faculty club, rathskeller, theater, and bowling alley. An athletics center houses an Olympic-size pool; indoor tennis, handball, and racquetball courts; basketball courts; exercise rooms; and an indoor track. A newly opened wellness center houses basketball courts, a lap pool, racquetball and squash courts, large fitness areas, and a multipurpose room.

Applying

Applicants for both tracks must have a bachelor's degree with a background in biological and chemical sciences (including courses in general and organic chemistry and 1 year of college physics) and a basic background in computer sciences and mathematics (including calculus and statistics). The focus of the applicant's background determines the appropriate biological or computer track of the M.S. program. Deficiencies in these courses may be made up after admission. An overall and advanced grade point average of at least a B and a combined score of at least 1100 on the verbal and quantitative portions of the General Test of the Graduate Record Examinations are recommended minimum admission requirements. The Columbian College operates on a semester system, with terms beginning in August and January. For admission to the M.S. program, application materials should be received by May 30 for fall admission and November 30 for spring admission. Deadlines are subject to change.

Correspondence and Information

For program information:
Fatah Kashanchi, Ph.D.
Graduate Advisor, Biology
Department of Biochemistry and
   Molecular Biology
School of Medicine and Health
   Sciences
The George Washington University
2300 Eye Street, NW
Washington, D.C. 20037
Phone: 202-994-1781
Fax: 202-994-1780
E-mail: bcmfxk@gwumc.edu

For program information:
Rahul Simha, Ph.D.
Graduate Advisor, Computer Science
Department of Computer Sciences
Columbian College of Arts and Sciences
801 22nd Street, NW
Washington, D.C. 20037
Phone: 202-994-7181
Fax: 202-994-4875
E-mail: csgrad@gwu.edu

For applications and catalogs:
Dean
Columbian College of Arts and Sciences
The George Washington University
Phone: 202-994-6210
Fax: 202-994-6213
E-mail: csgrad@gwu.edu
Web site: http://www.gwu.edu/~gradinfo/
   index2.htm

The George Washington University

THE FACULTY AND THEIR RESEARCH

The GWU School of Medicine, Ross Hall Laboratories
Mahnaz Badamchian, Ph.D., Associate Research Professor. Proteomics of parasitic nematodes.
Patricia Berg, Ph.D., Associate Professor. Breast cancer and leukemia.
Bernard Bouscarel, Ph.D., Associate Professor. Serum lipoproteins; albumin and related regulation.
Vincent Chiappinelli, Ph.D., Professor and Chair. Nicotinic receptors.
Stephanie Constant, Ph.D., Assistant Professor. Immunity to intracellular pathogens.
Sidong Fu, Ph.D., Assistant Research Professor. Disease-related gene mapping.
Allan Goldstein, Ph.D., Professor and Chair. Genomics education; molecular biology of immunopeptides.
Timothy Hales, Ph.D., Professor. Genomics and proteomics of neurotransmitter receptors.
John Hawdon, Ph.D., Associate Professor. Genomics of parasitic nematodes; genetics of drug resistance.
Peter Hotez, M.D., Ph.D., Professor and Chair. Vaccinology of tropical infectious diseases.
Valerie Hu, Ph.D., Associate Professor. Molecular and cellular responses to low-energy internal radiation.
Fatah Kashanchi, Ph.D., Associate Professor. Genomics and proteomics of HIV and other related viruses.
Ajit Kumar, Ph.D., Professor. RNA-protein interaction; transactivation of HIV gene expression.
Timothy McCaffrey, Ph.D., Associate Professor. Genomics and proteomics of cardiovascular disease.
Steven Patierno, Ph.D., Professor. Cancer genetics.
Anne Pumfery, Ph.D., Assistant Research Professor. Genomics and proteomics of HIV and HHV-8/KSHV.
Margaret Sutherland, Ph.D., Assistant Professor. Transgenic technologies.
Jack Vanderhoek, Ph.D., Professor. Major bioactive lipid metabolites related to cell proliferation, migration, and regulation of enzyme activities.
Thomas Wilke, Ph.D., Assistant Research Professor. Genomics of and evolution of snails and parasitic diseases.

Department of Computer Science
Abdelgahni Bellachia, D.Sc., Assistant Professor. Database systems; network computing; information systems.
Simon Berkovich, Ph.D., Professor. Information retrieval; algorithms and data structures; mathematical modeling.
Liliana Florea, Ph.D., Assistant Professor. Bioinformatics; algorithms.
Bhagirath Narahari, Ph.D., Professor. Algorithms; parallel computing; embedded systems.
Rahul Simha, Ph.D., Associate Professor. Bioinformatics; algorithms; simulation; distributed and parallel computing systems.
Jonathan Stanton, Ph.D., Assistant Professor. Distributed computing; cluster computing.
Abdou Youssef, Ph.D., Professor. Algorithms and data structures; multimedia information systems; data compression.

Columbian College of Arts and Sciences
John Balbach, Ph.D., Assistant Professor. Solid-state NMR.
Mark Reeves, Ph.D., Associate Professor. Optical microscopy.
Courtney Smith, Ph.D., Associate Professor. Molecular evolution.
Frank Turano, Ph.D., Associate Professor. Plant biology and biochemistry.
Akos Vertes, Ph.D., Professor. Mass spectrometry in life sciences.
Chen Zeng, Ph.D., Assistant Professor. Protein modeling.

Children's Hospital Laboratories
Eric Hoffman, Ph.D., Professor. Molecular basis of muscle disorders.
Christie Holland, Ph.D., Associate Professor. HIV pathogenesis.
Mary Rose, Ph.D., Associate Research Professor. Mucin genes.
Mendel Tuchman, M.D., Professor. Inborn errors of metabolism.

The Institute for Genomics Research
Najib El-Sayed, Ph.D., Assistant Professor of Microbiology and Tropical Medicine. Genomics of parasitic protozoa.
Claire M. Fraser, Ph.D., Professor of Pharmacology and President. Genomics.
Malcolm Gardner, Ph.D. Genomics of parasitic protozoa.
Elodie Ghedin, Ph.D. Parasite genomics; cell biology of trypanosomatids.
Ewen Kirkness, Ph.D., Associate Professor of Pharmacology. Neurotransmitter receptors.
Norman Lee, Ph.D. Genomics.
William Nierman, Ph.D., Professor of Biochemistry and Molecular Biology. Genomics.
Ian Paulsen, Ph.D. Membrane transporters; comparative genomics.
John Quackenbush, Ph.D., Associate Professor of Biochemistry and Molecular Biology. Bioinformatics; gene expression; mammalian microarray analysis.

Celera
Sorin Istreal, Ph.D. Bioinformatics of transcription and regulation of eukaryotic cells.

National Institutes of Health
John Brady, Ph.D., Chief, Laboratory of Molecular Virology. Molecular biology of HIV and HTLV-1.
Genoveffa Franchini, Ph.D., Chief, Laboratory of Immune Regulation. Molecular biology of human retroviruses and vaccines.
Donita Garland, Ph.D., Chief, Eye and Neurological Diseases. Proteomics of the eye and related diseases.
Kuan-The Jeang, Ph.D., Chief, Laboratory of Molecular Microbiology. Molecular biology of retroviruses; genomics of pathogens.
John Weinstein, Ph.D., Chief, Laboratory of Cancer and Drug Development. Genomics and bioinformatics of cancer.
Steve Zeichner, Ph.D., Staff Scientist, Pediatrics and Neurological Disorders. Molecular biology of retroviruses; genomics of pathogens.

Food and Drug Administration
Sufian F. Al-Khaldi, Ph.D., DNA Microarray Specialist. Genomics of cancer and microorganisms.

INDIANA UNIVERSITY–PURDUE UNIVERSITY INDIANAPOLIS

School of Medicine
Department of Medical and Molecular Genetics

Programs of Study

The Department of Medical and Molecular Genetics is one of the first, largest, and most renowned independent genetics departments in the nation. The Department offers several degrees in medical and molecular genetics, including the Doctor of Philosophy (Ph.D.), the Master of Science (M.S.), the combined M.D./Ph.D., and the M.S. with an emphasis in genetic counseling. Since its inception in 1966, the Department has trained more than 200 students. Faculty members are primarily in the Department of Medical and Molecular Genetics, and some have appointments in other School of Medicine departments, including Biochemistry, Medicine, Neurology, Oral-Facial Development, Pathology, Pediatrics, and Psychiatry, thereby providing the opportunity for a wide range of training and research opportunities in the rapidly changing field of human genetics.

The program offers course work in molecular, biochemical, cytogenetic, population, and clinical genetics. Research training of doctoral candidates is initiated in the first year through a series of laboratory rotations. Departmental course work is completed during the first two years of study, after which the student takes a preliminary examination. The student must defend a research proposal before advancing to candidacy.

Established in 1991, the Genetic Counseling Program is a two-year training program accredited by the American Board of Genetic Counseling. Students in the program obtain a unique and extensive clinical experience in more than twenty local genetic counseling clinics. The genetic counseling student takes either 36 credit hours of course work or, alternatively, 30 credit hours of course work including a thesis.

Research Facilities

The molecular genetics division faculty members and their laboratories offer opportunities for study and research on the mechanisms, diagnosis, and treatment of a wide variety of diseases. These include a number of neurogenetic, hematologic, and metabolic disorders. Most faculty laboratories are located in the modern Medical Research and Library Building and the recently completed Cancer Research Institute. The Department also has DNA diagnostic services and a lymphocyte cell bank. Departmental facilities include state-of-the-art cytogenetic laboratories with instrumentation and fluorescent microscopy for molecular cytogenetic techniques, including comparative genomic hybridization and multicolor fluorescent in situ hybridization (FISH) analysis. Data from large studies of genetic diseases, such as Huntington's disease, familial dementia syndromes, alcoholism, bipolar manic-depressive illness, osteoporosis, and primary pulmonary hypertension, are available for genetic analysis. Extensive computing systems are located within the Department for genetic mapping of complex diseases. The clinical genetics group provides genetic evaluation and counseling in medical genetics, bone dysplasia, Huntington's disease, and metabolism as well as outreach clinics. Genetic counseling students participate in more than twenty additional clinics, including prenatal diagnosis and numerous specialty clinics.

Financial Aid

Financial support, including an annual stipend of $22,000 for the 2005–06 academic year and a waiver of tuition and health insurance, is available to qualified applicants seeking the Ph.D. degree. Aid is awarded in the form of University fellowships (some of which are awarded under the Graduate Minority Program) and research assistantships. Funds are available for doctoral students to attend national scientific meetings. While assistance is offered competitively, the Department makes every possible effort to address the financial needs of its doctoral students during their period of training. Master's students often pursue financial aid and local job opportunities that are related to genetic counseling.

Cost of Study

Tuition for 2005–06 was $214.95 per credit for in-state students and $620.40 per credit for out-of-state students.

Living and Housing Costs

Off-campus facilities are readily available for students. The cost of living is below the national average. Additional information can be obtained from the Housing Department of Indiana University–Purdue University at Indianapolis via telephone at 317-274-7200 or via their Web site at http://www.housing.iupui.edu.

Student Group

The total student enrollment of Indiana University–Purdue University at Indianapolis is about 27,000, with nearly 7,000 students engaged in graduate study and another 1,500 enrolled in the medical and dental schools. The Department has approximately 45 graduate and postdoctoral students, 12 of whom are in the Genetic Counseling Program.

More than 300 students have been trained as part of the program and now hold outstanding positions in various areas of human genetics at medical schools, universities, hospitals, and research institutes and in industry. Trainees are actively involved in research and teaching and hold academic, research, and clinical positions as professors, laboratory directors, division directors, and deans. The genetic counseling graduates are working in prenatal, pediatric, cancer, and other specialty clinics and as state genetic coordinators.

Location

Indianapolis is the nation's twelfth-largest city, with more than 1 million residents. The city's cultural activities include programs supported by the Indianapolis Museum of Art, Eiteljorg Museum of Native American and Western Art, and the Indianapolis Symphony Orchestra. A diverse year-round program of theater, ballet, opera, and popular entertainment is presented at the Indiana Repertory Theatre, Clowes Hall, Conseco Fieldhouse, the Indianapolis Convention Center, and the RCA Dome. The city has many activities for children, including the world's largest children's museum and the Indianapolis Zoo.

The Department and The Medical Center

The Department of Medical and Molecular Genetics is an integral part of the Indiana University Medical Center, with faculty members drawn from many departments. The Medical Center is one of the nation's largest health-care centers. Located on 85 acres, 1 mile west of downtown Indianapolis, it is home to six hospitals, including the largest intensive care children's hospital in the country, James Whitcomb Riley Hospital for Children. The main campuses of Indiana University and Purdue University are within an hour's drive of the Medical Center campus.

Applying

Applicants are required to apply electronically. The link to the electronic application can be found at the departmental Web site. Prospective doctoral and noncounseling M.S. students should file applications for admission before January 15. Prospective genetic counseling students must submit their applications by February 1.

Applicants are expected to have had courses in genetics and biochemistry, with genetic counseling applicants also completing a course in psychology. Typically, a grade point average of at least 3.0 on a 4.0 scale is required for admission. Some prior research or related experience is an advantage. Scores on the General Test of the Graduate Record Examinations (GRE) must be available for review at the time of application. A Subject Test is not required. A minimum TOEFL score of 250 (computer-based score) or 600 (paper-based score) is required for international students whose native language is not English.

Correspondence and Information

Graduate Secretary
Department of Medical and Molecular Genetics
Indiana University School of Medicine, IB 130
975 West Walnut Street
Indianapolis, Indiana 46202-5251

Phone: 317-274-2238
Fax: 317-274-2293
E-mail: medgen@iupui.edu
Web site: http://www.iupui.edu/~medgen

Indiana University–Purdue University Indianapolis

THE FACULTY AND THEIR RESEARCH

Gary D. Bellus, Assistant Professor; Ph.D., 1985; M.D., 1989. Molecular mechanisms of epidermal differentiation; molecular genetics of human skeletal dysplasias; genetic skin disorders; neurofibromatosis.

Yan Chen, Associate Professor; M.D., 1983; Ph.D., 1994. Molecular genetics; signal transduction and animal studies related to receptor serine kinases and early development; cancer genetics.

Joe C. Christian, Professor Emeritus; Ph.D., 1960; M.D., 1964. Quantitative genetics of aging; alcoholism; cardiovascular disease risk factors; twin studies; clinical genetics.

P. Michael Conneally, Distinguished Professor Emeritus; Ph.D., 1962. Population genetics; gene mapping; genetic heterogeneity; genetic studies of Huntington's, Parkinson's, and Alzheimer's diseases; gene mapping of complex diseases.

Kenneth Cornetta, Professor and Chairman; M.D., 1982. Experimental hematopoiesis; chronic myeloid leukemia; use of retroviral gene transfer (gene therapy) in hematologic malignancies.

Stephen R. Dlouhy, Associate Research Professor; Ph.D., 1980. Molecular and biochemical genetics; inherited human neurologic disease; DNA and cell banking; mouse model systems; X-chromosome proteolipid protein gene.

Tatiana Foroud, P. Michael Conneally Professor; Ph.D., 1994. Population genetics; complex disease gene mapping.

David Gilley, Assistant Professor; Ph.D., 1992. Molecular genetics; telomeres in normal and cancer cells.

Brenda Grimes, Assistant Professor; Ph.D., 1990. Molecular genetics; human artificial chromosomes; gene therapy; human centromere function.

Brittney-Shea Herbert, Assistant Professor; Ph.D., 1998. Molecular genetics; breast cancer; telomerase and senescence.

Daniel Koller, Assistant Research Professor; Ph.D., 2001. Complex disease gene mapping; genetics of osteoporosis; linkage and linkage disequilibrium analysis methods.

Sean Mooney, Assistant Professor; Ph.D., 2001. Molecular function of phenotypically annotated genomic variation; bioinformatics.

Nuria Morral, Assistant Professor; Ph.D., 1993. Molecular genetics; diabetes.

Catherine G. Palmer, Professor Emeritus; Ph.D., 1953. Cytogenetics; cytogenetics of leukemia; clinical cytogenetics; prenatal diagnosis; molecular cytogenetics.

Kimberly A. Quaid, Professor; Ph.D., 1986. Psychosocial assessment of genetic technology; presymptomatic testing for genetic disorders.

Terry E. Reed, Professor; Ph.D., 1971; M.P.H., 1982. Dermatoglyphics; twin studies; epidemiology.

Elliot Rosen, Associate Professor; Ph.D., 1980. Molecular genetics; gene targeting; transgenic technology.

Virginia C. Thurston, Clinical Associate Professor; Ph.D., 1997. Molecular cytogenetics; clinical cytogenetics; chromosome structure; genetics of germ cell tumors; undergraduate medical education; graduate education in genetics.

Wilfredo Torres-Martinez, Clinical Associate Professor; M.D., 1986. Clinical genetics; dysmorphology; prenatal diagnosis; teratology.

Gail H. Vance, Professor; M.D., 1980. Molecular cytogenetics of germ cell tumors and thymomas; clinical cytogenetics; cancer genetics; familial cancer syndromes: risk assessment, genetic testing, and research database.

Laurence E. Walsh, Assistant Professor; M.D., 1987. Neurogenetics; pediatric neurology.

David D. Weaver, Professor Emeritus; M.D., 1966. Clinical genetics; dysmorphology; syndromology; birth defects; inherited disorders; prenatal diagnosis; embryology; teratology.

Kenneth White, Assistant Professor; Ph.D., 1997. Molecular genetics; metabolic bone diseases.

Syed-Adeel Zaidi, Clinical Assistant Professor; M.B.B.S., 1998. Clinical management of adults with genetic disorders.

Xin Zhang, Assistant Professor; Ph.D., 1998. Molecular genetics of vertebrate development.

Genetic Counselors

Jessica Claybrook, M.S., 2005; Board-Eligible Genetic Counselor and Indiana Teratogen Information Service Coordinator. Adult and cancer genetics.

Lisa Cushman, Student Project Coordinator; Ph.D., 2001; Certified Genetic Counselor. Pediatrics; fetal alcohol syndrome.

Paula Delk, Director, Genetic Counseling Program; M.S., 1994; Certified Genetic Counselor. Genetics education; craniofacial and skeletal dysplasia syndromes.

Cindy Hunter, M.S., 1996; Certified Genetic Counselor. Cancer genetic counseling.

Ellen Kucharski, M.S., 2004; Board-Eligible Genetic Counselor. Pediatric genetic counseling; clinic management.

Susan S. Romie, M.S., 1992; Certified Genetic Counselor. Genetics education; adult genetic counseling.

Kerry White, M.S., 2002; Certified Genetic Counselor. Pediatrics; research.

Associate Research Faculty

Simon J. Conway, Associate Professor; Ph.D., 1993. Congenital heart defects; pathogenesis of conotruncal and valvular heart defects.

Mary C. Dinauer, Professor; Ph.D., 1979; M.D., 1981. Molecular genetics; inherited disorders of phagocytes; molecular biology of phagocyte oxidant production; gene transfer in hematopoietic stem cells; Rho family GTPases.

John R. Duguid, Associate Professor; Ph.D., 1975; M.D., 1981. Neurology.

Michael J. Econs, Professor; M.D., 1983. Genetic aspects of metabolic bone disease.

Howard J. Edenberg, Chancellor's Professor; Ph.D., 1973. Genomics and bioinformatics; transcriptome analysis; regulation of gene expression; genetics of complex diseases; molecular biology of alcohol metabolism and alcoholism.

Luis F. Escobar, Assistant Professor; M.D., 1984. Clinical genetics.

Anthony B. Firulli, Associate Professor; Ph.D., 1993. Transcription bHLH proteins; cardiac development.

David A. Flockhart, Professor; Ph.D., 1976; M.D., 1987. Pharmacogenetics.

Lawrence M. Gelbert, Adjunct Professor; Ph.D., 1989. Pharmacogenomics and drug discovery; analysis of drug mechanism of action using microarray technology; analysis of genetic variation affecting therapeutic response and disease susceptibility; cancer genetics.

Bernardino Ghetti, Distinguished Professor; M.D., 1966. Degenerative disease of the nervous system in humans and animals; Alzheimer's disease; tauopathies; prion diseases; neurologic mutant mice.

Alan M. Golichowski, Professor; M.D., 1974; Ph.D., 1976. Obstetrics and gynecology.

Bryan E. Hainline, Clinical Associate Professor of Pediatrics, Assistant Professor of Medical and Molecular Genetics, and Director, Molecular Genetics Services and DNA Diagnostic Laboratory; M.D./Ph.D., 1981. Molecular genetics; inherited disorders of mitochondrial metabolism and biogenesis; molecular biology of fatty acid metabolism; trace-metal and cofactor metabolism; treatment of Prader-Willi syndrome.

James K. Hartsfield Jr., Professor; D.M.D., 1981; Ph.D., 1993. Clinical genetics; dysmorphology; the interaction of genetic and environmental factors that influence nonsyndromic craniofacial growth and development and the response to orthodontic treatment.

Robert Konrad, Adjunct Assistant Professor; M.D., 1991. Pancreatic beta cells; insulin secretion; type 2 diabetes; O-linked protein glycosylation.

Debomoy K. Lahiri, Professor; Ph.D., 1980. Molecular neurobiology, genetics, and biochemistry; mechanisms of aging and neuroprotection; cellular and molecular biological studies of Alzheimer's disease; gene regulation and expression of ApoE, beta-amyloid precursor protein, and beta-secretase; biochemical and genetic studies of autism and bipolar affective disorder.

Sheila P. Little, Adjunct Professor; Ph.D., 1978. Neurodegenerative disorders; Alzheimer's disease; bioinformatics.

Marc S. Mendonca, Associate Professor and Director; Ph.D., 1987. Radiation-induced carcinogenesis and tumor suppression gene deletion; characterization of a novel HeLa/cervical cancer tumor suppressor gene on 11q13; endothelial stem cell radiobiology; radiosensitization and noninvasive imaging of prostate, breast, and ovarian tumors.

Eric M. Meslin, Professor; Ph.D., 1989. Bioethics; health policy; research ethics.

Jill R. Murrell, Assistant Professor; Ph.D., 1992. Molecular genetics of neurodegenerative diseases, genetic risk factors for Alzheimer's disease, and tauopathies.

John I. Nurnberger Jr., Joyce and Iver Small Professor of Psychiatry and Professor of Medical and Molecular Genetics and Medical Neurobiology; M.D., 1975; Ph.D., 1983. Psychiatric genetics; affective disorders; addictive disorders; genetics of complex disease.

Lillie-Mae Padilla, Associate Professor; M.D., 1968. Prenatal diagnosis.

William H. Schneider, Adjunct Professor; Ph.D., 1976. History of human genetics and eugenics; social history of science.

Weinian Shou, Associate Professor; Ph.D., 1991. Molecular mechanisms of mammalian development and congenital birth defects.

Rebecca S. Wappner, Professor; M.D., 1970. Inborn errors of metabolism; newborn screening.

Ronald C. Wek, Professor; Ph.D., 1987. Cellular stress responses and disease; mechanisms controlling translation in response to diverse stress conditions.

Peter C. M. Young, Associate Professor Emeritus; Ph.D., 1967. Biochemistry of steroid hormones.

IOWA STATE UNIVERSITY

Genetics Program

Program of Study

The interdepartmental genetics program offers vigorous training in genetics leading to a Ph.D. or M.S. degree in one of thirteen departments: Agronomy; Animal Science; Biochemistry, Biophysics, and Molecular Biology; Biomedical Sciences; Ecology, Evolution, and Organismal Biology; Entomology; Food Science and Human Nutrition; Genetics, Development and Cell Biology; Horticulture; Natural Resource Ecology and Management; Plant Pathology; Veterinary Microbiology and Preventive Medicine; and Veterinary Pathology. The curriculum provides a thorough grounding in fundamental genetic principles, allows flexibility in the choice of research area, and emphasizes interdisciplinary research interactions.

Students are required to complete core courses in molecular genetics; transmission genetics; quantitative, population, or evolutionary genetics; and biochemistry. Students participate in an annual workshop on a current topic in genetics research (featuring internationally renowned researchers), attend the annual retreat, and participate in seminars. Additional courses in genetics and the life sciences may be selected to meet the specific interests of each student.

Research is the cornerstone of the training in genetics program. Faculty laboratories provide opportunities for research in many areas: regulation of gene expression; plant and animal genome mapping; developmental genetics; DNA replication; host-parasite interactions; genetic engineering of plants, animals, and microbes; molecular genetics; transposable elements; computational molecular biology; evolutionary genetics; population genetics; quantitative genetics; and selection theory. To choose a research program, major professor, and home department, genetics research assistants rotate through three faculty laboratories during their first year.

Research Facilities

Excellent facilities are available to support the graduate research program. The laboratories of program faculty members are located in a variety of modern, well-equipped buildings. On-campus instrumentation centers (protein chemistry, animal gene transfer, microscopy, plant transformation, nucleic acids, flow cytometry, and macromolecular structure determination) provide specialized expertise critical for molecular genetics. Computers are readily accessible. The University library maintains extensive collections of journals in all biological sciences, including online journals.

Financial Aid

All applicants are considered for nomination for four-year $25,000 Iowa State University (ISU) Plant Sciences Institute Fellowships, available for all areas of plant research. Applicants are considered for nomination for ISU Biotechnology Fellowships, interdepartmental genetics research assistantships, and NSF and USDA Fellowships. All domestic applicants are considered for need-based Miller Fellowships.

Cost of Study

Beginning fall 2006, Ph.D. students who are on half-time assistantships and maintain a 3.0 GPA and progress well toward their degree receive full tuition scholarships. M.S. students who are on half-time assistantships and maintain a 3.0 GPA and progress well toward their degree will be responsible for 50 percent of in-state tuition. Computer, activity and service fees, health facility, and health fees are assessed each semester. Information on tuition and fees is available at http://www.iastate.edu/~registrar/fees/. All students on assistantships receive paid single health insurance.

Living and Housing Costs

For on-campus housing, Fredriksen Court offers the best of both worlds: the freedom that comes with independent apartment living, and the convenience of living on campus. Thirty-two apartments in University Village are reserved for single graduate students and other single students who are at least 23 years of age. Additional information and costs can be found at http://www.housing.iastate.edu. For off-campus housing information in the Ames area, students should visit http://www.ames.ia.us/ and click the Relocation Information option. Upper-division Buchanan Resident Hall offers single or double suites, nearby parking, is close to campus, and offers a mature environment. CyRide, the city of Ames bus service, is free of charge for all ISU students.

Student Group

The current enrollment at ISU is 26,380 students, including 4,618 graduate students. There are 92 genetics majors. Forty percent are domestic; 60 percent are international; 52 percent are women; 6 percent are underrepresented; 67 percent are Ph.D. students, and 33 percent are M.S. students. During the 2004–05 school year, 25 genetics graduate students completed their training. For the last five years, the Genetics Program has recruited about 25 new graduate students each year.

Student Outcomes

Most students awarded doctoral degrees continue their training as postdoctoral associates at major research institutions in the United States or abroad in preparation for research and/or teaching positions in academia, industry, or government. A few go directly to permanent research positions in industry. Many students awarded master's degrees continue their training as doctoral students; however, some choose research support positions in academia, industry, or government. A complete list of outcomes is available at the Web site listed in the Correspondence and Information section.

Location

Ames is a city of 50,000 residents, including the student body of Iowa State University. The University community is located in a pleasant rural area 30 miles north of Des Moines, the capital of Iowa. An active cultural life is provided in Ames by local musical and theatrical groups and by the many international artists who perform regularly. Nearby Des Moines has a variety of activities to offer, and Minneapolis or Kansas City can be reached easily by a few hours' drive on interstate highways.

The University

Iowa State University was founded more than 130 years ago as Iowa State College, one of the first land-grant institutions in the nation. Besides the Graduate College, the University has six undergraduate colleges: Agriculture, Design, Engineering, Human Sciences, Liberal Arts and Sciences, and Veterinary Medicine. The scenic campus offers a good environment for the varied activities of those who work and study at the University. Additional information is available on the Internet at http://www.iastate.edu and through the ISU Web cam at http://www.iastate.edu/webcam/.

Applying

Applications for admission are accepted year-round. For the best funding opportunities for fall, all application materials should arrive by January 31; for spring, by September 1. Applicants may use University forms requested through the e-mail address listed in the Correspondence and Information section or may send the following materials to the address listed in the Correspondence and Information section: resume; statement of research interests and experience; transcripts; GRE scores; TOEFL scores, if applicable; and three letters of recommendation on letterhead, addressing academic and laboratory skills.

Correspondence and Information

Dr. Patrick Schnable, Chair
Interdepartmental Genetics
2102 Molecular Biology Building
Iowa State University
Ames, Iowa 50011-3260
Phone: 515-294-7697
     800-499-1972 (toll-free)
Fax: 515-294-6669
E-mail: genetics@iastate.edu
Web site: http://www.genetics.iastate.edu

Iowa State University

THE FACULTY AND THEIR RESEARCH

Senior faculty members associated with the Genetics Program are listed. Departmental affiliation (and concurrent affiliation where noted) appears in parentheses following each name.

Linda Ambrosio (Genetics, Development and Cell Biology), Ph.D., Princeton, 1985. Developmental and molecular genetics of *Drosophila;* signal transduction by D-*raf;* cell determination.

Diane Bassham (Genetics, Development and Cell Biology), Ph.D., Warwick (England), 1994. Protein trafficking in plants; plant vacuole biogenesis and function; environmental stress responses.

Thomas J. Baum (Plant Pathology), Ph.D., Clemson, 1993. Molecular analysis of the interactions between cyst nematodes and their hosts; signal transduction during nematode infection.

Gwyn A. Beattie (Plant Pathology), Ph.D., Wisconsin, 1991. Molecular ecology, genetics, and physiology of plant-associated bacteria; molecular mechanisms of plant-bacterial interactions.

Philip W. Becraft (Agronomy; Genetics, Development and Cell Biology), Ph.D., Berkeley, 1992. Molecular genetics of plant development; cell interactions; signal transduction.

Jeffrey Beetham (Veterinary Pathology; Entomology), Ph.D., California, Davis, 1994. Molecular parasitology; vector-parasite-host interactions; gene expression in protozoan parasite development.

P. Jeffrey Berger (Animal Science), Ph.D., Ohio State, 1970. Statistical approaches for prediction of genetic merit in animals.

Madan Bhattacharyya (Agronomy), Ph.D., Western Ontario, 1987. Recognition and signal transduction events in soybean–*Phytophthora sojae* interaction and phosphoinositide signal transduction pathway in *Arabidopsis.*

Diane Birt (Food Science and Human Nutrition), Ph.D., Purdue, 1975. Identification and study of cancer prevention by novel dietary constituents and cancer enhancement by overeating and obesity.

Tom Bobik (Biochemistry, Biophysics, and Molecular Biology), Ph.D., Illinois at Urbana-Champaign, 1990. Genetics and biochemistry of vitamin B12.

Adam Bogdanove (Plant Pathology), Ph.D., Cornell. Molecular mechanisms of bacterial plant pathogenesis and disease resistance.

Bryony C. Bonning (Entomology), Ph.D., London, 1989. Virus-based strategies for management of insect pests; molecular and physiological analysis of insect-virus interactions.

Volker Brendel (Genetics, Development and Cell Biology), Ph.D., Weizmann (Israel), 1986. Computational molecular biology; genome informatics.

Anne M. Bronikowski (Ecology, Evolution and Organismal Biology), Ph.D., Chicago, 1997. Genetics of aging; evolutionary genetics; adaptation of survival and reproductive schedules; mathematical models of population dynamics.

E. Charles Brummer (Agronomy), Ph.D., Georgia, 1993. Breeding and genetics of alfalfa and other forage crops.

Hui-Hsien Chou (Genetics, Development and Cell Biology; Computer Science), Ph.D., Maryland, College Park, 1996. Bioinformatics; computational biology; cellular automata; self-organization phenomena.

Clark R. Coffman (Genetics, Development and Cell Biology), Ph.D., California, San Diego, 1993. Cell migration and programmed cell death; developmental genetics of *Drosophila* germ cells.

Gloria Culver (Biochemistry, Biophysics, and Molecular Biology), Ph.D., Rochester, 1994. Ribosome assembly; RNA-protein recognition; RNA folding.

Jack C. Dekkers (Animal Science), Ph.D., Wisconsin, 1989. Integration of molecular and quantitative genetics for genetic improvement of livestock.

Drena L. Dobbs (Genetics, Development and Cell Biology), Ph.D., Oregon, 1983. Computational molecular biology; bioinformatics; macromolecular structure–function.

N. Matthew Ellinwood (Animal Science), D.V.M., 1997, Ph.D., 2000, Colorado State. Medical and molecular genetics; animal models of human disease; inborn errors of metabolism; neuropathic lysosomal storage diseases; gene therapy.

Rohan L. Fernando (Animal Science), Ph.D., Illinois at Urbana-Champaign, 1984. Population and quantitative genetics; statistical methods for mapping QTL; marker-assisted selection; genetic evaluation in crossbred populations.

Clark F. Ford (Food Science and Human Nutrition), Ph.D., Iowa, 1981. Molecular genetic analysis of enzyme structure and function; engineering of industrial proteins.

Jack R. Girton (Biochemistry, Biophysics, and Molecular Biology), Ph.D., Alberta, 1979. Developmental genetics of *Drosophila melanogaster,* genetics of neural development and programmed cell death.

Xun Gu (Agronomy; Genetics, Development and Cell Biology), Ph.D., Texas–Houston Health Science Center, 1996. Computational molecular biology; molecular evolution and comparative genomics; functional divergence of gene families; gene network regulation.

Richard B. Hall (Natural Resource Ecology and Management), Ph.D., Wisconsin, 1974. Genetics of biomass production; manipulations of phytochrome response; pest resistance in trees; genecology of natural and managed systems.

Larry J. Halverson (Agronomy), Ph.D., Wisconsin, 1991. Molecular genetics, physiology, and ecology of soil bacteria; water stress tolerance; biodegradation of pollutants.

David Hannapel (Horticulture), Ph.D., Purdue, 1985. Regulation of gene expression during plant growth and development.

Thomas C. Harrington (Plant Pathology), Ph.D., Berkeley, 1983. Fungal genetics, systematics, and evolution; forest pathology.

Eric R. Henderson (Genetics, Development and Cell Biology), Ph.D., UCLA, 1984. Telomere molecular biology; AFM; nanotechnology.

Martha James (Biochemistry, Biophysics, and Molecular Biology), Ph.D., Iowa State, 1989. Genetic, molecular, and biochemical analysis of starch synthesis in higher plants.

Jean-Luc Jannink (Agronomy), Ph.D., Minnesota, 1999. Quantitative and ecological genetics of small grain improvement; statistical analysis of DNA marker information in selection.

Fredric J. Janzen (Ecology, Evolution, and Organismal Biology), Ph.D., Chicago, 1992. Evolutionary genetics of animals; molecular evolution; quantitative genetics; theoretical and empirical studies of selection.

Kristen M. Johansen (Biochemistry, Biophysics, and Molecular Biology), Ph.D., Yale, 1989. Nuclear architectural remodeling; chromatin structure; spindle matrix.

Anumantha Kanthasamy (Biomedical Sciences), Ph.D., Madras (India), 1989. Cellular and molecular mechanisms of neurodegenerative processes in Parkinson's disease; environmental risk factors of neurodegenerative diseases; oxidative stress and apoptotic signaling; transgenic animals.

Susan J. Lamont (Animal Science), Ph.D., Illinois Medical Center, 1980. Immunogenetics, marker-assisted selection, molecular genetics of poultry.

Coralie Lashbrook (Horticulture), Ph.D., California, Davis, 1995. Molecular biology and functional genomics of abscission; ethylene and auxin control of cell separation; modulation of cell separation for crop improvement.

Dennis Lavrov (Ecology, Evolution and Organismal Biology), Ph.D., Michigan, 2000. Animal evolution, phylogenetics, and mitochondrial genomics; use of gene order data for the analysis of ancient relationships; comparative and population mitochondrial genomics of lower (nonbilaterian) animals.

Gustavo MacIntosh (Biochemistry, Biophysics, and Molecular Biology), Ph.D., Buenos Aires, 1977. Genetics and proteomics of secreted ribonucleases in plants; functional characterization of plant ribonucleases in relation to defense, development, and stress; signal transduction during wounding and stress responses.

W. Allen Miller (Plant Pathology; Biochemistry, Biophysics, and Molecular Biology), Ph.D., Wisconsin, 1984. Translation mechanisms; RNA plant virus replication and genomics; translation mechanisms; RNA plant virus replication and genomics; virus-aphid interactions.

F. Chris Minion (Veterinary Microbiology and Preventive Medicine), Ph.D., Alabama at Birmingham, 1983. Functional genomics and virulence mechanisms of microbial pathogens.

Diane Moody (Animal Science), Ph.D., Nebraska, 1998. Genetic improvement of dairy cattle; genetic regulation of osteoporosis in chickens.

John Nason (Ecology, Evolution, and Organismal Biology), Ph.D., California, Riverside, 1991. Analyses of habitat fragmentation effects on and coevolution of plants and insects.

Basil J. Nikolau (Biochemistry, Biophysics, and Molecular Biology), Ph.D., Massey (New Zealand), 1981. Metabolomics; functional genomics of metabolism; biochemical genetics; metabolic engineering.

Marit Nilsen-Hamilton (Biochemistry, Biophysics, and Molecular Biology), Ph.D., Cornell, 1973. Regulation of gene expression; growth factors; cancer; regulated aptamers for imaging gene expression and detection/treatment of cancer.

David J. Oliver (Genetics, Development and Cell Biology), Ph.D., Cornell, 1975. Control of gene expression by light and environmental factors; mitochondrial function; signal transduction in plants; heavy metals in plants.

Reuben Peters (Biochemistry, Biophysics, and Molecular Biology), Ph.D., California, San Francisco, 1998. Enzymatic mechanisms/engineering and metabolic pathway identification/engineering of plant terpenoid natural products.

Thomas Peterson (Genetics, Development and Cell Biology; Agronomy), Ph.D., California, Santa Barbara, 1984. Maize molecular genetics; gene regulation; effects of transposable elements on gene expression and recombination.

Gregory J. Phillips (Veterinary Microbiology and Preventative Medicine), Ph.D., Georgia, 1987. Bacterial genetics; genetic analysis of protein export and membrane protein insertion in bacteria.

Jo Anne Powell-Coffman (Genetics, Development and Cell Biology), Ph.D., California, San Diego, 1993. *C. elegans* development; aryl hydrocarbon receptor function and regulation; hypoxia signaling and response.

Stephen Proulx (Ecology, Evolution, and Organismal Biology), Ph.D., Utah, 2002. Evolution of genetic networks and application of mathematical and computational tools to these problems; theoretical and experimental genetics.

James Reecy (Animal Science), Ph.D., Purdue, 1995. Gene expression, mammalian growth and development, and molecular genetics of beef cattle.

Steven R. Rodermel (Genetics, Development and Cell Biology), Ph.D., Harvard, 1986. Nuclear-chloroplast interactions; regulatory mechanisms of photosynthesis.

Max F. Rothschild (Animal Science), Ph.D., Cornell, 1978. Gene identification; marker-assisted selection; structural and functional genomics in pigs, dogs, and shrimp.

Patrick S. Schnable (Agronomy), Ph.D., Iowa State, 1986. Plant genomics, molecular biology, and genetics.

Paul Scott (Agronomy), Ph.D., Purdue, 1992. Plant biochemistry; molecular genetics, crop improvement, and grain quality.

Randy C. Shoemaker (Agronomy; USDA-ARS), Ph.D., Iowa State, 1984. Genome mapping, organization, evolution and soybean gene regulation.

Martin H. Spalding (Genetics, Development and Cell Biology), Ph.D., Wisconsin, 1979. Metabolic regulation of gene expression in plants and algae.

Chad Stahl (Animal Science), Ph.D., Cornell, 2001. Recombinant protein expression; nutrient-gene interactions; nutritional biochemistry.

Christopher K. Tuggle (Animal Science), Ph.D., Minnesota, 1986. Gene expression in mammalian development; comparative mapping in mammals; structural and functional genomics; transgenic models.

Nicole Valenzuela (Ecology, Evolution and Organismal Biology), Ph.D., SUNY at Stony Brook, 1999. Sex determining mechanisms in vertebrates; gene discovery of factors involved in sex determination/differentiation cascade of reptiles; ecological genomics, and comparative genome evolution.

Erik Vollbrecht (Genetics, Development and Cell Biology), Ph.D., Berkeley, 1997. Molecular genetics and evolution of plant development; functional genomics; maize genetics; plant morphology.

Daniel F. Voytas (Genetics, Development and Cell Biology), Ph.D., Harvard, 1989. Retrotransposable elements of *Arabidopsis* and yeast.

Kan Wang (Agronomy), Ph.D., Ghent (Belgium), 1987. Efficient transformation systems for corn and soybeans; corn-based edible vaccines; molecular mechanisms in plant responses to environmental stress.

Jonathan Wendel (Ecology, Evolution, and Organismal Biology), Ph.D., North Carolina, 1983. Evolutionary genetics of plants; molecular evolution; evolution of crop plants.

Steven A. Whitham (Plant Pathology), Ph.D., Berkeley, 1995. Genetics and molecular genetics of plant-virus interactions; signaling in plant defense and stress responses.

Roger P. Wise (Plant Pathology, USDA-ARS), Ph.D., Michigan State, 1983. Plant genomics; molecular signaling in cereal-host/fungal-pathogen interactions; maize nuclear-mitochondrial interactions.

Eve Syrkin Wurtele (Genetics, Development and Cell Biology), Ph.D., UCLA, 1980. Molecular interactions between development and metabolism; biotin enzymes; acetyl-CoA; computational analysis of metabolic networks.

Yanhai Yin (Genetics, Development and Cell Biology), Ph.D., Scripps Research Institute, 1997. Molecular mechanisms of plant steroid hormone signal transduction in genetic model system *Arabidopsis.*

Qijing Zhang (Veterinary Microbiology and Preventive Medicine), Ph.D., Iowa State, 1994. Molecular biology of foodborne bacterial pathogens; genetic basis of antibiotic resistance; molecular mechanisms of pathogen-host interaction.

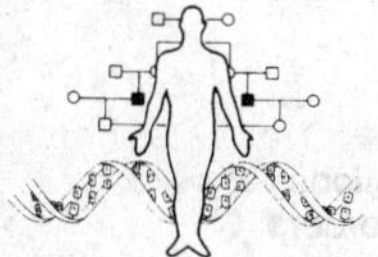

THE JOHNS HOPKINS UNIVERSITY

Predoctoral Training Program in Human Genetics and Molecular Biology

Program of Study	The University-wide Committee on Human Genetics, under the aegis of the Institute of Genetic Medicine, offers an interdivisional program, based in the School of Medicine, to train Ph.D. candidates for research careers in human genetics and molecular biology.

The Predoctoral Training Program in Human Genetics and Molecular Biology has been designed as an alternative to the combined M.D./Ph.D. program for those who are primarily interested in research and who want knowledge of human biology along with in-depth research training within a period of five years. This program is predicated on the belief that research progress is enhanced by detailed knowledge of the experimental organism. Therefore, the program offers a solid background in human biology as well as rigorous training in research strategies and techniques of molecular biology and genetics. Required courses include molecular and cell biology, biochemistry, human pathology, and pathophysiology, which provide insight into the human phenotype, as well as comprehensive courses in the genetics of humans and model organisms.

The cornerstone of the program is the thesis project. Research training is initiated as the student enters the program by a series of laboratory rotations. The faculty members come from clinical as well as basic science departments, have a wide range of research interests pertinent to human genetics, and carry out well-supported research programs. Students choose from a variety of projects, ranging from studies of molecular mechanisms of disease, genetic control of cell differentiation, and molecular cytogenetics to investigations into the pathogenesis of inborn errors of metabolism and genetic factors that predispose to disease. A wide choice of elective courses in diverse areas of genetics and molecular biology provides the means to achieve individual career goals. Students gain valuable experience as teaching assistants in such undergraduate courses as genetics and molecular biology. Supporting activities include journal clubs, the short summer course in mammalian genetics at the Jackson Laboratory, faculty and student research conferences, and seminars by visiting scientists.

The combination of research training in molecular biology and genetics with knowledge of human biology qualifies graduates to carry out sophisticated genetic studies of important biological problems relevant to genetic disease and to teach human biology as well as human genetics.

Research Facilities

Research laboratories are well equipped to carry out sophisticated research in all areas of genetics. The proximity of the renowned clinical facilities of the Johns Hopkins Hospital and Oncology Center provides access to a wealth of material for study. Computer and library facilities are excellent. Because the program in human genetics is a University-wide activity, supporting facilities are extensive.

Financial Aid

The program is supported by a limited number of teaching assistantships and predoctoral training funds from the National Institutes of Health. These fellowships, which are restricted to U.S. citizens and permanent U.S. residents, cover tuition, provide monthly stipends, and are awarded to essentially all students in the program. Students are encouraged, however, to apply for fellowships from outside sources (e.g., the National Science Foundation and Howard Hughes Medical Institute) before entering the program.

Cost of Study

Tuition ($29,500 in 2004–05) for students accepted into the program is provided by fellowship. Each student is charged a one-time matriculation fee of $660 upon arrival, and health insurance is required.

Living and Housing Costs

Single rooms and suites are available in the School of Medicine at costs ranging from $300 to $350 per month. Many students prefer to live in housing available near the Homewood (main) campus, and there is convenient, free transportation from that campus to the medical institutions. Information on monthly rates at University and local residences can be obtained from the Housing Office, Wolman Hall, Charles and 34th Streets, Baltimore, Maryland 21218.

Student Group

Students in the program interact with graduate students in the Biochemistry, Cell and Molecular Biology Program and in the many other graduate programs on the School of Medicine campus, as well as medical students and postdoctoral fellows in the laboratories.

Location

Baltimore is a vibrant port city with approximately 2.5 million inhabitants in the metropolitan area, which has been revitalized by extensive urban renewal. It is a center for culture, with outstanding museums, galleries, and restaurants of all types. There is a resident theater company as well as several experimental theaters, and the celebrated Baltimore Symphony and other musical groups perform frequently. In addition, Baltimore is an hour from Washington's museums and the Kennedy Center for the Performing Arts. The scenic Maryland countryside and nearby Chesapeake Bay provide many opportunities for recreational activities.

The University

Founded in 1876, the University has always been a leader in graduate education and biomedical research. Faculty members of the School of Medicine, the Department of Biology, the Carnegie Institution of Washington (Department of Embryology), and the School of Hygiene and Public Health provide aggregate strength in human genetics.

Applying

Applicants should have competence in genetics, chemistry (including organic), mathematics, and physics and should have some research experience. Early application is encouraged to ensure that all application materials are received by January 10. All materials should be addressed to the Predoctoral Training Program in Human Genetics and Molecular Biology. Credentials must include scores from the Graduate Record Examinations (to ensure that the program receives their GRE scores, applicants should specify the code for human genetics, 5316, on the Educational Testing Service's score report form). All applicants are notified of the admissions decision by April 15.

Correspondence and Information

Requests for information and applications should specify the Predoctoral Training Program in Human Genetics and Molecular Biology and be sent to the director of the program.

Dr. David Valle, Director
Predoctoral Training Program in Human Genetics and Molecular Biology
BRB 515
The Johns Hopkins University School of Medicine
733 North Broadway
Baltimore, Maryland 21205
E-mail: muscelli@jhmi.edu
Web site: http://www.hopkinsmedicine.org/humangenetics

The Johns Hopkins University

THE FACULTY AND THEIR RESEARCH

Ruben Adler, M.D., Professor of Ophthalmology. Cellular and molecular factors of retinal photoreceptors and neurons; retinal degenerations.
Dimitrios Avramopoulos, M.D., Ph.D., Assistant Professor of Psychiatry. Genetic basis of complex disorders, particularly psychiatric disorders.
Joel Bader, Ph.D., Assistant Professor of Biomedical Engineering. Computational biology and bioinformatics; statistical genetics.
Stephen B. Baylin, M.D., Professor of Oncology. Molecular determinants of tumor progression.
Michael Beer, Ph.D., Assistant Professor of Biomedical Engineering. Computational molecular biology and genomics.
Seth Blackshaw, Ph.D., Assistant Professor of Neuroscience. Developmental neurobiology.
Jef D. Boeke, Ph.D., Professor of Molecular Biology and Genetics. Transposition and gene silencing, mechanisms and regulation.
Simeon Boyadjiev, M.D., Assistant Professor of Pediatrics, Institute of Genetic Medicine. Clinical and molecular dysmorphology.
Nancy Braverman, M.D., Assistant Professor of Pediatrics, Institute of Genetic Medicine. Molecular biology and pathophysiology of peroxisome assembly disease.
Aravinda Chakravarti, Ph.D., Professor and Director, Institute of Genetic Medicine. Genetics of common, multifactorial disease.
Linzhao Cheng, Ph.D., Associate Professor of Gynecology/Obstetrics. Molecular genetics of human stem cells.
Janice Clements, Ph.D., Professor of Comparative Medicine. Molecular basis of lentivirus disease: HIV, SIV, and visna virus.
Ronald Cohn, M.D., Assistant Professor of Pediatrics. Molecular mechanisms of muscle regeneration.
Pierre A. Coulombe, Ph.D., Professor of Biological Chemistry. Cellular and molecular biology of wound healing in complex epithelia.
David Cutler, Ph.D., Assistant Professor, Institute of Genetic Medicine. Computational biology; genomics; evolution.
Garry R. Cutting, M.D., Professor of Pediatrics, Institute of Genetic Medicine. Molecular genetics of cystic fibrosis and ion channels.
Chi Dang, M.D., Ph.D., Professor of Medicine. Role of oncogenes in cell proliferation, neoplasia, and programmed cell death.
Hal Dietz, M.D., Professor of Pediatrics, Institute of Genetic Medicine. Molecular biology of heart development and of heritable disorders affecting the heart; influence of premature termination codons on RNA splicing and processing.
Andrew P. Feinberg, M.D., M.P.H., Professor of Medicine. Genomic imprinting in normal development and disease; genetics of childhood cancer; novel genomic technology.
Shannon Fisher, M.D., Ph.D., Assistant Professor, Institute of Genetic Medicine. Study of skeletal development, using the zebrafish as a model system.
John D. Gearhart, Ph.D., Professor of Physiology, Institute of Genetic Medicine. Mammalian developmental genetics; human stem cells.
Gregory Germino, M.D., Professor of Medicine. Molecular basis of renal cyst disease; renal tubular morphogenesis.
Stephen J. Gould, Ph.D., Professor of Cell Biology and Anatomy. Identification of genetic basis of peroxisome assembly disorders.
William B. Guggino, Ph.D., Professor of Physiology. Gene therapy of cystic fibrosis; chloride and water channels.
Ethylin W. Jabs, M.D., Professor of Pediatrics, Institute of Genetic Medicine. Genetics of craniofacial development; characterization of human centromeric regions; repetitive DNA; molecular cytogenetics; gene mapping.
Nicholas Katsanis, Ph.D., Associate Professor of Genetic Medicine, Institute of Genetic Medicine. Identification of responsible genes and interaction of the protein products to produce the pleiotrophic pathophysiology.
Scott Kern, M.D., Associate Professor of Pathology. Genetic alterations in pancreatic carcinoma and colorectal dysplasia.
Gary Ketner, Ph.D., Professor of Immunology. Virology; use of viruses for gene delivery.
Kenneth Kinzler, Ph.D., Professor of Oncology. Molecular genetics of cancer.
Lori Kotch, Ph.D., Assistant Professor of Pediatrics, Institute of Genetic Medicine. Cellular and molecular determinants for alcohol-induced birth defects.
Se-Jin Lee, M.D., Ph.D., Professor of Molecular Biology and Genetics. Growth and differentiation in mammalian development.
Nick Marsh-Armstrong, Ph.D., Assistant Professor of Neuroscience. Transgenic studies of CNS developmental gene regulation.
Debra Mathews, Ph.D., Research Scientist, Phoebe R. Berman Bioethics Institute. Intersection of bench science, public policy, and society.
Andrew McCallion, Ph.D., Assistant Professor of Comparative Medicine, Institute of Genetic Medicine. Functional genomics; complex disease; gene regulation.
Joshua Mendell, M.D., Ph.D., Assistant Professor, Institute of Genetic Medicine. Posttranscriptional regulation of gene expression.
Susan Michaelis, Ph.D., Professor of Cell Biology and Anatomy. Molecular mechanisms of signal transduction; protein targeting and secretion; yeast molecular genetics.
Akhilesh Pandey, M.D., Ph.D., Assistant Professor, Institute of Genetic Medicine. Phosphorylation and signal transduction; proteomics; mass spectrometry; bioinformatics.
Jonathan Pevsner, Ph.D., Associate Professor of Neurology. Molecular basis of neurological disorders; bioinformatics.
Randall Reed, Ph.D., Professor of Molecular Biology and Genetics. Molecular mechanisms of neuronal development and signal transduction.
Roger H. Reeves, Ph.D., Professor of Physiology, Institute of Genetic Medicine. Genetic and physical mapping; developmental consequences of aneuploidy/Down syndrome; manipulation of yeast artificial chromosomes; transgenic and knockout mice.
David Schlessinger, M.D., National Institute of Aging. Aging; development; X-chromosome genetics.
Alan F. Scott, Ph.D., Associate Professor of Medicine, Institute of Genetic Medicine. Studies of human transposable elements; mechanisms of genomic evolution; methods for DNA sequencing and analysis.
Gregg L. Semenza, M.D., Ph.D., Professor of Pediatrics, Institute of Genetic Medicine. Mechanisms of transcription regulation involved in the maintenance of cellular and systemic oxygen homeostasis; analysis of transgenic mice, transfected cells, and cloned transcription factors.
Geraldine Seydoux, Ph.D., Professor of Molecular Biology and Genetics. Molecular genetics of germ cell development.
Kirby D. Smith, Ph.D., Professor of Pediatrics, Institute of Genetic Medicine. Chromosome DNA organization; genomic evolution; regulation of gene function during development.
Hongjun Song, Ph.D, Assistant Professor of Neurology. Biology of neural stem cells and neurogenesis in the adult CNS.
Forrest Spencer, Ph.D., Associate Professor of Medicine, Institute of Genetic Medicine. Centromere function and cell cycle control in yeast; molecular genetics of chromosome segregation; yeast artificial chromosomes.
Sara Sukumar, Ph.D., Professor of Oncology. Breast cancer.
Catherine Thompson, Ph.D., Assistant Professor of Neurology. Molecular mechanisms of CNS development.
David Valle, M.D., Professor of Pediatrics, Institute of Genetic Medicine. Clinical biochemical and molecular studies of inborn errors of amino acid metabolism; disorders of peroxisome biogenesis; photoreceptor biology and inherited retinal degenerations.
Victor Velculescu, M.D., Ph.D., Assistant Professor of Oncology. Molecular genetics of human cancer.
Bert Vogelstein, M.D., Professor of Oncology. Molecular genetics of human cancer.
Tao Wang, M.D., Ph.D., Assistant Professor of Pediatrics, Institute of Genetic Medicine. X-linked mental retardation.
Don Zack, M.D., Ph.D., Professor of Ophthalmology. Molecular approaches to the study of retinal development and function.

NORTH CAROLINA STATE UNIVERSITY

Graduate Program in Genomic Sciences

Programs of Study

Genomic sciences is a dynamic field based on new technologies that facilitate the deciphering and analysis of the genome of organisms. Advances in genomic sciences are expected to lead to improvements in agriculture, health, and the environment. The Genomic Sciences Graduate Program at North Carolina State University (NCSU) is designed to provide the academic foundation in genomic sciences for the next generation of scientists and provides a strong and well-balanced graduate training program in two degree tracks: bioinformatics and functional genomics. Students majoring in either track receive a solid grounding in each field through a core of courses common to both programs. Bioinformatics is concerned with the management and interpretation of genomic data. The bioinformatics degree track leads to a course work–based Master of Bioinformatics and a research-oriented Ph.D. degree. The functional genomics degree track, which includes structural genomics, focuses on the application of high-throughput molecular biology to the study of gene sequence and expression, and macromolecular structure. The program offers the Master of Functional Genomics (nonthesis), and the Master of Science (M.S.) and Ph.D. degrees. For more information, prospective students should visit the Web site.

Research Facilities

The Genome Research Laboratory (GRL) provides access to the latest equipment for DNA fragment analysis, high-throughput automated DNA sequencing, and gene-expression analysis using DNA microarray technology. The facility provides training and guidance in the use of the equipment, allowing students in genomic sciences to gain practical experiences as part of their education. The GRL is adjacent to the Bioinformatics Research Center, the Forestry Biotechnology Laboratory, the Fungal Genomics Laboratory, and the Nematode Research Center on the University's Centennial Campus. In addition, a number of core research facilities on the NCSU campus provide state-of-the-art equipment and services at cost, including the Biological Resource Center, the Center for Electron Microscopy, the Monoclonal Antibody Facility, the Mass Spectrometry Facility, the Nuclear Magnetic Resource Facility, and the Cell and Molecular Imaging Facility. A number of experimental farms and greenhouses are available for plant research, as well as the Southeastern Plant Environmental Laboratory. Research equipment is also available in individual faculty research programs. Students work in or closely with the Bioinformatics Research Center for data management and analysis. Relationships have been established with several companies in the Research Triangle that are involved in genomics research.

Financial Aid

Graduate students in genomic sciences receive financial support in the form of fellowships, research assistantships, and internships from a number of sources.

Cost of Study

Tuition for a full program of graduate study in 2006–07 is estimated to be $2019 per semester for residents of North Carolina and $8043 per semester for nonresidents. Fees are estimated to be $632 per semester. Students with stipends are granted waivers of resident or nonresident tuition and health insurance, but they remain responsible for student fees.

Living and Housing Costs

Housing facilities are available on campus and in the community. Accommodations in the graduate dormitory are estimated to be $1995 per person per semester for double occupancy in 2006–07. Apartments in the E. S. King Village (married-student housing) are estimated to be $515–$630 per month. Most graduate students live off campus.

Student Group

The University enrolls approximately 30,000 students, including more than 5,000 graduate students. The University Graduate Student Association (UGSA) plays an active role in matters relating to graduate education.

Location

Raleigh, the capital of North Carolina, is a cosmopolitan city of more than 300,000 residents. It forms one corner of the renowned Research Triangle, which is a community of three major universities, several small colleges, and numerous government and private research laboratories. Considerable intellectual exchange occurs among North Carolina State University, Duke University, the University of North Carolina at Chapel Hill, and area companies and research institutions.

The University and The Program

North Carolina State University has served as a major center for scientific and technological education since its founding in 1887. The Genomic Sciences Graduate Program is a University-wide program initiated in 1999.

Applying

Candidates should have a bachelor's degree in mathematical, computer, physical, or biological sciences. Candidates should submit official transcripts of all college and university course work, Graduate Record Examinations (GRE) scores, and three letters of recommendation. International applicants whose native language is not English are required to take the Test of English as a Foreign Language (TOEFL). Application information can also be found on the Web at http://www2.ncsu.edu/grad/.

Correspondence and Information

Dr. David Bird
Co-Director of Genomic Sciences Graduate Program
Box 7253
North Carolina State University
Raleigh, North Carolina 27695

Phone: 919-515-6813
Fax: 919-515-9500
E-mail: david_bird@ncsu.edu
Web site: http://genomics.ncsu.edu

Dr. Zhao-Bang Zeng
Co-Director of Genomic Sciences Graduate Program
Box 7566
North Carolina State University
Raleigh, North Carolina 27695

Phone: 919-515-1942
Fax: 919-515-7315
E-mail: zeng@stat.ncsu.edu
Web site: http://genomics.ncsu.edu

North Carolina State University

THE FACULTY AND THEIR RESEARCH

College of Agriculture and Life Sciences
P. A. Agris: Nucleic acid design, biophysics.
G. Allen: Chromatin structure, gene expression.
J. Alonso: Signal transduction in *Arabidopsis.*
R. R. H. Anholt: Neurobiology of olfaction.
M. Ashwell: Quantitative genetics in livestock.
W. R. Atchley: Quantitative genetics, evolution.
P. Awadalla: Evolutionary genomics (humans, *drosophila*).
D. M. Bird: Genomics of nematode parasites.
R. Borski: Endrocrinology in vertebrates.
W. Boss: Plant responses to environment stimuli.
R. S. Boston: Protein interactions in maize.
B. Brizuela: Epithelial morphogenesis.
D. T. Brown: Sindbis virus structure-function.
J. W. Brown: RNA structure-function, evolution.
I. Carbone: Fungal population evolution.
J. Cassady: Food animal production.
J. Cavanagh: Bacterial protein structure (NMR).
A. C. Clark: Protein folding/assembly.
S. D. Clouse: Plant gene expression, development.
S. E. Curtis: Genes in cyanobacterial development.
M. E. Daub: Fungal molecular genetics, toxins.
R. A. Dean: Fungal genomics, development.
R. E. Dewey: Agronomically important crop genes.
E. Eisen: Mammalian complex traits, transgenes.
P. Estes: Central nervous system development.
C. Farin: Control of meiotic maturation of oocytes.
R. Franks: Development in *Arabidopsis.*
G. Gibson: Genomics of development, behavior.
J. Godwin: Molecular endocrinology, neurobiology.
M. M. Goodman: Crop plant evolution, isozymes.
M. B. Goshe: Proteomics of protein interactions (MS).
C. Grozinger: Neurogenomics in honey bees.
A. Grunden: Hyperthermophilic bacteria.
L. K. Hanley-Bowdoin: Geminivirus replication/host.
C. L. Hemenway: Plant virus RNA replication.
J. Holland: Genomics-assisted maize breeding.
S. Kathariou: Bacterial food-borne pathogens.
T. R. Klaenhammer: Bacteria in dairy fermentation.
S. Leath: Wheat disease resistance.
H.-C. Liu: Viral-host interactions, resistance.
S. Lommel: Plant virus-host interactions.
J. Lubischer: Developmental neurobiology.
T. F. C. Mackay: Quantitative variation, *Drosophila.*
J. W. Mahaffey: Development, body patterns.
L. Mathies: Animal development, *C. elegans.*
C. Mattos: Structural analyses of GTPases.
E. S. Maxwell: Structure/function of snoRNAs.
E. S. Miller: RNA-protein interactions, microbes.
P. Modziak: Skeletal muscle growth, myonuclei.
J. W. Moyer: Evolution of plant bunyaviruses.
J. Odle: Lipid digestion and metabolism.
J. Olson: Bacterial physiology.
C. H. Opperman: Genomics of nematode parasites.
G. A. Payne: Fungi and plant resistance, genetics.
J. Petitte: Germ cells, avian transgenics.
R. M. Petters: Reproductive physiology, embryos.
M. D. Purugganan: Flower development evolution.
J. B. Ristaino: Fungal species molecular markers.
D. Robertson: Plant virus gene silencing.
M. Sikes: Antigen receptor gene recombination.
R. C. Smart: Molecular chemical carcinogenesis.

C. V. Sullivan: Reproductive physiology of fish.
W. F. Thompson: Gene transfer in plants.
J. L. Thorne: Statistical analyses of DNA sequence.
B. M. Wiegmann: Molecular biosystematics.
P. Wollenzien: RNA structure-function.
Q. Xiang: Plant systematics.

College of Engineering
D. L. Bitzer: Information theory, genetic sequences.
J. M. Haugh: Cell engineering, signal transduction.
S. Heber: Splicing graphs, EST analysis.
R. M. Kelly: Hyperthermophilic microorganisms.
D. Lalush: In vivo molecular imaging.
M. A. Vlouk: Information theory and genetics.

College of Natural Resources
H. V. Amerson: Forest pathology, resistance.
V. Chiang: Transgenic trees, wood formation.
J. Frampton: Forest tree genetics, resistance.
B. Goldfarb: Plant development, root biology.
B. Li: Forest genetics, tree physiology.
L. Li: Wood formation.
B. H. Lui: Biochips, genome informatics.
S. McKeand: Breeding, ecophysiology.
R. Sederoff: Forest tree genomics.
R. Whetten: Genetic signals for xylem formation.

College of Physical and Mathematical Sciences
M. Davidian: Nonlinear models, covariate error.
S. Ghosh: Bayesian Markov Chain Monte Carlo.
J. Hughes-Oliver: Nonlinear models, spatial stats.
E. Kaltofen: Symbolic computation.
J. S. Lindsey: Molecular photonics, automation.
D. Muddiman: Proteomic approaches using mass spectrometry.
S. V. Muse: DNA sequence analysis, evolution.
D. Nielsen: Statistical genetics.
C. Sagui: Biomolecule dynamic simulations.
E. Stone: Statistical analyses of evolution.
A. Tsiatis: Statistical analyses, surrogate markers.
J-Y. Tzeng: Statistical analyses of complex traits.
B. Wang: Peptide mimetics, drug delivery.
B. Weir: Statistical tools for genetic markers.
R. Wolfinger: Mixed models for data assessment.
Z.-B. Zeng: Statistics/computer genetic models.
D. Zhang: Biostatistics, epidemiology, genetics.

College of Veterinary Medicine
K. B. Adler: Inflammatory signals in lung disease.
P. Arasu: Molecular host-parasite interactions.
J. Barnes: Cell growth, cytoprotection.
M. Breen: Canine genome, cancer.
G. A. Dean: Viral vaccines (FIV).
F. J. Fuller: Equine infectious anemia virus/host.
J. E. Gadsby: Growth factor regulation of ovaries.
J. Horowitz: Transcription and cell-cycle control.
D. H. Ley: Pathophysiology of avian mycoplasma.
L. Martin: Inflammatory signals in lung disease.
P. E. Orndorff: Genetics of virulence in bacteria.
J. Piedrahita: Transgenic animal development.
M. L. Rodriguez-Puebla: Cell-cycle control.
B. Sherry: Cardiac response to viral infection.
W. A. Tompkins: Viral immunology (FIV).
J. Yoder: Zebrafish immunogenomics.

Research Facilities | The Genome Research Laboratory (GRL), the Forestry Biotechnology Laboratory, the Fungal Genomics Laboratory and the Nematode Research Center on the University's Centennial Campus provide access to the latest equipment for DNA fragment analysis, high-throughput automated DNA sequencing, and gene expression analysis using DNA microarray technology. The GRL provides training and guidance in the use of the equipment, allowing students in genomic sciences to gain practical experience as part of their education. In addition, a number of core research facilities on the NCSU campus provide state-of-the-art equipment and services at cost, including the Biological Resource Center, the Center for Electron Microscopy, the Monoclonal Antibody Facility, the Mass Spectrometry Facility, the Nuclear Magnetic Resonance Facility, and the Cell and Molecular Imaging Facility. A number of experimental farms and greenhouses are available for plant research, as well as the Southeastern Plant Environmental Laboratory. Research equipment also is available in individual faculty research programs. Finally, students work closely with the Bioinformatics Research Center for data management and analysis. |
Financial Aid	Graduate students in functional genomics receive financial support in the form of fellowships and research assistantships from a number of sources.
Cost of Study	Tuition for a full program of graduate study in 2006–07 is estimated to be $2019 per semester for residents of North Carolina and $8043 per semester for nonresidents. Fees are estimated to be $632 per semester. Students with stipends are granted resident and nonresident tuition and health insurance but are responsible for student fees.
Living and Housing Costs	Housing facilities are available on campus and in the community. Accommodations in the graduate dormitory are estimated to be $1995 per person per semester for double occupancy in 2006–07. Apartments in the E. S. King Village (married student housing) are estimated to be $515–$630 per month. Most graduate students live off campus.
Student Group	The University enrolls approximately 30,000 students, including more than 5,000 graduate students. The University Graduate Student Association (UGSA) plays an active role in matters relating to graduate education.
Location	Raleigh, the capital of North Carolina, is a cosmopolitan city of more than 300,000 residents. It forms one corner of the renowned Research Triangle, a community of three major universities, several small colleges, and numerous government and private research laboratories. Considerable intellectual exchange occurs among North Carolina State University, Duke University, the University of North Carolina at Chapel Hill, and area companies and research institutions.
The University and The Program	North Carolina State University has served as a major center for scientific and technological education since its founding in 1887. The Genomic Sciences Graduate Program is a University-wide program initiated in 1999.
Applying	Candidates for graduate work should have a bachelor's degree in biological, physical, or mathematical sciences and a basic understanding of biological principles and processes. Candidates should submit official transcripts of all college and university course work, Graduate Record Examinations (GRE) scores, and three letters of recommendation. International applicants whose native language is not English are required to take the Test of English as a Foreign Language (TOEFL). Application information can be obtained at the Graduate School Web site (http://www2.acs.ncsu.edu/grad/).
Correspondence and Information	Dr. David Bird Co-Director of Genomic Sciences Graduate Program Box 7253 North Carolina State University Raleigh, North Carolina 27695 Phone: 919-515-6813 Fax: 919-515-9500 E-mail: david_bird@ncsu.edu Web site: http://genomics.ncsu.edu

North Carolina State University

THE FACULTY AND THEIR RESEARCH

College of Agriculture and Life Sciences
P. A. Agris: Nucleic acid design, biophysics.
G. Allen: Chromatin structure, gene expression.
J. Alonso: Signal transduction in *Arabidopsis.*
R. R. H. Anholt: Neurobiology of olfaction.
M. Ashwell: Quantitative genetics in livestock.
W. R. Atchley: Quantitative genetics, evolution.
P. Awadalla: Evolutionary genomics (humans, *drosophila*).
D. M. Bird: Genomics of nematode parasites.
R. Borski: Endocrinology in vertebrates.
W. Boss: Plant responses to environment stimuli.
R. S. Boston: Protein interactions in maize.
B. Brizuela: Epithelial morphogenesis.
D. T. Brown: Sindbis virus structure-function.
J. W. Brown: RNA structure-function, evolution.
I. Carbone: Fungal population evolution.
J. Cassady: Food animal production.
J. Cavanagh: Bacterial protein structure (NMR).
A. C. Clark: Protein folding/assembly.
S. D. Clouse: Plant gene expression, development.
S. E. Curtis: Genes in cyanobacterial development.
M. E. Daub: Fungal molecular genetics, toxins.
R. A. Dean: Fungal genomics, development.
R. E. Dewey: Agronomically important crop genes.
E. Eisen: Mammalian complex traits, transgenes.
P. Estes: Central nervous system development.
C. Farin: Control of meiotic maturation of oocytes.
R. Franks: Development in *Arabidopsis.*
G. Gibson: Genomics of development, behavior.
J. Godwin: Molecular endocrinology, neurobiology.
M. M. Goodman: Crop plant evolution, isozymes.
M. B. Goshe: Proteomics of protein interactions (MS).
C. Grozinger: Neurogenomics in honey bees.
A. Grunden: Hyperthermophilic bacteria.
L. K. Hanley-Bowdoin: Geminivirus replication/host.
C. L. Hemenway: Plant virus RNA replication.
J. Holland: Genomics-assisted maize breeding.
S. Kathariou: Bacterial food-borne pathogens.
T. R. Klaenhammer: Bacteria in dairy fermentation.
S. Leath: Wheat disease resistance.
H.-C. Liu: Viral-host interactions, resistance.
S. Lommel: Plant virus-host interactions.
J. Lubischer: Developmental neurobiology.
T. F. C. Mackay: Quantitative variation, *Drosophila.*
J. W. Mahaffey: Development, body patterns.
L. Mathies: Animal development, *C. elegans.*
C. Mattos: Structural analyses of GTPases.
E. S. Maxwell: Structure/function of snoRNAs.
E. S. Miller: RNA-protein interactions, microbes.
P. Modziak: Skeletal muscle growth, myonuclei.
J. W. Moyer: Evolution of plant bunyaviruses.
J. Odle: Lipid digestion and metabolism.
J. Olson: Bacterial physiology.
C. H. Opperman: Genomics of nematode parasites.
G. A. Payne: Fungi and plant resistance, genetics.
J. Petitte: Germ cells, avian transgenics.
R. M. Petters: Reproductive physiology, embryos.
M. D. Purugganan: Flower development evolution.
J. B. Ristaino: Fungal species molecular markers.
D. Robertson: Plant virus gene silencing.
M. Sikes: Antigen receptor gene recombination.
R. C. Smart: Molecular chemical carcinogenesis.

C. V. Sullivan: Reproductive physiology of fish.
W. F. Thompson: Gene transfer in plants.
J. L. Thorne: Statistical analyses of DNA sequence.
B. M. Wiegmann: Molecular biosystematics.
P. Wollenzien: RNA structure-function.
Q. Xiang: Plant systematics.

College of Engineering
D. L. Bitzer: Information theory, genetic sequences.
J. M. Haugh: Cell engineering, signal transduction.
S. Heber: Splicing graphs, EST analysis.
R. M. Kelly: Hyperthermophilic microorganisms.
D. Lalush: In vivo molecular imaging.
M. A. Vlouk: Information theory and genetics.

College of Natural Resources
H. V. Amerson: Forest pathology, resistance.
V. Chiang: Transgenic trees, wood formation.
J. Frampton: Forest tree genetics, resistance.
B. Goldfarb: Plant development, root biology.
B. Li: Forest genetics, tree physiology.
L. Li: Wood formation.
B. H. Lui: Biochips, genome informatics.
S. McKeand: Breeding, ecophysiology.
R. Sederoff: Forest tree genomics.
R. Whetten: Genetic signals for xylem formation.

College of Physical and Mathematical Sciences
M. Davidian: Nonlinear models, covariate error.
S. Ghosh: Bayesian Markov Chain Monte Carlo.
J. Hughes-Oliver: Nonlinear models, spatial stats.
E. Kaltofen: Symbolic computation.
J. S. Lindsey: Molecular photonics, automation.
D. Muddiman: Proteomic approaches using mass spectrometry.
S. V. Muse: DNA sequence analysis, evolution.
D. Nielsen: Statistical genetics.
C. Sagui: Biomolecule dynamic simulations.
E. Stone: Statistical analyses of evolution.
A. Tsiatis: Statistical analyses, surrogate markers.
J-Y. Tzeng: Statistical analyses of complex traits.
B. Wang: Peptide mimetics, drug delivery.
B. Weir: Statistical tools for genetic markers.
R. Wolfinger: Mixed models for data assessment.
Z.-B. Zeng: Statistics/computer genetic models.
D. Zhang: Biostatistics, epidemiology, genetics.

College of Veterinary Medicine
K. B. Adler: Inflammatory signals in lung disease.
P. Arasu: Molecular host-parasite interactions.
J. Barnes: Cell growth, cytoprotection.
M. Breen: Canine genome, cancer.
G. A. Dean: Viral vaccines (FIV).
F. J. Fuller: Equine infectious anemia virus/host.
J. E. Gadsby: Growth factor regulation of ovaries.
J. Horowitz: Transcription and cell-cycle control.
D. H. Ley: Pathophysiology of avian mycoplasma.
L. Martin: Inflammatory signals in lung disease.
P. E. Orndorff: Genetics of virulence in bacteria.
J. Piedrahita: Transgenic animal development.
M. L. Rodriguez-Puebla: Cell-cycle control.
B. Sherry: Cardiac response to viral infection.
W. A. Tompkins: Viral immunology (FIV).
J. Yoder: Zebrafish immunogenomics.

THE OHIO STATE UNIVERSITY

Department of Molecular Genetics

Program of Study

This department offers a graduate program leading to the M.S. and Ph.D. degrees in numerous areas of genetics, molecular biology, and cell and developmental biology. The graduate program consists of a broadly based core of courses in areas emphasizing the four-part nature of the department: genetics, cell biology, developmental biology, and molecular biology. In addition, a strong foundation in biochemistry is expected of all program graduates. Graduate students participate actively in the seminar and teaching programs of the department.

Areas of research emphasis include nuclear gene expression and RNA splicing, neurogenesis and muscle development, microbial development, molecular evolution, organelle genetics (both mitochondrial and chloroplast), plant molecular biology, and cytogenetics. A wide variety of model systems are studied, including fungi, algae, and protists, including flowering plants, *Drosophila*, cultured mammalian cells, transgenic mice, and man. Bacterial studies emphasize host-parasite interactions and include both plant and animal pathogens.

The faculty participating in the molecular genetics graduate program consists of 18 members with appointments in the Department of Molecular Genetics and 6 faculty members drawn from five other departments in two colleges (Biological Sciences and Medicine). The faculty currently supervises 40 graduate students in the areas of research training offered within the department.

For details of requirements for M.S. and Ph.D. degrees, students should write to the Graduate Committee Chairman named below.

Research Facilities

In addition to the standard laboratory equipment used in molecular biology, special research facilities available to graduate students include controlled-environment chambers, laboratories for the culture of *Drosophila*, greenhouse and farm field space, an electron microscope, transgenic plant and animal biotechnology facilities, two computer centers, and radiation equipment. The Bioinstrumentation Facility includes automated peptide and oligonucleotide synthesis and automated sequencing of amino acids and nucleic acids. The Biological Sciences/Pharmacy Library houses 300 journals and 120,000 volumes and provides access to 8,000 online journals . Substantial resources are also available in the Agriculture, Science and Engineering, and Health Sciences Libraries. The libraries are members of OhioLINK, a statewide network providing access to databases, electronic journals, and the library catalogs of more than eighty colleges and universities.

Financial Aid

Prospective graduate students may apply for graduate associateships. In 2004–05, the normal stipend was $22,140 for twelve months of half-time service. In addition to the stipend, tuition and out-of-state fees are paid by the University. Various forms of University fellowship aid, as listed in the Ohio State *Graduate School Bulletin,* are awarded each year to successful applicants.

Cost of Study

In 2004–05, graduate students carrying a full academic load were charged $2750 per quarter if they were residents of Ohio and an additional $3961 per quarter if they were nonresidents. Tuition and out-of-state fees are paid by the University for graduate students who have been awarded associateships in molecular genetics.

Living and Housing Costs

The cost of a single room in the graduate residence halls was approximately $1556 per quarter in 2004–05; charges for meals vary according to meal plans and accommodations. Off-campus housing varies in price; both rooms and apartments are available within walking distance of the campus.

Student Group

There are currently 50,995 students at the Columbus campus of Ohio State. Of these, 10,210 are enrolled in the Graduate School. There are 40 graduate students associated with the molecular genetics program at the University.

Location

The Ohio State University is located in Columbus, the capital of Ohio. The main campus is located 2½ miles from downtown Columbus. Musical, dramatic, and other cultural programs are presented in Columbus. The city supports a gallery of fine arts, a symphony orchestra, the Ohio Historical Society's State Museum and reconstructed nineteenth-century Ohio village, the Museum of Science and Industry, and other institutions and activities usually associated with a large city. A wide range of sporting and outdoor recreational facilities is available in the Columbus area. The Columbus research community, of which Ohio State is a part, includes Battelle Memorial Institute (the world's largest nonprofit scientific research institute), the national headquarters for Chemical Abstracts, Children's Hospital, and the Riverside Hospital (a large, modern hospital complex). All of these institutions are located close to the Columbus campus.

The University

In many ways, Ohio State functions as a city within the metropolitan area of Columbus. The University is organized into twenty-six colleges and schools, including the Graduate School. Collectively, the colleges cover essentially all tertiary education, including such professional areas as law and medicine. The University is the major center for graduate studies in Ohio, thus providing a stimulating and challenging intellectual atmosphere.

Applying

Application forms for admission and financial aid may be obtained by accessing the Web site at http://gradapply.osu.edu or via e-mail at domestic.grad@osu.edu (international.grad@osu.edu for international students). An applicant must have a baccalaureate degree and should have adequate preparation for graduate study in molecular genetics.

Correspondence and Information

Dr. Mark A. Seeger
Graduate Committee Chairman
Department of Molecular Genetics
The Ohio State University
484 West 12th Avenue
Columbus, Ohio 43210-1292

Phone: 614-292-5106
Fax: 614-292-4466
Web site: http://www.osumolgen.org

The Ohio State University

THE FACULTY AND THEIR RESEARCH

David M. Bisaro, Professor and Director of the Molecular, Cellular and Developmental Biology Program; Ph.D., Wayne State. Geminivirus replication, gene expression, and pathogenesis.

Arthur H. M. Burghes, Professor; Ph.D., London. Human molecular genetics; neurologic diseases.

Helen M. Chamberlin, Associate Professor; Ph.D., Caltech. Cell differentiation during organogenesis.

Tien-Hsien Chang, Associate Professor; Ph.D., SUNY at Buffalo. Yeast RNA helicases; mRNA metabolism.

Albert de la Chapelle, Professor; Ph.D., Helsinki (Finland). Analysis of human cancer genes.

Susan E. Cole, Assistant Professor; Ph.D., Johns Hopkins. Analysis of cyclic mRNA expression during somitogenesis: linking the Notch pathway and the segmentation clock in vertebrate development.

Donald H. Dean, Professor; Ph.D., Michigan. Insecticidal proteins from *Bacillus thuringiensis.*

Andrea I. Doseff, Assistant Professor; Ph.D., SUNY at Stony Brook. Signal transduction pathways and regulation of the apoptotic machinery during innate immune response; cell differentiation and cancer.

Harold A. Fisk, Assistant Professor; Ph.D., California, San Diego. Cell cycle regulation of centrosome duplication and the Mps1 protein kinase.

Paul A. Fuerst, Professor; Ph.D., Brown. Molecular evolutionary genetics.

Paul K. Herman, Associate Professor; Ph.D., Caltech. Regulation of growth and cell cycle in yeast.

Russell J. Hill, Assistant Professor; Ph.D., Caltech. Specification of cell type in early embryogenesis.

Gustavo W. Leone, Assistant Professor; Ph.D., Calgary. Signal transduction, cell proliferation, apoptosis, and tumor models.

Berl R. Oakley, Professor and Interim Chair; Ph.D., London. Mitosis; microtubules; organellar movement.

Stephen A. Osmani, Professor and Ohio Eminent Scholar; Ph.D., King's. Cell cycle; nuclear migration; fungal development.

Michael C. Ostrowski, Professor; Ph.D., South Carolina. Signaling pathways and gene transcription.

Hay-Oak Park, Associate Professor; Ph.D., Wisconsin–Madison. Regulation of oriented cell division and cell cycle in yeast.

Deborah S. Parris, Professor; Ph.D., Virginia Tech. Structure-function analysis of herpes simplex virus DNA replication proteins.

Christoph Plass, Associate Professor; Ph.D., Medizinische Universität zu Lübeck (Germany). DNA methylation; development; cancer.

Mark Seeger, Associate Professor; Ph.D., Indiana. Neuronal development in *Drosophila.*

Amanda Simcox, Professor; Ph.D., Sussex (England). Pattern formation in *Drosophila,* including genomic approaches.

Harald Vaessin, Professor; Ph.D., Cologne (Germany). Neuronal lineage differentiation in *Drosophila.*

Desh Pal S. Verma, Professor; Ph.D., Western Ontario. Molecular genetics of symbiotic nitrogen fixation.

Michael Weinstein, Associate Professor; Ph.D., California, San Diego. TGF and FGF signaling in mammalian development and tumorigenesis.

STONY BROOK UNIVERSITY, STATE UNIVERSITY OF NEW YORK

with Cold Spring Harbor Laboratory and Brookhaven National Laboratory
Graduate Studies in Cellular and Developmental Biology

Program of Study

Cellular and developmental biology is a graduate specialization within the umbrella program of molecular and cellular biology. Emphasis is placed on the control mechanisms that define and regulate growing and developing systems. Students have the opportunity to perform Ph.D. thesis research in a number of laboratories at three collaborative institutions: the State University of New York at Stony Brook, Cold Spring Harbor Laboratory, or Brookhaven National Laboratory. The Ph.D. is granted by the State University of New York at Stony Brook.

The integrated nature of the program offers the student an extraordinary range of experimental science. Students are involved in ongoing research as soon as they arrive on campus. During the first academic year, they experience research in four different laboratories of program faculty members. Following the first academic year, a member of the program faculty is selected as a Ph.D. mentor. Diverse biological systems are available to investigate a variety of biological questions, including cancer, neurobiology, gene expression, infectious disease, protein trafficking, DNA replication, structural biology, signal transduction, cell cycle, apoptosis, virology, development, biological membranes, and immune defense.

The graduate program offers the student a unique opportunity to choose among three academic subspecialties: biochemistry and molecular biology, cellular and developmental biology, or immunology and pathology. During the first academic year, students take core courses in biochemistry, molecular genetics, and cell biology. In addition, students participate in a journal club/readings course, in which they learn to critically evaluate original research articles. Subsequent specialized courses and electives are then designed to enhance the student's knowledge in their selected academic subspecialty.

In the second and third semesters, graduate students develop teaching skills by serving as a teaching assistant with the guidance of the faculty instructor of the course. In the third and subsequent years, graduate students present their research progress to other students and faculty members in graduate student seminars. These provide opportunities to gain communication skills and learn about ongoing research of students in other laboratories.

In the third year, students prepare a written Ph.D. thesis proposal in consultation with their research faculty adviser. Following successful defense of the proposal, the student advances to candidacy, and the proposal committee and the faculty adviser become the student's Ph.D. thesis committee. This committee meets at least once a year to support the research efforts of the student.

Research Facilities

Research is supported by state-of-the-art facilities at Stony Brook. General computing services are available to students, including an account for e-mail and Internet access. The new Proteomics Center is a core facility at Stony Brook whose services include protein sequencing, peptide synthesis, analytical HPLC, preparative HPLC, MALDI-TOF mass spectrometry, and DNA sequencing. The University recently opened a DNA Microarray Center that offers researchers the ability to simultaneously analyze the expression of thousands of genes in humans, rodents, yeast, plants, bacteria, or *Drosophila*. The Center for Structural Biology has advanced instrumentation for NMR spectroscopy and X-ray crystallography. The University Microscopy Imaging Center assists in research projects requiring advanced light, electronic light, and electron-microscopy techniques.

Financial Aid

All students accepted into the graduate program receive full financial support. Students are provided with a stipend of $25,000 per year for cost-of-living expenses. In addition, tuition is waived for all graduate students accepted into the program. Health insurance is provided.

Cost of Study

University fees are approximately $750 per year.

Living and Housing Costs

Room rent for a single student living in an on-campus apartment ranges from $325 to $1500 per month, depending on size and number of occupants. Privately owned housing is available close to the University.

Student Group

There are approximately 100 graduate students in the program. Many enter the program as recent college graduates, and others have had previous graduate school or research experience. Generally, students have undergraduate majors in biology, biochemistry, or chemistry, but some students come from other disciplines. Diverse cultural and ethnical interests are supported by the campus community.

Location

Stony Brook is located on the wooded north shore of Long Island, New York, an area of picturesque villages, harbors, and beaches. It is about 50 miles east of New York City with easy access provided by the Long Island Railroad or major parkways so that students can take advantage of New York City's commercial, scientific, and cultural resources.

The University

The University was founded in 1957 in Oyster Bay, Long Island. In 1962 it moved to Stony Brook and now occupies more than 1,000 acres with more than 100 buildings. Stony Brook has exceptional strength in the sciences, mathematics, humanities, fine arts, social sciences, engineering, and health professions for both undergraduate and graduate students. It was classified by the Carnegie Foundation as one of the nation's leading research institutions. The University Hospital and Medical Center supports the Schools of Medicine, Dentistry, Nursing, and Health Technology and Management.

An indoor sports complex supports the campus's Division I athletics and Seawolf fans. The Staller Center for the Arts offers a wide variety of professional performances in music, film, dance, theater, and fine art exhibitions in its five theaters and University Art Gallery.

Applying

Applications should be submitted by January 15 for admission for the following September, although applications are accepted after that date. Applicants must have a baccalaureate degree from an accredited college or university and submit school transcripts. The General Test of the Graduate Record Examinations (GRE) is required, and a Subject Test in an appropriate field is optional. Three letters of recommendation are required.

Prospective candidates are invited to visit Stony Brook whenever possible. This affords students an opportunity to meet with program faculty members and students. To apply online, students should visit http://www.grad.sunysb.edu/applying/applying.htm.

Correspondence and Information

Director, Graduate Studies in Cellular and Developmental Biology
Life Sciences Room 336/338
State University of New York at Stony Brook
Stony Brook, New York 11794-5215

Phone: 631-632-8533
Fax: 631-632-9730
E-mail: mcbprog@life.bio.sunysb.edu
Web site: http://life.bio.sunysb.edu/mcb

Stony Brook University, State University of New York

THE FACULTY AND THEIR RESEARCH

Department of Biochemistry and Cell Biology
Paul M. Bingham, Ph.D., 1979. Genetic control of development and gene expression in animals. Deborah Brown, Ph.D., 1987. Structure and function of cholesterol/sphingolipid-rich membrane rafts and caveolae. Vitaly Citovsky, Ph.D., 1987. Nuclear targeting and intercellular communication in plants. Neta Dean, Ph.D., 1988. Protein glycosylation; fungal cell wall biosynthesis; fungal pathogenesis. Dale Deutsch, Ph.D., 1972. Molecular neurobiology of anandamide (the endogenous marijuana) regulation. J. Peter Gergen, Ph.D., 1982. Gene expression and development in *Drosophila*. Robert Haltiwanger, Ph.D., 1986. Role of protein glycosylation in signal transduction; Notch signaling. Bernadette Holdener, Ph.D., 1990. Role of protein folding in WNT signal transduction and development. Nancy Hollingsworth, Ph.D., 1988. Chromosome structure and function during meiosis in yeast. Jen-Chih Hsieh, Ph.D., 1994. Molecular mechanism of Wnt signaling. A. Wali Karzai, Ph.D., 1995. Structure and function of RNA-binding proteins and biochemical studies of the SmpB·SsrA quality control system for protein tagging, directed degradation, and ribosome rescue. William J. Lennarz, Ph.D., 1959. Biosynthesis and function of glycoproteins in cell-cell interactions. Erwin London, Ph.D., 1979. Membrane protein structure/translocation/folding; structure and function of sphingolipid/cholesterol rafts in membranes. Harvard Lyman, Ph.D., 1960. Photocontrol of chloroplast development. Kenneth B. Marcu, Ph.D., 1975. Molecular control of innate and adaptive immunity: NF-kappaB kinases and inflammatory response; antibody gene class switch recombination and somatic hypermutation. Aaron Neiman, Ph.D., 1994. Vesicle trafficking and intracellular signaling in yeast. Nisson Schechter, Ph.D., 1971. Homeobox and filament proteins in neuronal differentiation, growth, and regeneration. Sanford R. Simon, Ph.D., 1967. Extracellular matrix degradation by inflammatory and tumor cell proteases. Steven O. Smith, Ph.D., 1985. Structure and function of membrane proteins. James Staros, Ph.D., 1974. Biochemical and biophysical approaches to signal transduction by ErbB-family receptors. Rolf Sternglanz, Ph.D., 1967. Chromatin structure and function; histone modifying enzymes; gene expression. Gerald H. Thomsen, Ph.D., 1988. Regulation of early vertebrate development by growth factor signals, ubiquitin modification, and T-box family transcription factors.

Department of Chemistry
Carlos Simmerling, Ph.D., 1994. Computational studies of biomolecular structure and dynamics. Peter J. Tonge, Ph.D., 1986. Tuberculosis pathogenesis and drug discovery; enzyme mechanisms and rational inhibitor design; fluorescent proteins.

Department of Molecular Genetics and Microbiology
Jorge L. Benach, Ph.D., 1972. Pathogenesis of spirochetal infections; utilization of host macromolecules. Allen Bruce Futcher, Ph.D., 1983. Cell-cycle control in yeast and on analysis of gene expression networks using microarrays and bioinformatics. Michael Hayman, Ph.D., 1973. Signal transduction pathways in cell growth, differentiation, and cancer. Patrick Hearing, Ph.D., 1984. Adenovirus–host cell interactions; adenovirus assembly and vectors for gene therapy. Eugene Katz, Ph.D., 1969. Genetics/development in cellular slime molds. James Konopka, Ph.D., 1985. Hormone signal transduction/yeast cell morphogenesis. Janet Leatherwood, Ph.D., 1993. Cell-cycle control of DNA replication/microarray analyses of fission yeast. Nancy Reich, Ph.D., 1983. Signal transduction and gene expression in response to cytokines and virus. David D. Thanassi, Ph.D., 1995. Virulence factors of pathogenic bacteria. Eckard Wimmer, Ph.D., 1962. RNA virus genetics, replication, and pathogenicity; cellular virus receptors. Wei-Xing Zong, Ph.D., 1999. Molecular regulation of cell-death pathways.

Department of Neurobiology and Behavior
Simon Halegoua, Ph.D., 1978. Molecular control of the neuronal phenotype. Maurice Kernan, Ph.D., 1990. Differentiation and signal transduction in *Drosophila* mechanosensory cilia and sperm. Joel M. Levine, Ph.D., 1980. Cell-surface molecules of the developing nervous system. Gail Mandel, Ph.D., 1977. Gene expression in the nervous system. David McKinnon, Ph.D., 1987. Molecular physiology of sympathetic neurons and cardiac muscle. Howard Sirotkin, Ph.D., 1996. Genetic and molecular analysis of early vertebrate development. Lonnie P. Wollmuth, Ph.D., 1992. Molecular and biophysical mechanisms of excitatory synaptic transmission.

Department of Oral Biology and Pathology
Soosan Ghazizadeh, Ph.D., 1994. Keratinocyte stem cell lineage and use in cell and gene therapy.

Department of Pathology
Howard B. Fleit, Ph.D., 1980. Leukocyte Fc receptors; macrophage differentiation. Martha Furie, Ph.D., 1980. Interactions among pathogenic bacteria, endothelium, and leukocytes. Berhane Ghebrehiwet, D.V.M./D.Sc., 1974. Biochemistry; function of the complement system. Richard R. Kew, Ph.D., 1986. Leukocyte chemotaxis/inflammation. Ute M. Moll, M.D., 1985. The p53 and Rb/E2F tumor suppressor gene family: function/regulation in normal cells and tumor-associated inactivation. Eric D. Spitzer, M.D./Ph.D., 1985. Molecular biology of *Cryptococcus neoformans*. Gary Zieve, Ph.D., 1977. Assembly/transport of snRNP particles.

Department of Pharmacological Sciences
Miguel Berrios, Ph.D., 1983. Nuclear structure and function; cell biology of DNA damage and repair. Daniel Bogenhagen, M.D., 1977. Mitochondrial DNA; mitochondrial proteomics. Holly Colognato, Ph.D., 2000. Extracellular matrix in the brain; roles during development and during neurodegeneration. Michael A. Frohman, M.D./Ph.D., 1986. Signal transduction; membrane vesicle trafficking; regulated exocytosis; diabetes; phototransduction. Arthur P. Grollman, M.D., 1959. Molecular carcinogenesis and DNA repair. Craig C. Malbon, Ph.D., 1976. Wnt-Frizzled signaling, G-proteins, and development. Masaaki Moriya, Ph.D., 1981. Cellular response to DNA damage. Jeffery Pessin, Ph.D., 1980. Identification of insulin-mediated signal cascades regulation of intracellular GLUT4 vesicle trafficking and biogenesis. Joav Prives, Ph.D., 1968. Cytoskeletal membrane interactions in muscle cells. Styliani Anna Tsirka, Ph.D., 1989. Neuronal microglial interactions in the physiology and pathology of the central nervous system. Orlando D. Schärer, Ph.D., 1996. Chemical biology of mammalian DNA repair.

Department of Physiology and Biophysics
Roger A. Johnson, Ph.D., 1968. Adenine nucleoside 3'-polyphosphates, adenylyl cyclases, and signal transduction. Stuart McLaughlin, Ph.D., 1968. Calcium/ phospholipid second messenger system. W. Todd Miller, Ph.D., 1989. Tyrosine phosphorylation and signal transduction. Nicholas Nassar, Ph.D., 1992. Crystallographic and biochemical studies of signal proteins. Suzanne Scarlata, Ph.D., 1984. Structure/oligomerization of membrane proteins. Ilan Spector, Ph.D., 1967. Actin cytoskeleton in normal and cancer cells. Hsien-yu Wang, Ph.D., 1989. Wnt/Frizzled and G-protein signal transduction in development. Thomas White, Ph.D., 1994. Molecular biology and physiology of gap-junction channels.

School of Medicine and Other Departments
Wen-Tien Chen, Ph.D., 1979. Proteases and integrins in cancer invasion, metastasis, and angiogenesis. Yaacov Hod, Ph.D., 1977. Hormonal control of gene expression in prostate cancer cells. Jolyon Jesty, D.Phil., 1972. Mechanisms of thrombogenesis. Richard Lin, M.D., 1988. Signal transduction and cell growth. Erich R. Mackow, Ph.D., 1984. Hantavirus and rotavirus pathogenesis; viral regulations of cell signaling pathways and responses; viral attachment and entry; reverse genetics. Wolfgang Quitschke, Ph.D., 1983. Gene regulation of proteins associated with neurodegenerative diseases. Mario J. Rebecchi, Ph.D., 1984. Phospholipases and signal transduction. Roy T. Steigbigel, M.D., 1966. Immune dysfunction induced by HIV infection. William E. VanNostrand, Ph.D., 1985. Physiologic and pathophysiologic vascular functions of the Alzheimer's disease amyloid beta-protein precursor.

Brookhaven National Laboratory
John Dunn, Ph.D., 1970. Genomics and gene expression. Richard Setlow, Ph.D., 1947. DNA damage and repair; carcinogenesis and mutagenesis in fish. F. William Studier, Ph.D., 1963. Molecular genetics of phage T7; structural genomics.

Cold Spring Harbor Laboratory
Gregory Hannon, Ph.D., 1991. Genetics of growth in mammalian cells and dsRNA-induced gene silencing. Leemor Joshua-Tor, Ph.D., 1991. X-ray crystallography; molecular recognition; nucleic acid regulation; RNAi. Adrian Krainer, Ph.D., 1986. mRNA splicing; gene expression; RNA-protein interaction. Yuri Lazebnik, Ph.D., 1986. Molecular mechanisms of cancer and apoptosis. Scott Lowe, Ph.D., 1994. Modulation of apoptosis; chemosensitivity; senescence by cancer genes. Vivek Mittal, Ph.D., 1994. Id transcription factors; tumor-mediated neovascularization. K. Muthuswamy Senthil, Ph.D., 1995. Understanding cancer initiation using three-dimensional epithelial structures. David L. Spector, Ph.D., 1980. Spatial organization of gene expression. Arne Stenlund, Ph.D., 1984. DNA replication of papillomaviruses. Bruce Stillman, Ph.D., 1979. DNA replication and chromatin assembly in human and yeast cells. William P. Tansey, Ph.D., 1991. Regulation of oncoprotein stability. Nicholas Tonks, Ph.D., 1985. Characterization of protein tyrosine phosphatases. Linda Van Aelst, Ph.D., 1991. Role of Ras in mammalian cell transformation. Michael Wigler, Ph.D., 1978. Genomics and cancer.

STONY BROOK UNIVERSITY, STATE UNIVERSITY OF NEW YORK

with Cold Spring Harbor Laboratory and Brookhaven National Laboratory
Graduate Program in Genetics

Program of Study

This interinstitutional program, which leads to the Ph.D. degree, is designed to provide training in all major areas of genetics. It offers graduate training in molecular genetics, developmental genetics, computational biology and genomics, evolutionary genetics, and human genetics. All students, no matter what their particular interest, are exposed to all the areas of specialization offered within the program. This experience ensures that the students are prepared to cope with the broad range of challenges they may meet after leaving the program. All trainees take core courses in genetics, molecular biology, biochemistry, and cell biology. The first-year experience also includes laboratory rotations in which the student works in the laboratories of 4 different program faculty members. These rotations allow the student to gain firsthand knowledge of the methods and approaches taken by each laboratory and provide a basis for selecting a thesis research adviser. Students have the opportunity to further broaden their knowledge by participating in journal clubs and by taking elective courses from offerings both within and outside the program. The specific elective course or courses taken by a student are determined in conjunction with a faculty adviser to best meet the student's particular needs. Trainees participate in two ongoing research seminar series. A student research seminar provides trainees with a regular opportunity to present their work to their colleagues and genetics program faculty members. Students also attend research seminars given by internal and visiting faculty members in order to keep abreast of the latest developments and potential areas of future excitement in the field of genetics. The granting of degrees is vested in Stony Brook University, State University of New York.

Research Facilities

The primary training facilities are Stony Brook University, State University of New York, Cold Spring Harbor Laboratory, and Brookhaven National Laboratory. The Genetics Program faculty at Stony Brook is primarily drawn from departments within the College of Arts and Sciences and the School of Medicine but also includes faculty members from many other disciplines. The Life Sciences Building, which houses the Genetics Program office, is the home of the Departments of Molecular Genetics and Microbiology, Biochemistry and Cell Biology, Neurobiology and Behavior, and Ecology and Evolution, all of which are represented in the Genetics Program. The University Health Sciences Center, located across the street from the Life Sciences Building, is the primary home for departments in the School of Medicine, including program faculty members in the Departments of Medicine, Molecular Pathology and Immunology, Pharmacological Sciences, and Physiology and Biophysics. An important new facility at Stony Brook is the Center for Molecular Medicine, a state-of-the-art research building constructed adjacent to the Life Sciences Building. This building houses interdepartmental thematic research Centers in Brain and Spinal Cord Research, Developmental Genetics, Infectious Diseases, and Structural Biology, each of which involves Genetics Program faculty members. The Center for Molecular Medicine provides both an intellectual and physical catalyst for facilitating interactions among scientists with common interests in these areas of modern biology, irrespective of their departmental affiliation. Cold Spring Harbor Laboratory is a world-renowned research institute that has internationally recognized strengths in the areas of cancer biology, neurobiology, plant genetics, structural biology and bioinformatics. The world-class facilities that are available at Brookhaven National Laboratory, including access to the National Synchrotron Light Source, provide additional unique resources for trainees in the Genetics Program. Research faculty members at Brookhaven have widely recognized programs in the fields of molecular biology of microbial, plant, and animal systems and have a leading role in the emerging field of proteomics.

Financial Aid

Students are admitted with full financial support. The stipend for the year 2006–07 was $25,000. All students in the program also receive health insurance benefits from the University.

Cost of Study

All admitted students receive full tuition scholarships. There are miscellaneous University fees, including transportation, information technology, and student government fees, amounting to roughly $350 per semester.

Living and Housing Costs

Dormitory accommodations are available for single students. Prepaid meal plans and à la carte food service are available. Residence halls have limited cooking facilities. There are accommodations for married graduate students in campus apartment complexes. Privately owned housing is available off campus. Information may be obtained from the University's Off-Campus Housing Office or the Health Sciences Center's Office of Student Services.

Student Group

Each entering class in the Genetics Program consists of approximately 10 students. There are currently 65 full-time predoctoral students in the Program. The Program is one of several graduate programs in the biological and biomedical sciences at Stony Brook. Altogether, these various programs bring approximately 50 new predoctoral students to Stony Brook each year. Entering students come from a variety of backgrounds and form a diverse group, both nationally and internationally.

Student Outcomes

Graduates typically go on to postdoctoral research appointments followed by academic careers in teaching and research; alternatively, graduates will seek research careers in government or private industry.

Location

Stony Brook is located in a region of coves, beaches, and small historic villages on the North Shore of Long Island, approximately 60 miles east of New York City. The area has retained its distinctive New England flavor and combines the charm of a rural setting minutes from the waters of Long Island Sound with proximity to the cultural, scientific, and industrial resources of the nation's largest city. Cold Spring Harbor Laboratory is about 25 miles west, and Brookhaven National Laboratory about 18 miles east, of Stony Brook.

The University

Stony Brook University, State University of New York was founded in 1957 at Oyster Bay, Long Island. In 1960, it was designated the State University's fourth University Center with a mandate to develop undergraduate and graduate programs in the humanities, sciences, social sciences, and engineering. In 1962, it moved to its present location at Stony Brook, where it has grown to a campus with 105 buildings on 1,100 acres and a student body of 17,623, including 6,220 graduate students. Cold Spring Harbor Laboratory has grown into one of the most important independent centers for biological research in the world. Brookhaven National Laboratory, one of the major government research centers in the country, carries out important research in all areas of science and engineering.

Applying

Applicants normally apply by January 15 for financial support and admission in the following September. The General Test of the Graduate Record Examinations is required, with the Subject Test in either biology or biochemistry strongly recommended. There is an application fee of $60. Applications are considered on a rolling basis. Whenever possible, online applications are preferred. The Application for Admission, along with detailed instructions on how to apply, may be found on the Graduate School's Web site at http://www.grad.sunysb.edu.

Correspondence and Information

Chairman, Graduate Admissions Committee
Graduate Program in Genetics
Life Sciences Building, #130
Stony Brook University, State University of New York
Stony Brook, New York 11794-5222
Phone: 631-632-8812
E-mail: kbell@notes.cc.sunysb.edu
Web site: http://life.bio.sunysb.edu/gen

Stony Brook University, State University of New York

THE FACULTY AND THEIR RESEARCH

Cancer Biology
Dafna Bar-Sagi, Wen-Tien Chen, Howard Crawford, Greg Hannon, Michael Hayman, Yuri Lazebnik, Christopher Lee, Scott Lowe, Kenneth Marcu, Ute Moll, Sentil Muthuswamy, Scott Powers, Linda VanAelst, Michael Wigler, Wei-Xing Zong.

Cell Cycle, DNA Replication, Recombination, and Repair
Dan Bogenhagen, Paul Fisher, A. Bruce Futcher, Arther Grollman, Nancy Hollingsworth, Janet Leatherwood, Arne Stenlund, Bruce Stillman.

Development
Michael Frohman, J. Peter Gergen, Michael Hadjiargyrou, Jen-Chih Hsieh, Bernadette Holdener, Alea Mills, Howard Sirotkin, Ken-Ichi Takemaru, Gerald Thomsen.

Eukaryotic Cell Biology and Cell Signaling
Carl Anderson, Debbie Brown, Nicholas Carpino, Neta Dean, Tatsuya Hirano, James Konopka, William Lennarz, Craig Malbon, Aaron Neiman, Joav Prives, Clint Rubin, David Spector, William Tansey, Thomas White.

Genomics, Computational and Structural Biology
Dax Fu, Leemor Joshua-Tor, Richard McCombie, Vivek Mittal, Andy Neuwald, John Reinitz, Steve Skiena, Steven O. Smith, Michael Zhang.

Human Genetics and Gene Therapy
Wadie Bahou, J. Craig Cohen, Paul Freimuth, Eli Hatchwell, Patrick Hearing, Nancy Mendell, Jonathan Sebat.

Microbial Pathogenesis and Gene Therapy
Jorge Benach, James Bliska, Carol Carter, Martha Furie, Erich Mackow, Jacek Skowronski, David Thanassi, Eckard Wimmer.

Neurobiology and Behavior
Turhan Canli, Hollis Cline, Holly Colognato, Joshua Dubnau, Grigori Enikolopov, Z. Josh Huang, Maurice Kernan, Roberto Malinow, Mirjana Maletic-Savatic, Styliani-Anna Tsirka, Timothy Tully, Yi Zhong.

Plant Biology
Vitaly Citovsky, David Jackson, Wolfgang Lukowitz, Robert Martienssen, Marja Timmermans.

Population Genetics and Evolution
Michael Bell, Geeta Bharathan, Walter Eanes, John True.

Regulation of Gene Expression
Paul Bingham, John Dunn, Wali Karzai, Adrian Krainer, Nancy Reich, Rolf Sternglanz, F. William Studier.

THOMAS JEFFERSON UNIVERSITY

Jefferson College of Graduate Studies
Kimmel Cancer Center
Department of Microbiology and Immunology
Graduate Program in Genetics

Program of Study

The Graduate Program in Genetics is an interdepartmental program that focuses on training in the rapidly expanding field of molecular genetics and functional genomics. The program of study leading to a Ph.D. degree is designed to provide graduating students with training and research experience to pursue careers as independent scientific investigators in academic, government, or industrial settings. Students entering with a baccalaureate degree take core courses in genetics, biochemistry, molecular biology, cell biology, bioinformatics, and immunology. Advanced courses and special-topics courses in the genetic basis of human disease; the genetics of yeast, fruit flies, and mice; and the genetic basis of cancer are offered and can be taken in both the first and second years of study. Course work during the first year is supplemented by three rotations in the laboratories of faculty preceptors who conduct active research programs on diverse problems in molecular genetics and the genetics of cancer and disease. During this time, students learn state-of-the-art research techniques and the principles of sound methods of scientific investigation. Besides frequent laboratory seminars where faculty preceptors, postdoctoral fellows, and graduate students present their current research, a wide variety of joint-program and departmental seminars and lectureships by eminent scientists from outside the Jefferson academic community form an integral part of this program. A thesis laboratory is chosen after the first year of study, and formal course work is completed by the end of the second year of study. The curriculum is flexible enough to accommodate the diverse backgrounds and the special interests of individual students. The Graduate Program in Genetics, with the Graduate Programs in Immunology and Microbial Pathogenesis, Molecular Pharmacology and Structural Biology, and Biochemistry and Molecular Biology, make up the Joint Graduate Programs of the Kimmel Cancer Center. Students applying to the Joint Graduate Programs may perform research rotations or work with faculty members in any of these programs.

Current research interests of the genetics program faculty include functional genomics, genetics of cancer susceptibility, genetics of human disease, genetics of the immune system, developmental genetics, molecular genetics of animal models of human disease, molecular genetics of hematopoietic neoplasias and solid tumors, molecular mechanisms of altered growth regulation by oncogenes and tumor suppressor genes, transcriptional regulation, chromatin organization, and the control of gene expression, and mechanisms of ionizing and nonionizing radiation damage to cells.

Research Facilities

Research laboratories are primarily located in the Kimmel Cancer Center, which occupies 70,000 square feet in the Bluemle Life Sciences Building and 25,000 square feet in the adjacent Jefferson Alumni Hall. In addition to having extensive basic equipment and facilities, the program has access to numerous specialized resources. These include peptide synthesis and sequencing facilities, microarray and proteomics facilities, a flow cytometry facility, a pathogen-free animal facility, a facility for generating transgenic and knockout mice, a bioimaging facility, a pathology facility and tumor bank, extensive computer facilities, and an X-ray crystallography facility.

Financial Aid

Financial support is available to full-time Ph.D. students in the form of University fellowships and training grants. In 2006–07, students granted full fellowship support receive funds for payment of tuition and fees, along with a stipend of $24,500. Also available to students demonstrating financial need are Title IV funds. University loan programs are also available to qualifying students.

Cost of Study

Tuition and fees for full-time Ph.D. students are $15,640 per year in 2006–07.

Living and Housing Costs

Campus housing is available to graduate students in the program. In addition, there is reasonable alternative housing located near Thomas Jefferson University.

Student Group

Each year, approximately 15 students are admitted to the Joint Graduate Programs of the Kimmel Cancer Center. Students must choose one of the Joint Ph.D. Programs for graduate study by the end of the first year, so the number of students who enter the Genetics Ph.D. Program can vary. The program currently has 21 Ph.D. students. The University enrolls about 2,500 students. The College of Graduate Studies enrolls approximately 630 students, about half of whom are women.

Student Outcomes

Recent Ph.D. graduates have accepted postdoctoral positions in prestigious academic institutions, such as Harvard, Johns Hopkins, Tufts, Baylor, UC Davis, the University of Texas M. D. Anderson Cancer Center, and the National Institutes of Health, and in industry, including Bristol-Meyers Squibb, Merck, and Pfizer.

Location

The 13-acre campus of Thomas Jefferson University is located in the historic downtown area of Philadelphia within walking distance of many places of cultural interest, including concert halls, theaters, museums, art galleries, and historic sites. There are numerous intercollegiate and professional sports events. Convenient bus and subway lines connect the University with other local universities and colleges and with several outstanding libraries. The New Jersey shore and the Pennsylvania mountains offer year-round recreational opportunities, and New York City and Washington, D.C., are just a few hours away.

The University

Thomas Jefferson University is an academic health center. It evolved from the Jefferson Medical College, which was founded in 1824, and besides the Medical College, includes the College of Graduate Studies, the College of Health Professions, the University Hospital, and various affiliated hospitals and institutions. Jefferson Alumni Hall houses the TJU Fitness and Recreation Facility, study lounges, an indoor swimming pool, a sauna, a gymnasium, a handball/squash court, and volleyball courts.

Applying

Applications can be submitted at any time but should be received before March 1 for optimal consideration. Scores on the General Test of the Graduate Record Examinations (GRE), three letters of recommendation, and academic transcripts are required of all applicants. For those students whose native language is not English, scores on the Test of English as a Foreign Language (TOEFL) are required. Scores on an appropriate Subject Test of the GRE are strongly recommended. Prospective students are encouraged to visit the Department and to discuss the graduate program with members of the faculty.

Correspondence and Information

For applications:
Jessie Pervall
Director of Admissions and Recruitment
College of Graduate Studies
Thomas Jefferson University
1020 Locust Street, M-46
Philadelphia, Pennsylvania 19107-6799

Phone: 215-503-4400
Fax: 215-503-3433
E-mail: cgs-info@jefferson.edu
Web site: http://www.jefferson.edu/cgs

For program information:
Joanne Balitzky
Training Programs Coordinator
Kimmel Cancer Center
Thomas Jefferson University
233 South 10th Street, 910 BLSB
Philadelphia, Pennsylvania 19107-5541

Phone: 215-503-6687
Fax: 215-503-0622
E-mail: joanne.balitzky@jefferson.edu
Web site: http://www.jefferson.edu/jcgs/phd/genetics/

Thomas Jefferson University

THE FACULTY AND THEIR RESEARCH

David Abraham, Professor; Ph.D., Pennsylvania, 1983. Parasite immunology; role of eosinophils and neutrophils in innate and adaptive immunity to nematode parasite infections; mechanisms of immune-mediated resistance and development of vaccines against nematode infections; chemotherapy of leishmaniasis.

Emad S. Alnemri, Professor; Ph.D., Temple, 1991. Molecular mechanisms of programmed cell death (apoptosis); regulation of caspase activation during apoptosis and inflammation; role of mitochondria in cell death and survival.

Raffaele Baffa, Associate Professor; M.D., Padua (Italy), 1987. Urology; cancer biology; bladder cancer; prostate cancer; kidney cancer; carcinogenesis; molecular oncology; molecular pathology; tumor suppressor genes.

Renato Baserga, Professor; M.D., Milan, 1949. Genetic analysis of G1-phase and control of cell proliferation; growth factors and their receptors; apoptosis; anticancer therapy.

Jeffrey L. Benovic, Professor; Ph.D., Duke, 1986. Molecular and regulatory properties of G-protein–coupled receptors; role of receptor dysregulation in cancer and cardiovascular and neurological disorders.

Bruce M. Boman, Professor; M.D., Minnesota, 1976; Ph.D., Mayo, 1982. Colon cancer genetics; integration of molecular testing into diagnosis and management of colorectal cancer; mechanisms of action of the encoded product of the colon cancer susceptibility gene, adenomatous polyposis coli (APC).

Arthur M. Buchberg, Associate Professor; Ph.D., SUNY at Buffalo, 1983. Identification of genetic modifiers of cancer; mouse models of cancer; determining the role of homeobox genes in oncogenesis and apoptosis.

Bruno Calabretta, Professor; M.D., Modena (Italy), 1977; Ph.D., Rome, 1987. Molecular mechanisms of normal hematopoiesis and BCR/ABL-dependent leukemogenesis.

Marcella Devoto, Research Associate Professor; Ph.D., Bologna (Italy), 1983. Identification of Mendelian and complex disease genes by means of statistical genetic analysis, such as linkage and association studies.

Andrea D. Eckhart, Associate Professor; Ph.D., North Carolina at Chapel Hill, 1997. Molecular mechanisms underlying hypertension; understanding the role of regulation of 7-transmembrane spanning receptor signaling in cardiovascular disease in vivo and in vitro.

Ya Ming Hou, Professor; Ph.D., Berkeley, 1986. Genetic and biochemical studies of tRNA, including structure and function mechanism of tRNA aminoacylation, maintenance of the tRNA 3' end, maturation and processing, editing and repair, decoding on the ribosome, and development of bacterial pathogenesis in infectious diseases; targeting tRNAs as strategies against metabolic and neurodegenerative disorders.

James B. Jaynes, Associate Professor; Ph.D., Washington (Seattle), 1980. Developmental genetics and molecular biology of processes regulated by homeodomain transcription factors and higher order chromatin structure.

Erica S. Johnson, Assistant Professor; Ph.D., MIT, 1992. Mechanisms for regulating protein function; conjugation pathway and function of the ubiquitin-like protein SUMO.

James H. Keen, Professor; Ph.D., Cornell, 1976. Molecular mechanisms coupling signal transduction with receptor-mediated endocytosis and exocytosis; membrane transport studied by biochemical, molecular biological, and morphological approaches.

Michael P. King, Associate Professor; Ph.D., Caltech, 1987. Mammalian mitochondrial biogenesis; molecular genetics of mtDNA mutations in human disease; mitochondrial transformation; posttranscriptional modification of mitochondrial RNAs; motor-neuron development.

Carlisle Landel, Assistant Professor; Ph.D., Colorado at Boulder, 1982. Mouse genetics; using the laboratory mouse as a model system to examine the genetics of sperm cryopreservation.

Alexander Mazo, Associate Professor; Ph.D., Russian Academy of Sciences, 1976. Chromatin modifications by epigenetic factors and nuclear hormone receptors.

Steven McKenzie, Professor; M.D./Ph.D., Pennsylvania, 1985. Murine models of immune-mediated thrombocytopenia and thrombosis syndromes; molecular genetic events that underlie progression to myelodysplasia and leukemia; investigation in murine embryonic stem cell models of hematopoiesis and leukemogenesis coupled to cDNA microarray analysis of the molecular changes in gene expression.

W. Edward Mercer, Professor; Ph.D., Penn State, 1980. Molecular mechanisms of the p53 tumor suppressor protein in cell-cycle checkpoint control and apoptotic cell death; molecular mechanisms of cell-cycle–dependent kinases, protein kinase inhibitors, and protein phosphatases in cell-cycle checkpoint control and apoptosis.

Diane E. Merry, Associate Professor; Ph.D., Pennsylvania, 1991. Molecular pathogenesis of Kennedy's disease and other polyglutamine expansion diseases; protein misfolding and aggregation in neurodegenerative disease; mouse models of neurologic disease; domain function in trafficking and degradation of the androgen receptor; gene and pharmacological therapies for neurodegenerative diseases.

Andrea Morrione, Research Assistant Professor; Ph.D., Milan, 1992. Characterization of the role of adaptor protein Grb10 in regulation of IGF-I receptor ubiquitination, trafficking, and signaling; role of growth factor proepithelin and its receptor in bladder cancer.

Jay L. Rothstein, Professor; Ph.D., Chicago, 1988. Mechanisms controlling tumor immunity and chronic inflammation; function of oncogenes in promoting cancer and autoimmune disease; use of transgenic and conditional animal models to study thyroid cancer and thyroid autoimmune disease.

Linda D. Siracusa, Associate Professor and Program Director; Ph.D., SUNY at Buffalo, 1985. Identification and characterization of genes affecting cancer susceptibility; molecular genomics and biology of the gastrointestinal tract, with special emphasis on pathways leading to colorectal cancer.

Saul Surrey, Professor; Ph.D., Temple, 1970. Molecular basis for regulated expression of human globin genes in normal and disease states; determinants for normal hemoglobin assembly and degradation as well as polymerization of sickle hemoglobin; molecular analysis of developing human erythroid progenitors; array-based strategies for SNP/mutation detection.

Jouni Uitto, Professor; Ph.D./M.D., Helsinki, 1970. Molecular genetics of the cutaneous basement membrane zone; regulation of collagen and elastin gene expression and the molecular basis of heritable and acquired connective tissue diseases.

Scott A. Waldman, Professor; Ph.D., Thomas Jefferson, 1980; M.D., Stanford, 1987. Molecular mechanisms of signal transduction, with emphasis on receptor-effector coupling and postreceptor signaling mechanisms; molecular mechanisms underlying tissue-specific transcriptional regulation; translation of molecular signaling mechanisms to novel diagnostic and therapeutic approaches to patients with cancer.

Ya Wang, Associate Professor; Ph.D., Academy of Medical Science (Beijing), 1994. Role of Fhit in maintenance of genomic integrity following low-dose radiation; molecular mechanism by which checkpoint reduces cell sensitivity to high-energy particle–induced killing; the new role of MEPE/OF45 as a cofactor of CHK1 in DNA damage response.

Charlene J. Williams, Professor; Ph.D., Rutgers, 1983. Molecular genetics and biochemistry of heritable osteoarthropathies and inflammatory arthropathies.

Edward P. Winter, Associate Professor; Ph.D., SUNY at Stony Brook, 1984. Analysis of meiotic development using molecular/genetics in yeast; signal transduction and protein kinases; transcriptional regulation.

UNIVERSITY OF CINCINNATI

Graduate Program in Molecular and Developmental Biology
at the Cincinnati Children's Medical Center

Program of Study

The Graduate Program in Molecular and Developmental Biology (MDB) at the University of Cincinnati offers superb Ph.D. training. Based in the Cincinnati Children's Medical Center, one of the nation's three largest children's hospitals, the MDB program is in a unique position to offer courses and research training in the basic sciences underlying congenital and acquired diseases of children. These include fundamental mechanisms of development, organogenesis, organ dysgenesis as the basis of childhood congenital disease, infectious diseases, asthma, childhood cancer, reproductive sciences, human and molecular genetics, biomedical engineering, and neuroscience. Basic sciences are studied in a highly integrated manner with clinical research, and there are many opportunities for translational applications. Core courses in developmental biology, development and disease, molecular genetics, and cell biology provide a strong foundation for dissertation research. The combination of training in both basic science and translational research dramatically enhances the career opportunities for graduates. Typically, students in the program go on to postdoctoral positions in major academic or biotechnology laboratories. There are approximately 70 faculty members from the Children's Medical Center and other departments in the University of Cincinnati College of Medicine. They publish actively in major journals, and speak regularly at international meetings. The Children's Medical Center has outstanding research facilities and cutting-edge core technologies are available. Students take at least two laboratory rotations before deciding on a dissertation laboratory and must pass a qualifying examination in the second year for admission to candidacy.

Research Facilities

MDB faculty members are located at Cincinnati Children's Hospital Medical Center and the University of Cincinnati College of Medicine. Research laboratories at Cincinnati Children's Hospital Medical Center occupy 500,000 square feet of research space, much of which has been constructed or renovated since 1999. Additional laboratories are located across the street in basic science departments at the University of Cincinnati College of Medicine. State-of-the-art facilities for molecular, developmental, and cell biology studies, and core facilities for creating transgenic mice (genetically modified mice), imaging, DNA analysis, genomics, proteomics, and bioinformatics facilitate student research. MDB research programs are well-supported by multiple funding agencies, including the National Institutes of Health (NIH), the American Heart Association, the American Lung Association, and the March of Dimes.

Financial Aid

Students holding baccalaureate degrees who are accepted by the program are provided with a full-tuition scholarship, health insurance, and a base stipend of $20,500 per year. Student stipends are supported by research grants from a variety of sources in addition to NIH training grants in teratology, endocrinology, and cardiac and pulmonary biology. Increased student stipends of $24,000 are available to highly qualified applicants and include the Cincinnati Children's Distinguished Research Scholar Awards, Ryan Fellowships, and externally funded predoctoral awards.

Cost of Study

All students are awarded a full-tuition scholarship, health insurance, and a stipend.

Living and Housing Costs

Furnished apartments for single and married students, maintained by the University, rented for $510 to $730 per month in 2004–05, including utilities and telephone. These apartments are limited in number, and most graduate students find less expensive apartments in adjacent areas. The cost of living in Cincinnati is relatively low.

Student Group

In 2004–05, there were 50 students in the MDB program, part of a larger graduate student population of approximately 400 graduate students in biomedical sciences at the University of Cincinnati College of Medicine.

Location

Located in the southwest corner of Ohio, Cincinnati is nestled among seven hills that overlook the Ohio River, with a population of more than 1.5 million in the greater Cincinnati area. Music, art, theater, and ballet provide an impressive array of cultural activities. Cincinnati strongly supports professional baseball (the Reds) and football (the Bengals) as well as college basketball. City and county parks provide attractive and well-maintained settings for a variety of recreational activities.

The University

Founded in 1819 as the Cincinnati College and the Medical College of Ohio, the University of Cincinnati became a municipal university in 1870; in 1977, it joined the state university system. Located on five campuses, the University consists of eighteen component colleges and divisions that provide a wide range of graduate and professional programs. It is accredited by the North Central Association of Colleges and Schools. Over the years, the University has achieved a national and international reputation in many areas of research and graduate training.

Applying

Applicants are required to take the GRE General Test and provide undergraduate transcripts (GPA score results on a 4.0 scale or equivalent). Undergraduate training in biology is required, with recommended courses in biochemistry, cell biology, molecular biology, developmental biology, and genetics. For biology majors, courses in general and organic chemistry, general physics, and algebra or calculus are recommended. For nonbiology majors, a minor in biology is recommended. Research experience is a major asset to applicants. In addition, a statement of career goals and three letters of recommendation are required. Rolling admissions are granted after February 1, and it is recommended that applicants apply before March 1 to ensure the best chance for admission. Admitted students are encouraged to enter the program by July 1 of the year in which they are admitted. Early admission requests are considered, providing all of the admission items described above are submitted before February 1. The most promising applicants are invited to visit for an interview, for which all costs and arrangements are paid for by the program.

Correspondence and Information

Tim Le Cras, Ph.D.
Director of Recruiting
Division of Developmental Biology
Cincinnati Research Foundation
3333 Burnet Avenue, ML 7007
Cincinnati, Ohio 45229-3039

Phone: 513-636-4545
E-mail: mdbprog@cchmc.org
Web site: http://www.chmcc.org/dbprog/

University of Cincinnati

THE FACULTY AND THEIR RESEARCH

Cancer Biology
Susan Waltz. Growth factors and receptor tyrosine kinases in tumorigenesis.
Susanne Wells. Papilloma virus and cervical cancer.
Yi Zheng. Molecular mechanisms of Rho GTPase signal transduction.

Cardiovascular Biology
D. Woodrow Benson. Genetic analysis of congenital heart malformations.
Jay Degen. Hemostatic factors and signaling systems in development and disease.
Sandra Degen. Biology of blood coagulation proteins and related growth factors.
Jeffery Molkentin. Molecular control of cardiac development and hypertrophy.
Jeffrey Robbins. Molecular basis of heart performance and hypertrophic cardiomyopathy.
Stephanie Ware. Genetics of cardiovascular development.
David Wieczorek. Contractile protein isoform function.
Katherine Yutzey. Cardiac morphogenesis and gene expression.

Developmental Biology and Stem-Cell Biology
Michael Bates. Digestive system differentiation and gene expression.
Nadean Brown. Vertebrate eye development.
Kenneth Campbell. Vertebrate forebrain development.
Daniel Choo. Mouse inner ear differentiation and patterning.
Tiffany Cook. Genes controlling color vision in *Drosophila*.
Brian Gebelein. Homeobox genes and body patterning in *Drosophila*.
Janet Heasman. Xenopus germ layer determination.
Rashmi Hegde. Structural analysis of transcription factors.
Chia-Yi (Alex) Kuan. Programmed cell death in the developing nervous system.
Richard Lang. Early development of the vertebrate eye.
James Lessard. Cell and molecular biology of muscle development and function.
Xinhua Lin. Cell-cell signaling in *Drosophila*.
Jun Ma. Transcriptional regulation and development.
S. Steven Potter. Homeobox gene regulation in mammalian development.
James Wells. Early endoderm patterning and pancreas development.
Daniel Wiginton. Development and differentiation of the intestine.
Aaron Zorn. Endoderm development and organ function.

Human Genetics, Gene Therapy, and Molecular Medicine
Bruce Aronow. Chromatin structure in T cells; leukemia and gene therapy vectors.
Mitchell Cohen. Intestinal secretion; activators of intestinal guanylate cyclase.
Timothy Cripe. Human gene therapy vectors.
Timothy Crombleholme. Fetal tissue repair and scarless healing.
Gregory Grabowski. Pathogenesis and therapy of human genetic disease.
Min-Xin Guan. Human mitochondrial genetic disease.
David Williams. Blood stem cell biology and gene therapy.

Immunobiology
Robert Colbert. Antigen processing and presentation in autoimmunity.
Hartmut Geiger. Hematopoietic stem cells: DNA integrity and aging; hematopoietic stem cell mobilization.
Gurjit Khurana Hershey. Genetics and pathogenesis of atopic disorders.
David Hildeman. Molecular biology of antigen-specific T cells.
Christopher Karp. Cytokine-mediated dysregulation of cellular immunity in human disease.
Jonathan Katz. Immunologic aspects of type 1 diabetes mellitus.
Marc Rothenberg. Eosinophil biology; chemokine receptor signaling pathways.

Neuroscience
Masato Nakafuku. Molecular and cellular control of neural stem cells.
Sarah Pixley. Olfactory signal transduction.
Nancy Ratner. Schwann cell-neuron interactions in development and disease.
David Repaske. Molecular basis of endocrine disorders.
Randy Sallee. Pharmacology and pharmacogenetics of neuropsychiatric disorders.
Michael T. Williams. Drugs of abuse, neurotoxicology, and behavioral teratology.

Pulmonary Biology
Ann Akeson. Lung vascular development.
Tim Le Cras. Newborn lung development and disease.
Ward Rice. Cell signaling in lung development and differentiation.
John Shannon. Lung developmental biology.
Timothy Weaver. Structure-function analysis of proteins critical for lung homeostasis.
Jeffrey Whitsett. Lung morphogenesis, gene regulation, and surfactant biology.

Reproductive Biology and Teratology
George Daston. Mechanisms of chemical teratogenesis.
Stuart Handwerger. Placental and decidual gene expression.
Daniel Nebert. Molecular basis of environmental toxicity and teratogenesis.
Charles Vorhees. Drug-induced developmental neurotoxicity.
Christopher Wylie. Development of the germ line.

UNIVERSITY OF CONNECTICUT HEALTH CENTER

Graduate Program in Genetics and Developmental Biology

Program of Study

The genetics and developmental biology graduate program provides students with fundamental interdisciplinary training in modern molecular genetics and developmental biology, emphasizing cellular and molecular aspects as well as tissue interactions. The program is intended for students pursuing a Ph.D. degree and prepares students to compete for job opportunities in traditional medical and dental school departments as well as a productive research career in either academia or industry. A combined M.D./Ph.D. program is also available. Students are encouraged to obtain in-depth training through research and courses in biochemistry, molecular biology, cell biology, developmental biology, and genetics. Faculty members are from several basic science and clinical departments and study a wide range of organisms including yeast, worms, fruit flies, mice, and humans. Areas of research include the biology of human embryonic stem cells, mapping and cloning of genes responsible for human disease, RNA processing (including RNA editing, alternative splicing, antisense regulation, and RNA interference), the molecular mechanisms of aging, signal transduction pathways, microbial pathogenesis, developmental neurobiology, cell differentiation, musculoskeletal development, morphogenesis and pattern formation, reproductive biology, and endocrinology.

Research Facilities

The Department of Genetics and Developmental Biology is the academic home of the genetics and developmental biology graduate program. The Department of Genetics and Developmental Biology occupies three floors of the state-of-the-art Academic Research Building, which opened in 1999, as well as laboratory space in adjacent buildings. The department houses equipment and facilities for mouse transgenics, ES cell manipulation, DNA microarrays, nucleic acid sequencing, fluorescence microscopy, and digital imaging. Students also have ready access to first-rate flow cytometry and confocal microscopy facilities. Other institutional resources include a computer center and a library containing approximately 100,000 volumes and subscribing to more than 3,600 current periodicals. Students of the program therefore have a world-class opportunity for research and training in cutting-edge areas of genetics and developmental biology.

Financial Aid

Support for doctoral students engaged in full-time degree programs at the Health Center is provided on a competitive basis. Graduate research assistantships for 2006–07 provide a stipend of $26,000 per year, which includes a waiver of tuition/University fees for the fall and spring semesters and a student health insurance plan. While financial aid is offered competitively, the Health Center makes every possible effort to address the financial needs of all students during their period of training.

Cost of Study

For 2006–07, tuition is $3996 per semester ($7992 per year) for full-time students who are Connecticut residents and $10,368 per semester ($20,772 per year) for full-time out-of-state residents. General University fees are added to the cost of tuition for students who do not receive a tuition waiver. These costs are usually met by traineeships or research assistantships for doctoral students.

Living and Housing Costs

There is a wide range of affordable housing options in the greater Hartford area within easy commuting distance of the campus, including an extensive complex that is adjacent to the Health Center. Costs range from $600 to $800 per month for a one-bedroom unit; 2 or more students sharing an apartment usually pay less. University housing is not available at the Health Center.

Student Group

At Farmington, there are about 500 students in the Schools of Medicine and Dental Medicine, 150 Ph.D. students, and 50 postdoctoral fellows. There are no restrictions on the admission of out-of-state graduate students.

Location

The Health Center is located in the historic town of Farmington, Connecticut. Set in the beautiful New England countryside on a hill overlooking the Farmington Valley, it is close to ski areas, hiking trails, and facilities for boating, fishing, and swimming. Connecticut's capital city of Hartford, 7 miles east of Farmington, is the center of an urban region of approximately 800,000 people. The beaches of the Long Island Sound are about 50 minutes away to the south, and the beautiful Berkshires are a short drive to the northwest. New York City and Boston can be reached within 2½ hours by car. Hartford is the home of the acclaimed Hartford Stage Company, TheatreWorks, the Hartford Symphony and Chamber orchestras, two ballet companies, an opera company, the Wadsworth Atheneum (the oldest public art museum in the nation), the Mark Twain house, the Hartford Civic Center, and many other interesting cultural and recreational facilities. The area is also home to several branches of the University of Connecticut, Trinity College, and the University of Hartford, which includes the Hartt School of Music. Bradley International Airport (about 20 minutes from campus) serves the Hartford/Springfield area with frequent airline connections to major cities in this country and abroad. Frequent bus and rail service is also available from Hartford.

The Health Center

The 200-acre Health Center campus at Farmington houses a division of the University of Connecticut Graduate School, as well as the School of Medicine and Dental Medicine. The campus also includes the John Dempsey Hospital, associated clinics, and extensive medical research facilities, all in a centralized facility with more than 1 million square feet of floor space. The Health Center's newest research addition, the Academic Research Building, was opened in 1999. This impressive eleven-story structure provides 170,000 square feet of state-of-the-art laboratory space. The faculty at the center includes more than 260 full-time members. The institution has a strong commitment to graduate study within an environment that promotes social and intellectual interaction among the various educational programs. Graduate students are represented on various administrative committees concerned with curricular affairs, and the Graduate Student Organization (GSO) represents graduate students' needs and concerns to the faculty and administration, in addition to fostering social contact among graduate students in the Health Center.

Applying

Applications for admission should be submitted on standard forms obtained from the Graduate Admissions Office at the UConn Health Center or the Web site listed below. The application should be filed together with transcripts, three letters of recommendation, a personal statement, and recent results from the General Test of the Graduate Record Examinations. International students must take the Test of English as a Foreign Language (TOEFL) to satisfy Graduate School requirements. The deadline for completed applications and receipt of all supplemental materials is December 15. In accordance with the laws of the state of Connecticut and of the United States, the University of Connecticut Health Center does not discriminate against any person in its educational and employment activities on the grounds of race, color, creed, national origin, sex, age, or physical disability.

Correspondence and Information

Dr. William Mohler
Program Director
Department of Genetics and Developmental Biology
University of Connecticut Health Center
Farmington, Connecticut 06030-3301
Phone: 860-679-1833
E-mail: wmohler@neuron.uchc.edu
Web site: http://grad.uchc.edu

University of Connecticut Health Center

THE FACULTY AND THEIR RESEARCH

Andrew Arnold, M.D., Professor of Medicine and Murray-Heilig Chair in Molecular Medicine. Molecular genetic underpinnings of tumors of the endocrine glands; role of the cyclin D1 oncogene.

Choukri Ben Mamoun, Ph.D., Associate Professor of Genetics and Developmental Biology. Cell signaling and transport of nutrients in the human malaria parasite *Plasmodium falciparum.*

Gordon Carmichael, Ph.D., Professor of Genetics and Developmental Biology. Regulation of gene expression in eukaryotes.

Kevin Claffey, Ph.D., Assistant Professor of Cell Biology. Angiogenesis in cancer progression and metastasis; vascular endothelial growth factor (VEGF) expression; hypoxia-mediated gene regulation.

Stephen Clark, Ph.D., Associate Professor of Medicine. Characterization of mutations affecting connective tissues; molecular genetic mapping; generation and analysis of transgenic mice.

Asis Das, Ph.D., Professor of Microbiology. Basic genetic and biomechanical mechanisms that govern the elongation-termination and decision in transcription.

Caroline N. Dealy, Ph.D., Assistant Professor of Anatomy. Roles of various growth factors and signaling molecules, particularly IGF-I and insulin, in the regulation of chick limb development.

Paul Epstein, Ph.D., Associate Professor of Pharmacology. Receptor signal transduction, second messengers, and protein phosphorylation in control of cell growth and regulation; purification and regulation of cyclic nucleotide phosphodiesterases; role of calmodulin in mediating Ca^{2+}-dependent cell processes.

Guo-Hua Fong, Ph.D., Assistant Professor of Cell Biology. Developmental biology of the vascular system, VEGF-A receptor signal transduction, embryonic stem cells, and gene knock-out in mice.

Brenton R. Graveley, Ph.D., Associate Professor of Genetics and Developmental Biology. Regulation of alternative splicing in the mammalian nervous system and mechanisms of alternative splicing.

Arthur Günzl, Ph.D., Associate Professor, Center for Microbial Pathogenesis. Transcription and antigenic variation in the mammalian parasite *Trypanosoma brucei.*

Marc Hansen, Ph.D., Professor of Medicine. Molecular genetics of osteosarcoma and related bone diseases.

Timothy Hla, Ph.D., Professor of Cell Biology. Molecular mechanisms of angiogenesis; G-protein–coupled receptor signaling; cyclooxygenase (Cox-2) and cancer; lipid mediators.

Laurinda Jaffe, Ph.D., Professor of Cell Biology. Physiology of fertilization.

Robert A. Kosher, Ph.D., Professor of Anatomy. Limb development; role of extracellular matrix, cytoskeleton, and cyclic nucleotides in chondrogenesis; molecular regulation of gene activity during cartilage differentiation.

Barbara Kream, Ph.D., Professor of Medicine. Hormonal regulation of collagen gene expression in bone.

Marc Lalande, Ph.D., Professor of Genetics and Developmental Biology. Genomic imprinting; Angelman syndrome; mechanism of tissue-specific silencing of the Angelman ubiquitin ligase in mouse and human.

James Li, Ph.D., Assistant Professor of Genetics and Developmental Biology. Identifying the molecular mechanisms underlying formation of the mammalian cerebellum.

Alexander Lichtler, Ph.D., Associate Professor of Pediatrics. Regulation of collagen gene transcription; retrovirus vectors; role of homeobox genes in limb development.

Bruce Mayer, Ph.D., Associate Professor of Genetics and Developmental Biology. Biologically relevant Nck-interacting proteins.

Mina Mina, D.M.D., Ph.D., Associate Professor of Pediatric Dentistry. Characterization of genetic and epigenetic influences involved in pattern formation and skeletogenesis of the chick mandible and mouse tooth germ.

William Mohler, Ph.D., Assistant Professor of Genetics and Developmental Biology. Molecular and cellular mechanisms of cell fusion.

D. Kent Morest, M.D., Professor of Neuroscience. Role of cell and tissue interactions in the migration and differentiation of neurons; structure and function of neurons during development and synapse formation.

Achilles Pappano, Ph.D., Professor of Pharmacology. Development of electrical properties of the heart; maturation of cardiac autonomic receptors; neurogenesis in the embryonic heart.

John Peluso, Ph.D., Professor of Cell Biology. Control of ovarian follicle growth steroidogenesis in vitro; proto-oncogene expression and ovarian follicular growth.

Justin D. Radolf, M.D., Professor of Medicine and Genetics and Developmental Biology. Molecular pathogenesis and immunobiology of spirochetal infections.

Robert Reenan, Ph.D., Associate Professor of Genetics and Developmental Biology. Molecular mechanisms underlying electrical signaling in the *Drosophila* nervous system.

Blanka Rogina, Ph.D., Assistant Professor of Genetics and Developmental Biology. Molecular mechanism underlying aging process in *Drosophila melanogaster.*

Edward F. Rossomando, D.D.S., Ph.D., Professor of BioStructure and Function. Control of gene expression in tumor and nontumor cell lines in response to stimulation by monokines; coding, transmission, and processing of environmental signals in normal and abnormal development.

David Rowe, M.D., Professor of Pediatrics. Hormonal regulation of Type I collagen in mature and developing bone; heritable disorders of bone formation.

Mansoor Sarfarazi, Ph.D., Associate Professor of Surgery. Positional mapping and mutation analysis of human genetic disorders; primary open angle glaucoma; primary congenital glaucoma; synpolydactyly; dyslexia; mitral valve prolapse and ascending aortic aneurysm.

Marvin Tanzer, M.D., Professor of BioStructure and Function. Role of the extracellular matrix in developing systems; regulation of the expression of collagens, proteoglycans, laminin, and fibronectin.

Petros Tsipouras, M.D., Professor of Pediatrics. Heritable disorders of connective tissue, nosology, and genetics; genetic linkage studies; molecular mechanisms of mutations in human collagen genes.

William B. Upholt, Ph.D., Professor of BioStructure and Function. Regulation of gene expression during embryonic development; procollagen gene expression and regulation in limb chondrogenesis and skeletogenesis; pattern formation; homeobox genes.

Bruce White, Ph.D., Professor of Physiology. Control of prolactin gene expression at pretranslational level in GH3 cells; control of aromatase gene expression in ovarian and testicular tissues.

Stephen K. Wikel, Ph.D., Professor of Cell Biology. Characterization of the complex cellular and molecular immunology of the tick-host-pathogen interface.

Dianqing Wu, Ph.D., Associate Professor of Genetics and Developmental Biology. Signaling mechanisms of Wnt proteins; molecular basis and function of signal transduction.

Ren-He Xu, Ph.D., Associate Professor of Genetics and Developmental Biology. Biology of human embryonic stem cells.

University of Connecticut Health Center.

UNIVERSITY OF MASSACHUSETTS WORCESTER

Department of Molecular Genetics and Microbiology

Program of Study

The Department of Molecular Genetics and Microbiology at the University of Massachusetts Medical School offers a program of study and research that leads to a Ph.D. degree in medical sciences. The faculty of the Department is composed of 21 members whose research interests include molecular and cellular immunology, molecular biology, genetics, virology, bacteriology, developmental biology, and neurobiology. The Department is dedicated to graduate training, and its faculty members teach in the Graduate School Core Course as well as advanced graduate courses in immunology, development, nucleic acids, genetics, neuroendocrinology, virology, and microbial pathogenesis.

Incoming students should have good preparation in the fundamentals of biochemistry, genetics, and molecular or cell biology. Formal course requirements for the Ph.D. include the Graduate School Core Course, a two-semester interdisciplinary course integrating biochemistry, molecular biology, genetics, and cell biology, as well as two advanced topics courses. Students are required to undertake two or more laboratory rotations prior to selecting a laboratory for thesis research. In addition to the course work and research, students are expected to attend Departmental and medical school seminars and to participate in a weekly Departmental seminar program. Admission to candidacy is granted following successful completion of a qualifying exam administered at the conclusion of the second year. The Ph.D. degree generally requires about five years to obtain and is awarded following completion and successful defense of thesis research. The training received through this program is intended to provide students with the academic background, experimental tools, and analytical skills necessary to conduct independent research and teach modern biology.

Research Facilities

The Medical Center is a large, modern teaching hospital with extensive laboratory space for both basic and clinical research. The ten-story, 360,000-square-foot Aaron Lazare Medical Research Building opened in 2002. University departments and programs are also housed in the adjacent Massachusetts Biotechnology Research Park. The Department of Molecular Genetics and Microbiology occupies one floor in the Basic Sciences Wing of the Medical Center and laboratories in the Biotechnology Park.

Facilities and equipment are the most modern available for biomedical research and include core facilities for protein sequencing, nucleic acid sequencing, electron microscopy, tissue culture, fluorescence cell sorting, mass spectroscopy, and biomedical imaging, as well as a large and well-maintained experimental animal care facility.

Financial Aid

All beginning full-time students are offered a graduate assistantship with an annual stipend ($25,000). Research assistantships paying $25,000 are available to all students starting thesis research.

Cost of Study

Tuition fees are waived for all graduate students. Students pay fees for student activities and equipment of $190 per semester. Health insurance is provided by the University to all students. These figures are subject to change.

Living and Housing Costs

Apartments and rooms are available in the Worcester area and nearby communities. The large number of students in the area makes the sharing of apartments feasible, at costs that can be borne by students.

Student Group

The Department has 20 graduate students enrolled in the Ph.D. program. The total enrollment of the Graduate School of Biomedical Sciences is approximately 350 students.

Student Outcomes

Most graduates pursue postdoctoral positions at outstanding institutions and eventually take staff positions at biotechnology companies or faculty positions in academic institutions throughout the U.S. The greatest number of postdoctoral positions secured by recent graduates are at Harvard Medical School, Yale University, University of Washington, and the National Institutes of Health.

Location

Worcester is located 37 miles west of Boston, and the two cities share a wealth of scientific and cultural opportunities. The University maintains close ties with other local educational institutions, including Clark University, Holy Cross College, Worcester Polytechnic Institute, and the Worcester Foundation for Experimental Biology. For recreation, the beaches of Cape Cod and the ski slopes of Vermont and New Hampshire are just 1 to 2 hours away.

The Medical Center

The Medical Center is an integral part of the state university system, providing advanced degree programs in the Graduate School of Biomedical Sciences (GSBS), the Medical School (UMMS), and the Graduate School of Nursing (GSN). The Medical Center has a well-established research and teaching community composed of faculty members from seven basic science departments and thirteen clinical science departments. The GSBS offers Ph.D. and M.D./Ph.D. programs and a D.V.M./Ph.D. program with Tufts University Veterinary School in the nearby town of Grafton.

Applying

The Department accepts graduate students through the Graduate School of Biomedical Sciences. Applicants should have a bachelor's degree in a physical or biological science. Specific course requirements include a year each of calculus, organic chemistry, physics, and biology. Applicants must take the General Test of the Graduate Record Examinations. Three letters of recommendation and an application fee, $25 in state and $50 out of state, are required. Applications should be received as soon as possible and not later than January 2.

The University of Massachusetts Medical Center is an Equal Opportunity/Affirmative Action employer. The University of Massachusetts Medical School recruits minority and women candidates, considers all applicants, and, once students are accepted, ensures that they will not be discriminated against in any area.

The graduate catalog and application forms may be obtained from the address listed in this Close-Up. Potential applicants may wish to contact Dr. Richard Baker, Graduate Director, at 508-856-6046 for additional information about the Molecular Genetics and Microbiology Program.

Correspondence and Information

Graduate School of Biomedical Sciences
University of Massachusetts Worcester
55 Lake Avenue, North
Worcester, Massachusetts 01655
Phone: 508-856-4135
E-mail: gsbs@umassmed.edu
Web site: http://www.umassmed.edu/gsbs/

University of Massachusetts Worcester

THE FACULTY AND THEIR RESEARCH

Brian J. Akerley, Assistant Professor; Ph.D., UCLA, 1995. Genetics of bacterial pathogenesis. Environmental and genetic regulation of the phosphorylcholine epitope of *H. influenzae. Mol. Microbiol.* 55(3):724–38, 2005 (with Wong). Position-based scanning for comparative genomics and identification of genetic islands in *Haemophilus influenzae* type b. *Infect. Immun.* 71:1098–108, 2003 (with Bergman). A genome-scale analysis of genes required for growth or survival of *Haemophilus influenzae. Proc. Natl. Acad. Sci. U.S.A.* 99:966–71, 2002 (with Rubin et al.).

Richard E. Baker, Associate Professor; Ph.D., Penn State, 1980. Chromatin structure and centromere function. The histone fold domain of Cse4 is sufficient for *CEN* targeting and propagation of active centromeres in budding yeast. *Eukaryotic Cell* 3:1533–43, 2004 (with Morey et al.). The N terminus of the centromere H3-like protein Cse4p provides a function distinct from that of the histone fold domain. *Mol. Cell. Biol.* 20:7037–48, 2000 (with Chen et al.).

Paul R. Dobner, Associate Professor; Ph.D., Columbia, 1981. Neuropeptide function and regulation. Endogenous neurotensin facilitates visceral nociception and is required for stress-induced antinociception in mice and rats. *Neuroscience*, in press (with Gui and Carraway). Neurotensin-deficient mice show altered responses to antipsychotic drugs. *Proc. Natl. Acad. Sci. U.S.A.* 98:8048–53, 2001 (with Fadel et al.).

Rachel M. Gerstein, Associate Professor; Ph.D., Brandeis, 1991. Developmental regulation of V(D)J recombination and B-cell development. Pairs of violet-excited fluorophores for flow cytometric analysis. *Cytometry* 33:435–44, 1999 (with Anderson et al.). 8 color, 10 parameter flow cytometry to elucidate complex leukocyte heterogeneity. *Cytometry* 29:323–39, 1997 (with Roederer et al.). Defective B-cell development and function in Btk deficient mice. *Immunity* 3:283–300, 1995 (with Sideras et al.).

Jon Goguen, Associate Professor; Ph.D., Massachusetts, 1980. Mechanisms of bacterial metastasis; function and regulation of virulence determinants in *Yersinia pestis*, the causative agent of plague. Temperature sensing in *Yersinia pestis:* Translation of the LcrF activator protein is thermally regulated. *J. Bacteriol.* 175:7901–9, 1993 (with Hoe). A surface protease and the invasion character of plague. *Science* 258:1004–7, 1992 (with Sodeinde et al.). Temperature sensing in *Yersinia pestis:* Regulation of *yopE* transcription by *lcrF. J. Bacteriol.* 174:4275–86, 1992 (with Hoe and Minion).

Ronald M. Iorio, Associate Professor; Ph.D., Boston College, 1979. Viral pathogenesis and virulence; virus entry and the mechanism of virus-induced membrane fusion; viral glycoproteins and receptors. Decreased dependence on receptor recognition for the fusion promotion activity of L289A-mutated Newcastle disease virus fusion protein correlates with a monoclonal antibody-detected conformational change. *J. Virol.* 79:1180–90, 2005 (with Li et al.). Amino acid substitutions in the F-specific domain in the stalk of the Newcastle disease virus HN protein modulate fusion and interfere with its interaction with the F protein. *J. Virol.* 78:13053–61, 2004 (with Melanson).

Allan Jacobson, Professor and Chair; Ph.D., Brandeis, 1971. Cytoplasmic aspects of the posttranscriptional regulation of gene expression. Positive and negative regulation of poly(A) nuclease. *Mol. Cell. Biol.* 24:5521–33, 2004 (with Mangus et al.). Regulation of mRNA decay: Decapping goes solo. *Mol. Cell* 15:1–2, 2004. A faux 3'-UTR promotes aberrant termination and triggers nonsense-mediated mRNA decay. *Nature* 432:112–8, 2004 (with Amrani et al.).

Duane D. Jenness, Associate Professor; Ph.D., Berkeley, 1980. Genetic control of signal transduction in yeast. The C terminus of the *Saccharomyces cerevisiae* α-factor receptor contributes to the formation of preactivation complexes with its cognate G protein. *Mol. Cell. Biol.* 20:5321–9, 2000 (with Dosil et al.). Homo-oligomeric complexes of the yeast α-factor pheromone receptor are functional units of endocytosis. *Mol. Biol. Cell.* 11:2873–84, 2000 (with Yesilaltay).

Timothy F. Kowalik, Associate Professor; Ph.D., Utah State, 1989. Cell proliferation and DNA viruses; relationship between proliferation and apoptosis. Apoptosis associated with deregulated E2F activity is dependent on E2F1 and Atm/Nbs1/Chk2. *Mol. Cell. Biol.* 24:2968–77, 2004 (with Rogoff et al.). E2F1 transcription factor links proliferation to apoptosis. *Mol. Cell. Biol.* 22:5308–18, 2002 (with Rogoff et al.). Cytomegalovirus IE proteins in growth control. *J. Virol.* 74:8028–37, 2000 (with Castillo and Yurochko).

John M. Leong, Professor; Ph.D., 1985, M.D., 1987, Brown. Colonization by the Lyme disease and relapsing fever spirochetes; host-cell signaling by pathogenic *E. coli.* EspF$_U$ is a translocated EHEC effector that interacts with Tir and N-WASP and promotes Nck-independent actin assembly. *Dev. Cell* 7(2):217–28, 2004 (with Campellone et al.). B1b lymphocytes confer T cell–independent long-lasting immunity. *Immunity* 3:379–90, 2004. (with Alugupalli et al.).

Dannel McCollum, Associate Professor; Ph.D., California, San Diego, 1994. Cell-cycle control. The *S. pombe* Cdc14-like phosphatase Clp1p regulates chromosome bi-orientation and interacts with aurora kinase. *Dev. Cell* 7:755–62, 2004 (with Trautmann et al.). Initiation of cytokinesis is controlled through multiple modes of regulation of the Sid2p-Mob1p kinase complex. *Mol. Cell. Biol.* 24:3262–76, 2004 (with Hou et al.). Dma1p prevents mitotic exit and cytokinesis by inhibiting the septation initiation network (SIN). *Dev. Cell* 3:779–90, 2003 (with Geurtin et al.).

Trudy G. Morrison, Professor; Ph.D., Tufts, 1972. Enveloped virus entry; assembly; virus glycoprotein structure and function. The role of the cytoplasmic domain of the NDV fusion protein in association with lipid rafts. *J. Virol.* 77:12968–79, 2003 (with Dolganiuc et al). Interacting domains of the HN and F proteins of NDV. *J. Virol.* 77:11040–9, 2003 (with Gravel).

Kenan C. Murphy, Assistant Professor; Ph.D., Maryland, 1983. λ Red recombineering of pathogenic bacteria; mechanistic analysis of recombination and repair proteins. Lambda Red–mediated recombinogenic engineering of enterohemorrhagic and enteropathogenic *E. coli. BMC Mol. Biol.* 4:11, 2003 (with Campellone). Bacteriophage P22 Abc2 protein binds to RecC, increases the 5' strand nicking activity of RecBCD and together with λ Bet, promotes Chi-independent recombination. *J. Mol. Biol.* 296:385–401, 2000.

Anthony R. Poteete, Professor; Ph.D., MIT, 1977. Bacteriophage-promoted homologous recombination. Modulation of DNA repair and recombination by the bacteriophage λ Orf function in *Escherichia coli* K-12. *J. Bacteriol.*, in press. Phage λ Red–mediated adaptive mutation. *J. Bacteriol.* 184:3753–5, 2002 (with Wang and Foster). Recombination-promoting activity of the bacteriophage λ Rap protein in *Escherichia coli* K-12. *J. Bacteriol.* 184:4626–9, 2002 (with Fenton and Wang).

Peter M. Pryciak, Associate Professor; Ph.D., California, San Francisco, 1992. Yeast signal transduction and cell polarity. Interaction with the SH3 domain protein Bem1 regulates signaling by the *Saccharomyces cerevisiae* p21-activated kinase Ste20. *Mol. Cell. Biol.* 25:2177–90, 2005 (with Winters). Role of scaffolds in MAP kinase pathway specificity revealed by custom design of pathway-dedicated signaling proteins. *Curr. Biol.* 11:1815–24, 2001 (with Harris et al.).

Christopher M. Sassetti, Assistant Professor; Ph.D., California, San Francisco, 2000. Pathogenesis of tuberculosis. Genetic requirements for mycobacterial survival during infection. *Proc. Natl. Acad. Sci. U.S.A.* 100(22):12989–94, 2003 (with Rubin). Genes required for mycobacterial growth defined by high-density mutagenesis. *Mol. Microbiol.* 48:77–84, 2003 (with Rubin and Boyd). Comprehensive identification of conditionally essential genes in mycobacteria. *Proc. Natl. Acad. Sci. U.S.A.* 98:12712–7, 2001 (with Rubin and Boyd).

Janet Stavnezer, Professor; Ph.D., Johns Hopkins, 1971. Molecular mechanism of antibody class switching: role of DNA repair genes. Deletion of the nucleotide excision repair gene *Ercc1* reduces immunoglobulin class switching and alters mutations near switch recombination junctions. *J. Exp. Med.* 200:321–30, 2004 (with Schrader et al.). Mutations occur in the Ig Sμ region but rarely in Sγ regions prior to class switch recombination. *EMBO J.* 22:5893–903, 2003 (with Schrader et al.).

Donald J. Tipper, Emeritus Professor; Ph.D., Birmingham (England), 1959. Analysis of transmembrane insertion of mammalian prion proteins in yeast. Yeast genes controlling responses to topogenic signals in a model transmembrane protein. *Mol. Biol. Cell* 13:1158–74, 2002 (with Harley).

Michael R. Volkert, Professor; Ph.D., Rutgers, 1977. Human DNA repair and protection genes. Stress induction and mitochondrial localization of OXR1 proteins in yeast and humans. *Mol. Cell. Biol.* 24:3180–7, 2004 (with Elliott). The single-strand DNA binding activity of human PC4 functions to prevent mutagenesis and killing by oxidative DNA damage. *Mol. Cell. Biol.* 24:6084–93, 2004 (with Wang et al.).

Raymond M. Welsh, Professor; Ph.D., Massachusetts Amherst, 1972. Viral immunology and pathogenesis. Cutting edge: MHC class II–restricted killing in vivo during viral infection. *J. Immunol.* 174: 614–8, 2005 (with Jellison and Kim). Comprehensive early and lasting loss of memory CD8 T cells and functional memory during acute and persistent viral infections. *J. Immunol.* 172:3139–50, 2004 (with Kim). Immunological memory to viral infections. *Ann. Rev. Immunol.* 22:711–43, 2003 (with Selin and Szomolanyi-Tsuda).

Robert T. Woodland, Associate Professor; Ph.D., Pennsylvania, 1974. Regulation of B-cell survival and homeostasis, gene transfer into lymphocytes. Expression of a human coxsackie/adenovirus receptor transgene permits adenovirus infection of primary lymphocytes. *J. Immunol.* 165:4112–9, 2000 (with Schmidt et al.).

Affiliated Faculty
Elliot Androphy, M.D. (Medicine). Victor Boyartchuk, Ph.D. (Molecular Medicine). Paul R. Clapham, Ph.D. (Molecular Medicine). Ronald C. Desrosiers, Ph.D. (New England Regional Primate Center). Richard Ellison III, M.D. (Medicine). Francis A. Ennis, M.D. (Medicine). Robert W. Finberg, M.D. (Medicine). Douglas T. Golenbock, M.D. (Medicine). R. Paul Johnson, Ph.D. (New England Regional Primate Center). Jae Jung, Ph.D. (New England Regional Primate Center). Peter E. Newburger, M.D. (Medicine). Andrea Pereira, Ph.D. (Medicine). Ravindra Singh, Ph.D. (Medicine). Mario Stevenson, Ph.D. (Molecular Medicine). John L. Sullivan, M.D. (Medicine). Doyle Ward, Ph.D. (Molecular Medicine). Maria L. Zapp, Ph.D. (Molecular Medicine).

THE UNIVERSITY OF TENNESSEE AND OAK RIDGE NATIONAL LABORATORY

Graduate School of Genome Science and Technology

Program of Study

The University of Tennessee–Oak Ridge National Laboratory Graduate School of Genome Science and Technology (GST) has a focus on new developments in the biological and computational sciences that stem from genome sequencing efforts. The program is designed to take advantage of the unique opportunity that the University of Tennessee and Oak Ridge National Laboratory (ORNL) have for interaction and collaboration. Students are trained in emerging areas of genome science, with emphasis on mammalian genomics, structural biology, proteomics, computational biology/bioinformatics, and bioanalytical technologies. Scientists from both campuses participate in teaching. Students work on either Ph.D. or M.S. programs that integrate research projects mentored jointly by a faculty member from each campus. A yearlong introductory course in genome science and technology focuses on scientific inquiry that is being conducted on a genome-wide scale. Laboratory rotations during the first year offer students hands-on experiences in a variety of the focus areas in the GST program.

Research Facilities

GST facilities include excellent classrooms, laboratories, and administrative offices. These facilities include those of the ORNL Life Sciences Division, which houses well-equipped research areas and an animal-care and experimental area of 36,000 square feet. The library systems of Oak Ridge Associated Universities, Oak Ridge National Laboratory, and the University of Tennessee house excellent technical libraries.

Financial Aid

The stipend level is determined on a yearly basis and is set at a competitive level with other regional graduate programs. The stipend level for the 2006–07 academic year is $20,000. In addition, the program offers remission of tuition. Retention of support is contingent on a student's progress and adequate performance in the program.

Cost of Study

In 2006–07, the semester fees are $3183 for Tennessee residents. Tuition and fees for nonresident students are $8966 per semester. Tuition is waived for students with financial aid.

Living and Housing Costs

Reasonably priced apartments are available in the Oak Ridge and Knoxville areas, and low-cost University apartments are available in Knoxville. An automobile is a practical necessity for School and general transportation. Limited public transportation exists.

Student Group

The University of Tennessee Knoxville campus, located approximately 25 miles east of Oak Ridge, has more than 6,000 graduate students in various fields, including 300 students in the biological and chemical sciences and several hundred graduate students and postdoctoral fellows at ORNL. In its seventh year, this program has an enrollment of 40 graduate students and 2 postdoctoral fellows. It anticipates a significant expansion of the student body.

Location

The Knoxville/Oak Ridge area, with a population of nearly 400,000, has one of the highest concentrations of Ph.D. holders in the United States. Light industry and education are the predominant types of jobs in the region, and a relatively low cost of living is enjoyed. The area boasts an excellent climate and access to many lakes for fishing, boating, and water sports. Hiking, camping, and nature tours are available year-round in the nearby Great Smoky Mountains National Park, Big South Fork National River and Recreation Area, and Obed Wild and Scenic River.

The School

In 1998, the Graduate School of Biomedical Sciences was renamed the Graduate School of Genome Science and Technology, with emphases on new developments in the biological and computational sciences that stem from genome sequencing efforts.

Applying

The GST program welcomes applicants who have majored in the sciences. The application process includes the GST application form, the application form for the University of Tennessee Graduate School, three reference letters from qualified individuals, Graduate Record Examinations (GRE) scores, and official transcripts from all undergraduate and graduate institutions attended. A $35 fee must accompany the application to the University of Tennessee Graduate School. The deadline for application for fall admission is May 1 for U.S. citizens. For international applicants, the deadline for application for fall admission is February 1.

Correspondence and Information

For application material:
Dr. Cynthia B. Peterson, Director
The University of Tennessee–Oak Ridge National Laboratory
Graduate School of Genome Science and Technology
1060 Commerce Park
Oak Ridge, Tennessee 37830-8026
Phone: 865-574-1227
Fax: 865-574-4812 or 974-0361
Web site: http://gst.ornl.gov/

For Graduate School information:
Graduate Student Services
218 Student Services Building
The University of Tennessee
Knoxville, Tennessee 37996-0220
Phone: 865-974-3251
Fax: 865-974-6541
Web site: http://web.utk.edu/~gsinfo/

The University of Tennessee and Oak Ridge National Laboratory

THE FACULTY AND THEIR RESEARCH

GENETICS/GENOMICS
Cymbeline Culiat, Ph.D., Tennessee. DNA repair; genetic susceptibility; gene expression; embryonic stem cells; mutagenesis.
Ranjan Ganguly, Ph.D., Nebraska. Gene structure; gene regulation.
Dabney Johnson, Ph.D., Tennessee. Phenotype; screening; mouse genetics; mouse mutations; deletion complexes; mouse models.
Jung Han Kim, Ph.D., Tennessee. Etiology and mechanisms underlying non-insulin-dependent diabetes mellitus (NIDDM) and obesity; concomitantly related diseases.
Mitchell L. Klebig, Ph.D., Tennessee. Intracellular transport; endocytosis; fit1; Picalm; hematopoiesis; anemia; iron metabolism; scoliosis.
John Koontz, Ph.D., Kentucky. Role of insulin and glucocorticoids in controlling gene expression.
Kurt H. Lamour, Ph.D., Michigan State. Plant pathogens in the genus *Phytophthora;* functional genomics; molecular marker discovery.
Bruce McKee, Ph.D., Michigan State. Genetics of chromosome pairing and recombination, sperm development, and chromosome structure.
Ed Michaud, Ph.D., Tennessee. Transgenics and targeted mutagenesis in mice.
Naima Moustaid-Moussa, Ph.D., Paris. Nutritional and hormonal regulation of adipocyte gene expression and role in obesity and comorbidity; paracrine function of adipocytes.
Rebecca Prosser, Ph.D., Illinois. Cellular basis of mammalian circadian rhythms.
Todd B. Reynolds, Ph.D. Vanderbilt. Biofilm; yeast; Candida albicans; fungal pathogenesis; cell wall; phospholipids; protein trafficking.
Arnold M. Saxton, Ph.D., North Carolina State. Statistics; data analysis; metabolic pathways; complex traits; quantitative genetics; animal breeding.
Pamela Small, Ph.D., Stanford. Bacterial pathogenesis; host-bacterial interactions; mycobacterial pathogenesis.
C. Neal Stewart Jr., Ph.D., Virginia Polytechnic. Transgenic plants; introgression; functional genomics; environmental sensing; plant-insect interactions; plant-pathogen interactions; biophotonics; phytoremediation; bio/ag security.
Gary Truett, Ph.D., Georgia. Developmental genetics of growth, energy balance, and obesity.
Brynn Jones Voy, Ph.D., Tennessee. Using cDNA microarrays and molecular phenotypes to understand mouse models of human disease.
Hwa-Chain Robert Wang, Ph.D., Virginia. Cellular transformation to cell quiescence or apoptosis.
Yisong Wang, Ph.D., Karolinska (Stockholm). Cell cycle; cancer; cell cycle regulatory proteins; oncogenes; tumor suppressor genes; protein ubiquitination and degradation; mass spectrometry; ENU-induced mutant mice.
Jay Wimalasena, Ph.D., Colorado Health Sciences Center. Estrogens; breast; ovarian cancer; cell cycle; apoptosis; cyclins; Cdk; signal transduction.
Stan D. Wullschleger, Ph.D., Arkansas. Physiological response of plants to global environmental change.
Xuemin Xu, Ph.D., Tokyo Institute of Technology. Molecular and cellular mechanism of Alzheimer's disease and related neuronal degeneration; apoptosis.
Yun You, Ph.D., Purdue. Deletion complexes in mouse embryonic stem cells.

STRUCTURAL BIOLOGY/PROTEOMICS
Jeffrey Becker, Ph.D., Cincinnati. Peptides; hormones; receptors; membrane transport; fungal virulence; yeast.
Barry Bruce, Ph.D., Berkeley. Protein transport; membrane biogenesis; organelle evolution; photosynthesis.
Gerard Bunick, Ph.D., Pennsylvania. Protein cryocrystallography; nucleosome structure.
Chris Dealwis, Ph.D., London. Protein crystallography; rational design of renin inhibitors.
Elias Fernandez, Ph.D., Loyola Chicago. Three-dimensional structures of proteins and the relevance of these structures in biology.
Eli Greenbaum, Ph.D., Columbia. Photosynthesis; renewable hydrogen production; nanoscale photosynthesis; nanobiotechnology; molecular electronics; artificial sight; biomedical research.
Elizabeth Howell, Ph.D., Lehigh. Structure, function, assembly, and folding of chromosomal and plasmid forms of dihydrofolate reductase.
Nitin Jain, Ph.D., Brandeis. Structure and functional dynamics of proteins involved in biological electron transfer and immune response using NMR; protein-protein interactions; structure-based drug design.
Steve Kennel, Ph.D., California, San Diego. Radioimmunotherapy; vasculature; phage display; inflammation; cytokines.
Peter Mazur, Ph.D., Harvard. Cryobiology; cryopreservation; cell physiology; cell biophysics.
Bruce McKee, Ph.D., Michigan State. Genetics of chromosome pairing and recombination, sperm development, and chromosome structure.
Chamindrani Mendis-Handagama, D.V.M., Sri Lanka; Ph.D., Australia. Testis; Leydig cells; differentiation; thyroid hormone; PCBs; androgens; steroidogenesis.
Beth Mullin, Ph.D., North Carolina State. Plant-microbe interactions; nitrogen fixation; molecular evolution.
Andreas Nebenfuehr, Ph.D., Oregon State. Intracellular motility; protein targeting; secretion; Golgi apparatus.
Cynthia Peterson, Ph.D., LSU. Blood coagulation; heparin; serine protease inhibitors (serpins); fibrinolysis; extracellular matrix.
Dan Roberts, Ph.D., California, Davis. Structure and function of calcium regulatory proteins in plants.
Gary S. Sayler, Ph.D., Idaho. Ecological and toxicological impact of environmental contaminants on microbial communities; biodegradative mechanisms; plasmids and transposons.
Hildegard Schuller, Ph.D., Hannover School of Veterinary Medicine (Germany). Lung cancer; pancreatic cancer; growth regulation; neurotransmitter receptors; arachidonic acid cascade.
Engin Serpersu, Ph.D., Hacettepe (Turkey). Enzymatic catalysis at the molecular/structural level using NMR; computer modeling.
Pamela Small, Ph.D., Stanford. Host-pathogen interactions; cellular microbiology; comparative microbial genomics; mycobacterial polyketides.
Alan Solomon, M.D., Mount Sinai. Amyloid-associated diseases and multiple myeloma.
Albrecht von Arnim, Ph.D., East Anglia (England). Light signal transduction; *Arabidopsis;* gene silencing.
Jonathan S. Wall, Ph.D., Essex. Amyloid; Light chain; imaging; SPECT; antibody; immunotherapy; protein folding; protein aggregation.
Steven Wilhelm, Ph.D., Western Ontario. Gene regulation; functional genomics in natural systems.
Xuemin Xu, Ph.D., Tokyo Institute of Technology. Alzheimer's disease; neurodegenerative disorder; presenilin; PSAP; apoptosis.
Michael Zemel, Ph.D., Wisconsin. Cellular and molecular biology of nutrition and chronic disease.

COMPUTATIONAL BIOLOGY AND BIOINFORMATICS
Michael Berry, Ph.D., Illinois. Scientific computing; parallel numerical algorithms; information retrieval; data mining; bioinformatics; computational science and performance evaluation.
Hong Guo, Ph.D., Harvard. Computational molecular and structural biology.
Loren Hauser, Ph.D., California, Irvine. Comparative genomic sequences.
Brian Hingerty, Ph.D., Princeton. DNA structure; DNA-carcinogen adducts; NMR computer modeling.
Michael A. Langston, Ph.D., Texas A&M. Combinatorial optimization; computational biology; graph algorithms; parallel and distributed processing; reconfigurable computing.
Frank Larimer, Ph.D., Florida State. Genome structure; enzyme evolution; protein structure.
Dale A. Pelletier, Ph.D., Iowa. Genomics; proteomics; bacterial physiology; methods of isolation/characterization of protein complexes; anaerobic metabolism of aromatic compounds.
Nagiza F. Samatova, Ph.D., Russian Academy of Sciences. Bioinformatics; functional genomics and proteomics; protein-protein interactions; protein docking; data mining; algorithms; graph theory.
Ed Uberbacher, Ph.D., Pennsylvania. Computer analysis of genome sequences.
Dong Xu, Ph.D., Illinois at Urbana-Champaign. Computational biology; bioinformatics; protein modeling.
Ying Xu, Ph.D., Colorado. Computational structural biology; combinatorial optimization algorithms for molecular biology; data mining and modeling.

BIOANALYTICAL TECHNOLOGIES
Michelle Buchanan, Ph.D., Wisconsin–Madison. Mass spectroscopy.
Kelsey D. Cook, Ph.D., Wisconsin. Mass spectrometry; ionization mechanisms; reaction monitoring; polymer and biopolymer analysis; analytical chemistry.
Mitch Doktycz, Ph.D., Illinois at Chicago. Proteins and RNA technologies.
Robert Hettich, Ph.D., Purdue. Mass spectrometry for the molecular characterization of biomolecules.
Greg Hurst, Ph.D., Wisconsin–Madison. Development of new mass spectrometric techniques for analysis of DNA and proteins.
George W. Kabalka, Ph.D., Purdue. Radiopharmaceutical chemistry; organometallic chemistry.
Kurt Lamour, Ph.D., Michigan State. Functional genomics of plant pathogens in the genus *Phytophthora;* population biology and epidemiology.
Michael Paulus, Ph.D., Tennessee. Medical imaging; novel semiconductor detector development; analog integrated circuit design.
Michael J. Sepaniak, Ph.D., Iowa State. Capillary scale separations; applied laser spectroscopy and detection; chemical sensors; molecular modeling.
Gary Van Berkel, Ph.D., Washington State. Mass spectrometry fundamentals and applications.

uvm

UNIVERSITY OF VERMONT

Department of Microbiology and Molecular Genetics
Markey Center for Molecular Genetics

Program of Study

Graduate studies in the Department of Microbiology and Molecular Genetics at the University of Vermont provide training in research directed toward understanding fundamental issues in biology at the molecular level. Faculty members and students work together on important research problems in a wide variety of areas, emphasizing molecular aspects of gene expression, host-pathogen interaction, and structural biology.

Students begin independent research early in their program of study, through a series of laboratory rotations. At the same time, they pursue a program of courses designed to provide the background in concepts and techniques that a successful research scientist must have. The program is designed to fit the individual's needs and includes work in molecular genetics, microbiology, cell biology, molecular biology, biochemistry, and structural biology. An array of seminars, student research forums, journal clubs, and interdepartmental research meetings focusing on such areas as nucleic acid biochemistry, DNA repair, microbial pathogenesis, and signal transduction provide a stimulating and important part of students' training.

Research Facilities

The department occupies 3½ floors of the Stafford Biotechnology Center, which offers state-of-the-art research and teaching facilities. Specialized facilities and equipment are available for the latest techniques in molecular and cellular biology, including cell imaging, DNA synthesis and sequencing, molecular graphics, X-ray crystallography, phosphor imaging, electron microscopy, fluorescence-activated cell sorting, and a variety of biochemical, genomic, and proteomic techniques. The Dana Biomedical Library is an excellent research library, with ready access to all computerized databases. The Lucille P. Markey Charitable Trust awarded the department a $2.3-million grant to support the Center for Molecular Genetics.

Financial Aid

A stipend is provided to all students accepted into the program. The stipend ($20,772 per year in 2005–06) may take the form of a teaching assistantship or a research fellowship. Health insurance benefits are also provided to students.

Cost of Study

Tuition is covered by traineeships and fellowships, which, in addition to the annual stipend, are provided to all students.

Living and Housing Costs

The University offers modern off-campus town-house apartments for married graduate students on a space-available basis. A wide variety of off-campus rental housing is available. Students may choose from attractive neighborhoods near the University or a more rural environment within a short commuting distance. The best time for apartment-hunting is May 15 through August 1. Living expenses for a single student are estimated at $1000 per month.

Student Group

Current enrollment in the Graduate College is approximately 1,200. Students come from throughout the United States and many other countries. There are currently 33 students enrolled in the department's graduate program, which continues to expand. Graduates of the program pursue postdoctoral research before beginning careers in research and teaching. For a list of program alumni and their first positions after leaving the program, prospective students should visit the department's Web site.

Location

Burlington is located on the shore of Lake Champlain, nestled between the Green Mountains and the Adirondacks. It is a small, lively city that serves as the cultural, educational, and economic hub of a wide area. Cultural and recreational opportunities abound. Superb opportunities for downhill and cross-country skiing, skating, boating, hiking, swimming, and cycling are available in or near the city. Annual festivals celebrate Mozart, reggae, folk music, and jazz. Burlington is only 90 minutes by car or bus from Montreal, one of the most cosmopolitan cities in North America.

The University

The University of Vermont, founded in 1791, enrolls approximately 10,000 students. It is an unusual institution, serving a wide variety of functions, including those of a traditional New England liberal arts college, high-technology center, medical school, and agricultural college. All of these functions take place within a single campus, providing a diverse and stimulating environment.

Applying

Applications for admission are reviewed on an individual basis. Students should have the equivalent of one year of biology, two years of chemistry (including one year of organic), one year of physics, and mathematics through calculus. Students with good records who have not completed undergraduate requirements may be permitted to make up any deficiencies early in the program. Scores on the General Test of the Graduate Record Examinations are required, and a Subject Test in biochemistry, cell and molecular biology; biology; or chemistry is optional.

Correspondence and Information

Chair, Graduate Admissions Committee
Department of Microbiology and Molecular Genetics
Stafford Hall
University of Vermont
Burlington, Vermont 05405
Phone: 802-656-2164
Fax: 802-656-8749
E-mail: helen.brunelle@uvm.edu
Web site: http://www.uvm.edu/microbiology

University of Vermont

THE FACULTY AND THEIR RESEARCH

Erik Bateman, Research Associate Professor; Ph.D., Reading (England), 1983. Molecular mechanisms of eukaryotic RNA polymerase II transcription and regulation in *Acanthamoeba.*

Jeffrey P. Bond, Research Associate Professor; Ph.D., Rochester, 1989. Computational analysis of biomolecular sequences and structures; antibody engineering.

John M. Burke, Professor; Ph.D., MIT, 1983. Ribozymes, RNA processing, and RNA-protein interactions; in vitro selection; antiviral applications of ribozymes; targeted RNA cleavage.

Sylvie Doublié, Associate Professor; Ph.D., North Carolina, 1993. X-ray crystallography of protein–nucleic acid complexes.

Barry A. Finette, Professor; Ph.D., Texas at Austin, 1984; M.D., Texas Southwestern Medical Center at Dallas, 1988. Mutagenic mechanisms in the pediatric population; individual/population susceptibility to environmental/iatrogenic exposure to mutagenic and carcinogenic agents.

Christopher Francklyn, Professor; Ph.D., California, Santa Barbara, 1988. Protein–nucleic acid recognition; structure and function of RNA-binding proteins, design of nucleic acid–binding proteins.

Gregory Gilmartin, Associate Professor; Ph.D., Virginia, 1983. Molecular aspects of mRNA processing and its role in control of gene expression.

Joyce E. Heckman, Research Assistant Professor; Ph.D., MIT, 1976. Catalytic RNA; structure and function of the hairpin ribozyme.

Nicholas H. Heintz, Professor; Ph.D., Vermont, 1979. Initiation of DNA synthesis in mammalian cell; control of gene expression during the cell cycle; induction of cell proliferation by environmental factor.

Christopher D. Huston, Assistant Professor; M.D., Cornell, 1994. Host-pathogen interactions with intestinal protozoa; molecular mechanisms of *Entamoeba histolytica* phagocytosis.

Douglas I. Johnson, Professor; Ph.D., Purdue, 1983. Cell-cycle control of Cdc42-dependent signal transduction pathways regulating cell polarity in yeast.

Mariana Matrajt, Assistant Professor; Ph.D., Buenos Aires, 1997. Genetic approaches to development and differentiation of *Toxoplasma gondii.*

Keith P. Mintz, Research Assistant Professor; Ph.D., Vermont, 1990. Molecular mechanisms of adhesion; interaction of periodontopathogens with extracellular matrix proteins.

Scott W. Morrical, Professor; Ph.D., Wisconsin, 1987. Enzymology of DNA replication and homologous genetic recombination; biochemistry of protein-DNA interactions.

David S. Pederson, Associate Professor; Ph.D., Rochester, 1983. Cell-cycle control of DNA replication and transcription in yeast.

Markus Thali, Associate Professor; Ph.D., Zürich, 1990. Cell and molecular biology of virus-host relationships, e.g., lentivirus-host interactions during assembly and release.

Mary L. Tierney, Associate Professor; Ph.D., Michigan State, 1983. Molecular analysis of plant extracellular matrix structure and function; molecular genetic characterization of genes essential for early stages of plant embryogenesis.

Susan S. Wallace, Professor and Chairperson; Ph.D., Cornell, 1965. Molecular approaches to the study of DNA damage and repair; biological consequences of oxidative DNA base damage.

Gary Ward, Associate Professor; Ph.D., California, San Diego, 1985. Cell biology of parasitic protozoa and host-parasite interaction; mechanisms of host-cell invasion by *Plasmodium* and *Toxoplasma.*

Cedric Wesley, Assistant Professor; Ph.D., SUNY at Stony Brook, 1991. Cell-surface interactions regulating metazoan development; cell-surface notch signaling in *Drosophila melanogaster.*

Umadevi V. Wesley, Research Assistant Professor; Ph.D., SUNY at Stony Brook, 1994. Roles of cell-surface peptidases, dipeptidyl peptidase IV, and fibroblast activating protein in suppressing the malignant phenotype of human cancers.

Recent Publications from the Department

Peng, Z., and **E. Bateman.** Analysis of the 5S rRNA gene promoter from *Acanthamoeba castellanii. Mol. Microbiol.* 52:1123–32, 2004.

Han, J., and **J. M. Burke.** Model for general acid-base catalysis by the hammerhead ribozyme: pH-activity relationships of G8 and G12 variants at the putative active site. *Biochemistry* 44:7864–70, 2005.

Doublié, S., V. Bandaru, **J. P. Bond,** and **S. S. Wallace.** The crystal structure of human endonuclease VIII-like 1 (NEIL1) reveals a zincless finger motif required for glycosylase activity. *Proc. Natl. Acad. Sci. U.S.A.* 101:10284–9, 2004.

Martin, G., A. Möglich, W. Keller, and **S. Doublié.** Biochemical and structural insights into substrate binding and catalytic mechanism of mammalian Poly(A) polymerase. *J. Mol. Biol.* 341:911–25, 2004.

Rice S., et al. **(B. A. Finette).** Genotoxicity of therapeutic intervention in children with acute lymphocytic leukemia. *Cancer Res.* 64:4464–71, 2004.

Guth, E., S. H. Connolly, M. Bovee, and **C. S. Francklyn.** A substrate-assisted concerted mechanism for aminoacylation by a class II aminoacyl-tRNA synthetase. *Biochemistry* 44:3785–94, 2005.

Venkataraman, K., K. M. Brown, and **G. M. Gilmartin.** Analysis of a noncanonical poly(A) site reveals a tripartite mechanism for vertebrate poly(A) site recognition. *Genes Dev.* 19:1315–27, 2005.

Heckman, J. E., D. Lambert, and **J. M. Burke.** Photocrosslinking detects a compact, active structure of the hammerhead ribozyme. *Biochemistry* 44:4148–56, 2005.

Burch, P. M., Z. Yuan, A. Loonen, and **N. H. Heintz.** An ERK1,2-dependent program of chromatin trafficking of c-Fos and Fra-1 is required for cyclin D1 expression during cell cycle re-entry. *Mol. Cell. Biol.* 24:4696–709, 2004.

Huston, C. D., D. R. Boettner, V. Miller-Sims, and W. A. Petri. Apoptotic killing and phagocytosis of host cells by the parasite *Entamoeba histolytica. Infect. Immun.* 71:964–72, 2003.

Richman, T. J., et al. **(D. I. Johnson).** Analysis of cell-cycle specific localization of the Rdi1p RhoGDI and the structural determinants required for Cdc42p membrane localization and clustering at sites of polarized growth. *Curr. Genet.* 45:339–49, 2004.

Esheverria, P. C., and **M. Matrajt** et al. *Toxoplasma gondii* Hsp90 is a potential drug target whose expression and subcellular localization are developmentally regulated. *J. Mol Biol.* 350:723–34, 2005.

Mintz, K. P. Identification of an extracellular matrix protein adhesin, EmaA, which mediates the adhesion of *Actinobacillus actinomycetemcomitans* to collagen. *Microbiology* 150:2677–88, 2004.

Ma, Y., et al. **(S. W. Morrical).** Dual functions of single-stranded DNA-binding protein in helicase loading at the bacteriophage T4 DNA replication fork. *J. Biol. Chem.* 279:1035–45, 2004.

Zhu, W., et al. **(D. S. Pederson).** Evidence that the pre-mRNA splicing factor Clf1p plays a role in DNA replication in *Saccharomyces cerevisiae. Genetics* 160:1319–33, 2002.

Nydegger, S., et al. **(M. Thali).** HIV-1 egress is gated through late endosomal membranes. *Traffic* 4:902–11, 2003.

Bernhardt, C., and **M. L. Tierney.** Proline-rich cell wall proteins—building blocks for an expanding cell wall? *Plant Cell Walls,* ed. Taka Hayashi. Springer-Verlag Press, in press.

Watanabe, T., J. O. Blaisdell, **S. S. Wallace,** and **J. P. Bond.** Engineering functional changes in *Escherichia coli* endonuclease III based on phylogenetic and structural analyses. *J. Biol. Chem.* 280:34378–84, 2005.

Mital, J., M. Meissner, D. Soldati, and **G. E. Ward.** Conditional expression of *Toxoplasma gondii* apical membrane antigen-1 (TgAMA1) demonstrates that TgAMA1 plays a critical role in host cell invasion. *Mol. Biol. Cell.* 16:4341–9, 2005.

Ahimou, F., L.-P. Mok, B. Bardot, and **C. S. Wesley.** The adhesion force of notch with delta and the rate of notch signaling. *J. Cell Biol.* 167:1217–29, 2004.

Wesley, U. V., M. McGroarty, and A. Homoyouni. Dipeptidyl peptidase inhibits malignant phenotype of prostate cancer cells by blocking basic fibroblast growth factor signaling pathway. *Cancer Res.* 65:1325–34, 2005.

UNIVERSITY OF WISCONSIN–MADISON

Genetics Training Program

Program of Study

The program in genetics is offered by the Departments of Genetics and Medical Genetics. It offers a broad spectrum of graduate work in genetics, including many aspects of molecular and developmental genetics, immunogenetics, cytogenetics, and viral, bacterial, mammalian, plant, human, behavioral, population, and clinical genetics.

The faculty for the program comes primarily from the Laboratory of Genetics but includes members of other departments with allied interests. Students have access to more than 70 trainers in many different biological departments at the University of Wisconsin–Madison (UW-Madison).

Applicants are selected for their strong interest in and aptitude for genetic research. Students should plan to earn the Ph.D. degree, which normally takes five to six years. The first year is devoted to both research and course work. Formal Ph.D. degree requirements are minimal and can be chosen from a wide range of specialized courses as well as from the fundamental background curriculum. Students are expected to initiate research on their thesis projects during the second semester of their first year and to devote more time to independent research as their careers progress. To fulfill minor requirements, the student may take courses in a single department, or from several departments chosen to meet individual interests. Seminars, journal clubs, and experience in teaching are also an integral part of the formal training. Each student's course program is devised by a certifying committee consisting of the student, the major professor, and 3 other professors from the program.

Research time in the first semester is devoted to a series of research experiences in various laboratories that last approximately three to four weeks each. This rotation through laboratories of the student's choice helps the student decide on a major professor.

In addition to the Ph.D. program, a specialized master's degree is offered in medical genetics, with training in genetics counseling. Admission is independent of the Ph.D. program.

Research Facilities

The laboratory location for thesis research depends upon the choice of thesis adviser. Since the program is administratively centered in the Genetics/Biotechnology Building, many students conduct their thesis research in this facility. The Genetics Building and the Genetics/Biotechnology Building house laboratories provided with all the equipment necessary for state-of-the-art genetic research. Students who choose professors with laboratories outside the Genetics Buildings use comparable facilities. Ancillary facilities such as greenhouses, constant-environment laboratories, a high-voltage electron microscope, and animal quarters are also available.

Financial Aid

All students accepted into the Ph.D. degree program receive financial aid from either an NIH training grant or graduate school fellowships. In later years, support may be derived from a research assistantship. The net stipend in 2006–07 is $22,700 per year after payment of tuition and fees. Support is awarded on a continuing basis, provided that satisfactory progress is maintained toward completion of Ph.D. requirements, and is contingent upon availability of federal government grants. Limited financial aid is available to international students. Currently, there is no support for the M.S. students in genetics counseling.

Cost of Study

Graduate students appointed as research assistants are considered residents for purposes of tuition, and the cost of resident tuition is paid for students accepted into the Ph.D. program.

Living and Housing Costs

A wide variety of rooms and apartments can be found in the areas surrounding the campus and elsewhere in Madison. Most students live off campus, where rates vary considerably. Rooms for graduate students on campus were available from $4000 for the 2005–06 academic year, not including board; unfurnished apartments for married students (Eagle Heights) were $600–$700 per month, including heat and hot water.

Student Group

The present enrollment at UW-Madison is approximately 41,000, of whom about 8,700 are graduate students. The standard enrollment in the genetics program is 40–50 students. About 12 new students are usually admitted to the program each year. Students are drawn from all regions of the United States and a number of other countries. Students interact with numerous postdoctoral fellows, who have received their training from a large cross section of outstanding educational institutions.

Location

The city of Madison has a population of about 200,000. It is the capital city of the state and is surrounded by four attractive lakes. Recreational opportunities are excellent all year. Camping, canoeing, backpacking, fishing, hunting, skiing, and bicycling are popular activities in the area. The cosmopolitan population enjoys both city- and University-promoted concerts, recitals, plays and theater productions and excellent dining. There are numerous opportunities for both appreciation of and participation in the arts. In the campus mall area, impromptu art exhibits, musical performances, and local vendors of crafts can be found.

The University

Founded in 1848, the University of Wisconsin has consistently been ranked among the top universities in the country. The community of biologists is one of the largest, most diverse, and best of any educational institution in the country. It is characterized by a dynamic spirit of inquiry and an atmosphere of academic freedom. The University has unusual breadth in its offerings and has pioneered in placing its facilities and staff at the disposal of the state and the nation.

Applying

Most students admitted are selected during January and February. Applications must be complete by December 15. Midyear admissions are not considered. Admission is based mainly on demonstrated ability and interest in biology, mathematics, and the physical sciences. Minimum course prerequisites are mathematics through calculus, 1 year of physics, chemistry through organic chemistry, and three of the following four biology courses—cell biology, physiology, development, and population biology—as well as introductory courses in genetics and biochemistry. Scores on the GRE General Test and a Subject Test are required.

Correspondence and Information

Genetics Ph.D. Admissions Secretary
1420 Genetics/Biotechnology Building
University of Wisconsin–Madison
425 Henry Mall
Madison, Wisconsin 53706-1580
Phone: 608-262-1069
Web site: http://www.genetics.wisc.edu

University of Wisconsin–Madison

THE FACULTY AND THEIR RESEARCH

Philip Anderson, Professor. Molecular genetics of the nematode, *Caenorhabditis elegans.*

Frederick Blattner, Professor. Mechanisms of DNA replication, transcription, and recombination; cloning of eukaryotic DNA segments in prokaryotic carriers; immunoglobulins; DNA sequencing; phage lambda.

Anthony B. Bleecker, Professor (also with Botany). Genetic and hormonal control of growth and development in plants.

Sean B. Carroll, Professor (also with Molecular Biology). Developmental genetics of pattern formation; molecular evolution.

Nansi Jo Colley, Associate Professor (also with Ophthalmology). Molecular and genetic basis of photoreceptor cell function in *Drosophila melanogaster,* with particular emphasis on intracellular targeting and trafficking.

James F. Crow, Emeritus Professor. Population/evolution; *Drosophila;* human and mammalian research.

Michael R. Culbertson, Professor and Chair (also with Molecular Biology). Molecular genetics of yeast.

John Doebley, Professor. Evolutionary genetics; development and evolution.

William F. Dove, Professor (also with Oncology). Cellular and developmental genetics of *Physarum polycephalum* and of the mouse.

Albert H. Ellingboe, Professor (also with Plant Pathology). Mendelian and molecular genetics of host-parasite interactions.

William E. Engels, Professor. Genetic and molecular behavior of transposable elements in *Drosophila* chromosomes.

Barry S. Ganetzky, Professor. Neurogenetics; genetic, electrophysiological, and molecular analysis of mutants affecting the *Drosophila* nervous system.

Audrey Gasch, Assistant Professor. The role, regulation, and evolution of fungal genomic expression responses to stress.

F. Michael Hoffmann, Professor (also with Oncology). Roles of proto-oncogenes and growth factors in *Drosophila* development.

Akihiro Ikeda, Assistant Professor. Mouse genetics; development and function of the synapse; sensory neuronal disease.

Judith E. Kimble, Professor (also with Biochemistry and Molecular Biology). Genetics and molecular biology of development in *C. elegans.*

Ching Kung, Professor (also with Molecular Biology). Ion channels and sensory transduction in *Paramecium,* yeast, and *Escherichia coli.*

Allen Laughon, Professor. Molecular genetics; homeotic and segmentation gene interactions in *Drosophila.*

Patrick Masson, Professor. Maize transposable elements; mutational analysis of gravitropism and phototropism in *Arabidopsis.*

Bret Payseur, Assistant Professor. Population genetics; genetics of speciation; genetics of morphological evolution.

Francisco Pelegri, Assistant Professor. Genetic analysis of vertebrate development.

Tómas Prolla, Associate Professor. Molecular basis of cancer and DNA repair; generation of mouse models of human disease through gene targeting.

David Schwartz, Professor. Single molecule approaches to whole genome analysis.

Ahna Skop, Assistant Professor. Cytokinesis and cell-cycle proteomics.

Paul Sondel, Professor (also with Human Oncology). Immunogenetics; transplantation immunology; cancer immunology.

Xin Sun, Assistant Professor. Molecular genetics of vertebrate organogenesis.

Richard Vierstra, Professor. Molecular, genetic, and biochemical analysis of ubiquitin-dependent protein degradation and photomorphogenesis in plants, especially *Arabidopsis.*

Jon Wolff, Professor (also with Pediatrics). The treatment of genetic diseases by gene transfer.

Jerry C. P. Yin, Professor. Molecular genetics of learning and memory formation in *Drosophila* and mice.

The following members of other departments serve as trainers in the Genetics Training Program:

Caroline Alexander, Oncology.
Richard Amasino, Biochemistry.
Alan Attie, Veterinary Medicine and Biochemistry.
Maureen Barr, Pharmacy.
David Baum, Botany.
Andrew Bent, Plant Pathology.
Seth Blair, Zoology.
Grace Boekhoff-Falk, Anatomy.
Christopher Bradfield, Oncology.
David Brow, Biomolecular Chemistry.
Richard Burgess, Oncology.
Glenn H. Chambliss, Bacteriology.
Elizabeth C. Craig, Physiological Chemistry.
James E. Dahlberg, Physiological Chemistry.
Christopher Day, Botany.
Timothy Donohue, Bacteriology.
Diana Downs, Bacteriology.
Norman Drinkwater, Oncology.
David Eide, Nutritional Sciences and Biochemistry.
Marcin Filutowicz, Bacteriology.
Catherine Fox, Biomolecular Chemistry.
Michael Gould, Oncology.
Richard L. Gourse, Bacteriology.

Daniel Greenspan, Pathology.
Anne Griep, Anatomy.
Yevgenya Grinblat, Zoology and Anatomy.
Mary Halloran, Zoology and Anatomy.
Jeffrey Hardin, Zoology.
Colleen Hayes, Biochemistry.
Nancy Keller, Plant Pathology.
Patrick Krysan, Horticulture.
Robert Landick, Bacteriology.
Carol E. Lee, Zoology.
Sally Leong, Plant Pathology.
Janet E. Mertz, Oncology.
Amy Moser, Human Oncology.
William S. Reznikoff, Biochemistry.
Gary Roberts, Bacteriology.
Michael R. Sussman, Biochemistry.
Donald Waller, Botany.
Karen Wassarman, Bacteriology.
Bernard Weisblum, Pharmacology.
John White, Anatomy and Molecular Biology.
Marvin Wickens, Biochemistry.
Jaehyuk Yu, Food Microbiology and Toxicology.

WAKE FOREST UNIVERSITY

School of Medicine
Molecular Genetics Program

Program of Study

The interdisciplinary program in molecular genetics offers the degree of Doctor of Philosophy (Ph.D.) through the Graduate School of Wake Forest University. The program provides individualized training and research opportunities using molecular approaches to a wide variety of basic and biomedical research problems. The faculty members of the Molecular Genetics Program represent the Departments of Biochemistry, Medicine, Microbiology and Immunology, Neurobiology and Anatomy, Pathology, Pediatrics, Physiology and Pharmacology, and Cancer Biology. Many of the participating faculty members are also members of the Comprehensive Cancer Center of Wake Forest University. Students can select program faculty members from any of the represented departments. This affords students the opportunity to apply the fundamental techniques of molecular biology to diverse areas of research and prepares them for independent research and teaching careers.

During the first year, students participate in a core course curriculum in molecular biology, biochemistry, and virology designed to build a strong foundation for their specialized laboratory research. During the second year, advanced courses focus on current literature topics, scientific writing, and bioinformatics. Specialty courses are offered through the various departments of the program faculty. Students perform research rotations during their first year in laboratories of 2 or 3 program faculty members to prepare them to select their thesis research laboratory. Students are admitted to candidacy upon satisfactory completion of courses and an oral defense of a written research proposal at the end of the second year. They are expected to complete their degree in four to five years.

Research Facilities

Resources supporting the Molecular Genetics Program are centrally located on the fifth floor of the Hanes Research Building. In addition, each program faculty member has well-equipped laboratory space tailored to his or her specific research needs. Computer facilities give students and faculty members access to DNA and protein sequence databases and sequence analysis software. A molecular graphics facility is available for structural modeling of proteins and nucleic acids. There are many core laboratory facilities providing services that include automated DNA sequencing and genotyping, oligonucleotide synthesis, peptide sequencing and synthesis, proteomics, electron microscopy, nuclear magnetic resonance, cytogenetics, fluorescence-activated cell sorting, and tissue culture. The Carpenter Library of the medical center maintains an extensive collection of basic science and clinical journals and has extensive access to journals online.

Financial Aid

The Wake Forest University Graduate School of Arts and Sciences provides financial support in the form of dean's fellowships or graduate fellowships for first-year students ($46,772 yearly in 2005–06). In addition, all students are awarded tuition scholarships by the graduate school ($26,000 yearly in 2005–06). After the first year, students receive full support from their thesis adviser's research funds or NIH-sponsored training grants.

Cost of Study

Tuition costs ($26,000 yearly in 2005–06) are met in full by Wake Forest University Graduate School of Arts and Sciences for all students holding fellowships.

Living and Housing Costs

Most students live in apartments or rental houses within walking distance of the Medical Center. Rent ranges from $300 to $500 per month for one-bedroom apartments and from $500 to $700 per month for two-bedroom apartments. There are University-owned efficiency apartments for single and married graduate students on a space-available basis.

Student Group

In 2004–05, there were 675 graduate students enrolled on both campuses of Wake Forest University, of whom 233 were biomedical sciences students. The total enrollment in all schools at Wake Forest University in fall 2004 was 6,404.

Student Outcomes

The interdisciplinary program in molecular genetics graduated its first Ph.D. student in 1996. Program faculty members have broad graduate training experience. Graduates typically obtain excellent postdoctoral positions at universities, research institutes, biotechnology research laboratories, or the National Institutes of Health. Following their postdoctoral training, many continue with academic careers as professors and independent biomedical researchers. Others secure research and staff positions in private industry or government laboratories. Some graduates now hold leadership positions in both academia and industry.

Location

The Wake Forest University School of Medicine is located in a residential area of Winston-Salem on the Bowman Gray campus of Wake Forest University, 3 miles from the main campus. The city is in the northwestern region of North Carolina in the foothills of the Blue Ridge Mountains. A population of 185,000 enjoys a mild climate, with early springs and late autumns allowing outdoor activities during most of the year. Cultural attractions include the Southeast Center for Contemporary Art, the Reynolda House Museum of American Art, and performances presented by the nationally recognized North Carolina School of the Arts, the Winston-Salem Symphony, and the Little Theatre. The Sci Works Center and Old Salem area are educational and historic attractions. The city is home to a minor-league baseball team and has excellent recreational facilities.

The University and The Program

Wake Forest is a private university founded in 1834. The Wake Forest University School of Medicine, founded in 1902, has major research programs in genomics, cancer, neurological disorders, hypertension, aging, structural biology, and atherosclerosis. The research facilities at the Medical Center complex are extensive. The establishment of the Molecular Genetics Program exemplifies the commitment of this institution to remain in the forefront of scientific research. New faculty members have been and continue to be recruited in an ongoing effort to strengthen the School's capabilities in molecular genetics research.

Applying

The program invites applications from students who have completed undergraduate degrees with majors in the biological sciences and chemistry. Complete applications should be received by January 15, 2006, for admission in the fall semester. Applications should include scores from the Graduate Record Examinations, including the Subject Test in biology; biochemistry, cell and molecular biology; or chemistry. Also required are official undergraduate transcripts, three letters of reference, and a statement of personal interests. Application forms and additional information about the Molecular Genetics Program may be obtained upon request.

Correspondence and Information

Dr. Timothy Howard
Molecular Genetics Program
Wake Forest University School of Medicine
Medical Center Boulevard
Winston-Salem, North Carolina 27157
Phone: 336-713-7509
Fax: 336-716-7200
E-mail: tdhoward@wfubmc.edu
Web site: http://www.wfubmc.edu/molecular_genetics/

Wake Forest University

THE FACULTY AND THEIR RESEARCH

Steven Akman, Professor of Internal Medicine and Cancer Biology; M.D., Yeshiva (Einstein), 1975. DNA damage, mutagenesis, and genomic instability.

Martha A. Alexander-Miller, Associate Professor of Microbiology and Immunology; Ph.D., Washington (St. Louis), 1993. Regulation of CD8+ cytotoxic T lymphocytes; control of functional avidity.

Donald Bowden, Professor of Biochemistry and Medicine; Ph.D., Berkeley, 1978. Genetics of common diseases, with emphasis on type 2 diabetes, cardiovascular disease, and renal disease.

Bernard Brown II, Assistant Professor of Chemistry; Ph.D., North Carolina State, 1997. Chemical, physical, and structural properties of proteins and nucleic acids using a variety of spectroscopic and biochemical techniques, in addition to X-ray crystallography.

Bao-Li Chang, Assistant Professor of Pediatrics; Ph.D., Maryland, Baltimore, 2001. Genetic contributions to prostate cancer through family-based linkage and case-control association studies.

Yong Chen, Associate Professor of Cancer Biology; Ph.D., Free University of Brussels, 1989. Tumor suppressor genes; apoptosis; cell-cycle regulation; prostate cancer.

Al Claiborne, Professor of Biochemistry; Ph.D., Duke, 1979. Structural and mechanistic studies of flavoproteins in streptococcal oxygen metabolism; catalytic functions of flavin coenzymes.

Scott Cramer, Associate Professor of Cancer Biology; Ph.D., California, Santa Cruz, 1992. Molecular tools to understand and cure prostate cancer growth and metastasis.

Zheng Cui, Associate Professor of Pathology; M.D., Tsuenyi Medical (China), 1979; Ph.D., Massachusetts Amherst, 1988. Lipid signaling in apoptosis and carcinogenesis.

Paul Dawson, Associate Professor of Internal Medicine–Gastroenterology; Ph.D., SUNY at Stony Brook, 1986. Regulation and genetics of hepatic and intestinal cholesterol and bile acid metabolism.

Rajendar Deora, Assistant Professor of Microbiology and Immunology; Ph.D., Illinois at Chicago, 1997. *Bordetella* pathogenesis.

Karin Drotschmann, Assistant Professor of Cancer Biology; Ph.D., Göttingen (Germany), 1997. Defects in the mismatch repair pathway and implications for cancer predisposition, progression, and chemotherapy.

Barry I. Freedman, Professor of Internal Medicine and Nephrology; M.D., SUNY Downstate Medical Center, 1984. Molecular genetics of human renal disease, diabetes mellitus, and hypertension.

Roy Hantgan, Associate Professor of Biochemistry; Ph.D., Cornell, 1974. Molecular mechanisms of blood coagulation and fibrinolysis; conformation of proteins in solution.

Gregory Hawkins, Assistant Professor of Medicine, Section on Pulmonary/Critical Care, and Member, Center for Human Genomics; Ph.D., Maryland, College Park, 1990. Molecular genetics of complex diseases.

Ashok Hegde, Assistant Professor of Neurobiology and Anatomy; Ph.D., Centre for Cellular and Molecular Biology, Hyderabad, India, 1990. Molecular biological, electrophysiological, and behavioral approaches to study questions on long-term memory.

Elizabeth Hiltbold, Assistant Professor of Microbiology and Immunology; Ph.D., Emory, 1996. Antigen-presenting cell responses to intracellular bacterial pathogens; regulation of T-cell responses.

Thomas Hollis, Assistant Professor of Biochemistry; Ph.D., Texas at Austin, 1997. X-ray crystallographic studies of DNA repair proteins and Fanconi anemia–associated proteins.

Ross Holmes, Professor of Surgical Sciences–Urology; Ph.D., Australian National, 1978. Molecular and biochemical approaches to investigate intermediary metabolism in cells related to the synthesis of glycolate and oxalate.

David Horita, Assistant Professor of Biochemistry; Ph.D., Wisconsin–Madison, 1994. Structural biology underlying two systems: DNA repair and chromatin-mediated transcription regulation.

Timothy Howard, Assistant Professor of Pediatrics, Center for Human Genomics; Ph.D., Wake Forest, 1996. Identification of genes for complex diseases; asthma and allergy.

Susan Hutson, Professor of Biochemistry; Ph.D., Wisconsin–Madison, 1976. Structure-function of vitamin B6–dependent enzymes; nutrient gene regulation; excitatory amino acids and seizure control.

Islam Khan, Research Assistant Professor of Internal Medicine–Rheumatology; Ph. D., Lucknow (India), 1989. T-cell effector function in health and autoimmune diseases.

Constantinos Koumenis, Assistant Professor of Radiation Oncology; Ph.D., Houston, 1994. Effects of the tumor microenvironment on gene expression; development of radiosensitizing agents; stress-inducible gene therapy for brain tumors.

Mark O. Lively, Professor and Director of the Molecular Genetics Program and Biomolecular Resource Laboratories; Ph.D., Georgia Tech, 1978. Analysis of protein structure and function using mass spectrometry, proteomics, and bioinformatics; proteolytic enzymes; cell biology and biochemistry of laryngopharyngeal reflux and gastroesophageal reflux disease.

W. Todd Lowther, Assistant Professor of Biochemistry; Ph.D., Florida, 1994. X-ray crystallographic and biochemical analyses of enzymes that repair the oxidative damage to free and protein-incorporated methionine.

Doug Lyles, Professor of Biochemistry and Chairman; Ph.D., Mississippi Medical Center, 1975. Virus assembly; molecular pathogenesis of virus infection.

Charles McCall, Professor and Chairman, Medicine/Infectious Diseases; M.D., Wake Forest, 1961. Neutrophil biology and biochemistry; inflammation of lung; signal transduction; phospholipid metabolism; expression of IL-1, TNFα, and cyclooxygenase genes in phagocytes.

Deborah Meyers, Professor of Pediatrics and Public Health and of Medicine; Ph.D., Indiana, 1976. Family studies and association studies on the genetics of common respiratory diseases.

Mark Miller, Professor of Cancer Biology; Ph.D., Columbia, 1983. Molecular pathogenesis and chemoprevention of lung cancer.

Carol Milligan, Associate Professor of Neurobiology and Anatomy; Ph.D., Medical College of Pennsylvania, 1992. Neuronal development, degeneration, and plasticity after injury: identification and characterization of differentially expressed genes in motoneuron cell death.

Nilamadhab Mishra, Assistant Professor of Internal Medicine–Rheumatology; M.D., Berhampur (India), 1989. New therapeutic targets for the treatment and prevention of SLE.

Steven Mizel, Professor and Chairman, Microbiology and Immunology; Ph.D., Stanford, 1973. Vaccines against agents of bioterrorism; flagellin signal transduction.

Charles S. Morrow, Associate Professor of Biochemistry; Ph.D., Saint Louis, 1980; M.D., Missouri–Columbia, 1983. Transcriptional and posttranscriptional regulation of expression of genes associated with antineoplastic drug resistance.

Ronald Oppenheim, Professor of Neurobiology and Anatomy; Ph.D., Washington (St. Louis), 1967. Developmental neurobiology; programmed cell death; growth factors and neurotrophic molecules.

David Ornelles, Associate Professor of Microbiology and Immunology; Ph.D., MIT, 1987. Molecular virology of adenovirus; oncolytic and oncogenic potential of human adenovirus.

Griffith D. Parks, Associate Professor of Microbiology and Immunology; Ph.D., Wisconsin–Madison, 1987. Molecular biology of paramyxoviruses.

John Parks, Professor of Pathology; Ph.D., Wake Forest, 1979. Structure-function relationships of lecithin:cholesterol acyltransferase and high-density lipoproteins.

Derek Parsonage, Assistant Professor of Biochemistry; Ph.D., Birmingham (England), 1984. Enzymology of bacterial enzymes involved in defending against oxidative stress: enzymes that utilize flavin and cysteine residues at the catalytic site.

Fred Perrino, Associate Professor of Biochemistry; Ph.D., Cincinnati, 1986. Molecular mechanisms of mutagenesis during DNA replication in animal cells; use of in vitro model systems.

Mark Pettenati, Associate Professor of Pediatrics–Medical Genetics; Ph.D., West Virginia, 1983. Molecular cytogenetics.

Leslie Poole, Associate Professor of Biochemistry; Ph.D., Wake Forest, 1988. Mechanistic enzymology of bacterial enzymes involved in protection against oxidative stress; novel roles of catalytic cysteine residues.

Gaddamanugu Prasad, Associate Professor of General Surgery and Cancer Biology; Ph.D., Indian Institute of Science (Bangalore), 1985. Cytoskeletal proteins in tumor biology, mammary gland differentiation and breast cancer, DNA damage responses, and cancer markers.

Thomas Register, Associate Professor of Pathology; Ph.D., South Carolina, 1985. Modulation of the expression of cytokines and connective tissue genes during early atherogenesis.

Michele M. Sale, Assistant Professor of Internal Medicine, Section on Endocrinology and Metabolism, Center for Human Genomics; Ph.D., Tasmania, 1995. Identifying and evaluating genes for type 2 diabetes and stroke.

Yolanda Sánchez, Instructor, Internal Medicine–Molecular Medicine; Ph.D., Oviedo (Spain), 1988. Host-pathogen interactions in the lung.

David Sane, Associate Professor of Internal Medicine; M.D., Duke, 1983. Role of extracellular matrix proteins and integrin receptors in vascular pathobiology.

Fernando Segade, Assistant Professor of Internal Medicine–Endocrinology and Metabolism, Center for Human Genomics; Ph.D., Santiago (Spain), 1989. Functional genomic analysis of complex diseases.

Gregory Shelness, Associate Professor of Pathology; Ph.D., SUNY at Stony Brook, 1985. Intracellular protein targeting and transport; lipoprotein assembly and secretion.

Peter Smith, Professor of Biochemistry; Ph.D., Tennessee, 1971. Viral vectors bearing specific cytochrome P-450 genes.

William Sonntag, Professor of Physiology and Pharmacology; Ph.D., Tulane, 1979. Growth factors and aging.

Mary Sorci-Thomas, Associate Professor of Pathology; Ph.D., Wake Forest, 1984. Regulation of apolipoprotein gene expression; structure-function relationships of apoprotein A-I.

W. Edward Swords, Assistant Professor of Microbiology and Immunology; Ph.D., Alabama at Birmingham, 1996. Cellular microbiology; bacterial genetics and pathogenesis; innate host defenses and host-pathogen interactions.

Ann Tallant, Associate Professor of Physiology and Pharmacology; Ph.D., Tennessee, Memphis, 1983. Signal transduction; regulation of growth; hypertension; cancer treatment/prevention

Suzy V. Torti, Associate Professor of Biochemistry; Ph.D., Tufts, 1977. Regulation of iron metabolism in normal and transformed cells.

Alan Townsend, Professor of Biochemistry; Ph.D., North Carolina at Chapel Hill, 1986. Mechanisms of resistance to cytotoxic and mutagenic agents; enzymes of glutathione metabolism; chemoprevention of cancer; oxidative stress and antioxidant defenses.

Richard Weinberg, Professor of Internal Medicine, Associate in Biochemistry, and Associate Director, General Clinical Research Center; M.D., Johns Hopkins, 1975. Structure and function of human apolipoprotein A-IV (apo A-IV), an intestinal protein synthesized during lipid absorption and incorporated into the surface of nascent chylomicrons.

Daniel J. Wozniak, Associate Professor of Microbiology and Immunology; Ph.D., Ohio State, 1989. Pathogenesis of *Pseudomonas aeruginosa;* genetic and environmental regulation of capsule and toxin synthesis.

Jianfeng Xu, Professor of Pediatrics and Public Health and of Medicine; M.D., Shanghai Medical, 1984; Dr.P.H., Johns Hopkins, 1997. Identifying genetic and epigenetic changes at both germline and somatic levels that increase cancer risk and/or modify disease progression.

Barbara K. Yoza, Research Assistant Professor of Medicine, Section on Molecular Medicine; Ph.D., SUNY at Stony Brook, 1980. Molecular biology of inflammation, particularly signal transduction and regulation of gene expression in human monocytes during infection.

Liqing Yu, Assistant Professor of Pathology; M.D., Hubei Medical (China), 1985; Ph.D., Peking Union Medical, 1995. Molecular and cellular basis of cholesterol trafficking in vivo and in vitro.

Section 11
Marine Biology

This section contains a directory of institutions offering graduate work in marine biology, followed by in-depth entries submitted by institutions that chose to prepare detailed program descriptions. Additional information about programs listed in the directory but not augmented by an in-depth entry may be obtained by writing directly to the dean of a graduate school or chair of a department at the address given in the directory.

For programs offering related work, see also in this book Biological and Biomedical Sciences and Zoology. In Book 4, see Marine Sciences and Oceanography.

CONTENTS

Program Directory

Close-Ups

Marine Biology

California State University, Stanislaus, Graduate School, College of Arts, Letters, and Sciences, Department of Biological Sciences, Turlock, CA 95382. Offers marine sciences (MS). Part-time programs available. *Faculty:* 2 full-time (1 woman). *Students:* Average age 26. 1 applicant, 100% accepted, 0 enrolled. In 2005, 2 degrees awarded. *Degree requirements:* For master's, thesis. *Entrance requirements:* For master's, GRE General Test, GRE Subject Test, minimum GPA of 3.0. *Application deadline:* For fall admission, 2/15 for domestic students, 2/15 for international students; for spring admission, 9/15 for domestic students, 9/15 for international students. Application fee: $55. Electronic applications accepted. *Expenses:* Tuition, nonresident: part-time $339 per unit. Required fees: $999 per term. *Financial support:* In 2005–06, 2 students received support; fellowships, career-related internships or fieldwork, Federal Work-Study, and scholarships/grants available. Support available to part-time students. Financial award application deadline: 3/2; financial award applicants required to submit FAFSA. *Faculty research:* Marine mollusks, kelp forest ecology, marine fish, trace metals, paleoceanography and sedimentology. *Unit head:* Dr. Pamela Roe, Graduate Program Coordinator, 209-667-3476, E-mail: pam@science.csustan.edu.

College of Charleston, Graduate School, School of Sciences and Mathematics, Program in Marine Biology, Charleston, SC 29424-0001. Offers MS. Includes cooperative program faculty and laboratories from the Charleston Laboratory of the National Ocean Service; South Carolina Marine Resources Research Institute (MRRI); Medical University of South Carolina; and The Citadel, The Military College of South Carolina. *Faculty:* 31 full-time (9 women), 13 part-time/adjunct (7 women). *Students:* 28 full-time (23 women), 29 part-time (20 women); includes 4 minority (1 Asian American or Pacific Islander, 3 Hispanic Americans), 5 international. Average age 25. 76 applicants, 45% accepted, 18 enrolled. In 2005, 13 degrees awarded. *Degree requirements:* For master's, thesis, comprehensive exam. *Entrance requirements:* For master's, GRE General Test, GRE Subject Test in biology. Additional exam requirements/recommendations for international students: Required—TOEFL. *Application deadline:* For fall admission, 2/1 for domestic students; for spring admission, 11/1 for domestic students. Application fee: $35. Electronic applications accepted. *Expenses:* Tuition, state resident: full-time $6,668; part-time $256 per semester hour. Tuition, nonresident: full-time $15,342; part-time $587 per semester hour. Tuition and fees vary according to course load. *Financial support:* In 2005–06, 2 fellowships, 20 research assistantships (averaging $12,000 per year), 24 teaching assistantships (averaging $9,900 per year) were awarded; career-related internships or fieldwork, Federal Work-Study, and institutionally sponsored loans also available. Financial award application deadline: 4/1. *Faculty research:* Ecology, biological oceanography, environmental physiology, toxicology, cellular-molecular developmental biology. *Unit head:* Dr. David W. Owens, Director, 843-953-9200, Fax: 843-953-9199, E-mail: owensd@cofc.edu. *Application contact:* Susan Hallatt, Assistant Director of Graduate Admissions, 843-953-5614, Fax: 843-953-1434, E-mail: hallatts@cofc.edu.

See Close-Up on page 811.

Florida Institute of Technology, Graduate Programs, College of Science, Department of Biological Sciences, Program in Marine Biology, Melbourne, FL 32901-6975. Offers MS. Part-time programs available. *Students:* Average age 28. *Degree requirements:* For master's, thesis. *Entrance requirements:* For master's, GRE General Test, minimum GPA of 3.0. *Application deadline:* Applications are processed on a rolling basis. Electronic applications accepted. *Expenses:* Tuition: Part-time $825 per credit. *Financial support:* In 2005–06, 8 students received support; research assistantships with full and partial tuition reimbursements available, teaching assistantships with full and partial tuition reimbursements available, career-related internships or fieldwork and tuition remissions available. Financial award application deadline: 3/1; financial award applicants required to submit FAFSA. *Faculty research:* Ecology of coral reef fish communities and biology of *Foraminiferida*; ecology, physiology, reproduction, and morphology of sea stars, sea urchins, and other echinoderms. *Application contact:* Carolyn P. Farrior, Director of Graduate Admissions, 321-674-7118, Fax: 321-723-9468, E-mail: cfarrior@fit.edu.

See Close-Up on page 813.

Florida State University, Graduate Studies, College of Arts and Sciences, Department of Biological Science, Program in Marine Biology, Tallahassee, FL 32306. Offers MS, PhD. *Faculty:* 6 full-time (1 woman). *Students:* 16 full-time (6 women), 2 international. *Degree requirements:* For master's and doctorate, thesis/dissertation, teaching experience, seminar presentation, comprehensive exam, registration. *Entrance requirements:* For master's and doctorate, GRE General Test (minimum 1100: U-500, G-500), minimum upper division GPA of 3.0. Additional exam requirements/recommendations for international students: Required—TOEFL (minimum score 600 paper-based; 250 computer-based), IB 100. *Application deadline:* For fall admission, 1/15 for domestic students, 12/1 for international students; for spring admission, 10/15 for domestic students, 9/1 for international students. Application fee: $30. *Financial support:* In 2005–06, fellowships with full tuition reimbursements (averaging $19,000 per year), research assistantships with full tuition reimbursements (averaging $19,000 per year), teaching assistantships with full tuition reimbursements (averaging $17,600 per year) were awarded. Financial award application deadline: 1/15; financial award applicants required to submit FAFSA. *Faculty research:* Community ecology, physiological and biochemical adaptation, marine algae, behavioral ecology, marine ecology. *Application contact:* Judy Bowers, Coordinator, Graduate Affairs, 850-644-3023, Fax: 850-644-9829, E-mail: gradinfo@bio.fsu.edu.

Memorial University of Newfoundland, School of Graduate Studies, Department of Biology, St. John's, NL A1C 5S7, Canada. Offers biology (M Sc, PhD); marine biology (M Sc, PhD). Part-time programs available. *Students:* 74 full-time (46 women), 6 part-time (2 women), 25 international. 37 applicants, 22% accepted, 6 enrolled. In 2005, 15 master's, 5 doctorates awarded. *Degree requirements:* For master's, thesis; for doctorate, thesis/dissertation, oral defense of thesis, comprehensive exam. *Entrance requirements:* For master's, honors degree (minimum 2nd class standing) in related field. *Application deadline:* Applications are processed on a rolling basis. Application fee: $40 Canadian dollars. Electronic applications accepted. Students attend 2 or 3 semesters per year. *Expenses:* Tuition, nonresident: full-time $1,466; part-time $733 per semester. International tuition: $1,906 full-time. Tuition and fees vary according to degree level and program. *Financial support:* Fellowships, research assistantships, teaching assistantships, institutionally sponsored loans available. *Faculty research:* Northern flora and fauna, especially cold ocean and boreal environments. *Unit head:* Dr. Chris Parrish, Chair, 709-737-3225, Fax: 709-737-3018, E-mail: pmarino@mun.ca. *Application contact:* Dr. Paul Snelgrove, Graduate Officer, 709-737-3440, E-mail: graduate.biology@mun.ca.

Murray State University, College of Science, Engineering and Technology, Department of Water Science, Murray, KY 42071-0009. Offers MS. Part-time programs available.

Nicholls State University, Graduate Studies, College of Arts and Sciences, Department of Biological Sciences, Thibodaux, LA 70310. Offers marine and environmental biology (MS). Part-time programs available. *Faculty:* 10 full-time (3 women), 1 part-time/adjunct (0 women). *Students:* 21 full-time (5 women), 3 part-time (1 woman); includes 2 minority (1 African American, 1 Asian American or Pacific Islander), 1 international. 21 applicants, 71% accepted, 13 enrolled. In 2005, 6 master's awarded. *Degree requirements:* For master's, thesis, comprehensive exam, registration. *Entrance requirements:* For master's, GRE. Additional exam requirements/recommendations for international students: Required—TOEFL (minimum score 600 paper-based). *Application deadline:* For fall admission, 6/17 for domestic students; for spring admission, 11/15 priority date for domestic students. Application fee: $25. *Expenses:* Tuition, state resident: full-time $3,240. Tuition, nonresident: full-time $8,868. *Financial support:* In 2005–06, 6 students received support, including research assistantships (averaging $12,000 per year), teaching assistantships with tuition reimbursements available (averaging $8,000 per

year) *Faculty research:* Biovemediation, ecology, public health, biotechnology, physiology. Total annual research expenditures: $1.2 million. *Unit head:* Dr. Marilyn B. Kilgen, Head, 985-448-4710, E-mail: biol-mbk@nicholls.edu. *Application contact:* Dr. Raj Boopathy, Graduate Coordinator, 985-448-4716, E-mail: biol-rrb@nicholls.edu.

Northeastern University, College of Arts and Sciences, Department of Biology, Boston, MA 02115-5096. Offers bioinformatics (PMS); biology (MS, PhD); biotechnology (MS); marine biology (MS). Part-time programs available. *Faculty:* 26 full-time (10 women), 2 part-time/adjunct. *Students:* 79 full-time (43 women), 11 part-time (9 women); includes 7 minority (1 African American, 5 Asian Americans or Pacific Islanders, 1 Hispanic American), 22 international. Average age 28. 148 applicants, 25% accepted. In 2005, 20 master's, 3 doctorates awarded. Terminal master's awarded for partial completion of doctoral program. *Degree requirements:* For master's, thesis; for doctorate, thesis/dissertation, qualifying exam. *Entrance requirements:* For master's, GRE General Test, GRE Subject Test; for doctorate, GRE General Test. Additional exam requirements/recommendations for international students: Required—TOEFL (minimum score 250 computer-based). *Application deadline:* For fall admission, 1/15 for domestic students, 2/1 for international students. Application fee: $50. *Financial support:* In 2005–06, 34 teaching assistantships with tuition reimbursements (averaging $16,475 per year) were awarded; fellowships with tuition reimbursements, research assistantships with tuition reimbursements, career-related internships or fieldwork, Federal Work-Study, and tuition waivers (full and partial) also available. Financial award application deadline: 3/1; financial award applicants required to submit FAFSA. *Faculty research:* Biochemistry, cell and systems physiology, ecology, marine sciences, molecular biology. *Unit head:* Dr. Susan Powers-Lee, Chair, 617-373-2260, Fax: 617-373-3724, E-mail: gradbio@neu.edu. *Application contact:* Janeen Greene, Administrative Assistant, 617-373-2262, Fax: 617-373-3724, E-mail: gradbio@neu.edu.

See Close-Up on page 185.

Nova Southeastern University, Oceanographic Center, Program in Marine Biology, Fort Lauderdale, FL 33314-7796. Offers MS. *Faculty:* 15 full-time (1 woman), 5 part-time/adjunct (0 women). *Students:* 7 full-time (4 women), 89 part-time (60 women); includes 6 minority (1 African American, 1 American Indian/Alaska Native, 2 Asian Americans or Pacific Islanders, 2 Hispanic Americans), 2 international. In 2005, 15 degrees awarded. *Degree requirements:* For master's, thesis. *Entrance requirements:* For master's, GRE. Additional exam requirements/recommendations for international students: Required—TOEFL (minimum score 550 paper-based). *Application deadline:* Applications are processed on a rolling basis. Application fee: $50. *Expenses:* Contact institution. *Financial support:* In 2005–06, 6 research assistantships (averaging $4,000 per year), 3 teaching assistantships (averaging $3,500 per year) were awarded; career-related internships or fieldwork, Federal Work-Study, scholarships/grants, and unspecified assistantships also available. Support available to part-time students. Financial award applicants required to submit FAFSA. *Unit head:* Dr. Andrew Rogerson, Associate Dean, Director of Graduate Programs, 954-262-3600, Fax: 954-262-4020, E-mail: arogerso@nsu.nova.edu.

Nova Southeastern University, Oceanographic Center, Program in Marine Biology and Oceanography, Fort Lauderdale, FL 33314-7796. Offers marine biology (PhD); oceanography (PhD). *Faculty:* 15 full-time (1 woman), 5 part-time/adjunct (0 women). *Students:* 4 applicants, 75% accepted, 3 enrolled. *Degree requirements:* For doctorate, thesis/dissertation, comprehensive exam. *Entrance requirements:* For doctorate, GRE, master's degree. Application fee: $50. *Application contact:* Dr. Andrew Rogerson, Associate Dean, Director of Graduate Programs, 954-262-3600, Fax: 954-262-4020, E-mail: arogerso@nsu.nova.edu.

Rutgers, The State University of New Jersey, New Brunswick/Piscataway, Graduate School, Program in Environmental Sciences, New Brunswick, NJ 08901-1281. Offers air resources (MS, PhD); aquatic biology (MS, PhD); aquatic chemistry (MS, PhD); atmospheric science (MS, PhD); chemistry and physics of aerosol and hydrosol systems (MS, PhD); environmental chemistry (MS, PhD); environmental microbiology (MS, PhD); environmental toxicology (PhD); exposure assessment (PhD); fate and effects of pollutants (MS, PhD); pollution prevention and control (MS, PhD); water and wastewater treatment (MS, PhD); water resources (MS, PhD). *Faculty:* 81 full-time, 7 part-time/adjunct. *Students:* 49 full-time (27 women), 48 part-time (19 women); includes 10 minority (3 African Americans, 6 Asian Americans or Pacific Islanders, 1 Hispanic American), 24 international. Average age 32. 79 applicants, 41% accepted, 15 enrolled. In 2005, 8 master's, 10 doctorates awarded. Terminal master's awarded for partial completion of doctoral program. *Degree requirements:* For master's, thesis or alternative, oral final exam, comprehensive exam; for doctorate, thesis/dissertation, thesis defense, qualifying exam, comprehensive exam. *Entrance requirements:* For master's and doctorate, GRE General Test. Additional exam requirements/recommendations for international students: Required—TOEFL. *Application deadline:* For fall admission, 3/1 for domestic students; for spring admission, 11/1 for domestic students. Applications are processed on a rolling basis. Application fee: $50. Electronic applications accepted. *Expenses:* Tuition, state resident: full-time $10,440; part-time $435 per credit. Tuition, nonresident: full-time $15,520; part-time $647 per credit. Required fees: $129 per credit. Tuition and fees vary according to program. *Financial support:* In 2005–06, 10 fellowships with full tuition reimbursements (averaging $21,887 per year), 34 research assistantships with full tuition reimbursements (averaging $19,367 per year), 3 teaching assistantships with full tuition reimbursements (averaging $17,583 per year) were awarded; career-related internships or fieldwork and Federal Work-Study also available. Financial award application deadline: 1/15; financial award applicants required to submit FAFSA. *Faculty research:* Atmospheric sciences; biological waste treatment; contaminant fate and transport; exposure assessment; air, soil and water quality. Total annual research expenditures: $5.7 million. *Unit head:* John Reinfelder, Director, 732-932-8013, Fax: 732-932-8644, E-mail: reinfelder@envsci.rutgers.edu. *Application contact:* Dr. Paul J. Lioy, Graduate Admissions Committee, 732-932-0150, Fax: 732-445-0116, E-mail: ploy@eohsi.rutgers.edu.

San Francisco State University, Division of Graduate Studies, College of Science and Engineering, Department of Biology, Program in Marine Science, San Francisco, CA 94132-1722. Offers MS. *Entrance requirements:* For master's, minimum GPA of 2.5 in last 60 units.

Texas State University–San Marcos, Graduate School, College of Science, Department of Biology, Program in Aquatic Biology, San Marcos, TX 78666. Offers MS. *Students:* 16 full-time (7 women), 15 part-time (6 women); includes 5 Hispanic Americans. Average age 31. 9 applicants, 100% accepted, 6 enrolled. In 2005, 4 degrees awarded. *Degree requirements:* For master's, thesis, 3 seminars, comprehensive exam. *Entrance requirements:* For master's, GRE General Test, previous course work in biology, minimum GPA of 2.75 in last 60 hours of course work. Additional exam requirements/recommendations for international students: Required—TOEFL. *Application deadline:* For fall admission, 6/15 priority date for domestic students, 6/1 priority date for international students; for spring admission, 10/15 priority date for domestic students, 10/1 priority date for international students. Applications are processed on a rolling basis. Application fee: $40 ($90 for international students). *Expenses:* Tuition, area resident: Part-time $116 per credit. Tuition, state resident: full-time $3,168; part-time $176 per credit. Tuition, nonresident: full-time $8,136; part-time $452 per credit. Required fees: $1,112; $74 per credit. Full-time tuition and fees vary according to course load. *Financial support:* In 2005–06, 17 students received support; research assistantships, teaching assistantships available. Financial award application deadline: 4/1; financial award applicants required to submit FAFSA. *Unit head:* Dr. Walter Rast, Director, Aquatic Station, 512-245-8713, E-mail: wr10@txstate.edu.

University of Alaska Fairbanks, School of Fisheries and Ocean Sciences, Department of Marine Sciences and Limnology, Fairbanks, AK 99775-7520. Offers marine biology (MS, PhD); oceanography (MS, PhD), including biological oceanography (PhD), chemical oceanography (PhD), fisheries (PhD), geological oceanography (PhD), physical oceanography (PhD). Part-time programs available. Terminal master's awarded for partial completion of doctoral program.

Marine Biology

Degree requirements: For master's and doctorate, thesis/dissertation, comprehensive exam, registration. *Entrance requirements:* For master's and doctorate, GRE General Test. Additional exam requirements/recommendations for international students: Required—TOEFL. Electronic applications accepted. *Expenses:* Tuition, state resident: full-time $4,392; part-time $244 per credit. Tuition, nonresident: full-time $8,964; part-time $498 per credit. Required fees: $800; $5 per credit. $48 per contact hour. Tuition and fees vary according to course level, course load, campus/location and reciprocity agreements. *Faculty research:* Seafood science and nutrition, sustainable harvesting, chemical oceanography, marine biology, physical oceanography.

University of California, San Diego, Graduate Studies and Research, Scripps Institution of Oceanography, La Jolla, CA 92093. Offers earth sciences (PhD); marine biodiversity and conservation (MAS); marine biology (PhD); oceanography (PhD). Postbaccalaureate distance learning degree programs offered (minimal on-campus study). *Faculty:* 97. *Students:* 243 (126 women); includes 21 minority (2 African Americans, 2 American Indian/Alaska Native, 11 Asian Americans or Pacific Islanders, 6 Hispanic Americans) 63 international. 311 applicants, 24% accepted, 37 enrolled. In 2005, 9 master's, 25 doctorates awarded. *Median time to degree:* Of those who began their doctoral program in fall 1997, 100% received their degree in 8 years or less. *Entrance requirements:* For doctorate, GRE General Test, GRE Subject Test. Additional exam requirements/recommendations for international students: Required—TOEFL (minimum score 550 paper-based; 213 computer-based). *Application deadline:* For fall admission, 1/4 for domestic students. Application fee: $60 ($80 for international students). Electronic applications accepted. *Financial support:* Fellowships, research assistantships, health care benefits available. *Unit head:* Myrl C. Hendershott, Chair, 858-534-3206, E-mail: siodept@sio.ucsd.edu. *Application contact:* Dawn Huffman, Graduate Coordinator, 858-534-3206.

University of California, Santa Barbara, Graduate Division, College of Letters and Sciences, Division of Mathematics, Life, and Physical Sciences, Department of Ecology, Evolution, and Marine Biology, Santa Barbara, CA 93106. Offers MA, PhD. *Faculty:* 36 full-time (7 women), 6 part-time/adjunct (1 woman). *Students:* 61 full-time (29 women); includes 10 minority (1 African American, 6 Asian Americans or Pacific Islanders, 3 Hispanic Americans), 3 international. Average age 29. 150 applicants, 9% accepted, 4 enrolled. In 2005, 3 master's, 6 doctorates awarded. Terminal master's awarded for partial completion of doctoral program. *Median time to degree:* Of those who began their doctoral program in fall 1997, 90% received their degree in 8 years or less. *Degree requirements:* For master's, thesis (for some programs), comprehensive exam (for some programs), registration; for doctorate, thesis/dissertation, comprehensive exam, registration. *Entrance requirements:* For master's and doctorate, GRE General Test. Additional exam requirements/recommendations for international students: Required—TOEFL (minimum score 550 paper-based; 213 computer-based). *Application deadline:* For fall admission, 12/15 for domestic students, 12/15 for international students. Application fee: $60. Electronic applications accepted. *Financial support:* In 2005–06, 30 students received support, including 44 fellowships with full and partial tuition reimbursements available, 90 teaching assistantships with partial tuition reimbursements available (averaging $5,025 per year); research assistantships with full and partial tuition reimbursements available, career-related internships or fieldwork, Federal Work-Study, institutionally sponsored loans, scholarships/grants, traineeships, health care benefits, tuition waivers (full and partial), and unspecified assistantships also available. Financial award application deadline: 12/15; financial award applicants required to submit FAFSA. *Faculty research:* Ecology, population genetics, stream ecology, evolution, marine biology. *Unit head:* Alice Alldredge, Chair, 805-893-2415, Fax: 805-893-4724. *Application contact:* Jennifer Chater, Staff Graduate Advisor, 805-893-3023, Fax: 805-893-4724, E-mail: eemb-info@lifesci.ucsb.edu.

University of Colorado at Boulder, Graduate School, College of Arts and Sciences, Department of Ecology and Evolutionary Biology, Boulder, CO 80309. Offers animal behavior (MA); biology (MA, PhD); environmental biology (MA, PhD); evolutionary biology (MA, PhD); neurobiology (MA); population biology (MA); population genetics (PhD). *Faculty:* 26 full-time (5 women). *Students:* 44 full-time (24 women), 22 part-time (14 women); includes 16 minority (1 American Indian/Alaska Native, 3 Asian Americans or Pacific Islanders, 12 Hispanic Americans), 2 international. Average age 30. 20 applicants, 90% accepted. In 2005, 7 master's, 7 doctorates awarded. Terminal master's awarded for partial completion of doctoral program. *Degree requirements:* For master's, thesis or alternative, comprehensive exam; for doctorate, thesis/dissertation, comprehensive exam. *Entrance requirements:* For master's, GRE General Test, GRE Subject Test, minimum undergraduate GPA of 3.0; for doctorate, GRE General Test, GRE Subject Test. *Application deadline:* For fall admission, 1/2 priority date for domestic students, 12/1 priority date for international students. Application fee: $50 ($60 for international students). *Financial support:* In 2005–06, fellowships (averaging $7,844 per year), research assistantships (averaging $15,663 per year), teaching assistantships (averaging $13,829 per year) were awarded; Federal Work-Study, institutionally sponsored loans, and tuition waivers (full) also available. Financial award application deadline: 3/1. *Faculty research:* Behavior, ecology, genetics, morphology, endocrinology. Total annual research expenditures: $5.2 million. *Unit head:* Jeffry Mitton, Chair, 303-492-0505, Fax: 303-492-8699, E-mail: mitton@colorado.edu. *Application contact:* Jill Skarstadt, Graduate Coordinator, 303-492-7654, Fax: 303-492-8699, E-mail: skarstad@colorado.edu.

University of Guam, Graduate School and Research, College of Arts and Sciences, Program in Biology, Mangilao, GU 96923. Offers tropical marine biology (MS). *Degree requirements:* For master's, thesis, comprehensive exam. *Entrance requirements:* For master's, GRE General Test, GRE Subject Test. Additional exam requirements/recommendations for international students: Required—TOEFL. *Faculty research:* Maintenance and ecology of coral reefs.

University of Hawaii at Manoa, Graduate Division, Specialization in Marine Biology, Honolulu, HI 96822. Offers MS, PhD. Program is interdisciplinary; degree is in specific discipline with the specialization in Marine Biology. *Degree requirements:* For master's and doctorate, thesis/dissertation, research project. *Entrance requirements:* For master's, GRE Subject Test. Additional exam requirements/recommendations for international students: Required—TOEFL. Application fee: $50. *Expenses:* Tuition, state resident: full-time $8,400; part-time $200 per credit hour. Tuition, nonresident: full-time $11,088; part-time $462 per credit hour. Tuition and fees vary according to program. *Financial support:* Research assistantships, teaching assistantships available. *Faculty research:* Ecology, ichthyology, behavior of marine animals, developmental biology. *Unit head:* Dr. David Karl, Chair, 808-956-8964, Fax: 808-956-5059, E-mail: dkarl@hawaii.edu.

University of Maine, Graduate School, College of Natural Sciences, Forestry, and Agriculture, School of Marine Sciences, Program in Marine Biology, Orono, ME 04469. Offers MS, PhD. *Students:* 18 full-time (9 women), 13 part-time (10 women); includes 1 minority (Asian American or Pacific Islander), 5 international. Average age 29. 35 applicants, 9% accepted, 2 enrolled. In 2005, 6 master's, 1 doctorate awarded. *Degree requirements:* For master's and doctorate, thesis/dissertation. *Entrance requirements:* For master's and doctorate, GRE General Test. Additional exam requirements/recommendations for international students: Required—TOEFL. *Application deadline:* For fall admission, 2/1 for domestic students. Applications are processed on a rolling basis. Application fee: $50. Electronic applications accepted. *Financial support:* Fellowships with tuition reimbursements, research assistantships with tuition reimbursements, teaching assistantships with tuition reimbursements, career-related internships or fieldwork, Federal Work-Study, and tuition waivers (full and partial) available. Support available to part-time students. Financial award application deadline: 3/1. *Unit head:* Dr. Malcolm Snick, Coordinator, 207-581-2562. *Application contact:* Scott G. Delcourt, Associate Dean of the Graduate School, 207-581-3219, Fax: 207-581-3232, E-mail: graduate@maine.edu.

University of Massachusetts Dartmouth, Graduate School, College of Arts and Sciences, Department of Biology, North Dartmouth, MA 02747-2300. Offers biology (MS); marine biology (MS). Part-time programs available. *Faculty:* 17 full-time (6 women), 4 part-time/adjunct (3 women). *Students:* 7 full-time (4 women), 7 part-time (4 women), 1 international. Average age 27. 18 applicants, 33% accepted, 6 enrolled. In 2005, 2 degrees awarded. *Degree requirements:* For master's, thesis. *Entrance requirements:* For master's, GRE General Test, GRE Subject

Test. Additional exam requirements/recommendations for international students: Required—TOEFL (minimum score 500 paper-based). *Application deadline:* For fall admission, 5/7 for domestic students, 3/7 for international students; for spring admission, 11/15 priority date for domestic students, 9/15 priority date for international students. Application fee: $35 ($55 for international students). Electronic applications accepted. *Expenses:* Tuition, state resident: full-time $2,071; part-time $86 per credit. Tuition, nonresident: full-time $8,099; part-time $337 per credit. Required fees: $7,282; $393 per credit. *Financial support:* In 2005–06, 7 teaching assistantships with full tuition reimbursements (averaging $10,340 per year) were awarded; research assistantships with full tuition reimbursements, Federal Work-Study and unspecified assistantships also available. Support available to part-time students. Financial award application deadline: 3/1; financial award applicants required to submit FAFSA. *Faculty research:* Marine life and quality analysis, environmental impact to cranberry soil, oceanography, aquaculture. Total annual research expenditures: $190,000. *Unit head:* Dr. Nancy O'Connor, Director, 508-999-8217, Fax: 508-999-8196, E-mail: noconnor@umassd.edu. *Application contact:* Carol Novo, Graduate Admissions Officer, 508-999-8604, Fax: 508-999-8183, E-mail: graduate@umassd.edu.

University of Miami, Graduate School, Rosenstiel School of Marine and Atmospheric Science, Division of Marine Biology and Fisheries, Coral Gables, FL 33124. Offers MA, MS, PhD. *Faculty:* 27 full-time (6 women), 24 part-time/adjunct (7 women). *Students:* 54 full-time (32 women); includes 5 minority (3 African Americans, 2 Hispanic Americans), 9 international. Average age 28. 80 applicants, 13% accepted, 8 enrolled. In 2005, 1 master's, 6 doctorates awarded. Terminal master's awarded for partial completion of doctoral program. *Median time to degree:* Of those who began their doctoral program in fall 1997, 83% received their degree in 8 years or less. *Degree requirements:* For master's and doctorate, thesis/dissertation, comprehensive exam, registration. *Entrance requirements:* For master's and doctorate, GRE General Test. Additional exam requirements/recommendations for international students: Required—TOEFL (minimum score 550 paper-based; 213 computer-based). *Application deadline:* For fall admission, 1/1 for domestic students. Applications are processed on a rolling basis. Application fee: $50. Electronic applications accepted. *Financial support:* In 2005–06, 49 students received support, including 15 fellowships with tuition reimbursements available (averaging $22,380 per year), 30 research assistantships with tuition reimbursements available (averaging $22,380 per year), 4 teaching assistantships with tuition reimbursements available (averaging $22,380 per year); institutionally sponsored loans and scholarships/grants also available. Financial award application deadline: 3/1; financial award applicants required to submit FAFSA. *Faculty research:* Biochemistry, physiology, plankton, coral, biology. *Unit head:* Dr. Robert Cowen, Chairperson, 305-421-4177, E-mail: rcowen@rsmas.miami.edu. *Application contact:* Dr. Larry Peterson, Associate Dean, 305-421-4155, Fax: 305-421-4771, E-mail: gso@rsmas.miami.edu.

The University of North Carolina Wilmington, College of Arts and Sciences, Department of Biological Sciences, Wilmington, NC 28403-3297. Offers biology (MS); marine biology (MS, PhD). Part-time programs available. *Faculty:* 28 full-time (5 women). *Students:* 4 full-time (2 women), 53 part-time (31 women); includes 4 minority (2 African Americans, 2 Hispanic Americans), 2 international. Average age 29. 92 applicants, 15% accepted, 14 enrolled. In 2005, 13 degrees awarded. *Degree requirements:* For master's, thesis/dissertation, comprehensive exam; for doctorate, thesis/dissertation. *Entrance requirements:* For master's, GRE General Test, GRE Subject Test, minimum B average in undergraduate major. *Application deadline:* For fall admission, 3/15 for domestic students. Applications are processed on a rolling basis. Application fee: $45. *Financial support:* In 2005–06, 31 teaching assistantships were awarded; career-related internships or fieldwork and Federal Work-Study also available. Support available to part-time students. Financial award application deadline: 3/15. *Faculty research:* Marine processes, estuaries studies, biotechnology, underwater research, acid rain. *Unit head:* Dr. Martin H. Posey, Chairman, 910-962-3470, E-mail: poseym@uncw.edu. *Application contact:* Dr. Robert D. Roer, Dean, Graduate School, 910-962-4117, Fax: 910-962-3787, E-mail: roer@uncw.edu.

See Close-Up on page 815.

University of Oregon, Graduate School, College of Arts and Sciences, Department of Biology, Eugene, OR 97403. Offers ecology and evolution (MA, MS, PhD); marine biology (MA, MS, PhD); molecular, cellular and genetic biology (PhD); neuroscience and development (PhD). *Faculty:* 42 full-time (14 women), 5 part-time/adjunct (2 women). *Students:* 92 full-time (43 women), 2 part-time (both women); includes 9 minority (2 African Americans, 1 American Indian/Alaska Native, 4 Asian Americans or Pacific Islanders, 2 Hispanic Americans), 6 international. 18 applicants, 83% accepted. In 2005, 11 master's, 7 doctorates awarded. Terminal master's awarded for partial completion of doctoral program. *Degree requirements:* For master's, thesis (for some programs); for doctorate, thesis/dissertation. *Entrance requirements:* For master's and doctorate, GRE General Test, minimum GPA of 3.2. Additional exam requirements/recommendations for international students: Required—TOEFL. *Application deadline:* For fall admission, 12/15 for domestic students. *Financial support:* In 2005–06, 36 teaching assistantships were awarded; research assistantships, Federal Work-Study, institutionally sponsored loans, and scholarships/grants also available. Financial award application deadline: 2/1. *Faculty research:* Developmental neurobiology; evolution, population biology, and quantitative genetics; regulation of gene expression; biochemistry of marine organisms. *Unit head:* George Sprague, Head, 541-346-6051, Fax: 541-346-6056. *Application contact:* Lynne Romans, Admissions Contact, 541-346-4252, Fax: 541-346-6056, E-mail: lromans@uoregon.edu.

University of Southern California, Graduate School, College of Letters, Arts and Sciences, Department of Biological Sciences, Program in Marine Environmental Biology, Los Angeles, CA 90089. Offers PhD. *Degree requirements:* For doctorate, thesis/dissertation. *Entrance requirements:* For doctorate, GRE General Test. Additional exam requirements/recommendations for international students: Required—TOEFL. *Expenses:* Tuition: Full-time $25,416; part-time $1,059 per unit. Required fees: $484; $484 per year. Tuition and fees vary according to course load and program. *Faculty research:* Microbial ecology, physiology of larval development, biological community structure, Cambrian radiation.

University of Southern Mississippi, Graduate School, College of Science and Technology, Department of Biological Sciences, Hattiesburg, MS 39406-0001. Offers environmental biology (MS, PhD); marine biology (MS, PhD); microbiology (MS, PhD); molecular biology (MS, PhD). *Degree requirements:* For master's and doctorate, thesis/dissertation, comprehensive exam. *Entrance requirements:* For master's, GRE General Test, minimum GPA of 3.0; for doctorate, GRE General Test, minimum GPA of 3.5. Additional exam requirements/recommendations for international students: Required—TOEFL.

See Close-Up on page 297.

Western Illinois University, School of Graduate Studies, College of Arts and Sciences, Department of Biological Sciences, Macomb, IL 61455-1390. Offers biological sciences (MS); zoo and aquarium studies (Certificate). Part-time programs available. *Students:* 43 full-time (27 women), 20 part-time (16 women); includes 6 minority (2 African Americans, 1 American Indian/Alaska Native, 1 Asian American or Pacific Islander, 2 Hispanic Americans), 3 international. Average age 26. 38 applicants, 79% accepted. In 2005, 14 master's, 6 other advanced degrees awarded. *Degree requirements:* For master's, thesis or alternative. *Entrance requirements:* Additional exam requirements/recommendations for international students: Required—TOEFL (minimum score 550 paper-based; 213 computer-based). *Application deadline:* Applications are processed on a rolling basis. Application fee: $30. Electronic applications accepted. *Expenses:* Tuition, state resident: full-time $3,599; part-time $200 per semester hour. Tuition, nonresident: full-time $7,198; part-time $400 per semester hour. Required fees: $890; $49 per semester hour. Tuition and fees vary according to campus/location. *Financial support:* In 2005–06, 21 students received support, including 7 research assistantships with full tuition reimbursements available (averaging $6,288 per year), 14 teaching assistantships (averaging $6,288 per year) Financial award applicants required to submit FAFSA.

Marine Biology

Western Illinois University *(continued)*
Unit head: Dr. Richard V. Anderson, Chairperson, 309-298-2408. *Application contact:* Dr. Barbara Baily, Director of Graduate Studies/Associate Provost, 309-298-1806, Fax: 309-298-2345, E-mail: grad-office@wiu.edu.

Woods Hole Oceanographic Institution, MIT/WHOI Joint Program in Oceanography/Applied Ocean Science and Engineering, Woods Hole, MA 02543-1541. Offers applied ocean sciences (PhD); biological oceanography (PhD, Sc D); chemical oceanography (PhD, Sc D); civil and environmental and oceanographic engineering (PhD); electrical and oceanographic engineering (PhD); geochemistry (PhD); geophysics (PhD); marine biology (PhD); marine geochemistry (PhD, Sc D); marine geology (PhD, Sc D); marine geophysics (PhD); mechanical and oceanographic engineering (PhD); ocean engineering (PhD); oceanographic engineering (M Eng, MS, PhD, Sc D, Eng); paleoceanography (PhD); physical oceanography (PhD, Sc D). MS, PhD, and Sc D offered jointly with MIT. Terminal master's awarded for partial completion of doctoral program. *Degree requirements:* For master's and Eng, thesis (for some programs); for doctorate, thesis/dissertation. *Entrance requirements:* For master's, GRE General Test; for doctorate, GRE General Test, GRE Subject Test. Additional exam requirements/recommendations for international students: Required—TOEFL. Electronic applications accepted.

COLLEGE OF CHARLESTON, THE GRADUATE SCHOOL

Graduate Program in Marine Biology

Program of Study

The Graduate School of the College of Charleston offers a graduate program leading to a Master of Science degree in marine biology. The program specifically seeks to provide knowledge and skills that will allow graduates to pursue further graduate study and/or successfully pursue professional employment in the marine science field. Students in the program have the opportunity to be in daily contact with individuals involved in virtually the entire spectrum of research in coastal, ocean, and estuarine systems from molecules to ecosystems. The Graduate School benefits from its close associations with other academic and research institutions, which include The Citadel, the Marine Resources Research Institute of the South Carolina Department of Natural Resources and the associated Waddell Mariculture Center, the Medical University of South Carolina and its Marine Biomedicine and Environmental Science Program, and the Charleston Laboratory of the National Ocean Service. Faculty and staff at these institutions serve actively as members of the marine biology graduate faculty, working closely with students. The broad scope of interests of the graduate faculty members and facilities provide students with a wide variety of research and training opportunities in such areas as aquaculture, benthic ecology, biogeography, ecotoxicology, fisheries, fisheries management, ichthyology, immunology of marine organisms, invertebrate zoology, marine biodiversity, marine biomedical sciences, marine biotechnology, marine ecology, marine environmental sciences, marine genomics, marine mammal biology, molecular biology, oceanic productivity, oceanography, ornithology, physiology, physiological ecology, resource management, and systematics.

Research Facilities

The George D. Grice Marine Laboratory (GML) houses the main research and educational facilities of the Graduate Program in Marine Biology. The location of the Grice Laboratory near the mouth of Charleston Harbor provides an ideal setting for research and study. Research vessels, small boats, and kayaks provide students with immediate access to the relatively unspoiled and biologically rich South Carolina coast.

The Grice Laboratory is located at Fort Johnson on James Island, across the harbor from the College's main campus in historic downtown Charleston. GML is a part of a larger community at Fort Johnson, including laboratories of the South Carolina Marine Resources Research Institute, the Marine Biomedical and Environmental Sciences Program of the Medical University of South Carolina, the Charleston Laboratory of the National Ocean Service, and the Hollings Marine Laboratory. The College's main downtown campus supports the Graduate Program in Marine Biology with the Addlestone Library, modern academic computing areas, and a Science Center housing the Departments of Biology, Chemistry, Geology, and Physics. The Medical University of South Carolina, situated downtown within blocks of the College houses numerous teaching and research laboratories and offers graduate courses in several ancillary areas.

Financial Aid

The program has available teaching, research, and summer assistantships in which all out-of-state fees are waived. A number of scholarships are available for new and current students, including a new scholarship program in marine genomics with a current stipend of $20,000 per year.

Cost of Study

Tuition is $2709 for full-time in-state students and $6300 for out-of-state students for the 2006–07 academic year.

Living and Housing Costs

A limited number of single-occupancy dormitory rooms are available at the Grice Marine Laboratory during the academic year for $1595 per semester, but off-campus housing in Charleston is abundant. The cost of housing, food, and moderate extras totals approximately $800 per month for a single graduate student.

Student Group

There is an average enrollment of about 50 graduate students in the marine biology program; almost all of these students receive financial assistance in the form of scholarships, teaching assistantships, or research assistantships.

Student Outcomes

The program graduates 10–12 students annually. Approximately half of these go into Ph.D. programs; the other half begin career employment. Follow-up studies indicate that virtually all graduates are in a Ph.D. program three to twelve months after graduation or are employed within six months of graduation (most students find employment within three months of graduation). Employment positions include federal and state agencies, private firms, nongovernmental organizations, and a broad variety of teaching positions. In recent years, graduates have entered Ph.D. programs at Scripps, Virginia Institute of Marine Science (VIMS), George Washington, the University of South Carolina, Florida State University, University of Georgia, Old Dominion, UCLA, University of Delaware, University of Colorado, North Carolina State University, Clemson, Rutgers, University of Washington, and Duke.

Location

The College is located in the heart of one of America's most historic cities. Charleston and the surrounding area offer many interesting and attractive places to visit, including numerous antebellum homes and buildings, internationally acclaimed gardens, Fort Sumter, Cape Romain Wildlife Refuge, Francis Marion National Forest, and the well-known and undeveloped sea islands. The city also offers a wide variety of cultural and sports activities. Regional points of interest include the Appalachian Mountains and Atlanta, both 5 hours away, and Savannah, Georgia, and the popular Myrtle Beach area, both 2 hours away.

The College

The College of Charleston was founded in 1770 as the first municipal college in the United States. In 1970, the College of Charleston became part of the South Carolina state college system. In 1992, the graduate programs of the College of Charleston were reorganized as the University of Charleston, South Carolina. In 2002, the programs were renamed the Graduate School of the College of Charleston.

Applying

Prospective graduate students must apply by February 1 for admission the following fall. Students can also be admitted for spring enrollment (application deadline is November 1). Applicants must take the Graduate Record Examinations General Test and Subject Test in biology and include complete transcripts and at least three letters of recommendation in the College application packet.

Correspondence and Information

Professor David W. Owens, Director
Graduate Program in Marine Biology
Grice Marine Laboratory
205 Fort Johnson Road
Charleston, South Carolina 29412

Phone: 843-953-9200
Fax: 843-953-9199
E-mail: owensd@cofc.edu
Web site: http://www.cofc.edu/~marine

College of Charleston, The Graduate School

THE FACULTY AND THEIR RESEARCH

Dennis Allen. Estuarine ecology.

Agnes J. Ayme-Southgate. The assembly and function of muscle cells, using *Drosophila melanogaster* (fruit fly) as a model system; formation of the complex protein system known as the myofibril during development.

John E. Baatz. Mammalian lung biochemistry and molecular biology.

W. Leonard Balthis. Coastal ecosystem health, with an emphasis on the condition and distributions of benthic fauna in relation to human and natural disturbances.

Daniel W. Bearden. Physical and environmental chemistry; nuclear magnetic resonance; mass spectrometry; computational chemistry.

Paul R. Becker. Transport and fate of contaminants in Arctic ecosystems; geographic and species-specific patterns of contaminants in mammals; biological and chemical factors affecting the transport of contaminants through food webs.

Joseph Bernardo. Life history evolution; population genetic structure; life history and community structure; character displacement; amphibian ecology; population regulation; maternal effects; experimental design.

Craig L. Browdy. Shrimp reproduction and mariculture.

Karen G. Burnett. Marine biomedicine; immunology; molecular biology of marine organisms.

Louis E. Burnett Jr. Environmental physiology; respiration and transport processes in animals.

Christopher P. Buzzelli. Investigation of physical versus biological mechanisms and how they regulate estuarine and wetland habitat biotic structure and function.

Robert W. Chapman. Fisheries; genetics; population biology.

Steven J. Christopher. Development and application of high-accuracy analytical methodologies for the determination of trace element contaminants in marine biological matrices.

Loren D. Coen. Marine benthic ecology; plant-animal interactions; tropical ecology; crustacean biology.

Mark R. Collins. Fish biology and ecology; parasites of fishes.

Stacie E. Crowe. Benthic ecology; taxonomy of marine invertebrates.

Margaret A. Davidson. Coastal resource management and research.

Russell D. Day. Mercury toxicology in sea turtles and seabirds.

Isaure de Buron. Host-parasite interactions at the ecological, cellular, and molecular levels.

Richard H. Defran. Population characteristics and dynamics of coastal bottlenose dolphins.

Marie E. DeLorenzo. Environmental toxicology.

M. Richard DeVoe. Aquaculture policy; marine/coastal policy and management; science management.

Guy T. DiDonato. Relationship between tidal creek condition and land-use changes in coastal South Carolina.

Robert T. Dillon Jr. Biology of mollusks; genetics of gastropods and bivalves.

Giacomo R. DiTullio. Phytoplankton physiology and ecology; biogeochemical cycling.

Eric L. Dobson. Geochemical information systems; remote sensing.

Gregory J. Doucette. Physiological ecology of marine phytoplankton, marine biotoxins, and harmful algae.

Phillip Dustan. Marine ecology; coral reef ecology; biological oceanography.

Patricia A. Fair. Marine mammal health assessment and impacts of environmental stressors; toxicological effects of contaminants.

Wayne R. Fitzgibbon. Applying microphysiological techniques to the study of hormonal regulation of mammalian renal physiology and pathophysiology.

Michael H. Fulton. Environmental health; aquatic toxicology.

Sylvia B. Galloway. Coral health/disease characterization using genomic/proteomic approaches.

Thomas W. Greig. Fisheries population genetics; molecular marine forensics; evolutionary ecotoxicology.

Paul S. Gross. Genomics; shrimp immunity and sea urchin complement genes.

Danny J. Gustafson. Plant conservation genetics and restoration ecology.

Nancy H. Hadley. Molluscan mariculture.

Antony S. Harold. Phylogenetic systematics and biogeography of fishes.

Patrick J. Harris. Population biology of fishes; fisheries biology.

Willem J. Hillenius. Comparative anatomy of tetrapods, particularly, mammals, reptiles, and dinosaurs.

A. Frederick Holland. Environmental assessments; resource management; benthic ecology.

Melissa Hughes. Animal behavior, in particular, communication in song birds and crustaceans.

Jeffrey L. Hyland. Environmental monitoring and assessments; benthic ecology; ecotoxicology.

Eric R. James. The host-parasite interaction: immunity, biochemistry, apoptosis; cryopreservation of cells and organisms.

Michael Janech. Physiology of marine organisms; molecular and proteomic applications.

Pamela C. Jutte. Benthic ecology; invertebrate behavioral biology.

Karl J. Karnaky Jr. Cell biology of epithelial salt transport in fishes.

Jennifer M. Keller. Effects of environmental contaminants on marine wildlife health.

Peter B. Key. Aquatic toxicology of insecticides.

Rachael A. King. Systematics research on various local peracarid crustacean groups.

David M. Knott. Taxonomy and ecology of benthic and planktonic invertebrates.

Christopher Korey. *Drosophila* genetics; molecular genetics of human neurological disease using *Drosophila* as a model system.

Laura M. Kracker. GIS and spatial analysis of fish distribution, species diversity, and aquatic habitats; landscape ecology methodologies for large lake and marine ecosystems; underwater acoustics and remote sensing; bioinformatics applied to coral health.

John R. Kucklick. Analytical chemistry; aquatic ecotoxicology.

Eric R. Lacy. Biology of epithelial cells of osmoregulatory and digestive organs in fishes and mammals.

Mark D. Lazzaro. Cell biology: cytoskeletal function in pollen tube development; structure and function of plant secretory hairs, including salt glands of marine plants; digital and fluorescent microscopy.

Joshua Loefer. Fisheries biology; life history and remote tracking of large pelagic predators.

Philip P. Maier. Fisheries research.

Robert M. Martore. Marine artificial reefs; fisheries; benthic ecology.

Wayne E. McFee. Marine mammal strandings; marine mammal life history; dolphin/human interactions.

Elizabeth Meyer-Bernstein. Physiological mechanisms underlying the circadian timing system, using *Drosophila* and mouse model systems; studies at the molecular, cellular, system and behavioral levels.

Donald H. Miller. Mechanisms of osmoregulation in elasmobranchs.

Pamela J. Morris. Environmental microbiology.

Susan J. Morrison. Ecology of estuarine and marine microbes.

Duncan R. Munro. Mammalian physiology; normal and pathological gastric physiology.

Courtney Murren. Plant ecology.

Eugene J. Olmi. Fisheries recruitment and estuarine ecology; population biology of decapod crustaceans.

David W. Owens. Sea turtle behavior, physiology, and ecology.

Margie M. Peden-Adams. Sublethal toxicological effects of environmental contaminants.

Paul L. Pennington. Marine and estuarine ecotoxicology.

John S. Peters. Age and growth of fishes.

Craig J. Plante. Microbial ecology; benthic ecology; the influence of animal-microbe interactions on biogeochemical processes; the role of autoinduction in the development of marine biofilms.

Robert D. Podolsky. Functional biology and evolutionary ecology of marine invertebrates; larval ecology and life-history evolution; fertilization ecology; physiological ecology; phenotypic plasticity.

William Post. Ornithology; coastal avian ecology.

Seth Pritchard. Plant physiological ecology; physiological responses of plants to ongoing global environmental changes, including rising atmospheric carbon dioxide and ozone concentrations, warming, and soil salinization; implications for ecosystem function and food production.

John S. Ramsdell. Toxicology of algal derived toxins; mechanism of toxin action.

Marcel J. M. Reichert. Fish ecology; fisheries science.

Paul M. Rosenblum. Fish endocrinology, reproduction, and nutrition.

William A. Roumillat. Biology of fishes.

Gorka Sancho. Behavioral ecology of fishes; fisheries conservation.

Paul A. Sandifer. Biology of decapod crustaceans; aquaculture; coastal ecology.

Denise M. Sanger. Impacts of human land use; benthic ecology; water quality; sediment chemistry; toxicology.

Leslie R. Sautter. Biological oceanography; marine phytoplankton ecology; marine geology.

Brian G. Scholtens. Ecological models of plant-insect interactions.

Lori H. Schwacke. Development and application of mathematical and computer models for the analysis of marine mammal health data.

Geoffrey I. Scott. Aquatic toxicology.

George R. Sedberry. Community population and trophic ecology of marine fishes; coral reef biology; fisheries biology.

Al Segars. Health/population assessment in marine turtles.

Thomas Siewicki. Environmental toxicology; environmental modeling; risk assessment.

Theodore I. J. Smith. Aquaculture of crustaceans and fish; fisheries biology.

Erik Sotka. Ecology and evolution of marine biotic interactions; larval dispersal; molecular ecology; chemical ecology.

Jill R. Stewart. Water-quality research, with a concentration on detecting and tracking microbial pollution in coastal environments.

Allan E. Strand. Molecular ecology, evolution, and demography of plants.

Frances M. Van Dolah. Functional genomics of toxic dinoflagellates; effects of algal toxins on marine mammals and human consumers.

Robert F. Van Dolah. Benthic ecology; toxicology; environmental assessment; invertebrate community structure; population dynamics.

Keith Walters. Marine ecology; habitat restoration; marine snow dynamics; plant-animal interactions; meiofauna.

Gregory W. Warr. Structure and expression of fish antibody genes.

John E. Weinstein. Environmental toxicology; physiological ecology and toxicology of invertebrates and fish.

Charles A. Wenner. Ichthyology; ecology of deep water fishes; fisheries biology.

Elizabeth L. Wenner. Crustacean biology; marine and estuarine invertebrate and fish communities.

J. David Whitaker. Crustacean fisheries resource research.

Dara H. Wilber. Ecological impact assessment in the marine and estuarine environment.

Pace Wilber. Geographical information systems.

Edward F. Wirth. Effects of pesticides on crustaceans, particularly reproduction and physiology.

D. Reid Wiseman. Systematics and ecology of marine algae.

Cheryl M. Woodley. Application of biochemistry, molecular, and cellular biology to understanding the effects of biotic and abiotic stressors on ecosystem health.

David M. Wyanski. Life history and taxonomy of marine fishes; fisheries biology.

John D. Zardus. Evolution and ecology of commensal barnacles.

Anastasia M. Zimmerman. Molecular evolution of the vertebrate immune system; genome-wide analyses of innate and adaptive immune loci in fishes; use of the zebrafish as an immunological model.

FLORIDA INSTITUTE OF TECHNOLOGY

College of Science and Liberal Arts
Department of Biological Sciences
Graduate Program in Marine Biology and Ecology

Programs of Study

The marine biology and ecology section of the Department of Biological Sciences offers programs of study leading to the degrees of Master of Science and Doctor of Philosophy. Programs focus on the applied and theoretical biology of the organism and its relationship to its environment. Major areas of study and research include marine biology, aquaculture, biological conservation, and population ecology. Departmental research focuses on the biology of threatened and commercially exploited species and communities, ranging from life history analysis, taxonomic studies, and ecological description to techniques for habitat management and methods for culture and propagation of species.

The programs consist of fundamental studies in the biological sciences, with emphasis on individual research under the supervision of graduate faculty members. Programs of study are designed to meet individual needs and are prepared by the student in consultation with a committee of faculty members. Courses include those that contribute to the professional and scientific development of the student, those that correct deficiencies in the student's undergraduate preparation, and those required by the department. Seminars by visiting scientists, reviews of current research in the department and elsewhere, and close contact with faculty members and their continuing research help to prepare the student for an active program of independent study and research. Students are encouraged to pursue interdisciplinary studies, using cellular and molecular methods to solve problems in organismic biology. Modern biochemical and cell biology techniques can often provide sophisticated solutions to marine and aquatic biology problems, and students who develop interdisciplinary research projects often gain a competitive edge in career development. Advisory committee members include faculty members from the cellular and molecular biology section of the Department of Biological Sciences and members of other departments.

Research Facilities

Florida Tech is located in east-central Florida, an area with many opportunities for field research, with access to subtropical and tropical marine and terrestrial communities. There is a wide array of ecosystems, ranging from temperate salt marshes to subtropical upland forests, many of which support populations of endangered and threatened species. The Indian River lagoon provides a high-diversity marine ecosystem with many unusual features. Opportunities exist for fieldwork on state and federal wildlife refuges and other protected areas.

A major strength of the graduate program is the interaction that occurs between that department and several associated research institutions. Scientists from the Harbor Branch Oceanographic Institution (HBOI) in Fort Pierce, Florida; from the Caribbean Marine Research Center (CMRC) at Lee Stocking Island, Bahamas; from the Florida Department of Environmental Protection (FDEP); and from Holmes Regional Medical Center (HRMC) in Melbourne are members of the graduate faculty. These associations provide additional research resources as well as a high degree of flexibility in graduate programs in areas of course work and research.

Financial Aid

Graduate teaching and research assistantships are available to qualified students. For 2006–07, stipends range from $7600 to $12,500 for nine months. Computer-based information on scholarships, loan funds, and other student assistance may be obtained from the Financial Aid Office. A limited number of assistantships providing tuition remission only are also available.

Cost of Study

The 2006–07 tuition is $900 per semester credit hour for all students. As noted above, however, tuition is remitted for graduate assistants.

Living and Housing Costs

Room and board on campus cost approximately $3000 per semester in 2006–07. On-campus housing (dormitories and apartments) is available for full-time single and married graduate students, but priority for dormitory rooms is given to undergraduate students. Many apartment complexes and rental houses are available near the campus.

Student Group

The department currently has 65 graduate students enrolled from colleges throughout the United States. Approximately half of the graduate students are women, and approximately one fourth are married. Most graduate students receive financial support.

Student Outcomes

Graduates of the Department of Biological Sciences are employed by such facilities as the Florida Department of Environmental Protection; St. John's Water Management District; Brevard County; Sea World; Walt Disney World–The Living Seas; Dynamac; Bionetics; Nantucket Marine Lab; ETT Environmental, Inc.; Brigham and Women's Hospital; Dartmouth Medical School; University of Miami Medical School; Pfizer Foundation; Sri International; Autec; Kistler-Morse; Great Lakes Environmental Center; DuPont Pharmaceuticals; and the South Atlantic Fishery Management.

Location

Florida Tech's main campus is located in Melbourne, a residential community on Florida's east coast. Melbourne is the key city in south Brevard County, which also encompasses nine smaller communities on the mainland and the barrier island. The John F. Kennedy Space Center and Disney World are within a 90-minute drive of the University. The area's economy is supported by a well-balanced mix of industries in electronics, aviation, light manufacturing, opticals, communications, agriculture, and tourism.

The Institute

Florida Tech was founded in 1958 and has developed rapidly into a university providing both undergraduate and graduate education in the sciences and engineering for students from throughout the United States and many other countries. Current enrollment on the Melbourne campus is about 4,000. In addition to the marine biology and ecology program, Florida Tech offers graduate programs in applied mathematics, cell and molecular biology, chemical engineering, chemistry, civil engineering, computer science, electrical engineering, environmental engineering, mechanical engineering, ocean engineering, oceanography, operations research, physics, science education, space sciences, and systems engineering.

Applying

Further information and application forms for admission to the Graduate School may be obtained from the Graduate Admissions Office. Applicants must take the Graduate Record Examinations (General Test) and arrange to have the scores sent to the Graduate Admissions Office. Separate application for financial aid must be made on forms available from the department or the Graduate School and must be submitted to the department by March 1.

Correspondence and Information

Graduate Admissions Office
Florida Institute of Technology
150 West University Boulevard
Melbourne, Florida 32901

Phone: 321-674-8027
    800-944-4348 (toll-free in the U.S.)
Fax: 321-723-9468
E-mail: grad-admissions@fit.edu
Web site: http://www.fit.edu/grad

Dr. Gary N. Wells, Head
Department of Biological Sciences
Florida Institute of Technology
150 West University Boulevard
Melbourne, Florida 32901

Phone: 321-674-8034
E-mail: gwells@fit.edu
Web site: http://www.bio.fit.edu

Florida Institute of Technology

THE FACULTY AND THEIR RESEARCH

Mark B. Bush, Professor; Ph.D., Hull (England), 1986. Conservation biology, with an emphasis on the restoration and creation of coastal wetlands and on wetland conservation in general; paleoecology; refuge design and management. Distributional change and conservation on the Andean flank: A paleoecological perspective. *Global Ecol. Biogeogr.* 11:463–73, 2002.

Junda Lin, Professor; Ph.D., North Carolina at Chapel Hill, 1989. Molluscan and crustacean aquaculture; population and community ecology of marine benthos. A rearing system for the culture of ornamental decapod larvae. *Aquaculture* 218:329–39, 2003 (with Narciso et al.).

John G. Morris, Associate Professor; Ph.D., Illinois at Urbana-Champaign, 1974. Population ecology and critical habitat requirements of rare, threatened, and endangered species of Florida reptiles, birds, and mammals, focusing on the gopher tortoise, scrub jay, Atlantic bottlenose dolphin, and West Indian manatee; development of habitat suitability models employing GIS methodology.

Jonathan M. Shenker, Associate Professor; Ph.D., Oregon State, 1986. Finfish aquaculture in freshwater and marine systems; larval and juvenile fish biology and recruitment; Indian River and coastal ecosystems; assessment of the effects of pollutants on aquatic organisms; nuclear magnetic resonance spectrographic examination of the physiology of living organisms. Recruitment of tarpon *(Megalops atlanticus)*, leptocephali into the Indian River Lagoon, Florida. *Contrib. Mar. Sci.* 35:55–69, 2002 (with Cowie-Mojica et al.).

Richard A. Tankersley, Professor; Ph.D., Wake Forest, 1992. Behavioral processes control and modify the distribution, transport, and reproduction of estuarine dependent organisms. Settlement of blue crab *(Callinectes sapidus)* postlarvae during selective tidal-stream transport. *Mar. Biol.* 141:863–75, 2002 (with Welch and Forward).

Ralph G. Turingan, Associate Professor; Ph.D., Puerto Rico, 1993. Environmental biology of fishes; vertebrate functional morphology; ecological morphology of feeding systems; evolution of organismal design in vertebrates. Intraspecific variation in gape-prey size relationships and feeding success during early ontogeny in red drum, *Sciaenops ocellatus. Environ. Biol. Fishes* 66(1):75–84, 2002 (with Krebs).

Richard L. Turner, Associate Professor; Ph.D., South Florida, 1977. Echinoderm biology, including biochemistry, physiology, functional anatomy, reproduction, systematics, and ecology; reproduction and ecology of the Florida applesnail; physiological ecology of crustaceans. Occurrence and significance of the Atlantic ghost crab *Ocypode quadrata* from the upper Pleistocene to Holocene Anastasia formation of Florida. *J. Crustacean Biol.* 23(3):712–22, 2003 (with Portell and Beerensson).

Robert van Woesik, Professor; Ph.D., James Cook (Australia), 1993. Population and community ecology of scleractinian corals; spatial and temporal assessment of coral communities and the application of this ecology to the management of coral reefs. Processes regulating coral communities. *Comments Theor. Biol.* 7:199–212, 2002.

Associated Graduate Faculty

Megan Davis-Hodgkins, Director, Aquaculture Division, Harbor Branch Oceanographic Institution; Ph.D., Florida Tech., 1998. Early life history of conch: larval nutrition, larval dispersal and recruitment, metamorphosis, predator-prey interactions, and stock enhancement. The combined effects of temperature and salinity on growth, development, and survival for tropical gastropod veligers of *Strombus gigas. J. Shellfish Res.* 19:883–9, 2000.

Tamara Frank, Director, Department of Visual Ecology, Harbor Branch Oceanographic Institution; Ph.D., California, Santa Barbara, 1987. Effects of a decrease in downwelling irradiance on the daytime vertical distribution patterns of zooplankton and micronekton. *Mar. Biol.* 140:1181–93, 2002 (with Widder).

M. Dennis Hanisak, Director, Division of Marine Science, Harbor Branch Oceanographic Institution; Ph.D., Rhode Island, 1977. Marine botany; physiological ecology of marine plants; biology of deep-water macroalgae; nutrient dynamics; coral reef ecology; aquaculture, particularly marine plant cultivation. Physiological responses of *Vallisneria americana* transplants along a salinity gradient in the Callosahatchee Estuary (Southwest Florida). *Estuaries* 22:138–48, 1999.

Jose V. Lopez, Assistant Scientist, Division of Biomedical Research, Harbor Branch Oceanographic Institution; Ph.D., George Mason, 1995. Molecular evolution; conservation and biodiversity of marine invertebrates and their microbial associates. Characterization of genetic markers for *in vitro* cell line identification of the marine sponge *Axinella corrugata. J. Heredity* 93:27–36, 2002 (with Peterson et al.).

Paula M. Mikkelsen, Assistant Curator, Department of Invertebrates, American Museum of Natural History, New York; Ph.D., Florida Tech, 1994. Molluscan biodiversity, systematics, life histories, and phylogeny of shelled opisthobranch mollusks. Anatomy and biology of *Mysella pedroana* (Mollusca: Bivalvia: Galeommatoidea), and its commensal relationship with *Blepharipoda occidentalis* (Crustacea: Anomura: Albuneidae). *Zoologischer Anzeiger* 241:149–60, 2002 (with Boyko).

Richard Paperno, Research Administrator I, Florida Fish and Wildlife Conservation Commission, Florida Marine Research Institute; Ph.D., Delaware, 1991. Early life history of fishes; marine ecology. Age-0 spot *(leiostomus xanthurus)* from two estuaries along central Florida's East Coast: Comparisons of the timing of recruitment, seasonal changes in abundance, and rates of growth and mortality. *Fla. Sci.* 65:85–99, 2002.

Valerie J. Paul, Director, Smithsonian Marine Station at Fort Pierce; Ph.D., California, San Diego (Scripps), 1985. Marine chemical ecology; marine plant-herbivore interactions; coral reef ecology; marine natural products. The co-occurrence of chemical and structural defenses in the gorgonian corals of Guam. *Mar. Ecol. Prog. Ser.* 239:105–14, 2002 (with Puglisi et al.).

Shirley A. Pomponi, Director, Division of Biomedical Marine Research, Harbor Branch Oceanographic Institution; Ph.D., Miami (Florida), 1977. Systematics and chemotaxonomy of marine sponges; cell culture of marine sponges, including culture optimization studies. The oceans and human health: The discovery and development of marine-derived drugs. *Oceanography* 14:78–87, 2001.

Marty A. Riche, Research Fishery Biologist, USDA; Ph.D., Michigan State, 2000. Sustainable marine aquaculture technologies; spawning, nutrition, and physiology of southern flounder and black sea bass. Incorporation of plant protein feedstuffs into fishmeal diets for rainbow trout increases phosphorus availability. *Aquaculture Nutr.* 5:101–6, 1999 (with Brown).

John Scarpa, Associate Research Scientist, Aquaculture Division, Harbor Branch Oceanographic Institution; Ph.D., Texas A&M, 1989. Bivalve genetics. Method for the production of genetically modified bivalve trochophore larvae as a feed for marine fish larvae. *J. Appl. Aqua.* 12:1–11, 2002.

Hilary M. Swain, Executive Director, Archbold Biological Station, Lake Placid, Florida; Ph.D., Newcastle, 1981. Conservation biology; Florida ecosystems; geographic information systems (GIS); biodiversity; agroecology; restoration ecology. Characterization of scrub ecosystems of Brevard County, Florida. *Fla. Sci.* 62:13–47, 1999 (with Schmalzer and Boyle).

Bjorn G. Tunberg, Research Biologist, Smithsonian Marine Station, Fort Pierce, Florida; Ph.D., Göteborg (Sweden), 1984. Crustacean population ecology; ecosystem trophodynamics; mangrove systems; salt marsh restoration techniques. Studies on the covariation between physical factors and the long-term variation of the marine soft bottom macrofauna in western Sweden. *Estuarine, Coastal, Shelf Sci.* 50:373–85, 2000 (with Hagberg).

Edith A. Widder, Director, Bioluminescence Department, Harbor Branch Oceanographic Institution; Ph.D., California, Santa Barbara, 1982. Photophysiology and photoecology of marine organisms, with emphasis on bioluminescence and vision; instrumentation development and utilization of research submersibles for oceanic in situ radiometry. Ultraviolet absorption in transparent zooplankton and its implications for depth distribution and visual predation. *Mar. Biol.* 138:717–30, 2001 (with Johnsen).

UNIVERSITY OF NORTH CAROLINA WILMINGTON

Department of Biology and Marine Biology
Ph.D. in Marine Biology

Program of Study

The major emphasis of the Ph.D. program is to provide doctoral training in the areas that encompass modern marine biology. Faculty strengths include coastal and estuarine biology, crustacean biology, marine mammalogy, and molecular biology and systematics of marine organisms. The Ph.D. program in marine biology represents a thirty-year investment by the University of North Carolina Wilmington (UNCW) in marine biology. UNCW has immediate access to the marine environment, and the Department of Biology and Marine Biology has an undergraduate program in marine biology that is ranked fifth in the nation. The Department has awarded more than 190 M.S. marine biology degrees over the last twenty years and six Ph.D. degrees in cooperative programs. Students applying with a B.S. degree are admitted to the M.S. in marine biology program, complete all of the course requirements and electives, and conduct their thesis research. After preparing their research for publication, a student may then apply for admission to the Ph.D. program with the sponsorship of a graduate faculty member in the Department of Biology and Marine Biology. If admitted, students proceed through the curriculum in parallel with students admitted with M.S. degrees from other institutions. During the first two semesters, students enroll in the Graduate Seminars and, upon completion, take written and oral candidacy examinations based upon their research area, course work, and general knowledge of the biological sciences. Students then prepare a doctoral prospectus and pursue their research.

Research Facilities

The Department of Biology and Marine Biology is located in Dobo and Friday Halls, which house modern research and teaching facilities. Additional faculty research laboratories are located in the nearby Center for Marine Science (CMS), an integral part of the University that promotes basic and applied research in the fields of oceanography, coastal and wetlands studies, marine biomedical and environmental physiology, and marine biotechnology and aquaculture. The University's laboratories contain modern equipment that enables study in disciplines as diverse as molecular and cellular biology, light and electron microscopy, organismal biology and behavior, and ecology. Boats from 13 to 25 feet are available through CMS, and CMS also operates the 63-foot R/V *Cape Fear*. Undeveloped barrier islands are located a short distance from the University, as are two estuarine sanctuaries that provide broad expanses of coastal marsh. Locally, Wilmington and the surrounding area contain large wetland ecosystems and an expansive tidal swamp associated with the Cape Fear estuary.

Financial Aid

The Department provides full financial aid for all predoctoral graduate students in good standing. State-funded teaching assistantships and grant-funded research assistantships constitute the major sources of support. In 2004, the nine-month stipend was $16,000, paid over ten months. Additional summer support is frequently available.

Cost of Study

Tuition and fees are paid for all predoctoral graduate students. For 2005–06, this award amounted to $2766 per academic year for North Carolina residents and $10,144 per academic year for nonresidents.

Living and Housing Costs

The University is convenient to a variety of rental properties, both apartments and single-family housing. Prices vary considerably, but one-bedroom units are available in the $500 to $750 per month price range. An off-campus housing list can be obtained through the University housing office.

Student Group

This recently established Ph.D. program has admitted 11 students to date. These students join more than 60 M.S. biology and marine biology students who are currently in the Department. Many students are from out of state, and, over the past decade, 94 percent of the M.S. graduates have gone on to positions within their discipline or to Ph.D. programs. The University has an undergraduate population of approximately 10,600 and a graduate student population of 800.

Location

The University is located in southeastern North Carolina on a 650-acre campus midway between the Atlantic Ocean and the Cape Fear River. The city of Wilmington, which was founded in 1732, is on the east bank of the Cape Fear River and is 6 miles from Wrightsville Beach and 15 miles from Carolina Beach.

The University and The Department

The University was founded in 1947 as Wilmington College, a locally supported institution, and subsequently became part of the 16-campus University of North Carolina system. The Department of Biology and Marine Biology is located in Dobo and Friday Halls and has more than 30 graduate faculty members. Some faculty members have offices and laboratories at the Center for Marine Science. Interaction among biology faculty members is extensive and often includes faculty members in the Departments of Chemistry, Mathematics and Statistics, Earth Sciences, and Physics and Physical Oceanography.

Applying

Applicants should submit an application for graduate admission, official transcripts of all college work, official scores on the Graduate Record Examinations (General Test and Subject Test in biology), and three letters of reference. Applicants with an M.S. degree should submit a curriculum vitae, summary of M.S. thesis research (three pages maximum), and a statement of interest for Ph.D. research. Applicants should contact prospective mentors prior to applying, because admission to the program is dependent upon sponsorship by a faculty member.

Correspondence and Information

Dr. Ann Pabst, Graduate Coordinator
Ms. Tracie J. Chadwick, Graduate Secretary (contact for program information)
Department of Biology and Marine Biology
601 South College Road
Wilmington, North Carolina 28403-5915
Phone: 910-962-3536
E-mail: chadwickt@uncw.edu
Web site: http://www.uncw.edu/bio/index.html

University of North Carolina Wilmington

THE FACULTY AND THEIR RESEARCH

Daniel G. Baden, Ph.D., Miami (Florida), 1977. Marine toxicology: receptor site interactions with natural toxins. badend@uncw.edu

J. Craig Bailey, Ph.D., LSU, 1996. Molecular phylogeny and phycology: molecular and morphological analyses to infer relationships among marine and freshwater algal species. baileyc@uncw.edu

Stephen W. Brewer, Ph.D., California, Davis, 2000. Plant ecology. brewers@uncw.edu

Lawrence B. Cahoon, Ph.D., Duke, 1981. Biological oceanography and limnology: primary production, grazing, and nutrient dynamics, and water quality analysis and remediation. cahoon@uncw.edu

Gregory Chandler, Ph.D., Australian National, 2001. Plant systematics: evolution, biogeography, and systematics of the angiosperms. chandlerg@uncw.edu

Ileana E. Clavijo, Ph.D., Puerto Rico, 1982. Fisheries biology: ecological and behavioral interactions of coastal shelf fishes. clavijo@uncw.edu

Richard M. Dillaman, Ph.D., South Carolina, 1974. Morphology: light and electron microscopy investigations of mineral deposition. dillamanr@ucwil.edu

Michael J. Durako, Ph.D., South Florida, 1991. Wetlands plant ecology: restoration, physiological ecology, reproductive biology, and demographics of sea grasses. durakom@uncw.edu

Steven D. Emslie, Ph.D., Florida, 1987. Ornithology and paleobiology: fossil record of birds in the Pliocene/Pleistocene and evaluation of Royal and Sandwich tern breeding colonies. emslies@uncw.edu

Courtney T. Hackney, Ph.D., Mississippi State, 1977. Estuarine and wetlands ecology: linkages of physical, chemical, and biotic components. hackney@uncw.edu

Neil F. Hadley, Ph.D., Colorado, 1966. Entomology and physiology: comparative physiology of terrestrial arthropods. hadleyn@uncw.edu

Paul E. Hosier, Ph.D., Duke, 1973. Coastal plant ecology: ecological and geologic processes on barrier islands and beaches. hosier@uncw.edu

Donald F. Kapraun, Ph.D., Texas at Austin, 1969. Phycology: quantification of nuclear genomes in species of red seaweeds. kapraund@uncw.edu

Stephen T. Kinsey, Ph.D., Florida State, 1996. Biochemistry: use of nuclear magnetic resonance to probe metabolism inside living cells of marine organisms. kinseys@uncw.edu

Heather N. Koopman, Ph.D., Duke, 2001. Marine lipid physiology: Examine three facets of the physiology of marine animals—metabolism/health, specialized adaptations, and phylogenetic lineage. koopmanh@uncw.edu

Thomas E. Lankford Jr., Ph.D., Delaware, 1997. Ichthyology: growth and survival of fishes during their early life history stages. lankfordt@uncw.edu

Michael A. McCartney, Ph.D., SUNY at Stony Brook, 1994. Molecular ecology and evolution: evolutionary biology and ecology of marine invertebrates and fishes. mccartneym@uncw.edu

Joel J. Mintzes, Ph.D., Northwestern, 1974. Biological education: investigations of human learning and conceptual development. mintzes@uncw.edu

D. Ann Pabst, Ph.D., Duke, 1989. Cetacean biomechanics: musculoskeletal design and thermoregulatory function in whales, dolphins, and porpoises. pabsta@uncw.edu

David E. Padgett, Ph.D., Ohio State, 1975. Mycology: ecology of fungi in natural waters, particularly estuaries. padgett@uncw.edu

Joseph R. Pawlik, Ph.D., California, San Diego (Scripps), 1988. Chemical ecology: ecological functions of secondary metabolites from marine invertebrates and control of settlement of marine invertebrate larvae. pawlikj@uncw.edu

Martin H. Posey, Ph.D., Oregon, 1985. Benthic and coastal ecology: effects of predation, competition, biological disturbance, introduced species, and eutrophication on the structure of bottom communities. poseym@uncw.edu

L. Scott Quackenbush, Ph.D., Florida State, 1981. Crustacean endocrinology: mechanisms crustaceans use to control growth and molting. quackenbushs@uncw.edu

Robert D. Roer, Ph.D., Duke, 1979. Physiology and mineralization: mechanisms of ion translocation and calcification in the Crustacea. roer@uncw.edu

Richard Satterlie, Ph.D., California, Santa Barbara, 1978. Neurophysiology. satterlier@uncw.edu

Laela S. Sayigh, Ph.D., MIT/Woods Hole Oceanographic Institution, 1992. Dolphin vocalizations: development and functions of signature whistles in free-ranging bottlenose dolphins. sayighl@uncw.edu

Frederick Scharf, Ph.D., Massachusetts Amherst, 2001. Fisheries biology: role of ecological processes in structuring aquatic communities; population dynamics of marine and estuarine fishes. scharff@uncw.edu

Thomas H. Shafer, Ph.D., Ohio State, 1975. Developmental biology: gene expression of marine organisms throughout their life cycle. shafert@uncw.edu

Ronald K. Sizemore, Ph.D., Maryland, 1975. Marine microbiology: ecology, taxonomy, and molecular biology of the genus *Vibrio.* sizemorer@uncw.edu

Bongkeun Song, Ph.D., Rutgers, 2000. Marine microbiology. songb@uncw.edu

Ann E. Stapleton, Ph.D., Chicago, 1990. Plant molecular biology: physiological analysis of how plants are affected by UV. stapletona@uncw.edu

Alina M. Szmant, Ph.D., Rhode Island, 1980. Coral reef biology: physiological ecology and nutrient dynamics of reef corals. szmanta@uncw.edu

Carmelo R. Tomas, Ph.D., Rhode Island, 1977. Phytoplankton biology: physiological ecology of harmful algal bloom species. tomasc@uncw.edu

Wm. David Webster, Ph.D., Texas Tech, 1985. Mammalogy: terrestrial vertebrates and nesting ecology of loggerhead sea turtles. webste@uncw.edu

Ami E. Wilbur, Ph.D., Delaware, 1995. Shellfish aquaculture: development of genetic markers in marine bivalve mollusks as tools for evaluation of culture efforts. wilbura@uncw.edu

Section 12
Microbiological Sciences

This section contains a directory of institutions offering graduate work in microbiological sciences, followed by in-depth entries submitted by institutions that chose to prepare detailed program descriptions. Additional information about programs listed in the directory but not augmented by an in-depth entry may be obtained by writing directly to the dean of a graduate school or chair of a department at the address given in the directory.

For programs offering related work, see also in this book Biochemistry; Biological and Biomedical Sciences; Botany and Plant Biology; Cell, Molecular, and Structural Biology; Ecology, Environmental Biology, and Evolutionary Biology; Entomology; Genetics, Developmental Biology, and Reproductive Biology; Parasitology; Pathology and Pathobiology; Physiology; and Zoology. In Book 4, see Agricultural and Food Sciences and Chemistry; in Book 5, see Agricultural Engineering and Bioengineering and Biomedical Engineering and Biotechnology; and in Book 6, see Allied Health, Dentistry and Dental Sciences, Pharmacy and Pharmaceutical Sciences, Public Health, and Veterinary Medicine and Sciences.

CONTENTS

Bacteriology

The University of Iowa, Roy J. and Lucille A. Carver College of Medicine and Graduate College, Graduate Programs in Medicine, Department of Microbiology, Iowa City, IA 52242-1316. Offers general microbiology and microbial physiology (MS, PhD); immunology (MS, PhD); microbial genetics (MS, PhD); pathogenic bacteriology (MS, PhD); virology (MS, PhD). *Faculty:* 23 full-time (3 women), 11 part-time/adjunct (2 women). *Students:* 54 full-time (25 women); includes 4 minority (3 Asian Americans or Pacific Islanders, 1 Hispanic American), 8 international. 93 applicants, 12% accepted, 4 enrolled. In 2005, 7 degrees awarded. *Median time to degree:* Of those who began their doctoral program in fall 1997, 80% received their degree in 8 years or less. *Degree requirements:* For master's, thesis/dissertation; for doctorate, thesis/dissertation, comprehensive exam. *Entrance requirements:* For master's and doctorate, GRE General Test. Additional exam requirements/recommendations for international students: Required—TOEFL. *Application deadline:* For fall admission, 2/1 for domestic students, 2/1 for international students. Application fee: $50 ($75 for international students). Electronic applications accepted. *Expenses:* Tuition, state resident: part-time $1,882 per term. Tuition, nonresident: full-time $17,338; part-time $4,907 per term. Tuition and fees vary according to course load and program. *Financial support:* In 2005–06, 63 research assistantships with full tuition reimbursements (averaging $22,000 per year) were awarded; institutionally sponsored loans, scholarships/grants, traineeships, and health care benefits also available. *Faculty research:* Biocatalysis and blue jeans, gene regulation, processing and transport of HIV, retroviral pathogenesis, biodegradation. Total annual research expenditures: $9 million. *Unit head:* Dr. Michael A. Apicella, Head, 319-335-7810, E-mail: grad-micro-info@uiowa.edu.

See Close-Up on page 923.

University of Prince Edward Island, Atlantic Veterinary College, Graduate Program in Veterinary Medicine, Charlottetown, PE C1A 4P3, Canada. Offers anatomy (M Sc, PhD); bacteriology (M Sc, PhD); clinical pharmacology (M Sc, PhD); clinical sciences (M Sc, PhD); epidemiology (M Sc, PhD), including reproduction; fish health (M Sc, PhD); food animal nutrition (M Sc, PhD); immunology (M Sc, PhD); microanatomy (M Sc, PhD); parasitology (M Sc, PhD); pathology (M Sc, PhD); pharmacology (M Sc, PhD); physiology (M Sc, PhD); toxicology (M Sc, PhD); veterinary science (M Vet Sc); virology (M Sc, PhD). Part-time programs available. *Faculty:* 76 full-time (25 women), 49 part-time/adjunct (8 women). *Students:* 54 full-time (32 women), 2 part-time. Average age 30. In 2005, 7 master's, 6 doctorates awarded. *Degree requirements:* For master's and doctorate, thesis/dissertation. *Entrance requirements:* For master's, DVM, B Sc honors degree, or equivalent; for doctorate, M Sc. *Application deadline:* Applications are processed on a rolling basis. Application fee: $50. *Expenses: Contact institution.* Tuition charges are reported in Canadian dollars. Part-time tuition and fees vary according to course level, degree level, campus/location, program and student level. *Financial support:* In 2005–06, 4 fellowships (averaging $25,000 Canadian dollars per year), 4 research assistantships (averaging $16,500 Canadian dollars per year) were awarded; career-related internships or fieldwork also available. *Faculty research:* Animal health management, infectious diseases, fin fish and shellfish health, basic biomedical sciences, ecosystem health. Total annual research expenditures: $1.2 million Canadian dollars. *Unit head:* Dr. James Bellamy, Associate Dean of Graduate Studies and Research, 902-566-0856, E-mail: bellamy@upei.ca. *Application contact:* Cheryl Gaudet, Registrar's Office, 902-566-0781, Fax: 902-566-0795, E-mail: registrar@upei.ca.

The University of Tennessee Health Science Center, College of Graduate Health Sciences, Department of Molecular Sciences, Memphis, TN 38163-0002. Offers bacterial pathogenesis (PhD); biochemistry (PhD); immunology (PhD); microbiology (PhD); molecular and cell biology-pathology (PhD); molecular biology (PhD); signal transduction (PhD); structural biology (PhD); virology (PhD). *Faculty:* 17 full-time (4 women). *Students:* 25 full-time (11 women); includes 14 minority (1 African American, 13 Asian Americans or Pacific Islanders). Average age 24. 83 applicants, 12% accepted. In 2005, 6 degrees awarded. *Degree requirements:* For doctorate, thesis/dissertation, oral and written preliminary and comprehensive exams. *Entrance requirements:* For doctorate, GRE General Test, GRE Subject Test, minimum GPA of 3.0. Additional exam requirements/recommendations for international students: Required—TOEFL. *Application deadline:* For fall admission, 5/15 for domestic students. Application fee: $0. *Financial support:* In 2005–06, 8 fellowships, 21 research assistantships were awarded; teaching assistantships, traineeships and tuition waivers (full) also available. *Unit head:* Dr. David Hasty, Chairman, 901-448-6150, Fax: 901-448-7360. *Application contact:* Ida W. Mosby, Director, Enrollment Services, 901-448-5560, E-mail: imosby@utmem.edu.

The University of Texas Medical Branch, Graduate School of Biomedical Sciences, Center for Biodefense and Emerging Infectious Diseases, Galveston, TX 77555. Offers biodefense training (PhD). *Entrance requirements:* For doctorate, GRE, an overall and advanced grade point average of 3.0; personal interviews encouraged. *Expenses:* Tuition, state resident: full-time $8,350; part-time $90 per credit hour. Tuition, nonresident: full-time $21,450; part-time $366 per credit hour. Required fees: $1,027; $11 per credit hour. $60 per term.

The University of Texas Medical Branch, Graduate School of Biomedical Sciences, Program in Emerging and Tropical Infectious Diseases, Galveston, TX 77555. Offers PhD, MD/PhD. *Degree requirements:* For doctorate, thesis/dissertation. *Entrance requirements:* For doctorate, GRE General Test. *Application deadline:* Applications are processed on a rolling basis. Application fee: $25 ($50 for international students). *Expenses:* Tuition, state resident: full-time $8,350; part-time $90 per credit hour. Tuition, nonresident: full-time $21,450; part-time $366 per credit hour. Required fees: $1,027; $11 per credit hour. $60 per term. *Financial support:* In 2005–06, fellowships (averaging $23,000 per year), research assistantships with full tuition reimbursements (averaging $23,000 per year) were awarded; traineeships and unspecified assistantships also available. *Faculty research:* Emerging diseases, tropical diseases, parasitology, vitology and bacteriology.

See Close-Up on page 949.

University of Virginia, College and Graduate School of Arts and Sciences, Department of Microbiology, Charlottesville, VA 22903.

University of Washington, Graduate School, School of Public Health and Community Medicine, Graduate Program in Pathobiology, Seattle, WA 98195. Offers MS, PhD. *Faculty:* 22 full-time (8 women), 11 part-time/adjunct (6 women). *Students:* 47 full-time (31 women); includes 7 minority (1 African American, 4 Asian Americans or Pacific Islanders, 2 Hispanic Americans), 6 international. Average age 47. 55 applicants, 31% accepted, 12 enrolled. In 2005, 1 master's, 2 doctorates awarded. Terminal master's awarded for partial completion of doctoral program. *Median time to degree:* Of those who began their doctoral program in fall 1997, 99% received their degree in 8 years or less. *Degree requirements:* For master's, thesis/dissertation, registration; for doctorate, thesis/dissertation, comprehensive exam, registration. *Entrance requirements:* For master's and doctorate, GRE General Test, minimum GPA of 3.0. Additional exam requirements/recommendations for international students: Required—TOEFL. *Application deadline:* For fall admission, 12/15 for domestic students, 11/15 for international students. Application fee: $50. *Financial support:* In 2005–06, 2 fellowships with tuition reimbursements (averaging $22,560 per year), 18 research assistantships with tuition reimbursements (averaging $22,560 per year) were awarded; career-related internships or fieldwork, institutionally sponsored loans, scholarships/grants, traineeships, tuition waivers (full and partial), and unspecified assistantships also available. Financial award application deadline: 3/1; financial award applicants required to submit FAFSA. *Faculty research:* Pathogenesis of chlamydiae, molecular biology of parasites, signal transduction, antigenic analysis, molecular biology of tumor viruses. *Unit head:* , Dr. Andreas Stergachis, Acting Chair, 206-543-8350, Fax: 206-543-3873, E-mail: stergach@u.washington.edu. *Application contact:* Joseph A. Daniels, Manager of Student Services, 206-543-4338, Fax: 206-543-3873, E-mail: pathobio@u.washington.edu.

See Close-Up on page 1129.

University of Wisconsin–Madison, Graduate School, College of Agricultural and Life Sciences, Department of Bacteriology, Madison, WI 53706-1380. Offers MS. Part-time programs available. *Faculty:* 100 full-time (27 women). *Students:* 19 full-time (11 women), 13 part-time (9 women); includes 7 minority (2 African Americans, 1 American Indian/Alaska Native, 3 Asian Americans or Pacific Islanders, 1 Hispanic American), 4 international. Average age 28. 36 applicants, 36% accepted, 10 enrolled. In 2005, 21 degrees awarded. *Entrance requirements:* Additional exam requirements/recommendations for international students: Required—TOEFL. *Application deadline:* For fall admission, 7/5 for domestic students, 6/1 for international students; for spring admission, 11/30 for domestic students, 10/15 for international students. Applications are processed on a rolling basis. Application fee: $45. Electronic applications accepted. *Financial support:* In 2005–06, 6 students received support, including 1 fellowship with tuition reimbursement available (averaging $18,000 per year), 2 research assistantships with tuition reimbursements available (averaging $18,500 per year) *Faculty research:* Microbial physiology, gene regulation, microbial ecology, plant-microbe interactions, symbiosis. Total annual research expenditures: $8.8 million. *Unit head:* Dr. Glenn Chambliss, Chair, 608-262-2914, Fax: 608-262-9865. *Application contact:* Diana M. Downs, Professor, Program Director, 608-265-4630, Fax: 608-262-9865, E-mail: bactms@bact.wisc.edu.

Immunology

Albany Medical College, Graduate Programs in the Biological Sciences, Center for Immunology and Microbial Disease, Albany, NY 12208-3479. Offers MS, PhD. Part-time programs available. *Faculty:* 17 full-time (2 women), 11 part-time/adjunct (6 women). *Students:* 16 full-time (10 women). Average age 25. 26 applicants, 46% accepted, 5 enrolled. In 2005, 7 master's, 2 doctorates awarded. Terminal master's awarded for partial completion of doctoral program. *Degree requirements:* For master's, thesis; for doctorate, thesis/dissertation, oral qualifying exam, written preliminary exam, 1 published paper-peer review, comprehensive exam. *Entrance requirements:* For master's and doctorate, GRE General Test. Additional exam requirements/recommendations for international students: Required—TOEFL. *Application deadline:* For fall admission, 3/15 for domestic students. Applications are processed on a rolling basis. Application fee: $0 ($60 for international students). *Financial support:* In 2005–06, 10 research assistantships (averaging $23,000 per year) were awarded; Federal Work-Study, scholarships/grants, and tuition waivers (full) also available. Financial award applicants required to submit FAFSA. *Faculty research:* Microbial and viral pathogenesis, cancer development and cell transformation, biochemical and genetic mechanisms responsible for human disease. *Unit head:* Dr. Thomas D. Friedrich, Graduate Director, 518-262-6750, Fax: 518-262-6161, E-mail: dgs_cimd@mail.amc.edu.

Albert Einstein College of Medicine, Sue Golding Graduate Division of Medical Sciences, Department of Microbiology and Immunology, Bronx, NY 10461. Offers PhD, MD/PhD. *Degree requirements:* For doctorate, thesis/dissertation. *Entrance requirements:* For doctorate, GRE General Test. Additional exam requirements/recommendations for international students: Required—TOEFL. *Faculty research:* Nature of histocompatibility antigens, lymphoid cell receptors, regulation of immune responses and mechanisms of resistance to infection.

Baylor College of Medicine, Graduate School of Biomedical Sciences, Department of Immunology, Houston, TX 77030-3498. Offers PhD, MD/PhD. *Faculty:* 29 full-time (9 women). *Students:* 24 full-time (12 women); includes 9 minority (2 African Americans, 3 Asian Americans or Pacific Islanders, 4 Hispanic Americans), 12 international. Average age 29. 68 applicants, 18% accepted, 5 enrolled. In 2005, 4 degrees awarded. *Median time to degree:* Of those who began their doctoral program in fall 1997, 100% received their degree in 8 years or less. *Degree requirements:* For doctorate, thesis/dissertation, public defense. *Entrance requirements:* For doctorate, GRE General Test, GRE Subject Test (strongly recommended), minimum GPA of 3.0. Additional exam requirements/recommendations for international students: Required—TOEFL. *Application deadline:* For fall admission, 1/1 for domestic students. Application fee: $30. Electronic applications accepted. *Expenses:* Tuition: Full-time $8,200. Full-time tuition and fees vary according to program. *Financial support:* In 2005–06, 23 students received support, including 9 fellowships (averaging $20,000 per year), 15 research assistantships (averaging $20,000 per year); teaching assistantships, career-related internships or fieldwork, Federal Work-Study, institutionally sponsored loans, health care benefits, and tuition waivers (full) also available. Financial award applicants required to submit FAFSA. *Faculty research:* Structure and function of major histocompatibility antigens, induction and regulation of T-cell immune responses, microbial genetics, pathophysiology of bacterial and viral infections, control and epidemiology of respiratory viruses. *Unit head:* Dr. Dorothy Lewis, Director, 713-798-6427, Fax: 713-798-7949. *Application contact:* Kelly Levitt, Graduate Program Administrator, 713-798-3921, Fax: 713-798-3900, E-mail: klevitt@bcm.tcm.edu.

See Close-Up on page 849.

Baylor College of Medicine, Graduate School of Biomedical Sciences, Interdepartmental Program in Cell and Molecular Biology, Houston, TX 77030-3498. Offers biochemistry (PhD); cell and molecular biology (PhD); genetics (PhD); human genetics (PhD); immunology (PhD); microbiology (PhD); virology (PhD). *Faculty:* 99 full-time (21 women). *Students:* 56 full-time (28 women); includes 20 minority (4 African Americans, 1 American Indian/Alaska Native, 6 Asian Americans or Pacific Islanders, 9 Hispanic Americans), 5 international. Average age 28. 164 applicants, 18% accepted, 12 enrolled. In 2005, 3 doctorates awarded. *Median time to degree:* Of those who began their doctoral program in fall 1997, 63% received their degree in 8 years or less. *Degree requirements:* For doctorate, thesis/dissertation, public defense. *Entrance requirements:* For doctorate, GRE General Test, GRE Subject Test (strongly recommended), minimum GPA of 3.0. Additional exam requirements/recommendations for international students: Required—TOEFL. *Application deadline:* For fall admission, 1/1 for domestic students. Applications are processed on a rolling basis. Application fee: $30. Electronic applications accepted. *Expenses:* Tuition: Full-time $8,200. Full-time tuition and fees vary according to program. *Financial support:* In 2005–06, 52 students received support, including 20 fellowships (averaging $23,000 per year), 36 research assistantships (averaging $23,000 per year); teaching assistantships, Federal Work-Study, institutionally sponsored loans, health care benefits, and tuition waivers (full) also available. Financial award applicants required to submit FAFSA. *Faculty research:* Gene expression and regulation, developmental biology and genetics, signal transduction and membrane biology, aging process, molecular virology. *Unit head:* Dr. Tom

Immunology

Baylor College of Medicine (continued)
Cooper, Director, 713-798-6557. *Application contact:* Lourdes Fernandez, Graduate Program Administrator, 713-798-6557, Fax: 713-798-6325, E-mail: cmbprog@bcm.edu.
See Close-Up on page 545.

Boston University, School of Medicine, Division of Graduate Medical Sciences, Department of Pathology and Laboratory Medicine, Immunology Training Program, Boston, MA 02215. Offers PhD, MD/PhD. *Degree requirements:* For doctorate, thesis/dissertation, qualifying exam. *Entrance requirements:* For doctorate, GRE General Test, GRE Subject Test. Additional exam requirements/recommendations for international students: Required—TOEFL. *Application deadline:* For fall admission, 1/15 for domestic students; for spring admission, 10/15 priority date for domestic students. Electronic applications accepted. *Expenses:* Tuition: Full-time $31,530; part-time $985 per credit. Required fees: $316; $40 per semester. Tuition and fees vary according to course level and program. *Financial support:* Fellowships with tuition reimbursements, research assistantships with tuition reimbursements, Federal Work-Study, scholarships/grants, and traineeships available. *Unit head:* Dr. Ann Marshak-Rothstein, Director, 617-638-4284, Fax: 617-638-4286, E-mail: itp@bu.edu.
See Close-Up on page 853.

Brown University, Graduate School, Division of Biology and Medicine, Program in Molecular Biology, Cell Biology, and Biochemistry, Providence, RI 02912. Offers biochemistry (M Med Sc, Sc M, PhD), including biochemistry (Sc M, PhD), biology (Sc M, PhD), medical science (M Med Sc, PhD); biology (MA); cell biology (M Med Sc, Sc M, PhD), including biochemistry (Sc M, PhD), biology (Sc M, PhD), medical science (M Med Sc, PhD); developmental biology (M Med Sc, Sc M, PhD), including biochemistry (Sc M, PhD), biology (Sc M, PhD), medical science (M Med Sc, PhD); immunology (M Med Sc, Sc M, PhD), including biochemistry (Sc M, PhD), biology (Sc M, PhD), medical science (M Med Sc, PhD); molecular microbiology (M Med Sc, Sc M, PhD), including biochemistry (Sc M, PhD), biology (Sc M, PhD), medical science (M Med Sc, PhD). Part-time programs available. Terminal master's awarded for partial completion of doctoral program. *Degree requirements:* For master's, thesis (for some programs); for doctorate, one foreign language, thesis/dissertation, preliminary exam. *Entrance requirements:* For master's and doctorate, GRE General Test, GRE Subject Test. Additional exam requirements/recommendations for international students: Required—TOEFL. Electronic applications accepted. *Faculty research:* Molecular genetics, gene regulation.
See Close-Up on page 555.

Brown University, Graduate School, Division of Biology and Medicine, Program in Pathology and Laboratory Medicine, Providence, RI 02912. Offers biology (PhD); cancer biology (PhD); immunology and infection (PhD); medical science (PhD); pathobiology (Sc M); toxicology and environmental pathology (PhD). Terminal master's awarded for partial completion of doctoral program. *Degree requirements:* For doctorate, thesis/dissertation, preliminary exam. *Entrance requirements:* For master's and doctorate, GRE General Test, GRE Subject Test. Additional exam requirements/recommendations for international students: Required—TOEFL. Electronic applications accepted. *Faculty research:* Environmental pathology, carcinogenesis, immunopathology, signal transduction, innate immunity.

California Institute of Technology, Division of Biology, Program in Immunology, Pasadena, CA 91125-0001. Offers PhD. *Degree requirements:* For doctorate, thesis/dissertation, qualifying exam. *Entrance requirements:* For doctorate, GRE General Test. *Application deadline:* For fall admission, 1/1 for domestic students. Application fee: $0. *Financial support:* Application deadline: 1/1. *Application contact:* Elizabeth Ayala, Graduate Program Coordinator, 626-395-4497, Fax: 626-683-3343, E-mail: biograd@cco.caltech.edu.

Case Western Reserve University, School of Medicine and School of Graduate Studies, Graduate Programs in Medicine, Programs in Molecular and Cellular Basis of Disease, Program in Immunology, Cleveland, OH 44106. Offers MS, PhD, MD/PhD. *Degree requirements:* For doctorate, thesis/dissertation. *Entrance requirements:* For doctorate, GRE General Test, GRE Subject Test. Additional exam requirements/recommendations for international students: Required—TOEFL. *Faculty research:* Immunopathology, immunochemistry.

Colorado State University, College of Veterinary Medicine and Biomedical Sciences, Department of Microbiology, Immunology and Pathology, Fort Collins, CO 80523-0015. Offers immunology (MS); microbiology (MS, PhD); pathology (PhD). *Faculty:* 40 full-time (11 women), 2 part-time/adjunct (0 women). *Students:* 64 full-time (37 women), 37 part-time (21 women); includes 9 minority (1 American Indian/Alaska Native, 5 Asian Americans or Pacific Islanders, 3 Hispanic Americans), 17 international. Average age 30. 81 applicants, 22% accepted, 18 enrolled. In 2005, 6 master's, 7 doctorates awarded. *Degree requirements:* For master's, thesis/dissertation; for doctorate, thesis/dissertation, comprehensive exam. *Entrance requirements:* For master's and doctorate, GRE General Test, minimum GPA of 3.0. Additional exam requirements/recommendations for international students: Required—TOEFL. *Application deadline:* For fall admission, 1/1 for domestic students; for spring admission, 10/1 priority date for domestic students. Applications are processed on a rolling basis. Application fee: $50. Electronic applications accepted. *Expenses:* Tuition, state resident: full-time $3,690; part-time $205 per credit. Tuition, nonresident: full-time $14,958; part-time $831 per credit. Required fees: $1,061. *Financial support:* In 2005–06, 17 fellowships with tuition reimbursements (averaging $28,475 per year), 35 research assistantships with tuition reimbursements (averaging $21,715 per year), 7 teaching assistantships with tuition reimbursements (averaging $20,772 per year) were awarded; traineeships and unspecified assistantships also available. *Faculty research:* Medical and veterinary microbiology, pathology of disease, microbial pathogenesis, industrial and environmental microbiology, vector-borne disease. Total annual research expenditures: $21.1 million. *Unit head:* Dr. Jeffrey Wilusz, Head, 970-491-0652, Fax: 970-491-0603, E-mail: jeffrey.wilusz@colostate.edu. *Application contact:* Dr. Herbert Schweizer, Graduate Program Coordinator, 970-491-6136, Fax: 970-491-1815, E-mail: microbio@colostate.edu.
See Close-Up on page 861.

Cornell University, College of Veterinary Medicine, Ithaca, NY 14853-0001. Offers comparative biomedical science (PhD); immunology (MS, PhD); pharmacology (PhD); physiology (PhD); veterinary medicine (DVM); zoology (MS, PhD). *Accreditation:* AVMA. *Faculty:* 155 full-time (53 women). *Students:* 334 full-time (267 women); includes 71 minority (25 African Americans, 2 American Indian/Alaska Native, 18 Asian Americans or Pacific Islanders, 26 Hispanic Americans), 2 international. Average age 26. 871 applicants, 11% accepted, 78 enrolled. In 2005, 81 first professional degrees, 15 doctorates awarded. *Degree requirements:* For first-professional, thesis or alternative, on-site clinical training. *Entrance requirements:* GRE General Test or MCAT, undergraduate pre-medical science program, animal or veterinary experience, letter of recommendation. *Application deadline:* For fall admission, 10/1 for domestic students, 10/1 for international students. Application fee: $40. Electronic applications accepted. *Expenses:* Contact institution. *Financial support:* In 2005–06, 303 students received support, including 30 fellowships (averaging $24,854 per year), 88 research assistantships with tuition reimbursements available (averaging $24,854 per year); Federal Work-Study, institutionally sponsored loans, scholarships/grants, and unspecified assistantships also available. Financial award application deadline: 2/1; financial award applicants required to submit CSS PROFILE or FAFSA. *Faculty research:* Extensive biomedical research, comparative cancer, food safety. Total annual research expenditures: $49.3 million. *Unit head:* Dr. Donald F. Smith, Dean, 607-253-3771. *Application contact:* Jennifer A Mailey, Director of Admissions, 607-253-3700, Fax: 607-253-3709, E-mail: vet_admissions@cornell.edu.

Cornell University, Graduate School, Graduate Fields of Comparative Biomedical Sciences, Field of Immunology, Ithaca, NY 14853-0001. Offers cellular immunology (MS, PhD); immunochemistry (MS, PhD); immunogenetics (MS, PhD); immunopathology (MS, PhD); infection and immunity (MS, PhD). *Faculty:* 27 full-time (11 women). *Students:* 9 full-time (5 women), 3 international. 36 applicants, 14% accepted, 2 enrolled. In 2005, 1 degree awarded.

Terminal master's awarded for partial completion of doctoral program. *Degree requirements:* For master's, thesis/dissertation; for doctorate, thesis/dissertation, comprehensive exam. *Entrance requirements:* For master's and doctorate, GRE General Test, 2 letters of recommendation. Additional exam requirements/recommendations for international students: Required—TOEFL (minimum score 550 paper-based; 213 computer-based). *Application deadline:* For fall admission, 12/15 for domestic students. Application fee: $60. Electronic applications accepted. *Financial support:* In 2005–06, 8 students received support, including 4 fellowships with full tuition reimbursements available, 4 research assistantships with full tuition reimbursements available; teaching assistantships with full tuition reimbursements available, institutionally sponsored loans, scholarships/grants, health care benefits, tuition waivers (full and partial), and unspecified assistantships also available. Financial award applicants required to submit FAFSA. *Faculty research:* Avian immunology, mucosal immunity, anti-parasite and anti-viral immunity, neutrophil function, reproductive immunology. *Unit head:* Director of Graduate Studies, 607-253-3276, Fax: 607-253-3756. *Application contact:* Graduate Field Assistant, 607-253-3276, Fax: 607-253-3756, E-mail: graduate_edcvm@cornell.edu.

Cornell University, Joan and Sanford I. Weill Medical College and Graduate School of Medical Sciences, Weill Graduate School of Medical Sciences, Program in Immunology, New York, NY 10021-4896. Offers immunology (MS, PhD), including immunology, microbiology, pathology. *Faculty:* 35 full-time (10 women). *Students:* 38 full-time (22 women); includes 5 minority (1 African American, 4 Asian Americans or Pacific Islanders), 23 international. Average age 22. 60 applicants. In 2005, 6 degrees awarded. *Median time to degree:* Of those who began their doctoral program in fall 1997, 100% received their degree in 8 years or less. *Degree requirements:* For doctorate, thesis/dissertation, final exam. *Entrance requirements:* For doctorate, GRE General Test, GRE Subject Test, laboratory research experience, course work in biological sciences. Additional exam requirements/recommendations for international students: Required—TOEFL. *Application deadline:* For fall admission, 12/15 for domestic students. Application fee: $60. *Expenses:* Tuition: Full-time $32,320. Required fees: $1,025. *Financial support:* Fellowships, stipends available. *Unit head:* Selina Chen-Kiang, Director, 212-746-6440, E-mail: sckiang@med.cornell.edu.

Creighton University, School of Medicine and Graduate School, Graduate Programs in Medicine, Department of Medical Microbiology and Immunology, Omaha, NE 68178-0001. Offers MS, PhD. Terminal master's awarded for partial completion of doctoral program. *Degree requirements:* For master's, thesis, comprehensive exam; for doctorate, thesis/dissertation, preliminary exams. *Entrance requirements:* For master's and doctorate, GRE General Test. *Faculty research:* Infectious diseases, molecular biology, genetics, antimicrobial agents and chemotherapy, virology.
See Close-Up on page 863.

Dalhousie University, Faculty of Graduate Studies and Faculty of Medicine, Graduate Programs in Medicine, Department of Microbiology and Immunology, Halifax, NS B3H 4R2, Canada. Offers M Sc, PhD, MD/PhD. Part-time programs available. *Degree requirements:* For master's, thesis/dissertation; for doctorate, thesis/dissertation, comprehensive exam. *Entrance requirements:* For master's, GRE General Test, honors B Sc; for doctorate, GRE General Test, honors B Sc in microbiology, M Sc in discipline or transfer after 1 year in master's program. Additional exam requirements/recommendations for international students: Required—TOEFL. *Faculty research:* Virology, molecular genetics, pathogenesis, bacteriology, immunology.

Dartmouth College, Program in Immunology, Hanover, NH 03755.
See Close-Up on page 865.

Drexel University, College of Medicine, Biomedical Graduate Programs, Program in Microbiology and Immunology, Philadelphia, PA 19104-2875. Offers MS, PhD. Terminal master's awarded for partial completion of doctoral program. *Degree requirements:* For master's, thesis, comprehensive exam; for doctorate, thesis/dissertation, qualifying exam. *Entrance requirements:* For master's, GRE General Test, minimum GPA of 2.75; for doctorate, GRE General Test, minimum GPA of 3.0. Additional exam requirements/recommendations for international students: Required—TOEFL. Electronic applications accepted. *Faculty research:* Immunology of malarial parasites, virology, bacteriology, molecular biology, parasitology.

Duke University, Graduate School, Department of Immunology, Durham, NC 27710. Offers PhD. *Faculty:* 32 full-time. *Students:* 37 full-time (23 women); includes 5 minority (1 African American, 2 Asian Americans or Pacific Islanders, 2 Hispanic Americans), 14 international. 61 applicants, 20% accepted, 6 enrolled. In 2005, 2 doctorates awarded. *Degree requirements:* For doctorate, thesis/dissertation. *Entrance requirements:* For doctorate, GRE General Test, GRE Subject Test (recommended). Additional exam requirements/recommendations for international students: Required—IELT (preferred) or TOEFL. *Application deadline:* For fall admission, 12/31 for domestic students, 12/31 for international students. Application fee: $75. Electronic applications accepted. *Financial support:* Fellowships, research assistantships available. Financial award application deadline: 12/31. *Unit head:* , Michael Krangel, Director of Graduate Studies, 919-684-4985, Fax: 919-613-7878, E-mail: dgs-immunology@duke.edu.
See Close-Up on page 869.

East Carolina University, Brody School of Medicine, Department of Microbiology and Immunology, Greenville, NC 27858-4353. Offers PhD. *Faculty:* 15 full-time (2 women), 6 part-time/adjunct (0 women). *Students:* 5 full-time (4 women), 9 part-time (4 women), 3 international. Average age 28. 16 applicants, 38% accepted. In 2005, 3 degrees awarded. *Median time to degree:* Of those who began their doctoral program in fall 1997, 67% received their degree in 8 years or less. *Degree requirements:* For doctorate, thesis/dissertation, comprehensive exam, registration. *Entrance requirements:* For doctorate, GRE General Test. Additional exam requirements/recommendations for international students: Required—TOEFL. *Application deadline:* For fall admission, 4/15 for domestic students. Applications are processed on a rolling basis. Application fee: $50. *Expenses:* Tuition, state resident: full-time $2,516. Tuition, nonresident: full-time $12,832. *Financial support:* In 2005–06, 17 fellowships with tuition reimbursements (averaging $21,500 per year) were awarded Financial award application deadline: 6/1. *Faculty research:* Molecular virology, genetics of bacteria, yeast and somatic cells, bacterial physiology and metabolism, bioterrorism. Total annual research expenditures: $1.6 million. *Unit head:* Dr. Charles J. Smith, Interim Chair, 252-744-2700, Fax: 252-744-3104, E-mail: smithcha@ecu.edu. *Application contact:* Dr. Richard A. Franklin, Director of Graduate Studies, 252-744-2705, Fax: 252-744-3104, E-mail: franklinra@ecu.edu.

Emory University, Graduate School of Arts and Sciences, Division of Biological and Biomedical Sciences, Program in Immunology and Molecular Pathogenesis, Atlanta, GA 30322-1100. Offers PhD. *Faculty:* 46 full-time (9 women). *Students:* 62 full-time (29 women); includes 11 minority (5 African Americans, 5 Asian Americans or Pacific Islanders, 1 Hispanic American), 8 international. Average age 27. 136 applicants, 23% accepted, 17 enrolled. In 2005, 12 degrees awarded. *Median time to degree:* Of those who began their doctoral program in fall 1997, 100% received their degree in 8 years or less. *Degree requirements:* For doctorate, thesis/dissertation, comprehensive exam, registration. *Entrance requirements:* For doctorate, GRE General Test, minimum GPA of 3.0 in science course work. Additional exam requirements/recommendations for international students: Required—TOEFL. *Application deadline:* For fall admission, 1/3 for domestic students, 1/3 for international students. Application fee: $50. Electronic applications accepted. *Expenses:* Tuition: Full-time $14,400. Required fees: $217. *Financial support:* In 2005–06, 30 students received support, including 30 fellowships with full tuition reimbursements available (averaging $23,000 per year); institutionally sponsored loans, health care benefits, and tuition waivers (full) also available. *Faculty research:* Transplantation immunology, autoimmunity, microbial pathogenesis. *Unit head:* Dr. Aron Lukacher, Director, 404-727-1896, Fax: 404-727-5764, E-mail: alukach@emory.edu. *Application contact:* 404-727-2545, Fax: 404-727-3322, E-mail: gdbbs@emory.edu.

Immunology

Florida State University, Graduate Studies, College of Arts and Sciences, Department of Biological Science, Program in Immunology, Tallahassee, FL 32306. Offers MS, PhD. *Faculty:* 2 full-time (0 women). *Students:* 8 full-time (4 women), 3 international. *Degree requirements:* For master's and doctorate, thesis/dissertation, teaching experience, seminar presentation, comprehensive exam, registration. *Entrance requirements:* For master's, GRE General Test (minimum 1100: U-500, G-500), minimum upper division GPA of 3.0; for doctorate, GGRE General Test (minimum 1100: U-500, G-500), minimum upper division GPA of 3.0. Additional exam requirements/recommendations for international students: Required—TOEFL (minimum score 600 paper-based; 250 computer-based), IB 100. *Application deadline:* For fall admission, 1/15 for domestic students, 12/1 for international students; for spring admission, 10/15 for domestic students, 9/1 for international students. Application fee: $30. *Financial support:* In 2005–06, fellowships with full tuition reimbursements (averaging $19,000 per year), research assistantships with full tuition reimbursements (averaging $19,000 per year), teaching assistantships with full tuition reimbursements (averaging $17,600 per year) were awarded. Financial award application deadline: 1/15; financial award applicants required to submit FAFSA. *Faculty research:* Immunogenetics. *Application contact:* Judy Bowers, Coordinator, Graduate Affairs, 850-644-3023, Fax: 850-644-9829, E-mail: gradinfo@bio.fsu.edu.

Georgetown University, Graduate School of Arts and Sciences, Programs in Biomedical Sciences, Department of Microbiology and Immunology, Washington, DC 20057. Offers biohazardous threat agents and emerging infectious diseases (MS); microbiology and immunology research (PhD); science policy and advocacy for the healthcare arena in a global setting (MS); teaching microbiology and immunology (MS). *Degree requirements:* For doctorate, thesis/dissertation, comprehensive exam. *Entrance requirements:* For master's, GRE General Test, 3 letters of reference, bachelor's degree in related field; for doctorate, GRE General Test, 3 letters of reference, graduate degree in related field. Additional exam requirements/recommendations for international students: Required—TOEFL (minimum score 505 paper-based; 213 computer-based). Electronic applications accepted. *Faculty research:* Pathogenesis and basic biology of the fungus Candida albicans, molecular biology of viral hepatitis, molecular and cellular biology of the Hepatitis B and Delta viruses, dengue virus, immunopathological mechanisms in Multiple Sclerosis.

See Close-Up on page 871.

The George Washington University, Columbian College of Arts and Sciences, Institute for Biomedical Sciences, Program in Microbiology and Immunology, Washington, DC 20052. Offers PhD. *Students:* 3 full-time (1 woman), 5 part-time; includes 1 minority (Asian American or Pacific Islander), 2 international. Average age 27. *Degree requirements:* For doctorate, thesis/dissertation. *Entrance requirements:* For doctorate, GRE General Test, minimum GPA of 3.0. Additional exam requirements/recommendations for international students: Required—TOEFL (minimum score 600 paper-based; 250 computer-based). *Application deadline:* For fall admission, 1/2 priority date for domestic students, 1/2 priority date for international students. Applications are processed on a rolling basis. Application fee: $60. Electronic applications accepted. *Financial support:* Fellowships with tuition reimbursements available. *Unit head:* Dr. D. Leitenberg, Head, 202-994-3532, Fax: 202-994-2913. *Application contact:* Information Contact, 202-994-3532, Fax: 202-994-2913, E-mail: mtmjxl@gwumc.edu.

See Close-Up on page 873.

Harvard University, School of Public Health, Department of Immunology and Infectious Diseases, Boston, MA 02115-6096. Offers PhD, SD. Part-time programs available. *Degree requirements:* For doctorate, thesis/dissertation, qualifying exam. *Entrance requirements:* For doctorate, GRE. Additional exam requirements/recommendations for international students: Required—TOEFL (minimum score 560 paper-based; 220 computer-based); Recommended—IELT (minimum score 7). Electronic applications accepted. *Faculty research:* Infectious disease epidemiology and tropical public health, vector biology, ecology and control, virology.

Indiana University–Purdue University Indianapolis, Indiana University School of Medicine, Department of Microbiology and Immunology, Indianapolis, IN 46202-2896. Offers MS, PhD, MD/MS, MD/PhD. *Faculty:* 20 full-time (2 women). *Students:* 18 full-time (15 women), 10 part-time (6 women); includes 5 minority (4 African Americans, 1 Hispanic American), 10 international. Average age 28. In 2005, 1 master's, 10 doctorates awarded. Terminal master's awarded for partial completion of doctoral program. *Degree requirements:* For master's and doctorate, thesis/dissertation. *Entrance requirements:* For master's and doctorate, GRE General Test, previous course work in calculus, cell biology, chemistry, genetics, physics, and biochemistry. *Application deadline:* For fall admission, 3/1 for domestic students. Applications are processed on a rolling basis. Application fee: $50 ($60 for international students). *Expenses:* Tuition, state resident: full-time $5,159; part-time $215 per credit hour. Tuition, nonresident: full-time $14,890; part-time $620 per credit hour. Required fees: $614. Tuition and fees vary according to campus/location and program. *Financial support:* Fellowships with full tuition reimbursements, research assistantships with full tuition reimbursements, teaching assistantships with full tuition reimbursements, Federal Work-Study, institutionally sponsored loans, scholarships/grants, traineeships, and tuition waivers (partial) available. Financial award application deadline: 2/1. *Faculty research:* Host-parasite interactions, molecular biology, cellular and molecular immunology and hematology, viral and bacterial pathogenesis, cancer research. Total annual research expenditures: $4.2 million. *Unit head:* Dr. Hal E. Broxmeyer, Chairman, 317-274-7672, Fax: 317-274-4090, E-mail: hbroxmey@iupui.edu. *Application contact:* 317-274-7671, Fax: 317-274-4090.

See Close-Up on page 875.

Iowa State University of Science and Technology, Graduate College, Interdisciplinary Programs, Program in Immunobiology, Ames, IA 50011. Offers MS, PhD. *Students:* 17 full-time (12 women), 2 part-time (both women); includes 2 minority (both Hispanic Americans), 5 international. 22 applicants, 14% accepted, 2 enrolled. In 2005, 2 master's, 3 doctorates awarded. *Degree requirements:* For master's and doctorate, one foreign language, thesis/dissertation. *Entrance requirements:* For master's and doctorate, GRE General Test, resumé. Additional exam requirements/recommendations for international students: Required—IELTS (paper score 600; computer score 250) or TOEFL (score 7.1). *Application deadline:* For fall admission, 1/15 priority date for domestic students, 1/15 priority date for international students. Applications are processed on a rolling basis. Application fee: $30 ($70 for international students). Electronic applications accepted. *Expenses:* Tuition, state resident: full-time $6,410. Tuition, nonresident: full-time $16,422. Tuition and fees vary according to program. *Financial support:* In 2005–06, 16 research assistantships with full and partial tuition reimbursements (averaging $16,125 per year) were awarded; fellowships, teaching assistantships, scholarships/grants, health care benefits, and unspecified assistantships also available. *Faculty research:* Immunogenetics, cellular and molecular immunology, infectious disease, neuroimmunology. *Unit head:* Dr. Doug E Jones, Supervisory Committee Chair, 515-294-7252, E-mail: idgp@iastate.edu. *Application contact:* Mary Jordison, Information Contact, 515-294-7252, Fax: 515-924-6790, E-mail: idgp@iastate.edu.

The Johns Hopkins University, Bloomberg School of Public Health, The W. Harry Feinstone Department of Molecular Microbiology and Immunology, Baltimore, MD 21218-2699. Offers MHS, Sc M, PhD. Part-time programs available. *Faculty:* 38 full-time (9 women), 17 part-time/adjunct (4 women). *Students:* 64 full-time (39 women), 3 part-time (1 woman); includes 18 minority (5 African Americans, 1 American Indian/Alaska Native, 8 Asian Americans or Pacific Islanders, 4 Hispanic Americans), 20 international. Average age 28. 143 applicants, 38% accepted, 25 enrolled. In 2005, 9 master's, 5 doctorates awarded. Terminal master's awarded for partial completion of doctoral program. *Median time to degree:* Of those who began their doctoral program in fall 1997, 50% received their degree in 8 years or less. *Degree requirements:* For master's, thesis (for some programs), master's essay, written exams; for doctorate, thesis/dissertation, 1 year full-time residency, oral and written exams. *Entrance requirements:* For master's, GRE General Test or MCAT, 3 letters of recommendation, curriculum vitae; for doctorate, GRE General Test, 3 letters of recommendation, curriculum vitae. Additional exam

requirements/recommendations for international students: Required—TOEFL (minimum score 600 paper-based; 250 computer-based). *Application deadline:* For fall admission, 11/10 priority date for domestic students, 11/10 priority date for international students. Applications are processed on a rolling basis. Application fee: $45. Electronic applications accepted. *Expenses:* Tuition: Full-time $30,960. Tuition and fees vary according to degree level and program. *Financial support:* In 2005–06, 101 students received support, including 3 fellowships (averaging $23,600 per year); Federal Work-Study, institutionally sponsored loans, scholarships/grants, and stipends also available. Support available to part-time students. Financial award application deadline: 3/15; financial award applicants required to submit FAFSA. *Faculty research:* Immunological disorders, viral and bacterial infections, parasitic diseases, vector-borne diseases, pathogenesis, parasite immunology, biochemistry of parasitic protozoa, vector biology, ecology of infectious diseases. Total annual research expenditures: $22.4 million. *Unit head:* Dr. Diane E. Griffin, Chair, 410-955-3459, Fax: 410-955-0105, E-mail: dgriffin@jhsph.edu. *Application contact:* E-mail: mmi@jhsph.edu.

The Johns Hopkins University, School of Medicine, Graduate Programs in Medicine, Program in Immunology, Baltimore, MD 21218-2699. Offers PhD. *Degree requirements:* For doctorate, thesis/dissertation, oral exam, final thesis seminar, comprehensive exam, registration. *Entrance requirements:* For doctorate, GRE General Test, GRE Subject Test. Additional exam requirements/recommendations for international students: Required—TOEFL. *Expenses:* Tuition: Full-time $30,960. Tuition and fees vary according to degree level and program. *Faculty research:* HIV immunity, tumor immunity, major histocompatibility complex, transplantation, genetics of antibodies and T-cell receptors.

See Close-Up on page 877.

Long Island University, C.W. Post Campus, School of Health Professions and Nursing, Department of Biomedical Sciences, Program in Medical Biology, Brookville, NY 11548-1300. Offers hematology (MS); immunology (MS); medical biology (MS); medical chemistry (MS); medical microbiology (MS). Part-time and evening/weekend programs available. *Degree requirements:* For master's, thesis. *Entrance requirements:* For master's, minimum GPA of 2.75 in major. Electronic applications accepted. *Faculty research:* Hematopoiesis, growth factors in cancer, interleukins in allergy, PCR techniques.

Louisiana State University Health Sciences Center, School of Graduate Studies in New Orleans, Department of Microbiology, Immunology, and Parasitology, New Orleans, LA 70112-1393. Offers microbiology and immunology (MS, PhD). Terminal master's awarded for partial completion of doctoral program. *Degree requirements:* For master's, thesis; for doctorate, thesis/dissertation, preliminary exam, qualifying exam. *Entrance requirements:* For master's and doctorate, GRE General Test. Additional exam requirements/recommendations for international students: Required—TOEFL. *Faculty research:* Microbial physiology, animal virology, vaccine development, AIDS drug studies, pathogenic mechanisms, molecular immunology.

Louisiana State University Health Sciences Center at Shreveport, Department of Microbiology and Immunology, Shreveport, LA 71130-3932. Offers microbiology/immunology (MS). *Faculty:* 21 full-time (5 women). *Students:* 22 full-time (10 women); includes 5 minority (2 African Americans, 1 American Indian/Alaska Native, 1 Asian American or Pacific Islander, 1 Hispanic American). Average age 23. 44 applicants, 16% accepted. In 2005, 3 master's awarded. Terminal master's awarded for partial completion of doctoral program. *Degree requirements:* For master's, thesis/dissertation. *Entrance requirements:* For master's, GRE General Test. Additional exam requirements/recommendations for international students: Required—TOEFL. Application fee: $30. *Financial support:* In 2005–06, 2 fellowships (averaging $20,000 per year), 20 research assistantships (averaging $20,000 per year) were awarded; institutionally sponsored loans also available. Financial award application deadline: 7/1. *Faculty research:* Infectious disease, pathogenesis, molecular virology and biology. *Unit head:* Dr. Dennis J. O'Callaghan, Head, 318-675-5750, Fax: 318-675-5764.

See Close-Up on page 163.

Loyola University Chicago, Graduate School, Department of Microbiology and Immunology, Maywood, IL 60153. Offers immunology (MS, PhD); microbiology (MS, PhD); virology (MS, PhD). *Faculty:* 11 full-time (3 women). *Students:* 26 full-time (18 women), 1 (woman) part-time; includes 4 minority (1 African American, 3 Asian Americans or Pacific Islanders), 9 international. Average age 28. 74 applicants, 15% accepted, 4 enrolled. In 2005, 1 master's, 3 doctorates awarded. Terminal master's awarded for partial completion of doctoral program. *Degree requirements:* For master's, thesis/dissertation; for doctorate, thesis/dissertation, comprehensive exam. *Entrance requirements:* For master's and doctorate, GRE General Test. Additional exam requirements/recommendations for international students: Required—TOEFL. *Application deadline:* Applications are processed on a rolling basis. Application fee: $40. Electronic applications accepted. *Expenses:* Tuition: Full-time $11,610; part-time $645 per credit. Required fees: $55 per semester. *Financial support:* In 2005–06, 5 fellowships with tuition reimbursements (averaging $22,000 per year), 24 research assistantships with tuition reimbursements (averaging $22,000 per year) were awarded; institutionally sponsored loans and scholarships/grants also available. Financial award application deadline: 2/15. *Faculty research:* Viral pathogenesis, microbial physiology and genetics, immunoglobulin genetics and differentiation of the immune response, signal transduction and host-parasite interactions. *Unit head:* Dr. Katherine L. Knight, Chair, 708-216-3385, Fax: 708-216-9574, E-mail: kknight@lumc.edu. *Application contact:* Dr. Karen Visick, Graduate Program Director, 708-216-0869, Fax: 708-216-9574, E-mail: kvisick@lumc.edu.

See Close-Up on page 879.

Mayo Graduate School, Graduate Programs in Biomedical Sciences, Program in Immunology, Rochester, MN 55905. Offers PhD. *Degree requirements:* For doctorate, oral defense of dissertation, qualifying oral and written exam. *Entrance requirements:* For master's, GRE, 1 year of chemistry, biology, calculus, and physics. Additional exam requirements/recommendations for international students: Required—TOEFL. Electronic applications accepted. *Faculty research:* Immunogenetics, autoimmunity, receptor signal transduction, T lymphocyte activation, transplantation.

See Close-Up on page 881.

McGill University, Faculty of Graduate and Postdoctoral Studies, Faculty of Medicine, Department of Microbiology and Immunology, Montréal, QC H3A 2T5, Canada. Offers M Sc, M Sc A, PhD. *Degree requirements:* For master's and doctorate, thesis/dissertation. *Entrance requirements:* For master's, minimum GPA of 3.0. *Faculty research:* Virology, molecular genetics, immunology, molecular biology, microbial physiology.

McMaster University, Faculty of Health Sciences and School of Graduate Studies, Program in Medical Sciences, Molecular Immunology, Virology, and Inflammation Area, Hamilton, ON L8S 4M2, Canada. Offers M Sc, PhD. *Students:* 66 full-time, 2 part-time. In 2005, 4 master's, 2 doctorates awarded. *Degree requirements:* For master's, thesis/dissertation; for doctorate, thesis/dissertation, comprehensive exam. *Entrance requirements:* For master's, honors B Sc, B+ average in related field; for doctorate, M Sc, minimum B+ average, students with proven research experience and an A average may be admitted with a B Sc degree. Additional exam requirements/recommendations for international students: Required—TOEFL (minimum score 580 paper-based; 237 computer-based). *Application deadline:* For fall admission, 3/31 for domestic students. Applications are processed on a rolling basis. Application fee: $85. *Financial support:* Teaching assistantships available. *Unit head:* Dr. Mark McDermott, Coordinator, 905-525-9140 Ext. 22874. *Application contact:* Dr. Carl Richards, Associate Dean, 905-525-9140 Ext. 22983, Fax: 905-546-1129.

Medical University of South Carolina, College of Graduate Studies, Program in Microbiology and Immunology, Charleston, SC 29425-0002. Offers Pharm D, MS, PhD, DMD/PhD, MD/PhD. *Faculty:* 18 full-time (7 women). *Students:* 23 full-time (12 women); includes 4 minority (2 Asian Americans or Pacific Islanders, 2 Hispanic Americans), 1 international.

Immunology

Medical University of South Carolina (continued)

Average age 28. 181 applicants, 34% accepted, 43 enrolled. In 2005, 1 master's, 6 doctorates awarded. Terminal master's awarded for partial completion of doctoral program. *Degree requirements:* For master's, thesis, research seminar; for doctorate, thesis/dissertation, teaching and research seminar, oral and written exams. *Entrance requirements:* For master's and doctorate, GRE General Test, interview. Additional exam requirements/recommendations for international students: Required—TOEFL (minimum score 600 paper-based; 250 computer-based). *Application deadline:* For fall admission, 1/15 priority date for domestic students, 1/15 priority date for international students. Applications are processed on a rolling basis. Application fee: $0 ($75 for international students). Electronic applications accepted. *Financial support:* In 2005–06, 15 students received support, including fellowships with partial tuition reimbursements available (averaging $21,000 per year); Federal Work-Study and scholarships/grants also available. Financial award application deadline: 3/15; financial award applicants required to submit FAFSA. *Faculty research:* Inmate and adoptive immunology, gene therapy/vector development, vaccinology, protemonics of biowarfare agents, bacterial and fungal pathogenesis. *Unit head:* Dr. James S. Norris, Chair, 843-792-7915, Fax: 843-792-6590, E-mail: norrisjs@musc.edu. *Application contact:* Cheryl Brown, 843-792-4620, Fax: 843-792-4645, E-mail: brownche@musc.edu.

See Close-Up on page 885.

New York Medical College, Graduate School of Basic Medical Sciences, Microbiology and Immunology Department, Valhalla, NY 10595-1691. Offers MS, PhD, MD/PhD. Part-time and evening/weekend programs available. Terminal master's awarded for partial completion of doctoral program. *Degree requirements:* For master's, thesis/dissertation; for doctorate, thesis/dissertation, comprehensive exam. *Entrance requirements:* For master's and doctorate, GRE General Test. Additional exam requirements/recommendations for international students: Required—TOEFL. *Faculty research:* Tumor and transplantation immunology, molecular mechanisms of DNA repair, virus-host interactions.

New York University, Graduate School of Arts and Science, Department of Biology, New York, NY 10012-1019. Offers biology (PhD); biomedical journalism (MS); cancer and molecular biology (PhD); computational biology (PhD); computers in biological research (MS); developmental genetics (PhD); general biology (MS); immunology and microbiology (PhD); molecular genetics (PhD); neurobiology (PhD); oral biology (MS); plant biology (PhD); recombinant DNA technology (MS). Part-time programs available. *Faculty:* 24 full-time (5 women), 8 part-time/adjunct. *Students:* 104 full-time (52 women), 41 part-time (23 women); includes 28 minority (2 African Americans, 20 Asian Americans or Pacific Islanders, 6 Hispanic Americans), 47 international. Average age 27. 349 applicants, 56% accepted, 39 enrolled. In 2005, 59 master's, 4 doctorates awarded. Terminal master's awarded for partial completion of doctoral program. *Degree requirements:* For master's, thesis or alternative, qualifying paper; for doctorate, thesis/dissertation, comprehensive exam. *Entrance requirements:* For master's, GRE General Test; for doctorate, GRE General Test, GRE Subject Test. Additional exam requirements/recommendations for international students: Required—TOEFL. *Application deadline:* For fall admission, 1/4 for domestic students. Application fee: $80. *Financial support:* Fellowships with tuition reimbursements, research assistantships with tuition reimbursements, teaching assistantships with tuition reimbursements, career-related internships or fieldwork, Federal Work-Study, institutionally sponsored loans, scholarships/grants, health care benefits, and unspecified assistantships available. Financial award application deadline: 1/4; financial award applicants required to submit FAFSA. *Faculty research:* Genomics, molecular and cell biology, development and molecular genetics, molecular evolution of plants and animals. *Unit head:* Gloria Coruzzi, Chairman, 212-998-8200, Fax: 212-995-4015, E-mail: biology@nyu.edu. *Application contact:* Stephen Small, Director of Graduate Studies, 212-998-8200, Fax: 212-995-4015, E-mail: biology@nyu.edu.

New York University, School of Medicine and Graduate School of Arts and Science, Sackler Institute of Graduate Biomedical Sciences, Graduate Programs in Molecular Oncology and Immunology, New York, NY 10012-1019. Offers immunology (PhD); molecular oncology (PhD). *Degree requirements:* For doctorate, one foreign language, thesis/dissertation, qualifying exam. *Entrance requirements:* For doctorate, GRE General Test, GRE Subject Test. Additional exam requirements/recommendations for international students: Required—TOEFL. Electronic applications accepted.

See Close-Up on page 891.

North Carolina State University, College of Veterinary Medicine, Program in Comparative Biomedical Sciences, Raleigh, NC 27695. Offers cell biology and morphology (MS, PhD); epidemiology and population medicine (MS, PhD); immunology (MS, PhD); microbiology and immunology (MS, PhD); pathology (MS, PhD); pharmacology (MS, PhD); specialized veterinary medicine (MS). Part-time programs available. *Degree requirements:* For master's and doctorate, thesis/dissertation. *Entrance requirements:* For master's and doctorate, GRE General Test. Additional exam requirements/recommendations for international students: Required—TOEFL (minimum score 550 paper-based; 213 computer-based). Electronic applications accepted. Expenses: Contact institution. *Faculty research:* Infectious diseases, cell biology, pharmacology and toxicology, genomics, pathology and population medicine.

North Carolina State University, Graduate School, College of Agriculture and Life Sciences and College of Veterinary Medicine, Program in Immunology, Raleigh, NC 27695. Offers MS, PhD. *Degree requirements:* For master's and doctorate, thesis/dissertation. *Entrance requirements:* For master's and doctorate, GRE General Test. Additional exam requirements/recommendations for international students: Required—TOEFL (minimum score 550 paper-based; 213 computer-based). Electronic applications accepted. *Faculty research:* Immunogenetics, immunopathology, immunotoxicology, immunoparasitology, molecular and infectious disease immunology.

Northwestern University, Northwestern University Feinberg School of Medicine, Department of Microbiology-Immunology, Chicago, IL 60611-3008.

Northwestern University, Northwestern University Feinberg School of Medicine and Interdepartmental Degree Programs, Integrated Graduate Programs in the Life Sciences, Chicago, IL 60611. Offers cancer biology (PhD); cell biology (PhD); developmental biology (PhD); evolutionary biology (PhD); immunology and microbial pathogenesis (PhD); molecular biology and genetics (PhD); neurobiology (PhD); pharmacology and toxicology (PhD); structural biology and biochemistry (PhD). *Degree requirements:* For doctorate, thesis/dissertation, written and oral qualifying exams, comprehensive exam. *Entrance requirements:* For doctorate, GRE General Test. Additional exam requirements/recommendations for international students: Required—TOEFL (minimum score 600 paper-based; 250 computer-based). Electronic applications accepted.

See Close-Up on page 189.

The Ohio State University, College of Medicine and Public Health and Graduate School, Graduate Programs in the Basic Medical Sciences, Integrated Biomedical Science Graduate Program, Columbus, OH 43210. Offers immunology (MS, PhD); medical genetics (MS, PhD); molecular virology (MS, PhD); pharmacology (MS, PhD). *Degree requirements:* For doctorate, thesis/dissertation. *Entrance requirements:* For master's, GRE General Test; for doctorate, GRE. Additional exam requirements/recommendations for international students: Required—TOEFL (minimum score 600 paper-based; 250 computer-based), TSE. Electronic applications accepted.

Oregon Health & Science University, School of Medicine, Graduate Programs in Medicine, Department of Molecular Microbiology and Immunology, Portland, OR 97239-3098. Offers PhD. *Degree requirements:* For doctorate, thesis/dissertation. *Entrance requirements:* For doctorate, GRE General Test. *Faculty research:* Molecular biology of bacterial and viral pathogens, cellular and humoral immunology, molecular biology of microbes.

The Pennsylvania State University Milton S. Hershey Medical Center, Graduate School Programs in the Biomedical Sciences, Graduate Program in Microbiology and Immunology, Hershey, PA 17033-2360. Offers genetics (PhD); immunology (MS, PhD); microbiology (MS); microbiology/virology (PhD); molecular biology (PhD). *Students:* 23 full-time (11 women); includes 2 minority (both Asian Americans or Pacific Islanders), 2 international. Average age 27. Terminal master's awarded for partial completion of doctoral program. *Median time to degree:* Of those who began their doctoral program in fall 1997, 100% received their degree in 8 years or less. *Degree requirements:* For master's, thesis or alternative, registration; for doctorate, thesis/dissertation, oral exam, comprehensive exam, registration. *Entrance requirements:* For master's, GRE or MCAT; for doctorate, GRE General Test or MCAT, minimum GPA of 3.0. Additional exam requirements/recommendations for international students: Required—TOEFL. *Application deadline:* Applications are processed on a rolling basis. Application fee: $45. Electronic applications accepted. *Financial support:* In 2005–06, 23 research assistantships with full tuition reimbursements were awarded; fellowships with full tuition reimbursements, scholarships/grants, health care benefits, and unspecified assistantships also available. Financial award applicants required to submit FAFSA. *Faculty research:* Virus replication and assembly, oncogenesis, interactions of viruses with host cells and animal model systems. *Unit head:* , Dr. Richard J. Courtney, Chair, 717-531-7659, Fax: 717-531-6522, E-mail: micro-grad-hmc@psu.edu. *Application contact:* Billie Burns, Secretary, 717-531-7659, Fax: 717-531-6522, E-mail: micro-grad-hmc@psu.edu.

Purdue University, School of Veterinary Medicine and Graduate School, Graduate Programs in Veterinary Medicine, Department of Veterinary Pathobiology, West Lafayette, IN 47907. Offers biochemistry and molecular biology (MS, PhD); comparative epidemiology (MS, PhD); epidemiology (MS, PhD); immunology (MS, PhD); infectious diseases (MS, PhD); interdisciplinary genetics (PhD); laboratory animal medicine (MS, PhD); microbiology (MS, PhD); molecular virology (MS, PhD); parasitology (MS, PhD); pathobiology (MS, PhD); public health epidemiology (MS, PhD); toxicology (MS, PhD); veterinary anatomic pathology (MS, PhD); veterinary clinical pathology (MS, PhD); virology (MS, PhD). *Faculty:* 32 full-time (7 women). *Students:* 49 full-time (20 women), 3 part-time (1 woman); includes 2 minority (both African Americans), 31 international. Average age 35. In 2005, 3 master's, 8 doctorates awarded. Terminal master's awarded for partial completion of doctoral program. *Degree requirements:* For master's, thesis (for some programs); for doctorate, thesis/dissertation. *Entrance requirements:* For master's and doctorate, GRE General Test. Additional exam requirements/recommendations for international students: Required—TOEFL (minimum score 575 paper-based), TWE (minimum score 4). *Application deadline:* For fall admission, 8/12 for domestic students, 6/15 for international students; for spring admission, 1/12 for domestic students, 10/15 for international students. Application fee: $55. *Financial support:* Fellowships, research assistantships, teaching assistantships available. Financial award application deadline: 3/1; financial award applicants required to submit FAFSA. *Unit head:* Dr. H. Hogenesch, Head, 765-494-7543.

Queen's University at Kingston, School of Graduate Studies and Research, Faculty of Health Sciences, Department of Microbiology and Immunology, Kingston, ON K7L 3N6, Canada. Offers M Sc, PhD. Part-time programs available. *Degree requirements:* For master's, thesis/dissertation, registration; for doctorate, thesis/dissertation, comprehensive exam, registration. *Entrance requirements:* For master's and doctorate, minimum B+ average. Additional exam requirements/recommendations for international students: Required—TOEFL (minimum score 600 paper-based; 250 computer-based). Electronic applications accepted. *Faculty research:* Bacteriology, virology, immunology, education in microbiology and immunology, microbial pathogenesis.

Rosalind Franklin University of Medicine and Science, School of Graduate and Postdoctoral Studies, Department of Microbiology and Immunology, North Chicago, IL 60064-3095. Offers medical microbiology (MS, PhD); microbiology and immunology (MS, PhD). Part-time programs available. Terminal master's awarded for partial completion of doctoral program. *Degree requirements:* For master's and doctorate, thesis/dissertation. *Entrance requirements:* For master's and doctorate, GRE General Test. Additional exam requirements/recommendations for international students: Required—TOEFL, TWE. *Faculty research:* Molecular biology, parasitology, virology.

Rush University, Graduate College, Division of Immunology and Microbiology, Program in Immunology/Microbiology, Chicago, IL 60612-3832. Offers immunology (MS, PhD); virology (MS, PhD). Part-time programs available. *Faculty:* 8 full-time (4 women). *Students:* 15 full-time (7 women); includes 1 minority (African American), 4 international. Average age 30. 62 applicants, 6% accepted, 4 enrolled. In 2005, 2 degrees awarded. Terminal master's awarded for partial completion of doctoral program. *Median time to degree:* Of those who began their doctoral program in fall 1997, 99% received their degree in 8 years or less. *Degree requirements:* For master's, thesis; for doctorate, thesis/dissertation, comprehensive preliminary exam. *Entrance requirements:* For master's, GRE General Test; for doctorate, GRE General Test, interview, minimum GPA of 3.0. Additional exam requirements/recommendations for international students: Required—TOEFL. *Application deadline:* For fall admission, 4/1 for domestic students. Applications are processed on a rolling basis. Application fee: $25. Electronic applications accepted. *Financial support:* In 2005–06, 3 students received support, including 6 research assistantships with full tuition reimbursements available (averaging $20,000 per year); Federal Work-Study, institutionally sponsored loans, scholarships/grants, and tuition waivers (full and partial) also available. Support available to part-time students. Financial award application deadline: 4/15. *Faculty research:* Human genetics, autoimmunity, tumor biology, complement, HIV immunopathology genesis. Total annual research expenditures: $800,000. *Application contact:* Connie M. Lambert, Department Administrator, 312-563-2563, Fax: 312-563-3552, E-mail: connie_m_lambert@rush.edu.

Rutgers, The State University of New Jersey, New Brunswick/Piscataway, Graduate School, Program in Cell and Developmental Biology, New Brunswick, NJ 08901-1281. Offers cell biology (MS, PhD); developmental biology (MS, PhD); immunology (MS). Part-time programs available. *Faculty:* 121 full-time. *Students:* 37 full-time (17 women), 9 part-time (8 women); includes 12 minority (2 African Americans, 1 American Indian/Alaska Native, 6 Asian Americans or Pacific Islanders, 3 Hispanic Americans), 20 international. Average age 31. 74 applicants, 36% accepted, 13 enrolled. In 2005, 8 master's, 7 doctorates awarded. Terminal master's awarded for partial completion of doctoral program. *Degree requirements:* For master's, thesis or alternative; for doctorate, thesis/dissertation, written qualifying exam. *Entrance requirements:* For master's, GRE General Test; for doctorate, GRE General Test, GRE Subject Test (recommended), minimum GPA of 3.0. Additional exam requirements/recommendations for international students: Required—TOEFL. *Application deadline:* For fall admission, 1/5 for domestic students. Applications are processed on a rolling basis. Application fee: $50. Electronic applications accepted. *Expenses:* Tuition: state resident: full-time $10,440; part-time $435 per credit. Tuition, nonresident: full-time $15,520; part-time $647 per credit. Required fees: $129 per credit. Tuition and fees vary according to program. *Financial support:* In 2005–06, 36 students received support, including 16 fellowships with full tuition reimbursements available (averaging $24,000 per year), 14 research assistantships with full tuition reimbursements available (averaging $21,500 per year), 6 teaching assistantships with full tuition reimbursements available (averaging $16,988 per year) Financial award application deadline: 1/15; financial award applicants required to submit FAFSA. *Faculty research:* Signal transduction, developmental biology, cell biology, developmental genetics, developmental neurobiology. *Unit head:* Dr. Richard Padgett, Acting Director, 732-445-0251, Fax: 732-445-6370, E-mail: padgett@waksman.rutgers.edu. *Application contact:* Carolyn J. Ambrose, Administrative Assistant, 732-445-3430, Fax: 732-445-6370, E-mail: ambrose@biology.rutgers.edu.

Rutgers, The State University of New Jersey, New Brunswick/Piscataway, Graduate School, Program in Microbiology and Molecular Genetics, New Brunswick, NJ 08901-1281. Offers applied microbiology (MS, PhD); clinical microbiology (MS, PhD); computational molecular biology (PhD); immunology (MS, PhD); microbial biochemistry (MS, PhD); molecular genetics (MS, PhD); virology (MS, PhD). Part-time programs available. *Faculty:* 132 full-time. *Students:* 57 full-time (31 women), 18 part-time (8 women); includes 16 minority (9 Asian Americans or

Pacific Islanders, 7 Hispanic Americans), 20 international. Average age 29. 104 applicants, 32% accepted, 19 enrolled. In 2005, 9 master's, 10 doctorates awarded. Terminal master's awarded for partial completion of doctoral program. *Median time to degree:* Of those who began their doctoral program in fall 1997, 100% received their degree in 8 years or less. *Degree requirements:* For master's, thesis or alternative, comprehensive exam, registration; for doctorate, thesis/dissertation, written qualifying exam, comprehensive exam, registration. *Entrance requirements:* For master's, GRE General Test, minimum GPA of 3.0; for doctorate, GRE General Test, GRE Subject Test (recommended), minimum GPA of 3.0. Additional exam requirements/recommendations for international students: Required—TOEFL. *Application deadline:* For fall admission, 1/5 priority date for domestic students, 11/1 priority date for international students. Applications are processed on a rolling basis. Application fee: $50. Electronic applications accepted. *Expenses:* Tuition, state resident: full-time $10,440; part-time $435 per credit. Tuition, nonresident: full-time $15,520; part-time $647 per credit. Required fees: $129 per credit. Tuition and fees vary according to program. *Financial support:* In 2005–06, 48 students received support, including 14 fellowships with full tuition reimbursements available (averaging $24,000 per year), 25 research assistantships with full tuition reimbursements available (averaging $24,000 per year), 9 teaching assistantships with full tuition reimbursements available (averaging $16,988 per year); Federal Work-Study, institutionally sponsored loans, scholarships/grants, and unspecified assistantships also available. Financial award application deadline: 1/5; financial award applicants required to submit FAFSA. *Faculty research:* Molecular genetics and microbial physiology; virology and pathogenic microbiology; applied, environmental and industrial microbiology; computers in molecular biology. *Unit head:* Dr. Andrew K. Vershon, Director, 732-445-2905, Fax: 732-445-6370, E-mail: vershon@waksman.rutgers.edu. *Application contact:* Diane Murano, Administrative Assistant, 732-445-5086, Fax: 732-445-6370, E-mail: murano@biology.rutgers.edu.

Saint Louis University, Graduate School and School of Medicine, Graduate Program in Biomedical Sciences and Graduate School, Department of Molecular Microbiology and Immunology, St. Louis, MO 63103-2097. Offers PhD. *Faculty:* 20 full-time (4 women), 1 (woman) part-time/adjunct. *Students:* 22 full-time (10 women); includes 1 minority (Hispanic American), 1 international. Average age 27. 4 applicants, 50% accepted, 2 enrolled. In 2005, 1 degree awarded. *Degree requirements:* For doctorate, thesis/dissertation, qualifying exams, comprehensive exam. *Entrance requirements:* For doctorate, GRE General Test, letters of recommendation, resumé, interview. Additional exam requirements/recommendations for international students: Required—TOEFL (minimum score 550 paper-based; 213 computer-based). *Application deadline:* For fall admission, 7/1 for domestic students, 7/1 for international students; for spring admission, 11/1 for domestic students, 11/1 for international students. Applications are processed on a rolling basis. Application fee: $40. *Expenses:* Tuition: Part-time $760 per credit hour. Required fees: $55 per semester. *Financial support:* In 2005–06, 12 students received support. Health care benefits and unspecified assistantships available. Support available to part-time students. Financial award application deadline: 6/1; financial award applicants required to submit FAFSA. *Faculty research:* Biodefense, cancer gene therapy, rheumatoid arthritis. Total annual research expenditures: $3.6 million. *Unit head:* Dr. William Wold, Chairperson, 314-977-8857, Fax: 314-773-3403, E-mail: woldws@slu.edu. *Application contact:* Gary Behrman, Associate Dean of the Graduate School, 314-977-3827, E-mail: behrmang@slu.edu.

Stanford University, School of Medicine, Graduate Programs in Medicine, Department of Microbiology and Immunology, Stanford, CA 94305-9991. Offers PhD. *Degree requirements:* For doctorate, thesis/dissertation, 2 quarters teaching assistantship, comprehensive exam. *Entrance requirements:* For doctorate, GRE General Test, GRE Subject Test (biology or biochemistry). Additional exam requirements/recommendations for international students: Required—TOEFL. Electronic applications accepted. *Faculty research:* Molecular pathogenesis of bacteria viruses and parasites, immune system function, autoimmunity, molecular biology.

See Close-Up on page 893.

Stanford University, School of Medicine, Graduate Programs in Medicine, Program in Immunology, Stanford, CA 94305-9991. Offers PhD. *Degree requirements:* For doctorate, thesis/dissertation, qualifying examination. *Entrance requirements:* For doctorate, GRE General Test, GRE Subject Test. Additional exam requirements/recommendations for international students: Required—TOEFL. Electronic applications accepted.

State University of New York at Buffalo, Graduate School, Graduate Programs in Cancer Research and Biomedical Sciences at Roswell Park Cancer Institute, Department of Immunology at Roswell Park Cancer Institute, Buffalo, NY 14263. Offers PhD. *Faculty:* 7 part-time/adjunct (2 women). *Students:* 24 full-time (10 women), 5 part-time (3 women); includes 1 minority (Asian American or Pacific Islander), 15 international. Average age 27. 32 applicants, 38% accepted. In 2005, 3 degrees awarded. *Median time to degree:* Of those who began their doctoral program in fall 1997, 80% received their degree in 8 years or less. *Degree requirements:* For doctorate, thesis/dissertation. *Entrance requirements:* For doctorate, GRE General Test, GRE Subject Test. Additional exam requirements/recommendations for international students: Required—TOEFL, TWE, TSE. *Application deadline:* For fall admission, 2/1 for domestic students. Applications are processed on a rolling basis. Application fee: $35. Electronic applications accepted. *Financial support:* In 2005–06, 10 students received support, including 4 fellowships with full tuition reimbursements available (averaging $21,000 per year), 6 research assistantships with full tuition reimbursements available (averaging $21,000 per year); Federal Work-Study also available. Financial award application deadline: 6/1; financial award applicants required to submit FAFSA. *Faculty research:* Immunochemistry, immunobiology, molecular immunology, hybridoma studies, recombinant DNA studies. Total annual research expenditures: $2 million. *Unit head:* Dr. Soldano Ferrone, Chair, 716-845-8534, E-mail: soldano.ferrone@roswellpark.org. *Application contact:* Craig R. Johnson, Director of Admissions, 716-845-2339, Fax: 716-845-8178, E-mail: craig.johnson@roswellpark.edu.

State University of New York at Buffalo, Graduate School, School of Medicine and Biomedical Sciences, Graduate Programs in Medicine and Biomedical Sciences, Department of Microbiology and Immunology, Buffalo, NY 14260. Offers MA, PhD. *Faculty:* 16 full-time (4 women), 2 part-time/adjunct (0 women). *Students:* 39 full-time (19 women), 1 (woman) part-time; includes 4 minority (3 Asian Americans or Pacific Islanders, 1 Hispanic American), 13 international. Average age 29. 12 applicants, 17% accepted, 0 enrolled. In 2005, 5 master's, 6 doctorates awarded. *Median time to degree:* Of those who began their doctoral program in fall 1997, 100% received their degree in 8 years or less. *Degree requirements:* For master's, comprehensive exam; for doctorate, thesis/dissertation, departmental qualifying exam. *Entrance requirements:* For master's, GRE General Test; for doctorate, GRE General Test, 3 letters of recommendation. Additional exam requirements/recommendations for international students: Required—TOEFL. *Application deadline:* For fall admission, 2/1 priority date for domestic students, 2/1 priority date for international students. Applications are processed on a rolling basis. Application fee: $35. Electronic applications accepted. *Financial support:* In 2005–06, 7 fellowships with tuition reimbursements (averaging $23,500 per year), 25 research assistantships with tuition reimbursements (averaging $21,000 per year), teaching assistantships with tuition reimbursements (averaging $19,000 per year) were awarded; Federal Work-Study, institutionally sponsored loans, traineeships, health care benefits, and unspecified assistantships also available. Financial award application deadline: 2/1; financial award applicants required to submit FAFSA. *Faculty research:* Bacteriology, immunology, parasitology, virology, microbial pathogenesis. Total annual research expenditures: $6.1 million. *Unit head:* Dr. John Hay, Chairman, 716-829-2907, Fax: 716-829-2158. *Application contact:* Dr. Laurie K. Read, Director of Graduate Studies, 716-829-2176, Fax: 716-829-2158.

State University of New York Upstate Medical University, College of Graduate Studies, Department of Microbiology and Immunology, Syracuse, NY 13210-2334. Offers MS, PhD, MD/PhD. Terminal master's awarded for partial completion of doctoral program. *Degree requirements:* For master's, thesis/dissertation; for doctorate, thesis/dissertation, comprehensive exam. *Entrance requirements:* For master's, GRE General Test, GRE Subject Test, interview; for

doctorate, GRE General Test, interview. Additional exam requirements/recommendations for international students: Required—TOEFL. *Faculty research:* Virology, molecular biology, microbial genetics, parasitology.

Stony Brook University, State University of New York, Graduate School, College of Arts and Sciences, Department of Biochemistry and Cell Biology, Molecular and Cellular Biology Program, Stony Brook, NY 11794. Offers biochemistry and molecular biology (PhD); biological sciences (MA); cellular and developmental biology (PhD); immunology and pathology (PhD); molecular and cellular biology (PhD). *Students:* 116 full-time (69 women), 2 part-time (1 woman); includes 16 minority (3 African Americans, 10 Asian Americans or Pacific Islanders, 3 Hispanic Americans), 62 international. Average age 30. 342 applicants, 14% accepted. In 2005, 7 degrees awarded. *Degree requirements:* For doctorate, thesis/dissertation, teaching experience, comprehensive exam. *Entrance requirements:* For doctorate, GRE General Test, GRE Subject Test. Additional exam requirements/recommendations for international students: Required—TOEFL. *Application deadline:* For fall admission, 1/15 for domestic students. Application fee: $50. *Expenses:* Tuition, state resident: full-time $6,900; part-time $288 per credit. Tuition, nonresident: full-time $10,920; part-time $455 per credit. Required fees: $704. *Financial support:* Fellowships, research assistantships, teaching assistantships, Federal Work-Study available. *Application contact:* Information Contact, 631-632-8533, Fax: 631-632-9730.

See Close-Up on page 597.

Temple University, Health Sciences Center, School of Medicine and Graduate School, Graduate Programs in Medicine, Department of Microbiology and Immunology, Philadelphia, PA 19140-5104. Offers MS, PhD, MD/PhD. *Degree requirements:* For master's, thesis; for doctorate, thesis/dissertation, research seminars. *Entrance requirements:* For master's and doctorate, GRE General Test, GRE Subject Test, minimum GPA of 3.0. Additional exam requirements/recommendations for international students: Required—TOEFL (minimum score 600 paper-based; 250 computer-based). Electronic applications accepted. *Expenses:* Tuition, state resident: full-time $8,694; part-time $483 per credit. Tuition, nonresident: full-time $12,672; part-time $704 per credit. Required fees: $500; $122 per semester. Tuition and fees vary according to course level, campus/location and program. *Faculty research:* Molecular and cellular immunology, molecular and biochemical microbiology, molecular genetics.

See Close-Up on page 897.

Texas A&M University System Health Science Center, Graduate School of Biomedical Sciences, Department of Medical Microbiology and Immunology, College Station, TX 77840. Offers immunology (PhD); microbiology (PhD); molecular biology (PhD); virology (PhD). *Degree requirements:* For doctorate, thesis/dissertation. *Entrance requirements:* For doctorate, GRE General Test, minimum GPA of 3.0. *Faculty research:* Molecular pathogenesis, microbial therapeutics.

Thomas Jefferson University, Jefferson College of Graduate Studies, Program in Immunology and Microbial Pathogenesis, Philadelphia, PA 19107. Offers PhD. *Students:* 32 full-time (16 women); includes 2 minority (1 Asian American or Pacific Islander, 1 Hispanic American), 7 international. 63 applicants, 13% accepted, 5 enrolled. In 2005, 1 doctorate awarded. *Degree requirements:* For doctorate, thesis/dissertation, comprehensive exam, registration. *Entrance requirements:* For doctorate, GRE General Test, minimum GPA of 3.2. Additional exam requirements/recommendations for international students: Required—TOEFL (minimum score 213 computer-based). *Application deadline:* For fall admission, 3/1 priority date for domestic students, 1/1 priority date for international students. Applications are processed on a rolling basis. Application fee: $50. Electronic applications accepted. *Expenses:* Tuition: Full-time $14,894; part-time $800 per credit. *Financial support:* In 2005–06, 9 students received support, including 32 fellowships with full tuition reimbursements available; research assistantships, Federal Work-Study, institutionally sponsored loans, scholarships/grants, and traineeships also available. Support available to part-time students. Financial award application deadline: 5/1; financial award applicants required to submit FAFSA. *Unit head:* Dr. Jay Rothstein, Program Director, 215-503-4622, E-mail: jay.rothstein@jefferson.edu. *Application contact:* Jessie F. Pervall, Director of Admissions, 215-503-0155, Fax: 215-503-9920, E-mail: jessie.pervall@jefferson.edu.

See Close-Up on page 899.

Tufts University, Sackler School of Graduate Biomedical Sciences, Program in Immunology, Boston, MA 02155. Offers PhD. *Faculty:* 23 full-time (8 women). *Students:* 38 full-time (24 women); includes 7 minority (6 Asian Americans or Pacific Islanders, 1 Hispanic American), 14 international. Average age 28. 96 applicants, 13% accepted, 5 enrolled. In 2005, 7 degrees awarded. *Degree requirements:* For doctorate, thesis/dissertation. *Entrance requirements:* For doctorate, GRE General Test, 3 letters of reference. Additional exam requirements/recommendations for international students: Required—TOEFL. *Application deadline:* For fall admission, 1/15 priority date for domestic students, 1/15 priority date for international students. Applications are processed on a rolling basis. Application fee: $65. Electronic applications accepted. *Financial support:* In 2005–06, 38 students received support, including 38 research assistantships with full tuition reimbursements available (averaging $29,000 per year); scholarships/grants and tuition waivers (full) also available. Financial award application deadline: 1/15. *Faculty research:* Genetic analysis of lymphocyte function, ontogeny and activation, transformation of hematopoietic cells, autoimmunity, the immune response to infection. *Unit head:* Dr. Henry H. Wortis, Director, 617-636-6836, Fax: 617-636-2990, E-mail: henry.wortis@tufts.edu. *Application contact:* 617-636-6767, Fax: 617-636-0375, E-mail: sackler-school@tufts.edu.

Tulane University, School of Medicine and Graduate School, Graduate Programs in Medicine, Department of Microbiology and Immunology, New Orleans, LA 70118-5669. Offers MS, PhD, MD/PhD. MS and PhD offered through the Graduate School. *Degree requirements:* For master's, thesis; for doctorate, 2 foreign languages, thesis/dissertation. *Entrance requirements:* For master's, GRE General Test, minimum B average in undergraduate course work; for doctorate, GRE General Test, GRE Subject Test. Additional exam requirements/recommendations for international students: Required—TOEFL or TSE. Electronic applications accepted. *Faculty research:* Vaccine development, viral pathogenesis, molecular virology, bacterial pathogenesis, fungal pathogenesis.

Uniformed Services University of the Health Sciences, School of Medicine, Programs in Biomedical Sciences, Department of Microbiology and Immunology, Bethesda, MD 20814-4799. Offers PhD. Students referred to program in Emerging Infectious Diseases, Microbiology Track. *Faculty:* 13 full-time (7 women), 8 part-time/adjunct (4 women). *Students:* 1 (woman) full-time. Average age 28. In 2005, 1 degree awarded. *Degree requirements:* For doctorate, one foreign language, thesis/dissertation, qualifying exam, comprehensive exam. *Application deadline:* For fall admission, 1/15 for domestic students. Applications are processed on a rolling basis. Application fee: $0. *Financial support:* Tuition waivers (full) available. *Faculty research:* Infections diseases, virology, bacteriology, microbial pathogenesis. *Unit head:* Dr. Alison O'Brien, Chair, 301-295-3400, E-mail: aobrien@usuhs.mil. *Application contact:* Janet M. Anastasi, Graduate Program Coordinator, 301-295-9474, Fax: 301-295-6772, E-mail: janastasi@usuhs.mil.

See Close-Up on page 903.

Uniformed Services University of the Health Sciences, School of Medicine, Programs in Biomedical Sciences, Graduate Program in Emerging Infectious Diseases, Bethesda, MD 20814-4799. Offers PhD. *Faculty:* 13 full-time (3 women), 7 part-time/adjunct (1 woman). *Students:* 39 full-time (27 women); includes 8 minority (3 African Americans, 5 Asian Americans or Pacific Islanders). Average age 26. 58 applicants, 21% accepted, 8 enrolled. In 2005, 2 degrees awarded. *Median time to degree:* Of those who began their doctoral program in fall 1997, 100% received their degree in 8 years or less. *Degree requirements:* For doctorate, thesis/dissertation, qualifying exam, comprehensive exam. *Entrance requirements:* For doctorate, GRE General Test. Additional exam requirements/recommendations for international

Immunology

Uniformed Services University of the Health Sciences *(continued)*
students: Required—TOEFL. *Application deadline:* For fall admission, 1/15 for domestic students. Applications are processed on a rolling basis. Application fee: $0. *Financial support:* In 2005–06, fellowships with full tuition reimbursements (averaging $23,000 per year); scholarships/grants and tuition waivers (full) also available. *Unit head:* Dr. Christopher Broder, Director, 301-295-3401, E-mail: cbroder@usuhs.mil. *Application contact:* Janet M. Anastasi, Graduate Program Coordinator, 301-295-9474, Fax: 301-295-6772, E-mail: janastasi@usuhs.mil.

See Close-Up on page 901.

Université de Montréal, Faculty of Medicine and Faculty of Graduate Studies, Graduate Programs in Medicine, Department of Microbiology and Immunology, Montréal, QC H3C 3J7, Canada. Offers M Sc, PhD. *Faculty:* 39 full-time (10 women), 29 part-time/adjunct (8 women). *Students:* 88 full-time (57 women). 29 applicants, 69% accepted, 19 enrolled. In 2005, 16 master's, 4 doctorates awarded. Terminal master's awarded for partial completion of doctoral program. *Degree requirements:* For master's, thesis; for doctorate, thesis/dissertation, general exam. *Entrance requirements:* For master's and doctorate, proficiency in French, knowledge of English. *Application deadline:* For fall and spring admission, 2/1. For winter admission, 11/1 for domestic students. Application fee: $30. Electronic applications accepted. *Unit head:* Pierre Belhumeur, Director, 514-343-6273, Fax: 514-343-5701. *Application contact:* George Szatmari, Chairperson, 514-343-5796, Fax: 514-343-5701.

Université de Montréal, Faculty of Medicine and Faculty of Graduate Studies, Graduate Programs in Medicine, Program in Virology and Immunology, Montréal, QC H3C 3J7, Canada. Offers PhD. *Students:* 15 full-time (9 women). 3 applicants, 33% accepted, 1 enrolled. In 2005, 1 degree awarded. *Degree requirements:* For doctorate, thesis/dissertation, general exam. *Entrance requirements:* For doctorate, proficiency in French, knowledge of English. *Application deadline:* For fall and spring admission, 2/1. For winter admission, 11/1 for domestic students. Application fee: $30. Electronic applications accepted. *Unit head:* Pierre Belhumeur, Director, 514-343-6273, Fax: 514-343-5701. *Application contact:* Silvie Jauvim, Information Contact, 514-343-3129.

Université de Sherbrooke, Faculty of Medicine and Health Sciences, Graduate Programs in Medicine, Program in Immunology, Sherbrooke, QC J1K 2R1, Canada. Offers M Sc, PhD. *Students:* 10 full-time (5 women), 15 part-time (6 women). 7 applicants, 86% accepted, 5 enrolled. In 2005, 1 master's, 1 doctorate awarded. *Application deadline:* For fall admission, 6/30 for domestic students. For winter admission, 10/31 for domestic students; for spring admission, 2/28 for domestic students. Application fee: $50. Electronic applications accepted. *Unit head:* Dr. Jana Stankova, Director, 819-564-5268, E-mail: stankova@usherbrooke.ca.

Université du Québec, Institut National de la Recherche Scientifique, Graduate Programs, Research Center—INRS—Institut Armand-Frappier—Human Health, Québec, QC G1K 9A9, Canada. Offers applied microbiology (M Sc); biology (PhD); experimental health sciences (M Sc); virology and immunology (M Sc, PhD). Programs given in French. Part-time programs available. *Faculty:* 46. *Students:* 170 full-time (100 women), 25 international. Average age 28. In 2005, 24 master's, 5 doctorates awarded. *Degree requirements:* For doctorate, thesis/dissertation. *Entrance requirements:* For master's and doctorate, appropriate bachelor's degree, proficiency in French. *Application deadline:* For fall admission, 3/30 for domestic students, 3/30 for international students. For winter admission, 11/1 for domestic students. Application fee: $30 Canadian dollars. *Financial support:* Fellowships, research assistantships, teaching assistantships available. *Faculty research:* Immunity, infection and cancer; toxicology and environmental biotechnology; molecular pharmacochemistry. *Unit head:* Pierre Talbot, Director, 450-681-5010 Ext. 4406, E-mail: pierre.talbot@iaf.inrs.ca. *Application contact:* Michel Barbeau, Registrar, 418-654-2518, Fax: 418-654-3858, E-mail: michel.barbeau@adm.inrs.ca.

Université Laval, Faculty of Medicine, Graduate Programs in Medicine, Programs in Microbiology-Immunology, Québec, QC G1K 7P4, Canada. Offers M Sc, PhD. Terminal master's awarded for partial completion of doctoral program. *Degree requirements:* For master's, thesis/dissertation; for doctorate, thesis/dissertation, comprehensive exam. *Entrance requirements:* For master's and doctorate, knowledge of French, comprehension of written English. Electronic applications accepted.

University at Albany, State University of New York, School of Public Health, Department of Biomedical Sciences, Program in Immunobiology and Immunochemistry, Albany, NY 12222-0001. Offers MS, PhD. *Degree requirements:* For master's and doctorate, thesis/dissertation. *Entrance requirements:* For master's and doctorate, GRE General Test, GRE Subject Test. Application fee: $60. *Financial support:* Application deadline: 2/1. *Unit head:* Dr. James Dias, Chair, Department of Biomedical Sciences, 518-474-2662.

University of Alberta, Faculty of Medicine and Dentistry and Faculty of Graduate Studies and Research, Graduate Programs in Medicine, Department of Medical Microbiology and Immunology, Edmonton, AB T6G 2E1, Canada. Offers M Sc, PhD. Terminal master's awarded for partial completion of doctoral program. *Degree requirements:* For master's and doctorate, thesis/dissertation. *Entrance requirements:* For master's and doctorate, GRE. Additional exam requirements/recommendations for international students: Required—TOEFL (paper score 600; computer score 250) or Michigan English Language Assessment Battery. Tuition and fees charges are reported in Canadian dollars. *Expenses:* Tuition, state resident: part-time $562 Canadian dollars per term. Tuition, nonresident: full-time $3,375 Canadian dollars. Required fees: $573 Canadian dollars; $84 Canadian dollars per term. *Faculty research:* Cellular and reproductive immunology, microbial pathogenesis, mechanisms of antibiotic resistance, molecular biology of mammalian viruses, antiviral chemotherapy.

The University of Arizona, College of Medicine, Graduate Programs in Medicine, Department of Microbiology and Immunology, Tucson, AZ 85721. Offers MS, PhD. *Degree requirements:* For master's and doctorate, thesis/dissertation. *Entrance requirements:* For master's and doctorate, GRE General Test, minimum GPA of 3.0. *Faculty research:* Environmental and pathogenic microbiology, molecular biology.

University of Arkansas for Medical Sciences, College of Medicine and Graduate School, Graduate Programs in Medicine, Department of Microbiology and Immunology, Little Rock, AR 72205-7199. Offers MS, PhD, MD/PhD. *Faculty:* 20 full-time (8 women). *Students:* 17 full-time, 2 part-time. *Degree requirements:* For master's and doctorate, thesis/dissertation. *Entrance requirements:* For master's and doctorate, GRE General Test. Additional exam requirements/recommendations for international students: Required—TOEFL. Application fee: $0. *Financial support:* Research assistantships available. Support available to part-time students. *Unit head:* Dr. Roger G. Rank, Chairman, 501-686-5144. *Application contact:* Dr. Lee Soderberg, Co-Coordinator of Graduate Studies, 501-686-6368, E-mail: soderberglees@uams.edu.

The University of British Columbia, Faculty of Graduate Studies, Faculty of Science, Department of Microbiology and Immunology, Vancouver, BC V6T 1Z1, Canada. Offers M Sc, PhD. *Degree requirements:* For master's, thesis/dissertation; for doctorate, thesis/dissertation, comprehensive exam. *Entrance requirements:* For master's and doctorate, GRE General Test. Additional exam requirements/recommendations for international students: Required—TOEFL (minimum score 590 paper-based; 243 computer-based). Electronic applications accepted. *Faculty research:* Bacterial genetics, metabolism, pathogenic bacteriology, virology.

University of Calgary, Faculty of Medicine and Faculty of Graduate Studies, Department of Medical Science, Calgary, AB T2N 1N4, Canada. Offers cancer biology (M Sc, PhD); immunology (M Sc, PhD); joint injury and arthritis research (M Sc, PhD); medical education (M Sc, PhD); medical science (M Sc, PhD); mountain medicine and high altitude physiology (M Sc). *Faculty:* 114 full-time (47 women), 5 part-time/adjunct (0 women). *Students:* 121 full-time (69 women), 1 part-time. 68 applicants, 29% accepted, 19 enrolled. In 2005, 19 master's, 7 doctorates awarded. *Median time to degree:* Of those who began their doctoral program in fall 1997, 100% received their degree in 8 years or less. *Degree requirements:* For master's,

thesis; for doctorate, thesis/dissertation, candidacy exam. *Entrance requirements:* For master's, minimum undergraduate GPA of 3.2; for doctorate, minimum graduate GPA of 3.2. Additional exam requirements/recommendations for international students: Required—TOEFL (minimum score 600 paper-based; 250 computer-based). *Application deadline:* For fall admission, 6/15 priority date for domestic students, 5/15 priority date for international students. For winter admission, 10/15 for domestic students; for spring admission, 3/15 for domestic students. Applications are processed on a rolling basis. Application fee: $100 ($130 for international students). Electronic applications accepted. *Financial support:* In 2005–06, 30 students received support, including 22 research assistantships, 2 teaching assistantships; scholarships/grants and tuition waivers (partial) also available. *Faculty research:* Cancer biology, immunology, joint injury and arthritis, medical education, population genomics. *Unit head:* Dr. Francine Smith, Graduate Coordinator, 403-220-6852, Fax: 403-210-8109, E-mail: fsmith@ucalgary.ca. *Application contact:* Christine Szefer, Graduate Program Administrator, 403-220-6852, Fax: 403-210-8109, E-mail: cszefer@ucalgary.ca.

University of California, Berkeley, Graduate Division, School of Public Health, Group in Infectious Diseases and Immunity, Berkeley, CA 94720-1500. Offers PhD. *Entrance requirements:* For doctorate, GRE General Test, minimum GPA of 3.0.

University of California, Davis, Graduate Studies, Graduate Group in Immunology, Davis, CA 95616. Offers MS, PhD. *Faculty:* 46 full-time. *Students:* 48 full-time (31 women); includes 13 minority (1 African American, 9 Asian Americans or Pacific Islanders, 3 Hispanic Americans), 13 international. Average age 29. 67 applicants, 36% accepted, 13 enrolled. In 2005, 7 master's, 1 doctorate awarded. Terminal master's awarded for partial completion of doctoral program. *Median time to degree:* Of those who began their doctoral program in fall 1997, 66.7% received their degree in 8 years or less. *Degree requirements:* For master's, thesis (for some programs), comprehensive exam (for some programs); for doctorate, thesis/dissertation. *Entrance requirements:* For master's and doctorate, GRE General Test. Additional exam requirements/recommendations for international students: Required—TOEFL (minimum score 550 paper-based; 213 computer-based). *Application deadline:* For fall admission, 1/15 for domestic students, 1/15 for international students. Application fee: $60. Electronic applications accepted. *Financial support:* In 2005–06, 39 students received support, including 3 fellowships with full and partial tuition reimbursements available (averaging $11,364 per year), 17 research assistantships with full and partial tuition reimbursements available (averaging $15,391 per year), 8 teaching assistantships with partial tuition reimbursements available (averaging $15,082 per year); Federal Work-Study, institutionally sponsored loans, scholarships/grants, tuition waivers (full and partial), and unspecified assistantships also available. Financial award application deadline: 1/15; financial award applicants required to submit FAFSA. *Faculty research:* Immune regulation in autoimmunity, immunopathology, immunotoxicology, tumor immunology, avian immunology. *Unit head:* Hilary Benton, Graduate Program Chair, 530-752-3720, E-mail: hpbenton@ucdavis.edu. *Application contact:* Tania Heta, Administrative Assistant, 530-752-8551, E-mail: taheta@ucdavis.edu.

University of California, Los Angeles, School of Medicine and Graduate Division, Graduate Programs in Medicine, Department of Microbiology, Immunology and Molecular Genetics, Los Angeles, CA 90095. Offers MS, PhD. *Degree requirements:* For doctorate, thesis/dissertation, oral and written qualifying exams. *Entrance requirements:* For doctorate, GRE General Test, GRE Subject Test. Additional exam requirements/recommendations for international students: Required—TOEFL.

University of California, San Diego, Graduate Studies and Research, Division of Biology, Program in Immunology, Virology, and Cancer Biology, La Jolla, CA 92093. Offers PhD. Offered in association with the Salk Institute. *Degree requirements:* For doctorate, thesis/dissertation, qualifying exam. Electronic applications accepted.

University of California, San Francisco, Graduate Division, Department of Microbiology and Immunology, San Francisco, CA 94143. Offers PhD. *Degree requirements:* For doctorate, thesis/dissertation. *Entrance requirements:* For doctorate, GRE General Test.

University of Chicago, Division of the Biological Sciences, Biomedical Sciences: Cancer, Immunology, Nutrition, Pathology, and Microbiology, Committee on Immunology, Chicago, IL 60637-1513. Offers PhD. *Faculty:* 37 full-time (11 women). *Students:* 38 full-time (15 women); includes 10 minority (3 African Americans, 6 Asian Americans or Pacific Islanders, 1 Hispanic American), 4 international. Average age 27. In 2005, 3 degrees awarded. *Degree requirements:* For doctorate, thesis/dissertation, registration. *Entrance requirements:* For doctorate, GRE General Test. Additional exam requirements/recommendations for international students: Required—TOEFL. *Application deadline:* For fall admission, 12/28 priority date for domestic students, 12/28 priority date for international students. Application fee: $55. Electronic applications accepted. *Financial support:* In 2005–06, 27 students received support, including 9 fellowships with full tuition reimbursements available (averaging $26,301 per year), 16 research assistantships with full tuition reimbursements available (averaging $26,301 per year); institutionally sponsored loans and traineeships also available. *Faculty research:* Molecular immunology, transplantation, autoimmunology, neuroimmunology, tumor immunology. Total annual research expenditures: $15 million. *Unit head:* Dr. Albert Bendelac, Chairman, 773-834-8646, Fax: 773-702-4634, E-mail: abendelac@bsd.uchicago.edu. *Application contact:* Rebecca Levine, Administrative Assistant, Student Services, 773-834-3899, Fax: 773-702-4634, E-mail: rlevine@huggins.bsd.uchicago.edu.

University of Cincinnati, Division of Research and Advanced Studies, College of Medicine, Graduate Programs in Biomedical Sciences, Immunobiology Training Program, Cincinnati, OH 45221. Offers MS, PhD. *Degree requirements:* For master's, seminar, thesis with oral defense; for doctorate, seminar, dissertation with oral defense, written and oral candidacy exams.

See Close-Up on page 909.

University of Colorado at Denver and Health Sciences Center, Graduate School, Program in Biomedical Sciences, Department of Immunology, Denver, CO 80262. Offers PhD. In 2005, 5 degrees awarded. *Degree requirements:* For doctorate, thesis/dissertation, 4 laboratory rotations, comprehensive exam. *Entrance requirements:* For doctorate, 4 letters of recommendation. Additional exam requirements/recommendations for international students: Required—TOEFL (minimum score 550 paper-based; 213 computer-based). *Application deadline:* For fall admission, 1/15 for domestic students. Application fee: $50. *Expenses:* Tuition, state resident: full-time $11,730. Tuition, nonresident: full-time $22,980. Tuition and fees vary according to degree level and program. *Financial support:* Applicants required to submit FAFSA. *Unit head:* Dr. John C. Cambier, Chair, 303-398-1325, E-mail: cambierj@njc.org. *Application contact:* Jane Lanners, Administrative Assistant, 303-398-1305, E-mail: lannersj@njc.org.

See Close-Up on page 911.

University of Connecticut Health Center, Graduate School, Programs in Biomedical Sciences, Program in Immunology, Farmington, CT 06030. Offers PhD, DMD/PhD, MD/PhD. *Degree requirements:* For doctorate, thesis/dissertation, comprehensive exam, registration. *Entrance requirements:* For doctorate, GRE General Test. Additional exam requirements/recommendations for international students: Required—TOEFL (minimum score 600 paper-based; 250 computer-based). Electronic applications accepted.

See Close-Up on page 915.

University of Florida, College of Medicine, Department of Pathology, Immunology and Laboratory Medicine, Gainesville, FL 32611. Offers immunology and molecular pathology (PhD). *Faculty:* 36 full-time (7 women), 1 (woman) part-time/adjunct. *Degree requirements:* For doctorate, thesis/dissertation. *Entrance requirements:* For doctorate, GRE General Test, minimum GPA of 3.0. Additional exam requirements/recommendations for international students: Required—TOEFL. *Application deadline:* For fall admission, 2/15 for domestic students. Application fee: $30. Electronic applications accepted. *Expenses:* Tuition, state resident: full-time $6,234. Tuition, nonresident: full-time $21,359. Tuition and fees vary according to program.

Financial support: In 2005–06, research assistantships with full tuition reimbursements (averaging $24,341 per year); fellowships with full tuition reimbursements, teaching assistantships, institutionally sponsored loans and traineeships also available. *Faculty research:* Molecular immunology, autoimmunity and transplantation, tumor biology, oncogenic viruses, human immunodeficiency viruses. *Unit head:* Dr. James M. Crawford, Chairman, 352-392-6840, Fax: 352-392-6249, E-mail: crawford@pathology.ufl.edu. *Application contact:* Dr. Wayne McCormack, Associate Dean of Graduate Education, 352-392-7413, Fax: 352-846-3466, E-mail: mccormac@pathology.ufl.edu.

University of Florida, College of Medicine and Graduate School, Interdisciplinary Program in Biomedical Sciences, Concentration in Immunology and Microbiology, Gainesville, FL 32611. Offers PhD. *Faculty:* 66. *Students:* 33 full-time (20 women); includes 9 minority (1 African American, 7 Asian Americans or Pacific Islanders, 1 Hispanic American). In 2005, 9 degrees awarded. *Degree requirements:* For doctorate, thesis/dissertation. *Entrance requirements:* For doctorate, GRE General Test, minimum GPA of 3.0. Additional exam requirements/recommendations for international students: Required—TOEFL. *Application deadline:* For fall admission, 2/15 for domestic students. Applications are processed on a rolling basis. Application fee: $30. Electronic applications accepted. *Expenses:* Tuition, state resident: full-time $6,234. Tuition, nonresident: full-time $21,359. Tuition and fees vary according to program. *Financial support:* Fellowships with full tuition reimbursements, research assistantships with full tuition reimbursements, teaching assistantships, institutionally sponsored loans, scholarships/grants, and traineeships available. *Unit head:* Dr. Paul Gulig, Director, 352-392-0050, E-mail: gulig@ufl.edu. *Application contact:* Dr. Wayne McCormack, Associate Dean of Graduate Education, 352-392-7413, Fax: 352-846-3466, E-mail: mccormac@pathology.ufl.edu.

University of Guelph, Ontario Veterinary College and Graduate Program Services, Graduate Programs in Veterinary Sciences, Department of Pathobiology, Guelph, ON N1G 2W1, Canada. Offers anatomic pathology (DV Sc, Diploma); clinical pathology (Diploma); comparative pathology (M Sc, PhD); immunology (M Sc, PhD); laboratory animal science (DV Sc); pathology (M Sc, PhD, Diploma); veterinary infectious diseases (M Sc, PhD); zoo animal/wildlife medicine (DV Sc). *Faculty:* 26. *Students:* 66 (22 women). 47 applicants, 32% accepted. In 2005, 6 master's, 7 doctorates, 1 other advanced degree awarded. *Degree requirements:* For master's and doctorate, thesis/dissertation. *Entrance requirements:* For master's, DVM with B average or an honours degree in biological sciences; for doctorate, DVM or MSC degree, minimum B+ average. Additional exam requirements/recommendations for international students: Required—TOEFL (minimum score 550 paper-based; 213 computer-based). *Application deadline:* For fall admission, 2/1 for domestic students. Applications are processed on a rolling basis. Application fee: $75. *Financial support:* In 2005–06, 40 students received support, including 20 fellowships, research assistantships (averaging $13,500 per year), 13 teaching assistantships (averaging $1,600 per year); career-related internships or fieldwork also available. *Faculty research:* Pathogenesis; diseases of animals, wildlife, fish, and laboratory animals; parasitology; immunology; veterinary infectious diseases; laboratory animal science. Total annual research expenditures: $2.5 million. *Unit head:* Dr. John Prescott, Chair, Fax: 519-824-5930, E-mail: jprescott@ovc.uoguelph.ca. *Application contact:* Dr. J. MacInnes, Graduate Coordinator, 519-824-4120 Ext. 54731, Fax: 519-767-0809, E-mail: macinnes@uoguelph.ca.

University of Illinois at Chicago, College of Medicine and Graduate College, Graduate Programs in Medicine, Department of Microbiology and Immunology, Chicago, IL 60607-7128. Offers PhD, MD/PhD. *Degree requirements:* For doctorate, thesis/dissertation. *Entrance requirements:* For doctorate, GRE General Test, minimum GPA of 2.75. Additional exam requirements/recommendations for international students: Required—TOEFL.

See Close-Up on page 919.

The University of Iowa, Graduate College, Program in Immunology, Iowa City, IA 52242-1316. Offers PhD, MD/PhD. *Students:* 12 full-time (6 women), 14 part-time (7 women), 5 international. 19 applicants, 47% accepted, 5 enrolled. In 2005, 5 degrees awarded. *Degree requirements:* For doctorate, thesis/dissertation, comprehensive exam, registration. *Entrance requirements:* For doctorate, GRE General Test, minimum GPA of 3.0. Additional exam requirements/recommendations for international students: Required—TOEFL (minimum score 600 paper-based; 250 computer-based). *Application deadline:* For fall admission, 2/1 priority date for domestic students, 2/1 priority date for international students. Application fee: $60 ($85 for international students). Electronic applications accepted. *Expenses:* Tuition, state resident: part-time $1,882 per term. Tuition, nonresident: full-time $17,338; part-time $4,907 per term. Tuition and fees vary according to course load and program. *Financial support:* In 2005–06, 1 fellowship, 25 research assistantships with partial tuition reimbursements were awarded; teaching assistantships with partial tuition reimbursements, scholarships/grants also available. Financial award applicants required to submit FAFSA. *Unit head:* Dr. Gail Bishop, Director, 319-335-7748, E-mail: gail-bishop@uiowa.edu. *Application contact:* Paulette Scheler, Program Associate, 800-551-7748.

The University of Iowa, Roy J. and Lucille A. Carver College of Medicine and Graduate College, Graduate Programs in Medicine, Department of Microbiology, Iowa City, IA 52242-1316. Offers general microbiology and microbial physiology (MS, PhD); immunology (MS, PhD); microbial genetics (MS, PhD); pathogenic bacteriology (MS, PhD); virology (MS, PhD). *Faculty:* 23 full-time (3 women), 11 part-time/adjunct (2 women). *Students:* 54 full-time (25 women); includes 4 minority (3 Asian Americans or Pacific Islanders, 1 Hispanic American), 8 international. 93 applicants, 12% accepted, 4 enrolled. In 2005, 7 degrees awarded. *Median time to degree:* Of those who began their doctoral program in fall 1997, 80% received their degree in 8 years or less. *Degree requirements:* For master's, thesis/dissertation; for doctorate, thesis/dissertation, comprehensive exam. *Entrance requirements:* For master's and doctorate, GRE General Test. Additional exam requirements/recommendations for international students: Required—TOEFL. *Application deadline:* For fall admission, 2/1 for domestic students, 2/1 for international students. Application fee: $50 ($75 for international students). Electronic applications accepted. *Expenses:* Tuition, state resident: part-time $1,882 per term. Tuition, nonresident: full-time $17,338; part-time $4,907 per term. Tuition and fees vary according to course load and program. *Financial support:* In 2005–06, 63 research assistantships with full tuition reimbursements (averaging $22,000 per year) were awarded; institutionally sponsored loans, scholarships/grants, traineeships, and health care benefits also available. *Faculty research:* Biocatalysis and blue jeans, gene regulation, processing and transport of HIV, retroviral pathogenesis, biodegradation. Total annual research expenditures: $9 million. *Unit head:* Dr. Michael A. Apicella, Head, 319-335-7810, E-mail: grad-micro-info@uiowa.edu.

See Close-Up on page 923.

University of Kansas, Graduate Studies Medical Center, Interdisciplinary Graduate Program in Biomedical Sciences, Department of Microbiology, Molecular Genetics and Immunology, Lawrence, KS 66045. Offers PhD, MD/PhD. *Faculty:* 8. *Students:* 2 full-time (both women), 5 part-time (3 women), 1 international. Average age 27. In 2005, 2 degrees awarded. *Degree requirements:* For doctorate, thesis/dissertation, comprehensive exam. *Entrance requirements:* For doctorate, GRE General Test. *Expenses:* Tuition, state resident: full-time $4,859. Tuition, nonresident: full-time $12,000. Required fees: $589. Tuition and fees vary according to program. *Financial support:* Fellowships with tuition reimbursements, research assistantships with partial tuition reimbursements, teaching assistantships with full and partial tuition reimbursements available. *Faculty research:* Bacterial and viral pathogenesis, molecular virology, microbial molecular biology and genetics, immunochemistry. *Unit head:* Dr. Opendra Narayan, Chairman, 913-588-7010, Fax: 913-588-7295, E-mail: bnarayan@kume.edu. *Application contact:* Dr. Joe Lutkenhaus, Director of Graduate Studies, 913-588-7054, Fax: 913-588-7295, E-mail: jlutkenh@kumc.edu.

University of Louisville, School of Medicine, Department of Microbiology and Immunology, Louisville, KY 40292-0001. Offers MS, PhD. *Students:* 33 full-time (15 women), 4 part-time (3 women); includes 5 minority (4 African Americans, 1 Asian American or Pacific Islander), 23 international. Average age 29. In 2005, 11 master's, 6 doctorates awarded. *Degree requirements:* For master's and doctorate, thesis/dissertation. *Entrance requirements:* For master's and doctorate, GRE General Test, 1 year of course work in biology, organic chemistry, physics; 1 semester of course work in calculus and quantitative analysis, biochemistry, or molecular biology. *Application deadline:* For fall admission, 2/1 for domestic students. Application fee: $50. Electronic applications accepted. *Expenses:* Tuition, state resident: full-time $6,006; part-time $334 per credit hour. Tuition, nonresident: full-time $16,554; part-time $920 per credit hour. Tuition and fees vary according to course load, degree level and program. *Financial support:* Fellowships with tuition reimbursements, research assistantships with tuition reimbursements available. *Unit head:* Dr. Robert D. Stout, Chair, 502-852-5351, Fax: 502-852-7531, E-mail: bobstout@louisville.edu. *Application contact:* Dr. Richard D Miller, Director of Admissions, 502-852-5630, Fax: 502-852-7531, E-mail: rdmill01@louisville.edu.

University of Manitoba, Faculty of Medicine and Faculty of Graduate Studies, Graduate Programs in Medicine, Department of Immunology, Winnipeg, MB R3T 2N2, Canada. Offers M Sc, PhD. Terminal master's awarded for partial completion of doctoral program. *Degree requirements:* For master's, thesis; for doctorate, one foreign language, thesis/dissertation. *Faculty research:* Immediate hypersensitivity, regulation of the immune response, natural immunity, cytokines, inflammation.

University of Maryland, Graduate School, Graduate Programs in Medicine, Department of Microbiology and Immunology, Baltimore, MD 21201. Offers MS, PhD, MD/PhD. Part-time programs available. *Faculty:* 21 full-time (4 women). *Students:* 37 full-time (21 women), 1 part-time; includes 8 minority (2 African Americans, 4 Asian Americans or Pacific Islanders, 2 Hispanic Americans), 15 international. Average age 29. 86 applicants, 8% accepted, 6 enrolled. In 2005, 1 master's, 4 doctorates awarded. *Degree requirements:* For master's, thesis; for doctorate, thesis/dissertation, oral exam. *Entrance requirements:* For master's and doctorate, GRE General Test, minimum GPA of 3.0. Additional exam requirements/recommendations for international students: Required—TOEFL, TOEFL or IELTS; Recommended—IELT. *Application deadline:* For fall admission, 7/1 for domestic students, 1/15 for international students. Application fee: $50. Electronic applications accepted. *Expenses:* Tuition, state resident: full-time $8,079; part-time $409 per credit hour. Tuition, nonresident: full-time $18,384; part-time $731 per credit hour. Required fees: $695; $10 per credit hour. Tuition and fees vary according to degree level and program. *Financial support:* Fellowships, research assistantships, teaching assistantships, career-related internships or fieldwork available. Support available to part-time students. Financial award application deadline: 2/15. *Faculty research:* Epidemiology, ecology of infectious microorganisms, electron microscopy, medical microbiology, molecular biology. *Unit head:* Dr. Jan Cerny, Chairman, 410-706-7114, Fax: 410-706-2129. *Application contact:* Dr. Harry Mobley, Program Director, 410-706-0466, Fax: 410-706-1617, E-mail: hmobley@umaryland.edu.

See Close-Up on page 929.

University of Maryland, School of Medicine, Graduate Program in Life Sciences, Program in Molecular Microbiology and Immunology, Baltimore, MD 21201. Offers PhD. *Expenses:* Tuition, state resident: full-time $8,079; part-time $409 per credit hour. Tuition, nonresident: full-time $18,384; part-time $731 per credit hour. Required fees: $695; $10 per credit hour. Tuition and fees vary according to degree level and program.

University of Massachusetts Worcester, Graduate School of Biomedical Sciences, Program in Immunology-Virology, Worcester, MA 01655-0115. Offers medical sciences (PhD). *Faculty:* 45 full-time (16 women). *Degree requirements:* For doctorate, thesis/dissertation. *Entrance requirements:* For doctorate, GRE General Test. Additional exam requirements/recommendations for international students: Required—TOEFL (minimum score 600 paper-based; 250 computer-based). *Application deadline:* For fall admission, 12/15 for domestic students, 12/15 for international students. Applications are processed on a rolling basis. Application fee: $25 ($50 for international students). *Expenses:* Tuition, state resident: full-time $2,640. Tuition, nonresident: full-time $9,856. Required fees: $5,685. *Financial support:* In 2005–06, research assistantships with full tuition reimbursements (averaging $25,235 per year); unspecified assistantships also available. *Faculty research:* Molecular and immunological studies on viral pathogenesis and oncology, AIDS viruses and other retroviruses, herpes viruses, influenza, Dengue and Newcastle disease viruses. *Unit head:* Dr. Alan Rothman, Director, 508-856-6976. *Application contact:* Michael Cole, Director of Admissions and Recruitment, 508-856-4779, Fax: 508-856-3659, E-mail: michael.cole@umassmed.edu.

See Close-Up on page 933.

University of Medicine and Dentistry of New Jersey, Graduate School of Biomedical Sciences, Graduate Programs in Biomedical Sciences–Newark, Program in Experimental Pathology, Newark, NJ 07107. Offers PhD. *Entrance requirements:* Additional exam requirements/recommendations for international students: Required—TOEFL. *Application deadline:* For fall admission, 2/1 for domestic students. *Financial support:* Fellowships, research assistantships, Federal Work-Study, institutionally sponsored loans, and tuition waivers (full and partial) available. *Unit head:* Dr. Muriel Lambert, Program Director, 973-972-4405, Fax: 973-972-7293, E-mail: mlambert@umdnj.edu.

University of Medicine and Dentistry of New Jersey, Graduate School of Biomedical Sciences, Graduate Programs in Biomedical Sciences–Piscataway, Program in Molecular Genetics, Microbiology and Immunology, Piscataway, NJ 08854-5635. Offers MS, PhD, MD/PhD. Terminal master's awarded for partial completion of doctoral program. *Degree requirements:* For master's and doctorate, thesis/dissertation, qualifying exam. *Entrance requirements:* For master's and doctorate, GRE General Test. Additional exam requirements/recommendations for international students: Required—TOEFL. *Application deadline:* For fall admission, 1/5 for domestic students. Applications are processed on a rolling basis. Application fee: $40. *Financial support:* Fellowships, research assistantships, teaching assistantships available. Financial award application deadline: 5/1. *Faculty research:* Interferon, receptors, retrovirus evolution, Arbo virus/host cell interactions. *Unit head:* Dr. Joseph P. Dougherty, Director, 732-235-4588, Fax: 732-235-5223, E-mail: doughejp@umdnj.edu.

University of Miami, Graduate School, Miller School of Medicine, Graduate Programs in Medicine, Department of Microbiology and Immunology, Coral Gables, FL 33124. Offers PhD, MD/PhD. *Faculty:* 38 full-time (8 women). *Students:* 29 full-time (21 women); includes 4 minority (1 African American, 3 Hispanic Americans), 10 international. Average age 29. 84 applicants, 21% accepted, 7 enrolled. In 2005, 9 degrees awarded. *Median time to degree:* Of those who began their doctoral program in fall 1997, 100% received their degree in 8 years or less. *Degree requirements:* For doctorate, thesis/dissertation, oral and written qualifying exams. *Entrance requirements:* For doctorate, GRE General Test. Additional exam requirements/recommendations for international students: Required—TOEFL. *Application deadline:* For fall admission, 12/31 for domestic students. Applications are processed on a rolling basis. Application fee: $50. Electronic applications accepted. *Financial support:* In 2005–06, fellowships with full tuition reimbursements (averaging $22,000 per year), research assistantships with full tuition reimbursements (averaging $22,000 per year) were awarded; institutionally sponsored loans also available. *Faculty research:* Cellular and molecular immunology, molecular and pathogenic virology, pathogenic bacteriology and gene therapy of cancer. Total annual research expenditures: $5.8 million. *Unit head:* Dr. Eckhard R. Podack, Chairman, 305-243-6694, Fax: 305-243-5522, E-mail: epodack@med.miami.edu. *Application contact:* Karen T. Del Rio, Administrator, Graduate Program, 305-243-5682, Fax: 305-243-6903, E-mail: kdelrio@med.miami.edu.

See Close-Up on page 937.

University of Michigan, Medical School and Horace H. Rackham School of Graduate Studies, Program in Biomedical Sciences (PIBS), Department of Microbiology and Immunology, Ann Arbor, MI 48109. Offers PhD. *Degree requirements:* For doctorate, oral defense of dissertation, preliminary exam. *Entrance requirements:* For doctorate, GRE General Test.

Immunology

University of Michigan (continued)
Additional exam requirements/recommendations for international students: Required—TOEFL (minimum score 600 paper-based), TWE; Recommended—TSE. Electronic applications accepted. *Expenses:* Tuition, state resident: full-time $14,082; part-time $894 per credit hour. Tuition, nonresident: full-time $28,500; part-time $1,675 per credit hour. Required fees: $189; $189 per unit. *Faculty research:* Gene regulation, molecular biology of animal and bacterial viruses, molecular and cellular networks, pathogenesis and microbial genetics.

University of Michigan, Medical School and Horace H. Rackham School of Graduate Studies, Program in Biomedical Sciences (PIBS), Program in Immunology, Ann Arbor, MI 48109-0619. Offers PhD. *Degree requirements:* For doctorate, thesis/dissertation. *Expenses:* Tuition, state resident: full-time $14,082; part-time $894 per credit hour. Tuition, nonresident: full-time $28,500; part-time $1,675 per credit hour. Required fees: $189; $189 per unit. *Faculty research:* Tumor immunology, aging, tolerance, autoimmunity, antigen presentation.

University of Minnesota, Duluth, Medical School, Microbiology, Immunology and Molecular Pathobiology Section, Duluth, MN 55812-2496. Offers MS, PhD. *Faculty:* 5 full-time (2 women), 3 part-time/adjunct (1 woman). *Students:* 18 applicants, 0% accepted.Terminal master's awarded for partial completion of doctoral program. *Degree requirements:* For master's, thesis, final oral exam; for doctorate, thesis/dissertation, final exam, oral and written preliminary exams. *Entrance requirements:* For master's and doctorate, GRE General Test. Additional exam requirements/recommendations for international students: Required—TOEFL. *Application deadline:* For fall admission, 5/1 for domestic students. Applications are processed on a rolling basis. Application fee: $30. *Financial support:* Research assistantships available. *Faculty research:* Immunomodulation, molecular diagnosis of rabies, cytokines, cancer immunology, cytomegalovirus infection. Total annual research expenditures: $207,223. *Unit head:* Dr. Arthur Johnson, Head, 218-726-7561, Fax: 218-726-6235, E-mail: ajohnso1@d.umn.edu.

University of Missouri–Columbia, School of Medicine and Graduate School, Graduate Programs in Medicine, Department of Molecular Microbiology and Immunology, Columbia, MO 65211. Offers MS, PhD. *Faculty:* 14 full-time (3 women), 1 (woman) part-time/adjunct. *Students:* 24 full-time (15 women), 13 part-time (6 women); includes 2 minority (both African Americans), 10 international. Average age 25. In 2005, 2 master's, 3 doctorates awarded. Terminal master's awarded for partial completion of doctoral program. *Degree requirements:* For master's and doctorate, thesis/dissertation. *Entrance requirements:* For master's and doctorate, GRE General Test, minimum GPA of 3.0. *Application deadline:* For fall admission, 1/31 for domestic students. Application fee: $45 ($60 for international students). *Financial support:* Fellowships, research assistantships, teaching assistantships, institutionally sponsored loans available. Financial award application deadline: 3/1. *Faculty research:* Molecular biology, host-parasite interactions. *Unit head:* Dr. David R. Lee, Director of Graduate Studies, 573-882-7893, E-mail: leedr@missouri.edu.

The University of North Carolina at Chapel Hill, School of Medicine and Graduate School, Graduate Programs in Medicine, Department of Microbiology and Immunology, Chapel Hill, NC 27599. Offers immunology (MS, PhD); microbiology (MS, PhD). *Faculty:* 38 full-time (15 women), 26 part-time/adjunct (5 women). *Students:* 69 full-time (31 women), 2 part-time (both women); includes 9 minority (2 African Americans, 5 Asian Americans or Pacific Islanders, 2 Hispanic Americans), 11 international. Average age 29. 119 applicants, 21% accepted, 7 enrolled. In 2005, 3 master's, 9 doctorates awarded. Terminal master's awarded for partial completion of doctoral program. *Median time to degree:* Of those who began their doctoral program in fall 1997, 100% received their degree in 8 years or less. *Degree requirements:* For master's and doctorate, thesis/dissertation, comprehensive exam, registration. *Entrance requirements:* For master's and doctorate, GRE General Test, minimum GPA of 3.0. Additional exam requirements/recommendations for international students: Required—TOEFL (minimum score 600 paper-based; 250 computer-based). *Application deadline:* For fall admission, 1/1 priority date for domestic students, 1/1 priority date for international students. Applications are processed on a rolling basis. Application fee: $70. Electronic applications accepted. *Financial support:* In 2005–06, 3 fellowships with full tuition reimbursements (averaging $22,000 per year), 66 research assistantships with full tuition reimbursements (averaging $22,000 per year) were awarded; scholarships/grants, traineeships, health care benefits, and unspecified assistantships also available. *Faculty research:* HIV pathogenesis, immune response, t-cell mediated autoimmunity, alpha-viruses, bacterial chemotaxis. Total annual research expenditures: $8.1 million. *Unit head:* Jeffrey Frelinger, Chairman, 919-966-1191, Fax: 919-962-8103, E-mail: jfrelin@med.unc.edu. *Application contact:* Dixie Flannery, Student Services Manager, 919-966-1191, Fax: 919-962-8103, E-mail: microimm@listserv.med.unc.edu.

University of North Dakota, School of Medicine and Graduate School, Graduate Programs in Medicine, Department of Microbiology and Immunology, Grand Forks, ND 58202. Offers MS, PhD. *Faculty:* 6 full-time (1 woman). *Students:* 1 (woman) full-time, 12 part-time (7 women). 25 applicants, 12% accepted, 3 enrolled. In 2005, 1 master's, 2 doctorates awarded. *Degree requirements:* For master's, thesis or alternative, comprehensive exam; for doctorate, thesis/dissertation, final examination, comprehensive exam. *Entrance requirements:* For master's and doctorate, GRE General Test, minimum GPA of 3.0. Additional exam requirements/recommendations for international students: Required—TOEFL (minimum score 550 paper-based; 213 computer-based). *Application deadline:* For fall admission, 2/15 priority date for domestic students, 2/15 priority date for international students; for spring admission, 10/15 priority date for domestic students, 10/15 priority date for international students. Applications are processed on a rolling basis. Application fee: $35. Electronic applications accepted. *Financial support:* In 2005–06, 8 students received support, including research assistantships (averaging $13,997 per year), teaching assistantships with full tuition reimbursements available (averaging $13,997 per year); fellowships, Federal Work-Study, institutionally sponsored loans, scholarships/grants, and tuition waivers (full and partial) also available. Support available to part-time students. Financial award application deadline: 3/15; financial award applicants required to submit FAFSA. *Faculty research:* Genetic and immunological aspects of a murine model of human multiple sclerosis, termination of DNA replication, cell division in bacteria, yersinia pestis. *Unit head:* Dr. David Bradley, Director, 701-777-2214, Fax: 701-777-3527, E-mail: dbradley@medicine.nodak.edu.

University of North Texas Health Science Center at Fort Worth, Graduate School of Biomedical Sciences, Fort Worth, TX 76107-2699. Offers anatomy and cell biology (MS, PhD); biochemistry and molecular biology (MS, PhD); biomedical sciences (MS, PhD); biotechnology (MS); forensic genetics (MS); integrative physiology (MS, PhD); medical science (MS); microbiology and immunology (MS, PhD); pharmacology (MS, PhD); science education (MS). *Faculty:* 57 full-time (9 women), 2 part-time/adjunct (0 women). *Students:* 177 full-time (98 women), 42 part-time (33 women); includes 61 minority (15 African Americans, 3 American Indian/Alaska Native, 27 Asian Americans or Pacific Islanders, 16 Hispanic Americans), 53 international. Average age 28. 237 applicants, 62% accepted, 91 enrolled. In 2005, 37 master's, 15 doctorates awarded. Terminal master's awarded for partial completion of doctoral program. *Degree requirements:* For master's and doctorate, thesis/dissertation. *Entrance requirements:* For master's and doctorate, GRE General Test. Additional exam requirements/recommendations for international students: Required—TOEFL. *Application deadline:* For fall admission, 5/1 for domestic students. Application fee: $25 ($50 for international students). *Expenses:* Contact institution. *Financial support:* In 2005–06, 80 research assistantships (averaging $16,000 per year) were awarded; fellowships, teaching assistantships, career-related internships or fieldwork, Federal Work-Study, institutionally sponsored loans, scholarships/grants, and traineeships also available. Support available to part-time students. Financial award application deadline: 4/1; financial award applicants required to submit FAFSA. *Faculty research:* Alzheimer's disease, aging, eye diseases, cancer, cardiovascular disease. Total annual research expenditures: $21 million. *Unit head:* Dr. Thomas Yorio, Dean, 817-735-2560, Fax: 817-735-0243, E-mail: yoriot@hsc.unt.edu. *Application contact:* Carla Lee, Director of Graduate Admissions and Services, 817-735-2560, Fax: 817-735-0243, E-mail: gsbs@hsc.unt.edu.

See Close-Up on page 279.

University of Oklahoma Health Sciences Center, College of Medicine and Graduate College, Graduate Programs in Medicine, Department of Microbiology and Immunology, Oklahoma City, OK 73190. Offers immunology (MS, PhD); microbiology (MS, PhD). Part-time programs available. Terminal master's awarded for partial completion of doctoral program. *Degree requirements:* For master's, thesis or alternative; for doctorate, one foreign language, thesis/dissertation. *Entrance requirements:* For doctorate, GRE General Test, 3 letters of recommendation. Additional exam requirements/recommendations for international students: Required—TOEFL. *Faculty research:* Molecular genetics, pathogenesis, streptococcal infections, gram-positive virulence, monoclonal antibodies.

University of Ottawa, Faculty of Graduate and Postdoctoral Studies, Faculty of Medicine, Department of Biochemistry, Microbiology and Immunology, Ottawa, ON K1N 6N5, Canada. Offers biochemistry (M Sc, PhD); microbiology and immunology (M Sc, PhD). *Faculty:* 72 full-time (19 women), 22 part-time/adjunct (8 women). *Students:* 163 full-time, 22 part-time. 35 applicants, 69% accepted, 16 enrolled. In 2005, 16 master's, 17 doctorates awarded. *Degree requirements:* For master's, thesis; for doctorate, thesis/dissertation, seminar, comprehensive exam. *Entrance requirements:* For master's, honors degree or equivalent, minimum B average; for doctorate, master's degree, minimum B+ average. *Application deadline:* For fall admission, 3/1 priority date for domestic students, 2/15 priority date for international students. For winter admission, 11/15 for domestic students; for spring admission, 4/1 for domestic students. Application fee: $75. Electronic applications accepted. *Expenses:* Tuition: Part-time $260 per credit. Tuition and fees vary according to course load and program. *Financial support:* Fellowships, research assistantships with full tuition reimbursements, teaching assistantships with full tuition reimbursements, career-related internships or fieldwork, Federal Work-Study, scholarships/grants, traineeships, tuition waivers (full and partial), and unspecified assistantships available. Financial award application deadline: 2/15. *Faculty research:* General biochemistry, molecular biology, microbiology, host biology, nutrition and metabolism. *Unit head:* Dr. Zemin Yao, Chair, 613-562-5800 Ext. 5459, Fax: 613-562-5440. *Application contact:* Carol Ann Kelly, Academic Assistant, 613-562-5800 Ext. 5424, Fax: 613-562-5440, E-mail: grads@uottawa.ca.

University of Pennsylvania, National Institutes of Health Sponsored Programs, Graduate Group in Immunology, Philadelphia, PA 19104. Offers PhD, MD/PhD, VMD/PhD. *Faculty:* 100. *Students:* 83 full-time (46 women); includes 22 minority (2 African Americans, 1 American Indian/Alaska Native, 15 Asian Americans or Pacific Islanders, 4 Hispanic Americans), 4 international. 104 applicants, 25% accepted, 10 enrolled. In 2005, 6 doctorates awarded. *Degree requirements:* For doctorate, thesis/dissertation, 2 preliminary examinations. *Entrance requirements:* For doctorate, GRE General Test, undergraduate major in natural or physical science. Additional exam requirements/recommendations for international students: Required—TOEFL. *Application deadline:* For fall admission, 12/15 priority date for domestic students, 12/1 priority date for international students. Applications are processed on a rolling basis. Application fee: $70. Electronic applications accepted. *Financial support:* In 2005–06, 34 students received support; fellowships, scholarships/grants, traineeships, and unspecified assistantships available. *Faculty research:* Immunoglobulin structure and function, cell surface receptors, lymphocyte functional transplantation immunology, cellular immunology, molecular biology of immunoglobulins. *Unit head:* Dr. Steven L. Reiner, Chairman, 215-746-5536. *Application contact:* Jennifer Laverty, Graduate Coordinator, 215-746-5536, E-mail: jlaverty@mail.med.upenn.edu.

See Close-Up on page 941.

University of Pennsylvania, School of Medicine, Biomedical Graduate Studies, Graduate Group in Immunology, Philadelphia, PA 19104. Offers PhD, MD/PhD, VMD/PhD. *Faculty:* 100. *Students:* 83 full-time (46 women); includes 22 minority (2 African Americans, 1 American Indian/Alaska Native, 15 Asian Americans or Pacific Islanders, 4 Hispanic Americans), 4 international. 104 applicants, 25% accepted, 10 enrolled. In 2005, 6 doctorates awarded. *Degree requirements:* For doctorate, thesis/dissertation, 2 preliminary exams. *Entrance requirements:* For doctorate, GRE General Test, undergraduate major in natural or physical science. Additional exam requirements/recommendations for international students: Required—TOEFL. *Application deadline:* For fall admission, 12/15 priority date for domestic students, 12/1 priority date for international students. Applications are processed on a rolling basis. Application fee: $70. Electronic applications accepted. *Financial support:* In 2005–06, 34 students received support; fellowships, research assistantships, scholarships/grants, traineeships, and unspecified assistantships available. *Faculty research:* Immunoglobulin structure and function, cell surface receptors, lymphocyte functional transplantation immunology, cellular immunology, molecular biology of immunoglobulins. *Unit head:* Dr. Steven L. Reiner, Chairman, 215-746-5536. *Application contact:* Jennifer Laverty, Graduate Coordinator, 215-746-5536, E-mail: jlaverty@mail.med.upenn.edu.

See Close-Up on page 941.

University of Pittsburgh, School of Medicine, Graduate Programs in Medicine, Program in Immunology, Pittsburgh, PA 15260. Offers MS, PhD. *Faculty:* 54 full-time (15 women). *Students:* 31 full-time (17 women); includes 3 minority (1 African American, 2 Asian Americans or Pacific Islanders), 10 international. Average age 28. 415 applicants, 22% accepted, 42 enrolled. In 2005, 10 degrees awarded. *Median time to degree:* Of those who began their doctoral program in fall 1997, 95% received their degree in 8 years or less. *Degree requirements:* For doctorate, thesis/dissertation, comprehensive exam, registration. *Entrance requirements:* For doctorate, GRE General Test, GRE Subject Test, minimum QPA of 3.0. Additional exam requirements/recommendations for international students: Required—TOEFL (minimum score 600 paper-based; 250 computer-based), IELT (minimum score 7). *Application deadline:* For fall admission, 12/15 priority date for domestic students, 12/15 priority date for international students. Application fee: $40. Electronic applications accepted. *Expenses:* Tuition, state resident: full-time $13,194; part-time $537 per credit. Tuition, nonresident: full-time $25,012; part-time $1,026 per credit. Required fees: $700; $164 per term. Tuition and fees vary according to campus/location and program. *Financial support:* In 2005–06, research assistantships with full tuition reimbursements (averaging $21,500 per year); teaching assistantships with full tuition reimbursements, institutionally sponsored loans, scholarships/grants, traineeships, health care benefits, and unspecified assistantships also available. *Faculty research:* Human T-cell biology, opportunistic infections associated with AIDS, autoimmunity, immunoglobin gene expression, tumor immunology. *Unit head:* Dr. Russell D. Salter, Program Director, 412-648-7471, Fax: 412-624-7736, E-mail: rds@pitt.edu. *Application contact:* Graduate Studies Administrator, 412-648-8957, Fax: 412-648-1007, E-mail: gradstudies@medschool.pitt.edu.

University of Prince Edward Island, Atlantic Veterinary College, Graduate Program in Veterinary Medicine, Charlottetown, PE C1A 4P3, Canada. Offers anatomy (M Sc, PhD); bacteriology (M Sc, PhD); clinical pharmacology (M Sc, PhD); clinical sciences (M Sc, PhD); epidemiology (M Sc, PhD), including reproduction; fish health (M Sc, PhD); food animal nutrition (M Sc, PhD); immunology (M Sc, PhD); microanatomy (M Sc, PhD); parasitology (M Sc, PhD); pathology (M Sc, PhD); pharmacology (M Sc, PhD); physiology (M Sc, PhD); toxicology (M Sc, PhD); veterinary science (M Vet Sc); virology (M Sc, PhD). Part-time programs available. *Faculty:* 76 full-time (25 women), 49 part-time/adjunct (8 women). *Students:* 54 full-time (32 women), 2 part-time. Average age 30. In 2005, 7 master's, 6 doctorates awarded. *Degree requirements:* For master's and doctorate, thesis/dissertation. *Entrance requirements:* For master's, DVM, B Sc honors degree, or equivalent; for doctorate, M Sc. *Application deadline:* Applications are processed on a rolling basis. Application fee: $50. *Expenses:* Contact institution. Tuition charges are reported in Canadian dollars. Part-time tuition and fees vary according to course level, degree level, campus/location, program and student level. *Financial support:* In 2005–06, 4 fellowships (averaging $25,000 Canadian dollars per year), 4 research assistantships (averaging $16,500 Canadian dollars per year) were awarded; career-related internships or fieldwork also available. *Faculty research:* Animal health management, infectious diseases, fin fish and shellfish health, basic biomedical sciences, ecosystem health. Total annual research expenditures: $1.2 million Canadian dollars.

Unit head: Dr. James Bellamy, Associate Dean of Graduate Studies and Research, 902-566-0856, E-mail: bellamy@upei.ca. *Application contact:* Cheryl Gaudet, Registrar's Office, 902-566-0781, Fax: 902-566-0795, E-mail: registrar@upei.ca.

University of Rochester, School of Medicine and Dentistry, Graduate Programs in Medicine and Dentistry, Department of Microbiology and Immunology, Rochester, NY 14627-0250. Offers microbiology (MS, PhD). *Degree requirements:* For doctorate, thesis/dissertation, qualifying exam. *Entrance requirements:* For master's and doctorate, GRE General Test.

University of South Alabama, College of Medicine and Graduate School, Program in Basic Medical Sciences, Specialization in Microbiology and Immunology, Mobile, AL 36688-0002. Offers PhD. *Faculty:* 10 full-time (0 women). *Degree requirements:* For doctorate, thesis/dissertation. *Entrance requirements:* For doctorate, GRE General Test or MCAT. *Application deadline:* For fall admission, 4/1 for domestic students. Applications are processed on a rolling basis. *Expenses:* Tuition, state resident: full-time $4,008. Tuition, nonresident: full-time $8,016. Required fees: $692. *Financial support:* Fellowships, research assistantships, institutionally sponsored loans available. Financial award application deadline: 4/1. *Faculty research:* Mechanisms of tumor immunity, host response to infectious agents, virus replication, immune regulation, mechanisms of resistance to viruses and bacteria. *Unit head:* Dr. Joseph H. Coggin, Chair, 251-460-6339. *Application contact:* Lanette Flagge, Academic Advisor, 251-460-6153.

The University of South Dakota, School of Medicine and Health Sciences and Graduate School, Biomedical Sciences Graduate Program, Molecular Microbiology and Immunology Group, Vermillion, SD 57069-2390. Offers MA, PhD. *Faculty:* 6 full-time (1 woman). *Students:* 4 full-time (1 woman), 1 international. Average age 27. 11 applicants, 36% accepted, 2 enrolled.Terminal master's awarded for partial completion of doctoral program. *Degree requirements:* For master's, thesis/dissertation; for doctorate, thesis/dissertation, comprehensive exam, registration. *Entrance requirements:* For master's and doctorate, GRE General Test, minimum GPA of 3.0. Additional exam requirements/recommendations for international students: Required—TOEFL (minimum score 550 paper-based; 213 computer-based). *Application deadline:* For fall admission, 4/15 priority date for domestic students, 4/15 priority date for international students. Applications are processed on a rolling basis. Application fee: $35. *Expenses: Contact institution.* Tuition and fees vary according to course load, program and reciprocity agreements. *Financial support:* In 2005–06, 4 students received support, including 4 fellowships with partial tuition reimbursements available (averaging $20,772 per year), 1 research assistantship with partial tuition reimbursement available (averaging $10,386 per year); Federal Work-Study and unspecified assistantships also available. Financial award application deadline: 4/15; financial award applicants required to submit FAFSA. *Faculty research:* Structure-function membranes, plasmids, immunology, virology, pathogenesis. Total annual research expenditures: $325,000.

University of Southern California, Keck School of Medicine and Graduate School, Graduate Programs in Medicine, Department of Molecular Microbiology and Immunology, Los Angeles, CA 90033. Offers MS, PhD. *Faculty:* 9 full-time (1 woman), 1 (woman) part-time/adjunct. *Students:* 33 full-time (19 women), 1 part-time; includes 8 minority (7 Asian Americans or Pacific Islanders, 1 Hispanic American), 17 international. Average age 28. 68 applicants, 12% accepted, 5 enrolled. In 2005, 6 master's, 5 doctorates awarded. *Median time to degree:* Of those who began their doctoral program in fall 1997, 80% received their degree in 8 years or less. *Degree requirements:* For master's, thesis optional; for doctorate, thesis/dissertation. *Entrance requirements:* For master's, GRE General Test, minimum GPA of 3.0; for doctorate, GRE General Test, GRE Subject Test, minimum GPA of 3.0. Additional exam requirements/ recommendations for international students: Required—TOEFL (minimum score 250 paper-based). *Application deadline:* For fall admission, 2/15 priority date for domestic students, 2/15 priority date for international students; for spring admission, 7/15 for domestic students, 2/15 for international students. Applications are processed on a rolling basis. Application fee: $65 ($75 for international students). Electronic applications accepted. *Expenses:* Tuition: Full-time $25,416; part-time $1,059 per unit. Required fees: $484; $484 per year. Tuition and fees vary according to course load and program. *Financial support:* In 2005–06, 22 students received support, including 1 fellowship with full tuition reimbursement available (averaging $21,000 per year), 20 research assistantships with full tuition reimbursements available (averaging $21,000 per year), 1 teaching assistantship with full tuition reimbursement available (averaging $21,000 per year); institutionally sponsored loans, traineeships, and tuition waivers (partial) also available. Financial award application deadline: 6/1. *Faculty research:* Animal virology, microbial genetics, molecular and cellular immunology, cellular differentiation control of protein synthesis. Total annual research expenditures: $2.5 million. *Unit head:* Dr. Stanley M. Tahara, Director, Graduate Studies, 323-442-1722, Fax: 323-442-1721, E-mail: stahara@usc.edu. *Application contact:* Laura C. Steel, Graduate Administrator, 323-442-2337, Fax: 323-442-1721, E-mail: lsteel@usc.edu.

University of Southern Maine, School of Applied Science, Engineering, and Technology, Program in Applied Immunology and Molecular Biology, Portland, ME 04104-9300. Offers MS. Part-time programs available. *Degree requirements:* For master's, thesis. *Entrance requirements:* For master's, GRE General Test, GRE Subject Test, minimum GPA of 3.0. Additional exam requirements/recommendations for international students: Required—TOEFL. Electronic applications accepted. *Faculty research:* Flow cytometry, cancer, epidemiology, monoclonal antibodies, DNA diagnostics.

University of South Florida, College of Medicine and College of Graduate Studies, Graduate Programs in Medical Sciences, Tampa, FL 33620-9951. Offers anatomy (PhD); biochemistry and molecular biology (MS, PhD), including biochemistry and molecular biology (PhD), bioinformatics and computational biology (MS); medical microbiology and immunology (PhD); pathology (PhD); pharmacology and therapeutics (PhD), including medical sciences; physiology and biophysics (PhD). *Students:* 108 full-time (53 women), 34 part-time (26 women); includes 31 minority (9 African Americans, 9 Asian Americans or Pacific Islanders, 13 Hispanic Americans), 30 international. 117 applicants, 99% accepted, 116 enrolled. In 2005, 9 master's, 4 doctorates awarded. *Degree requirements:* For doctorate, thesis/dissertation. *Entrance requirements:* For doctorate, GRE General Test, minimum GPA of 3.0. Application fee: $30. *Expenses: Contact institution. Financial support:* Institutionally sponsored loans and scholarships/ grants available. Financial award application deadline: 4/1; financial award applicants required to submit FAFSA. *Unit head:* Dr. Joseph J. Krzanowski, Associate Dean for Research and Graduate Affairs, 813-974-4181, Fax: 813-974-4317, E-mail: jkrzanow@com1.med.usf.edu.

The University of Tennessee Health Science Center, College of Graduate Health Sciences, Department of Molecular Sciences, Memphis, TN 38163-0002. Offers bacterial pathogenesis (PhD); biochemistry (PhD); immunology (PhD); microbiology (PhD); molecular and cell biology-pathology (PhD); molecular biology (PhD); signal transduction (PhD); structural biology (PhD); virology (PhD). *Faculty:* 17 full-time (4 women). *Students:* 25 full-time (11 women); includes 14 minority (1 African American, 13 Asian Americans or Pacific Islanders). Average age 24. 83 applicants, 12% accepted. In 2005, 6 degrees awarded. *Degree requirements:* For doctorate, thesis/dissertation, oral and written preliminary and comprehensive exams. *Entrance requirements:* For doctorate, GRE General Test, GRE Subject Test, minimum GPA of 3.0. Additional exam requirements/recommendations for international students: Required—TOEFL. *Application deadline:* For fall admission, 5/15 for domestic students. Application fee: $0. *Financial support:* In 2005–06, 8 fellowships, 21 research assistantships were awarded; teaching assistantships, traineeships and tuition waivers (full) also available. *Unit head:* Dr. David Hasty, Chairman, 901-448-6150, Fax: 901-448-7360. *Application contact:* Ida W. Mosby, Director, Enrollment Services, 901-448-5560, E-mail: imosby@utmem.edu.

The University of Texas at Austin, Graduate School, College of Natural Sciences, Institute for Cellular and Molecular Biology, Program in Microbiology and Immunology, Austin, TX 78712-1111. Offers PhD. *Degree requirements:* For doctorate, thesis/dissertation. *Entrance requirements:* For doctorate, GRE General Test. Additional exam requirements/recommendations for inter-national students: Required—TOEFL. Electronic applications accepted. *Faculty research:* Pathogenesis, virology, cell biology, molecular genetics.

The University of Texas Health Science Center at Houston, Graduate School of Biomedical Sciences, Program in Immunology, Houston, TX 77225-0036. Offers MS, PhD, MD/PhD. *Faculty:* 53 full-time (9 women). *Students:* 28 full-time (16 women); includes 9 minority (1 American Indian/Alaska Native, 7 Asian Americans or Pacific Islanders, 1 Hispanic American), 6 international. Average age 25. 37 applicants, 30% accepted, 6 enrolled. In 2005, 5 degrees awarded. Terminal master's awarded for partial completion of doctoral program. *Degree requirements:* For master's and doctorate, thesis/dissertation. *Entrance requirements:* For master's and doctorate, GRE General Test. Additional exam requirements/recommendations for international students: Required—TOEFL, TWE. *Application deadline:* For fall admission, 1/15 for domestic students; for spring admission, 11/1 for domestic students. Applications are processed on a rolling basis. Application fee: $10. Electronic applications accepted. *Financial support:* Fellowships with full tuition reimbursements, research assistantships with full tuition reimbursements, teaching assistantships, institutionally sponsored loans, scholarships/grants, and health care benefits available. Financial award application deadline: 1/15. *Faculty research:* Dendritic cells and innate/adaptive immune responses, lymphocyte activation and differentiation, host-pathogen relationships, tumor immunobiology, cytokine and chemokine biology. *Unit head:* Dr. Stephanie S. Watowich, Director, 713-563-3262, Fax: 713-563-3357, E-mail: swatowic@mdanderson.org. *Application contact:* Dr. Victoria P. Knutson, Assistant Dean of Admissions, 713-500-9860, Fax: 713-500-9877, E-mail: victoria.p.knutson@uth.tmc.edu.

The University of Texas Health Science Center at San Antonio, Graduate School of Biomedical Sciences, Department of Microbiology and Immunology, San Antonio, TX 78229-3900. Offers microbiology and immunology (PhD); microbiology and immunology primary and secondary science teacher track (MS). Terminal master's awarded for partial completion of doctoral program. *Degree requirements:* For master's and doctorate, thesis/dissertation. *Entrance requirements:* For master's and doctorate, GRE General Test, minimum GPA of 3.0. *Faculty research:* Molecular immunology, mechanisms of pathogenesis, molecular genetics, vaccine and immunodiagnostic development.

See Close-Up on page 947.

The University of Texas Medical Branch, Graduate School of Biomedical Sciences, Program in Microbiology and Immunology, Galveston, TX 77555. Offers MS, PhD. *Students:* 29 full-time (13 women), 2 part-time (both women); includes 5 minority (3 African Americans, 2 Hispanic Americans), 9 international. Average age 30. In 2005, 3 doctorates awarded. *Degree requirements:* For master's, thesis or alternative; for doctorate, thesis/dissertation. *Entrance requirements:* For doctorate, GRE General Test, minimum GPA of 3.0. Additional exam requirements/recommendations for international students: Required—TOEFL (minimum score 550 paper-based; 213 computer-based). *Application deadline:* Applications are processed on a rolling basis. Application fee: $30 ($75 for international students). Electronic applications accepted. *Expenses:* Tuition, state resident: full-time $8,350; part-time $90 per credit hour. Tuition, nonresident: full-time $21,450; part-time $366 per credit hour. Required fees: $1,027; $11 per credit hour. $60 per term. *Financial support:* In 2005–06, research assistantships with full tuition reimbursements (averaging $23,000 per year) Financial award applicants required to submit FAFSA. *Unit head:* Dr. Thomas K. Hughes, Director, 409-772-6660, Fax: 409-772-2295, E-mail: tkhughes@utmb.edu. *Application contact:* Martha J. Lewis, Coordinator II Special Programs, 409-772-2322, Fax: 409-772-2295, E-mail: mlewis@utmb.edu.

See Close-Up on page 951.

The University of Texas Southwestern Medical Center at Dallas, Southwestern Graduate School of Biomedical Sciences, Division of Basic Science, Program in Immunology, Dallas, TX 75390. Offers PhD. *Faculty:* 47 full-time (12 women), 2 part-time/adjunct (0 women). *Students:* 51 full-time (29 women); includes 13 minority (1 African American, 1 American Indian/Alaska Native, 6 Asian Americans or Pacific Islanders, 5 Hispanic Americans), 9 international. Average age 28. In 2005, 8 doctorates awarded. *Degree requirements:* For doctorate, thesis/dissertation, qualifying exam. *Entrance requirements:* For doctorate, GRE General Test, minimum GPA of 3.0. Additional exam requirements/recommendations for international students: Required—TOEFL. *Application deadline:* For fall admission, 1/5 for domestic students. Applications are processed on a rolling basis. Application fee: $0. Electronic applications accepted. *Expenses:* Tuition, state resident: full-time $6,550; part-time $50 per credit hour. Tuition, nonresident: full-time $19,650; part-time $326 per credit hour. Required fees: $42 per credit hour. Tuition and fees vary according to degree level and program. *Financial support:* Fellowships, research assistantships available. *Faculty research:* Antibody diversity and idiotype, cytotoxic effector mechanisms, natural killer cells, biology of immunoglobulins, oncogenes. *Unit head:* Dr. Nicolai Van Oers, Chair, 214-648-1236, Fax: 214-648-1902, E-mail: nicolai.vanoers@utsouthwestern.edu. *Application contact:* Dr. Nancy E. Street, Associate Dean, 214-648-6708, Fax: 214-648-2102, E-mail: nancy.street@utsouthwestern.edu.

University of Toronto, School of Graduate Studies, Life Sciences Division, Department of Immunology, Toronto, ON M5S 1A1, Canada. Offers M Sc, PhD. *Degree requirements:* For master's and doctorate, thesis/dissertation, thesis defense. *Entrance requirements:* For master's, GRE (international applicants), resumé, 3 letters of reference; for doctorate, GRE (international applicants). Additional exam requirements/recommendations for international students: Required—TOEFL, TWE.

University of Utah, The Graduate School, College of Science and Graduate Programs in Medicine, Program in Molecular Biology, Salt Lake City, UT 84112-1107.

University of Virginia, College and Graduate School of Arts and Sciences, Immunology Training Program, Charlottesville, VA 22903. Offers PhD, MD/PhD. *Entrance requirements:* For doctorate, GRE General Test, GRE Subject Test. Additional exam requirements/ recommendations for international students: Required—TOEFL. *Application deadline:* For fall admission, 2/15 for domestic students. Application fee: $60. *Expenses:* Tuition, state resident: full-time $7,731. Tuition, nonresident: full-time $18,672. Required fees: $1,479. Full-time tuition and fees vary according to degree level and program. *Application contact:* Glenn Glover, Graduate Program Administrator, 434-924-2412, Fax: 434-924-1221, E-mail: gmg6n@virginia.edu.

See Close-Up on page 953.

University of Washington, School of Medicine and Graduate School, Graduate Programs in Medicine, Department of Immunology, Seattle, WA 98195. Offers PhD. *Degree requirements:* For doctorate, thesis/dissertation. *Entrance requirements:* For doctorate, GRE General Test, BA or BS in related field. Additional exam requirements/recommendations for international students: Required—TOEFL (minimum score 600 paper-based; 250 computer-based). Electronic applications accepted. *Faculty research:* Molecular and cellular immunology, regulation of lymphocyte differentiation and responses, genetics of immune recognition genetics and pathogenesis of autoimmune diseases, signal transduction.

The University of Western Ontario, Faculty of Graduate Studies, Biosciences Division, Department of Microbiology and Immunology, London, ON N6A 5B8, Canada. Offers M Sc, PhD. *Degree requirements:* For master's and doctorate, thesis/dissertation, oral and written exam. *Entrance requirements:* For master's, honors degree or equivalent in microbiology, immunology, or other biological science; minimum B average; for doctorate, M Sc in microbiology and immunology. *Faculty research:* Virology, molecular pathogenesis, cellular immunology, molecular biology.

Vanderbilt University, Graduate School and School of Medicine, Department of Microbiology and Immunology, Nashville, TN 37232-2363. Offers MS, PhD, MD/PhD. *Faculty:* 40 full-time (6

Immunology

Vanderbilt University *(continued)*
women). *Students:* 50 full-time (27 women); includes 6 minority (2 African Americans, 1 Asian American or Pacific Islander, 3 Hispanic Americans), 9 international. In 2005, 1 master's, 11 doctorates awarded. *Degree requirements:* For master's, thesis; for doctorate, thesis/dissertation, final and qualifying exams. *Entrance requirements:* For master's and doctorate, GRE General Test, GRE Subject Test (recommended). *Application deadline:* For fall admission, 1/15 for domestic students, 1/15 for international students. Application fee: $0. Electronic applications accepted. *Expenses:* Tuition: Part-time $1,283 per semester hour. Required fees: $2,202; $1,101 per semester. One-time fee: $30. Tuition and fees vary according to course load, program and student level. *Financial support:* Fellowships with full tuition reimbursements, research assistantships with full tuition reimbursements, Federal Work-Study, institutionally sponsored loans, traineeships, and tuition waivers (partial) available. Financial award application deadline: 1/15. *Faculty research:* Molecular and cellular immunology, molecular genetics and immunogenetics, cellular microbiology of pathogen-host interaction, virology, biotechnology. *Unit head:* Jacek Hawiger, Chair, 615-343-3435, Fax: 615-343-7392, E-mail: jacek.hawiger@vanderbilt.edu. *Application contact:* Eugene M. Oltz, Director of Graduate Studies, 615-343-0453, Fax: 615-343-7392, E-mail: eugene.oltz@vanderbilt.edu.

Virginia Commonwealth University, Medical College of Virginia-Professional Programs, School of Medicine and Graduate Programs, School of Medicine Graduate Programs, Department of Microbiology and Immunology, Richmond, VA 23284-9005. Offers microbiology and genetics (MS); microbiology and immunology (MS, PhD, CBHS); molecular biology and genetics (PhD). *Faculty:* 33 full-time (8 women). *Students:* 66 full-time (43 women), 9 part-time (6 women); includes 21 minority (12 African Americans, 5 Asian Americans or Pacific Islanders, 4 Hispanic Americans), 11 international. 95 applicants, 48% accepted. In 2005, 25 master's, 7 doctorates, 3 other advanced degrees awarded. *Degree requirements:* For master's, thesis; for doctorate, thesis/dissertation, comprehensive oral and written exams. *Entrance requirements:* For master's, GRE General Test or MCAT; for doctorate, GRE General Test, MCAT. *Application deadline:* For fall admission, 2/15 for domestic students. Application fee: $50. *Expenses:* Tuition, state resident: full-time $6,268; part-time $405 per credit. Tuition, nonresident: full-time $15,904; part-time $940 per credit. Required fees: $751 per semester hour. Tuition and fees vary according to course load and program. *Financial support:* Fellowships, research assistantships, teaching assistantships available. *Faculty research:* Microbial physiology and genetics, molecular biology, crystallography of biological molecules, antibiotics and chemotherapy, membrane transport. *Unit head:* Dr. Dennis E. Ohman, Chair, 804-828-9728, Fax: 804-828-9946, E-mail: deohman@vcu.edu. *Application contact:* Dr. Guy A. Cabral, Chair, Graduate Program, 804-828-2306, E-mail: gacabral@vcu.edu.

Wake Forest University, School of Medicine and Graduate School, Graduate Programs in Medicine, Department of Microbiology and Immunology, Winston-Salem, NC 27109. Offers PhD. *Degree requirements:* For doctorate, thesis/dissertation. *Entrance requirements:* For doctorate, GRE General Test. Additional exam requirements/recommendations for international students: Required—TOEFL. Electronic applications accepted. *Faculty research:* Molecular immunology, bacterial pathogenesis and molecular genetics, viral pathogenesis, regulation of mRNA metabolism, leukocyte biology.

See Close-Up on page 959.

Washington University in St. Louis, Graduate School of Arts and Sciences, Division of Biology and Biomedical Sciences, Program in Immunology, St. Louis, MO 63130-4899. Offers PhD. *Degree requirements:* For doctorate, thesis/dissertation. *Entrance requirements:* For doctorate, GRE General Test, GRE Subject Test. Electronic applications accepted.

Wayne State University, School of Medicine and Graduate School, Graduate Programs in Medicine, Department of Immunology and Microbiology, Detroit, MI 48202. Offers MS, PhD, MD/PhD. *Faculty:* 12 full-time. *Students:* 15 full-time (10 women), 1 (woman) part-time; includes 4 minority (1 African American, 2 Asian Americans or Pacific Islanders, 1 Hispanic American), 2 international. Average age 28. 37 applicants, 16% accepted, 3 enrolled. In 2005, 3 degrees awarded. Terminal master's awarded for partial completion of doctoral program. *Degree requirements:* For master's and doctorate, thesis/dissertation. *Entrance requirements:* For master's, GRE, minimum GPA of 2.5; for doctorate, GRE, minimum GPA of 3.0. Additional exam requirements/recommendations for international students: Required—TOEFL (minimum score 550 paper-based; 213 computer-based); Recommended—TWE (minimum score 6). *Application deadline:* For fall admission, 7/15 for domestic students, 6/1 for international students. Applications are processed on a rolling basis. Application fee: $30 ($50 for international students). Electronic applications accepted. *Expenses:* Tuition, state resident: part-time $338 per credit hour. Tuition, nonresident: part-time $746 per credit hour. Required fees: $24 per credit hour. Full-time tuition and fees vary according to program. *Financial support:* In 2005–06, 1 fellowship with tuition reimbursement, 12 research assistantships with tuition reimbursements (averaging $20,167 per year) were awarded; teaching assistantships, career-related internships or fieldwork and Federal Work-Study also available. *Faculty research:*

Immune regulation, bacterial pathophysiology, molecular biology/viruses/bacteria, cellular and molecular immunology, microbial pathogenesis. Total annual research expenditures: $2.5 million. *Unit head:* Dr. Paul Montgomery, Chair, 313-577-1591, Fax: 313-577-1155, E-mail: aa2411@wayne.edu. *Application contact:* Harley Tse, Associate Professor, 313-577-1564, Fax: 313-577-1155, E-mail: ad6056@wayne.edu.

West Virginia University, Davis College of Agriculture, Forestry and Consumer Sciences, Interdisciplinary Program in Genetics and Developmental Biology, Morgantown, WV 26506. Offers animal breeding (MS, PhD); biochemical and molecular genetics (MS, PhD); cytogenetics (MS, PhD); descriptive embryology (MS, PhD); developmental genetics (MS); experimental morphogenesis teratology (MS); human genetics (MS, PhD); immunogenetics (MS, PhD); life cycles of animals and plants (MS, PhD); molecular aspects of development (MS, PhD); mutagenesis (MS, PhD); oncology (MS, PhD); plant genetics (MS, PhD); population and quantitative genetics (MS, PhD); regeneration (MS, PhD); teratology (PhD); toxicology (MS, PhD). *Students:* 13 full-time (7 women), 6 part-time (4 women), 11 international. Average age 28. In 2005, 2 master's, 4 doctorates awarded. *Degree requirements:* For master's, thesis/dissertation; for doctorate, thesis/dissertation, comprehensive exam. *Entrance requirements:* For master's, GRE or MCAT, minimum GPA of 2.75. Additional exam requirements/recommendations for international students: Required—TOEFL. Application fee: $45. *Expenses:* Tuition, state resident: full-time $4,582; part-time $258 per credit hour. Tuition, nonresident: full-time $13,820; part-time $741 per credit hour. *Financial support:* In 2005–06, 3 research assistantships with tuition reimbursements (averaging $9,936 per year), 4 teaching assistantships with tuition reimbursements (averaging $9,936 per year) were awarded; fellowships, Federal Work-Study, institutionally sponsored loans, and tuition waivers (full and partial) also available. Financial award application deadline: 2/1; financial award applicants required to submit FAFSA. Total annual research expenditures: $1 million. *Unit head:* Dr. J. Nath, Chairman, 304-293-6023 Ext. 4333, Fax: 304-293-2960, E-mail: joginder.nath@mail.wvu.edu.

West Virginia University, School of Medicine, Graduate Programs at the Health Science Center, Biomedical Sciences Graduate Program, Program in Immunology and Microbial Pathogenesis, Morgantown, WV 26506. Offers MS, PhD. *Faculty:* 28 full-time (4 women). *Students:* 19 full-time (9 women); includes 2 minority (1 African American, 1 Asian American or Pacific Islander), 1 international. Average age 26. In 2005, 5 degrees awarded. *Median time to degree:* Of those who began their doctoral program in fall 1997, 97% received their degree in 8 years or less. *Degree requirements:* For doctorate, thesis/dissertation, comprehensive exam. *Entrance requirements:* For doctorate, GRE General Test, minimum GPA of 3.0. Additional exam requirements/recommendations for international students: Required—TOEFL. *Application deadline:* For fall admission, 3/1 priority date for domestic students, 1/15 priority date for international students. Applications are processed on a rolling basis. Application fee: $0. Electronic applications accepted. *Financial support:* In 2005–06, research assistantships with full tuition reimbursements (averaging $20,000 per year); institutionally sponsored loans, traineeships, and health care benefits also available. *Faculty research:* Regulation of signal transduction in immune responses, immune responses in bacterial and viral diseases, peptide and DNA vaccines for contraception, inflammatory bowel disease, physiology of pathogenic microbes. Total annual research expenditures: $2.5 million. *Unit head:* Dr. John Barnett, Chair, 304-293-2649, Fax: 304-293-7823, E-mail: john.barnett@hsc.wvu.edu. *Application contact:* Dr. Jia Luo, Graduate Coordinator, 304-293-7208, Fax: 304-293-7823, E-mail: jia.luo@hsc.wvu.edu.

See Close-Up on page 961.

Wright State University, School of Graduate Studies, College of Science and Mathematics, Program in Microbiology and Immunology, Dayton, OH 45435. Offers MS. Part-time programs available. *Degree requirements:* For master's, thesis. *Entrance requirements:* Additional exam requirements/recommendations for international students: Required—TOEFL. *Faculty research:* Reproductive immunology, viral pathogenesis, virus-host cell interactions.

Yale University, Graduate School of Arts and Sciences, Department of Immunobiology, New Haven, CT 06520. Offers PhD. *Degree requirements:* For doctorate, thesis/dissertation. *Entrance requirements:* For doctorate, GRE General Test.

Yale University, School of Medicine and Graduate School of Arts and Sciences, Combined Program in Biological and Biomedical Sciences (BBS), Immunology Track, New Haven, CT 06520. Offers PhD, MD/PhD. *Students:* 5 full-time. *Degree requirements:* For doctorate, thesis/dissertation. *Entrance requirements:* For doctorate, GRE General Test. Additional exam requirements/recommendations for international students: Required—TOEFL. *Application deadline:* For fall admission, 12/8 for domestic students, 12/8 for international students. Electronic applications accepted. *Financial support:* Fellowships, research assistantships available. *Unit head:* Dr. Al Bothwell, Director of Graduate Studies, 203-785-3857, Fax: 203-737-1764. *Application contact:* Barbara Giamattei, Registrar, 203-785-3857, E-mail: barbara.giamattei@yale.edu.

Infectious Diseases

Cornell University, Graduate School, Graduate Fields of Comparative Biomedical Sciences, Field of Comparative Biomedical Sciences, Ithaca, NY 14853-0001. Offers cellular and molecular medicine (MS, PhD); developmental and reproductive biology (MS, PhD); infectious diseases (MS, PhD); population medicine and epidemiology (MS); population medicine and epidemiology sciences (PhD); structural and functional biology (MS, PhD). *Faculty:* 135 full-time (38 women). *Students:* 41 full-time (23 women); includes 4 minority (1 African American, 3 Asian Americans or Pacific Islanders), 19 international. 58 applicants, 60% accepted, 33 enrolled. In 2005, 4 degrees awarded. *Degree requirements:* For master's, thesis/dissertation; for doctorate, thesis/dissertation, comprehensive exam. *Entrance requirements:* For master's and doctorate, GRE General Test, 2 letters of recommendation. Additional exam requirements/recommendations for international students: Required—TOEFL (minimum score 550 paper-based; 213 computer-based). *Application deadline:* For fall admission, 12/15 for domestic students. Application fee: $60. Electronic applications accepted. *Financial support:* In 2005–06, 41 students received support, including 13 fellowships with full tuition reimbursements available, 28 research assistantships with full tuition reimbursements available; teaching assistantships with full tuition reimbursements available, institutionally sponsored loans, scholarships/grants, health care benefits, tuition waivers (full and partial), and unspecified assistantships also available. Financial award applicants required to submit FAFSA. *Faculty research:* Receptors and signal transduction, viral and bacterial infectious diseases, tumor metastasis, clinical sciences/nutritional disease, development/neurologic disorders. *Unit head:* Director of Graduate Studies, 607-253-3276, Fax: 607-253-3756. *Application contact:* Graduate Field Assistant, 607-253-3276, Fax: 607-253-3756, E-mail: graduate_edcvm@cornell.edu.

Georgetown University, Graduate School of Arts and Sciences, Programs in Biomedical Sciences, Department of Microbiology and Immunology, Washington, DC 20057. Offers biohazardous threat agents and emerging infectious diseases (MS); microbiology and immunology research (PhD); science policy and advocacy for the healthcare arena in a global setting (MS); teaching microbiology and immunology (MS). *Degree requirements:* For doctorate, thesis/dissertation, comprehensive exam. *Entrance requirements:* For master's, GRE General Test, 3 letters of reference, bachelor's degree in related field; for doctorate, GRE General Test, 3 letters of reference, graduate degree in related field. Additional exam requirements/

recommendations for international students: Required—TOEFL (minimum score 505 paper-based; 213 computer-based). Electronic applications accepted. *Faculty research:* Pathogenesis and basic biology of the fungus Candida albicans, molecular biology of viral hepatitis, molecular and cellular biology of the Hepatitis B and Delta viruses, dengue virus, immunopathological mechanisms in Multiple Sclerosis.

See Close-Up on page 871.

The George Washington University, School of Public Health and Health Services, Department of Epidemiology and Biostatistics, Washington, DC 20052. Offers biostatistics (MPH); epidemiology (MPH); health information systems (MPH); microbiology and emerging infectious diseases (MSPH). *Accreditation:* CEPH. *Degree requirements:* For master's, case study or special project. *Entrance requirements:* For master's, GMAT, GRE General Test, or MCAT. Additional exam requirements/recommendations for international students: Required—TOEFL.

Harvard University, School of Public Health, Department of Immunology and Infectious Diseases, Boston, MA 02115-6096. Offers PhD, SD. Part-time programs available. *Degree requirements:* For doctorate, thesis/dissertation, qualifying exam. *Entrance requirements:* For doctorate, GRE. Additional exam requirements/recommendations for international students: Required—TOEFL (minimum score 560 paper-based; 220 computer-based); Recommended—IELT (minimum score 7). Electronic applications accepted. *Faculty research:* Infectious disease epidemiology and tropical public health, vector biology, ecology and control, virology.

The Johns Hopkins University, Bloomberg School of Public Health, Department of Epidemiology, Baltimore, MD 21205. Offers cancer epidemiology (MHS, Sc M, PhD, Sc D); cardiovascular disease epidemiology (MHS, Sc M, PhD, Sc D); clinical epidemiology (MHS, Sc M, PhD, Sc D); clinical trials (PhD, Sc D); epidemiology (Dr PH, Sc D); epidemiology (general) (MHS, Sc M, PhD); epidemiology of aging (MHS, Sc M, PhD, Sc D); human genetics/genetic epidemiology (MHS, Sc M, PhD, Sc D); infectious disease epidemiology (MHS, Sc M, PhD, Sc D); occupational/environmental epidemiology (MHS, Sc M, PhD, Sc D). *Faculty:* 84 full-time (45 women), 81 part-time/adjunct (33 women). *Students:* 148 full-time (111 women), 52 part-time (37 women); includes 48 minority (14 African Americans, 30 Asian Americans or Pacific

Islanders, 4 Hispanic Americans), 36 international. Average age 31. 293 applicants, 37% accepted, 56 enrolled. In 2005, 27 master's, 22 doctorates awarded. *Median time to degree:* Of those who began their doctoral program in fall 1997, 72% received their degree in 8 years or less. *Degree requirements:* For master's, thesis, 1 year full-time residency, comprehensive exam, registration; for doctorate, thesis/dissertation, 1 year full-time residency, oral and written exams, teaching requirement, comprehensive exam, registration. *Entrance requirements:* For master's, GRE General Test, 3 letters of recommendation, curriculum vitae; for doctorate, GRE General Test or MCAT, minimum 1 year work experience, 3 letters of recommendation, curriculum vitae. Additional exam requirements/recommendations for international students: Required—TOEFL (minimum score 600 paper-based; 250 computer-based). *Application deadline:* For fall admission, 12/1 for domestic students. Applications are processed on a rolling basis. Application fee: $45. Electronic applications accepted. *Expenses:* Tuition: Full-time $30,960. Tuition and fees vary according to degree level and program. *Financial support:* In 2005–06, 212 students received support, including 2 fellowships (averaging $28,859 per year); Federal Work-Study, institutionally sponsored loans, scholarships/grants, traineeships, tuition waivers (partial), and stipends also available. Support available to part-time students. Financial award application deadline: 3/15; financial award applicants required to submit FAFSA. *Faculty research:* Cancer and congenital malformations, nutritional epidemiology, AIDS, tuberculosis, cardiovascular disease, risk assessment. Total annual research expenditures: $47.8 million. *Unit head:* Dr. Jonathan M. Samet, Chairman, 410-955-3286, Fax: 410-955-0863, E-mail: jsamet@jhsph.edu. *Application contact:* Frances S. Burman, Senior Academic Coordinator, 410-955-3926, Fax: 410-955-0863, E-mail: fburman@jhsph.edu.

Loyola University Chicago, Graduate School, Marcella Niehoff School of Nursing, Population-Based Infection Control and Environmental Safety Program, Chicago, IL 60611-2196. Offers MSN. Part-time and evening/weekend programs available. *Entrance requirements:* For master's, Illinois license, 3 letters of recommendation. *Expenses:* Tuition: Full-time $11,610; part-time $645 per credit. Required fees: $55 per semester. *Application contact:* Dr. Vicki A. Keough, Associate Dean, 773-508-3263, Fax: 773-508-3241, E-mail: vkeough@luc.edu.

Purdue University, School of Veterinary Medicine and Graduate School, Graduate Programs in Veterinary Medicine, Department of Veterinary Pathobiology, West Lafayette, IN 47907. Offers biochemistry and molecular biology (MS, PhD); comparative epidemiology (MS, PhD); epidemiology (MS, PhD); immunology (MS, PhD); infectious diseases (MS, PhD); interdisciplinary genetics (PhD); laboratory animal medicine (MS, PhD); microbiology (MS, PhD); molecular virology (MS, PhD); parasitology (MS, PhD); pathobiology (MS, PhD); public health epidemiology (MS, PhD); toxicology (MS, PhD); veterinary anatomic pathology (MS, PhD); veterinary clinical pathology (MS, PhD); virology (MS, PhD). *Faculty:* 32 full-time (7 women). *Students:* 49 full-time (20 women), 3 part-time (1 woman); includes 2 minority (both African Americans), 31 international. Average age 35. In 2005, 3 master's, 8 doctorates awarded. Terminal master's awarded for partial completion of doctoral program. *Degree requirements:* For master's, thesis (for some programs); for doctorate, thesis/dissertation. *Entrance requirements:* For master's and doctorate, GRE General Test. Additional exam requirements/recommendations for international students: Required—TOEFL (minimum score 575 paper-based), TWE (minimum score 4). *Application deadline:* For fall admission, 8/12 for domestic students, 6/15 for international students; for spring admission, 1/12 for domestic students, 10/15 for international students. Application fee: $55. *Financial support:* Fellowships, research assistantships, teaching assistantships available. Financial award application deadline: 3/1; financial award applicants required to submit FAFSA. *Unit head:* Dr. H. Hogenesch, Head, 765-494-7543.

Uniformed Services University of the Health Sciences, School of Medicine, Programs in Biomedical Sciences, Graduate Program in Emerging Infectious Diseases, Bethesda, MD 20814-4799. Offers PhD. *Faculty:* 13 full-time (3 women), 7 part-time/adjunct (1 woman). *Students:* 39 full-time (27 women); includes 8 minority (3 African Americans, 5 Asian Americans or Pacific Islanders). Average age 26. 58 applicants, 21% accepted, 8 enrolled. In 2005, 2 degrees awarded. *Median time to degree:* Of those who began their doctoral program in fall 1997, 100% received their degree in 8 years or less. *Degree requirements:* For doctorate, thesis/dissertation, qualifying exam, comprehensive exam. *Entrance requirements:* For doctorate, GRE General Test. Additional exam requirements/recommendations for international students: Required—TOEFL. *Application deadline:* For fall admission, 1/15 for domestic students. Applications are processed on a rolling basis. Application fee: $0. *Financial support:* In 2005–06, fellowships with full tuition reimbursements (averaging $23,000 per year); scholarships/grants and tuition waivers (full) also available. *Unit head:* Dr. Christopher Broder, Director, 301-295-3401, E-mail: cbroder@usuhs.mil. *Application contact:* Janet M. Anastasi, Graduate Program Coordinator, 301-295-9474, Fax: 301-295-6772, E-mail: janastasi@usuhs.mil.

See Close-Up on page 901.

Université de Montréal, Faculty of Medicine and Faculty of Graduate Studies, Graduate Programs in Medicine, Program in Specialized Studies, Montréal, QC H3C 3J7, Canada. Offers anesthesia (DESS); diagnostic radiology (DESS); family medicine (DESS); medical biochemistry (DESS); medical genetics (DESS); medicine (DESS); microbiology and infectious diseases (DESS); nuclear medicine (DESS); obstetrics and gynecology (DESS); ophthalmology (DESS); pediatrics (DESS); psychiatry (DESS); radiology-oncology (DESS); surgery (DESS). *Faculty:* 159 full-time (37 women), 345 part-time/adjunct (102 women). *Entrance requirements:* For degree, proficiency in French. *Application deadline:* For fall and spring admission, 2/1. For winter admission, 11/1 for domestic students. Application fee: $30. Electronic applications accepted. *Unit head:* Renée Roy, Vice Dean, 514-343-7798.

Université Laval, Faculty of Medicine, Post-Professional Programs in Medical Studies, Québec, QC G1K 7P4, Canada. Offers anatomy–pathology (DESS); anesthesia–resuscitation (DESS); cardiology (DESS); care of older people (Diploma); clinical research (DESS); community health (DESS); dermatology (DESS); diagnostic radiology (DESS); emergency medicine (Diploma); family medicine (DESS); general surgery (DESS); geriatrics (DESS); hematology (DESS); internal medicine (DESS); maternal and fetal medicine (Diploma); medical biochemistry (DESS); medical microbiology and infectious diseases (DESS); medical oncology (DESS); nephrology (DESS); neurology (DESS); neurosurgery (DESS); obstetrics and gynecology (DESS); ophthalmology (DESS); orthopedic surgery (DESS); oto-rhino-laryngology (DESS); palliative medicine (Diploma); pediatrics (DESS); plastic surgery (DESS); psychiatry (DESS); pulmonary medicine (DESS); radiology–oncology (DESS); thoracic surgery (DESS); urology (DESS). *Degree requirements:* For other advanced degree, comprehensive exam. *Entrance requirements:* For degree, knowledge of French. Electronic applications accepted.

University of Calgary, Faculty of Medicine and Faculty of Graduate Studies, Department of Microbiology and Infectious Diseases, Calgary, AB T2N 1N4, Canada. Offers M Sc, PhD. *Faculty:* 14 full-time (3 women). *Students:* 26 full-time (13 women). Average age 29. 25 applicants, 36% accepted. In 2005, 2 master's, 5 doctorates awarded. *Degree requirements:* For master's, thesis, oral thesis exam; for doctorate, thesis/dissertation, candidacy exam, oral thesis exam. *Entrance requirements:* For master's and doctorate, minimum GPA of 3.2. Additional exam requirements/recommendations for international students: Required—TOEFL (minimum score 580 paper-based; 237 computer-based). *Application deadline:* For fall admission, 5/15 for domestic students, 4/15 for international students. For winter admission, 9/15 for domestic students; for spring admission, 4/15 for domestic students. Applications are processed on a rolling basis. Application fee: $100 ($130 for international students). Electronic applications accepted. *Financial support:* In 2005–06, 7 students received support, including 7 fellowships (averaging $13,370 per year), 24 research assistantships (averaging $51,600 per year) Financial award application deadline: 2/1. *Faculty research:* Bacteriology, virology, parasitology, immunology. Total annual research expenditures: $5.6 million. *Unit head:* Dr. Donald E. Woods, Graduate Coordinator, E-mail: midgrad@ucalgary.ca. *Application contact:* Julie Boyd, Graduate Program Administrator, 403-220-2558, Fax: 403-210-8109, E-mail: boydj@ucalgary.ca.

University of California, Berkeley, Graduate Division, School of Public Health, Division of Public Health Biology and Epidemiology, Berkeley, CA 94720-1500. Offers epidemiology (MPH,

MS, PhD), including epidemiology (MS, PhD), epidemiology and biostatistics (MPH); infectious diseases (MPH, PhD). *Accreditation:* CEPH (one or more programs are accredited). *Degree requirements:* For master's, comprehensive exam, thesis/dissertation, oral and written exam. *Entrance requirements:* For master's, GRE General Test, minimum GPA of 3.0; MD, DDS, DVM, or PhD in biomedical science (MPH); for doctorate, GRE General Test, minimum GPA of 3.0.

University of California, Berkeley, Graduate Division, School of Public Health, Group in Infectious Diseases and Immunity, Berkeley, CA 94720-1500. Offers PhD. *Entrance requirements:* For doctorate, GRE General Test, minimum GPA of 3.0.

University of Georgia, College of Veterinary Medicine and Graduate School, Graduate Programs in Veterinary Medicine, Athens, GA 30602. Offers infectious diseases (MS, PhD), including infectious diseases, medical microbiology; pathology (MS, PhD); physiology and pharmacology (MS, PhD), including pharmacology, physiology; population health (MAM); toxicology (MS, PhD); veterinary anatomy and radiology (MS), including veterinary anatomy. *Faculty:* 65 full-time (22 women). *Students:* 64 full-time, 15 part-time; includes 7 minority (6 African Americans, 1 Hispanic American), 29 international. Average age 27. 88 applicants, 39% accepted, 24 enrolled. In 2005, 6 master's, 13 doctorates awarded. *Degree requirements:* For master's and doctorate, thesis/dissertation, comprehensive exam, registration. *Entrance requirements:* For master's and doctorate, GRE General Test, 3 letters of recommendation. Additional exam requirements/recommendations for international students: Required—TOEFL (minimum score 550 paper-based; 213 computer-based). *Application deadline:* For fall admission, 7/1 for domestic students, 4/15 for international students; for spring admission, 11/15 for domestic students, 10/15 for international students. Applications are processed on a rolling basis. Application fee: $50. *Expenses: Contact institution. Financial support:* In 2005–06, 102 research assistantships with full tuition reimbursements (averaging $23,000 per year) were awarded; unspecified assistantships also available. Financial award application deadline: 3/1; financial award applicants required to submit FAFSA. *Faculty research:* Vascular biomedicine, environmental toxicology, food safety, vaccines and emergency diseases. Total annual research expenditures: $11.3 million. *Unit head:* Dr. Harry W. Dickerson, Associate Dean for Research and Graduate Affairs, 706-542-5734, Fax: 706-542-8254, E-mail: hwd@vet.uga.edu.

University of Georgia, College of Veterinary Medicine and Graduate School, Graduate Programs in Veterinary Medicine, Department of Infectious Diseases, Program in Infectious Diseases, Athens, GA 30602. Offers MS, PhD. *Faculty:* 9 full-time (3 women). *Students:* 21 full-time, 2 part-time; includes 2 minority (both African Americans), 6 international. 34 applicants, 41% accepted, 9 enrolled. In 2005, 2 degrees awarded. *Degree requirements:* For master's, thesis; for doctorate, one foreign language, thesis/dissertation. *Entrance requirements:* For master's and doctorate, GRE General Test. *Application deadline:* For fall admission, 7/1 for domestic students; for spring admission, 11/15 for domestic students. Application fee: $50. Electronic applications accepted. *Financial support:* Fellowships, research assistantships, teaching assistantships, unspecified assistantships available. *Application contact:* Dr. Susan Little, Graduate Coordinator, 706-542-5792, Fax: 706-542-5771, E-mail: slittle@vet.uga.edu.

University of Guelph, Ontario Veterinary College and Graduate Program Services, Graduate Programs in Veterinary Sciences, Department of Pathobiology, Guelph, ON N1G 2W1, Canada. Offers anatomic pathology (DV Sc, Diploma); clinical pathology (Diploma); comparative pathology (M Sc, PhD); immunology (M Sc, PhD); laboratory animal science (DV Sc); pathology (M Sc, PhD, Diploma); veterinary infectious diseases (M Sc, PhD); zoo animal/wildlife medicine (DV Sc). *Faculty:* 26. *Students:* 66 (22 women). 47 applicants, 32% accepted. In 2005, 6 master's, 7 doctorates, 1 other advanced degree awarded. *Degree requirements:* For master's and doctorate, thesis/dissertation. *Entrance requirements:* For master's, DVM with B average or an honours degree in biological sciences; for doctorate, DVM or MSC degree, minimum B+ average. Additional exam requirements/recommendations for international students: Required—TOEFL (minimum score 550 paper-based; 213 computer-based). *Application deadline:* For fall admission, 2/1 for domestic students. Applications are processed on a rolling basis. Application fee: $75. *Financial support:* In 2005–06, 40 students received support, including 20 fellowships, research assistantships (averaging $13,500 per year), 13 teaching assistantships (averaging $1,600 per year); career-related internships or fieldwork also available. *Faculty research:* Pathogenesis; diseases of animals, wildlife, fish, and laboratory animals; parasitology; immunology; veterinary infectious diseases; laboratory animal science. Total annual research expenditures: $2.5 million. *Unit head:* Dr. John Prescott, Chair, Fax: 519-824-5930, E-mail: jprescott@ovc.uoguelph.ca. *Application contact:* Dr. J. MacInnes, Graduate Coordinator, 519-824-4120 Ext. 54731, Fax: 519-767-0809, E-mail: macinnes@uoguelph.ca.

University of Minnesota, Twin Cities Campus, School of Public Health, Division of Environmental Health Sciences, Area in Environmental Infectious Diseases, Minneapolis, MN 55455-0213. Offers MPH, MS, PhD. *Degree requirements:* For doctorate, thesis/dissertation. *Entrance requirements:* For master's and doctorate, GRE General Test. *Application deadline:* For fall admission, 3/1 for domestic students. Applications are processed on a rolling basis. Application fee: $55 ($75 for international students). *Expenses:* Tuition, state resident: full-time $8,748; part-time $729 per credit. Tuition, nonresident: full-time $15,848; part-time $1,321 per credit. Full-time tuition and fees vary according to class time, course load, program and reciprocity agreements. *Financial support:* Application deadline: 3/1. *Application contact:* Kathy Soupir, Major Coordinator, 612-625-0622, Fax: 612-626-4837, E-mail: soupi001@umn.edu.

University of Pittsburgh, Graduate School of Public Health, Department of Infectious Diseases and Microbiology, Pittsburgh, PA 15260. Offers MPH, MS, Dr PH, PhD, MD/PhD. *Faculty:* 18 full-time (6 women), 3 part-time/adjunct (0 women). *Students:* 36 full-time (24 women), 11 part-time (8 women); includes 6 minority (3 African Americans, 1 Asian American or Pacific Islander, 2 Hispanic Americans), 13 international. Average age 28. 57 applicants, 49% accepted, 12 enrolled. In 2005, 6 master's, 8 doctorates awarded. Terminal master's awarded for partial completion of doctoral program. *Median time to degree:* Of those who began their doctoral program in fall 1997, 66% received their degree in 8 years or less. *Degree requirements:* For master's, one foreign language, thesis/dissertation, comprehensive exam (for some programs), registration; for doctorate, one foreign language, thesis/dissertation, comprehensive exam, registration. *Entrance requirements:* For master's and doctorate, GRE General Test. Additional exam requirements/recommendations for international students: Required—TOEFL (minimum score 550 paper-based; 213 computer-based). *Application deadline:* For fall admission, 1/15 for domestic students, 5/1 for international students. Applications are processed on a rolling basis. Application fee: $50 ($60 for international students). Electronic applications accepted. *Expenses:* Tuition, state resident: full-time $13,194; part-time $537 per credit. Tuition, nonresident: full-time $25,012; part-time $1,026 per credit. Required fees: $700; $164 per term. Tuition and fees vary according to campus/location and program. *Financial support:* In 2005–06, 32 students received support, including 31 research assistantships with tuition reimbursements available (averaging $20,333 per year); career-related internships or fieldwork, traineeships, health care benefits, tuition waivers (partial), and unspecified assistantships also available. Financial award applicants required to submit FAFSA. *Faculty research:* HIV, Epstein-Barr virus, virology, immunology, malaria. Total annual research expenditures: $13.5 million. *Unit head:* Dr. Charles Rinaldo, Chairman, 412-624-3928, Fax: 412-624-4953, E-mail: rinaldo@pitt.edu. *Application contact:* Dr. Phalguni Gupta, Associate Professor, 412-624-7998, Fax: 412-624-4953, E-mail: pgupta1@pitt.edu.

See Close-Up on page 943.

The University of Texas Medical Branch, Graduate School of Biomedical Sciences, Center for Biodefense and Emerging Infectious Diseases, Galveston, TX 77555. Offers biodefense training (PhD). *Entrance requirements:* For doctorate, GRE, an overall and advanced grade point average of 3.0; personal interviews encouraged. *Expenses:* Tuition, state resident: full-time $8,350; part-time $90 per credit hour. Tuition, nonresident: full-time $21,450; part-time $366 per credit hour. Required fees: $1,027; $11 per credit hour. $60 per term.

Infectious Diseases

The University of Texas Medical Branch, Graduate School of Biomedical Sciences, Program in Emerging and Tropical Infectious Diseases, Galveston, TX 77555. Offers PhD, MD/PhD. *Degree requirements:* For doctorate, thesis/dissertation. *Entrance requirements:* For doctorate, GRE General Test. *Application deadline:* Applications are processed on a rolling basis. Application fee: $25 ($50 for international students). *Expenses:* Tuition, state resident: full-time $8,350; part-time $90 per credit hour. Tuition, nonresident: full-time $21,450; part-time $366 per credit hour. Required fees: $1,027; $11 per credit hour. $60 per term. *Financial support:* In 2005–06, fellowships (averaging $23,000 per year), research assistantships with full tuition reimbursements (averaging $23,000 per year) were awarded; traineeships and unspecified assistantships also available. *Faculty research:* Emerging diseases, tropical diseases, parasitology, vitology and bacteriology.

See Close-Up on page 949.

Yale University, School of Medicine and Graduate School of Arts and Sciences, Combined Program in Biological and Biomedical Sciences (BBS), Microbiology Track, New Haven, CT 06520. Offers PhD, MD/PhD. *Students:* 5 full-time. *Degree requirements:* For doctorate, thesis/dissertation. *Entrance requirements:* For doctorate, GRE General Test, GRE Subject Test. Additional exam requirements/recommendations for international students: Required—TOEFL. *Application deadline:* For fall admission, 12/8 for domestic students, 12/8 for international students. Electronic applications accepted. *Financial support:* Fellowships, research assistantships available. *Unit head:* Dr. Joann Sweasy, Director of Graduate Studies, 203-737-2626, E-mail: darlene.a.smith@yale.edu. *Application contact:* Darlene Smith, Registrar, 203-737-2404, E-mail: darlene.a.smith@yale.edu.

Medical Microbiology

Creighton University, School of Medicine and Graduate School, Graduate Programs in Medicine, Department of Medical Microbiology and Immunology, Omaha, NE 68178-0001. Offers MS, PhD. Terminal master's awarded for partial completion of doctoral program. *Degree requirements:* For master's, thesis, comprehensive exam; for doctorate, thesis/dissertation, preliminary exams. *Entrance requirements:* For master's and doctorate, GRE General Test. *Faculty research:* Infectious diseases, molecular biology, genetics, antimicrobial agents and chemotherapy, virology.

See Close-Up on page 863.

Idaho State University, Office of Graduate Studies, College of Arts and Sciences, Department of Biological Sciences, Pocatello, ID 83209. Offers biology (MNS, MS, DA, PhD); clinical laboratory science (MS); microbiology (MS). *Accreditation:* NAACLS. *Degree requirements:* For master's, one foreign language, thesis, comprehensive exam, registration (for some programs); for doctorate, 2 foreign languages, thesis/dissertation, comprehensive exam, registration. *Entrance requirements:* For master's, GRE General Test, minimum GPA of 3.0 in all upper division classes; for doctorate, GRE General Test, GRE Subject Test, minimum GPA of 3.0 in all upper division classes. Additional exam requirements/recommendations for international students: Required—TOEFL (minimum score 550 paper-based; 213 computer-based). *Faculty research:* Ecology and evolutionary biology, plant and animal physiology, plant and animal developmental biology, immunology, molecular biology.

Rosalind Franklin University of Medicine and Science, School of Graduate and Postdoctoral Studies, Department of Microbiology and Immunology, Program in Medical Microbiology, North Chicago, IL 60064-3095. Offers MS, PhD. *Degree requirements:* For master's and doctorate, thesis/dissertation. *Entrance requirements:* For master's and doctorate, GRE General Test. Additional exam requirements/recommendations for international students: Required—TOEFL, TWE.

Rutgers, The State University of New Jersey, New Brunswick/Piscataway, Graduate School, Program in Microbiology and Molecular Genetics, New Brunswick, NJ 08901-1281. Offers applied microbiology (MS, PhD); clinical microbiology (MS, PhD); computational molecular biology (PhD); immunology (MS, PhD); microbial biochemistry (MS, PhD); molecular genetics (MS, PhD); virology (MS, PhD). Part-time programs available. *Faculty:* 132 full-time. *Students:* 57 full-time (31 women), 18 part-time (8 women); includes 16 minority (9 Asian Americans or Pacific Islanders, 7 Hispanic Americans), 20 international. Average age 29. 104 applicants, 32% accepted, 19 enrolled. In 2005, 9 master's, 10 doctorates awarded. Terminal master's awarded for partial completion of doctoral program. *Median time to degree:* Of those who began their doctoral program in fall 1997, 100% received their degree in 8 years or less. *Degree requirements:* For master's, thesis or alternative, comprehensive exam, registration; for doctorate, thesis/dissertation, written qualifying exam, comprehensive exam, registration. *Entrance requirements:* For master's, GRE General Test, minimum GPA of 3.0; for doctorate, GRE General Test, GRE Subject Test (recommended), minimum GPA of 3.0. Additional exam requirements/recommendations for international students: Required—TOEFL. *Application deadline:* For fall admission, 1/5 priority date for domestic students, 11/1 priority date for international students. Applications are processed on a rolling basis. Application fee: $50. Electronic applications accepted. *Expenses:* Tuition, state resident: full-time $10,440; part-time $435 per credit. Tuition, nonresident: full-time $15,520; part-time $647 per credit. Required fees: $129 per credit. Tuition and fees vary according to program. *Financial support:* In 2005–06, 48 students received support, including 14 fellowships with full tuition reimbursements available (averaging $24,000 per year), 25 research assistantships with full tuition reimbursements available (averaging $24,000 per year), 9 teaching assistantships with full tuition reimbursements available (averaging $16,988 per year); Federal Work-Study, institutionally sponsored loans, scholarships/grants, and unspecified assistantships also available. Financial award application deadline: 1/5; financial award applicants required to submit FAFSA. *Faculty research:* Molecular genetics and microbial physiology; virology and pathogenic microbiology; applied, environmental and industrial microbiology; computers in molecular biology. *Unit head:* Dr. Andrew K. Vershon, Director, 732-445-2905, Fax: 732-445-6370, E-mail: vershon@waksman.rutgers.edu. *Application contact:* Diane Murano, Administrative Assistant, 732-445-5086, Fax: 732-445-6370, E-mail: murano@biology.rutgers.edu.

Texas A&M University System Health Science Center, Graduate School of Biomedical Sciences, Department of Medical Microbiology and Immunology, College Station, TX 77840. Offers immunology (PhD); microbiology (PhD); molecular biology (PhD); virology (PhD). *Degree requirements:* For doctorate, thesis/dissertation. *Entrance requirements:* For doctorate, GRE General Test, minimum GPA of 3.0. *Faculty research:* Molecular pathogenesis, microbial therapeutics.

Texas Tech University Health Sciences Center, Graduate School of Biomedical Sciences, Department of Microbiology and Immunology, Lubbock, TX 79430. Offers medical microbiology (MS, PhD). *Faculty:* 10 full-time (2 women), 13 part-time/adjunct (1 woman). *Students:* 16 full-time (6 women); includes 6 minority (5 Asian Americans or Pacific Islanders, 1 Hispanic American). Average age 25. 8 applicants, 25% accepted, 2 enrolled. Terminal master's awarded for partial completion of doctoral program. *Degree requirements:* For master's and doctorate, thesis/dissertation. *Entrance requirements:* For master's and doctorate, GRE General Test, minimum GPA of 3.0. Additional exam requirements/recommendations for international students: Required—TOEFL (minimum score 550 paper-based; 213 computer-based). *Application deadline:* For fall admission, 5/15 priority date for domestic students, 4/15 priority date for international students; for spring admission, 11/15 for domestic students, 10/15 for international students. Applications are processed on a rolling basis. Application fee: $45. Electronic applications accepted. *Financial support:* In 2005–06, 15 research assistantships with tuition reimbursements (averaging $20,500 per year) were awarded; fellowships, scholarships/grants, health care benefits, and unspecified assistantships also available. *Faculty research:* Genetics, pathogenic bacteriology, molecular biology, virology, medical mycology. *Unit head:* Dr. Ronald Curtis Kennedy, Chair, 806-743-2545, Fax: 806-743-2334, E-mail: ronald.kennedy@ttuhsc.edu. *Application contact:* Dr. Abdul N. Hamood, Graduate Adviser, 806-743-1707, Fax: 806-743-2334, E-mail: abdul.hamood@ttuhsc.edu.

Université du Québec, Institut National de la Recherche Scientifique, Graduate Programs, Research Center—INRS—Institut Armand-Frappier—Human Health, Québec, QC G1K

9A9, Canada. Offers applied microbiology (M Sc); biology (PhD); experimental health sciences (M Sc); virology and immunology (M Sc, PhD). Programs given in French. Part-time programs available. *Faculty:* 46. *Students:* 170 full-time (100 women), 25 international. Average age 28. In 2005, 24 master's, 5 doctorates awarded. *Degree requirements:* For doctorate, thesis/dissertation. *Entrance requirements:* For master's and doctorate, appropriate bachelor's degree, proficiency in French. *Application deadline:* For fall admission, 3/30 for domestic students. For winter admission, 11/1 for domestic students. Application fee: $30 Canadian dollars. *Financial support:* Fellowships, research assistantships, teaching assistantships available. *Faculty research:* Immunity, infection and cancer; toxicology and environmental biotechnology; molecular pharmacochemistry. *Unit head:* Pierre Talbot, Director, 450-681-5010 Ext. 4406, E-mail: pierre.talbot@iaf.inrs.ca. *Application contact:* Michel Barbeau, Registrar, 418-654-2518, Fax: 418-654-3858, E-mail: michel.barbeau@adm.inrs.ca.

University of Alberta, Faculty of Medicine and Dentistry and Faculty of Graduate Studies and Research, Graduate Programs in Medicine, Department of Medical Microbiology and Immunology, Edmonton, AB T6G 2E1, Canada. Offers M Sc, PhD. Terminal master's awarded for partial completion of doctoral program. *Degree requirements:* For master's and doctorate, thesis/dissertation. *Entrance requirements:* For master's and doctorate, GRE. Additional exam requirements/recommendations for international students: Required—TOEFL (paper score 600; computer score 250) or Michigan English Language Assessment Battery. Tuition and fees charges are reported in Canadian dollars. *Expenses:* Tuition, state resident: part-time $562 Canadian dollars per term. Tuition, nonresident: full-time $3,375 Canadian dollars. Required fees: $573 Canadian dollars; $84 Canadian dollars per term. *Faculty research:* Cellular and reproductive immunology, microbial pathogenesis, mechanisms of antibiotic resistance, molecular biology of mammalian viruses, antiviral chemotherapy.

University of Georgia, College of Veterinary Medicine and Graduate School, Graduate Programs in Veterinary Medicine, Athens, GA 30602. Offers infectious diseases (MS, PhD), including infectious diseases, medical microbiology; pathology (MS, PhD); physiology and pharmacology (MS, PhD), including pharmacology, physiology; population health (MAM); toxicology (MS, PhD); veterinary anatomy and radiology (MS), including veterinary anatomy. *Faculty:* 65 full-time (22 women). *Students:* 64 full-time, 15 part-time; includes 7 minority (6 African Americans, 1 Hispanic American), 29 international. Average age 27. 88 applicants, 39% accepted, 24 enrolled. In 2005, 6 master's, 13 doctorates awarded. *Degree requirements:* For master's and doctorate, thesis/dissertation, comprehensive exam, registration. *Entrance requirements:* For master's and doctorate, GRE General Test, 3 letters of recommendation. Additional exam requirements/recommendations for international students: Required—TOEFL (minimum score 550 paper-based; 213 computer-based). *Application deadline:* For fall admission, 7/1 for domestic students, 4/15 for international students; for spring admission, 11/15 for domestic students, 10/15 for international students. Applications are processed on a rolling basis. Application fee: $50. *Expenses:* Contact institution. *Financial support:* In 2005–06, 102 research assistantships with full tuition reimbursements (averaging $23,000 per year) were awarded; unspecified assistantships also available. Financial award application deadline: 3/1; financial award applicants required to submit FAFSA. *Faculty research:* Vascular biomedicine, environmental toxicology, food safety, vaccines and emergency diseases. Total annual research expenditures: $11.3 million. *Unit head:* Dr. Harry W. Dickerson, Associate Dean for Research and Graduate Affairs, 706-542-5734, Fax: 706-542-8254, E-mail: hwd@vet.uga.edu.

University of Georgia, College of Veterinary Medicine and Graduate School, Graduate Programs in Veterinary Medicine, Department of Infectious Diseases, Program in Medical Microbiology, Athens, GA 30602. Offers MS, PhD. *Students:* 5 full-time, 4 part-time; includes 2 minority (1 African American, 1 Hispanic American), 2 international. In 2005, 3 master's, 2 doctorates awarded. *Degree requirements:* For master's, thesis; for doctorate, one foreign language, thesis/dissertation. *Entrance requirements:* For master's and doctorate, GRE General Test. *Application deadline:* For fall admission, 7/1 for domestic students; for spring admission, 11/15 for domestic students. Application fee: $50. Electronic applications accepted. *Financial support:* Fellowships, research assistantships, teaching assistantships, unspecified assistantships available. *Application contact:* Dr. Susan Little, Graduate Coordinator, 706-542-5792, Fax: 706-542-5771, E-mail: slittle@vet.uga.edu.

University of Hawaii at Manoa, John A. Burns School of Medicine and Graduate Division, Graduate Programs in Biomedical Sciences, Department of Tropical Medicine, Medical Microbiology and Pharmacology, Honolulu, HI 96822. Offers tropical medicine (MS, PhD). Part-time programs available. *Faculty:* 25 full-time (10 women), 2 part-time/adjunct (0 women). *Students:* 12 full-time (3 women), 2 part-time (1 woman); includes 4 minority (all Asian Americans or Pacific Islanders), 5 international. Average age 29. 34 applicants, 32% accepted, 8 enrolled. Terminal master's awarded for partial completion of doctoral program. *Degree requirements:* For master's and doctorate, thesis/dissertation. *Entrance requirements:* For master's and doctorate, GRE. *Application deadline:* For fall admission, 3/1 for domestic students, 3/1 for international students; for spring admission, 9/1 for domestic students, 9/1 for international students. Application fee: $50. *Expenses:* Tuition, state resident: full-time $8,400; part-time $200 per credit hour. Tuition, nonresident: full-time $11,088; part-time $462 per credit hour. Tuition and fees vary according to program. *Financial support:* Fellowships, tuition waivers (full) available. *Faculty research:* Immunological studies of dengue, malaria, Kawasaki's disease, lupus erythematosus, and rheumatoid disease. *Unit head:* Dr. Sandra Chang, Graduate Chair, 808-732-1477, Fax: 808-732-1483, E-mail: sandrac@hawaii.edu.

University of Manitoba, Faculty of Medicine and Faculty of Graduate Studies, Graduate Programs in Medicine, Department of Medical Microbiology, Winnipeg, MB R3T 2N2, Canada. Offers M Sc, PhD. Part-time programs available. Terminal master's awarded for partial completion of doctoral program. *Degree requirements:* For master's, thesis; for doctorate, one foreign language, thesis/dissertation. *Entrance requirements:* For master's and doctorate, minimum GPA of 3.0. Electronic applications accepted. *Faculty research:* HIV, bacterial adhesion, sexually transmitted diseases, virus structure/function and assembly.

Announcement: The Department of Medical Microbiology offers studies leading to MSc and PhD degrees with research/academic experience suitable for a career in basic microbiology or infectious diseases. Scientific interests and research projects range from understanding gene regulation and molecular basis of cellular functions to development of vaccines and diagnostics for human health and veterinary diseases.

See Close-Up on page 927.

University of Minnesota, Duluth, Medical School, Microbiology, Immunology and Molecular Pathobiology Section, Duluth, MN 55812-2496. Offers MS, PhD. *Faculty:* 5 full-time (2 women), 3 part-time/adjunct (1 woman). *Students:* 18 applicants, 0% accepted.Terminal master's awarded for partial completion of doctoral program. *Degree requirements:* For master's, thesis, final oral exam; for doctorate, thesis/dissertation, final exam, oral and written preliminary exams. *Entrance requirements:* For master's and doctorate, GRE General Test. Additional exam requirements/recommendations for international students: Required—TOEFL. *Application deadline:* For fall admission, 5/1 for domestic students. Applications are processed on a rolling basis. Application fee: $30. *Financial support:* Research assistantships available. *Faculty research:* Immunomodulation, molecular diagnosis of rabies, cytokines, cancer immunology, cytomegalovirus infection. Total annual research expenditures: $207,223. *Unit head:* Dr. Arthur Johnson, Head, 218-726-7561, Fax: 218-726-6235, E-mail: ajohnso1@d.umn.edu.

University of South Florida, College of Medicine and College of Graduate Studies, Graduate Programs in Medical Sciences, Tampa, FL 33620-9951. Offers anatomy (PhD); biochemistry and molecular biology (MS, PhD), including biochemistry and molecular biology (PhD); bioinformatics and computational biology (MS); medical microbiology and immunology (PhD); pathology (PhD); pharmacology and therapeutics (PhD), including medical sciences; physiology and biophysics (PhD). *Students:* 108 full-time (53 women), 34 part-time (26 women); includes 31 minority (9 African Americans, 9 Asian Americans or Pacific Islanders, 13 Hispanic Americans), 30 international. 117 applicants, 99% accepted, 116 enrolled. In 2005, 9 master's, 4 doctorates awarded. *Degree requirements:* For doctorate, thesis/dissertation. *Entrance requirements:* For doctorate, GRE General Test, minimum GPA of 3.0. Application fee: $30. *Expenses: Contact institution. Financial support:* Institutionally sponsored loans and scholarships/grants available. Financial award application deadline: 4/1; financial award applicants required to submit FAFSA. *Unit head:* Dr. Joseph J. Krzanowski, Associate Dean for Research and Graduate Affairs, 813-974-4181, Fax: 813-974-4317, E-mail: jkrzanow@com1.med.usf.edu.

University of Wisconsin–La Crosse, Office of University Graduate Studies, College of Science and Health, Department of Biology, Program in Clinical Microbiology, La Crosse, WI 54601-3742. Offers MS. *Faculty:* 8 full-time (3 women), 1 (woman) part-time/adjunct. *Students:* 8 full-time (5 women), 8 part-time (7 women); includes 1 minority (Asian American or Pacific Islander), 2 international. Average age 25. 9 applicants, 89% accepted, 5 enrolled. In 2005, 2 degrees awarded. *Degree requirements:* For master's, thesis, registration. *Entrance requirements:* For master's, GRE General Test, minimum GPA of 2.85. Additional exam requirements/recommendations for international students: Required—TOEFL (minimum score 550 paper-based; 213 computer-based). Application fee: $45. *Expenses:* Tuition, state resident: part-time $354 per credit. Tuition, nonresident: part-time $943 per credit. Tuition and fees vary according to course load, program and reciprocity agreements. *Faculty research:* Bacterial pathogenesis, drug discovery, virus replication, influenza vaccines, molecular epidemiology. *Unit head:* Dr. Mike Hoffman, Head, 608-785-6984. *Application contact:* Kathryn Kiefer, Associate Director of Admissions, 608-785-8939, E-mail: admissions@uwlax.edu.

University of Wisconsin–Madison, Medical School and Graduate School, Graduate Programs in Medicine and College of Agricultural and Life Sciences, Microbiology Doctoral Training Program, Madison, WI 53706-1380. Offers PhD. *Faculty:* 90 full-time (25 women). *Students:* 112 full-time (67 women); includes 23 minority (4 African Americans, 1 American Indian/Alaska Native, 5 Asian Americans or Pacific Islanders, 13 Hispanic Americans), 7 international. Average age 24. 213 applicants, 21% accepted, 21 enrolled. In 2005, 12 degrees awarded. *Degree requirements:* For doctorate, thesis/dissertation, preliminary exam, 2 semesters of teaching. *Entrance requirements:* For doctorate, GRE. Additional exam requirements/recommendations for international students: Required—TOEFL (minimum score 580 paper-based; 237 computer-based). *Application deadline:* For fall admission, 12/1 for domestic students, 12/1 for international students. Application fee: $45. Electronic applications accepted. *Financial support:* In 2005–06, 112 students received support, including 38 fellowships with tuition reimbursements available (averaging $22,500 per year), 74 research assistantships with tuition reimbursements available (averaging $22,500 per year); career-related internships or fieldwork, scholarships/grants, traineeships, health care benefits, and tuition waivers (full) also available. *Faculty research:* Microbial pathogenesis, gene regulation, immunology, virology, cell biology. Total annual research expenditures: $15.1 million. *Unit head:* Dr. Heidi Goodrich-Blair, Director, 608-265-4537, Fax: 608-262-9865, E-mail: hgblair@bact.wisc.edu. *Application contact:* Kathryn A. Holtgraver, Program Coordinator, 608-265-0689, Fax: 608-262-9865, E-mail: kathyh@bact.wisc.edu.

See Close-Up on page 957.

Microbiology

Albany Medical College, Graduate Programs in the Biological Sciences, Center for Immunology and Microbial Disease, Albany, NY 12208-3479. Offers MS, PhD. Part-time programs available. *Faculty:* 17 full-time (2 women), 11 part-time/adjunct (6 women). *Students:* 16 full-time (10 women). Average age 25. 26 applicants, 46% accepted, 5 enrolled. In 2005, 7 master's, 2 doctorates awarded. Terminal master's awarded for partial completion of doctoral program. *Degree requirements:* For master's, thesis; for doctorate, thesis/dissertation, oral qualifying exam, written preliminary exam, 1 published paper-peer review, comprehensive exam. *Entrance requirements:* For master's and doctorate, GRE General Test. Additional exam requirements/recommendations for international students: Required—TOEFL. *Application deadline:* For fall admission, 3/15 for domestic students. Applications are processed on a rolling basis. Application fee: $0 ($60 for international students). *Financial support:* In 2005–06, 10 research assistantships (averaging $23,000 per year) were awarded; Federal Work-Study, scholarships/grants, and tuition waivers (full) also available. Financial award applicants required to submit FAFSA. *Faculty research:* Microbial and viral pathogenesis, cancer development and cell transformation, biochemical and genetic mechanisms responsible for human disease. *Unit head:* Dr. Thomas D. Friedrich, Graduate Director, 518-262-6750, Fax: 518-262-6161, E-mail: dgs_cimd@mail.amc.edu.

Albert Einstein College of Medicine, Sue Golding Graduate Division of Medical Sciences, Department of Microbiology and Immunology, Bronx, NY 10461. Offers PhD, MD/PhD. *Degree requirements:* For doctorate, thesis/dissertation. *Entrance requirements:* For doctorate, GRE General Test. Additional exam requirements/recommendations for international students: Required—TOEFL. *Faculty research:* Nature of histocompatibility antigens, lymphoid cell receptors, regulation of immune responses and mechanisms of resistance to infection.

American University of Beirut, Graduate Programs, Faculty of Medicine, Beirut, Lebanon. Offers biochemistry (MS); human morphology (MS); microbiology and immunology (MS); neuroscience (MS); pharmacology and therapeutics (MS); physiology (MS). *Degree requirements:* For master's, one foreign language, thesis (for some programs), comprehensive exam, registration. *Entrance requirements:* For master's, GRE, letter of recommendation.

Arizona State University, Division of Graduate Studies, College of Liberal Arts and Sciences, Division of Natural Sciences and Mathematics, School of Life Sciences, Tempe, AZ 85287. Offers biology (MNS); cell and developmental biology (MS, PhD); microbiology (MNS, MS, PhD). *Accreditation:* NAACLS. *Degree requirements:* For master's, thesis optional for MNS, required for MS; for doctorate, one foreign language, thesis/dissertation. *Entrance requirements:* For master's and doctorate, GRE.

Auburn University, Graduate School, College of Sciences and Mathematics, Department of Biological Sciences, Auburn University, AL 36849. Offers botany (MS, PhD); microbiology (MS, PhD); zoology (MS, PhD). *Faculty:* 28 full-time (5 women). *Students:* 44 full-time (23 women), 49 part-time (27 women); includes 8 minority (5 African Americans, 3 Hispanic Americans), 17 international. 65 applicants, 51% accepted, 22 enrolled. In 2005, 11 master's, 1 doctorate awarded. *Entrance requirements:* For master's and doctorate, GRE General Test. Additional exam requirements/recommendations for international students: Required—TOEFL. *Application deadline:* For fall admission, 7/7 for domestic students; for spring admission, 11/24 for domestic students. Electronic applications accepted. *Financial support:* Research assistantships, teaching assistantships available. *Unit head:* Dr. James M Barbaree, Chair, 334-844-1647, Fax: 334-844-1645. *Application contact:* Dr. Stephen L. McFarland, Acting Dean of the Graduate School, 334-844-4700.

Baylor College of Medicine, Graduate School of Biomedical Sciences, Department of Molecular Virology and Microbiology, Houston, TX 77030-3498. Offers PhD, MD/PhD. *Faculty:* 45 full-time (13 women). *Students:* 44 full-time (22 women); includes 7 minority (3 African Americans, 1 Asian American or Pacific Islander, 3 Hispanic Americans), 12 international. Average age 28. 64 applicants, 22% accepted, 7 enrolled. In 2005, 3 doctorates awarded. *Median time to degree:* Of those who began their doctoral program in fall 1997, 67% received their degree in 8 years or less. *Degree requirements:* For doctorate, thesis/dissertation, public defense. *Entrance requirements:* For doctorate, GRE General Test, GRE Subject Test (strongly recommended), minimum GPA of 3.0. Additional exam requirements/recommendations for international students: Required—TOEFL. *Application deadline:* For fall admission, 1/1 for domestic students. Applications are processed on a rolling basis. Application fee: $30. Electronic applications accepted. *Expenses:* Tuition: Full-time $8,200. Full-time tuition and fees vary according to program. *Financial support:* In 2005–06, 14 fellowships (averaging $23,000 per year), 30 research assistantships (averaging $23,000 per year) were awarded; teaching assistantships, career-related internships or fieldwork, Federal Work-Study, institutionally sponsored loans, health care benefits, and tuition waivers (full) also available. Financial award applicants

required to submit FAFSA. *Faculty research:* Molecular biology of virus replication, viruses and cancer, viral genetics, viral infectious diseases, environmental virology. *Unit head:* Dr. Frank Ramig, Director, 713-798-4830, Fax: 713-798-5075, E-mail: rramig@bcm.edu. *Application contact:* Rosa Banegas, Graduate Program Administrator, 713-798-4472, Fax: 713-798-5075, E-mail: rbanegas@bcm.edu.

See Close-Up on page 851.

Baylor College of Medicine, Graduate School of Biomedical Sciences, Interdepartmental Program in Cell and Molecular Biology, Houston, TX 77030-3498. Offers biochemistry (PhD); cell and molecular biology (PhD); genetics (PhD); human genetics (PhD); immunology (PhD); microbiology (PhD); virology (PhD). *Faculty:* 99 full-time (41 women). *Students:* 56 full-time (28 women); includes 20 minority (4 African Americans, 1 American Indian/Alaska Native, 6 Asian Americans or Pacific Islanders, 9 Hispanic Americans), 5 international. Average age 28. 164 applicants, 18% accepted, 12 enrolled. In 2005, 3 doctorates awarded. *Median time to degree:* Of those who began their doctoral program in fall 1997, 63% received their degree in 8 years or less. *Degree requirements:* For doctorate, thesis/dissertation, public defense. *Entrance requirements:* For doctorate, GRE General Test, GRE Subject Test (strongly recommended), minimum GPA of 3.0. Additional exam requirements/recommendations for international students: Required—TOEFL. *Application deadline:* For fall admission, 1/1 for domestic students. Applications are processed on a rolling basis. Application fee: $30. Electronic applications accepted. *Expenses:* Full-time tuition and fees vary according to program. *Financial support:* In 2005–06, 52 students received support, including 20 fellowships (averaging $23,000 per year), 36 research assistantships (averaging $23,000 per year); teaching assistantships, Federal Work-Study, institutionally sponsored loans, health care benefits, and tuition waivers (full) also available. Financial award applicants required to submit FAFSA. *Faculty research:* Gene expression and regulation, developmental biology and genetics, signal transduction and membrane biology, aging process, molecular virology. *Unit head:* Dr. Tom Cooper, Director, 713-798-6557. *Application contact:* Lourdes Fernandez, Graduate Program Administrator, 713-798-6557, Fax: 713-798-6325, E-mail: cmbprog@bcm.edu.

See Close-Up on page 545.

Boston University, School of Medicine, Division of Graduate Medical Sciences, Department of Microbiology, Boston, MA 02118. Offers MA, PhD, MD/PhD. *Faculty:* 12 full-time (3 women), 12 part-time/adjunct (2 women). *Students:* 9 full-time (6 women); includes 1 minority (Asian American or Pacific Islander), 1 international. Average age 28.Terminal master's awarded for partial completion of doctoral program. *Degree requirements:* For master's, thesis/dissertation; for doctorate, thesis/dissertation, comprehensive exam. *Entrance requirements:* For master's and doctorate, GRE General Test, GRE Subject Test. Additional exam requirements/recommendations for international students: Required—TOEFL. *Application deadline:* For fall admission, 1/15 for domestic students; for spring admission, 10/15 for domestic students. Electronic applications accepted. *Expenses:* Tuition: Full-time $31,530; part-time $985 per credit. Required fees: $316; $40 per semester. Tuition and fees vary according to course level and program. *Financial support:* Fellowships, research assistantships, Federal Work-Study, scholarships/grants, and traineeships available. *Faculty research:* Eukaryotic cell biology, tumor cell biology, nutrition and cancer, experimental tumor therapy, photobiology. *Unit head:* Dr. Ronald B. Corley, Chairman, 617-638-4284, Fax: 617-638-4286, E-mail: rbcorley@bu.edu. *Application contact:* Dr. Gregory Viglianti, Graduate Director, 617-638-7790, Fax: 617-638-4286, E-mail: gviglian@bu.edu.

See Close-Up on page 855.

Brandeis University, Graduate School of Arts and Sciences, Programs in Life Sciences, Program in Molecular and Cell Biology, Waltham, MA 02454-9110. Offers genetics (PhD); microbiology (PhD); molecular and cell biology (MS, PhD); molecular biology (PhD); neurobiology (PhD). *Faculty:* 18 full-time (8 women). *Students:* 43 full-time (21 women); includes 2 minority (both Asian Americans or Pacific Islanders), 11 international. Average age 27. 173 applicants, 8% accepted, 6 enrolled. In 2005, 2 master's, 4 doctorates awarded. Terminal master's awarded for partial completion of doctoral program. *Median time to degree:* Of those who began their doctoral program in fall 1997, 100% received their degree in 8 years or less. *Degree requirements:* For master's, research project, thesis optional; for doctorate, thesis/dissertation, teaching assistant experience, comprehensive exam. *Entrance requirements:* For master's and doctorate, GRE General Test, resumé, 3 letters of recommendation. Additional exam requirements/recommendations for international students: Required—TOEFL (minimum score 600 paper-based; 250 computer-based). *Application deadline:* For fall admission, 1/15 for domestic students. Applications are processed on a rolling basis. Application fee: $55. Electronic applications accepted. *Financial support:* In

Microbiology

Brandeis University *(continued)*
2005–06, 20 fellowships with full tuition reimbursements (averaging $26,500 per year), 23 research assistantships with full tuition reimbursements (averaging $26,500 per year), 5 teaching assistantships with full tuition reimbursements (averaging $3,000 per year) were awarded; scholarships/grants, traineeships, health care benefits, and tuition waivers (full and partial) also available. Financial award application deadline: 4/15; financial award applicants required to submit CSS PROFILE or FAFSA. *Faculty research:* Regulation of gene expression by transcription factors, molecular neurobiology, immunology, molecular mechanisms of genetic recombination, and cell differentiation. *Unit head:* Dr. Piali Sengupta, Chair, 781-736-2686, Fax: 781-736-3107, E-mail: piali@bradeis.edu. *Application contact:* Marcia Cabral, Information Officer, 781-736-3100, Fax: 781-736-3107, E-mail: cabral@brandeis.edu.

Brigham Young University, Graduate Studies, College of Biological and Agricultural Sciences, Department of Microbiology and Molecular Biology, Provo, UT 84602-1001. Offers microbiology (MS, PhD); molecular biology (MS, PhD). *Faculty:* 16 full-time (2 women). *Students:* 19 full-time (15 women); includes 1 minority (Asian American or Pacific Islander), 4 international. Average age 23. 16 applicants, 19% accepted, 2 enrolled. In 2005, 7 master's, 1 doctorate awarded. *Degree requirements:* For master's and doctorate, thesis/dissertation, comprehensive exam. *Entrance requirements:* For master's, GRE General Test, minimum GPA of 3.0 during previous 2 years; for doctorate, GRE General Test, minimum GPA of 3.0. Additional exam requirements/recommendations for international students: Required—TOEFL, IELT. *Application deadline:* For fall admission, 2/1 priority date for domestic students, 2/1 priority date for international students. For winter admission, 6/30 for domestic students. Application fee: $50. Electronic applications accepted. *Financial support:* In 2005–06, 14 students received support, including 7 research assistantships with full tuition reimbursements available (averaging $14,100 per year), 7 teaching assistantships with full tuition reimbursements available (averaging $14,100 per year); institutionally sponsored loans and tuition waivers (partial) also available. Financial award application deadline: 2/1. *Faculty research:* Immunology, environmental microbiology, molecular genetics, molecular virology, tumor biology. Total annual research expenditures: $399,094. *Unit head:* Dr. Brent L. Nielsen, Chair, 801-422-1102, Fax: 801-422-0519, E-mail: brent_nielsen@byu.edu. *Application contact:* Dr. Laura Bridgewater, Graduate Coordinator, 801-422-2434, Fax: 801-422-0519, E-mail: laura_bridgewater@byu.edu.

Brown University, Graduate School, Division of Biology and Medicine, Program in Molecular Biology, Cell Biology, and Biochemistry, Providence, RI 02912. Offers biochemistry (M Med Sc, Sc M, PhD), including biochemistry (Sc M, PhD), biology (Sc M, PhD), medical science (M Med Sc, PhD); biology (MA); cell biology (M Med Sc, Sc M, PhD), including biochemistry (Sc M, PhD), biology (Sc M, PhD), medical science (M Med Sc, PhD); developmental biology (M Med Sc, Sc M, PhD), including biochemistry (Sc M, PhD), biology (Sc M, PhD), medical science (M Med Sc, PhD); immunology (M Med Sc, Sc M, PhD), including biochemistry (Sc M, PhD), biology (Sc M, PhD), medical science (M Med Sc, PhD); molecular microbiology (M Med Sc, Sc M, PhD), including biochemistry (Sc M, PhD), biology (Sc M, PhD), medical science (M Med Sc, PhD). Part-time programs available. Terminal master's awarded for partial completion of doctoral program. *Degree requirements:* For master's, thesis (for some programs); for doctorate, one foreign language, thesis/dissertation, preliminary exam. *Entrance requirements:* For master's and doctorate, GRE General Test, GRE Subject Test. Additional exam requirements/recommendations for international students: Required—TOEFL. Electronic applications accepted. *Faculty research:* Molecular genetics, gene regulation.

See Close-Up on page 555.

California State University, Fullerton, Graduate Studies, College of Natural Science and Mathematics, Department of Biological Science, Fullerton, CA 92834-9480. Offers biological science (MS); botany (MS); microbiology (MS). Part-time programs available. *Students:* 18 full-time (10 women), 41 part-time (26 women); includes 23 minority (3 African Americans, 9 Asian Americans or Pacific Islanders, 11 Hispanic Americans), 3 international. Average age 27. 61 applicants, 39% accepted, 19 enrolled. In 2005, 10 degrees awarded. *Degree requirements:* For master's, thesis. *Entrance requirements:* For master's, DAT, GRE General Test and GRE Subject Test, or MCAT, minimum GPA of 3.0 in biology. Application fee: $55. *Expenses:* Tuition, nonresident: part-time $339 per unit. *Financial support:* Teaching assistantships, career-related internships or fieldwork, Federal Work-Study, institutionally sponsored loans, and scholarships/grants available. Support available to part-time students. Financial award application deadline: 3/1. *Faculty research:* Glycosidase release and the block to polyspermy in ascidian eggs. *Unit head:* Dr. Robert Koch, Chair, 714-278-3614. *Application contact:* Dr. Michael Horn, Adviser, 714-278-3707.

California State University, Long Beach, Graduate Studies, College of Natural Sciences and Mathematics, Department of Biological Sciences, Long Beach, CA 90840. Offers microbiology (MPH, MS), including medical technology, microbiology (MS); nurse epidemiology. Part-time programs available. *Faculty:* 18 full-time (2 women). *Students:* 9 full-time (8 women), 42 part-time (24 women); includes 14 minority (2 American Indian/Alaska Native, 4 Asian Americans or Pacific Islanders, 8 Hispanic Americans), 1 international. Average age 29. 68 applicants, 37% accepted, 15 enrolled. In 2005, 6 degrees awarded. *Entrance requirements:* For master's, GRE Subject Test, minimum GPA of 3.0. *Application deadline:* For fall admission, 7/1 for domestic students; for spring admission, 12/1 for domestic students. Applications are processed on a rolling basis. Application fee: $55. Electronic applications accepted. *Expenses:* Tuition, nonresident: part-time $339 per semester hour. *Financial support:* Teaching assistantships, Federal Work-Study, institutionally sponsored loans, scholarships/grants, traineeships, and unspecified assistantships available. Financial award application deadline: 3/2. *Unit head:* Dr. Editte Gharakhanian, Chair, 562-985-4806, Fax: 562-985-8878, E-mail: eghara@csulb.edu. *Application contact:* Dr. Judith A Brusslan, Graduate Advisor, 562-985-4806, Fax: 562-985-8878, E-mail: bruss@csulb.edu.

Case Western Reserve University, School of Medicine and School of Graduate Studies, Graduate Programs in Medicine, Department of Molecular Biology and Microbiology, Cleveland, OH 44106-4960. Offers cellular biology (PhD); microbiology (PhD); molecular biology (PhD); molecular virology (MD/PhD). Students are admitted to an integrated Biomedical Sciences Training Program involving 11 basic science programs at Case Western Reserve University. *Degree requirements:* For doctorate, thesis/dissertation. *Entrance requirements:* For doctorate, GRE General Test, GRE Subject Test. Additional exam requirements/recommendations for international students: Required—TOEFL. *Application deadline:* Applications are processed on a rolling basis. Application fee: $25. Electronic applications accepted. *Financial support:* In 2005–06, 24 students received support, including fellowships with full tuition reimbursements available (averaging $23,000 per year); Federal Work-Study, scholarships/grants, traineeships, and tuition waivers (full) also available. *Faculty research:* Gene expression in eukaryotic and prokaryotic systems; microbial physiology; intracellular transport and signaling; mechanisms of oncogenesis; molecular mechanisms of RNA processing, editing, and catalysis. Total annual research expenditures: $4.8 million. *Unit head:* Dr. Jonathan Karn, Chairman, 216-368-3420, Fax: 216-368-3055, E-mail: jonathan.karn@case.edu. *Application contact:* Dr. Patrick Viollier, Admissions Coordinator, 216-368-3420, Fax: 216-368-3055, E-mail: patrick.viollier@case.edu.

See Close-Up on page 557.

The Catholic University of America, School of Arts and Sciences, Department of Biology, Program in Cell and Microbial Biology, Washington, DC 20064. Offers cell biology (MS, PhD); microbiology (MS, PhD). Part-time programs available. *Students:* 3 full-time (1 woman), 8 part-time (6 women); includes 1 Asian American or Pacific Islander, 5 international. Average age 30. 14 applicants, 64% accepted, 3 enrolled. Terminal master's awarded for partial completion of doctoral program. *Degree requirements:* For master's, thesis or alternative, comprehensive exam; for doctorate, thesis/dissertation, comprehensive exam. *Entrance requirements:* For master's and doctorate, GRE General Test, GRE Subject Test, 3 letters of recommendation.

Additional exam requirements/recommendations for international students: Required—TOEFL (minimum score 580 paper-based; 237 computer-based). *Application deadline:* For fall admission, 2/1 for domestic students; for spring admission, 11/15 priority date for domestic students. Applications are processed on a rolling basis. Application fee: $55. Electronic applications accepted. *Expenses:* Tuition: Full-time $24,800; part-time $940 per credit. Required fees: $1,090; $285 per term. Part-time tuition and fees vary according to course load and program. *Financial support:* Fellowships, research assistantships, teaching assistantships, career-related internships or fieldwork, scholarships/grants, tuition waivers (full and partial), and unspecified assistantships available. Support available to part-time students. Financial award application deadline: 2/1; financial award applicants required to submit FAFSA. *Faculty research:* Cell differentiation, regulation of cell growth, drug resistance, gene cloning and sequencing, developmental biology and neurobiology. *Unit head:* Dr. Venigalla Rao, Chair, Department of Biology, 202-319-5267, Fax: 202-319-5721, E-mail: rao@cua.edu.

Clemson University, Graduate School, College of Agriculture, Forestry and Life Sciences, Department of Biological Sciences, Program in Microbiology, Clemson, SC 29634. Offers MS, PhD. *Students:* 23 full-time (13 women), 5 part-time (2 women); includes 2 minority (both African Americans), 8 international. Average age 26. 29 applicants, 14% accepted, 3 enrolled. In 2005, 5 master's awarded. *Degree requirements:* For master's and doctorate, thesis/dissertation. *Entrance requirements:* For master's and doctorate, GRE General Test. Additional exam requirements/recommendations for international students: Required—TOEFL. *Application deadline:* For fall admission, 6/1 for domestic students, 4/15 for international students. Application fee: $50. *Financial support:* Research assistantships, teaching assistantships available. Financial award application deadline: 3/1; financial award applicants required to submit FAFSA. *Faculty research:* Anaerobic microbiology, microbiology and ecology of soil and aquatic systems, genetic engineering, monoclonal antibodies and immunomodulation. *Unit head:* Dr. Malcolm J. B. Paynter, Chair, 864-656-3581, Fax: 864-656-1127, E-mail: pmalcol@clemson.edu. *Application contact:* Dr. Margaret Ptacek, Graduate Coordinator, 864-656-6964, Fax: 864-656-0435, E-mail: mptacek@clemson.edu.

See Close-Ups on pages 857 and 859.

Colorado State University, College of Veterinary Medicine and Biomedical Sciences, Department of Microbiology, Immunology and Pathology, Fort Collins, CO 80523-0015. Offers immunology (MS); microbiology (MS, PhD); pathology (PhD). *Faculty:* 40 full-time (11 women), 2 part-time/adjunct (0 women). *Students:* 64 full-time (37 women), 37 part-time (21 women); includes 9 minority (1 American Indian/Alaska Native, 5 Asian Americans or Pacific Islanders, 3 Hispanic Americans), 17 international. Average age 30. 81 applicants, 22% accepted, 18 enrolled. In 2005, 6 master's, 7 doctorates awarded. *Degree requirements:* For master's, thesis/dissertation; for doctorate, thesis/dissertation, comprehensive exam. *Entrance requirements:* For master's and doctorate, GRE General Test, minimum GPA of 3.0. Additional exam requirements/recommendations for international students: Required—TOEFL. *Application deadline:* For fall admission, 1/1 for domestic students; for spring admission, 10/1 priority date for domestic students. Applications are processed on a rolling basis. Application fee: $50. Electronic applications accepted. *Expenses:* Tuition, state resident: full-time $3,690; part-time $205 per credit. Tuition, nonresident: full-time $14,958; part-time $831 per credit. Required fees: $1,061. *Financial support:* In 2005–06, 17 fellowships with tuition reimbursements (averaging $28,475 per year), 35 research assistantships with tuition reimbursements (averaging $21,715 per year), 7 teaching assistantships with tuition reimbursements (averaging $20,772 per year) were awarded; traineeships and unspecified assistantships also available. *Faculty research:* Medical and veterinary microbiology, pathology of disease, microbial pathogenesis, industrial and environmental microbiology, vector-borne disease. Total annual research expenditures: $21.1 million. *Unit head:* Dr. Jeffrey Wilusz, Head, 970-491-0652, Fax: 970-491-0603, E-mail: jeffrey.wilusz@colostate.edu. *Application contact:* Dr. Herbert Schweizer, Graduate Program Coordinator, 970-491-6136, Fax: 970-491-1815, E-mail: microbio@colostate.edu.

See Close-Up on page 861.

Columbia University, College of Physicians and Surgeons and Graduate School of Arts and Sciences, Graduate School of Arts and Sciences at the College of Physicians and Surgeons, Department of Microbiology, New York, NY 10032. Offers biomedical sciences (M Phil, MA, PhD). Only candidates for the PhD are admitted. Terminal master's awarded for partial completion of doctoral program. *Degree requirements:* For doctorate, thesis/dissertation. *Entrance requirements:* For master's, GRE General Test; for doctorate, GRE. Additional exam requirements/recommendations for international students: Required—TOEFL. *Expenses:* Tuition: Full-time $31,448. Tuition and fees vary according to course level, course load, campus/location and program. *Faculty research:* Prokaryotic molecular biology, immunology, virology, yeast molecular genetics, regulation of gene expression.

Cornell University, Graduate School, Graduate Fields of Agriculture and Life Sciences, Field of Microbiology, Ithaca, NY 14853-0001. Offers PhD. *Faculty:* 48 full-time (10 women). *Students:* 63 applicants, 33% accepted, 7 enrolled. In 2005, 8 doctorates awarded. *Degree requirements:* For doctorate, thesis/dissertation, 2 semesters of teaching experience, comprehensive exam. *Entrance requirements:* For doctorate, GRE General Test, 3 letters of recommendation. Additional exam requirements/recommendations for international students: Required—TOEFL (minimum score 550 paper-based; 213 computer-based). *Application deadline:* For fall admission, 1/15 for domestic students. Application fee: $60. Electronic applications accepted. *Financial support:* In 2005–06, 56 students received support, including 5 fellowships with full tuition reimbursements available, 44 research assistantships with full tuition reimbursements available, 7 teaching assistantships with full tuition reimbursements available; institutionally sponsored loans, scholarships/grants, health care benefits, tuition waivers (full and partial), and unspecified assistantships also available. Financial award applicants required to submit FAFSA. *Faculty research:* Microbial diversity, molecular biology, biotechnology, microbial ecology, phytobacteriology. *Unit head:* Director of Graduate Studies, 607-255-3088. *Application contact:* Graduate Field Assistant, 607-255-3088, E-mail: microfield@cornell.edu.

Dalhousie University, Faculty of Graduate Studies and Faculty of Medicine, Graduate Programs in Medicine, Department of Microbiology and Immunology, Halifax, NS B3H 4R2, Canada. Offers M Sc, PhD, MD/PhD. Part-time programs available. *Degree requirements:* For master's, thesis/dissertation; for doctorate, thesis/dissertation, comprehensive exam. *Entrance requirements:* For master's, GRE General Test, honors B Sc; for doctorate, GRE General Test, honors B Sc in microbiology, M Sc in discipline or transfer after 1 year in master's program. Additional exam requirements/recommendations for international students: Required—TOEFL. *Faculty research:* Virology, molecular genetics, pathogenesis, bacteriology, immunology.

Dartmouth College, Program in Microbiology and Molecular Pathogenesis, Hanover, NH 03755.

See Close-Up on page 867.

Dartmouth College, School of Arts and Sciences, Department of Microbiology, Hanover, NH 03755. Offers PhD. *Faculty:* 24 full-time (10 women), 4 part-time/adjunct (2 women). *Students:* 38 full-time (16 women); includes 1 minority (Hispanic American), 11 international. Average age 28. 336 applicants, 20% accepted, 32 enrolled. In 2005, 3 doctorates awarded. Terminal master's awarded for partial completion of doctoral program. *Degree requirements:* For doctorate, thesis/dissertation, teaching experience. *Entrance requirements:* For doctorate, GRE General Test, GRE Subject Test. *Application deadline:* For fall admission, 1/7 for domestic students, 1/7 for international students. Application fee: $40. *Expenses:* Tuition: Full-time $31,770. *Financial support:* In 2005–06, 41 students received support, including fellowships with full tuition reimbursements available (averaging $23,000 per year), research assistantships with full tuition reimbursements available (averaging $23,000 per year); Federal Work-Study, institutionally sponsored loans, scholarships/grants, traineeships, tuition waivers (full and partial), and unspecified assistantships also available. *Faculty research:* Immune regulation and monitoring molecular parasitology, targeted tumor cell vaccines, HIV infection, viral immunology. Total annual research expenditures: $11.6 million. *Unit head:* , Dr. Ron K. Taylor, Director, 603-650-

Microbiology

1632, Fax: 603-650-6223, E-mail: ron.k.taylor@dartmouth.edu. *Application contact:* Marcia L. Ingalls, Administrative Assistant, 608-650-4522, Fax: 608-650-6223, E-mail: marcia.l.ingalls@dartmouth.edu.

Drexel University, College of Medicine, Biomedical Graduate Programs, Program in Microbiology and Immunology, Philadelphia, PA 19104-2875. Offers MS, PhD. Terminal master's awarded for partial completion of doctoral program. *Degree requirements:* For master's, thesis, comprehensive exam; for doctorate, thesis/dissertation, qualifying exam. *Entrance requirements:* For master's, GRE General Test, minimum GPA of 2.75; for doctorate, GRE General Test, minimum GPA of 3.0. Additional exam requirements/recommendations for international students: Required—TOEFL. Electronic applications accepted. *Faculty research:* Immunology of malarial parasites, virology, bacteriology, molecular biology, parasitology.

Duke University, Graduate School, Department of Molecular Genetics and Microbiology, Durham, NC 27710. Offers PhD. *Faculty:* 22 full-time. *Students:* 36 full-time (22 women); includes 4 minority (2 African Americans, 2 Asian Americans or Pacific Islanders), 10 international. Average age 22. 77 applicants, 26% accepted, 9 enrolled. In 2005, 1 degree awarded. *Degree requirements:* For doctorate, thesis/dissertation. *Entrance requirements:* For doctorate, GRE General Test, GRE Subject Test (recommended). Additional exam requirements/recommendations for international students: Required—IELT (preferred) or TOEFL. *Application deadline:* For fall admission, 12/31 for domestic students, 12/31 for international students. Application fee: $75. Electronic applications accepted. *Financial support:* In 2005–06, fellowships with full tuition reimbursements (averaging $19,350 per year), research assistantships with full tuition reimbursements (averaging $19,350 per year) were awarded; Federal Work-Study also available. Financial award application deadline: 12/31. *Unit head:* Dr. Hubert Amrein, Director of Graduate Studies, 919-681-1518, Fax: 919-684-2790, E-mail: dgsmicro@mc.duke.edu.

See Close-Up on page 771.

East Carolina University, Brody School of Medicine, Department of Microbiology and Immunology, Greenville, NC 27858-4353. Offers PhD. *Faculty:* 15 full-time (2 women), 6 part-time/adjunct (0 women). *Students:* 5 full-time (4 women), 9 part-time (4 women), 3 international. Average age 28. 16 applicants, 38% accepted. In 2005, 3 degrees awarded. *Median time to degree:* Of those who began their doctoral program in fall 1997, 67% received their degree in 8 years or less. *Degree requirements:* For doctorate, thesis/dissertation, comprehensive exam, registration. *Entrance requirements:* For doctorate, GRE General Test. Additional exam requirements/recommendations for international students: Required—TOEFL. *Application deadline:* For fall admission, 4/15 for domestic students. Applications are processed on a rolling basis. Application fee: $50. *Expenses:* Tuition, state resident: full-time $2,516. Tuition, nonresident: full-time $12,832. *Financial support:* In 2005–06, 17 fellowships with tuition reimbursements (averaging $21,500 per year) were awarded Financial award application deadline: 6/1. *Faculty research:* Molecular virology, genetics of bacteria, yeast and somatic cells, bacterial physiology and metabolism, bioterrorism. Total annual research expenditures: $1.6 million. *Unit head:* Dr. Charles J. Smith, Interim Chair, 252-744-2700, Fax: 252-744-3104, E-mail: smithcha@ecu.edu. *Application contact:* Dr. Richard A. Franklin, Director of Graduate Studies, 252-744-2705, Fax: 252-744-3104, E-mail: franklinra@ecu.edu.

East Tennessee State University, James H. Quillen College of Medicine, Biomedical Science Graduate Program, Johnson City, TN 37614. Offers anatomy (MS, PhD); biochemistry (MS, PhD); biophysics (MS, PhD); microbiology (MS, PhD); pharmacology (MS, PhD); physiology (MS, PhD). Part-time programs available. *Faculty:* 49 full-time (12 women), 1 (woman) part-time/adjunct. *Students:* 30 full-time (19 women), 5 part-time (4 women); includes 2 minority (1 African American, 1 Asian American or Pacific Islander), 11 international. Average age 31. 78 applicants, 13% accepted, 9 enrolled. In 2005, 1 master's, 4 doctorates awarded. Terminal master's awarded for partial completion of doctoral program. *Degree requirements:* For master's, one foreign language, thesis, comprehensive qualifying exam; for doctorate, 2 foreign languages, thesis/dissertation. *Entrance requirements:* For master's, GRE General Test, minimum GPA of 3.0, bachelor's degree in biological or related science; for doctorate, GRE General Test, GRE Subject Test. Additional exam requirements/recommendations for international students: Required—TOEFL (minimum score 550 paper-based; 213 computer-based). *Application deadline:* For fall admission, 3/15 for domestic students; for spring admission, 3/1 for domestic students. Application fee: $25 ($35 for international students). *Expenses:* Contact institution. *Financial support:* In 2005–06, 7 research assistantships with full tuition reimbursements (averaging $15,000 per year) were awarded; teaching assistantships with full tuition reimbursements, career-related internships or fieldwork, Federal Work-Study, institutionally sponsored loans, scholarships/grants, and tuition waivers (full) also available. Financial award application deadline: 7/1; financial award applicants required to submit FAFSA. Total annual research expenditures: $2.1 million. *Unit head:* Dr. Mitchell E. Robinson, Assistant Dean, Director, 423-439-4658, E-mail: robinson@etsu.edu.

East Tennessee State University, School of Graduate Studies, College of Arts and Sciences, Department of Biological Sciences, Johnson City, TN 37614. Offers biology (MS); microbiology (MS). *Faculty:* 12 full-time (0 women). *Students:* 18 full-time (11 women), 1 (woman) part-time; includes 1 minority (African American), 3 international. Average age 27. 22 applicants, 73% accepted, 8 enrolled. In 2005, 6 degrees awarded. *Degree requirements:* For master's, thesis or alternative, comprehensive exam. *Entrance requirements:* For master's, GRE General Test or GRE Subject Test, minimum GPA of 3.0. Additional exam requirements/recommendations for international students: Required—TOEFL (minimum score 550 paper-based; 213 computer-based). *Application deadline:* For fall admission, 7/15 for domestic students; for spring admission, 11/1 for domestic students. Applications are processed on a rolling basis. Application fee: $25 ($35 for international students). *Expenses:* Tuition, nonresident: full-time $9,312; part-time $404 per hour. Required fees: $261 per hour. *Financial support:* In 2005–06, 1 research assistantship with full tuition reimbursement (averaging $6,000 per year), 15 teaching assistantships with full tuition reimbursement (averaging $6,000 per year) were awarded; career-related internships or fieldwork and institutionally sponsored loans also available. Financial award application deadline: 7/1; financial award applicants required to submit FAFSA. *Faculty research:* Vertebrate natural history, mutation rates in fruit flies, regulation of plant secondary metabolism, plant biochemistry, timekeeping in honeybees, gene expression in diapausing flies. Total annual research expenditures: $226,807. *Unit head:* Dr. Dan M. Johnson, Chair, 423-439-4329, Fax: 423-439-5958, E-mail: johnsodm@etsu.edu.

Emory University, Graduate School of Arts and Sciences, Division of Biological and Biomedical Sciences, Program in Microbiology and Molecular Genetics, Atlanta, GA 30322-1100. Offers PhD. *Faculty:* 25 full-time (4 women). *Students:* 25 full-time (12 women); includes 6 minority (2 African American, 2 Asian Americans or Pacific Islanders, 2 Hispanic Americans), 3 international. Average age 27. 52 applicants, 17% accepted, 4 enrolled. In 2005, 5 doctorates awarded. *Median time to degree:* Of those who began their doctoral program in fall 1997, 100% received their degree in 8 years or less. *Degree requirements:* For doctorate, thesis/dissertation, comprehensive exam, registration. *Entrance requirements:* For doctorate, GRE General Test, minimum GPA of 3.0 in science course work. Additional exam requirements/recommendations for international students: Required—TOEFL. *Application deadline:* For fall admission, 1/3 for domestic students, 1/3 for international students. Application fee: $50. Electronic applications accepted. *Expenses:* Tuition: Full-time $14,400. Required fees: $217. *Financial support:* In 2005–06, 11 students received support, including 11 fellowships with full tuition reimbursements available (averaging $23,000 per year); scholarships/grants, health care benefits, and tuition waivers (full) also available. *Faculty research:* Bacterial genetics and physiology, microbial development, molecular biology of viruses and bacterial pathogens, DNA recombination. *Unit head:* Dr. Bill Shafer, Director, 404-728-7688, Fax: 404-329-2210, E-mail: wshafer@emory.edu. *Application contact:* 404-727-2545, Fax: 404-727-3322, E-mail: gdbbs@emory.edu.

Emporia State University, School of Graduate Studies, College of Liberal Arts and Sciences, Department of Biological Sciences, Emporia, KS 66801-5087. Offers botany (MS); environmental biology (MS); general biology (MS); microbial and cellular biology (MS); zoology (MS). Part-time programs available. *Faculty:* 13 full-time (2 women), 4 part-time/adjunct (2 women). *Students:* 3 full-time (1 woman), 17 part-time (7 women), 3 international. 10 applicants, 90% accepted, 7 enrolled. In 2005, 5 degrees awarded. *Degree requirements:* For master's, comprehensive exam or thesis. *Entrance requirements:* For master's, GRE, appropriate undergraduate degree, interview, letters of reference. Additional exam requirements/recommendations for international students: Required—TOEFL (minimum score 450 paper-based; 133 computer-based). *Application deadline:* For fall admission, 8/15 for domestic students. Applications are processed on a rolling basis. Application fee: $30 ($75 for international students). Electronic applications accepted. *Expenses:* Tuition, state resident: full-time $2,890; part-time $132 per credit. Tuition, nonresident: full-time $9,258; part-time $422 per credit. Required fees: $626; $41 per credit. Tuition and fees vary according to degree level. *Financial support:* In 2005–06, 4 research assistantships with full tuition reimbursements (averaging $6,492 per year), 9 teaching assistantships with full tuition reimbursements (averaging $6,492 per year) were awarded; fellowships, career-related internships or fieldwork, Federal Work-Study, institutionally sponsored loans, health care benefits, and unspecified assistantships also available. Financial award application deadline: 3/15; financial award applicants required to submit FAFSA. *Faculty research:* Fisheries, range, and wildlife management; aquatic, plant, grassland, vertebrate, and invertebrate ecology; mammalian and plant systematics, taxonomy, and evolution; immunology, virology, and molecular biology. *Unit head:* Dr. J. Richard Schrock, Chair, 620-341-5311, Fax: 620-341-5607, E-mail: jschrock@emporia.edu. *Application contact:* Dr. Derek Zelmer, Graduate Coordinator, 620-341-5623, Fax: 620-341-5607, E-mail: dzelmer@emporia.edu.

Florida State University, Graduate Studies, College of Arts and Sciences, Department of Biological Science, Program in Microbiology, Tallahassee, FL 32306. Offers MS, PhD. *Faculty:* 2 full-time (0 women). *Students:* 5 full-time (4 women), 2 international. *Degree requirements:* For master's and doctorate, thesis/dissertation, teaching experience, seminar presentation, comprehensive exam, registration. *Entrance requirements:* For master's and doctorate, GRE General Test (minimum 1100: U-500, G-500), minimum upper division GPA of 3.0. Additional exam requirements/recommendations for international students: Required—TOEFL (minimum score 600 paper-based; 250 computer-based), IB 100. *Application deadline:* For fall admission, 1/15 for domestic students, 12/1 for international students; for spring admission, 10/15 for domestic students, 9/1 for international students. Application fee: $30. *Financial support:* In 2005–06, fellowships with full tuition reimbursements (averaging $19,000 per year), research assistantships with full tuition reimbursements (averaging $19,000 per year), teaching assistantships with full tuition reimbursements (averaging $17,600 per year) were awarded. Financial award application deadline: 1/15; financial award applicants required to submit FAFSA. *Faculty research:* Prokaryotes and eukaryotes in biotechnology, recombinant DNA technology, microbioecology, biochemistry. *Application contact:* Judy Bowers, Coordinator, Graduate Affairs, 850-644-3023, Fax: 850-644-9829, E-mail: gradinfo@bio.fsu.edu.

George Mason University, College of Science, Fairfax, VA 22030. Offers bioinformatics (MS, PhD); climate dynamics (PhD); computational sciences (MS); computational sciences and informatics (PhD); computational social science (PhD); computational techniques and applications (Certificate); earth systems and geoinformation science (PhD); earth systems science (MS); nanotechnology and nanoscience (Certificate); neuroscience (PhD); physical sciences (PhD); remote sensing and earth image processing (Certificate). Part-time and evening/weekend programs available. *Degree requirements:* For doctorate, thesis/dissertation, comprehensive exam, registration. *Entrance requirements:* For master's and doctorate, GRE General Test, minimum GPA of 3.0 in last 60 hours. Additional exam requirements/recommendations for international students: Required—TOEFL. Electronic applications accepted. *Expenses:* Tuition, area resident: Full-time $5,244; part-time $219 per credit. Tuition, state resident: part-time $651 per credit. Tuition, nonresident: full-time $15,636. Required fees: $1,524; $65 per credit. *Faculty research:* Space sciences and astrophysics, fluid dynamics, materials modeling and simulation, bioinformatics, global changes and statistics.

George Mason University, College of Arts and Sciences, Department of Biology, Program in Biology, Fairfax, VA 22030. Offers bioinformatics (MS); ecology, systematics and evolution (MS); interpretive biology (MS); molecular and microbiology (PhD); molecular, microbial, and cellular biology (MS); organismal biology (MS). Part-time programs available. *Degree requirements:* For master's, thesis or alternative. *Entrance requirements:* For master's, GRE General Test, GRE Subject Test, bachelor's degree in biology or equivalent. Electronic applications accepted. *Expenses:* Tuition, area resident: Full-time $5,244; part-time $219 per credit. Tuition, state resident: part-time $651 per credit. Tuition, nonresident: full-time $15,636. Required fees: $1,524; $65 per credit.

Georgetown University, Graduate School of Arts and Sciences, Programs in Biomedical Sciences, Department of Microbiology and Immunology, Washington, DC 20057. Offers biohazardous threat agents and emerging infectious diseases (MS); microbiology and immunology research (PhD); science policy and advocacy for the healthcare arena in a global setting (MS); teaching microbiology and immunology (MS). *Degree requirements:* For doctorate, thesis/dissertation, comprehensive exam. *Entrance requirements:* For master's, GRE General Test, 3 letters of reference, bachelor's degree in related field; for doctorate, GRE General Test, 3 letters of reference, graduate degree in related field. Additional exam requirements/recommendations for international students: Required—TOEFL (minimum score 505 paper-based; 213 computer-based). Electronic applications accepted. *Faculty research:* Pathogenesis and basic biology of the fungus Candida albicans, molecular biology of viral hepatitis, molecular and cellular biology of the Hepatitis B and Delta viruses, dengue virus, immunopathological mechanisms in Multiple Sclerosis.

See Close-Up on page 871.

The George Washington University, Columbian College of Arts and Sciences, Institute for Biomedical Sciences, Program in Microbiology and Immunology, Washington, DC 20052. Offers PhD. *Students:* 3 full-time (1 woman), 5 part-time; includes 1 minority (Asian American or Pacific Islander), 2 international. Average age 27. *Degree requirements:* For doctorate, thesis/dissertation. *Entrance requirements:* For doctorate, GRE General Test, minimum GPA of 3.0. Additional exam requirements/recommendations for international students: Required—TOEFL (minimum score 600 paper-based; 250 computer-based). *Application deadline:* For fall admission, 1/2 priority date for domestic students, 1/2 priority date for international students. Applications are processed on a rolling basis. Application fee: $60. Electronic applications accepted. *Financial support:* Fellowships with tuition reimbursements available. *Unit head:* Dr. D. Leitenberg, Head, 202-994-3532, Fax: 202-994-2913. *Application contact:* Information Contact, 202-994-3532, Fax: 202-994-2913, E-mail: mtmjxl@gwumc.edu.

See Close-Up on page 873.

The George Washington University, School of Public Health and Health Services, Department of Epidemiology and Biostatistics, Washington, DC 20052. Offers biostatistics (MPH); epidemiology (MPH); health information systems (MPH); microbiology and emerging infectious diseases (MSPH). *Accreditation:* CEPH. *Degree requirements:* For master's, case study or special project. *Entrance requirements:* For master's, GMAT, GRE General Test, or MCAT. Additional exam requirements/recommendations for international students: Required—TOEFL.

Georgia State University, College of Arts and Sciences, Department of Biology, Program in Applied and Environmental Microbiology, Atlanta, GA 30303-3083. Offers MS, PhD. *Degree requirements:* For master's, thesis or alternative, exam; for doctorate, thesis/dissertation, exam. *Entrance requirements:* For master's and doctorate, GRE General Test. Additional exam requirements/recommendations for international students: Required—TOEFL. Electronic applications accepted. *Expenses:* Tuition, state resident: full-time $4,368; part-time $182 per semester hour. Tuition, nonresident: full-time $8,732; part-time $728 per semester hour. Required fees: $46 per semester hour.

Microbiology

Harvard University, Graduate School of Arts and Sciences, Division of Medical Sciences, Boston, MA 02115. Offers biological chemistry and molecular pharmacology (PhD); cell biology (PhD); genetics (PhD); microbiology and molecular genetics (PhD); pathology (PhD), including experimental pathology. *Students:* 433 full-time (210 women). In 2005, 83 doctorates awarded. *Degree requirements:* For doctorate, thesis/dissertation. *Entrance requirements:* For doctorate, GRE General Test, GRE Subject Test. Additional exam requirements/recommendations for international students: Required—TOEFL. Application fee: $60. *Expenses:* Tuition: Full-time $28,752. Full-time tuition and fees vary according to program and student level. *Financial support:* Fellowships, research assistantships, teaching assistantships, institutionally sponsored loans and tuition waivers (full) available. Financial award application deadline: 1/1. *Unit head:* Administrator, 617-432-2029. *Application contact:* Administrator, 617-432-2029.

Howard University, Graduate School of Arts and Sciences, Department of Microbiology, Washington, DC 20059-0002. Offers PhD. *Degree requirements:* For doctorate, one foreign language, thesis/dissertation, qualifying exam, teaching experience, comprehensive exam. *Entrance requirements:* For doctorate, GRE General Test, minimum GPA of 3.0 in sciences. Additional exam requirements/recommendations for international students: Required—TOEFL. *Faculty research:* Immunology, molecular and cellular microbiology, microbial genetics, microbial physiology, pathogenic bacteriology, medical mycology, medical parasitology, virology.

Idaho State University, Office of Graduate Studies, College of Arts and Sciences, Department of Biological Sciences, Pocatello, ID 83209. Offers biology (MNS, MS, DA, PhD); clinical laboratory science (MS); microbiology (MS). *Accreditation:* NAACLS. *Degree requirements:* For master's, one foreign language, thesis, comprehensive exam, registration (for some programs); for doctorate, 2 foreign languages, thesis/dissertation, comprehensive exam, registration. *Entrance requirements:* For master's, GRE General Test, minimum GPA of 3.0 in all upper division classes; for doctorate, GRE General Test, GRE Subject Test, minimum GPA of 3.0 in all upper division classes. Additional exam requirements/recommendations for international students: Required—TOEFL (minimum score 550 paper-based; 213 computer-based). *Faculty research:* Ecology and evolutionary biology, plant and animal physiology, plant and animal developmental biology, immunology, molecular biology.

Illinois Institute of Technology, Graduate College, College of Science and Letters, Department of Biological, Chemical and Physical Sciences, Biology Division, Chicago, IL 60616-3793. Offers biochemistry (MS); biology (MBS, PhD); biotechnology (MS); cell biology (MS); microbiology (MS); molecular biochemistry and biophysics (MS, PhD). Part-time and evening/weekend programs offered (no on-campus study). Postbaccalaureate distance learning degree programs offered (no on-campus study). Terminal master's awarded for partial completion of doctoral program. *Degree requirements:* For master's, thesis (for some programs), comprehensive exam; for doctorate, thesis/dissertation, comprehensive exam. *Entrance requirements:* For master's and doctorate, GRE General Test, minimum undergraduate GPA of 3.0. Additional exam requirements/recommendations for international students: Required—TOEFL (minimum score 550 paper-based; 213 computer-based). Electronic applications accepted. *Faculty research:* Protein crystallography, small angle x-ray diffraction of muscle, spectroscopy of multidomain proteins, structure and function of cell cycle proteins, development of anticancer drugs.

Illinois State University, Graduate School, College of Arts and Sciences, Department of Biological Sciences, Normal, IL 61790-2200. Offers biological sciences (MS); biology (PhD); biotechnology (MS); botany (PhD); ecology (PhD); genetics (PhD); microbiology (PhD); physiology (PhD); zoology (PhD). Part-time programs available. *Faculty:* 27 full-time (6 women). *Students:* 36 full-time (28 women), 25 part-time (17 women); includes 1 minority (Asian American or Pacific Islander), 23 international. 80 applicants, 21% accepted. In 2005, 14 master's, 5 doctorates awarded. *Degree requirements:* For master's, thesis or alternative; for doctorate, variable foreign language requirement, thesis/dissertation, 2 terms of residency. *Entrance requirements:* For master's, GRE General Test, minimum GPA of 2.6 in last 60 hours of course work; for doctorate, GRE General Test. *Application deadline:* Applications are processed on a rolling basis. Application fee: $30. *Expenses:* Tuition, state resident: full-time $3,060; part-time $170 per credit hour. Tuition, nonresident: full-time $6,390; part-time $355 per credit hour. Required fees: $1,411; $47 per credit hour. *Financial support:* In 2005–06, 22 research assistantships (averaging $13,909 per year), 38 teaching assistantships (averaging $12,617 per year) were awarded; Federal Work-Study, tuition waivers (full), and unspecified assistantships also available. Financial award application deadline: 4/1. *Faculty research:* CRUI: osmoregulation in euryhaline fish: physiology, ecology and molecular biology; the PRISM project: enhancing science and math education; cell structure—functions of Na pump assembly. *Unit head:* Dr. Hou Tak Takucheung, Acting Chairperson, 309-438-3669. *Application contact:* Derek A. McCracken, Graduate Adviser, 309-438-3664.

See Close-Up on page 147.

Indiana State University, School of Graduate Studies, College of Arts and Sciences, Department of Life Sciences, Terre Haute, IN 47809-1401. Offers ecology (PhD); life sciences (MS); microbiology (PhD); physiology (PhD); science education (MS); sports medicine (PhD). *Faculty:* 24 full-time (8 women), 12 part-time/adjunct (2 women). *Students:* 51 full-time (24 women), 24 part-time (10 women); includes 3 minority (all Asian Americans or Pacific Islanders), 11 international. Average age 26. 45 applicants, 73% accepted, 29 enrolled. In 2005, 9 master's, 2 doctorates awarded. *Degree requirements:* For master's, thesis (for some programs); for doctorate, thesis/dissertation, comprehensive exam. *Entrance requirements:* For master's and doctorate, GRE General Test. *Application deadline:* For fall admission, 7/1 for domestic students; for spring admission, 11/1 priority date for domestic students. Applications are processed on a rolling basis. Application fee: $35. Electronic applications accepted. *Expenses:* Tuition, state resident: full-time $6,288; part-time $262 per credit hour. Tuition, nonresident: full-time $12,504; part-time $521 per credit hour. *Financial support:* In 2005–06, 26 teaching assistantships with partial tuition reimbursements (averaging $8,005 per year) were awarded; research assistantships with partial tuition reimbursements, Federal Work-Study, institutionally sponsored loans, and tuition waivers (partial) also available. Financial award application deadline: 3/1; financial award applicants required to submit FAFSA. *Unit head:* Dr. Swapan Ghosh, Interim Chairperson, 812-237-2400.

See Close-Up on page 149.

Indiana University Bloomington, Graduate School, College of Arts and Sciences, Department of Biology, Program in Microbiology, Bloomington, IN 47405-7000. Offers MA, PhD. PhD offered through the University Graduate School. Part-time programs available. *Faculty:* 15 full-time (2 women). *Students:* 22 full-time (11 women), 1 (woman) part-time; includes 3 minority (1 African American, 2 Hispanic Americans), 5 international. In 2005, 4 degrees awarded. *Degree requirements:* For master's and doctorate, thesis/dissertation. *Entrance requirements/recommendations for international students:* Required—TOEFL. *Application deadline:* For fall admission, 1/5 priority date for domestic students, 12/1 priority date for international students; for spring admission, 9/1 priority date for domestic students. Applications are processed on a rolling basis. Application fee: $45. Electronic applications accepted. *Expenses:* Tuition, state resident: full-time $5,437; part-time $227 per credit hour. Tuition, nonresident: full-time $15,836; part-time $660 per credit hour. Required fees: $821. Tuition and fees vary according to campus/location and program. *Financial support:* In 2005–06, 21 students received support, including fellowships with tuition reimbursements available (averaging $17,000 per year), research assistantships with tuition reimbursements available (averaging $17,000 per year), teaching assistantships with tuition reimbursements available (averaging $17,000 per year). Financial award application deadline: 1/15. *Faculty research:* Fungal ecology, bacterial photogenesis, microbial genetics. *Unit head:* Dr. Yves Brun, Head, 812-855-8860, Fax: 812-855-6705, E-mail: ybrun@bio.indiana.edu. *Application contact:* Gretchen Clearwater, Adviser for Graduate Affairs, 812-855-1861, Fax: 812-855-6705, E-mail: biograd@bio.indiana.edu.

Indiana University–Purdue University Indianapolis, Indiana University School of Medicine, Department of Microbiology and Immunology, Indianapolis, IN 46202-2896. Offers MS, PhD,

MD/MS, MD/PhD. *Faculty:* 20 full-time (2 women). *Students:* 18 full-time (15 women), 10 part-time (6 women); includes 5 minority (4 African Americans, 1 Hispanic American), 10 international. Average age 28. In 2005, 1 master's, 10 doctorates awarded. Terminal master's awarded for partial completion of doctoral program. *Degree requirements:* For master's and doctorate, thesis/dissertation. *Entrance requirements:* For master's and doctorate, GRE General Test, previous course work in calculus, cell biology, chemistry, genetics, physics, and biochemistry. *Application deadline:* For fall admission, 3/1 for domestic students. Applications are processed on a rolling basis. Application fee: $50 ($60 for international students). *Expenses:* Tuition, state resident: full-time $5,159; part-time $215 per credit hour. Tuition, nonresident: full-time $14,890; part-time $620 per credit hour. Required fees: $614. Tuition and fees vary according to campus/location and program. *Financial support:* Fellowships with full tuition reimbursements, research assistantships with full tuition reimbursements, teaching assistantships with full tuition reimbursements, Federal Work-Study, institutionally sponsored loans, scholarships/grants, traineeships, and tuition waivers (partial) available. Financial award application deadline: 2/1. *Faculty research:* Host-parasite interactions, molecular biology, cellular and molecular immunology and hematology, viral and bacterial pathogenesis, cancer research. Total annual research expenditures: $4.2 million. *Unit head:* Dr. Hal E. Broxmeyer, Chairman, 317-274-7672, Fax: 317-274-4090, E-mail: hbroxmey@iupui.edu. *Application contact:* 317-274-7671, Fax: 317-274-4090.

See Close-Up on page 875.

Iowa State University of Science and Technology, College of Veterinary Medicine and Graduate College, Graduate Programs in Veterinary Medicine, Department of Veterinary Microbiology and Preventive Medicine, Ames, IA 50011. Offers veterinary microbiology (MS, PhD). *Faculty:* 21 full-time, 9 part-time/adjunct. *Students:* 29 full-time (14 women), 14 international (4 women); includes 3 minority (1 African American, 2 Hispanic Americans), 14 international. 18 applicants, 33% accepted, 4 enrolled. In 2005, 9 master's, 7 doctorates awarded. *Degree requirements:* For master's, thesis or alternative; for doctorate, thesis/dissertation. *Entrance requirements:* For master's and doctorate, GRE General Test. Additional exam requirements/recommendations for international students: Required—TOEFL (paper score 550; computer score 213) or IELTS (score 6.5). *Application deadline:* For fall admission, 2/1 priority date for domestic students, 2/1 priority date for international students. Applications are processed on a rolling basis. Application fee: $30 ($70 for international students). Electronic applications accepted. *Financial support:* In 2005–06, 23 research assistantships with partial tuition reimbursements (averaging $15,645 per year), 1 teaching assistantship with partial tuition reimbursement (averaging $15,440 per year) were awarded; fellowships, scholarships/grants, health care benefits, and unspecified assistantships also available. *Faculty research:* Bacteriology, immunology, virology, public health and food safety. *Unit head:* Dr. Lisa Nolan, Chair, 515-294-3534, E-mail: vetmicro@iastate.edu. *Application contact:* Dr. Eileen Thacker, Director of Graduate Education, 515-294-5097, E-mail: vetmicro@instate.edu.

Iowa State University of Science and Technology, Graduate College, Interdisciplinary Programs, Program in Microbiology, Ames, IA 50011. Offers MS, PhD. *Students:* 16 full-time (9 women), 7 part-time (3 women); includes 3 minority (2 African Americans, 1 Hispanic American), 4 international. 35 applicants, 40% accepted, 6 enrolled. In 2005, 1 master's, 4 doctorates awarded. *Degree requirements:* For master's, thesis or alternative; for doctorate, thesis/dissertation. *Entrance requirements:* For master's and doctorate, GRE General Test. Additional exam requirements/recommendations for international students: Required—TOEFL (paper score 550; computer score 213 or IELTS (score 6.5). *Application deadline:* For fall admission, 1/15 priority date for domestic students, 1/15 priority date for international students. Application fee: $30 ($70 for international students). Electronic applications accepted. *Expenses:* Tuition, state resident: full-time $6,410. Tuition, nonresident: full-time $16,422. Tuition and fees vary according to program. *Financial support:* In 2005–06, 11 research assistantships with partial tuition reimbursements (averaging $15,247 per year), 5 teaching assistantships with partial tuition reimbursements (averaging $15,230 per year) were awarded; scholarships/grants, health care benefits, and unspecified assistantships also available. *Unit head:* Dr. Gwyn Beattie, Chair, Supervising Committee, 515-294-9050, E-mail: gbeattie@iastate.edu.

The Johns Hopkins University, Bloomberg School of Public Health, The W. Harry Feinstone, Department of Molecular Microbiology and Immunology, Baltimore, MD 21218-2699. Offers MHS, Sc M, PhD. Part-time programs available. *Faculty:* 38 full-time (9 women), 17 part-time/adjunct (4 women). *Students:* 64 full-time (39 women), 3 part-time (1 woman); includes 18 minority (5 African Americans, 1 American Indian/Alaska Native, 8 Asian Americans or Pacific Islanders, 4 Hispanic Americans), 20 international. Average age 28. 143 applicants, 38% accepted, 25 enrolled. In 2005, 9 master's, 5 doctorates awarded. Terminal master's awarded for partial completion of doctoral program. *Median time to degree:* Of those who began their doctoral program in fall 1997, 50% received their degree in 8 years or less. *Degree requirements:* For master's, thesis (for some programs), master's essay, written exams; for doctorate, thesis/dissertation, 1 year full-time residency, oral and written exams. *Entrance requirements:* For master's, GRE General Test or MCAT, 3 letters of recommendation, curriculum vitae; for doctorate, GRE General Test, 3 letters of recommendation, curriculum vitae. Additional exam requirements/recommendations for international students: Required—TOEFL (minimum score 600 paper-based; 250 computer-based). *Application deadline:* For fall admission, 11/10 priority date for domestic students, 11/10 priority date for international students. Applications are processed on a rolling basis. Application fee: $45. Electronic applications accepted. *Expenses:* Tuition: Full-time $30,960. Tuition and fees vary according to degree level and program. *Financial support:* In 2005–06, 101 students received support, including 3 fellowships (averaging $23,600 per year); Federal Work-Study, institutionally sponsored loans, scholarships/grants, and stipends also available. Support available to part-time students. Financial award application deadline: 3/15; financial award applicants required to submit FAFSA. *Faculty research:* Immunological disorders, viral and bacterial infections, parasitic diseases, vector-borne diseases, pathogenesis, parasite immunology, biochemistry of parasitic protozoa, vector biology, ecology of infectious diseases. Total annual research expenditures: $22.4 million. *Unit head:* Dr. Diane E. Griffin, Chair, 410-955-3459, Fax: 410-955-0105, E-mail: dgriffin@jhsph.edu. *Application contact:* E-mail: mmi@jhsph.edu.

Kansas State University, Graduate School, College of Arts and Sciences, Division of Biology, Program in Microbiology, Manhattan, KS 66506. Offers PhD. *Students:* 4 full-time (1 woman), 1 international. Average age 24. In 2005, 3 degrees awarded. *Degree requirements:* For doctorate, thesis/dissertation. *Entrance requirements:* Additional exam requirements/recommendations for international students: Required—TOEFL (minimum score 550 paper-based; 213 computer-based). *Application deadline:* For fall admission, 1/15 priority date for domestic students, 1/15 priority date for international students; for spring admission, 3/1 for domestic students, 8/1 for international students. Applications are processed on a rolling basis. Application fee: $30 ($55 for international students). *Expenses:* Tuition, state resident: full-time $5,160; part-time $215 per credit hour. Tuition, nonresident: full-time $12,816; part-time $534 per credit hour. Required fees: $564. *Financial support:* Research assistantships, teaching assistantships, institutionally sponsored loans and scholarships/grants available. Support available to part-time students. Financial award application deadline: 3/1; financial award applicants required to submit FAFSA. *Faculty research:* Immune cell function, virology, cell viability, water quality. Total annual research expenditures: $3 million. *Application contact:* David Rintoul, Director, 785-532-6663, Fax: 785-532-6653, E-mail: drintoul@ksu.edu.

Loma Linda University, School of Medicine, Department of Biochemistry/Microbiology, Loma Linda, CA 92350. Offers MS, PhD. Part-time programs available. *Faculty:* 37 full-time (7 women), 5 part-time/adjunct (1 woman). *Students:* 6 full-time (4 women), 6 part-time (3 women); includes 4 Asian Americans or Pacific Islanders, 2 Hispanic Americans, 3 international. *Degree requirements:* For master's, thesis or alternative; for doctorate, thesis/dissertation. *Entrance requirements:* For master's and doctorate, GRE General Test. *Application deadline:* Applications are processed on a rolling basis. Application fee: $40. *Financial support:* Tuition waivers (full and partial) available. Support available to part-time students. *Faculty research:*

Physical chemistry of macromolecules, biochemistry of endocrine system, biochemical mechanism of bone volume regulation. *Unit head:* Dr. Lawrence C. Sowers, Coordinator, 909-824-4527.

Long Island University, C.W. Post Campus, School of Health Professions and Nursing, Department of Biomedical Sciences, Program in Medical Biology, Brookville, NY 11548-1300. Offers hematology (MS); immunology (MS); medical biology (MS); medical chemistry (MS); medical microbiology (MS). Part-time and evening/weekend programs available. *Degree requirements:* For master's, thesis. *Entrance requirements:* For master's, minimum GPA of 2.75 in major. Electronic applications accepted. *Faculty research:* Hematopoiesis, growth factors in cancer, interleukins in allergy, PCR techniques.

Louisiana State University Health Sciences Center, School of Graduate Studies in New Orleans, Department of Microbiology, Immunology, and Parasitology, New Orleans, LA 70112-1393. Offers microbiology and immunology (MS, PhD). Terminal master's awarded for partial completion of doctoral program. *Degree requirements:* For master's, thesis; for doctorate, thesis/dissertation, preliminary exam, qualifying exam. *Entrance requirements:* For master's and doctorate, GRE General Test. Additional exam requirements/recommendations for international students: Required—TOEFL. *Faculty research:* Microbial physiology, animal virology, vaccine development, AIDS drug studies, pathogenic mechanisms, molecular immunology.

Louisiana State University Health Sciences Center at Shreveport, Department of Microbiology and Immunology, Shreveport, LA 71130-3932. Offers microbiology/immunology (MS). *Faculty:* 21 full-time (5 women). *Students:* 22 full-time (10 women); includes 5 minority (2 African Americans, 1 American Indian/Alaska Native, 1 Asian American or Pacific Islander, 1 Hispanic American). Average age 23. 44 applicants, 16% accepted. In 2005, 3 master's awarded. Terminal master's awarded for partial completion of doctoral program. *Degree requirements:* For master's, thesis/dissertation. *Entrance requirements:* For master's, GRE General Test. Additional exam requirements/recommendations for international students: Required—TOEFL. Application fee: $30. *Financial support:* In 2005–06, 2 fellowships (averaging $20,000 per year), 20 research assistantships (averaging $20,000 per year) were awarded; institutionally sponsored loans also available. Financial award application deadline: 7/1. *Faculty research:* Infectious disease, pathogenesis, molecular virology and biology. *Unit head:* Dr. Dennis J. O'Callaghan, Head, 318-675-5750, Fax: 318-675-5764.

See Close-Up on page 163.

Loyola University Chicago, Graduate School, Department of Microbiology and Immunology, Maywood, IL 60153. Offers immunology (MS, PhD); microbiology (MS, PhD); virology (MS, PhD). *Faculty:* 11 full-time (3 women). *Students:* 26 full-time (18 women), 1 (woman) part-time; includes 4 minority (1 African American, 3 Asian Americans or Pacific Islanders), 9 international. Average age 28. 74 applicants, 15% accepted, 4 enrolled. In 2005, 1 master's, 3 doctorates awarded. Terminal master's awarded for partial completion of doctoral program. *Degree requirements:* For master's, thesis/dissertation; for doctorate, thesis/dissertation, comprehensive exam. *Entrance requirements:* For master's and doctorate, GRE General Test. Additional exam requirements/recommendations for international students: Required—TOEFL. *Application deadline:* Applications are processed on a rolling basis. Application fee: $40. Electronic applications accepted. *Expenses:* Tuition: Full-time $11,610; part-time $645 per credit. Required fees: $55 per semester. *Financial support:* In 2005–06, 5 fellowships with tuition reimbursements (averaging $22,000 per year), 24 research assistantships with tuition reimbursements (averaging $22,000 per year) were awarded; institutionally sponsored loans and scholarships/grants also available. Financial award application deadline: 2/15. *Faculty research:* Viral pathogenesis, microbial physiology and genetics, immunoglobulin genetics and differentiation of the immune response, signal transduction and host-parasite interactions. *Unit head:* Dr. Katherine L. Knight, Chair, 708-216-3385, Fax: 708-216-9574, E-mail: kknight@lumc.edu. *Application contact:* Dr. Karen Visick, Graduate Program Director, 708-216-0869, Fax: 708-216-9574, E-mail: kvisick@lumc.edu.

See Close-Up on page 879.

Marquette University, Graduate School, College of Arts and Sciences, Department of Biology, Milwaukee, WI 53201-1881. Offers cell biology (MS, PhD); developmental biology (MS, PhD); ecology (MS, PhD); endocrinology (MS, PhD); evolutionary biology (MS, PhD); genetics (MS, PhD); microbiology (MS, PhD); molecular biology (MS, PhD); muscle and exercise physiology (MS, PhD); neurobiology (MS, PhD); reproductive physiology (MS, PhD). Terminal master's awarded for partial completion of doctoral program. *Degree requirements:* For master's, thesis, 1 year of teaching experience or equivalent, comprehensive exam; for doctorate, thesis/dissertation, 1 year of teaching experience or equivalent, qualifying exam. *Entrance requirements:* For master's and doctorate, GRE General Test, GRE Subject Test. Additional exam requirements/recommendations for international students: Required—TOEFL. *Faculty research:* Microbial and invertebrate ecology, evolution of gene function, DNA methylation, DNA arrangement.

McGill University, Faculty of Graduate and Postdoctoral Studies, Faculty of Agricultural and Environmental Sciences, Department of Natural Resource Sciences, Montréal, QC H3A 2T5, Canada. Offers agrometeorology (M Sc, PhD); entomology (M Sc, PhD); forest science (M Sc, PhD); microbiology (M Sc, PhD); neotropical environment (M Sc, PhD); soil science (M Sc, PhD); wildlife biology (M Sc, PhD). *Degree requirements:* For master's and doctorate, thesis/dissertation, registration. *Entrance requirements:* For master's, minimum GPA of 3.0 or 3.2 in the last 2 years of university study. Additional exam requirements/recommendations for international students: Required—TOEFL (minimum score 550 paper-based; 213 computer-based), IELT (minimum score 7). Electronic applications accepted. *Faculty research:* Toxicology, reproductive physiology, parasites, wildlife management, genetics.

McGill University, Faculty of Graduate and Postdoctoral Studies, Faculty of Medicine, Department of Microbiology and Immunology, Montréal, QC H3A 2T5, Canada. Offers M Sc, M Sc A, PhD. *Degree requirements:* For master's and doctorate, thesis/dissertation. *Entrance requirements:* For master's, minimum GPA of 3.0. *Faculty research:* Virology, molecular genetics, immunology, molecular biology, microbial physiology.

Medical College of Wisconsin, Graduate School of Biomedical Sciences, Department of Microbiology and Molecular Genetics, Milwaukee, WI 53226-0509. Offers MS, PhD, MD/MS, MD/PhD. *Degree requirements:* For doctorate, thesis/dissertation, comprehensive exam, registration. *Entrance requirements:* For doctorate, GRE General Test. Additional exam requirements/recommendations for international students: Required—TOEFL. *Faculty research:* Virology, immunology, bacterial toxins, regulation of gene expression.

Medical University of South Carolina, College of Graduate Studies, Program in Microbiology and Immunology, Charleston, SC 29425-0002. Offers Pharm D, MS, PhD, DMD/PhD, MD/PhD. *Faculty:* 18 full-time (7 women). *Students:* 23 full-time (12 women); includes 4 minority (2 Asian Americans or Pacific Islanders, 2 Hispanic Americans), 1 international. Average age 28. 181 applicants, 34% accepted, 43 enrolled. In 2005, 1 master's, 6 doctorates awarded. Terminal master's awarded for partial completion of doctoral program. *Degree requirements:* For master's, thesis, research seminar; for doctorate, thesis/dissertation, teaching and research seminar, oral and written exams. *Entrance requirements:* For master's and doctorate, GRE General Test, interview. Additional exam requirements/recommendations for international students: Required—TOEFL (minimum score 600 paper-based; 250 computer-based). *Application deadline:* For fall admission, 1/15 priority date for domestic students, 1/15 priority date for international students. Applications are processed on a rolling basis. Application fee: $0 ($75 for international students). Electronic applications accepted. *Financial support:* In 2005–06, 15 students received support, including fellowships with partial tuition reimbursements available (averaging $21,000 per year); Federal Work-Study and scholarships/grants also available. Financial award application deadline: 3/15; financial award applicants required to submit FAFSA. *Faculty research:* Inmate and adoptive immunology, gene therapy/vector development, vaccinology, protemonics of biowarfare agents, bacterial and fungal pathogenesis.

Unit head: Dr. James S. Norris, Chair, 843-792-7915, Fax: 843-792-6590, E-mail: norrisjs@musc.edu. *Application contact:* Cheryl Brown, 843-792-4620, Fax: 843-792-4645, E-mail: brownche@musc.edu.

See Close-Up on page 885.

Meharry Medical College, School of Graduate Studies, Department of Microbiology, Nashville, TN 37208-9989. Offers PhD. *Degree requirements:* For doctorate, thesis/dissertation, comprehensive exam. *Entrance requirements:* For doctorate, GRE General Test, GRE Subject Test, undergraduate degree in related science. *Faculty research:* Microbial and bacterial pathogenesis, viral transcription, immune response to viruses and parasites.

Miami University, Graduate School, College of Arts and Sciences, Department of Microbiology, Oxford, OH 45056. Offers MS, PhD. Part-time programs available. *Degree requirements:* For master's, thesis, final exam; for doctorate, thesis/dissertation, final exams, comprehensive exam. *Entrance requirements:* For master's, GRE General Test, minimum undergraduate GPA of 3.0 during previous 2 years or 2.75 overall; for doctorate, GRE General Test, minimum undergraduate GPA of 2.75, 3.0 graduate. Additional exam requirements/recommendations for international students: Required—TOEFL (minimum score 550 paper-based; 213 computer-based), TWE (minimum score 4). Electronic applications accepted.

Michigan State University, College of Osteopathic Medicine and The Graduate School, Graduate Studies in Osteopathic Medicine, East Lansing, MI 48824. Offers biochemistry and molecular biology (MS, PhD); microbiology (MS); microbiology and molecular genetics (MS, PhD); pharmacology and toxicology (MS, PhD), including pharmacology and toxicology, pharmacology and toxicology-environmental toxicology (PhD); physiology (MS, PhD). *Students:* 2 full-time (1 woman), 1 international. Average age 27. *Expenses:* Tuition, state resident: part-time $330 per credit hour. Tuition, nonresident: part-time $685 per credit hour. Tuition and fees vary according to program. *Unit head:* Dr. Veronica M. Maher, Associate Dean, Graduate Studies, 517-353-7785, Fax: 517-353-9004, E-mail: maher@msu.edu. *Application contact:* Kathie Schafer, Director of Admissions, 517-353-7740, Fax: 517-355-3296, E-mail: comadm@com.msu.edu.

Michigan State University, College of Veterinary Medicine and The Graduate School, Graduate Program in Veterinary Medicine and College of Natural Science and Graduate Programs in Human Medicine, Department of Microbiology and Molecular Genetics, East Lansing, MI 48824. Offers industrial microbiology (MS); microbiology (MS, PhD), including environmental toxicology (MS); microbiology and molecular genetics (MS, PhD); microbiology—environmental toxicology (PhD). *Faculty:* 34 full-time (10 women). *Students:* 47 full-time (28 women), 2 part-time (1 woman); includes 3 minority (2 African Americans, 1 Hispanic American), 21 international. Average age 26. 104 applicants, 13% accepted. In 2005, 3 master's, 9 doctorates awarded. *Degree requirements:* For master's, thesis (for some programs); for doctorate, thesis/dissertation, comprehensive exam. *Entrance requirements:* For master's and doctorate, GRE General Test, minimum GPA of 3.0, 3 letters of recommendation. Additional exam requirements/recommendations for international students: Required—TOEFL (minimum score 550 paper-based; 213 computer-based), Michigan State University ELT (85), Michigan ELAB (83). *Application deadline:* For fall admission, 8/27 for domestic students. Application fee: $50. Electronic applications accepted. *Expenses:* Tuition, state resident: part-time $330 per credit hour. Tuition, nonresident: part-time $685 per credit hour. Tuition and fees vary according to program. *Financial support:* In 2005–06, 12 fellowships with tuition reimbursements (averaging $6,638 per year), 29 research assistantships with tuition reimbursements (averaging $15,427 per year), 3 teaching assistantships with tuition reimbursements (averaging $14,823 per year) were awarded; scholarships/grants, traineeships, health care benefits and unspecified assistantships also available. *Faculty research:* Microbial physiology, ecology and evolution: molecular pathogenesis and infectious diseases; genomics and genetics; cancer, differentiation, immunology and cell biology. Total annual research expenditures: $5.2 million. *Unit head:* Dr. Walter Esselman, Chairperson, 517-355-6463 Ext. 1510, Fax: 517-353-8957, E-mail: mmgchair@msu.edu. *Application contact:* Suzanne Peacock, Graduate Program Coordinator, 517-432-2288, Fax: 517-353-8957, E-mail: micgrad@msu.edu.

See Close-Up on page 887.

Michigan State University, The Graduate School, College of Agriculture and Natural Resources, MSU-DOE Plant Research Laboratory, East Lansing, MI 48824. Offers biochemistry and molecular biology (PhD); cellular and molecular biology (PhD); crop and soil sciences (PhD); genetics (PhD); microbiology and molecular genetics (PhD); plant biology (PhD); plant physiology (PhD). Offered jointly with the Department of Energy. *Faculty:* 9 full-time (2 women). *Degree requirements:* For doctorate, thesis/dissertation, laboratory rotation, defence of dissertation, comprehensive exam. *Entrance requirements:* For doctorate, GRE General Test, acceptance into one of the affiliated department programs; 3 letters of recommendation; bachelor's degree or equivalent in life sciences, chemistry, biochemistry, or biophysics; research experience. Application fee: $50. Electronic applications accepted. *Expenses:* Tuition, state resident: part-time $330 per credit hour. Tuition, nonresident: part-time $685 per credit hour. Tuition and fees vary according to program. *Faculty research:* Role of hormones in the regulation of plant development and physiology, molecular mechanisms associated with signal recognition, development and application of genetic methods and materials, protein routing and function. Total annual research expenditures: $7.4 million. *Unit head:* Dr. Kenneth Keegstra, Director, 517-353-2270, Fax: 517-353-9168, E-mail: keegstra@msu.edu. *Application contact:* Janet Taylor, Graduate Program Secretary, 517-353-2270, Fax: 517-353-9168, E-mail: prl@msu.edu.

Montana State University, College of Graduate Studies, College of Letters and Science, Department of Microbiology, Bozeman, MT 59717. Offers MS, PhD. Part-time programs available. *Faculty:* 10 full-time (3 women), 2 part-time/adjunct (1 woman). *Students:* 5 full-time (4 women), 18 part-time (9 women); includes 2 minority (1 Asian American or Pacific Islander, 1 Hispanic American), 5 international. Average age 30. 25 applicants, 32% accepted, 8 enrolled. In 2005, 2 master's, 2 doctorates awarded. *Degree requirements:* For master's, comprehensive exam, registration; for doctorate, thesis/dissertation, comprehensive exam, registration. *Entrance requirements:* For master's and doctorate, GRE General Test. Additional exam requirements/recommendations for international students: Required—TOEFL (minimum score 550 paper-based; 213 computer-based). *Application deadline:* For fall admission, 7/15 priority date for domestic students, 5/15 priority date for international students; for spring admission, 12/1 priority date for domestic students, 10/1 priority date for international students. Applications are processed on a rolling basis. Application fee: $30. Electronic applications accepted. *Expenses:* Tuition, state resident: full-time $4,132. Tuition, nonresident: full-time $11,332. *Financial support:* In 2005–06, 22 students received support, including 5 fellowships with full tuition reimbursements available (averaging $18,000 per year), 12 research assistantships with full tuition reimbursements available (averaging $18,000 per year), 5 teaching assistantships with full tuition reimbursements available (averaging $15,000 per year); health care benefits also available. Financial award application deadline: 3/1; financial award applicants required to submit FAFSA. *Faculty research:* Environmental health, infectious disease, immunology, bioinformatics, environmental microbiology. Total annual research expenditures: $2.5 million. *Unit head:* Dr. Timothy Ford, Department Head, 406-994-2903, Fax: 406-994-4926, E-mail: tford@montana.edu.

Mount Sinai School of Medicine of New York University, Graduate School of Biological Sciences, New York, NY 10029-6504. Offers biophysics, structural biology and biomathematics (PhD); community medicine (MPH); genetic counseling (MS); genetics and genomic sciences (PhD); mechanisms of disease and therapy (PhD); microbiology (PhD); molecular, cellular, biochemical and developmental sciences (PhD); neurosciences (PhD). *Students:* 218 full-time (109 women). 4,208 applicants, 7% accepted, 117 enrolled. Terminal master's awarded for partial completion of doctoral program. *Degree requirements:* For master's, registration; for doctorate, thesis/dissertation, registration. *Entrance requirements:* For doctorate, GRE General Test, GRE Subject Test, 3 years of college pre-med course work. Additional exam requirements/

Microbiology

Mount Sinai School of Medicine of New York University (continued)
recommendations for international students: Required—TOEFL. *Application deadline:* For fall admission, 1/15 for domestic students. Application fee: $75. Electronic applications accepted. *Expenses:* Tuition: Full-time $33,250. Required fees: $1,600. Full-time tuition and fees vary according to degree level, program and reciprocity agreements. *Financial support:* In 2005–06, fellowships with full tuition reimbursements (averaging $26,000 per year), research assistantships with full tuition reimbursements (averaging $26,000 per year) were awarded; Federal Work-Study, institutionally sponsored loans, scholarships/grants, health care benefits, and unspecified assistantships also available. Financial award application deadline: 4/5; financial award applicants required to submit FAFSA. *Faculty research:* Cancer, gene therapy, minimally invasive surgery, cardiac translational research. Total annual research expenditures: $162.2 million. *Unit head:* Dr. Diomedes Logothetis, Dean, 212-241-6546, Fax: 212-241-0651, E-mail: diomedes.logothetis@mssm.edu. *Application contact:* Lily Recanati, Manager, 212-241-3267, Fax: 212-241-0651, E-mail: lily.recanati@mssm.edu.

See Close-Up on page 177.

New York Medical College, Graduate School of Basic Medical Sciences, Microbiology and Immunology Department, Valhalla, NY 10595-1691. Offers MS, PhD, MD/PhD. Part-time and evening/weekend programs available. Terminal master's awarded for partial completion of doctoral program. *Degree requirements:* For master's, thesis/dissertation; for doctorate, thesis/dissertation, comprehensive exam. *Entrance requirements:* For master's and doctorate, GRE General Test. Additional exam requirements/recommendations for international students: Required—TOEFL. *Faculty research:* Tumor and transplantation immunology, molecular mechanisms of DNA repair, virus-host interactions.

New York University, Graduate School of Arts and Science, Department of Biology, New York, NY 10012-1019. Offers biology (PhD); biomedical journalism (MS); cancer and molecular biology (PhD); computational biology (PhD); computers in biological research (MS); developmental genetics (PhD); general biology (MS); immunology and microbiology (MS); molecular genetics (PhD); neurobiology (PhD); oral biology (MS); plant biology (PhD); recombinant DNA technology (MS). Part-time programs available. *Faculty:* 24 full-time (5 women), 8 part-time/adjunct. *Students:* 104 full-time (52 women), 41 part-time (23 women); includes 28 minority (2 African Americans, 20 Asian Americans or Pacific Islanders, 6 Hispanic Americans), 47 international. Average age 27. 349 applicants, 56% accepted, 39 enrolled. In 2005, 59 master's, 4 doctorates awarded. Terminal master's awarded for partial completion of doctoral program. *Degree requirements:* For master's, thesis or alternative, qualifying paper; for doctorate, thesis/dissertation, comprehensive exam. *Entrance requirements:* For master's, GRE General Test; for doctorate, GRE General Test, GRE Subject Test. Additional exam requirements/recommendations for international students: Required—TOEFL. *Application deadline:* For fall admission, 1/4 for domestic students. Application fee: $80. *Financial support:* Fellowships with tuition reimbursements, research assistantships with tuition reimbursements, teaching assistantships with tuition reimbursements, career-related internships or fieldwork, Federal Work-Study, institutionally sponsored loans, scholarships/grants, health care benefits, and unspecified assistantships available. Financial award application deadline: 1/4; financial award applicants required to submit FAFSA. *Faculty research:* Genomics, molecular and cell biology, development and molecular genetics, molecular evolution of plants and animals. *Unit head:* Gloria Coruzzi, Chairman, 212-998-8200, Fax: 212-995-4015, E-mail: biology@nyu.edu. *Application contact:* Stephen Small, Director of Graduate Studies, 212-998-8200, Fax: 212-995-4015, E-mail: biology@nyu.edu.

New York University, School of Medicine and Graduate School of Arts and Science, Sackler Institute of Graduate Biomedical Sciences, Department of Microbiology, New York, NY 10012-1019. Offers PhD, MD/PhD. *Degree requirements:* For doctorate, one foreign language, thesis/dissertation, qualifying exam, comprehensive exam. *Entrance requirements:* For doctorate, GRE General Test, GRE Subject Test. Additional exam requirements/recommendations for international students: Required—TOEFL. *Faculty research:* Aspects of microbiology, parasitology, and genetics; virology.

See Close-Up on page 889.

North Carolina State University, College of Veterinary Medicine, Program in Comparative Biomedical Sciences, Raleigh, NC 27695. Offers cell biology and morphology (MS, PhD); epidemiology and population medicine (MS, PhD); immunology (MS, PhD); microbiology and immunology (MS, PhD); pathology (MS, PhD); pharmacology (MS, PhD); specialized veterinary medicine (MS). Part-time programs available. *Degree requirements:* For master's and doctorate, thesis/dissertation. *Entrance requirements:* For master's and doctorate, GRE General Test. Additional exam requirements/recommendations for international students: Required—TOEFL (minimum score 550 paper-based; 213 computer-based). Electronic applications accepted. Expenses: Contact institution. *Faculty research:* Infectious diseases, cell biology, pharmacology and toxicology, genomics, pathology and population medicine.

North Carolina State University, Graduate School, College of Agriculture and Life Sciences, Department of Microbiology, Program in Microbiology, Raleigh, NC 27695. Offers MS, PhD. *Degree requirements:* For master's, thesis (for some programs); for doctorate, thesis/dissertation. *Entrance requirements:* For master's and doctorate, GRE. Electronic applications accepted.

North Dakota State University, The Graduate School, College of Agriculture, Food Systems, and Natural Resources, Department of Veterinary and Microbiological Sciences, Fargo, ND 58105. Offers microbiology (MS); molecular pathogenesis (PhD); natural resource management (MS). Part-time programs available. *Degree requirements:* For master's, thesis; for doctorate, thesis/dissertation, oral and written preliminary exams. *Entrance requirements:* For master's and doctorate, GRE. Additional exam requirements/recommendations for international students: Required—TOEFL. *Faculty research:* Bacterial gene regulation, antibiotic resistance, molecular virology, mechanisms of bacterial pathogenesis, immunology of animals.

Northwestern University, Northwestern University Feinberg School of Medicine, Department of Microbiology-Immunology, Chicago, IL 60611-3008.

Northwestern University, Northwestern University Feinberg School of Medicine and Interdepartmental Degree Programs, Integrated Graduate Programs in the Life Sciences, Chicago, IL 60611. Offers cancer biology (PhD); cell biology (PhD); developmental biology (PhD); evolutionary biology (PhD); immunology and microbial pathogenesis (PhD); molecular biology and genetics (PhD); neurobiology (PhD); pharmacology and toxicology (PhD); structural biology and biochemistry (PhD). *Degree requirements:* For doctorate, thesis/dissertation, written and oral qualifying exams, comprehensive exam. *Entrance requirements:* For doctorate, GRE General Test. Additional exam requirements/recommendations for international students: Required—TOEFL (minimum score 600 paper-based; 250 computer-based). Electronic applications accepted.

See Close-Up on page 189.

The Ohio State University, Graduate School, College of Biological Sciences, Department of Microbiology, Columbus, OH 43210. Offers MS, PhD. *Degree requirements:* For master's, thesis optional; for doctorate, thesis/dissertation. *Entrance requirements:* For master's and doctorate, GRE General Test, GRE Subject Test in biology or biochemistry (recommended). Additional exam requirements/recommendations for international students: Required—TOEFL (minimum score 600 paper-based; 250 computer-based), TSE. Electronic applications accepted.

Ohio University, Graduate Studies, College of Arts and Sciences, Department of Biological Sciences, Program in Microbiology, Athens, OH 45701-2979. Offers MS, PhD. *Faculty:* 11 full-time (2 women). *Students:* 5 full-time (2 women), 2 international. Average age 24. 8 applicants, 25% accepted, 2 enrolled. In 2005, 2 doctorates awarded. *Median time to degree:* Of those who began their doctoral program in fall 1997, 90% received their degree in 8

years or less. *Degree requirements:* For master's, thesis, 1 quarter of teaching experience; for doctorate, thesis/dissertation, 2 quarters of teaching experience, comprehensive exam. *Entrance requirements:* For master's and doctorate, GRE General Test. Additional exam requirements/recommendations for international students: Required—TOEFL (minimum score 620 paper-based; 260 computer-based). *Application deadline:* For fall admission, 1/15 for domestic students, 1/15 for international students. Application fee: $45. Electronic applications accepted. *Financial support:* In 2005–06, 4 students received support, including research assistantships with full tuition reimbursements available (averaging $15,500 per year), 4 teaching assistantships with full tuition reimbursements available (averaging $15,500 per year); fellowships with full tuition reimbursements available, Federal Work-Study, institutionally sponsored loans, and tuition waivers (full) also available. Financial award application deadline: 1/15. *Faculty research:* Bacteriology, virology, microbial genetics, cell and molecular biology, molecular virology. Total annual research expenditures: $1 million. *Unit head:* Dr. Donald B. Miles, Graduate Chair, 740-593-2317, Fax: 740-593-0300, E-mail: milesd@ohio.edu. *Application contact:* Dr. Donald B. Miles, Graduate Chair, 740-593-2317, Fax: 740-593-0300, E-mail: milesd@ohio.edu.

Oklahoma State University, College of Arts and Sciences, Department of Microbiology and Molecular Genetics, Stillwater, OK 74078. Offers MS, PhD. *Faculty:* 14 full-time (0 women). *Students:* 10 full-time (2 women), 15 part-time (11 women); includes 2 minority (1 African American, 1 American Indian/Alaska Native), 15 international. Average age 28. 55 applicants, 15% accepted, 4 enrolled. In 2005, 2 degrees awarded. *Degree requirements:* For master's, thesis/dissertation; for doctorate, thesis/dissertation, comprehensive exam. *Entrance requirements:* For master's and doctorate, GRE General Test. Additional exam requirements/recommendations for international students: Required—TOEFL. *Application deadline:* For fall admission, 6/1 for domestic students, 3/1 for international students. Applications are processed on a rolling basis. Application fee: $40 ($75 for international students). Electronic applications accepted. *Expenses:* Tuition, state resident: full-time $4,253; part-time $139 per credit hour. Tuition, nonresident: full-time $12,569; part-time $485 per credit hour. Required fees: $43 per credit hour. One-time fee: $20 part-time. Tuition and fees vary according to course load and program. *Financial support:* In 2005–06, 9 research assistantships (averaging $14,198 per year), 13 teaching assistantships (averaging $14,042 per year) were awarded; scholarships/grants, health care benefits, tuition waivers (full), and unspecified assistantships also available. Financial award application deadline: 3/1. *Faculty research:* Bioinformatics, genomics-genetics, virology, environmental microbiology, development-molecular mechanisms. *Unit head:* Dr. Robert V. Miller, Head, 405-744-7180.

Oregon Health & Science University, School of Medicine, Graduate Programs in Medicine, Department of Molecular Microbiology and Immunology, Portland, OR 97239-3098. Offers PhD. *Degree requirements:* For doctorate, thesis/dissertation. *Entrance requirements:* For doctorate, GRE General Test. *Faculty research:* Molecular biology of bacterial and viral pathogens, cellular and humoral immunology, molecular biology of microbes.

Oregon State University, College of Veterinary Medicine, Program in Veterinary Science, Corvallis, OR 97331. Offers microbiology (MS); pathology (MS); toxicology (MS). Part-time programs available. *Students:* 3 full-time (2 women), 2 international. Average age 27. In 2005, 1 degree awarded. *Degree requirements:* For master's, thesis. *Entrance requirements:* For master's, minimum GPA of 3.0 in last 90 hours. Additional exam requirements/recommendations for international students: Required—TOEFL. *Application deadline:* For fall admission, 11/1 for domestic students. Application fee: $50. *Expenses:* Contact institution. *Financial support:* Research assistantships, Federal Work-Study, institutionally sponsored loans, and scholarships/grants available. Support available to part-time students. Financial award application deadline: 2/1. *Faculty research:* Calf diseases, bovine foot rot, caliciviruses, effects of toxic agents on immune systems. *Unit head:* Dr. Linda L. Blythe, Associate Dean, 541-737-2098, Fax: 541-737-4245, E-mail: linda.blythe@orst.edu.

Oregon State University, Graduate School, College of Science, Department of Microbiology, Corvallis, OR 97331. Offers MA, MAIS, MS, PhD. Part-time programs available. *Faculty:* 9 full-time (5 women), 2 part-time/adjunct (0 women). *Students:* 27 full-time (15 women), 3 part-time (1 woman); includes 3 minority (2 Asian Americans or Pacific Islanders, 1 Hispanic American), 5 international. Average age 29. In 2005, 3 master's, 2 doctorates awarded. Terminal master's awarded for partial completion of doctoral program. *Degree requirements:* For master's, thesis; for doctorate, one foreign language, thesis/dissertation. *Entrance requirements:* For master's and doctorate, GRE General Test, minimum GPA of 3.0 in last 90 hours. Additional exam requirements/recommendations for international students: Required—TOEFL. *Application deadline:* For fall admission, 3/1 for domestic students. Applications are processed on a rolling basis. Application fee: $50. *Expenses:* Tuition, area resident: Part-time $301 per credit. Tuition, state resident: full-time $8,139; part-time $501 per credit. Tuition, nonresident: full-time $14,376; part-time $532 per credit. Required fees: $1,266. *Financial support:* Fellowships, research assistantships, teaching assistantships, career-related internships or fieldwork, Federal Work-Study, and institutionally sponsored loans available. Support available to part-time students. Financial award application deadline: 2/1. *Faculty research:* Genetics, physiology, biotechnology, pathogenic microbiology, plant virology. *Unit head:* Dr. Theo W Dreher, Chair, 541-737-4441, Fax: 541-737-0496. *Application contact:* Sharon Jansen, Graduate Admissions Clerk, 541-737-4441, Fax: 541-737-0496, E-mail: jansens@orst.edu.

The Pennsylvania State University Milton S. Hershey Medical Center, Graduate School Programs in the Biomedical Sciences, Graduate Program in Microbiology and Immunology, Hershey, PA 17033-2360. Offers genetics (PhD); immunology (MS, PhD); microbiology (MS); microbiology/virology (PhD); molecular biology (PhD). *Students:* 23 full-time (11 women); includes 2 minority (both Asian Americans or Pacific Islanders), 2 international. Average age 27. Terminal master's awarded for partial completion of doctoral program. *Median time to degree:* Of those who began their doctoral program in fall 1997, 100% received their degree in 8 years or less. *Degree requirements:* For master's, thesis or alternative, registration; for doctorate, thesis/dissertation, oral exam, comprehensive exam, registration. *Entrance requirements:* For master's, GRE or MCAT; for doctorate, GRE General Test or MCAT, minimum GPA of 3.0. Additional exam requirements/recommendations for international students: Required—TOEFL. *Application deadline:* Applications are processed on a rolling basis. Application fee: $45. Electronic applications accepted. *Financial support:* In 2005–06, 23 research assistantships with full tuition reimbursements were awarded; fellowships with full tuition reimbursements, scholarships/grants, health care benefits, and unspecified assistantships also available. Financial award applicants required to submit FAFSA. *Faculty research:* Virus replication and assembly, oncogenesis, interactions of viruses with host cells and animal model systems. *Unit head:* Dr. Richard J. Courtney, Chair, 717-531-7659, Fax: 717-531-6522, E-mail: micro-grad-hmc@psu.edu. *Application contact:* Billie Burns, Secretary, 717-531-7659, Fax: 717-531-6522, E-mail: micro-grad-hmc@psu.edu.

The Pennsylvania State University University Park Campus, Graduate School, Eberly College of Science, Department of Biochemistry and Molecular Biology, Program in Biochemistry, Microbiology, and Molecular Biology, State College, University Park, PA 16802-1503. Offers MS, PhD. *Degree requirements:* For master's, thesis/dissertation; for doctorate, thesis/dissertation, comprehensive exam. *Entrance requirements:* For master's and doctorate, GRE General Test, GRE Subject Test. Application fee: $45. *Expenses:* Tuition, state resident: full-time $12,518; part-time $522 per credit. Tuition, nonresident: full-time $23,004; part-time $959 per credit. Required fees: $484. Tuition and fees vary according to course load, campus/location and program. *Financial support:* Fellowships, unspecified assistantships available.

See Close-Up on page 407.

Purdue University, Graduate School, School of Science, Department of Biological Sciences, West Lafayette, IN 47907. Offers biochemistry (PhD); biophysics (PhD); cell and developmental biology (PhD); ecology, evolutionary and population biology (MS, PhD), including ecology, evolutionary biology, population biology; genetics (MS, PhD); microbiology (MS, PhD); molecular biology (PhD); neurobiology (MS, PhD); plant physiology (PhD). *Faculty:* 47 full-time (9 women), 4 part-time/adjunct (1 woman). *Students:* 97 full-time (53 women), 8 part-time (4 women);

includes 13 minority (3 African Americans, 1 American Indian/Alaska Native, 4 Asian Americans or Pacific Islanders, 5 Hispanic Americans), 50 international. Average age 28. 168 applicants, 29% accepted, 23 enrolled. In 2005, 18 master's, 9 doctorates awarded. Terminal master's awarded for partial completion of doctoral program. *Degree requirements:* For master's, thesis (for some programs); for doctorate, thesis/dissertation, seminars, teaching experience. *Entrance requirements:* For master's and doctorate, GRE General Test. Additional exam requirements/recommendations for international students: Required—TOEFL, TSE. *Application deadline:* For fall admission, 2/15 for domestic students, 1/31 for international students. Applications are processed on a rolling basis. Application fee: $55. Electronic applications accepted. *Financial support:* In 2005–06, 15 fellowships, 60 research assistantships, 53 teaching assistantships were awarded. Support available to part-time students. Financial award application deadline: 2/15; financial award applicants required to submit FAFSA. *Unit head:* Dr. Richard J Kuhn, Head, 765-494-4407. *Application contact:* Nancy Konopka, Graduate Studies Office Manager, 765-494-8142, Fax: 765-494-0876, E-mail: njk@bilbo.bio.purdue.edu.

Purdue University, School of Veterinary Medicine and Graduate School, Graduate Programs in Veterinary Medicine, Department of Veterinary Pathobiology, West Lafayette, IN 47907. Offers biochemistry and molecular biology (MS, PhD); comparative epidemiology (MS, PhD); epidemiology (MS, PhD); immunology (MS, PhD); infectious diseases (MS, PhD); interdisciplinary genetics (PhD); laboratory animal medicine (MS, PhD); microbiology (MS, PhD); molecular virology (MS, PhD); parasitology (MS, PhD); pathobiology (MS, PhD); public health epidemiology (MS, PhD); toxicology (MS, PhD); veterinary anatomic pathology (MS, PhD); veterinary clinical pathology (MS, PhD); virology (MS, PhD). *Faculty:* 32 full-time (7 women). *Students:* 49 full-time (20 women), 3 part-time (1 woman); includes 2 minority (both African Americans), 31 international. Average age 35. In 2005, 3 master's, 8 doctorates awarded. Terminal master's awarded for partial completion of doctoral program. *Degree requirements:* For master's, thesis (for some programs); for doctorate, thesis/dissertation. *Entrance requirements:* For master's and doctorate, GRE General Test. Additional exam requirements/recommendations for international students: Required—TOEFL (minimum score 575 paper-based), TWE (minimum score 4). *Application deadline:* For fall admission, 8/12 for domestic students, 6/15 for international students; for spring admission, 1/12 for domestic students, 10/15 for international students. Application fee: $55. *Financial support:* Fellowships, research assistantships, teaching assistantships available. Financial award application deadline: 3/1; financial award applicants required to submit FAFSA. *Unit head:* Dr. H. Hogenesch, Head, 765-494-7543.

Queen's University at Kingston, School of Graduate Studies and Research, Faculty of Health Sciences, Department of Microbiology and Immunology, Kingston, ON K7L 3N6, Canada. Offers M Sc, PhD. Part-time programs available. *Degree requirements:* For master's, thesis/dissertation, registration; for doctorate, thesis/dissertation, comprehensive exam, registration. *Entrance requirements:* For master's and doctorate, minimum B+ average. Additional exam requirements/recommendations for international students: Required—TOEFL (minimum score 600 paper-based; 250 computer-based). Electronic applications accepted. *Faculty research:* Bacteriology, virology, immunology, education in microbiology and immunology, microbial pathogenesis.

Quinnipiac University, School of Health Sciences, Programs in Medical Laboratory Sciences, Hamden, CT 06518-1940. Offers biomedical sciences (MHS); laboratory management (MHS); microbiology (MHS). *Accreditation:* NAACLS. Part-time and evening/weekend programs available. *Faculty:* 2 full-time (0 women), 3 part-time/adjunct (2 women). *Students:* 21 full-time (15 women), 26 part-time (20 women); includes 6 minority (3 African Americans, 2 Asian Americans or Pacific Islanders, 1 Hispanic American), 2 international. Average age 29. 31 applicants, 77% accepted, 17 enrolled. In 2005, 10 degrees awarded. *Degree requirements:* For master's, thesis optional. *Entrance requirements:* For master's, minimum GPA of 2.5; bachelor's degree in biological, medical, or health sciences. Additional exam requirements/recommendations for international students: Required—TOEFL (minimum score 575 paper-based; 233 computer-based). *Application deadline:* For fall admission, 7/30 priority date for domestic students, 5/30 priority date for international students; for spring admission, 12/15 priority date for domestic students, 10/15 priority date for international students. Applications are processed on a rolling basis. Application fee: $45. Electronic applications accepted. *Expenses:* Tuition: Part-time $570 per credit. *Financial support:* Tuition waivers (partial) and unspecified assistantships available. Support available to part-time students. Financial award application deadline: 4/15; financial award applicants required to submit FAFSA. *Faculty research:* Microbial physiology, fermentation technology. *Unit head:* Dr. Kenneth Kaloustian, Director, 203-582-8676, Fax: 203-582-3443, E-mail: ken.kaloustian@quinnipiac.edu. *Application contact:* Louise Howe, Associate Director of Graduate Admissions, 800-462-1944, Fax: 203-582-3443, E-mail: graduate@quinnipiac.edu.

Rensselaer Polytechnic Institute, Graduate School, School of Science, Department of Biology, Troy, NY 12180-3590. Offers biochemistry (MS, PhD); biophysics (MS, PhD); cell biology (MS, PhD); developmental biology (MS, PhD); microbiology (MS, PhD); molecular biology (MS, PhD). Part-time programs available. Terminal master's awarded for partial completion of doctoral program. *Degree requirements:* For master's and doctorate, thesis/dissertation, comprehensive exam, registration. *Entrance requirements:* For master's and doctorate, GRE General Test. Additional exam requirements/recommendations for international students: Required—TOEFL. Electronic applications accepted. *Expenses:* Tuition: Full-time $31,000; part-time $1,320 per credit. Required fees: $1,623. *Faculty research:* Bioinformatics, molecular biology/biochemistry, cell and tissue biology, environment, ecology.

See Close-Up on page 199.

Rosalind Franklin University of Medicine and Science, School of Graduate and Postdoctoral Studies, Department of Microbiology and Immunology, North Chicago, IL 60064-3095. Offers medical microbiology (MS, PhD); microbiology and immunology (MS, PhD). Part-time programs available. Terminal master's awarded for partial completion of doctoral program. *Degree requirements:* For master's and doctorate, thesis/dissertation. *Entrance requirements:* For master's and doctorate, GRE General Test. Additional exam requirements/recommendations for international students: Required—TOEFL, TWE. *Faculty research:* Molecular biology, parasitology, virology.

Rush University, Graduate College, Division of Immunology and Microbiology, Chicago, IL 60612-3832. Offers microbiology (PhD); virology (MS, PhD), including immunology, virology. *Degree requirements:* For doctorate, thesis/dissertation, comprehensive preliminary exam. *Entrance requirements:* For doctorate, GRE General Test, interview, minimum GPA of 3.0. Additional exam requirements/recommendations for international students: Required—TOEFL. *Faculty research:* Immune interactions of cells and membranes, HIV immunopathogenesis, autoimmunity, tumor biology.

Rutgers, The State University of New Jersey, New Brunswick/Piscataway, Graduate School, Program in Microbiology and Molecular Genetics, New Brunswick, NJ 08901-1281. Offers applied microbiology (MS, PhD); clinical microbiology (MS, PhD); computational molecular biology (PhD); immunology (MS, PhD); microbial biochemistry (MS, PhD); molecular genetics (MS, PhD); virology (MS, PhD). Part-time programs available. *Faculty:* 132 full-time. *Students:* 57 full-time (31 women), 18 part-time (8 women); includes 16 minority (9 Asian Americans or Pacific Islanders, 7 Hispanic Americans), 20 international. Average age 29. 104 applicants, 32% accepted, 19 enrolled. In 2005, 9 master's, 10 doctorates awarded. Terminal master's awarded for partial completion of doctoral program. *Median time to degree:* Of those who began their doctoral program in fall 1997, 100% received their degree in 8 years or less. *Degree requirements:* For master's, thesis or alternative, comprehensive exam, registration; for doctorate, thesis/dissertation, written qualifying exam, comprehensive exam, registration. *Entrance requirements:* For master's, GRE General Test, minimum GPA of 3.0; for doctorate, GRE General Test, GRE Subject Test (recommended), minimum GPA of 3.0. Additional exam requirements/recommendations for international students: Required—TOEFL. *Application deadline:* For fall admission, 1/5 priority date for domestic students, 11/1 priority date for international students. Applications are processed on a rolling basis. Application fee: $50.

Electronic applications accepted. *Expenses:* Tuition, state resident: full-time $10,440; part-time $435 per credit. Tuition, nonresident: full-time $15,520; part-time $647 per credit. Required fees: $129 per credit. Tuition and fees vary according to program. *Financial support:* In 2005–06, 48 students received support, including 14 fellowships with full tuition reimbursements available (averaging $24,000 per year), 25 research assistantships with full tuition reimbursements available (averaging $24,000 per year), 9 teaching assistantships with full tuition reimbursements available (averaging $16,988 per year); Federal Work-Study, institutionally sponsored loans, scholarships/grants, and unspecified assistantships also available. Financial award application deadline: 1/5; financial award applicants required to submit FAFSA. *Faculty research:* Molecular genetics and microbial physiology; virology and pathogenic microbiology; applied, environmental and industrial microbiology; computers in molecular biology. *Unit head:* Dr. Andrew K. Vershon, Director, 732-445-2905, Fax: 732-445-6370, E-mail: vershon@waksman.rutgers.edu. *Application contact:* Diane Murano, Administrative Assistant, 732-445-5086, Fax: 732-445-6370, E-mail: murano@biology.rutgers.edu.

Saint Louis University, Graduate School and School of Medicine, Graduate Program in Biomedical Sciences and Graduate School, Department of Molecular Microbiology and Immunology, St. Louis, MO 63103-2097. Offers PhD. *Faculty:* 20 full-time (4 women), 1 (woman) part-time/adjunct. *Students:* 22 full-time (10 women); includes 1 minority (Hispanic American), 1 international. Average age 27. 4 applicants, 50% accepted, 2 enrolled. In 2005, 1 degree awarded. *Degree requirements:* For doctorate, thesis/dissertation, qualifying exams, comprehensive exam. *Entrance requirements:* For doctorate, GRE General Test, letters of recommendation, resumé, interview. Additional exam requirements/recommendations for international students: Required—TOEFL (minimum score 550 paper-based; 213 computer-based). *Application deadline:* For fall admission, 7/1 for domestic students, 7/1 for international students; for spring admission, 11/1 for domestic students, 11/1 for international students. Applications are processed on a rolling basis. Application fee: $40. *Expenses:* Tuition: Part-time $760 per credit hour. Required fees: $55 per semester. *Financial support:* In 2005–06, 12 students received support. Health care benefits and unspecified assistantships available. Support available to part-time students. Financial award application deadline: 6/1; financial award applicants required to submit FAFSA. *Faculty research:* Biodefense, cancer gene therapy, rheumatoid arthritis. Total annual research expenditures: $3.6 million. *Unit head:* Dr. William Wold, Chairperson, 314-977-8857, Fax: 314-773-3403, E-mail: woldws@slu.edu. *Application contact:* Gary Behrman, Associate Dean of the Graduate School, 314-977-3827, E-mail: behrmang@slu.edu.

San Diego State University, Graduate and Research Affairs, College of Sciences, Department of Biological Sciences, Program in Microbiology, San Diego, CA 92182. Offers MS. *Students:* 3 full-time (1 woman), 3 part-time (all women); includes 1 minority (Hispanic American), 3 international. Average age 30. 12 applicants, 8% accepted, 1 enrolled. In 2005, 1 degree awarded. *Degree requirements:* For master's, thesis, oral exam. *Entrance requirements:* For master's, GRE General Test, GRE Subject Test, resumé or curriculum vitae, 2 letters of recommendation.. Additional exam requirements/recommendations for international students: Required—TOEFL. *Application deadline:* For fall admission, 5/1 for domestic students; for spring admission, 11/1 for domestic students, 10/1 for international students. Applications are processed on a rolling basis. Application fee: $55. Electronic applications accepted. *Financial support:* Fellowships, research assistantships, teaching assistantships, unspecified assistantships available. Financial award applicants required to submit FAFSA. *Unit head:* Terry Frey, Head, 619-594-6756, Fax: 619-594-5676, E-mail: gradcoor@sciences.sdsu.edu. *Application contact:* Terry Frey, Graduate Coordinator, 619-594-6756, Fax: 619-594-5676, E-mail: gradcoor@sciences.sdsu.edu.

San Francisco State University, Division of Graduate Studies, College of Science and Engineering, Department of Biology, Program in Microbiology, San Francisco, CA 94132-1722. Offers MA. *Entrance requirements:* For master's, minimum GPA of 2.5 in last 60 units.

San Jose State University, Graduate Studies and Research, College of Science, Department of Biological Sciences, San Jose, CA 95192-0001. Offers biological sciences (MA, MS); molecular biology and microbiology (MS); organismal biology, conservation and ecology (MS); physiology (MS). Part-time programs available. *Students:* 44 full-time (35 women), 33 part-time (24 women); includes 39 minority (33 Asian Americans or Pacific Islanders, 6 Hispanic Americans), 12 international. Average age 31. 140 applicants, 40% accepted, 29 enrolled. In 2005, 39 degrees awarded. *Entrance requirements:* For master's, GRE. *Application deadline:* For fall admission, 6/29 for domestic students; for spring admission, 11/30 for domestic students. Applications are processed on a rolling basis. Application fee: $59. Electronic applications accepted. *Expenses:* Tuition, nonresident: part-time $339 per unit. Required fees: $1,286 per semester. Tuition and fees vary according to course load and degree level. *Financial support:* In 2005–06, 13 teaching assistantships were awarded; Federal Work-Study also available. Financial award applicants required to submit FAFSA. *Faculty research:* Systemic physiology, molecular genetics, SEM studies, toxicology, large mammal ecology. *Unit head:* Dr. Sally Veregge, Chair, 408-924-4900, Fax: 408-924-4840. *Application contact:* Dr. Howard Shellhammer, Graduate Coordinator, 408-924-4897.

Seton Hall University, College of Arts and Sciences, Department of Biological Sciences, South Orange, NJ 07079-2697. Offers biology (MS); microbiology (MS); molecular bioscience (PhD). Part-time and evening/weekend programs available. *Students:* 17 full-time (14 women), 35 part-time (18 women). Average age 29. 33 applicants, 70% accepted, 14 enrolled. In 2005, 8 degrees awarded. *Degree requirements:* For master's, research paper or thesis, seminar. *Application deadline:* For fall admission, 7/1 priority date for domestic students, 7/1 priority date for international students; for spring admission, 11/1 priority date for domestic students, 11/1 priority date for international students. Applications are processed on a rolling basis. Application fee: $50. Electronic applications accepted. *Financial support:* Teaching assistantships, career-related internships or fieldwork and Federal Work-Study available. *Faculty research:* Neurobiology, genetics, immunology, molecular biology, cellular physiology, toxicology, microbiology, bioinformatics. *Unit head:* Dr. Carolyn Bentivegna, Chair, 973-761-9044, Fax: 973-761-9596, E-mail: bentivca@shu.edu. *Application contact:* Dr. Carroll D. Rawn, Director of Graduate Studies, 973-761-9054, Fax: 973-761-9596, E-mail: rawncarr@shu.edu.

See Close-Up on page 593.

South Dakota State University, Graduate School, College of Agriculture and Biological Sciences, Department of Biology/Microbiology, Brookings, SD 57007. Offers biology (MS); microbiology (MS). *Degree requirements:* For master's, thesis, oral exam. *Entrance requirements:* For master's, GRE. Additional exam requirements/recommendations for international students: Required—TOEFL. *Faculty research:* Plant tissue culture, molecular biology studies of metabolic regulation in plants, mechanisms of mammalian gene expression, aquatic-wetland ecosystem ecology, stress-induced immunosuppression on parasite-induced pathology.

Southern Illinois University Carbondale, Graduate School, College of Science, Program in Molecular Biology, Microbiology, and Biochemistry, Carbondale, IL 62901-4701. Offers MS, PhD. *Faculty:* 16 full-time (2 women). *Students:* 39 full-time (23 women), 37 part-time (19 women); includes 11 minority (9 African Americans, 2 Asian Americans or Pacific Islanders), 40 international. Average age 25. 134 applicants, 16% accepted, 17 enrolled. In 2005, 7 master's, 4 doctorates awarded. *Degree requirements:* For master's and doctorate, thesis/dissertation. *Entrance requirements:* For master's, GRE, minimum GPA of 2.7; for doctorate, GRE, minimum GPA of 3.25. Additional exam requirements/recommendations for international students: Required—TOEFL. *Application deadline:* Applications are processed on a rolling basis. Application fee: $20. *Financial support:* In 2005–06, 40 students received support, including 3 fellowships with full tuition reimbursements available, 24 research assistantships with full tuition reimbursements available, 12 teaching assistantships with full tuition reimbursements available; Federal Work-Study and institutionally sponsored loans also available. Support available to part-time students. Financial award application deadline: 3/1. *Faculty research:* Prokaryotic gene regulation and expression; eukaryotic gene regulation; microbial, phylogenetic, and metabolic diversity; immune responses to tumors, pathogens, and autoantigens; protein fold-

Microbiology

Southern Illinois University Carbondale (continued)

ing and structure. *Unit head:* Dr. John Martinko, Director, 618-453-8116, Fax: 618-453-8036, E-mail: martinko.mbmb@science.siu.edu. *Application contact:* Donna Mueller, Office Systems Specialist II, 618-536-2349, Fax: 618-453-8036, E-mail: mueller@micro.siu.edu.

See Close-Up on page 595.

Southwestern Oklahoma State University, College of Professional and Graduate Studies, School of Behavioral Sciences and Education, Specialization in Health Sciences and Microbiology, Weatherford, OK 73096-3098. Offers M Ed. *Faculty:* 7 full-time (4 women), 11 part-time/adjunct (5 women). *Students:* 1 (woman) full-time; minority (African American) Average age 24. *Expenses:* Tuition, state resident: part-time $130 per credit hour. Tuition, nonresident: part-time $305 per credit hour. Required fees: $25 per credit hour. Tuition and fees vary according to course level and course load. *Unit head:* Dr. Gary Wolgamott, Associate Dean, 580-774-3079, Fax: 530-774-3795, E-mail: gary.wolgamott@swosu.edu.

Stanford University, School of Medicine, Graduate Programs in Medicine, Department of Microbiology and Immunology, Stanford, CA 94305-9991. Offers PhD. *Degree requirements:* For doctorate, thesis/dissertation, 2 quarters teaching assistantship, comprehensive exam. *Entrance requirements:* For doctorate, GRE General Test, GRE Subject Test (biology or biochemistry). Additional exam requirements/recommendations for international students: Required—TOEFL. Electronic applications accepted. *Faculty research:* Molecular pathogenesis of bacteria viruses and parasites, immune system function, autoimmunity, molecular biology.

See Close-Up on page 893.

State University of New York at Buffalo, Graduate School, School of Medicine and Biomedical Sciences, Graduate Programs in Medicine and Biomedical Sciences, Department of Microbiology and Immunology, Buffalo, NY 14260. Offers MA, PhD. *Faculty:* 16 full-time (4 women), 2 part-time/adjunct (0 women). *Students:* 39 full-time (19 women), 1 (woman) part-time; includes 4 minority (3 Asian Americans or Pacific Islanders, 1 Hispanic American), 13 international. Average age 29. 12 applicants, 17% accepted, 0 enrolled. In 2005, 5 master's, 6 doctorates awarded. *Median time to degree:* Of those who began their doctoral program in fall 1997, 100% received their degree in 8 years or less. *Degree requirements:* For master's, comprehensive exam; for doctorate, thesis/dissertation, departmental qualifying exam. *Entrance requirements:* For master's, GRE General Test; for doctorate, GRE General Test, 3 letters of recommendation. Additional exam requirements/recommendations for international students: Required—TOEFL. *Application deadline:* For fall admission, 2/1 priority date for domestic students, 2/1 priority date for international students. Applications are processed on a rolling basis. Application fee: $35. Electronic applications accepted. *Financial support:* In 2005–06, 7 fellowships with tuition reimbursements (averaging $23,500 per year), 25 research assistantships with tuition reimbursements (averaging $21,000 per year), teaching assistantships with tuition reimbursements (averaging $19,000 per year) were awarded; Federal Work-Study, institutionally sponsored loans, traineeships, health care benefits, and unspecified assistantships also available. Financial award application deadline: 2/1; financial award applicants required to submit FAFSA. *Faculty research:* Bacteriology, immunology, parasitology, virology, microbial pathogenesis. Total annual research expenditures: $6.1 million. *Unit head:* Dr. John Hay, Chairman, 716-829-2907, Fax: 716-829-2158. *Application contact:* Dr. Laurie K. Read, Director of Graduate Studies, 716-829-2176, Fax: 716-829-2158.

State University of New York Upstate Medical University, College of Graduate Studies, Department of Microbiology and Immunology, Syracuse, NY 13210-2334. Offers MS, PhD, MD/PhD. Terminal master's awarded for partial completion of doctoral program. *Degree requirements:* For master's, thesis/dissertation; for doctorate, thesis/dissertation, comprehensive exam. *Entrance requirements:* For master's, GRE General Test, GRE Subject Test, interview; for doctorate, GRE General Test, interview. Additional exam requirements/recommendations for international students: Required—TOEFL. *Faculty research:* Virology, molecular biology, microbial genetics, parasitology.

Stony Brook University, State University of New York, Health Sciences Center, School of Medicine and Graduate School, Graduate Programs in Medicine, Department of Molecular Genetics and Microbiology, Stony Brook, NY 11794. Offers molecular microbiology (PhD). *Faculty:* 18 full-time (6 women). *Students:* 33 full-time (22 women); includes 5 minority (2 African Americans, 1 Asian American or Pacific Islander, 2 Hispanic Americans), 7 international. Average age 27. 44 applicants, 16% accepted. In 2005, 7 degrees awarded. *Degree requirements:* For doctorate, thesis/dissertation, comprehensive exam. *Entrance requirements:* For doctorate, GRE General Test, GRE Subject Test. Additional exam requirements/recommendations for international students: Required—TOEFL. *Application deadline:* For fall admission, 1/15 for domestic students. Application fee: $50. *Expenses:* Tuition, state resident: full-time $6,900; part-time $288 per credit. Tuition, nonresident: full-time $10,920; part-time $455 per credit. Required fees: $704. *Financial support:* In 2005–06, 12 fellowships, 29 research assistantships, 2 teaching assistantships were awarded; Federal Work-Study also available. Financial award application deadline: 3/15. *Faculty research:* Adenovirus molecular genetics, molecular biology of tumors, virus SV40, mechanism of tumor infection by SAV virus. Total annual research expenditures: $8.2 million. *Unit head:* Dr. Dafna Bar-Sagi, Chair, 631-632-8812, Fax: 631-632-9797. *Application contact:* Dr. Patrick Hearing, Director, 631-632-8813, Fax: 631-632-9797, E-mail: hearing@asterix.bio.sunysb.edu.

See Close-Up on page 895.

Temple University, Health Sciences Center, School of Medicine and Graduate School, Graduate Programs in Medicine, Department of Microbiology and Immunology, Philadelphia, PA 19140-5104. Offers MS, PhD, MD/PhD. *Degree requirements:* For master's, thesis; for doctorate, thesis/dissertation, research seminars. *Entrance requirements:* For master's and doctorate, GRE General Test, GRE Subject Test, minimum GPA of 3.0. Additional exam requirements/recommendations for international students: Required—TOEFL (minimum score 600 paper-based; 250 computer-based). Electronic applications accepted. *Expenses:* Tuition, state resident: full-time $8,694; part-time $483 per credit. Tuition, nonresident: full-time $12,672; part-time $704 per credit. Required fees: $500; $122 per semester. Tuition and fees vary according to course level, campus/location and program. *Faculty research:* Molecular and cellular immunology, molecular and biochemical microbiology, molecular genetics.

See Close-Up on page 897.

Texas A&M University, College of Science, Department of Biology, Program in Microbiology, College Station, TX 77843. Offers MS, PhD. *Students:* Average age 28. *Degree requirements:* For master's, thesis or alternative, registration; for doctorate, thesis/dissertation, comprehensive exam, registration. *Entrance requirements:* For master's and doctorate, GRE General Test. Additional exam requirements/recommendations for international students: Required—TOEFL. Application fee: $50 ($75 for international students). *Expenses:* Tuition, state resident: full-time $4,488; part-time $187 per credit hour. Tuition, nonresident: full-time $11,112; part-time $463 per credit hour. Required fees: $1,974. *Financial support:* Application deadline: 4/1; *Unit head:* Dr. Mark Zoran, Head, 979-845-7755.

Texas A&M University, College of Veterinary Medicine, Graduate Programs in Veterinary Medicine, Department of Veterinary Pathobiology, College Station, TX 77843. Offers genetics (MS, PhD); veterinary microbiology (MS, PhD); veterinary parasitology (MS); veterinary pathology (MS, PhD). Part-time programs available. Postbaccalaureate distance learning degree programs offered. *Faculty:* 16 full-time (3 women), 1 part-time/adjunct (1 woman). *Students:* 45 full-time (28 women), 20 part-time (13 women); includes 7 minority (2 African Americans, 5 Hispanic Americans), 19 international. Average age 33. 25 applicants, 76% accepted, 19 enrolled. In 2005, 5 master's, 5 doctorates awarded. Terminal master's awarded for partial completion of doctoral program. *Degree requirements:* For master's and doctorate, thesis/dissertation, seminars. *Entrance requirements:* For master's and doctorate, GRE General Test, minimum GPA of 3.0 in last 60 hours. Additional exam requirements/recommendations

for international students: Required—TOEFL. *Application deadline:* For fall admission, 3/1 for domestic students; for spring admission, 8/1 priority date for domestic students. Applications are processed on a rolling basis. Application fee: $50 ($75 for international students). Electronic applications accepted. *Expenses:* Tuition, state resident: full-time $4,488; part-time $187 per credit hour. Tuition, nonresident: full-time $11,112; part-time $463 per credit hour. Required fees: $1,974. *Financial support:* In 2005–06, fellowships with partial tuition reimbursements (averaging $16,000 per year), research assistantships with partial tuition reimbursements (averaging $15,400 per year), teaching assistantships with partial tuition reimbursements (averaging $16,000 per year) were awarded; Federal Work-Study, institutionally sponsored loans, scholarships/grants, traineeships, health care benefits, and unspecified assistantships also available. Support available to part-time students. Financial award applicants required to submit FAFSA. *Faculty research:* Infectious and noninfectious diseases of animals and birds, animal genetics, molecular biology, immunology, virology. *Unit head:* Dr. Ann B. Kier, Head, 979-845-5941, Fax: 979-845-9231, E-mail: akier@cvm.tamu.edu. *Application contact:* Dr. G. G. Wagner, Graduate Advisor, 979-845-2851, Fax: 979-862-1147, E-mail: gwagner@cvm.tamu.edu.

Texas A&M University System Health Science Center, Graduate School of Biomedical Sciences, Department of Medical Microbiology and Immunology, College Station, TX 77840. Offers immunology (PhD); microbiology (PhD); molecular biology (PhD); virology (PhD). *Degree requirements:* For doctorate, thesis/dissertation. *Entrance requirements:* For doctorate, GRE General Test, minimum GPA of 3.0. *Faculty research:* Molecular pathogenesis, microbial therapeutics.

Texas Tech University, Graduate School, College of Arts and Sciences, Department of Biological Sciences, Lubbock, TX 79409. Offers biological informatics (MS); biology (MS, PhD); microbiology (MS); zoology (MS, PhD). Part-time programs available. *Faculty:* 29 full-time (4 women). *Students:* 101 full-time (47 women), 9 part-time (2 women); includes 7 minority (2 Asian Americans or Pacific Islanders, 5 Hispanic Americans), 42 international. Average age 30. 74 applicants, 50% accepted, 16 enrolled. In 2005, 8 master's, 5 doctorates awarded. *Degree requirements:* For master's, thesis (for some programs); for doctorate, thesis/dissertation. *Entrance requirements:* For master's and doctorate, GRE General Test. Additional exam requirements/recommendations for international students: Required—TOEFL (minimum score 550 paper-based; 213 computer-based). *Application deadline:* Applications are processed on a rolling basis. Application fee: $50 ($60 for international students). Electronic applications accepted. *Expenses:* Tuition, state resident: full-time $4,296. Tuition, nonresident: full-time $10,920. Required fees: $1,992. Tuition and fees vary according to program. *Financial support:* In 2005–06, 51 students received support, including 22 research assistantships with partial tuition reimbursements available (averaging $13,849 per year), 74 teaching assistantships with partial tuition reimbursements available (averaging $13,899 per year); career-related internships or fieldwork, Federal Work-Study, and institutionally sponsored loans also available. Support available to part-time students. Financial award application deadline: 4/15; financial award applicants required to submit FAFSA. *Faculty research:* Genome organization and evolution, plant stress response, climate change on arid ecosystems, cell signaling hormone regulation. Total annual research expenditures: $3.1 million. *Unit head:* Dr. John C. Zak, Chair, 806-742-2715, Fax: 806-742-2963, E-mail: john.zak@ttu.edu. *Application contact:* Dr. Randall M. Jeter, Graduate Adviser, 806-742-2710, Fax: 806-742-2963, E-mail: randall.jeter@ttu.edu.

Thomas Jefferson University, Jefferson College of Graduate Studies, Program in Immunology and Microbial Pathogenesis, Philadelphia, PA 19107. Offers PhD. *Students:* 32 full-time (16 women); includes 2 minority (1 Asian American or Pacific Islander, 1 Hispanic American), 7 international. 63 applicants, 13% accepted, 5 enrolled. In 2005, 1 doctorate awarded. *Degree requirements:* For doctorate, thesis/dissertation, comprehensive exam, registration. *Entrance requirements:* For doctorate, GRE General Test, minimum GPA of 3.2. Additional exam requirements/recommendations for international students: Required—TOEFL (minimum score 213 computer-based). *Application deadline:* For fall admission, 3/1 priority date for domestic students, 1/1 priority date for international students. Applications are processed on a rolling basis. Application fee: $50. Electronic applications accepted. *Expenses:* Tuition: Full-time $14,894; part-time $800 per credit. *Financial support:* In 2005–06, 9 students received support, including 32 fellowships with full tuition reimbursements available; research assistantships, Federal Work-Study, institutionally sponsored loans, scholarships/grants, and traineeships also available. Support available to part-time students. Financial award application deadline: 5/1; financial award applicants required to submit FAFSA. *Unit head:* Dr. Jay Rothstein, Program Director, 215-503-4622, E-mail: jay.rothstein@jefferson.edu. *Application contact:* Jessie F. Pervall, Director of Admissions, 215-503-0155, Fax: 215-503-9920, E-mail: jessie.pervall@jefferson.edu.

See Close-Up on page 899.

Thomas Jefferson University, Jefferson College of Graduate Studies, Program in Microbiology, Philadelphia, PA 19107. Offers MS. Part-time and evening/weekend programs available. *Faculty:* 11 full-time (2 women), 2 part-time/adjunct (0 women). *Students:* 28 applicants, 68% accepted, 16 enrolled. In 2005, 17 degrees awarded. *Degree requirements:* For master's, thesis, registration. *Entrance requirements:* For master's, GRE General Test, minimum GPA of 3.0. Additional exam requirements/recommendations for international students: Required—TOEFL (minimum score 213 computer-based). *Application deadline:* For fall admission, 8/1 priority date for domestic students, 3/1 priority date for international students. For winter admission, 12/1 for domestic students; for spring admission, 4/1 for domestic students. Applications are processed on a rolling basis. Application fee: $50. Electronic applications accepted. *Expenses:* Contact institution. *Financial support:* In 2005–06, 24 students received support. Federal Work-Study and institutionally sponsored loans available. Support available to part-time students. Financial award application deadline: 5/1; financial award applicants required to submit FAFSA. *Faculty research:* Virological procedures for characterization of HIV isolates, epidemiological studies of nosocomial infections, immunogenetics, molecular microbiology for identification of infectious diseases. *Unit head:* Dr. Dennis M. Gross, Associate Dean, 215-503-0156, Fax: 215-503-3433, E-mail: dennis.gross@jefferson.edu. *Application contact:* Dr. Dennis M. Gross, Associate Dean, 215-503-0156, Fax: 215-503-3433, E-mail: dennis.gross@jefferson.edu.

Tufts University, Sackler School of Graduate Biomedical Sciences, Department of Molecular Biology and Microbiology, Boston, MA 02155. Offers molecular microbiology (PhD), including microbiology, molecular biology, molecular microbiology. *Faculty:* 18 full-time (7 women). *Students:* 48 full-time (33 women); includes 12 minority (1 American Indian/Alaska Native, 5 Asian Americans or Pacific Islanders, 6 Hispanic Americans), 12 international. Average age 28. 100 applicants, 12% accepted, 6 enrolled. In 2005, 5 degrees awarded. *Degree requirements:* For doctorate, thesis/dissertation, graduate seminar, journal club, graduate research, comprehensive exam, registration. *Entrance requirements:* For doctorate, GRE General Test, 3 letters of reference. Additional exam requirements/recommendations for international students: Required—TOEFL. *Application deadline:* For fall admission, 1/15 priority date for domestic students, 1/15 priority date for international students. Applications are processed on a rolling basis. Application fee: $65. Electronic applications accepted. *Financial support:* In 2005–06, 48 students received support, including 48 research assistantships with full tuition reimbursements available (averaging $29,000 per year); scholarships/grants, health care benefits, and tuition waivers (full) also available. Financial award application deadline: 1/15. *Faculty research:* Fundamental problems of molecular biology of prokaryotes, eukaryotes and their viruses. *Unit head:* Dr. Abraham L. Sonenshein, Director, 617-636-6761, Fax: 617-636-0337, E-mail: linc.sonenshein@tufts.edu. *Application contact:* 617-636-6767, Fax: 617-633-0375, E-mail: sackler-school@tufts.edu.

Tulane University, School of Medicine and Graduate School, Graduate Programs in Medicine, Department of Microbiology and Immunology, New Orleans, LA 70118-5669. Offers MS, PhD, MD/PhD. MS and PhD offered through the Graduate School. *Degree requirements:* For master's, thesis; for doctorate, 2 foreign languages, thesis/dissertation. *Entrance requirements:* For master's, GRE General Test, minimum B average in undergraduate course work; for doctorate, GRE General Test, GRE Subject Test. Additional exam requirements/recommendations for international students: Required—TOEFL or TSE. Electronic applications accepted. *Faculty research:* Vaccine development, viral pathogenesis, molecular virology, bacterial pathogenesis, fungal pathogenesis.

Uniformed Services University of the Health Sciences, School of Medicine, Programs in Biomedical Sciences, Department of Microbiology and Immunology, Bethesda, MD 20814-4799. Offers PhD. Students referred to program in Emerging Infectious Diseases, Microbiology Track. *Faculty:* 13 full-time (7 women), 8 part-time/adjunct (4 women). *Students:* 1 (woman) full-time. Average age 28. In 2005, 1 degree awarded. *Degree requirements:* For doctorate, one foreign language, thesis/dissertation, qualifying exam, comprehensive exam. *Application deadline:* For fall admission, 1/15 for domestic students. Applications are processed on a rolling basis. Application fee: $0. *Financial support:* Tuition waivers (full) available. *Faculty research:* Infections diseases, virology, bacteriology, microbial pathogenesis. *Unit head:* Dr. Alison O'Brien, Chair, 301-295-3400, E-mail: aobrien@usuhs.mil. *Application contact:* Janet M. Anastasi, Graduate Program Coordinator, 301-295-9474, Fax: 301-295-6772, E-mail: janastasi@usuhs.mil.

See Close-Up on page 903.

Université de Montréal, Faculty of Medicine and Faculty of Graduate Studies, Graduate Programs in Medicine, Department of Microbiology and Immunology, Montréal, QC H3C 3J7, Canada. Offers M Sc, PhD. *Faculty:* 39 full-time (10 women), 29 part-time/adjunct (8 women). *Students:* 88 full-time (57 women). 29 applicants, 69% accepted, 19 enrolled. In 2005, 16 master's, 4 doctorates awarded. Terminal master's awarded for partial completion of doctoral program. *Degree requirements:* For master's, thesis; for doctorate, thesis/dissertation, general exam. *Entrance requirements:* For master's and doctorate, proficiency in French, knowledge of English. *Application deadline:* For fall and spring admission, 2/1. For winter admission, 11/1 for domestic students. Application fee: $30. Electronic applications accepted. *Unit head:* Pierre Belhumeur, Director, 514-343-6273, Fax: 514-343-5701. *Application contact:* George Szatmari, Chairperson, 514-343-5796, Fax: 514-343-5701.

Université de Montréal, Faculty of Medicine and Faculty of Graduate Studies, Graduate Programs in Medicine, Program in Specialized Studies, Montréal, QC H3C 3J7, Canada. Offers anesthesia (DESS); diagnostic radiology (DESS); family medicine (DESS); medical biochemistry (DESS); medical genetics (DESS); medicine (DESS); microbiology and infectious diseases (DESS); nuclear medicine (DESS); obstetrics and gynecology (DESS); ophthalmology (DESS); pediatrics (DESS); psychiatry (DESS); radiology-oncology (DESS); surgery (DESS). *Faculty:* 159 full-time (37 women), 345 part-time/adjunct (102 women). *Entrance requirements:* For degree, proficiency in French. *Application deadline:* For fall and spring admission, 2/1. For winter admission, 11/1 for domestic students. Application fee: $30. Electronic applications accepted. *Unit head:* Renée Roy, Vice Dean, 514-343-7798.

Université de Sherbrooke, Faculty of Medicine and Health Sciences, Graduate Programs in Medicine, Program in Microbiology, Sherbrooke, QC J1K 2R1, Canada. Offers M Sc, PhD. *Faculty:* 7 full-time (1 woman). *Students:* 15 full-time (7 women), 11 part-time (6 women). Average age 23. 9 applicants, 67% accepted, 5 enrolled. In 2005, 5 master's, 3 doctorates awarded. *Degree requirements:* For master's and doctorate, thesis/dissertation. *Application deadline:* For fall admission, 6/30 for domestic students. For winter admission, 10/31 for domestic students; for spring admission, 2/28 for domestic students. Application fee: $50. Electronic applications accepted. *Faculty research:* Oncogenes, study of splicing mechanism. *Unit head:* Dr. Claudine Rancourt, Director, 819-820-6868 Ext. 16808, E-mail: claudine.rancourt@usherbrooke.ca.

Université du Québec, Institut National de la Recherche Scientifique, Graduate Programs, Research Center—INRS—Institut Armand-Frappier—Human Health, Québec, QC G1K 9A9, Canada. Offers applied microbiology (M Sc); biology (PhD); experimental health sciences (M Sc); virology and immunology (M Sc, PhD). Programs given in French. Part-time programs available. *Faculty:* 46. *Students:* 170 full-time (100 women), 25 international. Average age 28. In 2005, 24 master's, 5 doctorates awarded. *Degree requirements:* For doctorate, thesis/dissertation. *Entrance requirements:* For master's and doctorate, appropriate bachelor's degree, proficiency in French. *Application deadline:* For fall admission, 3/30 for domestic students, 3/30 for international students. For winter admission, 11/1 for domestic students. Application fee: $30 Canadian dollars. *Financial support:* Fellowships, research assistantships, teaching assistantships available. *Faculty research:* Immunity, infection and cancer; toxicology and environmental biotechnology; molecular pharmacochemistry. *Unit head:* Pierre Talbot, Director, 450-681-5010 Ext. 4406, E-mail: pierre.talbot@iaf.inrs.ca. *Application contact:* Michel Barbeau, Registrar, 418-654-2518, Fax: 418-654-3858, E-mail: michel.barbeau@adm.inrs.ca.

Université Laval, Faculty of Agricultural and Food Sciences, Program in Agricultural Microbiology, Québec, QC G1K 7P4, Canada. Offers M Sc, PhD. Terminal master's awarded for partial completion of doctoral program. *Degree requirements:* For master's, thesis/dissertation; for doctorate, thesis/dissertation, comprehensive exam. *Entrance requirements:* For master's and doctorate, knowledge of French and English. Electronic applications accepted.

Université Laval, Faculty of Medicine, Graduate Programs in Medicine, Programs in Microbiology-Immunology, Québec, QC G1K 7P4, Canada. Offers M Sc, PhD. Terminal master's awarded for partial completion of doctoral program. *Degree requirements:* For master's, thesis/dissertation; for doctorate, thesis/dissertation, comprehensive exam. *Entrance requirements:* For master's and doctorate, knowledge of French, comprehension of written English. Electronic applications accepted.

Université Laval, Faculty of Sciences and Engineering, Department of Biochemistry and Microbiology, Programs in Microbiology, Québec, QC G1K 7P4, Canada. Offers M Sc, PhD. Terminal master's awarded for partial completion of doctoral program. *Degree requirements:* For master's, thesis/dissertation; for doctorate, thesis/dissertation, comprehensive exam. *Entrance requirements:* For master's and doctorate, knowledge of French, comprehension of written English. Electronic applications accepted.

The University of Alabama at Birmingham, Graduate Programs in Joint Health Sciences, Department of Cell Biology, Graduate Program in Cellular and Molecular Biology, Birmingham, AL 35294.

See Close-Ups on pages 601 and 603.

The University of Alabama at Birmingham, Graduate Programs in Joint Health Sciences, Department of Microbiology, Birmingham, AL 35294. Offers PhD. The department participates in the Cellular and Molecular Biology Graduate Program. *Students:* 82 full-time (47 women), 1 part-time; includes 17 minority (7 African Americans, 2 American Indian/Alaska Native, 8 Asian Americans or Pacific Islanders), 22 international. 56 applicants, 13% accepted. In 2005, 12 degrees awarded. *Degree requirements:* For doctorate, thesis/dissertation. *Entrance requirements:* For doctorate, GRE General Test, interview. *Application deadline:* Applications are processed on a rolling basis. Electronic applications accepted. *Expenses:* Tuition, state resident: part-time $170 per credit hour. Tuition, nonresident: full-time $4,612; part-time $425 per credit hour. International tuition: $10,732 full-time. Required fees: $11 per credit hour. $124 per term. Tuition and fees vary according to course load, degree level and program. *Financial support:* Fellowships available. *Unit head:* , Dr. David D. Chaplin, Chair, 205-934-3470, Fax: 205-934-1426. *Application contact:* Information Contact, 205-934-0621, Fax: 205-975-2536.

See Close-Up on page 905.

The University of Alabama at Birmingham, Graduate Studies and School of Medicine, Cellular and Molecular Biology Graduate Program, Birmingham, AL 35294.

University of Alberta, Faculty of Graduate Studies and Research, Department of Biological Sciences, Edmonton, AB T6G 2E1, Canada. Offers environmental biology and ecology (M Sc, PhD); microbiology and biotechnology (M Sc, PhD); molecular biology and genetics (M Sc, PhD); physiology and cell biology (M Sc, PhD); plant biology (M Sc, PhD); systematics and evolution (M Sc, PhD). *Faculty:* 72 full-time (15 women), 15 part-time/adjunct (4 women). *Students:* 238 full-time (117 women), 32 part-time (15 women), 31 international. 206 applicants, 42% accepted. In 2005, 29 master's, 31 doctorates awarded. Terminal master's awarded for partial completion of doctoral program. *Degree requirements:* For master's and doctorate, thesis/dissertation, registration. *Entrance requirements:* Additional exam requirements/recommendations for international students: Required—TOEFL. *Application deadline:* For fall admission, 3/1 for domestic students. Applications are processed on a rolling basis. Application fee: $0. Tuition and fees charges are reported in Canadian dollars. *Expenses:* Tuition, state resident: part-time $562 Canadian dollars per term. Tuition, nonresident: full-time $3,375 Canadian dollars. Required fees: $573 Canadian dollars; $84 Canadian dollars per term. *Financial support:* In 2005–06, 4 research assistantships with partial tuition reimbursements (averaging $12,000 per year), 103 teaching assistantships with partial tuition reimbursements (averaging $12,300 per year) were awarded; career-related internships or fieldwork and scholarships/grants also available. *Unit head:* , Laura Frost, Chair, 780-492-1904. *Application contact:* Dr. John P. Chang, Associate Chair for Graduate Studies, 780-492-1257, Fax: 780-492-9457, E-mail: bio.grad.coordinator@ualberta.ca.

The University of Arizona, College of Medicine, Graduate Programs in Medicine, Department of Microbiology and Immunology, Tucson, AZ 85721. Offers MS, PhD. *Degree requirements:* For master's and doctorate, thesis/dissertation. *Entrance requirements:* For master's and doctorate, GRE General Test, minimum GPA of 3.0. *Faculty research:* Environmental and pathogenic microbiology, molecular biology.

University of Arkansas for Medical Sciences, College of Medicine and Graduate School, Graduate Programs in Medicine, Department of Microbiology and Immunology, Little Rock, AR 72205-7199. Offers MS, PhD, MD/PhD. *Faculty:* 20 full-time (8 women). *Students:* 17 full-time, 2 part-time. *Degree requirements:* For master's and doctorate, thesis/dissertation. *Entrance requirements:* For master's and doctorate, GRE General Test. Additional exam requirements/recommendations for international students: Required—TOEFL. Application fee: $0. *Financial support:* Research assistantships available. Support available to part-time students. *Unit head:* Dr. Roger G. Rank, Chairman, 501-686-5144. *Application contact:* Dr. Lee Soderberg, Co-Coordinator of Graduate Studies, 501-686-6368, E-mail: soderberglees@uams.edu.

The University of British Columbia, Faculty of Graduate Studies, Faculty of Science, Department of Microbiology and Immunology, Vancouver, BC V6T 1Z1, Canada. Offers M Sc, PhD. *Degree requirements:* For master's, thesis/dissertation; for doctorate, thesis/dissertation, comprehensive exam. *Entrance requirements:* For master's and doctorate, GRE General Test. Additional exam requirements/recommendations for international students: Required—TOEFL (minimum score 590 paper-based; 243 computer-based). Electronic applications accepted. *Faculty research:* Bacterial genetics, metabolism, pathogenic bacteriology, virology.

University of Calgary, Faculty of Medicine and Faculty of Graduate Studies, Department of Microbiology and Infectious Diseases, Calgary, AB T2N 1N4, Canada. Offers M Sc, PhD. *Faculty:* 14 full-time (3 women). *Students:* 26 full-time (13 women). Average age 29. 25 applicants, 36% accepted. In 2005, 2 master's, 5 doctorates awarded. *Degree requirements:* For master's, thesis, oral thesis exam; for doctorate, thesis/dissertation, candidacy exam, oral thesis exam. *Entrance requirements:* For master's and doctorate, minimum GPA of 3.2. Additional exam requirements/recommendations for international students: Required—TOEFL (minimum score 580 paper-based; 237 computer-based). *Application deadline:* For fall admission, 5/15 for domestic students, 4/15 for international students. For winter admission, 9/15 for domestic students; for spring admission, 4/15 for domestic students. Applications are processed on a rolling basis. Application fee: $100 ($130 for international students). Electronic applications accepted. *Financial support:* In 2005–06, 7 students received support, including 7 fellowships (averaging $13,370 per year), 24 research assistantships (averaging $51,600 per year) Financial award application deadline: 2/1. *Faculty research:* Bacteriology, virology, parasitology, immunology. Total annual research expenditures: $5.6 million. *Unit head:* Dr. Donald E. Woods, Graduate Coordinator, E-mail: midgrad@ucalgary.ca. *Application contact:* Julie Boyd, Graduate Program Administrator, 403-220-2558, Fax: 403-210-8109, E-mail: boydj@ucalgary.ca.

University of California, Berkeley, Graduate Division, Group in Microbiology, Berkeley, CA 94720-1500. Offers PhD. *Degree requirements:* For doctorate, thesis/dissertation. *Entrance requirements:* For doctorate, GRE General Test, minimum GPA of 3.0.

University of California, Davis, Graduate Studies, Graduate Group in Microbiology, Davis, CA 95616. Offers MS, PhD. *Faculty:* 80 full-time. *Students:* 55 full-time (33 women); includes 12 minority (8 Asian Americans or Pacific Islanders, 4 Hispanic Americans), 8 international. Average age 30. 109 applicants, 31% accepted, 10 enrolled. In 2005, 5 degrees awarded. Terminal master's awarded for partial completion of doctoral program. *Median time to degree:* Of those who began their doctoral program in fall 1997, 5% received their degree in 8 years or less. *Degree requirements:* For master's and doctorate, thesis/dissertation. *Entrance requirements:* For master's and doctorate, GRE General Test, minimum GPA of 3.0. Additional exam requirements/recommendations for international students: Required—TOEFL (minimum score 550 paper-based; 213 computer-based). *Application deadline:* For fall admission, 1/15 for domestic students, 1/15 for international students. Application fee: $60. Electronic applications accepted. *Financial support:* In 2005–06, 55 students received support, including 10 fellowships with full and partial tuition reimbursements available (averaging $12,906 per year), 40 research assistantships with full and partial tuition reimbursements available (averaging $15,104 per year), 5 teaching assistantships with partial tuition reimbursements available (averaging $15,082 per year); Federal Work-Study, institutionally sponsored loans, scholarships/grants, and tuition waivers (full and partial) also available. Financial award application deadline: 1/15; financial award applicants required to submit FAFSA. *Faculty research:* Microbial physiology and genetics, microbial molecular and cellular biology, microbial ecology, microbial pathogenesis and immunology, urology. *Unit head:* Linda Bisson, Chair, 530-752-3835, E-mail: lfbisson@ucdavis.edu. *Application contact:* Lewanna Archer, Administrative Assistant, 530-752-0262, Fax: 530-752-0914, E-mail: leacher@ucdavis.edu.

See Close-Up on page 907.

University of California, Irvine, College of Medicine and School of Biological Sciences, Department of Microbiology and Molecular Genetics, Irvine, CA 92697. Offers biological sciences (MS, PhD). Students apply through the Graduate Program in Molecular Biology, Genetics, and Biochemistry. *Degree requirements:* For doctorate, thesis/dissertation. *Entrance requirements:* For doctorate, GRE General Test, GRE Subject Test, minimum GPA of 3.0. Additional exam requirements/recommendations for international students: Required—TOEFL (minimum score 550 paper-based; 213 computer-based), TSE. Electronic applications accepted. *Faculty research:* Molecular biology and genetics of viruses, bacteria, and yeast; immune response; molecular biology of cultured animal cells; genetic basis of cancer; genetics and physiology of infectious agents.

University of California, Los Angeles, School of Medicine and Graduate Division, Graduate Programs in Medicine, Department of Microbiology, Immunology and Molecular Genetics, Los Angeles, CA 90095. Offers MS, PhD. *Degree requirements:* For doctorate, thesis/dissertation, oral and written qualifying exams. *Entrance requirements:* For doctorate, GRE General Test, GRE Subject Test. Additional exam requirements/recommendations for international students: Required—TOEFL.

University of California, Riverside, Graduate Division, Program in Microbiology, Riverside, CA 92521-0102. Offers MS, PhD. Part-time programs available. *Faculty:* 27 full-time (4 women). *Students:* 12 full-time (8 women); includes 2 minority (both Asian Americans or Pacific Island-

Microbiology

University of California, Riverside (continued)

ers), 7 international. Average age 29. In 2005, 2 degrees awarded. Terminal master's awarded for partial completion of doctoral program. *Degree requirements:* For master's, thesis; for doctorate, thesis/dissertation, qualifying exams. *Entrance requirements:* For master's and doctorate, GRE General Test, minimum GPA of 3.2. Additional exam requirements/recommendations for international students: Required—TOEFL (minimum score 550 paper-based; 213 computer-based); Recommended—TSE (minimum score 50). *Application deadline:* For fall admission, 5/1 for domestic students, 2/1 for international students. For winter admission, 9/1 for domestic students; for spring admission, 12/1 for domestic students. Applications are processed on a rolling basis. Application fee: $60 ($75 for international students). Electronic applications accepted. *Expenses:* Tuition, nonresident: full-time $14,694. Required fees:$9,009. Full-time tuition and fees vary according to program. *Financial support:* In 2005–06, research assistantships (averaging $14,000 per year), teaching assistantships (averaging $15,000 per year) were awarded; fellowships, career-related internships or fieldwork, Federal Work-Study, institutionally sponsored loans, and tuition waivers (full and partial) also available. Financial award application deadline: 2/1; financial award applicants required to submit FAFSA. *Faculty research:* Host-pathogen interactions; environmental microbiology; bioremediation; molecular microbiology; microbial genetics, physiology, and pathogenesis. *Unit head:* Dr. James Borneman, Director, 951-827-4115. *Application contact:* Zina Romero, Graduate Program Assistant, 800-735-0717, Fax: 951-827-5913, E-mail: microbio@ucr.edu.

University of California, San Diego, School of Medicine and Graduate Studies and Research, Molecular Pathology Program, La Jolla, CA 92093. Offers bioinformatics (PhD); cancer biology/oncology (PhD); cardiovascular sciences and disease (PhD); microbiology (PhD); molecular pathology (PhD); neurological disease (PhD); stem cell and developmental biology (PhD); structural biology/drug design (PhD). *Entrance requirements:* For doctorate, GRE General Test, GRE Subject Test. Additional exam requirements/recommendations for international students: Required—TOEFL. Electronic applications accepted.

See Close-Up on page 1119.

University of California, San Francisco, Graduate Division, Department of Microbiology and Immunology, San Francisco, CA 94143. Offers PhD. *Degree requirements:* For doctorate, thesis/dissertation. *Entrance requirements:* For doctorate, GRE General Test.

University of Central Florida, Burnett College of Biomedical Sciences, Program in Molecular Biology and Microbiology, Orlando, FL 32816. Offers microbiology (MS); molecular biology (MS). Part-time and evening/weekend programs available. *Faculty:* 21. *Students:* 16 full-time (11 women), 12 part-time (9 women); includes 9 minority (1 African American, 5 Asian Americans or Pacific Islanders, 3 Hispanic Americans), 7 international. Average age 28. 22 applicants, 32% accepted, 6 enrolled. In 2005, 7 degrees awarded. *Degree requirements:* For master's, thesis, comprehensive exam. *Entrance requirements:* For master's, GRE General Test, minimum GPA of 3.0 in last 60 hours. Additional exam requirements/recommendations for international students: Required—TOEFL. *Application deadline:* For fall admission, 3/15 for domestic students; for spring admission, 12/1 for domestic students. Application fee: $30. Electronic applications accepted. *Expenses:* Tuition, state resident: full-time $5,788. Tuition, nonresident: full-time $21,927. Required fees: $241 per credit hour. *Financial support:* In 2005–06, fellowships with partial tuition reimbursements (averaging $2,800 per year), research assistantships with partial tuition reimbursements (averaging $4,700 per year), teaching assistantships with partial tuition reimbursements (averaging $4,400 per year) were awarded; career-related internships or fieldwork, Federal Work-Study, institutionally sponsored loans, tuition waivers (partial), and unspecified assistantships also available. Financial award application deadline: 3/1; financial award applicants required to submit FAFSA. *Application contact:* Dr. Karl X. Chai, Coordinator, 407-823-4846, E-mail: kxchai@mail.ucf.edu.

University of Chicago, Division of the Biological Sciences, Biomedical Sciences: Cancer, Immunology, Nutrition, Pathology, and Microbiology, Committee on Microbiology, Chicago, IL 60637-1513. Offers PhD. *Faculty:* 17 full-time (4 women). *Students:* 19 full-time (10 women); includes 5 minority (1 African American, 2 Asian Americans or Pacific Islanders, 2 Hispanic Americans), 1 international. Average age 27. In 2005, 3 degrees awarded. *Degree requirements:* For doctorate, thesis/dissertation, registration. *Entrance requirements:* For doctorate, GRE General Test. Additional exam requirements/recommendations for international students: Required—TOEFL. *Application deadline:* For fall admission, 12/28 priority date for domestic students, 12/28 priority date for international students. Application fee: $55. Electronic applications accepted. *Financial support:* In 2005–06, 5 students received support, including fellowships with full tuition reimbursements available (averaging $26,301 per year), research assistantships with full tuition reimbursements available (averaging $26,301 per year); institutionally sponsored loans and traineeships also available. Financial award application deadline: 1/5; financial award applicants required to submit FAFSA. *Faculty research:* Molecular genetics, herpes virus, adipoviruses, Picarna viruses, ENS viruses. *Unit head:* Dr. Olaf Schneewind, Chairman, 773-834-9061, Fax: 773-702-1631, E-mail: oschnee@delphi.bsd.uchicago.edu. *Application contact:* Rebecca Levine, Administrative Assistant, Student Services, 773-834-3899, Fax: 773-702-4634, E-mail: rlevine@huggins.bsd.uchicago.edu.

University of Cincinnati, Division of Research and Advanced Studies, College of Medicine, Graduate Programs in Biomedical Sciences, Department of Molecular Genetics, Biochemistry and Microbiology, Cincinnati, OH 45267. Offers MS, PhD. Terminal master's awarded for partial completion of doctoral program. *Degree requirements:* For master's, thesis or alternative; for doctorate, thesis/dissertation, qualifying exam. *Entrance requirements:* For master's, GRE General Test; for doctorate, GRE General Test, GRE Subject Test. Additional exam requirements/recommendations for international students: Required—TOEFL (minimum score 590 paper-based; 243 computer-based), TWE. Electronic applications accepted. *Faculty research:* Cancer biology and developmental genetics, gene regulation and chromosome structure, microbiology and pathogenic mechanismis, structural biology, membrane biochemistry and signal transduction.

University of Colorado at Boulder, Graduate School, College of Arts and Sciences, Department of Ecology and Evolutionary Biology, Boulder, CO 80309. Offers animal behavior (MA); biology (MA, PhD); environmental biology (MA, PhD); evolutionary biology (MA, PhD); neurobiology (MA); population biology (MA); population genetics (PhD). *Faculty:* 26 full-time (5 women). *Students:* 44 full-time (24 women), 22 part-time (14 women); includes 16 minority (1 American Indian/Alaska Native, 3 Asian Americans or Pacific Islanders, 12 Hispanic Americans), 2 international. Average age 30. 20 applicants, 90% accepted. In 2005, 7 master's, 7 doctorates awarded. Terminal master's awarded for partial completion of doctoral program. *Degree requirements:* For master's, thesis or alternative, comprehensive exam; for doctorate, thesis/dissertation, comprehensive exam. *Entrance requirements:* For master's, GRE General Test, GRE Subject Test, minimum undergraduate GPA of 3.0; for doctorate, GRE General Test, GRE Subject Test. *Application deadline:* For fall admission, 1/2 priority date for domestic students, 12/1 priority date for international students. Application fee: $50 ($60 for international students). *Financial support:* In 2005–06, fellowships (averaging $7,844 per year), research assistantships (averaging $15,663 per year), teaching assistantships (averaging $13,829 per year) were awarded; Federal Work-Study, institutionally sponsored loans, and tuition waivers (full) also available. Financial award application deadline: 3/1. *Faculty research:* Behavior, ecology, genetics, morphology, endocrinology. Total annual research expenditures: $5.2 million. *Unit head:* Jeffry Mitton, Chair, 303-492-0505, Fax: 303-492-8699, E-mail: mitton@colorado.edu. *Application contact:* Jill Skarstadt, Graduate Coordinator, 303-492-7654, Fax: 303-492-8699, E-mail: skarstad@colorado.edu.

University of Colorado at Denver and Health Sciences Center, Graduate School, Program in Biomedical Sciences, Program in Microbiology, Denver, CO 80262. Offers PhD. In 2005, 5 degrees awarded. *Degree requirements:* For doctorate, thesis/dissertation, comprehensive exam. *Entrance requirements:* For doctorate, GRE, minimum GPA of 3.0, 4 letters of recommendation. Additional exam requirements/recommendations for international students: Required—TOEFL

(minimum score 550 paper-based; 213 computer-based). *Application deadline:* For fall admission, 1/1 for domestic students. Application fee: $50. Electronic applications accepted. *Expenses:* Tuition, state resident: full-time $11,730. Tuition, nonresident: full-time $22,980. Tuition and fees vary according to degree level and program. *Financial support:* Fellowships, research assistantships, teaching assistantships, Federal Work-Study and institutionally sponsored loans available. Support available to part-time students. Financial award application deadline: 3/15; financial award applicants required to submit FAFSA. *Faculty research:* Molecular mechanisms of picornavirus replication, mechanisms of papovavirus assembly, human immune response in multiple sclerosis. Total annual research expenditures: $3.5 million. *Unit head:* Dr. Randall K. Holmes, Chair, 303-315-7903. *Application contact:* Dr. Ronald E. Gill, Director, 303-724-4227.

See Close-Up on page 913.

University of Connecticut, Graduate School, College of Liberal Arts and Sciences, Department of Molecular and Cell Biology, Field of Microbiology, Storrs, CT 06269. Offers MS, PhD. *Faculty:* 20 full-time (5 women). *Students:* 18 full-time (11 women); includes 2 minority (1 African American, 1 Hispanic American). Average age 28. 36 applicants, 19% accepted, 7 enrolled. In 2005, 3 master's, 4 doctorates awarded. Terminal master's awarded for partial completion of doctoral program. *Degree requirements:* For master's, comprehensive exam; for doctorate, thesis/dissertation. *Entrance requirements:* For master's and doctorate, GRE General Test, GRE Subject Test. Additional exam requirements/recommendations for international students: Required—TOEFL (minimum score 550 paper-based; 213 computer-based). *Application deadline:* For fall admission, 2/1 priority date for domestic students, 2/1 priority date for international students; for spring admission, 11/1 for domestic students, 10/1 for international students. Applications are processed on a rolling basis. Application fee: $55. Electronic applications accepted. *Expenses:* Tuition, state resident: part-time $444 per credit hour. Tuition, nonresident: part-time $1,154 per credit hour. Tuition and fees vary according to course load. *Financial support:* In 2005–06, 5 research assistantships with full tuition reimbursements, 11 teaching assistantships with full tuition reimbursements were awarded; fellowships, Federal Work-Study, scholarships/grants, health care benefits, and unspecified assistantships also available. Financial award application deadline: 2/1; financial award applicants required to submit FAFSA. *Application contact:* Anne St. Onje, Graduate Coordinator, 860-486-4314, Fax: 860-486-3943, E-mail: ann.st_onje@uconn.edu.

See Close-Up on page 611.

University of Delaware, College of Arts and Sciences, Department of Biological Sciences, Newark, DE 19716. Offers biotechnology (MS); cancer biology (MS, PhD); cell and extracellular matrix biology (MS, PhD); cell and systems physiology (MS, PhD); developmental biology (MS, PhD); ecology and evolution (MS, PhD); microbiology (MS, PhD); molecular biology and genetics (MS, PhD). *Faculty:* 39 full-time (11 women). *Students:* 64 full-time (46 women), 2 part-time; includes 7 minority (4 African Americans, 2 Asian Americans or Pacific Islanders, 1 Hispanic American), 18 international. Average age 26. 113 applicants, 31% accepted, 19 enrolled. In 2005, 5 master's, 4 doctorates awarded. Terminal master's awarded for partial completion of doctoral program. *Median time to degree:* Of those who began their doctoral program in fall 1997, 100% received their degree in 8 years or less. *Degree requirements:* For master's, thesis/dissertation, preliminary exam; for doctorate, thesis/dissertation, preliminary exam, comprehensive exam. *Entrance requirements:* For master's and doctorate, GRE General Test. Additional exam requirements/recommendations for international students: Required—TOEFL (minimum score 600 paper-based; 250 computer-based); Recommended—TWE, TSE. *Application deadline:* For fall admission, 4/15 for domestic students, 1/15 for international students; for spring admission, 10/1 for domestic students. Applications are processed on a rolling basis. Application fee: $60. Electronic applications accepted. *Financial support:* In 2005–06, 26 students received support, including fellowships with full tuition reimbursements available (averaging $19,000 per year), 19 research assistantships with full tuition reimbursements available (averaging $19,000 per year), 26 teaching assistantships with full tuition reimbursements available (averaging $19,000 per year); tuition waivers (partial) also available. Financial award application deadline: 4/15. *Faculty research:* Microorganisms, bone, cancer metastasis, developmental biology, cell biology, DNA. Total annual research expenditures: $8.3 million. *Unit head:* Dr. Daniel D. Carson, 302-831-6977, Fax: 302-831-2281, E-mail: dcarson@udel.edu. *Application contact:* Dr. Melinda K. Duncan, Graduate Coordinator, 302-831-1841, Fax: 302-831-2281, E-mail: danders@udel.edu.

University of Florida, College of Medicine, Department of Molecular Genetics and Microbiology, Gainesville, FL 32611. Offers MS, PhD. *Faculty:* 26 full-time (7 women). *Students:* 145 (76 women); includes 20 minority (5 African Americans, 9 Asian Americans or Pacific Islanders, 6 Hispanic Americans) 34 international. In 2005, 26 degrees awarded. Terminal master's awarded for partial completion of doctoral program. *Degree requirements:* For master's and doctorate, GRE General Test, minimum GPA of 3.0. Additional exam requirements/recommendations for international students: Required—TOEFL. *Application deadline:* For fall admission, 2/15 for domestic students. Applications are processed on a rolling basis. Application fee: $30. Electronic applications accepted. *Expenses:* Tuition, state resident: full-time $6,234. Tuition, nonresident: full-time $21,359. Tuition and fees vary according to program. *Financial support:* In 2005–06, 38 research assistantships with full tuition reimbursements (averaging $24,464 per year) were awarded; fellowships with full tuition reimbursements, institutionally sponsored loans, traineeships, and unspecified assistantships also available. *Unit head:* Dr. Henry Baker, Chairman, 352-392-0680, E-mail: hvbaker@ufl.edu. *Application contact:* Dr. Wayne McCormack, Associate Dean of Graduate Education, 352-392-7413, Fax: 352-846-3466, E-mail: mccormac@pathology.ufl.edu.

University of Florida, College of Medicine and Graduate School, Interdisciplinary Program in Biomedical Sciences, Concentration in Immunology and Microbiology, Gainesville, FL 32611. Offers PhD. *Faculty:* 66. *Students:* 33 full-time (20 women); includes 9 minority (1 African American, 7 Asian Americans or Pacific Islanders, 1 Hispanic American). In 2005, 9 degrees awarded. *Degree requirements:* For doctorate, thesis/dissertation. *Entrance requirements:* For doctorate, GRE General Test, minimum GPA of 3.0. Additional exam requirements/recommendations for international students: Required—TOEFL. *Application deadline:* For fall admission, 2/15 for domestic students. Applications are processed on a rolling basis. Application fee: $30. Electronic applications accepted. *Expenses:* Tuition, state resident: full-time $6,234. Tuition, nonresident: full-time $21,359. Tuition and fees vary according to program. *Financial support:* Fellowships with full tuition reimbursements, research assistantships with full tuition reimbursements, teaching assistantships, institutionally sponsored loans, scholarships/grants, and traineeships available. *Unit head:* Dr. Paul Gulig, Director, 352-392-0050, E-mail: gulig@ufl.edu. *Application contact:* Dr. Wayne McCormack, Associate Dean of Graduate Education, 352-392-7413, Fax: 352-846-3466, E-mail: mccormac@pathology.ufl.edu.

University of Florida, Graduate School, College of Agricultural and Life Sciences, Department of Microbiology and Cell Science, Gainesville, FL 32611. Offers biochemistry and molecular biology (MS, PhD); microbiology and cell science (MS, PhD). *Faculty:* 18 full-time (5 women). *Students:* 25 (9 women); includes 4 minority (1 African American, 2 Asian Americans or Pacific Islanders, 1 Hispanic American) 6 international. 31 applicants, 26% accepted. In 2005, 1 master's, 4 doctorates awarded. *Degree requirements:* For doctorate, thesis/dissertation. *Entrance requirements:* For master's and doctorate, GRE General Test, minimum GPA of 3.0. *Application deadline:* For fall admission, 6/1 for domestic students. Applications are processed on a rolling basis. Application fee: $20. Electronic applications accepted. *Expenses:* Tuition, state resident: full-time $6,234. Tuition, nonresident: full-time $21,359. Tuition and fees vary according to program. *Financial support:* In 2005–06, 20 students received support, including 14 research assistantships (averaging $15,696 per year), 5 teaching assistantships (averaging $14,677 per year); fellowships *Faculty research:* Biomass conversion, membrane and cell wall chemistry, plant biochemistry and genetics. *Unit head:* Dr. Eric Triplett, Chair, 352-392-1906,

Fax: 352-392-5922, E-mail: ewt@ufl.edu. *Application contact:* Dr. James F. Preston, Coordinator, 352-392-5923, Fax: 352-392-5922, E-mail: jpreston@ufl.edu.

University of Georgia, Graduate School, College of Arts and Sciences, Department of Microbiology, Athens, GA 30602. Offers MS, PhD. *Faculty:* 15 full-time (4 women). *Students:* 49 full-time, 1 part-time; includes 5 minority (2 African Americans, 3 Asian Americans or Pacific Islanders), 7 international. 63 applicants, 25% accepted, 6 enrolled. In 2005, 1 master's, 6 doctorates awarded. *Degree requirements:* For master's, thesis; for doctorate, one foreign language, thesis/dissertation. *Entrance requirements:* For master's and doctorate, GRE General Test. Additional exam requirements/recommendations for international students: Required—TOEFL (minimum score 550 paper-based; 213 computer-based), TSE (minimum score 50). *Application deadline:* For fall admission, 7/1 for domestic students; for spring admission, 11/15 for domestic students. Application fee: $50. Electronic applications accepted. *Financial support:* In 2005–06, 9 fellowships (averaging $20,000 per year), 20 research assistantships (averaging $18,461 per year), 12 teaching assistantships (averaging $18,461 per year) were awarded; unspecified assistantships also available. Financial award application deadline: 12/15. *Unit head:* Dr. Duncan Krause, Head, 706-542-2671, Fax: 706-542-2674, E-mail: dkrause@uga.edu. *Application contact:* Dr. Anne Summers, Information Contact, 706-542-2669, Fax: 706-542-2674, E-mail: summers@uga.edu.

See Close-Up on page 917.

University of Guelph, Graduate Program Services, College of Biological Science, Department of Molecular and Cellular Biology, Guelph, ON N1G 2W1, Canada. Offers biochemistry (M Sc, PhD); biophysics (M Sc, PhD); botany (M Sc, PhD); microbiology (M Sc, PhD); molecular biology and genetics (M Sc, PhD). *Faculty:* 43 full-time (6 women). *Students:* 136 full-time (52 women). Average age 26. 56 applicants, 30% accepted, 17 enrolled. In 2005, 4 master's, 2 doctorates awarded. *Degree requirements:* For master's, thesis/dissertation, research proposal; for doctorate, thesis/dissertation, research proposal, comprehensive exam, registration. *Entrance requirements:* Additional exam requirements/recommendations for international students: Required—TOEFL (minimum score 550 paper-based; 213 computer-based), IELT (minimum score 7). *Application deadline:* For fall admission, 7/1 for domestic students, 1/1 for international students. For winter admission, 11/1 for domestic students; for spring admission, 3/1 for domestic students. Applications are processed on a rolling basis. Application fee: $75. Electronic applications accepted. *Financial support:* Fellowships, research assistantships, teaching assistantships available. Support available to part-time students. *Faculty research:* Physiology, structure, genetics, and ecology of microbes; virology and microbial technology. *Unit head:* Dr. Chris Whitfield, Chair, 519-824-4120 Ext. 53361, Fax: 519-827-1802, E-mail: cwhitfie@uoguelph.ca. *Application contact:* Laurie Winn, Graduate Admissions Secretary, 519-824-4320 Ext. 52730, Fax: 519-767-1656, E-mail: lwinn@uoguelph.ca.

University of Hawaii at Manoa, Graduate Division, Colleges of Arts and Sciences, College of Natural Sciences, Department of Microbiology, Honolulu, HI 96822. Offers MS, PhD. *Faculty:* 20 full-time (5 women). *Students:* 36 full-time (22 women), 3 part-time (1 woman); includes 13 minority (12 Asian Americans or Pacific Islanders, 1 Hispanic American), 15 international. Average age 29. 42 applicants, 38% accepted, 6 enrolled. In 2005, 9 master's awarded. *Median time to degree:* Of those who began their doctoral program in fall 1997, 50% received their degree in 8 years or less. *Degree requirements:* For doctorate, thesis/dissertation. *Entrance requirements:* For master's and doctorate, GRE. *Application deadline:* For fall admission, 3/1 for domestic students, 2/1 for international students; for spring admission, 9/1 for domestic students, 8/1 for international students. Application fee: $50. *Expenses:* Tuition, state resident: full-time $8,400; part-time $200 per credit hour. Tuition, nonresident: full-time $11,088; part-time $462 per credit hour. Tuition and fees vary according to program. *Financial support:* In 2005–06, 14 research assistantships (averaging $16,996 per year), 17 teaching assistantships (averaging $14,657 per year) were awarded. *Faculty research:* Virology, immunology, microbial physiology, medical microbiology, bacterial genetics. *Unit head:* Dr. Paul Q. Patek, Chairperson, 808-956-8553, Fax: 808-956-5339, E-mail: patek@hawaii.edu. *Application contact:* Dr. Paul Q. Patek, Chairperson, 808-956-8553, Fax: 808-956-5339, E-mail: patek@hawaii.edu.

University of Idaho, College of Graduate Studies, College of Agricultural and Life Sciences, Department of Microbiology, Molecular Biology and Biochemistry, Moscow, ID 83844-2282. Offers MS, PhD. *Students:* 30 full-time (15 women), 8 part-time (3 women), 19 international. Average age 30. In 2005, 5 master's, 4 doctorates awarded. *Degree requirements:* For master's, thesis; for doctorate, one foreign language, thesis/dissertation. *Entrance requirements:* For master's, minimum GPA of 2.8; for doctorate, minimum undergraduate GPA of 2.8, 3.0 graduate. *Application deadline:* For fall admission, 8/1 for domestic students; for spring admission, 12/16 for domestic students. Application fee: $55 ($60 for international students). *Expenses:* Tuition, state resident: full-time $4,508. Tuition, nonresident: full-time $8,770; part-time $130 per credit. Required fees: $217 per credit. *Financial support:* Research assistantships, teaching assistantships available. Financial award application deadline: 2/15. *Faculty research:* Environmental fields, food, immunology. *Unit head:* Dr. Patricia Hartzell, Head, 208-885-0572.

University of Illinois at Chicago, College of Medicine and Graduate College, Graduate Programs in Medicine, Department of Microbiology and Immunology, Chicago, IL 60607-7128. Offers PhD, MD/PhD. *Degree requirements:* For doctorate, thesis/dissertation. *Entrance requirements:* For doctorate, GRE General Test, minimum GPA of 2.75. Additional exam requirements/recommendations for international students: Required—TOEFL.

See Close-Up on page 919.

University of Illinois at Urbana–Champaign, Graduate College, College of Liberal Arts and Sciences, School of Molecular and Cellular Biology, Department of Microbiology, Champaign, IL 61820. Offers MS, PhD. *Faculty:* 13 full-time (2 women), 2 part-time/adjunct (0 women). *Students:* 83 full-time (47 women), 2 part-time (both women); includes 10 minority (1 African American, 5 Asian Americans or Pacific Islanders, 4 Hispanic Americans), 30 international. 108 applicants, 23% accepted, 17 enrolled. In 2005, 7 master's, 5 doctorates awarded. *Degree requirements:* For doctorate, thesis/dissertation. *Entrance requirements:* For master's, GRE, minimum GPA of 3.0. *Application deadline:* Applications are processed on a rolling basis. Application fee: $50 ($60 for international students). Electronic applications accepted. *Financial support:* In 2005–06, 26 fellowships, 49 research assistantships, 58 teaching assistantships were awarded. Financial award application deadline: 2/15. *Faculty research:* Bacterial physiology and genetics, bacterial pathogenesis, host-pathogen interaction, molecular immunology. *Unit head:* Dr. John E. Cronan, Head, 217-333-1737, Fax: 217-244-6697, E-mail: j-cronan@uiuc.edu. *Application contact:* Marsha Dunlap, Secretary, 217-333-6472, Fax: 217-244-6697, E-mail: mldunlap@uiuc.edu.

See Close-Up on page 921.

The University of Iowa, Roy J. and Lucille A. Carver College of Medicine and Graduate College, Graduate Programs in Medicine, Department of Microbiology, Iowa City, IA 52242-1316. Offers general microbiology and microbial physiology (MS, PhD); immunology (MS, PhD); microbial genetics (MS, PhD); pathogenic bacteriology (MS, PhD); virology (MS, PhD). *Faculty:* 23 full-time (3 women), 11 part-time/adjunct (2 women). *Students:* 54 full-time (25 women); includes 4 minority (3 Asian Americans or Pacific Islanders, 1 Hispanic American), 8 international. 93 applicants, 12% accepted, 4 enrolled. In 2005, 7 degrees awarded. *Median time to degree:* Of those who began their doctoral program in fall 1997, 80% received their degree in 8 years or less. *Degree requirements:* For master's, thesis/dissertation; for doctorate, thesis/dissertation, comprehensive exam. *Entrance requirements:* For master's and doctorate, GRE General Test. Additional exam requirements/recommendations for international students: Required—TOEFL. *Application deadline:* For fall admission, 2/1 for domestic students, 2/1 for international students. Application fee: $50 ($75 for international students). Electronic applications accepted. *Expenses:* Tuition, state resident: part-time $1,882 per term. Tuition, nonresident: full-time $17,338; part-time $4,907 per term. Tuition and fees vary according to

course load and program. *Financial support:* In 2005–06, 63 research assistantships with full tuition reimbursements (averaging $22,000 per year) were awarded; institutionally sponsored loans, scholarships/grants, traineeships, and health care benefits also available. *Faculty research:* Biocatalysis and blue jeans, gene regulation, processing and transport of HIV, retroviral pathogenesis, biodegradation. Total annual research expenditures: $9 million. *Unit head:* Dr. Michael A. Apicella, Head, 319-335-7810, E-mail: grad-micro-info@uiowa.edu.

See Close-Up on page 923.

University of Kansas, Graduate School, College of Liberal Arts and Sciences, Division of Biological Sciences, Department of Molecular Biosciences, Lawrence, KS 66045. Offers biochemistry and biophysics (MA, PhD); microbiology (MA, PhD); molecular, cellular, and developmental biology (MA, PhD). *Faculty:* 44. *Students:* 48 full-time (28 women), 5 part-time (3 women); includes 1 minority (Asian American or Pacific Islander), 26 international. Average age 27. 111 applicants, 19% accepted. In 2005, 5 master's, 5 doctorates awarded. Terminal master's awarded for partial completion of doctoral program. *Degree requirements:* For master's and doctorate, thesis/dissertation, comprehensive exam. *Entrance requirements:* For master's and doctorate, GRE General Test. Additional exam requirements/recommendations for international students: Required—TOEFL, IELT, TOEFL (paper 570; computer 230) or IELT; Recommended—TWE, TSE. *Application deadline:* For fall admission, 1/15 priority date for domestic students, 1/15 priority date for international students. Applications are processed on a rolling basis. Application fee: $55 ($60 for international students). Electronic applications accepted. *Expenses:* Tuition, state resident: full-time $4,859. Tuition, nonresident: full-time $12,000. Required fees: $589. Tuition and fees vary according to program. *Financial support:* Fellowships, research assistantships with tuition reimbursements, teaching assistantships with tuition reimbursements available. Financial award application deadline: 3/1. *Faculty research:* Structure and function of proteins, genetics of organism development, molecular genetics, neurophysiology, molecular virology and pathogenics, developmental biology, cell biology. *Unit head:* Kathy Suprenant, Chair, 785-864-4631, Fax: 785-864-5294, E-mail: ksupre@ku.edu. *Application contact:* John P. Connolly, Information Contact, 785-864-4311, Fax: 785-864-5924, E-mail: jconnolly@ku.edu.

University of Kansas, Graduate Studies Medical Center, Interdisciplinary Graduate Program in Biomedical Sciences, Department of Microbiology, Molecular Genetics and Immunology, Lawrence, KS 66045. Offers PhD, MD/PhD. *Faculty:* 8. *Students:* 2 full-time (both women), 5 part-time (3 women), 1 international. Average age 27. In 2005, 2 degrees awarded. *Degree requirements:* For doctorate, thesis/dissertation, comprehensive exam. *Entrance requirements:* For doctorate, GRE General Test. *Expenses:* Tuition, state resident: full-time $4,859. Tuition, nonresident: full-time $12,000. Required fees: $589. Tuition and fees vary according to program. *Financial support:* Fellowships with tuition reimbursements, research assistantships with partial tuition reimbursements, teaching assistantships with full and partial tuition reimbursements available. *Faculty research:* Bacterial and viral pathogenesis, molecular virology, microbial molecular biology and genetics, immunochemistry. *Unit head:* Dr. Opendra Narayan, Chairman, 913-588-7010, Fax: 913-588-7295, E-mail: bnarayan@kume.edu. *Application contact:* Dr. Joe Lutkenhaus, Director of Graduate Studies, 913-588-7054, Fax: 913-588-7295, E-mail: jlutkenh@kumc.edu.

University of Kentucky, Graduate School, Graduate School Programs from the College of Medicine, Program in Microbiology and Immunology, Lexington, KY 40506-0032. Offers microbiology (PhD). *Faculty:* 34 full-time (7 women), 1 part-time/adjunct (0 women). *Students:* 26 full-time (13 women), 9 part-time (6 women); includes 2 minority (1 African American, 1 Asian American or Pacific Islander), 13 international. Average age 27. 19 applicants, 100% accepted, 18 enrolled. In 2005, 4 degrees awarded. *Median time to degree:* Of those who began their doctoral program in fall 1997, 96.3% received their degree in 8 years or less. *Degree requirements:* For doctorate, thesis/dissertation, comprehensive exam. *Entrance requirements:* For doctorate, GRE General Test, minimum undergraduate GPA of 3.0. Additional exam requirements/recommendations for international students: Required—TOEFL (minimum score 550 paper-based; 213 computer-based). *Application deadline:* For fall admission, 7/17 priority date for domestic students, 2/1 priority date for international students; for spring admission, 12/13 for domestic students, 6/15 for international students. Applications are processed on a rolling basis. Application fee: $40 ($55 for international students). Electronic applications accepted. *Expenses:* Tuition, state resident: full-time $6,308; part-time $331 per credit hour. Tuition, nonresident: full-time $13,968; part-time $756 per credit hour. Tuition and fees vary according to course load, degree level and program. *Financial support:* In 2005–06, 6 fellowships with full tuition reimbursements (averaging $4,589 per year), 29 research assistantships with full tuition reimbursements (averaging $21,000 per year) were awarded; teaching assistantships with full tuition reimbursements, Federal Work-Study, scholarships/grants, traineeships, health care benefits, and unspecified assistantships also available. Support available to part-time students. Financial award application deadline: 3/15. *Unit head:* Dr. Brian Stevenson, Head, 859-257-9358, Fax: 859-257-8994, E-mail: brian.stevenson@uky.edu. *Application contact:* Dr. Brian Jackson, Senior Associate Dean, 859-257-8176, Fax: 859-323-1928.

See Close-Up on page 925.

University of Louisville, School of Medicine, Department of Microbiology and Immunology, Louisville, KY 40292-0001. Offers MS, PhD. *Students:* 33 full-time (15 women), 4 part-time (3 women); includes 5 minority (4 African Americans, 1 Asian or Pacific Islander), 23 international. Average age 29. In 2005, 11 master's, 6 doctorates awarded. *Degree requirements:* For master's and doctorate, thesis/dissertation. *Entrance requirements:* For master's and doctorate, GRE General Test, 1 year of course work in biology, organic chemistry, physics; 1 semester of course work in calculus and quantitative analysis, biochemistry, or molecular biology. *Application deadline:* For fall admission, 2/1 for domestic students. Application fee: $50. Electronic applications accepted. *Expenses:* Tuition, state resident: full-time $6,006; part-time $334 per credit hour. Tuition, nonresident: full-time $16,554; part-time $920 per credit hour. Tuition and fees vary according to course load, degree level and program. *Financial support:* Fellowships with tuition reimbursements, research assistantships with tuition reimbursements available. *Unit head:* Dr. Robert D. Stout, Chair, 502-852-5351, Fax: 502-852-7531, E-mail: bobstout@louisville.edu. *Application contact:* Dr. Richard D Miller, Director of Admissions, 502-852-5630, Fax: 502-852-7531, E-mail: rdmill01@louisville.edu.

University of Maine, Graduate School, College of Natural Sciences, Forestry, and Agriculture, Department of Biochemistry, Molecular Biology, and Microbiology, Orono, ME 04469. Offers biochemistry (MPS, MS); biochemistry and molecular biology (PhD); microbiology (MPS, MS, PhD). *Students:* 40 full-time (23 women), 17 part-time (11 women); includes 2 minority (1 Asian American or Pacific Islander, 1 Hispanic American), 13 international. Average age 31. 42 applicants, 48% accepted, 11 enrolled. In 2005, 5 master's awarded. *Degree requirements:* For doctorate, thesis/dissertation. *Entrance requirements:* For master's and doctorate, GRE General Test. Additional exam requirements/recommendations for international students: Required—TOEFL. *Application deadline:* For fall admission, 2/1 for domestic students. Applications are processed on a rolling basis. Application fee: $50. Electronic applications accepted. *Financial support:* In 2005–06, 5 research assistantships with tuition reimbursements (averaging $18,000 per year), 12 teaching assistantships with tuition reimbursements (averaging $16,000 per year) were awarded; tuition waivers (full and partial) also available. Financial award application deadline: 3/1. *Unit head:* Dr. John Singer, Chair, 207-581-2810, Fax: 207-581-2801. *Application contact:* Scott G. Delcourt, Associate Dean of the Graduate School, 207-581-3219, Fax: 207-581-3232, E-mail: graduate@maine.edu.

University of Manitoba, Faculty of Graduate Studies, Faculty of Science, Biological Sciences Division, Department of Microbiology, Winnipeg, MB R3T 2N2, Canada. Offers M Sc, PhD. *Degree requirements:* For master's, thesis; for doctorate, one foreign language, thesis/dissertation.

University of Maryland, Graduate School, Graduate Programs in Medicine, Department of Microbiology and Immunology, Baltimore, MD 21201. Offers MS, PhD, MD/PhD. Part-time

Microbiology

University of Maryland (continued)
programs available. *Faculty:* 21 full-time (4 women). *Students:* 37 full-time (21 women), 1 part-time; includes 8 minority (2 African Americans, 4 Asian Americans or Pacific Islanders, 2 Hispanic Americans), 15 international. Average age 29. 86 applicants, 8% accepted, 6 enrolled. In 2005, 1 master's, 4 doctorates awarded. *Degree requirements:* For master's, thesis; for doctorate, thesis/dissertation, oral exam. *Entrance requirements:* For master's and doctorate, GRE General Test, minimum GPA of 3.0. Additional exam requirements/recommendations for international students: Required—TOEFL, TOEFL or IELTS; Recommended—IELT. *Application deadline:* For fall admission, 7/1 for domestic students, 1/15 for international students. Application fee: $50. Electronic applications accepted. *Expenses:* Tuition, state resident: full-time $8,079; part-time $409 per credit hour. Tuition, nonresident: full-time $18,384; part-time $731 per credit hour. Required fees: $695; $10 per credit hour. Tuition and fees vary according to degree level and program. *Financial support:* Fellowships, research assistantships, teaching assistantships, career-related internships or fieldwork available. Support available to part-time students. Financial award application deadline: 2/15. *Faculty research:* Epidemiology, ecology of infectious microorganisms, electron microscopy, medical microbiology, molecular biology. *Unit head:* Dr. Jan Cerny, Chairman, 410-706-7114, Fax: 410-706-2129. *Application contact:* Dr. Harry Mobley, Program Director, 410-706-0466, Fax: 410-706-1617, E-mail: hmobley@umaryland.edu.

See Close-Up on page 929.

University of Maryland, School of Medicine, Graduate Program in Life Sciences, Program in Molecular Microbiology and Immunology, Baltimore, MD 21201. Offers PhD. *Expenses:* Tuition, state resident: full-time $8,079; part-time $409 per credit hour. Tuition, nonresident: full-time $18,384; part-time $731 per credit hour. Required fees: $695; $10 per credit hour. Tuition and fees vary according to degree level and program.

University of Massachusetts Amherst, Graduate School, College of Natural Resources and the Environment, Department of Microbiology, Amherst, MA 01003. Offers MS, PhD. Part-time programs available. *Faculty:* 20 full-time (6 women). *Students:* 34 full-time (24 women), 4 part-time (2 women); includes 9 minority (1 African American, 5 Asian Americans or Pacific Islanders, 3 Hispanic Americans), 6 international. Average age 27. 48 applicants, 50% accepted, 10 enrolled. In 2005, 10 master's, 1 doctorate awarded. Terminal master's awarded for partial completion of doctoral program. *Degree requirements:* For master's, thesis or alternative; for doctorate, thesis/dissertation. *Entrance requirements:* For master's and doctorate, GRE General Test. Additional exam requirements/recommendations for international students: Required—TOEFL (minimum score 530 paper-based; 197 computer-based). *Application deadline:* For fall admission, 12/1 priority date for domestic students, 12/1 priority date for international students; for spring admission, 10/1 for domestic students, 10/1 for international students. Applications are processed on a rolling basis. Application fee: $40 ($65 for international students). Electronic applications accepted. *Expenses:* Tuition, state resident: part-time $110 per credit. Tuition, nonresident: part-time $414 per credit. Required fees: $2,824 per term. One-time fee: $250 part-time. Full-time tuition and fees vary according to course load, campus/location, program and reciprocity agreements. *Financial support:* In 2005–06, research assistantships with full tuition reimbursements (averaging $13,657 per year), teaching assistantships with full tuition reimbursements (averaging $13,657 per year) were awarded; fellowships with full tuition reimbursements, career-related internships or fieldwork, Federal Work-Study, scholarships/grants, traineeships, and unspecified assistantships also available. Support available to part-time students. Financial award application deadline: 2/1. *Unit head:* Dr. Steve Sandler, Director, 413-577-4391, Fax: 413-545-1578.

See Close-Up on page 931.

University of Massachusetts Worcester, Graduate School of Biomedical Sciences, Department of Molecular Genetics and Microbiology, Worcester, MA 01655-0115. Offers PhD. *Faculty:* 21 full-time (3 women). *Degree requirements:* For doctorate, thesis/dissertation. *Entrance requirements:* For doctorate, GRE General Test. Additional exam requirements/recommendations for international students: Required—TOEFL (minimum score 600 paper-based; 250 computer-based). *Application deadline:* For fall admission, 12/15 for domestic students, 12/15 for international students. Applications are processed on a rolling basis. Application fee: $25 ($50 for international students). *Expenses:* Tuition, state resident: full-time $2,640. Tuition, nonresident: full-time $9,856. Required fees: $5,685. *Financial support:* In 2005–06, research assistantships with full tuition reimbursements (averaging $25,235 per year); unspecified assistantships also available. *Faculty research:* Gene structure, regulation of gene expression, pathogenesis. *Unit head:* Dr. Allan Jacobson, Chair, 508-856-2442. *Application contact:* Michael Cole, Director of Admissions and Recruitment, 508-856-4779, Fax: 508-856-3659, E-mail: michael.cole@umassmed.edu.

See Close-Up on page 797.

University of Medicine and Dentistry of New Jersey, Graduate School of Biomedical Sciences, Graduate Programs in Biomedical Sciences–Newark, Department of Microbiology and Molecular Genetics, Newark, NJ 07107. Offers PhD. *Degree requirements:* For doctorate, thesis/dissertation, qualifying exam. *Entrance requirements:* For doctorate, GRE General Test. Additional exam requirements/recommendations for international students: Required—TOEFL. *Application deadline:* For fall admission, 2/1 for domestic students. Application fee: $40. *Financial support:* Fellowships, research assistantships, Federal Work-Study, institutionally sponsored loans, and tuition waivers (full and partial) available. Financial award application deadline: 5/1. *Faculty research:* Molecular genetics of yeast, mutagenesis and carcinogenesis of DNA, bacterial protein synthesis, mammalian cell genetics, adenovirus gene expression. *Unit head:* Dr. Vivian Bellofatto, Program Director, 973-972-4483 Ext. 4406, Fax: 973-972-8981.

Announcement: This program is designed to provide training in molecular genetics and biochemistry of microbial systems. The department pioneers research in many areas of molecular biology involving bacteria, yeast, parasites, and mammalian systems. Major areas of interest include gene expression, growth control, host-pathogen interactions, and DNA replication, mutagenesis, and recombination.

See Close-Up on page 935.

University of Medicine and Dentistry of New Jersey, Graduate School of Biomedical Sciences, Graduate Programs in Biomedical Sciences–Piscataway, Program in Molecular Genetics, Microbiology and Immunology, Piscataway, NJ 08854-5635. Offers MS, PhD, MD/PhD. Terminal master's awarded for partial completion of doctoral program. *Degree requirements:* For master's and doctorate, thesis/dissertation, qualifying exam. *Entrance requirements:* For master's and doctorate, GRE General Test. Additional exam requirements/recommendations for international students: Required—TOEFL. *Application deadline:* For fall admission, 1/5 for domestic students. Applications are processed on a rolling basis. Application fee: $40. *Financial support:* Fellowships, research assistantships, teaching assistantships available. Financial award application deadline: 5/1. *Faculty research:* Interferon, receptors, retrovirus evolution, Arbo virus/host cell interactions. *Unit head:* Dr. Joseph P. Dougherty, Director, 732-235-4588, Fax: 732-235-5223, E-mail: doughejp@umdnj.edu.

University of Miami, Graduate School, Miller School of Medicine, Graduate Programs in Medicine, Department of Microbiology and Immunology, Coral Gables, FL 33124. Offers PhD, MD/PhD. *Faculty:* 38 full-time (8 women). *Students:* 29 full-time (21 women); includes 4 minority (1 African American, 3 Hispanic Americans), 10 international. Average age 24. 94 applicants, 21% accepted, 7 enrolled. In 2005, 9 degrees awarded. *Median time to degree:* Of those who began their doctoral program in fall 1997, 100% received their degree in 8 years or less. *Degree requirements:* For doctorate, thesis/dissertation, oral and written qualifying exams. *Entrance requirements:* For doctorate, GRE General Test. Additional requirements/recommendations for international students: Required—TOEFL. *Application deadline:* For fall admission, 12/31 for domestic students. Applications are processed on a

rolling basis. Application fee: $50. Electronic applications accepted. *Financial support:* In 2005–06, fellowships with full tuition reimbursements (averaging $22,000 per year), research assistantships with full tuition reimbursements (averaging $22,000 per year) were awarded; institutionally sponsored loans also available. *Faculty research:* Cellular and molecular immunology, molecular and pathogenic virology, pathogenic bacteriology and gene therapy of cancer. Total annual research expenditures: $5.8 million. *Unit head:* Dr. Eckhard R. Podack, Chairman, 305-243-6694, Fax: 305-243-5522, E-mail: epodack@med.miami.edu. *Application contact:* Karen T. Del Rio, Administrator, Graduate Program, 305-243-5682, Fax: 305-243-6903, E-mail: kdelrio@med.miami.edu.

See Close-Up on page 937.

University of Michigan, Medical School and Horace H. Rackham School of Graduate Studies, Program in Biomedical Sciences (PIBS), Department of Microbiology and Immunology, Ann Arbor, MI 48109. Offers PhD. *Degree requirements:* For doctorate, oral defense of dissertation, preliminary exam. *Entrance requirements:* For doctorate, GRE General Test. Additional exam requirements/recommendations for international students: Required—TOEFL (minimum score 600 paper-based), TWE; Recommended—TSE. Electronic applications accepted. *Expenses:* Tuition, state resident: full-time $14,082; part-time $894 per credit hour. Tuition, nonresident: full-time $28,500; part-time $1,675 per credit hour. Required fees: $189; $189 per unit. *Faculty research:* Gene regulation, molecular biology of animal and bacterial viruses, molecular and cellular networks, pathogenesis and microbial genetics.

University of Minnesota, Twin Cities Campus, Medical School and Graduate School, Graduate Programs in Medicine, PhD Program in Microbiology, Immunology and Cancer Biology, Minneapolis, MN 55455-0213. Offers PhD. *Degree requirements:* For doctorate, thesis/dissertation. *Entrance requirements:* For doctorate, GRE General Test. Additional exam requirements/recommendations for international students: Required—TOEFL (minimum score 600 paper-based; 250 computer-based). Electronic applications accepted. *Expenses:* Tuition, state resident: full-time $8,748; part-time $729 per credit. Tuition, nonresident: full-time $15,848; part-time $1,321 per credit. Full-time tuition and fees vary according to class time, course load, program and reciprocity agreements. *Faculty research:* Virology, microbiology, cancer biology, immunology.

See Close-Up on page 939.

University of Mississippi Medical Center, School of Graduate Studies in the Health Sciences, Department of Microbiology, Jackson, MS 39216-4505. Offers MS, PhD, MD/PhD. *Faculty:* 17 full-time (6 women). *Students:* 29 full-time (17 women), 1 (woman) part-time; includes 4 minority (all African Americans), 10 international. Average age 27. 20 applicants, 60% accepted, 5 enrolled. In 2005, 2 degrees awarded. Terminal master's awarded for partial completion of doctoral program. *Median time to degree:* Of those who began their doctoral program in fall 1997, 100% received their degree in 8 years or less. *Degree requirements:* For master's, thesis; for doctorate, thesis/dissertation, first authored publication. *Entrance requirements:* For master's and doctorate, GRE General Test, minimum GPA of 3.0. *Application deadline:* For fall admission, 7/1 for domestic students, 3/1 for international students. Applications are processed on a rolling basis. Application fee: $10. *Financial support:* In 2005–06, 23 students received support, including 23 research assistantships (averaging $16,559 per year); scholarships/grants also available. Financial award application deadline: 4/1. *Faculty research:* Immunology, virology, microbial physiology/genetics, parasitology. Total annual research expenditures: $1 million. *Unit head:* Dr. Richard O'Callghan, Chairman, 601-984-1700, Fax: 601-984-1708. *Application contact:* Dr. V. Gregory Chinchar, Director, Graduate Program in Microbiology, 601-984-1743, Fax: 601-984-1708, E-mail: vchinchar@microbio.umsmed.edu.

University of Missouri–Columbia, School of Medicine and Graduate School, Graduate Programs in Medicine, Department of Molecular Microbiology and Immunology, Columbia, MO 65211. Offers MS, PhD. *Faculty:* 14 full-time (3 women), 1 (woman) part-time/adjunct. *Students:* 24 full-time (15 women), 13 part-time (6 women); includes 2 minority (both African Americans), 10 international. Average age 25. In 2005, 2 master's, 3 doctorates awarded. Terminal master's awarded for partial completion of doctoral program. *Degree requirements:* For master's and doctorate, thesis/dissertation. *Entrance requirements:* For master's and doctorate, GRE General Test, minimum GPA of 3.0. *Application deadline:* For fall admission, 1/31 for domestic students. Application fee: $45 ($60 for international students). *Financial support:* Fellowships, research assistantships, teaching assistantships, institutionally sponsored loans available. Financial award application deadline: 3/1. *Faculty research:* Molecular biology, host-parasite interactions. *Unit head:* Dr. David R. Lee, Director of Graduate Studies, 573-882-7893, E-mail: leedr@missouri.edu.

The University of Montana, Graduate School, College of Arts and Sciences, Division of Biological Sciences, Program in Biochemistry and Microbiology, Missoula, MT 59812-0002. Offers biochemistry (MS); integrative microbiology and biochemistry (PhD); microbial ecology (MS, PhD); microbiology (MS). Terminal master's awarded for partial completion of doctoral program. *Degree requirements:* For master's, thesis; for doctorate, variable foreign language requirement, thesis/dissertation. *Entrance requirements:* For master's and doctorate, GRE General Test. *Application deadline:* For fall admission, 2/1 for domestic students. Application fee: $45. *Expenses:* Tuition, state resident: part-time $267 per credit. Tuition, nonresident: part-time $665 per credit. Part-time tuition and fees vary according to course load and degree level. *Financial support:* In 2005–06, research assistantships with tuition reimbursements (averaging $9,400 per year), teaching assistantships with full tuition reimbursements (averaging $9,400 per year) were awarded; Federal Work-Study and tuition waivers (full and partial) also available. Financial award application deadline: 3/1; financial award applicants required to submit FAFSA. *Faculty research:* Ribosome structure, medical microbiology/pathogenesis, microbial ecology/environmental microbiology. *Application contact:* Janean Clark, Graduate Programs Secretary, 406-243-5222, Fax: 406-243-4184, E-mail: jmclark@selway.umt.edu.

University of Nebraska Medical Center, Graduate Studies, Department of Pathology and Microbiology, Omaha, NE 68198. Offers MS, PhD. Part-time programs available. *Faculty:* 36 full-time (6 women), 50 part-time/adjunct (4 women). *Students:* 31 full-time (15 women), 3 part-time (2 women); includes 15 minority (1 African American, 13 Asian Americans or Pacific Islanders, 1 Hispanic American). Average age 28. 23 applicants, 43% accepted, 9 enrolled. In 2005, 6 doctorates awarded. Terminal master's awarded for partial completion of doctoral program. *Degree requirements:* For master's and doctorate, thesis/dissertation, comprehensive exam. *Entrance requirements:* For master's, previous course work in biology, chemistry, mathematics, and physics; for doctorate, GRE General Test, previous course work in biology, chemistry, mathematics, and physics. Additional exam requirements/recommendations for international students: Required—TOEFL. *Application deadline:* Applications are processed on a rolling basis. Application fee: $45. *Expenses:* Tuition, area resident: Part-time $200 per hour. Tuition, nonresident: part-time $538 per hour. Required fees: $308; $59 per term. *Financial support:* In 2005–06, 12 fellowships with tuition reimbursements (averaging $21,000 per year), 19 research assistantships with tuition reimbursements (averaging $21,000 per year) were awarded; institutionally sponsored loans and tuition waivers (full) also available. Support available to part-time students. Financial award application deadline: 3/1. *Faculty research:* Carcinogenesis, cancer biology, immunobiology, molecular virology, molecular genetics. *Unit head:* Dr. Rakesh K. Siagh, Chair, Graduate Committee, 402-559-9948, Fax: 402-559-4077, E-mail: rsingh@unmc.edu.

University of New Hampshire, Graduate School, College of Life Sciences and Agriculture, Department of Microbiology, Durham, NH 03824. Offers MS, PhD. Part-time programs available. *Faculty:* 7 full-time. *Students:* 7 full-time (5 women), 10 part-time (5 women), 1 international. Average age 28. 28 applicants, 25% accepted, 3 enrolled. In 2005, 3 degrees awarded. Terminal master's awarded for partial completion of doctoral program. *Degree requirements:* For master's and doctorate, thesis/dissertation. *Entrance requirements:* For master's and doctorate, GRE General Test. Additional exam requirements/recommendations for international students: Required—TOEFL (minimum score 550 paper-based; 213 computer-based); Recommended—TSE. *Application deadline:* For fall admission, 4/1 priority date for domestic students, 4/1 priority date for international students. For winter admission, 12/1 for domestic students. Applications are processed on a rolling basis. Application fee: $60. Electronic applications accepted. *Expenses:* Tuition, state resident: full-time $8,010; part-time $445 per credit hour. Tuition, nonresident: full-time $19,730; part-time $810 per credit hour. Required fees: $322 per semester. Tuition and fees vary according to course load and program. *Financial support:* In 2005–06, 1 fellowship, 1 research assistantship, 10 teaching assistantships were awarded; career-related internships or fieldwork, Federal Work-Study, scholarships/grants, and tuition waivers (full and partial) also available. Support available to part-time students. Financial award application deadline: 2/15. *Faculty research:* Bacterial host-parasite interactions, immunology, microbial structures, bacterial and bacteriophage genetics, virology. *Unit head:* Dr. Aaron Mangolin, Chairperson, 603-862-0211. *Application contact:* Flora Joyal, Administrative Assistant, 603-862-4095, E-mail: flora.joyal@unh.edu.

University of New Mexico, School of Medicine, Biomedical Sciences Graduate Program, Albuquerque, NM 87131-5196. Offers biochemistry and molecular biology (MS, PhD); cell biology and physiology (MS, PhD); molecular genetics and microbiology (MS, PhD); neuroscience (MS, PhD); pathology (MS, PhD); toxicology (MS, PhD). Part-time programs available. Terminal master's awarded for partial completion of doctoral program. *Degree requirements:* For master's, thesis/dissertation; for doctorate, thesis/dissertation, comprehensive exam. *Entrance requirements:* For master's and doctorate, GRE General Test, minimum undergraduate GPA of 3.0. Additional exam requirements/recommendations for international students: Required—TOEFL. Electronic applications accepted. *Expenses:* Tuition, state resident: full-time $5,676. Tuition, nonresident: full-time $14,974; part-time $238 per credit hour. Required fees: $385 per term. Tuition and fees vary according to course load and program. *Faculty research:* Signal transduction, infectious disease, biology of cancer, structural biology, neuroscience.

The University of North Carolina at Chapel Hill, School of Medicine and Graduate School, Graduate Programs in Medicine, Department of Microbiology and Immunology, Chapel Hill, NC 27599. Offers immunology (MS, PhD); microbiology (MS, PhD). *Faculty:* 38 full-time (15 women), 26 part-time/adjunct (5 women). *Students:* 69 full-time (31 women), 2 part-time (both women); includes 9 minority (2 African Americans, 5 Asian Americans or Pacific Islanders, 2 Hispanic Americans), 11 international. Average age 29. 119 applicants, 21% accepted, 7 enrolled. In 2005, 3 master's, 9 doctorates awarded. Terminal master's awarded for partial completion of doctoral program. *Median time to degree:* Of those who began their doctoral program in fall 1997, 100% received their degree in 8 years or less. *Degree requirements:* For master's and doctorate, thesis/dissertation, comprehensive exam, registration. *Entrance requirements:* For master's and doctorate, GRE General Test, minimum GPA of 3.0. Additional exam requirements/recommendations for international students: Required—TOEFL (minimum score 600 paper-based; 250 computer-based). *Application deadline:* For fall admission, 1/1 priority date for domestic students, 1/1 priority date for international students. Applications are processed on a rolling basis. Application fee: $70. Electronic applications accepted. *Financial support:* In 2005–06, 3 fellowships with full tuition reimbursements (averaging $22,000 per year), 66 research assistantships with full tuition reimbursements (averaging $22,000 per year) were awarded; scholarships/grants, traineeships, health care benefits, and unspecified assistantships also available. *Faculty research:* HIV pathogenesis, immune response, t-cell mediated autoimmunity, alpha-viruses, bacterial chemotaxis. Total annual research expenditures: $8.1 million. *Unit head:* Jeffrey Frelinger, Chairman, 919-966-1191, Fax: 919-962-8103, E-mail: jfrelin@med.unc.edu. *Application contact:* Dixie Flannery, Student Services Manager, 919-966-1191, Fax: 919-962-8103, E-mail: microimm@listserv.med.unc.edu.

University of North Dakota, School of Medicine and Graduate School, Graduate Programs in Medicine, Department of Microbiology and Immunology, Grand Forks, ND 58202. Offers MS, PhD. *Faculty:* 6 full-time (1 woman). *Students:* 1 (woman) full-time, 12 part-time (7 women). 25 applicants, 12% accepted, 3 enrolled. In 2005, 1 master's, 2 doctorates awarded. *Degree requirements:* For master's, thesis or alternative, comprehensive exam; for doctorate, thesis/dissertation, final examination, comprehensive exam. *Entrance requirements:* For master's and doctorate, GRE General Test, minimum GPA of 3.0. Additional exam requirements/recommendations for international students: Required—TOEFL (minimum score 550 paper-based; 213 computer-based). *Application deadline:* For fall admission, 2/15 priority date for domestic students, 2/15 priority date for international students; for spring admission, 10/15 priority date for domestic students, 10/15 priority date for international students. Applications are processed on a rolling basis. Application fee: $35. Electronic applications accepted. *Financial support:* In 2005–06, 8 students received support, including research assistantships (averaging $13,997 per year), teaching assistantships with full tuition reimbursements available (averaging $13,997 per year); fellowships, Federal Work-Study, institutionally sponsored loans, scholarships/grants, and tuition waivers (full and partial) also available. Support available to part-time students. Financial award application deadline: 3/15; financial award applicants required to submit FAFSA. *Faculty research:* Genetic and immunological aspects of a murine model of human multiple sclerosis, termination of DNA replication, cell division in bacteria, yersinia pestis. *Unit head:* Dr. David Bradley, Director, 701-777-2214, Fax: 701-777-3527, E-mail: dbradley@medicine.nodak.edu.

University of North Texas Health Science Center at Fort Worth, Graduate School of Biomedical Sciences, Fort Worth, TX 76107-2699. Offers anatomy and cell biology (MS, PhD); biochemistry and molecular biology (MS, PhD); biomedical sciences (MS, PhD); biotechnology (MS); forensic genetics (MS); integrative physiology (MS, PhD); medical science (MS); microbiology and immunology (MS, PhD); pharmacology (MS, PhD); science education (MS, PhD). *Faculty:* 57 full-time (9 women), 2 part-time/adjunct (0 women). *Students:* 177 full-time (98 women), 42 part-time (33 women); includes 61 minority (15 African Americans, 3 American Indian/Alaska Native, 27 Asian Americans or Pacific Islanders, 16 Hispanic Americans), 53 international. Average age 28. 237 applicants, 62% accepted, 91 enrolled. In 2005, 37 master's, 15 doctorates awarded. Terminal master's awarded for partial completion of doctoral program. *Degree requirements:* For master's and doctorate, thesis/dissertation. *Entrance requirements:* For master's and doctorate, GRE General Test. Additional exam requirements/recommendations for international students: Required—TOEFL. *Application deadline:* For fall admission, 5/1 for domestic students. Application fee: $25 ($50 for international students). *Expenses:* Contact institution. *Financial support:* In 2005–06, 80 research assistantships (averaging $16,000 per year) were awarded; fellowships, teaching assistantships, career-related internships or fieldwork, Federal Work-Study, institutionally sponsored loans, scholarships/grants, and traineeships also available. Support available to part-time students. Financial award application deadline: 4/1; financial award applicants required to submit FAFSA. *Faculty research:* Alzheimer's disease, aging, eye diseases, cancer, cardiovascular disease. Total annual research expenditures: $21 million. *Unit head:* Dr. Thomas Yorio, Dean, 817-735-2560, Fax: 817-735-0243, E-mail: yoriot@hsc.unt.edu. *Application contact:* Carla Lee, Director of Graduate Admissions and Services, 817-735-2560, Fax: 817-735-0243, E-mail: gsbs@hsc.unt.edu.

See Close-Up on page 279.

University of Oklahoma, Graduate College, College of Arts and Sciences, Department of Botany and Microbiology, Program in Microbiology, Norman, OK 73019-0390. Offers MS, PhD. *Students:* 30 full-time (19 women), 6 part-time (3 women); includes 2 minority (1 American Indian/Alaska Native, 1 Hispanic American), 9 international. 19 applicants, 53% accepted, 3 enrolled. In 2005, 3 master's, 5 doctorates awarded. Terminal master's awarded for partial completion of doctoral program. *Degree requirements:* For master's, thesis, oral exam; for doctorate, one foreign language, thesis/dissertation, general exam. *Entrance requirements:* For master's and doctorate, GRE. Additional exam requirements/recommendations for international students: Required—TOEFL (minimum score 550 paper-based; 213 computer-based). *Application deadline:* For fall admission, 6/1 for domestic students, 4/1 for international students; for spring admission, 12/1 for domestic students, 9/1 for international students. Applications are processed on a rolling basis. Application fee: $40 ($90 for international students).

Expenses: Tuition, state resident: full-time $3,029; part-time $126 per credit hour. Tuition, nonresident: full-time $10,807; part-time $450 per credit hour. Required fees: $1,231; $44 per credit hour. Tuition and fees vary according to course load and program. *Financial support:* In 2005–06, 8 students received support; research assistantships with full tuition reimbursements available, teaching assistantships with full tuition reimbursements available, scholarships/grants and unspecified assistantships available. Financial award applicants required to submit FAFSA. *Faculty research:* Bioinformatics, molecular ecology, anaerobic microbiology, biomediation, environmental microbiology. *Application contact:* Adell Hopper, Staff Assistant, 405-325-4322, Fax: 405-325-7619, E-mail: ahopper@ou.edu.

University of Oklahoma Health Sciences Center, College of Medicine and Graduate College, Graduate Programs in Medicine, Department of Microbiology and Immunology, Oklahoma City, OK 73190. Offers immunology (MS, PhD); microbiology (MS, PhD). Part-time programs available. Terminal master's awarded for partial completion of doctoral program. *Degree requirements:* For master's, thesis or alternative; for doctorate, one foreign language, thesis/dissertation. *Entrance requirements:* For doctorate, GRE General Test, 3 letters of recommendation. Additional exam requirements/recommendations for international students: Required—TOEFL. *Faculty research:* Molecular genetics, pathogenesis, streptococcal infections, gram-positive virulence, monoclonal antibodies.

University of Ottawa, Faculty of Graduate and Postdoctoral Studies, Faculty of Medicine, Department of Biochemistry, Microbiology and Immunology, Ottawa, ON K1N 6N5, Canada. Offers biochemistry (M Sc, PhD); microbiology and immunology (M Sc, PhD). *Faculty:* 72 full-time (19 women), 22 part-time/adjunct (8 women). *Students:* 163 full-time, 22 part-time. 35 applicants, 69% accepted, 16 enrolled. In 2005, 16 master's, 17 doctorates awarded. *Degree requirements:* For master's, thesis; for doctorate, thesis/dissertation, seminar, comprehensive exam. *Entrance requirements:* For master's, honors degree or equivalent, minimum B average; for doctorate, master's degree, minimum B+ average. *Application deadline:* For fall admission, 3/1 priority date for domestic students, 2/15 priority date for international students. For winter admission, 11/15 for domestic students; for spring admission, 4/1 for domestic students. Application fee: $75. Electronic applications accepted. *Expenses:* Tuition: Part-time $260 per credit. Tuition and fees vary according to course load and program. *Financial support:* Fellowships, research assistantships with full tuition reimbursements, teaching assistantships with full tuition reimbursements, career-related internships or fieldwork, Federal Work-Study, scholarships/grants, traineeships, tuition waivers (full and partial), and unspecified assistantships available. Financial award application deadline: 2/15. *Faculty research:* General biochemistry, molecular biology, microbiology, host biology, nutrition and metabolism. *Unit head:* Dr. Zemin Yao, Chair, 613-562-5800 Ext. 5459, Fax: 613-562-5440. *Application contact:* Carol Ann Kelly, Academic Assistant, 613-562-5800 Ext. 5424, Fax: 613-562-5440, E-mail: grads@uottawa.ca.

University of Pennsylvania, School of Medicine, Biomedical Graduate Studies, Graduate Group in Cell and Molecular Biology, Program in Microbiology, Virology, and Parasitology, Philadelphia, PA 19104. Offers PhD, MD/PhD, VMD/PhD. *Degree requirements:* For doctorate, thesis/dissertation. *Entrance requirements:* For doctorate, GRE General Test, previous course work in science. Additional exam requirements/recommendations for international students: Required—TOEFL. *Application deadline:* For fall admission, 12/15 priority date for domestic students, 12/1 priority date for international students. Applications are processed on a rolling basis. Application fee: $70. Electronic applications accepted. *Financial support:* Fellowships, research assistantships, scholarships/grants, traineeships, and unspecified assistantships available. *Unit head:* Dr. Paul Bates, Chair, 215-573-3509. *Application contact:* Emily D. Geib, Coordinator, 215-898-3918, Fax: 215-573-2104, E-mail: eddz@mail.med.upenn.edu.

University of Pittsburgh, Graduate School of Public Health, Department of Infectious Diseases and Microbiology, Pittsburgh, PA 15260. Offers MPH, MS, Dr PH, MD/PhD. *Faculty:* 18 full-time (6 women), 3 part-time/adjunct (0 women). *Students:* 36 full-time (24 women), 11 part-time (8 women); includes 6 minority (3 African Americans, 1 Asian American or Pacific Islander, 2 Hispanic Americans), 13 international. Average age 28. 57 applicants, 49% accepted, 12 enrolled. In 2005, 6 master's, 8 doctorates awarded. Terminal master's awarded for partial completion of doctoral program. *Median time to degree:* Of those who began their doctoral program in fall 1997, 66% received their degree in 8 years or less. *Degree requirements:* For master's, one foreign language, thesis/dissertation, comprehensive exam (for some programs), registration; for doctorate, one foreign language, thesis/dissertation, comprehensive exam, registration. *Entrance requirements:* For master's and doctorate, GRE General Test. Additional exam requirements/recommendations for international students: Required—TOEFL (minimum score 550 paper-based, 213 computer-based). *Application deadline:* For fall admission, 1/15 for domestic students, 5/1 for international students. Applications are processed on a rolling basis. Application fee: $50 ($60 for international students). Electronic applications accepted. *Expenses:* Tuition, state resident: full-time $13,194; part-time $537 per credit. Tuition, nonresident: full-time $25,012; part-time $1,026 per credit. Required fees: $700; $164 per term. Tuition and fees vary according to campus/location and program. *Financial support:* In 2005–06, 32 students received support, including 31 research assistantships with tuition reimbursements available (averaging $20,333 per year); career-related internships or fieldwork, traineeships, health care benefits, tuition waivers (partial), and unspecified assistantships also available. Financial award applicants required to submit FAFSA. *Faculty research:* HIV, Epstein-Barr virus, virology, immunology, malaria. Total annual research expenditures: $13.5 million. *Unit head:* Dr. Charles Rinaldo, Chairman, 412-624-3928, Fax: 412-624-4953, E-mail: rinaldo@pitt.edu. *Application contact:* Dr. Phalguni Gupta, Associate Professor, 412-624-7998, Fax: 412-624-4953, E-mail: pgupta1@pitt.edu.

See Close-Up on page 943.

University of Pittsburgh, School of Medicine, Graduate Programs in Medicine, Program in Molecular Virology and Microbiology, Pittsburgh, PA 15260. Offers MS, PhD. *Faculty:* 36 full-time (6 women). *Students:* 33 full-time (20 women); includes 7 minority (2 African Americans, 2 Asian Americans or Pacific Islanders, 3 Hispanic Americans), 3 international. Average age 28. 415 applicants, 22% accepted, 42 enrolled. In 2005, 4 degrees awarded. *Median time to degree:* Of those who began their doctoral program in fall 1997, 95% received their degree in 8 years or less. *Degree requirements:* For doctorate, thesis/dissertation, comprehensive exam, registration. *Entrance requirements:* For doctorate, GRE General Test, GRE Subject Test, minimum QPA of 3.0. Additional exam requirements/recommendations for international students: Required—TOEFL (minimum score 600 paper-based; 250 computer-based), IELT (minimum score 7). *Application deadline:* For fall admission, 12/15 priority date for domestic students, 12/15 priority date for international students. Application fee: $40. Electronic applications accepted. *Expenses:* Tuition, state resident: full-time $13,194; part-time $537 per credit. Tuition, nonresident: full-time $25,012; part-time $1,026 per credit. Required fees: $700; $164 per term. Tuition and fees vary according to campus/location and program. *Financial support:* In 2005–06, 23 research assistantships with full tuition reimbursements (averaging $21,500 per year) were awarded; fellowships with full tuition reimbursements, teaching assistantships with full tuition reimbursements, institutionally sponsored loans, scholarships/grants, traineeships, health care benefits, and unspecified assistantships also available. *Faculty research:* Host-pathogen interactions, persistent microbial infections, microbial genetics and gene expression, microbial pathogenesis, anti-bacterial therapeutics. *Unit head:* Dr. Jo Anne Flynn, Program Director, 412-624-7743, Fax: 412-648-3394, E-mail: joanne@pitt.edu. *Application contact:* Graduate Studies Administrator, 412-648-8957, Fax: 412-648-1077, E-mail: gradstudies@medschool.pitt.edu.

University of Puerto Rico, Medical Sciences Campus, School of Medicine, Division of Graduate Studies, Department of Microbiology and Medical Zoology, San Juan, PR 00936-5067. Offers medical zoology (MS, PhD); microbiology and medical zoology (MS, PhD). *Faculty:* 15 full-time (8 women), 2 part-time/adjunct (both women). *Students:* 32 full-time (26 women); includes 28 minority (all Hispanic Americans) Average age 23. 20 applicants, 25%

Microbiology

University of Puerto Rico, Medical Sciences Campus (continued)
accepted, 5 enrolled. In 2005, 2 master's, 1 doctorate awarded. *Median time to degree:* Of those who began their doctoral program in fall 1997, 50% received their degree in 8 years or less. *Degree requirements:* For master's, one foreign language, thesis/dissertation, registration; for doctorate, one foreign language, thesis/dissertation, comprehensive exam, registration. *Entrance requirements:* For master's and doctorate, GRE General Test, GRE Subject Test, interview, minimum GPA of 3.0, 3 letters of recommendation. *Application deadline:* For fall admission, 2/15 for domestic students, 2/15 for international students; for spring admission, 9/15 for domestic students, 9/15 for international students. Application fee: $15. *Expenses:* Tuition, state resident: full-time $3,600; part-time $100 per credit hour. Tuition, nonresident: full-time $4,655. Required fees: $1,734. Tuition and fees vary according to class time, degree level and program. *Financial support:* In 2005–06, 9 fellowships with full tuition reimbursements, 17 research assistantships with full tuition reimbursements, 17 teaching assistantships with full tuition reimbursements were awarded; institutionally sponsored loans and tuition waivers (full) also available. Financial award application deadline: 4/30. *Faculty research:* Molecular and general parasitology, immunology, development of viral vaccines and antiviral agents, fungal dimorphism, AIDS, pathogenesis. Total annual research expenditures: $5.1 million. *Unit head:* Dr. Guillermo J. Vázquez, Director, 787-758-2525 Ext. 1309, Fax: 787-758-4808, E-mail: gvazquez@rcm.upr.edu. *Application contact:* Dr. Iraida E. Robledo, Graduate Program Coordinator, 787-758-2525 Ext. 1311, Fax: 787-758-4808, E-mail: irobledo@rcm.upr.edu.

University of Rhode Island, Graduate School, College of the Environment and Life Sciences, Department of Cell and Molecular Biology, Kingston, RI 02881. Offers biochemistry (MS, PhD); microbiology (MS, PhD), including biodegradation (MS), cellular development (MS), electron microscopy and ultrastructure (MS), genetics and molecular biology (MS), immunology (MS), marine and freshwater ecosystems (MS), microbial pathogenesis (MS), microbial physiology (MS), protozoology (MS), virology (MS), water-pollution microbiology (MS); molecular genetics (MS, PhD). In 2005, 2 degrees awarded. *Degree requirements:* For master's and doctorate, thesis/dissertation. *Entrance requirements:* For master's and doctorate, GRE General Test. Additional exam requirements/recommendations for international students: Required—TOEFL. *Expenses:* Tuition, state resident: full-time $5,522; part-time $307 per credit. Tuition, nonresident: full-time $15,992; part-time $888 per credit. Required fees: $1,786; $73 per credit. One-time fee: $80 part-time. *Financial support:* Fellowships, research assistantships, teaching assistantships available. *Unit head:* Dr. Jay Sperry, Chairperson, 401-874-5900.

University of Rochester, School of Medicine and Dentistry, Graduate Programs in Medicine and Dentistry, Department of Microbiology and Immunology, Rochester, NY 14627-0250. Offers microbiology (MS, PhD). *Degree requirements:* For doctorate, thesis/dissertation, qualifying exam. *Entrance requirements:* For master's and doctorate, GRE General Test.

University of Saskatchewan, College of Medicine, Department of Microbiology and Immunology, Saskatoon, SK S7N 5A2, Canada. Offers M Sc, PhD. *Faculty:* 12. *Students:* 19. *Degree requirements:* For master's and doctorate, thesis/dissertation, registration. *Entrance requirements:* Additional exam requirements/recommendations for international students: Required—TOEFL. *Application deadline:* For fall admission, 7/1 for domestic students. Applications are processed on a rolling basis. Application fee: $50. *Financial support:* Fellowships, research assistantships, teaching assistantships available. Financial award application deadline: 1/31. *Unit head:* Dr. W. Xiao, Graduate Chair, 306-966-4306, Fax: 306-966-4311. *Application contact:* Dr. P. Bretscher, Graduate Chair, 306-966-4306, Fax: 306-966-4311, E-mail: wei.xiao@usask.ca.

University of Saskatchewan, Western College of Veterinary Medicine and College of Graduate Studies and Research, Graduate Programs in Veterinary Medicine, Department of Veterinary Microbiology, Saskatoon, SK S7N 5A2, Canada. Offers M Sc, M Vet Sc, PhD. *Faculty:* 12 full-time (1 woman). *Students:* 31 full-time (16 women); includes 4 minority (all African Americans) In 2005, 2 master's, 3 doctorates awarded. *Degree requirements:* For master's, thesis/dissertation, registration (for some programs); for doctorate, thesis/dissertation, registration. *Entrance requirements:* Additional exam requirements/recommendations for international students: Required—IELT or TOEFL. *Application deadline:* For fall admission, 7/1 for domestic students. Applications are processed on a rolling basis. Application fee: $50. *Financial support:* Fellowships, teaching assistantships available. Financial award application deadline: 1/31. *Faculty research:* Immunology, vaccinology, epidemiology, virology, parasitology. *Unit head:* Dr. Vikram Misra, Head, 306-966-4307, Fax: 306-966-4311, E-mail: vikram.misra@usask.ca. *Application contact:* Dr. J. Gordan, Graduate Chair, 306-966-7214, E-mail: gordan@usask.ca.

University of South Alabama, College of Medicine and Graduate School, Program in Basic Medical Sciences, Specialization in Microbiology and Immunology, Mobile, AL 36688-0002. Offers PhD. *Faculty:* 10 full-time (0 women). *Degree requirements:* For doctorate, thesis/dissertation. *Application deadline:* For fall admission, 4/1 for domestic students. Applications are processed on a rolling basis. Application fee: $25. *Expenses:* Tuition, state resident: full-time $4,008. Tuition, nonresident: full-time $8,016. Required fees: $692. *Financial support:* Fellowships, research assistantships, institutionally sponsored loans available. Financial award application deadline: 4/1. *Faculty research:* Mechanisms of tumor immunity, host response to infectious agents, virus replication, immune regulation, mechanisms of resistance to viruses and bacteria. *Unit head:* Dr. Joseph H. Coggin, Chair, 251-460-6339. *Application contact:* Lanette Flagge, Academic Advisor, 251-460-6153.

The University of South Dakota, School of Medicine and Health Sciences and Graduate School, Biomedical Sciences Graduate Program, Molecular Microbiology and Immunology Group, Vermillion, SD 57069-2390. Offers MA, PhD. *Faculty:* 6 full-time (1 woman). *Students:* 4 full-time (1 woman), 1 international. Average age 27. 11 applicants, 36% accepted, 2 enrolled.Terminal master's awarded for partial completion of doctoral program. *Degree requirements:* For master's, thesis/dissertation; for doctorate, thesis/dissertation, comprehensive exam, registration. *Entrance requirements:* For master's and doctorate, GRE General Test, minimum GPA of 3.0. Additional exam requirements/recommendations for international students: Required—TOEFL (minimum score 550 paper-based; 213 computer-based). *Application deadline:* For fall admission, 4/15 priority date for domestic students, 4/15 priority date for international students. Applications are processed on a rolling basis. Application fee: $35. *Expenses:* Contact institution. Tuition and fees vary according to course load, program and reciprocity agreements. *Financial support:* In 2005–06, 4 students received support, including 4 fellowships with partial tuition reimbursements available (averaging $20,772 per year), 1 research assistantship with partial tuition reimbursement available (averaging $10,386 per year); Federal Work-Study and unspecified assistantships also available. Financial award application deadline: 4/15; financial award applicants required to submit FAFSA. *Faculty research:* Structure-function membranes, plasmids, immunology, virology, pathogenesis. Total annual research expenditures: $325,000.

University of Southern California, Keck School of Medicine and Graduate School, Graduate Programs in Medicine, Department of Molecular Microbiology and Immunology, Los Angeles, CA 90033. Offers MS, PhD. *Faculty:* 9 full-time (1 woman), 1 (woman) part-time/adjunct. *Students:* 33 full-time (19 women), 1 part-time; includes 8 minority (7 Asian Americans or Pacific Islanders, 1 Hispanic American), 17 international. Average age 28. 68 applicants, 12% accepted, 5 enrolled. In 2005, 6 master's, 5 doctorates awarded. *Median time to degree:* Of those who began their doctoral program in fall 1997, 80% received their degree in 8 years or less. *Degree requirements:* For master's, thesis optional; for doctorate, thesis/dissertation. *Entrance requirements:* For master's, GRE General Test, minimum GPA of 3.0; for doctorate, GRE General Test, GRE Subject Test, minimum GPA of 3.0. Additional exam requirements/recommendations for international students: Required—TOEFL (minimum score 250 paper-based). *Application deadline:* For fall admission, 2/15 priority date for domestic students, 2/15 priority date for international students; for spring admission, 7/15 for domestic students, 2/15 for international students. Applications are processed on a rolling basis. Application fee: $65 ($75 for international students). Electronic applications accepted. *Expenses:* Tuition: Full-time $25,416; part-time $1,059 per unit. Required fees: $484; $484 per year. Tuition and fees vary according to course load and program. *Financial support:* In 2005–06, 22 students received support, including 1 fellowship with full tuition reimbursement available (averaging $21,000 per year), 20 research assistantships with full tuition reimbursements available (averaging $21,000 per year), 1 teaching assistantship with full tuition reimbursement available (averaging $21,000 per year); institutionally sponsored loans, traineeships, and tuition waivers (partial) also available. Financial award application deadline: 6/1. *Faculty research:* Animal virology, microbial genetics, molecular and cellular immunology, cellular differentiation control of protein synthesis. Total annual research expenditures: $2.5 million. *Unit head:* Dr. Stanley M. Tahara, Director, Graduate Studies, 323-442-1722, Fax: 323-442-1721, E-mail: stahara@usc.edu. *Application contact:* Laura C. Steel, Graduate Administrator, 323-442-2337, Fax: 323-442-1721, E-mail: lsteel@usc.edu.

University of Southern Mississippi, Graduate School, College of Science and Technology, Department of Biological Sciences, Hattiesburg, MS 39406-0001. Offers environmental biology (MS, PhD); marine biology (MS, PhD); microbiology (MS, PhD); molecular biology (MS, PhD). *Degree requirements:* For master's and doctorate, thesis/dissertation, comprehensive exam. *Entrance requirements:* For master's, GRE General Test, minimum GPA of 3.0; for doctorate, GRE General Test, minimum GPA of 3.5. Additional exam requirements/recommendations for international students: Required—TOEFL.

See Close-Up on page 297.

University of South Florida, College of Graduate Studies, College of Arts and Sciences, Department of Biology, Tampa, FL 33620-9951. Offers biology (PhD); botany (MS); ecology (PhD); microbiology (MS); physiology (PhD); zoology (MS). Part-time programs available. *Faculty:* 21. *Students:* 54 full-time (32 women), 26 part-time (17 women); includes 8 minority (2 African Americans, 1 American Indian/Alaska Native, 1 Asian American or Pacific Islander, 4 Hispanic Americans), 14 international. 76 applicants, 34% accepted, 13 enrolled. In 2005, 3 master's, 3 doctorates awarded. *Degree requirements:* For master's, thesis (for some programs), graduate seminar in biology; for doctorate, 2 foreign languages, thesis/dissertation, essay of research interest, comprehensive exam. *Entrance requirements:* For master's, GRE General Test, minimum undergraduate GPA of 3.0 in last 60 hours of course work; for doctorate, GRE General Test, GRE Subject Test in biology, minimum undergraduate GPA of 3.0 in last 60 hours of course work. Additional exam requirements/recommendations for international students: Required—TOEFL (minimum score 570 paper-based), TSE (minimum score 50). *Application deadline:* For fall admission, 2/1 priority date for domestic students, 3/1 priority date for international students; for spring admission, 10/1 for domestic students, 8/1 for international students. Application fee: $30. Electronic applications accepted. *Financial support:* Fellowships with full tuition reimbursements, research assistantships with full tuition reimbursements, teaching assistantships with full tuition reimbursements, Federal Work-Study and unspecified assistantships available. Financial award application deadline: 6/30. *Unit head:* Sydney Pierce, Chairperson, 813-974-3250, Fax: 813-974-3263. *Application contact:* Christine Smith, Graduate Advisor, 813-974-4747, Fax: 813-974-3263, E-mail: csmith2@chuma1.cas.usf.edu.

The University of Tennessee, Graduate School, College of Arts and Sciences, Department of Microbiology, Knoxville, TN 37996. Offers MS, PhD. Part-time programs available. *Degree requirements:* For master's and doctorate, thesis/dissertation. *Entrance requirements:* For master's and doctorate, GRE General Test, minimum GPA of 2.7. Additional exam requirements/recommendations for international students: Required—TOEFL. Electronic applications accepted.

The University of Tennessee Health Science Center, College of Graduate Health Sciences, Department of Molecular Sciences, Memphis, TN 38163-0002. Offers bacterial pathogenesis (PhD); biochemistry (PhD); immunology (PhD); microbiology (PhD); molecular and cell biology–pathology (PhD); molecular biology (PhD); signal transduction (PhD); structural biology (PhD); virology (PhD). *Faculty:* 17 full-time (4 women). *Students:* 25 full-time (11 women); includes 14 minority (1 African American, 13 Asian Americans or Pacific Islanders). Average age 24. 83 applicants, 12% accepted. In 2005, 6 degrees awarded. *Degree requirements:* For doctorate, thesis/dissertation, oral and written preliminary and comprehensive exams. *Entrance requirements:* For doctorate, GRE General Test, GRE Subject Test, minimum GPA of 3.0. Additional exam requirements/recommendations for international students: Required—TOEFL. *Application deadline:* For fall admission, 5/15 for domestic students. Application fee: $0. *Financial support:* In 2005–06, 8 fellowships, 21 research assistantships were awarded; teaching assistantships, traineeships and tuition waivers (full) also available. *Unit head:* Dr. David Hasty, Chairman, 901-448-6150, Fax: 901-448-7360. *Application contact:* Ida W. Mosby, Director, Enrollment Services, 901-448-5560, E-mail: imosby@utmem.edu.

The University of Texas at Austin, Graduate School, College of Natural Sciences, Institute for Cellular and Molecular Biology, Program in Microbiology and Immunology, Austin, TX 78712-1111. Offers PhD. *Degree requirements:* For doctorate, thesis/dissertation. *Entrance requirements:* For doctorate, GRE General Test. Additional exam requirements/recommendations for international students: Required—TOEFL. Electronic applications accepted. *Faculty research:* Pathogenesis, virology, cell biology, molecular genetics.

The University of Texas at Austin, Graduate School, College of Natural Sciences, School of Biological Sciences, Program in Microbiology, Austin, TX 78712-1111. Offers MA, PhD. *Entrance requirements:* For master's and doctorate, GRE General Test. Electronic applications accepted.

See Close-Up on page 945.

The University of Texas Health Science Center at Houston, Graduate School of Biomedical Sciences, Program in Microbiology and Molecular Genetics, Houston, TX 77225-0036. Offers MS, PhD, MD/PhD. *Faculty:* 18 full-time (5 women). *Students:* 28 full-time (16 women); includes 7 minority (1 African American, 1 Asian American or Pacific Islander, 5 Hispanic Americans), 4 international. Average age 26. 32 applicants, 63% accepted, 8 enrolled. In 2005, 1 degree awarded. Terminal master's awarded for partial completion of doctoral program. *Degree requirements:* For master's and doctorate, thesis/dissertation. *Entrance requirements:* For master's and doctorate, GRE General Test. Additional exam requirements/recommendations for international students: Required—TOEFL, TWE. *Application deadline:* For fall admission, 1/15 for domestic students; for spring admission, 11/1 for domestic students. Applications are processed on a rolling basis. Application fee: $10. Electronic applications accepted. *Financial support:* Fellowships with full tuition reimbursements, research assistantships with full tuition reimbursements, teaching assistantships, institutionally sponsored loans, scholarships/grants, and health care benefits available. Financial award application deadline: 1/15. *Faculty research:* Microbial genomics, microbial diversity, gene regulation, molecular pathogenesis, sensory transduction. *Unit head:* Dr. Samuel Kaplan, Director, 713-500-5502, Fax: 713-500-5499, E-mail: samuel.kaplan@uth.tmc.edu. *Application contact:* Dr. Victoria P. Knutson, Assistant Dean of Admissions, 713-500-9860, Fax: 713-500-9877, E-mail: victoria.p.knutson@uth.tmc.edu.

The University of Texas Health Science Center at San Antonio, Graduate School of Biomedical Sciences, Department of Microbiology and Immunology, San Antonio, TX 78229-3900. Offers microbiology and immunology (PhD); microbiology and immunology primary and secondary science teacher track (MS). Terminal master's awarded for partial completion of doctoral program. *Degree requirements:* For master's and doctorate, thesis/dissertation. *Entrance requirements:* For master's and doctorate, GRE General Test, minimum GPA of 3.0. *Faculty research:* Molecular immunology, mechanisms of pathogenesis, molecular genetics, vaccine and immunodiagnostic development.

See Close-Up on page 947.

The University of Texas Medical Branch, Graduate School of Biomedical Sciences, Program in Microbiology and Immunology, Galveston, TX 77555. Offers MS, PhD. *Students:* 29 full-time (13 women), 2 part-time (both women); includes 5 minority (3 African Americans, 2 Hispanic

Americans), 9 international. Average age 30. In 2005, 3 doctorates awarded. *Degree requirements:* For master's, thesis or alternative; for doctorate, thesis/dissertation. *Entrance requirements:* For doctorate, GRE General Test, minimum GPA of 3.0. Additional exam requirements/recommendations for international students: Required—TOEFL (minimum score 550 paper-based; 213 computer-based). *Application deadline:* Applications are processed on a rolling basis. Application fee: $30 ($75 for international students). Electronic applications accepted. *Expenses:* Tuition, state resident: full-time $8,350; part-time $90 per credit hour. Tuition, nonresident: full-time $21,450; part-time $366 per credit hour. Required fees: $1,027; $11 per credit hour. $60 per term. *Financial support:* In 2005–06, research assistantships with full tuition reimbursements (averaging $23,000 per year) Financial award applicants required to submit FAFSA. *Unit head:* Dr. Thomas K. Hughes, Director, 409-772-6660, Fax: 409-772-2295, E-mail: tkhughes@utmb.edu. *Application contact:* Martha J. Lewis, Coordinator II Special Programs, 409-772-2322, Fax: 409-772-2295, E-mail: mlewis@utmb.edu.

See Close-Up on page 951.

The University of Texas Southwestern Medical Center at Dallas, Southwestern Graduate School of Biomedical Sciences, Division of Basic Science, Program in Molecular Microbiology, Dallas, TX 75390. Offers PhD. *Faculty:* 34 full-time (10 women), 2 part-time/adjunct (0 women). *Students:* 28 full-time (17 women), 1 part-time; includes 3 minority (2 African Americans, 1 Hispanic American), 1 international. Average age 28. In 2005, 2 doctorates awarded. *Degree requirements:* For doctorate, thesis/dissertation, oral and written exams. *Entrance requirements:* For doctorate, GRE General Test, minimum GPA of 3.0. Additional exam requirements/recommendations for international students: Required—TOEFL. *Application deadline:* For fall admission, 1/5 for domestic students. Applications are processed on a rolling basis. Application fee: $0. Electronic applications accepted. *Expenses:* Tuition, state resident: full-time $6,550; part-time $50 per credit hour. Tuition, nonresident: full-time $19,650; part-time $326 per credit hour. Required fees: $42 per credit hour. Tuition and fees vary according to degree level and program. *Financial support:* Fellowships, research assistantships, institutionally sponsored loans available. *Faculty research:* Cell and molecular immunology, molecular pathogenesis of infectious disease, virology. *Unit head:* Dr. Michael Gale, Chair, 214-648-5940, Fax: 214-648-1899, E-mail: michael.gale@utsouthwestern.edu. *Application contact:* Dr. Nancy E. Street, Associate Dean, 214-648-6708, Fax: 214-648-2102, E-mail: nancy.street@utsouthwestern.edu.

The University of Toledo, College of Graduate Studies, Department of Microbiology and Immunology, Toledo, OH 43606-3390. Offers MS. Part-time programs available. *Degree requirements:* For master's, thesis, qualifying exam. *Entrance requirements:* For master's, GRE General Test, minimum undergraduate GPA of 3.0. *Expenses:* Tuition, state resident: full-time $6,623; part-time $308 per credit hour. Tuition, nonresident: full-time $13,232; part-time $735 per credit hour. *Faculty research:* Gene regulation, bacterial and fungal genetics, viral replication, immunology, microbial ecology.

University of Utah, The Graduate School, College of Science and Graduate Programs in Medicine, Program in Molecular Biology, Salt Lake City, UT 84112-1107.

University of Vermont, College of Medicine and Graduate College, Graduate Programs in Medicine, Department of Microbiology and Molecular Genetics, Burlington, VT 05405. Offers MS, PhD, MD/MS, MD/PhD. *Faculty:* 18 full-time (5 women). *Students:* 23 (16 women); includes 1 minority (Hispanic American) 13 international. 56 applicants, 32% accepted, 5 enrolled. In 2005, 4 doctorates awarded. *Degree requirements:* For master's and doctorate, thesis/dissertation. *Entrance requirements:* For master's and doctorate, GRE General Test. Additional exam requirements/recommendations for international students: Required—TOEFL (minimum score 550 paper-based; 213 computer-based). *Application deadline:* For fall admission, 2/1 for domestic students. Applications are processed on a rolling basis. Application fee: $40. Electronic applications accepted. *Expenses:* Tuition, area resident: Part-time $410 per credit hour. Tuition, nonresident: part-time $1,034 per credit hour. *Financial support:* Fellowships, research assistantships, teaching assistantships available. Financial award application deadline: 3/1. *Unit head:* , Dr. Susan S. Wallace, Chairperson, 802-656-2164. *Application contact:* Dr. S. Doublie, Coordinator, 802-656-2164.

See Close-Up on page 801.

University of Victoria, Faculty of Graduate Studies, Faculty of Science, Department of Biochemistry and Microbiology, Victoria, BC V8W 2Y2, Canada. Offers biochemistry (M Sc, PhD); microbiology (M Sc, PhD). *Faculty:* 14 full-time (3 women), 1 part-time/adjunct (0 women). *Students:* Average age 22. 37 applicants, 43% accepted, 5 enrolled. In 2005, 3 master's, 1 doctorate awarded. *Degree requirements:* For master's, thesis, seminar; for doctorate, thesis/dissertation, seminar, candidacy exam. *Entrance requirements:* For master's, GRE General Test (75th percentile), minimum B+ average; for doctorate, GRE General Test (75th percentile), minimum B+ average, M Sc. Additional exam requirements/recommendations for international students: Required—TOEFL (minimum score 600 paper-based; 250 computer-based). *Application deadline:* For fall admission, 5/31 for domestic students, 12/15 for international students. Applications are processed on a rolling basis. Application fee: $75 ($125 for international students). Electronic applications accepted. Tuition charges are reported in Canadian dollars. *Expenses:* Tuition, nonresident: full-time $4,492 Canadian dollars; part-time $749 Canadian dollars per term. International tuition: $5,346 Canadian dollars full-time. Tuition and fees vary according to course load, campus/location and program. *Financial support:* In 2005–06, 5 fellowships (averaging $14,250 per year), 16 research assistantships (averaging $11,000 per year), 20 teaching assistantships (averaging $2,000 per year) were awarded; career-related internships or fieldwork, institutionally sponsored loans, and awards, graduate teaching fellowships also available. Financial award application deadline: 2/15. *Faculty research:* Molecular pathogenesis, prokaryotic, eukaryotic, macromolecular interactions, microbial surfaces, virology, molecular genetics. Total annual research expenditures: $2.2 million. *Unit head:* Dr. Claire Cupples, Chair, 250-721-7077, Fax: 250-721-8855, E-mail: ccupples@uvic.ca. *Application contact:* Melinda Powell, Graduate Secretary, 250-721-8861, Fax: 250-721-8855, E-mail: biocgsec@uvic.ca.

University of Virginia, School of Medicine, Department of Microbiology, Charlottesville, VA 22903. Offers PhD, MD/PhD. *Students:* 93 full-time (57 women); includes 11 minority (2 African Americans, 5 Asian Americans or Pacific Islanders, 4 Hispanic Americans), 16 international. Average age 27. In 2005, 11 degrees awarded. *Degree requirements:* For doctorate, thesis/dissertation. *Entrance requirements:* For doctorate, GRE. Additional exam requirements/recommendations for international students: Required—TOEFL (minimum score 600 paper-based; 250 computer-based). *Application deadline:* Applications are processed on a rolling basis. Application fee: $60. Electronic applications accepted. *Expenses:* Tuition, state resident: full-time $7,731. Tuition, nonresident: full-time $18,672. Required fees: $1,479. Full-time tuition and fees vary according to degree level and program. *Financial support:* Fellowships, traineeships and unspecified assistantships available. Financial award applicants required to submit FAFSA. *Faculty research:* Virology, membrane biology and molecular genetics. *Unit head:* J. Thomas Parsons, Chairman, 434-924-1071, Fax: 434-982-1071. *Application contact:* Peter C. Brunjes, Associate Dean for Graduate Programs and Research, 434-924-7184, Fax: 434-924-6737, E-mail: grad-a-s@virginia.edu.

See Close-Up on page 955.

University of Washington, School of Medicine and Graduate School, Graduate Programs in Medicine, Department of Microbiology, Seattle, WA 98195. Offers PhD. *Degree requirements:* For doctorate, thesis/dissertation. *Entrance requirements:* For doctorate, GRE General Test, GRE Subject Test (recommended). Electronic applications accepted. *Faculty research:* Bacterial genetics and physiology, mechanisms of bacterial and viral pathogenesis, bacterial-plant interaction.

The University of Western Ontario, Faculty of Graduate Studies, Biosciences Division, Department of Microbiology and Immunology, London, ON N6A 5B8, Canada. Offers M Sc,

PhD. *Degree requirements:* For master's and doctorate, thesis/dissertation, oral and written exam. *Entrance requirements:* For master's, honors degree or equivalent in microbiology, immunology, or other biological science; minimum B average; for doctorate, M Sc in microbiology and immunology. *Faculty research:* Virology, molecular pathogenesis, cellular immunology, molecular biology.

University of Wisconsin–La Crosse, Office of University Graduate Studies, College of Science and Health, Department of Biology, La Crosse, WI 54601-3742. Offers aquatic sciences (MS); biology (MS); cellular and molecular biology (MS); clinical microbiology (MS); microbiology (MS); nurse anesthesia (MS); physiology (MS). *Accreditation:* AANA/CANAEP. Part-time programs available. *Faculty:* 18 full-time (5 women), 1 part-time/adjunct (0 women). *Students:* 23 full-time (10 women), 52 part-time (25 women); includes 5 minority (1 American Indian/Alaska Native, 3 Asian Americans or Pacific Islanders, 1 Hispanic American), 3 international. Average age 26. 61 applicants, 44% accepted, 23 enrolled. In 2005, 14 degrees awarded. *Degree requirements:* For master's, thesis, comprehensive exam, registration. *Entrance requirements:* For master's, GRE General Test, minimum GPA of 2.85. Additional exam requirements/recommendations for international students: Required—TOEFL (minimum score 550 paper-based; 213 computer-based). *Application deadline:* For fall admission, 3/1 for domestic students. Applications are processed on a rolling basis. Application fee: $45. Electronic applications accepted. *Expenses:* Tuition, state resident: part-time $354 per credit. Tuition, nonresident: part-time $943 per credit. Tuition and fees vary according to course load, program and reciprocity agreements. *Financial support:* In 2005–06, 10 students received support, including 4 research assistantships with partial tuition reimbursements available (averaging $10,000 per year), 10 teaching assistantships with partial tuition reimbursements available (averaging $9,600 per year); career-related internships or fieldwork, Federal Work-Study, health care benefits, unspecified assistantships, and grant-funded positions also available. Support available to part-time students. Financial award application deadline: 3/15; financial award applicants required to submit FAFSA. *Faculty research:* Cell and molecular biology, physiology, environmental sciences, mycology, biomedical general. Total annual research expenditures: $700,000. *Unit head:* Dr. Tom Volk, Program Director, 608-785-6972, Fax: 608-785-6959, E-mail: volk.thom@uwlax.edu. *Application contact:* Kathryn Kiefer, Associate Director of Admissions, 608-785-8939, E-mail: admissions@uwlax.edu.

University of Wisconsin–Madison, Medical School and Graduate School, Graduate Programs in Medicine and College of Agricultural and Life Sciences, Microbiology Doctoral Training Program, Madison, WI 53706-1380. Offers PhD. *Faculty:* 90 full-time (25 women). *Students:* 112 full-time (67 women); includes 23 minority (4 African Americans, 1 American Indian/Alaska Native, 5 Asian Americans or Pacific Islanders, 13 Hispanic Americans), 7 international. Average age 24. 213 applicants, 21% accepted, 21 enrolled. In 2005, 12 degrees awarded. *Degree requirements:* For doctorate, thesis/dissertation, preliminary exam, 2 semesters of teaching. *Entrance requirements:* For doctorate, GRE. Additional exam requirements/recommendations for international students: Required—TOEFL (minimum score 580 paper-based; 237 computer-based). *Application deadline:* For fall admission, 12/1 for domestic students, 12/1 for international students. Application fee: $45. Electronic applications accepted. *Financial support:* In 2005–06, 112 students received support, including 38 fellowships with tuition reimbursements available (averaging $22,500 per year), 74 research assistantships with tuition reimbursements available (averaging $22,500 per year); career-related internships or fieldwork, scholarships/grants, traineeships, health care benefits, and tuition waivers (full) also available. *Faculty research:* Microbial pathogenesis, gene regulation, immunology, virology, cell biology. Total annual research expenditures: $15.1 million. *Unit head:* Dr. Heidi Goodrich-Blair, Director, 608-265-4537, Fax: 608-262-9865, E-mail: hgblair@bact.wisc.edu. *Application contact:* Kathryn A. Holtgraver, Program Coordinator, 608-265-0689, Fax: 608-262-9865, E-mail: kathyh@bact.wisc.edu.

See Close-Up on page 957.

University of Wisconsin–Oshkosh, The School of Graduate Studies, College of Letters and Science, Department of Biology and Microbiology, Oshkosh, WI 54901. Offers biology (MS), including botany, microbiology, zoology. *Degree requirements:* For master's, thesis, comprehensive exam, registration. *Entrance requirements:* For master's, GRE General Test, minimum GPA of 3.0, BS in biology. Additional exam requirements/recommendations for international students: Required—TOEFL (minimum score 550 paper-based; 213 computer-based). Electronic applications accepted.

Utah State University, School of Graduate Studies, College of Agriculture, Department of Nutrition and Food Sciences, Logan, UT 84322. Offers dietetic administration (MDA); food microbiology and safety (MFMS); nutrition and food sciences (MS, PhD); nutrition science (MS, PhD), including molecular biology. Postbaccalaureate distance learning degree programs offered. *Faculty:* 13 full-time (6 women), 1 (woman) part-time/adjunct. *Students:* 55 full-time (26 women), 13 part-time (12 women); includes 4 minority (all African Americans), 15 international. Average age 27. 14 applicants, 57% accepted, 4 enrolled. In 2005, 8 master's, 3 doctorates awarded. *Degree requirements:* For master's, thesis, BS core competency courses BS core competency courses; for doctorate, thesis/dissertation, teaching experience, inst 7920 and BS core competency courses, comprehensive exam, registration. *Entrance requirements:* For master's, GRE General Test, minimum GPA of 3.0, course work in chemistry; for doctorate, GRE General Test, minimum GPA of 3.2, course work in chemistry, MS or manuscript in referred journal. Additional exam requirements/recommendations for international students: Required—TOEFL (minimum score 550 paper-based). *Application deadline:* For fall admission, 6/15 priority date for domestic students, 6/15 priority date for international students; for spring admission, 10/15 priority date for domestic students, 10/15 priority date for international students. Applications are processed on a rolling basis. Application fee: $50 ($60 for international students). Electronic applications accepted. *Financial support:* In 2005–06, 19 students received support, including fellowships with partial tuition reimbursements available (averaging $14,484 per year), 22 research assistantships with partial tuition reimbursements available (averaging $12,600 per year), 3 teaching assistantships with partial tuition reimbursements available (averaging $6,000 per year); Federal Work-Study, institutionally sponsored loans, scholarships/grants, tuition waivers (full and partial), unspecified assistantships, and fellowships, international students get support from their countries also available. Financial award application deadline: 3/1. *Faculty research:* Mineral balance, meat microbiology and nitrate interactions, milk ultrafiltration, lactic culture, milk coagulation. Total annual research expenditures: $319,280. *Unit head:* Dr. Charles C. Carpenter, Head, 435-797-2126, Fax: 435-797-2379, E-mail: chuck@cc.usu.edu. *Application contact:* Pam Zetterquist, Staff Assistant II, 435-797-4041, Fax: 435-797-2379, E-mail: pzett@cc.usu.edu.

Vanderbilt University, Graduate School and School of Medicine, Department of Microbiology and Immunology, Nashville, TN 37232-2363. Offers MS, PhD, MD/PhD. *Faculty:* 40 full-time (6 women). *Students:* 50 full-time (27 women); includes 6 minority (2 African Americans, 1 Asian American or Pacific Islander, 3 Hispanic Americans), 9 international. In 2005, 1 master's, 11 doctorates awarded. *Degree requirements:* For master's, thesis; for doctorate, thesis/dissertation, final and qualifying exams. *Entrance requirements:* For master's and doctorate, GRE General Test, GRE Subject Test (recommended). *Application deadline:* For fall admission, 1/15 for domestic students, 1/15 for international students. Application fee: $0. Electronic applications accepted. *Expenses:* Tuition: Part-time $1,283 per semester hour. Required fees: $2,202; $1,101 per semester. One-time fee: $30. Tuition and fees vary according to course load, program and student level. *Financial support:* Fellowships with full tuition reimbursements, research assistantships with full tuition reimbursements, Federal Work-Study, institutionally sponsored loans, traineeships, and tuition waivers (partial) available. Financial award application deadline: 1/15. *Faculty research:* Molecular and cellular immunology, molecular genetics and immunogenetics, cellular microbiology of pathogen-host interaction, virology, biotechnology. *Unit head:* Jacek Hawiger, Chair, 615-343-3435, Fax: 615-343-7392, E-mail: jacek.hawiger@vanderbilt.edu. *Application contact:* Eugene M. Oltz, Director of Graduate Studies, 615-343-0453, Fax: 615-343-7392, E-mail: eugene.oltz@vanderbilt.edu.

Microbiology

Virginia Commonwealth University, Medical College of Virginia-Professional Programs, School of Medicine and Graduate Programs, School of Medicine Graduate Programs, Department of Microbiology and Immunology, Richmond, VA 23284-9005. Offers microbiology and genetics (MS); microbiology and immunology (MS, PhD, CBHS); molecular biology and genetics (PhD). *Faculty:* 33 full-time (8 women). *Students:* 66 full-time (43 women), 9 part-time (6 women); includes 21 minority (12 African Americans, 5 Asian Americans or Pacific Islanders, 4 Hispanic Americans), 11 international. 95 applicants, 48% accepted. In 2005, 25 master's, 7 doctorates, 3 other advanced degrees awarded. *Degree requirements:* For master's, thesis; for doctorate, thesis/dissertation, comprehensive oral and written exams. *Entrance requirements:* For master's, GRE General Test or MCAT; for doctorate, GRE General Test, MCAT. *Application deadline:* For fall admission, 2/15 for domestic students. Application fee: $50. *Expenses:* Tuition, state resident: full-time $6,268; part-time $405 per credit. Tuition, nonresident: full-time $15,904; part-time $940 per credit. Required fees: $751 per semester hour. Tuition and fees vary according to course load and program. *Financial support:* Fellowships, research assistantships, teaching assistantships available. *Faculty research:* Microbial physiology and genetics, molecular biology, crystallography of biological molecules, antibiotics and chemotherapy, membrane transport. *Unit head:* Dr. Dennis E. Ohman, Chair, 804-828-9728, Fax: 804-828-9946, E-mail: deohman@vcu.edu. *Application contact:* Dr. Guy A. Cabral, Chair, Graduate Program, 804-828-2306, E-mail: gacabral@vcu.edu.

Virginia Polytechnic Institute and State University, Graduate School, College of Science, Department of biological Sciences, Blacksburg, VA 24061. Offers botany (MS, PhD); ecology and evolutionary biology (MS, PhD); genetics and developmental biology (MS, PhD); microbiology (MS, PhD); zoology (MS, PhD). *Faculty:* 38 full-time (9 women). *Students:* 70 full-time (30 women), 4 part-time (1 woman); includes 6 minority (2 African Americans, 1 American Indian/Alaska Native, 2 Asian Americans or Pacific Islanders, 1 Hispanic American), 12 international. Average age 27. 81 applicants, 22% accepted, 14 enrolled. In 2005, 9 master's, 8 doctorates awarded. *Entrance requirements:* For master's and doctorate, GRE General Test. Additional exam requirements/recommendations for international students: Required—TOEFL (minimum score 550 paper-based; 213 computer-based). *Application deadline:* Applications are processed on a rolling basis. Application fee: $45. Electronic applications accepted. *Expenses:* Tuition, state resident: full-time $6,558; part-time $364 per credit. Tuition, nonresident: full-time $11,296; part-time $628 per credit. Required fees: $1,419; $468 per credit. $234 per term. *Financial support:* In 2005–06, 28 research assistantships with full tuition reimbursements (averaging $16,689 per year), 37 teaching assistantships with full tuition reimbursements (averaging $13,902 per year) were awarded; career-related internships or fieldwork, Federal Work-Study, scholarships/grants, and unspecified assistantships also available. *Faculty research:* Freshwater ecology, cell cycle regulation, behavioral ecology, motor proteins. *Unit head:* Dr. Bob Jones, Chairman, 540-231-9514, Fax: 540-231-9307, E-mail: rhjones@vt.edu. *Application contact:* Sue Rasmussen, Graduate Secretary, 540-231-8929, Fax: 540-231-9307, E-mail: sueras@vt.edu.

Wagner College, Division of Graduate Studies, Department of Biological Sciences, Program in Microbiology, Staten Island, NY 10301-4495. Offers MS. Part-time and evening/weekend programs available. *Students:* 5 full-time (1 woman), 4 part-time (2 women); includes 3 African Americans, 2 Asian Americans or Pacific Islanders. 6 applicants, 100% accepted, 5 enrolled. In 2005, 5 degrees awarded. *Degree requirements:* For master's, comprehensive exam or thesis. *Entrance requirements:* For master's, minimum GPA of 2.6, proficiency in statistics, undergraduate major in science. Additional exam requirements/recommendations for international students: Required—TOEFL (minimum score 550 paper-based; 217 computer-based). *Application deadline:* For fall admission, 8/1 priority date for domestic students, 6/30 priority date for international students; for spring admission, 12/10 for domestic students, 11/15 for international students. Applications are processed on a rolling basis. Application fee: $50 ($85 for international students). *Expenses:* Tuition: Full-time $14,760; part-time $820 per credit. *Financial support:* Career-related internships or fieldwork, tuition waivers (partial), unspecified assistantships, and alumni fellowships available. Financial award applicants required to submit FAFSA. *Unit head:* Dr. Roy Mosher, Director, 718-420-4072, E-mail: rmosher@wagner.edu. *Application contact:* Kristina Muller, Admissions Office, Senior Associate Director, 718-390-3411, Fax: 718-390-3105, E-mail: kmuller@wagner.edu.

Wake Forest University, School of Medicine and Graduate School, Graduate Programs in Medicine, Department of Microbiology and Immunology, Winston-Salem, NC 27109. Offers PhD. *Degree requirements:* For doctorate, thesis/dissertation. *Entrance requirements:* For doctorate, GRE General Test. Additional exam requirements/recommendations for international students: Required—TOEFL. Electronic applications accepted. *Faculty research:* Molecular immunology, bacterial pathogenesis and molecular genetics, viral pathogenesis, regulation of mRNA metabolism, leukocyte biology.

See Close-Up on page 959.

Washington State University, College of Veterinary Medicine and Graduate School, Graduate Programs in Veterinary Science, Pullman, WA 99164. Offers veterinary and comparative anatomy, pharmacology, and physiology (MS, PhD), including neuroscience, veterinary science; veterinary clinical sciences (MS, PhD); veterinary microbiology and pathology (MS, PhD), including veterinary science. Part-time programs available. *Faculty:* 72 full-time (13 women), 6 part-time/adjunct (4 women). *Students:* 50 full-time (24 women). Average age 30. In 2005, 3 master's, 4 doctorates awarded. Terminal master's awarded for partial completion of doctoral program. *Degree requirements:* For master's and doctorate, thesis/dissertation, oral exam. *Entrance requirements:* For master's and doctorate, GRE General Test, minimum GPA of 3.0. *Application deadline:* For fall admission, 12/31 for domestic students. Applications are processed on a rolling basis. Application fee: $35. Electronic applications accepted. *Expenses:* Contact institution. Part-time tuition and fees vary according to campus/location and program. *Financial support:* In 2005–06, 21 research assistantships with partial tuition reimbursements, 8 teaching assistantships with partial tuition reimbursements were awarded; fellowships, career-related internships or fieldwork, Federal Work-Study, institutionally sponsored loans, scholarships/grants, traineeships, tuition waivers (partial), and teaching associateships also available. Financial award application deadline: 12/1; financial award applicants required to submit FAFSA. *Application contact:* Julie K. Smith, Principal Assistant, 509-335-3064, Fax: 509-335-0160, E-mail: jksmith@vetmed.wsu.edu.

Washington State University, Graduate School, College of Sciences, School of Molecular Biosciences, Program in Microbiology, Pullman, WA 99164. Offers MS, PhD. *Faculty:* 23 full-time (5 women), 21 part-time/adjunct (4 women). *Students:* 12 full-time (9 women); includes 2 minority (both Hispanic Americans), 4 international. Average age 25. 231 applicants, 18% accepted, 13 enrolled. In 2005, 1 master's, 1 doctorate awarded. Terminal master's awarded for partial completion of doctoral program. *Degree requirements:* For master's, thesis/dissertation, oral exam; for doctorate, thesis/dissertation, oral exam, comprehensive exam, registration. *Entrance requirements:* For master's and doctorate, GRE General Test, minimum GPA of 3.0. Additional exam requirements/recommendations for international students: Required—TOEFL (minimum score 550 paper-based; 213 computer-based). *Application deadline:* For fall admission, 12/15 for domestic students, 12/15 for international students. Application fee: $35. Electronic applications accepted. *Expenses:* Tuition, state resident: full-time $6,295; part-time $336 per credit. Tuition, nonresident: full-time $15,949; part-time $819 per credit. Required fees: $933. Part-time tuition and fees vary according to campus/location and program. *Financial support:* In 2005–06, fellowships with full tuition reimbursements (averaging $18,852 per year), 2 research assistantships with full and partial tuition reimbursements (averaging $18,852 per year), 10 teaching assistantships with full and partial tuition reimbursements (averaging $18,852 per year) were awarded; Federal Work-Study, institutionally sponsored loans, health care benefits, and unspecified assistantships also available. Financial award application deadline: 4/1; financial award applicants required to submit FAFSA. *Faculty research:* Viral-host interaction, bacterial-host interaction, microbial medicine, microbial pathogenesis, cancer biology. Total annual research expenditures: $5.8 million. *Application contact:* Kelly G. McGovern, Academic Coordinator, 509-335-4566, Fax: 509-335-1907, E-mail: smbgrad@wsu.edu.

Washington University in St. Louis, Graduate School of Arts and Sciences, Division of Biology and Biomedical Sciences, Program in Molecular Microbiology and Microbial Pathogenesis, St. Louis, MO 63110. Offers PhD. *Degree requirements:* For doctorate, thesis/dissertation. *Entrance requirements:* For doctorate, GRE General Test, GRE Subject Test. Electronic applications accepted.

Wayne State University, School of Medicine and Graduate School, Graduate Programs in Medicine, Department of Immunology and Microbiology, Detroit, MI 48202. Offers MS, PhD, MD/PhD. *Faculty:* 12 full-time. *Students:* 15 full-time (10 women), 1 (woman) part-time; includes 4 minority (1 African American, 2 Asian Americans or Pacific Islanders, 1 Hispanic American), 2 international. Average age 28. 37 applicants, 16% accepted, 3 enrolled. In 2005, 3 degrees awarded. Terminal master's awarded for partial completion of doctoral program. *Degree requirements:* For master's and doctorate, thesis/dissertation. *Entrance requirements:* For master's, GRE, minimum GPA of 2.5; for doctorate, GRE, minimum GPA of 3.0. Additional exam requirements/recommendations for international students: Required—TOEFL (minimum score 550 paper-based; 213 computer-based); Recommended—TWE (minimum score 6). *Application deadline:* For fall admission, 7/15 for domestic students, 6/1 for international students. Applications are processed on a rolling basis. Application fee: $30 ($50 for international students). Electronic applications accepted. *Expenses:* Tuition, state resident: part-time $338 per credit hour. Tuition, nonresident: part-time $746 per credit hour. Required fees: $24 per credit hour. Full-time tuition and fees vary according to program. *Financial support:* In 2005–06, 1 fellowship with tuition reimbursement, 12 research assistantships with tuition reimbursements (averaging $20,167 per year) were awarded; teaching assistantships, career-related internships or fieldwork and Federal Work-Study also available. *Faculty research:* Immune regulation, bacterial pathophysiology, molecular biology/viruses/bacteria, cellular and molecular immunology, microbial pathogenesis. Total annual research expenditures: $2.5 million. *Unit head:* Dr. Paul Montgomery, Chair, 313-577-1591, Fax: 313-577-1155, E-mail: aa2411@wayne.edu. *Application contact:* Harley Tse, Associate Professor, 313-577-1564, Fax: 313-577-1155, E-mail: ad6056@wayne.edu.

West Virginia University, School of Medicine, Graduate Programs at the Health Science Center, Biomedical Sciences Graduate Program, Program in Immunology and Microbial Pathogenesis, Morgantown, WV 26506. Offers MS, PhD. *Faculty:* 28 full-time (4 women). *Students:* 19 full-time (9 women); includes 2 minority (1 African American, 1 Asian American or Pacific Islander), 1 international. Average age 26. In 2005, 5 degrees awarded. *Median time to degree:* Of those who began their doctoral program in fall 1997, 97% received their degree in 8 years or less. *Degree requirements:* For doctorate, thesis/dissertation, comprehensive exam. *Entrance requirements:* For doctorate, GRE General Test, minimum GPA of 3.0. Additional exam requirements/recommendations for international students: Required—TOEFL. *Application deadline:* For fall admission, 3/1 priority date for domestic students, 1/15 priority date for international students. Applications are processed on a rolling basis. Application fee: $0. Electronic applications accepted. *Financial support:* In 2005–06, research assistantships with full tuition reimbursements (averaging $20,000 per year); institutionally sponsored loans, traineeships, and health care benefits also available. *Faculty research:* Regulation of signal transduction in immune responses, immune responses in bacterial and viral diseases, peptide and DNA vaccines for contraception, inflammatory bowel disease, physiology of pathogenic microbes. Total annual research expenditures: $2.5 million. *Unit head:* Dr. John Barnett, Chair, 304-293-2649, Fax: 304-293-7823, E-mail: john.barnett@hsc.wvu.edu. *Application contact:* Dr. Jia Luo, Graduate Coordinator, 304-293-7208, Fax: 304-293-7823, E-mail: jia.luo@hsc.wvu.edu.

See Close-Up on page 961.

Wright State University, School of Graduate Studies, College of Science and Mathematics, Program in Microbiology and Immunology, Dayton, OH 45435. Offers MS. Part-time programs available. *Degree requirements:* For master's, thesis. *Entrance requirements:* Additional exam requirements/recommendations for international students: Required—TOEFL. *Faculty research:* Reproductive immunology, viral pathogenesis, virus-host cell interactions.

Yale University, School of Medicine and Graduate School of Arts and Sciences, Combined Program in Biological and Biomedical Sciences (BBS), Microbiology Track, New Haven, CT 06520. Offers PhD, MD/PhD. *Students:* 5 full-time. *Degree requirements:* For doctorate, thesis/dissertation. *Entrance requirements:* For doctorate, GRE General Test, GRE Subject Test. Additional exam requirements/recommendations for international students: Required—TOEFL. *Application deadline:* For fall admission, 12/8 for domestic students, 12/8 for international students. Electronic applications accepted. *Financial support:* Fellowships, research assistantships available. *Unit head:* Dr. Joann Sweasy, Director of Graduate Studies, 203-737-2626, E-mail: darlene.a.smith@yale.edu. *Application contact:* Darlene Smith, Registrar, 203-737-2404, E-mail: darlene.a.smith@yale.edu.

Virology

Baylor College of Medicine, Graduate School of Biomedical Sciences, Department of Molecular Virology and Microbiology, Houston, TX 77030-3498. Offers PhD, MD/PhD. *Faculty:* 45 full-time (13 women). *Students:* 44 full-time (22 women); includes 7 minority (3 African Americans, 1 Asian American or Pacific Islander, 3 Hispanic Americans), 12 international. Average age 28. 64 applicants, 22% accepted, 7 enrolled. In 2005, 3 doctorates awarded. *Median time to degree:* Of those who began their doctoral program in fall 1997, 67% received their degree in 8 years or less. *Degree requirements:* For doctorate, thesis/dissertation, public defense. *Entrance requirements:* For doctorate, GRE General Test, GRE Subject Test (strongly recommended), minimum GPA of 3.0. Additional exam requirements/recommendations for international students: Required—TOEFL. *Application deadline:* For fall admission, 1/1 for domestic students. Applications are processed on a rolling basis. Application fee: $30. Electronic applications accepted. *Expenses:* Tuition: Full-time $8,200. Full-time tuition and fees vary according to program. *Financial support:* In 2005–06, 14 fellowships (averaging $23,000 per year), 30 research assistantships (averaging $23,000 per year) were awarded; teaching assistantships, career-related internships or fieldwork, Federal Work-Study, institutionally sponsored loans, health care benefits, and tuition waivers (full) also available. Financial award applicants required to submit FAFSA. *Faculty research:* Molecular biology of virus replication, viruses and cancer, viral genetics, viral infectious diseases, environmental virology. *Unit head:* Dr. Frank Ramig, Director, 713-798-4830, Fax: 713-798-5075, E-mail: rramig@bcm.edu. *Application* *contact:* Rosa Banegas, Graduate Program Administrator, 713-798-4472, Fax: 713-798-5075, E-mail: rbanegas@bcm.edu.

See Close-Up on page 851.

Baylor College of Medicine, Graduate School of Biomedical Sciences, Interdepartmental Program in Cell and Molecular Biology, Houston, TX 77030-3498. Offers biochemistry (PhD); cell and molecular biology (PhD); genetics (PhD); human genetics (PhD); immunology (PhD); microbiology (PhD); virology (PhD). *Faculty:* 99 full-time (21 women). *Students:* 56 full-time (28 women); includes 20 minority (4 African Americans, 1 American Indian/Alaska Native, 6 Asian Americans or Pacific Islanders, 9 Hispanic Americans), 5 international. Average age 28. 164 applicants, 18% accepted, 12 enrolled. In 2005, 3 doctorates awarded. *Median time to degree:* Of those who began their doctoral program in fall 1997, 63% received their degree in 8 years or less. *Degree requirements:* For doctorate, thesis/dissertation, public defense. *Entrance requirements:* For doctorate, GRE General Test, GRE Subject Test (strongly recommended), minimum GPA of 3.0. Additional exam requirements/recommendations for international students: Required—TOEFL. *Application deadline:* For fall admission, 1/1 for domestic students. Applications are processed on a rolling basis. Application fee: $30. Electronic applications accepted. *Expenses:* Tuition: Full-time $8,200. Full-time tuition and fees vary according to program. *Financial support:* In 2005–06, 52 students received support, including 20 fellowships (averaging $23,000 per year), 36 research assistantships (averaging $23,000 per year); teaching assistantships, Federal Work-Study, institutionally sponsored loans, health care benefits, and tuition waivers (full) also available. Financial award applicants required to submit FAFSA. *Faculty research:* Gene expression and regulation, developmental biology and genetics, signal transduction and membrane biology, aging process, molecular virology. *Unit head:* Dr. Tom Cooper, Director, 713-798-6557. *Application contact:* Lourdes Fernandez, Graduate Program Administrator, 713-798-6557, Fax: 713-798-6325, E-mail: cmbprog@bcm.edu.

See Close-Up on page 545.

Loyola University Chicago, Graduate School, Department of Microbiology and Immunology, Maywood, IL 60153. Offers immunology (MS, PhD); microbiology (MS, PhD); virology (MS, PhD). *Faculty:* 11 full-time (3 women). *Students:* 26 full-time (18 women), 1 (woman) part-time; includes 4 minority (1 African American, 3 Asian Americans or Pacific Islanders), 9 international. Average age 28. 74 applicants, 15% accepted, 4 enrolled. In 2005, 1 master's, 3 doctorates awarded. Terminal master's awarded for partial completion of doctoral program. *Degree requirements:* For master's, thesis/dissertation; for doctorate, thesis/dissertation, comprehensive exam. *Entrance requirements:* For master's and doctorate, GRE General Test. Additional exam requirements/recommendations for international students: Required—TOEFL. *Application deadline:* Applications are processed on a rolling basis. Application fee: $40. Electronic applications accepted. *Expenses:* Tuition: Full-time $11,610; part-time $645 per credit. Required fees: $55 per semester. *Financial support:* In 2005–06, 5 fellowships with tuition reimbursements (averaging $22,000 per year), 24 research assistantships with tuition reimbursements (averaging $22,000 per year) were awarded; institutionally sponsored loans and scholarships/grants also available. Financial award application deadline: 2/15. *Faculty research:* Viral pathogenesis, microbial physiology and genetics, immunoglobulin genetics and differentiation of the immune response, signal transduction and host-parasite interactions. *Unit head:* Dr. Katherine L. Knight, Chair, 708-216-3385, Fax: 708-216-9574, E-mail: kknight@lumc.edu. *Application contact:* Dr. Karen Visick, Graduate Program Director, 708-216-0869, Fax: 708-216-9574, E-mail: kvisick@lumc.edu.

See Close-Up on page 879.

Mayo Graduate School, Graduate Programs in Biomedical Sciences, Program in Virology and Gene Therapy, Rochester, MN 55905. Offers PhD.

See Close-Up on page 883.

McMaster University, Faculty of Health Sciences and School of Graduate Studies, Program in Medical Sciences, Molecular Immunology, Virology, and Inflammation Area, Hamilton, ON L8S 4M2, Canada. Offers M Sc, PhD. *Students:* 66 full-time, 2 part-time. In 2005, 4 master's, 2 doctorates awarded. *Degree requirements:* For master's, thesis/dissertation; for doctorate, thesis/dissertation, comprehensive exam. *Entrance requirements:* For master's, honors B Sc, B+ average in related field; for doctorate, M Sc, minimum B+ average, students with proven research experience and an A average may be admitted with a B Sc degree. Additional exam requirements/recommendations for international students: Required—TOEFL (minimum score 580 paper-based; 237 computer-based). *Application deadline:* For fall admission, 3/31 for domestic students. Applications are processed on a rolling basis. Application fee: $85. *Financial support:* Teaching assistantships available. *Unit head:* Dr. Mark McDermott, Coordinator, 905-525-9140 Ext. 22874. *Application contact:* Dr. Carl Richards, Associate Dean, 905-525-9140 Ext. 22983, Fax: 905-546-1129.

The Ohio State University, College of Medicine and Public Health and Graduate School, Graduate Programs in the Basic Medical Sciences, Integrated Biomedical Science Graduate Program, Columbus, OH 43210. Offers immunology (MS, PhD); medical genetics (MS, PhD); molecular virology (MS, PhD); pharmacology (MS, PhD). *Degree requirements:* For doctorate, thesis/dissertation. *Entrance requirements:* For master's, GRE General Test; for doctorate, GRE. Additional exam requirements/recommendations for international students: Required—TOEFL (minimum score 600 paper-based; 250 computer-based), TSE. Electronic applications accepted.

The Pennsylvania State University Milton S. Hershey Medical Center, Graduate School Programs in the Biomedical Sciences, Graduate Program in Microbiology and Immunology, Hershey, PA 17033-2360. Offers genetics (PhD); immunology (MS, PhD); microbiology (MS); microbiology/virology (MS, PhD); molecular biology (PhD). *Students:* 23 full-time (11 women); includes 2 minority (both Asian Americans or Pacific Islanders), 2 international. Average age 27. Terminal master's awarded for partial completion of doctoral program. *Median time to degree:* Of those who began their doctoral program in fall 1997, 100% received their degree in 8 years or less. *Degree requirements:* For master's, thesis or alternative, registration; for doctorate, thesis/dissertation, oral exam, comprehensive exam, registration. *Entrance requirements:* For master's, GRE or MCAT; for doctorate, GRE General Test or MCAT, minimum GPA of 3.0. Additional exam requirements/recommendations for international students: Required—TOEFL. *Application deadline:* Applications are processed on a rolling basis. Application fee: $45. Electronic applications accepted. *Financial support:* In 2005–06, 23 research assistantships with full tuition reimbursements were awarded; fellowships with full tuition reimbursements, scholarships/grants, health care benefits, and unspecified assistantships also available. Financial award applicants required to submit FAFSA. *Faculty research:* Virus replication and assembly, oncogenesis, interactions of viruses with host cells and animal model systems. *Unit head:* Dr. Richard J. Courtney, Chair, 717-531-7659, Fax: 717-531-6522, E-mail: micro-grad-hmc@psu.edu. *Application contact:* Billie Burns, Secretary, 717-531-7659, Fax: 717-531-6522, E-mail: micro-grad-hmc@psu.edu.

Purdue University, School of Veterinary Medicine and Graduate School, Graduate Programs in Veterinary Medicine, Department of Veterinary Pathobiology, West Lafayette, IN 47907. Offers biochemistry and molecular biology (MS, PhD); comparative epidemiology (MS, PhD); epidemiology (MS, PhD); immunology (MS, PhD); infectious diseases (MS, PhD); interdisciplinary genetics (PhD); laboratory animal medicine (MS, PhD); microbiology (MS, PhD); molecular virology (MS, PhD); parasitology (MS, PhD); pathobiology (MS, PhD); public health epidemiology (MS, PhD); toxicology (MS, PhD); veterinary anatomic pathology (MS, PhD); veterinary clinical pathology (MS, PhD); virology (MS, PhD). *Faculty:* 32 full-time (7 women). *Students:* 49 full-time (20 women), 3 part-time (1 woman); includes 2 minority (both African Americans), 31 international. Average age 35. In 2005, 3 master's, 8 doctorates awarded. Terminal master's awarded for partial completion of doctoral program. *Degree requirements:* For master's, thesis (for some programs); for doctorate, thesis/dissertation. *Entrance requirements:* For master's and doctorate, GRE General Test. Additional exam requirements/recommendations for inter-

national students: Required—TOEFL (minimum score 575 paper-based), TWE (minimum score 4). *Application deadline:* For fall admission, 8/12 for domestic students, 6/15 for international students; for spring admission, 1/12 for domestic students, 10/15 for international students. Application fee: $55. *Financial support:* Fellowships, research assistantships, teaching assistantships available. Financial award application deadline: 3/1; financial award applicants required to submit FAFSA. *Unit head:* Dr. H. Hogenesch, Head, 765-494-7543.

Rush University, Graduate College, Division of Immunology and Microbiology, Program in Immunology/Microbiology, Chicago, IL 60612-3832. Offers immunology (MS, PhD); virology (MS, PhD). Part-time programs available. *Faculty:* 8 full-time (4 women). *Students:* 15 full-time (7 women); includes 1 minority (African American), 4 international. Average age 30. 62 applicants, 6% accepted, 4 enrolled. In 2005, 2 degrees awarded. Terminal master's awarded for partial completion of doctoral program. *Median time to degree:* Of those who began their doctoral program in fall 1997, 99% received their degree in 8 years or less. *Degree requirements:* For master's, thesis; for doctorate, thesis/dissertation, comprehensive preliminary exam. *Entrance requirements:* For master's, GRE General Test; for doctorate, GRE General Test, interview, minimum GPA of 3.0. Additional exam requirements/recommendations for international students: Required—TOEFL. *Application deadline:* For fall admission, 4/1 for domestic students. Applications are processed on a rolling basis. Application fee: $25. Electronic applications accepted. *Financial support:* In 2005–06, 3 students received support, including 6 research assistantships with full tuition reimbursements available (averaging $20,000 per year); Federal Work-Study, institutionally sponsored loans, scholarships/grants, and tuition waivers (full and partial) also available. Support available to part-time students. Financial award application deadline: 4/15. *Faculty research:* Human genetics, autoimmunity, tumor biology, complement, HIV immunopathology genesis. Total annual research expenditures: $800,000. *Application contact:* Connie M. Lambert, Department Administrator, 312-563-2563, Fax: 312-563-3552, E-mail: connie_m_lambert@rush.edu.

Rutgers, The State University of New Jersey, New Brunswick/Piscataway, Graduate School, Program in Microbiology and Molecular Genetics, New Brunswick, NJ 08901-1281. Offers applied microbiology (MS, PhD); clinical microbiology (MS, PhD); computational molecular biology (PhD); immunology (MS, PhD); microbial biochemistry (MS, PhD); molecular genetics (MS, PhD); virology (MS, PhD). Part-time programs available. *Faculty:* 132 full-time. *Students:* 57 full-time (31 women), 18 part-time (8 women); includes 16 minority (9 Asian Americans or Pacific Islanders, 7 Hispanic Americans), 20 international. Average age 29. 104 applicants, 32% accepted, 19 enrolled. In 2005, 9 master's, 10 doctorates awarded. Terminal master's awarded for partial completion of doctoral program. *Median time to degree:* Of those who began their doctoral program in fall 1997, 100% received their degree in 8 years or less. *Degree requirements:* For master's, thesis or alternative, comprehensive exam, registration; for doctorate, thesis/dissertation, written qualifying exam, comprehensive exam, registration. *Entrance requirements:* For master's, GRE General Test, minimum GPA of 3.0; for doctorate, GRE General Test, GRE Subject Test (recommended), minimum GPA of 3.0. Additional exam requirements/recommendations for international students: Required—TOEFL. *Application deadline:* For fall admission, 1/5 priority date for domestic students, 11/1 priority date for international students. Applications are processed on a rolling basis. Application fee: $50. Electronic applications accepted. *Expenses:* Tuition, state resident: full-time $10,440; part-time $435 per credit. Tuition, nonresident: full-time $15,520; part-time $647 per credit. Required fees: $129 per credit. Tuition and fees vary according to program. *Financial support:* In 2005–06, 48 students received support, including 14 fellowships with full tuition reimbursements available (averaging $24,000 per year), 25 research assistantships with full tuition reimbursements available (averaging $24,000 per year), 9 teaching assistantships with full tuition reimbursements available (averaging $16,988 per year); Federal Work-Study, institutionally sponsored loans, scholarships/grants, and unspecified assistantships also available. Financial award application deadline: 1/5; financial award applicants required to submit FAFSA. *Faculty research:* Molecular genetics and microbial physiology; virology and pathogenic microbiology; applied, environmental and industrial microbiology; computers in molecular science. *Unit head:* Dr. Andrew K. Vershon, Director, 732-445-2905, Fax: 732-445-6370, E-mail: vershon@waksman.rutgers.edu. *Application contact:* Diane Murano, Administrative Assistant, 732-445-5086, Fax: 732-445-6370, E-mail: murano@biology.rutgers.edu.

Texas A&M University System Health Science Center, Graduate School of Biomedical Sciences, Department of Medical Microbiology and Immunology, College Station, TX 77840. Offers immunology (PhD); microbiology (PhD); molecular biology (PhD); virology (PhD). *Degree requirements:* For doctorate, thesis/dissertation. *Entrance requirements:* For doctorate, GRE General Test, minimum GPA of 3.0. *Faculty research:* Molecular pathogenesis, microbial therapeutics.

Université de Montréal, Faculty of Medicine and Faculty of Graduate Studies, Graduate Programs in Medicine, Program in Virology and Immunology, Montréal, QC H3C 3J7, Canada. Offers PhD. *Students:* 15 full-time (9 women). 3 applicants, 33% accepted, 1 enrolled. In 2005, 1 degree awarded. *Degree requirements:* For doctorate, thesis/dissertation, general exam. *Entrance requirements:* For doctorate, proficiency in French, knowledge of English. *Application deadline:* For fall and spring admission, 2/1. For winter admission, 11/1 for domestic students. Application fee: $30. Electronic applications accepted. *Unit head:* Pierre Belhumeur, Director, 514-343-6273, Fax: 514-343-5701. *Application contact:* Silvie Jauvim, Information Contact, 514-343-3129.

Université du Québec, Institut National de la Recherche Scientifique, Graduate Programs, Research Center—INRS—Institut Armand-Frappier—Human Health, Québec, QC G1K 9A9, Canada. Offers applied microbiology (M Sc); biology (PhD); experimental health sciences (M Sc); virology and immunology (M Sc, PhD). Programs given in French. Part-time programs available. *Faculty:* 46. *Students:* 170 full-time (100 women), 25 international. Average age 28. In 2005, 24 master's, 5 doctorates awarded. *Degree requirements:* For doctorate, thesis/dissertation. *Entrance requirements:* For master's and doctorate, appropriate bachelor's degree, proficiency in French. *Application deadline:* For fall admission, 3/30 for domestic students, 3/30 for.international students. For winter admission, 11/1 for domestic students. Application fee: $30 Canadian dollars. *Financial support:* Fellowships, research assistantships, teaching assistantships available. *Faculty research:* Immunity, infection and cancer; toxicology and environmental biotechnology; molecular pharmacochemistry. *Unit head:* Pierre Talbot, Director, 450-681-5010 Ext. 4406, E-mail: pierre.talbot@iaf.inrs.ca. *Application contact:* Michel Barbeau, Registrar, 418-654-2518, Fax: 418-654-3858, E-mail: michel.barbeau@adm.inrs.ca.

University of California, San Diego, Graduate Studies and Research, Division of Biology, Program in Immunology, Virology, and Cancer Biology, La Jolla, CA 92093. Offers PhD. Offered in association with the Salk Institute. *Degree requirements:* For doctorate, thesis/dissertation, qualifying exam. Electronic applications accepted.

The University of Iowa, Roy J. and Lucille A. Carver College of Medicine and Graduate College, Graduate Programs in Medicine, Department of Microbiology, Iowa City, IA 52242-1316. Offers general microbiology and microbial physiology (MS, PhD); immunology (MS, PhD); microbial genetics (MS, PhD); pathogenic bacteriology (MS, PhD); virology (MS, PhD). *Faculty:* 23 full-time (3 women), 11 part-time/adjunct (2 women). *Students:* 54 full-time (25 women); includes 4 minority (3 Asian Americans or Pacific Islanders, 1 Hispanic American), 8 international. 93 applicants, 12% accepted, 4 enrolled. In 2005, 7 degrees awarded. *Median time to degree:* Of those who began their doctoral program in fall 1997, 80% received their degree in 8 years or less. *Degree requirements:* For master's, thesis/dissertation; for doctorate, thesis/dissertation, comprehensive exam. *Entrance requirements:* For master's and doctorate, GRE General Test. Additional exam requirements/recommendations for international students: Required—TOEFL. *Application deadline:* For fall admission, 2/1 for domestic students, 2/1 for international students. Application fee: $50 ($75 for international students). Electronic applications accepted. *Expenses:* Tuition, state resident: part-time $1,882 per term. Tuition, nonresident: full-time $17,338; part-time $4,907 per term. Tuition and fees vary according to

Virology

The University of Iowa (continued)
course load and program. *Financial support:* In 2005–06, 63 research assistantships with full tuition reimbursements (averaging $22,000 per year) were awarded; institutionally sponsored loans, scholarships/grants, traineeships, and health care benefits also available. *Faculty research:* Biocatalysis and blue jeans, gene regulation, processing and transport of HIV, retroviral pathogenesis, biodegradation. Total annual research expenditures: $9 million. *Unit head:* Dr. Michael A. Apicella, Head, 319-335-7810, E-mail: grad-micro-info@uiowa.edu.

See Close-Up on page 923.

University of Massachusetts Worcester, Graduate School of Biomedical Sciences, Program in Immunology-Virology, Worcester, MA 01655-0115. Offers medical sciences (PhD). *Faculty:* 45 full-time (16 women). *Degree requirements:* For doctorate, thesis/dissertation. *Entrance requirements:* For doctorate, GRE General Test. Additional exam requirements/recommendations for international students: Required—TOEFL (minimum score 600 paper-based; 250 computer-based). *Application deadline:* For fall admission, 12/15 for domestic students, 12/15 for international students. Applications are processed on a rolling basis. Application fee: $25 ($50 for international students). *Expenses:* Tuition, state resident: full-time $2,640. Tuition, nonresident: full-time $9,856. Required fees: $5,685. *Financial support:* In 2005–06, research assistantships with full tuition reimbursements (averaging $25,235 per year); unspecified assistantships also available. *Faculty research:* Molecular and immunological studies on viral pathogenesis and oncology, AIDS viruses and other retroviruses, herpes viruses, influenza, Dengue and Newcastle disease viruses. *Unit head:* Dr. Alan Rothman, Director, 508-856-6976. *Application contact:* Michael Cole, Director of Admissions and Recruitment, 508-856-4779, Fax: 508-856-3659, E-mail: michael.cole@umassmed.edu.

See Close-Up on page 933.

University of Pennsylvania, School of Medicine, Biomedical Graduate Studies, Graduate Group in Cell and Molecular Biology, Program in Microbiology, Virology, and Parasitology, Philadelphia, PA 19104. Offers PhD, MD/PhD, VMD/PhD. *Degree requirements:* For doctorate, thesis/dissertation. *Entrance requirements:* For doctorate, GRE General Test, previous course work in science. Additional exam requirements/recommendations for international students: Required—TOEFL. *Application deadline:* For fall admission, 12/15 priority date for domestic students, 12/1 priority date for international students. Applications are processed on a rolling basis. Application fee: $70. Electronic applications accepted. *Financial support:* Fellowships, research assistantships, scholarships/grants, traineeships, and unspecified assistantships available. *Unit head:* Dr. Paul Bates, Chair, 215-573-3509. *Application contact:* Emily D. Geib, Coordinator, 215-898-3918, Fax: 215-573-2104, E-mail: eddz@mail.med.upenn.edu.

University of Pittsburgh, School of Medicine, Graduate Programs in Medicine, Program in Molecular Virology and Microbiology, Pittsburgh, PA 15260. Offers MS, PhD. *Faculty:* 36 full-time (6 women). *Students:* 33 full-time (20 women); includes 7 minority (2 African Americans, 2 Asian Americans or Pacific Islanders, 3 Hispanic Americans), 3 international. Average age 28. 415 applicants, 22% accepted, 42 enrolled. In 2005, 4 degrees awarded. *Median time to degree:* Of those who began their doctoral program in fall 1997, 95% received their degree in 8 years or less. *Degree requirements:* For doctorate, thesis/dissertation, comprehensive exam, registration. *Entrance requirements:* For doctorate, GRE General Test, GRE Subject Test, minimum QPA of 3.0. Additional exam requirements/recommendations for international students: Required—TOEFL (minimum score 600 paper-based; 250 computer-based), IELT (minimum score 7). *Application deadline:* For fall admission, 12/15 priority date for domestic students, 12/15 priority date for international students. Application fee: $40. Electronic applications accepted. *Expenses:* Tuition, state resident: full-time $13,194; part-time $537 per credit. Tuition, nonresident: full-time $25,012; part-time $1,026 per credit. Required fees: $700; $164 per term. Tuition and fees vary according to campus/location and program. *Financial support:* In 2005–06, 23 research assistantships with full tuition reimbursements (averaging $21,500 per year) were awarded; fellowships with full tuition reimbursements, teaching assistantships with full tuition reimbursements, institutionally sponsored loans, scholarships/grants, traineeships, health care benefits, and unspecified assistantships also available. *Faculty research:* Host-pathogen interactions, persistent microbial infections, microbial genetics and gene expression, microbial pathogenesis, anti-bacterial therapeutics. *Unit head:* Dr. Jo Anne Flynn, Program Director, 412-624-7743, Fax: 412-648-3394, E-mail: joanne@pitt.edu. *Application contact:* Graduate Studies Administrator, 412-648-8957, Fax: 412-648-1077, E-mail: gradstudies@medschool.pitt.edu.

University of Prince Edward Island, Atlantic Veterinary College, Graduate Program in Veterinary Medicine, Charlottetown, PE C1A 4P3, Canada. Offers anatomy (M Sc, PhD); bacteriology (M Sc, PhD); clinical pharmacology (M Sc, PhD); clinical sciences (M Sc, PhD); epidemiology (M Sc, PhD), including reproduction; fish health (M Sc, PhD); food animal nutrition (M Sc, PhD); immunology (M Sc, PhD); microanatomy (M Sc, PhD); parasitology (M Sc, PhD); pathology (M Sc, PhD); pharmacology (M Sc, PhD); physiology (M Sc, PhD); toxicology (M Sc, PhD); veterinary science (M Vet Sc); virology (M Sc, PhD). Part-time programs available. *Faculty:* 76 full-time (25 women), 49 part-time/adjunct (8 women). *Students:* 54 full-time (32 women), 2 part-time. Average age 30. In 2005, 7 master's, 6 doctorates awarded. *Degree requirements:* For master's and doctorate, thesis/dissertation. *Entrance requirements:* For master's, DVM, B Sc honors degree, or equivalent; for doctorate, M Sc. *Application deadline:* Applications are processed on a rolling basis. Application fee: $50.

Expenses: Contact institution. Tuition charges are reported in Canadian dollars. Part-time tuition and fees vary according to course level, degree level, campus/location, program and student level. *Financial support:* In 2005–06, 4 fellowships (averaging $25,000 Canadian dollars per year), 4 research assistantships (averaging $16,500 Canadian dollars per year) were awarded; career-related internships or fieldwork also available. *Faculty research:* Animal health management, infectious diseases, fin fish and shellfish health, basic biomedical sciences, ecosystem health. Total annual research expenditures: $1.2 million Canadian dollars. *Unit head:* Dr. James Bellamy, Associate Dean of Graduate Studies and Research, 902-566-0856, E-mail: bellamy@upei.ca. *Application contact:* Cheryl Gaudet, Registrar's Office, 902-566-0781, Fax: 902-566-0795, E-mail: registrar@upei.ca.

The University of Tennessee Health Science Center, College of Graduate Health Sciences, Department of Molecular Sciences, Memphis, TN 38163-0002. Offers bacterial pathogenesis (PhD); biochemistry (PhD); immunology (PhD); microbiology (PhD); molecular and cell biology-pathology (PhD); molecular biology (PhD); signal transduction (PhD); structural biology (PhD); virology (PhD). *Faculty:* 17 full-time (4 women). *Students:* 25 full-time (11 women); includes 14 minority (1 African American, 13 Asian Americans or Pacific Islanders). Average age 24. 83 applicants, 12% accepted. In 2005, 6 degrees awarded. *Degree requirements:* For doctorate, thesis/dissertation, oral and written preliminary and comprehensive exams. *Entrance requirements:* For doctorate, GRE General Test, GRE Subject Test, minimum GPA of 3.0. Additional exam requirements/recommendations for international students: Required—TOEFL. *Application deadline:* For fall admission, 5/15 for domestic students. Application fee: $0. *Financial support:* In 2005–06, 8 fellowships, 21 research assistantships were awarded; teaching assistantships, traineeships and tuition waivers (full) also available. *Unit head:* Dr. David Hasty, Chairman, 901-448-6150, Fax: 901-448-7360. *Application contact:* Ida W. Mosby, Director, Enrollment Services, 901-448-5560, E-mail: imosby@utmem.edu.

The University of Texas Health Science Center at Houston, Graduate School of Biomedical Sciences, Program in Virology and Gene Therapy, Houston, TX 77225-0036. Offers MS, PhD, MD/PhD. *Faculty:* 29 full-time (8 women). *Students:* 8 full-time (3 women); includes 3 minority (1 Asian American or Pacific Islander, 2 Hispanic Americans). Average age 23. 17 applicants, 94% accepted, 11 enrolled. In 2005, 2 degrees awarded. Terminal master's awarded for partial completion of doctoral program. *Degree requirements:* For master's and doctorate, thesis/dissertation. *Entrance requirements:* For master's and doctorate, GRE General Test. Additional exam requirements/recommendations for international students: Required—TOEFL, TWE. *Application deadline:* For fall admission, 1/15 for domestic students; for spring admission, 11/1 for domestic students. Applications are processed on a rolling basis. Application fee: $10. Electronic applications accepted. *Financial support:* Fellowships with full tuition reimbursements, research assistantships with full tuition reimbursements, teaching assistantships, institutionally sponsored loans, scholarships/grants, and health care benefits available. Financial award application deadline: 1/15. *Faculty research:* Molecular virology and biology, viral vaccines and therapeutics, viral immunology, viral pathogenesis, viral nucleic acid protein biochemistry. *Unit head:* Dr. Jagannadha K. Sastry, Director, 713-563-3304, Fax: 713-563-3276, E-mail: jsastry@mdanerson.org. *Application contact:* Dr. Victoria P. Knutson, Assistant Dean of Admissions, 713-500-9860, Fax: 713-500-9877, E-mail: victoria.p.knutson@uth.tmc.edu.

The University of Texas Medical Branch, Graduate School of Biomedical Sciences, Center for Biodefense and Emerging Infectious Diseases, Galveston, TX 77555. Offers biodefense training (PhD). *Entrance requirements:* For doctorate, GRE, an overall and advanced grade point average of 3.0; personal interviews encouraged. *Expenses:* Tuition, state resident: full-time $8,350; part-time $90 per credit hour. Tuition, nonresident: full-time $21,450; part-time $366 per credit hour. Required fees: $1,027; $11 per credit hour. $60 per term.

The University of Texas Medical Branch, Graduate School of Biomedical Sciences, Program in Emerging and Tropical Infectious Diseases, Galveston, TX 77555. Offers PhD, MD/PhD. *Degree requirements:* For doctorate, thesis/dissertation. *Entrance requirements:* For doctorate, GRE General Test. *Application deadline:* Applications are processed on a rolling basis. Application fee: $25 ($50 for international students). *Expenses:* Tuition, state resident: full-time $8,350; part-time $90 per credit hour. Tuition, nonresident: full-time $21,450; part-time $366 per credit hour. Required fees: $1,027; $11 per credit hour. $60 per term. *Financial support:* In 2005–06, fellowships (averaging $23,000 per year), research assistantships with full tuition reimbursements (averaging $23,000 per year) were awarded; traineeships and unspecified assistantships also available. *Faculty research:* Emerging diseases, tropical diseases, parasitology, vitology and bacteriology.

See Close-Up on page 949.

Yale University, School of Medicine and Graduate School of Arts and Sciences, Combined Program in Biological and Biomedical Sciences (BBS), Microbiology Track, New Haven, CT 06520. Offers PhD, MD/PhD. *Students:* 5 full-time (4 women). *Degree requirements:* For doctorate, thesis/dissertation. *Entrance requirements:* For doctorate, GRE General Test, GRE Subject Test. Additional exam requirements/recommendations for international students: Required—TOEFL. *Application deadline:* For fall admission, 12/8 for domestic students, 12/8 for international students. Electronic applications accepted. *Financial support:* Fellowships, research assistantships available. *Unit head:* , Dr. Joann Sweasy, Director of Graduate Studies, 203-737-2626, E-mail: darlene.a.smith@yale.edu. *Application contact:* Darlene Smith, Registrar, 203-737-2404, E-mail: darlene.a.smith@yale.edu.

Cross-Discipline Announcements

Massachusetts Institute of Technology, School of Science, Department of Biology, Cambridge, MA 02139-4307.

Graduate work in the Department of Biology at MIT leads to the PhD degree. Research opportunities include many areas of modern biology: biochemistry, biophysics, cellular and developmental biology, immunology, microbiology, and neurobiology. Students come from a wide variety of backgrounds. Previous experience in the biological sciences, although desirable, is not a prerequisite for admission. Formal courses in biochemistry, genetics, and the method and logic of molecular biology are required of all students early in their graduate program. Each student is encouraged to follow the program of study that best meets his or her educational goals. Special emphasis is placed on research leading to the PhD thesis.

Princeton University, Graduate School, Department of Molecular Biology, Princeton, NJ 08544-1019.

Graduate studies in the Department of Molecular Biology at Princeton emphasize training in research and encourage students to apply molecular, biochemical, structural, genetic, and computational approaches to biological problems. Faculty members and students pursue research in a wide variety of areas of biochemistry and cell biology, biological dynamics and computational biology, biomedicine and societal issues, biophysics and structural biology, developmental biology, genetics and genomics, microbiology, neurobiology, oncology, and virology.

University of Pittsburgh, School of Arts and Sciences, Department of Biological Sciences, Program in Molecular, Cellular, and Developmental Biology, Pittsburgh, PA 15260.

In the department's graduate program in molecular, cellular, and developmental biology, faculty research programs use prokaryotic and eukaryotic experimental systems and molecular and genetic approaches to understand the structure and function of genes and proteins, macromolecular interactions, cell-specific gene regulation, regulation of cell proliferation, and embryogenesis.

BAYLOR COLLEGE OF MEDICINE

Department of Immunology

Program of Study	The Department of Immunology has ongoing research programs in molecular aspects of immunology, including MHC evolution, structure, and function; antigen presentation; T- and B-cell lymphocyte differentiation; dendritic cell and regulatory T-cell function; signal transduction mechanisms of lymphocyte activation and apoptosis; immunosenescence; immunotherapy, gene therapy, and genetic and cancer vaccine development; inflammation; cytokines; immune-related diseases, including asthma, autoimmunity, AIDS, and inherited immunodeficiencies; and stem cell biology.

The Ph.D. degree usually requires four to six years of full-time study and research. Beginning in the first term, an exceptional program provides students with an interactive course in the problem-solving skills needed for graduate-level work. During the first year, all students complete a core curriculum that includes immunology, biochemistry, genetics, and molecular and cell biology. After three or four research rotations, a major adviser is selected and a supporting committee is established to help the student set up a program in his or her major field of interest. In subsequent years, the student concentrates on laboratory research; participates in advanced seminars; passes a qualifying exam, which consists of writing and presenting an NIH-style grant; and prepares a dissertation. Small classes and seminars permit close interaction with faculty members. Reciprocal agreements enable graduate students to receive credit for courses taken at other Houston-area institutions of higher learning.

Research Facilities

The Department of Immunology has state-of-the-art research facilities that offer the open-space design popular in laboratories today, with the north windows offering a terrific view of the Houston skyline. There are spaces for shared equipment and core facilities for flow cytometry and protein chemistry within the department that feature the latest instruments and technologies. Baylor College of Medicine has a new animal housing facility and is home to the human genome project.

Financial Aid

The department currently provides generous assistantships of $23,000 and health insurance to qualified first-year students. Support for subsequent years is provided from funds of the faculty members. Students able to procure extramural funding are rewarded with financial benefits.

Cost of Study

The cost of tuition, which is typically waived, is $8200 for the 2007–08 academic year.

Living and Housing Costs

Houston is the fourth-largest city in the United States, with the greater Houston area containing 4 million inhabitants. This population base includes a wide variety of racial and ethnic groups that give Houston a rich diversity and cosmopolitan feel but without the high cost of living. Costs of housing, health care, utilities, groceries, and transportation are all below the national average.

Student Group

The Department of Immunology has 22 students from throughout the United States and abroad pursuing Ph.D. degrees. The graduate school enrollment at the College of Medicine is more than 400. Each year, 4 to 6 new students are accepted into the Graduate Program in Immunology.

Student Outcomes

Graduates of the Department of Immunology Ph.D. programs are located in different academic, government, and industrial institutions throughout the world.

Location

Houston is a dynamic, diverse city with a rapidly growing population of approximately 4 million. For recreation, the Gulf of Mexico is only an hour's ride by car. The semitropical climate offers pleasant days throughout the year. Symphony, opera, ballet, live theater, a thriving museum district, a revitalized downtown, year-round major-league sports, and a large number of diverse ethnic groups help to make Houston an entertaining and exciting city in which to live. A new light rail system, linking the medical center with downtown, makes it easier than ever to enjoy all Houston has to offer.

The College

Baylor College of Medicine is a private institution that ranks in the top ten nationally, according to *U.S. News & World Report*, and first in Texas for federal research funding. The College is part of the Texas Medical Center complex, which is located in southwest Houston on 200 acres of land bordered by Hermann Park and Zoo, residential areas, and Rice University. The Medical Center is world renowned and represents a model for the concept of the central concentration of medical facilities. It contains two medical schools and seven hospitals with more than 4,000 beds. The Texas Medical Center library houses approximately 175,000 books, subscribes to more than 4,500 journals, has nearly 4,000 items in its audiovisual collection (including computer software), and subscribes to several major computerized information-retrieval services.

Applying

The application deadline for the 2007 school year is January 1, 2007. The preferred entry time is August (first term). Prospective students must take either the Graduate Record Examinations (GRE) General Test or Medical College Admission Test (MCAT), and a GRE Subject Test is highly recommended in biology, biochemistry, or molecular biology.

Correspondence and Information

Dorothy E. Lewis, Director of Graduate Studies
Department of Immunology
Baylor College of Medicine
One Baylor Plaza, M929
Houston, Texas 77030
Phone: 713-798-6054
Fax: 713-798-3700
E-mail: immuno@bcm.edu
Web site: http://www.bcm.edu/immuno

Baylor College of Medicine

THE FACULTY AND THEIR RESEARCH

Professors
M. Zouhair Atassi, Ph.D. Protein structure and function; molecular and cellular immunology.
Richard G. Cook, Ph.D. Structure and function of class I MHC molecules.
Mark Entman, M.D. Cardiovascular inflammation.
David P. Huston, M.D. Molecular and cellular mechanisms of allergic inflammation.
Dorothy E. Lewis, Ph.D. Mechanisms of CD8 T-cell dysfunction in HIV infection.
Cliona Rooney, Ph.D. Immunology and pathogenesis of Epstein-Barr virus.
Roger D. Rossen, M.D. Molecular mechanisms that regulate monocyte macrophage trafficking and that modulate their function and differentiation.
William T. Shearer, M.D., Ph.D. Immunopathogenesis of pediatric HIV-1 infection.
C. Wayne Smith, M.D. Adhesive mechanisms of neutrophils.
Tse-Hua Tan, Ph.D. Signal transduction by stress-activated MAP kinases and phosphatases in lymphocyte activation and apoptosis.
Rongfu Wang, Ph.D. Identification of tumor antigens recognized by CD4+ and CD8+ T cells; development of novel cancer vaccine strategies; identification of cancer-specific genes.
Li-Yuan Yu-Lee, Ph.D. Role of prolactin in the growth, differentiation, and maturation of cells of the immune system.
Jingwu Zang, M.D., Ph.D. Autoimmune mechanisms involved in the pathogenesis of multiple sclerosis.

Associate Professors
John W. Belmont, M.D., Ph.D. Hematopoietic and cardiovascular development; JNK pathway signal transduction; mammalian condensin.
Holly H. Birdsall, M.D., Ph.D. Leukocyte transendothelial migration and inflammation.
Si-Yi Chen, M.D., Ph.D. Genetic modification of immune cells for tumor and HIV therapy.
Margaret Conner, Ph.D. Pathogenesis of enteric virus infection.
David Corry, M.D. T-cell differentiation and mechanisms of T-cell–mediated allergic airway disease.
Margaret Goodell, Ph.D. Hematopoietic stem cell biology.
Barry Myones, M.D. Function of complement receptors on macrophages and neutrophils in mediation of phagocytosis and cell-cell adhesion, and on erythrocytes in immune complex clearance.
Frank M. Orson, M.D. Genetic immunization for infectious disease (HIV), cancer (prostate), and inflammation (allergy).
David M. Spencer, Ph.D. Tumor immunobiology and gene therapy.
A. Clinton White, M.D. Interactions between human pathogenic parasites and the host cytokine and chemokine responses.

Assistant Professors
Stuart L. Abramson, M.D., Ph.D. Asthma, cytokine regulation of myeloid cell superoxide production; phagocyte dysfunction in HIV infection.
Catherine Bollard, M.B.B.S. Augmentation of antitumor immunity against leukemic cells.
Min Chen, Ph.D. Molecular regulation of apoptosis in the immune system.
Stephen Gottschalk, M.D. Immunotherapy for virus-associated malignancies.
Shuhua Han, M.D. Biology of germinal center reactions and mechanisms of the humoral immune response.
Farrah Kheradmand, M.D. Pathogenesis of obstructive lung disease.
Daniel Lacorazza, Ph.D. Molecular events that control hemopoiesis and development of the immune system.
Margarita Martinez Moczygemba, Ph.D. Allergic inflammation.
John R. Rodgers, Ph.D. Immune responses to gene therapy.
Jin Wang, Ph.D. Molecular regulation of lymphocyte apoptosis and immunity.
Biao Zheng, M.D., Ph.D. Development of in vivo immune responses and germinal center biology.

BAYLOR COLLEGE OF MEDICINE

Department of Molecular Virology and Microbiology

Program of Study

The Department of Molecular Virology and Microbiology offers a program leading to the Ph.D. degree, with specialization in molecular virology or molecular microbiology. Doctoral and postdoctoral research training opportunities in molecular virology focus on general animal virology, viral genetics, viral infectious diseases, transcriptional and translational regulation of viral gene expression, viral pathogenesis, structural virology, and cancer virology, with an emphasis on molecular mechanisms. Additional virology programs focus on respiratory viruses, including application and development of various assays and animal models; T- and B-cell immune responses; mediators of immunity; mucosal immunology; pathogenesis; and antivirals. Doctoral and postdoctoral research training opportunities in molecular microbiology focus on microbial genomics, molecular genetics, physiology and regulation of bacterial pathogenesis, genetic and physical analysis of protein structure-function relationships, mechanisms of antibiotic resistance, biochemical analysis of membrane transport, DNA topology, and site-specific recombination.

The graduate program leading to the Ph.D. degree generally requires four or five years of full-time study. In the first year, a graduate student takes courses in general virology and microbiology, molecular biology, biochemistry, immunology, cell biology, and advanced topics in virology or microbiology. Students also perform laboratory research rotations. In the second year, students select a research mentor, complete the qualifying examination, and begin thesis research on an original research problem. In subsequent years, the focus is on laboratory research. Combined M.D./Ph.D., Ph.D./M.P.H., and Ph.D./M.B.A. programs are also available. There is no program leading to the M.S. degree.

Research Facilities

The research accomplishments of the Department's faculty members place them at the forefront of a number of modern research areas, providing an excellent research environment. Through the faculty members, who are keenly interested in teaching, modern research facilities are available to predoctoral and postdoctoral trainees at all levels.

The Department occupies approximately 53,000 square feet of modern, recently renovated, air-conditioned laboratories. All equipment and facilities essential for modern virologic, microbiologic, and molecular research are available, including electron, fluorescence, and confocal microscopes; oligonucleotide and peptide synthesizers; nucleic acid and protein sequencing cores; phosphorimaging core; clean labs dedicated to PCR; and animal facilities. Separate facilities for DNA chip production and analysis and for generation of transgenic mice are available. Individual laboratories contain modern equipment necessary for the ongoing research projects in those labs. The Department also provides research libraries and sophisticated computer support.

Financial Aid

The stipend for all Ph.D. students is $23,000 for the 2006–07 academic year. Tuition scholarships are available to all students. Health insurance is provided for all Ph.D. students; additional financial aid is also available.

Cost of Study

All students in the Department are awarded full tuition scholarships.

Living and Housing Costs

A large number of modern apartment complexes are located within a few miles of the Texas Medical Center; rent for one bedroom begins at approximately $600 per month. Two dormitories provide housing on the Texas Medical Center campus for single and married students, but early reservations are advised. Automobile ownership increases access to the range of available housing.

Student Group

The Department maintains a student body of about 50, and there are about 25 postdoctoral trainees. Schoolwide, the Graduate School has a student body of 534, and there are 850 postdoctoral trainees. This student body is drawn from both American and international applicants. In addition, there are about 650 medical students at the College.

Student Outcomes

Recent graduates have pursued postdoctoral studies in virology, microbiology, or other fields, at a wide variety of institutions throughout the United States and abroad. Graduates of the Department hold appointments at universities and medical schools throughout the world, as well as staff scientist positions in biotechnology and pharmaceutical companies and government agencies.

Location

The Houston metropolitan area, with a population of more than 4 million, offers urban amenities, including symphony, opera, ballet, live theater, art museums, year-round major league sports, and a large number of diverse ethnic groups, which help to make Houston an entertaining and exciting city in which to live. The semitropical climate offers pleasant days throughout the year, and summer comfort is ensured in this "most air-conditioned city in the world." Houston is within an hour of Gulf Coast beaches.

The College

Baylor College of Medicine is a private institution that ranks among the top twelve nationally and first in Texas for federal research funding. Baylor has more than 1,000 faculty members, including more than 250 engaged in full-time research. There is a high degree of interdisciplinary cooperation, not only among the basic sciences, but also with clinical investigators and investigators at other institutions in the Medical Center and elsewhere in Houston. The Texas Medical Center is located in southwest Houston and is bounded by Hermann Park and Zoo, residential areas, and Rice University. The Medical Center contains two medical schools, thirteen renowned hospitals, and other associated institutions. It has an excellent library shared by all institutions and provides an environment that places a wealth of resources at the disposal of students and the faculty.

Applying

The application deadline is January 1. The preferred entry time is August (first term). Prospective students must take the Graduate Record Exam (GRE). The Subject Test in either biology, chemistry, or molecular biology is highly recommended. An undergraduate GPA of 3.0 (on a 4-point scale) and GRE scores at or above the 70th percentile are expected for favorable consideration.

Correspondence and Information

Dr. Frank Ramig
Director of Graduate Studies
Department of Molecular Virology and
  Microbiology
Baylor College of Medicine
Houston, Texas 77030
Phone: 713-798-4830
Fax: 713-798-3586
E-mail: rramig@bcm.edu

Dr. Sarah Highlander
Assistant Director
Department of Molecular Virology and
  Microbiology
Baylor College of Medicine
Houston, Texas 77030
Phone: 713-798-6311
Fax: 713-798-7375
E-mail: sarahh@bcm.edu

Dr. Susan Marriott
Assistant Director
Department of Molecular Virology and
  Microbiology
Baylor College of Medicine
Houston, Texas 77030
Phone: 713-798-4440
Fax: 713-798-4435
E-mail: susanm@bcm.edu

Baylor College of Medicine

THE FACULTY AND THEIR RESEARCH

Ervin Adam, M.D., Professor. Epidemiology of human malignancies associated with viral infections. *Am. J. Obstet. Gynecol.* 182:257, 2000.
*Robert L. Atmar, M.D., Associate Professor (Medicine). Epidemiology/pathogenesis of respiratory and enteric viruses. *J. Virol.* 71:405, 2003.
*Carol J. Baker, M.D., Professor (Pediatrics). Pathogenesis and prevention of streptococcal infections; neonatal immunity. *J. Infect. Dis.* 179:576, 1999.
 Sarah H. Blutt, Ph.D., Assistant Professor. Intestinal immunity, B cells, and IgA. *J. Virol.* 78:6974, 2004.
 Janet S. Butel, Ph.D., Distinguished Service Professor and Chair. Polyomaviruses and pathogenesis of human disease. *J. Virol.* 78:9306, 2004.
*Thomas R. Cate, M.D., Professor (Medicine). Influenza virus: Host defenses and vaccines. *Semin. Respir. Infect.*13:17, 1998.
*Wah Chiu, Ph.D., Professor (Biochemistry). Electron cryomicroscopy of viruses. *Nature* 439:612, 2006.
 Margaret E. Conner, Ph.D., Associate Professor. Pathogenesis, immunity, and vaccines in enteric viruses. *Lancet* 362(9394):1445, 2003.
 Robert B. Couch, M.D., Professor. Viral respiratory diseases: vaccines, chemotherapy, immunity. *Dev. Biol.* 115:25, 2003.
 Lawrence A. Donehower, Ph.D., Professor. p53 tumor suppressor gene in cancer and aging. *Genes Dev.* 19:1162, 2005.
*Herbert L. DuPont, M.D., Professor (Internal Medicine). Travelers' diarrhea and travel medicine. *Am. J. Gastroenterol.* 99:383, 2004.
 Hanaa El Sahli, M.D., Assistant Professor. Tuberculosis epidemiology; vaccine evaluation. *Scand. J. Infect. Dis.* 36:106, 2004.
 Mary K. Estes, Ph.D., Professor. Molecular biology pathogenesis and vaccines of gastrointestinal viruses. *Trends Microbiol.* 12:279, 2004.
 Brian E. Gilbert, Ph.D., Associate Professor. Aerosol drug delivery for viruses and cancer. *Clin. Cancer Res.* 10:2319, 2004.
 W. Paul Glezen, M.D., Professor. Respiratory virus infections: Epidemiology and immunity. *Vaccine* 23:1540, 2005.
*David Y. Graham, M.D., Professor (Medicine). *Helicobacter pylori* infection; pathogenesis; therapy. *Gut* 53:1235, 2004.
*Stephen B. Greenberg, M.D., Professor (Medicine). Respiratory viral infections: interferon, antivirals, and immunity. *J. Am. Med. Assoc.* 283:499, 2000.
*Richard J. Hamill, M.D., Professor (Medicine). Epidemiology and treatment of fungal infections; antifungals. *AIDS* 19:399, 2005.
*Christophe Herman, Ph.D., Assistant Professor (Molecular and Human Genetics). Proteins degradation; stress response; membrane quality control
   mechanism. *Mol. Cell* 11:659, 2003.
 Christine H. Herrmann, Ph.D., Assistant Professor. HIV; viral gene expression; nuclear structure. *Curr. HIV Res.* 1:395, 2003.
 Sarah K. Highlander, Ph.D., Associate Professor. Microbial genomics and genetics of bacterial pathogens. *Front Biosci.* 6:1128, 2001.
 F. Blaine Hollinger, M.D., Professor. Viral hepatitis; hepatitis A virus. In *Fields Virology*, 5th edition, Philadelphia: Lippincott Williams & Wilkins, in press.
 Ronald T. Javier, Ph.D., Associate Professor. Viral tumorigenesis and adenoviruses. *EMBO J.*, 2006, doi:10.1038/sj.emboj.7601030.
 Wendy A. Keitel, M.D., Associate Professor. Vaccine development; pathogenesis and immune responses to viral infections. *Vaccine* 18:531, 1999.
 Jason T. Kimata, Ph.D., Assistant Professor. Retroviral replication and pathogenesis. *J. Virol.* 76:6425, 2002.
 Paul D. Ling, Ph.D., Associate Professor. Molecular biology of Epstein-Barr virus latency. *J. Virol.* 78:3919, 2004.
 Richard E. Lloyd, Ph.D., Associate Professor. Picornaviruses: translation control; translation control during apoptosis. *Mol. Cell Biol.* 24:1779, 2004.
 Susan J. Marriott, Ph.D., Associate Professor. Transcription and transformation of human retroviruses. *Virology* 324:540, 2004.
*Edward O. Mason Jr., Ph.D., Professor (Pediatrics). Epidemiology and pathogenesis of *S. pneumoniae* and *S. aureus*. *Pediatr. Infect. Dis. J.* 21:141, 2002.
 Innocent Mbawuike, Ph.D., Assistant Professor. T-cell function and age-related immunodeficiency in virus infection. *J. Immunol.* 162: 2530, 1999.
 Bradley M. Mitchell, Ph.D., Assistant Professor. (Ophthalmology). Viral and fungal ocular infections and pathogenesis. *J. Infect. Dis.* 190:192, 2004.
 Flor M. Muñoz, M.D., Assistant Professor. Maternal and infant immunization; epidemiology of respiratory viruses. *Am. J. Obstet. Gynecol.*, 192(4):1098, 2005.
*Daniel M. Musher, M.D., Professor (Medicine). Bacterial infection and host response. *N. Engl. J. Med.* 348:1256, 2003.
*Frank M. Orson, M.D., Associate Professor (Medicine). Genetic immunization for HIV and allergic diseases. *J. Immunol.* 164:6313, 2000.
 Timothy G. Palzkill, Ph.D., Professor. Protein structure-function; functional genomics. *J. Biol. Chem.* 280:17786, 2005.
*Pyong Woo Park, Ph.D., Assistant Professor (Medicine). Extracellular matrix; lung inflammatory diseases; host defense. *Nature* 411:98, 2001.
 Joseph F. Petrosino, Ph.D., Assistant Professor. Functional genomics and vaccine discovery for biodefense and infectious disease pathogens.
 Pedro A. Piedra, M.D., Associate Professor. Respiratory virus vaccine research: Influenza, RSV, hMPV, and adenovirus. *Pediatrics* 116(3):397, 2005.
*Bidadi V. Venkataram Prasad, Ph.D., Professor (Biochemistry). Structure-function of gastroenteric viruses. *Nature* 417:311, 2002.
 Frank Ramig, Ph.D., Professor. Reoviridae: genetics, pathogenesis, and RNA replication. *J. Virol.* 78:10213, 2004.
 Andrew P. Rice, Ph.D., Professor. HIV gene expression. *J. Virol.* 78:8114, 2004.
*Cliona M. Rooney, Ph.D., Professor (Pediatrics). Epstein-Barr virus and tumor immunology, pathology, and immunotherapy. *Blood* 105:4247, 2005.
*Susan M. Rosenberg, Ph.D., Professor (Molecular and Human Genetics). Genetic instability; mutation; DNA repair and recombination; antibiotic resistance.
   Mol. Cell 7:571, 2001.
 Edward B. Siwak, Ph.D., Assistant Professor. Blood-borne viruses: HIV-1 and HCV. *Transfusion* 44:1179, 2004.
 Betty L. Slagle, Ph.D., Associate Professor. Hepatitis viruses and liver cancer. *J. Virol.* 76:11770, 2002.
 Richard E. Sutton, M.D., Ph.D., Assistant Professor. Lentiviral vectors for gene transfer and the study of HIV. *J. Virol.* 80:3406, 2006.
 John A. Vanchiere, M.D., Ph.D., Assistant Professor (Pediatrics). Pathogenesis of BK virus disease; natural history of human polyomavirus infection. *J. Infect. Dis.* 192:658, 2005.
*James Versalovic, M.D., Ph.D., Assistant Professor (Pathology). Probiotics; *Lactobacillus;* immunomodulation. *Infect. Immun.* 73:912, 2005.
*George M. Weinstock, Ph.D., Professor (Molecular and Human Genetics). Infectious disease genomics and molecular genetics. *Nature* 440:7082, 2006.
*A. Clinton White, M.D., Professor (Medicine). Host-parasite responses in cysticercosis and cryptosporidiosis. *J. Infect. Dis.* 192:1294–302, 2005.
 Philip R. Wyde, Ph.D., Professor. Animal models for studying influenza, measles, respiratory syncytial human metapneumovirus and parainfluenza virus
   pathogenesis, prevention and/or treatment. *Antiviral Res.* 66:57, 2005.
*Qizhi Cathy Yao, M.D., Ph.D., Associate Professor (Surgery). HIV and tumor immunology and vaccine development. *J. Immunol.* 173:1951, 2004.
*Boris Yoffe, M.D., Professor (Medicine). Hepatitis viruses and pathogenesis of liver diseases. *Hepatology* 36:592, 2002.
*Edward J. Young, M.D., Professor (Medicine). *Brucella* infections; pathogenesis of the Jarisch-Herxheimer reaction. *Clin. Infect. Dis.* 31:904, 2000.
 E. Lynn Zechiedrich, Ph.D., Associate Professor. DNA topoisomerases; genomic instability; multidrug resistance. *Proc. Natl. Acad. Sci. U.S.A.* 103:2386, 2006.
*Xiaoliu (Shaun) Zhang, M.D., Ph.D., Associate Professor (Pediatrics). Conditionally replicating herpes simplex virus. *Cancer Res.* 62:2306, 2002.

Joint appointees (primary appointment)

BOSTON UNIVERSITY

Immunology Training Program

Program of Study	The Immunology Training Program administers a multidisciplinary course of study that leads to the Ph.D. degree. Students are admitted jointly to the program and to a collaborating department (microbiology or pathology). During their first two years, students take courses in cellular and molecular immunology, as well as courses in other areas of modern cellular and molecular biology appropriate to their departmental specialization. During this period, students also do research rotations in the laboratories of immunology faculty members to introduce them to both research methodologies and potential dissertation projects. After a qualifying examination, students formulate a dissertation project with the advice of their sponsor and a committee of other faculty members and then embark upon full-time research work. Students are expected to present their work in public seminars and eventually to describe their findings in one or more research publications in addition to the dissertation. The goal of the program is to prepare students for postdoctoral work in immunology and for careers as independent research scientists. Joint M.D./Ph.D. students may also apply for admission to the Immunology Training Program.
Research Facilities	The immunology research laboratories are well equipped for state-of-the-art work in molecular and cellular immunology. The Medical Center maintains extensive animal-care facilities, including transgenic mouse and rat facilities; core protein and nucleic acid synthesis and analysis facilities; confocal deconvolution and electron microscopy facilities; core facilities for genomics and proteomics analysis; and facilities for structural biology, together with excellent computer and library facilities. All central facilities are within a short distance from the research laboratories.
Financial Aid	Students admitted to the program are offered full tuition support, a stipend ($27,000 for the 2006–07 academic year) for the first two years of study, and coverage of health insurance and activity fees, through an NIH-funded training grant. Advanced students are supported by the laboratories in which they are engaged in dissertation research.
Cost of Study	Tuition and fees for the 2006–07 academic year are approximately $33,500 for a full-time student.
Living and Housing Costs	Many students share houses or apartments in Boston and surrounding areas at a total cost of $900 to $1200 per month. The University housing office can assist in rental arrangements.
Student Group	There are 50 to 60 graduate students and 20 to 30 postdoctoral fellows in the twenty research laboratories participating in the Immunology Training Program. The program directly supports 8 graduate students and 4 fellows each year, and other students are supported by departmental or institutional funds or other training grants. There are about 200 students enrolled at any one time in the seven departments that make up the Division of Graduate Medical Sciences.
Location	The Boston/Cambridge area provides one of the richest academic and scientific environments in the country. Immunology research programs at other local universities provide useful opportunities for collaborative work, and distinguished visitors are continually brought to the Boston area by way of the Immunology Seminar series at each institution. The University and the community offer first-rate athletic and recreational facilities, as well as many activities and interest groups in which graduate students are welcome to participate.
The University and The Program	Boston University is a private institution founded in 1839. The School of Medicine is located at the Boston University Medical Center, about 2 miles from the main Charles River Campus of Boston University. It is among the top fifteen institutions in the country in terms of the amount of NIH-derived research support per faculty member. The Immunology Training Program involves the coordinated efforts of faculty members in the Departments of Microbiology, Pathology, Medicine, and Biochemistry. Its weekly seminar series provides the focus not only for the exchange of ideas with visiting scientists but also for an unusually high degree of collaborative interaction.
Applying	Applicants should have completed or be about to complete a bachelor's degree program that includes a broad and rigorous background in cell and molecular biology, genetics, and immunology as well as chemistry, physics, and biochemistry. Scores from the GRE General Test are required. Applications must be received by January 1 for admission in the fall semester.
Correspondence and Information	Ann Marshak-Rothstein, Ph.D., ITP Director Department of Microbiology School of Medicine Boston University 715 Albany Street Boston, Massachusetts 02118 Phone: 617-638-4284 Fax: 617-638-4286 E-mail: itp@bu.edu Web site: http://www.bumc.bu.edu/immunology

Boston University

THE FACULTY AND THEIR RESEARCH

David Center, Professor of Medicine; M.D., Boston University. Mechanisms and role of CD4-transduced signals in interleukin 16-stimulation of CD4+ T cells; T regulatory cells; nuclear factors associated with control of the CD4+ T-cell cycle; functions of pro-interleukin-16 and its dysregulation in T-cell leukemias; effects of interleukin 16 stimulation of CD4+ T cells. *J. Allergy Clin. Immunol.* 117:86–91, 2006; *J. Immunol.* 172:1654, 2004.

Ronald B. Corley, Professor of Microbiology; Ph.D., Duke. B-lymphocyte development and function; IgM antibody structure and function in innate and adaptive immunity; role of Toll-like receptors in autoimmunity. *BMC Immunol.* 6:8, 2005; *J. Immunol.* 174:983, 2005; *Cell. Immunol.* 229:68, 2004; *Int. Immunol.* 16:1411, 2004; *J. Autoimmunity* 23:333, 2004; *Eur. J. Immunol.* 32:2328, 2002.

William Cruikshank, Professor of Medicine; Ph.D., Boston University. Characterization, process of synthesis and secretion, and clinical relevance to inflammation of the CD4+ T-cell-specific cytokine Interleukin 16 (IL-16). *J. Immunol.* 176:2337–45, 2006; *Endocrinology* 147:1941–9, 2006; *J. Allergy Clin. Immunol.* 117:86–91, 2006; *Cell Immunol.* 237:17–27, 2005; *J. Neurosci. Res.* 79:680, 2005.

Douglas V. Faller, Professor of Medicine; M.D., Harvard; Ph.D., MIT. Effects of oncogenes and oncogenic viruses on signal transduction, cellular growth and differentiation, apoptosis, and the cellular immune system. *Chem. Biol. Drug Design*, in press; *Virology* 345:390–403, 2006; *J. Biol. Chem.* 202:87–99, 2005; *EMBO J.* 23:2293–303, 2004; *J. Cell Physiol.* 198:277–94, 2004.

Caroline A. Genco, Professor of Medicine; Ph.D., Rochester. Microbial pathogenesis; vaccine development; host-parasite interactions; innate immune response to bacterial pathogens; bacterial gene regulation. *Mol. Microbiol.* 60:963, 2006; *Cell. Microbiol.* 8:738, 2006; *Circulation* 109:2801, 2004; *Proc. Natl. Acad. Sci.* 100:9542, 2003.

Dana Graves, Professor of Oral Biology; D.D.S., Columbia; D.M.Sc., Harvard. Mechanisms through which diabetes alters the host response to bacteria, focusing on apoptotic pathways that are induced by pathogens and how they are altered by diabetes. *Am. J. Pathol.*, in press; *Diabetes* 55:487–95, 2006; *J. Biol. Chem.* 280:12096, 2005.

Adam Lerner, Associate Professor of Medicine; M.D., Yale. Cyclic nucleotide phosphodiesterase inhibitors in lymphoid malignancies; role of glucocorticoids in cAMP-mediated apoptosis; role of AND-34 in breast cancer and B-cell signal transduction. *Biochem. J.* 393:21, 2006; *Biochem. Pharm.* 69:473, 2005; *Mol. Cancer Res.* 3:32, 2005; *J. Immunol.* 170:969, 2003; *Blood* 101:4122, 2003; *Blood* 103:2661, 2003.

Stuart M. Levitz, Adjunct Professor of Medicine; M.D., NYU. The interplay between the host immune system and medically important fungi. *J. Immunol.* 176:3053–61, 2006; *J. Immunol.* 175:7496–503, 2005; *Proc. Natl. Acad. Sci. U.S.A.* 98:10422, 2001.

Ann Marshak-Rothstein, Professor of Microbiology; Ph.D., Pennsylvania. Activation of autoreactive B cells by BCR/TLR-coengagement; role of exogenous and endogenous Fas-ligand in lymphocyte apoptosis, inflammation, and autoimmunity. *Immunol. Rev.* 204:27, 2005; *J. Immunol.* 172:6598, 2004; *Immunity* 19:837, 2003; *Nature* 416:603, 2002; *Exp. Med.* 191:1209, 2000.

John Murphy, Professor of Medicine; Ph.D., Connecticut. Molecular genetic analysis of the diphtheria toxin repressor (DtxR); molecular mechanism of diphtheria toxin catalytic domain translocation from early endosomes to the eukaryotic cell cytosol. *J. Cell Biol.* 160:1139, 2003; *J. Bacteriol.* 185:2251, 2003; *Proc. Natl. Acad. Sci. U.S.A.* 100:3808, 2003; *J. Microbiol. Methods* 51:63, 2002.

Barbara S. Nikolajczyk, Assistant Professor of Microbiology; Ph.D., North Carolina at Chapel Hill, 1991. Transcriptional regulation of immunoglobulin and cytokine genes by changes in chromatin structure; inflammatory diseases. *Mol. Immunol.* 43:1541–8, 2006; *J. Biol. Chem.* 281:9227–37, 2006; *J. Immunol.* 174:2834, 2005.

David Seldin, Professor of Medicine; M.D., Ph.D., Harvard/MIT Health Sciences. Application of transgenic and knockout mouse models to study serine-threonine kinases in lymphoproliferative diseases and other malignancies; immunoglobulin light chain amyloidosis. *Cancer Res.* 65(13):5792–801, 2005; *Toxicol. Pathol.* 33:726–37, 2005; *Oncogene* 21:5280–8, 2002.

Jacqueline Sharon, Professor of Pathology and Laboratory Medicine; Ph.D., Columbia. Generation and characterization of recombinant antibodies and T-cell receptors for the treatment and diagnosis of cancer and infectious diseases. *J. Cell. Biochem.* 96:305, 2005; *Immunol. Lett.* 91:179, 2004; *Int. J. Parasitol.* 33:281, 2003.

David Sherr, Professor of Environmental Health; Ph.D., Cornell. Mechanisms of immunosuppression mediated by environmental pollutants; molecular mechanisms of mammary carcinogenesis; targeting tumor-associated peptides for cancer immunotherapy. *Mol. Pharmacol.* 67:1740, 2005; *Int. J. Cancer* 115:333, 2005; *J. Immunol.* 173:3165, 2004; *Toxicol. Sci.* 79:211, 2004.

Gail E. Sonenshein, Professor of Biochemistry; Ph.D., MIT. Roles of NF-kappa B, forkhead, and myc transcription factors in control of cell proliferation; survival and neoplastic transformation of B cells and breast cancer. *Cancer Res.* 66:2570, 2006; *J. Virol.* 79:95, 2005; *J. Immunol.* 172:5522, 2004; *J. Biol. Chem.* 279:40593, 2004; *Mol. Cell. Biol.* 24:8681, 2004; *Mol. Cell. Biol.* 23:5738, 2003.

Gregory Viglianti, Associate Professor of Microbiology; Ph.D., Minnesota. Molecular biology of HIV-1; role of TLRs in autoimmunity. *J. Exp. Med.* 202:1171, 2005; *J. Virol.* 78:2819, 2004; *J. Biol. Chem.* 279:43604, 2004; *J. Endocrinology Res.* 10:247, 2004; *Immunity* 19:837, 2003; *AIDS Res. Hum. Retroviruses* 18:649, 2002; *J. Biol. Chem.* 277:50579, 2002.

Lee Wetzler, Professor of Medicine; M.D., SUNY Upstate Medical Center. Vaccine development; immune adjuvants; innate immunity and innate immune receptors; microbial immunity and pathogenesis; effect of bacterial products on eukaryotic cells. *J. Immunol.* 174:3345, 2005; *Cell. Microbiol.* 5:99, 2003; *Trends Microbiol.* 11:87, 2003; *J. Immunol.* 168:1533, 2002; *Infect. Immunol.* 69:5031, 2001; *Proc. Natl. Acad. Sci. U.S.A.* 97:9097, 2000.

BOSTON UNIVERSITY

School of Medicine
Department of Microbiology

Programs of Study

The Department of Microbiology of the Boston University School of Medicine offers Ph.D., M.D./Ph.D., and M.A. programs in microbiology and molecular pathogenesis and immunology. These programs of study are offered as interdepartmental study programs. The department also participates in interdepartmental programs in cellular and molecular biology and in hematology/oncology.

During the first two years, students complete formal course work in an individualized curriculum consisting of course work in microbiology, microbial pathogenesis virology, genetics, microbial and eukaryotic immunology, cell biology, biochemistry, or other related topics of special interest to the student. Special seminars, journal clubs, and research in progress seminars complement the formal course work. During the first year, students are required to do short rotation projects in three laboratories, culminating in the selection of a laboratory for a dissertation project. At the end of the second year, students take qualifying examinations. For the remainder of the graduate program, students are engaged in full-time original research that provides the basis for their dissertation.

Research Facilities

The research facilities of the department are in contiguous laboratories in recently renovated space that includes all equipment necessary to carry out research in molecular and cellular biology. The Medical Center maintains extensive animal-care facilities, including transgenic mouse and rat facilities; core protein and nucleic acid synthesis and analysis facilities; confocal deconvolution and electron microscopy facilities; core facilities for genomics and proteomics analysis; and facilities for structural biology, together with excellent computer and library facilities. All central facilities are within a short distance of the research laboratories.

Financial Aid

Students admitted to the Ph.D. program are granted full financial aid, including stipend, tuition, health insurance, and activity fees. For the 2006–07 academic year, the initial stipend for all trainees is $27,000. Tuition remission includes activity fees and health insurance. Students are encouraged to apply for fellowships from outside sources.

Cost of Study

Full-time tuition for the 2006–07 academic year is $33,500 per year.

Living and Housing Costs

Most students live in shared housing in Boston and surrounding areas. The total cost of living in the Boston area ranges from $900 to $1200 per month.

Student Group

The department has approximately 30 graduate students. There are a total of 180 graduate students enrolled in the Division of Graduate Medical Sciences at the medical school. Graduates currently hold posts in major universities, research institutions, and industry.

Location

Boston University School of Medicine is part of a large medical complex that includes Boston Medical Center and several research institutes. It is situated in a city known for its excellence in scientific and medical education as well as its historic, cultural, recreational, and sports activities. As the hub of New England, Boston is close to the countryside, mountain ranges, and seashore.

The University

Boston University is an independent, coeducational, nonsectarian university. It was incorporated in 1869 and has been a leader in the areas of health care and science. Numerous interdisciplinary programs offer broad possibilities for combining career goals with personal interests.

Applying

Candidates for admission must have a strong background in biological and physical sciences. The applicant's academic record, references, GRE test scores, and related work experience are considered in the admissions process. The GRE General Test is required. International students must demonstrate competence in English. Applications must be received by January 1 for admission in the fall semester.

Correspondence and Information

Director of Graduate Studies
Department of Microbiology
Boston University School of Medicine
715 Albany Street
Boston, Massachusetts 02118
Phone: 617-638-4284
Fax: 617-638-4286
E-mail: microbio@bu.edu
Web site: http://www.bumc.bu.edu/microbiology

For applications:

Division of Graduate Medical Sciences
Boston University School of Medicine
715 Albany Street
Boston, Massachusetts 02118
Phone: 617-638-5120
Fax: 617-638-4842
Web site: http://www.bumc.bu.edu/gms

Boston University

THE FACULTY AND THEIR RESEARCH

Selwyn A. Broitman, Professor of Microbiology; Ph.D., Michigan State, 1956; FACG. Cholesterol-lowering drugs and hepatic metastases.

Iih-Nan (George) Chou, Professor of Microbiology; Ph.D., Illinois, 1971. Cell biology, cytoskeleton, and cancer research; mechanisms of cell injury by carcinogenic metals. *Toxicol. Appl. Pharmacol.* 193:202, 2003; *J. Cell Biochem.* 88:152, 2003; *J. Cell Biochem.* 78:550, 2000; *Toxicol. Appl. Pharmacol.* 140:461, 1996.

Ronald B. Corley, Professor of Microbiology; Ph.D., Duke, 1975. B-lymphocyte development and function; IgM antibody structure and function; secreted IgM antibody in innate and adaptive immunity; role of Toll-like receptors in autoimmunity. *J. Immunol.* 174:983, 2005; *BMC Immunol.* 6:8, 2005; *Int. Immunol.* 16:1411, 2004; *J. Autoimmunity* 23:333, 2004.

Susan H. Fisher, Professor of Microbiology; Ph.D., Tufts, 1978. Regulation of nitrogen metabolism in *B. subtilis;* DNA-protein interactions. *J. Biol. Chem.* 280:33298, 2005; *Cell* 107:427, 2002; *J. Bacteriol.* 182:5939, 2000; *J. Mol. Biol.* 300:29, 2000; *Mol. Microbiol.* 32:223, 1999; *J. Bacteriol.* 180:2943, 1998; *Proc. Natl. Acad. Sci. U.S.A.* 93:8841, 1996.

Caroline A. Genco, Professor of Medicine; Ph.D., Rochester, 1987. Microbial pathogenesis; vaccine development; host-parasite interactions; innate immune response to bacterial pathogens and chronic inflammation; bacterial gene regulation. *Mol. Microbiol.* 60:963, 2006; *Cellular Microbiol.* 8:738, 2006; *Circulation* 109:2801, 2004; *Proc. Natl. Acad. Sci.* 100:9542, 2003.

Suryaram (Rahm) Gummuluru, Assistant Professor of Microbiology; Ph.D., Rochester, 1996. Virus-host interactions in HIV-1 pathogenesis. *Proc. Natl. Acad. Sci.* 103:738, 2006; *J. Virol.* 77:12865, 2003; *AIDS* 16:517, 2002; *J. Virol.* 76:10692, 2002; *J. Virol.* 74:10822, 2000; *J. Virol.* 73:5422, 1999.

Mark S. Klempner, Professor of Medicine; M. D., Cornell, 1973. Basic molecular biology and pathogenic mechanisms of the Lyme disease spirochete, *Borrelia burgdorferi*, and novel molecular methods for detecting, identifying, and quantifying microorganisms. *J. Phys. Chem. B* 109:312–20, 2005; *J. Infect. Dis.* 192:1010–3, 2005; *J. Infect. Dis.* 192:2020–6, 2005.

Stuart M. Levitz, Adjunct Professor of Medicine; M.D., NYU, 1979. The interplay between host immune system and medically important fungi. *J. Immunol.* 176:3053–61, 2006; *J. Immunol.* 175:7496–503, 2005; *J. Immunol.* 172:3059, 2004; *Infect. Immun.* 72:1746, 2004; *J. Biol. Chem.* 278:37561, 2003; *J. Immunol.* 168:5303–9, 2002; *J. Biol. Chem.* 277:39320, 2002; *Proc. Natl. Acad. Sci. U.S.A.* 98:10422, 2001.

Ann Marshak-Rothstein, Professor of Microbiology; Ph.D., Pennsylvania, 1977. Activation of autoreactive B cells by BCR/TLR-coengagement; role of exogenous and endogenous Fas-ligand in lymphocyte apoptosis, inflammation, and autoimmunity. *Immunol. Rev.* 204:27, 2005; *J. Immunol.* 172:6598, 2004; *Immunity* 19:837, 2003; *Nature* 416:603, 2002; *Exp. Med.* 191:1209, 2000.

John R. Murphy, Research Professor of Medicine; Ph.D., Connecticut, 1972. Molecular genetic analysis of the diphtheria toxin repressor (DtxR); molecular mechanism of diphtheria toxin catalytic domain translocation from early endosomes to the eukaryotic cell cytosol. *J. Cell Biol.* 160:1139, 2003; *J. Bacteriol.* 185:2251, 2003; *Proc. Natl. Acad. Sci. U.S.A.* 100:3808, 2003; *J. Microbiol. Methods* 51:63, 2002.

Barbara S. Nikolajczyk, Assistant Professor of Microbiology; Ph.D., North Carolina at Chapel Hill, 1991. Transcriptional regulation of immunoglobulin and cytokine genes by changes in chromatin structure and inflammatory diseases. *Mol. Immunol.* 43:1541–8, 2006; *J. Biol. Chem.* 281:9227–37, 2006; *J. Immunol.* 174:2834, 2005.

John C. Samuelson, Professor of Molecular and Cell Biology; M.D./Ph.D., Harvard, 1984. The study of the biochemistry, cell biology, and evolution of amebae and giardias using molecular biological methods. *Eukaryotic Cell* 5:836–48, 2006; *Proc. Natl. Acad. Sci. U.S.A.* 102:1548–53, 2005.

David Seldin, Professor of Medicine; M.D., Ph.D., Harvard/MIT Health Sciences, 1980. Application of transgenic and knockout mouse models to study serine-threonine kinases in lymphoproliferative diseases and other malignancies; immunoglobulin light chain amyloidosis. *Cancer Res.* 65(13):5792–801, 2005; *Toxicol. Pathol.* 33:726–37, 2005; *Oncogene* 21:5280, 2002.

Guillermo E. Taccioli, Assistant Professor; Ph.D., Buenos Aires, 1989. V(D)J recombination; DNA repair: Chromosomal instability; ionizing radiation response. *Cancer Res.* 65:10223, 2005; *J. Cell. Biol.* 167:627, 2004; *J. Biol. Chem.* 278:22136, 2003; *EMBO J.* 21:1, 2002; *Nat. Genet.* 31:159, 2002; *Immunity* 9:355, 1998.

Gregory Viglianti, Associate Professor of Microbiology; Ph.D., Minnesota, 1984. Molecular biology of HIV-1; role of TLRs in autoimmunity. *J. Exp. Med.* 202:1171, 2005; *J. Virol.* 78:2819, 2004; *J. Biol. Chem.* 279:43604, 2004; *J. Endocrinology Res.* 10:247, 2004; *Immunity* 19:837, 2003; *AIDS Res. Hum. Retroviruses* 18:649, 2002; *J. Biol. Chem.* 277:50579, 2002.

Lee M. Wetzler, Professor of Medicine; M.D., SUNY Upstate Medical Center, 1982. Vaccine development; immune adjuvants; innate immunity and innate immune receptors; microbial immunity and pathogenesis; effect of bacterial products on eukaryotic cells. *J. Immunol.* 174:3545, 2005; *Cell. Microbiol.* 5:99, 2003; *Trends Microbiol.* 11:87, 2003; *J. Immunol.* 168:1533, 2002; *Infect. Immun.* 69:5031, 2001; *Proc. Natl. Acad. Sci. U.S.A.* 97:9097, 2000.

Glen B. Zamansky, Associate Professor of Microbiology; Ph.D., Harvard, 1978. Keratinocyte cell biology and cellular responses to ultraviolet radiation.

CLEMSON UNIVERSITY

Master of Science in Microbiology

Program of Study	The graduate program in microbiology, offered through the Department of Biological Sciences, offers the Master of Science (M.S.) degree. The microbiology program encompasses a wide variety of disciplines, including bacterial physiology and metabolism, pathogenic microbiology, carcinogenesis and cancer cell biology, immunology, microbial genetics, microbial ecology, and applied and environmental microbiology. The overall goals of the program are to develop scientists with strong interdisciplinary skills in research design, critical thinking, and communication in microbiology, as well as expertise in a specific research area. The M.S. degree is usually completed within two years and requires 24 semester credit hours of formal course work and 6 semester credit hours of thesis research. Students complete a curriculum that includes course work in cellular and physiological microbiology, microbial genetics, and environmental microbiology. The first year of the program emphasizes general course work and research topic and proposal development. The second year focuses on research and thesis preparation and defense. The M.S. degree in microbiology prepares students for careers in industry, hospital and public health laboratories, and state and federal agencies.
Research Facilities	The Department of Biological Sciences houses research instrumentation and maintains affiliations with several campuswide facilities. Imaging laboratories contain confocal, scanning, and transmission electron microscopes. Additional electron microscopes are available through the University's off-campus microscopy facility. The Clemson University DNA sequencing facility provides ABI and LiCor instrumentation for DNA sequencing and fragment analysis. A CT scanning facility is available through the Greenville Hospital System. Specific instruments include a real-time PCR machine, a fluorimeter plate reader, and spectrophotometers. Walk-in incubators, cold rooms, special laboratories equipped for tissue culture, analytical ultracentrifugation, isotope analysis facilities, and various chromatography facilities are also available. The Biotechnology Greenhouse has a number of environmental chambers for research use. Students have opportunities to work with physicians and biomedical research faculty members at nearby facilities, including the research laboratory of the Greenville Hospital System in Greenville, South Carolina, and the Greenwood Genetics Center in Greenwood, South Carolina. The Robert Muldrow Cooper Library, Clemson's main library, has more than 2 million volumes and 6,000 serial and online journals.
Financial Aid	Graduate teaching assistantships are available through the Department of Biological Sciences to support teaching in undergraduate laboratories. These assistantships may be renewed annually as long as a student is making satisfactory progress toward the degree, and they are limited to three years for M.S. candidates. The twelve-month stipend is currently $12,000 for M.S. students. Graduate research assistantships are also available through extramurally funded research programs with individual faculty members within the department. A limited number of graduate teaching assistantships are currently available for the upcoming academic year.
Cost of Study	Tuition for 2006–07 is $4643 per semester for in-state students and $9255 per semester for nonresidents. Off-campus rates are $535 per hour for in-state students and $918 per hour for nonresidents. Graduate assistants pay a flat fee of $1079 per semester and $348 per summer session. Graduate fellows pay South Carolina resident fees.
Living and Housing Costs	On-campus housing is available. For information, students should visit http://www.housing.clemson.edu. The cost of living in Clemson is quite low compared to the national average. Students who choose to live off the campus typically spend $300–$400 per month for rent, depending on location, amenities, roommates, etc.
Student Group	The microbiology program seeks students with strong backgrounds in the biological sciences whose interests complement existing emphasis areas in biomedical research, food safety, and bioremediation. The program has approximately 17 students. Of these, 65 percent are women and 35 percent, men; 82 percent are full-time and 18 percent, part-time; 76 percent are U.S. residents and 24 percent, international students.
Student Outcomes	Recent graduates are employed in basic research; in industrial, hospital, public health, and pollution control laboratories; with state and federal agencies; and in teaching positions in colleges and universities. Others are pursuing continuing graduate or postdoctoral studies in various institutions throughout the country.
Location	Clemson is a small, beautiful college town near the Blue Ridge Mountains and Lake Hartwell in upstate South Carolina. The Upstate is one of the country's fastest-growing areas and is the midpoint of the Charlotte-to-Atlanta I-85 corridor, a multistate area along Interstate 85 that runs from metro Atlanta to Richmond, Virginia, and encompasses Charlotte, North Carolina, and North Carolina's Research Triangle. Atlanta and Charlotte are each a 2-hour's drive away. Many financial institutions and other industries have national headquarters for a major presence in the Upstate, including Wachovia, Bank of America, BMW, Bon Secours St. Francis Health System, Bosch North America, Bowater, Charter Communications, Ernst and Young, Fluor Corporation, IBM, Microsoft, Michelin of North America, and many others.
The University	Clemson is classified by the Carnegie Foundation as Doctoral/Research University–Extensive, a category comprising less than 4 percent of all universities in America. The University's mission is to fulfill the covenant between its founder and the people of South Carolina to establish a "high seminary of learning" through its responsibilities of teaching, research, and extended public service. The University has identified eight areas of academic emphasis that create collaborations that, in turn, help fulfill the University's mission.
Applying	Applicants may apply on the Web at http://www.grad.clemson.edu/p_apply.html. Applications, along with a $50 nonrefundable fee, should be received no later than five weeks prior to registration. Every required item in support of the application must be on file by that date. Students are advised to contact the department for the deadlines of the program of proposed study.
Correspondence and Information	Dr. Margaret B. Ptacek Graduate Programs Coordinator Department of Biological Sciences 132 Long Hall Clemson University Clemson, South Carolina 29634-0314 Phone: 864-656-2328 Fax: 864-656-0435 E-mail: mptacek@clemson.edu Web site: http://www.clemson.edu/biosci

Clemson University

THE FACULTY AND THEIR RESEARCH

Min Cao, Assistant Professor; Ph.D., Cornell, 2002. Microbial genetics, microbial pathogenesis, and genomics; using the food-borne pathogen *Listeria monocytogenes* as a model to study bacterial stress response (especially oxidative and nitrosative stress), identify novel virulence factors, and develop new genetic tools (e.g., mariner-based transposon system). (E-mail: mcao@clemson.edu)

Wen Chen, Associate Professor; Ph.D., Ohio, 1991. Prolactin receptor antagonists for anti–human breast cancer therapy; development of protein-based therapeutics; molecular cloning of novel genes related to breast cancer formation. (E-mail: wenc@clemson.edu)

Steven S. Hayasaka, Professor; Ph.D., Oregon State, 1975. Development of bacteriocins for use in preventing growth of food-borne pathogens; growth and activity of a *Sphingomonas paucimobilis* strain during bioremediation of fluoranthene-polluted soil (gene organization and biochemical pathway). (E-mail: hayasas@clemson.edu)

Thomas A. Hughes, Professor; Ph.D., North Carolina State, 1981. Cloning, sequencing, identification, and, ultimately, the regulation of an enhancer of the bacteriocin lactacin B; gene organization and biochemical pathway of *Sphingomonas paucimobilis* for polyaromatic hydrocarbon degradation. (E-mail: T020509@clemson.edu)

Xiuping Jiang, Assistant Professor; Ph.D., Maryland, 1996. Developing nanoparticles-based methods for rapid detection of food-borne pathogens such as *Escherichia coli* O157:H7, *Listeria monocytogenes, Campylobacter jejuni,* and *Salmonella* from food and environmental samples; investigating pathogen movement from the preharvest environment to fresh food; determining emergence and spread of ceftriaxone resistance in *Salmonella* due to the treatment of calves with ceftiofur on the farm. (E-mail: xiuping@clemson.edu)

Harry D. Kurtz, Assistant Professor; Ph.D., Idaho, 1989. Examining microbial ecosystems living in the deserts of southeastern Utah; developing management tools for use by the Bureau of Land Management and National Park Service for maintenance and care of parks and monuments in this area of Utah; developing methods to aid efforts to stabilize coastal dunes in South Carolina. (E-mail: hkurtz@clemson.edu)

Lyndon L. Larcom, Professor; Ph.D., Pittsburgh, 1968. Mechanisms of carcinogenesis; effects of damage to DNA and mechanisms for repair of the damage; cellular defects in leukemias; techniques for quantitative molecular analysis. (E-mail: lllrcm@clemson.edu)

Tamara L. McNealy, Assistant Professor; Ph.D., Heidelberg, 2003. Virulence mechanisms of intracellular respiratory pathogens, in particular *Francisella tularensis* and *Legionella pneumophila,* and how the natural environment of these pathogens has adapted them for pathogenicity in humans; host-pathogen interactions; gene regulation/expression in response to environmental signals; expression of bacterial membrane proteins in regulating immune response of host. (E-mail: tmcneal@clemson.edu)

Andrew S. Mount, Lecturer; Ph.D., Clemson, 1999. Study of cellular and molecular biology of biomineralization in mollusks, using the Eastern oyster *Crassostrea virginica* as a model: understanding how the organism nucleates calcium carbonate crystals, the mantle as a shell-forming organ, the role of collagen, and investigation of the role of the immune system in shell formation; developing novel functional genomic approaches that will enable transcriptome analysis of specific cell types related to the secretion of organic matrix proteins. (E-mail: mount@clemson.edu)

Kimberly S. Paul, Assistant Professor; Ph.D., Princeton, 1998. Parasite-host adaptation in African Trypanosomes; cell biology, biochemistry and molecular biology of fatty acid metabolism; environmental sensing and regulation of lipid uptake and metabolism; lipid trafficking; mitochondrial biology. (E-mail: kpaul@clemson.edu)

Charles D. Rice, Professor; Ph.D., William and Mary, 1989. Comparative marine immunobiology, with a special interest in immunobiology of fishes; veterinary immunology; molecular and cellular aspects of neuroendocrine-immune interactions; ontogeny and phylogeny of tumor immunology; immunotoxicology. (E-mail: cdrice@clemson.edu)

Thomas R. Scott, Adjunct Professor; Ph.D., Georgia, 1983. Immunology, with a concentration on cellular immunity and cancer immunology; identification of cell surface markers on immunologically important cells; isolation of important cytokines regulating cell growth and differentiation. (E-mail: trscott@clemson.edu)

Lesly A. Temesvari, Associate Professor; Ph.D., Windsor, 1987. Molecular and cellular mechanisms that govern biogenesis and function of endosomes and lysosomes; cellular and molecular biological approaches to investigate the role of several small-molecular-weight Rab GTPases in endosomal and lysosomal membrane and protein trafficking and in pathogenicity of the protozoan parasite *Entamoeba histolytica.* (E-mail: ltemesv@clemson.edu)

T. R. Jeremy Tzeng, Assistant Professor; Ph.D., Clemson, 1998. Evaluation of nanoparticle compositions for their ability to neutralize microbial pathogens; evaluation of phytochemical compounds for antimicrobial and antitumor activities. (E-mail: tzuenrt@clemson.edu)

Yanzhang Wei, Assistant Professor; Ph.D., Ohio, 1996. Cancer immunotherapy: dendritic cell–mediated cancer immunotherapy; cancer gene therapy; novel approaches for targeted cancer therapy. (E-mail: ywei@clemson.edu)

Xianzhone Yu, Assistant Professor; Ph.D., Ohio, 1998. Molecular and cellular mechanisms of tumor angiogenesis; gene therapy targeting on tumor angiogenesis; establishing tumor models through transgenic technique; tumor therapeutic agent screening. (E-mail: xyu@clemson.edu)

CLEMSON UNIVERSITY

Ph.D. in Microbiology

Program of Study	The graduate program in microbiology, offered through the Department of Biological Sciences, includes the Ph.D. degree. The microbiology program encompasses a wide variety of disciplines, including bacterial physiology and metabolism, pathogenic microbiology, carcinogenesis and cancer cell biology, immunology, microbial genetics, microbial ecology, and applied and environmental microbiology. The overall goals of the program are to develop scientists with strong interdisciplinary skills in research design, critical thinking, and communication in microbiology, as well as expertise in a specific research area. The Ph.D. degree is usually completed within four years. There is no minimum credit hour requirement; instead, a program is established by each student in consultation with his or her advisory committee. The program requires an approved plan of study (second year), successful completion of the qualifying exam for candidate status (third year), and research and dissertation preparation and defense (second through fourth years). The Ph.D. in microbiology prepares students for careers as research scientists in academic institutions, in research laboratories, and with state, federal, and private agencies.
Research Facilities	The Department of Biological Sciences houses research instrumentation and maintains affiliations with several campuswide facilities. Imaging laboratories contain confocal, scanning, and transmission electron microscopes. Additional electron microscopes are available through the University's off-campus microscopy facility. The Clemson University DNA sequencing facility provides ABI and LiCor instrumentation for DNA sequencing and fragment analysis. A CT scanning facility is available through the Greenville Hospital System. Specific instruments include a real-time PCR machine, a fluorimeter plate reader, and spectrophotometers. Walk-in incubators, cold rooms, special laboratories equipped for tissue culture, analytical ultracentrifugation, isotope analysis facilities, and various chromatography facilities are also available. The Biotechnology Greenhouse has a number of environmental chambers for research use. Students have opportunities to work with physicians and biomedical research faculty members at nearby facilities, including the research laboratory of the Greenville Hospital System in Greenville, South Carolina, and the Greenwood Genetics Center in Greenwood, South Carolina. The Robert Muldrow Cooper Library, Clemson's main library, has more than 2 million volumes and 6,000 serial and online journals.
Financial Aid	Graduate teaching assistantships are available through the Department of Biological Sciences to support teaching in undergraduate laboratories. These assistantships may be renewed annually as long as a student is making satisfactory progress toward the degree, and they are limited to five years for Ph.D. candidates. The twelve-month stipends are currently $15,000 for Ph.D. students. Graduate research assistantships are also available through extramurally funded research programs with individual faculty members within the department. A limited number of graduate teaching assistantships are currently available for the upcoming academic year.
Cost of Study	Tuition for 2006–07 is $4643 per semester for in-state students and $9255 per semester for nonresidents. Off-campus rates are $535 per hour for in-state students and $918 per hour for nonresidents. Graduate assistants pay a flat fee of $1079 per semester and $348 per summer session. Graduate fellows pay South Carolina resident fees.
Living and Housing Costs	On-campus housing is available. For information, students should visit http://www.housing.clemson.edu. The cost of living in Clemson is quite low compared to the national average. Students who choose to live off the campus typically spend $300–$400 per month for rent, depending on location, amenities, roommates, etc.
Student Group	The microbiology program seeks students with strong backgrounds in the biological sciences whose interests complement existing emphasis areas in biomedical research, food safety, and bioremediation. The program has approximately 17 students. Of these, 53 percent are women and 47 percent, men; 71 percent are full-time and 29 percent, part-time; 71 percent are U.S. residents and 29 percent, international students.
Student Outcomes	Recent graduates are employed in basic research; in industrial, hospital, public health, and pollution control laboratories; with state and federal agencies; and in teaching positions in colleges and universities. Others are pursuing continuing graduate or postdoctoral studies in various institutions throughout the country.
Location	Clemson is a small, beautiful college town near the Blue Ridge Mountains and Lake Hartwell in upstate South Carolina. The Upstate is one of the country's fastest-growing areas and is the midpoint of the Charlotte-to-Atlanta I-85 corridor, a multistate area along Interstate 85 that runs from metro Atlanta to Richmond, Virginia, and encompasses Charlotte, North Carolina, and North Carolina's Research Triangle. Atlanta and Charlotte are each a 2-hour's drive away. Many financial institutions and other industries have national headquarters for a major presence in the Upstate, including Wachovia, Bank of America, BMW, Bon Secours St. Francis Health System, Bosch North America, Bowater, Charter Communications, Ernst and Young, Fluor Corporation, IBM, Microsoft, Michelin of North America, and many others.
The University	Clemson is classified by the Carnegie Foundation as Doctoral/Research University–Extensive, a category comprising less than 4 percent of all universities in America. The University's mission is to fulfill the covenant between its founder and the people of South Carolina to establish a "high seminary of learning" through its responsibilities of teaching, research, and extended public service. The University has identified eight areas of academic emphasis that create collaborations that, in turn, help fulfill the University's mission.
Applying	Applicants may apply on the Web at http://www.grad.clemson.edu/p_apply.html. Applications, along with a $50 nonrefundable fee, should be received no later than five weeks prior to registration. Every required item in support of the application must be on file by that date. Students are advised to contact the department for the deadlines of the program of proposed study.
Correspondence and Information	Dr. Margaret B. Ptacek Graduate Programs Coordinator Department of Biological Sciences 132 Long Hall Clemson University Clemson, South Carolina 29634-0314 Phone: 864-656-2328 Fax: 864-656-0435 E-mail: mptacek@clemson.edu Web site: http://www.clemson.edu/biosci

Clemson University

THE FACULTY AND THEIR RESEARCH

Min Cao, Assistant Professor; Ph.D., Cornell, 2002. Microbial genetics, microbial pathogenesis, and genomics; using the food-borne pathogen *Listeria monocytogenes* as a model to study bacterial stress response (especially oxidative and nitrosative stress), identify novel virulence factors, and develop new genetic tools (e.g., mariner-based transposon system). (E-mail: mcao@clemson.edu)

Wen Chen, Associate Professor; Ph.D., Ohio, 1991. Prolactin receptor antagonists for anti–human breast cancer therapy; development of protein-based therapeutics; molecular cloning of novel genes related to breast cancer formation. (E-mail: wenc@clemson.edu)

Steven S. Hayasaka, Professor; Ph.D., Oregon State, 1975. Development of bacteriocins for use in preventing growth of food-borne pathogens; growth and activity of a *Sphingomonas paucimobilis* strain during bioremediation of fluoranthene-polluted soil (gene organization and biochemical pathway). (E-mail: hayasas@clemson.edu)

Thomas A. Hughes, Professor; Ph.D., North Carolina State, 1981. Cloning, sequencing, identification, and, ultimately, the regulation of an enhancer of the bacteriocin lactacin B; gene organization and biochemical pathway of *Sphingomonas paucimobilis* for polyaromatic hydrocarbon degradation. (E-mail: T020509@clemson.edu)

Xiuping Jiang, Assistant Professor; Ph.D., Maryland, 1996. Developing nanoparticles-based methods for rapid detection of food-borne pathogens such as *Escherichia coli* O157:H7, *Listeria monocytogenes, Campylobacter jejuni,* and *Salmonella* from food and environmental samples; investigating pathogen movement from the preharvest environment to fresh food; determining emergence and spread of ceftriaxone resistance in *Salmonella* due to the treatment of calves with ceftiofur on the farm. (E-mail: xiuping@clemson.edu)

Harry D. Kurtz, Assistant Professor; Ph.D., Idaho, 1989. Examining microbial ecosystems living in the deserts of southeastern Utah; developing management tools for use by the Bureau of Land Management and National Park Service for maintenance and care of parks and monuments in this area of Utah; developing methods to aid efforts to stabilize coastal dunes in South Carolina. (E-mail: hkurtz@clemson.edu)

Lyndon L. Larcom, Professor; Ph.D., Pittsburgh, 1968. Mechanisms of carcinogenesis; effects of damage to DNA and mechanisms for repair of the damage; cellular defects in leukemias; techniques for quantitative molecular analysis. (E-mail: Illrcm@clemson.edu)

Tamara L. McNealy, Assistant Professor; Ph.D., Heidelberg, 2003. Virulence mechanisms of intracellular respiratory pathogens, in particular *Francisella tularensis* and *Legionella pneumophila,* and how the natural environment of these pathogens has adapted them for pathogenicity in humans; host-pathogen interactions; gene regulation/expression in response to environmental signals; expression of bacterial membrane proteins in regulating immune response of host. (E-mail: tmcneal@clemson.edu)

Andrew S. Mount, Lecturer; Ph.D., Clemson, 1999. Study of cellular and molecular biology of biomineralization in mollusks, using the Eastern oyster *Crassostrea virginica* as a model: understanding how the organism nucleates calcium carbonate crystals, the mantle as a shell-forming organ, the role of collagen, and investigation of the role of the immune system in shell formation; developing novel functional genomic approaches that will enable transcriptome analysis of specific cell types related to the secretion of organic matrix proteins. (E-mail: mount@clemson.edu)

Kimberly S. Paul, Assistant Professor; Ph.D., Princeton, 1998. Parasite-host adaptation in African Trypanosomes; cell biology, biochemistry and molecular biology of fatty acid metabolism; environmental sensing and regulation of lipid uptake and metabolism; lipid trafficking; mitochondrial biology. (E-mail: kpaul@clemson.edu)

Charles D. Rice, Professor; Ph.D., William and Mary, 1989. Comparative marine immunobiology, with a special interest in immunobiology of fishes; veterinary immunology; molecular and cellular aspects of neuroendocrine-immune interactions; ontogeny and phylogeny of tumor immunology; immunotoxicology. (E-mail: cdrice@clemson.edu)

Thomas R. Scott, Adjunct Professor; Ph.D., Georgia, 1983. Immunology, with a concentration on cellular immunity and cancer immunology; identification of cell surface markers on immunologically important cells; isolation of important cytokines regulating cell growth and differentiation. (E-mail: trscott@clemson.edu)

Lesly A. Temesvari, Associate Professor; Ph.D., Windsor, 1987. Molecular and cellular mechanisms that govern biogenesis and function of endosomes and lysosomes; cellular and molecular biological approaches to investigate the role of several small-molecular-weight Rab GTPases in endosomal and lysosomal membrane and protein trafficking and in pathogenicity of the protozoan parasite *Entamoeba histolytica.* (E-mail: ltemesv@clemson.edu)

T. R. Jeremy Tzeng, Assistant Professor; Ph.D., Clemson, 1998. Evaluation of nanoparticle compositions for their ability to neutralize microbial pathogens; evaluation of phytochemical compounds for antimicrobial and antitumor activities. (E-mail: tzuenrt@clemson.edu)

Yanzhang Wei, Assistant Professor; Ph.D., Ohio, 1996. Cancer immunotherapy: dendritic cell–mediated cancer immunotherapy; cancer gene therapy; novel approaches for targeted cancer therapy. (E-mail: ywei@clemson.edu)

Xianzhone Yu, Assistant Professor; Ph.D., Ohio, 1998. Molecular and cellular mechanisms of tumor angiogenesis; gene therapy targeting on tumor angiogenesis; establishing tumor models through transgenic technique; tumor therapeutic agent screening. (E-mail: xyu@clemson.edu)

COLORADO STATE UNIVERSITY

College of Veterinary Medicine and Biomedical Sciences
Department of Microbiology, Immunology and Pathology

Programs of Study

The Department of Microbiology, Immunology and Pathology offers programs leading to the M.S. and Ph.D. degrees in selected areas of Microbiology, Immunology and Pathology. Areas of research emphasis are infectious diseases, including medical/veterinary bacteriology, virology, and immunology; biotechnology, including industrial microbiology and microbial and molecular genetics; and environmental microbiology/microbial ecology. Among the currently active projects in medical microbiology are the following: investigation of mycobacterial surface structures and their roles in pathogenesis and immunity; development of novel diagnostic and immunization technologies; mechanisms of pathogenesis, antibiotic action and drug resistance in various bacteria; molecular mechanisms of animal-virus genome replication, expression, and latency; replication and transmission of viruses in vertebrate and arthropod hosts; pathogenic mechanisms in avian and ovine retroviruses; genetic mapping of microbial and arthropod genomes; and construction of virus and plasmid vectors for genetic alteration of prokaryotic and eukaryotic organisms. Projects in environmental microbiology include transformation and degradation of organic compounds by soil microorganisms and evaluation of the effects of disturbances and pollutants on microbial components of ecosystems.

The M.S. degree program usually requires two to three years for completion, and satisfactory completion of the Ph.D. degree requires at least four or five years. Students devote their first year of study to course work in core areas of microbiology, immunology and pathology, and related disciplines, laboratory rotations, and identification of a research project, an adviser, and a graduate committee. Students with a strong background in microbiology may serve as teaching assistants during the first year. The master's degree program culminates with an original thesis, describing results of the research, which is defended before the faculty committee in an oral examination. Ph.D. students are admitted to candidacy by passing written and oral qualifying examinations, usually during their second year of study. Their research results are presented in a dissertation and orally defended before the faculty committee for completion of the degree.

Student and faculty development is enriched by participation in interdisciplinary programs such as the Program of Excellence in Infectious Diseases, and Programs in Cell and Molecular Biology and Ecology. The department also participates in the joint D.V.M./Ph.D. degree program of the College of Veterinary Medicine and Biomedical Sciences. The post-D.V.M. combined microbiology residency/Ph.D. graduate training program is sponsored through the College of Veterinary Medicine and Biomedical Sciences.

Research Facilities

The Department of Microbiology, Immunology and Pathology occupies research, faculty, graduate student, and administrative office space in three buildings, including the microbiology and pathology facilities on the main campus and the Arthropod-Borne Infectious Diseases Laboratory on the Foothills campus. The total space comprises 158,000 square feet. In addition, members of the department work in the Diagnostic Laboratory of the Veterinary Teaching Hospital and adjunct members work at the CDC facility located on the foothills campus. A broad range of instrumentation is available, including ultracentrifuges, spectrophotometers, real-time PCR machines, gas chromatographs, mass spectrometers, HPLC, electrophoresis equipment, microcomputers, flow cytometers, and growth chambers for a variety of culture conditions. Biocontainment level-3 is available for both microbial cultures and animals, including a 12,000-square-foot state-of-the-art BL3 facility on the Foothills campus. Adjoining facilities provide DNA sequencing, oligonucleotide and peptide synthesis and sequencing, electron and confocal microscopy, and laboratory-animal and livestock housing.

Financial Aid

Graduate students are supported by teaching and research assistantships, federally and privately funded fellowships, and traineeships and scholarships. Total support includes tuition plus a stipend, which is currently about $20,772 per year. Stipend levels for post-D.V.M. students are higher and are determined by the College of Veterinary Medicine and Biomedical Sciences.

Cost of Study

For 2006–07, graduate student tuition and fees for two semesters are approximately $5500 for Colorado residents and $16,895 for nonresidents.

Living and Housing Costs

A limited number of University apartments are available on campus for single students, couples, or families. A wide variety of living accommodations at moderate rents can also be found off campus. An estimate of minimum but adequate living expenses for a single student is $17,000 to $20,000 per academic year.

Student Group

There are about 80 graduate students in the graduate program in microbiology, immunology and pathology. More than half of these are Ph.D. candidates, and about half are women. Microbiology also has about 300 undergraduate majors. Of the 25,000 students at Colorado State, about 4,100 are graduate and professional students.

Location

Fort Collins, located at the base of the Rocky Mountains and the mouth of Poudre Canyon, is at an elevation of 5,000 feet. The air is clear and dry, and year-round temperatures are moderate. There are more than 300 days of sunshine each year. The city has a population of 145,000 with museums, a public library, a civic symphony, galleries, and dance and drama groups, and is a regional shopping center. The mountains afford easily accessible winter and summer recreation. Denver is 65 miles to the south, and Laramie, Wyoming, is 65 miles to the north.

The University

Colorado State is a land-grant university founded in 1870. It has eight colleges and a graduate school. The Department of Microbiology, Immunology and Pathology is one of five departments in the College of Veterinary Medicine and Biomedical Sciences. The Microbiology and the Pathology Building are on the 883-acre main campus in the heart of Fort Collins, and the Arthropod-borne and Infectious Diseases Laboratory, the Centers for Disease Control, and other research facilities are 3 miles west on the 1,700-acre Foothills campus.

Applying

Admission to the graduate program is open to holders of bachelor's, master's, and D.V.M. degrees or the equivalent. Applicants should have backgrounds in chemistry, biochemistry, physics, general biology, and microbiology. The complete application includes undergraduate and graduate transcripts, letters of recommendation, a referee's evaluation form, and scores from the General Test of the Graduate Record Examinations. For application details and forms, students should consult the School's Web site at http://www.cvmbs.colostate.edu/mip/gradprogs/gradindex.html. Applications should be complete by January 1 for admission in the fall semester, which is when most students who are accepted matriculate. In addition, applications received throughout the year for unforeseen funding opportunities are considered. Notifications of admission and stipend awards are made soon after those dates.

Correspondence and Information

Graduate Assistant
Graduate Programs in Microbiology, Immunology and Pathology
Colorado State University
1862 Campus Delivery
Fort Collins, Colorado 80523-1682

Phone: 970-491-3228
E-mail: microbio@colostate.edu
Web site: http://www.cvmbs.colostate.edu/mip/

Colorado State University

THE FACULTY AND THEIR RESEARCH

Additional faculty members are listed on the department's Web site.

Ramesh K. Akkina, Professor; Ph.D., Minnesota, 1982. Molecular virology; viral pathogenesis; gene therapy; molecular studies on gene structure; evolution of veterinary viral pathogens.

Anne C. Avery, Assistant Professor; D.V.M., Pennsylvania, 1990; Ph.D., Cornell, 1991. Autoimmune and immune-mediated disease and infectious diseases of the hematopoietic system.

Paul R. Avery, Assistant Professor; D.V.M., Pennsylvania, 1991; Ph.D., Colorado State, 2002. AIDS; FIV.

Barry J. Beaty, University Distinguished Professor; Ph.D., Wisconsin–Madison, 1976. Arbovirology; vector-virus interactions; epidemiology of arbovirus disease; virus gene structure–biological function relationships; rapid virus diagnosis; mosquito molecular biology.

John T. Belisle, Associate Professor; Ph.D., Colorado State, 1992. Bacterial physiology and molecular genetics; molecular analysis of the posttranslational modifications of mycobacterial proteins; definition of mycobacterial proteins involved in pathogenesis and immunity; quorum sensing in mycobacteria.

William C. Black IV, Associate Professor; Ph.D., Iowa State, 1985. Molecular genetics and systematics of medically important arthropods; genome mapping in insects, with emphasis on genetics of vector competence for pathogens.

Carol D. Blair, Professor; Ph.D., Berkeley, 1968. Molecular virology; replication of arboviruses in vectors and vertebrates; regulation of viral gene expression; virus genetics; molecular mechanisms of pathogenesis; persistent infections.

Patrick J. Brennan, University Distinguished Professor; Ph.D., Dublin, 1965. Mycobacteria and their diseases (tuberculosis, leprosy, opportunistic mycobacterioses, bovine tuberculosis); chemical definition, immunoreactivity, and role in pathogenesis of components of cell walls and membranes; new skin test antigens for diagnosis of leprosy and TB; biosynthesis of cell wall and discovery of new drug targets; biochemical and genetic basis of morphological variations.

Jonathan O. Carlson, Professor; Ph.D., Berkeley, 1974. Molecular biology of viruses; gene expression in prokaryotic and eukaryotic cells.

Delphi Chatterjee, Research Associate Professor; Ph.D., London, 1981. Structural biology and glycomics; definition of macromolecules in mycobacteria; use of state-of-the-art NMR and MS as tools for macromolecular analysis and also to understand the roles of these molecules in pathogenesis and antibiotic resistance.

Dean C. Crick, Research Assistant Professor; Ph.D., Western Ontario, 1989. Biochemistry and physiology of *Mycobacterium* species; cell-wall biosynthesis; lipid biosynthesis; isoprenoid metabolism; drug development.

James C. DeMartini, Professor; D.V.M., 1966, Ph.D., 1972, California, Davis. Immunology and pathogenesis of viral diseases of ruminants, especially retroviruses and herpesviruses; localization of viral nucleic acids and proteins; immune responses; molecular mechanisms of retroviral oncogenesis; ecology of infectious diseases.

Steven W. Dow, Associate Professor; D.V.M., Georgia, 1982; Ph.D., Colorado State, 1992. Immune vaccines in cancer.

Nancy DuTeau, Research Assistant Professor; Ph.D., Iowa State, 1985. Molecular environmental genetics; genetic markers for population studies and microbial diversity; biomarkers for assessing exposure to environmental hazards.

E. J. Ehrhart, Associate Professor; D.V.M., Missouri, 1987; Ph.D., Colorado State, 1996. Cancer; tumor markers; diagnostic molecular pathology.

Brian D. Foy, Assistant Professor; Ph.D., Tulane, 2001. Vector biology; arbovirology; malaria; immunological interactions at the interface of hosts, vectors, and pathogens.

Claudia Gentry-Weeks, Associate Professor; Ph.D., Oklahoma Health Sciences Center, 1985. Bacterial pathogenesis and genetic analysis of infectious agents of humans and animals; genetic methods used to identify virulence factors of bacterial pathogens; antibiotic resistance of *Salmonella enterica;* pathogenesis of *Enterococcus faecalis.*

Daniel H. Gould, Professor; D.V.M., Colorado State, 1968; Ph.D., California, Davis, 1972. Veterinary diagnostic pathology; toxic, metabolic, and nutritional diseases of ruminants; neuropathology and neurotoxicology.

Edward A. Hoover, Professor; D.V.M., Illinois at Urbana-Champaign, 1967; Ph.D., Ohio State, 1970. Pathogenesis/intervention for retrovirus infections (feline and simian immunodeficiency viruses (FIV, SIV) and feline leukemia virus (FeLV)); pathogenesis of prion disease.

Doreene R. Hyatt, Assistant Professor, Ph.D., Arizona, 1996. Foodborne enteric bacteria and antimicrobial resistance.

Julia M. Inamine, Associate Professor; Ph.D., Duke, 1982. Microbial and molecular genetics; molecular genetic analysis of cell envelope biosynthesis in mycobacteria.

Angelo Izzo, Assistant Professor; Ph.D., Adelaide, 1992. Association between the immune response and pathogenesis during pulmonary mycobacterial infection.

Mercedes Gonzales Juarrero, Assistant Professor; Ph.D., Universidad Auonoma de Madrid (Spain), 1990. Role and dynamics of macrophage and dendritic cell populations during the early and late stages of infection with *Mycobacterium tuberculosis.*

Anne Lenaerts, Assistant Professor; Ph.D., Ghent. Establishing new models to improve and/or facilitate screening of experimental compounds against Mtb.

James C. Linden, Professor; Ph.D., Iowa State, 1969. Industrial microbiology; bioprocess engineering; bioconversion of cellulose; ethanol and SCP fermentations; microbial production of enzymes; plant-cell culture secondary metabolite production; plant elicitation of defense responses.

Michael McNeil, Research Professor; Ph.D., Colorado at Boulder, 1984. Development of new drugs against tuberculosis and other mycobacteria targeting cell wall biosynthesis; elucidation of mycobacterial cell wall biosynthetic pathways.

Robert W. Norrdin, Professor; D.V.M., Cornell, 1962; Ph.D., Cornell, 1969. Bone remodeling; metabolic bone diseases; nutritional and orthopedic pathology.

Kenneth E. Olson, Associate Professor; Ph.D., Colorado State, 1992. Arbovirology; molecular biology; alphavirus gene expression in mosquitoes.

Christine S. Olver, Assistant Professor; D.V.M., 1987, Ph.D., 1994, Ohio State. Red blood-cell receptor interactions in murine malaria infections; role of the Duffy glycoprotein in *Plasmodium yoelii* infections.

Ian M. Orme, Professor; Ph.D., London, 1981. Cellular immunology; nature and characteristics of T-lymphocytes involved in immunity to mycobacteria; immunosenescence and tuberculosis disease; animal models for TB vaccine and drug screening.

Sandra L. Quackenbush, Assistant Professor; Ph.D., Colorado State, 1994. Viral pathogenesis, particularly viral-induced oncogenesis.

Joel Rovnak, Assistant Professor; Ph.D., Cornell, 1999. Characterization of molecular mechanisms of oncogenesis, particularly where these mechanisms involve viral control of transcription and splicing.

Alan R. Schenkel, Assistant Professor; Ph.D., Wisconsin–Madison, 1998. Roles of adhesion molecules in leukocyte extravasation (exit from blood into tissues) in response to inflammation.

Herbert P. Schweizer, Professor; Ph.D., Konstanz (Germany), 1983. Molecular microbiology, genetics, physiology, and efflux pumps of *Pseudomonas aeruginosa;* drug discovery and resistance; development of molecular and genetic tools for analysis of pathogenic bacteria.

Richard A. Slayden, Assistant Professor; Ph.D., Colorado State, 1997. Bacterial differentiation into distinct cellular populations; phenotypically optimization throughout the disease process enhance bacterial survival and tolerance to chemotherapeutics.

Mary Anna Thrall, Professor; D.V.M., Purdue, 1970. Pathophysiology and experimental treatment, specifically bone marrow transplantation, therapy of ethylene glycol toxicoses, clinical cytology.

Richard G. Titus, Professor; Ph.D., Washington (Seattle), 1978. Immunology, molecular biology, biochemistry: vaccine development and drug discovery for pathogens *(Leishmania, Borrelia)* transmitted by insects (arthropods).

Sue VandeWoude, Associate Professor of Comparative Medicine; D.V.M., Virginia-Maryland Regional College of Veterinary Medicine, 1986. Feline lentivirus biology, immunology, and receptor usage; host-lentivirus adaptation; laboratory animal/comparative medicine; environmental enrichment.

Varalakshmi D. Vissa, Assistant Professor; Ph.D., Maryland, Baltimore, 1991. Basic and applied post-genome leprosy research, with focus on epidemiology (strain typing and detection of drug resistance); development of diagnostic reagents; studies on physiology of *Mycobacterium leprae* and genetics of cell wall synthesis; multiple polymorphic loci for molecular typing of strains of *Mycobacterium.*

Jeffrey Wilusz, Professor and Department Head; Ph.D., Duke, 1985. Posttranscriptional control of gene expression, including mRNA stability in mammalian cells, mRNA-protein interactions, and 3' end processing.

CREIGHTON UNIVERSITY

School of Medicine
Department of Medical Microbiology and Immunology

Programs of Study

Programs are offered leading to M.S. and Ph.D. degrees. Students are encouraged to acquire a breadth of knowledge in the major disciplines of microbiology and then concentrate in a research area such as virology and HIV pathogens; cellular, molecular, and clinical immunology; microbial pathogenicity; infectious diseases; antimicrobial agents and chemotherapy; microbial and molecular genetics; microbial physiology; and epidemiology, prions, and multiple sclerosis.

The minimum required curriculum for the M.S. degree, which is usually completed in two full years, consists of 30 semester hours of graduate credit, including both formal courses and thesis research. For the Ph.D., an additional 60 hours are required. Postdoctoral positions are also available.

Research Facilities

The department occupies 12,000 square feet of space in modern buildings. More than three quarters of the available space is used for research laboratories, which are completely equipped with all essential facilities, including a BSL-3 biocontainment facility. There is also a centrally located departmental pool of major equipment. Ample space is provided for each student. Well-appointed animal quarters and an animal operating room are available. The Clinical Microbiology Laboratory at St. Joseph Hospital is well equipped for extensive instruction in diagnostic microbiology, immunopathology, and viral serology. The Health Sciences Library has more than 210,000 volumes and currently receives 1,600 serials.

Financial Aid

Financial aid is generally available in the form of stipends and/or remission of tuition. Graduate assistantships that are financed by the University and other related sources are available. The assistantship appointee assists in clerical and other duties performed in the student's major department. These appointments require no less than 10 and no more than 20 clock hours of the student's time per week.

Cost of Study

Graduate tuition is $595 per semester hour for 2006–07. Students typically enroll for 8 to 12 hours of course work each semester.

Living and Housing Costs

Creighton University offers on-campus housing facilities for unmarried students who are not residents of Omaha. In addition, students can find adequate low-rent housing off campus. Information on housing may be obtained from the Housing Office.

Student Group

All states and fifty-four countries are represented in the Creighton student body. The University enrolls nearly 6,500 students, of whom more than 500 are members of the graduate division. There are usually between 10 and 15 graduate students in the Department of Medical Microbiology and Immunology.

Student Outcomes

The spectrum of graduate alumni extends into all facets of society and the professions, including scientists who have entered the fields of medicine, pharmacy, and allied health and individuals who have joined microbiology departments, pharmaceutical companies, industrial corporations, and clinical microbiology laboratories. The list includes, but is not limited to, former students and postdoctoral fellows employed at Pfizer, Inc.; Abbott Laboratories; National Aeronautics and Space Administration; Yale University; Wyeth-Ayerst Pharmaceutical Laboratories; Medical College of Virginia; Rocky Mountain National Laboratory; the U.S. Food and Drug Administration; the University of Nebraska Medical Center; and the United States Centers for Disease Control.

Location

Creighton University is located in Omaha, Nebraska, a dynamic city in the heart of the nation. Omaha has grown to be a modern city of commerce and industry, and it is a major distribution center serving the expanding markets of mid-America. The population of Omaha and the urban areas surrounding it is more than 650,000. Educational and cultural institutions of a high order include the Joslyn Art Museum and the Omaha Symphony Orchestra. Omaha is also well represented in both professional and collegiate athletics and is the home of the College World Series. Public recreation opportunities are provided by Omaha's parks, golf courses, swimming pools, and the Henry Doorly Zoo. Omaha has outstanding religious, educational, and medical facilities and is truly a place for all people.

The University

As Creighton University enters its second century of service as a national resource, the health sciences are undergoing many changes. Medicine, dentistry, pharmacy, and nursing have been organized under one administrative structure at the University, directed by a vice president for health sciences. With renovation of the multimillion-dollar Criss Health Sciences Center and completion of the new Beirne Tower, educational programs are being implemented, research capabilities expanded, and health-care delivery systems studied. The Health Sciences Center includes the Schools of Medicine, Pharmacy and Allied Health Professions, Nursing, and Dentistry and is supported by a teaching hospital with a connecting Health Professions Center, the Boys Town Institute for Communication Disorders in Children, and the well-equipped Bio-Information Center.

Applying

Completed applications must be on file with the Graduate School on or before February 1 for consideration for admission in the fall semester. Applications should include transcripts, scores on the General Test of the Graduate Record Examinations, and three letters of recommendation.

Correspondence and Information

Philip D. Lister, Ph.D.
Director of Graduate Program
Department of Medical Microbiology and Immunology
School of Medicine
Creighton University
Omaha, Nebraska 68178
Phone: 402-280-2921
Fax: 402-280-1875
E-mail: cumedmicro@creighton.edu
Web site: http://www.mmi.creighton.edu/index.html

Creighton University

THE FACULTY AND THEIR RESEARCH

Devendra K. Agrawal, Professor; Ph.D. (biochemistry), Lucknow (India), 1978; Ph.D. (medical sciences), McMaster, 1984. Transcription/translation of cell-adhesion molecules; mechanisms of restenosis/intimal hyperplasia; apoptosis of vascular smooth-muscle cells; allergy/asthma.

Jason C. Bartz. Assistant Professor; Ph.D. (veterinary science), Wisconsin–Madison, 1998. Prion diseases; pathogenesis of neurodegenerative disorders; neurovirology.

Michael Belshan, Assistant Professor; Ph.D. (molecular, cellular, and developmental biology), Iowa State, 1999. Virus-host cell interactions; virus replication and pathogenesis in human immunodeficiency virus (HIV) replication.

Marvin J. Bittner, Associate Professor; M.D. (infectious diseases), Harvard, 1976. Clinical infectious diseases; travel medicine; hospital epidemiology.

Stephen J. Cavalieri, Associate Professor; Ph.D. (medical microbiology), West Virginia, 1981. Clinical microbiology; antimicrobial susceptibility testing; mycobacteriology; rapid diagnostic testing for infectious diseases; clinical virology.

Edward A. Chaperon, Associate Professor; Ph.D. (immunology), Wisconsin, 1965. Cellular immunology.

Archana Chatterjee, Associate Professor; M.D., Ph.D. (pediatric infectious disease), Nebraska Medical Center, 1993. Maternal-fetal transmission of cytomegalovirus infection; the role of cytomegalovirus in atherosclerosis; antibiotic utilization by physicians; antibiotic resistance; vaccine development.

Kristen M. Drescher, Associate Professor; Ph.D. (molecular microbiology and immunology), Johns Hopkins, 1996. Pathogenesis of demyelinating diseases; multiple sclerosis.

Paul Fey, Assistant Clinical Professor of Medicine; Ph.D. (medical microbiology), Creighton, 1995. Molecular epidemiology and characterization of *Staphylococcus* species.

Martha Gentry-Nielsen, Professor; Ph.D. (infectious disease), Oklahoma State, 1984. The effects of ethanol ingestion and liver cirrhosis on susceptibility to pneumococcal infection.

Richard V. Goering, Professor and Chair; Ph.D. (microbiology), Iowa State, 1972. Molecular techniques for the epidemiological analysis of nosocomial pathogens; genetics of antibiotic resistance.

Gary L. Gorby, Associate Professor; M.D. (infectious disease), Northeastern Ohio College of Medicine, 1983. Pathogenesis of *Neisseria gonorrhoeae* infections.

Venkatesh Govindarajan, Assistant Professor; Ph.D. (developmental/molecular biology), Houston, 1997. Fibroblast growth factor (FGF) signaling during ocular and skeletal development.

Nancy D. Hanson, Associate Professor; Ph.D. (medical microbiology), Nebraska Medical Center, 1991. Regulation of inducible beta-lactamase expression in *Serratia;* molecular mechanisms of antimicrobial resistance; antivirals.

Edward A. Horowitz, Associate Professor; M.D. (infectious disease), Creighton, 1978. Clinical aspects of antimicrobial efficiency and resistance.

Floyd C. Knoop, Professor; Ph.D. (medical microbiology/biochemistry), Tennessee, 1974. Mechanisms of NAD-dependent ADP-ribosylation and receptor-mediated transmembrane signaling by microbial toxins; basis of adenyl and guanyl cyclase/protein kinase activation.

Philip D. Lister, Associate Professor; Ph.D. (medical microbiology), Creighton, 1992. In vitro and in vivo models of infections to study new antibiotics, drug combinations, new dosing strategies, and basic bacterial-antibiotic interactions relating to development of antibiotic resistance and treatment of drug-resistant bacteria.

Laurel C. Preheim, Professor; M.D. (infectious disease), Northwestern, 1973. Host-parasite interactions; microbial virulence factors; host defense mechanisms; effect of alcohol ingestion and alcoholic liver disease on resistance to pneumococcal pneumonia.

Mark E. Rupp, Associate Clinical Professor; M.D. (infectious disease), Baylor College of Medicine, 1984. Pathogenesis of prosthetic device infections; adherence of coagulase-negative staphylococci; infection control/hospital epidemiology.

Patrick C. Swanson, Associate Professor; Ph.D. (chemistry), Michigan, 1995. V(D)J recombination and other processes underlying antigen receptor diversity.

Kenneth S. Thomson, Professor; Ph.D. (microbiology), Tasmania (Australia), 1988. Antibiotic resistance mechanisms in bacteria; clinical relevance of antibiotic susceptibility tests; beta-lactam antibiotics; quinolone antibiotics.

Zhaoyi Wang, Assistant Professor; Ph.D. (molecular genetics), Washington (St. Louis), 1994. Tumorigenesis of breast cancer; estrogen signaling.

DARTMOUTH COLLEGE

Dartmouth Medical School
Program in Immunology

Program of Study

The Immunology Program at Dartmouth Medical School (DMS) is an interdepartmental program that trains both postdoctoral fellows and predoctoral students in cellular and molecular immunology. Most graduate students pursuing the Ph.D. degree in immunology train in labs in the umbrella Program in Molecular and Cellular Biology (MCB) and receive their degree from the Microbiology and Immunology Department. Other immunology students matriculate through the Ph.D. programs of the Department of Physiology or the Department of Pharmacology and Toxicology. Ph.D. degrees are granted through these departments for those students. Individuals interested in applying for admission to the Program in Immunology may choose to direct their application, through the Immunology Program, to either the Department of Physiology, the Department of Pharmacology and Toxicology, or the Program in Molecular and Cellular Biology. Qualified students are notified of acceptance through the admission committees from the respective graduate Ph.D. programs. Course requirements and curriculum for students vary somewhat depending on the graduate program into which they are enrolled. As an example, during the first year, students enrolled in the Program in Molecular and Cellular Biology (http://www.dartmouth.edu/~mcb/) take an intensive three-term core course in biochemistry, cell and molecular biology genetics, immunology, and molecular pathogenesis. Students also need to complete three additional advanced-level elective courses. After the first year of three rotations in student-chosen laboratories, a thesis laboratory is determined. The MCB graduate programs utilize laboratory rotations in the first year as a means to introduce students to different investigators and their model systems. By the end of the second year, it is generally expected that a student has taken the required advanced-level courses, completed their teaching experience, and passed the qualifying exam. Prospective students should visit the individual Web sites for program requirements for the Department of Physiology (http://www.dartmouth.edu/~physiol/) and the Department of Pharmacology and Toxicology (http://dms.dartmouth.edu/pharmtox/).

Research Facilities

Most of the immunology laboratories are located at the Dartmouth-Hitchcock Medical Center (DHMC) in Lebanon, New Hampshire, on 225 acres, 2 miles from the undergraduate college and the Hanover-based elements of the Medical School. The DHMC facility has fully integrated patient care, medical education, and research activities. The immunology labs at DMS are state-of-the-art and well equipped. In addition to the standard biochemical and immunological instrumentation, specialized equipment and facilities, such as electron microscopes, an automatic protein sequenator, tissue culture facilities, fluorescence-activated cell sorters, a confocal image analysis facility, irradiation facilities, nucleic acid and peptide synthesizers, and nuclear magnetic resonance spectrometers are available. The Matthews Fuller Health Sciences and Dana Biomedical Libraries house a complete collection of journals in the biomedical sciences. Dartmouth is known for its outstanding Computational Center, which offers both regular and short courses, as well as access to terminals throughout the Medical School and the College. The facility is linked to the computer networks at Dartmouth College, including the Internet, electronic mail service, literature searching, and other network capabilities.

Financial Aid

The Program in Immunology has a training grant from the National Institutes of Health that supports predoctoral students and postdoctoral fellows. Each student enrolled in the program can be supported by the training grant or by a Dartmouth fellowship that provides a full tuition scholarship, a prepaid health insurance plan, and a competitive student stipend. The stipend for 2006–07 is $23,500 for incoming students and $24,000 for those who have completed their qualifying exam.

Cost of Study

All students currently enrolled in the program receive a scholarship for the full cost of tuition, which is $44,396 for 2006–07.

Living and Housing Costs

Dartmouth assists graduate students in arranging for housing. Single-student housing (usually off campus) costs $580 to $855 per month. Married students can rent College-owned duplex apartments for $620 to $780 per month plus heat and utilities.

Student Group

The Program in Immunology is composed of faculty participants, graduate students, and postdoctoral fellows from the Departments of Biochemistry, Medicine, Microbiology and Immunology, Pathology, Physiology, Pharmacology and Toxicology, and Surgery. The student population at DMS in the six graduate programs totals 149, large enough to allow for a variety of educational experiences but small enough to permit close faculty-student interactions.

Student Outcomes

Immunology Program graduate students receiving the Ph.D. degree from Dartmouth and postdoctoral fellows are prepared for postdoctoral or faculty positions in academia, including both primarily teaching or research-oriented colleges and universities; in research institutes; in biotech and other companies; or positions in business or law. The placement of both graduate students and postdoctoral fellows has been very successful and includes positions at the most prestigious institutes.

Location

Life in the Hanover-Lebanon area offers an attractive combination of cultural activities in a rural setting. Concerts and dramatic productions are held the year round, particularly at Dartmouth's Hopkins Center. Alpine and Nordic skiing, hiking, and lake and white-water canoeing opportunities are outstanding, as are those for running, biking, and other seasonal sports. In addition, the ocean beaches of New Hampshire and Maine are about 2 hours away, and Boston, Montreal, and New York City can be reached by car in 2, 3, and 5 hours, respectively.

The College and The Program

Dartmouth was founded in 1769 as a college committed to liberal learning. The Medical School was established in 1797. The smallest of the Ivy League institutions, Dartmouth has a long-standing tradition of close student-faculty ties, a tradition strongly endorsed by the Program in Immunology. The program is quite diversified, bringing expertise to bear on immunological questions from the areas of biochemistry, physiology, medicine, molecular biology, cell biology, and structural biology. The strength of the program lies in its commitment to the training of students by developing close personal interactions between students and faculty and staff members.

Applying

Application forms from the Program in Molecular and Cellular Biology, Department of Physiology, or Department of Pharmacology and Toxicology can be obtained from the MCB Web site at http://www.dartmouth.edu/~mcb/. Applications are reviewed beginning January 7. All applications and supporting documentation (GRE scores, transcripts, and letters of recommendation) should be received as soon as possible but not later than January 7. The Program in Immunology welcomes applications from students from minority groups who, at present, make up approximately 10 percent of the combined graduate and medical school student population. General information about admission and application forms and a list of faculty publications can be obtained from the address listed in this description.

Correspondence and Information

Marcia Ingalls, Administrative Coordinator
Postdoctoral and Graduate Student Education and Recruitment, Immunology Program
Department of Microbiology and Immunology–Borwell Building
Dartmouth Medical School
One Medical Center Drive
Lebanon, New Hampshire 03756
E-mail: immunology@dartmouth.edu
Web site: http://www.dms.dartmouth.edu/immuno

Dartmouth College

THE FACULTY AND THEIR RESEARCH

Susana N. Asin, Research Assistant Professor of Microbiology and Immunology; Ph.D., Cordoba, 1992. Cytokine and chemokine regulation of HIV-1 infection and replication in mucosal and submucosal cell populations from the female reproductive tract and in peripheral blood cells; coinfection by sexually transmitted pathogens and HIV infection in the female reproductive tract.

Richard J. Barth Jr., Associate Professor of Surgery; M.D., Harvard, 1985. Tumor immunotherapy.

Brent L. Berwin, Assistant Professor of Microbiology and Immunology; Ph.D., Wisconsin–Madison, 1999. Immune regulation and antigen trafficking by molecular chaperones and scavenger receptors.

Constance E. Brinckerhoff, Professor of Medicine and of Biochemistry; Ph.D., SUNY at Buffalo, 1968. Cell and molecular biology of connective tissue degradation in rheumatoid arthritis and cancer.

Christopher M. Burns, Assistant Professor of Medicine; M.D., Albany Medical College, 1982. Murine lupus; CD40-CD40L interactions in autoimmunity.

David J. Bzik, Professor of Microbiology and Immunology; Ph.D., Penn State, 1983. Molecular basis of pathogenesis in Apicomplexa; drug targets in pyrimidine and purine metabolism.

Jacqueline Y. Channon-Smith, Research Assistant Professor of Microbiology and Immunology; Ph.D., London, 1984. Monitoring of the immune response of human peripheral blood leukocytes to specific and nonspecific stimuli, especially to determine the accuracy of treatment during clinical trials.

Jose R. Conejo-Garcia, Assistant Professor of Microbiology and Immunology and of Medicine; M.D., Zaragoza (Spain), 1990; Ph.D., Alcala (Spain), 1998. Contribution of leukocytes to tumor vascularization and growth; tumor immunotherapy.

Ruth I. Conner, Research Assistant Professor in Microbiology and Immunology; Ph.D., Ohio State, 1988. HIV-1 mother-to-child transmission through breastfeeding; mechanisms of HIV-1 transmission in the neonatal gastrointestinal tract; inhibition of HIV-1 by lactic acid bacteria; innate antiviral immunity in human breast milk and the neonatal gastrointestinal mucosa; transmission of drug-resistant HIV-1 breast milk.

Ruth W. Craig, Professor of Pharmacology and Toxicology; Ph.D., SUNY Roswell Park, 1984. Molecular mechanisms involved in the induction of differentiation and apoptosis in myelomonocytic cells.

Joyce A. DeLeo, Professor of Anesthesiology and of Pharmacology and Toxicology; Ph.D., Oklahoma, 1988. Neuropharmacology; neuroimmunology; mechanisms leading to chronic pain with a focus on spinal neuroimmune responses, central neuroimmune activation, and neuroinflammation, using molecular, cellular, and in vivo behavior pharmacological approaches.

Marc S. Ernstoff, Professor of Medicine; M.D., NYU, 1978. Tumor immunology of prostate cancer, melanoma, and renal-cell carcinoma.

Camilo E. Fadul, Associate Professor of Medicine; M.D., Rosario (Colombia), 1979. Immunology and immunotherapy of tumors of the central nervous system.

Michael W. Fanger, Professor of Microbiology and Immunology and of Medicine; Ph.D., Yale, 1967. Immunology and immunotherapy; HIV infection of mucosal cells; tumor-associated antigens and targeted vaccines.

Roy A. Fava, Research Associate Professor of Medicine; Ph.D., Vermont, 1982. Angiogenesis; TGF-β; VEGF; arthritis; inflammation; cytokines; lymphotoxin-beta; bone growth and repair.

Steven N. Fiering, Associate Professor of Microbiology and Immunology; Ph.D., Stanford, 1990. Chromatin-based regulation of the mammalian genome; transcriptional regulation of the β-globin locus; transgenic mice.

James D. Gorham, Associate Professor of Pathology and of Microbiology and Immunology; Ph.D., 1991, M.D., 1992, NYU. Immunology of the liver; immune regulation by TGF-β; hepatic immunology; regulation of Th1 development.

William R. Green, Professor and Chair of Microbiology and Immunology; Ph.D., Case Western Reserve, 1977. Cell-mediated immunity, especially cytolytic T cells, to retrovirus-induced leukemia or immunodeficiency; viral pathogenesis; escape mechanisms from viral and tumor immunity; smallpox vaccine design.

Paul M. Guyre, Professor of Physiology and of Microbiology and Immunology; Ph.D., New Hampshire, 1979. Mechanisms of steroid regulation of immunity and inflammation; human macrophage activation; dendritic cell antigen presentation; Fc receptors; sepsis; atherosclerosis; autoimmune disease.

William F. Hickey, Professor of Pathology; M.D., Vermont, 1977. Development of inflammation in the nervous system; autoimmune diseases; immunology of the central nervous system.

Henry N. Higgs, Assistant Professor of Biochemistry; Ph.D., Washington (Seattle), 1996. Roles of lymphocyte microvilli and actin cytoskeleton in immune function and metastasis.

Alexandra L. Howell, Research Associate Professor of Medicine and of Microbiology and Immunology; Ph.D., Texas Health Science Center, 1983. HIV-1 infection of cells and tissues in the female reproductive tract; sex hormone regulation of HIV-1 infection in blood cells and cells from the reproductive tract; innate antiviral immune mechanisms in the female reproductive tract.

Lloyd H. Kasper, Professor of Medicine and of Microbiology and Immunology; M.D., Rush, 1975. Autoimmunity; experimental inflammatory bowel disease and multiple sclerosis.

John A. Kelly, Assistant Professor of Medicine; M.D., Ireland, 2003. Stat5 role in T-cell homeostasis and oncogenesis.

Mark L. Lang, Research Assistant Professor of Microbiology and Immunology; Ph.D., Dundee (Scotland), 1998. Humoral immunity and CD1d-restricted NKT cells; CD1d intracellular trafficking and antigen presentation.

John F. Modlin, Professor and Chair of Pediatrics and Professor of Medicine; M.D., Duke, 1971. Virology; vaccines; immunization policy.

Peter M. Morganelli, Research Associate Professor of Microbiology and Immunology; Ph.D., Dartmouth, 1989. Immune mechanisms in atherosclerosis; macrophage and lipoprotein metabolism; human IgG Fc receptors.

Ralph C. Nichols, Research Assistant Professor of Medicine and of Microbiology and Immunology; Ph.D., George Washington, 1991. Molecular biology; regulation of gene expression by transcriptional and posttranscriptional mechanisms; arthritis and inflammatory disease; toxicological diseases of the liver; dys-regulation of cytokines in cancer.

Randolph J. Noelle, Professor of Microbiology and Immunology; Ph.D., Albany Medical College, 1980. B-lymphocyte activation; T-helper-cell action; CD40 signaling; B-cell memory and plasma development autoimmunity.

George A. O'Toole, Assistant Professor of Microbiology and Immunology; Ph.D., Wisconsin–Madison, 1994. Biofilm formation by the bacterial pathogens *Pseudomonas aeruginosa* and *Staphylococcus aureus;* biofilm formation in soil microbes; biofilm antibiotic resistance; role of biofilms on implant infections and disease.

William F. C. Rigby, Professor of Medicine and of Microbiology and Immunology; M.D., Harvard, 1979. Posttranscriptional regulation of cytokine and CD154 (CD40 ligand) gene expression; tristetraprolin function in mRNA turnover; Von Hippel-Lindau regulation of mRNA stability.

Charles L. Sentman, Assistant Professor of Microbiology and Immunology; Ph.D., Texas Southwestern Medical Center at Dallas, 1990. Natural killer (NK) cells; role of NK receptors on NK cells and cytolytic T cells; innate immunity; immunotherapy.

Nancy A. Speck, Professor of Biochemistry; Ph.D., Northwestern, 1983. Transcriptional regulation of hematopoiesis and leukemia.

Ronald K. Taylor, Professor of Microbiology and Immunology; Ph.D., Maryland, 1984. Bacterial pathogenesis; colonization mechanisms; virulence gene regulation; vaccine and antimicrobial design.

Mary Jo Turk, Assistant Professor of Microbiology and Immunology; Ph.D., Purdue, 2001. Mechanisms of concomitant tumor immunity.

Edward J. Usherwood, Assistant Professor of Microbiology and Immunology; Ph.D., Cambridge, 1995. Interaction between murine gammaherpesvirus 68 and the immune system; viral vaccines; immunological memory.

Matthew P. Vincenti, Research Associate Professor of Medicine; Ph.D., SUNY Health Science Center at Syracuse, 1992. Interleukin-1-dependent signal transduction and activation of matrix metalloproteinase gene expression.

C. Fordham von Reyn, Professor of Medicine and of Infectious Diseases; M.D., Harvard, 1971. Vaccines for HIV-associated tuberculosis, mycobacterial skin testing; AIDS; nontuberculous mycobacteria.

Hillary D. White, Research Associate Professor of Microbiology and Immunology; Ph.D., California, Santa Barbara, 1979. Mucosal and tumor immunology in the human and murine female reproductive tract, with a focus on the regulation of cytotoxic T lymphocytes by reproduction tract sex steroid and stress hormones.

Charles R. Wira, Professor of Physiology; Ph.D., Dartmouth, 1970. Role of sex hormones in the regulation of innate and adaptive immunity in the human and rodent female reproductive tract.

Mark P. Yeager, Professor of Anesthesiology and of Medicine; M.D., McGill, 1974. Glucocorticoid control of the human systemic inflammatory response; nonlinear modeling of systemic sepsis in humans.

DARTMOUTH COLLEGE

Dartmouth Medical School
Program in Microbiology and Molecular Pathogenesis

Program of Study	Enrollment in the Program in Microbiology and Molecular Pathogenesis at Dartmouth offers the opportunity to pursue a Ph.D. degree at a leading academic institution that is located in an outstanding rural setting. The recently expanded multidisciplinary program includes faculty members from the Departments of Biochemistry, Biological Sciences, Chemistry, Microbiology and Immunology, Physiology, and Pharmacology and Toxicology in both the Medical School and the College. This provides a wide range of opportunity for study and interaction that is generally administered through the Graduate Program in Molecular and Cellular Biology (MCB). Individuals interested in applying for admission may choose to direct their application through the Graduate Program in Molecular and Cellular Biology or through a specific department. Course requirements and curriculum for students vary somewhat depending on the graduate program into which they are enrolled. As an example, during the first year, students enrolled in the MCB Program normally take a comprehensive three-term course in molecular and cellular biology, immunology, and molecular genetics. By the end of the second year, it is generally expected that a student will have taken advanced-level courses in areas such as biochemistry, biology, microbial systems and pathogenesis, molecular genetics, and immunology; three such elective courses are required. The graduate program utilizes laboratory rotations in the first year as a means to introduce students to laboratories in which they may do their thesis research.
Research Facilities	The laboratories at Dartmouth are state-of-the-art and well equipped. In addition to the College and Medical School laboratories located on the main campus, the Dartmouth-Hitchcock Medical Center has recently constructed a campus on 225 acres in Lebanon, New Hampshire, about 2 miles from the main campus. The facility has fully integrated patient care, medical education, and research activities and is one of the few entirely new medical centers in the country. In addition to standard modern instrumentation available in individual laboratories, core facilities are maintained that include specialized equipment such as electron microscopes, automatic protein and DNA sequencers, tissue culture and hybridoma facilities, fluorescence-activated cell sorter, confocal image analysis, deoxyoligonucleotide synthesizer, peptide synthesizer, DNA microarrayer and scanner, DNA chip reader, nuclear magnetic resonance spectrometers, and equipment for state-of-the-art genomic and proteomic analyses. The Dana Biomedical Library houses a complete collection of journals in the biomedical sciences. Dartmouth is known for its outstanding Computation Center, which develops and licenses cutting-edge software and offers both regular and short courses and terminals that are readily accessible throughout the Medical School and the College, including all of the laboratories. The facility provides e-mail service, literature and database searching, extensive electronic journal access, and other network facilities.
Financial Aid	Each student enrolled in the program receives a Dartmouth Fellowship that provides a full-tuition scholarship, a prepaid health insurance plan, and a student stipend. The stipend for 2006–07 is approximately $23,500 for incoming students and $24,000 for those who have completed their qualifying exams.
Cost of Study	The tuition for 2006–07 is $44,396.
Living and Housing Costs	Dartmouth assists graduate students in arranging for housing. Single-student housing (usually off campus) costs $580–$855 per month. Married students can rent college-owned duplexes. Duplex apartments rent for $620 to $780 per month plus heat and utilities.
Student Group	The graduate student population in the MCB and related graduate programs totals about 149 students, 25 admitted each year. Dartmouth has approximately 4,200 undergraduates and 1,520 students in the graduate and professional schools.
Student Outcomes	Students completing the program generally progress to postdoctoral positions in similar academic or corporate research laboratories around the country. Eventually, the majority accept faculty positions at colleges, universities, or medical schools; direct research in biotechnology laboratories; or take positions in business or law.
Location	Life in the Hanover-Lebanon area offers an attractive combination of cultural activities in a rural setting. Concerts and dramatic productions are held on a year-round basis. Alpine and Nordic skiing, hiking, and lake and white-water canoeing are outstanding, as are running, biking, and other seasonal sports. In addition, the ocean beaches of New Hampshire and Maine are about 2 hours away. Finally, when the urge to spend time in a metropolitan area arises, Boston, Montreal, and New York can be reached by car in 2, 3, and 5 hours, respectively.
The College and Medical School	Dartmouth was founded in 1769 as a college committed to liberal learning. The Medical School was established in 1797. The smallest of the Ivy League institutions, Dartmouth has a long-standing tradition of close student-faculty ties, a tradition strongly endorsed by the Graduate Program. The Hopkins Center for the Performing Arts serves as the cultural focus of Dartmouth. The center incorporates concert halls, theaters, art galleries, art studios, and crafts workshops. A wide range of subsidized musical, dance, and theatrical productions by visiting artists are sponsored by the center.
Applying	Application forms to study microbiology and molecular pathogenesis through the Graduate Program in Molecular and Cellular Biology and/or specific departments can be obtained at the address given in this description. All applications and supporting documentation (GRE scores, transcripts, and letters of recommendation) should be submitted so as to be received no later than January 7. It is the long-standing policy of Dartmouth to actively support equality for all persons regardless of race or ethnic background. No student is denied or otherwise discriminated against because of race, color, sex, religion, handicap, or national or ethnic origin.
Correspondence and Information	Dr. Ronald K. Taylor Department of Microbiology and Immunology Vail Building Dartmouth Medical School Hanover, New Hampshire 03755 Phone: 603-650-1632 Web site: http://www.dartmouth.edu/~molpath

Dartmouth College

THE FACULTY AND THEIR RESEARCH

David J. Bzik, Professor of Microbiology and Immunology; Ph.D., Penn State, 1983. Molecular basis of pathogenesis in apicomplexa; drug targets in pyrimidine and purine metabolism; attenuation of parasite virulence and vaccine design.

Ambrose L. Cheung, Professor of Microbiology and Immunology; M.D., Northwestern, 1980. Regulation of virulence determinants; stress-induced response; in vivo gene expression and interactions with host cells by *Staphylococcus aureus*.

Kathryn L. Cottingham, Associate Professor of Biological Sciences; Ph.D., Wisconsin–Madison, 1996. Aquatic community and ecosystem ecology; quantitative analysis of ecological and genomic data.

Michael W. Fanger, Professor of Microbiology and Immunology and Professor of Medicine; Ph.D., Yale, 1967. Immunology and immunotherapy; Fc receptors; HIV infection of mucosal cells; tumor-associated antigens and targeted vaccines.

Steven N. Fiering, Assistant Professor of Microbiology and Immunology; Ph.D., Stanford, 1990. Chromatin-based regulation of the mammalian genome; transcriptional regulation of the β-globin locus; transgenic mice.

James D. Gorham, Associate Professor of Pathology and of Microbiology and Immunology; Ph.D., 1991, M.D., 1992, NYU. Immunology of the liver; immune regulation by TGF-β1; regulation of Th1 development.

William R. Green, Professor and Chair of Microbiology and Immunology; Ph.D., Case Western Reserve, 1977. Cell-mediated immunity, especially cytolytic T cells, to retrovirus-induced leukemia or immunodeficiency; viral pathogenesis; escape mechanisms from viral and tumor immunity.

Mary Lou Guerinot, Professor of Biological Sciences; Ph.D., Dalhousie, 1979. Genetic regulation of the nitrogen-fixing symbiosis between rhizobia and legumes; iron regulation and gene expression in plants *(Arabidopsis thaliana)* and bacteria.

Paul M. Guyre, Professor of Physiology and of Microbiology and Immunology; Ph.D., New Hampshire, 1979. Mechanisms of steroid regulation of immunity and inflammation; human macrophage activation; dendritic cell antigen presentation; Fc receptors; sepsis; atherosclerosis; autoimmune disease.

Deborah A. Hogan, Assistant Professor of Microbiology and Immunology; Ph.D., Michigan State, 1999. Molecular interactions between bacteria and fungi; host-associated microbial communities; bacterial and fungal pathogenesis.

Alexandra Howell, Research Associate Professor of Microbiology and Immunology and of Medicine; Ph.D., Texas, 1983. Interaction between myeloid cells (monocytes, macrophages, and neutrophils) and HIV-1; targeted antigen delivery to myeloid antigen presenting cells as a novel vaccine approach; HIV-1 infection of cells and tissues of the female reproductive tract and the effect of sex hormones on infectivity.

Lloyd H. Kasper, Professor of Medicine and of Microbiology and Immunology; M.D., Rush, 1975. Autoimmunity, experimental inflammatory bowel disease, and multiple sclerosis.

Randolph J. Noelle, Professor of Microbiology and Immunology; Ph.D., Albany Medical College, 1980. β-lymphocyte activation; T–helper cell action; CD40 signaling, β-cell memory, and plasma development autoimmunity.

George A. O'Toole, Associate Professor of Microbiology and Immunology; Ph.D., Wisconsin–Madison, 1994. Biofilm formation by the bacterial pathogens *Pseudomonas aeruginosa* and *Staphylococcus aureus;* biofilm antibiotic in soil microbes; biofilm antibiotic resistance; role of biofilms on implant infections and disease.

Jeffrey Parsonnet, Associate Professor of Medicine and of Microbiology and Immunology; M.D., NYU, 1979. Immunology and pathogenesis of toxic shock syndrome and septic shock and factors influencing production of toxic shock syndrome toxin-1.

William F. C. Rigby, Professor of Medicine and of Microbiology and Immunology; M.D., Harvard, 1979. Posttranscriptional regulation of cytokine and CD154 (CD40 ligand) gene expression; tristetraprolin function in mRNA turnover; polypyrimidine tract binding protein function; Von Hippel-Lindau regulation of mRNA stability.

Karen A. Skorupski, Research Associate Professor of Microbiology and Immunology; Ph.D., Rutgers, 1988. Transcriptional regulation of bacterial virulence gene expression; identification of new targets for antimicrobial design.

Nancy A. Speck, Professor of Biochemistry; Ph.D., Northwestern, 1983. Transcriptional regulation of hematopoiesis and leukemia.

Paula Sundstrom, Professor of Microbiology and Immunology; Ph.D., Washington (Seattle), 1986. Molecular pathogenesis of the AIDS-related fungal pathogen, *Candida albicans;* adherence; morphological transitions; virulence gene regulation; antifungal drug design.

Surachai Supattapone, Assistant Professor of Biochemistry; Ph.D., John Hopkins, 1988; D.Phil., Oxford, 1991; M.D., Johns Hopkins, 1992. Molecular pathogenesis of prion diseases.

Ronald K. Taylor, Professor of Microbiology and Immunology; Ph.D., Maryland, Baltimore County, 1984. Bacterial pathogenesis; colonization mechanisms; virulence gene regulation; vaccine and antimicrobial design.

Edward J. Usherwood, Assistant Professor of Microbiology and Immunology; Ph.D., Cambridge, 1995. Interaction between murine gammaherpesvirus 68 and the immune system; viral vaccines; immunological memory.

C. Fordham von Reyn, Professor of Medicine; M.D., Harvard, 1971. Vaccines against HIV-associated tuberculosis; tuberculin skin testing; nontuberculosis mycobacteria; AIDS in Africa.

Charles R. Wira, Professor of Physiology; Ph.D., Dartmouth, 1970. Sex hormone regulation of mucosal immunity in the female reproductive tract.

DUKE UNIVERSITY

Graduate Program in Immunology

Program of Study	The Graduate School offers a program in immunology through the Department of Immunology leading to the degree of Doctor of Philosophy. Research opportunities are available in various areas of molecular and cellular immunology, including immunogenetics, immunochemistry, and tumor immunology. Formal course work is concentrated in the first three semesters and includes courses in biochemistry, cell biology, molecular biology, and genetics as well as comprehensive training in immunology. During the first year, students rotate in various laboratories of program faculty members before selecting the laboratory in which they will pursue original research. Following their formal course work, students take a qualifying examination and then concentrate on research, leading to a dissertation. Seminars, journal clubs, laboratory meetings, and an annual retreat are additional, important components of the program.
Research Facilities	The Department of Immunology is housed in modern, well-equipped research facilities. Faculty members are committed to providing students with the opportunity to use state-of-the-art techniques for molecular and cellular immunology and biomedical research. Core facilities available for research include a fully equipped fluorescence flow cytometry facility, micromolecular facilities for oligonucleotide and peptide synthesis and sequencing, a transgenic facility, a microscopy facility with access to confocal and high-resolution transmission and scanning electron microscopes, and computing facilities.
Financial Aid	All applicants for admission are automatically considered for institutional and departmental fellowships and traineeships. All students are supported by the Basic Immunology Training Grant, Duke Graduate School, research grants, and private foundations.
Cost of Study	Tuition, fees, and health insurance for 2005–06 were approximately $34,000. A stipend of $23,000 was provided to defray the cost of living expenses.
Living and Housing Costs	Off-campus housing, including single rooms, apartments, and houses, vary in price from $500 to $1000 per month, depending on location and furnishings. Many apartments are available within walking distance of the University.
Student Group	At Duke there are more than 12,000 students enrolled in degree programs, including approximately 6,000 students in graduate and professional programs. Of the 34 students currently enrolled in the Graduate Program in Immunology, 3 are members of underrepresented minority groups, 13 are international students, and 20 are women.
Student Outcomes	The majority of Ph.D. graduates have elected to do postdoctoral work in academic centers. A smaller group of recent graduates have found postdoctoral work in industrial laboratories. After postdoctoral training, most graduates enter academic positions; a few have found positions in industry and in government.
Location	With a population of 227,000, Durham is located midway between the Atlantic Ocean and the Appalachian Mountains, in a metropolitan area of more than 1 million people. The climate is temperate, and the cultural life of the community is well developed. Twelve miles away is the noted Research Triangle Park, where the North Carolina Biotechnology Center, the Environmental Protection Agency, the National Institute of Environmental Health Sciences, the Triangle Universities Computation Center, the Microelectronics Center of North Carolina, and numerous private enterprises are located. Other major universities in the Triangle area include the University of North Carolina at Chapel Hill and North Carolina State University at Raleigh.
The University	Duke University, with its Gothic quadrangles and 8,500 acres of pine forest, encompasses a complex of modern research facilities. Unlike medical centers in many other areas, Duke Medical Center, Duke Clinic, and the Duke Hospital are located on the main college campus, which encourages interdisciplinary interaction at many levels. The Department of Immunology has a dual role as a member department of the Graduate School and as a preclinical basic science department in the Medical School. Thus, research and educational opportunities are available in both basic academic and clinical areas. The Medical Center is the site of the Comprehensive Cancer Center.
Applying	Applicants should have strong undergraduate training in biology, genetics, and biochemistry. Graduate Record Examinations scores (General and Subject Tests) are required from all applicants. The TOEFL is required of all international applicants whose native language is not English. Applications should be received by December 31. Information describing the program in detail is available.
Correspondence and Information	Michael Krangel Director of Graduate Studies Department of Immunology Box 3010 Duke University Medical Center Durham, North Carolina 27710 Phone: 919-684-4985 Fax: 919-684-8982 E-mail: dgs-immunology@duke.edu Web site: http://immunology.mc.duke.edu

Duke University

THE FACULTY AND THEIR RESEARCH

Soman Abraham, Professor; Ph.D., Newcastle upon Tyne (England), 1981. Mast cell modulation of innate and adaptive immunity.
Rebecca H. Buckley, Professor; M.D., North Carolina at Chapel Hill, 1958. Genetic disorders of the immune system; bone marrow transplantation; human T- and B-cell ontogeny.
Lindsay Cowell, Assistant Professor; Ph.D., North Carolina State, 2000. Computational methods for study of the immune system.
Michael M. Frank, Professor; M.D., Harvard, 1960. Effector function of antibody and complement in immune-mediated damage and inflammation.
Eli Gilboa, Professor; Ph.D., Weizmann (Israel), 1977. Biology of dendritic cells and dendritic cell–based cancer vaccines.
Michael Dee Gunn, Associate Professor; M.D., Texas Health Science Center at Dallas, 1983. Lymphoid chemokines, dendritic cells, and the regulation of immune responses.
Barton F. Haynes, Professor; M.D., Baylor College of Medicine, 1973. Study of human T-cell maturation and thymus biology; biology of human retroviruses; autoimmune disease.
You-Wen He, Assistant Professor; M.D., Ph.D., Miami, 1996. T-lymphocyte development, apoptosis, and tumorigenesis.
Garnett Kelsoe, Professor; D.Sc., Harvard, 1979. Lymphocyte activation and selection by antigen; V(D)J rearrangement and hypermutation.
Thomas B. Kepler, Professor; Ph.D., Brandeis, 1989. Computational immunology, with emphasis on antigen receptor diversification; leukocyte communication and induced functional reorganization of the immune system.
Motonari Kondo, Assistant Professor; M.D., Ph.D., Tohoku (Japan), 1995. Molecular mechanisms of lineage commitment in lymphocyte development.
Michael S. Krangel, Associate Professor; Ph.D., Harvard, 1982. Rearrangement and expression of T-cell receptor genes; chemotactic cytokines.
Robert J. Lefkowitz, Professor; M.D., Columbia, 1966. Isolation and identification of neural receptor sites; interaction of neurotransmitters and receptors.
David S. Pisetsky, Professor; M.D., Ph.D., Yeshiva (Einstein), 1973. Genetics of autoimmune diseases; immunoglobulin V-region genetics; B-cell activation.
Jonathan Poe, Assistant Research Professor; Ph.D., East Tennessee State, 1997. B-cell–restricted coreceptors CD19 and CD22 influence on B-cell development; function and malignant transformation in vivo using mouse models of autoimmunity and cancer.
Jeff Rathmell, Assistant Professor, Ph.D., Stanford, 1996. Study of lymphocyte cell death and metabolism.
Marcella Sarzotti-Kelsoe, Associate Research Professor; Ph.D., Torino (Italy), 1980. Neonatal development of T-cell responses to virus and DNA vaccines; stem-cell development of T and B cells in humans with SCID.
Thomas F. Tedder, Professor and Chair; Ph.D., Alabama at Birmingham. Structure and function of human leukocyte adhesion molecules; B-lymphocyte activation.
J. Brice Weinberg, Professor; M.D., Arkansas Medical Sciences, 1969. Inflammation; host defense; nitric oxide.
Kent J. Weinhold, Professor; Ph.D., Thomas Jefferson, 1979. Vaccine strategies and CD8 effector-cell functions in AIDS and cancer.
Yiping Yang, Assistant Professor; M.D., Zhejiang Medical (China), 1985; Ph.D., Michigan, 1993. T-cell tolerance, T-cell memory and antigen-defined cancer immunotherapy.
Weiguo Zhang, Associate Professor; Ph.D., Yeshiva (Einstein), 1994. T-cell receptor signal transduction.
Xiaoping Zhong, Assistant Professor, Ph.D., Duke (1997). T-cell receptor signaling in T-cell development and function; innate immunity.
Minghua Zhu, Assistant Research Professor; Ph.D., Yeshiva (Einstein), 1995. T-cell receptor signal transduction.
Yuan Zhuang, Associate Professor; Ph.D., Yale, 1989. Gene regulation in the development and differentiation of B and T lymphocytes.

GEORGETOWN UNIVERSITY

Medical Center
Department of Microbiology and Immunology

Programs of Study	The Department of Microbiology and Immunology offers several degrees and certificates, including a Ph.D. or a general Master of Science degree in microbiology and immunology, a Master of Science degree in biomedical science policy and advocacy, a Master of Science degree in biohazardous threat agents and emerging infectious diseases, an interdisciplinary graduate certificate in biodefense and public policy, and a new online graduate certificate in biohazardous threat agents and emerging infectious diseases. Both certificates are graduate programs fully recognized by the University and operating on a semester schedule. All M.S. and certificate programs are available part-time.

Predoctoral students can specialize in various areas of contemporary biomedical microbiology and immunology. Research activities of the Department span a broad spectrum of subdisciplines and utilize cellular, biochemical, and molecular approaches to current problems in microbiology and immunology.

The general M.S. allows students to enhance their background in microbiology, immunology, and various aspects of modern molecular genetics. Laboratory experience in specialized techniques is also available. The program is intended to enhance career advancement in academia and industry or prepare a student for a future in research.

The M.S. in biomedical science policy and advocacy provides core courses in the biomedical sciences (microbiology, immunology, virology, and biochemistry), introduces the topics of science policy and advocacy, and presents students with in-depth studies in various science policy issues via the Science, Technology, and International Affairs (STIA) Division of the Edmund E. Walsh School of Foreign Service. This is an interdisciplinary program that invites applicants from all undergraduate backgrounds.

The M.S. in biohazardous threat agents and emerging infectious diseases provides training in basic sciences and its application to the field of biohazardous threat agents and emerging infectious diseases. The program curriculum focuses on the scientific background, potential dissemination, successful detection, and treatment approaches to biohazardous threat agents and emerging infectious diseases.

The graduate certificate in biodefense and public policy is a joint effort between the Georgetown University (GU) Medical Center and the Georgetown Public Policy Institute. The certificate's combination of scientific and administrative disciplines enables graduates to provide leadership not only in the technical aspects of emergency response but also in the crucial policy and administrative challenges that arise in planning response strategies and responding to incidents. Courses are given for credit; students who wish to continue their studies in compatible M.S. or Ph.D. programs at Georgetown are eligible to receive 15 credits toward the full graduate degree.

The online graduate certificate in threat agents and emerging infectious diseases is an outgrowth of the Master of Science program, having the same focus and consisting of about 50 percent of the M.S. program's course work taught by the same faculty members. The certificate aims to prepare current and future professionals in defense, security, and health industries by enhancing their understanding of the top biohazardous threat agents and presenting essential information about the microbiological and human impact, as well as biothreat surveillance and homeland security. Courses are given for credit; students who wish to continue their studies in compatible M.S. or Ph.D. programs at Georgetown are eligible to receive 12 credits toward the full graduate degree.

Research Facilities	The Department occupies 12,000 square feet in the Medical-Dental Building and has access to additional facilities in the Medical Center. In addition, the Department has a library/conference room, general purpose laboratories, and offices, as well as well-equipped research laboratories that contain laminar-flow hoods, biosafety cabinets, beta and gamma spectrometers, spectrophotometers, plate readers, ultracentrifuges, controlled-environment incubators, cold rooms, tissue culture facilities, luminescent microscopes, and more. The Department is adjacent to the Dahlgren Medical Library and the Research Resource Facility.
Financial Aid	Some Ph.D. students qualify for a stipend of $23,772 for the 2006 fiscal year; on average, 2 students are awarded stipends annually. Information about loans and scholarships from external sources is available from the Graduate School's Office of Financial Aid (http://finaid. georgetown.edu). Georgetown University also collaborates with the CONACYT organization to bring Mexican citizens to the University and provides financial assistance to cover tuition, housing, and medical insurance during their stay. Interested students must meet the usual admission requirements for the program.
Cost of Study	Full-time tuition for 2006–07 is $31,512 per year plus fees. Students enrolled for fewer than 12 credits per semester are charged $1313 per credit. Students who have received approval to enroll for more than 12 credits in a semester are charged no more than the full-time tuition rate.
Living and Housing Costs	On-campus housing is primarily for undergraduate students. Exceptions are made for graduate students with disabilities through Disability Support Services. The Georgetown University Auxiliary Services provides off-campus housing referral services to all students. The Georgetown University Office of International Student Services assists all international students with finding housing when they arrive.
Student Group	The Department had 111 graduate students, including 16 Ph.D. candidates and 95 M.S. students, in 2005–06. There are nearly 13,000 students enrolled in the University; about 6,000 are candidates for advanced degrees.
Student Outcomes	Recent graduates of the Department of Microbiology and Immunology are employed by Duke University, Georgetown University, Stanford University, the National Institutes of Health, Northrop Grumman Corporation, and Science Applications International Corporation (SAIC), among numerous others. Graduates have also gained admission to several competitive medical schools and Ph.D. programs.
Location	Beyond Washington's most familiar vistas on Capitol Hill, the city includes a lively urban center. Casual cafés and upscale bistros line the trendy streets of Georgetown, while the downtown district offers a host of new restaurants. Kayakers tackle the Potomac River as it winds past the elegant marble tributes to America's great leaders.
The University and The Department	Begun in 1789, Georgetown today is a thriving student-centered, international research university. Distinguished faculty members combine scholarship and professional experience as diplomats, public health officials, economists, authors, and scientists to foster every student's full potential.

The Department of Microbiology and Immunology is concerned with the study of the immunological and biological properties of bacteria, viruses, and fungi. Its research activities span a broad spectrum of subdisciplines and utilize cellular, biochemical, and molecular approaches to study current problems in microbiology and immunology.

Applying	The deadlines for M.S. and certificate programs are July 1 for the fall semester and November 1 for the spring semester. The only deadline for Ph.D. programs for the summer is February 2. Exceptional applications are considered after these dates on a case-by-case basis. Students are selected based on their undergraduate record, GRE scores, personal statement, recommendations, and, when possible, interviews with members of the Department. Applicants to the master's program in biomedical science policy and advocacy need not have a strong scientific background, as the program provides an opportunity for students to take introductory courses in microbiology and immunology.
Correspondence and Information	Eugenia Pyntikova, Program Coordinator Department of Microbiology and Immunology Medical-Dental Building, Room SW311 Georgetown University 3900 Reservoir Road Washington, D.C. 20057-1440 Phone: 202-687-3422 Fax: 202-687-1800 E-mail: ep72@georgetown.edu Web site: http://microbiology.georgetown.edu

Georgetown University

THE FACULTY AND THEIR RESEARCH

Richard A. Calderone, Professor and Chair; Ph.D., West Virginia. Recognition of mammalian cells and signaling events by the human pathogen *Candida albicans*.

John L. Casey, Associate Professor; Ph.D., Berkeley. Molecular biology of viral hepatitis.

Ronald L. Cihlar, Professor; Ph.D., Massachusetts. Pathogenesis and basic biology of the fungus *Candida albicans*.

Michael F. Cole, Professor; Ph.D., London. Regulation of commensal and exogenous pathogenic bacteria at the mucosal surface by the secretory immune response and the ontogeny of the secretory immune system.

Paul J. Cote, Professor; Ph.D., Georgetown. Immunobiology and pathogenesis of chronic WHV infection and disease; virologic, serologic, hepatic, and cellular and molecular immunologic responses in neonatal WHV infection; antiviral and immunotherapy of chronic WHV infection.

William Fonzi, Associate Professor; Ph.D., Texas A&M. Identification and characterization of the genes and gene products that contribute to the virulence of *Candida albicans*, using molecular genetic techniques.

John L Gerin, Professor; Ph.D., Tennessee. Molecular and cellular biology of the hepatitis B and delta viruses.

Herbert Herscowitz, Professor; Ph.D., Hahnemann. Utilization of the immune system to augment novel therapeutic approaches against various forms of cancer.

Maja Maric, Assistant Professor; Ph.D., NYU. Antigen processing and presentation of antigens in the context of MHC class II and its effects on autoimmune diseases, tumors, and infections; role of gamma interferon inducible–lysosomal thiol reductase (GILT) in antigen processing.

R. Pad Padmanabhan, Professor; Ph.D., Wayne State. Molecular mechanisms of RNA replication and polyprotein processing in Dengue viral life cycle.

Leonard Rosenthal, Professor; Ph.D., Kansas State. Transforming and anti-transforming activities of human herpesviruses HCMV, HHV-6, and HHV-8 and their association with AIDS/KS.

Non–Tenure Track Faculty

Ana Albors, Assistant Professor; Ph.D., Valencia (Spain). *Candida albicans*.

Gail Feser, Research Associate. *Candida albicans*.

Brent E. Korba, Professor; Ph.D., Maryland. Antiviral and immunotherapy of chronic WHV infection; hepadnavirus infection of non-hepatic cells and tissues.

Dongmei Li, Assistant Professor; Ph.D., Beijing. *Candida albicans*.

Philip Posch, Instructor; Ph.D., Georgetown. Impact of cytokine gene polymorphisms on cytokine expression levels; hematopoietic stem cell transplantation and cancer.

Secondary Faculty

Joseph Bellanti, M.D., SUNY at Buffalo.

Armead Johnson, Ph.D., Baylor.

Stephen Peters, Ph.D., George Washington.

Richard Schlegel, Professor and Interim Chair, Pathology; M.D., Ph.D., Northwestern. Biology of oncogenic human papillomaviruses, ranging from analysis of viral oncoproteins to development of molecular vaccines and therapeutics.

Adjunct Faculty

Michael P. Bray, M.D., Dartmouth; M.P.H., Johns Hopkins.

Sheldon Brodel, Ph.D., Maryland, Baltimore.

Jeffrey Collmann, Ph.D., Adelaide (Australia).

William Daddio, Ph.D., Notre Dame.

Lawrence Kerr, Ph.D., Vanderbilt.

James Kvach, Ph.D., Miami (Ohio).

Daniel Lucey, M.D., M.P.H., Dartmouth; FACP.

JoAnn Rinaudo, Ph.D., Toronto.

Norman Schaad, Ph.D., California, Davis.

Joseph Timpone, M.D., Georgetown.

James Wilson, M.D., Cincinnati.

Catherine Woytowicz, Ph.D., California, Riverside.

THE GEORGE WASHINGTON UNIVERSITY

Graduate Program in Microbiology and Immunology

Program of Study	The mission of the Graduate Program in Microbiology and Immunology within the Institute for Biomedical Sciences at The George Washington University is to provide a flexible, rigorous training program, so that its graduates may become outstanding independent research scientists and teachers. It is a multidisciplinary program comprising faculty members at The George Washington University as well as at academically affiliated institutions such as Children's National Medical Center, The Institute for Genomic Research, and the National Institutes of Health. Current research strengths and training opportunities include the study of host-pathogen relationships, inflammation, vaccine development, T lymphocyte development and activation, cancer immunology, molecular parasitology, immunoparasitology, molecular retrovirology (HIV/AIDS), and microbial genomics and proteomics. The Department of Microbiology, Immunology and Tropical Medicine at The George Washington University is also the home of the Human Hookworm Vaccine Initiative, a public private partnership with the Sabin Vaccine Institute, and several investigators maintain active international collaborations with research laboratories in Brazil, Honduras, Panama, and China.

The program leading to a Ph.D. degree is designed to permit flexibility. During the first year, students take the core curriculum required of all students, including Macromolecular Interactions: Proteins, Macromolecular Interactions: Nucleic Acids and Information Processing, Cell Biology, and two electives offered by the graduate programs within the Institute for Biomedical Sciences. In order to gain expertise in experimental research and to familiarize themselves with the research interests of the faculty members, students are required to rotate through three laboratories. There is also a series of mini-courses to develop skills for careers in science.

Upon completion of core courses and laboratory rotations, the student selects a degree program as well as a research mentor. A research advisory committee, consisting of the research adviser and two or three additional faculty members, guides the student through the completion of the dissertation. Students who choose the microbiology and immunology program have the option to take a variety of advanced courses in microbiology and/or molecular and cellular immunology, as well as electives offered by other programs within the University, during the second year of study. Typically during the summer after the second year, students undertake their comprehensive exams, are advanced to Ph.D. candidacy, and focus on completing their thesis research project.

Research Facilities Extensive research and core facilities are available in well-equipped faculty laboratories housed in GW's Medical Center, Children's Research Institute of Children's National Medical Center, The Institute for Genomic Research, and selected laboratories at the National Institutes of Health (NIH). The University's Gelman Library and Himmelfarb Health Sciences Library, the National Library of Medicine, the Library of Congress, and numerous government agencies and other university libraries are available to students. Innovative research collaborations are available with the laboratories of NIH investigators affiliated with the program, and students are invited to attend the numerous research seminars and colloquia conducted throughout the area.

Financial Aid Individual fellowships for $23,000 plus tuition award waivers are available to entering students. Advanced students are supported by research grants and/or other funding sources.

Cost of Study The cost of tuition for the 2006–07 academic year is $970 per credit hour and the cost of student association fees is $1 per credit hour. Tuition and fees are typically provided through the Institute for Biological Sciences and/or participating institutions.

Living and Housing Costs University housing is not generally available to graduate students. Information on off-campus housing is available through the Office of Campus Life, which hosts several apartment-hunting weekends during the summer. The cost of living in the Washington area is comparable to that of other major metropolitan areas.

Student Group The total on-campus student body includes 10,274 full-time and 8,738 part-time students. The Columbian College and Graduate School of Arts and Sciences enrolls 1,806 students (579 in the Ph.D. program), including approximately 170 in biomedical sciences.

Location The University benefits from the abundant cultural, historical, and educational offerings of metropolitan Washington, D.C. The University is close to the National Institutes of Health (including the National Cancer Institute), the National Library of Medicine, the Smithsonian Institution, the Food and Drug Administration, and the Environmental Protection Agency.

The University The George Washington University, chartered by Congress in 1821, is private and nonsectarian. It holds regional accreditation from the Middle States Association of Colleges and Schools and has received professional recognition for specific programs. Diversified offerings at all levels of the University associate it with the people and activities of many organizations that are exclusive to the Washington area.

The campus, located four blocks west of the White House, is a reflection of the varied character of the surrounding area, with a mixture of large modern buildings, traditional town houses, and classroom and dormitory buildings. All the major facilities of the University are located on campus, including the University hospital and medical school complex. A safe, clean, and modern Metro-rail system connects the Medical Center with urban Washington, D.C.; suburban Virginia and Maryland; and the NIH.

Applying Interested students should write to the University for a brochure detailing the program and for applications for admission and financial aid. Undergraduate requirements include at least a B average and courses in general and organic chemistry, general biology, general physics, and calculus. Minor deficiencies may be removed during the summer and first semester after acceptance. Acceptable scores on the Graduate Record Examinations must be submitted. A personal interview is highly recommended. The deadline for applications is March 1. Applicants interested in fellowship support must submit a completed application for admission by January 15. International students must apply by June 1; TOEFL scores are required.

Correspondence and Information

Microbiology and Immunology Program
The Institute For Biomedical Sciences
Ross Hall, Suite 605
The George Washington University
2300 Eye Street, NW
Washington, D.C. 20037
Phone: 202-994-2179
Fax: 202-994-0967
E-mail: gwibs@gwu.edu
Web site: http://www.gwumc.edu/ibs/

For application and University bulletins:

Graduate Admissions
Columbian College of Arts and Sciences
Phillips Hall, Room #107
The George Washington University
801 22nd Street NW
Washington, D.C. 20052
Phone: 202-994-6210
Fax: 202-994-6213
E-mail: askccas@gwu.edu
Web site: http://columbian.gwu.edu

The George Washington University

THE FACULTY AND THEIR RESEARCH

Jeff Bethony; Assistant Professor of Microbiology, Immunology, and Tropical Medicine; Ph.D. SUNY at Buffalo, 2000. Genetic and parasite epidemiology.

Maria Elena Bottazzi, Associate Research Professor of Microbiology, Immunology, and Tropical Medicine; Ph.D., Florida, 1995. Host-parasite relationships during hookworm disease; project manager human hookworm vaccine initiative.

Michael Bukrinsky, Professor of Microbiology, Immunology, and Tropical Medicine; M.D., Ph.D., Moscow Medical, 1984. Regulation of HIV nuclear importation; HIV and innate immunity.

Stephanie Constant, Assistant Professor of Microbiology, Immunology, and Tropical Medicine; Ph.D., York (England), 1991. Host-pathogen interactions and regulation of inflammation.

Edward DeFabo, Research Professor of Environment and Occupational Health; Ph.D., George Washington, 1974. Photoreceptor for immunosuppression; UV radiation carcinogenesis; UV effects on cellular immunity.

Ben Dickens, Research Associate Professor of Microbiology, Immunology, and Tropical Medicine; Ph.D., Florida, 1978. Inflammatory complications resulting in developmental pathology in extremely premature infants.

John Hawdon, Associate Professor of Microbiology, Immunology, and Tropical Medicine; Ph.D., Pennsylvania, 1991. Hookworm infective process, nematode growth and development, hookworm population genetics.

Robert Hawley, Professor of Anatomy and Cell Biology; Ph.D., Toronto, 1984. Regulation of hematopoietic cell development; gene therapy.

Peter J. Hotez, Professor and Chair of Microbiology, Immunology, and Tropical Medicine; M.D., Ph.D., Cornell-Rockefeller, 1987. Development of hookworm vaccine; control of neglected tropical diseases.

Fatah Kashanchi, Associate Professor of Biochemistry and Molecular Biology; Ph.D., Kansas , 1991. Genomics and proteomics of HIV-1 and HTLV-1 infected cells.

Imtiaz Khan, Professor, Microbiology, Immunology and Tropical Medicine; Ph.D., Banaras Hindu (India), 1983. Immune responses to infections by opportunistic pathogens.

Ajit Kumar; Professor of Biochemistry and Molecular Biology; Ph.D., Chicago, 1968. Regulation of viral gene trans-activation; role of cellular factors; RNA protein interactions.

David Leitenberg, Assistant Professor of Microbiology, Immunology, and Tropical Medicine, Pediatrics and Pathology; M.D, Ph.D., Iowa, 1990. Regulation of T-cell activation and differentiation; modulation of signal transduction during T cell development.

Nancy Noben-Trauth, Assistant Professor of Microbiology, Immunology, and Tropical Medicine; Ph.D., Iowa, 1992. Cytokine function in the immune response to parasites; cytokine regulation of inflammatory bowel disease.

Frances P. Noonan, Professor in Environment and Occupational Health; Ph.D., Queensland (Australia), 1977. Skin cancer; ultraviolet radiation regulation of immunity; genetic control of susceptibility to UV immunosuppression; UV effects on autoimmunity and infectious disease.

Gary L. Simon, Professor of Medicine, of Microbiology, Immunology, and Tropical Medicine, and of Biochemistry and Molecular Biology; M.D., Maryland, 1975; Ph.D., Wisconsin, 1972. HIV/AIDS pathogenesis.

L. Courtney Smith, Associate Professor of Biology; Ph.D., UCLA, 1985. Origins and evolution of the vertebrate immune system in sea urchins.

Children's National Medical Center

Anamaris M. Colberg-Poley; Professor of Pediatrics and of Biochemistry and Molecular Biology; Ph.D., Penn State Hershey Medical Center, 1980. Regulation of gene expression and protein trafficking of cytomegalovirus.

Stephan Ladish, Professor of Pediatrics and of Biochemistry and Molecular Biology; M.D., Pennsylvania, 1973. Tumor immunosuppression by gangliosides; gangliosides and metabolism.

Sasa Radoja, Assistant Professor of Pediatrics and of Microbiology, Immunology, and Tropical Medicine; Ph.D., NYU, 2001. Development and lytic function of CD8+ cytotoxic T lymphocytes.

Mary Rose, Professor of Pediatrics, and of Biochemistry and Molecular Biology; Ph.D., Case Western, 1970, Lung inflammation, asthma, and genetic regulation of mucin production.

Stanislav Vukmanovic, Associate Professor of Pediatrics and of Microbiology, Immunology and Tropical Medicine; M.D., Ph.D., Belgrade (Yugoslavia), 1991. T-cell repertoire selection; maintenance and survival of peripheral T cells.

Kanneboyina Nagaraju, Associate Professor of Pediatrics; D.V.M., Andhra (India), 1986; Ph.D., Sanjay Gandhi Postgraduate Institute of Medical Sciences (India), 1995. Mechanisms of initiation and perpetuation of autoimmune and inflammatory responses in systemic autoimmune rheumatic diseases.

The Institute for Genomic Research

Claire Fraser, Professor of Pharmacology and Physiology and Investigator, The Institute for Genomic Research; Ph.D., SUNY at Buffalo, 1981. Microbial genomics.

William Nierman; Investigator, The Institute for Genomic Research; Ph.D., Berkeley, 1979. Microbial pathogen genomics.

National Institute of Health

B. J. Fowlkes, Adjunct Associate Professor of Genetics and Immunology and Senior Investigator, Laboratory of Cellular and Molecular Immunology, NIAID; Ph.D., George Washington, 1985. T-cell differentiation in the thymus; thymus selection.

Andy Hurwitz; Principle Investigator, National Cancer Institute, Frederick, Ph.D., Yeshiva, 1994; Tumor immunology and tumor vaccine development.

Ligia Pinto, Adjunct Assistant Professor of Immunology; Ph.D., Lisbon, 1995. Immune response to HPV and HIV.

Jeffrey Schlom, Adjunct Professor of Genetics and of Immunology; Ph.D., Rutgers, 1969. Tumor immunology; monoclonal antibodies.

Pam Schwartzberg, Senior Investigator, Genetic Disease Research Branch, NHGRI; M.D., Ph.D., Columbia University, 1992. T lymphocyte signal transduction; T lymphocyte activation and development.

INDIANA UNIVERSITY–PURDUE UNIVERSITY INDIANAPOLIS

School of Medicine
Department of Microbiology and Immunology

Programs of Study

The Department offers a comprehensive program of study and research leading to the Ph.D. degree. The program emphasizes an interdisciplinary approach of molecular, cellular, and biochemical techniques as a means to solve current problems in molecular and cellular immunology, microbial pathogenesis, virology, and cancer biology. The primary goal of the Department is to prepare students for a career in basic research.

The Department participates in the Indiana University School of Medicine Biomedical Sciences Program, which offers an open-enrollment/gateway system that provides a shared first-year experience for all Ph.D. students. Students have the freedom to explore research areas through three rotations in laboratories across programs and choose entry into any of the ten Ph.D. programs at the conclusion of the first academic year. The open-enrollment system enhances the community of graduate students by offering a shared collaborative culture, a vital component of today's interdisciplinary nature of biomedical science research. Information on the gateway program can be found at http://www.medicine.iu.edu/~gradschl/.

Formal course work is typically completed by the end of the second year. At this time, students take a written and oral qualifying examination to be admitted to candidacy for an advanced degree. The remaining time in the program is spent toward developing both research and problem-solving skills. Progression to the Ph.D. normally takes four to five years.

In addition to course work and research, students further develop their scientific and communication skills through their participation in supervised teaching, student seminars, journal clubs, and representation of their own research at local and national meetings. The Department sponsors a weekly seminar series that, in conjunction with campuswide seminars, covers a diverse set of topics exposing students to national and international research investigators.

Within the Department, faculty interests are diverse, creating opportunities for a wide variety of research projects. Current topics of investigation include cancer (cancer cell biology, DNA repair, gene therapy, hematopoiesis, stem cell transplantation, tumor immunity, viral oncogenesis); immunology (autoimmunity, innate immunity, development and differentiation of immune cell function, transplantation biology); and pathogenesis (gene therapy and viral vectors, viral gene regulation, mechanisms of host-cell invasion, viral and bacterial pathogenesis). Faculty members also have affiliations with the NCI-designated Indiana University Cancer Center, Walther Oncology Center, Wells Center for Pediatric Research, and Sexually Transmitted Disease Center and training programs in cancer, diabetes, aging, and gene transfer/gene therapy.

Research Facilities

New laboratories throughout the Department are well equipped with the necessary tools for current multidisciplinary approaches to scientific investigation in microbiology and immunology. Computer facilities within the Department and on campus connect students to the Internet. Students have access to the Biochemistry Biotechnology Facility, which contains the latest equipment for DNA and protein synthesis and sequencing, image analysis, and computer sequence analysis. Genomic, proteomic, and flow cytometry facilities are available to all researchers. Transgenic and knockout mice can be generated in a modern animal-care facility located within the School of Medicine.

Financial Aid

Financial support includes a stipend, currently $23,000 annually; health insurance; and tuition. This support is provided by University fellowships, research grants, Departmental ships, and work-study programs. Numerous opportunities are available for spousal employment on campus and within the metropolitan area.

Cost of Study

Normally, all students in the doctoral program receive a full stipend, health insurance, and payment of tuition costs. Students are expected to purchase any required textbooks and pay for parking and the preparation of the thesis. Travel costs and expenses for students to represent their research at national meetings are generally available.

Living and Housing Costs

Indianapolis is a very affordable place to live and study. Apartments are available within walking distance of or a short drive from the campus. Major streets and interstate highways make the campus easily accessible.

Student Group

The Department maintains a group of approximately 35 graduate students. Typically, between 6 and 8 students are admitted each year. Their undergraduate backgrounds include degrees in chemistry, biochemistry, microbiology, and molecular and cellular biology. Recent graduates have moved on to postdoctoral research positions in academic and industrial settings. Others are primary faculty members teaching at liberal arts colleges.

Location

Indianapolis, a rapidly developing metropolitan area of more than 1 million people, has gained nationwide recognition as an attractive place to live. It is the capital city of Indiana, and the campus is located within walking distance of a vibrant downtown. The Indiana University–Purdue University Indianapolis (IUPUI) campus contains world-class athletics facilities for swimming and diving as well as for track and field. Many Olympic qualifying and collegiate championships continue to be held on campus at these facilities. Professional sports include baseball, basketball, football, hockey, and soccer. The city has one of the nation's finest orchestras and art museums. Cultural activities include touring plays, theater, and art festivals. Indianapolis is home to two recreational reservoirs, large parks, and many fine public golf courses. The popular Indianapolis 500 auto race takes place in the city every May.

The University and The Department

Indiana University is a comprehensive statewide university system, with core campuses located at Indianapolis and Bloomington. The Department is located in the Indiana University School of Medicine on the campus of IUPUI in Indianapolis. The campus also includes law, allied health, dental, and nursing schools; four hospitals; and the IUPUI undergraduate schools.

Applying

A strong background in molecular and/or cellular biology, biochemistry, microbiology, or related fields is recommended for applicants. Undergraduate courses should include basic biology, genetics, general and organic chemistry, physics, calculus, and biochemistry. The General Test of the GRE is required. Applications should be submitted by January 15 for matriculation in the fall semester. Students are selected on the basis of undergraduate grades, test scores, the personal statement, research interests, letters of recommendation, and interviews with faculty members. Information on applying to the gateway program can be found at http://www.medicine.iu.edu/~gradschl/.

Correspondence and Information

Chair—Graduate Recruitment Committee
Department of Microbiology and Immunology
Indiana University School of Medicine
635 Barnhill Drive, MS420
Indianapolis, Indiana 46202-5120

Phone: 317-274-7671
Fax: 317-274-4090
E-mail: rhaak@iupui.edu
Web site: http://www.iupui.edu/~micro/

Indiana University–Purdue University Indianapolis

THE FACULTY AND THEIR RESEARCH

Ghalib Alkhatib, Ph.D. Role of human host factors, including chemokine receptors, in HIV membrane fusion and HIV pathogenesis.

Byron E. Batteiger, M.D. Examine the fine specificity of the human humoral immunity to *Chlamydia trachomatis* genital infections.

Margaret Bauer, Ph.D. Focus on the pathogenesis of bacterial sexually transmitted diseases, with an emphasis on the host-pathogen interactions of *Haemophilus ducreyi* and *Chlamydia trachomatis* with the human host.

Janice S. Blum, Ph.D. Elucidation of the intracellular processing events and the steps that regulate antigen presentation.

Darron R. Brown, M.D. Human papillomaviruses; epidemiology of HPV infections.

Hal E. Broxmeyer, Ph.D. Hematopoiesis; factors regulating the self-renewal, proliferation, and differentiation of stem and progenitor cells.

Randy R. Brutkiewicz, Ph.D. Assembly, trafficking, and function of nonclassical MHC class I molecules.

David W. Clapp, M.D. Hematopoiesis; in utero marking of fetal hematopoietic stem cells; gene transfer.

Kenneth Cornetta, M.D. Retroviral vectors; gene transfer/gene therapy.

Alexander L. Dent, Ph.D. Role of transcription factors in the development of the immune system and lymphoid malignancies.

Mary C. Dinauer, Ph.D. Structure and regulation of phagocyte NADPH oxidase and role in host defense; regulation of neutrophil function by Rac GTPases; gene therapy of inherited hematopoietic disorders.

Xin-Yuan Fu, Ph.D. PTK/STAT signaling pathway and its roles in cell-cycle control, apoptosis, immune responses, development, and human diseases; roles of STAT factor in signal transduction and cancer; molecular mechanisms of FGFR-related skeletal disorders.

Thomas A. Gardner, M.D. Cancer gene therapy: discovery, development, modification, and testing of tumor-specific promoters and novel delivery mechanism.

Laura Haneline, M.D. Hematopoiesis; role of Fanconi anemia proteins in protecting hematopoietic stem cells from apoptosis and malignant transformation; gene transfer/gene therapy.

Johnny J. He, Ph.D. HIV pathogenesis in the central nervous system; regulation of HIV gene expression; HIV interaction with hematopoietic progenitor cells.

Meei-Huey Jeng, Ph.D. Hormonal regulation of gene expression in breast cancer and normal breast.

Chinghai Kao, Ph.D. Tissue/tumor-specific promoter-based gene therapy.

Mark H. Kaplan, Ph.D. Function of STAT proteins in the immune system.

Michael J. Klemsz, Ph.D. Regulation of gene expression in the immune system; molecular biology of ETS-domain transcription factors; antigen presentation via MHC class I molecules.

Louis M. Pelus, Ph.D. Mechanisms associated with how a novel chemokine rapidly induces mobilization of hematopoietic stem cells from bone marrow to peripheral blood; role of inhibitor of apoptosis proteins in cancer cell growth; progression of cells through the cell cycle.

Ann Roman, Ph.D. Pathobiology of human papillomaviruses; differentiation-dependent regulation of viral gene expression; viral perturbation of cellular function; viral transformation.

Martin L. Smith, Ph.D. p53 and cancer; role of p53 in nucleotide excision repair.

Stanley M. Spinola, M.D. Pathogenesis and host response of *Haemophilus ducreyi* in an experimental model of human infection.

Edward F. Srour, Ph.D. Characterization and biology of human hematopoietic stem cells; transplantation.

David S. Wilkes, M.D. Role of accessory cells and humoral immunity in lung allograft rejection.

Frank Yang, Ph.D. Use *Borrelia* transmission between ticks and mammals as a model system to study vector-pathogen and host-pathogen interactions.

THE JOHNS HOPKINS UNIVERSITY

School of Medicine
Graduate Program in Immunology

Program of Study	The Graduate Program in Immunology, in cooperation with the Department of Molecular Biology and Genetics in the School of Medicine, offers a program of study leading to the Ph.D. degree in immunology. Candidates are not accepted to work for the M.A. degree. Formal course work is concentrated in the first year and includes several required courses in the School of Medicine's Graduate Program in Biochemistry, Cellular and Molecular Biology, emphasizing the relevance of basic molecular and cellular biology to immunology. The immunology section of the curriculum includes basic immunology and an advanced course in molecular immunology in which leaders in specific areas of research are invited to discuss topics of current interest with students. There is also a weekly seminar series that attracts immunologists of international stature. After students have completed the first year of formal course work, original research leading to a dissertation is the major feature of the training program. For the dissertation, students can select from a wide range of problems under study, including the mechanisms of antigen recognition by T cells expressing the alpha-beta or gamma-delta receptors; the pathogenesis of AIDS and AIDS vaccine development; the structure and function of proteins encoded by genes from the major histocompatibility complex; the biochemistry of lymphocyte activation; T-cell development and T-cell activation; leukocyte chemotaxis and the mechanisms of tumor-cell destruction by T cells, macrophages, and macrophage-derived products; immunoglobulin gene rearrangement and mutation and B-lymphocyte development; immunologic approaches to the treatment of cancer; dendritic cells and their role in antigen presentation; mechanisms of transplant rejection; programmed cell death and its relationship to autoimmune disease; antigen processing; and mechanisms of immunologic tolerance. Most students complete their studies in four to six years.
Research Facilities	The program is committed to providing students with the opportunity to use state-of-the-art techniques in molecular immunology. These facilities include core units for HPLC/FPLC, peptide sequencing and synthesis, oligonucleotide synthesis, PCR instrumentation, a fluorescence polarization photometer, a laser fluorescence photobleaching recovery instrument, a surface plasmon resonance instrument, standard biochemical and tissue culture facilities, and a fully equipped fluorescence flow cytometry core facility.
Financial Aid	Candidates accepted into the program are offered support that provides payment of tuition, medical insurance, and a stipend; in 2006–07, the stipend is $25,200 per year.
Cost of Study	Tuition for 2006–07 is $32,000 per year. Applicants accepted to the program have complete tuition coverage.
Living and Housing Costs	Single rooms and suites are available through the School of Medicine at costs that range from $600 to $700 per month in 2006–07. Off-campus housing is available within a few miles of the School at generally reasonable rates.
Student Group	The goal of the immunology training program is to train the next generation of immunologists to contribute to the generation of new knowledge on the basic mechanisms of the immune system and apply this knowledge to the understanding and treatment of disease. This is accomplished by selecting and supporting qualified trainees and providing relevant course work, with the participation of highly interactive faculty members who are accomplished researchers in the field of immunology. Students are an integral part of the general graduate medical environment, which includes nearly 450 Ph.D. students enrolled in the basic science departments of the School of Medicine. Five hundred medical students are also enrolled in the School of Medicine, and a comparable number of graduate students are enrolled in the adjacent School of Hygiene and Public Health.
Location	Baltimore and its suburbs have a population of approximately 1.5 million people. The city, which is located on an arm of Chesapeake Bay, has gained national attention for the success of its urban revitalization programs. The city is the site of a beautiful waterfront development that includes many charming shops, restaurants, and hotels; a science center; an extremely popular baseball stadium; and the National Aquarium. The city's cultural attractions are extensive and include a theater for traveling Broadway shows, a repertory theater, an opera company, a symphony orchestra of national reputation, and several museums and major art galleries. In addition, Baltimore is 1 hour from Washington, D.C., and 3 hours from New York City. Year-round opportunities for outdoor recreation are provided by the Chesapeake Bay; the surrounding countryside; the mountains of western Maryland, Pennsylvania, and Virginia; and the Atlantic Coast.
The University and The School	Johns Hopkins University School of Medicine was opened in 1893. The University proper (founded in 1876) was the first American institution to place primary emphasis upon graduate study. As a result of this tradition, there is a very high proportion of graduate students in all divisions of the University.
Applying	Students are normally admitted in September. The bachelor's degree from a qualified college or university is required. Applicants are expected to have had training in organic and inorganic chemistry, general biology, physics, and calculus. Courses in immunology, biochemistry, and molecular biology are recommended but are not required for admission. Underrepresented minority students are encouraged to apply. Applications should include all scores from the Graduate Record Examinations (including a Subject Test in biology or chemistry), transcripts of undergraduate grades, at least two letters of recommendation, and a statement of personal career objectives. Personal interviews at the University are recommended. The deadline for completed applications is January 10.
Correspondence and Information	For an application: Office of Graduate Student Affairs Johns Hopkins University School of Medicine 1830 Building, Suite 2-107 Baltimore, Maryland 21205 Phone: 410-614-3385 E-mail: grad_study@jhmi.edu For specific information about the Graduate Program in Immunology: Graduate Program in Immunology Johns Hopkins University School of Medicine Broadway Research Building 733 North Broadway, Suite 631 Baltimore, Maryland 21205 Phone: 410-955-2709 Fax: 410-955-0964 E-mail: ajames@jhmi.edu Web site: http://www.med.jhu.edu/gradweb/immunology/

The Johns Hopkins University

THE FACULTY AND THEIR RESEARCH

Mario Amzel, Ph.D., Professor (Biophysics). Recognition of flexible peptide; mimicry of antigen by anti-idiotypic antibodies; affinity maturation.

William Baldwin, M.D., Ph.D., Professor (Pathology). Complement mediated injury in allograft rejection.

Bruce Bochner, Professor (Medicine). Mechanisms of recruitment, activation, and survival of eosinophils, basophils, and mast cells in allergic inflammation; Siglec immunobiology; role of IgE in human allergic and anaphylactic responses.

Lieping Chen, M.D., Ph.D., Professor (Dermatology). T-lymphocyte costimulation and coinhibition; tumor immunology; immunotherapy.

Linzhao Cheng, Ph.D., Associate Professor (OB-GYN, Institute for Cell Engineering). Immune properties of human stem cells and transplantation.

Andrea Cox, M.D., Ph.D., Assistant Professor (Oncology). Cellular immune response to hepatitis C virus infection in order to better understand viral mechanisms of immune evasion and to reverse those mechanisms to enhance immunity to the virus.

Stephen Desiderio, M.D., Ph.D., Professor (Molecular Biology and Genetics, Institute for Cell Engineering). Molecular mechanisms of lymphocyte differentiation and activation; immunoglobulin and T-cell receptor gene assembly.

Charles Drake, M.D., Ph.D., Assistant Professor (Oncology). Immune response to cancer at a cellular and transcriptional level; using these data to augment the immune response to tumors in both experimental models and patients with cancer.

Michael Edidin, Ph.D., Professor (Biology, Institute for Cell Engineering). Biophysical characteristics and immunologic functions of MHC molecules; scanning near-field microscopy; analysis of membrane domains.

Ephraim Fuchs, M.D., Associate Professor (Oncology). Developing clinically feasible strategies for breaking immunologic to cancer, resulting in the regression of advanced disease.

Patricia Gearhart, Ph.D., Senior Investigator (Laboratory of Gerontology). Somatic mutation in immunoglobulin variable genes; DNA repair; B-cell differentiation.

Diane Griffin, M.D., Ph.D., Professor and Chair (Molecular Microbiology and Immunology). Pathogenesis of viral infections.

Abdel Hamad, D.V.M., Ph.D., Assistant Professor (Pathology). Role of the Fas pathway in regulating the cellular and molecular basis of autoimmune diabetes; development and function of intraepithelial double negative alpha/beta T cells and their role in mucosal tolerance.

Alan Hess, Ph.D., Professor (Oncology). Immunology of bone-marrow transplantation and graft-versus-host disease.

Elizabeth Jaffee, M.D., Professor (Oncology). Analysis of antitumor immune responses against human tumors; identification of the targets of tumor-specific cytotoxic T cells.

Abraham Kupfer, Ph.D., Professor (Cell Biology, Institute for Cell Engineering). Cell activation by antigen presenting cells: The structure and function of the immunological synapse and its roles in the induction of activation or tolerance.

Daniel Leahy, Ph.D., Professor (Biophysics and Biophysical Chemistry). Three-dimensional structure of proteins involved in cell-cell and cell-matrix interaction and signaling.

Hyam Levitsky, M.D., Professor (Oncology). Vaccine development; mechanisms of antigen-specific tolerance; identification of tumors.

Li Lin, Ph.D., Principal Investigator (Laboratory of Cardiovascular Science). NF-κB and angiotensin II signaling in cardiovascular system.

Drew Pardoll, M.D., Ph.D., Professor (Oncology). Dissection of the antitumor immune responses; cancer gene therapy vaccine engineering; T-cell development and tolerance.

Joel Pomerantz, Ph.D., Assistant Professor (Biological Chemistry, Institute for Cell Engineering). Signal transduction in normal and cancer cells of the immune system.

Jonathan Powell, M.D., Ph.D., Assistant Professor (Medicine). Signaling molecules and induction of T-cell anergy.

Stuart Ray, M.D., Associate Professor (Medicine). Immune response to hepatitis C virus; identification of T-cell epitopes.

Noel Rose, M.D., Professor (Pathology). Cellular and molecular mechanisms of apoptosis; relevance to autoimmunity.

Antony Rosen, M.D., Professor (Medicine). Mechanisms of autoimmunity with particular emphasis on the roles of apoptosis.

Scheherazade Sadegh-Nasseri, Ph.D., Associate Professor (Pathology). Molecular mechanisms involved in presentation of antigens to T cells.

Jonathan Schneck, M.D., Ph.D., Professor (Pathology). Design of novel compounds that regulate T-cell responses and structural analysis of proteins central in the generation of immune response.

Ranjan Sen, Ph.D., Chief (Laboratory of Cellular and Molecular Biology). Regulation of active and passive cell death; the role of NF-kB in linking innate and adaptive immune responses.

Robert Siliciano, M.D., Ph.D., Professor (Medicine, Howard Hughes Medical Institute). T-cell recognition of HIV; mechanisms of cytotoxicity and antigen processing in AIDS vaccine development; analysis of virus load, latency, and mechanisms of CD4 depletion in HIV infection.

Mark J. Soloski, Ph.D., Professor (Medicine). Structure and function of mouse and human MHC gene products; immune response to stress proteins; role of MHC gene products in regulating T-cell immune responses during infection and autoimmunity; function of GPI-anchored surface structures in T-cell differentiation and function.

T. C. Wu, M.D., Ph.D., Professor (Pathology). Antigen-specific vaccine development.

Fidel Zavala, M.D., Professor (Molecular Microbiology and Immunology). Characterization of mechanisms involved in the development of effector and memory CD8+ T cells against infectious pathogens, particularly malaria parasites.

Zhou Zhu, M.D., Ph.D., Associate Professor (Medicine). Immune regulation of allergic inflammation.

LOYOLA UNIVERSITY CHICAGO

Stritch School of Medicine
Department of Microbiology and Immunology

Programs of Study	The Department of Microbiology and Immunology emphasizes a program of study leading to the Ph.D. degree in the principal fields of microbiology, virology, and immunology. The core curriculum consists of courses in microbiology, virology, immunology, molecular biology, and biochemistry and is designed to train students for a career in research and teaching. Advanced course offerings include cellular and molecular immunology, microbial genetics and physiology, molecular virology, and molecular biology of oncogenesis. Major emphasis is also placed on developing communication skills. All Ph.D. candidates must pass a comprehensive qualifying examination, demonstrate achievement in independent research, and prepare and successfully defend a dissertation. Major research is being conducted in molecular biology of the immune system, immune mechanisms of host resistance, molecular and cellular aspects of lymphocyte differentiation, regulation of gene expression, virus entry, replication and pathogenesis, antiviral immunity, regulation of transcription and translation, bacterial physiology and metabolism, sporulation, microbial pathogenesis, signal transduction, microbial genetics, diagnostic microbiology, biochemistry of membrane proteins, neuroimmunology, cancer immunology, mucosal immunology, and immunodermatology.
Research Facilities	The Department of Microbiology and Immunology is modern and well equipped, containing a wide range of instrumentation and all facilities required for research. These include a flow cytometry facility, a transgenic animal facility, an animal-care facility, a macromolecular analysis facility, an electron microscopy facility, a fluorescence microscope facility, a cell-culture facility, a phosphoimager, and a spectrofluorometer, circular dichroism spectrometer. The Medical Center library contains more than 187,500 volumes relating to the biological and health sciences and receives more than 2,200 periodicals, as well as 1,400 e-journals, annually. Microfilm, computer access and search programs, and an interlibrary loan system are also available.
Financial Aid	All students accepted into the Ph.D. program receive a fellowship that provides a stipend of $22,000 and tuition remission.
Cost of Study	As covered in the Financial Aid section, all students receive a stipend and remission of tuition.
Living and Housing Costs	The cost of housing near Loyola Medical Center varies from $450 to $550 per month for a one-bedroom apartment and from $550 to $650 per month for a two-bedroom apartment.
Student Group	Loyola University has a total enrollment of approximately 15,000 students. At the Medical Center, there are about 100 graduate students and 520 medical students. Approximately 32 graduate students are enrolled in the Department of Microbiology and Immunology.
Student Outcomes	Recent program graduates have accepted positions as postdoctoral fellows at various research universities and at the National Institutes of Health (NIH). Other graduates have gone on to become research scientists in pharmaceutical companies, and still others have become educators at small colleges and universities.
Location	The Loyola University Medical Center campus is located in suburban Maywood, approximately 13 miles (20 minutes) west of downtown Chicago. The well-known museums, planetarium, aquarium, and lakeshore of Chicago are all within a half-hour drive of the campus. Chicago is the home of the Chicago Symphony, the Lyric Opera, and numerous professional sports teams. The Chicago area is surrounded by many parks and forest preserves; ski areas, lakes, and camping are available within a 3-hour drive.
The University	Loyola University Chicago is one of the oldest private universities in Illinois. The Medical Center in Maywood was opened in 1968, and the graduate and medical education programs have enjoyed vibrant growth. Graduates of the microbiology and immunology program and the programs of the other basic science departments hold prominent academic, industrial, and clinical positions throughout the nation.
Applying	Students may complete the application process online at https://applyyourself.com/?id=loyolamc. Students must submit applications by May 1 in order to be considered for financial aid. The Department requires that students submit scores from the Graduate Record Examinations prior to acceptance. Students whose native language is not English are required to submit scores from the Test of English as a Foreign Language (TOEFL) as well.
Correspondence and Information	Dr. Thomas M. Gallagher Admissions Committee Chairman Department of Microbiology and Immunology Stritch School of Medicine Loyola University Chicago 2160 South First Avenue Maywood, Illinois 60153 Phone: 708-216-3385 Web site: http://www.lumc.edu/microimmuno

Loyola University Chicago

THE FACULTY AND THEIR RESEARCH

Katherine L. Knight, Professor and Chairperson; Ph.D., Indiana, 1966. Interactions between host and commensal bacteria; molecular regulation of B-cell development and generation of antibody diversity; mucosal immune responses.

Susan C. Baker, Professor; Ph.D., Vanderbilt, 1986. Molecular virology; mechanisms of coronavirus RNA synthesis and RNA recombination; functions of proteases in viral replication.

Joseph W. Brewer, Associate Professor; Ph.D., Duke, 1995. B-cell differentiation; biology of endoplasmic reticulum (ER); signaling pathways that regulate ER homeostasis.

Manuel O. Díaz, Professor of Medicine, M.D., Uruguay, 1976. Mutations, chromosome rearrangements, and epigenetic changes involved in cell transformation and immortalization in leukemia and lymphoma.

Luisa A. DiPietro, Professor of Surgery and of Microbiology and Immunology; D.D.S., 1980, Ph.D., 1989, Illinois. Macrophage function in wound repair; the macrophage as a regulator of physiologic and pathologic angiogenesis.

Adam Driks, Associate Professor of Microbiology and Immunology; Ph.D., Brandeis, 1989. Cellular differentiation in bacteria; structural and genetic analysis of *Bacillus anthracis* and *B. subtilis* spore assembly; development of tools to combat biological weapons.

Kimberly E. Foreman, Associate Professor; Ph.D., Cincinnati, 1992. Pathogenesis of virally induced tumors; molecular interactions between viruses in dual viral infections; mechanisms of tumor resistance to apoptosis.

Thomas M. Gallagher, Associate Professor; Ph.D., Wisconsin, 1987. Mechanisms of enveloped virus entry into host cells; structure and function of viral glycoproteins.

David W. Hecht, Professor of Medicine and of Microbiology and Immunology; M.D., Loyola of Chicago, 1982. DNA transfer, antibiotic resistance, and virulence factors of pathogenic bacteria.

David H. Keating, Assistant Professor; Ph.D., Illinois, 1996. Bacterial-plant communication mechanisms required for symbiotic nitrogen fixation; structure and function of bacterial cell surface polysaccharides.

Herbert L. Mathews, Professor of Microbiology and Immunology; Ph.D., West Virginia, 1977. Cellular and molecular mechanisms of host resistance; effect of stress upon the human immune response.

Brian J. Nickoloff, Professor of Pathology, Microbiology and Immunology; M.D., 1979, Ph.D., 1983, Wayne State. Immunobiology of benign and malignant skin diseases.

Liang Qiao, Associate Professor; M.D., Lausanne (Switzerland), 1992. Mechanisms of mucosal tolerance to unharmful foreign antigens and induction of mucosal immunity to pathogens; development of vaccines against HIV, SARS-cov, HPV, and cancers; mechanisms of tumor resistance to immunotherapy.

John A. Robinson, Professor of Medicine and of Microbiology and Immunology; Chief, Section of Therapeutic Apheresis; and Associate Dean for Research; M.D., Illinois Medical Center, 1964. Clinical evaluations of experimental protocols to induce tolerance in thoracic organ implementation; development of experimental protocols to enable transplantation of sensitized patients.

Gayatri Vedantam, Assistant Professor of Medicine; Ph.D., Illinois, 1996. Antibiotic resistance gene transfer in human commensal bacteria; mobile DNA; bacterial virulence.

Karen L. Visick, Associate Professor; Ph.D., Washington (Seattle), 1993. Colonization of animal host epithelium by symbiotic bacteria; microbial genetics and molecular biology.

Christopher M. Wiethoff, Assistant Professor; Ph.D., Kansas, 2002. Mechanisms of nonenveloped virus cell entry; mechanisms of adenovirus capsid disassembly; characterization of viral proteins involved in cell membrane penetration.

Pamela L. Witte, Professor of Cell Biology, Neurobiology and Anatomy, and Microbiology and Immunology; Ph.D., Texas, 1984. Cellular and molecular regulation of B-cell development; hematopoietic microenvironments; immunodeficiency of aging.

Alan J. Wolfe, Professor; Ph.D., Arizona, 1985. Genomic and genetic mapping of global signaling networks; modulation of gene expression by nucleoprotein complexes; effect of ions on cellular behavior.

A graduate student explains his hypothesis to a colleague.

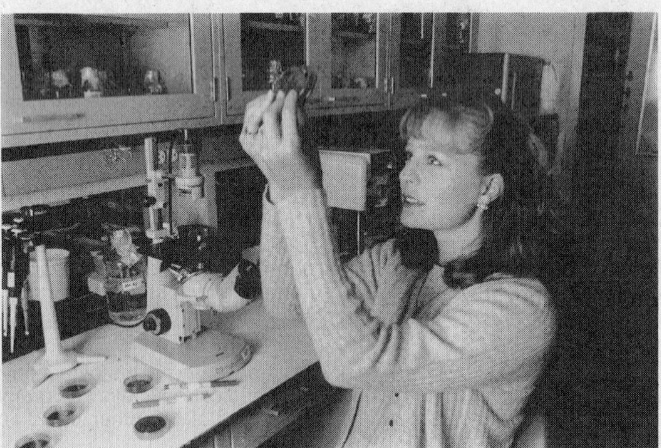

A graduate student examines the results of a recent experiment.

MAYO GRADUATE SCHOOL

Graduate Program in Immunology

Program of Study

The Department of Immunology, Mayo Graduate School, offers graduate training leading to the Ph.D. and M.D./Ph.D. degrees in biomedical sciences (immunology). Students have the opportunity to obtain a comprehensive education in the biomedical sciences by taking a broad range of formal course work in various departments within the Mayo Graduate School. In addition, they carry out advanced research, leading to the dissertation, in any one of a wide variety of areas within the rapidly expanding field of immunology. These areas encompass the fields of biochemistry, cell biology, genetics, and molecular biology and a variety of aspects of immunopathology.

Research Facilities

Departmental research laboratories are well equipped for the most modern cellular, biochemical, and molecular studies in immunology. The Department of Immunology has access to specialized research facilities that include hybridoma production, DNA synthesis/sequencing, protein synthesis/sequencing, fluorescence-activated cell sorting, electron microscopy, gene transfer, and gene knockout capability.

Financial Aid

Full-time graduate students are provided with Mayo Graduate School predoctoral research fellowships. They provide a yearly stipend ($22,900 in 2005–06) in addition to medical insurance and other benefits. Receipt of fellowship stipends does not obligate students to work as laboratory assistants or teaching assistants. Since stipends are provided by the Graduate School, students are not constrained by funding limitations in their choice of research advisers.

Cost of Study

Tuition costs for course work taken at Mayo's campuses in Rochester, Minnesota; Jacksonville, Florida; and Scottsdale, Arizona, are provided in addition to the yearly stipend. Students pay no tuition or ancillary fees.

Living and Housing Costs

Living costs in Rochester are comparable to those in cities of similar size within the Upper Midwest and generally lower than those in urban areas of the East and West Coasts. A single student can live comfortably in Rochester on the stipend provided.

Student Group

The Department of Immunology's graduate program approximately (25–30 students) is specifically organized to educate a select group of students. The high ratio of faculty to students provides each student with an unusual opportunity for close collaboration with internationally recognized researchers.

Student Outcomes

Approximately three fourths of the students enrolled in Mayo Ph.D. programs graduate with a Ph.D. degree. Students have gone on to competitive postdoctoral positions and have advanced to the following institutions: University of Texas Southwestern Medical Center at Dallas; University of California, San Diego; Mount Sinai Medical Center; University of Wisconsin–Madison; University of Minnesota; and Eppley Cancer Research Institute.

Location

The city of Rochester combines the best of two worlds—the warmth and friendliness of a small town with the bustling commerce, entertainment, and conveniences of a metropolis. Lectures, symphony concerts, art exhibits, and a civic theater contribute to a cosmopolitan atmosphere, unusual for a city this size.

The Graduate School

The Mayo Graduate School is a division of Mayo Foundation, which includes Mayo Clinic, Mayo Graduate School of Medicine, Mayo Medical School, Mayo School of Health-Related Sciences, and affiliated hospitals. Mayo Foundation is accredited by the Commission on Institutions of Higher Education of the North Central Association of Colleges and Schools.

Applying

Scores on the GRE General Test are required and must indicate strong academic ability. A GRE Subject Test is highly recommended. Specific course prerequisites include two years of college chemistry, one year of calculus, one year of biology, and one year of physics, with evidence of superior performance in all courses. Official transcripts from schools attended, three letters of recommendation, and a summary of the student's research experience, scientific interests, and career goals are also required. Interviews are strongly recommended; virtually all students accepted have been interviewed. Applications to the M.D./Ph.D. program are initiated via the AMCAS system and must be submitted by November 1. The Ph.D. portion of the M.D./Ph.D. application should be submitted by December 15. Applications for the Ph.D. program should be submitted by December 15. Later applications may be considered on a space-available basis. The application fee is $30. Mayo Foundation is an affirmative action and equal opportunity educator and employer.

Correspondence and Information

Hirohito Kita, M.D.
Graduate Program Director
Department of Immunology
Mayo Clinic College of Medicine
Mayo Graduate School
200 First Street, SW
Rochester, Minnesota 55905
Phone: 507-284-9891
E-mail: felmlee.terri@mayo.edu

Mayo Graduate School

THE FACULTY AND THEIR RESEARCH

Daniel D. Billadeau, Ph.D. Molecular mechanisms controlling lymphocyte activation. Research is focused on the identification and characterization of intracellular signaling cascades that link extracellular stimuli to cellular activation events that control actin cytoskeletal changes. Using a combination of biochemical and genetic approaches, the ongoing research projects focus on the molecular mechanisms by which several intracellular signaling molecules, including dynamin2, Vav1, HS1, and WAVE2, modulate T-cell activation by regulating de novo actin polymerization at the immune synapse in response to TCR engagement.

Richard J. Bram, M.D., Ph.D. Signal transduction. Research stems from an interest in molecular signaling events that control cellular behavior in activation of the immune system and in neoplastic transformation. Ongoing projects include studies on the mechanism of action of CAML, an intracellular signal transduction protein; characterization of TACI, a lymphocyte-specific TNFR family member; and identification of other novel proteins capable of regulating signals important in cell growth, function, or transformation.

Marilia Cascalho, M.D., Ph.D. Molecular and cellular immunology. One central theme is the study of the generation of B-lymphocyte memory and somatic hypermutation. These studies focus on mechanisms of survival of B cells undergoing somatic hypermutation and class switch recombination. Another theme concerns B-cell–dependent T-cell development and function. These studies investigate how B cells and immunoglobulin promote thymocyte development and positive selection of T lymphocytes, studying gene-targeted mice with various constitutive and regulatable B-cell deficiencies.

Chella S. David, Ph.D. Immunogenetics. Research centers on immunogenetic aspects of immune response, with emphasis on the major histocompatibility complex (MHC) HLA class II genes. Transgenic mice expressing human MHC genes are used in structure-function analysis, disease correlation, and thymic regulation of T-cell development. Major ongoing studies on HLA-linked disease models are collagen-induced arthritis, experimental autoimmune encephalomyelitis, diabetes, lupus, celiac disease, myocarditis, relapsing polychondritis, toxic shock syndrome, and dermatitis herpetiformis.

Karen M. Hedin, Ph.D. Molecular immunology and signal transduction. Research is focused on the molecular mechanisms and biological impact of chemokine receptor signaling in lymphocytes and endothelial cells. Chemokines are hormones that regulate cell growth, movement, and gene expression by binding and activating cell-surface G-protein–coupled receptors. Projects include uncovering the mechanisms that permit chemokines to regulate the migration, activation, and cytokine expression patterns of T lymphocytes; defining the roles of chemokines in lymphoid tumor formation and dissemination; and characterizing chemokine actions during the rejection of transplanted organs.

Diane F. Jelinek, Ph.D. Molecular and cellular immunology; tumor biology. Research centers on normal and malignant human B lymphocytes. Molecular strategies are used to understand how various stimuli elicit their cellular effects on normal B lymphocytes and how these pathways become altered in the B-cell malignancies, multiple myeloma, and B-cell chronic lymphocytic leukemia. Projects include dissecting cytokine-mediated signaling pathways in normal versus transformed B cells and the elucidation of genes that are transcriptionally activated in response to specific stimulation.

Hirohito Kita, M.D. Allergy, inflammation, and host defense. Research is focused on elucidating the mechanisms of allergic diseases, with a special interest in eosinophilic leukocytes and innate immunity. Ongoing studies include exploring the roles of adhesion molecules, GPI-anchored proteins, Fc receptors, and protease-activated receptors in activation and mediator release by eosinophils; characterizing the intracellular signaling mechanisms coupling serpentine receptors and the effector functions of granulocytes; and dissecting the mechanisms of allergic inflammation and its outcomes by analyzing and manipulating murine models of diseases. Results from these in vitro and animal models are applied to humans. Several clinical research projects are ongoing to understand the immunological mechanisms of patients with asthma and chronic rhinosinusitus and to provide better and safer treatment for these patients.

Keith L. Knutson, Ph.D. Research focuses on the development of immune-based therapies for cancer, with an emphasis on T-cell therapy. The program includes identification of T-cell peptide epitopes from known tumor antigens, reversing immunosuppression, overcoming immunological tolerance to self, improving engraftment of transferred T cells, rapid ex vivo expansion of antigen-specific T cells, cancer vaccines, and novel in vivo immunologic monitoring techniques. An emphasis is placed on breast and ovarian cancers.

Eugene D. Kwon, M.D. Tumor immunology. Research centers on methods to evoke a potent immune response to treat relatively advanced forms of malignancy. Specific areas of research pertain to the preclinical and clinical use of novel vaccines and antibodies to activate antitumoral T cells, the use of hormone manipulations to boost or rebuild host immunity, and the treatment of patients with immunotherapy in order to induce clinical tumor regression. A special emphasis is placed on developing highly state-of-the-art immunotherapies to be tested in clinical phase I or II trials to treat patients with prostate, kidney, or bladder cancer.

Paul J. Leibson, M.D., Ph.D. Mechanisms of lymphocyte activation and antitumor immunity. Research is focused on exploring human cell-mediated antitumor immunity, with a special interest in natural killer (NK) cell activation. Projects include characterizing the intracellular signaling events that positively or negatively regulate the ability of NK cells to mediate antitumor cytotoxicity, evaluating the cell regulation initiated by newly identified receptors expressed on the surface of cytotoxic lymphocytes, identifying and characterizing mechanisms that regulate critical cytoskeletal changes needed for effective antitumor immunity, and assessing novel therapeutic strategies that can be used to enhance tumor killing.

Vanda A. Lennon, M.D., Ph.D. Neuroimmunology. Specific interests: autoimmunity and its overlap with cancer, the antigenicity of plasma membrane channels and neurotransmitter receptors expressed in neurons, glia, and muscle, as well as in carcinomas of lung, ovary, and breast, and in thymic epithelial neoplasms. These antigens include voltage-gated cation channels (P/Q-type and N-type calcium channels and potassium channels), acetylcholine receptors, novel channels, and other synapse-related molecules. The tumor proteins activate helper, effector, and regulatory T cells and initiate production of autoantibodies that cause patients to present with paraneoplastic disorders of the central and peripheral nervous systems. The counterpart proteins in neurons and muscle are the unintended targets of otherwise highly effective tumor immune responses. The research laboratory is developing instructive new animal models of neurological autoimmunity based on insights gathered from the clinical laboratory's immunological studies of thousands of patients with neurological autoimmunity related to cancer. This activity in turn translates contemporary immunological and neurobiological knowledge to the diagnosis and treatment of autoimmunity and cancer.

David J. McKean, Ph.D. Molecular immunology and signal transduction. Research is focused on the molecular characterization of the signals needed to initiate activation and differentiation in double positive thymocytes and T lymphocytes. Intracellular second-messenger signaling and gene transcription control mechanisms are characterized to identify the positive and negative regulatory roles of costimulatory receptor signals in antigen receptor–activated cells.

Kay L. Medina, Ph.D. Regulation of B-cell development in mouse and man. Research is focused on understanding the cellular and molecular mechanisms that control the differentiation of hematopoietic stem cells into B lymphocytes. Projects include studies investigating the role of transcription factors and signaling molecules in B cell-fate specification from stem cells, the role of systemic factors in regulating steady-state B lymphocyte production, and the impact of cellular oncogenes on B-cell development.

Larry R. Pease, Ph.D. Molecular and cellular immunology. Studies address various aspects of immune recognition of viral and tumor antigens. Current focus is on dendritic cells and T-cell components of the immune response. Interests span a variety of topics, including repertoire development, antigen presentation, activation of dendritic cells, activation of naïve T cells, and mobilization of the immune response to viral pathogens and tumors. A variety of mouse mutant, transgenic, and knockout models are employed experimentally to address questions in vitro and in vivo.

Jeffrey L. Platt, M.D. Transplantation biology. Research addresses novel approaches to replacing organs, including stem-cell biology, organogenesis, regeneration, and xenotransplantation. Another theme concerns the means by which the functions of toll-like receptors are controlled and the origin of sepsis and the sepsis syndrome. Two other themes concern the rebuilding of the immune system, particularly the T-cell compartment and control of T-cell allogeneic responses.

Eric M. Poeschla, M.D. Human immunodeficiency virus (HIV/AIDS) pathogenesis; human gene therapy with lentiviral vectors (HIV, FIV). Molecular virology is the basic orientation of the laboratory. Work currently focuses on the function and trafficking of the viral integrase protein, viral integration, and the role of HIV-1 integrase interactor LEDGF/p75 in the HIV-1 life cycle. Lentiviral RNA encapsidation and postentry restriction factors (e.g., TRIM5alpha) are further interests. Human gene therapy research involves novel lentiviral vectors and therapeutic transgenes. One project develops vectors, animal models, and therapeutic transgenes for glaucoma.

Moses Rodriguez, M.D. Viral immunopathology. The laboratory is concerned with the role of the immune response and virus persistence in demyelination of the central nervous system. Studies use Theiler's virus, a picornavirus that, when inoculated into susceptive mice, produces immune-mediated demyelination that is similar to multiple sclerosis. The group is investigating the role of host genes in the control of susceptibility to infection in order to understand the effect of class I and class II MHC-restricted effector T cells limiting virus infection or promoting demyelination. Experiments are also underway to investigate how antibodies promote remyelination in the central nervous system.

Richard G. Vile, Ph.D. Gene and immunotherapy of cancer. Research interests are based on development of efficient and targeted viral vectors for in vivo delivery of therapeutic genes to tumor cells and development of effective ways to stimulate antitumor immune responses through gene expression in tumor cells. The aim of the laboratory is to develop treatments for neoplastic disease that are more gentle and effective than current treatments. There is also an increasing interest in the use of stem cells and immune cells as carriers of viral constructs so that the major hurdle of systemic delivery of vectors to tumors can be overcome in patients.

Peter J. Wettstein, Ph.D. Immunogenetics. Principal investigations center on the genetic control of the immune response to minor histocompatibility antigens that provide a strong obstacle to tissue transplantation. Experimental studies are directed toward the roles played by both specific polymorphic major histocompatibility complex molecules that present minor antigens as well as the polymorphic genes that encode the target antigens. The experimental approach to identifying major antigens involves the purification and sequencing by mass spectrometry on antigenic peptides eluted from major histocompatibility complex molecules.

MAYO GRADUATE SCHOOL

Graduate Program in Virology and Gene Therapy

Program of Study

The Molecular Medicine Program at Mayo Graduate School offers graduate training leading to the Ph.D. and M.D./Ph.D. degrees in biomedical sciences (virology and gene therapy). Students have the opportunity to obtain a comprehensive education in the biomedical sciences by taking a broad range of formal course work in various departments within the Mayo Graduate School. In addition, they carry out advanced research, leading to the dissertation, in any one of a wide variety of areas within the fields of virology or gene therapy. These areas encompass the fields of biochemistry, cell biology, genetics, and molecular biology.

Research Facilities

Research laboratories are well equipped for the most modern cellular, biochemical, and molecular studies in virology and gene therapy. Mayo has outstanding core research facilities that include DNA synthesis/sequencing, protein synthesis/sequencing, fluorescence-activated cell sorting, electron microscopy, gene transfer, and gene knockout capability. The Molecular Medicine Program operates a vector production facility and a toxicology core for preclinical and clinical studies. Additional information can be found in Mayo Graduate School's Graduate Program in Biomedical Sciences entry in Section 1 of *Peterson's Graduate Programs in the Biological Sciences.*

Financial Aid

Full-time graduate students are provided with Mayo Graduate School predoctoral research fellowships. They provide a yearly stipend (approximately $22,900 in 2005–06) in addition to low-cost medical insurance and other benefits. Receipt of fellowship stipends does not obligate students to work as laboratory assistants or teaching assistants. Since stipends are provided by the Graduate School, students are not constrained by funding limitations in their choice of research advisers.

Cost of Study

Tuition costs for course work taken at Mayo and off campus are provided in addition to the yearly stipend; students pay no tuition or ancillary fees.

Living and Housing Costs

Living costs in Rochester are comparable to those in cities of similar size within the Upper Midwest. A single student can live comfortably on the stipend provided.

Student Group

Approximately three fourths of the students enrolled in Mayo Ph.D. programs graduate with a Ph.D. degree. Students take approximately five years to complete a Ph.D. Most students have authored several publications by the time they graduate.

Location

The city of Rochester combines the best of two worlds—the warmth and friendliness of a small town and the bustling commerce, entertainment, and conveniences of nearby Minneapolis. Lectures, symphony concerts, art exhibits, and a civic theater contribute to a cosmopolitan atmosphere, unusual for a city of its size.

The Graduate School

The Mayo Graduate School is a division of Mayo Foundation, which includes Mayo Clinic, Mayo Graduate School of Medicine, Mayo Medical School, Mayo School of Health-Related Sciences, and affiliated hospitals. Mayo Foundation is accredited by the Higher Learning Commission of the North Central Association of Colleges and Schools.

Applying

Scores on the GRE General Test are required and must indicate strong academic ability. A GRE Subject Test is highly recommended. Specific course prerequisites include 2 years of college chemistry, 1 year of calculus, 1 year of biology, and 1 year of physics, with evidence of superior performance in all courses. Official transcripts from schools attended, three letters of recommendation, and a summary of the student's research experience, scientific interests, and career goals are also required. Interviews are strongly recommended; virtually all students accepted have been interviewed. Applications to the M.D./Ph.D. program are initiated via the AMCAS system and must be submitted by November 1. The Ph.D. portion of the M.D./Ph.D. application should be submitted by December 31. Applications for the Ph.D. program should be submitted by December 31. Later applications may be considered on a space-available basis. Mayo Foundation is an affirmative action and equal opportunity educator and employer.

Correspondence and Information

Roberto Cattaneo, Ph.D.
Graduate Program Director
Molecular Medicine Program
Mayo Graduate School
Guggenheim 18
200 First Street, SW
Rochester, Minnesota 55905
Phone: 507-538-1188
E-mail: sanford.becky@mayo.edu

Mayo Graduate School

THE FACULTY AND THEIR RESEARCH

Andrew D. Badley, M.D. Regulation of cell death induced by different stimuli, including infections, drugs, and ischemia; how viral proteins, such as those made by HIV, impact and alter the cell death response; evaluation and development of novel therapeutic approaches to modifying the host cell death response, i.e., ways of selectively inhibiting or inducing cell death in targeted cell populations.

Roberto Cattaneo, Ph.D. Replication and transcription, protein synthesis and transport, particle assembly and release, and entry into cells of morbilliviruses, which are animal viruses with six structural proteins and a negative-strand RNA genome; main model system is measles virus (MV), an important human pathogen for which a safe and effective live attenuated vaccine strain exists; developing new divalent MV-based vaccines; transforming MV into a vector for targeting and eliminating cancer cells.

Mark J. Federspiel, Ph.D. Elucidation of mechanisms of retrovirus entry; major model uses avian leukosis virus (ALV) envelope glycoproteins and several cloned cellular receptors of ALV; several novel genetic approaches used to force ALV to evolve its receptor usage and identify important regions of these proteins; molecular, cellular, biochemical, and biophysical analyses to characterize the viral glycoprotein-receptor interactions.

Evanthia Galanis, M.D. Translational applications of gene transfer/virotherapy in the treatment of solid tumors; targeting cytotoxicity of fusogenic membrane glycoproteins by exploiting the overexpression of matrix metalloproteases in tumor microenvironment; developing targeted attenuated strains of measles virus with selectivity for gliomas; testing combinations of replicating adenoviral vectors and small molecules in breast cancer.

Yasuhiro Ikeda, Ph.D., D.V.M. Modification of human immunodeficiency virus–based vectors yielding higher titers with increased safety; characterization of anti-HIV-1 cellular factors for a novel anti-AIDS therapy; making a modified HIV-1, which is replication competent in rhesus monkeys for the development of a better AIDS animal model.

Zvonimir S. Katusic, M.D., Ph.D. Mechanisms of endothelial dysfunction and the role of nitric oxide in pathogenesis of cerebrovascular disease; pharmacologic, genetic, and cell-based therapies studied in animal models of vascular injury and cerebral vasospasm.

Christopher G. A. McGregor, M.D., FRCS. Gene therapy; xenotransplantation; new physical delivery systems and various viral vectors for targeted gene therapy to the heart; effects of various biologically relevant genes on the development of cardiac allograft vasculopathy; prevention, diagnosis, mechanisms, and treatment of delayed xenograft rejection, using transgenic porcine hearts, pursued in a preclinical transplant model.

John C. Morris III, M.D. Sodium-iodide symporter (NIS)–mediated gene therapy of cancer; feasibility and efficacy of NIS gene transfer in vivo using mouse models, mechanisms of maximizing NIS protein expression and activity, and cell killing effect of 131 I in several cancer models in vitro and in vivo; preclinical and toxicity studies with a replication-deficient adenovirus containing the NIS gene; phase I clinical trial in men with locally recurrent prostate cancer after external beam radiotherapy.

Kah-Whye Peng, Ph.D. Measles virotherapy for ovarian cancer; mechanisms of measles tumor selectivity; vector targeting; receptor density and affinity; noninvasive monitoring of viruses; nanotechnology for imaging and cancer therapy.

Eric M. Poeschla, M.D. Human gene therapy; lentiviral vectors; lentiviral pathogenesis (HIV infection/AIDS); investigation of basic and translational aspects, particularly human gene therapy with vectors derived from feline immunodeficiency virus (FIV), with focus on postmitotic therapeutic gene therapy targets in eye and brain; molecular virological studies of HIV disease involving comparative lentivirology: genome encapsidation, lentivirus-specific aspects of reverse transcription, and the enabling of infection of nondividing cells by nuclear import of the reverse-transcribed genome, a critical property for gene therapy and AIDS pathogenesis.

Gregory A. Poland, M.D. Vaccines; vaccine-preventable diseases; vaccine immunogenetics; biodefense vaccines; vaccine development; mechanisms of immune response development to vaccines; vaccine-preventable infectious diseases, particularly, vaccines against agents of bioterrorism, predictors of vaccine response, antigen processing, and HLA presentation; using novel mass spectrometry techniques to develop new measles and vaccinia vaccines; understanding the immunogenetics of measles and rubella vaccine responses; basic mechanisms of adoptive immunity to vaccines.

Moses Rodriguez, M.D. Viral immunopathology; role of the immune response and virus persistence in demyelination of the central nervous system; studies using Theiler's virus, a picornavirus that, when inoculated into susceptive mice, produces immune-mediated demyelination that is similar to multiple sclerosis; role of host genes in the control of susceptibility to infection in order to understand the effect of class I and class II MHC–restricted effector T cells limiting virus infection or promoting demyelination; investigation of how antibodies promote remyelination in the central nervous system.

Stephen J. Russell, M.D., Ph.D. Gene therapy and oncolytic virotherapy for multiple myeloma, ovarian cancer, pancreatic cancer, and glioma, using measles, retrovirus, lentivirus, adenovirus, Sindbis, and nonviral vectors; engineering viral membrane glycoproteins; targeting virus entry; pharmacokinetic and pharmacodynamic studies of gene therapy; nonimmune in vivo monitoring and imaging of viral gene expression; radiovirotherapy.

David B. Schowalter, M.D., Ph.D. Development of durable, nontoxic gene therapy systems capable of targeting the liver, including adenovirus, helper-dependent adenovirus, adeno-associated virus, and plasmid-based systems; model diseases, including two fatty acid oxidation disorders: medium and very long chain acyl-CoA dehydrogenase deficiencies; use of hepatic gene therapies as a preventative health tool.

Robert D. Simari, M.D. Expression and modulation of factors that regulate the development of atherosclerosis and its sequelae; emphasis on regulation of the tissue factor pathway in vascular disease and the integrative roles of proliferation and thrombosis within the vasculature; development of transgenic models for atherosclerosis as well as somatic gene transfer applications of gene therapies. (lab grants from the National Institutes of Health, American Heart Association, and the Miami Heart Research Institute)

Richard G. Vile, Ph.D. Gene and immunotherapy of cancer; development of efficient and targeted viral vectors for in vivo delivery of therapeutic genes to tumor cells; development of effective ways to stimulate antitumor immune responses through gene expression in tumor cells; development of treatments for neoplastic disease that are more gentle and effective than current treatments; use of stem cells and immune cells as carriers of viral constructs, so that systemic delivery of vectors to tumors can be overcome in patients.

Joseph D. Yao, M.D. Development and clinical application of molecular diagnostic methods for infective agents, including viral hepatitis, HIV, pathogenic mycobacteria/fungi, and bacterial pathogens.

MEDICAL UNIVERSITY OF SOUTH CAROLINA

College of Graduate Studies
Department of Microbiology and Immunology

Programs of Study

The microbiology and immunology graduate program is the largest discipline-based graduate program at the Medical University of South Carolina (MUSC). The program of study for each student is planned jointly by the student and his or her advisory committee and is highly individualized according to the student's needs and goals.

All Ph.D. students participate in the first-year common curriculum of the College of Graduate Studies and take 12 credit hours in advanced and elective courses after the first year. Requirements for graduation include satisfactory completion of all required course work and the qualifying examination, selection of a dissertation adviser and advisory committee, and submission and oral defense of the dissertation proposal. All requirements must be completed at least one year prior to the student's final dissertation/thesis defense.

Master's students may complete the first-year core curriculum as their course requirement or take electives offered through other departments. Requirements for graduation include satisfactory completion of all required course work, selection of a thesis adviser and advisory committee, and submission and oral defense of the thesis proposal. All requirements must be completed at least six months prior to the student's final thesis defense.

Students must maintain a minimum GPA of 3.0 in all courses. Students may retake courses once at the discretion of the course director and program committee if this grade is not achieved. Failure to achieve the required grade at that time usually results in dismissal from the program, as does a grade of less than 2.0 in any course. In addition, all students are required to attend and present yearly at the Microbiology and Immunology Departmental Seminar Series.

Research Facilities

As a research institution, the University has numerous research facilities. The Biomolecular Computing Resource is available to graduate students and faculty and staff members. The Protein Sequence Facility provides protein sequence analysis for the MUSC research community as well as for outside academic and corporate researchers. The University also has several resources where students can get information about applying for research grants. The library has access to numerous electronic databases, journals, e-books, and biomedical resources.

Financial Aid

Graduate students are eligible for scholarships ranging from $21,000 to $24,000 and loan programs. Most students are guaranteed loans with attractive interest rates and repayment options. Repayment of most loans and interest can be deferred until completion of the program.

Cost of Study

Tuition for the master's program in 2005–06 was $398 per credit or $4587 per semester for a full course load ($521 and $6162 for out-of-state students, respectively). Ph.D. candidates paid $397 per credit or $4308 for the semester ($518 and $5487 for out-of-state students, respectively). Students must also pay a University fee and library fee of $70. Ph.D. students are eligible for reduced tuition, which is approximately $1210 per semester.

Living and Housing Costs

On average, the rent for one-bedroom apartments in Charleston is $500–$850 per month and two bedrooms average $900–$1100.

Student Group

The majority of the University's student body is native to South Carolina, although students come from many other states and countries. Approximately 60 percent of the students in the College of Graduate Studies are from out of state.

Location

Charleston's residents enjoy cultural programs and exhibits by the Charleston Symphony, the Charleston Opera Company, the Choral Society, the Charleston Museum, and the Gibbes Museum of Art. Charleston is also the site of the annual Spoleto Arts Festival. Outdoor activities include skiing and windsurfing as well as camping on the barrier islands in the adjacent wildlife sanctuaries. Charleston is a 3-hour drive from the Appalachian Mountains.

The University and The College

The Medical University of South Carolina's College of Graduate Studies evolved from a program of graduate education in the basic medical sciences and became a division in 1965. It offers opportunities for advanced learning and research and for promoting the extension of knowledge in the biomedical sciences. The University pursues three interrelated missions: education, research, and clinical service. The University's cooperative and visionary leadership has the support of a diverse professional workforce, the strength of accountable allocation of resources, and expanded community involvement.

Applying

To apply, a strong undergraduate background in basic science is recommended, as are scores on the General Test of the GRE. Scores more than five years old are unacceptable. Applications may be submitted at any time; however, to ensure consideration for fellowship support, the application process must be completed by February 1. Applicants are required to apply online. U.S. applicants for the Ph.D. program can apply without a fee. Master's degree applicants and international applicants pay a $75 nonrefundable processing fee.

Financial aid applications are available after February 1. Students must apply annually to be considered for the next academic year. Eligibility is based upon the completion of all required forms, specific program regulations, and satisfactory academic progress toward completion of a degree. Once an application has been reviewed, the student is notified in writing of the aid that has been awarded.

Correspondence and Information

Dr. Lucille London
Graduate Program Coordinator
Department of Microbiology and Immunology
Medical University of South Carolina
173 Ashley Avenue
P.O. Box 250504
Charleston, South Carolina 29425
Phone: 843-792-5014
Fax: 843-792-2464
E-mail: londonl@musc.edu
Web site: http://www2.musc.edu/mic/micro4803/index.html

Medical University of South Carolina

THE FACULTY

Prabhakar Baliga, Professor and Chief, Division of Transplant Surgery; M.B.B.S., Madras (India), 1982.
Edward Balish, Professor of Microbiology and Immunology; Ph.D., Syracuse, 1963.
Narendra L. Banik, Professor (also with Neurology); Ph.D., London, 1970.
Craig C. Beeson, Associate Professor of Pharmaceutical Sciences; Ph.D., California, Irvine, 1993.
Narayan R. Bhat, Professor of Neurology; Ph.D., Indian Institute of Science, 1978.
Robert J. Boackle, Professor (also with Stomatology); Ph.D., Alabama, 1974.
Kenneth D. Chavin, M.D., Assistant Professor (also with Surgery); Ph.D., Medical University of South Carolina, 1993.
David J. Cole, Professor of Surgery; M.D., Cornell, 1986.
James A. Cook, Professor of Physiology and Neurosciences; Ph.D., Tulane, 1975.
Maurizio Del Poeta, Assistant Professor of Biochemistry and Molecular Biology; M.D., Ancona (Italy), 1982.
Jian-yun Dong, Professor; M.D., Capital Medical Institute (Beijing), 1984; Ph.D., Alabama, 1992.
Sebastiano Gattoni-Celli, Associate Professor of Radiation Oncology; M.D., Naples (Italy), 1972.
Gary S. Gilkeson, Professor of Medicine; M.D., Southwestern Medical College, 1979.
William E. Gillanders, Assistant Professor of Surgery; M.D., Duke, 1991.
Hiroko Hama, Assistant Professor of Biochemistry and Molecular Biology; Ph.D., Okayama (Japan), 1990.
Azizul Haque, Assistant Professor of Microbiology and Immunology; Ph.D., Saga Medical (Japan), 1997.
Eric R. James, Associate Professor of Ophthalmology; Ph.D., London, 1974.
Laura M. Kasman, Assistant Professor; Ph.D., Harvard, 1992.
Lucille London, Associate Professor; Ph.D., Pennsylvania, 1986.
Steven London, D.D.S., Associate Professor and Associate Dean for Research, College of Dental Medicine; Ph.D., Pennsylvania, 1987.
Maria F. Lopes-Virella, M.D., Professor of Medicine and of Pathology and Laboratory Medicine; Ph.D., Lisbon, 1990.
Harold D. May, Associate Professor; Ph.D., Virginia Tech, 1987.
James S. Norris, Professor and Chair; Ph.D., Colorado, 1971.
Jim Oates, Assistant Professor of Rheumatology and Immunology; M.D., Johns Hopkins, 1991.
Janardan P. Pandey, Professor; Ph.D., Wisconsin, 1972.
Swapan K. Ray, Assistant Professor of Neurology; Ph.D., Calcutta, 1989.
Titus Reaves, Assistant Professor of Cell Biology and Anatomy; Ph.D., South Carolina, 1998.
Semyon Rubinchik, Research Assistant Professor.
Michael G. Schmidt, Professor and Vice Chairman; Ph.D., Indiana, 1985.
Inderjit Singh, Professor of Pediatrics, of Cell Biology and Anatomy, and of Biochemistry and Molecular Biology; Ph.D., Iowa State, 1974.
Natalie Sutkowski, Assistant Professor of Microbiology and Immunology; Ph.D., Rutgers, 1994.
Stephen Tomlinson, Associate Professor (also with Stomatology); Ph.D., Cambridge, 1989.
Maria Trojanowska, Associate Professor of Medicine; Ph.D., Polish Academy of Science, 1980.
William R. Tyor, Associate Professor (also with Neurology); M.D., Duke, 1981.
Gabriel T. Virella, M.D., Professor; Ph.D., Lisbon, 1974.
Isabel Virella-Lowell, Assistant Professor; M.D., Medical University of South Carolina, 1995.
Christinia Voelkel-Johnson, Assistant Professor; Ph.D., North Carolina State, 1995.
Gregory W. Warr, Professor of Biochemistry and Molecular Biology; Ph.D., London, 1973.
Caroline Westwater, Research Assistant Professor; Ph.D., Aberdeen, 1996.
Cynthia F. Wright, Associate Professor of Pathology and Laboratory Medicine; SUNY at Albany, 1985.

MICHIGAN STATE UNIVERSITY

Department of Microbiology and Molecular Genetics

Programs of Study

More than 50 full-time faculty members of the Department of Microbiology and Molecular Genetics offer research expertise over a wide range of subdisciplines, including traditional areas of microbiology as well as molecular and cellular biology. The Department has four areas of research emphasis: microbial physiology, ecology, and evolution; cancer, differentiation, immunology, and cell biology; molecular pathogenesis and virology; and functional and comparative genomics and genetics. The Department of Microbiology and Molecular Genetics participates in the integrated Graduate Program in Cell, Molecular, and Structural Biology; the Genetics Program; Cell and Molecular Biology Program; Center for Emerging and Infectious Diseases; National Food Safety and Toxicology Center; Astrobiology Institute; and Center for Microbial Ecology. The Doctor of Philosophy degree in microbiology is granted in the Colleges of Natural Science, Human Medicine, Osteopathic Medicine, and Veterinary Medicine. Postdoctoral study is offered and encouraged.

Graduate education is coordinated by the Director of Graduate Studies, who provides guidance in initial enrollment and counsels students about their education and future careers. Programs are planned on an individual basis with the candidate's selected major professor and a guidance committee. In addition to the course credits required by the guidance committee, 24 semester credits of research and one year of residence are required for the Ph.D. degree. All graduate students are trained in teaching and assist in one laboratory course. In addition, Departmental seminars and campuswide journal clubs provide a valuable opportunity to broaden a student's background in diverse research areas. Four to five years are usually needed to complete all requirements.

Research Facilities

The Department occupies about 70,000 square feet of space situated on three floors of the Biomedical Physical Sciences Building, along with additional space in other locations on campus. Individual faculty laboratories are well equipped and are supplemented with common Departmental facilities and equipment and central University facilities. The Department maintains PCs for use by graduate students, which can be directly linked to Michigan State University (MSU) mainframes. The Department is also home to the publication operations of the world-famous Bergey's Manual Trust and the Ribosomal Database Project.

Financial Aid

Financial subsidy, in the form of various assistantships, traineeships, fellowships, and scholarships, is available for qualified applicants on a competitive basis. In 2006–07, the total package begins at $29,977 per year and includes a stipend, tuition and fee waiver equivalent to 22 credits, and individual health coverage. Generally, students who are accepted receive support.

Cost of Study

As noted in the Financial Aid section, costs are separately covered by fellowships or waived for students receiving assistantships.

Living and Housing Costs

The University provides convenient and economical housing for both married and single graduate students. Single men and women may be housed in Owen Graduate Center. There are 1,800 apartments, owned and operated by the University, that are used primarily as rentals for married students. Monthly rents are $559 to $660. Most graduate students live in privately owned, off-campus rooms and apartments, which are available at competitive rates.

Student Group

More than 8,300 of the University's enrollment of almost 45,000 are graduate and professional students. The Department averages close to 60 graduate students, who come from throughout the United States and from other countries. Equal opportunity is provided, regardless of sex or ethnic origin. About 25 postdoctoral students are in residence. A Microbiology Graduate Student Workshop exists as a forum to discuss ongoing doctoral research and also functions as an organizational and social group. Graduate students participate actively in Departmental committees and elect members who maintain liaison with the Department chairperson.

Student Outcomes

The majority of graduates enter postdoctoral training programs in preparation for research and teaching careers in academe and industry.

Location

The Red Cedar River flows through the center of the gently rolling campus of 2,010 acres, which is graced by more than 5,700 species of trees, shrubs, and vines. Adjacent to the campus is the residential city of East Lansing, which adjoins Lansing, the state capital. Graduate students and their families may enjoy the many opportunities for cultural and social development offered by the University and neighboring civic groups. A great variety of recreational facilities are available at the University and in the greater Lansing area.

The University

Michigan State, the pioneer land-grant university, was founded in 1855 with the philosophy that the opportunity for learning should not be restricted to any one group or class but should be made freely available to all who can make good use of it.

Applying

Applicants for entry into the program in microbiology usually have had preparation in physics, inorganic chemistry, organic chemistry, and mathematics through the calculus level in addition to courses in the biological sciences. Preparation in the fundamentals of microbiology is desirable but not necessary. Applicants should have a minimum grade point average of 3.0 on a 4.0 scale and grades of 3.0 or above in quantitative science courses, including mathematics. Interested students with degrees in any of the physical or biological sciences are invited to apply.

Application is completed online, with accompanying documents, including transcripts of all prior college education, a personal letter of intent and objective, scores on the General Test of the Graduate Record Examinations, and three letters of recommendation from academics, sent to the Graduate Secretary by December 31 (absolute deadline of February 1) for consideration for admission in the fall semester of the same year. Admission decisions are made by the Department's graduate committee, Graduate Recruiting Officer, and Chairperson.

Correspondence and Information

Graduate Secretary
Department of Microbiology and Molecular Genetics
2215 Biomedical Physical Sciences
Michigan State University
East Lansing, Michigan 48824-4320

Phone: 517-432-2288
Fax: 517-353-8957
E-mail: micgrad@msu.edu
Web site: http://www.mmg.msu.edu

Michigan State University

THE FACULTY AND THEIR RESEARCH

Cancer, Differentiation, Immunology, and Cell Biology
Kathryn H. Brooks, Associate Professor; Ph.D., Iowa. Immunology; autoimmunity; biology of CD5[+] B cells.

Wendy C. Champness, Professor; Ph.D., Michigan State. Microbial molecular genetics and biotechnology; genetic regulation of *Streptomyces* antibiotics; developmental genetics.

Susan E. Conrad, Professor; Ph.D., Caltech. Control of proliferation in normal and tumorigenic cells; control of gene expression during the cell cycle; estrogen regulation of human breast cancer cell proliferation.

Paul M. Coussens, Professor; Ph.D., Penn State. Molecular virology; viral gene expression; oncogenes.

Walter J. Esselman, Professor and Chair; Ph.D., Penn State. Expression and function of lymphocyte CD45 protein tyrosine phosphatase in the regulation of signaling, the cell cycle, and transformation.

Michele M. Fluck, University Distinguished Professor; Ph.D., Geneva. Viral oncology; oncogene activation of transcription factors and their effect on chromatin structure and DNA replication; mammary gland oncogenesis in mice and its dependence upon estrogen.

Steven R. Heidemann, Professor; Ph.D., Princeton. Mechanisms of neural growth; role of mechanical force and the cytoskeleton.

Felipe Kierszenbaum, Professor; Ph.D., Buenos Aires. Immunology, biochemistry, and molecular biology of parasitic diseases.

Donna J. Koslowsky, Associate Professor; Ph.D., Washington (Seattle). RNA editing; organellar gene expression in trypanosomes.

Lee Kroos, Professor; Ph.D., Stanford. Regulation of gene expression during development; prokaryotic RNA polymerases and regulatory proteins; *Bacillus* sporulation; cell-cell interactions during *Myxococcus* fruiting-body formation.

John E. Linz, Professor; Ph.D., LSU. Genetics and control of mycotoxin biosynthesis; fungal development; food-borne microbial pathogens.

Ronald J. Patterson, Professor; Ph.D., Northwestern. Cell biology; nucleocytoplasmic RNA transport; role of nuclear galectins in pre-mRNA splicing.

Rosetta N. Reusch, Professor; Ph.D., Columbia. Metabolism and functions of poly-3-hydroxybutyrate and inorganic polyphosphates; polyhydroxybutyrate-associated proteins; genetic competence; gene transfer; ion channels.

Richard C. Schwartz, Professor; Ph.D., MIT. Regulation of proinflammatory cytokines by C/EBP transcription factors; regulation of progesterone receptor expression in mammary tissue; hematopoietic growth and differentiation.

Animesh A. Sinha, Associate Professor and Chief, Dermatology and Cutaneous Sciences; M.D., Ph.D., Alberta. Genetic and molecular basis of autoimmune disease; systems biology of complex disease.

Ian A. York, Assistant Professor; Ph.D., McMaster. Biology of antigen presentation.

Functional and Comparative Genomics and Genetics
Andrea Amalfitano, Osteopathic Heritage Foundation Professor; Ph.D., Michigan State. Virus-mediated gene transfer for the treatment of genetic and acquired disease; virology, innate immunity, vaccine development, and gene therapy for inborn errors of metabolism.

Robert A. Britton, Assistant Professor; Ph.D., Baylor College of Medicine. Microbial genomics and genetics; chromosome structure and function; regulation of gene expression; genomics of probiotic bacteria.

Jerry B. Dodgson, Professor; Ph.D., Wisconsin–Madison. Genome mapping in birds; RNA interference in chickens; transgenics.

Karen Friderici, Professor; Ph.D., Michigan State. Molecular pathology of human genetic disease; genetics of hearing loss.

John C. Fyfe, Associate Professor; Ph.D., Pennsylvania; D.V.M., Oregon State. Molecular, cellular, and clinical biology of inherited disease; animal models of human genetic disease.

Julius H. Jackson, Professor; Ph.D., Kansas. Microbial physiology and genetics; genomic evolution and theory; gene and chromosome evolution.

Kazem Kashefi, Assistant Professor; Ph.D., London. Microbial diversity, redox transformations, and community dynamics of extreme environments; biotechnological applications of extremophilic organisms.

Michael F. Thomashow, Professor; Ph.D., UCLA. Regulation and function of genes induced in response to environmental stresses.

Patrick J. Venta, Associate Professor; Ph.D., Michigan. Genome mapping in the dog; diagnostic tests for genetic disease.

Vilma Yuzbasiyan-Gurkan, Associate Professor; Ph.D., Istanbul. Comparative mammalian genetics and genomics; cancer genetics; prion diseases.

Microbial Physiology, Ecology, and Evolution
Thomas R. Corner, Professor and Associate Chair (Undergraduate); Ph.D., Rochester. Bacterial membranes; endospore heat resistance.

Frank B. Dazzo, Professor; Ph.D., Florida. Microbial ecology and physiology; beneficial plant-microbe interactions; computer-assisted microscopy.

George M. Garrity, Professor; Sc.D., Pittsburgh. Bacterial systematics; ecology; bioinformatics; computational biology.

Robert P. Hausinger, Professor and Associate Chair (Graduate); Ph.D., Minnesota. Enzymology and microbial physiology; metallocenter assembly mechanisms; biodegradation; DNA repair enzymes.

Jay T. Lennon, Assistant Professor; Ph.D., Dartmouth. Microbial diversity; food webs; ecosystem functioning; ecological and evolutionary roles of viruses in ecosystems.

Richard E. Lenski, Hannah Professor; Ph.D., North Carolina. Ecology, genetics, and evolution of microbial populations.

John E. Merrill, Assistant Professor; Ph.D., Washington (Seattle). Biotechnology of algae; algal photosynthetic pigments; bioremediation and aquaculture of algae.

C. Adinarayana Reddy, Professor; Ph.D., Illinois. Physiology, ecology, and molecular biology of lignin biodegradation by wood-rotting fungi; fungal degradation of xenobiotics: bioremediation, veterinary bacteriology.

Gemma Reguera, Assistant Professor; Ph.D., Massachusetts Amherst. Electron transfer and biofilm physiology; microbial nanowires and biotechnological applications.

Thomas M. Schmidt, Professor; Ph.D., Ohio State. Microbial ecology and evolution; analysis of patterns and extent of microbial diversity in the environment.

James M. Tiedje, University Distinguished Professor and Director, Center for Microbial Ecology; Ph.D., Cornell. Molecular microbial ecology; biodegradation of soil and sediments.

Barry L. Williams, Assistant Professor; Ph.D., Illinois at Urbana-Champaign. Evolutionary genetics; the molecular genetic basis of evolutionary diversification.

Molecular Pathogenesis and Virology
Cindy Grove Arvidson, Assistant Professor; Ph.D., UCLA. Bacteriology; protein targeting, virulence, and regulation of gene expression in the pathogenic *Neisseriae*.

Michael Bagdasarian, Professor; M.D., Ph.D., Warsaw. Molecular biology; molecular pathogenesis; protein targeting in prokaryotes.

Todd Ciche, Assistant Professor; Ph.D., Wisconsin–Madison. Symbiosis in the insect pathogenic nematode *Heterorhabditis bacteriophora* and the enteric bioluminescent bacterium *Photorhabdus luminescens;* genes involving the symbiont specific colonization of infective juvenile-stage nematode intestine and nematode growth and reproduction.

Sheng-Yang He, Professor; Ph.D., Cornell. Molecular genetics of plant-pathogen interactions; type III protein secretion.

Roger K. Maes, Associate Professor; D.V.M., Ghent (Belgium); Ph.D., Purdue. Clinical virology; natural host models of herpesvirus latency; molecular biology and immunity involving feline herpesvirus-1; mucosal immunization.

Veronica M. Maher, University Distinguished Professor; Ph.D., Wisconsin–Madison. Mechanisms of mutagenesis and recombination in human cells; DNA-induced damage by chemical carcinogens or radiation; DNA repair; molecular mechanisms of carcinogenesis.

Linda S. Mansfield, Professor; V.M.D., Ph.D., Pennsylvania. Parasite immunobiology; host bacterial-parasite interactions; molecular parasitology.

J. Justin McCormick, University Distinguished Professor; Ph.D., Catholic University. Mechanisms of malignant transformation of human cells; role of oncogenes in transformation; chemical- and radiation-induced carcinogenesis.

Leonel Mendoza, Associate Professor; Ph.D., Texas. Cloning and characterization of the genes in *Pythium* vaccines.

Martha H. Mulks, Professor; Ph.D., Rensselaer. Medical and veterinary microbiology; genetics and virulence factors of pathogenic bacteria, especially *Actinobacillus pleuropneumoniae* and *Neisseria gonorrhoeae;* microbial IgA proteases; vaccine development.

Maria J. Patterson, Professor; Ph.D., Northwestern; M.D., Michigan State. Infectious disease; medical and clinical microbiology.

James J. Pestka, Professor; Ph.D., Cornell. Modulation of immune function through diet; immunotoxicology.

Suzanne M. Thiem, Associate Professor; Ph.D., Idaho. Molecular biology and genetics of baculoviruses; control of virus gene expression; viral pathogenesis and host range.

Thomas S. Whittam, Hannah Professor; Ph.D., Arizona. Population genetics and molecular evolution of pathogenic bacteria; emerging infectious diseases; host-parasite coevolution.

Vincent B. Young, Assistant Professor; M.D., Ph.D., Stanford. Molecular pathogenesis, microbial ecology, and immunology of enteric bacterial infections.

Yong-Hui Zheng, Assistant Professor; Ph.D., Hokkaido (Japan). Innate immunity to HIV-1; restriction factors to retroviruses.

NEW YORK UNIVERSITY

Sackler Institute of Graduate Biomedical Sciences
School of Medicine
Department of Microbiology

Program of Study

The Department of Microbiology of the Sackler Institute of Graduate Sciences at New York University School of Medicine offers a comprehensive graduate training program leading to the Ph.D. The Department of Microbiology promotes a modern, integrated approach to the study of microbiology that includes research opportunities in prokaryotic and eukaryotic cells and viruses that infect them. The research interests of faculty members range from the basic mechanisms of gene expression to microbial pathogenesis and the host response to infection. The purpose of the graduate program is to provide students with a strong foundation in the principles and techniques of modern cellular and molecular biology as applied to the investigation of microbial pathogenesis and gene regulation. To achieve this goal, students take a core curriculum that includes courses in microbiology, biochemistry, and cellular and molecular biology and are offered elective courses in microbial pathogenesis, genetics, immunology, virology, and computer applications.

The doctoral program usually takes four to five years to complete. Students spend their first year taking courses and familiarizing themselves, via a rotation program, with research directed by participating faculty members. At the end of the first year, students select an area of research they wish to pursue and a research adviser. In their second year, students continue course work and begin organized doctoral research. Preliminary examinations are given at the end of the second year. Successful candidates then devote their full time to research and advanced elective courses until they complete their doctoral project.

Research Facilities

Extensive and modern research facilities and laboratory equipment are available within the Department of Microbiology, located in the Medical Science Building and the Skirball Institute of the New York University School of Medicine. Molecular biology and biotechnology research is supported by state-of-the-art equipment that includes complete computer facilities; instrumentation for protein and DNA sequencing and synthesis; and extensive tissue culture, electron microscopy, gene cloning, and P3 virology laboratories. Additional specialized research facilities are available through the NYU Cancer Institute, located at the School of Medicine.

Financial Aid

All graduate students are supported by either assistantships or traineeships, which carry stipends of $26,000 per year and also cover all tuition fees and health insurance costs. Financial support is provided for the entire duration of study. Housing loans of $1500 a year are available through the Sackler Institute.

Cost of Study

Full tuition scholarships cover most education-related expenses.

Living and Housing Costs

University-supplied apartments at the Medical Center are available and guaranteed to all. Rents range from about $575 to $1390 per month depending on the types of accommodations. Privately owned rooms and apartments are available at a wide range of rents throughout New York City.

Student Group

The graduate enrollment in the Department averages about 30, with equal numbers of men and women, and includes M.D./Ph.D. students. The graduate program at the School of Medicine (the Sackler Institute) has approximately 150 Ph.D. and 70 M.D./Ph.D. students.

Location

The School of Medicine is located in the center of Manhattan. The city's cultural opportunities, such as its offerings in music, theater, art, and dance and the lectures at its many universities and scholarly societies are unparalleled. There are lovely scenic areas and excellent recreational facilities near the city.

The University

New York University, a private institution founded in 1831, has a total enrollment of more than 40,000 students. With its campus at Washington Square, the University includes sixteen colleges, schools, and divisions at six major centers in Manhattan. The School of Medicine is located in one of the largest medical centers in the world, at 31st Street and First Avenue in the heart of New York City.

Applying

Admission is strictly limited to students with strong undergraduate preparation in both biological and physical sciences, including organic chemistry, calculus, and physics. The Department strongly encourages applications from underrepresented minority students.

Application by January 1 is recommended. Students are selected on the basis of college records, recommendations, and interviews with members of the departmental graduate committee. All applicants must hold a Bachelor of Science degree or its equivalent and submit scores on both the General and Subject tests of the Graduate Record Examinations.

Correspondence and Information

Dr. Michael Garabedian
Graduate Adviser for Microbiology
New York University School of Medicine
550 First Avenue
New York, New York 10016
Phone: 212-263-7662
E-mail: garabm01@med.nyu.edu
Web site: http://microbiology.med.nyu.edu/docs/home.shtml

New York University

THE FACULTY AND THEIR RESEARCH

The faculty pursues a broad spectrum of research in modern biology. Research interests include the following topics: microbial and molecular genetics; mechanisms of pathogenicity and host resistance to infectious agents; AIDS, retrovirology, and oncogenic viruses; growth factors, cytokines, and mechanisms of signal transduction. The entire Departmental faculty is listed below.

Claudio Basilico, M.D., Professor of Microbiology and Chairman. Control of proliferation of normal and cancer cells; growth factors and oncogenes.

Joel Belasco, Ph.D., Professor of Microbiology. mRNA degradation; protein-RNA interactions; structure and assembly of HIV-1 Rev; microRNA function.

Martin J. Blaser, M.D., Professor of Medicine and Microbiology and Chairman of Medicine. Biology of bacterial persistence in mammalian hosts.

Andrew Darwin, Ph.D., Assistant Professor of Microbiology. Molecular pathogenesis of Yersinia.

Heran Darwin, Ph.D., Assistant Professor of Microbiology. Molecular mechanisms of mycobacterium tuberculosis pathogenesis.

Irina Derkatch, Ph.D., Assistant Professor of Microbiology. Biogenesis and physiology of prions.

Joel Ernst, M.D., Professor of Medicine and Microbiology, Director of Infectious Diseases. Immunity to tuberculosis; pathogenesis of tuberculosis.

Alvin E. Friedman-Kien, M.D., Professor of Dermatology and Microbiology. Pathogenesis and viral etiology of AIDS-related Kaposi's sarcoma.

Michael J. Garabedian, Ph.D., Associate Professor of Microbiology. Mechanism of steroid hormone receptor action.

Nathaniel Landau, Ph.D., Professor of Microbiology. Mechanisms of intracellular resistance to HIV.

David E. Levy, Ph.D., Professor of Molecular Pathology and Microbiology. Regulation of gene expression by viruses and cytokines; mechanisms regulating cell growth and malignancy of animal cells.

Alka Mansukhani, Ph.D., Research Assistant Professor. The role of growth factors and their receptors in development and oncogenesis.

Ian Mohr, Ph.D., Assistant Professor of Microbiology. Virus-host interactions; regulation of translation of herpesvirus-infected cells; oncolytic viruses.

Martin S. Nachbar, M.D., Associate Professor of Medicine and Microbiology. Computer applications for teaching in the sciences.

Richard P. Novick, M.D., Professor of Microbiology. Mobile genes, quorum sensing, and the regulation of bacterial virulence.

Guillermo I. Perez–Perez, Ph.D., Assistant Professor Microbiology and Medicine. Epidemiology and pathophysiology of enteric bacterial infections.

Robert J. Schneider, Ph.D., Professor of Microbiology. Regulation of gene expression in normal and transformed cells.

Naoko Tanese, Ph.D., Associate Professor of Microbiology. Mechanisms of transcriptional regulation in eukaryotes.

Derya Unutmaz, M.D., Associate Professor of Microbiology. Immune response to HIV infection.

Jan T. Vilcek, M.D., Professor of Microbiology. Molecular mechanisms of cytokine actions.

Angus Wilson, Ph.D., Assistant Professor of Microbiology. Regulation of cellular and viral gene expression during herpesvirus infection.

Hans-Georg Wisniewski, Ph.D., Research Assistant Professor of Microbiology. The role of TSG-6 in inflammation.

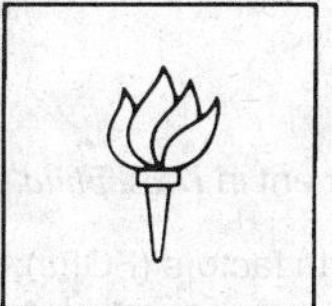

NEW YORK UNIVERSITY

School of Medicine
Sackler Institute of Graduate Biomedical Sciences
Graduate Programs in Molecular Oncology and in Immunology

Programs of Study

The program is intended to train predoctoral and postdoctoral individuals in the areas of molecular oncology, humoral and cellular immunology, and innate immunity. Particular emphasis is placed on a combined education in these disciplines for a better understanding of the mechanisms by which normal cells are transformed into tumor cells and of the host factors, particularly the immunological factors, which regulate the progressive growth of transformed cells. A unique aspect of the program is that trainees become proficient in the basics of both immunology and molecular oncology and then focus on cancer-related topics in one of these two areas for their research. The participating faculty members have expertise in the molecular biology of oncogenes and tumor suppressor genes; signal transduction; genetic alteration in human cancer and the molecular aspects of tumor development; regulation of the mammalian cell cycle; mechanisms of transcriptional and epigenetic gene regulation, with an emphasis on the role of chromatin structure; the pathogenesis of AIDS; cytokine and chemokine biology; molecular genetics, as studied by gene transfer into eukaryotic cells; the molecular biology of immunoglobulin genes; the chemistry and biology of the complement cascade and of complement receptors; the interfacing of the hematopoietic and lymphopoietic systems; and many aspects of cellular and tumor immunology, including immunosuppression, cytotoxic T-lymphocyte and macrophage responses, lymphocyte differentiation, the migration and regulation of lymphocytes, and the development of cancer vaccines and cancer gene therapy.. The programs interrelate wherever possible and aim to prepare students for research and teaching careers in molecular oncology and immunology. Training is coordinated with related Sackler Institute programs in cellular and molecular biology and microbiology, and students take many of the same courses as students in these programs. Extensive use of animal models is emphasized in many research projects.

Research Facilities

The New York University Medical Center is one of the largest teaching and research complexes in the country. Research laboratories are located in the Basic Medical Science Building, the new Smilow Research Center, the Skirball Institute of Biomolecular Medicine, and Tisch, Bellevue, and VA Hospitals. The participating faculty members occupy modern research laboratories equipped for research in their areas of interest. There are centralized departmental facilities as well as Medical Center–wide resources available through the NYU Cancer Institute and the Skirball Institute. These facilities and resources include the latest nucleic acid and protein sequencing and analysis instrumentation; structural biology laboratories with state-of-the-art X-ray crystallography equipment; P-2 and P-3 laboratories for studying highly infectious agents; modern microscopy equipment such as laser confocal microscopes, transmission and scanning electron microscopes, and image-processing facilities; and major animal-care centers, which include a transgenic mouse unit, an animal histology and pathology laboratory, and an animal imaging facility. Excellent libraries and state-of-the-art computer facilities (ranging from individual personal computers to access to mainframe computers and minicomputers) are available to all students.

Financial Aid

All graduate students are supported by either assistantships or traineeships, which carried stipends of $26,000 for the 2005–06 academic year, in addition to all tuition fees and health insurance costs. Financial support is provided for the duration of study. No teaching or laboratory assisting is required for the receipt of financial aid. Low-interest housing loans of $1500 per year are also available for qualified students, as are loans for the purchase of personal computers.

Cost of Study

The cost of study is usually met by the financial aid awards, as described above.

Living and Housing Costs

University housing in dormitory facilities and apartment complexes within the immediate vicinity of the Medical Center is guaranteed for all incoming graduate students. Depending on the type of accommodation, rents range from $600 per month for a dormitory room to $730 to $840 per month for apartments. The services of the Medical Center's housing office are also available to students who have special needs or wish to live off campus.

Student Group

There are approximately 190 Ph.D. and 80 M.D./Ph.D. candidates in the Institute. On average, 40 to 50 new students matriculate each year. Students are drawn from a pool of highly qualified national and international applicants. About 16 percent of the students are from underrepresented minority groups, 60 percent are women, and 30 percent are international students. In addition to the graduate students, there are 676 M.D. candidates enrolled in the School of Medicine.

Location

The New York University Medical Center, where the Sackler Institute is located, is in the heart of midtown Manhattan, on First Avenue between 23rd and 34th Streets. The location provides easy access to the myriad cultural, social, and educational opportunities of New York City with its world-class museums, libraries, theater, dance, music, and sporting events. To facilitate access to these activities, the Medical School has established a student activities office, which supplies information and discount tickets to many of these events.

The University

New York University, which was founded in 1831, is the largest private university in the country, with an enrollment of more than 50,000 students. The University includes thirteen colleges and schools at five major centers in Manhattan. The main campus is located around Washington Square in the historic Greenwich Village area of the city. University-sponsored bus service connects the main campus to the Medical Center. The Medical Center is composed of the School of Medicine and affiliated hospitals and research institutes and has a history of more than 150 years of training nationally and internationally recognized physicians and scientists.

Applying

Only full-time students are admitted into the Sackler Institute to the Ph.D. or M.D./Ph.D. programs. Applicants must have at least a bachelor's degree or its equivalent from a college or university of recognized standing and have a strong background in the biological, chemical, and physical sciences. Evaluation for admission to the different programs is processed by the individual program admissions committees and is based on previous academic achievement, letters of recommendation, assessment of the applicant's scientific potential, and scores on the Graduate Record Examinations (GRE). It is highly recommended that applicants also take a GRE Subject Test in biology or chemistry. The Test of English as a Foreign Language (TOEFL) is required for international students. After an initial screening, the selected candidates living in the United States are invited to NYU for interviews. Applicants are encouraged to submit application forms and all supporting material by January 4. Applications received after this date are considered at the discretion of the admissions committee of the desired program.

Correspondence and Information

Dr. David Levy
Graduate Student Advisor
Department of Pathology
New York University School of Medicine
550 First Avenue
New York, New York 10016
Web site: http://www.med.nyu.edu/sackler/training/path/

New York University

THE FACULTY AND THEIR RESEARCH

Erika A. Bach, Ph.D., Assistant Professor of Pharmacology. Role of JAK-STAT and receptor tyrosine kinase signaling pathways in development in *Drosophila*.

Ross Basch, M.D., Professor of Pathology. Developmental hematology; differentiation of immunologically competent cells.

Claudio Basilico, M.D., Professor and Chairman of Microbiology. Oncogenes and growth factors; expression and function of fibroblast growth factors (FGFs); control of cell-cycle progression.

Nina Bhardwaj, M.D., Ph.D., Professor of Medicine. Immunobiology of antigen-presenting cells called dendritic cells (DCs).

Peter Brooks, Ph.D., Associate Professor. Role of integrins and cryptic ECM epitopes in angiogenesis and tumor metastasis.

Steven Burakoff, M.D., Laura and Isaac Perlmutter Professor; Director, Skirball Institute of Biomolecular Medicine; and Director, NYU Cancer Institute. Negative regulatory mechanisms that control signal transduction in lymphocytes.

Moses V. Chao, Ph.D., Professor of Cell Biology, Physiology, and Neuroscience. Mechanisms of action of trophic factor receptors; glial cell differentiation.

Pamela Cowin, Ph.D., Professor of Cell Biology. Breast cancer; mammary development and stem cells; Wnt signaling and cell adhesion.

Gregory David, Ph.D., Assistant Professor of Pharmacology. Regulation of chromatin structure and chromosomal dynamics in normal and transformed cells.

Michael L. Dustin, Ph.D., Associate Professor of Immunology. T-cell activation, migration, and adhesion; formation of the immunological synapse.

Brian D. Dynlacht, M.D., Ph.D., Professor of Pathology. Control of cell-cycle progression; transcriptional regulatory mechanisms in mammalian cells.

Adrian Erlebacher, M.D., Ph.D., Assistant Professor of Pathology. Immunobiology of pregnancy.

Alan B. Frey, Ph.D., Associate Professor of Cell Biology. Tumor immunology.

Hodaka Fujii, M.D., Ph.D., Assistant Professor of Pathology. Mechanisms of signal transduction from receptors of immunoregulatory cytokines.

Michael Garabedian, Ph.D., Associate Professor of Microbiology. Mechanisms of signal transduction and transcriptional regulation by steroid hormone receptors.

Leslie I. Gold, Ph.D., Associate Professor. Mechanisms of loss of growth regulation in hormone regulated cancers cancer (TGF-β/cell cycle); steroid–growth factor interactions and stromal-epithelial interactions in normal and malignant growth.

Stevan R. Hubbard, Ph.D., Associate Professor of Pharmacology. Structural and mechanistic aspects of signaling by receptor tyrosine kinases.

Hannah L. Klein, Ph.D., Professor of Biochemistry and Medicine. DNA damage avoidance pathways and maintenance of genomic stability; roles of repair factors, DNA damage checkpoints, and sister chromatid cohesion.

Juan Lafaille, Ph.D., Associate Professor of Pathology and Medicine. Molecular pathogenesis of autoimmune and allergic diseases.

Peng Lee, M.D., Ph.D., Assistant Professor of Pathology. Androgen receptor, its cofactors and target genes in endocrine malignancies, including prostate and breast cancers; stromal microenvironment in cancer development and progression.

David E. Levy, Ph.D., Professor of Pathology and Microbiology. Regulation of gene expression by interferons, viruses, growth factors, and cytokines; molecular biology of signal transduction; mechanisms regulating cell growth and malignancy of animal cells.

Daniel Meruelo, Ph.D., Professor of Pathology. Immunogenetics; virology; molecular biology; genetic control of leukemia susceptibility; AIDS research; gene therapy.

Moosa Mohammadi, Ph.D., Associate Professor of Pharmacology. Molecular mechanisms of fibroblast growth factor (FGF) receptor regulation and signaling.

Elizabeth W. Newcomb, Ph.D., Associate Professor of Pathology. Mechanisms of action of anticancer drugs on cell death and angiogenesis; radiotherapy as an immune modulator; experimental glioma animal model to test novel therapeutic approaches.

Michele Pagano, M.D., Professor of Pathology. Cell-cycle control; ubiquitin-mediated proteolysis; cancer biology.

Angel Pellicer, M.D., Ph.D., Professor of Pathology. Oncogenes: mechanism of activation and function using transgenic technology.

Mark R. Philips, M.D., Professor of Medicine, Cell Biology, and Pharmacology. Posttranslational processing and membrane targeting of CAAX proteins, such as the oncogene product *ras*.

Daniel B. Rifkin, Ph.D., Professor of Cell Biology. Mechanisms of growth factor action; studies on TGF-β signaling.

David Ron, M.D., Professor of Medicine and Cell Biology. Cellular response to malfolded proteins.

David B. Roth, M.D., Ph.D., Irene Diamond Professor and Chairman of Pathology. Mechanisms of antigen receptor gene rearrangement, maintenance of genomic stability, and lymphomagenesis.

Robert Schneider, Ph.D., Professor of Microbiology. Mechanisms of cellular transformation and pathogenesis in human carcinoma.

Jane Skok, Ph.D., Assistant Professor of Pathology. Nuclear organization of immunoglobulin genes.

Susan Smith, Ph.D., Associate Professor of Pathology. Molecular mechanisms of telomere function.

George Teebor, M.D., Professor of Pathology. DNA repair in mammalian cells; radiation and oxidative damage to DNA; DNA repair enzymology.

Daniel Turnbull, Ph.D., Associate Professor of Radiology and Pathology. Noninvasive imaging methods to analyze mouse models of disease.

David Zagzag, M.D., Ph.D., Associate Professor of Pathology and Neurosurgery. Molecular biology of angiogenesis and invasion in central nervous system embryogenesis and oncogenesis.

Edward Ziff, Ph.D., Professor of Biochemistry. Relationship of synaptic organization to neurotransmitter receptor function at excitatory synapses; mechanisms of regulation of glutamate receptors and synaptic transmission.

Susan Zolla-Pazner, Ph.D., Professor of Pathology. Acquired immune deficiency syndrome (AIDS) and the effects of HIV infection on the regulation of the immune response.

STANFORD UNIVERSITY

School of Medicine
Department of Microbiology and Immunology

Programs of Study

The Department of Microbiology and Immunology offers programs leading to the Ph.D. degree in microbiology and immunology. Graduate students spend part of the first year in formal courses, which, depending on their background, may include offerings in medical microbiology, immunology, virology, genetics, molecular parasitology, molecular biology, and biochemistry, as well as stress response and biostatistics. Courses in subjects pertinent to the students' particular interests, such as medical parasitology, histology, and general pathology are also available in other departments of the University. A Master of Science degree is awarded only in exceptional circumstances.

In their first year, students usually do at least two rotations in different laboratories. In the spring quarter of the second year, each student orally defends a formal research proposal on a topic outside the intended thesis project. In the autumn quarter of the third year, a research proposal based on the student's own thesis topic is defended to his or her thesis committee. This research occupies the remainder of their stay in the department. All students are also required to obtain some teaching experience, usually by serving as a teaching assistant in at least two department courses.

Research Facilities

The facilities of the department are located in the modern Fairchild Building, adjacent to the Stanford Medical Center. The department occupies more than 20,000 square feet of space, including labs well equipped for research in microbiology and immunology. Additional facilities are available in the Medical Center and in the Beckman Center for Molecular and Genetic Medicine.

Financial Aid

Students are encouraged to apply for non-Stanford fellowships or scholarships (e.g., NSF, Howard Hughes, California State, and others). Teaching and research assistantships are also available that provide a net stipend of $25,500 for twelve months and a waiver of tuition fees. Since these positions are predicated on half-time assistantships/half-time studies (10 units per quarter), a student with a bachelor's degree could complete the Ph.D. requirement of 135 units in a minimum time of 3¾ calendar years.

Cost of Study

Tuition was $38,084 for the 2003–04 academic year, or $9521 per quarter, for students not supported by grants. However, tuition costs are normally covered for all students admitted to the Ph.D. program.

Living and Housing Costs

On-campus housing is available for single graduate students, couples, and families with children. Other suitable housing can be found off campus.

Student Group

There is an average enrollment of approximately 100 graduate students and postdoctoral fellows in the department. All parts of the United States and many other countries are usually represented.

Location

The city of Palo Alto (population approximately 60,000) and the adjoining campus are located between the hills bordering the Pacific Ocean and San Francisco Bay. The climate is mild, at midday averaging 60 degrees in winter and 74 degrees in summer. Humidity is low, nights are cool, and rainfall is rare from May through September. The campus area is contiguous with a series of peninsular towns, many of which provide cultural events, and it is 35 miles from San Francisco, which is reached by freeway or public conveyance in three quarters of an hour. The University of California, Berkeley, campus is about an hour away by car.

The University

The University is situated in the central portion of the San Francisco Peninsula, with easy access to the urban centers of the Bay Area. The University was deeded more than 8,000 acres of land at its founding in 1885. Of this land, the 5,000 acres set aside for academic purposes provide a spacious campus area. The student body numbers between 11,000 and 12,000, about half of whom are graduate students.

Applying

Applications are made to the Office of Graduate Admissions, with a deadline of December 15, for admission and financial support for the autumn quarter. Under certain circumstances a student may begin study in another quarter. Candidates for admission to the department should have completed all of the essential premedical sciences. The department requires the results of both the General Test and a Subject Test, preferably in biology, of the Graduate Record Examinations. The department strongly advises prospective students to take the GRE in October so that scores are available when applications are evaluated. Applicants are considered without regard to race, color, creed, religion, sex, marital status, age, physical handicap, or national origin.

Correspondence and Information

Julia A. Wong
Microbiology and Immunology Student Services Coordinator
Department of Microbiology and Immunology
Stanford University School of Medicine
299 Campus Drive
Stanford, California 94305-5124
E-mail: juwong@stanford.edu
Web site: http://cmgm.stanford.edu/micro/index.html

Stanford University

THE FACULTY AND THEIR RESEARCH

Ann M. Arvin, M.D., Professor. Molecular virology of and immunity to varicella-zoster virus (VZV); virus–target cell interactions in the pathogenesis of VZV infection in the natural human host and in the SCID-hu model; analysis of the T-cell-mediated host response. *J. Virol.* 77:5964, 5607, 1268, 489, 2003; *J. Virol.* 74:11425, 76:8468, 2002; *Proc. Natl. Acad. Sci. U.S.A.* 95:11969, 1998.

Helen M. Blau, Ph.D., Professor. Plasticity of stem cells; immune responses in tissue regeneration; RNAi and signal transduction pathways; blood vessel growth and inhibition; tumor biology. *Nat. Cell Biol.,* in press; *PNAS* 100:2088–93, 2003; *Cell* 111:589–601, 2002; *Science* 298:361–2, 2002; *Nature* 419:437, 2002.

John C. Boothroyd, Ph.D., Professor. Pathogenesis of the protozoan parasite *Toxoplasma*, with emphasis on population biology, gene regulation, protein trafficking, invasion, and developmental biology. *Mol. Biochem. Parasitol.* 125:155–8, 2002; *Eukaryotic Cell* 1:329–40, 2002; *Nature Structural Biol.* 9:606–11, 2002; *Science* 294:161–5, 2001.

Yueh-hsiu Chien, Ph.D., Professor. Development and function of gd T cells. Genetic and molecular analyses of bacterial infection on host adaptive immune response. *Proc. Natl. Acad. Sci. U.S.A.* 98:10261–6, 2001; *Science* 287:314–6, 2000; *J. Exp. Med.* 190:1343–50, 1999; *Immunity* 8:667–73, 1998; *Ann. Rev. Immunol.* 14:511–32, 1996.

Christopher H. Contag, Ph.D., Assistant Professor. Molecular mechanisms of pathogenesis and host response to insult using in vivo cellular and molecular imaging to reveal biological processes in living animal models of disease. Emphases on gene expression and cell migration in vivo. *Nature Med.* 4:245–7, 1999; *Proc. Natl. Acad. Sci. U.S.A.* 96:12044–9, 1998.

Mark M. Davis, Ph.D., Professor and Chair. Mechanisms of T-lymphocyte recognition and differentiation; thymic selection; dynamics of cell surface molecules during T-cell recognition. *Nature* 418:552–6, 419:845–9, 2002; *Science* 289:1349–52, 2000; *Ann. Rev. Immunol.* 16:523–44, 1998; *Science* 274:94–6, 1996.

Stanley Falkow, Ph.D., Professor. Genetic and molecular mechanisms of microbial pathogenicity; disruption of the epithelial apical-junctional complex by *Helicobacter pylori* CagA. *Science* 300:1430–4, 2003.

Stephen J. Galli, M.D., Professor. Regulation of mast cell and basophil development, phenotype, and function; roles of mast cells in innate and acquired immunity and inflammation; pathogenesis of asthma and allergic inflammation. *Nature Rev. Immunol.* 2:773, 2002; *Curr. Opin. Immunol.* 11:53–9, 1999; *J. Exp. Med.* 188:234–8, 1998; *Nature* 392:90–3, 1998; 390:172–5, 1997.

K. Christopher Garcia, Ph.D., Assistant Professor. Structural biology of cell surface receptor activation in autoimmunity and host-pathogen interactions. *Ann. Rev. Immunol.* 17:369–97, 1999; *Science* 274:209–19, 1996.

Harry B. Greenberg, M.D., Professor. Gene coding assignments for rotavirus, with specific emphasis on the molecular determinants of virulence, host range, and immunity; immunity to hepatitis C and influenza viruses. *J. Virol.* 76:4467–82, 2002; *J. Clin. Inv.* 109:1203–13, 2002; *J. Immunol.* 166:877–84, 2001; *J. Clin. Inv.* 106:1541–52, 2000; *Proc. Natl. Acad. Sci. U.S.A.* 96:5692–7, 1999.

Peter Jackson, Ph.D., Assistant Professor. Control of the mitotic cell cycle by cyclin-dependent kinases, dual specificity phosphatases, and ubiquitin ligases; biochemical, enzymological, and structural studies of substrate recognition and ubiquitin chain formation mechanisms for E3 ubiquitin ligases; chemical biology of small molecule inhibitors of mitosis. *Dev. Cell* 4:813–26, 799–812, 2003; *Nature* 416:850–4, 2002; *Cell* 105:645–55, 2001; *Nature Cell Biol.* 3:715–22, 2001; *Genes Dev.* 13:2242–57, 2001; *Trends Cell Biol.* 10:429–39, 2000.

Karla Kirkegaard, Ph.D., Professor. Molecular genetics and biochemistry of RNA viral genome replication, including the mechanism of RNA-dependent RNA polymerases and of cellular membrane rearrangements during viral infection. *J. Virol.* 77:7843–55, 2003; *Proc. Natl. Acad. Sci. U.S.A.* 100:7289–94, 2003; *J. Biol. Chem.* 277:16324–31, 2002; *Science* 296:2218–22, 2002.

A. C. Matin, Ph.D., Professor. Drug resistance in bacteria and their biofilms; stress response under zero gravity; molecular approaches to bioremediation. *Antimicrob. Agents Chemother.* 46:2458, 2002; In *Encyclopedia of Environmental Microbiology*, vol. 6, pp. 3034–6. New York: John Wiley & Sons, 2001; *Mol. Microbiol.* 36:414–23, 2000.

Hugh O. McDevitt, M.D., Professor. Structure-function analysis of H2 and HLA-D region gene products; how these proteins regulate the immune response and predispose to autoimmunity; development of methods to selectively alter Ia/DR function in vivo. *J. Exp. Med.* 196:481–92, 2002; *Proc. Natl. Acad. Sci. U.S.A.* 99:12287–92, 2002; *J. Exp. Med.* 193:1333–40, 2001.

Edward S. Mocarski, Ph.D., Professor. Molecular biology and pathogenesis of cytomegaloviruses; identification of genes involved in replication, latency, and pathogenesis, including those that modulate the host immune response. *J. Virol.* 77:7563–74, 77:631–41, 2003; *Science* 296:1323–6, 2002.

Garry P. Nolan, Ph.D., Associate Professor. Multidimensional proteomics of phosphoproteins by FACS, HIV-1 host protein parasitism; retroviral libraries for complementation cloning of signaling molecules in T-cell biology, angiogenesis. *Immunity* 16:51–65, 2002; *Nature Biotechnol.* 20:155–62, 2002; *Nature Genetics* 27:23–9, 2001; *Immunity* 15:687–90, 1999; *Cell* 95:595–604, 1998.

Peter Parham, Ph.D., Professor. Regulation of cytotoxic T cells and natural killer cells by MHC class I polymorphisms, with an emphasis on the diversity of MHC class I in human populations and comparisons in other species. *Immunity* 10:439–49, 1999; 7:739–51, 753–63, 1997.

Charles G. Prober, M.D., Professor. Conduct of clinical research focused on infections in children with an emphasis on viral infections, especially those caused by herpes viruses. Neonatal herpes infection: Diagnosis, treatment and prevention. *Semin. Neonatol.* 7:283–91, 2002.

David A. Relman, M.D., Associate Professor. Molecular, genome-wide analysis of pathogen-host cell interactions; *Bordetella sp.* pathogenesis and genomics; pathogen discovery using cultivation-independent molecular methods; exploration of human microbial ecology. *Appl. Environ. Microbiol.* 69:1687–94, 2003; *Proc. Natl. Acad. Sci. U.S.A.* 100:1896–901, 2003; 99:972–7, 2002; *Cell Microbiol.* 4:825–33, 2002.

Peter Sarnow, Ph.D., Professor. Molecular biology of RNA virus-host interactions; mechanism of internal initiation of protein synthesis in viral and cellular mRNA. *J. Virol.* 77:2801–6, 2003; *J. Mol. Biol.* 324:889–2, 2002; *EMBO J.* 20:240–9, 2001; *Proc. Natl. Acad. Sci. U.S.A.* 98:12972–7, 2001; *Cell* 102:511–20, 2000; *Mol. Cell. Biol.* 20:4490–9, 2000; *Proc. Natl. Acad. Sci.* 13:13118–23, 1999.

David S. Schneider, Ph.D., Assistant Professor. Using the fruit fly *Drosophila* as a model vector to study the cell biology and genetics of malaria transmission. *Science* 288:2376–9, 2000.

Gary K. Schoolnik, M.D., Professor. Pathogenesis of *M. tuberculosis*; pathogenesis of enteropathogenic *E. coli*; environmental persistence of *V. cholerae* in environmental habitats. *Science* 284:1520–3, 1999; *Proc. Natl. Acad. Sci. U.S.A.* 96:4028–33, 1999; *Science* 280:2114–8, 1998.

Upinder Singh, M.D., Assistant Professor. Genetic and genomic studies of protozoan parasites *Entamoeba histolytica* and *Toxoplasma gondii* with an emphasis on host-pathogen interactions, pathogenesis, and developmental biology. *J. Biosciences* 27 (supplement 3):595–601, 2002; *Mol. Biochem. Parasitol.* 120:107–16, 2002; *Eukaryotic Cell* 1:329–40, 2002; *Mol. Microbiol.* 44:721–33, 2002.

Bruce A. D. Stocker, M.D., Professor Emeritus. *Salmonella* genetics; auxotrophic *Salmonella* and *Shigella* as live vaccines and live-vaccine presenters of protective antigens or epitopes of other pathogens. Evidence for an interrupted-chromosome intermediate in a phage-mediated transduction. *Vaccine* 16:24, 1998; 14:251, 1996; *J. Bacteriol.* 178:1113, 1996.

Man-Wah Tan, Ph.D., Assistant Professor. Genetics and genomics dissection of host-pathogen interactions using a multipathogen *C. elegans* pathogenesis system; genetics and molecular analyses of signaling pathways in host innate immune response. *Curr. Opin. Microbiol.* 3:29–34, 2000; *Cell* 96:47–56, 1999; *Proc. Natl. Acad. Sci. U.S.A.* 96:715–20, 2408–99, 1999.

Julie A. Theriot, Ph.D., Assistant Professor. Cell biology of interactions between infectious bacteria and the human host cell actin cytoskeleton; actin-based motility of bacterial pathogens. *J. Cell Biol.* 135:647–60, 1996; *Ann. Rev. Cell Dev. Biol.* 11:213–39, 1995; *Proc. Natl. Acad. Sci. U.S.A.* 92:6572–6, 1995; *Mol. Microbiol.* 17:945–51, 1995.

Lucy S. Tompkins, M.D., Ph.D., Professor. Molecular pathogenic studies of *Helicobacter* and *Campylobacter*; genetic and molecular studies to characterize the host-parasite interaction. *Helicobacter pylori* enter and survive within multivesicular vacuoles of epithelial cells. *Cell Microbiol.* 4:667–90, 2002.

STONY BROOK UNIVERSITY, STATE UNIVERSITY OF NEW YORK

Studies in Molecular Genetics and Microbiology

Program of Study

The Department of Molecular Genetics and Microbiology offers a program of study leading to the Ph.D. degree in molecular genetics and microbiology, which is awarded through the Graduate School of Stony Brook University.

Studies are designed to prepare a student to become an effective basic research scientist. The major areas of study are cancer, infectious disease, molecular genetics, signal transduction, and cell cycle regulation. Throughout the period of training, the student is mentored by an advisory committee composed of faculty members who are active in several research areas and a research adviser who is selected by the student at the end of the first year of study. The individualized program aims to develop breadth of understanding in the basic disciplines through active participation in laboratory research, course work, and seminars. At the end of the third semester of study, candidates for the Ph.D. must pass a written comprehensive examination. At the end of the second year, they submit and orally defend a research proposal in the form of a research grant application. Students must also submit and defend a dissertation based on an original research project and be the first author of a publication in the literature. Typically, this requires five years. The program faculty members are in the Department of Molecular Genetics and Microbiology, in other departments at Stony Brook University, at Cold Spring Harbor Laboratory, at The Feinstein Institute for Medical Research, and at Brookhaven National Laboratory.

Research Facilities

The Department of Molecular Genetics and Microbiology is housed in the Life Sciences Building and the Centers for Molecular Medicine, a new state-of-the-art research and teaching facility that serves as a physical and intellectual bridge between investigators in the adjacent Life Sciences Building and the nearby University Health Sciences Center. Centralized facilities include cell culture, monoclonal antibody production, DNA microarray analysis, DNA sequencing, a computer facility that includes a graphics computer workstation and analysis software, liquid scintillation counters, a phosphoimager, a laser densitometer, an Odyssey Infrared Imaging System, a fermentation apparatus, fluorescence microscopes, and dishwashing services. The research facilities at Cold Spring Harbor Laboratory and Brookhaven National Laboratory are world-renowned and are available to members of the department.

The library of the Health Sciences Center holds 250,510 volumes, including more than 4,375 journal titles and an expanding array of electronic journal access. Other campus libraries include the Frank Melville, Jr. Memorial Library and other satellite libraries in the physics and chemistry buildings.

Financial Aid

Students are admitted with full financial support, which is provided in 2006–07 at the rate of $25,000 for the twelve-month period. In addition to the stipend, full-tuition scholarships are provided for all students receiving full-time support. Health insurance is provided by the University for all graduate students in the program.

Cost of Study

Miscellaneous University fees amount to approximately $290 each semester. All admitted students receive tuition scholarships.

Living and Housing Costs

Dormitory accommodations are available for single students. Residence halls have limited cooking facilities. The University apartments are designed to house graduate, married, and health sciences students. There are accommodations for married graduate students in campus apartment complexes. Privately owned housing is widely available off campus and is used by most graduate students. Information may be obtained from the University's Off Campus Housing Office.

Student Group

The present student enrollment at Stony Brook is 17,763 students, including 6,668 graduate students enrolled in a wide range of programs. Currently, there are approximately 300 graduate students enrolled in various biomedical graduate programs.

Student Outcomes

Graduates typically go on to postdoctoral research appointments, followed by academic careers in teaching and research; alternatively, graduates seek research careers in government or private industry.

Location

Stony Brook is located in a region of coves, beaches, and small historic villages on the North Shore of Long Island, approximately 60 miles east of New York City. The area has retained its distinctive New England flavor and is just minutes from the waters of Long Island Sound and the Atlantic Ocean, with proximity to the cultural, scientific, and industrial resources of New York City. The nearby biological research institutes at Cold Spring Harbor and Brookhaven National Laboratories provide additional advantages for the scientific community.

The University

Stony Brook is a 1,100-acre universe where world-renowned faculty members have created a stimulating, highly interactive environment for graduate studies, especially in the sciences and health professions. Established in 1957 as part of the SUNY system, Stony Brook is now recognized by the Association of American Universities (AAU) as one of the nation's finest universities, emphasizing research in a variety of disciplines.

Applying

Application forms and further information may be obtained by writing to the Department of Molecular Genetics and Microbiology or downloaded from the Internet. The General Test of the Graduate Record Examinations is required, with the Subject Test in either biology or biochemistry preferred. To ensure that the scores are available to meet the application deadline, the tests should be taken no later than December. The deadline for receipt of applications for the fall semester is January 15, but all applications are considered on a rolling basis. There is an application fee of $60 that must accompany the application.

Correspondence and Information

Chairman, Graduate Admissions Committee
Department of Molecular Genetics and Microbiology
State University of New York
Stony Brook, New York 11794-5222

Phone: 631-632-8812
E-mail: molecular_genetics@notes.cc.sunysb.edu
Web site: http://www.mgm.stonybrook.edu/index.shtml

Stony Brook University, State University of New York

THE FACULTY AND THEIR RESEARCH

Carl W. Anderson, Brookhaven National Laboratory; Ph.D., Washington (St. Louis), 1970. Cell cycle control and cellular response to DNA damage.

Wadie Bahou, Medicine/Hematology; M.D., Massachusetts, 1980. Vascular biology and thrombosis; molecular genetics.

Jorge Benach, Molecular Genetics and Microbiology; Ph.D., Rutgers, 1971. Infectious disease immunology and pathogenesis.

James Bliska, Molecular Genetics and Microbiology; Ph.D., Berkeley, 1988. Molecular and cellular basis of bacterial-host cell interactions; pathogenesis.

Nicholas Carpino, Molecular Genetics and Microbiology; Ph.D., SUNY at Stony Brook, 1997. Hematopoietic cell development, function, and intracellular signaling pathways; T-cell signaling pathways and immune system biology.

Carol A. Carter, Molecular Genetics and Microbiology; Ph.D., Yale, 1972. Replication of HIV; control mechanisms in protein assembly; viral pathogenesis.

Edward Chan, Pediatrics and Molecular Genetics and Microbiology; M.D., SUNY at Buffalo, 1997. Growth factor receptors and cancer.

Christopher Cutler, Periodontics; D.D.S., Emory, 1986. Periodontal disease.

Neta Dean, Biochemistry and Cell Biology; Ph.D., UCLA, 1988. Protein trafficking in yeast.

Nicholas Delihas, Molecular Genetics and Microbiology; Ph.D., Yale, 1961. Control of gene expression by regulatory RNAs; structure, function, and evolution of small RNAs.

John J. Dunn, Brookhaven National Laboratory; Ph.D., Rutgers, 1970. Transcription, processing, and translation of RNA.

Paul Freimuth, Brookhaven National Laboratory; Ph.D., Columbia, 1986. Adenovirus reproduction; virus-cellular receptor binding.

Martha Furie, Pathology and Molecular Genetics and Microbiology; Ph.D., Rockefeller, 1980. Interactions among endothelial cells, leukocytes, and pathogenic bacteria.

A. Bruce Futcher, Molecular Genetics and Microbiology; D.Phil., Oxford, 1981. Cell cycle; cyclins; yeast genetics.

Christine Ginocchio, North Shore University Hospital; Ph.D., SUNY at Stony Brook, 1993. HIV pathogenesis.

Gregory Hannon, Cold Spring Harbor Laboratory; Ph.D., Case Western Reserve, 1992. Growth control in mammalian cells; posttranscriptional gene silencing.

Michael J. Hayman, Molecular Genetics and Microbiology; Ph.D., National Institute for Medical Research (London), 1973. Mechanism of transformation by retroviral oncogenes; erythroid differentiation.

Janet C. Hearing, Molecular Genetics and Microbiology; Ph.D., SUNY at Stony Brook, 1984. Molecular analysis of latent Epstein-Barr virus DNA replication.

Patrick Hearing, Molecular Genetics and Microbiology; Ph.D., Northwestern, 1980. Viral molecular genetics; eukaryotic transcriptional regulation; adenovirus and hepatitis B virus replication; gene therapy; cancer.

Leemor Joshua-Torr, Cold Spring Harbor Laboratory; Ph.D., Weizmann (Israel), 1991. Structural biology and molecular recognition.

Wali Karzai, Ph.D., Johns Hopkins, 1995. Structure and function of RNA-binding proteins and biochemical studies of the SmpB-SsrA quality control system.

Eugene R. Katz, Molecular Genetics and Microbiology; Ph.D., Cambridge, 1969. Developmental genetic studies on *Dictyostelium discoideum;* role of membrane sterols in cell growth and development.

Richard Kew, Pathology; Ph.D., SUNY at Stony Brook, 1986. Leukocyte chemotaxis; inflammation; pulmonary immunopathology.

James Konopka, Molecular Genetics and Microbiology; Ph.D., UCLA, 1985. Cell growth and development in yeast; G protein-coupled receptor signal transduction; *C. albicans* pathogenesis.

Janet Leatherwood, Molecular Genetics and Microbiology; Ph.D., Johns Hopkins, 1993. Cell-cycle control of DNA replication.

Christopher Lee, Urology and Molecular Genetics and Microbiology; M.D., University of Medicine and Dentistry of New Jersey–Robert Wood Johnson Medical School/Rutgers. Cancer vaccine development.

Huilin Li, Brookhaven National Laboratory; Ph.D., University of Sciences and Technology (China), 1994. Structural biology of macromolecular assemblies and membrane proteins by cryo-electron microscopy.

Erwin London, Biochemistry and Cell Biology; Ph.D., Cornell, 1980. Membrane protein folding and lipid interaction; protein translocation across membranes.

Scott Lowe, Cold Spring Harbor Laboratory; Ph.D., MIT, 1994. Apoptosis; anticancer therapy resistance.

Benjamin Luft, Medicine; M.D., Yeshiva (Einstein), 1976. Pathobiology of *Borellia* and toxoplasma.

Erich R. Mackow, Medicine; Ph.D., Temple, 1984. Rotavirus molecular genetics; determinants of viral neutralization and pathogenesis; reverse genetics; recombinant vectors and mucosal immunity; biochemistry of cell fusion; viral pathogenesis.

Kenneth B. Marcu, Biochemistry and Cell Biology; Ph.D., SUNY at Stony Brook, 1975. Immunoglobulin gene expression and recombination.

Todd Miller, Physiology and Biophysics; Ph.D., Rockefeller, 1988. Regulation and substrate specificity of tyrosine kinases.

Ute Moll, Pathology; M.D., Ulm (Germany), 1985. Function and regulation of the p53 family in apoptosis and cancer.

Senthil Muthuswamy, Ph.D., McMaster, 1995. Understanding cancer initiation using three-dimensional epithelial structures.

Aaron Neiman, Biochemistry and Cell Biology; Ph.D., California, San Francisco, 1994. Developmental regulation of the secretory pathway.

Kepal Patel, Surgery and Molecular Genetics and Microbiology; M.D., University of Medicine and Dentistry of New Jersey–Robert Wood Johnson Medical School/Rutgers, 1996. Genetic profiling in the progression of thyroid cancer.

Aniko Paul, Molecular Genetics and Microbiology; Ph.D., Stanford, 1966. Biochemical and genetic studies of poliovirus replication.

Nancy Reich, Molecular Genetics and Microbiology; Ph.D., SUNY at Stony Brook, 1983. Control of interferon-induced gene expression.

Jacek Skowronski, Cold Spring Harbor Laboratory; Ph.D., Lodz (Poland), 1981. HIV genes and signal transduction in T cells.

Eric Spitzer, Pathology; M.D., Johns Hopkins, 1985. Molecular biology of microbial pathogens.

Roy T. Steigbigel, Medicine and Molecular Genetics and Microbiology; M.D., Rochester, 1966. Treatment of HIV infection.

Bettie M. Steinberg, Long Island Jewish Hospital; Ph.D., SUNY at Stony Brook, 1976. Molecular biology of human papillomaviruses.

Arne Stenlund, Cold Spring Harbor Laboratory; Ph.D., Uppsala (Sweden), 1984. DNA replication of bovine papillomavirus.

Bruce W. Stillman, Cold Spring Harbor Laboratory; Ph.D., Australian National, 1979. Eukaryotic DNA replication and its control: DNA tumor viruses and *Saccharomyces cerevisiae.*

F. William Studier, Brookhaven National Laboratory; Ph.D., Caltech, 1963. Bacteriophage T7 genetics.

William Tansey, Cold Spring Harbor Laboratory; Ph.D., Sydney, 1991. Oncogene regulation; transcription; protein destruction.

David Thanassi, Molecular Genetics and Microbiology; Ph.D., Berkeley, 1995. Biogenesis of bacterial adhesion organelles; bacterial pathogenesis.

Gerald Thomsen, Biochemistry and Cell Biology; Ph.D., Rockefeller, 1988. Embryonic induction in Xenopus.

Peter Tonge, Chemistry; Ph.D., Birmingham (England), 1986. Enzyme mechanisms and rational drug design.

Nicholas Tonks, Cold Spring Harbor Laboratory; Ph.D., Dundee, 1985. Posttranslational modification; phosphorylation; phosphatases; signal transduction; protein structure and function.

Kevin J. Tracey, The Feinstein Institute for Medical Research; M.D., Boston University, 1983. The cholinergic anti-inflammatory pathway.

Michael Wigler, Cold Spring Harbor Laboratory; Ph.D., Columbia, 1978. Signal transduction and growth control in eukaryotes; cancer.

Eckard Wimmer, Molecular Genetics and Microbiology; Dr.rer.nat., Göttingen (Germany), 1962. Molecular biology of picornavirus replication and molecular basis of pathogenesis; translational regulation of gene expression.

Wei-Xing Zong, Molecular Genetics and Microbiology; Ph.D., University of Medicine and Dentistry of New Jersey–Robert Wood Johnson Medical School/Rutgers, 1999. Molecular regulation of apoptotic and necrotic cell death.

TEMPLE UNIVERSITY
of the Commonwealth System of Higher Education

Health Sciences Center
Department of Microbiology and Immunology

Programs of Study	The Department of Microbiology and Immunology in the School of Medicine offers programs leading to the Ph.D. and M.S. degrees in microbiology and immunology. The areas of concentration include molecular and cellular immunology, molecular and biochemical microbiology, and eukaryotic and prokaryotic molecular genetics. Students enter the program in late August. Research training begins soon after matriculation with a research orientation program. Students perform two months of rotations usually in three selected laboratories in the Department, becoming familiar with the research being done in each. They then select an adviser and a graduate research advisory committee to assist in the development and assessment of their research project. The student's research is reviewed semiannually by the research advisory committee.

Students generally take two years of formal classwork, which includes basic microbiology, immunology, biochemistry, and molecular genetics. These courses are part of an interdisciplinary program, and a variety of courses offered by other departments at the Health Sciences Center are also available. Elective courses, taken in the second year, include advanced courses in immunology, microbial physiology, and molecular genetics. Temple also offers an M.D./Ph.D. program; interested students should apply through the medical school.

Current active research areas include the following: (1) Molecular and cellular immunology: molecular and cellular immunology of T and B lymphocytes; molecular aspects of macrophage physiology; molecular genetics of T-cell antigen receptors; immune response to neoplasia; molecular basis for the superantigen activity of certain microbial agents and analysis of the immunosuppressive activity of these and other microbial agents; cellular and molecular analysis of the role of opioid receptor expression in the function of the cells of the immune system; molecular aspects of autoimmune disease; molecular basis for the function of cytokines; microbial immunity. (2) Molecular and biochemical microbiology: role of peptide signals in plasmid maintenance; analysis of cell division; mechanisms of long-term survival and asymptomatic carriage of bacteria; molecular biology of DNA and RNA viruses, including human immunodeficiency virus and Epstein-Barr virus. (3) Eukaryotic and prokaryotic molecular genetics: molecular genetics of sporulation in *Bacillus;* cellular and viral oncogenes and their role in cell growth and transformation; molecular modeling and protein structure-function relationships; molecular genetics of the immune response.

Teaching experience is an important part of the training program. To develop confidence in group teaching, each graduate student takes a course in the presentation of scientific information. In the second and third years, the student has the opportunity to assist in teaching. |
Research Facilities	The Department occupies approximately 30,000 square feet. The program is housed in state-of-the-art facilities. These include the construction of a BSL3 facility, with the support of a construction grant from the National Institutes of Health (NIH). Instrumentation that is available for training and research use includes liquid-scintillation and gamma spectrometers; a confocal microscope; infrared, ultraviolet, and visible spectrophotometers; high-speed centrifuges and ultracentrifuges; laminar-flow hoods; cold and warm rooms; computers for data analysis; facilities for image analysis; hybridoma production and electrophoresis equipment; and chromatographs, including HPLC, FPLC, and GLC. Available facilities include a fluorescence activated cell sorter/analyzer; an automated DNA sequencer; a protein/amino acid sequencer; an oligonucleotide synthesizer; a peptide synthesizer; electron spin resonance, nuclear magnetic resonance, and mass spectrometers; and a transgenic mouse facility.
Financial Aid	Full and partial fellowships and scholarships are awarded based on scholastic ability and need. The University deadline for some of these applications is March 15. The Department recently renewed an NIH training grant that supports a number of students. Additional graduate student support is provided by research grants. All students presently in the program receive financial support, including tuition. All faculty members in the Department are funded, primarily by the NIH. The average peer-reviewed grant support is in excess of $300,000 per faculty member per year.
Cost of Study	Temple is a state-affiliated university. In 2006–07, tuition for graduate study for out-of-state students is $746 per semester hour. Pennsylvania residents are allowed to matriculate at a reduced cost, which is set at $511 per semester hour in 2006–07.
Living and Housing Costs	Graduate students usually live in apartments or rooms in various sections of Philadelphia. Rents vary considerably, but one or two students are often able to occupy an apartment with rent in the range of $550 to $700 per month.
Student Group	There are 154 graduate students at the Health Sciences Center, of whom 41 are enrolled in the Department of Microbiology and Immunology. The total student population of the Health Sciences Center, including professional students, is about 2,000.
Student Outcomes	Since 1965, more than 100 students have graduated with a Ph.D. These individuals have been very successful in seeking and obtaining positions. Approximately 68 percent of graduates hold academic positions, 20 percent have industrial research positions, and 10 percent are directors of clinical microbiology or immunology laboratories.
Location	Philadelphia has much to offer in art, music, and other cultural pursuits. It is a major center for medical research, and most of the seminars, colloquia, and similar events sponsored by other educational and research institutions in the city are open to Temple graduate students.
The Health Sciences Center	The Health Sciences Center includes the Schools of Medicine, Dentistry, and Pharmacy; the College of Allied Health Professions; and the Temple University Hospital. It is located on Broad Street, five blocks from the North Philadelphia railroad station and about 1½ miles north of the main campus of Temple University. A subway station for the Broad Street line is located adjacent to campus.
Applying	There is no fixed list of courses that are prerequisites for admission, but understanding of, and success in, present-day microbiology and immunology is based on good preparation in the biological, chemical, and physical sciences. Each applicant must take the General Test of the Graduate Record Examinations. International applicants should take the TOEFL. The University is an Equal Opportunity/Affirmative Action employer.
Correspondence and Information	Dr. Patrick J. Piggot Professor and Acting Chair Department of Microbiology and Immunology Temple University School of Medicine 3400 North Broad Street Philadelphia, Pennsylvania 19140 Phone: 215-707-3207 Fax: 215-707-7788 E-mail: drykard@temple.edu Web site: http://www.temple.edu/medicine/departments_centers/basic_science/microbiology.htm

Temple University

THE FACULTY AND THEIR RESEARCH

Bettina A. Buttaro, Associate Professor; Ph.D., Minnesota. Molecular basis of pathogenicity of streptococci and enterococci; long-term persistence and human carriage of Group A streptococci; peptide-mediated stabilization of plasmids in enterococci; *S. mutans* persistence in biofilms.

Marion M. Chan, Assistant Professor; Ph.D., Cornell. Innate immunity; roles of proinflammatory cytokines and oxidants (nitric oxide) in microbial infections, autoimmune diseases, and tumorigenesis and their modulation by anti-inflammatories and antioxidants.

Toby K. Eisenstein, Professor; Ph.D., Bryn Mawr. Immunomodulation by microbes and opioids; mechanisms of immunity and immunosuppression induced by *Salmonella* and other intracellular pathogens of macrophages; role of cytokines and nitric oxide in immunosuppression; neuroimmunopharmacology, mechanisms of opioid-mediated immunosuppression.

Earl E. Henderson, Professor; Ph.D., Chicago. Lymphocyte transformation by the Epstein-Barr virus (EBV); DNA repair and host-cell reactivation in eukaryotic and prokaryotic systems utilizing T4 endonuclease V for genetic engineering in human lymphoblastoid cell lines and fibroblasts; regulation of HIV expression and interaction with Epstein-Barr virus.

Walter K. Long, Associate Professor; Ph.D., Illinois. Role of DNA methylation in latency, reactivation, and oncogenesis of herpesviruses; regulation of viral gene expression and virus replication and role of viral and cellular genes; interactions between herpesviruses and human immunodeficiency virus.

Marc Monestier, Professor; M.D., Ph.D., Lyon. Molecular basis of autoimmune diseases; structure and properties of antichromatin antibodies; animal models of chemically induced autoimmunity; antiphospholipid auto-antibodies.

Patrick J. Piggot, Professor and Acting Chair; Ph.D., London. Regulation of gene expression during cell differentiation, using as an experimental system sporulation of *Bacillus subtilis;* mechanisms for establishing cell-type-specific gene expression; postexponential metabolism and survival of *S. mutans.*

Chris D. Platsoucas, L. H. Carnell Professor; Ph.D., MIT. Molecular genetics of human T cells; trimolecular complex (T-cell receptor/MHC/peptide) studies in human antitumor responses (melanoma, ovarian carcinoma, breast carcinoma); autoimmune diseases; rejection of cardiac allografts.

Arthur G. Schwartz, Professor; Ph.D., Harvard. Biological role of the adrenal steroid dehydroepiandrosterone and its role as a natural protector against the development of cancer and other age-related diseases.

Tomasz Skorski, Professor; M.D., Ph.D., Medical Academy (Warsaw). Leukemogenic mechanisms activated by BCR/ABL oncogenic tyrosine kinase.

Kenneth J. Soprano, Professor; Ph.D., Rutgers. Molecular biology and recombinant DNA approaches to understanding the mechanism of action of the *c-jun* and *c-myc* oncogenes during the mammalian cell cycle; identification of molecular events that regulate cell-cycle progression in normal human cells and how the regulation and function of these events are altered in human breast and ovarian tumor cells; use of retinoids to identify growth suppressors of human breast and ovarian tumor cells.

Alexander Y. Tsygankov, Associate Professor; Ph.D., Russian Academy of Sciences. Molecular mechanisms of signal transduction in T cells; interactions of tyrosine protein kinases with their substrates in the course of T-cell activation.

THOMAS JEFFERSON UNIVERSITY

Jefferson College of Graduate Studies
Department of Microbiology and Immunology
Kimmel Cancer Center
Graduate Program in Immunology and Microbial Pathogenesis

Programs of Study

The Graduate Program in Immunology and Microbial Pathogenesis, leading to the Doctor of Philosophy degree, allows students to concentrate within this field under the structure of the Department of Microbiology and Immunology and in connection with the Kimmel Cancer Center. Students entering with a baccalaureate degree take a core biomedical course in the fall semester, incorporating biochemistry, molecular biology, cell biology, and genetics. With this foundation, students continue on with core courses in immunology, virology, and microbiology. Advanced courses are taken in both the first and second years of study, and the curriculum is flexible in order to accommodate the individual student's background and interests, including advanced topics in virology, microbial pathogenesis, and tumor immunology. Laboratory rotations allow students to experience different areas of research before choosing a preceptor and direction for thesis work by the end of the first year. Concentrated thesis research continues from then on as the predominant activity for the remainder of the student's graduate experience. Guidance for students is provided by the research adviser and a thesis faculty committee. Students are given considerable opportunity to continue to develop their scientific knowledge base in immunology through weekly journal clubs and research seminars. The Graduate Program in Immunology and Microbial Pathogenesis, with the Graduate Programs in Biochemistry and Molecular Biology, Genetics, and Molecular Pharmacology and Structural Biology, make up the Joint Graduate Programs of the Kimmel Cancer Center. Students applying to the Joint Graduate Programs may perform research rotations or work with faculty members in any of these programs.

Current research interests within the program include antigen presentation, antiviral agents, autoimmunity, B-cell development, cancer immunology, cancer inflammation, cell biology of malaria, cell growth regulation and differentiation, cellular immunology, cell activation and signal transduction, chemical and antigenic structure of virus particles, cytokines, developmental immunology, immunochemistry, immunogenetics, immunoparasitology, immunoregulation, latent-virus infections, microbial immunology and pathogenesis, molecular immunology, neuroimmunology, neurovirology, regulation of viral gene expression, reproductive immunology, transplantation immunology, viral oncogenes, viral replication, and virus-cell interactions.

Research Facilities

Research laboratories are primarily located in the Kimmel Cancer Center, occupying 70,000 square feet of the Bluemle Life Sciences Building and 25,000 square feet in the adjacent Jefferson Alumni Hall. The Center is exceptionally well equipped and provides a rich environment for interaction among research disciplines. Facilities are available for cell sorting by flow cytometry, peptide and oligonucleotide synthesis and sequencing, proteomics, microarray, scanning and transmission electron microscopy, transgenic mouse production, and X-ray crystallography. State-of-the-art, high-speed, Internet-ready computer systems are also available.

Financial Aid

Financial support is available to full-time Ph.D. students in the form of University fellowships. In 2006–07, students granted full fellowship support receive funds for payment of tuition and a stipend of $24,500. Also available to students demonstrating financial need are Title IV funds. University loan programs are available to qualifying students.

Cost of Study

Tuition and fees for full-time Ph.D. students are $15,640 per year in 2006–07.

Living and Housing Costs

Campus housing is available to graduate students in the program; there is also reasonably priced alternative housing near the University.

Student Group

Each year, approximately 15 students are admitted to the Joint Graduate Programs of the Kimmel Cancer Center. First-year students must choose one of the programs for graduate study by the end of the year; therefore, the actual number of students entering the Graduate Program in Immunology and Microbial Pathogenesis varies.

Student Outcomes

Recent Ph.D. graduates have accepted postdoctoral positions in prestigious academic institutions, such as Harvard, Baylor, Johns Hopkins, UC Davis, and M. D. Anderson Cancer Center at the University of Texas; at the Trudeau Institute and the NIH; and in industry, including Bristol-Meyers Squibb, Centocor, and Pfizer.

Location

The 13-acre campus of Thomas Jefferson University is located in the vibrant historical downtown area of Philadelphia within walking distance of many places of cultural interest. Numerous intercollegiate and professional athletics events take place nearby, within easy access by bus and subway lines. Philadelphia is located a few hours away from New York City and Washington, D.C., and the nearby Jersey shore and Pennsylvania mountains offer year-round recreational activities.

The University

Thomas Jefferson University is an academic health center. It evolved from the Jefferson Medical College, which was founded in 1824. In addition to the medical college, the University includes the Jefferson College of Graduate Studies, the Jefferson College of Health Professions, the University hospital, and various affiliated hospitals and institutions. Jefferson Alumni Hall houses the TJU Fitness and Recreation Facility, study lounges, a swimming pool, a sauna, a gymnasium, and a handball/squash court.

At present, the University enrolls about 2,500 students. The College of Graduate Studies enrolls approximately 630 students, about half of whom are women.

Applying

Applications may be submitted at any time but should be received before March 1 for optimal consideration. Scores on the General Test of the Graduate Record Examinations (GRE), three letters of recommendation, and academic transcripts are required of all applicants. For those students whose native language is not English, scores on the Test of English as a Foreign Language (TOEFL) are required. Scores on an appropriate Subject Test of the GRE are strongly recommended. Prospective students are encouraged to visit the Department and discuss the graduate program with members of the faculty.

Correspondence and Information

For applications:
Jessie Pervall
Director of Admissions and Recruitment
College of Graduate Studies
Thomas Jefferson University
1020 Locust Street, M-46
Philadelphia, Pennsylvania 19107-6799

Phone: 215-503-4400
Fax: 215-503-3433
E-mail: cgs-info@jefferson.edu
Web site: http://www.jefferson.edu/cgs

For program information:
Joanne Balitzky
Training Programs Coordinator
Kimmel Cancer Center
Thomas Jefferson University
233 South 10th Street, 910 BLSB
Philadelphia, Pennsylvania 19107-5541

Phone: 215-503-6687
Fax: 215-503-0622
E-mail: joanne.balitzky@jefferson.edu
Web site: http://www.jefferson.edu/jcgs/phd/imp/

Thomas Jefferson University

THE FACULTY AND THEIR RESEARCH

David Abraham, Professor; Ph.D., Pennsylvania, 1983. Parasite immunology; role of eosinophils and neutrophils in innate and adaptive immunity to nematode parasite infections; mechanisms of immune-mediated resistance and development of vaccines against nematode infections; chemotherapy of leishmaniasis.

Kishore R. Alugupalli, Assistant Professor; Ph.D., Lund (Sweden), 1996. Molecular basis of bacterial pathogenesis using murine infection models; B 1 b lymphocytes in T cell–independent IgM memory; role of Toll-like receptors in innate and adaptive immune responses.

Melvin Bosma, Professor; Ph.D., Illinois at Urbana-Champaign, 1966. How autoreactive B lymphocytes change the specificity of their antigen receptor.

Arthur M. Buchberg, Associate Professor; Ph.D., SUNY at Buffalo, 1983. Immunogenetics: steps involved in murine myeloid and B-cell leukemogenesis; study of the involvement of homeobox genes in transformation and apoptosis; identification of genes involved in altering susceptibility to colon cancer.

Bruno Calabretta, Professor; M.D./Ph.D., Rome, 1987. Regulation of granulocytic differentiation by members of the CCAAT family of transcription factors.

Catherine E. Calkins, Professor; Ph.D., Purdue, 1972. Cellular immunology; immunoregulation of antiself reactivity; autoimmune disease; antiviral immunity; immunopathogenesis of hepatitis B virus.

Bernhard Dietzschold, Professor; D.V.M., Giessen (Germany), 1967. Delineation of molecular mechanisms involved in the pathogenesis of rabies; development of novel recombinant virus vaccines against rabies and other emerging viral diseases such as SARS and Nipah virus infection.

Laurence C. Eisenlohr, Professor; V.M.D., 1983, Ph.D., 1988, Pennsylvania. Cell biology of MHC class I- and class II-restricted antigen processing of viral antigens; T-cell responses and memory to viral antigens; antitumor immunity.

Mark A. Feitelson, Professor; Ph.D., UCLA, 1979. Cellular and molecular mechanisms responsible for pathogenesis of chronic hepatitis B virus infection, especially in development of hepatocellular carcinoma; animal models of chronic hepatitis B virus infection; structure and function of the hepatitis B X gene product; role of virus genetic variation in the outcome of infection; contribution of alcohol and underlying viral infection to alcoholic liver diseases; pathogenesis of hepatitis C virus.

Neal Flomenberg, Professor; M.D., Thomas Jefferson, 1976. Marrow transplantation and graft-host interactions; characterization and manipulation of the immune response to malignant disease in an allogeneic setting (such as graft-versus-leukemia responses); approaches to promote graft-host tolerance after marrow or solid organ transplantation and to accelerate immune reconstitution.

Phyllis R. Flomenberg, Associate Professor; M.D., Yeshiva (Einstein), 1980. Adenoviruses; *Vaccinia virus;* viral immunology; tumor immunology; immunotherapy.

Richard R. Hardy, Professor; Ph.D., Caltech, 1980. Developmental immunology of B cells, including their generation from stem cells in fetal liver and bone marrow; regulation of B-cell development by pre-BCR and BCR signaling; development of B-cell subsets, including B-1 B cells, with relevance to autoimmunity and leukemia.

D. Craig Hooper, Associate Professor; Ph.D., McGill, 1983. Neuroimmunology; contribution of free radicals to immunity; central nervous system (CNS) inflammation; autoimmunity; virus clearance from the CNS; CNS tumor immunity; immunopathogenesis; T cell–dependent inflammatory mechanisms; mechanisms of cell invasion into the CNS; CD4 T-cell priming; passive immunization; vaccination.

Donald L. Jungkind, Professor; Ph.D., Texas Medical Branch, 1972. Molecular diagnosis and automation in clinical microbiology; computer-based training in infectious diseases.

Tim Manser, Professor and Chairperson; Ph.D., Utah, 1982. Role of antigen receptor specificity in B-cell selection, tolerance, and memory; germinal center reaction; regulation of B-cell responses by inhibitory Fc receptors.

W. Edward Mercer, Professor; Ph.D., Penn State, 1980. Molecular mechanisms of DNA tumor virus host-cell interactions in cell-cycle deregulation and resistance to apoptosis.

Glenn F. Rall, Associate Professor; Ph.D., Vanderbilt, 1990. Viral spread within neurons; recruitment and function of the immune response to neurotropic viral challenges; transgenic models of viral pathogenesis.

Ulrich Rodeck, Professor; M.D., Germany, 1981. Regulation of epithelial cell survival by the epidermal growth factor receptor; coordinate regulation of cell-cycle progression and cell survival by growth factor– and adhesion-dependent signal transduction; signal transduction events controlling cell fate in the anchorage-independent state; zebrafish as a new in vivo model system for the genotoxic stress response.

Michael J. Root, Assistant Professor; M.D./Ph.D., Harvard, 1997. Structure and function of glycoproteins involved in viral and cell-cell membrane fusion; design of viral entry inhibitors and immunogens for vaccine development.

Jay L. Rothstein, Professor and Program Director; Ph.D., Chicago, 1988. Mechanisms controlling tumor immunity and chronic inflammation; the function of oncogenes in promoting cancer and autoimmune disease; the use of transgenic and conditional animal models to study thyroid cancer and thyroid autoimmune disease.

Matthias J. Schnell, Associate Professor; Ph.D., Hohenheim-Stuttgart (Germany), 1994. Recombinant rhabdovirus-based vectors as potential live and killed vaccines against HIV-1, other infectious diseases, and cancer; molecular biology and pathogenesis of negative-stranded RNA viruses; molecular mechanism of neurotropism of rabies virus.

Charles P. Scott, Assistant Professor; Ph.D., Pennsylvania, 1997. Combinatorial drug discovery; rational drug design; chemical genomics of host-pathogen interactions and cancer.

Linda D. Siracusa, Associate Professor; Ph.D., SUNY at Buffalo, 1985. Immunogenetics: identification and characterization of genes, allelic variants, and mutations involved in mammalian disease processes.

Alagarsamy Srinivasan, Professor; Ph.D., Jawaharlal Nehru (New Delhi), 1977. Retroviruses: structure, replication, and virion morphogenesis; HIV molecular biology; role of HIV genes in AIDS pathogenesis; genetic heterogeneity of HIV.

David S. Strayer, Professor; M.D./Ph.D., Chicago, 1976. Gene delivery and gene therapy; use of viral gene transfer vectors for immunization against lentiviral antigens; application of immunostimulatory cytokines to immunization; development of strategies to protect from and to treat infection by HIV, HBV, and HCV, using transgenes designed to improve cellular resistance to these viruses, including such transgenes as single chain Fv antibodies, interferons alpha and gamma, and inhibitory RNA species (siRNAs, ribozymes, antisense).

Yuri Sykulev, Associate Professor; M.D./Ph.D., Pyrogov Moscow State Medical, 1982. Mechanisms of lymphocyte activation; factors limiting responses of T cells that recognize viral-infected and cancer cells; implication for vaccine development; mechanism of T- and NK-cell cytolytic activities.

Theodore F. Taraschi, Professor; Ph.D., Rutgers, 1980. Elucidation of mechanism(s) responsible for the transport of parasite proteins to the host cell membrane in human erythrocytes infected with the malaria parasite *Plasmodium falciparum;* characterization of parasite DNA base excision and mismatch repair pathways; identification of the structures and mechanisms responsible for hemoglobin internalization and transport to the parasite food vacuole.

John L. Wagner, Associate Professor; M.D., Temple, 1989. Canine immunogenetics and allogeneic peripheral blood stem-cell transplantation; gene expression in graft-versus-host disease post–bone marrow transplantation.

Scott A. Waldman, Professor; Ph.D., Thomas Jefferson, 1980; M.D., Stanford, 1987. Molecular mechanisms of signal transduction, with emphasis on receptor-effector coupling and post-receptor signaling mechanisms; translation of molecular signaling mechanisms to novel diagnostic and therapeutic approaches to patients with cancer; defining cross talk between systemic and mucosal immune systems; exploring the relationship between immune-mediated inflammation and neoplasia; identifying unique antigenic targets to serve as effective antitumor vaccines.

David Wiest, Associate Professor; Ph.D., Duke, 1991. Understanding the molecular basis for control of T lineage development by molecular effectors differentially elicited by distinct TCR isotypes.

Hui Zhang, Associate Professor; M.D., Sun Yatsen Medical Sciences, 1982; Ph.D., SUNY Health Science Center at Syracuse, 1994. Function of virion infectivity factor (Vif) encoded by HIV-1; HIV-1 pathogenesis; proteomics; RNA/DNA editing; microRNA; high-throughput assay to screen antiviral drugs.

Jianke Zhang, Assistant Professor; Ph.D., Purdue, 1993. Immune tolerance and homeostasis; lymphocyte apoptosis; cytokine-receptor signal transduction; lymphocyte development; signal transduction mechanisms by the Fas/TNF receptor family members that mediate apoptosis and growth/proliferation; transgenic and gene targeting (knockout) analyses of protein functions in mice.

UNIFORMED SERVICES UNIVERSITY OF THE HEALTH SCIENCES

F. Edward Hébert School of Medicine
Graduate Program in Emerging Infectious Diseases

Program of Study

One of the missions of the Uniformed Services University (USU) is to provide both civilians and military students with high-quality training leading to advanced degrees in the biomedical sciences. The Graduate Program in Emerging Infectious Diseases (EID) is designed for applicants who wish to pursue an interdisciplinary program of study leading to the Ph.D. degree and was created for students who are primarily interested in the pathogenesis, host response, and epidemiology of infectious diseases. No M.S. degree program is currently offered. A broadly based core program of formal training is combined with an intensive laboratory research experience in the different disciplines encompassed by the field of infectious diseases. Courses are taught by an interdisciplinary EID faculty who hold primary appointments in the Departments of Microbiology and Immunology, Pathology, Preventive Medicine and Biometrics, Pediatrics, and Medicine. Research training emphasizes modern methods in molecular biology and cell biology, as well as interdisciplinary approaches.

During the first two years, all students are required to complete a series of broadly based core courses and laboratory rotations. Students also select one of three academic tracks in which to focus the remainder of their course work. The three tracks are microbiology and immunology, pathology, and preventive medicine/parasitology. Advanced course work is required in each academic track. In addition, each student selects a faculty member with whom he or she would like to carry out a thesis research project. By the end of the second year, the student must complete all requirements for advancement to candidacy for the Ph.D. degree, which includes satisfactory completion of formal course work and passage of the qualifying examination. After advancement to candidacy, the student must complete an original research project and prepare and defend a written dissertation under the supervision of his or her faculty adviser and an advisory committee.

Research Facilities

Each academic department of the University is provided with laboratories for the support of a variety of research projects. Laboratories are available in most areas of study that encompass the interdisciplinary field of emerging infectious diseases, including both basic and medical aspects of bacteriology, bacterial genetics, virology, cellular and molecular immunology, parasitology, pathogenic mechanisms of disease, pathology of infectious disease, and epidemiology of infectious diseases. Resources available to students within the University include a phosophorimager, a real-time PCR machine, FACSCAN and fluorescent cell sorters, automated oligonucleotide and peptide synthesizers, high-resolution transmission and scanning electron microscopes, confocal microscopes, a certified central animal facility, and state-of-the-art computer facilities. In addition, a BSL-3 biohazard containment laboratory suite is available. The library/learning resources center houses more than 521,000 bound volumes, subscribes to nearly 3,000 journals (print and online), and maintains 100 IBM and Macintosh personal computers for use by students, faculty members, and staff members. Biostatisticians serve as a resource for students and faculty members.

Financial Aid

Stipends are available for civilian applicants. Awards of stipends are competitive, are for one-year periods, and may be renewed. The 2006–07 stipend level begins at $24,000 per year. Special fellowships are also available.

Cost of Study

Graduate students in the Emerging Infectious Diseases Program are not required to pay tuition or fees. Civilian students do not incur obligations to the United States government for service after completion of their graduate training programs.

Living and Housing Costs

There is a reasonable supply of affordable rental housing in the area. The University does not have housing for graduate students. Living costs in the greater Washington, D.C., area are comparable to those of other East Coast metropolitan areas.

Student Group

The first full-time graduate students were admitted to the EID program in fall 2000. There are currently 34 full-time students enrolled in the EID graduate program. The University also has Ph.D. programs in departmentally based basic biomedical sciences, as well as interdisciplinary graduate programs in molecular and cell biology and in neurosciences.

Location

The greater Washington metropolitan area has a population of about 3 million that includes the District of Columbia and the surrounding areas of Maryland and Virginia. The region is a center of education and research and is home to five major universities, four medical schools, and numerous other internationally recognized private and government research centers. In addition, multiple cultural advantages exist in the area and include theaters, a major symphony orchestra, major-league sports, and world-famous museums. The Metro subway system has a station adjacent to the campus and provides a convenient connection from the University to cultural attractions and activities in downtown Washington. The international community in Washington is the source of many diverse cuisines and international cultural events. For a wide variety of outdoor activities, the Blue Ridge Mountains, Chesapeake Bay, and Atlantic coast beaches are all within a 1- to 3-hours' drive. Many national and local parks serve the area for weekend hikes, bicycling, and picnics.

The University

USU is located just outside Washington, D.C., in Bethesda, Maryland. The campus is situated in an attractive, park-like setting on the grounds of the National Naval Medical Center (NNMC) and across the street from the National Institutes of Health (NIH). Wooded areas with jogging and biking trails surround the University. NIH, NNMC, and other research institutes in the area provide additional resources to enhance the education experience of graduate students at USU. Students can visit the USUHS Web site at http://www.usuhs.mil.eid.

Applying

The Admissions Committee, in consultation with other faculty members, evaluates applications to the program. Each applicant must submit an application form, complete academic transcripts of postsecondary education, and results of the Graduate Record Examinations. No GRE Subject Test is required. In addition, three letters of recommendation from individuals familiar with the academic achievements and/or research experience of the applicant are required, as well as a personal statement that expresses the applicant's career objectives. USU subscribes fully to the policy of equal educational opportunity and selects students on a competitive basis without regard to race, color, gender, creed, or national origin. Application forms may be obtained from the University Web site (available at http://www.usuhs.mil/geo/Application.pdf). Completed applications should be received on or before January 15.

Both civilians and military personnel are eligible to apply. Prior to acceptance, each applicant must complete a baccalaureate degree that includes required courses in mathematics, biology, physics, and chemistry (inorganic, organic, and biochemistry). Advanced-level courses in microbiology, molecular biology, genetics, and cell biology are very strongly recommended. All students are expected to have a reasonable level of computer literacy. Active-duty military applicants must obtain the approval and sponsorship of their parent military service, in addition to acceptance into the EID graduate program.

Correspondence and Information

Dr. Eleanor S. Metcalf
Associate Dean for Graduate Education
Uniformed Services University
4301 Jones Bridge Road
Bethesda, Maryland 20814-4755

Phone: 800-772-1747 (toll-free)
Web site: http://www.usuhs.mil

Dr. Christopher C. Broder, Director
Graduate Program in Emerging Infectious Diseases
Uniformed Services University
4301 Jones Bridge Road
Bethesda, Maryland 20814-4755

Phone: 301-295-5749
Fax: 301-295-9861
E-mail: cbrodercbroder@usuhs.mil
Web site: http://www.usuhs.mil/mic/eid

Uniformed Services University of the Health Sciences

THE FACULTY

The interdisciplinary graduate programs at USU are superimposed on the departmental structure. Therefore, all faculty members in the interdisciplinary Graduate Program in Emerging Infectious Diseases (EID) have primary appointments in either a basic science or a clinical department and secondary appointments in EID. The faculty is derived primarily from the Departments of Microbiology and Immunology, Pathology, Preventive Medicine and Biometrics, Pediatrics, and Medicine. Thus, the faculty in EID includes the experts in infectious diseases, regardless of department. For additional information, students should visit the USU Academic Department Web site at http://www.usuhs.mil/acad/aca_dept.html. To address e-mail to specific faculty members at USU, students should use the first letter of their first name plus the last name and @usuhs.mil as the address; for example, to send e-mail to John Doe, the address would be jdoe@usuhs.mil.

Richard G. Andre, Ph.D.; Professor, Preventive Medicine and Biometrics.
Naomi E. Aronson, M.D.; Associate Professor, Medicine.
Christopher C. Broder, Ph.D.; Professor, Microbiology and Immunology.
W. Patrick Carney, Ph.D.; Professor, Preventive Medicine and Biometrics.
Wei-Mei Ching, M.D.; Adjunct Associate Professor, Rickettsial Diseases, NMRC.
Richard M. Conran, M.D., Ph.D.; Professor, Pathology.
John H. Cross, Ph.D.; Professor, Preventive Medicine and Biometrics.
David F. Cruess, Ph.D.; Professor, Preventive Medicine and Biometrics.
Stephen J. Davies, Ph.D.; Assistant Professor, Microbiology and Immunology.
Andre T. Dubois, M.D., Ph.D.; Research Professor, Medicine.
Maryna C. Eichelberger, Ph.D.; Adjunct Assistant Professor, Microbiology and Immunology.
Chou-Zen Giam, Ph.D.; Professor, Microbiology and Immunology.
Scott W. Gordon, Ph.D.; Assistant Professor, Preventive Medicine and Biometrics.
Patricia Guerry, Ph.D.; Adjunct Associate Professor, Immunology, NMRC.
Val G. Hemming, M.D.; Professor, Pediatrics.
Tomoko I. Hooper, M.D.; Associate Professor, Preventive Medicine and Biometrics.
Ann E. Jerse, Ph.D.; Associate Professor, Microbiology and Immunology.
Elliott Kagan, M.D.; Professor, Pathology.
Niranjan Kanesa-Thasan, M.D.; Adjunct Assistant Professor, Preventive Medicine and Biometrics.
Tadeusz J. Kochel, Ph.D.; Adjunct Assistant Professor, Immunology, NMRC.
Philip R. Krause, M.D.; Adjunct Assistant Professor, CBER, FDA.
Larry W. Laughlin, M.D., Ph.D.; Professor, Preventive Medicine and Biometrics.
Radha K. Maheshwari, Ph.D.; Professor, Pathology.
Ernest L. Maynard, Ph.D.; Assistant Professor, Biochemistry and Molecular Biology.
Anthony T. Maurelli, Ph.D.; Professor, Microbiology and Immunology.
D. Scotty Merrell, Ph.D.; Assistant Professor, Microbiology and Immunology.
Eleanor S. Metcalf, Ph.D.; Professor, Microbiology and Immunology.
Edward Mitre, M.D.; Assistant Professor, Microbiology and Immunology.
Alison D. O'Brien, Ph.D.; Professor, Microbiology and Immunology.
Christian F. Ockenhouse, M.D., Ph.D.; Adjunct Assistant Professor, Infectious Diseases, WRAIR.
Martin G. Ottolini, M.D.; Associate Professor, Pediatrics.
Gerald V. Quinnan Jr., M.D.; Professor, Preventive Medicine and Biometrics.
Allen L. Richards, Ph.D.; Associate Professor, Preventive Medicine and Biometrics.
Donald R. Roberts, Ph.D.; Professor, Preventive Medicine and Biometrics.
Capt. Stephen J. Savarino, USN; Adjunct Assistant Professor, Infectious Diseases, NMRC.
Brian C. Schaefer, Ph.D.; Assistant Professor, Microbiology and Immunology.
Clifford M. Snapper, M.D.; Professor, Pathology.
V. Ann Stewart, D.V.M., Ph.D.; Adjunct Assistant Professor, Infectious Diseases, WRAIR.
John D. Stocker, M.D.; Professor, Pathology.
Col. Jose A. Stoute, USA; Associate Professor, Medicine.
Jesus G. Valenzuela, Ph.D.; Adjunct Assistant Professor, Vector Biology, NIAID, NIH.
Lt. Col. Glenn W. Wortmann, USA; Adjunct Assistant Professor, Infectious Diseases, WRAIR.
Shuenn-Jue L. Wu, Ph.D.; Adjunct Associate Professor, Immunology, NMRC.
Pengfei Zhang, Ph.D.; Research Associate Professor, Preventive Medicine and Biometrics.

UNIFORMED SERVICES UNIVERSITY OF THE HEALTH SCIENCES

F. Edward Hébert School of Medicine
Graduate Program in Microbiology and Immunology

Programs of Study

One of the missions of the University is to provide both civilian and military students with high-quality training leading to advanced degrees in the biomedical sciences. The Department of Microbiology and Immunology offers a graduate program leading to the Ph.D. degree. This program, a component of the Interdisciplinary Graduate Program in Emerging Infectious Diseases (EID), is designed for full-time students who wish to pursue professional research and academic careers in the various disciplines encompassed within microbiology and immunology. A broadly based core program of formal training is combined with an intensive laboratory research experience in microbiology and immunology. Research training emphasizes modern methods in molecular biology and cell biology, as well as interdisciplinary approaches. A wide range of interests is represented in the department, including both basic and medical aspects of bacteriology, genetics, virology, immunology, parasitology, infectious diseases, and pathogenic mechanisms. For more information, students can access the department's Web site at http://www.usuhs.mil/mic.

During the first two years, students are required to complete a series of broadly based formal core courses and laboratory rotations. Students in microbiology and immunology focus the remainder of their course work on advanced courses in this area. In addition, students select a faculty member with whom they wish to carry out their thesis research project. By the end of the second year, the student must complete all requirements for advancement to candidacy for the Ph.D. degree, which includes satisfactory completion of formal course work and passage of the qualifying examination. Since teaching is an integral part of a graduate training program, all students who have faculty advisers within the department are required to act as teaching assistants, at least during the first two years. After advancement to candidacy, students complete an original research project and prepare and defend a written dissertation under the supervision of their faculty adviser and an advisory committee.

Research Facilities

The Department of Microbiology and Immunology occupies more than 15,500 square feet of space in the modern and well-equipped research wing of the Hébert School of Medicine. The diverse research projects directed by the department's 10 faculty members are supported by special resources within the University, including a phosophorimager, FACSCAN and fluorescent cell sorters, automated oligonucleotide and peptide synthesizers, large-volume fermentors, high-resolution transmission and scanning electron microscopes, animal resource facilities, and up-to-date computer hardware and software. In addition, BSL-3 biohazard containment laboratories are available. The library/learning resources center houses more than 521,000 bound volumes, subscribes to nearly 3,000 journals (print and online), and contains 150 IBM and Macintosh personal computers for use by students, faculty, and staff.

Financial Aid

Stipends are available for civilian applicants. Awards of stipends are competitive, are for one-year periods, and may be renewed. The 2006–07 stipend level begins at $24,000 per year. Special fellowships are also available.

Cost of Study

Graduate students are not required to pay tuition. Civilian graduate students do not incur obligations to the United States government for service after completion of their graduate training programs. Active-duty military personnel incur an obligation for additional military service in accordance with Department of Defense regulations that govern sponsored graduate education.

Living and Housing Costs

The University does not have housing for graduate students; however, many different types of rental housing exist in the area. Living costs in the greater Washington, D.C., area are comparable to those of other East Coast metropolitan areas.

Student Group

The first graduate students in the Department of Microbiology and Immunology were admitted in 1978. The students are now part of the interdisciplinary EID program. Currently, there are 35 students in the EID program.

Location

The greater Washington metropolitan area, which includes the District of Columbia and the surrounding areas of Maryland and Virginia, has a population of about 3 million. The region is a center of education and research and is home to five major universities, four medical schools, and numerous other internationally recognized private and government research centers. In addition, multiple cultural advantages exist in the area, including theaters, a major symphony orchestra, major-league sports, and world-famous museums. The Metro subway system has a station adjacent to the campus and provides a convenient connection from the University to cultural attractions in downtown Washington. The international community in Washington is the source of many diverse cuisines and international cultural events. For a wide variety of outdoor activities, the Blue Ridge Mountains, Chesapeake Bay, and Atlantic coast beaches are all within a 1–3 hour drive. Many national and local parks serve the area and are accessible for afternoon hikes, bikes, and picnics.

The University

USUHS is located just outside Washington, D.C., in Bethesda, Maryland. The campus is situated on an attractive, wooded site at the National Naval Medical Center and is across the street from the National Institutes of Health. These institutes can provide additional resources to enhance the educational experience of graduate students at USUHS. Students can visit the USUHS Web site at http://www.usuhs.mil.

Applying

Both civilians and military personnel are eligible to apply for admission into the EID Ph.D. program. Applicants are accepted only as full-time students into the EID Ph.D. program, with a concentration in microbiology and immunology. No M.S. degree program is offered. Prior to acceptance, each applicant must complete a baccalaureate degree that includes college-level courses in mathematics, biology, physics, inorganic and organic chemistry, and biochemistry. Advanced-level courses in microbiology, molecular biology, genetics, and cell biology are very strongly recommended. Laboratory experience is valued, and courses in molecular biology and molecular genetics are exceptionally valuable. Active-duty military applicants must obtain the approval and sponsorship of their parent military service, in addition to acceptance into the EID graduate program.

Each applicant must submit a USUHS graduate training application form, complete academic transcripts of postsecondary education, and results of the Graduate Record Examinations. In addition, three letters of recommendation from individuals familiar with the academic achievements and/or research experience of the applicant and a personal statement that expresses the applicant's career objectives are required. USUHS subscribes fully to the policy of equal educational opportunity and selects students on a competitive basis without regard to race, color, sex, creed, or national origin. Application forms may be obtained online at http://www.usuhs.mil/geo/Application.pdf or the Department of Microbiology and Immunology Web site (listed in the Correspondence and Information section). Completed applications should be received on or before January 15 for matriculation in late August.

Correspondence and Information

Dr. Eleanor S. Metcalf
Associate Dean for Graduate Education
Uniformed Services University of the Health Sciences
4301 Jones Bridge Road
Bethesda, Maryland 20814-4799
Phone: 800-772-1747 (toll-free)

Dr. Christopher C. Broder, Director
Graduate Program in Emerging Infectious Diseases
Uniformed Services University of the Health Sciences
4301 Jones Bridge Road
Bethesda, Maryland 20814-4799
Phone: 301-295-5749
Fax: 301-295-1545
E-mail: cbroder@usuhs.mil
Web site: http://www.usuhs.mil/mic

Uniformed Services University of the Health Sciences

THE FACULTY AND THEIR RESEARCH

A comprehensive list of faculty publications may be obtained by writing to the department or by visiting the Web site at http://www.usuhs.mil/mic.

Christopher C. Broder, Professor; Ph.D., Florida, 1989. Structural and functional studies on viral envelope glycoproteins and the interactions between enveloped viruses and their cellular receptors. Current studies concern HIV-1 envelope/CD4/coreceptor interactions and membrane fusion; immunogenicity of oligomeric HIV-1 Env; and the fusion and attachment glycoproteins of Hendra and Nipah viruses, newly identified paramyxoviruses. Potent neutralization of Hendra and Nipah viruses by human monoclonal antibodies. *J. Virol.* 80:891, 2006 (with Zhu et al.). Hendra and Nipah viruses: Different and dangerous. *Nat. Rev. Microbiol.* 4:23, 2006 (with Eaton et al.). From the Cover: Ephrin-B2 ligand is a functional receptor for Hendra virus and Nipah virus. *Proc. Natl. Acad. Sci. U.S.A.* 102:10652, 2005 (with Bonaparte et al.). Protection of rhesus monkeys against infection with minimally pathogenic, simian-human immunodeficiency virus: Correlations with neutralizing antibodies and cytotoxic T cells. *J. Virol.* 79:3358, 2005 (with Quinnan et al.). Receptor binding, fusion inhibition, and induction of cross-reactive neutralizing antibodies by a soluble G glycoprotein of Hendra virus. *J. Virol.* 79:6690, 2005 (with Bossart et al.).

Stephen J. Davies, Assistant Professor; B.V.Sc., Bristol (England) 1993; Ph.D., Cornell, 1998. Molecular biology, biochemistry, and developmental biology of helminth parasites and immunobiology of helminth infections. Current focus is on schistosome infections and mechanisms by which these parasites activate CD4+ T cells, modulate CD4+ T-cell function, and exploit signals from CD4+ T cells to facilitate transmission. In vivo imaging of tissue eosinophilia and eosinopoietic responses to schistosome worms and eggs. *Int. J. Parasitol.* 35:851–9, 2005 (with S. J. Smith et al.). *Schistosoma mansoni:* Sex-specific modulation of parasite growth by host immune signals. *Exp. Parasitol.* 106:59–61, 2004 (with Hernandez, Lim, and McKerrow). Developmental plasticity in schistosomes and other helminths. *Int. J. Parasitol.* 33:1277–84, 2003 (with McKerrow). Modulation of blood fluke development in the liver by hepatic CD4+ lymphocytes. *Science* 294:1358–61, 2001 (with Grogan et al.).

Chou-Zen Giam, Professor and Vice Chair; Ph.D., Connecticut Health Center, 1983. Molecular biology and pathogenesis of human retroviruses and human herpesvirus type 8; cell cycle; anaphase promoting complex; ubiquitination; chromosome instability; cancer; transcriptional regulation; signal transduction; protein-protein and protein-DNA interactions. Current studies concern the human T-lymphotropic virus type 1 (HTLV-1), with emphasis on elucidating the mechanisms of action of the HTLV-1 transcriptional activator, Tax, and understanding the replicative processes of HTLV-1 and the lymphoproliferative diseases (adult T-cell leukemia and tropical spastic paraparesis) it causes. Activation of the anaphase promoting complex by HTLV-1 Tax leads to senescence. *EMBO J.,* 2006 (with Kuo). HTLV-1 Tax directly binds the Cdc20-associated anaphase promoting complex and activates it ahead of schedule. *Proc. Natl. Acad. Sci. U.S.A.* 102:63, 2005 (with Liu et al.). Human T-lymphotropic virus type 1 oncoprotein Tax promotes unscheduled degradation of Pds1p/securin and Clb2p/cyclin B1 and causes chromosomal instability. *Mol. Cell. Biol.* 23(15):5269–81, 2003 (with Liu et al.).

Ann E. Jerse, Associate Professor; Ph.D., Maryland, Baltimore, 1991. Adaptation mechanisms utilized by *Neisseria gonorrhoeae* in the urogenital tract. Current studies focus on testing the role of selected genes in gonococcal evasion of host innate defenses and testing candidate vaccines and microbicidal agents for the capacity to prevent experimental gonococcal infection in female mice. *Neisseria gonorrhoeae* α 2,3-sialyltransferase protects against opsonophagocytic killing by mouse neutrophils and enhances survival during experimental murine genital tract infection. *Infect. Immun.,* in press (with Wu). In vivo selection for *Neisseria gonorrhoeae* opacity protein expression in the absence of human CEACAM receptors. *Infect. Immun.* 74:2965, 2006 (with Simms). A *Neisseria gonorrhoeae* catalase mutant is more sensitive to hydrogen peroxide and paraquat, an inducer of toxic oxygen radicals. *Microb. Pathol.* 37:55, 2004 (with Soler-Garcia). Inhibition of *Neisseria gonorrhoeae* genital tract infection by leading candidate topical microbicides in a mouse model. *J. Infect. Dis.* 189:410, 2004 (with Spencer et al.).

Anthony T. Maurelli, Professor; Ph.D., Alabama at Birmingham, 1983. Molecular genetics of intracellular bacterial pathogens. Current studies involve secreted virulence products of *Shigella* and how these proteins are transported out of the bacterium. Additional projects involve molecular pathogenesis and metabolic pathways of *Chlamydia* and the evolution of bacterial pathogens. Characterization of *Chlamydia* MurC-DdlA, a fusion protein exhibiting D-alanyl-D-alanine ligase activity involved in peptidoglycan synthesis and D-cycloserine sensitivity. *Mol. Microbiol.* 57:41–52, 2005 (with McCoy). Fitness cost due to mutations in the 16S rRNA associated with spectinomycin resistance in *Chlamydia psittaci* 6BC. *Antimicrob. Agents Chemother.* 49:4455–64, 2005 (with Binet). MxiE regulates intracellular expression of factors secreted by the *Shigella flexneri* 2a type III secretion system. *J. Bacteriol.* 184:4409–19, 2002 (with Kane et al.). Pathoadaptive mutations that enhance virulence: Genetic organization of the *cadA* regions of *Shigella* spp. *Infect. Immun.* 69:7471–80, 2001 (with Day et al.).

D. Scott Merrell, Assistant Professor; Ph.D., Tufts, 2001. Host-pathogen interactions using *Helicobacter pylori* as a model system. Current studies concern utilization of genomic tools to elucidate the role of *H. pylori* regulatory factors in adaptation to survival in the human stomach, bacterial control of virulence factor expression, and identification and characterization of host signaling pathways modulated by infection. Iron and pH homeostasis intersect at the level of Fur regulation in the gastric pathogen *Helicobacter pylori. Infect. Immun.* 74(1):602–14, 2006 (with Gancz et al.). Profiling of microdissected gastric epithelial cells reveals a cell type specific response to *Helicobacter pylori* infection. *Gastroenterology* 127(5):1446–62, 2004 (with Mueller et al.). Phosphorylation-independent effects of CagA during interaction of *H. pylori* with polarized monolayers. *J. Infect. Dis.* 190(8):1516–23, 2004 (with El-etr et al.).

Eleanor S. Metcalf, Professor; Ph.D., Pennsylvania, 1976. Host-parasite interactions in *Salmonella* infections; understanding the mechanisms that regulate the host responses to gram-negative bacteria and the virulence factors of the bacteria that permit the organism's survival in the host. Current studies focus on identification of immunocompetent host cells that are responsive to *Salmonella typhimurium*. A novel intraepithelial T-cell subset induced after oral infection with *S. typhimurium.,* in press (with Lopez-Briones et al.). Infection-induced expansion of a MHC class Ib–dependent intestinal intraepithelial gammadelta T cell subset. *J. Immunol.* 172:6828–37, 2004 (with Davies et al.). Bacterial and host factors involved in the major histocompatibility complex class Ib–restricted presentation of *Salmonella* Hsp 60: A novel pathway. *Infect. Immun.* 72:2843–9:2004 (with Lo et al.). Molecular mimicry mediated by MHC class Ib molecules after infection with gram negative pathogens. *Nature Med.* 6:215–8, 2000 (with Lo et al.).

Edward Mitre, Assistant Professor; M.D., Johns Hopkins, 1995. Immune responses to helminth infections. Current studies focus on the role of IgE in filarial infections. IgE memory: Persistence of antigen specific IgE responses years after treatment of human filarial infections. *J. Allergy Clin. Immunol.,* in press (with Nutman). Saturation of immunoglobulin E (IgE) binding sites by polyclonal IgE does not explain the protective effect of helminth infections on atopy. *Infect. Immun.* 73:4106–11, 2005 (with Norwood and Nutman). Parasite antigen-driven basophils are a major source of IL-4 in human filarial infections. *J. Immunol.* 172:2439–45, 2004 (with Taylor et al.). Lack of basophilia in human parasitic infections. *Am. J. Trop. Med. Hyg.* 69(1):87–91, 2003 (with Nutman).

Alison D. O'Brien, Professor and Chair; Ph.D., Ohio State, 1976. Molecular mechanisms of enteric bacterial pathogenesis. Current studies concern the role of *E. coli* Shiga toxins in the pathogenesis of hemorrhagic colitis and the hemolytic uremic syndrome, analysis of the mode of action of the Rho-modifying cytotoxic necrotizing factor and its role in the pathogenesis of *E. coli*–mediated urinary tract infections, and identification of *B. anthracis* exosporium antigens as potential targets for incorporation into a second-generation protective antigen (PA)–based vaccine. A single amino acid substitution in the enzymatic domain of cytotoxic necrotizing factor type 1 of *Escherichia coli* alters the tissue culture phenotype to that of the dermonecrotic toxin of *Bordetella* spp. *Mol. Microbiol.,* in press (with McNichol et al.). Genetic toxoids of Shiga toxin types 1 and 2 protect mice against homologous but not heterologous toxin challenge. *Vaccine* 24:1142–8, 2006 (with Wen et al.). The established intimin-receptor Tir and the putative eucaryotic intimin-receptors nucleolin and 1 integrin localize at or near the site of enterohemorrhagic *Escherichia coli* O157:H7 adherence to enterocytes in vivo. *Infect. Immun.* 74:1255–65, 2006 (with Sinclair et al.). Cytotoxic necrotizing factor type 1 production by uropathogenic *Escherichia coli* modulates polymorphonuclear leukocyte function. *Infect. Immun.* 73:5301–10, 2005 (with Davis et al.).

Brian C. Schaefer, Assistant Professor; Ph.D., Harvard, 1995. Molecular mechanisms of signal transmission from lymphocyte antigen receptors to NF-κB. Current studies focus on the role of changes in protein-protein association and in protein localization for signal transmission to NF-κB, the mechanism of CTLA-4 inhibition of NF-κB activation, the role of translocations of the Bcl10 and MALT1 genes in MALT lymphoma, and the role of the TCR-directed NF-κB signaling pathway in T-cell development and the host response to pathogen infection. POLKADOTS are foci of functional interactions between cytosolic intermediates in T cell receptor–induced activation of NF-κB. *Mol. Biol. Cell.,* 2006 (online; with Rossman et al.). Complex and dynamic redistribution of NF-κB signaling intermediates in response to T cell receptor stimulation. *Proc. Natl. Acad. Sci. U.S.A.* 101:1004–9, 2004 (with Kappler et al.).

THE UNIVERSITY OF ALABAMA AT BIRMINGHAM

Department of Microbiology

Program of Study

The Department of Microbiology, in conjunction with the Departments of Biochemistry and Molecular Genetics, Cell Biology, and Neurobiology, offers course work and individual laboratory research leading to the Ph.D. degree. The program is designed to provide high-quality interdisciplinary training in cell and molecular biology to a selected group of predoctoral students, preparing them to become independent investigators in these disciplines. Students are immersed in research at the forefront of scientific endeavor and provided with sufficient guidance and course work to place their research in the proper perspective.

This interdisciplinary program includes faculty members from Anesthesiology, the Arthritis Center, Biochemistry and Molecular Genetics, Biology, the Cancer Center, Cell Biology, Endocrinology, Geographic Medicine, Hematology/Oncology, Infectious Disease, Medicine, Microbiology, Neurobiology, Neurology, Neuropsychiatry, Oral Biology, Pathology, Pediatrics, Pharmacology, Physiology and Biophysics, Psychiatry, and Public Health.

The first-year curriculum emphasizes three areas: acquisition of a working knowledge of contemporary cellular and molecular biology through an intensive, integrated course in biochemistry, genetics, cell biology, virology, immunology, and neurobiology; involvement in a diversity of laboratory research training experiences; and the development of skills in reading, writing, and speaking. Advanced students are engaged primarily in research but also take some advanced courses and tutorials in specialized areas of interest and participate in seminars. Completion of requirements for the Ph.D. usually takes six years. No foreign language is required.

Areas of specialization for dissertation research include fundamental molecular biology; biochemistry of nucleic acids; prokaryotic and eukaryotic molecular and cell biology; molecular virology; viral, microbial, and mammalian genetics; immunogenetics; cellular, developmental, and tumor immunology; immunochemistry; biological macromolecules and membranes; host-parasite relationships; infectious diseases; biochemistry of connective tissues; X-ray crystallography; NMR spectroscopy; molecular biophysics; ion-channel structure and biophysics; synaptic function and plasticity; neuronal development; and glial-neuronal signaling. Students are encouraged to present research results at scientific meetings and publish in scientific journals.

Research Facilities

Faculty members participating in the program have more than 400,000 square feet of laboratory space. In addition to well-equipped labs, a number of core facilities are available. These include automated DNA sequencing, DNA chip analysis, NMR spectroscopy, electron microscopy, protein and nucleic acid synthesis and analysis, mass spectrometry, confocal microscopy, fluorescence-activated cell sorting, large-scale bacterial fermentation, X-ray diffraction, a P3 containment facility, computer facilities, and a hybridoma facility.

Financial Aid

All students admitted to the program receive support through national or state granting agencies in the amount of $23,000 plus student health insurance and full payment of tuition and fees.

Cost of Study

Tuition and fees for the 2004–05 academic year were $5500 for in-state students and $11,000 for out-of-state students. As indicated above, tuition and fees are paid for all students.

Living and Housing Costs

The cost of living in Birmingham is slightly lower than the U.S. average. Housing is readily available in the Medical Center area.

Student Group

The graduate program of the Departments of Biochemistry and Molecular Genetics, Cell Biology, Neurobiology, and Microbiology consists of more than 130 faculty members, 190 full-time graduate students, and about 150 postdoctoral fellows and visiting faculty members. The total enrollment at the University of Alabama at Birmingham is more than 15,000.

Student Outcomes

Graduates typically go on to postdoctoral research appointments, followed by careers in academic research and teaching, research in the biotechnology industry, or other science-related professions.

Location

Birmingham is located in the lovely rolling foothills of the Appalachian Mountain range in central Alabama. A metropolitan area that includes 1 million people, Birmingham is only a few hours' drive from Atlanta, Nashville, New Orleans, and the Gulf Coast. The city has excellent art and historical museums, theaters, libraries, a symphony orchestra, a ballet, a zoo, and botanical gardens. A host of recreational opportunities, including camping, swimming, fishing, hiking, golf, tennis, and boating, are available the year round in numerous local and state parks.

The University

The University of Alabama at Birmingham consists of University College, the Graduate School, and the Medical Center. It is located in the largest population center of the state and is heavily engaged in research, instruction, and service programs. Its dedication to excellence in biomedical research is exemplified by its consistent ranking in the top twenty institutions in receipt of federal research funds.

Applying

Applications to the program are evaluated by the Admissions Committee in consultation with other faculty members. The admission decision is based on scores achieved on the Graduate Record Examinations (a combined score of at least 1200, nominally, on the verbal and quantitative portions of the General Test), undergraduate grade point average (consideration is given to the curriculum completed), letters of evaluation, prior research experience, and a personal interview.

To be accepted into the program, the student should have completed a B.S. degree that includes the following undergraduate course work by the time of entrance: calculus, general and organic chemistry, and at least one introductory course in zoology or biology. Courses in physical chemistry, genetics, cell biology, and related topics are also beneficial to the candidate. Any remedial course work must be completed with a grade of B or better before the end of the first full year of doctoral study.

Applications are strongly encouraged from individuals with prior research experience, an M.S. degree in a related area, or a professional degree such as the M.D., D.M.D., D.V.M., or O.D.

The program anticipates admitting 35 to 40 students each year.

Correspondence and Information

Admissions Committee
Cellular and Molecular Biology Graduate Program
Suite 260, Bevill Biomedical Research Building
The University of Alabama at Birmingham
Birmingham, Alabama 35294-2170

Phone: 800-262-7764 (toll-free)
E-mail: cmb@uab.edu
Web site: http://www.cmb.uab.edu

The University of Alabama at Birmingham

THE FACULTY AND THEIR RESEARCH

Cell Physiology, Adhesion, and Signaling
S. Barnes, Ph.D.: bile acids, polyphenols, and metabolism and their effects on protein expression and function in chronic disease.
Z. Bebok, M.D.: membrane protein biogenesis in epithelial cells (CFTR as model); unfolded protein response.
R. Carter, M.D.: immune and autoimmune lymphocytes; structure and function in vivo.
C. Chang, Ph.D.: signal pathways in frog development.
D. Chaplin, M.D., Ph.D.: control of inflammation and lymphoid organ formation by LT and TNF.
J. Chatham, Ph.D.: cardiomyocyte function and metabolism in diabetes and ischemic heart disease.
J. Collawn, Ph.D.: intracellular protein sorting.
C. Gladson; M.D.: molecular mechanisms promoting tumor-cell proliferation; invasion and angiogenesis in gliomas.
J. Hagood, M.D.: fibroblast signaling in lung remodeling and fibrogenesis.
J. Kearney, Ph.D.: B cells; B-cell development; hybridomas; transgenic and knockout mice; immunoregulation; *B. anthracis*.
N. Kedishvili, Ph.D.: regulation of retinoic acid homeostasis.
F. Lin, M.D., Ph.D.: regulation of cell growth by G-protein-coupled receptor signaling.
R. Marchase, Ph.D.: calcium regulation and its impairment in diabetes.
G. Marques, Ph.D.: TGF-ß signaling in nervous system development and function.
M. Miller, Ph.D.: function and evolution of intercellular communication mechanisms.
E. Schwiebert, Ph.D.: extracellular nucleotide signaling and epithelial cell biology and physiology.
L. Schwiebert, Ph.D.: airway inflammation.
E. Sztul, Ph.D.: membrane traffic; protein degradation.
J. Thompson, Ph.D.: molecular mechanisms of angiogenesis.
A. Woods, Ph.D.: *Syndecan proteoglycans* in cell adhesion and matrix assembly.

Gene Regulation and Expression
A. Agarwal, M.D.: regulation of heme oxygenase gene expression in kidney and vascular injury.
L. Bridges, M.D., Ph.D.: genetic influences on treatment response in rheumatoid arthritis.
C. Chen, Ph.D.: mechanism and regulation of mammalian mRNA turnover.
X. Chen, Ph.D.: the p53 tumor-suppressor gene family and transcriptional regulation.
D. Crawford, M.D., Ph.D.: the role of G2/M-specific genes in mitosis and G2 DNA damage checkpoint regulation.
P. Higgins, Ph.D.: genetic and biochemical studies of chromosome dynamics.
C. Klug, Ph.D.: hematopoietic stem-cell biology and acute leukemias.
W.-C. Lin, M.D., Ph.D.: cell-cycle control and DNA damage response.
S. Lobo-Ruppert, Ph.D.: role of transcription factor oncogenes in tumor initiation.
R. Mayne, Ph.D.: collagen and collagen-binding proteins in health and disease.
J. McDonald, M.D.: cellular life and death signals in cancer; AIDS and bone disease.
M. Ruppert, M.D., Ph.D.: role of zinc finger transcription factors in tumor progression.
T. Ryan, Ph.D.: gene regulation; stem cells; mouse models; mutagenesis; cell therapies.
C. Turnbough, Ph.D.: *B. anthracis* spore structure-function/bacterial gene regulation.
H. Wang, Ph.D.: role of histone modification in chromatin function.

Immunology
S. Barnum, Ph.D.: role of complement, acute phase proteins, and adhesion molecules in acute and chronic inflammation in the central nervous system.
T. Benveniste, Ph.D.: immune/nervous system interactions.
O. Branch, Ph.D.: malaria molecular epidemiology and immunology.
P. Bucy, M.D., Ph.D.: T-cell regulation of immune responses in vivo.
R. Davis, M.D.: lymphocyte development and mechanisms of lymphomagenesis.
C. Elson, Ph.D.: chronic intestinal inflammation.
K. Fujihashi, D.D.S., Ph.D.: mucosal immunology; regulation of S-igA antibody responses; mucosal vaccine development.
V. Ghanta, Ph.D.: tumor immunology; CNS and immune system interactions.
Z. Hel, Ph.D.: development and testing of novel HIV/AIDS vaccine strategies.
L. Justement, Ph.D.: analysis of molecular mechanisms regulating lymphocyte biology.
J. Kabarowski, Ph.D.: regulation of inflammation and chronic inflammatory disease by lysophospholipid receptors.
J. Kapp, Ph.D.: immune regulation and transplantation.
J. Katz, Ph.D., D.D.S.: vaccine delivery systems; inflammation; innate immunity.
R. Kimberly, M.D.: autoimmunity, molecular mechanisms, and genetic risk.
W. Koopman, M.D.: pathogenesis of immune disease.
H. Kubagawa, M.D.: molecular genetics and immunopathology of host defense.
J. Mestecky, M.D.: mucosal immunity; vaccines.
M. Nahm, M.D.: adaptive and innate immune responses to vaccines against bacteria.
T. Strong, Ph.D.: identification of tumor antigens and development of cancer vaccines.
A. Szalai, Ph.D.: inflammation innate immunity and the acute phase proteins in health and disease.
L. Timares, Ph.D.: engineering dendritic cells for immunotherapy.
M. Walter, Ph.D.: structure and function of cytokines and their signaling molecules.
C. Weaver, M.D.: T-cell development.
Z. Zhang, Ph.D.: molecular regulation of early B-cell development and antibody repertoire formation.
T. Zhou, M.D.: specific induction of apoptosis in autoimmune and inflammatory cells.

Macromolecular Structure and Function
J. Blalock, Ph.D.: rational drug and vaccine design.
C. Brouillette, Ph.D.: protein structural cooperativity and energetics.
D. Chattopadhyay, Ph.D.: structure-function analysis of proteins.
I. Chesnokov, Ph.D.: DNA replication and cell cycle in eukaryotes.
H. Cheung, Ph.D.: regulatory mechanism in cardiac muscle.
L. DeLucas, Ph.D.: protein crystallography/protein crystal growth.
G. Elgavish, Ph.D.: NMR studies of intact hearts.
S. Frank, M.D.: growth hormone action and GH receptor structure and function.
B. Freeman, Ph.D.: tissue metabolism of reactive inflammatory mediators.
R. Krishna, Ph.D.: structural biology and biomolecular NMR spectroscopy.
J. Murphy-Ullrich, Ph.D.: extracellular matrix control of cell and growth-factor function.

C. Raman, Ph.D.: lymphocyte activation, immune tolerance and autoimmunity tolerance, and autoimmunity.
J. Segrest, M.D., Ph.D.: structural biology of supramolecular assemblies, particularly lipoproteins and membranes.
B. Sha, Ph.D.: structure and function of molecular chaperones.
N. Sthanam, Ph.D.: bacterial surface protein anchoring.

Molecular Genetics and Disease
S. Abulkadir, Ph.D.: molecular genetics of prostate cancer.
P. Atkinson, M.D., Ph.D.: primary immunodeficiency/role of infection in chronic diseases.
D. Bedwell, Ph.D.: translation termination; treatment of genetic diseases.
P. Burrows, Ph.D.: B-lymphocyte development and function.
P. Detloff, Ph.D.: mouse models of human genetic disorders.
K. Dybvig, Ph.D.: pathogenic mechanisms of mycoplasmas.
L. Guay-Woodford, M.D.: characterizing molecular determinants involved in PKD pathogenesis.
K. Jiao, M.D., Ph.D.: TGF-ß/BMP signaling during cardiogenesis.
K. Kirk, Ph.D.: the CFTR chloride channel.
J. Kudlow, M.D.: nucleocytoplasmic O-glycosylation in the control of cell function.
J. Mountz, M.D.: gene therapy; T-cell aging; immunogenetics; T-cell imaging.
H. Schroeder, M.D., Ph.D.: development/function of lymphocyte antigen receptors.
R. Serra, Ph.D.: mechanism of TGF-ß action in developmental and disease processes.
T. Townes, Ph.D.: developmental regulation of gene expression.
T. Unnasch, Ph.D.: molecular epidemiology and ecology of vector-borne diseases.
D. Welch, Ph.D.: molecular basis of tumor progression and metastasis.
B. Yoder, Ph.D.: cilia signaling and dysfunction in development and disease.

Molecular Pathogenesis
D. Balkovetz, M.D., Ph.D.: epithelial cell biology; epithelial cell signal regulation; regulation of paracellular transport across epithelial cell tight junctions.
W. Benjamin, Ph.D.: molecular typing of *Streptococcus pneumoniae*.
D. Briles, Ph.D.: bacterial pathogenesis; virulence; immunity; pneumococcus.
N. Childers, D.D.S., Ph.D.: development of a vaccine for the prevention of dental caries.
M. Cooper, M.D.: developmental immunobiology, with emphasis on B-cell and T-cell differentiation; clinical immunology, with emphasis on immunodeficiency diseases and lymphoid malignancies.
S. Hollingshead, Ph.D.: mechanisms of variation in microbial pathogenesis.
R. Lorenz, M.D., Ph.D.: cellular and molecular immunology of the gastrointestinal tract.
S. Michalek, Ph.D.: mucosal vaccines and host mechanisms involved in inflammation.
R. Morrison, Ph.D.: immunobiology of Chlamydia infection.
C. Morrow, Ph.D.: understanding, at the molecular level, virus–host cell interactions.
M. Niederweis, Ph.D.: the role of porins in outer-membrane permeability and drug resistance of mycobacteria.
D. Pritchard, Ph.D.: roles of glycoconjugates in pathogenesis.
J. Rayner, Ph.D.: cell biology of the malaria parasite *Plasmodium falciparum*.
A. Steyn, Ph.D.: mechanism of mycobacterium tuberculosis virulence.
K. Waites, M.D.: diagnostic microbiology; epidemiology and mechanisms of antimicrobial resistance; mycoplasma and ureaplasma diseases; staphylococcal diseases.
H. Wu, Ph.D.: bacteria-host interaction.
J. Yother, Ph.D.: capsular polysaccharides of *Streptococcus pneumoniae*.

Neurobiology
D. Benos, Ph.D.: molecular basis of operation on ion channels and transporters.
M. Brenner, Ph.D.: molecular studies of astrocytes in health and disease.
L. Dobrunz, Ph.D.: synaptic transmission and plasticity in hippocampus.
J. Hablitz, Ph.D.: cellular mechanisms of neurotransmission.
G. Johnson, Ph.D.: molecular mechanisms of neurodegeneration.
R. Jope, Ph.D.: neuronal signaling mechanisms regulating gene expression and cell death.
R. Lester, Ph.D.: molecular and cellular mechanisms of nicotine addiction.
L. Pozzo-Miller, Ph.D.: neurotrophins on Ca2+ signaling, synapse development, and plasticity.
K. Roth, M.D., Ph.D.: molecular regulation of neuronal cell death.
D. Ruden, Ph.D.: environmental and developmental toxicology.
H. Sontheimer, Ph.D.: the role of neuroglia in brain function and disease.
D. Sweatt, Ph.D.: signal transduction mechanisms in learning and memory.
A. Theibert, Ph.D.: role of phosphoinositides in developmental neurobiology.
S. Wilson, Ph.D.: mouse models of neurodegeneration.
M. Wyss, Ph.D.: control of the autonomic nervous system.
Y. Zhou, Ph.D.: modulation of ion channels; regulation of neuronal excitability and synaptic transmission.

Virology
W. Britt, M.D.: human herpesviruses, molecular virology, and pathogenesis.
L. Chow, Ph.D.: human papillomavirus DNA replication and pathogenesis.
T. T. Dokland, Ph.D.: cryoelectron microscopy and X-ray crystallography of virus assembly processes.
J. Engler, Ph.D.: identifying novel diagnostic peptides and cell-specific ligands.
B. Hahn, M.D.: origin and evolution of primate lentiviruses.
J. Kappes, Ph.D.: HIV, molecular virology, and pathogenesis.
R. Kaslow, M.D.: immunogenetic determinants in AIDS and other diseases.
O. Kutsch, Ph.D.: HIV-1 latency and drug screening.
E. Lefkowitz, Ph.D.: bioinformatics; biodefense; microbial genomics and evolution.
M. Luo, Ph.D.: structure-based approaches to anti-infectious agents.
P. Prevelige, Ph.D.: structural biology of viral assembly and infection.
G. Shaw, M.D., Ph.D.: evolution and persistence of HIV-1.
W. Sullender, M.D.: respiratory syncytial virus; antigenic diversity.
S. Thompson, Ph.D.: translation initiation and replication of RNA viruses.
R. Whitley, M.D.: herpesvirus; varicella zoster virus.
A. Zajac, Ph.D.: antiviral immunity; T-cell responses; immunological memory.

UNIVERSITY OF CALIFORNIA, DAVIS

Graduate Group in Microbiology

Programs of Study

The Graduate Group in Microbiology offers programs leading to the M.S. and Ph.D. degrees in microbiology. The Ph.D. track is a research program that emphasizes training in modern approaches to microbiological problems. The M.S. track is designed as a terminal degree program requiring submission of a research thesis. The Group is recognized for its strength in microbial physiology, genetics, molecular biology, and virology and for the interdisciplinary breadth of its faculty. The 70 faculty members of the Group conduct research in fundamental, applied, and pathogenic microbiology, and they represent departments affiliated with the Division of Biological Sciences, the College of Agricultural and Environmental Sciences, and the professional Schools of Medicine and Veterinary Medicine.

The formal course work for the Ph.D. degree is based on a one-year core sequence that includes microbial diversity and physiology, molecular biology, and pathogenic microbiology. First-year Ph.D. students also enroll in a course in recombinant DNA methodology and conduct research during laboratory rotations with faculty mentors selected by mutual consent. Advanced graduate courses and seminars appropriate to microbiological research are offered through affiliated departments, including Microbiology, Medical Microbiology and Immunology, Veterinary Microbiology and Immunology, Molecular and Cellular Biology, and Food Science and Technology. The Ph.D. qualifying examination consists of a research-based oral examination at the end of the second year. The Ph.D. program takes, on average, five years to complete, including the dissertation research.

Research Facilities

The University of California, Davis, is a leading research institution and is committed to providing modern research facilities and state-of-the-art equipment. In addition to well-equipped faculty and departmental laboratories for microbiological research, campuswide facilities include laboratories for electron microscopy, nuclear magnetic resonance, gas and liquid chromatography, mass and UV/visible/IR spectrometry, amino acid sequencing, and peptide and oligonucleotide synthesis. Access to controlled microbial, plant, and animal growth facilities is readily available to researchers in the Group.

Financial Aid

All domestic Ph.D. students are supported by a first-year stipend ($1770 per month in 2005–06) and provided with a nonresident tuition fellowship. In subsequent years, all Ph.D. students receive support from a variety of sources, including University fellowships, teaching assistantships, and research assistantships. Students in the M.S. program can be supported through fellowships and assistantships.

Cost of Study

Registration fees were $2987 per quarter in 2004–05, including health insurance. Nonresident tuition was $4898 per quarter. Out-of-state domestic students establish California residence in their first year. All fees are paid for students supported as research assistants, and partial fees are paid for students supported as teaching assistants.

Living and Housing Costs

Single graduate students typically pay about $1000 per month for on-campus or off-campus housing, food, transportation, and miscellaneous expenses. Expenses for married graduate students are generally $800–$1000 more than for single students.

Student Group

There are approximately 65 full-time students in the Graduate Group in Microbiology. The Microbiology Graduate Students Association provides academic and social orientation, organizes a biennial student research conference, and nominates members to serve on administrative committees of the Group.

Location

Davis is an environmentally conscious town of about 50,000 located in the agriculturally diverse Sacramento Valley, 15 miles west of Sacramento and 70 miles northeast of San Francisco. Davis provides ready access to cultural events and to aquatic and mountain sports and is celebrated for its bicycling residents.

The University

The University of California, Davis, is one of nine campuses of the University of California. There are about 24,000 students enrolled in the undergraduate colleges, the professional schools, and the graduate division. The students are served by a faculty of more than 1,500. The Davis campus is renowned for teaching and research in biological and agricultural sciences and for one of the most diversified teaching faculties of the University of California system. The Davis campus libraries rank in the top twenty-five academic libraries in North America.

Applying

Application forms may be obtained from the Admissions address or applicants may apply online at http://som.ucdavis.edu/departments/microbiology/graduate/mgg. The completed application form, a $60 application fee, and all official university/college/community college transcripts should be returned to the Group address in this In-Depth Description by January 15. Strong preference is given to doctoral applicants. Students interested in a master's degree should contact the chair of the Graduate Group prior to submitting an application. International students from non-English-speaking countries must also include their TOEFL scores (at least 625 or above). All prospective students must have the results of the General Test of the Graduate Record Examinations (subject test in biology or biochemistry is helpful but not required), and three letters of recommendation forwarded to the Group address in this In-Depth Description.

Correspondence and Information

For program information:

Graduate Group in Microbiology
Department of Medical Microbiology and Immunology
University of California, Davis
One Shields Avenue
Davis, California 95616
Phone: 530-752-0262
Web site: http://som.ucdavis.edu/departments/microbiology/
      graduate/mgg

For an application:

Admissions
Graduate Studies and Research
University of California, Davis
One Shields Avenue
Davis, California 95616
Web site: http://gradstudies.ucdavis.edu/b4apply.htm

University of California, Davis

THE FACULTY AND THEIR RESEARCH

E. P. Baldwin, Assistant Professor of Molecular and Cellular Biology; Ph.D., Berkeley. Analysis and engineering of protein structure and function using molecular genetics, biochemistry, and X-ray crystallography.

P. A. Barry, Assistant Professor, Center for Comparative Medicine; Ph.D., Utah. Mechanisms of viral persistence, pathogenesis, and vaccine strategies.

S. W. Barthold, Professor of Veterinary Microbiology and Immunology; D.V.M., California, Davis; Ph.D., Wisconsin. Host-agent interactions during persistent infections, with emphasis on Lyme disease and granulocytic ehrlichiosis.

N. Baumgarth, Associate Professor, Center for Comparative Medicine; Ph.D., Hannover (Germany). Regulation of immune responses to infectious agents; focusing on immunity to influenza virus, HIV and *Borrelia burgdorfei*.

A. J. Bäumler, Professor of Medical Microbiology and Immunology; Ph.D., Turbingen (Germany). Molecular mechanisms involved during the interaction of *Salmonella* with the intestinal mucosa.

B. L. Beaman, Professor of Medical Microbiology and Immunology; Ph.D., Kansas. Host-parasite interactions; pathogenesis.

C. L. Bevins, Professor of Medical Microbiology and Immunology; M.D., Ph.D., Maryland. Innate immunity; mucosal immunology; antimicrobial peptides.

L. F. Bisson, Professor of Viticulture and Enology; Ph.D., Berkeley. Regulation of hexose transport in yeasts.

A. C. Brault, Assistant Professor of Pathology, Microbiology, and Immunology; Ph.D., Texas. Molecular epidemiology and pathogenic determinants of arthropod-borne viruses such as West Nile virus.

G. Bruening, Professor of Plant Pathology; Ph.D., Wisconsin. Replication of plant viruses and satellite RNAs.

S. M. Burgess, Assistant Professor of Molecular and Cellular Biology; Ph.D., California, San Francisco. Chromosome organization and nuclear function in *S. cerevisiae*.

B. A. Byrne, Assistant Professor of Pathology, Microbiology, and Immunology; Ph.D., Washington State. Host-pathogen interactions and clinical microbiology; plasmid-encoded virulence factors of *Rhodococcus equi* (a horse and human pulmonary pathogen).

C. J. Cardona, Assistant Professor of Population Health and Reproduction; D.V.M., Purdue; Ph.D., Michigan State. Pathogenesis of avian viruses; immune responses; viral pathogenicity; viral transmission.

R. H. Cheng, Professor of Molecular and Cellular Biology; Ph.D., Purdue. Virus assembly; model systems of macromolecules; proteome imaging; supermolecule symmetry; cryoEM and structural modeling.

A. T. W. Cheung, Professor of Pathology; Ph.D., UCLA. Host defense in human neonates and nonhuman primates.

B. B. Chomel, Professor of Population Health and Reproduction; D.V.M., Lyons (France). Epidemiological and diagnostic aspects of some major infectious zoonoses, with emphasis on cat-scratch disease and related *Bartonella* infections; rabies and pet bites.

R. Y. Chuang, Professor of Pharmacology and Toxicology; Ph.D., California, Davis. Gene regulation of virus-infected cells.

D. O. Cliver, Professor of Population Health and Reproduction; Ph.D., Ohio State. Microbiologic safety of food and water.

P. A. Conrad, Professor of Pathology, Microbiology, and Immunology; D.V.M., Colorado State; Ph.D., Edinburgh. Human and veterinary parasitology.

J. S. Cullor, Associate Professor of Population Health and Reproduction; Ph.D.; California, Davis. Protection and prevention of endotoxic shock.

M. E. Dahmus, Professor of Molecular and Cellular Biology; Ph.D., Caltech. Interactions of RNA polymerase with the transcription apparatus.

S. Dandekar, Professor and Chair, Medical Microbiology and Immunology; Professor of Internal Medicine; Ph.D., Baroda (India). Mucosal infections and immunity; molecular pathogenesis of AIDS and co-infections; functional genomics of infectious diseases.

A. Gelli, Assistant Professor of Medical Pharmacology and Toxicology; Ph.D., Toronto. Cell surface signaling interactions between fungal pathogens and host cells that promote the pathogenesis of human fungal infections.

L. J. Gershwin, Professor of Veterinary Microbiology and Immunology; D.V.M., Ph.D., California, Davis. Microbiology and immunology of bovine respiratory disease.

R. P. Hedrick, Professor of Veterinary Medicine and Epidemiology; Ph.D., Oregon State. Infectious diseases of fish and shellfish.

J. W. B. Hershey, Professor of Biological Chemistry; Ph.D., Rockefeller. Translational control mechanisms and initiation factors.

W. D. Heyer, Professor of Microbiology; Ph.D., Bern (Switzerland). Molecular mechanisms and regulation of DNA repair; meiosis; application of homologous recombination.

M. J. Holland, Professor of Biological Chemistry; Ph.D., Massachusetts. Regulation of gene expression in yeasts.

N. Hunter, Assistant Professor of Microbiology; Ph.D., Oxford. Crossing-over during meiosis; molecular, genetic, and cytological approaches to understanding the mechanism and regulation of crossing-over between homologous chromosomes during meiosis.

M. M. Igo, Associate Professor of Microbiology; Ph.D., Harvard. Bacterial signal transduction.

K. B. Kaplan, Assistant Professor of Molecular and Cellular Biology; Ph.D., California, San Francisco. Molecular genetics and biochemistry of chromosome segregation in *S. cerevisiae*.

S. C. Kowalczykowski, Professor of Microbiology; Ph.D., Georgetown. Molecular mechanisms of genetic recombination: DNA helicases, DNA binding proteins, recA protein, and recBCD enzyme.

H.-J. Kung, Professor, UC Davis Cancer Center; Ph.D., Caltech. Tyrosine kinases; oncogenic signals; molecular genetics of oncogenic viruses.

R. B. LeFebvre, Professor of Pathology, Microbiology, and Immunology; Ph.D., Colorado State. Development of diagnostic immunologic and genetic probes for pathogens.

P. S. C. Leung, Associate Adjunct Professor of Medicine; Ph.D., California, Davis. Molecular basis of human autoimmune diseases and allergy; microbiological aspects of primary biliary cirrhosis.

S. J. Lin, Assistant Professor of Microbiology; Ph.D., Johns Hopkins. Molecular mechanisms of calorie restriction mediated life span extension in y yeast *Saccharomyces cerevisiae*.

P. Luciw, Professor of Pathology; Ph.D., Pennsylvania. Regulation of gene expression in HIV and SIV; mechanisms of retroviral pathogenesis.

S. Luckhart, Associate Professor of Medical Microbiology and Immunology; Ph.D., Rutgers. Innate immunity; signal transduction; vector-borne diseases; malaria.

N. J. MacLachlan, Professor of Pathology, Microbiology, and Immunology; Ph.D., California, Davis. Molecular biology and pathogenesis of bluetongue virus and equine arteritis virus infections.

M. L. Marthas, Associate Adjunct Professor of the Primate Center and Veterinary Pathology; Ph.D., California, Davis. Molecular mechanisms of viral diseases, especially SIV and HIV.

K. A. McDonald, Associate Professor of Chemical Engineering and Materials Science; Ph.D., Maryland. Bioprocess engineering.

J. C. Meeks, Professor of Microbiology; Ph.D., Oregon. Physiology and genetics of free-living and symbiotic cyanobacteria.

C. J. Miller, Professor, California Regional Primate Research Center; Ph.D., California, Davis. Pathogenesis and host immune response to viral infections; AIDS.

D. A. Mills, Assistant Professor of Viticulture and Enology; Ph.D., Minnesota. Microbial ecology and molecular genetics of lactic acid bacteria (LAB).

D. C. Nelson, Professor of Microbiology; Ph.D., Oregon. Physiology and ecology of colorless sulfur bacteria.

T. W. North, Professor, Center for Comparative Medicine; Ph.D., Arizona. Antiviral drugs and mechanisms of viral drug resistance.

J. Nunnari, Associate Professor of Molecular and Cellular Biology; Ph.D., Vanderbilt. Mitochondrial organization and transmission; mechanisms of mitochondrial DNA inheritance.

D. M. Ogrydziak, Professor of Food Science and Technology; Ph.D., MIT. Protein processing; the genetics of *Yarrowia*.

B. I. Osburn, Professor of Veterinary Pathology; Ph.D., California, Davis. Immunology of bluetongue virus infection.

D. Pappagianis, Professor of Medical Microbiology and Immunology; M.D., Stanford; Ph.D., Berkeley. Biology of coccidioidomycosis.

R. E. Parales, Assistant Professor of Microbiology; Ph.D., Cornell. Bacterial degradation of environmental pollutants; chemotaxis of bacteria to organic pollutants.

N. C. Pedersen, Professor of Veterinary Medicine; Ph.D., Australian National. Feline and simian models for human AIDS research; pathogenesis and antiviral drugs.

C. Pomeroy, Professor of Medicine and Infectious Diseases; M.D., Michigan. Investigation of pathogenesis of murine (mouse) cytomegalovirus (MCMV) infection, specifically the role played by host cytokines in the defense against MCMV.

E. R. (Ted) Powers, Assistant Professor of Molecular and Cellular Biology; Ph.D., California, Santa Cruz. Nutrient sensing and control of cell growth in *S. cerevisiae*.

M. L. Privalsky, Professor of Microbiology; Ph.D., Berkeley. Retroviral oncogenes; transcriptional regulation; control of cellular growth and differentiation.

K. Radke, Associate Professor of Animal Sciences; Ph.D., California, Davis. Retroviral gene expression; oncogenic viruses.

G. H. Rhodes, Associate Adjunct Professor of Pathology; Ph.D., Berkeley. Development of vaccines against SIV; immune mechanisms of protection.

J. R. Roth, Professor of Microbiology; Ph.D., Johns Hopkins. Genetic approaches to the physiology and evolution of bacteria.

D. D. Y. Ryu, Professor of Chemical Engineering; Ph.D., MIT. Biochemical, biomolecular, and metabolic engineering.

M. A. Savageau, Professor of Biomedical Engineering; Ph.D., Stamford. Function, design, and evolution of bacteria.

E. T. Sawai, Associate Adjunct Professor of Medical Pathology; Ph.D., Baylor. Retroviral pathogenesis, AIDS, HIV/SIV, signal transduction, oncogenes.

J. M. Scholey, Professor of Molecular and Cellular Biology; Ph.D., Cambridge. Cell biology; functions of microtubule-based motor proteins and intracellular transport systems; mechanisms of mitosis in *Drosophila* embryo; microtubule-based transport and IFT-motors in *C. elegans* neurons.

K. M. Scow, Professor of Land, Air and Water Resources; Ph.D., Cornell. Soil microbial ecology; metabolism of organic pollutants.

I. H. Segel, Professor of Biochemistry and Biophysics; Ph.D., Wisconsin. Enzymology of microbial sulfur and nitrogen metabolism.

B. L. Shacklett, Assistant Professor of Medical Microbiology and Immunology; Ph.D., California, Davis. Cell-mediated immune responses to HIV-1 and other viruses in mucosal lymphoid tissues; trafficking of lymphoid cells to mucosal tissues and the central nervous system.

K. Shiozaki, Associate Professor of Microbiology; Ph.D., Kyoto (Japan). Molecular mechanism of stress response in eukaryotic cells, using fission yeast *Schizosaccharomyces pombe*.

M. H. Singer, Associate Professor of Microbiology; Ph.D., Wisconsin. Prokaryotic development and differentiation; bacterial signal transduction.

J. V. Solnick, Associate Professor of Infectious and Immunological Diseases; Ph.D., North Carolina. Molecular pathogenesis of *Helicobacter pylori* and related organisms.

E. E. Sparger, Adjunct Professor of Veterinary Medicine; D.V.M., Georgia; Ph.D., California, Davis. FIV pathogenesis and replication.

J. L. Stott, Professor of Pathology, Microbiology, and Immunology; Ph.D., California, Davis. Viral immunology.

M. Syvanen, Professor of Medical Microbiology and Immunology; Ph.D., Berkeley. Genetic control of biological detoxification mechanisms.

J. V. Torres, Assistant Professor of Medical Microbiology and Immunology; Ph.D., Baylor. Vaccine development; AIDS; retroviral immunology; HIV/SIV.

R. Tsolis, Assistant Professor of Medical Microbiology and Immunology; Ph.D., Oregon Health Sciences. Molecular biology of host-pathogen interactions during infection by *Brucella* spp.

S. Wuertz, Professor of Civil and Environmental Engineering; Ph.D., Massachusetts. Horizontal gene transfer in biofilms, molecular detection of environmental pathogens, microbial community analysis.

T. Yilma, Professor of Veterinary Microbiology and Immunology; Ph.D., California, Davis. Molecular virology: expression of heterologous genes.

G. M. Young, Assistant Professor of Food Science and Technology; Ph.D., Washington State. Genetics and molecular biology of virulence factor secretion by the pathogenic bacteria *Salmonella typhimurium* and *Yersinia enterocolitica*.

UNIVERSITY OF CINCINNATI

Cincinnati Children's Hospital Medical Center
Immunobiology Graduate Training Program

Programs of Study

Students can enter the Immunobiology Graduate Training Program in pursuit of either a Master of Science (M.S.) degree or a doctoral (Ph.D.) degree. The typical course of study requires two years for the master's degree and four to five years for the doctoral degree.

The program provides a sound foundation in molecular, genetic, and cellular approaches to immunological problems. The curriculum consists of one year of didactic course work followed by advancement to candidacy early in the second year. Throughout the program students engage in stimulating discussion of current immunologic advancements through participation in several research seminars and journal clubs.

In the second through fourth years, Ph.D. students focus entirely on their thesis research. The highly accomplished faculty members provide a wide variety of research experiences for the student to choose from, ranging from basic immunological mechanisms to the study of immune dysfunction in human disease. At the completion of this program, the student has a solid foundation in state-of-the-art immunologic, molecular, genetic, and genomic approaches to conducting medical research and is well prepared to meet the biomedical challenges of the future.

Research Facilities

In addition to a world-class faculty, graduate students in the Immunobiology Graduate Training Program enjoy access to world-class facilities while conducting their research at the Children's Hospital Research Foundation. All researchers and students have access to the hospital's core support facilities and capabilities.

Cincinnati Children's Hospital offers more research space than any other pediatric facility in the nation: 520,000 square feet in five buildings, including a dedicated research building that was most recently expanded in fall 2000. The hospital is close to patient-care facilities and the University of Cincinnati (UC) College of Medicine.

Since the Immunobiology Graduate Training Program is administered by the University of Cincinnati's College of Medicine, students have access to resources at UC. UC's Medical Sciences Building, one of the country's largest research-teaching structures under one roof, houses extensive state-of-the-art research facilities.

Financial Aid

Financial support is generous for graduate students in the program. All doctoral candidates accepted into the program receive full tuition remission, a nationally competitive stipend, and a health insurance plan. Eligible students can receive financial support from numerous scientific investigator and institutional research funds.

Cost of Study

Tuition costs per quarter are $2867 for in-state students and $5677 for out-of-state students. Other per-quarter costs are a general fee of $234 and campus life fee of $131.

Living and Housing Costs

The cost of living remains relatively low in comparison to similar and larger metropolitan areas across the country. While housing and gas prices are on the rise, students have found the city to be reasonably affordable. Current graduate students are generally able to find safe and comfortable apartments in the $400–$600 range.

Student Group

Students involved in the program have a strong interest in scientific research and have backgrounds in biology, chemistry, or premedicine.

Location

The greater Cincinnati metropolitan area consists of nearly 2 million people and is considered to be one of the best areas in the U.S. in which to live. Across the Ohio River is northern Kentucky, with an active riverfront shopping area, the Newport Aquarium, and opportunity for numerous fine dining experiences. Cincinnati has many attractions of its own, including the Cincinnati Chamber Orchestra, the Cincinnati Zoo, the Contemporary Arts Center, and professional baseball and football teams.

The Medical Center

Cincinnati Children's Hospital Medical Center serves the medical needs of infants, children, and adolescents with family-centered care, innovative research, and outstanding teaching programs.

Clinical procedures and treatments pioneered at Cincinnati Children's are now used throughout the world. The impact of its medical research breakthroughs has improved pediatric health today and will do so for generations to come.

Applying

Applicants to the Immunobiology Graduate Training Program at Cincinnati Children's Research Foundation and the University of Cincinnati are strongly urged to apply for admission before February 1 of the year in which they desire admission. First-year students are enrolled in July of the academic year. For admission to the program, a student should have at least one baccalaureate degree from an accredited institution, scores from the General Test and Subject Test of the Graduate Record Examinations, a GPA of 3.0 on a 4.0 scale, and a strong undergraduate background in biology.

Students may apply online at http://www.uc.edu/admissions/app/Program. Application packets should include the University of Cincinnati Application for Graduate Study; a $40 nonrefundable application fee; personal data sheet; personal background statement; three letters of recommendation; an official copy of GRE scores (and TOEFL scores, if applicable); and official undergraduate transcripts.

Correspondence and Information

Sonya Shields
Immunobiology Graduate Training Program
Cincinnati Children's Hospital Medical Center
3333 Burnet Avenue
Cincinnati, Ohio 45229-3039
Phone: 513-636-1339
E-mail: sonya.shields@cchmc.org
Web site: http://www.cchmc.org

University of Cincinnati

THE FACULTY AND THEIR RESEARCH

Yasmine Belkaid, Assistant Professor of Pediatrics; Ph.D., Paris XI (South), 1996. Role of regulatory T cells in pathogen persistence, disease reactivation, and immunity to persistent infection.

Charles Caldwell, Assistant Professor of Medicine; Ph.D., San Diego State. Modulating inflammation in a variety of disease states.

Rhonda Cardin, Assistant Professor of Pediatrics; Ph.D., Louisiana State, 1989. Cytomegalovirus pathogenesis and latency; mechanisms of immunological control of persistent CMV infection; immune evasion mechanisms and CMV-encoded chemokine receptors.

Sunil Chatterjee, Professor of Medicine; Ph.D., Calcutta, 1966. Construction and testing of cancer vaccines.

Monica Chiaramonte, Assistant Professor of Pediatrics; Ph.D., Buenos Aires, 1993. Immunological mechanisms underlying the pathogenesis of lung fibrosis; identification components of both innate and adaptive immunity involved in fibrotic processes.

Claire A. Chougnet, Assistant Professor of Pediatrics; Ph.D., Paris V, 1991. Interactions between antigen-presenting cells and T cells, with special focus on pathogenesis of infection by human immunodeficiency virus.

Robert Colbert, Associate Professor of Pediatrics; M.D./Ph.D., Rochester, 1987. Immunological mechanisms underlying pathogenesis of autoinflammatory diseases, in particular, the role of HLA-B27 in susceptibility to spondyloarthropathies.

Joan Cook-Mills, Associate Professor of Medicine; Ph.D., Michigan State, 1987. Endothelial cell regulation of leukocyte migration into inflammatory sites, such as in allergic asthma.

George Deepe, Professor of Medicine and Chief, Division of Infectious Diseases; M.D., Cincinnati, 1976. Analysis of protective immune response to the pathogenic fungus *Histoplasma capsulatum;* determining influence of cytokines and T-cell subpopulations on host control of the fungus.

Jay Degen, Assistant Professor of Pediatrics; Ph.D., Washington (Seattle), 1982. Role of growth factors and blood coagulation proteins in various biological processes.

Rodney DeKoter, Assistant Professor of Medicine; Ph.D., University of Western Ontario, Canada, 1996. Transcription factors that regulate the development and function of the immune system.

Lisa Filipovich, Professor of Pediatrics; M.D., Minnesota, 1974. Clinical research in BMT for immunodeficiencies and histiocytic disorders; immune reconstitution following alternative donor transplantation.

Fred Finkelman, McDonald Professor of Medicine; M.D., Yale, 1971. Cytokine biology, including regulation of cytokine responses and cytokine roles in allergy, asthma, autoimmunity, and infectious diseases.

David Glass, Professor of Pediatrics; M.D., Birmingham (England), 1965. Autoimmunity, especially of chronic rheumatic diseases of childhood, with application of high-throughput genomic and functional genomic methodologies.

Thomas A. Griffin, Resident Assistant Professor of Pediatrics; M.D./Ph.D., Case Western Reserve, 1991. Cellular and molecular biology of immunoproteasomes; molecular mechanisms regulating immunoproteasome biosynthesis; nonantigen-presenting functions of immunoproteasomes in T cells.

Alexei Grom, Resident Assistant Professor of Pediatrics; M.D., Leningrad Pediatric Medical, 1986. Systemic JRA and macrophage activation syndrome; NK cell function.

Gurjit Hershey, Associate Professor of Pediatrics; M.D./Ph.D., Washington (St. Louis), 1992. Genetics and development of atopic diseases, including asthma.

David Hildeman, Assistant Professor of Pediatrics; Ph.D., Wisconsin–Madison, 1997; Molecular biology of antigen-specific T cells; mechanisms involved in T-cell homeostasis, immunity, and autoimmunity.

Simon Hogan, Assistant Professor of Pediatrics; Ph.D., Australian National University, 1997. Characterizing the molecular mechanisms that regulate the accumulation and activation of eosinophils in the allergic gastrointestinal tract.

Jason Jiang, Associate Professor of Pediatrics; Ph.D., Baylor College of Medicine, 1988. Enteric viruses causing acute gastroenteritis in humans, particularly, human caliciviruses, including norovirus and sapovirus, and human rotaviruses.

Michael Jordan, Assistant Professor of Pediatrics; M.D., Texas Southwestern Medical Center at Dallas, 1993. Immunoregulation; immunodeficiency states; cancer immunotherapy.

Christopher Karp, Professor of Pediatrics and Director, Division of Molecular Immunology; M.D., North Carolina at Chapel Hill, 1986. Understanding molecular mechanisms responsible for regulation and dysregulation of immune responses in infectious and autoimmune human diseases.

Jonathan D. Katz, Associate Professor and Director, Diabetes Research Center; Ph.D., UCLA, 1990. Studies on role of T cells in autoimmune diabetes through application of standard immunological methods to transgenic and gene-targeted NOD mice and through use of functional genomics and proteomics.

Joerg Kohl, Professor of Pediatrics; M.D., Mainz, 1988. Role of complement system in bridging innate to adaptive immunity.

Alex B. Lentsch, Associate Professor of Medicine; Ph.D., Louisville, 1996. Regulation of inflammatory responses by cytokines, chemokines, and adhesion molecules; mechanisms governing angiogenesis in prostate tumors.

John Monaco Jr., Professor of Medicine; Ph.D., Stanford, 1983. Molecular mechanisms of antigen processing and presentation to T cells, with focus on specificity and biochemistry of proteasomes and role of interferon-inducible proteasome subunits in the major histocompatibility complex (MHC) class I antigen processing pathway.

Suzanne Morris, Resident Associate Professor of Medicine; Ph.D., North Carolina, 1989. B-cell activation and lifespan; role of membrane immunoglobulin isotypes in B-cell activation, survival, and tolerance; cytokine regulation of the immune response, including cytokine signaling and the mechanisms that regulate T-cell homeostasis.

Simon L. Newman, Professor of Medicine; Ph.D., Alabama, 1978. Innate immunity to fungi, particularly, *Histoplasma capsulatum;* understanding biology and biochemistry of interaction between *Histoplasma* yeasts, macrophages, and dendritic cells.

Judith Rhodes, Associate Professor of Medicine; Ph.D., UCLA, 1980. Pathways in the opportunistic fungus *Aspergillus fumigatus* that are involved in pathogenesis of invasive aspergillosis.

Marc E. Rothenberg, Professor of Pediatrics and Director, Division of Allergy/Immunology; M.D./Ph.D., Harvard, 1990. Molecular and cellular basis for allergic responses; role of chemokines in inflammation; novel therapeutic intervention strategies in patients with allergic disorders.

Nancy M. Sawtell, Associate Professor of Medicine; Ph.D., Cincinnati, 1986. Molecular mechanisms of herpes simplex virus latency and reactivation.

Susan Thompson, Associate Professor of Pediatrics; Ph.D., Tennessee, 1988. Genetic and functional genomic studies of juvenile rheumatoid arthritis (JRA) to advance understanding of the causes and mechanisms of disease pathogenesis.

Sherry Thornton, Assistant Professor of Pediatrics; Ph.D., Cincinnati, 1997. Molecular and cellular mechanisms underlying the pathogenesis of autoimmune arthritis, with a focus on angiogenesis in arthritis.

Peter D. Walzer, Professor of Anatomy and Cell Biology; M.D., Albany Medical College, 1968. The fungal opportunistic pathogen *Pneumocystis carinii,* a major cause of pneumonia in HIV patients and other immunocompromised hosts.

Bo Wang, Assistant Professor of Pediatrics; D.V.M., Shandong Agricultural (China), 1982; Ph.D., Louis Pasteur (Strasbourg), 1997. Immune-pathogenesis of type 1 diabetes.

Marc Wathelet, Assistant Professor of Medicine; Ph.D., Free University of Brussels, 1988. Innate immunity; antiviral defenses; interferons; virulence mechanisms; Ebola virus; SARS coronavirus.

Alison Weiss, Professor of Genetics and Biochemistry; Ph.D., Stanford, 1983. Bacterial toxins, including pertussis, adenylate cyclase, shiga, and anthrax toxins, with an emphasis on investigating vaccine efficacy and novel therapeutic approaches to counter toxin-mediated diseases.

David A. Williams, Professor of Pediatrics and Director, Division of Experimental Hematology; M.D., Indiana–Purdue at Indianapolis, 1979. Function of Ras-related Rho GTPases in hematopoiesis and immune cell function.

Marsha Wills-Karp, Professor of Pediatrics and Director, Division of Immunobiology; Ph.D., California, Santa Barbara, 1986. Asthma; allergy; T-cell immunology; genetics of asthma; cytokines.

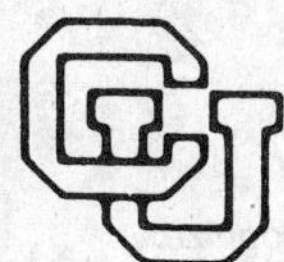

UNIVERSITY OF COLORADO AT DENVER
AND HEALTH SCIENCES CENTER

Department of Immunology

Program of Study

The Department of Immunology offers a Doctor of Philosophy degree program through the Graduate School of the University of Colorado. The Department was established in 1994 to bring together the very large Denver immunology community in one integrated training program. Immunology has a long history in Colorado, beginning in the days when patients with tuberculosis were sent to live at a high altitude and were cared for in research-oriented institutions. Dr. Webb, founder and first president of the American Association of Immunologists (1913–15), worked in Colorado. Many significant advances, from IgE and the helper T cell to fundamental studies on T- and B-cell antigen recognition and repertoire development, have been made in the University's laboratories.

The school is on the semester system, and the program begins in the last week of August. Students are required to take 60 credit hours of course work, which includes a three-semester series in immunology and molecular biology, as well as four laboratory rotations. A choice of electives completes the course requirements. In the second year, the student chooses a lab in which to perform dissertation research. During that year a comprehensive examination is scheduled, consisting of a written research proposal that is defended orally. The student then assembles a thesis committee of appropriate experts who meet with him or her at least every six months to guide the project. This committee makes the final decision about when the thesis is ready to be written, passes on it, and conducts the oral defense. Students generally graduate after four or five years.

The Department consists of 23 primary-appointment and 27 secondary-appointment faculty members, all of whom hold a Ph.D. or M.D., whose research interests cover almost every aspect of immunity, from the most fundamental studies of lymphocyte activation to allergy, autoimmunity, and tumor immunology. In addition, students have the opportunity to study in other areas of the biological sciences, including biochemistry, molecular biology, microbiology, virology, and cell biology.

Research Facilities

The program is centered at three institutions in proximity to each other: the University of Colorado Health Sciences Center (UCHSC), the Barbara Davis Center for Childhood Diabetes (BDCCD), and the National Jewish Medical and Research Center (NJMRC). Each of these institutions has modern facilities and instrumentation for research in contemporary immunology. The group has superb animal facilities, including barrier housing for transgenics; molecular biology cores; immunology and monoclonal antibody cores; nine flow cytometers; and very advanced instrumentation for confocal microscopy and image analysis and molecular studies. Excellent library and Internet connection facilities are also available.

Financial Aid

The Department provides a stipend of $22,500 (2006–07) plus tuition, fees, and student insurance for all accepted students who are citizens or permanent residents of the United States.

Cost of Study

Educational costs for 2006–07 are $123 per credit hour for state residents and $575 for nonresidents. Tuition is paid by the Department.

Living and Housing Costs

The campus is located in a pleasant residential neighborhood, and affordable housing is available within biking or walking distance of the campus.

Student Group

Five or 6 students per year are accepted into the immunology program. They come from diverse academic and cultural backgrounds. There are 43 students in the program, including 6 students who are enrolled in the combined M.D./Ph.D. program and 6 international students. Graduates have been very successful in securing academic appointments or positions in industry.

Location

Set at the foot of the Rocky Mountains at 1 mile high (1,620 meters), Denver is a sprawling, dynamic city with mild winters, dry summers, more than 300 days of sunshine per year, and the largest city park system in the nation. With a metropolitan population just under 2 million, the city offers a variety of cultural pursuits, such as museums, concert halls (including the famous Red Rocks amphitheater), a zoo, a botanical garden, and a superb public library. There are major-league sports, while the nearby mountains create a year-round mecca of outdoor recreational opportunities, such as skiing, hiking, rock climbing, camping, fishing, kayaking, and rafting.

The Health Sciences Center

Located in the residential heart of Aurora, the CU Health Sciences Center encompasses the Schools of Medicine, Nursing, Dentistry, and Pharmacy and the Graduate School. This widely renowned cluster of health sciences institutions includes two teaching hospitals—University Hospital and Colorado Psychiatric Hospital—the NCI-designated CU Comprehensive Cancer Center, and a constellation of research and treatment institutions listed among the most prestigious in the country. The CU Health Sciences Center is the major health research base in Colorado, currently attracting $113 million in research and training grants annually to advance the knowledge of biomedical sciences through basic and applied research. Approximately 400 graduate students and 600 full-time faculty members at the CU Health Sciences Center participate in a wide variety of academic and research activities associated with seventeen graduate programs.

Applying

A bachelor's degree or the equivalent is required, as is proven scientific ability indicated through performance in a college-level science program or performance in a research laboratory as a student or research technician. GRE scores and GPAs are expected to be high, and letters of recommendation and the student's personal statement are considered seriously. Except in extenuating circumstances, students are expected to have successfully completed courses in organic chemistry, general biology, and biochemistry. The application deadline is January 1. Approximately 15 students are interviewed each year; the program pays the domestic portion of an interviewee's airfare as well as other associated expenses. No student can be accepted without an interview.

Correspondence and Information

Immunology Graduate Coordinator
National Jewish Medical and Research Center
University of Colorado at Denver and Health Sciences Center
1400 Jackson Street, K520
Denver, Colorado 80206
Phone: 303-398-1305
Fax: 303-398-1396
E-mail: lannersj@njc.org
Web site: http://www.uchsc.edu/immuno

University of Colorado at Denver and Health Sciences Center

THE FACULTY AND THEIR RESEARCH

Donald Bellgrau, Professor; Ph.D., Pennsylvania. T-cell signaling and autoimmunity; vaccine development; immunology of privileged sites.
Willi Born, Professor; Ph.D., Max Planck Institute for Immunobiology (Germany). Biological importance of $\gamma\delta$ T cells and the ligands they recognize.
John Cambier, Professor; Ph.D., Iowa. Molecular and biochemical mechanisms of transmembrane signaling and activation of B and T cells.
J. John Cohen, Professor; M.D., Ph.D., McGill. Signal and effector pathways in hematopoietic and immune cell apoptosis.
James DeGregori, Assistant Professor; Ph.D., MIT. Role of E2F in thymocyte death.
George Eisenbarth, Professor; M.D., Ph.D., Duke. Immunotherapy of type 1 diabetes.
Brian M. Freed, Associate Professor; Ph.D., Albany Medical College. Mechanisms of immune suppression by cigarette smoke.
Laurent Gapin, Assistant Professor; Ph.D., Paris. Function of a novel class of T cells (NKT cells).
Ronald G. Gill, Associate Professor; Ph.D., UCLA. Immunobiology of pancreatic islet transplantation.
James Hagman, Assistant Professor; Ph.D., Washington (Seattle). Regulation of gene expression early in B-cell development.
Kathryn Haskins, Professor; Ph.D., Kansas. Immunoregulation in autoimmune diabetes; Th1 and Th2 T-cell function.
Peter Henson, Professor; Ph.D., Cambridge. Regulation of the inflammatory response.
V. Michael Holers, Professor; M.D., Washington (St. Louis). Signal transduction through complement receptors.
John Kappler, Professor; Ph.D., Brandeis. Structure and function of the T-cell receptor and its ligands.
Ross Kedl, Assistant Professor; Ph.D., Minnesota. Determining how toll-like receptors influence the balance between innate and adaptive immunity.
Laurel Lenz, Assistant Professor; Ph.D., Kansas State. Bacterial pathogenesis and immunity.
Philippa Marrack, Professor; Ph.D., Cambridge. T-cell development; T-cell responses and death.
Rebecca O'Brien, Associate Professor; Ph.D., Washington (Seattle). Antigen recognition by $\gamma\delta$ T cells.
Roberta Pelanda, Assistant Professor; Ph.D., Milan. B-cell development and selection.
Anne-Laure Perraud, Assistant Professor; Ph.D., Würzburg. Functional characterization of the TRPM cation-channels in the immune context.
Terry Potter, Associate Professor; Ph.D., Melbourne. Recognition of MHC class I molecules by CD8$^+$ T cells.
Christopher L. Reardon, Associate Professor; M.D., Ph.D., Arizona. Functions of $\gamma\delta$-T lymphocytes and their roles in neoplasia.
David W. H. Riches, Professor; Ph.D., Birmingham. Death receptor signal transduction and macrophage involvement in chronic inflammation.
Hong-Bing Shu, Assistant Professor; Ph.D., Emory. Signal transduction by TNF family members.
Jill E. Slansky, Assistant Professor; Ph.D., Wisconsin–Madison. T-cell response to tumors.
Raul M. Torres, Assistant Professor; Ph.D., Washington (Seattle). Signals guiding development and immune response of B lymphocytes.
Kenneth Tyler, Professor; M.D., Johns Hopkins. Mechanisms of virus-induced apoptosis and alterations in cellular signal transduction pathways.
Linda van Dyk, Assistant Professor; Ph.D., Texas. Pathogenesis and latency of herpesvirus.
Cara Wilson, Assistant Professor; M.D., Duke. Cell-mediated responses to HIV.
Lawrence Wysocki, Professor; Ph.D., Harvard. Somatic hypermutation during the B-cell response to antigen.
Gongyi Zhang, Assistant Professor; Ph.D., Academy of Science of China. X-ray crystallographic studies of macromolecules.

UNIVERSITY OF COLORADO AT DENVER AND HEALTH SCIENCES CENTER

Graduate Program in Microbiology

Program of Study

The University of Colorado at Denver and Health Sciences Center (UCDHSC) Graduate Program in Microbiology offers a comprehensive program of graduate courses and research training leading to a Ph.D. degree. The program prepares students for careers in teaching and research in academic institutions as well as research in the biotechnology industry. Entering students begin their graduate work by taking a common core curriculum that includes topics in biochemistry, genetics, and molecular cell biology. During the first year, students also take two or more electives chosen from courses in bacterial genetics and biochemistry, bacterial pathogenesis, molecular virology, viral pathogenesis, immunology, biochemistry, molecular structure, and pathology. During the first nine months, students do three research rotations in laboratories of graduate program faculty members. Students participate in research seminars and journal clubs throughout their graduate training. Dissertation research usually begins in the summer of the first year, and the research leading to the Ph.D. degree is generally completed within four to five years.

The graduate program faculty is a diverse and productive group, which has well-funded research programs and includes members of the Department of Microbiology and some members of the Departments of Pediatrics, Medicine, and Neurology. Areas of research interest include molecular mechanisms of bacterial and viral pathogenesis; the molecular biology of bacterial, viral, and cellular gene expression; and host-parasite interactions in infectious diseases.

Research Facilities

The UCDHSC Basic Science programs, including the microbiology department, are located in the newly constructed world-class research facilities on the new UCDHSC Fitzsimons Campus. The department occupies approximately 17,000 square feet of contiguous space, including laboratory research and support space; administrative, faculty, and graduate student offices; and conference rooms. Excellent specialized research equipment is available to support departmental research, including a deconvoluting fluorescence microscope, automated chromatography equipment, and facilities for fluorescence-activated cell sorting and real-time quantitative PCR. A spacious BSL3 containment facility is immediately adjacent to departmental space for experiments requiring a heightened level of safety and physical security. Also contained in Research Complex I are several core facilities available to support research, including a world-class NMR and X-ray crystallography facility; core facilities for proteomics, protein expression, DNA sequencing, and DNA microarray development and analysis; and facilities for producing monoclonal antibodies and transgenic and knockout mice.

Financial Aid

Funds available to support students include an NIH/NIAID predoctoral training grant, fellowships, and research grants. Students in their first year are supported by departmental and Graduate School funds. Students admitted into the program receive financial aid, including stipend, individual student health and dental insurance, and tuition and student fees. The 2006–07 stipend is $22,500.

Cost of Study

The costs of tuition, student fees, and individual student health and dental insurance are paid by the department in the first year and by the thesis mentor in subsequent years. The stipend award is renewable, contingent upon satisfactory progress in the graduate program.

Living and Housing Costs

Students can find convenient and affordable housing in the residential communities that surround the Fitzsimons Campus, often within walking or biking distance. Several local metro-Denver bus routes serve the campus, and there is convenient access to major freeways and bus routes for those preferring to reside in suburban communities.

Student Group

There are 20 to 25 students enrolled in the program. The students have diverse academic backgrounds and have degrees in such fields as biology, microbiology, biochemistry, and other related disciplines. Approximately half of the students studying in the program are women, and minority student enrollment is encouraged.

Location

The new UCDHSC Fitzsimons Campus is located in a predominantly residential neighborhood in the Denver suburb of Aurora, with convenient access to downtown shopping and business districts. Recreational facilities, shops, restaurants, conference facilities, and other amenities will be located within the Town Center on the campus grounds. Greater metropolitan Denver has a population of approximately 2 million and offers diverse cultural opportunities, including the fine arts, major-league sports, concert halls, and outdoor amphitheaters. The city enjoys a delightful climate, with mild winters, warm dry summers, and more than 300 days of sunshine each year. The Rocky Mountains, located nearby, offer an extraordinary year-round selection of outdoor recreational activities, including skiing, snowboarding, camping, hiking, fishing, hunting, and water sports.

The Center and The Program

The UCDHSC Fitzsimons Campus is a 240-acre complex on the site of the recently decommissioned Fitzsimons Army Medical Garrison, conveniently located in the Denver suburb of Aurora. This new campus will be the home of the Health Sciences Center, the Colorado Biosciences Park Aurora, the University of Colorado Hospital (UCH) and associated clinics, and the Children's and Veteran's Administration Hospitals and associated research activities. The University of Colorado at Denver and Health Sciences Center is the largest comprehensive health care and research facility in the region. It has gained both national and international recognition for accomplishments in biomedical research and patient care. The HSC includes the Schools of Medicine, Nursing, Dentistry, and Pharmacy and the UCH teaching hospital and clinics. The Graduate Program in Microbiology is a component of a strong and growing graduate program in the basic sciences associated with the School of Medicine.

Applying

Students enter the program in late August at the beginning of the fall semester. Students are selected on the basis of past performance and, where possible, individual interviews. Each year the program recruits a small number of highly qualified students who have the intellectual competence and strong motivation to become successful scientists. The University requires that applicants take the Graduate Record Examinations and achieve a minimum cumulative undergraduate GPA of 3.0 (on a 4.0 scale). The program recognizes that students who are attracted to a career in microbiology come from highly varied backgrounds. It is suggested that applicants have completed courses in biology, chemistry (general and organic), biochemistry, physics, genetics, and calculus before entering the program. Applications and supporting documents should be received by January 1. Applications are considered after this date on a space-available basis. Links to the online application are found on the department's Web site at http://www.uchsc.edu/sm/microbio.

Correspondence and Information

Graduate Program Director
Department of Microbiology
University of Colorado at Denver and Health Sciences Center
Mail Stop 8333
P.O. Box 6511
Aurora, Colorado 80045
Phone: 303-724-4224
Fax: 303-724-4226
E-mail: micrograd.program@uchsc.edu
Web site: http://www.uchsc.edu/sm/microbio

University of Colorado at Denver and Health Sciences Center

THE FACULTY AND THEIR RESEARCH

Bruce W. Banfield, Assistant Professor; Ph.D., British Columbia. Molecular mechanisms of alphaherpesvirus neuronal spread and virulence.

David J. Barton, Assistant Professor; Ph.D., Medical College of Ohio. Picornavirus RNA translation and replication; hepatitis C virus and innate antiviral immunity.

Thomas B. Campbell, Associate Professor; M.D., Texas Southwestern Medical Center at Dallas. Human herpesvirus 8 and Kaposi's sarcoma; HIV-1 replication fitness; clinical trials of antiretroviral therapy.

Sonia C. Flores, Assistant Professor; Ph.D., South Alabama. Mechanisms of HIV-1 Tat and Nef-dependent vascular endothelial cell phenotypic changes.

Robert L. Garcea, Professor; M.D., California, San Francisco. Mechanisms of papovavirus assembly.

Donald H. Gilden, Professor; M.D., Maryland. Varicella zoster virus latency in the human nervous system; human immune response in multiple sclerosis.

Ronald E. Gill, Associate Professor; Ph.D., Washington (Seattle). Regulation of gene expression by intercellular interactions during the development of *Myxococcus xanthus*.

Kathryn V. Holmes, Professor; Ph.D., Rockefeller. Molecular biology and pathogenesis of coronavirus infections, including SARS; virus-receptor interactions.

Randall K. Holmes, Professor and Chair; M.D., Ph.D., NYU. Structure-function relationships of bacterial toxins.

Dirk Homann, Assistant Professor; M.D., M.A., Free University of Berlin. Specific T-cell responses under conditions of pathogen control, persistent viral disease, and autoimmunity.

Edward N. Janoff, Professor; M.D., Arizona. Mucosal immunity; HIV transmission and vaccine; pneumococcal infections and vaccine; B cell regulation.

Laurel L. Lenz Jr., Assistant Professor, Ph.D., Washington (Seattle). Molecular mechanisms of *Listeria* pathogenesis; bacterial subversion of innate immunity.

Martin L. Pato, Associate Professor; Ph.D., Berkeley. Mechanism of transposition in prokaryotes; role of supercoiling and organization of DNA structure; biology of bacteriophage Mu.

Jerome Schaack, Associate Professor; Ph.D., Yale. Gene therapy vectors; adenovirus preterminal protein function in the regulation of the infectious cycle.

Kenneth L. Tyler, Professor; M.D., Johns Hopkins. Molecular and genetic basis of virus-induced cell death (apoptosis) using reovirus infection in cell culture and animal models (encephalitis, myocarditis, hepatitis).

Linda F. van Dyk, Assistant Professor; Ph.D., Texas Southwestern Medical Center at Dallas. Genetic and molecular approaches to pathogenesis and latent infection by lymphotropic herpesviruses.

Michael L. Vasil, Professor; Ph.D., Texas Southwestern Medical Center at Dallas. Mechanisms of bacterial pathogenesis with emphasis on genetic regulation and biochemistry of virulence factors.

F. Andres Vazquez-Torres, Assistant Professor; D.V.M., Ph.D., Wisconsin. Molecular and cellular biology of the microbial pathogenesis of the intracellular bacterium *Salmonella*.

Martin Voskuil, Assistant Professor; Ph.D., Wisconsin. Role of the *Mycobacterium tuberculosis* dormant state in latent disease utilizing genetic, DNA microarray, and biochemical techniques.

Looking west toward downtown and the Rocky Mountains.

UNIVERSITY OF CONNECTICUT HEALTH CENTER

Graduate Program in Immunology

Program of Study

A Ph.D. in immunology is offered through an interdepartmental program consisting of 22 faculty members. The immunology faculty members also participate in training students in the combined M.D./Ph.D. and D.M.D./Ph.D. programs. The central focus of the program is to train students to become independent investigators who will provide meaningful research and educational contributions to the areas of basic, applied, or clinical immunology. This goal is achieved by lectures, seminars, laboratory rotations, research presentations, and a concentration on laboratory research. In addition to basic and advanced immunology courses, students are given a strong foundation in biomedical sciences through the core curriculum in biochemistry, genetics, molecular biology, and cell biology. Research laboratory training aims to provide a foundation in modern laboratory techniques and concentrates on hypothesis-based analysis of problems. Research in the program is focused on the cellular and molecular aspects of immune system structure and function in animal models and in humans. Areas of emphasis include molecular immunology (mechanisms of antigen presentation, major histocompatibility complex genetics and function, cytokines and cytokine receptors, and tumor antigens), cellular immunology (biochemical mechanisms and biological aspects of signal transduction of lymphocytes and granulocytes; cellular and molecular requirements for thymic T-lymphocyte development, selection, and activation; cytokines in B- and T-cell development; regulation of antitumor immunity; immunoparasitology, including parasite genetics and immune recognition of parasite antigens; and mechanisms of inflammation), organ-based immunology (immune effector mechanisms of the intestine, lymphocyte interactions in the lung, and immune regulation of the eye), immunity to infectious agents (viruses, bacteria, parasites, including vector-borne organisms), and autoimmunity (animal models of autoimmune disease and effector mechanisms in human autoimmunity).

Research Facilities

The Graduate Program in Immunology is interdepartmental, and therefore provides a broad base of training possibilities as well as ample shared facilities. State-of-the-art equipment is available in individual laboratories for analysis of molecular and cellular parameters of immune system structure and function. In addition, Health Center–supported facilities provide equipment and expertise in areas of advanced data acquisition and analysis. These facilities include the Center for Cell Analysis and Modeling, the Fluorescence Flow Cytometry Facility, the Gene Targeting and Transgenic Facility, the Molecular Core Facility, the Microarray Facility, the Gregory P. Mullen Structural Biology Facility, and the Electron Microscopy Facility. The Health Center Library is well equipped with extensive journal and book holdings and rapid electronic access to database searching, the World Wide Web, and library holdings. A computer center is also housed in the library for student use and training.

Financial Aid

Support for doctoral students engaged in full-time degree programs at the Health Center is provided on a competitive basis. Graduate research assistantships for 2006–07 provide a stipend of $26,000 per year, which includes a waiver of tuition/University fees for the fall and spring semesters and a student health insurance plan. While financial aid is offered competitively, the Health Center makes every possible effort to address the financial needs of all students.

Cost of Study

For 2006–07, tuition is $3996 per semester ($7992 per year) for full-time students who are Connecticut residents and $10,368 per semester ($20,772 per year) for full-time out-of-state residents. General University fees are added to the cost of tuition for students who do not receive a tuition waiver. These costs are usually met by traineeships or research assistantships for doctoral students.

Living and Housing Costs

There is a wide range of affordable housing options in the greater Hartford area within easy commuting distance of the campus, including an extensive complex that is adjacent to the Health Center. Costs range from $600 to $800 per month for a one-bedroom unit; 2 or more students sharing an apartment usually pay less. University housing is not available.

Student Group

At present, there are 30 students in the Graduate Program in Immunology. There are 150 students in the various Ph.D. programs on the Health Center campus.

Student Outcomes

Graduates have traditionally been accepted into high-quality laboratories for postdoctoral training. Following their training, graduates have accepted a wide range of positions in research in universities, colleges, research institutes, and industry, including the biotechnology sector.

Location

The Health Center is located in the historic town of Farmington, Connecticut. Set in the beautiful New England countryside on a hill overlooking the Farmington Valley, it is close to ski areas, hiking trails, and facilities for boating, fishing, and swimming. Connecticut's capital city of Hartford, 7 miles east of Farmington, is the center of an urban region of approximately 800,000 people. The beaches of the Long Island Sound are about 50 minutes away to the south, and the beautiful Berkshires are a short drive to the northwest. New York City and Boston can be reached within 2½ hours by car. Hartford is the home of the acclaimed Hartford Stage Company, TheatreWorks, the Hartford Symphony and Chamber orchestras, two ballet companies, an opera company, the Wadsworth Atheneum (the oldest public art museum in the nation), the Mark Twain house, the Hartford Civic Center, and many other interesting cultural and recreational facilities. The area is also home to several branches of the University of Connecticut, Trinity College, and the University of Hartford, which includes the Hartt School of Music. Bradley International Airport (about 20 minutes from campus) serves the Hartford/Springfield area with frequent airline connections to major cities in this country and abroad. Frequent bus and rail service is also available from Hartford.

The Health Center

The 200-acre Health Center campus at Farmington houses a division of the University of Connecticut Graduate School, as well as the School of Medicine and Dental Medicine. The campus also includes the John Dempsey Hospital, associated clinics, and extensive medical research facilities, all in a centralized facility with more than 1 million square feet of floor space. The Health Center's newest research addition, the Academic Research Building, was opened in 1999. This impressive eleven-story structure provides 170,000 square feet of state-of-the-art laboratory space. The faculty at the center includes more than 260 full-time members. The institution has a strong commitment to graduate study within an environment that promotes social and intellectual interaction among the various educational programs. Graduate students are represented on various administrative committees concerned with curricular affairs, and the Graduate Student Organization (GSO) represents graduate students' needs and concerns to the faculty and administration, in addition to fostering social contact among graduate students in the Health Center.

Applying

Applications for admission should be submitted on standard forms obtained from the Graduate Admissions Office at the UConn Health Center or on the Web site. The application should be filed together with transcripts, three letters of recommendation, a personal statement, and recent results from the General Test of the Graduate Record Examinations. International students must take the Test of English as a Foreign Language (TOEFL) to satisfy Graduate School requirements. The deadline for completed applications and receipt of all supplemental materials is December 15. In accordance with the laws of the state of Connecticut and of the United States, the University of Connecticut Health Center does not discriminate against any person in its educational and employment activities on the grounds of race, color, creed, national origin, sex, age, or physical disability.

Correspondence and Information

Dr. Lynn Puddington, Program Director
Graduate Program in Immunology
Department of Immunology
CIIVR, MC 1319
University of Connecticut Health Center
Farmington, Connecticut 06030-1319

Phone: 860-679-4655
Fax: 860-679-1868
E-mail: puddington@nso1.uchc.edu
Web site: http://www.uchc.edu

University of Connecticut Health Center

THE FACULTY AND THEIR RESEARCH

Adam J. Adler, Associate Professor of Immunology; Ph.D., Columbia. Mechanisms of T-cell tolerance induction to peripheral self- and tumor-antigens; immunological properties of prostate cancer.

Hector L. Aguila, Assistant Professor of Immunology; Ph.D., Yeshiva (Einstein). Hematopoiesis and bone marrow microenvironment; lymphoid cell development; stem cell biology.

Pierluigi E. Bigazzi, Professor of Pathology; M.D., Florence. Anti-idiotypic immunity in autoimmunity; spontaneous and experimental animal models of autoimmune diseases.

Robert B. Clark, Associate Professor of Immunology; M.D., Stanford. Autoimmunity; immune regulation; regulatory T cells.

Robert Cone, Professor of Immunology; Ph.D., Michigan. Ocular immunology; regulatory T cells; neuroimmunology.

Irving Goldschneider, Professor of Immunology; M.D., Pennsylvania. T- and B-cell development; acquired thymic tolerance; cytokines.

Chi-Kuang Huang, Associate Professor of Immunology; Ph.D., Connecticut. Signal transduction in stimulated neutrophil and lymphocytes; roles of protein kinase and phosphoproteins in cell activation; chemotaxis.

Donald L. Kreutzer, Professor of Pathology and Surgery; Ph.D., Kansas. Immunopathology and molecular mechanisms of inflammation; mediators and regulators of leukocyte chemotaxis; modulation of inflammatory reactions by the vascular endothelium.

Leo Lefrancois, Professor of Immunology; Ph.D., Wake Forest. T-cell memory; immune response to infection; tolerance; vaccines.

Zihai (Zack) Li, Associate Professor of Immunology; M.D., Henan (China); Ph.D., Mount Sinai. Role of heat shock proteins (HSP) in the early phase of T-cell priming.

Joseph A. Lorenzo, Professor of Medicine; M.D., SUNY Downstate Medical Center. Relationships between bone-absorbing osteoclasts and immune cells.

Bijay Mukherji, Professor of Medicine; M.D., Calcutta (India). Tumor immunology and cancer vaccines; tumor-specific antigens.

James O'Rourke, Professor of Immunology and Surgery; M.D., Georgetown. Vascular biology; tissue plasminogen activator synthesis, transport, and release.

Lynn Puddington, Associate Professor of Immunology; Ph.D., Wake Forest. Allergic asthma; neonatal immunity and tolerance; developmental immunology.

Justin D. Radolf, Professor of Medicine and of Genetics and Developmental Biology; M.D., California, San Francisco. Molecular pathogenesis and immunobiology of spirochetal infections.

T. V. Rajan, Professor of Immunology; Ph.D., M.D., All-India Institute of Medical Sciences. Parasitology; filariasis; molecular immunoparasitology.

Pramod K. Srivastava, Professor of Immunology; Ph.D., Hyderabad (India). Heat shock proteins as peptide chaperones; roles in antigen presentation and applications in immunotherapy of cancer, infectious diseases, and autoimmune disorders.

Roger S. Thrall, Professor of Immunology and Surgery; Ph.D., Marquette. Immune cells; pulmonary inflammation.

Anthony T. Vella, Assistant Professor of Immunology; Ph.D., Cornell. T-cell immunity; costimulation; adjuvants and cytokines.

Stephen K. Wikel, Professor of Cell Biology and Immunology; Ph.D., Saskatchewan. Characterization of the complex cellular and molecular immunology of the arthropod vector-host-pathogen interface.

Carol A. Wu, Assistant Professor of Immunology; Ph.D., Vanderbilt. Viral respiratory infection and asthma.

Richard A. Zeff, Associate Professor of Immunology; Ph.D., Rush. Major histocompatibility complex; antigen processing and presentation.

The University of Connecticut Health Center.

UNIVERSITY OF GEORGIA

Department of Microbiology

Programs of Study	The Department of Microbiology offers M.S. and Ph.D. degrees. Students with a B.S. may apply directly for either program. Degree requirements include the successful completion of course work and comprehensive exams, fulfillment of the Graduate School residency requirement, and submission of a thesis or dissertation consisting of original, scholarly research in the field of microbiology. Students pursue a variety of disciplines, including microbial physiology and genetics, cell biology, microbial pathogenesis, ecology, evolution, population biology, biotechnology, and bioinformatics. All students take a core curriculum in microbial physiology and metabolism, molecular biology, and microbial diversity. Advanced courses in specialized areas are also available through the Department of Microbiology and the Division of Biological Sciences. The M.S. program is generally completed in two to three years, while the Ph.D. requires five to six years. The program of study is designed by the student and his or her advisory committee to provide a broad foundation in microbiology, preparing the student for a career in research and/or teaching in academia, industry, or the government.
Research Facilities	The Department of Microbiology occupies more than 34,000 square feet of renovated space in the Biological Sciences Building. Facilities are also available for environmental studies at the University of Georgia Marine Institute on Sapelo Island and the Savannah River Ecology Laboratory. Departmental laboratories are equipped for state-of-the-art research in microbiology. Students may also use professionally staffed University research facilities for computational analysis, electron and laser-scanning confocal microscopy, flow cytometry, proteomics, mass spectrometry, DNA and protein sequencing and synthesis, large-scale fermentations, monoclonal antibody production, and research animal care. Most of these facilities are in the same building, while others are just a few blocks away.
Financial Aid	Graduate students in the Department of Microbiology are supported by assistantships or fellowships, with awards starting at $20,000 per year for Ph.D. candidates and $18,000 per year for M.S. candidates. Several Ph.D. candidates have been awarded the University's highly competitive Presidential Graduate Fellowship, which has an annual stipend of $22,000. In addition to these awards, full annual tuition is waived for every graduate student on any kind of fellowship or assistantship; this amounts to an additional value of $19,792. Thus, the total financial package (assistantship/fellowship plus tuition waiver) ranges from $37,792 to $41,792, depending on the type of assistantship/fellowship.
Cost of Study	Student activity fees covering registration, health service, campus transportation, and computer use are $495 per semester.
Living and Housing Costs	The cost of living (including food, clothing, housing, and utilities) is generally quite reasonable in Athens (students can compare the cost of living at various locations at http://homefair.com/homefair/calc/salcalc.html). Dormitory rooms are available. In addition, apartments are available in the University's family housing unit (http://www.uga.edu/housing/gradfam/index.html). A variety of apartments, duplexes, and rental homes can be found off campus.
Student Group	In 2005–06, 50 students were enrolled in graduate studies in the Department of Microbiology, representing Europe, Asia, South America, the Middle East, and fifteen U.S. states. The Microbiology Graduate Student Association promotes communication between students and faculty members, sponsors visiting seminar speakers and social events, and participates in the campuswide Graduate Student Association.
Location	The University of Georgia is located in Athens, one of America's great college towns. Athens has a community of 98,000 and is situated in a rolling, wooded area in the Piedmont of northeastern Georgia. The climate is moderate, with mean temperatures ranging from 33°F to 53°F (1°C to 12°C) in January and 68°F to 89°F (20°C to 32°C) in July. The Appalachian Trail begins only 70 miles to the north, and Atlanta lies about the same distance to the southwest. Within a modest drive are the barrier islands of the Georgia coast, the Okefenokee Swamp, and the Chattahoochee National Forest. Numerous recreational and cultural activities are available in the Athens area and nearby.
The University and The Department	Chartered in 1785, the University of Georgia was the nation's first state-supported university. It is composed of thirteen schools and has a total enrollment in of 32,500 students. The University campus covers approximately 3,500 acres. Established more than fifty years ago in the Franklin College of Arts and Sciences, the Department of Microbiology, among the top five non–medical school microbiology departments in the nation, is highly regarded for its mastery of a broad range of microbial systems. Among the 15 faculty members, there are 3 recipients of prestigious NSF Presidential Young Investigator Awards, 1 Guggenheim Fellow, and 3 recipients of NIH Research Career Development Awards. Research in the Department of Microbiology usually generates more than $2 million annually from sources such as the NIH, NSF, EPA, USDA, Department of Energy, Office of Naval Research, and several foundations and corporate sponsors. The Department members usually generate more than seventy publications annually in scientific journals.
Applying	Application forms are available online at the Web site. To ensure consideration for any assistantship, duplicate copies of the entire application, including related material, should be sent to both the Graduate School and the Graduate Coordinator of the Department of Microbiology before December 15. All recent successful applicants had a baccalaureate in the biological sciences with previous research experience either as an undergraduate or in employment. Undergraduate GPAs are usually about 3.5, and GRE scores (verbal and quantitative) are approximately 1250. International students whose native language is not English must have a TSE score of at least 50 for acceptance. A TOEFL Academic Speaking Test (TAST) score of at least 26 may be substituted for the TSE score.
Correspondence and Information	Graduate Coordinator Department of Microbiology 527 Biological Sciences Building University of Georgia Athens, Georgia 30602-2605 Phone: 706-542-2045 E-mail: mibcoord@uga.edu Web site: http://www.uga.edu/mib

University of Georgia

THE FACULTY AND THEIR RESEARCH

Daniel G. Colley, Professor; Ph.D., Tulane, 1968. Immunology of schistosomiasis in mice (at the University of Georgia) and people (at the Kenya Medical Research Institute); public health interventions for parasitic diseases.
PD-L2+ dendritic cells and PD-1+ CD4+ T cells in *Schistosomiasis* correlate with morbidity. *Parasite Immunol.* 27:45, 2005. Resistant and susceptible people in Kenya. *Lancet* 360:592, 2002.

Harry A. Dailey, Professor; Ph.D., UCLA, 1976.
Examination of mitochondrial protein targeting of haem synthetic enzymes: In vivo identification of three functional haem-responsive motifs in 5-aminolaevulinate synthase. *Biochem. J.* 385:1–6, 2005. Production and characterization of erythropoietic protoporphyric heterodimeric ferrochelatases. *Blood* 106:1098–104, 2005.

Timothy R. Hoover, Associate Professor; Ph.D., Wisconsin, 1988. Transcriptional activation in bacteria; virulence factors in *Helicobacter pylori.*
Stable accumulation of sigma[54] in *Helicobacter pylori* requires the novel protein HP0958. *J. Bacteriol.* 187:4463, 2005. *Helicobacter pylori* FlgR is an enhancer-independent activator of sigma[54]–RNA polymerase holoenzyme. *J. Bacteriol.* 186:4535, 2004.

Anna C. Glasgow Karls, Associate Professor; Ph.D., Wisconsin, 1986.
Analysis of the Piv recombinase-related gene family of *Neisseria gonorrhoeae. J. Bacteriol.* 187:1276–86, 2005. Piv site-specific invertase requires a DEED motif analogous to the catalytic center of the RuvC Holliday junction resolvases. *J. Bacteriol.* 187:3431–7, 2005.

Duncan C. Krause, Professor; Ph.D., North Carolina at Chapel Hill, 1982.
Mutant analysis reveals specific requirement for protein P30 in *Mycoplasma pneumoniae* gliding motility. *J. Bacteriol.* 187:6281–9, 2005. Cellular engineering in a minimal microbe. *Mol. Microbiol.* 51:917–24, 2004. HMW1 is required for stability and localization of HMW2 to the attachment organelle of *Mycoplasma pneumoniae. J. Bacteriol.* 186:8221–8, 2004.

Robert Maier, Professor and Eminent Scholar; Ph.D., Wisconsin, 1977. Roles of metalloproteins such as hydrogenases and oxidative stress resistance enzymes in virulence by pathogenic bacteria.
The *Helicobacter pylori* MutS protein confers protection from oxidative DNA damage. *Mol. Microbiol.* 58:166–76, 2005.

Jan Mrázek, Assistant Professor; Ph.D., Czechoslovak Academy of Sciences, 1992. DNA sequence analysis; bioinformatics; comparative genomics.
Genomic comparisons among gamma-proteobacteria. *Env. Microbiol.*, in press. Frequent oligonucleotide motifs in genomes of three streptococci. *Nucleic Acids Res.* 30:4216, 2002. Highly expressed and alien genes of the Synechocystis genome. *Nucleic Acids Res.* 29:1590, 2001.

Ellen L. Neidle, Professor; Ph.D., Yale, 1987. Bacterial degradation of aromatic compounds; bioremediation; transcriptional regulation; gene amplification. Gene amplification involves site-specific short homology-independent illegitimate recombination in *Acinetobacter* sp. strain ADP1. *J. Mol. Biol.* 338:643–56, 2004. Selection for gene clustering by tandem amplification. *Annu. Rev. Microbiol.* 58:119–42, 2004.

Joy Doran Peterson, Assistant Professor; Ph.D., Florida, 1994.
Pectin-rich biorefinery for production of ethanol and specialty chemicals. *Ethanol Producer* 26–9, 2003. Fermentations of pectin rich biomass with recombinant bacteria to produce fuel ethanol. *Appl. Biochem. Biotechnol.* 84–86:141–52, 2000. With Cripe, Sutton, and Foster.

Mark A. Schell, Professor; Ph.D., Cornell, 1979. Host-pathogen interactions; microbial genomics; bioinformatics; gene regulation.
The genome sequence of the probiotic intestinal bacterium *Lactobacillus johnsonii* NCC533. *Proc. Natl. Acad. Sci. U.S.A.* 101:2512–7, 2004. Physiological insight into *Bifidobacterium longum* from genome sequence analysis. *Proc. Natl. Acad. Sci. U.S.A.* 99:14422–7, 2002.

Lawrence J. Shimkets, Professor; Ph.D., Minnesota, 1980. Regulation of cell-cell interactions in *Myxococcus* gene regulation, surface motility, and chemotaxis.
The Dif chemosensory system is directly involved in phosphatidylethanolamine sensory transduction in *Myxococcus xanthus. Mol. Microbiol.* 57:1466–508, 2005.

Eric V. Stabb, Assistant Professor; Ph.D., Wisconsin–Madison, 1997. *Vibrio fischeri–Euprymna scolopes* symbiosis; bioluminescence; LPS; genetics.
Characterization of pES213, a small mobilizable plasmid from *Vibrio fischeri. Plasmid* 54:114, 2005. Correlation between osmolarity and luminescence of symbiotic *Vibrio fischeri* strain ES114. *J. Bacteriol.* 186:2906, 2004.

Anne O. Summers, Professor; Ph.D., Washington (St. Louis), 1973. Resistance to toxic metals and antibiotics; lateral gene transfer; integrons and plasmid genomics.
Mobile genetic elements: The agents of open source evolution. *Nature Rev. Microbiol.* 3:722–32, 2005. Bacterial mercury resistance from atoms to ecosystems. *FEMS Microbiol. Rev.* 27:355, 2003.

William B. Whitman, Professor; Ph.D., Texas at Austin, 1978. Bacterial physiology/evolution; carbon metabolism in methanogens; prokaryotic systematics.
RNA-dependent cysteine biosynthesis in archaea. *Science* 307:1969–72, 2005. With Sauerwald et al. The importance of porE and porF in the anabolic pyruvate oxidoreductase of *Methanococcus maripaludis. Arch. Microbiol.* 181:68–73, 2004. With Lin et al.

Juergen K. W. Wiegel, Professor; Ph.D., Göttingen (Germany), 1973.
Diversity of aerobic and anaerobic alkalithermophiles. *Biochem. Soc. Trans.*, in press (with V. Kevbrin). *Caloramator viterbensis* sp. nov.: A novel thermophilic glycerol-fermenting bacterium isolated from a hot spring in Italy. *Int. J. Syst. Evol. Microbiol.* 52:1177, 2002.

Joint and Adjunct Faculty

Michael W. Adams, Research Professor, Department of Biochemistry and Molecular Biology; Ph.D., London, 1979. Physiology and enzymology of microorganisms growing near 100°C, using structural and functional genomic approaches.

Michael J. Adang, Professor, Department of Entomology; Ph.D., Washington State, 1981. Genetics and toxicology of *Bacillus thuringiensis* insecticidal proteins.

Russell W. Carlson, Professor, Department of Biochemistry and Molecular Biology, and Technical Director, Complex Carbohydrate Research Center; Ph.D., Colorado at Boulder, 1976. Molecular bases for bacterial-plant and bacterial-animal interactions.

Timothy P. Denny, Professor, Department of Plant Pathology; Ph.D., Cornell, 1983. Genetics and biochemistry of bacterial-plant interactions; mechanisms of systemic colonization of tomato by *Ralstonia solanacearum.*

Mark A. Eiteman, Professor, Department of Biological and Agricultural Engineering; Ph.D., Virginia, 1991. Fermentation technology and metabolic engineering of the production of biochemicals.

Marcus Fechheimer, Josiah Meigs Professor, Department of Cellular Biology; Ph.D., Johns Hopkins, 1980. Cell biology, molecular biology, and biochemistry of the actin cytoskeleton in nonmuscle cells and in neurodegenerative disease.

Joseph Frank, Professor, Department of Food Science and Technology; Ph.D., Wisconsin–Madison, 1977. Pathogen-food interactions, with an emphasis on direct observation using confocal scanning laser microscopy; food microbiology; food fermentations.

Robert E. Hodson, Professor, Department of Marine Sciences; Ph.D., California, San Diego (Scripps), 1977. Marine microbial ecology.

Sidney R. Kushner, Professor, Department of Genetics; Ph.D., Brandeis, 1970. Analysis of the mechanism of posttranscriptional genetic regulation in *Escherichia coli;* molecular analysis of mRNA decay and polyadenylation; structure-function analysis of DNA helicase II.

John J. Maurer, Associate Professor, Department of Avian Medicine; Ph.D., Texas Health Science Center at San Antonio, 1990. Antibiotic resistance, food safety, and molecular epidemiology of veterinary and foodborne pathogens.

Cory Momany, Associate Professor, Department of Pharmaceutical and Biomedical Sciences; Ph.D., Texas at Austin, 1990. Structural biology of prokaryotic transcriptional regulation, drug discovery, and bioremediation.

Mary Ann Moran, Professor, Department of Marine Sciences; Ph.D., Georgia, 1987. Microbial ecology and biogeochemistry in coastal marine environments; bacterial community structure; ecological genomics.

Andrew L. Neal, Assistant Research Professor, Savannah River Ecology Laboratory; Ph.D., Wales, Bangor, 1995. Molecular basis and physicochemistry of bacterial-mineral interactions; bioremediation of heavy metals and radionuclides; bacterial metal respiration.

Boris Striepen, Associate Professor, Department of Cellular Biology; Ph.D., Marburg (West Germany), 1995. Cell and molecular biology of intracellular protozoan parasites (Toxoplasma and Cryptosporidium).

Rick L. Tarleton, Distinguished Research Professor, Department of Cellular Biology; Ph.D., Wake Forest, 1983. Immunology of *Trypanosoma cruzi* infection and Chagas disease; vaccine development; proteomics and bioinformatics for trypanosomes.

UNIVERSITY OF ILLINOIS AT CHICAGO

College of Medicine
Department of Microbiology and Immunology

Programs of Study	Research within the Department of Microbiology and Immunology incorporates a wide range of modern biological studies and approaches. As part of the College of Medicine Graduate Education in Medical Sciences (GEMS) program, the Department's graduate program similarly affords a multidisciplinary and integrated approach to biological research. GEMS students apply to a preferred department, but they attend classes with students from all departments of the College of Medicine, and they are free to explore research opportunities in any laboratory in the GEMS program. The program is designed to provide students with a rigorous and thorough scientific education in core courses as well as in-depth study in microbiology, immunology, and virology. Critical skills are developed and fostered through participation in journal clubs, research talks, and regular Departmental seminars. Students undertake independent research in a chosen area of interest under the supervision of a faculty member. The Ph.D. degree is expected to require four or five years of study.

During their first year, graduate students complete fundamental, didactic course work in core GEMS courses including molecular biology, biochemistry, cell biology, genetics, integrative biology, and research methods. During their second year, students pursue advanced course work in microbiology, immunology, and virology taught by the Department of Microbiology and Immunology involving more extensive student presentation and discussion. In addition, throughout their graduate careers, students are required to participate in weekly seminar series (presented by guest speakers and Department faculty members), in Departmental journal clubs, and in the student research seminar series.

First-year students are provided with the opportunity to rotate through three research laboratories in any laboratory in the GEMS program. These three rotations allow students to gain firsthand knowledge of the research projects and also to learn various techniques. After completing the three rotations, students select their permanent thesis adviser.

By the end of the second year, all Ph.D. students are required to successfully pass an oral preliminary exam. The purpose of the preliminary examination is to ascertain the student's ability to present and defend a sound scientific proposal and also to test the student's general knowledge in the same as well as other related disciplines. Prior to graduation, all students are required to orally defend their dissertations.

Research Facilities   The research facilities and resources at the University of Illinois at Chicago (UIC) and in the Department of Microbiology and Immunology are those of a first-rate, research-oriented institution. The Departmental research laboratories occupy more than 23,000 square feet of space. The labs are well equipped for all aspects of research in microbiology, immunology, cell and molecular biology, biochemistry, and structural biology. Equipment is available for light and fluorescence microscopy, fluorometry and spectrophotometry, electrophoresis, cell culture, protein purification, and X-ray crystallography. Many larger facilities and specialized items are shared within the Department, fostering the interactive and communal research environment.

The University Research Resources Center includes facilities for fluorescence-activated cell sorting and analysis, genomics and proteomics, and for the synthesis and analysis of DNA, proteins, and peptides. A central electron microscope laboratory is equipped with transmission and scanning instruments, and instrumentation is also available for small molecule and protein structural analysis by nuclear magnetic resonance, circular dichroism, and mass spectroscopy. A modern biological resources laboratory houses and cares for animals used in research. The Library of Health Sciences, one of the largest of its type in the U.S., houses more than 400,000 books and journals; the library subscribes to more than 6,400 journal titles. The campus computer network, the Academic Data Network, provides each laboratory with high-speed Ethernet connections for University computer services, online library searches, and Internet access.

Financial Aid   Financial aid is offered to all incoming Ph.D. candidates. Tuition-free teaching and research assistantships support the students with an annual stipend of $24,000 in 2006–07.

Cost of Study   Students receiving stipends are exempt from tuition and other major fees. For other students, tuition and fees in 2006–07 are $3774 per semester for Illinois residents and $7965 per semester for out-of-state students. These costs are subject to change without notice.

Living and Housing Costs   Students may live in University residence halls or in off-campus apartments. Food and housing cost about $15,000 per year.

Student Group   UIC enrolls approximately 25,000 students from throughout the U.S. and overseas as well as from Illinois. Twenty percent of the total are engaged in studies in health science fields, and nearly 5,000 are enrolled at the graduate level. Approximately 40 students are enrolled in the doctoral program in the Department of Microbiology and Immunology.

Student Outcomes   After completing their doctoral work, many recent graduates have pursued postdoctoral training at some of the best academic institutions throughout the U.S. Other graduates have taken scientific positions in federal laboratories and regulatory agencies, and in the pharmaceutical and biotechnology industries.

Location   UIC is located on the Near West Side, 5 minutes by public transit from Chicago's downtown center, the Loop. It is part of a neighborhood that includes two historical landmark residential areas as well as the West Side Medical Center District, one of the world's largest and most advanced concentrations of health-care facilities. The campus has a variety of recreational facilities, and excellent cultural and entertainment programs are available. The city offers a wealth of concert halls, theaters, galleries, museums, and parks; several distinguished educational institutions; restaurants serving a variety of ethnic foods; and a lovely lakefront setting. Chicago is frequently host to major scientific meetings.

The University   One of two campuses of the University of Illinois, UIC was established in 1982 by merger of the University's Chicago Circle and Medical Center campuses. The University is the largest research and teaching institution in the Chicago area. It includes the Colleges of Art, Architecture and Urban Planning; Associated Health Professions; Business Administration; Dentistry; Education; Engineering; Health, Physical Education and Recreation; Liberal Arts and Sciences; Medicine; Nursing; and Pharmacy. It also includes the Graduate College, Honors College, Jane Addams College of Social Work, and the School of Public Health. UIC offers bachelor's degrees in ninety-seven fields, master's in seventy-nine areas, and doctorates in forty-four specializations. The University of Illinois Hospital and Clinics provide health care to more than a quarter of a million people annually.

Applying   Additional information and application forms may be obtained at the addresses listed in the contact section.

Correspondence and Information
Graduate Affairs Committee
Department of Microbiology and Immunology (M/C 790)
College of Medicine
University of Illinois at Chicago
835 South Wolcott, E-704
Chicago, Illinois 60612-7344
Phone: 312-996-9477
Fax: 312-996-6415
Web site: http://www.uic.edu/depts/mcmi/index2.html

University of Illinois at Chicago

THE FACULTY AND THEIR RESEARCH

Ananda Chakrabarty, Professor; Ph.D., Calcutta, 1965. Microorganisms and cancer; interaction of microbial redox proteins with the tumor suppressor protein p53; development of anticancer drugs from microbial sources.

Zheng Wei Chen, Professor and Director; M.D./Ph.D., Peking Union Medical College (Beijing), 1988. Immunity to bacterial and viral infections; immunology of the biological interaction between AIDS virus and *Mycobacterium tuberculosis* in primate model systems; HIV, TB, and malaria vaccine; nanomedicine; nanobiotechnology.

Edward Cohen, Professor; M.D., Washington (St. Louis), 1957. Cancer therapy with gene-modified cells; induction of immunity to breast cancer in mouse models of human disease.

Bin He, Associate Professor; Ph.D., Purdue, 1993. Biology and pathogenesis of herpes simplex viruses; virus-host interactions; virus strategies to escape cellular responses.

William Hendrickson, Associate Professor; Director, Research Resources Center; and Co-Director, Graduate Education in Medical Sciences Program; Ph.D., Tufts, 1981.

Prasad Kanteti, Associate Professor; Ph.D. Pune (India), 1982. Signaling pathways mediated by TNFR family members and T-cell coreceptors; interplay between costimulatory and death pathways.

Linda J. Kenney, Associate Professor; Ph.D., Pennsylvania, 1987. Two-component signal transduction; regulation of *Salmonella* pathogenicity island 2 gene expression; regulation of outer membrane proteins; EnvZ/OmpR.

Amy Kenter, Associate Professor; Ph.D., Yeshiva (Einstein), 1982. Genetic mechanisms of immunoglobulin gene recombination; DNA-binding proteins; B-cell-specific nucleic acid enzymes; plasmid recombination assays.

Howard Lipton, Professor; M.D., Nebraska at Omaha, 1964. Molecular pathogenesis of Theiler's virus–induced demyelinating disease.

Philip Matsumura, Professor; Ph.D., Rochester, 1975. Prokaryotic molecular biology; bacterial chemotaxis; protein structure and function.

Alan McLachlan, Professor; Ph.D., Aberdeen (Scotland), 1980. Regulation of hepatitis B virus transcription and replication.

Tapan Misra, Associate Professor; Ph.D., Calcutta, 1972. Bacterial genetics and molecular biology; protein purification; biochemical interactions of regulatory proteins with the regulatory elements of the DNA.

Bellur Prabhakar, Professor and Head; Ph.D., Johns Hopkins, 1980. Cloning, expression, and characterization of antigens involved in, and animal models of, human autoimmune diseases.

Lijun Rong, Associate Professor; Ph.D., Purdue, 1991. Molecular mechanisms of viral entry and pathogenesis of retroviruses, hepatitis C virus, and Ebola viruses.

Deepak Shukla, Assistant Professor; Ph.D., Illinois, 1996. Viral and cellular mediators of herpes simplex virus entry, molecular mechanism of viral entry, and viral pathogenesis.

Simon Silver, Professor; Ph.D., MIT, 1962. Bacterial molecular biology; environmental microbiology; toxic heavy metals; membrane transport; plasmid molecular biology.

Zuoming Sun, Assistant Professor; Ph.D., Duke, 1995. Immunology; signaling mechanisms regulating thymocyte survival and T-cell activation.

David Ucker, Professor; Ph.D., California, San Francisco, 1981. Molecular mechanism of physiological cell death; innate immune recognition and anti-inflammatory clearance of apoptotic cell corpses.

Karl Volz, Associate Professor; Ph.D., California, San Diego, 1981. Protein engineering; X-ray crystallography of biologically active proteins.

William Walden, Professor; Ph.D., Washington (St. Louis), 1983. Molecular mechanisms of eukaryotic iron regulation; translational regulation; protein-RNA interactions in gene regulation.

Affiliated Faculty

Lawrence Chan, Associate Professor; M.D., Pennsylvania. Molecular mechanism of chronic skin inflammation, atopic dermatitis.

James L. Cook, Professor; M.D., Baylor, 1972. Viral gene regulation of mammalian cell susceptibility to immunological injury.

Andrei Gartel, Assistant Professor; Institute of Virology (Moscow). Regulation of CDK inhibitor p21 in normal and cancer cells; anticancer drugs that target antiapoptotic pathways.

Gail Hecht, Associate Professor; M.D., Loyola, 1982. Interactions between enteric bacterial pathogens and host intestinal epithelial cells.

Ramaswamy Kalyanasundaram, Associate Professor; Ph.D., Calgary. Molecular basis of host evasion by pathogens; identify novel antigens for vaccine or drug development.

Jianxun Li, Associate Professor; Ph.D., CUNY Graduate Center, 1990. Regulation of cytoskeleton and integrin-mediated cell adhesion by protein kinases in the immune system.

Phillip Marucha, Professor; Ph.D., Connecticut. Stress-impaired oxygen dependent transcription in wounds; distress, saliva composition, and saliva functions; interdisciplinary research in cancer from an oral perspective; modulation of inflammation by stress and social factors; mechanisms of neuroendocrine regulation of wound healing during aging.

Lin Tao, Associate Professor; D.D.S./Ph.D., Connecticut, 1989. Microbial genetics of streptococci and lactobacilli.

Richard D. Ye, Associate Professor; M.D., Shanghai, 1982; Ph.D., Washington (St. Louis), 1988. Molecular mechanisms of leukocyte activation; immunopharmacology of G-protein–coupled receptors.

UNIVERSITY OF ILLINOIS AT URBANA–CHAMPAIGN

Department of Microbiology

Programs of Study

The Department of Microbiology offers a graduate program leading to a Ph.D. degree. The Ph.D. program emphasizes molecular microbiology, including molecular genetics, microbial physiology, microbial ecology and evolution, pathogenic microbiology, cell biology, and immunology. A Medical Scholars Program leading to an M.D./Ph.D. is also offered.

The graduate program is tailored to each student's previous studies, interests, and background. Students generally complete suggested course work in the first and second years. During the first semester, students rotate through several laboratories in order to learn experimental techniques and to aid in choosing a research project. Students choose a research adviser and propose a thesis project during the spring semester of the first year. An oral qualifying exam is given at the end of the second year. Students generally complete their Ph.D. in four to five years.

All students teach for at least two semesters as part of their graduate training.

Research Facilities

The department has excellent research facilities. All the major equipment and technical expertise needed for modern research in microbiology, molecular biology, and cell biology (including recombinant DNA technology, nucleic acid and protein sequencing, hybridoma isolation and monoclonal antibody production, protein purification and characterization, spectroscopy, fluorescent and electron microscopy, and computer analysis) are available in the department. In addition, many centralized facilities with specialized research equipment and services are available. The University has the third-largest university library in the United States.

Financial Aid

All students admitted to the Ph.D. program receive financial support, including tuition waivers. Several fellowships are awarded on a competitive basis to outstanding applicants. The starting stipend in 2005–06 was $22,660 per year plus a tuition waiver. Upper-level graduate students are generally supported by research assistantships or interdisciplinary training grants.

Cost of Study

Tuition waivers are provided to all students admitted to the Ph.D. program.

Living and Housing Costs

A variety of reasonably priced housing is available. The University maintains housing for single and married graduate students, and a large number of private apartments and houses are available near campus. More information can be obtained from the Housing Information Office, 2 Fred Turner Student Services Building, University of Illinois, 610 East John Street, Champaign, Illinois 61820.

Student Group

There are approximately 65 full-time graduate students in the Ph.D. program and 130 undergraduates majoring in microbiology. There are also many postdoctoral fellows in the department.

Student Outcomes

Graduates are employed in universities, industry, and government. Most Ph.D. students accept postdoctoral positions at major research institutions, after which they take research positions in academia or industry. Recent graduates have taken jobs at Johns Hopkins University, Yale University, Washington University in St. Louis, University of Pennsylvania, University of Wisconsin, Rockefeller University, Harvard Medical School, Upjohn Company, DuPont, Abbott Lab, Mayo Clinic, the U.S. Department of Agriculture, and Archer Daniels Midland.

Location

The University is located in the twin cities of Urbana and Champaign, which have a combined population of about 100,000. The community lies about 130 miles south of Chicago, 120 miles west of Indianapolis, and 180 miles northeast of St. Louis. There are numerous entertainment and recreational opportunities in and near Urbana-Champaign. A wide variety of musical, artistic, and sporting events take place on campus and in the community. The campus also has excellent recreation facilities.

The University and The Department

The University offers a wide variety of graduate and undergraduate degrees. About 34,000 students are enrolled, including about 7,000 graduate students.

The Department of Microbiology at the University of Illinois at Urbana-Champaign has been consistently ranked by the National Science Foundation and *U.S. News & World Report* as one of the best microbiology departments in the United States. The department encourages close interaction among faculty members and graduate students working in different laboratories. The department is part of an umbrella program in molecular and cellular biology that encompasses eighty research laboratories. Many laboratories collaborate closely with other departments in the School of Chemical Sciences and the Colleges of Medical Sciences and Veterinary Medicine.

Applying

Admission to the Ph.D. program in microbiology requires a bachelor's degree in biological or physical sciences. Applicants should have at least a B average in the last 60 hours of undergraduate work and any graduate work completed. The GRE General Test is also required. MCAT scores are accepted in lieu of GRE scores for applicants to the Medical Scholars Program. Both a minimum score of 590 (paper-based test) or 241 (computer-based test) on the TOEFL and a minimum score of 55 on the TSE (revised scoring) are required of applicants from countries where English is not the primary language.

Generally, students are only admitted in the fall semester. Although there is no absolute deadline for application, complete applications should be received by early January for full consideration.

Correspondence and Information

For additional information and application forms:

Department of Microbiology
B103 Chemical and Life Sciences Laboratory
University of Illinois at Urbana-Champaign
601 South Goodwin Avenue
Urbana, Illinois 61801-3709

E-mail: microinfo@life.uiuc.edu
Web site: http://www.life.uiuc.edu/micro

University of Illinois at Urbana–Champaign

THE FACULTY AND THEIR RESEARCH

Steven R. Blanke, Ph.D., Illinois at Urbana-Champaign, 1989. Molecular and cellular basis of infection biology and bacterial pathogenesis; bacterial persistence during infection; intracellular infection; complex systems; bacterial toxins; *Helicobacter pylori*; gastric ulcer disease and stomach cancer; *Bacillus anthracis. Helicobacter pylori* VacA: A paradigm for toxin multifunctionality. *Nat. Rev. Microbiol.* 3(4):320–32, 2005 (with Cover). Micro-managing the executioner: Pathogen targeting of mitochondria. *Trends Microbiol.* 13(2):64–71, 2005. *Helicobacter pylori* vacuolating cytotoxin enters cells, localizes to the mitochondria, and induces mitochondrial membrane permeability changes correlated to toxin channel activity. *Cell Microbiol.* 6(2):143–54, 2004 (with Willhite).

John E. Cronan Jr., Ph.D., California, Irvine, 1968. Regulation of lipid metabolism; protein-lipid interactions; synthesis of biotin and lipoic acid. A new *Escherichia coli* metabolic competency: Growth on fatty acids by a novel anaerobic β-oxidation pathway. *Mol. Microbiol.* 47:793–805, 2003 (with Campbell and Morgan-Kiss). The *Escherichia coli lipB* gene encodes lipoyl-ACP: Protein transferase. *J. Bacteriol.* 185:1582–9, 2003 (with Jordan).

Stephen K. Farrand, Ph.D., Rochester, 1973. Molecular signaling between bacteria and between bacteria and their hosts; quorum sensing; mechanisms of plasmid transfer. Mutational analysis of TraR: Correlating function with molecular structure of a quorum-sensing transcriptional activator. *J. Biol. Chem.* 278:13173–82, 2003 (with Luo, Smyth, and Qin). The quorum-sensing system of *Agrobacterium* plasmids: Its analysis and its utility. *Methods Enzymol.* 358:452–84, 2002 (with Qin and Oger). Co-evolution of the agrocinopine opines and the agrocinopine-mediated control of TraR, the quorum-sensing activator of the Ti plasmid conjugation system. *Mol. Microbiol.* 41:1173–86, 2001 (with Oger). The antiactivator TraM interferes with the autoinducer-dependent binding of TraR to DNA by interacting with the C-terminal region of the quorum-sensing activator. *J. Biol. Chem.* 275:7713–22, 2000 (with Luo and Qin). Quorum-sensing signal binding results in dimerization of TraR and its release from membranes into the cytoplasm. *EMBO J.* 19:5212–21, 2000 (with Qin et al.). TraG from RP4 and TraG and VirD4 from Ti plasmids confer relaxosome specificity to the conjugal transfer system of pTiC58. *J. Bacteriol.* 182:1541–8, 2000 (with Hamilton et al.). Signal-dependent DNA binding and functional domains of the quorum-sensing activator TraR as defined by repressor activity. *Proc. Natl. Acad. Sci. U.S.A.* 96:9009–14, 1999 (with Luo).

Bruce Fouke, Ph.D., SUNY at Stony Brook, 1993. Geomicrobiology: Microbe-water-mineral interactions. Phylogenetic diversity and distribution of bacteria in travertine depositional facies (Angel Terrace, Mammoth Hot Springs, Yellowstone National Park, U.S.A.). *J. Sedimentary Res.,* in press (with Bonheyo, Sanzenbacher, and Frias-Lopez). Cyanobacterial diversity associated with coral black band disease in Caribbean and Indo-Pacific Reefs. *Appl. Environ. Microbiol.,* in press (with Frias-Lopez, Bonheyo, and Jin).

Jeffrey F. Gardner, Ph.D., Marquette, 1975. Mechanism of site-specific recombination by bacteriophage λ and conjugative transposons; mechanisms of protein-DNA and protein-RNA interactions; characterization of bacteriophage λ excisionase mutants defective in DNA binding. Role of nucleoid proteins in the structure and function of the *E. coli* chromosome, in press (with R. C. Johnson, L. Johnson, and Schmidt). Integration host factor: Putting a twist on protein-DNA recognition. *J. Mol. Biol.* 330:493–502, 2003 (with Lynch, Mattis, and Rice). The molecular basis of cooperative DNA binding between λ integrase and excisionase. *Mol. Microbiol.* 50:89–99, 2003 (with Swalla, Cho, and Gumport). Conservation of structure and function among tyrosine recombinases: Homology-based modeling of the λ integrase core-binding domain. *Nucleic Acids Res.* 131:805–18, 2003 (with Swalla and Gumport). Regulation of site-specific recombination by the carboxyl terminus of λ integrase. *Nucleic Acids Res.* 30:5193–204, 2002 (with Kazmierczak, Swalla, Burgin, and Gumport).

James A. Imlay, Ph.D., Berkeley, 1987. Substantial DNA damage from submicromolar intracellular hydrogen peroxide detected in Hpx-mutants of *Escherichia coli. Proc. Natl. Acad. Sci. USA* 102:9317–22, 2005. Repair of oxidized iron-sulfur clusters in *Escherichia coli. J. Biol. Chem.* 279:44590–9, 2004 (with Djaman). Are respiratory enzymes the primary sources of intracellular hydrogen peroxide? *J. Biol. Chem.* 279:48742–50, 2004 (with Seaver). Pathways of oxidative damage. *Ann. Rev. Microbiol.* 57:395–418, 2003.

Andrei Kuzminov, Ph.D., Novosibirsk (Russia), 1990. Chromosomal lesions: mechanisms of formation, repair via homologous strand exchange, predisposition, and avoidance in *Escherichia coli* and *Deinococcus radiodurans.* Chromosomal fragmentation of dUTPase-deficient mutants of *Escherichia coli* and its recombinational repair. *Mol. Microbiol.* 51:1279–95, 2004 (with Kouzminova). RdgB acts to avoid chromosome fragmentation in *Escherichia coli. Mol. Microbiol.* 48:1711–25, 2003 (with Bradshaw). Single-strand interruptions in replicating chromosomes cause double-strand breaks. *Proc. Natl. Acad. Sci. U.S.A.* 98(15):8241–6, 2001.

William W. Metcalf, Ph.D., Purdue, 1991. Molecular, genetic, and biochemical analysis of microbial processes involved in the global cycling of carbon and phosphorus. Genetic analysis of the archaeon *Methanosarcina barkeri* Fusaro reveals a central role for Ech hydrogenase and ferredoxin in methanogenesis and carbon fixation. *Proc. Natl. Acad. Sci. U.S.A.* 99:5632–7, 2002 (with Meuer, Kuettner, Zhang, and Hedderich). Isolation and characterization of hypophosphite/2-oxoglutarate dioxygenase: A novel phosphorus oxidizing enzyme from *Pseudomonas stutzeri* WM88. *J. Biol. Chem.* 277:38262–71, 2002 (with White).

Charles G. Miller, Ph.D., Northwestern, 1968. Mechanisms of intracellular protein breakdown and proteolytic modification. Structure of peptidase T from *Salmonella typhimurium. Eur. J. Biochem.* 269:443–50, 2002 (with Håkansson). Aspartic peptide hydrolases in *Salmonella enterica* serovar Typhimurium. *J. Bacteriol.* 183:3089–97, 2001 (with Larsen and Knox).

Gary J. Olsen, Ph.D., Colorado, 1983. Genome analysis; gene expression in the Archaea; comparative analysis of molecules to study their evolution and structure. Evolution of eukaryotic transcription: Insights from the genome of *Giardia lamblia. Genome Res.* 14:1537–47, 2004 (with Best et al.). Evidence for similar subunit architectures between archaeal and eukaryal RNA polymerases. *FEMS Microbiol. Lett.* 195:85–90, 2001 (with Best). Archaeal RecA homologs: Different response to DNA-damaging agents in mesophilic and thermophilic Archaea. *Extremophiles* 5:265–75, 2001 (with Reich et al.). Thermal adaptation analyzed by comparison of protein sequences from mesophilic and extremely thermophilic *Methanococcus* species. *Proc. Natl. Acad. Sci. U.S.A.* 96:3578–83, 1999 (with Haney et al.).

Abigail A. Salyers, Ph.D., George Washington, 1969. Interaction of colonic bacteria with host; molecular microbial ecology; genetics of obligate anaerobes; polysaccharide uptake and catabolism by *Bacteroides;* conjugative transposons of *Bacteroides.* Multiple gene products and sequences required for the excision of the mobilizable integrated *Bacteroides* element NBU1. *J. Bacteriol.* 182:928–36, 2000 (with Shoemaker and Wang). Conjugative transposons: Transmissible resistance islands. In *Pathogenicity Islands and Other Mobile Virulence Elements,* pp. 331–46, eds. J. B. Kaper and J. Hacker. Washington, D.C.: ASM Press, 1999 (with Shoemaker and Frias). Physiological characterization of SusG: An outer membrane protein essential for starch utilization by *Bacteroides thetaiotaomicron. J. Bacteriol.* 181:7206–11, 1999 (with Shipman).

Joanna L. Shisler, Ph.D., Emory, 1996. Poxvirus interaction with host cells; regulation of the NF-kappaB transcription factor by viral proteins. The vaccinia virus K1L gene product inhibits host NF-kappaB activation by preventing I kappa B alpha degradation. *J. Virol.* 78:3553, 2004 (with Jin). *Molluscum contagiosum* virus inhibitors of apoptosis: The MC159 v-FLIP protein blocks Fas-induced activation of procaspases and degradation of the related MC160 protein. *Virology* 282:14, 2001 (with Moss).

James M. Slauch, Ph.D., Princeton, 1990. Molecular mechanisms of *Salmonella* pathogenesis. HilD, HilC, and RtsA constitute a feed forward loop that controls expression of the SPI1 type three secretion system regulator *hilA* in *Salmonella enterica* serovar Typhimurium. *Mol. Microbiol.* 57:691–705, 2005 (with Ellermeier). Physical properties allow SodCI but not SodCII to contribute to virulence in *Salmonella enterica* serovar Typhimurium strain 14028. *J. Bacteriol.* 186:5230–8, 2004 (with Krishnakumar, Craig, and Imlay).

Richard I. Tapping, Ph.D., McMaster, 1995. Innate immunity: Role of Toll-like receptors in host defense against microbial pathogens. Mycobacterial lipoarabinomannan mediates physical interactions between TLR1 and TLR2 to induce signaling. *J. Endotoxin Res.* 9(4):264–8, 2003 (with Tobias). Leptospiral lipopolysaccharide activates cells through a TLR2-dependent mechanism. *Nature Immunol.* 2:346–52, 2001 (with Ulevitch).

Carin K. Vanderpool, Ph.D., Minnesota, 2003. Small noncoding RNA regulation of bacterial stress responses; physiology and genetics of sugar phosphate stress in *Escherichia coli.*

Brenda A. Wilson, Ph.D., Johns Hopkins, 1989. Anti-toxin therapeutics. *Curr. Opin. Biotech.* 13:267–74, 2002. Bacterial toxin effects on host-cell signal transduction pathways. *Circ. Res.* 90:850–7, 2002 (with Sabri); *J. Biol. Chem.* 275:2239–45, 2000 (with Seo); and *J. Biol. Chem.* 272:1268–75, 1997. Bacterial toxin interaction with host cells. *Infect. Immun.* 68:4531–8, 2000. Molecular mechanisms of *Pasteurella, Bordetella, E. coli,* and *Clostridia* pathogenesis: Structure-function of bacterial protein toxins. *Infect. Immun.* 67:80–7, 1999.

Carl R. Woese, Ph.D., Yale, 1953. Molecular evolution of prokaryotes; structure and function of protein translation apparatus. On the nature of global classification. *Proc. Natl. Acad. Sci. U.S.A.* 89:2930–4, 1992 (with Whellis and Kandler). The ribosomal RNA database project: Updated description. *Nucleic Acids Res.* 19:4817, 1991 (with Olsen, Overbeek, and Larsen).

L. John Xu, Ph.D., Harvard, 1996. Molecular mechanisms underlying host-pathogen interactions; genetic basis of host resistance/susceptibility to bacterial infection. Diversity in the CDR3 region of V_H is sufficient for most antibody specificities. *Immunity* 13:37–45, 2000 (with Davis).

THE UNIVERSITY OF IOWA

College of Medicine
Department of Microbiology

Programs of Study

The Department of Microbiology, through the Graduate College, offers programs leading to the M.S. and Ph.D. degrees in the following five areas: pathogenic bacteriology, microbial genetics, immunology, general microbiology and microbial physiology, and virology.

Combined M.D./Ph.D. and D.D.S./Ph.D. programs are also available, and the Department has space for postdoctoral fellows.

Graduate students in the Department are aided by a faculty sponsor and a 3-member or 5-member advisory committee. Although every student must fulfill certain general requirements, each program is designed to satisfy the special needs of the individual student. Requirements for the Ph.D. degree include completing the necessary formal courses, successfully passing a comprehensive examination, and writing a scholarly thesis based on the results of original and significant research. The progress of each student is reviewed annually by the faculty. Candidates normally complete the Ph.D. degree in five or six years.

Research Facilities

The Department occupies approximately 38,000 square feet of space in the Bowen Science Building. Specialized modern equipment is available for research in each of the five program areas. In addition, the Department has a walk-in incubator and cold rooms, rooms for isotope work, an excellent electron microscope suite, and specially equipped isolation rooms under negative pressure for research on highly pathogenic microorganisms. The Department has a small library containing essential periodicals, and the excellent Hardin Library for the Health Sciences is nearby.

Financial Aid

All applicants for admission are considered for assistantships, which are funded through the Graduate College, the Department, training grants, and research grants. The annual stipend amount for 2005–06 was $22,000. In addition, tuition and fees are paid for students who hold assistantships.

Cost of Study

Tuition and fees for residents of Iowa is $3354 per semester in 2006–07. Nonresident tuition and fees are $9051 per semester. Students with assistantships of 25 percent or more are considered residents for tuition purposes. Research equipment and supplies are provided by the Department or through grant funds.

Living and Housing Costs

Full board and a single or shared double room are available for single students in residence halls at costs ranging from approximately $7000 to $9500 per year. Married student housing is available through the University (one- or two-bedroom apartments renting for $400 to $550 per month) or off campus at a wide range of prices.

Student Group

The Department has between 50 and 60 graduate students enrolled full-time. Each student is provided with study and bench space for research. The total student enrollment at the University of Iowa is approximately 29,600, including 5,400 graduate students in various disciplines.

Student Outcomes

Over the past five years, graduates of the Department have taken academic positions at such institutions as Texas A&M, South Florida, and Pittsburgh and postdoctoral positions at nationally recognized research centers such as Stanford, Washington (Seattle), Rochester, Illinois, and California, San Francisco. A number of recent graduates have obtained industrial positions at major pharmaceutical, biotechnological, and environmental corporations.

Location

Iowa City is an actively growing university community located on the Iowa River in the eastern part of the state. In addition to the many scientific programs, the community has a variety of artistic, cultural, educational, and recreational activities.

Interstate Highway 80 passes nearby, and several airlines operate out of the Cedar Rapids/Iowa City airport.

The University and The Department

The University of Iowa is more than 150 years old and now consists of eleven colleges. The Department of Microbiology offers courses to students in seven of these colleges and also operates an undergraduate program. The University is a member of the Big Ten Conference and takes part in many cooperative teaching and research programs.

Applying

To be considered for admission, applicants must have a bachelor's degree from an accredited institution, with a good academic record in biology, chemistry, mathematics, and physics and a minimum grade point average of 3.0 on a 4.0 scale. Creditable scores on the Graduate Record Examinations are also required.

Official transcripts from each undergraduate and graduate institution attended must be submitted to the Director of Admissions, together with the completed application form. Scores on the Graduate Record Examinations should be included, if available at the time of applying; otherwise they may be sent later to the head of the department. Three letters of recommendation should be sent directly to the head of the department.

Correspondence and Information

For information about the program:
M. A. Apicella, M.D., Professor and Head
Department of Microbiology
3-403 Bowen Science Building
The University of Iowa
Iowa City, Iowa 52242-1109
E-mail: grad-micro-info@uiowa.edu
Web site: http://www.medicine.uiowa.edu/microbiology/

For application forms and information:
Office of Admissions
Calvin Hall
The University of Iowa
Iowa City, Iowa 52242-1396
Web site: http://www.uiowa.edu/admissions/graduate/
programs/program-details/microbiology-phd.html

The University of Iowa

THE FACULTY AND THEIR RESEARCH

L.-A. H. Allen, Associate Professor of Internal Medicine and Microbiology; Ph.D., Wisconsin–Madison, 1990. Perturbation of phagocyte function by bacterial pathogens.

M. A. Apicella, Professor and Head; M.D., SUNY Downstate Medical Center, 1963. Pathogenesis of human infection by pathogenic *Neisseria* and *Haemophilus.*

R. F. Ashman, Professor of Internal Medicine and Microbiology; M.D., Columbia, 1966. Mechanism of action of inhibitory CpG oligonucleotides.

G. A. Bishop, Professor of Microbiology and Internal Medicine; Ph.D., Michigan, 1983. Molecular mechanisms of B-lymphocyte activation by normal and viral oncogenic receptors.

J. E. Butler, Professor; Ph.D., Kansas, 1965. Antibody and T-cell repertoire development in neonates.

S. Clegg, Professor; Ph.D., Dundee (Scotland), 1978. The fimbriae of enterobacteria.

C. D. Cox, Professor; Ph.D., Georgia, 1971. Iron metabolism and pathogenicity of *Pseudomonas aeruginosa;* new products from corn.

M. O. Dailey, Associate Professor of Pathology and Microbiology; Ph.D., 1976, M.D., 1977, Chicago. Regulation of lymphocyte surface-adhesion molecules and lymphocyte migration.

M. G. Feiss, Professor; Ph.D., Washington (Seattle), 1969. Virus DNA packaging.

J. T. Harty, Professor; Ph.D., Minnesota, 1989. T-cell responses to infection.

A. R. Horswill, Assistant Professor; Ph.D., Wisconsin–Madison, 2001. Bacterial cell-to-cell communication using autoinducing peptides.

J. C. D. Houtman, Assistant Professor; Ph.D., Wisconsin–Madison, 1999. Role of adaptor proteins in T-cell activation.

W. Johnson, Professor and Vice Chairman; Ph.D., Rutgers, 1968. Mechanism of pathogenesis of infections caused by *Legionella pneumophila.*

B. D. Jones, Associate Professor; Ph.D., Maryland, 1989. Interactions between pathogenic bacteria and mammalian cells.

A. J. Klingelhutz, Assistant Professor; Ph.D., Wisconsin–Madison, 1991. Molecular mechanisms of immortalization and transformation by human papillomavirus.

D. M. Lubaroff, Professor of Urology and Microbiology; Ph.D., Yale, 1967. Tumor immunology and immunotherapy.

B. K. Martin, Assistant Professor; Ph.D., Utah, 1993. The molecular biology of immune complement and its role in autoimmune disease.

W. J. Maury, Assistant Professor; Ph.D., Virginia, 1988. Regulation of lentiviral cell tropism.

L. L. McCarter, Associate Professor; Ph.D., California, Davis, 1983. Signal transduction and regulation of gene expression in bacteria.

W. M. Nauseef, Professor of Internal Medicine and Microbiology; M.D., SUNY Upstate Medical Center, 1976. The molecular and cell biology of human neutrophils.

S. Perlman, Professor of Microbiology and Pediatrics; Ph.D., MIT, 1972; M.D., Miami (Florida), 1979. Study of virus-induced demyelination.

T. L. Ratliff, Professor of Urology and Microbiology; Ph.D., Arkansas, 1977. Signal transduction and tumor immunity via complement component C5a.

R. J. Roller, Associate Professor; Ph.D., Harvard, 1987. Molecular biology of herpes simplex infections.

P. B. Rothman, Professor and Head of Internal Medicine and Professor of Microbiology; M.D., Yale, 1985. Regulation of cytokine signaling.

G. V. Stauffer, Professor; Ph.D., Penn State, 1976. Molecular mechanisms controlling gene expression and how regulatory events serve the needs of the cell.

M. F. Stinski, Professor; Ph.D., Michigan State, 1969. Human cytomegalovirus pathogenesis and regulation of gene expression.

C. M. Stoltzfus, Professor; Ph.D., Wisconsin, 1971. HIV RNA splicing regulation.

S. M. Varga, Assistant Professor; Ph.D., Massachusetts, 1999. Viral immunology and immunopathology.

D. S. Weiss, Associate Professor; Ph.D., Berkeley, 1991. Cell division and protein localization in bacteria.

J. P. Weiss, Professor of Internal Medicine and Microbiology; Ph.D., NYU, 1981. Inflammatory response to invading bacteria.

M. E. Wilson, Professor of Internal Medicine and Microbiology; M.D., Rochester, 1980. Immunobiology of infection with the protozoan parasite *Leishmania chagasi.*

T. L. Yahr, Assistant Professor; Ph.D., Medical College of Wisconsin, 1998. Role of the type III secretion system in the pathogensis of *Pseudomonas aeruginosa.*

UNIVERSITY OF KENTUCKY

College of Medicine
Graduate Program in Microbiology, Immunology, and Molecular Genetics

Program of Study

The Graduate Program in Microbiology, Immunology, and Molecular Genetics offers study leading to the Doctor of Philosophy degree. Research opportunities are available in bacterial pathogenesis, physiology, and genetics; molecular and cellular immunology; tumor immunobiology; eukaryotic cell biology and genetics; and molecular virology. Students have multidisciplinary biomedical research opportunities and utilize modern research technology in a strong intellectual environment. A broad range of courses provides a sound didactic foundation for careers in research and teaching. All students pursuing degrees in the biomedical sciences at the University of Kentucky College of Medicine are admitted through the Integrated Biomedical Sciences curriculum (IBS). This first-year core curriculum provides broad-based exposure to biochemistry, cell biology, molecular biology, genetics, cell signaling, and integrated physiology and provides flexibility in selecting a research emphasis among the 125 faculty members in the biomedical sciences. Students select their doctoral degree program at the completion of the first-year core curriculum. During the first year, students rotate among various faculty laboratories and then select the laboratories for their independent research. During the second year, students take electives in microbiology, immunology, and biochemistry. Following the completion of four full-time semesters, students must pass a qualifying examination. Each student's program is tailored to the individual's interests and career goals and includes participation in one of several journals clubs, as well as training in the use of general equipment, oral presentations, and the multifaceted uses of computers in modern biology. The Ph.D. degree in microbiology, immunology, and molecular genetics generally requires four to five years to complete. A combined M.D./Ph.D. degree is also offered. Information regarding the microbiology, immunology, and molecular genetics program and admission through the Integrated Biomedical Sciences curriculum may be obtained from the Director of Graduate Studies at the address below.

Research Facilities

The faculty laboratory facilities are located in several adjacent buildings that are well equipped for all aspects of modern biomedical research. The department directs core facilities in flow cytometry and cell sorting, electron microscopy, and transgenic mice. In addition, core facilities are available for macromolecular structure, nuclear magnetic resonance spectroscopy, microarrays, genomics, animal resources, and biomedical image processing. Frequently used equipment, facilities, and techniques include an automated peptide synthesizer, a BL-3 facility, high-performance liquid chromatographs, analytic and preparative ultracentrifuges, a cesium irradiator, beta and gamma spectrophotometers, a flow cytometer, automated ELISA spectrometers, and facilities for optical imaging; digitized video microscopy; DNA cloning, sequencing, and transfection; and RNA and protein analyses, plus other techniques used in modern tissue, cell, and molecular biology. Students have access to the Health Sciences Library and an extensive computer network supported by department computer staff members.

Financial Aid

Financial aid includes full tuition and fees plus a stipend of $20,000.

Cost of Study

The 2006–07 in-state tuition for full-time graduate students is $3518 per semester; out-of-state tuition is $7577.

Living and Housing Costs

In 2006–07, the monthly cost of on-campus graduate student housing, including utilities, is $500 for an efficiency apartment, $618 for a one-bedroom apartment, and $672 for a two-bedroom apartment. Off-campus housing, including single rooms, apartments, and some houses, is available for $400 to $1000 per month, depending on location and furnishings. Housing is available within walking distance of the University.

Student Group

The University student body is composed of approximately 22,000 undergraduates and 5,000 graduate and professional students. In 2005–06, there were 36 students in the Graduate Program in Microbiology, Immunology, and Molecular Genetics.

Location

The University is located in Lexington, a metropolitan area of approximately 250,000. Lexington has numerous theaters, concert halls, and restaurants and more than 400 horse farms. There are many opportunities for outdoor recreation in the Bluegrass region, including a readily accessible network of state parks with opportunities for hiking and canoeing. In addition, the University has recreational facilities and offers a variety of artistic presentations. Lexington provides a safe and pleasant lifestyle; it has a comfortable climate, convenient shopping and entertainment, and diversity, such as ethnic foods and repertory movie theaters.

The University

The University of Kentucky was founded in 1865 as a land-grant college. Graduate training was begun in 1870, and graduate degrees were first awarded in 1876. The main campus of 1,800 acres is located south of downtown Lexington. The location of the Medical Center on the main University campus allows for teaching and research interactions with the Departments of Biological Sciences and Chemistry and the College of Agriculture. The University is committed to a broad range of programs, particularly those designed for advanced study and research. On the Lexington campus, students have the opportunity for specialization in ten colleges, including Arts and Sciences, Medicine, Dentistry, Law, Agriculture, Pharmacy, and Engineering.

Applying

Admission to IBS is based upon academic background, GPA, recommendations, GRE scores, and prior experience. Personal interviews provide critical perspectives for both the student and the Admissions Committee and are strongly encouraged. Students must meet the admissions requirements set by the Graduate School, including a bachelor's degree from a fully accredited institution of higher learning. The deadline for submission of international applications is February 1; domestic applications are accepted on a rolling basis.

Correspondence and Information

Program Director
Graduate Program in Microbiology, Immunology, and Molecular Genetics
Department of Microbiology and Immunology
University of Kentucky Medical Center
Lexington, Kentucky 40536-0298
Phone: 859-323-5257
    800-462-5257 (toll-free)
Web site: http://www.mc.uky.edu/microbiology

University of Kentucky

THE FACULTY AND THEIR RESEARCH

Subbarao Bondada, Professor; Ph.D., Tata Institute of Fundamental Research (India), 1976. Growth regulation of B lymphocytes in health, autoimmunity, and cancer; antibody responses to polysaccharide vaccines. (bondada@uky.edu)

James S. Bryson, Associate Professor; Ph.D., Miami (Ohio), 1985. Bone marrow transplantation; graft-versus-host disease. (jsbrys@uky.edu)

Jason A. Carlyon, Assistant Professor; Ph.D., Virginia Commonwealth, 1999. Pathogenic mechanisms of *Anaplasma phagocytophilum;* adherence, invasion, intracellular survival in neutrophils. (jason.carlyon@uky.edu)

Donald A. Cohen, Professor; Ph.D., Cincinnati, 1979. Mechanisms of induction of interstitial pneumonitis after bone marrow transplantation and AIDS. (dcohen@uky.edu)

Leslie Crofford, Chief, Division of Rheumatology; M.D., Tennessee, Memphis, 1984. The role of lipid mediators in immune/inflammatory arthritis. (lcrofford@uky.edu)

Jeffrey N. Davidson, Professor; Ph.D., Harvard, 1976. Regulation, expression, and organization of a multienzymatic protein crucial to cellular proliferation and pyrimidine homeostasis. (jndavid@uky.edu)

Edward L. DeMoll, Associate Professor; Ph.D., Texas at Austin, 1982. Strategies that microbes use to meet environmental challenges. (eldemol@uky.edu)

Willem DeVillers, Professor; Ph.D., Oxford (UK), 1995. Autoimmune Crohn's disease and ulcerated colitis and macrophange receptors. (willem.devillers@uky.edu)

Sarah E. F. D'Orazio, Assistant Professor; Ph.D., Miami (Florida), 1995. Protective immunity against intracellular bacterial pathogens; bacterial antigens recognized by CD8+ T cells; *Listeria monocytogenes* infection. (sarah.dorazio@uky.edu)

Jeffrey L. Ebersole, Professor; Ph.D., Pittsburgh, 1975. Microbial pathogens and immunobiology of the oral cavity. (jleber2@uky.edu)

Jacqueline Fetherston, Assistant Professor; Ph.D., Washington (St. Louis), 1981. Pathogenic mechanisms in *Yersinia pestis:* the role of iron transport systems in survival in mammals and fleas. (jdfeth01@uky.edu)

Beth Garvy, Assistant Professor; Ph.D., Michigan State, 1991. Host defense against *Pneumocystis carinii;* neonatal immunity; bone marrow transplantation; cellular immunology. (bgarv0@uky.edu)

Robert Geraghty, Assistant Professor; Ph.D., Wisconsin–Madison, 1995. Mechanisms by which alpha herpes viruses, such as herpes simplex virus 1 and 2, enter cells. (rgeragh@uky.edu)

Robert J. Jacob, Associate Professor; Ph.D., Syracuse, 1974. Molecular pathogenesis of herpes virus infection and latency; morphology and ultrastructure of biological/synthetic structures. (rjjaco00@uky.edu)

Charlotte Kaetzel, Associate Professor; Ph.D., Maryland, 1979. Gene regulation in the mucosal immune system; intestinal immunity. (cskaet@uky.edu)

Alan M. Kaplan, Professor and Chairman; Ph.D., Purdue, 1969. Graft-versus-host disease; molecular mechanisms of induction of pulmonary fibrosis. (akaplan@uky.edu)

Heinz Köhler, Professor; M.D., Munich, 1967. Development of superantibodies for immunotherapy of cancer and in vivo diagnosis of cell-cycle proteins. (hkohl00@uky.edu)

Guangxiang Luo, Associate Professor; M.D., Hunan Medical University (China), 1983. Molecular genetic analysis of hepatitis C virus replication, regulation, and host-cell interaction. (gluo0@uky.edu)

Charles Lutz, Professor; M.D., Ph.D., Chicago, 1982. Natural killer (NK) cell recognition of tumor cells; expression of NK receptor genes and ligands. (ctlutz2@uky.edu)

Kevin T. McDonagh, Professor; M.D., Columbia, 1984. Tumor immunology and gene therapy. (mcdonaugh@uky.edu)

Joseph McGillis, Associate Professor; Ph.D., George Washington, 1985. Mechanisms of nervous system influence on immunity and inflammation. (jpmcgi01@uky.edu)

Craig Miller, Professor; D.D.S., Kentucky, 1982. Oncogenic role of human papillomavirus and molecular mechanisms of reactivation of herpes simplex. (cmiller@uky.edu)

Bin-Tao Pan, Associate Professor; Ph.D., McGill, 1983. Mechanisms by which oncogenic Ras induces tumor formation. (btpan@uky.edu)

Robert D. Perry, Professor; Ph.D., Michigan State, 1978. Regulation and role of iron and heme transport systems in the pathogenesis of *Yersina pestis,* the cause of plague. (rperry@uky.edu)

Martha Peterson, Associate Professor; Ph.D., Wisconsin–Madison, 1984. Posttranscriptional gene regulation, splicing, and cleavage-polyadenylation regulation of the IgM gene during B-cell development. (mlpete01@uky.edu)

Carol Pickett, Associate Professor; Ph.D., Texas at Austin, 1983. Mechanisms of bacterial pathogenesis; interactions of bacterial toxins with eukaryotic cells. (cpicket@uky.edu)

Andrew Pierce, Assistant Professor; Ph.D., North Carolina–Chapel Hill, 1995. DNA double-strand-break repair. (apierce@uky.edu)

Vivek Rangnekar, Professor; Ph.D., Bombay (India), 1983. Transcription factors in apoptosis and tumor-growth suppression. (vmrang01@uky.edu)

Thomas L. Roszman, Professor; Ph.D., Michigan State, 1966. Role of calpain in regulating integrated-mediated interaction of T cells with the extracellular matrix. (tlrosz00@uky.edu)

Chongsuk Ryou, Assistant Professor; Ph.D., Wayne State (Michigan), 1998. Molecular mechanisms of prion pathogenesis. (cryou2@uky.edu)

Anthony Sinai, Assistant Professor; Ph.D., Rochester, 1994. Molecular and cell biology of *Toxoplasma gondii.* (sinai@uky.edu)

Charles Snow, Professor; Ph.D., Iowa, 1978. B-cell fate determination, B-cell–T-cell interactions; regulation of apoptotic pathways. (ecsnow01@uky.edu)

Brett Spear, Professor; Ph.D., Pennsylvania, 1985. Regulation of liver gene expression during development and disease. (bspear@uky.edu)

Marion Steiner, Associate Professor; Ph.D., Kentucky, 1968. Inflammatory mediators of Alzheimer's disease. (mrstei01@uky.edu)

Brian Stevenson, Assistant Professor; Ph.D., SUNY at Stony Brook, 1989. Bacterial pathogenesis; biology of *Borrelia burgdorferi,* the cause of Lyme disease. (bstev01@uky.edu)

Susan C. Straley, Professor; Ph.D., Cornell, 1972. *Yersina pestis* adherence, invasion, and virulence; protein expression, targeting, and mechanisms of action. (scstra01@uky.edu)

Glenn Telling, Associate Professor; Ph.D., Carnegie Mellon, 1990. Molecular and cellular mechanisms of prion pathogenesis. (gtell2@uky.edu)

Jerold G. Woodward, Professor; Ph.D., Utah, 1979. Mechanisms of autoimmune disease and the control of peripheral tolerance in the immune system. (jwood1@uky.edu)

Jianyou Zhang, Associate Professor; Ph.D., Texas at Austin, 1989. Mutation and recombination of retroviruses. (jzhan1@uky.edu)

UNIVERSITY OF MANITOBA

Faculty of Medicine
Department of Medical Microbiology

Program of Study

Graduates in honours or general science, medicine, dentistry, and veterinary medicine may apply for entry into the program, but students with a three-year B.Sc. degree must normally enroll in a pre-master's course arranged in consultation with the Graduate Studies Committee and the head of the Department. Course work requirements are determined on an individual basis, but a minimum of 12 credit hours is required for both the M.Sc. and Ph.D. degrees. In addition, all students are expected to participate in a seminar program, which is composed of journal club, special topics review, and research seminars that also include faculty members and visiting scientists. Each graduate program is arranged by the supervising professor and an advisory committee.

Scientific interests of the faculty members are broad and research projects range from the basic understanding of gene regulation and molecular basis of cellular functions to the development of vaccines and diagnostics for human health and veterinary diseases. The Department has active research programs in cell and molecular biology, immunology, virology, bacterial genetics, microbial pathogenicity, chlamydial biology, and clinical microbiology.

Research Facilities

The Department of Medical Microbiology occupies the fifth floor of the Basic Medical Sciences Building on the downtown campus of the Faculty of Medicine, including modern research laboratories. Teaching and research are also conducted within the Clinical Microbiology Laboratories of the Health Sciences Centre and within the infectious diseases programs of the Health Sciences Centre, Cadham Provincial Laboratory, St. Boniface Hospital, and Canadian Science Centre for Human and Animal Health.

The Department's equipment, much of which is shared, supports research ranging from molecular biology to clinical microbiology. It includes ample biohazard containment facilities; controlled environment equipment; ultracentrifugation, spectrophotometric, chromatographic, and electrophoretic equipment; a transmission electron microscope; fluorescent microscopes; liquid scintillation counters; personal computers; and computer terminals for direct access to the mainframe computer. A library and a number of other ancillary facilities are available.

Financial Aid

Graduate studies in the Department are usually supported by awards or funds from one or a combination of national granting agencies, provincial granting agencies, University of Manitoba Graduate Fellowships, Departmental graduate student stipends, and supervisor's research funds.

Aid programs involving national granting agencies, provincial granting agencies, and University of Manitoba Graduate Fellowships are open competitions that require students to submit applications and have sponsorship from the Department. University of Manitoba Graduate Fellowships also requires rating from the Department. Departmental graduate student stipend support requires application by the proposed supervisor. The award is based on the priorities set up for the funding program. Supervisor's research fund support is awarded solely at the discretion of the supervisor. The level of funding under national and provincial granting agencies and University of Manitoba Graduate Fellowships is determined according to the published schedule of each agency and adjusted periodically by the agency. Students in the Department are eligible to apply for national training programs in infectious diseases and allergy/asthma.

The Department is responsible for the administration of the graduate training program as well as the administration of the funds. It does not assume any responsibility regarding the availability or amount of funds from the source or the continued availability of funds from the source. The Graduate Studies Committee does not encourage self-funding, due to the increased probability of instability of funding that may affect the student's performance and his or her ability to complete a program.

Supervisors or potential supervisors generally recommend the appropriate studentship program to apply to, based on the project and the qualifications of the applicant. Upon submission of an application for a studentship that requires Departmental sponsorship, it is understood that the student has made a commitment to register in the Department if a studentship is awarded. Application deadlines are different for each program and submission of an application for admission should be made nine months prior to the expected registration whenever possible in order to have the greatest options in application for funding. The Department deadline to apply for admission is one month prior to the deadline set by the Faculty of Graduate Studies.

Cost of Study

In 2004–05, tuition was $8354 for the M.Sc. program, $8354 for the Ph.D. from M.Sc. program, and $12,531 for the Ph.D. from honours program, which were exclusive of Endowment, Student Organization, and other applicable fees. Students should note that these fees are based on the minimum time allowed for completion of the programs (two years for the M.Sc. degree and four years for the Ph.D.). Continuing fees are assessed annually beyond these minimum time periods; the maximum time to complete the programs are five years for the M.Sc. and seven years for the Ph.D.

Living and Housing Costs

Based on the approximate cost of living for 2004–05, estimates of some annual basic expenses were as follows: $4900 for accommodations; $700 for transportation (bus pass); $2880 food ($60 per week); $900 for clothing; $1200 for books, supplies, typing, and copying; and $900 for general expenses. The estimated total was $11,480.

Student Group

For the academic year 2004–05, there were 29 Ph.D., 31 M.Sc., 1 pre-M.Sc., and 2 M.D./Ph.D. students registered in the Department. Of those students, 43 percent received financial aid and 17 percent are international students.

Student Outcomes

Over the past five years, Ph.D. graduates of the Department have gone on to postdoctoral research at Focus Technologies (United States); Northwestern University (Chicago), University of Nairobi (Kenya), University of Kentucky (United States); Zycos, Inc. (Boston); and the Canadian Science Centre for Human and Animal Health (Winnipeg).

Location

The University is located in Winnipeg, Manitoba, a city of 650,000 who celebrate their ethnic richness in a unique summer festival called Folkorama. The city has been host to many international events, including the 1999 Pan American Games.

Spring, summer, fall and winter are four distinct seasons in Winnipeg. Recreational activities vary from swimming to skating and include music and theatre festivals, professional sports teams, and numerous restaurants.

The University and The Department

The University of Manitoba (U of M) has been the home university for many generations of international students. U of M graduates are found in every continent around the world. The University offers a full range of undergraduate and graduate programs in social sciences, humanities, and the natural sciences. In addition, there are programs in the major professions that are accredited by Canadian professional associations.

The Department of Medical Microbiology offers programs of study leading to the M.Sc. and Ph.D. degrees with research and academic experience suitable for a career in basic microbiology or infectious diseases.

Applying

Application forms for admission to the graduate program can be obtained from the Department of Medical Microbiology, Suite 510, 730 William Avenue, Winnipeg, Manitoba, Canada, R3E 0W3.

Correspondence and Information

Dr. Xi Yang, Graduate Studies Chair
Department of Medical Microbiology
523-730 William Avenue
Winnipeg, Manitoba R3E 0W3
Canada
Phone: 204-789-3481
Fax: 204-789-3926
E-mail: yangxi@ms.umanitoba.ca
Web site: http://www.umanitoba.ca/faculties/medicine/medical_microbiology

University of Manitoba

THE FACULTY AND THEIR RESEARCH

Michelle Alfa, Professor; Ph.D., Alberta. Nosocomial infections, including *Clostridium difficile*–associated diarrhea and infection transmission due to medical device biofilm buildup.

Anton Andonov, Adjunct Professor; M.D., Leningrad. Molecular biology; epidemiology; diagnostics and prevention (edible vaccines) of human hepatitis (A, B, C, D, E) and other blood-borne viruses (HGV, TTV, TLMV).

Fred Y. Aoki, M.D., Manitoba. Antiviral chemotherapy of herpes simplex virus, HIV, and influenza A virus infection.

Harvey Artsob, Adjunct Professor; Ph.D., McGill. Zoonotic disease agents and level 4 viruses; epidemiology, epizootiology, and diagnostic applications.

Kathryn Bernard, Adjunct Professor; M.Sc., McGill. Systematics of nonenteric bacterial pathogens; characterizing and development of laboratory response to bacterial agents of bioterrorism.

Jody Berry, Adjunct Professor; Ph.D., Manitoba. Immunogenetics of antibody responses to infectious agents; development of improved methods for deriving monoclonal antibody for diagnostic and therapeutic (human) monoclonal antibody to bacterium, viruses (HIV-1, foot-and-mouth disease virus), and toxins.

Jamie Blanchard, Associate Professor; M.D., Manitoba; Ph.D., Johns Hopkins. Epidemiology of HIV and STDs; infectious causes of chronic diseases.

Stephanie Booth, Adjunct Professor; Ph.D., Oxford. Transmissible spongiform encephalopathy pathogenesis and diagnosis; pathogen/host interactions by microarray analysis.

Timothy Booth, Adjunct Professor. Virus structure and morphogenesis; biological transmission by arthropod vectors; pathology and diagnosis of animal viruses; transmissible spongiform encephalopathies, including chronic wasting disease of deer and elk.

Eric J. Bow, Professor; M.D., Calgary. Pathogenesis of infections in immunocompromised hosts.

Jingxin Cao, Adjunct Professor; Ph.D., Surrey (England). Hepatitis B and C immunopathogenesis, with particular interests in those viral proteins with potential immunoregulatory functions; development of novel therapeutic vaccines for chronic viral infections.

Michael Carpenter, Adjunct Professor; Ph.D., Manitoba. Molecular mechanism of Hepatitis C virus replication and cellular pathogenesis; host-cell regulators of viral gene expression.

Clifford Clark, Adjunct Professor; Ph.D., Alberta. Virulence factors, including protein exotoxins, of *E. coli, Salmonella, Aeromonas; Campylobacter* species: population structure of, and gene transfer within, enteric bacteria; parasites: development and evaluation of molecular typing and fingerprinting methods.

Kevin M. Coombs, Professor and Associate Department Head; Ph.D., Texas. Structure-function aspects of virus assembly and pathogenesis.

Michael Coulthart, Adjunct Professor; Ph.D., McMaster. Molecular, population, and evolutionary genetics of host-pathogen interaction; molecular pathogenesis and molecular diagnostics of prion diseases; human population genetics; microbial population genetics.

Markus Czub, Adjunct Professor; D.V.M., Berlin; Dr.med.vet., Giessen (Germany). Pathogenesis of emerging paramyxoviruses *(Hendra, Nipah, Menangle)* and of transmissible spongiform encephalopathies (retroviruses, prions).

Magdy R. Dawood, Assistant Professor; Ph.D., Purdue. HIV detection; role of immune system in protection against HIV infection.

Michael A. Drebot, Adjunct Professor; Ph.D., Dalhousie. Development of molecular diagnostics for infectious agents associated with zoonotic disease.

John Embil, Associate Professor; M.D., Dalhousie. Epidemiology of hospital-acquired infections, particularly those caused by methicillin-resistant *Staphylococcus aureus* and vancomycin-resistant *Enterococcus;* epidemiology and pharmacogenomic factors in infections in feet of diabetics; evaluation of new antibiotics.

Joanne E. Embree, Professor and Department Head; M.D., Halifax. Perinatal infections: Effect on prematurity, pregnancy, and neonatal complications; STDs in pediatrics; pediatric AIDS.

Margaret Fast, Assistant Professor; M.D., Manitoba. Epidemiology of communicable disorders.

Heinz Feldmann, Associate Professor; M.D., Giessen (Germany). Molecular biology and pathogenesis of level-4 viruses, such as filoviruses and arenaviruses; development of diagnostics for level-4 agents.

Alan Ferguson, Assistant Professor; Ph.D., Bristol (England). Reproductive health, particularly the sociodemographic aspects of HIV/AIDS and family planning in sub-Saharan Africa and development of elicitation methods in these fields.

Keith Fowke, Associate Professor; Ph.D., Manitoba. Cellular immune responses to viral infections (i.e., HIV); role of cell death (apoptosis) in HIV infection.

Larry Gelmon, Assistant Professor; M.D., Saskatchewan. AIDS/STD training and capacity-building in Eastern and Southern Africa; tropical public health and international development.

Greg W. Hammond, Professor; M.D.C.M., McGill. Clinical virology and development of diagnostics for viral infections in humans; use of molecular probes for diagnosis of enteric adenovirus and HIV infections.

Godfrey Harding, Professor; M.D., Manitoba. Urinary tract infections: Epidemiology, pathogenesis, and management strategies.

Kent HayGlass, Professor; Ph.D., Western Ontario. Role of Th1- versus Th2-like populations in vivo; cytokine-mediated regulation of immune disorders (i.e., allergy, HIV) in humans and experimental animals.

Runtao He, Adjunct Professor; M.D., Beijing; Ph.D., Dalhousie. Molecular virology and cell biology; molecular pathogenesis of SARS coronavirus and hepatitis C virus.

Daryl J. Hoban, Professor; Ph.D., Manitoba. Antimicrobial and antifungal resistance; new-drug development; molecular biology of bacterial pathogenesis and diagnostics applications.

Michael Jackson, Adjunct Professor; Ph.D., London. Development and application of advanced in vivo imaging (MRI, fluorescence, etc.) and spectroscopic (infrared, Raman, etc.) techniques to study infectious diseases, with particular emphasis on link between infection and chronic disease; hepatitis and TSEs.

Francis T. Jay, Professor; Ph.D., Liverpool. Mechanisms of antiviral and antitumour activities of interferons; structure-function studies of interferons by gene modification and monoclonal antibody analysis.

Wendy Johnson, Adjunct Professor; Ph.D., Alberta. Molecular diagnostics and epidemiology; bacterial pathogenesis and virulence mechanisms; antibody-based therapeutics; recombinant therapeutic proteins.

Amin Kabani, Associate Professor; M.D., Bristol. Molecular diagnostics; antimicrobial resistance.

David Knox, Adjunct Professor; Ph.D., Ottawa. Prion or transmissible spongiform encephalopathy (TSE) diseases; comparative genomics used to identify differentially expressed host genes' characteristic of TSE infection, potential therapeutic targets, susceptibility markers, and evaluation of therapeutic interventions.

Yan Li, Adjunct Professor; Ph.D., Ottawa. Molecular biology and genetics of respiratory viruses, such as influenza, RSV, and parainfluenza virus; development of molecular diagnostics for respiratory viruses.

Bruce R. Light, Professor; M.D., Calgary. Pathophysiology of acute bacterial pneumonia in patients and animal models.

Evelyn Lo, Assistant Professor; M.D., Toronto. Nosocomial infections: epidemiology, prevention, and transmission; quality improvement in reducing ventilator-associated pneumonia.

Grant A. McClarty, Professor; Ph.D., Manitoba. Biochemical studies on cell-cell interactions that occur between the obligate intracellular bacterial pathogen *Chlamydia* and the host cell.

Chester Morris, Assistant Professor; M.D., British Columbia. Targeted HIV interventions in vulnerable groups; impact assessment of HIV in Africa and models of HIV service delivery in private sector in resource-limited settings.

Stephen Moses, Professor; M.D., Toronto; M.P.H., Johns Hopkins. Epidemiology and control of communicable diseases, with particular emphasis on STDs and HIV infection.

Michael Mulvey, Assistant/Adjunct Professor; Ph.D., Manitoba. Bacterial pathogenesis and novel antimicrobial resistance mechanisms.

Nicolaas Nagelkerke, Associate Professor; Ph.D., Amsterdam. Biostatistics and epidemiology, with a focus on methodology of infectious disease research; mathematical modelling of infectious diseases.

Lai-King Ng, Adjunct Professor; Ph.D., Edmonton. Surveillance and mechanisms of antibiotic resistance in *Neisseria gonorrhoeae* and enteric pathogens *(Salmonella, Shigella, Campylobacter,* and *Vibrio cholerae;)* pathogenesis of *Campylobacter jejuni.*

Lindsay E. Nicolle, Professor; M.D., Manitoba. Epidemiology and pathogenesis of urinary tract infection.

Ayman Noreddin, Assistant Professor; Ph.D., University of the Pacific. Population pharmacokinetics; Monte Carlo analysis and clinical simulations.

Pamela H. Orr, Professor; M.D., Manitoba. Epidemiology of *Chlamydia trachomatis;* epidemiology of infections in elderly and aboriginal populations.

Carla Osiowy, Adjunct Professor; Ph.D., Calgary. Molecular characterization of hepatitis B virus (HBV): investigation of drug-resistant and vaccine-escape mutants; molecular evolution studies; functional analysis of hepatitis C (HCV) protein; molecular interactions between HCV, HBV, and HIV.

Rosanna W. Peeling, Adjunct Professor; Ph.D., Manitoba. Pathogenesis of chlamydial infections, sexually transmitted diseases, and reproductive health.

Pierre Plourde, Associate Professor; M.D., Ottawa. Assessment of immigrants from developing countries; epidemiology of social, sexual, and IDU networks.

Francis A. Plummer, Professor; M.D., Manitoba. Ecology and epidemiology of *Neisseria gonorrhoeae;* immune response to *N. gonorrhoeae;* transmission dynamics, national history, and perinatal transmission of HIV-1 in Africa.

Stuart Rosser, Assistant Professor; M.D., Alberta. Clinical trials in HIV therapeutics and critical care; quality improvement in HIV care.

Allan R. Ronald, Professor Emeritus; M.D., Manitoba.

Alberto Severini, Adjunct Professor; M.D., Parma (Italy). Molecular mechanisms of herpesvirus replication; assays for DNA-binding proteins; molecular diagnostics of HSV-1, HSV-2, papilloma, and polioma viruses.

Meenu Sharma, Adjunct Professor; Ph.D., Panjab (India). Mycobacterial pathogenicity, with special interest in mycobacterial cell wall and host-immune response interaction; characterizing differences in protein expression profiles between prevalent strains of *M. tuberculosis* to provide better understanding of role of mycobacterial virulence determinants.

Graham A. Tipples, Adjunct Professor; Ph.D., Manitoba. Viral *Exanthemata* (i.e., measles, rubella, human herpes virus 6): Molecular and serological diagnostics; molecular epidemiology.

Paul Van Caeseele, Assistant Professor; M.D., Manitoba. Blood cultures; detection of resistant microorganisms; influenza; necrotising enterocolitis; First Nations/Inuit Health.

Stefan Wagener, Adjunct Professor; Ph.D., Germany. Survival and inactivation of infectious agents on environmental surfaces; assessing personal protective equipment, engineering controls, and chemical agents in their ability to control biohazards; developing new approaches to biological safety.

Gehua Wang, Adjunct Professor; M.D., Shandong (China). Molecular biology and pathogenesis of enterobacteria; identification and characterization of *Campylobacter* virulence factors.

Hana Weingartl, Adjunct Professor; Ph.D., Guelph. Veterinary virology; viral pathogenesis on molecular/cellular level; diagnostic applications.

John L. Wylie, Assistant Professor; Ph.D., Manitoba. Molecular diagnostics; molecular epidemiology of sexually transmitted bacterial pathogens within sexual networks.

Xi Yang, Associate Professor; M.D., Shandong (China); Ph.D., Manitoba. Host defense against intracellular bacterial pathogens: role of T cell and cytokine.

Xiao-Jian Yao, Assistant Professor; M.D., China; Ph.D., Montreal. Molecular biology of HIV-1 replication; investigation of cellular and viral factors (including integrase and DNA flap element) contributing to HIV-1 PIC nuclear import and nondividing cell infection; development of prevention strategies against HIV-1 infection.

George Zhanel, Professor; Pharm.D., Minnesota; Ph.D., Manitoba. Antibiotic resistance, including epidemiology, treatment, and study of cellular and molecular mechanisms of resistance.

UNIVERSITY OF MARYLAND

School of Medicine
Graduate Program in Molecular Microbiology and Immunology

Program of Study

The Graduate Program in Molecular Microbiology and Immunology offers a program leading to the Ph.D. degree that is designed to provide training in bacteriology, virology, parasitology, and immunology. The department actively pursues studies of cellular oncogenes, mechanisms of biological signal transduction, bacterial and parasite genetics, bacterial pathogenesis, retrovirology, and cellular and molecular immunology. The aim of the program is to prepare students for careers in research and teaching.

The required core curriculum includes courses in immunology, virology, bacteriology, parasitology, molecular biology, cell biology, and biochemistry. Students elect additional advanced courses in their particular areas of interest. A minimum of 38 semester hours of course work, with a minimum grade point average of 3.0, is required. During the first three semesters, a series of laboratory rotations familiarizes students with faculty research interests. Following completion of the core curriculum in the third semester of study, students must pass a qualifying examination prior to admission to candidacy for the Ph.D. degree. Upon the successful completion of the qualifying examination, students are accepted into the laboratory of a faculty adviser to begin dissertation research and develop independent benchtop skills. Completion of the Ph.D. program usually requires five years.

Research Facilities

The laboratories are housed in several modern, spacious buildings that are well-equipped for all phases of program research, including cell culture and the handling of infectious microorganisms. Laboratory facilities include a high-containment laboratory dedicated to AIDS-related research, a self-contained animal facility, and laboratories for biopolymer synthesis, hybridoma production, and cell sorting. Equipment includes automated peptide and oligonucleotide synthesizers, high-performance liquid chromatographs, a cesium irradiator, beta and gamma spectrometers, ultracentrifuges, automated enzyme-linked immunosorbent assay spectrophotometers, a fluorescence flow cytometer, and fluorescence microscopes. The department is an integral part of the Medical Biotechnology Center. The program includes members of the Medical Biotechnology Center, the Cancer Center, and the Center for Vaccine Development. In addition to housing a large number of bound volumes and current periodicals, it has access to more than thirty computer databases in the sciences and offers a computerized reference and bibliographic service.

Financial Aid

Financial aid is available in the form of a limited number of graduate student fellowships, which offer a stipend and remission of tuition and health insurance; research assistantships; and student loans.

Cost of Study

In 2006–07, tuition is $363 per credit hour for residents of the state of Maryland and $646 per credit hour for nonresidents.

Living and Housing Costs

Limited University housing is available. Most students live in apartments in the city or within a 30-minute commute. Costs ranged from $700 per month for a one-bedroom apartment to $1200 per month for two bedrooms in 2006.

Student Group

There are about 1,000 graduate health and professional students on the Baltimore campus. Currently, 43 are enrolled in the Ph.D. or M.D./Ph.D. program in microbiology and immunology.

Location

The School of Medicine is located on the University of Maryland, Baltimore campus. Situated in downtown Baltimore, the campus is close to a wide range of educational, cultural, and athletic activities. An extensive revitalization program for the city has made Baltimore an attractive tourist center. Most notable is the Inner Harbor area, which includes the National Aquarium, the Maryland Science Center, and numerous shops and restaurants. In addition, the Peabody Music Institute, the Walters Art Gallery, the Baltimore Symphony, and the Baltimore Museum of Art provide a rich cultural environment. The Chesapeake Bay is within easy driving distance of the campus.

The University

The University of Maryland School of Medicine, founded in 1807, is the fifth-oldest medical school in the country, and it represented the beginning of the development of the professional-school campus in Baltimore. Over the next 194 years, five other professional schools joined the Baltimore campus, including the Graduate School, which was established in 1918. More than 5,000 students are enrolled in various degree programs on campus. In 1984, consolidation of the Graduate Schools on the University of Maryland, Baltimore and the University of Maryland, Baltimore County campuses occurred. This consolidation was, in part, intended to facilitate interaction between faculty and students on both campuses.

Applying

Applicants to the program should have a strong undergraduate background in biology, chemistry, and mathematics. In addition to official transcripts of grades and three letters of recommendation, the program requires an undergraduate GPA of at least 3.0 (3.5 for graduate studies), a combined score of at least 1200 for the verbal and quantitative portions of the GRE General Test, and a 4.5 or higher score on the written portion of the GRE. For applicants whose native language is not English, a TOEFL score of at least 280 on the computer-based test is required. Students lacking certain prerequisites are sometimes admitted on a provisional basis. Applications must be completed by January 15 for admission to the following fall term.

Correspondence and Information

June Green
Graduate Program in Molecular Microbiology and Immunology
University of Maryland School of Medicine
660 West Redwood Street, HH Room 326
Baltimore, Maryland 21201-1559

Phone: 410-706-7126
Fax: 410-706-2129
E-mail: jgreen@umaryland.edu
Web site: http://microbiology.umaryland.edu

University of Maryland

THE FACULTY AND THEIR RESEARCH

Sergei P. Atamas, Assistant Professor of Medicine; M.D., Crimea State Medical (Ukraine), 1984; Ph.D., Moscow Medical (Russia), 1989. Inflammatory and immune mechanisms of tissue fibrosis; role of T cells and macrophages in pulmonary fibrosis; profibrotic and antifibrotic cytokines and chemokines.

Abdu F. Azad, Professor of Microbiology and Immunology; Ph.D., Johns Hopkins, 1976. Immunological and molecular aspects of host-parasite interactions.

Eileen Barry, Associate Professor of Medicine and of Microbiology and Immunology, Ph.D., Virginia Commonwealth, 1991. Development of live attenuated strains of bacterial pathogens, including *Salmonella* and *Shigella*, as vaccines and their use as live vectors for the expression of heterologous antigens.

Patrik Bavoil, Associate Professor of Oral and Craniofacial Biological Sciences; Ph.D., Berkeley, 1982. Pathogenesis of the obligate intracellular pathogen, *Chlamydia*, and its bacteriophages.

Nicholas H. Carbonetti, Associate Professor of Microbiology and Immunology; Ph.D., Leicester (England), 1985. Gene regulation of virulence factors of *Bordetella pertussis*; bacterial signal transduction; molecular pathogenesis; vaccine development.

Gregory B. Carey, Assistant Professor of Microbiology and Immunology; Ph.D., Virginia Commonwealth, 1995. Understanding the nature of B-cell receptor propagated signals that determine life and death outcomes of both normal and cancerous B cells.

Jan Cerny, Professor and Chairman of Microbiology and Immunology; M.D., Karlova (Prague), 1962; Ph.D., Czechoslovak Academy of Sciences, 1967. Regulation of immune responses; functions and the repertoire of anti-idiotypic T cells; immunological changes in aging.

Alan Cross, Professor of Medicine; M.D., Pennsylvania, 1970.

Shiladitya DasSarma, Professor, University of Maryland Biotechnology Institute; Ph.D., MIT, 1984.

Wendy Davidson, Associate Professor of Microbiology and Immunology; Ph.D., Australian National, 1977. Mechanisms of systemic autoimmunity and B-cell lymphomagenesis.

Anthony DeVico, Associate Professor, Institute of Human Virology and of Microbiology and Immunology; Ph.D., George Washington, 1992. HIV pathogenesis and chemokine systems; virus–host cell interactions; HIV vaccines.

Michael S. Donnenberg, Professor of Medicine and of Microbiology and Immunology; M.D., Columbia, 1983. Molecular basis of cellular invasion by enteropathogenic and uropathogenic *Escherichia coli*.

Donna Farber, Associate Professor of Department of Surgery and Microbiology and Immunology; Ph.D., California, 1990. Immunological memory: mechanisms for the generation, function, and maintenance of T-cell memory; signal transduction pathways in naive, effector, and memory CD4 T-cell subsets.

Ricardo Feldman, Associate Professor of Microbiology and Immunology; Ph.D., NYU, 1979. Role of cellular and viral oncogenes in the control of cell proliferation and differentiation; RNA tumor viruses.

Martin F. Flajnik, Professor of Microbiology and Immunology; Ph.D., Rochester, 1983. Evolution of the immune system.

Ashraf Fouad, Chairman, Department of Endodontics, Prothodontics and Operative Dentistry, and Director of the Postgraduate Endodontics Program; D.D.S., Iowa, 1992.

Robert Gallo, Professor, Medical Biotechnology Center, Medicine, and Microbiology and Immunology; Director, Institute for Human Virology; and Head of Tumor Biology, Greenbaum Cancer Center; M.D., Jefferson Medical, 1963. Pathogenesis of HIV-induced AIDS and AIDS malignancies, especially Kaposi's sarcoma; HIV molecular biology; biological approaches to the control of HIV infection and replication; herpes viruses 6, 7, and 8; other human retroviruses, including HTLV-1 and HTLV-2; discovery of novel human viruses.

Alfredo Garzino-Demo, Assistant Professor, Institute of Human Virology and Microbiology and Immunology; Ph.D., Torino (Italy), 1988. Mechanism of HIV inhibition by chemokines; role of chemokines in Th1/Th2 responses; structural studies on chemokines; role of chemokines in inflammatory diseases of the central nervous system.

Anthony Gaspari, Professor and Chair of Dermatology; M.D., Thomas Jefferson.

Patricia Gearhart, Associate Professor of Microbiology and Immunology; Ph.D., Pennsylvania, 1974. Mechanisms that introduce mutations into the variable gene and identifying DNA sequences near the variable gene.

Bret Hassel, Associate Professor of Microbiology and Immunology; Ph.D., Johns Hopkins, 1989. Molecular mechanisms responsible for biological activities of interferons.

Dhan V. Kalvakolanu, Professor of Microbiology and Immunology; Ph.D., Bangalore (India), 1988. Molecular mechanisms of actions of cytokines such as interferons and interferon-inducible gene products; synergistic antiproliferation action of retinoids, tamoxifen, and interferons on human tumor cells.

Roberta M. Kamin-Lewis, Associate Professor of Microbiology and Immunology; Ph.D., Berkeley, 1974. Regulation of T-cell activation; pathways of T-cell infection by HIV; HIV vaccine development.

James B. Kaper, Professor of Microbiology and Immunology, Medicine, and Biological Chemistry; Ph.D., Maryland, 1979. Molecular pathogenesis of diarrheagenic bacteria; development of vaccines and DNA probes.

Achsah Keegan, Professor of Microbiology and Immunology; Ph.D., Johns Hopkins, 1989. Cytokine receptor signaling: protection from apoptosis; regulation of allergy and asthma by IL-4; control of myeloid development and function.

Myron M. Levine, Professor of Medicine, Pediatrics, and Microbiology and Immunology; M.D., Medical College, Virginia, 1967; D.T.P.H., London, 1974. Pathogenesis of bacterial diarrheas; enteric vaccine development; field epidemiological studies of bacterial enteric infections.

George K. Lewis, Professor of the Institute of Human Virology and of Microbiology and Immunology; Ph.D., Mississippi, 1974. Immunogenicity of lentivirus envelope proteins; mechanisms of virus entry and neutralization.

Ferenc Livák, Assistant Professor of Microbiology and Immunology; M.D., Semmelweis (Hungary), 1988. Mechanism of antigen receptor repertoire formation; control of T-cell receptor gene rearrangement; developmental regulation of T-lymphocyte differentiation.

Volker Mai, Assistant Professor of Epidemiology and Preventive Medicine; Ph.D., Georgia, 1999; M.P.H., Harvard, 2000. Diet, composition of the microflora, and carcinogenesis; animal and human cancer prevention studies; microbial causes of diarrhea; transmission of antibiotic resistance genes in the gut.

Dean Mann, Professor of Pathology; M.D., St. Louis, 1963. Dendritic cell biology; antigen processing and presentation; tolerance induction; cancer vaccines.

Andrei Medvedev, Assistant Professor of Microbiology and Immunology; Ph.D., Moscow, 1989. Mechanisms of tolerance to LPS and mycobacterial components.

Gregory Melikian, Associate Professor, Institute of Human Virology; Ph.D., Moscow State, 1984.

Kamal Moudgil, Associate Professor of Microbiology and Immunology; M.D., 1983, Ph.D., 1989, New Delhi (India). Antigen processing and presentation, self-tolerance, and shaping of the self-directed T-cell repertoire; pathogenesis and regulation of autoimmune arthritis.

James P. Nataro, Professor of Pediatrics, Medicine, and Microbiology and Immunology; M.D./Ph.D., Maryland, 1987. Molecular pathogenesis of diarrheagenic bacteria; molecular diagnostics; vaccine development.

Diana Oram, Assistant Professor of Biomedical Sciences; Ph.D., Emory, 1999.

Yuko Ota, Assistant Professor, Department of Microbiology and Immunology; Ph.D.

Zeev Pancer, Assistant Professor, Center of Marine Biotechnology, UMBI; Ph.D.

Macela Pasetti, Assistant Professor and Chief, Applied Immunology Unit of Pediatrics; Ph.D.

C. David Pauza, Professor of the Institute of the Human Virology and Microbiology and Immunology, Ph.D.; Berkeley, 1981. Role of chemokines in mucosal immunity; biology of gamma delta T cells.

Christopher Plowe, Associate Professor, Department of Medicine and of Microbiology and Immunology; M.D., Cornell, 1986; M.P.H., Columbia, 1991. Malaria.

Marvin S. Reitz, Professor of the Institute of Human Virology and of Microbiology and Immunology; Ph.D., Purdue, 1970. Human herpes virus 8 (HHV-8) and its role in human diseases.

Suzana Rudulovic, Associate Professor of Microbiology and Immunology; M.D., 1986, Ph.D., 1992, Ljubljana (Slovenia). Molecular pathogenesis and genetics of the obligate intracellular *Rickettsia;* identification and characterization of virulence-associated genes and hemoysins; DNA vaccine development against *R. prowazekii*.

Abdul Ruknudin, Assistant Professor of Microbiology and Immunology; Ph.D., Madras (India), 1982. Role calcium ions play in the regulation of cell physiology.

John B. Sacci, Assistant Professor of Microbiology and Immunology; Ph.D., Maryland, 1989. Immunoparasitology and molecular parasitology; aspects of host-parasite interactions.

Maria S. Salvato, Professor, Institute of Human Virology; Ph.D., Berkeley, 1979. Pathogenesis of arenavirus hemorrhagic fever and arenavirus vaccines; mechanisms of virus-mediated cell death in AIDS; use of animal models and genomic/proteomic approaches to analyze virus/host interactions.

Dan H. Schulze, Associate Professor of Microbiology and Immunology; Ph.D., Texas at Austin, 1978. Ca^{2+} regulation in lymphocytes; gene expression during B-cell development.

David Scott, Professor of Surgery; Ph.D., Yale, 1969. Gene therapy for tolerance in autoimmune diseases: mechanisms of tolerance.

Mark Shirtliff, Assistant Professor of Biomedical Sciences; Ph.D., Texas Medical Branch, 2001. Bacterial biofilm study through differential gene expression (via microarrays) and proteomic (via 2-D gel electrophoresis and subsequent MALDI-TOF MS) analyses to identify vaccine candidates, novel diagnostic strategies, and antimicrobial targets that are tested in animal models of biofilm infection, including osteomyelitis, implant infection, and abscess.

Mark Strauch, Associate Professor of Biomedical Sciences; Ph.D.

Scott Strome, Chair of Otorhinolaryngology–Head and Neck Surgery; M.D., Harvard, 1991.

Marcelo B. Sztein, Professor of Pediatrics and Medicine and Associate Professor of Microbiology and Immunology; M.D., Buenos Aires, 1976. Immune-mediated mechanisms of protection against microbial infections; vaccine development; host-parasite interactions.

Stefanie N. Vogel, Professor of Microbiology and Immunology; Ph.D., Maryland, 1977. Cellular and molecular mechanisms of macrophage activation; initiation of innate immunity by microbial structures; molecular biology of host responsiveness to endotoxin.

Mark Williams, Assistant Professor of Microbiology and Immunology; Ph.D., Michigan, 1991. Redox regulation of signal transduction and gene expression.

Richard Zhao, Associate Professor of Pathology; Ph.D. Oregon State, 1991. Molecular, cellular, and virologic characterizations of HIV-1 viral protein R (Vpr) interactions with host cellular responses.

UNIVERSITY OF MASSACHUSETTS AMHERST

Department of Microbiology

Programs of Study

The Department of Microbiology at the University of Massachusetts Amherst offers programs of graduate study leading to the M.S. and Ph.D. degrees in microbiology. Postdoctoral training is also available. Courses covering various areas in the field of microbiology are offered by the departmental faculty, which is listed on the reverse side of this page.

In the Ph.D. program, formal course work is generally completed during the first two years. After the first year, an increasing proportion of time is dedicated to research. Students select dissertation problems from a wide spectrum of research areas pursued by the faculty. The following research fields are represented: bioinformatics, physiology, genetics, immunology, parasitology, pathogenic bacteriology, molecular biology, and microbial ecology with special opportunities in environmental microbiology. In addition, close ties are maintained with the Departments of Biochemistry, Biology, and Chemistry through the interdepartmental program for doctoral training in molecular and cellular biology. In the second year, Ph.D. candidates must pass a comprehensive preliminary examination. Degree requirements are completed by submission and defense of a dissertation. There is no foreign language requirement. Completion of the Ph.D. program generally takes four years beyond the bachelor's degree.

Research Facilities

The Department of Microbiology occupies space in the Morrill Science Center. Air-conditioned laboratories are spacious and well equipped for research and teaching. The modern apparatus necessary for investigation into all aspects of microbiology is available within departmental space. The department's facilities include tissue- and cell-culture laboratories, animal quarters, and various instrument rooms containing preparative and analytical ultracentrifuges, scintillation counters, fermentors, anaerobic chambers, equipment for DNA sequence analysis and chromatographic and electrophoretic procedures, photography, and other standard laboratory procedures. Centralized facilities provide state-of-the-art equipment and expertise to support research projects, such as the Central Microscopy Facility, Phosphorimager, Genomics and Bioinformatics Facility, High Field NMR Facility, and the Mass Spectrometry Facility.

Financial Aid

Financial aid is available in the form of University fellowships and teaching assistantships. Research assistantships are available for advanced graduate students. All assistantships include a waiver of tuition.

Cost of Study

For the academic year 2005–06, annual tuition for in-state residents was $110 per credit; nonresident tuition was $414 per credit. Full-time students register for at least 9 credits per semester. The mandatory fee assessed for full-time graduate students was $3458 per semester for in-state residents and $4034 for nonresidents. Fees are subject to change.

Living and Housing Costs

Graduate student housing is available in several twelve-month campus residence halls through University Housing Services. The University owns and manages 345 unfurnished apartments of various sizes for family housing on or near the campus. Off-campus housing is available; rents vary widely and depend on factors such as size and location.

Student Group

The department has approximately 35 graduate and 100 undergraduate students as well as 35 postdoctoral fellows. Enrollment at the Amherst campus is about 24,000, including 6,000 graduate students.

Location

The 1,450-acre campus of the University provides a rich cultural environment in a rural setting. Amherst is situated in the picturesque Pioneer Valley in historic western Massachusetts. The area is renowned for its natural beauty. Green open land framed by the outline of the Holyoke Range, clear streams, country roads, forests, grazing cattle, and shade trees are characteristic of the region. A broad spectrum of cultural activities and extensive recreational facilities are available within the University and at four neighboring colleges—Smith, Amherst, Mount Holyoke, and Hampshire. Opportunities for outdoor winter sports are exceptional. Amherst is 90 miles west of Boston and 175 miles north of New York City, and Cape Cod is a 3½-hour drive away.

The University

The University of Massachusetts is the state university of the Commonwealth of Massachusetts and is the flagship campus of the five-campus UMass system. Departments affiliated with the ten colleges and schools of the University offer a variety of graduate degrees through the Graduate School. The Amherst campus consists of approximately 150 buildings, including a twenty-eight-story library that is the largest at a state-supported institution in New England, with more than 5.7 million items

Applying

Application forms may be obtained from the Graduate Admissions Office, 530 Goodell Building, University of Massachusetts, 140 Hicks Way, Amherst, Massachusetts 01003-9333, or online at http://www.umass.edu/gradschool. Prospective students are required to take the Graduate Record Examinations. Applications for admission should be received by the Graduate Admissions Office by December 1 for September enrollment and by October 1 for January enrollment. Applications received after these dates are considered only if space is available.

Correspondence and Information

Graduate Program Director
Department of Microbiology
Morrill IV, N203
639 North Pleasant Street
University of Massachusetts
Amherst, Massachusetts 01003-9298

Phone: 413-545-2051
Fax: 413-545-1578
E-mail: microbio-dept@microbio.umass.edu
Web site: http://www.bio.umass.edu/micro/

University of Massachusetts Amherst

THE FACULTY AND THEIR RESEARCH

S. J. Sandler, Associate Professor and Interim Department Head; Ph.D., Berkeley. Molecular genetics of recombination; DNA replication and DNA repair in bacteria. *Mol. Microbiol.* 57:1074, 2005. *Mol. Microbiol.* 53:1343, 2004.

C. L. Baldwin, Adjunct Professor; Ph.D., Cornell. Cellular immunity to intracellular microbial parasites, including *Brucella abortus:* particular interest in the interaction of the microbe with macrophages and the control of infection by T-cell cytokines; stimulation and control of gamma/delta T-cell responses. In *Brucella: Molecular and Cellular Biology,* Norfolk, United Kingdom: Horizon Scientific Press. *J. Immun.* 174:3386–93.

J. Blanchard, Assistant Professor; Ph.D., Georgia. Evolution of cellular networks; Modeling cellular processes using genomic data; graphical representation of biological pathway data. *Evolution* 57:1959–72, 2003. *Trends Genet.* 16:315–20, 2000. *Science* 283:404–06, 1999.

J. P. Burand, Adjunct Associate Professor; Ph.D., Washington State. Biology and molecular biology of insect pathogenic viruses, particularly baculoviruses and nonoccluded insect viruses, with emphasis on virus-host interactions that affect the virulence and persistence of these viruses in insects. *J. Invertebr. Pathol.* 85:128–31, 2004. *J. Invertebr. Pathol.* 81:33–44, 2002. *J. Invertebr. Pathol.* 80:81–9, 2002. *Virus Genes* 23:17–25, 2001.

D. R. Cooley, Adjunct Associate Professor; Ph.D., Massachusetts. Ecology of diseases; plant pathogenic fungi and bacteria; plant disease management; integrated pest management; development of sustainable agricultural systems.

S. Goodwin, Associate Dean, College of Natural Resources and the Environment; Ph.D., Wisconsin.

J. F. Holden, Assistant Professor; Ph.D., Washington (Seattle). Physiology and ecology of hyperthermophilic archaea; geomicrobiology of deep-sea hydrothermal environments. *AGU Geophys. Monogr.* 144:13–24, 2004. *J. Am. Soc. Mass Spectrom.* 14:957–70, 2003. *Curr. Opin. Chem. Biol.* 7:160–5, 2003.

L. A. Katz, Adjunct Professor; Ph.D., Cornell. Evolution of eukaryotic microbes, particularly genome evolution, molecular systematics, and phylogeography; specific taxa include ciliates and amoeboid eukaryotes. *Trends Ecol. Evol.* 19:32–8, 2004. *J. Eukaryotes Microbiol.* 51:441–50, 2004. *Mol. Biol. Evolution* 21:555–62, 2004.

M. M. Klingbeil, Assistant Professor; Ph.D., Toledo. Molecular and biochemical parasitology, replication and repair of mitochondrial DNA (kinetoplast DNA network) in African *typanosomes*. *Science* 309:409–15, 2005. *Proc. Natl. Acad. Sci. U.S.A.* 101:4333–4, 2004. *J. Biol. Chem.* 278:49095–101, 2003. *Mol. Cell* 10:175–86, 2002. *Protist* 152:255–62, 2001.

G. R. Lanza, Professor and Director, Environmental Sciences Program; Ph.D., Virginia Tech. Microbial ecology and physiology; bioremediation/phytoremediation.

S. B. Leschine, Professor; Ph.D., Pittsburgh. Microbial physiology and ecology; cellulose decomposition; biofilms on natural polymers; enzymology of cellulose and chitin hydrolysis; fuels from biomass. *Handbook on Clostridia,* pp. 101–31, ed. P. Durre., CRC Press, 2005. *Arch. Microbiol.* 180:434–43, 2003 (with G. Reguera). *Int. J. Syst. Evol. Microbiol.* 52:1155–60, 2002 (with Warnick). *FEMS Microbiol. Lett.* 204:367–74, 2001 (with Reguera). *Int. J. Syst. Evol. Microbiol.* 51:123–32, 2001 (with Monserrate and Canale-Parola).

S. C. Long, Adjunct Professor of Microbiology and Associate Professor of Civil Environmental Engineering; Ph.D., North Carolina at Chapel Hill. Development and field testing of microbial source tracking tools for watershed and source water protection applications; assessment of land use impacts on microbial risk in drinking waters; evaluation of microbial safety of biosolids; evaluation of novel disinfection processes for application to small systems and/or the developing world. *Appl. Environ. Microbiol.* 70(10):5996–6004, 2004. *J. Water Supply: Res. Tech.-AQUA* 52:565–76, 2003.

D. R. Lovley, Distinguished University Professor; Ph.D., Michigan State. Genome-enabled study of the physiology and ecology of novel anaerobic microorganisms; environmental genomics; bioremediation of metal and organic contamination; microbial fuel cells; in silico cell modeling; life in extreme environments. *Nature* 435:1098–101, 2005. *Nature Microbiol. Rev.* 1:35–44, 2003. *Science* 301:934, 2003. *Nature* 416:767–9, 2002. *Science* 295:483–5, 2002. *Nature* 415:312-6, 2002.

W. J. Manning, Adjunct Professor; Ph.D., Delaware. Effects of ozone on plants and associated mycoflora; plants as bioindicators of ozone; effects of ozone and other air pollutants on plants in urban environments; managing invasive plants with fungal pathogens. *Environ. Pollut.* 126:73–81, 2003.

L. Margulis, Adjunct Professor and Distinguished University Professor. Evolution of microbes; symbiogenesis and the origin of the nucleus; evolution of cells. *Proc. Natl. Acad. Sci. U.S.A.* 31:175–91, 2005. *Acquiring Genomes: A theory of the origins of species,* 240 pp., 2002. *Proc. Natl. Acad. Sci. U.S.A.* 99:1410–3, 2002. *Proc. Natl. Acad. Sci. U.S.A.* 93:1071–6, 1996. *Symbiosis in Cell Evolution: Microbial communities in the Archean and Proterozoic eons,* 2nd edition, 452 pp., 1993.

L. C. Norkin, Professor; Ph.D., Columbia. Virology and obligate intracellular bacterial parasites: entry and intracellular trafficking and biology of the DNA tumor virus SV40 and the chlamydiae; interactions of microbial pathogens with caveolae and lipid "rafts." *BMC Virol. J.* 2:38, 2005. *BMC Infect. Dis.* 4:23, 2004. *Exp. Cell Res.* 287:67–78, 2003. *J. Virol.* 76:5156–66, 2002. *Exp. Cell Res.* 226:229–308, 2001. *Adv. Drug Deliv. Rev.* 49:301, 2001. *Exp. Cell Res.* 246:83–90, 1999. *Immunol. Rev.* 168:1322, 1999. *J. Gen. Virol.* 79:1469–77, 1998. *Mol. Biol. Cell* 7:1825–34, 1996.

K. Nüsslein, Assistant Professor; Ph.D., Michigan State. Microbial ecology of terrestrial and aquatic environments; relating the stress of environmental influences to community structure and function, with emphasis on understanding interactions among bacterial communities. *Appl. Environ. Microbiol.* 71:2484–92, 2005. *J. Polym. Sci.* 42:3860–4, 2004. *Langmuir* 19:6226–9, 2003. *Appl. Environ. Microbiol.* 65:3622–6, 1999. *Appl. Soil Ecol.* 13:109–22, 1999.

S. T. Petsch, Adjunct Assistant Professor; Ph.D., Yale. Transport, transformation, and biodegradation of natural organic matter in sediments, soils, and sedimentary rocks. *Palaeogeog. Palaeoclim. Palaeoecol.* 219:157–70, 2005. *Mar. Chem.* 92:353–66, 2004. *Org. Geochem.* 34:731–43, 2003.

P. J. Pomposiello, Assistant Professor; Ph.D., Michigan. Global transcriptional regulation in response to environmental stress. *Protein Sci.* 14(6):1673–8, 2005 (with Osborne, Siddiqui, Landgraf, and Gehring). *J. Bacteriol.* 185(22):6624–32, 2003 (with Carrasco, Koutsolioutsou, and Demple).

M. A. Riley, Adjunct Professor; Ph.D., Harvard. Microbial ecology and evolution and experimental evolution. *J. Evolutionary Biol.* 16:1236–8, 2004. *Nature* 428:412–4, 2004. *Nature* 418:171–4, 2002.

E. Stuart, Associate Professor; Ph.D., Illinois. Infection, immunity, and pathogenesis with especial interest in processes used by chlamydial species for entry into host cells and replication; generation, display, and utility of recombinant vaccine components targeting viral or bacterial peptides; vaccine components for use with *Chlamydia. Curr. Microbiol.* 49(1):13–21, 2004. *BMC Infect. Dis.* 4(1):23, 2004.

L. Sun, Assistant Adjunct Professor; Ph.D., Caltech. Protein engineering, biosynthesis of natural products, and complex biological systems. *ChemBioChem.* 3:781–3, 2002. *Protein Engineering* 14:699–704, 2001.

W. Webley, Assistant Research Professor; Ph.D., Massachusetts. Immunology and pathogenic bacteriology; elucidating entry pathways, survival, and host range of the obligate intracellular pathogens, *Chlamydiaceae;* developing novel antigen display and vaccine delivery systems for *Chlamydia* and other pathogenic bacteria of interest to community health; elucidating the role of *Chlamydia* in asthma and Thrombotic Thrombocytopenic Purpura (TTP). *Am. J. Respir. Crit. Care Med.* 171(10):1083–8, 2005. *BMC Infect. Dis.* 4(1):23, 2004 (with Stuart and Norkin). *Curr. Microbiol.* 49(1):13–21, 2004. *Am. J. Respir. Crit. Care Med.* 169(7):A586, 2004. *J. Clin. Apheresis* 18(2), 2003. *Exp. Cell Res.* 287(1):67–78, 2003 (with Stuart and Norkin).

R. M. Weis, Adjunct Professor; Ph.D., Stanford. Signaling transduction in the chemotaxis system of *E. coli. J. Bacteriol.* 186:7556–63, 2004. *J. Bacteriol.* 185:3636–43, 2003. *Biochim. Biophys. Acta* 1596:28–35, 2002. *Cell* 100:357–65, 2000.

R. L. Wick, Adjunct Professor; Ph.D., Virginia Tech. Disease diagnostics; diseases caused by Pythium and phytophthora; nematode diseases of turfgrasses. 2005–06 Fulbright Scholar to teach plant pathology in Bangladesh. In *Northeast Pepper Integrated Pest Management Manual,* pp. 51–7, 2001. In *Encyclopedia of Plant Pathology,* vols. 1 and 2. New York: John Wiley & Sons, 2001 (six entries). In *Diseases of Woody Ornamentals and Trees in Nurseries,* APS Press, 2001 (six chapters). *HortScience* 31:851–4, 1996 (with Boyle). *Compendium of Flowering Potted Plant Diseases,* 90 pp., 1995 (with Daughtrey and Peterson). *Adv. Plant Pathol.* 10:63–126, 1993 (with Hansen).

UNIVERSITY OF MASSACHUSETTS WORCESTER

Graduate School of Biomedical Sciences
Program in Immunology-Virology

Program of Study

The Program in Immunology-Virology (IVP) in the Graduate School of Biomedical Sciences at the University of Massachusetts Medical School at Worcester offers a program of study and research leading to a Ph.D. degree in medical sciences. The faculty members of the program have varied research interests, including molecular and cellular immunology, viral immunology, molecular virology, rheumatology, autoimmune diabetes, transplantation, cytokine induction, and signal transduction. The program teaches molecular and cellular immunology as well as virology and strongly emphasizes basic research and the training of graduate students and postdoctoral fellows.

Incoming students should have good preparation in the fundamentals of biochemistry, genetics, and biology. Formal course requirements for the Ph.D. include a core curriculum encompassing the disciplines of biochemistry, molecular biology, and cell biology. The IVP offers several specialization courses in immunology or virology. Students should undertake three to four laboratory rotations prior to selecting a laboratory for thesis research. In addition to taking formal course work and conducting research, students are expected to attend departmental and medical school seminars and to participate in a weekly seminar program. Admission to candidacy occurs following successful completion of a qualifying exam administered at the conclusion of the second year. The Ph.D. degree is awarded following completion and successful defense of thesis research. The training received through this program is intended to provide students with the academic background, experimental tools, and analytical skills necessary to conduct independent research and teach modern immunology and virology.

Research Facilities

The Medical School was constructed in 1974 and is generously endowed with office, classroom, and laboratory space. The program's research facilities are extensive, and the most recent addition was a 360,000-square-foot, ten-story research laboratory building, which opened in 2001. Facilities and equipment are the most modern available for biomedical research and include core facilities for protein sequencing, nucleic acid synthesis, electron microscopy, tissue culture, fluorescence cell sorting, BL3 virus laboratories, and transgenic mouse technology. A large and well-maintained experimental animal-care facility and modern computer and electronics services are available to all investigators.

Financial Aid

In 2004–05, beginning full-time students were offered a graduate assistantship at an annual stipend of $24,500. Students are granted similar stipends as research assistantships after they choose a lab for dissertation research. Tuition is waived for both graduate and research assistants.

Cost of Study

All full-time students are supported with assistantships and tuition waivers. Student health, dental, and hospitalization insurance is provided by the University, as is a disability insurance plan.

Living and Housing Costs

Apartments and rooms are available in the Worcester area and nearby communities. The large number of students in the area makes the sharing of apartments feasible at reasonable costs. The Housing Office at the medical school assists students in finding appropriate places to live.

Student Group

The Program in Immunology-Virology has 39 graduate students enrolled in the Ph.D. program plus 84 postdoctoral fellows. The total enrollment of the Graduate School is 342 students, of whom 32 are Ph.D./M.D. candidates.

Location

Worcester is the second-largest city in Massachusetts and is 40 miles west of Boston. Worcester and the surrounding towns offer varied living conditions and a wide range of educational and cultural activities. Located in the city are numerous other educational institutions, including Assumption College, Clark University, Holy Cross College, Worcester Polytechnic Institute, and Worcester State College.

The University and The School

The University of Massachusetts has five campuses: the main one at Amherst, about 40 miles west of Worcester; the Medical School in Worcester; and campuses in Boston, Dartmouth, and Lowell. The Medical School opened in 1970; there are now 100 medical students in each class. The Medical School joins regional institutions and the University of Massachusetts at Amherst in the training of other health-care professionals. The Ph.D. program began in 1979 and became the Graduate School of Biomedical Sciences in 1986.

Applying

The program accepts graduate students through the Graduate School of Biomedical Sciences. Applicants should have a bachelor's degree in a physical or biological science. Specific course requirements for admission include a year each of calculus, organic chemistry, physics, and biology. Applicants must take the General Test of the Graduate Record Examinations. Three letters of recommendation, original transcripts, and an application fee ($25 for Massachusetts residents, $50 for non-Massachusetts residents) are required. The deadline for fall applications is January 2. The University of Massachusetts Medical School is an Equal Opportunity/Affirmative Action employer. The University of Massachusetts Medical School recruits minority and women candidates, considers all applicants, and, once students are accepted, ensures that they will not be discriminated against in any area.

The graduate catalog and application forms may be obtained from the address listed in this In-Depth Description. Potential applicants may wish to contact Susan Foley at 508-856-4230 for additional information about the program.

Correspondence and Information

Graduate School of Biomedical Sciences
University of Massachusetts Worcester
Worcester, Massachusetts 01655
Phone: 508-856-4135
E-mail: susan.foley@umassmed.edu (for immunology-virology specialization)
Web site: http://www.umassmed.edu/ivp/

University of Massachusetts Worcester

THE FACULTY AND THEIR RESEARCH

Elliot J. Androphy, M.D., Professor. Papillomavirus transcription and oncogenic transformation. *Mol. Cell. Biol.* 24:2144–52, 2004.

Leslie J. Berg, Ph.D., Professor. Signal transduction proteins involved in T-cell development and activation. *Immunity* 21:67–80, 2004.

Rita Bortell, Ph.D., Associate Professor. Mechanisms of immunoregulation in modulation of autoimmune diabetes. *Proc. Natl. Acad. Sci. U.S.A.* 100:10382–7, 2003.

Cynthia A. Chambers, Ph.D., Assistant Professor. Regulation of T-cell activation by costimulatory molecules. *Ann. Rev. Immunol.* 19:565–94, 2001.

Francis Ka-Ming Chan, Ph.D., Assistant Professor. Programmed cell death; tumor necrosis factor signaling in healthy and diseased immune systems. *Methods Mol. Biol.* 261:371–82, 2004; *J. Biol. Chem.* 278(51): 51613–21, 2003.

Jason J. Chen, Ph.D., Assistant Professor. Regulation of cell-cycle checkpoint and apoptosis by the papillomavirus oncogenes. *J. Biol. Chem.* 278:43163–8, 2003.

Paul R. Clapham, Ph.D., Associate Professor. HIV envelope, receptors, and cell tropism. *J. Virol.* 78:6915–26, 2004.

Ronald C. Desrosiers, Ph.D., Professor. Strategies of immune evasion and vaccine development for AIDS. *Nature Med.* 5:723–5, 1999.

Stephen J. Doxsey, Ph.D., Associate Professor. Mechanisms of autoimmune disease development; immune aspects of scleroderma pathogenesis; intracellular pathogens in autoimmunity. *Clin. Exp. Immunol.* 137:288–97, 2004.

Francis A. Ennis, M.D., Professor. Human immune responses to virus infections; identifying structure of epitopes recognized by human CD4+ and CD8+ CTL and function of these CTL in clearance of infected cells and in immunopathology. *J. Exp. Med.* 197:927–32, 2003.

Robert W. Finberg, M.D., Professor. Virus receptor proteins and host-cell pathogenesis; host-cell surface proteins and the basis of cellular activation events. *Proc. Natl. Acad. Sci. U.S.A.* 101(5):1315–20, 2004.

Kate A. Fitzgerald, Ph.D., Assistant Professor. IKKε and TBK1, which are essential components of the IRF-3 signaling pathway. *Nat. Immunol.* 4(5):491–6, 2003.

Rachel M. Gerstein, Ph.D., Assistant Professor. Developmental regulation of V(D)J recombination and lymphocyte development. *J. Exp. Med.* 196:705–11, 2002; *J. Biol. Chem.* 10.1074/jbc.M200305200, 2002 (online).

Douglas T. Golenbock, M.D., Professor. Toll receptors; pathophysiology of sepsis and pelvic inflammatory disease. *J. Exp. Med.*, in press.

Dale L. Greiner, Ph.D., Professor. Pathogenesis of autoimmune type 1 diabetes and induction of islet transplantation tolerance. *J. Clin. Invest.* 112:795–808, 2003.

Steven Grossman, M.D., Ph.D., Assistant Professor. The ubiquitin/proteasome system and cancer. *Oncogene* 23:4121, 2004; *Science* 300:342–4, 2003.

Ronald M. Iorio, Ph.D., Associate Professor. Viral pathogenesis; virus-induced fusion; viral glycoproteins and receptors. *J. Virol.* 78:5299, 2004.

Jae U. Jung, Ph.D., Professor. Herpesvirus immune evasion and pathogenesis. *Mol. Cell. Biol.* 24:5369–82, 2004; *J. Exp. Med.* 200:681–7, 2004.

Joonsoo Kang, Ph.D., Assistant Professor. Molecular circuits controlling T-lymphocyte development and lineage commitment; cytokines essential for lymphocyte development and function. *J. Immunol.* 15:2307, 2004; *Curr. Opin. Immunol.* 16:180, 2004.

Michelle A. Kelliher, Ph.D., Assistant Professor. Mechanism of tal-1-induced leukemogenesis; death receptor signaling in apoptosis. *J. Exp. Med.* 196:15–17, 2002; *Oncongene* 20:3897–905, 2001.

Hardy Kornfeld, M.D., Professor. Host defense against respiratory pathogens. *Infect. Immun.* 71:254–9, 2003.

Timothy F. Kowalik, Ph.D., Associate Professor. Cell growth control, DNA viruses; RNAi and virus infection. *Mol. Cell. Biol.* 24:2968–77, 2004; *Cell Cycle* 3:40–1, 2004.

Evelyn A. Kurt-Jones, Ph.D., Associate Professor. Pattern recognition receptors TLR2, TLR4, and CD14 and their role in innate immunity to viral and bacterial infection. *Proc. Natl. Acad. Sci. U.S.A.* 101(5):1315–20, 2004; *Infect. Immun.* 72(11):6446–54, 2004.

John M. Leong, M.D., Ph.D., Professor. Host-cell signaling and tissue colonization by bacterial pathogens. *Dev. Cell* 7:217–28, 2004; *Immunity* 3:379–90, 2004.

Egil Lien, Ph.D., Assistant Professor. Mechanisms for microbial activation and evasion of innate immune responses via toll-like receptors. *J. Biol. Chem.* 279:39727–35, 2004; *Nature Immunol.* 4:1162–4, 2003.

Shan Lu, M.D., Ph.D., Associate Professor. DNA immunization and its applications; immunogenicity of protein and peptide antigens. *Vaccine* 22:3348–57, 2004; *Vaccine* 22:1764–72, 2004.

Elizabeth J. Luna, Ph.D., Professor. Membrane-cytoskeleton interactions in motile cells, e.g., neutrophils and tumor cells. *J. Cell Sci.* 117:5043–57, 2004; *J. Cell Sci.* 116:2261–75, 2003.

Katherine Luzuriaga, M.D., Professor. Immunopathogenesis of persistent viral infections (CMV, EBV, HIV-1); HIV-1 vaccine development; SARS monoclonal antibodies. *New Engl. J. Med.* 350(24):2471–80, 2004; *J. Immunol.* 170(11):5786–92, 2003; *J. Immunol.* 170(5):2590–8, 2003.

Trudy G. Morrison, Ph.D., Professor. Enveloped, RNA virus entry and assembly; structure and function of viral glycoproteins. *J. Virol.* 77:12968, 2003; *J. Virol.* 77:11040, 2003.

Peter E. Newburger, M.D., Professor. Regulation of functionally important genes in phagocytic leukocytes; molecular mechanisms of translation of selenium-containing proteins. *J. Leukocyte Biol.* 75:358–72, 2004; *Proc. Natl. Acad. Sci. U.S.A.* 101:6508–13, 2004.

Kenneth L. Rock, M.D., Professor. Antigen presentation on MHC class I and class II molecules; development of vaccine technology. *Immunity* 21(2):155–65, 2004; *Nature* 425:516–21, 2003.

Aldo A. Rossini, M.D., Professor. Animal models used to study tolerance; loss of tolerance that occurs in autoimmunity and induction of tolerance for islet transplantation. *Diabetologia*, in press; *Proc. Natl. Acad. Sci. U.S.A.* 100:10382–7, 2003.

Alan L. Rothman, M.D., Professor. Immunity and pathogenesis of flaviviral infections; T-lymphocyte responses to dengue and hepatitis C viruses. *J. Clin. Invest.* 113:946–51, 2004; *J. Immunol.* 168:5959, 2002.

Madelyn R. Schmidt, Ph.D., Assistant Professor. Study of lymphocyte homeostasis and gene therapy for immunodeficiency diseases and cancer. *J. Immunol.* 169:6765, 2002.

Carol E. Schrader, Ph.D., Assistant Professor. Role of DNA repair pathways in antibody class switch recombination. *J. Exp. Med.* 200:321–30, 2004; *EMBO J.* 22:5893–903, 2003.

Liisa K. Selin, M.D., Ph.D., Associate Professor. Acute and memory CD8 T-cell responses to viruses in mice and humans. *Immunity* 20:5–16, 2004; *Nature Immunol.* 3:627–34, 2002.

Neal Silverman, Ph.D., Assistant Professor. Innate immunity: recognition and signaling in *Drosophila. Immunity* 20:637–49, 2004.

Merav Socolovsky, M.D., Ph.D., Assistant Professor. Molecular mechanisms regulating the homeostasis of hematopoietic and red-cell progenitors in vivo, using novel flow-cytometric methodology. *Blood* 98:3261–73, 2001; *Cell* 98:181–91, 1999.

Janet Stavnezer, Ph.D., Professor. Antibody class switch recombination and DNA repair; mutations that occur in the Ig Sμ region but rarely in Sγ regions prior to class switch recombination. *EMBO J.* 22:5893–903, 2003.

Lawrence J. Stern, Ph.D., Associate Professor. Molecular recognition in the immune system; antigen presentation by MHC proteins; T-cell activation mechanisms. *Proc. Natl. Acad. Sci. U.S.A.* 101(36):13279–84, 2004 (online).

Mario Stevenson, Ph.D., Professor. Studies of HIV/SIV pathogenesis; new targets for inhibitors of HIV-1 replication. *Nature* 424:213–9, 2003; *Nature Med.* 9:853–60, 2003.

John L. Sullivan, M.D., Professor. Virological and immunological studies of HIV-1 and Epstein-Barr virus and SARS. *N. Engl. J. Med.* 350(24):2471–80, 2004; *J. Virol.* 78(10):5194–204, 2004; *J. Virol.* 78(20):11429–33, 2004.

Gyongyi Szabo, M.D., Ph.D., Professor. Innate immune functions in hepatitis C viral infection and liver disease and its modulation by alcohol. *Hepatology* 40:555–64, 2004; *Hepatology* 40:376–85, 2004; *J. Immunol.* 173:1198–407, 2004.

Eva Szomolanyi-Tsuda, M.D., Assistant Professor. Pathogenesis of polyomavirus infection in mice; T-cell-independent and T-cell-dependent antiviral antibody responses. *Nature Immunol.* 3:112–4, 2002; *Nature* 412:300–7, 2001.

Masanori Terajima, M.D., Ph.D., Assistant Professor. Immunopathogenesis of human hantavirus infections, hantavirus pulmonary syndrome (HPS), and hemorrhagic fever renal syndrome (HFRS); host immune responses against vaccinia virus. *J. Gen. Virol.* 85(7):1909–19, 2004.

Raymond M. Welsh, Ph.D., Professor. Viral immunology and pathogenesis. *Immunity* 20:5–16, 2004; *J. Immunol.* 172:3139–50, 2004.

Robert T. Woodland, Ph.D., Associate Professor. B-cell development and homeostasis. *Int. Immunol.* 17:3819–27, 2003; *J. Immunol.* 169:6795–805, 2002; *J. Immunol.* 168:2712–19, 2002.

Maria L. Zapp, Ph.D., Assistant Professor. Molecular mechanisms of cellular and viral nuclear RNA export; small molecule inhibitors of HIV-1 RNA-protein interactions. *Genes Dev.* 18:23–34, 2004; *J. Immunol.* 170:5615–24, 2003.

Robert B. Zurier, M.D., Professor. Influence of prostaglandins, essential fatty acids, and cannabinoids on inflammation and immune responses. *Biochem. Pharmacol.* 65:649–55, 2003.

UNIVERSITY OF MEDICINE AND DENTISTRY OF NEW JERSEY

*Graduate School of Biomedical Sciences
at New Jersey Medical School, Newark
Department of Microbiology and Molecular Genetics*

Programs of Study

The Department of Microbiology and Molecular Genetics, associated with New Jersey Medical School at the University of Medicine and Dentistry of New Jersey (UMDNJ) campus in Newark, offers a program of graduate studies that leads to the Doctor of Philosophy degree through the Graduate School of Biomedical Sciences. Combined-degree programs (M.D./Ph.D. or D.M.D./Ph.D.) are also available in conjunction with the New Jersey Medical School or the New Jersey Dental School. Research opportunities are available in such fields as molecular biology of yeast, parasites, mammalian cells, and bacteria; genetic engineering; virology; mammalian cell genetics; and mechanisms of mutagenesis and carcinogenesis. The program emphasizes course work in molecular and cellular genetics, microbial pathogenesis, and advanced topics in molecular biology. Students undertake laboratory research in the first year. Full-time thesis research is begun in the second year, after students successfully complete qualifying examinations and submit a research proposal. Students broaden their educational experience by taking courses in other departments and by attending departmental seminars and journal clubs. They also participate in teaching a short medical school laboratory course. Upon completion of the program, the graduate is well prepared for a career combining research and teaching in molecular biology in an academic environment, a research institute, or industry.

Students can enter this program after completing the requirements for a bachelor's or a master's degree. Students rotate through several faculty laboratories during the first year and select a faculty adviser by the start of the second year. Normally, five years of work are needed to complete all the requirements for the Ph.D. degree.

Research Facilities

The Department of Microbiology and Molecular Genetics is located in the International Center for Public Health, close to the UMDNJ health sciences complex in Newark. The department is equipped with state-of-the-art tissue culture facilities, warm and cold rooms, and all the necessary equipment and services for modern molecular biology and genetic research. Core facilities are available for studies with transgenic animals, confocal imaging, proteomics, and genomics. In addition, the Molecular Resource Center provides automated DNA sequencing, oligonucleotide synthesis, peptide synthesis, protein sequencing, and imaging. The department offers extensive networked computing facilities. Individual laboratories are equipped with centrifuges, micromanipulators, and equipment for electrophoresis, chromatography, and autoradiography. The Central Animal Care Facility, George F. Smith Library of the Health Sciences, and Scientific Data Processing Center all are conveniently located near the department.

Financial Aid

The UMDNJ Graduate School of Biomedical Sciences offers merit scholarships and graduate fellowships that provide tuition, health insurance, and a twelve-month stipend ($24,000 in 2005–06). A special allocation from the New Jersey Department of Higher Education is available for ethnic minority students who require financial assistance.

Cost of Study

Modest fees are charged for such items as membership in the Graduate Student Association. (Students receive a stipend, tuition waiver, and health insurance).

Living and Housing Costs

UMDNJ has no on-campus housing. Many graduate students share apartments in neighboring communities from which UMDNJ is easily accessible by automobile or public transportation. On-campus parking is available. The Graduate School office may provide assistance in finding housing for students.

Student Group

About 60 students, approximately half of them women, are currently enrolled in Ph.D. programs in the UMDNJ Graduate School at Newark. Faculty members of the Department of Microbiology and Molecular Genetics provide considerable attention to each graduate student.

Location

The Graduate School is located in the University Heights section of Newark. A city of 400,000 people, Newark is a short distance from New York City. Diverse cultural, sports, and outdoor activities are available within the metropolitan area, in the nearby countryside, or at the seashore. Rutgers University (Newark campus) and the New Jersey Institute of Technology are situated only a few blocks away and offer cross-registration for graduate courses. Many other major graduate and professional institutions are within easy driving distance.

The University

The University of Medicine and Dentistry of New Jersey is a multicampus state institution under the administration of a Board of Trustees appointed by the governor. The Newark health sciences campus includes the Graduate School of Biomedical Sciences, the New Jersey Medical School, University Hospital, the New Jersey Dental School, the School of Allied Health Professions, the Public Health Research Institute, and the National TB Center at UMDNJ. Graduate students in microbiology and molecular genetics have opportunities to collaborate with faculty members in other basic science departments, in the Public Health Research Institute, and in clinical departments of the medical and dental schools.

Applying

Requirements for admission include a degree from an accredited college or university and a satisfactory academic record, scores on the GRE General Test, and letters of reference from former instructors. A personal interview may also be required. Early application is encouraged. Prospective applicants are invited to consult appropriate faculty members concerning their professional interests and the specific opportunities for graduate study.

Correspondence and Information

For program information:
Graduate Program Director
Department of Microbiology and Molecular Genetics
UMDNJ—New Jersey Medical School
225 Warren Street
Newark, New Jersey 07103
Phone: 973-972-4483 Ext. 20698
Web site: http://njmsmicro.umdnj.edu/

For application forms:
Admissions Office
UMDNJ—Graduate School of Biomedical Sciences
185 South Orange Avenue
Newark, New Jersey 07103
Phone: 973-972-4511
Web site: http://www.umdnj.edu/gsbsnweb/

University of Medicine and Dentistry of New Jersey

THE FACULTY AND THEIR RESEARCH

Vivian Bellofatto, Professor; Ph.D., Yeshiva (Einstein), 1984. Regulation of mRNA synthesis and turnover in parasitic protozoa.

Purnima Bhanot, Assistant Professor; Ph.D., Johns Hopkins, 2000. Biology of the malaria parasite, *Plasmodium.*

Marjorie C. Brandriss, Professor; Ph.D., MIT, 1975. Regulation of transcriptional activation of gene expression in eukaryotes; regulation of proline metabolism.

Nancy D. Connell, Associate Professor; Ph.D., Harvard, 1989. Mycobacterial physiology; mechanisms of nutrient transport and multidrug resistance in mycobacteria; mycobacterial-macrophage interaction.

Karl Drlica, Professor; Ph.D., Berkeley, 1971. Mechanisms of antibiotic resistance in bacteria; antisense RNA and gene expression.

David Dubnau, Professor; Ph.D., Columbia, 1961. Molecular genetic analysis of global control of cellular competence and DNA uptake.

Stephen Garrett, Assistant Professor; Ph.D., Johns Hopkins, 1986. Molecular genetics of cell signaling and protein trafficking.

Emanuel Goldman, Professor; Ph.D., MIT, 1972. Control of translational efficiency and accuracy by tRNAs; codon bias; programmed translational frameshifts; regulation of gene expression in bacteria and phage.

M. Zafri Humayun, Professor; Ph.D., Indian Institute of Science, 1975. Molecular biology and genetic regulation of DNA replication fidelity; transient mutator responses.

David B. Kaback, Professor; Ph.D., Brandeis, 1976. Structure of eukaryotic chromosomes; control of meiotic recombination and chromosome segregation.

Fred Kramer, Professor; Ph.D., Rockefeller, 1969. Structure, function, and biotechnological utility of nucleic acids; molecular beacons.

David M. Lukac, Assistant Professor; Ph.D., Pennsylvania, 1997. Transcription regulation and cellular signaling in herpesvirus reactivation; virus-host interactions.

Leonard Mindich, Professor; Ph.D., Rockefeller, 1962. Molecular genetic analysis of virus assembly.

Carol S. Newlon, Professor and Chair; Ph.D., MIT, 1971. Replication and segregation of eukaryotic chromosomes; chromosomal DNA replication origin structure, regulation, and interaction.

J. Patrick O'Connor, Assistant Professor; Ph.D., Pittsburgh, 1990. Genetics of mammalian skeletal development and function.

Harvey L. Ozer, Professor and Senior Associate Dean for Research; M.D., Stanford, 1965. Regulation of cell proliferation, including carcinogenesis and cellular basis of aging.

Nikhat Parveen, Assistant Professor, Ph.D., Hawaii, 1995. Virulence factors of *Pseudomonas aeruginosa* and Lyme disease spirochete, *Borrelia burgdorferi.*

David Perlin, Professor and Scientific Director, Public Health Research Institute; Ph.D., Cornell, 1980. Drug resistance in fungal infections; energy coupling by plasma membrane proton pump.

Nicholas M. Ponzio, Professor; SUNY Downstate Medical Center, 1973. Cellular and molecular approaches to etiology and treatment of B-cell lymphomas.

Lynn S. Ripley, Professor; Ph.D., Illinois at Urbana-Champaign, 1974. Enzymological mechanisms of frame shifts, spontaneous mutagenesis, and molecular evolution.

Emilia Vitale, Associate Professor; Ph.D., Naples, 1982. Genetic and physical mapping of genes involved in neurodevelopmental disorders.

Ian Whitehead, Associate Professor; Ph.D., British Columbia, 1993. Mammalian signal transduction and oncogenesis.

Robert Wieder, Associate Professor; Ph.D., 1982, M.D., 1983, CUNY, Mount Sinai. Cell-cycle checkpoint control and programmed cell death in breast carcinogenesis.

Hua Zhu, Associate Professor; Ph.D. Columbia, 1993. Human cytomegalovirus and host-cell interactions.

The International Center of Public Health, which houses the Department of Microbiology and Molecular Genetics, Public Health Research Institute, and National Tuberculosis Center.

UNIVERSITY OF MIAMI

Miller School of Medicine
Department of Microbiology and Immunology

Programs of Study	The objective of the Department's graduate program is to prepare students for careers in microbiology and immunology. The program is designed to expose students to the central issues and cutting-edge research in the interdisciplinary biomedical sciences, with emphases on bacteriology, parasitology, virology, genetics, immunology, and molecular and cellular biology, together with intensive laboratory research training experience in their chosen area of specialization.

The Department offers training leading to the Ph.D. degree. An acceptable Ph.D. dissertation embodying original investigative findings suitable for publication must be presented and defended. The doctoral degree typically requires five years beyond the baccalaureate. The Department does not offer an M.S. degree program.

Since the Department aims to prepare its graduates for careers in research and teaching, all students gain valuable experience in the latter through participation in the teaching program.

The Department participates in the University's combined M.D./Ph.D. program. Medical students interested in advanced research in microbiology and immunology should consult the Director of the Graduate Program.

Each entering student is assigned to the First Year Advisor. This faculty member functions as the student's general adviser, providing information and advice regarding laboratory rotations and research programs until the selection of a mentor for the dissertation research. The student usually selects a research mentor by the end of the first year of study. After conducting initial research and formulating a dissertation proposal, a general Qualifying Examination is administered during the second year. Other requirements for the Ph.D. degree include 36 credits at the graduate level (exclusive of dissertation research), an overall average of B or better (no grades below C are accepted as part of the 36-credit requirement), a dissertation embodying original research and encompassing at least 24 credits, and an oral defense of the dissertation before the committee, which includes an invited examiner with an international reputation from another university.

Research Facilities

The teaching and research facilities of the Department are located in the Miller School of Medicine in four buildings that are well equipped for training and investigative work in all the disciplines of microbiology and immunology at current state-of-the-art levels. These include a transgene/gene knock-out mouse facility and a flow cytometry resource center with a tri-laser multispectral BD/LSR.

Financial Aid

Full stipends of $22,000 per annum are awarded to all accepted students.

Cost of Study

All students are awarded full-tuition fellowships.

Living and Housing Costs

There is no on-campus housing available for graduate students.

Student Group

Graduate enrollment in the Department of Microbiology and Immunology was 37 in 2005–06. The student body includes individuals from all parts of the United States as well as from other countries.

Location

From Fort Lauderdale to the Florida Keys, an exciting cosmopolitan community offers substantial cultural and recreational attractions. The suburb of Coral Gables, located about 14 miles from the School of Medicine campus, is one of the attractive municipalities that make up the Miami metropolitan area. The Coconut Grove area, with its art galleries and theaters, is another interesting component of the community. The University supports a full calendar of musical, dramatic, artistic, athletic, and scientific events during the year. Most departments and schools sponsor weekly seminars that are open to graduate students. For those interested in outdoor recreation, the Florida Keys and Everglades National Park are nearby.

The Department

The Department of Microbiology and Immunology was founded in 1953 and is housed in the School of Medicine at 1600 Northwest 10th Avenue in Miami. Its teaching responsibilities include predoctoral, postdoctoral, medical, and undergraduate major and minor programs. The Department has one of the highest levels of research funding at the University of Miami.

Applying

Prospective students entering in the fall should have their complete application, including letters of recommendation, submitted before December 31. Strong applicants should have a combined score (verbal and quantitative) equal to or greater than 1200 on the Graduate Record Examinations General Test and a minimum grade point average of 3.0 (on a 4.0 scale) to be strongly considered for admission. A GRE Subject Test is not required. Students should have a degree in a natural science and have completed courses in general microbiology and in inorganic and organic chemistry, as well as a year of biological science. Courses in genetics, biochemistry, physics, calculus, and a suitable foreign language (French, German, or Russian) are recommended.

The cost of a formal application is $50.

Correspondence and Information

Karen Del Rio
Administrator, Graduate Program
Department of Microbiology and Immunology (R138)
Miller School of Medicine
University of Miami
P.O. Box 016960
Miami, Florida 33101
Phone: 305-243-5682
Fax: 305-243-6903
E-mail: bugsandimmunity@miami.edu
Web site: http://chroma.med.miami.edu/micro

University of Miami

THE FACULTY AND THEIR RESEARCH

*B. Adkins, Associate Professor; Ph.D., California, San Diego. Maturation and function of T lymphocytes during fetal and neonatal life.

*A. L. Ager Jr., Research Associate Professor; Ph.D., Georgia. Chemotherapy of parasitic protozoa; prophylactic and suppressive effects of chemicals in mouse malaria; African and American trypanosomiasis.

N. H. Altman, Professor; V.M.D., Pennsylvania. Animal models of human disease.

*G. Barber, Professor; Ph.D., London. Molecular and genetic approaches toward elucidating the mechanisms of host defense against viral and malignant disease.

*A. Baur, Associate Professor; M.D., Erlangen-Nürnberg. Molecular function of the Nef protein of HIV-1.

*B. Blomberg, Professor; Ph.D., California, San Diego. Molecular regulation of young and aged B-lymphocyte development; tolerance induction in tissue transplantation; immune response during psychosocial intervention and cancer.

*L. H. Boise, Associate Professor; Ph.D., Virginia Commonwealth. Regulation of programmed cell death (apoptosis).

T. J. Cleary, Professor; Ph.D., Cincinnati. Clinical microbiology; amplification techniques for the detection of microbial pathogens; epidemiology of hospital-acquired infections.

V. Esquenazi, Research Professor; Ph.D., Miami (Florida). Transplantation immunobiology; genetics of transplantation immunity; serology and genetics of the major histocompatibility complex; HLA and disease associations.

*K. A. Fields, Assistant Professor; Ph.D., Kentucky. Molecular mechanisms of *Chlamydia* pathogenesis.

*M. A. Fletcher, Professor; Ph.D., Baylor. Immunopathology of HIV disease, cancer, and chronic fatigue syndrome; clinical immunology; psychoneuroimmunology.

*L. Fuller, Professor; Ph.D., Miami (Florida). Transplantation immunology; bone marrow regulatory cells; complement.

*S. B. Greer, Professor; Ph.D., Columbia. Radiation and repair; tumor radiosensitization with 5-halogenated nucleoside analogs; DNA hypomethylation of tumors as an approach to therapy; determining relative levels of pyrimidine metabolizing enzymes as a guide to therapy or selective rescue of tumors; gene therapy coupled with radiosensitization by halogenated pyrimidine analogs.

*E. W. Harhaj, Assistant Professor; Ph.D., Penn State Hershey Medical Center. Mechanisms of HTLV-I–mediated T-cell transformation; regulation of the NF-kB signaling pathway.

*P. A. J. Haslett, Assistant Professor; M.D., London. Clinical and laboratory approaches to experimental and therapeutic cellular immunology of human infectious diseases.

*L. Inverardi, Professor; M.D., Milano (Italy). Immunobiology of islet of Langerhans transplantation; xenotransplantation.

*R. Jurecic, Associate Professor; Ph.D., Zagreb (Croatia). Stem cell self-renewal and differentiation; stem cell therapy; mouse models of stem cell maintenance; cancer stem cells; hematological malignancies.

*N. S. Kenyon, Professor of Surgery, Medicine, Microbiology, and Immunology and Martin Kleiman Chair in Diabetes Research; Ph.D., Richmond. Immunology (transplantation and type 1 diabetes); preclinical/translational models of islet cell transplantation and tolerance induction; immunological studies of clinical islet transplant recipients.

N. Klimas, Professor; M.D., Miami (Florida). Immunologic abnormalities in AIDS and chronic fatigue syndrome.

*M. A. Kolber, Associate Professor; Ph.D., Illinois at Urbana-Champaign; M.D., Miami (Florida). Immunologic T-cell defects in HIV infection; clinical immunologic studies involving HIV seropositive individuals.

*K. Lee, Professor; M.D., Michigan. Dendritic cell biology and function.

*R. Levy, Professor; Ph.D., Ohio State. Cellular and molecular approaches for regulating engraftment and tolerance following allogeneic hematopoietic stem cell transplantation in cancer therapy.

*W. Li, Assistant Professor; Ph.D., Nebraska Medical Center. Autoimmunity and autoantigens.

*M. G. Lichtenheld, Associate Professor; M.D., Mainz. Genetics and epigenetics of cytotoxic lymphocytes; immunobiology of cell death.

*D. M. Lopez, Professor; Ph.D., Miami (Florida). Tumor immunology; breast cancer; cytokine regulation; cancer immunotherapy.

*T. R. Malek, Professor and Vice Chairman; Ph.D., Illinois at Chicago. Cytokine regulation of T-lymphocyte development and function; immunobiology of T regulatory cells; tumor immunology.

O. V. Martinez, Research Associate Professor; Ph.D., Miami (Florida). Microbiology: microbiological testing of tissues and organs for transplantation; antimicrobial susceptibility tests.

*J. M. Mathew, Research Associate Professor; Ph.D., Madurai (India). Immune responses and their regulation in human organ and cell transplantations.

M. McCarthy, Associate Professor of Neurology; M.D., Johns Hopkins; Ph.D., Harvard. Virology; neurovirology; pathogenesis of viral diseases of the nervous system; molecular basis of viral cytopathogenicity; neural cell biology of differentiation as it pertains to virus-cell interactions.

E. Mesri, Associate Professor; Ph.D., Buenos Aires. Molecular mechanisms of carcinogenesis and angiogenesis activation induced by the Kaposi's sarcoma–associated herpesvirus (KSHV).

*J. Miller, Professor; M.D., Yeshiva (Einstein). Bone marrow induction of donor specific tolerance in clinical solid organ transplantation.

*G. P. Munson, Assistant Professor; Ph.D., Northwestern. Bacterial pathogenesis; molecular, biochemical, and genetic investigations of virulence genes, with particular emphasis on transcriptional regulation of virulence.

*S. Pahwa, Professor; M.D., New Delhi (India). Immunopathogenesis of HIV infection for understanding the nature and mechanisms of immune dysfunction in HIV infected pediatric and adult patients.

*R. L. Pastori, Research Associate Professor; Ph.D., Buenos Aires. Molecular strategies for phenotype modulation in transplantation and autoimmunity; antisense ribozymes and protein transduction domains.

*G. V. Plano, Associate Professor; Ph.D., South Alabama. Molecular pathogenesis of *Yersinia pestis*.

*L. R. W. Plano, Associate Professor of Clinical Pediatrics; M.D., Ph.D., South Alabama. Molecular analysis of staphylococcal exfoliative toxin A and the interaction with desmoglein 1, responsible for staphylococcal scalded skin syndrome.

*E. R. Podack, Professor and Chairman; M.D., Frankfurt; Ph.D., Göttingen. Molecular immunology; mechanism of lymphocyte-mediated cytolysis and apoptosis; CTL regulation; heat-shock fusion proteins in immune regulation; immunobiology of TL1A, April, CD30-L and their receptors.

*A. Pugliese, Associate Professor; M.D., Palermo (Italy). Immunogenetics of type 1 diabetes; regulation of tolerance to diabetes-associated autoantigens; genetics of type 1 diabetes, in particular, HLA; prevention of diabetes.

*C. Ricordi, Professor and Chief (Cellular Transplantation); M.D., Milan (Italy). Biological replacement strategies for type 1 diabetes; isolation, purification, and transplantation of pancreatic islet cells; mechanisms of tolerance induction and chimerism; cellular transplantation and gene therapy; cell separation technologies; xenotransplantation; tissue engineering.

*R. L. Riley, Professor; Ph.D., Virginia. B-lymphocyte development and function; B-lymphocyte subsets; development of B-cell antibody repertoires; immunodeficiency of aging; autoimmunity.

*J. Rosenblatt; Professor; Chief, Hematology-Oncology and Medicine; and Associate Director for Clinical and Transplantation Research, Sylvester Comprehensive Cancer Center; M.D., UCLA. Human retroviruses/HTLV-I/II; human gene therapy and immunotherapy for cancer and HIV-1 infection; novel therapies for breast cancer and lymphoma.

*K. Schesser, Assistant Professor; Ph.D., Oregon State. Molecular and cellular analysis of host-pathogen interactions.

*W. A. Scott, Professor; Ph.D., Wisconsin–Madison. Molecular mechanisms of drug resistance in HIV reverse transcriptase.

A. Shatry, Research Assistant Professor; Ph.D., Montana State. Immunobiology of hematopoietic stem cell engraftment.

*G. R. Thomas, Assistant Professor; M.D., Rochester, Georgetown; FACS. Tumor immunology of head and neck cancer.

M. Torroella-Kouri, Research Assistant Professor; Ph.D., Institute of Molecular Genetics (Prague). Tumor immunology; role of cytokines in the host's tumor-induced immunodepressed status.

*V. Vincek, Associate Professor; M.D., Ph.D., Rijeka (Croatia). Immunogenetics; effects of UVB light on mammals.

**Currently participating graduate faculty members who mentor graduate students' dissertation research.*

UNIVERSITY OF MINNESOTA, TWIN CITIES CAMPUS

Medical School
Ph.D. Program in Microbiology, Immunology, and Cancer Biology

Program of Study

The University of Minnesota, Twin Cities Campus, has formed a new graduate training program entitled Microbiology, Immunology, and Cancer Biology (MICaB). This exciting development allows students to undertake research and specialized course work in one of the three areas (microbiology, immunology, or cancer biology) and to also receive exposure to the other two areas. This training prepares students for independent research careers in areas including, but not limited to, genetic engineering of microorganisms for biotechnology; viral, fungal, and bacterial pathogenesis; environmental sensing and development in microbes; microbial genomics; lymphocyte activation and development; transplantation immunology and autoimmunity; superantigens and the immune system; cell biology and metastasis; cancer genetics; and tumor implantology.

Students begin study in the fall and spend their first year doing course work, identifying an adviser by doing laboratory rotations, selecting a track, and initiating their thesis research project. During the first two years, all MICaB students take courses on the structure, function, and metabolism of microorganisms; molecular immunology; and pathobiology as well as specialized courses in their chosen track. In addition to course work and research, students have weekly opportunities to participate in lab meetings, journal clubs, and student research seminars. Students gain valuable teaching experience by assisting in laboratory courses for one semester during their graduate career. Most students complete their Ph.D. in four to five years.

Research Facilities

The laboratories of program faculty members are located within modern research facilities. On the Minneapolis campus, these include the state-of-the-art Nils Hasselmo Hall, Molecular and Cellular Biology Building, and Cancer Center Research Building. The Center for Microbial Physiology and Metabolic Engineering is located on the St. Paul campus. Students on both campuses have access to core facilities for oligonucleotide and peptide synthesis, flow cytometry, production of knockout and transgenic mice, confocal and electron microscopy, and biomedical image and molecular biology data analysis. Students also have free access to the Internet and the Web. Research facilities are supported by many departmental and collegiate libraries, including the Diehl Hall Biomedical Library.

Financial Aid

All Ph.D. students in good standing receive complete financial support, including a stipend of $23,000 (for 2006–07), a full tuition waiver, and one of the most comprehensive health-care benefit packages available in graduate education. Support sources are program funds, one of the six NIH training grants that support the program, University or nationally competitive fellowships, or faculty research grants.

Cost of Study

Students receive full-tuition scholarships and 95 percent coverage for health insurance. Members of the student's immediate family also qualify for tuition benefits. There is a student service fee each semester of approximately $280 for residents and nonresidents.

Living and Housing Costs

Most students live in off-campus rental housing. Rents range from $470 to $490 per month for a single room, $630 to $670 per month for a one-bedroom apartment, and $700 to $950 per month for a two-bedroom apartment. University housing for families and residence halls are available but often have waiting lists. For details, students should contact the Housing Office at 612-624-2994 or access the Web site at http://www.umn.edu/housing.

Student Group

Approximately 80 full-time graduate students are enrolled in the MICaB program, 2 of whom are elected as student representatives to plan social and professional development functions for the student group. Like the student population as a whole, MICaB students come from diverse backgrounds, giving the University a cosmopolitan character.

Location

The Minneapolis–St. Paul area is the cultural and industrial center of the upper Midwest. Both are highly attractive cities that offer flourishing downtowns, entertainment, sports, parks, lakes, and open spaces. The cold winter, cool spring, warm summer, and colorful fall provide excellent opportunities for year-round recreational activities.

The University

A land-grant institution established in 1869, the University has campuses in Minneapolis–St. Paul, Crookston, Duluth, and Morris as well as research stations around the state. The Twin Cities campus enrolls more than 37,000 students from fifty states and 100 countries, including 9,000 students in 163 graduate programs.

Applying

Applicants must have research experience and a bachelor's degree in an area of science that includes the following course work: a year of calculus, general chemistry, organic chemistry, physics, and one academic year or the equivalent of courses in the biological sciences, supplemented by courses in biochemistry and genetics. A course in histology, immunology, or microbiology is highly recommended. Required materials are the Graduate School application form, GRE scores (General Test only), the application fee, transcripts, and three letters of recommendation (preferably from supervisors of lab experience). The deadline for application is December 15 for entry the following fall semester.

Correspondence and Information

MICaB Graduate Program Admissions
University of Minnesota, Twin Cities Campus
Mayo Mail Code 196
1460 Mayo, 420 Delaware Street, SE
Minneapolis, Minnesota 55455-0312

Phone: 612-624-5947
Fax: 612-626-0623
E-mail: micab@mail.ahc.umn.edu
Web site: http://www.micab.umn.edu

University of Minnesota, Twin Cities Campus

THE FACULTY AND THEIR RESEARCH

Microbiology Track
Dwight L. Anderson, Professor; Ph.D., Minnesota, Twin Cities, 1961. *Bacillus subtilis* bacteriophage φ29 morphogenesis.
Sandra K. Armstrong, Associate Professor; Ph.D., Missouri–Columbia, 1986. Iron acquisition and gene regulation in *Bordetella pertussis*.
Judith G. Berman, Professor; Ph.D., Weizmann (Israel), 1984. Yeast telomeres and chromatin; *Candida albicans* morphogenesis.
Daniel R. Bond, Assistant Professor; Ph.D., Cornell, 1999. Physiology and functional genomics of dissimilatory Fe(III)-reducing bacteria; microbial fuel cells.
Wade A. Bresnahan, Assistant Professor; Ph.D., Texas Medical Branch, 1997. Molecular mechanisms of human cytomegalovirus replication and pathogenesis.
P. Patrick Cleary, Professor; Ph.D., Rochester, 1971. Molecular genetics of streptococcal cell-surface antigens.
Kathleen F. Conklin, Associate Professor; Ph.D., Tufts, 1982. Retroviruses; oncogenesis and gene regulation.
Anath Das, Professor; Ph.D., Nebraska–Lincoln, 1980. Molecular mechanism of DNA transport across cell membranes.
Dana A. Davis, Assistant Professor; Ph.D., Arizona, 1998. *Candida albicans* genetics and pathogenesis.
Gary M. Dunny, Professor; Ph.D., Michigan, 1978. Molecular biology of conjugative gene transfer in gram-positive bacteria.
Lynda B. M. Ellis, Professor; Ph.D., Brandeis, 1971. Computational biology and protein structure prediction.
Jeffrey A. Gralnick, Assistant Professor; Ph.D., Wisconsin–Madison, 2003. Physiology of *Shewanella*; geomicrobiology.
Ashley T. Haase, Regents Professor; M.D., Columbia College of Physicians and Surgeons, 1965. HIV pathogenesis.
Vivek Kapur, Professor; Ph.D., Penn State, 1991. Molecular mechanisms of bacterial pathogenicity and evolution.
Paul T. Magee, Professor; Ph.D., Berkeley, 1964. Genetics and molecular biology of *Candida albicans*.
Louis M. Mansky, Associate Professor; Ph.D., Iowa State, 1990. Cell and molecular biology of HIV and HTLV; HIV drug resistance; HTLV particle assembly and release; evolution of emerging viruses.
Larry L. McKay, Professor; Ph.D., Oregon State, 1969. Plasmid biology; genetics; applications of lactic acid bacteria.
Christian D. Mohr, Assistant Professor; Ph.D., Texas at San Antonio, 1993. Flagellar export and assembly in *C. crescentus* and *B. cepacia*.
Daniel J. O'Sullivan, Associate Professor; Ph.D., University College Cork (Ireland), 1984. Molecular analysis of lactic acid bacteria.
Stephen A. Rice, Assistant Professor; Ph.D., Utah, 1985. Molecular biology of herpes simplex virus.
Michael J. Sadowsky, Professor; Ph.D., Hawaii, 1983. Soil microbiology; biodegradation; *Rhizobium*- and *Bradyrhizobium*-host interactions.
Leslie Schiff, Professor; Ph.D., Tufts, 1985. Virus–host cell interactions; viral protein structure-function.
Patrick M. Schlievert, Professor; Ph.D., Iowa, 1976. Immunobiology and genetic control of staphylococcal and streptococcal pyrogenic toxins.
Janet L. Schottel, Professor; Ph.D., Washington (St. Louis), 1977. Plant pathogenesis; regulation of mRNA turnover.
John R. Schreiber, Professor; M.D., Tulane, 1980. Immune response to bacterial polysaccharides.
Pamela J. Skinner, Assistant Professor; Ph.D., Minnesota, Twin Cities, 1998. Neurodegenerative disease and HIV pathogenesis.
Peter J. Southern, Associate Professor; Ph.D., Edinburgh, 1978. Molecular basis of persistent virus infection and virus-induced disease.
Kenneth D. Vernick, Associate Professor; Ph.D., Maryland, 1988. Malaria-mosquito interactions.
Larry P. Wackett, Professor; Ph.D., Texas, 1984. Biodegradation and biocatalysis.
Carol L. Wells, Professor; Ph.D., Wisconsin, 1978. Virulence factors of normal flora intestinal bacteria.

Immunology Track
Mitchell S. Abrahamsen, Professor; Ph.D., Washington (Seattle), 1990. Molecular parasitology; molecular immunology.
Bruce R. Blazar, Professor; M.D., Albany Medical College, 1978. Immunobiology of transplantation.
Paul R. Bohjanen, Associate Professor; M.D./Ph.D., Michigan, 1993. T-lymphocyte mRNA stability.
Agustin P. Dalmasso, Professor; M.D., Cordoba (Argentina), 1958. Complement and transplantation immunology.
Michael A. Farrar, Associate Professor; Ph.D., Washington (St. Louis), 1993. Signal transduction and lymphocyte development.
Dale S. Gregerson, Professor; Ph.D., British Columbia, 1976. Tolerance and pathogenesis in autoimmune diseases of nervous tissues.
Reuben S. Harris, Assistant Professor; Ph.D., Alberta, 1997. Mechanisms of purposeful mutation.
Kristin A. Hogquist, Professor; Ph.D., Washington (St. Louis), 1991. T-cell development.
Koho Iizuka, Assistant Professor; M.D., Hirosaki, 1987. Natural killer cell biology.
Elizabeth Ingulli, Assistant Professor; M.D., SUNY Health Science Center, 1990. Transplantation immunology.
Stephen C. Jameson, Professor; Ph.D., Cambridge, 1988. Activation and development of CD8 T cells.
Ronald Jemmerson, Professor; Ph.D., Northwestern, 1978. B-cell and antibody recognition of protein antigens.
Marc K. Jenkins, Professor; Ph.D., Northwestern, 1985. Activation requirements for helper T lymphocytes.
Dan S. Kaufman, Assistant Professor; M.D./Ph.D., Mayo, 1996. Hematopoietic and endothelial cell development from embryonic cells.
Alexander Khoruts, Assistant Professor; M.D., Minnesota, Twin Cities, 1989. CD4 and T-cell-mediated immunopathogenesis and tolerance.
Stephen J. McSorley, Assistant Professor; Ph.D., Glasgow, 1995. Immunity to *Salmonella*.
Matthew F. Mescher, Professor; Ph.D., Harvard, 1976. Cytotoxic T-lymphocyte activation; signal transduction; immunotherapy.
Jeffrey S. Miller, Professor; Northwestern, 1985. Natural killer cell development.
Daniel L. Mueller, Professor; M.D., Wisconsin, 1983. Regulation of growth and effector function in T cells.
Erik J. Peterson, Assistant Professor; M.D., Minnesota, Twin Cities, 1990. Leukocyte activation and development.
Yoji Shimizu, Professor; Ph.D., Wisconsin, 1987. Cell adhesion and the immune response.
Bruce K. Walcheck, Associate Professor; Ph.D., Montana State, 1994. Cell adhesion and signaling; leukocyte recruitment.
Xianzheng Zhou, Assistant Professor; M.D., Jiangxi Medical College, 1983; Ph.D., Karolinska Institute, 1994. Human cancer immunology and immunotherapy; transplantation immunology.

Cancer Biology Track
Khalil Ahmed, Professor; Ph.D., McGill, 1960. Molecular mechanisms of cell growth and proliferation; kinase signaling in cancer.
Vivian J. Bardwell, Associate Professor; Ph.D., Wisconsin–Madison, 1990. Cancer biology.
Peter B. Bitterman, Professor; M.D., Yale, 1976. Mechanisms of fibroblast population control; proliferation and apoptosis in tissue repair.
Denis R. Clohisy, Professor; M.D., Northwestern, 1983. Cell biology of bone resorption.
Patrick Gaffney, Assistant Professor; M.D., Minnesota, Twin Cities, 1991. Cancer/autoimmunity genetics and genomics.
Jennifer L. Hall, Assistant Professor; Ph.D., Berkeley, 1995. Vascular biology.
Ameeta Kelekar, Assistant Professor; Ph.D., Princeton, 1987. Mechanisms of apoptosis.
Nobuaki Kikyo, Assistant Professor; Ph.D., Tokyo, 1993. Nuclear remodeling and cancer.
Carol A. Lange, Associate Professor; Ph.D., Colorado at Boulder, 1991. Signal transduction in breast cancer.
David A. Largaespada, Associate Professor; Ph.D., Wisconsin–Madison, 1992. Myeloid leukemia; mouse transgenesis.
Tucker W. LeBien, Professor; Ph.D., Nebraska, 1977. Normal and abnormal human B-cell development.
Brett K. Levay-Young, Research Associate; Ph.D., Berkeley, 1987. Wound healing; lung development.
Walter C. Low, Professor; Ph.D., Michigan, 1979. Cell and gene therapies.
Kim Mansky, Assistant Professor; Ph.D., Wisconsin–Madison, 1997. RANKL signaling and MITF activation of genes.
Patrick Mantyh, Professor; Ph.D., California, San Francisco, 1981; J.D., William Mitchell Law, 1994. Cancer pain biology.
James B. McCarthy, Professor; Ph.D., Catholic University, 1981. Cell adhesion; tumor invasion.
R. Scott McIvor, Professor; Ph.D., Minnesota, Twin Cities, 1982. Gene therapy.
Christopher A. Pennell, Associate Professor; Ph.D., North Carolina, 1984. Tumor immunology and immunotherapy.
Vitaly Polunovsky, Professor; Ph.D., Moscow State, 1975. Translational control for cancer.
Sundaram Ramakrishnan, Professor; Ph.D., All-India Institute of Medical Sciences, 1980. Tumor angiogenesis and experimental therapy.
Michel M. Sanders, Professor; Ph.D., Michigan, 1981. Hormonal regulation of gene expression.
Robert Sheaff, Assistant Professor; Ph.D., Colorado, 1994. Cell-cycle control.
Amy P. N. Skubitz, Associate Professor; Ph.D., Johns Hopkins, 1984. Cell-cell and cell-matrix interactions.
Daniel A. Vallera, Professor; Ph.D., Ohio State, 1978. Tumor immunology.
Brian G. Van Ness, Professor; Ph.D., Minnesota, Twin Cities, 1979. Molecular immunology.
Catherine M. Verfaillie, Professor; M.D., Leuven (Belgium), 1982. Cell biology; inflammation.
Jennifer A. Westendorf, Associate Professor; Ph.D., Mayo, 1996. Transcription factors in bone biology and cancer.
Douglas Yee, Professor; M.D., Chicago, 1981. Growth regulation of breast cancer.

UNIVERSITY OF PENNSYLVANIA / NATIONAL INSTITUTES OF HEALTH

Graduate Program in Immunology

Program of Study

The University of Pennsylvania/National Institutes of Health Graduate Training Program in Immunology is directed toward the acquisition of knowledge of molecular and basic immunology as well as the necessary foundation in the appropriate basic disciplines. These basic areas include biochemistry, molecular genetics, cell biology, virology, developmental biology, and physical chemistry, and all are related to understanding the mechanisms of disease. Based on a long history of excellent research in immunology at both the University of Pennsylvania (UP) and the National Institutes of Health (NIH) intramural laboratories, this combined program assures students of training with outstanding investigators in the most current laboratory facilities.

After taking basic courses in the first year of study and working part-time in a laboratory, students take advanced courses in their area of interest in the second year. Generally by the end of the second year, students are expected to pass a preliminary examination that tests their knowledge and potential for conducting independent immunology research. During the second year, students select a research supervisor at either the NIH or the University of Pennsylvania after exploring thesis research options through various short laboratory experiences during the first and second years. Opportunities are provided for students to explore NIH research labs in the summer before beginning the program and the summer following the first year of classes. Depending on the thesis research area, students may spend research time at either or both institutions and have an advisory committee composed of both NIH and University of Pennsylvania faculty members.

Research Facilities

At the University of Pennsylvania, the Graduate Program in Immunology utilizes the faculty and facilities of the many departments of the Schools of Arts and Science, Medicine, and Veterinary Medicine and laboratories of the Children's Hospital of Philadelphia. The research pursued in these various laboratories includes work in autoimmunity, clinical immunology, cytokines and growth factors, immune disorders, immunogenetics, immunohematology, immunological tolerance, immunoreceptors, immunoregulation, lymphocyte physiology, microbial immunology, molecular immunology, transplantation, and tumor immunology. As the federal government's primary agency for the support and conduct of biomedical research, the NIH's many institutes and centers offer state-of-the-art research equipment and employ nearly 1,200 tenured or tenure-track investigators and more than 3,700 postdoctoral scientists with either medical, dental, or graduate degrees. The NIH intramural research program, located on a 300-acre campus in Bethesda, Maryland, contains more than thirty biomedical research buildings housing a broad spectrum of biomedical and related scientific research. Four Nobel Laureates made their prize-winning discoveries in NIH laboratories, and more than 100 received training at NIH. Basic research in the biomedical sciences at the NIH is complemented by an active clinical research program at the unique 250-bed research hospital and laboratory complex, the Warren Grant Magnuson Clinical Center. The NIH campus is home to the National Institute for Allergy and Infectious Disease research and a Vaccine Research Development Center containing hundreds of researchers dedicated to immunology research. The National Library of Medicine, the world's largest medical library, is located next to the NIH campus.

Financial Aid

All University of Pennsylvania/NIH graduate students are supported through NIH Intramural Research Training Awards and receive support for stipend, tuition, and health benefits throughout their years of training. Stipends for first-year students are $24,000 and increase yearly based on the student's performance.

Cost of Study

All students accepted into the program are guaranteed full funding throughout their studies and research by training awards.

Living and Housing Costs

While at UP, the University offers graduate students on-campus living in two graduate towers, apartments, and dormitories. Most students find this the simplest solution when entering UP. Alternatively, there are many rooms, apartments, and houses available in the University City area. When at NIH, apartments, houses, and rooms in private residences are available for rent near the NIH campus. For information on available housing, applicants should visit the Web sites of local newspapers or contact the Graduate Partnerships Program (GPP) on the Web at http://gpp.nih.gov.

Student Group

The Graduate Student Group in Immunology is one of thirty-six graduate groups in the School of Arts and Sciences, which has a total of approximately 3,300 graduate students and is one of several groups in the Biomedical Graduate Studies Division, which contains approximately 400 students. The Graduate Group in Immunology has about 55 students. At NIH, there are more than 400 graduate students from more than 100 universities. While at NIH, graduate students enjoy services and activities sponsored by the Graduate Partnerships Program that are similar to those of a university and ensure student success while creating a strong NIH graduate student community.

Location

The University is located in Philadelphia, a historic shrine and cultural center. Philadelphia has several professional sports teams and is a few hours away from New York City and Washington, D.C. The 300-acre campus of NIH in Bethesda, Maryland, is close to Washington, D.C., affording a spectacular cultural and community environment and easy access to the Atlantic Ocean and the Allegheny and Appalachian Mountains for recreational activities. Other satellite campuses of NIH are in Maryland, Montana, North Carolina, and Arizona.

The University and The National Institutes of Health

The University of Pennsylvania is one of the oldest private universities in the U.S. Penn was founded in 1740 by Benjamin Franklin and became the nation's first university with a medical school in 1765. Since then, Penn has grown to include four undergraduate schools and twelve graduate schools. The NIH, like Penn, has a long history of training scientists and physicians. Thousands of scientists have completed postdoctoral training in NIH laboratories, and many NIH-trained scientists have received international recognition for their work. The NIH environment is rich in scientific exchange, and it provides opportunities for a broad biomedical research experience. Many graduate students have received their graduate training at the NIH through arrangements between the NIH and universities.

Applying

To apply, prospective students must be U.S. citizens or noncitizen nationals of the United States or must have been lawfully admitted for permanent residence (i.e., possess a valid Alien Registration Receipt Card or some other verification of such status). An entering class typically includes a variety of different undergraduate majors in the biological, chemical, or physical sciences. Information on submitting an application to the University of Pennsylvania/NIH Graduate Partnership Program in Immunology or joining an NIH laboratory for dissertation research may be found on the Web at http://gpp.nih.gov.

Correspondence and Information

Graduate Partnerships Program
Building 2, Room 2E06
National Institutes of Health–DHHS
2 Center Drive
Bethesda, Maryland 20892-0234
Phone: 301-594-9605
Fax: 301-594-9606
E-mail: gpp@nih.gov
Web site: http://gpp.nih.gov

NIH Partnership Director
Al Singer, Ph.D.
Phone: 301-496-5461
E-mail: singera@nih.gov

University Partnership Director
Steven Reiner, M.D.
BRB II/III, Room 414
University of Pennsylvania
421 Curie Boulevard
Philadelphia, Pennsylvania 19104
Phone: 215-746-5536
E-mail: sreiner@mail.med.upenn.edu
Web site: http://www.med.upenn.edu/
immun/

University of Pennsylvania / National Institutes of Health

THE FACULTY AND THEIR RESEARCH

NIH Faculty

Remy Bosselut, M.D., Ph.D. Genetic analysis of intrathymic T-cell development.

Rachel Caspi, Ph.D. Self-tolerance and autoimmunity with a focus on immunologically privileged retinal antigens.

Daniel Douek, M.D., Ph.D. Analysis of T-cell immune responses in humans in order to develop strategies to amplify, broaden, or focus T-cell immunity to HIV infection; vaccine development.

David N. Garboczi, Ph.D. Structural biology of membrane proteins and the T-cell receptor.

Ronald Germain, M.D., Ph.D. Peptide complexes with MHC molecules; self-peptide:MHC molecule recognition in development of thymic T cells; biochemical mechanisms of TCR discrimination between self and foreign peptide associated MHC molecules.

Michael Lenardo, M.D. Programmed cell death in immunological tolerance, autoimmunity, and viral pathogenesis by HIV.

John O'Shea, M.D. Cytokine signal transduction and the molecular basis of immunoregulation.

Sue Pierce, Ph.D. B-cell signaling; vaccine development; malaria.

William E. Paul, M.D. Mechanisms of action of type I cytokines, principally IL-4; biologic functions of cytokines; differentiation of naive T cells.

Larry Samelson, M.D. Signal transduction events mediated by the T-cell antigen receptor.

Pamela Schwartzberg, M.D., Ph.D. Signal transduction pathways of cytoplasmic tyrosine kinases; defects in these pathways contributing to disease processes.

Bob Seder, M.D. Development of vaccines for infectious diseases; cellular and molecular mechanisms by which various cytokines and costimulatory molecules regulate cellular immunity in vivo.

Alan Sher, Ph.D. Immune regulation in parasitic and bacterial infections.

Richard Siegel, M.D., Ph.D. TNF-receptor family signaling in normal and pathological immune responses; crosstalk between cell-death and survival signals.

Alfred Singer, M.D. Lymphocyte recognition and development; molecular and cellular recognition signals in self/nonself discrimination in thymic development.

Louis Staudt, M.D., Ph.D. Molecular pathogenesis of human leukemias; BCL-6 oncogene in large-cell lymphoma; micro-arrays of gene expression in human lymphomas and leukemias.

Peter Sun, Ph.D. Structural biology of NK cell receptors and TGF-β receptors.

Thomas A. Waldmann, M.D. Role of IL-2R and related receptors on growth and differentiation of normal and neoplastic T cells; IL-15 and IL-15 receptors.

University of Pennsylvania Faculty

Hydar Ali, Ph.D. Chemoattractant receptors in innate immunity and inflammatory diseases.

David Allman, Ph.D. Immune system development: ontogeny of B cells and dendritic cells; B-cell homeostasis.

Michael Atchison, Ph.D. Control of gene expression; differentiation; embryonic and hematopoietic development.

Gerd Blobel, M.D., Ph.D. Hematopoiesis; gene expression; transcription factors; chromatin.

Janis Burkhardt, Ph.D. Regulation and function of the T-cell cytoskeleton.

Michael Cancro, Ph.D. B-cell homeostasis and selection.

Martin Carroll, M.D. Mechanisms of transformation by tyrosine kinase fusion proteins.

Andrew Caton, Ph.D. Molecular and genetic basis of immune recognition.

John J. Cebra, Ph.D. Development and maintenance of the cellular system for mucosal cellular humoral immunity.

Youhai Chen, M.D., Ph.D. Autoimmunity, apoptosis, and immune tolerance.

John Kim Choi, M.D., Ph.D. Transcription factors and cell differentiation.

Yongwon Choi, Ph.D. Molecular mechanisms by which the TNF/TNFR superfamily regulates the immune system and bone.

Charles Clevenger, M.D., Ph.D. Prolactin receptor signal transduction; neuroendocrine-immune interactions.

Philip L. Cohen, M.D. Autoimmunity; programmed cell death.

Randy Q. Cron, M.D., Ph.D. Transcriptional regulation of cytokine genes and HIV-1 in primary CD4 T lymphocytes.

Steven D. Douglas, M.D. Human monocyte-macrophage and HIV; microglia; macrophage-neuropeptides.

Robert Eisenberg, M.D. Basis mechanisms of autoimmune disease.

Stephen G. Emerson, M.D., Ph.D. Molecular regulation of stem cell division.

Jan Erikson, Ph.D. B-cell development; tolerance induction and autoimmunity.

Hildegund Ertl, M.D. Immune response to viruses.

Jay P. Farrell, Ph.D. Immunological aspects of cutaneous and visceral leishmaniasis.

Peter J. Felsburg, V.M.D., Ph.D. Role of cytokines in T-cell development; X-linked SCID; gene therapy.

Bruce Freedman, V.M.D., Ph.D. Signal transduction mechanisms in immune cells.

David Gasser, Ph.D. Genetic factors controlling immunity; class III genes in the MHC.

Glen Gaulton, Ph.D. Molecular biology of T-cell growth and differentiation.

Francisco Gonzalez-Scarano, M.D. Pathogenesis of HIV and other neurotropic viruses.

Mark I. Greene, M.D., Ph.D., F.R.C.P. Receptors; reovirus, neu-growth factor receptors; T-cell receptors.

Stephan Grupp, M.D., Ph.D. Role of the B-cell receptor complex in B-cell signaling and lymphoid development.

Wayne W. Hancock, Ph.D. Transplant immunobiology; inflammation and mechanisms of disease.

Dorothee Herlyn, D.V.M. Cancer immunotherapy.

James Hoxie, M.D. Pathogenesis of human immunodeficiency viruses.

Christopher A. Hunter, Ph.D. Factors that regulate protective immunity to infection.

Carl H. June, M.D. Human T-cell adoptive immunotherapy; T-cell biology and mechanisms of costimulation; HIV pathogenesis mechanisms.

Thomas Kadesch, Ph.D. B-cell-specific transcription; hematopoiesis.

Malek Kamoun, M.D., Ph.D. T-cell receptors and regulatory pathways of T-cell activation; HLA molecules and regulation of immune response.

Gary Koretzky, M.D., Ph.D. Molecular mechanisms of lymphocyte activation.

John Lambris, Ph.D. Protein-protein interactions in the complement system.

Terri Laufer, M.D. T-cell development; autoimmunity; transgenesis; MHC class II function.

Arnold Levinson, M.D. Cellular and molecular basis of human autoimmune disease.

Nina Luning Prak, M.D., Ph.D. Rearrangement of antibody genes in normal and autoimmune states; development of a mouse model of L1 retrotransposition.

Michael Madaio, M.D. Humoral and cellular mechanisms of autoimmunity and nephritis.

Lionel Manson, Ph.D. Immunological responses during progressive tumor growth; MMTV.

Michael Marks, Ph.D. Protein trafficking and antigen presentation.

John Monroe, Ph.D. B-lymphocyte development, activation, and tolerance; HIV-associated B lymphoma.

Luis J. Montaner, D.V.M., Ph.D. Macrophage biology and cytokines; AIDS immunopathology and therapy.

Jonni S. Moore, Ph.D. Failure of normal growth regulatory controls; development of hematologic malignancies.

Ruth Muschel, Ph.D. Mechanisms of metastasis and tumor progression.

Ali Naji, M.D., Ph.D. Transplantation immunobiology and immune pathogenesis of diabetes mellitus.

Hooman Noorchashm, M.D., Ph.D. T-lymphocyte homeostasis; transplantation immunology; autoimmunity.

Paul Offit, M.D. Immunologic mechanisms by which adjuvants enhance immune response.

Jordan Orange, M.D., Ph.D. Natural killer cell biology and the innate immunological synapse.

Yvonne Paterson, Ph.D. Structural immunology; vaccine development; regulation of autoimmune disease.

Warren S. Pear, M.D., Ph.D. Hematopoietic development and transformation.

Edward Pearce, Ph.D. Immunobiology of schistosomiasis; CD4+ T helper response polarization; dendritic cell biology.

Ellen Puré, Ph.D. Cellular and molecular inflammation in chronic inflammation; cell adhesion molecules and extracellular matrix in tumor growth and metastasis.

Steven L. Reiner, M.D., Chairman. Host response to infectious diseases; lymphocyte differentiation; gene silencing; chromatin structure and DNA methylation.

Bruce R. Rosengard, M.D. Immunological tolerance; chronic rejection; xenotransplantation.

Susan Ross, Ph.D. Viral superantigens; immune response to viruses.

Milton Rossman, M.D. Molecular and cellular immunology of lymphocyte and macrophages in granulomatous lung disorders.

Alan Schreiber, M.D. Molecular and cell biology of Fc receptors; platelet immunology.

Phil Scott, Ph.D. T cells and cytokines in infectious disease, particularly leishmania.

Abraham Shaked, M.D., Ph.D. Organ transplantation.

Virginia Smith Shapiro, Ph.D. Role of cell surface protein CD28 in T-cell activation and initiation of a response; contributions to progression of multiple myeloma.

Hao Shen, Ph.D. Antigenic repertoire in bacterial pathogens.

Donald L. Siegel, Ph.D., M.D. Immune modulation of pathogenic autoantibodies and alloantibodies using phage display repertoire cloning.

Kathleen Sullivan, M.D., Ph.D. Immunodeficiency and autoimmunity; regulation of inflammation.

Laurence Turka, M.D. Transplantation immunology; tolerance; T-cell development.

Mark L. Tykocinski, M.D. Proteins with immunotherapeutic potential.

M. A. Wasik, M.D. Identification and characterization of oncoproteins contributing to transformation of T lymphocytes and development of malignant lymphoma.

Wilfried T. Weber, D.V.M., Ph.D. Lymphocyte biology; ontogeny.

David B. Weiner, Ph.D. DNA immunization.

Susan Weiss, Ph.D. Murine coronavirus; mouse hepatitis virus (MHV).

Drew Weissman, M.D., Ph.D. Immunopathogenesis of HIV infection; initiation of infection and control of viral replication; immunobiology of dendritic cells in HIV.

Andrew Wells, Ph.D. Cell-cycle control of T-cell responses.

Xiaolu Yang, Ph.D. Molecular mechanism of programmed cell death; signaling pathways of members of the tumor necrosis factor receptor (TNFR) family.

Sally Zigmond, Ph.D. Cell locomotion and chemotaxis.

UNIVERSITY OF PITTSBURGH

Graduate School of Public Health
Department of Infectious Diseases and Microbiology

Programs of Study

The Department of Infectious Diseases and Microbiology of the University of Pittsburgh Graduate School of Public Health (GSPH) was established in 1948. It offers a unique program of study in a public health and medical environment that leads to either the Ph.D., M.S., M.P.H., or Dr.P.H. degree. It merges the disciplines of molecular biology, immunology, epidemiology, disease prevention, and medicine to create a unique basic and applied research environment in which to study the pathogenesis of human infectious diseases and their prevention. The state-of-the-art research programs meet the challenge of new infectious disease problems and focus on understanding the mechanisms of pathogenesis of microbial infections at the cellular and molecular levels. These basic research programs relate directly to developing methods for disease prevention and treatment.

Major programmatic emphases in basic research within the Department include acquired immune deficiency syndrome (AIDS), herpesvirus infections in transplant recipients, cellular immunity against herpesviruses and retroviruses, dendritic cell biology, regulation of viral gene expression, antiviral drug resistance, *Legionella*, malaria, host genetic susceptibility, and diagnostic microbiology. Recently added programs of research include the vaccinology of emerging infections, including SARS coronavirus, avian influenza virus, West Nile virus and *Coxiella burnetii* (the agent of Q fever), and the cellular immune response to hepatitis C virus. Many of the programs are interdisciplinary and involve collaboration among basic scientists, clinicians, clinical microbiologists, and epidemiologists. The Department is also the home of the local site of the internationally recognized Multicenter AIDS Cohort Study (MACS) on the natural history of HIV infection. Departmental research is translated directly into clinical trials, as exemplified by the NIH integrated clinical and preclinical program that conducts clinical trials on immunotherapy of HIV infection.

The Department is renowned for its public health education programs for health-care providers (The Pennsylvania/MidAtlantic AIDS Education and Training Center) and for the control and prevention of AIDS in various population groups (Pennsylvania Prevention Project). The Department also is the home of the University of Pittsburgh's Center for Research on Health and Sexual Orientation, a unique program that focuses on the health problems of sexual minorities.

Research Facilities

The main portion of the Department is housed in the two contiguous GSPH buildings, Parran and Crabtree Halls. State-of-the-art immunology, molecular biology, and histopathology laboratories are housed throughout four floors of Parran Hall, many of which have recently been renovated as part of an ongoing expansion process. Central Departmental facilities include cold rooms, warm rooms, ultracentrifuges and super-speed centrifuges, a flow cytometry facility, and a fluorescence microscopy and imaging suite. The University of Pittsburgh has a microchemical facility that provides key services such as DNA and peptide synthesis and DNA and protein sequencing, and has genomics and proteomics core laboratories to aid researchers in joining the genomics revolution. Additional facilities include the AIDS prevention offices and MACS clinic in the Keystone Building, a few blocks from Parran Hall, and the Department's newly renovated administrative offices in Crabtree Hall.

Financial Aid

Financial support is provided to every doctoral student, for the full time of their education, in the form of graduate student research (GSR) fellowships. This support includes full tuition, a stipend, health insurance, and student fees and is provided during the entire course of study for all doctoral candidates.

Master's students are eligible for two IDM Public Health Scholarships of $1000 each, which are awarded annually based on academic performance. Awards for travel to scientific meetings are available to graduate students through the Aurelia Koros Travel Scholarship.

Cost of Study

Tuition and fees for the 2006–07 academic year total $17,400 for in-state students and $30,850 for out-of-state students.

Living and Housing Costs

A large number of apartments are available within a 3-mile radius of the Graduate School of Public Health. It is estimated that monthly living expenses total $500–$700 for rent, $250 for food, and $100 for miscellaneous expenses. All University of Pittsburgh students have free use of the Pittsburgh Port Authority bus service for transportation.

Student Group

The Department has approximately 40 graduate students from both the United States and numerous other countries. Each year, approximately 4 to 6 students are admitted to the Ph.D. and Dr.P.H. programs and 6 to 10 students are admitted to the M.S. and M.P.H. programs.

Student Outcomes

Graduates of the Department of Infectious Diseases and Microbiology have obtained positions in academic, government, and profit and nonprofit institutions throughout the world. Alumni are employed at such organizations as GlaxoSmithKline, Centers for Disease Control, Pharmacia/UpJohn, Merck, National Institutes of Health, the University of Pittsburgh Medical Center, the University of California at Los Angeles, Dartmouth Medical School, the University of Maryland, Yale University, Mahidol University (Thailand), and King Faisal Hospital (Saudi Arabia). Graduates of the program have also gone on to pursue medical and other health-care professional degrees.

Location

The Graduate School of Public Health is located in the heart of the University of Pittsburgh campus in the historical Oakland area of Pittsburgh, which is within short walking distance of the Carnegie Library; Carnegie Natural History and Fine Art Museums; the 456-acre Schenley Park, which includes the Phipps Conservatory and Botanical Gardens; Carnegie-Mellon University; Carlow University; and the University of Pittsburgh Medical Center complex. Situated in scenic western Pennsylvania, Pittsburgh has undergone a tremendous rebirth in the technological and banking industries following its reign as a center of steel and coal production. Coupled with the numerous cultural and outdoor activities available, Pittsburgh is one of the most livable and affordable cities in the United States.

The Department

The Department of Infectious Diseases and Microbiology currently has 18 full-time faculty members, including 14 in the biosciences and 4 in public health education and prevention. The Department is housed in 28,713-square-feet of recently renovated state-of-the-art research facilities, of which 11,838 square feet are devoted to laboratories in Parran and Crabtree Halls. Currently, the Department receives more than $11 million annually in extramural research funds.

Applying

Admissions are made for the fall term only. To be considered for admission, completed applications must be received on or before February 1. Candidates are evaluated for admission based upon their academic record, GRE scores, previous work and/or research experience, and letters of recommendation. Additional information can be found on the Department's Web site.

Correspondence and Information

Robin Leaf
Academic Coordinator
A419 Crabtree Hall
Graduate School of Public Health
University of Pittsburgh
130 DeSoto Street
Pittsburgh, Pennsylvania 15261

Phone: 412-624-3331
Fax: 412-383-8926
E-mail: ral9@pitt.edu
Web site: http://www.idm.pitt.edu

Dr. Todd Reinhart
Associate Professor
606 Parran Hall
Graduate School of Public Health
University of Pittsburgh
130 DeSoto Street
Pittsburgh, Pennsylvania 15261

Phone: 412-648-2341
Fax: 412-624-4873
E-mail: reinhar@pitt.edu

University of Pittsburgh

THE FACULTY AND THEIR RESEARCH

Professors
Charles R. Rinaldo Jr., Chairman; Ph.D., Utah, 1973. T-cell immunity to human immunodeficiency virus, human herpesvirus 8 (Kaposi's sarcoma–associated herpesvirus), and hepatitis C virus; vaccines for emerging infections; natural history of HIV infection; clinical virology.
Phalguni Gupta, Assistant Chairman; Ph.D., Wisconsin, 1972. Genetic variation of HIV; cellular and molecular basis of HIV pathogenesis; development of novel vaccine vectors against HIV; biochemical and molecular epidemiology of AIDS; molecular mechanisms of sexual transmission of HIV.

Associate Professors
Simon Barratt-Boyes, Ph.D., California, Davis, 1993. Dendritic cell biology and therapy; primate models of infectious disease; vaccines for emerging infections.
Lawrence A. Kingsley, Dr.P.H., Pittsburgh, 1983. Epidemiology of HIV/AIDS; metabolic/morphologic toxicities (lipodystrophy) of antiretroviral therapy; KSHV, HBV, HCV, and other STDs.
Todd A. Reinhart, Sc.D., Harvard, 1992. In vivo pathogenesis of simian immunodeficiency virus; roles of chemokines in viral immunodeficiencies and tuberculosis.
David T. Rowe, Ph.D., McMaster, 1983. Epstein-Barr virus latent gene expression; relation of viral load to EBV disease progression.
Anthony J. Silvestre, Ph.D., Pittsburgh, 1992. HIV/AIDS, sexuality, human diversity, public health, and sexual orientation.

Assistant Professors
Velpandi Ayyavoo, Ph.D., Madurai Kamaraj (India), 1991. Development of DNA-based vaccine for HIV-1; immunopathogenesis of HIV-1 accessory genes.
Rodger L. Beatty, Ph.D., Pittsburgh, 1997. HIV/AIDS and substance abuse behavioral research; HIV/AIDS education and community-based public health practice.
Linda Frank, Ph.D., M.S.N., Pittsburgh, 1990. AIDS education of health-care professionals; HIV treatment and adherence; mental health issues and HIV; prison health care; educational evaluation and outcome.
Emilia L. Lombardi, Ph.D., Akron, 1997. Discrimination and prejudice in access to health care; HIV prevention/treatment and substance use; gender/transgender/transsexual issues.
Jeremy J. Martinson, Ph.D., Oxford, 1991. Human genomic variation and resistance to HIV infection; genetic mutation and causes of neural tube closure defects; distribution of mutations in different regions of the human genome.
Douglas J. Perkins, Ph.D., Ohio State, 1997. Tropical and international medicine; protective immunity in malaria; genetic basis of disease susceptibility to malaria; impact of HIV/malaria coinfections on the immune response.
Tianyi Wang, Ph.D., Ohio State, 2001. Molecular biology and quantitative proteomics in the study of host defense mechanisms.

Research Assistant Professors
Yue Chen, M.D., Ph.D., Pittsburgh, 1998. Mucosal vaccine development against SIV infection.
Zheng Fan, M.D., Shanghai (China), 1970. Cellular immunity to human and simian immunodeficiency viruses.
Xiao-Li Huang, M.D., Shanghai (China), 1970. Cellular immunity to human immunodeficiency and herpesviruses.
Giovanna Rappocciolo, Ph.D., Milan, 1983. Cellular immunity to herpes viruses and immunodeficiency virus.

FACULTY WITH SECONDARY APPOINTMENTS

Professors
Albert D. Donnenberg, Ph.D., Johns Hopkins, 1980. T-cell lymphopoiesis.
Lee H. Harrison, M.D., Emory, 1982. Epidemiology and molecular epidemiology of bacterial pathogens, including *Escherichia coli* O157:H7, *Streptococcus pneumoniae*, Group B streptococcus, and *Neisseria meningitides;* epidemiology and vaccine prevention of HIV infection in Brazil; enhanced surveillance methods for detection of naturally occurring and bioterrorism-related outbreaks.
Donald A. Henderson, M.D., Rochester, 1954. Biodefense and public health preparedness.
John W. Mellors, M.D., Dartmouth, 1978. HIV antiviral drugs; antiretroviral drug resistance; HIV clinical trials.
Ronald C. Montelaro, Ph.D., Wisconsin, 1975. Molecular basis of HIV-1 and related animal lentivirus persistence and pathogenesis; AIDS vaccine development; antimicrobial peptides.
Patrick S. Moore, M.D., M.P.H., Utah, 1985. Investigation of basic interactions between KSHV and host-cell signaling pathways that contribute to cancer cell development; discovery of new pathogens in chronic diseases (cancers, autoimmune disorders), using molecular methods.
Michael Murphey-Corb, Ph.D., Louisiana State, 1980. Characterization of protective immune responses to SIV; development of SIV vaccines, with an emphasis on the mucosae; pathogenesis of SIV; SIV genetic evolution; development of nonhuman primate model for gene therapy.
Michael A. Parniak, Ph.D., Waterloo, 1978. Mechanisms of action and resistance of HIV-1 to reverse transcriptase inhibitors; RNase H as a potential antiviral, antifungal, and antibacterial target; nonnucleoside reverse transcriptase inhibitors as anti-HIV-1 microbicides; structural and biological properties of modified arabino antisense oligonucleotides.
Robert M. Wadowsky, Sc.D., Pittsburgh, 1983. Real-time PCR for measuring Epstein-Barr virus DNA load in transplant recipients; *Streptococcus pyogenes* and viruses, including rhinovirus, as agents of the common cold.
Victor L. Yu, M.D., Minnesota, 1970. Legionellosis; antibiotic-resistant bacteria.

Associate Professors
Frank J. Jenkins, Ph.D., Pennsylvania, 1984. Herpes simplex virus type 1 latency and reactivation; role of human herpesvirus 8 in Kaposi's sarcoma and other diseases; role of HSV-1 proteins in viral replication.
Deborah K. McMahon, M.D., Temple, 1981. Treatment of HIV/AIDS; HIV clinical trials.
A. William Pasculle, Sc.D., Pittsburgh, 1976. Host defenses and virulence factors of *Legionella* infections; clinical microbiology; bioterrorism; *Clostridium difficile.*

Assistant Professors
Kelly Stefano Cole, Ph.D., Pennsylvania, 1992. Protective immune responses and vaccines for HIV-1.
Richard Day, Ph.D., Berkeley, 1973; M.Sc. (Hyg.), Pittsburgh, 1988. Environmental and occupational epidemiology; psychiatric epidemiology; statistical computing; statistical consulting.
Pawel Kalinski, M.D., Warsaw, 1991; Ph.D., Amsterdam, 1998. Immune polarization; dendritic cell biology; therapeutic uses of dendritic cells.
Sharon Riddler, M.D., Wisconsin, 1986. Treatment of HIV/AIDS; HIV clinical trials; complications of antiretroviral therapy.
Ted Ross, Ph.D., Vanderbilt, 1996. Role of the tax gene of the human T-cell leukemia virus type-2 in transformation of human T lymphocytes.
Raj Shankarappa, M.V.Sc., Mysore (Bangalore); Ph.D., Maryland College Park, 1990. Molecular evolution and pathogenesis of HIV-1 and HCV.

THE UNIVERSITY OF TEXAS AT AUSTIN

School of Biological Sciences
Graduate Program in Microbiology

Programs of Study	The Graduate Program in Microbiology of the School of Biological Sciences offers a program leading to a Ph.D. in microbiology. Direct admission to a master's program is not currently offered. The focus of the graduate program is on solving fundamental problems of biology through modern molecular, biochemical, genetic, biochemical, and immunological approaches. The program encompasses topics concerning the cell and molecular biology, genetics, physiology, and biochemistry of microorganisms and cells of higher organisms. The program of study is designed to provide excellent training and research opportunities individually tailored to each student's needs. Students are offered a broad choice of formal course work, and the section has an extensive seminar program on contemporary topics. Requirements for the doctorate, in addition to formal courses, include passing the preliminary examination (usually administered during the second year of study), participation in student seminars, teaching experience, and original laboratory research culminating in writing a dissertation and, finally, in an oral defense. A laboratory rotation program and a series of introductory lectures by each professor facilitate selection of a suitable research mentor.
Research Facilities	The Graduate Program in Microbiology is housed primarily in the Experimental Science Building but will soon relocate to the recently completed Neural and Molecular Science Building. The program also occupies facilities in the Molecular Biology Building, the Patterson Laboratories Building, and Painter Hall. The facilities for graduate research are state-of-the-art. Laboratories are large, modern, and well equipped and are supported by excellent core facilities. Faculty research covers cellular and molecular immunology; microbial, cell, and developmental biology; phage genetics; molecular virology; molecular approaches to human pathogenesis; and genetic and biochemical analyses of genome transactions, including DNA replication, chromosome segregation, transcription, recombination, and transposition. The research programs of departmental faculty members are well funded, and faculty members publish in outstanding journals. The Program enjoys a very close relationship with the new Institute for Cellular and Molecular Biology and this is providing added vigor to the section's already strong research environment.
Financial Aid	Support may be in the form of fellowships and teaching or research assistantships. The basic stipend is competitive with other institutions and reflects the prevailing salary for a teaching assistant position in the College of Natural Sciences. For the 2004–05 academic year, stipends for teaching and research assistants were about $22,000 for twelve months. Tuition remission is available to teaching assistants. Graduate students are encouraged to apply for scholarships and/or fellowships. As part of the academic requirement for the Ph.D. degree, a student serves as a teaching assistant for two semesters. Teaching assistants and graduate research assistants may be considered for reappointment if they continue to meet the scholastic requirements for eligibility established by the Office of Graduate Studies.
Cost of Study	Tuition and fees for most graduate students during the 2004–05 academic year totaled $6200 (resident tuition for fall and spring semesters combined). Teaching assistants, research assistants, and University fellowship recipients qualify for resident tuition, and the University or research grants pay resident tuition costs for virtually all Ph.D. students.
Living and Housing Costs	The estimated cost of living for students living off campus for 2004–05 for one year was $14,200. Off-campus housing in the city of Austin ranged in price from $400 to $800 per month. Furnished and unfurnished University apartments for married graduate students rented for $408 to $601 per month. The estimated cost of attendance, including living expenses and tuition and fees, ranged from $15,500 to $22,000 for the year.
Student Group	Student enrollment in the Graduate Program in Microbiology for the 2004–05 year totaled more than 65, equally divided between men and women.
Location	Austin, the state capital, with a population of about 570,000, is a clean city with a rapidly growing high-technology industrial community. It is situated on the Colorado River on the fringe of the hill country. The Colorado River has been dammed to create seven scenic lakes near Austin. Outdoor recreational facilities for swimming, boating, and fishing are excellent. The Gulf of Mexico is less than 200 miles away. The climate is moderately dry and warm. Winters are very mild and the summer months are hot, but accommodations are air conditioned. The University sponsors many cultural and sporting events, and the city has a symphony orchestra, a lyric opera, a ballet company, and a repertory theater as well as a wide variety of live music venues.
The University	The University of Texas at Austin is a major research university that is home to 51,000 students, 3,000 faculty members, and 19,000 staff members. From teaching to research to public service, the University's activities support its core purpose: to transform lives for the benefit of society through the core values of learning, discovery, freedom, leadership, individual opportunity, and responsibility.
Applying	Students may begin a program of graduate study in the fall session. The priority deadline for application is January 15. The following materials are required to be submitted for full consideration: application for graduate study, statement of purpose, official college and university transcripts, official test scores, three letters of recommendation, and processing fees. Materials should be submitted to the Graduate and International Admissions Center (GIAC) and to the Microbiology Graduate Affairs Office. Successful applicants typically average a score on the Graduate Record Examinations of at least 1200 (combined quantitative and verbal) and have a GPA of 3.5 (on a 4.0 scale). International students are required to submit TOEFL test scores to the GIAC. The program requires a minimum score of 550 (paper-based) or the equivalent of 213 (computer-based) on the TOEFL. Training and laboratory experience in microbiology, general biology, genetics, molecular biology, and inorganic, organic, and biochemistry is desirable.
Correspondence and Information	Graduate Program in Microbiology The University of Texas at Austin 1 University Station, A5000 Austin, Texas 78712-0162 Phone: 512-471-4181 Fax: 512-471-7088 Web site: http://www.biosci.utexas.edu/graduate/micro

The University of Texas at Austin

THE FACULTY AND THEIR RESEARCH

Karen Artzt, Professor; Ph.D., Cornell. Mammalian developmental genetics.

Henry R. Bose Jr., Professor; Ph.D., Indiana. Malignant transformation of cells by reticuloendotheliosis virus (REV-T).

R. Malcolm Brown Jr., Professor; Ph.D., Texas at Austin. Biosynthesis of cellulose; high-resolution electron and light microscopy; Golgi structure-function.

Clarence S. M. Chan, Associate Professor; Ph.D., Cornell. Mechanism and control of chromosome segregation and polarized cell growth in yeast.

Arturo De Lozanne, Associate Professor; Ph.D., Stanford. Molecular basis of cytokinesis.

Jaquelin P. Dudley, Professor; Ph.D., Baylor College of Medicine. Transcription regulation and oncogenesis.

Charles F. Earhart, Professor; Ph.D., Purdue. Microbial iron assimilation; cell envelope phenomena.

Andrew Ellington, Associate Professor; Ph.D., Harvard. Evolutionary techniques in biopolymer and cell engineering; therapeutic and diagnostic applications of aptamers; protein and metabolic engineering.

George Georgiou, Professor; Ph.D., Cornell. Antibody engineering; protein folding in vivo; bacterial adhesion; biodegradation of recalcitrant pollutants.

Ellen Gottlieb, Assistant Professor; Ph.D., Yale. Regulation of eukaryotic gene expression; molecular mechanisms of mRNA transport and localization in development and in human disease.

David Graham, Assistant Professor; Ph.D., Illinois at Urbana-Champaign. Evolution and enzymology of biosynthetic pathways.

Rasika M. Harshey, Professor; Ph.D., Indian Institute of Science. DNA transposition and signal transduction.

David Herrin, Professor; Ph.D., South Florida. Splicing, mobility, and evolution of group-I introns; circadian clock regulation of transcription.

Jon M. Huibregtse, Associate Professor; Ph.D., Michigan. Ubiquitin proteolysis system.

Vishwanath R. Iyer, Assistant Professor; Ph.D., Harvard. Genomics; genome-wide transcriptional programs and mechanisms; microarray technology.

Makkuni Jayaram, Professor; Ph.D., Indian Institute of Science. Site-specific DNA recombination; molecular symbiosis in yeast.

Arlen W. Johnson, Associate Professor; Ph.D., Harvard. Mechanisms of mRNA degradation and ribosome biogenesis in yeast.

Robert M. Krug, Professor; Ph.D., Rochester. Viral gene expression and replication.

Alan M. Lambowitz, Professor; Ph.D., Yale. Gene expression; ribozymes; retroelements and reverse transcriptases; functional genomics; gene therapy.

Richard J. Meyer, Professor; Ph.D., Pennsylvania. Replication and conjugal transfer of broad host-range plasmids.

Ian J. Molineux, Professor; D. Phil., Oxford. DNA translocation mechanisms; phage therapy; phage-host interactions; evolutionary genetics.

Terry J. O'Halloran, Associate Professor; Ph.D., North Carolina at Chapel Hill. Regulation of membrane traffic in eukaryotic cells.

Tanya T. Paull, Assistant Professor; Ph.D., UCLA. DNA repair and genomic stability.

Shelley M. Payne, Professor; Ph.D., Texas Southwestern Medical Center. Pathogenic mechanisms of *Shigella, E. coli,* and *Vibrio cholerae;* genetics of bacterial iron transport systems.

Bob G. Sanders, Professor; Ph.D., Penn State. The role of vitamin E as a biological response modifier; cellular-molecular mechanisms of vitamin E's actions as an immunomodulator on tumor cell growth.

Scott W. Stevens, Assistant Professor; Ph.D., North Carolina at Chapel Hill. Processing and metabolism of RNA in eukaryotes.

Paul J. Szaniszlo, Professor; Ph.D., North Carolina at Chapel Hill. Fungal cellular and molecular biology; medical mycology.

Ming Tian, Assistant Professor; Ph.D., Harvard. Control and mechanism of DNA recombination required for generating antibody diversity in B cells.

Philip W. Tucker, Professor; Ph.D., MIT. Molecular mechanisms controlling gene expression.

James R. Walker, Professor; Ph.D., Texas at Austin. Chromosome replication; control of initiation and mechanism of polymerization.

THE UNIVERSITY OF TEXAS
HEALTH SCIENCE CENTER AT SAN ANTONIO
Department of Microbiology and Immunology

Program of Study

The Department offers a research-oriented program of graduate study leading to the Ph.D. degree that comprises emerging and established areas of microbiology and immunology. The program promotes a stimulating, supportive, and collegial environment for research training to enable students to pursue professional careers in academic or industrial research and teaching. Students have a high degree of flexibility in identifying research interests and selecting a mentor. Areas of specialization include bacterial, fungal, protozoan, and viral disease pathogenesis; biodefense; immunology; microbial physiology and genetics; molecular and cell biology; morphogenesis; and regulatory biochemistry. In the first year, students pursue laboratory research with several faculty members of their choosing to assist in selecting a mentor for thesis research. Formal course work comprises microbial physiology and pathogenesis, immunology, biochemistry, cell and molecular biology, and virology in the first year and a specialized course in a subsequent year. Students serve as teaching assistants for medical and dental student courses and participate in a seminar program to develop scientific communication skills. The oral qualifying examination is taken by the end of the second year. Students advance to doctoral candidacy after passing the qualifying examination and concentrate on their thesis research while participating in seminars and journal clubs. The mentor supervises the doctoral candidate's research and study, with a student-selected advisory committee periodically reviewing progress. The Ph.D. degree is awarded upon completion of a comprehensive dissertation research project and successful defense before an examining advisory committee. The Ph.D. degree program usually requires five to six years of full-time graduate study and research to complete.

Research Facilities

Laboratories are equipped with state-of-the-art instrumentation necessary for carrying out a broad program of research. Expertise from scientists and clinicians in departments in the Graduate School of Biomedical Sciences and Schools of Medicine or Dentistry is readily accessible and provides additional opportunities for collaborative investigations. Extensive core facilities are available for nucleic acid sequencing or synthesis; fluorescent, confocal, or electron microscopy; microarray technology; bioinformatics; transgenic mouse or embryonic stem cell technology; fluorescence-activated cell sorting; surface plasmon resonance; magnetic resonance imaging; mass spectrometry; and X-ray crystallography. The biomedical library maintains more than 3,000 subscriptions to scientific journals and over 84,000 books. Electronic access to nearly 300 journals and the Internet is available through an extensive computer network. Other facilities include modern animal laboratories and machine, glass, and electrical shops. Adjunct faculty members at the Southwest Foundation for Biomedical Research enable access to research facilities that include one of a select few primate research colonies and biosafety level 4 laboratories in the U.S.

Financial Aid

Students in good academic standing receive full financial support for the entire program of graduate study. Current assistantships include a twelve-month stipend, currently $21,500, and health insurance allowance. Funds originate from several sources, which include NIH training grants, research grants, and teaching assistantships. There are employment opportunities in the scientific, health-care, and business communities for spouses.

Cost of Study

The 2004–05 tuition and fees for the first year were $2942 ($1087 per semester and $768 for the summer term). Most students receive a waiver of nonresident tuition. The fee for student parking was $72 to $336 per year.

Living and Housing Costs

Overall living costs in San Antonio are about 10 percent lower than U.S. national averages. There is no state personal income tax. Private apartments near the University are plentiful, with an average monthly rent of about $500 to $600 for a one-bedroom apartment and $650 to $750 for a two-bedroom apartment. Amounts vary based on amenities offered. Most require six- to twelve-month lease contracts and security deposits.

Student Group

Approximately 5 students are admitted each year. Total current enrollment includes 36 full-time pre-Ph.D. students pursuing careers mainly in scientific research or teaching. Eighteen are women, 8 are members of underrepresented minorities, and 13 are international students. The program seeks ambitious and energetic individuals with the potential for innovation and discovery. The program strives to increase the diversity of the scientific profession in all respects and encourages applications from individuals who are members of groups underrepresented in the sciences.

Student Outcomes

Recent graduates of the Ph.D. program in microbiology and immunology have attained postdoctoral positions nationwide at prestigious academic and research institutions, such as Princeton University; Stanford University; Harvard Medical School; University of California, San Francisco; University of Wisconsin–Madison; M. D. Anderson Cancer Center; and the National Institutes of Health. Previous graduates have attained faculty and academic or industrial positions and have prominent and productive research careers.

Location

San Antonio is a vibrant metropolis bordering the Texas Hill Country with a rich Mexican and American cultural heritage. The city and surrounding area offers many cultural and entertainment opportunities, including the Riverwalk, historic Alamo, NBA champion Spurs, museums, municipal zoo, and Schlitterbahn Waterpark, Sea World, and Fiesta Texas theme parks. Its dry sunny climate, warm winters, and nearby parks, lakes, and rivers provide for outdoor activities throughout the year.

The University

Located in the South Texas Medical Center, The University of Texas Health Science Center at San Antonio comprises the Graduate School of Biomedical Sciences and the Medical, Dental, Nursing, and Allied Health Schools. The Department of Microbiology and Immunology is one of seven biomedical sciences departments created in 1972. With more than 700 faculty members and more than 2,500 students, the University continues to expand its commitment to educational excellence, scholarly research, the highest-quality health care, and public service.

Applying

Applications and supporting information should be submitted to the Office of the Registrar by January 10 for fall admission. Early consideration is given to applications received by December 15. Applicants must have a bachelor's degree or equivalent, satisfactory scores on the Graduate Record Examinations, a strong academic record, and letters of reference from individuals in academic positions. Application forms can be downloaded from the Department's Web site.

Correspondence and Information

Michael T. Berton
Director of Admissions
Department of Microbiology and Immunology
The University of Texas Health Science Center at San Antonio
7703 Floyd Curl Drive–MC 7758
San Antonio, Texas 78229-3900
Phone: 210-567-3952
E-mail: gradadm@uthscsa.edu
Web site: http://www.uthscsa.edu/micro

The University of Texas Health Science Center at San Antonio

THE FACULTY AND THEIR RESEARCH

Primary Faculty

John F. Alderete, Professor; Ph.D., Kansas, 1978. Host-parasite interactions; mechanisms of disease pathogenesis; STDs and trichomonosis; molecular biology of cytoadherence and phenotypic variation.

Joel B. Baseman, Professor and Chairman; Ph.D., Massachusetts, 1968. Mechanisms of host-parasite interplay; identification of novel mycoplasma virulence determinants.

Michael T. Berton, Associate Professor; Ph.D., Tennessee Health Science Center, 1985. Immunoglobulin gene expression and class switch recombination; cytokine and Toll-like receptor signaling; biodefense research; *Francisella tularensis* pathogenesis and immunity.

Santanu Bose, Assistant Professor; Ph.D., Medical College of Wisconsin, 1998. Molecular biology of viral virulence and innate immune defense to parainfluenza virus and respiratory syncytial virus infection.

Subramamanian Dhandayuthapani, Assistant Professor; Ph.D., Madras (India), 1987. Molecular genetics of *Mycobacterium tuberculosis* pathogenesis; two-component regulatory systems.

Peter H. Dube, Assistant Professor; Ph.D., SUNY at Stony Brook, 1999. Bacterial host-pathogen interaction in vivo; inflammation; dendritic cells; *Yersinia enterocolitica;* biodefense research; *Yersinia pestis.*

William G. Haldenwang, Professor; Ph.D., Texas at Austin, 1977. Molecular genetics; bacterial sporulation; RNA polymerase biochemistry.

Kenneth M. Izumi, Assistant Professor; Ph.D., UCLA, 1990. Molecular biology of Epstein-Barr virus transformation of B-lymphocyte growth.

David Kadosh, Assistant Professor; Ph.D., Harvard, 1998. Fungal pathogenesis; *Candida albicans;* filamentous growth; transcriptional regulation; genomics; virulence mechanisms.

David Kolodrubetz, Professor; Ph.D., Brandeis, 1980. Genetics of gram-negative periodontal pathogens; virulence factor expression and function; mechanisms of aerobic/anaerobic transcriptional regulation.

Keith A. Krolick, Professor; Ph.D., UCLA, 1977. Studies of autoimmune disease, with emphasis on myasthenia gravis in both human and animal systems; two-way communication between the immune system and muscle that determines disease severity.

Stephen J. Mattingly, Professor; Ph.D., Medical College of Georgia, 1972. Molecular mechanisms of pathogenicity of group B streptococci and *Pseudomonas aeruginosa.*

Carlos Orihuela, Assistant Professor; Ph.D., Texas Medical Branch at Galveston, 2001. *Streptococcus pneumoniae* virulence gene expression.

Virginia L. Thomas, Associate Professor; Ph.D., Texas Health Science Center at San Antonio, 1973. Pathogenic bacteriology; host-parasite interactions and bacterial biofilm formation.

Brian L. Wickes, Associate Professor; Ph.D., Catholic University, 1992. Medical mycology; molecular mechanisms of fungal pathogenesis; the association of mating type with virulence in *Cryptococcus neoformans.*

Wendell D. Winters, Associate Professor; Ph.D., Illinois at Chicago, 1968. Immune response to virus and tumor cell antigens during therapy in canine and human cancer patients; mechanisms of biological responses to electromagnetic field exposures; antimicrobial herbal medicines.

Yan Xiang, Assistant Professor; Ph.D., Case Western Reserve, 1997. Biology of poxviruses.

Guangming Zhong, Professor; Ph.D., Manitoba, 1991. *Chlamydia* pathogenesis and immune evasion; immune response to *Chlamydia; Chlamydia* and atherosclerosis.

Affiliated Faculty

Sunil K. Ahuja, Professor and Director, Veterans Administration Center for AIDS and HIV infection; M.D., Armed Forces Medical College (India), 1983. Molecular immunology and immunogenetic mechanisms of human immunodeficiency virus (HIV) pathogenesis.

Shou-Jiang Gao, Associate Professor; Ph.D., Bordeaux (France), 1993. Kaposi's sarcoma: associated herpesvirus (KSHV) epidemiology and pathogenesis.

Luis D. Giavedoni, Adjunct Professor; Ph.D., National (Buenos Aires), 1988. Pathogenesis of SIV infection of rhesus macaques.

M. Neal Guentzel, Professor; Ph.D., Texas at Austin, 1972. Public health microbiology; enteric bacteriology.

Anthony J. Infante, Professor; Ph.D., 1977, M.D., 1978, Indiana. Immunology; cellular cooperation in immune responses; immunodeficiency states.

James H. Jorgensen, Professor; Ph.D., Texas Medical Branch at Galveston, 1973. Clinical microbiology; automated and rapid methods of diagnosis; new antimicrobials.

Karl E. Klose, Associate Professor; Ph.D., Berkeley, 1993. Molecular mechanisms of *Vibrio cholerae, Francisella tularensis,* and *Salmonella* pathogenesis; prokaryotic transcription regulation.

Ellen Kraig, Professor; Ph.D., Brandeis, 1981. Molecular immunology; autoimmunity; bacterial pathogenesis.

Robert E. Lanford, Adjunct Professor; Ph.D., Baylor College of Medicine, 1979. Molecular biology of hepatitis B virus and hepatitis C virus replication and pathogenesis.

Peter C. Melby, Associate Professor; M.D., Colorado School of Medicine, 1983. Immunopathogenic mechanisms and protective immunity in leishmaniasis.

Jean L. Patterson, Adjunct Professor; Ph.D., Notre Dame, 1979. RNA viruses.

Judy M. Teale, Professor; Ph.D., Virginia, 1976. Immunobiology, immunology, and pathogenesis of neurocysticercosis; immune response to *Francisella tularensis.*

THE UNIVERSITY OF TEXAS MEDICAL BRANCH

Graduate School of Biomedical Sciences at Galveston
Emerging and Tropical Infectious Diseases Training Program

Program of Study

A program leading to the Ph.D. degree is offered. The program is open to those students holding a B.A., B.S., or an M.S. degree as well as an M.D. or D.V.M. degree. A combined M.D./Ph.D. program is also offered. The goal of the program is to prepare students for a career investigating the mechanisms of human disease. Most students enrolling in the University of Texas Medical Branch (UTMB) Graduate School of Biomedical Sciences (GSBS) enter and take courses within the basic biomedical science curriculum that include biochemistry, cell biology and molecular biology, and genetics as well as choices among various short modular courses on a variety of topics. First-year students are not committed to an individual graduate program and are free to examine the research and academic opportunities within all the GSBS programs. Students select a graduate program at the end of the first year. Upon entering the Emerging and Tropical Infectious Diseases Training Program, three different tracks are available depending on a student's basic interests in infectious diseases, toxicology, or pathobiology. Students are required to take courses in pathobiology of human diseases, ethics in science, teaching in pathology and experimental design, and grant writing. Students then develop a customized course of study depending on their academic background and interest by selecting among a variety of elective courses that fall into groups along the basic three educational tracks. Electives include courses in the evolution and pathogenesis of infectious diseases, tropical diseases, virology, viral phylogenetics, emerging infectious diseases, vector biology, toxicology, cardiovascular toxicology, and neuropathology. Course requirements are flexible and depend on the needs and interests of the student; students are encouraged to take courses in other programs. A preliminary examination and a dissertation are required.

Research Facilities

Research and clinical laboratories in the Department of Pathology are well equipped to support research in the areas of biodefense, tropical and emerging infectious diseases, toxicology, and pathobiology. Laboratories and core facilities are equipped for research that includes molecular biology; DNA sequencing; real-time PCR; imaging; protein chemistry; conventional, fluorescent, confocal, and electron microscopy; histochemistry; whole-body autoradiography; pharmacokinetic studies; enzymology; and proteomics and genomics. Special resources for infectious disease research include biosafety level (BSL) 2 and level 3 containment laboratories and insectaries and a recently completed BSL4 laboratory. The BSL4 facility is the only one of its kind on a U.S. university campus and enhances the University's research programs in biodefense and emerging infectious diseases. UTMB is one of eight institutions nationwide receiving grants to establish a Regional Center of Excellence (RCE) for Biodefense and Emerging Infectious Diseases Research from the National Institute of Allergy and Infectious Diseases (NIAID). The University has an exceptionally distinguished team of scientists in emerging and infectious diseases and biodefense. NIAID recently announced that UTMB is going to be the site of a $167-million Galveston National Laboratory (GNL), one of two large-scale national research facilities focusing on new and emerging disease threats. The GNL will serve scientists from all over the country, supplying much-needed space for research aimed at fighting microbes that could be used in a biological terrorist attack, as well as those that constitute emerging natural threats such as influenza and West Nile virus. UTMB is the only institution in the nation to achieve both the GNL and RCE designations. UTMB is also home to three World Health Organization collaborating centers, including one in tropical and emerging infectious diseases. Special resources for toxicology research include a National Institute of Environmental Health and Safety Research Center, a facility for inhalation exposures, and core facilities for custom microarrays, protein chemistry, and mass spectrometry. There is more than 385,000 square feet of space dedicated to research that is supported by more than $249 million in research grants. The Medical Branch has one of the nation's largest and most comprehensive medical libraries, which contains advanced information retrieval systems and subscribes to more than 4,600 biomedical periodicals in print or electronic forms.

Financial Aid

Graduate assistantships are available. In 2006–07, initial graduate assistant salaries are approximately $23,000 per year for Ph.D. students and $16,456 for M.S. students. Departmental graduate assistants qualify for in-state tuition fees.

Cost of Study

In the 2006–07 academic year, tuition is $50 per credit hour, and full-time students are required to take at least 9 units per term. The student service fee is $10.99 an hour (up to 12 hours), with a maximum of $150 per term (up to 18 hours). These tuition rates are subject to change. Student health fees are $55 per semester.

Living and Housing Costs

Living costs in Galveston vary according to personal lifestyles. Costs for rooms in University dormitories and fraternities begin at approximately $323 per month, per person, double occupancy, and the accommodations are within walking distance of the campus. Private rooms are $570 per month. Apartments are also available for single students beginning at approximately $670 per month.

Student Group

In 2004, total student enrollment at the Medical Branch was 2,335; 320 were graduate students. Approximately 30 students were enrolled in the experimental pathology program.

Location

Galveston, the oldest city and port in Texas, is located on Galveston Island, which is connected to the mainland by causeways and a ferry. The city, with a population of 70,000, is a major port and resort area on the Gulf of Mexico. The scientific facilities of the NASA Manned Spacecraft Center are nearby.

In addition to the usual recreational facilities of a city, Galveston Island has 32 miles of Gulf-front beach and numerous bay areas for water-related activities, including fishing, sailing, surfing, and waterskiing. Just an hour's drive away is the nation's fourth-largest city, Houston, with major shopping malls; acclaimed symphony, ballet, opera, and theater companies; music of all kinds; and professional sports teams.

The Medical Branch

The University of Texas Medical Branch is one of four Health Science Centers in the University of Texas System and includes, in addition to the Graduate School of Biomedical Sciences, the School of Allied Health Sciences, the School of Nursing, the School of Medicine, the World Health Organization Collaborating Center for Tropical Diseases, the Institute for the Medical Humanities, and seven hospitals. The Medical Branch is spread over 91 acres and is one of the largest centers for biomedical education and research in the Southwest. It has a faculty of 2,335 members, with 330 graduate faculty members. The Medical Branch is currently undergoing the largest physical expansion program in its 113-year history.

Applying

Applicants should hold a bachelor's degree or the equivalent, should have appropriate course preparation for the proposed area of study, and must have taken courses in biology, general chemistry, organic chemistry, and physics. An overall and advanced grade point average of 3.0 and the Graduate Record Examinations General Test are minimum admission requirements.

The Graduate School operates on a trimester system, with terms beginning in September, January, and April. Application materials should be received as early as possible and no later than ninety days before the expected date of enrollment. In general, students begin graduate school only in the fall semester. Personal interviews are encouraged.

Correspondence and Information

Dr. Alan T. Barrett
Emerging and Tropical Infectious Diseases Training Program
Center for Tropical Diseases
The University of Texas Medical Branch
1.116 Keiller Building
301 University Boulevard
Galveston, Texas 77555-0609
Phone: 409-772-6662
Fax: 409-747-2500
E-mail: hediaz@utmb.edu
Web site: http://www.utmb.edu/predocetid

The University of Texas Medical Branch

THE FACULTY AND THEIR RESEARCH

Program Faculty
Alan D. T. Barrett, Ph.D. Molecular basis of attenuation, virulence, and immunogenicity of flaviviruses.
Ashkok Chopra, Ph.D. Virulence factors from gram-negative pathogens.
Stephen Higgs, Ph.D. Interactions between mosquito vectors, the viruses they transmit, and their vertebrate hosts.
Gary R. Klimpel, Ph.D. Mucosal immune response to bacterial and viral infections.
Stanley M. Lemon, M.D. Molecular virology and mechanisms of pathogenesis of hepatitis C virus.
David W. Niesel, Ph.D. Molecular basis of pathogenicity of mycobacteria and enteric bacteria.
C. J. Peters, M.D. Pathogenesis of viral hemorrhagic fevers and related conditions.
Johnny W. Peterson, Ph.D. Molecular mechanisms of secretory and inflammatory diarrheal diseases.
Norbert J. Roberts Jr., M.D. Viral pathogenesis and antiviral immunity.
Chiaho Shih, Ph.D. Molecular mechanisms of chronicity of viral hepatitis and liver pathogenesis.
Lynn Soong, M.D., Ph.D. Molecular mechanisms of protective immunity and pathogenesis associated with *Leishmania* infections.
Robert B. Tesh, M.D. Epidemiology of rodent-associated and arthropod-borne viral diseases.
David H. Walker, M.D. Molecular studies of immunity to and pathogenesis of rickettsial and ehrlichial diseases.
Scott C. Weaver, Ph.D. Molecular genetics, evolution, and ecology of Venezuelan equine encephalitis emergence and virus-vector interactions.

Adjunct Program Faculty
Thomas Albrecht, Ph.D. Role of cellular proliferative signaling responses in the pathogenesis of herpesviruses.
Judith F. Aronson, M.D. Pathogenesis of arenavirus hemorrhagic fever in a guinea pig model.
Charles F. Fulhorst, D.V.M., Dr.P.H. Epidemiology and ecology of rodent-borne arenaviruses and hantaviruses.
Nisha Garg, Ph.D. Understanding the molecular mechanisms essential for growth; development of differentiation of *Trypanosoma cruzi*.
David G. Gorenstein, Ph.D. Structure-based design and structural biology of antiviral agents.
Norbert K. Herzog, Ph.D. Cellular signal transduction and oncogenesis.
Clifford W. Houston, Ph.D. Role of cytolytic enterotoxins in *Aeromonas*-mediated diseases.
S. David Hudnall, M.D. Molecular epidemiology and pathogenesis of human gammaherpesviruses (Epstein-Barr and HHV-8) infections.
Bruce Luxon, Ph.D. Structural biology of proteins and DNA using NMR and computational biology.
Peter W. Mason, Ph.D. Understanding and controlling arthropod-borne viral diseases.
William A. O'Brien, M.D. Mechanism of HIV entry into primary cells and virologic correlates of clinical progression of AIDS.
Vsevolod L. Popov, D.Sc. Cellular mechanisms of pathogenesis of rickettsial diseases at the ultrastructural level.
Victor E. Reyes, Ph.D. Regulation of antigen presentation during mucosal immune responses.
Dennis M. Walling, M.D. Molecular mechanisms of Epstein-Barr virus pathogenesis in epithelial cells, especially oral hairy leukoplakia.
Stanley J. Watowich, Ph.D. Structural studies of viral proteins and structure-based antiviral drug design.
Douglas Watts, Ph.D. Ecology and epidemiology of arbovirus diseases.
Shu-Yuan Xiao, M.D., Ph.D. Pathogenesis of arboviruses that cause hepatitis, encephalitis, and hemorrhagic fever.

THE UNIVERSITY OF TEXAS MEDICAL BRANCH

Graduate School of Biomedical Sciences at Galveston
Microbiology and Immunology Program

Programs of Study

An interdepartmental program leading to the Ph.D. degree is offered in microbiology and immunology. Subspecialty areas in which candidates can obtain training include bacterial and viral pathogenesis, parasitology, microbial genetics, molecular virology, host defense, neuroimmunoendocrinology, autoimmunity, and structural and molecular biology.

Students are engaged in laboratory research throughout the program and rotate through as many as three laboratories during the first year. Students are encouraged to begin their dissertation research early in their graduate program and should identify a major professor by the end of the first year. Formal course work is generally completed by the end of the second year. After passing qualifying examinations and submitting an approved research proposal, students are admitted to candidacy with the appointment of a research supervisory committee.

Although graduate training is directed primarily toward research, students have the opportunity to obtain teaching experience during their graduate program.

Research Facilities

Students have access to state-of-the-art research facilities within faculty laboratories and may utilize extensive core facilities at the University. The Department of Microbiology and Immunology is well equipped for modern microbiological research, with more than 34,000 square feet of space in the Medical Research Building. In addition, other program faculty members have laboratories located within the Departments of Internal Medicine, Obstetrics and Gynecology, Pediatrics, and Pathology. Numerous core facilities are available in the department and on the campus and provide excellent flow cytometry, NMR, and X-ray crystallography services, and routine peptide and nucleic acid sequencing and synthesis, among others. A modern medical library with extensive journal collections and online services is available.

Financial Aid

Graduate assistantships are available with a stipend of $23,000 when the student enters the first year Basic Biomedical Sciences Curriculum. A variety of funding sources (grants and fellowships) support students in subsequent years of study. The James W. McLaughlin Fellowship Fund awards competitive fellowships to UTMB predoctoral students working in the area of infection and immunity. Departmental graduate assistants pay tuition at the Texas resident rate.

Cost of Study

Basic tuition in 2005–06 was $50 per hour, with a minimum of $450 per term for Texas residents, and $325 per hour for nonresidents. Certain courses require minimal lab fees. The student activity fee was $10.99 per hour, with a maximum of $98.91 per term. These tuition rates are subject to change.

Living and Housing Costs

Rooms in University dormitories and fraternities begin at approximately $195 per month per person, double occupancy, and are within walking distance of the campus. Private rooms are $260 per month. Apartments are also available for single and married students, with rents beginning at approximately $245 per month per person, double occupancy.

Student Group

The department has an average enrollment of approximately 30 graduate students. About half are married.

Location

The city of Galveston, with a population of approximately 70,000, is a pleasant and historic island community that is the commercial center of Galveston County and a popular resort on the Gulf of Mexico. In addition to UTMB, the island is home to Galveston College and the Galveston campus of Texas A&M University, which are located nearby. The scientific facilities of the NASA Johnson Space Center are also located nearby. Houston, with a wide variety of academic, scientific, and entertainment resources, is located about 45 miles north of Galveston.

The University and The Medical Branch

The University of Texas Medical Branch is one of four Health Science Centers in the University of Texas System and includes the Graduate School of Biomedical Sciences, the School of Allied Health Sciences, the School of Nursing, the School of Medicine, the Marine Biomedical Institute, and the Institute for the Medical Humanities. The Medical Branch is one of the largest centers for biomedical education and research in the Southwest. It has approximately 1,000 faculty members. The graduate faculty numbers more than 300.

Applying

Applicants should have an undergraduate record that demonstrates a strong preparation in the sciences. Commonly, a successful applicant will have an overall grade point average of more than 3.0 (B). The applicant must take the verbal, quantitative, and writing sections of the Graduate Record Examinations (GRE). No minimum score is required; however, scores are useful to the Admissions Committee during their deliberations. Information pertaining to the GRE can be found at http://www.gre.org. The institution and department codes for UTMB-Biomedical Sciences are R6887-0299, respectively. A TOEFL score of no less than 600 is required for international students. However, students with lower scores but having research experience are frequently considered. The Graduate School operates on a trimester system, with terms beginning in September, January, and April. Application materials should be received by March 15 for September enrollment. New students are not enrolled in the spring (January) and summer (April) terms. The application fee is $30 for U.S. citizens and permanent residents and $75 for international applicants.

Correspondence and Information

Thomas K. Hughes Jr., Ph.D., Program Director
Microbiology and Immunology Graduate Program
The University of Texas Medical Branch
301 University Boulevard
Galveston, Texas 77555-1019
E-mail: micro-immuno.gsbs@utmb.edu

Martha Lewis, Program Secretary
Microbiology and Immunology Graduate Program
The University of Texas Medical Branch
301 University Boulevard
Galveston, Texas 77555-1019
Phone: 409-772-2322
E-mail: mlewis@utmb.edu
Web site: http://gsbs.utmb.edu
　　　　http://microbiology.utmb.edu

The University of Texas Medical Branch

THE FACULTY AND THEIR RESEARCH

For information on current publications, students should visit http://microbiology.utmb.edu.

T. Albrecht, Professor (Microbiology and Immunology); Ph.D., Penn State Hershey Medical Center, 1973. Cellular and molecular pathogenesis of herpesviruses; signal transduction; regulation of viral and cellular gene expression; cell-cycle perturbation; induction of cellular damage.

J. Aronson, Associate Professor (Pathology); M.D., North Carolina at Chapel Hill, 1985. Arenavirus pathogenesis.

S. Baron, Professor (Microbiology and Immunology, Internal Medicine); M.D., NYU, 1952. Interferon action; innate immunity; HIV; smallpox; cancer.

A. Barrett, Professor (Pathology); Ph.D., Warwick (England), 1984. Molecular basis of genetics of flaviviruses; bunyaviruses; tropical virus diseases; vaccine development.

D. Beasley, Assistant Professor (Microbiology and Immunology); Ph.D., Queensland (Australia), 1999. Molecular determinants of arbovirus virulence and antigenicity; emerging diseases; diagnostic assays.

I. Boldogh, Associate Professor (Microbiology and Immunology); D.M.&B., Ph.D., Debrecen (Hungary), 1974. Oxidative stress; functional changes of mitochondria in allergic inflammation and aging processes.

N. Bourne. Associate Professor (Pediatrics, Microbiology and Immunology) Ph.D., Wales, 1987. Genital herpes; hepatitis C virus; vaccines and antiviral agents.

A. Casola, Associate Professor (Pediatrics); M.D., Naples (Italy), 1989. Paramixovirus immunopathogenesis; inflammation; epithelial cell signaling.

T. Chan, Professor (Microbiology and Immunology); M.D., Taiwan, 1963; Ph.D., Yale, 1969. Transgenic mouse models for SARS and hepatitis C; asthma research using a mouse model; gene therapy.

A. K. Chopra, Professor (Microbiology and Immunology); Ph.D., Kurukshetra (India), 1982; C.Sc., Czechoslovak Academy of Sciences, 1984. Virulence gene regulation; pathogenic mechanisms; cell signaling; vaccine development; inflammation.

P. Christadoss, Professor (Microbiology and Immunology); M.D., Madras (India), 1977. Autoimmunity; cellular and molecular mechanisms of immunopathogenesis; new therapy for autoimmune diseases, i.e., myasthenia gravis.

M. W. Cloyd, Professor (Microbiology and Immunology); Ph.D., Duke, 1977. Biological, genetic, and molecular mechanisms of HIV and mouse leukemia virus pathogenesis.

D. H. Coppenhaver, Associate Professor (Microbiology and Immunology); Ph.D., Duke, 1976. Antimicrobial host defenses; nutrition in cancer and infection; molecular evolution.

R. A. Davey, Assistant Professor (Microbiology and Immunology); Ph.D., Adelaide (Australia), 1993. Murine leukemia viruses; gene therapy; viral entry.

T. Eaves-Pyles, Assistant Professor (Microbiology and Immunology); Ph.D., Cincinnati, 1998. Pathogenic bacteria and their products induce innate immune response in the human host.

D. M. Estes, Professor (Pathology, Microbiology and Immunology); Ph.D., Texas A&M, 1988. Immunoregulation and host defense.

I. V. Frolov, Associate Professor (Microbiology and Immunology); Ph.D., Institute of Molecular Biology (Russia), 1988. Virus replication; alphaviruses; flaviviruses; virus-host interactions.

C. F. Fulhorst, Associate Professor (Pathology); D.V.M., California, Davis, 1978; Dr. P.H., Berkeley, 1994. Epidemiology, ecology, and pathogenesis of rodent-borne viral zoonoses.

N. Garg, Assistant Professor (Microbiology and Immunology); Ph.D., Haryana Agricultural (India), 1988. Host pathomechanisms and mitochondrial abnormalities in progression of chagasic myocarditis; identification and testing of *T. cruzi* genes as vaccine candidates.

R. P. Garofalo, Professor (Pediatrics); M.D., Milan (Italy), 1985. Respiratory syncytial virus immunopathogenesis; airway inflammation; cytokine transcription; dendritic cell function; epithelial cell signaling.

J. A. Grant, Professor (Internal Medicine); M.D., Duke, 1966. Potency of antihistamines.

N. K. Herzog, Associate Professor (Pathology) and Assistant Dean, Graduate School of Biomedical Sciences; Ph.D., Texas at Austin, 1985. Molecular biology of cellular signal transduction in viral pathogenesis, using genomics, proteomics, and kinomics; antivirals; hemorrhagic fever viruses.

C. W. Houston, Herman Barnett Distinguished Professor (Microbiology and Immunology) and Associate Vice President, Educational Outreach; Ph.D., Oklahoma Health Sciences Center, 1979. *Aeromonas hydrophila*, cytotoxic enterotoxin.

T. K. Hughes Jr., Professor; Vice Chair, Graduate and Postgraduate Education; and Program Director (Microbiology and Immunology); Ph.D., Texas Medical Branch, 1982. Cytokine mechanisms of action; neuroimmunoendocrinology.

G. Klimpel, Professor (Microbiology and Immunology); Ph.D., Arizona, 1976. Innate and adaptive immune responses to different bacterial pathogens; leptospirosis; bioterrorist agents such as *Francisella tularensis*; bacterial pathogenesis.

R. König, Associate Professor (Microbiology and Immunology, Sealy Center for Molecular Science); Ph.D., Bern (Switzerland), 1984. Molecular and cellular mechanisms of T-cell activation and differentiation; signal transduction in T cells leading to activation or apoptosis; functional genomics of inflammation and autoimmune diseases; antitumor and biodefense vaccinology.

S. M. Lemon, Professor (Microbiology and Immunology) and Director, Institute of Human Infections and Immunity; M.D., Rochester, 1972. Molecular mechanisms involved with replication of positive-strand RNA viruses and the pathogenesis of human hepatitis viruses, particularly hepatitis C virus; antiviral drug development; mechanisms of immune evasion of hepatitis C virus.

M. A. Lett-Brown, Associate Professor (Internal Medicine); Ph.D., London, 1973. New sources of IL-4 in allergic/immune reactions.

K. Li, Assistant Professor (Microbiology and Immunology); M.D., 1994, Ph.D., 1999, Beijing Medical. Virus-host interactions in pathogenesis of hepatitis C virus and related viruses; innate immunity to viral infections.

S. Makino, Professor (Microbiology and Immunology); Ph.D., Tokyo, 1984. Molecular virology and pathogenesis of coronavirus, bunyavirus, and hepatitis C virus.

P. W. Mason, Professor; Ph.D., Massachusetts Amherst. Flavivirus (West Nile virus and dengue virus) and picornavirus (foot-and-mouth disease virus) pathogenesis and control.

J. W. McBride, Assistant Professor (Pathology, Microbiology and Immunology, Sealy Center for Vaccine Development); Ph.D., California, Davis. Obligate intracellular bacteria; pathobiology, immunity, molecular pathogenesis, vaccine development and diagnostics.

T. McNearney, Associate Professor (Anatomy and Neurosciences, Internal Medicine, Microbiology and Immunology); M.D., Saint Louis, 1981. Viral interference of host immunosurveillance; autoimmunity; contribution of neurotransmitters to generation and persistence of inflammatory arthropathies; outcome studies in systemic sclerosis.

G. N. Milligan, Associate Professor (Pediatrics, Microbiology and Immunology); Ph. D., Missouri–Columbia, 1988. Viral immunology; immunity to herpes simplex virus; neuroimmunology; mucosal immunology.

V. L. Motin, Associate Professor (Pathology, Microbiology and Immunology); Ph.D., Gamaleya Research Institute (Moscow), 1988. *Yersinia pestis* pathogenesis; vaccine; therapeutics.

J. E. Nichols, Associate Professor (Internal Medicine/Division of Infectious Diseases, Anesthesiology, Microbiology and Immunology); Ph.D., Texas Medical Branch, 1999. Viral pathogenesis and influence of aging on host response to pathogens (immunosenescence) and development of human model systems using tissue engineering.

D. W. Niesel, Professor (Microbiology and Immunology), Chair, and Vice Dean, Graduate School of Biomedical Sciences; Ph.D., North Carolina State, 1980. In vivo virulence gene expression; response to low-shear environments; bacterial pathogenesis; functional genomics.

W. A. O'Brien, Professor (Internal Medicine, Pathology, Microbiology and Immunology); M.D., Minnesota, 1982. Pathogenesis and treatment of HIV, including mechanism of HIV entry; antiretroviral resistance.

J. W. Peterson, Professor (Microbiology and Immunology); Ph.D., Texas Health Science Center at Dallas, 1972. Toxins; inflammation; molecular pathogenesis of bacterial infections of the intestine; therapeutics against diarrheal disease and anthrax.

R. B. Pyles, Assistant Professor (Pediatrics, Microbiology and Immunology) and Scientist, Sealy Center for Vaccine Development; Ph.D., Cincinnati, 1993. Herpesvirology; molecular virology; viral pathogenesis; innate immunology of mucosa.

S. Rajaraman, Professor (Pathology); M.D., Madras (India), 1966. Renal immunopathology; gut-renal peptides.

V. E. Reyes, Professor (Pediatrics, Microbiology and Immunology); Ph.D., Texas Medical Branch, 1986. Antigen processing; T-cell presentation; mucosal immunity; *H. pylori*; pathogenesis.

N. J. Roberts Jr., Professor (Internal Medicine) and Director, Division of Infectious Disease; M.D., NYU, 1971. Viral pathogenesis; respiratory syncytial virus; influenza virus.

E. R. Sherwood, Professor (Anesthesiology, Microbiology and Immunology); Ph.D., Tulane, 1986; M.D., Chicago, 1994. T cells and natural killer cells in systemic inflammation innate immunology; immunomodulation.

C. Shih, Professor (Pathology); Ph.D., MIT, 1982. Molecular virology and pathogenesis of human hepatitis B virus and hepatoma; vaccine development for tropical infectious agents.

E. M. Smith, Professor (Psychiatry and Behavioral Sciences); Ph.D., Baylor College of Medicine, 1980. Psychoneuroimmunology; immune and neuroendocrine systems; hormone production in immune responses.

L. Soong, Associate Professor (Microbiology and Immunology, Pathology); M.D., Shanghai (China), 1983; Ph.D., Georgia, 1992. Protective immunity and pathogenesis of intracellular pathogens; dendritic cell–pathogen interactions; cytokines/chemokines.

L. R. Stanberry, Professor (Pediatrics, Microbiology and Immunology); Chairman, Department of Pediatrics; and Director, Sealy Center for Vaccine Development; M.D., 1977, Ph.D., 1979, Illinois at Chicago. Herpes simplex virus pathogenesis and control; prophylactic and therapeutic vaccines.

J. Sun, Assistant Professor (Microbiology and Immunology, Sealy Center for Vaccine Development); M.D., Shanghai, 1983; Ph.D., Georgia, 1992. Protective immunity and pathogenesis of hepatitis C virus infection.

R. B. Tesh, George Dock Distinguished Professor (Pathology); M.D., Jefferson Medical, 1961. Epidemiology of arthropod-borne and other zoonotic viral diseases.

A. G. Torres, Assistant Professor (Microbiology and Immunology, Pathology, Sealy Center for Vaccine Development); Ph.D., Texas at Austin, 1999. Diarrheal disease, bacterial genetics, mechanisms of adhesion.

D. H. Walker, Professor and Chairman (Pathology); M.D., Vanderbilt, 1969. Immunity to and pathogenesis of arthropod-transmitted obligate intracellular bacterial infections (*Rickettsia* and *Ehrlichia*).

D. M. Walling, Associate Professor (Internal Medicine, Microbiology and Immunology); M.D., Johns Hopkins, 1987. Pathogenesis and oncogenesis of Epstein-Barr virus.

S. C. Weaver, Professor (Pathology); Ph.D., California, San Diego, 1993. Ecology, epidemiology, evolution, and pathogenesis of arboviral diseases; antiviral and vaccine development.

E. B. Whorton Jr., Professor and Director, Molecular Epidemiology Research Program; Ph.D., Oklahoma, 1968. Biostatistics and epidemiology methods of environmental and genetic toxicology and infectious diseases.

UNIVERSITY OF VIRGINIA

School of Medicine
Immunology Training Program

Program of Study

The Immunology Training Program emphasizes the use of contemporary techniques in cellular and molecular biology to address fundamental questions in immunology. The program places a major emphasis on a multidisciplinary approach to science that crosses traditional departmental boundaries. Reflecting this, the program faculty members are also members of the Departments of Microbiology, Biochemistry, Biomedical Engineering, Cell Biology, Pharmacology, Internal Medicine, Pathology, and Chemistry. Collectively, the faculty has a proven track record of research productivity, collegiality, and effective training of graduate students. A sample of the current research areas in the program include antigen processing and presentation, lymphocyte development, various aspects of signal transduction in cells of the immune system, lymphocyte migration, apoptosis, autoimmunity, viral immunology (including AIDS), and cancer immunology. Individuals become Ph.D. students in immunology when they choose a Ph.D. mentor with an appointment in an immunology program affiliated with the department (e.g., microbiology).

The course work in the interdisciplinary Immunology Training Program favors an emphasis on contemporary immunological issues coupled with a broad exposure to cell and molecular biology. The course work is generally completed by the middle of the second year. In addition, there are a number of formal and informal seminars and student colloquia for both learning and presenting scientific information. The real substance of the Ph.D. comes with the thesis research each student undertakes to present an original and significant contribution to the general field of immunology. In the first year, students generally do three laboratory rotations, at the end of which they pick their thesis lab. The development of the student as an independent scientist is fostered by a close working relationship between the student and his or her research adviser as well as by a thesis committee. Students receive individual attention in designing a course of study, completing their course work and laboratory rotations, and devising and conducting their thesis research. Most students complete their studies in four to six years. The dynamic interaction of various disciplines enables the students to develop the broad knowledge and special techniques necessary for a rewarding and successful career in the biomedical sciences.

Research Facilities

The research laboratories of the Immunology Training Program occupy lab space in several floors of adjoining buildings in the School of Medicine at the University of Virginia (UVA). These modern laboratory facilities provide an outstanding environment for conducting cutting-edge research in immunology. Immunology students have access to comprehensive research support services and equipment to meet their needs, such as flow cytometry and single-cell sorting, generation of transgenic and knockout mice, generation of hybridomas, specialized tissue culture, confocal microscopy/imaging technology, electron microscopy, and magnetic resonance imaging of small animals. In addition, a comprehensive biomolecular core facility performs DNA sequencing, oligonucleotide synthesis and chip microarray analysis, peptide synthesis, and various types of protein analysis. The latter is complemented by a world-renowned mass spectrometry facility. The newly renovated Claude Moore Health Sciences Library houses a huge collection of books and 3,375 journals, many of which are also available online in full-text form to UVA students and faculty. In addition, the Carter Immunology Center library also hosts a specific collection of immunology-related journals. The Academic Computing–Health Sciences Office provides expert assistance on computer-related issues, from data acquisition and imaging to broad area networking and bioinformatics.

Financial Aid

All students accepted into the program are given financial support in the form of an annual stipend. In 2006–07, the stipends were $23,100 to $26,100. Sources of financial support include national and University fellowships, National Research Service Awards of the NIH, teaching fellowships, and research assistantships.

Cost of Study

Tuition, fees, and health-care costs are covered by traineeships and fellowships.

Living and Housing Costs

University housing is available for single and married students, with rents that ranged from $400 to $600 a month in 2006–07. In addition, the Charlottesville community has extensive accommodations in a wide range of rents. Housing opportunities range from downtown apartments and condominiums to secluded country cottages.

Student Group

There are approximately 19,200 students enrolled at the University of Virginia, including 6,067 graduate and professional students. Approximately 280 students are currently enrolled in graduate programs of the School of Medicine (not including medical students). Training opportunities are provided by 235 faculty members from twenty-nine basic science, clinical, and engineering departments, offices, and centers. The small size of the institution and the compact campus further serve to foster dynamic interaction between all disciplines.

Location

Charlottesville is a university city located within historic Albemarle County, with a population of about 120,000. Charlottesville has been consistently ranked as one of America's best places to live by magazines such as *Money, Outside,* and *Kiplinger's Personal Finance.* Theaters, concert halls, and other recreational and athletics facilities are within easy access. The city is located about 20 miles from the Blue Ridge Mountains and Shenandoah National Park and 110 miles from Washington, D.C.

The University

The University of Virginia was founded in 1819 by Thomas Jefferson and is the major research university of the state of Virginia, attracting scholars and students from around the world. It is routinely ranked as one of the top two public universities in the country. The grounds of the University, which are exceptionally attractive, encompass about 1,600 green acres, and all facilities are within easy walking distance of each other.

Applying

Students interested in a Ph.D. in immunology should apply via the Biomedical Sciences Program at UVA at http://www.healthsystem. virginia.edu/internet/bims/. Applicants may choose to apply to one of seven Graduate Groups in Biomedical Sciences. Most students enter through the Microbiology, Immunology, and Infectious Disease Graduate Group, while others enter through such groups as Molecular Medicine. The Biomedical Sciences Web site describes the application process, done at http://artsandsciences. virginia.edu/admissions/apply.html.

Correspondence and Information

For program information:
Glenn Glover
Graduate Program Administrator
Carter Immunology Center
University of Virginia
P.O. Box 801386
Charlottesville, Virginia, 22908
Phone: 434-924-2412
E-mail: gmg6n@virginia.edu
Web site: http://www.healthsystem.virginia.edu/internet/CIC/
home.cfm

For application materials:
Office of the Dean
Graduate School of Arts and Sciences
422 Cabell Hall
University of Virginia
P.O. Box 400772
Charlottesville, Virginia 22904-4772
Web site: http://www.virginia.edu/~resadm/gsas/gsas.html

University of Virginia

THE FACULTY AND THEIR RESEARCH

Timothy P. Bender, Ph.D., Professor of Microbiology. B-cell development: regulation of the c-myb expression, modulation of its transcription activating properties, and its role in lymphocyte development and function.

W. Kline Bolton, M.D., Professor of Internal Medicine (Nephrology). Study of the mechanisms of autoimmune glomerulonephritis in man, using experimental animal models.

Larry Borish, M.D., Associate Professor of Internal Medicine (Allergy and Clinical Immunology). Molecular genetic studies in allergic disease and asthma.

Thomas J. Braciale, M.D., Ph.D., Professor of Microbiology and of Pathology. Host immune response to viral infection; virus and host interactions in the respiratory tract; T-cell effector activity.

Michael G. Brown, Ph.D., Assistant Professor of Internal Medicine (Rheumatology, Immunology, and Microbiology). NK cells and viral immunity.

Timothy N. J. Bullock, Ph.D., Assistant Professor of Research in Microbiology. CD4+ T-cell responses to melanoma; antigen processing and presentation by dendritic cells.

Steven M. Cohn, M.D., Ph.D., Associate Professor of Internal Medicine (Gastroenterology and Hepatology). Communication between the immune system and epithelium in chronic, inflammatory intestinal diseases.

Fabio Cominelli, M.D., Ph.D., Professor of Internal Medicine (Gastroenterology and Hepatology). Cytokine network in the regulation of mucosal immune responses and its alteration in association with intestinal inflammation.

Victor H. Engelhard, Ph.D., Professor of Microbiology. Identification of peptide antigens presented by MHC molecules to T lymphocytes; tumor immunology and self-tolerance; development of cancer vaccines.

Peter B. Ernst, D.V.M., Ph.D., Professor of Internal Medicine. Gastrointestinal inflammation, with specific interests in lymphoepithelial cell interactions in *H. pylori* infection and inflammatory bowel disease.

Shu Man Fu, M.D., Ph.D., Professor of Internal Medicine (Rheumatology) and of Microbiology. Lymphocyte activation and differentiation in normal and autoimmune states in man.

Joanna Goldberg, Ph.D., Associate Professor of Microbiology. Molecular mechanisms of bacterial pathogenesis; biosynthesis and regulation of bacterial polysaccharides and their role in disease.

Lloyd S. Gray, M.D., Associate Professor of Pathology and of Biophysics. Events constituting the antigen receptor–initiated signal transduction sequence in T lymphocytes.

Young S. Hahn, Ph.D., Associate Professor of Microbiology and of Pathology. Mechanism of immune evasion by hepatitis C virus.

M. L. Hammarskjöld, M.D., Ph.D., Professor of Microbiology. Regulation of human retrovirus gene expression.

Doris M. Haverstick, Ph.D., Assistant Professor of Pathology. Biochemistry and regulation of receptor-associated calcium influx in T lymphocytes leading to cytokine production and cellular proliferation.

Catherine C. Hedrick, Ph.D., Assistant Professor of Internal Medicine. Regulation of monocyte and T-cell adhesion to endothelium in atherogenesis; role of glucose and oxidized lipids in angiogenesis in diabetes.

John Christian Herr, Ph.D., Professor of Cell Biology. Sperm surface antigens, their roles in capacitation and fertilization, and their clinical application as components of contraceptive vaccines.

Peter Heymann, M.D., Associate Professor of Pediatrics. Viral infection, IgE antibody, and eosinophil responses in wheezing infants and children.

Kevin T. Hogan, Ph.D., Associate Professor of Surgery. Identification of tumor antigens recognized by T lymphocytes; cancer immunotherapy.

Donald F. Hunt, Ph.D., Professor of Chemistry and of Pathology. Development of new methods and instrumentation for characterizing the primary structures of peptides and proteins; application of these new methods to structural problems in immunology.

Dean H. Kedes, M.D., Ph.D., Assistant Professor of Internal Medicine (Infectious Diseases) and of Microbiology. Molecular pathogenesis and structural analysis of Kaposi's sarcoma–associated herpes virus.

Klaus Ley, M.D., Associate Professor of Biomedical Engineering and of Molecular Physiology and Biological Physics. Selectin- and integrin-mediated leukocyte adhesion in inflammation, lymphocyte and cancer cell trafficking, and atherosclerosis.

Peter Lobo, M.D., Associate Professor of Medicine and Director, Renal Transplant Medicine. Natural IgM anti-lymphocyte autoantibodies; HLA expression.

Ulrike Lorenz, Ph.D., Assistant Professor of Microbiology. Involvement of the protein tyrosine phosphatase SHP1 in hematopoietic signal transduction.

Marcia McDuffie, M.D., Associate Professor of Microbiology, of Internal Medicine (Rheumatology), and of Pediatrics. Identification of genes controlling the development of autoreactive T lymphocytes; characterization of the T-cell population regulating autoimmune activity; role of adhesion molecules in early thymocyte differentiation.

David Mullins, Ph.D., Assistant Professor of Microbiology, Human Immune Therapy Center. Cancer immunotherapy and tumor immunology.

Jerry L. Nadler, M.D., Professor of Medicine. Role of lipid pathways and oxidative stress in pathogenesis and complications of diabetes; viral vector approaches to diabetes treatment.

J. Thomas Parsons, Ph.D., Professor of Microbiology. Role of protein tyrosine kinases in mediating mechanisms of cell adhesion and cytoskeletal-directed intracellular signaling.

Sarah J. Parsons, Ph.D., Professor of Microbiology. Function and regulation of the proto-oncogene product c-Src; tyrosine kinase signaling in normal and neoplastic growth.

William A. Petri Jr., M.D., Ph.D., Professor of Internal Medicine (Infectious Diseases), of Microbiology, and of Pathology. Molecular mechanisms of pathogenesis of infectious diseases.

Theresa Pizarro, Ph.D., Assistant Professor of Internal Medicine (Gastroenterology and Hepatology). Establishment of efficient gene therapy models of enhanced anti-inflammatory peptide expression for the treatment of chronic inflammatory diseases.

Thomas A. E. Platts-Mills, M.D., Ph.D., Professor of Internal Medicine (Allergy and Clinical Immunology) and of Microbiology. Immune response to common environmental allergens, and the role these allergens play in chronic allergic diseases such as asthma and atopic dermatitis.

Timothy L. Pruett, M.D., Professor of Surgery (Transplant). Characterization of the effect of *E. coli* upon CD4 lymphocyte function and activation.

Kodi Ravichandran, Ph.D., Professor of Microbiology. Intracellular signaling in lymphocytes and phagocytes.

David Rekosh, Ph.D., Professor of Microbiology. Human immunodeficiency virus assembly and vector development.

Margo Roberts, Ph.D., Associate Professor of Microbiology and of Internal Medicine. Elucidation of the mechanisms governing T-cell development and thymic function; induction of therapeutically relevant immune responses via genetic manipulation.

James K. Roche, M.D., Ph.D., Associate Professor of Internal Medicine (Gastroenterology and Hepatology). Role of T lymphocytes in epithelial cell injury; mechanisms for damage to the intestinal mucosa during inflammation.

C. Edward Rose Jr., M.D., Professor of Internal Medicine (Pulmonary and Critical Care). Role of chemokines in asthma and other forms of lung inflammation.

Julianne J. Sando, Ph.D., Professor of Anesthesiology. Signal transduction pathways; activation of protein kinase C and its role in pathways leading to transcription of lymphokines.

Craig L. Slingluff Jr., M.D., Professor of Surgery (Oncology). Tumor immunology; identification of epitopes for tumor-specific T lymphocytes; development of human tumor vaccines.

Michael F. Smith, Ph.D., Assistant Professor of Internal Medicine (Gastroenterology and Hepatology). Regulation of gene expression in macrophages.

John W. Steinke, Ph.D., Assistant Professor of Research Asthma and Allergic Disease. Molecular genetic studies in allergic disease and asthma.

Ronald P. Taylor, Ph.D., Professor of Biochemistry. Complement activation; role of complement and complement receptors in host defense.

Kenneth S. K. Tung, M.D., Professor of Pathology and of Microbiology. Tolerance mechanism and events that activate pathogenic autoreactive T and B cells.

Judith A. Woodfolk, M.D., Ph.D., Assistant Professor of Internal Medicine (Allergy and Clinical Immunology). Characterization of antigen-specific T-cell repertoires and analysis of their regulatory role in the human immune response to extrinsic and intrinsic allergens.

UNIVERSITY OF VIRGINIA

Department of Microbiology

Program of Study

The Department of Microbiology offers a program of study and research leading to the Ph.D. degree from the Graduate School of Arts and Sciences. The program emphasizes molecular, cellular, and biochemical approaches to current microbiological problems. Students begin the program with a one-year period of instruction with core courses in cell and molecular biology and biochemistry and with elective courses in topics such as genetics, infectious disease, carcinogenesis, virology, immunology, and cell imaging. After three research rotations in faculty laboratories, a Ph.D. lab is chosen. Research to be incorporated into the doctoral dissertation is undertaken. There are excellent opportunities for research in infectious disease, biodefense, immunology, cancer research, cell and molecular biology, bacteriology, and virology. Students are mentored throughout the program of study and are trained to be independent, creative, and productive scientists. Those completing the program are prepared for careers in research and teaching at academic, government, or industrial institutions.

Research Facilities

Members of the Department have modern research laboratories in one of several adjoining buildings in the Health Sciences Center at the University of Virginia (UVA). The Department of Microbiology has close affiliations and shares some facilities with the Thaler Center for HIV and AIDS Research, the Beirne Carter Center for Immunology Research, and the Cancer Center. Departmental facilities include ample conference rooms, office space for graduate students, and a Departmental library of current periodicals. The newly renovated Health Sciences Library houses a large collection of books and periodicals and provides online full-text access to thousands of scientific journals.

Laboratories are well equipped with state-of-the-art instrumentation and computers for individual research programs. In addition, there are many central support facilities available to all members of the research community, including facilities for confocal and electron microscopy, transgenic mouse production, hybridoma production, DNA sequencing, oligonucleotide and peptide synthesis, microarray analysis, mass spectrometry, flow cytometry and single-cell sorting, image processing, and BL-3 laboratories for research on bioterrorism organisms. The Health Sciences arm of the Academic Computing Center provides assistance with data acquisition, computer graphics, bioinformatics, and networking.

Financial Aid

Grant support in the form of traineeships, assistantships, and fellowships is available to all students accepted into the program. These provide tuition, fees, and medical and dental insurance, as well as an annual stipend. In 2004–05, the stipends were $22,000 to $24,000.

Cost of Study

Tuition and fees are covered by traineeships and fellowships.

Living and Housing Costs

The Charlottesville community has extensive housing accommodations at a wide range of rents, from downtown apartments to University-area rentals to country cottages. University housing is also available for single and married students. Useful Web sites are the Blue Ridge Apartment Council at http://www.brac.com or the University's housing office at http://www.virginia.edu/residencelife/.

Student Group

The Department typically has 90 graduate students, including 15 M.D./Ph.D. candidates. There are 19,000 students enrolled in the University; about 4,500 are candidates for advanced degrees from the Graduate School of Arts and Sciences.

Location

Charlottesville is a university city with a population of more than 40,000 and a larger community of about 120,000. The city is located about 20 miles from the Blue Ridge Mountains, Skyline Drive, and the Shenandoah Valley and is 110 miles from Washington, D.C. It has excellent theaters, concert halls, sports arenas, recreational and athletic facilities, and restaurants. Students can take advantage of opportunities for hiking, canoeing, horseback riding, biking, and skiing. Charlottesville is unusually cosmopolitan for its size, with residents from around the world participating in the life of the University and the community. It routinely appears on lists of the best places to live in the United States.

The University and The Department

The University of Virginia, founded in 1819 by Thomas Jefferson, is a highly selective university and is the major research university in Virginia. As one of the premier pubic universities in the country, it attracts scholars and students from around the world. The grounds of the University, which are exceptionally attractive, encompass about 1,600 green acres, and all facilities are within easy walking distance of each other.

The Department of Microbiology shares buildings with the other basic science departments of the Medical School, including the Departments of Biochemistry and Molecular Genetics, Molecular Physiology and Biological Physics, Cell Biology, and Pharmacology. This provides a center for biomedical research and education and facilitates interdisciplinary studies. There are close affiliations among all departments of biological and physical sciences, whose facilities are available to graduate students in the various disciplines. Seminars and courses are planned jointly, as are research endeavors when appropriate.

The University of Virginia does not discriminate on the basis of age, sex, race, color, national origin, or religion in admission, education, or employment.

Applying

Students interested in a Ph.D. in microbiology should apply via the Biomedical Sciences program at UVA at http://www.healthsystem.virginia.edu/internet/bims/. Applicants may choose to apply to one of seven graduate groups in biomedical sciences. Most of UVA's students enter through the microbiology, immunology and infectious disease graduate group, while others choose to enter through another group, such as molecular medicine. One becomes a Ph.D. student in the Department of Microbiology when one chooses a Ph.D. mentor with an appointment in microbiology. The biomedical sciences Web site, listed in this Close-Up, describes the application process. Prospective students may apply at http://artsandsciences.virginia.edu/admissions/apply.html.

Correspondence and Information

Dr. Ann L. Beyer, Graduate Adviser
Department of Microbiology
Box 800734
University of Virginia
Charlottesville, Virginia 22908
Phone: 434-924-5930
Web site: http://www.healthsystem.virginia.edu/internet/microbiology/

University of Virginia

THE FACULTY AND THEIR RESEARCH

Timothy P. Bender, Professor; Ph.D., Michigan. Gene expression during lymphocyte differentiation.
Ann L. Beyer, Professor; Ph.D., Vanderbilt. Ribosomal RNA synthesis and processing; ribosome biogenesis.
Amy H. Bouton, Associate Professor; Ph.D., Virginia. The adaptor molecule p130^{cas} and signal transduction.
Thomas J. Braciale, Professor and Director, Beirne Carter Center for Immunology Research; M.D., Ph.D., Pennsylvania. Antigen recognition by T lymphocytes.
David L. Brautigan, Professor and Director, Markey Center for Cell Signaling and Molecular Biology Institute; Ph.D., Northwestern. Signaling, cycling, and protein phosphatases.
Jay C. Brown, Professor; Ph.D., Harvard. Virus structure and assembly.
Michael G. Brown, Assistant Professor; Ph.D., Virginia Commonwealth. NK cells and viral immunity.
James E. Casanova, Associate Professor; Ph.D., Wesleyan. Signal transduction and *Salmonella* pathogenesis.
Daniel A. Engel, Associate Professor; Ph.D., Yale. Transcriptional regulation in cancer cells.
Victor H. Engelhard, Professor; Ph.D., Illinois at Urbana-Champaign. Antigen recognition by T lymphocytes; antigen processing.
Peter B. Ernst, Professor, Ph.D., McMaster. Gastrointestinal inflammation.
Jay W. Fox, Professor and Director, Protein and Nucleic Acid Sequencing Facility; Ph.D., Colorado State. Basement membrane proteins and cell-binding mechanisms; protease protein chemistry.
Shu Man Fu, Professor; M.D., Stanford; Ph.D., Rockefeller. Lymphocyte activation and autoimmunity.
Joanna B. Goldberg, Professor; Ph.D., Berkeley. Bacterial polysaccharides and pathogenesis.
Young S. Hahn, Associate Professor; Ph.D., Caltech. Immune suppression by microorganisms, hepatitis C virus.
Marie-Louise Hammarskjold, Professor; M.D., Ph.D., Karolinska (Stockholm). Molecular biology of human immunodeficiency virus and Epstein-Barr virus.
Kevin C. Hazen, Associate Professor; Ph.D., Montana State. Fungal cell wall protein molecular biology and biochemistry; fungal adhesion mechanisms.
Paul S. Hoffman, Professor, Ph.D., Virginia Tech. Host-parasite interaction; *Legionella pneumophila* and *Helicobacter pylori*.
Alan F. Horwitz, Professor; Ph.D., Stanford. Cell adhesion and signaling in developmental and pathobiology; mechanisms of cell migration.
Robert J. Kadner, Professor; Ph.D., UCLA. Bacterial genetics; mechanisms of active transport; gene regulation.
Dean H. Kedes, Associate Professor; Ph.D., M.D., Yale. Molecular biology and pathogenesis of Kaposi's sarcoma–associated herpesvirus (KSHV/HHV-8).
Gary M. Kupfer, Associate Professor; M.D., Johns Hopkins. Molecular basis of inherited cancer syndromes; genomic instability.
Deborah A. Lannigan, Assistant Professor; Ph.D., Rochester. Kinases, drugs, and cancer.
Ulrike M. A. Lorenz, Assistant Professor; Ph.D., Berlin and Max Planck Institute. Involvement of tyrosine phosphatase in signal transduction.
Ian G. Macara, Professor; Ph.D., Sheffield (England). Understanding asymmetry: cell polarity and nucleocytoplasmic transport.
Barbara J. Mann, Associate Professor; Ph.D., Virginia. Molecular parasitology; pathogenesis of *Entamoeba histolytica* and *Francisella tularensis*.
Marcia J. McDuffie, Associate Professor; M.D., North Carolina at Chapel Hill. Immunogenetics of autoimmunity; T-cell differentiation.
Tom G. Obrig, Professor; Ph.D., Illinois. Bacterial pathogenesis.
J. Thomas Parsons, Professor and Chairman; Ph.D., Duke. Molecular genetics of tumor viruses.
Sarah J. Parsons, Professor; Ph.D., Duke. Proto-oncogenes in normal and neoplastic processes.
Lucy W. Pemberton, Associate Professor; Ph.D., Imperial Cancer Research Fund. Nucleocytoplasmic transport; nuclear transport receptors; import of protein complexes involved in transcription regulation.
William Petri, Professor; M.D., Ph.D., Virginia. Molecular parasitology.
Thomas Platts-Mills, Professor; M.D., Oxford. Allergens and allergic diseases.
Kodimangalam S. Ravichandran, Professor; Ph.D., Massachusetts. Intracellular signaling in T lymphocytes.
David M. Rekosh, Professor and Director, Thaler Center for HIV and AIDS Research; Ph.D., MIT. Molecular biology of human immunodeficiency virus and adenovirus.
Margo R. Roberts, Associate Professor; Ph.D., Leeds (England). T-cell development and immune therapy.
Martin A. Schwartz, Professor; Ph.D., Stanford. Signal transduction by integrins.
Corinne M. Silva, Assistant Professor; Ph.D., North Carolina at Chapel Hill. Signal transducers (STATs) in proliferation, differentiation, transformation, and tumorigenesis.
M. Mitchell Smith, Professor; Ph.D., Johns Hopkins. Chromatin structure and function; cell division cycle; yeast molecular genetics.
Michael F. Smith, Associate Professor; Ph.D., Temple. Regulation of the innate immune response; signal transduction in gastric cancer.
Kenneth S. K. Tung, Professor; M.D., Melbourne (Australia). Tolerance; regulatory T cells; mechanism of neonatal and adult autoimmune diseases.
Julie Davis Turner, Assistant Professor; Ph.D., Pennsylvania. Pathogenesis of *Mycobacterium tuberculosis* and HIV synergy; correlates of protection by mycobacteria.
Michael J. Weber, Professor and Director, Cancer Center; Ph.D., California, San Diego. Signal transduction by oncogenes; receptors and kinases.
Judith M. White, Professor; Ph.D., Harvard. Viral and cellular membrane fusion proteins.

The Rotunda, designed by Thomas Jefferson, on the Central Grounds of the University.

UNIVERSITY OF WISCONSIN–MADISON

Department of Bacteriology
Department of Medical Microbiology and Immunology
Microbiology Doctoral Training Program

Program of Study

The Departments of Bacteriology and Medical Microbiology and Immunology at the University of Wisconsin–Madison (UW–Madison) offer a microbiology Ph.D. degree for studies that deal with the genetics, biochemistry, physiology, cell biology, and pathogenesis of prokaryotic and lower eukaryotic microorganisms and diverse aspects of host-microbe interactions. The Microbiology Doctoral Training Program provides outstanding, diverse opportunities for graduate training. Areas of current research emphasize fundamental aspects of gene expression and regulation, microbial physiology, host defenses and host-pathogen interactions, and structural analysis of proteins and nucleic acids. Some work in these areas is fundamental, with the goal of understanding one of these areas in greater detail, or directed basic research to provide both insight and practical use of new information. Research programs include aspects of metabolic diversity and regulation, protein and nucleic acid biosynthesis, pathogenesis, immunology, cell biology, bioremediation, environmental microbiology, biotechnology and industrial microbiology, and food microbiology. The program encourages the fullest development of technical and intellectual skills. It reflects the interdisciplinary nature of both the field and the Madison campus by allowing the student to choose from among approximately 90 doctoral thesis advisers drawn from the Departments of Bacteriology, Biochemistry, Biomolecular Chemistry, Chemistry, Food Microbiology and Toxicology, Food Science, Genetics, Medical Microbiology and Immunology, Molecular Biology, Oncology, Pathobiological Sciences, Pharmacy, Pharmacology, and Plant Pathology. For more information about the program, students should visit the Web site listed in the Correspondence and Information section.

Research Facilities

Program laboratories are based in various locations during the construction of the new Microbial Sciences Building. The research laboratories in these locations are fully equipped for all aspects of contemporary research in microbiology, immunology, and cellular and molecular biology studies; there also are containment facilities and biohazard equipment, such as laminar flow hoods, for work with human pathogenic microbes. Additional facilities include ultracentrifuges, liquid scintillation counters, computer-assisted imagers of beta-emitting isotopes, a variety of low- and high-pressure chromatography systems, HPLCs with detectors, a scanning spectrofluorometer, anaerobic chambers, fermentors, a DNA-sequencing room, a graphics workstation for molecular modeling of proteins and nucleic acids, numerous cold- and constant-temperature rooms, a staffed photography studio, staffed electronics and machine shops, a microcomputer suite, and a research reading room. A large number of University-based facilities provide equipment and exercise for genetic and biochemical analysis of proteins and nucleic acids.

Financial Aid

Financial aid is available in the form of research assistantships, predoctoral traineeships, and fellowships. The 2005–06 research assistantship stipend is $22,500 for twelve months.

Cost of Study

Tuition is waived for graduate students in the biological sciences. Graduate students must pay fees, however, which are estimated to be approximately $700 for 2005–06. Health-care benefits for students are the same as those for the faculty and staff.

Living and Housing Costs

Rents for off-campus efficiencies range from $450 to $625 per month, one-bedroom apartments range from $550 to $900 per month, and two-bedroom apartments range from $650 to $1550 per month. Single graduate student housing is available for $525 to $610 per month for a one-bedroom apartment and $665 to $765 per month for a two-bedroom apartment. For more information, students should visit the Web site at http://www.housing.wisc.edu.

Student Group

UW–Madison's fall 2004 enrollment was 41,588, of whom 11,354 were graduate or professional students. There are currently about 120 graduate students enrolled in the program in microbiology.

Student Outcomes

Ph.D. degree recipients from the program are in demand at major food, pharmaceutical, biotechnology, and clinical settings. Ph.D. recipients who are considering tenure-track teaching/research positions or advanced positions in industry complete postdoctoral training at UW–Madison or elsewhere.

Location

Madison, the state's capital, is in a beautiful physical setting surrounded by four lakes and rolling countryside. Madison's exceptional cultural environment includes a civic/performing arts center, repertory theater companies, art museums and galleries, a symphony, and a chamber orchestra. The state capitol and the surrounding square is a center for outdoor concerts, art fairs, and farmers' markets. The Frank Lloyd Wright–designed Monona Terrace Convention Center is a jewel of the city. Madison's 215,697 residents enjoy many parks and recreational facilities, including a 1,200-acre arboretum. State Street, Madison's historic main street, is a unique and lively pedestrian shopping mall. Sports enthusiasts enjoy University and city intramural sports, sailing, canoeing, cross-country and downhill skiing, camping, bicycling, hunting, and fishing.

The University

The University, founded in 1848, is located 1 mile from the state capitol on a series of hills overlooking Lake Mendota. Few campuses can rival the beauty of the Madison campus, with its hills, woods, gardens, and almost 2 miles of lake shoreline. There are more than 115 doctoral programs in eleven major schools and colleges (many of them are near or at the top of many national rankings). In 2004, research and development spending was $714 million, which ranked UW–Madison first in the Big Ten and fifth among all universities in total research volume. With more than 2,000 professors who offer more than 8,000 courses, the range of academic opportunities at UW is unparalleled.

Applying

The deadline for application to the Ph.D. program in microbiology is December 1 for admission in the fall. The GRE is required; the Subject Test is not required but is recommended.

Correspondence and Information

Kathryn A. Holtgraver, Coordinator
Microbiology Doctoral Training Program
University of Wisconsin–Madison
420 Henry Mall
Madison, Wisconsin 53706
Phone: 608-265-0689
Fax: 608-262-9865
E-mail: micro@bact.wisc.edu
Web site: http://www.microbiology.wisc.edu

University of Wisconsin–Madison

THE FACULTY AND THEIR RESEARCH

Paul G. Ahlquist, Molecular Virology, Oncology, and Plant Pathology. RNA virus replication; gene expression; virus-host interactions.

Mark Albertini, Medicine. Host immunity against cancer.

Caitilyn Allen, Plant Pathology. Physiology and genetics of bacterial plant pathogens.

Philip Anderson, Genetics. Molecular genetics of *C. elegans.*

Jean-Michelle Ané, Agronomy. Plant-microbe symbiotic associations.

Aseem Ansari, Biochemistry. Chemical approaches to dissecting and directing the genome decoding machinery.

Teri C. Balser, Soil Science. Soil microbial community, ecology, and ecophysiology.

James Bangs, Medical Microbiology and Immunology. Molecular parasitology.

Stephen Barclay, Bacteriology. Developmental microbiology of *Dictyostelium, Pneumocystis,* and *Cryptosporidium.*

Helen E. Blackwell, Chemistry. Chemical approaches to modulate quorum sensing in animal- and plant-associated bacteria.

Frederick Blattner, Genetics. Molecular genetics, immunology, and genome studies.

Curtis Brandt, Medical Microbiology and Immunology. Virology; pathogenesis of herpes simplex virus; gene therapy; antiviral drugs.

David Brow, Biomolecular Chemistry. Mechanism of DNA transcription and RNA splicing.

Richard Burgess, Oncology. Biochemistry and molecular biology of RNA polymerase and transcription factors.

Glenn Chambliss, Bacteriology. Global regulatory controls in gram-positive bacteria and microbial degradation of energetic compounds.

Bruce Christensen, Animal Health and Biomedical Sciences. Cellular, molecular, genetic, and biochemical factors influencing the competence of mosquitoes that transmit diseases such as lymphatic filariasis, malaria, and specific arboviral diseases.

Elizabeth Craig, Biochemistry. Genetics of environmental stress; protein translocation and folding.

Michael Culbertson, Genetics. Yeast genetics.

Cameron Currie, Bacteriology. Microbial ecology and evolution, symbiosis, evolutionary genomics, and host-pathogen interactions.

Charles Czuprynski, Pathobiological Sciences. Immunology and pathogenic bacteria.

Joseph P. Dillard, Medical Microbiology and Immunology. Molecular mechanisms of disease caused by *Neisseria gonorrhoeae.*

Timothy Donohue, Bacteriology. Process and regulation of cytochrome biosynthesis.

Diana Downs, Bacteriology. Metabolic regulation and pathway integration.

Jorge Escalante-Semerena, Bacteriology. Genetics and biochemistry of anaerobic metabolism.

Robert Fillingame, Biomolecular Chemistry. Molecular mechanism of ATP synthesis during oxidative phosphorylation.

Marcin Filutowicz, Bacteriology. Regulation of DNA replication.

Katrina Forest, Bacteriology. Structural biology of pathogenesis; bacterial type IV pili and the critical bacterial cell division protein FtsZ.

Brian Fox, Biochemistry. Biochemical, catalytic, and spectroscopic studies of redox active enzymes; protein engineering.

Catherine Fox, Biomolecular Chemistry. The role of chromosome structure in gene expression and DNA replication.

Paul D. Friesen, Biochemistry. Molecular biology of baculoviruses; virus–host cell interactions and apoptosis.

Audrey Gasch, Genetics. Genomic expression of fungi.

Heidi Goodrich-Blair, Bacteriology. Nematode-bacteria symbiosis and stationary-phase *E. coli:* molecular signals in microbial communication.

Richard Gourse, Bacteriology. Prokaryotic transcription/regulation of ribosome synthesis.

Jenny Gumperz, Medical Microbiology and Immunology. Molecular and functional analyses of T lymphocytes in innate immunity.

Kenneth E. Hammel, Bacteriology (Forest Products Laboratory). Lignocellulose degradation by filamentous fungi.

Jo Handelsman, Plant Pathology. Molecular and ecological basis of plant-bacterial interactions.

Colleen E. Hayes, Biochemistry. Regulation of lymphocyte development and function.

William Hickey, Soil Science. Soil microbiology; bioremediation; physiology and molecular ecology; functional genomics.

Christina Hull, Biomolecular Chemistry. Molecular biology of human fungal pathogens.

Anna Huttenlocher, Pharmacology and Pediatrics. Molecular mechanisms of leukocyte chemotaxis and inflammation.

Thomas Jeffries, Bacteriology (Forest Products Laboratory). Bioconversion of renewable resources.

Eric Johnson, Food Microbiology and Toxicology and Bacteriology. *C. botulinum* and its toxicology; food and industrial microbiology.

Robert Kalejta, Molecular Virology. Viral modulation of the cell cycle.

Yoshihiro Kawaoka, Pathobiological Sciences. Molecular biology and pathogenesis of influenza and Ebola viruses.

James Keck, Biomolecular Chemistry. Structure and function of proteins involved in multiple nucleic acid metabolic pathways.

Nancy Keller, Plant Pathology. Host and pathogen signals governing toxin formation, spore development, and pathogenesis of seed by fungi.

Patricia Kiley, Biomolecular Chemistry. Transcriptional control of anaerobic gene expression in *E. coli.*

Bruce Klein, Pediatrics. Microbial-host interactions and pathogenesis.

Laura Knoll, Medical Microbiology and Immunology. Molecular mechanisms of development and pathogenesis of *Toxoplasma gondii.*

Ching Kung, Genetics. Microbial ion-channel proteins.

Robert Landick, Bacteriology. Transcriptional regulation/RNA polymerase structure and function.

Miroslav Malkovsky, Medical Microbiology and Immunology. Cellular genes regulating expression of primate immunodeficiency viruses.

John Mansfield, Bacteriology. Immunology and molecular biology of infectious disease.

William McClain, Bacteriology. tRNA structure and function.

Margaret McFall-Ngai, Medical Microbiology and Immunology. Host animal response interactions with beneficial bacterial symbionts.

Katherine McMahon, Civil and Environmental Engineering. Environmental microbiology and molecular microbial ecology.

Rebecca I. Montgomery, Biochemistry. Molecular mechanisms of virus-cell interactions that lead to pathogenesis.

Sean Palecek, Chemical and Biological Engineering. Genetics of fungal adhesion and biofilm formation; yeast dimorphism.

Ann Palmenberg, Biochemistry. Molecular biology of RNA picornaviruses; protein translation, proteolytic processing; RNA synthesis; viral pathogenesis; viral vaccines and vectors; computer-based sequence analysis.

Donna Paulnock, Medical Microbiology and Immunology. Cellular and molecular biology of the innate immune response.

Nicole Perna, Animal Health and Biomedical Sciences. Genomic and organismal evolution of plant- and animal-associated enterobacteria.

Richard Proctor, Medical Microbiology and Immunology. Bacterial endotoxins; mechanisms of bacterial inhibition of neutrophil function.

William Reznikoff, Biochemistry. Molecular biology of gene expression and transposition.

Gary Roberts, Bacteriology. Maturation and regulation of metalloproteins.

Eric Roden, Geology and Geophysics. Biogeochemistry and microbial ecology of soils and sediments.

Edward Ruby, Medical Microbiology and Immunology. Bacterial determinants of host-microbe interactions.

Matyas Sandor, Pathology and Laboratory Medicine. Role of T cells in infectious diseases.

Ronald Schell, Medical Microbiology and Immunology. Immune responses to *Borrelia burgdorferi.*

Stacey Schultz-Cherry, Medical Microbiology and Immunology. Basic cellular, biochemical, and molecular biology of human influenza virus and astrovirus pathogenesis.

David Schwartz, Chemistry. New, genomic science–based approaches to important biological problems.

Ben Shen, Pharmacy. Secondary metabolite biosynthesis and drug discovery.

Paul M. Sondel, Pediatrics, Human Oncology, and Genetics. Immunogenetics; tumor immunology; clinical oncology.

Gary Splitter, Animal Health and Biomedical Sciences. Immunology of infectious diseases.

James Steele, Food Science. Genetics of lactic acid bacteria.

Rob Striker, Medical Microbiology and Immunology. Nucleotide selectivity, error rate, and supramolecular structure of the hepatitis C viral (HCV) polymerase.

M. Suresh, Pathobiological Sciences. Regulation of T-cell proliferation, apoptosis, and memory.

Adel Talaat, Animal Health and Biomedical Sciences. Functional genomics for understanding the molecular pathogenesis of tuberculosis and paratuberculosis.

Michael Thomas, Bacteriology. Biochemical analysis and manipulation of natural product biosynthetic pathways.

Jon Thorson, Pharmacy. Understanding and exploiting biosynthetic pathways and enzymes in an effort to generate novel therapeutics.

Karen Wassarman, Bacteriology. Small RNAs in bacteria.

David I. Watkins, Pathology, Primate Research Center. Function and evolution of major histocompatibility complex class I molecules.

William Weidanz, Medical Microbiology and Immunology. Immunoparasitology; role of T cells in immunity to malaria.

Paul Weimer, Bacteriology. Physiology and ecology of fiber digestion by the anaerobic bacteria of the bovine rumen.

Rodney Welch, Medical Microbiology and Immunology. Molecular analysis of bacteria pathogens.

Jon Woods, Medical Microbiology and Immunology. Pathogenic mechanisms and host-parasite interactions of *Histoplasma capsulatum.*

John Yin, Chemical and Biomedical Engineering. Systems biology; virus dynamics; virus-host interactions.

Timothy Yoshino, Animal Health and Biomedical Sciences. Cellular and molecular biology of schistosome blood fluke–snail host interactions.

Jaehyuk Yu, Food Microbiology and Toxicology. Fungi and mycotoxins: molecular genetics of asexual sporulation and mycotoxin biosynthesis in filamentous fungi.

WAKE FOREST UNIVERSITY

School of Medicine
Department of Microbiology and Immunology

Programs of Study

The department offers the degree of Doctor of Philosophy (Ph.D.) through the Graduate School of Wake Forest University. Research programs are available in the areas of molecular pathogenesis (bacterial and viral), microbial genetics, immunology (cellular and molecular), and virology.

In the first year of residence, students take a core curriculum of courses consisting of biochemistry, molecular and cellular biology, virology, immunology, and bacteriology; these courses are specifically designed for graduate students and are not part of the medical student curriculum. Students also gain knowledge of a variety of experimental techniques by rotating through three faculty laboratories during the first year. In the second year, students enroll in Advanced Topics in Microbiology and Immunology—a course devoted to in-depth analysis of the research literature in contemporary areas of microbiology, virology, and immunology. After the first year of residence, students are actively engaged in thesis research. Advanced students present findings in the yearly Research Seminar Series.

Research Facilities

The department is housed in modern facilities totaling approximately 25,000 square feet and containing twelve well-equipped research laboratories. The department has a range of equipment for use in biochemical and molecular studies. Major equipment shared in the department includes scintillation counters, flow cytometers, high-speed cell sorters, ultracentrifuges, and PCR thermocyclers. An institutional computer facility used for gene analysis and molecular graphics is networked to most faculty labs. There are also core laboratories at the medical center. Supported by a grant from the National Cancer Institute, these laboratories include facilities for oligonucleotide synthesis, protein sequencing and peptide synthesis, tissue culture, and electron microscopy.

Financial Aid

Wake Forest University School of Medicine provides full tuition scholarships for all full-time graduate students in microbiology and immunology. In addition, all students accepted into the program receive fellowships of $20,772 (2005–06 level) and a health insurance supplement. Additional support can include travel to national scientific meetings to present recent work.

Cost of Study

All students accepted into the program are awarded fellowships that fully cover tuition fees and provide a stipend for living expenses.

Living and Housing Costs

Each student secures his or her own living quarters, usually from the wide selection of rooms, apartments, and houses in the residential neighborhood surrounding the medical center. Rents are considerably lower than the national mean, with a one-bedroom apartment averaging $600 per month. Other living costs are generally at or below those in other metropolitan areas.

Student Group

Thirty-four full-time students are currently enrolled in the department. All are working toward the Ph.D. degree. All of these students receive equal support, which is guaranteed during their graduate tenure (contingent on reasonable progress toward the degree).

Location

Wake Forest University School of Medicine is located in a residential area 3 miles from the undergraduate campus. Winston-Salem, which has a population of 180,000, is located in the foothills of the Blue Ridge Mountains. The area provides many job opportunities for spouses and continues to be an economically thriving region. Cultural events are extremely accessible, as Winston-Salem, in addition to being the home of Wake Forest University, is also the home of the North Carolina School of the Arts, the Winston-Salem Symphony, Salem College, and Winston-Salem State University.

A mild winter climate, without an oppressive summer climate, permits outdoor activities nearly year-round. Recreational opportunities abound, as there are more than sixty county and state parks nearby, including the 1,100-acre Tanglewood Park. The Blue Ridge Parkway and Great Smoky Mountain National Park are only 1 and 4 hours away, respectively.

The Department

The Department of Microbiology and Immunology is a well-established and well-recognized research department; the Ph.D. degree was first offered in 1964; more than 100 graduate degrees have been awarded by the department faculty. The teaching and research facilities are relatively new and spacious. The department is currently in the midst of a period of growth that has strengthened and will continue to strengthen its faculty in the areas of biodefense, bacterial and viral pathogenesis, and immunology. Members of the department are associated with the Comprehensive Cancer Center, the Biology of Inflammation Program, and the Molecular Genetics Research Program.

Applying

Applicants for the academic year beginning in mid-August should submit applications for admission and all supporting documents before January 15. Personal interviews are generally requested of all applicants prior to acceptance. Applicants are required to submit scores on the General Test of the Graduate Record Examinations (GRE) by January 15. GRE Subject Test scores are not required.

Correspondence and Information

Dr. Elizabeth Hiltbold
Chair, Graduate Admissions Committee
Department of Microbiology and Immunology
Wake Forest University Medical Center
Medical Center Boulevard
Winston-Salem, North Carolina 27157-1064

Phone: 336-716-8246
E-mail: bhiltbold@wfubmc.edu
Web site: http://www.wfubmc.edu/microbio

Wake Forest University

THE FACULTY AND THEIR RESEARCH

Griffith D. Parks, Associate Professor and Interim Chair; Ph.D., Wisconsin. Molecular biology and replication of RNA viruses; structure-function of viral proteins and viral RNA genome; viral vectors for therapy.

Martha Alexander-Miller, Associate Professor; Ph.D., Washington (St. Louis). Activation and regulation of CD8$^+$ cytotoxic T lymphocytes following viral infection; role of CTL avidity.

Subhashini Arimilli, Instructor; Ph.D., Osmania (India). Visual infection on human dendrite cells; regulation of antiviral cytotoxic T lymphocytes following respiratory infection in mice.

David A. Bass, Professor; M.D., Johns Hopkins; D.Phil., Oxford. Biochemistry of the oxidative burst response in neutrophils; signal transduction of neutrophil activation.

Rajendar Deora, Assistant Professor; Ph.D., Illinois at Chicago. Microbial pathogenesis; transcriptional regulation and functional genomics of host-pathogen interactions.

Purnima Dubey, Assistant Professor; Ph.D., Chicago. Noninvasive molecular imaging of the cellular antitumor immune response; prostate cancer biology; tumor suppressor genes controlling prostate cancer development and progression.

Jason M. Grayson, Assistant Professor; Ph.D., Texas Southwestern Medical Center at Dallas. Apoptosis of antigen-specific CD8$^+$ T cells during viral infection.

Elizabeth M. Hiltbold, Assistant Professor; Ph.D., Emory. Regulation of antigen presenting cell function by intracellular bacterial infection; visualizing T-cell activation through immunofluorescent microscopy.

Charles E. McCall, Professor; M.D., Wake Forest. Molecular basis of alterations in innate immunity during human sepsis; modulation of transcription factors; translation efficiency and proximal signaling responses to microbial toxins.

Steven B. Mizel, Professor; Ph.D., Stanford. Effect of bacterial flagellin on inflammatory cells; adjuvant properties of flagellin.

David A. Ornelles, Associate Professor; Ph.D., MIT. Cell-cycle regulation of mRNA transport and translation during adenovirus infection; genetic and molecular approaches to identify cellular factors required by adenovirus; replicating oncolytic adenoviruses.

Sean D. Reid, Assistant Professor; Ph.D., Penn State. Molecular biology and pathogenesis of group A *Streptococcus;* molecular, genetic, and genomic approaches to investigate bacterial pathogen-host interactions.

Stephen H. Richardson, Professor Emeritus; Ph.D., USC.

W. Edward Swords, Assistant Professor; Ph.D., Alabama at Birmingham. Cellular microbiology of airway infections; molecular basis for persistent colonization and inflammation by nontypeable *Haemophilus influenzae*.

Daniel J. Wozniak, Associate Professor; Ph.D., Ohio State. Molecular biology and pathogenesis of *Pseudomonas aeruginosa;* genetic and biochemical approaches to characterize environmental and transcriptional control of virulence factors.

Recent Graduates and Current Postdoctoral Positions

Erica Brown, Ph.D., National Institute of Allergy and Infectious Diseases (NIAID), National Institutes of Health (NIH); Bethesda, MD.
Andrew Cawthon, Ph.D., Children's Research Institute, Ohio State University; Columbus, OH.
Amy Chastain, Ph.D., Vanderbilt University School of Medicine; Nashville, TN.
Dinene Crater, Ph.D., Emory University School of Medicine; Atlanta, GA.
Michael Elliot, Ph.D., University of Virginia; Charlottesville, VA.
Peter Gray, Ph.D., University of Pennsylvania; Philadelphia, PA.
Michael Keller, Ph.D., U.S. Army Medical Research Institute of Infectious Diseases (USAMRIID); Fort Detrick, MD.
Sarah Kopecky, Ph.D., Mount Sinai School of Medicine; New York, NY.
Hemantha Don Kulasekara, Ph.D., University of Washington; Seattle, WA.
Karen LaRue, Ph.D., Los Alamos National Laboratories; Los Alamos, NM.
Susan Murphy, Ph.D., Duke University School of Medicine; Durham, NC.
Joseph Orlando, Ph.D., Harvard Medical School at the Beth Israel Deaconess Medical Center; Boston, MA.
Tina Peachman, Ph.D., Walter Reed Army Institute of Research; Silver Spring, MD.
Deborah Ramsey, Ph.D., Washington University in St. Louis; St. Louis, MO.
John C. Rassa, Ph.D., University of Pennsylvania; Philadelphia, PA.
Paula Roesch, Ph.D., University of Wisconsin–Madison; Madison, WI.
Samuel C. Woolwine III, Ph.D., Johns Hopkins University; Baltimore, MD.
Hang Yuan, Ph.D., Georgetown University School of Medicine; Washington, D.C.

WEST VIRGINIA UNIVERSITY

Graduate Program in Immunology and Microbial Pathogenesis

Programs of Study

The Graduate Program in Immunology and Microbial Pathogenesis is one of seven graduate programs in West Virginia University (WVU) Schools of Medicine and Pharmacy offering interdisciplinary biomedical research training leading to the Ph.D. or M.D./Ph.D. degree. Faculty members and students explore diverse areas of inquiry related to the medical implications of microbes and the human body's response to them. Research areas include the physiology of pathogenic microbes; microbial virulence factors; the biochemistry of inflammatory cytokines; interactions between microbes and their hosts; immune response in bacterial and viral diseases; effects of man-made pesticides and herbicides on the immune system; molecular aspects of cell signaling as it relates to cancer chemotherapy and cell growth control; cytotoxic T cells in transplant rejection and disease pathogenesis; bacterial diseases, including Lyme disease, cystic fibrosis, and streptococcal infections; and peptide and DNA vaccines for contraception.

Students benefit from individual attention by faculty members, within a research environment that is dynamic, collaborative, and interdisciplinary. In addition to course work and laboratory research, students participate in seminars, journal clubs, and research conferences. They also attend national scientific meetings and obtain valuable speaking and teaching experience. In addition to course work and laboratory research, students participate in seminars, journal clubs, and research conferences. Graduate trainees also have the opportunity to obtain teaching experience and attend national scientific meetings. The Ph.D. typically takes five years to complete. During Year 1, students matriculate in a common integrated core curriculum. This integrated first year allows students to build competence in key areas of contemporary science, gain exposure to the various training program options, meet potential dissertation advisers, and network scientifically and socially. In the second semester, students customize their course work by selecting from an array of program-specific electives. At the end of Year 1, students select a research adviser and can select Immunology and Microbial Pathogenesis as their training program. Year 2 consists of advanced course work, research, teaching, and the candidacy examination. Years 3 to 5 are devoted to dissertation research. The Graduate Program in Immunology and Microbial Pathogenesis also participates in the combined M.D./Ph.D. Scholars Program. As M.D./Ph.D. Scholars, students take the first two years of the medical curriculum, followed typically by three years of research as required for the Ph.D. degree before returning to the M.D. program.

Research Facilities

Institutional facilities include a computer-based learning center, a centralized animal facility with a transgenic barrier, and a library housing more than 205,000 volumes and 2,400 journals. Core facilities are available for examining gene expression or genetic variation (Affymetrix platform), image analysis, confocal and electron microscopy and laser capture microdissection, live-cell imaging, mass spectrometry, flow cytometry and high-speed cell sorting, proteomics, recombinant DNA technology, transgenic rodent biology, and functional neuroimaging (fMRI, PET/CT). Affiliated research centers include the National Institute for Occupational Safety and Health (NIOSH), Center for Advanced Imaging, Blanchette Rockefeller Neurosciences Institute, Sensory Neuroscience Research Center, and Mary Babb Randolph Cancer Center.

Financial Aid

Ph.D. and M.D./Ph.D. students in the biomedical sciences receive financial support during their training, provided they remain in good academic standing and excel in research. Such support includes full tuition, health insurance, and an annual stipend of $22,000. Combined M.D./Ph.D. students also receive medical tuition waivers.

Cost of Study

Students' tuition costs are covered.

Living and Housing Costs

The cost of an efficiency apartment in University-owned housing is approximately $400 per month. A limited number of University apartments are available for married students. Privately owned apartments in Morgantown cost $400 to $600 per month. In general, the cost of living is lower compared to larger cities.

Student Group

Total University enrollment is approximately 26,000 students, which includes 6,500 graduate and professional students. Graduate students come from all parts of the United States and many other countries.

Location

With an appealing balance to life, Morgantown is a vibrant university community of 80,000 residents in northern West Virginia. Located near the Pennsylvania border at the western edge of the Appalachian Mountains, abundant opportunities exist for activities such as world-class white-water rafting and kayaking, hiking and camping, mountain biking, fishing, and skiing. Morgantown has a cosmopolitan atmosphere with a range of activities usually found in much larger cities. It also enjoys proximity to major metropolitan centers: Pittsburgh is a 90-minute drive north and Washington, D.C., is a 3-hour drive east.

The University

West Virginia University is a comprehensive, land-grant, Carnegie-designated Doctoral/Research University–Extensive public institution. The University's academic Health Sciences Center includes the Schools of Medicine, Dentistry, Nursing, and Pharmacy, all of which offer graduate degree programs. There are seven Ph.D. biomedical research training programs in the Schools of Medicine and Pharmacy that benefit from a common, undifferentiated first year: Biochemistry and Molecular Biology, Cancer Cell Biology, Cellular and Integrative Physiology, Exercise Physiology, Immunology and Microbial Pathogenesis, Neuroscience, and Pharmaceutical and Pharmacological Sciences. Graduate faculty members in these programs are from various basic science and clinical departments throughout WVU and are members of interdisciplinary research centers in five health-related areas: cancer cell biology, cardiovascular sciences, immunopathology and microbial pathogenesis, neuroscience, and respiratory biology and lung diseases.

As a member of the Big East Conference, WVU participates in NCAA Division I sports. WVU also offers a wide variety of creative arts, theater, and entertainment opportunities.

Applying

Applicants must have a bachelor's degree and excellent GPA and GRE scores. Three letters of recommendation and a personal statement are required. Students are invited in groups of 10 for a paid, two-day visit/interview from January through March. Prospective students should visit the Web site at http://www.hsc.wvu.edu/som/resoff/gradprograms/PhD.asp for more information and an online application.

Correspondence and Information

Slawomir Lukomski, Ph.D., Graduate Director
Graduate Program in Immunology and Microbial Pathogenesis
West Virginia University
P.O. Box 9177
Morgantown, West Virginia 26506-9177
Phone: 304-293-6405
E-mail: slukomski@hsc.wvu.edu
Web site: http://www.hsc.wvu.edu/som/resoff/gradprograms/
    PhD.asp

Office of Research and Graduate Education
Health Sciences Center
West Virginia University
P.O. Box 9104
Morgantown, West Virginia 26506
Phone: 304-293-7116
E-mail: cnoel@hsc.wvu.edu

West Virginia University

THE FACULTY AND THEIR RESEARCH

John Barnett, Professor and Chair; Ph.D., Louisville. Consequences to the immune system of exposure to xenobiotic compounds.

Nyles Charon, Professor; Ph.D., Minnesota. Understanding spirochete motility and how this attribute allows the organisms to invade host tissue.

Christopher Cuff, Associate Professor; Ph.D., Temple. Cellular and molecular events that result in the development of antigen-specific immune responses in the gastrointestinal tract.

Thomas Elliott, Professor; Ph.D., California, San Diego. Novel mechanisms controlling gene expression in the enteric bacteria *Salmonella typhimurium* and *Escherichia coli.*

Solveig Ericson, Associate Professor; M.D., Ph.D., Dartmouth. Functional recovery of myeloid cells following bone marrow transplantation.

Daniel Flynn, Professor; Ph.D., North Carolina State. Signaling networks that regulate cellular motility and invasion in breast cancer.

Laura Gibson, Associate Professor; Ph.D., West Virginia. Effects of dose-intensive chemotherapy on the bone marrow microenvironment and the subsequent effect on hematopoietic recovery.

Bing-Hua Jiang, Assistant Professor; Ph.D., Mississippi. PI 3-kinase and angiogenesis.

Kenneth Landreth, Professor; Ph.D., Washington (Seattle). Mechanism regulating the production of immunocompetent lymphocytes in primary lymphoid tissues.

Slawomir Lukomski, Associate Professor; Ph.D., Lodz (Poland). Microbial pathogenesis; Group A streptococcus; collagen-like protein.

Jia Luo, Associate Professor; Ph.D., Iowa. Cell signaling during brain development; teratogenic effects of ethanol on developing central nervous systems; ErbB protein-mediated signal transduction in breast cancer.

Michael Luster, Adjunct Professor; Ph.D., Loyola of Chicago. Role of growth and proinflammatory cytokines in toxic damage and repair.

Rajesh Naz, Professor; Ph.D., All-India Institute of Medical Sciences. Peptide and DNA vaccines for contraception; mechanisms of immune infertility.

Joan Olson, Associate Professor; Ph.D., Oregon Health Sciences. Microbial pathogenesis and microbial-host interactions.

Giovanni Piedimonte, Professor; M.D., Rome (Italy). Viral respiratory infections; neuroimmunomodulation; asthma pathogenesis; lung development; airway neurobiology; cystic fibrosis.

Rosana Schafer, Associate Professor; Ph.D., Temple. Effects of in vivo exposure to commonly used herbicides on the immune system, specifically on the response to a bacterial infection.

James Sheil, Associate Professor; Ph.D., Kentucky. Structure-function relationships involved in the cell-mediated immune mechanisms underlying disease pathogenesis.

Uma Sundaram, Professor; M.D., Medical College of Ohio. Immune regulation of electrolyte and nutrient transport in chronically inflamed intestine (Crohn's disease).

Weixin Wang, Assistant Professor; Ph.D., Shanghai Institute for Biochemistry. DNA damage response; MAP kinase; apoptosis.

Section 13
Neuroscience and Neurobiology

This section contains a directory of institutions offering graduate work in neuroscience and neurobiology, followed by in-depth entries submitted by institutions that chose to prepare detailed program descriptions. Additional information about programs listed in the directory but not augmented by an in-depth entry may be obtained by writing directly to the dean of a graduate school or chair of a department at the address given in the directory.

For programs offering related work, see also in this book Anatomy; Biochemistry; Biological and Biomedical Sciences; Biophysics; Cell, Molecular, and Structural Biology; Genetics, Developmental Biology, and Reproductive Biology; Pathology and Pathobiology; Pharmacology and Toxicology; Physiology; and Zoology. In Book 2, see Psychology and Counseling, and in Book 6, see Optometry and Vision Sciences.

CONTENTS

Program Directories

Announcements

Cross-Discipline Announcements

Close-Ups

See also:

Biopsychology

American University, College of Arts and Sciences, Department of Psychology, Program in Psychology, Washington, DC 20016-8001. Offers experimental/biological psychology (MA); general psychology (MA); personality/social psychology (MA). Part-time programs available. *Students:* 33 full-time (31 women), 12 part-time (10 women); includes 7 minority (4 African Americans, 1 Asian American or Pacific Islander, 2 Hispanic Americans). Average age 25. In 2005, 25 master's awarded. *Degree requirements:* For master's, thesis or alternative, comprehensive exam. *Entrance requirements:* For master's, GRE General Test, GRE Subject Test. *Application deadline:* For fall admission, 4/30 for domestic students. Applications are processed on a rolling basis. Application fee: $50. *Expenses:* Tuition: Full-time $17,802; part-time $989 per credit. Required fees: $380. *Financial support:* Research assistantships, teaching assistantships available. Financial award application deadline: 2/1. *Faculty research:* Behavior therapy, cognitive behavior modification, pro-social behavior, conditioning and learning, olfaction.

Boston University, School of Medicine, Division of Graduate Medical Sciences, Program in Mental Health Counseling and Behavioral Medicine, Boston, MA 02215. Offers mental health and behavioral medicine (MA). *Entrance requirements:* For master's, GRE General Test. Additional exam requirements/recommendations for international students: Required—TOEFL. *Expenses:* Tuition: Full-time $31,530; part-time $985 per credit. Required fees: $316; $40 per semester. Tuition and fees vary according to course level and program. *Faculty research:* HIV/AIDS, trauma, behavioral medicine (obesity, breast cancer), neurosciences, autism, serious mental illness, sports psychology.

Carnegie Mellon University, College of Humanities and Social Sciences, Department of Psychology, Area of Cognitive Neuropsychology, Pittsburgh, PA 15213-3891. Offers PhD. *Degree requirements:* For doctorate, thesis/dissertation, comprehensive exam. *Entrance requirements:* For doctorate, GRE General Test. Additional exam requirements/recommendations for international students: Required—TOEFL, TSE.

Columbia University, Graduate School of Arts and Sciences, Division of Natural Sciences, Department of Psychology, New York, NY 10027. Offers experimental psychology (M Phil, MA, PhD); psychobiology (M Phil, MA, PhD); social psychology (M Phil, MA, PhD). *Faculty:* 23 full-time. *Students:* 35 full-time (19 women), 2 part-time. Average age 28. 156 applicants, 7% accepted. In 2005, 7 master's, 6 doctorates awarded. *Degree requirements:* For master's and doctorate, thesis/dissertation. *Entrance requirements:* For master's and doctorate, GRE General Test. Additional exam requirements/recommendations for international students: Required—TOEFL. Application fee: $75. *Expenses:* Tuition: Full-time $31,448. Tuition and fees vary according to course level, course load, campus/location and program. *Financial support:* Fellowships, teaching assistantships, Federal Work-Study and institutionally sponsored loans available. Support available to part-time students. Financial award application deadline: 1/5; financial award applicants required to submit FAFSA. *Unit head:* Geraldine Downey, Chair, 212-854-8718, Fax: 212-854-3609, E-mail: gd20@columbia.edu.

Cornell University, Graduate School, Graduate Fields of Arts and Sciences, Field of Psychology, Ithaca, NY 14853-0001. Offers biopsychology (PhD); human experimental psychology (PhD); personality and social psychology (PhD). *Faculty:* 49 full-time (19 women). *Students:* 143 applicants, 8% accepted, 6 enrolled. In 2005, 3 degrees awarded. *Degree requirements:* For doctorate, thesis/dissertation, 2 semesters of teaching experience, comprehensive exam. *Entrance requirements:* For doctorate, GRE General Test, 3 letters of recommendation. Additional exam requirements/recommendations for international students: Required—TOEFL (minimum score 550 paper-based; 213 computer-based). *Application deadline:* For fall admission, 12/15 for domestic students. Application fee: $60. Electronic applications accepted. *Financial support:* In 2005–06, 32 students received support, including 12 fellowships with full tuition reimbursements available, 20 teaching assistantships with full tuition reimbursements available; research assistantships with full tuition reimbursements available, institutionally sponsored loans, scholarships/grants, health care benefits, tuition waivers (full and partial), and unspecified assistantships also available. Financial award applicants required to submit FAFSA. *Faculty research:* Sensory and perceptual systems, social cognition, cognitive development, quantitative and computational modeling, behavioral neuroscience. *Unit head:* Director of Graduate Studies, 607-255-6364, Fax: 607-255-8433. *Application contact:* Graduate Field Assistant, 607-255-6364, Fax: 607-255-8433, E-mail: psychapp@cornell.edu.

Drexel University, College of Arts and Sciences, Department of Psychology, Philadelphia, PA 19104-2875. Offers clinical psychology (MA, MS, PhD), including clinical psychology (MA, MS), forensic psychology (PhD), health psychology (PhD), neuropsychology (PhD); law-psychology (PhD). *Accreditation:* APA (one or more programs are accredited). *Degree requirements:* For doctorate, thesis/dissertation, internship. *Entrance requirements:* For doctorate, GRE General Test. Additional exam requirements/recommendations for international students: Required—TOEFL. Electronic applications accepted. Expenses: Contact institution. *Faculty research:* Neurosciences, rehabilitation psychology, cognitive science, neurological assessment.

Duke University, Graduate School, Department of Psychology, Durham, NC 27708-0586. Offers biological psychology (PhD); clinical psychology (PhD); cognitive psychology (PhD); developmental psychology (PhD); experimental psychology (PhD); health psychology (PhD); human social development (PhD). *Accreditation:* APA (one or more programs are accredited). *Faculty:* 32 full-time. *Students:* 99 full-time (71 women); includes 16 minority (8 African Americans, 3 Asian Americans or Pacific Islanders, 5 Hispanic Americans), 18 international. 442 applicants, 7% accepted, 22 enrolled. In 2005, 7 doctorates awarded. *Degree requirements:* For doctorate, thesis/dissertation. *Entrance requirements:* For doctorate, GRE General Test. Additional exam requirements/recommendations for international students: Required—IELT (preferred) or TOEFL. *Application deadline:* For fall admission, 12/31 for domestic students, 12/31 for international students. Application fee: $75. Electronic applications accepted. *Financial support:* Fellowships, research assistantships, teaching assistantships, career-related internships or fieldwork and Federal Work-Study available. Financial award application deadline: 12/31. *Unit head:* Amy Needham, Co-Director of Graduate Studies, 919-660-5714, Fax: 919-660-5726, E-mail: amy.needham@duke.edu.

Graduate School and University Center of the City University of New York, Graduate Studies, Program in Psychology, New York, NY 10016-4039. Offers basic applied neurocognition (PhD); biopsychology (PhD); clinical psychology (PhD); developmental psychology (PhD); environmental psychology (PhD); experimental psychology (PhD); industrial psychology (PhD); learning processes (PhD); neuropsychology (PhD); psychology (PhD); social personality (PhD). *Faculty:* 119 full-time (40 women). *Students:* 488 full-time (361 women), 2 part-time (1 woman); includes 91 minority (34 African Americans, 1 American Indian/Alaska Native, 22 Asian Americans or Pacific Islanders, 34 Hispanic Americans), 58 international. Average age 32. 773 applicants, 18% accepted, 80 enrolled. In 2005, 50 degrees awarded. *Degree requirements:* For doctorate, one foreign language, thesis/dissertation. *Entrance requirements:* For doctorate, GRE General Test. Additional exam requirements/recommendations for international students: Required—TOEFL. *Application deadline:* For fall admission, 1/1 for domestic students. Application fee: $125. Electronic applications accepted. *Financial support:* In 2005–06, 272 students received support, including 190 fellowships, 34 research assistantships, 33 teaching assistantships; career-related internships or fieldwork, Federal Work-Study, institutionally sponsored loans, and tuition waivers (full and partial) also available. Financial award application deadline: 2/1; financial award applicants required to submit FAFSA. *Unit head:* Dr. Joseph Glick, Executive Officer, 212-817-8706, Fax: 212-817-1533, E-mail: jglick@gc.cuny.edu.

Harvard University, Graduate School of Arts and Sciences, Department of Psychology, Cambridge, MA 02138. Offers psychology (PhD), including behavior and decision analysis, cognition, developmental psychology, experimental psychology, personality, psychobiology,

psychopathology; social psychology (PhD). *Faculty:* 25 full-time, 4 part-time/adjunct. *Students:* 71 full-time (37 women). 166 applicants, 10% accepted. In 2005, 7 doctorates awarded. *Degree requirements:* For doctorate, thesis/dissertation, general exams. *Entrance requirements:* For doctorate, GRE General Test. Additional exam requirements/recommendations for international students: Required—TOEFL. *Application deadline:* For fall admission, 12/30 for domestic students. Application fee: $60. *Expenses:* Tuition: Full-time $28,752. Full-time tuition and fees vary according to program and student level. *Financial support:* Fellowships, research assistantships, teaching assistantships, career-related internships or fieldwork, Federal Work-Study, and institutionally sponsored loans available. Financial award application deadline: 12/30. *Unit head:* Michele Biscoe, Administrator, 617-495-9978. *Application contact:* Office of Admissions and Financial Aid, 617-495-5315.

Howard University, Graduate School of Arts and Sciences, Department of Psychology, Washington, DC 20059-0002. Offers clinical psychology (PhD); developmental psychology (PhD); experimental psychology (PhD); neuropsychology (PhD); personality psychology (PhD); psychology (MS); social psychology (PhD). *Accreditation:* APA (one or more programs are accredited). Part-time programs available. *Degree requirements:* For master's, thesis; for doctorate, thesis/dissertation, qualifying exam, comprehensive exam. *Entrance requirements:* For master's, GRE General Test, minimum GPA of 2.5, bachelor's degree in psychology or related field; for doctorate, GRE General Test, minimum GPA of 3.0. *Faculty research:* Personality and psychophysiology, educational and social development of African-American children, child and adult psychopathology.

Hunter College of the City University of New York, Graduate School, School of Arts and Sciences, Department of Psychology, New York, NY 10021-5085. Offers applied and evaluative psychology (MA); biopsychology and comparative psychology (MA); social, cognitive, and developmental psychology (MA). Part-time and evening/weekend programs available. *Faculty:* 14 full-time (8 women), 1 part-time/adjunct (0 women). *Students:* 7 full-time (all women), 47 part-time (38 women); includes 2 Asian Americans or Pacific Islanders, 5 Hispanic Americans. Average age 28. 64 applicants, 41% accepted, 15 enrolled. In 2005, 23 degrees awarded. *Degree requirements:* For master's, thesis, comprehensive exam. *Entrance requirements:* For master's, GRE General Test, minimum 12 credits of course work in psychology, including statistics and experimental psychology; 2 letters of recommendation. Additional exam requirements/recommendations for international students: Required—TOEFL. *Application deadline:* For fall admission, 4/1 for domestic students, 2/1 for international students; for spring admission, 11/1 for domestic students, 9/1 for international students. Applications are processed on a rolling basis. Application fee: $125. *Expenses:* Tuition, state resident: full-time $6,400; part-time $270 per credit. Tuition, nonresident: part-time $500 per credit. International tuition: $12,000 full-time. Required fees: $50 per term. Part-time tuition and fees vary according to course load and program. *Financial support:* Federal Work-Study, scholarships/grants, and tuition waivers (partial) available. Support available to part-time students. *Faculty research:* Personality, cognitive and linguistic development, hormonal and neural control of behavior, gender and culture, social cognition of health and attitudes. *Unit head:* Dr. Gordon A. Barr, Acting Chair, 212-772-5550. *Application contact:* William Zlata, Director for Graduate Admissions, 212-772-4482, Fax: 212-650-3336, E-mail: admissions@hunter.cuny.edu.

Indiana University–Purdue University Indianapolis, School of Science, Department of Psychology, Indianapolis, IN 46202-3275. Offers clinical rehabilitation psychology (MS, PhD); industrial/organizational psychology (MS); psychobiology of addictions (PhD). *Accreditation:* APA (one or more programs are accredited). Part-time programs available. *Faculty:* 13 full-time (3 women). *Students:* 30 full-time (25 women), 18 part-time (15 women); includes 2 minority (1 African American, 1 Asian American or Pacific Islander), 4 international. Average age 32. In 2005, 13 master's, 2 doctorates awarded. Terminal master's awarded for partial completion of doctoral program. *Degree requirements:* For master's and doctorate, thesis/dissertation. *Entrance requirements:* For master's, GRE General Test, minimum undergraduate GPA of 3.0; for doctorate, GRE General Test, GRE Subject Test (clinical rehabilitation psychology), minimum undergraduate GPA of 3.2. *Application deadline:* For fall admission, 2/1 for domestic students. Application fee: $50 ($60 for international students). *Expenses:* Tuition, state resident: full-time $5,159; part-time $215 per credit hour. Tuition, nonresident: full-time $14,890; part-time $620 per credit hour. Required fees: $614. Tuition and fees vary according to campus/location and program. *Financial support:* In 2005–06, 21 students received support; fellowships with partial tuition reimbursements available, research assistantships with partial tuition reimbursements available, teaching assistantships with partial tuition reimbursements available, career-related internships or fieldwork, Federal Work-Study, and institutionally sponsored loans available. Financial award application deadline: 3/1; financial award applicants required to submit FAFSA. *Faculty research:* Psychiatric rehabilitation, health psychology, psychopharmacology of drugs of abuse, personnel psychology, organizational decision making. *Unit head:* Dr. J. Gregor Fetterman, Chairman, 317-274-6943, Fax: 317-274-6756, E-mail: gfetter@iupui.edu. *Application contact:* Susie Wiesinger, Graduate Program Administrative Assistant, 317-274-6945, Fax: 317-274-6756, E-mail: swiesing@iupui.edu.

Louisiana State University and Agricultural and Mechanical College, Graduate School, College of Arts and Sciences, Department of Psychology, Baton Rouge, LA 70803. Offers biological psychology (MA, PhD); clinical psychology (MA, PhD); cognitive psychology (MA, PhD); developmental psychology (MA, PhD); industrial/organizational psychology (MA, PhD); school psychology (MA, PhD). *Accreditation:* APA (one or more programs are accredited). *Faculty:* 21 full-time (8 women). *Students:* 74 full-time (50 women), 27 part-time (21 women); includes 13 minority (6 African Americans, 2 American Indian/Alaska Natives, 3 Asian Americans or Pacific Islanders, 2 Hispanic Americans), 1 international. Average age 28. 253 applicants, 13% accepted, 221 enrolled. In 2005, 14 master's, 10 doctorates awarded. Terminal master's awarded for partial completion of doctoral program. *Degree requirements:* For master's, thesis; for doctorate, thesis/dissertation, 1 year internship. *Entrance requirements:* For master's and doctorate, GRE General Test, minimum GPA of 3.0. Additional exam requirements/recommendations for international students: Required—TOEFL (minimum score 550 paper-based; 213 computer-based). *Application deadline:* For fall admission, 1/25 priority date for domestic students, 5/15 priority date for international students. Applications are processed on a rolling basis. Application fee: $25. Electronic applications accepted. *Financial support:* In 2005–06, 74 students received support, including 1 fellowship (averaging $16,500 per year), 7 research assistantships with partial tuition reimbursements available (averaging $20,250 per year), 49 teaching assistantships with partial tuition reimbursements available (averaging $12,178 per year); career-related internships or fieldwork, Federal Work-Study, institutionally sponsored loans, scholarships/grants, tuition waivers (full and partial), and contracts also available. Financial award applicants required to submit FAFSA. Total annual research expenditures: $195,733. *Unit head:* Dr. Alan Baumeister, Chair, 225-578-4099, Fax: 225-578-4125, E-mail: abaumei@lau.edu. *Application contact:* Dr. Janet McDonald, Coordinator of Graduate Studies, 225-578-4116, Fax: 225-578-4125, E-mail: psmcdo@lsu.edu.

Memorial University of Newfoundland, School of Graduate Studies, Interdisciplinary Program in Cognitive and Behavioral Ecology, St. John's, NL A1C 5S7, Canada. Offers M Sc, PhD. *Students:* 16 full-time (8 women), 3 part-time (2 women), 3 international. 7 applicants, 29% accepted, 2 enrolled. In 2005, 1 doctorate awarded. *Degree requirements:* For master's, thesis, public lecture; for doctorate, thesis/dissertation, oral defense of dissertation, comprehensive exam. *Entrance requirements:* For master's, honors degree (minimum 2nd class standing) in related field; for doctorate, master's degree. *Application deadline:* For fall admission, 1/31 for domestic students. Applications are processed on a rolling basis. Application fee: $40 Canadian dollars. Electronic applications accepted. Students attend 2 or 3 semesters per year. *Expenses:* Tuition, nonresident: full-time $1,466; part-time $733 per semester. International tuition: $1,906 full-time. Tuition and fees vary according to degree level and program. *Financial support:* Fellowships, research assistantships, teaching assistantships, institutionally sponsored loans and scholarships/grants available. *Faculty research:*

Biopsychology

Memorial University of Newfoundland (continued)
Seabird feeding ecology, marine mammal and seabird energetics, systems of fish, seabird/seal/fisheries interaction. *Unit head:* Dr. Anne Storey, Chair, 709-778-7665, Fax: 709-737-3316, E-mail: astorey@play.psych.mun.ca. *Application contact:* Gail Kenny, Secretary, 709-737-8154, Fax: 709-737-3316, E-mail: gkenny@mun.ca.

Northwestern University, The Graduate School, Judd A. and Marjorie Weinberg College of Arts and Sciences, Department of Psychology, Evanston, IL 60208. Offers brain, behavior and cognition (PhD); clinical psychology (PhD); cognitive psychology (PhD); personality (PhD); social psychology (PhD). Admissions and degrees offered through The Graduate School. *Accreditation:* APA (one or more programs are accredited). Part-time programs available. *Faculty:* 23 full-time (6 women). *Students:* 77 (46 women); includes 6 minority (all Asian Americans or Pacific Islanders) 20 international. Average age 25. 348 applicants, 7% accepted, 13 enrolled. In 2005, 3 doctorates awarded. *Degree requirements:* For doctorate, thesis/dissertation. *Entrance requirements:* For doctorate, GRE General Test, GRE Subject Test. Additional exam requirements/recommendations for international students: Required—TOEFL, TSE. *Application deadline:* For fall admission, 12/31 for domestic students. Application fee: $60 ($75 for international students). Electronic applications accepted. *Financial support:* In 2005–06, 12 fellowships with full tuition reimbursements (averaging $17,052 per year), 3 research assistantships with partial tuition reimbursements (averaging $13,617 per year), 22 teaching assistantships with full tuition reimbursements (averaging $13,617 per year) were awarded; career-related internships or fieldwork, Federal Work-Study, institutionally sponsored loans, and tuition waivers (full and partial) also available. Financial award application deadline: 12/31; financial award applicants required to submit FAFSA. *Faculty research:* Memory and higher order cognition, anxiety and depression, effectiveness of psychotherapy, social cognition, molecular basis of memory. Total annual research expenditures: $3.4 million. *Unit head:* Alice Eagly, Interim Chair, 847-491-7859, Fax: 847-491-7859, E-mail: psych-chair@northwestern.edu. *Application contact:* Dan McAdams, Admission Officer, 847-491-4174, Fax: 847-491-7859, E-mail: dmca@northwestern.edu.

Northwestern University, The Graduate School and Northwestern University Feinberg School of Medicine, Program in Clinical Psychology, Evanston, IL 60208. Offers clinical psychology (PhD), including clinical neuropsychology, general clinical. PhD admissions and degree offered through The Graduate School. *Accreditation:* APA. *Faculty:* 17 full-time (10 women), 61 part-time/adjunct (35 women). *Students:* 29 full-time (24 women); includes 8 minority (5 African Americans, 2 Asian Americans or Pacific Islanders, 1 Hispanic American), 2 international. Average age 34. 159 applicants, 4% accepted, 5 enrolled. In 2005, 5 degrees awarded. *Degree requirements:* For doctorate, thesis/dissertation, clinical internship. *Entrance requirements:* For doctorate, GRE General Test, GRE Subject Test, minimum GPA of 3.2, course work in psychology. Additional exam requirements/recommendations for international students: Required—TOEFL. *Application deadline:* For fall admission, 1/12 for domestic students. Application fee: $60 ($75 for international students). *Financial support:* In 2005–06, 1 fellowship with full and partial tuition reimbursement (averaging $11,673 per year), 6 research assistantships with partial tuition reimbursements (averaging $16,285 per year) were awarded; career-related internships or fieldwork, institutionally sponsored loans, scholarships/grants, and tuition waivers (full) also available. Financial award application deadline: 1/12; financial award applicants required to submit FAFSA. *Faculty research:* Cancer and cardiovascular risk reduction, evaluation of mental health services and policy, neuropsychological assessment, outcome of psychotherapy, cognitive therapy, pediatric and clinical child psychology. *Unit head:* Dr. Mark A. Reinecke, Director, 312-908-1465, Fax: 312-908-5070, E-mail: m-reinecke@northwestern.edu. *Application contact:* Peter Zeldow, Admission Officer, 312-908-8733, E-mail: pbz080@northwestern.edu.

The Ohio State University, Graduate School, College of Social and Behavioral Sciences, Department of Psychology, Columbus, OH 43210. Offers clinical psychology (PhD); cognitive/experimental psychology (PhD); counseling psychology (PhD); developmental psychology (PhD); mental retardation and developmental disabilities (PhD); psychobiology (PhD); psychology (MA); quantitative psychology (PhD); social psychology (PhD). *Accreditation:* APA (one or more programs ɛ ə accredited). *Degree requirements:* For doctorate, thesis/dissertation. *Entrance requirements:* For master's and doctorate, GRE General Test. Additional exam requirements/recommendations for international students: Required—TOEFL (minimum score 600 paper-based; 250 computer-based). Electronic applications accepted.

Oregon Health & Science University, School of Medicine, Graduate Programs in Medicine, Department of Behavioral Neuroscience, Portland, OR 97239-3098. Offers MS, PhD, MD/PhD. *Degree requirements:* For master's, thesis; for doctorate, thesis/dissertation, written exam. *Entrance requirements:* For master's and doctorate, GRE General Test. *Faculty research:* Neural basis of behavior, behavioral pharmacology, behavioral genetics, neuropharmacology and neuroendocrinology, biological basis of drug seeking and addiction.

See Close-Up on page 1011.

Pacific Graduate School of Psychology, PGSP-Stanford Psy D Consortium Program, Palo Alto, CA 94303-4232. Offers Psy D. *Degree requirements:* For doctorate, thesis/dissertation. *Entrance requirements:* For doctorate, GRE, BA or MA in psychology or related area, minimum undergraduate GPA of 3.0, minimum graduate GPA of 3.3. Additional exam requirements/recommendations for international students: Required—TOEFL. Electronic applications accepted. *Faculty research:* Biopsychosocial research, neurobiology, psychopharmacology.

The Pennsylvania State University University Park Campus, Graduate School, College of Health and Human Development, Department of Biobehavioral Health, State College, University Park, PA 16802-1503. Offers MS, PhD. *Students:* 26 full-time (20 women); includes 6 minority (4 African Americans, 2 Hispanic Americans), 7 international. *Entrance requirements:* For doctorate, GRE General Test or MCAT. *Expenses:* Tuition, state resident: full-time $12,518; part-time $522 per credit. Tuition, nonresident: full-time $23,004; part-time $959 per credit. Required fees: $484. Tuition and fees vary according to course load, campus/location and program. *Unit head:* Dr. Lynn T. Kozlowski, Head, 814-863-7256, Fax: 814-863-7525, E-mail: ltk1@psu.edu.

Announcement: The Department of Biobehavioral Health at Penn State University seeks applicants for its PhD program. Department provides interdisciplinary training emphasizing research methodology and targeted toward improving human health across the lifespan. Students develop understanding of biological and behavioral factors in health and disease using human or animal models. Faculty members have expertise in health psychology, behavioral medicine, addiction, epidemiology, social psychology, medical sociology, and cognition. Health promotion topics include international and minority health, women's health, sexual health, and substance abuse. Biological topics include cardiovascular disease, neuropharmacology, endocrinology, physiology, stress, genetics, psychopharmacology, and psychoneuroimmunology. Call 814-863-7256 or visit www.bbh.hhdev.psu.edu/grad/.

The Pennsylvania State University University Park Campus, Graduate School, College of the Liberal Arts, Department of Psychology, State College, University Park, PA 16802-1503. Offers clinical psychology (MS, PhD); cognitive psychology (MS, PhD); developmental psychology (MS, PhD); industrial/organizational psychology (MS, PhD); psychobiology (MS, PhD); social psychology (MS, PhD). *Accreditation:* APA (one or more programs are accredited). *Students:* 113 full-time (84 women), 15 part-time (13 women); includes 21 minority (6 African Americans, 9 Asian Americans or Pacific Islanders, 6 Hispanic Americans), 18 international. *Degree requirements:* For master's and doctorate, thesis/dissertation. *Entrance requirements:* For master's and doctorate, GRE General Test, GRE Subject Test. *Expenses:* Tuition, state resident: full-time $12,518; part-time $522 per credit. Tuition, nonresident: full-time $23,004; part-time $959 per credit. Required fees: $484. Tuition and fees vary according to course load, campus/location and program. *Financial support:* Fellowships, research assistantships, teaching assistantships available. *Unit head:* Dr. Kevin R. Murphy, Head, 814-865-9515, Fax:

814-863-7002, E-mail: krmurphy@psu.edu. *Application contact:* Dr. Kevin R. Murphy, Head, 814-865-9515, Fax: 814-863-7002, E-mail: krmurphy@psu.edu.

Rutgers, The State University of New Jersey, Newark, Graduate School, Program in Integrative Neuroscience, Newark, NJ 07102. Offers PhD. Part-time programs available. *Faculty:* 27 full-time (11 women), 1 part-time/adjunct (0 women). *Students:* 19 full-time (10 women), 14 part-time (8 women); includes 7 minority (1 African American, 5 Asian Americans or Pacific Islanders, 1 Hispanic American). 68 applicants, 29% accepted, 7 enrolled. In 2005, 2 degrees awarded. *Degree requirements:* For doctorate, thesis/dissertation. *Entrance requirements:* For doctorate, GRE, minimum GPA of 3.0. *Application deadline:* For fall admission, 2/1 for domestic students. Applications are processed on a rolling basis. Application fee: $50. Electronic applications accepted. *Expenses:* Tuition, state resident: full-time $10,440; part-time $435 per credit. Tuition, nonresident: full-time $15,520; part-time $637 per credit. *Financial support:* In 2005–06, 21 teaching assistantships with full and partial tuition reimbursements (averaging $19,367 per year) were awarded Financial award application deadline: 3/1. *Faculty research:* Systems neuroscience, cognitive neuroscience, molecular neuroscience, behavioral neuroscience. Total annual research expenditures: $3 million. *Unit head:* Dr. Ian Creese, Director, 973-353-1080 Ext. 3300, Fax: 973-353-1272, E-mail: creese@axon.rutgers.edu.

Rutgers, The State University of New Jersey, Newark, Graduate School, Program in Psychology, Newark, NJ 07102. Offers cognitive neuroscience (PhD); cognitive science (PhD); perception (PhD); psychobiology (PhD); social cognition (PhD). *Faculty:* 15 full-time (5 women), 3 part-time/adjunct (2 women). *Students:* 17 full-time (12 women), 11 part-time (5 women); includes 7 minority (1 African American, 4 Asian Americans or Pacific Islanders, 2 Hispanic Americans). 42 applicants, 12% accepted, 4 enrolled. In 2005, 2 doctorates awarded. *Degree requirements:* For doctorate, thesis/dissertation, comprehensive exam, registration. *Entrance requirements:* For doctorate, GRE General Test, GRE Subject Test, minimum undergraduate B average. *Application deadline:* For fall admission, 2/1 for domestic students; for spring admission, 11/1 for domestic students. Application fee: $50. Electronic applications accepted. *Expenses:* Tuition, state resident: full-time $10,440; part-time $435 per credit. Tuition, nonresident: full-time $15,520; part-time $637 per credit. *Financial support:* In 2005–06, 16 students received support, including 11 fellowships with full and partial tuition reimbursements available (averaging $18,000 per year), 9 teaching assistantships with full tuition reimbursements available (averaging $16,988 per year); research assistantships, career-related internships or fieldwork, health care benefits, and minority scholarships also available. Financial award application deadline: 2/1. *Faculty research:* Visual perception (luminance, motion), neuroendocrine mechanisms in behavior (reproduction, pain), attachment theory, connectionist modeling of cognition. *Unit head:* Dr. Harold Siegel, Director, 973-353-5440 Ext. 236, Fax: 973-353-1171, E-mail: hisiegal@andromeda.rutgers.edu.

Rutgers, The State University of New Jersey, New Brunswick/Piscataway, Graduate School, Program in Psychology, New Brunswick, NJ 08901-1281. Offers biopsychology and behavioral neuroscience (PhD); clinical psychology (PhD); cognitive psychology (PhD); interdisciplinary developmental psychology (PhD); interdisciplinary health psychology (PhD); social psychology (PhD). *Accreditation:* APA. *Faculty:* 112 full-time, 5 part-time/adjunct. *Students:* 97 full-time (74 women), 2 part-time (1 woman); includes 18 minority (4 African Americans, 8 Asian Americans or Pacific Islanders, 6 Hispanic Americans), 6 international. Average age 28. 396 applicants, 6% accepted, 16 enrolled. In 2005, 9 doctorates awarded. *Median time to degree:* Of those who began their doctoral program in fall 1997, 98% received their degree in 8 years or less. *Degree requirements:* For doctorate, thesis/dissertation. *Entrance requirements:* For doctorate, GRE General Test. Additional exam requirements/recommendations for international students: Required—TOEFL (minimum score 577 paper-based; 233 computer-based). *Application deadline:* For fall admission, 12/15 for domestic students, 12/15 for international students. Application fee: $50. Electronic applications accepted. *Expenses:* Tuition, state resident: full-time $10,440; part-time $435 per credit. Tuition, nonresident: full-time $15,520; part-time $647 per credit. Required fees: $129 per credit. Tuition and fees vary according to program. *Financial support:* In 2005–06, 85 students received support, including 13 fellowships with full tuition reimbursements available (averaging $18,000 per year), 21 research assistantships with full tuition reimbursements available (averaging $16,988 per year), 45 teaching assistantships with full tuition reimbursements available (averaging $16,988 per year); career-related internships or fieldwork, Federal Work-Study, scholarships/grants, traineeships, and unspecified assistantships also available. Financial award application deadline: 12/15. *Faculty research:* Learning and memory, behavioral ecology, hormones and behavior, psychopharmacology, anxiety disorders. Total annual research expenditures: $4 million. *Unit head:* Barbara McCrady, Director, 732-445-2556, Fax: 732-445-2263, E-mail: bmccrady@rci.rutgers.edu. *Application contact:* Anne Sokolowski, Administrative Assistant, 732-445-2555, Fax: 732-445-2263, E-mail: anne@psych-b.rutgers.edu.

State University of New York at Binghamton, Graduate School, School of Arts and Sciences, Department of Psychology, Specialization in Behavioral Neuroscience, Binghamton, NY 13902-6000. Offers MA, PhD. *Degree requirements:* For master's, thesis; for doctorate, thesis/dissertation, departmental qualifying exam. *Entrance requirements:* For master's and doctorate, GRE General Test, GRE Subject Test. Additional exam requirements/recommendations for international students: Required—TOEFL. Electronic applications accepted.

Stony Brook University, State University of New York, Graduate School, College of Arts and Sciences, Department of Psychology, Program in Biopsychology, Stony Brook, NY 11794. Offers PhD. *Students:* 14 full-time (10 women), 1 part-time; includes 2 minority (1 African American, 1 Hispanic American), 4 international. Average age 28. 33 applicants, 12% accepted. In 2005, 1 degree awarded. *Degree requirements:* For doctorate, thesis/dissertation. *Entrance requirements:* For doctorate, GRE General Test, GRE Subject Test. Additional exam requirements/recommendations for international students: Required—TOEFL. *Application deadline:* For fall admission, 1/15 for domestic students. Application fee: $50. *Expenses:* Tuition, state resident: full-time $6,900; part-time $288 per credit. Tuition, nonresident: full-time $10,920; part-time $455 per credit. Required fees: $704. *Application contact:* Director, 631-632-7855, Fax: 631-632-7876.

Announcement: The research interests of the faculty and students in biopsychology span a broad range of topics in cognitive and behavioral neuroscience, including gene-behavior interactions, psychopharmacology, anatomical plasticity, human brain imaging (fMRI, TMS, ERP), rodent models of psychopathology, clinical neuroscience, and neuroethics. The Biopsychology Research Groups have numerous collaborations with other researchers at Stony Brook, nearby Brookhaven National Laboratory, and at other institutions in the United States and abroad.

See Close-Up on page 1017.

Texas A&M University, College of Liberal Arts, Department of Psychology, Program in Behavioral and Cellular Neuroscience, College Station, TX 77843. Offers MS, PhD. *Degree requirements:* For master's and doctorate, thesis/dissertation. *Entrance requirements:* For master's and doctorate, GRE. Additional exam requirements/recommendations for international students: Required—TOEFL. *Application deadline:* For fall admission, 1/5 for domestic students, 1/5 for international students. Application fee: $50 ($75 for international students). *Expenses:* Tuition, state resident: full-time $4,488; part-time $187 per credit hour. Tuition, nonresident: full-time $11,112; part-time $463 per credit hour. Required fees: $1,974. *Financial support:* Fellowships with partial tuition reimbursements, research assistantships with partial tuition reimbursements, teaching assistantships with partial tuition reimbursements, career-related internships or fieldwork, institutionally sponsored loans, health care benefits, and unspecified assistantships available. Financial award application deadline: 1/5; financial award applicants required to submit FAFSA. *Unit head:* James Grau, Area Coordinator, 979-845-2584, Fax: 979-845-4727, E-mail: j-grau@tamu.edu. *Application contact:* Sharon Starr, Graduate Admissions Supervisor, 979-458-1710, Fax: 979-845-4727, E-mail: gradadv@psyc.tamu.edu.

Biopsychology

University at Albany, State University of New York, College of Arts and Sciences, Department of Psychology, Albany, NY 12222-0001. Offers autism (Certificate); biopsychology (PhD); clinical psychology (PhD); general/experimental psychology (PhD); industrial/organizational psychology (PhD); psychology (MA); social/personality psychology (PhD). *Accreditation:* APA (one or more programs are accredited). *Students:* 57 full-time (42 women), 63 part-time (50 women). Average age 31. In 2005, 16 master's, 14 doctorates, 21 other advanced degrees awarded. *Degree requirements:* For doctorate, thesis/dissertation. *Entrance requirements:* For doctorate, GRE General Test, GRE Subject Test. Additional exam requirements/recommendations for international students: Required—TOEFL (minimum score 550 paper-based; 213 computer-based). *Application deadline:* For fall admission, 1/15 for domestic students, 1/15 for international students. Application fee: $60. Electronic applications accepted. *Financial support:* Fellowships, research assistantships, teaching assistantships, career-related internships or fieldwork available. Financial award application deadline: 2/1. *Unit head:* Edelgard Wulfert, Chair, 518-442-4820.

The University of British Columbia, Faculty of Arts and Faculty of Graduate Studies, Department of Psychology, Vancouver, BC V6T 1Z1, Canada. Offers behavioral neuroscience (MA, PhD); clinical psychology (MA, PhD); cognitive science (MA, PhD); developmental psychology (MA, PhD); forensic psychology (MA, PhD); psychometrics (MA, PhD); social/personality psychology (MA, PhD). *Accreditation:* APA (one or more programs are accredited). *Faculty:* 49 full-time (17 women), 19 part-time/adjunct (12 women). *Students:* 112 full-time (77 women). Average age 29. 280 applicants, 12% accepted, 15 enrolled. In 2005, 16 master's, 17 doctorates awarded. Terminal master's awarded for partial completion of doctoral program. *Median time to degree:* Of those who began their doctoral program in fall 1997, 60% received their degree in 8 years or less. *Degree requirements:* For master's, thesis/dissertation; for doctorate, thesis/dissertation, comprehensive exam, registration (for some programs). *Entrance requirements:* For master's and doctorate, GRE General Test, GRE Subject Test. Additional exam requirements/recommendations for international students: Required—TOEFL (minimum score 550 paper-based; 230 computer-based). *Application deadline:* For fall admission, 1/15 for domestic students, 1/15 for international students. Applications are processed on a rolling basis. Application fee: $90 Canadian dollars ($150 Canadian dollars for international students). Electronic applications accepted. *Financial support:* In 2005–06, 112 students received support, including fellowships with full and partial tuition reimbursements available (averaging $16,000 per year), research assistantships with full and partial tuition reimbursements available (averaging $5,000 per year), teaching assistantships with full and partial tuition reimbursements available (averaging $10,000 per year); career-related internships or fieldwork, Federal Work-Study, institutionally sponsored loans, scholarships/grants, health care benefits, tuition waivers (full and partial), and unspecified assistantships also available. Financial award application deadline: 1/15. *Faculty research:* Clinical, developmental, social/personality, cognition, behavioral neuroscience. Total annual research expenditures: $5 million Canadian dollars. *Unit head:* Dr. Eric Eich, Head, 604-822-3078, Fax: 604-822-6923, E-mail: ee@psych.ubc.ca. *Application contact:* Rose Tam, Graduate Secretary, 604-822-3144, Fax: 604-822-6923, E-mail: gradsec@psych.ubc.ca.

University of Connecticut, Graduate School, College of Liberal Arts and Sciences, Department of Psychology, Field of Psychology, Storrs, CT 06269. Offers behavioral neuroscience (PhD); biopsychology (PhD); clinical psychology (MA, PhD); cognition and instruction (PhD); developmental psychology (MA, PhD); ecological psychology (PhD); experimental psychology (PhD); general psychology (MA, PhD); industrial/organizational psychology (PhD); language and cognition (PhD); neuroscience (PhD); social psychology (MA, PhD). *Accreditation:* APA (one or more programs are accredited). *Faculty:* 51 full-time (23 women). *Students:* 134 full-time (90 women), 21 part-time (10 women); includes 13 minority (7 African Americans, 2 Asian Americans or Pacific Islanders, 4 Hispanic Americans), 21 international. Average age 28. 510 applicants, 10% accepted, 49 enrolled. In 2005, 22 master's, 15 doctorates awarded. Terminal master's awarded for partial completion of doctoral program. *Degree requirements:* For master's, comprehensive exam; for doctorate, thesis/dissertation. *Entrance requirements:* For master's and doctorate, GRE General Test, GRE Subject Test. Additional exam requirements/recommendations for international students: Required—TOEFL (minimum score 550 paper-based; 213 computer-based). *Application deadline:* For fall admission, 2/1 priority date for domestic students, 2/1 priority date for international students; for spring admission, 11/1 for domestic students, 10/1 for international students. Applications are processed on a rolling basis. Application fee: $55. Electronic applications accepted. *Expenses:* Tuition, state resident: part-time $444 per credit hour. Tuition, nonresident: part-time $1,154 per credit hour. Tuition and fees vary according to course load. *Financial support:* In 2005–06, 67 research assistantships with full tuition reimbursements, 58 teaching assistantships with full tuition reimbursements were awarded; fellowships, career-related internships or fieldwork, Federal Work-Study, scholarships/grants, health care benefits, and unspecified assistantships also available. Financial award application deadline: 2/1; financial award applicants required to submit FAFSA. *Application contact:* Deborah Doucette, Administrative Assistant, 860-486-2057, Fax: 860-486-2760, E-mail: futuregr@psych.psy.uconn.edu.

University of Illinois at Urbana–Champaign, Graduate College, College of Liberal Arts and Sciences, Department of Psychology, Champaign, IL 61820. Offers applied measurement (MS); biological psychology (MA, PhD); brain and cognition (MA, PhD); clinical psychology (MA, PhD); cognitive psychology (MA, PhD); developmental psychology (MA, PhD); personnel psychology (MS); quantitative psychology (MA, PhD); social-personality-organizational (MA, PhD); visual cognition and human performance (MA, PhD). *Accreditation:* APA (one or more programs are accredited). *Faculty:* 56 full-time (19 women), 3 part-time/adjunct (all women). *Students:* 163 full-time (105 women), 14 part-time (10 women); includes 28 minority (8 African Americans, 11 Asian Americans or Pacific Islanders, 9 Hispanic Americans), 47 international. 346 applicants, 18% accepted, 29 enrolled. In 2005, 19 master's, 24 doctorates awarded. *Degree requirements:* For doctorate, thesis/dissertation. *Entrance requirements:* For master's, GRE General Test, GRE Subject Test (recommended), minimum GPA of 3.0; for doctorate, GRE General Test, GRE Subject Test (recommended). Additional exam requirements/recommendations for international students: Required—TOEFL, TSE. *Application deadline:* For fall admission, 1/4 for domestic students. Application fee: $50 ($60 for international students). Electronic applications accepted. *Financial support:* In 2005–06, 32 fellowships with full tuition reimbursements (averaging $15,000 per year), 87 research assistantships with full tuition reimbursements (averaging $14,412 per year), 98 teaching assistantships with full tuition reimbursements (averaging $14,412 per year) were awarded; career-related internships or fieldwork, traineeships, and tuition waivers (full) also available. Financial award application deadline: 12/1. Total annual research expenditures: $5.7 million. *Unit head:* Dr. Lawrence Hubert, Interim Head, 217-333-0632, Fax: 217-244-5876, E-mail: lhubert@uiuc.edu. *Application contact:* Cheryl Berger, Assistant Head for Graduate Affairs, 217-333-0631, Fax: 217-244-5876, E-mail: gradstdy@cyrus.psych.uiuc.edu.

University of Michigan, Horace H. Rackham School of Graduate Studies, College of Literature, Science, and the Arts, Department of Psychology, Ann Arbor, MI 48109. Offers biopsychology (PhD); clinical psychology (PhD); cognition and perception (PhD); developmental psychology (PhD); organizational psychology (PhD); personality psychology (PhD); social psychology (PhD). *Accreditation:* APA. *Faculty:* 86 full-time, 69 part-time/adjunct. *Students:* 157 full-time (112 women); includes 51 minority (21 African Americans, 1 American Indian/Alaska Native, 17 Asian Americans or Pacific Islanders, 12 Hispanic Americans), 32 international. Average age 25. 698 applicants, 9% accepted, 29 enrolled. In 2005, 25 degrees awarded. *Degree requirements:* For doctorate, thesis/dissertation, oral defense of dissertation, preliminary exam, comprehensive exam. *Entrance requirements:* For doctorate, GRE General Test (optional), GRE Subject Test (optional). Additional exam requirements/recommendations for international students: Required—TOEFL. *Application deadline:* For fall admission, 12/15 for domestic students. Application fee: $60 ($75 for international students). Electronic applications accepted. *Expenses:* Tuition, state resident: full-time $14,082; part-time $894 per credit hour. Tuition, nonresident: full-time $28,500; part-time $1,675 per credit hour. Required fees: $189; $189 per unit. *Financial support:* Fellowships with full tuition reimbursements, research assistantships with full tuition reimbursements, teaching assistantships with full tuition reimbursements, career-related internships or fieldwork available. Financial award application deadline: 4/15. *Unit head:* Richard Gonzalez, Chair, 734-764-7429. *Application contact:* Lesley A. Newton, Psychology Graduate Office, 731-763-2131, Fax: 734-615-7584, E-mail: psych.grad.office@umich.edu.

University of Minnesota, Twin Cities Campus, Graduate School, College of Liberal Arts, Department of Psychology, Program in Cognitive and Biological Psychology, Minneapolis, MN 55455-0213. Offers PhD. *Students:* 30 full-time (15 women); includes 5 minority (1 African American, 4 Asian Americans or Pacific Islanders), 6 international. 38 applicants, 18% accepted, 5 enrolled. In 2005, 5 degrees awarded. *Median time to degree:* Of those who began their doctoral program in fall 1997, 75% received their degree in 8 years or less. *Degree requirements:* For doctorate, thesis/dissertation, comprehensive exam. *Entrance requirements:* For doctorate, GRE General Test, GRE Subject Test (recommended), 12 credits of upper-level psychology courses, including a course in statistics or psychological measurement. Additional exam requirements/recommendations for international students: Required—TOEFL (minimum score 550 paper-based; 213 computer-based); Recommended—TSE (minimum score 50). *Application deadline:* For fall admission, 12/20 for domestic students, 12/20 for international students. Application fee: $55 ($75 for international students). *Expenses:* Tuition, state resident: full-time $8,748; part-time $729 per credit. Tuition, nonresident: full-time $15,848; part-time $1,321 per credit. Full-time tuition and fees vary according to class time, course load, program and reciprocity agreements. *Financial support:* In 2005–06, fellowships with full tuition reimbursements (averaging $17,500 per year), research assistantships with full tuition reimbursements (averaging $11,895 per year), teaching assistantships with full tuition reimbursements (averaging $11,895 per year) were awarded; career-related internships or fieldwork, traineeships, and tuition waivers (partial) also available. Financial award application deadline: 12/20. *Unit head:* Sheng He, Area Director, 612-626-0752. *Application contact:* Coordinator, 612-624-4181, Fax: 612-626-2079, E-mail: psyapply@tc.umn.edu.

University of Missouri–Columbia, Graduate School, College of Arts and Sciences, Division of Biological Sciences, Columbia, MO 65211. Offers cellular, molecular and developmental biology (MA, PhD); evolutionary biology and ecology (MA, PhD); neurobiology and behavior (MA, PhD). *Faculty:* 43 full-time (12 women). *Students:* 48 full-time (26 women), 28 part-time (10 women); includes 7 minority (2 African Americans, 3 Asian Americans or Pacific Islanders, 2 Hispanic Americans), 20 international. In 2005, 3 master's, 8 doctorates awarded. Terminal master's awarded for partial completion of doctoral program. *Degree requirements:* For master's, thesis/dissertation; for doctorate, thesis/dissertation, comprehensive exam. *Entrance requirements:* For master's and doctorate, GRE General Test, minimum GPA of 3.0. *Application deadline:* For fall admission, 1/15 for domestic students. Applications are processed on a rolling basis. Application fee: $45 ($60 for international students). *Financial support:* Fellowships, research assistantships, teaching assistantships, institutionally sponsored loans available. *Unit head:* Dr. Ray Semlitsch, Director of Graduate Studies, 573-884-6396, E-mail: semlitschr@missouri.edu. *Application contact:* Nila Emerich, Application Contact, 800-553-5698.

University of Nebraska at Omaha, Graduate Studies and Research, College of Arts and Sciences, Department of Psychology, Omaha, NE 68182. Offers developmental psychology (PhD); industrial/organizational psychology (MS, PhD); psychobiology (MA); school psychology (MS, Ed S). Part-time programs available. *Faculty:* 18 full-time (8 women). *Students:* 43 full-time (28 women), 24 part-time (20 women); includes 2 minority (both Hispanic Americans), 3 international. Average age 27. 104 applicants, 45% accepted, 28 enrolled. In 2005, 15 master's, 6 other advanced degrees awarded. *Degree requirements:* For master's, thesis (for some programs), comprehensive exam. *Entrance requirements:* For master's, GRE General Test, GRE Subject Test, previous course work in psychology, including statistics and a laboratory course; minimum GPA of 3.0; for doctorate, GRE General Test. Additional exam requirements/recommendations for international students: Required—TOEFL (minimum score 500 paper-based; 173 computer-based). *Application deadline:* For fall admission, 2/1 for domestic students. Application fee: $45. Electronic applications accepted. *Expenses:* Tuition, state resident: part-time $172 per credit. Tuition, nonresident: part-time $452 per credit. Part-time tuition and fees vary according to campus/location. *Financial support:* In 2005–06, 43 students received support; fellowships, research assistantships with tuition reimbursements available, teaching assistantships with tuition reimbursements available, career-related internships or fieldwork, Federal Work-Study, institutionally sponsored loans, scholarships/grants, tuition waivers (partial), and unspecified assistantships available. Support available to part-time students. Financial award application deadline: 3/1; financial award applicants required to submit FAFSA. *Unit head:* Dr. Kenneth Deffenbacher, Chairperson, 402-554-2592.

University of Oklahoma Health Sciences Center, College of Medicine and Graduate College, Graduate Programs in Medicine, Department of Psychiatry and Behavioral Sciences, Oklahoma City, OK 73190. Offers biological psychology (MS, PhD). *Degree requirements:* For master's and doctorate, thesis/dissertation. *Entrance requirements:* For doctorate, GRE General Test, 3 letters of recommendation. Additional exam requirements/recommendations for international students: Required—TOEFL. *Faculty research:* Behavioral neuroscience, human neuropsychology, psychophysiology, behavioral medicine, health psychology.

University of Oregon, Graduate School, College of Arts and Sciences, Department of Psychology, Eugene, OR 97403. Offers clinical psychology (PhD); cognitive psychology (MA, MS, PhD); developmental psychology (MA, MS, PhD); physiological psychology (MA, MS, PhD); psychology (MA, MS, PhD); social/personality psychology (MA, MS, PhD). *Accreditation:* APA (one or more programs are accredited). *Faculty:* 27 full-time (11 women), 11 part-time/adjunct (8 women). *Students:* 85 full-time (53 women), 12 part-time (9 women); includes 10 minority (7 Asian Americans or Pacific Islanders, 3 Hispanic Americans), 15 international. 176 applicants, 11% accepted. In 2005, 21 master's, 11 doctorates awarded. Terminal master's awarded for partial completion of doctoral program. *Degree requirements:* For doctorate, thesis/dissertation. *Entrance requirements:* For master's, GRE General Test, minimum GPA of 3.0; for doctorate, GRE General Test. Additional exam requirements/recommendations for international students: Required—TOEFL. *Application deadline:* For fall admission, 12/1 for domestic students. Application fee: $50. *Financial support:* In 2005–06, 48 teaching assistantships were awarded; research assistantships, career-related internships or fieldwork also available. *Unit head:* Marjorie Taylor, Head, 541-346-4921, Fax: 541-346-4911. *Application contact:* Lori Olsen, Admissions Contact, 541-346-5060, Fax: 541-346-4911, E-mail: gradsec@psych.uoregon.edu.

The University of Texas at Austin, Graduate School, The Institute for Neuroscience, Austin, TX 78712-1111. Offers MA, PhD, MD/PhD. Terminal master's awarded for partial completion of doctoral program. *Degree requirements:* For master's and doctorate, thesis/dissertation. *Entrance requirements:* For master's and doctorate, GRE. Electronic applications accepted. *Faculty research:* Cellular/molecular biology, neurobiology, pharmacology, behavioral neuroscience.

See Close-Up on page 1059.

University of Windsor, Faculty of Graduate Studies and Research, Faculty of Arts and Social Sciences, Department of Psychology, Windsor, ON N9B 3P4, Canada. Offers adult clinical (MA, PhD); applied social psychology (MA, PhD); child clinical (MA, PhD); clinical neuropsychology (MA, PhD). *Accreditation:* APA (one or more programs are accredited). *Faculty:* 26 full-time (10 women). *Students:* 108 full-time (88 women), 1 (woman) part-time. 182 applicants, 23% accepted. In 2005, 17 master's, 12 doctorates awarded. *Degree requirements:* For master's, thesis/dissertation; for doctorate, thesis/dissertation, comprehensive exam. *Entrance requirements:* For master's, GRE General Test, GRE Advanced Test in Psychology, minimum B average; for doctorate, GRE General Test, GRE Advanced Test in Psychology, master's degree. Additional exam requirements/recommendations for international students: Required—TOEFL (minimum score 600 paper-based; 250 computer-based). *Application deadline:* For fall admission, 1/15 for domestic students. Application fee: $55. Electronic applications accepted. *Financial support:* In 2005–06, 66 teaching assistantships (averaging $8,956 per year) were awarded; Federal Work-Study, scholarships/grants,

Biopsychology

University of Windsor (continued)
tuition waivers (full and partial), unspecified assistantships, and bursaries also available. Financial award application deadline: 2/15. *Faculty research:* Gambling, suicidology, emotional competence, psychotherapy and trauma. *Unit head:* Dr. Shelagh Towson, Head, 519-253-3000 Ext. 2215, Fax: 519-973-7021, E-mail: towson@uwindsor.ca. *Application contact:* Applicant Services, 519-253-3000 Ext. 6459, Fax: 519-971-3653, E-mail: gradadmit@uwindsor.ca.

University of Wisconsin–Madison, Graduate School, College of Letters and Science, Department of Psychology, Program in Biology of Brain and Behavior, Madison, WI 53706-1380. Offers PhD. *Degree requirements:* For doctorate, thesis/dissertation, comprehensive exam, registration. *Entrance requirements:* For doctorate, GRE General Test, minimum undergradu-ate GPA of 3.0. Additional exam requirements/recommendations for international students: Required—TOEFL. Electronic applications accepted.

Wayne State University, School of Medicine and Graduate School, Graduate Programs in Medicine, Department of Psychiatry and Behavioral Neurosciences, Detroit, MI 48202. Offers MS. *Faculty:* 13 full-time. *Expenses:* Tuition, state resident: part-time $338 per credit hour. Tuition, nonresident: part-time $746 per credit hour. Required fees: $24 per credit hour. Full-time tuition and fees vary according to program. *Financial support:* In 2005–06, 6 research assistantships with tuition reimbursements (averaging $22,533 per year) were awarded Total annual research expenditures: $9.7 million. *Unit head:* Manuel Tancer, Chair, 313-577-1673, E-mail: aa2656@wayne.edu. *Application contact:* Dr. Michael J. Bannon, Director, 313-577-4271, Fax: 313-993-4269, E-mail: mbannon@med.wayne.edu.

Neurobiology

Albert Einstein College of Medicine, Sue Golding Graduate Division of Medical Sciences, Department of Neuroscience, Bronx, NY 10461. Offers PhD, MD/PhD. *Degree requirements:* For doctorate, thesis/dissertation. *Entrance requirements:* For doctorate, GRE General Test. Additional exam requirements/recommendations for international students: Required—TOEFL. *Faculty research:* Structure-function relations at chemical and electrical synapses, mechanisms of electrogenesis, analysis of neuronal subsystems.

Boston University, School of Medicine, Division of Graduate Medical Sciences, Department of Anatomy and Neurobiology, Boston, MA 02118. Offers MA, PhD, MD/PhD. Part-time programs available. *Faculty:* 13 full-time (3 women), 6 part-time/adjunct (0 women). *Students:* 32 full-time (19 women); includes 7 minority (2 African Americans, 3 Asian Americans or Pacific Islanders, 2 Hispanic Americans), 4 international. Average age 28.Terminal master's awarded for partial completion of doctoral program. *Degree requirements:* For master's and doctorate, thesis/dissertation, qualifying exam. *Entrance requirements:* For master's and doctorate, GRE General Test, GRE Subject Test. Additional exam requirements/recommendations for international students: Required—TOEFL. *Application deadline:* For spring admission, 10/15 priority date for domestic students. Electronic applications accepted. *Expenses:* Tuition: Full-time $31,530; part-time $985 per credit. Required fees: $316; $40 per semester. Tuition and fees vary according to course level and program. *Financial support:* Fellowships with tuition reimbursements, research assistantships with tuition reimbursements, Federal Work-Study, scholarships/grants, and traineeships available. *Faculty research:* Neuroanatomy, development of the nervous system, aging, respiratory system, reproductive system. *Unit head:* Dr. Mark Moss, Chairman, 617-638-4200, Fax: 617-638-4216.

Brandeis University, Graduate School of Arts and Sciences, Programs in Life Sciences, Program in Molecular and Cell Biology, Waltham, MA 02454-9110. Offers genetics (PhD); microbiology (PhD); molecular and cell biology (MS, PhD); molecular biology (PhD); neurobiology (PhD). *Faculty:* 18 full-time (8 women). *Students:* 43 full-time (21 women); includes 2 minority (both Asian Americans or Pacific Islanders), 11 international. Average age 27. 173 applicants, 8% accepted, 6 enrolled. In 2005, 2 master's, 4 doctorates awarded. Terminal master's awarded for partial completion of doctoral program. *Median time to degree:* Of those who began their doctoral program in fall 1997, 100% received their degree in 8 years or less. *Degree requirements:* For master's, research project, thesis optional; for doctorate, thesis/dissertation, teaching assistant experience, comprehensive exam, registration. *Entrance requirements:* For master's and doctorate, GRE General Test, resumé, 3 letters of recommendation. Additional exam requirements/recommendations for international students: Required—TOEFL (minimum score 600 paper-based; 250 computer-based). *Application deadline:* For fall admission, 1/15 for domestic students. Applications are processed on a rolling basis. Application fee: $55. Electronic applications accepted. *Financial support:* In 2005–06, 20 fellowships with full tuition reimbursements (averaging $26,500 per year), 23 research assistantships with full tuition reimbursements (averaging $26,500 per year), 5 teaching assistantships with full tuition reimbursements (averaging $3,000 per year) were awarded; scholarships/grants, traineeships, health care benefits, and tuition waivers (full and partial) also available. Financial award application deadline: 4/15; financial award applicants required to submit CSS PROFILE or FAFSA. *Faculty research:* Regulation of gene expression by transcription factors, molecular neurobiology, immunology, molecular mechanisms of genetic recombination, and cell differentiation. *Unit head:* Dr. Piali Sengupta, Chair, 781-736-2686, Fax: 781-736-3107, E-mail: piali@bradeis.edu. *Application contact:* Marcia Cabral, Information Officer, 781-736-3100, Fax: 781-736-3107, E-mail: cabral@brandeis.edu.

California Institute of Technology, Division of Biology, Program in Neurobiology, Pasadena, CA 91125-0001. Offers PhD. *Degree requirements:* For doctorate, thesis/dissertation, qualifying exam. *Entrance requirements:* For doctorate, GRE General Test. *Application deadline:* For fall admission, 1/1 for domestic students. Application fee: $0. *Financial support:* Application deadline: 1/1. *Application contact:* Elizabeth Ayala, Graduate Program Coordinator, 626-395-4497, Fax: 626-683-3343, E-mail: biograd@cco.caltech.edu.

Carnegie Mellon University, Mellon College of Science, Department of Biological Sciences, Pittsburgh, PA 15213-3891. Offers biochemistry (PhD); biophysics (PhD); cell biology (PhD); computational biology (MS, PhD); developmental biology (PhD); genetics (PhD); molecular biology (PhD); neurobiology (PhD). *Degree requirements:* For doctorate, thesis/dissertation, comprehensive exam. *Entrance requirements:* For doctorate, GRE General Test, GRE Subject Test, interview. Electronic applications accepted. *Faculty research:* Genetic structure, function, and regulation; protein structure and function; biological membranes; biological spectroscopy.

See Close-Up on page 93.

Case Western Reserve University, School of Medicine and School of Graduate Studies, Graduate Programs in Medicine, Department of Neurosciences, Cleveland, OH 44106. Offers neurobiology (PhD); neuroscience (PhD). *Faculty:* 19 full-time (5 women). *Students:* 51 full-time (33 women); includes 11 minority (1 African American, 6 Asian Americans or Pacific Islanders, 4 Hispanic Americans), 15 international. Average age 30. 59 applicants, 10% accepted, 3 enrolled. In 2005, 5 doctorates awarded. *Median time to degree:* Of those who began their doctoral program in fall 1997, 100% received their degree in 8 years or less. *Degree requirements:* For doctorate, thesis/dissertation. *Entrance requirements:* For doctorate, GRE General Test. Additional exam requirements/recommendations for international students: Required—TOEFL. *Application deadline:* For fall admission, 3/15 for domestic students. Applications are processed on a rolling basis. Application fee: $50. Electronic applications accepted. *Financial support:* In 2005–06, 51 students received support, including 9 fellowships (averaging $23,000 per year), 42 research assistantships (averaging $23,000 per year); traineeships, health care benefits, and unspecified assistantships also available. Financial award application deadline: 4/1. *Faculty research:* Neurotropic factors, synapse formation, regeneration, determination of cell fate, cellular neuroscience. Total annual research expenditures: $5 million. *Unit head:* Dr. Lynn Landmesser, Chair, 216-368-3996, Fax: 216-368-4650, E-mail: ltl@po.cwru.edu. *Application contact:* Gina Schon, Graduate Student Coordinator, 216-368-6253, Fax: 216-368-4650, E-mail: glv@po.cwru.edu.

Announcement: Neurobiology is a multidisciplinary program based in the Department of Neurosciences. Research programs utilize a variety of modern approaches, including morphological, molecular, biochemical, genetic, and physiological techniques to study various aspects of neurobiology.

See Close-Up on page 993.

Columbia University, College of Physicians and Surgeons and Graduate School of Arts and Sciences, Graduate School of Arts and Sciences at the College of Physicians and Surgeons, Program in Neurobiology and Behavior, New York, NY 10032. Offers M Phil, PhD, MD/PhD. Only candidates for the PhD are admitted. *Degree requirements:* For doctorate, thesis/dissertation. *Entrance requirements:* For master's and doctorate, GRE General Test. Additional exam requirements/recommendations for international students: Required—TOEFL. *Expenses:* Contact institution. Tuition and fees vary according to course level, course load, campus/location and program. *Faculty research:* Cellular and molecular mechanisms of neural development, neuropathology, neuropharmacology.

Cornell University, Graduate School, Graduate Fields of Agriculture and Life Sciences, Field of Neurobiology and Behavior, Ithaca, NY 14853-0001. Offers behavioral biology (MS, PhD), including behavioral ecology, chemical ecology, ethology, neuroethology, sociobiology; neurobiology (MS, PhD), including cellular and molecular neurobiology, neuroanatomy, neurochemistry, neuropharmacology, neurophysiology, sensory physiology. *Faculty:* 40 full-time (6 women). *Students:* 35 full-time (15 women); includes 6 minority (1 African American, 1 American Indian/Alaska Native, 2 Asian Americans or Pacific Islanders, 2 Hispanic Americans), 7 international. 44 applicants, 25% accepted, 4 enrolled. In 2005, 1 master's, 6 doctorates awarded. *Degree requirements:* For doctorate, thesis/dissertation, 1 year of teaching experience, seminar presentation, comprehensive exam. *Entrance requirements:* For doctorate, GRE General Test, GRE Subject Test (biology), 3 letters of recommendation. Additional exam requirements/recommendations for international students: Required—TOEFL (minimum score 550 paper-based; 213 computer-based). *Application deadline:* For fall admission, 12/1 for domestic students. Application fee: $60. Electronic applications accepted. *Financial support:* In 2005–06, 32 students received support, including 21 fellowships with full tuition reimbursements available, 5 research assistantships with full tuition reimbursements available, 6 teaching assistantships with full tuition reimbursements available; institutionally sponsored loans, scholarships/grants, health care benefits, tuition waivers (full and partial), and unspecified assistantships also available. Financial award applicants required to submit FAFSA. *Faculty research:* Cellular neurobiology and neuropharmacology, integrative neurobiology, social behavior, chemical ecology, neuroethology. *Unit head:* Director of Graduate Studies, 607-254-4340, Fax: 607-254-4340. *Application contact:* Graduate Field Assistant, 607-254-4340, Fax: 607-254-4340, E-mail: nbb_field@cornell.edu.

See Close-Up on page 995.

Dalhousie University, Faculty of Graduate Studies and Faculty of Medicine, Graduate Programs in Medicine, Department of Anatomy and Neurobiology, Halifax, NS B3H 4R2, Canada. Offers M Sc, PhD. *Degree requirements:* For master's and doctorate, thesis/dissertation. *Entrance requirements:* For master's and doctorate, GRE (recommended), minimum A- average. Additional exam requirements/recommendations for international students: Required—TOEFL. *Faculty research:* Neuroscience histology, cell biology, neuroendocrinology, evolutionary biology.

Duke University, Graduate School, Department of Biological Anthropology and Anatomy, Durham, NC 27710. Offers cellular and molecular biology (PhD); gross anatomy and physical anthropology (PhD), including comparative morphology of human and non-human primates, primate social behavior, vertebrate paleontology; neuroanatomy (PhD). *Faculty:* 6 full-time. *Students:* 14 full-time (9 women); includes 1 minority (African American), 2 international. 39 applicants, 18% accepted, 3 enrolled. In 2005, 3 doctorates awarded. *Degree requirements:* For doctorate, one foreign language, thesis/dissertation. *Entrance requirements:* For doctorate, GRE General Test. Additional exam requirements/recommendations for international students: Required—IELT (preferred) or TOEFL. *Application deadline:* For fall admission, 12/31 for domestic students, 12/31 for international students. Application fee: $75. Electronic applications accepted. *Financial support:* Fellowships, teaching assistantships, Federal Work-Study available. Financial award application deadline: 12/31. *Unit head:* Daniel Schmitt, Director of Graduate Studies, 919-684-5664, Fax: 919-684-8034, E-mail: l.squires@baa.mc.duke.edu.

Duke University, Graduate School, Department of Neurobiology, Durham, NC 27708-0586. Offers PhD. *Faculty:* 61 full-time. *Students:* 44 full-time (25 women); includes 6 minority (3 African Americans, 2 Asian Americans or Pacific Islanders, 1 Hispanic American), 20 international. 122 applicants, 14% accepted, 5 enrolled. In 2005, 2 doctorates awarded. *Degree requirements:* For doctorate, variable foreign language requirement, thesis/dissertation. *Entrance requirements:* For doctorate, GRE General Test. Additional exam requirements/recommendations for international students: Required—IELT (preferred) or TOEFL. *Application deadline:* For fall admission, 12/31 for domestic students, 12/31 for international students. Application fee: $75. Electronic applications accepted. *Financial support:* Fellowships, research assistantships, teaching assistantships, Federal Work-Study available. Financial award application deadline: 12/31. *Unit head:* Dona Chikaraishi, Director, 919-681-4269, Fax: 919-684-4431, E-mail: sink@neuro.duke.edu.

Georgia State University, College of Arts and Sciences, Department of Biology, Program in Neurobiology and Behavior, Atlanta, GA 30303-3083. Offers MS, PhD. *Degree requirements:* For master's, thesis or alternative, exam; for doctorate, thesis/dissertation, exam. *Entrance requirements:* For master's and doctorate, GRE General Test. Additional exam requirements/recommendations for international students: Required—TOEFL. Electronic applications accepted. *Expenses:* Tuition, state resident: full-time $4,368; part-time $182 per semester hour. Tuition, nonresident: full-time $8,732; part-time $728 per semester hour. Required fees: $46 per semester hour.

See Close-Up on page 999.

Harvard University, Graduate School of Arts and Sciences, Program in Neuroscience, Boston, MA 02115. Offers neurobiology (PhD). *Degree requirements:* For doctorate, thesis/dissertation, qualifying exam. *Entrance requirements:* For doctorate, GRE General Test, GRE Subject Test. Additional exam requirements/recommendations for international students: Required—TOEFL. *Application deadline:* For fall admission, 12/15 for domestic students. *Expenses:* Tuition: Full-time $28,752. Full-time tuition and fees vary according to program and student level. *Financial support:* Fellowships, research assistantships, teaching assistantships, institutionally sponsored loans, scholarships/grants, tuition waivers (full), and stipends available.

Financial award application deadline: 1/1. *Faculty research:* Relationship between diseases of the nervous system and basic science. *Unit head:* Dr. Jonathan Cohen, Chair, 617-432-1728. *Application contact:* Dr. Jonathan Cohen, Chair, 617-432-1728.

See Close-Up on page 1001.

Indiana University–Purdue University Indianapolis, Indiana University School of Medicine, Program in Medical Neurobiology, Indianapolis, IN 46202-2896. Offers MS, PhD. *Students:* 9 full-time (4 women), 12 part-time (8 women); includes 1 minority (Asian American or Pacific Islander), 1 international. Average age 33. In 2005, 4 degrees awarded. Terminal master's awarded for partial completion of doctoral program. *Degree requirements:* For master's and doctorate, thesis/dissertation. *Entrance requirements:* For master's and doctorate, GRE General Test, previous course work in calculus, organic chemistry, and physics. *Application deadline:* For fall admission, 2/1 for domestic students; for spring admission, 9/15 priority date for domestic students. Applications are processed on a rolling basis. Application fee: $50 ($60 for international students). *Expenses:* Tuition, state resident: full-time $5,159; part-time $215 per credit hour. Tuition, nonresident: full-time $14,890; part-time $620 per credit hour. Required fees: $614. Tuition and fees vary according to campus/location and program. *Financial support:* Fellowships with full tuition reimbursements, research assistantships with full tuition reimbursements, teaching assistantships with full tuition reimbursements, career-related internships or fieldwork, Federal Work-Study, institutionally sponsored loans, scholarships/grants, traineeships, tuition waivers (partial), and unspecified assistantships available. Financial award application deadline: 2/1. *Faculty research:* Neurobiology from molecular level to complex behavioral interactions. Total annual research expenditures: $5.3 million. *Unit head:* Dr. Jay Simon, Director, 317-274-5848, Fax: 317-274-1365.

Louisiana State University Health Sciences Center, School of Graduate Studies in New Orleans, Department of Cell Biology and Anatomy, New Orleans, LA 70112-2223. Offers cell biology and anatomy (MS, PhD), including cell biology, developmental biology, neurobiology and anatomy. *Degree requirements:* For master's and doctorate, thesis/dissertation. *Entrance requirements:* For master's and doctorate, GRE General Test, GRE Subject Test, minimum undergraduate GPA of 3.0. Additional exam requirements/recommendations for international students: Required—TOEFL. *Faculty research:* Visual system organization, neural development, plasticity of sensory systems, information processing through the nervous system, visuomotor integration.

Loyola University Chicago, Graduate School, Department of Cell Biology, Neurobiology and Anatomy, Maywood, IL 60153. Offers MS, PhD, MD/PhD. Part-time programs available. *Faculty:* 17 full-time (6 women), 5 part-time/adjunct (1 woman). *Students:* 17 full-time (8 women), 1 part-time; includes 5 minority (1 African American, 3 Asian Americans or Pacific Islanders, 1 Hispanic American), 3 international. Average age 27. 30 applicants, 23% accepted, 7 enrolled. In 2005, 4 master's, 4 doctorates awarded. Terminal master's awarded for partial completion of doctoral program. *Median time to degree:* Of those who began their doctoral program in fall 1997, 100% received their degree in 8 years or less. *Degree requirements:* For master's, thesis/dissertation; for doctorate, thesis/dissertation, comprehensive exam. *Entrance requirements:* For master's, GRE General Test, minimum GPA of 3.0; for doctorate, GRE General Test, GRE Subject Test (biology), minimum GPA of 3.0. Additional exam requirements/recommendations for international students: Required—TOEFL (minimum score 600 paper-based; 250 computer-based). *Application deadline:* For fall admission, 5/1 priority date for domestic students, 5/1 priority date for international students. Applications are processed on a rolling basis. Application fee: $0. Electronic applications accepted. *Expenses:* Tuition: Full-time $11,610; part-time $645 per credit. Required fees: $55 per semester. *Financial support:* In 2005–06, 8 fellowships with full tuition reimbursements (averaging $22,000 per year), 5 research assistantships with full tuition reimbursements (averaging $22,000 per year) were awarded; Federal Work-Study and unspecified assistantships also available. Financial award application deadline: 5/1; financial award applicants required to submit FAFSA. *Faculty research:* Brain steroids, immunology, neuroregeneration, cytokines. Total annual research expenditures: $1 million. *Unit head:* Dr. John Clancy, Chair, 708-216-3352. *Application contact:* Dr. Tharkery S. Gray, Graduate Program Director, 708-216-3345.

See Close-Up on page 575.

Marquette University, Graduate School, College of Arts and Sciences, Department of Biology, Milwaukee, WI 53201-1881. Offers cell biology (MS, PhD); developmental biology (MS, PhD); ecology (MS, PhD); endocrinology (MS, PhD); evolutionary biology (MS, PhD); genetics (MS, PhD); microbiology (MS, PhD); molecular biology (MS, PhD); muscle and exercise physiology (MS, PhD); neurobiology (MS, PhD); reproductive physiology (MS, PhD). Terminal master's awarded for partial completion of doctoral program. *Degree requirements:* For master's, thesis, 1 year of teaching experience or equivalent, comprehensive exam; for doctorate, thesis/dissertation, 1 year of teaching experience or equivalent, qualifying exam. *Entrance requirements:* For master's and doctorate, GRE General Test, GRE Subject Test. Additional exam requirements/recommendations for international students: Required—TOEFL. *Faculty research:* Microbial and invertebrate ecology, evolution of gene function, DNA methylation, DNA arrangement.

New York University, Graduate School of Arts and Science, Department of Biology, New York, NY 10012-1019. Offers biology (PhD); biomedical journalism (MS); cancer and molecular biology (PhD); computational biology (PhD); computers in biological research (MS); developmental genetics (PhD); general biology (MS); immunology and microbiology (PhD); molecular genetics (PhD); neurobiology (PhD); oral biology (MS); plant biology (PhD); recombinant DNA technology (MS). Part-time programs available. *Faculty:* 24 full-time (5 women), 8 part-time/adjunct. *Students:* 104 full-time (52 women), 41 part-time (23 women); includes 28 minority (2 African Americans, 20 Asian Americans or Pacific Islanders, 6 Hispanic Americans), 47 international. Average age 27. 349 applicants, 56% accepted, 39 enrolled. In 2005, 59 master's, 4 doctorates awarded. Terminal master's awarded for partial completion of doctoral program. *Degree requirements:* For master's, thesis or alternative, qualifying paper; for doctorate, thesis/dissertation, comprehensive exam. *Entrance requirements:* For master's, GRE General Test; for doctorate, GRE General Test, GRE Subject Test. Additional exam requirements/recommendations for international students: Required—TOEFL. *Application deadline:* For fall admission, 1/4 for domestic students. Application fee: $80. *Financial support:* Fellowships with tuition reimbursements, research assistantships with tuition reimbursements, teaching assistantships with tuition reimbursements, career-related internships or fieldwork, Federal Work-Study, institutionally sponsored loans, scholarships/grants, health care benefits, and unspecified assistantships available. Financial award application deadline: 1/4; financial award applicants required to submit FAFSA. *Faculty research:* Genomics, molecular and cell biology, development and molecular genetics, molecular evolution of plants and animals. *Unit head:* Gloria Coruzzi, Chairman, 212-998-8200, Fax: 212-995-4015, E-mail: biology@nyu.edu. *Application contact:* Stephen Small, Director of Graduate Studies, 212-998-8200, Fax: 212-995-4015, E-mail: biology@nyu.edu.

Northwestern University, The Graduate School, Judd A. and Marjorie Weinberg College of Arts and Sciences, Department of Neurobiology and Physiology, Evanston, IL 60208. Offers MS. Admissions and degrees offered through The Graduate School. Part-time programs available. *Faculty:* 14 full-time (4 women), 1 part-time/adjunct (0 women). *Students:* 11 (6 women); includes 5 minority (4 Asian Americans or Pacific Islanders, 1 Hispanic American) 1 international. Average age 24. 45 applicants, 16% accepted, 4 enrolled. In 2005, 6 degrees awarded. *Degree requirements:* For master's, thesis. *Entrance requirements:* For master's, GRE General Test and MCAT (strongly recommended). Additional exam requirements/recommendations for international students: Required—TOEFL. *Application deadline:* For fall admission, 3/3 for domestic students, 3/31 for international students. Applications are processed on a rolling basis. Application fee: $60 ($75 for international students). Electronic applications accepted. *Expenses:* Contact institution. *Financial support:* In 2005–06, 5 students received support. Institutionally sponsored loans available. Financial award application deadline: 1/15; financial award applicants required to submit FAFSA. *Faculty research:* Sensory neurobiology and neuroendocrinology, reproductive biology, vision physiology and psychophysics, cell and developmental biology. Total annual research expenditures: $5.5 million. *Unit head:* David Ferster, Chair, 847-491-4137, Fax: 847-491-5211, E-mail: ferster@northwestern.edu. *Application contact:* Dr. Michael Kennedy, Associate Chair, 847-491-5521, Fax: 847-491-5211, E-mail: m-kennedy2@northwestern.edu.

Northwestern University, Northwestern University Feinberg School of Medicine and Interdepartmental Degree Programs, Integrated Graduate Programs in the Life Sciences, Chicago, IL 60611. Offers cancer biology (PhD); cell biology (PhD); developmental biology (PhD); evolutionary biology (PhD); immunology and microbial pathogenesis (PhD); molecular biology and genetics (PhD); neurobiology (PhD); pharmacology and toxicology (PhD); structural biology and biochemistry (PhD). *Degree requirements:* For doctorate, thesis/dissertation, written and oral qualifying exams, comprehensive exam. *Entrance requirements:* For doctorate, GRE General Test. Additional exam requirements/recommendations for international students: Required—TOEFL (minimum score 600 paper-based; 250 computer-based). Electronic applications accepted.

See Close-Up on page 189.

Purdue University, Graduate School, School of Science, Department of Biological Sciences, West Lafayette, IN 47907. Offers biochemistry (PhD); biophysics (PhD); cell and developmental biology (PhD); ecology, evolutionary and population biology (MS, PhD), including ecology, evolutionary biology, population biology; genetics (MS, PhD); microbiology (MS, PhD); molecular biology (PhD); neurobiology (MS, PhD); plant physiology (PhD). *Faculty:* 47 full-time (9 women), 4 part-time/adjunct (1 woman). *Students:* 97 full-time (53 women), 8 part-time (4 women); includes 13 minority (3 African Americans, 1 American Indian/Alaska Native, 4 Asian Americans or Pacific Islanders, 5 Hispanic Americans), 50 international. Average age 28. 168 applicants, 29% accepted, 23 enrolled. In 2005, 18 master's, 9 doctorates awarded. Terminal master's awarded for partial completion of doctoral program. *Degree requirements:* For master's, thesis (for some programs); for doctorate, thesis/dissertation, seminars, teaching experience. *Entrance requirements:* For master's and doctorate, GRE General Test. Additional exam requirements/recommendations for international students: Required—TOEFL, TSE. *Application deadline:* For fall admission, 2/15 for domestic students, 1/31 for international students. Applications are processed on a rolling basis. Application fee: $55. Electronic applications accepted. *Financial support:* In 2005–06, 15 fellowships, 60 research assistantships, 53 teaching assistantships were awarded. Support available to part-time students. Financial award application deadline: 2/15; financial award applicants required to submit FAFSA. *Unit head:* Dr. Richard J Kuhn, Head, 765-494-4407. *Application contact:* Nancy Konopka, Graduate Studies Office Manager, 765-494-8142, Fax: 765-494-0876, E-mail: njk@bilbo.bio.purdue.edu.

Rutgers, The State University of New Jersey, New Brunswick/Piscataway, Graduate School, Program in Physiology and Neurobiology, New Brunswick, NJ 08901-1281. Offers PhD. *Faculty:* 76 full-time. *Students:* 4 full-time (3 women); includes 2 minority (1 African American, 1 Asian American or Pacific Islander). Average age 26. 7 applicants, 29% accepted, 2 enrolled. *Degree requirements:* For doctorate, thesis/dissertation, qualifying exam. *Entrance requirements:* For master's and doctorate, GRE General Test, GRE Subject Test (biology or chemistry), minimum undergraduate GPA of 3.0. Additional exam requirements/recommendations for international students: Required—TOEFL. *Application deadline:* For fall admission, 3/1 for domestic students. Applications are processed on a rolling basis. Application fee: $50. *Expenses:* Tuition, state resident: full-time $10,440; part-time $435 per credit. Tuition, nonresident: full-time $15,520; part-time $647 per credit. Required fees: $129 per credit. Tuition and fees vary according to program. *Financial support:* In 2005–06, fellowships with full tuition reimbursements (averaging $17,000 per year), research assistantships with full tuition reimbursements (averaging $15,500 per year), teaching assistantships with full tuition reimbursements (averaging $16,988 per year) were awarded; tuition waivers (full) also available. Financial award application deadline: 3/1; financial award applicants required to submit FAFSA. *Faculty research:* Neuronal growth factors, neuronal gene expression, neurogenetics, circulation controls, reproduction. Total annual research expenditures: $3.4 million. *Unit head:* Jianjie Ma, Graduate Program Director, 732-235-4494, Fax: 732-235-5885, E-mail: maj2@umdnj.edu. *Application contact:* David Egger, Director of Admissions, 732-235-4522, Fax: 732-235-4029.

Université Laval, Faculty of Medicine, Graduate Programs in Medicine, Programs in Neurobiology, Québec, QC G1K 7P4, Canada. Offers M Sc, PhD. Terminal master's awarded for partial completion of doctoral program. *Degree requirements:* For master's, thesis/dissertation; for doctorate, thesis/dissertation, comprehensive exam. *Entrance requirements:* For master's and doctorate, knowledge of French and English. Electronic applications accepted.

University at Albany, State University of New York, College of Arts and Sciences, Department of Biological Sciences, Specialization in Molecular, Cellular, Developmental, and Neural Biology, Albany, NY 12222-0001. Offers MS, PhD. *Degree requirements:* For master's, one foreign language; for doctorate, one foreign language, thesis/dissertation. *Entrance requirements:* For master's and doctorate, GRE General Test. Application fee: $60. *Financial support:* Minority assistantships available. *Unit head:* Dr. Albert Millis, Chair, Department of Biological Sciences, 518-442-4300.

The University of Alabama at Birmingham, Graduate Programs in Joint Health Sciences, Department of Cell Biology, Graduate Program in Cellular and Molecular Biology, Birmingham, AL 35294.

See Close-Ups on pages 601 and 603.

The University of Alabama at Birmingham, Graduate Programs in Joint Health Sciences, Department of Neurobiology, Birmingham, AL 35294. Offers PhD. The department participates in the Cellular and Molecular Biology Graduate Program. *Students:* 35 full-time (16 women), 1 (woman) part-time; includes 4 minority (2 African Americans, 1 Asian American or Pacific Islander, 1 Hispanic American), 9 international. 10 applicants, 80% accepted. In 2005, 5 degrees awarded. *Degree requirements:* For doctorate, thesis/dissertation. *Entrance requirements:* For doctorate, GRE, interview. *Application deadline:* Applications are processed on a rolling basis. Electronic applications accepted. *Expenses:* Tuition, state resident: part-time $170 per credit hour. Tuition, nonresident: full-time $4,612; part-time $425 per credit hour. International tuition: $10,732 full-time. Required fees: $11 per credit hour. $124 per term. Tuition and fees vary according to course load, degree level and program. *Unit head:* , Dr. J. David Sweatt, Chair, Fax: 205-934-6571. *Application contact:* Information Contact, 205-975-5573, Fax: 205-934-6571.

See Close-Ups on pages 1029 and 1027.

University of Arkansas for Medical Sciences, College of Medicine and Graduate School, Graduate Programs in Medicine, Department of Neurobiology and Developmental Sciences, Little Rock, AR 72205-7199. Offers MS, PhD, MD/PhD. *Faculty:* 31 full-time (10 women), 7 part-time/adjunct (4 women). *Students:* 14 full-time, 5 part-time. *Degree requirements:* For master's and doctorate, thesis/dissertation. *Entrance requirements:* For master's, GRE General Test; for doctorate, GRE General Test, GRE Subject Test. Additional exam requirements/recommendations for international students: Required—TOEFL. Application fee: $0. *Financial support:* Research assistantships available. Support available to part-time students. *Unit head:* Dr. Gwen Childs, Chair, 501-686-5180. *Application contact:* Dr. David Davies, Graduate Coordinator, 501-686-5184, E-mail: dldavies@uams.edu.

See Close-Up on page 1031.

University of California, Irvine, College of Medicine and School of Biological Sciences, Department of Anatomy and Neurobiology, Irvine, CA 92697. Offers biological sciences (MS, PhD). *Degree requirements:* For doctorate, thesis/dissertation. *Entrance requirements:* For master's and doctorate, GRE General Test, GRE Subject Test. Additional exam requirements/recommendations for international students: Required—TOEFL (minimum score 550 paper-based; 213 computer-based), TSE. Electronic applications accepted. *Faculty research:*

Neurobiology

University of California, Irvine (continued)
Neurotransmitter immunocytochemistry, intracellular physiology, molecular neurobiology, forebrain organization and development, structure and function of sensory and motor systems.

University of California, Irvine, Office of Graduate Studies, School of Biological Sciences, Department of Neurobiology and Behavior, Irvine, CA 92697. Offers biological sciences (MS, PhD). *Degree requirements:* For doctorate, thesis/dissertation. *Entrance requirements:* For master's and doctorate, GRE General Test, GRE Subject Test, minimum GPA of 3.0. Additional exam requirements/recommendations for international students: Required—TOEFL (minimum score 550 paper-based; 213 computer-based). Electronic applications accepted. *Faculty research:* Synaptic processes, neurophysiology, neuroendocrinology, neuroanatomy, molecular neurobiology.

University of California, Los Angeles, School of Medicine and Graduate Division, Graduate Programs in Medicine, Department of Neurobiology, Los Angeles, CA 90095. Offers anatomy and cell biology (PhD). *Degree requirements:* For doctorate, thesis/dissertation, oral and written qualifying exams. *Entrance requirements:* For doctorate, GRE General Test, GRE Subject Test, bachelor's degree in physical or biological science. *Faculty research:* Neuroendocrinology, neurophysiology.

University of California, San Diego, Graduate Studies and Research, Division of Biology, Program in Computational Neurobiology, La Jolla, CA 92093-0348. Offers PhD. Offered in association with the Salk Institute. *Degree requirements:* For doctorate, thesis/dissertation, qualifying exam. Electronic applications accepted.

Announcement: Computational neurobiology is an interdisciplinary PhD program in the Division of Biological Sciences designed to train young scientists with the broad range of scientific and technical skills that are essential to understanding the computational resources of neural systems. This program welcomes students with backgrounds in physics, chemistry, biology, psychology, computer science, and mathematics. For more information, visit http://www.biology.ucsd.edu/grad/CN_overview.html.

University of California, San Diego, Graduate Studies and Research, Division of Biology, Program in Neurobiology, La Jolla, CA 92093. Offers PhD. Offered in association with the Salk Institute. *Degree requirements:* For doctorate, thesis/dissertation, qualifying exam. Electronic applications accepted.

University of Chicago, Division of the Biological Sciences, Department of Neurobiology, Pharmacology, and Cell Physiology, Committee on Neurobiology, Chicago, IL 60637-1513. Offers PhD. *Faculty:* 45 full-time (14 women). *Students:* 40 full-time (21 women); includes 11 minority (2 African Americans, 4 Asian Americans or Pacific Islanders, 5 Hispanic Americans), 5 international. Average age 28. In 2005, 6 degrees awarded. *Degree requirements:* For doctorate, thesis/dissertation, preliminary exam. *Entrance requirements:* For doctorate, GRE General Test. Additional exam requirements/recommendations for international students: Required—TOEFL. *Application deadline:* For fall admission, 12/28 priority date for domestic students, 12/28 priority date for international students. Application fee: $55. Electronic applications accepted. *Financial support:* In 2005–06, 25 students received support, including fellowships with tuition reimbursements available (averaging $26,301 per year), research assistantships with tuition reimbursements available (averaging $26,301 per year); institutionally sponsored loans, scholarships/grants, traineeships, and health care benefits also available. Financial award applicants required to submit FAFSA. *Faculty research:* Immunogenetic aspects of neurologic disease. *Unit head:* Peggy Mason, Chairman, 773-705-3849, E-mail: neurobiology@chicago.edu. *Application contact:* Diane J. Hall, Graduate Administrative Director, 773-702-6371, Fax: 773-702-1216, E-mail: d-hall@uchicago.edu.

University of Colorado at Boulder, Graduate School, College of Arts and Sciences, Department of Ecology and Evolutionary Biology, Boulder, CO 80309. Offers animal behavior (MA); biology (MA, PhD); environmental biology (MA, PhD); evolutionary biology (MA, PhD); neurobiology (MA); population biology (MA); population genetics (PhD). *Faculty:* 26 full-time (5 women). *Students:* 44 full-time (24 women), 22 part-time (14 women); includes 16 minority (1 American Indian/Alaska Native, 3 Asian Americans or Pacific Islanders, 12 Hispanic Americans), 2 international. Average age 30. 20 applicants, 90% accepted. In 2005, 7 master's, 7 doctorates awarded. Terminal master's awarded for partial completion of doctoral program. *Degree requirements:* For master's, thesis or alternative, comprehensive exam; for doctorate, thesis/dissertation, comprehensive exam. *Entrance requirements:* For master's, GRE General Test, GRE Subject Test, minimum undergraduate GPA of 3.0; for doctorate, GRE General Test, GRE Subject Test. *Application deadline:* For fall admission, 1/2 priority date for domestic students, 12/1 priority date for international students. Application fee: $50 ($60 for international students). *Financial support:* In 2005–06, fellowships (averaging $7,844 per year), research assistantships (averaging $15,663 per year), teaching assistantships (averaging $13,829 per year) were awarded; Federal Work-Study, institutionally sponsored loans, and tuition waivers (full) also available. Financial award application deadline: 3/1. *Faculty research:* Behavior, ecology, genetics, morphology, endocrinology. Total annual research expenditures: $5.2 million. *Unit head:* Jeffry Mitton, Chair, 303-492-0505, Fax: 303-492-8699, E-mail: mitton@colorado.edu. *Application contact:* Jill Skarstadt, Graduate Coordinator, 303-492-7654, Fax: 303-492-8699, E-mail: skarstad@colorado.edu.

University of Connecticut, Graduate School, College of Liberal Arts and Sciences, Department of Physiology and Neurobiology, Storrs, CT 06269. Offers physiology and neurobiology (MS, PhD), including comparative physiology, endocrinology, neurobiology. *Faculty:* 16 full-time (5 women). *Students:* 35 full-time (18 women), 2 part-time (1 woman); includes 5 minority (1 African American, 4 Asian Americans or Pacific Islanders), 16 international. Average age 27. 44 applicants, 32% accepted, 13 enrolled. In 2005, 5 master's, 6 doctorates awarded. Terminal master's awarded for partial completion of doctoral program. *Degree requirements:* For master's, comprehensive exam; for doctorate, thesis/dissertation. *Entrance requirements:* For master's and doctorate, GRE General Test, GRE Subject Test. Additional exam requirements/recommendations for international students: Required—TOEFL (minimum score 550 paper-based; 213 computer-based). *Application deadline:* For fall admission, 2/1 priority date for domestic students, 2/1 priority date for international students; for spring admission, 11/1 for domestic students, 10/1 for international students. Applications are processed on a rolling basis. Application fee: $55. Electronic applications accepted. *Expenses:* Tuition, state resident: part-time $444 per credit hour. Tuition, nonresident: part-time $1,154 per credit hour. Tuition and fees vary according to course load. *Financial support:* In 2005–06, 15 research assistantships with full tuition reimbursements, 14 teaching assistantships with full tuition reimbursements were awarded; fellowships, Federal Work-Study, scholarships/grants, health care benefits, and unspecified assistantships also available. Financial award application deadline: 2/1. *Unit head:* Angel L. de Blas, Head, 860-486-3285, Fax: 860-486-3303, E-mail: angel.deblas@uconn.edu. *Application contact:* J. Larry Renfro, Chairperson, 860-486-4119, Fax: 860-486-3303, E-mail: larry.renfro@uconn.edu.

See Close-Up on page 1261.

University of Illinois at Chicago, Graduate College, College of Liberal Arts and Sciences, Department of Biological Sciences, Chicago, IL 60607-7128. Offers cell and developmental biology (PhD); ecology and evolution (MS, DA, PhD); genetics and development (PhD); molecular biology (MS, PhD); neurobiology (MS, PhD); plant biology (MS, DA, PhD). *Degree requirements:* For master's, thesis; for doctorate, thesis/dissertation, preliminary exam. *Entrance requirements:* For master's and doctorate, GRE General Test, GRE Subject Test, previous course work in physics, calculus, and organic chemistry; minimum GPA of 2.75. Additional exam requirements/recommendations for international students: Required—TOEFL. Electronic applications accepted.

University of Illinois at Chicago, Graduate College, Program in Neuroscience, Chicago, IL 60607-7128. Offers PhD, MD/PhD. Admissions and degrees offered through participating Departments of Anatomy and Cell Biology, Biochemistry, Bioengineering, Biological Sciences, Chemistry, Pathology, Pharmacology, Physiology and Biophysics, and Psychology. *Degree requirements:* For doctorate, thesis/dissertation. *Entrance requirements:* For doctorate, GRE General Test, minimum GPA of 3.75 on a 5.0 scale. Additional exam requirements/recommendations for international students: Required—TOEFL. *Faculty research:* Neurobiology and behavior.

See Close-Up on page 1045.

The University of Iowa, Graduate College, College of Liberal Arts and Sciences, Department of Biological Sciences, Iowa City, IA 52242-1316.

See Close-Up on page 257.

The University of Iowa, Graduate College, College of Liberal Arts and Sciences, Department of Psychology, Iowa City, IA 52242-1316. Offers neural and behavioral sciences (PhD); psychology (MA, PhD). *Faculty:* 31 full-time, 10 part-time/adjunct. *Students:* 48 full-time (29 women), 48 part-time (31 women); includes 13 minority (3 African Americans, 3 Asian Americans or Pacific Islanders, 7 Hispanic Americans), 11 international. 264 applicants, 11% accepted, 21 enrolled. In 2005, 9 master's, 11 doctorates awarded. *Degree requirements:* For master's, exam, thesis optional; for doctorate, thesis/dissertation, comprehensive exam, registration. *Entrance requirements:* For master's and doctorate, GRE General Test, minimum GPA of 3.0. Additional exam requirements/recommendations for international students: Required—TOEFL (minimum score 550 paper-based; 213 computer-based). *Application deadline:* For fall admission, 1/1 for domestic students, 1/1 for international students. Application fee: $60 ($85 for international students). Electronic applications accepted. *Expenses:* Tuition, state resident: part-time $1,882 per term. Tuition, nonresident: full-time $17,338; part-time $4,907 per term. Tuition and fees vary according to course load and program. *Financial support:* In 2005–06, 9 fellowships, 43 research assistantships with partial tuition reimbursements, 35 teaching assistantships with partial tuition reimbursements were awarded. Financial award applicants required to submit FAFSA. *Unit head:* , Gregg C. Oden, Chair, 319-335-2405, Fax: 319-335-0191.

University of Kentucky, Graduate School, Graduate School Programs from the College of Medicine, Program in Anatomy and Neurobiology, Lexington, KY 40506-0032. Offers anatomy (PhD). *Faculty:* 32 full-time (8 women). *Students:* 21 full-time (6 women), 3 part-time (all women); includes 4 minority (1 African American, 1 Asian American or Pacific Islander, 2 Hispanic Americans), 8 international. Average age 27. 13 applicants, 92% accepted, 10 enrolled. In 2005, 4 degrees awarded. *Median time to degree:* Of those who began their doctoral program in fall 1997, 83.3% received their degree in 8 years or less. *Degree requirements:* For doctorate, thesis/dissertation, comprehensive exam. *Entrance requirements:* For doctorate, GRE General Test, minimum undergraduate GPA of 3.0. Additional exam requirements/recommendations for international students: Required—TOEFL (minimum score 550 paper-based; 213 computer-based). *Application deadline:* For fall admission, 7/17 priority date for domestic students, 2/1 priority date for international students; for spring admission, 12/13 priority date for domestic students, 6/15 priority date for international students. Applications are processed on a rolling basis. Application fee: $40 ($55 for international students). Electronic applications accepted. *Expenses:* Tuition, state resident: full-time $6,308; part-time $331 per credit hour. Tuition, nonresident: full-time $13,968; part-time $756 per credit hour. Tuition and fees vary according to course load, degree level and program. *Financial support:* In 2005–06, 6 fellowships with full tuition reimbursements (averaging $5,000 per year), 22 research assistantships with full tuition reimbursements (averaging $21,000 per year), 22 teaching assistantships with full tuition reimbursements (averaging $10,500 per year) were awarded; tuition waivers (full) also available. Financial award application deadline: 3/15. *Faculty research:* Neuroendocrinology, developmental neurobiology, neurotrophic substances, neural plasticity and trauma, neurobiology of aging. Total annual research expenditures: $1.4 million. *Unit head:* Dr. Douglas Gould, Director of Graduate Studies, 859-323-5484, Fax: 859-323-5946, E-mail: djgould@uky.edu. *Application contact:* Dr. Brian Jackson, Senior Associate Dean, 859-257-8176, Fax: 859-323-1928.

University of Louisville, School of Medicine, Department of Anatomical Sciences and Neurobiology, Louisville, KY 40292-0001. Offers MS, PhD. *Students:* 26 full-time (9 women), 7 part-time (3 women); includes 1 minority (African American), 17 international. Average age 28. In 2005, 13 master's, 4 doctorates awarded. *Degree requirements:* For master's and doctorate, thesis/dissertation. *Entrance requirements:* For master's and doctorate, GRE General Test. Additional exam requirements/recommendations for international students: Required—TOEFL. *Application deadline:* For fall admission, 1/15 for domestic students; for spring admission, 4/15 for domestic students. Application fee: $50. Electronic applications accepted. *Expenses:* Tuition, state resident: full-time $6,006; part-time $334 per credit hour. Tuition, nonresident: full-time $16,554; part-time $920 per credit hour. Tuition and fees vary according to course load, degree level and program. *Financial support:* Fellowships with full tuition reimbursements, research assistantships with full tuition reimbursements, teaching assistantships with full tuition reimbursements available. *Unit head:* Dr. Fred J. Roisen, Chair, 502-852-5165, Fax: 502-852-6228, E-mail: fjrois01@gwise.louisville.edu. *Application contact:* Dr. Kathleen Klueber, Director of Graduate Studies, 502-852-5172, Fax: 502-852-6228, E-mail: kmklue01@gwise.louisville.edu.

University of Maryland, Graduate School, Graduate Programs in Medicine, Department of Anatomy and Neurobiology, Baltimore, MD 21201. Offers MS, MD/PhD. Part-time and evening/weekend programs available. *Faculty:* 16 full-time (2 women). *Students:* 7 full-time (4 women); includes 1 minority (African American), 3 international. Average age 29. 1 applicant, 0% accepted, 0 enrolled. *Degree requirements:* For master's, thesis optional; for doctorate, one foreign language, thesis/dissertation. *Entrance requirements:* For master's, GRE General Test, minimum GPA of 3.0; for doctorate, GRE General Test, GRE Subject Test (recommended), minimum GPA of 3.0. Additional exam requirements/recommendations for international students: Required—TOEFL, TOEFL or IELTS; Recommended—IELT. *Application deadline:* For fall admission, 5/31 for domestic students, 1/15 for international students. Application fee: $50. Electronic applications accepted. *Expenses:* Tuition, state resident: full-time $8,079; part-time $409 per credit hour. Tuition, nonresident: full-time $18,384; part-time $731 per credit hour. Required fees: $695; $10 per credit hour. Tuition and fees vary according to degree level and program. *Financial support:* Research assistantships available. Support available to part-time students. Financial award application deadline: 2/15. *Faculty research:* Neural networks, chemical sensory pathways, electrophysiology, developmental neurobiology. *Unit head:* Dr. Michael T. Shipley, Chair, 410-706-3590, Fax: 410-706-2512, E-mail: mshipley@umaryland.edu. *Application contact:* Dr. George Markelonis, Program Director, 410-706-3713, Fax: 410-706-2512, E-mail: gmarkel@umaryland.edu.

University of Missouri–Columbia, Graduate School, College of Arts and Sciences, Division of Biological Sciences, Columbia, MO 65211. Offers cellular, molecular and developmental biology (MA, PhD); evolutionary biology and ecology (MA, PhD); neurobiology and behavior (MA, PhD). *Faculty:* 43 full-time (12 women). *Students:* 48 full-time (26 women), 28 part-time (10 women); includes 7 minority (2 African Americans, 3 Asian Americans or Pacific Islanders, 2 Hispanic Americans), 20 international. In 2005, 3 master's, 8 doctorates awarded. Terminal master's awarded for partial completion of doctoral program. *Degree requirements:* For master's, thesis/dissertation; for doctorate, thesis/dissertation, comprehensive exam. *Entrance requirements:* For master's and doctorate, GRE General Test, minimum GPA of 3.0. *Application deadline:* For fall admission, 1/15 for domestic students. Applications are processed on a rolling basis. Application fee: $45 ($60 for international students). *Financial support:* Fellowships, research assistantships, teaching assistantships, institutionally sponsored loans available. *Unit head:* Dr. Ray Semlitsch, Director of Graduate Studies, 573-884-6396, E-mail: semlitschr@missouri.edu. *Application contact:* Nila Emerich, Application Contact, 800-553-5698.

The University of North Carolina at Chapel Hill, School of Medicine and Graduate School, Graduate Programs in Medicine, Curriculum in Neurobiology, Chapel Hill, NC 27599. Offers PhD. *Faculty:* 68 full-time (19 women), 12 part-time/adjunct (0 women). *Students:* 32 full-time (23

Neuroscience

women); includes 6 minority (3 African Americans, 1 Asian American or Pacific Islander, 2 Hispanic Americans), 2 international. 49 applicants, 22% accepted, 4 enrolled. In 2005, 6 degrees awarded. *Degree requirements:* For doctorate, thesis/dissertation, comprehensive exam. *Entrance requirements:* For doctorate, GRE General Test, minimum GPA of 3.0. *Application deadline:* For fall admission, 3/1 for domestic students. Application fee: $55. Electronic applications accepted. *Financial support:* In 2005–06, 12 fellowships with full tuition reimbursements (averaging $22,000 per year), 20 research assistantships with full tuition reimbursements (averaging $22,000 per year) were awarded; teaching assistantships *Unit head:* Dr. Paul B. Manis, Director, 919-966-8926, Fax: 919-966-4348, E-mail: paul_manis@med.unc.edu. *Application contact:* Lori Blalock, Administrative Assistant, 919-966-1260, Fax: 919-966-4348, E-mail: lori_blalock@med.unc.edu.

See Close-Up on page 1051.

University of Pennsylvania, School of Arts and Sciences, Program in Neurobiology and Physiology, Philadelphia, PA 19104. Offers PhD. *Entrance requirements:* For doctorate, GRE General Test, GRE Subject Test. Additional exam requirements/recommendations for international students: Required—TOEFL. Electronic applications accepted.

University of Pittsburgh, School of Medicine, Graduate Programs in Medicine, Program in Neurobiology, Pittsburgh, PA 15260. Offers MS, PhD. *Degree requirements:* For doctorate, thesis/dissertation. *Entrance requirements:* For doctorate, GRE General Test, GRE Subject Test, minimum QPA of 3.0. Additional exam requirements/recommendations for international students: Required—TOEFL. Application fee: $30 ($40 for international students). *Expenses:* Tuition, state resident: full-time $13,194; part-time $537 per credit. Tuition, nonresident: full-time $25,012; part-time $1,026 per credit. Required fees: $700; $164 per term. Tuition and fees vary according to campus/location and program. *Faculty research:* Development and plasticity, biophysics and signal transduction, neural systems and computational modeling. *Application contact:* Graduate Studies Administrator, 412-648-8957, Fax: 412-648-1236, E-mail: biomed_phd@fs1.dean-med.pitt.edu.

University of Rochester, School of Medicine and Dentistry, Graduate Programs in Medicine and Dentistry, Department of Neurobiology and Anatomy, Program in Neurobiology and Anatomy, Rochester, NY 14627-0250. Offers MS, PhD. *Degree requirements:* For doctorate, thesis/dissertation, qualifying exam. *Entrance requirements:* For master's and doctorate, GRE General Test.

University of Southern California, Keck School of Medicine and Graduate School, Graduate Programs in Medicine, Department of Cell and Neurobiology, Los Angeles, CA 90089. Offers MS, PhD. *Faculty:* 23 full-time (4 women), 8 part-time/adjunct (2 women). *Students:* 9 full-time (3 women); includes 3 minority (all Asian Americans or Pacific Islanders) Average age 25. 15 applicants, 0% accepted, 0 enrolled.Terminal master's awarded for partial completion of doctoral program. *Degree requirements:* For master's, thesis or alternative; for doctorate, thesis/dissertation. *Entrance requirements:* For master's, GRE General Test, minimum GPA of 3.0; for doctorate, GRE General Test. *Application deadline:* For fall admission, 3/5 priority date for domestic students, 3/5 priority date for international students. Application fee: $65 ($75 for international students). Electronic applications accepted. *Expenses:* Tuition: Full-time $25,416; part-time $1,059 per unit. Required fees: $484; $484 per year. Tuition and fees vary according to course load and program. *Financial support:* In 2005–06, 3 research assistantships (averaging $24,876 per year), teaching assistantships (averaging $9,056 per year) were awarded; fellowships, Federal Work-Study, institutionally sponsored loans, and tuition waivers (partial) also available. Support available to part-time students. *Faculty research:* Neurobiology and development, gene therapy in vision, lacrimal glands, neuroendocrinology, signal transduction mechanisms. *Unit head:* Dr. Mikel Henry Snow, Chair, 323-442-1881, Fax: 323-442-3466. *Application contact:* Darlene Marie Campbell, Project Specialist, 323-442-1881, Fax: 323-442-0466, E-mail: dmc@usc.edu.

The University of Tennessee Health Science Center, College of Graduate Health Sciences, Department of Anatomy and Neurobiology, Memphis, TN 38163-0002. Offers PhD. *Faculty:* 26 full-time (4 women), 1 (woman) part-time/adjunct. *Students:* 13 full-time (8 women); includes 9 minority (8 Asian Americans or Pacific Islanders, 1 Hispanic American). Average age 24. 49 applicants, 12% accepted. In 2005, 3 degrees awarded. *Degree requirements:* For doctorate, thesis/dissertation, oral and written preliminary and comprehensive exams. *Entrance requirements:* For doctorate, GRE General Test, minimum GPA of 3.0. *Application deadline:* For fall admission, 5/15 for domestic students. Application fee: $0. Electronic applications accepted. *Financial support:* Fellowships, research assistantships, teaching assistantships, Federal Work-Study, traineeships, and tuition waivers (full) available. Financial award application deadline: 2/25. *Unit head:* Dr. Stephen T. Kitai, Chairman, 901-440-5957, Fax: 901-448-7193. *Application contact:* Ida W. Mosby, Director, Enrollment Services, 901-448-5560, E-mail: imosby@utmem.edu.

See Close-Up on page 1057.

The University of Texas at Austin, Graduate School, The Institute for Neuroscience, Austin, TX 78712-1111. Offers MA, PhD, MD/PhD. Terminal master's awarded for partial completion of doctoral program. *Degree requirements:* For master's and doctorate, thesis/dissertation. *Entrance requirements:* For master's and doctorate, GRE. Electronic applications accepted. *Faculty research:* Cellular/molecular biology, neurobiology, pharmacology, behavioral neuroscience.

See Close-Up on page 1059.

The University of Texas at San Antonio, College of Sciences, Department of Biology, Program in Biology, San Antonio, TX 78249-0617. Offers cell and molecular biology (PhD); neurobiology (PhD). *Degree requirements:* For doctorate, thesis/dissertation, comprehensive exam. *Entrance requirements:* For doctorate, GRE General Test, minimum GPA of 3.0. Additional exam requirements/recommendations for international students: Required—TOEFL. Expenses: Contact institution. *Faculty research:* Neurophysiology, neurotoxicology, neural circuit analysis, neuroendocrinology, development of biosensors for the detection of toxins.

University of Utah, School of Medicine and The Graduate School, Graduate Programs in Medicine, Department of Neurobiology and Anatomy, Salt Lake City, UT 84112-1107. Offers M Phil, MS, PhD. Part-time programs available. Terminal master's awarded for partial completion of doctoral program. *Degree requirements:* For master's and doctorate, one foreign language, thesis/dissertation. *Entrance requirements:* For master's and doctorate, GRE General Test. *Expenses:* Tuition, state resident: full-time $2,932; part-time $369 per credit. Tuition, nonresident: full-time $10,350; part-time $1,302 per credit. Required fees: $516 per

term. Tuition and fees vary according to course load and program. *Faculty research:* Neuroscience, neuroanatomy, developmental neurobiology, neurogenetics.

University of Vermont, College of Medicine and Graduate College, Graduate Programs in Medicine, Department of Anatomy and Neurobiology, Burlington, VT 05405. Offers PhD, MD/PhD. *Students:* 15 (6 women) 4 international. 14 applicants, 43% accepted, 1 enrolled. *Degree requirements:* For doctorate, thesis/dissertation. *Entrance requirements:* For doctorate, GRE General Test. Additional exam requirements/recommendations for international students: Required—TOEFL (minimum score 550 paper-based; 213 computer-based). *Application deadline:* For fall admission, 1/15 for domestic students. Applications are processed on a rolling basis. Application fee: $40. Electronic applications accepted. *Expenses:* Tuition, area resident: Part-time $410 per credit hour. Tuition, nonresident: part-time $1,034 per credit hour. *Financial support:* Teaching assistantships available. Financial award application deadline:3/1. *Faculty research:* Autonomic neurobiology, developmental neurobiology, neurotransmitter expression and release, plasticity and regeneration. *Unit head:* Dr. Rodney L. Parsons, Chairperson, 802-656-2230. *Application contact:* Dr. Rae Nishi, Coordinator, 802-656-2230, E-mail: rae.nishi@uvm.edu.

University of Washington, School of Medicine and Graduate School, Graduate Programs in Medicine, Graduate Program in Neurobiology and Behavior, Seattle, WA 98195. Offers neurobiology (PhD); neuroscience (PhD). *Degree requirements:* For doctorate, thesis/dissertation. *Entrance requirements:* For doctorate, GRE. Additional exam requirements/recommendations for international students: Required—TOEFL, TSE. Electronic applications accepted. *Faculty research:* Motor, sensory systems, neuroplasticity, animal behavior, neuroendocrinology, computational neuroscience.

University of Wisconsin–Madison, Medical School and Graduate School, Graduate Programs in Medicine, Department of Physiology, Madison, WI 53706-1380. Offers PhD. *Faculty:* 22 full-time (6 women). *Students:* 28 full-time (14 women); includes 10 minority (1 African American, 8 Asian Americans or Pacific Islanders, 1 Hispanic American). Average age 22. 72 applicants, 14% accepted, 6 enrolled. In 2005, 3 degrees awarded. *Degree requirements:* For doctorate, thesis/dissertation, written exams. *Entrance requirements:* For doctorate, GRE, minimum GPA of 3.0. Additional exam requirements/recommendations for international students: Required—TOEFL (minimum score 580 paper-based; 237 computer-based). *Application deadline:* For fall admission, 1/15 for domestic students. Applications are processed on a rolling basis. Application fee: $45. Electronic applications accepted. *Financial support:* In 2005–06, fellowships with tuition reimbursements (averaging $22,500 per year), research assistantships with tuition reimbursements (averaging $22,500 per year), teaching assistantships with tuition reimbursements (averaging $22,500 per year) were awarded. *Faculty research:* Studies in molecular cellular systems, cardiovascular, neuroscience. *Unit head:* Dr. Richard L. Moss, Chair, 608-262-1939, Fax: 608-265-5072, E-mail: rlmoss@physiology.wisc.edu. *Application contact:* Sue S. Krey, Program Assistant 4, 608-262-9114, Fax: 608-265-5512, E-mail: krey@physiology.wisc.edu.

Wake Forest University, School of Medicine and Graduate School, Graduate Programs in Medicine, Department of Neurobiology and Anatomy, Winston-Salem, NC 27109. Offers PhD. *Degree requirements:* For doctorate, thesis/dissertation. *Entrance requirements:* For doctorate, GRE General Test. Additional exam requirements/recommendations for international students: Required—TOEFL. Electronic applications accepted. *Faculty research:* Sensory neurobiology, reproductive endocrinology, regulatory processes in cell biology.

Wayne State University, School of Medicine and Graduate School, Graduate Programs in Medicine, Cellular and Clinical Neurobiology Program, Detroit, MI 48202. Offers PhD. *Faculty:* 29. *Students:* 7 full-time (6 women); includes 1 minority (Asian American or Pacific Islander), 2 international. Average age 26. 6 applicants, 33% accepted. In 2005, 3 degrees awarded. *Degree requirements:* For doctorate, thesis/dissertation. *Entrance requirements:* For doctorate, GRE General Test, GRE Subject Test. Additional exam requirements/recommendations for international students: Required—TOEFL (minimum score 550 paper-based; 213 computer-based); Recommended—TWE (minimum score 6). *Application deadline:* For fall admission, 4/1 for domestic students, 6/1 for international students. Applications are processed on a rolling basis. Application fee: $30 ($50 for international students). Electronic applications accepted. *Expenses:* Tuition, state resident: part-time $338 per credit hour. Tuition, nonresident: part-time $746 per credit hour. Required fees: $24 per credit hour. Full-time tuition and fees vary according to program. *Financial support:* In 2005–06, 4 fellowships were awarded; research assistantships, Federal Work-Study also available. Financial award application deadline: 4/1. *Faculty research:* Cellular and molecular properties of dopamine and serotonin neurons, molecular and cellular neurobiology, neuronal and glial signal transduction, biological bases of neurological and psychiatric disorders. *Unit head:* Dr. Michael J. Bannon, Director, 313-577-1471, Fax: 313-993-4269, E-mail: mbannon@med.wayne.edu. *Application contact:* Gregory Kapatos, Graduate Director, 313-577-5965.

Wesleyan University, Graduate Programs, Department of Biology, Middletown, CT 06459-0260. Offers cell biology (PhD); comparative physiology (PhD); developmental biology (PhD); genetics (PhD); neurophysiology (PhD); population biology (PhD). *Faculty:* 12 full-time (3 women). *Students:* 29 full-time (15 women), 10 international. Average age 26. 131 applicants. In 2005, 2 doctorates awarded. *Degree requirements:* For doctorate, one foreign language, thesis/dissertation. *Entrance requirements:* For doctorate, GRE Subject Test. *Application deadline:* For fall admission, 2/15 for domestic students. Applications are processed on a rolling basis. Application fee: $0. *Expenses:* Tuition: Full-time $24,732. One-time fee: $20 full-time. *Financial support:* Research assistantships, teaching assistantships, stipends available. *Faculty research:* Microbial population genetics, genetic basis of evolutionary adaptation, genetic regulation of differentiation and pattern formation in *drosophila*. *Unit head:* Dr. Michael Weir, Chairman, 860-685-2402, E-mail: mweir@wesleyan.edu. *Application contact:* Marjorie Fitzgibbons, Information Contact, 860-685-2157, E-mail: mfitzgibbons@wesleyan.edu.

See Close-Up on page 323.

Yale University, Graduate School of Arts and Sciences, Department of Molecular, Cellular, and Developmental Biology, Program in Neurobiology, New Haven, CT 06520. Offers PhD. *Degree requirements:* For doctorate, thesis/dissertation. *Entrance requirements:* For doctorate, GRE General Test, GRE Subject Test.

Yale University, Graduate School of Arts and Sciences, Department of Neurobiology, New Haven, CT 06520. Offers PhD. *Degree requirements:* For doctorate, thesis/dissertation. *Entrance requirements:* For doctorate, GRE General Test, GRE Subject Test.

Neuroscience

Albany Medical College, Graduate Programs in the Biological Sciences, Center for Neuropharmacology and Neuroscience, Albany, NY 12208-3479. Offers MS, PhD. Part-time programs available. *Faculty:* 21 full-time (6 women). *Students:* 16 full-time (8 women); includes 2 minority (1 African American, 1 Asian American or Pacific Islander), 3 international. Average age 26. 22 applicants, 64% accepted, 4 enrolled. In 2005, 5 master's, 3 doctorates awarded. Terminal master's awarded for partial completion of doctoral program. *Median time to degree:* Of those who began their doctoral program in fall 1997, 100% received their degree in 8 years or less. *Degree requirements:* For master's, thesis/dissertation; for doctorate, thesis/dissertation, comprehensive exam. *Entrance requirements:* For master's and doctorate, GRE General Test. Additional exam requirements/recommendations for international students: Required—TOEFL. *Application deadline:* For fall admission, 3/15 for domestic students. Applications are processed on a rolling basis. Application fee: $0 ($60 for international students). *Financial support:* In 2005–06, 16 students received support, including 6 fellowships (averaging $23,000 per year), 10 research assistantships (averaging $23,000 per year); Federal Work-Study, scholarships/grants, and tuition waivers (full) also available. Financial award applicants required to submit FAFSA. *Faculty research:* Molecular and cellular neuroscience,

Neuroscience

Albany Medical College (continued)
neuronal development. *Unit head:* Dr. Stanley D. Glick, Director, 518-262-5303, Fax: 518-262-5799, E-mail: cnninfo@mail.amc.edu. *Application contact:* Dr. Richard Keller, Graduate Director, 518-262-5303, Fax: 518-262-5799, E-mail: cnninfo@mail.amc.edu.

American University, College of Arts and Sciences, Department of Psychology, Program in Behavior, Cognition, and Neuroscience, Washington, DC 20016-8001. Offers PhD. *Students:* 9 full-time (5 women), 7 part-time (4 women); includes 1 minority (African American), 1 international. Average age 26. In 2005, 1 degree awarded. *Degree requirements:* For doctorate, thesis/dissertation, comprehensive exam. *Entrance requirements:* For doctorate, GRE General Test, GRE Subject Test. *Application deadline:* For fall admission, 2/1 for domestic students. Application fee: $50. *Expenses:* Tuition: Full-time $17,802; part-time $989 per credit. Required fees: $380. *Financial support:* Fellowships, research assistantships, teaching assistantships, career-related internships or fieldwork, Federal Work-Study, institutionally sponsored loans, and tuition waivers (full and partial) available. Support available to part-time students. Financial award application deadline: 2/1. *Faculty research:* Psychophysics, drug discrimination learning, choice behavior, conditioning and learning, olfaction and taste. *Application contact:* Sara Holland, Senior Administrative Assistant, 202-885-1717, Fax: 202-885-1023.

American University of Beirut, Graduate Programs, Faculty of Medicine, Beirut, Lebanon. Offers biochemistry (MS); human morphology (MS); microbiology and immunology (MS); neuroscience (MS); pharmacology and therapeutics (MS); physiology (MS). *Degree requirements:* For master's, one foreign language, thesis (for some programs), comprehensive exam, registration. *Entrance requirements:* For master's, GRE, letter of recommendation.

Arizona State University, Division of Graduate Studies, College of Liberal Arts and Sciences, Department of Biology, Program in Neuroscience, Tempe, AZ 85287. Offers MS, PhD. Terminal master's awarded for partial completion of doctoral program. *Degree requirements:* For master's, thesis; for doctorate, thesis/dissertation, oral exam. *Entrance requirements:* For master's and doctorate, GRE General Test, GRE Subject Test. Additional exam requirements/recommendations for international students: Required—TOEFL (minimum score 600 paper-based; 250 computer-based); Recommended—TSE.

Arizona State University, Division of Graduate Studies, College of Liberal Arts and Sciences, Division of Natural Sciences and Mathematics, Department of Psychology, Tempe, AZ 85287. Offers behavioral neuroscience (PhD); clinical psychology (PhD); cognitive/behavioral systems (PhD); developmental psychology (PhD); environmental psychology (PhD); quantitative research methods (PhD); social psychology (PhD). *Accreditation:* APA. *Degree requirements:* For doctorate, thesis/dissertation. *Entrance requirements:* For doctorate, GRE General Test, GRE Subject Test.

Baylor College of Medicine, Graduate School of Biomedical Sciences, Department of Neuroscience, Houston, TX 77030-3498. Offers PhD, MD/PhD. *Faculty:* 38 full-time (5 women). *Students:* 54 full-time (24 women); includes 13 minority (1 African American, 7 Asian Americans or Pacific Islanders, 5 Hispanic Americans), 13 international. Average age 29. 92 applicants, 16% accepted, 11 enrolled. In 2005, 3 degrees awarded. *Median time to degree:* Of those who began their doctoral program in fall 1997, 75% received their degree in 8 years or less. *Degree requirements:* For doctorate, thesis/dissertation, public defense. *Entrance requirements:* For doctorate, GRE General Test, GRE Subject Test (strongly recommended), minimum GPA of 3.0. Additional exam requirements/recommendations for international students: Required—TOEFL. *Application deadline:* For fall admission, 2/1 for domestic students. Application fee: $30. Electronic applications accepted. *Expenses:* Tuition: Full-time $8,200. Full-time tuition and fees vary according to program. *Financial support:* In 2005–06, 48 students received support, including 15 fellowships (averaging $23,000 per year), 39 research assistantships (averaging $23,000 per year); Federal Work-Study, institutionally sponsored loans, health care benefits, and tuition waivers (full) also available. Financial award applicants required to submit FAFSA. *Faculty research:* Molecular and developmental neurobiology, neurobiology of disease, neuroanatomy, neurophysiology, neural systems analysis. *Unit head:* Dr. Mariella DeBiasi, Director, 713-798-7270. *Application contact:* Krista Delfaco, Graduate Program Administrator, 713-798-7270, Fax: 713-798-3946, E-mail: kdefalco@bcm.edu.

See Close-Up on page 985.

Baylor College of Medicine, Graduate School of Biomedical Sciences, Program in Developmental Biology, Houston, TX 77030-3498. Offers PhD, MD/PhD. *Faculty:* 42 full-time (10 women). *Students:* 38 full-time (17 women); includes 7 minority (1 American Indian/Alaska Native, 5 Asian Americans or Pacific Islanders, 1 Hispanic American), 21 international. Average age 27. 67 applicants, 25% accepted, 7 enrolled. In 2005, 2 doctorates awarded. *Median time to degree:* Of those who began their doctoral program in fall 1997, 100% received their degree in 8 years or less. *Degree requirements:* For doctorate, thesis/dissertation, public defense. *Entrance requirements:* For doctorate, GRE General Test, GRE Subject Test (strongly recommended), minimum GPA of 3.0. Additional exam requirements/recommendations for international students: Required—TOEFL. *Application deadline:* For fall admission, 1/1 for domestic students. Application fee: $30. Electronic applications accepted. *Expenses:* Tuition: Full-time $8,200. Full-time tuition and fees vary according to program. *Financial support:* In 2005–06, 37 students received support, including 10 fellowships (averaging $23,000 per year), 28 research assistantships (averaging $23,000 per year); career-related internships or fieldwork, Federal Work-Study, institutionally sponsored loans, health care benefits, tuition waivers (full), and stipends also available. *Faculty research:* Molecular and genetic approaches to study pattern formation in *Dictyostelium, Drosophila, C.elegans,* mouse, *Xenopus,* and zebrafish; cross-species approach. *Unit head:* Dr. Hugo Bellen, Director, 713-798-6410. *Application contact:* Catherine Tasnier, Graduate Program Administrator, 713-798-6410, Fax: 713-798-5386, E-mail: cat@bcm.edu.

See Close-Up on page 757.

Baylor University, Graduate School, College of Arts and Sciences, Department of Psychology and Neuroscience, Program in Neuroscience, Waco, TX 76798. Offers MA, PhD. *Students:* 10 full-time (7 women); includes 2 minority (both Hispanic Americans) In 2005, 3 master's, 2 doctorates awarded. *Degree requirements:* For doctorate, comprehensive exam. *Entrance requirements:* For master's and doctorate, GRE General Test. *Application deadline:* Applications are processed on a rolling basis. Application fee: $25. *Unit head:* Dr. Jaime Diaz-Granados, Graduate Program Director, 254-710-2961, Fax: 254-710-3033, E-mail: jim_diaz-granados@baylor.edu. *Application contact:* Bettye Keel, Graduate Coordinator, 254-710-2811, Fax: 254-710-3033, E-mail: bettye_keel@baylor.edu.

Boston University, Graduate School, College of Arts and Sciences, Department of Cognitive and Neural Systems, Boston, MA 02215. Offers MA, PhD. *Students:* 61 full-time (9 women), 10 part-time (2 women); includes 4 minority (1 African American, 2 Asian Americans or Pacific Islanders, 1 Hispanic American), 31 international. Average age 30. 70 applicants, 59% accepted, 15 enrolled. In 2005, 2 degrees awarded. Terminal master's awarded for partial completion of doctoral program. *Degree requirements:* For master's, one foreign language, comprehensive exam, registration; for doctorate, one foreign language, thesis/dissertation, comprehensive exam, registration. *Entrance requirements:* For master's and doctorate, GRE General Test, GRE Subject Test (recommended), 3 letters of recommendation. Additional exam requirements/recommendations for international students: Required—TOEFL (minimum score 550 paper-based; 213 computer-based). *Application deadline:* For fall admission, 5/1 for domestic students, 5/1 for international students; for spring admission, 11/15 for domestic students, 11/15 for international students. Application fee: $60. *Expenses:* Tuition: Full-time $31,530; part-time $985 per credit. Required fees: $316; $40 per semester. Tuition and fees vary according to course level and program. *Financial support:* In 2005–06, 3 fellowships with full tuition reimbursements (averaging $16,500 per year), 48 research assistantships with full tuition reimbursements (averaging $16,000 per year), 2 teaching assistantships with full tuition reimbursements were awarded; Federal Work-Study and unspecified assistantships also available. Support

available to part-time students. Financial award application deadline: 1/15; financial award applicants required to submit FAFSA. *Unit head:* Stephen Grossberg, Chairman, 617-353-7858, Fax: 617-353-7755, E-mail: steve@bu.edu. *Application contact:* Carol Y. Jefferson, Administrative Assistant, 617-353-7676, Fax: 617-353-7755, E-mail: caroly@bu.edu.

Boston University, Graduate School of Arts and Sciences, Interdepartmental Program in Neuroscience, Boston, MA 02215. Offers MA, PhD. *Students:* 21 full-time (12 women), 1 part-time; includes 3 minority (1 African American, 1 Asian American or Pacific Islander, 1 Hispanic American), 3 international. Average age 28. 120 applicants, 11% accepted, 6 enrolled. In 2005, 2 master's, 7 doctorates awarded. Terminal master's awarded for partial completion of doctoral program. *Degree requirements:* For master's, one foreign language, thesis/dissertation, registration; for doctorate, one foreign language, thesis/dissertation, comprehensive exam, registration. *Entrance requirements:* For master's and doctorate, GRE General Test, 3 letters of recommendation. Additional exam requirements/recommendations for international students: Required—TOEFL (minimum score 550 paper-based; 213 computer-based). *Application deadline:* For fall admission, 1/15 for domestic students, 1/15 for international students. Application fee: $60. *Expenses:* Tuition: Full-time $31,530; part-time $985 per credit. Required fees: $316; $40 per semester. Tuition and fees vary according to course level and program. *Financial support:* In 2005–06, 8 students received support, including 1 fellowship with full tuition reimbursement available (averaging $16,500 per year), 5 research assistantships with full tuition reimbursements available (averaging $16,000 per year), 2 teaching assistantships with full tuition reimbursements available (averaging $16,000 per year) Financial award application deadline: 1/15; financial award applicants required to submit FAFSA. *Unit head:* Dr. William Eldred, Director, 617-353-2439, Fax: 617-358-1857, E-mail: eldred@bu.edu. *Application contact:* Lori Lamport, Program Administrator, 617-358-1123, Fax: 617-358-1857, E-mail: neurosci@bu.edu.

See Close-Up on page 987.

Boston University, School of Medicine, Division of Graduate Medical Sciences, Program in Behavioral Neuroscience, Boston, MA 02215. Offers PhD, MD/PhD. Part-time programs available. *Faculty:* 14 part-time/adjunct (3 women). *Students:* 6 full-time (5 women), 3 part-time (all women), 1 international. Average age 41. *Degree requirements:* For doctorate, thesis/dissertation. *Entrance requirements:* For doctorate, GRE General Test, GRE Subject Test. Additional exam requirements/recommendations for international students: Required—TOEFL. *Application deadline:* For fall admission, 1/15 for domestic students; for spring admission, 10/15 priority date for domestic students. Electronic applications accepted. *Expenses:* Tuition: Full-time $31,530; part-time $985 per credit. Required fees: $316; $40 per semester. Tuition and fees vary according to course level and program. *Financial support:* In 2005–06, 4 fellowships, 5 research assistantships were awarded; Federal Work-Study, scholarships/grants, and traineeships also available. *Faculty research:* Human brain dysfunction, language disorders, disorders of purposeful movement, path of learning. *Unit head:* Dr. Marlene Oscar Berman, Director, 617-638-4803, Fax: 617-638-4806, E-mail: oscar@bu.edu.

Brandeis University, Graduate School of Arts and Sciences, Department of Psychology, Waltham, MA 02454-9110. Offers cognitive neuroscience (PhD); general psychology (MA); social/developmental psychology (PhD). MA program offered to students enrolled in PhD program only. *Faculty:* 15 full-time (3 women), 11 part-time/adjunct (7 women). *Students:* 21 full-time (17 women); includes 7 minority (5 Asian Americans or Pacific Islanders, 2 Hispanic Americans). Average age 27. 55 applicants, 16% accepted, 5 enrolled. In 2005, 4 master's, 10 doctorates awarded. *Median time to degree:* Of those who began their doctoral program in fall 1997, 80% received their degree in 8 years or less. *Degree requirements:* For doctorate, thesis/dissertation, comprehensive exam. *Entrance requirements:* For doctorate, GRE General Test, GRE Subject Test (recommended), 3 letters of recommendation. Additional exam requirements/recommendations for international students: Required—TOEFL (minimum score 600 paper-based; 250 computer-based). *Application deadline:* For fall admission, 1/15 for domestic students, 1/15 for international students. Application fee: $55. Electronic applications accepted. *Financial support:* In 2005–06, 16 students received support, including 16 fellowships with full tuition reimbursements available (averaging $17,500 per year); institutionally sponsored loans, scholarships/grants, traineeships, health care benefits, tuition waivers (full), and unspecified assistantships also available. *Faculty research:* Development, cognition, social aging, perception. Total annual research expenditures: $5.1 million. *Unit head:* Paul DiZio, Director of Graduate Studies, 781-736-3300, Fax: 781-736-3291, E-mail: dizio@brandeis.edu. *Application contact:* Donna J. Coletti, Graduate Admissions Coordinator, 781-736-3303, Fax: 781-736-3291, E-mail: coletti@brandeis.edu.

Brandeis University, Graduate School of Arts and Sciences, Programs in Life Sciences, Program in Neuroscience, Waltham, MA 02454-9110. Offers MS, PhD. *Faculty:* 19 full-time (6 women). *Students:* 40 full-time (18 women); includes 12 minority (6 African Americans, 3 Asian Americans or Pacific Islanders, 3 Hispanic Americans). Average age 24. 119 applicants, 13% accepted, 8 enrolled. In 2005, 4 master's, 8 doctorates awarded. Terminal master's awarded for partial completion of doctoral program. *Median time to degree:* Of those who began their doctoral program in fall 1997, 100% received their degree in 8 years or less. *Degree requirements:* For master's, research project, thesis optional; for doctorate, thesis/dissertation, qualifying exams, teaching experience, journal club, comprehensive exam, registration. *Entrance requirements:* For master's and doctorate, GRE General Test, resumé, 3 letters of recommendation. Additional exam requirements/recommendations for international students: Required—TOEFL (minimum score 600 paper-based; 250 computer-based). *Application deadline:* For fall admission, 1/15 for domestic students. Applications are processed on a rolling basis. Application fee: $55. Electronic applications accepted. *Financial support:* In 2005–06, 22 fellowships with tuition reimbursements (averaging $26,500 per year), 18 research assistantships with tuition reimbursements (averaging $26,500 per year), 6 teaching assistantships with tuition reimbursements (averaging $3,000 per year) were awarded; scholarships/grants, traineeships, health care benefits, and tuition waivers (full and partial) also available. Support available to part-time students. Financial award application deadline: 4/15; financial award applicants required to submit CSS PROFILE or FAFSA. *Faculty research:* Behavioral neuroscience, cellular and molecular neuroscience, computational and integrative neuroscience. *Unit head:* Dr. Leslie Griffith, Chair, Fax: 781-736-3107, E-mail: griffith@brandeis.edu. *Application contact:* Marcia Cabral, Information Officer, 781-736-3100, Fax: 781-736-3107, E-mail: cabral@brandeis.edu.

Brigham Young University, Graduate Studies, College of Biological and Agricultural Sciences, Department of Physiology and Developmental Biology, Provo, UT 84602-1001. Offers neuroscience (MS, PhD); physiology and developmental biology (MS, PhD). Part-time programs available. *Faculty:* 17 full-time (0 women). *Students:* 25 full-time (11 women); includes 1 minority (Asian American or Pacific Islander), 1 international. Average age 26. 26 applicants, 38% accepted, 9 enrolled. In 2005, 5 master's, 2 doctorates awarded. Terminal master's awarded for partial completion of doctoral program. *Degree requirements:* For master's and doctorate, thesis/dissertation. *Entrance requirements:* For master's, GRE General Test, minimum GPA of 3.0 during previous 2 years; for doctorate, GRE General Test, minimum GPA of 3.0 overall. Additional exam requirements/recommendations for international students: Required—TOEFL. *Application deadline:* For fall admission, 2/1 priority date for domestic students, 2/1 priority date for international students. For winter admission, 9/10 for domestic students. Application fee: $50. Electronic applications accepted. *Financial support:* In 2005–06, 26 students received support, including 13 research assistantships with full tuition reimbursements available (averaging $15,500 per year), 13 teaching assistantships with partial tuition reimbursements available (averaging $14,900 per year); fellowships with partial tuition reimbursements available, career-related internships or fieldwork, institutionally sponsored loans, scholarships/grants, tuition waivers (full and partial), unspecified assistantships, and tuition awards also available. Financial award application deadline: 2/1. *Faculty research:* Sex differentiation of the brain, exercise physiology, developmental biology, membrane biophysics, neuroscience. Total annual research expenditures: $892,386. *Unit head:* Dr. James P. Porter, Chair, 801-422-9160, Fax: 801-422-0700, E-mail: james_porter@byu.edu. *Application contact:*

Neuroscience

Dr. Dixon J. Woodbury, Graduate Coordinator, 801-422-7562, Fax: 801-422-0700, E-mail: dixon_woodbury@byu.edu.

See Close-Up on page 1243.

Brock University, Graduate Studies, Faculty of Social Sciences, Program in Psychology, St. Catharines, ON L2S 3A1, Canada. Offers behavioral neuroscience (MA, PhD); life span development (MA, PhD); social personality (MA, PhD). Part-time programs available. *Faculty:* 26 full-time (13 women), 7 part-time/adjunct (3 women\. *Students:* 36 full-time (26 women), 2 part-time (1 woman), 4 international. 69 applicants, 38% accepted. In 2005, 8 degrees awarded. *Degree requirements:* For master's, thesis. *Entrance requirements:* For master's, GRE, honors degree. Additional exam requirements for international students: Required—TOEFL. *Application deadline:* For fall admission, 1/15 for domestic students. Applications are processed on a rolling basis. Application fee: $75. Electronic applications accepted. *Financial support:* Fellowships, research assistantships, teaching assistantships, career-related internships or fieldwork, scholarships/grants, and unspecified assistantships available. Support available to part-time students. *Faculty research:* Social personality, behavioral neuroscience, life-span development. *Unit head:* Dr. Carolyn Hafer, Graduate Program Director, 905-688-5550 Ext. 4297, Fax: 905-688-6922, E-mail: chafer@brocku.ca. *Application contact:* Dr. Carolyn Hafer, Graduate Program Director, 905-688-5550 Ext. 4297, Fax: 905-688-6922, E-mail: chafer@brocku.ca.

Brown University, Graduate School, Department of Neuroscience, Providence, RI 02912. Offers PhD. *Degree requirements:* For doctorate, thesis/dissertation, comprehensive exam. *Entrance requirements:* For doctorate, GRE.

See Close-Up on page 989.

Brown University, Graduate School, Division of Biology and Medicine, Department of Neuroscience, Providence, RI 02912. Offers PhD. *Degree requirements:* For doctorate, thesis/dissertation, preliminary exam. *Entrance requirements:* For doctorate, GRE General Test, GRE Subject Test. Additional exam requirements/recommendations for international students: Required—TOEFL. Electronic applications accepted. *Faculty research:* Neurophysiology, systems neuroscience, membrane biophysics, neuropharmacology, sensory systems.

Brown University, National Institutes of Health Sponsored Programs, Department of Neuroscience, Providence, RI 02912. Offers PhD. *Degree requirements:* For doctorate, thesis/dissertation, comprehensive exam.

See Close-Up on page 989.

California Institute of Technology, Division of Engineering and Applied Science, Option in Computation and Neural Systems, Pasadena, CA 91125-0001. Offers MS, PhD. *Faculty:* 3 full-time (0 women). *Students:* 45 full-time (11 women); includes 4 minority (1 African American, 1 Asian American or Pacific Islander, 2 Hispanic Americans), 19 international. 113 applicants, 13% accepted, 8 enrolled. In 2005, 1 master's, 7 doctorates awarded. Terminal master's awarded for partial completion of doctoral program. *Degree requirements:* For doctorate, thesis/dissertation, qualifying exam. *Entrance requirements:* For doctorate, GRE General Test. *Application deadline:* For fall admission, 1/15 for domestic students. Application fee: $0. *Financial support:* In 2005–06, 2 research assistantships were awarded; fellowships, teaching assistantships, Federal Work-Study and institutionally sponsored loans also available. Financial award application deadline: 1/15. *Faculty research:* Biological and artificial computational devices, modeling of sensory processes and learning, theory of collective computation. *Unit head:* Dr. Pietro Perona, Executive Officer, 626-395-4867.

Carleton University, Faculty of Graduate Studies, Faculty of Arts and Social Sciences, Department of Psychology, Ottawa, ON K1S 5B6, Canada. Offers psychology (M Sc, MA, PhD), including neuroscience (M Sc). Part-time programs available. *Degree requirements:* For master's, thesis/dissertation; for doctorate, thesis/dissertation, comprehensive exam. *Entrance requirements:* For master's, honors degree; for doctorate, GRE, master's degree. Additional exam requirements/recommendations for international students: Required—TOEFL. Application fee: $75. *Financial support:* Fellowships, research assistantships, teaching assistantships, institutionally sponsored loans, scholarships/grants, and unspecified assistantships available. *Faculty research:* Behavioral neuroscience, social and personality psychology, cognitive/perception, developmental psychology, computer user research and evaluation, forensic psychology, health psychology. *Unit head:* Mary Gick, Chair, 613-520-2600 Ext. 2644, Fax: 613-520-3667. *Application contact:* Etelle Bourassa, Graduate Studies Assistant, 613-520-2600 Ext. 2644, Fax: 613-520-3667.

Carnegie Mellon University, Center for the Neural Basis of Cognition, Pittsburgh, PA 15213-3891.

See Close-Up on page 991.

Case Western Reserve University, School of Medicine and School of Graduate Studies, Graduate Programs in Medicine, Department of Neurosciences, Cleveland, OH 44106. Offers neurobiology (PhD); neuroscience (PhD). *Faculty:* 19 full-time (5 women). *Students:* 51 full-time (33 women); includes 11 minority (1 African American, 6 Asian Americans or Pacific Islanders, 4 Hispanic Americans), 15 international. Average age 30. 59 applicants, 10% accepted, 3 enrolled. In 2005, 5 doctorates awarded. *Median time to degree:* Of those who began their doctoral program in fall 1997, 100% received their degree in 8 years or less. *Degree requirements:* For doctorate, thesis/dissertation. *Entrance requirements:* For doctorate, GRE General Test. Additional exam requirements/recommendations for international students: Required—TOEFL. *Application deadline:* For fall admission, 3/15 for domestic students. Applications are processed on a rolling basis. Application fee: $50. Electronic applications accepted. *Financial support:* In 2005–06, 51 students received support, including 9 fellowships (averaging $23,000 per year), 42 research assistantships (averaging $23,000 per year); traineeships, health care benefits, and unspecified assistantships also available. Financial award application deadline: 4/1. *Faculty research:* Neurotropic factors, synapse formation, regeneration, determination of cell fate, cellular neuroscience. Total annual research expenditures: $5 million. *Unit head:* , Dr. Lynn Landmesser, Chair, 216-368-3996, Fax: 216-368-4650, E-mail: ltl@po.cwru.edu. *Application contact:* Gina Schon, Graduate Student Coordinator, 216-368-6253, Fax: 216-368-4650, E-mail: glv@po.cwru.edu.

See Close-Up on page 993.

College of Staten Island of the City University of New York, Graduate Programs, Center for Developmental Neuroscience and Developmental Disabilities, Staten Island, NY 10314-6600. Offers neuroscience (PhD); neuroscience, mental retardation and developmental disabilities (MS). *Faculty:* 2 full-time (0 women), 2 part-time/adjunct (1 woman). *Students:* Average age 34. 14 applicants, 79% accepted, 11 enrolled. In 2005, 1 degree awarded. *Degree requirements:* For master's, thesis, oral preliminary exam, thesis defense. *Entrance requirements:* For master's, GRE, General Aptitude Test and Advanced test in biology, psychology or by permission of the program coordinator in other fields, 3 letters of recommendation, minimum GPA of 3.0, 2 semesters of course work in biology and psychology, 1 semester of course work in calculus, 1 semester of course work statistics. Additional exam requirements/recommendations for international students: Required—TOEFL (minimum score 550 paper-based; 213 computer-based). *Application deadline:* Applications are processed on a rolling basis. Application fee: $125. *Expenses:* Tuition, state resident: full-time $6,400; part-time $270 per credit. Tuition, nonresident: part-time $500 per credit. Required fees: $328; $101 per semester. *Financial support:* Applicants required to submit FAFSA. *Faculty research:* Depression, 5-HT1A receptor and neuroplasticity, the 5HT1A receptor and brain development, the influence of rapid transcranial magnetic stimulation on the neurotransmission in the nervous system. Total annual research expenditures: $158,285. *Unit head:* Dr. Probal Banerjee, Head, 718-982-3938, Fax: 718-982-3944, E-mail: banerjee@mail.csi.cuny.edu. *Application contact:* Emmanuel Esperance, Deputy

Director of Office of Recruitment and Admissions, 718-982-2190, Fax: 718-982-2500, E-mail: admissions@mail.csi.cuny.edu.

Colorado State University, Graduate School, College of Natural Sciences, Department of Psychology, Fort Collins, CO 80523-0015. Offers applied social psychology (MS, PhD); behavioral neuroscience (MS, PhD); cognitive psychology (MS, PhD); counseling psychology (MS, PhD); industrial-organizational psychology (MS, PhD). *Accreditation:* APA. *Faculty:* 24 full-time (11 women). *Students:* 67 full-time (48 women), 30 part-time (27 women); includes 22 minority (1 African American, 5 American Indian/Alaska Native, 6 Asian Americans or Pacific Islanders, 10 Hispanic Americans), 3 international. Average age 28. 334 applicants, 7% accepted, 23 enrolled. In 2005, 9 master's, 13 doctorates awarded. *Median time to degree:* Of those who began their doctoral program in fall 1997, 100% received their degree in 8 years or less. *Degree requirements:* For master's, thesis/dissertation; for doctorate, thesis/dissertation, comprehensive exam, registration. *Entrance requirements:* For master's and doctorate, GRE General Test, GRE Subject Test, minimum GPA of 3.5. Additional exam requirements/recommendations for international students: Required—TOEFL. *Application deadline:* For fall admission, 12/15 for domestic students, 2/15 for international students. Application fee: $50. Electronic applications accepted. *Expenses:* Tuition, state resident: full-time $3,690; part-time $205 per credit. Tuition, nonresident: full-time $14,958; part-time $831 per credit. Required fees: $1,061. *Financial support:* In 2005–06, 10 research assistantships with full tuition reimbursements (averaging $863 per year), 52 teaching assistantships with full tuition reimbursements (averaging $863 per year) were awarded; career-related internships or fieldwork, Federal Work-Study, institutionally sponsored loans, scholarships/grants, and unspecified assistantships also available. *Faculty research:* Environmental psychology, cognitive learning, health psychology, counseling and clinical issues. Total annual research expenditures: $4.3 million. *Unit head:* Ernest L. Chavez, Chair, 970-491-6364, Fax: 970-491-1032, E-mail: ernest.chavez@colostate.edu. *Application contact:* Joanne Moran, Program Assistant I, 970-491-7298, Fax: 970-491-1032, E-mail: joanne.moran@colostate.edu.

Colorado State University, Graduate School, Program in Molecular, Cellular and Integrative Neurosciences, Fort Collins, CO 80523-0015.

Cornell University, Joan and Sanford I. Weill Medical College and Graduate School of Medical Sciences, Weill Graduate School of Medical Sciences, Program in Neuroscience, New York, NY 10021-4896. Offers PhD, MD/PhD. *Faculty:* 36 full-time (9 women). *Students:* 43 full-time (22 women); includes 8 minority (4 African Americans, 2 Asian Americans or Pacific Islanders, 2 Hispanic Americans), 12 international. Average age 22. 69 applicants. In 2005, 7 doctorates awarded. *Median time to degree:* Of those who began their doctoral program in fall 1997, 100% received their degree in 8 years or less. *Degree requirements:* For doctorate, thesis/dissertation, final exam. *Entrance requirements:* For doctorate, GRE General Test, GRE Subject Test, undergraduate training in biology, organic chemistry, physics, and mathematics. Additional exam requirements/recommendations for international students: Required—TOEFL. *Application deadline:* For fall admission, 12/15 for domestic students. Application fee: $60. *Expenses:* Tuition: Full-time $32,320. Required fees: $1,025. *Financial support:* Fellowships, stipends available. *Unit head:* , Betty J. Casey, Co-Director, 212-746-5832, E-mail: bjc2002@med.cornell.edu.

Dalhousie University, Faculty of Graduate Studies, College of Arts and Science, Faculty of Science, Department of Psychology, Halifax, NS B3H 4R2, Canada. Offers clinical psychology (PhD); psychology (M Sc, PhD); psychology/neuroscience (M Sc, PhD). *Accreditation:* APA (one or more programs are accredited). *Degree requirements:* For master's and doctorate, thesis/dissertation. *Entrance requirements:* For doctorate, GRE General Test. Additional exam requirements/recommendations for international students: Required—TOEFL. *Faculty research:* Physiological psychology, psychology of learning, learning and behavior, forensic clinical health psychology, development perception and cognition.

Dalhousie University, Faculty of Graduate Studies, Neuroscience Institute, Halifax, NS B3H 4R2, Canada. Offers M Sc, PhD. Offered in conjunction with the Departments of Anatomy and Neurobiology, Biochemistry, Pharmacology, Physiology and Biophysics, and Psychology. *Degree requirements:* For doctorate, thesis/dissertation. *Entrance requirements:* For master's and doctorate, 4 year honors degree or equivalent, minimum B+ average. Additional exam requirements/recommendations for international students: Required—TOEFL. *Faculty research:* Molecular, cellular, systems, behavioral and clinical neuroscience.

Dartmouth College, School of Arts and Sciences, Department of Psychological and Brain Sciences, Hanover, NH 03755. Offers cognitive neuroscience (PhD); psychology (PhD). *Faculty:* 18 full-time (4 women), 2 part-time/adjunct (1 woman). *Students:* 34 full-time (17 women); includes 2 minority (1 Asian American or Pacific Islander, 1 Hispanic American), 8 international. Average age 26. 95 applicants, 21% accepted, 9 enrolled. In 2005, 5 doctorates awarded. *Degree requirements:* For doctorate, thesis/dissertation. *Entrance requirements:* For doctorate, GRE General Test, GRE Subject Test. Additional exam requirements/recommendations for international students: Required—TOEFL. *Application deadline:* For fall admission, 1/15 for domestic students. Application fee: $40. *Expenses:* Tuition: Full-time $31,770. *Financial support:* In 2005–06, 31 students received support, including fellowships with full tuition reimbursements available (averaging $21,000 per year), research assistantships with full tuition reimbursements available (averaging $21,000 per year); Federal Work-Study, institutionally sponsored loans, and tuition waivers (full) also available. *Faculty research:* Behavioral neuroscience, cognitive neuroscience, cognitive science, social/personality psychology. Total annual research expenditures: $5.1 million. *Unit head:* Dr. Howard C. Hughes, Chair, 603-646-3181, Fax: 603-646-1419, E-mail: howard.hughes@dartmouth.edu. *Application contact:* Nancy Tenney, Departmental Administrative, 603-646-3181, E-mail: nancy.4.tenney@dartmouth.edu.

Dartmouth College, School of Arts and Sciences, The Neuroscience Center, Hanover, NH 03755. Offers PhD, MD/PhD. Degrees awarded through participating programs. In 2005, 3 degrees awarded. *Expenses:* Tuition: Full-time $31,770. *Application contact:* Information Contact, 603-650-8561, Fax: 603-650-8449, E-mail: NeuroscienceCenter@dartmouth.edu.

See Close-Up on page 997.

Drexel University, College of Medicine, Biomedical Graduate Programs, Program in Neuroscience, Philadelphia, PA 19104-2875. Offers PhD. *Degree requirements:* For doctorate, thesis/dissertation, qualifying exam. *Entrance requirements:* For doctorate, GRE General Test, or MCAT, minimum GPA of 2.75. Additional exam requirements/recommendations for international students: Required—TOEFL. Electronic applications accepted. *Faculty research:* Central monoamine systems, drugs of abuse, anatomy/physiology of sensory systems, neurodegenerative disorders and recovery of function, neuromodulation and synaptic plasticity.

Duke University, Graduate School, Department of Cognitive Neuroscience, Durham, NC 27708-0586. Offers PhD, Certificate. *Faculty:* 25 full-time. *Students:* 4 full-time (3 women). 76 applicants, 13% accepted, 4 enrolled. *Entrance requirements:* Additional exam requirements/recommendations for international students: Required—IELT (preferred) or TOEFL. *Application deadline:* For fall admission, 12/31 for domestic students, 12/31 for international students. *Unit head:* Marty Woldorff, Director of Graduate Studies, 919-681-0604, Fax: 919-681-0815, E-mail: woldorff@duke.edu.

Emory University, Graduate School of Arts and Sciences, Department of Psychology, Atlanta, GA 30322-1100. Offers clinical psychology (PhD); cognition and development (PhD); neuroscience and animal behavior (PhD). *Accreditation:* APA. *Faculty:* 32 full-time (13 women). *Students:* 71 full-time (53 women); includes 7 minority (4 African Americans, 2 Asian Americans or Pacific Islanders, 1 Hispanic American), 3 international. Average age 25. 312 applicants, 6% accepted, 11 enrolled. In 2005, 11 doctorates awarded. *Median time to degree:* Of those who began their doctoral program in fall 1997, 78% received their degree in 8 years or less. *Degree requirements:* For doctorate, thesis/dissertation, comprehensive exam, registration. *Entrance requirements:* For doctorate, GRE General Test, minimum GPA of 3.25. Additional exam requirements/recommendations for international students: Required—TOEFL. Applica-

Neuroscience

Emory University (continued)
tion deadline: For fall admission, 1/3 for domestic students, 1/3 for international students. Application fee: $50. Electronic applications accepted. Expenses: Tuition: Full-time $14,400. Required fees: $217. Financial support: In 2005–06, 47 students received support, including 52 fellowships with full tuition reimbursements available (averaging $16,796 per year); research assistantships, teaching assistantships, career-related internships or fieldwork, Federal Work-Study, institutionally sponsored loans, scholarships/grants, and tuition waivers also available. Financial award application deadline: 4/15. Faculty research: Neuroscience and animal behavior; adult and child psychopathology, cognition development assessment. Unit head: Dr. Elaine Walker, Chair, 404-727-7438, Fax: 404-727-0372. Application contact: Terry Legge, Graduate Program Specialist, 404-727-7456, Fax: 404-727-0372, E-mail: tlegge@emory.edu.

Emory University, Graduate School of Arts and Sciences, Division of Biological and Biomedical Sciences, Program in Neuroscience, Atlanta, GA 30322-1100. Offers PhD. Faculty: 104 full-time (15 women). Students: 90 full-time (57 women); includes 20 minority (7 African Americans, 1 American Indian/Alaska Native, 7 Asian Americans or Pacific Islanders, 5 Hispanic Americans), 8 international. Average age 27. 125 applicants, 28% accepted, 19 enrolled. In 2005, 9 degrees awarded. Median time to degree: Of those who began their doctoral program in fall 1997, 100% received their degree in 8 years or less. Degree requirements: For doctorate, thesis/dissertation, comprehensive exam, registration. Entrance requirements: For doctorate, GRE General Test, minimum GPA of 3.0 in science course work. Additional exam requirements/recommendations for international students: Required—TOEFL. Application deadline: For fall admission, 1/3 for domestic students, 1/3 for international students. Application fee: $50. Electronic applications accepted. Expenses: Tuition: Full-time $14,400. Required fees: $217. Financial support: In 2005–06, 30 students received support, including 30 fellowships with full tuition reimbursements available (averaging $23,000 per year); institutionally sponsored loans, health care benefits, and tuition waivers (full) also available. Faculty research: Cell and molecular biology, development, behavior, neurodegenerative disease. Unit head: Dr. Yoland Smith, Director, 404-727-7519, Fax: 404-727-3278, E-mail: yolands@rmy.emory.edu. Application contact: 404-727-2545, Fax: 404-727-3322, E-mail: gdbbs@emory.edu.

Florida Atlantic University, Charles E. Schmidt College of Science, Center for Complex Systems and Brain Sciences, Boca Raton, FL 33431-0991. Offers PhD. Faculty: 15 full-time (1 woman). Students: 20 full-time (5 women), 3 part-time (1 woman); includes 2 minority (1 American Indian/Alaska Native, 1 Hispanic American), 9 international. Average age 31. 11 applicants, 18% accepted, 2 enrolled. Degree requirements: For doctorate, thesis/dissertation. Entrance requirements: For doctorate, GRE General Test, minimum GPA of 3.0 in last 60 hours of undergraduate course work. Additional exam requirements/recommendations for international students: Required—TOEFL. Application deadline: For fall admission, 2/15 priority date for domestic students, 2/15 priority date for international students; for spring admission, 11/1 for domestic students, 8/15 for international students. Application fee: $30. Expenses: Tuition, state resident: full-time $4,394; part-time $244 per credit. Tuition, nonresident: full-time $16,441; part-time $912 per credit. Financial support: In 2005–06, 2 fellowships with full tuition reimbursements (averaging $16,060 per year), 5 research assistantships with partial tuition reimbursements (averaging $16,060 per year), 9 teaching assistantships with partial tuition reimbursements (averaging $16,060 per year) were awarded; Federal Work-Study, traineeships, and unspecified assistantships also available. Faculty research: Motor behavior, speech perception, nonlinear dynamics and fractals, behavioral neuroscience, cellular and molecular neuroscience. Total annual research expenditures: $1.5 million. Unit head: Dr. J. A. Scott Kelso, Director, 561-297-2230, Fax: 561-297-3634, E-mail: kelso@ccs.fau.edu. Application contact: Dr. Betty Tuller, Associate Director, 561-297-2227, Fax: 561-297-3634, E-mail: tuller@ccs.fau.edu.

Florida State University, Graduate Studies, College of Arts and Sciences, Department of Biological Science and Department of Psychology and Department of Nutrition, Food, and Exercise Sciences, Program in Neuroscience, Tallahassee, FL 32306. Offers PhD. Faculty: 24 full-time (6 women). Students: 37 full-time (22 women); includes 3 minority (2 African Americans, 1 Hispanic American), 2 international. Average age 24. 56 applicants, 21% accepted, 12 enrolled. In 2005, 3 degrees awarded. Median time to degree: Of those who began their doctoral program in fall 1997, 100% received their degree in 8 years or less. Degree requirements: For doctorate, thesis/dissertation, teaching experience, comprehensive exam, registration. Entrance requirements: For doctorate, GRE General Test. Additional exam requirements/recommendations for international students: Required—TOEFL (minimum score 600 paper-based; 250 computer-based). Application deadline: For fall admission, 12/15 for domestic students, 12/15 for international students. Application fee: $30. Financial support: In 2005–06, fellowships with full tuition reimbursements (averaging $16,500 per year), research assistantships with full tuition reimbursements (averaging $18,000 per year), teaching assistantships with full tuition reimbursements (averaging $18,000 per year) were awarded; scholarships/grants, traineeships, and unspecified assistantships also available. Financial award application deadline: 12/15; financial award applicants required to submit FAFSA. Faculty research: Behavioral ecology, neurophysiology, sensory physiology, neuroendocrinology, neurotransmission. Unit head: Dr. Robert Contreras, Director, 850-644-1751, Fax: 850-644-7739, E-mail: contreras@psy.fsu.edu. Application contact: Janice Parker, Office Manager, 850-644-3076, Fax: 850-644-0349, E-mail: parker@neuro.fsu.edu.

Florida State University, Graduate Studies, College of Arts and Sciences, Department of Psychology, Interdisciplinary Program in Neuroscience, Tallahassee, FL 32306. Offers PhD. Faculty: 10 full-time (4 women). Students: 22 full-time (12 women); includes 3 minority (1 African American, 2 Hispanic Americans). Average age 24. 26 applicants, 35% accepted, 9 enrolled. In 2005, 3 doctorates awarded. Median time to degree: Of those who began their doctoral program in fall 1997, 100% received their degree in 8 years or less. Degree requirements: For doctorate, thesis/dissertation, preliminary exam. Entrance requirements: For doctorate, GRE General Test (minimum 1100: verbal 500, quantitative 500), minimum GPA of 3.0, research experience, letters of recommendation. Additional exam requirements/recommendations for international students: Required—TOEFL (minimum score 550 paper-based; 213 computer-based). Application deadline: For fall admission, 12/15 for domestic students, 12/15 for international students. Application fee: $30. Electronic applications accepted. Financial support: In 2005–06, 2 fellowships with full tuition reimbursements (averaging $18,000 per year), 17 research assistantships with full tuition reimbursements (averaging $20,700 per year), 1 teaching assistantship with full tuition reimbursement (averaging $17,000 per year) were awarded; Federal Work-Study, institutionally sponsored loans, scholarships/grants, traineeships, and unspecified assistantships also available. Financial award applicants required to submit FAFSA. Faculty research: Sensory processes, neural development and plasticity, circadian rhythms, behavioral and molecular genetics, hormonal control of behavior. Total annual research expenditures: $1.6 million. Unit head: Dr. Robert Contreras, Director, 850-644-1751, Fax: 850-644-7739, E-mail: contreras@psy.fsu.edu. Application contact: Cherie P. Miller, Graduate Program Assistant, 850-644-2499, Fax: 850-644-7739, E-mail: grad-info@psy.fsu.edu.

George Mason University, College of Science, Fairfax, VA 22030. Offers bioinformatics (MS, PhD); climate dynamics (PhD); computational sciences (MS); computational sciences and informatics (PhD); computational social science (PhD); computational techniques and applications (Certificate); earth systems and geoinformation science (PhD); earth systems science (MS); nanotechnology and nanoscience (Certificate); neuroscience (PhD); physical sciences (PhD); remote sensing and earth image processing (Certificate). Part-time and evening/weekend programs available. Degree requirements: For doctorate, thesis/dissertation, comprehensive exam, registration. Entrance requirements: For master's and doctorate, GRE General Test, minimum GPA of 3.0 in last 60 hours. Additional exam requirements/recommendations for international students: Required—TOEFL. Electronic applications accepted. Expenses: Tuition, area resident: Full-time $5,244; part-time $219 per credit. Tuition, state resident: part-time $651 per credit. Tuition, nonresident: full-time $15,636. Required fees:

$1,524; $65 per credit. Faculty research: Space sciences and astrophysics, fluid dynamics, materials modeling and simulation, bioinformatics, global changes and statistics.

Georgetown University, Graduate School of Arts and Sciences, Programs in Biomedical Sciences, Program in Neuroscience, Washington, DC 20057. Offers PhD, MD/PhD. Degree requirements: For doctorate, thesis/dissertation. Entrance requirements: For doctorate, GRE General Test. Additional exam requirements/recommendations for international students: Required—TOEFL.

The George Washington University, Columbian College of Arts and Sciences, Institute for Biomedical Sciences, Program in Neuroscience, Washington, DC 20052. Offers PhD. Degree requirements: For doctorate, thesis/dissertation, general exam. Entrance requirements: For doctorate, GRE General Test, interview, minimum GPA of 3.0. Additional exam requirements/recommendations for international students: Required—TOEFL (minimum score 600 paper-based; 250 computer-based). Electronic applications accepted. Faculty research: Cerebral ischemia, neural transplantation, molecular mechanisms of action, neurotransmitter systems, psychobiology of learning and memory.

Graduate School and University Center of the City University of New York, Graduate Studies, Program in Psychology, New York, NY 10016-4039. Offers basic applied neurocognition (PhD); biopsychology (PhD); clinical psychology (PhD); developmental psychology (PhD); environmental psychology (PhD); experimental psychology (PhD); industrial psychology (PhD); learning processes (PhD); neuropsychology (PhD); psychology (PhD); social personality (PhD). Faculty: 119 full-time (40 women). Students: 488 full-time (361 women), 2 part-time (1 woman); includes 91 minority (34 African Americans, 1 American Indian/Alaska Native, 22 Asian Americans or Pacific Islanders, 34 Hispanic Americans), 58 international. Average age 32. 773 applicants, 18% accepted, 80 enrolled. In 2005, 50 degrees awarded. Degree requirements: For doctorate, one foreign language, thesis/dissertation. Entrance requirements: For doctorate, GRE General Test. Additional exam requirements/recommendations for international students: Required—TOEFL. Application deadline: For fall admission, 1/1 for domestic students. Application fee: $125. Electronic applications accepted. Financial support: In 2005–06, 272 students received support, including 190 fellowships, 34 research assistantships, 33 teaching assistantships; career-related internships or fieldwork, Federal Work-Study, institutionally sponsored loans, and tuition waivers (full and partial) also available. Financial award application deadline: 2/1; financial award applicants required to submit FAFSA. Unit head: Dr. Joseph Glick, Executive Officer, 212-817-8706, Fax: 212-817-1533, E-mail: jglick@gc.cuny.edu.

Harvard University, Graduate School of Arts and Sciences, Program in Neuroscience, Boston, MA 02115. Offers neurobiology (PhD). Degree requirements: For doctorate, thesis/dissertation, qualifying exam. Entrance requirements: For doctorate, GRE General Test, GRE Subject Test. Additional exam requirements/recommendations for international students: Required—TOEFL. Application deadline: For fall admission, 12/15 for domestic students. Expenses: Tuition: Full-time $28,752. Full-time tuition and fees vary according to program and student level. Financial support: Fellowships, research assistantships, teaching assistantships, institutionally sponsored loans, scholarships/grants, tuition waivers (full), and stipends available. Financial award application deadline: 1/1. Faculty research: Relationship between diseases of the nervous system and basic science. Unit head: Dr. Jonathan Cohen, Chair, 617-432-1728. Application contact: Dr. Jonathan Cohen, Chair, 617-432-1728.

See Close-Up on page 1001.

Indiana University Bloomington, Graduate School, College of Arts and Sciences, Program in Neuroscience, Bloomington, IN 47405-7000. Offers PhD. Offered through the University Graduate School. Students: 6 full-time (1 woman), 6 part-time (2 women), 2 international. Average age 30. In 2005, 1 degree awarded. Degree requirements: For doctorate, thesis/dissertation, oral exam. Entrance requirements: For doctorate, GRE General Test. Additional exam requirements/recommendations for international students: Required—TOEFL. Application deadline: For fall admission, 1/15 for domestic students; for spring admission, 9/1 priority date for domestic students. Applications are processed on a rolling basis. Application fee: $50 ($60 for international students). Electronic applications accepted. Expenses: Tuition, state resident: full-time $5,437; part-time $227 per credit hour. Tuition, nonresident: full-time $15,836; part-time $660 per credit hour. Required fees: $821. Tuition and fees vary according to campus/location and program. Financial support: Fellowships with tuition reimbursements, teaching assistantships with full tuition reimbursements, institutionally sponsored loans available. Financial award application deadline: 1/15. Faculty research: Synaptic transmitter systems, neurophysiology, neuropharmacology, neurochemistry, somatosensory and sensorimotor functions. Unit head: Dr. J. Michael Walker, Director, 812-855-7756. Application contact: Faye Caylor, Administrative Assistant, 812-855-7756, Fax: 812-855-4520, E-mail: fcaylor@indiana.edu.

Iowa State University of Science and Technology, Graduate College, Interdisciplinary Programs, Program in Neuroscience, Ames, IA 50011. Offers MS, PhD. Students: 24 full-time (12 women), 3 part-time (1 woman); includes 2 minority (both Asian Americans or Pacific Islanders), 11 international. 10 applicants, 10% accepted, 0 enrolled. In 2005, 1 doctorate awarded. Terminal master's awarded for partial completion of doctoral program. Degree requirements: For master's and doctorate, thesis/dissertation. Entrance requirements: For master's and doctorate, GRE General Test, resumé. Additional exam requirements/recommendations for international students: Required—TOEFL (paper score 550; computer score 220) or IELTS (score 6.5). Application deadline: For fall admission, 1/15 priority date for domestic students, 1/15 priority date for international students. Application fee: $30 ($70 for international students). Expenses: Tuition, state resident: full-time $6,410. Tuition, nonresident: full-time $16,422. Tuition and fees vary according to program. Financial support: In 2005–06, 18 research assistantships with full and partial tuition reimbursements (averaging $22,000 per year), 6 teaching assistantships with full and partial tuition reimbursements were awarded; scholarships/grants, health care benefits, and unspecified assistantships also available. Faculty research: Behavioral pharmacology and · immunology, developmental neurobiology, neuroendocrinology, neuroregulatory mechanisms at the cellular level, signal transduction in neurons. Unit head: Dr. Timothy Day, Chair, Supervising Committee, 515-294-7252, E-mail: idgp@iastate.edu. Application contact: Mary Jordison, Information Contact, 515-294-7252, Fax: 515-924-6790, E-mail: idgp@iastate.edu.

Announcement: The neuroscience graduate program involves 31 faculty members across 9 disciplines. The program's philosophy is based on the combination of mentoring, curriculum, and collaborative research. Research areas include neurobiology, neural degenerative diseases, neural stem cell research, animal models, learning and memory, and neuromuscular system research. World Wide Web: http://www.neuroscience.iastate.edu.

The Johns Hopkins University, School of Medicine, Graduate Programs in Medicine, Department of Neuroscience, Baltimore, MD 21218-2699. Offers PhD. Faculty: 97 full-time (18 women). Students: 82 full-time (30 women); includes 30 minority (2 African Americans, 27 Asian Americans or Pacific Islanders, 1 Hispanic American), 29 international. Average age 28. 187 applicants, 13% accepted, 9 enrolled. In 2005, 13 doctorates awarded. Degree requirements: For doctorate, thesis/dissertation, thesis defense, comprehensive exam, registration. Entrance requirements: For doctorate, GRE General Test, GRE Subject Test, bachelor's degree in science or mathematics. Additional exam requirements/recommendations for international students: Required—TOEFL. Application deadline: For winter admission, 12/15 for domestic students. Application fee: $75. Expenses: Tuition: Full-time $30,960. Tuition and fees vary according to degree level and program. Financial support: In 2005–06, 78 fellowships (averaging $24,000 per year) were awarded; tuition waivers (full) also available. Financial award application deadline: 1/1. Faculty research: Neurophysiology, neurochemistry, neuroanatomy, pharmacology, development. Unit head: Dr. Solomon H. Snyder, Chairman, 410-955-3024, Fax: 410-955-3623. Application contact: Dr. David J. Linden, Director of Graduate Studies, 410-955-7947, Fax: 410-955-3623, E-mail: dlinden@jhmi.edu.

Neuroscience

Kent State University, School of Biomedical Sciences, Program in Neuroscience, Kent, OH 44242-0001. Offers MS, PhD. Offered in cooperation with Northeastern Ohio Universities College of Medicine. Terminal master's awarded for partial completion of doctoral program. *Degree requirements:* For master's and doctorate, thesis/dissertation. *Entrance requirements:* For master's and doctorate, GRE General Test. *Faculty research:* Plasticity of the nervous system, learning and memory processes–neural correlates, neuroendocrinology of cyclic behavior, synaptic neurochemistry.

Louisiana State University Health Sciences Center, School of Graduate Studies in New Orleans, Interdisciplinary Neuroscience PhD Training Program, New Orleans, LA 70112-2223. Offers PhD, MD/PhD. *Degree requirements:* For doctorate, thesis/dissertation. *Entrance requirements:* For doctorate, GRE General Test, GRE Subject Test, previous course work in chemistry, mathematics, physics, and computer science. Additional exam requirements/recommendations for international students: Required—TOEFL. *Faculty research:* Visual system, second messengers, drugs and behavior, signal transduction, plasticity and development.

See Close-Up on page 1003.

Loyola University Chicago, Graduate School, Program in Neuroscience, Maywood, IL 60153. Offers MS, PhD, MD/PhD. *Faculty:* 32 full-time (9 women). *Students:* 19 full-time (12 women), 1 (woman) part-time, 8 international. Average age 29. 25 applicants, 12% accepted, 3 enrolled. Terminal master's awarded for partial completion of doctoral program. *Median time to degree:* Of those who began their doctoral program in fall 1997, 20% received their degree in 8 years or less. *Degree requirements:* For master's and doctorate, thesis/dissertation, comprehensive exam. *Entrance requirements:* For master's, GRE or MCAT; for doctorate, GRE General Test. Additional exam requirements/recommendations for international students: Required—TOEFL. *Application deadline:* For fall admission, 3/15 priority date for domestic students, 3/15 priority date for international students. Applications are processed on a rolling basis. Application fee: $0. *Expenses:* Tuition: Full-time $11,610; part-time $645 per credit. Required fees: $55 per semester. *Financial support:* In 2005–06, 20 students received support, including 6 fellowships with full tuition reimbursements available (averaging $22,000 per year); Federal Work-Study also available. Financial award application deadline: 3/15; financial award applicants required to submit FAFSA. *Faculty research:* Alzheimer's and Parkinson's disease, drugs of abuse, neuroendocrinology, neuroimmunology, brain cancer. Total annual research expenditures: $3.5 million. *Unit head:* Dr. Edward J. Neafsey, Director, 708-216-3355, Fax: 708-216-6823, E-mail: eneafse@lumc.edu. *Application contact:* Peggy Richied, Administrative Secretary, 708-216-4841, Fax: 708-216-6823, E-mail: prichie@lumc.edu.

Massachusetts Institute of Technology, School of Science, Department of Brain and Cognitive Sciences, Cambridge, MA 02139-4307. Offers cognitive science (PhD); neuroscience (PhD). *Faculty:* 33 full-time (8 women). *Students:* 79 full-time (34 women); includes 21 minority (5 African Americans, 1 American Indian/Alaska Native, 11 Asian Americans or Pacific Islanders, 4 Hispanic Americans), 12 international. Average age 26. 313 applicants, 8% accepted, 12 enrolled. In 2005, 10 degrees awarded. *Degree requirements:* For doctorate, thesis/dissertation, comprehensive exam. *Entrance requirements:* For doctorate, GRE General Test. Additional exam requirements/recommendations for international students: Required—TOEFL (minimum score 577 paper-based; 233 computer-based). *Application deadline:* For fall admission, 12/15 for domestic students, 12/15 for international students. Application fee: $70. Electronic applications accepted. *Expenses:* Tuition: Full-time $32,100. Required fees: $200. Part-time tuition and fees vary according to course load. *Financial support:* In 2005–06, 76 students received support, including 53 fellowships with tuition reimbursements available (averaging $23,817 per year), 22 research assistantships with tuition reimbursements available (averaging $24,476 per year), 2 teaching assistantships with tuition reimbursements available (averaging $24,600 per year); Federal Work-Study, institutionally sponsored loans, scholarships/grants, traineeships, health care benefits, and unspecified assistantships also available. *Faculty research:* Vision, learning and memory, motor control, plasticity. Total annual research expenditures: $18.5 million. *Unit head:* Prof. Mriganka Sur, Department Head, 617-253-8784, E-mail: bcs-info@mit.edu. *Application contact:* Brain and Cognitive Sciences Academic, 617-253-7403, E-mail: bcs-admissions@mit.edu.

See Close-Up on page 1005.

Mayo Graduate School, Graduate Programs in Biomedical Sciences, Program in Molecular Neuroscience, Rochester, MN 55905. Offers PhD. Program also offered in Jacksonville, FL. *Degree requirements:* For doctorate, oral defense of dissertation, qualifying oral and written exam. *Entrance requirements:* For doctorate, GRE, 1 year of chemistry, biology, calculus, and physics. Additional exam requirements/recommendations for international students: Required—TOEFL. Electronic applications accepted. *Faculty research:* Cholinergic receptor/Alzheimer's; molecular biology, channels, receptors, and mental disease; neuronal cytoskeleton; growth factors; gene regulation.

McGill University, Faculty of Graduate and Postdoctoral Studies, Faculty of Medicine, Department of Neurology and Neurosurgery, Montréal, QC H3A 2T5, Canada. Offers M Sc, PhD. *Degree requirements:* For master's and doctorate, thesis/dissertation. *Entrance requirements:* For master's, B Sc; for doctorate, M Sc or MD. *Faculty research:* Neuroimmunology, molecular neuroscience, CNS regeneration, neuroanatomy, neuropsychology.

McMaster University, Faculty of Health Sciences and School of Graduate Studies, Program in Medical Sciences, Neurosciences and Behavioral Sciences Area, Hamilton, ON L8S 4M2, Canada. Offers M Sc, PhD. *Students:* 36 full-time, 1 part-time. In 2005, 6 master's, 2 doctorates awarded. *Degree requirements:* For master's, thesis/dissertation; for doctorate, thesis/dissertation, comprehensive exam. *Entrance requirements:* For master's, honors B Sc, B+ average in related field; for doctorate, M Sc, minimum B+ average, students with proven research experience and an A average may be admitted with a B Sc degree. Additional exam requirements/recommendations for international students: Required—TOEFL (minimum score 580 paper-based; 237 computer-based). *Application deadline:* For fall admission, 3/31 for domestic students. Applications are processed on a rolling basis. Application fee: $85. *Financial support:* Teaching assistantships available. *Unit head:* Dr. Laurie Doering, Coordinator, 905-525-9140 Ext. 22913. *Application contact:* Dr. Carl Richards, Associate Dean, 905-525-9140 Ext. 22983, Fax: 905-546-1129.

Medical College of Georgia, School of Graduate Studies, Program in Neuroscience, Augusta, GA 30912. Offers PhD. *Students:* 7 full-time (3 women); includes 1 Hispanic American, 2 international. *Financial support:* In 2005–06, 7 research assistantships with partial tuition reimbursements (averaging $22,500 per year) were awarded *Faculty research:* Learning and memory, neuronal migration, development of dendrites, synapse formation, regeneration. *Unit head:* Dr. Robert Yu, Director of Institute of Neuroscience, 706-721-0699, Fax: 706-721-7915, E-mail: ryu@mail.mcg.edu. *Application contact:* Dr. Deborah Lewis, Program Director, 706-721-6345, E-mail: dlewis@mail.mcg.edu.

Medical University of South Carolina, College of Graduate Studies, Program in Neurosciences, Charleston, SC 29425-0002. Offers Pharm D, MS, PhD, DMD/PhD, MD/PhD. *Faculty:* 25 full-time (6 women). *Students:* 27 full-time (14 women); includes 1 minority (Hispanic American), 3 international. Average age 28. 181 applicants, 34% accepted, 43 enrolled. In 2005, 2 degrees awarded. Terminal master's awarded for partial completion of doctoral program. *Degree requirements:* For master's, thesis, research seminar; for doctorate, thesis/dissertation, teaching and research seminar, oral and written exams. *Entrance requirements:* For master's and doctorate, GRE General Test, interview. Additional exam requirements/recommendations for international students: Required—TOEFL (minimum score 600 paper-based; 250 computer-based). *Application deadline:* For fall admission, 1/15 priority date for domestic students, 1/15 priority date for international students. Applications are processed on a rolling basis. Application fee: $0 ($75 for international students). Electronic applications accepted. *Financial support:* In 2005–06, 6 students received support, including fellowships with partial tuition reimbursements available (averaging $21,000 per year); Federal Work-Study and scholarships/grants also available. Financial award application deadline: 3/15; financial award applicants required

to submit FAFSA. *Faculty research:* Addiction, aging, movement disorders, membrane physiology, neurotransmission and behavior. *Unit head:* Dr. Peter Kalivas, Chair, 843-792-4400, Fax: 843-792-6590, E-mail: kalivasp@musc.edu. *Application contact:* Cheryl Brown, 843-792-4620, Fax: 843-792-4645, E-mail: brownche@musc.edu.

Michigan State University, The Graduate School, College of Natural Science, Program in Neuroscience, East Lansing, MI 48824. Offers MS, PhD. *Faculty:* 38 full-time (15 women). *Students:* 26 full-time (16 women), 1 (woman) part-time; includes 6 minority (1 African American, 3 Asian Americans or Pacific Islanders, 2 Hispanic Americans), 4 international. Average age 27. 39 applicants, 8% accepted. In 2005, 1 degree awarded. *Degree requirements:* For doctorate, thesis/dissertation, laboratory rotation, defense of dissertation, comprehensive exam. *Entrance requirements:* For doctorate, GRE General Test, minimum GPA of 3.0, 3 letters of recommendation, bachelor's degree in a biological/physical science or related field. Additional exam requirements/recommendations for international students: Required—TOEFL (minimum score 550 paper-based; 213 computer-based), Michigan State University ELT (85), Michigan ELAB (83). *Application deadline:* For fall admission, 12/27 for domestic students. Application fee: $50. Electronic applications accepted. *Expenses:* Tuition, state resident: part-time $330 per credit hour. Tuition, nonresident: part-time $685 per credit hour. Tuition and fees vary according to program. *Financial support:* In 2005–06, 16 fellowships with tuition reimbursements (averaging $10,730 per year), 11 research assistantships with tuition reimbursements (averaging $15,480 per year) were awarded; scholarships/grants and unspecified assistantships also available. *Faculty research:* Autonomic nervous system function, neural development and plasticity, neuroendocrine mechanisms of behavior, neurodegenerative and neuromuscular disease, synaptic transmission/signal transduction. Total annual research expenditures: $881,598. *Unit head:* Dr. Cheryl L. Sisk, Director, 517-353-8947, Fax: 517-432-2744, E-mail: sisk@msu.edu. *Application contact:* Adrianna Felpausch, Graduate Admissions Associate, 517-353-8947, Fax: 517-432-2744, E-mail: feldpa15@msu.edu.

Montana State University, College of Graduate Studies, College of Letters and Science, Department of Cell Biology and Neuroscience, Bozeman, MT 59717. Offers neuroscience (MS, PhD). Part-time programs available. *Faculty:* 7 full-time (3 women). *Students:* 2 full-time (1 woman), 9 part-time (5 women), 2 international. Average age 28. 7 applicants, 43% accepted, 3 enrolled. In 2005, 3 degrees awarded. *Degree requirements:* For master's, comprehensive exam, registration; for doctorate, thesis/dissertation, comprehensive exam, registration. *Entrance requirements:* For master's and doctorate, GRE General Test. Additional exam requirements/recommendations for international students: Required—TOEFL (minimum score 550 paper-based; 213 computer-based). *Application deadline:* For fall admission, 7/15 priority date for domestic students, 5/15 priority date for international students; for spring admission, 12/1 priority date for domestic students, 10/1 priority date for international students. Applications are processed on a rolling basis. Application fee: $30. Electronic applications accepted. *Expenses:* Tuition, state resident: full-time $4,132. Tuition, nonresident: full-time $11,332. *Financial support:* In 2005–06, 3 fellowships with full tuition reimbursements (averaging $20,000 per year), 8 research assistantships with full tuition reimbursements (averaging $15,000 per year), 2 teaching assistantships with full tuition reimbursements (averaging $10,000 per year) were awarded; scholarships/grants, traineeships, and unspecified assistantships also available. Financial award application deadline: 3/1; financial award applicants required to submit FAFSA. *Faculty research:* Neural development, axonal sprouting, visual processing, neural information coding, membrane signaling proteins. Total annual research expenditures: $7.2 million. *Unit head:* Dr. Gwen Jacobs, Department Head, 406-994-5120, Fax: 406-994-7077, E-mail: gwen@cns.montana.edu.

Mount Sinai School of Medicine of New York University, Graduate School of Biological Sciences, New York, NY 10029-6504. Offers biophysics, structural biology and biomathematics (PhD); community medicine (MPH); genetic counseling (MS); genetics and genomic sciences (PhD); mechanisms of disease and therapy (PhD); microbiology (PhD); molecular, cellular, biochemical and developmental sciences (PhD); neurosciences (PhD). *Students:* 218 full-time (109 women). 4,208 applicants, 7% accepted, 117 enrolled. Terminal master's awarded for partial completion of doctoral program. *Degree requirements:* For master's, registration; for doctorate, thesis/dissertation, registration. *Entrance requirements:* For doctorate, GRE General Test, GRE Subject Test, 3 years of college pre-med course work. Additional exam requirements/recommendations for international students: Required—TOEFL. *Application deadline:* For fall admission, 1/15 for domestic students. Application fee: $75. Electronic applications accepted. *Expenses:* Tuition: Full-time $33,250. Required fees: $1,600. Full-time tuition and fees vary according to degree level, program and reciprocity agreements. *Financial support:* In 2005–06, fellowships with full tuition reimbursements (averaging $26,000 per year), research assistantships with full tuition reimbursements (averaging $26,000 per year) were awarded; Federal Work-Study, institutionally sponsored loans, scholarships/grants, health care benefits, and unspecified assistantships also available. Financial award application deadline: 4/5; financial award applicants required to submit FAFSA. *Faculty research:* Cancer, gene therapy, minimally invasive surgery, cardiac translational research. Total annual research expenditures: $162.2 million. *Unit head:* Dr. Diomedes Logothetis, Dean, 212-241-6546, Fax: 212-241-0651, E-mail: diomedes.logothetis@mssm.edu. *Application contact:* Lily Recanati, Manager, 212-241-3267, Fax: 212-241-0651, E-mail: lily.recantati@mssm.edu.

See Close-Up on page 177.

New York Medical College, Graduate School of Basic Medical Sciences, Department of Cell Biology, Valhalla, NY 10595-1691. Offers cell biology and neuroscience (MS, PhD). Part-time and evening/weekend programs available. Terminal master's awarded for partial completion of doctoral program. *Degree requirements:* For master's, thesis/dissertation; for doctorate, thesis/dissertation, comprehensive exam. *Entrance requirements:* For master's and doctorate, GRE General Test. Additional exam requirements/recommendations for international students: Required—TOEFL. *Faculty research:* Mechanisms of growth control in skeletal muscle, cartilage differentiation, cytoskeletal functions, signal transduction pathways, neuronal development and plasticity.

See Close-Up on page 577.

New York University, Graduate School of Arts and Science, Center for Neural Science, New York, NY 10012-1019. Offers PhD. *Faculty:* 15 full-time (3 women), 4 part-time/adjunct. *Students:* 36 full-time (15 women); includes 7 minority (1 African American, 1 American Indian/Alaska Native, 3 Asian Americans or Pacific Islanders, 2 Hispanic Americans), 11 international. Average age 28. 120 applicants, 17% accepted, 6 enrolled. In 2005, 5 degrees awarded. *Degree requirements:* For doctorate, one foreign language, thesis/dissertation. *Entrance requirements:* For doctorate, GRE, interview. Additional exam requirements/recommendations for international students: Required—TOEFL. *Application deadline:* For fall admission, 1/4 for domestic students. Application fee: $80. *Financial support:* Fellowships with tuition reimbursements, research assistantships with tuition reimbursements, career-related internships or fieldwork, Federal Work-Study, scholarships/grants, and health care benefits available. Financial award application deadline: 1/4; financial award applicants required to submit FAFSA. *Faculty research:* Systems and integrative neuroscience; combining biology, cognition, computation, and theory. *Unit head:* J. Anthony Movshon, Chair, 212-998-7780, Fax: 212-995-4011, E-mail: cns@nyu.edu. *Application contact:* Lynne Kiorpes, Director of Graduate Studies, 212-998-7780, Fax: 212-995-4011, E-mail: cns@nyu.edu.

See Close-Up on page 1007.

New York University, School of Medicine and Graduate School of Arts and Science, Sackler Institute of Graduate Biomedical Sciences, Department of Neuroscience and Physiology, New York, NY 10012-1019. Offers neuroscience (PhD); physiology (PhD). *Degree requirements:* For doctorate, one foreign language, thesis/dissertation, qualifying exam, comprehensive exam. *Entrance requirements:* For doctorate, GRE General Test. Additional exam requirements/recommendations for international students: Required—TOEFL. *Faculty research:* Synaptic transmission, retinal physiology, signal transduction, CNS intrinsic properties, cerebellar function.

Neuroscience

Northwestern University, The Graduate School, Institute for Neuroscience, Evanston, IL 60208. Offers PhD. Admissions and degree offered through The Graduate School. *Faculty:* 150 full-time (37 women). *Students:* 111 (63 women); includes 20 minority (3 African Americans, 11 Asian Americans or Pacific Islanders, 6 Hispanic Americans) 25 international. 169 applicants, 35% accepted, 23 enrolled. In 2005, 13 degrees awarded. *Degree requirements:* For doctorate, thesis/dissertation. *Entrance requirements:* For doctorate, GRE General Test. Additional exam requirements/recommendations for international students: Required—TOEFL. *Application deadline:* For fall admission, 12/30 priority date for domestic students, 12/30 priority date for international students. Application fee: $65 ($75 for international students). *Financial support:* In 2005–06, 16 fellowships with full tuition reimbursements (averaging $22,500 per year), 35 research assistantships with full tuition reimbursements (averaging $22,500 per year), 10 teaching assistantships with full tuition reimbursements (averaging $22,500 per year) were awarded; scholarships/grants, traineeships, and tuition waivers (full) also available. Financial award application deadline: 12/31; financial award applicants required to submit FAFSA. *Faculty research:* Circadian rhythms, synaptic neurotransmissions, cognitive neuroscience, sensory/motor systems, cell biology and structure/function, neurobiology of disease. Total annual research expenditures: $1.4 million. *Unit head:* Enrico Mugnaini, Director, 312-503-4333, Fax: 312-503-7345, E-mail: e-mugnaini@northwestern.edu. *Application contact:* Robert Harper-Mangels, Assistant Director, NUIN, 312-503-1873, Fax: 312-503-7345, E-mail: r-mangels@northwestern.edu.

Announcement: One of the nation's largest neuroscience programs. More than 130 faculty members, internationally known for state-of-the-art research, provide doctoral training noted for its depth and breadth. Interdisciplinary research training is available in all areas of basic and clinical neuroscience.

See Close-Up on page 1009.

Northwestern University, The Graduate School and Judd A. and Marjorie Weinberg College of Arts and Sciences, Interdepartmental Biological Sciences Program (IBiS), Evanston, IL 60208. Offers biochemistry, molecular biology, and cell biology (PhD), including biochemistry, cell and molecular biology, molecular biophysics, structural biology; biotechnology (PhD); cell and molecular biology (PhD); developmental biology and genetics (PhD); hormone action and signal transduction (PhD); neuroscience (PhD); structural biology, biochemistry, and biophysics (PhD). Participants in the Interdepartmental Biological Sciences Program include the Departments of Biochemistry, Molecular Biology, and Cell Biology; Chemistry; Neurobiology and Physiology; Chemical Engineering; Civil Engineering; and Evanston Hospital. *Entrance requirements:* For doctorate, GRE General Test. Additional exam requirements/recommendations for international students: Required—TOEFL (minimum score 600 paper-based), TSE(minimum score 50). Electronic applications accepted. *Faculty research:* Developmental genetics, gene regulation, DNA-protein interactions, biological clocks, bioremediation.

See Close-Up on page 187.

The Ohio State University, College of Medicine and Public Health and Graduate School, Graduate Programs in the Basic Medical Sciences, Neuroscience Graduate Studies Program, Columbus, OH 43210. Offers PhD. *Degree requirements:* For doctorate, thesis/dissertation, registration. *Entrance requirements:* For doctorate, GRE General Test, GRE Subject Test. Additional exam requirements/recommendations for international students: Required—TOEFL (minimum score 600 paper-based; 250 computer-based). Electronic applications accepted. *Faculty research:* Spinal cord injury, behavioral neuroscience, molecular cellular neurobiology, neurophysiology, neuroimmunology.

Ohio University, Graduate Studies, College of Arts and Sciences, Department of Biological Sciences, Program in Neuroscience, Athens, OH 45701-2979. Offers MS, PhD. *Degree requirements:* For master's and doctorate, thesis/dissertation, comprehensive exam, registration. *Entrance requirements:* For master's, GRE General Test, minimum GPA of 3.2 for last degree earned; for doctorate, GRE General Test, minimum GPA of 3.2 for last degree earned. Additional exam requirements/recommendations for international students: Required—TOEFL (minimum score 620 paper-based; 260 computer-based). *Faculty research:* Computational neuroscience, neurobiology of aging, vestibular neuroscience, ion channels and electrogenic pumps.

Oregon Health & Science University, School of Medicine, Graduate Programs in Medicine, Department of Behavioral Neuroscience, Portland, OR 97239-3098. Offers MS, PhD, MD/PhD. *Degree requirements:* For master's, thesis; for doctorate, thesis/dissertation, written exam. *Entrance requirements:* For master's and doctorate, GRE General Test. *Faculty research:* Neural basis of behavior, behavioral pharmacology, behavioral genetics, neuropharmacology and neuroendocrinology, biological basis of drug seeking and addiction.

See Close-Up on page 1011.

Oregon Health & Science University, School of Medicine, Graduate Programs in Medicine, Neuroscience Graduate Program, Portland, OR 97239-3098. Offers PhD, MD/PhD. *Degree requirements:* For doctorate, thesis/dissertation. *Entrance requirements:* For doctorate, GRE General Test. Additional exam requirements/recommendations for international students: Required—TOEFL. *Faculty research:* Signal transduction, receptors and ion channels, transcriptional regulation, drug abuse, systems, behavioral neuroscience.

See Close-Up on page 1013.

The Pennsylvania State University Milton S. Hershey Medical Center, Graduate School Programs in the Biomedical Sciences, The Huck Institutes of the Life Sciences, Intercollege Graduate Program in Neuroscience, Hershey, PA 17033-2360. Offers MS, PhD, MD/PhD. *Students:* 18 full-time (7 women), 1 part-time; includes 4 minority (1 African American, 3 Asian Americans or Pacific Islanders), 1 international. Average age 27.Terminal master's awarded for partial completion of doctoral program. *Degree requirements:* For master's, thesis or alternative, registration; for doctorate, thesis/dissertation, oral exam, comprehensive exam, registration. *Entrance requirements:* For master's, GRE General Test; for doctorate, GRE General Test, minimum GPA of 3.0. Additional exam requirements/recommendations for international students: Required—TOEFL (minimum score 500 paper-based; 213 computer-based). *Application deadline:* Applications are processed on a rolling basis. Application fee: $45. Electronic applications accepted. *Financial support:* In 2005–06, 6 fellowships with full tuition reimbursements, 18 research assistantships with full tuition reimbursements were awarded; institutionally sponsored loans, scholarships/grants, health care benefits, and unspecified assistantships also available. Financial award applicants required to submit FAFSA. *Faculty research:* Behavioral neuroscience, growth factors and neuropeptides, molecular neurobiology and neurogenetics, neuronal aging and brain metabolism, neuronal and glial development. *Unit head:* Dr. Robert Milner, Program Director, 717-531-1045, Fax: 717-531-4139, E-mail: neuro-grad-hmc@psu.edu. *Application contact:* Sue Myers, Program Assistant, 717-531-1045, Fax: 717-531-4139, E-mail: neuro-grad-hmc@psu.edu.

The Pennsylvania State University University Park Campus, Graduate School, Intercollege Graduate Programs, Intercollege Graduate Program in Integrative Biosciences, State College, University Park, PA 16802-1503. Offers integrative biosciences (PhD), including biomolecular transport dynamics, cell and developmental biology, cellular and molecular mechanisms of toxicity, chemical biology, ecological and molecular plant physiology, immunobiology, molecular medicine, neuroscience, nutrition science. *Students:* 110 full-time (58 women), 2 part-time (both women); includes 6 minority (3 African Americans, 3 Asian Americans or Pacific Islanders), 61 international. *Entrance requirements:* For master's and doctorate, GRE General Test. Application fee: $45. *Expenses:* Tuition, state resident: full-time 12,518; part-time $522 per credit. Tuition, nonresident: full-time $23,004; part-time $959 per credit. Required fees: $484. Tuition and fees vary according to course load, campus/location and program. *Financial support:* Fellowships available. *Unit head:* Dr. Richard J. Frisque, Co-Director, 814-863-3523, Fax: 814-863-1357, E-mail: rjf6@psu.edu.

See Close-Up on page 197.

Princeton University, Graduate School, Department of Ecology and Evolutionary Biology, Princeton, NJ 08544-1019. Offers biology (PhD); neuroscience (PhD). *Degree requirements:* For doctorate, thesis/dissertation. *Entrance requirements:* For doctorate, GRE General Test, GRE Subject Test. Additional exam requirements/recommendations for international students: Required—TOEFL (minimum score 600 paper-based; 250 computer-based). Electronic applications accepted.

See Close-Up on page 705.

Princeton University, Graduate School, Department of Psychology, Princeton, NJ 08544-1019. Offers neuroscience (PhD); psychology (PhD). *Degree requirements:* For doctorate, thesis/dissertation. *Entrance requirements:* For doctorate, GRE General Test, GRE Subject Test. Additional exam requirements/recommendations for international students: Required—TOEFL (minimum score 550 paper-based). Electronic applications accepted.

Rosalind Franklin University of Medicine and Science, School of Graduate and Postdoctoral Studies, Department of Neuroscience, North Chicago, IL 60064-3095. Offers PhD, MD/PhD. *Degree requirements:* For doctorate, thesis/dissertation, original research project, comprehensive exam. *Entrance requirements:* For doctorate, GRE General Test. Additional exam requirements/recommendations for international students: Required—TOEFL, TWE.

See Close-Up on page 1015.

Rush University, Graduate College, Division of Neuroscience, Chicago, IL 60612-3832. Offers MS, PhD. *Faculty:* 11 full-time (5 women). *Students:* 11 full-time (5 women); includes 2 minority (both Asian Americans or Pacific Islanders) In 2005, 4 degrees awarded. Terminal master's awarded for partial completion of doctoral program. *Degree requirements:* For master's and doctorate, thesis/dissertation. *Entrance requirements:* For master's and doctorate, GRE General Test. Additional exam requirements/recommendations for international students: Required—TOEFL. *Application deadline:* For winter admission, 2/28 for domestic students. Applications are processed on a rolling basis. Application fee: $40. Electronic applications accepted. *Financial support:* In 2005–06, 11 students received support, including fellowships with full tuition reimbursements available (averaging $20,772 per year); research assistantships with full tuition reimbursements available, teaching assistantships, tuition waivers (full) also available. Financial award application deadline: 2/28. *Faculty research:* Neurodegenerative disorders, neurobiology of memory, aging, pathology and genetics of Alzheimer's disease. *Unit head:* Dr. Leyla deToledo-Morrell, Director.

Rutgers, The State University of New Jersey, Newark, Graduate School, Program in Integrative Neuroscience, Newark, NJ 07102. Offers PhD. Part-time programs available. *Faculty:* 27 full-time (11 women), 1 part-time/adjunct (0 women). *Students:* 19 full-time (10 women), 14 part-time (8 women); includes 7 minority (1 African American, 5 Asian Americans or Pacific Islanders, 1 Hispanic American). 68 applicants, 29% accepted, 7 enrolled. In 2005, 2 degrees awarded. *Degree requirements:* For doctorate, thesis/dissertation. *Entrance requirements:* For doctorate, GRE, minimum GPA of 3.0. *Application deadline:* For fall admission, 2/1 for domestic students. Applications are processed on a rolling basis. Application fee: $50. Electronic applications accepted. *Expenses:* Tuition, state resident: full-time $10,440; part-time $435 per credit. Tuition, nonresident: full-time $15,520; part-time $637 per credit. *Financial support:* In 2005–06, 21 teaching assistantships with full and partial tuition reimbursements (averaging $19,367 per year) were awarded Financial award application deadline: 3/1. *Faculty research:* Systems neuroscience, cognitive neuroscience, molecular neuroscience, behavioral neuroscience. Total annual research expenditures: $3 million. *Unit head:* Dr. Ian Creese, Director, 973-353-1080 Ext. 3300, Fax: 973-353-1272, E-mail: creese@axon.rutgers.edu.

Rutgers, The State University of New Jersey, Newark, Graduate School, Program in Psychology, Newark, NJ 07102. Offers cognitive neuroscience (PhD); cognitive science (PhD); perception (PhD); psychobiology (PhD); social cognition (PhD). *Faculty:* 15 full-time (5 women), 3 part-time/adjunct (2 women). *Students:* 17 full-time (12 women), 11 part-time (5 women); includes 7 minority (1 African American, 4 Asian Americans or Pacific Islanders, 2 Hispanic Americans). 42 applicants, 12% accepted, 4 enrolled. In 2005, 2 doctorates awarded. *Degree requirements:* For doctorate, thesis/dissertation, comprehensive exam, registration. *Entrance requirements:* For doctorate, GRE General Test, GRE Subject Test, minimum undergraduate B average. *Application deadline:* For fall admission, 2/1 for domestic students; for spring admission, 11/1 for domestic students. Application fee: $50. Electronic applications accepted. *Expenses:* Tuition, state resident: full-time $10,440; part-time $435 per credit. Tuition, nonresident: full-time $15,520; part-time $637 per credit. *Financial support:* In 2005–06, 16 students received support, including 11 fellowships with full and partial tuition reimbursements available (averaging $18,000 per year), 9 teaching assistantships with full tuition reimbursements available (averaging $16,988 per year); research assistantships, career-related internships or fieldwork, health care benefits, and minority scholarships also available. Financial award application deadline: 2/1. *Faculty research:* Visual perception (luminance, motion), neuroendocrine mechanisms in behavior (reproduction, pain), attachment theory, connectionist modeling of cognition. *Unit head:* Dr. Harold Siegel, Director, 973-353-5440 Ext. 236, Fax: 973-353-1171, E-mail: hisiegal@andromeda.rutgers.edu.

Rutgers, The State University of New Jersey, New Brunswick/Piscataway, Graduate School, Programs in the Molecular Biosciences, Program in Neuroscience, New Brunswick, NJ 08901-1281. Offers PhD. *Students:* 48 full-time (25 women); includes 32 minority (1 African American, 30 Asian Americans or Pacific Islanders, 1 Hispanic American), 25 international. *Degree requirements:* For doctorate, thesis/dissertation, qualifying exam. *Entrance requirements:* For doctorate, GRE, minimum B average, letters of reference. Additional exam requirements/recommendations for international students: Required—TOEFL (minimum score 550 paper-based; 213 computer-based). *Application deadline:* For fall admission, 3/1 priority date for domestic students, 3/1 priority date for international students. Applications are processed on a rolling basis. Application fee: $50. Electronic applications accepted. *Expenses:* Tuition, state resident: full-time $10,440; part-time $435 per credit. Tuition, nonresident: full-time $15,520; part-time $647 per credit. Required fees: $129 per credit. Tuition and fees vary according to program. *Financial support:* In 2005–06, 7 students received support, including 3 fellowships with full tuition reimbursements available (averaging $17,000 per year), 3 teaching assistantships with full tuition reimbursements available (averaging $16,988 per year); unspecified assistantships also available. *Faculty research:* Growth; trophic factors; neurogenetic, molecular and cellular neuroscience; developmental neurobiology. *Unit head:* David Egger, Graduate Program Director, 732-235-5388, Fax: 732-235-5885, E-mail: egger@umdnj.edu. *Application contact:* Joan Mordes, Program Assistant, 732-235-5390, Fax: 732-235-5885, E-mail: mordesja@umdnj.edu.

Seton Hall University, College of Arts and Sciences, Department of Psychology, South Orange, NJ 07079-2697. Offers experimental psychology (MS), including behavioral neuroscience. *Students:* 7 full-time (3 women), 3 part-time (2 women). Average age 25. 15 applicants, 67% accepted, 6 enrolled. *Entrance requirements:* Additional exam requirements/recommendations for international students: Required—TOEFL. *Application deadline:* For fall admission, 7/1 priority date for domestic students, 7/1 priority date for international students. Applications are processed on a rolling basis. Electronic applications accepted. *Financial support:* Federal Work-Study, scholarships/grants, and unspecified assistantships available. *Faculty research:* Behavioral neuroscience, cognitive psychology, social psychology, perception/motor skills, memory. *Unit head:* Dr. Jeffrey Levy, Chair, 973-761-9484, Fax: 973-275-5829, E-mail: levyjeff@shu.edu. *Application contact:* Dr. Michael Vigorito, Director of Graduate Studies, 973-761-9484, Fax: 973-275-5829, E-mail: vigorimi@shu.edu.

Stanford University, School of Medicine, Graduate Programs in Medicine, Neurosciences Program, Stanford, CA 94305-9991. Offers PhD. *Degree requirements:* For doctorate, thesis/dissertation. *Entrance requirements:* For doctorate, GRE General Test, GRE Subject Test. Additional exam requirements/recommendations for international students: Required—TOEFL. Electronic applications accepted.

Neuroscience

State University of New York at Buffalo, Graduate School, College of Arts and Sciences, Department of Psychology, Buffalo, NY 14260. Offers behavioral neuroscience (PhD); clinical psychology (PhD); cognitive psychology (PhD); general psychology (MA); social-personality (PhD). *Accreditation:* APA (one or more programs are accredited). *Faculty:* 30 full-time (11 women), 8 part-time/adjunct (6 women). *Students:* 74 full-time (48 women), 9 part-time (2 women); includes 8 minority (3 African Americans, 2 Asian Americans or Pacific Islanders, 3 Hispanic Americans), 14 international. Average age 25. 332 applicants, 15% accepted, 23 enrolled. In 2005, 23 master's, 7 doctorates awarded. Terminal master's awarded for partial completion of doctoral program. *Median time to degree:* Of those who began their doctoral program in fall 1997, 50% received their degree in 8 years or less. *Degree requirements:* For master's, project; for doctorate, thesis/dissertation. *Entrance requirements:* For master's and doctorate, GRE General Test. Additional exam requirements/recommendations for international students: Required—TOEFL (minimum score 550 paper-based; 213 computer-based). *Application deadline:* For fall admission, 1/5 for domestic students, 1/5 for international students. Application fee: $35. Electronic applications accepted. *Financial support:* In 2005–06, 81 students received support, including 20 fellowships with full tuition reimbursements available (averaging $14,400 per year), 10 research assistantships with full tuition reimbursements available (averaging $10,400 per year), 37 teaching assistantships with full tuition reimbursements available (averaging $10,400 per year); career-related internships or fieldwork, Federal Work-Study, institutionally sponsored loans, tuition waivers (partial), and minority fellowships also available. Financial award application deadline: 1/5; financial award applicants required to submit FAFSA. *Faculty research:* Neural, endocrine, and molecular bases of behavior; adult mood and anxiety disorders; relationship dysfunction; attention deficit/hyperactivity disorder; psycho-linguistics. Total annual research expenditures: $7.9 million. *Unit head:* Dr. Paul A. Luce, Chair, 716-645-3650 Ext. 203, Fax: 716-645-3801, E-mail: psychair@acsu.buffalo.edu. *Application contact:* Michele Nowacki, Coordinator of Admissions, 716-645-3650 Ext. 209, Fax: 716-645-3801, E-mail: psych@acsu.buffalo.edu.

State University of New York at Buffalo, Graduate School, School of Medicine and Biomedical Sciences, Graduate Programs in Medicine and Biomedical Sciences, Program in Neuroscience, Buffalo, NY 14260. Offers MS, PhD. Part-time programs available. *Students:* 26 full-time (14 women), 4 part-time, 22 international. In 2005, 2 master's, 8 doctorates awarded. Terminal master's awarded for partial completion of doctoral program. *Degree requirements:* For master's, thesis or alternative, registration; for doctorate, thesis/dissertation, comprehensive exam, registration. *Entrance requirements:* For master's, GRE General Test; for doctorate, GRE General Test, 3 letters of recommendation. Additional exam requirements/recommendations for international students: Required—TOEFL. *Application deadline:* For fall admission, 2/1 priority date for domestic students, 2/1 priority date for international students. Application fee: $35. *Financial support:* In 2005–06, 1 student received support, including 19 research assistantships with full tuition reimbursements available (averaging $21,000 per year) *Faculty research:* Neural plasticity, development, synapse, neurodisease, genetics of neuropathology. Total annual research expenditures: $6 million. *Unit head:* Dr. Malcolm Slaughter, Professor, 716-829-3240, Fax: 716-829-2364, E-mail: mslaught@buffalo.edu. *Application contact:* Patricia R. Raiff, Program Administrator, 716-827-2419, Fax: 716-829-3849, E-mail: praiff@buffalo.edu.

State University of New York Downstate Medical Center, School of Graduate Studies, Program in Neural and Behavioral Science, Brooklyn, NY 11203-2098. Offers PhD, MD/PhD. *Faculty:* 44 full-time (11 women). *Students:* 28 full-time (12 women); includes 4 Asian Americans or Pacific Islanders, 15 international. Average age 31. 23 applicants, 17% accepted, 3 enrolled. In 2005, 5 degrees awarded. *Degree requirements:* For doctorate, thesis/dissertation, comprehensive exam, registration. *Entrance requirements:* For doctorate, GRE. *Application deadline:* For fall admission, 4/1 for domestic students. Application fee: $35. *Financial support:* In 2005–06, 28 students received support, including teaching assistantships with tuition reimbursements available (averaging $23,500 per year); fellowships, career-related internships or fieldwork, Federal Work-Study, and health care benefits also available. *Faculty research:* Molecular neuroscience, cellular neuroscience, systems neuroscience, behavioral neuroscience, behavior. *Unit head:* Mark Stewart, Director, 718-270-1167. *Application contact:* Denise Sheares, Admissions Officer, 718-270-2738, Fax: 718-270-3378, E-mail: dsheares@downstate.edu.

State University of New York Upstate Medical University, College of Graduate Studies, Program in Neuroscience, Syracuse, NY 13210-2334. Offers PhD. *Degree requirements:* For doctorate, thesis/dissertation, comprehensive exam. *Entrance requirements:* For doctorate, GRE General Test. Additional exam requirements/recommendations for international students: Required—TOEFL.

Stony Brook University, State University of New York, Graduate School, College of Arts and Sciences, Department of Neuroscience, Stony Brook, NY 11794. Offers PhD. *Faculty:* 14 full-time (1 woman), 1 part-time/adjunct (0 women). *Students:* 34 full-time (21 women); includes 2 minority (both Asian Americans or Pacific Islanders), 14 international. Average age 27. 131 applicants, 17% accepted. In 2005, 8 degrees awarded. *Degree requirements:* For doctorate, thesis/dissertation, teaching experience, comprehensive exam. *Entrance requirements:* For doctorate, GRE General Test, GRE Subject Test, minimum GPA of 3.0. Additional exam requirements/recommendations for international students: Required—TOEFL. *Application deadline:* For fall admission, 1/15 for domestic students. Application fee: $50. *Expenses:* Tuition, state resident: full-time $6,900; part-time $288 per credit. Tuition, nonresident: full-time $10,920; part-time $455 per credit. Required fees: $704. *Financial support:* In 2005–06, 3 fellowships, 27 research assistantships, 5 teaching assistantships were awarded; Federal Work-Study also available. *Faculty research:* Biophysics; neurochemistry; cellular, developmental, and integrative neurobiology. Total annual research expenditures: $7.1 million. *Unit head:* Dr. Lorne Mendell, Chair, 631-632-8616, Fax: 631-632-6661, E-mail: lmendell@nstep.cc.sunysb.edu. *Application contact:* Dr. Gary Matthews, Director, 631-632-8616, Fax: 631-632-6661, E-mail: ggmatthews@notes.cc.sunysb.edu.

Announcement: The doctoral program in neuroscience prepares students for research careers in modern neuroscience. With approximately 40 faculty members, the program offers opportunities for world-class research on all aspects of neuroscience, from molecules to behavior. All students receive full tuition waivers and stipends to cover living costs.

See Close-Up on page 1019.

Syracuse University, Graduate School, L. C. Smith College of Engineering and Computer Science, Department of Biomedical and Chemical Engineering, Syracuse, NY 13244. Offers bioengineering (ME, MS, PhD); chemical engineering (MS, PhD); neuroscience (MS). Evening/weekend programs available. *Students:* 58 full-time (19 women), 12 part-time (3 women); includes 4 minority (2 African Americans, 2 Asian Americans or Pacific Islanders), 38 international. 105 applicants, 50% accepted, 19 enrolled. *Entrance requirements:* For master's, GRE General Test, GRE Subject Test. Additional exam requirements/recommendations for international students: Required—TOEFL. *Application deadline:* For fall admission, 3/1 for domestic students. Applications are processed on a rolling basis. Application fee: $65. Electronic applications accepted. *Financial support:* Fellowships with full tuition reimbursements, research assistantships with full tuition reimbursements, teaching assistantships with full tuition reimbursements available. *Faculty research:* Audition, taction, multisensory, computational neuroscience. *Unit head:* Dr. Gustav Engbretson, Chair, 315-443-1931, Fax: 315-443-1184, E-mail: gengbret@syr.edu. *Application contact:* Dawn Long, 315-443-1931, E-mail: dmlong@syr.edu.

Syracuse University, Graduate School, L. C. Smith College of Engineering and Computer Science, Institute for Sensory Research, Syracuse, NY 13244. Offers neuroscience (MS). Part-time and evening/weekend programs available. *Faculty:* 13. *Students:* 9 full-time (4 women), 2 part-time (1 woman); includes 1 minority (Asian American or Pacific Islander), 1 international. 6 applicants, 67% accepted, 3 enrolled. *Entrance requirements:* For master's, GRE General Test, GRE Subject Test. *Application deadline:* For fall admission, 3/1 for domestic students. Applications are processed on a rolling basis. Application fee: $65. Electronic applications accepted. *Financial support:* Fellowships with full tuition reimbursements, research

assistantships with full tuition reimbursements, teaching assistantships with full tuition reimbursements, tuition waivers (partial) and full support awards available. *Faculty research:* Psychophysics of auditory, somatosensory, and visual systems; psychophysics; neuroanatomy; neurochemistry. *Unit head:* Dr. Robert Smith, Director, 315-443-9702, Fax: 315-443-1184, E-mail: bob_smith@isr.syr.edu.

Teachers College Columbia University, Graduate Faculty of Education, Department of Biobehavioral Studies, Program in Neuroscience and Education, New York, NY 10027-6696. Offers Ed M, Ed D. *Students:* 5 full-time (all women), 8 part-time (all women); includes 2 minority (1 Asian American or Pacific Islander, 1 Hispanic American), 4 international. Average age 25. 18 applicants, 78% accepted, 7 enrolled. In 2005, 1 degree awarded. *Application deadline:* For fall admission, 5/15 for domestic students. Application fee: $50. *Expenses:* Tuition: Full-time $22,440; part-time $935 per credit. Required fees: $240 per term. *Financial support:* Career-related internships or fieldwork, Federal Work-Study, institutionally sponsored loans, and tuition waivers (full and partial) available. Support available to part-time students. Financial award application deadline: 2/1. *Faculty research:* Neuropsychological diagnosis and intervention. *Application contact:* Debbie Lesperance, Assistant Director of Admission, 212-678-3710, Fax: 212-678-4171.

Temple University, Health Sciences Center, School of Medicine and Graduate School, Graduate Programs in Medicine, Department of Neuroscience, Philadelphia, PA 19122-6096. Offers MS, PhD. *Entrance requirements:* For master's and doctorate, GRE or MCAT, minimum GPA of 3.0. Additional exam requirements/recommendations for international students: Required—TOEFL. *Application deadline:* For fall admission, 7/15 for domestic students, 12/15 for international students; for spring admission, 1/1 for domestic students, 8/1 for international students. Applications are processed on a rolling basis. Application fee: $50. Electronic applications accepted. *Expenses:* Tuition, state resident: full-time $8,694; part-time $483 per credit. Tuition, nonresident: full-time $12,672; part-time $704 per credit. Required fees: $500; $122 per semester. Tuition and fees vary according to course level, campus/location and program. *Financial support:* Application deadline: 1/15; *Unit head:* Dr. Kamel Khalili, Chair, 215-204-0678, Fax: 215-204-0679, E-mail: kamel.khalili@temple.edu.

Texas A&M University, College of Liberal Arts, Department of Psychology, Program in Behavioral and Cellular Neuroscience, College Station, TX 77843. Offers MS, PhD. *Degree requirements:* For master's and doctorate, thesis/dissertation. *Entrance requirements:* For master's and doctorate, GRE. Additional exam requirements/recommendations for international students: Required—TOEFL. *Application deadline:* For fall admission, 1/5 for domestic students, 1/5 for international students. Application fee: $50 ($75 for international students). *Expenses:* Tuition, state resident: full-time $4,488; part-time $187 per credit hour. Tuition, nonresident: full-time $11,112; part-time $463 per credit hour. Required fees: $1,974. *Financial support:* Fellowships with partial tuition reimbursements, research assistantships with partial tuition reimbursements, teaching assistantships with partial tuition reimbursements, career-related internships or fieldwork, institutionally sponsored loans, health care benefits, and unspecified assistantships available. Financial award application deadline: 1/5; financial award applicants required to submit FAFSA. *Unit head:* James Grau, Area Coordinator, 979-845-2584, Fax: 979-845-4727, E-mail: j-grau@tamu.edu. *Application contact:* Sharon Starr, Graduate Admissions Supervisor, 979-458-1710, Fax: 979-845-4727, E-mail: gradadv@psyc.tamu.edu.

Texas A&M University, College of Science, Department of Biology, Faculty of Neuroscience, College Station, TX 77843. Offers MS, PhD. *Degree requirements:* For doctorate, thesis/dissertation, qualifying exam. *Entrance requirements:* For doctorate, GRE General Test. Additional exam requirements/recommendations for international students: Required—TOEFL. *Application deadline:* For fall admission, 2/1 for domestic students. Applications are processed on a rolling basis. Application fee: $50 ($75 for international students). *Expenses:* Tuition, state resident: full-time $4,488; part-time $187 per credit hour. Tuition, nonresident: full-time $11,112; part-time $463 per credit hour. Required fees: $1,974. *Financial support:* Fellowships, teaching assistantships available. Financial award application deadline: 4/1. *Faculty research:* Learning and memory, circadian rhythms, neural development and function, neuroendocrinology, neuro-pharmacology and neurotoxicology. Total annual research expenditures: $1.6 million.

See Close-Up on page 1021.

Texas A&M University System Health Science Center, Graduate School of Biomedical Sciences, Department of Neuroscience, College Station, TX 77840. Offers PhD. *Degree requirements:* For doctorate, thesis/dissertation. *Entrance requirements:* For doctorate, GRE General Test. Electronic applications accepted. *Faculty research:* Fetal alcohol syndrome, circadian rhythms, neuroendocrinology, bone and joint diseases, molecular neurobiology.

Texas A&M University System Health Science Center, Graduate School of Biomedical Sciences, Department of Neuroscience and Experimental Therapeutics, College Station, TX 77840. Offers PhD.

See Close-Up on page 1023.

Texas Tech University Health Sciences Center, Graduate School of Biomedical Sciences, Department of Pharmacology and Neuroscience, Lubbock, TX 79430. Offers MS, PhD, MD/PhD, MS/PhD. *Faculty:* 13 full-time (4 women). *Students:* 6 full-time (2 women); includes 1 minority (Asian American or Pacific Islander), 3 international. Average age 24. 17 applicants, 0% accepted, 0 enrolled. In 2005, 1 master's, 1 doctorate awarded. Terminal master's awarded for partial completion of doctoral program. *Degree requirements:* For master's and doctorate, thesis/dissertation. *Entrance requirements:* For master's and doctorate, GRE General Test, minimum GPA of 3.0. Additional exam requirements/recommendations for international students: Required—TOEFL. *Application deadline:* For fall admission, 5/15 priority date for domestic students, 4/15 priority date for international students; for spring admission, 11/15 for domestic students, 10/15 for international students. Applications are processed on a rolling basis. Application fee: $45. Electronic applications accepted. *Financial support:* In 2005–06, 3 students received support, including 5 research assistantships (averaging $20,500 per year); Federal Work-Study, institutionally sponsored loans, scholarships/grants, and health care benefits also available. Financial award applicants required to submit FAFSA. *Faculty research:* Neuroscience, neuropsychopharmacology, autonomic pharmacology, cardiovascular pharmacology, molecular pharmacology. Total annual research expenditures: $677,422. *Unit head:* Dr. Reid L. Norman, Chair, 806-743-2425 Ext. 222, Fax: 806-743-2744, E-mail: reid.norman@ttuhsc.edu. *Application contact:* Dr. Michael P. Blanton, Associate Professor, 806-743-2425 Ext. 229, Fax: 806-743-2744, E-mail: michael.blanton@ttuhsc.edu.

Thomas Jefferson University, Jefferson College of Graduate Studies, Graduate Program in Neuroscience, Philadelphia, PA 19107. Offers PhD. Offered in conjunction with the Farber Institute for Neuroscience. *Students:* 7 full-time (5 women); includes 1 African American, 1 Asian American or Pacific Islander. 31 applicants, 19% accepted, 6 enrolled. *Degree requirements:* For doctorate, thesis/dissertation. *Entrance requirements:* For doctorate, GRE General Test, strong background in the sciences, interview, previous research experience. *Application deadline:* For fall admission, 3/1 priority date for domestic students, 3/1 priority date for international students. Application fee: $50. *Expenses:* Tuition: Full-time $14,894; part-time $800 per credit. *Financial support:* In 2005–06, 7 fellowships were awarded; research assistantships, teaching assistantships, institutionally sponsored loans, scholarships/grants, and stipend, Title IV funds also available. *Unit head:* Dr. Elisabeth Van Bockstaele, Program Director, 215-503-1245, Fax: 215-955-4949, E-mail: elisabeth.vanbockstaele@jefferson.edu. *Application contact:* Jessie F. Pervall, Director of Admissions and Recruitment, 215-503-4400, Fax: 215-503-9920, E-mail: cgs-info@jefferson.edu.

Tufts University, Sackler School of Graduate Biomedical Sciences, Program in Neuroscience, Boston, MA 02155. Offers PhD. *Faculty:* 28 full-time (8 women). *Students:* 17 full-time (13 women); includes 2 minority (both Asian Americans or Pacific Islanders), 2 international. Average age 27. 74 applicants, 12% accepted, 3 enrolled. In 2005, 5 degrees awarded.

Neuroscience

Tufts University (continued)
Degree requirements: For doctorate, thesis/dissertation. *Entrance requirements:* For doctorate, GRE General Test, 3 letters of reference. Additional exam requirements/recommendations for international students: Required—TOEFL. *Application deadline:* For fall admission, 1/15 priority date for domestic students, 1/15 priority date for international students. Applications are processed on a rolling basis. Application fee: $65. Electronic applications accepted. *Financial support:* In 2005–06, 17 students received support, including 17 research assistantships with full tuition reimbursements available (averaging $29,000 per year); fellowships, scholarships/grants, health care benefits, and tuition waivers (full) also available. Financial award application deadline: 1/15. *Faculty research:* Electrophysiology, molecular neurobiology, structure and function of sensory systems, neural regulation of development. *Unit head:* Dr. Dale Hunter, Director, 617-636-3646, Fax: 617-636-2413. *Application contact:* 617-636-6767, Fax: 617-636-0375, E-mail: sackler-school@tufts.edu.

Tulane University, School of Medicine and Graduate School, Graduate Programs in Medicine, Program in Neuroscience, New Orleans, LA 70118-5669. Offers MS, PhD, MD/PhD. MS and PhD offered through the Graduate School. *Degree requirements:* For doctorate, thesis/dissertation, qualifying exam. *Entrance requirements:* For doctorate, GRE General Test. Additional exam requirements/recommendations for international students: Required—TOEFL or TSE. Electronic applications accepted. *Faculty research:* Neuroendocrinology, ion channels, neuropeptides.

Uniformed Services University of the Health Sciences, School of Medicine, Programs in Biomedical Sciences, Graduate Program in Neuroscience, Bethesda, MD 20814-4799. Offers PhD. *Faculty:* 31 full-time (13 women), 9 part-time/adjunct (2 women). *Students:* 23 full-time (11 women); includes 2 minority (1 Asian American or Pacific Islander, 1 Hispanic American), 4 international. Average age 26. 30 applicants, 47% accepted, 10 enrolled. In 2005, 2 degrees awarded. *Median time to degree:* Of those who began their doctoral program in fall 1997, 100% received their degree in 8 years or less. *Degree requirements:* For doctorate, thesis/dissertation, qualifying exams, comprehensive exam. *Entrance requirements:* For doctorate, GRE General Test, minimum GPA of 3.0; course work in biology, general chemistry, organic chemistry. Additional exam requirements/recommendations for international students: Required—TOEFL. *Application deadline:* For fall admission, 1/15 for domestic students. Applications are processed on a rolling basis. Application fee: $0. *Financial support:* In 2005–06, fellowships with full tuition reimbursements (averaging $23,000 per year); tuition waivers (full) also available. *Faculty research:* Neuronal development and plasticity, molecular neurobiology, environmental adaptations, stress and injury. *Unit head:* Dr. Regina Armstrong, Director, 301-295-3205, Fax: 301-295-1996, E-mail: rarmstrong@usuhs.mil. *Application contact:* Janet M. Anastasi, Graduate Program Coordinator, 301-295-9474, Fax: 301-295-6772, E-mail: janastasi@usuhs.mil.

See Close-Up on page 1025.

Université de Montréal, Faculty of Medicine and Faculty of Graduate Studies, Graduate Programs in Medicine, Montréal, QC H3C 3J7, Canada. Offers biochemistry (M Sc, PhD, DEPD), including biochemistry (M Sc, PhD), clinical biochemistry (DEPD); biomedical engineering (M Sc A, PhD, DESS); biomedical sciences (M Sc, PhD); communal and public health (M Sc, PhD, DESS), including community health (M Sc, DESS), public health (PhD); environmental and occupational health (M Sc, DESS); health administration (M Sc, DESS); health sciences (DESS), including anesthesia, diagnostic radiology, family medicine, medical biochemistry, medical genetics, medicine, microbiology and infectious diseases, nuclear medicine, obstetrics and gynecology, ophthalmology, pediatrics, psychiatry, radiology-oncology, surgery; microbiology and immunology (M Sc, PhD); nutrition (M Sc, PhD); pathology and cellular biology (M Sc, PhD); pharmacology (M Sc, PhD); physiology (M Sc, PhD), including biophysics and molecular physiology, neurological sciences, physiology; speech therapy and audiology (MOA, DESS), including speech therapy (DESS), speech-language pathology and audiology (MOA); virology and immunology (PhD). Terminal master's awarded for partial completion of doctoral program. *Degree requirements:* For master's, thesis; for doctorate, thesis/dissertation, general exam. *Entrance requirements:* For master's and doctorate, proficiency in French, knowledge of English; for other advanced degree, proficiency in French. *Application deadline:* For fall and spring admission, 2/1. For winter admission, 11/1 for domestic students. Application fee: $30. Electronic applications accepted. *Financial support:* Fellowships, research assistantships, teaching assistantships, career-related internships or fieldwork, institutionally sponsored loans, and tuition waivers (full) available. *Application contact:* Dr. Pierre Boyle, Vice Dean of Studies, 514-343-6300, Fax: 514-343-5751, E-mail: pierre.boyle@umontreal.ca.

Université de Montréal, Faculty of Medicine and Faculty of Graduate Studies, Graduate Programs in Medicine, Department of Physiology, Program in Neurological Sciences, Montréal, QC H3C 3J7, Canada. Offers M Sc, PhD. *Students:* 67 full-time (39 women). 30 applicants, 47% accepted, 13 enrolled. In 2005, 14 master's, 11 doctorates awarded. Terminal master's awarded for partial completion of doctoral program. *Degree requirements:* For master's, thesis; for doctorate, thesis/dissertation, general exam. *Entrance requirements:* For master's and doctorate, proficiency in French, knowledge of English. *Application deadline:* For fall and spring admission, 2/1. For winter admission, 11/1 for domestic students. Application fee: $30. Electronic applications accepted. *Financial support:* Fellowships, research assistantships available. *Unit head:* John Kalaska, Head, 514-343-6349, Fax: 514-343-7072.

University at Albany, State University of New York, School of Public Health, Department of Biomedical Sciences, Program in Neuroscience, Albany, NY 12222-0001. Offers MS, PhD. *Degree requirements:* For master's and doctorate, thesis/dissertation. *Entrance requirements:* For master's and doctorate, GRE General Test, GRE Subject Test. Application fee: $60. *Financial support:* Application deadline: 2/1. *Unit head:* Dr. James Dias, Chair, Department of Biomedical Sciences, 518-474-2662.

The University of Alabama at Birmingham, Graduate Programs in Joint Health Sciences, Department of Cell Biology, Graduate Program in Cellular and Molecular Biology, Birmingham, AL 35294.

See Close-Ups on pages 601 and 603.

The University of Alabama at Birmingham, Graduate Programs in Joint Health Sciences, Department of Cell Biology, Graduate Program in Neuroscience, Birmingham, AL 35294. Offers PhD. *Students:* 6 full-time (2 women); includes 1 minority (African American), 2 international. 30 applicants, 40% accepted. Application fee: $35 ($60 for international students). *Expenses:* Tuition, state resident: part-time $170 per credit hour. Tuition, nonresident: full-time $4,612; part-time $425 per credit hour. International tuition: $10,732 full-time. Required fees: $11 per credit hour. $124 per term. Tuition and fees vary according to course load, degree level and program. *Unit head:* , Dr. John Joseph Hablitz, Head, 205-934-1550, Fax: 205-934-6571, E-mail: hablitz@uab.edu.

See Close-Up on page 601.

The University of Alabama at Birmingham, Graduate Programs in Joint Health Sciences, Department of Neurobiology, Birmingham, AL 35294. Offers PhD. The department participates in the Cellular and Molecular Biology Graduate Program. *Students:* 35 full-time (16 women), 1 (woman) part-time; includes 4 minority (2 African Americans, 1 Asian American or Pacific Islander, 1 Hispanic American), 9 international. 10 applicants, 80% accepted. In 2005, 5 degrees awarded. *Degree requirements:* For doctorate, thesis/dissertation. *Entrance requirements:* For doctorate, GRE, interview. *Application deadline:* Applications are processed on a rolling basis. Electronic applications accepted. *Expenses:* Tuition, state resident: part-time $170 per credit hour. Tuition, nonresident: full-time $4,612; part-time $425 per credit hour. Tuition, international: $10,732 full-time. Required fees: $11 per credit hour. $124 per term. Tuition and fees vary according to course load, degree level and program. *Unit head:* Dr. J. David

Sweatt, Chair, Fax: 205-934-6571. *Application contact:* Information Contact, 205-975-5573, Fax: 205-934-6571.

See Close-Ups on pages 1029 and 1027.

The University of Alabama at Birmingham, School of Social and Behavioral Sciences, Department of Psychology, Birmingham, AL 35294. Offers behavioral neuroscience (PhD); clinical psychology (PhD); developmental psychology (PhD); medical psychology (PhD); psychology (MA, PhD). *Accreditation:* APA (one or more programs are accredited). *Students:* 61 full-time (47 women), 3 part-time (1 woman); includes 12 minority (11 African Americans, 1 Asian American or Pacific Islander), 6 international. 178 applicants, 10% accepted. In 2005, 8 master's, 8 doctorates awarded. *Application deadline:* Applications are processed on a rolling basis. Application fee: $35 ($60 for international students). Electronic applications accepted. *Expenses:* Tuition, state resident: part-time $170 per credit hour. Tuition, nonresident: full-time $4,612; part-time $425 per credit hour. International tuition: $10,732 full-time. Required fees: $11 per credit hour. $124 per term. Tuition and fees vary according to course load, degree level and program. *Financial support:* Career-related internships or fieldwork available. *Faculty research:* Biological basis of behavior structure, function of the nervous system. *Unit head:* Dr. Carl E. McFarland, Chair, 205-934-3850, E-mail: cmcfarla@uab.edu.

University of Alberta, Faculty of Medicine and Dentistry and Faculty of Graduate Studies and Research, Graduate Programs in Medicine, Centre for Neuroscience, Edmonton, AB T6G 2E1, Canada. Offers M Sc, PhD. Terminal master's awarded for partial completion of doctoral program. *Degree requirements:* For master's and doctorate, thesis/dissertation. *Entrance requirements:* Additional exam requirements/recommendations for international students: Required—TOEFL (minimum score 600 paper-based; 250 computer-based). Tuition and fees charges are reported in Canadian dollars. *Expenses:* Tuition, state resident: part-time $562 Canadian dollars per term. Tuition, nonresident: full-time $3,375 Canadian dollars. Required fees: $573 Canadian dollars; $84 Canadian dollars per term. *Faculty research:* Sensory and motor mechanisms, neural growth and regeneration, molecular neurobiology, synaptic mechanisms, behavioral and psychiatric neuroscience.

The University of Arizona, Graduate College, Graduate Interdisciplinary Programs, Graduate Interdisciplinary Program in Neuroscience, Tucson, AZ 85721. Offers PhD. *Degree requirements:* For doctorate, thesis/dissertation. *Entrance requirements:* For doctorate, GRE. Additional exam requirements/recommendations for international students: Required—TOEFL. *Faculty research:* Cognitive neuroscience, developmental neurobiology, speech and hearing, motor control, insect neurobiology.

The University of British Columbia, Faculty of Arts and Faculty of Graduate Studies, Department of Psychology, Vancouver, BC V6T 1Z1, Canada. Offers behavioral neuroscience (MA, PhD); clinical psychology (MA, PhD); cognitive science (MA, PhD); developmental psychology (MA, PhD); forensic psychology (MA, PhD); psychometrics (MA, PhD); social/personality psychology (MA, PhD). *Accreditation:* APA (one or more programs are accredited). *Faculty:* 49 full-time (17 women), 19 part-time/adjunct (12 women). *Students:* 112 full-time (77 women). Average age 29. 280 applicants, 12% accepted, 15 enrolled. In 2005, 16 master's, 17 doctorates awarded. Terminal master's awarded for partial completion of doctoral program. *Median time to degree:* Of those who began their doctoral program in fall 1997, 60% received their degree in 8 years or less. *Degree requirements:* For master's, thesis/dissertation; for doctorate, thesis/dissertation, comprehensive exam, registration (for some programs). *Entrance requirements:* For master's and doctorate, GRE General Test, GRE Subject Test. Additional exam requirements/recommendations for international students: Required—TOEFL (minimum score 550 paper-based; 230 computer-based). *Application deadline:* For fall admission, 1/15 for domestic students, 1/15 for international students. Applications are processed on a rolling basis. Application fee: $90 Canadian dollars ($150 Canadian dollars for international students). Electronic applications accepted. *Financial support:* In 2005–06, 112 students received support, including fellowships with full and partial tuition reimbursements available (averaging $16,000 per year), research assistantships with full and partial tuition reimbursements available (averaging $5,000 per year), teaching assistantships with full and partial tuition reimbursements available (averaging $10,000 per year); career-related internships or fieldwork, Federal Work-Study, institutionally sponsored loans, scholarships/grants, health care benefits, tuition waivers (full and partial), and unspecified assistantships also available. Financial award application deadline: 1/15. *Faculty research:* Clinical, developmental, social/personality, cognition, behavioral neuroscience. Total annual research expenditures: $5 million Canadian dollars. *Unit head:* Dr. Eric Eich, Head, 604-822-3078, Fax: 604-822-6923, E-mail: ee@psych.ubc.ca. *Application contact:* Rose Tam, Graduate Secretary, 604-822-3144, Fax: 604-822-6923, E-mail: gradsec@psych.ubc.ca.

The University of British Columbia, Faculty of Medicine and Faculty of Graduate Studies, Graduate Programs in Medicine, Program in Neuroscience, Vancouver, BC V6T 1Z1, Canada. Offers M Sc, PhD. *Faculty:* 96 full-time (24 women). *Students:* 114 full-time (53 women). Average age 24. 103 applicants, 26% accepted, 27 enrolled. In 2005, 15 master's, 5 doctorates awarded. *Degree requirements:* For master's, thesis/dissertation, registration; for doctorate, thesis/dissertation, comprehensive exam, registration. *Entrance requirements:* For master's and doctorate, GRE General Test. Additional exam requirements/recommendations for international students: Required—TOEFL (minimum score 600 paper-based; 250 computer-based). *Application deadline:* For fall admission, 12/31 for domestic students, 12/31 for international students. Applications are processed on a rolling basis. Application fee: $90 Canadian dollars ($150 Canadian dollars for international students). Electronic applications accepted. *Financial support:* In 2005–06, 114 students received support, including 4 teaching assistantships; fellowships, research assistantships, institutionally sponsored loans, scholarships/grants, traineeships, and tuition waivers (full and partial) also available. *Faculty research:* Behavioral neurobiology, molecular biology of ion channels and receptors, physiology of synaptic transmission role of calcium in brain, neuroanatomy and imaging. *Unit head:* Dr. Steven R. Vincent, Chairman, 604-822-7375, Fax: 604-822-7981, E-mail: neurosci@interchange.ubc.ca. *Application contact:* Liz Wong, Secretary, 604-822-7375, Fax: 604-822-7981, E-mail: neurosci@interchange.ubc.ca.

University of Calgary, Faculty of Medicine and Faculty of Graduate Studies, Department of Neuroscience, Calgary, AB T2N 1N4, Canada. Offers M Sc, PhD. *Faculty:* 22 full-time (3 women). *Students:* 73 full-time (38 women). Average age 28. 43 applicants, 21% accepted, 9 enrolled. In 2005, 10 master's, 5 doctorates awarded. *Degree requirements:* For master's, thesis, oral thesis exam; for doctorate, thesis/dissertation, candidacy exam, oral thesis exam. *Entrance requirements:* For master's and doctorate, minimum GPA of 3.2 during previous 2 years. Additional exam requirements/recommendations for international students: Required—TOEFL (minimum score 580 paper-based; 237 computer-based). *Application deadline:* For fall admission, 5/15 for domestic students, 4/15 for international students. For winter admission, 9/15 for domestic students; for spring admission, 2/15 for domestic students. Applications are processed on a rolling basis. Application fee: $100 ($130 for international students). Electronic applications accepted. *Financial support:* In 2005–06, 38 fellowships, 28 research assistantships with partial tuition reimbursements (averaging $74,150 per year), 3 teaching assistantships with partial tuition reimbursements (averaging $9,000 per year) were awarded; scholarships/grants also available. *Faculty research:* Cellular pharmacology and neurotoxicology, developmental neurobiology, molecular basis of neurodegenerative diseases, neural systems, ion channels. *Unit head:* Dr. W. M. Stell, Graduate Coordinator, Fax: 403-283-2700, E-mail: wstell@ucalgary.ca. *Application contact:* Julie Boyd, Graduate Program Administrator, 403-220-2558, Fax: 403-210-8109, E-mail: neurosci@ucalgary.ca.

University of California, Berkeley, Graduate Division, Neuroscience Graduate Program, Berkeley, CA 94720-3190. Offers PhD. *Degree requirements:* For doctorate, thesis/dissertation, qualifying exam. *Entrance requirements:* For doctorate, GRE General Test, GRE Subject Test, minimum GPA of 3.0. Additional exam requirements/recommendations for international students:

Required—TOEFL. Electronic applications accepted. *Faculty research:* Ion channels, signal transduction, plasticity, neural networks, neural cell fate and pattern formation.

See Close-Up on page 1033.

University of California, Davis, Graduate Studies, Graduate Group in Neuroscience, Davis, CA 95616. Offers PhD. *Faculty:* 60 full-time. *Students:* 35 full-time (20 women); includes 8 minority (1 African American, 5 Asian Americans or Pacific Islanders, 2 Hispanic Americans), 3 international. Average age 28. 134 applicants, 13% accepted, 6 enrolled. In 2005, 3 degrees awarded. *Median time to degree:* Of those who began their doctoral program in fall 1997, 50% received their degree in 8 years or less. *Degree requirements:* For doctorate, thesis/dissertation. *Entrance requirements:* For doctorate, GRE General Test, GRE Subject Test. Additional exam requirements/recommendations for international students: Required—TOEFL (minimum score 550 paper-based; 213 computer-based). *Application deadline:* For fall admission, 1/5 for domestic students, 1/5 for international students. Application fee: $60. Electronic applications accepted. *Financial support:* In 2005–06, 33 students received support, including 6 fellowships with full and partial tuition reimbursements available (averaging $14,835 per year), 17 research assistantships with full and partial tuition reimbursements available (averaging $16,105 per year), 4 teaching assistantships with partial tuition reimbursements available (averaging $15,082 per year); Federal Work-Study, institutionally sponsored loans, scholarships/grants, tuition waivers (full and partial), and unspecified assistantships also available. Financial award application deadline: 1/15; financial award applicants required to submit FAFSA. *Faculty research:* Neuroethology, cognitive neurosciences, cortical neurophysics, cellular and molecular neurobiology. *Unit head:* Kenneth H. Britten, Graduate Program Chair, 530-754-5080, E-mail: khbritten@ucdavis.edu. *Application contact:* Troy Crowder, Graduate Program Assistant, 530-757-8845, Fax: 530-752-8832, E-mail: tacrowder@ucdavis.edu.

See Close-Up on page 1035.

University of California, Los Angeles, School of Medicine and Graduate Division, Graduate Programs in Medicine, Neuroscience Program, Los Angeles, CA 90095. Offers PhD. *Degree requirements:* For doctorate, thesis/dissertation, oral and written qualifying exams. *Entrance requirements:* For doctorate, GRE General Test.

University of California, Riverside, Graduate Division, Program in Neuroscience, Riverside, CA 92521-0102. Offers PhD. *Faculty:* 18 full-time (7 women). *Students:* 17 full-time (8 women); includes 2 minority (both Asian Americans or Pacific Islanders), 6 international. Average age 29. In 2005, 1 degree awarded. *Degree requirements:* For doctorate, thesis/dissertation, 2 quarters of teaching experience, qualifying exams. *Entrance requirements:* For doctorate, GRE General Test, minimum GPA of 3.2. Additional exam requirements/recommendations for international students: Required—TOEFL (minimum score 550 paper-based; 213 computer-based); Recommended—TSE. *Application deadline:* For fall admission, 5/1 for domestic students, 2/1 for international students. For winter admission, 9/1 for domestic students; for spring admission, 12/1 for domestic students. Applications are processed on a rolling basis. Application fee: $60 ($75 for international students). Electronic applications accepted. *Expenses:* Tuition, nonresident: full-time $14,694. Required fees: $9,009. Full-time tuition and fees vary according to program. *Financial support:* In 2005–06, research assistantships (averaging $14,000 per year), teaching assistantships (averaging $15,000 per year) were awarded; fellowships, tuition waivers (full and partial) also available. Financial award application deadline: 1/5; financial award applicants required to submit FAFSA. *Faculty research:* Modulation of ion channels, synaptic transmission, sensor-motor processing, plasticity in adult CNS, neural control of eating behaviors. *Unit head:* Dr. Sarjeet Gill, Director. *Application contact:* Perla Fabelo, Graduate Student Affairs Assistant, 800-735-0717, Fax: 951-827-5517, E-mail: neuro@ucr.edu.

University of California, San Diego, Graduate Studies and Research, Interdisciplinary Program in Cognitive Science, La Jolla, CA 92093. Offers cognitive science/anthropology (PhD); cognitive science/communication (PhD); cognitive science/computer science and engineering (PhD); cognitive science/linguistics (PhD); cognitive science/neuroscience (PhD); cognitive science/philosophy (PhD); cognitive science/psychology (PhD); cognitive science/sociology (PhD). Admissions through affiliated departments. *Faculty:* 65 full-time (14 women). *Students:* 7 full-time (2 women); includes 1 minority (Asian American or Pacific Islander) Average age 26. 2 applicants, 100% accepted. In 2005, 1 degree awarded. *Degree requirements:* For doctorate, thesis/dissertation, registration. *Entrance requirements:* For doctorate, GRE General Test, acceptance into one of the 8 participating departments. *Application deadline:* Applications are processed on a rolling basis. Application fee: $0. *Faculty research:* Language and cognition, philosophy of mind, visual perception, biological anthropology, sociolinguistics. *Unit head:* Gary Cottrell, Director, 858-534-7141, Fax: 858-534-1128, E-mail: gcottrell@ucsd.edu. *Application contact:* Beverley Walton, Coordinator, 858-534-4387, E-mail: bwalton@ucsd.edu.

University of California, San Diego, School of Medicine and Graduate Studies and Research, Molecular Pathology Program, La Jolla, CA 92093. Offers bioinformatics (PhD); cancer biology/oncology (PhD); cardiovascular sciences and disease (PhD); microbiology (PhD); molecular pathology (PhD); neurological disease (PhD); stem cell and developmental biology (PhD); structural biology/drug design (PhD). *Entrance requirements:* For doctorate, GRE General Test, GRE Subject Test. Additional exam requirements/recommendations for international students: Required—TOEFL. Electronic applications accepted.

See Close-Up on page 1119.

University of California, San Diego, School of Medicine and Graduate Studies and Research, Neurosciences Program, La Jolla, CA 92093. Offers PhD. *Degree requirements:* For doctorate, thesis/dissertation, qualifying exam. *Entrance requirements:* For doctorate, GRE General Test, GRE Subject Test. Additional exam requirements/recommendations for international students: Required—TOEFL. Electronic applications accepted. *Faculty research:* Neurophysiology, neuropharmacology, neurochemistry.

See Close-Up on page 1037.

University of California, San Francisco, Graduate Division, Program in Neuroscience, San Francisco, CA 94143. Offers PhD. *Degree requirements:* For doctorate, thesis/dissertation. *Entrance requirements:* For doctorate, GRE General Test, GRE Subject Test. *Faculty research:* Molecular neurobiology, synaptic plasticity, mechanisms of motor learning.

University of Chicago, Division of the Biological Sciences, Department of Neurobiology, Pharmacology, and Cell Physiology, Committee on Computational Neuroscience, Chicago, IL 60637-1513. Offers PhD. *Faculty:* 27 full-time (5 women), 2 part-time/adjunct (1 woman). *Students:* 17 full-time (7 women); includes 2 minority (1 African American, 1 Asian American or Pacific Islander), 2 international. Average age 27. *Degree requirements:* For doctorate, thesis/dissertation, preliminary exam. *Entrance requirements:* For doctorate, GRE General Test. *Application deadline:* For fall admission, 12/28 for domestic students, 12/28 for international students. Application fee: $55. Electronic applications accepted. *Financial support:* In 2005–06, fellowships with tuition reimbursements (averaging $26,301 per year), research assistantships with tuition reimbursements (averaging $26,301 per year) were awarded; institutionally sponsored loans, scholarships/grants, traineeships, and health care benefits also available. Financial award applicants required to submit FAFSA. *Unit head:* Dr. Philip Ulinski, Chair, 773-702-8081. *Application contact:* Diane J. Hall, Graduate Administrative Director, 773-702-6371, Fax: 773-702-1216, E-mail: d-hall@uchicago.edu.

University of Cincinnati, Division of Research and Advanced Studies, Interdisciplinary PhD Study Program in Neuroscience, Cincinnati, OH 45221. Offers PhD. *Degree requirements:* For doctorate, thesis/dissertation, qualifying exam. *Entrance requirements:* For doctorate, GRE General Test. Additional exam requirements/recommendations for international students: Required—TOEFL. Electronic applications accepted. *Faculty research:* Developmental neurobiology, membrane and channel biophysics, molecular neurobiology, neuroendocrinology, neuronal cell biology.

University of Colorado at Denver and Health Sciences Center, Graduate School, Program in Biomedical Sciences, Program in Neuroscience, Denver, CO 80262. Offers PhD, PhD/MD. *Faculty:* 41 full-time (12 women). *Students:* 24 full-time (12 women); includes 2 minority (both Hispanic Americans) Average age 26. 21 applicants, 48% accepted, 6 enrolled. In 2005, 4 degrees awarded. *Median time to degree:* Of those who began their doctoral program in fall 1997, 8% received their degree in 8 years or less. *Degree requirements:* For doctorate, thesis/dissertation, lab rotations, comprehensive exam. *Entrance requirements:* For doctorate, GRE, minimum GPA of 3.0. Additional exam requirements/recommendations for international students: Required—TOEFL (minimum score 570 paper-based; 230 computer-based). *Application deadline:* For fall admission, 1/15 for domestic students, 1/15 for international students. Applications are processed on a rolling basis. Application fee: $50. Electronic applications accepted. *Expenses:* Tuition, state resident: full-time $11,730. Tuition, nonresident: full-time $22,980. Tuition and fees vary according to degree level and program. *Financial support:* In 2005–06, 3 students received support, including 2 fellowships with full tuition reimbursements available (averaging $1,200 per year); institutionally sponsored loans, scholarships/grants, traineeships, and health care benefits also available. Financial award application deadline: 3/15; financial award applicants required to submit FAFSA. *Faculty research:* Neurobiology of olfaction, ion channels, schizophrenia, spinal cord regeneration, neurotransplantation. *Unit head:* Dr. Diego Restrepo, Director, 303-724-3405, Fax: 303-724-3420, E-mail: diego.restrepo@uchsc.edu. *Application contact:* Mellodee Phillips, Administrator, 303-315-7872, Fax: 303-724-3121, E-mail: mellodee.phillips@uchsc.edu.

See Close-Up on page 1039.

University of Connecticut, Graduate School, College of Liberal Arts and Sciences, Department of Psychology, Field of Psychology, Storrs, CT 06269. Offers behavioral neuroscience (PhD); biopsychology (PhD); clinical psychology (MA, PhD); cognition and instruction (PhD); developmental psychology (MA, PhD); ecological psychology (PhD); experimental psychology (PhD); general psychology (MA, PhD); industrial/organizational psychology (PhD); language acquisition (PhD); neuroscience (PhD); social psychology (MA, PhD). *Accreditation:* APA (one or more programs are accredited). *Faculty:* 51 full-time (23 women). *Students:* 134 full-time (90 women), 21 part-time (10 women); includes 13 minority (7 African Americans, 2 Asian Americans or Pacific Islanders, 4 Hispanic Americans), 21 international. Average age 28. 510 applicants, 10% accepted, 49 enrolled. In 2005, 22 master's, 15 doctorates awarded. Terminal master's awarded for partial completion of doctoral program. *Degree requirements:* For master's, comprehensive exam; for doctorate, thesis/dissertation. *Entrance requirements:* For master's and doctorate, GRE General Test, GRE Subject Test. Additional exam requirements/recommendations for international students: Required—TOEFL (minimum score 550 paper-based; 213 computer-based). *Application deadline:* For fall admission, 2/1 priority date for domestic students, 2/1 priority date for international students; for spring admission, 11/1 for domestic students, 10/1 for international students. Applications are processed on a rolling basis. Application fee: $55. Electronic applications accepted. *Expenses:* Tuition, state resident: part-time $444 per credit hour. Tuition, nonresident: part-time $1,154 per credit hour. Tuition and fees vary according to course load. *Financial support:* In 2005–06, 67 research assistantships with full tuition reimbursements, 58 teaching assistantships with full tuition reimbursements were awarded; fellowships, career-related internships or fieldwork, Federal Work-Study, scholarships/grants, health care benefits, and unspecified assistantships also available. Financial award application deadline: 2/1; financial award applicants required to submit FAFSA. *Application contact:* Deborah Doucette, Administrative Assistant, 860-486-2057, Fax: 860-486-2760, E-mail: futuregr@psych.psy.uconn.edu.

University of Connecticut Health Center, Graduate School, Programs in Biomedical Sciences, Program in Neuroscience, Farmington, CT 06030. Offers PhD, DMD/PhD, MD/PhD. *Degree requirements:* For doctorate, thesis/dissertation, comprehensive exam, registration. *Entrance requirements:* For doctorate, GRE General Test, interview (recommended). Additional exam requirements/recommendations for international students: Required—TOEFL (minimum score 600 paper-based; 250 computer-based). Electronic applications accepted. *Faculty research:* Biology of degeneration, regeneration, plasticity, and transplantation; biology of neurotransmission in the nervous system; cell structure of neural tissue and neural development.

See Close-Up on page 1041.

University of Delaware, College of Arts and Sciences, Department of Psychology, Newark, DE 19716. Offers behavioral neuroscience (PhD); clinical psychology (PhD); cognitive psychology (PhD); social psychology (PhD). *Accreditation:* APA. *Faculty:* 28 full-time (10 women), 2 part-time/adjunct (1 woman). *Students:* 62 full-time (36 women); includes 9 minority (6 African Americans, 1 Asian American or Pacific Islander, 2 Hispanic Americans), 1 international. Average age 28. 230 applicants, 8% accepted. 15 enrolled. In 2005, 3 doctorates awarded. *Degree requirements:* For doctorate, thesis/dissertation. *Entrance requirements:* For doctorate, GRE General Test. Additional exam requirements/recommendations for international students: Required—TOEFL (minimum score 600 paper-based; 250 computer-based). *Application deadline:* For fall admission, 1/7 for domestic students. Application fee: $60. Electronic applications accepted. *Financial support:* In 2005–06, 41 students received support, including 6 fellowships with full tuition reimbursements available (averaging $14,675 per year), 11 research assistantships with full tuition reimbursements available (averaging $14,675 per year), 20 teaching assistantships with full tuition reimbursements available (averaging $14,675 per year); career-related internships or fieldwork, Federal Work-Study, institutionally sponsored loans, scholarships/grants, tuition waivers (full and partial), and minority fellowships also available. Financial award application deadline: 1/7. *Faculty research:* Emotion development, neural and cognitive aspects of memory, neural control of feeding, intergroup relations, social cognition and communication. Total annual research expenditures: $3.2 million. *Unit head:* Dr. Thomas M. DiLorenzo, Chair, 302-831-2271, Fax: 302-831-3645. *Application contact:* Linda Scarpitti, Information Contact, 302-831-2271, Fax: 302-831-3645, E-mail: linnie@udel.edu.

University of Florida, College of Medicine, Department of Neuroscience, Gainesville, FL 32611. Offers MS, PhD. *Faculty:* 52. *Students:* 1 full-time (0 women); minority (African American) Terminal master's awarded for partial completion of doctoral program. *Degree requirements:* For master's and doctorate, thesis/dissertation. *Entrance requirements:* For master's and doctorate, GRE General Test, minimum GPA of 3.0. Additional exam requirements/recommendations for international students: Required—TOEFL. *Application deadline:* For fall admission, 2/15 for domestic students. Application fee: $30. Electronic applications accepted. *Expenses:* Tuition, state resident: full-time $6,234. Tuition, nonresident: full-time $21,359. Tuition and fees vary according to program. *Financial support:* In 2005–06, research assistantships with full tuition reimbursements (averaging $23,528 per year); fellowships with full tuition reimbursements, traineeships and unspecified assistantships also available. *Faculty research:* Neural injury and repair, neuroimmunology and endocrinology, neurophysiology, neurotoxicology, cellular and molecular neurobiology. *Unit head:* Dr. Douglas Anderson, Chairman, 352-392-6641, E-mail: anderson@mbi.ufl.edu. *Application contact:* Dr. Wayne McCormack, Associate Dean of Graduate Education, 352-392-7413, Fax: 352-846-3466, E-mail: mccormac@pathology.ufl.edu.

University of Florida, College of Medicine and Graduate School, Interdisciplinary Program in Biomedical Sciences, Concentration in Neuroscience, Gainesville, FL 32611. Offers PhD. *Faculty:* 24 full-time (5 women), 2 part-time/adjunct (1 woman). *Students:* 44 (24 women); includes 8 minority (2 African Americans, 5 Asian Americans or Pacific Islanders, 1 Hispanic American) 8 international. In 2005, 7 degrees awarded. *Degree requirements:* For doctorate, thesis/dissertation. *Entrance requirements:* For doctorate, GRE General Test, minimum GPA of 3.0. Additional exam requirements/recommendations for international students: Required—TOEFL. *Application deadline:* For fall admission, 2/15 for domestic students. Application fee: $30. Electronic applications accepted. *Expenses:* Tuition, state resident: full-time $6,234. Tuition, nonresident: full-time $21,359. Tuition and fees vary according to program. *Financial support:* In 2005–06, 35 research assistantships (averaging $23,528 per year) were awarded; fellowships *Faculty research:* Neural injury and repair, neurophysiology, neurotoxicology, cellular and molecular neurobiology, neuroimmunology and endocrinology. *Unit head:* Dr. Sue

Neuroscience

University of Florida (continued)

Semple-Rowland, Director, 352-392-3598, E-mail: rowland@mbi.ufl.edu. *Application contact:* Dr. Wayne McCormack, Associate Dean of Graduate Education, 352-392-7413, Fax: 352-846-3466, E-mail: mccormac@pathology.ufl.edu.

University of Georgia, Graduate School, Biomedical and Health Sciences Institute, Athens, GA 30602. Offers neuroscience (PhD). *Students:* 1 applicant, 100% accepted, 1 enrolled. *Application contact:* Dr. Jan Sandor, Director of Graduate Admissions, 706-542-1787, Fax: 706-542-9480, E-mail: gradadm@uga.edu.

University of Guelph, Ontario Veterinary College and Graduate Program Services, Graduate Programs in Veterinary Sciences, Department of Clinical Studies, Guelph, ON N1G 2W1, Canada. Offers anesthesiology (M Sc, DV Sc); cardiology (Diploma); clinical studies (Diploma); emergency/critical care (Diploma); medicine (M Sc, DV Sc); neurology (M Sc, DV Sc); ophthalmology (M Sc, DV Sc); surgery (M Sc, DV Sc). *Faculty:* 37. *Students:* 32 (20 women). *Degree requirements:* For master's, thesis/dissertation; for doctorate, thesis/dissertation, comprehensive exam. *Entrance requirements:* Additional exam requirements/recommendations for international students: Required—TOEFL (minimum score 550 paper-based; 213 computer-based), IELT (minimum score 7). *Application deadline:* For fall admission, 12/6 for domestic students. For winter admission, 10/30 for domestic students; for spring admission, 2/28 for domestic students. Applications are processed on a rolling basis. Application fee: $75. Electronic applications accepted. *Financial support:* Fellowships, research assistantships, teaching assistantships, career-related internships or fieldwork and scholarships/grants available. *Faculty research:* Orthopedics, respirology, oncology, exercise physiology, cardiology. Total annual research expenditures: $1.5 million. *Unit head:* Dr. Chris M. Brown, Chair, 519-823-8800 Ext. 54080, Fax: 519-767-0311, E-mail: cmbrown@ovc.uoguelph.ca. *Application contact:* Dr. J. Scott Weese, Graduate Coordinator, 519-823-8800 Ext. 54064, Fax: 519-767-0311, E-mail: jsweese@uoguelph.ca.

University of Hartford, College of Arts and Sciences, Department of Biology, Program in Neuroscience, West Hartford, CT 06117-1599. Offers MS. Part-time and evening/weekend programs available. *Faculty:* 2 full-time (1 woman), 3 part-time/adjunct (1 woman). *Students:* 10 full-time (6 women), 6 part-time (5 women); includes 1 minority (Hispanic American), 7 international. Average age 29. 17 applicants, 65% accepted. In 2005, 4 degrees awarded. *Degree requirements:* For master's, oral exams, thesis optional. *Entrance requirements:* For master's, GRE General Test, GRE Subject Test, MCAT. Additional exam requirements/recommendations for international students: Required—TOEFL (minimum score 550 paper-based; 213 computer-based). *Application deadline:* Applications are processed on a rolling basis. Application fee: $40 ($55 for international students). Electronic applications accepted. *Expenses:* Tuition: Part-time $515 per credit. Required fees: $200 per term. Tuition and fees vary according to program. *Financial support:* In 2005–06, 2 research assistantships with partial tuition reimbursements (averaging $6,000 per year), 5 teaching assistantships with partial tuition reimbursements (averaging $5,000 per year) were awarded. Financial award application deadline: 6/1. *Faculty research:* Neurobiology of aging, central actions of neural steroids, neuroendocrine control of reproduction, retinopathies in sharks, plasticity in the central nervous system. Total annual research expenditures: $15,900. *Unit head:* Dr. Jacob Harney, Head, 860-768-5372, Fax: 860-768-5002, E-mail: harney@hartford.edu.

See Close-Up on page 1043.

University of Illinois at Urbana–Champaign, Graduate College, College of Liberal Arts and Sciences, School of Molecular and Cellular Biology, Neuroscience Program, Champaign, IL 61820. Offers PhD. *Students:* 57 full-time (33 women), 3 part-time; includes 13 minority (1 American Indian/Alaska Native, 6 Asian Americans or Pacific Islanders, 6 Hispanic Americans), 18 international. 112 applicants, 7% accepted, 6 enrolled. In 2005, 5 degrees awarded. *Degree requirements:* For doctorate, thesis/dissertation. *Application deadline:* Applications are processed on a rolling basis. Application fee: $50 ($60 for international students). Electronic applications accepted. *Financial support:* In 2005–06, 13 fellowships, 42 research assistantships, 21 teaching assistantships were awarded. Financial award application deadline: 2/15. *Unit head:* Gene Robinson, Director, 217-333-4971, Fax: 217-244-3499, E-mail: generobi@uiuc.edu. *Application contact:* Sam Beshers, Coordinator, 217-333-4971, Fax: 217-244-3499, E-mail: beshers@uiuc.edu.

The University of Iowa, Graduate College, Program in Neuroscience, Iowa City, IA 52242-1316. Offers PhD, MD/PhD. *Students:* 12 full-time (7 women), 24 part-time (8 women); includes 3 minority (1 African American, 1 Asian American or Pacific Islander, 1 Hispanic American), 14 international. 13 applicants, 31% accepted, 3 enrolled. In 2005, 7 degrees awarded. *Degree requirements:* For doctorate, thesis/dissertation, comprehensive exam, registration. *Entrance requirements:* For doctorate, GRE General Test, minimum GPA of 3.0. Additional exam requirements/recommendations for international students: Required—TOEFL (minimum score 550 paper-based; 213 computer-based). *Application deadline:* For fall admission, 1/1 priority date for domestic students, 1/1 priority date for international students. Applications are processed on a rolling basis. Application fee: $60 ($85 for international students). Electronic applications accepted. *Expenses:* Tuition, state resident: part-time $1,882 per term. Tuition, nonresident: full-time $17,338; part-time $4,907 per term. Tuition and fees vary according to course load and program. *Financial support:* In 2005–06, 4 fellowships, 28 research assistantships with partial tuition reimbursements, 2 teaching assistantships with partial tuition reimbursements were awarded. Financial award applicants required to submit FAFSA. *Faculty research:* Molecular, cellular, and developmental systems; behavioral neurosciences. *Unit head:* Dr. Daniel Tranel, Director, 319-335-9968, E-mail: daniel-tranel@uiowa.edu. *Application contact:* Anita Kafer, Program Associate, 319-335-9968, E-mail: phd@genetics.uiowa.edu.

Announcement: The Neuroscience PhD Program emphasizes interdisciplinary training in all levels of brain research: molecular, cellular, systems, and cognitive/behavior. More than 90 faculty members have active research programs in specialties that include developmental neurobiology, firing dynamics, growth factors, ion channels, neuroimaging, and neural bases of memory, language, and emotion. Strong lab rotation system guides students' choice of a dissertation sponsor. For further information, visit http://neuroscience.grad.uiowa.edu or e-mail gen-neuro@uiowa.edu.

University of Kansas, Graduate School, School of Pharmacy, Department of Pharmacology and Toxicology, Program in Neurosciences, Lawrence, KS 66045. Offers MS, PhD. *Students:* 6 full-time (2 women), 1 (woman) part-time, 1 international. Average age 27. 9 applicants, 56% accepted. In 2005, 1 degree awarded. *Degree requirements:* For master's and doctorate, GRE. Additional exam requirements/recommendations for international students: Required—TOEFL. *Application deadline:* For fall admission, 1/15 priority date for domestic students, 1/15 priority date for international students. Applications are processed on a rolling basis. Electronic applications accepted. *Expenses:* Tuition, state resident: full-time $4,859. Tuition, nonresident: full-time $12,000. Required fees: $589. Tuition and fees vary according to program. *Financial support:* Fellowships, research assistantships, teaching assistantships available. Financial award application deadline: 1/15.

University of Lethbridge, School of Graduate Studies, Lethbridge, AB T1K 3M4, Canada. Offers accounting (MScM); addictions counseling (M Sc); agricultural biotechnology (M Sc); agricultural studies (M Sc, MA); anthropology (MA); archaeology (MA); art (MA); biochemistry (M Sc); biological sciences (M Sc); biomolecular science (PhD); biosystems and biodiversity (PhD); Canadian studies (MA); chemistry (M Sc); computer science (M Sc); computer science and geographical information science (M Sc); counseling psychology (M Ed); dramatic arts (MA); earth, space, and physical science (PhD); economics (MA); educational leadership (M Ed); English (MA); environmental science (M Sc); evolution and behavior (PhD); exercise science (M Sc); finance (MScM); French (MA); French/German (MA); French/Spanish (MA); general education (M Ed); general management (MScM); geography (M Sc, MA); German

(MA); health sciences (M Sc, MA); history (MA); human resource management and labour relations (MScM); individualized multidisciplinary (M Sc, MA); information systems (MScM); international management (MScM); kinesiology (M Sc, MA); management (M Sc, MA); marketing (MScM); mathematics (M Sc); music (MA); Native American studies (MA); neuroscience (M Sc, PhD); new media (MA); nursing (M Sc); philosophy (MA); physics (M Sc); policy and strategy (MScM); political science (MA); psychology (M Sc, MA); religious studies (MA); sociology (MA); theoretical and computational science (PhD); urban and regional studies (MA). Part-time and evening/weekend programs available. *Faculty:* 250. *Students:* 193 full-time, 145 part-time. 35 applicants, 100% accepted, 35 enrolled. In 2005, 40 degrees awarded. *Degree requirements:* For doctorate, thesis/dissertation, comprehensive exam. *Entrance requirements:* For master's, GMAT (M Sc management), bachelor's degree in related field, minimum GPA of 3.0 during previous 20 graded semester courses, 2 years teaching or related experience (M Ed); for doctorate, master's degree, minimum graduate GPA of 3.5. Additional exam requirements/recommendations for international students: Required—TOEFL. Application fee: $60 Canadian dollars. *Expenses:* Tuition, nonresident: part-time $531 per course. Required fees: $83 per year. Tuition and fees vary according to degree level and program. *Financial support:* Fellowships, research assistantships, teaching assistantships, scholarships/grants, health care benefits, and unspecified assistantships available. *Faculty research:* Movement and brain plasticity, gibberellin physiology, photosynthesis, carbon cycling, molecular properties of main-group ring components. *Unit head:* Dr. Shamsul Alam, Dean, 403-329-2121, Fax: 403-329-2097, E-mail: inquiries@uleth.ca. *Application contact:* Kathy Schrage, Administrative Assistant, Office of the Academic Vice President, 403-329-2121, Fax: 403-329-2097, E-mail: inquiries@uleth.ca.

University of Maryland, Graduate School, Graduate Programs in Medicine, Department of Physiology, Baltimore, MD 21201. Offers neuroscience (PhD); physiology (MS, PhD); reproductive endocrinology (PhD). *Faculty:* 28 full-time (6 women), 21 part-time/adjunct. *Students:* 28 full-time (15 women), 3 part-time (1 woman); includes 7 minority (2 African Americans, 5 Asian Americans or Pacific Islanders), 6 international. Average age 28. 41 applicants, 29% accepted, 6 enrolled. In 2005, 3 master's, 2 doctorates awarded. *Degree requirements:* For doctorate, thesis/dissertation. *Entrance requirements:* For doctorate, GRE General Test, GRE Subject Test, minimum GPA of 3.0. Additional exam requirements/recommendations for international students: Required—TOEFL, TOEFL or IELTS; Recommended—IELT. *Application deadline:* For fall admission, 7/1 for domestic students, 1/15 for international students. Application fee: $50. Electronic applications accepted. *Expenses:* Tuition, state resident: full-time $8,079; part-time $409 per credit hour. Tuition, nonresident: full-time $18,384; part-time $731 per credit hour. Required fees: $695; $10 per credit hour. Tuition and fees vary according to degree level and program. *Financial support:* Fellowships, research assistantships available. Financial award application deadline: 2/15. *Faculty research:* Membrane physiology, biophysics and morphology, central nervous system physiology, EEG analysis, information theory. *Unit head:* Dr. Mordecai P. Blaustein, Chairman, 410-706-3345, Fax: 410-706-3345, E-mail: mblaust@umaryland.edu. *Application contact:* Dr. Robert Koos, Director of Graduate Studies, 410-706-8033, Fax: 410-706-8341, E-mail: rkoos@umaryland.edu.

University of Maryland, Graduate School, Graduate Programs in Medicine, Program in Neuroscience and Cognitive Sciences, Baltimore, MD 21201. Offers MS, PhD, MD/PhD. Part-time programs available. *Faculty:* 55 full-time. *Students:* 59 full-time (23 women); includes 5 minority (2 African Americans, 3 Asian Americans or Pacific Islanders), 1 international. Average age 28. 62 applicants, 29% accepted, 8 enrolled. In 2005, 6 doctorates awarded. Terminal master's awarded for partial completion of doctoral program. *Degree requirements:* For master's and doctorate, thesis/dissertation. *Entrance requirements:* For master's and doctorate, GRE General Test, minimum GPA of 3.0. Additional exam requirements/recommendations for international students: Required—TOEFL, TOEFL or IELTS; Recommended—IELT. *Application deadline:* For fall admission, 7/1 for domestic students, 1/15 for international students. Application fee: $50. Electronic applications accepted. *Expenses:* Tuition, state resident: full-time $8,079; part-time $409 per credit hour. Tuition, nonresident: full-time $18,384; part-time $731 per credit hour. Required fees: $695; $10 per credit hour. Tuition and fees vary according to degree level and program. *Financial support:* Fellowships, research assistantships available. Financial award application deadline: 2/15. *Faculty research:* Molecular, biochemical, and cellular pharmacology; membrane biophysics; synaptology; developmental neurobiology. *Unit head:* Dr. Michael T. Shipley, Chair, 410-706-3590, Fax: 410-706-2512, E-mail: mshipley@umaryland.edu. *Application contact:* Dr. Peg McCarthy, Director, 410-706-8912, E-mail: mmccarth@umaryland.edu.

University of Maryland, School of Medicine, Graduate Program in Life Sciences, Program in Neuroscience, Baltimore, MD 21201. Offers PhD. *Expenses:* Tuition, state resident: full-time $8,079; part-time $409 per credit hour. Tuition, nonresident: full-time $18,384; part-time $731 per credit hour. Required fees: $695; $10 per credit hour. Tuition and fees vary according to degree level and program.

University of Maryland, Baltimore County, Graduate School, College of Natural Sciences and Mathematics, Department of Biological Sciences and Department of Psychology, Program in Neurosciences and Cognitive Sciences, Baltimore, MD 21250. Offers PhD. *Faculty:* 18 full-time (8 women). *Students:* 7 full-time (5 women); includes 4 minority (1 African American, 2 Asian Americans or Pacific Islanders, 1 Hispanic American). 8 applicants, 50% accepted, 0 enrolled. *Degree requirements:* For doctorate, thesis/dissertation, comprehensive exam (for some programs). *Entrance requirements:* For doctorate, GRE General Test, minimum GPA of 3.0. Additional exam requirements/recommendations for international students: Required—TOEFL. *Application deadline:* For fall admission, 2/1 for domestic students, 1/1 for international students. Applications are processed on a rolling basis. Application fee: $50. Electronic applications accepted. *Expenses:* Tuition, state resident: part-time $395 per credit. Tuition, nonresident: part-time $652 per credit. Required fees: $82 per credit. Tuition and fees vary according to course load, program and reciprocity agreements. *Financial support:* In 2005–06, 7 students received support, including 1 fellowship with tuition reimbursement available (averaging $22,500 per year), research assistantships with tuition reimbursements available (averaging $21,500 per year), teaching assistantships with tuition reimbursements available (averaging $20,500 per year) *Unit head:* Dr. Phyllis Robinson, Director, 410-455-3669, Fax: 410-455-3875, E-mail: biograd@umbc.edu.

University of Maryland, Baltimore County, Graduate School, College of Natural and Mathematical Sciences, Department of Chemistry and Biochemistry, Program in Biochemistry, Baltimore, MD 21250. Offers biochemistry (PhD); neuroscience (PhD). *Students:* Average age 25. *Degree requirements:* For doctorate, thesis/dissertation, comprehensive exam. *Entrance requirements:* For doctorate, GRE General Test, minimum GPA of 3.0. Additional exam requirements/recommendations for international students: Required—TOEFL (minimum score 550 paper-based; 213 computer-based). *Application deadline:* For fall admission, 7/1 priority date for domestic students, 1/1 priority date for international students; for spring admission, 12/1 for domestic students. Applications are processed on a rolling basis. Application fee: $45. Electronic applications accepted. *Expenses:* Tuition, state resident: part-time $395 per credit. Tuition, nonresident: part-time $652 per credit. Required fees: $82 per credit. Tuition and fees vary according to course load, program and reciprocity agreements. *Financial support:* Fellowships with full tuition reimbursements, research assistantships with full tuition reimbursements, teaching assistantships with tuition reimbursements, tuition waivers (full) available. *Unit head:* Dr. Michael F. Summers, Graduate Coordinator, 410-455-2527. *Application contact:* Kathleen Reinecke, Biochemistry Coordinator, 410-706-8417, Fax: 410-706-8297, E-mail: kreineck@umaryland.edu.

See Close-Up on page 433.

University of Maryland, College Park, Graduate Studies, College of Behavioral and Social Sciences, Department of Hearing and Speech Sciences, College Park, MD 20742. Offers audiology (MA, PhD); hearing and speech sciences (Au D); language pathology (MA, PhD); neuroscience (PhD); speech (MA, PhD). *Accreditation:* ASHA (one or more programs are

Neuroscience

accredited). *Faculty:* 17 full-time (all women), 15 part-time/adjunct (12 women). *Students:* 70 full-time (69 women), 16 part-time (all women). 234 applicants, 45% accepted, 36 enrolled. In 2005, 13 master's, 1 doctorate awarded. *Degree requirements:* For master's, thesis optional; for doctorate, thesis/dissertation, written and oral exams. *Entrance requirements:* For master's, GRE General Test, minimum GPA of 3.5, 3 letters of recommendation; for doctorate, GRE General Test, minimum GPA of 3.5. Additional exam requirements/recommendations for international students: Required—TOEFL. *Application deadline:* For fall admission, 2/1 for domestic students, 2/1 for international students. Applications are processed on a rolling basis. Application fee: $60. Electronic applications accepted. *Financial support:* In 2005–06, 21 fellowships with full tuition reimbursements (averaging $9,144 per year), 8 research assistantships (averaging $7,870 per year), 14 teaching assistantships with tuition reimbursements (averaging $6,756 per year) were awarded; career-related internships or fieldwork, Federal Work-Study, and scholarships/grants also available. Support available to part-time students. Financial award applicants required to submit FAFSA. *Faculty research:* Speech perception, language acquisition, bilingualism, hearing loss. Total annual research expenditures: $404,799. *Unit head:* Dr. Nan B. Ratner, Chair, 301-405-4217, Fax: 301-314-2023, E-mail: nratner@hesp.umd.edu. *Application contact:* Dean of Graduate School, 301-405-4190, Fax: 301-314-9305.

See Close-Up on page 1047.

University of Maryland, College Park, Graduate Studies, Interdepartmental Programs, Program in Neurosciences and Cognitive Sciences, College Park, MD 20742. Offers PhD. *Students:* 36 full-time (19 women), 4 part-time; includes 6 minority (3 African Americans, 1 American Indian/Alaska Native, 2 Asian Americans or Pacific Islanders), 17 international. 73 applicants, 7% accepted, 5 enrolled. In 2005, 6 degrees awarded. *Degree requirements:* For doctorate, thesis/dissertation, comprehensive exam. *Entrance requirements:* For doctorate, GRE General Test, 3 letters of recommendation. Additional exam requirements/recommendations for international students: Required—TOEFL. *Application deadline:* For fall admission, 1/7 for domestic students, 1/7 for international students. Applications are processed on a rolling basis. Application fee: $60. Electronic applications accepted. *Financial support:* In 2005–06, 10 fellowships (averaging $11,747 per year) were awarded; teaching assistantships, Federal Work-Study and scholarships/grants also available. Support available to part-time students. Financial award applicants required to submit FAFSA. *Faculty research:* Molecular neurobiology, cognition, neural and behavioral systems language, memory, human development. *Unit head:* Dr. Norbert Hornstein, Chairman, 301-405-7002, Fax: 301-314-7104, E-mail: nhorste@umd.edu.

University of Maryland, College Park, National Institutes of Health Sponsored Programs, Department of Hearing and Speech Sciences, College Park, MD 20742. Offers audiology (MA, PhD); hearing and speech sciences (Au D); language pathology (MA, PhD); neuroscience (PhD); speech (MA, PhD). *Faculty:* 17 full-time (all women), 15 part-time/adjunct (12 women). *Students:* 70 full-time (69 women), 16 part-time (all women). 234 applicants, 45% accepted, 36 enrolled. In 2005, 13 master's, 1 doctorate awarded. *Degree requirements:* For master's, thesis optional; for doctorate, thesis/dissertation, written and oral examinations. *Entrance requirements:* For master's, GRE General Test, minimum GPA of 3.5, 3 letters of recommendation; for doctorate, GRE General Test, minimum GPA of 3.5. Additional exam requirements/recommendations for international students: Required—TOEFL. *Application deadline:* For fall admission, 2/1 for domestic students, 2/1 for international students. Applications are processed on a rolling basis. Application fee: $60. Electronic applications accepted. *Financial support:* In 2005–06, 21 fellowships with full tuition reimbursements (averaging $9,144 per year), 8 research assistantships (averaging $7,870 per year), 14 teaching assistantships with tuition reimbursements (averaging $6,756 per year) were awarded; career-related internships or fieldwork, Federal Work-Study, and scholarships/grants also available. Support available to part-time students. Financial award applicants required to submit FAFSA. *Faculty research:* Speech, perception, language acquisition, bilingualism, hearing loss. Total annual research expenditures: $404,799. *Application contact:* Dean of Graduate School, 301-405-4190, Fax: 301-314-9305.

See Close-Up on page 1047.

University of Massachusetts Amherst, Graduate School, Interdisciplinary Programs, Program in Neuroscience and Behavior, Amherst, MA 01003. Offers MS, PhD. *Students:* 42 full-time (30 women), 1 (woman) part-time; includes 6 minority (2 African Americans, 2 Asian Americans or Pacific Islanders, 2 Hispanic Americans), 4 international. Average age 27. 65 applicants, 26% accepted, 13 enrolled. In 2005, 3 degrees awarded. Terminal master's awarded for partial completion of doctoral program. *Degree requirements:* For master's and doctorate, thesis/dissertation. *Entrance requirements:* For master's and doctorate, GRE General Test, GRE Subject Test. Additional exam requirements/recommendations for international students: Required—TOEFL (minimum score 530 paper-based; 197 computer-based). *Application deadline:* For fall admission, 1/2 priority date for domestic students, 1/2 priority date for international students. Applications are processed on a rolling basis. Application fee: $40 ($65 for international students). Electronic applications accepted. *Expenses:* Tuition, state resident: part-time $110 per credit. Tuition, nonresident: part-time $414 per credit. Required fees: $2,824 per term. One-time fee: $250 part-time. Full-time tuition and fees vary according to course load, campus/location, program and reciprocity agreements. *Financial support:* In 2005–06, 22 fellowships with full tuition reimbursements (averaging $5,798 per year), 1 research assistantship with full tuition reimbursement (averaging $5,936 per year) were awarded; teaching assistantships with full tuition reimbursements, career-related internships or fieldwork, Federal Work-Study, scholarships/grants, traineeships, and unspecified assistantships also available. Support available to part-time students. Financial award application deadline: 1/12. *Unit head:* Dr. Jerrold S. Meyer, Acting Director, 413-545-2046, Fax: 413-545-3243, E-mail: jmeyer@psych.umass.edu. *Application contact:* 413-545-0721.

University of Massachusetts Worcester, Graduate School of Biomedical Sciences, Program in Neuroscience, Worcester, MA 01655-0115. Offers PhD. *Faculty:* 50 full-time (9 women). *Degree requirements:* For doctorate, thesis/dissertation. *Entrance requirements:* For doctorate, GRE General Test. Additional exam requirements/recommendations for international students: Required—TOEFL (minimum score 600 paper-based; 250 computer-based). *Application deadline:* For fall admission, 12/15 for domestic students, 12/15 for international students. Application fee: $25 ($50 for international students). *Expenses:* Tuition, state resident: full-time $2,640. Tuition, nonresident: full-time $9,856. Required fees: $5,685. *Financial support:* In 2005–06, research assistantships with full tuition reimbursements (averaging $25,235 per year); unspecified assistantships also available. *Faculty research:* Molecular, biophysical, and cellular techniques to investigate cell function with emphasis on signal transduction processes, cell growth, and proliferation. *Unit head:* Dr. Dvid Weaver, Director, 508-856-2495. *Application contact:* Michael Cole, Director of Admissions and Recruitment, 508-856-4779, Fax: 508-856-3659, E-mail: michael.cole@umassmed.edu.

University of Medicine and Dentistry of New Jersey, Graduate School of Biomedical Sciences, Graduate Programs in Biomedical Sciences–Newark, Program in Integrative Neuroscience, Newark, NJ 07107. Offers PhD. *Degree requirements:* For doctorate, thesis/dissertation, qualifying exam. *Entrance requirements:* For doctorate, GRE General Test, minimum GPA of 3.5. Additional exam requirements/recommendations for international students: Required—TOEFL. *Application deadline:* For fall admission, 2/1 for domestic students. Application fee: $40. *Financial support:* Fellowships, research assistantships available. Financial award application deadline: 5/1. *Unit head:* Dr. Steven Levison, Program Director, 973-676-1000 Ext. 11155, Fax: 973-972-5059, E-mail: levisost@umdnj.edu.

University of Medicine and Dentistry of New Jersey, Graduate School of Biomedical Sciences, Graduate Programs in Biomedical Sciences–Piscataway, Program in Neuroscience, Piscataway, NJ 08854-5635. Offers MS, PhD, MD/PhD. *Degree requirements:* For master's and doctorate, thesis/dissertation, qualifying exam. *Application deadline:* For fall admission, 1/5 for domestic students. Application fee: $40. *Unit head:* Dr. David Egger, Acting Co-Director, 732-235-4822, Fax: 732-235-4029, E-mail: egger@umdnj.edu.

University of Miami, Graduate School, College of Arts and Sciences, Department of Psychology, Coral Gables, FL 33124. Offers adult clinical (PhD); applied developmental psychology (PhD); behavioral neuroscience (PhD); child clinical (PhD); health clinical (PhD); psychology (MS). *Accreditation:* APA (one or more programs are accredited). *Faculty:* 29 full-time (15 women). *Students:* 92 full-time (67 women); includes 26 minority (6 African Americans, 5 Asian Americans or Pacific Islanders, 15 Hispanic Americans). Average age 25. 280 applicants, 10% accepted, 21 enrolled. In 2005, 12 degrees awarded. *Median time to degree:* Of those who began their doctoral program in fall 1997, 75% received their degree in 8 years or less. *Degree requirements:* For doctorate, thesis/dissertation, comprehensive exam. *Entrance requirements:* For doctorate, GRE General Test, minimum GPA of 3.5. Additional exam requirements/recommendations for international students: Required—TOEFL. *Application deadline:* For fall admission, 12/1 for domestic students, 12/1 for international students. Application fee: $50. Electronic applications accepted. *Financial support:* In 2005–06, 11 fellowships with full tuition reimbursements (averaging $20,772 per year), 47 research assistantships with full tuition reimbursements (averaging $20,772 per year), 21 teaching assistantships with full tuition reimbursements (averaging $15,579 per year) were awarded; career-related internships or fieldwork, institutionally sponsored loans, scholarships/grants, and traineeships also available. Financial award applicants required to submit FAFSA. *Faculty research:* Behavioral factors in cardiovascular disease and cancer adult psychopathology, developmental disabilities, social and emotional development, mechanisms of coping. Total annual research expenditures: $13.2 million. *Unit head:* Dr. A. Rodney Wellens, Chairman, 305-284-2814, Fax: 305-284-3402. *Application contact:* Patricia L. Perreira, Staff Associate, 305-284-2814, Fax: 305-284-3402, E-mail: inquire@psy.miami.edu.

University of Miami, Graduate School, Miller School of Medicine, Graduate Programs in Medicine, Neuroscience Program, Miami, FL 33101. Offers PhD, MD/PhD. *Students:* 32 (13 women); includes 4 minority (1 American Indian/Alaska Native, 1 Asian American or Pacific Islander, 2 Hispanic Americans) 7 international. Average age 28. 67 applicants, 15% accepted, 5 enrolled. In 2005, 4 degrees awarded. *Degree requirements:* For doctorate, thesis/dissertation, qualifying exam. *Entrance requirements:* For doctorate, GRE General Test. Additional exam requirements/recommendations for international students: Required—TOEFL (minimum score 550 paper-based; 213 computer-based). *Application deadline:* For fall admission, 1/15 priority date for domestic students, 1/15 priority date for international students. Applications are processed on a rolling basis. Application fee: $50. Electronic applications accepted. *Financial support:* In 2005–06, 17 fellowships with tuition reimbursements (averaging $22,000 per year), 4 research assistantships with tuition reimbursements (averaging $22,000 per year) were awarded; institutionally sponsored loans also available. *Faculty research:* Cellular and molecular biology, transduction, nerve regeneration and embryonic development, membrane biophysics. Total annual research expenditures: $6 million. *Unit head:* Dr. John L. Bixby, Chairman, Neuroscience Program Steering Committee, 305-243-4874, Fax: 305-243-2970, E-mail: jbixby@miami.edu. *Application contact:* Samone Welch, Graduate Coordinator, 305-243-3368, Fax: 305-243-2970, E-mail: neurosci@med.miami.edu.

See Close-Up on page 1049.

University of Michigan, Horace H. Rackham School of Graduate Studies, Neuroscience Program, Ann Arbor, MI 48109. Offers PhD. *Faculty:* 117 full-time (33 women). *Students:* 74 full-time (41 women); includes 25 minority (3 African Americans, 15 Asian Americans or Pacific Islanders, 7 Hispanic Americans). Average age 26. 88 applicants, 25% accepted, 10 enrolled. In 2005, 11 doctorates awarded. *Median time to degree:* Of those who began their doctoral program in fall 1997, 100% received their degree in 8 years or less. *Degree requirements:* For doctorate, oral defense of dissertation, preliminary exam. *Entrance requirements:* For doctorate, GRE General Test, GRE Subject Test. Additional exam requirements/recommendations for international students: Required—TOEFL. *Application deadline:* For fall admission, 12/15 for domestic students. Application fee: $55. Electronic applications accepted. *Expenses:* Tuition, state resident: full-time $14,082; part-time $894 per credit hour. Tuition, nonresident: full-time $28,500; part-time $1,675 per credit hour. Required fees: $189; $189 per unit. *Financial support:* In 2005–06, 20 fellowships with full tuition reimbursements (averaging $23,500 per year), 54 research assistantships with full tuition reimbursements (averaging $23,500 per year), teaching assistantships with full tuition reimbursements (averaging $23,500 per year) were awarded; traineeships and health care benefits also available. *Faculty research:* Developmental neurobiology, molecular neurobiology, systems neurobiology, hearing and chemical senses. *Unit head:* Dr. Peter Hitchcock, Director, 734-763-8169, Fax: 734-746-0884, E-mail: peterh@umich.edu. *Application contact:* Student Services, 734-763-9638, Fax: 734-647-0717, E-mail: neuroscience.program@umich.edu.

University of Minnesota, Twin Cities Campus, Graduate School, Graduate Program in Neuroscience, Minneapolis, MN 55455-0213. Offers MS, PhD. Terminal master's awarded for partial completion of doctoral program. *Degree requirements:* For master's and doctorate, thesis/dissertation. *Entrance requirements:* For doctorate, GRE. Electronic applications accepted. *Expenses:* Tuition, state resident: full-time $8,748; part-time $729 per credit. Tuition, nonresident: full-time $15,848; part-time $1,321 per credit. Full-time tuition and fees vary according to class time, course load, program and reciprocity agreements. *Faculty research:* Cellular and molecular neuroscience, behavioral neuroscience, developmental neuroscience, neurodegenerative diseases, pain, addiction, motor control.

University of Nebraska Medical Center, Graduate Studies, Department of Pharmacology and Experimental Neuroscience, Omaha, NE 68198. Offers neuroscience (MS, PhD); pharmacology (MS, PhD). *Faculty:* 20 full-time, 9 part-time/adjunct. *Students:* 21 full-time (13 women), 4 international. Average age 27. 18 applicants, 22% accepted, 4 enrolled. In 2005, 1 master's, 2 doctorates awarded. Terminal master's awarded for partial completion of doctoral program. *Median time to degree:* Of those who began their doctoral program in fall 1997, 70% received their degree in 8 years or less. *Degree requirements:* For master's and doctorate, thesis/dissertation, comprehensive exam, registration. *Entrance requirements:* For master's and doctorate, GRE General Test. Additional exam requirements/recommendations for international students: Required—TOEFL (minimum score 600 paper-based; 250 computer-based). *Application deadline:* Applications are processed on a rolling basis. Application fee: $45. Electronic applications accepted. *Expenses:* Tuition, area resident: Part-time $200 per hour. Tuition, nonresident: part-time $538 per hour. Required fees: $308; $59 per term. *Financial support:* In 2005–06, 1 fellowship with full tuition reimbursement (averaging $21,000 per year), 9 research assistantships with full tuition reimbursements (averaging $21,000 per year) were awarded; career-related internships or fieldwork, institutionally sponsored loans, scholarships/grants, and unspecified assistantships also available. Support available to part-time students. *Faculty research:* Neuropharmacology, molecular pharmacology, toxicology, molecular biology, neuroscience. Total annual research expenditures: $1.8 million. *Unit head:* Dr. Jialin Zheng, Program Chair, 402-559-5656, Fax: 402-559-7495, E-mail: jzheng@unmc.edu.

University of New Mexico, School of Medicine, Biomedical Sciences Graduate Program, Albuquerque, NM 87131-5196. Offers biochemistry and molecular biology (MS, PhD); cell biology and physiology (MS, PhD); molecular genetics and microbiology (MS, PhD); neuroscience (MS, PhD); pathology (MS, PhD); toxicology (MS, PhD). Part-time programs available. Terminal master's awarded for partial completion of doctoral program. *Degree requirements:* For master's, thesis/dissertation; for doctorate, thesis/dissertation, comprehensive exam. *Entrance requirements:* For master's and doctorate, GRE General Test, minimum undergraduate GPA of 3.0. Additional exam requirements/recommendations for international students: Required—TOEFL. Electronic applications accepted. *Expenses:* Tuition, state resident: full-time $5,676. Tuition, nonresident: full-time $14,974; part-time $238 per credit hour. Required fees: $385 per term. Tuition and fees vary according to course load and program. *Faculty research:* Signal transduction, infectious disease, biology of cancer, structural biology, neuroscience.

University of Oklahoma Health Sciences Center, College of Medicine and Graduate College, Graduate Programs in Medicine, Department of Neuroscience, Oklahoma City, OK

Neuroscience

University of Oklahoma Health Sciences Center (continued)
73190. Offers MS, PhD. *Degree requirements:* For doctorate, thesis/dissertation. *Entrance requirements:* For master's and doctorate, GRE General Test, 3 letters of recommendation. Additional exam requirements/recommendations for international students: Required—TOEFL.

University of Oregon, Graduate School, College of Arts and Sciences, Department of Biology, Eugene, OR 97403. Offers ecology and evolution (MA, MS, PhD); marine biology (MA, MS, PhD); molecular, cellular and genetic biology (PhD); neuroscience and development (PhD). *Faculty:* 42 full-time (14 women), 5 part-time/adjunct (2 women). *Students:* 92 full-time (43 women), 2 part-time (both women); includes 9 minority (2 African Americans, 1 American Indian/Alaska Native, 4 Asian Americans or Pacific Islanders, 2 Hispanic Americans), 6 international. 18 applicants, 83% accepted. In 2005, 11 master's, 7 doctorates awarded. Terminal master's awarded for partial completion of doctoral program. *Degree requirements:* For master's, thesis (for some programs); for doctorate, thesis/dissertation. *Entrance requirements:* For master's and doctorate, GRE General Test, minimum GPA of 3.2. Additional exam requirements/recommendations for international students: Required—TOEFL. *Application deadline:* For fall admission, 12/15 for domestic students. *Financial support:* In 2005–06, 36 teaching assistantships were awarded; research assistantships, Federal Work-Study, institutionally sponsored loans, and scholarships/grants also available. Financial award application deadline: 2/1. *Faculty research:* Developmental neurobiology; evolution, population biology, and quantitative genetics; regulation of gene expression; biochemistry of marine organisms. *Unit head:* George Sprague, Head, 541-346-6051, Fax: 541-346-6056. *Application contact:* Lynne Romans, Admissions Contact, 541-346-4252, Fax: 541-346-6056, E-mail: lromans@uoregon.edu.

University of Pennsylvania, School of Medicine, Biomedical Graduate Studies, Graduate Group in Neuroscience, Philadelphia, PA 19104. Offers PhD, MD/PhD. *Faculty:* 95. *Students:* 129 full-time (60 women); includes 26 minority (4 African Americans, 2 American Indian/Alaska Native, 16 Asian Americans or Pacific Islanders, 4 Hispanic Americans), 16 international. 193 applicants, 22% accepted, 10 enrolled. In 2005, 8 doctorates awarded. *Degree requirements:* For doctorate, thesis/dissertation, research project. *Entrance requirements:* For doctorate, GRE General Test. Additional exam requirements/recommendations for international students: Required—TOEFL. *Application deadline:* For fall admission, 12/15 priority date for domestic students, 12/1 priority date for international students. Applications are processed on a rolling basis. Application fee: $70. Electronic applications accepted. *Financial support:* Fellowships, research assistantships, teaching assistantships, scholarships/grants, traineeships, and unspecified assistantships available. *Faculty research:* Molecular and cellular neuroscience, behavioral neuroscience, developmental neurobiology, systems neuroscience and neurophysiology, neurochemistry. *Unit head:* Dr. Michael P. Nusbaum, Chairperson, 215-898-1585. *Application contact:* Fiona Cowan, Coordinator, 215-898-8048, Fax: 215-573-2248, E-mail: fiona@mail.med.upenn.edu.

University of Pittsburgh, Center for Neuroscience, Pittsburgh, PA 15260. Offers neurobiology (PhD); neuroscience (PhD). *Faculty:* 88 full-time (20 women). *Students:* 81 full-time (41 women); includes 12 minority (2 African Americans, 1 American Indian/Alaska Native, 6 Asian Americans or Pacific Islanders, 3 Hispanic Americans), 17 international. Average age 25. 138 applicants, 31% accepted, 17 enrolled. In 2005, 7 doctorates awarded. *Median time to degree:* Of those who began their doctoral program in fall 1997, 100% received their degree in 8 years or less. *Degree requirements:* For doctorate, thesis/dissertation, comprehensive exam, registration. *Entrance requirements:* For doctorate, GRE, interview. Additional exam requirements/recommendations for international students: Required—TOEFL (minimum score 550 paper-based; 213 computer-based). *Application deadline:* For fall admission, 1/2 priority date for domestic students, 1/2 priority date for international students. Application fee: $50. Electronic applications accepted. *Expenses:* Contact institution. Tuition and fees vary according to campus/location and program. *Financial support:* In 2005–06, 38 fellowships with full tuition reimbursements (averaging $21,500 per year), 40 research assistantships with full tuition reimbursements (averaging $21,500 per year), 3 teaching assistantships with full tuition reimbursements (averaging $21,500 per year) were awarded. Financial award application deadline: 1/2. *Faculty research:* Behavioral/systems/cognitive, cell and molecular, development/plasticity/repair, neurobiology of disease. *Unit head:* Dr. Alan Sved, Co-Director, Graduate Program, 412-624-6996, Fax: 412-624-9188. *Application contact:* Joan M. Blaney, Administrator, 412-624-5043, Fax: 412-624-9198, E-mail: jblaney@pitt.edu.

See Close-Up on page 1053.

University of Rochester, School of Medicine and Dentistry, Graduate Programs in Medicine and Dentistry, Department of Neurobiology and Anatomy, Interdepartmental Program in Neuroscience, Rochester, NY 14627-0250. Offers MS, PhD. Terminal master's awarded for partial completion of doctoral program. *Degree requirements:* For doctorate, one foreign language, thesis/dissertation, qualifying exam. *Entrance requirements:* For master's and doctorate, GRE General Test.

University of South Alabama, College of Medicine and Graduate School, Program in Basic Medical Sciences, Specialization in Cell Biology and Neuroscience, Mobile, AL 36688-0002. Offers PhD. *Faculty:* 13 full-time (2 women). In 2005, 1 degree awarded. *Degree requirements:* For doctorate, thesis/dissertation, oral and written preliminary exams, research proposal, qualifying exam. *Entrance requirements:* For doctorate, GRE General Test, minimum GPA of 3.0. Additional exam requirements/recommendations for international students: Required—TOEFL. *Application deadline:* For fall admission, 4/30 for domestic students. Applications are processed on a rolling basis. Application fee: $25. *Expenses:* Tuition, state resident: full-time $4,008. Tuition, nonresident: full-time $8,016. Required fees: $692. *Financial support:* Fellowships, institutionally sponsored loans available. Financial award application deadline:4/1. *Faculty research:* Cytoskeleton-membrane interactions, neural basis of oral motor behavior, microtubule organizing centers, molecular biology of human chromosomes, mechanisms of synaptic transmissions. *Unit head:* Dr. Glen Wilson, Chair, 251-460-6490. *Application contact:* Lanette Flagge, Academic Advisor, 251-460-6153.

The University of South Dakota, School of Medicine and Health Sciences and Graduate School, Biomedical Sciences Graduate Program, Program in Neuroscience, Vermillion, SD 57069-2390. Offers MA, PhD. *Faculty:* 7 full-time (2 women), 2 part-time/adjunct (1 woman). *Students:* 11 full-time (6 women), 3 international. Average age 26. 11 applicants, 45% accepted, 3 enrolled. In 2005, 1 degree awarded. Terminal master's awarded for partial completion of doctoral program. *Degree requirements:* For master's, thesis/dissertation, registration; for doctorate, thesis/dissertation, comprehensive exam, registration. *Entrance requirements:* For master's and doctorate, GRE General Test, minimum GPA of 3.0. Additional exam requirements/recommendations for international students: Required—TOEFL (minimum score 550 paper-based; 213 computer-based). *Application deadline:* For fall admission, 4/15 for domestic students, 4/15 for international students. Applications are processed on a rolling basis. Application fee: $35. *Expenses:* Contact institution. *Financial support:* In 2005–06, 11 students received support, including 10 fellowships with partial tuition reimbursements available (averaging $20,772 per year), 1 research assistantship with partial tuition reimbursement available (averaging $10,386 per year); traineeships and unspecified assistantships also available. Financial award application deadline: 4/15; financial award applicants required to submit FAFSA. *Faculty research:* Central nervous system learning, neural plasticity, respiratory control. Total annual research expenditures: $1.3 million.

University of Southern California, Graduate School, College of Letters, Arts and Sciences, Neuroscience Graduate Program, Los Angeles, CA 90089. Offers PhD. *Degree requirements:* For doctorate, thesis/dissertation. *Entrance requirements:* For doctorate, GRE General Test. Additional exam requirements/recommendations for international students: Required—TOEFL. *Expenses:* Tuition: Full-time $25,416; part-time $1,059 per unit. Required fees: $484; $484 per year. Tuition and fees vary according to course load and program. *Faculty research:* Cellular

and molecular neurobiology, behavioral and systems neurobiology, cognitive neuroscience, computation neuroscience and neural engineering, neuroscience of aging.

See Close-Up on page 1055.

The University of Texas at Austin, Graduate School, The Institute for Neuroscience, Austin, TX 78712-1111. Offers MA, PhD, MD/PhD. Terminal master's awarded for partial completion of doctoral program. *Degree requirements:* For master's and doctorate, thesis/dissertation. *Entrance requirements:* For master's and doctorate, GRE. Electronic applications accepted. *Faculty research:* Cellular/molecular biology, neurobiology, pharmacology, behavioral neuroscience.

See Close-Up on page 1059.

The University of Texas at Dallas, School of Behavioral and Brain Sciences, Program in Cognition and Neuroscience, Richardson, TX 75083-0688. Offers MS, PhD. Part-time and evening/weekend programs available. *Faculty:* 11 full-time (1 woman). *Students:* 35 full-time (19 women), 28 part-time (20 women); includes 16 minority (2 African Americans, 10 Asian Americans or Pacific Islanders, 4 Hispanic Americans), 9 international. Average age 32. 54 applicants, 59% accepted, 20 enrolled. In 2005, 22 master's, 3 doctorates awarded. *Degree requirements:* For master's, internship. *Entrance requirements:* For master's and doctorate, GRE General Test, minimum GPA of 3.0 in upper-level coursework in field. Additional exam requirements/recommendations for international students: Required—TOEFL (minimum score 550 paper-based; 213 computer-based). *Application deadline:* For fall admission, 7/15 for domestic students; for spring admission, 11/15 for domestic students. Applications are processed on a rolling basis. Application fee: $50 ($100 for international students). Electronic applications accepted. *Expenses:* Tuition, state resident: part-time $5,450; part-time $303 per credit. Tuition, nonresident: full-time $12,648; part-time $703 per credit. Tuition and fees vary according to program. *Financial support:* In 2005–06, 6 research assistantships with tuition reimbursements (averaging $10,576 per year), 12 teaching assistantships with tuition reimbursements (averaging $9,586 per year) were awarded; fellowships, Federal Work-Study also available. Support available to part-time students. Financial award application deadline: 4/30; financial award applicants required to submit FAFSA. *Faculty research:* Combination of biological, behavioral, and computational approaches for evaluating biological and artificial information processing systems. *Unit head:* Dr. Jim Bartlett, Head, PHD Program, 972-83-2355, Fax: 972-883-2491, E-mail: jbartlet@utdallas.edu. *Application contact:* Dr. Robert D. Stillman, Head, 972-883-3106, Fax: 972-883-3022, E-mail: stillman@utdallas.edu.

The University of Texas Health Science Center at Houston, Graduate School of Biomedical Sciences, Program in Neuroscience, Houston, TX 77225-0036. Offers MS, PhD, MD/PhD. *Faculty:* 40 full-time (8 women). *Students:* 30 full-time (16 women); includes 4 minority (1 American Indian/Alaska Native, 3 Asian Americans or Pacific Islanders), 6 international. Average age 26. 34 applicants, 97% accepted, 13 enrolled. In 2005, 4 degrees awarded. Terminal master's awarded for partial completion of doctoral program. *Degree requirements:* For master's and doctorate, thesis/dissertation. *Entrance requirements:* For master's and doctorate, GRE General Test. Additional exam requirements/recommendations for international students: Required—TOEFL, TWE. *Application deadline:* For fall admission, 1/15 for domestic students; for spring admission, 11/1 for domestic students. Applications are processed on a rolling basis. Application fee: $10. Electronic applications accepted. *Financial support:* Fellowships with full tuition reimbursements, research assistantships with full tuition reimbursements, teaching assistantships, institutionally sponsored loans, scholarships/grants, and health care benefits available. Financial award application deadline: 1/15. *Faculty research:* Neuroplasticity and memory, visual sciences, neurotrauma and stroke, neurotransmitter vesicle trafficking and secretion, cognition and autism. *Unit head:* Dr. Michael D Mauk, Director, 713-500-5619, Fax: 713-500-0621, E-mail: michael.d.mauk@uth.tmc.edu. *Application contact:* Dr. Victoria P. Knutson, Assistant Dean of Admissions, 713-500-9860, Fax: 713-500-9877, E-mail: victoria.p.knutson@uth.tmc.edu.

The University of Texas Medical Branch, Graduate School of Biomedical Sciences, Program in Neuroscience, Galveston, TX 77555. Offers PhD. *Students:* 18 full-time (7 women), 1 (woman) part-time; includes 4 minority (3 Asian Americans or Pacific Islanders, 1 Hispanic American), 6 international. Average age 28. In 2005, 3 degrees awarded. *Degree requirements:* For doctorate, thesis/dissertation. *Entrance requirements:* For doctorate, GRE General Test. Additional exam requirements/recommendations for international students: Required—TOEFL (minimum score 550 paper-based; 213 computer-based). *Application deadline:* Applications are processed on a rolling basis. Application fee: $30 ($75 for international students). Electronic applications accepted. *Expenses:* Tuition, state resident: full-time $8,350; part-time $90 per credit hour. Tuition, nonresident: full-time $21,450; part-time $366 per credit hour. Required fees: $1,027; $11 per credit hour. $60 per term. *Financial support:* In 2005–06, fellowships (averaging $23,000 per year), research assistantships with full tuition reimbursements (averaging $23,000 per year) were awarded. Financial award applicants required to submit FAFSA. *Unit head:* Dr. James E. Blankenship, Director, 409-772-2107, Fax: 409-762-4687, E-mail: jeblanke@utmb.edu. *Application contact:* Cindy Cheatham, Coordinator II Special Programs, 409-772-2107, Fax: 409-762-9382, E-mail: cacheath@utmb.edu.

The University of Texas Southwestern Medical Center at Dallas, Southwestern Graduate School of Biomedical Sciences, Division of Basic Science, Program in Neuroscience, Dallas, TX 75390. Offers PhD. *Faculty:* 35 full-time (5 women). *Students:* 59 full-time (27 women); includes 14 minority (1 African American, 9 Asian Americans or Pacific Islanders, 4 Hispanic Americans), 23 international. Average age 27. In 2005, 6 doctorates awarded. *Degree requirements:* For doctorate, thesis/dissertation, qualifying exam. *Entrance requirements:* For doctorate, GRE General Test, minimum GPA of 3.0. Additional exam requirements/recommendations for international students: Required—TOEFL. *Application deadline:* For fall admission, 1/5 for domestic students. Applications are processed on a rolling basis. Application fee: $0. Electronic applications accepted. *Expenses:* Tuition, state resident: full-time $6,650; part-time $50 per credit hour. Tuition, nonresident: full-time $19,650; part-time $366 per credit hour. Required fees: $42 per credit hour. Tuition and fees vary according to degree level and program. *Financial support:* Fellowships, research assistantships available. *Faculty research:* Ion channels, sensory transduction, membrane excitability and biophysics, synaptic transmission, developmental neurogenetics. *Unit head:* Dr. Jane E. Johnson, Chair, 214-648-1870, Fax: 214-648-1801, E-mail: jane.johnson@utsouthwestern.edu. *Application contact:* Dr. Nancy E. Street, Associate Dean, 214-648-6708, Fax: 214-648-2102, E-mail: nancy.street@utsouthwestern.edu.

The University of Toledo, College of Graduate Studies, Program in Cellular and Molecular Neurobiology, Toledo, OH 43606-3390. Offers MS, PhD. Part-time programs available. Terminal master's awarded for partial completion of doctoral program. *Degree requirements:* For master's and doctorate, thesis/dissertation, qualifying exam. *Entrance requirements:* For master's and doctorate, GRE General Test, minimum undergraduate GPA of 3.0. *Expenses:* Tuition, state resident: full-time $6,623; part-time $308 per credit hour. Tuition, nonresident: full-time $13,232; part-time $735 per credit hour. *Faculty research:* Developmental neuroscience, sensory systems, neuropharmacology, neurophysiology, substances.

See Close-Up on page 311.

University of Utah, School of Medicine and The Graduate School, Graduate Programs in Medicine, Program in Neuroscience, Salt Lake City, UT 84112-1107. Offers PhD. *Degree requirements:* For doctorate, thesis/dissertation. *Entrance requirements:* For doctorate, GRE General Test, minimum GPA of 3.0. Additional exam requirements/recommendations for international students: Required—TOEFL (minimum score 500 paper-based; 173 computer-based); Recommended—TWE (minimum score 6). Electronic applications accepted. *Expenses:* Tuition, state resident: full-time $2,932; part-time $369 per credit. Tuition, nonresident: full-time $10,350; part-time $1,302 per credit. Required fees: $516 per term. Tuition and fees vary according to course load and program. *Faculty research:* Brain and behavioral neuroscience,

Neuroscience

cellular neuroscience, molecular neuroscience, neurobiology of disease, developmental neuroscience.

See Close-Up on page 1061.

University of Vermont, College of Medicine and Graduate College, Graduate Programs in Medicine, Department of Anatomy and Neurobiology, Burlington, VT 05405. Offers PhD, MD/PhD. *Students:* 15 (6 women) 4 international. 14 applicants, 43% accepted, 1 enrolled. *Degree requirements:* For doctorate, thesis/dissertation. *Entrance requirements:* For doctorate, GRE General Test. Additional exam requirements/recommendations for international students: Required—TOEFL (minimum score 550 paper-based; 213 computer-based). *Application deadline:* For fall admission, 1/15 for domestic students. Applications are processed on a rolling basis. Application fee: $40. Electronic applications accepted. *Expenses:* Tuition, area resident: Part-time $410 per credit hour. Tuition, nonresident: part-time $1,034 per credit hour. *Financial support:* Teaching assistantships available. Financial award application deadline:3/1. *Faculty research:* Autonomic neurobiology, developmental neurobiology, neurotransmitter expression and release, plasticity and regeneration. *Unit head:* Dr. Rodney L. Parsons, Chairperson, 802-656-2230. *Application contact:* Dr. Rae Nishi, Coordinator, 802-656-2230, E-mail: rae.nishi@uvm.edu.

University of Virginia, School of Medicine, Department of Neuroscience, Charlottesville, VA 22904. Offers PhD, MD/PhD. *Students:* 26 full-time (15 women); includes 4 minority (2 African Americans, 1 Asian American or Pacific Islander, 1 Hispanic American), 2 international. Average age 27. In 2005, 5 degrees awarded. *Degree requirements:* For doctorate, thesis/dissertation. *Entrance requirements:* Additional exam requirements/recommendations for international students: Required—TOEFL. *Application deadline:* Applications are processed on a rolling basis. Application fee: $60. Electronic applications accepted. *Expenses:* Tuition, state resident: full-time $7,731. Tuition, nonresident: full-time $18,672. Required fees: $1,479. Full-time tuition and fees vary according to degree level and program. *Financial support:* Applicants required to submit FAFSA. *Unit head:* Kevin S. Lee, Director, 434-924-9111, Fax: 434-982-4380. *Application contact:* Information Contact, 434-982-4285, Fax: 434-982-4785, E-mail: neurograd@virginia.edu.

University of Washington, School of Medicine and Graduate School, Graduate Programs in Medicine, Graduate Program in Neurobiology and Behavior, Seattle, WA 98195. Offers neurobiology (PhD); neuroscience (PhD). *Degree requirements:* For doctorate, thesis/dissertation. *Entrance requirements:* For doctorate, GRE. Additional exam requirements/recommendations for international students: Required—TOEFL, TSE. Electronic applications accepted. *Faculty research:* Motor, sensory systems, neuroplasticity, animal behavior, neuroendocrinology, computational neuroscience.

The University of Western Ontario, Faculty of Graduate Studies, Biosciences Division, Clinical Neurological Sciences, London, ON N6A 5B8, Canada. Offers M Sc, PhD. Terminal master's awarded for partial completion of doctoral program. *Degree requirements:* For master's and doctorate, thesis/dissertation. *Entrance requirements:* For master's, honors degree or equivalent, minimum B+ average; for doctorate, master's degree, minimum B+ average. *Faculty research:* Behavioral neuroscience, neural regeneration and degeneration, visual development, human motor function.

University of Wisconsin–Madison, Graduate School, College of Letters and Science, Department of Psychology, Program in Cognitive Neurosciences, Madison, WI 53706-1380. Offers PhD. *Degree requirements:* For doctorate, thesis/dissertation, comprehensive exam, registration. *Entrance requirements:* For doctorate, GRE General Test, minimum undergraduate GPA of 3.0. Additional exam requirements/recommendations for international students: Required—TOEFL. Electronic applications accepted.

University of Wisconsin–Madison, Graduate School, Neuroscience Training Program, Madison, WI 53706-1380. Offers MS, PhD. *Degree requirements:* For doctorate, thesis/dissertation. *Entrance requirements:* For doctorate, GRE General Test. Electronic applications accepted.

University of Wisconsin–Madison, Medical School, Training Program in Hearing, Madison, WI 53706-1380.

University of Wisconsin–Madison, School of Veterinary Medicine, Department of Animal Health and Biomedical Sciences, Program in Comparative Biosciences, Madison, WI 53706-1380. Offers anatomy (MS, PhD); biochemistry (MS, PhD); cellular and molecular biology (MS, PhD); environmental toxicology (MS, PhD); neurosciences (MS, PhD); pharmacology (MS, PhD); physiology (MS, PhD). *Degree requirements:* For doctorate, thesis/dissertation.

University of Wyoming, Graduate School, College of Health Sciences, Division of Communication Disorders, Program in Neuroscience, Laramie, WY 82070. Offers audiology (PhD). *Accreditation:* ASHA. Part-time programs available. *Faculty:* 5 full-time (3 women). *Students:* 2 full-time (1 woman), 2 part-time (both women), 1 international. 1 applicant, 100% accepted, 1 enrolled. *Entrance requirements:* Additional exam requirements/recommendations for international students: Required—TOEFL. *Application deadline:* For fall admission, 2/15 for domestic students; for spring admission, 11/1 for domestic students. Application fee: $50. *Expenses:* Tuition, state resident: full-time $3,720; part-time $155 per credit hour. Tuition, nonresident: full-time $10,704; part-time $446 per credit hour. Required fees: $666; $162 per semester. Tuition and fees vary according to course load and program. *Financial support:* In 2005–06, 1 student received support, including 1 research assistantship with partial tuition reimbursement available (averaging $10,062 per year); Federal Work-Study, institutionally sponsored loans, and scholarships/grants also available. Financial award application deadline: 2/15. *Faculty research:* Audiometric techniques with infants, applications of insert earphones, auditory evoked potentials and neurodevelopment. Total annual research expenditures: $117,000. *Application contact:* Claoma T. Woodall, Office Associate, Senior, 307-766-6427, Fax: 307-766-5584, E-mail: woodall@uwyo.edu.

Vanderbilt University, Graduate School and School of Medicine, Program in Neuroscience, Nashville, TN 37240-1001. Offers PhD. *Faculty:* 57 full-time (11 women). *Students:* 61 full-time (32 women); includes 7 minority (4 African Americans, 2 Asian Americans or Pacific Islanders, 1 Hispanic American), 13 international. 40 applicants, 10% accepted, 1 enrolled. In 2005, 7 degrees awarded. *Entrance requirements:* For doctorate, GRE General Test. *Application deadline:* For fall admission, 1/15 for domestic students, 1/15 for international students. Application fee: $0. Electronic applications accepted. *Expenses:* Tuition: Part-time $1,283 per semester hour. Required fees: $2,202; $1,101 per semester. One-time fee: $30. Tuition and fees vary according to course load, program and student level. *Financial support:* Fellowships with full tuition reimbursements, research assistantships with full tuition reimbursements, institutionally sponsored loans, traineeships, and tuition waivers (partial) available. Financial award application deadline: 1/15. *Faculty research:* Molecular neuroscience, neural development, synaptic and systems plasticity, neuropharmacology, synaptic transmission. *Unit head:* Elaine Sanders-Bush, Director, 615-936-3736, Fax: 615-936-0212. *Application contact:* Mary Early-Zald, Director of Graduate Studies, 615-936-2610, Fax: 615-936-0212, E-mail: mary.early-zald@vanderbilt.edu.

Virginia Commonwealth University, Medical College of Virginia-Professional Programs, School of Medicine, Graduate Program in Neuroscience, Richmond, VA 23284-9005.

See Close-Up on page 1063.

Virginia Commonwealth University, Medical College of Virginia-Professional Programs, School of Medicine and Graduate Programs, School of Medicine Graduate Programs, Department of Anatomy, Richmond, VA 23284-9005. Offers anatomy (MS, PhD, CBHS); anatomy and physical therapy (PhD); neuroscience (MS, PhD). *Faculty:* 31 full-time (6 women). *Students:* 53 full-time (23 women), 3 part-time (all women). 97 applicants, 52% accepted. In 2005, 3 master's, 3 doctorates, 16 other advanced degrees awarded. *Degree requirements:* For master's, thesis; for doctorate, thesis/dissertation, comprehensive oral and written exams. *Entrance requirements:* For master's, DAT, GRE General Test, or MCAT; for doctorate, DAT, GRE General Test, MCAT. *Application deadline:* For fall admission, 2/15 for domestic students. Application fee: $50. *Expenses:* Tuition, state resident: full-time $6,268; part-time $405 per credit. Tuition, nonresident: full-time $15,904; part-time $940 per credit. Required fees: $751 per semester hour. Tuition and fees vary according to course load and program. *Financial support:* Fellowships available. *Unit head:* , Dr. John T. Povlishock, Chair, 804-828-9623, Fax: 804-828-9477, E-mail: jtpovlis@vcu.edu. *Application contact:* Dr. George R. Leichnetz, Director, Graduate Programs in Anatomy, 804-828-9512, Fax: 804-828-9477, E-mail: grleichn@vcu.edu.

See Close-Up on page 349.

Virginia Commonwealth University, Medical College of Virginia-Professional Programs, School of Medicine and Graduate Programs, School of Medicine Graduate Programs, Department of Biochemistry, Richmond, VA 23284-9005. Offers biochemistry (MS, PhD, CBHS); bioinformatics (MS); molecular biology and genetics (MS, PhD); neurosciences (PhD). *Faculty:* 26 full-time (6 women). *Students:* 48 full-time (24 women), 9 part-time (8 women); includes 12 minority (2 African Americans, 8 Asian Americans or Pacific Islanders, 2 Hispanic Americans), 8 international. 61 applicants, 75% accepted. In 2005, 1 master's, 4 doctorates, 2 other advanced degrees awarded. *Degree requirements:* For master's, thesis; for doctorate, thesis/dissertation, comprehensive oral and written exams. *Entrance requirements:* For master's and doctorate, DAT, GRE General Test, MCAT. *Application deadline:* For fall admission, 2/15 for domestic students. Application fee: $50. *Expenses:* Tuition, state resident: full-time $3,185; part-time $405 per credit. Tuition, nonresident: full-time $7,952; part-time $940 per credit. Required fees: $751 per semester hour. Tuition and fees vary according to course load and program. *Financial support:* Fellowships, research assistantships available. *Faculty research:* Molecular biology, peptide/protein chemistry, neurochemistry, enzyme mechanisms, macromolecular structure determination. Total annual research expenditures: $3.5 million. *Unit head:* Dr. Sarah Spiegel, Chair, 804-828-9762, Fax: 804-828-1473, E-mail: sspiegel@vcu.edu. *Application contact:* Dr. Keith R. Shelton, Program Director, 804-828-9886, Fax: 804-828-1473, E-mail: krshelto@vcu.edu.

See Close-Up on page 447.

Virginia Commonwealth University, Medical College of Virginia-Professional Programs, School of Medicine and Graduate Programs, School of Medicine Graduate Programs, Department of Pharmacology and Toxicology, Richmond, VA 23284-9005. Offers molecular biology and genetics (PhD); neurosciences (PhD); pharmacology (PhD, CBHS); pharmacology and toxicology (MS). *Faculty:* 36 full-time (9 women). *Students:* 35 full-time (18 women), 11 part-time (all women); includes 7 minority (2 African Americans, 4 Asian Americans or Pacific Islanders, 1 Hispanic American), 5 international. 90 applicants, 33% accepted. In 2005, 6 master's, 2 doctorates awarded. Terminal master's awarded for partial completion of doctoral program. *Degree requirements:* For master's, thesis; for doctorate, thesis/dissertation, comprehensive oral and written exams. *Entrance requirements:* For master's, DAT, GRE General Test or MCAT; for doctorate, GRE General Test, MCAT, DAT. *Application deadline:* For fall admission, 4/1 for domestic students. Application fee: $50. *Expenses:* Tuition, state resident: full-time $6,268; part-time $405 per credit. Tuition, nonresident: full-time $15,904; part-time $940 per credit. Required fees: $751 per semester hour. Tuition and fees vary according to course load and program. *Financial support:* Fellowships, teaching assistantships available. *Faculty research:* Drug abuse, drug metabolism, pharmacodynamics, peptide synthesis, receptor mechanisms. *Unit head:* Dr. Billy R. Martin, Chair, 804-828-8407, Fax: 804-828-2117, E-mail: brmartin@vcu.edu. *Application contact:* Sheryol Cox, Graduate Program Coordinator, 804-828-8400, Fax: 804-828-2117, E-mail: swcox@vcu.edu.

See Close-Up on page 1219.

Virginia Commonwealth University, Medical College of Virginia-Professional Programs, School of Medicine and Graduate Programs, School of Medicine Graduate Programs, Department of Physiology, Richmond, VA 23284-9005. Offers neurosciences (PhD); physiology (MS, PhD, CBHS). *Faculty:* 30 full-time (3 women). *Students:* 67 full-time (36 women), 5 part-time (3 women); includes 26 minority (7 African Americans, 16 Asian Americans or Pacific Islanders, 3 Hispanic Americans), 2 international. 144 applicants, 69% accepted. In 2005, 7 master's, 5 doctorates, 18 other advanced degrees awarded. Terminal master's awarded for partial completion of doctoral program. *Degree requirements:* For master's, thesis; for doctorate, thesis/dissertation, comprehensive oral and written exams. *Entrance requirements:* For master's, DAT, GRE General Test, or MCAT; for doctorate, GRE General Test, MCAT, DAT. *Application deadline:* For fall admission, 2/15 for domestic students. Application fee: $50. *Expenses:* Tuition, state resident: full-time $6,268; part-time $405 per credit. Tuition, nonresident: full-time $15,904; part-time $940 per credit. Required fees: $751 per semester hour. Tuition and fees vary according to course load and program. *Financial support:* Fellowships, research assistantships, teaching assistantships, career-related internships or fieldwork and tuition waivers (full) available. *Unit head:* Dr. Margaret C. Biber, Chair, 804-828-9756, Fax: 804-828-7382, E-mail: mcbiber@vcu.edu. *Application contact:* Dr. George D. Ford, Director, 804-828-9501, Fax: 804-828-7382, E-mail: gdford@vcu.edu.

Wake Forest University, School of Medicine and Graduate School, Graduate Programs in Medicine, Interdisciplinary Program in Neuroscience, Winston-Salem, NC 27109. Offers PhD. *Degree requirements:* For doctorate, thesis/dissertation, oral and written qualifying exams. *Entrance requirements:* For doctorate, GRE General Test. Additional exam requirements/recommendations for international students: Required—TOEFL. Electronic applications accepted. *Faculty research:* Neurobiology of substance abuse, learning and memory, aging, sensory neurobiology, nervous system development.

Washington State University, College of Veterinary Medicine and Graduate School, Graduate Programs in Veterinary Science, Pullman, WA 99164. Offers veterinary and comparative anatomy, pharmacology, and physiology (MS, PhD), including neuroscience, veterinary science; veterinary clinical sciences (MS, PhD); veterinary microbiology and pathology (MS, PhD), including veterinary science. Part-time programs available. *Faculty:* 72 full-time (13 women), 6 part-time/adjunct (4 women). *Students:* 50 full-time (24 women). Average age 30. In 2005, 3 master's, 4 doctorates awarded. Terminal master's awarded for partial completion of doctoral program. *Degree requirements:* For master's and doctorate, thesis/dissertation, oral exam. *Entrance requirements:* For master's and doctorate, GRE General Test, minimum GPA of 3.0. *Application deadline:* For fall admission, 12/31 for domestic students. Applications are processed on a rolling basis. Application fee: $35. Electronic applications accepted. *Expenses:* Contact institution. Part-time tuition and fees vary according to campus/location and program. *Financial support:* In 2005–06, 21 research assistantships with partial tuition reimbursements, 8 teaching assistantships with partial tuition reimbursements were awarded; fellowships, career-related internships or fieldwork, Federal Work-Study, institutionally sponsored loans, scholarships/grants, traineeships, tuition waivers (partial), and teaching associateships also available. Financial award application deadline: 12/1; financial award applicants required to submit FAFSA. *Application contact:* Julie K. Smith, Principal Assistant, 509-335-3064, Fax: 509-335-0160, E-mail: jksmith@vetmed.wsu.edu.

Washington State University, College of Veterinary Medicine and Graduate School, Graduate Programs in Veterinary Science, Department of Veterinary and Comparative Anatomy, Pharmacology, and Physiology, Program in Neuroscience, Pullman, WA 99164. Offers MS, PhD. Part-time programs available. *Faculty:* 42 full-time (16 women). *Students:* 26 full-time (13 women); includes 1 minority (Hispanic American), 10 international. Average age 30. 40 applicants, 25% accepted, 5 enrolled. In 2005, 1 master's, 6 doctorates awarded. Terminal master's awarded for partial completion of doctoral program. *Degree requirements:* For master's, thesis, written exam; for doctorate, thesis/dissertation, written exam, oral exam. *Entrance requirements:* For master's and doctorate, GRE General Test, minimum GPA of 3.0. Additional exam requirements/recommendations for international students: Required—TOEFL. *Applica-*

Neuroscience

Washington State University *(continued)*
tion deadline: For fall admission, 12/31 for domestic students. Applications are processed on a rolling basis. Application fee: $35. Electronic applications accepted. *Expenses:* Tuition, state resident: full-time $6,295; part-time $336 per credit. Tuition, nonresident: full-time $15,949; part-time $819 per credit. Required fees: $933. Part-time tuition and fees vary according to campus/location and program. *Financial support:* In 2005–06, 4 research assistantships with partial tuition reimbursements (averaging $19,500 per year), 8 teaching assistantships with partial tuition reimbursements (averaging $19,500 per year) were awarded; career-related internships or fieldwork, Federal Work-Study, institutionally sponsored loans, and scholarships/grants also available. Financial award application deadline: 12/31. *Faculty research:* Neural mechanisms of substance abuse, molecular mechanisms of sleep regulations, neuroendocrinology of pituitary cells, behavioral aspects of sexual differentiation. Total annual research expenditures: $4.8 million. *Application contact:* Pam Colbert, Coordinator, 509-335-0986, Fax: 509-335-4650, E-mail: colbertp@vetmed.wsu.edu.

Washington University in St. Louis, Graduate School of Arts and Sciences, Department of Philosophy, Program in Philosophy/Neuroscience/Psychology, St. Louis, MO 63130-4899. Offers PhD. *Degree requirements:* For doctorate, thesis/dissertation. *Entrance requirements:* For doctorate, GRE General Test, sample of written work. Electronic applications accepted.

Washington University in St. Louis, Graduate School of Arts and Sciences, Division of Biology and Biomedical Sciences, Program in Neurosciences, St. Louis, MO 63130-4899. Offers PhD. *Degree requirements:* For doctorate, thesis/dissertation. *Entrance requirements:* For doctorate, GRE General Test, GRE Subject Test. Electronic applications accepted.

Wayne State University, Graduate School, College of Liberal Arts and Sciences, Department of Psychology, Detroit, MI 48202. Offers human development (MA); psychology (MA, MS, PhD), including behavioral and cognitive neuroscience (PhD), clinical psychology (PhD), cognitive and social psychology (PhD), industrial/organizational psychology (PhD), psychology (MA, MS). *Accreditation:* APA (one or more programs are accredited). *Faculty:* 33 full-time (8 women). *Students:* 108 full-time (85 women), 21 part-time (18 women); includes 20 minority (12 African Americans, 5 Asian Americans or Pacific Islanders, 3 Hispanic Americans), 13 international. Average age 28. 214 applicants, 20% accepted, 17 enrolled. In 2005, 25 master's, 13 doctorates awarded. *Degree requirements:* For doctorate, thesis/dissertation. *Entrance requirements:* For doctorate, GRE General Test, GRE Subject Test. Additional exam requirements/recommendations for international students: Required—TOEFL (minimum score 550 paper-based; 213 computer-based); Recommended—TWE (minimum score 6). *Application deadline:* Applications are processed on a rolling basis. Application fee: $30 ($50 for international students). Electronic applications accepted. *Expenses:* Tuition, state resident: part-time $338 per credit hour. Tuition, nonresident: part-time $746 per credit hour. Required fees: $24 per credit hour. Full-time tuition and fees vary according to program. *Financial support:* In 2005–06, 1 fellowship with tuition reimbursement (averaging $11,663 per year), 17 research assistantships with tuition reimbursements (averaging $16,398 per year), 47 teaching assistantships with tuition reimbursements (averaging $12,797 per year) were awarded; career-related internships or fieldwork also available. Financial award application deadline: 2/1. *Faculty research:* Clinical neuropsychology, cognitive, emotional and neurological aspects of aging; industrial/organizational psychology, health psychology, alcohol and drug abuse. Total annual research expenditures: $4.7 million. *Unit head:* Douglas Whitman, Chair, 313-577-2803, Fax: 313-577-7636, E-mail: dwhitman@wayne.edu. *Application contact:* Dr. Melissa Kaplan-Estrin, Graduate Director, 313-577-2824, Fax: 313-577-7636, E-mail: mkestrin@sun.science.wayne.edu:

West Virginia University, School of Medicine, Graduate Programs at the Health Science Center, Biomedical Sciences Graduate Program, Program in Neuroscience, Morgantown, WV 26506. Offers MS, PhD. *Faculty:* 20 full-time (5 women). *Students:* 17 full-time (5 women); includes 6 minority (4 Asian Americans or Pacific Islanders, 2 Hispanic Americans), 2 international. Average age 25. In 2005, 1 master's awarded. *Median time to degree:* Of those who began their doctoral program in fall 1997, 96% received their degree in 8 years or less. *Degree requirements:* For doctorate, thesis/dissertation, comprehensive exam. *Entrance requirements:* For doctorate, GRE General Test, minimum GPA of 3.0. Additional exam requirements/recommendations for international students: Required—TOEFL. *Application deadline:* For fall admission, 3/1 priority date for domestic students, 1/15 priority date for international students. Applications are processed on a rolling basis. Application fee: $0. Electronic applications accepted. *Financial support:* In 2005–06, research assistantships with full tuition reimbursements (averaging $20,000 per year); institutionally sponsored loans, traineeships, and health care benefits also available. *Faculty research:* Sensory neuroscience, cognitive neuroscience, neural injury, homeostasis, behavioral neuroscience. Total annual research expenditures: $5 million. *Unit head:* Dr. Richard D. Dey, Chair, 304-293-5979, Fax: 304-293-8159, E-mail: richard.dey@hsc.wvu.edu. *Application contact:* Dr. Albert S. Berrebi, Program Coordinator, 304-293-2357, Fax: 304-293-2902, E-mail: albert.berrebi@hsc.wvu.edu.

See Close-Up on page 1065.

Yale University, Graduate School of Arts and Sciences, Interdepartmental Neuroscience Program, New Haven, CT 06520. Offers PhD. *Degree requirements:* For doctorate, thesis/dissertation. *Entrance requirements:* For doctorate, GRE General Test. Expenses: Contact institution.

Yale University, School of Medicine and Graduate School of Arts and Sciences, Combined Program in Biological and Biomedical Sciences (BBS), Neuroscience Track, New Haven, CT 06520. Offers PhD, MD/PhD. *Students:* 8 full-time. *Degree requirements:* For doctorate, thesis/dissertation. *Entrance requirements:* For doctorate, GRE General Test. Additional exam requirements/recommendations for international students: Required—TOEFL. *Application deadline:* For fall admission, 12/8 for domestic students, 12/8 for international students. Electronic applications accepted. *Financial support:* Fellowships, research assistantships available. *Unit head:* Dr. Charles Greer, Co-Director, 203-785-5932. *Application contact:* Carol Russo, Graduate Registrar, 203-785-5932, Fax: 203-785-5971, E-mail: carol.russo@yale.edu.

Cross-Discipline Announcements

Massachusetts Institute of Technology, School of Science, Department of Biology, Cambridge, MA 02139-4307.

Graduate work in the Department of Biology at MIT leads to the PhD degree. Research opportunities include many areas of modern biology: biochemistry, biophysics, cellular and developmental biology, immunology, microbiology, and neurobiology. Students come from a wide variety of backgrounds. Previous experience in the biological sciences, although desirable, is not a prerequisite for admission. Formal courses in biochemistry, genetics, and the method and logic of molecular biology are required of all students early in their graduate program. Each student is encouraged to follow the program of study that best meets his or her educational goals. Special emphasis is placed on research leading to the PhD thesis.

Princeton University, Graduate School, Department of Molecular Biology, Princeton, NJ 08544-1019.

Graduate studies in the Department of Molecular Biology at Princeton emphasize training in research and encourage students to apply molecular, biochemical, structural, genetic, and computational approaches to biological problems. Faculty members and students pursue research in a wide variety of areas of biochemistry and cell biology, biological dynamics and computational biology, biomedicine and societal issues, biophysics and structural biology, developmental biology, genetics and genomics, microbiology, neurobiology, oncology, and virology.

University of Washington, School of Medicine and Graduate School, Graduate Programs in Medicine, Department of Pharmacology, Seattle, WA 98195.

The department is engaged in research in molecular genetic approaches to dissecting signal transduction systems and ion transport, regulation of gene expression, molecular pharmacology, and genetic analyses of vertebrate development. Emphasis is on in vivo models, including cultured cells and transgenic animals.

BAYLOR COLLEGE OF MEDICINE

Department of Neuroscience

Program of Study	The Department of Neuroscience offers students interested in the nervous system a course of study leading to the Ph.D. in neuroscience or to a combined M.D./Ph.D. in conjunction with Baylor's Medical School. The goal of this program is to provide students with intensive education and training in neuroscience and an opportunity to excel in the laboratory. This educational experience is provided by the faculty members in the Department, each of whom is actively involved in research designed to provide insight into the function of the central nervous system. These faculty members, in combination with Baylor's extensive programs in other related disciplines, offer graduate students both a sound education in neuroscience and a remarkably broad graduate experience.

Graduate students are offered the opportunity to devote all of their time to research and study, and entering students are encouraged to pursue both research and course work. With the counsel of a faculty adviser, students select from among graduate-level courses and research topics in molecular neurobiology, neurophysiology, neurobiology of disease, neuroanatomy, systems neuroscience, computational neuroscience, biochemistry, and physiology, as well as participate in the neuroscience seminar program and student journal club. Research rotations in faculty laboratories are expected to lead to the definition of a field of interest and the selection of a thesis research project. Students are admitted to candidacy upon successful completion of a written and an oral qualifying exam that is given in the second year of graduate school. Subsequent progress in the program is dependent upon original research leading to a formal dissertation. Progress in both research and course work is followed closely by the major thesis adviser, a thesis advisory committee, and an overall graduate student advisory committee.

Research Facilities

The Department of Neuroscience, which is based in the Smith Research Building, has outstanding modern research facilities for molecular neurobiology, neurochemistry, neuroanatomy, neurophysiology, biophysics, behavioral neuroscience, optical imaging, and computer science. The Department has also undergone a major expansion of the faculty, which has resulted in new research opportunities for students. The local environment is research intensive and adjoins other active research programs ongoing at Baylor College of Medicine.

Financial Aid

All students receive stipends from the Department. For the 2005–06 academic year, the stipend for first-year students was $23,000 plus medical insurance. Students are also encouraged, however, to apply for NSF and NRSA predoctoral fellowships as well as fellowships from various private foundations.

Cost of Study

All tuition costs are covered by tuition scholarships. A matriculation fee of $25 and a graduation fee of $140 are the responsibility of the student.

Living and Housing Costs

There are a limited number of on-campus apartments of various types, costing $250 to $350 per month. Apartments in a similar price range are also available near the Medical Center. Many students share apartments or houses, and there are many employment opportunities for students' spouses.

Student Group

Created in 1978, the graduate program in neuroscience has 47 full-time graduate students. A federally funded training program for predoctoral and postdoctoral trainees in neuroscience funds a dynamic and highly interactive training environment. The Graduate School has 500 students and the Medical School, 682. In its admissions policies, Baylor places no restrictions on sex, age, religion, race, or country of origin.

Location

Baylor College of Medicine is located within the Texas Medical Center, a large and vigorous research community that also includes the University of Texas Health Science Center, eight teaching hospitals, and the M. D. Anderson Cancer Center. Rice University is within two blocks of the Medical Center. Houston is also the home of the University of Houston and numerous other colleges and universities. The fourth-largest city in the nation, Houston is an exciting cultural and metropolitan center. Ballet, opera, symphony, and theater are excellent. Many fine museums and parks enhance life in the city. Professional and amateur sports are very popular. The climate permits participation in a wide variety of outdoor activities, and the Gulf Coast beaches are only a short drive from the city.

The College

Baylor College of Medicine is an independent, private institution dedicated to excellence in research and to both graduate and medical education. Both the research activities and graduate education at Baylor are characterized by a high level of interdisciplinary cooperation among all Baylor departments. Baylor facilitates research activities by supporting extensive resources such as the Baylor Information Network, major symposia, and numerous core research facilities. There is also ample opportunity for extensive interaction with other institutions in the Texas Medical Center, and Rice University has close ties with Baylor and provides important complementary research strengths. A reciprocity agreement among Rice University, the University of Texas Health Science Center, the University of Houston, and Baylor also allows graduate students to take courses at any of the participating institutions.

Applying

Applicants should have a baccalaureate degree and a sound preparation in the life and physical sciences. Prospective students are required to submit the results of the General Test of the Graduate Record Examinations and are encouraged to take GRE Subject Tests in scientific areas in which they have expertise. GRE scores that are more than two years old at the time of application are not acceptable.

The completed application should be accompanied by transcripts from all schools attended and by three letters of recommendation. Applications must be complete in all respects by June 1, but submission of applications by January 1 is strongly encouraged. Applications are evaluated and ranked by a faculty committee, and decisions regarding admission and financial support are made as soon as possible after the receipt of a completed application. The academic year begins on August 1.

Correspondence and Information

For information:
Dr. Mariella De Biasi
Director of Graduate Studies
Department of Neuroscience
Baylor College of Medicine
One Baylor Plaza
Houston, Texas 77030
Phone: 800-367-9049 (toll-free)
Fax: 713-798-3946
E-mail: nsc_ask@cns.neusc.bcm.tmc.edu
Web site: http://neuro.neusc.bcm.tmc.edu

For an application:
Graduate Admissions Office
Baylor College of Medicine
Houston, Texas 77030
Phone: 713-798-4060
Web site: http://www.bcm.tmc.edu/gradschool

Baylor College of Medicine

THE FACULTY AND THEIR RESEARCH

Department of Neuroscience faculty members, drawn from several basic science and clinical departments, strive to promote collaborative research in the different disciplines.

A. Anderson, Assistant Professor, Neurology; M.D., Texas Medical Branch. Role of cellular signaling mechanisms in the development of epilepsy.

S. F. Basinger, Assistant Dean, Ophthalmology; Ph.D., Syracuse. Anatomy and biochemistry of the retina; molecular biology of photoreceptors; peptide transmitters.

H. J. Bellen, Professor, Institute for Molecular Genetics, and Investigator, Howard Hughes Medical Institute; Ph.D., California, Davis. Development of the nervous system in the fruit fly *Drosophila melanogaster.*

W. E. Brownell, Professor, Otorhinolaryngology; Ph.D., Chicago. Biophysics and cell biology of cochlear transduction.

G. D. Clark, Associate Professor, Pediatrics, Neurology, Neuroscience, and Cain Foundation Laboratories; M.D., LSU Medical Center. Lipid receptors in brain.

M. C. Crair, Assistant Professor, Neuroscience; Ph.D., Berkeley. Cortical development and plasticity.

J. A. Dani, Professor, Neuroscience; Ph.D., Minnesota. Synaptic communication; molecular excitability.

G. D'Arcangelo, Assistant Professor, Neurology and Neuroscience; Ph.D., NYU. Molecular mechanisms of neuronal proliferation, differentiation, and migration.

S. Davies, Assistant Professor, Neurosurgery and Neuroscience; Ph.D., London. Molecular and cell biology of axon regeneration and glial scar formation in the injured adult mammalian central nervous system.

R. Davis, Professor, Cell Biology; Ph.D., California, Davis. Molecular and cellular biology of learning and memory.

M. De Biasi, Associate Professor, Neuroscience; Ph.D., Padova (Italy). Autonomic nervous system: analysis from molecules to wake behavior.

Michael J. Friedlander, Ph.D., Professor and Chair, Neuroscience; Illinois at Urbana-Champaign. Synapse function; role of nitric oxide in neural signaling; molecular basis of learning.

F. Gabbiani, Assistant Professor; Ph.D., Swiss Federal Institute of Technology. Information processing in the nervous system.

G. Mardon, Professor, Pathology and Neuroscience; Ph.D., MIT. Function of retinal determination genes in *Drosophila* and vertebrates.

P. R. Montague, Professor, Neuroscience; Ph.D., Alabama at Birmingham. Computational neuroscience and molecular mechanisms of cortical plasticity.

J. L. Noebels, Professor, Neurology; M.D., Yale; Ph.D., Stanford. Developmental neurogenetics; gene control of neuronal excitability; cellular neurophysiology of epilepsy.

P. A. Overbeek, Professor, Cell Biology; Ph.D., Michigan. Transgenic mice; retinal development and neural communication; insertional mutagenesis.

R. E. Paylor, Associate Professor, Molecular and Human Genetics; Ph.D., Colorado. Genetic basis of complex behavioral traits.

P. Pfaffinger, Associate Professor, Neuroscience; Ph.D., Washington (Seattle). Molecular biology and biophysics of potassium channels.

C. Rosenmund, Associate Professor, Molecular and Human Genetics and Neuroscience; Ph.D., Oregon Health Sciences. Molecular mechanisms of neurotransmitter release and short-term plasticity at central synapses; development of synapses; impact of heterogeneous synapse function on information processing.

P. Saggau, Professor, Neuroscience; Ph.D., Munich. Cellular mechanisms of learning and memory; associative mechanisms; optical recording of neural activity.

P. Schulz, Associate Professor, Neurology; M.D., Boston University. Cellular mechanisms underlying memory and dementia.

D. Shine, Associate Professor, Neurosurgery; Ph.D., Texas Medical Branch. Molecular and cellular neurobiology.

J. Stringer, Associate Professor, Pharmacology; M.D., Ph.D., Virginia. Mechanisms that control seizure onset and termination in limbic circuits of the brain.

J. W. Swann, Professor, Pediatrics; Ph.D., Maryland. Development of hippocampal local circuits; synaptic transmission; epilepsy.

R. H. Thalmann, Professor, Cell Biology; Ph.D., Michigan. Physiology of inhibitory receptor systems in mammalian brain.

S. M. Wu, Professor, Ophthalmology; Ph.D., Harvard. Electrophysiology of the vertebrate retina; electrical and pharmacological properties of retinal synapses; retinal circuitry and visual perception.

H. Zheng, Associate Professor; Ph.D., Peking. Alzheimer's disease.

H. Y. Zoghbi, Professor, Pediatrics, and Investigator, Howard Hughes Medical Institute; M.D., American (Beirut). Molecular basis of neurodegenerative diseases.

BOSTON UNIVERSITY

Program in Neuroscience

Program of Study

The Program in Neuroscience is an interdisciplinary program, administered through the Graduate School of Arts and Sciences, with strong affiliations with the following departments: Anatomy and Neurobiology, Biology, Biomedical Engineering, Cognitive and Neural Systems, Health Sciences, Mathematics, Pharmacology, Physiology and Biophysics, and Psychology. The neuroscience faculty supervises graduate study leading to the M.A. and Ph.D. degrees in neuroscience. This program takes advantage of the wide range of neuroscience expertise among many departments at Boston University to provide students with a broad-based multidisciplinary program, which is unique in its focus on both experimental and computational approaches to the understanding of neural systems. The faculty approaches all disciplines of neuroscience, from the molecular and cellular bases of neurobiological processes through the clinical diagnosis of neurological diseases. A broad range of research areas are represented, including cellular and molecular neurobiology, cellular and systems plasticity, structural neurobiology, neurophysiology, sensorimotor integration, learning and memory, cognitive sciences, computational modeling, biomolecular engineering, and neuropharmacology.

In the first two years, students take required courses and complete laboratory rotations. Students are required to complete survey courses that cover the broad areas of knowledge in both experimental and computational neuroscience, and they take elective courses offered through the various participating departments. Students complete a qualifying exam at the end of their second year, propose and defend their thesis topic, and engage in full-time research toward their dissertation. After completing their research, students defend their thesis in front of a 5-person committee. It is anticipated that it will take students five years to complete the program.

Research Facilities

The neuroscience faculty members are housed in various areas on the Charles River campus and the Medical and Woods Hole campuses. The laboratories comprising the program provide faculty members and students with cutting-edge technology for the experimental investigation of the nervous system using approaches such as molecular neurobiology, neuroanatomy at the microscopic and systems levels, neurophysiology in a broad range of systems and levels of analysis, behavioral and neuropsychological analyses of humans and animals, and cognitive neuroscience, including human brain imaging. In addition, there are many faculty members who pursue understanding of the nervous system through computational analyses, including modeling at a broad range of levels, from biophysical properties of cells and cell systems to neural network models of behavioral and cognitive functions.

Financial Aid

All Ph.D. graduate students accepted into the program have financial support from an NIH-supported training grant, competitive teaching and University fellowships, or research assistantships available from grants or contracts held by faculty members. The graduate stipend for 2006–07 is $24,000 for twelve months, plus tuition scholarships.

Cost of Study

Tuition in 2006–07 was $33,330 for a full-time student (eight courses per year). These costs, along with health insurance, are covered by fellowships.

Living and Housing Costs

The cost of living in Boston is comparable with other major cities in the United States. Most graduate students live in off-campus housing, although University housing is available for interested graduate students.

Student Group

Currently the program has 26 students, of whom 15 are women and 7 are members of minority groups or are international students. All of these students are funded through research assistantships or fellowships.

Student Outcomes

Graduates of the Program in Neuroscience are prepared for careers in teaching, research, and public service. Potential employers of program graduates include colleges and universities, pharmaceutical companies, private industry, and government agencies.

Location

Because Boston has a large concentration of major universities and other institutions, students enjoy an unusually stimulating intellectual life, enhanced by a variety of seminars, colloquiums, and lectures. Boston University is located in proximity to museums, parks, and attractions such as Fenway Park, the Museum of Fine Arts, and downtown Boston.

The University

Boston University's main campus is located along the Charles River and extends more than 45 acres. There are sixteen schools and colleges with a wide range of graduate and professional programs. The Graduate School of Arts and Sciences has been offering advanced programs leading to the M.A. and Ph.D. degrees since its founding in 1974. The Medical Campus is located adjacent to Boston Medical Center in the South End of downtown Boston.

Applying

Applications for fall semester admission should be filed by December 15. A personal statement, three letters of recommendation, and GRE scores are required. International students must submit scores on the Test of English as a Foreign Language (TOEFL). Students are encouraged to list their research interests in the personal statement. Applications are available through the Graduate School of Arts and Sciences at http://www.bu.edu/cas/graduate/, and questions may be directed to the Program in Neuroscience Administrator.

Correspondence and Information

Program in Neuroscience Administrator
Boston University
5 Cummington Street
Boston, Massachusetts 02215

Phone: 617-358-1123
Fax: 617-358-1124
E-mail: neurosci@bu.edu
Web site: http://www.bu.edu/neuro/pin

Boston University

THE FACULTY AND THEIR RESEARCH

Jelle Atema, Professor of Biology; Ph.D., Michigan. Sensory biology.
Helen Barbas, Professor of Health Sciences; Ph.D., McGill. Prefrontal cortex in primates.
Michael J. Baum, Professor of Biology; Ph.D., McGill. Behavioral endocrinology.
Gene Blatt, Associate Professor of Anatomy and Neurobiology; Ph.D., Thomas Jefferson. Learning and memory.
Daniel Bullock, Associate Professor of Cognitive and Neural Systems; Ph.D., Stanford. Sensorimotor learning/control.
Gloria V. Callard, Professor of Biology, College of Arts and Sciences; Ph.D., Rutgers. Neuroendocrinology.
Gail Carpenter, Professor of Cognitive and Neural Systems; Ph.D., Wisconsin–Madison. Learning and memory.
James Cherry, Associate Professor of Psychology; Ph.D., North Carolina State. Olfaction; cognition; behavior.
H. Steven Coburn, Professor of Biomedical Engineering; Ph.D., MIT. Audition.
Michael Cohen, Associate Professor of Cognitive and Neural Systems; Ph.D., Harvard. Speech and language processing.
James Collins, Professor of Biomedical Engineering; Ph.D., Oxford. Posture and locomotion.
Paul B. Cook, Assistant Professor of Biology; Ph.D., Berkeley. Modulation of retinal neural circuits.
Geoffrey Cooper, Professor of Biology; Ph.D., Miami (Florida). Cellular growth control; cancer.
M. Carter Cornwall, Professor of Physiology; Ph.D., Utah. Visual transduction and adaptation.
Alice Cronin-Golomb, Professor of Psychology; Ph.D., Caltech. Vision and neurodegenerative disease.
Charles DeLisi, Professor of Biomedical Engineering; Ph.D., NYU. Optimization algorithms.
Carlo DeLuca, Professor of Biomedical Engineering; Ph.D., Queens at Kingston. Neuromuscular signals and controls.
Vincent Dionne, Professor of Biology; Ph.D., Arizona. Olfactory transduction mechanisms.
Howard Eichenbaum, Professor of Psychology; Ph.D., Michigan. Hippocampal mediation of declarative memory.
William D. Eldred, Professor of Biology; Ph.D., Colorado Health Sciences Center. Nitric oxide/cGMP signal transduction in retina.
Mary S. Erskine, Professor of Biology; Ph.D., Connecticut. Reproductive endocrinology.
David H. Farb, Professor of Pharmacology; Ph.D., Brandeis. Amino acid receptors.
J. Fernando Garcia-Diaz, Associate Professor of Physiology; Ph.D., Malaga (Spain). Expression and modulation of ion channels.
Terrell T. Gibbs, Assistant Professor of Pharmacology; Ph.D., Harvard. Neuromodulators.
Gerald Gottlieb, Research Professor of NeuroMuscular Research Center; Ph.D., Illinois Medical Center.
Stephen Grossberg, Professor of Cognitive and Neural Systems; Ph.D., Rockefeller. Visual and auditory perception.
Frank Guenther, Associate Professor of Cognitive and Neural Systems; Ph.D., Boston University. Speech production and perception; biological sensorimotor control; functional brain imaging.
Catherine L. Harris, Associate Professor of Psychology; Ph.D., California, San Diego. Representation of language.
Michael Hasselmo, Professor of Psychology; D.Phil., Oxford. Neuromodulators.
Robert E. Hausman, Professor of Biology; Ph.D., Northwestern. Developmental biology.
Allyn Hubbard, Professor of Electrical and Computer Engineering; Ph.D., Wisconsin. Auditory physiology.
Kathleen Kantak, Professor of Psychology; Ph.D., Syracuse. Behavioral pharmacology of drug abuse.
Thomas Kemper, Professor of Anatomy and Neurobiology, School of Medicine; M.D., Illinois Medical Center. Neuropathology; development.
Dae-Shik Kim, Associate Professor of Anatomy and Neurobiology, School of Medicine; Ph.D., Max Plank Institute. Functional MRI.
Nancy Kopell, Professor of Mathematics; Ph.D., Berkeley. Dynamics of oscillatory networks.
Thomas H. Kunz, Professor of Biology; Ph.D., Kansas. Mammal physiology and behavior.
Susan Leeman, Professor of Pharmacology; Ph.D., Radcliffe. Functions of neuropeptides.
Simon Levy, Associate Professor of Physiology; Ph.D., Boston University. Calcium signaling in nerve cells.
Jacqueline Liederman, Associate Professor of Psychology; Ph.D., Rochester. Developmental neuropsychology.
Jen-Wei Lin, Associate Professor of Biology; Ph.D., SUNY at Buffalo. Electrophysiology; neurotransmitter secretion.
Jennifer Luebke, Research Assistant Professor of Anatomy and Neurobiology; Ph.D., Boston University. Electrophysiology.
Henry Marcucella, Professor of Psychology; Ph.D., Boston University. Animal behavior.
Ennio Mingolla, Professor of Cognitive and Neural Systems; Ph.D., Connecticut. Network models of visual processes.
Mark Moss, Professor of Anatomy and Neurobiology; Ph.D., Northeastern. Age-related cognitive dysfunction.
David Mostofsky, Professor of Psychology; Ph.D., Boston University. Operant conditioning; behavioral medicine.
David Mountain, Professor of Biomedical Engineering; Ph.D., Wisconsin–Madison. Auditory information processing.
Enrico Nasi, Professor of Physiology; Ph.D., Bryn Mawr. Light-dependent ionic channels and phototransduction.
Christopher L. Passaglia, Assistant Professor of Biomedical Engineering; Ph.D., Syracuse. Visual processing; retinal physiology.
Alan Peters, Professor of Anatomy and Neurobiology; Ph.D., Bristol (England). Mammalian cerebral cortex.
Douglas Rosene, Associate Professor of Anatomy and Neurobiology; Ph.D., Rochester. Learning and memory.
Michele Rucci, Assistant Professor of Cognitive and Neural Systems; Ph.D., Scuola Superiore (Italy). Vision; sensory-motor control and learning; computational neuroscience.
Shelley Russek, Assistant Professor of Pharmacology; Ph.D., Boston University. Gene regulation in nervous system.
Julie Sandell, Associate Professor of Anatomy and Neurobiology; Ph.D., MIT. Primate retina in cerebrovascular disease.
Judith Schotland, Assistant Professor of Health Sciences; Ph.D., Northwestern. Spinal neural networks and movement.
Eric Schwartz, Professor of Cognitive and Neural Systems; Ph.D., Columbia. Computational neuroscience.
Kamal Sen, Assistant Professor of Biomedical Engineering; Ph.D., Brandeis. Neural coding of natural sounds.
Barbara Shinn-Cummingham, Associate Professor of Cognitive and Neural Systems; Ph.D., MIT. Binaural hearing; psychoacoustics.
Jean-Jacques Soghomonian, Associate Professor of Anatomy and Neurobiology; Ph.D., Montreal. Motor control and sensorimotor integration.
David Somers, Assistant Professor of Psychology; Ph.D., Boston University. Visual perception, attention, and awareness using fMRI and modeling.
Chantal Stern, Associate Professor of Psychology; D.Phil., Oxford. Functional MRI cognition.
Malvin Teich, Professor of Biomedical Engineering; Ph.D., Cornell. Wavelet analysis of fractal signals.
James F. A. Traniello, Professor of Biology; Ph.D., Harvard. Behavioral ecology; insects.
Susan Tsunoda, Assistant Professor of Biology; Ph.D., Washington (St. Louis). Signal transduction; protein tagging; *Drosophila* phototransduction.
Lucia Vaina, Professor of Biomedical Engineering; Ph.D., Paris IV (Sorbonne). Computational learning; vision.
Deborah Vaughan, Professor of Anatomy and Neurobiology; Ph.D., Boston University. Morphology of central nervous system and aging.
Herbert F. Voigt, Professor of Biomedical Engineering; Ph.D., Johns Hopkins. Auditory neurophysiology.
Matt Wachowiak, Assistant Professor of Biology; Ph.D., Florida. Olfactory coding and synaptic processing; imaging; neurophysiology.
Takeo Watanabe, Professor of Psychology; Ph.D., Tokyo. Visual perception and cognition.
John White, Associate Professor of Biomedical Engineering; Ph.D., Johns Hopkins. Membrane conductance; ion channels.
Eric Widmaier, Professor of Biology; Ph.D., California, San Francisco. Neuroendocrinology; cell signaling and gene regulation.
Ayako Yamaguichi, Assistant Professor of Biology; Ph.D., California, Davis. Behavioral neurobiology; neuroendocrinology; animal communication.

BROWN UNIVERSITY /
NATIONAL INSTITUTES OF HEALTH
Graduate Program in Neuroscience

Program of Study	The goal of the Graduate Program in Neuroscience is to prepare students for a professional career devoted to scientific investigation of the nervous system. Research in the program encompasses genes, molecules, cells, networks, systems, and behavior. Students enrolled in the Partnership Program between Brown and the National Institutes of Health (NIH) complete their course work at Brown and demonstrate general academic competence in neuroscience at the end of the first year. This unique alliance between Brown and NIH investigators creates a very broad range of research opportunities. Laboratory rotations are taken at Brown during the first academic year, and summer rotations are done at NIH in the summers before and after the first academic year. By the end of the first academic year, students choose either an NIH research mentor with whom to do their dissertation research or establish a collaborative dissertation between a Brown and an NIH investigator. Faculty committees advise students throughout their studies. Students propose and defend a thesis topic at the end of the second year. Students receive their Ph.D.'s from Brown after satisfying program requirements and completing a significant body of original research. The training program also emphasizes the conceptual foundations of neuroscience, teaching skills, oral presentation of research results, clear writing of scientific reports, and proposals for research support. Successful completion of the program leads to the Ph.D. in neuroscience.

Neuroscience is an interdisciplinary field, and entering students have varied backgrounds. However, training in biology and related basic sciences and mathematics is essential. The program's requirements include graduate courses in the component disciplines of neuroscience, participation in weekly journal clubs and research seminars, a comprehensive examination, a written thesis proposal, a substantial body of original research, and a doctoral dissertation and oral defense. Interdisciplinary work is encouraged, and participating departments include Neuroscience, Cell and Molecular Biology, Pharmacology, Physiology, Psychology, Cognitive and Linguistic Sciences, Physics, Computer Science, and Applied Math. Typically, five years are required to complete the Ph.D. The core of the training involves close interaction with faculty members to develop expertise in biological, behavioral, and theoretical aspects of neuroscience.

Research Facilities
Graduate research and training are carried out in neuroscience program laboratories. A wide variety of experimental approaches are represented in this graduate program, which is further enhanced by the alliance of Brown and NIH investigators. Faculty trainers employ diverse modern technologies, including structural NMR, genetic engineering, electrophysiology, molecular and cell biological methods, two-photon microscopy, high-dimensional simultaneous microelectrode recording, behavioral neurophysiology, psychophysical and behavioral analyses, functional MRI, mathematical modeling and computer simulation of neural systems, and brain-machine interfaces. Most of the laboratories at Brown are scheduled to be located in a new state-of-the-art research building linked to the Biomedical Research Center.

As the federal government's primary agency for the support and conduct of biomedical research, the NIH offers state-of-the-art research equipment. NIH employs more than 1,200 tenured or tenure-track investigators and nearly 3,700 postdoctoral scientists with medical, dental, or graduate degrees. The NIH intramural research program, located on a 300-acre campus in Bethesda, Maryland, contains more than thirty biomedical research buildings housing twenty-seven institutes and centers in a broad spectrum of biomedical and related scientific research. Four Nobel laureates made their prize-winning discoveries in NIH laboratories, and more than 100 received training at NIH. Basic research in the biomedical sciences at the NIH is complemented by an active clinical research program at the unique 250-bed research hospital and laboratory complex, the Warren Grant Magnuson Clinical Center. The National Library of Medicine, the world's largest medical library, is located on the NIH campus.

Financial Aid
In 2006–07, stipends are $25,000 for twelve months. In addition, tuition, health insurance, and health services fees are paid by the program and the University.

Cost of Study
All students accepted to the program receive full support, including tuition, stipend, and health insurance fees. Full tuition for academic year 2006–07 is $33,888; NIH tuition is $21,533. This amount decreases to $2118 when students complete 24 graduate course credits. All students are assessed a health service fee of $526.

Living and Housing Costs
Apartments near Brown are available in a pleasant residential area that surrounds the University, with rents ranging from $600 to $1200 per month. Dormitory housing for graduate students is also available. Near NIH, there are various houses, apartments, and rooms in private residences available for rent. The cost of living in the Washington, D.C., area is comparable to other major cities in the United States. For information on available housing near NIH, applicants should visit the Web sites of local newspapers or contact the Graduate Partnerships Program office at the NIH Web site.

Student Group
There are more than 5,700 undergraduates, 1,500 graduate students, and 329 medical students at Brown University. Students come from all regions of the United States and more than fifty other countries. There are approximately 135 graduate students within the Division of Biology and Medicine and 39 full-time students in the Graduate Program in Neuroscience. At NIH, there are over 400 graduate students from more than 100 universities. While at NIH, graduate students enjoy services and activities sponsored by the Graduate Partnerships Program, similar to a university campus. These services and an active Graduate Student Council help build a strong graduate student community.

Location
Brown University is in a historic district, within walking distance of the center of the city of Providence, the capital of Rhode Island. The Brown campus itself is a 133-acre complex of architecturally diverse old and new buildings around the central College Green. Many cultural resources are available, including concerts, theater, museums, and art galleries, in addition to the seminars, colloquia, and other events presented by the University. Students may use the University's recreational facilities for both indoor and outdoor exercise. The state is a mecca for ocean sports and many other forms of recreation. Rhode Island is centrally located for easy travel to other parts of New England, including Cape Cod, Martha's Vineyard, Nantucket, and the ski areas of Vermont, New Hampshire, and Maine. Boston is within a 1-hour drive of Brown, and New York City is 150 miles away. The National Institutes of Health is in a northwest suburban area, approximately 10 miles from the cultural life of Washington, D.C. The Metro transit system provides easy access to museums, theaters, and events within the Washington metropolitan area.

The University and The Institutes
Founded in 1764 as the seventh college in America and the third in New England, Brown University began offering graduate courses in 1850. The first Ph.D. was awarded in 1889. The Graduate School was formally established in 1927. The Division of Biology and Medicine provides undergraduate and graduate education and research in the biological sciences and medicine. It includes the School of Medicine, which awards the M.D. degree. Faculty members on campus and in the eight affiliated hospitals participate in the graduate programs offering research degrees. Brown was one of the first universities in the United States to offer an undergraduate major in neuroscience. The multidisciplinary Graduate Program in Neuroscience was created in 1985. The NIH, like Brown, has a long history of training scientists. Thousands have completed postdoctoral training in NIH laboratories, and many NIH-trained scientists have received international recognition for their work. The NIH environment is rich in scientific exchange and provides opportunities for a broad biomedical research experience. Graduate students have been receiving their research training at the NIH for more than fifty years.

Applying
To apply for the Partnership Program, prospective students must be U.S. citizens or noncitizen nationals of the United States or must have been lawfully admitted for permanent residence (i.e., possess a valid Alien Registration Receipt Card or some other verification of such status). Completed applications are due in the Office of the Dean of the Graduate School by December 15 to receive maximum consideration. Applications received after this date are considered on a case-by-case basis, but no application for admission in the fall semester is considered after April 15. The GRE General Test is required. The October GRE is the latest examination that permits test results to be delivered in January. Students whose primary language is not English must submit scores on the Test of English as a Foreign Language (TOEFL). Applicants to the Brown/NIH Partnership must also submit the online GPP application by December 15, which can be found at http://gpp.nih.gov.

Correspondence and Information

Graduate Partnerships Program
Building 2, Room 2E06
National Institutes of Health
2 Center Drive
Bethesda, Maryland 20892-0234

Phone: 301-594-9605
Fax: 301-594-9606
E-mail: gpp@nih.gov
Web site: http://gpp.nih.gov

Neuroscience Graduate Program Director
Department of Neuroscience
Brown University
Box 1953
Providence, Rhode Island 02912

Phone: 401-863-3440
Fax: 401-863-1074
E-mail: Neuroscience@brown.edu
Web site: http://neuroscience.brown.edu

Brown University / National Institutes of Health

THE FACULTY AND THEIR RESEARCH

The National Institutes of Health (NIH)

Below is a listing of neuroscientists directly associated with the NIH/Brown Partnership Program in Neuroscience. However, there are more than 200 neuroscience labs that students can consider. A full listing of neuroscience faculty at NIH can be found at http://neuroscience.nih.gov/.

Robert Adelstein, M.D., Harvard. Role of cytoplasmic myosin II in cell and organ function and in mouse development. NHLBI

Leonardo Belluscio, Ph.D., Columbia. Development and modification of neural circuitry in the mammalian olfactory bulb. NINDS

Perry Blackshear, M.D.; Harvard; D.Phil., Oxford. Signal transduction in response to polypeptide hormones and other agonists. NIEHS

Andres Buonanno, Ph.D., Washington (St. Louis). Regulation of neuronal plasticity by the neuregulin/ErbB signaling pathway: implications for neurological disorders. NICHD

Matthew Daniels, Ph.D., Chicago. Regulation of neuronal outgrowth and synaptogenesis in the central nervous system. NHLBI

Raymond Dionne, D.D.S., Georgetown; Ph.D., Virginia Commonwealth. Clinical studies evaluating novel therapeutic agents and the examination of molecular mechanisms related to pain and analgesia in humans. NIDCR

Marilyn Huestis, Ph.D., Maryland, Baltimore. Pharmacokinetics and pharmacodynamics of drugs of abuse; cannabinoids. NIDA

Donald Ingram, Ph.D., Georgia. Behavioral neuroscience of aging. NINDS

Bechara Kachar, M.D., São Paulo. Cellular and molecular basis of auditory sensory transduction. NIDCD

Jennifer Lippincott-Schwartz, Ph.D., Johns Hopkins. Distribution, dynamics, and biogenesis of eukaryotic organelles. NICHD

Mortimer Mishkin, Ph.D., McGill. Neural mechanisms of learning and memory in primates. NIMH

Anil Mukherjee, M.D., SUNY at Buffalo; Ph.D., Utah. Heritable neurodegenerative and autoimmune disorders. NICHD

John Newman, M.D., Rochester. Mechanisms underlying auditory communication in primates and other mammals. NICHD

Michael Rogawski, M.D., Ph.D., Yale. Epilepsy mechanisms and treatment approaches. NINDS

Norman Salem, Ph.D., Rochester. Polyunsaturated lipid metabolism and function in the nervous system. NIAAA

Toni Shippenberg, Ph.D., Baylor College of Medicine. Neuropeptides and drug addiction. NIDA

David Sibley, Ph.D., California, San Diego. Molecular neuropharmacology of dopamine receptors. NINDS

Kenton Swartz, Ph.D., Harvard. Structure and mechanics of ion channels. NINDS

George Uhl, M.D., Ph.D., Johns Hopkins. Genetics and molecular neurobiology of addiction. NIDA

Ted Usdin, M.D., Ph.D., Washington (St. Louis). Biological role of neurotransmitters and neuromodulators. NIMH

Ling-Gang Wu, M.D., Second Military Medical College (Shanghai); Ph.D., Baylor College of Medicine. Cellular and molecular mechanisms mediating and regulating endocytosis. NINDS.

Richard Youle, Ph.D., South Carolina. Molecular and cellular mechanisms of cell death during development and disease; engineering therapeutic proteins to promote apoptosis of cancer cells and prevent the apoptosis of neurons. NINDS.

Brown University

Carlos Aizenman, Assistant Professor; Ph.D., Johns Hopkins. Developmental and experience-driven mechanisms underlying cellular excitability, network activity, and synaptic function in the developing visual system.

David M. Berson, Professor of Neuroscience; Ph.D., MIT. Anatomy and physiology of the mammalian visual system.

Elie Bienenstock, Associate Professor of Applied Mathematics and Neuroscience; Ph.D., Brown. Mathematical and computational models of neural coding, motor control, and vision.

Michael Black, Professor of Computer Science; Ph.D., Yale. Neural engineering; computational neuroscience; brain-machine interface.

Sheila E. Blumstein, Professor of Cognitive and Linguistic Science; Ph.D., Harvard. Neural basis of speech and language processing; acoustic properties of speech.

Wayne Bowen, Professor of Biology; Ph.D., Cornell. Biochemical mechanisms involved in the neuronal action of opiates and related compounds; Sigma-2 receptor-mediated apoptosis; mechanisms and role in cell proliferation and survival.

Rebecca D. Burwell, Associate Professor of Psychology; Ph.D., North Carolina at Chapel Hill. Neural mechanisms of memory and attention.

Barry W. Connors, Professor of Neuroscience; Ph.D., Duke. Cellular physiology and anatomy of cerebral cortex; basic mechanisms of epilepsy.

Leon N Cooper, Professor of Physics and Neuroscience; Ph.D., Columbia. Neural modeling of learning and memory.

John P. Donoghue, Professor and Chairman of Neuroscience; Ph.D., Brown. Neural control of skilled motor learning and performance; neural plasticity.

Anna Dunaevsky, Assistant Professor; Ph.D., Massachusetts Amherst. Cellular and molecular mechanisms underlying formation, maintenance, and modification of synapses.

Justin R. Fallon, Professor; Ph.D., Pennsylvania. Molecular basis of synapse formation and plasticity.

Stuart Geman, Professor of Applied Mathematics; Ph.D., MIT. Compositional vision; neural representation; neural modeling.

Edward Hawrot, Professor and Chairman of Pharmacology; Ph.D., Harvard. Structure and function of acetylcholine receptors.

Julie A. Kauer, Associate Professor of Molecular Pharmacology and Physiology; Ph. D., Yale. Mechanisms underlying synaptic plasticity related to memory formation and drug addiction.

Diane Lipscombe, Associate Professor of Neuroscience; Ph.D., London. Molecular biology and biophysics of neuronal ion channels.

John Marshall, Associate Professor; Ph.D., Cambridge. Regulation of neurotransmitter receptors and ion channels.

Mayank Mehta, Assistant Professor; Ph.D., Indian Institute of Science. Mechanisms of learning and memory: computational modeling and in vivo parallel recording in hippocampus and cortex.

Michael A. Paradiso, Professor of Neuroscience; Ph.D., Brown. Information processing and plasticity in visual cortex; visual perception.

Robert Reenan, Associate Professor; Ph.D., Harvard. Identification and characterization of genes involved in mismatch repair in *Saccharomyces cerevisiae*.

David Ress, Associate Professor of Neuroscience; Ph.D., Stanford. Use of functional magnetic-resonance imaging to obtain similar information directly from the awake, behaving human brain.

Jerome N. Sanes, Professor of Neuroscience (Research); Ph.D., Rochester. Brain mechanisms underlying voluntary movement and motor skill learning.

David L. Sheinberg, Assistant Professor; Ph.D., Brown. Physiological basis of high-level vision: processing and plasticity in extrastriate cortex.

Andrea M. Simmons, Professor of Psychology and Neuroscience; Ph.D., Harvard. Auditory physiology and behavior; acoustic communication in anuran amphibians.

James A. Simmons, Professor of Neuroscience; Ph.D., Princeton. Behavioral and physiological analysis of bat echolocation.

Michael J. Tarr, Professor; Ph.D., MIT. Behavioral, computational, and neural basis of high-level vision; object and face recognition; visual categorization; perceptual expertise.

Mark Zervas, Assistant Professor; Ph.D., Yeshiva (Einstein). Ganglioside expression and function in development and disease.

Anita L. Zimmerman, Associate Professor of Molecular Pharmacology and Physiology; Ph.D., Miami (Florida). Molecular biology and biophysics of ion channels from neurons and heart cells.

The Dale and Betty Bumpers Vaccine Research Center (VRC), one of more than thirty major research buildings on the NIH campus, was established to facilitate research in vaccine development.

The National Human Genome Research Institute (NHGRI) led the Human Genome Project for the NIH, which culminated in the completion of the full human genome sequence in April 2003.

CARNEGIE MELLON UNIVERSITY
UNIVERSITY OF PITTSBURGH
Center for the Neural Basis of Cognition

Role of the Center

The Center for the Neural Basis of Cognition (CNBC) offers interdisciplinary Ph.D. and postdoctoral training programs that are operated jointly with affiliated programs at Carnegie Mellon University (CMU) and the University of Pittsburgh (PITT). Affiliated departments from Carnegie Mellon include Biological Sciences, Computer Science, Psychology, Robotics, the Center for Automated Learning and Discovery (CALD), and Statistics and, from the University of Pittsburgh, the Departments of Mathematics, Neurobiology, Neuroscience, and Psychology. Mechanisms exist for students in other departments to participate.

The CNBC training program brings together several of the strongest programs of each of the two universities to train interdisciplinary scientists who are interested in understanding how cognitive processes arise from neural mechanisms in the brain. Students combine intensive training in their chosen specialty with broad exposure to other disciplines that touch on neural computation and problems of higher brain function. Members of the CNBC community have access to a wide range of resources, including positron emission tomography (PET) and magnetic resonance imaging (MRI) scanners for functional brain imaging; neurophysiology laboratories for recording from awake, behaving animals; electron and confocal microscopes for structural imaging; high-performance computing facilities, including an in-house supercomputer for neural simulation and image analysis; and patient populations for neuropsychological studies.

The CNBC graduate program is carried out in conjunction with the departmental training programs and requires two to four additional courses beyond the standard departmental training requirements. Students are encouraged to develop a program that includes interdisciplinary training within the CNBC as well as within their department. The CNBC membership typically provides the student with special educational opportunities (e.g., seminar series), a CNBC fellowship that provides travel support, a computer, and a CNBC certificate upon completion of the program.

Facilities

The CNBC offices are located in Mellon Institute at CMU and in the Biomedical Science Tower at PITT. The CNBC includes state-of-the-art neurophysiology laboratories, office space for faculty members and postdoctoral and graduate students, computer facilities, and conference rooms where research meetings and seminars are held. Laboratory facilities of CNBC faculty members are located in the affiliated departments. The Pittsburgh NMR Center and the Pittsburgh Supercomputing Center are located in the Mellon Institute. There is also a Brain Imaging Research Center jointly managed by the University of Pittsburgh and Carnegie Mellon, housing a Siemens 3T scanner.

Carnegie Mellon University is a private institution devoted to liberal professional education. The University has an endowment in excess of $400 million, a total enrollment of about 7,300, and approximately 550 teaching faculty members. Almost 50 percent of the day students live on campus. The University of Pittsburgh is one of the oldest institutions of higher education in the U.S. and is one of the nation's most distinguished comprehensive universities. Its medical school is ranked in the top tier of medical research institutions. PITT has an enrollment of approximately 32,000 students and employs approximately 10,000 faculty and staff members. CMU and PITT are members of the Association of American Universities (AAU), an organization composed of sixty-three eminent doctorate-granting research institutions in the U.S. and Canada.

Funding

The Center for the Neural Basis of Cognition was initiated in 1994 by a major gift from the R. K. Mellon Foundation. It builds on several government-funded collaborations between the two universities. Recent grants include a National Institute of Mental Health Education Grant and a National Science Foundation Integrative Graduate Education and Research Training Program Grant.

Entrance

To be admitted to the CNBC Training Program, students must be accepted into the doctoral program of a home department at either of the two universities. Applications to the home department and the CNBC program may be submitted simultaneously.

Application to the CNBC training program is made by completing the short application that is available at the CNBC Web site. The application must be accompanied by a brief, two-page essay that describes the student's interest in the field and his or her goals in entering the program. Applications must be received by January 1. The following additional materials are obtained from the home department to which the student is applying, if not supplied by the student directly: an undergraduate transcript, GRE scores, and three letters of recommendation. Students who send copies of these items directly to the CNBC admissions office speed up processing of the application. Prior research experience is considered in the evaluation of applications. Where possible, finalists for admission are invited to Pittsburgh for personal interviews before the final admission decision.

The CNBC works with the degree-granting programs throughout the admissions process, and a final decision on admission to the CNBC program is made after the student is admitted to one of these programs.

Student Support

Admitted students are eligible to receive a computer (if not supplied by their department) and support for travel to meetings and conferences. Admitted students may also apply for CNBC fellowship awards that include one year of stipend and tuition support.

Student Profile

In 2005, the CNBC had 83 Ph.D. trainees and 53 postdoctoral fellows from CMU Departments of Biology, Psychology, Computer Science, Robotics, and Statistics and from the PITT Departments of Neuroscience, Neurobiology, Mathematics, Psychology, and Bioengineering.

Forty-three students have completed the CNBC training program. Most have secured postdoctoral or junior faculty positions at excellent institutions, including Yale, Princeton, UC-Davis, UC-Berkeley, Max Planck Institute, Albert Einstein College of Medicine, NIH, and University College London.

Correspondence and Information

Center for the Neural Basis of Cognition
115 Mellon Institute
4400 Fifth Avenue
Pittsburgh, Pennsylvania 15213

Phone: 412-268-4000
Fax: 412-268-5060
Web site: http://www.cnbc.cmu.edu

Carnegie Mellon University / University of Pittsburgh

THE FACULTY AND THEIR RESEARCH

E. T. Ahrens, Assistant Professor, Department of Biological Sciences, CMU. Advancing the state of the art of magnetic resonance imaging (MRI) and using these techniques to visualize development, connectivity, function, and pathology of the vertebrate nervous system; extremely high-resolution MRI, or magnetic resonance microscopy (MRM), an emerging technique that can observe the inner workings of intact living animals in three dimensions at near-cellular resolution.

S. G. Amara, Professor and Chair, Department of Neurobiology, PITT. Molecular genetic, electrophysiological, and cell biological approaches to explore the relationships among neurotransmitter transporter structure, substrate transport, inhibitor binding, and ion permeation.

J. Anderson, Professor, Department of Psychology and Computer Science, CMU. ACT-R: Computer simulation used to model a variety of cognitive phenomena in the domains of attention, learning and memory, problem solving, and language.

P. Arenth, Assistant Professor, Department of Physical Medicine and Rehabilitation, PITT. Issues related to rehabilitation of cognitive deficits associated with traumatic brain injury.

G. Barrionuevo, Professor, Department of Neuroscience, PITT. Biophysical analysis of synaptic integration and plasticity in cells of the hippocampus and prefrontal cortex.

A. Barth, Assistant Professor, Department of Biological Sciences, CMU. Identifying the molecules and pathways involved in developmental and adult plasticity using both in vivo manipulations to induce changes in synaptic strength and whole-cell electrophysiological recordings.

J. T. Becker, Professor, Department of Psychiatry and Neurology, PITT. Information processing defects that can produce human memory disorders and the nature and extent of the neuroanatomical damage that produces these defects.

G. Bi, Assistant Professor, Department of Neurobiology, PITT. Cellular characterization of a complete set of rules of activity-dependent synaptic modification to elucidate the underlying cellular mechanisms; at circuitry level, to investigate how cellular rules may influence the activity-dependent development and remodeling of neural circuits.

C. W. Bradberry, Associate Professor, Psychiatry and Laboratory Medicine. Influence of D1 receptor activation in prefrontal cortex on other neurotransmitter systems; nature of serotonergic dysfunction in aggressive/impulsive macaques; behavioral/neurochemical comparisons between cocaine and psychoactive metabolites of cocaine being studied in humans in other laboratories.

J. P. Card, Associate Professor, Center for Neuroscience, PITT. Defining the functional organization of neural circuits involved in control of behavioral state and autonomic function; developing tools for transneuronal tracing of neural circuits.

P. Carpenter, Professor, Department of Psychology, CMU. Functional characteristics and neural processes that underlie complex cognitive skills (fMRI [functional magnetic resonance imaging]).

R. Cho, Assistant Professor, Department of Psychiatry, PITT. Taking cognitive neuroscience approaches (imaging and computational methods) to elucidate the architecture and dynamics of cognitive control and how they are disturbed in serious mental illness, such as schizophrenia.

C. Cidis-Meltzer, Associate Professor, Department of Radiology, PITT. Using imaging to better understand brain structure-function relationships in aging and age-related neuropsychiatric disorders.

C. Colby, Associate Professor, Department of Neuroscience and CNBC, PITT. Neural mechanisms of attention, memory, and spatial representation (single-unit recording in primates and functional imaging in humans).

J. C. Crowley, Assistant Professor, Department of Biological Sciences, CMU. Formation of circuitry in the visual system, specifically development of neural processing modules in primary visual cortex.

X. T. Cui, Assistant Professor, Bioengineering Department, PITT. Interfacing nervous system via electroactive and bioactive conducting polymer complex; surface modification and characterization of biosensors, bioMEMS, and implantable medical devices; conducting polymer actuators and drug release for biomedical application; polymer-based biomaterial and tissue engineering.

S. T. DeKosky, Professor, Departments of Psychiatry, Neurology, and Neurobiology, PITT. Neurodegenerative human disease, traumatic brain injury, and lesion-induced synaptogenesis and axonal sprouting.

W. F. Eddy, Professor, Department of Statistics, CMU. Analysis of fMRI data.

G. B. Ermentrout, Professor, Department of Mathematics, PITT. Biological oscillations and mathematical modeling.

H. Feldman, Ronald L. and Patricia M. Violi Professor of Child Development, Pediatrics, PITT. Biological and environmental basis of language development.

J. Fiez, Associate Professor, Psychology and Neuroscience and the CNBC, PITT. Understanding neural basis of cognitive functions using neuroimaging as the primary methodology; other methodologies, such as behavioral studies of normal brain-damaged subjects, are used to provide converging evidence to support neuroimaging results and to provide an interdisciplinary perspective.

N. J. Gandhi, Assistant Professor, Otolaryngology and Neuroscience, PITT. Neural control of coordinated movements of the eyes and head as well as integration of different types of eye movements as in saccades and smooth pursuit.

C. R. Genovese, Assistant Professor, Department of Statistics, CMU. Statistical methods for making effective inferences from data in complex scientific problems, including fMRI.

P. Gianaros, Assistant Professor, Departments of Psychiatry and Psychology, PITT. Using cardiovascular behavioral neuroimaging to define how the human brain regulates cardiovascular and autonomic nervous system responses to stress and contributes to the risk for cardiovascular disease.

A. R. Hariri, Director, Developmental Imaging Genomics Program, Department of Psychiatry, PITT. Identifying biological mechanisms that contribute to complex cognitive and emotional behaviors to understand how individual differences in these behaviors emerge and how such differences may confer vulnerability to psychiatric disease.

L. Holt, Assistant Professor, Department of Psychology, CMU. Empirical data collected in perceptual and learning paradigms with human participants.

J. P. Horn, Associate Professor, Department of Neurobiology, PITT. Interactions between metabotropic and nicotinic receptors give rise to synaptic plasticity in sympathetic ganglia.

S. Iyengar, Professor and Chair, Department of Statistics, PITT. Spike train data analysis for both single trains and multiple simultaneously recorded neurons.

J. Johnson, Associate Professor, Neuroscience and Psychiatry, PITT. Ion channels, which are fundamental to the movement and processing of information in all nervous systems.

M. A. Just, Professor, Department of Psychology, CMU. Visual thinking, language comprehension, and problem-solving processes with fMRI and cognitive methods.

K. Kandler, Associate Professor, Department of Neurobiology, PITT. How immature inhibitory pathways develop, become reorganized, and become functionally aligned with excitatory connections.

R. E. Kass, Professor, Department of Statistics, CMU. Transmission of information by collections of neurons; Bayesian statistics.

S.-G. Kim, Professor, Department of Neurobiology, PITT. Development of novel MR techniques to measure physiological changes induced by neural activity; investigation of biophysics of MRI signals; measurement of physiological changes induced by neural activity; exploration of limits of spatial and temporal resolution of functional MRI (fMRI); application of MR techniques to brain research.

R. L. Klatzky, Professor, Department of Psychology, HCII, CMU. Human perception and cognition (haptic perception and spatial cognition).

H. R. Koerber, Associate Professor, Department of Neurobiology, PITT. Development and adult plasticity of spinal networks involved in sensory information processing.

T. S. Lee, Associate Professor, Department of Computer Science and CNBC, CMU. Computational and neural processes involved in vision.

M. S. Lewicki, Assistant Professor, Department of Computer Science and CNBC, CMU. Computational principles underlying how the brain represents and learns higher-level representations.

D. Lewis, Professor, Departments of Psychiatry and Neuroscience, PITT. Anatomical analysis of the functional architecture of primate neocortex and cortical circuitry in pathophysiology of human neuropsychiatric disorders.

B. Luna, Associate Professor, Department of Psychiatry, PITT. Characterizing changes that occur in brain function that subserve the maturation of cognition by bridging developmental psychology and neuroscience.

P. K. Machamer, Associate Professor and Director, History, Philosophy of Science, Center for Philosophy of Science (CPS), PITT. How knowledge requires actions that have associated social criteria that warrant their counting as knowledge.

E. Machery, Assistant Professor, Department of Philosophy, PITT. Exploring the theoretical issues that are raised by psychology and cognitive science—particularly the themes of concepts, simple heuristics, evolution/culture/cognition, and philosophy.

B. MacWhinney, Professor, Department of Psychology, CMU. Language acquisition: brain structure and developments.

J. L. McClelland, Professor, Departments of Psychology and Computer Science, CMU; Adjunct Professor, Department of Neuroscience, PITT; and Co-Director, CNBC. Computational models of perception, memory, and language; mechanisms of learning and cognitive development.

S. Meriney, Associate Professor, Department of Neuroscience, PITT. Studying mechanisms that control synaptic plasticity in the nervous system.

N. Minshew, Associate Professor, Department of Psychiatry and Neurology, PITT. Investigating the cognitive and brain basis of autism and the genetics of this disorder.

T. Mitchell, Fredkin Professor of AI Learning and Director, Computer Science, Center for Automated Learning and Discovery (CALD).

B. Moghaddam, Professor of Neuroscience, PITT. Mechanisms that underlie cognitive control of goal-directed behavior and how these mechanisms are disrupted in animal models of psychiatric disorders.

R. Y. Moore, Professor, Departments of Neurology, Neuroscience, and Psychiatry, PITT. Functional organization of the mammalian hypothalamus, with particular emphasis on the circadian timing system.

P. Munro, Faculty, SIS Information Science and Telecommunications, PITT. Investigations to abstract mathematical and computational principles underlying learning at the synaptic, neuronal, and systems levels.

A. P. Monaghan Nichols, Assistant Professor, Department of Neurobiology, PITT. Genetic analysis of vertebrate CNS development.

C. Olson, Professor and Director, Primate Physiology Laboratory, CNBC, CMU. Centered spatial awareness and frontal cortex; studies of posterior cingulate cortex (single unit recording in primates).

C. A. Perfetti, Professor, Department of Psychology and Linguistics, PITT. Cognitive science of language and reading processes, including lower- and higher-level processes and the nature of reading ability.

M. Phillips, Professor, Department of Psychiatry, PITT. Examining neural response to emotional stimuli in normal and psychiatric populations using fMRI.

M. F. Pogue-Geile, Associate Professor, Departments of Psychology and Psychiatry, PITT. Neuropsychology and genetics of schizophrenia.

L. Reder, Professor, Department of Psychology, CMU. How information is organized and stored in memory and how it is accessed and used.

E. Reichle, Assistant Professor, PITT. Understanding how language, attention, and vision guide eye movements during normal reading.

J. Ricker, Associate Professor of Physical Medicine and Rehabilitation, PITT. Examination of neurobiological correlates and predictors of cognitive impairment, recovery, and rehabilitation following brain trauma in humans.

J. Rubin, Assistant Professor, Department of Mathematics, PITT. Geometric dynamical analysis of model networks of coupled neurons.

W. Schneider, Professor, Department of Psychology, PITT. Dynamic cortical processing and brain imaging.

A. B. Schwartz, Professor, Department of Neurobiology, PITT. Brain processes that subserve volitional movement; basic cortical mechanisms subserving arm movement; applied bioengineering of cortical output.

S. Sesack, Associate Professor, Neuroscience and Psychiatry. Dopamine neurons and their prefrontal cortical targets, which play an important role in the regulation of cognitive functions and have been implicated in the pathophysiology of schizophrenia.

G. J. Siegle, Assistant Professor, Department of Psychiatry, PITT. Understanding how individual differences in information processing (e.g., attending to, remembering, and interpreting information) are related to affective psychopathologies, particularly unipolar depression; ultimate goals of this research include better understanding of the nature of affective disorders and development of interventions tailored to account for individual differences in information processing styles.

D. J. Simons, Professor, Department of Neurobiology, PITT. Sensory information processing in mammalian cerebral cortex.

M. Sommer, Assistant Professor, Department of Neuroscience and CNBC, PITT. How brain areas interact as circuits, with specific focus on one, critical experimentally tractable network—that which mediates our ability to see the world around us.

P. Strick, VA Senior Research Career Scientist; Professor, Departments of Neurobiology, Psychiatry, and Neurological Surgery, PITT; and Co-Director, CNBC. Central control of movement; motor skill acquisition and retention; basal ganglia and cerebellar involvement in motor and cognitive function.

F. Thiels, Professor of Neuroscience, PITT. How animals acquire information from the environment and use that information to guide their behavior; understanding biological substrates of learning and memory, one of the most sought-after goals of neuroscience.

N. Tokowicz, Assistant Professor, Department of Psychology, PITT. Cognitive processes related to language learning and use.

C. A. Tompkins, Professor, Communication Science and Disorders, PITT. Understanding cognitive underpinnings of neurologic communication disorders in adults.

D. S. Touretzky, Research Professor, Departments of Computer Science and Robotics and CNBC, CMU. Neural representations in rodent navigation; hippocampal processing; computational models of animal learning.

N. Urban, Assistant Professor, Department of Biological Sciences, CMU. Understanding physiological mechanisms underlying the functional and computational properties of brain neuronal networks through detailed studies of physiological properties of synapses, cells, and circuits involved in the performance of a given task.

V. Ventura, Research Scientist, Statistics, CMU. Modeling neuronal firing patterns.

K. Verdolini, Associate Professor, Department of Communication Science and Disorders, PITT. Influence of specified cognitive manipulations on the outcome of voice training and therapy; role of exercise in laryngeal wound healing; effectiveness of selected treatment programs for teachers and other professional voice users with phonotraumatic injury.

M. Wheeler, Assistant Professor, Psychology, PITT. Using fMRI to identify neural correlates of memory retrieval and recognition memory; neural bases of decision in memory tasks and how attentional control influences fMRI activity during retrieval.

CASE WESTERN RESERVE UNIVERSITY

Case School of Medicine
Department of Neurosciences

Program of Study

The Department of Neurosciences at Case Western Reserve University offers graduate training in a number of areas of neuroscience research leading to a Ph.D. degree in the neurosciences. The intent is to provide interdisciplinary training in modern neurobiology through a combination of course work, seminars, and research experience. Faculty members conduct research on mechanisms of development of sensory and motor systems, regeneration, axon pathfinding, synaptic function and plasticity, neurotrophin gene expression and trophic regulation, systems neuroscience, neuropharmacology, aging, neuron-glial interactions, neural lineages, simple neural networks, expression and regulation of neurotransmitters and receptors, and neurogenetics. Individuals are trained in basic research with a view toward an academic profession.

In the first year, students enroll in a comprehensive graduate course in cell and molecular biology that involves faculty members from most of the biological sciences departments. This is complemented by a two-semester course in cellular and molecular neuroscience. During the first and second years, students select from a number of graduate-level courses in neurobiology, including neuroethology and behavior, developmental neurobiology, molecular neurobiology, synaptic neurochemistry and neuropharmacology, and computational neuroscience. The graduate program is flexible, allowing students to tailor course work to their interests (molecular, physiological, pharmacological, developmental, computational) and goals. Three research rotations are required. Advancement to candidacy is based on the successful completion of course work, laboratory performance, and a two-part qualifying exam. Most students complete their Ph.D. training within four to five years.

The Department participates in the integrated Biomedical Science Training Program. This interdepartmental program comprises many related training areas within the Case School of Medicine, with provisions for the student to explore different areas during the first year.

Research Facilities

Participating faculty members are drawn from a number of departments in the medical school, including neurosciences, neurology, pharmacology, and physiology, as well as from the biology department. Thesis research is often interdisciplinary, involving techniques of biochemistry, biophysics, molecular biology, and morphology. The Department and associated laboratories are well equipped for such analyses. Available facilities include a wide array of specially equipped light microscopes, oligonucleotide synthesizers, thermal cyclers for PCR, a confocal microscope facility, transmission and scanning electron microscopes, darkrooms, constant-temperature rooms, and a dual-laser flow cytometry facility. Several laboratories have well-developed cell- and tissue-culture facilities, including those needed for hybridoma work, and facilities for image analysis.

Financial Aid

Financial aid includes a complete tuition scholarship plus an annual stipend ($23,000 in 2005–06). This is derived from NIH institutional training grants, individual fellowships, and Department resources.

Cost of Study

In 2005–06, tuition at Case Western Reserve University Graduate School was $20,000 per year and fees were $874 per year. These costs are covered by the Department.

Living and Housing Costs

Single rooms are available on campus starting from $5400 per year. Within a 2-mile radius of the campus, numerous apartments are available for married and single students at rents ranging from $500 to $800 per month.

Student Group

The University's total enrollment is approximately 9,500. Of this number, 5,900 are enrolled in the Graduate School and the professional schools, and 3,600 in the undergraduate college. Forty-five graduate students are currently enrolled in the program.

Location

Case Western Reserve University is located in the University Circle area on the eastern outskirts of the city. Recognized as the cultural center of Cleveland, the area contains Severance Hall (home of the Cleveland Orchestra), the Museums of Art and Natural History, Garden Center, Institute of Art, Institute of Music, Western Reserve Historical Society, Cleveland Playhouse, and Crawford Auto-Aviation Museum. The Cleveland Hopkins International Airport is 40 minutes away by rapid transit.

The University

In 1967, Case Western Reserve University was formed by a federation of Western Reserve University and Case Institute of Technology. The Case School of Medicine ranks among the top twenty schools in the United States with respect to federal funding of research. Graduate education in biomedical sciences has been strengthened by the addition of interdisciplinary programs offered by the basic science departments in the medical school and other University departments.

Applying

There are two options for applying to the Ph.D. program. Option one allows an applicant to apply to the Biomedical Sciences Training Program. Option two allows an applicant to apply to the Neuroscience Program directly. Applications should be submitted by early fall or winter for admission the following July. Late applications are accepted. Additional information and applications may be obtained by writing to the addresses below. The completed application should be accompanied by official transcripts from each undergraduate and graduate institution attended and results of the GRE General Test (verbal, quantitative, and analytical portions) and a Subject Test.

Correspondence and Information

When applying to the Biomedical Sciences Training Program:
Biomedical Sciences Training Program Coordinator
Case School of Medicine, WG46
Case Western Reserve University
10900 Euclid Avenue
Cleveland, Ohio 44106-4934

When applying to the Neuroscience Program directly:
Graduate Program Coordinator
Case School of Medicine
Department of Neurosciences
Case Western Reserve University
10900 Euclid Avenue
Cleveland, Ohio 44106-4975
E-mail: nrb@po.cwru.edu
Web site: http://neurowww.cwru.edu

Case Western Reserve University

THE FACULTY AND THEIR RESEARCH

Listed below are faculty members participating in the graduate training program in the neurosciences.

Heather Broihier, Assistant Professor of Neurosciences; Ph.D., MIT. Genetic and molecular analysis of motoneuron specification and differentiation in *Drosophila.*

Hillel Chiel, Professor of Biology and Neurosciences; Ph.D., MIT. Feeding behavior in *Aplysia;* analysis of neural networks in vivo and in vitro; computational neuroscience.

Evan S. Deneris, Professor of Neurosciences; Ph.D., UCLA. Transcriptional mechanisms of gene expression in neurons; neuronal nicotinic acetylcholine receptor genes.

Thomas Dick, Associate Professor of Medicine and Neurosciences; Ph.D., Texas Tech. Neural regulation of respiration; central pattern generators; sensory modulation of the breathing pattern and motoneuronal activity.

Dominique M. Durand, Professor of Biomedical Engineering and Neurosciences; Ph.D., Toronto. Computational neuroscience; cellular mechanisms of hypoxia and epilepsy; neuroprostheses; electric and magnetic field effects on neurons.

David Friel, Associate Professor of Neurosciences; Ph.D., Chicago. Subcellular calcium signaling cascades triggered by membrane depolarization and their modulation by intracellular ion channels.

Alison Hall, Associate Professor of Neurosciences and Pharmacology; Ph.D., Case Western Reserve. Influence of cell lineage and environmental cues on peripheral nervous system differentiation.

Stefan Herlitze, Assistant Professor of Neurosciences; Ph.D., Heidelberg. Regulation of Ca^{2+} channel function in *C. elegans.*

Stephen W. Jones, Associate Professor of Physiology and Biophysics; Ph.D., Cornell. Characterization of neuronal ion channels and their regulation by neurotransmitters and second messengers.

David M. Katz, Professor of Neurosciences; Ph.D., SUNY at Stony Brook. How neuronal growth factors and neural activity regulate development of functional neural circuits; developmental neurobiology of breathing.

Diana Kunze, Professor of Neurosciences; Ph.D., Utah. Electrophysiological and immunocytochemical studies of peripheral and central neurons involved in autonomic reflexes.

Joseph LaManna, Professor of Neurology, Physiology, Neurosciences, and Anatomy; Ph.D., Duke. Regulation of cerebral metabolism and blood flow.

Lynn Landmesser, Professor of Neurosciences; Ph.D., UCLA. Molecular mechanisms of axon guidance and synapse formation; synaptic plasticity.

Gary Landreth, Professor of Neurosciences and Neurology; Ph.D., Michigan. Mechanism of action of nerve growth factor; molecular mechanisms of Alzheimer's disease.

John Leigh, Professor of Neurology; M.D., Newcastle upon Tyne. Control of ocular motor systems.

David Lust, Professor of Neurosurgery; Ph.D., SUNY Upstate Medical Center. Age-dependent changes in the evolution of cellular damage induced by reperfusion following brain ischemia.

Alfred T. Malouf, Assistant Professor of Pediatrics and Neurosciences; Ph.D., Johns Hopkins. Neurophysiology and development of axon pathways in the hippocampus.

Robert H. Miller, Professor of Neurosciences; Ph.D., University College (London). Lineage and function of glial cells.

Stephen O'Gorman, Associate Professor of Neurosciences; Ph.D., Harvard. Analysis of hindbrain and craniofacial development in mammals using homologous and site-specific recombination.

Guillermo Pilar, Professor of Neurosciences; M.D., Buenos Aires (Argentina). Mechanisms of neuronal homeostasis during developmental neuronal death; membrane and intracellular pathways mediating modulation of transmitter release.

Robert Petersen, Associate Professor of Pathology and Neurosciences; Ph.D., Minnesota. The cell and molecular biology of prion and other neurodegenerative diseases.

Roy E. Ritzmann, Professor of Biology and Neurosciences; Ph.D., Virginia. Neural circuits that control behavior and implications for robotic design.

Iain Robinson, Assistant Professor of Neurosciences; Ph.D., Liverpool (England). The genetically tractable organism *Drosophila melanogaster,* in combination with electrophysiology, imaging, biochemistry, and molecular biology, is being used to examine the molecular basis of exocytosis and endocytosis.

Bryan Roth, Associate Professor of Psychiatry and Neurosciences; M.D., Ph.D., St. Louis. Serotonin receptor expression and regulation.

Ruth E. Siegel, Associate Professor of Pharmacology; Ph.D., Harvard. Development and expression of the $GABA_A$ receptor gene family.

Jerry Silver, Professor of Neurosciences; Ph.D., Case Western Reserve. Role of glial cells in the development and regeneration of neural circuits.

Corey Smith, Assistant Professor of Physiology and Biophysics; Ph.D., Colorado. Understanding how chromaffin cells meet secretory demands.

Ben W. Strowbridge, Associate Professor of Neurosciences; Ph.D., Yale. Synaptic plasticity in the hippocampus; cellular mechanisms of epilepsy; synaptic organization of the olfactory bulb.

Bruce Trapp, Professor of Neurosciences; Ph.D., Loyola Chicago. Cellular and molecular mechanisms of myelination during normal development; cellular and molecular mechanisms of demyelination and inflammation in multiple sclerosis brain.

Christopher Wilson, Assistant Professor of Pediatrics and Neurosciences; Ph.D., California, Davis. Neural control of respiration, specifically striving to understand how respiratory rhythm is generated and modified in mammalian brainstem neural networks.

Richard E. Zigmond, Professor of Neurosciences; Ph.D., Rockefeller. Changes in neuronal gene expression related to regeneration.

Case Western Reserve University campus.

The University provides a pleasant living environment for students.

CORNELL UNIVERSITY

Field of Neurobiology and Behavior

Program of Study

The graduate Field of Neurobiology and Behavior offers a program that leads primarily to a Ph.D. degree; an M.S. degree may be obtained, but applications are not encouraged from students who wish to pursue the master's as a terminal degree. In addition to working in areas of specific interest to members of the field, students may benefit from experience in the related fields of physiology, biochemistry, ecology and evolutionary biology, entomology, genetics, microbiology, psychology, and zoology as well as in other areas in the physical and natural sciences.

The student's major and minor programs are under the supervision of a special graduate committee, 3 members of which are chosen by the student in consultation with his or her adviser. A fourth member is assigned by the Director of Graduate Studies in Neurobiology and Behavior.

Research Facilities

Modern biochemical, neurophysiological, electron microscopic, acoustical, and behavioral research facilities are available. A fully equipped and staffed electronics shop can be used by graduate students. Cornell possesses one of the world's finest library collections of biological literature.

In addition, Cornell has available for research near the campus extensive tracts of natural, wild land representing a wide range of ecological conditions. The Field possesses a variety of acoustical and optical field research equipment. There are opportunities for summer study at the Shoals Marine Laboratory, off the coast of New Hampshire.

Financial Aid

Cornell fellowships and teaching and research assistantships in biology are available to qualified students. Additional graduate stipends are available from two NIH training grants: one funds students in molecular and cellular neurobiology; the other funds students in integrative neurobiology and behavior, with special emphasis on projects that bridge physiological and behavioral studies. To be eligible for support from the training grant, students must hold U.S. citizenship or a permanent visa.

Cost of Study

Tuition and fees are ordinarily paid by traineeships, fellowships, or teaching assistantships.

Living and Housing Costs

For the 2005–06 academic year, the cost of living for a single student ranged from $16,096 to $23,467. This figure included the cost of books, room and board, and personal expenses. Costs of tuition and travel should be added to arrive at an approximate cost of attendance for the nine-month academic year.

Student Group

There are nearly 6,000 students enrolled in Cornell's various graduate programs. The Field of Neurobiology and Behavior enrolls about 40 graduate students.

Location

Cornell University is located near Ithaca above Cayuga Lake, one of the Finger Lakes of central New York. The entire region is noted for its scenery, fishing, water sports, gorges, parks, and wineries. Ithaca is accessible by automobile, bus, and airplane.

The University

Cornell University, chartered in 1865 as the land-grant institution of the state of New York, includes nine endowed and four state-supported schools and colleges as well as the Graduate School. The opportunities for interaction between scientists of various disciplines, the extensive holdings of Cornell's libraries, and the other advantages provided by a major university contribute to a productive and stimulating experience for the researcher, teacher, and scholar. Cornell is an Equal Opportunity/Affirmative Action educator.

Applying

Interested individuals should visit the Cornell University Graduate School Web site for application information (http://www.gradschool. cornell.edu/). Completed forms are due December 1 for admission the following fall. Decisions are usually reached by March 1. Applicants must present scores on the Graduate Record Examinations (the General Test and a Subject Test in one of the sciences). Students are encouraged to apply early and take the GRE in the fall. Application for midyear admission is not encouraged. Correspondence in advance with individual faculty members is appropriate.

Prospective candidates are encouraged to correspond with individual faculty members. They should, however, first contact the Graduate Field Assistant of Neurobiology and Behavior at W363 Seeley G. Mudd Hall or telephone 607-254-4340 to obtain additional information.

Correspondence and Information

Director of Graduate Studies
Field of Neurobiology and Behavior
Seeley Mudd Hall
Cornell University
Ithaca, New York 14853
E-mail: nbb_field@cornell.edu
Web site: http://www.nbb.cornell.edu/neurobio/field/index.html

Cornell University

THE FACULTY AND THEIR RESEARCH

Elizabeth Adkins-Regan, Ph.D., Professor. Mechanisms of behavior; hormonal and neural mechanisms of avian social behavior. Hormonal mechanisms of mate choice. *Am. Zool.* 38(1):166–78, 1998.

Kraig Adler, Ph.D., Professor. Photoreception and orientation of vertebrates; behavior, ecology, and evolution of amphibians and reptiles. True navigation by an amphibian. *Anim. Behav.* 50:855–8, 1995 (with Phillips and Borland).

Andrew H. Bass, Ph.D., Professor. Animal communication, especially production and encoding of vocal signals; evolution and sexual differentiation of brain and behavior. Steroid-dependent auditory plasticity leads to adaptive coupling of sender and receiver. *Science* 305:404–7, 2004 (with Sisneros, Forlano, and Deitcher).

Ronald Booker, Ph.D., Associate Professor. Control of neurodevelopment and postembryonic neurogenesis in the CNS of insects. Effects of parasitism and octopamine on the frontal ganglion and foregut of the moth, *Manduca sexta. J. Exp. Biol.*, in press (with C. I. Miles).

Jack Bradbury, Ph.D., Professor. Evolution of animal social behavior and communication, with current projects on functions of vocal learning in wild parrots, evolution of signal repertoire size, and role of sound field shape in vocal communication. *Principles of Animal Communication.* Sunderland, Mass.: Sinauer Associates, 1998 (with Vehrencamp).

Christopher W. Clark, Ph.D., Senior Scientist and Imogene Johnson Director, Bioacoustics Research Program. Abundances and acoustic behaviors of birds, fish, elephants, and whales; evolution and function of acoustic signaling behavior in animals. Potential use of sounds by whales for probing the environment: Evidence from models and empirical measurements. In *Echolocation in Bats and Dolphins*, eds. Thomas et al. University of Chicago Press, 2004 (with Ellison).

David L. Deitcher, Ph.D., Associate Professor. Mechanisms of neurotransmitter and neuropeptide release; synaptic development. A *Drosophila* SNAP-25 null mutant reveals context-dependent redundancy with SNAP-24 in neurotransmission. *Genetics* 162:259–71, 2002 (with Vilinsky et al.).

Timothy J. DeVoogd, Ph.D., Professor. Animal communication; neuroanatomical plasticity; quantitative neurobiology; neurobiology of avian caching; hormonal influences on brain structure; avian song system. Neural constraints on the complexity of avian song. *Brain Behav. Evol.* 63:221–32, 2004.

Janis L. Dickinson, Ph.D., Associate Professor. Behavioral ecology of birds and insects, with a focus on dispersal, mating behavior, and cooperative breeding; relationship between resource wealth, nepotism, and family group living. *Proc. R. Soc. London B* 272:2423–8, 2005.

Thomas Eisner, Ph.D., Professor. Chemical communication, ecology, and evolution; behavioral ecology, particularly of insects; insect-plant interactions; conservation. *For Love of Insects,* 464 pp. Harvard University Press, 2003.

Stephen T. Emlen, Ph.D., Professor. Evolution of animal social behavior, with current emphasis on behavioral ecology of cooperative nesting in birds. Predicting family dynamics in social vertebrates. In *Behavioural Ecology: An Evolutionary Approach,* 4th ed., pp. 228–53, eds. J. R. Krebs and N. B. Davies. London: Blackwell Scientific, 1997.

John Ewer, Ph.D., Associate Professor. Genetic and neuroendocrine control of postembryonic insect development; insect behavior. Neuroendocrine control of larval ecdysis behavior in *Drosophila*: Complex regulation by partially redundant neuropeptides. *J. Neurosci.* 24:4283–92, 2004 (with Clark and Del Campo).

Joseph R. Fetcho, Ph.D., Professor. Motor control with a focus on hindbrain and spinal cord; optical and genetic approaches for studying neuronal structure and function; regeneration. Cyclic amp–induced repair of zebrafish spinal circuits. *Science* 305:254–8, 2004 (with Bhatt, Otto, and Depoister).

Barbara Finlay, Ph.D., Professor. Development and evolution of neural systems in mammals. Developmental structure of brain evolution. *Behav. Brain Sci.* 24:263–308, 2001 (with Darlington and Nicastro).

Cole Gilbert, Ph.D., Associate Professor. Invertebrate neuroethology; sensory guidance; neural evolution in arthropods. Visual control of cursorial prey pursuit by tiger beetles (Cicindelidae). *J. Comp. Physiol. A* 181:217–30, 1997.

Bruce P. Halpern, Ph.D., Professor. Sensory physiology; chemoreception. Identification of air-phase retronasal and orthonasal odorant pairs. *Chem. Senses* 30:1–14, 2005.

Ronald M. Harris-Warrick, Ph.D., Professor. Neuromodulation of neural networks in lobster and mouse spinal cord; molecular manipulation of ion channels in neurons. Activity-independent homeostasis in rhythmically active neurons. *Neuron* 37:109–20, 2003 (with MacLean).

Carl D. Hopkins, Ph.D., Professor. Neuroethology of animal communication; origins and evolution of electrogenesis in fishes. Design features for electric communication. *J. Exp. Biol.* 202:1217–28, 1999.

Howard C. Howland, Ph.D., Professor. Sensory physiology, especially vision; development of focusing and refractive state in humans; animal vision, models of myopia and emmetropization; optical quality of the eye. The effects of constant and diurnal illumination of the pineal gland and the eyes on ocular growth in chicks. *Invest. Ophthalmol. Vis. Sci.* 44:3692–7, 2003 (with Li).

Ronald R. Hoy, Ph.D., Professor. Animal communication; behavior genetics of invertebrates; regeneration and development in invertebrate nervous systems. Categorical perception of sound frequency by crickets. *Science* 273:1542–4, 1996 (with May and Wyttenbach).

Robert E. Johnston, Ph.D., Professor. Animal communication, social recognition, and memory, mostly by chemical signals; neural mechanisms of social recognition and memory; mechanisms of vertebrate social behavior (behavioral, olfactory, neural, hormonal). Kin recognition and the "armpit effect": Evidence of self-referent phenotype matching. *Proc. R. Soc. London B* 267:695–700, 2000 (with Mateo).

Michael I. Kotlikoff, Ph.D., Professor. Cellular signaling; ion channels; genetic cell sensors; conditional gene targeting. Gene trap and gene inversion methods for conditional gene inactivation in the mouse. *Nucleic Acids Res.* 33:14–9, 2005 (with Xin et al.).

David Lin, Ph.D., Assistant Professor. Axon guidance and target formation during the development of the mouse olfactory system, using genetic, in vitro, and genomic approaches. Spatial patterns of gene expression in the olfactory bulb. *Proc. Natl. Acad. Sci. U.S.A.* 101:12718–23, 2004.

Manfred Lindau, Dr.rer.nat., Professor. Biophysics; exocytosis; mechanisms of vesicle fusion and transmitter release. Electrochemical imaging of fusion pore openings by electrochemical detector arrays. *Proc. Natl. Acad. Sci. U.S.A.* 102:13879–84, 2005 (with Hafez et al.).

Christiane Linster, Ph.D., Associate Professor. Neural coding and memory in the olfactory system; emphasis on neuromodulatory influences; combined approach using behavior, electrophysiology, and computational modeling. Computation in the olfactory system. *Chem. Senses* 30:801, 2005 (with Cleland).

Ellis R. Loew, Ph.D., Associate Professor. Sensory physiology; sensory ecology; retinal physiology and biochemistry; clinical electrophysiology; visual sciences. The underwater visual environment. In *The Visual System of Fish*, pp. 1–44, eds. R. Douglas and M. Djamgoz. London: Chapman and Hall, 1990 (with McFarland).

David P. McCobb, Ph.D., Associate Professor. Ion channels, cellular excitability, and steroid hormones; integrative neuroendocrinology of adrenal and pituitary function, stress, social behavior, and emotion. Acute modulation of adrenal chromaffin cell BK channel gating and cell excitability by glucocorticoids. *J. Neurophysiol.* 91:561–70, 2004 (with Lovell and King).

Linda M. Nowak, Ph.D., Associate Professor. Molecular and electrophysiological analysis of glutamate receptors, ion channels, and response mechanisms in synaptic transmission. Extracellular 3', 5' cyclic guanosine monophosphate inhibits kainate-activated responses in cultured mouse cerebellar neurons. *J. Pharmacol. Exp. Ther.* 286:99–109, 1998.

Robert E. Oswald, Ph.D., Professor. Molecular neurobiology and biophysics; neurotransmitter receptors and signal transduction. Structural mobility of the extracellular ligand-binding domain of an ionotropic glutamate receptor. Analysis of NMR relaxation dynamics. *Biochemistry* 41:10472–81, 2002 (with McFeeters).

H. Kern Reeve, Professor. Theoretical sociobiology; evolution of cooperation and conflict in animals and plants; behavioral ecology; plasticity in insect social behavior. Genetic support for the evolutionary theory of reproductive transactions in social wasps. *Proc. R. Soc. London B* 267:75–9, 2000.

Wendell L. Roelofs, Ph.D., Professor. Insect chemical communication: production and perception; behavior, biosynthesis, and genetics. Chemistry of sex attraction. *Proc. Natl. Acad. Sci. U.S.A.* 92:44–9, 1995.

Thomas D. Seeley, Ph.D., Professor. Sociobiology: physiology, ecology, and evolution of animal societies, especially the societies of insects. How does an informed minority of scouts guide a honeybee swarm as it flies to its new home? *Anim. Behav.* 71:161–71, 2006 (with Beekman and Fathke).

Paul W. Sherman, Ph.D., Professor. Animal social behavior. Reflections on animal selves. *Trends Ecol. Evol.* 19:176–80, 2004 (with Bekoff).

Michael Spivey, Ph.D., Associate Professor. Eye movements and neural networks. Integration of visual and linguistic information during spoken-language comprehension. *Science* 268:1632–4, 1995 (with Tanenhaus).

Sandra Vehrencamp, Ph.D., Professor. Evolution of animal communication systems, especially bird song; behavioral ecology of cooperative nesting in birds. Is song-type matching a conventional signal of aggressive intentions? *Proc. R. Soc. London B* 268:1637–42, 2001.

Charles Walcott, Ph.D., Professor. Animal orientation and navigation; animal acoustic communication. Multi-modal orientation cues in homing pigeons. *Integr. Comp. Biol.* 45:574–81, 2005.

Gregory A. Weiland, Ph.D., Associate Professor. Molecular neurobiology and neuropharmacology; receptor and ion channel mechanisms. Two domains of the *beta* subunit of neuronal nicotinic acetylcholine receptors contribute to the affinity of substance P. *J. Pharmacol. Exp. Ther.* 286:619–26, 1998 (with Oswald).

DARTMOUTH COLLEGE

Dartmouth Medical School
Neuroscience Center at Dartmouth

Program of Study

The Neuroscience Center at Dartmouth (NCD) provides opportunities for qualified Ph.D. and M.D./Ph.D. students to be educated and highly trained for productive careers in translational neuroscience research through the Neuroscience Track of the Program in Experimental and Molecular Medicine (PEMM) at Dartmouth. Students utilize innovative approaches to investigate the neurobiology of disease to advance the understanding of the mechanisms of nervous system disorders to discover novel treatments. The interdisciplinary nature of the neuroscience track allows for a broad range of interactions with faculty members, residents, fellows, and other students. The NCD's faculty members represent departments that span the arts and sciences, Medical School, Thayer School of Engineering, and the Veteran Affairs Hospital. These members and the environment provide an exciting, interactive atmosphere that fosters the exchange of ideas and approaches, cultivates collaborations, enhances research activities, and promotes bi-directional translational research opportunities for both basic science and clinical faculty members. Translational research, which is emphasized within the neuroscience track of PEMM, is the process of applying ideas, insights, and information generated through basic scientific inquiry to the modification, cure, or prevention of human disease. As understanding of the biology of neurological and psychiatric disorders continues to grow, there are incredible opportunities for developing early biomarkers of disorders and novel disease-specific therapies. There is an enormous breadth of neuroscience research interests represented among the faculty members, so that students are able to pursue research in a wide range of clinically relevant neuroscience areas using state-of-the-art methodology. This multidisciplinary effort provides opportunities in areas including, but not limited to, addiction, chronic pain, epilepsy, language development, social cognition and brain science, mechanisms and treatment of neurodegeneration, and schizophrenia.

Overall, there is a thriving community of students engaged in graduate education and research at Dartmouth, and interaction among students in all the graduate programs is common and encouraged. Dartmouth, one of the Ivy League institutions, has a long-standing tradition of close student-faculty ties—one that is strongly endorsed by the Neuroscience Center at Dartmouth. Each student works closely with a faculty adviser and has the opportunity to interact daily with other members of the research community. Research rotations during the first year provide students with the opportunity to select a thesis lab from a faculty with diverse neuroscience research interests. The NCD sponsors speakers drawn from a host of nationally and internationally recognized research institutions to present a variety of seminars in a series held on campus. Student research-in-progress seminars are held on a regular basis, providing an opportunity both to learn about each other's research and to obtain feedback from faculty members and students. In addition, an integrated neuroscience journal club provides insights into the broad scope of the major and further interaction with other neuroscience graduate students. The general requirements for the Ph.D. degree include research rotations, successful completion of a series of core and elective courses, student research presentations, and a thesis. In addition to the written component, the thesis includes a public seminar and a closed defense.

Research Facilities

The Neuroscience Center at Dartmouth's faculty members occupy state-of-the-art laboratories housed in the Gilman, Moore, Remsen, and Vail buildings in Hanover and at the Dartmouth-Hitchcock Medical Center in Lebanon, northern New England's largest health-care facility. Frequent shuttle bus service connects the two research campuses. There are also Center faculty members affiliated with the Veterans Administration Hospital in White River Junction, Vermont (approximately 5 miles from the Hanover campus). Laboratories contain specialized equipment, including an automated protein sequencer, peptide synthesizer, oligonucleotide synthesizers, DNA sequencers, microarray gene core, phosphorimagers, electron microscopes, confocal and video microscopes, fluorescence activated-cell sorter, and nuclear magnetic resonance spectrometers. Moore Hall, home of psychological and brain sciences, offers a dedicated fMRI; small-animal laboratories; awake, behaving primate laboratories; computer labs; neurophysiological and psychophysical recording facilities; and video production and editing equipment.

Financial Aid

Graduate students in good standing are supported by fellowships or research grants that provide full tuition costs and a stipend. All qualified college seniors are encouraged to apply to the National Science Foundation, National Institutes of Health (NIH), or other agencies for graduate fellowship support.

Cost of Study

Students in the program receive a scholarship for the full cost of tuition.

Living and Housing Costs

Dartmouth College assists graduate students in arranging for appropriate housing in College facilities or in privately owned accommodations in the area. Unfurnished one- to three-bedroom apartments, equipped with a refrigerator and stove, are available for married graduate students. Rents for off-campus housing vary widely, but accommodations can often be found at rates similar to those for College housing.

Student Group

Dartmouth's undergraduate student body numbers approximately 4,000, and the graduate and professional school enrollment is about 1,375, including approximately 500 Ph.D. candidates in graduate programs of Arts and Sciences. Approximately 200 graduate students are enrolled in various Ph.D. programs in the Medical School. Graduate students seeking a neuroscience concentration benefit from the close interactions the program provides with other graduate students, faculty members, postdoctoral fellows, and staff members with varying experiences, interests, and backgrounds.

Student Outcomes

Graduate students receiving the Ph.D. degree from Dartmouth and postdoctoral fellows are prepared for postdoctoral or faculty positions in academia, including both primarily teaching and research-oriented colleges and universities; in research institutes; and in biotechnology and other companies or positions in business or law. The placement of both graduate students and postdoctoral fellows has been very successful and includes positions at the most prestigious institutions.

Location

Life in the Hanover/Lebanon area offers an attractive combination of academic, cultural, and recreational features in a rural setting. Concerts and dramatic productions are held on a year-round basis at the Hopkins Center for the Creative and Performing Arts. The Hood Art Museum houses the College's art collection. As full-time Dartmouth students, graduate students have access to Dartmouth College athletic and other facilities. Excellent hiking and mountaineering opportunities exist in the nearby White Mountains of New Hampshire and the Green Mountains of Vermont. The Appalachian Trail that passes through Hanover traverses both mountain ranges. Outstanding river and lake canoeing are available, and opportunities for fishing and hunting also abound. The beaches of New Hampshire and Maine are about 2½ hours away, and Boston, Montreal, and New York City are accessible by car in 2, 3, and 5 hours, respectively.

The College and The Center

Dartmouth was founded in 1769 as a college committed to liberal learning. Dartmouth Medical School (DMS), founded in 1797, is the fourth-oldest medical school in the country, one of only 125 medical schools in the country, and is a leader in medical education and biomedical research. Dartmouth College, Dartmouth Medical School, and the Dartmouth-Hitchcock Medical Center offer a rich culture of learning, scholarship, and research—a culture enhanced by its commitment to educational, scientific, and clinical excellence as well as to community and collegiality. Dartmouth has a long-standing tradition of close student-faculty ties, a tradition that is strongly endorsed by the Neuroscience Center at Dartmouth. The Neuroscience Center is diversified, bringing expertise to bear on neurological questions from areas such as biochemistry, physiology, medicine, molecular biology, cell biology, structural biology, and cognitive and behavioral psychology.

Applying

Students interested in graduate training in neurosciences need to apply through PEMM.

Correspondence and Information

Neuroscience Center at Dartmouth
7790 Borwell Building, Room 506E
Dartmouth-Hitchcock Medical Center
One Medical Center Drive
Lebanon, New Hampshire 03756
E-mail: NeuroscienceCenter@dartmouth.edu
Web site: http://dms.dartmouth.edu/ncd

Dartmouth College

THE FACULTY AND THEIR RESEARCH

Anatomy/Medicine (Neurology)
Rand Swenson: neuroanatomy; effect of manual procedures on chronic neck pain.

Anesthesiology
Gilbert Fanciullo: pain management therapies. Mark Yeager: neuroendocrine response to trauma; mathematical modeling of sepsis-induced immune dysfunction.

Anesthesiology/Pharmacology
Joyce De Leo: neuroimmunology; CNS neuroimmune responses in chronic pain.

Biochemistry
Dean Madden: structural biology of the glutamate receptor ion channel implicated in learning and memory and diseases such as epilepsy. Surachai Supattapone: pathogenesis of prion disease.

Biology
Patrick Dolph: molecular mechanisms of retinal degeneration in *Drosophila*. Roger Sloboda: microtubule-based motility in neurons.

Education
Donna Coch: developmental cognitive neuroscience; investigation of the development of reading skills using event-related potentials (ERPs).

Genetics
Charles Brenner: aprataxin and ataxiaoculomotor apraxia. Jay Dunlap: molecular biology of the biological clock. Claudio Pikielny: role of pheromones in complex stereotyped courtship and mating behaviors in the fruit fly *Drosophila melanogaster.*

Medicine
Stephen Berman: neuronal apoptosis. Jay Buckey: hyperbaric oxygen treatment. Jeffrey Cohen: neuromuscular, diabetic peripheral neuropathy; axonal transport in diabetic neuropathy. Camilo Fadul: neuro-oncology; immune mechanisms in brain tumor therapy. Gregory Holmes: epilepsy: how seizures affect brain development. Barbara Jobst: clinical characterization and electrical propagation pathways of epileptic seizures. Lloyd Kasper: neuroimmunology; immune modulation in multiple sclerosis. Stephen Lee: molecular pathogenesis of Parkinson's disease. Elija Stommel: diabetic neuropathy; CNS *Toxoplasma gondii* infection and cryptogenic epilepsy. Vijay Thadani: epilepsy; functional brain imaging; localization of seizure foci for the purpose of epilepsy surgery. Peter Williamson: surgical treatment of epilepsy.

Microbiology and Immunology
Randy Noelle: CD40 ligand in multiple sclerosis.

Music
Jon Appleton: how music is processed in the brain; alternative ways music reaches people with various disorders, such as dyslexia.

Orthopedic Surgery
James Weinstein: mechanisms of low back pain; radicular pain.

Pathology
Sherry Bursztajn: glutamate signaling and neuronal death. Brent Harris: neuropathology; CNS development and glial cell biology. William F. Hickey: neuropathology, neuroimmunology, and inflammatory diseases of the CNS. Miguel Marin-Padilla: epilepsy: structural alterations of the cerebral cortex of both clinical and experimental models of epilepsy. C. Harker Rhodes: neuropathology; glial neoplasms and genes involved in their development.

Pediatrics
Robert Darnall: neurobiology of cardiorespiratory control; role of the ventral medulla in sleep. James Filiano: pediatric neurology; mitochondrial disorders; brain abnormalities in SIDS. Richard Morse: epilepsy in children; functional brain imaging; prevention of seizure-induced damage on brain development. Richard Nordgren: attitudes in epilepsy, status epilepticus in children.

Physiology
Andrew Daubenspeck: respiratory-related cortical evoked responses. Leslie Henderson: sex steroid modulation of GABA receptor function. Jay Leiter: neural mechanisms of CO_2 chemosensitivity and pH regulation; respiratory neurobiology. Robert Maue: characterization of sodium channel function. Eugene Nattie: central CO_2 chemosensitivity. William North: arginine vasopressin and learning. Walter St. John: brain stem mechanisms of respiratory rhythmogenesis. Hermes H. Yeh: synaptic dynamics and plasticity in developing and adult CNS.

Psychiatry
Robert Ferguson: rehabilitation of neurocognitive deficits after chemotherapy. Laura Flashman: neuropsychology and neuroimaging of schizophrenia and neurological disorders. Matthew Friedman: psychobiology and clinical psychopharmacology of anxiety, affective, and stress-related disorders. Alan Green: clinical and biological studies of patients with schizophrenia and related psychiatric disorders; action of atypical and novel antipsychotic drugs. Ronald Green: morphology of auditory association cortex in dyslexia and schizophrenia. Thomas McAllister: traumatic brain injury; neuropsychiatric disorders; neuropharmacology; functional MRI. Brenna McDonald: clinical neuropsychology; use of structural and functional neuroimaging to examine memory and executive functions in pediatric neurological and neurodevelopmental disorders. Robert Roth: structural and functional imaging of affective and anxiety disorders. Heather Wishart: neuropsychology and brain imaging of multiple sclerosis; early dementia.

Psychological and Brain Sciences
David Bucci: neurobiological mechanisms of learning, memory, and attention; functional anatomy of cortical attention systems. Ann Clark: steroid influences on female sexual behavior. Yale Cohen: representation and synthesis of auditory space; multimodal representation of extrapersonal space; multimodal mechanisms of attention; sensorimotor integration. Catherine Cramer: development of ingestive behavior, especially physiological and experiential mechanisms subserving feeding discontinuities at weaning; development of tolerance and withdrawal from opiates in neonates; neonatal experience. Howard Hughes: visual neuroscience, cognitive neuroscience, and psychophysics; eye-head coordinating during gaze shifts; polymodal sensory integration; pattern and form processing; binocular interactions. William Kelley: how different kinds of information like words (verbal) or unfamiliar faces (nonverbal) are encoded; whether damage to brain areas shown to be active produces memory impairment. Laura-Ann Petitto: cognitive neuroscience—investigations using PET and MRI technology of the neurological substrates in the brain underlying language representation and use. Jeffrey Taube: neurobiology of learning and memory; neural mechanisms underlying the processing of spatial information. Peter Tse: visual neuroscience, fMRI, and psychophysics—how information about 3-D form is processed by the brain. John Van Horn: motor control and plasticity measured using functional magnetic resonance imaging.

Radiology
Laurence Cromwell: neuroimaging and interventional neuroradiology. Jeffrey Dunn: metabolic and structural studies of human disease models, with emphasis on neurological disorders and NMR microscopy. Clifford Eskey: imaging of cerebral perfusion and metabolism. Harold Swartz: ischemia-reperfusion injury; pathophysiological and physiological changes in oxygenation; melanin and Parkinson's disease.

Surgery
Perry Ball: neuroscience/critical care; spine. Ann-Christine Duhaime: brain injury in immaturity and epilepsy mechanisms. Jack Hoopes: pathogenesis and sparing of radiation myelopathy; modeling brain deformation during surgery; brain tumor oxygen and hypofractionated radiation therapy. David Roberts: stereotaxy; epilepsy; movement disorders. Henry Schmidek: pituitary tumors; molecular genetics of brain tumors. Sujatha Sundaram: pathogenesis of radiation myelopathy.

Thayer School of Engineering
Metin Akay: neural engineering; neural informatics; neural imaging and biomedical informatics. Keith Paulsen: neuroimaging; image-guidance; model-drive image reconstruction. Brian Pogue: near-infrared optical monitoring and treatment of tumors. Stuart Trembly: therapeutic heating of tissue with microwave energy.

GEORGIA STATE UNIVERSITY

College of Arts and Sciences
Programs in the Neurosciences

Programs of Study

Georgia State University (GSU) offers four graduate programs in the neurosciences and also opportunities for interdisciplinary training in the neurosciences through the Brains & Behavior Program. The Department of Biology offers a Ph.D. program and an M.S. concentration in neurobiology and behavior, the Department of Psychology offers a Ph.D. program in neuropsychology and behavioral neuroscience, and an M.D./Ph.D. program is offered in conjunction with the Medical College of Georgia. The Brains & Behavior Program offers interdisciplinary training in neuroscience through these graduate programs and those of six other departments: Chemistry, Computer Science, Computer Information Systems, Mathematics and Statistics (M.S. only), Philosophy (M.A. only), and Physics and Astronomy.

The Neurobiology and Behavior Program (http://biology.gsu.edu/graduate/PhD/nb/program.html) offers neuroscience research training using a diverse set of vertebrate and invertebrate model systems to study neural development, sensory processing, motor systems integration, computational neuroscience, single-channel structure and physiology, gene expression, and neuroendocrinology. A wide variety of techniques are used, including electrophysiology, molecular biology, confocal imaging and MRI, computer simulation, and behavioral analysis.

The Neuropsychology and Behavioral Neuroscience Program (http://www2.gsu.edu/~wwwpsy/NBNProg.htm) is a broadly integrated, interdisciplinary program. Research and training ranges from basic behavioral neuroscience to clinical and developmental neuropsychology and is focused on the central nervous system foundations and correlates of behavior in humans and other vertebrates. There are two tracks in the program: one allows for training leading to clinical licensure, and the other is oriented toward students intending to pursue a research/academic career either in basic neuroscience or in neuropsychology.

The Brains & Behavior Program (http://brainsbehavior.gsu.edu/grad_training.html) is a GSU Area of Focus program that promotes interdisciplinary research and training in neuroscience through a consortium of 70 faculty members and their students across the eight departments listed above. Applicants to the neuroscience programs or to the other departmental graduate programs can indicate their interest in the Brains & Behavior Program in their application.

Applicants to GSU neuroscience programs may also participate in the interinstitutional Graduate Scholars Program through the Center for Behavioral Neuroscience (CBN) (http://www.cbn-atl.org/education/graduatescholars.html). The NSF-sponsored CBN is a consortium of more than 60 neuroscientists from Georgia State and other Atlanta-area universities, including the Atlanta University Center, Emory University, Georgia Tech, and Morehouse Medical School. The purpose of the CBN is to foster interdisciplinary collaborative approaches toward understanding the basic neural mechanisms underlying the regulation of complex social behaviors and emotions. The CBN is also designed to enhance minority participation in science, encourage technology transfer to industry, and aid in bringing research findings to the public.

Three other neuroscience research centers at GSU help sponsor and coordinate activities: the Center for Brain and Health Sciences, the Center for Neural Communication and Computation, and the Language Research Center.

Research Facilities

Research laboratories at Georgia State University are located in four buildings on campus, soon to be joined by four additional buildings that make up a $200-million University Science Park. Core facilities include multiphoton and laser scanning confocal microscopes, high-field NMR and MRI, gene-array technology, behavioral monitoring, and grid computing. The Language Research Center is located on 60 wooded acres outside Atlanta. The University Library has a full complement of neuroscience research journals in print and online.

Financial Aid

Most Ph.D. students receive full tuition and stipend support at nationally competitive levels. Additional support is available for teaching and research assistants.

Cost of Study

Current figures are available at http://www.gsu.edu/tuition_fees.html. Tuition waivers are provided for all Ph.D. students who receive research and teaching assistantships.

Living and Housing Costs

Graduate student housing is available on campus in downtown Atlanta. Many graduate students live in private housing elsewhere in the metropolitan area. The cost of living is well below that of most of the large cities on the East and West coasts. GSU is easily accessible by car or subway.

Student Group

Georgia State University is a diverse public institution with more than 25,000 students. There are more than 7,500 graduate students, making it one of the largest graduate institutions in the Southeast. Students from 120 countries and forty-eight states attend the University. Approximately 5–10 Ph.D. students join each of the neuroscience programs in biology and psychology every year. There is an active Biology Graduate Students Association (http://www2.gsu.edu/~wwwbgs/).

Location

Georgia State University is located in the heart of Atlanta, which is an exciting, vibrant city. Often considered the cultural center of the Southeast, Atlanta has a thriving arts community with the Atlanta Symphony Orchestra, the Atlanta Ballet, the Alliance Theatre Company, and the High Museum of Art. In addition, Georgia State University operates the Rialto Performing Arts Center. Atlanta is also the home of four professional sports teams. Located along the scenic Chattahoochee River, Atlanta is a short drive from the southern end of the Appalachian Mountains. The climate is quite moderate: the mean July temperature is 73.4°F (23°C) and the mean January temperature is 50°F (10°C). There is a luxuriously long spring with many fragrant flowering trees and a magnificently colorful autumn.

The University

Georgia State University is part of the state university system, which includes Georgia Tech, the University of Georgia, and the Medical College of Georgia. The University offers more than fifty undergraduate and graduate degree programs, covering some 200 fields of study, through the Colleges of Arts and Sciences, Business Administration, Education, Health and Human Sciences, and Law and the School of Policy Studies.

Applying

Application materials may be obtained from the addresses in the Correspondence and Information section and online at http://biology.gsu.edu/graduate/admissions.html. In addition to the application forms, the applicant must submit official copies of transcripts from each institution attended, GRE General Test scores, three letters of recommendation, a statement of educational and career goals, and a $25 application fee. International students are also required to submit TOEFL scores.

Correspondence and Information

For neurobiology and behavior:
Graduate Coordinator: Neurobiology and Behavior
Department of Biology
Georgia State University
P.O. Box 4010
Atlanta, Georgia 30302-4010
Phone: 404-651-2759
Fax: 404-651-2509
E-mail: biolxm@langate.gsu.edu
Web site: http://biology.gsu.edu/neuro

For neuropsychology and behavioral neuroscience:
Director of Graduate Studies: Neuropsychology and
  Behavioral Neuroscience
Department of Psychology
University Plaza
Georgia State University
Atlanta, Georgia 30303-3083
Phone: 404-651-2456
Fax: 404-651-1391
E-mail: psyadvise-g@langate.gsu.edu
Web site: http://www.gsu.edu/~wwwpsy/NBNProg.htm
  http://www.gsu.edu/~wwwpsy/HowToApply.htm

Georgia State University

THE FACULTY AND THEIR RESEARCH

H. Elliott Albers, Ph.D., Tulane, 1979. Neuroendocrinology and behavior; circadian rhythms; communicative behaviors; peptide physiology. (Biology, Psychology)

Deborah J. Baro, Ph.D., Illinois at Chicago, 1989. Molecular approaches to neuronal systems: identifying and analyzing gene networks, integrating. (Biology)

Timothy J. Bartness, Ph.D., Florida, 1981. Neuroendocrinology of metabolism; annual rhythms, obesity, and food intake. (Biolology, Psychology)

Laura Carruth, Ph.D., Colorado at Boulder, 1998. Sexual differentiation of the avian central nervous system, using both a molecular and cellular approach; science outreach and education. (Biology)

Andrew N. Clancy, Ph.D., Texas, 1978. Neuroendocrinology and behavior; reproductive behavior; hormonal actions on the brain; neuroanatomy. (Biology)

Marsha G. Clarkson, Ph.D., Florida, 1979. Developmental psychology; perceptual development in infants; psychoacoustics; developmental psychobiology. (Psychology)

Gennady Cymbalyuk, Ph.D., Moscow State, 1996. Neurodynamics: Dynamics of neurons and circuits assessed by electrophysiological methods, computer simulations, and hybrid systems. (Physics)

Charles D. Derby, Ph.D., Boston University, 1982. Sensory neurobiology and integrative biology of the chemical senses (smell and taste), including behavioral, electrophysiological, anatomical, biochemical, and molecular approaches. (Biology)

Donald H. Edwards, Ph.D., Yale, 1976. Neuroethology; role of neuromodulation in neural circuit function and behavior; computational neuroscience. (Biology)

Kyle Frantz, Ph.D., Florida, 1999. Neurobehavioral effects of drugs of abuse in developing animals; neuroscience education and outreach. (Biology)

Matthew S. Grober, Ph.D., UCLA, 1988. Neuroethology: role of social interactions in regulating reproductive behavior and its underlying neuroendocrine mechanisms across the vertebrates. (Biology)

Julia K. Hilliard, Ph.D., Baylor College of Medicine, 1975. Viral immunology: host-virus interactions, viral tropism involving central nervous system targets, viral zoonoses and xenozoonoses; special emphasis on herpes B virus: rapid, enhanced diagnosis. (Biology)

Kim L. Huhman, Ph.D., Georgia, 1988. Neuroendocrinology and behavior; circadian rhythms; hormonal responses to stress and aggression. (Psychology)

Chun Jiang, Ph.D., Chinese Academy of Sciences, 1988. Cellular and molecular physiology; function and modulation of neuronal membrane ion channels. (Biology)

Paul S. Katz, Ph.D., Cornell, 1989. Electrophysiology; calcium imaging; neuromics; neural basis and evolution of simple behaviors in invertebrates; neuromodulators and second messengers in neuronal circuits. (Biology)

Tricia Zawacki King, Ph.D., Florida, 2000. Developmental and clinical neuropsychology; acquired brain injury; emotion. (Psychology)

Erin B. McClure, Ph.D., Emory, 2001. Developmental and clinical neuropsychology; child and adolescent mood and anxiety disorders; gender; emotion processing (fMRI and behavioral studies). (Psychology)

Mary Morris, Ph.D., Florida, 1986. Developmental and clinical neuropsychology; developmental learning disabilities; acquired brain injury. (Psychology)

Robin Morris, Ph.D., Florida, 1982. Developmental and clinical neuropsychology; developmental and attention related disabilities; acquired brain injury; childhood psychopathology. (Psychology)

Anne Z. Murphy, Ph.D., Cincinnati, 1992. Behavioral neuroendocrinology; sex differences in pain and analgesia; male and female reproductive behavior. (Biology)

Sarah L. Pallas, Ph.D., Cornell, 1987. Developmental neuroscience; sensory physiology; neuroanatomy; synaptic plasticity; recovery from brain damage; learning. (Biology)

Marise B. Parent, Ph.D., California, Irvine, 1993. Neural mechanisms of learning and memory; roles of amygdala and glucose in emotional memory. (Psychology)

Aras Petrulis, Ph.D., Cornell, 1998. Neurobiology of social behavior and recognition; chemosensory regulation of behavior; behavioral neuroendocrinology. (Psychology)

Vincent Rehder, Ph.D., Free University (Berlin), 1987. Cellular and molecular neurobiology; second-messenger systems and neuronal growth; cytoskeleton and cell motility; nerve cell regeneration. (Biology)

Diana L. Robins, Ph.D., Connecticut, 2002. Developmental and clinical neuropsychology; autism and pervasive developmental disorders; fMRI studies of emotion processing. (Psychology)

Eric Vanman, Ph.D., USC, 1994. Social neuroscience; human brain imaging and electroencephalography; emotion and social processes. (Psychology)

W. William Walthall, Ph.D., SUNY at Albany, 1984. Developmental neurobiology: integrated approaches to investigate underlying genetic programs. (Biology)

David A. Washburn, Ph.D., Georgia State, 1991. Comparative cognition; functional cerebral asymmetries; transcranial magnetic stimulation. (Psychology)

Irene T. Weber, Ph.D., Oxford, 1978. Ion channel structure-function; crystallography; bioinformatics. (Biology, Chemistry)

Walter Wilczynski, Ph. D., Michigan, 1978. Neural basis of natural behavior; neural systems underlying animal communication and reproductive behavior and interaction of the sensory and hormonal systems that mediate these behaviors. (Psychology)

HARVARD UNIVERSITY

Harvard Medical School
Division of Medical Sciences
Program in Neuroscience

Program of Study

The Program in Neuroscience at Harvard Medical School is an interdepartmental program that links basic science and clinical faculties throughout the Harvard community. Established in 1981, the program is an outgrowth of the Department of Neurobiology, which was founded in 1966. The department, through its outstanding multidisciplinary research laboratories and a series of didactic core courses, helped to establish the field of modern neurobiology. The Program in Neuroscience was created in recognition of the fact that neuroscience at Harvard had grown far beyond the borders of the Department of Neurobiology.

The current program faculty consists of a diverse group of more than 90 investigators whose research interests include cellular, molecular, developmental, genetic, systems, behavioral, immunological, neurological, and psychiatric approaches to the nervous system. Ion channels, synaptic function, neuronal development and differentiation, neuronal aging and degeneration, and visual CNS pathways are areas of particular strength. Understanding the pathophysiology of diseases of the nervous system is an important focus of the program. Numerous interactions exist between basic science and clinical laboratories, which have led to significant advances in the understanding of several major neurological diseases, including muscular dystrophy, Huntington's disease, epilepsy, and Alzheimer's disease.

Applicants should have a solid background in general biology; cell biology; genetics; biochemistry; organic, physical, and inorganic chemistry; general physics; and calculus. During the first eighteen months, students take a series of graduate-level courses and complete three laboratory rotations that serve as the basis for the selection of a dissertation adviser. Dissertation research begins during the second year of study. A qualifying examination, which includes a dissertation proposal, is taken by all students by June 30 of the second year of study. Students meet with their Dissertation Advisory Committee at least once a year, until the dissertation defense is made before a committee of 3 faculty members. The goal of the program is completion of the course work and dissertation defense in four to five years. In the first years of study, students are in the Department of Neurobiology, but for their dissertation research they select from among the wide range of advisers and laboratory settings. After the dissertation defense has been completed through the Division of Medical Sciences, the program faculty recommends candidates to the Faculty of Arts and Sciences of Harvard University for the Ph.D. in neurobiology.

Research Facilities

Research facilities include research laboratories in the Department of Neurobiology and the other basic science departments at the Harvard Medical School Longwood Quadrangle, as well as the Massachusetts General Hospital, Children's Hospital, Brigham and Women's Hospital, Beth Israel Deaconess Medical Center, McLean Hospital, and the departments of the Harvard University Faculty of Arts and Sciences in Cambridge. The Countway Library of Medicine is located at the quadrangle, and all other Harvard libraries are available to students.

Financial Aid

Scholarships and nonteaching fellowships, which cover tuition and fees and provide a monthly stipend, are awarded to virtually all students in the program. Students are encouraged, however, to apply for fellowships from outside sources before they begin their studies at Harvard. Enrolled students who receive competitively funded externally sponsored fellowships may be eligible to receive an educational allowance from the Division. International applicants are urged to seek financial support from their national governments and fellowship agencies. The annual stipend for 2006–07 is $28,008, which is $2334 per month.

Cost of Study

Tuition and fees in 2006–07 are $30,276 for the first and second years of study and $10,476 for subsequent years. Health insurance is provided by the University and is included in the costs. As indicated above, these costs are usually covered by fellowships.

Living and Housing Costs

Students have a variety of housing options to choose from, ranging from University dormitories and apartments to Vanderbilt Hall, located at Harvard Medical School, to shared apartments in and around Boston. Housing costs are comparable with those of other metropolitan university areas.

Student Group

The program currently has 85 students, with an incoming class size of about 13 to 18 students per year. Students come from widely diverse backgrounds and have a range of research interests that parallels the breadth of the program faculty members. Approximately half the program students are women. Coherence of the student group is supported through retreats and other social activities.

Location

There are two other medical schools and many colleges and universities in Boston. The Museum of Fine Arts, the Gardner Museum, Symphony Hall, and Fenway Park are all within walking distance of Harvard Medical School. The School is close to two subway lines that allow convenient access to the rest of Boston and surrounding communities. A free shuttle bus links the Medical School, the Harvard campus in Cambridge, and Massachusetts General Hospital.

The University

The Harvard University community is a large and varied one, strongly represented in the humanities as well as in the sciences. The University is located in Cambridge, and the Medical School and School of Public Health are in Boston amidst a cluster of notable Harvard-affiliated hospitals.

Applying

The Program in Neuroscience is administered by the Division of Medical Sciences at Harvard Medical School. The requirement for admission is undergraduate training in the physical sciences (physics, mathematics, and chemistry, including organic and physical chemistry) and biology. The General Test of the Graduate Record Examinations (GRE) is required, and a personal interview may be requested. The General Test should be taken no later than November so that score reports arrive in time to be considered.

The application for admission should be submitted to the Graduate School of Arts and Sciences by December 8. The application fee and deadline are subject to change. Final decisions concerning admissions are made in February or early March. Applicants are notified by letter from the Dean of the Graduate School of Arts and Sciences. Applications are available online at http://www.gsas.harvard.edu and students may request a hard copy via e-mail at dms@hms.harvard.edu. For specific questions about the program, students should contact Gina Conquest, Administrator of the Program in Neuroscience, at 617-432-0912 or e-mail gina_conquest@hms.harvard.edu. Members of underrepresented groups are encouraged to apply.

Correspondence and Information

Admission Office
Division of Medical Sciences
Harvard Medical School
260 Longwood Avenue
Boston, Massachusetts 02115
Phone: 617-432-0162
E-mail: dms@hms.harvard.edu
Web site: http://www.hms.harvard.edu

Harvard University

THE FACULTY AND THEIR RESEARCH

A partial list of the faculty members in the Program in Neuroscience is given below.

J. Assad, Ph.D. Mechanisms of visual processing in the visual cortex of awake behaving monkeys.
B. Bean, Ph.D. Neurotransmitter control of ion channels.
L. Benowitz, Ph.D. Molecular and cellular studies on nerve development and regeneration.
V. Bolshakov, Ph.D. Synaptic mechanisms of learning and memory acquisition.
A. Bonni, Ph.D. Intracellular signaling mechanisms in the development of the mammalian CNS.
R. Born, M.D., Ph.D. Neural circuitry of primate visual cortex.
X. Breakfield, Ph.D. Molecular etiology and gene therapy of neurologic diseases.
E. Brown, M.D., Ph.D. Statistical modeling and stochastic dynamical systems and analysis of neurophysiologic systems.
R. Burstein, Ph.D. Spinohyphothalamic and spinotelencephalic tracts: a neural substrate for autonomic and affective responses to pain.
W. Carlezon, Ph.D. Causal relationships between gene expression in the brain and motivated behavior.
C. Cepko, Ph.D. Cell fate determination in the vertebrate CNS.
C. Chen, M.D., Ph.D. Synaptic plasticity in the CNS, using electrophysiology and calcium imaging techniques.
D. Clapham, M.D., Ph.D. Investigation of receptors, G-proteins, ion channels, and intracellular calcium.
J. Cohen, Ph.D. Structure and function of ligand-gated ion channels.
D. Corey, Ph.D. Physiology and cell biology of transduction and adaptation in the inner ear.
G. Corfas, Ph.D. Molecular mechanisms of neurogenesis and neuronal differentiation.
D. Cotanche, Ph.D. Hair cell regeneration in the avian cochlea; development of the cochlear sensory epithelium.
C. Dulac, Ph.D. Molecular and developmental biology of sensory systems.
F. Engert, Ph.D. Plasticity; cellular mechanisms; development of functional networks and links to behavior.
M. Feany, M.D., Ph.D. Modeling human neurodegenerative diseases in the fruit fly *Drosophila*.
J. Flanagan, Ph.D. Cell-cell signaling molecules in neural development, axon guidance, and neural patterning.
L. Goodrich, Ph.D. Development of neural circuits.
M. Greenberg, Ph.D. Neurotransmitter and neurotrophin regulation of immediate early gene transcription.
C. Gu, Ph.D. Molecular mechanisms of how neural and vascular networks are coordinately developed, communicate, and evolve to work in concert during normal and disease states.
A. Hart, Ph.D. Genetic, molecular, and cellular analysis of the *C. elegans* nervous system.
Z. He, Ph.D. Cellular and molecular mechanisms in axon guidance and regeneration.
B. Hyman, M.D., Ph.D. Pathophysiology of Alzheimer's disease.
O. Isacson, M.D., Ph.D. Neuroprotection and neuronal repair in animal models of neurodegenerative diseases.
J. Kaplan, Ph.D. Genetic analysis of synaptic transmission and synaptic analysis.
K.-S. Kim, Ph.D. Molecular mechanisms of catecholaminergic-specific gene regulation.
E. Kravitz, Ph.D. Hormonal orchestration of behavior.
S. Kunes, Ph.D. Nervous system construction and function.
L. Kunkel, Ph.D. Molecular basis of human neuromuscular disease.
J. Lichtman, Ph.D. Study of synaptic competition by visualizing synaptic rearrangements directly in living animals using modern optical techniques.
M. Livingstone, Ph.D. Anatomical and physiological studies of the primate visual system.
E. Lo, Ph.D. Investigation of mechanisms underlying neuronal injury after stroke and trauma.
Q. Ma, Ph.D. Developmental studies of the mammalian nervous system.
J. Macklis, M.D., Ph.D. Neocortical development and cellular transplantation.
J. Majzoub, M.D. Regulation of neuroendocrine peptide gene expression in the CNS.
R. Masland, Ph.D. Functional organization of the retina.
M. Meister, Ph.D. Function of neural circuits in visual and olfactory systems.
V. Murthy, Ph.D. Mechanisms of synaptic transmission and plasticity.
C. Nelson, Ph.D. Developmental cognitive neuroscience, focusing primarily on memory and face processing.
D. Paul, Ph.D. Function and control of intercellular communication through gap junction.
S. Pomeroy, M.D., Ph.D. Expression and function of growth factors in the nervous system and in brain tumors.
W. Regehr, Ph.D. Role of presynaptic and postsynaptic calcium dynamics in determining synaptic efficacy.
C. Reid, M.D., Ph.D. Electrophysiological and computational study of the mammalian visual system.
P. Rosenberg, M.D., Ph.D. Mechanisms of action in neurotransmitters, specifically glutamate and norepinephrine.
G. Ruvkun, Ph.D. Transcriptional control of *C. elegans* neurogenesis and neural function.
B. Sabatini, M.D., Ph.D. Biochemical signaling within spines and boutons.
J. Sanes, Ph.D. Cellular and molecular studies of synapse formation in the vertebrate CNS.
C. Saper, M.D., Ph.D. Chemical coding of synaptic connections in the limbic and central autonomic systems.
T. Schwarz, Ph.D. Neurogenetic and biophysical analysis of synaptic transmission, exocytosis, and K$^+$ channel function.
R. Segal, M.D., Ph.D. Role of neurotrophins in cerebellar development.
D. Selkoe, M.D. Molecular and cell biological studies of the pathogenetic mechanisms of Alzheimer's disease.
C. Shatz, Ph.D. Activity-dependent development of the mammalian visual system.
C. Stiles, Ph.D. Basic principles of signal transduction in nerve cells and the developing nervous system.
K. Sweadner, Ph.D. Structure, function, and biological roles of Na, K-ATPase isoforms in excitable tissue.
R. Tanzi, Ph.D. Identification and characterization of genes involved in neurodegeneration in Alzheimer's disease and autism.
D. Van Vactor, Ph.D. Molecular analysis of mechanisms controlling motor axon guidance and target specificity in *Drosophila* embryo.
C. Walsh, M.D., Ph.D. Cerebral cortical development studied by retroviral-mediated gene transfer techniques.
C. Weitz, M.D., Ph.D. Molecular mechanisms of the circadian clock localized in the mammalian hypothalamus.
R. Wilson, Ph.D. Neural coding of chemosensory stimuli.
M. Wolfe, Ph.D. Combining chemistry and biology in the study of pathogenesis and treatment of neurodegenerative diseases.
C. Woolf, Ph.D. Changes in the function, chemistry, and structure of sensory neurons that contribute to pain.
B. Yankner, M.D., Ph.D. Neurobiology and biochemistry of beta amyloid protein associated with Alzheimer's disease.
G. Yellen, Ph.D. Biophysical analyses of functional motions of neuronal ion channels.
A. Young, M.D., Ph.D. Anatomy of the basal ganglia and the role of excitatory amino acids in Alzheimer's disease.
J. Yuan, Ph.D. Genes that control the determination of developmental cell death in vertebrates.

LOUISIANA STATE UNIVERSITY HEALTH SCIENCES CENTER

School of Graduate Studies
Interdisciplinary Neuroscience Training Program

Programs of Study

The Interdisciplinary Neuroscience Training Program at Louisiana State University (LSU) Health Sciences Center in New Orleans is a multidisciplinary, interdepartmental program offering predoctoral research training in fundamental and clinical neuroscience. The training program, which requires about four years to complete, leads to a Ph.D. in neuroscience and is offered by faculty members from twelve departments of the Health Sciences Center and the University of New Orleans. A Master of Science program is also offered. The breadth of research programs of the faculty encompasses all major areas of neuroscience, ranging from electrophysiology to molecular neurobiology and from genes to cells to human behavior and the neurobiology of disease. Training is designed to provide Ph.D. neuroscientists with a broad general knowledge of neuroscience as well as more intensive training in a highly specialized topic for academic and industrial research positions.

The core program includes specialized neuroscience courses (neuroscience, investigative neuroscience, molecular neurobiology, and behavioral neuroscience) and related biomedical courses (cell biology, statistics, and biochemistry). Rotation through a number of faculty laboratories is required and provides the students with research experience from the beginning of their tenure. Students participate in the neuroscience seminar series, neuroscience colloquia, Neuroscience Center retreats, and the many other brain science activities sponsored by the LSU Health Sciences Center. Students are also encouraged to become involved in the very active greater New Orleans chapter of the Society for Neuroscience.

Research Facilities

The faculty members are all located in modern buildings with state-of-the-art laboratories. In addition, the Health Sciences Center Core Laboratories, which is a service facility, provides oligonucleotide and peptide synthesis and sequencing, amino acid analysis and purification, mass spectrometry, flow cytometry, and phosphoimaging. Specialized equipment, including HPLCs, PCR cyclers, ultracentrifuges, and spectrophotometers, is available in the laboratories of many of the faculty members. Other multidisciplinary facilities include ultrastructure laboratories and a Computer Imaging Center for three-dimensional modeling and reconstruction, deconvolution, confocal microscopy, and digital image analysis. Equipment for high-speed sequencing of DNA has recently been purchased by the Neuroscience Center. Extensive library facilities are available at the Health Sciences Library of the Health Sciences Center and the University Library at the University of New Orleans. A comprehensive Computer Services Center is available on the Health Sciences Center campus.

Financial Aid

Graduate stipends are available from the Neuroscience Center, individual departments, and the LSU Health Sciences Center Graduate School. Supplements can be obtained from faculty grants and other sources for those students who are highly qualified. Fellowships from NIH, NSF, and the Howard Hughes Foundation are also available on a competitive basis.

Cost of Study

Students who receive stipends are not required to pay tuition but are responsible for activity fees of about $135 per semester.

Living and Housing Costs

The cost of living in the New Orleans area is generally below the national average. A recently remodeled dormitory with an exercise area and health facility is available on the Health Sciences Center campus. Living accommodations are also available in the historic Garden District, Uptown, and the Warehouse District, all of which offer distinctive Greek revival and Victorian architecture. Student apartments can also be found in the French Quarter at reasonable cost. Health care is available on campus through the student Health Center of the LSU Medical School.

Student Group

Approximately 110 students are enrolled in the LSU Health Sciences Center graduate school. Between 20 and 25 of these students are working in the neuroscience program and train in graduate programs of the various basic science departments of the Health Sciences Center. The Interdisciplinary Neuroscience Program accepts 4 to 5 students per year.

Location

The Health Sciences Center is in the heart of historic downtown New Orleans, within 5 minutes of the French Quarter. Exciting yearly events in New Orleans include Mardi Gras and the Jazz and Heritage Festival. The city is famous for its jazz and cuisine, and cultural opportunities, such as the symphony and opera, abound in New Orleans. The city is located only 90 minutes from the Gulf of Mexico; the subtropical climate permits a variety of year-round activities. The New Orleans metropolitan area has a population of approximately 1.5 million and has five major universities.

The Center

The Interdisciplinary Neuroscience Training Program is closely linked to the LSU Neuroscience Center of Excellence. The Neuroscience Center is a multidisciplinary, interdepartmental unit of LSU Health Sciences Center that is funded by the State of Louisiana Centers of Excellence Program, the National Institutes of Health, and the Department of Defense. The Neuroscience Center fosters interactions and collaborations among neuroscientists and creates a rich and stimulating environment for graduate education. In addition to graduate training, postdoctoral training is also available with the Neuroscience Center faculty. LSU Health Sciences Center is part of Louisiana State University; the main LSU campus is located in Baton Rouge. LSU is an EEOAA employer.

Applying

Applicants must hold a bachelor's degree or the equivalent thereof. Students should have taken courses in biology, chemistry, mathematics, physics, and computer science. The General Test of the Graduate Record Examinations is required, and the Subject Test in any area of science is recommended. Admission is determined by test scores, grades, recommendations, a written statement of interest and goals, and a personal interview when possible.

Correspondence and Information

Nicolas G. Bazan, M.D., Ph.D., Co-Director
R. Ranney Mize, Ph.D., Co-Director
Interdisciplinary Neuroscience Training Program
Neuroscience Center of Excellence
Louisiana State University Health Sciences Center
2020 Gravier Street, Suite D
New Orleans, Louisiana 70112

Phone: 504-599-0909
Fax: 504-568-5801
E-mail: rmize@lsuhsc.edu
    nbazan@lsuhsc.edu
    hbazan1@lsuhsc.edu
Web site: http://www.neuroscience.lsuhsc.edu

Louisiana State University Health Sciences Center

THE FACULTY AND THEIR RESEARCH

Mark Alliegro, Ph.D.; Cell Biology and Anatomy. Mechanisms of cell differentiation.

Rene Anand, Ph.D.; Neuroscience and Neurology. Neurobiology of ligand-gated ion channels.

Haydee E. P. Bazan, Ph.D.; Ophthalmology. Lipids involved in signal transduction mechanisms in the eye.

Nicolas G. Bazan, M.D., Ph.D.; Ophthalmology, Biochemistry and Molecular Biology, and Neurology; Director, Neuroscience Center, and Co-Director, Neuroscience Training Program. Neurochemistry and molecular biology of cell signal transduction; neurobiology.

Roger W. Beuerman, Ph.D.; Ophthalmology. Pain systems and nerve regeneration.

Richard P. Bobbin, Ph.D.; Otorhinolaryngology. Pharmacology of sensory systems, particularly the cochlea.

Kevin Brown, Ph.D.; Biochemistry and Molecular Biology. Molecular basis of ataxia-telangectasia (a genetic disorder leading to progressive neuronal degeneration); understanding molecular pathways activated after genome damage.

Claude Burgoyne, M.D.; Biomedical Engineering. Glaucomatous damage to the neural and connective tissues of the optic nerve head.

Carmen Canavier, Ph.D.; Psychology, University of New Orleans. Computational models of single neurons and small networks.

Chu Chen, Ph.D.; Neuroscience and Otorhinolaryngology. Neuromodulation of ionic channels and synaptic plasticity in hippocampal neurons.

Julia L. Cook, Ph.D.; Ophthalmology and Biochemistry and Molecular Biology. Molecular mechanisms of growth and growth regulation, with an emphasis on therapeutics and intervention.

Peter Cserjesi, Ph.D.; Cell Biology and Anatomy. Genetic and molecular events leading to peripheral nervous system determination and development.

Jeffrey Erickson, Ph.D.; Pharmacology and Neuroscience. Molecular and cell biology of vesicular neurotransmitter transporters.

Reha Erzurumlu, Ph.D.; Cell Biology and Anatomy. Axon-target interactions and neuronal plasticity in developing sensory pathways.

Bruce J. Fisch, M.D.; Neurology; Director, EEG and Epilepsy Programs.

William C. Gordon, Ph.D.; Ophthalmology. Cell biology of the retina and hippocampus under normal and pathological conditions.

Harry Gould, Ph.D. Mechanisms of the development of chronic pain.

Kevin W. Greve, Ph.D.; Psychology, University of New Orleans. Cognitive and clinical neuropsychology.

William Guido, Ph.D.; Cell Biology and Anatomy. Visual system physiology; visual development; plasticity.

John W. Haycock, Ph.D.; Biochemistry and Molecular Biology. Signal transduction systems and neurotransmitter function.

James M. Hill, Ph.D.; Ophthalmology. Ocular herpes simplex virus (HSV).

S. Michal Jazwinski, Ph.D.; Biochemistry and Molecular Biology. Molecular genetics; cell-cycle control and aging; longevity assurance genes.

Daniel R. Kapusta, Ph.D.; Pharmacology. Central nervous system control of renal function.

Bronya J. B. Keats, Ph.D.; Biometry and Genetics. Positional cloning of genes for nervous system disorders.

Bruce M. King, Ph.D.; Psychology, University of New Orleans. Biological basis of feeding behavior and obesity.

David G. Kline, M.D.; Neurosurgery.

Gerald LaHoste, Ph.D.; Psychology, University of New Orleans. Molecular mechanisms of synaptic plasticity, behavior, and neurodegeneration.

Thomas Lallier, Ph.D.; Cell Biology and Anatomy. Cell-surface interactions involved in axon guidance and neural crest cell migration.

Iris Lindberg, Ph.D.; Biochemistry and Molecular Biology. Biochemistry of neuropeptide biosynthesis.

Walter J. Lukiw, Ph.D.; Neuroscience and Ophthalmology. Alzheimer's disease; control of gene transcription; neurotoxicology; molecular biology of sleep and wakefulness.

Victor Marcheselli, Ph.D.; Ophthalmology and Neuroscience.

James G. May, Ph.D.; Psychology, University of New Orleans. Visual perception and human electrophysiology.

Michele Meneray, Ph.D.; Physiology.

R. Ranney Mize, Ph.D.; Head, Cell Biology and Anatomy, and Co-Director, Neuroscience Training Program. Visual system neurotransmitter function.

Dennis J. Paul, Ph.D.; Pharmacology. Opioid receptor pharmacology and neuroanatomy.

Johnny R. Porter, Ph.D.; Physiology. Neuroendocrine mechanisms in obesity: monoamine function.

Chandan Prasad, Ph.D.; Medicine/Endocrinology. Regulation of dopaminergic neurons.

Moshe Solomonow, Ph.D.; Orthopedics and Physiology. Special designs of prosthetic and orthotic devices; electrical stimulation of muscles; spinal cord injury rehabilitation; electromyography.

Kurt J. Varner, Ph.D.; Pharmacology. Central nervous system control of sympathetic and cardiovascular function.

Hélenè Varoqui, Ph.D.; Ophthalmology and Neuroscience. System A glutamine transporters in retina.

Wayne Vedeckis, Ph.D.; Biochemistry and Molecular Biology. Molecular mechanism of glucorticoid hormone action.

Judith Venuti, Ph.D.; Cell Biology and Anatomy. Ontogeny of the neuromuscular junction and molecular mechanisms of neural tube signaling to the somite.

Theodore G. Weyand, Ph.D.; Cell Biology and Anatomy. Visuomotor integration.

Mary C. Williams, Ph.D.; Psychology, University of New Orleans.

Peter Winsauer, Ph.D.; Pharmacology. Behavioral pharmacology of serotonin receptor subtype agonists and antagonists.

Eugene A. Woltering, M.D.; Chief, Surgical Endocrinology. Trophic and inhibitory effects of somatostatin on tumor growth and angiogenesis.

Houhui Xia, Ph.D.; Cell Biology and Anatomy and Neuroscience. Molecular mechanisms of synaptic plasticity: NMDA receptor signaling and CREB-mediated gene transcription.

MASSACHUSETTS INSTITUTE OF TECHNOLOGY

Department of Brain and Cognitive Sciences

Programs of Study

The Department of Brain and Cognitive Sciences offers graduate work leading to the Ph.D. in neuroscience and in cognitive science.

Graduate students follow a flexible curriculum that provides the opportunity to study in a variety of relevant fields. In their second term, incoming students choose a faculty adviser in whose laboratory they carry out supervised research. Students may subsequently switch to one or more other laboratories. The program places heavy emphasis on research; students are expected to devote considerable time to laboratory work beginning in the first year. In addition to research and formal course work, students participate in advanced seminars and assist in teaching undergraduate brain and cognitive science classes for three terms. There is a written qualifying examination at the end of the second year followed by an oral exam in the third year. Thesis research typically takes two to three years of full-time effort after the qualifying exams.

Research Facilities

The Department of Brain and Cognitive Sciences is administratively a part of MIT's School of Science and is closely allied with its Whitaker College of Health Sciences and Technology. There are thirty-four active laboratories equipped for studies ranging from molecular neurobiology to electrophysiology, neurotransmitters, neuropsychology, psychophysics, human information processing, and psycholinguistics. Each research laboratory is equipped as needed with computational facilities, and, for academic computing, students have unlimited access to MIT's campuswide Athena network. The department maintains a library, a machine shop, an electronics shop, and a video and confocal fluorescence microscopy laboratory. In keeping with MIT's interdisciplinary traditions, students also use other campus resources such as the Computer Science and Artificial Intelligence Laboratory, the Media Lab, the Speech Communication Group in the Research Laboratory of Electronics, the Clinical Research Center, and the facilities of other departments, such as Linguistics and Philosophy, Biology, Mechanical Engineering, and Electrical Engineering and Computer Science.

Financial Aid

Students are encouraged to apply for individual fellowships, such as NSF and NDSEG, or for other grants from outside sources to cover all or part of the cost of their education. Students in the department may be supported by federal training grants, research assistantships, teaching assistantships, or MIT-administered fellowships. Support provided by such appointments included full tuition payment and a stipend of $24,600 in the 2005–06 academic year. International students are urged to apply for fellowships in their native countries because sources of support for non-U.S. citizens are severely limited.

Cost of Study

As explained above, full tuition coverage and a stipend are offered to nearly all graduate students accepted into the program.

Living and Housing Costs

Monthly living costs (housing, food, and personal expenses) for a single graduate student generally average about $2300; for a married graduate student, $3000; and for a married graduate student with one child, $3200. Campus housing is currently available by lottery for all first-year graduate students.

Student Group

At present, the Department of Brain and Cognitive Sciences has 78 graduate students; approximately 15 new students are admitted each fall. Currently, there are 44 men and 34 women in the department, 11 of whom are international students.

Student Outcomes

Ten students were granted the Ph.D. in 2004–05, 7 hold postdoctoral fellowships, 1 is founder and CEO of a project management and system design consulting firm, 1 is a medical student, and 1 is Education-Industry Partnership Manager at Minnesota State Colleges and Universities.

Location

MIT is located on the Charles River in Cambridge, Massachusetts, facing the city of Boston. The department's facilities are located near Kendall Square, convenient to public transportation. There is a path along the Charles River for jogging and biking, and the MIT Sailing Club has boats to take out on the river.

The Institute

Founded in 1865, the Massachusetts Institute of Technology is recognized as one of the world's great institutions of learning and exploration. It is broadly organized into five schools: Architecture and Planning; Engineering; Humanities, Arts, and Social Sciences; Management; and Science. The Department of Brain and Cognitive Sciences is one of six departments in the School of Science. MIT's total enrollment is more than 10,000 students, approximately 60 percent of whom are engaged in graduate study. Students hail from all fifty states, the District of Columbia, four territories, and 106 other countries. Women constitute 30 percent and members of minority groups represent 16 percent of the graduate students. There are 983 members of the faculty.

Applying

Selection of students is based primarily upon evidence of research interest and potential for work within the areas represented in the department. Applicants are expected to have outstanding records, particularly in the fields of mathematics and the sciences, and should have a minimum of one year of college-level work in four of the following areas: physics, chemistry, biology, mathematics, computer science, and psychology/cognitive science. Students who do not fulfill these requirements may be considered on a case-by-case basis. Graduate Record Examinations test scores are required with the completed application, which must be received by December 15 to be considered for admission the following September.

Correspondence and Information

Graduate Office
Department of Brain and Cognitive Sciences
Building 46-2005
Massachusetts Institute of Technology
77 Massachusetts Avenue
Cambridge, Massachusetts 02139
Phone: 617-253-7403
E-mail: bcs-admissions@mit.edu
Web site: http://web.mit.edu/bcs

Massachusetts Institute of Technology

THE FACULTY AND THEIR RESEARCH

Edward H. Adelson, Professor; Ph.D., Michigan, 1979. Visual psychophysics and computational vision.
Mark F. Bear, Professor; Ph.D., Brown, 1984. Experience-dependent synaptic plasticity in the cerebral cortex.
Robert C. Berwick, Professor; Ph.D., MIT, 1982. Computational linguistics; learning theory.
Emilio Bizzi, Institute Professor; M.D., Rome, 1958. Motor control and motor learning.
Emery Brown, Professor; M.D., Harvard, 1987; Ph.D., Harvard, 1988. Neural signal processing algorithms and mechanisms of anesthesia.
Stephan Chorover, Professor; Ph.D., NYU, 1959. Human systems: neurobiological, psychological, and sociocultural levels of organization.
Martha Constantine-Paton, Professor; Ph.D., Cornell, 1976. Control of developmental synaptic plasticity.
Suzanne Corkin, Professor; Ph.D., McGill, 1964. Neural basis of memory and related cognitive foundations.
Robert Desimone, Professor; Ph.D., Princeton, 1979. Neural mechanisms of perception, attention, and memory.
James J. DiCarlo, Assistant Professor; M.D., Ph.D., Johns Hopkins, 1998. Neuronal mechanisms underlying object recognition.
Michale S. Fee, Associate Professor; Ph.D., Stanford, 1992. Neural mechanisms of sequence generation and learning.
John Gabrieli, Professor; Ph.D., MIT, 1987. Human memory systems.
Edward A. F. Gibson, Professor; Ph.D., Carnegie Mellon, 1991. Human language processing.
Ann M. Graybiel, Professor; Ph.D., MIT, 1971. Neural basis of implicit learning and action strategies.
Yasunori Hayashi, Assistant Professor; M.D., 1990, Ph.D., 1994, Kyoto (Japan). Molecular mechanisms of synaptic plasticity.
Alan Hein, Professor; Ph.D., Brandeis, 1960. Development and maintenance of spatially coordinated behavior.
Richard Held, Professor Emeritus; Ph.D., Harvard, 1952. Visual development: chronology and mechanism.
Susan Hockfield, Professor; Ph.D., Georgetown, 1979. Molecular substrates of mammalian development.
Neville Hogan, Professor; Ph.D., MIT, 1977. Motor control and robotic rehabilitation.
Alan Jasanoff, Assistant Professor; Ph.D., Harvard, 1998. Systems-level neural plasticity in low-level learning and perceptual behavior.
Nancy Kanwisher, Professor; Ph.D., MIT, 1986. Cognitive neuroscience of high-level vision.
Troy Littleton, Associate Professor; M.D., Ph.D., Baylor, 1994. Mechanisms of synapse formation, function, and plasticity.
Carlos Lois, Assistant Professor; M.D., Valencia, 1991; Ph.D., Rockefeller, 1995. Neurogenesis in the adult brain and the assembly of neuronal circuits.
Earl K. Miller, Professor; Ph.D., Princeton, 1990. Neural basis of goal-directed behavior.
Christopher Moore, Assistant Professor; Ph.D., MIT, 1998. Neural mechanisms of touch perception.
Elly Nedivi, Assistant Professor; Ph.D., Stanford, 1991. Molecular genetic analysis of synaptic plasticity.
Aude Oliva, Assistant Professor; Ph.D., Grenoble Polytechnic (France), 1995. Computational visual cognition.
Tomaso Poggio, Professor; Ph.D., Genoa, 1970. Computational neuroscience.
Mary C. Potter, Professor; Ph.D., Radcliffe, 1961. Cognitive processes: attention, perception, comprehension, and memory.
William G. Quinn, Professor; Ph.D., Princeton, 1971. Genetic and molecular studies of learning and memory in *Drosophila*.
Whitman Richards, Professor; Ph.D., MIT, 1965. Computational approaches to perception.
Rebecca Saxe, Assistant Professor; Ph.D., MIT, 2003. Theory of mind.
Peter H. Schiller, Professor; Ph.D., Clark, 1962. Visual and oculomotor systems.
Gerald E. Schneider, Professor; Ph.D., MIT, 1966. Axon regeneration in the brain; visuomotor control.
Laura Schulz, Assistant Professor; Ph.D., Berkeley, 2004. Causal learning in children; domain-general mechanisms of causal inference.
Wolfram Schultz, Professor; M.D., Heidleberg, 1972. Reward, prediction, and uncertainty.
H. Sebastian Seung, Professor; Ph.D., Harvard, 1990. Dynamics of neural networks.
Morgan H. Sheng, Professor; M.B.B.S., London, 1982; Ph.D., Harvard, 1990. Molecular mechanisms of synaptic plasticity.
Pawan Sinha, Associate Professor; Ph.D., MIT, 1995. Computational vision.
Jean-Jacques Slotine, Professor; Ph.D., MIT, 1983. Adaptation and learning in robots; principles of biological control.
Mriganka Sur, Professor and Department Head; Ph.D., Vanderbilt, 1978. Cortical plasticity and dynamics.
Joshua Tenenbaum, Assistant Professor; Ph.D., MIT, 1999. Human and machine learning; computational cognitive science.
Susumu Tonegawa, Professor; Ph.D., California, San Diego, 1968. Studies of memory mechanisms with genetically engineered mice.
Li-Huei Tsai, Professor; Ph.D., Texas Southwestern Medical Center at Dallas, 1990. Synaptic plasticity and neurodegeneration.
Kenneth N. Wexler, Professor; Ph.D., Stanford, 1970. Language acquisition and psycholinguistic theory.
Matthew A. Wilson, Professor; Ph.D., Caltech, 1990. Hippocampal learning and memory.
Richard J. Wurtman, Professor; M.D., Harvard, 1960. Effects of drugs, foods, and diseases on brain neurotransmitters and behavior.

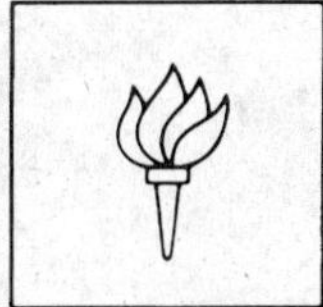

NEW YORK UNIVERSITY

Center for Neural Science

Program of Study

The Center for Neural Science at New York University (NYU) is an interdisciplinary department of the Faculty of Arts and Science, offering training leading to the Ph.D. degree in neural science. The program stresses an integrative approach to the organization and function of the nervous system. Designed to prepare students for careers in research at the forefront of neuroscience, the program emphasizes the application of diverse methodologies to problems in neuroscience to foster cooperation between experimental and theoretical aspects of the discipline. Available areas of research specialization include behavioral neuroscience, biophysics, brain imaging using fMRI, cognitive science, computational neuroscience, mathematical biology, cellular and molecular neurobiology, neurochemistry, neurophysiology, and visual neuroscience.

Entering students take a core curriculum that includes courses and a laboratory in neuroscience. In addition, a series of research laboratory rotations enables new students to sample several of the diverse approaches to the study of the nervous system that are available in the Center. After the first year, programs leading to the Ph.D. degree are designed to suit individual students' research goals and interests. An adviser and 2 other members of the faculty guide the student in the selection of courses and laboratory rotations. Students are encouraged to interact with other members of the Center, all of whom are available for advice and guidance. It is expected that the student fulfill the course requirements, complete the doctoral research, and defend a thesis within five years.

Research Facilities

The Center's core facilities and individual laboratories are located at the Washington Square Campus of New York University. Modern shared facilities and individual research laboratories provide an outstanding environment for biological and computational approaches to neuroscience.

Financial Aid

Financial support is provided for students in the neural science program through University fellowships, training grants, research assistantships, and teaching assistantships so that students will ordinarily devote full-time to their studies. Annual stipends for 2006–07 are $28,500 plus full tuition.

Cost of Study

In 2006–07, the cost is $971 (plus $292 in fees) per credit; usually, doctoral candidates register for 18 to 24 credits per year. As indicated above, recipients of fellowships and assistantships are awarded a waiver of tuition and fees.

Living and Housing Costs

Accommodations for qualified students are available in three graduate residence halls. These include 1- or 2-person studio apartments, private rooms with shared kitchen, and common areas in two- and three-bedroom apartments. Rents range from $854 to $1138 per month. For first-year doctoral candidates only, shared two-bedroom apartment suites are available in a pleasant apartment complex near the Washington Square campus. The rent for each student is $800 per month. In addition, the graduate school's referral service lists a wide variety of off-campus housing in all price ranges in the vicinity of the University.

Student Group

The doctoral program in neural science was established in 1989. There were 34 doctoral candidates in 2005–06. In addition, more than 25 doctoral candidates in other fields (e.g., biology, chemistry, computer science, mathematics) and 58 postdoctoral fellows work in the laboratories of faculty members of the Center for Neural Science.

Location

The Center for Neural Science is located at the Washington Square Campus of NYU. Situated in the heart of Greenwich Village, it is in one of the most desirable locations in New York City.

The University

New York University, a private, nonprofit institution, is the largest independent university in the country, with nearly 48,000 students. More than 2,500 full-time faculty members offer courses of study in NYU's fourteen undergraduate, graduate, and professional schools. As the cultural, artistic, and financial center of the nation, New York's limitless resources enrich both life in the city and the University's academic programs.

As part of the Graduate School of Arts and Science, the Center benefits from an outstanding academic environment and enjoys close affiliation with major research programs in the Courant Institute of Mathematical Sciences and in the Departments of Biology, Chemistry, Physics, and Psychology as well as research and clinical programs at the NYU School of Medicine.

Applying

Admission to the doctoral program in neural science is limited to qualified students, as documented by high scores on the Graduate Record Examinations, superior undergraduate grades, and excellent letters of recommendation. Students seeking admission should have a strong background in one or more of the academic areas related to neuroscience, such as biology, experimental psychology, chemistry, physics, mathematics, computer science, and engineering science. Individual candidates may make up deficiencies after admission by taking appropriate courses at the University. International students with suitable qualifications are welcome but must demonstrate their command of written and spoken English before admission to the Center. Qualified women and students who are members of minority groups are especially encouraged to apply.

Applications should be submitted before December 15 for admission in the fall.

Correspondence and Information

For admission, financial aid application forms, and the *Graduate School Bulletin:*

Office of Admissions
Graduate School of Arts and Science
New York University
P.O. Box 907
New York, New York 10276
Phone: 212-998-8050

For information about the Center and its academic and research programs:

Director of Graduate Studies
Center for Neural Science
809 Meyer Hall
New York University
New York, New York 10003
Phone: 212-998-7780
Web site: http://www.cns.nyu.edu

New York University

THE FACULTY AND THEIR RESEARCH

Core Faculty of the Center for Neural Science

Chiye Aoki, Professor of Neural Science and Biology; Ph.D., Rockefeller, 1985. Neuronal plasticity in neocortex.

Adam Carter, Assistant Professor of Neural Science; Ph.D., Harvard, 2002. Cellular mechanisms of synaptic integration and plasticity.

Nathaniel Daw, Assistant Professor of Neural Science and Psychology; Ph.D., Carnegie Mellon, 2003. Decision making and neuromodulation.

Samuel M. Feldman, Professor Emeritus of Neural Science and Psychology; Ph.D., McGill, 1959. Biological basis of behavior, with an emphasis on the regulation of movement.

Paul W. Glimcher, Associate Professor of Neural Science and Psychology; Ph.D., Pennsylvania, 1989. Planning and action in the oculomotor system.

Michael J. Hawken, Research Professor of Neural Science and Psychology; Ph.D., Otago (New Zealand), 1979. Vision and visual development; computational neuroscience.

David Heeger, Professor of Psychology and Neural Science; Ph.D., Pennsylvania, 1987. Functional imaging of the human brain (fMRI); computational neuroscience; vision; attention.

Souheil Inati, Assistant Professor of Neural Science and Psychology; Ph.D., MIT, 1999. Magnetic resonance imaging.

Lynne Kiorpes, Professor of Neural Science and Psychology; Ph.D., Washington (Seattle), 1982. Development of vision in primates and the effects of visual disorders on the development of visual system structure and function.

Eric Klann, Professor of Neural Science; Ph.D., Medical College of Virginia, 1989. Molecular mechanisms of learning and memory.

Joseph LeDoux, University Professor and Professor of Neural Science and Psychology; Ph.D., SUNY at Stony Brook, 1977. Memory and emotion.

Peter Lennie, Professor of Neural Science; Ph.D., Cambridge, 1972. Functional organization of vision.

Anthony Movshon, Silver Professor and Professor of Neural Science and Psychology; Ph.D., Cambridge, 1975. Vision and visual development.

Bijan Pesaran, Assistant Professor of Neural Science; Ph.D., CalTech, 2002. Neural activity underlying movement; planning and decision making and its application to brain-computer interfaces.

Alexander Reyes, Associate Professor of Neural Science; Ph.D., Washington (Seattle), 1990. Biophysical basis of information processing in single neurons; synaptic interaction of neurons in cortical networks.

John Rinzel, Professor of Neural Science and Mathematics; Ph.D., NYU, 1973. Theoretical neurobiology; properties of neurons and neural systems.

Nava Rubin, Associate Professor of Neural Science and Psychology; Ph.D., Hebrew (Jerusalem), 1993. Visual perception; neural basis of vision and cognition.

Dan Sanes, Professor of Neural Science and Biology; Ph.D., Princeton, 1984. Development and plasticity of the central auditory system.

Malcolm Semple, Associate Professor of Neural Science and Psychology; Ph.D., Monash (Australia), 1981. Auditory neurophysiology and psychoacoustics.

Robert Shapley, Natalie Clews Spencer Professor of the Sciences, and Professor of Neural Science, Biology, and Psychology; Ph.D., Rockefeller, 1970. Visual physiology and perception.

Eero Simoncelli, Associate Professor of Neural Science and Mathematics and Investigator, Howard Hughes Institute. Ph.D., MIT, 1993. Computational vision.

Wendy Suzuki, Associate Professor of Neural Science; Ph.D., California, San Diego, 1993. Neuroanatomical, electrophysiological, and behavioral studies of the organization of memory in the medial temporal lobe.

Associates of the Center for Neural Science

Karen Adolph, Associate Professor of Psychology and Neural Science; Ph.D., Emory, 1993. Infant learning and development; motor skill acquisition; perceptual exploration.

Efrain Azmitia, Professor of Biology and Neural Science; Ph.D., Rockefeller, 1973. Action of specific neuronotrophic and neurotoxic agents on the development and maintenance of selected neurotransmitter systems in the mammalian brain.

Justin Blau, Assistant Professor of Biology and Neural Science; Ph.D., King's College (London), 1992. Genetic analysis of circadian rhythm.

Marisa Carrasco, Professor of Psychology and Chair, Center for Neural Science; Ph.D., Princeton, 1989. Psychological and physiological processes in visual perception; models of visual attention.

Edgar Coons Jr., Professor of Psychology and Neural Science; Ph.D., Yale, 1964. Behavioral analysis of neuronal mechanisms mediating hunger, reward, and pain; aesthetics.

Clayton Curtis, Assistant Professor of Psychology and Neural Science; Ph.D., Minnesota, 1999. Cortical mechanisms of oculomotion and executive control using functional magnetic resonance imaging, psychophysiology, and pathological populations.

Lila Davachi, Assistant Professor of Psychology and Neural Science; Ph.D., Yale, 1999. Human memory formation.

Claude Desplan, Professor of Biology and Neural Science; Ph.D., Paris, 1983. Molecular genetic analysis of neural differentiation.

Davi Geiger, Associate Professor of Computer Science and Neural Science; Ph.D., MIT, 1990. Computer vision, learning, and applications.

Todd Holmes, Associate Professor of Biology and Neural Science; Ph.D., MIT, 1994. Signal transduction and ion-channel regulation.

Scott Johnson, Associate Professor of Psychology and Neural Science; Ph.D., Arizona State, 1992. Developmental cognition and perception.

Michael Landy, Professor of Psychology and Neural Science; Ph.D., Michigan, 1981. Psychophysics of spatial vision, motion, and depth processing; combination of visual cues; computational modeling of visual processes.

Laurence Maloney, Professor of Psychology and Neural Science; Ph.D., Stanford, 1985. Applications of mathematics and mathematical statistics to psychology and to the neurosciences.

James Matthews, Professor of Psychology and Neural Science; Ph.D., Brown, 1970. Motivation; conditioning; learning.

David McLaughlin, Professor of Mathematics and Neural Science and Provost, NYU; Ph.D., Indiana, 1971. Visual neuroscience; integrable waves; chaotic nonlinear waves; mathematical nonlinear optics.

Denis Pelli, Professor of Psychology and Neural Science; Ph.D., Cambridge, 1981. Psychophysics and computational modeling of visual perception, emphasizing the limitations imposed by the mediating neurons.

Charles Peskin, Professor of Mathematics and Neural Science; Ph.D., Yeshiva (Einstein), 1972. Application of mathematics and computing to the nervous system.

Elizabeth Phelps, Associate Professor of Psychology and Neural Science; Ph.D., Princeton, 1989. Interaction between human memory and emotion, combining behavioral and neural approaches.

Michael Shelley, Professor of Mathematics and Neural Science; Ph.D., Arizona, 1985. Applied mathematics, modeling, and large-scale computation; vision and computational neuroscience.

Daniel Tranchina, Professor of Biology, Mathematics, and Neural Science; Ph.D., Rockefeller, 1981. Experimental and theoretical studies of information processing in the retina.

Affiliates

Doris Aaronson, Professor of Psychology; Ph.D., Pennsylvania, 1966. Memory; mathematical models; speech perception; psycho-linguistics; reading; problem solving.

Ned Block, Professor of Philosophy; Ph.D., Harvard, 1971. Philosophy of mind; philosophy of science; cognitive science.

Adamantios I. Gafos, Professor of Linguistics; Ph.D., Johns Hopkins, 1996. Phonology; phonetics; morphology.

Murray Glanzer, Professor Emeritus of Psychology; Ph.D., Michigan, 1952. Memory; psycholinguistics.

Jerome Percus, Professor of Physics and Mathematics; Ph.D., Columbia, 1954. Chemical physics; mathematical biology; neural networks.

Carol Reiss, Professor of Biology; Ph.D., CUNY, Mount Sinai, 1978. Immune response to neurotropic viruses.

Edward Ziff, Professor of Biochemistry; Ph.D., Princeton, 1969. Mechanisms of growth control and gene regulation in animal cells and their DNA viruses; regulation and function of the *fos* gene.

NORTHWESTERN UNIVERSITY

Institute for Neuroscience

Programs of Study	The Institute for Neuroscience at Northwestern University provides a program of study leading to the Ph.D. in neuroscience. The Institute brings together a faculty of more than 160 members from more than twenty departments within four different schools at Northwestern University. The internationally recognized faculty provides state-of-the-art training in basic and clinical neuroscience at all levels of analysis, from molecular to behavioral. Some of the research interests of the faculty members include signal transduction, molecular structure and function, neurobiological basis of disease, circadian biology, movement and motor control, CNS physiology, development, sensory systems, learning and memory, cognition, and computational neuroscience.

Neuroscience students at Northwestern are admitted through the Institute, which offers a unified curriculum for all enrolled students. Students are required to take six courses, including three core courses and three electives. The core courses, Fundamentals of Neuroscience I, II, and III, are taken during the first year and provide students with an overview of the field. The core covers the following topics: molecular neuroscience, developmental neuroscience, sensory systems, motor systems, regulatory systems, and behavior and cognition. First-year students are also required to participate in three laboratory rotations. A dissertation laboratory is chosen by the end of the first year.

Graduate students further participate in an informal seminar program with their peers in which they discuss results and conclusions from their own experiments. Ongoing work is presented formally at an annual Neuroscience Symposium held on campus, and advanced students typically present their dissertation research at national meetings. Numerous seminars conducted each week by visiting scientists provide up-to-date details of research being done at other institutions. These various opportunities for rigorous exchange of ideas foster the primary goal of Ph.D. training—to help individuals become independent and innovative research scientists.

Research Facilities
As a major center of neuroscience research, Northwestern University has modern research facilities that cover all areas of contemporary investigation, from gene cloning to psychophysics. The laboratories of the faculty are found on both the Evanston and Chicago campuses of the University. New research buildings are currently being built in Evanston and Chicago.

Financial Aid
Entering students typically are supported as University Fellows. These fellowships for 2005–06 included a $23,750 stipend plus $500 in travel expenses and the full cost of tuition and health insurance. Second-year students are typically funded by teaching assistantships or training grants. Support in subsequent years is available to all students and comes from research assistantships or traineeships funded by NIH training grants. Financial aid applicants must submit the FAFSA. Fellowship support for members of underrepresented minority groups is available, and qualified individuals are encouraged to apply.

Cost of Study
Tuition is paid in full through fellowships, teaching assistantships, or grants.

Living and Housing Costs
Both University and private apartments and houses have rooms available starting at $550 per month.

Student Group
Last year, fourteen Ph.D. degrees were awarded. There are currently 120 students enrolled in the Institute, of whom 68 are women and 12 are underrepresented minorities; all attend full-time. Overall, there are 12,000 full-time students enrolled on the two campuses of the University; approximately 40 percent are in graduate and professional programs.

Student Outcomes
The majority of the Institute's graduates become postdoctoral fellows at prestigious research institutions. Many of the graduates are now junior faculty members at institutions such as the University of Washington, the University of Wisconsin, Bowdoin College, and Stony Brook, SUNY. Some graduates have pursued careers in industry (Abbott Labs, Unilabs Clinical Trials International, Structural GenomiX), while others have received medical degrees and gone on to do clinical research at places such as Children's Hospital in Boston and Johns Hopkins.

Location
The main campus of the University is located in Evanston, an attractive community along the shore of Lake Michigan. The Chicago campus, about 12 miles south of Evanston, is in an area known to visitors for its skyscrapers, fine shops, and restaurants. An immense variety of cultural, social, and recreational activities are to be found on and near both campuses.

The University
Northwestern University was founded in 1851 to create "a university of the highest order of excellence." It has grown to be one of the nation's largest and best-endowed private research universities. In breadth of top-ranked doctoral programs, Northwestern is among the top ten private universities in the country.

Applying
Outstanding students with a strong interest in neuroscience research are invited to apply. Each year the program intends to enroll about 16 new students, selected on the basis of letters of recommendation, GRE General Test scores, and past scholarly work. The application deadline is December 31 for the following fall. The application fee is $60 ($75 for international applications). The financial aid application deadline is December 31. Students can apply online at http://www.northwestern.edu/graduate/admission/onlineapp.html. Students who require a paper application can acquire one by contacting the Institute at the address listed in the Correspondence and Information section.

Correspondence and Information
Cheri Bell
Administrative Assistant
Northwestern University Institute for Neuroscience
Hogan 2-160
2205 Tech Drive
Evanston, Illinois 60208-3520
Phone: 847-491-2862
E-mail: neuro-info@northwestern.edu
Web site: http://www.northwestern.edu/nuin

Northwestern University

THE FACULTY AND THEIR RESEARCH

Sara Ahlgren, Ph.D.: Effect of ethanol on neural development.
Fraser Aird, Ph.D.: Molecular mechanisms involved in the effects of stress on behavior, endocrine, and immune functions.
George Alheid, Ph.D.: Neuroanatomy of the basal forebrain.
Ravi Allada, M.D.: Molecular basis of circadian behavioral rhythms.
Wayne Anderson, Ph.D.: Protein structure and biological function.
A. Vania Apkarian, Ph.D.: Sensation, pain perception, and management.
James Baker, Ph.D.: Brain circuitry that processes sensory information to produce movement; vestibular reflexes.
James Bartles, Ph.D.: Role of espins in actin cytoskeleton organization and function.
Joe Bass, Ph.D., M.D.: Cell and molecular regulation of body weight, feeding, and metabolism.
Susan Benloucif, Ph.D.: Human circadian rhythms and aging.
Mark Bevan, Ph.D.: Electrophysiological properties of the basal ganglia.
Eileen Bigio, M.D.: Neurodegeneration and Alzheimer's disease.
Lester Binder, Ph.D.: Microtubule proteins and neurodegenerative disease.
Martha Bohn, Ph.D.: Developmental neurobiology; neurotrophic factors; gene therapies for neurodegenerative diseases; steroid neurobiology.
James Booth, Ph.D.: Neural representations of cognitive processing during language acquisition and development.
Gary Borisy, Ph.D.: Cytoskeletal research; biochemistry, biophysics, and cell biology of microtubules.
Thomas Bozza, Ph.D.: Gene expression; axon guidance and odor mapping in the olfactory system.
Ann Bradlow, Ph.D.: Experimental phonetics; acoustic phonetics; speech perception; spoken word recognition.
David Brown, Ph.D.: Neuromusculoskeletal control during movement.
Richard Carthew, Ph.D.: Cell differentiation and morphogenesis during development.
Mary Ann Cheatham, Ph.D.: Genetic approaches to the study of cochlear physiology.
Anjen Chenn, Ph.D., M.D.: Developmental neurobiology; regulation of cell fate and cell proliferation.
Dane Chetkovich, M.D.: Molecular basis of synaptic plasticity.
Rex Chisholm, Ph.D.: Regulation of gene expression during development; molecular genetics of cell motility and learning.
Anis Contractor, Ph.D.: Molecular mechanism of synaptic plasticity.
Mauro Dal Canto, M.D.: Demyelinating diseases and viral infections of the nervous system.
Peter Dallos, Ph.D.: Biophysics and neurobiology of the cochlea; hair-cell physiology.
Julius DeWald, Ph.D.: Relationship between abnormal muscle coactivation patterns and cortical reorganization following stroke.
Steven DeVries, M.D., Ph.D.: Structure and function of the mammalian photoreceptor synapse.
Christine DiDonato, Ph.D.: Genetics of neuromuscular diseases.
John Disterhoft, Ph.D.: Cellular substrates of learning in the young and aging brain.
Margarita Dubocovich, Ph.D.: Cellular and molecular pharmacology of melatonin receptors.
Anne Duggan, Ph.D.: Macromolecular structure of neuronal acid-sensitive sodium channels.
Guy Edelman, M.D.: Physiology of neurosurgical disease.
Leon Epstein, M.D.: HIV neuoropathogenesis.
Yuanyi Feng, Ph.D.: Molecular mechanisms influencing cortical cevelopment.
Adriana Ferreira, M.D., Ph.D.: Mechanisms underlying establishment and maintenance of synaptic connections.
David Ferster, Ph.D.: Central mechanisms of vision in mammals; neuronal connectivity underlying receptive field properties.
Jaime Garcia-Anóveros, Ph.D.: Molecular mechanism of touch and pain sensation.
Yuri Geinisman, M.D., Ph.D.: Structural manifestations of synaptic plasticity.
Pablo Gejman, M.D.: Genetic basis of schizophrenia.
Vladimir Gelfand, Ph.D.: Cytoskeleton; molecular motors and regulation of organelle transport.
Dedre Gentner, Ph.D.: Cognition and language in learning and development.
Darren Gitelman. M.D.: Functional MRI.
Robert Goldman, Ph.D.: Intracellular organization of neurons; structure and function of neurofilaments.
Numa Gottardi-Littell, M.D.: Neuronal cell biology and the molecular pathogenesis of Alzheimer's.
Jay Gottfried, M.D., Ph.D.: Olfaction and functional imaging.
Dennis Groothuis, M.D.: Neuropathology of brain tumors; blood-brain barrier.
Adrian Gross, M.D.: Ion channel structure and function.
Jaime Grutzendler, M.D.: Molecular mechanisms regulating synaptic plasticity.
Timothy Hain, M.D.: Normal and abnormal vestibular function in humans.
Mitra Hartmann, Ph.D.: Neurobiology of active sensing behavior; sensory feedback.
Xiaolin He, Ph.D.: Cell-surface receptors in development and cancer.
C. J. Heckman, Ph.D.: Role of motor units and spinal circuits in normal and pathological motor function.
Afif Hentati, M.D.: Molecular genetics of motor neuron disease.
Laura Herzing, Ph.D.: Epigenetics and genome dynamics.
Philip Hockberger, Ph.D.: Intracellular messenger regulation of ion channel functions; synaptogenesis.
Robert Holmgren, Ph.D.: Genetic and molecular analysis of developmental fate in *Drosophila*.
Teresa Horton, Ph.D.: Circadian biology and aging.
James Houk, Ph.D.: Central nervous system control of motor behavior.
Philip Iannaccone, M.D., D.Phil.: Role of sonic hedgehog and environmental factors in neural development.
Kouichi Iwasaki, Ph.D.: Molecular biology and genetics of a rhythmic behavior.
J. Larry Jameson, M.D., Ph.D.: Transcriptional control of endocrine gene expression.
Mark Jung-Beeman, Ph.D.: Cognitive processing related to understanding complex language, "insight," and creative problem solving.
Richard Jungmann, Ph.D.: Gene expression regulation by second-messenger systems.
William Karpus, Ph.D.: Pathogenesis of demyelinating disease.
John Kessler, M.D.: Stem cell development, differentiation, and neuronal survival.
Ronald Kettner, Ph.D.: Neural and behavioral studies of eye movements and neural modeling.
Byung Kim, Ph.D.: Pathogenic mechanisms of Theiler's virus-induced demyelination.
William Klein, Ph.D.: Cell and molecular biology of brain development; growth cones and neural regeneration; Alzheimer's disease.
Jhumku Kohtz, Ph.D.: Role of signaling molecules (sonic hedgehog) in telencephalic development and neuronal differentiation.
Nina Kraus, Ph.D.: Auditory evoked potentials—neural generating systems and physiologic bases of hearing (auditory detection and discrimination).
Carole LaBonne, Ph.D.: Development of neural crest cells.
Charles Larson, Ph.D.: Neural control of vocalization.
Jon Levine, Ph.D.: Physiology of hypothalamic neuropeptides; regulation of pituitary hormone secretion.
Robert Levy, M.D., Ph.D.: Neuroscience of HIV infection; pain mechanisms; cerebral ischemia; brain imaging.
Honglin Li, Ph.D.: Molecular mechanisms regulating apoptosis.
Liming Li, Ph.D.: Prion proteins.
Robert Linsenmeier, Ph.D.: Neural microenvironment and function; sensory information processing in retina.
Howard Lipton, M.D.: Viral infections of the nervous system; multiple sclerosis.
Jerilyn Logemann, Ph.D.: Mechanisms of normal and abnormal swallowing.
Jon Lomasney, M.D.: Molecular basis for function and regulation of neurotransmitter and hormone receptors.

Gianmaria Maccaferri, Ph.D., M.D.: Synaptic communication; role of interneurons in the hippocampus.
Robert MacDonald, Ph.D.: Bilayer structure; membrane skeleton; fusion; neurosecretion.
Malcolm MacIver, Ph.D.: Mechanical and neural bases of goal-directed locomotor behavior.
Colum MacKinnon, Ph.D.: Spatial and temporal patterns of motor; cortical activity preceding voluntary movement.
Marco Martina, Ph.D., M.D.: Electrical properties of dendrites and their contribution to synaptic function.
Kelly Mayo, Ph.D.: Neuroendocrine peptides that influence the synthesis or secretion of pituitary hormones.
Donald McCrimmon, Ph.D.: Respiratory neurophysiology and pharmacology; motor control; homeostasis.
Kevin McKenna, Ph.D.: Spinal cord control of autonomic and sexual function.
William McKinney, M.D.: Depressive disorders.
David McLone, M.D., Ph.D.: Developmental neurobiology related to clinical disease.
Marsel Mesulam, M.D.: System-level neuroanatomy of the human brain; experimental neuropathology of Alzheimer's; mapping of cognitive domains.
Lee Miller, Ph.D.: Understanding signals that the brain creates to control limb movements.
Richard Miller, Ph.D.: Molecular aspects of synaptic communication.
Stephen Miller, Ph.D.: Immunologic effector mechanisms involved in mouse models of human multiple sclerosis.
Isabelle Mintz, Ph.D.: Ion channels; neuronal firing and control of dopamine release.
Richard Morimoto, Ph.D.: Neuronal protection; cellular and molecular response of neuronal cells to stress.
Jill Morris, Ph.D.: Biochemical basis of mental disorders.
Enrico Mugnaini, Director, M.D.: Cell class-specific gene expression in neuronal microcircuits.
Ferdinando Mussa-Ivaldi, Ph.D.: Motor learning and biorobotics.
Toshio Narahashi, Ph.D.: Electrophysiology, pharmacology, and toxicology of neuronal ion channels.
Maria-Grazia Nunzi, Ph.D.: Morphological and biochemical modifications of CNS structures during aging and Alzheimer's.
Zoltan Oltvai, M.D.: Apoptosis in the nervous system.
Puneet Opal, M.D., Ph.D.: Cellular processes in neurodegenerative diseases.
Ken Paller, Ph.D.: Human memory and cognitive neuroscience.
Peter Penzes, Ph.D.: Signal transduction mechanisms regulating development and structural plasticity of excitatory synapses.
Eric Perreault, Ph.D.: Spinal control of movement; muscle physiology; biomechanics.
Barry Peterson, Ph.D.: Neural circuitry underlying motor behavior and learning.
Lawrence Pinto, Ph.D.: Genetic and molecular neurobiology; biophysics of vision; retina.
Jelena Radvlica, M.D., Ph.D.: Cell biological mechanisms of memory.
Ann Ragin, Ph.D.: Neuroimaging.
Indira Raman, Ph.D.: Properties of ion channels and their role in neuronal specialization.
Yi Rao, M.D., Ph.D.: Molecular and cellular mechanisms of neuronal migration and axon guidance.
Paul Reber, Ph.D.: Functional neuroimaging of human declarative and nondeclarative memory; anatomical organization of memory systems throughout the brain.
Eva Redei, Ph.D.: Molecular mechanisms of the stress response.
Gary Robertson, M.D.: Regulation of water balance in health and disease.
Mark Rogers, Ph.D.: Neuromuscular control of posture, stance, and movement function.
J. Peter Rosenfeld, Ph.D.: Human event-related brain potential correlates of behavior; pain and opiate analgesia.
Aryeh Routtenberg, Ph.D.: Molecular biology and cellular physiology of synaptic plasticity.
Mario Ruggero, Ph.D.: Biophysics and neurobiology of the cochlea.
W. Zev Rymer, M.D., Ph.D.: Spinal regulation of movement; neural circuits; spinal cord disorders.
Vijay Sarthy, Ph.D.: Gene regulation development and functional organization of the vertebrate retina.
Mark Segraves, Ph.D.: Neuronal mechanisms underlying oculomotor behavior; contributions to the cerebral cortex to the generation of eye movements.
Gordon Shepherd, M.D., Ph.D.: Circuit organization of the neicortex.
Teepu Siddique, M.D.: Disorders of the human motor system.
Jonathon Siegel, Ph.D.: Synaptic physiology and ultrastructure of the cochlea; cochlear transduction.
Eugene Silinsky, Ph.D.: Biophysics and pharmacology of transmitter release.
Richard Silverman, Ph.D.: Mechanism of enzymes involved in neurotransmitter biosynthesis and degradation.
Joshua Singer, Ph.D.: Signal transduction in the visual system.
Greg Smith, Ph.D.: Role of infectious agents in disease.
Sara Solla, Ph.D.: Neural computation.
Patricia Spear, Ph.D.: Molecular biology of herpes simplex virus.
Nelson Spruston, Ph.D.: Synaptic integration in the central nervous system.
D. James Surmeier, Ph.D.: Neuromodulator control of cellular excitability.
Francis Szele, Ph.D.: Developmental neurobiology and mechanisms of plasticity.
Joseph Takahashi, Ph.D.: Cell and molecular biology of circadian rhythms; photoreception; signal transduction.
Edwin Taylor, Ph.D.: Mechanisms of molecular motors.
Robert Ten Eick, Ph.D.: Myocardial electrophysiology; cellular pathology; pharmacology.
Cynthia Thompson, Ph.D.: Neurolinguistics.
Jacek Topczewski, Ph.D.: Cellular differentiation; migration; organization during development.
Warren Tourtelotte, M.D., Ph.D.: Gene regulation in the CNS; nerve-muscle induction mechanisms; growth factor regulation.
Barbara Trommer, M.D.: Synaptic plasticity in developing hippocampus.
John Troy, Ph.D.: Neurophysiology of mammalian vision.
Fred Turek, Ph.D.: Photoperiodic control of neuroendocrine-gonadal activity; circadian rhythmicity.
Linda Van Eldik, Ph.D.: Glial-derived neurotrophic proteins; neurodegenerative disorders.
Robert Vasser, Ph.D.: Alzheimer's disease.
Alec Wang, Ph.D.: Architecture of neuronal circuitry in the mature vertebrate CNS.
Jack Waters, Ph.D.: Dendritic processing of synaptic input and relationship to learning.
Sandra Weintraub, Ph.D.: Neuropsychology of Alzheimer's and other dementias.
Donna Whitlon, Ph.D.: Mechanisms of neural development and regeneration in the mammalian cochlea.
Patrick Wong, Ph.D.: Functional neuroanatomy of speech perception, second language acquisition, aphasia recovery, and music cognition.
Theresa Woodruff, Ph.D.: Regulation of ovarian follicle growth during the mammalian reproductive cycle.
Catherine Woolley, Ph.D.: Structural plasticity; neuronal structure-function relationships; neuroendocrinology.
Beverly Wright, Ph.D.: Psychoacoustics in normal and impaired individuals; perceptual learning.
Chau Wu, Ph.D.: Electrophysiology and pharmacology of ion channels; repetitive discharges in myotonia and epilepsy.
Jane Wu, Ph.D.: Role of gene splicing in development and pathogenesis.
Alice Wyrwicz, Ph.D.: Neurobiologic applications of magnetic resonance imaging.
Phyllis Zee, M.D., Ph.D.: Age-related changes of the circadian system and sleep.
Jing Zheng, Ph.D.: Cochlear molecular biology and mechanisms of electromotility of outer hair cells.

OREGON HEALTH & SCIENCE UNIVERSITY

Department of Behavioral Neuroscience

Program of Study	The graduate program in the Department of Behavioral Neuroscience offers training that leads to the Doctor of Philosophy degree. The program emphasizes basic science training in behavioral neuroscience, with specialization in such areas as physiological psychology, behavioral and molecular genetics, behavioral pharmacology, and biological bases of addiction. The Department of Behavioral Neuroscience is one of six basic science departments in the School of Medicine and provides a unique environment especially suited for the education of multidisciplinary neuroscientists. Several faculty members maintain joint appointments in other basic science departments, such as the Departments of Cell and Developmental Biology, Physiology and Pharmacology, Psychiatry, and Neurology. Faculty research interests are diverse but concentrate on behavioral pharmacogenetics, addiction and drug conditioning, neuropharmacology, and neuroendocrinology. The program offers an opportunity to experience an integration of classical and modern techniques in neuroscience. The department also maintains ongoing research training programs funded by the National Institutes of Drug Abuse and the National Institute on Alcohol Abuse and Alcoholism. Additional research opportunities exist at the Vollum Institute for Advanced Biomedical Research, Portland State University, and the Veterans Affairs Medical Center (VAMC). Individual research, in close collaboration with members of the faculty, is considered an essential aspect of the doctoral program. During the first two years, students enrich their training by study in one or more of the allied basic science areas (e.g., pharmacology, biochemistry, molecular and cellular biology, molecular and medical genetics, neuroanatomy, neurochemistry, and neurophysiology) and complete core course work in statistics, research design, learning, and physiological psychology. Additional courses and seminars are selected in consultation with the academic adviser. A large portion of the student's effort during each academic term is devoted to training in research. Finally, all students participate each term in an active weekly seminar series (Issues in Behavioral Neuroscience), which is attended by the majority of the faculty members who are involved in the graduate training program. Participation in national and international meetings is encouraged and supported financially. Students are encouraged to rotate through three labs and must select a thesis adviser by the end of their first year. Students complete a research project requirement by the end of the second year and then advance to doctoral candidacy. The program usually requires an average of five years of study.
Research Facilities	The department is well equipped with the full range of behavioral, psychopharmacological, physiological, neurochemical, and molecular biological research facilities. Oregon Health & Science University (OHSU) and the VAMC facilities are on a shared campus that is reached by a 5-minute walk via the recently constructed VA Skybridge. The polymerase chain reaction genotyping laboratories are located on the OHSU campus. Complete American Association for Accreditation of Laboratory Animal Care (AAALAC)–accredited animal-care facilities are located both on the OHSU campus and at the VAMC. Other facilities include a comprehensive Behavioral Thermoregulation and Metabolism Laboratory at Portland State University as well as a third AAALAC-accredited on-site animal facility.
Financial Aid	Stipends and tuition support are generally available for all students. The stipend for first-year students in the 2006–07 academic year is $23,500. The stipend increases to $24,500 when students advance to doctoral candidacy. Tuition and health insurance are also provided for all students admitted to the program. Sources of support include departmental, grant-supported research assistantships and research traineeships on NRSA training grants from NIAAA and NIDA. All applications received by January 1 are considered for available support on a competitive basis.
Cost of Study	Resident tuition and fees for 2005–06 were $21,492. Nonresident tuition and fees for 2005–06 were $23,868.
Living and Housing Costs	Most students prefer to live in off-campus apartments, many of which are within walking distance of the University. Apartments in the campus area typically rent for $350 to $500 per month.
Student Group	In 2005–06, there were 28 graduate students in the program; on the average, between 3 and 5 students are admitted to the Ph.D. program each year. Oregon Health & Science University has approximately 1,500 students. Of these, 226 students are currently enrolled in the School of Medicine graduate programs.
Student Outcomes	From 1990 to 2005, the department has awarded twenty-nine Ph.D. degrees. Of these graduates, 9 hold faculty positions, 1 is an industrial research scientist, 2 are in clinical practice, 15 hold postdoctoral positions, and 2 are currently not involved in scientific endeavors.
Location	The University is located on 116 acres atop Marquam Hill in southwest Portland. The site allows views of downtown Portland, the Willamette and Columbia Rivers, and the Cascades, including Mount Hood and Mount St. Helens. The Portland metropolitan area is a major cultural center with a population of approximately 1.3 million. The proximity of recreation areas in the Columbia River Gorge, Cascades, High Desert, and the Oregon coast provide excellent access to skiing, sailing, windsurfing, hiking, biking, and mountain climbing, all within 1 to 2 hours' drive of the campus. Other colleges and universities in Portland include Reed College, Lewis & Clark College, Portland State University, and the University of Portland.
The University and The Department	Oregon Health & Science University is the state's only academic health center and is composed of the Schools of Medicine, Dentistry, and Nursing and the Oregon Graduate Institute School of Science and Engineering. Several affiliated institutes and clinics are located on the campus, including the VAMC, Vollum Institute for Advanced Biomedical Research, the Center for Research on Occupational and Environmental Toxicology (CROET), Shriners Hospital for Crippled Children, and three University hospitals. The department is administratively located in the medical school and includes faculty members from the VAMC, Vollum Institute, CROET, and several other University departments. Opportunities exist for interaction with Portland State University and the Oregon National Primate Research Center.
Applying	Applicants must supply a completed application form, transcripts of college or university work leading to a bachelor's degree (with a GPA of at least 3.0), scores on the Graduate Record Examinations General Test, and three letters of recommendation by January 1. A background in general biology and chemistry, statistics, and biopsychology is recommended. In addition, the highly competitive applicant should have completed course work in biochemistry, physiology, and neurobiology. Research experience (e.g., independent study, honors thesis, or laboratory work experience) is essential. Students are selected on the basis of research experience, undergraduate records, recommendations, and interviews with faculty members.
Correspondence and Information	Dr. Tamara Phillips Department of Behavioral Neuroscience, L470 Oregon Health & Science University 3181 Southwest Sam Jackson Park Road Portland, Oregon 97239-3098 Phone: 503-494-8464 Fax: 503-494-6877 E-mail: behavior@ohsu.edu Web site: http://www.ohsu.edu/behneuro/

Oregon Health & Science University

THE FACULTY AND THEIR RESEARCH

W. Kent Anger, Ph.D., Maine. Behavioral assessment of neurotoxicity; development of computerized methods for behavioral assessment and training/education.
Computer based instruction vs. traditional training in working adults: Information and quiz presentation and interactivity. *J. Org. Behav. Manage.*, in press. With Rohlman et al.

John K. Belknap, Ph.D., Colorado. Genetics of behavioral traits related to drug abuse.
The replicability of QTLs for murine alcohol preference behavior across eight independent studies. *Mammalian Genome* 12:893–9, 2001. With Atkins.

S. Paul Berger, M.D., Stanford. Psychopharmacology; behavioral sensitization; drug abuse; schizophrenia; translational preclinical and clinical research.
Altered behavioral response to dopamine D3 receptor agonists 7-OH-DPAT and PD 128907 following repetitive amphetamine administration. *Neuropsychopharmacology* 28:1422–32, 2003. With Richtand et al.

Kari J. Buck, Ph.D., Colorado. Biological (genetic) factors and neural circuits involved in complex behaviors, including risk factors for dependence on a variety of drugs of abuse.
Mpdz is a quantitative trait gene for drug withdrawal seizures. *Nature Neurosci.* 7:699–700, 2004. With Shirley, Walter, Reilley, and Fehr.

William E. Cameron, Ph.D., Arizona. Engaging biomedical researchers in K–12 and public science education.
A role for neuroscientists in engaging young minds. *Nature Rev.: Neurosci.* 4:763–8, 2003. With Chudler.

John C. Crabbe Jr., Ph.D., Colorado. Alcohol/drug mouse behavioral genetics; tolerance and dependence; mouse genetic animal models.
An analysis of the genetics of alcohol intoxication in inbred mice. *Neurosci. Biobehav. Rev.*, 28:785–802, 2205. With Metten et al.

Larry I. Crawshaw, Ph.D., California, Santa Barbara. Thermoregulation; ethanol; comparative physiology.
Effects of hypoxia, anoxia, and endogenous ethanol on thermoregulation in goldfish, *Carassius auratus*. *Am. J. Physiol.* 278:R545–55, 2000. With Rausch and Wallace.

Christopher L. Cunningham, Ph.D., Oregon. Behavioral pharmacology; learning and motivation; behavioral genetics.
Ethanol-induced conditioned place preference is expressed through a ventral tegmental area dependent mechanism. *Behav. Neurosci.* 119:213–23, 2005. With Bechtholt.

Deborah A. Finn, Ph.D., USC. Neuropharmacology; neuroendocrinology of ethanol intake and withdrawal.
The role of pregnane neurosteroids in ethanol withdrawal: Behavioral genetics approaches. *Pharmacol. Ther.* 101:91–112, 2004. With Ford, Wiren, Roselli, and Crabbe.

Kathleen A. Grant, Ph.D., Washington (Seattle). Behavioral pharmacology; behavioral endocrinology; neuroscience of addictive disorders.
Advances in nonhuman primate alcohol abuse and alcoholism research. *Pharmacol. Ther.* 100:235–55, 2003. With Bennett.

Daniel C. Hatton, Ph.D., Florida. Nutrition and behavior; mood disorders; blood pressure regulation.
The relationship between postpartum depression and breastfeeding. *J. Human Lactation* 21:444–50, 2005. With Harrison-Hohner et al.

Robert J. Hitzemann, Ph.D., California, San Francisco. Behavioral genetics; drug abuse; neuroimaging; psychopharmacology.
A strategy for the integration of QTL, gene expression, and sequence analyses. *Mammalian Genome* 14:733–47, 2003. With Malmanger et al.

Aaron Janowsky, Ph.D., Vanderbilt. Neurotransmitter transporters, receptors.
Effects of methamphetamine and lobeline on vesicular monoamine and dopamine transporter-mediated dopamine release in a cotransfected model system. *J. Pharmacol. Exp. Ther.* 310:1142–51, 2004. With Wilhelm et al.

Jeri S. Janowsky, Ph.D., Cornell. Cognitive neuroscience of aging and sex steroid effects; functional neuroimaging.
Androgen deprivation impairs memory in older men. *Behav. Neurosci.*, in press. With Bussiere, Beer, and Neuss.

K. Matthew Lattal, Ph.D., Pennsylvania. Behavioral and neurobiological mechanisms of learning and memory.
Behavioral impairments caused by injections of the protein synthesis inhibitor anisomycin after contextual retrieval reverse with time. *Proc. Natl. Acad. Sci. U.S.A.* 101 4667–72, 2004. With Abel.

Malcolm Low, M.D., Albany Medical College; Ph.D., Tufts. Regulation and function of hypothalamic/pituitary neuropeptides; transgenic mice.
Glucocorticoids exacerbate obesity and insulin resistance in neuronal-specific proopiomelanocortin deficient mice. *J. Clin. Invest.*, in press. With Smart et al.

Gregory P. Mark, Ph.D., Delaware. Neurochemistry of addiction; neurobiology of motivation and reinforcement.
Self-administration enhances excitatory synaptic transmission in the bed nucleus of the stria terminalis. *Nature Neurosci.* 8:413–4, 2005. With Dumont, Mader, and Williams.

Joseph D. Matarazzo, Ph.D., Northwestern. Behavior and health; behavioral cardiology; neuropsychology.
Health and behavior: The coming together of science and practice in psychology and medicine after a century of benign neglect. *J. Psychol. Med. Settings* 1:7–39, 1994.

Charles K. Meshul, Ph.D., Illinois. Parkinson's disease; immuno-electron microscopy; drug abuse; antipsychotic drugs; in vivo microdialysis.
L-Dopa-induced reversal in striatal glutamate following partial depletion of nigrostriatal dopamine with MPTP. *Neuroscience* 136:333–41, 2005. With Holmer et al.

Suzanne Mitchell, Ph.D., SUNY at Stony Brook. Human psychopharmacology; reinforcement and learning; impulse control.
Measures of impulsivity in cigarette smokers and nonsmokers. *Psychopharmacology* 146:455–64, 1999.

Kim Neve, Ph.D., California, Irvine. Dopamine receptors and transporters.
Preferential interaction between the dopamine D2 receptor and arrestin2 in neostriatal neurons. *Mol. Pharmacol.* 66:1635–42, 2004. With Macey and Gurevich.

Barry Oken, M.D., Medical College of Wisconsin. Human attention, alertness, and changes in visual processing related to aging and neurodegenerative diseases.
Randomized controlled trial of yoga and exercise in multiple sclerosis. *Neurology* 62:2058–64, 2004. With Kishiyama et al.

David S. Phillips, Ph.D., Purdue. Ototoxicity; research design; screening tests.
Early detection of ototoxicity using 1/6th octave steps. *J. Am. Acad. Audiology* 14(8):448–54, 2003. With Fausti et al.

Tamara J. Phillips, Ph.D., SUNY at Albany. Behavioral genetics; gene mapping for traits relevant to drug and alcohol addiction; behavioral pharmacology.
Gene expression differences in mice divergently selected for methamphetamine sensitivity. *Mammalian Genome* 16:291–305, 2005.

Jacob Raber, Ph.D., Weizmann (Israel). Aging; androgens; apolipoprotein E; Alzheimer's disease; androgens; histamine receptors; hypothalamic-pituitary-adrenal axis; methamphetamine.
Sex-differences in age-related cognitive decline in C57BL/6J mice associated with increased brain Microtubule-associated protein 2 and synaptophysin immunoreactivity. *Neuroscience*, in press. With Benice et al.

Andrey E. Ryabinin, Ph.D., Institute of Normal Physiology (Moscow). Stress-related neuropeptides: urocortin, CRF, vasopressin.
The Edinger-Westphal nucleus-lateral septum pathway and its relationship to alcohol consumption. *J. Neurosci.* 23:2477–87, 2003. With Bachtell et al.

Nathan Selden, M.D., Harvard; Ph.D., Cambridge. Behavioral pharmacology; pain; neuroanatomy; pediatric neurological surgery.
A single-pass tunneling technique for CSF shunting procedures. *Pediatr. Neurosurg.* 39:254–7, 2003. With Sandquist.

Alexander A. Stevens, Ph.D., New Hampshire. Neural basis of language and audition; perceptual and cortical reorganization in the blind.
Auditory perceptual consolidation in the blind. *Neuropsychologia* 43:1901–10, 2005. With Weaver.

Kristine M. Wiren, Ph.D., Connecticut. Molecular mechanisms mediating withdrawal; neuroendocrinology; steroid receptor regulation.
Molecular and genetic bases of the addictions. In *The Molecular and Genetic Basis of Neurologic and Psychiatric Disease*, pp. 779–89, eds. R. Rosenberg et al. Woburn, Mass.: Butterworth-Heinemann Medical Publishing, 2003 (with Crabbe).

OREGON HEALTH & SCIENCE UNIVERSITY

Neuroscience Graduate Program

Program of Study	The Neuroscience Graduate Program offers training leading to the Doctor of Philosophy (Ph.D.) degree. The program is intended to prepare students for a professional career in research and teaching. Selected medical students may pursue the M.D./Ph.D. degree.

The program currently offers opportunities for advanced study and research in molecular, cellular, developmental, endocrine, behavioral, and systems neuroscience. Research opportunities exist in basic and clinical departments, the Vollum Institute, the Center for Research on Occupational and Environmental Toxicology, the School of Dentistry, the Veterans Administration Medical Center, the Oregon Primate Center, and the Neurological Sciences Institute.

Required course work in the first year includes a three-term series in neuroscience (neurophysiology, molecular and cellular neuroscience, systems neuroscience) plus three laboratory rotations and seminars. Additional courses in biochemistry, molecular biology, genetics, cell biology, and advanced neuroscience topics make up the remainder of students' course work, typically over the first two years. Students identify their dissertation mentor at the end of the first year and spend considerable time in the laboratory during the second year defining their dissertation project. An oral qualifying exam, which incorporates a written thesis proposal, is taken after the second year. Students require an average of five to six years to complete the Ph.D.

There are more than 140 faculty members in the Neuroscience Graduate Program. They are well funded by extramural research grants and maintain modern laboratories for state-of-the-art biomedical research.

Research Facilities

The program faculty members are well equipped for their diverse research projects. Students of the program also have access to specialized resources within Oregon Health & Science University (OHSU) through their individual mentors.

Financial Aid

In 2005–06, all full-time Ph.D. students received stipends of at least $23,000 per year. In addition, fees such as health-care and hospitalization insurance are paid for students on stipends. After the first year of study, students are supported by funds provided by their thesis adviser or training grants. Students are encouraged to apply for extramural fellowships.

Cost of Study

Graduate students are enrolled as graduate research assistants and thus receive a full tuition waiver.

Living and Housing Costs

Many apartments are within walking distance of the campus or on a campus bus route and rent for $400 to $600 per month.

Student Group

The Oregon Health & Science University has approximately 1,300 students. Of these, 290 are enrolled in graduate programs. Currently, the Neuroscience Graduate Program has 64 graduate students.

Student Outcomes

Although the program has existed for only twelve years, 34 students have graduated, all have pursued biomedical research, and a number hold tenure-track faculty positions.

Location

The University is located on the hills of Sam Jackson Park near the center of the city of Portland. The site allows views of the Willamette River and the Cascades, including Mount Hood and Mount St. Helens. The Portland metropolitan area is a major cultural center with a population of about a million. The Willamette and Columbia rivers are popular areas for water sports. The driving time to the mountains and ocean beaches is 1–2 hours, and these areas provide superb recreational facilities. Portland's climate is mild, with summer temperatures between 70 and 80 degrees and winter temperatures between 40 and 50 degrees. Other colleges and universities located in Portland include Reed College, Lewis and Clark College, Portland State University, and the University of Portland.

The University

The Oregon Health & Science University is the only academic health center in the state of Oregon. It houses the Schools of Medicine, Dentistry, Nursing, and Engineering; the Vollum Institute; the Oregon Regional Primate Research Center; the Neurological Sciences Institute; the University Hospital and Clinics; and Doernbecher Children's Hospital. Affiliations exist with many other institutions, including Portland State University, the Veterans Administration Medical Center, and the Shriners Hospital for Crippled Children.

Applying

Students with a bachelor's degree in the biological or physical sciences and with a strong commitment to research are encouraged to apply. The deadline is January 1. Applicants must submit scores on the Graduate Record Examinations, and international applicants whose native language is not English must submit scores from the Test of English as a Foreign Language (TOEFL). Successful applicants generally have significant research experience. Applicants are notified of acceptance around April 1, and accepted students are strongly encouraged to begin research rotations on July 1.

Correspondence and Information

Neuroscience Graduate Program
Vollum Institute, L-474
Oregon Health & Science University
3181 Southwest Sam Jackson Park Road
Portland, Oregon 97239
Phone: 503-494-6932
Fax: 503-494-5518
E-mail: ngp@ohsu.edu
Web site: http://www.ohsu.edu/ngp

Oregon Health & Science University

THE FACULTY

Behavioral Neurobiology
John Belknap, Ph.D.
Kari Buck, Ph.D.
John Crabbe, Ph.D.
Christopher Cunningham, Ph.D.
Deborah Finn, Ph.D.
Edward Gallaher, Ph.D.
Daniel Hatton, Ph.D.
Gregory Mark, Ph.D.
Tamara Phillips, Ph.D.
Fred Risinger, Ph.D.
Robert Stackman, Ph.D.

Cell Neurobiology
Wolf Almers, Ph.D.
Michael Andresen, Ph.D.
Gary Banker, Ph.D.
Thomas Baumann, Ph.D.
R. Lane Brown, Ph.D.
David Dawson, Ph.D.
Matthew Frerking, Ph.D.
Craig Jahr, Ph.D.
Martin Kelly, Ph.D.
James Maylie, Ph.D.
Edwin McCleskey, Ph.D.
Charles Meshul, Ph.D.
Charles Roselli, Ph.D.
Larry Sherman, Ph.D.
Stephen Smith, Ph.D.
Gary Thomas, Ph.D.
Laurence Trussell, Ph.D.
Henrique vonGersdorff, Ph.D.
Gary Westbrook, M.D.
John Williams, Ph.D.

Developmental Neurobiology
Stephen Arch, Ph.D.
Stephen Back, M.D., Ph.D.
Agineszka Balkowiec, Ph.D.
Wenbiao Chen, Ph.D.
Jan Christian, Ph.D.
Philip Copenhaver, Ph.D.
Beth Habecker, Ph.D.
Karla Kent, Ph.D.
Steven Matsunoto, Ph.D.
David Morton, Ph.D.
Martha Neuringer, Ph.D.
Sergio Ojeda, D.V.M.
Bruce Patton, Ph.D.
Gary Reiness, Ph.D.
Oline Rønnekleiv, Ph.D.
Richard Simerly, Ph.D.
Henryk Urbanski, Ph.D.

Medical Neurobiology
Charles Allen, Ph.D.
Kent Anger, Ph.D.
Bruce Gold, Ph.D.
Steven Johnson, M.D., Ph.D.
Glen Kisby, Ph.D.
Alfred Lewy, M.D., Ph.D.
Daniel Marks, Ph.D.
Jay Nelson, Ph.D.
Edward Neuwelt, M.D.
John Nutt, Ph.D.
Halina Offner, M.D.
Mohammad Sabri, Ph.D.
Peter Spencer, Ph.D.
Dennis Trune, Ph.D.
Arthur Vandenbark, Ph.D.
Tania Vu, Ph.D.

Molecular Neurobiology
Grazyna Adamus, Ph.D.
John Adelman, Ph.D.
Cynthia Bethea, Ph.D.
Greg Burrows, Ph.D.
Vince Coghlan, Ph.D.
Roger Cone, Ph.D.
P. Michael Conn, Ph.D.
Caroline Enns, Ph.D.
Michael Forte, Ph.D.
Peter Gillespie, Ph.D.
Richard Goodman, M.D., Ph.D.
David Grandy, Ph.D.
Aaron Janowsky, Ph.D.
Michael Kapiloff, M.D.
Jeff Karpen, Ph.D.
Doris Kretzschmar, Ph.D.
Kristen Lampi, Ph.D.
Peter Larsson, Ph.D.
Elaine Lewis, Ph.D.
Malcolm Low, M.D., Ph.D.
Curtis Machida, Ph.D.
Bruce Magun, Ph.D.
Kim Neve, Ph.D.
Hemachandra Reddy, Ph.D.
David Robertson, Ph.D.
Peter Rotwein, Ph.D.
Andrey Ryabinin, Ph.D.
Bruce Schnapp, Ph.D.
John Scott, Ph.D.
Thomas Soderling, Ph.D.
Eliot Spindel, Ph.D.
Mary Stenzel-Poore, Ph.D.
Philip Stork, M.D.
Mathew Thayer, Ph.D.
Richard Walker, Ph.D.
Marcel Wehrli, Ph.D.
Kristine Wiren, Ph.D.

Systems Neurobiology
Sue Aicher, Ph.D.
Neal Barmack, Ph.D.
Curtis Bell, Ph.D.
Judy Cameron, Ph.D.
William Cameron, Ph.D.
Paul Cordo, Ph.D.
Michael Cowley, Ph.D.
Robert Duvoisin, Ph.D.
Alvin Eisner, Ph.D.
Mary Heinricher, Ph.D.
Fay Horak, Ph.D.
Jane Macpherson, Ph.D.
Claudio Mello, Ph.D.
Catherine Morgans, Ph.D.
Shaun Morrison, Ph.D.
Alfred Nuttall, Ph.D.
Robert Peterka, Ph.D.
Jacob Raber, Ph.D.
Patrick Roberts, Ph.D.
Lee Robertson, Ph.D.
David Rossi, Ph.D.
Show-Ling Shyng, Ph.D.
Susan Smith, Ph.D.
Harold Spies, Ph.D.
Peter Steyger, Ph.D.
Rowland Taylor, Ph.D.
Laurence Trussell, Ph.D.
John Welsh, Ph.D.
William Woodward, Ph.D.

ROSALIND FRANKLIN UNIVERSITY
OF MEDICINE AND SCIENCE
Interdepartmental Neuroscience Graduate Program

Programs of Study

The Interdepartmental Neuroscience Graduate Program coordinates training among neuroscientists in five departments of the Chicago Medical School: Biochemistry and Molecular Biology, Cell Biology and Anatomy, Cellular and Molecular Pharmacology, Neuroscience, and Physiology and Biophysics. The objective of the program is to prepare Ph.D. and M.D./Ph.D. students for careers in research and teaching. The course of study is guided by the integration of molecular, cellular, systems, and clinical perspectives and offers students training opportunities in the following areas: drug addiction and drug-induced neuroplasticity; basal ganglia; neurodegeneration; Alzheimer's and Parkinson's diseases; stem-cell and gene therapy strategies for neuronal repair; neurobiology of learning and behavior; growth factors and their receptors; regulation of neuronal gene expression; ion channels and ion transport; membrane biophysics; cell volume regulation; molecular biology of neurotransmitters, neuropeptides, and their receptors; neuroendocrinology; neuropharmacology; and neuroproteomics.

The Ph.D. degree requires successful completion of required course work, a qualifying examination, an original research project, and the writing and defense of an acceptable dissertation. The combined program leading to the M.D./Ph.D. degree requires students to apply to the program through the medical school; after acceptance, they enter the Graduate School and complete all requirements for the Ph.D. degree prior to entering the clinical training required for the M.D. The basic medical science courses required for the M.D. degree can be completed during the graduate phase of the combined program. Four to five years are usually required to complete the Ph.D. and six to seven years for the combined degrees.

Opportunities for postdoctoral training in neuroscience also are available.

Research Facilities

The University offers state-of-the-art equipment, including a confocal microscopy facility equipped for fixed specimen and live cell imaging and computer-assisted stereology; laser microdissection facility with quantitative "real-time" RT-PCR system; electron microscopy facility; two-photon microscopy; X-ray crystallography facility for determination of the complete three-dimensional structure of proteins, RNA, and DNA; Midwest Proteome Center-MALDI and electrospray MS/MS mass spectrometry; protein separation and bioinformatics facilities; molecular biology facilities with thermal cyclers, DNA oligonucleotide synthesizer, and a phosphorimager; equipment for in vivo and in vitro electrophysiological recordings; equipment for in vivo microdialysis and electrochemistry; HPLC equipment; fluorescence microscopes; cell and tissue culture facilities; computer modeling and graphics lab; and facilities for animal housing, surgery, behavioral testing, and monitoring of hormone secretions.

Financial Aid

Support for graduate students comes from University fellowships as well as research grants. Stipends are competitive with other Chicago-area health science graduate programs.

Cost of Study

Tuition was $20,496 in 2005–06, but most students received a full tuition waiver.

Living and Housing Costs

Housing is available both on and off campus. Current average housing costs range from $625 to $825 per month, depending on the choice of accommodations.

Student Group

During the past academic year, the University enrolled 1,687 students in its four schools: the Chicago Medical School, the School of Graduate and Postdoctoral Studies, Scholl College of Podiatric Medicine, and the College of Health Professions. The neuroscience program operates within the School of Graduate and Postdoctoral Studies and is currently training 16 students.

Student Outcomes

In recent years, matriculating students have obtained postdoctoral fellowships and residencies at top universities, including Duke, Northwestern, Stanford, Yale, and the University of Chicago. Neuroscience students have also been successful in obtaining their own NIH funding in the form of individual predoctoral National Research Service Awards. This success testifies to a small but high-quality program in which students receive extensive individualized mentoring.

Location

The University is located close to Lake Michigan in the beautiful North Shore suburbs of Chicago. Excellent public transportation provides ready access to the cultural and scientific resources of Chicago and many recreational opportunities in northern Illinois and nearby Wisconsin.

The University and The School

The graduate school of the University was established in 1968 as an expansion of the Chicago Medical School, which was founded in 1912. In 1981, the graduate school departments moved into a new research building on the North Chicago campus. In 2004, the University adopted the name Rosalind Franklin University of Medicine and Science in honor of the underrecognized pioneer of DNA research. In 2005, a new research building was opened. The University is committed to a period of rapid growth in basic science research.

Applying

Entrance requirements include a baccalaureate degree from an accredited college or university, satisfactory scores on the Graduate Record Examinations (GRE), and three letters of recommendation. The following undergraduate courses are recommended: biology, general and organic chemistry, physics, calculus, cell and molecular biology, physiology, biological psychology, and neuroscience. First priority is given to applications completed by February 1. Later applications are considered if space is available. Formal application materials may be obtained from the Office of Graduate Admissions, Rosalind Franklin University of Medicine and Science, 3333 Green Bay Road, North Chicago, Illinois 60064-3095.

Correspondence and Information

Lise Eliot, Ph.D.
Director, Interdepartmental Ph.D. Program in Neuroscience
Rosalind Franklin University of Medicine and Science
3333 Green Bay Road
North Chicago, Illinois 60064
Phone: 847-578-3416
Fax: 847-578-8515
E-mail: lise.eliot@rosalindfranklin.edu

Rosalind Franklin University of Medicine and Science

THE FACULTY AND THEIR RESEARCH

Participating faculty members are from the Departments of Biochemistry and Molecular Biology, Cell Biology and Anatomy, Cellular and Molecular Pharmacology, Neuroscience, and Physiology and Biophysics. More detailed information on each faculty member can be found on departmental Web pages at http://www.rosalindfranklin.edu.

Marjorie A. Ariano, Professor of Neuroscience; Ph.D., UCLA, 1977. Molecular, cellular, and behavioral studies of dopamine depletion in early stages of Parkinson's disease.

Christopher Brandon, Associate Professor of Cell Biology and Anatomy; Ph.D., Yeshiva (Einstein), 1974. Mechanisms of neural computation in the visual systems of vertebrates and invertebrates.

Pastor R. Couceyro, Assistant Professor of Cellular and Molecular Pharmacology; Oregon Health Sciences, 1994; Cocaine- and amphetamine-regulated transcript (CART) peptide and CART peptide receptor characterization; regulation of gene expression in brain by drugs of abuse.

Lisa Ebihara, Associate Professor of Physiology and Biophysics: M.D./Ph.D., Duke, 1981. Structure and function of gap junctional proteins.

Lise Eliot, Assistant Professor of Neuroscience; Ph.D., Columbia, 1991. Plasticity and sex differences in brain and cognitive development.

William N. Frost, Professor and Chair of Cell Biology and Anatomy; Ph.D., Columbia, 1987. Neural mechanisms of decision making, pattern generation, prepulse inhibition, and learning in two invertebrate model systems: the marine mollusks *Aplysia* and *Tritonia*.

Sarah S. Garber, Associate Professor of Physiology and Biophysics, Ph.D., Brandeis, 1987. Anion channel regulation; physiological regulation of cell volume under normal and pathological conditions.

Marc J. Glucksman, Associate Professor of Biochemistry and Molecular Biology; Ph.D., Columbia, 1990. Structural neurobiology of processing enzymes involved in neurodegenerative/neuropsychiatric disorders and reproduction; proteomics of the brain.

Richard A. Hawkins, Professor of Physiology and Biophysics; Ph.D., Harvard, 1969. Cerebral energy metabolism; cerebral nutrition and transport characteristics of the blood-brain barrier; metabolic encephalopathies; regional cerebral function.

Xiuti Hu, Research Associate Professor of Cellular and Molecular Pharmacology; M.D., Chongqing Medical College (China), 1983; Ph.D., Wayne State, 1993. Regulation of ion channel function by dopamine and Ca^{2+} signaling; neurobiological and molecular mechanisms of cocaine addiction.

Donghee Kim, Professor of Physiology and Biophysics; Ph.D., Michigan State, 1982. Molecular physiology of potassium channels in excitable cells.

Michela Marinelli, Assistant Professor of Cellular and Molecular Pharmacology; Ph.D., Bordeaux II, 1997. Neurophysiological mechanisms of addiction.

Robert A. Marr, Assistant Professor of Neuroscience; Ph.D., McMaster, 1999. Gene therapy approaches to the treatment of Alzheimer's disease and other neurodegenerative disorders.

Gloria E. Meredith, Professor and Chair of Cellular and Molecular Pharmacology; Ph.D., Georgetown, 1983. Neuronal plasticity, cell death and inflammation in Parkinson's disease; role of neurotrophins in addiction.

Kenneth E. Neet, Professor and Chair of Biochemistry and Molecular Biology; Ph.D., Florida, 1965. Growth factors and their receptors; signal transduction; apoptosis; neurobiology; protein structure-function.

Monica Oblinger, Professor of Cell Biology and Anatomy; Ph.D., Purdue, 1981. Molecular mechanisms involved in the recovery from neuronal injury; cytoskeletal genes in axonal regeneration; gonadal steroids and neuroprotection.

Daniel A. Peterson, Associate Professor of Neuroscience; Ph.D., Otago (New Zealand), 1991. Neuronal cell death mechanisms and therapeutic gene delivery; neurogenesis in the adult mammalian CNS.

Darryl R. Peterson, Professor of Physiology and Biophysics; Ph.D., Illinois, 1973. Blood-brain barrier and stroke; diabetic nephropathy.

Judith A. Potashkin, Associate Professor of Cellular and Molecular Pharmacology; Ph.D., SUNY at Buffalo, 1985. Changes in gene expression and the regulation of pre-mRNA splicing that occur with drug addiction and Parkinson's disease.

Hector Rasgado-Flores, Associate Professor of Physiology and Biophysics; Ph.D., Center for Research and Advanced Studies; CINVESTAV (Mexico), 1984. Volume regulatory mechanisms and transport of calcium and magnesium in muscle and nerve cells.

Barry Roberts, Research Professor of Cellular and Molecular Pharmacology; Ph.D., Cambridge, 1968. Regeneration and functional recovery of the injured spinal cord.

Henry Sackin, Professor of Physiology and Biophysics; Ph.D., Yale, 1978. Electrolyte transport; mechanotransduction; volume regulation; cloned K channels; structure-function of ion channels.

Ann Snyder, Research Associate Professor of Cellular and Molecular Pharmacology; Ph.D., Illinois at Urbana-Champaign, 1971. Glial and neuronal glucose transport and metabolism and neurotoxic effects of alcohol.

Heinz Steiner, Associate Professor of Cellular and Molecular Pharmacology; Ph.D., Düsseldorf (Germany), 1989. Dopamine and opioid regulation of basal ganglia circuits; molecular mechanisms of psychostimulant addiction.

Grace E. Stutzmann, Assistant Professor of Neuroscience; Ph.D., NYU, 1999. Neuronal calcium signaling and mechanisms of calcium dysregulation in disease states, particularly Alzheimer's disease.

Thanos Tzounopoulos, Assistant Professor of Cell Biology and Anatomy; Ph.D., Oregon Health & Science, 1997. Mechanisms and roles of synaptic plasticity on sensory processing; electrophysiological and imaging studies of auditory circuitries.

Janice H. Urban, Associate Professor of Physiology and Biophysics; Ph.D., Loyola of Chicago, 1987. Molecular and physiological aspects of hypothalamic neuroendocrine function; stress and reproductive hormones.

D. Eric Walters, Professor of Biochemistry and Molecular Biology; Ph.D., Kansas, 1978. Computer-aided drug design; modeling membrane proteins; chemosensory transduction mechanisms.

Anthony R. West, Assistant Professor of Neuroscience; Ph.D., Wayne State, 1997. Role of nitric oxide and dopamine interactions in Parkinson's disease and schizophrenia; electrophysiological and neurochemical studies of striatal neuron activity.

Francis J. White, Professor of Cellular and Molecular Pharmacology; Ph.D., South Carolina, 1981. Electrophysiology of dopamine systems; mechanisms of cocaine addiction.

Marina E. Wolf, Professor and Chair of Neuroscience; Ph.D., Yale, 1986. Role of neuronal plasticity in drug addiction; interactions between dopamine and glutamate receptors.

STONY BROOK UNIVERSITY, STATE UNIVERSITY OF NEW YORK

Department of Psychology
Graduate Training Program in Biopsychology

Programs of Study

The Department of Psychology includes four graduate training programs: biopsychology, clinical, cognitive/experimental, and social and health. Students are admitted to one of these areas. The Graduate Training Program in Biopsychology offers courses and research training leading to the degree of Doctor of Philosophy in biopsychology.

The biopsychology program provides students with a broad methodological and conceptual foundation in the biological bases of behavior and cognition. The full-time faculty members in biopsychology, together with participating faculty members from other areas in the psychology department and from other neuroscience programs around the campus, offer a wide selection of research opportunities for students. The curriculum emphasizes the importance of having a broad knowledge of the theories and techniques from both psychology and the neurosciences.

Students in the biopsychology program are trained in research through close collaboration with a faculty mentor. Required courses include statistics, cognitive and behavioral neuroscience I and II, and neuroanatomy, as well as two additional courses within area courses and three outside of area courses. In cooperation with the Department of Neurobiology and Behavior, the Department of Psychiatry, and Brookhaven National Laboratory's Medical and Chemistry Departments, interdisciplinary training is offered in behavioral neuroscience. In conjunction with the Cognitive/Experimental Area, interdisciplinary training is available in cognitive science. A quantitative minor is available. Students and faculty members present their research in the Cognitive and Behavioral Neuroscience Colloquium, which is offered each semester. Additional requirements include a second-year paper and a specialties examination, with advancement to candidacy at the end of the third year, followed by the doctoral dissertation proposal and defense. Additional information about requirements and training opportunities may be obtained by writing to the Graduate Program Coordinator.

Research Facilities

The Biopsychology Area maintains active laboratories with state-of-the-art equipment for research and graduate training. The Area has its own facilities for human electrophysiology (64 electrodes), transcranial magnetic stimulation, anatomical and histochemical analyses, image analysis, animal housing, surgery, underflow perfusion table, and animal behavioral testing. The biopsychology laboratories also have access to a neuron-tracing system and electron microscopy. The Department of Psychology has access to two fMRI facilities for neuroimaging studies at nearby Brookhaven National Laboratory and at Stony Brook University Hospital. Ample office and laboratory space is available for all graduate students.

Financial Aid

Ph.D. students are normally admitted with financial support, which was at least $13,276 for the 2006–07 academic (nine-month) year. This funding is associated with teaching or research responsibilities. Biopsychology students making good progress receive additional summer funding from sources such as summer teaching assignments, work-study, and faculty research grants; summer support currently averages more than $4000. A fifth year of support is available on a competitive basis. Some students receive competitive fellowships from the W. Burghardt Turner Foundation, Frederick E. G. Valergakis Research Fund, and the National Institutes of Health (National Research Service Award).

Cost of Study

As described in the Financial Aid section, Ph.D. students in good standing receive full tuition waivers. Miscellaneous fees (including health insurance) are approximately $700.

Living and Housing Costs

Housing is available on campus in residence halls or off campus in private housing. Information about campus residences can be obtained from http://www.sunysb.edu/stuaff/residence/. Off-campus housing information can be found at http://naples.cc.sunysb.edu/FSA/fsa.nsf/pages/housing.

Student Group

The Department of Psychology is one of Stony Brook's largest graduate departments. More than 600 Ph.D. degrees have been awarded since the program began more than forty years ago. Recent demographics show that 77 percent of the students are women, 7 percent are from underrepresented groups, and 11 percent are international students. During the 2006–07 academic year, 16 students were enrolled in the Graduate Training Program in Biopsychology.

Location

Stony Brook is located on the North Shore of Long Island, in a region of beaches and small historic villages. It is 60 miles east of New York City, conveniently connected by the Long Island Railroad (which stops at the edge of campus). Nearby research facilities at Cold Spring Harbor and Brookhaven National Laboratories provide additional advantages for the scientific community.

The University

The Stony Brook University, the flagship campus of the SUNY system, is a world-class, student-centered research university. Stony Brook is one of the top fifteen research institutions in the U.S. and one of the top three public research universities (Graham and Diamond, 1997). Stony Brook's new responsibility for the management of Brookhaven National Laboratory is testimony to the high quality of its programs in science, engineering, and health sciences. The University has more than 20,000 students, including nearly 7,000 graduate students.

Applying

The application deadline is January 15 for fall admission. The GRE General Test is required; the Subject Test in psychology is optional. For more information, prospective students can visit http://www.psychology.sunysb.edu. Online applications are required. They may be submitted to the Graduate School at http://www.grad.sunysb.edu/applying/applying.htm. Students should note that the Department of Psychology requires an additional application page, available on the Web site or via postal mail.

Correspondence and Information

Graduate Program Coordinator
Graduate Training Program in Biopsychology
Department of Psychology
Stony Brook University, State University of New York
Stony Brook, New York 11794-2500

E-mail: psychgradprogram@notes.cc.sunysb.edu
Web site: http://www.psychology.sunysb.edu

Stony Brook University, State University of New York

THE FACULTY AND THEIR RESEARCH

Brenda J. Anderson, Associate Professor; Ph.D., Illinois, 1993. How experience (negative or positive) alters the brain; measuring synapse numbers and size and metabolic capacity following hormone administration, stress, or exercise.

Turhan Canli, Assistant Professor; Ph.D., Yale, 1993. Neural basis of personality and emotion; genetics and brain function.

Hoi-Chung Leung, Assistant Professor; Ph.D., Northwestern, 1997. Prefrontal and parietal function in human cognition; neural mechanisms underlying spatial information processing and eye movement control; FMRI applications in cognitive neuroscience.

John Robinson, Associate Professor and Head, Biopsychology Area; Ph.D., New Hampshire, 1991. Rodent models of learning and memory disorders; behavioral actions of neuropeptides.

Nancy K. Squires, Chairperson and Professor; Ph.D., California, San Diego, 1972. Neuropsychology; neurophysiological measures of sensory and cognitive functions of the human brain, both in normal and clinical populations.

Patricia Whitaker-Azmitia, Professor; Ph.D., Toronto, 1979. Animal models of autism and Down syndrome; serotonin and its role in brain development.

Associated Faculty in Other Stony Brook Departments

Anat Biegon, Ph.D., Senior Scientist (Medical Department, Brookhaven National Laboratory).

S. John Gatley, Ph.D., Professor (Pharmacology, Northeastern University); Visiting Scientist (Medical Department, Brookhaven National Laboratory).

Rita Goldstein, Ph.D., Assistant Scientist (Medical Department, Brookhaven National Laboratory).

Mary Kritzer, Ph.D., Associate Professor (Neurobiology and Behavior, Stony Brook University).

Lauren Krupp, M.D., Associate Professor (Neurology, Stony Brook University).

Martas Maczaj, M.D., Associate Professor (Psychiatry, Stony Brook University).

Lawrence P. Morin, Ph.D., Professor (Psychiatry, Stony Brook University).

Panayotis (Peter) Thanos, Ph.D., Assistant Scientist (Medical Department, Brookhaven National Laboratory).

Rex Wang, Ph.D., Professor (Psychiatry, Stony Brook University).

STONY BROOK UNIVERSITY, STATE UNIVERSITY OF NEW YORK

Graduate Program in Neuroscience

Program of Study

The Department of Neurobiology and Behavior at Stony Brook University offers a University-wide doctoral training program for students interested in careers in the neurosciences. Students can enter the Graduate Program in Neuroscience from their undergraduate training or through the M.D./Ph.D. program at Stony Brook University.

The graduate program provides broad education in neuroscience, experience in teaching, and the opportunity to pursue original doctoral research in thirty-two laboratories. The program faculty includes neuroscientists at SUNY Stony Brook, recently ranked number two in the Graham/Diamond study of public research universities; at Cold Spring Harbor Laboratory, a private research institute; and at Brookhaven National Laboratory, a multidisciplinary research facility of the U.S. Department of Energy.

Research Facilities

Neuroscience program faculty members on the Stony Brook University campus are located in a complex formed by the Life Sciences Building, the adjacent Centers for Molecular Medicine, and the Health Sciences Center. Because the program crosses departmental and institutional boundaries, program members interact regularly with neuroscience faculty members and students from numerous other departments at the University. The additional ties with Cold Spring Harbor Laboratory and Brookhaven National Laboratory further enrich the academic environment of research and scholarship in neuroscience.

Molecular neuroscience facilities provide for analysis of protein and DNA biochemistry, including microsequencing, peptide mapping, synthesis of oligonucleotides and peptides, cellular transfection, and production of transgenic animals. Wide-ranging facilities for cellular and integrative electrophysiology exist for studies on dissociated neurons, brain slice preparations, neurons in situ, and genetically engineered cells in culture. Imaging facilities permit anatomical reconstruction using light and electron microscopy, fluorescence measurements on both conventional and confocal microscopes, and brain mapping with MRI and PET.

Financial Aid

All doctoral candidates in the Graduate Program in Neuroscience receive a full-tuition scholarship and, in 2006–07, an annual stipend of $25,000. The stipend compensates students for living costs such as University fees, medical insurance, books, and room and board. Sources of funding include teaching and research assistantships, predoctoral National Service Research Awards, and individual faculty research grants.

Cost of Study

Students are responsible for University fees totaling approximately $700 per year and for the cost of books.

Living and Housing Costs

On-campus housing costs are approximately $6000. Private housing is available in surrounding communities. An automobile is advisable for those students living off campus.

Student Group

The University currently enrolls more than 2,000 doctoral students, of whom about 350 are enrolled in programs in biological or biomedical sciences. At present, the Graduate Program in Neuroscience has 32 Ph.D. candidates.

Student Outcomes

New graduates typically accept postdoctoral research appointments in neuroscience at academic institutions. Program graduates pursue neuroscience-related professions in universities, medical schools, and government and industrial research institutions.

Location

Stony Brook and Cold Spring Harbor are located on the scenic North Shore of Long Island. A short distance southeast, at the island's geographic center, is Brookhaven National Laboratory. Recreational and cultural opportunities can be found at all three locations. Surrounding villages provide a variety of additional resources. To the south and east are beautiful ocean beaches. To the west is the metropolitan experience of New York City, which is easily accessible by train or car.

The University

Stony Brook was established in 1957 as part of the State University of New York system and is now recognized as one of the nation's finest public universities. It is classified by the Carnegie Foundation as a Doctoral/Research University–Extensive institution, which is the foundation's highest distinction; it is granted to fewer than 2 percent of all colleges and universities nationwide. Stony Brook is a member of the Association of American Universities.

Applying

Applications to the Graduate Program in Neuroscience can be considered at any time, but all admitted students begin in the fall semester. Application forms can be obtained by writing or calling the department or from the Graduate School's Web site. Entrance requirements are a bachelor's degree from an accredited college or university, one year of calculus and physics, inorganic chemistry, and organic chemistry, plus demonstrated proficiency in the biological sciences. Deficiencies in course requirements do not preclude admission. Strong candidates have undergraduate grades of B+ or better. Special consideration is given to promising applicants. Other requirements include satisfactory scores on the GRE and three letters of recommendation from people who are familiar with the applicant's academic background. A TOEFL score of 550 or better (215 or better on the computerized test) is required of international students. An application processing fee of $60 must be submitted with the application. Prospective students should apply at http://www.grad.sunysb.edu/applying/applying.htm.

Correspondence and Information

Graduate Program in Neuroscience
Department of Neurobiology and Behavior
Stony Brook University, State University of New York
Stony Brook, New York 11794-5230

Phone: 631-632-8630
Fax: 631-632-6661
Web site: http://www.hsc.stonybrook/som/neurobiology/

Stony Brook University, State University of New York

THE FACULTY AND THEIR RESEARCH

Biophysics and Cellular Neurobiology
Paul R. Adams, Professor; Ph.D., London. Models of synaptic learning; neocortical design.
Peter Brink, Professor; Ph.D., Illinois. Electrotonic synapses.
Gary G. Matthews, Professor; Ph.D., Pennsylvania. Cellular and molecular neurobiology of the retina.
Stuart McLaughlin, Professor; Ph.D., British Columbia. Biophysics of signal transduction.
Lonnie Wollmuth, Associate Professor; Ph.D., Washington (Seattle). Molecular mechanisms of synaptic transmission.

Molecular Neurobiology
Marian J. Evinger, Associate Professor; Ph.D., Washington (Seattle). Neural regulation of gene expression.
Michael Frohman, Professor; Ph.D., M.D., Pennsylvania. Regulation of exocytosis and cell shape by signaling proteins.
Simon Halegoua, Professor; Ph.D., SUNY at Stony Brook. Neuronal growth factor signaling and the control of phenotype and survival.
Maurice Kernan, Associate Professor; Ph.D., Wisconsin. *Drosophila* mechanosensory transduction; differentiation of sensory cilia and sperm.
Joel Levine, Professor; Ph.D., Washington (St. Louis). Molecular biology of nerve regeneration; nerve-glia interactions.
David McKinnon, Professor; Ph.D., Australian National. Molecular control of neuron firing properties.
Howard Sirotkin, Assistant Professor; Ph.D., Yeshiva (Einstein). Molecular genetics of vertebrate neural patterning.
Styliani-Anna Tsirka, Associate Professor; Ph.D., Thessaloniki. Neuronal-microglial interactions in the physiology and pathology of the central nervous system.

Integrative and Behavioral Neurobiology
Brenda J. Anderson, Associate Professor; Ph.D., Illinois. Neuroanatomical and metabolic plasticity.
Craig Evinger, Professor; Ph.D., Washington (Seattle). Motor control and learning; movement disorders.
Mary Kritzer, Associate Professor; Ph.D., Yale. Sex differences in cortical microcircuitry.
Lorne M. Mendell, Distinguished Professor; Ph.D., MIT. Functional effects of neurotrophins in pain and segmental reflex pathways.
Lawrence P. Morin, Professor; Ph.D., Rutgers. Neural control of mammalian circadian rhythms.
Irene C. Solomon, Associate Professor; Ph.D., California, Davis. Neural control of respiratory and cardiovascular function.
Stephen Yazulla, Professor; Ph.D., Delaware. Synaptic circuitry of the vertebrate retina.

Cold Spring Harbor Laboratories Faculty
Hollis Cline, Professor; Ph.D., Berkeley. Molecular control of neuronal plasticity.
Grigori Enikolopov, Associate Professor; Ph.D., USSR Academy of Sciences. Nitric oxide; neuron differentiation; survival.
Z. Josh Huang, Assistant Professor; Ph.D., Brandeis. Development and plasticity of the neocortical GABAergic circuits.
Zachary Mainen, Assistant Professor; Ph.D., California, San Diego. Neural coding and computations underlying rodent olfactory-guided behavior.
Roberto Malinow, Professor; M.D., NYU; Ph.D., Berkeley. Synaptic transmission and plasticity.
Timothy Tully, Professor; Ph.D., Illinois. Genetics of memory formation in *Drosophila*.
Anthony Zador, Assistant Professor; M.D., Ph.D., Yale. How the cortex solves the cocktail party problem.
Yi Zhong, Associate Professor; Ph.D., Berkeley. Molecular control of neuronal plasticity.

Brookhaven National Laboratory Faculty
Stephen L. Dewey, Senior Scientist; Ph.D., Iowa. Medical imaging and functional neurotransmitter interactions in substance abuse.
Andrew Gifford, Scientist; Ph.D., St. Andrews (Scotland). Pharmacology of brain receptors and neurotransmitter release.
Louis Peña, Scientist; Ph.D., UCLA. Cellular and molecular mechanisms of radiation sensitivity.
Peter Thanos, Neuroscientist; Ph.D., Eastern Virginia Medical School. CNS mechanisms of addiction.

TEXAS A&M UNIVERSITY

Faculty of Neuroscience

Programs of Study

The Faculty of Neuroscience offers a wide variety of interdisciplinary study programs that can be tailored to suit individual student needs in neuroscience training as part of the M.S. and Ph.D. degree programs of conventional departments. At Texas A&M, there is a special opportunity for collaboration among scientists studying different areas of neuroscience in five colleges (Agriculture, Liberal Arts, Medicine, Science, and Veterinary Medicine). Individual programs can be designed with emphasis on research training in areas such as molecular neurobiology, neuroanatomy, developmental neurobiology, neurochemistry, neuroendocrinology, neurogenetics, neuropharmacology, neurotoxicology, neurophysiology, physiological psychology, biological rhythms, and ethology.

Course requirements vary according to the admitting department, the degree program, the area of research specialization, and the experience and goals of the individual student. Each degree plan is developed by the student in consultation with a faculty supervisor and an advisory committee.

Research Facilities

The research facilities of each of the 34 neuroscientists in the program are available for use by students. These laboratories contain state-of-the-art equipment for neurophysiology, molecular biology, protein chemistry, chromatography, and behavioral monitoring; facilities of special value in neuroscience training include the Data Processing Center, the Gene Technologies Laboratory, the Laboratory Animal Resource and Research Facility, the Institute of Statistics, the Electron Microscopy Center, the Biological MRI Program, the Neuroendocrine/Neurochemical Core Facility, the Image Analysis Center, the Biotechnology Support Laboratory, and the Laboratories for Invertebrate Neuroendocrine Research.

Financial Aid

Fellowships, teaching assistantships, and research assistantships from $19,000 to $23,000 per year are available in the home departments of individual faculty members. Many appointments allow for a remission of out-of-state tuition for nonresident students.

Cost of Study

The 2005–06 tuition for Texas residents was $187 per credit hour, while nonresident tuition was $463 per credit hour. Students should expect additional expenses of $1000 to $1500 per semester for books and laboratory and miscellaneous fees (most of which are defrayed by fellowships).

Living and Housing Costs

Students living off campus typically pay between $500 and $800 per month for food and rent. The monthly rent for University-owned apartments, available for married students, ranges from $345 to $529. The University housing office can provide further information.

Student Group

Texas A&M University has a total enrollment of about 45,000 students, of whom 8,000 are graduate students. There are currently 30 graduate students in Faculty of Neuroscience laboratories.

Student Outcomes

Most recent graduates have accepted postdoctoral research positions at universities throughout the United States.

Location

College Station and the adjoining city of Bryan have a population of 120,000 and are located within 100 miles of both Austin and Houston and within 180 miles of Dallas, San Antonio, and the Gulf Coast. Waterskiing, fishing, hunting, boating, golf, tennis, softball, and horseback riding are all popular sports in the area. The Sunbelt location provides ample opportunity for students to participate in these activities. The community offers excellent schools, hospitals, numerous licensed day-care facilities, a regional shopping mall, and a range of restaurants and cultural activities. The two cities maintain forty parks, six swimming pools, four golf courses, and numerous tennis courts.

The University

Texas A&M University, the state's oldest public institution of higher learning, was founded in 1876 as a land-grant college. It is one of sixteen sea-grant universities in the nation and was designated a space-grant university in 1989. Texas A&M University is the home of the George Bush Presidential Library. In 1998, funded research at Texas A&M was $366.8 million, and the University ranked seventh in the National Science Foundation's ranking of colleges and universities based on total research expenditures. The University consists of ten academic colleges. The main campus of the University is situated on 5,200 acres. Its recreation facilities include an eighteen-hole golf course, tennis courts, indoor and outdoor swimming pools, and a polo field. Texas A&M University is a member of the Big Twelve Athletic Conference, and its intramural offerings constitute one of the best all-around sports programs in the country.

Applying

Applications and additional program information can be obtained from the address listed in the Correspondence and Information section or directly from the faculty. For each of the participating departments, applicants are accepted based on their undergraduate record, letters of recommendation, and Graduate Record Examinations (GRE) scores. The TOEFL is required of all international students whose native language is not English.

Prospective applicants are encouraged to contact members of the graduate faculty and to visit the campus to meet with faculty members and graduate students.

Correspondence and Information

Neuroscience Training Program
Faculty of Neuroscience/Graduate Advising Office
Department of Biology
Texas A&M University
College Station, Texas 77843-3258
E-mail: graduate@mail.bio.tamu.edu
Web site: http://www.bio.tamu.edu/neuro/

Texas A&M University

THE FACULTY AND THEIR RESEARCH

L. C. Abbott, Associate Professor of Veterinary Integrative Biosciences; Ph.D., Washington (Seattle), 1982; D.V.M., Washington State, 1988. Morphological, biochemical, molecular, and electrophysiological changes in the central nervous system during development of neurological disorders in tottering mice.

G. Alexander, Assistant Professor of Psychology; Ph.D., McGill, 1991. Neuropsychology; child development; human cognition.

R. Ballestero, Assistant Professor of Biology; Ph.D., Michigan, 1998. Study of nerve regeneration in the central nervous system of vertebrates; characterization of molecular mechanisms in the process of apoptosis.

L. R. Berghman, Professor of Poultry Science; Ph.D., Leuven (Belgium), 1988. Environmental psychophysiology; psychological responses to the stimulated environment.

J. Bizon, Assistant Professor of Psychology; Ph.D., California, Irvine, 1998. Neurobiological changes that contribute to individual differences in cognitive aging.

H. Bortfeld, Assistant Professor of Psychology; Ph.D., SUNY at Stony Brook, 1998. Development and learning of language and linguistics in infants.

P. Brandt, Assistant Professor of Neurosciences and Experimental Therapeutics; Ph.D., Kentucky, 1990. Calcium homeostasis and calcium-dependent regulation of steroid receptors.

G. R. Bratton, Professor of Veterinary Integrative Biosciences; D.V.M., 1966, Ph.D., 1977, Texas A&M. Mechanisms by which lead enters the body and produces toxicity; nutritional influences on lead intoxication; heavy-metal effects on reproductive function.

G. Carney, Assistant Professor of Biology; Ph.D., Georgia, 1998. Behavioral neurobiology; characterization of the target locus of a behavioral genetic hierarchy; *Drosophila* genetics.

V. M. Cassone, Professor of Biology; Ph.D., Oregon, 1983. Behavioral neurobiology; neurobiology of biological clocks; mechanisms of melatonin action.

A. Cepeda-Benito, Associate Professor of Psychology; Ph.D., Purdue, 1994. Theories and psychological constructs associated with substance use disorders, with emphasis on animal models of drug addiction, drug and food cravings, and prevention of drug use.

W.-J. A. Chen, Associate Professor of Neuroscience and Experimental Therapeutics; Ph.D., SUNY at Binghamton, 1992. Neurotoxicology; teratology; developmental psychobiology.

Y. Choe, Assistant Professor of Computer Science; Ph.D., Texas at Austin, 2001. Neural networks; self-organizing maps; influence of feature co-occurrence statistics, cortical structure, and functional performance; computational neuroscience.

S. Chopin, Professor of Biology; Ph.D., LSU Medical Center, 1974. Effects of endocrine disruptors during embryological development; microbial action of herbal remedies.

T. Cudd, Associate Professor of Veterinary Physiology and Pharmacology; D.V.M., Tennessee, 1982; Ph.D., Florida, 1992. Reflex control of endocrine and cardiovascular systems in the adult and fetus; control of the timing of parturition; fetal alcohol syndrome; eicosanoids in the brain.

S. Datta, Associate Professor of Biochemistry and Biophysics; Ph.D., California, San Diego, 1987. Development of the central nervous system in *Drosophila melanogaster.*

W. L. Dees, Professor of Veterinary Integrative Biosciences; Ph.D., Texas A&M, 1984. Molecular and physiological approaches for assessing neuroendocrine factors controlling and/or affecting the onset of female puberty.

D. P. Dohrman, Assistant Professor of Neuroscience and Experimental Therapeutics; Ph.D., Iowa, 1993. Cellular mechanisms of drug addiction; alcohol and nicotine interactions; signal transduction.

D. J. Earnest, Associate Professor of Neuroscience and Experimental Therapeutics; Ph.D., Northwestern, 1984. Neurobiology of mammalian circadian rhythms and their regulation by light-dark signals.

S. Eitan, Assistant Professor of Psychology; Ph.D., Weizmann (Israel), 1997. Opioid receptors and drugs of abuse.

R. Finnell, Professor and Director, Institute of Biosciences and Technology; Ph.D., Oregon Health Sciences, 1980. Interaction between specific genes and environmental toxicants as they influence normal embryonic development.

G. D. Frye, Professor of Neuroscience and Experimental Therapeutics; Ph.D., North Carolina, 1977. Neuropharmacology of ethanol intoxication, tolerance, and dependence; fetal alcohol syndrome; GABA receptor adaptation in the regulation of neuronal excitability; brain slice electrophysiology.

L. R. Garcia, Assistant Professor of Biology; Ph.D., Caltech, 1996. Male mating behavior in the *C. elegans* as a model to understand integration and regulation of chemosensory, mechanosensory, and motor outputs; genetic studies of sex-specific behavior.

J. B. Gelderd, Associate Professor of Neuroscience and Experimental Therapeutics; Ph.D., Florida, 1972. Pathophysiological processes occurring following spinal cord injury; methods to promote neural regeneration.

J. W. Grau, Professor of Psychology; Ph.D., Pennsylvania, 1985. Learning and memory, spinal plasticity, and pain modulation.

W. H. Griffith, Professor of Neuroscience and Experimental Therapeutics; Ph.D., Texas Medical Branch, 1980. Neurobiology of aging; cellular basis of epilepsy; brain slice electrophysiology and pharmacology; patch-clamp and single-channel recording of voltage and transmitter-activated currents.

M. Gonzales-Garcia, Associate Professor of Chemistry; Ph.D., Autonoma (Madrid), 1991. Regulation of apoptosis or programmed cell death (PCD) by members of the Bd-2 family of proteins; characterization of nerve regeneration–related proteins in the zebrafish's optic nerve.

P. Hardin, Professor of Biology; Ph.D., Indiana, 1987. Feedback loops in gene expression; circadian timekeeping mechanisms in *Drosophila;* neurobiology of animal behavior.

P. G. Harms, Professor of Animal Science; Ph.D., Purdue, 1969. Reproductive physiology; endocrine regulation of pregnancy.

C. Ketcham, Assistant Professor of Health and Kinesiology; Ph.D., Arizona, 2003. Motor control and the neuroscience of movement regulation.

G. Ko, Assistant Professor of Veterinary Integrative Biosciences; Ph.D., Northeastern Ohio College of Medicine, 1996. Circadian modulation of retinal cone ion channels.

K. Kotrla, Associate Professor of Psychiatry and Behavioral Sciences; M.D., Texas A&M, 1989. Utilizing neuroimaging techniques to investigate the functioning of neural networks in neuropsychiatric disorders.

J. Leibowitz, Professor of Microbial and Molecular Pathogenesis; M.D./Ph.D., Yeshiva (Einstein), 1975. Replication and gene expression of murine coronaviruses; molecular basis of pathogenesis of murine coronavirus-induced hepatitis and demyelinating disease.

J. Levine, Clinical Assistant Professor, Veterinary Small-Animal Clinical Sciences; D.V.M., Cornell, 2001. Neurology; neuropathy; central vestibular disease; granulomatous meningoencephalomyelitis.

J. Li, Assistant Professor of Veterinary Integrative Biosciences; Ph.D., Hawaii, 1997. Cellular and molecular mechanisms underlying cell death and survival; redox regulation of signaling molecules in cell death.

R. Lints, Assistant Professor of Biology; Ph.D., Melbourne, 1993. Developmental neurobiology of neurotransmitter systems in *C. elegans.*

T. Lints, Assistant Professor of Biology; Ph.D., Melbourne, 1993. Learning and memory; developmental neurobiology of birdsong.

E. Massa, Associate Professor of Biology; Ph.D., Michigan, 1985. *Drosophila* ion channel genetics; molecular determinants of membrane excitability; hippocampal neuron calcium-regulated gene expression.

B. McCormick, Professor Emeritus of Computer Science; Ph.D., Harvard, 1955. Construction of anatomically correct models of mouse brain networks.

M. W. Meagher, Professor of Psychology; Ph.D., North Carolina at Chapel Hill, 1989. Pain modulation; animal models of anxiety and mood disorders; learning and memory.

R. C. G. Miranda, Associate Professor of Neuroscience and Experimental Therapeutics; Ph.D., Rochester, 1989. Roles of gonadal estradiol-17β and neurotrophins in the GABAergic neurons of the cerebral cortex; development of circuits involved in cognitive and affective behaviors.

J. R. Nation, Professor of Psychology; Ph.D., Oklahoma, 1974. Neurochemical and neurobehavioral effects associated with environmental pollutants and drugs of abuse.

M. Packard, Professor of Psychology; Ph.D., McGill, 1991. Neurobiological bases of memory.

T. Pankiw, Assistant Professor of Entomology; Ph.D., Simon Fraser, 1996. Pheromone and genetic components of honeybee behavior.

B. Perkins, Assistant Professor of Biology; Ph.D., Baylor College of Medicine, 2000. Photoreceptor structure and development; vertebrate eye development; zebrafish genetics.

C. R. Reynolds, Professor of Educational Psychology; Ph.D., Georgia, 1978. Neuropsychology of memory in children; relationship of affective disorder to neuropsychological processing deficits.

C. A. Riccio, Professor of Educational Psychology; Ph.D., Georgia, 1993. Attention deficit hyperactivity disorder (ADHD); pediatric neuropsychology; individual assessment; language disorder.

I. S. Russell, Professor of Psychiatry and Behavioral Sciences; Ph.D., Indiana, 1961. Neural plasticity and mechanisms of learning and memory; recovery of visual function following cortical frontal eye-field lesions.

B. Setlow, Assistant Professor of Psychology; Ph.D. California, Irvine, 1998. Investigating role of nucleus accumbens and related brain systems in associative learning and goal-directed behavior; investigating interactions between drug addiction and learning/motivation.

M. Smotherman, Assistant Professor of Biology; Ph.D., UCLA, 1998. Comparative animal neurophysiology; auditory-feedback control of temporal call patterns in echolocating bats.

F. Sohrabji, Associate Professor of Professor of Neuroscience and Experimental Therapeutics; Ph.D., Rochester, 1991. Neurobiology of aging; role of gonadal hormones (estrogen) in neural repair; neural inflammation; maintenance of the blood-brain barrier.

G. Stoica, Professor of Veterinary Pathobiology; Ph.D., Michigan State, 1984. Mechanisms of retroviral-induced neurodegeneration.

R. Storts, Professor of Veterinary Pathobiology; D.V.M., Ohio State, 1957; Ph.D., Ohio State, 1966. Comparative neuropathology, with special interest in demyelinating and neurodegenerative diseases.

L. G. Tassinary, Professor of Architecture; Ph.D., Dartmouth, 1985; J.D., Boston College, 2003. Reproductive physiology; endocrine regulation of pregnancy.

E. Tiffany-Castiglioni, Professor of Veterinary Integrative Biosciences; Ph.D., Texas Medical Branch, 1979. Cellular mechanisms of neurotoxicity; functions of neuroglia; astroglial response to disease and trauma; cellular mechanisms of epilepsy.

P. J. Wellman, Professor of Psychology; Ph.D., Iowa State, 1980. Neurochemical basis for feeding and food intake.

G. B. Wells, Assistant Professor of Molecular and Cellular Medicine; Ph.D., Chicago, 1989. Role of protein structure in neurological diseases; clinical neuropathology.

R. D. Wells, Professor of Biosciences and Technology; Ph.D., Pittsburgh, 1964. DNA structure; triplet repeats; human hereditary neuromuscular diseases.

C. J. Welsh, Professor of Veterinary Integrative Biosciences; Ph.D., London, 1981. Neurotropic viruses; autoimmune diseases of nervous system cerebrovascular endothelial cells; Theiler's virus as a model for multiple sclerosis.

T. Wilcox, Associate Professor of Psychology; Ph.D., Arizona, 1993. Auditory information and object individuation in infancy.

U. Winzer-Serhan, Assistant Professor of Neuroscience and Experimental Therapeutics; Ph.D., Bremen (Germany), 1989. Nicotinic receptors and their influence on the developing brain.

K. Young, Associate Professor of Psychiatry and Behavioral Sciences; Ph.D., Texas, 1990. Anatomical and genetic studies of neurobasis of schizophrenia, Alzheimer's disease, and Parkinson disease.

D. Zimmer, Associate Professor of Veterinary Pathobiology; Ph.D., Baylor College of Medicine, 1983. Neurodegenerative diseases; oligodendrocyte pathobiology; interactions between p53 and S100 proteins.

M. J. Zoran, Associate Professor of Biology; Ph.D., Iowa State, 1987. Developmental neurobiology; molecular determinants of synaptogenesis and synaptic specificity; mechanisms of synaptic plasticity.

THE TEXAS A&M UNIVERSITY SYSTEM HEALTH SCIENCE CENTER

College of Medicine
Department of Neuroscience and Experimental Therapeutics

Programs of Study

The Department of Neuroscience and Experimental Therapeutics (NExT) participates in an interdisciplinary graduate program in the medical sciences that leads primarily to the Ph.D. degree. A distinctive feature of the program is broadly based instruction in the medical sciences with an emphasis in neuroscience and molecular mechanisms, including drug and toxin action. This interdisciplinary approach provides the strong conceptual framework in molecular, cellular, and systems-level biomedical sciences and cutting-edge research methods essential to the modern scientist. The Department's objective is to train students to become medical scientists who can make significant contributions to research in their chosen field, including pursuing professional careers as teachers and researchers in academic, industrial, or governmental positions.

The Ph.D. program in medical sciences requires a minimum of 96 semester hours composed of at least 32 hours of formal course work. The first-year program for each student is planned in consultation with a graduate adviser, who takes into account the student's previous training and interests. Typical graduate-level course work focuses on interesting topics in molecular, cellular, and systems biology related to medical sciences as well as laboratory rotations, statistics, and seminars related to the student's research interest. Courses in the second year include advanced topics in molecular, cellular, developmental, and structural-functional neuroscience and molecular mechanisms of drug and toxin action, as well as central nervous system pathobiology. Students are also encouraged, with the consent of their advisers, to enroll in several elective special-topics courses or audit courses that are related to their own developing interests. If interested, graduate students may also gain teaching experience by assisting in basic sciences courses for medical students. Cooperative programs with other departments in the College of Medicine, Texas A&M University, and the Texas A&M University System Health Science Center are also possible.

Research Facilities

The Department of Neuroscience and Experimental Therapeutics is located in the Joe H. Reynolds Medical Building on the Texas A&M University campus. The Medical Building is a modern, four-story structure adjacent to the Medical Sciences Library. Graduate students and postdoctoral fellows have the opportunity to work in faculty laboratories with specialized state-of-the-art equipment. Shared departmental resources include a quantitative image data acquisition and display system, gamma and beta scintillation counters, microplate readers, and other histological and molecular biology facilities. Also available is a state-of-the-art microarray facility for the fabrication of custom arrays and high-throughput data analysis.

Financial Aid

Assistantships and graduate student fellowships are available from the College of Medicine, the Graduate College, and private sources. Most full-time graduate students receive assistantships.

Cost of Study

In 2005–06, state minimum, graduate, and designate tuitions for Texas residents were expected to be $150 per semester credit hour. For nonresidents, state minimum tuition was expected to be $426 per semester credit hour. Nonresident students who hold graduate assistantships pay the same tuition as Texas residents. Students should expect additional expenses of $2500 to $3000 per semester for books and laboratory and miscellaneous fees. Students should visit the Web site at http://VPFN.tamu.edu/SFS/ for current information on the cost of study.

Living and Housing Costs

The cost of living is low, compared to the national average, and allows graduate students on limited incomes to maintain an increased standard of living. In 2004–05, students living off campus paid about $1000 each month for food, rent, and utilities. The monthly rent of the Texas A&M University–owned apartments for married students ranged from $345 to $529 plus the cost of electricity. Single graduate students may share a University apartment. The Texas A&M University Housing Office can supply additional information. Students can also visit the Web site at http://VPFN.tamu.edu/SFS for current information on the cost of living.

Student Group

The Texas A&M University System Health Science Center, College of Medicine, has 102 graduate students enrolled, with women representing 50 percent of the student body. Students come from all regions of the United States and a number of other countries. Graduates of the program have gone on to prestigious postdoctoral fellowships that have led to both academic and research positions.

Location

Texas A&M University System Health Science Center, College of Medicine, is located in College Station, Texas. The twin cities of Bryan and College Station have a combined population of approximately 120,000. The communities offer good school systems, a regional shopping mall, and a variety of other retail outlets. Students can take advantage of the cultural and intellectual activities associated with an academic community. The two cities maintain forty parks, six swimming pools, a golf course, and numerous tennis courts. Also, the Gulf Coast is approximately 150 miles from College Station. The nearly year-round warm climate of the area makes College Station a desirable place to live.

The University

The College of Medicine is located on the Texas A&M University campus. Texas A&M University is the state's oldest public institution of higher learning and was founded in 1876 as a land-grant college. It is one of sixteen sea-grant colleges in the nation and in 1989 became a space-grant college. It is also the home of the George Bush Presidential Library. The main campus of Texas A&M University is situated on 5,200 acres. Its recreation facilities include an eighteen-hole golf course, tennis courts, indoor and outdoor swimming pools, and a polo field. Texas A&M University is a member of the Big Twelve Athletic Conference, and its intramural offerings constitute one of the best all-around sports programs in the country.

Applying

Application forms for admission to the College of Medicine graduate program and financial support and additional information about the department can be obtained from the addresses in the Correspondence and Information section or online at http://www.gsbs.tamhsc.edu/admissions.html. Applicants are judged on the basis of their undergraduate record, scores on the verbal and quantitative portions of the Graduate Record Examinations (GRE) General Test, and letters of recommendation.

Correspondence and Information

Graduate Student Advisor
Department of Neuroscience and Experimental Therapeutics
College of Medicine
The Texas A&M University System Health Science Center
College Station, Texas 77843-1114
Phone: 979-845-4913
Fax: 979-845-0790
E-mail: sohrabji@medicine.tamhsc.edu
Web site: http://medicine.tamhsc.edu/basic_sciences/net

Office of Graduate Studies
College of Medicine
The Texas A&M University System Health Science Center
College Station, Texas 77843-1114
Phone: 979-845-0370

The Texas A&M University System Health Science Center

THE FACULTY AND THEIR RESEARCH

Research in the Department focuses on several broad areas in neuroscience: brain development and aging, neuroteratology, cellular and molecular basis of drug addiction, ocular pharmacology, neurodegeneration, and circadian rhythms.

Paul C. Brandt, Assistant Professor; Ph.D., Kentucky, 1990. Alterations of subcellular signal transduction in disease; senile dementia; autoimmune dry eye syndrome; plasma membrane Ca2+ pumps in calcium homeostasis and signal transduction; nitric oxide regulation and toxicity.

Wei-Jung A. Chen, Associate Professor; Ph.D., SUNY at Binghamton, 1992. Effects of substance abuse (alcohol, cocaine, and nicotine, for example) on the developing brain; polydrug interactions on the brain and cognitive development; fetal alcohol syndrome; use of three-dimensional stereological cell counting techniques, immunohistochemistry, radioimmunoassay, high-performance liquid chromatograph, gas chromatograph, and behavioral assessments.

George C. Y. Chiou, Professor; Ph.D., Vanderbilt, 1967. Treatment of uveitis and systemic inflammation with novel interleukin-1 blockers; systemic delivery of peptide drugs through the eyes; treatment of open-angle glaucoma and ischemic retinopathy with dopamine antagonists; action mechanisms of nitric oxide precursors for ischemic stroke and ocular hypertension; ocular pharmacology of natural products isolated from medicinal plants.

Douglas P. Dohrman, Assistant Professor; Ph.D., Iowa, 1993. Cellular and molecular mechanisms of alcohol and drug addiction; effects of alcohol, nicotine, and other drugs on brain development and cellular function; PKA and PKC signal transduction pathways; use of primary and immortalized cell cultures, immunocytochemistry, in vivo microdialysis, HPLC, and molecular biological techniques.

David J. Earnest, Professor; Ph.D., Northwestern, 1984. Neurobiology of the mammalian circadian clock in the suprachiasmatic nucleus (SCN) and its regulation by light-dark cycles; role of neurotrophins in use of multidisciplinary approaches to study the cellular and molecular mechanisms for normal SCN pacemaker function and the deterioration of circadian timekeeping that occurs in aging; Alzheimer's disease and an animal model for fetal alcohol syndrome (FAS); role of molecular elements of the circadian clockworks in toxin metabolism and mammary gland carcinogenesis.

Gerald D. Frye, Shelton Professor; Ph.D., North Carolina at Chapel Hill, 1977. Defining cellular and molecular mechanisms of ethanol neurotoxicity in fetal alcohol syndrome spectrum disorder; specifically addressing how intoxication disrupts synapse formation of GABAergic circuits to impair cognitive brain function, using patch clamp recordings in brain slices or in dispersed primary neuronal cell cultures.

John B. Gelderd, Professor and Interim Head; Ph.D., Florida, 1972. Pathophysiological processes occurring following spinal cord injury: exploration of methods to promote neural regeneration, reduce the destruction of spinal neuropil adjacent to a lesion, alleviate permanent paralysis, and promote integration and growth of transplanted embryonic nervous tissue.

William H. Griffith, Professor; Ph.D., Texas Medical Branch, 1980. Calcium homeostasis and brain aging; patch clamp recording of calcium currents and intracellular fluorescent calcium measurements; neuropharmacology of basal forebrain cholinergic neurons; behavioral characterization of cognitive status during aging; single-cell RT-PCR.

Rajesh C. Miranda, Associate Professor; Ph.D., Rochester, 1989. Role of cytokine-mediated intracellular signal transduction pathways and transcriptional regulation on the survival and differentiation of specific populations of neurons within the developing cerebral cortex; molecular biological, histological, tissue, and cell-culture techniques to elucidate mechanisms that regulate neuronal differentiation.

Joan E. Quarles, Assistant Professor; M.S., Michigan State, 1969. Transplantation of embryonic neural tissue; production of teaching materials using three-dimensional computer imaging.

Farida Sohrabji, Associate Professor; Ph.D., Rochester, 1991. Neurobiology of aging; steroid hormone influences on the aging brain and blood-brain barrier; elucidation of the role of estrogen and estrogen receptors; use of forebrain injury models, transgenics, and tissue-culture systems to determine the mechanisms of hormonal action on neurons, glia, and endothelial cells.

James R. West, Professor Emeritus; Ph.D., California, Irvine, 1975. Fetal alcohol syndrome and factors that influence the risk and severity of brain damage resulting from alcohol and other drugs of abuse during development; immunocytochemistry; stereological cell-counting techniques and other neurobiological measurements used to determine structural and functional deficits and their underlying mechanisms; rodent and ovine model systems.

Ursula Winzer-Serhan, Assistant Professor; Ph.D., Bremen (Germany), 1989. Neuropharmacology of nicotine; behavioral, molecular, and anatomical techniques to study the effects of developmental nicotine exposure in a postnatal rat model; specifically addressing how chronic developmental nicotine exposure translates into long-term behavioral and cognitive deficits.

Selected Faculty Publications

Brandt, P. C., and T. C. Vanaman. Studies of calmodulin-dependent regulation. *Methods Mol. Biol.* 284:111–27, 2004.

Beauregard, C., and **P. C. Brandt.** Peroxisome proliferator-activated receptor agonists inhibit interleukin-1 beta-mediated nitric oxide production in cultured lacrimal gland acinar cells. *J. Ocul. Pharm. Ther.* 19:579–87, 2003.

Grisel, J. J., and **W.-J. A. Chen.** Antioxidant pretreatment does not ameliorate alcohol-induced Purkinje cell loss in the developing rat cerebellum. *Alcohol. Clin. Exp. Res.* 29:1223–9, 2005.

Smith, A. M., D. R. Zeve, J. J. Grisel, and **W.-J. A. Chen.** Neonatal alcohol exposure increases malondialdehyde (MDA) and glutathione (GSH) levels in the developing cerebellum. *Brain Res. Dev. Brain Res.* 160:231–8, 2005.

Zou, Y., X. R. Xu, and **G. C. Chiou.** Effect of interleukin-1 blockers, CK112, and CK116 on rat experimental choroidal neovascularization in vivo and endothelial cell cultures in vitro. *J. Ocul. Pharmacol. Ther.* 22:19–25, 2006.

Park, Y. H., X. R. Xu, and **G. C. Chiou.** Structural requirements of flavonoids for increment of ocular blood flow in the rabbit and retinal function recovery in rat eyes. *J. Ocul. Pharmacol. Ther.* 20:189–200, 2004.

Dohrman, D. P., and C. K. Reiter. Ethanol modulates nicotine-induced up-regulation of nAChRs. *Brain Res.* 975:90–8, 2003.

Dohrman, D. P., and C. K. Reiter. Chronic ethanol reduces nicotine-induced dopamine release in PC12 cells. *Alcohol. Clin. Exp. Res.* 27:1846–51, 2003.

Menger, G. J., et al. **(D. J. Earnest).** Circadian profiling of the transcriptome in immortalized rat SCN cells. *Physiol. Genomics* 21: 370-381, 2005.

Bell-Pedersen, D., et al. **(D. J. Earnest).** Circadian rhythms from multiple oscillators: Lessons from diverse organisms. *Nature Rev. Genet.* 6:544–56, 2005.

Dubois, D. W., J. P. Trzeciakowski, A. R. Parrish, and **G. D. Frye.** GABAergic miniature postsynaptic currents in septal neurons show differential allosteric sensitivity after binge-like ethanol exposure. *Brain Res.,* in press.

DuBois, D. W., A. R. Parrish, J. P. Trzeciakowski, and **G. D. Frye.** Binge ethanol exposure delays development of GABAergic miniature postsynaptic currents in septal neurons. *Brain Res. Dev. Brain Res.* 152(2):199–212, 2004.

Han, S. H., D. Murchison, and **W. H. Griffith.** Low voltage–activated calcium and fast tetrodotoxin–resistant sodium currents define subtypes of cholinergic and noncholinergic neurons in rat basal forebrain. *Brain Res. Mol. Brain Res.* 134:226–38, 2006.

Murchison, D., D. C. Zawieja, and **W. H. Griffith.** Reduced mitochondrial buffering of voltage-gated calcium influx in aged rat basal forebrain neurons. *Cell Calcium* 36:61–75, 2004.

Santillano, D. R., et al. **(R. C. Miranda).** Ethanol induces cell-cycle activity and reduces stem cell diversity to alter both regenerative capacity and differentiation potential of cerebral cortical neuroepithelial precursors. *BMC Neurosci.* 6:59, 2005.

Cheema, Z. F., et al. **(R. C. Miranda).** The extracellular matrix, p53 and estrogen compete to regulate cell-surface Fas/Apo-1 suicide receptor expression in proliferating embryonic cerebral cortical precursors, and reciprocally, Fas-ligand modifies estrogen control of cell-cycle proteins. *BMC Neurosci.* 5:11, 2005.

Johnson, A. B., S. Bake, D. K. Lewis, and **F. Sohrabji.** Temporal expression of IL-1beta protein and mRNA in the brain after systemic LPS injection is affected by age and estrogen. *J. Neuroimmunol.* 174:82–91, 2006.

Sohrabji, F. Estrogen: A neuroprotective or proinflammatory hormone? Emerging evidence from reproductive aging models. *Ann. N.Y. Acad. Sci.* 1052:75–90, 2005.

Parnell, S. E., **J. R. West,** and **W.-J. A. Chen.** Nicotine decreases blood alcohol concentrations in adult rats: A phenomenon related to gastric function. *Alcohol. Clin. Exp. Res.,* in press.

Maier, S. E., and **J. R. West.** Alcohol and nutritional control treatments during neurogenesis in rat brain reduces total neuron number in locus ceoruleus but not in cerebellum or inferior olive. *Alcohol* 30:67–74, 2003.

West, J. R., and **J. B. Gelderd.** *The Cerebellum.* In *Neuroscience in Medicine,* 2nd ed., pp. 217–36, ed. P. M. Conn. Totowa, N.J.: Humana Press, 2003.

Winzer-Serhan, U. H., and F. M. Leslie. Expression of alpha5 nicotinic acetylcholine receptor subunit mRNA during hippocampal and cortical development. *J. Comp. Neurol.* 481:19–30, 2005.

UNIFORMED SERVICES UNIVERSITY OF THE HEALTH SCIENCES

F. Edward Hébert School of Medicine
Graduate Program in Neuroscience

Program of Study

The Uniformed Services University of the Health Sciences (USUHS) offers the Graduate Program in Neuroscience, a broadly based interdisciplinary program leading to the Ph.D. degree in neuroscience. Courses and research training are provided by the neuroscience faculty members, who hold primary appointments in the Departments of Anatomy, Physiology, and Genetics; Biochemistry; Medical and Clinical Psychology; Microbiology and Immunology; Neurology; Pathology; Pediatrics; Pharmacology; and Psychiatry at the University. The program permits considerable flexibility in the choice of courses and research areas; training programs are tailored to meet the individual requirements of each student. The program is designed for students with strong undergraduate training in the physical sciences, biology, or psychology who wish to pursue a professional career in neuroscience research. Integrated instruction in the development, structure, function, and pathology of the nervous system and its interaction with the environment is provided. Students in the program conduct their research under the direction of neuroscience faculty members in laboratories that are located in the medical school. During the first year of study, students begin formal course work. Each student is required to take laboratory training rotations in the research laboratories of program faculty members. By the end of the first year, students select a research area and a faculty thesis adviser. During the second year, students complete requirements for advancement to candidacy, including required course work and passage of the qualifying examination. After advancement to candidacy, each student develops an original research project and prepares and defends a written dissertation under the guidance of his or her faculty adviser and advisory committee.

Research Facilities

Each academic department at the University is provided with laboratories for the support of a variety of research projects. Neuroscience research laboratories available to students are suitable for research in most areas of neuroscience, including behavioral studies, electrophysiology, molecular and cellular neurobiology, neuroanatomy, neurochemistry, neuropathology, neuropharmacology, and neurophysiology. High-resolution electron microscopes, confocal microscopes, deconvolution wide-field fluorescence microscopes, a central resource facility providing custom synthesis of oligonucleotides and peptides and DNA sequencing, centralized animal facilities, computer support, a medical library, and a learning resources center are available within the University.

Financial Aid

Stipends are available on a competitive basis. Awards are made on a yearly basis and are renewable. For the 2006–07 academic year, stipends for entering students begin at $24,000. Outstanding students may be nominated for the Dean's Special Fellowship, which supports a stipend of $29,000.

Cost of Study

Graduate students in the neuroscience program are not required to pay tuition or fees. Civilian students incur no obligation to the United States government for service after completion of their graduate training program. Students are required to carry health insurance.

Living and Housing Costs

There is a reasonable supply of affordable rental housing in the area. The University does not have housing for graduate students. Students are responsible for making their own arrangements for accommodations. Costs in the Washington, D.C., area are comparable to those in other major metropolitan areas.

Student Group

The neuroscience graduate program is an active and growing graduate program; approximately 15 students are enrolled. The Uniformed Services University (USU) also has Ph.D. programs in departmentally based basic biomedical sciences, as well as interdisciplinary graduate programs in molecular and cell biology and in emerging infectious diseases. In addition to the graduate and medical programs in the medical school, the nursing school has graduate programs for nurse practitioners and nurse anesthetists.

Student Outcomes

Graduates hold faculty, research associate, postdoctoral, and other positions in universities, medical schools, government, and industrial research institutions.

Location

USU is located in Bethesda, Maryland, an immediately adjacent northern suburb of Washington, D.C. The University is located in a parklike setting at the National Naval Medical Center, which is across the street from the National Institutes of Health and the National Library of Medicine. A nearby Metro subway station provides convenient access to downtown Washington and surrounding areas. The D.C. metropolitan area has about 3 million people and offers seven major universities (including three other medical schools), numerous colleges, and internationally renowned research facilities. Wooded areas and jogging and biking trails surround the University. Cultural and recreational activities are plentiful, with a major park system, recreation facilities, museums, theaters, symphonies, opera, and major-league sports teams. The Chesapeake Bay, Atlantic Ocean, and Shenandoah Mountains are easily accessible.

The University

The University was established by Congress in 1972 to provide a comprehensive education in medicine to those who demonstrate potential for careers as Medical Corps officers in the uniformed services. Graduate programs in the basic medical sciences are offered to both civilian and military students and are an essential part of the academic environment at the University. The University is located in proximity to major research facilities, including the National Institutes of Health (NIH), the National Library of Medicine, Walter Reed Army Medical Center, Walter Reed Army Institute of Research, the Armed Forces Institute of Pathology, the Armed Forces Radiobiology Research Institute, the National Institute of Standards and Technology, and numerous biotechnology companies.

Uniformed Services University subscribes fully to the policy of equal educational opportunity and accepts students on a competitive basis without regard to race, color, sex, age, or creed.

Applying

Civilian applicants are accepted as full-time students only. Each applicant must have a bachelor's degree from an accredited academic institution. A strong background in science with courses in several of the following disciplines—biochemistry, biology, chemistry, mathematics, physics, physiology, and psychology—is desirable. Applicants must arrange for official transcripts of all prior college-level courses taken and their GRE scores (taken within the last two years) to be sent to the Office of Graduate Education. Students may elect to submit scores obtained in one or more GRE Subject Tests (from the subject areas listed above) in support of their application. Applicants must also arrange for letters of recommendation from 3 people who are familiar with their academic work to be sent to the University. For full consideration and evaluation for stipend support, completed applications should be received before January 15 for matriculation in late August. Late applications are evaluated on a space-available basis. There is no application fee. Application forms may be obtained from the Web site at http://cim.usuhs.mil/geo/application.htm.

Correspondence and Information

For applications:
Associate Dean for Graduate Education
Uniformed Services University
4301 Jones Bridge Road
Bethesda, Maryland 20814-4799
Phone: 301-295-3913
    800-772-1747 (toll-free)
E-mail: janastasi@usuhs.mil
Web site: http://cim.usuhs.mil/geo/

For information about the neuroscience program:
Regina C. Armstrong, Ph.D.
Director, Graduate Program in Neuroscience
Uniformed Services University
4301 Jones Bridge Road
Bethesda, Maryland 20814-4799
Phone: 301-295-3642
Fax: 301-295-1996
E-mail: rarmstrong@usuhs.mil
    nfinley@usuhs.mil
Web site: http://www.usuhs.mil/nes/home.html

Uniformed Services University of the Health Sciences

THE FACULTY AND THEIR RESEARCH

Denes V. Agoston, M.D., Ph.D., D.Sc., Associate Professor, Department of Anatomy, Physiology, and Genetics. Molecular and cellular mechanism of stem cell differentiation during development and following brain injury.

Juanita Anders, Ph.D., Associate Professor, Department of Anatomy, Physiology, and Genetics. Neurobiology and innovative therapies for neuronal regeneration of injured central nervous system: light as a therapy for nerve injury, transplantation of olfactory ensheathing cells (OECs), characterization of human OECs in vitro, and neural stem cells for repair of cerebral infarction.

Regina Armstrong, Ph.D., Professor, Department of Anatomy, Physiology, and Genetics, and Director, Graduate Program in Neuroscience. Cellular and molecular mechanisms of glial cell development and regeneration in demyelinating diseases.

Suzanne B. Bausch, Ph.D., Assistant Professor, Department of Pharmacology. Synaptic plasticity in epileptogenesis and seizure generation.

Rosemary C. Borke, Ph.D., Professor, Department of Anatomy, Physiology, and Genetics. Neuronal plasticity in development and regeneration.

Diane E. Borst, Ph.D., Research Assistant Professor, Department of Anatomy, Physiology, and Genetics. Molecular mechanisms of retinal gene regulation and function.

Maria F. Braga, D.D.S., Ph.D., Assistant Professor, Department of Anatomy, Physiology, and Genetics. Cellular and molecular mechanisms regulating neuronal excitability in the amygdala; pathophysiology of anxiety disorders and epilepsy.

Howard Bryant, Ph.D., Associate Professor, Department of Anatomy, Physiology, and Genetics. Electrophysiology of vascular smooth muscle.

Col. William W. Campbell, USAR, MC; M.D., M.S.H.A.; Professor and Chair, Department of Neurology, and Professor, Department of Neuroscience. Neuromuscular disease and clinical neurophysiology.

De-Maw Chuang, Ph.D., Adjunct Professor, Department of Psychiatry. Molecular and cellular of actions of mood stabilizers: neuroprotection against excitotoxicity-related neurodegeneration.

Thomas Côté, Ph.D., Associate Professor, Department of Pharmacology. Mu opioid receptor interaction with GTP-binding proteins and RGS proteins.

Brian Cox, Ph.D., Professor and Chair, Department of Pharmacology. Opiate drugs, endogenous opioids, and novel related peptides.

Patricia A. Deuster, Ph.D., M.P.H., Professor, Department of Military and Emergency Medicine. Mechanisms of neuroendocrine and immune activation with stress.

Martha M. Faraday, Ph.D., Assistant Professor, Department of Medical and Clinical Psychology. Psychobiology of stress vulnerability; biobehavioral assessment of stress-nicotine interactions.

Ying-Hong Feng, Associate Professor, M.D., Ph.D. Angiotensin receptor and signal transduction.

Zygmunt Galdzicki, Ph.D., Assistant Professor, Department of Anatomy, Physiology, and Genetics. Molecular and electrophysiological approach to understand mental retardation in Down syndrome; role of glutamate receptors in neurodegenerative disorders.

Neil Grunberg, Ph.D., Professor, Department of Medical and Clinical Psychology. Nicotine and tobacco; drug abuse; psychopharmacology; stress; environments and behavior.

Carl Gunderson, M.D., Professor, Department of Neurology. Education of medical students; history of military neurology.

Harry Holloway, M.D., Professor, Department of Psychiatry. Clinical psychiatry; alcohol and drug misuse; posttraumatic stress; neurobiology of psychiatric disorders; clinical psychopharmacology.

Christopher J. Hough, Ph.D., Research Assistant Professor, Department of Psychiatry. Affective and anxiety disorders; the mechanisms of action and mood stabilizers, including lithium and zinc.

Bahman Jabbari, M.D., Professor and Chair, Department of Neurology. Epilepsy and movement disorders; central electrophysiology (EEG, EP, sleep).

David Jacobowitz, Ph.D., Adjunct Professor, Department of Anatomy, Physiology, and Genetics. Gene and protein discovery in the diseased and developing brain.

Martha Johnson, Ph.D., Associate Professor, Department of Anatomy, Physiology, and Genetics. Education of first-year medical and graduate students in the anatomical and physiological sciences.

Sharon Juliano, Ph.D., Professor, Department of Anatomy, Physiology, and Genetics. Mechanisms of development and plasticity in the cerebral cortex, with particular emphasis on the migration of neurons into the cortical plate and factors maintaining the function and morphology of radial glia and Cajal-Retzius cells.

He Li, M.D., Ph.D., Assistant Professor, Department of Psychiatry. Neurobiological basis of post-traumatic stress disorder: synaptic plasticity and neuronal signaling in the amygdala circuitry.

Geoffrey Ling, M.D., Ph.D., Professor, Departments of Anesthesiology, Neurology, and Surgery. Novel therapeutics and diagnostic tools for traumatic brain injury and hemorrhagic shock; mechanisms of cellular injury and edema formation in traumatic brain injury.

Ann M. Marini, Ph.D., M.D., Associate Professor, Department of Neurology. Molecular and cellular mechanisms of intrinsic survival pathways through glutamate receptors to protect against neurodegenerative disorders.

Joseph McCabe, Ph.D., Professor and Vice Chair, Department of Anatomy, Physiology, and Genetics. Traumatic brain injury, hemorrhagic shock, and gene expression of neuroendocrine-related gene products.

J. Brian McCarthy, Assistant Professor, Department of Pharmacology. Receptor targeting and structural synaptic plasticity.

Leslie McKinney, Ph.D., Assistant Professor, Department of Anesthesiology. Ion transport in leukocytes and alteration of glial cell function by radiation.

Debra McLaughlin, Ph.D., Research Assistant Professor, Department of Anatomy, Physiology, and Genetics. Sensory processing and cortical neurophysiology.

David Mears, Ph.D., Assistant Professor, Department of Anatomy, Physiology, and Genetics. Electrophysiology and calcium signaling in neuroendocrine cells.

Gregory Mueller, Ph.D., Professor, Department of Anatomy, Physiology, and Genetics. Neuroendocrine regulation; neuropeptide gene expression; regulation of peptide biosynthesis and the proteomics of neuropeptide secretion.

Aryan Namboodiri, Ph.D., Assistant Professor, Department of Anatomy, Physiology, and Genetics. Neurobiology of N-acetylaspartate (NAA).

J. Timothy O'Neill, Ph.D., Research Assistant Professor, Departments of Pediatrics and Anatomy, Physiology, and Genetics. Mechanisms of control of newborn and developmental cerebral blood and oxygen supply.

Harvey B. Pollard, M.D., Ph.D., Professor and Chair, Department of Anatomy, Physiology, and Genetics. Molecular biology of secretory processes.

Merrily Poth, M.D., Professor, Department of Pediatrics. Neuroendocrinology and neuroimmunology; abnormalities of the HPA axis in depression and in obesity; growth in children and neuroendocrinology.

Christopher Reid, Ph.D., M.D., Research Assistant Professor, Department of Pharmacology. Cellular and genetic mechanisms governing the development of the mammalian forebrain.

Andrea Salazar, M.D., Professor, Department of Neurology. Clinical studies of head injuries resulting from trauma and viral infections.

Michael Schell, Ph.D., Assistant Professor, Department of Pharmacology. Synaptic mechanisms regulating the actin cytoskeleton in dendritic spines.

Aviva Symes, Ph.D., Associate Professor, Department of Pharmacology. Cytokine regulation of neuronal gene expression; mechanism of cytokine action after traumatic brain injury.

E. Fuller Torrey, M.D., Professor, Department of Psychiatry. Infectious agents as causes of schizophrenia and bipolar disorder.

Jack W. Tsao, M.D., D.Phil., Assistant Professor, Department of Neurology. Clinical studies on dementia, low back pain, and phantom limb pain; basic science studies on cellular and developmental mechanisms governing nerve degeneration.

Robert J. Ursano, M.D., Professor and Chairman, Department of Psychiatry. Post-traumatic stress disorders.

Ajay Verma, M.D., Ph.D., Assistant Professor, Department of Neurology. Regulation of brain metabolism by signal transduction.

Maree J. Webster, Ph.D., Research Assistant Professor, Department of Psychiatry. Neuropathology of severe mental illness; schizophrenia and bipolar disorder.

T. John Wu, Ph.D., Assistant Professor, Department of Obstetrics and Gynecology. Molecular mechanisms of peptide and steroid hormone action: regulation of fertility; behavior (e.g., depression and post-traumatic stress disorders).

THE UNIVERSITY OF ALABAMA
AT BIRMINGHAM
Department of Neurobiology

Program of Study

The Department of Neurobiology offers students who are interested in the nervous system a program of study that leads to the Ph.D. in neurobiology or a combined M.D./Ph.D. through the University of Alabama at Birmingham's (UAB) School of Medicine. The program has 39 faculty members, who are actively involved with research programs from a broad range of areas in contemporary neuroscience, with a major focus on topics of biomedical importance. Molecular, cellular, and biophysical approaches are emphasized to address fundamental problems of brain function, development, and disease. All students take a series of graduate courses in biochemistry and molecular biology, cell biology, ion-channel biophysics, synaptic function, developmental neurobiology, and medical neurobiology. In addition to the core curriculum, students perform a series of three laboratory rotations, which are selected from a series of advanced elective courses that include receptor and ion-channel kinetics, synaptic plasticity, intracellular signaling, and the biology of neurological and psychiatric disease. Students also participate in journal clubs, student presentations, and a visiting speakers program. Students are encouraged to attend national and international neuroscience meetings throughout their graduate training. A thesis adviser is selected by students at the end of the first year. Students take a departmental qualifying examination and present a thesis proposal in the second year. Upon successful completion of these requirements, the student is admitted to candidacy for the Ph.D. The Ph.D. is awarded after the defense of the final dissertation, usually in the fourth or fifth year.

Research Facilities

The Department of Neurobiology is housed in the Civitan International Research Center, which was built in 1992, with laboratories outfitted by major national foundations, which include the Lucille P. Markey Foundation and the W. M. Keck Foundation. The Department's primary faculty has 25,000 square feet of research space in the center, and the secondary faculty has an additional 25,000 square feet of research space in adjacent buildings. Facilities available to students include laser scanning confocal microscopy, molecular neurobiology core, neuronal cell culture suites, neuronal gene transfection core, electron microscopy core, fluorescent microscopy–digital imaging core, a large number of patch clamp electrophysiology suites, dye imaging facilities, and a neurobiology library that has current subscriptions to most major neuroscience journals. In addition, students have access to peptide and oligonucleotide sequencing facilities, hybridoma and transgenic animal facilities, and Lister Hill Library for the Health Sciences.

Financial Aid

All students receive a first-year stipend of $21,000 as well as tuition remission and health insurance. In the second year and each additional year of study, stipends, tuition, and fees are provided by training grants, mentor's research grants, or individual research fellowships and awards.

Cost of Study

The Department of Neurobiology and the mentor provide complete coverage for tuition and fees and full stipends for all students during all years in the program.

Living and Housing Costs

Students may choose to live in University dormitories or on-campus apartments or may select from a wide range of nearby off-campus apartments or houses. The cost of living in Birmingham is slightly below the U.S. average.

Student Group

The Department of Neurobiology admits between 5 and 6 students per year. Forty-one doctoral and postdoctoral students are currently training with primary faculty members in the Department, with another 40 training with affiliated neurobiology faculty members.

Location

Birmingham, a vibrant, rapidly growing community with a population of 1.2 million, is set along a series of beautiful mountain and valley ridge lines. The city has a rich cultural diversity, represented by UAB's Alys Stephens Center for the Performing Arts, the Alabama Symphony, the city's annual City Stages musical festival, and a variety of ethnic and heritage festivals. The climate is mild, providing opportunities for year-round outdoor recreation, such as hiking in nearby national forests and state parks, canoeing in the many rivers and lakes, and playing in numerous public tennis and golf facilities. Restaurants offer diverse cuisine and ambiance, with notable excellence in Chinese, Greek, Italian, Mexican, Thai, and French food. The nearby Florida beaches (a 4-hour drive) offer year-round coastal and aquatic amenities.

The University

The University of Alabama at Birmingham offers one of the leading environments in the U.S. to pursue graduate studies, particularly in the biomedical sciences. Doctoral programs are available in more than thirty disciplines. The University is a Carnegie Division I Research Institution and is consistently ranked in the top twenty institutions in the U.S. in federal funding for research.

Applying

Applications to the Department are evaluated by the Admission Committee. The admission decision is based on a combination of the undergraduate record, including curriculum taken and grade point average; scores on the Graduate Record Examinations (a combined score of 1200 on the verbal and quantitative portion of the General Test of the GRE is usually required); letters of recommendation; previous research experience; and a personal interview with members of the Admission Committee. TOEFL scores are required of students whose primary language is not English. The deadline for receipt of applications is February 1.

Correspondence and Information

Dr. Anne B. Theibert
Graduate Program Director
Department of Neurobiology
CIRC 576
The University of Alabama at Birmingham
1719 Sixth Avenue South
Birmingham, Alabama 35294-0021

Phone: 205-934-1550
Fax: 205-975-6330
E-mail: urthaler@nrc.uab.edu
Web site: http://www.neurobiology.uab.edu

The University of Alabama at Birmingham

THE FACULTY AND THEIR RESEARCH

Franklin R. Amthor, Professor of Psychology; Ph.D., Duke. Neurophysiology of the retina and neural modeling.

Dale J. Benos, Professor and Chairman of Physiology and Biophysics; Ph.D., Duke. Ion transport.

Etty Benveniste, Professor and Chair of Cell Biology; Ph.D., UCLA. Neuroimmunology; bidirectional communication between the immune and nervous systems.

J. Edwin Blalock, Professor of Physiology and Biophysics; Ph.D., Florida. Protein folding and interactions; immunotherapy; Ca channels and HIV.

Michael Brenner, Associate Professor of Neurobiology; Ph.D., Berkeley. Molecular control of transcription of cytoskeletal protein in glia; spinal injury.

William J. Britt, Professor of Pediatrics; M.D., Arizona. Molecular virology; pathogenesis of CNS virus infections; viral tropism for cells of CNS.

Steven L. Carroll, Associate Professor of Pathology; M.D., Ph.D., Baylor College of Medicine. Growth and differentiation factors in nervous system regeneration and neoplasia.

Peter J. Detloff, Associate Professor of Biochemistry and Molecular Genetics; Ph.D., Chicago. Mouse models of human neurological genetic disorders.

Lynn E. Dobrunz, Assistant Professor of Neurobiology; Ph.D., Johns Hopkins. Synaptic transmission, presynaptic properties of single synapses.

Leon S. Dure, Professor of Pediatrics and Neurology; M.D., Baylor College of Medicine. Inherited neurologic disease; movement disorders.

Paul D. R. Gamlin, Professor and Chairman of Physiological Optics; Ph.D., SUNY at Stony Brook. Neural control of eye movements.

Timothy J. Gawne, Associate Professor of Physiological Optics; Ph.D., Uniformed Services Health Sciences. Determining how information is encoded and processed in the brain and how it relates to vision.

Candece L. Gladson, Professor of Pathology; M.D., Loyola. Molecular mechanisms promoting tumor-cell proliferation, invasion, and angiogenesis in gliomas.

John J. Hablitz, Professor of Neurobiology; Ph.D., Houston. Development of ion channel gating and synaptic transmission by excitatory amino acids in the mammalian forebrain.

Gail V. W. Johnson, Professor of Psychiatry and Behavioral Neurobiology; Ph.D., Delaware. Metabolism and function of tau protein.

Kent Keyser, Professor of Physiological Optics; Ph.D., SUNY at Stony Brook. Neurotransmitters and receptors.

Kevin L. Kirk, Professor of Physiology and Biophysics; Ph.D., Iowa. Ion channel structure and function; cystic fibrosis.

Timothy W. Kraft, Associate Professor of Physiological Optics; Ph.D., Minnesota. Physiology and biophysics of photoreceptors, the retinal cells that convert light energy into electrochemical signals.

Robin A. J. Lester, Associate Professor of Neurobiology; Ph.D., Bristol. Molecular pharmacology of ligand- and voltage-gated ion channels in the central nervous system.

Michael S. Loop, Associate Professor of Physiological Optics; Ph.D., Florida State. Visual psychophysics; color vision in vertebrates.

Richard B. Marchase, Professor of Cell Biology and Associate Dean of Medicine; Ph.D., Johns Hopkins. Relationship between glucose metabolism and cytoplasmic calcium regulation.

Guillermo Marques, Assistant Professor of Cell Biology; Ph.D., Complutense (Madrid). Developmental and adult synaptic plasticity; regulation of gene expression during nervous system development; cell signaling and signal transduction by the TGF-beta/BMP pathway in neurons.

Lori L. McMahon, Assistant Professor of Physiology and Biophysics; Ph.D., St. Louis. Synaptic plasticity.

Anthony Nicholas, Associate Professor of Neurology; M.D., Ph.D. Texas Medical Branch. Monoamines and their receptors; Parkinson's disease; neurodegeneration.

Alan K. Percy, Professor of Pediatrics and Neurology; M.D., Stanford. Rett syndrome; myelination in developing brain; cocaine and neural development.

Lucas D. Pozzo-Miller, Assistant Professor of Neurobiology; Ph.D., Cordoba (Argentina). Role of microcompartmentalization of calcium in synaptic function and plasticity; role of brain-derived neurotrophic factor.

Kevin A. Roth, Professor of Pathology and Director of Neuropathology; M.D., Ph.D., Stanford. Molecular regulation of neuronal cell death role of Bc1-2 and caspase family members in programmed cell death.

Douglas M. Ruden, Associate Professor of Environmental Health Sciences; Ph.D., Caltech. Dominant mutations that affect EGF-R signal transduction.

Harald W. Sontheimer, Professor of Neurobiology; Ph.D., Heidelberg. Regulation and function of ion channels in glia; pathogenesis of gliomas.

J. David Sweatt, Professor and Chairman of Neurobiology; Ph.D., Vanderbilt. Signal transduction mechanisms in learning and memory.

Anne B. Theibert, Associate Professor of Neurobiology; Ph.D., Johns Hopkins. Molecular mechanisms of signal transduction in the CNS; role of phosphoinositides and intracellular calcium in neural development.

Scott M. Wilson, Assistant Professor of Neurobiology; Ph.D., Florida. Role of the ubiquitia-proteosome pathway in the nervous system.

J. Michael Wyss, Professor of Cell Biology; Ph.D., Washington (St. Louis). Plasticity in cerebral cortex; neuromodulation in the hypothalamus; the limbic cortex and neural cardiovascular control.

Jianhua Zhang, Assistant Professor of Pathology; Ph.D., Texas Southwestern Medical Center at Dallas. Glutamate receptor of activation, signaling, transcription regulation, and mechanisms of neuronal cell death.

Yi Zhou, Assistant Professor of Neurobiology; Ph.D., Minnesota. Modulation of ion channels; regulation of neuronal excitability and synaptic transmission.

THE UNIVERSITY OF ALABAMA AT BIRMINGHAM

Department of Neurobiology

Program of Study	The Department of Neurobiology, in conjunction with the Departments of Biochemistry and Molecular Genetics, Cell Biology, and Microbiology, offers course work and individual laboratory research leading to the Ph.D. degree. The program is designed to provide interdisciplinary training of high quality in cell and molecular biology to a select group of predoctoral students, preparing them to become independent investigators in these disciplines. Students are immersed in research at the forefront of scientific endeavor and are provided with sufficient guidance and course work to place their research in the proper perspective.

This interdisciplinary program includes faculty members from the following units: Anesthesiology, the Arthritis Center, Biochemistry and Molecular Genetics, Biology, the Cancer Center, Cell Biology, Endocrinology, Geographic Medicine, Hematology/Oncology, Infectious Disease, Medicine, Microbiology, Neurobiology, Neurology, Neuropsychiatry, Oral Biology, Pathology, Pediatrics, Pharmacology, Physiology and Biophysics, Psychiatry, and Public Health.

The first-year curriculum emphasizes three areas: acquisition of a working knowledge of contemporary cellular and molecular biology through an intensive, integrated course in biochemistry, genetics, cell biology, virology, immunology, and neurobiology; involvement in a diversity of laboratory research training experiences; and the development of skills in reading, writing, and speaking. Advanced students are engaged primarily in research, but they also take some advanced courses and tutorials in specialized areas of interest and participate in seminars. Completion of the requirements for the Ph.D. usually takes six years. No foreign language is required.

Areas of specialization for dissertation research include fundamental molecular biology; biochemistry of nucleic acids; prokaryotic and eukaryotic molecular and cell biology; molecular virology; viral, microbial, and mammalian genetics; immunogenetics; cellular, developmental, and tumor immunology; immunochemistry; biological macromolecules and membranes; host-parasite relationships; infectious diseases; biochemistry of connective tissues; X-ray crystallography; NMR spectroscopy; molecular biophysics; ion-channel structure and biophysics; synaptic function and plasticity; neuronal development; and glial-neuronal signaling. Students are encouraged to present research results at scientific meetings and publish in scientific journals.

Research Facilities Faculty members participating in the program have more than 400,000 square feet of laboratory space. In addition to well-equipped labs, a number of core facilities are available. These include automated DNA sequencing, DNA chip analysis, NMR spectroscopy, electron microscopy, protein and nucleic acid synthesis and analysis, mass spectrometry, confocal microscopy, fluorescence-activated cell sorting, large-scale bacterial fermentation, X-ray diffraction, a P3 containment facility, computer facilities, and a hybridoma facility.

Financial Aid All students admitted to the program receive support through national or state granting agencies in the amount of $23,000, plus student health insurance and full payment of tuition and fees.

Cost of Study Tuition and fees for the 2004–05 academic year were $5500 for in-state students and $11,000 for out-of-state students. As indicated above, tuition and fees are paid for all students.

Living and Housing Costs The cost of living in Birmingham is slightly below the U.S. average. Housing is readily available in the Medical Center area.

Student Group The graduate program of the Departments of Biochemistry and Molecular Genetics, Cell Biology, Neurobiology, and Microbiology has more than 130 faculty members, 190 full-time graduate students, and about 150 postdoctoral fellows and visiting faculty members. The total enrollment at the University of Alabama at Birmingham is more than 15,000.

Student Outcomes Graduates typically go on to postdoctoral research appointments, which are followed by careers in academic research and teaching, research in the biotechnology industry, or other science-related professions.

Location Birmingham is located in the lovely rolling foothills of the Appalachian Mountain range in central Alabama. A metropolitan area that includes 1 million people, Birmingham is only a few hours' drive from Atlanta, Nashville, New Orleans, and the Gulf Coast. The city has excellent art and historical museums, theaters, libraries, a symphony orchestra, a ballet, a zoo, and botanical gardens. A host of recreational opportunities, including camping, swimming, fishing, hiking, golf, tennis, and boating, are available year-round in numerous local and state parks.

The University The University of Alabama at Birmingham consists of University College, the Graduate School, and the Medical Center. It is located in the largest population center of the state and is heavily engaged in research, instruction, and service programs. Its dedication to excellence in biomedical research is exemplified by its consistent ranking among the top twenty institutions in receipt of federal research funds.

Applying Applications to the program are evaluated by the Admissions Committee, in consultation with other faculty members. The admission decision is based on scores achieved on the Graduate Record Examinations (with a combined score of at least 1200 on the verbal and quantitative portions of the General Test), undergraduate grade point average (with consideration given to the curriculum completed), letters of evaluation, prior research experience, and a personal interview.

To be accepted into the program, the student should have completed a B.S. degree that includes the following undergraduate course work by the time of entrance: calculus, general and organic chemistry, and at least one introductory course in zoology or biology. Courses in physical chemistry, genetics, cell biology, and related topics are also beneficial to the candidate. Any remedial course work must be completed with a grade of B or better before the end of the first full year of doctoral study.

Applications are strongly encouraged from individuals with prior research experience, including those who hold an M.S. degree in a related area or a professional degree, such as the M.D., D.M.D., D.V.M., or O.D. The program anticipates admitting 35 to 40 students each year.

Correspondence and Information
Admissions Committee
Cellular and Molecular Biology Graduate Program
Suite 260, Bevill Biomedical Research Building
The University of Alabama at Birmingham
Birmingham, Alabama 35294-2170
Phone: 800-262-7764 (toll-free)
E-mail: cmb@uab.edu
Web site: http://www.cmb.uab.edu

The University of Alabama at Birmingham

THE FACULTY AND THEIR RESEARCH

Cell Physiology, Adhesion, and Signaling
S. Barnes, Ph.D.: bile acids, polyphenols, and metabolism and their effects on protein expression and function in chronic disease.
Z. Bebok, M.D.: membrane protein biogenesis in epithelial cells (CFTR as model); unfolded protein response.
R. Carter, M.D.: immune and autoimmune lymphocytes; structure and function in vivo.
C. Chang, Ph.D.: signal pathways in frog development.
D. Chaplin, M.D., Ph.D.: control of inflammation and lymphoid organ formation by LT and TNF.
J. Chatham, Ph.D.: cardiomyocyte function and metabolism in diabetes and ischemic heart disease.
J. Collawn, Ph.D.: intracellular protein sorting.
C. Gladson, M.D.: molecular mechanisms promoting tumor-cell proliferation; invasion and angiogenesis in gliomas.
J. Hagood, M.D.: fibroblast signaling in lung remodeling and fibrogenesis.
J. Kearney, Ph.D.: B cells; B-cell development; hybridomas; transgenic and knockout mice; immunoregulation; *B. anthracis*.
N. Kedishvili, Ph.D.: regulation of retinoic acid homeostasis.
F. Lin, M.D., Ph.D.: regulation of cell growth by G-protein-coupled receptor signaling.
R. Marchase, Ph.D.: calcium regulation and its impairment in diabetes.
G. Marques, Ph.D.: TGF-ß signaling in nervous system development and function.
M. Miller, Ph.D.: function and evolution of intercellular communication mechanisms.
E. Schwiebert, Ph.D.: extracellular nucleotide signaling and epithelial cell biology and physiology.
L. Schwiebert, Ph.D.: airway inflammation.
E. Sztul, Ph.D.: membrane traffic; protein degradation.
J. Thompson, Ph.D.: molecular mechanisms of angiogenesis.
A. Woods, Ph.D.: *Syndecan proteoglycans* in cell adhesion and matrix assembly.

Gene Regulation and Expression
A. Agarwal, M.D.: regulation of heme oxygenase gene expression in kidney and vascular injury.
L. Bridges, M.D., Ph.D.: genetic influences on treatment response in rheumatoid arthritis.
C. Chen, Ph.D.: mechanism and regulation of mammalian mRNA turnover.
X. Chen, Ph.D.: the p53 tumor-suppressor gene family and transcriptional regulation.
D. Crawford, M.D., Ph.D.: the role of G2/M-specific genes in mitosis and G2 DNA damage checkpoint regulation.
P. Higgins, Ph.D.: genetic and biochemical studies of chromosome dynamics.
C. Klug, Ph.D.: hematopoietic stem-cell biology and acute leukemias.
W.-C. Lin, M.D., Ph.D.: cell-cycle control and DNA damage response.
S. Lobo-Ruppert, Ph.D.: role of transcription factor oncogenes in tumor initiation.
R. Mayne, Ph.D.: collagen and collagen-binding proteins in health and disease.
J. McDonald, M.D.: cellular life and death signals in cancer; AIDS and bone disease.
M. Ruppert, M.D., Ph.D.: role of zinc finger transcription factors in tumor progression.
T. Ryan, Ph.D.: gene regulation; stem cells; mouse models; mutagenesis; cell therapies.
C. Turnbough, Ph.D.: *B. anthracis* spore structure-function/bacterial gene regulation.
H. Wang, Ph.D.: role of histone modification in chromatin function.

Immunology
S. Barnum, Ph.D.: role of complement, acute phase proteins, and adhesion molecules in acute and chronic inflammation in the central nervous system.
T. Benveniste, Ph.D.: immune/nervous system interactions.
O. Branch, Ph.D.: malaria molecular epidemiology and immunology.
P. Bucy, M.D., Ph.D.: T-cell regulation of immune responses in vivo.
R. Davis, M.D.: lymphocyte development and mechanisms of lymphomagenesis.
C. Elson, Ph.D.: chronic intestinal inflammation.
K. Fujihashi, D.D.S., Ph.D.: mucosal immunology; regulation of S-igA antibody responses; mucosal vaccine development.
V. Ghanta, Ph.D.: tumor immunology; CNS and immune system interactions.
Z. Hel, Ph.D.: development and testing of novel HIV/AIDS vaccine strategies.
L. Justement, Ph.D.: analysis of molecular mechanisms regulating lymphocyte biology.
J. Kabarowski, Ph.D.: regulation of inflammation and chronic inflammatory disease by lysophospholipid receptors.
J. Kapp, Ph.D.: immune regulation and transplantation.
J. Katz, Ph.D., D.D.S.: vaccine delivery systems; inflammation; innate immunity.
R. Kimberly, M.D.: autoimmunity, molecular mechanisms, and genetic risk.
W. Koopman, M.D.: pathogenesis of immune disease.
H. Kubagawa, M.D.: molecular genetics and immunopathology of host defense.
J. Mestecky, M.D., Ph.D.: mucosal immunity; vaccines.
M. Nahm, M.D.: adaptive and innate immune responses to vaccines against bacteria.
T. Strong, Ph.D.: identification of tumor antigens and development of cancer vaccines.
A. Szalai, Ph.D.: inflammation innate immunity and the acute phase proteins in health and disease.
L. Timares, Ph.D.: engineering dendritic cells for immunotherapy.
M. Walter, Ph.D.: structure and function of cytokines and their signaling molecules.
C. Weaver, M.D.: T-cell development.
Z. Zhang, Ph.D.: molecular regulation of early B-cell development and antibody repertoire formation.
T. Zhou, M.D.: specific induction of apoptosis in autoimmune and inflammatory cells.

Macromolecular Structure and Function
J. Blalock, Ph.D.: rational drug and vaccine design.
C. Brouillette, Ph.D.: protein structural cooperativity and energetics.
D. Chattopadhyay, Ph.D.: structure-function analysis of proteins.
I. Chesnokov, Ph.D.: DNA replication and cell cycle in eukaryotes.
H. Cheung, Ph.D.: regulatory mechanism in cardiac muscle.
L. DeLucas, Ph.D.: protein crystallography/protein crystal growth.
G. Elgavish, Ph.D.: NMR studies of intact hearts.
S. Frank, M.D.: growth hormone action and GH receptor structure and function.
B. Freeman, Ph.D.: tissue metabolism of reactive inflammatory mediators.
R. Krishna, Ph.D.: structural biology and biomolecular NMR spectroscopy.
J. Murphy-Ullrich, Ph.D.: extracellular matrix control of cell and growth-factor function.

C. Raman, Ph.D.: lymphocyte activation, immune tolerance and autoimmunity tolerance, and autoimmunity.
J. Segrest, M.D., Ph.D.: structural biology of supramolecular assemblies, particularly lipoproteins and membranes.
B. Sha, Ph.D.: structure and function of molecular chaperones.
N. Sthanam, Ph.D.: bacterial surface protein anchoring.

Molecular Genetics and Disease
S. Abulkadir, Ph.D.: molecular genetics of prostate cancer.
P. Atkinson, M.D., Ph.D.: primary immunodeficiency/role of infection in chronic diseases.
D. Bedwell, Ph.D.: translation termination; treatment of genetic diseases.
P. Burrows, Ph.D.: B-lymphocyte development and function.
P. Detloff, Ph.D.: mouse models of human genetic disorders.
K. Dybvig, Ph.D.: pathogenic mechanisms of mycoplasmas.
L. Guay-Woodford, M.D.: characterizing molecular determinants involved in PKD pathogenesis.
K. Jiao, Ph.D.: TGF-ß/BMP signaling during cardiogenesis.
K. Kirk, Ph.D.: the CFTR chloride channel.
J. Kudlow, M.D.: nucleocytoplasmic O-glycosylation in the control of cell function.
J. Mountz, M.D.: gene therapy; T-cell aging; immunogenetics; T-cell imaging.
H. Schroeder, M.D., Ph.D.: development/function of lymphocyte antigen receptors.
R. Serra, Ph.D.: mechanism of TGF-ß action in developmental and disease processes.
T. Townes, Ph.D.: developmental regulation of gene expression.
T. Unnasch, Ph.D.: molecular epidemiology and ecology of vector-borne diseases.
D. Welch, Ph.D.: molecular basis of tumor progression and metastasis.
B. Yoder, Ph.D.: cilia signaling and dysfunction in development and disease.

Molecular Pathogenesis
D. Balkovetz, M.D., Ph.D.: epithelial cell biology; epithelial cell signal regulation; regulation of paracellular transport across epithelial cell tight junctions.
W. Benjamin, Ph.D.: molecular typing of *Streptococcus pneumoniae*.
D. Briles, Ph.D.: bacterial pathogenesis; virulence; immunity; pneumococcus.
N. Childers, D.D.S., Ph.D.: development of a vaccine for the prevention of dental caries.
M. Cooper, M.D.: developmental immunobiology, with emphasis on B-cell and T-cell differentiation; clinical immunology, with emphasis on immunodeficiency diseases and lymphoid malignancies.
S. Hollingshead, Ph.D.: mechanisms of variation in microbial pathogenesis.
R. Lorenz, M.D., Ph.D.: cellular and molecular immunology of the gastrointestinal tract.
S. Michalek, Ph.D.: mucosal vaccines and host mechanisms involved in inflammation.
R. Morrison, Ph.D.: immunobiology of Chlamydia infection.
C. Morrow, Ph.D.: understanding, at the molecular level, virus–host cell interactions.
M. Niederweis, Ph.D.: the role of porins in outer-membrane permeability and drug resistance of mycobacteria.
D. Pritchard, Ph.D.: roles of glycoconjugates in pathogenesis.
J. Rayner, Ph.D.: cell biology of the malaria parasite *Plasmodium falciparum*.
A. Steyn, Ph.D.: mechanism of mycobacterium tuberculosis virulence.
K. Waites, M.D.: diagnostic microbiology; epidemiology and mechanisms of antimicrobial resistance; mycoplasma and ureaplasma diseases; staphylococcal diseases.
H. Wu, Ph.D.: bacteria-host interaction.
J. Yother, Ph.D.: capsular polysaccharides of *Streptococcus pneumoniae*.

Neurobiology
D. Benos, Ph.D.: molecular basis of operation on ion channels and transporters.
M. Brenner, Ph.D.: molecular studies of astrocytes in health and disease.
L. Dobrunz, Ph.D.: synaptic transmission and plasticity in hippocampus.
J. Hablitz, Ph.D.: cellular mechanisms of neurotransmission.
G. Johnson, Ph.D.: molecular mechanisms of neurodegeneration.
R. Jope, Ph.D.: neuronal signaling mechanisms regulating gene expression and cell death.
R. Lester, Ph.D.: molecular and cellular mechanisms of nicotine addiction.
L. Pozzo-Miller, Ph.D.: neurotrophins on Ca2+ signaling, synapse development, and plasticity.
K. Roth, M.D., Ph.D.: molecular regulation of neuronal cell death.
D. Ruden, Ph.D.: environmental and developmental toxicology.
H. Sontheimer, Ph.D.: the role of neuroglia in brain function and disease.
D. Sweatt, Ph.D.: signal transduction mechanisms in learning and memory.
A. Theibert, Ph.D.: role of phosphoinositides in developmental neurobiology.
S. Wilson, Ph.D.: mouse models of neurodegeneration.
M. Wyss, Ph.D.: control of the autonomic nervous system.
Y. Zhou, Ph.D.: modulation of ion channels; regulation of neuronal excitability and synaptic transmission.

Virology
W. Britt, M.D.: human herpesviruses, molecular virology, and pathogenesis.
L. Chow, Ph.D.: human papillomavirus DNA replication and pathogenesis.
T. T. Dokland, Ph.D.: cryoelectron microscopy and X-ray crystallography of virus assembly processes.
J. Engler, Ph.D.: identifying novel diagnostic peptides and cell-specific ligands.
B. Hahn, M.D.: origin and evolution of primate lentiviruses.
J. Kappes, Ph.D.: HIV, molecular virology, and pathogenesis.
R. Kaslow, M.D.: immunogenetic determinants in AIDS and other diseases.
O. Kutsch, Ph.D.: HIV-1 latency and drug screening.
E. Lefkowitz, Ph.D.: bioinformatics; biodefense; microbial genomics and evolution.
M. Luo, Ph.D.: structure-based approaches to anti-infectious agents.
P. Prevelige, Ph.D.: structural biology of viral assembly and infection.
G. Shaw, M.D., Ph.D.: evolution and persistence of HIV-1.
W. Sullender, M.D.: respiratory syncytial virus; antigenic diversity.
S. Thompson, Ph.D.: translation initiation and replication of RNA viruses.
R. Whitley, M.D.: herpesvirus; varicella zoster virus.
A. Zajac, Ph.D.: antiviral immunity; T-cell responses; immunological memory.

UNIVERSITY OF ARKANSAS FOR MEDICAL SCIENCES

College of Medicine
Department of Neurobiology and Developmental Sciences

Programs of Study

The Department of Neurobiology and Developmental Sciences offers graduate training programs leading to the M.S. and Ph.D. degrees. The objective of these programs is to prepare graduates for careers in research and teaching. Areas of faculty research specialization include various aspects of neuroscience and developmental and cell biology. The Department's research-oriented M.S. and Ph.D. graduate programs offer access to a curriculum of multidepartmental core courses that include cell biology, physiology, biochemistry, gene expression, pharmacology, and ethics. A flexible menu of electives in specialized areas is offered to complement the research interests of students.

Research Facilities

The Department has enjoyed rapid growth in funding for its research programs. Extramural grant support for full-time faculty members in the Department totaled more than $5.8 million in 2004. The Department has two research divisions. The Division of Translational Neuroscience is exploring ways to eliminate neglect in stroke patients, comorbidities associated with back pain, spinal cord injuries and rehabilitation, long-term effects of neonatal pain, motion disorders, sleep and psychiatric disorders, depression and cardiac disease, and pain mechanisms. The Division of Cellular and Molecular Neuroscience is conducting research in the rapidly growing areas of neuroimmunology, neuroinflammation, and glial cell biology. Specific faculty interests include immune responses to brain infection, multiple sclerosis, Alzheimer's disease, fetal alcohol syndrome, and neuroendocrine regulation of growth and reproduction. In addition, developmental biology is an emerging research magnet area, with projects focused on pituitary cell differentiation, age-related changes in neuroendocrine function, and molecular regulation of protein synthesis in oocytes.

Laboratories equipped for in vivo and in vitro electrophysiology, tissue culture, biochemistry, molecular biology, laser microdissection, and image analysis allow for a breadth of research possibilities. The campus library has 3,463 journal subscriptions (online or in print), 43,000 volumes in its book collection, and large collections of audiovisual and computer-based instructional materials.

Financial Aid

Teaching and research assistantships are available on a competitive basis. The stipend is $21,500, and tuition is paid for graduate assistants.

Cost of Study

Tuition for the 2006–07 year is $252 per credit hour for Arkansas residents and $540 per credit hour for out-of-state students.

Living and Housing Costs

UAMS is located about 3 miles west of downtown Little Rock. The campus has two new (as of fall 2006) residency halls. Reasonably priced rooms, apartments, and houses are located near the University.

Student Group

There are 2,328 students enrolled in the various professional schools (Medicine, Pharmacy, Public Health, Health Related Professions, and Nursing) at the University of Arkansas for Medical Sciences. There are 129 students in the basic science departments of the Graduate School. A campuswide student association offers social and collegial relationships with other graduate students.

Location

Metropolitan Little Rock, which is located at the geographic center of Arkansas, has a population of approximately 300,000. It is the state capital and is served by several airlines and by the interstate highway system. The Ouachita, Ozark, and Boston Mountains are nearby. There are many lakes, rivers, and streams within driving distance. The city provides cultural opportunities, including the Arkansas Arts Center, the William J. Clinton Presidential Library, a symphony orchestra, several professional theater companies, and a zoo. There is a minor-league baseball team, and the University of Arkansas football team plays some of its home games in Little Rock.

The University

The University of Arkansas for Medical Sciences belongs to the University of Arkansas System. The Graduate School is part of the University of Arkansas for Medical Sciences, an institution that also includes colleges of medicine, public health, pharmacy, nursing, and health-related professions. Other units of the University of Arkansas system that are located in Little Rock include the University of Arkansas at Little Rock, the School of Law, and the Graduate School of Social Work.

Applying

The Department of Neurobiology and Developmental Sciences seeks applicants who have a strong undergraduate background in the biological and physical sciences. Score reports on the General Test of the Graduate Record Examinations are required of all applicants, as are transcripts of all academic records and three letters of recommendation. Interviews can be arranged at any time, and visits are encouraged. Prospective students should apply as early as possible in the academic year preceding the desired year of entrance. For further information and application forms for admission and financial aid, students should write to the address listed in this description.

Correspondence and Information

Dr. David L. Davies
Director of Graduate Studies
Department of Neurobiology and Developmental Sciences, #510
University of Arkansas for Medical Sciences
4301 West Markham Street
Little Rock, Arkansas 72205-7199
E-mail: daviesdavidl@uams.edu
Web site: http://www.uams.edu/neuroscience_cellbiology

University of Arkansas for Medical Sciences

THE FACULTY AND THEIR RESEARCH

E. D. Al-Chaer, Associate Professor (also with Pediatrics); J.D., Ph.D., Texas Medical Branch, 1996. Neural mechanisms associated with pain.

K. J. S. Anand, Professor (also with Pediatrics); M.D., Ph.D., Mahatma Gandhi (India), 1980. Neurobiology and behavioral effects of neonatal pain.

Steven W. Barger, Associate Professor (also with Geriatrics); Ph.D., Vanderbilt, 1992. Alzheimer's disease; neuroinflammation; aspects of signal transduction, transcription, and gene expression specific to neurons; neuronal cell death; excitotoxicity.

Helen Beneš, Research Associate Professor; Ph.D., Arkansas, 1983. Insect models to understand development and aging in multicellular organisms: the role of glutathione S-transferases (GSTs) in preventing oxidative damage during aging.

E. Robert Burns, Professor; Ph.D., Tulane, 1967. Experimental oncology/cell kinetics/chronochemotherapy; K–14 education.

M. Donald Cave, Professor; Ph.D., Illinois, 1965. Development of diagnostic molecular probes for microbial pathogens.

Jason Y. Chang, Associate Professor (also with Ophthalmology); Ph.D., Ohio State, 1988. Neuronal cell death; neurotoxicity; microglia; oxidative stress.

Amanda Charlesworth, Research Assistant Professor; Ph.D., University College, London, 1996. Molecular mechanisms that regulate mRNA translation in early development.

Yuzhi Chen, Assistant Professor (also with Geriatrics); Ph.D., Massachusetts Amherst, 1997. Neuroinflammatory response in Alzheimer's disease; neddylation and ubiquitination; adult brain stem cells.

Gwen V. Childs, Professor and Chairman; Ph.D., Iowa, 1972. Neuroendocrinology; pituitary regulation; reproductive system; cytochemistry and cell biology.

David L. Davies, Associate Professor; Ph.D., LSU Medical Center, 1982. Pathogenesis of brain damage following exposure to ethanol.

Maxim Dobretsov, Associate Professor (also with Anesthesiology); Ph.D., Saint Petersburg State (Russia), 1986. Neurobiology of pain; diabetic neuropathy.

John Dornhoffer, Professor (also with Otolaryngology); M.D., Kansas, 1988. Neurovestibular research; hearing loss and vertigo.

Paul D. Drew, Associate Professor; Ph.D., Maryland, 1989. Neuroimmunology and neurodegenerative disorders; regulation of gene expression; cytokine signaling: inflammation (multiple sclerosis and Alzheimer's disease).

Edgar Garcia-Rill, Professor; Ph.D., McGill, 1973. Disorders of reticular activating system; restitution locomotor function after spinal cord injury.

W. Sue T. Griffin, Professor (also with Geriatrics); Ph.D., Rochester, 1974. Developmental biology of central nervous system–immune system interactions in health and disease and age-related changes.

Abdallah Hayar, Assistant Professor; Ph.D., Louis Pasteur (France), 1996. Role of synchronized bursting activity in olfactory bulb processing of sensory input.

Cynthia J. M. Kane, Associate Professor (also with Pediatrics); Ph.D., Arkansas, 1986. Cytokine regulation of neurons and glia during development, injury, and neurodegenerative disease; fetal ethanol exposure in the developing brain.

Tammy L. Kielian, Associate Professor; Ph.D., Kansas, 1998. Role of pattern-recognition receptors in glial activation; effects of inflammation on CNS gap-junction communication; microbial pathogenesis in the CNS.

Robert E. Mrak, Professor (also with Pathology); M.D./Ph.D., California, Davis, 1976. Neuropathology, Alzheimer's disease.

Angus MacNicol, Associate Professor (also with Physiology and Biophysics and the Arkansas Cancer Research Center); Ph.D., University College (London), 1987. Molecular biology of early vertebrate development; signal transduction; neural differentiation and stem cell function.

Melanie C. MacNicol, Research Assistant Professor; Ph.D., UCLA, 1996. Cellular signal transduction in neuronal and oncogenic models.

Mark Mennemeier, Associate Professor (also with Psychiatry); Ph.D., Southern Illinois, 1988. Neural mechanisms of neglect.

Bruce W. Newton, Associate Professor; Ph.D., Kentucky, 1984. Hormone and target regulation of sexually dimorphic spinal nuclei; development of spinal autonomic nervous system; sex differences in pain.

Kevin D. Phelan, Associate Professor; Ph.D., Michigan State, 1988. Neurophysiology and neuropharmacology; regulation of synaptic transmission; neuronal-glial interactions; neuroimmunology.

Terry J. Sims, Associate Professor; Ph.D., Rochester, 1974. Development of the nervous system; regeneration following injury to the CNS.

Robert D. Skinner, Professor; Ph.D., Texas, 1969. Neurophysiology and neuroanatomy of vertebrate brain stem and spinal cord locomotor systems; rat model of anxiety; human auditory-evoked potentials in psychiatric diseases.

Patrick W. Tank, Professor; Ph.D., Michigan, 1976. Pattern formation during limb development and regeneration.

Yi-Hong Zhou, Assistant Professor; Ph.D., Dalhousie, 1994. Biology of brain tumors.

Recent Faculty Publications

Saab, C. Y., Y.-C. P. Arai, and **E. D. Al-Chaer.** Modulation of visceral nociceptive processing in the lumbar spinal cord following thalamic stimulation or inactivation and after dorsal column lesion in rats with neonatal colon irritation. *Brain Res.* 1008: 186–92, 2004.

Narsinghani, U., and **K. J. S. Anand.** Developmental neurobiology of pain in neonatal rats. *Nature (Lab Anim.)* 29:27–39, 2000.

Massa, P.T., et al. **(S. W. Barger).** NF-kappa-B in neurons? The uncertainty principle in neurobiology. *J. Neurochem.* 97:607–18, 2006.

Zakharkin, S. O., et al. **(H. Beneš).** Identification of two mariner-like elements in the genome of the autogenous mosquito, *Ochlerotatus atropalpus. Insect Biochem. Mol. Biol.* 34:377–86, 2004.

Burns, E. R. Clinical histology. *Clin. Anat.* 19:156–63, 2006.

Barnes, P. F., and **M. D. Cave.** Molecular epidemiology of tuberculosis: Relevance to disease control and pathogenesis. *N. Engl. J. Med.* 249:1149–56, 2003.

Garg, T. K., and **J. Y. Chang.** Methylmercury causes oxidative stress and cytotoxicity in microglia: Attenuation by 15-deoxy-delta 12, 14-prostaglandin J2. *J. Neuroimmunol.* 171:17–28, 2006.

Charlesworth, A., et al. **(A. M. MacNicol).** Musashi regulates the temporal order of mRNA translation during *Xenopus* oocyte maturation. *EMBO J.* 25:2792–801, 2006.

Chen, Y., et al. APP-BP1 mediates APP-induced apoptosis and DNA synthesis and is increased in Alzheimer's disease brain. *J. Cell Biol.* 163:27–33, 2003.

Davies, D. L., I. R. Niesman, F. A. Boop, and **K. D. Phelan.** Heterogeneity of astroglia cultured from adult human temporal lobe. *Int. J. Dev. Neurosci.* 18:151–60, 2000.

Dobretsov, M., and J. R. Stimers. Neuronal function and a3 isoform of the Na+, K+-ATPase. *Frontiers Biosci.* 10:2373–96, 2005.

Xu, J., and **P. D. Drew.** 9-cis retinoic acid suppresses inflammatory responses of microglia and astrocytes. *J. Neuroimmunol.* I171:135–44, 2006.

Garcia-Rill, E. Mechanisms of sleep and wakefulness. In *Sleep Medicine,* eds. Hawley and Belfus, Philadelphia, 2001.

Grimaldi, L. M. E., et al. **(W. S. T. Griffin).** Association of early onset Alzheimer's disease with an interlukin-1a gene polymorphism. *Ann. Neurol.* 47:361–5, 2000.

Carter, C. A., and **C. J. M. Kane.** Therapeutic potential of natural compounds that regulate the activity of protein kinase C. *Curr. Med. Chem.* 11:2883–902, 2004.

Esen, N., and **T. Kielian.** Central role for MyD88 in the responses of microglia to pathogen-associated molecular patterns (PAMPs). *J. Immunol.* 176:6802–11, 2006.

Mrak, R. E., and W. S. T. Griffin. Glia and their cytokines in progression of neurodegeneration. *Neurobiol. Aging* 349–54, 2005.

Phelan, K. D., and **B. W. Newton.** Sex differences in the response of rat lamina X neurons to exogenously applied galanin recorded in vitro. *Dev. Brain Res.* 122:157–63, 2000.

Sims, T., M. B. Durgan, and S. A. Gilmore. Transplantation of sciatic nerve segments into normal and glia-depleted spinal cords. *Exp. Brain Res.* 125:495–501, 1999.

Skinner, R. D., Y. Homma, and **E. Garcia-Rill.** Arousal mechanisms related to posture and locomotion. II. Ascending modulation. In *Brain Mechanisms for the Integration of Posture and Movement, Progress in Brain Research Series,* vol. 143, pp. 291–8, eds. S. Mori, D. Stuart, and M. Wiesendanger. Elsevier Science, 2004.

Iruthayanathan, M., **Y.-H. Zhou,** and **G. V. Childs.** DHEA restoration of GH gene expression in aging female rats, in vivo and in vitro: Evidence for actions via estrogen receptors. *Endocrinology* 146:5176–87, 2005.

Zhou Y.-H., et al. PAX6 suppresses growth of human glioblastoma cells. *J. Neurooncol.* 71: 223–9, 2005.

UNIVERSITY OF CALIFORNIA, BERKELEY

Neuroscience Graduate Program

Program of Study

The Neuroscience Graduate Program is a campuswide organization that includes more than 40 faculty members with state-of-the-art laboratories from the Departments of Molecular and Cell Biology, Psychology, and Integrative Biology in the College of Letters and Sciences; the Department of Chemical Engineering in the College of Chemistry; the Department of Environmental Science Policy and Management in the College of Natural Resources; the Helen Wills Neuroscience Institute; the School of Public Health; and the School of Optometry's Program in Vision Sciences. Faculty members participate in neuroscience graduate training and research from the molecular and genetic levels to the cognitive and computational levels. Areas of training and research include analysis of ion channels, receptors, and signal transduction mechanisms; formation, function, and plasticity of synapses; control of neural cell fate and pattern formation; neuronal growth cone guidance and target recognition; mechanisms of sensory processing in the visual, auditory, and olfactory systems; development and function of neural networks; motor control; and the neural basis of cognition. The preparations in use range from reductionist models to complex neural systems and include cells in culture, simple invertebrate and vertebrate organisms, model genetic systems, the mammalian cerebral cortex, and human brain imaging.

The graduate training includes a series of graduate-level courses, three laboratory rotations during the first year, placement in a thesis laboratory starting with the second year, a qualifying examination that includes a thesis proposal at the end of the second year, two semesters of teaching assistantship, and approximately three years of in-candidacy, full-time research ending with the thesis defense.

Research Facilities

In addition to the neuroscience faculty members' research laboratories, the Neuroscience Institute has three technology centers: the Brain Imaging Center, with a 4-Tesla fMRI and a networked computer system; the Functional Genomics Laboratory; and the Molecular Imaging Center. All facilities are used for research and training purposes. Faculty laboratories and the technology centers are housed in various buildings on the UC Berkeley main campus.

Financial Aid

All graduate students are fully supported by the program during their training. Financial aid packages include scholarships, teaching assistantships, and research assistantships. Applicants are strongly encouraged to apply for National Science Foundation fellowships in November during their senior year. University fellowships are also available. The 2006–07 minimum stipend level is set at $26,000 for twelve months.

Cost of Study

In 2006–07, University registration fees for legal California residents amount to approximately $8440 for the academic year, or two semesters. For nonresidents, tuition and fees for the same period of study amount to approximately $23,400. All admitted students have fees, tuition, and health and dental insurance paid in full by the program.

Living and Housing Costs

Most graduate students live in apartments or houses in the vicinity of the campus. Rents for shared housing range from $800 to $1000 per month. The University maintains a family student housing complex for married couples and single-parent families. The Student Housing Office has its own Web site (http://www.housing.berkeley.edu/) and assists students in locating housing both on and off campus.

Student Group

The Berkeley campus has approximately 35,000 students, including 9,500 graduate students. More than 2,800 international students from more than 100 countries add to the cosmopolitan flavor of the campus. The Neuroscience Graduate Program, with more than 100 graduate student affiliates from widely varied backgrounds and with wide-ranging interests and pursuits, is a vigorous, stimulating group within Berkeley's outstanding scientific community. Cohesion within this group is enhanced by weekly seminars, Journal Club activities, and annual off-site neuroscience retreats.

Student Outcomes

Neuroscience graduates find employment as postdoctoral fellows at peer institutions such as Caltech, Columbia, Harvard, MIT, Stanford, the University of California at San Francisco, the University of California at San Diego, and the NIH; as faculty members in academia; and as researchers in both private and public institutions. Some graduates work in biotechnology firms or choose to continue their studies at medical, veterinary, or law schools.

Location

The campus, at the foot of the Berkeley Hills and in the center of the city of Berkeley, is surrounded by business and residential districts. The setting retains a park atmosphere, with wooded glens, spacious plazas, and the picturesque Strawberry Creek running the length of the campus. Many areas and buildings on campus provide panoramic views of San Francisco and the entire nine-county Bay Area, a region renowned for its cultural and recreational activities, including public lectures, concerts, theater and other dramatic presentations, museums, galleries, exhibits, and a wide variety of restaurants, eateries, and cafés. Hiking and biking trails are in abundance, and sailing on the bay is nearby.

The University

The Berkeley campus, which was founded in 1868, is the oldest of the ten University of California campuses. Just to name a few of its outstanding and diverse faculty members, there are 6 out of a total of 19 Nobel laureates who are still active faculty members. There are 196 American Association for the Advancement of Science members, 230 American Academy of Arts and Sciences Fellows, 28 MacArthur Fellows (one of the most recent is Lu Chen, an active faculty member of the Neuroscience Graduate Program), 132 National Academy of Sciences members, 14 recipients of the National Medal of Science, 1 poet laureate, and 3 Pulitzer Prize winners. Generally considered the leading public university in the nation, with its graduate programs ranked first in the nation in both number (97 percent are in the top ten list) and scholarship (thirty-two "distinguished" programs) according to the National Research Council, Berkeley is continually developing new programs and redefining existing ones, such as the integrated Neuroscience Graduate Program, in an attempt to meet each new challenge and forge new paths.

Applying

The Neuroscience Graduate Program admits students starting in the fall semester only. Inquiries regarding application procedures should be directed to the Graduate Affairs Office. Applicants must use the online application method at http://www.grad.berkeley.edu/. The application deadline is December 15, but students should apply as early as possible. Graduate Record Examinations scores (Institute Code 4833; Department Code 0213) for the General Test and one Subject Test (same codes; tests in biochemistry and cell biology, chemistry, biology, psychology, computer science, or physics), a TOEFL (Institute Code 4833) for international applicants whose first language is not English (with a minimum score of 570 on the paper test or 233 on the CBT), and a minimum GPA of 3.0 are required. Admitted students have historically scored above the high 80th percentile on the tests and have higher average GPAs. The Neuroscience Graduate Program actively solicits applications from underrepresented groups.

Correspondence and Information

Graduate Affairs Office
Helen Wills Neuroscience Institute
132 Barker Hall, MC#3190
University of California, Berkeley
Berkeley, California 94720

Phone: 510-643-6438
Fax: 510-642-3192
E-mail: neurosci@uclink.berkeley.edu
Web site: http://neuroscience.berkeley.edu

University of California, Berkeley

THE FACULTY AND THEIR RESEARCH

Faculty members in the Neuroscience Institute are divided into five research areas: molecular neuroscience, cellular neuroscience, developmental neuroscience, systems neuroscience, and cognitive neuroscience. Individual faculty members may be involved in more than one research area.

Martin S. Banks, Professor. Visual space perception; psychophysics; modeling of vision; virtual reality.
Shaowen Bao, Assistant Adjunct Professor. Sensory processing.
George Bentley, Assistant Professor. Neural integration and transduction of environmental cues into endocrine signals.
Jose Carmena, Assistant Professor. Brain-machine interfaces; neural mechanisms of sensorimotor control and learning; neural ensemble computation; neuroprosthetics.
Lu Chen, Assistant Professor (MacArthur Fellow). Mechanisms of synapse formation during development and synapse modification in plasticity.
Yang Dan, Professor. Processing and computation of visual information in the thalamus and cortex.
Mark D'Esposito, Professor. Neural basis of working memory in humans; functions of human prefrontal cortex; *f*MRI.
Karen DeValois, Professor. Mechanisms of color and spatial vision.
Marian Cleeves Diamond, Professor. Mammalian forebrain structures and functions, especially the cerebral neocortex, including effects of enriched and impoverished experiential environments, aging, and immune responses.
John Flannery, Professor. Molecular genetics of transduction and retinal degeneration; gene therapy for retinal diseases.
Darlene Francis, Assistant Professor. Behavioral neuroscience; developmental psychobiology; animal models; stress; maternal care; gene-environment interaction.
Ralph Freeman, Professor. Development, plasticity, and function of central visual pathways.
Jack Gallant, Associate Professor. Neural mechanisms of visual form perception and attention.
Gian Garriga, Professor. Neuronal migration and axonal pathfinding in *C. elegans*.
Adam Gazzaley, Assistant Adjunct Professor. Neural mechanisms of attention and memory.
Donald Glaser, Professor and Nobel Laureate. Computational modeling of human vision.
Stephen Glickman, Professor. Comparative studies of species-characteristic behavior patterns in mammals.
Xiaohua Gong, Assistant Professor. Eye development and disease; cell-to-cell communication; intracellular signaling pathways in the lens.
Ehud Isacoff, Professor. Biophysical properties of ion channels; advanced optical methods.
Richard Ivry, Professor. Neural basis of motor control and motor learning in humans.
Lucia Jacobs, Associate Professor. Neural mechanisms of complex behaviors.
William Jagust, Professor. Structural and functional imaging of aging and dementia.
Daniela Kaufer, Assistant Professor. Molecular mechanisms underlying stress responses in the brain.
Stanley Klein, Professor. Modeling of spatial vision and its application in image compression.
Robert Knight, Professor. Role of human prefrontal cortex in attention and memory control.
Richard Kramer, Associate Professor. Mechanisms of cell signaling mediated by cyclic nucleotides.
Lance Kriegsfeld, Assistant Professor. Genetic, cellular, and hormonal mechanisms responsible for the temporal control of motivated behaviors.
Harold Lecar, Professor. Biophysical properties of ion channels; theoretical biophysics.
Dennis Levi, Professor. Peripheral and central mechanisms of amblyopia.
John Ngai, Professor. Cellular and molecular mechanisms of olfaction; functional genomics.
Bruno Olshausen, Associate Professor. Computational models of perception.
Mu-ming Poo, Professor. Mechanisms of axon guidance; synaptic plasticity.
Lynn Robertson, Adjunct Professor. Neural basis of human perception and attention.
David Schaffer, Associate Professor. Mechanisms of stem cell differentiation.
Kristin Scott, Assistant Professor (Merck Fellow). Taste recognition in *Drosophila*.
Arthur Shimamura, Professor. Biological basis of memory and cognitive functions in humans.
Michael Silver, Assistant Professor. Neural correlates of human visual perception and attention.
Noam Sobel, Associate Professor. Psychophysics of human olfaction; mechanisms of olfactory processing in humans; *f*MRI.
Friedrich Sommer, Associate Adjunct Professor. Models of associative memory; sensory processing.
Mark Tanouye, Professor. Molecular genetics and physiology of behavior in *Drosophila*.
Frédéric Theunissen, Associate Professor. Neural mechanisms underlying complex sound recognition.
Jonathan Wallis, Assistant Professor. Neuronal mechanisms of goal-directed behavior.
Frank Werblin, Professor. Processing of visual information in the retina.
Jeffery Winer, Professor. Neuroanatomy of the central auditory pathways.
Irving Zucker, Professor. Circadian rhythms; seasonal reproductive cycles; behavioral endocrinology; hibernation; daily torpor.
Robert Zucker, Professor. Role of calcium in transmitter release, synaptic transmission, and plasticity.

UNIVERSITY OF CALIFORNIA, DAVIS

Division of Biological Sciences
Neuroscience Doctoral Program

Program of Study	The Neuroscience Doctoral Program at the University of California, Davis (UC Davis), offers a program of graduate studies leading to the Ph.D. degree in neuroscience. The program is interdepartmental and has one of the largest faculties of any neuroscience program in the United States. The faculty bridges both basic science and clinical research, thereby offering uniquely broad research opportunities for students in the doctoral program. The research interests of the faculty include molecular, cellular, developmental, systems, behavioral, computational, cognitive, and neurological approaches to neuroscience. In order to understand the biology of learning, memory, language, behavior, disease, and sensory and motor systems, faculty members employ a diverse array of research techniques. These techniques include molecular genetics; biochemistry; sophisticated in vivo and in vitro electrophysiological techniques; optical, confocal, and multiphoton imaging; computational modeling; psychophysics; and fMRI and PET scanning.
	The Neuroscience Doctoral Program is designed to provide students with a thorough understanding of cellular, molecular, systems, cognitive, and developmental neuroscience. All students are required to take a core curriculum in these subjects along with neuroanatomy and quarterly journal clubs. The goal is to complete the program in five years. The first year entails course work and laboratory rotations. Students select a thesis mentor at the end of the first year and begin their research in the second year. The exam for advancement to candidacy consists of two parts: a combined written and oral qualifying exam and a thesis proposal. The qualifying examination is taken in the summer of the first year and the thesis proposal is completed by the end of the summer of the second year. The students' remaining training consists of successfully completing a research project and submitting the dissertation.
Research Facilities	The Neuroscience Doctoral Program is richly equipped for the full spectrum of modern neuroscience research and is among the best programs in the nation. Special research resources include the Center for Neuroscience, the M.I.N.D. Institute for the study of neurodevelopmental disorders, the Center for Mind and Brain, the California National Primate Center, the UC Davis Alzheimer's Research Center, the Medical School, the Veterinary School, the Institute of Theoretical Dynamics, and the Veterans Administration Medical Center. The Center for Neuroscience, in particular, is an important resource for the program. The center is an interdisciplinary unit that serves as a focal point for the study of neurosciences on the UC Davis campus. Through the center, students have access to cutting-edge equipment such as a multiphoton imaging system, several confocal microscopes, a gene chip array system, shared molecular equipment, a mass spectrometer, electron microscopes, and fMRI and PET scanners. The center also provides students with frequent exposure to scientists in all areas of neuroscience through its intense seminar series and emphasis on student interaction with the speakers. To provide cohesion among members of the program, the center also sponsors yearly retreats and other social gatherings.
Financial Aid	The UC Davis Neuroscience Doctoral Program guarantees full financial support, fees, and tuition for every student who is admitted to the program and who continues to make satisfactory progress. Support for graduate students comes from University fellowships, training grants, and research grants. Awards include tuition, fees, health insurance, and a stipend. For 2004–05, a stipend of $21,250 was provided for each student.
Cost of Study	All academic expenses, except books, are paid for every student admitted who continues to make satisfactory progress.
Living and Housing Costs	UC Davis offers University-operated on-campus housing for both single and married graduate students. In addition, in the city of Davis there are apartments, duplexes, and houses available to rent ($600 to $1100 for a one-bedroom apartment). All housing in Davis is within easy cycling distance of the campus and is close to campus bus stops.
Student Group	The University of California, Davis, has a total enrollment of nearly 25,000 full-time students. Of this number, approximately 3,215 are engaged in graduate studies, with approximately one third of those enrolled in a discipline relating to biological science. The Neuroscience Doctoral Program was founded in 1990 at the University of California, Davis. For the 2005–06 academic year, there were 42 students in the Neuroscience Doctoral Program. Students in the program have diverse backgrounds and a range of research interests that parallel the breadth of the faculty. Approximately 50 percent of the students are women.
Location	The city of Davis, with a population of approximately 52,000, is located in California's fertile Central Valley. It is situated 12 miles west of the capital city of Sacramento, 75 miles east of the San Francisco Bay Area, 120 miles west of Lake Tahoe, and 40 miles from Napa Valley. The 5,146-acre campus is surrounded by rich agricultural fields and includes a formal arboretum intertwined with a 2-mile-long footpath along the Putah Creek. Oak, redwood, and dryland trees and shrubs are predominant throughout the campus. As well as being a world-famous campus, Davis is known for its attention to the quality of college-town life and to the environment. In keeping with its environmental consciousness, Davis has the largest number of bicycles per capita of any city in the United States.
The University	UC Davis offers a full range of undergraduate and graduate programs, as well as professional schools in law, management, medicine, and veterinary medicine. It is the largest of the nine University of California campuses in acreage, second largest in budget, and third largest in enrollment.
Applying	Students are considered for admission for the fall quarter only. Applications can be obtained from the program's Web site. The application deadline is January 5. Late applications are considered as space and funding permit.
	Applicants sought are those who have a strong undergraduate background in the biological sciences, psychology, and/or the physical sciences. Students are selected for admission to the Neuroscience Doctoral Program on the basis of the quality of their previous academic work and their promise as creative scientists. Students should submit scores from the General Test and an appropriate Subject Test of the Graduate Record Examinations as well as official transcripts and three letters of recommendation. The program does not offer a separate master's degree.
Correspondence and Information	Graduate Program Coordinator Neuroscience Doctoral Program Center for Neuroscience University of California, Davis 1544 Newton Court Davis, California 95616 Phone: 530-757-8845 Fax: 530-757-8827 E-mail: tacrowder@ucdavis.edu Web site: http://neuroscience.ucdavis.edu/grad/

University of California, Davis

THE FACULTY AND THEIR RESEARCH

D. Amaral, Ph.D. Multidisciplinary analysis of neural systems involved in primate and human social behavior and memory.
K. Baynes, Ph.D. Mental representations of language.
R. Berman, Ph.D. Mechanisms of neuroplasticity; brain growth and repair.
A. Bonham, Ph.D. CNS control of the cardiovascular and respiratory systems.
K. Britten, Ph.D. Information processing in extrastriate visual cortex.
S. Bunge, Ph.D. Cognitive control; working memory; executive functions.
M. Burns, Ph.D. Temporal regulation of G-protein signaling in the phototransduction cascade; cellular neurophysiology.
E. Carstens, Ph.D. Sensory neurophysiology; pain and analgesia.
C. Carter, Ph.D. Neural mechanisms of attention and memory.
L. Chalupa, Ph.D. Development/plasticity of the mammalian visual system.
B. Chapman, Ph.D. Development of specific neuronal connections.
T.-Y. Chen, Ph.D. Structure and function of a class of transmembrane proteins; ion channels.
H.-J. Cheng, Ph.D. Molecular and cellular mechanisms of axon guidance.
B. Corbett, Ph.D. Autism spectrum disorders in children.
G. Cortopassi, Ph.D. Mitochondrial genetics and mitochondrial dysfunction in neurological disease.
W. DeBello, Ph.D. Cellular mechanisms of adaptive plasticity.
C. DeCarli, Ph.D. Relationship between brain structure and function in aging and disease.
E. Diaz, Ph.D. Molecular analysis of synapse formation.
E. Disbrow, Ph.D. Involvement of areas of the cerebral cortex in higher-order somatosensory processing.
J. Ditterich, Ph.D. Physiology of decision making and motor preparation in the cortex.
M. Ferns, Ph.D. Cellular and molecular basis of synapse formation in the development of the mammalian nervous system.
D. Gietzen, Ph.D. Neural control of food intake; responses to amino acid deficiency.
F. Gorin, M.D., Ph.D. Molecular neurobiology; neuroimmunology.
P. Hagerman, M.D. Molecular and clinical aspects of fragile-X mental retardation and autism.
R. Hagerman, M.D. Molecular clinical correlations in fragile X and treatment studies in autism and other neurodevelopmental disorders.
L. Hall, Ph.D. Molecular genetics and functional analysis of ion channels and G-protein-coupled receptors in *Drosophila*.
A. Ishida, Ph.D. Light adaptation in proximal retina.
P. Janata, Ph.D. Basic neural systems that underlie perception, attention, memory, action, and emotion interact in the context of natural behaviors, with an emphasis on music.
E. Jones, M.D., Ph.D. Structure, function, and development of the thalamus and cerebral cortex.
P. Knoepfler, Ph.D. Embryonic development, healing, and regeneration of stem cells.
L. Krubitzer, Ph.D. Anatomical connections and electrophysiological properties of neurons in the neocortex.
J. LaSalle, Ph.D. Human genetics; epigenetics; Rett syndrome; Angelman syndrome.
N. L'Etoile, Ph.D. Olfactory adaptation and discrimination using the model organism *C. elegans*.
B. Lyeth, Ph.D. Pathophysiology of traumatic brain injury.
R. Maddock, M.D. fMRI, cognitive, and psychophysiological studies of affective processes.
G. R. Mangun, Ph.D. Cognitive neuroscience of attention.
A. McAllister, Ph.D. Cellular and molecular mechanisms of synapse formation in the developing cerebral cortex.
L. Miller, Ph.D. Auditory neuroscience and speech recognition.
B. Mulloney, Ph.D. Neurobiology; neuroethology; computational neuroscience; education for careers in biological research.
P. Pappone, Ph.D. Function of extracellular ATP/P2Y receptor signaling in adipocytes.
I. Pessah, Ph.D. Molecular mechanisms by which environmental factors influence neurodevelopment.
D. Pleasure, M.D. Cellular and molecular biological techniques to study the development and regeneration of the nervous system.
C. Ranganath, Ph.D. Cognitive neuroscience of human memory; functions of the prefrontal cortex and medial temporal lobes; neuroimaging; human neuropsychology.
G. Recanzone, Ph.D. Neural correlates of auditory perception.
D. Richman, M.D. Structure-function relationships of the neuromuscular junction and the nicotinic acetylcholine receptor; human and animal model diseases of neuromuscular transmission.
S. Rivera, Ph.D. Origins and development of symbolic representation in infants and children.
P. Schwartzkroin, Ph.D. Cellular, synaptic, and network properties that give rise to and/or result from seizures and epilepsy.
F. Sharp, M.D. Molecular neurobiology; genomics; neural cell injury and death; blood genomics of neurological diseases.
K. Sigvardt, Ph.D. Neural mechanisms and modeling of Parkinson's disease.
M. Sutter, Ph.D. Auditory system performance through psychophysics; recording single neuron's responses to sound.
D. Swick, Ph.D. Cognitive neuropsychology of memory, language, and executive functions; human electrophysiology and functional neuroimaging.
J. Trimmer, Ph.D. Ion channel signaling complexes in mammalian neurons.
W. Usrey, Ph.D. Functional properties of neural circuits for vision.
A. Vazquez, Ph.D. Molecular basis of age-related hearing loss and hearing loss due to exposure to loud noise.
J. Werner, Ph.D. Color and spatial vision and their changes across the life span.
M. Wilson, Ph.D. Cellular mechanisms of signal processing in the retina, in particular, how retinal synapses work.
D. Woods, Ph.D. Cortical bases of attention and perception using psychophysics, event-related brain potentials, and functional MRI.
E. Yamoah, Ph.D. Molecular and cellular mechanisms of mechanoelectrical transduction in hair cells.
A. Yonelinas, Ph.D. Neural substrates and functional properties of human memory; neuroimaging.
C. J. Zhou, Ph.D. Molecular mechanism of pattern formation during embryonic neural development and pattern maintenance for postembryonic neural stem cell renewal and tissue regeneration.

UNIVERSITY OF CALIFORNIA, SAN DIEGO

School of Medicine
Neuroscience Graduate Program

Program of Study

The Neuroscience Graduate Program, established in 1969, is a Ph.D. program administered by the Neuroscience Group under the auspices of the Department of Neuroscience at the University of California, San Diego (UCSD) School of Medicine. This program offers an outstanding opportunity for graduate training in one of the most highly interactive scientific environments available in the United States. Recently, the graduate program was named by the National Research Council of the National Academy of Sciences as the top ranking neuroscience graduate program in the country. There is an enormous breadth of neuroscience research interests represented among the faculty members, so that students are able to pursue study in a wide range of neuroscience areas, and they often carry out their dissertation research in collaboration with more than one laboratory. The common purpose of all the participants in the Graduate Program in Neuroscience is to foster and maintain a community of excellence in neuroscience and to prepare students to develop creative and innovative scientific research in order to achieve productive and successful careers.

Research Facilities

In addition to on-campus laboratories, the program has laboratories located at the Salk Institute, the Scripps Research Institute, the Scripps Institution of Oceanography, the UCSD/VA Medical Centers, and the Burnham Institute.

The more than 130 faculty members in the Neuroscience Group provide instruction and diverse research opportunities to graduate students entering the program.

Financial Aid

The first year of funding is provided by the Graduate Program. Beginning in the second year, responsibility for funding students shifts to the thesis adviser, although the program provides funds when they are available. International students should note that because they do not qualify for residence, the tuition and fees paid by their research adviser are higher. In order to offset the higher fees, these students are encouraged to apply for transferable fellowships from their home governments.

Cost of Study

In 2005–06, the tuition and fees for nonresident students for three quarters was $24,000.

Living and Housing Costs

There is a surplus of housing available near campus. One- and two-bedroom apartments range from $800 to $1500 per month. UCSD also provides limited, subsidized on-campus housing. Although a car is helpful, it is not necessary, as there are many cycling paths and reliable public transportation within the vicinity of campus. Bus service is also free within a several-mile radius of campus, making access to groceries and shopping fairly easy.

Student Group

Currently, there are 84 graduate students working in the laboratories of the Neuroscience Group members. More than 800 graduate students are matriculated in the various biomedical sciences programs of the University. In addition, there are approximately 500 medical students on campus.

Location

San Diego has a Mediterranean climate. The average daily temperature during the summer months is in the mid-70s. In the winter, one can expect daytime temperatures to reach the mid-60s. Average annual rainfall is below 9 inches, and the abundant sunshine permits outdoor activities throughout the year.

San Diego's beaches are one of its greatest attractions; swimming, sunbathing, and surfing in February are common. During the winter months, one can within several hours' driving distance, enjoy skiing on the slopes of Big Bear.

San Diego is the second-largest city in California and the sixth largest in the United States. The local San Diego area has many attractions, including the world-famous San Diego Zoo, the Wild Animal Park, Old Town, Sea World, and Balboa Park, which houses many of San Diego's museums.

The University

UCSD is one of the newest of the ten campuses that make up the University of California system. Today UCSD is recognized throughout the academic world both for the eminence of its faculty and for the quality of its graduate and undergraduate programs. The history of its growth may help to explain how UCSD has been able to achieve a stature comparable to that of institutions that were founded a century or more ago.

Applying

Applicants should have a B.S. or B.A. with course work in neuroscience, psychology, physiological psychology, mathematics through calculus, general physics, general biology, general chemistry, organic chemistry, biochemistry, computer science, or engineering. Deficiencies in these areas can be corrected through appropriate course work in the first year of residence. Scores on Graduate Record Examinations, scores on the TOEFL for international students, three letters of recommendation, official transcripts of undergraduate grades, and a statement of purpose are required. December 19 is the deadline for the receipt of the application form and all application materials. Applicants are informed of the admissions decision by March 15.

Correspondence and Information

Department of Neuroscience, 0662
University of California, San Diego
9500 Gilman Drive
La Jolla, California 92093-0662
E-mail: neurograd@ucsd.edu
Web site: http://medicine.ucsd.edu/neurosci

University of California, San Diego

THE FACULTY AND THEIR RESEARCH

Prospective students are encouraged to visit the Web site at http://medicine.ucsd.edu/neurosci for updated faculty and research information.

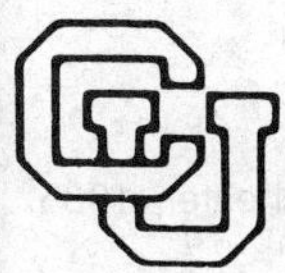

UNIVERSITY OF COLORADO AT DENVER AND HEALTH SCIENCES CENTER

Neuroscience Training Program

Program of Study

The University of Colorado at Denver and Health Sciences Center offers a multidisciplinary, interdepartmental training program in neuroscience. An internationally recognized faculty of more than 36 members from five basic science and four clinical departments provide training in molecular, cellular, systems, and developmental neuroscience leading to the Ph.D. or M.D./Ph.D. degree. Students are supported by NIH training grants.

During their first year in the program, students take core courses consisting of instruction in cellular and molecular neurobiology, systems neurobiology, developmental neurobiology, and pharmacology and anatomy of the central nervous system (CNS) as well as biochemistry and molecular biology. Students also participate in a faculty-student neuroscience seminar program and engage in laboratory research rotations. This training encompasses both a critical evaluation of the original literature and hands-on experience with state-of-the-art research techniques.

The laboratory rotations are designed to expose students to a variety of research areas and to help them decide on a specific area of neuroscience in which to concentrate and choose a thesis research project. By the second year, students select a faculty adviser to supervise their research and begin their thesis research. Upon satisfactory completion of a comprehensive examination, the student is a candidate for the Ph.D. degree in neuroscience.

Research Facilities

Faculty research activities span a broad spectrum from molecular to behavioral neuroscience. Faculty laboratories are well supported by NIH and other grants. A variety of shared facilities for confocal microscopy, 2-photon microscopy image analysis, oligonucleotide synthesis, peptide synthesis and microsequencing, fluorescence-activated cell sorting, generation of gene-targeted mice, gene chip, microarrays, and large-scale bacterial and mammalian cell culture are available in the Health Sciences Center. A research building has 20,000 square feet of research space for the neuroscience program as well as computer, library, and study rooms for student use. Associated neuroscience faculty members have an additional 44,000 square feet of laboratory space. The Denison Medical Library has more than 200,000 volumes and is online with several major libraries.

Financial Aid

Students admitted to the Neurobiology Training Program are eligible for fellowship support that pays tuition and fees and provides a stipend of $22,500 in 2007–08. Stipends increase on a regular basis.

Cost of Study

Tuition for 2007–08 is $123 per semester hour for Colorado residents and $575 per semester hour for nonresidents. Health insurance and fees are approximately $3082 annually. Students in the program receive complete support.

Living and Housing Costs

The Health Sciences Center is situated in an attractive residential neighborhood a few miles from downtown Denver. There are many apartments and houses within walking distance. Rents start at approximately $700 per month.

Student Group

The Health Sciences Center has about 2,500 students enrolled in its different programs, with about 360 students in its Ph.D. degree programs. The neuroscience program currently has 31 students, including 6 M.D./Ph.D. students and 15 women.

Location

Denver is a large metropolitan city of more than 2 million people, lying at the foot of the Rocky Mountains. Denver's cultural attractions include major museums, a symphony orchestra, numerous theaters, and professional sporting events, all within minutes of the Health Sciences Center. The high elevation, dry climate, and sunny days provide opportunities for year-round outdoor activities, including skiing, backpacking, climbing, fishing, and biking.

The Center

The University of Colorado at Denver and Health Sciences Center is the largest comprehensive health facility in Colorado and the Rocky Mountain region. It is composed of the University of Colorado Medical School, Dental School, Pharmacy School, Nursing School, Graduate School, and 400-bed University Hospital. This nationally recognized research facility is among the leaders in federal research support.

Applying

Students interested in pursuing a career in the neurosciences are invited to apply. Undergraduate course requirements include one year each of biology (including cell biology), calculus, organic chemistry, and physics. Some research experience and a course in biochemistry are strongly recommended. Applicants must take the GRE General Test, and a Subject Test is recommended. TOEFL scores are required of students whose first language is not English. Application materials are available online at the Graduate School Web site linked to the Neuroscience home page at http://www.uchsc.edu/neuroscience. For fall enrollment, completed applications must be received by December 15.

Correspondence and Information

Diego Restrepo, Ph.D., Director
Mellodee Phillips, Program Administrator
Neuroscience Training Program
University of Colorado at Denver and Health Sciences Center
P.O. Box 6511
Mail Stop 8315
Aurora, Colorado 80045
Phone: 303-724-3120
Fax: 303-724-3121
E-mail: mellodee.phillips@uchsc.edu
　　　　diego.restrepo@uchsc.edu
Web site: http://www.uchsc.edu/neuroscience

University of Colorado at Denver and Health Sciences Center

THE FACULTY AND THEIR RESEARCH

Diego Restrepo, Director, Neuroscience Program and Professor, Department of Cellular and Developmental Biology; Ph.D., Rochester, 1985. Neurobiology of olfaction at the systems and cellular levels; neural implants.

Kristin Artinger, Assistant Professor of Craniofacial Biology; Ph.D., California, Irvine, 1995. Developmental mechanisms involved during vertebrate embryogenesis.

Linda Barlow, Assistant Professor of Cellular and Developmental Biology; Ph.D., Washington (Seattle), 1991. Molecular-, cellular-, and tissue-level mechanisms that generate and pattern the taste system during embryogenesis.

Ulrich Bayer, Assistant Professor of Pharmacology; Ph.D., Heinrich-Pette-Institut, 1996. Molecular memory in signal transduction and neuronal function.

Kurt Beam, Professor of Physiology and Biophysics; Ph.D., Washington (Seattle), 1974. Biophysics of excitation-contraction coupling.

Tim Benke, Assistant Professor of Pediatric Neurology; M.D./Ph.D., Baylor, 1995. Impact of early-life seizures on synaptic plasticity.

William J. Betz, Professor and Chairman of Physiology; Ph.D., Yale, 1969. Exocytosis, endocytosis, and endosomal trafficking.

Steven G. Britt, Associate Professor of Cellular and Developmental Biology and Ophthalmology; M.D., Texas Medical Branch, 1986. Molecular genetics of sensory system development and discrimination.

Michael D. Browning, Professor of Pharmacology; Ph.D., California, Irvine, 1979. Molecular mechanisms of neuronal function with particular emphasis on protein phosphorylation.

John H. Caldwell, Professor of Cellular and Developmental Biology; Ph.D., Washington (St. Louis), 1977. Ion channels in the nervous system.

Mark Dell'Acqua, Assistant Professor of Pharmacology; Ph.D., Harvard, 1995. Postsynaptic protein kinase and phosphatase signaling complexes in synaptic plasticity.

Thomas E. Finger, Professor of Cellular and Developmental Biology; Ph.D., MIT, 1975. Chemical senses (taste, smell, and trigeminal); receptor cell development and differentiation; transmission and processing of primary sensory information.

Curt R. Freed, Professor of Medicine and Pharmacology; M.D., Harvard, 1969. Neurotransplantation for Parkinson's disease; differentiation of human embryonic stem cells; cellular repair of embryonic and newborn brain; protein pathology in neurodegenerative disease.

Robert Freedman, Professor of Psychiatry and Pharmacology; M.D., Harvard, 1972. Neurobiological studies of schizophrenia; basic and clinical psychopharmacology.

Kim A. Heidenreich, Professor of Pharmacology; Ph.D., Vermont, 1979. Molecular mechanisms of insulin and insulin-like growth factor action in the central nervous system.

Leslie Krushel, Assistant Professor of Pharmacology; Ph.D., Toronto, 1990. Mechanisms and regulation of protein synthesis in the nervous system.

Sherry Leonard, Associate Professor of Pharmacology and Psychiatry; Ph.D., Colorado at Denver, 1986. Molecular neurobiology of schizophrenia.

Simon Rock Levinson, Professor of Physiology; Ph.D., Cambridge, 1975. Molecular biology of ion channels.

Lee Niswander, Professor of Pediatrics; Ph.D., Case Western Reserve, 1990. Genetics and molecular mechanisms of neural patterning and closure of the neural tube during vertebrate embryogenesis.

Manisha Patel, Assistant Professor of Molecular Toxicology; Ph.D., Purdue, 1992. Oxidative stress in neurodegeneration.

Anne-Laure Perraud, Assistant Professor of Immunology; Ph.D., Würzburg (Germany), 1998. Biological role of the TRPM cationic channels; ion-channel function in the immune system.

Karl H. Pfenninger, Professor of Cellular and Developmental Biology; M.D., Zurich, 1971. Cell and molecular biology of neurite outgrowth, the nerve growth cone, and network formation during brain development.

Katie Rennie, Assistant Professor of Otolaryngology; Ph.D., Bristol (England), 1991. Ion channels in type I and II vestibular hair cells.

Angeles B. Ribera, Professor of Physiology; Ph.D., Columbia, 1982. Development of neuronal excitability; activity-dependent neuronal development.

William A. Sather, Associate Professor of Pharmacology; Ph.D., Washington (Seattle), 1988. Structural-basis of calcium channel function.

Nathan Schoppa, Assistant Professor of Physiology; Ph.D., Yale, 1995. Cellular and synaptic physiology of the olfactory bulb.

Nicholas W. Seeds, Professor of Biochemistry, Biophysics, and Genetics; Ph.D., Iowa, 1968. Developmental neurobiology; molecular aspects of neuronal migration and neurite outgrowth.

James M. Sikela, Associate Professor of Pharmacology; Ph.D., Case Western Reserve, 1983. Evolutionary neurogenomics; genome organization and human genetic disease.

Celia Sladek, Professor of Physiology and Biophysics and of Medicine; Ph.D., Northwestern, 1970. Neuroendocrinology; regulation of vasopressin and oxytocin secretion.

John M. Stewart, Professor of Biochemistry, Biophysics, and Genetics; Ph.D., Illinois, 1952. Neuropeptide biology and chemistry.

Dan Tollin, Assistant Professor of Physiology and Biophysics; D.Phil., Oxford, 1998. Sound localization in the auditory system.

Kenneth L. Tyler, Reuler-Lewin Family Professor of Neurology and Professor of Medicine, Microbiology, and Immunology; M.D., Johns Hopkins, 1978. Molecular and genetic basis of virus-induced cell death (apoptosis) and pathogenesis in neuronal and nonneuronal cells and tissues.

Sukumar Vijayaraghavan, Associate Professor of Physiology and Biophysics; Ph.D., Indian Institute of Science, 1987. Nicotinic acetylcholine receptor's role in normal and abnormal functions of the central nervous system.

Bruce Wallace, Professor of Physiology; Ph.D., Harvard, 1974. Mechanisms of postsynaptic differentiation.

Nancy R. Zahniser, Professor of Pharmacology; Ph.D., Pittsburgh, 1977. Regulation of neurotransmitter release; transporters and receptors in the brain; mechanisms underlying cocaine and ethanol actions.

W. Michael Zawada, Associate Professor of Clinical Pharmacology and Toxicology; Ph.D., Arkansas Medical Sciences, 1993. Cell-replacement therapies in the brain; signaling mechanisms of cell death in Parkinson's disease; effects of alcohol on stem cells.

UNIVERSITY OF CONNECTICUT HEALTH CENTER

Graduate Program in Neuroscience

Program of Study	The neuroscience program at the University of Connecticut Health Center is committed to fostering an interdisciplinary training environment that works towards understanding the normal function and disorders of the nervous system. Special emphasis is placed upon preparing students for research and teaching careers in both academic and industrial settings. The interdepartmental nature of the program offers comprehensive conceptual and experimental training in molecular, systems, and behavioral neuroscience. The faculty of the neuroscience program engages in research that involves cellular, molecular, and developmental neurobiology; neuroanatomy; neuroimaging; neurophysiology; neurochemistry; neuroendocrinology; neuropharmacology; and neuropathology. Specific examples of research topics include cellular and molecular bases of synaptic neurotransmission, including the structure and function of ligand-gated neurotransmitter receptors and voltage-sensitive ion channels; genetic and epigenetic regulation of membrane biogenesis in neurons and glia; electrophysiology of excitable tissue; development of the autonomic nervous system; development of neural tissue; stimulus coding, synaptic organization, and development of sensory systems; structure and function of auditory and gustatory systems; computational neuroscience; degeneration, regeneration, plasticity, and transplantation; and neurobiology of acute and chronic neurological disorders, notably, epilepsy, stroke, spinal cord injury, schizophrenia, multiple sclerosis, partial loss of hearing, and deafness. The curriculum instills a broad background in the neurosciences, with strong grounding in cell and molecular biology and systems-level neurobiology. The curriculum can be individualized to fit the diverse backgrounds and interests of students. Introductory core courses in the neurosciences cover cell and molecular neuroscience, electrophysiology, neuroanatomy, systems neuroscience, and developmental neurobiology. Elective courses are offered in computational neuroscience, physiology of excitable tissue, biochemistry, neuroimmunology, genetics, cell biology, and neuropharmacology. Students fulfill didactic course requirements in the first year and a half and complete three laboratory rotations by fall of the second year, at which time they select a laboratory for dissertation research. Participation in weekly journal clubs provides a broad perspective of the field.
Research Facilities	Because of the interdepartmental format, the students have access to all of the facilities of modern biomedical research at the University of Connecticut Health Center, including those in clinical and basic science departments. Most of the neuroscience faculty members are housed in the same building on adjoining floors, providing for a congenial atmosphere of informal scientific exchange and collaborations between laboratories. The Center for Cell Analysis and Modeling (CCAM) has state-of-the-art facilities for confocal and two-photon microscopy and image analysis and is available to members of the Program in Neuroscience. The Lyman Maynard Stowe Library has an extensive collection of periodicals and monographs as well as subscriptions to journals of current interest in the field of neuroscience.
Financial Aid	Support for doctoral students engaged in full-time degree programs at the Health Center is provided on a competitive basis. Graduate research assistantships for 2006–07 provide a stipend of $26,000 per year, which includes a waiver of tuition/University fees for the fall and spring semesters and a student health insurance plan. While financial aid is offered competitively, the Health Center makes every possible effort to address the financial needs of all students.
Cost of Study	For 2006–07, tuition is $3996 per semester ($7992 per year) for full-time students who are Connecticut residents and $10,368 per semester ($20,772 per year) for full-time out-of-state residents. General University fees are added to the cost of tuition for students who do not receive a tuition waiver. These costs are usually met by traineeships or research assistantships for doctoral students.
Living and Housing Costs	There is a wide range of affordable housing options in the greater Hartford area within easy commuting distance of the campus, including an extensive complex that is adjacent to the Health Center. Costs range from $600 to $800 per month for a one-bedroom unit; 2 or more students sharing an apartment usually pay less. University housing is not available.
Student Group	Twenty-two students are registered in the Ph.D. program (including combined-degree students). The total number of master's and Ph.D. students at the Health Center is approximately 400, and there are about 125 medical and dental students per class.
Location	The Health Center is located in the historic town of Farmington, Connecticut. Set in the beautiful New England countryside on a hill overlooking the Farmington Valley, it is close to ski areas, hiking trails, and facilities for boating, fishing, and swimming. Connecticut's capital city of Hartford, 7 miles east of Farmington, is the center of an urban region of approximately 800,000 people. The beaches of the Long Island Sound are about 50 minutes away to the south, and the beautiful Berkshires are a short drive to the northwest. New York City and Boston can be reached within 2½ hours by car. Hartford is the home of the acclaimed Hartford Stage Company, TheatreWorks, the Hartford Symphony and Chamber orchestras, two ballet companies, an opera company, the Wadsworth Atheneum (the oldest public art museum in the nation), the Mark Twain house, the Hartford Civic Center, and many other interesting cultural and recreational facilities. The area is also home to several branches of the University of Connecticut, Trinity College, and the University of Hartford, which includes the Hartt School of Music. Bradley International Airport (about 20 minutes from campus) serves the Hartford/Springfield area with frequent airline connections to major cities in this country and abroad. Frequent bus and rail service is also available from Hartford.
The Health Center	The 200-acre Health Center campus at Farmington houses a division of the University of Connecticut Graduate School, as well as the School of Medicine and Dental Medicine. The campus also includes the John Dempsey Hospital, associated clinics, and extensive medical research facilities, all in a centralized facility with more than 1 million square feet of floor space. The Health Center's newest research addition, the Academic Research Building, was opened in 1999. This impressive eleven-story structure provides 170,000 square feet of state-of-the-art laboratory space. The faculty at the center includes more than 260 full-time members. The institution has a strong commitment to graduate study within an environment that promotes social and intellectual interaction among the various educational programs. Graduate students are represented on various administrative committees concerned with curricular affairs, and the Graduate Student Organization (GSO) represents graduate students' needs and concerns to the faculty and administration, in addition to fostering social contact among graduate students in the Health Center.
Applying	Applications for admission should be submitted on standard forms obtained from the Graduate Admissions Office at the UConn Health Center or on the Web site. The application should be filed together with transcripts, three letters of recommendation, a personal statement, and recent results from the General Test of the Graduate Record Examinations. International students must take the Test of English as a Foreign Language (TOEFL) to satisfy Graduate School requirements. The deadline for completed applications and receipt of all supplemental materials is December 15. Earlier submission of applications is recommended, and interviews are considered highly desirable. Applicants should have had undergraduate instruction in chemistry and biology. In accordance with the laws of the state of Connecticut and of the United States, the University of Connecticut Health Center does not discriminate against any person in its educational and employment activities on the grounds of race, color, creed, national origin, sex, age, or physical disability.
Correspondence and Information	Dr. Eric Levine, Neuroscience Program Director Lori Capozzi, Neuroscience Graduate Program Coordinator University of Connecticut Health Center, E-4056 Farmington, Connecticut 06030-3401 Phone: 860-679-2658 or 2319 Fax: 860-679-8766 E-mail: capozzi@uchc.edu Web site: http://grad.uchc.edu

University of Connecticut Health Center

THE FACULTY AND THEIR RESEARCH

Srdjan Antic, Assistant Professor in Neuroscience; M.D., Belgrade. Dendritic integration of synaptic inputs; dopaminergic modulation of dendritic excitability.

Rashmi Bansal, Associate Professor in Neuroscience; Ph.D., Central Drug Research Institute, 1976. Developmental, cellular, and molecular biology of oligodendrocytes; growth-factor regulation of development and function and its relationship to neurodegenerative disease, including multiple sclerosis.

Elisa Barbarese, Professor in Neuroscience and Neurology; Ph.D. McGill, 1978. Molecular and cellular biology of neural cells, with emphasis on RNA trafficking.

Leslie R. Bernstein, Associate Professor in Neuroscience; Ph.D., Illinois, 1984. Behavioral neuroscience: psychoacoustics, binaural hearing.

John H. Carson, Professor in Molecular, Microbial, and Structural Biology; Ph.D., MIT, 1972. Molecular and developmental neurobiology; myelination; intracellular RNA trafficking.

Lisa Conti, Assistant Professor in Psychiatry; Ph.D., Vermont, 1986. Behavioral neuroscience: roles of stress and neuropeptides in animal models of psychiatric disorders.

Betty Eipper, Professor in Neuroscience; Ph.D., Harvard, 1973. Cell biology, biochemistry, and physiology of peptide synthesis, storage, and secretion in neurons and endocrine cells.

Paul M. Epstein, Associate Professor in Pharmacology; Ph.D., Yeshiva (Einstein), 1975. Receptor signal transduction, second messengers, and protein phosphorylation in control of cell growth and regulation.

Marion E. Frank, Professor in BioStructure and Function and Director, Connecticut Chemosensory Clinical Research Center; Ph.D., Brown, 1968. Gustatory neurophysiology, neuroanatomy, behavior, and disorders; chemosensory information processing; clinical testing of oral chemosensory function in humans.

James Hewett, Assistant Professor in Neuroscience; Ph.D., Michigan State, 1991. Mechanisms of cell injury and inflammation in the central nervous system.

Sandra Hewett, Associate Professor in Neuroscience; Ph.D., Michigan State, 1992. Mechanisms underlying acute and chronic cell death in the central nervous system.

Duck O. Kim, Professor in Neuroscience and Biological Engineering Program; D.Sc., Washington (St. Louis), 1972. Neurobiology and biophysics of the auditory system; computational neuroscience of single neurons and neural systems; experimental otolaryngology; biomedical engineering.

Shigeyuki Kuwada, Professor in Neuroscience; Ph.D., Cincinnati, 1973. Neurophysiology and anatomy of mammalian auditory system; principles of binaural signal processing, electrical audiometry in infants.

Eric S. Levine, Assistant Professor in Pharmacology; Ph.D., Princeton, 1992. Synapse plasticity and role of neuromodulators in brain development and learning.

Leslie Loew, Professor in Cell Biology; Ph.D., Cornell, 1974. Morphological determinants of cell physiology; image-based computational models of cellular biology; spatial variations of cell membrane electrophysiology; new optical methods for probing living cells.

Richard Mains, Professor and Chair in Neuroscience; Ph.D., Harvard, 1973. Pituitary; neuronal tissue culture; peptides, vesicles; enzymes; drug abuse; development.

Louise McCullough, Assistant Professor of Neurology and Neuroscience; M.D., Ph.D., Connecticut. Effects of estrogens on stroke.

D. Kent Morest, Professor in Neuroscience and Communication Sciences and Director of the Center for Neurological Sciences; M.D., Yale, 1960. Synaptic organization and fine structure of nervous system: plastic changes following activity changes; noise-induced hearing loss; development of synapses; tissue culture; neuronal transplantation.

Douglas L. Oliver, Professor in Neuroscience and Biomedical Engineering; Ph.D., Duke, 1977. Synaptic organization; parallel information processing in CNS; role of ionic currents, channel expression in information processing; neurocytology, morphology, cellular physiology of CNS sensory systems; biology of hearing and deafness.

Joel S. Pachter, Professor in Pharmacology; Ph.D., NYU, 1983. Mechanisms regulating pathogenesis of CNS infectious/inflammatory disease.

David S. Papermaster, Professor in Neuroscience; M.D., Harvard, 1963. Premature death (apoptosis) of retinal neurons; membrane biosynthesis and localization in photoreceptors of transgenic frogs and mice.

Achilles J. Pappano, Professor in Pharmacology; Ph.D., Pennsylvania, 1966. Regulation of excitation-contraction coupling in mammalian heart by autonomic transmitter signaling pathways.

Steven E. Pfeiffer, Professor in Neuroscience; Ph.D., Washington (St. Louis), 1967. Developmental, cell, and molecular biology of oligodendrocytes (OLs) and myelin membrane biogenesis and function; axon-glial interaction, lipid rafts; multiple sclerosis.

Steven J. Potashner, Professor in Neuroscience; Ph.D., McGill, 1971. Synaptic and transmitter biochemistry of neural connections in the auditory nervous system and their plasticity in the adult animal.

Matthew N. Rasband, Assistant Professor in Neuroscience; Ph.D., Rochester, 1999. Ion channels and neuroglial interactions at the node of Ranvier.

Martin R. Schiller, Assistant Professor in Neuroscience; Ph.D., Utah State, 1990. Neuronal regeneration; secretory pathway; cell signaling.

William J. Shoemaker, Associate Professor in Psychiatry; Ph.D., MIT, 1971. Neuropharmacology; CNS peptides and receptors; fetal alcohol syndrome; genetics of mental diseases.

Henry Smilowitz, Professor in Pharmacology; Ph.D., MIT, 1972. Development of novel brain tumor therapies for advanced, experimental malignant gliomas in rats and mice, using both radiation therapy and gene-mediated immunoprophylaxis.

David M. Waitzman, Associate Professor in Neurology; M.D./Ph.D., CUNY, Mount Sinai, 1982. Neurophysiology; oculomotor system; gaze control system; modeling of CNS.

Zhaowen Wang, Assistant Professor in Neuroscience; Ph.D., Michigan State, 1993. Molecular mechanisms of synaptic transmission, focusing on neurotransmitter release and mechanisms of potassium channel localization, using *C. elegans* as a model organism.

Nada Zecevic, Assistant Professor in Neuroscience; M.D., 1970, Ph.D., 1978, Belgrade. Cellular and molecular aspects of CNS development; primate cerebral cortex; oligodendrocyte progenitors, stem cells, microglia; multiple sclerosis.

UNIVERSITY OF HARTFORD

Department of Biology
Neuroscience Graduate Program

Program of Study

The Neuroscience Track of the Master's Program at the University of Hartford is interdisciplinary in nature, providing experiences at the interface of biology, psychology, and other natural sciences. The program is designed to meet the needs of both full-time and part-time nontraditional students, with most courses and laboratories offered in the evenings. In addition, the Neuroscience Master's Program offers two options. The thesis option requires a 6-credit thesis and is designed for students who are interested in research positions or who plan to pursue doctoral training in the future. The nonthesis option requires a 3-credit research paper where a particular applied/clinical interest is integrated with course work taken. This option is intended as an appropriate terminal degree for students who are pursuing advancement in nonresearch positions.

Research by faculty members in the Department of Biology and Psychology covers a broad range of areas, including neurodegenerative disorders, brain aging, mechanisms of steroid action on neurological function, neuroendocrine control of reproduction, cognitive psychology, developmental psychology, psychopathology, eating behavior, and stress. In addition to research opportunities at the University of Hartford, the Neuroscience Program has ties with top neuroscience researchers at nearby UConn Health Center, Trinity College, Wesleyan University, and the University of Connecticut.

Formal course work provides a solid foundation in neurophysiology, neuroanatomy, neuropharmacology, neuroendocrinology, molecular biology, clinical neurology, physiological psychology, advanced cell biology, biochemistry, vertebrate histology, sensory processes, advanced research methods, advanced animal physiology, and pathophysiology. Additional special topics courses designed for one-on-one interaction between the student and a faculty member provide specialized training for students who are focused in a particular area of interest. The Cooperative Education Program provides paid work experience for graduate students in a biological industry, under the supervision of a faculty member. The neuroscience master's program is designed to provide a versatile educational opportunity in scholarship and research.

Research Facilities

The laboratories at the University of Hartford are equipped to provide research opportunities in most areas of biological and neuroscience research. Collaborations with scientists at neighboring institutions give students access to a wide range of research facilities. The Department of Biology is equipped with a USDA/NIH-approved small animal facility, surgery suite, radioisotope laboratory, photographic dark room, computer lab, and laboratory space and equipment necessary for cell culture and molecular biological experimentation. The Harry Jack Gray Center houses the William H. Mortensen Library, the Mildred P. Allen Memorial Library, and a conference center that is available for regional meetings and conferences.

Financial Aid

A limited number of qualified students may receive a Regents Fellowship in Neuroscience, which provides a partial tuition remission and a $5000 stipend. Students who receive the fellowship serve as research assistants to a professor in the Department of Biology. Limited teaching assistantships are also available. Students who enroll in the thesis track of the program are encouraged to submit research grants for in-house and/or extramural funding opportunities. These grants provide invaluable grant writing experience for students who intend to pursue doctoral training.

Cost of Study

Tuition and fees in 2006–07 are as follows: tuition fee per graduate credit hour, $375; registration fee per semester, $30; technology fee per semester, $25; parking fee per semester, $45 (or $75 per year); graduation fee, $208. All fees are subject to change.

Living and Housing Costs

Graduate students attending the University of Hartford obtain off-campus housing. A listing of apartments in the area is available from the Chamber of Commerce. The cost of living in the greater Hartford area varies from town to town, with single bedroom apartments ranging from $450 to $750 and up.

Student Group

The program currently has 20 graduate students, split evenly between full-time and part-time students. Ten are women and 25 percent are from other countries. Forty percent of the graduate students are serving as either teaching or research assistants.

Student Outcomes

The Neuroscience Graduate Program has placed students into neurobiology doctoral programs throughout the United States, including the University of Connecticut, University of Florida, University of Minnesota, Florida State University, and Penn State University. Graduates of the program have also accepted positions with pharmaceutical companies, enrolled in medical school, or continued their jobs in nursing, health care, and physical therapy.

Location

The University maintains two campuses that are located in the residential suburb of West Hartford. The area provides an environment conducive to the development of the student's cultural and intellectual pursuits. Facilities include libraries, museums, theaters, the Hartford Civic Center and Coliseum, a symphony orchestra, several other colleges, modern shopping centers, fine restaurants, an international airport, surface transportation, and intercity highway systems.

The University

The University of Hartford offers a variety of academic programs that are available at few universities of its size, yet strives to foster individual attention. With approximately 7,000 full- and part-time students, the campus is large enough to achieve the goals of a university without becoming a massive, impersonal institution. Many opportunities for career preparation can be realized within the approximately eighty-five undergraduate majors and more than fifty-five graduate degree programs offered by the University. All degree programs carry regional and state accreditation or licensure.

The University combines the vitality of youth with a rich heritage of the past. It was founded in 1877, when the first of its three original schools was established. The Hartford Art School (1877), Hillyer College (1879), and the Hartt School (1920), all recognized institutions of higher education, joined in 1957 to form the University of Hartford. The University is an independent, coeducational, nonsectarian institution. The variety of its programs attracts a diverse student body from the urban and general metropolitan area, from about three fourths of the states of the Union, and currently from sixty-six other countries.

Applying

The academic year begins in early September. Ideal preparation includes a bachelor's degree in a science such as biology, psychobiology, physics, chemistry, psychology, or engineering. Optional undergraduate preparation includes courses in physics, chemistry, biology, psychology, and mathematics. An otherwise-well-qualified student whose background does not include all the requisite courses may be considered. All applicants must submit GRE General Test scores or MCAT scores. For health professionals, results of licensing and/or certification examinations are accepted. Official transcripts, three letters of recommendation from academic sources/supervisory personnel, and a letter of intent describing the student's professional and career goals are also required. Application fees are $40 for domestic applicants and $55 for international applicants.

Applications for admission to the Neuroscience Graduate Program are accepted year-round but applicants are encouraged to apply as early as possible and certainly by mid-July for admission in the fall.

Correspondence and Information

Jacob P. Harney, Ph.D.
Director, Neuroscience Graduate Program
Department of Biology, 369 Dana Hall
University of Hartford
200 Bloomfield Avenue
West Hartford, Connecticut 06117-1599
Phone: 860-768-5780 (Office)
      860-768-5928 (Laboratory)
Fax: 860-768-5002
E-mail: harney@hartford.edu
Web site: http://uhaweb.hartford.edu/biology/MNeuroscience.html

University of Hartford

THE FACULTY AND THEIR RESEARCH

Katherine Black, Assistant Professor of Psychology; Ph.D., New Hampshire. Developmental psychology; adolescent-parent relationships; peer-interactions; gender differences.

Joanna D. Borucinska, Associate Professor of Biology; Ph.D., Connecticut. Veterinary pathology; parasitology; zoonotic diseases; diseases of wildlife; environmentally-induced diseases.

Caryn Christensen, Associate Professor of Psychology; Ph.D., Ohio. Cognitive psychology; judgment and decision making; expert-novice differences in cognition.

Martin Cohen, Professor of Biology; Ph.D., Rhode Island. Plant physiology; science education technology.

William H. Coleman, Professor of Biology; Ph.D., Chicago. Microbiology; molecular biology of streptococcus resistance; PCR as a diagnostic tool of infection.

Jacob P. Harney, Associate Professor and Chair, Department of Biology; Ph.D., Florida. Mechanism of steroid action on neurological function; neurodegenerative disorders; epilepsy; calorie restriction and the ketogenic diet.

Mala L. Matacin, Associate Professor of Psychology; Ph.D., Cincinnati. Body image; behavioral medicine/health psychology; gender issues; eating behaviors and stress.

Linda F. Quenzer, Adjunct Assistant Professor of Biology; Ph.D., Massachusetts. Neuropharmacology and neuropsychopharmacology.

UNIVERSITY OF ILLINOIS AT CHICAGO

College of Medicine
Department of Anatomy and Cell Biology
Program in Neuroscience

Program of Study

The graduate program at the University of Illinois at Chicago (UIC) emphasizes training leading to the Ph.D. degree, emphasizing cellular and systems neurobiology as well as cell biology. Specific areas of research include axonal transport, neural development, Alzheimer's disease, neuroplasticity and regeneration, vestibulocochlear physiology and functional anatomy, ion channel regulation, neuroendocrine control of behavior, and functional interactions of the plasma membrane with cytoskeletal elements. Students are expected to have a clear desire to enter a research career. The training program provides an intensive experience in fundamental research and frequent opportunities to demonstrate appropriate forward progress. Basic courses in cell and molecular biology and biochemistry are taken in the first year, when students also do rotations through different research laboratories. The first-year curriculum is coordinated through an interdepartmental program in the College of Medicine and the Graduate College called Graduate Education in Medical Sciences (GEMS), thus allowing students from different programs to interact and explore different biomedical disciplines. A wide range of advanced courses in neuroscience, cell biology, and anatomy are available during the second year, which prepare the student to take advantage of the rapidly changing face of science. The course work, along with progress on a thesis research project, prepares the student for preliminary examinations. The average time needed to complete the degree in recent years has been about five years. The training prepares students for a research or teaching career in academia or industry.

Research Facilities

Cellular structure and function, biochemistry, and molecular biology are investigated with electron microscopes, confocal and video microscopy, and quantitative computer-assisted image analysis techniques. A new biochemical and molecular biology core provides a range of equipment for contemporary biology, including phosphorimaging, FPLC, real-time PCR, and cell culture. Electrophysiology is probed with patch clamp and microelectrode techniques, by using cells modified through molecular biology techniques or by using selected sensory systems to reveal the functions of receptors, mechanisms of intracellular trafficking and signal transduction, and characteristics of ion channels. The Library of the Health Sciences' collection of nearly 500,000 volumes and more than 5,100 journals supports education and research. A science library provides additional resources in terms of both books and journals. The libraries provide extensive free access to databases, and professional staff members provide training and computer-assisted searches.

Financial Aid

The successful applicant is anticipated to receive a research or teaching assistantship with a current stipend of $24,000 per year and a tuition waiver. The student is responsible for about $900 in fees each semester.

Cost of Study

For the 2005–06 year, in-state tuition and other required fees were nearly $12,000 per year. Out-of-state tuition and fees were approximately $28,000 per year. Costs are subject to change.

Living and Housing Costs

Nestled among the health sciences colleges, on-campus housing is available, including an apartment-style residence hall for graduate, professional, and older undergraduate students who are looking for convenience and an intensive study environment. The residence facilities connect directly to the Chicago Illini Union, the recently remodeled swimming pool, and the fitness center. Numerous privately owned rooms and apartments are available in the University area at a wide range of prices.

Student Group

The number of graduate students in the Department of Anatomy and Cell Biology is relatively small, allowing each student a great deal of individual attention and guidance in research training. However, the GEMS program includes 40 to 50 new students each year, which provides a critical mass of students for educational, scientific, and social activities. Thus the department combines the best of both large and small programs.

Location

The University of Illinois at Chicago is located in the heart of the cosmopolitan city of Chicago. The city combines the best of the arts, cuisine, entertainment, history, and sports. The campus is centrally located near The Loop and is surrounded by vibrant neighborhoods like Greek Town, Little Italy, Pilsen, and Chinatown. UIC is served by the Blue Line, with ready access to both airports, and is only minutes away from the famed Michigan Avenue and downtown. There is an abundance of cultural activities and institutions in the area, including the Art Institute of Chicago, art galleries, the Chicago Symphony Orchestra, the blues and jazz clubs of a lively music scene, and outstanding theaters that include the Chicago Shakespeare Theater, Steppenwolf Theatre Company, and The Second City.

The University

UIC is the largest institution of higher learning in the Chicago area and is one of only eighty-eight Research I institutions nationally. Its history began in the 1890s, when the Chicago College of Pharmacy and the College of Physicians and Surgeons of Chicago became part of the University of Illinois. Today, UIC is located on approximately 185 acres in an area that includes two historic landmark residential neighborhoods and the West Side Medical Center District, the largest concentration of advanced public and private health-care facilities in the world.

Applying

A general background in biology, chemistry, and physics is expected. Acceptance is based on the student's potential for scholarly research and teaching as shown by his or her undergraduate record, letters of recommendation, and Graduate Record Examinations General Test scores. The Graduate College application deadline for fall admission for degree candidates is May 15, but prospective students are strongly encouraged to apply by February 15 for the Anatomy and Cell Biology (GEMS) program. Applications from international students must be received by February 15 to allow processing. Applications received bearing a postmark later than the application deadline are returned to the applicant.

Correspondence and Information

Director, Graduate Studies Committee
Department of Anatomy and Cell Biology, M/C 512
University of Illinois at Chicago
808 South Wood Street, Room 578
Chicago, Illinois 60612-7308
Phone: 312-996-6791
Fax: 312-413-0354
E-mail: conwell@uic.edu
Web site: http://www.uic.edu/depts/mcan

University of Illinois at Chicago

THE FACULTY AND THEIR RESEARCH

Conwell H. Anderson, Associate Professor; Ph.D., Kansas, 1969. Neuroendocrinology; morphological aspects of the hypothalamus and its influence on reproduction.

Jonathan J. Art, Associate Professor; Ph.D., Chicago, 1979. Cochlear hair cell physiology to characterize the cellular mechanisms that contribute to sensory transduction and signal processing in the nervous system; confocal laser microscopy.

Scott T. Brady, Professor and Head; Ph.D., Southern California, 1978. Molecular mechanisms of axonal transport; specializations of the neuronal cytoskeleton; glial modulation of neuronal function; effects of physiological stress on neuron structure and function.

Rochelle S. Cohen, Professor; Ph.D., Connecticut, 1973. Estrogen effects on affective state and estrogen-induced neurochemical and structural changes in the limbic system; neural substrates of reproductive behavior; synaptic plasticity.

Christopher P. Fall, Assistant Professor; Ph.D., Virginia, 1998. Computational and experimental approaches to understanding the neuromodulation of cortical activity patterns; computational and experimental cell biology of intracellular Ca2+ signaling and related second messengers; mechanisms of cell death in neurodegenerative disease.

Robert Gould, Professor; Ph.D., Johns Hopkins, 1970. Mechanisms involved in myelination of rodent and cartilaginous fish central nervous systems during development; evolution of the developing myelin sheath.

Mary Jo LaDu, Assistant Professor, Ph.D. Illinois at Chicago, 1991. Alzheimer's disease (AD): structural and functional interactions between the amyloid-beta peptide (AB) and apolipoprotein E (apoE); in vitro and in vivo models to test the various hypotheses that arise from the study of AB and apoE4.

Orly Lazarov, Assistant Professor, Ph.D. Weizmann (Israel), 2000. Physiological roles of proteins associated with Alzheimer's disease and the mechanisms by which mutant forms of these proteins induce AD in vivo.

Anna Lysakowski, Associate Professor; Ph.D., Illinois at Chicago, 1984. Organization, physiology, and function of the vestibular sensory apparatus extending from the cellular to system level.

Yasuko Nakajima, Professor; M.D./Ph.D., Tokyo, 1962. Signal transduction mechanisms of neurotransmitter effects on brain neurons, using a unique method of culturing neurons from specific brain nuclei, such as cholinergic neurons from the basal forebrain and dopaminergic neurons from the substantia nigra (these neurons are related to Alzheimer's and Parkinson's diseases).

Adrienne A. Rogalski-Wilk, Associate Professor; Ph.D., Illinois, 1981. Cell and molecular biology of novel plasma membrane–actin cytoskeletal linkage.

James R. Unnerstall, Associate Professor; Ph.D., Johns Hopkins, 1984. Plasticity and regeneration in central catecholamine systems during the aging process, emphasizing the compensatory responses of specific neural systems following physiological stress, insult, or injury; neurotrophic factors.

Adjunct Research Faculty

Neelima Chauhan, Research Assistant Professor; Ph.D., Baroda (India). Preventing and reversing Alzheimer's pathogenesis in transgenic models of Alzheimer's disease by intracerebroventricular passive immunization combined with dietary satins, Propentofylline, or herbal alternatives in reducing cerebral amyloid burden and tau phosphorylation.

George De Vries, Adjunct Professor; Ph.D., Illinois at Chicago, 1979. Axonal oligodendrocyte signaling in multiple sclerosis.

Thomas Diekwisch, Professor and Head, Oral Biology; D.D.S., Ph.D., Marburg (West Germany), 1990. Craniofacial biology, development, and genetics.

Douglas Feinstein, Research Associate Professor; Ph.D., Johns Hopkins, 1984. Novel therapeutic means to reduce inflammatory damage and neuronal death in multiple sclerosis and Alzheimer's disease.

Anne George, Professor; Ph.D., Madras (India), 1983. Identification and characterization of acidic proteins involved in biomineralization.

Nalin Kumar, Professor; D.Phil., Oxford, 1979. Molecular structure and function of gap junctions; intercellular channels between adjacent cells.

Charles Laurito, Professor; M.D., Pittsburgh, 1975. Providing relief for patients suffering from acute, chronic, and cancer-related pain; basic understanding of how noxious stimuli are perceived as being painful; neurotransmitters released at the spinal cord level.

Deborah Little, Assistant Professor; Ph.D., Brandeis, 2002. Biological basis of compensatory processes in normal aging and in disease in language comprehension, attention, learning, and memory.

Jeremy Mao, Associate Professor; Ph.D., Alberta, 1992. Regeneration of tissues and organs by stem cells using cell biology, molecular biology, and mechanobiology tools.

Subhash Pandey, Associate Professor; Ph.D., Kanpur (India), 1987. Molecular and cellular neurobiology of alcoholism and drugs of abuse.

George Pappas, Professor; Ph.D., Ohio State, 1952. Structure and function of synapses; release chromaffin granule from adrenal chromaffin cells; control chronic pain with chromaffin cell transplant.

Harris Ripps, Professor; Ph.D., Columbia, 1959. Role of gap junctions in organizing the neuronal circuitry of the vertebrate retina and in the spread of death signals from cells undergoing apoptosis to their otherwise healthy neighbors.

Tapas Das Gupta, M.D./Ph.D., London, 1967. Cancer biology and chemotherapeutic agents.

William Wolf, Adjunct Associate Professor; Ph.D., George Washington, 1985. Monoaminergic modulation of neuronal plasticity in neurological disorders such as Parkinson's disease and stroke and psychiatric disorders that include drug abuse, schizophrenia, and affective disorders.

UNIVERSITY OF MARYLAND, COLLEGE PARK/ NATIONAL INSTITUTES OF HEALTH

Graduate Program in Sensory and Communication Neuroscience

Program of Study

The partnership between the Division of Intramural Research of the National Institute on Deafness and Other Communication Disorders (NIDCD) and the University of Maryland (UMD) Center for Comparative and Evolutionary Biology of Hearing Program (C-CEBH) creates the novel opportunity to work with two major groups of faculty members with distinguished research records.

Students in the University of Maryland, College Park/National Institutes of Health (UMD/NIH) program receive graduate training through the Neuroscience and Cognitive Science (NACS) doctoral program at the University of Maryland, College Park, and receive the Ph.D. degree from UMD and that program. Graduate students in NACS are trained in broad areas of cellular, molecular, systems, cognitive, and computational neuroscience. Students interested in working with UMD and/or NIDCD investigators in the combined program are strongly encouraged to contact those investigators who share their research interests. Students who are interested in collaborative research that involves investigators from both UMD and NIH are strongly encouraged to explore such possibilities.

The combined resources to support this UMD/NIH program are unparalleled. The C-CEBH at UMD is a campuswide program that is devoted to studies of auditory neuroscience in the broadest sense. The laboratories in C-CEBH cover research approaches from molecular biology to human psychoacoustics to computational modeling, and species from insects to fish to birds to mammals. The faculty members of the C-CEBH come from five departments: biology, electrical and computer engineering, hearing and speech sciences, linguistics, and psychology. The training program is open to students admitted to NACS or to the departments represented by any of the core faculty members in the program. The NIDCD is the home of nationally and internationally known scholars in the communication sciences who can serve as student dissertation advisers. Research approaches in the laboratories of NIDCD span molecular biology, cell biology, and genetics to human brain imaging and computational sciences. Students are able to select co-advisers from NIDCD or C-CEBH, although students working with NIDCD advisers must have a co-advisers at UMD. All faculty members from C-CEBH and NIDCD are in the NACS program.

Research Facilities

The NIH is the world's premier biomedical research institution. As the federal government's primary agency for the conduct of biomedical research, the NIH's many Institutes/Centers employ nearly 1,200 tenured or tenure-track investigators and more than 3,600 postdoctoral scientists with either medical, dental, or graduate degrees. The NIH intramural research program, located on a 300-acre campus in Bethesda, as well as Rockville, Maryland, 10 miles from downtown Washington, D.C., and 8 miles from the UMD campus, has been the scene of many exciting scientific advances. It houses more than thirty research buildings comprising twenty-seven Institutes/Centers in a broad spectrum of biomedical and related scientific research. Four Nobel Laureates made their prize-winning discoveries in NIH laboratories, and more than 100 received part of their training at the NIH. Basic research in the biomedical sciences at the NIH is complemented by an active clinical research program at the unique 250-bed research hospital and laboratory complex, the Warren Grant Magnuson Clinical Center. The NIH campus is also home to the National Library of Medicine, the world's largest medical library.

The University of Maryland is located in College Park, about 8 miles from NIH and 10 miles from downtown Washington, D.C. The campus has considerable strength in the biological and physical sciences and in engineering. The campus houses facilities for studies of biological sciences that range from molecular biology to systems-level studies of the nervous system. The NACS program, within which students in the joint NIH/C-CEBH program train, has more than 85 faculty members from fourteen academic departments. Course work is directed at providing trainees with a broad understanding of these areas and then allowing students to specialize in areas of particular interest to them. Programs are highly individualized.

Financial Aid

UMD/NIH graduate students are supported jointly in a variety of ways. Sources may include fellowships, training grant support, research assistantships, and/or teaching assistantships. The level of support is highly competitive with other leading institutions and always includes a stipend, health benefits, student fees, and tuition support. Stipends for first-year students in this program are $24,800 for 2006–07and increase yearly based on the student's performance. Support is guaranteed for the duration of satisfactory graduate education.

Cost of Study

Training awards cover tuition costs.

Living and Housing Costs

Apartments, houses, and rooms in private residences are available for rent near the UMD and NIH campuses. For information on available housing, prospective students should visit the Web sites of the University of Maryland (https://access12.umd.edu/ndconfig.nd/OchStudentSearch/Welcome), the Graduate Partnerships Program (GPP) (http://gpp.nih.gov), or local newspapers or contact the GPP.

Student Group

Students coming to NIH/C-CEBH join thousands of other graduate students who are doing their research in UMD/NIH laboratories. UMD/NIH continues to seek bright, creative, and industrious students to join in the quest for groundbreaking biomedical discoveries. While at UMD/NIH, graduate students enjoy services and activities sponsored by UMD and the Graduate Partnerships Program that ensure student success and create a strong graduate student community. A comfortable and stimulating academic setting is created for students in College Park at the University and by a new centrally located Graduate Student Lounge, bookstore, and auditoriums for research seminars. While at the NIH, students join more than 400 other graduate students from more than 100 universities who are doing their research in NIH laboratories. The Graduate Partnerships Program office at the NIH sponsors graduate student services and activities similar to those at the University to ensure student success and create a strong graduate student community.

Location

The 300-acre campus of NIH in Bethesda, Maryland, is close to Washington, D.C., affording a spectacular cultural and community environment. Other satellite campuses of NIH are in other Maryland locations (such as Rockville, for NIDCD), Montana, North Carolina, and Arizona. College Park is only 8 miles from Bethesda, which allows easy access to both institutions.

The University and The National Institutes of Health

The NIH and UMD have long histories of training scientists and physicians. Thousands of scientists have completed postdoctoral training in NIH laboratories, and many NIH-trained scientists have received international recognition for their work. The UMD/NIH collaborative environment is rich in scientific exchange, and it provides opportunities for a broad biomedical research experience. In times past, many graduate students have received their graduate training at the NIH through arrangements between the NIH and local universities. Numerous new graduate collaborations are being established by GPP.

Applying

To apply, prospective students must be U.S. citizens or eligible for lawful admission to the United States. An entering class typically includes a variety of different undergraduate majors in the biological, chemical, quantitative, or physical sciences. Information on submitting an application to the NIH/UMD is available on the Web at http://www.NACS.umd.edu/. Information about the Graduate Partnerships Program or joining an NIH laboratory for dissertation research can be found on the program's Web site.

Correspondence and Information

Graduate Partnerships Program
Building 2, Room 2E06
National Institutes of Health–DHHS
2 Center Drive
Bethesda, Maryland 20892-0234
Phone: 301-594-9605
Fax: 301-594-9606
E-mail: gpp@nih.gov
Web site: http://gpp.nih.gov

NIH Partnership Director
Dr. David Robinson
Phone: 301-496-1601
E-mail: robinsod@nidcd.nih.gov

University Partnership Director
Dr. Arthur N. Popper
Department of Biology
University of Maryland
College Park, Maryland 20742
Phone: 301-405-1940
E-mail: apopper@umd.edu

University of Maryland, College Park / National Institutes of Health

THE FACULTY AND THEIR RESEARCH

NIDCD Investigators
James F. Battey Jr., M.D., Ph.D. Structure, function, and regulation of bombesin receptors.
Allen Braun, M.D. PET and fMRI activation studies of language.
Richard Chadwick, Ph.D. Biomechanics of cochlear fine tuning.
Dennis Drayna, Ph.D. Genetic linkage and positional cloning in human communication disorders.
Thomas Friedman, Ph.D. Molecular genetics of hereditary deafness.
Andrew Griffith, M.D., Ph.D. Molecular mechanisms of genetic deafness.
Barry Horwitz, Ph.D. Neuromodeling, functional brain imaging, and language.
Kuni Iwasa, Ph.D. Biophysical mechanisms of sensory transduction.
Bechara Kachar, M.D. Molecular basis of transduction in auditory sensory organs.
Matthew Kelley, Ph.D. Development of the mammalian cochlea.
John Northup, Ph.D. Structure and function of signal-transducing G-proteins.
Robert J. Wenthold, Ph.D. Neurochemical mechanisms of hearing.
Doris Wu, Ph.D. Molecular basis for the morphogenesis of the inner ear.

University of Maryland, College Park, C-CEBH Professors
Catherine E. Carr, Professor of Biology. Development of hearing.
Monita Chatterjee, Assistant Professor of Hearing and Speech Sciences. Cochlear implants and auditory processes.
Robert J. Dooling, Professor of Psychology and C-CEBH Co-Director. Auditory psychoacoustics.
Sandra Gordon-Salant, Professor of Hearing and Speech Sciences. Speech perception, psychoacoustics, signal processing, hearing aids.
William Hall, Professor of Psychology. Anatomy and development of the auditory CNS.
Cynthia Moss, Professor of Psychology. Echolocation and signal processing.
David Poeppel, Assistant Professor of Linguistics and Biology. Speech processing in the brain.
Arthur N. Popper, Professor of Biology. Comparative auditory mechanisms.
Brenda Ryals, Professor of Communications Sciences and Disorders, James Madison University. Auditory properties, morphology and regeneration of the ear.
Shihab Shamma, Professor of Electrical and Computer Engineering. Computation in the auditory system.
Jonathan Simon, Assistant Professor of Electrical and Computer Engineering. Auditory computation.
David Yager, Associate Professor of Psychology and Affiliate Associate Professor of Biology. Physiological mechanisms of hearing.

The Dale and Betty Bumpers Vaccine Research Center (VRC), one of more than thirty major research buildings on the NIH campus, was established to facilitate research in vaccine development.

The National Human Genome Research Institute led the Human Genome Project for the National Institutes of Health, which culminated in the completion of the full human genome sequence in April 2003.

UNIVERSITY OF MIAMI

Neuroscience Program

Program of Study	The Neuroscience Program, established in 1988, is an interdisciplinary program leading to the Ph.D. degree. Its objective is to train highly qualified individuals for independent research and teaching careers in the neurosciences. Graduate and postgraduate training is the major goal of the program, with specific emphasis on cellular, molecular, and integrative approaches to neurobiology. During the first one to two years of the graduate program, students devote their time to course work and become acquainted with current research in their areas of interest. A core curriculum consisting of molecular/cell biology, membrane biophysics, neurophysiology, neuroanatomy, and integrative neuroscience is taken by all students; these courses can be supplemented with more than fifty elective courses. Students also attend and participate in research seminars and journal clubs. A series of laboratory rotations provides the opportunity for students to choose from among more than fifty research laboratories for their dissertation research. The Neuroscience Program Steering Committee oversees students' course work and laboratory rotations until they have passed the qualifying exam and have chosen a thesis adviser. From then on, the student's progress is followed by an individually tailored dissertation committee.

Faculty research interests focus on the mechanisms involved in cellular communication and signal transduction, gene expression in electrically excitable cells, synapse formation, neuronal growth and survival, neuroglia, and neuron–target cell recognition. Other areas of faculty research include neuroimmunology, autonomic control, cerebral metabolism and blood flow, and degenerative changes within specific neural pathways in neurological disorders and stroke. Program faculty members work in basic science departments as well as in such prominent research centers as the Miami Project to Cure Paralysis, the Neurotrauma Research Center, and the Cerebrovascular Disease Center.

The University of Miami School of Medicine offers a course of study leading to both the M.D. and Ph.D. degrees that can be completed in approximately seven years. Applicants interested in the combined-degree program should contact the M.D./Ph.D. Program Office.

Research Facilities

Faculty laboratories are well equipped for state-of-the-art research in neuroscience and include more than two dozen sophisticated electrophysiological setups, laser confocal and digital image analysis systems, and advanced molecular biology and biochemistry equipment. Additional shared facilities include those for DNA and peptide synthesis, protein sequencing and amino acid analysis, monoclonal antibody screening and production, fluorescence-activated cell sorting, scanning and transmission electron microscopy, and computers for the analysis of protein structure. The medical library receives more than 1,800 periodicals and holds more than 200,000 volumes; additional resources are available in the University libraries on the Coral Gables campus.

Financial Aid

Stipends, which start at $22,000 in 2006–07, are awarded to all students who are admitted into the program. Exceptional students who are chosen for the Lois Pope Life Fellows Program receive a discretionary educational allowance.

Cost of Study

All students accepted into the program receive a full tuition waiver.

Living and Housing Costs

The cost of living in the Miami area is typical for a medium-sized city in the Southeast.

Student Group

The University of Miami currently enrolls 14,000 students, including about 2,600 graduate students. Approximately 35 graduate students and 40 postdoctoral fellows are currently working in the laboratories of Neuroscience Program members.

Location

Miami is a rapidly growing, multiethnic, cosmopolitan community that stands at the gateway to Latin America and the Caribbean. It offers a wide variety of films, concerts, opera, ballet, and theatrical performances, in addition to having several fine museums. The area is a sports mecca, home to professional teams from the NFL, NBA, and NHL and has national-caliber college teams in football and baseball. Tennis, golf, and other outdoor sports can be enjoyed all year.

Because of the city's subtropical location, the climate is mild year round. Nearby parks, ocean beaches, tropical gardens, and wildlife sanctuaries offer abundant opportunities for outdoor recreation. The Florida Keys, coral reefs for snorkeling and diving, and the Everglades and Biscayne National Parks are all within an hour's drive of Miami. An additional benefit of the subtropical climate is that numerous scientific symposia and medical society meetings are held in the area.

The University and The Program

The University of Miami is a private, nondenominational, coeducational institution with more than 1,400 full-time faculty members. The Neuroscience Program is a broad interdisciplinary program that unites neuroscientists from three campuses and more than a dozen basic science and clinical departments. Its 60 faculty members are housed in the departments of Biochemistry and Molecular Biology, Biology, Cell Biology and Anatomy, Molecular and Cellular Pharmacology, Microbiology and Immunology, Neurological Surgery, Neurology, Otolaryngology, Pathology, Physiology and Biophysics, Psychiatry, and Psychology, and the Marine School.

Collectively, members of the Neuroscience Program hold more than $20 million in federal research support. Five members have received Jacob Javits Awards in recognition of their scientific contributions to the neurosciences. On the national level, members of the Neuroscience Program hold offices in the Society for Neurosciences and Society for Cell Biology, chair Gordon Conferences and study sections, and are on the editorial boards of the *Journal of Cell Biology, Journal of Neurobiology, Journal of Physiology, Journal of Biological Chemistry,* and *Proceedings of the Royal Society of London,* as well as other more specialized journals in their fields.

Applying

Applicants must have a bachelor's degree in one of the biological, behavioral, or physical sciences. Applicants are expected to have a strong quantitative background and should score in the 80th percentile or higher on the General Test of the GRE and have a GPA of at least 3.0 (on a 4.0 scale). Consideration for admission is limited to those who plan to pursue the doctoral degree. There is an application fee of $50. Application forms are available online at the University's Web site.

Correspondence and Information

Neuroscience Program (R-50)
School of Medicine
University of Miami
P.O. Box 011351
Miami, Florida 33101
Phone: 305-243-3368
      800-952-5386 (toll-free)
Fax: 305-243-2970
E-mail: neurosci@med.miami.edu
Web site: http://chroma.med.miami.edu/neuro/

University of Miami

THE FACULTY AND THEIR RESEARCH

Ellen F. Barrett. Electrophysiological and dye-imaging studies of myelinated axons and nerve terminals.

John N. Barrett. Neurotrophic factors; neuronal response to injury.

Antoni Barrientos. Yeast models of mitochondrial and neurodegenerative disorders.

John Bethea. CNS inflammation; cytokine signal transduction; neuronal response to injury.

Sanjoy K. Bhattacharya. Neurodegenerative ocular diseases; integration of sensory systems; neuronal signaling and neuroprotection strategies.

John L. Bixby. Neuronal signaling by tyrosine phosphatases and cell adhesion receptors; role of agrin in neuromuscular synaptogenesis.

Richard J. Bookman. Molecular mechanisms of neurotransmitter release; intracellular Ca^{2+} imaging.

Walter Bradley. Animal models of motor neuron diseases; clinical and new drug development research in amyotrophic lateral sclerosis.

Mary B. Bunge. Growth, differentiation, injury, and repair of nervous tissue.

Kermit L. Carraway. Modulation of signaling through ErbB receptors; anti-adhesion/protection mechanisms.

Nirupa Chaudhari. Sensory transduction: molecular biology of receptors, ion channels, signaling.

Gerhard P. Dahl. Biophysics and molecular biology of ion channels; cell-cell channels and membrane receptors.

Robert A. Davidoff. Spinal cord physiology and pharmacology; transmitters and messengers.

W. Dalton Dietrich. Pathophysiology and treatment of brain and spinal cord injury.

Mary J. Eaton. Cell and molecular therapies for the consequences of spinal cord injury.

Carl Eisdorfer. Biogenic amines; dementia and depression.

Lynne A. Fieber. Comparative physiology of single cells in the nervous system; ion channels in cellular communication.

M. Elizabeth Fini. Tissue response to stress; repair and regeneration; vision science.

Myron D. Ginsberg. Ischemic and dysmetabolic brain injury; cerebral blood flow and metabolism.

Jeffrey L. Goldberg. Development and regeneration in the visual system.

Edward J. Green. Electrophysiological correlates of learning in mammals and recovery of function following CNS insults.

Cristof T. Grewer. Neurotransmitter transporters; electrophysiology; transport kinetics.

Abigail S. Hackam. Cellular mechanisms of retinal development and degeneration.

John C. Hackman. Spinal cord neurophysiology and neuropharmacology; actions of transmitters and modulators.

Bingren Hu. Molecular mechanisms of cell death and survival after brain ischemia; novel therapeutic targets for anti-ischemic compounds.

George Inana. Molecular neurobiology of retinal function and diseases.

Yossef Itzhak. Animal models of drug addiction: neuropsychopharmacology, neurochemistry, and molecular biology of drugs and abuse.

Robert W. Keane. Neuroimmunology; developmental neurobiology.

W. Glenn Kerrick. Control and regulation of muscle contraction by Ca^{2+} and protein phosphorylation.

David Landowne. Nerve excitation; sodium channels; optical techniques.

Richard K. Lee. Neuroprotection and pathophysiology of glaucoma.

Vance P. Lemmon. Cell adhesion molecules; axon growth and guidance.

Wei Li. Autoimmunity in CNS and peripheral nerve system, including optic neuritis.

Daniel J. Liebl. Molecular, cellular, and developmental approach to identify the regulatory functions of axonal growth and regeneration following CNS injury.

Charles W. Luetje. Molecular biology of central nervous system nicotinic cholinergic receptors.

Karl L. Magleby. Ion-channel gating mechanisms.

Deborah C. Mash. Neurotransmitter (including cholinergic) receptors in brain; alterations in aging and Alzheimer's disease.

Philip M. McCabe. Neural substrates of learning; differential classically conditioned responses.

Carlos T. Moraes. Mitochondrial biology and genetics; neuromuscular diseases.

Vincent T. Moy. Cell-cell interaction; molecular recognition; immunology; model membrane systems; atomic force microscopy and reflection interference microscopy.

Kenneth J. Muller. Synaptic integration; axon growth and synapse formation; nerve repair.

Joseph T. Neary. Signal transduction and protein phosphorylation in astrocytes.

Wolfgang F. Nonner. Molecular basis of ionic selectivity and conduction in ionic channels.

Michael D. Norenberg. Role of astrocytes in neurologic disease.

Martin Oudega. Spinal cord injury; repair strategies; neurotrophic factors; biodegradable materials.

Ozcan Ozdamar. Analysis and clinical applications of auditory evoked potentials.

Miguel A. Perez-Pinzon. CNS injury: pathophysiological mechanisms of cell death and neuroprotective strategies; special emphasis on cerebral ischemia, mitochondrial physiology, and neurodegenerative diseases.

Vittorio Porciatti. Electrophysiology of the visual system, applied to neuroprotection of the optic nerve in glaucoma models.

James D. Potter. Molecular biology of the regulation of muscle contraction and Ca^{2+} binding proteins.

Lincoln T. Potter. Receptors; neuropharmacology and neurochemistry of cholinergic synapses.

Eugene Roberts. Ion regulation in brain tissue; age-related changes in the brain's response to anoxia, hypoxia, or ischemia.

Stephen D. Roper. Molecular and cellular biology of taste transduction; signal transduction.

Richard L. Rotundo. Regulation of gene expression in cells; biogenesis and localization of synaptic components.

Jacqueline Sagen. Cellular implants for the alleviation of chronic pain.

Michael C. Schmale. Marine neurobiology; neurogenic tumors.

Neil Schneiderman. Cardiovascular neurobiology.

Thomas J. Sick. Brain metabolism/electrophysiology.

Vladlen Z. Slepak. Molecular mechanisms of signal transduction in CNS.

Christine K. Thomas. Motor control and spinal cord injury.

Kathryn W. Tosney. Axonal guidance; muscle morphogenesis; neural crest migration; regulation of growth cone motility, adhesion, and cytoskeleton.

Pantelis Tsoulfas. CNS stem cells; mechanisms of stem cell maintenance and differentiation; neurotrophic factor signaling.

Brant D. Watson. Lipid peroxidation in experimental stroke induced by laser-driven photothrombotic occlusion of cerebral arteries.

Eva Widerström-Noga. Neuropathic pain and SCI: evaluation and treatment.

David L. Wilson. Neuroscience of mind and consciousness.

Patrick M. Wood. Neurobiology of human Schwann cells.

RESEARCH AREAS

Behavioral Neurobiology
Edward J. Green; Philip M. McCabe; Yossef Itzhak; Kenneth J. Muller; Eva Widerström-Noga.

Developmental Neurobiology
John N. Barrett; John Bethea; John L. Bixby; Richard J. Bookman; Mary B. Bunge; Kermit L. Carraway; Nirupa Chaudhari; W. Dalton Dietrich; Lynne A. Fieber; M. Elizabeth Fini; Jeffrey L. Goldberg; Abigail S. Hackam; Robert W. Keane; Richard K. Lee; Kenneth J. Muller; Martin Oudega; Richard L. Rotundo; Christine K. Thomas; Pantelis Tsoulfas; Patrick M. Wood.

Cell/Molecular Neurobiology
Ellen F. Barrett; John N. Barrett; Antoni Barrientos; John Bethea; John L. Bixby; Richard J. Bookman; Kermit L. Carraway; Nirupa Chaudhari; Gerhard P. Dahl; Mary J. Eaton; Lynne A. Fieber; Jeffrey L. Goldberg; Cristof T. Grewer; Abigail S. Hackam; George Inana; Robert W. Keane; Richard K. Lee; Vance P. Lemmon; Daniel J. Liebl; Charles W. Luetje; Brian A. Masters; Carlos T. Moraes; Vincent T. Moy; Kenneth J. Muller; Joseph T. Neary; Wolfgang F. Nonner; Michael D. Norenberg; Vittorio Porciatti; James D. Potter; Lincoln T. Potter; Richard L. Rotundo; Vladlen Z. Slepak; Pantelis Tsoulfas; David L. Wilson.

CNS Injury and Repair
Ellen F. Barrett; John L. Bixby; Richard J. Bookman; Walter Bradley (neuromuscular diseases, including ALS); Nirupa Chaudhari; Bingren Hu; W. Glenn Kerrick; David Landowne; Karl L. Magleby; Kenneth J. Muller; Wolfgang F. Nonner; James D. Potter; Richard L. Rotundo.

Neurological Disorders
John N. Barrett; John Bethea; Mary B. Bunge; W. Dalton Dietrich; Carl Eisdorfer; Lynne A. Fieber; Myron D. Ginsberg; Jeffrey L. Goldberg; Bingren Hu; George Inana; Robert W. Keane; Vance P. Lemmon; Wei Li; Deborah C. Mash; Carlos T. Moraes; Joseph T. Neary; Michael D. Norenberg; Martin Oudega; Miguel A. Perez-Pinzon; Vittorio Porciatti; Lincoln T. Potter; Eugene Roberts; Michael C. Schmale; Thomas J. Sick; Brant D. Watson; Eva Widerström-Noga.

Sensory Neurobiology
Nirupa Chaudhari; M. Elizabeth Fini; Abigail S. Hackam; George Inana; Vance P. Lemmon; Kenneth J. Muller; Ozcan Ozdamar; Vittorio Porciatti; Stephen D. Roper; Jacqueline Sagen; Eva Widerström-Noga.

Synapses
Ellen F. Barrett; John L. Bixby; Richard J. Bookman; George Inana; Karl L. Magleby; Kenneth J. Muller; Richard L. Rotundo; Thomas J. Sick.

Transmitters and Receptors
Richard J. Bookman; Nirupa Chaudhari; Robert A. Davidoff; Cristof T. Grewer; John C. Hackman; Yossef Itzhak; Charles W. Luetje; Karl L. Magleby; Joseph T. Neary; Wolfgang F. Nonner; Michael D. Norenberg; Miguel A. Perez-Pinzon; James D. Potter; Jacqueline Sagen.

UNIVERSITY OF NORTH CAROLINA AT CHAPEL HILL

Curriculum in Neurobiology Program

Program of Study

The Curriculum in Neurobiology Program at the University of North Carolina (UNC) at Chapel Hill is a multidisciplinary Ph.D. training program established in 1966, involving more than 70 faculty researchers in fourteen departments in the School of Medicine and College of Arts and Sciences. In addition, many faculty members have affiliations with specialized research centers in the University, including the UNC Neuroscience Center, the Bowles Center for Alcohol Studies, the Neurodevelopmental Disorders Research Center, and the Lineberger Cancer Center. The goal of this training program is to provide a structured set of experiences to endow trainees with a scholarly understanding of nervous system structure, function, and development. Courses also emphasize evaluation of the literature, the art of experimental design and analysis, exposure to and an appreciation of an extensive range of advanced methodological approaches to neuroscience, the "survival skills" that are necessary for future career development, the ethical and societal issues of basic science research as a career, and identification of the important questions and challenges for future neuroscience research.

Predoctoral trainees in the Curriculum in Neurobiology Program acquire a background in several major areas of neurobiology by taking required and elective courses. In addition, during the first year of their training, students learn the concepts and techniques of several disciplines, through research apprenticeships in three different laboratories. This background, combined with extensive interdepartmental collaborations among faculty members, trains students to use multidisciplinary approaches in solving problems of normal and disturbed nervous system function.

Courses required for the Ph.D. degree in neurobiology include Cellular and Molecular Neurobiology (six blocks: Introduction, Receptors, Electrical Signaling, Synaptic Transmission and Plasticity, Signal Transduction, and Neuroanatomy and Systems), Developmental Neurobiology Seminar in Neurobiology, and Biological Basis of Behavior; two elective specialty courses; and three research apprenticeships in different laboratories. The Cellular and Molecular Neurobiology series and the Seminar course are required during the first year. Students then take Developmental Neurobiology, Biological Basis of Behavior, and Seminar in Neurobiology in the second year. Additional elective courses in biochemistry, statistics, molecular biology, physiology, and others are available to enhance training. At the end of the second year, students take the written portion of the Graduate Qualifying Exam. By the end of the third year, students defend their dissertation proposal in an oral qualifying exam. The average time to complete the Ph.D. is 5.3 years.

Research Facilities

Faculty members are highly competitive in receiving extramural grant support, with nearly $36 million in research funding coming annually from the NIH, NSF, and private foundations. These funds support individual research laboratories. Also available are an impressive array of core facilities available to faculty members and students. Among these are state-of-the-art facilities for production of transgenic and knockout animal models, peptide and DNA synthesis, gene chip array analysis, mouse-behavior core, electron microscopy and laser confocal microscopy, molecular spectroscopy and proteomics, tissue culture, protein and DNA sequencing, cell sorting and harvesting, and functional imaging.

The offices of the Curriculum in Neurobiology are in the Neuroscience Research Building, which also houses the UNC Neuroscience Center. A central conference room and computing and student office space is conveniently located near the labs of many of the primary faculty members in the training program, creating a real neuroscience community. The Neuroscience Center also maintains specialized core facilities for neuroscience researchers, including an Affymetrix gene chip array instrument, a Taqman real-time PCR instrument, and multiphoton confocal microscopes.

Financial Aid

The program holds a Joint Predoctoral Training Grant from NINDS (with support from NIMH and NIDA), which funds most students during the first two years. In subsequent years, students are supported either by their mentor's research grants or individual fellowships. Students entering through different paths (M.D./Ph.D. or through the Interdisciplinary Biomedical Sciences Program affiliate immediately with a laboratory and are supported by their mentors.

Cost of Study

Full-time tuition and fees (12 credits or more) for the 2006–07 academic year are about $5008 for North Carolina residents and $19,006 for nonresidents. Students take 3 credits per semester typically after the second year, corresponding to about $2024 per year for residents and $5500 per year for nonresidents. Tuition and fees are usually met by appointment on a training grant or by support from the dissertation mentor.

Living and Housing Costs

The University owns residence halls for single students and apartments for married students. Room costs in 2005–06 average $3940 per academic year in the residence halls for double occupancy. Rents for apartments for married students range from $6000 to $8000 per year. Off-campus housing is more expensive.

Student Group

More than one third of the students at UNC Chapel Hill are in graduate or professional training. The Curriculum in Neurobiology Program averages 35 graduate students. Graduate students in the program are a diverse group of individuals. Their undergraduate majors range from experimental psychology, biology, and chemistry to physics, mathematics, engineering, music, and biotechnology. Many have prior research experience, and some have worked in research labs for a few years before returning to graduate school. There are several students enrolled in the dual M.D./Ph.D. program in the School of Medicine (for information on this program, students should visit http://www.med.unc.edu/mdphd). Currently, two thirds of the students are women.

Student Outcomes

Since the program began in 1966, the University has produced many neuroscientists, who continue to enjoy outstanding careers as department chairs; directors of neuroscience training programs; researchers/teachers in academia, industry, and government; and other positions. Currently, UNC neurobiology alumni hold faculty positions at Yale, the University of Minnesota, Emory, Vanderbilt, Harvard, and Cornell, among others. Each year, most of the graduates go on to their choice of many excellent postdoctoral positions in top neuroscience labs in the U.S. and abroad.

Location

Chapel Hill, with a population of approximately 90,000, is located in the center of the state, between the Atlantic Ocean and the Appalachian Mountains. The climate is mild, and myriad cultural, entertainment, sports, scientific, and business opportunities exist in the area, which is one of the most popular places to live in the United States. Many lakes, parks, and hiking and riding trails provide outdoor enjoyment, and moderate weather conditions make year-round outdoor activities possible. The mountains and the Atlantic coast are easily accessible, with the Great Smoky Mountains, the Blue Ridge Mountains, and the Carolina beaches within a few hours' drive. Duke University in Durham is 8 miles away, and the Raleigh and Greensboro campuses of the University of North Carolina are 30 and 50 miles away, respectively. Fifteen miles from Chapel Hill, the Research Triangle Park provides the setting for the National Institute of Environmental Health Sciences, the Environmental Protection Agency, and other public and private research facilities, including the giant pharmaceutical firm GlaxoSmithKline and a number of vibrant growing biotech firms. The proximity to Chapel Hill of these universities and Research Triangle Park creates a unique and stimulating scientific and cultural climate as well as employment opportunities.

The University

The University of North Carolina at Chapel Hill is the oldest state university in the United States. Approximately 27,000 students, of whom 9,000 are graduate and professional students, are enrolled annually. The cultural life of the University and community is rich and diverse. From the newly renovated Memorial Hall to the Ackland Art Musuem to the Morehead Planetarium and Science Center to the North Carolina Botanical Garden, Carolina offers a vast array of educational and cultural opportunities. Now in its third century, UNC belongs to the select group of American and Canadian campuses forming the Association of American Universities. UNC's academic offerings span more than 100 fields, including bachelor's, master's, and doctoral degrees as well as professional degrees in dentistry, medicine, pharmacy, and law. Five health schools—which, with UNC Hospitals, comprise one of the nation's most complete academic medical centers—are integrated with liberal arts, basic sciences, and high-tech academic programs. UNC Chapel Hill is rated among the twenty very "best buy" public universities in the U.S. and Canada as judged by the 2005 *Fiske Guide to Colleges,* based on the quality of the academic programs in relation to the cost of attendance. The Carolina academic community benefits from a library with more than 5.6 million volumes.

Applying

Application to the program is made through the Graduate School at UNC. Applications must be submitted online, and application instructions and materials can be found at the Graduate School application Web site at http://gradschool.unc.edu/forms.html. Students may contact the Graduate School by e-mail at gradinfo@unc.edu. The Curriculum in Neurobiology Program and the UNC Graduate School require the official Graduate Record Examinations (GRE) scores (Subject Tests are not required). The program also requires letters of recommendation from 3 people qualified to evaluate the applicant's academic and research experience. The program also requires a one- to two-page essay describing past research experience, future plans, unique factors and experiences that may contribute to diversity in the training program, special reasons for applying to the particular program at the University of North Carolina, specific research and training interests, and a list of the laboratories that the student wants to consider for dissertation research. The recommendation letters and personal statement can be either sent directly to the program or entered online.

Correspondence and Information

Lori Blalock
Administrative Assistant
UNC Curriculum in Neurobiology
CB 7320
The University of North Carolina at Chapel Hill
Chapel Hill, North Carolina 27599-7320
Phone: 919-966-1260
E-mail: neurobiology@med.unc.edu
Web site: http://www.med.unc.edu/neurobiology

University of North Carolina at Chapel Hill

THE FACULTY

Primary Training Faculty

Eva Anton, Assistant Professor of Cell and Molecular Physiology; Ph.D., Duke, 1994.

Steven L. Bachenheimer, Professor of Microbiology and Immunology; Ph.D., Tennessee, 1965.

Albert Baldwin, Professor of Biology; Ph.D., Virginia, 1984.

Aysenil Belger, Ph.D., Assistant Professor of Psychiatry and Psychology; Illinois at Urbana-Champaign, 1993.

Manzoor Bhat, Assistant Professor of Cell and Molecular Physiology; Ph.D., Indian Institute of Science, 1992.

George R. Breese, Professor of Psychiatry and Pharmacology; Ph.D., Tennessee, 1965.

Jay Brenman, Assistant Professor of Cell and Developmental Biology; Ph.D., California, San Francisco, 1997.

Regina Carelli, Associate Professor of Psychology; Ph.D., Rutgers, 1991.

Richard E. Cheney, Associate Professor of Cell and Molecular Physiology; Ph.D., Washington (St. Louis), 1989.

Jacquie Crawley, Professor of Psychology; Ph.D., Maryland College Park; 1976.

Fulton T. Crews, Professor of Pharmacology and Director, Center for Alcohol Studies; Ph.D., Michigan, 1978.

Stephen T. Crews, Professor of Biochemistry and Biophysics; Ph.D., Caltech, 1982.

Mohanish Deshmukh, Assistant Professor of Cell and Developmental Biology; Ph.D., Carnegie Mellon, 1994.

Luda Diatchenko, Associate Professor of Dentistry; M.D., 1990, Ph.D., Moscow, 1993.

Linda Dykstra, Professor of Psychology and Pharmacology and Dean, UNC Graduate School; Ph.D., Chicago, 1972.

Paul B. Farel, Professor of Cell and Molecular Physiology; Ph.D., UCLA, 1970.

Douglas C. Fitzpatrick, Assistant Professor of Otolaryngology/Head and Neck Surgery; Ph.D., North Carolina, 1987.

Rita Fuchs-Lokensgard, Assistant Professor of Psychology; Ph.D., Arizona State, 2000.

John H. Gilmore, Professor of Psychiatry; M.D., North Carolina, 1985.

Michael F. Goy, Associate Professor of Cell and Molecular Physiology; Ph.D., Wisconsin, 1977.

Chistina Grobin, Assistant Professor of Psychiatry; Ph.D., Yale, 1996.

T. Kendall Harden, Professor of Pharmacology; Ph.D., Mississippi, 1974.

Clyde Hodge, Associate Professor of Psychiatry, Center for Alcohol Studies; Ph.D., Auburn, 1991.

Joseph Hopfinger, Assistant Professor of Psychology; Ph.D., California, Davis, 1998.

Mark Hollins, Professor of Psychology; Ph.D., Brown, 1971.

Xuemei Huang, Assistant Professor of Neurology; M.D., Beijing, 1987; Ph.D., Purdue, 1994.

Fred Jarskog, Associate Professor of Psychiatry; M.D., Wake Forest, 1991.

Joey Johns, Associate Professor of Psychiatry; Ph.D., Georgia, 1988.

Anthony Lamantia, Associate Professor of Cell and Molecular Physiology; Ph.D., Yale, 1988.

Jean M. Lauder, Professor of Cell and Developmental Biology; Ph.D., Purdue, 1972.

Kenneth J. Lohmann, Associate Professor of Biology; Ph.D., Washington (Seattle), 1988.

Frank M. Longo, Professor and Chair, Neurology; M.D., 1981, Ph.D., 1983, California, San Diego.

P. Kay Lund, Professor of Cell and Molecular Physiology; Ph.D., Newcastle Upon Tyne, 1979.

Donald T. Lysle, Professor of Psychology; Ph.D., Pittsburgh, 1986.

Richard B. Mailman, Professor of Psychiatry and Pharmacology; Ph.D., North Carolina State, 1974.

William Maixner, Professor of Endodontics and Pharmacology; Ph.D., Iowa, 1982.

C. J. Malanga, Assistant Professor of Neurology; M.D., 1997, Ph.D., 1995, West Virginia.

Patricia F. Maness, Professor of Biochemistry and Biophysics; Ph.D., Texas at Houston, 1975.

Paul B. Manis, Professor of Otolaryngology and Interim Director, Neurobiology Curriculum; Ph.D., Florida, 1981.

Glenn K. Matsushima, Assistant Professor of Microbiology and Immunology; Ph.D., USC, 1988.

Ken D. McCarthy, Professor of Pharmacology; Ph.D., Utah, 1975.

Rick B. Meeker, Professor of Neurology; Ph.D., Bowling Green State, 1976.

Gerhard W. D. Meissner, Professor of Biochemistry and Biophysics and Physiology; Ph.D., Berlin, 1965.

Sharon L. Milgram, Associate Professor of Cell and Developmental Biology; Ph.D., Emory, 1991.

A. Leslie Morrow, Professor of Psychiatry; Ph.D., California, San Diego, 1985.

Robert A. Nicholas, Associate Professor of Pharmacology; Ph.D., California, San Diego, 1984.

Carol Otey, Assistant Professor of Cell and Molecular Physiology; Ph.D., North Carolina at Chapel Hill, 1986.

Edward R. Perl, Sarah Graham Kenan Professor of Cell and Molecular Physiology; M.D., Illinois, 1949.

Larysa H. Pevney, Assistant Professor of Genetics; Ph.D., Columbia, 1992.

Ben Philpot, Assistant Professor of Cell and Molecular Physiology; Ph.D., Virginia, 1997.

Mitchell J. Picker, Professor of Psychology; Ph.D., Western Michigan, 1984.

Joe Piven, Professor of Psychiatry and Director, Neurodevelopmental Disorders Research Center; M.D., Maryland, 1981.

Franck Polleux, Assistant Professor of Pharmacology; Ph.D., Claude Bernard (France), 1997.

Robert Rosenberg, Associate Professor of Pharmacology and Physiology; Ph.D., Yale, 1985.

Aldo Rustioni, Professor of Cell and Developmental Biology; M.D., Milan (Italy), 1965.

R. Jude Samulski, Professor of Pharmacology and Director, Gene Therapy Center; Ph.D., Florida, 1982.

Robert W. Sealock, Professor of Cell and Molecular Physiology; Ph.D., Purdue, 1972.

Richard Segal, Professor of Allied Health Sciences; Ph.D., Virginia, 1984.

David Siderovski, Assistant Professor of Pharmacology; Ph.D., Toronto, 1997.

William D. Snider, Professor of Neurology and Director, UNC Neuroscience Center; M.D., North Carolina at Chapel Hill, 1977.

Ann E. Stuart, Professor of Cell and Molecular Physiology and Ophthalmology; Ph.D., Yale, 1969.

Kathleen K. Sulik, Professor of Cell and Developmental Biology; Ph.D., Tennessee, 1974.

Kinuko I. Suzuki, Professor of Pathology; M.D., Osaka City, 1959.

Todd E. Thiele, Assistant Professor of Psychology; Ph.D., Kansas State, 1995.

Paul Tiesinga, Assistant Professor of Physics and Astronomy; Ph.D., Utrecht (Netherlands), 1996.

Jenny P. Ting, Professor of Microbiology and Immunology; Ph.D., Northwestern, 1979.

Terry VanDyke, Professor of Biochemistry and Biophysics; Ph.D., Florida, 1981.

Juli Valtschanoff, Assistant Professor of Cell and Developmental Biology; Ph.D./M.D., Sofia (Bulgaria), 1986.

Richard Weinberg, Associate Professor of Cell and Developmental Biology; Ph.D., Washington (Seattle), 1983.

Ellen Weiss, Associate Professor of Cell and Developmental Biology; Ph.D., Tennessee, 1985.

Barry L. Whitsel, Professor of Cell and Molecular Physiology; Ph.D., Illinois, 1965.

R. Mark Wightman, Kenan Professor of Chemistry; Ph.D., North Carolina at Chapel Hill, 1974.

Associate Training Faculty

Walter E. Bollenbacher, Professor of Biology; Ph.D., UCLA, 1974.

Thomas W. Bouldin, Professor of Pathology; M.D. North Carolina, 1974.

Gregory K. Essick, Associate Professor of Prosthodontics; D.D.S./Ph.D., North Carolina, 1983.

Karamarie Fecho, Assistant Professor of Anesthesiology; Ph.D., North Carolina, 1995.

Liesa Glantz, Assistant Professor of Psychiatry; Ph.D., Pittsburgh, 1999.

James C. Garbutt, Associate Professor of Psychiatry; M.D., Illinois at Chicago, 1975.

Lawrence I. Gilbert, William Rand Kenan Jr. Professor of Biology; Ph.D., Cornell, 1958.

Robert S. Greenwood, Professor of Neurology; M.D., Texas at Galveston, 1968.

John Hong, Adjunct Associate Professor of Pharmacology; Ph.D., Kansas, 1973.

James F. Howard Jr., Professor of Neurology; M.D., Vermont, 1974.

Henry S. Hsiao, Associate Professor of Biomedical Engineering; Ph.D., Berkeley, 1971.

David L. McIlwain, Professor of Cell and Molecular Physiology; M.D., Washington (St. Louis), 1964.

Cort A. Pedersen, Professor of Psychiatry; M.D., North Carolina, 1979.

Peter Petrusz, Professor of Cell Biology and Anatomy; M.D./Ph.D., Pecs (Hungary), 1963; Karolinska (Sweden), 1972.

Paul G. Shinkman, Professor of Psychology; Ph.D., Michigan, 1962.

Kunihiko Suzuki, Professor of Neurology and Psychiatry; M.D., Tokyo, 1959.

Hugh A. Tilson, Adjunct Professor of Neurobiology and Director, Neurotoxicology, EPA; Ph.D., Minnesota, 1972.

R. Haven Wiley, Professor of Biology; Ph.D., Rockefeller, 1970.

UNIVERSITY OF PITTSBURGH

Center for Neuroscience
Graduate Training Program

Program of Study

The Center for Neuroscience Graduate Training Program is a Ph.D. degree program offered cooperatively by the School of Medicine and the School of Arts and Sciences. The program introduces students to the fundamental issues and experimental approaches in neuroscience and trains them in the theory and practice of laboratory research. Research interests of the training faculty focus on several prominent themes, including behavioral/systems/cognitive, cell and molecular, development/plasticity/repair, and the neurobiology of disease. Major features of the program include the extensive collaborative interactions among its faculty members and its affiliation with the Center for the Neural Basis of Cognition, which is a joint program of the University of Pittsburgh and Carnegie Mellon University.

First-year students complete core course requirements in cellular, molecular, and systems neuroscience; participate in research in one to three laboratories; take the Preliminary Examination, which is based on presenting a journal article; and choose a laboratory for their thesis research. In the second year, students focus more intensively on laboratory research and, typically serve as a teaching assistant for one course. In the third year, students take a Comprehensive Exam modeled after a grant application and write a thesis proposal. Subsequent years are used to complete a dissertation based on their original research. Additional requirements include elective courses, a course in statistics, a course in professional ethics, journal clubs and research seminars, and obtaining research experience in at least two laboratories.

Research Facilities

Research in training laboratories is supported by a large number of grants funded by the federal government and private agencies. Thus, each faculty member operates and maintains a research laboratory containing state-of-the-art equipment, and excellent facilities are available for anatomical, behavioral, biochemical, computational, imaging, molecular, pharmacological, and physiological studies of the nervous system and its function. The University maintains a large central computer facility that is available to all members of the research community via widely distributed terminals. Several biomedical libraries are located on campus.

Financial Aid

All graduate students are supported by stipends from University fellowships and federally funded research and training grants. The stipend was $21,500 for 2005–06, and students receive a one-time $2000 educational allowance. In addition, tuition costs are remitted for full-time graduate students. A health insurance program is provided and paid for by the Center.

Cost of Study

See the Financial Aid section.

Living and Housing Costs

Living expenses in Pittsburgh are moderate. While many students live within walking or biking distance of the campus, an excellent public transportation system connects the University with all surrounding areas and is free to University students. Costs for private housing range from $400 to $525 per month for efficiency apartments and from $450 to $750 per month for multiple-bedroom apartments.

Student Group

Normally, there are 60–70 graduate students in the neuroscience training program; about half are women. Students generally have B.S. degrees in biology, chemistry, computer science, mathematics, neuroscience, or psychology, with a cumulative grade point average of at least 3.4 (on a 4.0 scale) and a cumulative GRE score typically greater than 1200 (verbal and quantitative) and a 4.5 in analytical writing.

Location

Pittsburgh is situated in the foothills of the Allegheny Mountains in western Pennsylvania, at the confluence of the Ohio, Allegheny, and Monongahela Rivers. Though originally a center of steel production, Pittsburgh is now largely a city of corporate headquarters, universities, and light industry. The University is located in the largely residential Oakland section of the city, 3 miles from downtown Pittsburgh.

The University

Founded in 1787, the University of Pittsburgh is the oldest institution of higher education west of the Allegheny Mountains. It is state-supported but governed and operated as a private, nonsectarian, coeducational university. Graduate programs are offered in fourteen professional schools as well as the School of Arts and Sciences. Enrollment exceeds 20,000 students, about 5,000 of whom are graduate and professional students.

Applying

Applications for graduate study and supporting documents should be submitted in time for evaluation in early January, and students usually begin graduate work in the summer or fall term. Members of underrepresented groups are encouraged to apply. Admission to the program is based on college transcripts, Graduate Record Examinations scores, letters of recommendation, research experience, and personal interviews.

Correspondence and Information

Center for Neuroscience
446 Crawford Hall
University of Pittsburgh
Pittsburgh, Pennsylvania 15260
Phone: 412-624-5043
Fax: 412-624-9198
E-mail: jblaney@pitt.edu
Web site: http://cnup.neurobio.pitt.edu

University of Pittsburgh

THE FACULTY AND THEIR RESEARCH

Behavioral/Systems/Cognitive

C. D. Balaban, Ph.D., Professor (Otolaryngology). Anatomy, neurophysiology, and neurochemistry of vestibular function.

M. Behrmann, Ph.D., Adjunct Professor (Neuroscience). Visual cognition in normal and brain-damaged humans.

G. Q. Bi, Ph.D., Assistant Professor (Neurobiology). Development and plasticity of neural circuits.

C. W. Bradberry, Ph.D., Associate Professor (Psychiatry). Neurochemical and cognitive traits in primates associated with chronic exposure to drugs of abuse or a predisposition to self-administer them.

J. P. Card, Ph.D., Associate Professor (Neuroscience). Functional organization of hypothalamus and related systems; viral neurotropism.

C. L. Colby, Ph.D., Associate Professor (Neuroscience). Cortical mechanisms of memory, attention, and spatial representation in primates.

J. C. Crowley, Ph.D., Adjunct Assistant Professor (Neurobiology). Development of visual system structure and function.

W. C. de Groat, Ph.D., Professor (Pharmacology). Neurochemistry and neuroanatomy of autonomic reflex control.

G. B. Ermentrout, Ph.D., Professor (Mathematics). Computational and theoretical models of neural and muscle physiology.

J. A. Fiez, Ph.D., Associate Professor (Psychology). Neuroimaging and behavioral studies of language, working memory, motivation, and learning.

N. J. Gandhi, Ph.D., Assistant Professor (Otolaryngology). Neural control of coordinated oculomotor and skeletomotor movements.

A. A. Grace, Ph.D., Professor (Neuroscience). Neurophysiology of basal ganglia system related to psychiatric disorders.

A. R. Hariri, Ph.D., Assistant Professor (Psychiatry). Functional neuroimaging of genetic effects on brain function.

J. P. Horn, Ph.D., Professor (Neurobiology). Synaptic computation in sympathetic ganglia.

F. J. Jenkins, Ph.D., Associate Professor (Pathology). Nervous and immune systems interactions in HSV latency.

S.-G. Kim, Ph.D., Professor (Neurobiology). Functional imaging methods and biophysics.

A. E. Kline, Ph.D., Assistant Professor (Physical Medicine and Rehabilitation). Pharmacological and environmental approaches for neurobehavioral and histological recovery after experimental traumatic brain injury.

T. S. Lee, Ph.D., Adjunct Associate Professor (Neuroscience). Computational and electrophysiological study of the neural basis of visual perception and recognition.

B. Luna, Ph.D., Associate Professor (Psychiatry). Brain basis of cognitive maturation through adolescence to adulthood.

J. L. McClelland, Ph.D., Adjunct Professor (Neuroscience). Computational models addressing neural mechanisms of perception, memory, language, and cognitive development.

B. Moghaddam, Ph.D., Professor (Neuroscience). Cellular basis of goal-directed behavior and animal models of schizophrenia.

P. W. Munro, Ph.D., Associate Professor (Information Science). Adaptation, learning, and plasticity in simulated neural networks.

C. R. Olson, Ph.D., Adjunct Professor (Neuroscience). Cortical mechanisms of cognition in primates.

T. M. Plant, Ph.D., Professor (Cell Biology and Physiology). Neurobiology of the onset of puberty in primates.

E. D. Reichle, Ph.D., Assistant Professor (Psychology). Computation models of eye-movement control during reading; the neural systems mediating the "eye-mind" link.

L. Rinaman, Ph.D., Associate Professor (Neuroscience). Neural circuits for anxiety, stress, and emotional learning; organization and postnatal development.

J. E. Rubin, Ph.D., Associate Professor (Mathematics). Theoretical and computational modeling of dynamics in neuronal networks.

W. Schneider, Ph.D., Professor (Psychology). Brain imaging and computational neuroscience in visual attention.

R. H. Schor, Ph.D., Associate Professor (Otolaryngology). Information processing in the vestibulospinal system.

A. B. Schwartz, Ph.D., Professor (Neurobiology). Cerebral basis for volitional movement and cortical neural prosthetics.

S. R. Sesack, Ph.D., Professor (Neuroscience). Functional neuroanatomy of cortical and monoamine systems.

D. J. Simons, Ph.D., Professor (Neurobiology). Sensory physiology of the cerebral cortex.

M. A. Sommer, Ph.D., Assistant Professor (Neuroscience). Neuronal circuits mediating perception, cognition, and action in primates.

P. L. Strick, Ph.D., Professor (Neurobiology). CNS circuits of motor and cognitive functions; motor skill acquisition and retention; functional imaging.

E. M. Stricker, Ph.D., University Professor (Neuroscience). Central control of homeostatic regulatory systems.

A. F. Sved, Ph.D., Professor (Neuroscience). Central neural control of the autonomic nervous system and cardiovascular function.

E. Thiels, Ph.D., Research Assistant (Neurobiology). Behavioral, biochemical, and physiological analysis of learning and memory.

B. J. Yates, Ph.D., Professor (Otolaryngology). Vestibular influences on autonomic control and navigation.

Cell and Molecular

E. Aizenman, Ph.D., Professor (Neurobiology). Cellular and molecular mechanisms of neurodegeneration.

K. Albers, Ph.D., Associate Professor (Medicine). Neurotrophic factors in sensory system development and function.

S. G. Amara, Ph.D., Professor (Neurobiology). Molecular and cellular biology of neurotransmitter transporters.

G. Barrionuevo, M.D., Professor (Neuroscience). Synaptic physiology in hippocampus and prefrontal cortex.

A. L. Barth, Ph.D., Assistant Professor (Neuroscience). Plasticity in developing and adult neocortex.

R. P. Bowser, Ph.D., Associate Professor (Pathology). Molecular mechanisms in neurodegeneration.

M. Cascio, Ph.D., Assistant Professor (Molecular Genetics and Biochemistry). Structure and function of neurotransmitter-gated channels.

C. T. Chu, M.D., Ph.D., Associate Professor (Pathology). Oxidative neuronal cell death: Signaling, autophagy, and Parkinson's disease.

P. J. Grabowski, Ph.D., Professor (Biological Sciences). Regulation of alternative splicing in the nervous system.

J. T. Greenamyre, M.D., Ph.D., Professor (Neurology). Mechanisms of neurodegeneration and neuroprotection in Parkinson's and Huntington's diseases.

A. R. Hariri, Ph.D., Assistant Professor (Psychiatry). Functional neuroimaging of genetic effects on brain function.

J. P. Horn, Ph.D., Professor (Neurobiology). Synaptic computation in sympathetic ganglia.

J. W. Johnson, Ph.D., Associate Professor (Neuroscience). Biophysics, pharmacology, and regulation of glutamate receptors.

E. S. Levitan, Ph.D., Professor (Pharmacology). Neurosecretion and channel expression.

Y. J. Liu, M.D., Assistant Professor (Neurology). Dopamine toxicity in Parkinson's disease.

S. D. Meriney, Ph.D., Associate Professor (Neuroscience). Regulation and modulation of presynaptic ion channels and transmitter release.

M. J. Palladino, Ph.D., Assistant Professor (Pharmacology). Molecular mechanisms of neurodegenerative diseases.

R. G. Perez, Ph.D., Assistant Professor (Neurology). Molecular mechanisms of proteins implicated in neurodegenerative disease.

G. E. Torres, Ph.D., Assistant Professor (Neurobiology). Cellular and molecular neuroscience.

N. N. Urban, Ph.D., Adjunct Assistant Professor (Neuroscience). Synaptic and dendritic physiology and imaging in the olfactory bulb.

D. C. Wood, Ph.D., Associate Professor (Neuroscience). Mechanical stimulus transduction and habituation.

Y. Xu, Ph.D., Professor (Anesthesiology). Gene and stem-cell therapy in brain ischemia/membrane protein structure by NMR.

Neurobiology of Disease

C. L. Achim, M.D., Ph.D., Associate Professor (Pathology). Neurotrophins and immune mediators in the human brain.

E. Aizenman, Ph.D., Professor (Neurobiology). Cellular and molecular mechanisms of neurodegeneration.

R. P. Bowser, Ph.D., Associate Professor (Pathology). Molecular mechanisms in neurodegeneration.

J. L. Cameron, Ph.D., Associate Professor (Psychiatry). Effects of stress on behavior and health.

J. Chen, M.D., Professor (Neurology). Mechanisms of cell death and neuroprotection in cerebral ischemia and Parkinson's disease.

C. T. Chu, M.D., Ph.D., Associate Professor (Pathology). Oxidative neuronal cell death: Signaling, autophagy, and Parkinson's disease.

D. B. DeFranco, Ph.D., Professor (Pharmacology). Signal transduction and neurodegeneration.

S. T. DeKosky, M.D., Professor and Chair (Neurology). Neurodegenerative dementias and brain injury.

A. A. Grace, Ph.D., Professor (Neuroscience). Neurophysiology of basal ganglia system related to psychiatric disorders.

S. H. Graham, M.D., Ph.D., Professor (Neurology). Neuronal cell death in ischemia and trauma.

J. T. Greenamyre, M.D., Ph.D., Professor (Neurology). Mechanisms of neurodegeneration and neuroprotection in Parkinson's and Huntington's diseases.

T. G. Hastings, Ph.D., Associate Professor (Neurology). Oxidative mechanisms associated with neurodegeneration.

D. A. Lewis, M.D., Professor (Psychiatry). Functional architecture of the prefrontal cortex and schizophrenia.

Y. J. Liu, M.D., Assistant Professor (Neurology). Dopamine toxicity in Parkinson's disease.

K. Mirnics, M.D., Ph.D., Assistant Professor (Psychiatry). Complex gene expression patterns in psychiatric disorders.

R. Y. Moore, M.D., Ph.D., Professor (Neurology). Neural mechanisms and disorders of circadian rhythms and sleep.

M. J. Palladino, Ph.D., Assistant Professor (Pharmacology). Molecular mechanisms of neurodegenerative diseases.

R. G. Perez, Ph.D., Assistant Professor (Neurology). Molecular mechanisms of proteins implicated in neurodegenerative disease.

N. F. Schor, M.D., Ph.D., Professor (Pediatrics). Neurochemical approaches to neuroblastoma therapy

S. R. Sesack, Ph.D., Professor (Neuroscience). Functional neuroanatomy of cortical and monoamine systems.

E. L. Sibille, Ph.D., Assistant Professor (Psychiatry). Molecular mechanisms of depression and aging.

P. L. Strick, Ph.D., Professor (Neurobiology). CNS circuits of motor and cognitive functions; motor skill acquisition and retention; functional imaging.

R. A. Sweet, M.D., Associate Professor (Psychiatry). Genetic and neuropathologic studies of psychosis in Alzheimer's disease; neuropathology of schizophrenia and depression.

G. E. Torres, Ph.D., Assistant Professor (Neurobiology). Cellular and molecular neuroscience.

C. A. Wiley, M.D., Ph.D., Professor (Pathology). Mechanisms of neurodegeneration during viral infection.

Y. Xu, Ph.D., Professor (Anesthesiology). Gene and stem-cell therapy in brain ischemia/membrane protein structure by NMR.

M. J. Zigmond, Ph.D., Professor (Neurology). Cell death and neuroprotection in neurodegenerative disease and stress.

Development/Plasticity/Repair

K. Albers, Ph.D., Associate Professor (Medicine). Neurotrophic factors in sensory system development and function.

G.-Q. Bi, Ph.D., Assistant Professor (Neurobiology). Development and plasticity of neural circuits.

J. C. Crowley, Ph.D., Adjunct Assistant Professor (Neurobiology). Development of visual system structure and function.

B. M. Davis, Ph.D., Associate Professor (Medicine). Cellular and molecular mechanisms of central and peripheral nervous system development and plasticity.

W. M. Halfter, Ph.D., Associate Professor (Neurobiology). Axonal pathfinding in the developing chick visual system.

K. Kandler, Ph.D., Associate Professor (Neurobiology). Development and plasticity of inhibitory circuits.

H. R. Koerber, Ph.D., Associate Professor (Neurobiology). Development, plasticity, and pain processing in sensory neurons and the spinal dorsal horn.

C. F. Lagenaur, Ph.D., Associate Professor (Neurobiology). Neurite outgrowth and synaptogenesis.

C. Lance-Jones, Ph.D., Associate Professor (Neurobiology). Motoneuron development and spinal cord patterning.

L. E. Lillien, Ph.D., Associate Professor (Neurobiology). CNS stem cells.

S. D. Meriney, Ph.D., Associate Professor (Neuroscience). Regulation and modulation of presynaptic ion channels and transmitter release.

A. P. Monaghan-Nichols, Ph.D., Assistant Professor (Neurobiology). Genetic analysis of vertebrate CNS development.

J. E. Rubin, Ph.D., Associate Professor (Mathematics). Theoretical and computational modeling of dynamics in neuronal networks.

E. Thiels, Ph.D., Research Assistant (Neurobiology). Behavioral, biochemical, and physiological analysis of learning and memory.

J. W. Yip, Ph.D., Associate Professor (Neurobiology). Molecular control of neuronal migration.

UNIVERSITY OF SOUTHERN CALIFORNIA

Neuroscience Graduate Program

Program of Study

The Neuroscience Graduate Program (NGP) at the University of Southern California (USC) represents a highly interdisciplinary approach to understanding neural function. An outstanding range of scientists drawn from many different departments across the University maintain state-of-the-art laboratories that offer diverse opportunities for graduate training leading to the Ph.D. Students gain a broad understanding of basic concepts in biological, cognitive, and modeling approaches to brain structure and function through shared interdisciplinary core courses. Thereafter, each student's specific program of study is tailored to individual interests, which enables the student to gain a deep understanding of selected topics during the Ph.D. thesis research. Areas of specialization include cellular and molecular neuroscience, behavioral and systems neuroscience, cognitive neuroscience, computational neuroscience and neural engineering, and the neurobiology of disease and aging.

The philosophy behind the curriculum is to get students into neuroscience research as quickly as possible, while providing them with the best tools to perform it. In total, the curriculum has six components—core courses, elective courses, neuroscience seminars, journal clubs, laboratory rotations, and dissertation research. During the first year, graduate students take two core courses that span the breadth of neuroscience, from molecular to cognitive and computational, and participate in a weekly chalk-talk seminar series entitled Neurolunch. The core courses cover basic experimental and theoretical concepts in neuroscience and prepare students for advanced study. Students typically choose three different laboratories for research rotations during their first year before selecting an adviser. Therefore, they have the opportunity to sample different approaches to neuroscience before choosing the subject of their dissertation work. A qualifying examination that includes a thesis proposal in the style of an actual grant proposal is completed by the end of the second year. An important cornerstone of the program is to provide students with a flexible and supportive environment. The most important part of the curriculum is hands-on experience in research, in which interaction with a mentor provides the student with the skills necessary to become a neuroscientist.

Research Facilities

The Neuroscience Graduate Program at USC includes 65 faculty members from approximately fifteen different departments throughout the University. Their research laboratories are located at the University Park Campus (UPC), the Health Sciences Campus (HSC), and Children's Hospital Los Angeles (CHLA). Facilities are available for all aspects of research and include complete resources for neuroanatomical, neurochemical, electrophysiological, cell-molecular, and behavioral studies. Labs occupy modern research space, with state-of-the-art facilities that include specialized equipment for cell culturing and sorting, protein and DNA synthesis and sequencing, confocal and two-photon microscopy, and more. In addition to faculty members' laboratories, shared core facilities are available for research in several areas, including genomics and proteomics, and provide imaging technologies for visualizing neurons at different levels of structure and function, including a 3T Magnetom Trio scanner for fMRI studies at the Cognitive Neuroscience Imaging Center. The USC libraries house more than 4.2 million volumes and maintain online subscriptions to more than 20,000 current journals.

Financial Aid

All students admitted to the program are fully supported during their graduate training. Support is provided in many forms, including teaching assistantships, research assistantships, training grants, and University fellowships. All awards include full tuition remission, a monthly stipend for living expenses (currently a minimum of $1906 per month or about $22,870 annually for first-year students), and payment of student health insurance and student health center fees. All students are encouraged to apply for funding from external sources, which include the National Institutes of Health and the National Science Foundation.

Cost of Study

All graduate students in the program are supported by training grants, individual fellowships, or research or teaching assistantships.

Living and Housing Costs

Most graduate students live in apartments or houses in the vicinity of the campus. Rents for shared housing range from $600 to $1000 per month. University apartments for graduate students and their families are available on a limited, first-come, first-served basis. All units are fully furnished, with estimated monthly rates ranging from $450 to $995 per month.

Student Group

The program includes 80 full-time students of widely varied backgrounds with broad-ranging interests and pursuits. Currently, 44 of the students are women, 5 are members of underrepresented minority groups, and 29 are international students. The graduate students constitute a vigorous group and are proactive in providing input to and leadership for the program through their student-run organization, Neuroscience Graduate Forum (NGF). Students and faculty members together constitute an engaged community of scholars whose interactions are enhanced by weekly seminars, journal club activities, and annual symposia and off-site neuroscience retreats. In addition, group social and recreational activities take advantage of the multicultural vibrancy and diversity of Los Angeles and its surroundings.

Student Outcomes

Neuroscience graduates have continued their careers as postdoctoral fellows at major research universities, including Caltech, Harvard, MIT, Stanford, the University of California at San Francisco, and numerous other institutions; as faculty members in academia; and as researchers in both private and public institutions. Some graduates have chosen to pursue careers in teaching or biotechnology.

Location

Situated 3 miles south of downtown Los Angeles, the 235-acre University Park Campus is adjacent to museums and recreational facilities of the Exposition Park complex. The 50-acre Health Sciences Campus, situated 7 miles from the University Park Campus, is 4 miles east of downtown Los Angeles. Cultural activities include outstanding museums, theaters, film, concerts, and clubs as well as the world-famous L. A. Philharmonic in Disney Hall. Professional sports teams include the Lakers, Kings, and Dodgers. Southern California's legendary climate provides outstanding recreational opportunities, and Los Angeles is close to both major mountain ranges and beaches.

The University

Founded in 1880, the University of Southern California is a private research university of international distinction. It ranks among the nation's top ten private research universities in terms of receiving government funds for research and development support. USC is one of sixty-two U.S. and Canadian research universities elected to the Association of American Universities.

Applying

Entry into the program requires a bachelor's degree in a related subject from an accredited four-year college. Most applicants have research experience and a strong undergraduate foundation in their chosen discipline, including a rigorous course presence in chemistry, physics, psychology, biology, engineering, or computational science. To apply, students must complete the application found on the program's Web site and submit official transcripts from all institutions attended, three letters of recommendation, a statement of purpose, and official GRE test scores. International students must also submit TOEFL scores. The deadline for complete applications, including all necessary documents, is January 1. Admissions decisions are made on a rolling basis beginning in late February, with final decisions made by the end of April.

Correspondence and Information

Linda Bazilian, Graduate Programs Manager
Graduate Program in Neuroscience
University of Southern California
3641 Watt Way, HNB 120, MC 2520
Los Angeles, California 90089-2520
Phone: 213-821-1088
Fax: 213-740-8123
E-mail: bazilian@usc.edu
Web site: http://www.usc.edu/neuroscience

University of Southern California

THE FACULTY AND THEIR RESEARCH

Ronald Alkana, Professor of Molecular Pharmacology and Toxicology; Pharm.D., USC, 1970; Ph.D., California, Irvine, 1975. Molecular mechanisms of drug action; psychopharmacology; behavioral pharmacology; ethanol; ligand-gated ion channels; allosteric modulation.

Elaine S. Andersen, Professor of Linguistics; Ph.D., Stanford, 1978. Aging; Alzheimer's disease; language acquisition; language and cognitive impairment; language loss.

Michelle Arbeitman, Assistant Professor of Biological Sciences; Ph.D., Stanford, 1998. How the sex-specific potential for courtship behavior is built into the nervous system and how the sex-specific deployment of a regulatory hierarchy acts on a molecular level to generate and maintain aspects of sex-specific differences in adults.

Michael A. Arbib, Fletcher Jones Professor of Computer Science; Ph.D., MIT, 1963. Computational and cognitive neuroscience; mirror neurons and action recognition; brain mechanisms of language and their evolution; epistemology; neural networks; simulation; schema theory; neuroinformatics.

Donald B. Arnold, Assistant Professor of Biological Sciences; Ph.D., Johns Hopkins, 1992. Ion channel; localization; neuron; confocal microscopy.

Laura A. Baker, Associate Professor of Psychology; Ph.D., Colorado, 1983. Behavioral genetics; childhood aggression; antisocial behavior.

Michel Baudry, Professor of Biological Sciences; Ph.D., Paris VII, 1977. Plasticity; neurodegeneration; glutamate receptors; oxygen free radicals; neuron/silicon interface; human brain project.

Theodore W. Berger, Professor of Biomedical Engineering; Ph.D., Harvard, 1976. Neurophysiology of memory and learning; nonlinear systems analysis of hippocampal network properties.

Irving Biederman, Harold W. Dornsife Professor of Neuroscience; Ph.D., Michigan, 1966. Shape recognition; cognitive neuroscience; perceptual and cognitive pleasure.

Sarah W. Bottjer, Professor of Biology; Ph.D., Indiana, 1979. Vocal learning in songbirds; neurogenesis and cell death; activity-dependent refinement of neural connectivity; synaptic plasticity and neurophysiology.

Nina S. Bradley, Associate Professor of Biokinesiology and Physical Therapy; Ph.D., UCLA, 1986. Motor control; sensorimotor development; embryonic motility.

Roberta Diaz Brinton, Professor of Molecular Pharmacology and Toxicology; Ph.D., Arizona, 1984. Learning and memory; Alzheimer's disease; learning disabilities; estrogen; neuroactive steroids; vasopressin; neurochips.

Samantha J. Butler, Assistant Professor of Neurobiology; Ph.D., Princeton, 1996. Mechanisms that establish connections between neurons.

Jonah R. Chan, Assistant Professor of Cell Biology and Neurobiology; Ph.D., Illinois at Urbana-Champaign, 2000. Role of neurotrophins and their receptors in the regulation of myelination; intrinsic cellular mechanisms influencing cell-fate decisions of the Schwann cell and the oligodendrocyte progenitor.

Jeannie Chen, Assistant Professor of Ophthalmology; Ph.D., USC, 1990. Signal transduction; photoreceptor function; retinal degeneration.

Robert Hsiu-Ping Chow, Associate Professor of Physiology and Biophysics; M.D./Ph.D., Pennsylvania, 1988. Exocytosis; synaptic transmission; SNARE proteins.

Cheryl M. Craft, Mary D. Allen Professor for Vision Research, Doheny Eye Institute, and Professor and Chair, Department of Cell and Neurobiology; Ph.D., Texas Health Science Center at San Antonio, 1984. Molecular neurogenetics; pineal gland; melatonin; G-protein–coupled receptor; photoreceptors; genetics of inherited blindness.

Robert A. Farley, Professor of Physiology and Biophysics, Biochemistry, and Molecular Biology; Ph.D., Rochester, 1975. NaK-ATPase; ion transport; protein structure; membrane proteins.

Caleb E. Finch, ARCO–William F. Kieschnick Professor in the Neurobiology of Aging; Ph.D., Rockefeller, 1969. Brain; aging; inflammation; demography; evolution; estrogen; Alzheimer's disease.

Ione Fine, Assistant Professor of Ophthalmology; Ph.D., Rochester, 1999. Perceptual learning and plasticity, with an emphasis on the effects of visual deprivation; measuring performance in patients implanted with electrode retinal prostheses.

Judy A. Garner, Associate Professor of Cell Biology and Neurobiology; Ph.D., Case Western Reserve, 1979. Axonal transport mechanisms; herpes simplex infection of the nervous system; intracellular trafficking; neuronal regeneration.

Andrew K. Groves, Adjunct Assistant Professor of Cell Biology and Neurobiology. Molecular mechanisms responsible for the development and regeneration of the inner ear.

Norberto M. Grzywacz, Professor of Biomedical Engineering; Ph.D., Hebrew (Jerusalem), 1984. Retina; ganglion cells; directional selectivity; motion perception; representation of visual information; receptive-field development; Bayesian vision.

Joseph B. Hellige, Professor of Psychology; Ph.D., Wisconsin–Madison, 1974. Perception; vision; attention; hemispheric asymmetry; laterality.

Judith A. Hirsch, Associate Professor of Biological Sciences; Ph.D., Wisconsin–Madison, 1986. Visual cortex; patch clamp; synaptic potential; interneuron.

Mark S. Humayun, Professor of Ophthalmology, Biomedical Engineering, and Cell Biology and Neurobiology; M.D., Duke, 1989; Ph.D., North Carolina at Chapel Hill, 1994. New therapies for severe retinal diseases.

Laurent Itti, Assistant Professor of Computer Science and Adjunct Assistant Professor of Psychology; Ph.D., Caltech, 2000. Computational modeling; biological vision; computer vision; functional neuroimaging; human psychophysics; neural networks; artificial intelligence; neuromorphic engineering; neuromimetic algorithms.

Chien-Ping Ko, Professor of Biological Sciences; Ph.D., Washington (St. Louis), 1975. Synapse formation; sprouting; remodeling and maintenance; synaptic transmission; synapse-glial interactions; neuromuscular junction; Schwann cells; acetylcholine receptor aggregation; nerve regeneration; neurodegenerative diseases.

Emily R. Liman, Assistant Professor of Biological Sciences; Ph.D., Harvard, 1992. Ion channel; patch clamp; olfactory transduction; pheromone; signal transduction.

Gerald E. Loeb, Professor of Biomedical Engineering; M.D., Johns Hopkins, 1972. Neural prosthesis; sensorimotor control; muscle; electromyography; spinal cord.

Zhong-Lin Lu, Associate Professor of Psychology; Ph.D., NYU, 1992. Computational and psychophysical study of visual motion, texture, and attention; visual memory systems; perceptual learning; computer image processing; neurophysiological and neuromagnetic study of sensory and attentional processes; visual neural networks; brain imaging.

Le Ma, Assistant Professor of Cell Biology and Neurobiology; Ph.D., Harvard, 1999. Axon growth, guidance, and branching; Slit/Robo signaling; live cell imaging.

David D. McKemy, Assistant Professor of Biology and Dentistry; Ph.D., Nevada, Reno, 1999. Somatosensation; nociception; ion channels; temperature; pain; signal transduction.

Thomas H. McNeill, Professor of Cell Biology and Neurobiology, Neurology, and Neurogerontology; Ph.D., Rochester, 1980. Parkinson's disease; Alzheimer's disease; plasticity; reactive synaptogenesis; growth factors.

Bartlett W. Mel, Associate Professor of Biomedical Engineering; Ph.D., Illinois at Urbana-Champaign, 1989. Computational neuroscience; neural engineering; dendritic trees; learning and memory; vision; object recognition.

Carol Ann Miller, Professor of Pathology and Neurology; M.D., Thomas Jefferson, 1965. Neuronal specificity; selective vulnerability neurodegeneration; Alzheimer's disease.

Toben H. Mintz, Assistant Professor of Psychology; Ph.D., Rochester, 1996. Language acquisition and language representation; computational modeling.

Christian J. Pike, Assistant Professor, Andrus Gerontology Center, and Hanson Family Assistant Professor of Gerontology; Ph.D., California, Irvine, 1994. Alzheimer's disease; apoptosis; neurodegeneration; neuroprotection; steroid hormones; reactive gliosis.

Michael W. Quick, Professor of Biology; Ph.D., Emory, 1992. Addiction; neurotransmitter transporters; nicotinic acetylcholine receptors; protein-protein interactions; regulation of synaptic proteins.

Adrian Raine, Robert G. Wright Professor of Psychology; D.Phil., York, 1982. Brain imaging; psychophysiology; neuropsychology; violence; schizotypal personality.

Stefan Schaal, Assistant Professor of Computer Science; Ph.D., Munich Technical, 1991. Statistical learning; motor control; computational neuroscience; biomimetic and humanoid robotics; nonlinear dynamics; motor psychophysics.

Paula Elyse Schauwecker, Associate Professor of Cell Biology and Neurobiology; Ph.D., USC, 1994. Neurodegenerative disease; mouse genetics; neuronal death; gene expression.

Nicolas Schweighofer, Assistant Professor of Biokinesiology and Physical Therapy; Ph.D., UCLA. Understanding the computational, neural, and psychological bases of the interactions between skill learning and motivation in humans, particularly the role of neuromodulatory systems (mainly dopaminergic and serotonergic).

Neil Segil, Research Associate Professor of Cell Biology and Neurobiology; Ph.D., Columbia, 1989. Development; ear; hearing; cell cycle; cyclin-dependent kinase; transcription.

Magdalene Seiler, Assistant Professor, Doheny Eye Institute; Ph.D., Munich, 1985. Transplantation of immature retinal cells (fetal retinal sheets or retinal progenitor cells) to rodent models of retinal degeneration with the aim of restoring vision.

Jean C. Shih, Boyd and Elsie Welin Professor of Pharmacology, Molecular Pharmacology, and Toxicology; Ph.D., California, Riverside, 1968. Molecular biology of aminergic neurotransmitters in development, aging, and disease states.

Larry W. Swanson, Milo Don and Lucille Appleman Professor of Biological Sciences; Ph.D., Washington (Seattle), 1972. Hunger and thirst; limbic system; motivation; structural neuroscience.

Armand R. Tanguay Jr., Associate Professor of Electrical Engineering–Electrophysics, Materials Science, and Biomedical Engineering; Ph.D., Yale, 1977. Neural networks; optics.

Richard F. Thompson, Keck Professor of Psychology and Biological Sciences; Ph.D., Wisconsin, 1956. Memory trace; engram; cerebellum; hippocampus; classical conditioning; procedural memory; declarative memory; long-term potentiation and long-term depression in the hippocampus.

Bosco S. Tjan, Assistant Professor of Psychology; Ph.D., Minnesota, 1996. Object recognition; scene perception; reading; spatial vision; ideal observers; signal-detection theory; information theory; neural networks; parallel and distributed computing; visual psychophysics; neural imaging.

Christoph von der Malsburg, Professor of Computer Science; Dr.rer.nat., Heidelberg, 1970. Functional modeling of vision processes; computer simulation; cognitive architecture; figure-ground separation; invariant object recognition.

Zuo-Zhong Wang, Associate Professor of Cell Biology and Neurobiology; Ph.D., Utah, 1993. Synapse formation; receptor and ion channel expression and trafficking; signal transduction; neuromuscular junction; cholinergic synapses; acetylcholine receptors; neurodegenerative diseases.

Alan G. Watts, Associate Professor of Neuroscience, Physiology, and Biophysics; D.Phil., Oxford, 1983. Ingestive behaviors; anorexia; neuropeptides; stress; hypothalamus; neuroanatomy; structure-function relationships.

Leslie Weiner, Professor of Neurology; M.D., Cincinnati, 1961. T-cell-receptor vaccines in multiple sclerosis; gene therapy in autoimmune disease; immunologic studies in Alzheimer's disease; neuroendocrine influences on the immune response; immune regulation in neurologic disease.

Ruth I. Wood, Associate Professor of Cell Biology and Neurobiology; Ph.D., Michigan, 1991. Behavioral neuroendocrinology; limbic anatomy; chemosensory; anabolic steroid.

Li I. Zhang, Assistant Professor of Physiology and Biophysics; Ph.D., California, San Diego. Functional microcircuitry; cortical representation; molecular and activity-dependent mechanisms; plasticity of cortical circuits; neural basis for various cortical pathologies.

THE UNIVERSITY OF TENNESSEE
HEALTH SCIENCE CENTER

College of Medicine
Department of Anatomy and Neurobiology

Program of Study	The Department of Anatomy and Neurobiology, in conjunction with the Integrated Program in Biomedical Science (IPBS) at the University of Tennessee (UT) Health Science Center and the UT Neuroscience Institute, offers a diversified research and training program in neuroscience leading to the Ph.D. degree. All students enter the Neuroscience track of the IPBS program and in their first year complete an interdisciplinary course of study that includes courses in cellular and molecular biology and systems biology. Beginning in the first year, neuroscience students also participate in the neuroscience seminar series and the neuroscience student symposium. In the second year, students complete advanced neuroscience course work by taking courses offered in the Department of Anatomy and Neurobiology. Individual instruction and small-group conferences are important parts of these courses. Students also begin independent research projects with a mentor of their choice in the second year. A large number of neuroscience faculty members are available to choose from with specialties in all areas of neurobiology. The Department of Anatomy and Neurobiology offers courses in development/molecular neuroscience, morphological neuroscience, cellular neuroscience, and behavioral neuroscience as well as the neuroscience seminar series and student symposium. The department also offers courses in developmental biology, gross anatomy, and microscopic anatomy. The usual time needed to attain the Ph.D. is four years beyond the bachelor's degree. The training is designed to equip graduates to assume productive roles in research and health science environments.
Research Facilities	The Department of Anatomy and Neurobiology is housed in a three-building complex that is conveniently located in the Health Science Center. Within the department, a wide variety of specialized research equipment is available. Students may use transmission electron microscopy; image analysis and densitometry; confocal microscopy; freeze fracture; cytochemistry and immunocytochemistry; computerized morphometry; extracellular and intracellular electrophysiology; various radioisotope techniques, including autoradiography and radioimmunoassay; HPLC; DNA microarray analysis; and many other modern biochemical techniques when pursuing problems in the neurobiological sciences. Other research facilities, such as the Molecular Resource Center, are also available on campus. Some faculty members and facilities are also located off campus at the Veterans Administration and St. Jude Children's Research Hospitals.
Financial Aid	Stipends for graduate doctoral fellowships are $21,000 per year with exemption from tuition and course fees. Stipends are provided by the program for the first two years and thereafter by their supporting mentor. Outstanding graduates of U.S. institutions are eligible for Alumni Endowment Scholarships that contribute up to $3,000 of additional annual support.
Cost of Study	Tuition and course fees for 2005–06 are $8124 per year for Tennessee residents and $20,600 per year for out-of-state students.
Living and Housing Costs	The cost for on-campus accommodations for single students was $4320 per year. A great variety of furnished and unfurnished apartments are available near the campus. The cost of living in Memphis is below the national average for cities of its size.
Student Group	The University of Tennessee Health Science Center is the largest center in the mid-South for the training of students in the health sciences. The Health Science Center, with a total enrollment of approximately 2,100 students, consists of the College of Medicine, College of Dentistry, College of Nursing, College of Pharmacy, College of Allied Health Professions, and Graduate School of Medical Sciences. Currently, there are 25 graduate students and 67 postdoctoral fellows participating in the Neuroscience Institute.
Location	Memphis, a progressive mid-Southern city of about 650,000 residents, lies on the Mississippi River and is known for its music heritage. There are many parks within the city and several state parks, lakes, and wooded areas nearby, offering recreational opportunities. Memphis has many cultural advantages, including a symphony orchestra, ballet, art academy, opera, professional and amateur theater, and excellent library and museum systems. There are several institutions of higher education in the area, including Rhodes College, University of Memphis, Christian Brothers University, and LeMoyne Owen College.
The University	Founded in 1794, the main campus of the University of Tennessee is in Knoxville. The Health Science Center, established in 1911, is one of several other campuses. This major subdivision is located in the Memphis Medical Center and is composed of the health science colleges of the University.
Applying	Applicants must have a baccalaureate degree, preferably with a major in a biological science. Scores on the General Test of the Graduate Record Examinations are required. Application is made through the Integrated Program in Biomedical Science and must be completed by March 1 if the candidate is to be considered for admission in the subsequent fall quarter. Detailed instructions for application are available on the department's Web site, which is listed in the Correspondence and Information section, and the IPBS Web site at http://www.utmem.edu/grad/IPBS/. The University of Tennessee does not discriminate on the basis of race, color, national origin, sex, religion, age, handicap, or veteran status. All programs and facilities of the University are accessible to the handicapped.
Correspondence and Information	Director of Graduate Studies Department of Anatomy and Neurobiology College of Medicine The University of Tennessee Health Science Center 855 Monroe Avenue Memphis, Tennessee 38163 Phone: 901-448-5965 Fax: 901-448-7193 E-mail: gradneuro@utmem.edu Web site: http://www.utmem.edu/anatomy-neurobiology/

The University of Tennessee Health Science Center

THE FACULTY AND THEIR RESEARCH

William A. Armstrong, Ph.D., Professor. Electrophysiological and morphological studies of the hypothalamo-neurohypophyseal system.

John D. Boughter Jr., Ph.D., Assistant Professor. Physiological, anatomical, and behavioral approaches to the study of taste in inbred strains of mice.

Joseph C. Callaway, Ph.D., Associate Professor. Integration of synaptic inputs in the dendrites of individual CNS neurons; the regional properties of neurons; the interaction between pairs of synaptically connected neurons; the plasticity of synaptic inputs.

Angela R. Cantrell, Ph.D., Assistant Professor. Ion channel function in neurodegenerative disorders and mental illness.

Edward Chaum, M.D., Ph.D., Associate Professor. Gene therapy for macular degeneration.

Tom Curran, Ph.D., Professor. Mechanisms of CNS development.

Allesandra d'Azzo, Ph.D., Professor. Role of the lysosomal system in normal cellular metabolism and in pathological conditions associated with lysosomal storage disorders (LSD).

Michael Dyer, Ph.D., Assistant Professor. Retinal physiology (ERG), electron microscopy, cell sorting, in vivo mouse models of retinoblastoma, and computational modeling of proliferation during development.

Andrea J. Elberger, Ph.D., Professor. Morphological, physiological, and behavioral analysis of the corpus callosum and the developing visual system.

Matthew Ennis, Ph.D., Professor. Neuroanatomy, neurophysiology and neurochemistry of brain circuits involved in pain/analgesia; taste and olfaction at the molecular, cellular, systems, and behavioral levels.

Malinda E. Fitzgerald, Ph.D., Adjunct Assistant Professor. Choroidal blood flow and retinal pathology in aging.

Robert C. Foehring, Ph.D., Professor. Influence of connections on motoneuron properties and ionic mechanisms and second messengers involved in modulation of neuron firing properties.

Eldon E. Geisert Jr., Ph.D., Professor. Adhesive interactions between astrocytes and CNS neurons following injury and during normal development.

Daniel Goldowitz, Ph.D., Professor. Developmental neurobiology, particularly how genetic information directs and instructs the formation of the mammalian central nervous system.

Kristin Hamre, Ph.D., Assistant Professor. Interaction of the mouse genotype with the effects of ethanol on the severity of ethanol-induced deficits following in utero alcohol and the response of the animal to adult ethanol exposure.

David L. Hasty, Ph.D., Professor. Immunological analysis of development; analysis of bacterial adherence mechanisms.

Karen A. Hasty, Ph.D., Professor. Normal and pathophysiological interaction of cells with extracellular matrices; biochemical and immunological definition, using monoclonal antibody technology.

Detlef Heck, Ph.D., Assistant Professor. Dynamics of network activity and the representation of behavior in neocortex and cerebellum; structure and function of the cerebellar cortical network; neuronal mechanisms of cerebro-cerebellar communication.

Paul Herron, Ph.D., Associate Professor. Neural basis of sensation and perception.

Ramin Homayouni, Ph.D., Assistant Professor. Molecular basis of cortical development.

Marcia G. Honig, Ph.D., Professor. Developmental neurobiology; mechanisms underlying pathway selection and the development of specific neuronal connections.

Monica M. Jablonski, Ph.D., Associate Professor. Interactions between the retina and the retinal pigment epithelium that are required for normal development; differentiation and organization of the light-sensitive membranes of the photoreceptor outer segments.

Michael Jacewicz, M.D., Professor. Mechanisms of ischemic brain injury; cerebral blood flow; neurologic complications of coronary bypass grafting.

Dianna A. Johnson, Ph.D., Professor. The role of neurotransmitters in the development of retinal circuitry through tropic, trophic, and excitotoxic actions.

Eldridge F. Johnson, Ph.D., Professor. Morphological and physiological characteristics of motor output cells in the neocortex and striatum.

Richard J. Kasser, Ph.D., Associate Professor. Changing extensibility of muscle by facilitation of antagonist muscles; improving joint range of motion using selective muscle strengthening; investigating joint motion and its effect on subjective pain relief.

Hitoshi Kita, Ph.D., Professor. Electrophysiological and neuroanatomical studies of the basal ganglia, using intracellular recording and staining techniques on in vivo and in vitro preparations and light- and electron-microscopic analysis.

Mark LeDoux, M.D., Associate Professor. Pathophysiology of motor systems emphasizing dystonia.

Douglas B. Matthews, Ph.D., Assistant Professor. Role of the hippocampal system in spatial cognitive processing; effects of acute and chronic ethanol on single-unit electrophysiology and gene expression in the brain.

Peter J. McKinnon, Ph.D., Associate Professor. Analysis of selected molecules that are important for the development and maintenance of the central nervous system, including astrocyte extracellular matrix proteins and PI3-kinases.

Guy Mittleman, Ph.D., Associate Professor. Neurophysiological basis of motivated behavior; addiction, schizophrenia, psychopharmacology.

James I. Morgan, Ph.D., Professor. Discovery of genes involved in the development and differentiation of specific neuronal cell types; utilizing novel experimental strategies to identify such genes and to characterize their structural properties.

Randall J. Nelson, Ph.D., Professor. Sensorimotor integration in primary somatosensory cortex of primates.

Thaddeus S. Nowak Jr., Ph.D., Professor. Neuropathology: changes in gene expression in response to injury after ischemia and other insults.

Akinniran Oladehin, Ph.D., Professor. Neural basis of physical activity: effects of exercise on aging brain.

Guillermo Oliver, Ph.D., Associate Professor. Role of homeobox genes during early murine pattern formation, with a particular emphasis in eye and central nervous system development.

Melburn R. Park, Ph.D., Associate Professor. Morphological and physiological characterization of the synaptic organization of the dorsal raphe nucleus of the mammalian brain.

William A. Pulsinelli, M.D., Ph.D., Professor. Cellular and molecular mechanisms leading to brain injury from cerebral hypoxia and ischemia.

Anton J. Reiner, Ph.D., Professor. Anatomy, function, and evolution of the basal ganglia, cerebral cortex, and visual system; neuronal control of the autonomic function of the eye.

James T. Robertson, M.D., Professor. Brain tumor registry; clinical trials of drugs; clinical research on the therapy of lower back pain and the sciatica factors.

Thomas Schikorski, Ph.D., Assistant Professor. Synaptic transmission and plasticity; long-term potentiation; learning and memory.

Reese S. Scroggs, Ph.D., Associate Professor. Voltage clamp studies on mechanisms of selective modulation of functional subpopulations of sensory neurons and amygdala neurons by serotonin.

Richard Smeyne, Ph.D., Associate Professor. Genetic analysis of Parkinson's disease; growth factors and development.

David V. Smith, Ph.D., Simon R. Bruesch Professor. Neurobiology of the gustatory system: extracellular recording in CNS, in vitro patch recording of receptor activity, immunocytochemistry, behavioral analyses, sensory coding, and multivariate analyses.

Robert S. Waters, Ph.D., Professor. Mechanisms underlying plasticity in neonatal and adult somatosensory cortices: in vivo intracellular recording and labeling of cortical neurons.

Robert W. Williams, Ph.D., Professor. Dynamics of neuron populations during development; processes and patterns of axonal growth in the primate visual system.

Jian Zuo, Ph.D., Associate Professor. Molecular genetics of development and degeneration of hair cells in mouse inner ear.

THE UNIVERSITY OF TEXAS AT AUSTIN

The Institute for Neuroscience
Graduate Program in Neuroscience

Programs of Study

The Graduate Program in Neuroscience is offered through the Institute for Neuroscience at the University of Texas at Austin. Using a multidisciplinary approach, the program involves the research and teaching activities related to the neurosciences through twelve different academic departments and more than 60 participating faculty members.

The program is designed to achieve three primary goals, including a broad-based exposure to the various areas of neuroscience, the opportunity to pursue research in the area and laboratory of the student's choice, and a supportive environment so that students can be successful in their graduate careers and enjoy themselves during this formative period of scientific study.

Training offered by the Graduate Program in Neuroscience consists of course work and laboratory experience. The course work required of all graduate students includes the two-semester Principles of Neuroscience course, taught by faculty members of the Institute for Neuroscience, and a graduate-level statistics course. Other elective courses are individually chosen by students in consultation with their mentor and the graduate adviser.

Students in the neuroscience graduate program take a qualifying examination near the end of their second year. Upon successful completion of the qualifying exam, students choose their dissertation committee, submit a proposal for dissertation research, and apply for Ph.D. candidacy. The focus is then nearly entirely on the research that culminates in earning the Ph.D.

Research Facilities

The University of Texas at Austin's Institute for Neuroscience has research facilities covering a variety of programs that support the study of neuroscience, including a DNA core facility, protein core facility, mouse genetic engineering facility, microscopy core facility, electron microscopy facility, animal resources center, and the Center for Computational Biology and Bioinformatics.

Financial Aid

Students admitted to the graduate program are initially provided with financial support for the first twelve months of study. Possible sources of this support can be Institute for Neuroscience fellowships, University of Texas fellowships, training grants, graduate research assistantships (RAs), or graduate teaching assistantships (TAs). After the first year of study, and assuming that the student is progressing successfully, each student is expected to receive funding from either their lab supervisor and/or a grant or fellowship for which they apply.

Cost of Study

Tuition and fees for 9 graduate credit hours (required for fall and spring semester enrollment) are approximately $3000 for Texas residents and approximately $6000 for non-Texas residents. Students should note that non-Texas residents usually receive the Texas residency rate because of their appointments as RAs, TAs, or fellowship recipients. Tuition and fees are updated once a year and, thus, are subject to change.

Living and Housing Costs

Living expenses in the Austin area are comparable to those of other large regional cities. For current information about housing on campus, students should visit the Housing and Food Service site at http://www.utexas.edu/student/housing/. To find information about renting apartments or houses, plus other interesting topics about living in Austin, they should visit the Web site at http://www.austin360.com/.

Student Group

There are approximately 60 students enrolled in the Institute for Neuroscience's graduate program.

Location

The University of Texas at Austin is located in Austin, Texas, the state capital, with a population of about 1 million within the city limits and more than 1.5 million in the greater metropolitan area. There is an abundance of outdoor activities in Austin, including two lakes for a variety of water activities. Austin offers museums, art galleries, and botanical gardens to explore, along with many fine restaurants, clubs, and movie theaters.

The University

The University of Texas at Austin, founded in 1883, is a co-ed institution with more than 50,000 students, including undergraduates and graduates. Its 350-acre urban campus in Austin also provides students with easy access to San Antonio. It is a state-supported university and part of the University of Texas System. The University operates on calendar semesters and offers bachelor's, master's, doctoral, and first professional degree programs.

Applying

Application for admission to the Graduate Program in Neuroscience requires the completion of two application packets. The primary application packet is sent to the Graduate and International Admission Center. A second packet containing supplementary information is sent to the Institute for Neuroscience. Detailed instructions about applying can be found at http://www.utexas.edu/neuroscience. Applicants must include all transcripts from each college or university attended, official GRE test scores, official TOEFL scores for international students, a resume, and a nonrefundable application fee of $50 for U.S. citizens/permanent residents or $75 for international students.

December 1 is the deadline for special fellowship consideration, and January 1 is the deadline for other financial assistance consideration. All materials must be received by those dates—incomplete applications will not be reviewed. New students are admitted for the fall semester only; there are no spring or summer admissions.

Correspondence and Information

Graduate Coordinator
The Institute for Neuroscience
The University of Texas at Austin
Austin, Texas 78712-0187
Phone: 512-471-3640
Fax: 512-471-0390
E-mail: neuroscience@mail.clm.utexas.edu
Web site: http://www.utexas.edu/neuroscience

The University of Texas at Austin

THE FACULTY AND THEIR RESEARCH

Creed W. Abell, Professor of Medicinal Chemistry; Ph.D., Wisconsin. Determining molecular properties of certain proteins in the nervous system; applying results of these studies to Parkinson's disease, Alzheimer's disease, and aging.

Lawrence Abraham, Professor of Kinesiology and Health Education; Ed.D., Columbia Teacher's College. Motor learning and biomechanics; physical education; kinesiology.

Seema Agarwala, Assistant Professor of Neurobiology; Ph.D., SUNY at Stony Brook. Patterning of the vertebrate brain.

Adriana Alcantara, Assistant Professor of Psychology; Ph.D., Illinois at Urbana-Champaign. The neuroplasticity that occurs within specific cellular microcircuits at the receptor-mediated neuronal signaling level; synaptic rewiring that may underlie addiction.

Richard Aldrich, Professor of Neurobiology and Chair; Ph.D., Stanford. Molecular mechanisms of ion channel function and their role in electrical signaling.

Karen Artzt, Professor of Molecular Genetics and Microbiology; Ph.D., Cornell. Early development of vertebrate nervous system, with goal of identifying genes involved in tissue organization and relevant to embryonal tumors.

Nigel Atkinson, Associate Professor of Neurobiology; Ph.D., Penn State Hershey Medical Center. How ion channel genes are regulated; consequences of this regulation.

Chadrajit Baja, Professor of Computer Science and Director; Ph.D., Cornell. Integrated approaches to computational modeling, simulations, mathematical analysis, and interrogative visualization, especially for dynamic scientific phenomena at multiple scales.

Adela Ben-Yakar, Assistant Professor of Mechanical Engineering; Ph.D., Stanford. Fundamentals of femtosecond-laser interaction with biological tissues and nanomaterials in the development of novel techniques, such as laser nanosurgery and 3-D microfabrication and nanofabrication techniques.

Susan Bergeson, Assistant Professor of Neurobiology; Ph.D., Oregon Health Sciences. Molecular biological, biochemical, genetic, pharmacological, and behavioral techniques to study alcohol/drug abuse–related physiological and behavioral traits.

George Bittner, Professor of Neurobiology; Ph.D., Stanford. How cellular and molecular mechanisms produce or induce long-term survival of the distal (anucleate) stump of severed axons and their eventual reconnection/regeneration.

Craig Champlin, Associate Professor of Speech Communication; Ph.D., Kansas. Hearing science; instrumentation; electrophysiological audiometry; hearing conservation.

Lawrence Cormack, Associate Professor of Psychology; Ph.D., Berkeley. Psychophysics of human visual system and its relation to perception and behavior.

David Crews, Professor of Integrative Biology; Ph.D., Rutgers. Reproductive biology; development and function of sex and individual differences.

Yvon Delville, Assistant Professor of Psychology; Ph.D., Massachusetts. Various aspects of behavioral neuroendocrinology.

Michael Domjan, Professor of Psychology and Director; Ph.D., McMaster. Animal learning; biological constraints on learning; learning mechanisms in reproductive behavior; comparative psychology.

Sharon Dormire, Assistant Professor of Nursing; Ph.D., Florida; RN. Changes in glucose delivery to the brain in women during the perimenopause and its effect on memory.

Christine Duvauchelle, Associate Professor of Pharmacology; Ph.D., California, Santa Barbara. Exploring neural mechanisms of reward, using behavioral, pharmacological, neurochemical, and immunological methodologies.

Roger Farrar, Professor of Kinesiology and Health Education; Ph.D., Massachusetts Amherst. Effects of exercise on motor systems of the body during the aging process, using the rat as an animal model.

Benito Fernández-Rodriguez, Associate Professor of Mechanical Engineering; Ph.D. Modeling, simulation, and control design of nonlinear systems; adaptive, robust, and intelligent control; system identification and diagnostics.

Wilson Geisler, Professor of Psychology and Director; Ph.D., Indiana. Vision and visual perception; motion perception; pattern vision.

Nace Golding, Assistant Professor of Neurobiology; Ph.D., Wisconsin–Madison. The influence of dendritic properties on the way neurons in the mammalian central auditory system process information about sound.

Ruben Gonzales, Professor of Pharmacology; Ph.D., Texas. Neurochemical basis for ethanol drinking behavior.

Francisco Gonzalez-Lima, Professor of Psychology and Pharmacology and Director; Ph.D., Puerto Rico, Medical Sciences Campus. Neuroscience; neuroanatomy; neurobiology; physiological psychology; psychobiology; learning and memory; neural mechanisms of behavior.

Andrea Gore, Associate Professor of Pharmacology; Ph.D., Wisconsin–Madison. Mechanisms by which the brain controls development/aging.

Lisa Griffin, Assistant Professor of Kinesiology and Health Education. How the central nervous system controls muscle force under various conditions in younger and older adult populations.

Jeffrey Gross, Assistant Professor of Molecular, Cell, and Developmental Biology; Ph.D., Duke. Vertebrate eye development and visual system function.

Adron Harris, Professor of Neurobiology and Director; Ph.D., North Carolina. Structure and function of ion channels, with emphasis on molecular mechanisms responsible for alcohol and drug actions.

Hans Hofmann, Assistant Professor of Integrative Biology; Ph.D., Max Planck Institute for Behavioral Physiology (Seewiesen). Molecular and hormonal mechanisms that underlie social behavior and its evolution.

Alex Huk, Assistant Professor of Neurobiology; Ph.D., Stanford. Decision making; how brains accrue, weigh, and remember evidence.

Jody Jensen, Associate Professor of Kinesiology and Health Education; Ph.D., Maryland. Developmental motor control.

Dan Johnston, Professor of Neurobiology and Director; Ph.D., Duke. Properties and mechanisms of long-term synaptic potentiation (LTP) and depression (LTD); synaptic substrates for aspects of memory.

Theresa Jones, Assistant Professor of Psychology; Ph.D., Texas at Austin. Plasticity of neural structure and synaptic connectivity in adult animals following brain damage and during learning.

Robert Josephs, Associate Professor of Psychology; Ph.D., Michigan. Testosterone and its importance for growth and functioning throughout the body; testosterone and social behavior; sex differences in cognitive function and test performance.

Swathi Kiran, Assistant Professor of Communication Sciences and Disorders; Ph.D., Northwestern. Effects of brain damage on communication.

Todd W. Maddox, Associate Professor of Psychology; Ph.D., California, Santa Barbara. The neurobiological underpinnings of category learning and attentional processes.

Dennis McFadden, Professor of Psychology; Ph.D., Indiana. Sensation and perception; audition; sex differences in the auditory system; prenatal hormonal effects and effects of drugs on the auditory system.

John Mihic, Associate Professor of Neurobiology; Ph.D., Toronto. Molecular mechanisms through which alcohol, inhaled solvents, sedatives, and anesthetics act on ligand-activated ion channels.

Risto Miikkulainen, Professor of Computer Science; Ph.D., UCLA. Biologically inspired computation, e.g., neural networks, genetic algorithms.

Hitoshi Morikawa, Assistant Professor of Neurobiology; Ph.D., Kyoto. Dopaminergic neurons in the ventral midbrain.

Richard Morrisett, Associate Professor of Pharmacology; Ph.D., Alabama at Birmingham. Synaptic alterations and their function in native intact systems underlying alcohol-related brain disorders, alcoholism, and other neurological disorders.

Martin Poenie, Associate Professor of Molecular Cell and Developmental Biology; Ph.D., Stanford. How T cells polarize their cytoskeleton in response to antigenic target cells.

George Pollak, Professor of Neurobiology and Chair; Ph.D., Maryland. Mammalian auditory system; how the auditory system processes communication signals; cues that enable animals to associate a sound with its location in space.

Mendell Rimer, Assistant Professor Neurobiology; Ph.D., Maryland, Baltimore. Synapses and how the nervous system works normally and under pathological situations.

Michael Ryan, Professor of Integrative Biology; Ph.D., Cornell. Evolution and function of animal behavior.

Tim Shallert, Professor of Psychology; Ph.D., Arizona State. How the central nervous system responds to loss of nerve cells; how behavior influences mechanisms of brain repair.

Christine Schmidt, Associate Professor of Biomedical Engineering; Ph.D., Illinois. Repair of tissues in nervous and cardiovascular systems.

Eyal Seidemann, Assistant Professor of Psychology; Ph.D., Stanford. How perceptual events and motor plans are represented and processed in primate cerebral cortex.

Jason Shear, Associate Professor of Chemistry and Biochemistry; Ph.D., Stanford. Bioanalytical, biomaterial, and biophysical studies at molecular and cellular levels, with emphases on high-speed and high-sensitivity chemical separations, laser-based studies of neuronal development and communication, and design of sensor arrays.

Harel Shouval, Assistant Professor of Biomedical Engineering; Ph.D., Brown. Rules by which changes in synaptic strength, the basis of learning, memory, and development in the cortex, take place.

Wesley Thompson, Professor of Neurobiology; Ph.D., Berkeley. How connections between neurons (synapses) are formed and maintained.

John Wallingford, Assistant Professor of Molecular Cell and Developmental Biology; Ph.D., Texas at Austin. Molecular and cellular basis of embryonic morphogenesis.

Richard Wilcox, Professor of Pharmacology and Chair; Ph.D., Southern Illinois. Computational chemistry; addiction science education; medical outcomes research.

Harold Zakon, Professor of Neurobiology and Chair; Ph.D., Cornell. Regulation of sodium and potassium currents by sex steroid hormones and phosphorylation in the electric organ of weakly electric fish.

Bing Zhang, Assistant Professor of Neurobiology; Ph.D., Cornell. Molecular mechanisms by which synapse structure and function are regulated.

UNIVERSITY OF UTAH

Program in Neuroscience

Program of Study	The Program in Neuroscience draws on 76 faculty members who are involved with neuroscience research and teaching; fourteen departments in the School of Medicine and the Colleges of Engineering, Pharmacy, Science, and Social and Behavioral Science are represented. The program offers a course of study leading to the Ph.D. in neuroscience, participates in the School of Medicine's M.D./Ph.D. program, and provides opportunities for postdoctoral training. Training is offered in neuroanatomy, neurophysiology, developmental neurobiology, behavioral neuroscience, and biochemical and pharmacological aspects of neurobiology.

The program of study for the Ph.D. begins with a required core of basic science courses, including neuroanatomy, neurophysiology, and systems in neuroscience. A second level of formal course work relating to the student's area of specialization and dissertation research is planned on an individual basis, and, if a student wishes, the doctoral degree can be taken with emphasis in one of the basic disciplines.

Since both interaction among members of the program and multidisciplinary training are primary goals, students are expected to participate throughout their training in several neuroscience seminar programs. They gain familiarity with problems and techniques through research apprenticeships in at least three laboratories early in their training. A supervisory committee composed of several faculty members meets regularly with the student to plan his or her training.

Research Facilities
The well-equipped modern research laboratories of participating faculty members are located in the Departments of Biology, Bioengineering, Human Genetics, Internal Medicine, Mathematics, Neurobiology and Anatomy, Neurology, Neurosurgery, Ophthalmology, Pharmacology and Toxicology, Physiology, Psychiatry, Psychology, and Radiology. The facilities and expertise necessary for applying most anatomical, behavioral, biochemical, and physiological techniques to study the nervous system are readily available. The proximity of strong research groups in cell biology, genetics, medicine, and molecular biology provides additional resources for neuroscience research.

Financial Aid
Support for graduate students is available from training and research grants, from University sources, and from teaching assistantships. The level of support is comparable to that at other major universities.

Cost of Study
Tuition and fees for all supported students are waived.

Living and Housing Costs
The cost of dormitory rooms for single students in the University residence halls ranges from $2886 to $3064 per academic year. Also available are double-room apartments for 2 single students (6 people sharing a bathroom, living room, and kitchen) that range in cost from $2432 to $2654 (including utilities) per academic year. A variety of optional meal plans are available through University Food Services. Rents for unfurnished University apartments range from $620 to $1280 per month, including all utilities except a telephone. Off-campus housing is available near the University. Costs are subject to increase for 2006–07.

Student Group
There are 5,382 graduate students at the University, of whom about 770 are working toward an advanced degree in one of a dozen departments in the biological sciences. Graduates of the program typically accept postdoctoral appointments, then go on to careers in teaching and/or research.

Location
Salt Lake City, with a metropolitan population of more than 800,000 residents, is the cultural and economic center of the intermountain region and was chosen to host the 2002 Winter Olympic Games. The Utah Symphony, Ballet West, and Repertory Dance Theatre are internationally known. For sports entertainment, Salt Lake supports the Utah Jazz as well as the Utah Grizzlies hockey team and Salt Lake Buzz baseball. The "Crossroads of the West," Salt Lake City is only a day's drive from a number of national parks, including Yellowstone, Grand Teton, Rocky Mountain, Mesa Verde, Grand Canyon, Zion, Bryce Canyon, and Canyonlands. The University is located at the base of the Wasatch Mountains. Within 15 miles of the campus are peaks that rise above 11,000 feet and provide the finest in year-round recreational opportunities, such as good fishing, hunting, mountain climbing, golfing, and some of the best skiing in the world.

The University
Founded in 1850, the University of Utah is the oldest state university west of the Missouri River. A vigorous institution with a teaching faculty of 1,500 and an adjunct, research, and clinical faculty of 1,800, it is a comprehensive university, including Colleges of Medicine, Law, Education, Business, Mines and Mineral Industries, Science, Humanities, Social and Behavioral Science, Health, Fine Arts, Engineering, Nursing, Pharmacy, and Social Work in addition to the Graduate School. Among its many assets are an excellent library and an outstanding computer center.

Applying
Inquiries and requests for application materials should be addressed to the program as indicated below. Application forms for admission will be supplied. Applicants are required to submit a completed application form, transcripts of their undergraduate record, GRE scores (scores on the verbal, quantitative, and analytical portions of the General Test), three letters of recommendation, and a statement of career interests. Applications are due by January 15 for entrance the following September, but earlier submission is encouraged. Final decisions on admission are normally made following personal interviews with prospective students, arranged at the program's expense.

Correspondence and Information
Program in Neuroscience
University of Utah
20 North 1900 East, 401 MREB
Salt Lake City, Utah 84132-3401

Phone: 801-581-4820
Fax: 801-581-4233
E-mail: tracy.marble@hsc.utah.edu
Web site: http://neuroscience.med.utah.edu

University of Utah

THE FACULTY AND THEIR RESEARCH

Neurobiology of Disease

Wolfgang Baehr, Professor of Ophthalmology; Ph.D., Heidelberg, 1967. Phototransduction; the retinoid cycle; inherited retinal disease.

Byron D. Bair, Associate Professor of Internal Medicine and Psychiatry; M.D., Utah, 1986. Disruptive behaviors in geriatric populations and impact on health.

Kevin M. Flanigan, Associate Professor of Neurology, Pathology, and Human Genetics; M.D., Rush, 1990. Genetic and molecular characterization of inherited neuromuscular diseases.

Robert S. Fujinami, Professor of Neurology; Ph.D., Northwestern, 1977. Neurovirology.

Daniel W. Fults, Professor of Neurosurgery; M.D., Texas at Austin, 1979. Genetic abnormalities that cause human brain tumors.

John E. Greenlee, Professor of Neurology; M.D., Rochester, 1969. Remote effects of cancer on the nervous system; viral infections of the nervous system.

Randy L. Jensen, Associate Professor of Neurosurgery; M.D., Utah, 1991. Neuro-oncology; hypoxia and angiogenesis in brain tumors.

Rajendra Kumar-Singh, Assistant Professor of Ophthalmology; Ph.D., Dublin, 1993. Gene therapy vectors and their application to the treatment of inherited diseases of the CNS.

Janet E. Lainhart, Associate Professor of Psychiatry; M.D., Wayne State, 1979. Neurobiology of autism.

Marina Myles-Worsley, Research Professor of Psychiatry; Ph.D., Utah, 1986. Neurobiological correlates of schizophrenia.

Aloisia Schmid, Assistant Professor of Human Genetics; Ph.D., Emory. *Drosophila* genetics in the pathologies of human disease.

Patrick A. Tresco, Professor of Bioengineering; Ph.D., Brown, 1991. Therapies for CNS degenerative disorders.

H. Steve White, Professor of Pharmacology and Toxicology; Ph.D., Utah, 1984. Neuropharmacology; anticonvulsant drug mechanisms; genetics of audiogenic seizures.

Molecular Neuroscience

Brenda L. Bass, Professor of Biochemistry; Ph.D., Colorado, 1985. Determining the roles of RNA editing in the nervous system, using *C. elegans* as a model system.

Jeanne Frederick, Research Assistant Professor of Ophthalmology; Ph.D., Wisconsin–Madison, 1979. Expression of mouse bipolar cell metabotropic glutamate receptor, mGluR6.

Erik M. Jorgensen, Professor of Biology; Ph.D., Washington (Seattle), 1989. Genetic analysis of neurotransmission in *C. elegans*.

Suzanne Mansour, Associate Professor of Human Genetics; Ph.D., Berkeley, 1985. Genes involved in inner ear development and function.

Andres Villu Maricq, Professor of Biology; Ph.D., Berkeley, 1987; M.D., California, San Francisco, 1990. Workings of synapses: molecular mechanisms regulating neurotransmission in *C. elegans*.

J. Michael McIntosh, Research Professor of Biology and Professor of Psychiatry; M.D., UCLA, 1987. Molecular pharmacology of CNS receptors and ion channels.

Shannon J. Odelberg, Research Assistant Professor of Internal Medicine; Ph.D., Virginia Commonwealth, 1987. Molecular basis of regeneration and cellular plasticity.

Baldomero M. Olivera, Distinguished Professor of Biology; Ph.D., Caltech, 1966. Ion channels, membrane receptors, and sensory transduction.

Scott W. Rogers, Associate Professor of Neurobiology and Anatomy; Ph.D., Utah, 1986. Molecular biology and neurobiology of ligand-activated receptors.

Alejandro Sánchez Alvarado, Associate Professor of Neurobiology and Anatomy; Ph.D., Cincinnati, 1992. Molecular basis of metazoan regeneration.

Michael C. Sanguinetti, Professor of Physiology; Ph.D., California, Davis, 1982. Ionic basis of cardiac excitability; mechanisms of ion channel gating.

Qiang Wu, Assistant Professor of Human Genetics; Ph.D., SUNY at Stony Brook, 1998. Characterization of neural protocadherin in gene clusters in the brain.

Doju Yoshikami, Professor of Biology; Ph.D., Cornell, 1970. Cellular and molecular physiology of ion channels.

Kang Zhang, Assistant Professor of Ophthalmology; Ph.D., 1991, M.D., 1995, Harvard. Molecular genetics and functional investigations of retinal diseases.

Developmental Neuroscience

Michael Bastiani, Professor of Biology; Ph.D., California, Davis, 1981. Development of neuronal connections.

Mario Capecchi, Distinguished Professor of Biology and Human Genetics; Ph.D., Harvard, 1967. Gene manipulation in mammalian and other eukaryotic cells.

Chi-Bin Chien, Associate Professor of Neurobiology and Anatomy; Ph.D., Caltech, 1990. Axonal pathfinding in zebrafish mutants.

Maureen Condic, Associate Professor; Ph.D., Berkeley, 1989. Interactions of sensory neurons with extracellular matrix.

Richard Dorsky, Assistant Professor of Neurobiology and Anatomy; Ph.D., California, San Diego, 1996. Signaling pathways in zebrafish neural cell fate determination.

Sabine Fuhrmann, Assistant Professor of Ophthalmology; Ph.D., Freiburg, 1996. Retinal development and cell-cell interactions.

David J. Grunwald, Professor of Human Genetics; Ph.D., Wisconsin–Madison, 1981. Tissue specification during zebrafish embryogenesis.

Edward Levine, Assistant Professor of Ophthalmology; Ph.D., SUNY at Stony Brook, 1994. Development and regeneration in the mammalian CNS.

Thomas N. Parks, George and Lorna Winder Professor of Neuroscience; Ph. D., Yale, 1978. Nerve cell development; neurobiology of the auditory system.

Tatjana Piotrowski, Assistant Professor of Neurobiology and Anatomy; Ph.D., Max Planck Institute (Germany), 1988. Sensory organ formation in zebrafish.

Yukio Saijoh, Assistant Professor of Neurobiology and Anatomy; Ph.D., Tohoku (Japan), 1991. Pattern formation of vertebrates.

Gary Schoenwolf, Professor of Neurobiology and Anatomy; Ph.D., Illinois at Urbana-Champaign, 1977. Pattern formation during early embryogenesis of vertebrates.

Sheryl A. Scott, Professor of Neurobiology and Anatomy; Ph.D., Yale, 1976. Development of sensory neurons.

Monica L. Vetter, Associate Professor of Neurobiology and Anatomy; Ph.D., California, San Francisco, 1994. Retinal neurogenesis.

Cellular Neuroscience

John F. Ash, Professor of Neurobiology and Anatomy; Ph.D., Stanford, 1974. Molecular physiology of membrane transport.

Bruce Bamber, Research Assistant Professor of Pharmacology and Toxicology; Ph.D., Washington (Seattle), 1994. Regulation of the $GABA_A$ receptor.

Gregory A. Clark, Associate Professor of Bioengineering; Ph.D., California, Irvine, 1982. Electrophysical and computational analyses of neuronal plasticity in simple systems (*Aplysia* and *Hermissenda*).

Annette E. Fleckenstein, Associate Professor of Pharmacology and Toxicology; Ph.D., Michigan, 1994. Neuropharmacology, neurochemistry, and aminergic transporters.

Colleen C. Hegg, Research Assistant Professor of Physiology; Ph.D., Wisconsin–Madison, 1996. Elucidating the mechanisms of injury-evoked regeneration in the mouse olfactory system.

Alan R. Light, Research Professor of Anesthesiology; Ph.D., SUNY Upstate Medical University, 1977. Neurobiology of pain enhancement after injury and inflammation.

Mary T. Lucero, Professor of Physiology; Ph.D., California, Davis, 1989. Olfactory physiology.

Robert E. Marc, Professor of Ophthalmology; Ph.D., Texas–Houston Health Science Center, 1975. Retinal circuitry and remodeling.

William C. Michel, Professor of Physiology; Ph.D., California, Santa Barbara, 1985. Olfactory and gustatory physiology.

Karen S. Wilcox, Research Associate Professor of Pharmacology and Toxicology; Ph.D., Pennsylvania, 1993. Electrophysiological mechanisms of action of novel anticonvulsant drugs.

Brain and Behavior

Alessandra Angelucci, Assistant Professor of Ophthalmology; M.D., Rome (Italy), 1990; Ph.D., MIT, 1996. Structure and function of the primate visual cerebral cortex.

Paul C. Bressloff, Professor of Mathematics; Ph.D., King's College, 1988. Theoretical neuroscience.

Donnell J. Creel, Research Professor of Ophthalmology; Ph.D., Utah, 1969. Sensory anomalies associated with hypopigmentation and diagnostic use of electrophysiological tests in patients.

Sarah Creem-Regehr, Assistant Professor of Psychology; Ph.D., Virginia, 2000. Cognitive neuroscience; visual perception; visual-motor control; spatial cognition.

William R. Crowley, Professor of Pharmacology and Toxicology; Ph.D., Rutgers, 1976. Neuroendocrine and neurochemical factors that regulate the secretion of anterior and posterior pituitary hormones.

Frances J. Friedrich, Associate Professor of Psychology; Ph.D., Kansas, 1980. Cognitive neuropsychology; language processes and spatial attention.

Franz Goller, Associate Professor of Biology, Ph.D., Notre Dame, 1992. Behavioral physiology of sound production and song learning in songbirds.

Glen R. Hanson, Professor of Pharmacology and Toxicology; D.D.S., UCLA, 1973; Ph.D., Utah, 1978. Interaction between neuropeptide and biogenic amine transmitter systems in the CNS.

Kenneth Horch, Professor of Bioengineering; Ph.D., Yale, 1971. Somatosensory physiology; neuroprosthetics.

Kristen A. Keefe, Associate Professor of Pharmacology and Toxicology; Ph.D., Pittsburgh, 1992. Pharmacology of neuropsychiatric disorders.

Raymond Kesner, Professor of Psychology; Ph.D., Illinois at Urbana-Champaign, 1965. Neurobiological basis of learning and memory.

Audie G. Leventhal, Professor of Neurobiology and Anatomy; Ph.D., SUNY at Albany, 1977. Development, structure, and function of mammalian visual pathways.

Richard A. Normann, Professor of Bioengineering; Ph.D., Berkeley, 1973. Information processing in the vertebrate retina; phototransduction.

Richard D. Rabbitt, Professor of Bioengineering; Ph.D., Rensselaer, 1986. Biophysics, biomechanics, microelectric impedance tomography of isolated hair cells.

Gary Rose, Professor of Biology; Ph.D., Cornell, 1983. Information processing in the auditory and electrosensory systems.

Bert N. Uchino, Associate Professor of Psychology; Ph.D., Ohio, 1993. Influence of social factors on the aging process.

Neil Vickers, Associate Professor of Biology; Ph.D., California, Riverside, 1992. Neurobiology of olfaction.

VIRGINIA COMMONWEALTH UNIVERSITY

Medical College of Virginia
School of Medicine
Graduate Program in Neuroscience

Program of Study

The Virginia Commonwealth University (VCU) Graduate Program in Neuroscience (GPN) promotes an integrated interdisciplinary course of graduate study in the neurosciences. The program is a joint effort involving four departments: Anatomy and Neurobiology, Biochemistry, Pharmacology and Toxicology, and Physiology. Once admitted into the GPN, students in their first year complete courses in physiology, biochemistry, molecular biology, and neuroanatomy. These core studies establish a foundation for future study. In addition, GPN students take a cellular and molecular neuroscience and methods course, attend research seminars, and carry out neuroscience laboratory rotations. At the end of the first year, GPN students select a faculty adviser and matriculate into that adviser's home department. The second year of study provides more specialized training. At the end of the second year, GPN students take written and oral qualifying exams for advancement toward candidacy. Each student must conduct an original research project, prepare a written dissertation, and defend his or her dissertation research in a final oral examination. Completion of the program, usually in four years, culminates in a Ph.D. degree from one of the four departments listed above. Although neuroscience research at VCU spans a wide range of topics, VCU has particular strengths in substance abuse, epilepsy, neural trauma, visuo/motor systems, and neurotransmission/signal transduction research.

Research Facilities

VCU has outstanding state-of-the-art research facilities and resources totaling more than 400,000 square feet of research space. These include a $28 million, 100,000-square-foot Medical Sciences Building and the Virginia Biotechnology Research Park, both of which opened in 1996; Sanger Hall; McGuire Hall; and the Blackwell-Smith Building. The campus is currently undergoing additional expansion of its facilities. Close interactions also exist between the program and the Medical College of Virginia (MCV) Clinical Neuroscience Center. Facilities are available for investigations involving behavioral, molecular, structural, biochemical, and physiological neuroscience. To facilitate an emphasis on cellular and molecular neuroscience, GPN students have access to recombinant DNA techniques, DNA arrays, laser-scanning confocal microscopy, fluorescence microscopy, high-pressure liquid chromatography, flow cytometry, electron microscopy, and patch clamp electrophysiology. Support services include excellent library facilities; cores in genomics, proteomics, and imaging; an extensive and up-to-date computer facility; high-speed connections to the Internet; biomedical machine shops; and other top-notch resources.

Financial Aid

Although the School has a number of NIH predoctoral training grants, students may also be supported by faculty members who provide grant funds for research assistantships. GPN students are supported by a stipend of $21,000 per year.

Cost of Study

Tuition and fees in 2006–07 are $10,808 per year for Virginia residents and $23,117 per year for out-of-state students. Tuition and fees are typically covered by fellowships or assistantships; residual costs are usually covered by the student's research adviser.

Living and Housing Costs

Virginia Commonwealth University offers limited on-campus housing for single students; however, most students prefer to live off campus. Richmond offers a wide variety of housing accommodations at a wide range of prices.

Student Group

Virginia Commonwealth University, the largest university in the state, has a total enrollment of approximately 22,000 students. On the VCU Medical Campus, more than 2,500 students undertake course work and research in medicine, dentistry, pharmacy, nursing, allied health, and various fields of graduate study.

Location

Richmond, the capital of Virginia, and its surrounding suburbs have a population of about 800,000. The area has a rich history, a moderate climate, and a reasonable cost of living. The city has a fine art and music scene, many cultural amenities, and excellent restaurants. Within a 90-mile radius are Washington, D.C.; Colonial Williamsburg; the Blue Ridge Mountains; and renowned Atlantic Coast beaches.

The University and The College

Virginia Commonwealth University comprises two campuses: the academic campus and the VCU Medical Campus, formerly the Medical College of Virginia (MCV). MCV was founded in 1838 and became an independent institution in 1854. The MCV campus consists of the Schools of Medicine, Dentistry, Nursing, Pharmacy, and Allied Health Professions. The academic campus is about 1 mile west of the Medical Campus and is primarily a liberal arts campus. While GPN is a relatively recent addition to the list of VCU's graduate programs, interdisciplinary efforts to recruit and train neuroscience graduate students at VCU have existed since 1992.

Applying

Admission to the School of Graduate Studies requires applicants to have earned a baccalaureate degree and have knowledge of biology, chemistry, mathematics, and physics. Applicants must have taken the General Test of the Graduate Record Examinations. Applicants must have evidence of superior scholarship and strong GRE scores. Three letters of recommendation and a letter summarizing the applicant's goals are also required. The selection of applicants begins in December, and applicants are encouraged to apply as early as possible.

Correspondence and Information

Dr. L. S. Satin
Director
Graduate Program in Neuroscience
Box 980524
Virginia Commonwealth University School of Medicine
Richmond, Virginia 23298
Phone: 804-828-7823
Fax: 804-828-1532
E-mail: lsatin@vcu.edu
Web site: http://www.vcu.edu/neurograd

Virginia Commonwealth University

THE FACULTY AND THEIR RESEARCH

Neural Injury

John W. Bigbee, Professor of Anatomy; Ph.D., Stanford, 1982. Neuronal development; nerve migration and neuritogenesis; regeneration of neurons; cellular basis of neurotrauma.

Earl F. Ellis, Professor of Pharmacology and Toxicology; Ph.D., Wake Forest, 1974. Biochemical mechanisms of traumatic brain injury.

Robert J. Hamm, Professor of Psychology; Ph.D., Southern Illinois, 1974. Pathobiology of traumatic brain injury, with an emphasis on receptor-mediated alterations in neuronal signaling due to injury; enhancement of neuronal regenerative capacity to reduce disabilities that follow traumatic brain injury.

Randall Merchant, Professor of Anatomy; Ph.D., North Dakota, 1978. Neuroimmunology; brain-tumor biology and cancer immunotherapy.

Linda Phillips, Associate Professor of Anatomy; Ph.D., Wake Forest, 1980. Changes in CNS protein and RNA during injury-induced synaptic reorganization; biochemical mechanisms of brain trauma.

John T. Povlishock, Professor and Chair of Anatomy; Ph.D., St. Louis, 1973. Pathobiology of traumatic brain injury in animals and man; immunocytological, ultrastructural, and computer-assisted analysis of cytoskeletal and axolemmal changes in neurotrauma; changes in brain microvasculature following brain trauma.

Thomas M. Reeves, Associate Professor of Anatomy; Ph.D., Southern Illinois, 1984. Electrophysiological correlates of functional recovery following brain injury; neural plasticity; synaptic reorganization.

Developmental Neurobiology/Aging/CNS Tumor Biology

William C. Broaddus, Associate Professor of Neurosurgery and Anatomy; M.D., Ph.D., Case Western, 1984. Molecular biology of brain tumors; stem cells.

Raymond J. Colello, Associate Professor of Anatomy; D.Phil., Oxford, 1990. Molecular mechanisms regulating myelination in the CNS; myelin gene transcription; demyelinating diseases.

Helen Fillmore, Assistant Professor of Neurosurgery and Anatomy; Ph.D., Memphis, 1993. Molecular biology of glioma; stem cells; metalloproteases.

Babette Fuss, Associate Professor of Anatomy and Neurobiology; Ph.D., Swiss Federal Institute of Technology, 1992. Oligodendrocyte behavior during normal development and under pathological conditions.

Martin R. Graf, Assistant Professor of Anatomy and Neurobiology; Ph.D., California, Irvine, 1996. Neuroimmunology associated with CNS neoplasms.

Michael S. Grotewiel, Associate Professor of Human Genetics; Ph.D., Vanderbilt, 1995. Molecular genetic basis of the nervous system and aging in *Drosophila*.

Scott C. Henderson, Associate Professor of Anatomy and Neurobiology; Ph.D., Western Ontario, 1990. Optical imaging; multiphoton confocal imaging; developmental neurobiology.

Carmen Sato-Bigbee, Associate Professor of Biochemistry and Molecular Biophysics; Ph. D., Buenos Aires, 1985. Signal transduction pathways and gene regulation in oligodendrocyte differentiation.

Glenn van Tuyle, Associate Professor of Biochemistry and Molecular Biophysics; Ph.D., Thomas Jefferson, 1971. Deletions of mitochondrial DNA as a function of aging; DNA-protein interactions.

Neurotransmission/Signal Transduction/Ion Channels

Hamid Akbarali, Professor of Pharmacology and Toxicology; Ph.D., St. John's Memorial (Canada), 1988. Ion channels in gastrointestinal smooth muscle; channel expression and modulation.

Galya Abdrakhmanova, Assistant Professor of Pharmacology and Toxicology; M.D., Ph.D., Charles (Prague). Molecular pharmacology of neuronal nicotinic receptors.

Clive M. Baumgarten, Professor of Physiology; Ph.D., Northwestern, 1976. Physiology and biophysics of ion channels of nerve and cardiac cell membranes; structure-function relations in channel proteins.

Margaret Boadle-Biber, Professor and Chair of Physiology; D.Phil., Oxford, 1967. Brain serotonergic systems.

Robert J. DeLorenzo, Professor of Neurology; Ph.D., 1973, M.D., Yale, 1974. Molecular neuroscience; neuropharmacology of neuroleptic drugs; biochemical basis of calcium effects in neurons.

John DeSimone, Professor of Physiology; Ph.D., Harvard, 1971. Transduction mechanisms in taste receptor cells; ion transport; optical methods.

John R. Grider, Professor of Physiology; Ph.D., Hahnemann, 1981. Neural circuitry of gut peristaltic reflexes; second messenger systems.

Kimberle M. Jacobs, Assistant Professor of Anatomy and Neurobiology; Ph.D., Brown, 1994. Neocortical organization; cellular mechanisms of cortical plasticity and epilepsy; GABA receptors.

A. Rory McQuiston, Associate Professor of Anatomy and Neurobiology; Ph.D., Alberta, 1995. Control of synaptic integration and network activity by the modulation of inhibitory interneurons.

Leslie S. Satin, Professor of Pharmacology and Toxicology and Director, Graduate Program in Neuroscience; Ph.D., UCLA, 1982. Physiology, biophysics, and pharmacology of endocrine and neuronal ion channels; pancreatic islet biology; role of calcium ions in neurohormone secretion; ion channel modulation; impact of CNS injury on synaptic transmission; GABA and glutamate receptor channels.

Sompong Sombati, Assistant Professor of Neurology; Ph.D., Oregon, 1983. Modulation of NMDA receptor using patch clamp techniques; epileptiform activity in neuronal culture; excitotoxicity; synaptic transmission.

Sarah Spiegel, Professor and Chair of Biochemistry; Ph.D., Weizmann (Israel), 1983. Biological functions of sphingosine 1-phosphate in apoptosis and neuronal development.

Gea-Ny Tseng, Associate Professor of Physiology; Ph.D., Columbia, 1985. Potassium channel structure and function; pharmacology of ion channels; mutagenesis.

Visuo/Motor-Special Senses

Richard M. Costanzo, Professor of Physiology; Ph.D., SUNY Upstate Medical Center, 1975. Neurophysiology; regeneration of sensory systems.

Stephen J. Goldberg, Professor of Anatomy; Ph.D., Clark, 1971. Neurophysiological and mechanical properties of brainstem motor units involved in motor control; tongue and eye movement control; basic properties of motor units and movement control.

George R. Leichnetz, Professor of Anatomy; Ph.D., Ohio State, 1970. Definition of neuroanatomical connections from eye movement–related cortical regions to brainstem; understanding basis of control and modulation of ocular motility.

J. Ross McClung, Professor of Anatomy; Ph.D., Texas Medical Branch, 1971. Neuronal organization of spinal cord; ocular motor nuclei and functional organization of the hypoglossal nucleus.

Alexandre E. Medina, Assistant Professor of Anatomy; Ph.D., Rio de Janeiro, 2000. Neuronal plasticity and effects of agents that alter brain development; cellular and molecular neurobiology.

M. Alex Meredith, Professor of Anatomy; Ph.D., Virginia Commonwealth, 1981. Neuroanatomy and electrophysiology of multisensory information processing in the CNS and translation of neuronal signals into motor commands.

Steven Price, Professor of Physiology; Ph.D., Princeton, 1961. Receptor mechanisms in taste and smell.

Mary S. Shall, Associate Professor of Anatomy; Ph.D., Virginia Commonwealth, 1991. Mechanical parameters and behavior of motor units in fine motor control; impact of vestibular system input on muscle development.

Steven Shapiro, Professor of Neurology; M.D., Pittsburgh, 1975. Neurophysiology; evoked potentials; auditory nervous system.

Substance Abuse

Robert L. Balster, Professor of Pharmacology and Toxicology; Ph.D., Houston, 1970. Animal models of drug dependence; behavioral pharmacology; excitatory amino acids; inhalation studies.

Patrick M. Beardsley, Professor of Pharmacology and Toxicology; Ph.D., Minnesota, 1982. Behavioral pharmacology; development of medications for drug-dependency disorders.

Jill C. Bettinger, Assistant Professor of Pharmacology and Toxicology; Ph.D., Minnesota, 1998. Neuroscience and genetics; molecular genetics of alcohol response and state-dependent learning in *C. elegans.*

Charles D. Cook, Assistant Professor of Pharmacology and Toxicology; Ph.D., North Carolina at Chapel Hill, 2000. Behavioral and pharmacological studies of CNS drugs; pain research.

M. Imad Damaj, Associate Professor of Pharmacology and Toxicology; Ph.D., Paris, 1991. CNS pharmacology; drugs of abuse; behavioral correlates to biochemical events; nicotine pharmacology and other drugs of abuse; transgenic animal models and drug abuse.

Andrew G. Davies, Ph.D., Assistant Professor of Pharmacology and Toxicology; Melbourne, 1993. *C. elegans* behavioral genetics and neuroscience; alcohol and drugs of abuse.

William L. Dewey, Professor of Pharmacology and Toxicology; Ph.D., Connecticut, 1967. Mechanisms of action of marijuana constituents; narcotic analgesics and their antagonists; role of endogenous substances in various diseases; sudden infant death; neurosciences.

Louis S. Harris, Harvey Haag Professor of Pharmacology and Toxicology; Ph.D., Harvard, 1958. Relationship between chemical and biochemical factors and pharmacological actions of drugs affecting the central nervous system.

Edward J. N. Ishac, Associate Professor of Pharmacology and Toxicology; Ph.D., Monash (Australia), 1983. Biochemical pharmacology; second messenger systems and modulation of neurotransmitters.

Kenneth S. Kendler, Rachel Banks Professor of Psychiatry and Human Genetics and Director, Psychiatric Genetics Research Program; M.D., Stanford, 1977. Genetics of psychiatric and drug abuse disorders; schizophrenia; depression.

Aron H. Lichtman, Associate Professor of Pharmacology and Toxicology; Ph.D., Dartmouth, 1989. Behavioral pharmacology; behavioral neuroscience; cannabinoid pharmacology and other drugs of abuse; endocannabinoids.

Billy R. Martin, Harris Professor of Pharmacology; Ph.D., North Carolina at Chapel Hill, 1974. Characterization of receptors and second messengers mediating the action of drugs of abuse.

Michael F. Miles, Professor of Pharmacology and Toxicology; Ph.D., Northwestern, 1983. Functional genomics and mechanisms of drug abuse; DNA arrays; ethanol and cocaine sensitization.

Katherine Nicholson, Assistant Professor of Pharmacology and Toxicology; D.V.M., Ph.D., Virginia Commonwealth, 1998. Behavioral pharmacology of drugs of abuse; glutamate receptors.

Susan E. Robinson, Professor of Pharmacology and Toxicology; Ph.D., Vanderbilt, 1976. Interaction of putative neurotransmitters with cholinergic and glutamatergic neurons; effect of prenatal exposure to drugs on developing neurotransmitters.

Dana E. Selley, Associate Professor of Pharmacology and Toxicology; Ph.D., Rochester, 1991. Biochemical neuropharmacology; mechanisms of drug efficacy; tolerance and dependence; G-protein–mediated signal transduction.

Keith L. Shelton, Assistant Professor of Pharmacology and Toxicology; Ph.D., Virginia Commonwealth. Behavioral pharmacology of abused drugs.

Laura Sim-Selley, Associate Professor of Pharmacology and Toxicology; Ph.D., Rochester, 1991. Functional neuroanatomy of G-protein–coupled receptors; neuropharmacology of psychoactive drugs: opiates, cannabinoids, ethanol, and cocaine.

Forrest L. Smith, Associate Professor of Pharmacology and Toxicology; Ph.D., Texas Tech, 1989. Neonatal opiate tolerance/dependence; behavioral and cellular consequences; intracellular mechanisms of opiate tolerance/dependence.

Stephen A. Varvel, Assistant Professor of Pharmacology and Toxicology; Ph.D., Virginia Commonwealth, 2000. Physiological role of endocannabinoid system; learning and memory mechanisms; novel pharmacological approaches for age-related illnesses.

Sandra P. Welch, Professor of Pharmacology and Toxicology; Ph.D., Virginia Commonwealth, 1986. Cellular pharmacology; role of free intracellular calcium in tolerance and physical dependence to drugs of abuse; endocrine modulation of tolerance; second messenger systems.

Jenny L. Wiley, Associate Professor of Pharmacology and Toxicology; Ph.D., Virginia Commonwealth, 1991. Behavior pharmacology; excitatory amino acids; cannabinoids; animal models of psychiatric disorders.

Neuroendocrine

Richard J. Krieg, Professor of Anatomy; Ph.D., UCLA, 1975. Central control of pituitary gland secretion by brain hormones and neurotransmitters.

WEST VIRGINIA UNIVERSITY

Graduate Program in Neuroscience

Programs of Study

The Graduate Program in Neuroscience is one of seven graduate programs in West Virginia University Schools of Medicine and Pharmacy offering interdisciplinary biomedical research training leading to the Ph.D. or M.D./Ph.D. degree. Research areas emphasize sensory neuroscience, cognitive neuroscience, behavioral neuroscience, neural injury, and homeostasis. Diseases studied include Alzheimer's and related dementias, asthma and airway innervation, blindness, depression and anxiety and stress-related disorders, obesity and eating disorders, and stroke and related cerebrovascular events. Advanced imaging techniques are also used to study areas of the brain involved in visual and auditory signal processing and cognition.

Students benefit from individual attention by faculty members in a research environment that is dynamic, collaborative, and interdisciplinary. In addition to course work and laboratory research, students participate in seminars, journal clubs, and research conferences. They also attend national scientific meetings and obtain valuable speaking and teaching experience. The Ph.D. typically takes five years to complete. During Year 1, students matriculate in a common integrated core curriculum. This integrated first year allows students to build competence in key areas of contemporary science, gain exposure to the various training program options, meet potential dissertation advisers, and network scientifically and socially. In the second semester, students customize their course work by selecting from an array of program-specific electives. At the end of Year 1, students select a research adviser and can select Neuroscience as their training program. Year 2 consists of advanced course work, research, teaching, and the candidacy examination. Years 3 to 5 are devoted to dissertation research. The Graduate Program in Neuroscience also participates in the combined M.D./Ph.D. Scholars Program. As an M.D./Ph.D. Scholar, students take the first two years of the medical curriculum, followed typically by three years of research as required for the Ph.D. degree before returning to the M.D. program.

Research Facilities

Institutional facilities include a computer-based learning center, a centralized animal facility with a transgenic barrier, and a library housing more than 205,000 volumes and 2,400 journals. Core facilities are available for examining gene expression or genetic variation (Affymetrix platform), image analysis, confocal and electron microscopy and laser-capture microdissection, live-cell imaging, mass spectrometry, flow cytometry and high-speed cell sorting, proteomics, recombinant DNA technology, transgenic rodent biology, and functional neuroimaging (fMRI, PET/CT). Affiliated research centers include the National Institute for Occupational Safety and Health (NIOSH), Center for Advanced Imaging, Blanchette Rockefeller Neurosciences Institute, Sensory Neuroscience Research Center, and Mary Babb Randolph Cancer Center.

Financial Aid

Students in the Ph.D. or M.D./Ph.D. biomedical sciences programs receive financial support during their training, provided they remain in good academic standing and excel in research. Such support includes full tuition, health insurance, and an annual stipend of $22,000. Combined M.D./Ph.D. students also receive medical tuition waivers.

Cost of Study

Students' tuition costs are covered.

Living and Housing Costs

The cost of an efficiency apartment in University-owned housing is approximately $400 per month. A limited number of University apartments are available for married students. Privately owned apartments in Morgantown cost $400 to $600 per month. In general, the cost of living is lower compared to larger cities.

Student Group

Total University enrollment is approximately 26,000 students, which includes 6,500 graduate and professional students. Graduate students come from all parts of the United States and many other countries.

Location

Morgantown is a vibrant university community, with an appealing balance to life, of 80,000 residents in northern West Virginia. Located near the Pennsylvania border at the western edge of the Appalachian Mountains, abundant opportunities exist for activities such as world-class white-water rafting and kayaking, hiking and camping, mountain biking, fishing, and skiing. Morgantown has a cosmopolitan atmosphere with a range of activities usually found in much larger cities. It also enjoys proximity to major metropolitan centers: Pittsburgh is a 90-minute drive north, and Washington, D.C., is a 3-hour drive east.

The University

West Virginia University is a comprehensive, land-grant, Carnegie-designated Doctoral/Research University–Extensive public institution. The University's academic Health Sciences Center includes the Schools of Medicine, Dentistry, Nursing, and Pharmacy, all of which offer graduate degree programs. There are seven Ph.D. biomedical research training programs in the Schools of Medicine and Pharmacy, which benefit from the common, undifferentiated first year: Biochemistry and Molecular Biology, Cancer Cell Biology, Cellular and Integrative Physiology, Exercise Physiology, Immunology and Microbial Pathogenesis, Neuroscience, and Pharmaceutical and Pharmacological Sciences. Graduate faculty members in these programs are from various basic science and clinical departments throughout WVU and are members of interdisciplinary research centers in five health-related areas: cancer cell biology, cardiovascular sciences, immunopathology and microbial pathogenesis, neuroscience, and respiratory biology and lung diseases.

As a member of the Big East Conference, WVU participates in NCAA Division I sports. WVU also offers a variety of creative arts, theater, and entertainment opportunities.

Applying

Applicants must have a bachelor's degree and excellent GPA and GRE scores. Three letters of recommendation and a personal statement are required. Applicants are invited in groups of 10 for a paid, two-day visit/interview January through March. Students should visit the Web site at http://www.hsc.wvu.edu/som/resoff/gradprograms/PhD.asp for more information and an online application.

Correspondence and Information

Albert Berrebi, Ph.D., Graduate Director
Graduate Program in Neuroscience
West Virginia University
P.O. Box 9128
Morgantown, West Virginia 26506
Phone: 304-293-2357
E-mail: aberrebi@hsc.wvu.edu
Web site: http://www.hsc.wvu.edu/som/resoff/gradprograms/
    PhD.asp

Office of Research and Graduate Education
Health Sciences Center
West Virginia University
P.O. Box 9104
Morgantown, West Virginia 26506
Phone: 304-293-7116
E-mail: cnoel@hsc.wvu.edu

West Virginia University

THE FACULTY AND THEIR RESEARCH

Ariel Agmon, Associate Professor; Ph.D., Stanford. Development of synaptic connections in neocortex.

Stephen Alway, Associate Professor and Chair; Ph.D., McMaster. Effects of innervation, peripheral neuropathies, and loading on apoptosis of aging skeletal muscle.

Albert Berrebi, Professor; Ph.D., Connecticut. Anatomy and physiology of auditory brain stem circuits.

Heather Billings, Research Assistant Professor; Ph.D., Rutgers. Ovulation and estrous behavior; estrogen-responsive neurons in preoptic area of hypothalamus; estrogen receptor transcriptional activity.

Cathrin Buetefisch, Assistant Professor; M.D., Free University of Berlin (Germany). Movement disorders; neuroplasticity and functional recovery of movement following stroke in humans; functional MRI.

Janet Cyr, Assistant Professor; Ph.D., Texas Southwestern. Molecular basis of hair cell transduction.

Richard Dey, Professor and Chair; Ph.D., Michigan State. Neuroanatomical organization and neuronal mediators in the lung.

Robert Goodman, Professor and Chair; Ph.D., Pittsburgh. Hypothalamic regulation of reproductive function; physiological infertility in seasonal breeders.

Stephen Graber, Associate Professor; Ph.D., Vermont. Molecular biology of G protein-mediated signal transduction.

Stan Hileman, Associate Professor; Ph.D., Kentucky. Neuronal pathways controlling food intake, nutrition, and fertility.

Jason Huber, Assistant Professor; Ph.D., Florida A&M. Pharmacology; blood brain barrier; diabetes.

Gregory Konat, Professor; Ph.D., Odense (Denmark). Molecular mechanisms of oxidative stress-induced neurodegeneration and neoplasia.

James Lewis, Assistant Professor; Ph.D., Caltech. Mapping and exploring brain regions responsible for sound recognition and for localizing sounds in three-dimensional space; integration of auditory information with the other senses (vision, motor/touch).

Jia Luo, Associate Professor; Ph.D., Iowa. Growth factor regulation and ethanol effects on brain development.

Peter Mathers, Associate Professor; Ph.D., Caltech. Molecular control of visual and auditory development.

Janine Mendola, Associate Professor; Ph.D., MIT. Neural basis of perception and memory in health and disease; functional organization of the primate visual system.

James O'Donnell, Professor; Ph.D., Chicago. Pharmacology; phosphodiesterases and depression.

Giovanni Piedimonte, Professor; M.D., Rome (Italy). Viral respiratory infections; neuroimmunomodulation; asthma pathogenesis; lung development; airway neurobiology; cystic fibrosis.

Aina Puce, Professor; Ph.D., Melbourne. Functional neuroimaging; electrophysiology; signal processing; neural modeling; cognitive neuroscience; face/object recognition; nonverbal communication.

Visvanathan Ramamurthy, Assistant Professor, Ph.D., Wesleyan. Biochemical mechanisms behind gene mutations that result in photoreceptor cell death; lipid modifications of proteins in photoreceptors; novel treatment for blinding diseases.

Benjamin Ramsden, Assistant Professor; Ph.D., Swinburne (Australia). Organization of cortical function, visual processing, and combined optical imaging and electrophysiology in mammalian sensory cortex.

Anthony Realini, Associate Professor; M.D., North Carolina at Chapel Hill. Clinical glaucoma management; intraocular pressure behavior; imaging of the optic nerve; functional visual assessment.

Charles Rosen, Assistant Professor; M.D., Ph.D., NYU. Neuroprotective agents for stroke; traumatic brain injury; neurodegenerative disease.

Adrienne Salm, Professor; Ph.D., Michigan State. Neurobiology of anxiety disorders: interactions of adverse early life experience and brain plasticity.

Bernard Schreurs, Professor; Ph.D., Iowa. Learning and memory; synaptic plasticity; functional imaging; psychoneuroimmunology.

Maxim Sokolov, Assistant Professor; Ph.D., Weizmann (Israel). Biochemistry of vision and molecular control of vision-related proteins.

George Spirou, Professor; Ph.D., Florida. Neural feedback circuits mediating selective attention to sounds.

William Wonderlin, Associate Professor; Ph.D., Johns Hopkins. Ion channel physiology and biophysics.

Sepideh Zareparsi, Assistant Professor; Ph.D., Oregon Health Sciences. Genetic analysis of complex ophthalmic disorders; gene-expression analysis of retinal aging and age-related retinal disorders.

Section 14
Nutrition

This section contains a directory of institutions offering graduate work in nutrition, followed by in-depth entries submitted by institutions that chose to prepare detailed program descriptions. Additional information about programs listed in the directory but not augmented by an in-depth entry may be obtained by writing directly to the dean of a graduate school or chair of a department at the address given in the directory.

For programs offering related work, see also in this book Biochemistry, Biological and Biomedical Sciences, Botany and Plant Biology, Microbiological Sciences, Pathology and Pathobiology, Pharmacology and Toxicology, and Physiology. In Book 2, see Economics (Agricultural Economics and Agribusiness) and Family and Consumer Sciences; in Book 4, see Agricultural and Food Sciences and Chemistry; in Book 5, see Agricultural Engineering and Bioengineering and Biomedical Engineering and Biotechnology; and in Book 6, see Allied Health, Public Health, and Veterinary Medicine and Sciences.

CONTENTS

Program Directory

Close-Ups

See also:

Nutrition

American Health Sciences University, Program in Nutrition Science, Aurora, CO 80012. Offers MS. Part-time and evening/weekend programs available. Postbaccalaureate distance learning degree programs offered (no on-campus study). *Faculty:* 4 part-time/adjunct (3 women). *Students:* Average age 35. *Entrance requirements:* For master's, course work in anatomy, physiology, and basic nutrition. Application fee: $100. Electronic applications accepted. *Expenses:* Tuition: Full-time $2,700; part-time $300 per credit. *Unit head:* Ann L Peterson, Academic Dean, 800-530-8079, Fax: 303-367-2577, E-mail: dean@ahsu.edu. *Application contact:* Dr. Chaitali Adhikari, Director of Graduate Program, 303-340-2054, Fax: 303-367-2577, E-mail: masters@ahsu.edu.

American University of Beirut, Graduate Programs, Faculty of Agricultural and Food Sciences, Beirut, Lebanon. Offers agricultural economics (MS); animal sciences (MS); ecosystem management (MSES); food technology (MS); irrigation (MS); mechanization (MS); nutrition (MS); plant protection (MS); plant science (MS); poultry science (MS); soils (MS). *Degree requirements:* For master's, one foreign language, thesis (for some programs), comprehensive exam, registration. *Entrance requirements:* For master's, GRE, letter of recommendation.

Andrews University, School of Graduate Studies, College of Arts and Sciences, Department of Nutrition, Berrien Springs, MI 49104. Offers MS. Part-time programs available. *Faculty:* 2 full-time (0 women), 2 part-time/adjunct (both women). *Students:* 7 full-time (6 women), 4 part-time (all women); includes 6 minority (4 African Americans, 1 Asian American or Pacific Islander, 1 Hispanic American), 3 international. Average age 32. In 2005, 3 degrees awarded. *Application deadline:* Applications are processed on a rolling basis. Application fee: $43. *Unit head:* Dr. Winston Craig, Chairperson, 269-471-3370. *Application contact:* Carolyn Hurst, Supervisor of Graduate Admission, 800-253-2874, Fax: 269-471-3228, E-mail: graduate@andrews.edu.

Arizona State University at the Polytechnic Campus, East College, Department of Nutrition, Mesa, AZ 85212. Offers MS. Part-time and evening/weekend programs available. *Degree requirements:* For master's, thesis. *Entrance requirements:* Additional exam requirements/recommendations for international students: Required—TOEFL (minimum score 550 paper-based; 213 computer-based); Recommended—TWE, TSE. Electronic applications accepted. *Faculty research:* Vitamin C metabolism, nutrition and exercise, nutrient intakes, nutritional status, metabolic feeding studies.

Auburn University, Graduate School, College of Human Sciences, Department of Nutrition and Food Science, Auburn University, AL 36849. Offers MS, PhD. Part-time programs available. *Faculty:* 13 full-time (6 women). *Students:* 11 full-time (6 women), 12 part-time (11 women); includes 2 minority (both African Americans), 6 international. 17 applicants, 53% accepted, 3 enrolled. In 2005, 9 master's, 2 doctorates awarded. *Degree requirements:* For master's, thesis (for some programs); for doctorate, thesis/dissertation. *Entrance requirements:* For master's and doctorate, GRE General Test. *Application deadline:* For fall admission, 7/7 for domestic students; for spring admission, 11/24 for domestic students. Applications are processed on a rolling basis. Application fee: $25 ($50 for international students). Electronic applications accepted. *Financial support:* Research assistantships, teaching assistantships, career-related internships or fieldwork and Federal Work-Study available. Support available to part-time students. Financial award application deadline: 3/15. *Faculty research:* Food quality and safety, diet, food supply, physical activity in maintenance of health, prevention of selected chronic disease states. *Unit head:* Dr. Douglas B White, Head, 334-844-4261. *Application contact:* Dr. Stephen L. McFarland, Acting Dean of the Graduate School, 334-844-4700.

Barnes-Jewish College of Nursing and Allied Health, Division of Allied Health, St. Louis, MO 63110-1091. Offers dietetic internship (Certificate); education (MSAH); management (MSAH); nutrition (MSAH). Part-time and evening/weekend programs available. *Faculty:* 1 part-time/adjunct (0 women). *Students:* 28 full-time (23 women), 9 part-time (8 women); includes 5 minority (all African Americans) Average age 35. 43 applicants, 53% accepted. In 2005, 2 master's, 14 Certificates awarded. *Degree requirements:* For master's, thesis or alternative, registration. *Entrance requirements:* For master's, minimum GPA of 3.0, 2 references, statistics course. Additional exam requirements/recommendations for international students: Required—TOEFL (minimum score 550 paper-based; 213 computer-based). *Application deadline:* Applications are processed on a rolling basis. Application fee: $25. *Expenses:* Tuition: Part-time $396 per hour. *Financial support:* In 2005–06, 1 student received support. Federal Work-Study and scholarships/grants available. Support available to part-time students. Financial award application deadline: 5/7. *Unit head:* Kelly Eiden, Interim Dean, 314-454-8987. *Application contact:* Christie N. Schneider, Admissions Director, 314-454-7538, Fax: 314-454-5239, E-mail: cns5347@bjc.org.

Bastyr University, Graduate and Professional Programs, School of Nutrition and Exercise Science, Kenmore, WA 98028-4966. Offers nutrition (MS); nutrition and clinical health psychology (MS). Part-time programs available. *Students:* 63 full-time (61 women), 18 part-time (15 women); includes 9 minority (6 African Americans, 2 Asian Americans or Pacific Islanders, 1 Hispanic American). Average age 30. 72 applicants, 74% accepted, 35 enrolled. In 2005, 33 degrees awarded. *Degree requirements:* For master's, thesis optional. *Entrance requirements:* For master's, BS with 1 year of course work in chemistry, biochemistry, physiology and nutrition. Additional exam requirements/recommendations for international students: Required—TOEFL (minimum score 550 paper-based; 213 computer-based). *Application deadline:* For fall admission, 3/15 priority date for domestic students, 3/15 priority date for international students. Applications are processed on a rolling basis. Application fee: $75. *Expenses:* Tuition: Full-time $18,030; part-time $306 per quarter hour. Required fees: $1,311. Tuition and fees vary according to course load. *Financial support:* Career-related internships or fieldwork, Federal Work-Study, and scholarships/grants available. Support available to part-time students. Financial award application deadline: 4/15; financial award applicants required to submit FAFSA. *Unit head:* Dr. Jennifer Lovejoy, Chair, 425-823-1300, Fax: 425-823-6222. *Application contact:* Admissions Office, 425-602-3330, Fax: 425-602-3090, E-mail: admiss@bastyr.edu.

See Close-Up on page 1083.

Baylor University, Graduate School, School of Education, Department of Health, Human Performance and Recreation, Waco, TX 76798. Offers exercise, nutrition and preventive health (PhD); health, human performance and recreation (MS Ed). *Accreditation:* NCATE. Part-time programs available. *Faculty:* 13 full-time (5 women), 3 part-time/adjunct (1 woman). *Students:* 52 full-time (26 women), 41 part-time (22 women); includes 17 minority (9 African Americans, 3 American Indian/Alaska Native, 2 Asian Americans or Pacific Islanders, 3 Hispanic Americans), 1 international. 30 applicants, 87% accepted. In 2005, 35 degrees awarded. *Degree requirements:* For master's, thesis optional. *Entrance requirements:* For master's, GRE General Test. *Application deadline:* For fall admission, 4/1 for domestic students; for spring admission, 10/1 for domestic students. Applications are processed on a rolling basis. Application fee: $25. Electronic applications accepted. *Financial support:* In 2005–06, 35 students received support, including 22 teaching assistantships; career-related internships or fieldwork, Federal Work-Study, institutionally sponsored loans, tuition waivers (partial), and recreation supplements also available. *Faculty research:* Behavior change theory, pedagogy, nutrition and enzyme therapy, exercise testing, health planning. *Unit head:* Dr. Mike Greenwood, Graduate Program Director, 254-710-3505, Fax: 254-710-3527, E-mail: mike_greenwood@baylor.edu. *Application contact:* Suzanne Keener, Administrative Assistant, 254-710-3588, Fax: 254-710-3870.

Benedictine University, Graduate Programs, Program in Nutrition and Wellness, Lisle, IL 60532-0900. Offers MS. *Students:* 12 (all women) 21 applicants, 90% accepted, 11 enrolled. *Entrance requirements:* Additional exam requirements/recommendations for international students: Required—TOEFL (minimum score 550 paper-based; 213 computer-based). *Application deadline:* Applications are processed on a rolling basis. Application fee: $40. Electronic

applications accepted. *Financial support:* Career-related internships or fieldwork and health care benefits available. Support available to part-time students. *Faculty research:* Community and corporate wellness risk assessment, health behavior change, self-efficacy, evaluation of health program impact and effectiveness. Total annual research expenditures: $8,335. *Unit head:* Catherine Arnold, Director, 630-829-6534, E-mail: carnold@ben.edu. *Application contact:* Kari Gibbons, Director, Admissions, 630-829-6200, Fax: 630-829-6584, E-mail: kgibbons@ben.edu.

Boston University, Sargent College of Health and Rehabilitation Sciences, Department of Health Sciences, Boston, MA 02215. Offers applied anatomy and physiology (MS, PhD); nutrition (MS). Part-time programs available. *Faculty:* 11 full-time (9 women), 6 part-time/adjunct (5 women). *Students:* 57 full-time (49 women), 5 part-time (all women); includes 7 minority (all Asian Americans or Pacific Islanders), 8 international. Average age 26. 67 applicants, 40% accepted, 20 enrolled. *Degree requirements:* For master's, thesis or alternative; for doctorate, one foreign language, thesis/dissertation. *Entrance requirements:* For master's, GRE General Test, minimum GPA of 3.0; for doctorate, GRE General Test. Additional exam requirements/recommendations for international students: Required—TOEFL (minimum score 550 paper-based). *Application deadline:* For fall admission, 3/1 for domestic students; for spring admission, 10/1 for domestic students. Applications are processed on a rolling basis. Application fee: $65. Electronic applications accepted. *Expenses:* Tuition: Full-time $31,530; part-time $985 per credit. Required fees: $316; $40 per semester. Tuition and fees vary according to course level and program. *Financial support:* In 2005–06, 20 fellowships with full tuition reimbursements, 7 research assistantships with full tuition reimbursements, 6 teaching assistantships with full tuition reimbursements were awarded; career-related internships or fieldwork, Federal Work-Study, institutionally sponsored loans, scholarships/grants, and tuition waivers (partial) also available. Support available to part-time students. Financial award application deadline: 4/15. *Faculty research:* Muscle metabolism, body acid-base balance, human performance, physical conditioning, diabetes. *Unit head:* Dr. Eileen O'Keefe, Chair, 617-353-7532, E-mail: ebokeefe@bu.edu. *Application contact:* Sharon Sankey, Director, Student Services, 617-353-2713, Fax: 617-353-7500, E-mail: ssankey@bu.edu.

Boston University, School of Medicine, Division of Graduate Medical Sciences, Program in Medical Nutrition Sciences, Boston, MA 02215. Offers MA, PhD. *Entrance requirements:* For master's and doctorate, GRE or MCAT. Additional exam requirements/recommendations for international students: Required—TOEFL. *Expenses:* Tuition: Full-time $31,530; part-time $985 per credit. Required fees: $316; $40 per semester. Tuition and fees vary according to course level and program.

See Close-Up on page 1085.

Bowling Green State University, Graduate College, College of Education and Human Development, School of Family and Consumer Sciences, Bowling Green, OH 43403. Offers food and nutrition (MFCS); human development and family studies (MFCS). Part-time programs available. *Faculty:* 22 full-time (17 women), 4 part-time/adjunct (2 women). *Students:* 13 full-time (all women), 5 part-time (all women), 6 international. Average age 27. 14 applicants, 71% accepted, 8 enrolled. In 2005, 3 degrees awarded. *Degree requirements:* For master's, thesis. *Entrance requirements:* For master's, GRE General Test, minimum GPA of 3.0. Additional exam requirements/recommendations for international students: Required—TOEFL. *Application deadline:* For fall admission, 2/1 for domestic students. Application fee: $30. *Financial support:* In 2005–06, 7 research assistantships with full tuition reimbursements (averaging $7,302 per year), 5 teaching assistantships with full tuition reimbursements (averaging $7,077 per year) were awarded; career-related internships or fieldwork and unspecified assistantships also available. Financial award applicants required to submit FAFSA. *Faculty research:* Public health, wellness, social issues and policies, ethnic foods, nutrition and aging. *Unit head:* Dr. Deborah Wooldridge, Director, 419-372-7823. *Application contact:* Dr. Rebecca Pobocik, Graduate Coordinator, 419-372-7849.

Brigham Young University, Graduate Studies, College of Biological and Agricultural Sciences, Department of Nutrition, Dietetics and Food Science, Provo, UT 84602-1001. Offers food science (MS); nutrition (MS). *Faculty:* 13 full-time (5 women). *Students:* 15 full-time (11 women); includes 3 minority (1 African American, 1 American Indian/Alaska Native, 1 Hispanic American). Average age 24. 5 applicants, 60% accepted, 3 enrolled. In 2005, 4 degrees awarded. *Degree requirements:* For master's, thesis, comprehensive exam, registration. *Entrance requirements:* For master's, GRE General Test. Additional exam requirements/recommendations for international students: Required—TOEFL (minimum score 550 paper-based; 213 computer-based). *Application deadline:* For fall admission, 2/1 for domestic students, 2/1 for international students. For winter admission, 6/30 for domestic students. Application fee: $50. Electronic applications accepted. *Financial support:* In 2005–06, 7 students received support, including 4 research assistantships (averaging $16,665 per year), 3 teaching assistantships (averaging $16,665 per year); career-related internships or fieldwork, institutionally sponsored loans, and scholarships/grants also available. Financial award application deadline: 4/1. *Faculty research:* Dairy foods, lipid oxidation, food processes, magnesium and selenium nutrition, nutrient effect on gene expression. Total annual research expenditures: $234,094. *Unit head:* Dr. Lynn V. Ogden, Chair, 801-422-3912, Fax: 801-422-0258, E-mail: lynn_ogden@byu.edu. *Application contact:* Dr. Merrill J. Christensen, Graduate Coordinator, 801-422-5255, Fax: 801-422-0258, E-mail: merrill_christensen@byu.edu.

Brooklyn College of the City University of New York, Division of Graduate Studies, Department of Health and Nutrition Science, Program in Nutrition Sciences, Brooklyn, NY 11210-2889. Offers nutrition (MS). Part-time programs available. *Students:* 3 full-time (all women), 51 part-time (47 women); includes 15 minority (7 African Americans, 7 Asian Americans or Pacific Islanders, 1 Hispanic American), 5 international. 31 applicants, 84% accepted, 11 enrolled. In 2005, 14 degrees awarded. *Degree requirements:* For master's, thesis or alternative. *Entrance requirements:* For master's, GRE, 18 credits in health-related areas, 2 letters of recommendation. Additional exam requirements/recommendations for international students: Required—TOEFL. *Application deadline:* For fall admission, 3/1 for domestic students, 2/1 for international students; for spring admission, 11/1 for domestic students, 10/1 for international students. Application fee: $50. *Expenses:* Tuition, state resident: full-time $6,400; part-time $270 per credit. Tuition, nonresident: full-time $12,000; part-time $500 per credit. Required fees: $118 per term. *Financial support:* Federal Work-Study, institutionally sponsored loans, and scholarships/grants available. Support available to part-time students. Financial award application deadline: 5/1; financial award applicants required to submit FAFSA. *Faculty research:* Medical ethics, AIDS, history of public health, diet restriction, palliative care, risk reduction/disease prevention, metabolism, diabetes. *Unit head:* Dr. Kathleen Axen, Graduate Deputy Chairperson, 718-951-5909, Fax: 718-951-4670, E-mail: kaxen@brooklyn.cuny.edu. *Application contact:* Marianne Booufall-Tynan, Director of Admissions, 718-951-5902, E-mail: adminqry@brooklyn.cuny.edu.

California State Polytechnic University, Pomona, Academic Affairs, College of Agriculture, Pomona, CA 91768-2557. Offers agricultural science (MS); animal science (MS); foods and nutrition (MS). Part-time programs available. *Faculty:* 43 full-time (13 women), 16 part-time/adjunct (9 women). *Students:* 20 full-time (16 women), 39 part-time (29 women); includes 14 minority (1 African American, 4 Asian Americans or Pacific Islanders, 9 Hispanic Americans), 4 international. Average age 30. 58 applicants, 76% accepted, 20 enrolled. In 2005, 13 degrees awarded. *Degree requirements:* For master's, thesis or alternative. *Application deadline:* For fall admission, 5/1 for domestic students. For winter admission, 10/15 for domestic students; for spring admission, 1/2 for domestic students. Applications are processed on a rolling basis. Application fee: $55. Electronic applications accepted. *Expenses:* Tuition, nonresident: full-time $9,021. Required fees: $3,597. *Financial support:* Career-related internships or fieldwork, Federal Work-Study, and institutionally sponsored loans available. Support available to part-time students. Financial award application deadline: 3/2; financial award applicants required to submit FAFSA. *Faculty research:* Equine nutrition, physiology, and reproduction; leadership development; bioartificial pancreas; plant science; ruminant and human nutrition. *Unit head:* Dr. Wayne R. Bidlack, Dean, 909-869-2200, E-mail: wrbidlack@csupomona.edu.

California State University, Chico, Graduate School, College of Natural Sciences, Department of Biological Sciences, Program in Nutritional and Food Science, Chico, CA 95929-0002. Offers nutrition education (MS). *Degree requirements:* For master's, thesis, seminar presentation. *Entrance requirements:* For master's, GRE General Test, 2 letters of recommendation. Additional exam requirements/recommendations for international students: Required—TOEFL (minimum score 550 paper-based; 213 computer-based). Electronic applications accepted.

California State University, Long Beach, Graduate Studies, College of Health and Human Services, Department of Family and Consumer Sciences, Program in Nutritional Sciences, Long Beach, CA 90840. Offers MS. Part-time programs available. *Faculty:* 8 full-time (6 women). *Students:* 20 full-time (19 women), 18 part-time (all women); includes 17 minority (11 Asian Americans or Pacific Islanders, 6 Hispanic Americans), 1 international. Average age 30. 40 applicants, 48% accepted, 17 enrolled. In 2005, 4 degrees awarded. *Degree requirements:* For master's, comprehensive exam or thesis. *Entrance requirements:* For master's, minimum GPA of 3.0. *Application deadline:* For fall admission, 7/1 for domestic students; for spring admission, 12/1 for domestic students. Applications are processed on a rolling basis. Application fee: $55. Electronic applications accepted. *Expenses:* Tuition, nonresident: part-time $339 per semester hour. *Financial support:* Federal Work-Study, institutionally sponsored loans, and scholarships/grants available. Financial award application deadline: 3/2. *Faculty research:* Protein and water-soluble vitamins, sensory evaluation of foods, mineral deficiencies in humans, child nutrition, minerals and blood pressure. *Unit head:* Dr. M Sue Stanley, Chair, 562-985-4484, Fax: 562-985-4414, E-mail: stanleym@csulb.edu.

California State University, Los Angeles, Graduate Studies, College of Health and Human Services, Department of Kinesiology and Nutritional Sciences, Major in Nutritional Science, Los Angeles, CA 90032-8530. Offers MS. *Accreditation:* ADtA. Part-time and evening/weekend programs available. *Students:* 60 full-time (55 women), 36 part-time (32 women); includes 49 minority (3 African Americans, 34 Asian Americans or Pacific Islanders, 12 Hispanic Americans). In 2005, 9 degrees awarded. *Degree requirements:* For master's, project or thesis. *Entrance requirements:* For master's, minimum GPA of 3.0. Additional exam requirements/recommendations for international students: Required—TOEFL. *Application deadline:* For fall admission, 6/30 for domestic students; for spring admission, 2/1 for domestic students. Applications are processed on a rolling basis. Application fee: $55. *Financial support:* Career-related internships or fieldwork and Federal Work-Study available. Support available to part-time students. Financial award application deadline: 3/1. *Faculty research:* Human nutrition, nutrition education, nutrition and fitness for the elderly. *Unit head:* Laura Calderon, Coordinator, 343-343-4650.

Case Western Reserve University, School of Medicine and School of Graduate Studies, Graduate Programs in Medicine, Department of Nutrition, Cleveland, OH 44106. Offers dietetics (MS); nutrition (MS, PhD), including nutrition and biochemistry (PhD); public health nutrition (MS). Part-time programs available. *Faculty:* 19 full-time (11 women), 31 part-time/adjunct (24 women). *Students:* 61 full-time (53 women), 15 part-time (13 women); includes 38 minority (2 African Americans, 31 Asian Americans or Pacific Islanders, 5 Hispanic Americans). Average age 25. 47 applicants, 53% accepted, 21 enrolled. In 2005, 17 master's, 3 doctorates awarded. Terminal master's awarded for partial completion of doctoral program. *Degree requirements:* For master's, thesis (for some programs); for doctorate, thesis/dissertation. *Entrance requirements:* For master's, GRE General Test; for doctorate, GRE General Test, GRE Subject Test. Additional exam requirements/recommendations for international students: Required—TOEFL. *Application deadline:* For fall admission, 3/1 for domestic students; for spring admission, 11/1 for domestic students. Applications are processed on a rolling basis. Application fee: $50. *Financial support:* In 2005–06, 11 students received support, including 3 fellowships with full tuition reimbursements available (averaging $23,000 per year), 9 research assistantships with full tuition reimbursements available (averaging $23,000 per year); career-related internships or fieldwork, Federal Work-Study, and tuition waivers (partial) also available. *Faculty research:* Fatty acid metabolism, application of gene therapy to nutritional problems, dietary intake methodology, nutrition and physical fitness, metabolism during infancy and pregnancy. Total annual research expenditures: $1.5 million. *Unit head:* Dr. Henri Brunengraber, Chairman, 216-368-2440, Fax: 216-368-6644, E-mail: hxb8@case.edu. *Application contact:* Pamela A. Woodruff, Department Assistant III, 216-368-2440, Fax: 216-368-6644, E-mail: paw5@case.edu.

See Close-Up on page 1087.

Central Michigan University, College of Graduate Studies, College of Education and Human Services, Department of Human Environmental Studies, Mount Pleasant, MI 48859. Offers human development and family studies (MA); nutrition and dietetics (MS). *Faculty:* 21 full-time (14 women). *Students:* 3 full-time (all women), 23 part-time (22 women). Average age 26. In 2005, 3 degrees awarded. *Degree requirements:* For master's, thesis or alternative, registration. *Entrance requirements:* For master's, GRE (MA), minimum GPA of 3.0 in last 60 hours, 15 credits of course work in human development and family studies or related area (MA). *Application deadline:* For fall admission, 2/1 for domestic students; for spring admission, 9/15 for domestic students. Application fee: $35 ($45 for international students). *Expenses:* Tuition, area resident: Part-time $325 per credit hour. Tuition, state resident: part-time $603 per credit hour. Tuition and fees vary according to degree level and reciprocity agreements. *Financial support:* In 2005–06, 2 research assistantships were awarded; fellowships with tuition reimbursements, career-related internships or fieldwork and Federal Work-Study also available. Financial award application deadline: 3/7. *Faculty research:* Human growth and development, family studies and human sexuality, nutritional food science/food services, apparel and textile retailing, computer-aided design for apparel and interior design. *Unit head:* Dr. Kathryn Koch, Chairperson, 989-774-3218, Fax: 989-774-2435, E-mail: kathryn.e.koch@cmich.edu. *Application contact:* Dr. Phame Camarena, Human Development Program Coordinator, 989-774-5600, E-mail: camar1pm@cmich.edu.

Central Washington University, Graduate Studies, Research and Continuing Education, College of Education and Professional Studies, Department of Family and Consumer Sciences, Ellensburg, WA 98926. Offers family and consumer sciences education (MS); family studies (MS); nutrition (MS). Part-time programs available. *Faculty:* 8 full-time (all women). *Students:* 8 full-time (7 women), 4 part-time (3 women); includes 2 minority (1 Asian American or Pacific Islander, 1 Hispanic American), 1 international. 5 applicants, 100% accepted, 5 enrolled. In 2005, 2 degrees awarded. *Degree requirements:* For master's, thesis or alternative. *Entrance requirements:* For master's, GRE General Test (nutrition), minimum GPA of 3.0. Additional exam requirements/recommendations for international students: Required—TOEFL (minimum score 550 paper-based; 213 computer-based). *Application deadline:* For fall admission, 4/1 for domestic students. For winter admission, 10/1 for domestic students; for spring admission, 1/1 for domestic students. Applications are processed on a rolling basis. Application fee: $50. Electronic applications accepted. *Expenses:* Tuition, state resident: full-time $1,968; part-time $197 per credit. Tuition, nonresident: full-time $4,320; part-time $432 per credit. Required fees: $623. Tuition and fees vary according to degree level. *Financial support:* In 2005–06, 1 teaching assistantship with partial tuition reimbursement (averaging $8,100 per year) was awarded; research assistantships, Federal Work-Study and unspecified assistantships also available. Financial award application deadline: 3/1; financial award applicants required to submit FAFSA. *Unit head:* Dr. Jan Bowers, Chair, 509-963-2766. *Application contact:* Justine Eason, Admissions Program Coordinator, 509-963-3103, Fax: 509-963-1799, E-mail: masters@cwu.edu.

Chapman University, Graduate Studies, Wilkinson College of Letters and Sciences, Department of Physical Sciences, Orange, CA 92866. Offers food science and nutrition (MS). Part-time and evening/weekend programs available. *Faculty:* 3 full-time (2 women), 2 part-

time/adjunct (1 woman). *Students:* 9 full-time (4 women), 8 part-time (5 women); includes 5 minority (2 African Americans, 3 Asian Americans or Pacific Islanders), 6 international. Average age 28. 12 applicants, 42% accepted, 0 enrolled. In 2005, 4 degrees awarded. *Degree requirements:* For master's, thesis, comprehensive exam, registration. *Entrance requirements:* For master's, GRE General Test, minimum undergraduate GPA of 3.0. Additional exam requirements/recommendations for international students: Required—TOEFL (minimum score 550 paper-based). *Application deadline:* Applications are processed on a rolling basis. Application fee: $55. Electronic applications accepted. *Expenses:* Contact institution. *Financial support:* In 2005–06, 11 students received support, including 3 fellowships (averaging $2,500 per year); Federal Work-Study also available. Financial award application deadline: 6/30; financial award applicants required to submit FAFSA. *Unit head:* Dr. Daniel Wellman, Chair, 714-744-7826, E-mail: prakash@chapman.edu. *Application contact:* Jojo Delfin, Information Contact, 714-997-6786, Fax: 714-997-6713, E-mail: delfin@chapman.edu.

Clemson University, Graduate School, College of Agriculture, Forestry and Life Sciences, Department of Food Science and Human Nutrition, Program in Food, Nutrition, and Culinary Science, Clemson, SC 29634. Offers MS. *Students:* 12 full-time (8 women), 6 part-time (3 women); includes 1 minority (African American), 5 international. 20 applicants, 60% accepted, 6 enrolled. In 2005, 4 degrees awarded. *Degree requirements:* For master's, thesis. *Entrance requirements:* For master's, GRE General Test. Additional exam requirements/recommendations for international students: Required—TOEFL. *Application deadline:* For fall admission, 6/1 for domestic students, 4/15 for international students. Applications are processed on a rolling basis. Application fee: $50. Electronic applications accepted. *Financial support:* Fellowships with partial tuition reimbursements, research assistantships with partial tuition reimbursements, teaching assistantships with partial tuition reimbursements available. Financial award applicants required to submit FAFSA. *Unit head:* Dr. Paul Dawson, Coordinator, 864-656-1138, Fax: 864-656-3131, E-mail: pdawson@clemson.edu.

College of Saint Elizabeth, Department of Foods and Nutrition, Morristown, NJ 07960-6989. Offers dietetic internship (Certificate); nutrition (MS). Part-time and evening/weekend programs available. *Faculty:* 3 full-time (all women), 2 part-time/adjunct (both women). *Students:* 16 full-time (13 women), 20 part-time (all women); includes 4 minority (1 African American, 3 Asian Americans or Pacific Islanders), 1 international. Average age 34. In 2005, 11 master's, 12 other advanced degrees awarded. *Application deadline:* Applications are processed on a rolling basis. Application fee: $35. Electronic applications accepted. *Expenses:* Tuition: Part-time $609 per credit. Required fees: $170 per term. One-time fee: $35 part-time. Tuition and fees vary according to course load, campus/location and reciprocity agreements. *Financial support:* Federal Work-Study, tuition waivers (partial), and unspecified assistantships available. Support available to part-time students. Financial award application deadline: 3/15; financial award applicants required to submit FAFSA. *Faculty research:* Medical nutrition intervention, public policy, obesity, hunger and food security, osteoporosis, nutrition and exercise. *Unit head:* Dr. Anne Boresma, Director of the Graduate Program in Nutrition, 973-290-4065, Fax: 973-290-4167, E-mail: nutrition@cse.edu. *Application contact:* Michael Szarek, Director of Enrollment Management, 973-290-4112, Fax: 973-290-4167, E-mail: mszarek@cse.edu.

Colorado State University, Graduate School, College of Applied Human Sciences, Department of Food Science and Human Nutrition, Fort Collins, CO 80523-0015. Offers food science (MS, PhD); nutrition (MS, PhD). Part-time programs available. *Faculty:* 14 full-time (6 women). *Students:* 35 full-time (29 women), 26 part-time (22 women); includes 2 minority (1 Asian American or Pacific Islander, 1 Hispanic American), 3 international. Average age 29. 70 applicants, 43% accepted, 20 enrolled. In 2005, 26 master's, 3 doctorates awarded. *Degree requirements:* For master's, thesis optional; for doctorate, thesis/dissertation, comprehensive exam, registration. *Entrance requirements:* For master's and doctorate, GRE General Test, minimum GPA of 3.0. Additional exam requirements/recommendations for international students: Required—TOEFL (minimum score 550 paper-based; 213 computer-based). *Application deadline:* For fall admission, 2/1 priority date for domestic students, 2/1 priority date for international students; for spring admission, 8/1 priority date for domestic students, 8/1 priority date for international students. Applications are processed on a rolling basis. Application fee: $50. Electronic applications accepted. *Expenses:* Tuition, state resident: full-time $3,690; part-time $205 per credit. Tuition, nonresident: full-time $14,958; part-time $831 per credit. Required fees: $1,061. *Financial support:* In 2005–06, 3 fellowships (averaging $3,730 per year), 15 research assistantships with full and partial tuition reimbursements (averaging $13,884 per year), 6 teaching assistantships with full and partial tuition reimbursements (averaging $10,863 per year) were awarded; career-related internships or fieldwork, Federal Work-Study, institutionally sponsored loans, scholarships/grants, traineeships, and unspecified assistantships also available. Financial award application deadline: 2/1. *Faculty research:* Metabolic regulation, nutrition education, food safety, obesity and diabetes. Total annual research expenditures: $3.4 million. *Unit head:* Dr. Christopher Melby, Head, 970-491-1944, Fax: 970-491-7252, E-mail: christopher.melby@colostate.edu. *Application contact:* Dr. Jennifer Anderson, Graduate Coordinator, 970-491-7334, Fax: 970-491-3875, E-mail: anderson@cahs.colostate.edu.

Columbia University, College of Physicians and Surgeons, Institute of Human Nutrition, MS Program in Nutrition, New York, NY 10032. Offers MS, MPH/MS. Part-time and evening/weekend programs available. *Degree requirements:* For master's, thesis. *Entrance requirements:* For master's, GRE General Test. Additional exam requirements/recommendations for international students: Required—TOEFL. *Expenses:* Tuition: Full-time $31,448. Tuition and fees vary according to course level, course load, campus/location and program.

See Close-Up on page 1089.

Columbia University, College of Physicians and Surgeons, Institute of Human Nutrition and Graduate School of Arts and Sciences at the College of Physicians and Surgeons, PhD Program in Nutrition, New York, NY 10032. Offers M Phil, MA, PhD, MD/PhD. Only candidates for the PhD are admitted. *Degree requirements:* For master's and doctorate, thesis/dissertation. *Entrance requirements:* For doctorate, GRE General Test. Additional exam requirements/recommendations for international students: Required—TOEFL. *Expenses:* Tuition: Full-time $31,448. Tuition and fees vary according to course level, course load, campus/location and program. *Faculty research:* Growth and development, nutrition and metabolism.

See Close-Up on page 1089.

Cornell University, Graduate School, Graduate Fields of Agriculture and Life Sciences and Graduate Fields of Human Ecology, Field of Nutrition, Ithaca, NY 14853-0001. Offers animal nutrition (MPS, MS, PhD); community nutrition (MPS, MS, PhD); human nutrition (MPS, MS, PhD); international nutrition (MPS, MS, PhD); nutritional biochemistry (MPS, MS, PhD). *Faculty:* 65 full-time (25 women). *Students:* 64 full-time (46 women); includes 12 minority (4 African Americans, 4 Asian Americans or Pacific Islanders, 4 Hispanic Americans), 24 international. 93 applicants, 25% accepted, 14 enrolled. In 2005, 5 master's, 10 doctorates awarded. *Degree requirements:* For master's, thesis (MS), project papers (MPS); for doctorate, thesis/dissertation, comprehensive exam. *Entrance requirements:* For master's and doctorate, GRE General Test, previous course work in organic chemistry with laboratory, and in biochemistry, 2 letters of recommendation. Additional exam requirements/recommendations for international students: Required—TOEFL (minimum score 550 paper-based; 213 computer-based). *Application deadline:* For fall admission, 1/10 for domestic students; for spring admission, 10/1 for domestic students. Application fee: $60. Electronic applications accepted. *Financial support:* In 2005–06, 58 students received support, including 29 fellowships with full tuition reimbursements available, 8 research assistantships with full tuition reimbursements available, 21 teaching assistantships with full tuition reimbursements available; institutionally sponsored loans, scholarships/grants, health care benefits, tuition waivers (full and partial), and unspecified assistantships also available. Financial award applicants required to submit FAFSA. *Faculty research:* Nutritional biochemistry, experimental human and animal nutrition, international nutrition, community nutrition. *Unit head:* Director of Graduate Studies, 607-255-2528,

Nutrition

Cornell University (continued)
Fax: 607-255-0178. *Application contact:* Graduate Field Assistant, 607-255-2528, Fax: 607-225-0178, E-mail: nutrition_gfr@cornell.edu.

Cornell University, Graduate School, Graduate Fields of Arts and Sciences, Field of International Development, Ithaca, NY 14853-0001. Offers development policy (MPS); international nutrition (MPS); international planning (MPS); international population (MPS); science and technology policy (MPS). *Faculty:* 59 full-time (20 women). *Students:* 34 applicants, 18% accepted, 5 enrolled. In 2005, 17 degrees awarded. *Degree requirements:* For master's, project paper. *Entrance requirements:* For master's, GRE General Test (recommended), 2 academic recommendations, 2 years of development experience. Additional exam requirements/recommendations for international students: Required—TOEFL. *Application deadline:* Applications are processed on a rolling basis. Application fee: $60. Electronic applications accepted. *Financial support:* In 2005–06, 9 students received support, including 6 fellowships with full tuition reimbursements available, 3 teaching assistantships with full tuition reimbursements available; research assistantships with full tuition reimbursements available, institutionally sponsored loans, scholarships/grants, health care benefits, tuition waivers (full and partial), and unspecified assistantships also available. Financial award applicants required to submit FAFSA. *Faculty research:* Development policy, international nutrition, international planning, science and technology policy, international population. *Unit head:* Director of Graduate Studies, 607-255-3037, Fax: 607-255-1005. *Application contact:* Graduate Field Assistant, 607-255-3037, Fax: 607-255-1005, E-mail: mpsid@cornell.edu.

Drexel University, College of Arts and Sciences, Department of Bioscience and Biotechnology, Program in Nutrition and Food Sciences, Philadelphia, PA 19104-2875. Offers food science (MS); nutrition science (PhD). Part-time programs available. Terminal master's awarded for partial completion of doctoral program. *Degree requirements:* For master's and doctorate, thesis/dissertation. *Entrance requirements:* For master's and doctorate, GRE General Test. Additional exam requirements/recommendations for international students: Required—TOEFL. Electronic applications accepted. *Faculty research:* Metabolism of lipids, W-3 fatty acids, obesity, diabetes and heart disease, mineral metabolism.

D'Youville College, Department of Dietetics, Buffalo, NY 14201-1084. Offers MS. (five year program that begins at freshman entry). *Accreditation:* ADtA. *Faculty:* 3 full-time (2 women), 3 part-time/adjunct (all women). *Students:* 32 full-time (30 women), 14 part-time (12 women); includes 3 minority (2 African Americans, 1 Hispanic American), 4 international. Average age 26. 35 applicants, 71% accepted, 10 enrolled. In 2005, 6 degrees awarded. *Degree requirements:* For master's, thesis, registration. *Entrance requirements:* Additional exam requirements/recommendations for international students: Required—TOEFL (minimum score 500 paper-based; 173 computer-based). *Application deadline:* Applications are processed on a rolling basis. Application fee: $25. Electronic applications accepted. *Financial support:* In 2005–06, 37 students received support. *Faculty research:* Nutrition education, clinical nutrition, herbal supplements, obesity. *Unit head:* Dr. Edward Weiss, Chair, 716-829-7832, Fax: 716-829-8137. *Application contact:* Ronald Dannecker, Director of Admissions, 716-829-7600, Fax: 716-829-7900, E-mail: admiss@dyc.edu.

East Carolina University, Graduate School, College of Human Ecology, Department of Nutrition and Hospitality Management, Greenville, NC 27858-4353. Offers nutrition (MS). Part-time programs available. *Students:* 8 full-time (6 women), 18 part-time (16 women), 1 international. Average age 29. 7 applicants, 43% accepted, 3 enrolled. In 2005, 8 degrees awarded. *Degree requirements:* For master's, thesis optional. *Entrance requirements:* For master's, GRE. Additional exam requirements/recommendations for international students: Required—TOEFL. *Application deadline:* For fall admission, 6/1 for domestic students. Applications are processed on a rolling basis. Application fee: $50. *Expenses:* Tuition, state resident: full-time $2,516. Tuition, nonresident: full-time $12,832. *Financial support:* In 2005–06, 7 students received support, including 2 fellowships (averaging $1,500 per year), 6 teaching assistantships with partial tuition reimbursements available (averaging $7,500 per year); Federal Work-Study, institutionally sponsored loans, scholarships/grants, and unspecified assistantships also available. Support available to part-time students. Financial award application deadline: 6/1. *Faculty research:* Lifecycle nutrition, nutrition and disease, nutrition for fish species, food service management. *Unit head:* Dr. William Forsythe, Chairperson, 252-328-6850, Fax: 252-328-4276, E-mail: forsythew@ecu.edu. *Application contact:* Dean of Graduate School, 252-328-6012, Fax: 252-328-6071, E-mail: gradschool@ecu.edu.

Eastern Illinois University, Graduate School, Lumpkin College of Business and Applied Sciences, School of Family and Consumer Sciences, Charleston, IL 61920-3099. Offers dietetics (MS); family and consumer sciences (MS). Part-time programs available. *Faculty:* 12 full-time (10 women). *Students:* 34 full-time (33 women), 29 part-time (all women). In 2005, 40 degrees awarded. *Degree requirements:* For master's, comprehensive exam. *Application deadline:* For fall admission, 7/31 for domestic students. Applications are processed on a rolling basis. Application fee: $30. *Expenses:* Tuition, state resident: part-time $150 per credit hour. Tuition, nonresident: part-time $452 per credit hour. Required fees: $738. *Financial support:* In 2005–06, 2 research assistantships with tuition reimbursements (averaging $7,200 per year), 6 teaching assistantships with tuition reimbursements (averaging $7,200 per year) were awarded; career-related internships or fieldwork also available. *Unit head:* Dr. James Painter, Chairperson, 217-581-6076, Fax: 217-581-6090, E-mail: jepainter@eiu.edu. *Application contact:* Dr. Frances Murphy, Coordinator, 217-581-6997, Fax: 217-581-6090, E-mail: flmurphy@eiu.edu.

Eastern Kentucky University, The Graduate School, College of Health Sciences, Department of Family and Consumer Sciences, Richmond, KY 40475-3102. Offers community nutrition (MS). Part-time programs available. *Entrance requirements:* For master's, GRE General Test, minimum GPA of 2.5.

East Tennessee State University, School of Graduate Studies, College of Business and Technology, Department of Family and Consumer Sciences, Johnson City, TN 37614. Offers clinical nutrition (MS). *Faculty:* 6 full-time (all women). *Students:* 13 full-time (11 women), 4 part-time (all women), 1 international. Average age 28. 18 applicants, 67% accepted. In 2005, 6 degrees awarded. *Degree requirements:* For master's, thesis, oral exam. *Entrance requirements:* For master's, GRE, ADA-approved undergraduate didactic program in dietetics. Additional exam requirements/recommendations for international students: Required—TOEFL (minimum score 550 paper-based; 213 computer-based). *Application deadline:* For fall admission, 2/15 for domestic students. Applications are processed on a rolling basis. Application fee: $25 ($35 for international students). *Expenses:* Tuition, nonresident: full-time $9,312; part-time $404 per hour. Required fees: $261 per hour. *Financial support:* In 2005–06, 4 research assistantships with full tuition reimbursements (averaging $5,500 per year), 2 teaching assistantships with full tuition reimbursements (averaging $5,500 per year) were awarded; career-related internships or fieldwork, Federal Work-Study, and scholarships/grants also available. Financial award application deadline: 7/1; financial award applicants required to submit FAFSA. *Faculty research:* Kindergarten students, measures of percent body fat during pregnancy, writing to read in preschool children, computer research in food systems management, students' knowledge of aging. Total research expenditures: $26,794. *Unit head:* Dr. Jamie Kridler, Chair, 423-439-7538, Fax: 423-439-7539, E-mail: kridler@etsu.edu.

Emory University, Graduate School of Arts and Sciences, Division of Biological and Biomedical Sciences, Program in Nutrition and Health Sciences, Atlanta, GA 30322-1100. Offers PhD. *Faculty:* 47 full-time (24 women). *Students:* 32 full-time (29 women); includes 4 minority (2 African Americans, 2 Asian Americans or Pacific Islanders), 11 international. Average age 27. 45 applicants, 20% accepted, 6 enrolled. In 2005, 1 degree awarded. *Median time to degree:* Of those who began their doctoral program in fall 1997, 100% received their degree in 8 years or less. *Degree requirements:* For doctorate, thesis/dissertation, comprehensive exam, registration. *Entrance requirements:* For doctorate, GRE General Test, minimum GPA of 3.0 in science course work. Additional exam requirements/recommendations for international students:

Required—TOEFL. *Application deadline:* For fall admission, 1/3 for domestic students, 1/3 for international students. Application fee: $50. Electronic applications accepted. *Expenses:* Tuition: Full-time $14,400. Required fees: $217. *Financial support:* In 2005–06, 12 students received support, including 12 fellowships with full tuition reimbursements available (averaging $23,000 per year); institutionally sponsored loans, health care benefits, and tuition waivers (full) also available. *Faculty research:* Biochemistry, molecular and cell biology, clinical nutrition, community and preventive health, nutritional epidemiology. *Unit head:* Dr. Aryeh Stein, Director, 404-727-4255, Fax: 404-727-1278, E-mail: astein2@sph.emory.edu. *Application contact:* 404-727-2545, Fax: 404-727-3322, E-mail: gdbbs@emory.edu.

Emory University, Rollins School of Public Health, Department of Global Health, Atlanta, GA 30322-1100. Offers global health (MPH); public nutrition (MSPH, PhD). *Accreditation:* CEPH. Part-time programs available. *Students:* 133 full-time (112 women), 8 part-time (6 women). Average age 27. 264 applicants, 65% accepted, 69 enrolled. In 2005, 77 degrees awarded. *Degree requirements:* For master's, thesis, practicum. *Entrance requirements:* For master's, GRE General Test. Additional exam requirements/recommendations for international students: Required—TOEFL (minimum score 550 paper-based; 213 computer-based). *Application deadline:* For fall admission, 1/15 priority date for domestic students, 1/1 priority date for international students. Application fee: $75. Electronic applications accepted. *Expenses:* Tuition: Full-time $14,400. Required fees: $217. *Financial support:* Fellowships with full and partial tuition reimbursements, career-related internships or fieldwork, Federal Work-Study, institutionally sponsored loans, and scholarships/grants available. Support available to part-time students. Financial award application deadline: 1/15. *Unit head:* Dr. Reynaldo Martorell, Chair, 404-727-9888, Fax: 404-727-1278, E-mail: rmart77@sph.emory.edu. *Application contact:* Suzanne Mason, Assistant Director of Academic Programs, 404-727-0263, Fax: 404-727-4590, E-mail: smason@sph.emory.edu.

Florida International University, College of Health and Urban Affairs, School of Public Health, Department of Dietetics and Nutrition, Miami, FL 33199. Offers MS, PhD. Part-time programs available. *Faculty:* 10 full-time (9 women). *Students:* 29 full-time (25 women), 11 part-time (8 women); includes 13 minority (4 African Americans, 1 Asian American or Pacific Islander, 8 Hispanic Americans), 11 international. Average age 31. 27 applicants, 59% accepted, 5 enrolled. In 2005, 12 master's, 3 doctorates awarded. *Degree requirements:* For master's and doctorate, thesis/dissertation. *Entrance requirements:* For master's, GRE General Test, minimum GPA of 3.0; for doctorate, GRE General Test. Additional exam requirements/recommendations for international students: Required—TOEFL. *Application deadline:* For fall admission, 4/1 for domestic students; for spring admission, 10/1 for domestic students. Applications are processed on a rolling basis. Application fee: $25. *Expenses:* Tuition, state resident: full-time $4,294; part-time $239 per credit. Tuition, nonresident: full-time $15,641; part-time $869 per credit. Required fees: $252; $126 per term. Tuition and fees vary according to program. *Financial support:* In 2005–06, 1 fellowship, 1 research assistantship, 1 teaching assistantship were awarded; career-related internships or fieldwork, Federal Work-Study, and institutionally sponsored loans also available. *Faculty research:* Clinical nutrition, cultural food habits, pediatric nutrition, diabetes, dietetic education. *Unit head:* Dr. Dian Weddle, Chairperson, 305-348-2878, Fax: 305-348-1996.

See Close-Up on page 1091.

Florida State University, Graduate Studies, College of Human Sciences, Department of Nutrition, Food, and Exercise Sciences, Tallahassee, FL 32306. Offers exercise science (PhD), including exercise physiology (MS, PhD), motor learning and control (MS, PhD); movement science (MS), including exercise physiology (MS, PhD), motor learning and control (MS, PhD); nutrition and food science (PhD); nutrition and food sciences (MS), including clinical nutrition, food science, nutrition and sport, nutrition science, nutrition education and health promotion. *Faculty:* 13 full-time (10 women). *Students:* 37 full-time (24 women), 28 part-time (20 women); includes 17 minority (9 African Americans, 4 Asian Americans or Pacific Islanders, 4 Hispanic Americans), 10 international. 67 applicants, 67% accepted, 23 enrolled. In 2005, 22 master's, 2 doctorates awarded. *Degree requirements:* For master's, thesis optional; for doctorate, thesis/dissertation, registration. *Entrance requirements:* For master's and doctorate, GRE General Test, minimum GPA of 3.0. Additional exam requirements/recommendations for international students: Required—TOEFL. *Application deadline:* For fall admission, 7/1 for domestic students, 5/1 for international students; for spring admission, 11/1 for domestic students, 12/1 for international students. Application fee: $30. Electronic applications accepted. *Financial support:* In 2005–06, 43 students received support, including 3 fellowships with partial tuition reimbursements available (averaging $10,000 per year), 9 research assistantships with partial tuition reimbursements available (averaging $8,000 per year), 22 teaching assistantships with partial tuition reimbursements available (averaging $8,000 per year); career-related internships or fieldwork, Federal Work-Study, institutionally sponsored loans, scholarships/grants, and unspecified assistantships also available. Financial award application deadline: 1/15; financial award applicants required to submit FAFSA. *Faculty research:* Nutrition and exercise, vitamin A deficiency, protein biochemistry, cardiovascular responses to exercises, physiological effects of cigarette smoking related to health and wellness. *Unit head:* Dr. Bahram Arjmandi, Chair, 850-644-1828, Fax: 850-645-5000. *Application contact:* Ursula Tate, Program Assistant, 850-644-4800, Fax: 850-645-5000, E-mail: utate@mailer.fsu.edu.

Framingham State College, Graduate Programs, Department of Chemistry and Food Science, Framingham, MA 01701-9101. Offers food science and nutrition science (MS). Part-time and evening/weekend programs available. *Entrance requirements:* For master's, GRE General Test.

Framingham State College, Graduate Programs, Program in Human Nutrition, Framingham, MA 01701-9101. Offers MS. *Accreditation:* ADtA.

Georgia State University, College of Health and Human Sciences, School of Health Professions, Division of Nutrition, Atlanta, GA 30303-3083. Offers MS, Certificate. Part-time and evening/weekend programs available. *Faculty:* 10 full-time (9 women), 1 (woman) part-time/adjunct. *Students:* 33 full-time (31 women), 3 part-time (all women); includes 6 minority (5 African Americans, 1 Hispanic American), 4 international. Average age 28. 61 applicants, 38% accepted, 18 enrolled. In 2005, 14 degrees awarded. *Degree requirements:* For master's, thesis optional. *Entrance requirements:* For master's, GRE or MAT. Additional exam requirements/recommendations for international students: Required—TOEFL (minimum score 550 paper-based; 213 computer-based). *Application deadline:* For fall admission, 5/15 for domestic students; for spring admission, 10/1 priority date for domestic students. Applications are processed on a rolling basis. Application fee: $50. Electronic applications accepted. *Expenses:* Tuition, state resident: full-time $4,368; part-time $182 per semester hour. Tuition, nonresident: full-time $8,732; part-time $728 per semester hour. Required fees: $46 per semester hour. *Financial support:* In 2005–06, research assistantships with full and partial tuition reimbursements (averaging $1,500 per year); Federal Work-Study, institutionally sponsored loans, scholarships/grants, tuition waivers (partial), and unspecified assistantships also available. Support available to part-time students. Financial award application deadline: 4/1; financial award applicants required to submit FAFSA. *Faculty research:* Food safety, nutrition education, sports nutrition, obesity, arthritis and nutrition. Total annual research expenditures: $56,000. *Unit head:* Dr. Mildred Cody, Director, 404-651-3085, E-mail: mcody@gsu.edu. *Application contact:* Bill Andrews, Senior Academic Advisor, 404-651-3064, Fax: 404-651-4871, E-mail: wandrews@gsu.edu.

Harvard University, School of Public Health, Department of Nutrition, Boston, MA 02115-6096. Offers nutrition (DPH, PhD, SD); nutritional epidemiology (DPH, SD); public health nutrition (DPH, SD). *Degree requirements:* For doctorate, thesis/dissertation, qualifying exam. *Entrance requirements:* For doctorate, GRE. Additional exam requirements/recommendations for international students: Required—TOEFL (minimum score 560 paper-based; 220 computer-based); Recommended—IELTS (minimum score 7). Electronic applications accepted.

Nutrition

Howard University, Graduate School of Arts and Sciences, Department of Nutritional Sciences, Washington, DC 20059-0002. Offers nutrition (MS, PhD). Part-time and evening/weekend programs available. *Faculty:* 4 full-time (2 women). *Students:* 6 full-time (5 women), 13 part-time (10 women); includes 14 minority (all African Americans), 3 international. In 2005, 3 master's, 3 doctorates awarded. *Median time to degree:* Of those who began their doctoral program in fall 1997, 89% received their degree in 8 years or less. *Degree requirements:* For master's and doctorate, thesis/dissertation, comprehensive exam. *Entrance requirements:* For master's and doctorate, GRE General Test, minimum GPA of 3.0. *Application deadline:* For fall admission, 2/15 for domestic students; for spring admission, 11/1 for domestic students. Applications are processed on a rolling basis. Application fee: $45. Electronic applications accepted. *Financial support:* In 2005–06, 2 fellowships with tuition reimbursements, 3 research assistantships with tuition reimbursements, 2 teaching assistantships were awarded; career-related internships or fieldwork, institutionally sponsored loans, scholarships/grants, traineeships, health care benefits, tuition waivers (full), and unspecified assistantships also available. Financial award application deadline: 4/1. *Faculty research:* Pregnancy outcome proteins, fiber, nutrition and cancer, nutrition education. Total annual research expenditures: $1.4 million. *Unit head:* Dr. Allan A. Johnson, Chair, 202-806-5666, Fax: 202-806-9233, E-mail: ajohnson@howard.edu. *Application contact:* Dr. Allan A. Johnson, Chair, 202-806-5666, Fax: 202-806-9233, E-mail: ajohnson@howard.edu.

Idaho State University, Office of Graduate Studies, Kasiska College of Health Professions, Department of Health and Nutrition Sciences, Pocatello, ID 83209. Offers dietetics (Certificate); health education (MHE); public health (MPH). Part-time programs available. *Degree requirements:* For master's, internship, thesis optional. *Entrance requirements:* For master's, GRE General Test (at least 30th percentile on each section), minimum GPA of 3.0 for upper division classes, 2 letters of recommendation. Additional exam requirements/recommendations for international students: Required—TOEFL (minimum score 600 paper-based; 213 computer-based). *Faculty research:* Epidemiology, environmental health, nutrition and aging, dietetics.

Immaculata University, College of Graduate Studies, Program in Nutrition Education, Immaculata, PA 19345. Offers nutrition education (MA); nutrition education/approved pre-professional practice program (MA). Part-time and evening/weekend programs available. *Students:* 17 full-time (all women), 39 part-time (38 women). Average age 34. 12 applicants, 83% accepted, 10 enrolled. In 2005, 8 degrees awarded. *Degree requirements:* For master's, thesis optional. *Entrance requirements:* For master's, GRE or MAT, minimum GPA of 3.0. *Application deadline:* Applications are processed on a rolling basis. Application fee: $30. Electronic applications accepted. *Expenses:* Tuition: Part-time $460 per credit hour. Tuition and fees vary according to course level and program. *Financial support:* Application deadline: 5/1. *Faculty research:* Sports nutrition, pediatric nutrition, changes in food consumption patterns in weight loss, nutritional counseling. *Unit head:* Dr. Laura Frank, Chair, 610-647-4400 Ext. 3482, E-mail: lfrank@immaculata.edu. *Application contact:* 610-647-4400 Ext. 3211, Fax: 610-993-8550, E-mail: graduate@immaculata.edu.

Indiana State University, School of Graduate Studies, College of Arts and Sciences, Department of Family and Consumer Sciences, Terre Haute, IN 47809-1401. Offers child development and family life (MS); clothing and textiles (MS); dietetics (MS); family and consumer sciences education (MS); nutrition and foods (MS). *Accreditation:* ADtA. Part-time programs available. *Faculty:* 7 full-time (6 women). *Students:* 9 full-time (all women), 14 part-time (13 women); includes 5 minority (3 African Americans, 2 Asian Americans or Pacific Islanders), 3 international. Average age 30. 13 applicants, 69% accepted, 6 enrolled. In 2005, 10 degrees awarded. *Degree requirements:* For master's, thesis optional. *Application deadline:* For fall admission, 7/1 for domestic students; for spring admission, 11/1 priority date for domestic students. Applications are processed on a rolling basis. Application fee: $20. Electronic applications accepted. *Expenses:* Tuition, state resident: full-time $6,288; part-time $262 per credit hour. Tuition, nonresident: full-time $12,504; part-time $521 per credit hour. *Financial support:* In 2005–06, 3 research assistantships with partial tuition reimbursements (averaging $2,869 per year) were awarded; teaching assistantships, tuition waivers (partial) also available. Financial award application deadline: 3/1; financial award applicants required to submit FAFSA. *Unit head:* Dr. Frederica Kramer, Chairperson, 812-237-3297.

Indiana University Bloomington, School of Health, Physical Education and Recreation, Department of Applied Health Science, Bloomington, IN 47405-7000. Offers health behavior (PhD); health promotion (MS); human development/family studies (MS); nutrition science (MS); public health (MPH); safety management (MS); school and college health education (MS). PhD offered through the University Graduate School. *Accreditation:* CEPH (one or more programs are accredited). Part-time programs available. *Faculty:* 16 full-time (7 women). *Students:* 25 full-time (18 women), 19 part-time (8 women); includes 1 minority (African American), 4 international. Average age 29.Terminal master's awarded for partial completion of doctoral program. *Degree requirements:* For master's, thesis optional; for doctorate, thesis/dissertation. *Entrance requirements:* For master's, GRE or minimum GPA of 2.8, 12 hours of course work in health education; for doctorate, GRE. *Application deadline:* Applications are processed on a rolling basis. Application fee: $50 ($60 for international students). *Expenses:* Tuition, state resident: full-time $5,437; part-time $227 per credit hour. Tuition, nonresident: full-time $15,836; part-time $660 per credit hour. Required fees: $821. Tuition and fees vary according to campus/location and program. *Financial support:* Fellowships, teaching assistantships with tuition reimbursements, career-related internships or fieldwork, Federal Work-Study, institutionally sponsored loans, scholarships/grants, tuition waivers (partial), and fee remissions available. Financial award application deadline: 3/1. *Faculty research:* Cancer education, HIV/AIDS and drug education, public health, parent-child interactions, safety education. *Unit head:* Dr. Mohammad R. Torabi, Chair, 812-855-4808. *Application contact:* William Yarber, Graduate Coordinator, 812-855-7974, Fax: 812-855-3936, E-mail: yarber@indiana.edu.

Indiana University of Pennsylvania, Graduate School and Research, College of Health and Human Services, Department of Food and Nutrition, Program in Food and Nutrition, Indiana, PA 15705-1087. Offers MS. Part-time programs available. *Degree requirements:* For master's, thesis optional. *Entrance requirements:* For master's, GRE General Test, 2 letters of recommendation. Additional exam requirements/recommendations for international students: Required—TOEFL.

Indiana University–Purdue University Indianapolis, Indiana University School of Medicine, School of Health and Rehabilitation Sciences, Program in Nutrition and Dietetics, Indianapolis, IN 46202-2896. Offers MS. *Students:* Average age 26. *Degree requirements:* For master's, thesis. *Entrance requirements:* For master's, GRE General Test, minimum GPA of 3.0. *Application deadline:* For fall admission, 1/15 for domestic students; for spring admission, 10/15 for domestic students. Applications are processed on a rolling basis. Application fee: $50 ($60 for international students). *Expenses:* Tuition, state resident: full-time $5,159; part-time $215 per credit hour. Tuition, nonresident: full-time $14,890; part-time $620 per credit hour. Required fees: $614. Tuition and fees vary according to campus/location and program. *Financial support:* Fellowships, research assistantships, career-related internships or fieldwork available. Support available to part-time students. *Faculty research:* Clinical nutrition–human applications. *Unit head:* Dr. Jacquelynn O'Palka, Department Chair, 317-278-0934.

Iowa State University of Science and Technology, Graduate College, College of Human Sciences and College of Agriculture, Department of Food Science and Human Nutrition, Ames, IA 50011. Offers food science and technology (MS, PhD); nutrition (MS, PhD). *Faculty:* 31 full-time, 1 part-time/adjunct. *Students:* 53 full-time (39 women), 6 part-time (3 women); includes 3 minority (all African Americans), 32 international. 57 applicants, 12% accepted, 7 enrolled. In 2005, 12 master's, 6 doctorates awarded. *Degree requirements:* For master's and doctorate, thesis/dissertation. *Entrance requirements:* For master's and doctorate, GRE General Test. Additional exam requirements/recommendations for international students: Required—TOEFL (paper score 550; computer score 213) or IELTS (score 6.5). *Application deadline:* For fall admission, 2/1 priority date for domestic students, 2/1 priority date for international students. Applications are processed on a rolling basis. Application fee: $50 ($70 for international students).

Electronic applications accepted. *Expenses:* Tuition, state resident: full-time $6,410. Tuition, nonresident: full-time $16,422. Tuition and fees vary according to program. *Financial support:* In 2005–06, 51 research assistantships with full and partial tuition reimbursements (averaging $15,144 per year) were awarded; fellowships, teaching assistantships, scholarships/grants also available. *Unit head:* Dr. Ruth S. MacDonald, Chair, 515-294-5991, Fax: 515-294-8181, E-mail: ruthmacd@iastate.edu. *Application contact:* Dr. Kevin Schalinske, Director of Graduate Education, 515-294-9230.

The Johns Hopkins University, Bloomberg School of Public Health, Department of International Health, Baltimore, MD 21218-2699. Offers disease prevention and control (MHS, PhD); health systems (MHS, PhD); human nutrition (MHS, PhD); international health (Dr PH); social and behavioral interventions (MHS, PhD). *Faculty:* 107 full-time (47 women), 254 part-time/adjunct. *Students:* 193 full-time (156 women), 9 part-time (4 women); includes 39 minority (3 African Americans, 1 American Indian/Alaska Native, 26 Asian Americans or Pacific Islanders, 9 Hispanic Americans), 58 international. Average age 27. 433 applicants, 42% accepted, 85 enrolled. In 2005, 47 master's, 9 doctorates awarded. *Median time to degree:* Of those who began their doctoral program in fall 1997, 71% received their degree in 8 years or less. *Degree requirements:* For master's, thesis (for some programs), internship, comprehensive exam, registration; for doctorate, thesis/dissertation, 1 year full-time residency, oral and written exams, comprehensive exam, registration. *Entrance requirements:* For master's, GRE General Test or MCAT, 3 letters of recommendation, resumé; for doctorate, GRE General Test, 3 letters of recommendation, resumé. Additional exam requirements/recommendations for international students: Required—TOEFL (minimum score 600 paper-based; 280 computer-based). *Application deadline:* For fall admission, 1/2 priority date for domestic students, 1/2 priority date for international students. Applications are processed on a rolling basis. Application fee: $45. Electronic applications accepted. *Expenses:* Tuition: Full-time $30,960. Tuition and fees vary according to degree level and program. *Financial support:* In 2005–06, 190 students received support, including 6 fellowships (averaging $30,000 per year); Federal Work-Study, institutionally sponsored loans, scholarships/grants, traineeships, and stipends also available. Support available to part-time students. Financial award application deadline: 3/15; financial award applicants required to submit FAFSA. *Faculty research:* Respiratory diseases, oral rehydration programs, vitamin A deficiency, infant malnutrition, health manpower planning. Total annual research expenditures: $59.4 million. *Unit head:* Dr. Robert E. Black, Chairman, 410-955-3934, Fax: 410-955-7159, E-mail: rblack@jhsph.edu. *Application contact:* Jennifer Shaffer, Academic Program Administrator, 410-955-3734, Fax: 410-955-7159, E-mail: jshaffer@jhsph.edu.

Kansas State University, Graduate School, College of Human Ecology, Department of Human Nutrition, Manhattan, KS 66506. Offers food science (MS, PhD); human nutrition (MS, PhD); public health (MS). Part-time programs available. *Faculty:* 11 full-time (5 women), 2 part-time/adjunct (both women). *Students:* 25 full-time (18 women), 13 part-time (10 women); includes 3 minority (1 African American, 1 American Indian/Alaska Native, 1 Asian American or Pacific Islander), 13 international. Average age 25. 19 applicants, 42% accepted, 2 enrolled. In 2005, 5 master's, 2 doctorates awarded. *Degree requirements:* For master's, thesis or alternative, residency; for doctorate, thesis/dissertation, residency. *Entrance requirements:* For master's, GRE General Test, minimum undergraduate GPA of 3.0; for doctorate, GRE General Test, minimum graduate GPA of 3.5, course work in biochemistry and statistics. Additional exam requirements/recommendations for international students: Required—TOEFL (minimum score 600 paper-based; 250 computer-based). *Application deadline:* For fall admission, 2/1 for domestic students, 2/1 for international students; for spring admission, 10/1 for domestic students, 8/1 for international students. Applications are processed on a rolling basis. Application fee: $30 ($55 for international students). Electronic applications accepted. *Expenses:* Tuition, state resident: full-time $5,160; part-time $215. Tuition, nonresident: full-time $12,816; part-time $534. Required fees: $564. *Financial support:* In 2005–06, 17 research assistantships (averaging $14,685 per year), 2 teaching assistantships with full tuition reimbursements (averaging $9,450 per year) were awarded; fellowships, career-related internships or fieldwork, Federal Work-Study, institutionally sponsored loans, scholarships/grants, and tuition waivers (full) also available. Support available to part-time students. Financial award application deadline: 3/1; financial award applicants required to submit FAFSA. *Faculty research:* Assessment of food portion size, bone mineral density and turnover in post menopausal women, weight maintenance in cancer prevention, dietary antioxidants and age related macular degeneration, food consumption and nutrient intakes of older women living alone. Total annual research expenditures: $940,352. *Unit head:* Dr. Denis Medeiros, Head, 785-532-0150, Fax: 785-532-3132, E-mail: medeiros@ksu.edu. *Application contact:* Janet Finney, Office Specialist, 785-532-5508, Fax: 785-532-3132, E-mail: nutrgrad@ksu.edu.

Kent State University, Graduate School of Education, Health, and Human Services, School of Family and Consumer Studies, Program in Nutrition, Kent, OH 44242-0001. Offers dietetic internship (MS); nutrition (MS). *Unit head:* Dr. Maureen Blankemeyer, Coordinator, 330-672-8610, E-mail: mblankem@kent.edu. *Application contact:* Nancy Miller, Academic Program Coordinator, Office of Graduate Student Services, 330-672-2576, Fax: 330-672-9162, E-mail: ogs@kent.edu.

Lehman College of the City University of New York, Division of Natural and Social Sciences, Department of Health Sciences, Program in Nutrition, Bronx, NY 10468-1589. Offers approved preprofessional practice (MS); clinical nutrition (MS); community nutrition (MS). *Degree requirements:* For master's, thesis or alternative.

Loma Linda University, School of Public Health, Department of Nutrition, Loma Linda, CA 92350. Offers public health nutrition (MPH, Dr PH). *Accreditation:* ADtA. *Faculty:* 12 full-time (6 women), 28 part-time/adjunct (23 women). *Students:* 3 full-time (all women), 2 part-time (both women); includes 1 Asian American or Pacific Islander, 3 international. *Degree requirements:* For doctorate, thesis/dissertation. *Entrance requirements:* For doctorate, GRE General Test. Additional exam requirements/recommendations for international students: Required—Michigan English Language Assessment Battery or TOEFL. *Application deadline:* Applications are processed on a rolling basis. Application fee: $100. *Financial support:* Application deadline: 5/15. *Faculty research:* Sports nutrition in minorities, dietary determinance of chronic disease, protein adequacy in vegetarian diets, relationship of dietary intake to hormone level. *Unit head:* Dr. Joan Sabaté, Chair, 909-824-4598, Fax: 909-824-4087. *Application contact:* Teri Tamayose, Director of Admissions and Academic Records, 909-824-4694, Fax: 909-824-4087, E-mail: ttamayose@sph.llu.edu.

Long Island University, C.W. Post Campus, School of Health Professions and Nursing, Department of Nutrition, Brookville, NY 11548-1300. Offers dietetic internship (Certificate); nutrition (MS). Part-time and evening/weekend programs available. *Degree requirements:* For master's, thesis. *Entrance requirements:* For master's, minimum GPA of 2.75 in major. Electronic applications accepted. *Faculty research:* Hematopoiesis, interleukins in allergy, growth factors effect in metastasis affecting behavioral change for nutrition.

Louisiana Tech University, Graduate School, College of Applied and Natural Sciences, School for Human Ecology, Ruston, LA 71272. Offers dietetics (MS); human ecology (MS). Part-time programs available. *Degree requirements:* For master's, thesis or alternative, Registered Dietician Exam eligibility. *Entrance requirements:* For master's, GRE General Test.

Marywood University, Academic Affairs, College of Health and Human Services, Department of Nutrition and Dietetics, Program in Nutrition, Scranton, PA 18509-1598. Offers MS. *Students:* 15 full-time (all women), 8 part-time (7 women). *Expenses:* Tuition: Part-time $643 per credit. Required fees: $370; $185 per term. $50 per term. *Application contact:* Deborah M. Flynn, Coordinator of Graduate Advising (Enrollment Management), 570-348-2322, E-mail: flynn@ac.marywood.edu.

See Close-Up on page 1093.

Marywood University, Academic Affairs, College of Health and Human Services, Department of Nutrition and Dietetics, Program in Sports Nutrition and Exercise Science, Scranton, PA 18509-1598. Offers MS. *Students:* 10 full-time (6 women), 5 part-time (3 women). *Expenses:* Tuition:

Nutrition

Marywood University *(continued)*
Part-time $643 per credit. Required fees: $370; $185 per term. $50 per term. *Application contact:* Deborah M. Flynn, Coordinator of Graduate Advising (Enrollment Management), 570-348-2322, E-mail: flynn@ac.marywood.edu.

See Close-Up on page 1093.

McGill University, Faculty of Graduate and Postdoctoral Studies, Faculty of Agricultural and Environmental Sciences, School of Dietetics and Human Nutrition, Montréal, QC H3A 2T5, Canada. Offers M Sc, M Sc A, PhD. Part-time programs available. Terminal master's awarded for partial completion of doctoral program. *Degree requirements:* For master's, thesis/dissertation, registration; for doctorate, thesis/dissertation, comprehensive exam, registration. *Entrance requirements:* For master's, GRE, minimum GPA of 3.2 during the last full-time terms of a completed bachelor's degree program in nutrition or a closely related field; for doctorate, M Sc, minimum GPA of 3.2. Additional exam requirements/recommendations for international students: Required—TOEFL (minimum score 560 paper-based; 220 computer-based), IELT (minimum score 7). Electronic applications accepted. *Faculty research:* Nutritional epidemiology, clinical nutrition, developmental nutrition, biochemistry and metabolism, study of indigenous foods.

Meredith College, John E. Weems Graduate School, Department of Human Environmental Sciences, Raleigh, NC 27607-5298. Offers nutrition (MS). *Faculty:* 4 full-time (3 women). *Students:* 8 full-time (all women), 16 part-time (15 women); includes 1 minority (African American), 1 international. Average age 28. 22 applicants, 68% accepted, 10 enrolled. In 2005, 4 degrees awarded. *Degree requirements:* For master's, thesis optional. *Entrance requirements:* For master's, GRE, recommendations, interview. Additional exam requirements/recommendations for international students: Required—TOEFL. *Application deadline:* For fall admission, 7/1 priority date for domestic students, 7/1 priority date for international students; for spring admission, 11/1 priority date for domestic students, 11/1 priority date for international students. Applications are processed on a rolling basis. Application fee: $50. Electronic applications accepted. *Expenses:* Contact institution. Part-time tuition and fees vary according to program. *Financial support:* Application deadline: 2/15; *Unit head:* Dr. Deborah Tippett, Head, 919-760-2355, Fax: 919-760-2819, E-mail: tipettd@meredith.edu. *Application contact:* Dr. William H. Landis, Director, 919-760-2355, Fax: 919-760-2819, E-mail: landisb@meredith.edu.

Michigan State University, The Graduate School, College of Agriculture and Natural Resources and College of Natural Science, Department of Food Science and Human Nutrition, East Lansing, MI 48824. Offers food science (MS, PhD); food science—environmental toxicology (PhD); human nutrition (MS, PhD); human nutrition-environmental toxicology (PhD). *Faculty:* 22 full-time (8 women). *Students:* 45 full-time (30 women), 6 part-time (all women); includes 6 minority (3 African Americans, 2 Asian Americans or Pacific Islanders, 1 Hispanic American), 23 international. Average age 30. 67 applicants, 12% accepted. In 2005, 12 master's, 7 doctorates awarded. *Degree requirements:* For master's, oral examination in defense of thesis, thesis optional; for doctorate, thesis/dissertation, 1 term assistant teaching, 1 year residency, oral presentation and defense of dissertation, comprehensive exam. *Entrance requirements:* For master's, GRE General Test, minimum GPA of 3.0 in last 2 years of undergraduate course work, 3 letters of recommendation; for doctorate, GRE General Test, minimum GPA of 3.0, 3 letters of recommendation. Additional exam requirements/recommendations for international students: Required—TOEFL (minimum score 550 paper-based; 213 computer-based), Michigan State University ELT (85), Michigan ELAB (83). *Application deadline:* For fall admission, 12/27 for domestic students. Application fee: $50. Electronic applications accepted. *Expenses:* Tuition, state resident: part-time $330 per credit hour. Tuition, nonresident: part-time $685 per credit hour. Tuition and fees vary according to program. *Financial support:* In 2005–06, 11 fellowships with tuition reimbursements (averaging $8,443 per year), 30 research assistantships with tuition reimbursements (averaging $12,420 per year), 2 teaching assistantships with tuition reimbursements (averaging $13,088 per year) were awarded; scholarships/grants and unspecified assistantships also available. *Faculty research:* Food safety and toxicology, food processing and quality enhancement, biochemical nutrition, community nutrition. Total annual research expenditures: $3.2 million. *Unit head:* Dr. Gale M. Strasburg, Chairperson, 517-355-8474 Ext. 100, Fax: 517-353-8963, E-mail: stragale@msu.edu. *Application contact:* Deborah Klein, Graduate Secretary, 517-355-8474 Ext. 118, Fax: 517-353-8963, E-mail: kleinde@msu.edu.

Middle Tennessee State University, College of Graduate Studies, College of Education and Behavioral Science, Department of Human Sciences, Murfreesboro, TN 37132. Offers child development and family studies (MS); nutrition and food science (MS). Part-time and evening/weekend programs available. Postbaccalaureate distance learning degree programs offered. *Degree requirements:* For master's, thesis, comprehensive exam. *Entrance requirements:* For master's, GRE or MAT. Additional exam requirements/recommendations for international students: Required—TOEFL (minimum score 525 paper-based; 195 computer-based). Electronic applications accepted. *Faculty research:* Courtship relationships, feminist methodology and epistemology in family studies, school uniforms, body fat in elderly, asynchronous distance education.

Mississippi State University, College of Agriculture and Life Sciences, Department of Food Science, Nutrition and Health Promotion, Mississippi State, MS 39762. Offers food science (PhD); food science, nutrition and health promotion (MS); nutrition (MS, PhD). *Faculty:* 13 full-time (5 women), 3 part-time/adjunct (all women). *Students:* 24 full-time (14 women), 14 part-time (8 women); includes 2 minority (1 African American, 1 Asian American or Pacific Islander), 17 international. Average age 28. 31 applicants, 29% accepted, 4 enrolled. In 2005, 10 master's, 2 doctorates awarded. *Degree requirements:* For master's and doctorate, thesis/dissertation, comprehensive exam, registration. *Entrance requirements:* For master's, GRE General Test, minimum GPA of 2.8; for doctorate, GRE General Test, minimum GPA of 3.0. Additional exam requirements/recommendations for international students: Required—TOEFL. *Application deadline:* For fall admission, 7/1 for domestic students; for spring admission, 11/1 for domestic students. Applications are processed on a rolling basis. Application fee: $30. Electronic applications accepted. *Expenses:* Tuition, state resident: full-time $4,312; part-time $240 per hour. Tuition, nonresident: full-time $9,772; part-time $543 per hour. International tuition: $10,102 full-time. Tuition and fees vary according to course load. *Financial support:* In 2005–06, 1 teaching assistantship with full tuition reimbursement (averaging $9,360 per year) was awarded; research assistantships with full tuition reimbursements, Federal Work-Study, institutionally sponsored loans, scholarships/grants, and unspecified assistantships also available. Financial award application deadline: 4/1; financial award applicants required to submit FAFSA. *Faculty research:* Food preservation, food chemistry, food safety, food processing, product development. Total annual research expenditures: $1.5 million. *Unit head:* Dr. Benjy Mikel, Head, 662-325-3200, Fax: 662-325-8728, E-mail: wbm50@ra.msstate.edu. *Application contact:* Philip G. Bonfanti, Director of Admissions, 662-325-4104, Fax: 662-325-8872, E-mail: admit@msstate.edu.

Montclair State University, The Graduate School, College of Education and Human Services, Department of Science and Physical Education Exercise, Montclair, NJ 07043-1624. Offers health and physical education (Certificate); health education (Certificate); nutrition and exercise science (Certificate); physical education (MA, Certificate), including coaching and sports administration (MA), exercise science (MA), physical education (MA), teaching and supervision of physical education (MA). Part-time and evening/weekend programs available. *Faculty:* 13 full-time (7 women), 12 part-time/adjunct (7 women). *Students:* 23 full-time (11 women), 35 part-time (8 women); includes 5 minority (2 African Americans, 1 Asian American or Pacific Islander, 2 Hispanic Americans). 29 applicants, 72% accepted, 14 enrolled. In 2005, 26 master's, 10 other advanced degrees awarded. *Degree requirements:* For master's, comprehensive exam. *Entrance requirements:* For master's, GRE General Test, 2 letters of recommendation; for Certificate, 2 letters of recommendation (nutrition and exercise science concentration). Additional exam requirements/recommendations for international students:

Required—TOEFL (minimum score 83 computer-based). *Application deadline:* Applications are processed on a rolling basis. Application fee: $60. Electronic applications accepted. *Expenses:* Tuition, state resident: part-time $409 per credit. Tuition, nonresident: part-time $604 per credit. Required fees: $56 per credit. Tuition and fees vary according to course load, degree level and program. *Financial support:* In 2005–06, 4 research assistantships with full tuition reimbursements (averaging $5,000 per year) were awarded; Federal Work-Study, scholarships/grants, and unspecified assistantships also available. Support available to part-time students. Financial award application deadline: 3/1; financial award applicants required to submit FAFSA. *Unit head:* Dr. Joseph Donnelly, Chairperson, 973-655-4154.

Montclair State University, The Graduate School, College of Education and Human Services, Department of Health and Nutrition Sciences, Montclair, NJ 07043-1624. Offers food safety instructor (Certificate); nutrition and food science (MS). *Faculty:* 11 full-time (6 women), 20 part-time/adjunct (14 women). *Students:* 6 full-time (all women), 14 part-time (13 women); includes 2 minority (both African Americans) 16 applicants, 75% accepted, 8 enrolled. In 2005, 3 master's, 2 other advanced degrees awarded. *Degree requirements:* For master's, thesis optional. *Entrance requirements:* For master's, GRE, 2 letters of recommendation. Additional exam requirements/recommendations for international students: Required—TOEFL (minimum score 83 computer-based). *Expenses:* Tuition: Full-time $3,001; part-time $409 per credit. Required fees: $56 per credit. Tuition and fees vary according to course load, degree level and program. *Financial support:* In 2005–06, 2 research assistantships (averaging $7,000 per year) were awarded *Faculty research:* Adolescent physical activity. Total annual research expenditures: $182,000. *Unit head:* Dr. Shahla Wunderlich, Chairperson, 973-655-6854, E-mail: wunderlichs@mail.montclair.edu.

Mount Mary College, Graduate Programs, Program in Dietetics, Milwaukee, WI 53222-4597. Offers administrative dietetics (MS); clinical dietetics (MS); nutrition education (MS). Part-time and evening/weekend programs available. *Faculty:* 1 (woman) full-time, 2 part-time/adjunct (1 woman). *Students:* 10 full-time (all women), 11 part-time (all women). Average age 23. 19 applicants, 53% accepted, 10 enrolled. In 2005, 1 degree awarded. *Degree requirements:* For master's, thesis. *Entrance requirements:* For master's, minimum GPA of 2.75, completion of ADA and DPD requirements. Additional exam requirements/recommendations for international students: Required—TOEFL (minimum score 500 paper-based; 173 computer-based). *Application deadline:* For fall admission, 2/15 for domestic students. Application fee: $35 ($75 for international students). *Expenses:* Tuition: Part-time $472 per credit. Required fees: $90; $45 per term. Tuition and fees vary according to program. *Financial support:* Career-related internships or fieldwork and Federal Work-Study available. Support available to part-time students. Financial award application deadline: 5/1; financial award applicants required to submit FAFSA. *Unit head:* Lisa Stark, Director, 414-258-4810 Ext. 398, E-mail: starkl@mtmary.edu.

Mount Saint Vincent University, Graduate Programs, Department of Applied Human Nutrition, Halifax, NS B3M 2J6, Canada. Offers M Sc AHN, MAHN. Part-time and evening/weekend programs available. *Faculty:* 6 full-time (5 women), 1 (woman) part-time/adjunct. *Students:* 7 full-time (all women), 14 part-time (all women). Average age 27. 63 applicants, 27% accepted, 6 enrolled. In 2005, 5 degrees awarded. *Degree requirements:* For master's, thesis (for some programs). *Entrance requirements:* For master's, bachelor's degree in related field, minimum GPA of 3.0, professional experience. *Application deadline:* For fall admission, 3/1 for domestic students. For winter admission, 11/1 for domestic students. Applications are processed on a rolling basis. Application fee: $50 Canadian dollars. Electronic applications accepted. Tuition and fees charges are reported in Canadian dollars. *Expenses:* Tuition, area resident: Part-time $1,466 Canadian dollars per credit. Required fees: $43 Canadian dollars per credit. One-time fee: $5 Canadian dollars. Tuition and fees vary according to course level, course load, campus/location and program. *Financial support:* In 2005–06, research assistantships (averaging $400 Canadian dollars per year); fellowships, career-related internships or fieldwork and scholarships/grants also available. Financial award application deadline: 5/1; financial award applicants required to submit FAFSA. *Unit head:* Linda L. Mann, Chairperson, 902-457-6146, Fax: 902-457-6134, E-mail: linda.mann@msvu.ca. *Application contact:* Karl Turner, Assistant Registrar/Admissions, 902-457-6117, Fax: 902-457-6498, E-mail: karl.turner@msvu.ca.

New York Institute of Technology, Graduate Division, School of Allied Health and Life Sciences, Program in Clinical Nutrition, Old Westbury, NY 11568-8000. Offers MS, DO/MS. Part-time and evening/weekend programs available. *Students:* 23 full-time (17 women), 12 part-time (9 women); includes 11 minority (5 African Americans, 4 Asian Americans or Pacific Islanders, 2 Hispanic Americans), 7 international. Average age 32. 51 applicants, 75% accepted, 16 enrolled. In 2005, 11 degrees awarded. *Degree requirements:* For master's, thesis (for some programs), comprehensive exam. *Entrance requirements:* For master's, minimum QPA of 2.85. Additional exam requirements/recommendations for international students: Required—TOEFL (minimum score 550 paper-based; 213 computer-based). *Application deadline:* For fall admission, 7/1 for domestic students; for spring admission, 12/1 priority date for domestic students. Applications are processed on a rolling basis. Application fee: $50. Electronic applications accepted. *Expenses:* Tuition: Full-time $33,654. Required fees: $600. Tuition and fees vary according to student level. *Financial support:* Fellowships, research assistantships with partial tuition reimbursements, career-related internships or fieldwork, institutionally sponsored loans, tuition waivers (full and partial), and unspecified assistantships available. Support available to part-time students. Financial award applicants required to submit FAFSA. *Faculty research:* Medical nutrition training. *Unit head:* Dr. Deborah Williams, Chair, 516-686-3803, E-mail: settinge@nyit.edu. *Application contact:* Jacquelyn Nealon, Dean of Admissions and Financial Aid, 516-686-7925, Fax: 516-686-7613, E-mail: jnealon@nyit.edu.

New York University, The Steinhardt School of Education, Department of Nutrition, Food Studies, and Public Health, Program in Nutrition and Dietetics, New York, NY 10012-1019. Offers clinical nutrition (MS); foods and nutrition (MS); nutrition and dietetics (PhD). Part-time and evening/weekend programs available. *Faculty:* 13 full-time (11 women). *Students:* 79 full-time (77 women), 105 part-time (103 women); includes 36 minority (11 African Americans, 17 Asian Americans or Pacific Islanders, 8 Hispanic Americans), 14 international. 111 applicants, 41% accepted, 33 enrolled. In 2005, 31 master's, 1 doctorate awarded. *Degree requirements:* For master's, thesis (for some programs); for doctorate, thesis/dissertation. *Entrance requirements:* For doctorate, GRE General Test, interview. Additional exam requirements/recommendations for international students: Required—TOEFL. *Application deadline:* For fall admission, 1/15 priority date for domestic students, 1/15 priority date for international students; for spring admission, 11/1 for domestic students, 11/1 for international students. Applications are processed on a rolling basis. Application fee: $50 ($60 for international students). *Financial support:* Fellowships with full and partial tuition reimbursements, career-related internships or fieldwork, Federal Work-Study, scholarships/grants, tuition waivers (partial), and unspecified assistantships available. Financial award application deadline: 2/1; financial award applicants required to submit FAFSA. *Faculty research:* Nutrition and race, childhood obesity and other eating disorders, nutritional epidemiology, nutrition policy, nutrition and health promotion. *Unit head:* Dr. Lisa Sasson, Director, 212-998-5580, Fax: 212-995-4194. *Application contact:* 212-998-5030, Fax: 212-995-4328, E-mail: grad.admissions@nyu.edu.

North Carolina Agricultural and Technical State University, Graduate School, School of Agriculture and Environmental and Allied Sciences, Department of Human Environment and Family Sciences/Food and Nutrition, Greensboro, NC 27411. Offers food and nutrition (MS). Part-time and evening/weekend programs available. *Degree requirements:* For master's, thesis or alternative, qualifying exam, comprehensive exam. *Entrance requirements:* For master's, GRE General Test, minimum GPA of 2.6.

North Carolina State University, Graduate School, College of Agriculture and Life Sciences and College of Veterinary Medicine, Program in Nutrition, Raleigh, NC 27695. Offers MN, MS, PhD. Part-time programs available. *Degree requirements:* For master's, thesis (for some programs); for doctorate, thesis/dissertation. *Entrance requirements:* For master's and doctorate, GRE General Test. Additional exam requirements/recommendations for international students: Required—TOEFL. Electronic applications accepted. *Faculty research:* Effects of food/feed ingredients and components on health and growth, community nutrition, waste management and reduction, experimental animal nutrition.

North Dakota State University, The Graduate School, College of Human Development and Education, Department of Health, Nutrition, and Exercise Sciences, Fargo, ND 58105. Offers entry level athletic training (MS); exercise science (MS); nutrition science (MS); public health (MS); sport pedagogy (MS); sports recreation management (MS). Part-time and evening/weekend programs available. *Faculty:* 12 full-time (6 women). *Students:* 28 full-time, 18 part-time; includes 6 minority (3 African Americans, 1 American Indian/Alaska Native, 2 Asian Americans or Pacific Islanders). 19 applicants, 100% accepted, 15 enrolled. In 2005, 4 degrees awarded. *Degree requirements:* For master's, thesis (for some programs), comprehensive exam (for some programs). *Entrance requirements:* For master's, minimum GPA of 3.0. *Application deadline:* Applications are processed on a rolling basis. Application fee: $45 ($60 for international students). Electronic applications accepted. *Financial support:* In 2005–06, 26 students received support, including 18 teaching assistantships with full tuition reimbursements available (averaging $6,500 per year) Financial award application deadline: 3/31. *Faculty research:* Biomechanics, sport specialization, recreation. Total annual research expenditures: $5,000. *Unit head:* Brad Strand, Head, 701-231-9718, Fax: 701-231-8872, E-mail: bradford.strand@ndsu.edu. *Application contact:* Pamela Hansen, 701-231-8093, Fax: 701-231-8872, E-mail: pamela.j.hansen@ndsu.edu.

Northern Illinois University, Graduate School, College of Health and Human Sciences, School of Family, Consumer and Nutrition Sciences, De Kalb, IL 60115-2854. Offers applied family and child studies (MS); nutrition and dietetics (MS). *Accreditation:* AAMFT/COAMFTE. Part-time programs available. *Faculty:* 16 full-time (14 women), 2 part-time/adjunct (1 woman). *Students:* 51 full-time (49 women), 36 part-time (35 women); includes 8 minority (3 African Americans, 2 Asian Americans or Pacific Islanders, 3 Hispanic Americans), 1 international. Average age 28. 91 applicants, 70% accepted, 38 enrolled. In 2005, 18 degrees awarded. *Degree requirements:* For master's, internship, thesis in nutrition and dietetics. *Entrance requirements:* For master's, GRE General Test, minimum GPA of 2.75. Additional exam requirements/recommendations for international students: Required—TOEFL (minimum score 550 paper-based; 213 computer-based). *Application deadline:* For fall admission, 6/1 for domestic students, 5/1 for international students; for spring admission, 11/1 for domestic students, 10/1 for international students. Applications are processed on a rolling basis. Application fee: $30. Electronic applications accepted. *Expenses:* Tuition, state resident: full-time $4,565; part-time $191 per credit hour. Tuition, nonresident: full-time $9,129; part-time $382 per credit hour. *Financial support:* In 2005–06, 17 research assistantships with full tuition reimbursements, 17 teaching assistantships with full tuition reimbursements were awarded; fellowships with full tuition reimbursements, career-related internships or fieldwork, Federal Work-Study, scholarships/grants, tuition waivers (full), and unspecified assistantships also available. Support available to part-time students. Financial award applicants required to submit FAFSA. *Faculty research:* Preliminary child development, hospitality administration in Asia, sports nutrition, eating disorders. *Unit head:* Dr. Laura Smart, Acting Chair, 815-753-1960, Fax: 815-753-1321, E-mail: lsmart@niu.edu.

The Ohio State University, Graduate School, College of Food, Agricultural, and Environmental Sciences, Program in Food Science and Nutrition, Columbus, OH 43210. Offers MS, PhD. *Degree requirements:* For master's, thesis optional; for doctorate, thesis/dissertation. *Entrance requirements:* For master's and doctorate, GRE General Test. Additional exam requirements/recommendations for international students: Required—TOEFL (paper 550; computer 213) or IELT (7) or Michigan English Language Assessment Battery (89). Electronic applications accepted.

The Ohio State University, Graduate School, College of Human Ecology, Department of Human Nutrition, Columbus, OH 43210. Offers food service management (MS, PhD); foods (MS, PhD); nutrition (MS, PhD). *Degree requirements:* For master's, thesis optional; for doctorate, thesis/dissertation. *Entrance requirements:* For master's and doctorate, GRE General Test. Additional exam requirements/recommendations for international students: Required—TOEFL (minimum score 577 paper-based; 233 computer-based). Electronic applications accepted.

The Ohio State University, Graduate School, College of Human Ecology, Program in Nutrition, Columbus, OH 43210. Offers PhD. *Degree requirements:* For doctorate, thesis/dissertation. *Entrance requirements:* For doctorate, GRE General Test. Additional exam requirements/recommendations for international students: Required—TOEFL (minimum score 573 paper-based; 230 computer-based). Electronic applications accepted.

Ohio University, Graduate Studies, College of Health and Human Services, School of Human and Consumer Sciences, Athens, OH 45701-2979. Offers child development and family life (MSHCS); food and nutrition (MSHCS). Part-time programs available. *Faculty:* 13 full-time (9 women), 5 part-time/adjunct (all women). *Students:* 11 full-time (9 women), 5 part-time (4 women); includes 5 minority (2 African Americans, 3 Asian Americans or Pacific Islanders). Average age 26. 14 applicants, 86% accepted, 5 enrolled. In 2005, 8 degrees awarded. *Degree requirements:* For master's, thesis. *Entrance requirements:* For master's, GRE. Additional exam requirements/recommendations for international students: Required—TOEFL. *Application deadline:* For fall admission, 8/30 for domestic students. Applications are processed on a rolling basis. Application fee: $45. *Financial support:* In 2005–06, 6 teaching assistantships were awarded; career-related internships or fieldwork, Federal Work-Study, and institutionally sponsored loans also available. Financial award application deadline: 3/15. *Faculty research:* Diversity, developmentally appropriate activities, death and dying, gerontology, sexuality education. *Unit head:* Dr. V. Ann Paulins, Director, 740-593-2880, Fax: 740-593-0289, E-mail: paulins@ohio.edu.

Oklahoma State University, College of Human Environmental Sciences, Department of Nutritional Sciences, Stillwater, OK 74078. Offers MS, PhD. *Faculty:* 19 full-time (16 women). *Students:* 33 full-time (29 women), 24 part-time (20 women); includes 9 minority (1 African American, 6 American Indian/Alaska Native, 1 Asian American or Pacific Islander, 1 Hispanic American), 18 international. Average age 30. 22 applicants, 59% accepted, 8 enrolled. In 2005, 9 degrees awarded. *Degree requirements:* For master's and doctorate, thesis/dissertation. *Entrance requirements:* For master's and doctorate, GRE. Additional exam requirements/recommendations for international students: Required—TOEFL. *Application deadline:* For fall admission, 7/1 priority date for domestic students, 3/1 priority date for international students. Applications are processed on a rolling basis. Application fee: $40 ($75 for international students). Electronic applications accepted. *Expenses:* Tuition, state resident: full-time $4,253; part-time $139 per credit hour. Tuition, nonresident: full-time $12,569; part-time $485 per credit hour. Required fees: $43 per credit hour. One-time fee: $20 part-time. Tuition and fees vary according to course load and program. *Financial support:* In 2005–06, 25 students received support, including 11 research assistantships (averaging 11,174 Albanian leks per year), 14 teaching assistantships (averaging 12,405 Albanian leks per year); career-related internships or fieldwork, Federal Work-Study, and tuition waivers (partial) also available. Support available to part-time students. Financial award application deadline: 3/1. *Faculty research:* Nutritional sciences, micronutrients and chronic disease, phytochemicals, nutrition education, osteoporosis, food service administration. *Unit head:* Dr. Nancy M. Betts, Head, 405-744-5040.

Oregon State University, Graduate School, College of Health and Human Sciences, Department of Nutrition and Food Management, Corvallis, OR 97331. Offers MAIS, MS, PhD. Part-time programs available. *Faculty:* 4 full-time (3 women), 4 part-time/adjunct (all women). *Students:* 14 full-time (13 women), 3 part-time (all women); includes 2 minority (both Asian Americans or Pacific Islanders), 2 international. Average age 38. In 2005, 3 master's awarded. *Degree requirements:* For master's and doctorate, thesis/dissertation. *Entrance requirements:* For master's and doctorate, minimum GPA of 3.0 in last 90 hours of course work. Additional exam requirements/recommendations for international students: Required—TOEFL. *Application deadline:* For fall admission, 3/1 for domestic students. Applications are processed on a

rolling basis. Application fee: $50. *Expenses:* Tuition, area resident: Part-time $301 per credit. Tuition, state resident: full-time $8,139; part-time $501 per credit. Tuition, nonresident: full-time $14,376; part-time $532 per credit. Required fees: $1,266. *Financial support:* Fellowships, research assistantships, teaching assistantships, career-related internships or fieldwork, Federal Work-Study, and institutionally sponsored loans available. Support available to part-time students. Financial award application deadline: 2/1. *Faculty research:* Human metabolic studies, trace minerals, food science, food management. *Unit head:* Dr. Anthony R. Wilcox, Head, 541-737-2643, Fax: 541-737-6914. *Application contact:* Laura Reid, Office Manager, 541-737-3561, Fax: 541-737-6914, E-mail: reidl@orst.edu.

The Pennsylvania State University University Park Campus, Graduate School, College of Health and Human Development, Department of Nutritional Sciences, State College, University Park, PA 16802-1503. Offers human nutrition (M Ed), including nutrition and public health (M Ed, MS), nutrition education (M Ed, MS); nutrition (MS, PhD), including applied human nutrition (MS), nutrition (PhD), nutrition and public health (M Ed, MS), nutrition education (M Ed, MS), nutrition science (MS). *Students:* 27 full-time (20 women), 2 part-time (1 woman); includes 1 minority (Asian American or Pacific Islander), 13 international. *Entrance requirements:* For master's and doctorate, GRE General Test or MCAT. *Expenses:* Tuition, state resident: full-time $12,518; part-time $522 per credit. Tuition, nonresident: full-time $23,004; part-time $959 per credit. Required fees: $484. Tuition and fees vary according to course load, campus/location and program. *Unit head:* Dr. Michael H. Green, Professor in Charge, 814-863-2914, Fax: 814-863-6103, E-mail: mhg@psu.edu.

The Pennsylvania State University University Park Campus, Graduate School, Intercollege Graduate Programs, Intercollege Graduate Program in Integrative Biosciences, State College, University Park, PA 16802-1503. Offers integrative biosciences (PhD), including biomolecular transport dynamics, cell and developmental biology, cellular and molecular mechanisms of toxicity, chemical biology, ecological and molecular plant physiology, immunobiology, molecular medicine, neuroscience, nutrition science. *Students:* 110 full-time (58 women), 2 part-time (both women); includes 6 minority (3 African Americans, 3 Asian Americans or Pacific Islanders), 61 international. *Entrance requirements:* For master's and doctorate, GRE General Test. Application fee: $45. *Expenses:* Tuition, state resident: full-time $12,518; part-time $522 per credit. Tuition, nonresident: full-time $23,004; part-time $959 per credit. Required fees: $484. Tuition and fees vary according to course load, campus/location and program. *Financial support:* Fellowships available. *Unit head:* Dr. Richard J. Frisque, Co-Director, 814-863-3523, Fax: 814-863-1357, E-mail: rjf6@psu.edu.

See Close-Up on page 197.

Purdue University, Graduate School, College of Consumer and Family Sciences, Department of Foods and Nutrition, West Lafayette, IN 47907. Offers nutrition (MS, PhD). *Faculty:* 15 full-time (7 women), 17 part-time/adjunct (14 women). *Students:* 39 full-time (32 women), 9 part-time (8 women); includes 1 minority (African American), 23 international. Average age 30. 49 applicants, 49% accepted, 10 enrolled. In 2005, 3 master's, 2 doctorates awarded. *Degree requirements:* For master's and doctorate, thesis/dissertation. *Entrance requirements:* For master's, GRE General Test (minimum 1000); for doctorate, GRE General Test. Additional exam requirements/recommendations for international students: Required—TOEFL (minimum score 600 paper-based). *Application deadline:* For fall admission, 1/16 for domestic students, 1/16 for international students. Applications are processed on a rolling basis. Application fee: $55. Electronic applications accepted. *Financial support:* Fellowships, research assistantships, teaching assistantships available. Support available to part-time students. Financial award applicants required to submit FAFSA. *Faculty research:* Nutrient requirements, nutrient metabolism, nutrition and disease prevention. *Unit head:* Dr. C. M. Weaver, Head, 765-494-8237, Fax: 765-494-0674. *Application contact:* Dawn Haan, Graduate Secretary, 765-494-8231, Fax: 765-494-0906, E-mail: mccammac@purdue.edu.

Rosalind Franklin University of Medicine and Science, College of Health Professions, Department of Nutrition, North Chicago, IL 60064-3095. Offers clinical education (MS); clinical nutrition (MS). Part-time and evening/weekend programs available. Postbaccalaureate distance learning degree programs offered (no on-campus study). *Degree requirements:* For master's, thesis optional. *Entrance requirements:* For master's, minimum GPA of 2.75, registered dietitian (RD), professional certificate or license. Expenses: Contact institution. *Faculty research:* Nutrition education, distance learning, computer-based graduate education, childhood obesity, nutrition medical education.

Rush University, College of Health Sciences, Department of Clinical Nutrition, Chicago, IL 60612-3832. Offers MS. Part-time programs available. *Faculty:* 33 part-time/adjunct (all women). *Students:* 18 full-time (all women), 3 part-time (all women). Average age 24. 37 applicants, 27% accepted, 10 enrolled. In 2005, 8 degrees awarded. *Degree requirements:* For master's, thesis. *Entrance requirements:* For master's, GRE General Test, minimum GPA of 3.0, course work in statistics, undergraduate didactic program approved by the American Dietetic Association. Additional exam requirements/recommendations for international students: Required—TOEFL. *Application deadline:* For winter admission, 2/15 for domestic students. Application fee: $40. *Financial support:* In 2005–06, 18 students received support. Career-related internships or fieldwork, Federal Work-Study, institutionally sponsored loans, scholarships/grants and stipends available. Support available to part-time students. Financial award applicants required to submit FAFSA. *Faculty research:* Food service management, chronic disease prevention/treatment, obesity, Alzheimer's. *Unit head:* Dr. Rebecca A. Dowling, Chairperson, 312-942-7075, Fax: 312-942-3075, E-mail: rebecca_dowling@rush.edu. *Application contact:* Dr. Linda Lafferty, Director, Dietetic Internship, 312-942-7845, Fax: 312-942-5203, E-mail: linda_lafferty@rush.edu.

Rutgers, The State University of New Jersey, New Brunswick/Piscataway, Graduate School, Program in Nutritional Sciences, New Brunswick, NJ 08901-1281. Offers MS, PhD. Part-time programs available. *Faculty:* 35 full-time. *Students:* 16 full-time (13 women), 13 part-time (all women); includes 3 minority (1 African American, 2 Hispanic Americans), 7 international. Average age 32. 55 applicants, 25% accepted, 6 enrolled. Terminal master's awarded for partial completion of doctoral program. *Degree requirements:* For master's and doctorate, thesis/dissertation. *Entrance requirements:* For master's and doctorate, GRE General Test. *Application deadline:* For fall admission, 4/1 for domestic students. Applications are processed on a rolling basis. Application fee: $50. *Expenses:* Tuition, state resident: full-time $10,440; part-time $435 per credit. Tuition, nonresident: full-time $15,520; part-time $647 per credit. Required fees: $129 per credit. Tuition and fees vary according to program. *Financial support:* In 2005–06, 13 students received support, including 2 fellowships with full tuition reimbursements available (averaging $17,000 per year), 5 research assistantships with full tuition reimbursements available (averaging $15,700 per year), 4 teaching assistantships with full tuition reimbursements available (averaging $16,988 per year); Federal Work-Study also available. Financial award application deadline: 1/15; financial award applicants required to submit FAFSA. *Faculty research:* Nutrition and gene expression, nutrition and disease (obesity, diabetes, cancer, osteoporosis, alcohol), community nutrition and nutrition education, cellular lipid transport and metabolism. *Unit head:* Dawn Brasaemle, Director, 732-932-6524, Fax: 732-932-6837, E-mail: brasaemle@aesop.rutgers.edu. *Application contact:* Suzy Kiefer, Graduate Program Secretary, 732-932-9439, Fax: 732-932-6837, E-mail: mkiefer@rci.rutgers.edu.

Sage Graduate School, Graduate School, Division of Health and Rehabilitation Sciences, Program in Nutrition, Troy, NY 12180-4115. Offers applied nutrition (MS). *Faculty:* 2 full-time (both women), 1 (woman) part-time/adjunct. *Students:* 6 full-time (5 women), 15 part-time (all women); includes 16 minority (2 Asian Americans or Pacific Islanders, 14 Hispanic Americans), 1 international. Average age 31. 18 applicants, 94% accepted, 17 enrolled. Application fee: $40. *Expenses:* Tuition: Full-time $9,270; part-time $515 per credit hour. *Unit head:* Rayane AbuSabha, Director of Didactic Program Dietetics, 518-244-2396, Fax: 518-244-4586, E-mail: abusar@sage.edu.

Nutrition

Saint Louis University, Graduate School, Doisy College of Health Sciences and Graduate School, Department of Nutrition and Dietetics, St. Louis, MO 63103-2097. Offers MS. Part-time programs available. *Faculty:* 7 full-time (5 women), 8 part-time/adjunct (7 women). *Students:* 31 full-time (28 women), 12 part-time (8 women); includes 4 minority (all African Americans), 4 international. Average age 27. 31 applicants, 90% accepted, 22 enrolled. In 2005, 7 degrees awarded. *Degree requirements:* For master's, comprehensive exam. *Entrance requirements:* For master's, GRE General Test, letters of recommendation, resumé. Additional exam requirements/recommendations for international students: Required—TOEFL (minimum score 550 paper-based; 213 computer-based). *Application deadline:* For fall admission, 7/1 for domestic students, 7/1 for international students; for spring admission, 11/1 for domestic students, 11/1 for international students. Applications are processed on a rolling basis. Application fee: $40. *Expenses:* Tuition: Part-time $760 per credit hour. Required fees: $55 per semester. *Financial support:* In 2005–06, 15 students received support; teaching assistantships with full tuition reimbursements available, career-related internships or fieldwork, health care benefits, and unspecified assistantships available. Financial award application deadline: 6/1; financial award applicants required to submit FAFSA. *Faculty research:* Hypertension prevention, weight management, geriatrics, sports nutrition, pediatrics. *Unit head:* Dr. Mildred Mattfeldt-Beman, Chairperson, 314-977-8523, Fax: 314-977-8520, E-mail: mattfield@slu.edu. *Application contact:* Gary Behrman, Associate Dean of the Graduate School, 314-977-3827, E-mail: behrmang@slu.edu.

San Diego State University, Graduate and Research Affairs, College of Professional Studies and Fine Arts, Department of Exercise and Nutritional Sciences, Program in Nutritional Science, San Diego, CA 92182. Offers nutritional sciences (MS). *Students:* 13 full-time (10 women), 5 part-time (all women); includes 3 minority (1 Asian American or Pacific Islander, 2 Hispanic Americans), 2 international. 54 applicants, 39% accepted, 11 enrolled. In 2005, 1 degree awarded. *Degree requirements:* For master's, thesis. *Entrance requirements:* For master's, GRE General Test, 2 letters of reference. Additional exam requirements/recommendations for international students: Required—TOEFL. *Application deadline:* For fall admission, 2/1 for domestic students, 2/1 for international students. Applications are processed on a rolling basis. Application fee: $55. Electronic applications accepted. *Financial support:* Teaching assistantships, career-related internships or fieldwork and unspecified assistantships available. Financial award applicants required to submit FAFSA. *Unit head:* Larry Verity, Head, 619-594-5979, Fax: 619-594-6553, E-mail: ensgrad@mail.sdsu.edu. *Application contact:* Larry Verity, Graduate Advisor, 619-594-6489, Fax: 619-594-6553, E-mail: ensgrad@mail.sdsu.edu.

San Jose State University, Graduate Studies and Research, College of Applied Sciences and Arts, Department of Nutrition and Food Science, San Jose, CA 95192-0001. Offers MS. *Students:* 72 full-time (66 women), 44 part-time (41 women); includes 30 minority (2 African Americans, 24 Asian Americans or Pacific Islanders, 4 Hispanic Americans), 19 international. Average age 32. 82 applicants, 63% accepted, 34 enrolled. In 2005, 11 degrees awarded. *Application deadline:* For fall admission, 6/27 for domestic students; for spring admission, 11/30 for domestic students. Applications are processed on a rolling basis. Application fee: $59. Electronic applications accepted. *Expenses:* Tuition, nonresident: part-time $339 per unit. Required fees: $1,286 per semester. Tuition and fees vary according to course load and degree level. *Financial support:* Applicants required to submit FAFSA. *Unit head:* Dr. Lucy McProud, Chair, 408-924-3100, Fax: 408-924-3114. *Application contact:* Dr. Pansilo Belo, Graduate Advisor, 408-924-3108.

Simmons College, School for Health Studies, Program in Nutrition and Health Promotion, Boston, MA 02115. Offers nutrition and health promotion (MS); sports nutrition (Certificate). Certificate program offered entirely online. Part-time and evening/weekend programs available. Postbaccalaureate distance learning degree programs offered. *Degree requirements:* For master's, research project. *Entrance requirements:* For master's, GRE, courses in community nutrition, human nutrition, introduction to nutrition, organic and inorganic chemistry, statistics, anatomy and physiology. Additional exam requirements/recommendations for international students: Required—TOEFL (minimum score 550 paper-based; 230 computer-based). Electronic applications accepted. Expenses: Contact institution. *Faculty research:* Nutrition supplements of athletes' nutrition and health-related behaviors of college-age women, glutamine supplementation in AIDS, cancer and vitamin A, dietary patterns and chronic disease development, gene interactions in obesity.

South Carolina State University, School of Graduate Studies, School of Applied Professional Sciences, Department of Family and Consumer Sciences, Orangeburg, SC 29117-0001. Offers individual and family development (MS); nutritional sciences (MS). Part-time and evening/weekend programs available. *Degree requirements:* For master's, departmental qualifying exam, thesis optional. *Entrance requirements:* For master's, GRE, MAT, or NTE, minimum GPA of 2.7. Electronic applications accepted. *Faculty research:* Societal competence, relationship of parent-child interaction to adult, quality of well-being of rural elders.

Southeast Missouri State University, School of Graduate Studies, Department of Health, Human Performance and Recreation, Cape Girardeau, MO 63701-4799. Offers community wellness and leisure services (MPA); nutrition and exercise science (MS). Part-time programs available. *Faculty:* 8 full-time (3 women). *Students:* 6 full-time (3 women), 10 part-time (6 women). Average age 28. 4 applicants, 100% accepted. In 2005, 4 degrees awarded. *Degree requirements:* For master's, thesis or alternative. *Entrance requirements:* For master's, GRE General Test (MS), minimum GPA of 3.0 (MS), minimum GPA of 2.7 (MPA). Additional exam requirements/recommendations for international students: Required—TOEFL (minimum score 550 paper-based; 213 computer-based). *Application deadline:* For fall admission, 8/1 for domestic students, 4/1 for international students; for spring admission, 11/21 for domestic students, 9/1 for international students. Applications are processed on a rolling basis. Application fee: $20 ($100 for international students). Electronic applications accepted. *Expenses:* Tuition, state resident: part-time $186 per hour. Tuition, nonresident: part-time $339 per hour. Required fees: $114; $13 per hour. Tuition and fees vary according to course load, degree level and campus/location. *Financial support:* In 2005–06, 9 students received support, including 1 research assistantship with full tuition reimbursement available (averaging $6,600 per year), 1 teaching assistantship with full tuition reimbursement available (averaging $6,600 per year); unspecified assistantships also available. Financial award applicants required to submit FAFSA. *Faculty research:* Health issues of athletes, body composition assessment, exercise training. *Unit head:* Dr. Joe Pujol, Chairperson, 573-651-2664, Fax: 573-651-5150, E-mail: jpujol@semo.edu. *Application contact:* Marsha L. Arant, Senior Administrative Assistant, Office of Graduate Studies, 573-651-2192, Fax: 573-651-2001, E-mail: marant@semo.edu.

Southern Illinois University Carbondale, Graduate School, College of Agriculture, Department of Animal Science, Food and Nutrition, Program in Food and Nutrition, Carbondale, IL 62901-4701. Offers MS. *Faculty:* 15 full-time (6 women). *Students:* 9 full-time (7 women), 10 part-time (7 women); includes 4 minority (1 African American, 2 Asian Americans or Pacific Islanders, 1 Hispanic American), 1 international. Average age 29. 9 applicants, 78% accepted, 1 enrolled. In 2005, 3 master's awarded. *Degree requirements:* For master's, thesis or alternative. *Entrance requirements:* For master's, minimum GPA of 2.7. Additional exam requirements/recommendations for international students: Required—TOEFL. *Application deadline:* Applications are processed on a rolling basis. Application fee: $0. *Financial support:* In 2005–06, 12 students received support; fellowships, research assistantships, teaching assistantships, career-related internships or fieldwork, Federal Work-Study, institutionally sponsored loans, and tuition waivers (full) available. Support available to part-time students. *Faculty research:* Public health nutrition, nutrition physiology, soybean utilization, nutrition education. Total annual research expenditures: $100,000. *Application contact:* Dr. Carol Boushey, Director, Dietetic Internship Program, 618-453-7514, Fax: 618-453-7517.

State University of New York at Buffalo, Graduate School, School of Public Health and Health Professions, Department of Exercise and Nutrition Sciences, Buffalo, NY 14260. Offers exercise science (MS, PhD); nutrition (MS). Part-time programs available. *Faculty:* 17 full-time (5 women), 7 part-time/adjunct (all women). *Students:* 61 full-time (46 women), 24 part-time (16 women); includes 13 minority (5 African Americans, 2 American Indian/Alaska Native, 5 Asian Americans or Pacific Islanders, 1 Hispanic American), 11 international. Average age 23. 110 applicants, 38% accepted, 30 enrolled. In 2005, 27 master's, 3 doctorates awarded. *Median time to degree:* Of those who began their doctoral program in fall 1997, 100% received their degree in 8 years or less. *Degree requirements:* For master's, comprehensive exam or thesis; for doctorate, thesis/dissertation, comprehensive exam. *Entrance requirements:* For master's, GRE General Test (nutrition), minimum GPA of 2.8; for doctorate, GRE General Test, minimum GPA of 2.8 (PhD). Additional exam requirements/recommendations for international students: Required—TOEFL (minimum score 550 paper-based; 213 computer-based). *Application deadline:* Applications are processed on a rolling basis. Application fee: $35. Electronic applications accepted. *Financial support:* In 2005–06, 10 students received support, including 10 teaching assistantships with full tuition reimbursements available (averaging $11,000 per year); research assistantships with tuition reimbursements available, career-related internships or fieldwork, Federal Work-Study, institutionally sponsored loans, scholarships/grants, health care benefits, tuition waivers (full and partial), unspecified assistantships, and stipends also available. Financial award application deadline: 3/15; financial award applicants required to submit FAFSA. *Faculty research:* Cardiovascular disease-diet and exercise, respiratory control and muscle function, plasticity of connective and neural tissue; exercise nutrition, diet and cancer. Total annual research expenditures: $369,000. *Unit head:* Dr. John X. Wilson, Chair, 716-829-2941 Ext. 208, Fax: 716-829-2428, E-mail: jxwilson@buffalo.edu. *Application contact:* Dr. Atif B. Awad, Director of Graduate Studies, 716-829-3680 Ext. 231, Fax: 716-829-3700, E-mail: awad@buffalo.edu.

Syracuse University, Graduate School, College of Human Services and Health Professions, Department of Nutrition Science and Food Management, Syracuse, NY 13244. Offers MA, MS. *Accreditation:* ADtA. Part-time programs available. *Students:* 21 full-time (20 women), 10 part-time (9 women); includes 3 minority (1 American Indian/Alaska Native, 2 Asian Americans or Pacific Islanders), 7 international. 26 applicants, 81% accepted, 9 enrolled. *Degree requirements:* For master's, thesis (for some programs). *Entrance requirements:* For master's, GRE General Test. Additional exam requirements/recommendations for international students: Required—TOEFL. *Application deadline:* For fall admission, 4/15 for domestic students; for spring admission, 11/1 for domestic students. Application fee: $65. Electronic applications accepted. *Financial support:* Fellowships with tuition reimbursements, research assistantships with full and partial tuition reimbursements, teaching assistantships with tuition reimbursements, tuition waivers (partial) available. *Unit head:* Dr. Norman Faiola, Chair, 315-443-4550, Fax: 315-443-2562, E-mail: nafaiola@syr.edu.

Teachers College Columbia University, Graduate Faculty of Education, Department of Health and Behavioral Studies, Program in Nutrition and Education, New York, NY 10027-6696. Offers nutrition education (Ed M, MS, Ed D); nutrition education and public health nutrition (Ed M, MS, Ed D), including community nutrition education (Ed M), nutrition and public health (MS, Ed D), nutrition education (MS, Ed D). Part-time and evening/weekend programs available. *Faculty:* 1 (woman) full-time, 3 part-time/adjunct. *Students:* 6 full-time (all women), 61 part-time (all women); includes 16 minority (1 African American, 1 American Indian/Alaska Native, 10 Asian Americans or Pacific Islanders, 4 Hispanic Americans), 4 international. Average age 32. 51 applicants, 76% accepted, 8 enrolled. In 2005, 8 master's, 1 doctorate awarded. Terminal master's awarded for partial completion of doctoral program. *Degree requirements:* For master's, integrative project, thesis optional; for doctorate, thesis/dissertation. *Entrance requirements:* For master's, GRE General Test or MAT, previous course work in science; for doctorate, GRE General Test, sample of written work, previous course work in science. *Application deadline:* For fall admission, 5/15 for domestic students; for spring admission, 12/1 for domestic students. Application fee: $50. *Expenses:* Tuition: Full-time $22,440; part-time $935 per credit. Required fees: $240 per term. *Financial support:* Fellowships, research assistantships, career-related internships or fieldwork, Federal Work-Study, institutionally sponsored loans, and tuition waivers (full and partial) available. Support available to part-time students. Financial award application deadline: 2/1. *Faculty research:* Psychosocial determinants of eating behavior, food supply and environmental education, development and evaluation of nutrition education. *Application contact:* Peter Shon, Assistant Director of Admission, 212-678-3305, Fax: 212-678-4171, E-mail: shon@exchange.tc.columbia.edu.

Texas A&M University, Intercollegiate Faculty of Nutrition, College Station, TX 77843. Offers MS, PhD. Part-time programs available. *Entrance requirements:* For master's and doctorate, GRE General Test. Additional exam requirements/recommendations for international students: Required—TOEFL. *Application deadline:* Applications are processed on a rolling basis. Application fee: $50 ($75 for international students). *Expenses:* Tuition, state resident: full-time $4,488; part-time $187 per credit hour. Tuition, nonresident: full-time $11,112; part-time $463 per credit hour. Required fees: $1,974. *Financial support:* Fellowships, research assistantships, teaching assistantships available. Financial award application deadline: 4/1; financial award applicants required to submit FAFSA. *Faculty research:* Cellular and molecular nutrition, domestic animal nutrition, applied human nutrition, food quality and nutrient composition. *Unit head:* Dr. Robert S. Chapkin, Chair, 979-845-1735, Fax: 979-862-2378, E-mail: r-chapkin@tamu.edu. *Application contact:* Joy C. McKenzie, Program Coordinator, 979-845-1735, Fax: 979-862-2378, E-mail: nutrsec@tamu.edu.

Texas Southern University, Graduate School, College of Liberal Arts and Behavioral Sciences, Department of Human Services and Consumer Sciences, Houston, TX 77004-4584. Offers human services and consumer sciences (MS), including child development, comprehensive human services and consumer sciences, foods and nutrition. Part-time and evening/weekend programs available. *Faculty:* 3 full-time (all women), 1 (woman) part-time/adjunct. *Students:* 9 full-time (7 women), 18 part-time (16 women); includes 26 minority (25 African Americans, 1 Hispanic American). Average age 33. 14 applicants, 93% accepted, 6 enrolled. In 2005, 6 degrees awarded. *Degree requirements:* For master's, thesis (for some programs), comprehensive exam. *Entrance requirements:* For master's, GRE General Test, minimum GPA of 2.5. Additional exam requirements/recommendations for international students: Required—TOEFL. *Application deadline:* For fall admission, 7/15 for domestic students. Applications are processed on a rolling basis. Application fee: $50 ($75 for international students). *Expenses:* Tuition, state resident: full-time $1,728; part-time $1,152 per credit hour. Tuition, nonresident: full-time $6,174; part-time $4,116 per credit hour. Required fees: $2,122. Tuition and fees vary according to course load and degree level. *Financial support:* Research assistantships, teaching assistantships, career-related internships or fieldwork and institutionally sponsored loans available. Financial award application deadline: 5/1. *Faculty research:* Food radiation/food for space travel, adolescent parenting, gerontology/grandparenting. *Unit head:* Dr. Shirley R. Nealy, Chair, 713-313-7638, Fax: 713-313-7228, E-mail: nealy_sr@tsu.edu.

Texas Tech University, Graduate School, College of Human Sciences, Department of Nutrition, Hospitality, and Retailing, Program in Food and Nutrition, Lubbock, TX 79409. Offers MS, PhD. Part-time programs available. *Students:* 16 full-time (all women), 6 part-time (all women); includes 1 minority (Hispanic American), 2 international. Average age 25. 24 applicants, 42% accepted, 4 enrolled. In 2005, 10 master's awarded. *Degree requirements:* For master's, thesis optional; for doctorate, thesis/dissertation. *Entrance requirements:* For master's and doctorate, GRE General Test. Additional exam requirements/recommendations for international students: Required—TOEFL (minimum score 550 paper-based; 213 computer-based). *Application deadline:* Applications are processed on a rolling basis. Application fee: $50 ($60 for international students). Electronic applications accepted. *Expenses:* Tuition, state resident: full-time $4,296. Tuition, nonresident: full-time $10,920. Required fees: $1,992. Tuition and fees vary according to program. *Financial support:* Research assistantships with partial tuition reimbursements, teaching assistantships with partial tuition reimbursements, career-related internships or fieldwork, Federal Work-Study, institutionally sponsored loans, and scholarships/grants available. Support available to part-time students. Financial award application deadline: 4/15; financial award applicants required to submit FAFSA. *Faculty research:* Assessment of nutritional status; nutritional status and health effects of selenium, vitamin E,

Nutrition

biotin, and iron; diabetes; obesity. *Unit head:* Dr. Brent Shriver, Graduate Advisor, 806-742-3068 Ext. 242, Fax: 806-742-3042, E-mail: brent.shriver@ttu.edu.

Texas Woman's University, Graduate School, College of Health Sciences, Department of Nutrition and Food Sciences, Denton, TX 76201. Offers exercise and sports nutrition (MS); food science (MS); institutional administration (MS); nutrition (MS, PhD). Part-time and evening/weekend programs available. *Students:* 65 full-time (63 women), 83 part-time (79 women); includes 40 minority (10 African Americans, 12 Asian Americans or Pacific Islanders, 18 Hispanic Americans), 19 international. Average age 29. In 2005, 30 master's, 3 doctorates awarded. *Degree requirements:* For master's, comprehensive exam; for doctorate, thesis/dissertation, qualifying exam, comprehensive exam. *Entrance requirements:* For master's, GRE General Test (verbal 350, quantitative 450), minimum GPA of 3.25, resumé; for doctorate, GRE General Test (verbal 450, quantitative 550), minimum GPA of 3.5, 2 letters of reference. Additional exam requirements/recommendations for international students: Required—TOEFL (minimum score 550 paper-based; 213 computer-based). *Application deadline:* Applications are processed on a rolling basis. Application fee: $30 ($50 for international students). Electronic applications accepted. *Expenses:* Tuition, state resident: full-time $2,934; part-time $163. Tuition, nonresident: full-time $7,974; part-time $152. *Financial support:* In 2005–06, 18 research assistantships (averaging $10,206 per year), 5 teaching assistantships (averaging $10,206 per year) were awarded; career-related internships or fieldwork, Federal Work-Study, institutionally sponsored loans, scholarships/grants, traineeships, health care benefits, and unspecified assistantships also available. Support available to part-time students. Financial award application deadline: 3/1; financial award applicants required to submit FAFSA. *Faculty research:* Food science, food safety, clinical nutrition, nutrition and cancer, weight management. *Unit head:* Dr. Carolyn M. Bednar, Interim Chair, 940-898-2636, Fax: 940-898-2634, E-mail: cbednar@twu.edu. *Application contact:* Samuel Wheeler, Coordinator of Graduate Admissions, 940-898-3188, Fax: 940-898-3081, E-mail: wheelersr@twu.edu.

Tufts University, The Gerald J. and Dorothy R. Friedman School of Nutrition Science and Policy, Medford, MA 02155. Offers humanitarian assistance (MAHA); nutrition (MS, PhD). Part-time programs available. *Faculty:* 12 full-time (8 women), 98 part-time/adjunct (45 women). *Students:* 194 full-time (165 women), 30 part-time (24 women); includes 21 minority (1 African American, 12 Asian Americans or Pacific Islanders, 8 Hispanic Americans), 24 international. Average age 29. 220 applicants, 55% accepted, 69 enrolled. In 2005, 42 master's, 9 doctorates awarded. *Median time to degree:* Of those who began their doctoral program in fall 1997, 96% received their degree in 8 years or less. *Degree requirements:* For doctorate, thesis/dissertation, comprehensive exam. *Entrance requirements:* For master's and doctorate, GRE General Test. Additional exam requirements/recommendations for international students: Required—TOEFL. *Application deadline:* For fall admission, 1/15 priority date for domestic students, 1/15 priority date for international students. Applications are processed on a rolling basis. Application fee: $65. Electronic applications accepted. *Expenses:* Contact institution. Tuition and fees vary according to program. *Financial support:* In 2005–06, 169 students received support, including 3 fellowships with full and partial tuition reimbursements available (averaging $20,000 per year), 36 research assistantships with full and partial tuition reimbursements available (averaging $20,000 per year), teaching assistantships with full and partial tuition reimbursements available (averaging $2,000 per year); career-related internships or fieldwork, Federal Work-Study, institutionally sponsored loans, scholarships/grants, traineeships, health care benefits, and unspecified assistantships also available. Support available to part-time students. Financial award applicants required to submit FAFSA. *Faculty research:* Nutritional biochemistry and metabolism, cell and molecular biochemistry, epidemiology, policy/planning, applied nutrition. Total annual research expenditures: $6.8 million. *Unit head:* Stacey M. Herman, Director of Student Affairs, 617-636-3711, Fax: 617-636-3600, E-mail: stacey.herman@tufts.edu. *Application contact:* Kristina S. Bonanno, Coordinator of Admissions and Recruitment, 617-636-3777, Fax: 617-636-3600, E-mail: nutritionadmissions@tufts.edu.

Tulane University, School of Public Health and Tropical Medicine, Department of Community Health Sciences, Program in Nutrition, New Orleans, LA 70118-5669. Offers MPH. *Degree requirements:* For master's, comprehensive exam. *Entrance requirements:* For master's, GRE General Test. Additional exam requirements/recommendations for international students: Required—TOEFL.

Tuskegee University, Graduate Programs, College of Agricultural, Environmental and Natural Sciences, Department of Food and Nutritional Sciences, Tuskegee, AL 36088. Offers MS. *Faculty:* 4 full-time (3 women). *Students:* 15 full-time (13 women), 2 part-time (1 woman); includes 14 minority (all African Americans), 1 international. Average age 30. In 2005, 8 degrees awarded. *Degree requirements:* For master's, thesis. *Entrance requirements:* For master's, GRE General Test. Additional exam requirements/recommendations for international students: Required—TOEFL (minimum score 500 paper-based; 173 computer-based). *Application deadline:* For fall admission, 7/15 for domestic students. Applications are processed on a rolling basis. Application fee: $25 ($35 for international students). *Expenses:* Tuition: Full-time $12,400. Required fees: $300; $490 per credit. *Financial support:* Application deadline: 4/15. *Unit head:* Dr. Ralphenia Pace, Head, 334-727-8162.

Université de Moncton, School of Food Science, Nutrition and Family Studies, Moncton, NB E1A 3E9, Canada. Offers foods/nutrition (M Sc). Part-time programs available. *Faculty:* 4 full-time (2 women). *Students:* 7 full-time (4 women). 3 applicants, 0% accepted. In 2005, 4 degrees awarded. *Degree requirements:* For master's, one foreign language, thesis. *Entrance requirements:* For master's, previous course work in statistics. *Application deadline:* For fall admission, 6/1 priority date for domestic students, 2/1 priority date for international students. For winter admission, 11/15 for domestic students; for spring admission, 3/31 for domestic students. Applications are processed on a rolling basis. Application fee: $39. Electronic applications accepted. *Financial support:* In 2005–06, 3 research assistantships were awarded; fellowships, career-related internships or fieldwork and scholarships/grants also available. Financial award application deadline: 5/23. *Faculty research:* Clinic nutrition (anemia, elderly, osteoporosis), applied nutrition, metabolic activities of lactic bacteria, solubility of low density lipoproteins, bile acids. *Unit head:* Regina M. Robichaud, Director, 506-858-4003, Fax: 506-858-4283, E-mail: robichr@umoncton.ca.

Université de Montréal, Faculty of Medicine and Faculty of Graduate Studies, Graduate Programs in Medicine, Department of Nutrition, Montréal, QC H3C 3J7, Canada. Offers M Sc, PhD. *Faculty:* 18 full-time (10 women), 1 (woman) part-time/adjunct. *Students:* 62 full-time (42 women), 10 part-time (9 women). 91 applicants, 20% accepted, 15 enrolled. In 2005, 7 master's, 2 doctorates awarded. Terminal master's awarded for partial completion of doctoral program. *Degree requirements:* For master's, thesis; for doctorate, thesis/dissertation, general exam. *Entrance requirements:* For master's, MD, B Sc in nutrition or equivalent, proficiency in French; for doctorate, M Sc in nutrition or equivalent, proficiency in French. *Application deadline:* For fall and spring admission, 2/1. For winter admission, 11/1 for domestic students. Application fee: $30. Electronic applications accepted. *Faculty research:* Nutritional aspects of diabetes, obesity, anorexia nervosa, lipid metabolism, hepatic function. *Unit head:* Dominique Garrel, Director, 514-343-6401, Fax: 514-343-7395. *Application contact:* Olivier Receveur, Assistant Director of Graduate Studies, 514-343-7023, Fax: 514-343-7395.

Université Laval, Faculty of Agricultural and Food Sciences, Department of Food Sciences and Nutrition, Programs in Nutrition, Québec, QC G1K 7P4, Canada. Offers M Sc, PhD. Terminal master's awarded for partial completion of doctoral program. *Degree requirements:* For master's, thesis/dissertation; for doctorate, thesis/dissertation, comprehensive exam. *Entrance requirements:* For master's and doctorate, knowledge of French and English. Electronic applications accepted.

The University of Akron, Graduate School, College of Fine and Applied Arts, School of Family and Consumer Sciences, Program in Nutrition and Dietetics, Akron, OH 44325. Offers MS. *Students:* 4 full-time (all women), 1 (woman) part-time, 2 international. Average age 26. 4 applicants, 75% accepted, 3 enrolled. In 2005, 1 degree awarded. *Degree requirements:* For master's, thesis or project, thesis optional. *Entrance requirements:* For master's, GRE General

Test, minimum GPA of 2.75, letters of recommendation. Additional exam requirements/recommendations for international students: Required—TOEFL (minimum score 550 paper-based; 213 computer-based), Michigan English Language Assessment Battery. *Application deadline:* For fall admission, 8/15 for domestic students. Applications are processed on a rolling basis. Application fee: $30 ($40 for international students). Electronic applications accepted. *Expenses:* Tuition, state resident: full-time $5,816; part-time $323 per credit. Tuition, nonresident: full-time $9,976; part-time $554 per credit. Required fees: $794; $43 per credit. $12 per term. Tuition and fees vary according to course load, degree level and program. *Unit head:* Dr. Deborah Marino, Associate Professor, 330-972-6322, E-mail: debora7@uakron.edu.

The University of Alabama, Graduate School, College of Human Environmental Sciences, Department of Human Nutrition and Hospitality Management, Tuscaloosa, AL 35487. Offers MSHES. Part-time programs available. Postbaccalaureate distance learning degree programs offered (no on-campus study). *Faculty:* 4 full-time (2 women), 1 (woman) part-time/adjunct. *Students:* 17 full-time (16 women), 31 part-time (28 women); includes 7 minority (2 African Americans, 2 Asian Americans or Pacific Islanders, 3 Hispanic Americans), 6 international. Average age 30. 27 applicants, 85% accepted, 18 enrolled. In 2005, 22 master's awarded. *Degree requirements:* For master's, thesis optional. *Entrance requirements:* For master's, minimum GPA of 3.0. Additional exam requirements/recommendations for international students: Required—TOEFL. *Application deadline:* For fall admission, 7/6 for domestic students. Applications are processed on a rolling basis. Application fee: $25. Electronic applications accepted. *Expenses:* Tuition, state resident: full-time $4,864; part-time $382 per hour. Tuition, nonresident: full-time $13,516; part-time $796 per hour. *Financial support:* In 2005–06, 4 students received support, including 2 research assistantships (averaging $8,100 per year), 2 teaching assistantships (averaging $8,100 per year); career-related internships or fieldwork also available. Financial award application deadline: 3/15. *Faculty research:* Fat determination of low-fat foods, maternal and child nutrition, obesity and eating disorders, community nutrition interventions. *Unit head:* Dr. Olivia W. Kendrick, Chair and Associate Professor, 205-348-6150, Fax: 205-348-3789, E-mail: okendric@ches.ua.edu.

The University of Alabama at Birmingham, School of Health Related Professions, Department of Nutrition Sciences, Division of Clinical Nutrition and Dietetics, Birmingham, AL 35294. Offers clinical nutrition (MS); dietetic internship (Certificate). Part-time programs available. *Students:* 24 full-time (23 women), 4 part-time (all women); includes 6 minority (1 African American, 3 Asian Americans or Pacific Islanders, 2 Hispanic Americans), 1 international. 45 applicants, 67% accepted. In 2005, 6 master's awarded. *Degree requirements:* For master's, thesis. *Entrance requirements:* For master's, GRE General Test or MAT, bachelor's degree in dietetics or related field; for Certificate, GRE General Test or MAT, bachelor's degree in dietetics. *Application deadline:* For fall admission, 7/1 for domestic students. Applications are processed on a rolling basis. Application fee: $35 ($60 for international students). Electronic applications accepted. *Expenses:* Tuition, state resident: part-time $170 per credit hour. Tuition, nonresident: full-time $4,612; part-time $425 per credit hour. International tuition: $10,732 full-time. Required fees: $11 per credit hour. $124 per term. Tuition and fees vary according to course load, degree level and program. *Financial support:* In 2005–06, 5 students received support, including 5 research assistantships; career-related internships or fieldwork also available. *Faculty research:* Clinical assessment, folic acid, energy metabolism, nutrition and cancer, nutrition for children and adolescents with special health care needs. *Unit head:* Dr. Sarah L. Morgan, Director, 205-934-3006, Fax: 205-934-7049, E-mail: smorgan@uab.edu.

The University of Alabama at Birmingham, School of Health Related Professions, Department of Nutrition Sciences, Program in Nutrition Sciences, Birmingham, AL 35294. Offers PhD. *Students:* 11 full-time (9 women); includes 1 minority (American Indian/Alaska Native), 3 international. 11 applicants, 45% accepted. In 2005, 3 doctorates awarded. *Degree requirements:* For doctorate, thesis/dissertation. *Entrance requirements:* For doctorate, GRE General Test. *Application deadline:* Applications are processed on a rolling basis. Application fee: $35 ($60 for international students). Electronic applications accepted. *Expenses:* Tuition, state resident: part-time $170 per credit hour. Tuition, nonresident: full-time $4,612; part-time $425 per credit hour. International tuition: $10,732 full-time. Required fees: $11 per credit hour. $124 per term. Tuition and fees vary according to course load, degree level and program. *Financial support:* In 2005–06, 6 fellowships with tuition reimbursements, 1 research assistantship with tuition reimbursement were awarded; career-related internships or fieldwork also available. *Faculty research:* Energy metabolism, obesity, body composition, cancer prevention, bone metabolism. Total annual research expenditures: $2 million. *Unit head:* Dr. Timothy R. Nagy, Interim Chair, 205-934-4088, E-mail: tnagy@uab.edu.

University of Alaska Anchorage, Community and Technical College, Dietetic Internship Program, Anchorage, AK 99508-8060. Offers Certificate. *Students:* 4 full-time (3 women). *Entrance requirements:* Additional exam requirements/recommendations for international students: Required—TOEFL (minimum score 550 paper-based; 213 computer-based). *Application deadline:* For fall admission, 2/15 for domestic students, 2/15 for international students. *Unit head:* Dr. Jon Gehler, Dean, 907-786-6400, Fax: 907-786-6401. *Application contact:* Elisa S. Mattison, Coordinator for Graduate Studies, 907-786-1096, Fax: 907-786-1021, E-mail: emattison@uaa.alaska.edu.

The University of Arizona, Graduate College, Graduate Interdisciplinary Programs, Graduate Interdisciplinary Program in Nutritional Sciences, Tucson, AZ 85721. Offers dietetics (MS); epidermalogical nutrition/public health nutrition (PhD); human/clinical nutrition (PhD); molecular nutrition (PhD); nutritional biochemistry (MS, PhD). Part-time programs available. *Degree requirements:* For master's and doctorate, thesis/dissertation, registration. *Entrance requirements:* For master's and doctorate, GRE, minimum GPA of 3.0. Additional exam requirements/recommendations for international students: Required—TOEFL. *Faculty research:* International nutrition, nutrition and aging, vitamin metabolism.

University of Arkansas for Medical Sciences, Graduate School, Program in Clinical Nutrition, Little Rock, AR 72205-7199. Offers MS. Part-time programs available. *Faculty:* 9 full-time (6 women), 1 (woman) part-time/adjunct. *Students:* 3 full-time, 9 part-time. *Degree requirements:* For master's, thesis. Application fee: $0. *Financial support:* Research assistantships available. Support available to part-time students. *Unit head:* Dr. Reza Hakkak, Director, 501-686-5715, E-mail: hakkakreza@uams.edu.

University of Bridgeport, Nutrition Institute, Bridgeport, CT 06604. Offers human nutrition (MS). Part-time and evening/weekend programs available. Postbaccalaureate distance learning degree programs offered (no on-campus study). *Faculty:* 1 full-time (0 women), 16 part-time/adjunct (7 women). *Students:* 6 full-time (all women), 127 part-time (92 women); includes 14 minority (2 African Americans, 5 Asian Americans or Pacific Islanders, 7 Hispanic Americans), 18 international. Average age 37. 283 applicants, 53% accepted, 132 enrolled. In 2005, 65 degrees awarded. *Degree requirements:* For master's, thesis, research project. *Entrance requirements:* For master's, previous course work in anatomy, biochemistry, organic chemistry, or physiology. *Application deadline:* For fall admission, 8/1 for domestic students; for spring admission, 12/1 priority date for domestic students. Applications are processed on a rolling basis. Application fee: $25 ($35 for international students). Electronic applications accepted. *Expenses:* Contact institution. Tuition and fees vary according to degree level and program. *Financial support:* In 2005–06, 33 students received support. Available to part-time students. Application deadline: 6/1; *Unit head:* Dr. David M. Brady, Director, 203-576-4667, Fax: 203-576-4591, E-mail: dbrady@bridgeport.edu.

The University of British Columbia, Faculty of Graduate Studies, Faculty of Land and Food Systems, Human Nutrition Program, Vancouver, BC V6T 1Z1, Canada. Offers M Sc, PhD. *Faculty:* 4 full-time (3 women), 1 (woman) part-time/adjunct. *Students:* 36 full-time (32 women). 26 applicants, 15% accepted, 3 enrolled. In 2005, 4 degrees awarded. *Degree requirements:* For master's, thesis/dissertation, registration; for doctorate, thesis/dissertation, comprehensive exam, registration. *Entrance requirements:* Additional exam requirements/recommendations for international students: Required—TOEFL (minimum score 577 paper-based; 233 computer-

Nutrition

The University of British Columbia (continued)
based). *Application deadline:* For fall admission, 2/1 for domestic students, 1/1 for international students. For winter admission, 8/1 for domestic students; for spring admission, 10/1 for domestic students. Applications are processed on a rolling basis. Application fee: $90 Canadian dollars ($150 Canadian dollars for international students). Electronic applications accepted. *Financial support:* In 2005–06, 7 fellowships (averaging $12,195 per year) were awarded; institutionally sponsored loans, scholarships/grants, and tuition waivers (partial) also available. *Faculty research:* Basic nutrition, clinical nutrition, community nutrition, women's health, pediatric nutrition. Total annual research expenditures: $655,645. *Unit head:* Dr. Susan Barr, Graduate Advisor, 604-822-6766, Fax: 604-822-4400, E-mail: gradapp@interchange.ubc.ca. *Application contact:* Graduate Programs Manager, 604-822-4593, Fax: 604-822-4400, E-mail: gradapp@interchange.ubc.ca.

University of California, Berkeley, Graduate Division, Group in Molecular and Biochemical Nutrition, Berkeley, CA 94720-1500. Offers PhD. *Degree requirements:* For doctorate, thesis/dissertation, qualifying exam. *Entrance requirements:* For doctorate, GRE General Test, minimum GPA of 3.0. Additional exam requirements/recommendations for international students: Required—TOEFL. Electronic applications accepted. *Faculty research:* Regulation of metabolism, nutritional genomics and nutrient-gene interactions, transport, metabolism and function of minerals, carcinogenesis and dietary anti-carcinogens.

See Close-Up on page 1095.

University of California, Berkeley, Graduate Division, School of Public Health, Master's Internationalist Program, Berkeley, CA 94720-1500. Offers community health education (MPH); epidemiology (MPH); interdisciplinary (MPH); maternal and child health (MPH); public health nutrition (MPH). *Accreditation:* CEPH. *Entrance requirements:* For master's, GRE General Test, minimum GPA of 3.0.

University of California, Davis, Graduate Studies, Graduate Group in Nutritional Biology, Davis, CA 95616. Offers MS, PhD. *Faculty:* 68 full-time. *Students:* 103 full-time (84 women); includes 16 minority (3 African Americans, 1 American Indian/Alaska Native, 9 Asian Americans or Pacific Islanders, 3 Hispanic Americans), 20 international. Average age 31. 110 applicants, 31% accepted, 23 enrolled. In 2005, 4 master's, 7 doctorates awarded. *Median time to degree:* Of those who began their doctoral program in fall 1997, 46.7% received their degree in 8 years or less. *Degree requirements:* For master's and doctorate, thesis/dissertation. *Entrance requirements:* For master's, GRE General Test, minimum GPA of 3.0; for doctorate, GRE General Test, minimum GPA of 30. Additional exam requirements/recommendations for international students: Required—TOEFL (minimum score 550 paper-based; 213 computer-based). *Application deadline:* 1/15 for domestic students, 1/15 for international students. Applications are processed on a rolling basis. Application fee: $60. Electronic applications accepted. *Financial support:* In 2005–06, 90 students received support, including 36 fellowships with full and partial tuition reimbursements available (averaging $8,959 per year), 37 research assistantships with full and partial tuition reimbursements available (averaging $13,214 per year), 13 teaching assistantships with partial tuition reimbursements available (averaging $15,082 per year); Federal Work-Study, scholarships/grants, tuition waivers (full and partial), and unspecified assistantships also available. Support available to part-time students. Financial award application deadline: 1/15; financial award applicants required to submit FAFSA. *Faculty research:* Human/animal nutrition. *Unit head:* Chris Calvert, Graduate Chair, 530-752-1269, E-mail: ccalvert@ucdavis.edu. *Application contact:* Karen Kornelly, Administrative Assistant, 530-754-7684, Fax: 530-752-8966, E-mail: klkornelly@ucdavis.edu.

University of California, Davis, Graduate Studies, Program in Maternal and Child Nutrition, Davis, CA 95616. Offers MAS. *Degree requirements:* For master's, comprehensive exam. *Entrance requirements:* Additional exam requirements/recommendations for international students: Required—TOEFL (minimum score 550 paper-based; 213 computer-based). *Application deadline:* For fall admission, 7/1 for domestic students, 7/1 for international students. *Application contact:* Sharoon Munowitch, Graduate Program Staff Coordinator, 530-757-8896.

University of Central Oklahoma, College of Graduate Studies and Research, College of Education, Department of Human Environmental Sciences, Edmond, OK 73034-5209. Offers family and child studies (MS); family and consumer science education (MS); interior design (MS); nutrition-food management (MS). Part-time programs available. *Faculty:* 6 full-time (all women), 9 part-time/adjunct (6 women). *Students:* 37 full-time (27 women), 37 part-time (35 women); includes 24 minority (14 African Americans, 5 American Indian/Alaska Native, 2 Asian Americans or Pacific Islanders, 3 Hispanic Americans), 4 international. Average age 31. 21 applicants, 95% accepted. In 2005, 28 degrees awarded. *Entrance requirements:* Additional exam requirements/recommendations for international students: Required—TOEFL (minimum score 550 paper-based; 213 computer-based). *Application deadline:* Applications are processed on a rolling basis. Application fee: $25. Electronic applications accepted. *Expenses:* Tuition, state resident: full-time $2,988; part-time $125 per credit hour. Tuition, nonresident: full-time $4,728; part-time $197 per credit hour. Required fees: $716; $16 per credit hour. *Financial support:* Career-related internships or fieldwork and unspecified assistantships available. Financial award application deadline: 3/31; financial award applicants required to submit FAFSA. *Faculty research:* Dietetics and food science. *Unit head:* Dr. Tana Stufflebean, Chairperson, 405-974-5787.

University of Chicago, Division of the Biological Sciences, Biomedical Sciences: Cancer, Immunology, Nutrition, Pathology, and Microbiology, Committee on Molecular Metabolism and Nutrition, Chicago, IL 60637-1513. Offers PhD. *Students:* 16 full-time (10 women); includes 5 minority (3 Asian Americans or Pacific Islanders, 2 Hispanic Americans), 1 international. Average age 27. In 2005, 2 degrees awarded. *Degree requirements:* For doctorate, thesis/dissertation, registration. *Entrance requirements:* For doctorate, GRE General Test. Additional exam requirements/recommendations for international students: Required—TOEFL. *Application deadline:* For fall admission, 12/28 priority date for domestic students, 12/28 priority date for international students. Application fee: $55. Electronic applications accepted. *Financial support:* In 2005–06, fellowships with full tuition reimbursements (averaging $26,301 per year), research assistantships with full tuition reimbursements (averaging $26,301 per year) were awarded; institutionally sponsored loans, scholarships/grants, traineeships, and health care benefits also available. Financial award applicants required to submit FAFSA. *Faculty research:* Regulation of lipoprotein metabolism, cellular vitamin metabolism, obesity and body composition, adipocyte differentiation. *Unit head:* Dr. Matthew Brady, Acting Chairman, 773-702-2346, Fax: 773-702-4634, E-mail: mbrady@medicine.bsd.uchicago.edu. *Application contact:* Rebecca Levine, Administrative Assistant, Student Services, 773-834-3899, Fax: 773-702-4634, E-mail: rlevine@huggins.bsd.uchicago.edu.

University of Cincinnati, Division of Research and Advanced Studies, College of Allied Health Sciences, Department of Nutritional Science, Cincinnati, OH 45221. Offers MS. Part-time programs available. *Degree requirements:* For master's, thesis. *Entrance requirements:* For master's, GRE General Test. Additional exam requirements/recommendations for international students: Required—TOEFL (minimum score 550 paper-based; 230 computer-based). Electronic applications accepted. *Faculty research:* Phytochemicals-osteoarthritis, pediatric hypertension and hypercholesterol, cancer prevention/Type II DB.

University of Connecticut, Graduate School, College of Agriculture and Natural Resources, Department of Nutritional Sciences, Field of Nutritional Sciences, Storrs, CT 06269. Offers MS, PhD. *Faculty:* 14 full-time (7 women). *Students:* 36 full-time (26 women), 9 part-time (8 women); includes 5 minority (2 African Americans, 3 Hispanic Americans), 14 international. Average age 30. 34 applicants, 26% accepted, 9 enrolled. In 2005, 5 master's, 6 doctorates awarded. Terminal master's awarded for partial completion of doctoral program. *Degree requirements:* For master's, thesis/dissertation, comprehensive exam; for doctorate, thesis/dissertation. *Entrance requirements:* For master's and doctorate, GRE General Test. Additional exam requirements/recommendations for international students: Required—TOEFL (minimum score 550 paper-based; 213 computer-based). *Application deadline:* For fall admission, 2/1

priority date for domestic students, 2/1 priority date for international students; for spring admission, 11/1 for domestic students, 10/1 for international students. Applications are processed on a rolling basis. Application fee: $55. Electronic applications accepted. *Expenses:* Tuition, state resident: part-time $444 per credit hour. Tuition, nonresident: part-time $1,154 per credit hour. Tuition and fees vary according to course load. *Financial support:* In 2005–06, 28 research assistantships with full tuition reimbursements, 5 teaching assistantships with full tuition reimbursements were awarded; fellowships, Federal Work-Study, scholarships/grants, health care benefits, and unspecified assistantships also available. Financial award application deadline: 2/1; financial award applicants required to submit FAFSA. *Application contact:* Maria-Luz Fernandez, Coordinator, 860-486-5547, Fax: 860-486-3674, E-mail: maria-luz.fernandez@uconn.edu.

University of Delaware, College of Health Sciences, Department of Health, Nutrition, and Exercise Sciences, Newark, DE 19716. Offers exercise science (MS), including biomechanics, exercise physiology, motor control; health promotion (MS); human nutrition (MS). Part-time programs available. *Faculty:* 17 full-time (6 women). *Students:* 41 full-time (27 women), 26 part-time (22 women); includes 6 minority (all African Americans), 4 international. Average age 27. 73 applicants, 60% accepted, 34 enrolled. In 2005, 20 degrees awarded. *Degree requirements:* For master's, thesis, registration. *Entrance requirements:* For master's, GRE General Test, interview, minimum GPA of 3.0. Additional exam requirements/recommendations for international students: Required—TOEFL (minimum score 550 paper-based; 213 computer-based), TSE (minimum score 50). *Application deadline:* For fall admission, 3/15 priority date for domestic students, 3/15 priority date for international students; for spring admission, 10/15 priority date for domestic students, 10/15 priority date for international students. Applications are processed on a rolling basis. Application fee: $60. Electronic applications accepted. *Financial support:* In 2005–06, 10 teaching assistantships with full tuition reimbursements (averaging $11,031 per year) were awarded; fellowships with full tuition reimbursements, research assistantships with full tuition reimbursements, career-related internships or fieldwork, Federal Work-Study, scholarships/grants, and tuition waivers (full and partial) also available. Support available to part-time students. Financial award application deadline: 4/15. *Faculty research:* Sport biomechanics, rehabilitation biomechanics, vascular dynamics. Total annual research expenditures: $407,241. *Unit head:* Dr. Susan J. Hall, Chair, 302-831-2625, Fax: 302-831-4261, E-mail: sjhall@udel.edu. *Application contact:* Dr. Michelle Provost-Craig, Graduate Coordinator, 302-831-6326, Fax: 302-831-3963, E-mail: provost@udel.edu.

University of Florida, Graduate School, College of Agricultural and Life Sciences, Department of Food Science and Human Nutrition, Gainesville, FL 32611. Offers MS, PhD. *Faculty:* 23 full-time (7 women), 1 (woman) part-time/adjunct. *Students:* 73 (44 women); includes 13 minority (3 African Americans, 4 Asian Americans or Pacific Islanders, 6 Hispanic Americans) 21 international. 84 applicants, 61% accepted. In 2005, 18 master's, 6 doctorates awarded. *Degree requirements:* For master's, thesis optional; for doctorate, thesis/dissertation. *Entrance requirements:* For master's and doctorate, GRE General Test, minimum GPA of 3.0. Additional exam requirements/recommendations for international students: Required—TOEFL. *Application deadline:* For fall admission, 6/1 for domestic students. Applications are processed on a rolling basis. Application fee: $20. Electronic applications accepted. *Expenses:* Tuition, state resident: full-time $6,234. Tuition, nonresident: full-time $21,359. Tuition and fees vary according to program. *Financial support:* In 2005–06, 13 research assistantships (averaging $13,562 per year), 13 teaching assistantships (averaging $12,990 per year) were awarded; fellowships, career-related internships or fieldwork also available. *Faculty research:* Pesticide research, nutritional biochemistry and microbiology, food safety and toxicology assessment and dietetics, food chemistry. *Application contact:* Dr. Harry Sitren, Coordinator, 352-392-1991 Ext. 216, Fax: 352-392-9467, E-mail: hssitren@ifas.ufl.edu.

University of Georgia, Graduate School, College of Family and Consumer Sciences, Department of Foods and Nutrition, Athens, GA 30602. Offers MFCS, MS, PhD. *Faculty:* 10 full-time (6 women). *Students:* 35 full-time, 2 part-time; includes 6 minority (4 African Americans, 2 Hispanic Americans), 7 international. 52 applicants, 60% accepted, 14 enrolled. In 2005, 18 master's, 2 doctorates awarded. *Degree requirements:* For master's, thesis (MS); for doctorate, thesis/dissertation. *Entrance requirements:* For master's, GRE General Test, minimum GPA of 3.0, course work in biochemistry and physiology; for doctorate, GRE General Test, master's degree, minimum GPA of 3.0. *Application deadline:* For fall admission, 7/1 for domestic students; for spring admission, 11/15 for domestic students. Application fee: $50. Electronic applications accepted. *Financial support:* Fellowships, research assistantships, teaching assistantships, unspecified assistantships available. *Unit head:* Dr. Rebecca M. Mullis, Head, 706-542-4875, Fax: 706-542-5059, E-mail: rmm@fcs.uga.edu. *Application contact:* Dr. Mary Ann Johnson, Graduate Coordinator, 706-542-2292, Fax: 706-542-5059, E-mail: mjohnson@fcs.uga.edu.

University of Guelph, Graduate Program Services, College of Biological Science, Department of Human Health and Nutritional Sciences, Guelph, ON N1G 2W1, Canada. Offers nutritional sciences (M Sc, PhD). *Faculty:* 17 full-time (8 women). *Students:* 67 full-time (48 women), 8 part-time (5 women). 83 applicants, 24% accepted. In 2005, 18 master's, 2 doctorates awarded. *Degree requirements:* For master's, thesis (for some programs); for doctorate, thesis/dissertation, comprehensive exam. *Entrance requirements:* Additional exam requirements/recommendations for international students: Required—TOEFL (minimum score 550 paper-based; 213 computer-based). *Application deadline:* For fall admission, 4/1 priority date for domestic students, 12/1 priority date for international students. Applications are processed on a rolling basis. Application fee: $75. Electronic applications accepted. *Financial support:* Fellowships, research assistantships, teaching assistantships, scholarships/grants and unspecified assistantships available. *Faculty research:* Nutrition and biochemistry, exercise metabolism and physiology, toxicology, gene expression, biomechanics and ergonomics. *Unit head:* Dr. T. E. Graham, Chair, 519-824-4120 Ext. 56168, Fax: 519-763-5902, E-mail: terrygra@uoguelph.ca. *Application contact:* Dr. D. J. Dyck, Graduate Coordinator, 519-824-4120 Ext. 56578, Fax: 519-763-5902, E-mail: ddyck@uoguelph.ca.

University of Guelph, Graduate Program Services, College of Social and Applied Human Sciences, Department of Family Relations and Applied Nutrition, Guelph, ON N1G 2W1, Canada. Offers applied nutrition (MAN); family relations and human development (M Sc, PhD), including applied human nutrition, couple and family therapy (M Sc), family relations and human development. *Accreditation:* AAMFT/COAMFTE (one or more programs are accredited). Part-time programs available. *Faculty:* 21 full-time (14 women). *Students:* 54 full-time (51 women), 8 part-time (7 women); includes 8 minority (2 African Americans, 5 Asian Americans or Pacific Islanders, 1 Hispanic American), 2 international. Average age 30. 108 applicants, 19% accepted, 21 enrolled. In 2005, 18 master's, 4 doctorates awarded. *Median time to degree:* Of those who began their doctoral program in fall 1997, 100% received their degree in 8 years or less. *Degree requirements:* For master's, thesis (for some programs); for doctorate, thesis/dissertation, comprehensive exam. *Entrance requirements:* For master's, minimum B+ average; for doctorate, master's degree in family relations and human development or related field with a minimum B+ average or master's degree in applied human nutrition. Additional exam requirements/recommendations for international students: Required—TOEFL (minimum score 600 paper-based, 250 computer-based). *Application deadline:* For fall admission, 2/1 priority date for domestic students, 2/1 priority date for international students. Application fee: $75. Electronic applications accepted. *Financial support:* In 2005–06, 50 students received support; fellowships, research assistantships, teaching assistantships, career-related internships or fieldwork and health care benefits available. *Faculty research:* Child and adolescent development, social gerontology, family roles and relations, couple and family therapy, applied human nutrition. Total annual research expenditures: $500,000. *Unit head:* Dr. Kerry Daly, Chair, 519-824-4120 Ext. 56326, Fax: 519-766-0691, E-mail: kdaly@uoguelph.ca. *Application contact:* Jo Anne Waechter, Graduate Secretary, 519-824-4120 Ext. 53968, Fax: 519-766-0691, E-mail: jwaechte@uoguelph.ca.

University of Hawaii at Manoa, Graduate Division, College of Tropical Agriculture and Human Resources, Department of Human Nutrition, Food and Animal Sciences, Program in

Nutrition

Nutritional Science, Honolulu, HI 96822. Offers MS. *Students:* 11 full-time (10 women), 2 part-time (1 woman); includes 4 minority (all Asian Americans or Pacific Islanders), 4 international. Average age 27. 21 applicants, 14% accepted, 3 enrolled. *Entrance requirements:* For master's, GRE General Test. *Application deadline:* For fall admission, 2/1 for domestic students, 2/1 for international students; for spring admission, 9/1 for domestic students, 9/1 for international students. Application fee: $50. *Expenses:* Tuition, state resident: full-time $8,400; part-time $200 per credit hour. Tuition, nonresident: full-time $11,088; part-time $462 per credit hour. Tuition and fees vary according to program. *Financial support:* Tuition waivers (full) available. Financial award application deadline: 3/1. *Faculty research:* Nutritional biochemistry, human nutrition, nutrition education, international nutrition, nutritional epidemiology. *Unit head:* Dr. Michael Dunn, Graduate Chairperson, 808-956-8356, Fax: 808-956-4024, E-mail: mdunn@hawaii.edu. *Application contact:* Dr. Michael Dunn, Graduate Chairperson, 808-956-8356, Fax: 808-956-4024, E-mail: mdunn@hawaii.edu.

University of Illinois at Chicago, Graduate College, College of Applied Health Sciences, Program in Human Nutrition and Dietetics, Chicago, IL 60607-7128. Offers MS, PhD. *Accreditation:* ADtA. *Degree requirements:* For master's and doctorate, thesis/dissertation. *Entrance requirements:* For master's and doctorate, GRE General Test, minimum GPA of 2.75. Additional exam requirements/recommendations for international students: Required—TOEFL. Electronic applications accepted. *Faculty research:* Nutrition for the elderly, inborn errors of metabolism, nutrition and cancer, lipid metabolism, dietary fat markers.

University of Illinois at Urbana–Champaign, Graduate College, College of Agricultural, Consumer and Environmental Sciences, Department of Food Science and Human Nutrition, Champaign, IL 61820. Offers MS, PhD. *Faculty:* 26 full-time (10 women), 1 (woman) part-time/adjunct. *Students:* 44 full-time (34 women), 14 part-time (9 women); includes 7 minority (1 African American, 5 Asian Americans or Pacific Islanders, 1 Hispanic American), 24 international. 89 applicants, 19% accepted, 14 enrolled. In 2005, 14 master's, 4 doctorates awarded. *Degree requirements:* For doctorate, one foreign language, thesis/dissertation. *Entrance requirements:* For master's, minimum GPA of 3.0. *Application deadline:* For fall admission, 5/15 for domestic students. Applications are processed on a rolling basis. Application fee: $50 ($60 for international students). Electronic applications accepted. *Financial support:* In 2005–06, 7 fellowships, 36 research assistantships, 11 teaching assistantships were awarded; tuition waivers (full and partial) also available. Financial award application deadline: 2/15. Total annual research expenditures: $4.4 million. *Unit head:* Faye Dong, Head, 217-244-4498, Fax: 217-265-0925, E-mail: fayedong@uiuc.edu. *Application contact:* Terri Cummings, Director of Student Services, 217-244-4405, Fax: 217-265-0925, E-mail: tcumming@uiuc.edu.

University of Illinois at Urbana–Champaign, Graduate College, College of Agricultural, Consumer and Environmental Sciences, Division of Nutritional Sciences, Champaign, IL 61820. Offers MS, PhD. *Students:* 51 full-time (41 women), 2 part-time (both women); includes 7 minority (1 African American, 3 Asian Americans or Pacific Islanders, 3 Hispanic Americans), 15 international. 44 applicants, 27% accepted, 10 enrolled. In 2005, 6 master's, 6 doctorates awarded. *Degree requirements:* For doctorate, thesis/dissertation. *Entrance requirements:* For master's, GRE, minimum GPA of 3.0. *Application deadline:* For fall admission, 7/15 for domestic students. Applications are processed on a rolling basis. Application fee: $50 ($60 for international students). Electronic applications accepted. *Financial support:* In 2005–06, 14 fellowships, 36 research assistantships, 7 teaching assistantships were awarded. Financial award application deadline: 2/15. *Unit head:* Sharon M. Donovan, Director, 217-333-4177, Fax: 217-333-9368, E-mail: sdonovan@uiuc.edu. *Application contact:* Linda L. Barenthin, Administrative Secretary, 217-333-4177, Fax: 217-333-9368, E-mail: lbarenth@uiuc.edu.

University of Kansas, Graduate Studies Medical Center, School of Allied Health, Department of Dietetics and Nutrition, Lawrence, KS 66045. Offers dietetic internship (Certificate); dietetics and nutrition (MS). Part-time programs available. *Faculty:* 4. *Students:* 26 full-time (25 women), 12 part-time (all women), 1 international. Average age 27. In 2005, 9 degrees awarded. *Degree requirements:* For master's, oral exam, thesis optional. *Application deadline:* For fall admission, 2/15 for domestic students; for spring admission, 10/31 priority date for domestic students. Application fee: $35. *Expenses:* Tuition, state resident: full-time $4,859. Tuition, nonresident: full-time $12,000. Required fees: $589. Tuition and fees vary according to program. *Financial support:* In 2005–06, 18 students received support; fellowships, research assistantships with partial tuition reimbursements available, teaching assistantships with full and partial tuition reimbursements available available. *Faculty research:* Weight management, docosahexaenoic acid (DHA) and infant health, gastrointestinal diseases, nutrition support. *Unit head:* Dr. Janice E. Harris, Chairperson, 913-588-7682, Fax: 913-588-7685, E-mail: jharris@kumc.edu. *Application contact:* Rachel Barkley, Dietetic Internship Director, 913-588-5359, Fax: 913-588-7685. E-mail: rbarkley@kumc.edu.

University of Kentucky, Graduate School, Graduate School Programs in the College of Agriculture, Program in Hospitality and Dietetic Administration, Lexington, KY 40506-0032. Offers MS. *Faculty:* 8 full-time (6 women). *Students:* 16 full-time (15 women), 1 (woman) part-time; includes 1 minority (Asian American or Pacific Islander), 5 international. Average age 29. 5 applicants, 100% accepted, 4 enrolled. In 2005, 4 degrees awarded. *Degree requirements:* For master's, thesis optional. *Entrance requirements:* For master's, GRE General Test, minimum undergraduate GPA of 2.5. Additional exam requirements/recommendations for international students: Required—TOEFL (minimum score 550 paper-based; 213 computer-based). *Application deadline:* For fall admission, 7/17 priority date for domestic students, 2/1 priority date for international students; for spring admission, 12/13 priority date for domestic students, 6/15 priority date for international students. Applications are processed on a rolling basis. Application fee: $40 ($55 for international students). Electronic applications accepted. *Expenses:* Tuition, state resident: full-time $6,308; part-time $331 per credit hour. Tuition, nonresident: full-time $13,968; part-time $756 per credit hour. Tuition and fees vary according to course load, degree level and program. *Financial support:* In 2005–06, 10 students received support, including 1 research assistantship with full tuition reimbursement available (averaging $5,000 per year), 9 teaching assistantships with full tuition reimbursement available (averaging $5,250 per year); fellowships with full tuition reimbursements available, Federal Work-Study, scholarships/grants, traineeships, health care benefits, tuition waivers (partial), and unspecified assistantships also available. Support available to part-time students. Financial award application deadline: 3/15. *Unit head:* Dr. Lisa Gaetke, Director of Graduate Studies, 859-257-1031, Fax: 859-257-3707, E-mail: lisa.gaetke@uky.edu. *Application contact:* Dr. Brian Jackson, Senior Associate Dean, 859-257-8176, Fax: 859-323-1928.

University of Kentucky, Graduate School, Program in Nutritional Sciences, Lexington, KY 40506-0032. Offers MSNS, PhD. *Faculty:* 33 full-time (14 women), 2 part-time/adjunct (0 women). *Students:* 29 full-time (20 women), 3 part-time (all women); includes 3 minority (all Asian Americans or Pacific Islanders), 17 international. Average age 31. 22 applicants, 64% accepted, 6 enrolled. In 2005, 3 master's, 6 doctorates awarded. *Median time to degree:* Of those who began their doctoral program in fall 1997, 89.5% received their degree in 8 years or less. *Degree requirements:* For doctorate, thesis/dissertation, comprehensive exam. *Entrance requirements:* For master's, GRE General Test, minimum undergraduate GPA of 2.5; for doctorate, GRE General Test, minimum graduate GPA of 3.0. Additional exam requirements/recommendations for international students: Required—TOEFL (minimum score 550 paper-based; 213 computer-based). *Application deadline:* For fall admission, 7/17 priority date for domestic students, 2/1 priority date for international students; for spring admission, 12/13 priority date for domestic students, 6/15 priority date for international students. Applications are processed on a rolling basis. Application fee: $40 ($55 for international students). Electronic applications accepted. *Expenses:* Tuition, state resident: full-time $6,308; part-time $331 per credit hour. Tuition, nonresident: full-time $13,968; part-time $756 per credit hour. Tuition and fees vary according to course load, degree level and program. *Financial support:* In 2005–06, 3 fellowships with full tuition reimbursements (averaging $2,000 per year), 21 research assistantships with full tuition reimbursements (averaging $15,400 per year) were awarded; teaching assistantships with full tuition reimbursements, Federal Work-Study, scholarships/grants, traineeships, health care benefits, tuition waivers (partial), and unspecified assistantships also available.

Support available to part-time students. Financial award application deadline: 3/15. *Faculty research:* Nutrition and AIDS, nutrition and alcoholism, nutrition and cardiovascular disease, nutrition and cancer, nutrition and diabetes. Total annual research expenditures: $6 million. *Unit head:* Dr. Steven R. Post, Director of Graduate Studies, 859-323-4933, Fax: 859-257-3646, E-mail: spost@uky.edu. *Application contact:* Dr. Brian Jackson, Senior Associate Dean, 859-257-8176, Fax: 859-323-1928.

See Close-Up on page 1097.

University of Maine, Graduate School, College of Natural Sciences, Forestry, and Agriculture, Department of Food Science and Human Nutrition, Orono, ME 04469. Offers food and nutritional sciences (PhD); food science and human nutrition (MS). Part-time programs available. *Faculty:* 4 full-time (1 woman). *Students:* 23 full-time (20 women), 7 part-time (6 women); includes 2 minority (1 African American, 1 Asian American or Pacific Islander), 4 international. Average age 29. 22 applicants, 41% accepted, 5 enrolled. In 2005, 9 master's, 1 doctorate awarded. *Degree requirements:* For master's and doctorate, thesis/dissertation. *Entrance requirements:* For master's, GRE General Test, minimum GPA of 3.0; for doctorate, GRE General Test. Additional exam requirements/recommendations for international students: Required—TOEFL. *Application deadline:* For fall admission, 2/1 for domestic students. Applications are processed on a rolling basis. Application fee: $50. Electronic applications accepted. *Financial support:* In 2005–06, 9 research assistantships with tuition reimbursements (averaging $13,500 per year), 4 teaching assistantships with tuition reimbursements (averaging $12,000 per year) were awarded; scholarships/grants and tuition waivers (full and partial) also available. Financial award application deadline: 3/1. *Faculty research:* Product development of fruit and vegetables, lipid oxidation in fish and meat, analytical methods development, metabolism of potato glycoalkaloids, seafood quality. *Unit head:* Dr. Rodney Bushway, Chair, 207-581-1626, Fax: 207-581-1636. *Application contact:* Scott G. Delcourt, Associate Dean of the Graduate School, 207-581-3219, Fax: 207-581-3232, E-mail: graduate@maine.edu.

University of Manitoba, Faculty of Graduate Studies, Faculty of Human Ecology, Department of Human Nutritional Sciences, Winnipeg, MB R3T 2N2, Canada. Offers M Sc. *Degree requirements:* For master's, thesis.

University of Manitoba, Faculty of Graduate Studies, Interdisciplinary Programs, Department of Foods and Nutritional Sciences, Winnipeg, MB R3T 2N2, Canada. Offers PhD.

University of Maryland, College Park, Graduate Studies, College of Agriculture and Natural Resources, Department of Nutrition and Food Science, Program in Nutrition, College Park, MD 20742. Offers MS, PhD. *Students:* 21 full-time (19 women), 2 part-time (both women); includes 3 minority (1 African American, 1 Asian American or Pacific Islander, 1 Hispanic American), 11 international. 23 applicants, 13% accepted, 2 enrolled. In 2005, 2 master's, 1 doctorate awarded. *Degree requirements:* For master's, thesis; for doctorate, thesis/dissertation, candidacy exam, comprehensive exam. *Entrance requirements:* For master's, GRE General Test, minimum GPA of 3.0, 3 letters of recommendation; for doctorate, GRE General Test, minimum GPA of 3.0. Additional exam requirements/recommendations for international students: Required—TOEFL. *Application deadline:* For fall admission, 1/10 for domestic students, 1/10 for international students; for spring admission, 9/10 for domestic students, 6/1 for international students. Applications are processed on a rolling basis. Application fee: $60. Electronic applications accepted. *Financial support:* In 2005–06, 2 fellowships (averaging $8,757 per year) were awarded; research assistantships, teaching assistantships with tuition reimbursements Financial award applicants required to submit FAFSA. *Faculty research:* Nutrition education, carbohydrates and physical activity. *Application contact:* Dean of Graduate School, 301-405-4190, Fax: 301-314-9305.

University of Massachusetts Amherst, Graduate School, School of Public Health and Health Sciences, Department of Nutrition, Amherst, MA 01003. Offers nutrition (MPH, MS); public health (PhD). *Faculty:* 8 full-time (4 women). *Students:* 12 full-time (all women), 2 part-time (both women), 10 international. Average age 26. 30 applicants, 60% accepted, 8 enrolled. In 2005, 3 master's, 1 doctorate awarded. *Degree requirements:* For master's, thesis or alternative. *Entrance requirements:* For master's, GRE General Test. Additional exam requirements/recommendations for international students: Required—TOEFL (minimum score 530 paper-based; 197 computer-based). *Application deadline:* For fall admission, 2/1 priority date for domestic students, 2/1 priority date for international students; for spring admission, 10/1 for domestic students, 10/1 for international students. Applications are processed on a rolling basis. Application fee: $40 ($65 for international students). Electronic applications accepted. *Expenses:* Tuition, state resident: part-time $110 per credit. Tuition, nonresident: part-time $414 per credit. Required fees: $2,824 per term. One-time fee: $250 part-time. Full-time tuition and fees vary according to course load, campus/location, program and reciprocity agreements. *Financial support:* In 2005–06, research assistantships with full tuition reimbursements (averaging $4,773 per year), teaching assistantships with full tuition reimbursements (averaging $6,789 per year) were awarded; fellowships with full tuition reimbursements, career-related internships or fieldwork, Federal Work-Study, scholarships/grants, traineeships, and unspecified assistantships also available. Support available to part-time students. Financial award application deadline: 2/1. *Unit head:* Dr. Nancy Cohen, Acting Head, 413-545-0470, Fax: 413-545-1074, E-mail: cohen@nutrition.umass.edu.

University of Medicine and Dentistry of New Jersey, School of Health Related Professions, Department of Interdisciplinary Studies, Program in Health Sciences, Newark, NJ 07107-1709. Offers cardiopulmonary sciences (PhD); clinical laboratory sciences (PhD); health sciences (MS); interdisciplinary studies (PhD); nutrition (PhD); physical therapy/movement science (PhD). *Degree requirements:* For doctorate, thesis/dissertation. *Entrance requirements:* For doctorate, interview, writing sample. Additional exam requirements/recommendations for international students: Required—TOEFL. *Application deadline:* Applications are processed on a rolling basis. Application fee: $50. *Unit head:* Dr. Margaret L. Kildoff, Director, 973-972-8576, Fax: 973-972-5258, E-mail: scanlan@umdnj.edu.

University of Medicine and Dentistry of New Jersey, School of Health Related Professions, Department of Primary Care, Dietetic Internship Program, Newark, NJ 07107-1709. Offers Certificate. *Entrance requirements:* For degree, bachelor's degree in dietetics, nutrition, or related field; interview; minimum GPA of 2.5. Additional exam requirements/recommendations for international students: Required—TOEFL. *Application deadline:* For fall admission, 2/5 for domestic students. Application fee: $50. *Unit head:* Denise Langevin, Director, 908-889-2488, Fax: 908-889-2487, E-mail: langevdd@umdnj.edu. *Application contact:* M. Geraldine McKay, Co-Director, 908-889-2494, Fax: 908-889-2487, E-mail: mckayge@umdnj.edu.

University of Medicine and Dentistry of New Jersey, School of Health Related Professions, Department of Primary Care, Program in Clinical Nutrition, Newark, NJ 07107-1709. Offers MS, DCN. *Entrance requirements:* For master's, minimum GPA of 3.0, proof of registered dietician status. *Application deadline:* For fall admission, 4/1 for domestic students; for spring admission, 10/15 for domestic students. Application fee: $50. *Unit head:* Dr. Riva Touger-Decker, Director, 973-972-6596, Fax: 973-972-7403, E-mail: decker@umdnj.edu. *Application contact:* Diane Rigassio, Assistant Professor, 973-972-6731, Fax: 973-972-7403, E-mail: rigassdl@umdnj.edu.

University of Memphis, Graduate School, College of Education, Department of Health and Sport Sciences, Memphis, TN 38152. Offers clinical nutrition (MS); exercise and sport science (MS); health promotion (MS); physical education teacher education (MS), including teacher education; sport and leisure commerce (MS). *Accreditation:* NCATE. Part-time and evening/weekend programs available. *Faculty:* 23 full-time (8 women), 11 part-time/adjunct (8 women). *Students:* Average age 28. 50 applicants, 62% accepted. In 2005, 14 degrees awarded. *Degree requirements:* For master's, thesis, comprehensive exam. *Entrance requirements:* For master's, GRE General Test or GMAT (sport and leisure commerce). *Application deadline:* For fall admission, 5/1 for domestic students; for spring admission, 11/1 for domestic students. Applications are processed on a rolling basis. Application fee: $25 ($50 for international students).

Nutrition

University of Memphis *(continued)*
Financial support: In 2005–06, 13 research assistantships with full tuition reimbursements (averaging $6,000 per year), 4 teaching assistantships with full tuition reimbursements (averaging $6,000 per year) were awarded; career-related internships or fieldwork, tuition waivers (partial), and community assistantships also available. *Faculty research:* Sport marketing and consumer analysis, health psychology, smoking cessation, psychosocial aspects of cardiovascular disease, global health promotion. Total annual research expenditures: $1.3 million. *Unit head:* Dr. Michael H. Hamrick, Chairman, 901-678-4165, Fax: 901-678-3591, E-mail: mhamrick@memphis.edu. *Application contact:* Christina Little, Academic Services Coordinator, 901-678-4316, Fax: 901-678-3591, E-mail: aclittle@memphis.edu.

University of Michigan, School of Public Health, Department of Environmental Health Sciences, Program in Human Nutrition, Ann Arbor, MI 48109. Offers MPH, MS. MS offered through the Horace H. Rackham School of Graduate Studies. *Accreditation:* ADtA. *Degree requirements:* For master's, thesis. *Entrance requirements:* For master's, GRE General Test. *Application deadline:* For fall admission, 2/1 for domestic students. Applications are processed on a rolling basis. Application fee: $60 ($75 for international students). *Expenses:* Tuition, state resident: full-time $14,082; part-time $894 per credit hour. Tuition, nonresident: full-time $28,500; part-time $1,675 per credit hour. Required fees: $189; $189 per unit. *Financial support:* Fellowships, teaching assistantships available. Financial award application deadline: 3/1. *Application contact:* Susan Crawford, Student Services Coordinator, E-mail: sph.ehs.inquiries@umich.edu.

University of Minnesota, Twin Cities Campus, Graduate School, College of Agricultural, Food, and Environmental Sciences and College of Human Ecology, Program in Nutrition, Minneapolis, MN 55455-0213. Offers MS, PhD. Part-time programs available. *Faculty:* 25 full-time (16 women), 15 part-time/adjunct (7 women). *Students:* 32 full-time (all women), 20 part-time (17 women); includes 5 minority (1 African American, 2 Asian Americans or Pacific Islanders, 2 Hispanic Americans), 13 international. Average age 30. 42 applicants, 45% accepted, 17 enrolled. In 2005, 7 master's, 1 doctorate awarded. Terminal master's awarded for partial completion of doctoral program. *Degree requirements:* For master's, thesis (for some programs); for doctorate, thesis/dissertation. *Entrance requirements:* For master's, GRE General Test, previous course work in general chemistry, organic chemistry, physiology, biology, biochemistry, statistics; minimum GPA of 3.0; for doctorate, GRE General Test, previous course work in general chemistry, organic chemistry, calculus, biology, physics, physiology, biochemistry, statistics; minimum GPA of 3.0. Additional exam requirements/recommendations for international students: Required—TOEFL (minimum score 550 paper-based; 213 computer-based). *Application deadline:* For fall admission, 6/15 for domestic students, 6/15 for international students, 10/15 for spring admission, 10/15 for domestic students, 10/15 for international students. Applications are processed on a rolling basis. Application fee: $55 ($75 for international students). Electronic applications accepted. *Expenses:* Tuition, state resident: full-time $8,748; part-time $729 per credit. Tuition, nonresident: full-time $15,848; part-time $1,321 per credit. Full-time tuition and fees vary according to class time, course load, program and reciprocity agreements. *Financial support:* In 2005–06, 27 students received support, including 2 fellowships with full tuition reimbursements available (averaging $16,000 per year), 18 research assistantships with full tuition reimbursements available (averaging $16,000 per year), 7 teaching assistantships with full tuition reimbursements available (averaging $16,000 per year); career-related internships or fieldwork, Federal Work-Study, institutionally sponsored loans, scholarships/grants, traineeships, health care benefits, and unspecified assistantships also available. Support available to part-time students. *Faculty research:* Diet and chronic disease: from basic biological and molecular biology approaches to a public health/intervention/epidemiology perspective. Total annual research expenditures: $2.3 million. *Unit head:* Dr. Mindy S. Kurzer, Director of Graduate Studies, 612-624-9789, Fax: 612-625-5272, E-mail: mkurzer@tc.umn.edu. *Application contact:* Susan K. Viker, Assistant Coordinator, 612-624-6753, Fax: 612-625-5272, E-mail: sviker@umn.edu.

University of Minnesota, Twin Cities Campus, School of Public Health, Major in Public Health Nutrition, Minneapolis, MN 55455-0213. Offers MPH. Part-time programs available. *Faculty:* 13 full-time (8 women), 2 part-time/adjunct (0 women). *Students:* 20 full-time (18 women), 7 part-time (all women). Average age 30. 31 applicants, 81% accepted, 13 enrolled. In 2005, 6 degrees awarded. *Degree requirements:* For master's, fieldwork, master's project. *Entrance requirements:* For master's, GRE General Test. Additional exam requirements/recommendations for international students: Required—TOEFL. *Application deadline:* For fall admission, 1/15 priority date for domestic students, 1/15 priority date for international students. Applications are processed on a rolling basis. Application fee: $55 ($75 for international students). Electronic applications accepted. *Expenses:* Contact institution. Full-time tuition and fees vary according to class time, course load, program and reciprocity agreements. *Financial support:* In 2005–06, 1 fellowship with partial tuition reimbursement, 3 research assistantships with partial tuition reimbursements, 1 teaching assistantship with partial tuition reimbursement were awarded; career-related internships or fieldwork, Federal Work-Study, institutionally sponsored loans, scholarships/grants, traineeships, and unspecified assistantships also available. Financial award application deadline: 1/15; financial award applicants required to submit FAFSA. *Faculty research:* Nutrition and pregnancy outcomes, nutrition and women's health, child growth and nutrition, child and adolescent nutrition and eating behaviors, obesity and eating disorder prevention. *Unit head:* Dr. Leslie Lytle, Chair, 612-624-1818, Fax: 612-624-0315, E-mail: lytle@epi.umn.edu. *Application contact:* Shelley Cooksey, Major Coordinator, 612-626-8802, Fax: 612-624-0315, E-mail: cooksey@epi.umn.edu.

University of Missouri–Columbia, Graduate School, College of Agriculture, Food and Natural Resources, Department of Food and Hospitality Systems, Columbia, MO 65211. Offers food science (MS, PhD); foods and food systems management (MS); human nutrition (MS). *Faculty:* 7 full-time (4 women). *Students:* 7 full-time (5 women), 14 part-time (6 women); includes 1 minority (Hispanic American), 13 international. In 2005, 6 master's, 2 doctorates awarded. Terminal master's awarded for partial completion of doctoral program. *Degree requirements:* For doctorate, thesis/dissertation. *Entrance requirements:* For master's and doctorate, GRE General Test, minimum GPA of 3.0. *Application deadline:* For fall admission, 4/1 for domestic students. Applications are processed on a rolling basis. Application fee: $45 ($60 for international students). *Financial support:* Research assistantships, teaching assistantships, institutionally sponsored loans available. *Unit head:* Dr. Andrew D. Clarke, Director of Graduate Studies, 573-882-2610, E-mail: clakrea@missouri.edu.

University of Missouri–Columbia, Graduate School, College of Agriculture, Food and Natural Resources and College of Human Environmental Science, Nutritional Sciences Program, Columbia, MO 65211. Offers nutrition (MS, PhD). *Degree requirements:* For master's, thesis, comprehensive exam; for doctorate, thesis/dissertation, comprehensive oral and written exam, qualifying exam. *Entrance requirements:* For master's and doctorate, GRE General Test, minimum GPA of 3.0. Additional exam requirements/recommendations for international students: Required—TOEFL. Electronic applications accepted. *Faculty research:* Whole-animal models, cultured-cell models, enzymology and protein characterization, radioisotope tracer methodology, lipid metabolism.

See Close-Up on page 1099.

University of Missouri–Columbia, Graduate School, College of Human Environmental Science, Department of Nutritional Sciences, Columbia, MO 65211. Offers exercise physiology (MA, PhD); nutritional sciences (MS, PhD). *Faculty:* 5 full-time (4 women), 1 part-time/adjunct (0 women). *Students:* 10 full-time (7 women), 6 part-time (4 women); includes 1 minority (African American), 5 international. In 2005, 7 master's awarded. *Degree requirements:* For doctorate, thesis/dissertation. *Entrance requirements:* For master's and doctorate, GRE General Test, minimum GPA of 3.0. Additional exam requirements/recommendations for international students: Required—TOEFL. *Application deadline:* Applications are processed on a rolling basis. Application fee: $45 ($60 for international students). *Financial support:* Fellow-

ships, research assistantships, teaching assistantships, institutionally sponsored loans available. *Unit head:* Dr. Kevin Fritsche, Director of Graduate Studies, 573-882-4288.

University of Nebraska–Lincoln, Graduate College, College of Agricultural Sciences and Natural Resources, Interdepartmental Area of Nutrition, Lincoln, NE 68588. Offers MS, PhD. *Degree requirements:* For master's, thesis optional; for doctorate, thesis/dissertation, comprehensive exam. *Entrance requirements:* For master's and doctorate, GRE General Test. Additional exam requirements/recommendations for international students: Required—TOEFL (minimum score 550 paper-based; 213 computer-based). Electronic applications accepted. *Faculty research:* Human nutrition and metabolism, animal nutrition and metabolism, biochemistry, community and clinical nutrition.

University of Nebraska–Lincoln, Graduate College, College of Education and Human Sciences, Department of Nutritional Science and Dietetics, Lincoln, NE 68588. Offers MS. *Degree requirements:* For master's, thesis optional. *Entrance requirements:* For master's, GRE General Test. Additional exam requirements/recommendations for international students: Required—TOEFL (minimum score 550 paper-based; 213 computer-based). Electronic applications accepted. *Faculty research:* Foods/food service administration, community nutrition science, diet-health relationships.

University of Nebraska Medical Center, School of Allied Health Professions and College of Medicine, Program in Medical Nutrition Education, Omaha, NE 68198. Offers dietetic internship (Certificate). *Faculty:* 1 (woman) full-time, 22 part-time/adjunct (all women). *Students:* 6 full-time. Average age 23. 36 applicants, 11% accepted, 4 enrolled. In 2005, 6 degrees awarded. *Application deadline:* For winter admission, 2/15 for domestic students. Application fee: $45. *Expenses:* Tuition, area resident: Part-time $200 per hour. Tuition, nonresident: part-time $538 per hour. Required fees: $308; $59 per term. *Faculty research:* Nutrition intervention outcomes. *Unit head:* Glenda R. Woscyna, Director, 402-559-7365, Fax: 402-559-6010, E-mail: gwoscyna@nebraskamed.com.

University of Nevada, Reno, Graduate School, College of Agriculture, Biotechnology and Natural Resources, Department of Nutrition, Reno, NV 89557. Offers MS. *Faculty:* 9. *Students:* 2 full-time (both women), 1 (woman) part-time. Average age 28. 5 applicants, 80% accepted, 3 enrolled. In 2005, 1 degree awarded. *Degree requirements:* For master's, thesis optional. *Entrance requirements:* For master's, GRE, minimum GPA of 3.0. Additional exam requirements/recommendations for international students: Required—TOEFL. *Application deadline:* For fall admission, 2/1 for domestic students; for spring admission, 11/1 for domestic students. Applications are processed on a rolling basis. Application fee: $60 ($95 for international students). *Expenses:* Tuition, area resident: Full-time $2,767; part-time $923 per semester. Tuition, state resident: full-time $5,733; part-time $1,911 per semester. Tuition, nonresident: full-time $12,679; part-time $1,911 per semester. International tuition: $13,878 full-time. Required fees: $404; $202 per term. One-time fee: $90. *Financial support:* In 2005–06, 5 research assistantships, 3 teaching assistantships were awarded; Federal Work-Study, institutionally sponsored loans, tuition waivers (full), and unspecified assistantships also available. Financial award application deadline: 3/1. *Faculty research:* Nutritional education, food technology, therapeutic human nutrition, human nutritional requirements, diet and disease. *Unit head:* Dr. Jamie Benedict, Graduate Program Director, 775-784-6440.

University of New Hampshire, Graduate School, College of Life Sciences and Agriculture, Department of Animal and Nutritional Sciences, Program in Animal and Nutritional Sciences, Durham, NH 03824. Offers PhD. *Faculty:* 23 full-time. *Students:* 5 full-time (3 women), 1 (woman) part-time. Average age 29. 5 applicants, 40% accepted, 2 enrolled. In 2005, 1 degree awarded. *Entrance requirements:* For doctorate, GRE. Additional exam requirements/recommendations for international students: Required—TOEFL (minimum score 550 paper-based; 213 computer-based). *Application deadline:* For fall admission, 4/1 priority date for domestic students, 4/1 priority date for international students. Application fee: $60. Electronic applications accepted. *Expenses:* Tuition, state resident: full-time $8,010; part-time $445 per credit hour. Tuition, nonresident: full-time $19,730; part-time $810 per credit hour. Required fees: $322 per semester. Tuition and fees vary according to course load and program. *Financial support:* In 2005–06, 1 fellowship, 1 research assistantship, 3 teaching assistantships were awarded; scholarships/grants, traineeships, and unspecified assistantships also available. Support available to part-time students. *Application contact:* Ann Barbarits, Administrative Assistant, 603-862-2178, E-mail: ansc.grad.program.info@unh.edu.

University of New Hampshire, Graduate School, College of Life Sciences and Agriculture, Department of Animal and Nutritional Sciences, Program in Nutritional Sciences, Durham, NH 03824. Offers MS. Part-time programs available. *Faculty:* 23 full-time. *Students:* 1 (woman) full-time, 12 part-time (10 women). Average age 33. 8 applicants, 50% accepted, 3 enrolled. In 2005, 7 degrees awarded. *Degree requirements:* For master's, thesis. *Entrance requirements:* Additional exam requirements/recommendations for international students: Required—TOEFL (minimum score 550 paper-based; 213 computer-based); Recommended—TSE. *Application deadline:* For fall admission, 4/1 priority date for domestic students, 4/1 priority date for international students. For winter admission, 12/1 for domestic students. Applications are processed on a rolling basis. Application fee: $60. Electronic applications accepted. *Expenses:* Tuition, state resident: full-time $8,010; part-time $445 per credit hour. Tuition, nonresident: full-time $19,730; part-time $810 per credit hour. Required fees: $322 per semester. Tuition and fees vary according to course load and program. *Financial support:* In 2005–06, 1 fellowship, 4 teaching assistantships were awarded; research assistantships, career-related internships or fieldwork, Federal Work-Study, and scholarships/grants also available. Support available to part-time students. Financial award application deadline: 2/15.

University of New Haven, Graduate School, College of Arts and Sciences, Program in Human Nutrition, West Haven, CT 06516-1916. Offers MS.

University of New Mexico, Graduate School, College of Education, Department of Individual, Family and Community Education, Program in Nutrition, Albuquerque, NM 87131-2039. Offers MS. Part-time programs available. *Students:* 18 full-time (all women), 8 part-time (7 women); includes 3 minority (2 Asian Americans or Pacific Islanders, 1 Hispanic American), 1 international. Average age 32. 8 applicants, 100% accepted, 7 enrolled. In 2005, 10 degrees awarded. *Degree requirements:* For master's, comprehensive exam or thesis. *Entrance requirements:* For master's, GRE, bachelor's degree in nutrition or prerequisites. *Application deadline:* For fall admission, 6/1 for domestic students; for spring admission, 11/1 for domestic students. Application fee: $40. Electronic applications accepted. *Expenses:* Tuition, state resident: full-time $5,676. Tuition, nonresident: full-time $14,974; part-time $238 per credit hour. Required fees: $385 per term. Tuition and fees vary according to course load and program. *Financial support:* In 2005–06, 9 students received support, including 1 research assistantship (averaging $4,619 per year), 4 teaching assistantships with full and partial tuition reimbursements available (averaging $3,219 per year); health care benefits also available. Financial award application deadline: 3/1; financial award applicants required to submit FAFSA. *Faculty research:* Nutritional needs of children, food safety in schools, obesity prevention, phytochemicals, international nutrition. *Unit head:* Dr. Karen Heller, Graduate Coordinator, 505-277-6430, E-mail: kheller@unm.edu. *Application contact:* Cynthia Salas, Information Contact, 505-277-4535, Fax: 505-277-8361, E-mail: casalas@unm.edu.

The University of North Carolina at Chapel Hill, Graduate School, School of Public Health, Department of Nutrition, Chapel Hill, NC 27599. Offers nutrition (MPH, Dr PH, PhD); nutritional biochemistry (MS); professional practice program (MPH). *Accreditation:* ADtA. *Faculty:* 19 full-time (13 women), 14 part-time/adjunct. *Students:* 91 full-time (76 women); includes 26 minority (5 African Americans, 1 American Indian/Alaska Native, 18 Asian Americans or Pacific Islanders, 2 Hispanic Americans). Average age 27. 113 applicants, 37% accepted, 31 enrolled. In 2005, 24 master's, 6 doctorates awarded. *Median time to degree:* Of those who began their doctoral program in fall 1997, 100% received their degree in 8 years or less. *Degree requirements:* For master's, thesis, major paper, comprehensive exam, registration; for doctorate, thesis/dissertation, comprehensive exam, registration. *Entrance requirements:* For master's

Nutrition

and doctorate, GRE General Test, minimum GPA of 3.0. Additional exam requirements/recommendations for international students: Required—TOEFL. *Application deadline:* For fall admission, 1/1 priority date for domestic students, 1/1 priority date for international students; for spring admission, 10/15 for domestic students. Applications are processed on a rolling basis. Application fee: $70. Electronic applications accepted. *Financial support:* In 2005–06, 58 students received support, including 26 fellowships with full and partial tuition reimbursements available (averaging $20,002 per year), 22 research assistantships with full and partial tuition reimbursements available (averaging $23,841 per year), 10 teaching assistantships with full and partial tuition reimbursements available (averaging $1,000 per year); career-related internships or fieldwork, Federal Work-Study, institutionally sponsored loans, scholarships/grants, traineeships, health care benefits, and unspecified assistantships also available. Financial award application deadline: 3/1; financial award applicants required to submit FAFSA. *Faculty research:* Nutrition policy, management and leadership development, lipid and carbohydrate metabolism, dietary trends and determinants, transmembrane signal transduction and carcinogenesis, maternal and child nutrition. Total annual research expenditures: $7.1 million. *Unit head:* Dr. June Stevens, Chair, 919-966-1065, Fax: 919-966-7216, E-mail: june_stevens@unc.edu. *Application contact:* Joanne O. Lee, Registrar, 919-966-7212, Fax: 919-966-7216, E-mail: joanne_lee@unc.edu.

The University of North Carolina at Greensboro, Graduate School, School of Human Environmental Sciences, Department of Nutrition, Greensboro, NC 27412-5001. Offers MS, PhD. *Students:* 19 full-time (15 women), 10 part-time (9 women). 28 applicants, 50% accepted. *Degree requirements:* For master's and doctorate, thesis/dissertation. *Entrance requirements:* For master's and doctorate, GRE General Test. Additional exam requirements/recommendations for international students: Required—TOEFL. *Application deadline:* For fall admission, 3/1 for domestic students; for spring admission, 11/1 for domestic students. Application fee: $45. *Expenses:* Tuition, state resident: part-time $302 per credit hour. Tuition, nonresident: part-time $1,683 per credit hour. Required fees: $51 per credit hour. Tuition and fees vary according to course load and program. *Financial support:* Fellowships with full tuition reimbursements, research assistantships with full tuition reimbursements, teaching assistantships with full tuition reimbursements, unspecified assistantships available. *Unit head:* Dr. Debbie E Kipp, Chair, 336-334-5332, Fax: 336-334-4129, E-mail: dekipp@uncg.edu. *Application contact:* Michelle Harkleroad, Director of Graduate Admissions, 336-334-4884, Fax: 336-334-4424, E-mail: mbharkle@uncg.edu.

University of North Florida, College of Health, Department of Public Health, Jacksonville, FL 32224-2645. Offers community health (MPH); geriatric management (MSH); health administration (MHA); health behavior research and evaluation (Certificate); nutrition (MSH); rehabilitation counseling (MS). *Accreditation:* CORE. Part-time and evening/weekend programs available. *Faculty:* 19 full-time (17 women). *Students:* 75 full-time (62 women), 63 part-time (47 women); includes 37 minority (18 African Americans, 8 Asian Americans or Pacific Islanders, 11 Hispanic Americans), 5 international. Average age 31. 136 applicants, 51% accepted, 43 enrolled. In 2005, 55 degrees awarded. *Degree requirements:* For master's, thesis optional. *Entrance requirements:* For master's, GRE General Test (MSH, MS, MPH), GMAT or GRE General Test (MHA), minimum GPA of 3.0 in last 60 hours. Additional exam requirements/recommendations for international students: Required—TOEFL (minimum score 500 paper-based; 173 computer-based). *Application deadline:* For fall admission, 7/1 priority date for domestic students, 5/1 priority date for international students; for spring admission, 11/10 priority date for domestic students, 10/1 priority date for international students. Applications are processed on a rolling basis. Application fee: $30. Electronic applications accepted. *Expenses:* Tuition, state resident: full-time $4,391; part-time $244 per semester hour. Tuition, nonresident: full-time $15,036; part-time $835 per semester hour. Required fees: $789; $44 per semester hour. *Financial support:* In 2005–06, 81 students received support, including 11 teaching assistantships (averaging $2,942 per year); research assistantships, career-related internships or fieldwork, Federal Work-Study, scholarships/grants, and tuition waivers (partial) also available. Support available to part-time students. Financial award application deadline: 4/1; financial award applicants required to submit FAFSA. *Faculty research:* Dietary supplements; alcohol, tobacco, and other drug use prevention; turnover among health professionals; aging; psychosocial aspects of disabilities. Total annual research expenditures: $360,730. *Unit head:* Dr. Judith Perkin, Chair, 904-620-2840, Fax: 904-620-2848, E-mail: jperkin@unf.edu. *Application contact:* Rachel Broderick, Director of Advising, 904-620-2817, Fax: 904-620-1770, E-mail: rbroderi@unf.edu.

University of Oklahoma Health Sciences Center, Graduate College, College of Allied Health, Department of Nutritional Sciences, Oklahoma City, OK 73190. Offers MS. *Accreditation:* ADtA. *Degree requirements:* For master's, thesis optional. *Entrance requirements:* For master's, GRE General Test, interview, 3 letters of reference. Additional exam requirements/recommendations for international students: Required—TOEFL (minimum score 550 paper-based).

University of Pittsburgh, School of Health and Rehabilitation Sciences, Program in Health and Rehabilitation Sciences, Pittsburgh, PA 15260. Offers dietetics (MS); health and rehabilitation sciences (MS), including clinical dietetics, health care supervision and management, health information systems, occupational therapy, physical therapy, rehabilitation counseling, rehabilitation science and technology, sports medicine; wellness and human performance (MS). *Accreditation:* APTA. Part-time and evening/weekend programs available. *Faculty:* 26 full-time (14 women), 13 part-time/adjunct (4 women). *Students:* 43 full-time (31 women), 49 part-time (33 women); includes 7 minority (5 African Americans, 1 Asian American or Pacific Islander, 1 Hispanic American), 16 international. Average age 31. 123 applicants, 79% accepted, 55 enrolled. In 2005, 63 degrees awarded. *Entrance requirements:* For master's, minimum GPA of 3.0. Additional exam requirements/recommendations for international students: Required—TOEFL. *Application deadline:* Applications are processed on a rolling basis. Application fee: $30 ($40 for international students). Electronic applications accepted. *Expenses:* Tuition, state resident: full-time $13,194; part-time $537 per credit. Tuition, nonresident: full-time $25,012; part-time $1,026 per credit. Required fees: $700; $164 per term. Tuition and fees vary according to campus/location and program. *Financial support:* In 2005–06, 6 research assistantships with full tuition reimbursements (averaging $17,647 per year) were awarded; teaching assistantships, Federal Work-Study, institutionally sponsored loans, traineeships, and unspecified assistantships also available. Support available to part-time students. Financial award applicants required to submit FAFSA. *Faculty research:* Assistive technology, seating and wheeled mobility, cellular neurophysiology, low back syndrome, augmentative communication. Total annual research expenditures: $1.1 million. *Application contact:* Shameem Gangjee, Director of Admissions, 412-383-6558, Fax: 412-383-6535, E-mail: admissions@shrs.pitt.edu.

University of Puerto Rico, Medical Sciences Campus, College of Health Related Professions, Program in Dietetics Internship, San Juan, PR 00936-5067. Offers Certificate. *Degree requirements:* For Certificate, one foreign language. *Entrance requirements:* For degree, minimum GPA of 2.50, interview, participation in the computer matching process by the American Dietetic Association. *Expenses:* Tuition, state resident: full-time $3,600; part-time $100 per credit hour. Tuition, nonresident: full-time $4,655. Required fees: $1,734. Tuition and fees vary according to class time, degree level and program.

University of Puerto Rico, Medical Sciences Campus, Graduate School of Public Health, Department of Human Development, Program in Health Science Nutrition, San Juan, PR 00936-5067. Offers MS. Part-time programs available. *Students:* 24 (22 women) 1 international. 17 applicants, 76% accepted. In 2005, 4 degrees awarded. *Degree requirements:* For master's, thesis. *Entrance requirements:* For master's, GRE, previous course work in algebra, biochemistry, biology, chemistry, and social sciences. *Application deadline:* For fall admission, 3/15 for domestic students. Application fee: $20. *Financial support:* Research assistantships, teaching assistantships, Federal Work-Study and institutionally sponsored loans available. Financial award application deadline: 4/30. *Unit head:* Dr. Winna Rivera, Coordinator, 787-758-2525 Ext. 1433, Fax: 787-759-6719, E-mail: winnarivera@hotmail.com. *Application contact:* Prof. Mayra E. Santiago-Vargas, Counselor, 787-756-5244, Fax: 787-759-6719, E-mail: msantiago@rcm.upr.edu.

University of Puerto Rico, Medical Sciences Campus, School of Medicine, Division of Graduate Studies, Department of Biochemistry, San Juan, PR 00936-5067. Offers MS, PhD. *Degree requirements:* For master's, one foreign language, thesis/dissertation; for doctorate, one foreign language, thesis/dissertation, comprehensive exam. *Entrance requirements:* For master's and doctorate, GRE General Test, GRE Subject Test, interview, minimum GPA of 3.0. *Expenses:* Tuition, state resident: full-time $3,600; part-time $100 per credit hour. Tuition, nonresident: full-time $4,655. Required fees: $1,734. Tuition and fees vary according to class time, degree level and program. *Faculty research:* Aging, plasma proteins, vitamins and minerals, protein structure/function, glycosilation of proteins.

University of Puerto Rico, Río Piedras, College of Education, Program in Family Ecology and Nutrition, San Juan, PR 00931-3300. Offers M Ed. Part-time programs available. *Students:* 12 full-time (all women), 6 part-time (all women); all minorities (all Hispanic Americans) Average age 25. 1 applicant, 100% accepted, 1 enrolled. In 2005, 2 degrees awarded. *Degree requirements:* For master's, thesis. *Entrance requirements:* For master's, EXADEP, minimum GPA of 3.0, letter of recommendation. *Application deadline:* For fall admission, 2/1 for domestic students, 2/1 for international students. Application fee: $17. *Expenses:* Tuition, state resident: part-time $100 per credit. Tuition, nonresident: part-time $294 per credit. Required fees: $72 per term. *Financial support:* Fellowships, research assistantships, teaching assistantships, career-related internships or fieldwork, Federal Work-Study, institutionally sponsored loans, and tuition waivers (partial) available. Financial award application deadline: 5/31. *Unit head:* Dr. Ivonne Pasarell, Coordinator, 787-764-0000 Ext. 2607, Fax: 787-764-0000 Ext.2374. *Application contact:* Information Contact, 787-764-0000 Ext. 4368, Fax: 787-763-4130.

University of Rhode Island, Graduate School, College of the Environment and Life Sciences, Department of Nutrition and Food Sciences, Program in Food Science and Nutrition, Kingston, RI 02881. Offers food science (MS, PhD); nutrition (MS, PhD). In 2005, 2 degrees awarded. *Entrance requirements:* For master's and doctorate, GRE General Test. Additional exam requirements/recommendations for international students: Required—TOEFL. *Application deadline:* For fall admission, 4/15 for domestic students. Applications are processed on a rolling basis. Application fee: $35. *Expenses:* Tuition, state resident: full-time $5,522; part-time $307 per credit. Tuition, nonresident: full-time $15,992; part-time $888 per credit. Required fees: $1,786; $73 per credit. One-time fee: $80 part-time. *Unit head:* Prof. Geoffrey Greene, Graduate Coordinator, 401-874-4028, E-mail: gwg@uri.edu.

University of Southern California, Keck School of Medicine and Graduate School, Graduate Programs in Medicine, Department of Preventive Medicine, Master of Public Health Program, Los Angeles, CA 90089. Offers biometry/epidemiology (MPH); health communication (MPH); health promotion (MPH); preventive nutrition (MPH). *Accreditation:* CEPH. Part-time programs available. *Faculty:* 16 full-time (9 women), 8 part-time/adjunct (2 women). *Students:* 113 full-time (81 women), 10 part-time (9 women); includes 61 minority (9 African Americans, 1 American Indian/Alaska Native, 39 Asian Americans or Pacific Islanders, 12 Hispanic Americans), 21 international. Average age 26. 268 applicants, 67% accepted, 55 enrolled. In 2005, 54 degrees awarded. *Degree requirements:* For master's, practicum, final report, oral presentation. *Entrance requirements:* For master's, GRE General Test, MCAT, GMAT, DAT, minimum GPA of 3.0. Additional exam requirements/recommendations for international students: Required—TOEFL (minimum score 600 paper-based; 250 computer-based). *Application deadline:* For fall admission, 6/1 priority date for domestic students, 6/1 priority date for international students; for spring admission, 11/15 priority date for domestic students, 10/1 priority date for international students. Applications are processed on a rolling basis. Application fee: $65 ($75 for international students). Electronic applications accepted. *Expenses:* Tuition: Full-time $25,416; part-time $1,059 per unit. Required fees: $484; $484 per year. Tuition and fees vary according to course load and program. *Financial support:* In 2005–06, 120 students received support, including 7 research assistantships with full tuition reimbursements available (averaging $24,876 per year), 11 teaching assistantships with partial tuition reimbursements available (averaging $24,876 per year); career-related internships or fieldwork, Federal Work-Study, institutionally sponsored loans, scholarships/grants, health care benefits, unspecified assistantships, and staff tuition remission also available. Support available to part-time students. Financial award application deadline: 2/1; financial award applicants required to submit CSS PROFILE or FAFSA. *Faculty research:* Substance abuse prevention, cancer and heart disease prevention, mass media and health communication research, health promotion, treatment compliance. Total annual research expenditures: $12 million. *Unit head:* Dr. Thomas W. Valente, Director, 626-457-6678, Fax: 626-457-6699, E-mail: tvalente@usc.edu. *Application contact:* Nemesia P. Kelly, Program Specialist, 626-457-6603, Fax: 626-457-6699, E-mail: nkelly@usc.edu.

University of Southern Mississippi, Graduate School, College of Health, Center for Community Health, Hattiesburg, MS 39406-0001. Offers health education (MPH); health policy/administration (MPH); occupational/environmental health (MPH); public health nutrition (MPH). *Accreditation:* CEPH. Part-time and evening/weekend programs available. *Degree requirements:* For master's, comprehensive exam. *Entrance requirements:* For master's, GRE General Test, minimum GPA of 2.75 in last 60 hours. Additional exam requirements/recommendations for international students: Required—TOEFL. *Faculty research:* Rural health care delivery, school health, nutrition of pregnant teens, risk factor reduction, sexually transmitted diseases.

University of Southern Mississippi, Graduate School, College of Health, Center for Nutrition and Food Systems, Hattiesburg, MS 39406-0001. Offers MS, PhD.

The University of Tennessee, Graduate School, College of Education, Health and Human Sciences, Department of Nutrition, Knoxville, TN 37996. Offers nutrition (MS), including nutrition science, public health nutrition. Part-time programs available. *Degree requirements:* For master's, thesis or alternative. *Entrance requirements:* For master's, GRE General Test, minimum GPA of 2.7. Additional exam requirements/recommendations for international students: Required—TOEFL. Electronic applications accepted.

The University of Tennessee at Martin, Graduate Programs, College of Agriculture and Applied Sciences, Department of Family and Consumer Sciences, Martin, TN 38238-1000. Offers dietetics (MSFCS); general family and consumer sciences (MSFCS). Part-time programs available. *Faculty:* 6. *Students:* 19 (all women) 19 applicants, 58% accepted, 6 enrolled. In 2005, 6 degrees awarded. *Degree requirements:* For master's, thesis optional. *Entrance requirements:* For master's, GRE General Test, minimum GPA of 2.5. Additional exam requirements/recommendations for international students: Required—TOEFL (minimum score 525 paper-based; 197 computer-based). *Application deadline:* For fall admission, 8/1 priority date for domestic students, 8/1 priority date for international students; for spring admission, 1/1 for domestic students, 1/1 for international students. Applications are processed on a rolling basis. Application fee: $25 ($50 for international students). Electronic applications accepted. *Expenses:* Tuition, state resident: full-time $5,200; part-time $291 per hour. Tuition, nonresident: full-time $9,000; part-time $794 per hour. International tuition: $14,200 full-time. *Financial support:* In 2005–06, 6 students received support. Scholarships/grants, tuition waivers (partial), and unspecified assistantships available. Financial award application deadline: 3/1. *Faculty research:* Children with developmental disabilities, regional food product development and marketing, parent education. *Unit head:* Dr. Lisa LeBleu, Coordinator, 731-881-7116, E-mail: llebleu@utm.edu. *Application contact:* Linda S. Arant, Student Service Specialist, 731-881-7012, Fax: 731-881-7499, E-mail: larant@utm.edu.

The University of Texas at Austin, Graduate School, College of Natural Sciences, Department of Human Ecology, Program in Nutritional Sciences, Austin, TX 78712-1111. Offers nutrition (MA); nutritional sciences (PhD). *Degree requirements:* For master's and doctorate, thesis/dissertation. *Entrance requirements:* For master's and doctorate, GRE General Test. Additional exam requirements/recommendations for international students: Required—TOEFL. Electronic applications accepted. *Faculty research:* Nutritional biochemistry, nutrient health assessment, obesity, nutrition education, molecular/cellular aspects of nutrient functions.

University of the Incarnate Word, School of Graduate Studies and Research, HEB School of Business and Administration, Programs in Administration, San Antonio, TX 78209-6397. Offers

Nutrition

University of the Incarnate Word *(continued)*
adult education (MAA); applied administration (MAA); communication arts (MAA); English (MAA); instructional tech (MAA); international business (Certificate); multidisciplinary sciences (MAA); nutrition (MAA); organizational development (MAA, Certificate); project management (Certificate); sports management (MAA); urban administration (MAA). *Students:* 2 full-time (1 woman), 166 part-time (99 women); includes 100 minority (27 African Americans, 1 American Indian/Alaska Native, 1 Asian American or Pacific Islander, 71 Hispanic Americans), 11 international. Average age 35. In 2005, 46 degrees awarded. *Entrance requirements:* For master's, GMAT, GRE, MAT. Additional exam requirements/recommendations for international students: Required—TOEFL. *Application deadline:* For fall admission, 8/15 for domestic students; for spring admission, 12/31 for domestic students. Applications are processed on a rolling basis. Application fee: $20. *Expenses:* Tuition: Full-time $9,810; part-time $545 per credit hour. Required fees: $828; $46 per credit. One-time fee: $30. *Financial support:* Federal Work-Study, scholarships/grants, and Federal Loans available. *Unit head:* Dr. Dan Dominguez, MAA Director, 210-829-3180, Fax: 210-805-3564, E-mail: dominigue@uiwtx.edu. *Application contact:* Andrea Cyterski-Acosta, Dean of Enrollment, 210-829-6005, Fax: 210-829-3921, E-mail: cyterski@uiwtx.edu.

University of the Incarnate Word, School of Graduate Studies and Research, School of Mathematics, Sciences, and Engineering, Program in Nutrition, San Antonio, TX 78209-6397. Offers MS. Part-time and evening/weekend programs available. *Students:* 4 full-time (all women), 19 part-time (18 women); includes 12 minority (all Hispanic Americans) Average age 27. In 2005, 14 degrees awarded. *Degree requirements:* For master's, thesis optional. *Entrance requirements:* For master's, GRE General Test, minimum GPA of 3.0. Additional exam requirements/recommendations for international students: Required—TOEFL. *Application deadline:* For fall admission, 8/15 for domestic students; for spring admission, 12/31 for domestic students. Applications are processed on a rolling basis. Application fee: $20. *Expenses:* Tuition: Full-time $9,810; part-time $545 per credit hour. Required fees: $828; $46 per credit. One-time fee: $30. *Financial support:* Research assistantships, teaching assistantships, and Federal Work-Study, scholarships/grants, and Federal Loans available. *Faculty research:* Diabetes assessment, modified diets, minority health issues, health promotion, outcomes research. *Unit head:* Dr. Joseph Bonilla, Chair, 210-829-3908, Fax: 210-829-3153, E-mail: josephb@universe.uiwtx.edu. *Application contact:* Andrea Cyterski-Acosta, Dean of Enrollment, 210-829-6005, Fax: 210-829-3921, E-mail: cyterski@uiwtx.edu.

University of Toronto, School of Graduate Studies, Life Sciences Division, Department of Nutritional Sciences, Toronto, ON M5S 1A1, Canada. Offers M Sc, PhD, M Sc/PhD. Part-time programs available. *Degree requirements:* For master's, thesis, oral thesis defense; for doctorate, thesis/dissertation, departmental examination, oral examination, comprehensive exam. *Entrance requirements:* For master's, minimum B average, background in nutrition or an area of biological or health sciences, 2 letters of reference; for doctorate, minimum B+ average in final 2 years, background in nutrition or an area of biological of health sciences, 2 letters of reference. Additional exam requirements/recommendations for international students: Required—TOEFL (580 paper; 237 computer), TWE (5), IELTS (7), MELAB (85), or COPE (4) required.

University of Utah, The Graduate School, College of Health, Division of Nutrition, Salt Lake City, UT 84112-1107. Offers MS. *Accreditation:* ADtA. *Faculty:* 3 full-time (0 women), 4 part-time/adjunct (3 women). *Students:* 13 full-time (11 women), 3 part-time (1 woman); includes 2 minority (both Asian Americans or Pacific Islanders), 1 international. Average age 28. 29 applicants, 55% accepted. In 2005, 10 degrees awarded. *Degree requirements:* For master's, thesis, comprehensive exam. *Entrance requirements:* For master's, GRE General Test, minimum undergraduate GPA of 3.0. Additional exam requirements/recommendations for international students: Required—TOEFL (minimum score 500 paper-based; 173 computer-based). *Application deadline:* For fall admission, 2/3 for domestic students, 2/3 for international students. Application fee: $45 ($65 for international students). Electronic applications accepted. *Expenses:* Contact institution. Tuition and fees vary according to course load and program. *Financial support:* In 2005–06, 20 students received support, including 1 fellowship (averaging $5,500 per year), 3 research assistantships with partial tuition reimbursements available (averaging $5,250 per year), 7 teaching assistantships with partial tuition reimbursements available (averaging $5,250 per year); career-related internships or fieldwork, Federal Work-Study, and institutionally sponsored loans also available. Financial award application deadline: 3/1; financial award applicants required to submit FAFSA. *Faculty research:* Cholesterol metabolism, sport nutrition education, metabolic and critical care, cardiovascular nutrition, wilderness nutrition. Total annual research expenditures: $58,843. *Unit head:* Dr. E. Wayne Askew, Director, 801-581-8240, Fax: 801-585-3874, E-mail: wayne.askew@health.utah.edu. *Application contact:* Jean Zancanella, Academic Adviser, 801-581-5280, Fax: 801-585-3874, E-mail: jean.zancanella@health.utah.edu.

University of Vermont, Graduate College, College of Agriculture and Life Sciences, Department of Nutrition and Food Sciences, Burlington, VT 05405. Offers family and consumer sciences (MAT); nutritional sciences (MS). *Faculty:* 8 full-time (5 women), 3 part-time/adjunct (2 women). *Students:* 13 (9 women); includes 1 minority (African American) 5 international. 17 applicants, 71% accepted, 7 enrolled. In 2005, 1 degree awarded. *Degree requirements:* For master's, thesis. *Entrance requirements:* For master's, GRE General Test. Additional exam requirements/recommendations for international students: Required—TOEFL (minimum score 550 paper-based; 213 computer-based). *Application deadline:* For fall admission, 4/1 for domestic students; for spring admission, 11/15 for domestic students. Applications are processed on a rolling basis. Application fee: $40. *Expenses:* Tuition, area resident: Part-time $410 per credit hour. Tuition, nonresident: part-time $1,034 per credit hour. *Financial support:* Research assistantships available. Financial award application deadline: 3/1. *Unit head:* Dr. J. Harvey-Berino, Director, 802-656-3374, Fax: 802-656-0407.

University of Washington, Graduate School, School of Public Health and Community Medicine, Department of Epidemiology, Interdisciplinary Graduate Program in Nutritional Sciences, Seattle, WA 98195. Offers MPH, MS, PhD. *Accreditation:* ADtA. Part-time programs available. *Faculty:* 11 full-time (7 women), 5 part-time/adjunct (all women). *Students:* 42 full-time (35 women), 12 part-time (11 women); includes 8 minority (6 Asian Americans or Pacific Islanders, 2 Hispanic Americans), 1 international. Average age 25. 71 applicants, 55% accepted, 15 enrolled. In 2005, 11 master's, 3 doctorates awarded. Terminal master's awarded for partial completion of doctoral program. *Median time to degree:* Of those who began their doctoral program in fall 1997, 98% received their degree in 8 years or less. *Degree requirements:* For master's, thesis, practicum (MPH); for doctorate, thesis/dissertation, comprehensive exam, registration. *Entrance requirements:* For master's and doctorate, GRE General Test, experience in health sciences (preferred), minimum GPA of 3.0. Additional exam requirements/recommendations for international students: Required—TOEFL (minimum score 580 paper-based; 237 computer-based). *Application deadline:* For fall admission, 2/1 priority date for domestic students, 2/1 priority date for international students. Applications are processed on a rolling basis. Application fee: $50. *Financial support:* In 2005–06, 19 students received support, including 8 research assistantships with full tuition reimbursements available (averaging $10,000 per year), 2 teaching assistantships with full tuition reimbursements available (averaging $10,000 per year); career-related internships or fieldwork also available. Financial award application deadline: 3/1. *Faculty research:* Lipids, trace elements, nutrition in disease prevention. *Unit head:* Dr. Adam Drewnowski, Director, 206-543-1730, Fax: 206-685-1696, E-mail: adamdrew@u.washington.edu. *Application contact:* Carey Purnell, Graduate Program Assistant, 206-543-1730, Fax: 206-685-1696, E-mail: nutr@u.washington.edu.

University of Wisconsin–Madison, Graduate School, College of Agricultural and Life Sciences, Department of Nutritional Sciences, Madison, WI 53706-1380. Offers MS, PhD. Part-time programs available. *Faculty:* 11 full-time (5 women). *Students:* 40 full-time (30 women), 11 international. Average age 27. 57 applicants, 25% accepted, 9 enrolled. In 2005, 2 master's, 5 doctorates awarded. Terminal master's awarded for partial completion of doctoral program. *Median time to degree:* Of those who began their doctoral program in fall 1997, 99% received

their degree in 8 years or less. *Degree requirements:* For master's, thesis or research report; for doctorate, thesis/dissertation. *Entrance requirements:* For master's and doctorate, GRE General Test. Additional exam requirements/recommendations for international students: Required—TOEFL. *Application deadline:* For fall admission, 1/5 priority date for domestic students, 1/5 priority date for international students. Applications are processed on a rolling basis. Application fee: $45. Electronic applications accepted. *Financial support:* In 2005–06, 4 fellowships with tuition reimbursements (averaging $20,772 per year), 32 research assistantships with tuition reimbursements (averaging $20,772 per year), 2 teaching assistantships with tuition reimbursements (averaging $20,772 per year) were awarded; Federal Work-Study, institutionally sponsored loans, and traineeships also available. Support available to part-time students. Financial award applicants required to submit FAFSA. *Faculty research:* Human and animal nutrition, nutrition epidemiology, nutrition education, biochemical and molecular nutrition. *Unit head:* Dr. Roger A. Sunde, Chair, 608-262-2727, Fax: 608-262-5860. *Application contact:* Cheri Bill-Mohoney, Program Assistant, 608-262-2513, Fax: 608-262-5860, E-mail: cmohoney@nutrisci.wisc.edu.

University of Wisconsin–Stevens Point, College of Professional Studies, School of Health Promotion and Human Development, Program in Nutritional Sciences, Stevens Point, WI 54481-3897. Offers MS. Part-time programs available. *Faculty:* 10 full-time (5 women). In 2005, 2 degrees awarded. *Degree requirements:* For master's, thesis or alternative. *Entrance requirements:* For master's, minimum GPA of 2.75. *Application deadline:* For fall admission, 5/1 for domestic students. Applications are processed on a rolling basis. Application fee: $45. *Expenses:* Tuition, state resident: full-time $5,619; part-time $312 per credit. Tuition, nonresident: full-time $16,229; part-time $902 per credit. Required fees: $651; $64 per credit. *Financial support:* Research assistantships, teaching assistantships, career-related internships or fieldwork and Federal Work-Study available. Support available to part-time students. Financial award application deadline: 5/1; financial award applicants required to submit FAFSA. *Unit head:* Marty Loy, Head, School of Health Promotion and Human Development, 715-346-2830, Fax: 715-346-3751.

University of Wisconsin–Stout, Graduate School, College of Human Development, Program in Food and Nutritional Sciences, Menomonie, WI 54751. Offers MS. Part-time programs available. *Faculty:* 33 full-time (18 women). *Students:* 17 full-time (14 women), 4 part-time (3 women); includes 2 minority (both African Americans), 5 international. Average age 30. 26 applicants, 77% accepted, 9 enrolled. In 2005, 1 degree awarded. *Degree requirements:* For master's, thesis, 40 credits, minimum of 20 credits, must be 700 level or higher. *Entrance requirements:* For master's, minimum GPA of 3.0. Additional exam requirements/recommendations for international students: Required—TOEFL (minimum score 500 paper-based; 173 computer-based). *Application deadline:* Applications are processed on a rolling basis. Application fee: $45. Electronic applications accepted. *Expenses:* Tuition, state resident: part-time $301 per credit hour. Tuition, nonresident: part-time $532 per credit hour. *Financial support:* In 2005–06, 4 research assistantships with partial tuition reimbursements (averaging $6,358 per year) were awarded; teaching assistantships, Federal Work-Study, scholarships/grants, health care benefits, tuition waivers (full and partial), and unspecified assistantships also available. Support available to part-time students. Financial award application deadline: 4/1; financial award applicants required to submit FAFSA. *Faculty research:* Nutritional biochemistry, nutritional aspects of degenerative diseases and carcinogenesis, community nutrition programs impacting chronic disease and malnutrition, innovations in nutrition education. *Unit head:* Dr. Carol Seaborn, Director, 715-232-2216, E-mail: seaborne@uwstout.edu. *Application contact:* Anne E. Johnson, Graduate Student Evaluator, 715-232-1322, Fax: 715-232-2413, E-mail: johnsona@uwstout.edu.

University of Wyoming, Graduate School, College of Agriculture, Department of Animal Sciences, Program in Food Science and Human Nutrition, Laramie, WY 82070. Offers MS. *Faculty:* 8 full-time (2 women). *Students:* 3 full-time (all women), 5 part-time (3 women), 2 international. Average age 33. 4 applicants, 50% accepted, 0 enrolled. *Degree requirements:* For master's, thesis. *Entrance requirements:* For master's, GRE General Test, minimum GPA of 3.0. Additional exam requirements/recommendations for international students: Required—TOEFL (minimum score 525 paper-based). *Application deadline:* For fall admission, 2/1 priority date for domestic students, 2/1 priority date for international students; for spring admission, 9/1 priority date for domestic students, 9/1 priority date for international students. Applications are processed on a rolling basis. Application fee: $50. Electronic applications accepted. *Expenses:* Tuition, state resident: full-time $3,720; part-time $155 per credit hour. Tuition, nonresident: full-time $10,704; part-time $446 per credit hour. Required fees: $666; $162 per semester. Tuition and fees vary according to course load and program. *Financial support:* In 2005–06, 1 student received support, including research assistantships with tuition reimbursements available (averaging $7,000 per year); career-related internships or fieldwork, Federal Work-Study, institutionally sponsored loans, scholarships/grants, and unspecified assistantships also available. Financial award application deadline: 3/1. *Faculty research:* Protein and lipid metabolism, food microbiology, food safety, meat science. *Unit head:* Dr. Warrie J. Means, Professor, 307-766-3404, Fax: 307-766-2355, E-mail: dcrule@uwyo.edu. *Application contact:* Office Assistant, Senior, 307-766-2224, Fax: 307-766-2355, E-mail: animalscience@uwyo.edu.

Utah State University, School of Graduate Studies, College of Agriculture, Department of Nutrition and Food Sciences, Logan, UT 84322. Offers dietetic administration (MDA); food microbiology and safety (MFMS); nutrition and food sciences (MS, PhD); nutrition science (MS, PhD), including molecular biology. Postbaccalaureate distance learning degree programs offered. *Faculty:* 13 full-time (6 women), 1 (woman) part-time/adjunct. *Students:* 55 full-time (26 women), 13 part-time (12 women); includes 4 minority (all African Americans), 15 international. Average age 27. 14 applicants, 57% accepted, 4 enrolled. In 2005, 8 master's, 3 doctorates awarded. *Degree requirements:* For master's, thesis, BS core competency courses BS core competency courses; for doctorate, thesis/dissertation, teaching experience, inst 7920 and BS core competency courses, comprehensive exam, registration. *Entrance requirements:* For master's, GRE General Test, minimum GPA of 3.0, course work in chemistry; for doctorate, GRE General Test, minimum GPA of 3.2, course work in chemistry, MS or manuscript in referred journal. Additional exam requirements/recommendations for international students: Required—TOEFL (minimum score 560 paper-based). *Application deadline:* For fall admission, 6/15 priority date for domestic students, 6/15 priority date for international students; for spring admission, 10/15 priority date for domestic students, 10/15 priority date for international students. Applications are processed on a rolling basis. Application fee: $50 ($60 for international students). Electronic applications accepted. *Financial support:* In 2005–06, 19 students received support, including fellowships with partial tuition reimbursements available (averaging $14,484 per year), 22 research assistantships with partial tuition reimbursements available (averaging $12,600 per year), 3 teaching assistantships with partial tuition reimbursements available (averaging $6,000 per year); Federal Work-Study, institutionally sponsored loans, scholarships/grants, tuition waivers (full and partial), unspecified assistantships, and fellowships, international students get support from their countries also available. Financial award application deadline: 3/1. *Faculty research:* Mineral balance, meat microbiology and nitrate interactions, milk ultrafiltration, lactic culture, milk coagulation. Total annual research expenditures: $319,280. *Unit head:* Dr. Charles C. Carpenter, Head, 435-797-2126, Fax: 435-797-2379, E-mail: chuck@cc.usu.edu. *Application contact:* Pam Zetterquist, Staff Assistant II, 435-797-4041, Fax: 435-797-2379, E-mail: pzett@cc.usu.edu.

Virginia Polytechnic Institute and State University, Graduate School, College of Agriculture and Life Sciences, Department of Human Nutrition, Foods and Exercise, Blacksburg, VA 24061. Offers MS, PhD. *Faculty:* 21 full-time (11 women). *Students:* 36 full-time (23 women), 3 part-time (all women); includes 5 minority (3 African Americans, 1 Asian American or Pacific Islander, 1 Hispanic American), 9 international. Average age 27. 33 applicants, 48% accepted, 10 enrolled. In 2005, 8 master's, 2 doctorates awarded. *Entrance requirements:* Additional exam requirements/recommendations for international students: Required—TOEFL (minimum score 550 paper-based; 213 computer-based). *Application deadline:* Applications are processed on a rolling basis. Application fee: $45. Electronic applications accepted. *Expenses:* Tuition,

Nutrition

state resident: full-time $6,558; part-time $364 per credit. Tuition, nonresident: full-time $11,296; part-time $628 per credit. Required fees: $1,419; $468 per credit. $234 per term. *Financial support:* In 2005–06, 2 fellowships with full tuition reimbursements (averaging $16,740 per year), 8 research assistantships with full tuition reimbursements (averaging $18,823 per year), 17 teaching assistantships with full tuition reimbursements (averaging $13,552 per year) were awarded; career-related internships or fieldwork, Federal Work-Study, scholarships/grants, and unspecified assistantships also available. Financial award application deadline: 4/1. *Faculty research:* Nutrition and food science research. *Unit head:* Dr. Gerald Jubb, Head, 540-231-4672, Fax: 540-231-3916. *Application contact:* Sherry Terry, Information Contact, 540-231-4640, Fax: 540-231-3916, E-mail: terrysd@vt.edu.

Washington State University, Graduate School, College of Agricultural, Human, and Natural Resource Sciences, Department of Food Science and Human Nutrition, Program in Human Nutrition, Pullman, WA 99164. Offers MS. *Faculty:* 13. *Students:* 4 full-time (3 women), 1 (woman) part-time; includes 1 minority (Asian American or Pacific Islander), 3 international. Average age 29. 17 applicants, 18% accepted, 3 enrolled. In 2005, 4 degrees awarded. *Degree requirements:* For master's, oral exam, written exam, thesis optional. *Entrance requirements:* For master's, GRE General Test, minimum GPA of 3.0, resumé, 3 letters of recommendation, letter of interest. Additional exam requirements/recommendations for international students: Required—TOEFL (minimum score 550 paper-based; 213 computer-based). *Application deadline:* For fall admission, 2/1 priority date for domestic students, 2/1 priority date for international students; for spring admission, 8/1 priority date for domestic students, 7/1 priority date for international students. Applications are processed on a rolling basis. Application fee: $35. Electronic applications accepted. *Expenses:* Tuition, state resident: full-time $6,295; part-time $336 per credit. Tuition, nonresident: full-time $15,949; part-time $819 per credit. Required fees: $933. Part-time tuition and fees vary according to campus/location and program. *Financial support:* In 2005–06, 3 students received support, including 1 research assistantship with full and partial tuition reimbursement available (averaging $16,317 per year), 2 teaching assistantships with full and partial tuition reimbursements available (averaging $11,637 per year); career-related internships or fieldwork, Federal Work-Study, scholarships/grants, tuition waivers (partial), and unspecified assistantships also available. Financial award application deadline: 2/1; financial award applicants required to submit FAFSA. *Faculty research:* Nutrition education programs, cultural issues in community nutrition programs. *Unit head:* Dr. Mariam Edlefsen, Director, 509-335-1395, E-mail: medlefsen@wsu.edu. *Application contact:* Jodi L. Anderson, Academic Coordinator, 509-335-4763, Fax: 509-335-4815, E-mail: fshn@wsu.edu.

Washington State University, Graduate School, College of Agricultural, Human, and Natural Resource Sciences, Department of Food Science and Human Nutrition, Program in Nutrition, Pullman, WA 99164. Offers PhD. *Faculty:* 13. *Students:* 1 (woman) full-time, 1 (woman) part-time, 1 international. Average age 29. 16 applicants, 6% accepted, 0 enrolled. In 2005, 2 degrees awarded. *Degree requirements:* For doctorate, thesis/dissertation, oral exam, written exam. *Entrance requirements:* For doctorate, GRE, minimum GPA of 3.0, resumé, 3 letters of recommendation. Additional exam requirements/recommendations for international students: Required—TOEFL (minimum score 550 paper-based; 213 computer-based). *Application deadline:* For fall admission, 2/1 priority date for domestic students, 2/1 priority date for international students; for spring admission, 8/1 priority date for domestic students, 7/1 priority date for international students. Applications are processed on a rolling basis. Application fee: $35. Electronic applications accepted. *Expenses:* Tuition, state resident: full-time $6,295; part-time $336 per credit. Tuition, nonresident: full-time $15,949; part-time $819 per credit. Required fees: $933. Part-time tuition and fees vary according to campus/location and program. *Financial support:* In 2005–06, 2 students received support, including 1 fellowship (averaging $3,250 per year), 1 research assistantship with full and partial tuition reimbursement available

(averaging $12,353 per year); teaching assistantships with full and partial tuition reimbursements available, career-related internships or fieldwork, Federal Work-Study, institutionally sponsored loans, scholarships/grants, health care benefits, tuition waivers (partial), and unspecified assistantships also available. Financial award application deadline: 2/1; financial award applicants required to submit FAFSA. *Faculty research:* Breastfeeding and lactation influence on maternal child wellbeing, sports anemia preschool nutrition. *Unit head:* Dr. Jill Shultz, Chair, 509-335-6181, Fax: 509-335-4815, E-mail: armstroj@wsu.edu. *Application contact:* Jodi L. Anderson, Academic Coordinator, 509-335-4763, Fax: 509-335-4815, E-mail: fshn@wsu.edu.

Wayne State University, Graduate School, College of Liberal Arts and Sciences, Department of Nutrition and Food Science, Detroit, MI 48202. Offers MA, MS, PhD. *Faculty:* 14 full-time (13 women). *Students:* 33 full-time (29 women), 5 part-time (4 women); includes 7 minority (3 African Americans, 4 Asian Americans or Pacific Islanders), 18 international. Average age 31. 26 applicants, 62% accepted, 8 enrolled. In 2005, 7 master's, 1 doctorate awarded. Terminal master's awarded for partial completion of doctoral program. *Degree requirements:* For master's, thesis (for some programs); for doctorate, thesis/dissertation. *Entrance requirements:* For master's and doctorate, GRE General Test, minimum GPA of 3.0. Additional exam requirements/recommendations for international students: Required—TOEFL (minimum score 550 paper-based; 213 computer-based); Recommended—TWE (minimum score 6). *Application deadline:* For fall admission, 7/1 for domestic students, 6/1 for international students. Applications are processed on a rolling basis. Application fee: $50 ($50 for international students). Electronic applications accepted. *Expenses:* Tuition, state resident: part-time $338 per credit hour. Tuition, nonresident: part-time $746 per credit hour. Required fees: $24 per credit hour. Full-time tuition and fees vary according to program. *Financial support:* In 2005–06, 10 students received support, including 1 fellowship with tuition reimbursement available, 7 teaching assistantships (averaging $14,997 per year); research assistantships, career-related internships or fieldwork and Federal Work-Study also available. Financial award application deadline: 4/1. *Faculty research:* Nutrition, cancer and gene expression, food micro-biology and food safety, lipids, lipoprotein and cholesterol metabolism, obesity and diabetes, metabolomics. Total annual research expenditures: $934,572. *Unit head:* Dr. K. L. Catherine Jen, Chair, 313-577-2500, E-mail: cjen@sun.science.wayne.edu. *Application contact:* Pramod Khosla, Assistant Professor, 313-577-9055, Fax: 313-577-8616, E-mail: pkhosla@sun.science.wayne.edu.

West Virginia University, Davis College of Agriculture, Forestry and Consumer Sciences, Division of Animal and Veterinary Sciences, Program in Animal and Veterinary Sciences, Morgantown, WV 26506. Offers breeding (MS); food sciences (MS); nutrition (MS); physiology (MS); production management (MS); reproduction (MS). Part-time programs available. *Degree requirements:* For master's, thesis, oral and written exams. *Entrance requirements:* For master's, minimum GPA of 2.5. Additional exam requirements/recommendations for international students: Required—TOEFL. Application fee: $45. *Expenses:* Tuition, state resident: full-time $4,582; part-time $258 per credit hour. Tuition, nonresident: full-time $13,820; part-time $741 per credit hour. *Financial support:* Research assistantships, teaching assistantships, Federal Work-Study, institutionally sponsored loans, and tuition waivers (full and partial) available. Financial award application deadline: 2/1; financial award applicants required to submit FAFSA. *Faculty research:* Animal nutrition, reproductive physiology, food science. *Unit head:* Dr. Hillar Klandorf, Coordinator, 304-293-4372 Ext. 2050, Fax: 304-293-3676, E-mail: hillar.klandorf@mail.wvu.edu.

Winthrop University, College of Arts and Sciences, Department of Human Nutrition, Rock Hill, SC 29733. Offers MS. Part-time programs available. *Degree requirements:* For master's, thesis optional. *Entrance requirements:* For master's, GRE General Test, or PRAXIS, interview, minimum GPA of 3.0. Electronic applications accepted.

BASTYR UNIVERSITY

School of Nutrition and Exercise Science

Program of Study

Bastyr University is a small private university that features a strong sense of community and fosters both academic and personal growth. Bastyr's School of Nutrition and Exercise Science offers a 65-quarter-credit program leading to a Master of Science in Nutrition. The courses at Bastyr balance a whole, natural foods philosophy with an understanding of human biochemistry and nutrient metabolism. The core curriculum includes course work in nutritional biochemistry, nutritional assessment, diet therapy, research methods, disease processes, counseling, eating disorders, and whole foods cooking. Electives include sports nutrition, vegetarian nutrition, and mental health. Experiential course work includes clinical shifts and practicum experiences in Bastyr Center for Natural Health and in the community. Students may choose a research emphasis that culminates in a thesis or pursue the new program in nutrition and clinical health psychology. This option, replacing the University's former clinical track in nutrition, is designed to enhance the nutritional counseling career field by also leading to licensure as a mental health counselor.

As nutrition consultants, graduates function both independently and in association with other professionals in a variety of settings such as private clinics, schools, and medical and dental offices as well as in the health and fitness industry. Graduates may become involved in the planning and implementation of nutrition programs for communities, businesses, or individuals. Graduates may also work in higher education or research, or they may proceed to doctoral studies. In addition, Bastyr University offers a Master of Science in Nutrition plus the Didactic Program in Dietetics (M.S.N./DPD), which is developmentally accredited by the Council on Accreditation for the American Dietetic Association. Bastyr University also offers a developmentally accredited dietetic internship that emphasizes community and public health nutrition. Completion of both the M.S.N./DPD and the dietetic internship prepares graduates to become registered dietitians (RDs).

Research Facilities

The University's mission includes the pursuit of scientific research on the use of nutrition and natural therapies in the management and treatment of health-care problems and in the prevention of chronic disease. Students can develop research projects in the community and the clinic. An active research department provides the opportunity for varied research experiences in nutrition. Bastyr nutrition students also have the opportunity of working with researchers in other institutions in the area, such as the Fred Hutchinson Cancer Research Center and Harborview Medical Center. The University maintains a medical library with current journals; special collections in the areas of nutrition, Oriental medicine, and naturopathic medicine; and audiovisual equipment and materials. Students also have access to the University of Washington Health Sciences Library.

Financial Aid

Students are eligible to participate in the University Scholarship Program, the Federal Stafford Loan Program, the Federal Perkins Loan Program, and the Federal Work-Study Program. Washington State residents are also eligible to participate in the Washington State Work-Study Program. In addition, there are private donor scholarships earmarked for students entering the Master of Science in Nutrition program. Applicants seeking financial aid should complete the application process by May 15. Complete financial aid information is available on the Bastyr University Web site, listed in this In-Depth Description. Students may request a financial aid packet by contacting the financial aid office at 425-602-3407.

Cost of Study

Tuition for 2004–05 was $292 per credit for the first 15 credits per quarter and $200 per credit for 16 or more credits. Total tuition and fees for a full-time student in the first year were approximately $13,680 for the academic year. Books and supplies cost approximately $820.

Living and Housing Costs

The University provides limited housing. Many students live in shared housing off campus; the average monthly rent ranges from $400 to $800 per person. The Student Services Office can direct students to listings of available housing. The Washington Association of Student Financial Aid Administrators estimates that living expenses, including transportation and personal expenses, average $1070 per month.

Student Group

The University's total enrollment in 2004–05 was 1,126; 749 of those students were enrolled in the graduate and professional programs. Twenty-five percent were Washington residents; twenty-six other states were represented. Five percent of the students were from other countries. Eighty-seven students were enrolled in the M.S. in Nutrition program. Their average age was 28.

Location

The academic and administrative campus is on a 47-acre site adjoining St. Edward's State Park in Kenmore, Washington, about 15 miles northeast of Seattle. Several hiking trails to Lake Washington begin at the edge of the campus; abundant opportunities for outdoor activities exist nearby in the Puget Sound area. The clinical campus is located in a Seattle neighborhood. The city offers a full range of activities and facilities, including museums, theaters, fine restaurants of many cuisines, a major opera and symphony orchestra, major-league sports, and outdoor activities. The Puget Sound area has a large number of academic institutions, including seven universities and many colleges, community colleges, and professional schools.

The University

Bastyr University was founded as a naturopathic medical college in 1978 to meet the growing need for scientifically trained naturopathic physicians, brought about by increased consumer interest in, and demand for, natural health-oriented care and preventive medicine. The University's mission includes serving as an effective leader in the improvement of the health and well-being of the human community. Since 1984, as a part of its mission to provide comprehensive education in natural medicine, the University has added graduate and undergraduate programs in nutrition, Oriental medicine, acupuncture, health psychology, exercise science and wellness, and herbal sciences as well as an M.A. program in applied behavioral sciences through the Leadership Institute of Seattle (LIOS). Bastyr University is accredited by the Northwest Commission on Colleges and Universities. A comprehensive administrative fee enables all students to visit Bastyr Center for Natural Health, the University's teaching clinic (a co-pay is due at each visit). Counseling services, acupuncture, physical medicine, and the herbal dispensary are all accessible to students.

Applying

Admission is based on academic achievement, personal and social development, relevant experience, and demonstrated humanistic qualities. Credentials to be submitted are all official transcripts, two letters of recommendation, a completed application form, and a $75 application fee. Minimum prerequisites are a baccalaureate degree with courses (including labs) in biochemistry, inorganic chemistry, organic chemistry, and physiology as well as courses in college-level algebra and nutrition. Prospective students with baccalaureate degrees and a solid history of academic performance may apply for admission to the postbaccalaureate program, which provides for one year of prerequisite preparation. Specific information is available from the Admissions Office. Applications should be received by the University by March 15 for admission the following fall, although late applications are considered if space is available.

Correspondence and Information

Admissions Office
Bastyr University
14500 Juanita Drive, NE
Kenmore, Washington 98028
Phone: 425-602-3330
Fax: 425-602-3090
E-mail: admissions@bastyr.edu
Web site: http://www.bastyr.edu/sub/adtrack.asp?adid=pe02

Bastyr University

THE FACULTY

Cheri Boekenoogen, M.S.N., Oklahoma.
Debra Boutin, M.S., Case Western Reserve.
Jeanne M. Cullen, M.S., Bastyr; RD, CD.
Ann Fiitante, M.S., Columbia; RD.
Mark Kestin, Ph.D., Flinders (Australia); M.P.H., Harvard.
Beverly Kindblade, M.S., Washington (Seattle); RD.
Samer Koutoubi, M.D., Institute of Medicine and Pharmacy (Romania); Ph.D., Florida International.
Cynthia Lair, B.A., Wichita State; CHN.
Buck Levin, Ph.D., North Carolina at Greensboro; RD.
Jennifer C. Lovejoy, Ph.D., Emory.
Scott Murdoch, Ph.D., North Carolina at Greensboro; RD.
Tiffany Reiss, Ph.D., Virginia Tech.

Students prepare organic vegetables in the University's demonstration nutrition kitchen.

BOSTON UNIVERSITY

School of Medicine
Division of Graduate Medical Sciences
Programs in Medical Nutrition Sciences

Programs of Study

The Programs in Medical Nutrition Sciences are the newest offerings in the Division of Graduate Medical Sciences at the Boston University School of Medicine, with tracks leading to the Master of Arts (M.A.) degree, the Doctor of Philosophy (Ph.D.) degree, and M.D./Ph.D. degree. These rigorous and interdisciplinary programs involve the full collaboration of the Boston University Schools of Medicine, Dental Medicine, and Public Health.

The main goal of the programs is to provide multidisciplinary training in medical nutrition sciences that prepares students for a wide variety of relevant career opportunities in four primary areas: basic, epidemiological, and clinical nutrition research; nutrition and public health promotion at the individual and population levels; nutrition policy and program planning; and nutrition and health communications. The core curriculums incorporate state-of-the-art, advanced training and a required thesis or dissertation in various areas of research and practice, including basic and laboratory nutritional sciences and related research methods and technologies; techniques for translating nutritional sciences into medical nutrition therapies, interventions, products, and services; epidemiological, clinical, and outcomes research methods used to evaluate relationships between nutritional status and health outcomes; methods to design and test the efficacy and impact of medical nutrition therapies and nutritional interventions; processes for the formulation and monitoring of nutrition-related health policies, programs, and campaigns; and strategies for effective public and professional communications of nutritional science.

The M.A. requires 32 credits, including 22 (6) core course credits and 10 (6) elective credits. Students generally complete their degrees in four semesters, including their thesis research experience. The M.A. program provides an excellent educational framework for those interested in pursuing wide-ranging nutrition careers in diverse settings such as major teaching hospitals; public and private health service organizations; local, state, and federal government; private research organizations; pharmaceutical companies; consumer product companies; electronic and print media; or public relations and health consulting firms. The M.A. program is also designed for those who desire a sound background in nutrition and preventive medicine prior to entering medical school or doctoral training. Ph.D. students who enter the program with an existing master's degree or another doctoral degree and professional experience in nutrition may complete the program in 32 credits. Those with a bachelor's degree in nutrition or a closely related science complete a 64-credit degree program that includes the 32-credit core sequence in the M.A. program. M.D./Ph.D. students complete dual-degree requirements, including the Ph.D. curriculum and dissertation research.

Research Facilities

Boston University School of Medicine ranks twelfth among U.S. medical schools for sponsored research. There are more than 600 funded research programs and 1,000 active clinical trials, providing an exceptional environment for students interested in advanced study and practice in a variety of medical, biomedical, and public health careers. The Boston University Medical Center Campus is home to the Clinical Research Center, Pulmonary Center, Arthritis Center, Cardiovascular Institute, Cancer Center, Center of Human Genetics, Obesity Center, Gerontology Center, and Center for Research in Women's Health as well as the Mallory Institute of Pathology, the Thorndike Memorial Laboratory, the Maxwell Finland Laboratory for Infectious Diseases, and the Sloan Epidemiology Unit. The prevalence of joint appointments between basic science and clinical departments attests to the high level of cooperation between basic scientists and clinicians in the conceptualization, discovery, and development of unusual research programs.

Financial Aid

Many forms of financial aid, including fellowships, assistantships, scholarships, loans, and work-study, are available to graduate students. The Division of Graduate Medical Sciences/Office of Financial Aid (617-638-5216) provides information and resources for applicants.

Cost of Study

For the 2005–06 academic year, full-time students (12–18 credits) paid $39,510 per year in tuition, and student fees were $450.

Living and Housing Costs

Boston University Medical Campus is situated in the South End neighborhood of the city of Boston. The Office of Housing provides information about on-campus residence options. The Office of Off-Campus Services (OCS) offers student guidance about off-campus housing in Boston's neighborhoods, average rental prices, and roommate matching. Public transportation in and around Boston is very good and readily accessible. The estimated cost for room and board in 2003–04 was $12,243 for twelve months.

Student Group

The Boston University Medical Campus is a unique environment with more than 2,000 students and 650 postgraduate trainees pursuing advanced study in medical, biomedical, dental, public health, and related disciplines. The student body represents the full range of cultural, educational, geographic, and ethnic diversity.

Location

The city of Boston, incorporated in 1822, is the economic and cultural hub of New England. The city is rich in history, old-world charm, and modern vitality. It is home to nearly 590,000 residents, many institutions of higher education, some of the world's finest hospitals, and numerous cultural and professional sports organizations. Boston-based jobs, primarily within the finance, health care, educational, and service areas numbered nearly 660,000 in 2002. A wide range of programs and services are available to meet the diverse needs of Boston's residents and visitors.

The University

Boston University is the fourth-largest, private research university in the U.S. It is a hub of intellectual, scientific, and cultural activity with 2,500 faculty members and 29,000 students from all fifty states and 143 countries. Boston University has two campuses—one situated along the Charles River in Boston, and another semi-autonomous medical campus located in the South End neighborhood of Boston. The medical campus consolidates the resources and activities of the School of Medicine, the School of Public Health, the Goldman School of Dental Medicine, Boston Medical Center Hospital, and Biosquare, a complex that houses specialized research facilities and biotechnology companies.

Applying

Applicants should be aware that admission to the Programs in Medical Nutrition Sciences is on a rolling basis. All applicants must have met the requirements for a baccalaureate degree with a strong background in the biological and physical sciences (a minimum of 28 credits or the equivalent). Applicants must submit the completed application, a nonrefundable application fee of $50 for the paper application ($60 for the online application), official transcripts or records of each college or university attended, letters of recommendation from 3 faculty members, and official results of the GRE test. MCAT results may be substituted with prior approval. In addition to these requirements, international applicants must submit the completed and signed International Student Data Form and a declaration of financial support, including supporting documentation; certified copies and certified English translations of all academic achievements in each college or university attended; and the results of the TOEFL. Students who matriculate at the doctoral level generally qualify for fellowship and scholarship awards. Applicants are encouraged to submit application forms and all supporting materials by May 1 for the fall and November 15 for the spring.

Correspondence and Information

Barbara E. Millen, Professor and Chair
Program in Medical Nutrition Sciences
Division of Graduate Medical Sciences
Boston University Medical Center
715 Albany Street L-317
Boston, Massachusetts 02118-2531
Phone: 617-638-4472
Fax: 617-638-4483
E-mail: bmillen@bu.edu
Web site: http://www.bumc.bu.edu/medicalnutrition

Boston University

THE FACULTY AND THEIR RESEARCH

Core Faculty Members

Caroline Apovian, M.D., Associate Professor of Medicine, Department of Endocrinology, Nutrition, and Diabetes in the Boston University School of Medicine; FACN. Dr. Apovian is currently the Director of the Nutrition and Weight Management Center and Co-director of the Nutrition and Metabolic Support Service at Boston Medical Center. She also serves as Director of Clinical Research at the Obesity Research Center of Boston University Medical Center. Research interests: obesity and co-morbidities; nutrition support in the critically ill patient; the elderly and physical function. As Director of the Nutrition and Weight Management Center at Boston Medical Center, Dr. Apovian's group investigates dietary and physical activity interventions in the obese, improvements in co-morbid conditions with lifestyle interventions, and health outcomes associated with gastric bypass surgery in morbidly obese individuals. Within the Nutrition and Metabolic Support Service, Dr. Apovian's team is involved in clinical research related to parenteral/enteral nutrition in the critically ill patient, home total parenteral nutrition, and the economics of nutrition support in the hospital. Dr. Apovian is the principal investigator for the Centers for Obesity Research and Education (C.O.R.E.) in New England and has been an active member of the Obesity Research Nutrition Center (BONRC) for more than three years. Dr. Apovian was recently awarded the Physician Nutrition Specialist Award given by the American Society of Clinical Nutrition to advance nutrition education to doctors, residents, and medical students.

Elizabeth Krall Kaye, Ph.D., M.P.H, Professor, Department of Health Policy and Health Services Research in the Goldman School of Dental Medicine. Dr. Krall Kaye is also an investigator with the VA Normative Aging and Dental Longitudinal studies at the Boston VA Medical Center and co-investigator in the Center for Research to Evaluate and Eliminate Dental Disparities. Research interests: the link between systemic and oral diseases, such as osteoporosis and periodontal disease; interaction of genetics and diet on risk of osteoporosis; osteoporosis in men; nutrition and oral health; oral health consequences of smoking. Dr. Krall Kaye has held grants from the National Institute of Dental and Craniofacial Research and U.S. Department of Veterans Affairs. Her work with Normative Aging and Dental Longitudinal studies has generated many collaborations, and these projects provide opportunities for student activity in nutritional epidemiology research.

Barbara E. Millen, D.P.H., Professor of Public Health and Socio-Medical Sciences in the Schools of Public Health and Medicine and Chair of M.A., Ph.D., and M.D./Ph.D. programs in Medical Nutrition Sciences; RD, FADA. Research interests: nutritional epidemiology of cardiovascular diseases; efficacy of preventive nutrition interventions for adult and minority populations; nutrition and health communications. Dr. Millen has led the NIH/NHLBI–funded Framingham Nutrition Studies for the past fifteen years and directs a research team that has examined the cross-sectional and prospective relationships between diet, nutritional status, and cardiovascular disease risk factors and related outcomes in male and female participants of the Framingham Heart study. Other collaborative investigations include prospective research on the relationship between nutritional risk and health outcomes in frail, minority elderly, 70–106 years of age; studies of nutrition and chronic disease risk reduction in a variety of adult female and male populations; the efficacy of community-based nutrition interventions; and food insecurity in urban populations. Dr. Millen is also involved in multidisciplinary investigations on innovative methods of nutrition and health communications with younger and older adult populations. Dr. Millen has supervised more than 200 master's and doctoral students and offers ongoing opportunities for students in graduate programs in the Schools of Public Health and Medicine.

Martha Nunn, D.D.S., Ph.D., Associate Professor, Department of Health Policy and Health Services Research in the Goldman School of Dental Medicine, and Director of the Biometry Core for Research to Reduce Oral Health Disparities. Research interests: statistical methods applied to dental research problems; generalized estimating equations; generalized linear mixed models; multiple endpoint survival analysis. Among the research topics she has explored are: the etiology of periodontitis; preventive practices and tooth retention; changes in oral health status with chronic diseases and aging; fluoride treatment and oral health status; and measures of nutritional risk and oral health status.

Rahul Ray, Ph.D., Associate Professor of Medicine, Department of Endocrinology, Nutrition, and Diabetes and Research Associate Professor, Department of Biophysics and Physiology in the Boston University School of Medicine. Research interests: structural biology of the vitamin D and estrogen endocrine systems (structure of hormone receptors, structure-activity relationship studies); proteomic/combinatorial approaches to develop drugs for cancers of prostate and breast, and novel approaches to site-specific delivery of cancer drugs. Dr. Ray has led several NIH/NIDDK–funded grants to study the structural basis for multiple biological properties of vitamin D hormone, including its anti-cancer activity. Dr. Ray is also the recipient of several technology-development grants for cancer drug-discovery, nuclear receptor–mediated drug delivery, and assay of biomolecules. Dr. Ray was a member of the Special Emphasis Panel of the NIDDK/NIH in 1999. More than 15 predoctoral, doctoral scientists, and junior faculty members have been trained in Dr. Ray's laboratory during the past years.

Participating Faculty

Krystoff Blusjtazn, Ph.D., Professor, Department of Pathology, BU School of Medicine.
Lewis Braverman, M.D., Chief, Section in Endocrinology, Boston Medical Center.
Tai Chen, Ph.D., Core Laboratory Director, BU General Clinical Research Center.
Barbara Corkey, M.D., Professor, Molecular Medicine/Obesity Department.
Diane Cullum-Dugan, Center for Nutrition and Weight Management; RD, LD.
Curtis Ellison, M.D., Professor and Chief, Department of Preventive Medicine, BU School of Medicine.
Susan Fish, Pharm.D., M.P.H., Associate Director, BU/Office of Clinical Research, and Director, BU Institutional Review Board.
Christine Gebeshian, M.S., Nutrition Research Manager, BU General Clinical Research Center; RD.
Nawfal Istfan, M.D., Ph.D., Associate Professor, Boston Medical Center.
Samer Karamohamed, M.D., Instructor, Neurology, BU School of Medicine.
Carine Lenders, M.D., Pediatrics, Boston Medical Center.
Robert Lowe, M.D., Boston Medical Center.
Alana Malabanan, M.D., Assistant Professor, Boston Medical Center.
Peter A. Merkel, M.D., M.P.H., Assistant Professor, BU School of Medicine; Associate Program Director, BU General Clinical Research Center.
Jeffrey Migneault, M.D., Assistant Professor, Information Systems Unit, BU School of Medicine.
Tom Moore, M.D., Professor, BU School of Medicine; Assistant Provost for Clinical Research.
Vivien Morris, M.P.H., M.S.; Nutrition and Fitness for Life Program, Boston Medical Center; RD, LDN.
Ashish Saha, Ph.D., Assistant Professor, BU School of Medicine.
Sibaji Sarkar, Ph.D., Instructor, BU School of Medicine.
Domenic Screnci, Ed.D., Director of Business Affairs, Educational Media Center, BU School of Medicine.
Keith Tornheim, Ph.D., Assistant Professor, Department of Biochemistry, BU School of Medicine.
Mary Walsh, Ph.D., Associate Professor, Department of Biophysics and Physiology, BU School of Medicine.
Lorrie Young, M.S., Boston Medical Center; RD.

CASE WESTERN RESERVE UNIVERSITY

Case School of Medicine
Department of Nutrition

Programs of Study	The Department of Nutrition offers programs that span the breadth of the discipline from applied nutrition and dietetics to basic nutritional sciences. The programs lead to the following degrees: Master of Science in Nutrition, Master of Science in Public Health Nutrition Internship (includes fieldwork experiences in public health and community-based agencies), Coordinated Dietetic Internship/Master's (with University Hospitals of Cleveland and Veterans Affairs Medical Center), and Doctor of Philosophy.

The master's degree programs require from one to two years of course work depending upon the student's undergraduate preparation and the specific program. A thesis or nonthesis option may be selected.

The Ph.D. program emphasizes nutritional biochemistry/metabolism, molecular nutrition, and human nutrition. It builds upon faculty expertise in the Departments of Nutrition, Biochemistry, Molecular Biology, Medicine, and Pediatrics. In recent years, investigators in nutritional biochemistry have used molecular biology techniques to enhance understanding of human metabolism. In the first year, nutrition students join graduate students from the other basic science departments in an integrated course that provides a broad introduction to cellular and molecular biology. The subsequent graduate program includes formal courses in human nutrition and nutritional biochemistry, seminars, and, most importantly, the performance of original research.

Research Facilities

Several well-equipped laboratories are housed in the Case School of Medicine Building. The department has access to clinical research units at University Hospitals of Cleveland and MetroHealth Medical Center for the conduct of human clinical nutrition and whole-body metabolism studies. The department has all general equipment necessary for conducting studies in nutritional biochemistry, including three gas chromatographs and mass spectrometers used for human investigation with stable isotopes. Facilities for isolated organ perfusion and a comprehensive organic chemistry laboratory for synthesis of new nutrients are also available.

Financial Aid

The University sponsors a federally funded work-study program as well as loan assistance. Students in the Coordinated Dietetic Internship/Master's Program are paid a stipend by the hospital for the first twelve months of the program. Ph.D. students in nutritional biochemistry or molecular nutrition received an annual stipend of $23,000 in 2005–06.

Cost of Study

Tuition for 2005–06 was $1112 per credit hour. This amounted to $13,350 per semester for full-time study. Partial tuition scholarships are available for some master's students. Full tuition scholarships are provided by the department for students in the Ph.D. nutritional biochemistry or molecular nutrition program. A limited number of teaching assistantships are available for Ph.D. students in human nutrition.

Living and Housing Costs

Graduate and professional students wishing on-campus housing may live in a University dormitory or, if they are married, in one of a limited number of University apartments for married students. Most graduate students find privately owned apartments near the campus. Costs are below average for large urban areas.

Student Group

The University has 9,228 students, of whom about 5,360 are enrolled in graduate and professional schools. About 305 students attend adjacent Institutes of music and art. Ninety-one graduate students are in residence in the department.

Location

Cleveland is an industrial and financial center. The city is richly endowed with cultural facilities, nearly all of them located in a single area known as the University Circle. In this area are the University, the Cleveland Orchestra, the Museum of Art, the Museum of Natural History, and several excellent repertory theaters. The city is also the home of the Cleveland Indians and Cavaliers and the Rock and Roll Hall of Fame. The camping, sailing, and skiing areas of Ohio, western Pennsylvania, and western New York are readily accessible.

The University and The Department

The strength of its combined science departments places the University in the top rank of institutions in this country. The Department of Nutrition has a number of associate members in several departments in the Case School of Medicine and in the University and has strong ties with the Departments of Chemistry, Biology, and Biomedical Engineering. A major emphasis is on interdisciplinary training programs in biological and chemical sciences.

Applying

Prerequisites for entrance into the Ph.D. program are organic chemistry, biology, and mathematics through calculus. Applications should be submitted in late autumn or early winter for an anticipated entrance in the following autumn semester. Application forms may be obtained from the department or online at http://www.applyweb.com/apply/cwrug/menu.html. Ph.D. applicants are required to take the Graduate Record Examinations, including the General Test and one Subject Test.

Information concerning application and prerequisites for specific master's programs should be obtained directly from the department.

Correspondence and Information

Graduate Admissions Committee
Department of Nutrition (M.S. or Ph.D.)
Case Western Reserve University
10900 Euclid Avenue
Cleveland, Ohio 44106-4906
Phone: 216-368-2440
E-mail: paw5@cwru.edu
Web site: http://www.cwru.edu/med/nutrition/home.html

Coordinator
Biomedical Science Training Program (Ph.D. only)
Case School of Medicine
West 452
Case Western Reserve University
Cleveland, Ohio 44106-4935
Phone: 216-368-3347

Case Western Reserve University

THE FACULTY AND THEIR RESEARCH

Hope Barkoukis, Assistant Professor; Ph.D., Case Western Reserve, 1997. Clinical nutrition, diet and liver cirrhosis.

Henri Brunengraber, Professor and Chairman; M.D., 1968, Ph.D., 1976, Brussels. Metabolic regulation; control of flux through metabolic pathways; design and testing of artificial nutrients; noninvasive probes of liver metabolism; markers of alcoholism; mass spectrometry.

Colleen Croniger, Assistant Professor; Ph.D., Case Western Reserve, 1990. Metabolic regulation of carbohydrate and lipid metabolism using genetically altered animal models.

Paul Ernsberger, Associate Professor; Ph.D., Northwestern, 1984. Genetic obesity and the role of nutrition in cardiovascular disease; novel signaling pathways.

Maria Hatzoglou, Professor; Ph.D., Athens, 1985. Identification of retroviral receptors and their use in gene therapy.

Takhir Kasumov, Assistant Professor; Ph.D., Moscow State, 1993. Synthesis of artificial nutrients.

Mary Beth Kavanagh, Instructor; M.S., Case Western Reserve, 1992; RD, LD. Nutritional education for nursing and dentistry; nutrition and the media.

Janos Kerner, Assistant Professor; Ph.D., Hungarian Academy, 1986. Mitochondrial fatty acid oxidation and the regulation of the pathway by a malonyl-CoA.

Jane Korsberg, Instructor and Private Practice; M.S., Case Western Reserve, 1977; RD, LD.

Edith Lerner, Associate Professor and Vice-Chairman; Ph.D., Wisconsin, 1971. Assessment of nutritional status during pregnancy; trace mineral metabolism during pregnancy; nutritional requirements for the preterm infant.

Duna Massillon, Assistant Professor; Ph.D., Montreal, 1994. Nutrient regulation of gene expression.

Laura Nagy, Professor; Ph.D., Berkeley, 1986. Effects of environmental factors such as diet and drugs on cellular signal transduction mechanisms.

Isabel Parraga, Associate Professor and Director, M.S. in Public Health Nutritional Internship Program; Ph.D., Case Western Reserve, 1992. Nutritional anthropology; maternal and child nutrition; public health nutrition; child growth and schistosomiasis.

Stephen Previs, Assistant Professor; Ph.D., Case Western Reserve, 1997. Application of stable isotopes for measuring carbohydrate, fat, and protein metabolism.

Tamara Kurtis Randall, Instructor; Public Health in Nutritional Internship Program, M.S., Case Western Reserve; RD, LD.

Alison Steiber, Assistant Professor and Director, Coordinated Dietetic Internship/M.S. Degree Program; Ph.D., Michigan State, 2005; RD. General nutrition assessment of patients, with emphasis in chronic renal failure.

James Swain, Assistant Professor and Director, Didactic Program in Dietetics; Ph.D., Iowa State, 2000; RD. Food chemistry; nutrition in management of chronic disease; absorption and efficacy in humans of different forms of iron used in food fortification.

Kou-Yi Tserng, Associate Professor; Ph.D., Illinois at Chicago, 1974. Fatty acid metabolism; beta-oxidation; unsaturated fatty acids; stable isotope tracers.

Jonathan Whittaker, Associate Professor; M.R.C.P., 1973, University College (London). Molecular basis of insulin binding and receptor activation.

Secondary Appointments

Saul Genuth, Professor in Medicine; M.D., Western Reserve, 1957. Diabetes mellitus; blood glucose control and complications.

Sharon Groh-Wargo, Assistant Professor in Pediatrics, Neonatal Nutritionist, MetroHealth Medical Center, Cleveland; Ph.D.; Case Western Reserve, 2002. Tolerance of a nutrient-enriched post-discharge formula for preterm infants.

Richard Hanson, Professor of Biochemistry; Ph.D., Brown, 1963. Hormonal control of gene expression.

Douglas Kerr, Associate Professor in Pediatrics; M.D./Ph.D., Western Reserve, 1965. Metabolic disorders in infants.

John Kirwan, Associate Professor in Reproductive Biology; Ph.D., Ball State, 1987. Aging; metabolism; endocrinology.

William Stanley, Associate Professor; Ph.D., Berkeley, 1986. Cardiac metabolism in health and disease.

Anthony Tavill, Professor in Medicine; M.D., Manchester (England), 1970. Hepatic metabolism in liver disease, iron metabolism and iron overload; protein metabolism and the liver; nutritional consequences of gastrointestinal disease; effects of alcohol on hepatic metabolism; circulatory disturbances in liver disease.

Adjunct Faculty

Adjunct faculty provide students with specialized clinical, research, and/or public health fieldwork experiences. Their current job position is listed directly after their name.

Phyllis Allen, Office of Public Health, Columbia, SC; M.S., Case Western Reserve, 1989; RD.

Janet Anselmo, Diabetes Self Management Program, Louis Stokes VA Medical Center, Cleveland, Ohio; M.S., Case Western Reserve, 1988; RD, LD.

Casey Atkinson, Cuyahoga County Board of Health, B.S., Youngstown State, 1999; RD, LD.

Johanna Asarian-Anderson, County of Los Angeles Department of Health Services; M.P.H., UCLA, 1972; RD.

Joan Atkinson, Maine Nutrition Network, Augusta, Maine; M.S., Case Western Reserve, 1999; RD, LD.

Anika Avery-Grant, Centers for Dialysis Care, Euclid, Ohio; M.S., Case Western Reserve, 1994; RD, LD.

Cynthia Bayerl, Department of Public Health; M.S., Boston University, 1976; RD, LDN.

Cynthia Blackburn, Consultant; M.S., Kent State, 1990; RD, LD.

Carmen Blakely-Adams, CCHS-Huron Hospital; B.S., Michigan State, 1986; RD, LD.

Elizabeth Boone, UH-Bedford Medical Center; B.S., Ohio, 1994; RD, LD.

Josephine Anne Cialone, North Carolina Division of Maternal and Child Health; M.S., Case Western Reserve, 1980; RD.

Nenita Clemente, MetroHealth Center for Community Health; M.S., Case Western Reserve, 1975; RD.

Cheri Collier, Clement Center, MetroHealth Centers for Community Health, Cleveland; M.S. Case Western Reserve, 1998; RD, LD.

Susan Comfort, St. Vincent's Charity Hospital; M.S., Case Western Reserve, 1988; RD, LD.

Janice Davis, Western Reserve Area on Aging, Cleveland; M.S., Case Western Reserve, 1986; RD.

Helen Dumski, Diabetes Association of Greater Cleveland; B.S., Ohio State; RD, LD.

Denise Ferris, West Virginia Office of Nutrition Services; Ph.D., Pittsburgh, 1989; RD.

Karen Fiedler, Adjunct Associate Professor; Ph.D., Tennessee, Knoxville, 1977.

Evangeline Fowler, Nutritional Services, St. John Westshore Hospital; M.S., Case Western Reserve, 1987.

Lorna Fuller, Sodexho Marriott Services, Huron Hospital; M.S., Kent State, 1995; RD, LD.

Deborah Gammell, MetroHealth Medical Center (WIC), Cleveland; M.S., Miami (Ohio), 1980; RD, LD.

Brenda Garritson, Baylor University Medical Center, Dallas, TX; M.S., Texas Woman's, 1993; RD, LD.

Peggy Gates, MetroHealth Medical Center (WIC), Cleveland; M.Ed., Cleveland State, 2001; RD, LD.

Melinda Gedeon, UHHS Bedford Medical Center; B.S., Ohio State; RD, LD.

Martha Halko, Cuyahoga County Board of Health; M.S., Akron, 2000; RD, LD.

Cathy Hastings, Pinellas County Health Department, St. Petersburg, Florida; M.P.H., South Florida, 1992; RD, LD.

Valerie Heimbach, MetroHealth Medical Center (WIC), Cleveland; M.S., Indiana of Pennsylvania, 1996; RD, LD.

Karen Horvath, UHHS-Bedford Medical Center; B.S., Akron, 1986; RD, LD.

Claire Hughes, Hawaii State Department of Health; Dr.P.H., Hawaii, 1998; RD.

Lisa Isham, Cuyahoga County Board of Health; M.S., Case Western Reserve, 1997; RD, LD.

Elvira Jarka, Health and Human Services, Kansas City, Missouri; M.P.H., Illinois, 1996; M.S., Southern Illinois.

Jan Kallio, Massachusetts WIC Program; M.S., Case Western Reserve, 1977; RD.

Jennifer Kernc, Centers for Dialysis Care, Cleveland, Ohio; B.S., Akron; RD, LD.

Natalia Kliszczuk-Smolio, Mt. Alverna Home, Inc., Parma, Ohio; B.S., Cincinnati, 1982; RD, LD.

Richard Koletsky, Adjunct Assistant Clinical Professor; M.D., Case Western Reserve, 1975.

Jennifer Kravec, MetroHealth Medical Center; B.S., Ohio State, 1997; RD, LD.

Perri Kushan, Menorah Park Center for Senior Living, Beachwood, Ohio; B.S., Akron, 1986; RD, LD.

Lois Lenard, Nutrition and Food Service, Louis Stokes Cleveland VA Medical Center, Cleveland, Ohio; B.S., Kent State, 1974; RD, LD.

Anita Martin, Duval County Public Health Unit, Jacksonville, Florida; M.P.H., North Carolina at Chapel Hill, 1980; RD, LD.

Mary Ann McGuckin, Akron Health Department; M.S., Case Western Reserve, 1974; RD.

Linda Novak-Eedy, Menorah Park Center for Senior Living, Beachwood, Ohio; B.S., Bowling Green State, 1983; RD, LD.

Lisa Ogg, Nutrition Health Professional, Cuyahoga County WIC Program; B.S., Kent State, 1996; RD, LD.

Michelle Ogurwale, MetroHealth Center for Community Health (WIC); M.S., Southern Illinois Carbondale; RD, LD.

Christine Polisena, Whole Health Management; M.S., Case Western Reserve, 1988; M.B.A. Cleveland State, 1992; RD, LD, FADA.

Barbara Pryor, Ohio Department of Health; M.S., Ohio State, 1994; RD, LD.

Anne Raguso, Director, Dietetic Internship, VA Medical Center, Cleveland; Ph.D., Case Western Reserve, 1985; RD, LD.

Anna Rostafinski, MetroHealth Medical Center (WIC); M.S., Case Western Reserve, 1988; RD, LD.

Jo Ann Ruggeri, CDC at Citiview; B.S., Ohio State, 1966; RD, LD.

Joanne Samuels, Solon City School District; B.S. SUNY, 1992; RD, LD.

Miriam Seidel, Jewish Health Care Foundation, Pittsburgh, Pennsylvania; M.S., Boston University, 1979; RD.

Najeeba Shine, Cuyahoga County Board of Health; M.S., Case Western Reserve, 1992; RD, LD.

Ruth Shrock, Ohio Department of Health, Columbus, Ohio; M.S., Ohio State, 1976, 1999; RD, LD.

Suzanne Silverstein, Virginia Department of Health; M.A., George Washington, 1999; RD.

Donna Skoda, Summit County General Health District, Stow, Ohio; M.S., Case Western Reserve, 1980; RD.

Sara Snow, MetroHealth Medical Center, Cleveland, Ohio; M.S., Case Western Reserve, 1997; RD, LD.

Lura Elizabeth Spinks, Berea City Schools; M.S., Ohio State; RD.

Margaret Tate, Arizona Department of Health; M.S., Colorado State, 1979; RD.

Norliza Tayag, WIC Program, Santa Clara County, California; B.S., San Diego State, 1997; RD.

Anita Ullman, Sodexho Marriott Services, Hillcrest Hospital; M.S., Case Western Reserve, 1985; RD, LD.

Karen Y. Warman, General Clinical Research Center, University Hospitals of Cleveland; M.S., Texas Woman's, 1979; RD.

Melissa Wilson, Centers for Dialysis Care, Cleveland, Ohio; B.S., Mercyhurst, 1997; RD, LD.

Diane Ohama Yates, Medical Nutritional Representative, Ross Products Division, Abbott Laboratories, Cleveland; B.S., Hawaii, 1978; RD, LD.

Mary Zyga, Clinical Specialties, Inc.; B.S., Ohio State, 1981; RD, LD.

COLUMBIA UNIVERSITY

Institute of Human Nutrition

Programs of Study

The Institute of Human Nutrition offers M.S. and Ph.D. programs in nutrition and nutrition-related sciences. The development of scientific curiosity and acquisition of knowledge and technical competence are primary goals of graduate education. The achievement of these goals is fostered at the Institute of Human Nutrition at Columbia by supervised research, guidance, and collaboration with distinguished laboratories and faculty members in an exciting scholarly and research-oriented environment.

The M.S. program is focused on basic science skills in nutrition. It requires one year of course work and the completion of a master's thesis.

The Institute also offers an M.S. program that is designed for practicing physicians and health professionals. The program is offered on a full- or part-time basis and is designed to meet the academic and scheduling needs of a practicing professional.

In the first year of study, the Ph.D. program is typically devoted to acquiring a broad base of scientific knowledge through course work and three research rotations. During the second year, the thesis adviser is chosen, and the student begins research in an area of special interest while taking additional courses, if needed, and completes the qualifying examination. Subsequent years involve the conduct and completion of a dissertation research project, followed by submission and defense of the dissertation. Although M.A. and M.Phil. degrees are awarded during the course of study as requirements are satisfied, only students who wish to obtain the Ph.D. are admitted to this program.

Research Facilities

The administrative offices of the Institute of Human Nutrition, part of the Columbia University Medical Center complex, are located on 168th Street in upper Manhattan. There are extensive research and clinical facilities at the Medical Center. Special facilities available for nutrition research and training include the Arteriosclerosis Research Center, the Herbert Irving Comprehensive Cancer Center, the General Clinical Research Center, and the Obesity Research Center, which is located at St. Luke's Roosevelt Hospital Center at 114th Street and Amsterdam Avenue. The Health Sciences Library contains more than 500,000 books and periodicals. It also includes a media center with audiovisual learning aids and computers with word processing, statistical, and other program capabilities for student use.

Financial Aid

Financial aid or loans are available to qualified M.S. students. All students, domestic or international, who are accepted into the Ph.D. program are awarded support that fully provides for the tuition and the medical insurance fees required by the University. These students also receive a stipend for their personal use that commences with registration and normally continues throughout graduate study. This stipend is approximately $27,335 for the 2006–07 school year.

Cost of Study

Full-time tuition for the 2006–07 year is approximately $37,680.

Living and Housing Costs

Housing is provided on the Health Sciences Campus and the Morningside Heights Campus of Columbia University. Accommodations include University Residence Halls, which consist of furnished 2- and 4-person suites, and Institutional Real Estate Apartments, which include studios and one-, two-, and three-bedroom apartments. Membership at the Bard Athletic Club is free to students. It contains a swimming pool, squash courts, a gymnasium, a sauna, and exercise equipment and is accessible to the handicapped. A full-time trainer is on staff. Programs offered include aerobics, weight training, and swimming lessons.

Student Group

There are approximately 50 students in the M.S. and 30 in the Ph.D. programs. There are 349 doctoral students enrolled in the various departments and programs at the Medical Center. Approximately 56 percent of students are women. Students are drawn from all parts of the United States and abroad.

Location

The Columbia University Medical Center complex is located at 168th Street and Broadway in upper Manhattan. New York's world-renowned cultural activities are all easily accessible by public transportation, as are sporting events and other recreational opportunities.

The University

Columbia University, a privately supported institution, is one of the world's leading educational, scholarly, and research centers. Founded by charter as King's College in 1754, it is one of the oldest universities in the country.

In addition to the Institute of Human Nutrition, components of the University that are situated on the Health Sciences Campus include the College of Physicians and Surgeons, the School of Nursing, the Mailman School of Public Health, and the School of Dental and Oral Surgery. The Center for Neurobiology and Behavior, the Hughes Institute for Structural Biology, and the Institute for Cancer Genetics are also based on the campus. Classrooms and laboratory facilities for graduate school programs are located in the Julius and Armand Hammer Health Sciences Center, the College of Physicians and Surgeons, the William Black Medical Research Building, the Russ Berrie Pavilion, the Irving Cancer Center Research Building, and the Research Annex of the New York State Psychiatric Institute.

Applying

Scores from the General Test of the Graduate Record Examinations (GRE) are required. An advanced Subject Test is recommended but not required. International applicants whose native language is not English are required to take the Test of English as a Foreign Language (TOEFL); applicants with a TOEFL score below 600 are reviewed with caution. Completed applications and all supporting material should be submitted by the beginning of January for consideration for the fall semester.

Correspondence and Information

For general program information and M.S. program applications:
Sharon R. Akabas, Ph.D.
Susan Vannucci, Ph.D.
M.S. Program Advisers
Office of Student Affairs
Columbia University Medical Center
630 West 168th Street
New York, New York 10032
Phone: 212-305-4808
Fax: 212-305-3079
E-mail: nutrition@columbia.edu

For Ph.D. program applications:
Admissions Office
Graduate School of Arts and Sciences
Columbia University Medical Center
701 West 168th Street, Room 406
New York, New York 10032
Phone: 212-305-8058
Fax: 212-305-1031
E-mail: gsasatpands@columbia.edu

For graduate program assistance:
Debra J. Wolgemuth, Ph.D.
Chair, Ph.D. Training Committee
Institute of Human Nutrition
PH 15 East, Room 1512
Columbia University Medical Center
630 West 168th Street
New York, New York 10032
Phone: 212-305-4808
Fax: 212-305-3079
E-mail: nutrition@columbia.edu

Columbia University

THE FACULTY

Domenico Accili, M.D., Professor of Medicine. Insulin resistance; mechanisms of insulin receptor signaling.

Richard J. Baer, Ph.D., Professor of Pathology. Function of the BRCA1 breast cancer susceptibility gene; retinoid signaling in mammary glands.

William S. Blaner, Ph.D., Professor of Nutritional Sciences. Retinoids and vitamin A metabolism.

Carol N. Boozer, D.Sc., Research Scientist, Department of Medicine. Energy expenditure and substrate oxidation in humans and rodent models; assessment of free-living physical activity in humans.

David A. Brenner, M.D., Professor of Medicine. Study of the regulation of gene expression in the liver and intestines under normal and pathological states.

Angela M. Christiano, Ph.D., Associate Professor of Dermatology and of Genetics and Development. Genetic basis of skin and hair disorders in humans; basic epidermal biology.

Jeanine M. D'Armiento, M.D., Assistant Professor of Medicine. Metalloproteases and lung pathophysiology.

Richard J. Deckelbaum, M.D., Robert R. Williams Professor of Nutrition; Professor of Pediatrics; Professor of Epidemiology; Director, Institute of Human Nutrition; and Chair. Lipoprotein-receptor-cell interaction; lipid emulsion metabolism; free fatty acids and cell lipid metabolism and gene expression.

Bernard F. Erlanger, Ph.D., Professor of Microbiology. Biologically significant receptors: the relationship of their structures to their metabolic and regulatory activities.

Dympna Gallagher, Ed.D., Assistant Professor of Nutritional Medicine. Energy expenditure and body composition at the organ-tissue level, both cross-sectionally and longitudinally, in growth, aging, and type 2 diabetes.

Anne A. Gershon, M.D., Professor of Pediatrics. Virus infectivity; infant immunity relevant to viral infections; varicella-zoster virus (VZV), the highly contagious etiologic agent of chickenpox (varicella) and shingles (zoster).

Michael D. Gershon, M.D., Professor of Anatomy and Cell Biology. Enteric nervous system and varicella-zoster virus.

Henry N. Ginsberg, M.D., Irving Professor of Medicine. Regulation of plasma lipoprotein metabolism; regulation of apoprotein B secretion from hepatocytes; relationship of insulin resistance to lipoprotein metabolism; mouse models of insulin resistance and dyslipidemia.

Ira J. Goldberg, M.D., Dickinson Richards Professor of Medicine. Lipoprotein metabolism; lipolytic enzymes; endothelial cell biology; atherosclerosis.

Maxwell E. Gottesman, M.D., Ph.D., Charles H. Revson Professor of Biochemistry and Molecular Biophysics and Professor of Microbiology. Gene regulation; cell-cycle control; vitamin A metabolism.

Lloyd A. Greene, Ph.D., Professor of Pathology. The mechanisms of neuronal differentiation and degeneration and the regulation thereof by external growth factors.

Geoffrey R. Howe, Ph.D., Professor of Public Health (Epidemiology). Relationship between diet and cancer and radiation and cancer; scientific support provided to NCI-funded studies conducted in the Ukraine and Belarus to study long-term risks of thyroid cancer and leukemia following the Chernobyl accident.

Li-Shin Huang, Ph.D., Research Scientist in Medicine. Mouse genetics; molecular biology of apolipoproteins; n-3 fatty acids and genes interaction; transgenic and knockout mouse models of dyslipidemia associated with diabetes.

Sudha Kashyap, M.D., Associate Professor of Clinical Pediatrics. Nutritional support of preterm infants.

Harry R. Kissileff, Ph.D., Associate Professor of Clinical Psychology (Psychiatry). Physiological control of eating in humans.

Rudolph L. Leibel, M.D., Professor of Pediatrics and of Medicine. Molecular physiology of energy homeostasis/weight regulation in rodents and humans; molecular genetics of obesity and type 2 diabetes in rodents and humans.

Cathy L. Mendelsohn, Ph.D., Assistant Professor of Urology. Retinoids and development of the urogenital tract.

Frederica P. Perera, Dr.P.H., Professor of Public Health (Environmental Sciences). Molecular epidemiology; risk assessment; carcinogenesis.

Francis Xavier Pi-Sunyer, M.D., Professor of Medicine. Carbohydrate and lipid metabolism; obesity; diabetes mellitus; food intake regulation.

Ravichandran Ramasamy, Ph.D., Assistant Professor of Medicine. Metabolic mechanisms of diabetic cardiomyopathy; mechanisms of ischemic injury; mitochondrial signaling and cardiac cell death.

Lorna W. Role, Ph.D., Professor of Anatomy and Cell Biology. The generation, plasticity, and maintenance of cholinergic and cholinoceptive synapses in the mammalian brain.

Neil S. Shachter, M.D., Assistant Professor of Medicine. Molecular mechanisms of hypertriglyceridemia using genetic epidemiologic and transgenic mouse techniques.

Lawrence S. Shapiro, Ph.D., Associate Professor of Biochemistry and Molecular Biophysics. Possible biochemical causes for adult-onset obesity.

Stephen L. Sturley, Ph.D., Associate Professor of Clinical Nutrition. Extracellular and intracellular sterol transport.

Ira A. Tabas, M.D., Ph.D., Richard J. Stock Professor and Vice-Chairman of Research of Medicine and of Anatomy and Cell Biology. Molecular and cellular biology of macrophages in the areas of vascular biology and atherosclerosis; cholesterol-induced cell signaling pathways involving apoptosis, phagocytosis, and intracellular lipid trafficking; in vivo correlations using a variety of transgenic and knockout mouse models.

Alan R. Tall, M.D., Tilden Weger Bieler Professor of Medicine. Plasma lipoprotein metabolism; atherosclerosis; protein structure-function and mutagenesis; regulation of gene expression; molecular nutrition.

David A. Talmage, Ph.D., Associate Professor of Clinical Nutrition (Pediatrics). Role of signal transduction pathways in regulating cellular proliferation, differentiation, and survival; mechanisms of retinoid action through affecting signal transduction pathways.

Timothy C. Wang, M.D., Dorothy L. and Daniel H. Silberberg Professor of Medicine and Chief, Gastroenterology Division. The role of inflammation, cytokines, and growth factors in the development of gastrointestinal cancers.

Sharon L. Wardlaw, M.D., Professor of Medicine, Division of Endocrinology. Neuroendocrine control of pituitary function and the hypothalamic regulation of energy balance.

I. Bernard Weinstein, M.D., Frode Jensen Professor of Medicine and Professor of Public Health and of Genetics and Development. Molecular mechanisms of carcinogenesis.

Christine L. Williams, M.D., Ph.D., Professor of Clinical Pediatrics. Child nutrition and pediatric preventive cardiology; in particular, issues related to lipids and cardiovascular disease risk as well as obesity in preschool children.

Debra J. Wolgemuth, Ph.D., Professor of Genetics and Development and Obstetrics and Gynecology. Role of retinoid signaling during development and differentiation in the mammalian reproductive system; cell-cycle control during mitosis and meiosis in the germ line.

FLORIDA INTERNATIONAL UNIVERSITY

College of Health and Urban Affairs
Robert Stempel School of Public Health
Dietetics and Nutrition Program

Programs of Study

The Dietetics and Nutrition Program at Florida International University (FIU) offers graduate programs leading to the M.S. and Ph.D. degrees. The M.S. degree requires a minimum of 37 credits beyond the B.S. degree; the Ph.D. requires a minimum of 56 credits beyond the M.S. degree. A Dietetic Internship Program is available to M.S. degree–seeking students. The Ph.D. program requires original research and the presentation of a dissertation. The M.S. program requires either a thesis or a project. Graduate students are guided by a major professor and a faculty advisory committee.

The graduate programs in dietetics and nutrition offer exceptional nutrition research opportunities, including an ethnically diverse population with recent immigrant groups from the Caribbean and Central and South America. Internationally recognized research in nutrition status and HIV/AIDS is ongoing. The National Policy and Resource Center on Nutrition and Aging addresses the burgeoning population of older adults and engages in a wide variety of activities, ranging from innovative long-term care and community-based demonstration projects to nationwide policy analysis and technical assistance.

Numerous affiliated community agencies and institutions work collaboratively with FIU's nationally recognized faculty to establish research funding and high-quality education for students. Experienced dietitians and nutritionists find the program, its faculty members, and the setting particularly relevant to their career advancement. Cooperative research with faculty members in public health and biological sciences, the Center on Aging, acute- and long-term-care facilities, and diverse community agencies broaden the scope of research opportunities for graduate students.

Research Facilities

The department has well-equipped research laboratories in the new Health and Life Science Building. Animal research laboratories are also available at the University.

The library is the largest in the state of Florida. A state-of-the-art computer system has been installed in the library and in School offices. Computer access to the University VAX, VMS, and SOLIX systems is possible from offices and labs.

FIU has met Carnegie classification criteria as a Research I public university with a continually growing student body and faculty.

Financial Aid

Stipends for graduate assistantships in teaching or research range from $1700 to $6000 per semester depending on the nature of the job and the number of hours worked. Full-time enrollment is a requirement.

The Department of Dietetics and Nutrition has scholarships for new and currently enrolled students. Teaching assistantships are available to qualified Ph.D. students and provide partial tuition fee waivers and stipends. For more information, students should contact the graduate program director at 305-348-2878. More generous scholarships for qualified students are available from the Financial Aid Office (http://www.fiu.edu/orgs/finaid).

Cost of Study

In 2006–07 tuition is $259.73 per credit hour for Florida residents and $763.80 per credit hour for out-of-state students. Additional per semester fees such as intercollegiate athletics, health services, and transportation access are not included in these figures.

Living and Housing Costs

Housing is limited and should be petitioned for with ample time. Campus housing includes apartment-style accommodations available in the form of studio apartments to shared units. Costs vary within the choice of housing and residential life. Off-campus housing near the University is also available. Information is available from the University Housing Office at http://www.fiu.edu/~housing.

Student Group

FIU has a student body of 35,000, including about 4,200 graduate students in 109 M.S. and thirty-one Ph.D. programs. The student body in dietetics and nutrition comprises students from diverse cultural backgrounds.

Location

The University is located in Miami, Florida's largest urban center and a major transportation and business hub of the southeastern U.S. Miami is an exciting, dynamic global marketplace. Miami Beach is known for its historic art deco district and the numerous hotels that line its beaches. Miami has been reinvented into a hemispheric crossroads for trade, travel, culture, and communications.

The University

FIU admitted its first class in 1972 and is one of the youngest institutions in Florida. However, it is the largest institution of higher learning in south Florida and the fourth-largest in the state. FIU is a metropolitan university that provides a rich cultural diversity on campus as well as in the community.

Applying

Admission to the M.S. Dietetics and Nutrition Program requires a minimum 3.0 GPA and a minimum score of 1000 (verbal and quantitative) on the General Test of the GRE. Ph.D. admission requires an M.S. degree in a related field with a minimum 3.3 GPA and a minimum score of 1120 on the GRE, three letters of reference, a statement of goals and research interests, and submission of a master's thesis or project report or a published manuscript resulting from master's work on which the applicant is among the first 3 authors. A major consideration in the decision to admit a student is the compatibility in research focus between the faculty member and the student. Students are encouraged to discuss research interests with faculty members prior to applying. International applicants need a TOEFL score of at least 550.

Correspondence and Information

Carrie C. Sanchez
Assistant Director of Academic Support
Florida International University
Miami, Florida 33199
Phone: 305-348-5609
Fax: 305-348-5980
E-mail: csanchez@fiu.edu

Florida International University

THE FACULTY AND THEIR RESEARCH

Marianna Baum, Professor; Ph.D. (biochemical nutrition), Florida State. Nutritional status and HIV.

Adriana Campa, Assistant Professor; Ph.D. (international relations) Miami (Florida); RD, American Dietetic Association. Nutrition education.

Victoria H. Castellanos, Associate Professor and Ph.D. Coordinator; Ph.D. (nutrition/physiological chemistry), California, Davis; RD, American Dietetic Association. Physiological and psychological control of food intake; obesity; physiological chemistry.

Michele Ciccazzo, Associate Professor; Ph.D. (food and nutrition), Florida State; RD, American Dietetic Association. Sports nutrition; nutrition education.

Zisca R. Dixon, Associate Professor; Ph.D. (food science and technology), Texas A&M; RD, American Dietetic Association. Antioxidant nutrients; food safety education.

Evelyn B. Enrione, Associate Professor and Associate Dean; Ph.D. (nutrition), Purdue; RD, American Dietetic Association. Clinical nutrition: TPN and enteral nutrition therapy in relation to disease.

Valerie George, Research Associate Professor; Ph.D. (nutrition), Laval. Epidemiology; nutrition education/behavior change in dietary and exercise intervention; nutrition education in exceptional children; world hunger.

Susan P. Himburg, Professor; Ph.D. (education), Miami; RD, American Dietetic Association; FADA, Charter Fellow American Dietetic Association. Education; cost-benefit analysis; program evaluation documentation of clinical outcomes/cost effectiveness.

Fatma G. Huffman, Professor; Ph.D. (biochemistry/nutrition), Auburn; RD, American Dietetic Association. Nutritional biochemistry; mineral bioavailability; nutrition and chronic disease; diabetes and coronary heart disease risk among minority populations; dietary, genetic, and biochemical risk assessment.

Marcia H. Magnus, Associate Professor; Ph.D. (nutrition), Cornell. Nutrition education.

Dian O. Weddle, Associate Professor and Co-Director of the National Policy and Resource Center on Nutrition and Aging; Ph.D. (public policy analysis), Illinois at Chicago; RD, American Dietetic Association; FADA, Charter Fellow American Dietetic Association. Public policy analysis related to nutrition and aging; home- and community-based nutrition and aging outcomes and program effectiveness.

Nancy S. Wellman, Professor and Director of National Policy and Resource Center on Nutrition and Aging; Ph.D. (education), Miami; RD, American Dietetic Association; FADA, Charter Fellow American Dietetic Association. Aging; consumer nutrition; nutrition communication.

Affiliated Research Faculty–Center on Aging

Burton D. Dunlop, Director of Research; Ph.D. (sociology), Illinois. Research and evaluation of issues and programs in aging; health and social services and long-term care.

Pamela Elfenbein, Director of Education and Training; Ph.D. (medical social work), Miami. Health promotion; disease prevention; healthy aging.

Max Rothman, Executive Director; J.D., Michigan. Policy, needs assessment, and evaluations of public programs concerning elders; long-term care in Florida; public policy and aging.

MARYWOOD UNIVERSITY

Department of Nutrition and Dietetics

Programs of Study

The Master of Science (M.S.) degree in nutrition provides many exciting career opportunities in a growing field and the option of continued graduate education in a doctoral program. Professional opportunities include the following: advocates who influence the development and interpretation of food- and nutrition-related legislation; clinicians who assess, plan, implement, and evaluate a client's nutrition care as a member of a health-care team; resource managers who administer food-service systems, community nutrition programs, and clinical practices; educators who teach clients, health-care professionals, employees, and the general public; researchers who directly conduct nutrition-related research and who interpret the research to the public; professionals with responsibilities for continuing education, involvement in professional societies, and compliance with the code of ethics for the Profession of Dietetics of the American Dietetic Association; and food and nutrition specialists who provide accurate, up-to-date information. Degree candidates are required to complete 36 credit hours of study, 18 of which must come from the core and research sequence, while the other 18 credit hours may be selected from electives offered.

The M.S. degree in sports nutrition and exercise science is an interdisciplinary program that combines and integrates the fields of nutrition and exercise science to prepare students professionally for careers that focus on maximizing athletic performance and/or the prevention and control of chronic diseases, such as obesity, diabetes, heart disease, and cancer. The program combines classroom learning and research opportunities in Marywood's state-of-the-art Human Performance Lab. A wide variety of professional career opportunities are available with professional, national, and collegiate sports teams; in hospitals and rehabilitation centers; in sports and fitness clubs; in corporate fitness and health centers; with the nutritional and sports supplement industry; and in sports training camps. There are also opportunities for teaching, academic or medical research, and positions in private practice. The degree also provides excellent preparation for further research or education, including medical school and doctoral programs. Degree candidates are required to complete 45 credit hours of study, 24 of which must come from the core of nutrition and exercise courses, 9 from the research sequence, and 12 from various electives.

Both programs require the student to successfully complete and defend an original research project as a master's thesis. To that end, after consultation with the Department chair, students are assigned a faculty member to mentor the candidate throughout the research project.

The Department of Nutrition and Dietetics also offers a dietician internship leading to achievement of the core competencies and general emphasis of the American Dietetic Association. The Dietetic Internship has been granted accreditation by the Commission on Accreditation for Dietetic Education (CADE).

Research Facilities

The Human Performance Laboratory includes equipment that measures and tests aerobic capacity, anaerobic power, strength, bone density, airway function, and body composition as well as a biochemistry lab and climate-control room. The food science laboratory at Marywood is a state-of-the-art facility that gives students a unique learning experience. In addition to traditional food-science equipment, the lab contains industrial equipment to train students for the food preparation in real-world commercial settings. The lab has eight fully functional work stations, food-science-testing equipment, a blast chiller and rethermalization unit, a metabolic kitchen, an ingredient room, a food-prep area, and a sensory evaluation laboratory, which enables students to conduct more accurate food-sensory testing.

The sensory evaluation laboratory allows students to test the products made in the food science laboratory. The products are evaluated, and the data can then be entered into a computer for further analysis. The healthy-demo classroom offers an area where demonstrations take place for the campus and the community.

Computer facilities, with the latest word processing and statistical software, are available within the Department and in the Learning Resources Center (LRC). The library, also located in the LRC, has holdings of more than 216,190 volumes, 338,190 microforms, 43,530 media items, and 965 current periodicals. The library also participates in a national and international interlibrary loan network.

Financial Aid

Assistantships, scholarships, and loans are available to graduate students. Students enrolled for at least 6 credits per semester can borrow under the Federal Stafford Student Loan. Information is available from the financial aid office

Cost of Study

Tuition is $672 per credit. The per-semester general fee is $415 for those enrolled in 12 or more credits.

Living and Housing Costs

Students can pursue off-campus housing. Marywood is located in a residential area, and rental apartments are available for graduate students.

Location

Marywood University is situated in a suburban area known as the Green Ridge section of Scranton, a city of about 75,000. Located a little more than 100 miles west of New York City and 100 miles north of Philadelphia, Scranton is served by the Scranton–Wilkes-Barre International Airport and is accessible by a network of superhighways. The Pocono Mountains resort areas and several beautiful lakes can be reached within 45 minutes or less. The Montage Mountain recreation area is only 15 minutes away.

The University and The Department

Marywood University, established 1915, is an independent, comprehensive Catholic university, owned and sponsored by the Congregation of Sisters, Servants of the Immaculate Heart of Mary. Graduate studies were inaugurated in 1921. The mission of the Department of Nutrition and Dietetics is to provide a dedicated faculty, high-quality facilities, and a diverse environment that supports nationally recognized undergraduate and graduate education, research, and service in the areas of nutrition, dietetics, family/consumer sciences, and sports nutrition/exercise science.

Applying

Nutrition applicants are required to have a Bachelor of Arts or Bachelor of Science degree in nutrition, foods, or dietetics from an accredited undergraduate institution. Sports nutrition and exercise science applicants must have at least one year of anatomy and physiology and one chemistry course from an accredited undergraduate institution. Other individuals are encouraged to apply with the understanding that they may be required to take undergraduate courses as prerequisites to satisfy the Departmental admissions committee.

Students should submit the completed application, the $30 application fee, and transcripts of prior academic work. International students should also submit TOEFL scores. The preferred application deadline is April 15 for the fall semester and November 15 for the spring semester. However, applications received after these dates are processed on a rolling basis.

Correspondence and Information

Dr. Alan Levine, Co-Chair
Department of Nutrition and Dietetics
Marywood University
2300 Adams Avenue
Scranton, Pennsylvania 18509
Phone: 866-279-9663 Ext. 6290 (toll-free)
E-mail: levine@marywood.edu
Web site: http://www.marywood.edu/departments/nutr_diet/home.html

Marywood University

THE FACULTY

The goal of the faculty of the Department of Nutrition and Dietetics is to integrate the theories, practices, and skills of nutrition and exercise science and bring them to bear on the challenges that both athletes and nonathletes face in choosing diet and exercise regimens to maintain optimum health and maximize performance, as well as the challenges of choosing appropriate diet and exercise patterns to minimize the chance of developing chronic diseases.

Alan Levine, Professor and Co-Chair, Department of Nutrition and Dietetics; Ph.D., NYU; RD, CMFC.
Kathleen McKee, Professor and Co-Chair, Department of Nutrition and Dietetics; Ph.D., Drexel; RD.
Marianne E. Borja, Director, Coordinated Program, Department of Nutrition and Dietetics; Ed.D., Temple; RD, FADA.
Maureen Dunne-Touhey, Director, Distance Education Dietetic Internship, and Clinical Coordinator, Dietetic Internship; M.S., CUNY, Lehman; RD.
Sandra Graham, Coordinator, Family and Consumer Sciences; M.S., Georgia State.
Lee Harrison, Professor, Department of Nutrition and Dietetics; Ph.D., NYU; RD, CDN, FADA.
Kenneth W. Rundell, Professor of Health Science and Director, Human Performance Laboratory; Ph.D., Syracuse; FACSM.

UNIVERSITY OF CALIFORNIA, BERKELEY

Department of Nutritional Sciences and Toxicology
Graduate Group in Molecular and Biochemical Nutrition

Program of Study

The Graduate Program in Molecular and Biochemical Nutrition at the University of California (UC), Berkeley, provides excellent research opportunities focused on the interactions among nutrients, phytochemicals, and metabolism, culminating in the Ph.D. Academic Senate faculty members in the Graduate Group supervise the programs of study/dissertation research of graduate students in Molecular and Biochemical Nutrition. Many members of the Molecular and Biochemical Nutrition faculty also belong to other interdisciplinary Graduate Groups, such as Comparative Biochemistry, Endocrinology, and Molecular Toxicology. Students from these other interdisciplinary Graduate Groups pursue their research in the Department of Nutritional Sciences and Toxicology, along with Ph.D. students enrolled in Molecular and Biochemical Nutrition.

A major strength of the Graduate Group program is close interaction between students and faculty members. Exposure to a variety of areas encompassed by modern metabolic and systems biology provides students with the outlook needed to plan research and apply research findings. Graduate students attend courses and seminars in many departments, including Molecular and Cell Biology, Integrative Biology, Chemistry, Statistics, and Public Health. Although course work is required, the program places a strong emphasis on the student's research training. The student's dissertation research is monitored by a faculty advisory committee, which provides advice and constructive criticism.

The Ph.D. degree in molecular and biochemical nutrition generally requires five years to complete. In addition to the diverse opportunities for research and academic study, graduate students gain teaching competency as teaching assistants for at least one semester. Ph.D. graduates are qualified to fill research positions in universities, industry, or government institutions and are prepared for teaching positions.

Research Facilities

Modern facilities for experimental biological nutrition research are available in the Department. In addition, the UC Berkeley campus offers a wide variety of research support facilities. The Berkeley campus library system is recognized as one of the finest in the world. The Department enjoys affiliations with the Western Regional Research Center/USDA–Albany, the Western Human Nutrition Research Center/USDA–Davis, the Bay Area Human Nutrition Center at San Francisco General Hospital, the University of California at San Francisco, Medical School/Moffitt Hospital, and the Children's Hospital of Oakland Research Institute.

Financial Aid

All graduate students working toward a Ph.D. receive support through a variety of sources. The funding package includes payment of fees (including nonresident tuition, if applicable) and a stipend for living expenses. The Department has training grants for predoctoral students, as well as research and teaching assistantships and fellowships. Applicants seeking financial aid must submit the Free Application for Federal Student Aid (FAFSA) by April 15 for maximum consideration.

Cost of Study

The 2004–05 registration fees were $2900.50 per semester. Tuition for nonresident students was an additional $7347 per semester. However, tuition and fees are covered by the financial aid package provided by this program.

Living and Housing Costs

There are many housing options in the Bay Area. Graduate students seeking housing should contact Residential and Student Service Programs, 2610 Channing Street, Berkeley, California 94720-2272, or telephone 510-642-3642, to receive information. The estimated budget for room and board for the 2004–05 academic year was $13,330.

Student Group

Approximately 33,000 students, including about 9,000 graduate students, are enrolled at UC Berkeley. Students in the Graduate Group in Molecular and Biochemical Nutrition work closely with students in other Graduate Groups hosted by the faculty housed in Morgan Hall and elsewhere.

Student Outcomes

The field of molecular and biochemical nutrition provides a host of challenging and intellectually stimulating problems of importance to human health and well-being. Graduates find employment with universities or research institutes, national and international biology programs, state extension services, and as private consultants.

Location

The campus is located in the city of Berkeley (population 106,500), spread across 1,232 acres overlooking San Francisco Bay. Although the campus has grown and developed considerably since its inception, great effort has been made to retain its original park-like atmosphere. San Francisco and many other Bay Area attractions are easily accessible through BART, the trans-county commuter rail system. The climate in the Bay Area is mild throughout the year. Temperatures range from about 40°F in the winter to about 77°F in the summer.

The University

The University of California was the product of a merger between the College of California (a private institution) and the Agricultural, Mining, and Mechanical Arts College (a Morrill Land Grant Act institution) in 1868. The Berkeley campus is the oldest of the ten-campus system. UC Berkeley is internationally renowned for its academic excellence. The faculty includes 8 Nobel laureates and 128 members of the National Academy of Sciences. In a recent national study, UC Berkeley was the only college or university that had every one of its departments ranked among the top five in the country.

UC Berkeley is also a cosmopolitan university. Its students present a rich diversity in their interests, backgrounds, cultures, and aspirations. Together, the campus and the San Francisco Bay Area provide an enormous array of intellectual, cultural, and recreational opportunities.

Applying

Admission is based on a variety of factors, including academic achievement and relevant experience. Applications must be received by January 5 and submitted online. The Graduate Group in Molecular and Biochemical Nutrition does not accept applications for a terminal master's or for studies beginning in spring. In addition, two official transcripts, GRE General Test scores, and three confidential letters of recommendation are also required by January 5. All applicants who are from countries where English is not an official language are required to submit TOEFL or TSE scores.

Recommended preparation includes a minimum of one course each in biochemistry, biochemistry lab, biology, quantitative analysis lab, physics with lab, calculus, statistics, physiology, and nutrition or another upper-division biological science and one year each of general chemistry with lab and organic chemistry with lab.

Correspondence and Information

Graduate Admissions
Graduate Group in Molecular and Biochemical Nutrition
117 Morgan Hall #3104
University of California, Berkeley
Berkeley, California 94720-3104

Phone: 510-643-2863
Fax: 510-642-0535
E-mail: mbn@nature.berkeley.edu
Web site: http://mbn.berkeley.edu

University of California, Berkeley

THE FACULTY AND THEIR RESEARCH

Professors
Bruce N. Ames, Ph.D. (biochemistry/genetics), Caltech, 1953. Aging mechanisms; DNA damage from nutritional imbalances; cancer prevention.
Leonard F. Bjeldanes, Ph.D. (organic chemistry), UCLA, 1970. Food toxicology; chemical carcinogenesis.
John E. Casida, Ph.D. (biochemistry/entomology), Wisconsin–Madison, 1954. Pesticide chemistry and toxicology; metabolism and mode of action of organic toxicants; insect biochemistry.
Benito O. de Lumen, Ph.D. (agricultural chemistry-biochemistry), California, Davis, 1971. Molecular biology/biochemistry of seeds as food sources.
Sharon E. Fleming, Ph.D. (food science and nutrition), Saskatchewan, 1975. Metabolism by intestinal epithelial cells.
John G. Forte, Ph.D. (physiology), Pennsylvania, 1961. Membrane transport and trafficking; pumps and transporters.
Marc Hellerstein, M.D., Yale, 1979; Ph.D. (nutritional biochemistry), MIT, 1986. Hepatic metabolic regulation, nutrition, and inflammation.
Isao Kubo, Ph.D. (organic chemistry), Osaka City (Japan), 1969. Natural products chemistry.
Joseph L. Napoli, Professor and Chair; Ph.D. (medicinal chemistry), Michigan, 1975. Retinoid endocrinology, metabolism, and function.
Barry Shane, Ph.D. (biochemistry), London, 1970. Metabolic and genetic regulation of vitamin metabolism.
Martyn Smith, Ph.D. (biochemistry), St. Bartholomew's Hospital Medical College (London), 1980. Development of biomarkers of early effect and susceptibility to leukemia and lymphoma; genetic toxicology of pesticides.
Hei Sook Sul, Ph.D. (nutritional sciences), Wisconsin–Madison. Lipid metabolism; adipose cell differentiation.

Associate Professors
Nancy K. Amy, Ph.D. (biology), Virginia, 1973. Regulation of trace-element metabolism.
Gregory W. Aponte, Ph.D. (nutrition), California, Davis, 1982. Gastrointestinal peptides and nutrient assimilation.
Chris Vulpe, Ph.D. (genetics), 1994, M.D., 1996, California, San Francisco. Genetic approaches to study mammalian copper and iron metabolism.
Jen (Wally) Wang, Ph.D. (molecular physiology), Vanderbilt, 1998. Mechanisms of glucocorticoid receptor-regulated metabolism.

Adjunct Professors
Ronald M. Krauss, M.D., Harvard, 1968. Genetic and nutrition regulation of lipoprotein metabolism.
Robert O. Ryan, Ph.D. (biochemistry), Nevada, Reno, 1982. Structure and function of exchangeable apolipoproteins; their influence on plasma lipid homeostasis and ability to reversibly associate with circulating lipoprotein particles.
Elizabeth C. Thiel, Ph.D. (biochemistry), Columbia. Genetic structure of ferritin protein and the role it plays in iron overload.

UNIVERSITY OF KENTUCKY

Graduate Center for Nutritional Sciences

Programs of Study

The University of Kentucky's Graduate Center for Nutritional Sciences is an interdisciplinary program that provides high-quality educational training and research experience leading to M.S. and Ph.D. degrees. Students have access to faculty expertise across twenty-eight departments and divisions at the University's Colleges of Medicine, Health Sciences, and Agriculture, preparing them for a variety of careers in academia, industry, and government. The program includes didactic course offerings in biochemistry, molecular biology, physiology, macronutrient and micronutrient metabolism, statistics, and nutrition and chronic disease. Electives are offered with a research emphasis in molecular metabolism, clinical nutrition, and behavioral aspects related to nutrition.

Students begin the program with one year of course work coupled with a nutritional sciences seminar and literature discussion course as well as research rotations in faculty laboratories. Research to be incorporated into the doctoral dissertation is pursued in the laboratory of one or more of the Center's faculty members. A primary area of research and training targets nutrition and chronic disease, with a focus on obesity and the associated disorders that include cardiovascular disease, diabetes, hypertension, and cancer. Other areas of specialty include nutrition and oxidative stress, clinical nutrition, and food science.

Research Facilities

The Graduate Center for Nutritional Sciences offers state-of-the-art equipment to study nearly every aspect of modern biology and medicine and assures students challenging and varied research opportunities. More than 36,000 square feet of new laboratory space have been dedicated to the Center that houses the newest equipment for cell culture and human and animal studies. The University of Kentucky (UK) also has in place multiple core facilities; among these are a microarray facility as well as cores for electron microscopy, confocal microscopy, flow cytometry, and magnetic resonance imaging.

University of Kentucky clinical facilities for training and research include the nationally recognized Chandler Medical Center, the Veteran's Administration Hospital, Sanders-Brown Center on Aging, Markey Cancer Center, and the Kentucky Clinic. Opportunities for community-based research exist locally and throughout the state. Research support services include agricultural and medical libraries, the Faculty Academic Computing and Technology Support (FACTS) Center, Statistics Consulting Laboratories, and media services. Access to the Internet is available in all major buildings.

Financial Aid

Traineeships, assistantships, and fellowships may be available to students accepted into the program. These provide tuition, health insurance, and fees as well as an annual stipend. In 2005, stipends for Ph.D. students ranged from $18,000 to $22,000.

Cost of Study

In 2006–07, Kentucky residents pay $4120 for full-time tuition per semester, and nonresidents pay $9654; per-credit-hour costs are $438.25 and $1053.25, respectively. Traineeships, assistantships, and fellowships cover tuition.

Living and Housing Costs

The University offers graduate and family housing for full-time graduate, doctoral, or professional students. Efficiency, one-bedroom, and two-bedroom apartments are available. They range in price from $500 to $672 per month, depending on size and location. Many students live in conveniently located off-campus housing, with one-bedroom apartments renting from $400 to $450 per month and two-bedroom apartments renting from $485 to $550 per month.

Student Group

The University of Kentucky enrolls more than 34,000 students, of whom more than 6,100 are enrolled in its eighty-six graduate programs. The University awards approximately 5,400 degrees annually, 1,300 of which are graduate degrees. The Center for Nutritional Sciences has 45 students.

Location

The University of Kentucky's 670-acre campus is located within walking distance of downtown Lexington, Kentucky, in the heart of the picturesque Bluegrass region. Lexington is a progressive community with a population of approximately 265,000, melding the amenities of a large city with the charm of a small town. Lexington features numerous theaters, concert halls, and restaurants, is surrounded by thoroughbred horse farms, and offers many opportunities for outdoor recreation, including a readily accessible network of state parks. Lexington is located about 70 miles from Louisville and 90 miles from Cincinnati, Ohio; it lies within a 500-mile radius of nearly three fourths of the manufacturing, retail sales, and population of the United States.

The University and The Center

The University of Kentucky is a public, research-extensive, land-grant university. Founded in 1865 as the Agricultural and Mechanical College, the University has grown into a comprehensive public institution of higher learning, with ninety-eight undergraduate programs, master's degrees in ninety-six fields, and doctoral degrees in sixty-two programs. The Chandler Medical Center, University Hospital, UK Children's Hospital, and several Centers of Excellence consistently receive national rankings for excellence in teaching, research, and patient care. Two thousand full-time faculty members and 9,000 full-time staff members are employed by the University.

The Graduate Center for Nutritional Sciences was initiated as a Ph.D.–granting program in 1989. The Center became a multidisciplinary graduate unit in 2000, offering an M.S. program with four emphases: clinical nutrition, community nutrition, molecular and biochemical nutrition, and sports and wellness nutrition. The Center is one of six interdisciplinary graduate training programs on campus, involving more than 60 faculty members from seven different colleges and/or schools. The mission of the Center is to train highly skilled nutritional scientists equipped to tackle critical nutrition-related disease and health issues and pursue promising careers in the rapidly expanding nutritional sciences field.

The University of Kentucky is an equal-opportunity university and encourages applications from all academically qualified people interested in educational opportunities.

Applying

Applications are accepted year-round. International students must submit their applications by February 1 for admission in the fall semester and June 15 for the spring semester. Admission to the graduate program requires a B.S. degree from a fully accredited institution of higher learning. Admission to the M.S. in Nutritional Sciences Program, with a clinical nutrition emphasis, is limited to those with a B.S. in dietetics or nutrition. Applicants to both the M.S. and Ph.D. programs must meet the Center's minimum grade point average requirements and General Test of the Graduate Record Examinations (GRE) score requirements. International applicants must achieve minimum TOEFL scores and demonstrate proficiency in verbal and written English communication.

For detailed admissions requirements, admission forms, and other program information, applicants should contact the Graduate Center for Nutritional Sciences or visit the Center's Web site at http://www.mc.uky.edu/nutrisci/.

The University of Kentucky is an equal opportunity university and encourages applications from all academically qualified people interested in educational opportunities.

Correspondence and Information

Steven Post, Ph.D., Director of Graduate Studies
Graduate Center for Nutritional Sciences
521 Charles T. Wethington Building
University of Kentucky
900 South Limestone
Lexington, Kentucky 40536-0200

Phone: 859-323-4933 Ext. 81366
E-mail: spost@uky.edu
Web site: http://www.mc.uky.edu/nutrisci/

University of Kentucky

THE FACULTY AND THEIR RESEARCH

K. Addo, Associate Professor of Food Science; Ph.D., Washington State, 1990. Developing analytical and rheological methods to evaluate the quality of cereal grains; development of fat replacers for use in baked products; nitrogen management and soft red winter wheat quality; wheat gluten and meat protein interactions using transglutaminase.

K. B. Ain, Professor of Internal Medicine; M.D., Brown, 1981. Thyroid cancer and disease.

J. W. Anderson, Professor of Medicine and Clinical Nutrition; M.D., Northwestern, 1961. Clinical research on nutrition management, obesity (adults/teens), and dyslipidemia.

D. Archbold, Associate Professor of Horticulture and Landscape Architecture; Ph.D., Michigan State, 1982. Interaction of production and harvest practices and postharvest handling techniques on fresh produce, especially fruit, quality, nutritional value, and safety.

G. A. Boissonneault, Professor of Clinical Nutrition; Ph.D., Illinois, 1982. Diet and cancer; diet and cardiovascular disease; diet and immunity.

J. A. Boling, Professor of Animal Nutrition; Ph.D., Wisconsin, 1967. Protein metabolism; nutrient-endocrine interrelationships.

M. G. Boosalis, Associate Professor of Clinical Nutrition; Ph.D., Minnesota, 1984. Acute phase response and their relationship to nutritional parameters.

G. Bruckner, Professor of Clinical Nutrition; Ph.D., Kentucky, 1979. Lipid metabolism, phytoestrogens, and obesity.

D. Bruemmer, Assistant Professor of Medicine; M.D., Hamburg. Role of nuclear hormone receptors in cardiovascular complications of obesity and diabetes.

A. Cantor, Associate Professor and Poultry Coordinator, Nutrition and Management; Ph.D., Cornell, 1974. Availability of organic sources of trace minerals and the use of exogenous enzymes to improve nutrient utilization in animal diets.

L. A. Cassis, Professor and Director of the Graduate Center for Nutritional Studies; Ph.D., West Virginia, 1984. The renin-angiotensin system as a link between obesity and cardiovascular disease.

L. H. Chen, Professor of Biochemical Nutrition and Associate Director; Ph.D., Louisville, 1964. Antioxidant defense systems, n-3 fatty acids, nutrition and aging.

C. K. Chow, Professor of Biochemical Nutrition; Ph.D., Illinois, 1969. Reactive oxygen species-induced degenerative disorders and its protection by dietary antioxidants.

J. L. Clasey, Associate Professor of Kinesiology and Health Promotion; Ph.D., Illinois at Urbana-Champaign, 1993. Human body composition and the relationship between body composition and health and disease.

A. Daugherty, Professor of Internal Medicine and Physiology and Director, Cardiovascular Research Center; Ph.D., Bath (England), 1981. Role of angiotensin peptides on the development of atherosclerosis and aortic aneurysms.

F. DeBeer, Professor and Chair, Department of Internal Medicine/Endocrinology; M.D., Pretoria (South Africa), 1983. Impact that inflammation has on lipoprotein metabolism and atherosclerosis.

M. DeBeer, Assistant Professor of Internal Medicine; Ph.D., Stellenbosch (South Africa), 1992; The influence of HDL size and composition during normal and inflammatory conditions on its interaction with HDL receptor SR-BI

W. J. S. de Villiers, Associate Professor of Internal Medicine and Chairman, Division of Gastroenterology; M.D., Stellenbosch (South Africa), 1991; D. Phil., Oxford, 1995. Pathophysiological role of macrophage scavenger receptors in animal models of chronic inflammation, such as inflammatory bowel disease and atherosclerosis.

N. D'Souza, Assistant Professor of Internal Medicine; Ph.D., Mumbai (India), 1985; M.P.H., Tulane, 1997. Effects of alcohol intake and nutrition on lung and liver immune defenses.

H. W. Forsythe, Associate Professor of Nutrition; Ph.D., Oklahoma State, 1987. Maternal and child nutrition; nutritional risk factors in pregnancy; nutrition for children with neurodevelopmental disorders: autism/PDD, Tourette's syndrome, and attention deficit hyperactivity disorder; nutritional assessment of multiethnic groups.

L. Gaetke, Associate Professor of Nutrition and Food Science; Ph.D., Kentucky, 1994. Therapeutic nutrition and dietetics; clinical outcomes of nutrition intervention; appetite regulation.

V. S. Gallicchio, Professor of Medicine and Clinical Sciences; Ph.D., NYU, 1976. Effects of hematopoietic growth factors/cytokines on eicosonoid metabolism; signal transduction mechanisms via activation of phospholipase activity and their clinical utility; nutritional and metal cation influences in immunology diseases.

H. P. Glauert, Professor of Biochemical Nutrition; Ph.D., Michigan State, 1982. Effect of nutrition on chemical carcinogenesis; nutritional toxicology; oxidative DNA damage; eicosanoids; transgenic animal models of metabolism and carcinogenesis.

M. C. Gong, Associate Professor of Physiology; Ph.D., Peking Union Medical College (China), 1994. Determining the effects of angiotensin peptides on the development of atherosclerosis and aortic aneurysms.

B. Hennig, Professor of Nutrition and Toxicology; Ph.D., Iowa State, 1982. Role of nutrition (e.g., lipids, antioxidants) and environmental contaminants on endothelial cell activation.

D. F. Hildebrand, Professor of Plant Biochemistry and Genetics; Ph.D., Illinois, 1982. Biochemistry and genetics of lipid metabolism in plants; characteristics and improvement of food nutritional quality; biotechnology.

D. Karounos, Associate Professor of Medicine; M.D., Kentucky, 1980. Immunology of type 1 diabetes; role of rubella virus in the pathogenesis of type 1 diabetes; prevention of diabetes with insulin or insulin analogs; intensive control of type 2 diabetes and macrovascular complications of diabetes.

E. J. Kasarskis Jr., Professor of Neurology; M.D., 1975, Ph.D., 1976, Wisconsin. Nutritional research in ALS (Lou Gehrig's disease) and other human neurodegenerative diseases.

T. Kelly, Associate Professor in Behavioral Science; Ph.D., Minnesota, 1983. Effects of methadone on the behavioral mechanisms of conditioned reinforcement and deprivation in the pigeon; behavioral aspects of alcoholism.

T. R. Kemp, Professor of Horticulture and Landscape Architecture; Ph.D., Kentucky, 1970. Identification of volatile chemical markers for bacteria such as *E. coli* and *Salmonella* to allow detection of these pathogenic bacteria on foods.

G.-M. Li, Associate Professor of Pathology; Ph.D., Wayne State, 1991. Role of DNA repair in cancer and folate deficiency–induced diseases.

M. Lindemann, Professor of Animal Sciences; Ph.D., Minnesota, 1981. Interaction of nutrition with reproduction (fecundity and milk production) and lean tissue growth, with emphasis on the minerals chromium, calcium, phosphorus, omega-3 fatty acids, and amino acids.

R. A. Lodder, Associate Professor of Pharmacy; Ph.D., Indiana, 1988. Nondestructive spectroscopic techniques to monitor the effects of nutrition on the stability of atherosclerotic plaques.

C. Mao, Assistant Professor of Nutritional Sciences; Ph.D., Rene Diderot (Paris), 1990. Role and mechanism of action of Wnt proteins and Wnt-signaling components in the control of endothelial cell behavior and communication with macrophages and smooth muscle cells.

J. Matthews, Assistant Professor of Animal Sciences; Ph.D., Virginia Tech, 1995. Characterization of amino acid and peptide transport proteins in mammals and regulation of pancreatic acinar and hepatic cell metabolites.

S. Ozcan, Assistant Professor of Biochemistry; Ph.D., Heinrich-Heine (Germany), 1993. Regulation of insulin production by glucose and how defects in this process lead to diabetes.

B. T. Pan, Associate Professor of Surgery; Ph.D., McGill, 1983. Role of nutrients in cancer development.

T. D. Porter, Associate Professor of Pharmaceutical Sciences; Ph.D., Illinois, 1981. Regulation of cholesterol synthesis.

S. Post, Associate Professor of Molecular and Biomedical Pharmacology, Nutritional Sciences, and Pharmaceutical Sciences and Director of Graduate Studies, Graduate Center for Nutritional Studies; Ph.D., Chicago, 1992. Cellular mechanisms that regulate the ability of a cell to sense and respond to changes in the extracellular environment.

D. K. St. Clair, Professor of Toxicology; Ph.D., Iowa, 1984. Structure and function of antioxidant enzymes and their role in aging and cancer.

W. St. Clair, Assistant Professor of Medicine; Ph.D., Iowa, 1985; M.D., Kentucky, 1995. Genitourinary cancer; central nervous system tumors; intensity-modulated radiation therapy (IMRT); gamma knife radiosurgery.

E. J. Smart, Professor of Pediatrics; Ph.D., Wisconsin–Madison, 1992. The role of caveolae and signal transduction in diabetes.

B. T. Spear, Associate Professor of Microbiology and Immunology; Ph.D., Pennsylvania, 1985. Control of liver gene expression during liver development and disease.

M. Toborek, Professor of Surgery; M.D., 1985, Ph.D., 1989, Silesian School of Medicine (Poland). Effect of nutrients on molecular/cellular metabolism of vascular endothelial cells.

D. R. van der Westhuyzen, Professor of Internal Medicine; Ph.D., Cape Town (South Africa), 1974. Role of lipoprotein receptors in the regulation of HDL metabolism, cellular cholesterol, and atherosclerosis.

N. Webb, Associate Professor of Internal Medicine; Ph.D., Kentucky, 1999. The role of inflammation in cardiovascular diseases, including atherosclerosis and abdominal aortic aneurysms.

T. Winter, Associate Professor of Internal Medicine; M.D., Godfrey Huggins (Zimbabwe), 1984; Ph.D., Cape Town (South Africa), 2001. Clinical nutrition and inflammatory bowel disease.

Y. L. Xiong, Professor and Coordinator of Food Science; Ph.D., Washington State, 1989. Antioxidant strategies to control protein and lipid oxidation in muscle foods and other functional food products.

J. W. Yates, Associate Professor, Department of Kinesiology and Health Promotion; Ph.D., Penn State, 1980. Strength and endurance training and muscle soreness.

H. Zhu, Assistant Professor of Molecular and Cellular Biochemistry; Ph.D., UCLA, 2000. Proteomic studies of oxidative stress and metalloproteins in human diseases.

UNIVERSITY OF MISSOURI–COLUMBIA

Nutritional Sciences Program

Programs of Study

The Nutritional Sciences Program at the University of Missouri–Columbia (MU) is an interdisciplinary group comprising the Nutritional Sciences Graduate Program (NSGP) and the Food for the 21st Century (F21C) Nutritional Sciences Cluster. The NSGP coordinates the core graduate nutrition curriculum and offers M.S. and Ph.D. degrees in nutrition, which are awarded through the Graduate School. M.A. and Ph.D. degrees in exercise physiology are also offered in conjunction with the Nutritional Sciences Departments. In addition, M.S. or Ph.D. degrees in animal science, biochemistry, and food science are offered through the respective departments to students studying nutrition. The F21C Nutrition Cluster is a research program consisting of more than 20 faculty members from nine departments in the Colleges of Agriculture, Foods and Natural Resources, Arts and Sciences, Human Environmental Sciences, and Medicine plus the Research Reactor. This rich environment offers a wide range of interdisciplinary research opportunities for the degree candidate. Entering graduate students are expected to have undergraduate training in chemistry and biology, including a two-semester course in biochemistry and an upper-level nutrition course. Some prerequisites can be met during the first year of graduate study.

The M.S. degree requires a minimum of 30 credit hours of course work beyond the bachelor's degree.

Requirements for the Ph.D. degree in nutrition include the course work required for the master's degree and additional course work such that the student attains a mastery of the broad fundamentals of metabolic integration, lipid and mineral biochemistry, signal transduction, and regulation of gene expression, as well as the demonstrated ability to conduct independent, innovative research. Students are admitted to full candidacy for the Ph.D. upon passing a comprehensive oral and written exam based on course work, research experience, and the ability to keep current with the latest trends in nutritional sciences.

Research opportunities include a broad range of approaches to the study of nutrition, including whole-animal models, cultured-cell models, enzymology and protein characterization, radioisotope tracer methodology, lipid metabolism, and receptor methodology. There is particular strength in nutritional biochemistry of macroelements and trace elements, in neurological sciences and signal transduction, and in nutraceuticals. The group is increasingly emphasizing application of molecular biology techniques to the study of nutrition.

Research Facilities

The research labs of the Nutritional Sciences faculty are housed in modern facilities in Eckles Hall, Gwynn Hall, McKee Gym, the Animal Sciences Research Center, and the School of Medicine. The F21C Nutrition Cluster provides funds for a series of programs that include seminars, strengthening mini-grants, graduate fellowships, undergraduate research internships, and mini-sabbaticals. MU has traditionally encouraged collaborative research among different campus units. This philosophy has continued with the F21C program, the molecular biology program, a series of campus core facilities, the Research Reactor, and the Life Sciences Enhancement program. Service centers provide monoclonal antibodies, custom oligonucleotides, protein sequence analysis, a transgenic animal facility, analytical chemistry labs, a modern NMR/mass spectroscopy facility, and an electron microscopy facility. MU has the largest university research reactor in the United States, providing unique radioisotopes and facilities for nutrition research. The 3.5-acre Animal Sciences Research Complex houses small- and large-animal facilities, surgical facilities, environmental chambers, metabolic facilities, and the whole-body counter. These programs give MU researchers an advantage in enlisting the multidisciplinary approaches necessary for today's cutting-edge research. MU libraries house more than 3.11 million volumes, and their online computer service eases access to 22,700 active serials.

Financial Aid

Graduate assistantships are available through the NSGP and from individual departments. Typical stipends range from \$16,500 to \$21,000 per year for Ph.D. candidates and from \$11,000 to \$12,000 per year for M.S. candidates. Outstanding students whose applications are received by February 1 are eligible for a number of fellowship opportunities, including those in Life Sciences, Molecular Biology, and the F21C Nutrition Cluster. These fellowships provide a higher level of support by waiving all fees and providing travel awards and can last for up to five years.

Cost of Study

Resident (\$209.60 per credit) and nonresident (\$402.40 per credit) tuition (2004–05) is waived for students on assistantships or fellowships.

Living and Housing Costs

On-campus apartments for single or married students can be readily obtained through the University Residential Life Office (573-882-7275). Rental units vary in cost from \$305 to \$500 per month. Most graduate students live off campus, and apartment, duplex, house, and mobile home rentals are readily available. In general, the cost of living in Columbia is low to moderate.

Student Group

The Columbia campus of the University of Missouri had an enrollment of 26,124 students in fall 2002, 7,256 of whom were in graduate or professional school. Of the 6,426 graduate students, 1,409 are from other countries. Currently, about 50 graduate students are enrolled in the nutritional sciences programs, including 20 from other countries.

Location

Columbia, a city of 84,000 people, is located in a rural area of wooded, rolling terrain and is equidistant (125 miles) from St. Louis and Kansas City. Besides the cultural and recreational outlets of these cities, Columbia has nationally recognized recreational programs, several excellent theater groups, and a lively series of concerts. The Missouri River and the Ozarks provide such outdoor activities as fishing, hunting (duck, turkey, deer), rock climbing, and canoeing. Missouri is a leading white-water state.

The University

The University of Missouri–Columbia was founded in 1839 and was the first state university west of the Mississippi. It was patterned after the ideals of Thomas Jefferson, a vigorous advocate of public higher education. In 1862, the University became a land-grant institution. The Columbia campus is the largest in a four-campus system that ranks among the nation's top twenty in enrollment. The pleasantly landscaped campus is within walking distance of downtown Columbia.

Applying

Applications should be submitted as early as possible (usually no later than March 1) for fall admission. University graduate application forms (http://web.missouri.edu/~gradschl/) are available online. While spring admissions are granted, assistantship support is usually more readily available for fall admission. Applicants must have an average of B or better in science (chemistry, physics, mathematics, and biology) from an accredited institution and must provide scores on the General Test of the Graduate Record Examinations. TOEFL scores are required for international students.

Correspondence and Information

Dr. Michael Petris
Nutritional Sciences Graduate Program
217 Gwynn Hall
University of Missouri–Columbia
Columbia, Missouri 65211
Phone: 573-882-4526
Fax: 573-882-0185
E-mail: Petrism@missouri.edu
Web site: http://web.missouri.edu/~nutsci/

University of Missouri–Columbia

THE FACULTY AND THEIR RESEARCH

Nutritional Biochemistry of Minerals

Richard P. Dowdy, Associate Professor and Associate Dean Emeritus; Ph.D., North Carolina State, 1967. Influence of aging on trace-element and macroelement nutrition.

David J. Eide, Professor, Nutritional Sciences/Biochemistry, and Director, Graduate Education; Ph.D., Wisconsin–Madison, 1987. The molecular biology of metal ion transport proteins and the regulation of metal ion homeostasis in eukaryotic cells.

Michael T. Henzl, Associate Professor, Biochemistry; Ph.D., Wisconsin–Madison, 1980. Biophysical characterization of calcium-binding proteins.

J. Steven Morris, Adjunct Assistant Professor, Research Reactor; Ph.D., Missouri–Columbia, 1973. Use of neutron activation analysis in nutrition.

Boyd L. O'Dell, Professor Emeritus, Biochemistry; Ph.D., Missouri–Columbia, 1943. First limiting metabolic defect in zinc deficiency.

Michael J. Petris, Assistant Professor, Nutritional Sciences/Biochemistry; Ph.D., Melbourne (Australia), 1998. Regulation of copper transport and copper-transporting P-type ATPases in eukaryotic cells and in prokaryotic models.

J. David Robertson, Professor, Chemistry/Research Reactor; Ph.D., Maryland, 1986. Radionuclear, analytical, and environmental chemistry; Alzheimer's disease and trace elements.

Elizabeth E. Rogers, Assistant Professor, Nutritional Sciences/Biochemistry; Ph.D., Harvard, 1997. The molecular biology of iron uptake and iron deficiency signaling in plants.

Neurological Sciences and Signal Transduction

Lane L. Clarke, Associate Professor, Veterinary Biomedical Sciences; D.V.M., Missouri–Columbia, 1982; Ph.D., North Carolina State, 1989. Electrolyte transport processes in airway and intestinal epithelia during health and disease.

Kevin L. Fritsche, Associate Professor, Animal Science; Ph.D., Illinois at Urbana-Champaign, 1988. Effect of dietary fats on immune functions; vitamin E requirements; cytokine-lipid mediator networks.

Grace Y. Sun, Professor, Biochemistry; Ph.D., Oregon State, 1966. Nutritional factors modulating brain signaling mechanisms: implication in neurodegenerative diseases related to stroke, alcoholism, and aging.

Gary A. Weisman, Professor, Biochemistry/Nutritional Sciences; Ph.D., Nebraska–Lincoln, 1982. Role of nucleotide receptors in the regulation of cardiovascular, neuronal, and neoplastic cell functions.

Peter A. Wilden, Associate Professor, Pharmacology; Ph.D., Iowa, 1988. Molecular mechanisms of insulin action; ATP-stimulated intracellular signaling in vascular smooth muscle; transmembrane and intracellular signaling processes.

Nutraceuticals

Dennis B. Lubahn, Associate Professor; Ph.D., Duke, 1983. Biochemistry of cancer; nutrition and reproduction.

Ruth S. MacDonald, Professor, Nutritional Sciences; Ph.D., Minnesota, 1985; RD. Role of nutrients on endocrine receptors that regulate cell growth and metabolism; nutraceuticals and cancer risk.

Human Nutrition and Exercise Physiology

Stephen D. Ball, Assistant Professor/State Nutrition Specialists, Nutritional Sciences; Ph.D., Arizona State, 2002. Body composition, obesity, and bone health.

Laura S. Hillman, Professor, Child Health; M.D., Yale, 1968. Calcium metabolism in the premature infant; chronic pediatric diseases.

Pamela S. Hinton, Assistant Professor, Nutritional Sciences; Ph.D., Wisconsin–Madison, 1997. Nutrition, physical activity, and women's health.

Thomas P. LaFontaine, Adjunct Instructor, Nutritional Sciences; Ph.D., Missouri–Columbia, 1983. Physiology of clinical exercise.

Tom R. Thomas, Professor, Nutritional Sciences; Ph.D., Missouri–Columbia, 1976. Exercise physiology; nutrition and exercise.

Animal Nutrition

Jeffre D. Firman, Professor, Animal Science; Ph.D., Maryland, 1987. Neural regulation of food and water intake in birds.

Monty S. Kerley, Professor, Animal Science; Ph.D., Illinois at Urbana-Champaign, 1986. Ruminant nutrition; nutritional effects of anaerobic fermentation in the rumen and large intestine.

David R. Ledoux, Associate Professor, Animal Science; Ph.D., Florida, 1987. Mineral and vitamin availability and requirements.

James N. Spain, Associate Professor, Animal Science; Ph.D., Virginia Tech, 1989. Nutrition of the periparturient dairy cow, with emphasis on the impact of energy metabolism on milk production and reproduction.

Trygve L. Veum, Professor, Animal Science; Ph.D., Cornell, 1968. Amino acid and mineral availability and nutrition for lean growth in swine.

James E. Williams, Professor, Animal Science; Ph.D., West Virginia, 1977. Ruminant nutrition, with emphasis on by-product utilization and metabolic regulation of growth.

<h1 style="text-align:center">Section 15
Parasitology</h1>

This section contains a directory of institutions offering graduate work in parasitology, followed by in-depth entries submitted by institutions that chose to prepare detailed program descriptions. Additional information about programs listed in the directory but not augmented by an in-depth entry may be obtained by writing directly to the dean of a graduate school or chair of a department at the address given in the directory.

For programs offering related work, see also in this book Biological and Biomedical Sciences and Microbiological Sciences. In Book 6, see Allied Health and Public Health.

CONTENTS

Program Directory

Close-Ups

Parasitology

Louisiana State University Health Sciences Center, School of Graduate Studies in New Orleans, Department of Microbiology, Immunology, and Parasitology, New Orleans, LA 70112-1393. Offers microbiology and immunology (MS, PhD). Terminal master's awarded for partial completion of doctoral program. *Degree requirements:* For master's, thesis; for doctorate, thesis/dissertation, preliminary exam, qualifying exam. *Entrance requirements:* For master's and doctorate, GRE General Test. Additional exam requirements/recommendations for international students: Required—TOEFL. *Faculty research:* Microbial physiology, animal virology, vaccine development, AIDS drug studies, pathogenic mechanisms, molecular immunology.

McGill University, Faculty of Graduate and Postdoctoral Studies, Faculty of Agricultural and Environmental Sciences, Institute of Parasitology, Montréal, QC H3A 2T5, Canada. Offers biotechnology (M Sc A, Certificate); parasitology (M Sc, PhD). Terminal master's awarded for partial completion of doctoral program. *Degree requirements:* For master's and doctorate, thesis/dissertation, registration. *Entrance requirements:* For master's, BSc, minimum GPA of 3.2; for doctorate, M Sc, minimum GPA of 3.2; for Certificate, minimum GPA of 3.0, B Sc in biological sciences. Additional exam requirements/recommendations for international students: Required—TOEFL (minimum score 550 paper-based; 213 computer-based), IELT (minimum score 7). Electronic applications accepted. *Faculty research:* Biochemistry, biochemical pharmacology, ecology, epidemiology, immunology.

New York University, School of Medicine and Graduate School of Arts and Science, Sackler Institute of Graduate Biomedical Sciences, Department of Medical and Molecular Parasitology, New York, NY 10012-1019. Offers PhD, MD/PhD. *Degree requirements:* For doctorate, one foreign language, thesis/dissertation, qualifying exam, comprehensive exam. *Entrance requirements:* For doctorate, GRE General Test. Additional exam requirements/recommendations for international students: Required—TOEFL. *Faculty research:* Immunoparasitology, cell biology of parasites, genetics of parasites, mode of action of antiparasitic drugs.

Purdue University, School of Veterinary Medicine and Graduate School, Graduate Programs in Veterinary Medicine, Department of Veterinary Pathobiology, West Lafayette, IN 47907. Offers biochemistry and molecular biology (MS, PhD); comparative epidemiology (MS, PhD); epidemiology (MS, PhD); immunology (MS, PhD); infectious diseases (MS, PhD); interdisciplinary genetics (PhD); laboratory animal medicine (MS, PhD); microbiology (MS, PhD); molecular virology (MS, PhD); parasitology (MS, PhD); pathobiology (MS, PhD); public health epidemiology (MS, PhD); toxicology (MS, PhD); veterinary anatomic pathology (MS, PhD); veterinary clinical pathology (MS, PhD); virology (MS, PhD). *Faculty:* 32 full-time (7 women). *Students:* 49 full-time (20 women), 3 part-time (1 woman); includes 2 minority (both African Americans), 31 international. Average age 35. In 2005, 3 master's, 8 doctorates awarded. Terminal master's awarded for partial completion of doctoral program. *Degree requirements:* For master's, thesis (for some programs); for doctorate, thesis/dissertation. *Entrance requirements:* For master's and doctorate, GRE General Test. Additional exam requirements/recommendations for international students: Required—TOEFL (minimum score 575 paper-based), TWE (minimum score 4). *Application deadline:* For fall admission, 8/12 for domestic students, 6/15 for international students; for spring admission, 1/12 for domestic students, 10/15 for international students. Application fee: $55. *Financial support:* Fellowships, research assistantships, teaching assistantships available. Financial award application deadline: 3/1; financial award applicants required to submit FAFSA. *Unit head:* Dr. H. Hogenesch, Head, 765-494-7543.

Texas A&M University, College of Veterinary Medicine, Graduate Programs in Veterinary Medicine, Department of Veterinary Pathobiology, College Station, TX 77843. Offers genetics (MS, PhD); veterinary microbiology (MS, PhD); veterinary parasitology (MS); veterinary pathology (MS, PhD). Part-time programs available. Postbaccalaureate distance learning degree programs offered. *Faculty:* 16 full-time (3 women), 4 part-time/adjunct (1 woman). *Students:* 45 full-time (28 women), 20 part-time (13 women); includes 7 minority (2 African Americans, 5 Hispanic Americans), 19 international. Average age 33. 25 applicants, 76% accepted, 19 enrolled. In 2005, 5 master's, 5 doctorates awarded. Terminal master's awarded for partial completion of doctoral program. *Degree requirements:* For master's and doctorate, thesis/dissertation, seminars. *Entrance requirements:* For master's and doctorate, GRE General Test, minimum GPA of 3.0 in last 60 hours. Additional exam requirements/recommendations for international students: Required—TOEFL. *Application deadline:* For fall admission, 3/1 for domestic students; for spring admission, 8/1 priority date for domestic students. Applications are processed on a rolling basis. Application fee: $50 ($75 for international students). *Expenses:* Tuition, state resident: full-time $4,488; part-time $187 per credit hour. Tuition, nonresident: full-time $11,112; part-time $463 per credit hour. Required fees: $1,974. *Financial support:* In 2005–06, fellowships with partial tuition reimbursements (averaging $16,000 per year), research assistantships with partial tuition reimbursements (averaging $15,400 per year), teaching assistantships with partial tuition reimbursements (averaging $16,000 per year) were awarded; Federal Work-Study, institutionally sponsored loans, scholarships/grants, traineeships, health care benefits, and unspecified assistantships also available. Support available to part-time students. Financial award applicants required to submit FAFSA. *Faculty research:* Infectious and noninfectious diseases of animals and birds, animal genetics, molecular biology, immunology, virology. *Unit head:* Dr. Ann B. Kier, Head, 979-845-5941, Fax: 979-845-9231, E-mail: akier@cvm.tamu.edu. *Application contact:* Dr. G. G. Wagner, Graduate Advisor, 979-845-2851, Fax: 979-862-1147, E-mail: gwagner@cvm.tamu.edu.

Tulane University, School of Public Health and Tropical Medicine, Department of Tropical Medicine, New Orleans, LA 70118-5669. Offers clinical tropical medicine and travelers health (Diploma); parasitology (MS, MSPH, PhD, Sc D); public health and tropical medicine (MPHTM). MS and PhD offered through the Graduate School. *Degree requirements:* For master's, thesis/dissertation; for doctorate, thesis/dissertation, comprehensive exam. *Entrance requirements:* For master's, GRE General Test, minimum B average in undergraduate course work; for doctorate, GRE General Test. Additional exam requirements/recommendations for international students: Required—TOEFL or TSE.

University of Notre Dame, Graduate School, College of Science, Department of Biological Sciences, Notre Dame, IN 46556. Offers aquatic ecology, evolution and environmental biology (MS, PhD); cellular and molecular biology (MS, PhD); genetics (MS, PhD); physiology (MS, PhD); vector biology and parasitology (MS, PhD). *Faculty:* 34 full-time (8 women), 3 part-time/adjunct (0 women). *Students:* 124 full-time (55 women); includes 11 minority (1 African American, 6 Asian Americans or Pacific Islanders, 4 Hispanic Americans), 39 international. 95 applicants, 34% accepted, 22 enrolled. In 2005, 4 master's, 11 doctorates awarded. Terminal master's awarded for partial completion of doctoral program. *Median time to degree:* Of those who began their doctoral program in fall 1997, 61% received their degree in 8 years or less. *Degree requirements:* For master's and doctorate, thesis/dissertation, comprehensive exam. *Entrance requirements:* For master's and doctorate, GRE General Test. Additional exam requirements/recommendations for international students: Required—TOEFL. *Application deadline:* For fall admission, 2/1 for domestic students; for spring admission, 11/1 for domestic students. Applications are processed on a rolling basis. Application fee: $50. Electronic applications accepted. *Financial support:* In 2005–06, 124 students received support, including 24 fellowships with full tuition reimbursements available (averaging $22,000 per year), 47 research assistantships with full tuition reimbursements available (averaging $15,250 per year), 45 teaching assistantships with full tuition reimbursements available (averaging $16,000 per year); traineeships and tuition waivers (full) also available. Financial award application deadline: 2/1. *Faculty research:* Tropical disease, molecular genetics, neurobiology, evolutionary biology, aquatic biology. Total annual research expenditures: $15.3 million. *Unit head:* Dr. Gary A. Lamberti, Director of Graduate Studies, 574-631-6552, Fax: 574-631-7413, E-mail: biology.biosadm.1@nd.edu. *Application contact:* Dr. Terrence J. Akai, Director of Graduate Admissions, 574-631-7706, Fax: 574-631-4183, E-mail: gradad@nd.edu.

See Close-Up on page 281.

University of Pennsylvania, School of Medicine, Biomedical Graduate Studies, Graduate Group in Cell and Molecular Biology, Program in Microbiology, Virology, and Parasitology, Philadelphia, PA 19104. Offers PhD, MD/PhD, VMD/PhD. *Degree requirements:* For doctorate, thesis/dissertation. *Entrance requirements:* For doctorate, GRE General Test, previous course work in science. Additional exam requirements/recommendations for international students: Required—TOEFL. *Application deadline:* For fall admission, 12/15 priority date for domestic students, 12/1 priority date for international students. Applications are processed on a rolling basis. Application fee: $70. Electronic applications accepted. *Financial support:* Fellowships, research assistantships, scholarships/grants, traineeships, and unspecified assistantships available. *Unit head:* Dr. Paul Bates, Chair, 215-573-3509. *Application contact:* Emily D. Geib, Coordinator, 215-898-3918, Fax: 215-573-2104, E-mail: eddz@mail.med.upenn.edu.

University of Prince Edward Island, Atlantic Veterinary College, Graduate Program in Veterinary Medicine, Charlottetown, PE C1A 4P3, Canada. Offers anatomy (M Sc, PhD); bacteriology (M Sc, PhD); clinical pharmacology (M Sc, PhD); clinical sciences (M Sc, PhD); epidemiology (M Sc, PhD), including reproduction; fish health (M Sc, PhD); food animal nutrition (M Sc, PhD); immunology (M Sc, PhD); microanatomy (M Sc, PhD); parasitology (M Sc, PhD); pathology (M Sc, PhD); pharmacology (M Sc, PhD); physiology (M Sc, PhD); toxicology (M Sc, PhD); veterinary science (M Vet Sc); virology (M Sc, PhD). Part-time programs available. *Faculty:* 76 full-time (25 women), 49 part-time/adjunct (8 women). *Students:* 54 full-time (32 women), 2 part-time. Average age 30. In 2005, 7 master's, 6 doctorates awarded. *Degree requirements:* For master's and doctorate, thesis/dissertation. *Entrance requirements:* For master's, DVM, B Sc honors degree, or equivalent; for doctorate, M Sc. *Application deadline:* Applications are processed on a rolling basis. Application fee: $50. *Expenses:* Contact institution. Tuition charges are reported in Canadian dollars. Part-time tuition and fees vary according to course level, degree level, campus/location, program and student level. *Financial support:* In 2005–06, 4 fellowships (averaging $25,000 Canadian dollars per year), 4 research assistantships (averaging $16,500 Canadian dollars per year) were awarded; career-related internships or fieldwork also available. *Faculty research:* Animal health management, infectious diseases, fin fish and shellfish health, basic biomedical sciences, ecosystem health. Total annual research expenditures: $1.2 million Canadian dollars. *Unit head:* Dr. James Bellamy, Associate Dean of Graduate Studies and Research, 902-566-0856, E-mail: bellamy@upei.ca. *Application contact:* Cheryl Gaudet, Registrar's Office, 902-566-0781, Fax: 902-566-0795, E-mail: registrar@upei.ca.

University of Washington, Graduate School, School of Public Health and Community Medicine, Graduate Program in Pathobiology, Seattle, WA 98195. Offers MS, PhD. *Faculty:* 22 full-time (8 women), 11 part-time/adjunct (6 women). *Students:* 47 full-time (31 women); includes 7 minority (1 African American, 4 Asian Americans or Pacific Islanders, 2 Hispanic Americans), 6 international. Average age 47. 55 applicants, 31% accepted, 12 enrolled. In 2005, 1 master's, 2 doctorates awarded. Terminal master's awarded for partial completion of doctoral program. *Median time to degree:* Of those who began their doctoral program in fall 1997, 99% received their degree in 8 years or less. *Degree requirements:* For master's, thesis/dissertation, registration; for doctorate, thesis/dissertation, comprehensive exam, registration. *Entrance requirements:* For master's and doctorate, GRE General Test, minimum GPA of 3.0. Additional exam requirements/recommendations for international students: Required—TOEFL. *Application deadline:* For fall admission, 12/15 for domestic students, 11/15 for international students. Application fee: $50. *Financial support:* In 2005–06, 2 fellowships with tuition reimbursements (averaging $22,560 per year), 18 research assistantships with tuition reimbursements (averaging $22,560 per year) were awarded; career-related internships or fieldwork, institutionally sponsored loans, scholarships/grants, traineeships, tuition waivers (full and partial), and unspecified assistantships also available. Financial award application deadline: 3/1; financial award applicants required to submit FAFSA. *Faculty research:* Pathogenesis of chlamydiae, molecular biology of parasites, signal transduction, antigenic analysis, molecular biology of tumor viruses. *Unit head:* Dr. Andreas Stergachis, Acting Chair, 206-543-8350, Fax: 206-543-3873, E-mail: stergach@u.washington.edu. *Application contact:* Joseph A. Daniels, Manager of Student Services, 206-543-4338, Fax: 206-543-3873, E-mail: pathobio@u.washington.edu.

See Close-Up on page 1129.

Yale University, School of Medicine, School of Public Health, Program in Parasitology, New Haven, CT 06520. Offers PhD. *Degree requirements:* For doctorate, thesis/dissertation, residency, comprehensive exam. *Entrance requirements:* For doctorate, GRE General Test. Additional exam requirements/recommendations for international students: Required—TOEFL.

See Close-Up on page 1103.

YALE UNIVERSITY
Parasitology Program

Program of Study	Parasitology at Yale University is an interdisciplinary graduate program of training and research in the study of parasites and their effects on their hosts. The faculty shares a commitment to understanding the biology of parasites and their vectors through employing cellular, molecular, genetic, and immunologic approaches. The faculty, primarily in the Department of Epidemiology and Public Health, consists of scientists in several academic departments at Yale University, including biology, immunobiology, cell biology and molecular biophysics, and biochemistry. Students are admitted into the Department of Epidemiology and Public Health. The program of graduate study is designed to provide individualized education in modern parasitology and to prepare students for independent careers in research and teaching. Course work generally occupies the first two years of study. Each student, together with a faculty advisory committee, outlines a course of study tailored to the individual's background and career goals. A program of course work may include parasitology, general microbiology, and/or epidemiology, as well as complementary courses offered in other programs such as cell biology, molecular biology, biochemistry, genetics, ecology, and statistics. The program also sponsors a journal club and seminars in parasitology. All students participate in three laboratory rotations, with different faculty members, in their areas of interest. Laboratory rotations ensure that students quickly become familiar with the variety of research opportunities available in the program. An individualized qualifying exam on topics selected by each student, in consultation with the faculty, is given before the end of the second year. Students then undertake an original research project under the direct supervision of a faculty member. The remaining degree requirements include submission of a dissertation prospectus, completion of a research project, and writing of a dissertation and its oral presentation.
Research Facilities	Faculty members occupy space at Science Hill and Yale Medical School. Facilities have the latest instruments necessary for research in modern parasitology. Faculty members and students have access to a number of special facilities, including a protein and nucleic acid facility for automated DNA sequencing, oligonucleotide synthesis, and peptide sequencing and synthesis; confocal and electron microscopic facilities; fluorescence-activated cell sorting; and computer facilities. Special laboratory facilities for the biocontainment of infectious agents and for rearing and studying insect vectors are also available. The Kline Science Library and Harvey Cushing/John Hay Whitney Medical Library have extensive collections of journals, books, and electronic data resources.
Financial Aid	All students admitted to the program receive a stipend, health insurance coverage, and tuition costs. The stipend for traineeships is approximately $26,750 in 2006–07.
Cost of Study	Tuition and fees are usually covered by financial aid.
Living and Housing Costs	Dormitory accommodations for women and men are available for the 2006–07 academic year; current costs range from $3510 to $5600. The University also maintains apartments for married students; rents range from $690 to $985 per month. For additional information about residential facilities in New Haven, students should contact the University Housing Bureau, Hendrie Hall, 165 Elm Street, New Haven, Connecticut 06520, or visit http://www.yale.edu/gradhousing/.
Student Group	About 300 graduate students are enrolled in the various programs in the biological sciences at Yale University. Approximately 10 graduate students and 40 postdoctoral fellows are associated with the parasitology program.
Location	New Haven is a city of 150,000, located on the north shore of Long Island Sound. The city is the birthplace of the American pizza and the American Society for Microbiology and also has an active cultural life. A wide spectrum of theater is available through Yale Repertory Theater, Yale School of Drama, and the Shubert and Long Wharf Theaters. New Haven is the home of three symphony orchestras; numerous classical, rock, and jazz groups; and several museums, including the Yale Art Gallery, Center for British Art, and the Peabody Museum of Natural History. Recreational facilities are available at the University's gymnasium, skating rink, playing fields, tennis courts, and sailing club. New Haven is close to state parks that offer a wide variety of hiking, biking, swimming, and rock-climbing opportunities.
The University	Yale University is a private university, founded in 1701. The University community is a large and diverse one, consisting of about 5,000 undergraduate students, 5,000 graduate and professional students, 1,500 postdoctoral fellows, and 2,000 faculty members.
Applying	Application forms and information about the graduate program and financial aid can be obtained at http://www.yale.edu/graduateschool/admissions/application.html. Students may apply either through the Biological and Biomedical Sciences (BBS)/Microbiology Graduate Program or through the Department of Epidemiology and Public Health. Completed forms as well as GRE General Test and Subject Test scores must be submitted by December 6, 2006. Applicants for whom English is not the native language are required to submit TOEFL scores.
Correspondence and Information	Dr. Diane McMahon-Pratt Department of Epidemiology and Public Health Yale University School of Medicine 60 College Street, LEPH 711 P.O. Box 208034 New Haven, Connecticut 06520-8034 E-mail: diane.mcmahon-pratt@yale.edu

Yale University

THE FACULTY AND THEIR RESEARCH.

Herve Agaisse, Assistant Professor of Microbial Pathogenesis; Ph.D., Pasteur Institute (Paris), 1996. Innate immune responses to pathogen infection in insects, using the *Drosophila* system.

Serap Aksoy, Professor of Epidemiology and Public Health; Ph.D., Columbia, 1982. Interaction of African trypanosomes with their vector, the tsetse fly; genetic modification of vector competence.

Karen Anderson, Professor of Pharmacology; Ph.D., Ohio State, 1982. Enzymatic and receptor-ligand interactions; structure-based drug design; antiparasitic drugs.

Norma Andrews, Professor of Microbial Pathogenesis and Cell Biology; Ph.D., São Paulo (Brazil), 1983. Cellular and molecular strategies used by the intracellular protozoan parasite *Trypanosoma cruzi* to invade and survive inside host cells.

Richard Bucala, Professor of Medicine and Pathology; Ph.D., Rockefeller, 1985; M.D., Cornell, 1986. Host response to tissue invasion and in the host-parasite interactions producing disease; role of the host cytokine MIF in the complications of malaria and *Leishmania* infection.

Michael Cappello, Professor of Pediatrics and of Epidemiology and Public Health; M.D., Georgetown, 1988. Tropical diseases and parasitology; hookworm; mechanisms of pathogenesis and vaccines.

Erol Fikrig, Professor of Medicine, Microbial Pathogenesis, and Epidemiology and Public Health; M.D., Cornell, 1985. Vectors and vector-borne diseases; mechanisms of pathogenesis and transmission; Lyme disease, human granulocytic ehrlichiosis, and West Nile virus.

Durland Fish, Professor of Epidemiology and Public Health; Ph.D., Florida, 1976. Epidemiology of vector-borne pathogens; landscape epidemiology of zoonoses; population regulation of arthropod vectors.

Richard Flavell, Professor of Immunobiology; Ph.D., Hull (England), 1970. Roles of TGF-β and IL-19 in the negative regulation of IFNγ production; control and regulation of *Leishmania major* infection.

Theodore Holford, Professor of Epidemiology and Public Health; Ph.D., Yale, 1973. Development and application of statistical methods in epidemiology; use of geographic information systems (GIS) to study factors that affect vector ecology and the subsequent risks that arise from vector-borne disease.

Diane McMahon-Pratt, Professor of Epidemiology and Public Health; Ph.D., Harvard, 1978. Immunology and developmental biochemistry of the parasitic protozoa *Leishmania*.

Leonard Munstermann, Senior Research Scientist in Epidemiology and Public Health; Ph.D., Notre Dame, 1979. Population genetics and taxonomy of medically important insects (mosquitoes and sand flies), focusing on the species *Aedes albopictus* and *Lutzomyia longipalpis,* respectively, and their role in disease transmission.

Curtis L. Patton, Professor of Epidemiology and Public Health; Ph.D., Michigan, 1966. Developmental biology of trypanosomes and malarial parasites.

Jeffrey Powell, Professor of Ecology and Evolutionary Biology; Ph.D., California, Davis, 1972. Molecular evolutionary genetics of vectors, especially the major vectors of malaria in Africa, the *Anopheles gambiae* complex.

Nancy H. Ruddle, Professor of Epidemiology and Public Health and of Immunobiology; Ph.D., Yale, 1968. Lymphotoxin (TNF-β) and tumor necrosis factor (TNF-α); regulation, mechanism of action, and biologic role in retroviral, parasitic, and autoimmune diseases.

Christian Tschudi, Associate Professor of Infectious Diseases; Ph.D., Basel, 1982. RNA modification and transcriptional mechanisms in African trypanosomes.

Elisabetta Ullu, Professor of Infectious Diseases and Cell Biology; Ph.D., Rome, 1973. Mechanisms and cofactors governing gene expression; pre-mRNA trans-splicing and polyadenylation in the protozoan parasite *Trypanosoma brucei.*

Graham Warren, Professor of Cell Biology; Ph.D., Cambridge, 1972. Golgi biogenesis; *Toxoplasma* and trypanosomes.

Liangbiao Zheng, Associate Professor of Epidemiology and Public Health; Ph.D., Harvard, 1990. Population genetics of anopheline mosquitoes.

Section 16
Pathology and Pathobiology

This section contains a directory of institutions offering graduate work in pathology and pathobiology, followed by in-depth entries submitted by institutions that chose to prepare detailed program descriptions. Additional information about programs listed in the directory but not augmented by an in-depth entry may be obtained by writing directly to the dean of a graduate school or chair of a department at the address given in the directory.

For programs offering related work, see also in this book Anatomy; Biochemistry; Biological and Biomedical Sciences; Cell, Molecular, and Structural Biology; Genetics, Developmental Biology, and Reproductive Biology; Microbiological Sciences; Pharmacology and Toxicology; and Physiology. In Book 6, see Allied Health, Public Health, and Veterinary Medicine and Sciences.

CONTENTS

Program Directories

Close-Ups

See also:

Molecular Pathology

Massachusetts Institute of Technology, School of Engineering, Biological Engineering Division, Cambridge, MA 02139-4307. Offers applied biosciences (PhD, Sc D); bioengineering (PhD, Sc D); biomedical engineering (M Eng); genetic toxicology (PhD, Sc D); molecular and systems bacterial pathogenesis (PhD, Sc D); molecular and systems toxicology and pharmacology (PhD, Sc D); molecular systems toxicology (PhD, Sc D); toxicology (SM, PhD, Sc D). *Faculty:* 16 full-time (4 women). *Students:* 126 full-time (56 women); includes 25 minority (3 African Americans, 17 Asian Americans or Pacific Islanders, 5 Hispanic Americans), 39 international. Average age 26. 349 applicants, 18% accepted, 28 enrolled. In 2005, 4 master's, 12 doctorates awarded. Terminal master's awarded for partial completion of doctoral program. *Degree requirements:* For master's, thesis; for doctorate, thesis/dissertation, qualifying exams, comprehensive exam. *Entrance requirements:* For master's and doctorate, GRE General Test. Additional exam requirements/recommendations for international students: Required—TOEFL (minimum score 600 paper-based; 250 computer-based). *Application deadline:* For fall admission, 12/31 for domestic students, 12/31 for international students. Application fee: $70. Electronic applications accepted. *Expenses:* Tuition: Full-time $32,100. Required fees: $200. Part-time tuition and fees vary according to course load. *Financial support:* In 2005–06, 112 students received support, including 57 fellowships with tuition reimbursements available (averaging $27,248 per year), 64 research assistantships with tuition reimbursements available (averaging $26,559 per year), 3 teaching assistantships with tuition reimbursements available (averaging $26,180 per year); Federal Work-Study, institutionally sponsored loans, scholarships/grants, traineeships, health care benefits, and unspecified assistantships also available. Total annual research expenditures: $22.5 million. *Unit head:* Prof. Douglas A. Lauffenburger, Director, 617-252-1629, Fax: 617-258-0204, E-mail: lauffen@mit.edu. *Application contact:* Biological Engineering Academic Office, 617-253-1712, Fax: 617-258-8676, E-mail: be-acad@mit.edu.

North Dakota State University, The Graduate School, College of Agriculture, Food Systems, and Natural Resources, Department of Veterinary and Microbiological Sciences, Fargo, ND 58105. Offers microbiology (MS); molecular pathogenesis (PhD); natural resource management (MS). Part-time programs available. *Degree requirements:* For master's, thesis; for doctorate, thesis/dissertation, oral and written preliminary exams. *Entrance requirements:* For master's and doctorate, GRE. Additional exam requirements/recommendations for international students: Required—TOEFL. *Faculty research:* Bacterial gene regulation, antibiotic resistance, molecular virology, mechanisms of bacterial pathogenesis, immunology of animals.

Texas Tech University Health Sciences Center, School of Allied Health Sciences, Program in Molecular Pathology, Lubbock, TX 79430. Offers MS. *Accreditation:* NAACLS. *Faculty:* 4 full-time (3 women). *Students:* 16 full-time (9 women); includes 7 minority (1 African American, 1 American Indian/Alaska Native, 1 Asian American or Pacific Islander, 4 Hispanic Americans). Average age 25. 23 applicants, 70% accepted, 16 enrolled. In 2005, 15 degrees awarded. *Entrance requirements:* Additional exam requirements/recommendations for international students: Required—TOEFL. *Application deadline:* For spring admission, 3/1 priority date for domestic students. Application fee: $35. Electronic applications accepted. *Financial support:* Career-related internships or fieldwork, institutionally sponsored loans, and scholarships/grants available. Financial award applicants required to submit FAFSA. *Unit head:* Dr. Hal Larsen, Chair, 806-743-3247, E-mail: hal.larsen@ttuhsc.edu.

University of California, San Diego, School of Medicine and Graduate Studies and Research, Molecular Pathology Program, La Jolla, CA 92093. Offers bioinformatics (PhD); cancer biology/oncology (PhD); cardiovascular sciences and disease (PhD); microbiology (PhD); molecular pathology (PhD); neurological disease (PhD); stem cell and developmental biology (PhD); structural biology/drug design (PhD). *Entrance requirements:* For doctorate, GRE General Test, GRE Subject Test. Additional exam requirements/recommendations for international students: Required—TOEFL. Electronic applications accepted.

See Close-Up on page 1119.

University of Medicine and Dentistry of New Jersey, Graduate School of Biomedical Sciences, Graduate Programs in Biomedical Sciences–Newark, Program in Experimental Pathology, Newark, NJ 07107. Offers PhD. *Entrance requirements:* Additional exam requirements/recommendations for international students: Required—TOEFL. *Application deadline:* For fall admission, 2/1 for domestic students. *Financial support:* Fellowships, research assistantships, Federal Work-Study, institutionally sponsored loans, and tuition waivers (full and partial) available. *Unit head:* Dr. Muriel Lambert, Program Director, 973-972-4405, Fax: 973-972-7293, E-mail: mlambert@umdnj.edu.

University of Pittsburgh, School of Medicine, Graduate Programs in Medicine, Program in Cellular and Molecular Pathology, Pittsburgh, PA 15260. Offers MS, PhD. *Faculty:* 45 full-time (11 women). *Students:* 27 full-time (14 women); includes 4 minority (1 African American, 2 Asian Americans or Pacific Islanders, 1 Hispanic American), 7 international. Average age 28. 415 applicants, 22% accepted, 42 enrolled. In 2005, 4 degrees awarded. *Median time to degree:* Of those who began their doctoral program in fall 1997, 95% received their degree in 8 years or less. *Degree requirements:* For doctorate, thesis/dissertation, comprehensive exam, registration. *Entrance requirements:* For doctorate, GRE General Test, GRE Subject Test, minimum QPA of 3.0. Additional exam requirements/recommendations for international students: Required—TOEFL (minimum score 600 paper-based; 250 computer-based), IELT (minimum score 7). *Application deadline:* For fall admission, 12/15 priority date for domestic students, 12/15 priority date for international students. Application fee: $40. Electronic applications accepted. *Expenses:* Tuition, state resident: full-time $13,194; part-time $537 per credit. Tuition, nonresident: full-time $25,012; part-time $1,026 per credit. Required fees: $700; $164 per term. Tuition and fees vary according to campus/location and program. *Financial support:* In 2005–06, fellowships with full tuition reimbursements (averaging $21,500 per year), research assistantships with full tuition reimbursements (averaging $21,500 per year) were awarded; teaching assistantships with full tuition reimbursements, institutionally sponsored loans, scholarships/grants, traineeships, health care benefits, and unspecified assistantships also available. *Faculty research:* Liver growth and differentiation, pathogenesis of neurodegeneration, cancer research. *Unit head:* Dr. Robert Bowser, Graduate Program Director, 412-383-7819, Fax: 412-383-7969, E-mail: bowserrp@upmc.edu. *Application contact:* Graduate Studies Administrator, 412-648-8957, Fax: 412-648-1077, E-mail: gradstudies@medschool.pitt.edu.

The University of Texas Health Science Center at Houston, Graduate School of Biomedical Sciences, Program in Molecular Pathology, Houston, TX 77225-0036. Offers MS, PhD, MD/PhD. *Faculty:* 33 full-time (8 women). *Students:* 9 full-time (7 women); includes 3 minority (2 Asian Americans or Pacific Islanders, 1 Hispanic American), 2 international. Average age 24. 21 applicants, 81% accepted, 10 enrolled.Terminal master's awarded for partial completion of doctoral program. *Degree requirements:* For master's and doctorate, thesis/dissertation. *Entrance requirements:* For master's and doctorate, GRE General Test. Additional exam requirements/recommendations for international students: Required—TOEFL, TWE. *Application deadline:* For fall admission, 1/15 for domestic students; for spring admission, 11/1 for domestic students. Applications are processed on a rolling basis. Application fee: $10. Electronic applications accepted. *Financial support:* Fellowships with full tuition reimbursements, research assistantships with full tuition reimbursements, teaching assistantships, institutionally sponsored loans, scholarships/grants, and health care benefits available. Financial award application deadline: 1/15. *Faculty research:* Cellular and molecular pathology, carcinogenesis and chemical pathology, immunopathology, structural and computational biology, pathogenesis of infectious disease. *Unit head:* Dr. Diane L. Hickson-Bick, Director, 713-500-5328, Fax: 713-500-0730, E-mail: diane.1.bick@uth.tmc.edu. *Application contact:* Dr. Victoria P. Knutson, Assistant Dean of Admissions, 713-500-9860, Fax: 713-500-9877, E-mail: victoria.p.knutson@uth.tmc.edu.

Yale University, School of Medicine and Graduate School of Arts and Sciences, Combined Program in Biological and Biomedical Sciences (BBS), Pharmacological Sciences and Molecular Medicine Track, New Haven, CT 06520. Offers PhD, MD/PhD. *Students:* 6 full-time. *Degree requirements:* For doctorate, thesis/dissertation. *Entrance requirements:* For doctorate, GRE General Test. Additional exam requirements/recommendations for international students: Required—TOEFL. *Application deadline:* For fall admission, 12/8 for domestic students, 12/8 for international students. Electronic applications accepted. *Financial support:* Fellowships, research assistantships available. *Unit head:* Dr. David F. Stern, Co-Director, 203-785-4832, Fax: 203-785-7467, E-mail: kathleen.fisher@yale.edu. *Application contact:* Kathy Fisher, Registrar, 203-785-4545, E-mail: kathleen.fisher@yale.edu.

Pathobiology

Auburn University, College of Veterinary Medicine and Graduate School, Graduate Programs in Veterinary Medicine, Auburn University, AL 36849. Offers biomedical sciences (MS, PhD), including anatomy, physiology and pharmacology (MS); biomedical sciences (PhD); clinical sciences (MS); large animal surgery and medicine (MS); pathobiology (MS); radiology (MS); small animal surgery and medicine (MS). Part-time programs available. *Faculty:* 76 full-time (24 women). *Students:* 11 full-time (8 women), 33 part-time (20 women); includes 3 minority (all Hispanic Americans), 14 international. 41 applicants, 37% accepted, 11 enrolled. In 2005, 9 master's, 3 doctorates awarded. *Degree requirements:* For doctorate, thesis/dissertation. *Entrance requirements:* For master's, GRE General Test; for doctorate, GRE General Test, GRE Subject Test. *Application deadline:* For fall admission, 7/7 for domestic students; for spring admission, 11/24 for domestic students. Applications are processed on a rolling basis. Application fee: $25 ($50 for international students). Electronic applications accepted. *Financial support:* Research assistantships, teaching assistantships, Federal Work-Study available. Support available to part-time students. Financial award application deadline: 3/15. *Application contact:* Dr. Stephen L. McFarland, Acting Dean of the Graduate School, 334-844-4700.

Brown University, Graduate School, Division of Biology and Medicine, Program in Pathology and Laboratory Medicine, Providence, RI 02912. Offers biology (PhD); cancer biology (PhD); immunology and infection (PhD); medical science (PhD); pathobiology (Sc M); toxicology and environmental pathology (PhD). Terminal master's awarded for partial completion of doctoral program. *Degree requirements:* For doctorate, thesis/dissertation, preliminary exam. *Entrance requirements:* For master's and doctorate, GRE General Test, GRE Subject Test. Additional exam requirements/recommendations for international students: Required—TOEFL. Electronic applications accepted. *Faculty research:* Environmental pathology, carcinogenesis, immunopathology, signal transduction, innate immunity.

Columbia University, College of Physicians and Surgeons and Graduate School of Arts and Sciences, Graduate School of Arts and Sciences at the College of Physicians and Surgeons, Department of Pathology, New York, NY 10032. Offers pathobiology (M Phil, MA, PhD). Only candidates for the PhD are admitted. Terminal master's awarded for partial completion of doctoral program. *Degree requirements:* For doctorate, thesis/dissertation. *Entrance requirements:* For master's and doctorate, GRE General Test. Additional exam requirements/recommendations for international students: Required—TOEFL. *Expenses:* Tuition: Full-time $31,448. Tuition and fees vary according to course level, course load, campus/location and program. *Faculty research:* Virology, molecular biology, cell biology, neurobiology, immunology.

Drexel University, College of Medicine, Biomedical Graduate Programs, Interdisciplinary Program in Molecular Pathobiology, Philadelphia, PA 19104-2875. Offers PhD, MD/PhD. *Degree requirements:* For doctorate, thesis/dissertation, qualifying exams, comprehensive exam. *Entrance requirements:* For doctorate, GRE General Test, minimum GPA of 3.0. Additional exam requirements/recommendations for international students: Required—TOEFL. Electronic applications accepted. *Faculty research:* Cell and molecular immunology, tumor immunology, molecular genetics, immunopathology, immunology of aging.

The Johns Hopkins University, School of Medicine, Graduate Programs in Medicine, Department of Pathology, Baltimore, MD 21218-2699. Offers pathobiology (PhD). *Faculty:* 59 full-time (10 women). *Students:* 27 full-time (16 women); includes 5 minority (2 African Americans, 2 Asian Americans or Pacific Islanders, 1 Hispanic American), 9 international. Average age 28. 107 applicants, 7% accepted, 7 enrolled. *Degree requirements:* For doctorate, thesis/dissertation, qualifying oral exam. *Entrance requirements:* For doctorate, GRE General Test, GRE Subject Test (biology or chemistry), previous course work with laboratory in organic and inorganic chemistry, general biology, calculus; interview. Additional exam requirements/recommendations for international students: Required—TOEFL. *Application deadline:* For winter admission, 1/10 for domestic students. Application fee: $90. Electronic applications accepted. *Expenses:* Tuition: Full-time $30,960. Tuition and fees vary according to degree level and program. *Financial support:* In 2005–06, 27 fellowships with full tuition reimbursements (averaging $24,600 per year) were awarded; scholarships/grants also available. *Faculty research:* Role of mutant proteins in Alzheimer's disease, nuclear protein function in breast and prostate cancer, medically important fungi, glycoproteins in HIV pathogenesis. Total annual research expenditures: $35.3 million. *Unit head:* Dr. J. Brooks Jackson, Chair, 410-955-9790, Fax: 410-955-0394. *Application contact:* Wilhelmena M. Braswell, Academic Program Coordinator, 443-287-3163, Fax: 410-955-9777, E-mail: wbraswel@jhmi.edu.

Kansas State University, College of Veterinary Medicine and Graduate School, Graduate Programs in Veterinary Medicine, Department of Diagnostic Medicine/Pathobiology, Manhattan, KS 66506. Offers biomedical science (MS); diagnostic medicine/pathobiology (PhD). *Faculty:* 25 full-time (4 women), 6 part-time/adjunct (2 women). *Students:* 20 full-time (9 women), 6 part-time (4 women), 14 international. Average age 30. 6 applicants, 0% accepted, 0 enrolled. In 2005, 4 master's, 3 doctorates awarded. Terminal master's awarded for partial completion of doctoral program. *Degree requirements:* For master's, thesis. *Entrance requirements:* For master's and doctorate, interviews. Additional exam requirements/recommendations for international students: Required—TOEFL (minimum score 550 paper-based; 213 computer-based). *Application deadline:* For fall admission, 2/1 for domestic students; for spring admission, 10/1 for domestic students. Applications are processed on a rolling basis. Application fee: $30 ($55 for international students). *Expenses:* Tuition, state resident: full-time $5,160; part-time $215 per credit hour. Tuition, nonresident: full-time $12,816; part-time $534 per credit hour. Required fees: $564. *Financial support:* In 2005–06, 20 research assistantships (averaging $16,257 per year) were awarded; teaching assistantships, Federal Work-

Study, institutionally sponsored loans, and scholarships/grants also available. Financial award application deadline: 3/1; financial award applicants required to submit FAFSA. *Faculty research:* Infectious disease of animals, food safety and security, epidemiology and public health, toxicology, and pathology. Total annual research expenditures: $2.8 million. *Unit head:* M. M. Chengappa, Head, 785-532-4403, E-mail: chengapa@vet.ksu.edu. *Application contact:* T. G. Nagaraja, Director, 785-532-1214, E-mail: tnagaraj@ksu.edu.

Medical University of South Carolina, College of Graduate Studies, Program in Molecular and Cellular Biology and Pathobiology, Charleston, SC 29425-0002. Offers cell biology and anatomy (PhD); marine biomedicine (PhD). *Faculty:* 173 full-time (41 women). *Students:* 62 full-time (32 women); includes 8 minority (3 African Americans, 1 American Indian/Alaska Native, 4 Asian Americans or Pacific Islanders), 7 international. Average age 28. 181 applicants, 34% accepted, 43 enrolled. In 2005, 20 doctorates awarded. *Degree requirements:* For doctorate, thesis/dissertation, teaching and research seminar, oral and written exams. *Entrance requirements:* For doctorate, GRE General Test, interview, minimum GPA of 3.2. Additional exam requirements/recommendations for international students: Required—TOEFL (minimum score 600 paper-based; 250 computer-based). *Application deadline:* For fall admission, 1/15 priority date for domestic students, 1/15 priority date for international students. Applications are processed on a rolling basis. Application fee: $0 ($75 for international students). Electronic applications accepted. *Financial support:* In 2005–06, 12 students received support, including fellowships with partial tuition reimbursements available (averaging $21,000 per year); Federal Work-Study and scholarships/grants also available. Financial award application deadline: 3/15; financial award applicants required to submit FAFSA. Total annual research expenditures: $10 million. *Unit head:* Dr. Donald R. Menick, Director, 843-856-5045, Fax: 843-792-6590, E-mail: menickd@musc.edu. *Application contact:* Cheryl Brown, 843-792-4620, Fax: 843-792-4645, E-mail: brownche@musc.edu.

Michigan State University, College of Veterinary Medicine and The Graduate School, Graduate Program in Veterinary Medicine, Department of Pathobiology and Diagnostic Investigation, East Lansing, MI 48824. Offers pathology (MS, PhD); pathology–environmental toxicology (PhD). *Faculty:* 18 full-time (5 women), 1 part-time/adjunct (0 women). *Students:* 10 full-time (6 women), 8 part-time (6 women); includes 6 minority (4 African Americans, 1 American Indian/Alaska Native, 1 Asian American or Pacific Islander), 4 international. Average age 30. 5 applicants, 40% accepted. In 2005, 4 degrees awarded. *Degree requirements:* For master's and doctorate, thesis/dissertation. *Entrance requirements:* For master's and doctorate, minimum GPA of 3, bachelor's and medical degrees. Additional exam requirements/recommendations for international students: Required—TOEFL (minimum score 550 paper-based; 213 computer-based), Michigan State University ELT (85), Michigan ELAB (83). *Application deadline:* For fall admission, 12/27 for domestic students. Application fee: $50. Electronic applications accepted. *Expenses:* Tuition, state resident: part-time $330 per credit hour. Tuition, nonresident: part-time $685 per credit hour. Tuition and fees vary according to program. *Financial support:* In 2005–06, 5 fellowships with tuition reimbursements (averaging $6,421 per year), 4 research assistantships with tuition reimbursements (averaging $14,036 per year) were awarded; scholarships/grants and unspecified assistantships also available. *Faculty research:* Microbiology and immunology, neurobiology and anatomy, toxicology. Total annual research expenditures: $2.2 million. *Unit head:* Dr. Willie M. Reed, Chairperson, 517-353-0635, Fax: 517-432-5836, E-mail: reed@dcpah.msu.edu. *Application contact:* Denise Harrison, Department Secretary, 517-432-4685, Fax: 517-432-5836, E-mail: harrison@dcpah.msu.edu.

The Ohio State University, College of Medicine and Public Health and Graduate School, Graduate Programs in the Basic Medical Sciences, Department of Pathology, Columbus, OH 43210. Offers experimental pathobiology (MS); pathology assistant (MS). *Accreditation:* NAACLS. *Degree requirements:* For master's, thesis, comprehensive exam (for some programs). *Entrance requirements:* For master's, GRE General Test. Additional exam requirements/recommendations for international students: Required—TOEFL (minimum score 550 paper-based; 213 computer-based). Electronic applications accepted. *Faculty research:* Clinical pathology, transplantation pathology, cancer research, neuropathology, vascular pathology.

The Ohio State University, College of Veterinary Medicine and Graduate School, Program in Veterinary Medicine, Department of Veterinary Biosciences, Columbus, OH 43210. Offers anatomy and cellular biology (MS, PhD); pathobiology (MS, PhD); pharmacology (MS, PhD); toxicology (MS, PhD); veterinary physiology (MS, PhD). *Faculty research:* Microvasculature, muscle biology, neonatal lung and bone development.

The Pennsylvania State University University Park Campus, Graduate School, College of Agricultural Sciences, Department of Veterinary and Biomedical Sciences, State College, University Park, PA 16802-1503. Offers pathobiology (MS, PhD). *Entrance requirements:* For master's and doctorate, GRE General Test. Additional exam requirements/recommendations for international students: Required—TOEFL. *Expenses:* Tuition, state resident: full-time $12,518; part-time $522 per credit. Tuition, nonresident: full-time $23,004; part-time $959 per credit. Required fees: $484. Tuition and fees vary according to course load, campus/location and program. *Unit head:* Dr. C. Channa Reddy, Head, 814-865-7696, Fax: 814-863-6140, E-mail: crrl@psu.edu. *Application contact:* Information Contact, E-mail: pathobiology@psu.edu.

Purdue University, School of Veterinary Medicine and Graduate School, Graduate Programs in Veterinary Medicine, Department of Veterinary Pathobiology, West Lafayette, IN 47907. Offers biochemistry and molecular biology (MS, PhD); comparative epidemiology (MS, PhD); epidemiology (MS, PhD); immunology (MS, PhD); infectious diseases (MS, PhD); interdisciplinary genetics (PhD); laboratory animal medicine (MS, PhD); microbiology (MS, PhD); molecular virology (MS, PhD); parasitology (MS, PhD); pathobiology (MS, PhD); public health epidemiology (MS, PhD); toxicology (MS, PhD); veterinary anatomic pathology (MS, PhD); veterinary clinical pathology (MS, PhD); virology (MS, PhD). *Faculty:* 32 full-time (7 women). *Students:* 49 full-time (20 women), 3 part-time (1 woman); includes 2 minority (both African Americans), 31 international. Average age 35. In 2005, 3 master's, 8 doctorates awarded. Terminal master's awarded for partial completion of doctoral program. *Degree requirements:* For master's, thesis (for some programs); for doctorate, thesis/dissertation. *Entrance requirements:* For master's and doctorate, GRE General Test. Additional exam requirements/recommendations for international students: Required—TOEFL (minimum score 575 paper-based), TWE (minimum score 4). *Application deadline:* For fall admission, 8/12 for domestic students, 6/15 for international students; for spring admission, 1/12 for domestic students, 10/15 for international students. Application fee: $55. *Financial support:* Fellowships, research assistantships, teaching assistantships available. Financial award application deadline: 3/1; financial award applicants required to submit FAFSA. *Unit head:* Dr. H. Hogenesch, Head, 765-494-7543.

Texas A&M University, College of Veterinary Medicine, Graduate Programs in Veterinary Medicine, Department of Veterinary Pathobiology, College Station, TX 77843. Offers genetics (MS, PhD); veterinary microbiology (MS, PhD); veterinary parasitology (MS); veterinary pathology (MS, PhD). Part-time programs available. Postbaccalaureate distance learning degree programs offered. *Faculty:* 16 full-time (3 women), 4 part-time/adjunct (1 woman). *Students:* 45 full-time (28 women), 20 part-time (13 women); includes 7 minority (2 African Americans, 5 Hispanic Americans), 19 international. Average age 33. 25 applicants, 76% accepted, 19 enrolled. In 2005, 5 master's, 5 doctorates awarded. Terminal master's awarded for partial completion of doctoral program. *Degree requirements:* For master's and doctorate, thesis/dissertation, seminars. *Entrance requirements:* For master's and doctorate, GRE General Test, minimum GPA of 3.0 in last 60 hours. Additional exam requirements/recommendations for international students: Required—TOEFL. *Application deadline:* For fall admission, 3/1 for domestic students; for spring admission, 8/1 priority date for domestic students. Applications are processed on a rolling basis. Application fee: $50 ($75 for international students). Electronic applications accepted. *Expenses:* Tuition, state resident: full-time $4,488; part-time $187 per credit hour. Tuition, nonresident: full-time $11,112; part-time $463 per credit hour. Required fees: $1,974. *Financial support:* In 2005–06, fellowships with partial tuition reimbursements (averaging $16,000 per year), research assistantships with partial tuition reimbursements

(averaging $15,400 per year), teaching assistantships with partial tuition reimbursements (averaging $16,000 per year) were awarded; Federal Work-Study, institutionally sponsored loans, scholarships/grants, traineeships, health care benefits, and unspecified assistantships also available. Support available to part-time students. Financial award applicants required to submit FAFSA. *Faculty research:* Infectious and noninfectious diseases of animals and birds, animal genetics, molecular biology, immunology, virology. *Unit head:* Dr. Ann B. Kier, Head, 979-845-5941, Fax: 979-845-9231, E-mail: akier@cvm.tamu.edu. *Application contact:* Dr. G. G. Wagner, Graduate Advisor, 979-845-2851, Fax: 979-862-1147, E-mail: gwagner@cvm.tamu.edu.

Uniformed Services University of the Health Sciences, School of Medicine, Programs in Biomedical Sciences, Department of Pathology, Bethesda, MD 20814-4799. Offers comparative pathology (PhD); molecular pathobiology (PhD). *Faculty:* 19 full-time (6 women), 80 part-time/adjunct (25 women). *Students:* 6 full-time (5 women); includes 1 minority (African American), 1 international. Average age 27. 11 applicants, 45% accepted, 1 enrolled. *Median time to degree:* Of those who began their doctoral program in fall 1997, 100% received their degree in 8 years or less. *Degree requirements:* For doctorate, one foreign language, thesis/dissertation, qualifying exam, comprehensive exam. *Entrance requirements:* For doctorate, GRE General Test, minimum GPA of 3.0. Additional exam requirements/recommendations for international students: Required—TOEFL. *Application deadline:* For fall admission, 1/15 for domestic students. Applications are processed on a rolling basis. Application fee: $0. *Financial support:* In 2005–06, fellowships with full tuition reimbursements (averaging $23,000 per year); tuition waivers (full) also available. *Faculty research:* Molecular pathology, genetics, virology. *Unit head:* Dr. Robert M. Friedman, Chair, 301-295-3450, E-mail: rfriedman@usuhs.mil. *Application contact:* Janet M. Anastasi, Graduate Program Coordinator, 301-295-9474, Fax: 301-295-6772, E-mail: janastasi@usuhs.mil.

The University of Arizona, Graduate College, College of Agriculture and Life Sciences, Department of Veterinary Science and Microbiology, Program in Pathobiology, Tucson, AZ 85721. Offers MS, PhD. Terminal master's awarded for partial completion of doctoral program. *Degree requirements:* For master's, thesis/dissertation, registration; for doctorate, thesis/dissertation, comprehensive exam, registration. *Entrance requirements:* For master's and doctorate, GRE. Additional exam requirements/recommendations for international students: Required—TOEFL (minimum score 550 paper-based; 213 computer-based). *Faculty research:* Antibiotic resistance, molecular pathogenesis of bacteria, food safety, diagnosis of animal disease, parasitology.

University of Cincinnati, Division of Research and Advanced Studies, College of Medicine, Graduate Programs in Biomedical Sciences, Program in Pathobiology and Molecular Medicine, Cincinnati, OH 45267. Offers pathology (PhD), including anatomic pathology, laboratory medicine, pathobiology and molecular medicine. *Degree requirements:* For doctorate, thesis/dissertation, qualifying exam. *Entrance requirements:* For doctorate, GRE General Test. Additional exam requirements/recommendations for international students: Required—TOEFL (minimum score 620 paper-based; 260 computer-based). Electronic applications accepted. *Faculty research:* Cardiovascular and lipid disorders, digestive and kidney disease, endocrine and metabolic disorders, hematologic and oncogenic, immunology and infectious disease.

University of Connecticut, Graduate School, College of Agriculture and Natural Resources, Department of Pathobiology and Veterinary Science, Field of Pathobiology and Veterinary Science, Storrs, CT 06269. Offers pathobiology (MS, PhD). *Faculty:* 12 full-time (4 women). *Students:* 20 full-time (12 women), 5 part-time (3 women); includes 2 minority (both Asian Americans or Pacific Islanders), 5 international. Average age 29. 19 applicants, 63% accepted, 11 enrolled. In 2005, 2 master's, 1 doctorate awarded. Terminal master's awarded for partial completion of doctoral program. *Degree requirements:* For master's, comprehensive exam; for doctorate, thesis/dissertation. *Entrance requirements:* For master's and doctorate, GRE General Test, GRE Subject Test. Additional exam requirements/recommendations for international students: Required—TOEFL (minimum score 550 paper-based; 213 computer-based). *Application deadline:* For fall admission, 2/1 priority date for domestic students, 2/1 priority date for international students; for spring admission, 11/1 for domestic students, 10/1 for international students. Applications are processed on a rolling basis. Application fee: $55. Electronic applications accepted. *Expenses:* Tuition, state resident: part-time $444 per credit hour. Tuition, nonresident: part-time $1,154 per credit hour. Tuition and fees vary according to course load. *Financial support:* In 2005–06, 17 research assistantships with full tuition reimbursements, 1 teaching assistantship with full tuition reimbursement were awarded; fellowships, Federal Work-Study, scholarships/grants, health care benefits, and unspecified assistantships also available. Financial award application deadline: 2/1; financial award applicants required to submit FAFSA. *Application contact:* Steven Geary, Chairperson, 860-486-4000, Fax: 860-486-2794, E-mail: steven.geary@uconn.edu.

University of Illinois at Urbana–Champaign, College of Veterinary Medicine and Graduate College, Graduate Programs in Veterinary Medicine, Department of Veterinary Pathobiology, Champaign, IL 61820. Offers MS, PhD. Part-time programs available. Postbaccalaureate distance learning degree programs offered (minimal on-campus study). *Faculty:* 22 full-time (6 women), 1 (woman) part-time/adjunct. *Students:* 22 full-time (16 women), 2 part-time; includes 7 minority (1 African American, 3 Asian Americans or Pacific Islanders, 3 Hispanic Americans), 8 international. Average age 26. 21 applicants, 19% accepted, 3 enrolled. In 2005, 4 master's, 4 doctorates awarded. Terminal master's awarded for partial completion of doctoral program. *Degree requirements:* For master's and doctorate, thesis/dissertation. *Entrance requirements:* For master's and doctorate, GRE, minimum GPA of 3.0. *Application deadline:* Applications are processed on a rolling basis. Application fee: $50 ($60 for international students). Electronic applications accepted. *Financial support:* In 2005–06, 5 fellowships, 6 research assistantships, 6 teaching assistantships were awarded; tuition waivers (full and partial) also available. *Faculty research:* Epidemiology, immunology, microbiology, parasitology, clinical pathology. *Unit head:* Daniel L. Rock, Head, 217-333-2449, Fax: 217-244-7421, E-mail: dlrock@uiuc.edu. *Application contact:* Tish Lehigh, Graduate Program Coordinator, 217-244-8924, Fax: 217-244-7421, E-mail: lehigh@uiuc.edu.

University of Missouri–Columbia, College of Veterinary Medicine and Graduate School, Graduate Programs in Veterinary Medicine, Department of Veterinary Pathobiology, Columbia, MO 65211. Offers laboratory animal medicine (MS); pathobiology (MS, PhD). *Faculty:* 36 full-time (10 women), 1 (woman) part-time/adjunct. *Students:* 25 full-time (13 women), 21 part-time (8 women); includes 11 minority (5 African Americans, 2 American Indian/Alaska Native, 2 Asian Americans or Pacific Islanders, 2 Hispanic Americans), 13 international. In 2005, 3 degrees awarded. *Degree requirements:* For master's, thesis; for doctorate, 2 foreign languages, thesis/dissertation. *Entrance requirements:* For master's and doctorate, GRE General Test, minimum GPA of 3.0. *Application deadline:* For fall admission, 5/1 for domestic students. Application fee: $45 ($60 for international students). *Financial support:* Research assistantships, teaching assistantships, institutionally sponsored loans available. *Unit head:* Dr. Catherine Vogelweid, Director of Graduate Studies, 573-882-5670, E-mail: vogelweidc@missouri.edu. *Application contact:* Dr. Ronald Terjung, Associate Dean for Research and Postdoctoral Studies, 573-882-2635, E-mail: terjungr@missouri.edu.

University of Rochester, School of Medicine and Dentistry, Program in Pathobiology and Molecular Medicine, Rochester, NY 14627-0250.

University of Southern California, Keck School of Medicine and Graduate School, Graduate Programs in Medicine, Department of Pathology, Los Angeles, CA 90089. Offers experimental and molecular pathology (MS); pathobiology (PhD). *Faculty:* 52 full-time (10 women). *Students:* 58 full-time (33 women); includes 20 minority (15 Asian Americans or Pacific Islanders, 5 Hispanic Americans), 23 international. Average age 26. 40 applicants, 55% accepted, 16 enrolled. In 2005, 6 master's, 8 doctorates awarded. *Degree requirements:* For master's, thesis or alternative; for doctorate, thesis/dissertation. *Entrance requirements:* For master's, GRE General Test, minimum GPA of 3.0; for doctorate, GRE General Test, minimum GPA of 3.0, BS in natural sciences. Additional exam requirements/recommendations for international

Pathobiology

University of Southern California *(continued)*
students: Required—TOEFL. *Application deadline:* For fall admission, 3/1 for domestic students. Application fee: $65. Electronic applications accepted. *Expenses:* Tuition: Full-time $25,416; part-time $1,059 per unit. Required fees: $484; $484 per year. Tuition and fees vary according to course load and program. *Financial support:* In 2005–06, 43 students received support, including 3 fellowships with tuition reimbursements available (averaging $24,876 per year), 38 research assistantships with tuition reimbursements available (averaging $24,876 per year), 2 teaching assistantships with tuition reimbursements available (averaging $24,876 per year); scholarships/grants also available. Financial award application deadline: 2/1. *Faculty research:* Immunology of lymphomas and leukemias, lung cancer, molecular basis of oncogenesis, central nervous system disease, organic chemical carcinogens and carcinogenic metal salts. Total annual research expenditures: $12.2 million. *Unit head:* , Dr. Clive R. Taylor, Chairman, 323-442-1180, Fax: 323-442-3314, E-mail: taylor@pathfinder.usc.edu. *Application contact:* Lisa A. Doumak, Student Services Assistant, 323-442-1168, Fax: 323-442-3049, E-mail: doumak@pathfinder.usc.edu.

See Close-Up on page 1125.

University of Toronto, School of Graduate Studies, Life Sciences Division, Department of Laboratory Medicine and Pathobiology, Toronto, ON M5S 1A1, Canada. Offers M Sc, PhD, MD/PhD. *Degree requirements:* For master's, thesis; for doctorate, thesis/dissertation, oral defense of thesis. *Entrance requirements:* For master's, minimum B+ average in final 2 years, research experience, 2 letters of recommendation, resumé, interview; for doctorate, minimum A– average, 2 letters of recommendation, research experience, resumé, interview. Additional exam requirements/recommendations for international students: Required—TOEFL (600 paper, 250 computer), TWE(5) or IELTS (7).

University of Washington, Graduate School, School of Public Health and Community Medicine, Graduate Program in Pathobiology, Seattle, WA 98195. Offers MS, PhD. *Faculty:* 22 full-time (8 women), 11 part-time/adjunct (6 women). *Students:* 47 full-time (31 women); includes 7 minority (1 African American, 4 Asian Americans or Pacific Islanders, 2 Hispanic Americans), 6 international. Average age 47. 55 applicants, 31% accepted, 12 enrolled. In 2005, 1 master's, 2 doctorates awarded. Terminal master's awarded for partial completion of doctoral program. *Median time to degree:* Of those who began their doctoral program in fall 1997, 99% received their degree in 8 years or less. *Degree requirements:* For master's, thesis/dissertation, registration; for doctorate, thesis/dissertation, comprehensive exam, registration. *Entrance requirements:* For master's and doctorate, GRE General Test, minimum GPA of 3.0. Additional exam requirements/recommendations for international students: Required—TOEFL. *Application deadline:* For fall admission, 12/15 for domestic students, 11/15 for international students. Application fee: $50. *Financial support:* In 2005–06, 2 fellowships with tuition reimbursements (averaging $22,560 per year), 18 research assistantships with tuition reimbursements (averaging $22,560 per year) were awarded; career-related internships or fieldwork, institutionally sponsored loans, scholarships/grants, traineeships, tuition waivers (full and partial), and unspecified assistantships also available. Financial award application deadline: 3/1; financial award applicants required to submit FAFSA. *Faculty research:* Pathogenesis of chlamydiae, molecular biology of parasites, signal transduction, antigenic analysis, molecular biology of tumor viruses. *Unit head:* Dr. Andreas Stergachis, Acting Chair, 206-543-8350, Fax: 206-543-3873, E-mail: stergach@u.washington.edu. *Application contact:* Joseph A. Daniels, Manager of Student Services, 206-543-4338, Fax: 206-543-3873, E-mail: pathobio@u.washington.edu.

See Close-Up on page 1129.

University of Wyoming, Graduate School, College of Agriculture, Department of Veterinary Sciences, Laramie, WY 82070. Offers pathobiology (MS). *Faculty:* 8 full-time (0 women), 8 part-time/adjunct (3 women). *Students:* 1 (woman) full-time. Average age 25. 2 applicants, 100% accepted. *Degree requirements:* For master's, thesis/dissertation. *Entrance requirements:* For master's, GRE General Test, minimum GPA of 3.0. Additional exam requirements/recommendations for international students: Required—TOEFL. *Application deadline:* For fall admission, 6/1 for domestic students. Applications are processed on a rolling basis. Application fee: $50. *Expenses:* Tuition, state resident: full-time $3,720; part-time $155 per credit hour. Tuition, nonresident: full-time $10,704; part-time $446 per credit hour. Required fees: $666; $162 per semester. Tuition and fees vary according to course load and program. *Financial support:* In 2005–06, research assistantships with full tuition reimbursements (averaging $10,062 per year), teaching assistantships with full tuition reimbursements (averaging $10,062 per year) were awarded; Federal Work-Study also available. Financial award application deadline: 3/1. *Faculty research:* Infectious diseases, pathology, toxicology, immunology, microbiology. Total annual research expenditures: $200,000. *Unit head:* Dr. Donal O'Toole, Head, 307-742-6638, Fax: 307-721-2051, E-mail: dot@uwyo.edu. *Application contact:* Beth Lee Howell, Office Associate Senior, 307-766-7638, Fax: 307-721-2051, E-mail: bethlee@uwyo.edu.

Wake Forest University, School of Medicine and Graduate School, Graduate Programs in Medicine, Program in Molecular and Cellular Pathobiology, Winston-Salem, NC 27103. Offers MS, PhD. *Degree requirements:* For master's and doctorate, thesis/dissertation. *Entrance requirements:* For master's and doctorate, GRE General Test. Additional exam requirements/recommendations for international students: Required—TOEFL. Electronic applications accepted. *Faculty research:* Atherosclerosis, lipoproteins, arterial wall metabolism.

See Close-Up on page 1135.

Yale University, School of Medicine and Graduate School of Arts and Sciences, Combined Program in Biological and Biomedical Sciences (BBS), Pharmacological Sciences and Molecular Medicine Track, New Haven, CT 06520. Offers PhD, MD/PhD. *Students:* 6 full-time. *Degree requirements:* For doctorate, thesis/dissertation. *Entrance requirements:* For doctorate, GRE General Test. Additional exam requirements/recommendations for international students: Required—TOEFL. *Application deadline:* For fall admission, 12/8 for domestic students, 12/8 for international students. Electronic applications accepted. *Financial support:* Fellowships, research assistantships available. *Unit head:* Dr. David F. Stern, Co-Director, 203-785-4832, Fax: 203-785-7467, E-mail: kathleen.fisher@yale.edu. *Application contact:* Kathy Fisher, Registrar, 203-785-4545, E-mail: kathleen.fisher@yale.edu.

Pathology

Albert Einstein College of Medicine, Sue Golding Graduate Division of Medical Sciences, Department of Pathology, Bronx, NY 10467. Offers PhD, MD/PhD. *Degree requirements:* For doctorate, thesis/dissertation. *Entrance requirements:* For doctorate, GRE General Test. Additional exam requirements/recommendations for international students: Required—TOEFL. *Faculty research:* Clinical and disease-related research at tissue, cellular, and subcellular levels; biochemistry and morphology of enzyme and lysosome disorders.

Baylor College of Medicine, Graduate School of Biomedical Sciences, Program in Developmental Biology, Houston, TX 77030-3498. Offers PhD, MD/PhD. *Faculty:* 42 full-time (10 women). *Students:* 38 full-time (17 women); includes 7 minority (1 American Indian/Alaska Native, 5 Asian Americans or Pacific Islanders, 1 Hispanic American), 21 international. Average age 27. 67 applicants, 25% accepted, 7 enrolled. In 2005, 2 doctorates awarded. *Median time to degree:* Of those who began their doctoral program in fall 1997, 100% received their degree in 8 years or less. *Degree requirements:* For doctorate, thesis/dissertation, public defense. *Entrance requirements:* For doctorate, GRE General Test, GRE Subject Test (strongly recommended), minimum GPA of 3.0. Additional exam requirements/recommendations for international students: Required—TOEFL. *Application deadline:* For fall admission, 1/1 for domestic students. Application fee: $30. Electronic applications accepted. *Expenses:* Tuition: Full-time $8,200. Full-time tuition and fees vary according to program. *Financial support:* In 2005–06, 37 students received support, including 10 fellowships (averaging $23,000 per year), 28 research assistantships (averaging $23,000 per year); career-related internships or fieldwork, Federal Work-Study, institutionally sponsored loans, health care benefits, tuition waivers (full), and stipends also available. *Faculty research:* Molecular and genetic approaches to study pattern formation in *Dictyostelium*, *Drosophila*, *C.elegans*, mouse, *Xenopus*, and zebrafish; cross-species approach. *Unit head:* Dr. Hugo Bellen, Director, 713-798-6410. *Application contact:* Catherine Tasnier, Graduate Program Administrator, 713-798-6410, Fax: 713-798-5386, E-mail: cat@bcm.edu.

See Close-Up on page 757.

Boston University, School of Medicine, Division of Graduate Medical Sciences, Department of Pathology and Laboratory Medicine, Boston, MA 02118. Offers cell and molecular biology (PhD); experimental pathology (PhD); immunology (PhD). Part-time programs available. *Faculty:* 12 full-time (4 women), 13 part-time/adjunct (2 women). *Students:* 19 full-time (10 women), 1 (woman) part-time; includes 4 minority (2 Asian Americans or Pacific Islanders, 2 Hispanic Americans), 6 international. Average age 29. *Degree requirements:* For doctorate, thesis/dissertation, qualifying exam. *Entrance requirements:* For doctorate, GRE General Test, GRE Subject Test. Additional exam requirements/recommendations for international students: Required—TOEFL. *Application deadline:* For fall admission, 1/15 for domestic students; for spring admission, 10/15 priority date for domestic students. Electronic applications accepted. *Expenses:* Tuition: Full-time $31,530; part-time $985 per credit. Required fees: $316; $40 per semester. Tuition and fees vary according to course level and program. *Financial support:* In 2005–06, 10 fellowships, 3 research assistantships were awarded; Federal Work-Study, scholarships/grants, and traineeships also available. *Faculty research:* Toxicology, carcinogenesis, endocytosis, cytogenetics. *Unit head:* Leonard Gottlieb, Chairman, 617-638-4500, Fax: 617-638-4085. *Application contact:* Dr. Adrianne Rogers, Associate Chairman, 617-638-4500, Fax: 617-638-4085, E-mail: aerogers@bu.edu.

Brown University, Graduate School, Division of Biology and Medicine, Program in Pathology and Laboratory Medicine, Providence, RI 02912. Offers biology (PhD); cancer biology (PhD); immunology and infection (PhD); medical science (PhD); pathobiology (Sc M); toxicology and environmental pathology (PhD). Terminal master's awarded for partial completion of doctoral program. *Degree requirements:* For doctorate, thesis/dissertation, preliminary exam. *Entrance requirements:* For master's and doctorate, GRE General Test, GRE Subject Test. Additional exam requirements/recommendations for international students: Required—TOEFL. Electronic applications accepted. *Faculty research:* Environmental pathology, carcinogenesis, immunopathology, signal transduction, innate immunity.

Case Western Reserve University, School of Medicine and School of Graduate Studies, Graduate Programs in Medicine, Programs in Molecular and Cellular Basis of Disease, Cleveland, OH 44106. Offers cell biology (MS, PhD); immunology (MS, PhD); pathology (MS, PhD). *Faculty:* 37 full-time (7 women). *Students:* 52 full-time (30 women); includes 16 minority (2 African Americans, 14 Asian Americans or Pacific Islanders). Average age 29. In 2005, 5 degrees awarded. Terminal master's awarded for partial completion of doctoral program. *Median time to degree:* Of those who began their doctoral program in fall 1997, 100% received their degree in 8 years or less. *Degree requirements:* For master's and doctorate, thesis/dissertation. *Entrance requirements:* For master's and doctorate, GRE General Test, GRE Subject Test. Additional exam requirements/recommendations for international students: Required—TOEFL (minimum score 550 paper-based; 213 computer-based). Application fee: $50. Electronic applications accepted. *Financial support:* In 2005–06, fellowships with full tuition reimbursements (averaging $23,000 per year); scholarships/grants, traineeships, and tuition waivers (full) also available. *Faculty research:* Neurobiology, molecular biology, cancer biology, biomaterials, biocompatibility. Total annual research expenditures: $11.4 million. *Unit head:* John B Lowe, Chairman, 216-368-0890, Fax: 216-368-0494, E-mail: jbl21@case.edu. *Application contact:* Tandalea Harney, Graduate Programs Director, 216-368-0890, Fax: 216-368-0494, E-mail: trh2@case.edu.

Colorado State University, College of Veterinary Medicine and Biomedical Sciences, Department of Microbiology, Immunology and Pathology, Fort Collins, CO 80523-0015. Offers immunology (MS); microbiology (MS, PhD); pathology (PhD). *Faculty:* 40 full-time (11 women), 2 part-time/adjunct (0 women). *Students:* 64 full-time (37 women), 37 part-time (21 women); includes 9 minority (1 American Indian/Alaska Native, 5 Asian Americans or Pacific Islanders, 3 Hispanic Americans), 17 international. Average age 30. 81 applicants, 22% accepted, 18 enrolled. In 2005, 6 master's, 7 doctorates awarded. *Degree requirements:* For master's, thesis/dissertation; for doctorate, thesis/dissertation, comprehensive exam. *Entrance requirements:* For master's and doctorate, GRE General Test, minimum GPA of 3.0. Additional exam requirements/recommendations for international students: Required—TOEFL. *Application deadline:* For fall admission, 1/1 for domestic students; for spring admission, 10/1 priority date for domestic students. Applications are processed on a rolling basis. Application fee: $50. Electronic applications accepted. *Expenses:* Tuition, state resident: full-time $3,690; part-time $205 per credit. Tuition, nonresident: full-time $14,958; part-time $831 per credit. Required fees: $1,061. *Financial support:* In 2005–06, 17 fellowships with tuition reimbursements (averaging $28,475 per year), 35 research assistantships with tuition reimbursements (averaging $21,715 per year), 7 teaching assistantships with tuition reimbursements (averaging $20,772 per year) were awarded; traineeships and unspecified assistantships also available. *Faculty research:* Medical and veterinary microbiology, pathology of disease, microbial pathogenesis, industrial and environmental microbiology, vector-borne disease. Total annual research expenditures: $21.1 million. *Unit head:* Dr. Jeffrey Wilusz, Head, 970-491-0652, Fax: 970-491-0603, E-mail: jeffrey.wilusz@colostate.edu. *Application contact:* Dr. Herbert Schweizer, Graduate Program Coordinator, 970-491-6136, Fax: 970-491-1815, E-mail: microbio@colostate.edu.

See Close-Up on page 861.

Columbia University, College of Physicians and Surgeons and Graduate School of Arts and Sciences, Graduate School of Arts and Sciences at the College of Physicians and Surgeons, Department of Pathology, New York, NY 10032. Offers pathobiology (M Phil, MA, PhD). Only candidates for the PhD are admitted. Terminal master's awarded for partial completion of doctoral program. *Degree requirements:* For doctorate, thesis/dissertation. *Entrance requirements:* For master's and doctorate, GRE General Test. Additional exam requirements/recommendations for international students: Required—TOEFL. *Expenses:* Tuition: Full-time $31,448. Tuition and fees vary according to course level, course load, campus/location and program. *Faculty research:* Virology, molecular biology, cell biology, neurobiology, immunology.

Dalhousie University, Faculty of Graduate Studies and Faculty of Medicine, Graduate Programs in Medicine, Department of Pathology, Halifax, NS B3H 4R2, Canada. Offers M Sc. Part-time programs available. *Degree requirements:* For master's, oral defense of thesis. *Faculty research:* Tumor immunology, molecular oncology, clinical chemistry, hematology, molecular genetics/oncology.

Duke University, Graduate School, Department of Pathology, Durham, NC 27710. Offers PhD. *Accreditation:* NAACLS. *Faculty:* 28 full-time. *Students:* 24 full-time (12 women); includes 6 minority (2 African Americans, 1 Asian American or Pacific Islander, 3 Hispanic Americans), 6 international. 23 applicants, 35% accepted, 4 enrolled. In 2005, 5 doctorates awarded. *Degree requirements:* For doctorate, thesis/dissertation. *Entrance requirements:* For doctorate, GRE General Test, GRE Subject Test (recommended). Additional exam requirements/recommendations for international students: Required—IELT (preferred) or TOEFL. *Application deadline:* For fall admission, 12/31 for domestic students, 12/31 for international students. Application fee: $75. Electronic applications accepted. *Financial support:* Fellowships, research assistantships, Federal Work-Study available. Financial award application deadline: 12/31. *Unit head:* Soman Abraham, Director of Graduate Studies, 919-684-3630, Fax: 919-684-8756, E-mail: hunr0033@mc.duke.edu.

See Close-Up on page 1115.

Duke University, School of Medicine, Pathologists' Assistant Program, Durham, NC 27708-0586. Offers MHS. *Accreditation:* NAACLS. *Students:* 12 full-time (10 women); includes 1 minority (Hispanic American) Average age 26. 41 applicants, 15% accepted, 6 enrolled. In 2005, 6 degrees awarded. *Degree requirements:* For master's, comprehensive exam. *Entrance requirements:* For master's, GRE. Additional exam requirements/recommendations for international students: Required—TOEFL, IELT. *Application deadline:* For fall admission, 2/28 for domestic students. Application fee: $50. *Expenses:* Contact institution. Financial support: In 2005–06, 11 students received support. Scholarships/grants available. Financial award application deadline: 5/1; financial award applicants required to submit FAFSA. *Unit head:* Dr. Kenneth R. Broda, Program Director, 919-681-4847, Fax: 919-681-7799, E-mail: broda001@mc.duke.edu. *Application contact:* Pamela Vollmer, Associate Director, 919-684-2159, E-mail: vollm003@mc.duke.edu.

Georgetown University, Graduate School of Arts and Sciences, Programs in Biomedical Sciences, Department of Pathology, Washington, DC 20057. Offers MS, PhD, MD/PhD, MS/PhD. *Degree requirements:* For master's, thesis/dissertation; for doctorate, thesis/dissertation, comprehensive exam. *Entrance requirements:* For master's and doctorate, GRE General Test. Additional exam requirements/recommendations for international students: Required—TOEFL. *Faculty research:* Virus-induced diabetes, viral oncology, renal pathophysiology.

Harvard University, Graduate School of Arts and Sciences, Division of Medical Sciences, Boston, MA 02115. Offers biological chemistry and molecular pharmacology (PhD); cell biology (PhD); genetics (PhD); microbiology and molecular genetics (PhD); pathology (PhD), including experimental pathology. *Students:* 433 full-time (210 women). In 2005, 83 doctorates awarded. *Degree requirements:* For doctorate, thesis/dissertation. *Entrance requirements:* For doctorate, GRE General Test, GRE Subject Test. Additional exam requirements/recommendations for international students: Required—TOEFL. Application fee: $60. *Expenses:* Tuition: Full-time $28,752. Full-time tuition and fees vary according to program and student level. *Financial support:* Fellowships, research assistantships, teaching assistantships, institutionally sponsored loans and tuition waivers (full) available. Financial award application deadline: 1/1. *Unit head:* Administrator, 617-432-2029. *Application contact:* Administrator, 617-432-2029.

Indiana University–Purdue University Indianapolis, Indiana University School of Medicine, Department of Pathology and Laboratory Medicine, Indianapolis, IN 46202-2896. Offers MS, PhD, MD/PhD. *Accreditation:* NAACLS. *Faculty:* 7 full-time (6 women), 13 part-time (11 women); includes 3 minority (2 African Americans, 1 Asian American or Pacific Islander), 3 international. Average age 30. In 2005, 1 degree awarded. *Degree requirements:* For master's and doctorate, thesis/dissertation. *Entrance requirements:* For master's and doctorate, GRE General Test. Additional exam requirements/recommendations for international students: Required—TOEFL. *Application deadline:* For fall admission, 1/15 for domestic students. Applications are processed on a rolling basis. Application fee: $50 ($60 for international students). *Expenses:* Tuition, state resident: full-time $5,159; part-time $215 per credit hour. Tuition, nonresident: full-time $14,890; part-time $620 per credit hour. Required fees: $614. Tuition and fees vary according to campus/location and program. *Financial support:* Fellowships, research assistantships with full tuition reimbursements, teaching assistantships with full tuition reimbursements, institutionally sponsored loans available. Financial award application deadline: 2/1. *Faculty research:* Intestinal microecology and anaerobes, molecular pathogenesis of infectious diseases, AIDS pneumocystis, sports medicine toxicology, neuropathology of aging. *Unit head:* Dr. John Eble, Chairman, 317-274-4806. *Application contact:* Dr. Diane S. Leland, Graduate Adviser, 317-274-0148.

Iowa State University of Science and Technology, College of Veterinary Medicine and Graduate College, Graduate Programs in Veterinary Medicine, Department of Veterinary Pathology, Ames, IA 50011. Offers MS, PhD. *Faculty:* 13 full-time, 6 part-time/adjunct. *Students:* 6 full-time (5 women), 6 part-time (3 women), 3 international. 6 applicants, 50% accepted, 3 enrolled. In 2005, 6 degrees awarded. *Degree requirements:* For master's, thesis or alternative; for doctorate, thesis/dissertation. *Entrance requirements:* Additional exam requirements/recommendations for international students: Required—GRE General Test (international applicants), TOEFL (paper score 550; computer score 213) or IELTS (score 6.5). *Application deadline:* For fall admission, 1/1 priority date for domestic students, 1/1 priority date for international students; for spring admission, 9/1 priority date for domestic students, 9/1 priority date for international students. Applications are processed on a rolling basis. Application fee: $30 ($70 for international students). Electronic applications accepted. *Financial support:* In 2005–06, 4 research assistantships with partial tuition reimbursements (averaging $15,842 per year) were awarded; teaching assistantships with partial tuition reimbursements, scholarships/grants, health care benefits, and unspecified assistantships also available. *Faculty research:* Veterinary clinical pathology, veterinary parasitology, veterinary toxicology. *Unit head:* Dr. Claire B. Andreasen, Chair, 515-294-3282. *Application contact:* Dr. Mark Ackerman, Director of Graduate Education, 515-294-3647, E-mail: mackerma@iastate.edu.

The Johns Hopkins University, School of Medicine, Graduate Programs in Medicine, Department of Pathology, Baltimore, MD 21218-2699. Offers pathobiology (PhD). *Faculty:* 59 full-time (10 women). *Students:* 27 full-time (16 women); includes 5 minority (2 African Americans, 2 Asian Americans or Pacific Islanders, 1 Hispanic American), 9 international. Average age 28. 107 applicants, 7% accepted, 7 enrolled. *Degree requirements:* For doctorate, thesis/dissertation, qualifying oral exam. *Entrance requirements:* For doctorate, GRE General Test, GRE Subject Test (biology or chemistry), previous course work with laboratory in organic and inorganic chemistry, general biology, calculus; interview. Additional exam requirements/recommendations for international students: Required—TOEFL. *Application deadline:* For winter admission, 1/10 for domestic students. Application fee: $90. Electronic applications accepted. *Expenses:* Tuition: Full-time $30,960. Tuition and fees vary according to degree level and program. *Financial support:* In 2005–06, 27 fellowships with full tuition reimbursements (averaging $24,600 per year) were awarded; scholarships/grants also available. *Faculty research:* Role of mutant proteins in Alzheimer's disease, nuclear protein function in breast and prostate cancer, medically important fungi, glycoproteins in HIV pathogenesis. Total annual research expenditures: $35.3 million. *Unit head:* Dr. J. Brooks Jackson, Chair, 410-955-9790, Fax: 410-955-0394. *Application contact:* Wilhelmena M. Braswell, Academic Program Coordinator, 443-287-3163, Fax: 410-955-9777, E-mail: wbraswel@jhmi.edu.

Loma Linda University, School of Medicine, Department of Pathology and Human Anatomy, Loma Linda, CA 92350. Offers MS, PhD. Part-time programs available. *Faculty:* 33 full-time (5 women), 26 part-time/adjunct (4 women). *Students:* 7 full-time (6 women), 3 part-time (1 woman); includes 2 African Americans, 1 Asian American or Pacific Islander, 3 Hispanic Americans. Terminal master's awarded for partial completion of doctoral program. *Degree requirements:* For master's, thesis; for doctorate, 2 foreign languages, thesis/dissertation. *Entrance requirements:* For master's and doctorate, GRE General Test. *Application deadline:* Applications are processed on a rolling basis. Application fee: $40. *Financial support:* Tuition waivers (full and partial) available. Support available to part-time students. *Faculty research:* Neuro-

endocrine system, histochemistry and image analysis, effect of age and diabetes on PNS, electron microscopy, histology. *Unit head:* Dr. Kenneth Wright, Coordinator, 909-824-301.

Louisiana State University Health Sciences Center, School of Graduate Studies in New Orleans, Department of Pathology, New Orleans, LA 70112-2223. Offers MS, PhD, MD/PhD. Part-time programs available. Terminal master's awarded for partial completion of doctoral program. *Degree requirements:* For master's and doctorate, thesis/dissertation. *Entrance requirements:* For master's and doctorate, GRE General Test. Additional exam requirements/recommendations for international students: Required—TOEFL. *Faculty research:* Immunohematology, hematology, experimental and epidemiological studies in atherosclerosis, epidemiology of cancer.

McGill University, Faculty of Graduate and Postdoctoral Studies, Faculty of Medicine, Department of Pathology, Montréal, QC H3A 2T5, Canada. Offers M Sc, PhD. *Degree requirements:* For master's and doctorate, thesis/dissertation. *Entrance requirements:* For master's, GRE, MD/DDS or bachelor's degree with high academic standing, minimum GPA of 3.0; for doctorate, GRE. Additional exam requirements/recommendations for international students: Required—TOEFL. *Faculty research:* Ultrastructural and physiological alterations in disease, immunopathology, transplantation biology and autoimmune disease, mechanisms of carcinogenesis, environmental pathology.

Medical College of Wisconsin, Graduate School of Biomedical Sciences, Department of Pathology, Milwaukee, WI 53226-0509. Offers MS, PhD, MD/MS, MD/PhD. *Degree requirements:* For doctorate, thesis/dissertation, comprehensive exam, registration. *Entrance requirements:* For master's, GRE General Test, minimum GPA of 3.0; previous course work in biochemistry, biology, calculus, chemistry, and physics; for doctorate, GRE General Test. Additional exam requirements/recommendations for international students: Required—TOEFL. *Faculty research:* Virology, immunology, metabolism, molecular pathology.

Medical University of South Carolina, College of Graduate Studies, Program in Pathology and Laboratory Medicine, Charleston, SC 29425-0002. Offers Pharm D, MS, PhD, DMD/MD/PhD. *Faculty:* 21 full-time (10 women). *Students:* 16 full-time (11 women); includes 2 minority (1 African American, 1 Hispanic American), 3 international. Average age 28. 181 applicants, 34% accepted, 43 enrolled. Terminal master's awarded for partial completion of doctoral program. *Degree requirements:* For master's, thesis, research seminar; for doctorate, thesis/dissertation, teaching and research seminar, oral and written exams. *Entrance requirements:* For master's and doctorate, GRE General Test, interview, minimum GPA of 3.2. Additional exam requirements/recommendations for international students: Required—TOEFL (minimum score 600 paper-based; 250 computer-based). *Application deadline:* For fall admission, 1/15 priority date for domestic students, 1/15 priority date for international students. Applications are processed on a rolling basis. Application fee: $0 ($75 for international students). Electronic applications accepted. *Financial support:* In 2005–06, 4 students received support, including fellowships with partial tuition reimbursements available (averaging $21,000 per year); Federal Work-Study and scholarships/grants also available. Financial award application deadline: 3/15; financial award applicants required to submit FAFSA. *Faculty research:* Neurobiology of hearing loss; inner ear ion homeostasis; cancer biology, genetics and radiation and chemotherapy, neural injury and repair, viral pathogenesis and cellular defense mechanisms. *Unit head:* Dr. Janice M. Lage, Chair, 843-792-3121, Fax: 843-792-6580, E-mail: lagejm@musc.edu. *Application contact:* Cheryl Brown, 843-792-4620, Fax: 843-792-4645, E-mail: brownche@musc.edu.

Michigan State University, College of Veterinary Medicine and The Graduate School, Graduate Program in Veterinary Medicine, Department of Pathobiology and Diagnostic Investigation, East Lansing, MI 48824. Offers pathology (MS, PhD); pathology–environmental toxicology (PhD). *Faculty:* 18 full-time (5 women), 1 part-time/adjunct (0 women). *Students:* 10 full-time (6 women), 8 part-time (6 women); includes 6 minority (4 African Americans, 1 American Indian/Alaska Native, 1 Asian American or Pacific Islander), 4 international. Average age 30. 5 applicants, 40% accepted. In 2005, 4 degrees awarded. *Degree requirements:* For master's and doctorate, thesis/dissertation. *Entrance requirements:* For master's and doctorate, minimum GPA of 3, bachelor's and medical degrees. Additional exam requirements/recommendations for international students: Required—TOEFL (minimum score 550 paper-based; 213 computer-based), Michigan State University ELT (85), Michigan ELAB (83). *Application deadline:* For fall admission, 12/27 for domestic students. Application fee: $50. Electronic applications accepted. *Expenses:* Tuition, state resident: part-time $330 per credit hour. Tuition, nonresident: part-time $685 per credit hour. Tuition and fees vary according to program. *Financial support:* In 2005–06, 5 fellowships with tuition reimbursements (averaging $6,421 per year), 4 research assistantships with tuition reimbursements (averaging $14,036 per year) were awarded; scholarships/grants and unspecified assistantships also available. *Faculty research:* Microbiology and immunology, neurobiology and anatomy, toxicology. Total annual research expenditures: $2.2 million. *Unit head:* Dr. Willie M. Reed, Chairperson, 517-353-0635, Fax: 517-432-5836, E-mail: reed@dcpah.msu.edu. *Application contact:* Denise Harrison, Department Secretary, 517-432-4685, Fax: 517-432-5836, E-mail: harrison@dcpah.msu.edu.

Mount Sinai School of Medicine of New York University, Graduate School of Biological Sciences, New York, NY 10029-6504. Offers biophysics, structural biology and biomathematics (PhD); community medicine (MPH); genetic counseling (MS); genetics and genomic sciences (PhD); mechanisms of disease and therapy (PhD); microbiology (PhD); molecular, cellular, biochemical and developmental sciences (PhD); neurosciences (PhD). *Students:* 218 full-time (109 women). 4,208 applicants, 7% accepted, 117 enrolled. Terminal master's awarded for partial completion of doctoral program. *Degree requirements:* For master's, registration; for doctorate, thesis/dissertation, registration. *Entrance requirements:* For doctorate, GRE General Test, GRE Subject Test, 3 years of college pre-med course work. Additional exam requirements/recommendations for international students: Required—TOEFL. *Application deadline:* For fall admission, 1/15 for domestic students. Application fee: $75. Electronic applications accepted. *Expenses:* Tuition: Full-time $33,250. Required fees: $1,600. Full-time tuition and fees vary according to degree level, program and reciprocity agreements. *Financial support:* In 2005–06, fellowships with full tuition reimbursements (averaging $26,000 per year), research assistantships with full tuition reimbursements (averaging $26,000 per year) were awarded; Federal Work-Study, institutionally sponsored loans, scholarships/grants, health care benefits, and unspecified assistantships also available. Financial award application deadline: 4/5; financial award applicants required to submit FAFSA. *Faculty research:* Cancer, gene therapy, minimally invasive surgery, cardiac translational research. Total annual research expenditures: $162.2 million. *Unit head:* Dr. Diomedes Logothetis, Dean, 212-241-6546, Fax: 212-241-0651, E-mail: diomedes.logothetis@mssm.edu. *Application contact:* Lily Recanati, Manager, 212-241-3267, Fax: 212-241-0651, E-mail: lily.recanati@mssm.edu.

See Close-Up on page 177.

New York Medical College, Graduate School of Basic Medical Sciences, Program in Experimental Pathology, Valhalla, NY 10595-1691. Offers MS, PhD, MD/PhD. Part-time and evening/weekend programs available. Terminal master's awarded for partial completion of doctoral program. *Degree requirements:* For master's, thesis/dissertation; for doctorate, thesis/dissertation, comprehensive exam. *Entrance requirements:* For master's and doctorate, GRE General Test. Additional exam requirements/recommendations for international students: Required—TOEFL. *Faculty research:* Atherogenesis and endothelial cell biology, immunology and inflammation, tumor biology and metastasis, mechanisms of chemical carcinogens, mechanisms of free radial tissue damage.

See Close-Up on page 1117.

North Carolina State University, College of Veterinary Medicine, Program in Comparative Biomedical Sciences, Raleigh, NC 27695. Offers cell biology and morphology (MS, PhD); epidemiology and population medicine (MS, PhD); immunology (MS, PhD); microbiology and immunology (MS, PhD); pathology (MS, PhD); pharmacology (MS, PhD); specialized veterinary medicine (MS). Part-time programs available. *Degree requirements:* For master's and doctorate, thesis/dissertation. *Entrance requirements:* For master's and doctorate, GRE General

Pathology

North Carolina State University (continued)
Test. Additional exam requirements/recommendations for international students: Required—TOEFL (minimum score 550 paper-based; 213 computer-based). Electronic applications accepted. Expenses: Contact institution. *Faculty research:* Infectious diseases, cell biology, pharmacology and toxicology, genomics, pathology and population medicine.

North Dakota State University, The Graduate School, College of Agriculture, Food Systems, and Natural Resources, Department of Veterinary and Microbiological Sciences, Fargo, ND 58105. Offers microbiology (MS); molecular pathogenesis (PhD); natural resource management (MS). Part-time programs available. *Degree requirements:* For master's, thesis; for doctorate, thesis/dissertation, oral and written preliminary exams. *Entrance requirements:* For master's and doctorate, GRE. Additional exam requirements/recommendations for international students: Required—TOEFL. *Faculty research:* Bacterial gene regulation, antibiotic resistance, molecular virology, mechanisms of bacterial pathogenesis, immunology of animals.

The Ohio State University, College of Medicine and Public Health and Graduate School, Graduate Programs in the Basic Medical Sciences, Department of Pathology, Columbus, OH 43210. Offers experimental pathobiology (MS); pathology assistant (MS). *Accreditation:* NAACLS. *Degree requirements:* For master's, thesis, comprehensive exam (for some programs). *Entrance requirements:* For master's, GRE General Test. Additional exam requirements/recommendations for international students: Required—TOEFL (minimum score 550 paper-based; 213 computer-based). Electronic applications accepted. *Faculty research:* Clinical pathology, transplantation pathology, cancer research, neuropathology, vascular pathology.

Oregon State University, College of Veterinary Medicine, Program in Veterinary Science, Corvallis, OR 97331. Offers microbiology (MS); pathology (MS); toxicology (MS). Part-time programs available. *Students:* 3 full-time (2 women), 2 international. Average age 27. In 2005, 1 degree awarded. *Degree requirements:* For master's, thesis. *Entrance requirements:* For master's, minimum GPA of 3.0 in last 90 hours. Additional exam requirements/recommendations for international students: Required—TOEFL. *Application deadline:* For fall admission, 11/1 for domestic students. Application fee: $50. *Expenses:* Contact institution. *Financial support:* Research assistantships, Federal Work-Study, institutionally sponsored loans, and scholarships/grants available. Support available to part-time students. Financial award application deadline: 2/1. *Faculty research:* Calf diseases, bovine foot rot, caliciviruses, effects of toxic agents on immune systems. *Unit head:* Dr. Linda L. Blythe, Associate Dean, 541-737-2098, Fax: 541-737-4245, E-mail: linda.blythe@orst.edu.

Purdue University, School of Veterinary Medicine and Graduate School, Graduate Programs in Veterinary Medicine, Department of Veterinary Pathobiology, West Lafayette, IN 47907. Offers biochemistry and molecular biology (MS, PhD); comparative epidemiology (MS, PhD); epidemiology (MS, PhD); immunology (MS, PhD); infectious diseases (MS, PhD); interdisciplinary genetics (PhD); laboratory animal medicine (MS, PhD); microbiology (MS, PhD); molecular virology (MS, PhD); parasitology (MS, PhD); pathobiology (MS, PhD); public health epidemiology (MS, PhD); toxicology (MS, PhD); veterinary anatomic pathology (MS, PhD); veterinary clinical pathology (MS, PhD); virology (MS, PhD). *Faculty:* 32 full-time (7 women). *Students:* 49 full-time (20 women), 3 part-time (1 woman); includes 2 minority (both African Americans), 31 international. Average age 35. In 2005, 3 master's, 8 doctorates awarded. Terminal master's awarded for partial completion of doctoral program. *Degree requirements:* For master's, thesis (for some programs); for doctorate, thesis/dissertation. *Entrance requirements:* For master's and doctorate, GRE General Test. Additional exam requirements/recommendations for international students: Required—TOEFL (minimum score 575 paper-based), TWE (minimum score 4). *Application deadline:* For fall admission, 8/12 for domestic students, 6/15 for international students; for spring admission, 1/12 for domestic students, 10/15 for international students. Application fee: $55. *Financial support:* Fellowships, research assistantships, teaching assistantships available. Financial award application deadline: 3/1; financial award applicants required to submit FAFSA. *Unit head:* Dr. H. Hogenesch, Head, 765-494-7543.

Queen's University at Kingston, School of Graduate Studies and Research, Faculty of Health Sciences, Department of Pathology, Kingston, ON K7L 3N6, Canada. Offers M Sc, PhD. Part-time programs available. *Degree requirements:* For master's, thesis/dissertation; for doctorate, thesis/dissertation, comprehensive exam. *Entrance requirements:* Additional exam requirements/recommendations for international students: Required—TOEFL. *Faculty research:* Immunopathology, cancer biology, immunology and metastases, cell differentiation, blood coagulation.

Quinnipiac University, School of Health Sciences, Program for Pathologists' Assistant, Hamden, CT 06518-1940. Offers MHS. *Accreditation:* NAACLS. *Faculty:* 2 full-time (0 women), 3 part-time/adjunct (0 women). *Students:* 34 full-time (26 women); includes 3 minority (1 African American, 1 Asian American or Pacific Islander, 1 Hispanic American). Average age 28. 74 applicants, 26% accepted, 18 enrolled. In 2005, 18 degrees awarded. *Degree requirements:* For master's, residency. *Entrance requirements:* For master's, interview, BS in biomedical science, minimum GPA of 2.8. *Application deadline:* For fall admission, 1/15 for domestic students. Applications are processed on a rolling basis. Application fee: $45. Electronic applications accepted. *Expenses:* Tuition: Part-time $570 per credit. *Financial support:* Career-related internships or fieldwork, tuition waivers (partial), and unspecified assistantships available. Financial award application deadline: 4/15; financial award applicants required to submit FAFSA. *Unit head:* Dr. Kenneth Kaloustian, Director, 203-582-8676, Fax: 203-582-3443, E-mail: ken.kaloustian@quinnipiac.edu. *Application contact:* Scott Farber, Director of Graduate Admissions, 800-462-1944, Fax: 203-582-3443, E-mail: graduate@quinnipiac.edu.

Rosalind Franklin University of Medicine and Science, College of Health Professions, Department of Clinical Laboratory Sciences, North Chicago, IL 60064-3095. Offers clinical laboratory science (MS); pathologist assistant (MS). Part-time programs available. *Degree requirements:* For master's, thesis (for some programs). *Entrance requirements:* For master's, minimum GPA of 2.8 in science, 2.5 overall. Additional exam requirements/recommendations for international students: Required—TOEFL. *Faculty research:* Clinical microbiology, hematology, and chemistry; pathology.

Rosalind Franklin University of Medicine and Science, School of Graduate and Postdoctoral Studies, Department of Pathology, North Chicago, IL 60064-3095. Offers MS, PhD, MD/MS, MD/PhD. *Accreditation:* NAACLS. Part-time programs available. Terminal master's awarded for partial completion of doctoral program. *Degree requirements:* For master's and doctorate, thesis/dissertation. *Entrance requirements:* For master's and doctorate, GRE General Test. Additional exam requirements/recommendations for international students: Required—TOEFL, TWE. *Faculty research:* Clinical chemistry, clinical microbiology, hepatopathology.

Saint Louis University, Graduate School and School of Medicine, Graduate Program in Biomedical Sciences and Graduate School, Department of Pathology, St. Louis, MO 63103-2097. Offers PhD. *Faculty:* 44 full-time (24 women), 3 part-time/adjunct (1 woman). *Students:* 2 full-time (both women), 2 part-time (both women); includes 1 minority (Asian American or Pacific Islander), 1 international. Average age 30. In 2005, 1 degree awarded. *Degree requirements:* For doctorate, oral and written preliminary exams, oral defense of dissertation. *Entrance requirements:* For doctorate, GRE General Test, letters of recommendation, resumé, interview. Additional exam requirements/recommendations for international students: Required—TOEFL (minimum score 550 paper-based; 213 computer-based). *Application deadline:* For fall admission, 7/1 for domestic students, 7/1 for international students; for spring admission, 11/1 for domestic students, 11/1 for international students. Applications are processed on a rolling basis. Application fee: $40. *Expenses:* Tuition: Part-time $760 per credit hour. Required fees: $55 per semester. *Financial support:* Health care benefits and tuition waivers (full) available. Financial award application deadline: 6/1; financial award applicants required to submit FAFSA. *Faculty research:* Gap junctions, mast cells, coagulation, pulmonary development, extracellular matrix. *Unit head:* Dr. Christine Janney, Chairperson, 314-977-8783, Fax:

314-268-5645, E-mail: janneycg@slu.edu. *Application contact:* Gary Behrman, Associate Dean of the Graduate School, 314-977-3827, E-mail: behrmang@slu.edu.

State University of New York at Buffalo, Graduate School, Graduate Programs in Cancer Research and Biomedical Sciences at Roswell Park Cancer Institute, Department of Cancer Pathology and Prevention at Roswell Park Cancer Institute, Buffalo, NY 14263. Offers PhD. *Faculty:* 8 part-time/adjunct (3 women). *Students:* 11 full-time (6 women); includes 2 minority (1 African American, 1 Asian American or Pacific Islander), 7 international. Average age 25. 14 applicants, 43% accepted, 4 enrolled. In 2005, 4 degrees awarded. *Degree requirements:* For doctorate, thesis/dissertation, dissertation defense, project, comprehensive exam. *Entrance requirements:* For doctorate, GRE General Test, GRE Subject Test (biochemistry, molecular biology), minimum GPA of 3.0, 3 letters of recommendation. Additional exam requirements/recommendations for international students: Required—TOEFL, TWE, TSE. *Application deadline:* For fall admission, 2/1 priority date for domestic students, 2/1 priority date for international students. Applications are processed on a rolling basis. Application fee: $35. Electronic applications accepted. *Financial support:* In 2005–06, 4 fellowships with full tuition reimbursements (averaging $21,000 per year), 5 research assistantships with full tuition reimbursements (averaging $21,000 per year) were awarded; Federal Work-Study also available. Financial award application deadline: 2/1; financial award applicants required to submit FAFSA. *Faculty research:* Molecular pathology of cancer, chemoprevention of cancer, genomic instability, molecular diagnosis and prognosis of cancer, molecular epidemiology. Total annual research expenditures: $5,400. *Unit head:* Dr. Clement Ip, Chairman, 716-845-8875, Fax: 716-845-8100, E-mail: clement.ip@roswellpark.org. *Application contact:* Craig R. Johnson, Director of Admissions, 716-845-2339, Fax: 716-845-8178, E-mail: craig.johnson@roswellpark.edu.

State University of New York at Buffalo, Graduate School, School of Medicine and Biomedical Sciences, Graduate Programs in Medicine and Biomedical Sciences, Department of Pathology and Anatomical Sciences, Buffalo, NY 14260. Offers anatomical sciences (MA, PhD); pathology (MA, PhD). *Faculty:* 15 full-time (2 women), 7 part-time/adjunct (2 women). *Students:* 7 full-time (5 women), 1 part-time, 1 international. Average age 25. In 2005, 1 master's, 1 doctorate awarded. *Entrance requirements:* For doctorate, GRE, 3 letters of recommendation. Additional exam requirements/recommendations for international students: Required—TOEFL. *Application deadline:* For fall admission, 2/1 priority date for domestic students, 2/1 priority date for international students. Application fee: $35. *Faculty research:* Immunopathology-immunobiology, experimental hypertension, neuromuscular disease, molecular pathology, cell motility and cytoskeleton. *Unit head:* Dr. Francisco Valazquez, Chairman, 716-829-2846, Fax: 716-829-2086, E-mail: fvelazquez@kaleidahealth.org. *Application contact:* Dr. Peter Nickerson, Director of Graduate Studies.

Stony Brook University, State University of New York, Graduate School, College of Arts and Sciences, Department of Biochemistry and Cell Biology, Molecular and Cellular Biology Program, Stony Brook, NY 11794. Offers biochemistry and molecular biology (PhD); biological sciences (MA); cellular and developmental biology (PhD); immunology and pathology (PhD); molecular and cellular biology (PhD). *Students:* 116 full-time (69 women), 2 part-time (1 woman); includes 16 minority (3 African Americans, 10 Asian Americans or Pacific Islanders, 3 Hispanic Americans), 62 international. Average age 30. 342 applicants, 14% accepted. In 2005, 7 degrees awarded. *Degree requirements:* For doctorate, thesis/dissertation, teaching experience, comprehensive exam. *Entrance requirements:* For doctorate, GRE General Test, GRE Subject Test. Additional exam requirements/recommendations for international students: Required—TOEFL. *Application deadline:* For fall admission, 1/15 for domestic students. Application fee: $50. *Expenses:* Tuition, state resident: full-time $6,900; part-time $288 per credit. Tuition, nonresident: full-time $10,920; part-time $455 per credit. Required fees: $704. *Financial support:* Fellowships, research assistantships, teaching assistantships, Federal Work-Study available. *Application contact:* Information Contact, 631-632-8533, Fax: 631-632-9730.

See Close-Up on page 597.

Temple University, Health Sciences Center, School of Medicine and Graduate School, Graduate Programs in Medicine, Department of Pathology and Laboratory Medicine, Philadelphia, PA 19122-6096. Offers PhD. *Faculty:* 9 full-time (1 woman). *Students:* 13 full-time (7 women), 17 part-time (11 women); includes 4 minority (3 Asian Americans or Pacific Islanders, 1 Hispanic American), 11 international. 45 applicants, 20% accepted, 7 enrolled. In 2005, 5 degrees awarded. *Degree requirements:* For doctorate, one foreign language, thesis/dissertation, research seminars. *Entrance requirements:* For doctorate, GRE General Test, GRE Subject Test, minimum GPA of 3.0. Additional exam requirements/recommendations for international students: Required—TOEFL (minimum score 600 paper-based; 250 computer-based). *Application deadline:* For fall admission, 1/15 priority date for domestic students, 12/15 priority date for international students; for spring admission, 11/1 priority date for domestic students, 8/1 priority date for international students. Applications are processed on a rolling basis. Application fee: $50. Electronic applications accepted. *Expenses:* Tuition, state resident: full-time $8,694; part-time $483 per credit. Tuition, nonresident: full-time $12,672; part-time $704 per credit. Required fees: $500; $122 per semester. Tuition and fees vary according to course level, campus/location and program. *Financial support:* Fellowships, research assistantships available. Financial award application deadline: 1/15; financial award applicants required to submit FAFSA. *Faculty research:* Molecular cloning, cell proliferation, cell cycle regulation, DNA repair, cytogenetics. *Unit head:* Dr. Henry Simpkins, Chair, 215-707-4353, Fax: 215-204-6864, E-mail: simpkins@temple.edu.

Texas A&M University, College of Veterinary Medicine, Graduate Programs in Veterinary Medicine, Department of Veterinary Pathobiology, College Station, TX 77843. Offers genetics (MS, PhD); veterinary microbiology (MS, PhD); veterinary parasitology (MS); veterinary pathology (MS, PhD). Part-time programs available. Postbaccalaureate distance learning degree programs offered. *Faculty:* 16 full-time (3 women), 4 part-time/adjunct (1 woman). *Students:* 45 full-time (28 women), 20 part-time (13 women); includes 7 minority (2 African Americans, 5 Hispanic Americans), 19 international. Average age 33. 25 applicants, 76% accepted, 19 enrolled. In 2005, 5 master's, 5 doctorates awarded. Terminal master's awarded for partial completion of doctoral program. *Degree requirements:* For master's and doctorate, thesis/dissertation, seminars. *Entrance requirements:* For master's and doctorate, GRE General Test, minimum GPA of 3.0 in last 60 hours. Additional exam requirements/recommendations for international students: Required—TOEFL. *Application deadline:* For fall admission, 3/1 for domestic students; for spring admission, 8/1 priority date for domestic students. Applications are processed on a rolling basis. Application fee: $50 ($75 for international students). Electronic applications accepted. *Expenses:* Tuition, state resident: full-time $4,488; part-time $187 per credit hour. Tuition, nonresident: full-time $11,112; part-time $463 per credit hour. Required fees: $1,974. *Financial support:* In 2005–06, fellowships with partial tuition reimbursements (averaging $16,000 per year), research assistantships with partial tuition reimbursements (averaging $15,400 per year), teaching assistantships with partial tuition reimbursements (averaging $16,000 per year) were awarded; Federal Work-Study, institutionally sponsored loans, scholarships/grants, traineeships, health care benefits, and unspecified assistantships also available. Support available to part-time students. Financial award applicants required to submit FAFSA. *Faculty research:* Infectious and noninfectious diseases of animals and birds, animal genetics, molecular biology, immunology, virology. *Unit head:* Dr. Ann B. Kier, Head, 979-845-5941, Fax: 979-845-9231, E-mail: akier@cvm.tamu.edu. *Application contact:* Dr. G. G. Wagner, Graduate Advisor, 979-845-2851, Fax: 979-862-1147, E-mail: gwagner@cvm.tamu.edu.

Texas A&M University System Health Science Center, Graduate School of Biomedical Sciences, Department of Pathology and Laboratory Medicine, College Station, TX 77840. Offers molecular pathology (PhD). *Degree requirements:* For doctorate, thesis/dissertation. *Entrance requirements:* For doctorate, GRE General Test. Electronic applications accepted. *Faculty research:* Oncogenes, tumor growth, angiogenesis, signal transduction, demyelinating, diseases.

Thomas Jefferson University, Jefferson College of Graduate Studies, Graduate Program in Molecular Cell Biology, Philadelphia, PA 19107. Offers developmental biology and teratology (PhD); pathology and cell biology (PhD). *Faculty:* 79 full-time. *Students:* 22 full-time (12 women), 1 part-time; includes 2 minority (1 African American, 1 Asian American or Pacific Islander), 3 international. 34 applicants, 26% accepted, 3 enrolled. In 2005, 2 degrees awarded. *Degree requirements:* For doctorate, thesis/dissertation. *Entrance requirements:* For doctorate, GRE General Test, minimum GPA of 3.2. Additional exam requirements/recommendations for international students: Required—TOEFL (minimum score 213 computer-based). *Application deadline:* For fall admission, 3/1 priority date for domestic students, 1/1 priority date for international students. Applications are processed on a rolling basis. Application fee: $50. Electronic applications accepted. *Expenses:* Tuition: Full-time $14,894; part-time $800 per credit. *Financial support:* In 2005–06, 2 students received support, including 222 fellowships with full tuition reimbursements available; research assistantships, Federal Work-Study, institutionally sponsored loans, scholarships/grants, and traineeships also available. Support available to part-time students. Financial award application deadline: 5/1; financial award applicants required to submit FAFSA. *Unit head:* Dr. Theodore Taraschi, Program Director, 215-503-5020, Fax: 215-503-0206, E-mail: theodore.taraschi@jefferson.edu. *Application contact:* Jessie F. Pervall, Director of Admissions, 215-503-0155, Fax: 215-503-9920, E-mail: jessie.pervall@jefferson.edu.

Uniformed Services University of the Health Sciences, School of Medicine, Programs in Biomedical Sciences, Department of Pathology, Bethesda, MD 20814-4799. Offers comparative pathology (PhD); molecular pathobiology (PhD). *Faculty:* 19 full-time (6 women), 80 part-time/adjunct (25 women). *Students:* 6 full-time (5 women); includes 1 minority (African American), 1 international. Average age 27. 11 applicants, 45% accepted, 1 enrolled. *Median time to degree:* Of those who began their doctoral program in fall 1997, 100% received their degree in 8 years or less. *Degree requirements:* For doctorate, one foreign language, thesis/dissertation, qualifying exam, comprehensive exam. *Entrance requirements:* For doctorate, GRE General Test, minimum GPA of 3.0. Additional exam requirements/recommendations for international students: Required—TOEFL. *Application deadline:* For fall admission, 1/15 for domestic students. Applications are processed on a rolling basis. Application fee: $0. *Financial support:* In 2005–06, fellowships with full tuition reimbursements (averaging $23,000 per year); tuition waivers (full) also available. *Faculty research:* Molecular pathology, genetics, virology. *Unit head:* Dr. Robert M. Friedman, Chair, 301-295-3450, E-mail: rfriedman@usuhs.mil. *Application contact:* Janet M. Anastasi, Graduate Program Coordinator, 301-295-9474, Fax: 301-295-6772, E-mail: janastasi@usuhs.mil.

Université de Montréal, Faculty of Medicine and Faculty of Graduate Studies, Graduate Programs in Medicine, Department of Pathology and Cellular Biology, Montréal, QC H3C 3J7, Canada. Offers M Sc, PhD. *Faculty:* 32 full-time (9 women), 1 part-time/adjunct (0 women). *Students:* 45 full-time (25 women). 26 applicants, 12% accepted, 3 enrolled. In 2005, 9 master's, 6 doctorates awarded. Terminal master's awarded for partial completion of doctoral program. *Degree requirements:* For master's, thesis; for doctorate, thesis/dissertation, general exam. *Entrance requirements:* For master's and doctorate, proficiency in French, knowledge of English. *Application deadline:* For fall and spring admission, 2/1. For winter admission, 11/1 for domestic students. Application fee: $30. Electronic applications accepted. *Financial support:* Tuition waivers (full) available. *Faculty research:* Immunopathology, cardiovascular pathology, oncogenetics, cellular neurocytology, muscular dystrophy. *Unit head:* Vincent F. Castellucci, Interim Director, 514-343-6294, Fax: 514-343-5755. *Application contact:* Roger Lippé, Information Contact, 514-343-5616, Fax: 514-343-5755.

Université Laval, Faculty of Medicine, Post-Professional Programs in Medical Studies, Québec, QC G1K 7P4, Canada. Offers anatomy–pathology (DESS); anesthesia–resuscitation (DESS); cardiology (DESS); care of older people (Diploma); clinical research (DESS); community health (DESS); dermatology (DESS); diagnostic radiology (DESS); emergency medicine (Diploma); family medicine (DESS); general surgery (DESS); geriatrics (DESS); hematology (DESS); internal medicine (DESS); maternal and fetal medicine (Diploma); medical biochemistry (DESS); medical microbiology and infectious diseases (DESS); medical oncology (DESS); nephrology (DESS); neurology (DESS); neurosurgery (DESS); obstetrics and gynecology (DESS); ophthalmology (DESS); orthopedic surgery (DESS); oto-rhino-laryngology (DESS); palliative medicine (Diploma); pediatrics (DESS); plastic surgery (DESS); psychiatry (DESS); pulmonary medicine (DESS); radiology–oncology (DESS); thoracic surgery (DESS); urology (DESS). *Degree requirements:* For other advanced degree, comprehensive exam. *Entrance requirements:* For degree, knowledge of French. Electronic applications accepted.

University at Albany, State University of New York, School of Public Health, Department of Biomedical Sciences, Program in Molecular Pathogenesis, Albany, NY 12222-0001. Offers MS, PhD. *Degree requirements:* For master's and doctorate, thesis/dissertation. *Entrance requirements:* For master's and doctorate, GRE General Test, GRE Subject Test. Application fee: $60. *Financial support:* Application deadline: 2/1. *Unit head:* Dr. James Dias, Chair, Department of Biomedical Sciences, 518-474-2662.

The University of Alabama at Birmingham, Graduate Programs in Joint Health Sciences, Department of Pathology, Birmingham, AL 35294. Offers PhD. *Students:* 49 full-time (19 women), 1 (woman) part-time; includes 7 minority (5 African Americans, 2 Asian Americans or Pacific Islanders), 23 international. 25 applicants, 68% accepted. In 2005, 10 degrees awarded. *Degree requirements:* For doctorate, thesis/dissertation. *Entrance requirements:* For doctorate, GRE General Test, interview. *Application deadline:* Applications are processed on a rolling basis. Application fee: $35 ($60 for international students). Electronic applications accepted. *Expenses:* Tuition, state resident: part-time $170 per credit hour. Tuition, nonresident: full-time $4,612; part-time $425 per credit hour. International tuition: $10,732 full-time. Required fees: $11 per credit hour. $124 per term. Tuition and fees vary according to course load, degree level and program. *Financial support:* In 2005–06, 15 fellowships were awarded; career-related internships or fieldwork also available. *Unit head:* Dr. Jay M. McDonald, Chairman, 205-934-4303, Fax: 205-934-5499, E-mail: mcdonald@uab.edu.

University of Alberta, Faculty of Medicine and Dentistry and Faculty of Graduate Studies and Research, Graduate Programs in Medicine, Department of Laboratory Medicine and Pathology, Edmonton, AB T6G 2E1, Canada. Offers medical sciences (M Sc, PhD). Part-time programs available. Terminal master's awarded for partial completion of doctoral program. *Degree requirements:* For master's, thesis; for doctorate, thesis/dissertation, candidacy exam. *Entrance requirements:* Additional exam requirements/recommendations for international students: Required—TOEFL. Tuition and fees charges are reported in Canadian dollars. *Expenses:* Tuition, state resident: part-time $562 Canadian dollars per term. Tuition, nonresident: full-time $3,375 Canadian dollars. Required fees: $573 Canadian dollars; $84 Canadian dollars per term. *Faculty research:* Transplantation, renal pathology, molecular mechanisms of diseases, cryobiology, immunodiagnostics, informatics/cybermedicine, neuroimmunology, microbiology.

University of Arkansas for Medical Sciences, College of Medicine and Graduate School, Graduate Programs in Medicine, Department of Pathology, Little Rock, AR 72205-7199. Offers MS. *Faculty:* 15 full-time (3 women). *Students:* 1 full-time. *Degree requirements:* For master's, thesis. *Entrance requirements:* For master's, GRE General Test. Additional exam requirements/recommendations for international students: Required—TOEFL. Application fee: $0. *Financial support:* Research assistantships available. Support available to part-time students. *Unit head:* Dr. Bruce Smoller, Chairman, 501-686-5170. *Application contact:* Dr. Kathleen D. Eisenach, Information Contact, 501-257-4827, E-mail: eisenachkathleend@uams.edu.

The University of British Columbia, Faculty of Medicine, Department of Pathology and Laboratory Medicine, Vancouver, BC V6T 1Z1, Canada. Offers experimental pathology (M Sc, PhD). *Faculty:* 139 full-time (47 women). *Students:* 87 full-time (41 women); includes 1 African American, 33 Asian Americans or Pacific Islanders. Average age 25. 40 applicants, 60% accepted, 20 enrolled. In 2005, 8 master's, 4 doctorates awarded. *Median time to degree:* Of those who began their doctoral program in fall 1997, 100% received their degree in 8 years or less. *Degree requirements:* For master's, thesis, registration; for doctorate, thesis/dissertation, internal oral defense, comprehensive exam, registration. *Entrance requirements:* For master's, GRE, upper-level course work in biochemistry and physiology; for doctorate, GRE. Additional exam requirements/recommendations for international students: Required—TOEFL (minimum score 570 paper-based; 230 computer-based), GRE General and Subject Test (international applicants). *Application deadline:* For fall admission, 3/1 for domestic students, 1/2 for international students. For winter admission, 8/1 for domestic students; for spring admission, 11/1 for domestic students. Applications are processed on a rolling basis. Application fee: $90 Canadian dollars ($150 Canadian dollars for international students). Electronic applications accepted. *Financial support:* In 2005–06, 12 students received support, including 24 fellowships with full tuition reimbursements available (averaging $20,000 per year), 87 research assistantships with full tuition reimbursements available (averaging $17,850 per year); teaching assistantships with full and partial tuition reimbursements available, institutionally sponsored loans, scholarships/grants, traineeships, health care benefits, tuition waivers (full and partial), and unspecified assistantships also available. *Faculty research:* Molecular biology of disease processes, cancer, hematopathology, atherosclerosis, pulmonary and cardiovascular pathophysiology. Total annual research expenditures: $13.5 million Canadian dollars. *Unit head:* Dr. Richard Hegele, Head, 604-822-7100, Fax: 604-822-7635, E-mail: rhegele@mrl.ubc.ca. *Application contact:* Dr. D. C. Walker, Graduate Adviser, 604-682-2344 Ext. 62705, Fax: 604-806-8351, E-mail: dwalker@mrl.ubc.ca.

University of California, Davis, Graduate Studies, Graduate Group in Comparative Pathology, Davis, CA 95616. Offers MS, PhD. *Accreditation:* NAACLS. *Faculty:* 108 full-time. *Students:* 89 full-time (52 women), 1 (woman) part-time; includes 20 minority (1 African American, 1 American Indian/Alaska Native, 12 Asian Americans or Pacific Islanders, 6 Hispanic Americans), 21 international. Average age 33. 40 applicants, 73% accepted, 22 enrolled. In 2005, 3 master's, 13 doctorates awarded. Terminal master's awarded for partial completion of doctoral program. *Median time to degree:* Of those who began their doctoral program in fall 1997, 73.3% received their degree in 8 years or less. *Degree requirements:* For master's, thesis (for some programs), comprehensive exam (for some programs); for doctorate, thesis/dissertation. *Entrance requirements:* For master's and doctorate, GRE General Test. Additional exam requirements/recommendations for international students: Required—TOEFL (minimum score 550 paper-based; 213 computer-based). *Application deadline:* For fall admission, 4/1 for domestic students, 3/1 for international students. Application fee: $60. Electronic applications accepted. *Financial support:* In 2005–06, 82 students received support, including 25 fellowships with full and partial tuition reimbursements available (averaging $11,336 per year), 39 research assistantships with full and partial tuition reimbursements available (averaging $18,505 per year); teaching assistantships with partial tuition reimbursements available, Federal Work-Study, institutionally sponsored loans, scholarships/grants, tuition waivers (full and partial), and unspecified assistantships also available. Financial award application deadline: 1/15; financial award applicants required to submit FAFSA. *Faculty research:* Immunopathology, toxicological and environmental pathology, reproductive pathology, pathology of infectious diseases. *Unit head:* Reen Lou, Chair, 530-752-2648, E-mail: rlou@ucdavis.edu. *Application contact:* Darlene Flemming, Administrative Assistant, 530-752-2657, Fax: 530-754-8124, E-mail: dhflemming@ucdavis.edu.

University of California, Los Angeles, School of Medicine and Graduate Division, Graduate Programs in Medicine, Program in Experimental Pathology, Los Angeles, CA 90095. Offers MS, PhD. *Degree requirements:* For doctorate, thesis/dissertation, oral and written qualifying exams. *Entrance requirements:* For master's, GRE General Test; for doctorate, GRE General Test, previous course work in physical chemistry and physics.

University of California, San Francisco, Graduate Division, Biomedical Sciences Graduate Group, Program in Experimental Pathology, San Francisco, CA 94143. Offers PhD. *Degree requirements:* For doctorate, thesis/dissertation. *Entrance requirements:* For doctorate, GRE General Test. *Faculty research:* Regulation of cell growth, mechanisms of immunological response, biology of parasitic diseases.

University of Chicago, Division of the Biological Sciences, Biomedical Sciences: Cancer, Immunology, Nutrition, Pathology, and Microbiology, Department of Pathology, Chicago, IL 60637-1513. Offers PhD. *Faculty:* 47 full-time (10 women). *Students:* 22 full-time (16 women); includes 4 minority (1 African American, 2 Asian Americans or Pacific Islanders, 1 Hispanic American), 6 international. Average age 28. In 2005, 4 degrees awarded. *Degree requirements:* For doctorate, thesis/dissertation, registration. *Entrance requirements:* For doctorate, GRE General Test. Additional exam requirements/recommendations for international students: Required—TOEFL. *Application deadline:* For fall admission, 12/28 for domestic students. Application fee: $55. Electronic applications accepted. *Financial support:* In 2005–06, 15 students received support, including fellowships with full tuition reimbursements available (averaging $26,301 per year), research assistantships with full tuition reimbursements available (averaging $26,301 per year); institutionally sponsored loans, scholarships/grants, traineeships, and health care benefits also available. Financial award applicants required to submit FAFSA. *Faculty research:* Vascular biology, apolipoproteins, cardiovascular disease, immunopathology. Total annual research expenditures: $18 million. *Unit head:* Dr. Stephen Meredith, Graduate Program Chair, 773-702-1267, Fax: 773-702-4634, E-mail: scmeredi@uchicago.edu. *Application contact:* Rebecca Levine, Administrative Assistant, Student Services, 773-834-3899, Fax: 773-702-4634, E-mail: rlevine@huggins.bsd.uchicago.edu.

University of Cincinnati, Division of Research and Advanced Studies, College of Medicine, Graduate Programs in Biomedical Sciences, Program in Pathobiology and Molecular Medicine, Cincinnati, OH 45267. Offers pathology (PhD), including anatomic pathology, laboratory medicine, pathobiology and molecular medicine. *Degree requirements:* For doctorate, thesis/dissertation, qualifying exam. *Entrance requirements:* For doctorate, GRE General Test. Additional exam requirements/recommendations for international students: Required—TOEFL (minimum score 620 paper-based; 260 computer-based). Electronic applications accepted. *Faculty research:* Cardiovascular and lipid disorders, digestive and kidney disease, endocrine and metabolic disorders, hematologic and oncogenic, immunology and infectious disease.

University of Florida, College of Medicine, Department of Pathology, Immunology and Laboratory Medicine, Gainesville, FL 32611. Offers immunology and molecular pathology (PhD). *Faculty:* 36 full-time (7 women), 1 (woman) part-time/adjunct. *Degree requirements:* For doctorate, thesis/dissertation. *Entrance requirements:* For doctorate, GRE General Test, minimum GPA of 3.0. Additional exam requirements/recommendations for international students: Required—TOEFL. *Application deadline:* For fall admission, 2/15 for domestic students. Application fee: $30. Electronic applications accepted. *Expenses:* Tuition, state resident: full-time $6,234. Tuition, nonresident: full-time $21,359. Tuition and fees vary according to program. *Financial support:* In 2005–06, research assistantships with full tuition reimbursements (averaging $24,341 per year); fellowships with full tuition reimbursements, teaching assistantships, institutionally sponsored loans and traineeships also available. *Faculty research:* Molecular immunology, autoimmunity and transplantation, tumor biology, oncogenic viruses, human immunodeficiency viruses. *Unit head:* Dr. James M. Crawford, Chairman, 352-392-6840, Fax: 352-392-6249, E-mail: crawford@pathology.ufl.edu. *Application contact:* Dr. Wayne McCormack, Associate Dean of Graduate Education, 352-392-7413, Fax: 352-846-3466, E-mail: mccormac@pathology.ufl.edu.

University of Georgia, College of Veterinary Medicine and Graduate School, Graduate Programs in Veterinary Medicine, Department of Pathology, Athens, GA 30602. Offers MS, PhD. *Faculty:* 10 full-time (5 women). *Students:* 17 full-time, 5 part-time; includes 1 minority (African American), 11 international. 19 applicants, 42% accepted, 6 enrolled. In 2005, 4 degrees awarded. *Degree requirements:* For master's, thesis; for doctorate, one foreign language, thesis/dissertation. *Entrance requirements:* For master's and doctorate, GRE General Test. *Application deadline:* For fall admission, 7/1 for domestic students; for spring admission, 11/15 for domestic students. Application fee: $50. Electronic applications accepted. *Financial support:* Fellowships, research assistantships, teaching assistantships, unspecified

Pathology

University of Georgia (continued)
assistantships available. *Unit head:* Dr. Barry G. Harmon, Head, 706-542-5831. *Application contact:* Dr. Jaroslava Halper, Graduate Coordinator, 706-542-5830, Fax: 706-542-5828, E-mail: jhalper@vet.uga.edu.

University of Guelph, Ontario Veterinary College and Graduate Program Services, Graduate Programs in Veterinary Sciences, Department of Pathobiology, Guelph, ON N1G 2W1, Canada. Offers anatomic pathology (DV Sc, Diploma); clinical pathology (Diploma); comparative pathology (M Sc, PhD); immunology (M Sc, PhD); laboratory animal science (DV Sc); pathology (M Sc, PhD, Diploma); veterinary infectious diseases (M Sc, PhD); zoo animal/wildlife medicine (DV Sc). *Faculty:* 26. *Students:* 66 (22 women). 47 applicants, 32% accepted. In 2005, 6 master's, 7 doctorates, 1 other advanced degree awarded. *Degree requirements:* For master's and doctorate, thesis/dissertation. *Entrance requirements:* For master's, DVM with B average or an honours degree in biological sciences; for doctorate, DVM or MSC degree, minimum B+ average. Additional exam requirements/recommendations for international students: Required—TOEFL (minimum score 550 paper-based; 213 computer-based). *Application deadline:* For fall admission, 2/1 for domestic students. Applications are processed on a rolling basis. Application fee: $75. *Financial support:* In 2005–06, 40 students received support, including 20 fellowships, research assistantships (averaging $13,500 per year), 13 teaching assistantships (averaging $1,600 per year); career-related internships or fieldwork also available. *Faculty research:* Pathogenesis; diseases of animals, wildlife, fish, and laboratory animals; parasitology; immunology; veterinary infectious diseases; laboratory animal science. Total annual research expenditures: $2.5 million. *Unit head:* Dr. John Prescott, Chair, Fax: 519-824-5930, E-mail: jprescott@ovc.uoguelph.ca. *Application contact:* Dr. J. MacInnes, Graduate Coordinator, 519-824-4120 Ext. 54731, Fax: 519-767-0809, E-mail: macinnes@uoguelph.ca.

The University of Iowa, Roy J. and Lucille A. Carver College of Medicine and Graduate College, Graduate Programs in Medicine, Department of Pathology, Iowa City, IA 52242-1316. Offers MS. *Faculty:* 24 full-time (5 women). *Students:* 9 full-time (7 women), 1 international. Average age 25. 6 applicants, 67% accepted, 4 enrolled. In 2005, 2 degrees awarded. *Degree requirements:* For master's, thesis. *Entrance requirements:* For master's, GRE. *Application deadline:* Applications are processed on a rolling basis. Application fee: $60 ($85 for international students). Electronic applications accepted. *Expenses:* Tuition, state resident: part-time $1,882 per term. Tuition, nonresident: full-time $17,338; part-time $4,907 per term. Tuition and fees vary according to course load and program. *Financial support:* In 2005–06, 3 research assistantships with full tuition reimbursements (averaging $22,000 per year) were awarded; tuition waivers (partial) also available. *Faculty research:* Oncology, microbiology, vascular biology, immunology, neuroscience. Total annual research expenditures: $5.2 million. *Unit head:* Michael Cohen, Head, 319-335-8232, Fax: 319-335-8348, E-mail: michael-cohen@uiowa.edu. *Application contact:* Thomas J. Waldschmidt, Graduate Program Director, 319-335-8223, E-mail: thomas-waldschmidt@uiowa.edu.

University of Kansas, Graduate Studies Medical Center, Interdisciplinary Graduate Program in Biomedical Sciences, Department of Pathology and Laboratory Medicine, Lawrence, KS 66045. Offers MA, PhD, MD/PhD. Part-time programs available. *Faculty:* 15. *Students:* 1 full-time (0 women), 4 part-time (3 women). Average age 27.Terminal master's awarded for partial completion of doctoral program. *Degree requirements:* For master's, thesis, comprehensive exam; for doctorate, variable foreign language requirement, thesis/dissertation, comprehensive exam. *Entrance requirements:* For master's and doctorate, GRE. Additional exam requirements/recommendations for international students: Required—TOEFL, TSE. *Application deadline:* For fall admission, 1/15 for domestic students. Applications are processed on a rolling basis. Application fee: $0. *Expenses:* Tuition, state resident: full-time $4,859. Tuition, nonresident: full-time $12,000. Required fees: $589. Tuition and fees vary according to program. *Financial support:* Fellowships, research assistantships, teaching assistantships with full and partial tuition reimbursements, Federal Work-Study, institutionally sponsored loans, and scholarships/grants available. Support available to part-time students. Financial award application deadline: 3/30; financial award applicants required to submit FAFSA. *Faculty research:* Calcification, CA neoplasia, cell signaling, cytogenetics. *Unit head:* Dr. Patricia Thomas, Acting Chair, 913-588-7070, Fax: 913-588-7073, E-mail: pthomas@kumc.edu. *Application contact:* Dr. Jill Pelling, Director of Graduate Studies, 913-588-7240, Fax: 913-588-7073, E-mail: jpelling@kumc.edu.

University of Manitoba, Faculty of Medicine and Faculty of Graduate Studies, Graduate Programs in Medicine, Department of Pathology, Winnipeg, MB R3T 2N2, Canada. Offers M Sc. *Degree requirements:* For master's, thesis. *Entrance requirements:* For master's, honours degree. *Faculty research:* Experimental pathology, cytogenetics, bone metastasis, cilial cell biology, breast cancer, neuropathology, neuroscience, neuroimmunology, immunobiology, transplantation.

University of Maryland, Graduate School, Graduate Programs in Medicine, Department of Pathology, Baltimore, MD 21201. Offers medical pathology (PhD); pathology (MS). *Accreditation:* NAACLS. Part-time programs available. *Faculty:* 51 full-time (11 women). *Students:* 21 full-time (16 women), 11 part-time (9 women); includes 2 African Americans, 7 international. Average age 29. 71 applicants, 28% accepted, 10 enrolled. In 2005, 8 master's, 4 doctorates awarded. *Degree requirements:* For doctorate, thesis/dissertation. *Entrance requirements:* For master's and doctorate, GRE General Test, minimum GPA of 3.0. Additional exam requirements/recommendations for international students: Required—TOEFL; Recommended—IELT. *Application deadline:* For fall admission, 7/1 for domestic students, 1/15 for international students. Application fee: $50. Electronic applications accepted. *Expenses:* Tuition, state resident: full-time $8,079; part-time $409 per credit hour. Tuition, nonresident: full-time $18,384; part-time $731 per credit hour. Required fees: $695; $10 per credit hour. Tuition and fees vary according to degree level and program. *Financial support:* Fellowships, research assistantships, teaching assistantships, career-related internships or fieldwork available. Support available to part-time students. Financial award application deadline: 2/15. *Faculty research:* Carcinogenesis, immunopathology, cell injury. *Unit head:* Dr. Sanford Stass, Chair, 410-328-1237, E-mail: pharris@umm.edu. *Application contact:* Dr. Anne Hamburger, Program Director, 410-706-7276, Fax: 410-706-8414, E-mail: ahamburg@umaryland.edu.

University of Medicine and Dentistry of New Jersey, Graduate School of Biomedical Sciences, Graduate Programs in Biomedical Sciences–Newark, Program in Experimental Pathology, Newark, NJ 07107. Offers PhD. *Entrance requirements:* Additional exam requirements/recommendations for international students: Required—TOEFL. *Application deadline:* For fall admission, 2/1 for domestic students. *Financial support:* Fellowships, research assistantships, Federal Work-Study, institutionally sponsored loans, and tuition waivers (full and partial) available. *Unit head:* Dr. Muriel Lambert, Program Director, 973-972-4405, Fax: 973-972-7293, E-mail: mlambert@umdnj.edu.

University of Michigan, Medical School and Horace H. Rackham School of Graduate Studies, Program in Biomedical Sciences (PIBS), Department of Pathology, Ann Arbor, MI 48109. Offers PhD. Offered through the Horace H. Rackham School of Graduate Studies. *Degree requirements:* For doctorate, thesis/dissertation, oral defense of dissertation, preliminary exam. *Entrance requirements:* For doctorate, GRE General Test, GRE Subject Test. Additional exam requirements/recommendations for international students: Required—TOEFL. Electronic applications accepted. *Expenses:* Tuition, state resident: full-time $14,082; part-time $894 per credit hour. Tuition, nonresident: full-time $28,500; part-time $1,675 per credit hour. Required fees: $189; $189 per unit. *Faculty research:* Regulation of cytokine gene expression, programmed cell death, pathogenesis of fibrosis, soluble mediators of inflammation.

See Close-Up on page 1121.

University of Mississippi Medical Center, School of Graduate Studies in the Health Sciences, Department of Pathology, Jackson, MS 39216-4505. Offers MS, PhD, MD/PhD. *Faculty:* 20 full-time (6 women), 1 part-time/adjunct (0 women). *Students:* 2 full-time (0 women). Average age 36. 7 applicants, 0% accepted, 0 enrolled. In 2005, 2 master's awarded. Terminal master's awarded for partial completion of doctoral program. *Degree requirements:* For master's, thesis; for doctorate, thesis/dissertation, first authored publication in peer-reviewed journal. *Entrance requirements:* For master's, GRE General Test, minimum GPA of 3.0; for doctorate, GRE General Test, GRE Subject Test, minimum GPA of 3.0. *Application deadline:* For fall admission, 7/1 for domestic students. Applications are processed on a rolling basis. Application fee: $10. *Financial support:* In 2005–06, 1 student received support, including 1 research assistantship (averaging $16,234 per year) Financial award application deadline: 3/31. *Faculty research:* Effects of rehabilitation therapy on immune system/hypothalamic/pituitary adrenal axis interaction; HLA, GC, CM, KM, and/or genetic factors in the pathogenesis of AIDS; stem cell research; renal disease. *Unit head:* Dr. Michael D. Hughson, Chairman, 601-984-1530, Fax: 601-984-1531, E-mail: mhughson@pathology.umsmed.edu. *Application contact:* Dr. Julius M. Cruse, Director, 601-984-1561, Fax: 601-984-1835, E-mail: jcruse@pathology.umsmed.edu.

University of Nebraska Medical Center, Graduate Studies, Department of Pathology and Microbiology, Omaha, NE 68198. Offers MS, PhD. Part-time programs available. *Faculty:* 36 full-time (6 women), 50 part-time/adjunct (4 women). *Students:* 31 full-time (15 women), 3 part-time (2 women); includes 15 minority (1 African American, 13 Asian Americans or Pacific Islanders, 1 Hispanic American). Average age 28. 23 applicants, 43% accepted, 9 enrolled. In 2005, 6 doctorates awarded. Terminal master's awarded for partial completion of doctoral program. *Degree requirements:* For master's and doctorate, thesis/dissertation, comprehensive exam. *Entrance requirements:* For master's, previous course work in biology, chemistry, mathematics, and physics; for doctorate, GRE General Test, previous course work in biology, chemistry, mathematics, and physics. Additional exam requirements/recommendations for international students: Required—TOEFL. *Application deadline:* Applications are processed on a rolling basis. Application fee: $45. *Expenses:* Tuition, area resident: Part-time $200 per hour. Tuition, nonresident: part-time $538 per hour. Required fees: $308; $59 per term. *Financial support:* In 2005–06, 12 fellowships with tuition reimbursements (averaging $21,000 per year), 19 research assistantships with tuition reimbursements (averaging $21,000 per year) were awarded; institutionally sponsored loans and tuition waivers (full) also available. Support available to part-time students. Financial award application deadline: 3/1. *Faculty research:* Carcinogenesis, cancer biology, immunobiology, molecular virology, molecular genetics. *Unit head:* Dr. Rakesh K. Siagh, Chair, Graduate Committee, 402-559-9948, Fax: 402-559-4077, E-mail: rsingh@unmc.edu.

University of New Mexico, School of Medicine, Biomedical Sciences Graduate Program, Albuquerque, NM 87131-5196. Offers biochemistry and molecular biology (MS, PhD); cell biology and physiology (MS, PhD); molecular genetics and microbiology (MS, PhD); neuroscience (MS, PhD); pathology (MS, PhD); toxicology (MS, PhD). Part-time programs available. Terminal master's awarded for partial completion of doctoral program. *Degree requirements:* For master's, thesis/dissertation; for doctorate, thesis/dissertation, comprehensive exam. *Entrance requirements:* For master's and doctorate, GRE General Test, minimum undergraduate GPA of 3.0. Additional exam requirements/recommendations for international students: Required—TOEFL. Electronic applications accepted. *Expenses:* Tuition, state resident: full-time $5,676. Tuition, nonresident: full-time $14,974; part-time $238 per credit hour. Required fees: $385 per term. Tuition and fees vary according to course load and program. *Faculty research:* Signal transduction, infectious disease, biology of cancer, structural biology, neuroscience.

The University of North Carolina at Chapel Hill, School of Medicine and Graduate School, Graduate Programs in Medicine, Department of Pathology and Laboratory Medicine, Chapel Hill, NC 27599. Offers experimental pathology (PhD). *Accreditation:* NAACLS. *Faculty:* 60 full-time (25 women), 5 part-time/adjunct (0 women). *Students:* 32 full-time (17 women); includes 1 African American, 1 American Indian/Alaska Native, 3 international. Average age 29. 40 applicants, 18% accepted, 4 enrolled. In 2005, 7 degrees awarded. *Median time to degree:* Of those who began their doctoral program in fall 1997, 100% received their degree in 8 years or less. *Degree requirements:* For doctorate, thesis/dissertation, oral exam, proposal defense, comprehensive exam. *Entrance requirements:* For doctorate, GRE General Test, GRE Subject Test (recommended). Additional exam requirements/recommendations for international students: Required—TOEFL (minimum score 550 paper-based; 213 computer-based). *Application deadline:* For fall admission, 4/15 for domestic students, 4/15 for international students. Applications are processed on a rolling basis. Application fee: $65. Electronic applications accepted. *Financial support:* In 2005–06, 14 fellowships with full and partial tuition reimbursements (averaging $21,500 per year), 16 research assistantships with full and partial tuition reimbursements (averaging $21,500 per year) were awarded; Federal Work-Study, institutionally sponsored loans, scholarships/grants, traineeships, tuition waivers (full), and unspecified assistantships also available. Financial award application deadline: 1/1; financial award applicants required to submit FAFSA. *Faculty research:* Carcinogenesis, blood coagulation, molecular biology and genetics and animal models of human disease, opportunistic infections in AIDS immunopathology. Total annual research expenditures: $16.2 million. *Unit head:* Dr. J. Charles Jennette, Brinkhous—Distinguished Professor and Chair, 919-966-4676, Fax: 919-966-6718, E-mail: charles.jennette@pathology.unc.edu. *Application contact:* Dr. William B. Coleman, Director of Graduate Admissions, 919-966-2699, Fax: 919-966-6718, E-mail: william.coleman@pathology.unc.edu.

University of Oklahoma Health Sciences Center, College of Medicine and Graduate College, Graduate Programs in Medicine, Department of Pathology, Oklahoma City, OK 73190. Offers PhD. *Degree requirements:* For doctorate, thesis/dissertation. *Entrance requirements:* For doctorate, GRE General Test, 3 letters of recommendation. Additional exam requirements/recommendations for international students: Required—TOEFL. *Faculty research:* Molecular pathology, tissue response in disease, anatomic pathology, immunopathology, histocytochemistry.

University of Pittsburgh, School of Medicine, Graduate Programs in Medicine, Program in Cellular and Molecular Pathology, Pittsburgh, PA 15260. Offers MS, PhD. *Faculty:* 45 full-time (11 women). *Students:* 27 full-time (14 women); includes 4 minority (1 African American, 2 Asian Americans or Pacific Islanders, 1 Hispanic American), 7 international. Average age 28. 415 applicants, 22% accepted, 42 enrolled. In 2005, 4 degrees awarded. *Median time to degree:* Of those who began their doctoral program in fall 1997, 95% received their degree in 8 years or less. *Degree requirements:* For doctorate, thesis/dissertation, comprehensive exam, registration. *Entrance requirements:* For doctorate, GRE General Test, GRE Subject Test, minimum QPA of 3.0. Additional exam requirements/recommendations for international students: Required—TOEFL (minimum score 600 paper-based; 250 computer-based), IELT (minimum score 7). *Application deadline:* For fall admission, 12/15 priority date for domestic students, 12/15 priority date for international students. Application fee: $40. Electronic applications accepted. *Expenses:* Tuition, state resident: full-time $13,194; part-time $537 per credit. Tuition, nonresident: full-time $25,012; part-time $1,026 per credit. Required fees: $700; $164 per term. Tuition and fees vary according to campus/location and program. *Financial support:* In 2005–06, fellowships with full tuition reimbursements (averaging $21,500 per year), research assistantships with full tuition reimbursements (averaging $21,500 per year) were awarded; teaching assistantships with full tuition reimbursements, institutionally sponsored loans, scholarships/grants, traineeships, health care benefits, and unspecified assistantships also available. *Faculty research:* Liver growth and differentiation, pathogenesis of neurodegeneration, cancer research. *Unit head:* Dr. Robert Bowser, Graduate Program Director, 412-383-7819, Fax: 412-383-7969, E-mail: bowserrp@upmc.edu. *Application contact:* Graduate Studies Administrator, 412-648-8957, Fax: 412-648-1077, E-mail: gradstudies@medschool.pitt.edu.

University of Prince Edward Island, Atlantic Veterinary College, Graduate Program in Veterinary Medicine, Charlottetown, PE C1A 4P3, Canada. Offers anatomy (M Sc, PhD); bacteriology (M Sc, PhD); clinical pharmacology (M Sc, PhD); clinical sciences (M Sc, PhD); epidemiology (M Sc, PhD), including reproduction; food animal nutrition (M Sc, PhD); immunology (M Sc, PhD); microanatomy (M Sc, PhD); parasitology (M Sc, PhD); pathology (M Sc, PhD); pharmacology (M Sc, PhD); physiology (M Sc, PhD); toxicology (M Sc, PhD); veterinary science (M Vet Sc); virology (M Sc, PhD). Part-time programs available. *Faculty:* 76 full-time (25 women), 49 part-time/adjunct (8 women). *Students:* 54 full-time (32 women), 2 part-time. Average age 30. In 2005, 7 master's, 6 doctorates awarded. *Degree requirements:* For master's and doctorate, thesis/dissertation. *Entrance requirements:* For master's, DVM, B Sc honors degree, or equivalent; for doctorate, M Sc. *Application deadline:* Applications are processed on a rolling basis. Application fee: $50. *Expenses:* Contact institution. Tuition charges are reported in Canadian dollars. Part-time tuition and fees vary according to course level, degree level, campus/location, program and student level. *Financial support:* In 2005–06, 4 fellowships (averaging $25,000 Canadian dollars per year), 4 research assistantships (averaging $16,500 Canadian dollars per year) were awarded; career-related internships or fieldwork also available. *Faculty research:* Animal health management, infectious diseases, fin fish and shellfish health, basic biomedical sciences, ecosystem health. Total annual research expenditures: $1.2 million Canadian dollars. *Unit head:* Dr. James Bellamy, Associate Dean of Graduate Studies and Research, 902-566-0856, E-mail: bellamy@upei.ca. *Application contact:* Cheryl Gaudet, Registrar's Office, 902-566-0781, Fax: 902-566-0795, E-mail: registrar@upei.ca.

University of Rochester, School of Medicine and Dentistry, Graduate Programs in Medicine and Dentistry, Department of Pathology and Laboratory Medicine, Rochester, NY 14642. Offers pathology (MS, PhD). *Degree requirements:* For doctorate, variable foreign language requirement, thesis/dissertation, qualifying exam. *Entrance requirements:* For doctorate, GRE General Test, GRE Subject Test.

See Close-Up on page 1123.

University of Saskatchewan, College of Medicine, Department of Pathology, Saskatoon, SK S7N 5A2, Canada. Offers M Sc, PhD. *Faculty:* 6. *Students:* 10. *Degree requirements:* For master's and doctorate, thesis/dissertation, registration. *Entrance requirements:* Additional exam requirements/recommendations for international students: Required—TOEFL. *Application deadline:* For fall admission, 7/1 for domestic students. Applications are processed on a rolling basis. Application fee: $50. *Financial support:* Fellowships, research assistantships, teaching assistantships available. Financial award application deadline: 1/31. *Unit head:* Dr. J. Krahn, Head, 306-966-2151, Fax: 306-966-2910, E-mail: john.krahn@usask.ca. *Application contact:* Dr. M. Qureshi, Graduate Chair, 306-655-6949, Fax: 306-655-2223, E-mail: mabood.queshi@saskatoonhealthregion.ca.

University of Saskatchewan, Western College of Veterinary Medicine and College of Graduate Studies and Research, Graduate Programs in Veterinary Medicine, Department of Veterinary Pathology, Saskatoon, SK S7N 5A2, Canada. Offers M Sc, M Vet Sc, PhD. *Faculty:* 12 full-time (4 women). *Students:* 20 full-time (11 women); includes 2 minority (both African Americans) In 2005, 3 master's, 1 doctorate awarded. *Degree requirements:* For master's, thesis/dissertation, registration (for some programs); for doctorate, thesis/dissertation, registration. *Entrance requirements:* Additional exam requirements/recommendations for international students: Required—IELT or TOEFL. *Application deadline:* For fall admission, 7/1 for domestic students. Applications are processed on a rolling basis. Application fee: $50. *Financial support:* Fellowships, teaching assistantships available. Financial award application deadline: 1/31. *Faculty research:* Thyroid, oncology, immunology/infectious diseases, vaccinology. *Unit head:* Dr. Marion Jackson, Head, 306-966-7332, Fax: 306-966-7439, E-mail: marion.jackson@usask.ca. *Application contact:* Dr. Andy Allen, Graduate Chair, 306-966-7492, Fax: 306-966-7439, E-mail: andrew.allen@usask.ca.

University of Southern California, Keck School of Medicine and Graduate School, Graduate Programs in Medicine, Department of Pathology, Los Angeles, CA 90089. Offers experimental and molecular pathology (MS); pathobiology (PhD). *Faculty:* 52 full-time (10 women). *Students:* 58 full-time (33 women); includes 20 minority (15 Asian Americans or Pacific Islanders, 5 Hispanic Americans), 23 international. Average age 26. 40 applicants, 55% accepted, 16 enrolled. In 2005, 6 master's, 8 doctorates awarded. *Degree requirements:* For master's, thesis or alternative; for doctorate, thesis/dissertation. *Entrance requirements:* For master's, GRE General Test, minimum GPA of 3.0; for doctorate, GRE General Test, minimum GPA of 3.0, BS in natural sciences. Additional exam requirements/recommendations for international students: Required—TOEFL. *Application deadline:* For fall admission, 3/1 for domestic students. Application fee: $65. Electronic applications accepted. *Expenses:* Tuition: Full-time $25,416; part-time $1,059 per unit. Required fees: $484; $484 per year. Tuition and fees vary according to course load and program. *Financial support:* In 2005–06, 43 students received support, including 3 fellowships with tuition reimbursements available (averaging $24,876 per year), 38 research assistantships with tuition reimbursements available (averaging $24,876 per year), 2 teaching assistantships with tuition reimbursements available (averaging $24,876 per year); scholarships/grants also available. Financial award application deadline: 2/1. *Faculty research:* Immunology of lymphomas and leukemias, lung cancer, molecular basis of oncogenesis, central nervous system disease, organic chemical carcinogens and carcinogenic metal salts. Total annual research expenditures: $12.2 million. *Unit head:* Dr. Clive R. Taylor, Chairman, 323-442-1180, Fax: 323-442-3314, E-mail: taylor@pathfinder.usc.edu. *Application contact:* Lisa A. Doumak, Student Services Assistant, 323-442-1168, Fax: 323-442-3049, E-mail: doumak@pathfinder.usc.edu.

See Close-Up on page 1125.

University of South Florida, College of Medicine and College of Graduate Studies, Graduate Programs in Medical Sciences, Tampa, FL 33620-9951. Offers anatomy (PhD); biochemistry and molecular biology (MS, PhD), including biochemistry and molecular biology (PhD), bioinformatics and computational biology (MS); medical microbiology and immunology (PhD); pathology (PhD); pharmacology and therapeutics (PhD), including medical sciences; physiology and biophysics (PhD). *Students:* 108 full-time (53 women), 34 part-time (26 women); includes 31 minority (9 African Americans, 9 Asian Americans or Pacific Islanders, 13 Hispanic Americans), 30 international. 117 applicants, 99% accepted, 116 enrolled. In 2005, 9 master's, 4 doctorates awarded. *Degree requirements:* For doctorate, thesis/dissertation. *Entrance requirements:* For doctorate, GRE General Test, minimum GPA of 3.0. Application fee: $30. *Expenses:* Contact institution. *Financial support:* Institutionally sponsored loans and scholarships/grants available. Financial award application deadline: 4/1; financial award applicants required to submit FAFSA. *Unit head:* Dr. Joseph J. Krzanowski, Associate Dean for Research and Graduate Affairs, 813-974-4181, Fax: 813-974-4317, E-mail: jkrzanow@com1.med.usf.edu.

The University of Tennessee Health Science Center, College of Graduate Health Sciences, Department of Pathology, Memphis, TN 38163-0002. Offers MS, PhD. Part-time programs available. *Faculty:* 17 full-time (5 women), 7 part-time/adjunct (2 women). *Students:* 3 full-time (1 woman); includes 2 minority (both Asian Americans or Pacific Islanders) Average age 26. 19 applicants, 5% accepted. In 2005, 2 doctorates awarded. Terminal master's awarded for partial completion of doctoral program. *Degree requirements:* For master's, thesis, comprehensive exam; for doctorate, thesis/dissertation, oral and written preliminary and comprehensive exams. *Entrance requirements:* For master's and doctorate, GRE General Test, minimum GPA of 3.0. Additional exam requirements/recommendations for international students: Required—TOEFL. *Application deadline:* For fall admission, 2/1 for domestic students. Application fee: $0. Electronic applications accepted. *Financial support:* In 2005–06, 4 teaching assistantships were awarded; research assistantships Financial award application deadline: 2/25. *Unit head:* Dr. Lawrence M. Pfeffer, Chairman, 901-448-7020, E-mail: lpfeffer@utmem.edu. *Application contact:* Ida W. Mosby, Director, Enrollment Services, 901-448-5560, E-mail: imosby@utmem.edu.

The University of Texas Medical Branch, Graduate School of Biomedical Sciences, Program in Experimental Pathology, Galveston, TX 77555. Offers PhD. *Students:* 33 full-time (18 women); includes 1 minority (Hispanic American), 3 international. In 2005, 2 doctorates awarded. *Degree requirements:* For doctorate, thesis/dissertation. *Entrance requirements:* For doctorate, GRE General Test. Additional exam requirements/recommendations for international students: Required—TOEFL (minimum score 550 paper-based; 213 computer-based). *Application deadline:* Applications are processed on a rolling basis. Application fee: $30 ($75 for international students). Electronic applications accepted. *Expenses:* Tuition, state resident:

full-time $8,350; part-time $90 per credit hour. Tuition, nonresident: full-time $21,450; part-time $366 per credit hour. Required fees: $1,027; $11 per credit hour. $60 per term. *Financial support:* In 2005–06, fellowships (averaging $23,000 per year), research assistantships with full tuition reimbursements (averaging $23,000 per year) were awarded. Financial award applicants required to submit FAFSA. *Unit head:* Dr. Stephen Higgs, Director, 409-747-2426, Fax: 409-747-2437, E-mail: sthiggs@utmb.edu. *Application contact:* Harriet Ross, 409-772-2521, Fax: 409-747-2400, E-mail: hross@utmb.edu.

See Close-Up on page 1127.

The University of Toledo, School of Graduate Studies, Department of Pathology, Toledo, OH 43606-3390. Offers MS. Part-time programs available. *Degree requirements:* For master's, thesis, qualifying exam. *Entrance requirements:* For master's, GRE General Test, minimum undergraduate GPA of 3.0. *Expenses:* Tuition, state resident: full-time $6,623; part-time $308 per credit hour. Tuition, nonresident: full-time $13,232; part-time $735 per credit hour. *Faculty research:* Cell injury, molecular and clinical molecular carcinogenesis, chemoprevention, hypertension.

University of Utah, School of Medicine and The Graduate School, Graduate Programs in Medicine, Department of Pathology, Salt Lake City, UT 84112-1107. Offers experimental pathology (PhD); laboratory medicine and biomedical science (MS). PhD offered after acceptance into the combined Program in Molecular Biology. *Degree requirements:* For doctorate, thesis/dissertation, comprehensive exam, registration. *Entrance requirements:* For doctorate, GRE, minimum GPA of 3.0. *Expenses:* Tuition, state resident: full-time $2,932; part-time $369 per credit. Tuition, nonresident: full-time $10,350; part-time $1,302 per credit. Required fees: $516 per term. Tuition and fees vary according to course load and program. *Faculty research:* Immunology, cell biology, signal transduction, gene regulation, receptor biology.

University of Vermont, College of Medicine and Graduate College, Graduate Programs in Medicine, Department of Pathology, Burlington, VT 05405. Offers MS, MD/MS. *Students:* 5 (3 women) 1 international. 6 applicants, 50% accepted, 3 enrolled. In 2005, 1 degree awarded. *Degree requirements:* For master's, thesis. *Entrance requirements:* For master's, GRE General Test. Additional exam requirements/recommendations for international students: Required—TOEFL (minimum score 550 paper-based; 213 computer-based). *Application deadline:* For fall admission, 4/1 for domestic students. Applications are processed on a rolling basis. Application fee: $40. *Expenses:* Tuition, area resident: Part-time $410 per credit hour. Tuition, nonresident: part-time $1,034 per credit hour. *Financial support:* Fellowships, research assistantships, traineeships available. Financial award application deadline: 3/1. *Unit head:* Dr. E. Bovill, Chairperson, 802-656-2210. *Application contact:* Dr. S. Huber, Coordinator, 802-656-2210.

University of Washington, School of Medicine and Graduate School, Graduate Programs in Medicine, Department of Pathology, Seattle, WA 98195. Offers molecular basis of disease (PhD); pathology (MS). *Degree requirements:* For doctorate, thesis/dissertation. *Entrance requirements:* For doctorate, GRE General Test. *Faculty research:* Viral oncogenesis, aging, mutagenesis and repair, extracellular matrix biology, vascular biology.

The University of Western Ontario, Faculty of Graduate Studies, Biosciences Division, Department of Pathology, London, ON N6A 5B8, Canada. Offers M Sc, PhD. *Degree requirements:* For master's, thesis/dissertation; for doctorate, thesis/dissertation, comprehensive exam. *Entrance requirements:* For master's and doctorate, minimum B+ average, honors degree. *Faculty research:* Heavy metal toxicology, transplant pathology, immunopathology, immunological cancers, neurochemistry, aging and dementia, cancer pathology.

University of Wisconsin–Madison, Medical School and Graduate School, Graduate Programs in Medicine, Department of Pathology and Laboratory Medicine, Madison, WI 53706-1380. Offers PhD. *Accreditation:* NAACLS. *Faculty:* 40 full-time (11 women). *Students:* 29 full-time (15 women); includes 3 minority (2 Asian Americans or Pacific Islanders, 1 Hispanic American), 14 international. Average age 28. 59 applicants, 17% accepted, 8 enrolled. In 2005, 1 degree awarded. *Median time to degree:* Of those who began their doctoral program in fall 1997, 50% received their degree in 8 years or less. *Degree requirements:* For doctorate, thesis/dissertation. *Entrance requirements:* For doctorate, GRE, minimum GPA of 3.0. Additional exam requirements/recommendations for international students: Required—TOEFL (minimum score 580 paper-based; 237 computer-based). *Application deadline:* For fall admission, 4/1 priority date for domestic students, 4/1 priority date for international students. Applications are processed on a rolling basis. Application fee: $45. Electronic applications accepted. *Financial support:* In 2005–06, 29 students received support, including fellowships with full tuition reimbursements available (averaging $22,500 per year), research assistantships with full tuition reimbursements available (averaging $22,500 per year); health care benefits also available. *Faculty research:* Immunology/immunopathology, cancer biology, neuroscience/neuropathology, growth factor/matrix biology, developmental pathology. *Unit head:* Dr. Michael N. Hart, Chair, 608-262-1189, Fax: 608-265-3301, E-mail: gharvey@facstaff.wisc.edu. *Application contact:* Gina King, Educational Support, 608-262-2665, Fax: 608-265-3301, E-mail: gradinfo@pathology.wisc.edu.

See Close-Up on page 1131.

Vanderbilt University, Graduate School and School of Medicine, Department of Cellular and Molecular Pathology, Nashville, TN 37240-1001. Offers PhD, MD/PhD. *Faculty:* 34 full-time (5 women). *Students:* 16 full-time (9 women); includes 4 minority (3 African Americans, 1 Hispanic American), 3 international. In 2005, 2 doctorates awarded. *Degree requirements:* For doctorate, thesis/dissertation, preliminary, qualifying, and final exams. *Entrance requirements:* For doctorate, GRE General Test. *Application deadline:* For fall admission, 1/15 for domestic students, 1/15 for international students. Application fee: $0. Electronic applications accepted. *Expenses:* Tuition: Part-time $1,283 per semester hour. Required fees: $2,202; $1,101 per semester. One-time fee: $30. Tuition and fees vary according to course load, program and student level. *Financial support:* Fellowships with full tuition reimbursements, research assistantships with full tuition reimbursements, Federal Work-Study, institutionally sponsored loans, traineeships, and tuition waivers (partial) available. Financial award application deadline: 1/15. *Faculty research:* Regulation of cell growth, vascular biology, tumor biology, immunology and retroviral pathology, angiogenesis. *Unit head:* Dr. Samuel A. Santoro, Chair, 615-322-2123, Fax: 615-343-7023. *Application contact:* Paul E. Bock, Director of Graduate Studies, 615-322-2123, Fax: 615-343-7023, E-mail: paul.e.bock@vanderbilt.edu.

Virginia Commonwealth University, Medical College of Virginia-Professional Programs, School of Medicine and Graduate Programs, School of Medicine Graduate Programs, Department of Pathology, Richmond, VA 23284-9005. Offers MS, PhD, MD/PhD. Part-time programs available. *Faculty:* 12 full-time (4 women). *Students:* 5 full-time (3 women), 1 part-time; includes 1 minority (Asian American or Pacific Islander), 1 international. 10 applicants, 30% accepted, 0 enrolled. Terminal master's awarded for partial completion of doctoral program. *Degree requirements:* For master's and doctorate, thesis/dissertation, comprehensive oral and written exams. *Entrance requirements:* For doctorate, GRE General Test, MCAT. *Application deadline:* For fall admission, 2/15 for domestic students. Application fee: $50. *Expenses:* Tuition, state resident: full-time $6,268; part-time $405 per credit. Tuition, nonresident: full-time $15,904; part-time $940 per credit. Required fees: $751 per semester hour. Tuition and fees vary according to course load and program. *Financial support:* Fellowships, teaching assistantships, tuition waivers (full) available. *Faculty research:* Biochemical and clinical applications of enzyme and protein immobilization, clinical enzymology. *Unit head:* Dr. David S. Wilkinson, Chair, 804-828-0183, Fax: 804-828-9749, E-mail: dswilkin@vcu.edu. *Application contact:* Dr. Alphonse Poklis, Director, Graduate Studies, 804-828-0272, Fax: 804-828-9749, E-mail: apoklis@vcu.edu.

See Close-Up on page 1133.

Pathology

Washington State University, College of Veterinary Medicine and Graduate School, Graduate Programs in Veterinary Science, Pullman, WA 99164. Offers veterinary and comparative anatomy, pharmacology, and physiology (MS, PhD), including neuroscience, veterinary science; veterinary clinical sciences (MS, PhD); veterinary microbiology and pathology (MS, PhD), including veterinary science. Part-time programs available. *Faculty:* 72 full-time (13 women), 6 part-time/adjunct (4 women). *Students:* 50 full-time (24 women). Average age 30. In 2005, 3 master's, 4 doctorates awarded. Terminal master's awarded for partial completion of doctoral program. *Degree requirements:* For master's and doctorate, thesis/dissertation, oral exam. *Entrance requirements:* For master's and doctorate, GRE General Test, minimum GPA of 3.0. *Application deadline:* For fall admission, 12/31 for domestic students. Applications are processed on a rolling basis. Application fee: $35. Electronic applications accepted. *Expenses: Contact institution.* Part-time tuition and fees vary according to campus/location and program. *Financial support:* In 2005–06, 21 research assistantships with partial tuition reimbursements, 8 teaching assistantships with partial tuition reimbursements were awarded; fellowships, career-related internships or fieldwork, Federal Work-Study, institutionally sponsored loans, scholarships/grants, traineeships, tuition waivers (partial), and teaching associateships also available. Financial award application deadline: 12/1; financial award applicants required to submit FAFSA. *Application contact:* Julie K. Smith, Principal Assistant, 509-335-3064, Fax: 509-335-0160, E-mail: jksmith@vetmed.wsu.edu.

Wayne State University, School of Medicine and Graduate School, Graduate Programs in Medicine, Department of Pathology, Detroit, MI 48202. Offers MS, PhD. *Accreditation:* NAACLS. *Faculty:* 15 full-time (2 women). *Students:* 11 full-time (8 women), 1 part-time; includes 2 minority (1 African American, 1 Asian American or Pacific Islander), 8 international. Average age 30. 28 applicants, 39% accepted, 3 enrolled. *Degree requirements:* For doctorate, thesis/dissertation. *Entrance requirements:* For doctorate, GRE General Test. Additional exam requirements/recommendations for international students: Required—TOEFL (minimum score 550 paper-based; 213 computer-based); Recommended—TWE (minimum score 6). *Application deadline:* For fall admission, 4/1 for domestic students, 6/1 for international students. Applications are processed on a rolling basis. Application fee: $30 ($50 for international students). Electronic applications accepted. *Expenses:* Tuition, state resident: part-time $338 per credit hour. Tuition, nonresident: part-time $746 per credit hour. Required fees: $24 per credit hour. Full-time tuition and fees vary according to program. *Financial support:* In 2005–06, 2 fellowships with tuition reimbursements, 8 research assistantships with tuition reimbursements (averaging $20,125 per year) were awarded; teaching assistantships, career-related internships or fieldwork and Federal Work-Study also available. Financial award application deadline: 4/30. *Faculty research:* Cardiovascular physiology, cancer biology, cellular and tissue proteases, cancer chemoprevention, lung development. Total annual research expenditures: $4.6 million. *Unit head:* David Grignon, Chair, 313-577-1102, Fax: 313-577-0057, E-mail: ad5033@wayne.edu. *Application contact:* Clement Diglio, Graduate Director, 313-577-1357, Fax: 313-577-0057, E-mail: cdiglio@med.wayne.edu.

Yale University, Graduate School of Arts and Sciences, Department of Experimental Pathology, New Haven, CT 06520. Offers PhD. *Degree requirements:* For doctorate, thesis/dissertation, qualifying exam. *Entrance requirements:* For doctorate, GRE General Test.

Yale University, School of Medicine and Graduate School of Arts and Sciences, Combined Program in Biological and Biomedical Sciences (BBS), Pharmacological Sciences and Molecular Medicine Track, New Haven, CT 06520. Offers PhD, MD/PhD. *Students:* 6 full-time. *Degree requirements:* For doctorate, thesis/dissertation. *Entrance requirements:* For doctorate, GRE General Test. Additional exam requirements/recommendations for international students: Required—TOEFL. *Application deadline:* For fall admission, 12/8 for domestic students, 12/8 for international students. Electronic applications accepted. *Financial support:* Fellowships, research assistantships available. *Unit head:* Dr. David F. Stern, Co-Director, 203-785-4832, Fax: 203-785-7467, E-mail: kathleen.fisher@yale.edu. *Application contact:* Kathy Fisher, Registrar, 203-785-4545, E-mail: kathleen.fisher@yale.edu.

Yale University, School of Medicine and Graduate School of Arts and Sciences, Pharmacological Sciences and Molecular Medicine Track, New Haven, CT 06520.

DUKE UNIVERSITY

Department of Pathology

Programs of Study	The Department of Pathology offers the degree of Doctor of Philosophy through the Graduate School of Arts and Sciences. The purpose of the Ph.D. program is to train individuals for careers as independent scientists investigating molecular mechanisms of disease. Flexible programs of training are available to fit individual requirements and goals. The program is open to those holding the B.A. or B.S. degree, as well as to individuals holding the M.S. or the M.D., D.V.M., D.D.S., or other doctoral degree. The Ph.D. program requires a minimum of six semesters of full-time course work and research, plus written and oral preliminary examinations, a dissertation, and a final oral examination. The average time for completion of degree requirements is four to five years. Holders of M.S. or M.D. degrees may obtain one semester of credit by transfer. Summer residence and participation in research programs throughout the course of study are required.

A core program of courses in cell and molecular biology, introductory pathology, and molecular aspects of disease is required. Other course requirements are flexible and depend on the interests and needs of the student. Course work in other departments is encouraged. |
Research Facilities	Research laboratories for the Department of Pathology are located primarily in the Medical Sciences Research Building (MSRB), a state-of-the-art research building located adjacent to Duke Hospital. Additional research laboratories are located in the Davison Building of Duke Clinic. Departmental research laboratories are fully equipped with necessary equipment and facilities to carry out research in modern cell and molecular biology, including tissue culture facilities; nucleic and amino acid sequencing; comparative genomic hybridization; flow cytometry; image analysis; immunohistochemistry; analytical and protein chemistry; and electron microscopy. Additional resources are available through the Duke University Comprehensive Cancer Center. The Clinical and Anatomic Pathology Laboratories within the Department of Pathology perform approximately 2.3 million laboratory tests, 45,000 cytologic examinations, and 350 autopsies per year, as well as collect and examine 30,000 surgical specimens. Thus, researchers within the department have ready access to tissue specimens and clinical colleagues. Informatics needs are served by a departmental library of more than 2,000 volumes, the Duke Medical Center and Duke University libraries, and an extensive departmental computer network, which is connected via fiber-optic cable to a variety of Duke servers and to the Internet.
Financial Aid	Fellowships are generally available for all accepted students. In 2005–06, fellowships consisted of a stipend of $20,800, plus tuition, fees, and health insurance.
Cost of Study	In 2005–06, yearly tuition was $25,650, plus $5630 in registration fees. A student health fee of $730 per year is required of full-time students. Tuition, fees, and health insurance are paid for fellowship recipients.
Living and Housing Costs	A limited number of on-campus apartments are available to graduate students; however, both furnished and unfurnished apartments can be rented in Durham through private citizens or real estate agencies. Prices vary considerably, and the following are approximations for apartments near the campus: one bedroom, $430 to $580; two bedroom, $550 to $700. Additional housing information and listings may be obtained from the Office of Housing Management at Duke University.
Student Group	Approximately 3,000 graduate and professional students are enrolled at Duke University, which has a total enrollment of about 9,000. The School of Medicine is an integral part of the University. About 18 graduate students are enrolled in the Department of Pathology.
Location	Durham is a city of nearly 200,000, located in the Piedmont region of North Carolina, with easy access to both seacoast and mountains. It is in the Research Triangle, within 10 to 15 miles of North Carolina State University, the University of North Carolina, and North Carolina Central University, all of which cooperate in teaching and research programs. It is close to a large number of government and private research institutes that enrich the scientific community. Duke University is situated at the southwest fringe of Durham on a tract of about 8,500 acres of rolling, wooded land.
The University	Duke University was founded in 1924. It was developed from Trinity College, a small liberal arts college with a history going back to 1839. From its beginning, Duke has had a substantial endowment, and it carries on a correspondingly ambitious program of research, instruction, and service.
Applying	Students are admitted in the fall. It is mandatory to complete the application process by December 31 for admission the following September. In unusual circumstances, candidates are considered for spring admission; however, it is not possible to grant financial aid to students who are admitted in the spring or summer. Applicants are favored who have good undergraduate backgrounds not only in biology but also in physics, mathematics, and chemistry. Scores on the GRE General Test are required; scores on an appropriate GRE Subject Test are recommended but not required. Students for whom English is not the native language must submit results of the TOEFL. In the case of applicants with advanced degrees (e.g., M.D., D.V.M., D.O., or D.D.S.), the department may petition the Graduate School to accept scores from other tests, such as the Medical College Admission Test or the Veterinary Aptitude Test, in lieu of the Graduate Record Examinations.
Correspondence and Information	Dr. Soman N. Abraham Director of Graduate Studies Department of Pathology Duke University Medical Center DUMC 3020 Durham, North Carolina 27710 Phone: 919-684-3630 E-mail: pathgrad@mc.duke.edu Web site: http://www.pathology.mc.duke.edu

Duke University

THE FACULTY AND THEIR RESEARCH

Experimental pathologists at Duke have their primary focus in the following three areas of research:

Inflammation research is aimed at investigating episodes that occur when white blood cells and their products interact with host cells and tissues, infectious agents, or environmental toxicants. Emphasis is placed on the events that initiate and regulate the inflammatory response and diseases that result from aberrant regulation.

Tumor biology involves analysis of neoplasia at all levels of involvement: molecular, nuclear, cell membrane, intracellular, and matrix. Current research includes the identification and analysis of neoplastic events and the application of novel therapies, including immunotherapy.

Vascular biology seeks to understand processes occurring at the interface between the vessel wall and blood. This includes studies of cell-cell, cell-matrix, and protein-cell interactions, particularly involving receptor-mediated events. Emphasis is also placed on the structure and function of blood proteins involved in these processes.

Inflammation
Soman Abraham, Professor; Ph.D., Newcastle, 1981. Bacterial interactions with inflammatory cells; mast cell biology.
Virginia Byers Kraus, Associate Professor; M.D., 1982, Ph.D., 1993, Duke. Molecular pathogenesis of arthritis and cartilage degradation.
Jeffrey H. Lawson, Assistant Professor; M.D., 1991, Ph.D., 1992, Vermont. Vascular surgery.
Salvatore Pizzo, Professor and Chairman of the Department; M.D./Ph.D., Duke, 1973. Biochemistry of proteinases and proteinase inhibitors.
Herman F. Staats, Assistant Research Professor; Ph.D., South Alabama. Vaccine strategies for mucosal immunity.
Mary E. Sunday, Professor; M.D./Ph.D., Harvard, 1982. Lung developmental biology; neuropeptides as mediators of altered immunity and lung injury.

Tumor Biology
Darell D. Bigner, Professor; M.D., 1965, Ph.D., 1971, Duke. Neuro-oncology; tumor immunology.
Henry S. Friedman, Assistant Professor; M.D., SUNY Upstate Medical Center, 1977. Tumor pharmacology; mechanisms of drug resistance.
Laura P. Hale, Assistant Research Professor; Ph.D., 1990, M.D., 1991, Duke. Immunopathology of breast cancer; immunotherapy.
Randy L. Jirtle, Associate Professor; Ph.D., Wisconsin, 1976. Liver carcinogenesis; radiation biology; genomic imprinting.
Jeffrey R. Marks, Assistant Professor; Ph.D., California, San Diego, 1985. Molecular genetics of breast cancer.
Hai Yan, Assistant Professor, Director of Molecular Oncogenomics Laboratory; M.D., Beijing Medical, 1991; Ph.D., Columbia, 1996. Molecular genetics of brain tumor.
Michael R. Zalutsky, Associate Professor; Ph.D., Washington (St. Louis), 1974. Radiation biology; nuclear chemistry.

Vascular Biology
Mark W. Dewhirst, Professor; D.V.M., 1975, Ph.D., 1979, Colorado State. Radiation biology; hyperthermia; tumor hypoxia.
Charles S. Greenberg, Professor; M.D., Hahnemann, 1976. Transglutaminase function; angiogenesis.
Maureane R. Hoffman, Professor; M.D./Ph.D., Iowa, 1982. Coagulation; platelet activation.
William H. Kane, Associate Professor; M.D./Ph.D., Washington (St. Louis), 1982. Molecular biology of coagulation factors.
Daniel J. Kenan, Assistant Professor; M.D./Ph.D., Duke, 1995. Combinatorial technologies for molecular phenotyping and recognition.
Thomas L. Ortel, Associate Professor; Ph.D., 1983, M.D., 1985, Indiana. Molecular mechanisms in hemorrhage and thrombosis.
Marilyn J. Telen, Associate Professor; M.D., NYU, 1977. Erythrocyte and tumor adhesion molecules; pathogenesis of sickle cell disease.
Miriam L. Wahl, Assistant Professor; Ph.D., Cincinnati, 1990. Manipulation of tumor intracellular pH and angiogenesis to potentiate cancer therapy.

Multidisciplinary
Gordon K. Klintworth, Professor; M.D., 1957, Ph.D., 1966, Witwatersrand (South Africa). Ophthalmic pathology and neuropathology.
Alan Proia, Professor; Ph.D., Rockefeller, 1979; M.D., Cornell, 1980. Ophthalmic pathology.
John Shelburne, Professor; Ph.D., 1971, M.D., 1972, Duke. Cell biology; electron microscopy.
Robin T. Vollmer, Associate Clinical Professor; M.D., Duke, 1967. Biomathematics; surgical pathology.

NEW YORK MEDICAL COLLEGE

Department of Pathology

Programs of Study

The Department of Pathology offers a vigorous multidisciplinary milieu for training that leads to the M.S. and Ph.D. degrees in experimental pathology. The programs focus on the comprehensive study of pathogenic mechanisms of human disease. This means that a multidisciplinary approach encompassing cellular and molecular biology, biochemistry, physiology, and immunology is essential for understanding the molecular basis of disease and its ultimate prevention and cure. Increasing awareness of environmental degradation has led to an increased need for people with training in environmental pathology in industry and government at both the master's and doctoral levels. A series of departmental courses examines the biological effects of environmental pollutants and their role in the pathogenesis of environmentally caused disease. These courses give students hands-on experience in environmental pathology and form the core of the program that leads to an M.S. degree in experimental pathology with an emphasis on environmental pathology or experimental pathology. A third track adds courses and practical experience related to evaluation and management of environmental pathogenic factors and leads to an M.S. degree in experimental pathology with an emphasis on environmental science and health. The Department also offers an advanced program leading to an M.S. in experimental pathology with an emphasis on toxicologic pathology for those already holding a doctoral degree (M.D., Ph.D., D.V.M., or equivalent) who require specialized training in this area. Admission to this program is strictly limited and highly competitive.

Graduates with an M.S. in experimental pathology are prepared for research, technical, and supervisory positions in scientific laboratories in academia, industry, and government. Completion of an M.S. degree in experimental pathology with an emphasis on environmental pathology or environmental science and health can be expected to take at least two years. An M.S. in experimental pathology with an emphasis on toxicologic pathology can be completed in one year of intense work. Graduates with a Ph.D. in experimental pathology are prepared to undertake careers as research scientists and teachers in academia, industry, and government. Students in the Ph.D. program are introduced to original laboratory research in the wide range of departmental research laboratories at the earliest stages of their training. Attainment of the Ph.D. requires formal course work, passing a comprehensive qualifying examination, and successfully defending a dissertation based on original research. Completion of a Ph.D. in experimental pathology can be expected to take five to six years.

Special areas of interest in the Department include biochemical toxicology, cell-cycle progression and apoptosis, chemical carcinogenesis, chronic mycobacterial and borrelial infectious disease, free-radical pathobiology, hypersensitivity and chronic inflammation, molecular genetics of hypertension, tissue engineering, tumor cell biology, and vascular, traumatic, and neoplastic neurologic disease.

Research Facilities

Departmental research facilities are located within the Basic Science Building and the neighboring Brander Cancer Research Institute and the Institute for Cancer Prevention. These facilities are fully equipped for research training in flow cytometry; biochemical, metabolic, and physiological studies; molecular genetics; tissue culture; isotope techniques; cell fractionation; and light, confocal, and electron microscopy and are supplemented by collaborative arrangements with faculty members of other basic science and clinical departments of the College. The Basic Science Building also houses the main College library of more than 180,000 volumes and the College's animal-care facilities.

Financial Aid

Graduate students admitted to the Ph.D. program normally receive stipends of $23,000 and tuition exemption. The minimum value of this support for Ph.D. students in 2005–06 was $40,320.

Cost of Study

Tuition for 2005–06 was approximately $590 per credit hour.

Living and Housing Costs

A limited number of one- and two-bedroom garden apartments are available on the Westchester Campus, and the College has a recently renovated dormitory. Students are advised to contact the Director of Housing, Office of Student Affairs, located in Sunshine Cottage, well in advance of their arrival in order to make arrangements for the rental of College or off-campus housing.

Student Group

At present, there are 12 M.S. students and 5 Ph.D. students in the graduate program in experimental pathology. There are also 7 postdoctoral research fellows associated with the program's laboratories. Graduate student enrollment in the six basic science departments at New York Medical College currently numbers 250. Approximately 360 first- and second-year medical students are also on campus during the academic year.

Location

New York Medical College is part of the Westchester Medical Center and is located in a scenic area of Westchester County, 28 miles north of New York City. This area also provides easy access to the abundant cultural and recreational resources of New York City and the Northeastern states.

The College

New York Medical College was chartered in 1860 by the New York State legislature. It has served the community continuously since that time as a center for medical education. The Graduate School of Basic Medical Sciences was established in 1963 as a unit of the College. The graduate program in the Department of Experimental Pathology was founded in 1971.

Applying

Prospective students should obtain application forms and apply as early as possible. Undergraduate preparation should include basic courses in biology, organic chemistry, mathematics, and physics. Applicants to the Ph.D. program require a bachelor's degree from an accredited college or university, preferably with a science major, and scores on the GRE General Test. Applicants from non-English-speaking countries must also submit scores on the Test of English as a Foreign Language (TOEFL).

Correspondence and Information

Henry P. Godfrey, M.D., Ph.D.
Ph.D. Graduate Program Director
Department of Pathology
New York Medical College
Valhalla, New York 10595-1690

Phone: 914-594-4160
Fax: 914-594-4163
E-mail: hgodfrey@nymc.edu
Web site: http://www.nymc.edu/pathology

Fred H. Moy, Ph.D.
M.S. Graduate Program Director
Department of Pathology
New York Medical College
Valhalla, New York 10595-1690

Phone: 914-594-4174
Fax: 914-594-4163
E-mail: moy@nymc.edu

New York Medical College

THE FACULTY AND THEIR RESEARCH

Iradge Argani, Professor and Interim Chairman; M.D., Tehran (Iran), 1955. Immunohematology and transplant immunology.

Praveen M. Chander, Associate Professor; M.D., All-India Institute of Medical Sciences, 1970. Pathogenesis of renal and vascular damage in stroke-prone spontaneously hypertensive rats; pathogenesis of HIV-associated nephropathy.

Wei Dai, Professor and Director, Molecular Carcinogenesis Division; Ph.D., Purdue, 1988. Cell-cycle regulation; checkpoint control; protein kinases.

Zbigniew Darzynikiewicz, Professor; M.D., 1960, Ph.D., 1966, Warsaw (Poland). Development of new methods of cell analysis applicable to flow cytometry; analysis of cell-cycle specificity of antitumor drugs.

Henry P. Godfrey, Professor and Ph.D. Graduate Program Director; M.D., Harvard, 1965; Ph.D., Birmingham (U.K.), 1980. Pathophysiology of tuberculosis and Lyme disease; diagnosis of chronic bacterial disease; biochemical mechanisms of delayed hypersensitivity.

Michael Iatropoulos, Research Professor; M.D., 1964, Ph.D., 1965, Tübingen (Germany). Comparative mechanisms of toxicity and carcinogenesis.

Alan M. Jeffrey, Research Professor; Ph.D., North Wales (U.K.), 1970. Toxicology and chemical carcinogenesis.

Meena Jhanwar-Uniyal, Clinical Associate Professor; Ph.D., Moscow, 1979. Involvement of tumor suppressor genes in tumorigenesis and metastasis; signal transduction pathways leading to cell-cycle regulation, differentiation, and apoptosis.

Ashok Kumar, Professor; Ph.D., Allahbad (India), 1963. Molecular genetics of renin-angiotensin system in hypertension and atherosclerosis.

Ellen M. Levee, Assistant Professor; Director, Department of Comparative Medicine; and Program Director, Fundamentals of Animal Research; D.V.M., Chihuahua (Mexico), 1985. Laboratory animal health care; analgesia in pigs and birds.

Jane H.-C. Lin, Assistant Professor; Ph.D., Illinois, 1986. Molecular mechanisms governing cell dysfunctions after neurologic trauma.

Paul A. Lucas, Associate Professor; Ph.D., Minnesota, 1981. Wound healing and tissue engineering.

Myron R. Melamed, Professor; M.D., Cincinnati, 1950. Flow and static cytometry of human cancer cells in conjunction with cytochemical, immunochemical, and in situ nucleic acid reactions for diagnostic and prognostic purposes.

Fred H. Moy, Associate Professor and M.S. Graduate Program Director; Ph.D., Sussex (U.K.), 1986. Biostatistics and epidemiology; methodology and applications in environmetrics; lead exposure; risk assessment; reproductive epidemiology; biomarkers and molecular epidemiology; evaluation and assessment in health sciences.

Frank Traganos, Professor; Ph.D., Cornell, 1979. Cell biology: study of mechanisms involved in control of cell-cycle progression (check points) and cell death (apoptosis) in model systems (cell cultures) and clinical material.

John H. Weisburger, Research Professor; Ph.D., Cincinnati, 1949; M.D. (hon.), Umeå (Sweden), 1980. Mechanisms of toxicity and carcinogenicity; mechanisms and role of promoters in major human cancers; role of nutrition in human carcinogenesis; rational means of prevention of cancer, coronary heart disease, and stroke.

Gary M. Williams, Research Professor; M.D., Pittsburgh, 1967. Mechanisms of carcinogenesis; metabolic and genetic effects of chemical carcinogens.

Reinhard E. Zachrau, Professor; M.D., Heidelberg (Germany), 1964. Spontaneous and induced tumor-specific, cell-mediated immunity in human breast cancer and its role in development of systemic metastasis and second primary cancers of breast and nonbreast origin.

UNIVERSITY OF CALIFORNIA, SAN DIEGO

Graduate Program in Molecular Pathology
USCD and the Burnham Institute

Program of Study

The Molecular Pathology Ph.D. program at the University of California, San Diego (UCSD), provides training in the biochemistry and molecular biology of human disease for students with undergraduate majors in biochemistry, molecular biology, and genetics. It is an interdepartmental and interinstitutional program jointly sponsored by the UCSD Department of Pathology and the Burnham Institute. The goal of the program is to prepare students for independent research careers in academia or the biotechnology sector, using a cooperative effort of academic and institutional faculty members in San Diego who are specifically interested in human disease. Faculty research is focused on human disease and is divided among six groups. The Cancer Biology group has combined faculty members from the independent National Cancer Centers at UCSD and the Burnham Institute. Their interests are divided among signal transduction, differentiation, cell death pathways, metastasis, and DNA-repair. The Stem Cell and Developmental Biology group focuses on the cellular and genetic events of differentiation and how the knowledge of such systems may be applied in the context of cellular and molecular therapeutics. The Neurobiology and Neurologic Disease group studies brain development, neurologic diseases, and pain research. The Structural Biology and Signal Transduction group investigates how protein-protein interactions and protein modifications control normal and disease processes. The Microbiology and Immunology group focuses on the pathogenic mechanisms of viruses, bacteria, and protozoa as well as the innate and acquired immune responses to infection. The Cardiovascular, Muscle, and Organ group is interested in diseases of the heart and skeletal muscles, the circulatory system, and the kidney. Faculty members reside at the UCSD School of Medicine, the Burnham Institute, the Scripps Research Institute, and the Salk Institute.

The goal of the program's academic training is to teach students how to identify important scientific questions and to solve them using modern research approaches. A molecular cell biology class focuses on recent advances and open questions in important areas of cell biology, focusing on those altered in human disease. A methods class describes the theory and application of research techniques to answering molecular questions underlying human disease. Three courses highlight known mechanisms and open questions in cancer, neurologic and muscle disease, and infectious disease. Lectures are accompanied by an original research paper exemplifying how to identify key questions, design good approaches, and use appropriate methods to derive clear answers. Educational tracks permit choice of a broad number of electives offered by any UCSD graduate program. To provide an understanding of how disease disrupts the structure and function of tissues and organs, molecular pathology students take medical school classes in histology and pathology. A course in mouse models of human disease empowers students to manipulate the mouse genome, to test how expression of genes thought to underlie human diseases alters the normal mouse biology, and to evaluate mutant phenotypes using their molecular biology and histology/pathology training. Students select a research laboratory based on three 6-week rotations in faculty laboratories. After their first year of classes and rotations, students pursue thesis research. Undergraduate teaching is optional but not required. The Annual Spring Research Symposium is held at Scripps Institute of Oceanography and provides a forum for students to present their recent work, discuss their ideas and hypotheses, and engage in some friendly scientific debate.

Research Facilities

UCSD research laboratories are located at the main campus in La Jolla and at the Medical Center in Hillcrest. The Burnham Institute is ½ mile north of UCSD. Both UCSD and Burnham locations have active seminar series and offer core technology facilities for imaging, protein mass spectrometry, and the production of transgenic or knockout mice.

Financial Aid

Students in good standing receive full support (tuition, fees, health insurance, plus a $25,000 stipend) throughout their training.

Cost of Study

As explained above, all graduate students receive full financial support.

Living and Housing Costs

A wide range of off-campus housing, which is listed in the Off-Campus Housing Office, is available in coastal communities within a 10- to 15-minute drive of the campus. On-campus housing is limited, and the waiting period ranges from two to three years.

Student Group

There are 17,000 undergraduate and 2,400 graduate students at UCSD, 51 of whom are in the Program in Molecular Pathology.

Location

UCSD is located in La Jolla, 10 miles north of San Diego and 2 miles from the ocean. The climate is one of the best in the United States, well suited for the recreational opportunities provided by the Pacific Ocean, the Laguna Mountains, and the Anza-Borrego desert.

The University

UCSD consistently ranks among the six best-funded research universities, and the broader scientific environment, which includes the Burnham, Salk, and Scripps Institutes, makes the La Jolla research community one of the best in the world.

Applying

Applications must be received by January 4. Applicants must be well grounded in biology and biochemistry, having undergraduate majors in biology, biochemistry, genetics, microbiology, or a similar field. GRE General and Subject Tests are required; however, the MCAT can replace the GRE. The admissions committee considers the composition of undergraduate courses, grade point averages, letters of recommendation, past research experiences, and GRE scores in selection of applicants. The ability to express oneself clearly in both oral and written English is essential. TOEFL scores are required when appropriate.

Correspondence and Information

Molecular Pathology Graduate Program
University of California, San Diego
9500 Gilman Drive, Mail Code 0612
La Jolla, California 92093-0612

Phone: 858-534-4324
E-mail: molpath@ucsd.edu
Web site: http://medicine.ucsd.edu/molpath/

University of California, San Diego

THE FACULTY AND THEIR RESEARCH

Robert Abraham, Ph.D., Professor, the Burnham Institute. Cell growth and signaling through DNA damage-induced cell-cycle checkpoints.
Katarina Akassoglou, Ph.D., Professor of Medicine. Autoimmune CNS disease; multiple sclerosis; new mouse models for multiple sclerosis.
Roland Blantz, M.D., Professor of Medicine. Regulation of glomerular and tubular function in the kidney; renal disease mechanisms.
Rolf Bodmer, Ph.D., Professor, the Burnham Institute. Cardiac development regulation by Notch, Polycomb, Hox, and GAT factors.
Ella Bossy-Wetsel, Ph.D., Assistant Professor, the Burnham Institute. Molecular biology of stroke.
Nigel Calcutt, Ph.D., Assistant Professor of Pathology. Peripheral nerve biochemistry; function and structure in diabetes.
Jue Chen, Ph.D., Assistant Professor of Medicine. Molecular bases of cardiac and skeletal muscle disease.
Kenneth Chien, M.D., Professor of Medicine. Molecular biology of cardiac hypertrophy.
Jacques Corbeil, Ph.D., Associate Adjunct Professor of Medicine. Genomic analysis of host-pathogen interactions; bioinformatics; HIV.
Lynette Corbeil, D.V.M., Ph.D., Adjunct Professor of Pathology. Immunity to natural and recombinant microbial antigens.
Marcia Dawson, Ph.D., Professor, the Burnham Institute. Drug discovery; bioorganic chemistry.
Jack Dixon, Ph.D., Professor of Pharmacology. Role of phosphatases in host-pathogen interactions and in cancer.
Danial Donoghue, Ph.D., Professor of Chemistry and Biochemistry. Tyrosine protein kinase biochemistry and function in diseases.
Kathryn R. Ely, Ph.D., Adjunct Professor of Pathology, the Burnham Institute. Protein crystallographic studies of macromolecular interactions.
Eva Engvall, Ph.D., Adjunct Professor of Pathology, the Burnham Institute. Basement membrane structure and function in development; cancer.
Sylvia Evans, Ph.D., Associate Adjunct Professor of Medicine. Heart lineage specification; stem cells; congenital heart disease.
Marilyn Farquhar, Ph.D., Professor of Pathology. Cell signaling and protein trafficking; cellular and molecular mechanisms of renal disease.
Gen-Sheng Feng, Ph.D., Adjunct Professor of Pathology, the Burnham Institute. Signaling by tyrosine phosphatases and adapter/scaffold proteins.
Joshua Fierer, M.D., Professor of Medicine. Polymorphonuclear leukocytes and innate immunity to *Salmonella* infections in mice.
Hudson Freeze, Ph.D., Adjunct Professor of Pathology, the Burnham Institute. Glycosylation defects; muscular dystrophy; Crohn's disease.
Stephen Frisch, Ph.D., Associate Adjunct Professor of Pathology. Tumor suppression, cell differentiation, and apoptosis.
Xiang-Dong Fu, Ph.D., Professor of Cellular and Molecular Medicine. Regulation of mRNA splicing in cell type-specificity; heart development.
Minoru Fukuda, Ph.D., Adjunct Professor of Pathology, the Burnham Institute. Carbohydrate-dependent cell recognition and tumor metastasis.
Fred Gage, Ph.D., Professor of Biology. Neuronal plasticity; neuronal replacement in Alzheimer's and spinal cord repair.
Richard Gallo, M.D., Ph.D., Associate Professor of Medicine. Antimicrobial defense mechanisms; innate immunity in skin and saliva.
Frances D. Gillin, Ph.D., Adjunct Professor of Pathology. Molecular and cell biology of host-parasite interactions.
Chris Glass, M.D., Ph.D., Professor of Medicine. Molecular biology of atherosclerosis and myeloid cell development.
Steve Gonias, M.D., Ph.D., Professor of Pathology. Regulation of cell physiology by proteases and their cell-surface receptors.
Roberta Gotleib, Ph.D., the Scripps Research Institute. Cardiomyocyte response to hypoxia; cardiomyocyte apoptosis; ischemic heart disease.
John Guatelli, M.D., Associate Professor of Medicine. Nef protein function in HIV infection.
Martin Haas, Ph.D., Professor of Biology. Stem cell differentiation; stem cell survival factors in amniotic fluid.
Dorit Hanein, Ph.D., Assistant Professor, the Burnham Institute. Modeling of cell contact and cell migration structures using high-resolution EM.
Steffan Ho, Ph.D., Assistant Professor of Pathology. Role of NFAT transcription factors in T-cell differentiation and in mature T-cell function.
Paul Insel, M.D., Ph.D., Professor of Pharmacology. G-protein-coupled receptors and G-protein-mediated signaling in health and disease.
Martin F. Kagnoff, M.D., Professor of Medicine. Host mucosal responses to microbial infection; inflammatory bowel disease and celiac disease.
Mark P. Kamps, Ph.D., Associate Professor of Pathology. Molecular mechanisms of oncogenes that cause leukemia.
Michael Karin, Ph.D., Professor of Pharmacology. Signal transduction in cancer and inflammation; microbial pathogenesis mechanisms.
Edward Koo, M.D., Ph.D., Professor of Neuroscience. Cellular mechanisms of neurodegenerative diseases, especially Alzheimer's disease.
Fred Levine, Ph.D., Professor of Medicine, the Burnham Institute. Pancreatic beta-cell development; stem cell applications in diabetes.
Robert Liddington, Ph.D., Professor, the Burnham Institute. Structural basis of integrin-mediated signaling; structure of virulence factors.
Stuart Lipton, M.D., Ph.D., Adjunct Professor of Pathology, the Burnham Institute. Neurodegeneration in stroke, dementia, and AIDS.
Paul Martin, Ph.D., Assistant Professor of Medicine. Glycosylation in muscular dystrophy; inclusion body myopathy; Alzheimer's.
Eliezer Masliah, M.D., Associate Professor of Pathology. Neurodegeneration in Alzheimer's, Parkinson's, and HIV.
Mark Mercola, Ph.D., Associate Professor of Pathology. Cardiac stem cells in development and regeneration; TGF-beta and Wnt signaling.
Andrew Mizisin, Ph.D., Assistant Professor of Pathology. Pathophysiology of toxic, metabolic, and immune-mediated neuropathies.
Tomas Mustelin, M.D., Ph.D., Professor, the Burnham Institute. Regulation of growth and survival by tyrosine protein phosphatases; leukemia.
Victor Nizet, M.D., Associate Professor of Pediatrics. Streptococcal virulence mechanisms; bacterial infections; meningitis; toxic shock.
Robert G. Oshima, Ph.D., Adjunct Professor of Pathology, the Burnham Institute. Ets transcription factors; tumor models; placenta development.
Elena B. Pasquale, Ph.D., Associate Adjunct Professor of Pathology, the Burnham Institute. Receptor tyrosine kinases of the Eph family in development and disease.
Manuel Perucho, Ph.D., Adjunct Professor of Pathology, the Burnham Institute. DNA-repair systems in normal cells and their mutation in cancer.
Henry C. Powell, M.D., Professor of Pathology. Diagnostic electron microscopy; peripheral and demyelinating neuropathies.
James Quigley, Ph.D., Adjunct Professor of Pathology, the Scripps Research Foundation. Metalloproteinases in tumor invasion.
John C. Reed, M.D., Ph.D., Adjunct Professor of Pathology, the Burnham Institute. Apoptosis regulation by Bc12 family proteins.
Sharon Reed, M.D., Professor of Pathology. Virulence mechanisms of the protozoan parasites, *Entamoeba histolytica* and *Toxoplasma gondii*.
Bing Ren, Ph.D., Assistant Professor of Cellular and Molecular Medicine. Genetic targets of oncoproteins and developmental signal pathways.
Douglas Richman, M.D., Professor of Pathology and Medicine. HIV neutralization; immune evasion.
David W. Rose, Ph.D., Associate Adjunct Professor of Medicine. Signal transduction; nuclear receptors; coactivator and corepressor exchange.
Michael G. Rosenfeld, M.D., Professor of Medicine. Transcription regulation of brain development; coactivator and corepressor mechanisms.
Guy S. Salvesen, Ph.D., Associate Adjunct Professor of Pathology, the Burnham Institute. Apoptosis; regulation of caspase proteolytic activity.
Diane Shelton, D.V.M., Adjunct Professor of Pathology. Cat and dog models of muscular dystrophy and other neuromuscular diseases.
Evan Snyder, M.D., Ph.D., Professor, the Burnham Institute. Stem cells and regenerative biology; applications for neurodegenerative disease.
Deborah H. Spector, Ph.D., Professor of Biology. Molecular biology and pathogenesis of Human Cytomegliovirus (HCMV).
David Tarin, M.D., Ph.D., Professor of Pathology. Tumor invasion and metastasis; molecular genetic programs driving metastatic spread.
Bruce Torbet, Ph.D., the Scripps Clinic and Research Foundation. Myeloid differentiation and HIV protease function.
Ajit Varki, M.D., Professor of Medicine. Sialic-acid-binding proteins in innate immunity and tumor metastasis.
Joe Vinetz, Ph.D., Associate Professor of Medicine. Strategies of blocking malaria transmission; mechanisms of leptospirosis pathogenesis.
Gernot Walter, Ph.D., Professor of Pathology. Biochemistry of protein phosphatase PP2a, a central enzyme in growth regulation/oncogenesis.
Ian Wilson, Ph.D., Adjunct Professor of Pathology, the Scripps Clinic and Research Foundation. Crystallography of cancer targets, HIV proteins, and immunoreceptors.
Tony Yaksh, Ph.D., Professor of Anesthesiology and Pharmacology. Mechanisms of pain and pain modulation; MAP kinases in pain signaling.
Dongxian Zhang, Ph.D., Assistant Professor, the Burnham Institute. Ligand-gated receptors in brain and neurodegenerative disease.

UNIVERSITY OF MICHIGAN

Department of Pathology

Program of Study

The Department of Pathology offers the Ph.D. degree through the Horace H. Rackham School of Graduate Studies. A special Ph.D./M.D. program is also open to students in the Medical Scientist Training Program at the University of Michigan. The primary goal of the doctorate in pathology program is to train individuals for careers as independent scientific investigators, with a focus on the study of the molecular and cellular mechanisms of disease processes. The course programs are formulated to meet the needs of individual students in consultation with their academic adviser and with approval of the Pathology Graduate Program Committee. Courses are chosen to provide each student with a background in basic areas of biochemistry, cell biology, immunology, and genetics in preparation for in-depth study of the cellular and molecular pathogenesis of disease. Additional course work depends on individual trainees' area of research specialization and may include courses in other basic science departments in the University. Students select a thesis adviser from the pathology faculty to guide their dissertation research. The research interests of the faculty are diverse and include investigative programs in tissue injury and repair, inflammation, aging, tumor biology, apoptosis, regulation of gene expression in disease processes, and the biology and pathobiology of cytokines, adhesion molecules, and extracellular matrix. An active program in molecular biology/genetics is present in the Department of Pathology in conjunction with the University of Michigan Howard Hughes Medical Institute for molecular genetics. Two National Institutes of Health training grants in lung immunopathology and experimental immunopathology support 4 graduate students and approximately 40 postdoctoral fellows in the department.

Research Facilities

The Department of Pathology is part of the University of Michigan Medical Center, which includes a hospital complex with a capacity of nearly 900 beds. The clinical activities of the department occupy approximately 40,000 gross square feet and the University Hospital, while the experimental programs are located in adjacent buildings, comprising approximately 23,000 square feet of recently renovated, modern, well-equipped laboratories. Several faculty members have appointments in the Howard Hughes Medical Institute for molecular genetics. In addition, diverse and excellent resources are available in other departments, both on the medical and adjacent main University campus. The medical campus maintains an outstanding medical library, and the University library system is among the largest in the nation.

Financial Aid

Financial aid for Ph.D students consists of research assistantships provided by the department and faculty research grants. This support covers tuition costs and includes a monthly living stipend.

Cost of Study

The estimated cost of tuition is $7013 per semester for Michigan residents and $14,190 per semester for nonresidents. Tuition is paid by the department for most students accepted into the program.

Living and Housing Costs

The University of Michigan provides a variety of on-campus housing for both single and married graduate students. However, many graduate students live in apartments or rooms in private homes. The cost of a one-bedroom unfurnished apartment close to campus is approximately $600 to $900 per month. Two University-operated apartment developments are available for married students; the approximate costs are $585 per month for efficiencies, $690 per month for one-bedroom apartments, $775 to $800 per month for two-bedroom apartments, and $992 to $1004 for three-bedroom apartments. Additional information is available on the University Housing Office's Web site at http://www.housing.umich.edu/info/.

Student Group

The University of Michigan's Ann Arbor campus enrolls approximately 35,000 students, including nearly 9,300 graduate students and 3,500 professional students. The majority of the graduate students are from out of state or from other countries, reflecting the cosmopolitan character of the University and of Ann Arbor.

Location

Located in southeastern Michigan, an hour's drive from Detroit, Ann Arbor combines the comfort and charm of small-city life with the vitality of a large academic community. As a home to people of diverse ethnic backgrounds, professions, and tastes, the city has something to offer everyone, including first-rate theater, concerts, numerous film societies, and Big Ten sports. Science and art museums feature special exhibits each year. Ann Arbor's summer street fairs have gained national recognition. Numerous parks and recreation facilities provide exercise and relaxation. Cross-country skiing in the winter and bicycling in the summer are popular. The Huron River, which flows through the city, offers opportunities for sailing, windsurfing, and canoeing. Detroit Metropolitan Airport, 30 minutes away, links Ann Arbor with U.S. and international cities. By car, Canada is only 1 hour away and Chicago, 4 hours.

The University

The University of Michigan emphasizes graduate and professional programs and has traditionally played a central role in American academic life.

Applying

Applications for fall term admission should be submitted to the Graduate Program in Biomedical Sciences (PIBS) by December 31. Students are not normally admitted for winter or summer terms. A bachelor's degree is required for admission, and a strong background in physical and biological sciences is recommended. The Graduate Record Examinations (the General Test and the Subject Test in biology or a physical science) and at least three letters of reference are required of all applicants. Detailed information on the department's faculty members, their research, and the University is available upon request and at the program's Web site, listed in the Correspondence and Information section.

Correspondence and Information

For program information:

Dr. Nicholas Lukacs
Graduate Program Committee
Department of Pathology
University of Michigan Medical School
4211 Medical Science Building I, 1301 Catherine Road
Ann Arbor, Michigan 48109-0602
Phone: 734-763-6454
Fax: 734-763-6476
E-mail: pathgradprog@path.med.umich.edu
Web site: http://www.pathology.med.umich.edu/gradprogram/
index.htm

For application materials and information:

Graduate Program in Biomedical Sciences (PIBS)
University of Michigan Medical School
1150 West Medical Center Drive
Ann Arbor, Michigan 48109-0619
Phone: 734-647-7005
877-294-0120 (toll-free)
Fax: 734-647-7022
E-mail: pibs@umich.edu
Web site: http://www.med.umich.edu/pibs/

University of Michigan

THE FACULTY AND THEIR RESEARCH

Anuska V. Andjelkovic-Zochowska, Assistant Professor; M.D., Ph.D., Belgrade, 1999. Molecular mechanism of leukocytes trafficking at the blood/brain barrier; chemokines and brain tumor metastasis.

James R. Baker Jr., Professor; M.D., Loyola Chicago, 1978. Autoimmune endocrine disease; nanotechnology for drug delivery; gene transfer and functional imaging.

Stephen W. Chensue, Professor; Ph.D., Wayne State, 1979; M.D., Michigan, 1983. Cellular and molecular mechanisms of leukocyte mobilization during innate and adaptive T-cell-mediated lung inflammation.

Arul Chinnaiyan, Associate Professor; M.D./Ph.D., Michigan, 1999. Using bioinformatics, integrative approaches, genomics, and proteomics to study cancer.

Kathleen R. Cho, Professor; M.D., Vanderbilt, 1984. Molecular pathogenesis of gynecological tumors.

Gregory Dressler, Associate Professor; Ph.D., Pennsylvania. Genetic basis of cellular differentiation and pattern formation in a complex multicellular tissue.

Colin S. Duckett, Assistant Professor and Associate Director, Molecular Mechanisms of Disease Program; Ph.D., London, 1994. Apoptosis and its deregulation in cancer and immunological disorders; intracellular signaling pathways regulating the activation of the transcription factor, nuclear factor kappa β (NF-$\kappa\beta$); analysis of inhibitor of apoptosis (IAP) protein function.

Barry G. England, Associate Professor; Ph.D., Wisconsin, 1971. Cellular and molecular regulation of the development and maturation of ovarian follicles; effects of stress on fertility and early fetal development.

Eric Fearon, Professor; M.D./Ph.D., Johns Hopkins, 1990. Functional analysis of Wnt and CDX2 pathways in normal and neoplastic cells; mouse gastrointestinal cancer models.

Jason E. Gestwicki, Assistant Professor; Ph.D., Wisconsin–Madison, 2002. Chemical biology; small molecule inhibitors of protein-protein interactions; controlling subcellular location of proteins with molecular machines; developing new therapeutic strategies.

Jay L. Hess, Carl V. Weller Professor and Chair; M.D., Ph.D., John Hopkins, 1989, Mechanisms of transcriptional regulation and transformation by MLL and HOX proteins.

Cory Hogaboam, Associate Professor; Ph.D., Calgary, 1993. Cytokine and chemokine involvement in chronic inflammatory and remodeling diseases of the lung.

Evan T. Keller, Professor; Ph.D., Wisconsin–Madison, 1996. Cancer metastasis; aging.

Celina G. Kleer, Associate Professor; M.D., Buenos Aires, 1993. Discovery and validation of novel tissue biomarkers in breast cancer and their molecular mechanism of action.

Steven L. Kunkel, Professor; Ph.D., Kansas, 1978. Regulation of cytokine gene expression; macrophage pathobiology.

Andrew P. Lieberman, Assistant Professor; M.D./Ph.D., Maryland, 1993. Study of hereditary neurological diseases, with particular interest in the polyglutamine expansion diseases and Niemann-Pick C.

Nicholas W. Lukacs, Associate Professor; Ph.D., Wayne State, 1992. Cytokine/chemokine responses associated with allergic airway responses (asthma), respiratory viral disease, and T-lymphocyte-mediated pulmonary inflammation.

Dervla Mellerick-Dressler, Assistant Professor; Ph.D., Pennsylvania, 1986. Molecular basis for cell-fate specification during central nervous system development; how the homeobox gene, vnd, and two of its transcriptional targets, Nk6 and ind, compartmentalize neuronal precursor cells, resulting in generation of distinct neuronal progeny; use of transgenic, cell biological, and molecular approaches to dissect regulatory activity of these homeodomain proteins in specifying neural stem cell identity.

Richard A. Miller, Professor; Ph.D., 1976, M.D. 1977, Yale. Aging T-cell activation; genetics of aging; long-lived mutant mice.

Hedwig Murphy, Assistant Professor; Ph.D., 1979, M.D., 1990, Wayne State. Role of vascular endothelial cells and dendritic cells in inflammation and autoimmunity; cellular and molecular mechanisms of hormonal modulation of inflammation and autoimmunity.

Gabriel Nunez, Associate Professor; M.D., Seville (Spain), 1977. Molecular and cellular regulation of programmed cell death (apoptosis) and innate immunity; microbial-host interactions; molecular mechanisms of inflammatory disorders, including Crohn's disease.

Sem H. Phan, Professor; Ph.D., 1975, M.D., 1976, Indiana. Molecular and cellular mechanisms of tissue repair and fibrosis; regulation of extracellular matrix and cytokine gene expression; myofibroblast differentiation.

Daniel G. Remick, Professor and Assistant Dean for Admissions; M.D., Mayo, 1982. Soluble mediators of inflammation, cellular and molecular mechanisms that control their production, and the effects these mediators exert.

Lloyd M. Stoolman, Professor; M.D., California, San Francisco, 1977. Lymphocyte recirculation and migration; functional characterization of the lymphocyte and endothelial receptors that mediate migration into lymphoid organs and sites of chronic inflammation; role of lymphocyte homing receptors in the hematogenous dissemination of lymphoid malignancies; role of lymphocyte–extracellular matrix interactions in lymphocyte migration.

James Varani, Professor; Ph.D., North Dakota, 1974. Cell-substrate adhesion and cell motility in normal mammalian cells and their malignant counterparts; biosynthesis and surface expression of extracellular matrix molecules, such as fibronectin, laminin, and thrombospondin and their involvement as endogenous regulators of adhesion, motility, invasion, and metastasis.

Peter A. Ward, Godfrey D. Stobbe Professor; M.D., Michigan, 1960. Mechanisms of the inflammatory response and its regulation; role of oxidants in tissue injury; mechanisms and participation of cytokines and chemokines; role of complement activation products in the inflammatory response and in sepsis.

Jeffrey S. Warren, Warthin Professor and Director, Division of Clinical Pathology; M.D., Ohio State, 1980. Role of cellular redox status in modulating chemokine expression; pathogenesis of granulomatous vasculitis; role of therapeutic erythropoietin in atherogenesis.

Thomas E. Wilson, Assistant Professor; M.D./Ph.D., Washington (St. Louis), 1994. Study of the normal functions of DNA double-strand break (DSB) repair pathways and the origins of chromosomal rearrangements associated with cancer, specifically translocations, deletions, and amplifications.

UNIVERSITY OF ROCHESTER

Department of Pathology and Laboratory Medicine
Pathways of Human Disease Program

Program of Study

Recent advances in the biological sciences present exciting opportunities for studying human disease and improving human health, thus requiring scientists with advanced training in biology and a solid foundation in the mechanisms of human disease.

The Pathways of Human Disease Program, offered through the Department of Pathology and Laboratory Medicine, trains scientists to understand the many challenges in curing human disease and to acquire the knowledge and skills to participate in these endeavors. The program emphasizes excellence in molecular biology, genetics, biochemistry, and cell biology to address fundamental research questions. Unlike other programs, however, the curriculum exposes the student to the entire spectrum of biomedical knowledge, from the molecular level to the whole-body level, in order to most effectively master the field of human disease. It combines core courses in biochemistry, cell biology, and molecular biology and genetics with unique, integrated courses covering anatomy, physiology, and human disease from a research perspective.

During the first year of study for the Ph.D. degree, students complete the core courses, the two-semester Pathways of Human Disease I and II course, and three laboratory rotations, one per calendar quarter, before choosing an adviser. Pathways of Human Disease Program graduate students have great flexibility in the selection of their thesis lab, since they are able to pursue research in any lab in the University of Rochester Medical Center (URMC) as long as its focus is on human disease research. During the second year, work begins in the adviser's laboratory, focusing more closely on advanced disease-oriented studies. Two electives, with at least one chosen from the program's offerings, are required, and students may take additional electives, depending on their background, areas of demonstrated competence, and research interests.

At the end of their second year, students take the Qualifying Examination, which includes a research proposal of no more than 25 pages written in the NIH format, as well as a 20- to 30-minute oral presentation of the proposal. Upon successful completion of the Qualifying Examination, students continue work on their research and dissertation, which ultimately leads to the thesis examination, consisting of a public seminar and a dissertation defense.

Research Facilities

The Department of Pathology and Laboratory Medicine is home to the Medical Center's Pathology/Morphology Imaging Core and the Electron Microscopy Core, with its PixCell laser capture microscope; Leica TCS SP spectral confocal microscope; Olympus AH-2 microscope linked to Image-Pro Plus v. 3 with a SPOT digital camera; Hitachi 7100 transmission electron microscope, with a MegaView III digital camera; and morphometric analysis software. Additional core facilities and equipment include a microarray core, liquid handling robotics, a computerized one- and two-dimensional gel analysis facility, a transgenic mouse facility, an animal tumor/xenograft facility, an APCI/electrospray mass spectrometer, a cell separation/flow cytometry facility, an NMR spectrometer, and a state-of-the-art X-ray crystallography facility.

The Medical School's Edward J. Miner Library provides the latest in remote computer access systems and, along with the Carlson Science Library on River Campus, contains approximately 2.75 million volumes and 12,000 serial publications.

Financial Aid

All students receive a full-tuition scholarship, paid health insurance, and an annual stipend. The stipend for 2006–07 is $23,000. The Offices for Graduate Education is responsible for student stipends for the first fifteen months of a student's enrollment. Thereafter, student stipends are funded by the adviser's research or training grants.

Cost of Study

In the 2006–07 academic year, the cost of the program is $23,000 per year, including all fees and health insurance. However, as outlined in the Financial Aid section, these costs are covered for all students.

Living and Housing Costs

A wide variety of housing accommodations is available for both married and single graduate students who are enrolled full-time. Goler House is a high-rise apartment building immediately adjacent to the Medical Center. Whipple Park, a University-owned town-house development, is about a mile away. Students living off campus can expect to spend $425 to $700 per month for a one-bedroom apartment or $500 to $1000 for a two-bedroom apartment, depending on size, location, and amenities. The University provides assistance through the Community Living Program (CLP) for those seeking housing in the Rochester community.

Student Group

In 2006–07, 44 students are enrolled in the program, comprising a mix of men and women from throughout the U.S. and around the world, including the Bahamas, Canada, China, Sri Lanka, and Taiwan.

Student Outcomes

Graduates of the program are qualified for positions in academia, industrial research, drug development for the pharmaceutical industry, and government agencies. Recent graduates are now working at Cold Spring Harbor Laboratory, Johnson & Johnson, Los Alamos National Laboratory, Yale University, Memorial Sloan-Kettering Cancer Center, and Harvard Medical School.

Location

Rochester is situated in upstate New York, where the Genesee River meets the southern shore of Lake Ontario. The metropolitan area has a population of nearly a million people. The city's commercial base is primarily technological. Major firms include Eastman Kodak Company, Xerox Corporation, and Bausch & Lomb. The area presents cultural opportunities unusual for its size and is especially noted for its musical activities, which center on the University's Eastman School of Music and the Rochester Philharmonic Orchestra. The city and five colleges and universities provide theater, dance, films, art galleries, and museums. Rochester is close to the Finger Lakes and Adirondack wilderness regions, and there are numerous parks and recreational areas.

The University

The University of Rochester, a private university founded in 1850, maintains a moderate size with an emphasis on graduate education and research, ranking as one of the nation's top research universities. Students participate in research programs that integrate cutting-edge, evidence-based medical science and clinical medical practice. The University's research portfolio has achieved top-fifteen rankings in NIH funding in neurology, neurosurgery, public health, and preventive medicine, and the University received the Outstanding Community Service Award in 2004 from the Association of American Medical Colleges. Currently, there are approximately 4,435 undergraduates and 2,310 full-time students enrolled in advanced degree programs at the University of Rochester.

Applying

Applicants must submit a complete online application, a personal and research statement, letters of recommendation, official college transcripts, and official GRE scores. International students must also submit TOEFL scores. Prospective students should have a baccalaureate degree that includes at least basic training in the biological sciences, chemistry, or physics and two to four semesters of mathematics. Applications must be received by January 1 for admission in the fall.

Correspondence and Information

Dr. Robert A. Mooney, Program Director
Department of Pathology and Laboratory Medicine
University of Rochester
601 Elmwood Avenue, Box 626
Rochester, New York 14642

Phone: 585-275-7811
Fax: 585-273-1101
E-mail: robert_mooney@urmc.rochester.edu
Web site: http://www.urmc.rochester.edu/GEBS/pwd/

University of Rochester

THE FACULTY AND THEIR RESEARCH

Bradford Berk, Professor of Medicine; Senior Vice President, Health Sciences; and CEO, URMC and Strong Health; M.D./Ph.D., Rochester, 1981. Molecular biology and genetics of cardiovascular disease.

Brendan Boyce, Professor of Pathology and Laboratory Medicine and Director, Surgical Pathology; M.B., Ch.B., Glasgow, 1972. Bone cell biology and bone pathology.

Laura Calvi, Assistant Professor of Medicine; M.D., Harvard, 1995. Microenvironment of hematopoietic stem cells, especially in osteoblasts, bone-forming cells; stem cell–microenvironmental interactions.

Chawnshang Chang, George Hoyt Whipple Professor of Pathology and Laboratory Medicine and Professor of Urology and of Radiation Oncology; Ph.D., Chicago, 1985. Molecular mechanisms of steroid hormone function in prostate.

Di Chen, Assistant Professor of Orthopaedics; M.D., Tianjin Medical (China), 1982; Ph.D., Louisville, 1992. Signal transduction mechanisms and function in bone development and bone remodeling.

P. Anthony di Sant'Agnese, Professor of Pathology and Laboratory Medicine; M.D., Columbia, 1975. Studies of neuroendocrine cells in the prostate; neuroendocrine differentiation of prostatic carcinoma.

Hicham Drissi, Assistant Professor of Orthopaedics; Ph.D., Paris V (Descartes), 1997. Regulatory mechanisms governing bone formation and skeletal development; role of regulatory proteins, particularly RUNX homology transcription factors, in development and pathogenesis.

Philip Fay, Professor of Medicine and of Biochemistry and Biophysics; Ph.D., Rochester, 1982. Human factor VIII structure and function; enzyme-cofactor interactions.

Charles Francis, Professor of Medicine and of Pathology and Laboratory Medicine; M.D., Pittsburgh, 1973. Role of fibroblast growth factors and vascular endothelial growth factor in endothelial cell responses; effect of ultrasound on the enzymatic process of blood clot dissolution.

Alan Friedman, Research Assistant Professor of Environmental Medicine; Ph.D., California, Santa Barbara, 1988. Proteomics and the biological basis for disease.

Jeffrey Hayes, Associate Professor of Biochemistry and Biophysics; Ph.D., Johns Hopkins, 1990. DNA structure; chromatin; protein-DNA interactions.

Wei Hsu, Assistant Professor of Biomedical Genetics, Oncology, and Dentistry; Ph.D., CUNY, Mount Sinai, 1994. Cellular signals and signal transduction mechanisms in mammalian development and disease progression.

Jiaoti Huang, Associate Professor of Pathology and Laboratory Medicine; Ph.D., NYU, 1991. Molecular mechanisms of oncogenesis; neuroendocrine differentiation in prostate cancer and its role in androgen-independent growth of prostate cancer cells.

Rulang Jiang, Associate Professor of Biomedical Genetics and Assistant Professor of Biology and Dentistry; Ph.D., Wesleyan, 1995. Molecular genetic basis of craniofacial development and birth defects.

Peter Keng, Professor of Radiation Oncology and of Biochemistry and Biophysics; Ph.D., Colorado State, 1978. Role of p53 in radiation sensitivity and cell-cycle delay; mechanisms by which interferon beta enhances radiation sensitivity in human tumor cells.

Yi-Fen Lee, Assistant Professor of Urology and of Pathology and Laboratory Medicine; Ph.D., Wisconsin–Madison, 1997. Non-genomic action of androgens and antiandrogens; vitamin D signaling in the prostate; physiological roles of the human testicular receptor 4.

Richard Libby, Assistant Professor of Ophthalmology; Boston College, 1997. Neurobiology of glaucoma, using mouse models of glaucoma and advanced mouse genetics to probe the pathophysiology of glaucoma; understanding the molecular processes that lead to retinal ganglion cell (RGC) death in glaucoma and why RGCs are more likely to die in some patients than in others.

Margot Mayer-Proschel, Associate Professor of Genetics; Ph.D., Würzberg (Germany), 1990. Biological and molecular mechanisms governing precursor cell division, differentiation, and survival in the brain.

Edward Messing, Professor and Chair of Urology; M.D., NYU, 1972. Early detection, prevention, and treatment of urologic malignancies.

Joseph Miano, Associate Professor of Medicine; Ph.D., New York Medical College, 1992. Smooth muscle cell differentiation control, with special emphasis on transgenic mouse models, genomics, and bioinformatics.

Laurie Milner, Associate Professor of Pediatrics; M.D., Texas Southwestern Medical Center at Dallas, 1986. Molecular mechanisms responsible for generation of diverse cell types from pluripotent stem cells; relation of aberrations in normal developmental pathways to pediatric malignancies.

Robert Mooney, Professor of Pathology and Laboratory Medicine; Ph.D., Johns Hopkins, 1980. Mechanisms of obesity-dependent insulin resistance in type 2 diabetes.

Mark D. Noble, Professor of Genetics; Ph.D., Stanford, 1977. Regeneration in the CNS; redox modulation of precursor cell function; developmental disorders; brain tumor biology and adverse effects of chemotherapy of neural stem cell populations.

Regis O'Keefe, Professor of Orthopaedics; M.D., Harvard, 1985; Ph.D., Rochester, 2000. Growth plate chondrocyte differentiation and maturation; biochemical mechanisms of prosthetic implant loosening; mechanisms of tumor-mediated bone loss.

James Palis, Professor of Pediatrics, Oncology, and Biomedical Genetics; M.D., Rochester, 1981. Development of the hematopoietic system in the mammalian embryo.

David A. Pearce, Associate Professor of Biochemistry and Biophysics; Ph.D., Bath (England), 1990. Yeast and mouse models for the study of human disease.

Robert Pierce, Assistant Professor of Pathology and Laboratory Medicine; M.D., Brown, 1993. Oxidative stress in tissue injury and neoplasia.

Carl Pinkert, Professor of Pathology and Laboratory Medicine and Director, Transgenic Facility; Ph.D., Georgia, 1993. Gene transfer, expression, and regulation, using transgenic animal modeling; mitochondria and mitochondrial gene transfer.

Chris Proschel, Research Assistant Professor of Biomedical Genetics; Ph.D. Astrocyte development and white matter disease.

J. Edward Puzas, Donald and Mary Clark Professor of Orthopaedics; Ph.D., Rochester, 1976. Molecular and cellular biology of the skeletal system.

Jay Reeder, Research Assistant Professor of Pathology and Laboratory Medicine; Ph.D., Rochester, 1997. Molecular mechanisms of bladder cancer.

William Ricke, Assistant Professor of Urology and of Pathology and Laboratory Medicine; Ph.D., Missouri–Columbia, 2000. Roles of hormones and microenvironment on development, treatment, and prevention of cancer.

Randy Rosier, Professor and Chair of Orthopaedics and Professor of Oncology and of Biochemistry and Biophysics; M.D., Ph.D. Regulation of endochondral ossification.

Paul Rothberg, Professor of Pathology and Laboratory Medicine and Director, Molecular Diagnostics Laboratory; Ph.D., SUNY at Stony Brook, 1981. Molecular genetics of cancer and of thrombophilia.

Daniel Ryan, Professor and Chair of Pathology and Laboratory Medicine; Ph.D., Johns Hopkins, 1975. B-cell development and differentiation.

Ignacio Sanz, Professor of Medicine and of Microbiology and Immunology; M.D., Santander (Spain), 1978. Human B-cell tolerance and autoimmunity, development, and function; innovative approaches to the study and treatment of autoimmune diseases.

Edward Schwarz, Associate Professor of Orthopaedics and of Microbiology and Immunology; Ph.D., Yeshiva (Einstein), 1993. Pro-inflammatory cytokine signal transduction and novel drug and gene therapies for rheumatoid arthritis.

Shey-Shing Sheu, Professor of Pharmacology and Physiology, of Medicine, and of Anesthesiology; Ph.D., Chicago, 1979. Mitochondrial Ca^{2+} transport mechanisms in heart and neuronal cells.

Patricia J. Simpson-Haidaris, Associate Professor of Medicine, of Microbiology and Immunology, and of Pathology and Laboratory Medicine; Ph.D., Notre Dame, 1991. Role of fibrinogen in acute-phase response in pneumonitis; fibrinogen gene expression in liver and lung.

Peter J. Sims, Professor of Pathology and Laboratory Medicine; M.D./Ph.D., Duke, 1980. Regulation of immune/inflammatory events and thromboembolic events by phospholipids and other components of the plasma membrane; structure and function of the phospholipid scramblase gene family: proteins implicated in control of the topology of plasma membrane phospholipids and in the regulation of cell proliferation, cell maturation, and apoptosis.

Harold Smith, Professor of Biochemistry and Biophysics; Ph.D. SUNY at Buffalo, 1982. Diversification of the proteome through messenger RNA editing and DNA mutations.

Charles Sparks, Professor of Pathology and Laboratory Medicine; M.D., Jefferson Medical, 1968. Biology of lipoprotein metabolism; lipoproteins as a risk factor in development of cardiovascular disease.

Janet Sparks, Associate Professor of Pathology and Laboratory Medicine; Ph.D., Pennsylvania, 1980. Insulin regulation of hepatic lipoprotein production through phosphatidylinositide-3-kinase activation via insulin receptor signaling pathways.

Nancy Wang, Professor of Pathology and Laboratory Medicine, of Pediatrics, and of Genetics; Ph.D., Minnesota, 1978. Cancer cytogenetics; chromosome microdissection and microcell-mediated chromosome transfer; tumor suppression gene identification for human ovarian cancer.

Therese Wiedmer, Associate Professor of Pathology and Laboratory Medicine; Ph.D., Bern (Switzerland), 1977. Role of phospholipid scramblases in growth factor signaling pathways underlying cell proliferation, maturation, and apoptosis; regulation of membrane phospholipid asymmetry.

Terry Wright, Associate Professor of Pediatrics and of Microbiology and Immunology; Ph.D., Rochester. Host immune response as a major component of lung injury associated with *Pneumocystis carinii* pneumonia; development of improved methods of treatment.

Lianping Xing, Assistant Professor of Pathology and Laboratory Medicine; Ph.D., Penn State, 1994. Molecular mechanisms by which TNF mediates the formation, activation, and survival of osteoclasts in inflammation-induced bone.

Haodong Xu, Assistant Professor of Pathology and Laboratory Medicine; M.D., 1987, Ph.D., 1993, Soowchow (China). Mechanisms in the regulation of voltage-gated potassium channels underlying cardiac repolarization; molecular mechanisms of lung tumor and fibrosing interstitial pneumonia.

Shuyuan Yeh, Assistant Professor of Urology and of Pathology and Laboratory Medicine; Ph.D., Wisconsin–Madison, 1996. Chemopreventive roles of vitamin E in prostate cancer; kinase pathways, androgen receptor, and coregulators.

Fay Young, Associate Professor of Medicine; M.D., Harvard, 1982. Normal and abnormal B-lymphocyte development.

Michael Zuscik, Assistant Professor of Orthopaedics; Ph.D., Rochester, 1993. Cellular and molecular processes that govern cartilage development and maturation.

UNIVERSITY OF SOUTHERN CALIFORNIA

Department of Pathology
Ph.D. Degree Program in Pathobiology
M.S. Degree Program in Experimental and Molecular Pathology

Programs of Study

The Department of Pathology offers a program leading to the M.S. or Ph.D. degree in pathobiology. The purpose of the program is to educate individuals who are interested in investigating mechanisms of disease and who will pursue careers in research, biotechnology, and teaching. Emphasis is placed on interdisciplinary approaches to the study of human disease. Areas currently under investigation include cellular and molecular biology of cancer, chemical carcinogenesis, virology, stem cell and developmental pathology, liver and pulmonary diseases, environmental pathology, and circulatory, endocrine, and neurodegenerative diseases.

Courses are offered in general pathology, eukaryotic gene expression, viral oncology, molecular biology of cancer, and biochemical and molecular bases of diseases as well as techniques in cellular and molecular pathology. Other graduate courses offered include biochemistry, physiology, histology, and microbiology. A wide range of courses geared to contribute to the individual's educational objectives are available.

The master's program in experimental and molecular pathology offers students the flexibility to prepare for a wide range of careers, including research, teaching, industry, or consulting.

The Ph.D. program in pathobiology may be pursued as part of the M.D./Ph.D. program offered by the University of Southern California (USC). Medical students, residents, and fellows may also apply to this program.

Postdoctoral positions are also available.

Research Facilities

The facilities of the Department include the latest equipment for research in cell biology, biochemistry, and molecular biology. Excellent facilities are also available in the neighboring Kenneth Norris Jr. Cancer Hospital Research Institute, Topping Research Tower, and Institute for Genetic Medicine. A separate containment level-3 facility is available for studies related to AIDS and human retroviruses and oncogenes. Library services are available at the Health Sciences Campus.

Financial Aid

Teaching assistantships are available for outstanding graduate students in their first year. Research assistantships are also available through faculty research grants. Special fellowships are available for qualified students. Most students are financially supported beginning in the second year by their mentors. No financial aid is available for the master's program at this time.

Cost of Study

Research assistantships and teaching assistantships include tuition. Full tuition for both the M.S. and Ph.D. programs is $38,124 for the 2006–07 academic year, plus mandatory fees of $1432 per year.

Living and Housing Costs

Housing arrangements are the responsibility of the students. Housing costs vary greatly, depending on location and type of accommodations. There are ample housing facilities in the many communities surrounding the medical school.

Student Group

Of the 32,836 students enrolled at the University of Southern California, 15,939 are engaged in graduate or professional study. The student body includes undergraduates from all parts of the United States and many other countries. There are 42 students enrolled in the Ph.D. pathobiology program and 16 enrolled in the M.S. in experimental and molecular pathology program. In addition, there are 3 students enrolled in the M.D./Ph.D. program.

Student Outcomes

Job opportunities for graduates are available in academic, hospital, and government institutions and in industry.

Location

The Health Sciences Campus of the University of Southern California is located in the center of Los Angeles, close to other universities and medical schools and within an hour's drive of both beaches and mountains. The city's excellent climate makes it possible to participate the year round in outdoor sports: camping and hiking at nearby mountain parks, skiing during the winter months, and swimming, surfing, and water sports. This exciting seaside city is rich in cultural, athletic, and scholastic opportunities.

The Health Sciences Campus

The graduate program in pathobiology is based at the Health Sciences Campus, which is approximately 10 miles from the main University campus. The Health Sciences Campus comprises the School of Medicine, School of Pharmacy, LAC-USC Healthcare Network, Doheny Eye Foundation, Norris Cancer Research Institute, Topping Research Tower, Institute for Genetic Medicine, and USC University Hospital.

Applying

Admission to the graduate program requires an undergraduate degree in the natural sciences and demonstration of competence and achievement in these studies. Scores on the General Test of the Graduate Record Examinations are required of all students, with a combined score of 1100 or better for the verbal and quantitative sections. The MCAT can substitute for the GRE upon request. A score of 600 or better on the Test of English as a Foreign Language (TOEFL) is required of international applicants. Applicants must have a minimum GPA of 3.0. Application forms and further information about this program and the requirements of the Graduate School can be obtained by writing to the Department.

The program is also integrated with the Ph.D. Programs in Biomedical and Biological Sciences (PIBBS), an integrated pan–Health Sciences Campus graduate program. Students can also enter the pathology program through the broader PIBBS program. For the PIBBS application, students can visit http://www.usc.edu/pibbs for directions on how to apply to their program.

Correspondence and Information

Chairman, Graduate Committee
Department of Pathology
School of Medicine
University of Southern California
2011 Zonal Avenue
Los Angeles, California 90089-9092
Phone: 323-442-1168
Fax: 323-442-3049
E-mail: doumak@pathfinder.usc.edu
Web site: http://www.usc.edu/medicine/pathology

University of Southern California

THE FACULTY AND THEIR RESEARCH

Michael J. Anderson, Assistant Professor; Ph.D., California, Irvine, 1993. Cell signaling, transformation, and inhibition of myogenic differentiation in childhood rhabdomyosarcoma. *Proc. Natl. Acad. Sci. U.S.A.* 98:1589, 2001. *Genomics* 47:187, 1998.

Saverio Bellusci, Assistant Professor; Ph.D., Paris, 1994. Molecular basis of lung development and repair. *Development* 132:2157–66, 2005; 124:4867–78, 1997. *Dev. Biol.* 277:316–31, 2005. *Dev. Cell* 4:11–8, 2003.

Norbert Berndt, Associate Professor; Ph.D., Düsseldorf, 1986. Control of the cell cycle and apoptosis by protein phosphatase 1. *Mol. Carcinog.*, in press. *Dev. Dynamics* 229(4):791–801, 2004. *Prog. Cell Cycle Res.* 5:497–510, 2003. *Oncogene* 20:6111, 2001. *J. Biol. Chem.* 274:29470, 1999. *Curr. Biol.* 7:375, 1997.

Thomas C. Chen, Associate Professor and Director, Neuro-Oncology; M.D., California, San Francisco, 1988; Ph.D., USC, 1996. Translational research for brain tumors. *Hum. Gene Ther.* 14:117, 2003. *Cancer* 97:2363, 2003. *Lab. Invest.* 78:165, 1998.

Cheng-Ming Chuong, Professor and Chairman, Graduate Committee; M.D., Taiwan, 1978; Ph.D., Rockefeller, 1983. Engineering stem cells into organs; morphoregulation; feather morphogenesis/stem cells. *Curr. Top. Dev. Biol.* 72:237, 2006. *Nature* 438:1026, 2005; 420:308, 2002. *Am. J. Pathol.* 164:1099, 2004.

Thomas D. Coates, Associate Professor; M.D., Michigan, 1975. Study of motility of human neutrophils and its relation to inherited increased propensity to infection; computer image analysis of biologic systems; role of inflammation in vascular damage in sickle cell disease. *J. Exp. Biol.* 199:741, 1996. *Biophys. J.* 67:2535, 1994. *Blood* 83:231, 1994.

Richard J. Cote, Professor; M.D., Chicago, 1980. Detection and characterization of micrometastases; molecular/antigenic changes in tumor progression; alterations of tumor suppressor gene p53. *Nature Biotechnol.* 19:856–60, 2001. *Cancer J. Sci. Am.* 5:2, 1999. *Lancet* 354:896, 1999. *J. Natl. Cancer Inst.* 91:1113, 1999; 90:1072, 1998. *Cancer Res.* 58:1090, 1998. *Nature* 385:123, 1997. *N. Engl. J. Med.* 331:1259, 1994.

Edward D. Crandall, Professor; Ph.D., Northwestern, 1964; M.D., Pennsylvania, 1972. Alveolar epithelial development, growth, differentiation, and pathobiology. *Am. J. Pathol.* 166:1321, 2005. *J. Histochem. Cytochem.* 2:759, 2004. *Cell Tissue Res.* 312:313, 2003. *J. Virol.* 75:11747, 2001.

J. A. Louis Dubeau, Professor; M.D., 1979, Ph.D., 1981, McGill. Biology, molecular genetics, and animal modeling of ovarian cancer. *Curr. Biol.* 15:561–5, 2005. *Gynecol. Oncol.* 72:437, 1999.

Alan L. Epstein, Professor; M.D./Ph.D., Stanford, 1978. Monoclonal antibody–based immunotherapy of cancer. *Clin. Cancer Res.* 11:8492, 2005; 11:3084, 2005; 5:51, 1999. *Blood* 101:4853, 2003. *Cancer Res.* 63:8384, 2003.

Parkash Gill, Professor; M.D., Punjabi (India), 1974. Molecular mechanisms of angiogenesis. *Blood*, in press. *Nature*, in press. *Proc. Natl. Acad. Sci. U.S.A.* 94:978, 1997. *N. Engl. J. Med.* 335:1261, 1996.

John Groffen, Professor; Ph.D., Groningen (Netherlands), 1981. Genetics of Ph-positive leukemias and innate immune system regulation. *Mol. Cell. Biol.* 25:5777–85, 2005. *Mol. Cancer* 3:10–22, 2004. *Oncogene* 22:8255–62, 2003; 21:8536–40, 2002. *Dev. Dynamics* 223:517, 2002.

Yuan-Ping Han, Assistant Professor; Ph.D., CUNY, Mount Sinai, 1996. Extracellular matrix (ECM) and MMP-mediated degradation in cell-fate determination and disease development. *J. Biol. Chem.* 279:4820, 2004; 277:27319, 2002.

Nora Heisterkamp, Professor; Ph.D., Groningen (Netherlands), 1981. Cell biology; signal transduction and treatment of Bcr-Abl-positive leukemia. *Mol. Cell. Biol.* 25:5777–85, 2005. *Blood* 106:1355–61, 2005. *Biochem. Biophys. Res. Commun.* 333:1276–83, 2005.

David R. Hinton, Professor and Vice Chairman, Department of Pathology; M.D., Toronto, 1978. Pathogenesis of age-related macular degeneration; role of retinal pigment epithelial cell in retinal disease; viral demyelination. *Invest. Ophthalmol. Vis. Sci.* 47:287–94, 2006; 46:4772–9, 2005.

Alan L. Hiti, Professor of Clinical; Ph.D., UCLA, 1981; M.D., Miami (Florida), 1983. Relationship of beta globin haplotypes and alpha gene status to sickle cell disease; applications of flow cytometry to diagnostic hematopathology and HIV disease. *Am. J. Dis. Child.* 147:1197, 1993.

Florence M. Hofman, Professor; Ph.D., Weizmann (Israel), 1976. Antiangiogenic therapy for cancer; signal transduction mechanisms in tumor-derived vasculature. *Am. J. Physiol.*, in press. *Cancer Res.* 65(22):10347–54, 2005.

Vesa Kaartinen, Assistant Professor; Ph.D., Kuopio (Finland), 1991. Molecular basis of craniofacial and cardiac development. *Dev. Biol.* 266:96–108, 2004. *Mech. Dev.* 121:173–82, 2004. *Development* 131:3481–90, 2004.

Linda Koss Kelly, Assistant Professor of Clinical; Ph.D., USC, 1985. Immunoelectron microscopy; renin and prorenin processing as related to nephropathy in diabetes and other disorders.

Anthony J. Keyser, Professor of Clinical; Ph.D., Emory, 1976. Connective tissue; clinical pathology; non-mucinous nature of the surface material of the bladder mucosa. *Clin. Biochem.* 30:613, 1997. *Am. J. Med. Sci.* 304:285, 1993.

Leslie A. Khawli, Professor of Clinical; Ph.D., USC, 1986. Generation of new methods for the immunotherapy and diagnosis of human cancer. *Clin. Cancer Res.* 11:8492, 2005; 11:5971, 2005; 11:3084, 2005. *Cancer Res.* 63:8384, 2003. *J. Immunother.* 26:320, 2003.

Tasneem A. Khwaja, Associate Professor; Ph.D., Cambridge, 1964. Experimental chemotherapy; evaluation of synthetic and natural products as anticancer agents in cellular and animal systems; studies of mechanisms of drug action and multidrug resistance. *J. Med. Chem.* 33:1914, 1990. *Oncology* 43:42, 1986.

Robert D. Ladner, Assistant Professor; Ph.D., Rutgers, 1996. Regulation and posttranslational modification of enzymes involved in thymidylate nucleotide metabolism. *Mol. Pharmacol.* 66:620, 2004. *Pharmacogenetics* 14:319, 2004. *Cancer Res.* 63:2898, 2003.

Joseph R. Landolph Jr., Associate Professor; Ph.D., Berkeley, 1976. Molecular biology of neoplastic transformation induced by carcinogenic Ni^{+2} and Cr^{+6} compounds; oncogene activation/tumor suppressor gene inactivation. *Toxicol. Sci.* 206:138, 2005. *Mol. Cell. Biochem.* 255:203, 2005.

Elizabeth R. Lawlor, Assistant Professor; M.D., McMaster, 1989; Ph.D., British Columbia, 2002. Cellular origins of pediatric sarcoma; cancer stem cells; molecular mechanisms of oncogene-induced transformation. *Oncogene* 21:307, 2002. *Cancer Res.* 58:2469, 1998.

Michael R. Lieber, Professor; Ph.D., 1981, M.D., 1983, Chicago. DNA repair in human cancer, aging, and the immune system. *Mol. Cell* 16:701, 2004; 2:477, 1998. *Nature* 428:88, 2004; 388:495, 1997; 388:492, 1997. *Nature Immunol.* 4:442, 2003. *Cell* 109:807, 2002; 108:781, 2002. *Curr. Biol.* 12:397, 2002.

Robert D. MacPhee Jr., Associate Professor of Clinical; Ph.D., USC, 1994. Development of novel methodologies in clinical laboratory medicine. In *Technical Digest of the Seventh International Meeting on Chemical Sensors*, pp. 461–4, 1998. *Clin. Exp. Immunol.* 107:21, 1997.

Carol A. Miller, Professor; M.D., Jefferson Medical, 1965. Molecular and cellular basis for degenerative and developmental neurologic diseases. *Proc. Natl. Acad. Sci. U.S.A.* 101(12):4210–5, 2004; 95:2586–91, 1998.

Kevin A. Nash, Assistant Professor; Ph.D., Aberdeen (England), 1987. Impact of antibiotics on microbial physiology; molecular basis of drug resistance and drug susceptibility in bacteria, especially *Mycobacterium tuberculosis*. *J. Antimicrob. Chemother.* 55:170, 2005. *Antimicrob. Agents Chemother.* 47:3053, 2003; 45:1982, 2001; 45:1607, 2001.

Paul K. Pattengale, Professor; M.D., NYU, 1970. Structure and function of human M-CSF; minimal residual disease in ALL; molecular pathogenesis of lymphoid cell neoplasms (mouse and human). *Am. J. Pathol.* 151:647, 1997. *Blood* 90:2901, 1997. *J. Biol. Chem.* 271:16388, 1996.

Michael F. Press, Professor; Ph.D., 1975, M.D., 1977, Chicago. Pathobiology of breast cancer. *Clin. Cancer Res.* 11:6598–607, 2005. *Clin. Breast Cancer* 6:240–6, 2005. *Breast Cancer Res. Treat.* 93:3–11, 2005.

Suraiya Rasheed, Professor and Director, Viral Oncology and Proteomics Research; Ph.D., Osmania (India), 1958; Ph.D., London, 1964. Application of proteomics technology to identify biomarkers in melanoma and neuronal stem cells; protein-protein interaction pathways in cancer and viral diseases. In *Thirteenth International Molecular Medicine Tri-Conference*, p. 90, 2006 (abstract book). *J. Translational Med.* 5:14, 2005 (doi: 10.1186/1479-5876-3-14). In *International Conference on Stem Cells Research and Therapeutics* vol. 4, p. 42, 2005.

C. Patrick Reynolds, Professor; M.D./Ph.D., Texas Health Science Center at Dallas, 1979. Laboratory and clinical development of anticancer drugs; mechanisms of drug resistance; ceramide as a therapeutic target; retinoids. *Clin. Cancer Res.* 11:2774–80, 2005. *Cancer Lett.* 206:169–80, 2004.

Pradip Roy-Burman, Professor; Ph.D., Calcutta, 1963. Modeling human prostate cancer in mice; mechanisms of prostate tumorigenesis and role of tissue microenvironment in disease progression. *Cancer Res.* 66:883, 2006; 65:5769, 2005. *Endocrine-Related Cancer* 11:225, 2004.

Russell P. Sherwin, Professor; M.D., Boston University, 1948. Environmental, occupational, and oncologic pathology; COPD; image analysis; immunopathology. *Mol. Cancer Ther.* 3:499–511, 2004. *Virchows Arch.* 437:422–8, 2000; 433:341–8, 1998. *Inhalation Toxicol.* 9:405–22, 1997; 7:1183–94, 1995; 7:1173, 1995.

Darryl Shibata, Professor; M.D., USC, 1983. Molecular clocks; investigating stem cells in human colon by using methylation patterns. *Proc. Natl. Acad. Sci. U.S.A.* 98:10839–44, 2001.

Michael R. Stallcup, Professor; Ph.D., Berkeley, 1974. Regulation of gene transcription by steroid hormone receptors and transcriptional coactivator proteins. *Genes Dev.* 19:1466, 2005. *Proc. Natl. Acad. Sci. U.S.A.* 102:3611, 2005. *Mol. Cell. Biol.* 24:2103, 2004. *Mol. Cell* 12:1537, 2003. *Science* 284:2174, 1999.

Clive R. Taylor, Professor and Chairman of the Department of Pathology; M.B.B.Chir., 1969; D.Phil., 1975, Oxford. Immunohistochemical methods applied to cell identification and tumor diagnosis; immunology of lymphomas and leukemias. *Adv. Pathol. Lab. Med.* 7:59, 1994.

Timothy J. Triche, Professor; M.D./Ph.D., Tulane, 1971. Molecular characterization of chimeric tumor genes, control of gene expression, and relationship to tumor phenotype, biologic aggressiveness, and treatment responsiveness. *Nat. Genet.* 6:146–51, 1994. *Diagn. Mol. Pathol.* 2:147–57, 1993. *EMBO J.* 12:4481–7, 1993.

Hide Tsukamoto, Professor; D.V.M., Tokyo, 1975; Ph.D., Kobe (Japan), 1988. Cellular and molecular mechanisms of liver cirrhosis; adipogenic transcriptional programs that are molecular targets for potential treatment of cirrhosis; iron-mediated intracellular signaling for activation of NF-kappaB in macrophages; development of a new prototype of bioartificial liver. *J. Biol. Chem.* 280:4959, 2005; 279:11392–401, 2004; 278:17646–54, 2003.

Randall Widelitz, Associate Professor; Ph.D., Arizona, 1986. Growth control in tumorigenesis and embryonic development; focus on roles of adhesion molecules, Wnts, and beta-catenin. *Curr. Top. Dev. Biol.* 72:237–74, 2005. *Mech. Dev.* 121:157–71, 2004. *Differentiation* 72:474–88, 2004.

THE UNIVERSITY OF TEXAS MEDICAL BRANCH

Graduate School of Biomedical Sciences at Galveston
Graduate Program in Experimental Pathology

Program of Study

A program leading to the Ph.D. degree is offered. The program is open to those students holding a B.A., B.S., or an M.S. degree as well as an M.D. or D.V.M. degree. A combined M.D./Ph.D. program is also offered. The goal of the program is to prepare students for a career investigating the mechanisms of human disease. Most students enrolling in the University of Texas Medical Branch (UTMB) Graduate School of Biomedical Sciences (GSBS) enter and take courses within the basic biomedical science curriculum that include biochemistry, cell biology and molecular biology, and genetics as well as choices among various short modular courses on a variety of topics. First-year students are not committed to an individual graduate program and are free to examine the research and academic opportunities within all the GSBS programs. Students select a graduate program at the end of the first year. Upon entering the Graduate Program in Experimental Pathology, three different tracks are available depending on a student's basic interests in infectious diseases, toxicology, or pathobiology. Students are required to take courses in pathobiology of human diseases, ethics in science, teaching in pathology and experimental design, and grant writing. Students then develop a customized course of study depending on their academic background and interest by selecting among a variety of elective courses that fall into groups along the basic three educational tracks. Electives include courses in the evolution and pathogenesis of infectious diseases, tropical diseases, virology, viral phylogenetics, emerging infectious diseases, vector biology, molecular toxicology, cardiovascular toxicology, and neuropathology, among others. Course requirements are flexible and depend on the needs and interests of the student; students are encouraged to take courses in other programs. A preliminary examination and a dissertation are required.

Research Facilities

Research and clinical laboratories in the Department of Pathology are well equipped to support research in the areas of biodefense, tropical and emerging infectious diseases, toxicology, and pathobiology. Laboratories and core facilities are equipped for research that includes molecular biology; DNA sequencing; real-time PCR; imaging; protein chemistry; conventional, fluorescent, confocal, and electron microscopy; histochemistry; whole-body autoradiography; pharmacokinetic studies; enzymology; and proteomics and genomics. Special resources for infectious disease research include biosafety level (BSL) 2 and level 3 containment laboratories, insectaries, and a recently completed BSL4 laboratory. The BSL4 facility is the only one of its kind on a U.S. university campus and enhances the University's research programs in biodefense and emerging infectious diseases. UTMB is one of eight institutions nationwide receiving grants to establish a Regional Center of Excellence (RCE) for Biodefense and Emerging Infectious Diseases Research from the National Institute of Allergy and Infectious Diseases (NIAID). The university has an exceptionally distinguished team of scientists in emerging infectious diseases and biodefense. NIAID recently announced that UTMB will be the site of the $150-million National Biocontainment Laboratory (NBL), one of two large-scale national research facilities focusing on new and emerging disease threats. The NBL will serve scientists from all over the country, supplying much-needed space for research aimed at fighting microbes that could be used in a biological terrorist attack as well as those that constitute emerging natural threats such as SARS and the West Nile virus. UTMB is the only institution in the nation to achieve both the NBL and the REC designations. UTMB is also home to three World Health Organization collaborating centers, including one in tropical and emerging infectious diseases.

Special resources for toxicology research include a National Institute of Environmental Health and Safety Research Center, a facility for inhalation exposures, and core facilities for custom microarrays, protein chemistry, and mass spectrometry. There is more than 385,000 square feet of space dedicated to research that is supported by more than $249 million in research grants. The Medical Branch has one of the nation's largest and most comprehensive medical libraries, which contains advanced information retrieval systems and subscribes to more than 4,600 biomedical periodicals in print or electronic forms.

Financial Aid

Graduate assistantships are available. In 2006–07, initial graduate assistant salaries are approximately $23,000 per year for Ph.D. students and $16,456 for M.S. students. Departmental graduate assistants qualify for in-state tuition fees.

Cost of Study

In the 2006–07 academic year, tuition is $50 per credit hour, and full-time students are required to take at least 9 units per term. The student service fee is $10.99 an hour (up to 12 hours), with a maximum of $150 per term (up to 18 hours). These tuition rates are subject to change. Student health fees are $55 per semester.

Living and Housing Costs

Living costs in Galveston vary according to personal lifestyles. Costs for rooms in University dormitories and fraternities begin at approximately $323 per month, per person, double occupancy, and the accommodations are within walking distance of the campus. Private rooms are $570 per month. Apartments are also available for single students beginning at approximately $670 per month.

Student Group

In 2004, total student enrollment at the Medical Branch was 2,121; 320 were graduate students. Approximately 30 students were enrolled in the experimental pathology program.

Location

Galveston, the oldest city and port in Texas, is located on Galveston Island, which is connected to the mainland by causeways and a ferry. The city, with a population of 70,000, is a major port and resort area on the Gulf of Mexico. The scientific facilities of the NASA Manned Spacecraft Center are nearby.

In addition to the usual recreational facilities of a city, Galveston Island has 32 miles of Gulf-front beach and numerous bay areas for water-related activities, including fishing, sailing, surfing, and waterskiing. Just an hour's drive away is the nation's fourth-largest city, Houston, with major shopping malls; acclaimed symphony; ballet, opera, and theater companies; music of all kinds; and professional sports teams.

The Medical Branch

The University of Texas Medical Branch is one of four Health Science Centers in the University of Texas System and includes, besides the Graduate School of Biomedical Sciences, the School of Allied Health Sciences, the School of Nursing, the School of Medicine, the Marine Biomedical Institute, the Institute for the Medical Humanities, and seven hospitals. The Medical Branch is spread over 90 acres and is one of the largest centers for biomedical education and research in the Southwest. It has a faculty of 2,335 members, with 330 graduate faculty members. The Medical Branch is currently undergoing the largest physical expansion program in its 113-year history.

Applying

Applicants should hold a bachelor's degree or the equivalent, should have the appropriate course preparation for the proposed area of study, and must have taken courses in biology, general chemistry, organic chemistry, and physics. An overall and advanced grade point average of 3.0 and the Graduate Record Examinations General Test are minimum admission requirements.

The Graduate School operates on a trimester system, with terms beginning in September, January, and April. Application materials should be received as early as possible but no more than ninety days before the expected date of enrollment. In general, students begin graduate school only in the fall semester. Personal interviews are encouraged.

Correspondence and Information

Dr. Stephen Higgs
Director, Graduate Program in Experimental Pathology
Department of Pathology
The University of Texas Medical Branch
1.104 Keiller Building
301 University Boulevard
Galveston, Texas 77555-0609

Phone: 409-772-2521
Fax: 409-747-2400
E-mail: hediaz@utmb.edu
Web site: http://www.utmb.edu/pathology

The University of Texas Medical Branch

THE FACULTY AND THEIR RESEARCH

Ahmed E. Ahmed, Ph.D., Professor. Pathogenesis of environmental insults; development of biomarkers for exposure and effects.
G. A. S. Ansari, Ph.D., Professor. Xenobiotic metabolism, conjugation, and toxicity; biomonitoring.
Judith F. Aronson, M.D., Associate Professor. Pathogenesis of arenavirus disease.
Alan D. T. Barrett, Ph.D., Professor. Molecular basis of attenuation, virulence, and immunogenicity of flaviviruses.
David Beasley, Ph.D., Assistant Professor.
Joel D. Bessman, M.D., Professor. Sickle cell anemia and other hematologic diseases.
Paul J. Boor, M.D., Professor. Influences of amines, oxidative stress, and aging on cardiovascular pathology.
Nigel Bourne, Ph.D., Associate Professor. In vivo models to evaluate strategies of disease prevention and treatment, with an emphasis on herpes simplex virus infections.
Donald Bouyer, Ph.D., Assistant Professor. Delineating the rickettsial mechanisms involved in the pathogenic process.
Edward G. Brooks, M.D., Associate Professor. Interaction of environmental agents with inflammatory reactions in respiratory mucosa as it pertains to allergy and asthma.
Gerald A. Campbell, M.D., Ph.D., Associate Professor. Pathologic mechanisms of blood-brain injury in systemic and central nervous system diseases.
Claudia Castro, M.D., Assistant Professor. Lung diseases (interstitial and neoplasms); cervical, endometrial, and ovarian neoplasms.
Miles W. Cloyd, Ph.D., Professor. Biological, genetic, and molecular biology of HIV and other retroviruses.
Daniel F. Cowan, M.D., Professor. Environmental pathology; comparative and marine pathology.
Robert A. Davey, Ph.D., Assistant Professor. Entry mechanism and receptor recognition of murine leukemia viruses and alphaviruses.
Tarek Elghetany, M.D., Associate Professor. The study of the relationship of bone marrow failure syndromes to myelodysplastic syndromes.
D. Mark Estes, Ph.D., Professor. Immunoregulation of effector cell function/development in infection and cancer.
Charles Fulhorst, D.V.M., Dr.P.H., Associate Professor. Epidemiology and ecology of rodent-borne viral zoonoses.
Nisha Garg, Ph.D., Assistant Professor. Pathomechanisms of chagasic myocarditis; vaccine development against *Trypanosoma cruzi* infection.
Benjamin Gelman, M.D., Ph.D., Professor. Mechanisms of HIV-associated dementia and peripheral neuropathy.
Hal K. Hawkins, M.D., Ph.D., Associate Professor. Acute inflammatory reaction to tissue injury; wound healing and smoke inhalation injury.
Norbert Herzog, Ph.D., Associate Professor and Assistant Dean, GSBS. Viral-host interactions; cellular signal transductions in viral pathogenesis; antiviral development.
Stephen Higgs, Ph.D., Associate Professor. Interactions between mosquito vectors, the viruses they transmit, and the vertebrate host.
Mike Holbrook, Ph.D., Assistant Professor. Flavivirus and viral hemorrhagic fever pathogenesis.
S. David Hudnall, M.D., Professor. Pathogenesis of herpesvirus-induced human cancer.
Nahed Ismail, M.D., Ph.D., Assistant Professor. Immunopathogenesis of Ehrlichiosis. Role of dendritic cells in susceptibility and resistance to *Rickettsia*.
Mary Kanz, Ph.D., Associate Professor. Effects of drugs and toxicants on biliary epithelial cell function
Bhupendra Kaphalia, Ph.D., Associate Professor. Alcoholic pancreatitis and liver disease.
M. Firoze Khan, Ph.D., Associate Professor. Xenobiotics, oxidative stress, and signal transduction.
George C. Kramer, Ph.D., Professor. Cardiovascular and respiratory physiology: microcirculation, bioengineering, and pathology.
Stanley Lemon, M.D., Professor. Molecular mechanisms of positive-strand RNA virus replication and the pathogenesis of human hepatitis viruses, particularly hepatitis C.
Kui Li, M.D., Ph.D., Assistant Professor.
Peter Mason, Ph.D., Professor. Development of a detailed understanding of the pathogenesis of West Nile and dengue disease for the development of new treatments.
Jere McBride, Ph.D., Assistant Professor. Obligate intracellular bacteria pathobiology, host response, diagnostics, and vaccine development.
Michael R. McGinnis, Ph.D., Professor. Yeast gene expression in microgravity.
Gregg Milligan, Ph.D., Associate Professor. Activation and effector function of innate and acquired immunity in the female genital tract and peripheral nervous system.
Mary Treinen Moslen, Ph.D., Professor. Pathophysiology of xenobiotic-mediated small intestinal ulceration.
Vladimir Motin, Ph.D., Associate Professor. Yersinia pestis pathogenesis, vaccines, and therapeutics.
William O'Brien, M.D., Professor. Pathogenesis of HIV infection, including mechanism of viral entry and replication, cellular cofactors, and HIV drug resistance.
Anthony O. Okorodudu, Ph.D., Professor. Intracellular cations with respect to cell membrane changes in aging.
Juan Olano, M.D., Associate Professor. Pathogenesis of rickettsial and ehrlichial infections.
Slobodan Paessler, D.V.M., Assistant Professor. Viral pathogenesis and vaccine development.
Clarence J. Peters, M.D., Professor. Emerging and tropical diseases; virology and viral epidemiology.
John R. Petersen, Ph.D., Professor. Development of clinical methodologies and analytical validation of these methods.
Johnny W. Peterson, Ph.D., Professor. Pathogenesis of toxin-mediated bacterial infections, with emphasis on development of therapeutics for cholera and anthrax.
Vsevolod Popov, Ph.D., Professor. Ultrastructure of intracellular pathogens and their interactions with host cells.
Rick Pyles, Ph.D., Assistant Professor. Development of immunomodulating interventions and vaccines against sexually transmitted infections.
Chiaho Shih, Ph.D., Professor. Human cancer with viral etiology.
Anthony Simmons, M.D., Ph.D., Professor. Herpes simplex; immunology; antibody engineering.
Lynn Soong, M.D., Ph.D., Associate Professor. Protective immunity and pathogenesis of leishmaniasis.
Robert E. Tesh, M.D., Professor. Arthropod-borne and rodent-associated viral diseases; vector biology.
Daniel L. Traber, Ph.D., Professor. Pathophysiology of sepsis and acute lung injury.
Smita Vaidya, Ph.D., Professor. Histocompatibility antigens: their structure and function in association with transplantation.
Roger Vertrees, Ph.D., Assistant Professor. Use of hyperthermia to study molecular characteristics in cancer cells.
David H. Walker, M.D., Professor and Chairman. Pathobiology of and immunity to obligately intracellular bacterial pathogens.
Jonathan B. Ward Jr., M.D., Professor. Genetic toxicology in human populations and animals.
Douglas M. Watts, Ph.D., Professor. Evaluation of candidate vaccines and therapeutics for selected viral disease pathogens in laboratory and feral animal models.
Scott C. Weaver, Ph.D., Professor. Ecology and evolution of arboviral diseases, vaccine development; arbovirus-vector interactions.
Steven Weinman, M.D., Ph.D., Professor. Cellular pathogenesis of liver diseases; mitochondrial effects of hepatitis viruses.
Shu-Yuan Xiao, M.D., Associate Professor. Pathogenesis and pathology of viral diseases, including flaviviruses and phleboviruses.
Xue-jie Yu, M.D., Ph.D., Associate Professor. Pathogenesis of obligately intercellular bacteria and the mechanism of persistent ehrlichial infection.

CURRENT RESEARCH PROGRAMS

The experimental pathology programs described below are independent but closely interconnected and multidisciplinary. Further information can be found under the University's listing for toxicology and emerging and tropical infectious diseases training program in this edition of *Peterson's Guide to Graduate Programs in the Biological Sciences*. The program offers students a unique opportunity for close interactions with nationally and internationally prominent faculty members in the area of experimental pathology. Faculty research programs are funded by peer-reviewed sources, including the NIH, EPA, DARPA, DTRA, WHO, NASA, and American Heart Association. Faculty members have active research programs in virology, parasitology, arthropod vector biology, pathobiology and immunobiology of infectious and tropical diseases, *Rickettsia* and other intracellular pathogenic organisms, signal transduction, mechanisms of chemical injury, environmental toxicology, neuropathology, and cardiovascular disease.

Molecular Biology and Infectious Disease. Specialized training and research are available in many areas, including emerging and tropical diseases, biodefense, virology, parasitology, molecular mechanisms of infectious disease and pathogen transmission, vector biology, cellular signal transduction, pathogenesis, and mechanisms of central and peripheral nervous system injury due to infectious agents. Research ranges from the molecular genetic level to clinical and epidemiologic questions. The Center for Biodefense and Emerging Infectious Diseases at the University of Texas Medical Branch includes research programs on infectious diseases found in the U.S. and in other regions around the world. A number of viral (arenavirus, flavivirus, hantavirus, hepadnavirus, filovirus, and alphavirus), bacterial (*Shigella, Rickettsia, Ehrlichia, Bacillus anthracis, Vibrio cholera,* and *Salmonella*), parasitic (leishmania and malaria), and fungal diseases are being studied. In addition, epidemiological and medical entomological studies on insect- and rodent-borne diseases are being investigated. Center members are also engaged in the development of novel therapeutics and vaccines against potential biowarfare agents and naturally emerging infectious agents. The center includes the World Arbovirus Reference Center and faculty members who are internationally renowned in the areas of arbovirology, medical entomology, and vector biology.

Pathobiology. Research emphasizes the cellular mechanisms of a variety of disease processes, including neoplasia, immunology, ischemia, and aging. Specific areas of interest include alterations in the blood-brain barrier in chronic inflammatory and infectious diseases, perinatal hypoxic-ischemic brain injury, the inflammatory response to smoke inhalation, alterations in intracellular cation balance in aging, and comparative pathology, particularly of the dolphin and other marine mammals.

Toxicology. Researchers examine mechanisms for adverse effects of environmental toxicants and drugs. Laboratories examine the roles of chemical bioactivation, adduct formation, oxidative stress, transporter dysfunction, and DNA damage in chemically induced diseases. Compounds of particular interest are allyamine, alcohols, aniline, methylene dianiline, haloacetonitriles, toluene, antineoplastic drugs, and nonsteroidal anti-inflammatory drugs. Experimental systems utilized include animal models that address systemic problems; pathophysiological, whole-organ preparations that characterize target-tissue injury; and cell systems that define specific molecular responses. Laboratories focus on atherosclerosis and aneurysm; cancer; hematological disorders; splenic toxicity, hepatobiliary injury, and pancreatic damage; and intestinal ulceration.

UNIVERSITY OF WASHINGTON

School of Public Health and Community Medicine
Program in Pathobiology

Programs of Study

The Program in Pathobiology is unique in its focus on host pathogen interactions and human disease. It has basic research programs in bacterial pathogens, host responses, human disease, parasites and fungi, and viruses (including HIV). Pathobiology is a basic science program that is part of the School of Public Health and Community Medicine because of its interest in disease and public health.

The Program in Pathobiology trains Ph.D. students in basic research with an emphasis on disease. In the first year, students rotate through three laboratories to become acquainted with various research programs, eventually identifying a faculty mentor with whom they complete their Ph.D. Course work is completed within the first two years. Courses emphasize basic mechanisms and experimental techniques used in the study of disease. Also, at the end of the first year, students complete preliminary examinations to qualify for further study. Entry into the Ph.D. program requires only a B.S. or B.A. degree. M.S. degrees can be awarded. Applicants should contact the Program for additional information if considering an M.S. degree.

Research Facilities

Faculty laboratories are located throughout Seattle, including concentrations of faculty members in the Health Science Complex at the University, the Fred Hutchinson Cancer Research Center, the Seattle Biomedical Research Institute, the Harborview Medical Center, and the Northwest Biological Research Center. Some additional faculty members are located in biotech firms throughout Seattle. Students in the Program in Pathobiology benefit from this diversity by having access to faculty members and equipment at these various locations. To encourage interactions between faculty members and students, members of the Program attend a fall retreat that includes talks by faculty members and student poster presentations. In addition, students present their work at a student symposium in the spring and attend Program seminars throughout the year.

Financial Aid

All students in the Program in Pathobiology are supported by research assistantships and/or fellowships that include a competitive stipend, health insurance, and tuition. Most funding is provided through faculty research grants. Additional fellowships are available through training grant programs in the Program in Pathobiology and other programs. Application materials are available from the Web site at http://depts.washington.edu/pathobio.

Cost of Study

Full-time tuition is approximately $11,740 per year for Washington residents and $27,524 per year for nonresidents in 2006–07. Tuition is paid for students receiving financial support through research assistantships or fellowships.

Living and Housing Costs

The University has residence halls for single students as well as housing units for full-time students who have a spouse/domestic partner and dependent children. Houses, private rooms, and apartments are available in the surrounding neighborhoods and within walking distance of the campus. There are a variety of food service plans and dining options available on campus, as well as food markets off-campus including co-ops.

Student Group

In the 2006–07 academic year, there are 45 graduate students in the Program in Pathobiology (34 women, 11 men). The Program is diverse and more than 20 percent of students are international. There are approximately 825 graduate students in the Health Sciences complex, and the University has a student body of more than 36,000 students.

Student Outcomes

After completing a Ph.D. degree, students usually advance to postdoctoral work in their chosen field. Some students who complete their Ph.D. degree accept positions in pharmaceutical or biotechnology companies or laboratories in public health departments.

Location

Located in Seattle, there are diverse cultural and recreational opportunities, including seasonal festivals and music venues within the Emerald City. Seattle is a short drive to the Cascade and Olympic Mountains and is located near both salt and fresh water.

The School

The School of Public Health and Community Medicine (SPHCM) at the University of Washington is a leader in public health sciences, with departments in Biostatistics, Environmental Health, Epidemiology, Global Health, and Health Services—with several interdepartmental and interschool programs, including Pathobiology.

Applying

The application deadline is December 15 for U.S. citizens and permanent residents (November 1 for international students). Applicants must submit the following to the Program in Pathobiology: a Program application, answers to six essay questions, a curriculum vitae, three letters of recommendation, transcripts, and official Graduate Record Examinations (GRE) scores. For further details, applicants should visit the Web site at http://depts.washington.edu/pathobio. International applicants must also submit scores on the Test of English as a Foreign Language (TOEFL). All applicants must also submit a Graduate School application, fee, and transcripts to the Graduate School of the University. Late applications will not be considered.

Correspondence and Information

Manager of Student Services
Program in Pathobiology
Box 357238
University of Washington
Seattle, Washington 98195-7238

Phone: 206-543-4338
Fax: 206-543-3873
E-mail: pathobio@u.washington.edu
Web site: http://depts.washington.edu/pathobio

University of Washington

THE FACULTY AND THEIR RESEARCH

The faculty of the Program in Pathobiology is a diverse set of researchers working with a variety of pathogens (bacteria, parasite, fungi, virus), host responses (cellular and humoral immune response, wounding, cell-cell interactions), and disease states (metabolism diet). The faculty uses a variety of techniques, including genetics, molecular biology, genomics, proteomics, cell biology, immunology, biochemistry, and microbiology.

Lee Ann Campbell, Professor and Graduate Program Director; Ph.D., Penn State, 1982. Molecular biology and pathogenic mechanisms of chlamydiae.

Gerard Cangelosi, Research Assistant Professor; Ph.D., California, Davis, 1984. Molecular biology; environmental monitoring; clinical detection of pathogenic mycobacteria.

William Carter, Professor; Ph.D., California, Davis, 1974. Elucidation of components in cell attachment and cell spreading in normal cells.

Arturo Centurion-Lara, Adjunct Research Associate Professor; M.D., San Marcos (Peru), 1986. *Treponema pallidum,* the etiologic agent of syphilis.

Patrick Duffy, Affiliate Associate Professor; Ph.D., M.D., Duke, 1986. Malaria pathogenesis and vaccine development.

Michael Emerman, Affiliate Professor; Ph.D., Wisconsin–Madison, 1986. Interactions of the human immunodeficiency virus (HIV) with its host cells.

Jean Feagin, Associate Professor; Ph.D., Stanford, 1982. Molecular parasitology, emphasizing organelle gene organization and expression in protozoans.

Nancy Freitag, Associate Professor; Ph.D., UCLA, 1989. *L. monocytogenes* as a model system for understanding intracellular parasitism.

Michal Fried, Affiliate Assistant Professor; Ph.D., Hebrew (Israel), 2000. Malaria.

Nancy Haigwood, Professor; Ph.D., North Carolina, 1980. Host immunity in the control and prevention of AIDS.

Sen-itiroh Hakomori, Professor Emeritus; M.D., 1951, D.Med.Sci., 1956, Tohoku (Japan). Role of glycophingolipids in defining antigenicity, cellular interaction, and signal transduction.

John Hansen, Affiliate Associate Professor; Ph.D., Oregon State, 1996. Host–pathogen interactions in trout; functional genomics.

Amanda Jones, Adjunct Research Assistant Professor; Ph.D., Calgary, 1997. Pathogenesis of group B streptococcal (GBS) disease.

Stefan Kappe, Affiliate Assistant Professor; Ph.D., Notre Dame, 1998. Biology of pre-erythrocytic stages of the malaria parasite *Plasmodium*.

George E. Kenny, Professor Emeritus; Ph.D., Minnesota, 1961. Human immune response to infectious diseases; detection and biology of mycoplasmas.

Elizabeth A. Kirk, Assistant Professor; Ph.D., R.D., Washington (Seattle), 1995. Nutrient–gene interactions and their impact on chronic diseases such as atherosclerosis and diabetes.

David Koelle, Adjunct Associate Professor; M.D., Washington (Seattle), 1985. Cellular immune response to herpes simplex virus type 2 (HSV-2).

Cho-Chou Kuo, Professor; M.D., National Taiwan, 1960; Ph.D., Washington (Seattle), 1970. Antigenic analysis; immunology and pathogenesis of chlamydiae.

Gael Kurath, Affiliate Associate Professor; Ph.D., Oregon State, 1984. Molecular biology and evolution of RNA viruses that infect fish.

Paul Lampe, Research Associate Professor; Ph.D., Minnesota, 1984. Regulation of intercellular communication via gap junctions.

Renee C. LeBoeuf, Adjunct Professor; Ph.D., SUNY at Buffalo, 1977. Genetic and nutritional regulation of proteins involved in lipid transport.

Jaisri R. Lingappa, Associate Professor; Ph.D., Harvard, 1985; M.D., Massachusetts, 1987. Cell biology of virus assembly; host proteins involved in assembly of HIV and other viruses.

Maxine Linial, Adjunct Research Professor; Ph.D., Tufts, 1970. Retroviral replication and host cell interactions.

Sheila A. Lukehart, Adjunct Research Professor; Ph.D., UCLA, 1978. Antigens of Treponema pallidum; immunity to treponemal diseases.

Julia McElrath, Adjunct Professor; Ph.D., Medical University of South Carolina, 1978. HIV-specific CD4^{++} and CD8^{++} T cells.

Peter Myler, Research Associate Professor; Ph.D., Queensland (Australia), 1982. Regulation of gene expression in protozoan parasites.

Julia Overbaugh, Affiliate Professor; Ph.D., Colorado, 1983. How retroviruses disrupt T-cell function and cause disease in the host.

Marilyn Parsons, Professor; Ph.D., Stanford, 1979. Molecular and cellular parasitology.

Steve Polyak, Adjunct Research Assistant Professor; Ph.D., McMaster (Canada), 1993. Molecular virology and immunology.

Pradip Rathod, Adjunct Professor; Ph.D., Oregon Health Sciences, 1982. Malaria.

Robert L. Rausch, Professor Emeritus, Ohio State, 1945; Ph.D., Wisconsin, 1949; DVM. Parasitology; helminthic zoonoses.

Steven Reed, Research Professor; Ph.D., Montana, 1979. Immunological response to mycobacteria infections.

Marilyn Roberts, Professor; Ph.D., Washington (Seattle), 1978. Antibiotic resistance genes; mercury resistance; oral disease.

Timothy Rose, Associate Professor; Ph.D., Geneva (Switzerland), 1981. Molecular biology of tumor viruses; cell growth, differentiation, and transformation.

Michael E. Rosenfeld, Professor; Ph.D., Wisconsin–Madison, 1981. Mechanisms of atherogenesis, inflammation, and vascular cell gene expression.

David Sherman, Associate Professor; Ph.D., Vanderbilt, 1987. Molecular genetics, microbiology, and biochemistry of pathogenic mycobacteria.

Peter Small, Affiliate Associate Professor; M.D., Florida, 1985. Nature and consequences of genetic diversity within the species *M. tuberculosis*.

Arnold Smith, Professor; M.D., Missouri, 1964. *Haemophilus influenzae*.

Joseph Smith, Research Assistant Professor; Ph.D., Washington (Seattle), 1994. Clonal antigenic variation and cytoadherence of infected erythrocytes caused by a variable family of *Plasmodium falciparum* adhesion ligand proteins.

Leo Stamatatos, Associate Professor; Ph.D, McGill (Canada), 1988. HIV-cell interactions and vaccine development.

Kenneth Stuart, Professor; Ph.D., Iowa, 1969. Molecular biology of protozoan pathogens.

Patricia Totten, Adjunct Associate Professor; Ph.D., Washington (Seattle), 1990. Pathogenesis of chancoid and the discovery of novel organisms associated with STD syndromes.

Wesley Van Voorhis, Adjunct Associate Professor; Ph.D., M.D., Rockefeller, 1983; M.D., Cornell, 1984. Trypanosome, leishmania, plasmodia, *Treponema pallidum,* and chlamydia infections.

Theodore White, Professor; Ph.D., Michigan, 1984. Molecular mechanisms of virulence and drug resistance in pathogenic yeasts.

UNIVERSITY OF WISCONSIN–MADISON

Department of Pathology and Laboratory Medicine

Program of Study

A student can plan to study just about any biological process but will always keep a central focus on understanding the pathogenesis of human disease while in a pathology Ph.D. program in the Department of Pathology and Laboratory Medicine at the University of Wisconsin–Madison (UW–Madison). Applicants should be interested in bridging the gap between basic discovery and its application to the treatment of human diseases and come to the program with a solid background in the biological sciences. The Ph.D. program in pathology does not include forensic pathology.

The cellular and molecular pathology program at UW–Madison features four main biologic foci: cancer biology, immunology, extracellular matrix, and neuroscience. These diverse research areas are tied together by the common goal of elucidating the mechanisms of human disease. The classroom aspect of the program is focused on disease pathogenesis. Graduate students are free to fulfill their remaining course work requirements in a focused minor, depending on their area of research, from courses in numerous different departments. The breadth of course work options and research interests of UW–Madison trainers allow for students to receive cutting-edge training in all aspects of modern cellular and molecular biology. Prospective students are welcome to explore the program's Web site to find more information on the program and the research being performed by members of the faculty.

Research Facilities

The Department is housed in the Medical School complex at the center of the Madison campus and at the Clinical Sciences Center on the west side of campus. The Department has modern research facilities including centralized support facilities. In addition, the Integrated Microscopy Resource on the Madison campus is a national microscopy center that provides equipment and support for scanning and transmission of electron microscopy and video-enhanced fluorescence microscopy. Extensive library facilities, including departmental collections, the Medical School Library, and extensive holdings at other science libraries on the Madison campus, are available.

Financial Aid

Support for students is from a variety of sources, including Department teaching assistantships, departmental research assistantships, and research assistantships from individual faculty research grants. In addition, students compete for University fellowships and research assistantships on several training grants on the Madison campus.

Cost of Study

If possible, students receive stipend support during the duration of their graduate study. Support is offered via research assistantships, project assistantships, fellowships, or training grant positions provided by the Department, a thesis adviser's research funds, or other campus sources. For the 2006–07 school year, the stipend rate for graduate students in the Department of Pathology and Laboratory Medicine is $22,725 with an increase of approximately 4 percent annually. In addition, assistantship recipients qualify for full remission of nonresident and resident tuition. In 2005–06, tuition and fees were approximately $4370 per semester for Wisconsin residents and $12,005 per semester for nonresidents. Assistantships also allow for subsidized enrollment in the Graduate Assistant Health Insurance Program.

Living and Housing Costs

For single graduate students, the University maintains graduate student apartments that offer one- and two-bedroom units; rent ranged from $525 to $700 per month in 2005–06. University Student Apartments, better known as Eagle Heights, offers one- to three-bedroom unfurnished apartments; rent ranged from $585 to $925 per month in 2005–06. There may be a waiting list for the Eagle Heights units, and priority is given to students with dependent children. For more details, students should visit the Division of University Housing Web site at http://www.housing.wisc.edu. Most students live off campus; costs for off-campus housing vary depending on size and location. Additional information about off-campus housing is available at the UW–Madison Campus Information, Assistance, and Orientation Web site at http://www.wisc.edu/cac/housing/.

Student Group

The Madison campus is the flagship of the University of Wisconsin with an enrollment of more than 42,000 students, including 10,000 graduate students. In the sciences, the graduate students belong to individual department graduate programs or are members of interdepartmental training programs. The Department of Pathology and Laboratory Medicine is seeking to expand the graduate program in cellular and molecular pathology.

Student Outcomes

Students in this program have successfully pursued a number of options after obtaining their Ph.D. degrees. They have received postdoctoral training and obtained faculty positions in academic institutions, taken research positions in industry, or continued on to medical school and secured faculty positions in medical institutions.

Location

Madison, recently ranked as one of the top cities of America to live in, is the capital of the state, with a metropolitan population of approximately 215,000. The city, situated on four picturesque lakes, is approximately 150 miles northwest of Chicago. The city and the University offer a wide variety of educational, cultural, and recreational opportunities. Superb facilities are available for summer and winter sports, such as sailing, camping, hiking, ice-skating, skiing, and bicycling.

The University

The University, founded in 1848, is one of the Big Ten schools and has a rich tradition of excellence in research. The University System includes the main campus in Madison plus twelve other comprehensive universities, thirteen freshman/sophomore campuses (UW Colleges), and the UW–Extension Program.

Applying

Applicants should have a bachelor's degree and an undergraduate minimum grade point average of 3.0 (on a 4.0 scale). Applicants should have a strong background in organic and physical chemistry, biochemistry, biology, (including genetics), and mathematics through calculus. Completed application forms, Graduate Record Examinations (GRE) scores, transcripts, and a minimum of three letters of recommendation are required for an admission decision. Applications for fall admission are considered from December through April. Priority consideration is given to completed applications that are received by January 1.

Correspondence and Information

Chair, Graduate Education Committee
Department of Pathology
6120 Medical Science Center
1300 University Avenue
University of Wisconsin
Madison, Wisconsin 53706
Phone: 608-262-2665
Fax: 608-265-3301
E-mail: gradinfo@pathology.wisc.edu
Web site: http://www.pathology.wisc.edu/gradprogram/

University of Wisconsin–Madison

THE FACULTY AND THEIR RESEARCH

Caroline Alexander, Assistant Professor, Oncology. Role of breast stem cells in tumor induction; multiple functions of Wnt signaling in the regulation of mammary epithelial cell growth; changes in tumor susceptibility that are effected by alterations of normal development.

B. Lynn Allen-Hoffmann, Professor, Pathology. Keratinocyte growth and differentiation; cancer biology; extracellular matrix; integrin receptors; apoptosis.

Craig S. Atwood, Assistant Professor, Medicine. Hormonal regulation of aging and neurodegenerative diseases.

Grace Boekhoff-Falk, Associate Professor, Anatomy. Molecular biology of limb development.

Emery H. Bresnick, Professor, Pharmacology. Regulation of transcription, hematopoiesis, and leukemogenesis.

William J. Burlingham, Professor, Surgery. Mechanisms of transplant tolerance and rejection.

Joshua Coon, Assistant Professor, Chemistry. Bioanalytical chemistry; mass spectrometry.

Arjang Djamali, Assistant Professor, Medicine. The mechanisms of disease progression in kidney disease.

Karen Downs, Associate Professor, Anatomy. Development of the murine umbilical cord; patterning of mesoderm; vasculogenesis.

Zsuzsa Fabry, Professor, Pathology. Inflammatory reactions of the central nervous system; vascular biology.

Andreas Friedl, Associate Professor, Pathology. Heparan sulfate proteoglycans as modulators of growth factors in human disease; tumor angiogenesis.

Michael Fritsch, Assistant Professor, Pathology. Molecular pathways leading to neoplastic transformation; pediatric germ cell tumors.

Daniel Greenspan, Professor, Pathology. Eukaryotic genes: structure and expression, roles in development and genetic diseases of humans.

Anne Griep, Professor, Anatomy. Role of tumor suppressor proteins; growth factors and caspases.

Jeff Hardin, Associate Professor, Zoology. Morphogenesis and pattern formation during early development.

Michael N. Hart, Professor and Chair, Pathology. Immunology of the blood-brain barrier.

Anna Huttenlocher, Associate Professor, Pharmacology. Tumor invasion/metastasis and the chemotaxis of inflammatory cells.

Nizar Jarjour, Associate Professor, Medicine. Asthma; circadian rhythm; investigative bronchoscopy.

Juan Jaume, Assistant Professor, Medicine. Endocrine autoimmunity, type 1 diabetes, and thyroid diseases.

Bruce Klein, Professor, Pediatrics. Microbial immunology and pathogenesis.

Youngsook Lee, Assistant Professor, Anatomy. Transcriptional control of cardiovascular development and mechanisms of cardiac-specific gene regulation.

Gary E. Lyons, Associate Professor, Anatomy. Developmental biology of the mammalian cardiovascular system.

James S. Malter, Professor, Pathology. Posttranscriptional gene regulation; gene therapy; cytokine biology.

Albee Messing, Professor, Veterinary Medicine. Transgenic mice; developmental neuropathology; molecular neurobiology.

Deane F. Mosher, Professor, Medicine. Extracellular matrix and cell adhesion.

Terry D. Oberley, Professor, Pathology. Role of reactive oxygen metabolism in vivo and in vitro.

David O'Connor, Assistant Professor, Pathology. Understanding the relationships between genetics and pathogenesis during AIDS virus infections.

Julie Olson, Assistant Professor, Neurological Surgery. Role of resident immune cells and infiltrating peripheral immune cells in the response to virus infection in the CNS.

Donna P. Peters, Professor, Pathology. Structure and regulation of extracellular matrix formation.

Erik Ranheim, Assistant Professor, Pathology. Interaction of the immune system with tumor cells; role of frizzled/wnt/beta-catenin pathway in normal and neoplastic lymphoid development.

Alan Rapraeger, Professor, Pathology. Cell surface proteoglycans; heparin-binding growth factors.

Matyas Sandor, Professor, Pathology. Role of T cells in infectious diseases; Fc receptors on lymphocytes.

Christine M. Seroogy, Assistant Professor, Pediatrics. Novel E3 ubiquitin ligase called GRAIL in T-cell function and hematopoietic tissue development.

John Sheehan, Assistant Professor, Medicine. Blood coagulation, intrinsic tenase regulation, coagulation factor IX.

Nader Sheibani, Assistant Professor, Ophthalmology and Visual Sciences and Pharmacology. Cell adhesion and signaling in vascular cells, diabetic retinopathy.

Igor Slukvin, Assistant Professor, Pathology. Hematopoietic differentiation of human embryonic stem cells; immune-privileged properties of embryonic and fetal tissues.

Paul M. Sondel, Professor, Pediatrics. Immune-mediated recognition and destruction of neoplasms.

Gary A. Splitter, Professor, Animal Health and Biomedical Science. Host-microbe interaction and pathogenesis, immunology.

Dandan Sun, Associate Professor, Neurosurgery. The role of Na+—K+—2Cl- cotransporter in ion homeostasis and cell volume regulation in the central nervous system.

M. Suresh, Associate Professor, Pathobiological Sciences. Regulation of antiviral T cell responses at the molecular level.

Adel Talaat, Assistant Professor, Animal Health and Biomedical Science. Genomic and functional analyses of tuberculosis and paratuberculosis to understand pathogenesis and develop novel vaccines.

David I. Watkins, Professor, Pathology. MHC class I structure, function, and evolution.

Weixiong Zhong, Assistant Professor, Pathology. Molecular mechanisms of selenium in human prostate cancer chemoprevention.

VIRGINIA COMMONWEALTH UNIVERSITY

Medical College of Virginia
Department of Pathology

Programs of Study

The Department of Pathology of the School of Medicine offers the Ph.D. degree. All candidates for this degree are trained to be both research scientists and teachers. Although emphasis is placed on experimental pathology, opportunities are provided for high-quality instruction with clinical correlation in specialized areas of laboratory medicine, such as analytical and environmental toxicology, the pathobiology of cancer, clinical chemistry, electron microscopy, functional genomics, immunopathology, molecular pathology, and neuropathology.

Research Facilities

The department is fully equipped to carry out research in all aspects of pathology. In addition to a core laboratory, facilities and sophisticated instruments are available for such processes as light, fluorescence, and electron microscopy (transmission and scanning); chromatography; electrophoresis; analytical and preparative ultracentrifugation; immunochemistry; virology; cell and tissue culture; molecular biology, including PCR facilities; and radioisotope analysis. Modern computer facilities and other specialized types of equipment (e.g., a fluorescence-activated cell sorter and a laser capture microdissection instrument) are also readily available.

Financial Aid

Students are urged to compete for outside fellowships or scholarships. However, departmental fellowships and teaching assistantships are available on a competitive basis for the 2006–07 year (the amount being dependent upon the length of the award). Financial aid is awarded in special cases, according to eligibility.

Cost of Study

In 2006–07, tuition and fees are $3391 per semester for Virginia residents and $7952 semester for nonresidents. Summer 2006 tuition and fees were $2261 for Virginia residents and $5301 for nonresidents. A consolidated fee, included in the tuition cost, helps support student facilities, campus development, intercollegiate athletics, and the MCV Campus Student Health Service.

Living and Housing Costs

For most students living on or off campus, rent, food, transportation, and personal expenses average $10,000 for the academic year.

Student Group

The total enrollment at VCU is more than 20,645, of whom more than 4,690 are enrolled in the School of Graduate Studies. The Department of Pathology has approximately 10 graduate students in its Ph.D. program.

Student Outcomes

Graduate training in pathology offers a wide spectrum of employment options, ranging from research to laboratory management. Many program graduates are employed as university or government research scientists, directors of clinical laboratories, undergraduate teaching faculty members, and technical consultants for private industry.

Location

Richmond, the state capital of Virginia, has a population of 750,000. The city is a cosmopolitan environment in both classic and contemporary styles. Culturally, it is endowed with six museums, nine theaters, thirty-one municipal parks, and the first state-supported fine arts museum in the nation. It is so situated that Washington, D.C.; Colonial Williamsburg; Jamestown and the James River Plantations; the Atlantic Ocean; Monticello; the Blue Ridge Mountains; and the Skyline Drive can each be reached within a 2-hour drive.

The University and The Department

The Medical College of Virginia, founded in 1837, is among the four most influential health centers in the United States today, according to a recent study. In 1968, the Medical College of Virginia became the Health Sciences Division of VCU. The other component of VCU, known as the Academic Campus, was formerly the Richmond Professional Institute, founded in 1917. The Department of Pathology, located at the Medical College of Virginia on the downtown campus, is a department of the School of Medicine and an integral part of the School of Graduate Studies.

Applying

The Board of Visitors, the administration, and the faculty of VCU are committed to a policy of equal opportunity in education without regard to race, creed, sex, national origin, age, or physical handicap. To be considered for admission, applicants must hold a baccalaureate or an equivalent degree, with a major in the biological sciences or chemistry, prior to admission. VCU requires scores on the General Test of the Graduate Record Examinations. International applicants are required to take the Test of English as a Foreign Language (TOEFL). Applications are accepted until March 15.

Correspondence and Information

Dr. Shawn E. Holt
Department of Pathology
Medical College of Virginia
Box 980662
Richmond, Virginia 23298-0662
Phone: 804-828-4802
Fax: 804-828-9749
E-mail: seholt@hsc.vcu.edu
Web site: http://www.vcu.edu/gradweb/programs/medphdpath.html

Virginia Commonwealth University

THE FACULTY AND THEIR RESEARCH

F. P. Anderson, Assistant Professor (Clinical Chemistry); Ph.D., Virginia Commonwealth, 1988. Immunoassay development emphasizing reversible, biocompatible assay systems for low-concentration ligands of clinical importance.
A hospital laboratory's experience with TOSOH AIA-1200. *J. Clin. Immunol.* 15:222–28, 1992. Antibody properties for chemically reversible biosensor applications. *Anal. Chim. Acta* 227:135–43, 1989 (with Miller). Fiber optic immunochemical sensor for continuous, reversible measurement of phenytoin concentration. *Clin. Chem.* 34:1417–21, 1988 (with Miller).

A. Ferreira-Gonzalez, Associate Professor (Molecular Diagnostics); Ph.D., George Washington, 1994. Mechanism of carcinogenesis; application of molecular biology techniques to clinical diagnosis and management of disease; minimal residual disease; infectious disease.
Assessing clinical utility of hepatitis C virus quantitative RT-PCR data: Implications for identification of responders among α-interferon-treated patients. *Mol. Diagn.* 1(2):1–11, 1996 (with Shiffman et al.). *Ras* oncogene activation during hepatocarcinogenesis of the B6C3F1 male mice by dichloroacetic and trichloroacetic acid. *Carcinogenesis* 16(3):495–500, 1995 (with DeAngelo and Garrett).

C. T. Garrett, Professor and Division Chair (Molecular Diagnostics); M.D., Johns Hopkins, 1966; Ph.D., Wisconsin–Madison, 1977. Application of DNA-based molecular testing, primarily PCR, to clinical diagnosis and evaluation of human disease; mechanisms of carcinogenesis.
Molecular detection of resistance to activated protein C due to a common mutation in the coagulation factor V gene causing hereditary predisposition to thrombosis. *J. Clin. Lab. Analysis* 11:328–35, 1997 (with Ferreira-Gonzalez et al.). Correlations between p21 expression and clinicopathological findings, p53 gene and protein alterations, and survival in patients with endometrial carcinoma. *J. Pathol.* 183:318–24, 1997 (with Ito et al.). FDA regulation of analyte specific reagents (ASRs): Implications for nucleic acid–based molecular testing. *Diagn. Mol. Pathol.* 5:151–3, 1996 (with Ferreira-Gonzalez).

S. E. Holt, Associate Professor (Cellular and Molecular Pathogenesis); Ph.D., Texas A&M, 1994. Telomeres and telomerase in aging and cancer.
Extension of life-span by introduction of telomerase into normal human cells. *Science* 279:349–52, 1998 (with Bodnar et al.). Lack of cell cycle regulation of telomerase activity in human cell lines. *Proc. Natl. Acad. Sci. U.S.A.* 94:10687–92, 1997. Regulation of telomerase activity in immortal cell lines. *Mol. Cell. Biol.* 16:2932–9, 1996.

W. G. Miller, Professor (Information Systems and Clinical Chemistry); Ph.D., Arizona, 1973. Fiber-optic immunosensors; interlaboratory standardization, proficiency testing, and quality control; informatics.
The accuracy of laboratory measurements in clinical chemistry: A study of eleven routine analytes in the frozen serum, definitive methods, and reference methods. *Arch. Pathol. Med.* 122:587–608, 1998 (with Ross et al.). A fiber-optic immunosensor for myoglobin. *Clin. Chem.* 43:2128–36, 1997 (with Hanbury et al.). Matrix effects in the measurement and standardization of lipids and lipoproteins. In *Handbook of Lipoprotein Testing*, eds. N. Rifai, G. Warnick, and M. Dominiczac. AACC Press, 1997.

Y. Oh, Professor (Cellular and Molecular Pathogenesis); Ph.D., Stanford, 1993. Genomic and proteomic approaches to defining the pathophysiological role of IGFBP-related proteins in human cancer.
IGF-independent effects of the IGFBP superfamily. In *Insulin-Like Growth Factor,* eds. D. Le Roith, W. Zumkeller, and R. Baxter. Georgetown, Tex.: Landes Bioscience, 2003 (with Walker, Kim, and Yang). Butyrate, a histone deacetylase inhibitor, activates the human IGFBP-3 promoter in breast cancer cells: Molecular mechanism involves an Sp1/Sp3 multiprotein complex. *Endocrinology* 142:3817–27, 2001 (with Walker, Wilson, and Powell). Insulin-like growth factor binding proteins in endocrine-related neoplasia. In *Endocrine Oncology, Contemporary Endocrinology,* vol. 17, pp. 215–36, ed. S. P. Ethier. Totowa, N.J.: Humana Press, 2000 (with Minniti).

A. Poklis, Professor (Clinical Toxicology); Ph.D., Maryland, 1974. Development and application of new analytical methods for clinical and experimental pharmacodynamic and pharmacokinetic studies of drugs and intoxicants.
Urinary excretion of d-amphetamine following oral doses in man: Implications for urine drug testing. *J. Anal. Toxicol.* 22(6):481–6, 1998 (with Still et al.). The detection of acetylcodeine and 6-monoacetylmorphine in opiate positive urines. *Forensic Sci. Int.* 95(1):1–10, 1998 (with O'Neal). Forensic toxicology. In *An Introduction to Forensic Sciences*, pp. 107–32, ed. W. Eckert. Boca Raton, Fla.: CRC Press, 1997. Analytical/forensic toxicology. In *Casarett and Doull's Toxicology: The Basic Science of Poisons*, fifth edition, Chap. 31, pp. 951–68, eds. K. Klassen, M. Amdur, and J. Doull. New York: McGraw-Hill, 1995.

A. E. Sirica, Professor and Division Chair (Cellular and Molecular Pathogenesis); Ph.D., Connecticut, 1977. Hepatic and biliary cancer.
Immunohistochemical demonstration of MET overexpression in human intrahepatic cholangiocarcinoma and in hepatolithiasis. *Hum. Pathol.* 29:175–80, 1998 (with Terada et al.). A new rat bile ductular epithelial cell culture model characterized by the appearance of polarized bile ducts in vitro. *Hepatology* 26:537–49, 1997 (with Gainey). NEU overexpression in the furan rat model of cholangiocarcinogenesis compared with biliary ductal cell hyperplasia. *Am. J. Pathol.* 151:1685–94, 1997 (with Radaeva et al.).

J. L. Ware, Professor and Vice Chair for Research and Graduate Education (Cellular and Molecular Pathogenesis); Ph.D., North Carolina at Chapel Hill, 1979. Pathobiology of human prostate cancer and metastasis.
Metastatic sublines of an SV40 large T antigen immortalized human prostate epithelial cell line. *Prostate* 34:275–82, 1998 (with Bae et al.). Re-expression of the type 1 IGF receptor (IGF-1R) inhibits the malignant phenotype of SV40 T-immortalized human prostate epithelial cells. *Endocrinology* 138:1728–35, 1997 (with Plymate et al.). Prostate cancer progression: Implications of histopathology. *Am. J. Pathol.* 145:983–93, 1994.

WAKE FOREST UNIVERSITY

School of Medicine
Graduate School of Arts and Sciences
Program in Molecular and Cellular Pathobiology

Program of Study

The interdisciplinary Program in Molecular and Cellular Pathobiology, sponsored by the Department of Pathology, offers a course of study leading to the Ph.D. degree. The program is a research-intensive training program in experimental pathology designed to prepare students for research careers in academia or industry. Training for a career in independent research is emphasized. Through course work and seminars, students are given a firm background in biochemistry, physiology, pathobiology, immunology, cell and molecular biology, and statistics. Each student's curriculum is designed, however, according to his or her own educational background and research interests. Generally before the end of the second year, students must pass a preliminary examination designed to test their depth and breadth of knowledge as well as their research potential. Students are given the opportunity to gain research experience during the first year, and a final research project and an adviser are usually selected by the end of the second year. The average time for completion of the Ph.D. is five years. Graduates of this program have competed successfully for faculty positions in departments of pathology, biochemistry, medicine, physiology, and comparative medicine.

Research opportunities are available in the areas of cell and molecular biology, including gene discovery and translational genetics; macrophage and smooth-muscle cell biology; cancer biology; lipid and lipoprotein metabolism; arteriosclerosis; diabetes; hypertension; osteoporosis; nutrition; reproductive biology; nonhuman primate behavior; comparative genetics; and diseases of laboratory animals. Research is particularly strong in the area of cardiovascular disease, chiefly arteriosclerosis, with emphasis on the use of transgenic and gene-targeted mice and nonhuman primates as animal models for this disease.

Research Facilities

Resources of the Department of Pathology include facilities for confocal scanning and transmission electron microscopy (including an intermediate-voltage electron microscope) and tissue culture; modern and well-equipped biochemistry, molecular biology, and pathology laboratories; computer facilities; and a medical school library. A special facility, located on the Friedberg campus, includes a 200-acre research farm housing about 400 nonhuman primates, 200 transgenic mice, and a variety of other research animals. There are extensive research laboratories, including the nonhuman Comparative Medicine Clinical Research Center and the Center for Human Genomics.

Financial Aid

The Wake Forest University Graduate School of Arts and Sciences provides financial aid in the form of Graduate Fellowships to all matriculating students. Fellowships for the 2006–07 academic year pay a base stipend of $20,772 per year plus a full-tuition scholarship of $27,285, for a total of $48,057. An annual supplement that covers some of the health insurance costs (students pay $33.19 per month) is available to qualified students who enroll in the University student health-care plan. All matriculants receive a new ThinkPad computer from the Graduate School. The program also provides books, supplies, and travel for students to attend regional and national scientific meetings. After the first year, students are generally supported from other resources, such as nationally competitive training or research grants. Several students have been awarded competitive predoctoral grants from private agencies. Postdoctoral fellowships are also available to persons with Ph.D., M.D., or D.V.M. degrees, with stipends starting at $35,568, depending on postdoctoral experience.

Cost of Study

For 2006–07, tuition is $27,285 per twelve-month academic year.

Living and Housing Costs

The majority of the department's graduate students live off campus in houses or apartments, many within walking distance of the Medical Center, which is in a middle-class residential neighborhood. Rents start at $400 per month for one-bedroom apartments and range from $545 to $615 per month for two-bedroom apartments.

Student Group

The graduate programs of the Wake Forest University School of Medicine currently enroll 260 full-time students. Fifteen students are currently enrolled in the Program in Molecular and Cellular Pathobiology.

Location

Winston-Salem is a pleasant community of approximately 186,000 people located in the Piedmont area (rolling hills) of North Carolina. The climate is mild, with long and colorful spring and fall seasons. The Blue Ridge Parkway and the eastern Continental Divide are about 90 miles away, and the North Carolina beaches are within 200 miles. Recreational activities include skiing in the winter and biking, camping, and swimming in the summer. Wake Forest University offers the cultural and athletic events associated with a major university. In addition, Winston-Salem is the home of several other colleges, including the North Carolina School of the Arts, which provides a wide variety of cultural events.

The University and The School

Wake Forest University, founded in 1834, is one of the oldest institutions of higher learning in North Carolina. The School of Medicine was founded in 1902 as a two-year program and in 1941 became a four-year medical school renamed the Bowman Gray School of Medicine. The medical school is located on a 63-acre campus in a pleasant residential area of town. The Medical Center has 819 full-time faculty members, 438 medical students, and 581 residents. Facilities include an 830-bed teaching hospital. In 1997, the Medical Center was renamed Wake Forest University Baptist Medical Center. In addition to the Program in Molecular and Cellular Pathobiology, graduate programs are offered in the Departments of Biochemistry, Neurobiology and Anatomy, Physiology and Pharmacology, Microbiology and Immunology, Public Health Sciences, Neuroscience, Medical Engineering, Cancer Biology, Molecular Genetics, and Molecular Medicine.

Applying

Applications for admission are accepted from individuals with a bachelor's degree or its equivalent and an academic record that reflects achievement. The majority of applicants have majored in biology or chemistry. Applications should include transcripts, at least three letters of recommendation, Graduate Record Examinations scores, and a brief statement of career goals, including specific research interests. The deadline for applications is January 15. Application forms can be obtained by writing to the address given in this description or by visiting http://www.wfubmc.edu/graduate/.

Correspondence and Information

Dr. John S. Parks
Director of Molecular and Cellular Pathobiology Graduate Program
Department of Pathology, Section on Lipid Sciences
Wake Forest University School of Medicine
Winston-Salem, North Carolina 27157-1040
Phone: 336-716-2145
Fax: 336-716-6279
E-mail: jparks@wfubmc.edu
Web site: http://www1.wfubmc.edu/pathresearch/graduate_program/phd

Wake Forest University

THE FACULTY AND THEIR RESEARCH

Michael R. Adams, Professor of Pathology; D.V.M., Illinois at Urbana-Champaign, 1972. Atherosclerosis; medical primatology; reproductive medicine; women's health.

Isabelle M. Berquin, Assistant Professor of Pathology; Ph.D., Wayne State, 1996. Molecular pathogenesis of breast cancer; regulation of gene expression; epidermal growth factor receptor pathway; proteases in cancer.

David W. Busija, Professor of Physiology and Pharmacology; Ph.D., Kansas, 1978. Cerebral circulation; vascular biology; perinatal physiology.

Thomas B. Clarkson, Professor of Pathology and Director, Comparative Medicine Clinical Research Center; D.V.M., Georgia, 1954. Arteriosclerosis and osteoporosis; animal models; women's health; nutrition.

J. Mark Cline, Associate Professor of Pathology; D.V.M., 1986, Ph.D., 1992, North Carolina State. Hormonal carcinogenesis in mammary gland and endometrium; immunohistochemistry; comparative pathology.

John R. Crouse III, Professor of Internal Medicine; M.D., SUNY Downstate Medical Center, 1969. Cholesterol and lipoprotein metabolism in nonhuman primates and human beings.

Zheng Cui, Associate Professor of Pathology; M.D., Tsuenyi Medical School (China), 1979; Ph.D., Massachusetts Amherst, 1988. Lipid informatics; lipid metabolism; cancer immunology and cancer biology.

Paul A. Dawson, Associate Professor of Pathology and Internal Medicine; Ph.D., SUNY at Stony Brook, 1986. LDL gene expression and regulation; bile acids; cholesterol metabolism; molecular cloning; molecular genetics.

Purnima Dubey, Assistant Professor of Pathology and Microbiology/Immunology; Ph.D., Chicago, 1986. Mechanisms of antitumor immunity.

Iris J. Edwards, Associate Professor of Pathology; Ph.D., Wake Forest, 1989. Regulation of proteoglycan metabolism in atherogenesis and aging; growth factors; mechanisms of cell-cell communication.

Steven R. Feldman, Professor of Dermatology and Pathology; M.D./Ph.D., Duke, 1985. Control of cutaneous gene transcription and regulation of cutaneous proteolysis; molecular effects of ultraviolet light.

A. Julian Garvin, Professor of Pathology; M.D., 1972, Ph.D., 1975, Medical University of South Carolina. Molecular mechanisms of pediatric neoplasia.

Randolph L. Geary, Associate Professor of Pathology and Surgical Sciences; M.D., Washington (Seattle), 1986. Vascular surgery; biology of the artery wall: vascular gene transfer, arterial injury and repair, primate model of restenosis, and immunocytochemistry.

Kazushi Inoue, Assistant Professor of Pathology; M.D., Gifu University School of Medicine (Japan), 1984; Ph.D., Tokyo, 1990. Regulation of tumor suppressor genes; signal transduction; animal models of human cancer.

Jay R. Kaplan, Professor of Pathology and Anthropology; Ph.D., Northwestern, 1976. Primate behavior and biology; cardiovascular behavioral medicine; psychoimmunology; women's health; nutrition.

Nancy Kock, Associate Professor of Pathology; D.V.M., 1981, Ph.D., 1985, California, Davis. Animal models of infectious disease.

Timothy E. Kute, Associate Professor of Pathology; Ph.D., Louisville, 1975. Breast cancer characterization; flow cytometry studies; steroid hormones and their receptors and proteases in cancer.

Nilamadhab Mishra, Assistant Professor of Internal Medicine; M.D., M.K.C.G. Medical College, Berhampur (India), 1989. Epigenetics, chromatin remodeling, transcriptional regulation, gene expression, and identifying therapeutic targets by proteomics in lupus.

John S. Parks, Professor of Pathology; Ph.D., Wake Forest, 1979. High-density lipoprotein biogenesis, intravascular metabolism, and catabolism in gene-targeted and transgenic mice; gene-diet interactions in atherosclerosis.

Thomas C. Register, Assistant Professor of Pathology; Ph.D., South Carolina, 1985. Atherosclerosis; osteoporosis; cell and molecular biology of vascular and skeletal tissues.

Lawrence L. Rudel, Professor of Pathology and Biochemistry; Ph.D., Arkansas, 1969. Plasma lipoprotein metabolism in primate atherosclerosis and dietary effects on hepatic cholesterol metabolism.

Richard W. St. Clair, Professor of Pathology (Physiology); Ph.D., Colorado State, 1965. Arteriosclerosis; arterial lipid metabolism; lipid and lipoprotein metabolism of cells in culture; hormone effects on cellular lipoprotein metabolism; cellular cholesterol efflux.

Gregory S. Shelness, Professor of Pathology; Ph.D., SUNY at Stony Brook, 1984. Intracellular protein targeting and transport; lipoprotein assembly and secretion.

Carol A. Shively, Professor of Pathology; Ph.D., California, Davis, 1983. Comparative primate behavior; comparative atherosclerosis; behavioral, hormonal, and adipose influences on atherogenesis; women's health; alcohol abuse.

Mary Sorci-Thomas, Professor of Pathology; Ph.D., Wake Forest, 1984. Structure-function relationships of ApoA-I; regulation of apolipoprotein gene expression; control mechanisms at the transcriptional and posttranscriptional level.

Janice D. Wagner, Associate Professor of Pathology; D.V.M., Ohio State, 1985; Ph.D., Wake Forest, 1991. Atherosclerosis and diabetes mellitus; hormonal influences associated with atherogenesis.

William D. Wagner, Professor of Pathology; Ph.D., West Virginia, 1970. Arteriosclerosis; arterial connective tissue metabolism; proteoglycan structure and function.

Richard B. Weinberg, Professor of Internal Medicine (Gastroenterology) and of Physiology and Pharmacology and Associate in Biochemistry; M.D., Johns Hopkins, 1975. Structure and function of human apolipoprotein A-IV (apo A-IV); biophysics of protein-lipid interactions; lipoprotein assembly and metabolism; diet-gene interactions affecting human lipid metabolism.

J. Koudy Williams, Professor of Pathology and Surgical Sciences (General); D.V.M., Iowa State, 1983. Endothelial dysfunction and atherosclerosis; vasomotion; women's health.

Mark C. Willingham, Professor of Pathology; M.D., Medical College of South Carolina, 1969. Basic and translational research in tumor biology, including apoptosis; drug resistance in tumor cells and immunotoxin therapy of cancer.

Liqing Yu, Assistant Professor of Pathology (Lipid Sciences); M.D., Hubei Medical College (Xianning, China), 1985; Ph.D., Peking Union Medical College (China), 1995. Sterol trafficking; ABC transporters in lipid metabolism.

Section 17
Pharmacology and Toxicology

This section contains a directory of institutions offering graduate work in pharmacology and toxicology, followed by in-depth entries submitted by institutions that chose to prepare detailed program descriptions. Additional information about programs listed in the directory but not augmented by an in-depth entry may be obtained by writing directly to the dean of a graduate school or chair of a department at the address given in the directory.

For programs offering related work, see also in this book Biochemistry; Biological and Biomedical Sciences; Cell, Molecular, and Structural Biology; Ecology, Environmental Biology, and Evolutionary Biology; Genetics, Developmental Biology, and Reproductive Biology; Neuroscience and Neurobiology; Nutrition; Pathology and Pathobiology; and Physiology. In Book 2, see Psychology and Counseling; in Book 4, see Chemistry and Environmental Sciences and Management; in Book 5, see Chemical Engineering and Civil and Environmental Engineering; and in Book 6, see Pharmacy and Pharmaceutical Sciences, Public Health, and Veterinary Medicine and Sciences.

CONTENTS

Program Directories

Announcements

Close-Ups

See also:

Molecular Pharmacology

Albert Einstein College of Medicine, Sue Golding Graduate Division of Medical Sciences, Division of Biological Sciences, Department of Molecular Pharmacology, Bronx, NY 10461. Offers PhD, MD/PhD. *Degree requirements:* For doctorate, thesis/dissertation. *Entrance requirements:* For doctorate, GRE General Test. Additional exam requirements/recommendations for international students: Required—TOEFL. *Faculty research:* Effects of drugs on macromolecules, enzyme systems, cell morphology and function.

Brown University, Graduate School, Division of Biology and Medicine, Program in Molecular Pharmacology and Physiology, Providence, RI 02912. Offers MA, Sc M, PhD, MD/PhD. *Degree requirements:* For doctorate, thesis/dissertation, preliminary exam. *Entrance requirements:* For master's and doctorate, GRE General Test, GRE Subject Test. Additional exam requirements/recommendations for international students: Required—TOEFL. Electronic applications accepted. *Faculty research:* Structural biology, antiplatelet drugs, nicotinic receptor structure/function.

Harvard University, Graduate School of Arts and Sciences, Division of Medical Sciences, Boston, MA 02115. Offers biological chemistry and molecular pharmacology (PhD); cell biology (PhD); genetics (PhD); microbiology and molecular genetics (PhD); pathology (PhD); including experimental pathology. *Students:* 433 full-time (210 women). In 2005, 83 doctorates awarded. *Degree requirements:* For doctorate, thesis/dissertation. *Entrance requirements:* For doctorate, GRE General Test, GRE Subject Test. Additional exam requirements/recommendations for international students: Required—TOEFL. Application fee: $60. *Expenses:* Tuition: Full-time $28,752. Full-time tuition and fees vary according to program and student level. *Financial support:* Fellowships, research assistantships, teaching assistantships, institutionally sponsored loans and tuition waivers (full) available. Financial award application deadline: 1/1. *Unit head:* Administrator, 617-432-2029. *Application contact:* Administrator, 617-432-2029.

Massachusetts Institute of Technology, School of Engineering, Biological Engineering Division, Cambridge, MA 02139-4307. Offers applied biosciences (PhD, Sc D); bioengineering (PhD, Sc D); biomedical engineering (M Eng); genetic toxicology (PhD, Sc D); molecular and systems bacterial pathogenesis (PhD, Sc D); molecular and systems toxicology and pharmacology (PhD, Sc D); molecular systems toxicology (SM, Sc D); toxicology (SM, PhD, Sc D). *Faculty:* 16 full-time (4 women). *Students:* 126 full-time (56 women); includes 25 minority (3 African Americans, 17 Asian Americans or Pacific Islanders, 5 Hispanic Americans), 39 international. Average age 26. 349 applicants, 18% accepted, 28 enrolled. In 2005, 4 master's, 12 doctorates awarded. Terminal master's awarded for partial completion of doctoral program. *Degree requirements:* For master's, thesis; for doctorate, thesis/dissertation, qualifying exams, comprehensive exam. *Entrance requirements:* For master's and doctorate, GRE General Test. Additional exam requirements/recommendations for international students: Required—TOEFL (minimum score 600 paper-based; 250 computer-based). *Application deadline:* For fall admission, 12/31 for domestic students, 12/31 for international students. Application fee: $70. Electronic applications accepted. *Expenses:* Tuition: Full-time $32,100. Required fees: $200. Part-time tuition and fees vary according to course load. *Financial support:* In 2005–06, 112 students received support, including 57 fellowships with tuition reimbursements available (averaging $27,248 per year), 64 research assistantships with tuition reimbursements available (averaging $26,559 per year), 3 teaching assistantships with tuition reimbursements available (averaging $26,180 per year); Federal Work-Study, institutionally sponsored loans, scholarships/grants, traineeships, health care benefits, and unspecified assistantships also available. Total annual research expenditures: $22.5 million. *Unit head:* Prof. Douglas A. Lauffenburger, Director, 617-252-1629, Fax: 617-258-0204, E-mail: lauffen@mit.edu. *Application contact:* Biological Engineering Academic Office, 617-253-1712, Fax: 617-258-8676, E-mail: be-acad@mit.edu.

Mayo Graduate School, Graduate Programs in Biomedical Sciences, Program in Molecular Pharmacology and Experimental Therapeutics, Rochester, MN 55905. Offers PhD. *Degree requirements:* For doctorate, oral defense of dissertation, qualifying oral and written exam. *Entrance requirements:* For doctorate, GRE, 1 year of chemistry, biology, calculus, and physics. Additional exam requirements/recommendations for international students: Required—TOEFL. Electronic applications accepted. *Faculty research:* Patch clamping, G-proteins, pharmacogenetics, receptor-induced transcriptional events, cholinesterase biology.

See Close-Up on page 1173.

Medical University of South Carolina, College of Graduate Studies, Program in Cell and Molecular Pharmacology and Experimental Therapeutics, Charleston, SC 29425-0002. Offers Pharm D, MBS, MS, PhD, DMD/PhD, MD/PhD. *Faculty:* 17 full-time (3 women). *Students:* 7 full-time (5 women). Average age 28. 181 applicants, 34% accepted, 43 enrolled. In 2005, 1 master's, 1 doctorate awarded. *Degree requirements:* For doctorate, thesis/dissertation, teaching and research seminar, oral and written exams. *Entrance requirements:* For doctorate, GRE General Test, interview. Additional exam requirements/recommendations for international students: Required—TOEFL (minimum score 600 paper-based; 250 computer-based). *Application deadline:* For fall admission, 1/15 priority date for domestic students, 1/15 priority date for international students. Applications are processed on a rolling basis. Application fee: $0 ($75 for international students). Electronic applications accepted. *Financial support:* In 2005–06, 1 student received support, including fellowships with partial tuition reimbursements available (averaging $21,000 per year); Federal Work-Study and scholarships/grants also available. Financial award application deadline: 3/15; financial award applicants required to submit FAFSA. *Faculty research:* Hypertension, kallikrein-kinin, sodium/calcium exchange, thromboxane receptors, molecular toxicology. *Unit head:* Dr. Kenneth D. Tew, Chair, 843-792-2514, Fax: 843-792-2475, E-mail: tewk@musc.edu. *Application contact:* Cheryl Brown, 843-792-4620, Fax: 843-792-4645, E-mail: brownche@musc.edu.

New York University, School of Medicine and Graduate School of Arts and Science, Sackler Institute of Graduate Biomedical Sciences, Department of Pharmacology, New York, NY 10012-1019. Offers molecular pharmacology and signal transduction (PhD). *Degree requirements:* For doctorate, one foreign language, thesis/dissertation, qualifying exam, comprehensive exam. *Entrance requirements:* For doctorate, GRE General Test. Additional exam requirements/recommendations for international students: Required—TOEFL. *Faculty research:* Pharmacology and neurobiology, neuropeptides, receptor biochemistry, cytoskeleton, endocrinology.

Purdue University, College of Pharmacy and Pharmacal Sciences and Graduate School, Graduate Programs in Pharmacy and Pharmacal Sciences, Department of Medicinal Chemistry and Molecular Pharmacology, West Lafayette, IN 47907. Offers analytical medicinal chemistry (PhD); computational and biophysical medicinal chemistry (PhD); medicinal and bioorganic chemistry (PhD); medicinal biochemistry and molecular biology (PhD); molecular pharmacology and toxicology (PhD); natural products and pharmacognosy (PhD); nuclear pharmacy (MS); radiopharmaceutical chemistry and nuclear pharmacy (PhD). *Faculty:* 21 full-time (2 women), 2 part-time/adjunct (0 women). *Students:* 69 full-time (39 women), 2 part-time (1 woman); includes 8 minority (2 African Americans, 2 Asian Americans or Pacific Islanders, 4 Hispanic Americans), 21 international. Average age 26. 125 applicants, 25% accepted, 15 enrolled. In 2005, 2 master's, 11 doctorates awarded. Terminal master's awarded for partial completion of doctoral program. *Degree requirements:* For master's and doctorate, thesis/dissertation. *Entrance requirements:* For master's, GRE General Test, minimum B average; BS in biology, chemistry, or pharmacy; for doctorate, GRE General Test, minimum B average; BS in biology, chemistry, or pharmacology. Additional exam requirements/recommendations for international students: Required—TOEFL. *Application deadline:* Applications are processed on a rolling basis. Application fee: $55. Electronic applications accepted. *Financial support:* Fellowships, research assistantships, teaching assistantships, traineeships available. Support available to part-time students. Financial award applicants required to submit FAFSA. *Faculty research:* Drug design and development, cancer research, drug synthesis and analysis, chemical pharmacology, environmental toxicology. *Unit head:* Dr. R. F. Borch, Graduate Head,

765-494-1403. *Application contact:* Dr. D. E. Bergstrom, Graduate Committee, 765-494-6275, E-mail: bergstrom@pharmacy.purdue.edu.

Rosalind Franklin University of Medicine and Science, School of Graduate and Postdoctoral Studies, Department of Cellular and Molecular Pharmacology, North Chicago, IL 60064-3095. Offers MS, PhD, MD/MS, MD/PhD. Part-time programs available. Terminal master's awarded for partial completion of doctoral program. *Degree requirements:* For master's and doctorate, thesis/dissertation. *Entrance requirements:* For master's and doctorate, GRE General Test. Additional exam requirements/recommendations for international students: Required—TOEFL, TWE. *Faculty research:* Control of gene expression in higher organisms, molecular mechanism of action of growth factors and hormones, hormonal regulation in brain neuropsychopharmacology.

Rutgers, The State University of New Jersey, New Brunswick/Piscataway, Graduate School, Program in Cellular and Molecular Pharmacology, Piscataway, NJ 08854. Offers PhD. *Faculty:* 49 full-time. *Students:* 8 full-time (6 women); includes 2 minority (both Hispanic Americans), 4 international. Average age 26. 36 applicants, 31% accepted, 5 enrolled. *Degree requirements:* For doctorate, thesis/dissertation, qualifying exam. *Entrance requirements:* For doctorate, GRE General Test, GRE Subject Test. Additional exam requirements/recommendations for international students: Required—TOEFL. *Application deadline:* For fall admission, 3/15 for domestic students. Application fee: $50. *Expenses:* Tuition, state resident: full-time $10,440; part-time $435 per credit. Tuition, nonresident: full-time $15,520; part-time $647 per credit. Required fees: $129 per credit. Tuition and fees vary according to program. *Financial support:* In 2005–06, 1 fellowship with full tuition reimbursement (averaging $24,000 per year), 2 research assistantships with full tuition reimbursements (averaging $15,800 per year), teaching assistantships (averaging $16,988 per year) were awarded; traineeships also available. Financial award application deadline: 2/15; financial award applicants required to submit FAFSA. *Faculty research:* Molecular, cellular, and neuropharmacology; drug metabolism; intracellular signaling systems; protein synthesis and processing; carcinogenesis. *Unit head:* Dr. Marc Gartenberg, Director, 732-235-5800, Fax: 732-235-4073, E-mail: gartenbe@umdnj.edu.

Stanford University, School of Medicine, Graduate Programs in Medicine, Department of Molecular Pharmacology, Stanford, CA 94305-9991. Offers PhD. *Degree requirements:* For doctorate, thesis/dissertation, qualifying examination. *Entrance requirements:* For doctorate, GRE General Test, GRE Subject Test. Additional exam requirements/recommendations for international students: Required—TOEFL. Electronic applications accepted. *Faculty research:* Action of such drugs as epinephrine, cell differentiation and development, microsomal enzymes, neuropeptide gene expression.

State University of New York at Buffalo, Graduate School, Graduate Programs in Cancer Research and Biomedical Sciences at Roswell Park Cancer Institute, Department of Molecular Pharmacology and Cancer Therapeutics at Roswell Park Cancer Institute, Program in Molecular Pharmacology and Cancer Therapeutics, Buffalo, NY 14263. Offers PhD. *Faculty:* 28 full-time (8 women). *Students:* 25 full-time (16 women); includes 10 minority (2 African Americans, 8 Asian Americans or Pacific Islanders), 3 international. Average age 26. 26 applicants, 35% accepted. In 2005, 7 degrees awarded. *Degree requirements:* For doctorate, thesis/dissertation, departmental qualifying exam, grant proposal. *Entrance requirements:* For doctorate, GRE General Test (recommended). Additional exam requirements/recommendations for international students: Required—TOEFL; Recommended—TWE. *Application deadline:* For fall admission, 2/1 for domestic students. Applications are processed on a rolling basis. Application fee: $35. Electronic applications accepted. *Financial support:* In 2005–06, 25 students received support, including 6 fellowships with full tuition reimbursements available (averaging $20,772 per year), 19 research assistantships with full tuition reimbursements available (averaging $20,772 per year) *Faculty research:* Molecular pharmacology, cancer cell biology, molecular biology, biochemistry, chemotherapy. Total annual research expenditures: $6.5 million. *Unit head:* Dr. Enrico Mihich, Chair, 716-845-8226, Fax: 716-845-8857. *Application contact:* Dr. David W. Goodrich, Director of Graduate Studies, 716-845-4506, Fax: 716-845-8857, E-mail: david.goodrich@roswellpark.org.

Announcement: A PhD program in molecular pharmacology and cancer therapeutics at the Roswell Park Cancer Institute. Program faculty members perform cutting-edge research in a number of areas, including cell cycle regulation, differentiation, signal transduction, chromatin structure, cancer genes, and mouse models of cancer. Preclinical and clinical research opportunities are also available. Stipends and tuition waivers are provided.

See Close-Up on page 1181.

Thomas Jefferson University, Jefferson College of Graduate Studies, Program in Molecular Pharmacology and Structural Biology, Philadelphia, PA 19107. Offers PhD. *Faculty:* 48 full-time. *Students:* 15 full-time (9 women), 1 part-time; includes 3 minority (2 African Americans, 1 Asian American or Pacific Islander), 3 international. 15 applicants, 20% accepted, 0 enrolled. In 2005, 4 degrees awarded. *Degree requirements:* For doctorate, thesis/dissertation, comprehensive exam, registration. *Entrance requirements:* For doctorate, GRE General Test, minimum GPA of 3.2. Additional exam requirements/recommendations for international students: Required—TOEFL (minimum score 213 computer-based). *Application deadline:* For fall admission, 3/1 priority date for domestic students, 3/1 priority date for international students. Applications are processed on a rolling basis. Application fee: $50. Electronic applications accepted. *Expenses:* Tuition: Full-time $14,894; part-time $800 per credit. *Financial support:* In 2005–06, 1 student received support, including 15 fellowships with full tuition reimbursements available; research assistantships, Federal Work-Study, institutionally sponsored loans, scholarships/grants, and traineeships also available. Support available to part-time students. Financial award application deadline: 5/1; financial award applicants required to submit FAFSA. *Faculty research:* Biochemistry and cell, molecular and structural biology of cell-surface and intracellular receptors, molecular modeling, signal transduction. *Unit head:* Dr. Jeffrey L. Benovic, program Director, 215-503-4607, Fax: 215-923-1098, E-mail: jeff.benovic@jefferson.edu. *Application contact:* Jessie F. Pervall, Director of Admissions, 215-503-0155, Fax: 215-503-9920, E-mail: jessie.pervall@jefferson.edu.

See Close-Up on page 1185.

University of Connecticut Health Center, Graduate School, Programs in Biomedical Sciences, Program in Cellular and Molecular Pharmacology, Farmington, CT 06030. Offers PhD, DMD/PhD, MD/PhD. *Degree requirements:* For doctorate, thesis/dissertation, comprehensive exam, registration. *Entrance requirements:* For doctorate, GRE General Test. Additional exam requirements/recommendations for international students: Required—TOEFL (minimum score 600 paper-based; 250 computer-based). Electronic applications accepted. *Faculty research:* Neuropharmacology, cardiovascular-pulmonary pharmacology, endocrine-reproductive pharmacology, immunopharmacology and chemotherapy.

See Close-Up on page 1203.

University of Massachusetts Worcester, Graduate School of Biomedical Sciences, Department of Biochemistry and Molecular Pharmacology, Worcester, MA 01655-0115. Offers PhD. *Faculty:* 34 full-time (5 women). *Degree requirements:* For doctorate, thesis/dissertation. *Entrance requirements:* For doctorate, GRE General Test. Additional exam requirements/recommendations for international students: Required—TOEFL (minimum score 600 paper-based; 250 computer-based). *Application deadline:* For fall admission, 12/15 for domestic students, 12/15 for international students. Applications are processed on a rolling basis. Application fee: $25 ($50 for international students). *Expenses:* Tuition, state resident: full-time $2,640. Tuition, nonresident: full-time $9,856. Required fees: $5,685. *Financial support:* In 2005–06, research assistantships with full tuition reimbursements (averaging $25,235 per year); unspecified assistantships also available. *Faculty research:* Neuropharmacology, control of DNA metabolism, clinical pharmacology and toxicology, chemical biology. *Unit head:* Dr. C. Robert Matthews, Chair, 508-856-2251, Fax: 508-856-2151. *Application contact:* Michael Cole, Director of Admissions and Recruitment, 508-856-4779, Fax: 508-856-3659, E-mail: michael.cole@umassmed.edu.

University of Medicine and Dentistry of New Jersey, Graduate School of Biomedical Sciences, Graduate Programs in Biomedical Sciences–Piscataway, Program in Cellular and Molecular Pharmacology, Piscataway, NJ 08854-5635. Offers MS, PhD, MD/PhD. *Degree requirements:* For master's and doctorate, thesis/dissertation, qualifying exam. *Application deadline:* For fall admission, 1/5 for domestic students. Application fee: $40. *Financial support:* Application deadline: 5/1. *Unit head:* Dr. Marc R. Gartenberg, Director, 732-235-5800, Fax: 732-235-4073, E-mail: gartenbe@umdnj.edu.

University of Nevada, Reno, School of Medicine and Graduate School, Graduate Programs in Medicine, Interdisciplinary Program in Cellular and Molecular Pharmacology and Physiology, Reno, NV 89557. Offers MS, PhD. *Faculty:* 8. *Students:* 13 full-time (4 women), 11 part-time (7 women); includes 1 minority (Asian American or Pacific Islander), 19 international. Average age 27. 7 applicants, 100% accepted, 7 enrolled. In 2005, 1 degree awarded. Terminal master's awarded for partial completion of doctoral program. *Degree requirements:* For master's, thesis; for doctorate, one foreign language, thesis/dissertation. *Entrance requirements:* For master's, GRE General Test, minimum GPA of 2.75; for doctorate, GRE General Test, minimum GPA of 3.0. Additional exam requirements/recommendations for international students: Required—TOEFL. *Application deadline:* For fall admission, 3/1 for domestic students; for spring admission, 11/1 for domestic students. Applications are processed on a rolling basis. Application fee: $60 ($95 for international students). *Expenses:* Tuition, area resident: Full-time $2,767; part-time $923 per semester. Tuition, state resident: full-time $5,733; part-time $1,911 per semester. Tuition, nonresident: full-time $12,679; part-time $1,911 per semester. International tuition: $13,878 full-time. Required fees: $404; $202 per term. One-time fee: $90. *Financial support:* Research assistantships, teaching assistantships available. Support available to part-time students. Financial award application deadline: 3/1. *Faculty research:* Neuropharmacology, toxicology, cardiovascular pharmacology, neuromuscular pharmacology. *Unit head:* Dr. Norman LeBlanc, Graduate Program Director, 775-784-1462.

University of Pittsburgh, School of Medicine, Graduate Programs in Medicine, Program in Molecular Pharmacology, Pittsburgh, PA 15260. Offers PhD. *Faculty:* 45 full-time (9 women). *Students:* 24 full-time (10 women); includes 5 minority (2 African Americans, 3 Asian Americans or Pacific Islanders), 6 international. Average age 28. 415 applicants, 22% accepted, 42 enrolled. In 2005, 6 degrees awarded. *Median time to degree:* Of those who began their doctoral program in fall 1997, 95% received their degree in 8 years or less. *Degree requirements:* For doctorate, thesis/dissertation, comprehensive exam, registration. *Entrance requirements:* For doctorate, GRE General Test, GRE Subject Test, minimum QPA of 3.0. Additional exam requirements/recommendations for international students: Required—TOEFL (minimum score 600 paper-based; 250 computer-based), IELT (minimum score 7). *Application deadline:* For fall admission, 12/15 priority date for domestic students, 12/15 priority date for international students. Application fee: $40. Electronic applications accepted. *Expenses:* Tuition, state resident: full-time $13,194; part-time $537 per credit. Tuition, nonresident: full-time $25,012; part-time $1,026 per credit. Required fees: $700; $164 per term. Tuition and fees vary according to campus/location and program. *Financial support:* In 2005–06, 7 fellowships with full tuition reimbursements (averaging $21,500 per year), 15 research assistantships with full tuition reimbursements (averaging $21,500 per year) were awarded; teaching assistantships with partial tuition reimbursements, institutionally sponsored loans, scholarships/grants, traineeships, health care benefits, tuition waivers (full), and unspecified assistantships also available. *Faculty research:* Drug discovery, signal transduction, cancer therapeutics, neurophramacology, cardiovascular and renal pharmacology. *Unit head:* Dr. Donald DeFranco, Professor and Director of Pharmacology Program, 412-624-4259, Fax: 412-648-1945, E-mail: dod1@pitt.edu. *Application contact:* Graduate Studies Administrator, 412-648-8957, Fax: 412-648-1236, E-mail: gradstudies@medschool.pitt.edu.

University of Southern California, School of Pharmacy and Graduate School, Graduate Programs in Pharmacy, Graduate Program in Molecular Pharmacology and Toxicology, Los Angeles, CA 90089. Offers MS, PhD, Pharm D/PhD. *Degree requirements:* For master's and doctorate, thesis/dissertation. *Entrance requirements:* For master's and doctorate, GRE General Test. *Expenses:* Tuition: Full-time $25,416; part-time $1,059 per unit. Required fees: $484; $484 per year. Tuition and fees vary according to course load and program.

Announcement: Graduate study in the Department of Molecular Pharmacology and Toxicology integrates state-of-the-art investigations of drug-receptor interactions and the molecular signal transduction pathways that mediate drug and physiological function, with an emphasis on oxidative stress and free radical biochemistry and toxicology, gene expression, atherosclerosis, and neurobiology. The molecular pharmacology program emphasizes application of molecular approaches, including receptor cloning, mutagenesis, gene regulation, and transgenic (gene knockout) analyses. The toxicology program combines sophisticated quantitative analyses such as EPR, NMR, HPLC, and MS/GC with molecular analyses of functional consequences.

Molecular Toxicology

Massachusetts Institute of Technology, School of Engineering, Biological Engineering Division, Cambridge, MA 02139-4307. Offers applied biosciences (PhD, Sc D); bioengineering (PhD, Sc D); biomedical engineering (M Eng); genetic toxicology (PhD, Sc D); molecular and systems bacterial pathogenesis (PhD, Sc D); molecular and systems toxicology and pharmacology (PhD, Sc D); molecular systems toxicology (PhD, Sc D); toxicology (SM, PhD, Sc D). *Faculty:* 16 full-time (4 women). *Students:* 126 full-time (56 women); includes 25 minority (3 African Americans, 17 Asian Americans or Pacific Islanders, 5 Hispanic Americans), 39 international. Average age 26. 349 applicants, 18% accepted, 28 enrolled. In 2005, 4 master's, 12 doctorates awarded. Terminal master's awarded for partial completion of doctoral program. *Degree requirements:* For master's, thesis; for doctorate, thesis/dissertation, qualifying exams, comprehensive exam. *Entrance requirements:* For master's and doctorate, GRE General Test. Additional exam requirements/recommendations for international students: Required—TOEFL (minimum score 600 paper-based; 250 computer-based). *Application deadline:* For fall admission, 12/31 for domestic students, 12/31 for international students. Application fee: $70. Electronic applications accepted. *Expenses:* Tuition: Full-time $32,100. Required fees: $200. Part-time tuition and fees vary according to course load. *Financial support:* In 2005–06, 112 students received support, including 57 fellowships with tuition reimbursements available (averaging $27,248 per year), 64 research assistantships with tuition reimbursements available (averaging $26,559 per year), 3 teaching assistantships with tuition reimbursements available (averaging $26,180 per year); Federal Work-Study, institutionally sponsored loans, scholarships/grants, traineeships, health care benefits, and unspecified assistantships also available. Total annual research expenditures: $22.5 million. *Unit head:* Prof. Douglas A. Lauffenburger, Director, 617-252-1629, Fax: 617-258-0204, E-mail: lauffen@mit.edu. *Application contact:* Biological Engineering Academic Office, 617-253-1712, Fax: 617-258-8676, E-mail: be-acad@mit.edu.

New York University, Graduate School of Arts and Science, Department of Environmental Medicine, New York, NY 10012-1019. Offers environmental health sciences (MS, PhD), including biostatistics (PhD), environmental hygiene (MS), epidemiology (PhD), ergonomics and biomechanics (PhD), exposure assessment and health effects (PhD), molecular toxicology/carcinogenesis (PhD), toxicology. Part-time programs available. *Faculty:* 26 full-time (7 women). *Students:* 42 full-time (33 women), 21 part-time (10 women); includes 11 minority (2 African Americans, 6 Asian Americans or Pacific Islanders, 3 Hispanic Americans), 21 international. Average age 32. 56 applicants, 38% accepted, 9 enrolled. In 2005, 7 master's, 6 doctorates awarded. Terminal master's awarded for partial completion of doctoral program. *Degree requirements:* For master's, thesis or alternative; for doctorate, one foreign language, thesis/dissertation, oral and written exams. *Entrance requirements:* For master's and doctorate, GRE General Test, GRE Subject Test, minimum GPA of 3.0; bachelor's degree in biological, physical, or engineering science. Additional exam requirements/recommendations for international students: Required—TOEFL. *Application deadline:* For fall admission, 11/14 for domestic students. Application fee: $80. *Financial support:* Fellowships with tuition reimbursements, teaching assistantships with tuition reimbursements, career-related internships or fieldwork, Federal Work-Study, institutionally sponsored loans, and health care benefits available. Financial award application deadline: 2/1; financial award applicants required to submit FAFSA. *Unit head:* Dr. Max Costa, Chair, 845-731-3661, Fax: 845-351-3317, E-mail: ehs@env.med.nyu.edu. *Application contact:* Dr. Jerome J Solomon, Director of Graduate Studies, 845-731-3661, Fax: 845-351-3317, E-mail: ehs@env.med.nyu.edu.

See Close-Up on page 183.

North Carolina State University, Graduate School, College of Agriculture and Life Sciences and College of Veterinary Medicine, Department of Environmental and Molecular Toxicology, Raleigh, NC 27695. Offers M Tox, MS, PhD. Terminal master's awarded for partial completion of doctoral program. *Degree requirements:* For master's, thesis (for some programs); for doctorate, thesis/dissertation. *Entrance requirements:* For master's and doctorate, GRE General Test, minimum GPA of 3.0. Electronic applications accepted. *Faculty research:* Chemical fate, carcinogenesis, developmental and endocrine toxicity, xenobiotic metabolism, signal transduction.

See Close-Up on page 1175.

Oregon State University, Graduate School, College of Agricultural Sciences, Department of Environmental and Molecular Toxicology, Corvallis, OR 97331. Offers toxicology (MS, PhD). *Faculty:* 12 full-time (4 women), 5 part-time/adjunct (0 women). *Students:* 25 full-time (16 women), 3 part-time (1 woman); includes 6 minority (1 American Indian/Alaska Native, 2 Asian Americans or Pacific Islanders, 3 Hispanic Americans), 5 international. In 2005, 1 master's, 1 doctorate awarded. Application fee: $50. *Expenses:* Tuition, area resident: Part-time $301 per credit. Tuition, state resident: full-time $8,139; part-time $501 per credit. Tuition, nonresident: full-time $14,376; part-time $532 per credit. Required fees: $1,266. *Unit head:* Dr. Lawrence R. Curtis, Head, 541-737-1764, Fax: 541-737-0497, E-mail: larry.curtis@orst.edu.

University of California, Berkeley, Graduate Division, Group in Molecular Toxicology, Berkeley, CA 94720-1500. Offers PhD.

University of California, Los Angeles, Graduate Division, School of Public Health, Department of Environmental Health Sciences, Interdepartmental Program in Molecular Toxicology, Los Angeles, CA 90095. Offers PhD. *Degree requirements:* For doctorate, thesis/dissertation, oral and written qualifying exams. *Entrance requirements:* For doctorate, GRE General Test. Electronic applications accepted.

See Close-Up on page 1195.

University of Cincinnati, Division of Research and Advanced Studies, College of Medicine, Graduate Programs in Biomedical Sciences, Department of Environmental Health, Programs in Environmental Genetics and Molecular Toxicology, Cincinnati, OH 45221. Offers MS, PhD. *Degree requirements:* For doctorate, thesis/dissertation. *Entrance requirements:* For master's, GRE; a minimum grade-point average (GPA) of 3.00; 3 letters of recommendation. Additional exam requirements/recommendations for international students: Required—TOEFL (minimum score 520 paper-based; 190 computer-based).

University of Pittsburgh, Graduate School of Public Health, Department of Environmental and Occupational Health, Pittsburgh, PA 15260. Offers environmental and occupational health (MPH); molecular toxicology (MS, PhD); occupational medicine (MPH), including occupational and environmental medicine; risk assessment (Certificate). *Accreditation:* CEPH (one or more programs are accredited). Part-time programs available. *Faculty:* 23 full-time (7 women), 17 part-time/adjunct (2 women). *Students:* 8 full-time (7 women), 8 part-time (3 women); includes 4 minority (3 African Americans, 1 Hispanic American), 4 international. Average age 35. 22 applicants, 64% accepted, 4 enrolled. In 2005, 6 master's, 3 doctorates, 1 other advanced degree awarded. *Degree requirements:* For master's, thesis, registration; for doctorate, thesis/dissertation, preliminary exams, comprehensive exam, registration. *Entrance requirements:* For master's and Certificate, GRE General Test; for doctorate, GRE General Test, minimum GPA of 3.4; background in biology, physics, chemistry and calculus. Additional exam requirements/recommendations for international students: Required—TOEFL (minimum score 550 paper-based; 213 computer-based). *Application deadline:* For fall admission, 2/15 priority date for domestic students, 2/15 priority date for international students. Applications are processed on a rolling basis. Application fee: $50 ($60 for international students). Electronic applications accepted. *Expenses:* Tuition, state resident: full-time $13,194; part-time $537 per credit. Tuition, nonresident: full-time $25,012; part-time $1,026 per credit. Required fees: $700; $164 per term. Tuition and fees vary according to campus/location and program. *Financial support:* In 2005–06, 12 students received support, including 2 fellowships (averaging $21,150 per year), 10 research assistantships with full tuition reimbursements available (averaging $20,333 per year); scholarships/grants, traineeships, tuition waivers (partial), and unspecified assistantships also available. Support available to part-time students. Financial award applicants required to submit FAFSA. *Faculty research:* Molecular toxicology, redox signaling, gene environment interaction, progenitor-progency lineage, occupational and pulmonary medicine. Total annual research expenditures: $5.2 million. *Unit head:* Dr. Bruce R. Pitt, Professor and Department Chair, 412-624-8400, Fax: 412-383-7658, E-mail: brucep@pitt.edu. *Application contact:* Eileen Penny Weiss, Student Affairs Administrator, 412-383-7297, Fax: 412-383-7658, E-mail: pweiss@eoh.pitt.edu.

University of Pittsburgh, School of Medicine, Graduate Programs in Medicine, Molecular Toxicology, Pittsburgh, PA 15260. *Faculty:* 18 full-time (4 women). *Students:* 1 full-time (0 women). Average age 28. 415 applicants, 22% accepted, 42 enrolled. *Entrance requirements:* Additional exam requirements/recommendations for international students: Required—TOEFL (minimum score 600 paper-based; 250 computer-based), IELT (minimum score 7). *Application deadline:* For fall admission, 12/15 priority date for domestic students, 12/15 priority date for international students. Application fee: $40. Electronic applications accepted. *Expenses:* Tuition, state resident: full-time $13,194; part-time $537 per credit. Tuition, nonresident: full-time $25,012; part-time $1,026 per credit. Required fees: $700; $164 per term. Tuition and fees vary according to campus/location and program. *Financial support:* In 2005–06, 48 research assistantships with full tuition reimbursements (averaging $21,500 per year) were awarded;

Molecular Toxicology

University of Pittsburgh (continued)
teaching assistantships, institutionally sponsored loans, scholarships/grants, traineeships, and unspecified assistantships also available. *Faculty research:* Biochemistry and molecular genet-ics, cell biology and molecular physiology, cellular and molecular pathology, immunology, molecular pharmacology. *Application contact:* Graduate Studies Administrator, 412-648-8957, Fax: 412-648-1077, E-mail: gradstudies@medschool.pitt.edu.

Pharmacology

Albany Medical College, Graduate Programs in the Biological Sciences, Center for Neuro-pharmacology and Neuroscience, Albany, NY 12208-3479. Offers MS, PhD. Part-time programs available. *Faculty:* 21 full-time (6 women). *Students:* 16 full-time (8 women); includes 2 minority (1 African American, 1 Asian American or Pacific Islander), 3 international. Average age 26. 22 applicants, 64% accepted, 4 enrolled. In 2005, 5 master's, 3 doctorates awarded. Terminal master's awarded for partial completion of doctoral program. *Median time to degree:* Of those who began their doctoral program in fall 1997, 100% received their degree in 8 years or less. *Degree requirements:* For master's, thesis/dissertation; for doctorate, thesis/dissertation, comprehensive exam. *Entrance requirements:* For master's and doctorate, GRE General Test. Additional exam requirements/recommendations for international students: Required—TOEFL. *Application deadline:* For fall admission, 3/15 for domestic students. Applications are processed on a rolling basis. Application fee: $0 ($60 for international students). *Financial support:* In 2005–06, 16 students received support, including 6 fellowships (averaging $23,000 per year), 10 research assistantships (averaging $23,000 per year); Federal Work-Study, scholarships/grants, and tuition waivers (full) also available. Financial award applicants required to submit FAFSA. *Faculty research:* Molecular and cellular neuroscience, neuronal development. *Unit head:* Dr. Stanley D. Glick, Director, 518-262-5303, Fax: 518-262-5799, E-mail: cnninfo@mail.amc.edu. *Application contact:* Dr. Richard Keller, Graduate Director, 518-262-5303, Fax: 518-262-5799, E-mail: cnninfo@mail.amc.edu.

Alliant International University–San Francisco Bay, California School of Professional Psychology, Program in Psychopharmacology, San Francisco, CA 94133-1221. Offers Post-Doctoral MS. Part-time programs available. *Faculty:* 9 part-time/adjunct (1 woman). *Students:* Average age 45. In 2005, 35 degrees awarded. *Entrance requirements:* For master's, doctorate in clinical psychology, minimum GPA of 3.0 in psychology and overall. Additional exam requirements/recommendations for international students: Required—TOEFL. *Application deadline:* For fall admission, 4/1 priority date for domestic students, 4/1 priority date for international students. Applications are processed on a rolling basis. Application fee: $65. Electronic applications accepted. *Expenses:* Tuition: Part-time $795 per unit. Tuition and fees vary according to degree level and program. *Financial support:* Federal Work-Study and institutionally sponsored loans available. Financial award application deadline: 2/15; financial award applicants required to submit FAFSA. *Unit head:* Dr. Steven R. Tulkin, Director, 415-955-2162, Fax: 415-955-2178, E-mail: stulkin@alliant.edu. *Application contact:* Katie Rosenfeld, Program Assistant, 415-955-2163, Fax: 415-955-2178, E-mail: krosenfeld@alliant.edu.

American University of Beirut, Graduate Programs, Faculty of Medicine, Beirut, Lebanon. Offers biochemistry (MS); human morphology (MS); microbiology and immunology (MS); neuroscience (MS); pharmacology and therapeutics (MS); physiology (MS). *Degree requirements:* For master's, one foreign language, thesis (for some programs), comprehensive exam, registration. *Entrance requirements:* For master's, GRE, letter of recommendation.

Argosy University/Hawai'i, Graduate Studies, The American School of Professional Psychology, Postdoctoral Program in Clinical Psychopharmacology, Honolulu, HI 96813. Offers Certificate. *Accreditation:* APA. Evening/weekend programs available. *Faculty:* 1 full-time (0 women), 10 part-time/adjunct (5 women). *Students:* 4 full-time. Average age 40. 4 applicants, 100% accepted, 4 enrolled. *Entrance requirements:* For degree, doctoral degree, state license. Additional exam requirements/recommendations for international students: Required—TOEFL (minimum score 550 paper-based; 213 computer-based). *Application deadline:* For fall admission, 1/15 for domestic students; for spring admission, 10/1 for domestic students. Applications are processed on a rolling basis. Application fee: $50. *Expenses: Contact institution. Financial support:* Application deadline: 5/1; *Unit head:* Dr. Suzanne Anthony, Chair Graduate Psychology Department, 888-323-2777, Fax: 808-536-5505, E-mail: hawaii@argosyu.edu. *Application contact:* Cherie Andrade, Director of Admissions, 888-323-2777, Fax: 808-536-5505, E-mail: candrade@argosyu.edu.

Auburn University, College of Veterinary Medicine and Graduate School, Graduate Programs in Veterinary Medicine, Auburn University, AL 36849. Offers biomedical sciences (MS, PhD), including anatomy, physiology and pharmacology (MS), biomedical sciences (PhD), clinical sciences (MS), large animal surgery and medicine (MS), pathobiology (MS), radiology (MS), small animal surgery and medicine (MS). Part-time programs available. *Faculty:* 76 full-time (24 women). *Students:* 11 full-time (8 women), 33 part-time (20 women); includes 3 minority (all Hispanic Americans), 14 international. 41 applicants, 37% accepted, 11 enrolled. In 2005, 9 master's, 3 doctorates awarded. *Degree requirements:* For master's, thesis/dissertation. *Entrance requirements:* For master's, GRE General Test; for doctorate, GRE General Test, GRE Subject Test. *Application deadline:* For fall admission, 7/7 for domestic students; for spring admission, 11/24 for domestic students. Applications are processed on a rolling basis. Application fee: $25 ($50 for international students). Electronic applications accepted. *Financial support:* Research assistantships, teaching assistantships, Federal Work-Study available. Support available to part-time students. Financial award application deadline: 3/15. *Application contact:* Dr. Stephen L. McFarland, Acting Dean of the Graduate School, 334-844-4700.

Baylor College of Medicine, Graduate School of Biomedical Sciences, Department of Pharmacology, Houston, TX 77030-3498. Offers PhD, MD/PhD. *Faculty:* 10 full-time (2 women). *Students:* 4 full-time (1 woman), (all international). Average age 27. 17 applicants, 18% accepted, 3 enrolled. In 2005, 3 doctorates awarded. *Median time to degree:* Of those who began their doctoral program in fall 1997, 100% received their degree in 8 years or less. *Degree requirements:* For doctorate, thesis/dissertation, public defense. *Entrance requirements:* For doctorate, GRE General Test, GRE Subject Test (strongly recommended), minimum GPA of 3.0. Additional exam requirements/recommendations for international students: Required—TOEFL. *Application deadline:* For fall admission, 2/1 for domestic students. Application fee: $30. Electronic applications accepted. *Expenses:* Tuition: Full-time $8,200. Full-time tuition and fees vary according to program. *Financial support:* In 2005–06, 3 fellowships (averaging $23,000 per year), 1 research assistantship with tuition reimbursement (averaging $20,000 per year) were awarded; career-related internships or fieldwork, Federal Work-Study, institutionally sponsored loans, health care benefits, and tuition waivers (full) also available. Financial award applicants required to submit FAFSA. *Faculty research:* Cancer research, biochemistry, molecular proteins and U-RNA, gene cleaning, tumor markers. *Unit head:* Dr. P. K. Chan, Director, 713-798-7915, E-mail: pchan@bcm.edu.

Boston University, School of Medicine, Division of Graduate Medical Sciences, Department of Pharmacology and Experimental Therapeutics, Boston, MA 02118. Offers MA, PhD, MD/PhD. *Faculty:* 12 full-time (4 women). *Students:* 20 full-time (10 women), 6 part-time (3 women); includes 3 minority (1 African American, 2 Asian Americans or Pacific Islanders), 5 international. Average age 31. Terminal master's awarded for partial completion of doctoral program. *Degree requirements:* For master's and doctorate, thesis/dissertation. *Entrance requirements:* For master's and doctorate, GRE General Test, GRE Subject Test. Additional exam requirements/recommendations for international students: Required—TOEFL. *Application deadline:* For fall admission, 1/15 for domestic students; for spring admission, 10/15 priority date for domestic students. Electronic applications accepted. *Expenses:* Tuition: Full-time $31,530; part-time $985 per credit. Required fees: $316; $40 per semester. Tuition and fees vary according to course level and program. *Financial support:* In 2005–06, fellowships with tuition reimbursements (averaging $19,000 per year), research assistantships with tuition reimbursements (averaging $19,000 per year) were awarded; Federal Work-Study, scholarships/grants, traineeships, tuition waivers, and research stipends also available. *Faculty research:* Molecular pharmacology, neuropharmacology, peptide receptors, psychopharmacology. *Unit head:* Dr. David H. Farb, Chairman, 617-638-4300, Fax: 617-638-4329, E-mail: dfarb@bu.edu. *Application contact:* Dr. Carol T. Walsh, Graduate Director, 617-638-4326, Fax: 617-638-4329, E-mail: ctwalsh@bu.edu.

See Close-Up on page 1161.

Boston University, School of Medicine, Division of Graduate Medical Sciences, Program in Clinical Investigation, Boston, MA 02215. Offers MA. *Expenses:* Tuition: Full-time $31,530; part-time $985 per credit. Required fees: $316; $40 per semester. Tuition and fees vary according to course level and program. *Application contact:* Michelle Hall, Assistant Director of Admissions, 617-638-5121, Fax: 617-638-5740, E-mail: natashah@bu.edu.

Case Western Reserve University, School of Medicine and School of Graduate Studies, Graduate Programs in Medicine, Department of Pharmacology, Cleveland, OH 44106. Offers MS, PhD, MD/PhD. *Faculty:* 13 full-time (5 women), 12 part-time/adjunct (2 women). *Students:* 33 full-time (22 women); includes 7 minority (3 African Americans, 3 Asian Americans or Pacific Islanders, 1 Hispanic American), 9 international. Average age 26. 39 applicants, 15% accepted, 5 enrolled. In 2005, 4 degrees awarded. Terminal master's awarded for partial completion of doctoral program. *Median time to degree:* Of those who began their doctoral program in fall 1997, 100% received their degree in 8 years or less. *Degree requirements:* For master's, comprehensive exam, registration; for doctorate, thesis/dissertation, comprehensive exam, registration. *Entrance requirements:* For doctorate, GRE General Test, GRE Subject Test. Additional exam requirements/recommendations for international students: Required—TOEFL. *Application deadline:* For fall admission, 3/1 for domestic students; for winter admission, 8/1 for domestic students; for spring admission, 8/15 for domestic students. Applications are processed on a rolling basis. Application fee: $50. Electronic applications accepted. *Financial support:* In 2005–06, 4 fellowships with full tuition reimbursements (averaging $23,000 per year) were awarded; research assistantships with full tuition reimbursements, health care benefits and tuition waivers (full) also available. *Faculty research:* Aspects of cellular, molecular, and clinical pharmacology; neuroendocrine pharmacology; drug metabolism. Total annual research expenditures: $5.3 million. *Unit head:* Dr. Kris Palczewski, Acting Chair, 216-368-4497, Fax: 216-368-1300, E-mail: kxp65@case.edu. *Application contact:* Camala Jayne Thompson, Coordinator of Graduate Admissions in Pharmacology, 216-368-4617, Fax: 216-368-3395, E-mail: cami@case.edu.

Columbia University, College of Physicians and Surgeons and Graduate School of Arts and Sciences, Graduate School of Arts and Sciences at the College of Physicians and Surgeons, Department of Pharmacology, New York, NY 10032. Offers pharmacology (M Phil, MA, PhD); pharmacology-toxicology (M Phil, MA, PhD). Only candidates for the PhD are admitted. Terminal master's awarded for partial completion of doctoral program. *Degree requirements:* For doctorate, thesis/dissertation. *Entrance requirements:* For master's and doctorate, GRE General Test. Additional exam requirements/recommendations for international students: Required—TOEFL. *Expenses:* Tuition: Full-time $31,448. Tuition and fees vary according to course level, course load, campus/location and program. *Faculty research:* Cardiovascular pharmacology, receptor pharmacology, neuropharmacology, membrane biophysics, eicosanoids.

Cornell University, College of Veterinary Medicine, Ithaca, NY 14853-0001. Offers comparative biomedical science (PhD); immunology (MS, PhD); pharmacology (PhD); physiology (PhD); veterinary medicine (DVM); zoology (MS, PhD). *Accreditation:* AVMA. *Faculty:* 155 full-time (53 women). *Students:* 334 full-time (267 women); includes 71 minority (25 African Americans, 2 American Indian/Alaska Native, 18 Asian Americans or Pacific Islanders, 26 Hispanic Americans), 2 international. Average age 26. 871 applicants, 11% accepted, 78 enrolled. In 2005, 81 first professional degrees, 15 doctorates awarded. *Degree requirements:* For first-professional, thesis or alternative, on-site clinical training. *Entrance requirements:* GRE General Test or MCAT, undergraduate pre-medical science program, animal or veterinary experience, letter of recommendation. *Application deadline:* For fall admission, 10/1 for domestic students, 10/1 for international students. Application fee: $40. Electronic applications accepted. *Expenses: Contact institution. Financial support:* In 2005–06, 303 students received support, including 30 fellowships (averaging $24,854 per year), 88 research assistantships with tuition reimbursements available (averaging $24,854 per year); Federal Work-Study, institutionally sponsored loans, scholarships/grants, and unspecified assistantships also available. Financial award application deadline: 2/1; financial award applicants required to submit CSS PROFILE or FAFSA. *Faculty research:* Extensive biomedical research, comparative cancer, food safety. Total annual research expenditures: $49.3 million. *Unit head:* Dr. Donald F. Smith, Dean, 607-253-3771. *Application contact:* Jennifer A Mailey, Director of Admissions, 607-253-3700, Fax: 607-253-3709, E-mail: vet_admissions@cornell.edu.

Cornell University, Graduate School, Graduate Fields of Comparative Biomedical Sciences, Field of Pharmacology, Ithaca, NY 14853-0001. Offers MS, PhD. *Faculty:* 21 full-time (5 women). *Students:* 28 full-time, 11% accepted, 1 enrolled. In 2005, 4 degrees awarded. *Degree requirements:* For master's, thesis/dissertation; for doctorate, thesis/dissertation, comprehensive exam. *Entrance requirements:* For master's and doctorate, GRE General Test, 3 letters of recommendation. Additional exam requirements/recommendations for international students: Required—TOEFL (minimum score 550 paper-based; 213 computer-based). *Application deadline:* For fall admission, 12/5 for domestic students. Application fee: $60. Electronic applications accepted. *Financial support:* In 2005–06, 13 students received support, including 7 fellowships with full tuition reimbursements available, 5 research assistantships with full tuition reimbursements available, 1 teaching assistantship with full tuition reimbursement available; institutionally sponsored loans, scholarships/grants, health care benefits, tuition waivers (full and partial), and unspecified assistantships also available. Financial award applicants required to submit FAFSA. *Faculty research:* Signal transduction, ion channels, calcium signaling, G proteins, cancer cell biology. *Unit head:* Director of Graduate Studies, 607-253-3276, Fax: 607-253-3756. *Application contact:* Graduate Field Assistant, 607-253-3276, Fax: 607-253-3756, E-mail: graduate_edcvm@cornell.edu.

See Close-Up on page 1163.

Cornell University, Joan and Sanford I. Weill Medical College and Graduate School of Medical Sciences, Weill Graduate School of Medical Sciences, Program in Pharmacology, New York, NY 10021-4896. Offers PhD, MD/PhD. *Faculty:* 26 full-time (6 women). *Students:* 53 full-time (31 women); includes 10 minority (1 African American, 5 Asian Americans or

Pacific Islanders, 4 Hispanic Americans), 15 international. 45 applicants. In 2005, 8 doctorates awarded. *Median time to degree:* Of those who began their doctoral program in fall 1997, 100% received their degree in 8 years or less. *Degree requirements:* For doctorate, thesis/dissertation, final exam. *Entrance requirements:* For doctorate, GRE General Test, GRE Subject Test, previous course work in natural and/or health sciences. Additional exam requirements/recommendations for international students: Required—TOEFL. *Application deadline:* For fall admission, 12/15 for domestic students. Application fee: $60. *Expenses:* Tuition: Full-time $32,320. Required fees: $1,025. *Financial support:* In 2005–06, 6 fellowships (averaging $18,156 per year) were awarded; stipends also available. *Unit head:* Lonny Levin, Director, 212-746-6752, E-mail: llevin@med.cornell.edu.

Creighton University, School of Medicine and Graduate School, Graduate Programs in Medicine and College of Arts and Sciences, Department of Pharmacology, Omaha, NE 68178-0001. Offers pharmaceutical sciences (MS); pharmacology (MS, PhD). Terminal master's awarded for partial completion of doctoral program. *Degree requirements:* For master's, thesis, comprehensive exam; for doctorate, thesis/dissertation, oral and written preliminary exams, comprehensive exam. *Entrance requirements:* For master's and doctorate, GRE General Test, minimum GPA of 3.0, undergraduate degree in sciences. *Faculty research:* Pharmacology secretion, cardiovascular-renal pharmacology, adrenergic receptors, signal transduction, genetic regulation of receptors.

Dalhousie University, Faculty of Graduate Studies and Faculty of Medicine, Graduate Programs in Medicine, Department of Pharmacology, Halifax, NS B3H 4R2, Canada. Offers M Sc, PhD, MD/PhD. *Degree requirements:* For master's, thesis/dissertation; for doctorate, thesis/dissertation, comprehensive exam. *Entrance requirements:* Additional exam requirements/recommendations for international students: Required—TOEFL. *Faculty research:* Electrophysiology and neurochemistry; endocrinology, immunology and cancer research; molecular biology; cardiovascular and autonomic; drug biotransformation and metabolism; ocular pharmacology.

Dartmouth College, Dartmouth Medical School, Program in Experimental and Molecular Medicine, Hanover, NH 03755. Offers experimental and molecular medicine (PhD); pharmcology and toxicology (PhD); phychological and brain sciences (PhD); physiology (PhD). *Degree requirements:* For doctorate, thesis/dissertation, comprehensive exam, registration. *Entrance requirements:* For doctorate, GRE General Test, 3 letters of recommendation. Additional exam requirements/recommendations for international students: Required—TOEFL. Electronic applications accepted. *Expenses:* Tuition: Full-time $31,770.

Dartmouth College, School of Arts and Sciences, Department of Pharmacology and Toxicology, Hanover, NH 03755. Offers MD/PhD. *Faculty:* 14 full-time (2 women). *Students:* 32 full-time (17 women); includes 2 minority (1 African American, 1 Hispanic American), 11 international. Average age 28. 65 applicants, 29% accepted, 10 enrolled. In 2005, 5 degrees awarded. *Degree requirements:* For doctorate, thesis/dissertation. *Entrance requirements:* For doctorate, GRE General Test, GRE Subject Test, bachelor's degree in biological, chemical, or physical science. Additional exam requirements/recommendations for international students: Required—TOEFL. *Application deadline:* For fall admission, 1/15 for domestic students. Electronic applications accepted. *Expenses:* Tuition: Full-time $31,770. *Financial support:* In 2005–06, 29 students received support, including fellowships with full tuition reimbursements available (averaging $23,250 per year), research assistantships with full tuition reimbursements available (averaging $23,250 per year); Federal Work-Study, institutionally sponsored loans, traineeships, and tuition waivers (full and partial) also available. Financial award application deadline: 4/15. *Faculty research:* Molecular biology of carcinogenesis, DNA repair and gene expression, biochemical and environmental toxicology, protein receptor ligand interactions. Total annual research expenditures: $7.1 million. *Unit head:* Dr. Ethan Dmitrovsky, Chair, 603-650-1667, Fax: 603-650-1129, E-mail: pharmacology.and.toxicology@dartmouth.edu. *Application contact:* Ruth Bedor, Coordinator of Graduate Studies, 603-650-1667, Fax: 603-650-1129, E-mail: ruth.bedor@dartmouth.edu.

See Close-Up on page 1165.

Drexel University, College of Medicine, Biomedical Graduate Programs, Pharmacology and Physiology Program, Philadelphia, PA 19104-2875. Offers MS, PhD, MD/PhD. Part-time programs available. Terminal master's awarded for partial completion of doctoral program. *Degree requirements:* For master's, comprehensive exam; for doctorate, thesis/dissertation, qualifying exam. *Entrance requirements:* For master's, GRE General Test, minimum GPA of 2.75; for doctorate, GRE General Test, minimum GPA of 3.0. Additional exam requirements/recommendations for international students: Required—TOEFL. Electronic applications accepted. *Faculty research:* Cardiovascular pharmacology, drugs of abuse, neurotransmitter mechanisms.

Duke University, Graduate School, Department of Pharmacology and Cancer Biology, Durham, NC 27710. Offers pharmacology (PhD). *Faculty:* 39 full-time. *Students:* 57 full-time (35 women); includes 10 minority (4 African Americans, 6 Asian Americans or Pacific Islanders), 7 international. Average age 23. 57 applicants, 21% accepted, 4 enrolled. In 2005, 11 degrees awarded. *Median time to degree:* Of those who began their doctoral program in fall 1997, 100% received their degree in 8 years or less. *Degree requirements:* For doctorate, thesis/dissertation, registration. *Entrance requirements:* For doctorate, GRE General Test, minimum GPA of 2.8. Additional exam requirements/recommendations for international students: Required—TOEFL, IELT (preferred) or TOEFL. *Application deadline:* For fall admission, 12/31 for domestic students, 12/31 for international students. Application fee: $75. Electronic applications accepted. *Financial support:* In 2005–06, 15 students received support, including 11 fellowships with tuition reimbursements available (averaging $23,000 per year), 8 research assistantships with tuition reimbursements available (averaging $25,000 per year), 2 teaching assistantships with tuition reimbursements available (averaging $23,000 per year); Federal Work-Study, scholarships/grants, traineeships, health care benefits, and unspecified assistantships also available. Financial award application deadline: 12/31. *Faculty research:* Developmental pharmacology, neuropharmacology, molecular pharmacology, toxicology, cell growth. *Unit head:* Dr. Donald McDonnell, Director of Graduate Studies, 919-681-8086, Fax: 919-681-7767, E-mail: donald.mcdonnell@duke.edu. *Application contact:* Raylene Means, Assistant to Director of Graduate Studies, 919-618-8601, Fax: 919-681-7767, E-mail: means003@mc.duke.edu.

Duquesne University, School of Pharmacy, Graduate School of Pharmaceutical Sciences, Program in Pharmacology/Toxicology, Pittsburgh, PA 15282-0001. Offers MS, PhD. Part-time programs available. *Faculty:* 6 full-time (2 women). *Students:* 14 full-time (9 women), 4 part-time (1 woman); includes 2 minority (both African Americans), 4 international. 46 applicants, 2% accepted, 1 enrolled. In 2005, 1 degree awarded. *Degree requirements:* For master's and doctorate, thesis/dissertation; for doctorate, thesis/dissertation, comprehensive exam. *Entrance requirements:* For master's and doctorate, GRE General Test. Additional exam requirements/recommendations for international students: Required—TOEFL, TSE. *Application deadline:* For fall admission, 2/1 for domestic students. Applications are processed on a rolling basis. Application fee: $50. *Financial support:* In 2005–06, 13 students received support, including 1 research assistantship with full tuition reimbursement available (averaging $12 per year), 12 teaching assistantships with full tuition reimbursements available; fellowships Financial award applicants required to submit FAFSA. *Unit head:* Dr. Paula Witt-Enderby, Head, 412-396-4346.

East Carolina University, Brody School of Medicine, Department of Pharmacology, Greenville, NC 27858-4353. Offers PhD, MD/PhD. *Faculty:* 10 full-time (2 women). *Students:* 2 full-time (1 woman), 7 part-time (5 women); includes 2 minority (both Asian Americans or Pacific Islanders), 2 international. Average age 28. 13 applicants, 23% accepted. In 2005, 4 degrees awarded. *Median time to degree:* Of those who began their doctoral program in fall 1997, 34% received their degree in 8 years or less. *Degree requirements:* For doctorate, thesis/dissertation, comprehensive exam, registration. *Entrance requirements:* For doctorate, GRE General Test, GRE Subject Test. Additional exam requirements/recommendations for international students: Required—TOEFL. *Application deadline:* For fall admission, 6/15 for

domestic students. Applications are processed on a rolling basis. Application fee: $50. *Expenses:* Tuition, state resident: full-time $2,516. Tuition, nonresident: full-time $12,832. *Financial support:* In 2005–06, fellowships with full tuition reimbursements (averaging $21,500 per year) Financial award application deadline: 6/1. *Faculty research:* GNS/behavioral pharmacology, cardiovascular pharmacology, cell signaling and second messenger, effects of calcium channel blockers. Total annual research expenditures: $955,705. *Unit head:* Dr. David A. Taylor, Chairman, 252-744-2734, Fax: 252-744-3203, E-mail: taylorda@ecu.edu. *Application contact:* Dr. Saeed Dar, Director of Graduate Studies, 252-744-2758, Fax: 252-744-3203, E-mail: dars@ecu.edu.

East Tennessee State University, James H. Quillen College of Medicine, Biomedical Science Graduate Program, Johnson City, TN 37614. Offers anatomy (MS, PhD); biochemistry (MS, PhD); biophysics (MS, PhD); microbiology (MS, PhD); pharmacology (MS, PhD); physiology (MS, PhD). Part-time programs available. *Faculty:* 49 full-time (12 women), 1 (woman) part-time/adjunct. *Students:* 30 full-time (19 women), 5 part-time (4 women); includes 2 minority (1 African American, 1 Asian American or Pacific Islander), 11 international. Average age 31. 78 applicants, 13% accepted, 9 enrolled. In 2005, 1 master's, 4 doctorates awarded. Terminal master's awarded for partial completion of doctoral program. *Degree requirements:* For master's, one foreign language, thesis, comprehensive qualifying exam; for doctorate, 2 foreign languages, thesis/dissertation. *Entrance requirements:* For master's, GRE General Test, minimum GPA of 3.0, bachelor's degree in biological or related science; for doctorate, GRE General Test, GRE Subject Test. Additional exam requirements/recommendations for international students: Required—TOEFL (minimum score 550 paper-based; 213 computer-based). *Application deadline:* For fall admission, 3/15 for domestic students; for spring admission, 3/1 for domestic students. Application fee: $25 ($35 for international students). *Expenses:* Contact institution. *Financial support:* In 2005–06, 7 research assistantships with full tuition reimbursements (averaging $15,000 per year) were awarded; teaching assistantships with full tuition reimbursements, career-related internships or fieldwork, Federal Work-Study, institutionally sponsored loans, scholarships/grants, and tuition waivers (full) also available. Financial award application deadline: 7/1; financial award applicants required to submit FAFSA. Total annual research expenditures: $2.1 million. *Unit head:* Dr. Mitchell E. Robinson, Assistant Dean, Director, 423-439-4658, E-mail: robinson@etsu.edu.

Emory University, Graduate School of Arts and Sciences, Division of Biological and Biomedical Sciences, Program in Molecular and Systems Pharmacology, Atlanta, GA 30322-1100. Offers PhD. *Faculty:* 45 full-time (9 women). *Students:* 46 full-time (30 women); includes 14 minority (10 African Americans, 1 American Indian/Alaska Native, 2 Asian Americans or Pacific Islanders, 1 Hispanic American), 6 international. Average age 27. 48 applicants, 38% accepted, 9 enrolled. In 2005, 3 degrees awarded. *Median time to degree:* Of those who began their doctoral program in fall 1997, 100% received their degree in 8 years or less. *Degree requirements:* For doctorate, thesis/dissertation, comprehensive exam, registration. *Entrance requirements:* For doctorate, GRE General Test, minimum GPA of 3.0 in science course work. Additional exam requirements/recommendations for international students: Required—TOEFL. *Application deadline:* For fall admission, 1/3 for domestic students, 1/3 for international students. Application fee: $50. Electronic applications accepted. *Expenses:* Tuition: Full-time $14,400. Required fees: $217. *Financial support:* In 2005–06, 18 students received support, including 18 fellowships with full tuition reimbursements available (averaging $23,000 per year); institutionally sponsored loans, health care benefits, and tuition waivers (full) also available. *Faculty research:* Transmembrane signaling, neuropharmacology, neurophysiology and neurodegeneration, metabolism and molecular toxicology, cell and developmental biology. *Unit head:* Dr. Edward Morgan, Director, 404-727-5986, Fax: 404-727-0365. *Application contact:* 404-727-2545, Fax: 404-727-3322, E-mail: gdbbs@emory.edu.

Fairleigh Dickinson University, College at Florham, Silberman College of Business, Center for Healthcare Management Studies, Program in Pharmaceutical Studies, Madison, NJ 07940-1099. Offers MBA, Certificate. *Students:* 6 full-time (3 women), 24 part-time (8 women), 2 international. Average age 33. 15 applicants, 60% accepted, 4 enrolled. In 2005, 12 degrees awarded. *Application deadline:* Applications are processed on a rolling basis. Application fee: $40. *Unit head:* Dr. Donald Zimmerman, Director, Center for Healthcare Management Studies, 201-692-7224, Fax: 201-692-7216, E-mail: hemba@fdu.edu.

Florida Agricultural and Mechanical University, Division of Graduate Studies, Research, and Continuing Education, College of Pharmacy and Pharmaceutical Sciences, Graduate Programs in Pharmaceutical Sciences, Tallahassee, FL 32307-3200. Offers environmental toxicology (PhD); medicinal chemistry (MS, PhD); pharmaceutics (MS, PhD); pharmacology/toxicology (MS, PhD); pharmacy administration (MS). *Accreditation:* CEPH. *Degree requirements:* For master's and doctorate, thesis/dissertation, publishable paper, comprehensive exam. *Entrance requirements:* For master's and doctorate, GRE General Test, minimum GPA of 3.0 in last 60 hours. Additional exam requirements/recommendations for international students: Required—TOEFL. *Faculty research:* Anticancer agents, anti-inflammatory drugs, chronopharmacology, neuroendocrinology, microbiology.

Georgetown University, Graduate School of Arts and Sciences, Programs in Biomedical Sciences, Department of Pharmacology, Washington, DC 20057. Offers PhD, MD/PhD. *Degree requirements:* For doctorate, thesis/dissertation, comprehensive exam. *Entrance requirements:* For doctorate, GRE General Test, previous course work in biology and chemistry. Additional exam requirements/recommendations for international students: Required—TOEFL. *Faculty research:* Neuropharmacology, techniques in biochemistry and tissue culture.

The George Washington University, Columbian College of Arts and Sciences, Institute for Biomedical Sciences, Program in Pharmacology, Washington, DC 20052. Offers PhD. *Degree requirements:* For doctorate, thesis/dissertation, general exam. *Entrance requirements:* For doctorate, GRE General Test, interview, minimum GPA of 3.0. Additional exam requirements/recommendations for international students: Required—TOEFL (minimum score 600 paper-based; 250 computer-based). Electronic applications accepted. *Faculty research:* Biochemical pharmacology and toxicology, cancer chemotherapy, carcinogenesis, drug metabolism and disposition, pharmacokinetics.

Howard University, Graduate School of Arts and Sciences, Department of Pharmacology, Washington, DC 20059-0002. Offers MS, PhD, MD/PhD. Part-time programs available. *Faculty:* 8 full-time (2 women), 2 part-time/adjunct (0 women). *Students:* 5 full-time (3 women); includes 2 minority (both African Americans), 3 international. Average age 24. 20 applicants, 20% accepted. In 2005, 1 doctorate awarded. *Degree requirements:* For master's, thesis, comprehensive exam; for doctorate, one foreign language, thesis/dissertation, qualifying exam, comprehensive exam. *Entrance requirements:* For master's, GRE General Test, minimum GPA of 3.2, BS in chemistry, biology, pharmacy, psychology or related field; for doctorate, GRE General Test, minimum graduate GPA of 3.2. *Application deadline:* For fall admission, 4/1 for domestic students, 4/1 for international students; for spring admission, 11/1 for domestic students, 11/1 for international students. Application fee: $45. *Financial support:* In 2005–06, 2 fellowships with tuition reimbursements (averaging $14,500 per year), 2 research assistantships with tuition reimbursements (averaging $14,500 per year), 2 teaching assistantships with tuition reimbursements (averaging $14,500 per year) were awarded; career-related internships or fieldwork, institutionally sponsored loans, and scholarships/grants also available. Financial award application deadline: 4/1. *Faculty research:* Biochemical pharmacology, molecular pharmacology, neuropharmacology, drug metabolism, cancer research. Total annual research expenditures: $4 million. *Unit head:* Dr. Robert Taylor, Chairman, 202-806-6311, Fax: 202-806-4453, E-mail: rtaylor@howard.edu. *Application contact:* Dr. Martha I Davila-Garcia, Director of Graduate Studies, 202-806-7834, Fax: 202-806-7836, E-mail: mdavila-garcia@howard.edu.

Idaho State University, Office of Graduate Studies, College of Pharmacy, Department of Pharmaceutical Sciences, Pocatello, ID 83209. Offers biopharmaceutical analysis (PhD); biopharmaceutics (PhD); pharmaceutical chemistry (MS); pharmaceutical science (PhD); pharmaceutics (MS); pharmacognosy (MS); pharmacokinetics (PhD); pharmacology (MS, PhD). *Degree requirements:* For master's, one foreign language, thesis, thesis research, comprehensive exam, registration (for some programs); for doctorate, thesis/dissertation,

Pharmacology

Idaho State University (continued)
written and oral exams, comprehensive exam, registration. *Entrance requirements:* For master's, GRE General Test, minimum GPA of 3.0, 3 letters of recommendation; for doctorate, GRE General Test, BS degree in pharmacy or related field, minimum GPA of 3.0, 3 letters of recommendation. Additional exam requirements/recommendations for international students: Required—TOEFL (minimum score 550 paper-based; 213 computer-based). *Faculty research:* Metabolic toxicity of heavy metals, neuroendocrine pharmacology, cardiovascular pharmacology, cancer biology, immunopharmacology.

Indiana University Bloomington, Medical Sciences Program, Bloomington, IN 47405-7000. Offers anatomy and cell biology (MA, PhD); pharmacology (MS, PhD); physiology (MA, PhD). *Students:* 11 full-time (2 women), 9 part-time (5 women); includes 5 minority (1 African American, 4 Asian Americans or Pacific Islanders), 2 international. Average age 29. *Entrance requirements:* For master's, GRE, minimum GPA of 3.0; for doctorate, GRE. Additional exam requirements/recommendations for international students: Required—TOEFL. *Application deadline:* For fall admission, 1/15 for domestic students; for spring admission, 9/1 for domestic students. Application fee: $45 ($55 for international students). *Expenses:* Tuition, state resident: full-time $5,437; part-time $227 per credit hour. Tuition, nonresident: full-time $15,836; part-time $660 per credit hour. Required fees: $821. Tuition and fees vary according to campus/location and program. *Financial support:* Fellowships available. *Unit head:* Dr. John B. Watkins, Assistant Dean/Director, 812-855-0616. *Application contact:* Kimberly Bunch, Director of Graduate Admissions, 812-855-1119, E-mail: kbunch@indiana.edu.

Indiana University–Purdue University Indianapolis, Indiana University School of Medicine, Department of Pharmacology and Toxicology, Program in Pharmacology, Indianapolis, IN 46202-2896. Offers MS, PhD, MD/PhD. *Students:* 8 full-time (6 women), 9 part-time (5 women); includes 2 minority (both Hispanic Americans), 6 international. Average age 29.Terminal master's awarded for partial completion of doctoral program. *Degree requirements:* For master's and doctorate, thesis/dissertation. *Entrance requirements:* For master's and doctorate, GRE General Test, GRE Subject Test, minimum GPA of 3.0. *Application deadline:* For fall admission, 1/15 for domestic students. Applications are processed on a rolling basis. Application fee: $50 ($60 for international students). *Expenses:* Tuition, state resident: full-time $5,159; part-time $215 per credit hour. Tuition, nonresident: full-time $14,890; part-time $620 per credit hour. Required fees: $614. Tuition and fees vary according to campus/location and program. *Financial support:* In 2005–06, 9 students received support; fellowships with partial tuition reimbursements available, research assistantships with partial tuition reimbursements available, Federal Work-Study, institutionally sponsored loans, and tuition waivers (partial) available. Financial award application deadline: 1/15. *Faculty research:* Cardiovascular pharmacology, chemotherapy, pharmacokinetics, neuropharmacology. *Application contact:* Victor Elharrar, Director of Graduate Studies, 317-274-1564, Fax: 317-274-7714, E-mail: velharra@iupui.edu.

The Johns Hopkins University, School of Medicine, Graduate Programs in Medicine, Department of Pharmacology and Molecular Sciences, Baltimore, MD 21205. Offers PhD. *Faculty:* 42 full-time (10 women). *Students:* 56 full-time (29 women); includes 11 minority (4 African Americans, 7 Asian Americans or Pacific Islanders), 17 international. 151 applicants, 13% accepted, 11 enrolled. In 2005, 7 degrees awarded. *Degree requirements:* For doctorate, thesis/dissertation, departmental seminar, comprehensive exam. *Entrance requirements:* For doctorate, GRE General Test. Additional exam requirements/recommendations for international students: Required—TOEFL. *Application deadline:* For fall admission, 1/10 for domestic students. Application fee: $75. Electronic applications accepted. *Expenses:* Tuition: Full-time $30,960. Tuition and fees vary according to degree level and program. *Financial support:* In 2005–06, fellowships (averaging $24,600 per year) *Unit head:* Dr. Philip A. Cole, Chairman, 410-614-0540, Fax: 410-614-7717, E-mail: pcole@jhmi.edu. *Application contact:* Dr. James T. Stivers, Director of Admissions, 410-955-7117, Fax: 410-955-3023, E-mail: jstivers@jhmi.edu.

See Close-Up on page 1169.

Kent State University, School of Biomedical Sciences, Program in Pharmacology, Kent, OH 44242-0001. Offers MS, PhD. Offered in cooperation with Northeastern Ohio Universities College of Medicine. Terminal master's awarded for partial completion of doctoral program. *Degree requirements:* For master's and doctorate, thesis/dissertation. *Entrance requirements:* For master's and doctorate, GRE General Test. *Faculty research:* Neuropharmacology, psychotherapeutics and substance abuse, molecular biology of substance abuse, toxicology.

Loma Linda University, School of Medicine, Department of Physiology/Pharmacology, Loma Linda, CA 92350. Offers MS, PhD. Part-time programs available. *Faculty:* 28 full-time (5 women), 3 part-time/adjunct. *Students:* 8 full-time (4 women); includes 1 Hispanic American, 6 international. *Degree requirements:* For master's, thesis or alternative; for doctorate, 2 foreign languages, thesis/dissertation. *Entrance requirements:* For master's and doctorate, GRE General Test. *Application deadline:* Applications are processed on a rolling basis. Application fee: $40. *Financial support:* Tuition waivers (full and partial) available. Support available to part-time students. *Faculty research:* Drug metabolism, biochemical pharmacology, structure and function of cell membranes, neuropharmacology. *Unit head:* Dr. Cherng-ju Kim, 909-824-4564.

Long Island University, Brooklyn Campus, Arnold and Marie Schwartz College of Pharmacy and Health Sciences, Graduate Programs in Pharmacy, Division of Pharmaceutical Sciences, Brooklyn, NY 11201-8423. Offers cosmetic science (MS); industrial pharmacy (MS); pharmaceutics (PhD); pharmacology/toxicology (MS). Part-time and evening/weekend programs available. *Faculty:* 17 full-time (2 women), 3 part-time/adjunct (1 woman). *Students:* 94 full-time (46 women), 77 part-time (26 women); includes 22 minority (20 African Americans, 2 Hispanic Americans), 135 international. Average age 30. In 2005, 33 master's, 1 doctorate awarded. Terminal master's awarded for partial completion of doctoral program. *Degree requirements:* For master's, thesis optional; for doctorate, thesis/dissertation, candidacy exam. *Entrance requirements:* For master's and doctorate, minimum GPA of 3.0. *Application deadline:* Applications are processed on a rolling basis. Application fee: $30. *Financial support:* In 2005–06, 5 fellowships with full tuition reimbursements (averaging $13,500 per year), 18 teaching assistantships with full tuition reimbursements (averaging $6,000 per year) were awarded; Federal Work-Study also available. Financial award applicants required to submit FAFSA. *Unit head:* Dr. Fotios M. Plakogiannis, Director, 718-488-1101, E-mail: fotios. plakogiannis@liu.edu. *Application contact:* Edward Dettling, Director of Graduate Admissions, 718-488-1011, Fax: 718-797-2399, E-mail: admissions@brooklyn.liu.edu.

Louisiana State University Health Sciences Center, School of Graduate Studies in New Orleans, Department of Pharmacology and Experimental Therapeutics, New Orleans, LA 70112-2223. Offers MS, PhD, MD/PhD. Terminal master's awarded for partial completion of doctoral program. *Degree requirements:* For master's and doctorate, thesis/dissertation. *Entrance requirements:* For doctorate, GRE General Test. Additional exam requirements/recommendations for international students: Required—TOEFL. *Faculty research:* Neuropharmacology, gastrointestinal pharmacology, drug metabolism, behavioral pharmacology, cardiovascular pharmacology.

Louisiana State University Health Sciences Center at Shreveport, Department of Pharmacology, Toxicology and Neuroscience, Shreveport, LA 71130-3932. Offers pharmacology (PhD). *Faculty:* 11 full-time, 5 part-time/adjunct. *Students:* 14 full-time (8 women), 1 (woman) part-time; includes 5 minority (all Asian Americans or Pacific Islanders) Average age 25.Terminal master's awarded for partial completion of doctoral program. *Degree requirements:* For doctorate, thesis/dissertation. *Entrance requirements:* For doctorate, GRE General Test, minimum GPA of 3.0. Additional exam requirements/recommendations for international students: Required—TOEFL. *Application deadline:* For fall admission, 5/1 for domestic students. Application fee: $30. *Financial support:* Fellowships, research assistantships, teaching assistantships, institutionally sponsored loans available. Financial award application deadline: 5/1. *Faculty research:* Behavioral, cardiovascular, clinical, and gastrointestinal pharmacology; neuropharmacology; psychopharmacology; drug abuse; pharmacokinetics; neuroendocrinology,

psychoneuroimmunology, and stress; toxicology. *Unit head:* Dr. Nicholas E. Goeders, Head, 318-675-7850, Fax: 318-675-7857, E-mail: ngoede@lsuhsc.edu. *Application contact:* Dr. Tammy R. Dugas, Assistant Professor, 318-675-7867, Fax: 318-675-7857, E-mail: tdugas@lsuhsc. edu.

See Close-Up on page 163.

Loyola University Chicago, Graduate School, Department of Pharmacology and Experimental Therapeutics, Chicago, IL 60626. Offers MS, PhD, MBA/MS, MD/PhD. *Faculty:* 10 full-time (3 women), 7 part-time/adjunct (1 woman). *Students:* 17 full-time (11 women), 8 international. Average age 26. 30 applicants, 27% accepted. In 2005, 3 master's, 2 doctorates awarded. Terminal master's awarded for partial completion of doctoral program. *Degree requirements:* For master's and doctorate, thesis/dissertation, comprehensive exam. *Entrance requirements:* For master's and doctorate, GRE General Test, minimum GPA of 3.0. Additional exam requirements/recommendations for international students: Required—TOEFL. *Application deadline:* For fall admission, 2/1 for domestic students. Electronic applications accepted. *Expenses:* Tuition: Full-time $11,610; part-time $645 per credit. Required fees: $55 per semester. *Financial support:* In 2005–06, 2 fellowships with full tuition reimbursements (averaging $26,024 per year), 13 research assistantships with full tuition reimbursements (averaging $22,000 per year) were awarded; career-related internships or fieldwork and Federal Work-Study also available. Financial award application deadline: 2/1; financial award applicants required to submit FAFSA. *Faculty research:* Neuropharmacology, molecular pharmacology, neuroendocrinology, hematopharmacology, neurodegeneration. *Unit head:* Dr. Tarun Patel, Chair, 708-216-5773, Fax: 708-216-6956. *Application contact:* Dr. Kenneth L. Byron, Graduate Program Director, 708-327-2819, Fax: 708-216-6596, E-mail: kbyron@luc.edu.

See Close-Up on page 1171.

Massachusetts College of Pharmacy and Health Sciences, Graduate Studies, Program in Pharmacology, Boston, MA 02115-5896. Offers MS, PhD. *Accreditation:* ACPE (one or more programs are accredited). Terminal master's awarded for partial completion of doctoral program. *Degree requirements:* For master's, oral defense of thesis; for doctorate, one foreign language. *Entrance requirements:* For master's and doctorate, GRE General Test, minimum QPA of 3.0. Additional exam requirements/recommendations for international students: Required—TOEFL (minimum score 600 paper-based; 230 computer-based). *Faculty research:* Neuropharmacology, cardiovascular pharmacology, nutritional pharmacology, pulmonary physiology, drug metabolism.

McGill University, Faculty of Graduate and Postdoctoral Studies, Faculty of Medicine, Department of Pharmacology and Therapeutics, Montreal, QC H3G 1Y6, Canada. Offers M Sc, PhD. Terminal master's awarded for partial completion of doctoral program. *Degree requirements:* For master's and doctorate, thesis/dissertation. *Entrance requirements:* For master's and doctorate, minimum GPA of 3.2 during last 2 years of full-time study or 3.0 overall. Additional exam requirements/recommendations for international students: Required—TOEFL.

McMaster University, Faculty of Health Sciences and School of Graduate Studies, Program in Medical Sciences, Physiology/Pharmacology Area, Hamilton, ON L8S 4M2, Canada. Offers M Sc, PhD. *Students:* 40 full-time, 4 part-time. In 2005, 8 master's, 2 doctorates awarded. *Degree requirements:* For master's, thesis/dissertation; for doctorate, thesis/dissertation, comprehensive exam. *Entrance requirements:* For master's, honors B Sc, B+ average in related field; for doctorate, M Sc, minimum B+ average, students with proven research experience and an A average may be admitted with a B Sc degree. Additional exam requirements/recommendations for international students: Required—TOEFL (minimum score 580 paper-based; 237 computer-based). *Application deadline:* For fall admission, 3/31 for domestic students. Applications are processed on a rolling basis. Application fee: $85. *Financial support:* Teaching assistantships available. *Unit head:* Dr. Jan Huizinga, Coordinator, 905-525-9140 Ext. 22590. *Application contact:* Dr. Carl Richards, Associate Dean, 905-525-9140 Ext. 22983, Fax: 905-546-1129.

Medical College of Georgia, School of Graduate Studies, Department of Pharmacology, Augusta, GA 30912-1500. Offers PhD. *Faculty:* 17 full-time (4 women). *Students:* 15 full-time (7 women), 10 international. Average age 31. In 2005, 1 degree awarded. *Degree requirements:* For doctorate, thesis/dissertation. *Entrance requirements:* For doctorate, GRE General Test. Additional exam requirements/recommendations for international students: Required—TOEFL. *Application deadline:* For fall admission, 6/30 for domestic students, 4/11 for international students. Applications are processed on a rolling basis. Application fee: $30. *Financial support:* In 2005–06, 2 students received support, including 13 research assistantships with partial tuition reimbursements available (averaging $22,500 per year); fellowships with partial tuition reimbursements available, Federal Work-Study, institutionally sponsored loans, and scholarships/grants also available. Support available to part-time students. Financial award application deadline: 5/31; financial award applicants required to submit FAFSA. *Faculty research:* Protein signaling, neural development, cardiovascular pharmacology, endothelial cell function, neuropharmacology. *Unit head:* Dr. R. William Caldwell, Chair, 706-721-3383, Fax: 706-721-6059, E-mail: wcaldwell@mail.mcg.edu. *Application contact:* Dr. Richard E. White, Program Director, 706-721-7582, Fax: 706-721-2347, E-mail: rwhite@mail.mcg.edu.

Medical College of Wisconsin, Graduate School of Biomedical Sciences, Department of Pharmacology and Toxicology, Milwaukee, WI 53226-0509. Offers MS, PhD, MD/MS, MD/PhD. Terminal master's awarded for partial completion of doctoral program. *Degree requirements:* For master's, thesis; for doctorate, thesis/dissertation, oral and written qualifying exams, comprehensive exam, registration. *Entrance requirements:* For master's and doctorate, GRE General Test, minimum B average. Additional exam requirements/recommendations for international students: Required—TOEFL. *Faculty research:* Cardiovascular physiology and pharmacology, drugs of abuse, environmental and aquatic toxicology, central nervous system and biochemical pharmacology, signal transduction.

Meharry Medical College, School of Graduate Studies, Department of Pharmacology, Nashville, TN 37208-9989. Offers MS, PhD. *Degree requirements:* For master's, thesis/dissertation; for doctorate, thesis/dissertation, comprehensive exam. *Entrance requirements:* For master's and doctorate, GRE. *Faculty research:* Neuropharmacology, cardiovascular pharmacology, behavioral pharmacology, molecular pharmacology, drug metabolism, anticancer.

Michigan State University, College of Osteopathic Medicine and The Graduate School, Graduate Studies in Osteopathic Medicine and Graduate Programs in Human Medicine and Graduate Program in Veterinary Medicine, Department of Pharmacology and Toxicology, East Lansing, MI 48824. Offers pharmacology and toxicology (MS, PhD); pharmacology and toxicology-environmental toxicology (PhD). *Faculty:* 16 full-time (3 women). *Students:* 19 full-time (12 women), 1 (woman) part-time; includes 4 minority (1 American Indian/Alaska Native, 2 Asian Americans or Pacific Islanders, 1 Hispanic American), 6 international. Average age 27. 41 applicants, 20% accepted. In 2005, 1 master's, 6 doctorates awarded. *Degree requirements:* For master's, thesis; for doctorate, thesis/dissertation, oral presentation of dissertation proposal, oral exam in defense of dissertation, comprehensive exam. *Entrance requirements:* For master's, GRE General Test, minimum GPA of 3.4 in last 2 undergraduate years, 3 letters of recommendation; for doctorate, GRE General Test, minimum GPA of 3.0 in last 2 undergraduate years, 3 letters of recommendation. Additional exam requirements/recommendations for international students: Required—TOEFL (minimum score 600 paper-based; 220 computer-based). *Application deadline:* For fall admission, 12/27 for domestic students. Applications are processed on a rolling basis. Application fee: $50. Electronic applications accepted. *Expenses:* Tuition, state resident: part-time $330 per credit hour. Tuition, nonresident: part-time $685 per credit hour. Tuition and fees vary according to program. *Financial support:* In 2005–06, 12 fellowships with tuition reimbursements (averaging $9,713 per year), 8 research assistantships with tuition reimbursements (averaging $15,434 per year) were awarded; teaching assistantships, institutionally sponsored loans, scholarships/grants, and unspecified assistantships also available. Support available to part-time students. *Faculty research:* Effects of drugs and chemicals on macromolecules and actions in humans; neurological, cardiovascular, pulmonary and gastrointestinal pharmacology; environmental and analytical toxicology; endocrine pharmacology. Total annual research expenditures: $3.5 million. *Unit head:* Dr. Joseph R. Haywood, Chairperson, 517-353-7147, Fax: 517-353-8915, E-mail: haywoo12@msu.edu. *Application contact:* Diane Hummel, Graduate Secretary, 517-353-7146, Fax: 517-353-8915, E-mail: hummeld@msu.edu.

New York Medical College, Graduate School of Basic Medical Sciences, Program in Pharmacology, Valhalla, NY 10595-1691. Offers MS, PhD, MD/PhD. Part-time and evening/weekend programs available. Terminal master's awarded for partial completion of doctoral program. *Degree requirements:* For master's, thesis/dissertation; for doctorate, thesis/dissertation, comprehensive exam. *Entrance requirements:* For master's and doctorate, GRE General Test. Additional exam requirements/recommendations for international students: Required—TOEFL. *Faculty research:* Hypertension, neuroendocrine and renal physiology, metabolism of vasoactive peptides, neuroendocrine and hormonal control of circulation.

New York University, School of Medicine and Graduate School of Arts and Science, Sackler Institute of Graduate Biomedical Sciences, Department of Pharmacology, New York, NY 10012-1019. Offers molecular pharmacology and signal transduction (PhD). *Degree requirements:* For doctorate, one foreign language, thesis/dissertation, qualifying exam, comprehensive exam. *Entrance requirements:* For doctorate, GRE General Test. Additional exam requirements/recommendations for international students: Required—TOEFL. *Faculty research:* Pharmacology and neurobiology, neuropeptides, receptor biochemistry, cytoskeleton, endocrinology.

North Carolina State University, College of Veterinary Medicine, Program in Comparative Biomedical Sciences, Raleigh, NC 27695. Offers cell biology and morphology (MS, PhD); epidemiology and population medicine (MS, PhD); immunology (MS, PhD); microbiology and immunology (MS, PhD); pathology (MS, PhD); pharmacology (MS, PhD); specialized veterinary medicine (MS). Part-time programs available. *Degree requirements:* For master's and doctorate, thesis/dissertation. *Entrance requirements:* For master's and doctorate, GRE General Test. Additional exam requirements/recommendations for international students: Required—TOEFL (minimum score 550 paper-based; 213 computer-based). Electronic applications accepted. Expenses: Contact institution. *Faculty research:* Infectious diseases, cell biology, pharmacology and toxicology, genomics, pathology and population medicine.

Northeastern University, Bouvé College of Health Sciences Graduate School, Program in Pharmacology, Boston, MA 02115-5096. Offers pharmaceutical sciences (PhD); pharmacology (MS). Part-time and evening/weekend programs available. *Faculty:* 36 full-time (10 women). *Students:* 76 full-time (40 women), 26 part-time (19 women). Average age 30. 141 applicants, 48% accepted. In 2005, 15 master's, 8 doctorates awarded. *Degree requirements:* For master's, thesis optional. *Entrance requirements:* For master's, bachelor's degree in science, minimum GPA of 3.0. Additional exam requirements/recommendations for international students: Required—TOEFL. *Application deadline:* Applications are processed on a rolling basis. Application fee: $50. *Financial support:* In 2005–06, 18 research assistantships (averaging $16,575 per year) were awarded; Federal Work-Study and tuition waivers (partial) also available. Support available to part-time students. Financial award application deadline: 3/1; financial award applicants required to submit FAFSA. *Faculty research:* Nicotinic receptor subtypes, G-protein coupled receptors, dopamine receptor pharmacology, path-clamping of ion channel. Total annual research expenditures: $1.6 million. *Unit head:* Dr. Vladimir Torchilin, Director, 617-373-3216, Fax: 617-373-6756. *Application contact:* Margaret Schnabel, Director of Graduate Admissions, 617-373-2708, Fax: 617-373-4704, E-mail: bouvegrad@neu.edu.

See Close-Up on page 1177.

Northwestern University, Northwestern University Feinberg School of Medicine, Department of Molecular Pharmacology and Biological Chemistry, Chicago, IL 60611.

Northwestern University, Northwestern University Feinberg School of Medicine and Interdepartmental Degree Programs, Integrated Graduate Programs in the Life Sciences, Chicago, IL 60611. Offers cancer biology (PhD); cell biology (PhD); developmental biology (PhD); evolutionary biology (PhD); immunology and microbial pathogenesis (PhD); molecular biology and genetics (PhD); neurobiology (PhD); pharmacology and toxicology (PhD); structural biology and biochemistry (PhD). *Degree requirements:* For doctorate, thesis/dissertation, written and oral qualifying exams, comprehensive exam. *Entrance requirements:* For doctorate, GRE General Test. Additional exam requirements/recommendations for international students: Required—TOEFL (minimum score 600 paper-based; 250 computer-based). Electronic applications accepted.

See Close-Up on page 189.

Nova Southeastern University, Center for Psychological Studies, Master's Program in Counseling, Mental Health, School Guidance, and Clinical Pharmacology, Fort Lauderdale, FL 33314-7796. Offers clinical pharmacology (MS); mental health counseling (MS); school guidance and counseling (MS). Part-time and evening/weekend programs available. *Faculty:* 12 full-time (1 woman), 42 part-time/adjunct (10 women). *Students:* 218 full-time (189 women), 419 part-time (354 women); includes 338 minority (183 African Americans, 2 American Indian/Alaska Native, 10 Asian Americans or Pacific Islanders, 143 Hispanic Americans), 14 international. 380 applicants, 85% accepted, 167 enrolled. In 2005, 123 degrees awarded. *Degree requirements:* For master's, 3 practica. *Entrance requirements:* Additional exam requirements/recommendations for international students: Required—TOEFL (minimum score 550 paper-based; 213 computer-based). *Application deadline:* For fall admission, 7/29 for domestic students. For winter admission, 11/29 for domestic students; for spring admission, 3/29 for domestic students. Applications are processed on a rolling basis. Application fee: $50. Electronic applications accepted. *Financial support:* Career-related internships or fieldwork, Federal Work-Study, and institutionally sponsored loans available. Financial award application deadline: 4/1. *Faculty research:* Clinical and child clinical psychology, geriatrics, interpersonal violence. *Unit head:* , Karen S. Grosby, Interim Dean, 954-262-5701, Fax: 954-262-3859. *Application contact:* Carlos Perez, Director of Marketing, 954-262-5702, Fax: 954-262-3893, E-mail: perez@nova.edu.

The Ohio State University, College of Medicine and Public Health and Graduate School, Graduate Programs in the Basic Medical Sciences, Integrated Biomedical Science Graduate Program, Columbus, OH 43210. Offers immunology (MS, PhD); medical genetics (MS, PhD); molecular virology (MS, PhD); pharmacology (MS, PhD). *Degree requirements:* For doctorate, thesis/dissertation. *Entrance requirements:* For master's, GRE General Test; for doctorate, GRE. Additional exam requirements/recommendations for international students: Required—TOEFL (minimum score 600 paper-based; 250 computer-based), TSE. Electronic applications accepted.

The Ohio State University, College of Pharmacy and Graduate School, Graduate Programs in Pharmacy, Division of Pharmacology, Columbus, OH 43210. Offers MS, PhD. *Faculty:* 9 full-time (2 women). *Students:* 10 full-time (3 women); includes 1 minority (Asian American or Pacific Islander), 7 international. Average age 26. 21 applicants, 14% accepted, 1 enrolled. In 2005, 3 degrees awarded. *Median time to degree:* Of those who began their doctoral program in fall 1997, 100% received their degree in 8 years or less. *Degree requirements:* For doctorate, thesis/dissertation. *Entrance requirements:* For master's and doctorate, GRE General Test, minimum GPA of 3.0. Additional exam requirements/recommendations for international students: Required—TOEFL (minimum score 600 paper-based; 250 computer-based); Recommended—TSE. *Application deadline:* For fall admission, 2/1 for domestic students. For winter admission, 9/1 for domestic students; for spring admission, 11/1 for domestic students. Application fee: $40 ($50 for international students). Electronic applications accepted. *Financial support:* In 2005–06, 9 students received support, including 2 fellowships with full and partial tuition reimbursements available (averaging $22,000 per year), research assistantships with full and partial tuition reimbursements available (averaging $19,000 per year), 7 teaching assistantships with full and partial tuition reimbursements available (averaging $18,000 per year) Financial award application deadline: 2/1. *Faculty research:* Neuropharmacology, biochemical pharmacology, toxicology, drug receptor theory, molecular pharmacology. *Unit head:* Dr. Dale G. Hoyt, Coordinator, 614-292-6245, Fax: 614-292-9083, E-mail: hoyt.27@osu.edu.

Application contact: Kathy I. Brooks, Graduate Program Coordinator, 614-292-6822, Fax: 614-292-2588, E-mail: gadmbrks@dendrite.pharmacy.ohio-state.edu.

The Ohio State University, College of Veterinary Medicine and Graduate School, Program in Veterinary Medicine, Department of Veterinary Biosciences, Columbus, OH 43210. Offers anatomy and cellular biology (MS, PhD); pathobiology (MS, PhD); pharmacology (MS, PhD); toxicology (MS, PhD); veterinary physiology (MS, PhD). *Faculty research:* Microvasculature, muscle biology, neonatal lung and bone development.

Oregon Health & Science University, School of Medicine, Graduate Programs in Medicine, Department of Physiology and Pharmacology, Portland, OR 97239-3098. Offers pharmacology (PhD); physiology (PhD). *Degree requirements:* For doctorate, thesis/dissertation. *Entrance requirements:* For doctorate, GRE General Test, MCAT. Additional exam requirements/recommendations for international students: Required—TOEFL.

The Pennsylvania State University Milton S. Hershey Medical Center, Graduate School Programs in the Biomedical Sciences, Graduate Program in Pharmacology, Hershey, PA 17033-2360. Offers MS, PhD, MD/PhD, PhD/MBA. *Students:* 13 full-time (5 women), 1 (woman) part-time, 5 international. Average age 27.Terminal master's awarded for partial completion of doctoral program. *Degree requirements:* For master's, thesis or alternative, registration; for doctorate, thesis/dissertation, oral exam, comprehensive exam, registration. *Entrance requirements:* For master's, GRE General Test; for doctorate, GRE General Test, minimum GPA of 3.0. Additional exam requirements/recommendations for international students: Required—TOEFL (minimum score 550 paper-based; 213 computer-based). *Application deadline:* For fall admission, 2/1 priority date for domestic students, 2/1 priority date for international students. Applications are processed on a rolling basis. Application fee: $45. Electronic applications accepted. *Financial support:* In 2005–06, 14 research assistantships with full tuition reimbursements were awarded; fellowships with full tuition reimbursements, institutionally sponsored loans, scholarships/grants, health care benefits, and unspecified assistantships also available. Financial award applicants required to submit FAFSA. *Faculty research:* Ion pump structure and function, drug development and targeting, mechanisms of drug resistance, neuropharmacology and toxicology, breast cancer, identification of molecular targets for drug development in cancer and cardiovascular and neurological diseases. *Unit head:* Dr. Kent Vrana, Chair, 717-531-8285, Fax: 717-531-5013, E-mail: pharm-grad-hmc@psu.edu. *Application contact:* Elaine Neldigh, Program Secretary, 717-531-8285, Fax: 717-531-5013, E-mail: pharm-grad-hmc@psu.edu.

Purdue University, College of Pharmacy and Pharmacal Sciences and Graduate School, Graduate Programs in Pharmacy and Pharmacal Sciences, Department of Medicinal Chemistry and Molecular Pharmacology, West Lafayette, IN 47907. Offers analytical medicinal chemistry (PhD); computational and biophysical medicinal chemistry (PhD); medicinal and bioorganic chemistry (PhD); medicinal biochemistry and molecular biology (PhD); molecular pharmacology and toxicology (PhD); natural products and pharmacognosy (PhD); nuclear pharmacy (MS); radiopharmaceutical chemistry and nuclear pharmacy (PhD). *Faculty:* 21 full-time (2 women), 2 part-time/adjunct (0 women). *Students:* 69 full-time (39 women), 2 part-time (1 woman); includes 8 minority (2 African Americans, 2 Asian Americans or Pacific Islanders, 4 Hispanic Americans), 21 international. Average age 26. 125 applicants, 25% accepted, 15 enrolled. In 2005, 2 master's, 11 doctorates awarded. Terminal master's awarded for partial completion of doctoral program. *Degree requirements:* For master's and doctorate, thesis/dissertation. *Entrance requirements:* For master's, GRE General Test, minimum B average; BS in biology, chemistry, or pharmacy; for doctorate, GRE General Test, minimum B average; BS in biology, chemistry, or pharmacology. Additional exam requirements/recommendations for international students: Required—TOEFL. *Application deadline:* Applications are processed on a rolling basis. Application fee: $55. Electronic applications accepted. *Financial support:* Fellowships, research assistantships, teaching assistantships, traineeships available. Support available to part-time students. Financial award applicants required to submit FAFSA. *Faculty research:* Drug design and development, cancer research, drug synthesis and analysis, chemical pharmacology, environmental toxicology. *Unit head:* Dr. R. F. Borch, Graduate Head, 765-494-1403. *Application contact:* Dr. D. E. Bergstrom, Graduate Committee, 765-494-6275, E-mail: bergstrom@pharmacy.purdue.edu.

Purdue University, School of Veterinary Medicine and Graduate School, Graduate Programs in Veterinary Medicine, Department of Basic Medical Sciences, West Lafayette, IN 47907. Offers anatomy (MS, PhD); pharmacology (MS, PhD); physiology (MS, PhD). Part-time programs available. *Faculty:* 16 full-time (4 women), 4 part-time/adjunct (1 woman). *Students:* 24 full-time (12 women), 3 part-time (all women); includes 1 minority (Hispanic American), 18 international. Average age 27. 18 applicants, 28% accepted, 4 enrolled. In 2005, 1 master's, 3 doctorates awarded. Terminal master's awarded for partial completion of doctoral program. *Median time to degree:* Of those who began their doctoral program in fall 1997, 33% received their degree in 8 years or less. *Degree requirements:* For master's and doctorate, thesis/dissertation. *Entrance requirements:* For master's and doctorate, GRE General Test. Additional exam requirements/recommendations for international students: Required—TOEFL. *Application deadline:* For fall admission, 12/31 priority date for domestic students, 12/31 priority date for international students. Application fee: $55. Electronic applications accepted. *Financial support:* In 2005–06, 4 fellowships with partial tuition reimbursements (averaging $13,251 per year), 12 research assistantships with partial tuition reimbursements (averaging $15,012 per year), 2 teaching assistantships with partial tuition reimbursements (averaging $17,800 per year) were awarded. Financial award application deadline: 3/1; financial award applicants required to submit FAFSA. *Faculty research:* Development and regeneration, tissue injury and shock, biomedical engineering, ovarian function, bone and cartilage biology, cell and molecular biology. *Unit head:* Dr. Gordon L. Coppoc, Head, 765-494-8592, Fax: 765-494-0781, E-mail: coppoc@purdue.edu. *Application contact:* Dr. Kevin M. Hannon, Chairman, Graduate Committee, 765-494-5949, Fax: 765-494-0781, E-mail: bmsgrad@purdue.edu.

Queen's University at Kingston, School of Graduate Studies and Research, Faculty of Health Sciences, Department of Pharmacology and Toxicology, Kingston, ON K7L 3N6, Canada. Offers M Sc, PhD. *Degree requirements:* For master's, thesis/dissertation, registration; for doctorate, thesis/dissertation, comprehensive exam, registration. *Entrance requirements:* For master's, minimum 2nd class standing, honors bachelor of science degree (life sciences, health sciences, or equivalent); for doctorate, masters of science degree or outstanding performance in honors bachelor of science program. Additional exam requirements/recommendations for international students: Required—TOEFL (minimum score 600 paper-based; 250 computer-based). Electronic applications accepted. *Faculty research:* Biochemical toxicology, cardiovascular pharmacology and neuropharmacology.

Rush University, Graduate College, Division of Pharmacology, Chicago, IL 60612-3832. Offers clinical research (MS); pharmacology (MS, PhD). Terminal master's awarded for partial completion of doctoral program. *Degree requirements:* For master's and doctorate, thesis/dissertation. *Entrance requirements:* For master's and doctorate, GRE General Test, interview. Additional exam requirements/recommendations for international students: Required—TOEFL (minimum score 550 paper-based; 213 computer-based). *Faculty research:* Dopamine neurobiology and Parkinson's disease; cardiac electrophysiology and clinical pharmacology; neutrophil motility, apoptosis, and adhesion; angiogenesis; pulmonary vascular physiology.

Saint Louis University, Graduate School and School of Medicine, Graduate Program in Biomedical Sciences and Graduate School, Department of Pharmacological and Physiological Science, St. Louis, MO 63103-2097. Offers PhD. *Faculty:* 31 full-time (13 women), 1 part-time/adjunct (0 women). *Students:* 18 full-time (12 women), 1 (woman) part-time; includes 1 minority (African American), 4 international. Average age 27. 8 applicants, 88% accepted, 7 enrolled. In 2005, 5 degrees awarded. *Degree requirements:* For doctorate, thesis/dissertation, departmental qualifying exams, comprehensive exam. *Entrance requirements:* For doctorate, GRE General Test, letters of recommendation, resumé. Additional exam requirements/recommendations for international students: Required—TOEFL (minimum score 550 paper-based; 213 computer-based). *Application deadline:* For fall admission, 7/1 for domestic students,

Pharmacology

Saint Louis University (continued)

7/1 for international students; for spring admission, 11/1 for domestic students, 11/1 for international students. Applications are processed on a rolling basis. Application fee: $40. Electronic applications accepted. *Expenses:* Tuition: Part-time $760 per credit hour. Required fees: $55 per semester. *Financial support:* In 2005–06, 17 students received support. Traineeships and unspecified assistantships available. Financial award application deadline: 6/1; financial award applicants required to submit FAFSA. *Faculty research:* Molecular endocrinology, neuropharmacology, cardiovascular science, drug abuse, neurotransmitter and hormonal signaling mechanisms. *Unit head:* Dr. Thomas C. Westfall, Chairperson, 314-977-6400; Fax: 314-977-6411, E-mail: westfate@slu.edu. *Application contact:* Gary Behrman, Associate Dean of the Graduate School, 314-977-3827, E-mail: behrmang@slu.edu.

Southern Illinois University Carbondale, Graduate School, Graduate Program in Medicine, Program in Pharmacology, Carbondale, IL 62901-4701. Offers MS, PhD. *Faculty:* 13 full-time (1 woman). *Students:* 16 full-time (3 women), 4 part-time (3 women); includes 1 minority (Asian American or Pacific Islander), 14 international. Average age 30. 26 applicants, 31% accepted, 5 enrolled. In 2005, 2 doctorates awarded. *Degree requirements:* For master's and doctorate, thesis/dissertation. *Entrance requirements:* For master's, minimum GPA of 3.0; for doctorate, minimum GPA of 3.25. Additional exam requirements/recommendations for international students: Required—TOEFL. *Application deadline:* For fall admission, 2/15 for domestic students; for spring admission, 12/31 for domestic students. Applications are processed on a rolling basis. Application fee: $0. *Financial support:* Fellowships with full tuition reimbursements, tuition waivers (full) available. *Faculty research:* Autonomic nervous system pharmacology, biochemical pharmacology, neuropharmacology, toxicology, cardiovascular pharmacology. *Unit head:* Dr. Carl L. Faingold, Chairman, 217-545-2185, Fax: 217-524-0145. *Application contact:* Satu M. Somani, Director, 217-785-2196.

Announcement: The Department of Pharmacology at Southern Illinois University offers the degrees of MS and PhD in pharmacology. Since its inception, this unit has grown substantially in research capability, having acquired both research personnel and substantial extramural funding support. Although it is still relatively young, the department is highly energetic and rapidly achieving academic excellence.

See Close-Up on page 1179.

State University of New York at Buffalo, Graduate School, School of Medicine and Biomedical Sciences, Graduate Programs in Medicine and Biomedical Sciences, Department of Pharmacology and Toxicology, Buffalo, NY 14260. Offers biochemical pharmacology (MS); pharmacology (MA, PhD). *Faculty:* 18 full-time (1 woman), 1 part-time/adjunct (0 women). *Students:* 13 full-time (3 women), 4 part-time; includes 3 minority (2 Asian Americans or Pacific Islanders, 1 Hispanic American), 5 international. 12 applicants, 8% accepted, 1 enrolled. In 2005, 3 doctorates awarded. Terminal master's awarded for partial completion of doctoral program. *Median time to degree:* Of those who began their doctoral program in fall 1997, 100% received their degree in 8 years or less. *Degree requirements:* For master's and doctorate, thesis/dissertation. *Entrance requirements:* For master's and doctorate, GRE General Test, 3 letters of recommendation. Additional exam requirements/recommendations for international students: Required—TOEFL. *Application deadline:* For fall admission, 2/1 priority date for domestic students, 2/1 priority date for international students. Applications are processed on a rolling basis. Application fee: $35. Electronic applications accepted. *Financial support:* In 2005–06, 14 students received support, including 2 fellowships with full tuition reimbursements available (averaging $21,224 per year), 12 research assistantships with full tuition reimbursements available (averaging $21,000 per year); teaching assistantships, Federal Work-Study, scholarships/grants, health care benefits, and unspecified assistantships also available. Financial award application deadline: 2/1; financial award applicants required to submit FAFSA. *Faculty research:* Neuropharmacology, toxicology, signal transduction, molecular pharmacology, behavioral pharmacology. Total annual research expenditures: $1.6 million. *Unit head:* Dr. Ronald P. Rubin, Chairman, 716-829-2800, Fax: 716-829-2801, E-mail: rprubin@buffalo.edu. *Application contact:* Noreen A. Harbison, Information Contact, 716-829-2800, Fax: 716-829-2801, E-mail: harbison@buffalo.edu.

State University of New York Upstate Medical University, College of Graduate Studies, Department of Pharmacology, Syracuse, NY 13210-2334. Offers MS, PhD, MD/PhD. Terminal master's awarded for partial completion of doctoral program. *Degree requirements:* For master's, thesis/dissertation; for doctorate, thesis/dissertation, comprehensive exam. *Entrance requirements:* For master's and doctorate, GRE General Test, interview. Additional exam requirements/recommendations for international students: Required—TOEFL.

Stony Brook University, State University of New York, Health Sciences Center, School of Medicine and Graduate School, Graduate Programs in Medicine, Department of Pharmacological Sciences, Graduate Program in Molecular and Cellular Pharmacology, Stony Brook, NY 11794. Offers PhD. *Faculty:* 14 full-time (3 women). *Students:* 35 full-time (18 women); includes 9 minority (5 African Americans, 3 Asian Americans or Pacific Islanders, 1 Hispanic American), 13 international. Average age 28. 52 applicants, 31% accepted. In 2005, 4 degrees awarded. *Degree requirements:* For doctorate, thesis/dissertation, departmental qualifying exam. *Entrance requirements:* For doctorate, GRE General Test. Additional exam requirements/recommendations for international students: Required—TOEFL. *Application deadline:* For fall admission, 1/15 for domestic students. Applications are processed on a rolling basis. Application fee: $50. Electronic applications accepted. *Expenses:* Tuition: State resident: full-time $6,900; part-time $288 per credit. Tuition, nonresident: full-time $10,920; part-time $455 per credit. Required fees: $754. *Financial support:* In 2005–06, 19 fellowships, 21 research assistantships, 1 teaching assistantship were awarded; Federal Work-Study also available. Financial award application deadline: 3/15; financial award applicants required to submit FAFSA. *Faculty research:* Toxicology, molecular and cellular biochemistry. Total annual research expenditures: $12.1 million. *Application contact:* Beverly Ponte, Graduate Program Administrator, 631-444-3057, Fax: 631-444-3218, E-mail: bev@pharm.som.sunysb.edu.

Announcement: The Program in Molecular and Cellular Pharmacology, an interdisciplinary program administered by the Department of Pharmacological Sciences, offers broad research opportunities in cell and molecular biology. Advanced training in molecular genetics, neuropharmacology, genetic toxicology, biochemical pharmacology, medicinal chemistry, bioorganic chemistry, and membrane biology and biophysics and the mechanisms of action of drugs and hormones are emphasized.

See Close-Up on page 1183.

Temple University, Health Sciences Center, School of Medicine and Graduate School, Graduate Programs in Medicine, Department of Pharmacology, Philadelphia, PA 19122-6096. Offers PhD, MD/PhD. *Faculty:* 12 full-time (1 woman). *Students:* 5 full-time (2 women), 9 part-time (8 women); includes 4 minority (all African Americans), 2 international. 18 applicants, 11% accepted, 2 enrolled. In 2005, 3 degrees awarded. Terminal master's awarded for partial completion of doctoral program. *Degree requirements:* For doctorate, one foreign language, thesis/dissertation, research seminars, one foreign language. *Entrance requirements:* For doctorate, GRE General Test, minimum GPA of 3.0. Additional exam requirements/recommendations for international students: Required—TOEFL (minimum score 620 paper-based; 260 computer-based). *Application deadline:* For fall admission, 1/15 priority date for domestic students, 12/15 priority date for international students. Applications are processed on a rolling basis. Application fee: $50. Electronic applications accepted. *Expenses:* Tuition, state resident: full-time $8,694; part-time $483 per credit. Tuition, nonresident: full-time $12,672; part-time $704 per credit. Required fees: $500; $122 per semester. Tuition and fees vary according to course level, campus/location and program. *Financial support:* Fellowships, research assistantships, Federal Work-Study available. Financial award application deadline: 1/15; financial award applicants required to submit FAFSA. *Faculty research:* Cardiovascular and central nervous systems, biochemical pharmacology. *Unit head:* Dr. Nae Dun, Chair, 215-707-3498, Fax: 215-707-7068, E-mail: nae.dun@temple.edu.

Temple University, Health Sciences Center, School of Pharmacy, Department of Pharmaceutical Sciences, Program in Pharmacodynamics, Philadelphia, PA 19122-6096. Offers MS, PhD. *Entrance requirements:* For master's, GRE General Test, minimum undergraduate GPA of 3.0; for doctorate, GRE General Test, minimum GPA of 3.0. Additional exam requirements/recommendations for international students: Required—TOEFL (minimum score 600 paper-based; 250 computer-based). *Application deadline:* For fall admission, 1/15 for domestic students, 12/15 for international students. Application fee: $50. Electronic applications accepted. *Expenses:* Tuition, state resident: full-time $8,694; part-time $483 per credit. Tuition, nonresident: full-time $12,672; part-time $704 per credit. Required fees: $500; $122 per semester. Tuition and fees vary according to course level, campus/location and program. *Financial support:* Application deadline: 1/15; *Unit head:* Dr. Daniel Canney, Director of Graduate Studies, 215-707-4948, E-mail: canney@temple.edu.

Texas A&M University System Health Science Center, Graduate School of Biomedical Sciences, Department of Medical Pharmacology and Toxicology, College Station, TX 77840. Offers PhD. *Degree requirements:* For doctorate, thesis/dissertation. *Entrance requirements:* For doctorate, GRE General Test. *Faculty research:* Medical treatment of eye disease, fetal alcohol syndrome, Alzheimer's disease, glycme receptory steroids.

Texas Tech University Health Sciences Center, Graduate School of Biomedical Sciences, Department of Pharmacology and Neuroscience, Lubbock, TX 79430. Offers MS, PhD, MD/PhD, MS/PhD. *Faculty:* 13 full-time (4 women). *Students:* 6 full-time (2 women); includes 1 minority (Asian American or Pacific Islander), 3 international. Average age 24. 17 applicants, 0% accepted, 0 enrolled. In 2005, 1 master's, 1 doctorate awarded. Terminal master's awarded for partial completion of doctoral program. *Degree requirements:* For master's and doctorate, thesis/dissertation. *Entrance requirements:* For master's and doctorate, GRE General Test, minimum GPA of 3.0. Additional exam requirements/recommendations for international students: Required—TOEFL. *Application deadline:* For fall admission, 5/15 priority date for domestic students, 4/15 priority date for international students; for spring admission, 11/15 for domestic students, 10/15 for international students. Applications are processed on a rolling basis. Application fee: $45. Electronic applications accepted. *Financial support:* In 2005–06, 3 students received support, including 5 research assistantships (averaging $20,500 per year); Federal Work-Study, institutionally sponsored loans, scholarships/grants, and health care benefits also available. Financial award applicants required to submit FAFSA. *Faculty research:* Neuroscience, neuropsychopharmacology, autonomic pharmacology, cardiovascular pharmacology, molecular pharmacology. Total annual research expenditures: $677,422. *Unit head:* Dr. Reid L. Norman, Chair, 806-743-2425 Ext. 222, Fax: 806-743-2744, E-mail: reid.norman@ttuhsc.edu. *Application contact:* Dr. Michael P. Blanton, Associate Professor, 806-743-2425 Ext. 229, Fax: 806-743-2744, E-mail: michael.blanton@ttuhsc.edu.

Thomas Jefferson University, Jefferson College of Graduate Studies, Program in Pharmacology, Philadelphia, PA 19107. Offers MS. Part-time and evening/weekend programs available. *Faculty:* 17 full-time (2 women), 4 part-time/adjunct (1 woman). *Students:* Average age 41. 33 applicants, 85% accepted, 24 enrolled. In 2005, 18 degrees awarded. *Degree requirements:* For master's, thesis, registration. *Entrance requirements:* For master's, GRE General Test, minimum GPA of 3.0. Additional exam requirements/recommendations for international students: Required—TOEFL (minimum score 213 computer-based). *Application deadline:* For fall admission, 3/1 priority date for domestic students, 3/1 priority date for international students. For winter admission, 12/1 for domestic students; for spring admission, 4/1 for domestic students. Applications are processed on a rolling basis. Application fee: $50. Electronic applications accepted. *Expenses:* Contact institution. *Financial support:* In 2005–06, 15 students received support. Federal Work-Study and institutionally sponsored loans available. Support available to part-time students. Financial award application deadline: 5/1; financial award applicants required to submit FAFSA. *Faculty research:* Receptors, forensic toxicology, thrombosis and atherosclerosis, peptide receptors, signal transduction. *Unit head:* Dr. Dennis M. Gross, Associate Dean, 215-503-0156, Fax: 215-503-3433, E-mail: dennis.gross@jefferson.edu. *Application contact:* Jessie F. Pervall, Director of Admissions, 215-503-0155, Fax: 215-503-9920, E-mail: jessie.pervall@jefferson.edu.

See Close-Up on page 419.

Tufts University, Sackler School of Graduate Biomedical Sciences, Program in Pharmacology and Experimental Therapeutics, Boston, MA 02155. Offers PhD. *Faculty:* 21 full-time (5 women). *Students:* 17 full-time (11 women); includes 4 minority (2 African Americans, 1 Asian American or Pacific Islander, 1 Hispanic American), 2 international. Average age 25. 53 applicants, 9% accepted, 3 enrolled. In 2005, 2 degrees awarded. *Degree requirements:* For doctorate, thesis/dissertation, qualifying exam. *Entrance requirements:* For doctorate, GRE General Test, 3 letters of reference. Additional exam requirements/recommendations for international students: Required—TOEFL. *Application deadline:* For fall admission, 1/15 priority date for domestic students, 1/15 priority date for international students. Applications are processed on a rolling basis. Application fee: $65. Electronic applications accepted. *Financial support:* In 2005–06, 17 students received support, including 17 research assistantships with full tuition reimbursements available (averaging $29,000 per year); scholarships/grants, health care benefits, and tuition waivers (full) also available. Financial award application deadline: 1/15. *Faculty research:* Biochemical mechanisms of narcotic addiction, clinical psychopharmacology, pharmacokinetics, neurotransmitter receptors, neuropeptides. *Unit head:* Dr. Richard Shader, Director, 617-636-3856, Fax: 617-636-6738. *Application contact:* 617-636-6767, Fax: 617-636-0375, E-mail: sackler-school@tufts.edu.

Tulane University, School of Medicine and Graduate School, Graduate Programs in Medicine, Department of Pharmacology, New Orleans, LA 70118-5669. Offers MS, PhD, MD/MS, MD/PhD. MS and PhD offered through the Graduate School. *Degree requirements:* For master's, one foreign language, thesis; for doctorate, 2 foreign languages, thesis/dissertation. *Entrance requirements:* For master's, GRE General Test, minimum B average in undergraduate course work; for doctorate, GRE General Test. Additional exam requirements/recommendations for international students: Required—TOEFL or TSE. Electronic applications accepted.

Université de Montréal, Faculty of Medicine and Faculty of Graduate Studies, Graduate Programs in Medicine, Department of Pharmacology, Montréal, QC H3C 3J7, Canada. Offers M Sc, PhD. *Faculty:* 24 full-time (6 women), 6 part-time/adjunct (0 women). *Students:* 59 full-time (40 women). 57 applicants, 26% accepted, 14 enrolled. In 2005, 16 master's, 5 doctorates awarded. Terminal master's awarded for partial completion of doctoral program. *Degree requirements:* For master's, thesis; for doctorate, thesis/dissertation, general exam. *Entrance requirements:* For master's, proficiency in French; for doctorate, master's degree, proficiency in French. *Application deadline:* For fall and spring admission, 2/1. For winter admission, 11/1 for domestic students. Applications are processed on a rolling basis. Application fee: $30. Electronic applications accepted. *Financial support:* Institutionally sponsored loans available. *Faculty research:* Molecular, clinical, and cardiovascular pharmacology; pharmacokinetics; mechanisms of drug interactions and toxicity; neuropharmacology and receptology. *Unit head:* Patrick du Souich, Director, 514-343-6354, Fax: 514-343-2359. *Application contact:* René Cardinal, Graduate Chairman, 514-343-6111 Ext. 3083.

Université de Sherbrooke, Faculty of Medicine and Health Sciences, Graduate Programs in Medicine, Department of Pharmacology, Sherbrooke, QC J1K 2R1, Canada. Offers M Sc, PhD. *Faculty:* 6 full-time (0 women), 3 part-time/adjunct (0 women). *Students:* 21 full-time (8 women), 33 part-time (12 women). Average age 25. 15 applicants, 40% accepted, 6 enrolled. In 2005, 10 master's, 6 doctorates awarded. *Degree requirements:* For master's and doctorate, thesis/dissertation. *Application deadline:* For fall admission, 6/30 for domestic students. For winter admission, 10/31 for domestic students; for spring admission, 2/28 for domestic students. Application fee: $50. Electronic applications accepted. *Financial support:* Fellowships, research assistantships, tuition waivers (full) available. *Faculty research:* Pharmacology of peptide hormones, pharmacology of lipid mediators. *Unit head:* Dr. Guylain Boulay, Director, 819-561-5470, E-mail: guylain.boulay@usherbrooke.ca.

The University of Alabama at Birmingham, Graduate Programs in Joint Health Sciences, Department of Pharmacology and Toxicology, Birmingham, AL 35294. Offers pharmacology (PhD); toxicology (PhD). *Students:* 29 full-time (18 women), 1 part-time; includes 2 minority (both African Americans), 7 international. 56 applicants, 11% accepted. In 2005, 3 degrees awarded. *Degree requirements:* For doctorate, thesis/dissertation. *Entrance requirements:* For doctorate, GRE General Test, interview. *Application deadline:* Applications are processed on a rolling basis. Application fee: $35 ($60 for international students). Electronic applications accepted. *Expenses:* Contact institution. Tuition and fees vary according to course load, degree level and program. *Financial support:* In 2005–06, 6 fellowships were awarded *Faculty research:* Biochemical pharmacology, neuropharmacology, endocrine pharmacology. *Unit head:* Dr. Robert B. Diasio, Chair, 205-934-4578. *Application contact:* Graduate Coordinator, 205-934-4584, Fax: 205-934-4209, E-mail: rdiasio@uab.edu.

University of Alberta, Faculty of Graduate Studies and Research, Department of Pharmacology, Edmonton, AB T6G 2E1, Canada. Offers M Sc, PhD. *Faculty:* 20 full-time (7 women), 3 part-time/adjunct (0 women). *Students:* 22 full-time (10 women), 6 international. Average age 26. 19 applicants, 42% accepted. In 2005, 2 degrees awarded. Terminal master's awarded for partial completion of doctoral program. *Degree requirements:* For master's and doctorate, thesis/dissertation. *Entrance requirements:* For master's, B Sc, minimum GPA of 3.3 on a 4.0 scale or 7.0 on a 9.0 scale; for doctorate, M Sc in pharmacology or closely related field, honors bachelor's of science in pharmacology. *Application deadline:* For fall admission, 5/1 for domestic students. Applications are processed on a rolling basis. Tuition and fees charges are reported in Canadian dollars. *Expenses:* Tuition, state resident: part-time $562 Canadian dollars per term. Tuition, nonresident: full-time $3,375 Canadian dollars. Required fees: $573 Canadian dollars; $84 Canadian dollars per term. *Financial support:* In 2005–06, 22 students received support, including 2 research assistantships with partial tuition reimbursements available (averaging $15,428 per year), 2 teaching assistantships with partial tuition reimbursements available (averaging $15,428 per year); scholarships/grants also available. *Faculty research:* Cardiovascular pharmacology, neuropharmacology, cancer pharmacology, molecular pharmacology, toxicology. Total annual research expenditures: $3.6 million. *Unit head:* Dr. Susan P. Dunn, Graduate Coordinator, 780-492-0510, Fax: 780-492-4325.

The University of Arizona, Graduate College, College of Pharmacy, Department of Pharmacology and Toxicology and College of Medicine, Graduate Program in Medical Pharmacology, Tucson, AZ 85721. Offers MS, PhD. *Degree requirements:* For master's, thesis/dissertation; for doctorate, thesis/dissertation, comprehensive exam. *Entrance requirements:* Additional exam requirements/recommendations for international students: Required—TOEFL, GRE (score 600). *Faculty research:* Immunopharmacology, pharmacogenetics, pharma cogenomics, clinical pharmacology, ocularpharmacology and neuropharmacology.

The University of Arizona, Graduate College, Graduate Interdisciplinary Programs, Tucson, AZ 85721. Offers American Indian studies (MA, PhD); applied mathematics (MS, PMS, PhD), including applied mathematics (MS, PhD), mathematical sciences (PMS); arid land resource sciences (PhD); cancer biology (PhD); comparative cultural and literary studies (MA, PhD); epidemiology (MS, PhD); genetics (MS, PhD); gerontological studies (MS, Certificate); insect science (PhD); neuroscience (PhD); nutritional sciences (MS, PhD), including dietetics (MS), epidermalogical nutrition/public health nutrition (PhD), human/clinical nutrition (PhD), molecular nutrition (PhD), nutritional biochemistry; pharmacology and toxicology (PhD); physiological sciences (PhD); planning (MS); second language acquisition and teaching (PhD). Part-time programs available. *Entrance requirements:* Additional exam requirements/recommendations for international students: Required—TOEFL.

University of Arkansas for Medical Sciences, College of Medicine and Graduate School, Graduate Programs in Medicine, Department of Pharmacology and Toxicology, Little Rock, AR 72205-7199. Offers pharmacology (MS, PhD); toxicology (MS, PhD). *Faculty:* 18 full-time (2 women), 3 part-time/adjunct (0 women). *Students:* 13 full-time, 3 part-time. *Degree requirements:* For master's and doctorate, thesis/dissertation. *Entrance requirements:* For master's and doctorate, GRE General Test. Additional exam requirements/recommendations for international students: Required—TOEFL. Application fee: $0. *Financial support:* Research assistantships, teaching assistantships available. Support available to part-time students. *Unit head:* Dr. Nancy J. Rusch, Chair, 501-686-5510. *Application contact:* Dr. Philip R. Mayeux, Graduate Coordinator, 501-686-5510, E-mail: mayeuxphilipr@exchange.uams.edu.

See Close-Up on page 1189.

The University of British Columbia, Faculty of Medicine and Faculty of Graduate Studies, Graduate Programs in Medicine, Department of Anesthesiology, Pharmacology and Therapeutics, Vancouver, BC V6T 1Z3, Canada. Offers M Sc, PhD. *Faculty:* 13 full-time (1 woman), 3 part-time/adjunct (0 women). *Students:* 28 full-time (9 women); includes 11 Asian Americans or Pacific Islanders, 1 Hispanic American, 8 international. Average age 28. 37 applicants, 16% accepted, 5 enrolled. In 2005, 1 master's, 2 doctorates awarded. Terminal master's awarded for partial completion of doctoral program. *Median time to degree:* Of those who began their doctoral program in fall 1997, 100% received their degree in 8 years or less. *Degree requirements:* For master's, thesis/dissertation; for doctorate, thesis/dissertation, comprehensive exam. *Entrance requirements:* For master's, MD or appropriate bachelor's degree; for doctorate, MD or M Sc. Additional exam requirements/recommendations for international students: Required—TOEFL (minimum score 600 paper-based; 250 computer-based). *Application deadline:* For fall admission, 3/1 for domestic students; 3/1 for international students. For winter admission, 7/1 for domestic students; for spring admission, 11/1 for domestic students. Applications are processed on a rolling basis. Application fee: $90 Canadian dollars ($150 Canadian dollars for international students). Electronic applications accepted. *Financial support:* In 2005–06, 4 students received support, including 16 fellowships with full and partial tuition reimbursements available (averaging $17,500 Canadian dollars per year), 10 research assistantships with full and partial tuition reimbursements available (averaging $15,000 Canadian dollars per year); institutionally sponsored loans, scholarships/grants, tuition waivers (full and partial), and unspecified assistantships also available. Financial award application deadline: 1/15. *Faculty research:* Cellular, biochemical, autonomic, and cardiovascular pharmacology; neuropharmacology. Total annual research expenditures: $4.2 million Canadian dollars. *Unit head:* Dr. C. B. Warriner, Head, 604-827-3355, Fax: 604-822-2281, E-mail: brian.warriner@vch.ca. *Application contact:* Wynne Leung, Graduate Secretary, 604-822-2575, Fax: 604-822-6012, E-mail: wynnel@interchange.ubc.ca.

University of California, Davis, Graduate Studies, Graduate Group in Pharmacology and Toxicology, Davis, CA 95616. Offers MS, PhD. *Faculty:* 77 full-time. *Students:* 53 full-time (30 women); includes 15 minority (1 American Indian/Alaska Native, 12 Asian Americans or Pacific Islanders, 2 Hispanic Americans), 12 international. Average age 31. 69 applicants, 35% accepted, 13 enrolled. In 2005, 2 master's, 7 doctorates awarded. Terminal master's awarded for partial completion of doctoral program. *Median time to degree:* Of those who began their doctoral program in fall 1997, 33.3% received their degree in 8 years or less. *Degree requirements:* For master's, comprehensive exam or thesis; for doctorate, thesis/dissertation, qualifying exam. *Entrance requirements:* For master's and doctorate, GRE General Test, minimum GPA of 3.0, course work in biochemistry and/or physiology. Additional exam requirements/recommendations for international students: Required—TOEFL (minimum score 550 paper-based; 213 computer-based). *Application deadline:* For fall admission, 1/15 for domestic students, 1/15 for international students. Application fee: $60. Electronic applications accepted. *Financial support:* In 2005–06, 51 students received support, including 9 fellowships with full and partial tuition reimbursements available (averaging $13,536 per year), 21 research assistantships with full and partial tuition reimbursements available (averaging $16,611 per year), 1 teaching assistantship with partial tuition reimbursement available (averaging $15,082 per year); career-related internships or fieldwork, Federal Work-Study, institutionally sponsored loans, scholarships/grants, tuition waivers (full and partial), and unspecified assistantships also available. Financial award application deadline: 1/15; financial award applicants required to submit FAFSA. *Faculty research:* Respiratory, neurochemical, molecular,

genetic, and ecological toxicology. *Unit head:* Alan Buckpitt, Graduate Group Chair, 530-752-7874, E-mail: arbuckpitt@ucdavis.edu. *Application contact:* Judy Erwin, Graduate Administrative Assistant, 530-752-4516, Fax: 530-752-3394, E-mail: gjerwin@ucdavis.edu.

University of California, Irvine, College of Medicine, Department of Pharmacology, Irvine, CA 92697. Offers pharmacology and toxicology (MS, PhD). *Degree requirements:* For doctorate, thesis/dissertation. *Entrance requirements:* For master's, GRE, minimum GPA 3.0; for doctorate, GRE General Test, GRE Subject Test, minimum GPA of 3.0. Additional exam requirements/recommendations for international students: Required—TOEFL (minimum score 550 paper-based; 213 computer-based). Electronic applications accepted. *Faculty research:* Mechanisms of action and effects of drugs on the nervous system, behavior, skeletal muscle, heart, and blood vessels; basic processes in the nervous system, skeletal muscle, heart, and blood vessels.

Announcement: The PhD program in pharmacology integrates many aspects of pharmacology, from the discovery of novel molecules to their regulation of behavior. The program emphasizes studies on ligand-receptor interactions and offers research training in molecular and cellular pharmacology, neuropharmacology, and cardiovascular pharmacology.

See Close-Up on page 1191.

University of California, Los Angeles, School of Medicine and Graduate Division, Graduate Programs in Medicine, Department of Molecular and Medical Pharmacology, Los Angeles, CA 90095. Offers PhD. *Degree requirements:* For doctorate, thesis/dissertation, qualifying exams. *Entrance requirements:* For doctorate, GRE General Test. *Faculty research:* Cardiovascular pharmacology, chemical pharmacology, neuropharmacology, clinical pharmacology, molecular pharmacology.

See Close-Up on page 1193.

University of California, San Diego, School of Medicine and Graduate Studies and Research, Graduate Studies in Biomedical Sciences, Department of Pharmacology, La Jolla, CA 92093-0685. Offers PhD. *Degree requirements:* For doctorate, thesis/dissertation, qualifying exam. *Entrance requirements:* For doctorate, GRE General Test. Additional exam requirements/recommendations for international students: Required—TOEFL. Electronic applications accepted. *Faculty research:* Molecular and cellular pharmacology, cell and organ physiology, cellular and molecular biology.

See Close-Up on page 1197.

University of California, San Francisco, School of Pharmacy and Graduate Division, Pharmaceutical Sciences and Pharmacogenomics Graduate Group, San Francisco, CA 94143. Offers PhD. *Faculty:* 49 full-time (13 women). *Students:* 48 full-time (24 women). Average age 27. 100 applicants, 14% accepted. In 2005, 7 degrees awarded. *Degree requirements:* For doctorate, thesis/dissertation. *Entrance requirements:* For doctorate, GRE General Test, minimum GPA of 3.0. Additional exam requirements/recommendations for international students: Required—TOEFL. *Application deadline:* For fall admission, 1/15 for domestic students. Application fee: $60. *Financial support:* In 2005–06, 4 fellowships with full tuition reimbursements (averaging $25,000 per year), 29 research assistantships with full tuition reimbursements (averaging $23,000 per year), 8 teaching assistantships with full tuition reimbursements (averaging $23,000 per year) were awarded; career-related internships or fieldwork, institutionally sponsored loans, scholarships/grants, traineeships, tuition waivers (full), and unspecified assistantships also available. *Faculty research:* Drug development, drug delivery, molecular pharmacology. *Unit head:* Francis C. Szoka, Program Director, 415-476-3895, Fax: 415-476-0688, E-mail: szoka@cgl.ucsf.edu. *Application contact:* Debbie Acoba, Administrator, 415-476-1947, Fax: 415-476-4929, E-mail: pspg@itsa.ucsf.edu.

University of Chicago, Division of the Biological Sciences, Department of Neurobiology, Pharmacology, and Cell Physiology, Department of Neurobiology, Pharmacology and Physiology, Chicago, IL 60637-1513. Offers cell physiology (PhD); pharmacological and physiological sciences (PhD). *Faculty:* 33 full-time (8 women). *Students:* 4 full-time (0 women); includes 1 minority (Asian American or Pacific Islander) Average age 30. In 2005, 4 degrees awarded. *Degree requirements:* For doctorate, thesis/dissertation, preliminary exam. *Entrance requirements:* For doctorate, GRE General Test. Additional exam requirements/recommendations for international students: Required—TOEFL. *Application deadline:* For fall admission, 12/28 priority date for domestic students, 12/28 priority date for international students. Application fee: $55. Electronic applications accepted. *Financial support:* In 2005–06, fellowships with tuition reimbursements (averaging $26,301 per year), research assistantships with tuition reimbursements (averaging $26,301 per year) were awarded; institutionally sponsored loans, scholarships/grants, traineeships, and health care benefits also available. Financial award applicants required to submit FAFSA. *Faculty research:* Psychopharmacology, neuropharmacology. *Unit head:* Dr. J. Murray Sherman, Chairman, 773-834-2900, Fax: 773-702-3774, E-mail: msherman@bsd.uchicago.edu. *Application contact:* Diane J. Hall, Graduate Administrative Director, 773-702-6371, Fax: 773-702-1216, E-mail: d-hall@uchicago.edu.

University of Cincinnati, Division of Research and Advanced Studies, College of Medicine, Graduate Programs in Biomedical Sciences, Department of Pharmacology and Cell Biophysics, Cincinnati, OH 45221. Offers cell biophysics (PhD); pharmacology (PhD). *Degree requirements:* For doctorate, thesis/dissertation, qualifying exam. *Entrance requirements:* For doctorate, GRE General Test. Additional exam requirements/recommendations for international students: Required—TOEFL. Electronic applications accepted. *Faculty research:* Lipoprotein research, enzyme regulation, electrophysiology, gene actuation.

University of Colorado at Denver and Health Sciences Center, Graduate School, Program in Biomedical Sciences, Department of Pharmacology, Denver, CO 80262. Offers PhD. In 2005, 7 degrees awarded. *Degree requirements:* For doctorate, thesis/dissertation, major seminar, comprehensive exam. *Entrance requirements:* For doctorate, GRE General Test. *Application deadline:* For fall admission, 1/15 for domestic students. Application fee: $50. *Expenses:* Tuition, state resident: full-time $11,730. Tuition, nonresident: full-time $22,980. Tuition and fees vary according to degree level and program. *Financial support:* Fellowships, research assistantships, teaching assistantships, Federal Work-Study, institutionally sponsored loans, and traineeships available. Support available to part-time students. Financial award application deadline: 3/15; financial award applicants required to submit FAFSA. *Faculty research:* Genomics/bioinformatics, cellular biology, drugs of abuse, neuroscience, signal transduction. *Unit head:* Dr. Boris Tabakoff, Chair, 303-724-3668, Fax: 303-724-3663, E-mail: boris.tabakoff@uchsc.edu.

University of Connecticut, Graduate School, School of Pharmacy, Department of Pharmaceutical Sciences, Field of Pharmaceutical Sciences, Graduate Program in Pharmacology and Toxicology, Storrs, CT 06269. Offers pharmacology (MS, PhD); toxicology (MS, PhD). *Faculty:* 12 full-time (1 woman). *Students:* 5 full-time (4 women). Average age 25. 40 applicants, 8% accepted, 3 enrolled. In 2005, 1 degree awarded. Terminal master's awarded for partial completion of doctoral program. *Degree requirements:* For master's, thesis/dissertation, comprehensive exam; for doctorate, thesis/dissertation. *Entrance requirements:* For master's and doctorate, GRE General Test. Additional exam requirements/recommendations for international students: Required—TOEFL (minimum score 550 paper-based; 213 computer-based). *Application deadline:* For fall admission, 2/1 priority date for domestic students, 2/1 priority date for international students; for spring admission, 11/1 for domestic students, 10/1 for international students. Applications are processed on a rolling basis. Application fee: $55. Electronic applications accepted. *Expenses:* Tuition, state resident: part-time $444 per credit hour. Tuition, nonresident: part-time $1,154 per credit hour. Tuition and fees vary according to course load. *Financial support:* In 2005–06, 1 research assistantship with full tuition reimbursement, 3 teaching assistantships with full tuition reimbursements were awarded; fellowships, career-related internships or fieldwork, Federal Work-Study, scholarships/grants, traineeships, health care benefits, and unspecified assistantships also available. Financial award application

Pharmacology

University of Connecticut *(continued)*
deadline: 2/1; financial award applicants required to submit FAFSA. *Application contact:* Leslie Lebel, Administrative Assistant, 860-486-4066, Fax: 860-486-4998, E-mail: phrmacy8@uconnvm.uconn.edu.

See Close-Up on page 1201.

University of Florida, College of Medicine, Department of Pharmacology and Therapeutics, Gainesville, FL 32611. Offers PhD. *Faculty:* 18 full-time (5 women). *Degree requirements:* For doctorate, thesis/dissertation. *Entrance requirements:* For doctorate, GRE General Test, minimum GPA of 3.0. Additional exam requirements/recommendations for international students: Required—TOEFL. *Application deadline:* For fall admission, 2/15 for domestic students. Application fee: $30. Electronic applications accepted. *Expenses:* Tuition, state resident: full-time $6,234. Tuition, nonresident: full-time $21,359. Tuition and fees vary according to program. *Financial support:* In 2005–06, research assistantships with full tuition reimbursements (averaging $23,318 per year); fellowships with full tuition reimbursements, institutionally sponsored loans also available. *Faculty research:* Receptor and membrane pharmacology, autonomics, tetralogy, enzymes, opioid peptides. *Unit head:* Dr. Stephen P. Baker, Chairman, 352-392-3519, E-mail: sbaker@pharmacology.ufl.edu. *Application contact:* Dr. Wayne McCormack, Associate Dean of Graduate Education, 352-392-7413, Fax: 352-846-3466, E-mail: mccormac@pathology.ufl.edu.

University of Florida, College of Medicine and Graduate School, Interdisciplinary Program in Biomedical Sciences, Concentration in Physiology and Pharmacology, Gainesville, FL 32611. Offers PhD. *Faculty:* 33 full-time (9 women). *Students:* 26 full-time (16 women); includes 6 minority (all Asian Americans or Pacific Islanders) In 2005, 5 degrees awarded. *Degree requirements:* For doctorate, thesis/dissertation. *Entrance requirements:* For doctorate, GRE General Test, minimum GPA of 3.0. *Application deadline:* For fall admission, 2/15 for domestic students. Application fee: $30. Electronic applications accepted. *Expenses:* Tuition, state resident: full-time $6,234. Tuition, nonresident: full-time $21,359. Tuition and fees vary according to program. *Financial support:* In 2005–06, 24 research assistantships with full tuition reimbursements (averaging $23,320 per year) were awarded; fellowships with full tuition reimbursements, traineeships and unspecified assistantships also available. *Unit head:* Dr. Jeffrey Harrison, Director, 352-392-3227, E-mail: harrison@pharmacolog.ufl.edu. *Application contact:* Dr. Wayne McCormack, Associate Dean of Graduate Education, 352-392-7413, Fax: 352-846-3466, E-mail: mccormac@pathology.ufl.edu.

University of Florida, College of Pharmacy and Graduate School, Graduate Programs in Pharmacy, Department of Pharmacodynamics, Gainesville, FL 32611. Offers MSP, PhD, Pharm D/PhD. Part-time programs available. *Faculty:* 5 full-time (3 women), 2 part-time (both women), 5 international. Average age 25. 15 applicants, 20% accepted. In 2005, 2 degrees awarded. Terminal master's awarded for partial completion of doctoral program. *Degree requirements:* For master's and doctorate, thesis/dissertation. *Entrance requirements:* For master's and doctorate, GRE General Test, minimum GPA of 3.0. Additional exam requirements/recommendations for international students: Required—TOEFL. *Application deadline:* For fall admission, 6/1 for domestic students. Applications are processed on a rolling basis. Application fee: $30. Electronic applications accepted. *Expenses:* Tuition, state resident: full-time $6,234. Tuition, nonresident: full-time $21,359. Tuition and fees vary according to program. *Financial support:* In 2005–06, 2 research assistantships (averaging $24,967 per year), 4 teaching assistantships (averaging $19,936 per year) were awarded; fellowships, institutionally sponsored loans and unspecified assistantships also available. Support available to part-time students. Financial award application deadline: 4/1. *Faculty research:* Hypertension, aging, alcoholism, diabetes, toxicology. Total annual research expenditures: $641,489. *Unit head:* Dr. Maureen Keller-Wood, Chair, 352-392-8798, Fax: 352-392-9187, E-mail: kellerwd@cop.ufl.edu. *Application contact:* Dr. Joanna Peris, Graduate Coordinator, 352-392-9768, Fax: 352-392-9187, E-mail: peris@cop.ufl.edu.

University of Georgia, College of Pharmacy, Department of Pharmaceutical and Biomedical Sciences, Athens, GA 30602. Offers medicinal chemistry (MS, PhD); pharmaceutics (MS, PhD); pharmacology (MS, PhD); toxicology (MS, PhD). *Faculty:* 17 full-time (4 women). *Students:* 8 full-time, 3 part-time (all women); includes 3 minority (all African Americans), 3 international. 132 applicants, 11% accepted, 7 enrolled. In 2005, 1 degree awarded. *Median time to degree:* Of those who began their doctoral program in fall 1997, 100% received their degree in 8 years or less. *Degree requirements:* For master's, thesis; for doctorate, one foreign language, thesis/dissertation. *Entrance requirements:* For master's and doctorate, GRE General Test, minimum GPA of 3.0. Additional exam requirements/recommendations for international students: Required—TOEFL. *Application deadline:* For fall admission, 1/2 priority date for domestic students, 1/2 priority date for international students. Application fee: $50. Electronic applications accepted. *Financial support:* In 2005–06, fellowships with full tuition reimbursements (averaging $18,000 per year), 10 research assistantships with full tuition reimbursements (averaging $16,500 per year), 35 teaching assistantships with full tuition reimbursements (averaging $16,500 per year) were awarded; career-related internships or fieldwork, institutionally sponsored loans, tuition waivers (partial), and unspecified assistantships also available. Financial award application deadline: 1/1. *Faculty research:* Cancer and infectious diseases, drug delivery, neuropharmacology, cardiovascular pharmacology, bioanalytical chemistry, structural biology. *Unit head:* Dr. Vasu Nair, Head, 706-542-5610, Fax: 706-542-3398, E-mail: vnair@rx.uga.edu. *Application contact:* Dr. Anthony C. Capomacchia, Graduate Coordinator, 706-542-5403, Fax: 706-542-3398, E-mail: tcapomac@rx.uga.edu.

University of Georgia, College of Veterinary Medicine and Graduate School, Graduate Programs in Veterinary Medicine, Department of Physiology and Pharmacology, Athens, GA 30602. Offers pharmacology (MS, PhD); physiology (MS, PhD). *Faculty:* 11 full-time (3 women). *Students:* 17 full-time, 4 part-time; includes 1 minority (African American), 10 international. 22 applicants, 23% accepted, 4 enrolled. In 2005, 1 master's, 4 doctorates awarded. *Degree requirements:* For master's, thesis; for doctorate, one foreign language, thesis/dissertation. *Entrance requirements:* For master's and doctorate, GRE General Test. *Application deadline:* For fall admission, 7/1 for domestic students; for spring admission, 11/15 for domestic students. Application fee: $50. Electronic applications accepted. *Financial support:* Fellowships, research assistantships, teaching assistantships, unspecified assistantships available. *Unit head:* Dr. Thomas F. Murray, Head, 706-542-5859, Fax: 706-542-3015, E-mail: tmurray@vet.uga.edu. *Application contact:* Dr. Royal A. McGraw, Graduate Coordinator, 706-542-0661, Fax: 706-542-3015, E-mail: mcgraw@uga.edu.

University of Guelph, Ontario Veterinary College and Graduate Program Services, Graduate Programs in Veterinary Sciences, Department of Biomedical Sciences, Guelph, ON N1G 2W1, Canada. Offers morphology (M Sc, DV Sc, PhD); pharmacology (M Sc, DV Sc, PhD); physiology (M Sc, DV Sc, PhD); toxicology (M Sc, DV Sc, PhD). Part-time programs available. *Faculty:* 25. *Students:* 43 (27 women). In 2005, 6 master's, 2 doctorates awarded. *Median time to degree:* Of those who began their doctoral program in fall 1997, 100% received their degree in 8 years or less. *Degree requirements:* For master's, thesis/dissertation; for doctorate, thesis/dissertation, comprehensive exam. *Entrance requirements:* For master's, honors B Sc, minimum 75% average in last 20 courses; for doctorate, M Sc with thesis from accredited institution. Additional exam requirements/recommendations for international students: Required—TOEFL (minimum score 550 paper-based; 213 computer-based). *Application deadline:* Applications are processed on a rolling basis. Application fee: $75. Electronic applications accepted. *Financial support:* Fellowships, research assistantships, teaching assistantships available. *Faculty research:* Cellular morphology; endocrine, vascular and reproductive physiology; clinical pharmacology; veterinary toxicology; developmental biology. Total annual research expenditures: $3.5 million. *Unit head:* Dr. N. Maclusky, Chair, 519-824-4120 Ext. 54904, Fax: 511-767-1450. *Application contact:* Dr. G. Kirby, Graduate Coordinator, 519-824-4120 Ext. 54948, Fax: 519-767-1450, E-mail: gkirby@uoguelph.ca.

University of Houston, College of Pharmacy, Houston, TX 77204. Offers hospital pharmacy (MSPHR); medical chemistry and pharmacology (MS); pharmaceutics (MS, PhD); pharmacology (MS, PhD); pharmacy (Pharm D); pharmacy administration (MSPHR). *Accreditation:* ACPE. Part-time programs available. *Faculty:* 18 full-time (3 women), 17 part-time/adjunct (9 women). *Students:* 495 full-time (342 women), 23 part-time (13 women); includes 279 minority (26 African Americans, 198 Asian Americans or Pacific Islanders, 55 Hispanic Americans), 38 international. Average age 25. 599 applicants, 24% accepted, 139 enrolled. In 2005, 114 first professional degrees, 12 master's, 8 doctorates awarded. Terminal master's awarded for partial completion of doctoral program. *Degree requirements:* For master's and doctorate, thesis/dissertation. *Entrance requirements:* For Pharm D, PCAT, interview; for master's and doctorate, GRE General Test. Additional exam requirements/recommendations for international students: Required—TOEFL. *Application deadline:* For spring admission, 3/1 for domestic students. Applications are processed on a rolling basis. Application fee: $25 ($75 for international students). *Financial support:* In 2005–06, 3 fellowships with full tuition reimbursements (averaging $17,200 per year), 6 research assistantships with full tuition reimbursements (averaging $14,200 per year), 28 teaching assistantships with full tuition reimbursements (averaging $14,200 per year) were awarded; career-related internships or fieldwork, Federal Work-Study, institutionally sponsored loans, scholarships/grants, health care benefits, and unspecified assistantships also available. Support available to part-time students. Financial award application deadline: 3/10. *Faculty research:* Cardiovascular and renal pharmacology, cellular pharmacology, drug delivery systems, geriatrics, pharmacokinetics, pharmacoeconomics, pharmaceutical marketing, pharmaceutical management, behavioral pharmacy, and signal transduction. Total annual research expenditures: $1.3 million. *Unit head:* Dr. Sunny Ohia, Dean, 713-743-1253, Fax: 713-743-1259, E-mail: seohia@uh.edu. *Application contact:* Shara Zatopek, Assistant Dean for Admissions, 713-743-1262, Fax: 713-743-1259, E-mail: szatopek@uh.edu.

University of Illinois at Chicago, College of Medicine and Graduate College, Graduate Programs in Medicine, Department of Pharmacology, Chicago, IL 60607-7128. Offers PhD, MD/PhD. Part-time programs available. *Degree requirements:* For doctorate, thesis/dissertation. *Entrance requirements:* For doctorate, GRE General Test. Additional exam requirements/recommendations for international students: Required—TOEFL. *Faculty research:* Neuropharmacology; platelet, molecular, and behavioral pharmacology.

See Close-Up on page 1205.

The University of Iowa, Roy J. and Lucille A. Carver College of Medicine and Graduate College, Graduate Programs in Medicine, Department of Pharmacology, Iowa City, IA 52242-1316. Offers MS, PhD. *Faculty:* 16 full-time (4 women), 9 part-time/adjunct (4 women). *Students:* 31 full-time (11 women); includes 4 minority (1 African American, 1 Asian American or Pacific Islander, 2 Hispanic Americans), 10 international. Average age 27. 98 applicants, 5% accepted, 3 enrolled. In 2005, 1 master's, 2 doctorates awarded. Terminal master's awarded for partial completion of doctoral program. *Median time to degree:* Of those who began their doctoral program in fall 1997, 100% received their degree in 8 years or less. *Degree requirements:* For master's, thesis/dissertation; for doctorate, thesis/dissertation, comprehensive exam. *Entrance requirements:* For master's, GRE General Test; for doctorate, GRE General Test, minimum GPA of 3.0, undergraduate course work in biochemistry. Additional exam requirements/recommendations for international students: Required—TOEFL (minimum score 600 paper-based; 250 computer-based). *Application deadline:* For fall admission, 2/1 priority date for domestic students, 2/1 priority date for international students. Applications are processed on a rolling basis. Application fee: $60 ($85 for international students). Electronic applications accepted. *Expenses:* Tuition, state resident: part-time $1,882 per term. Tuition, nonresident: full-time $17,338; part-time $4,907 per term. Tuition and fees vary according to course load and program. *Financial support:* In 2005–06, 31 research assistantships with full tuition reimbursements (averaging $22,000 per year) were awarded; scholarships/grants, traineeships, and unspecified assistantships also available. *Faculty research:* Cancer and cell cycle, hormones and growth factors, nervous system function and dysfunction, receptors and signal transduction, stroke and hypertension. Total annual research expenditures: $5 million. *Unit head:* Dr. Donna L Hammond, Interim Head, 319-335-7946, Fax: 319-335-8930, E-mail: donna-hammond@uiowa.edu. *Application contact:* Dr. Dawn Quelle, Director, Graduate Admissions, 319-353-5749, Fax: 319-335-8930, E-mail: pharmacology-admissions@uiowa.edu.

University of Kansas, Graduate School, School of Pharmacy, Department of Pharmacology and Toxicology, Program in Pharmacology and Toxicology, Lawrence, KS 66045. Offers MS, PhD. *Students:* 12 full-time (6 women), 2 part-time (both women); includes 1 minority (Hispanic American), 6 international. Average age 28. 49 applicants, 10% accepted. In 2005, 2 degrees awarded. *Degree requirements:* For doctorate, thesis/dissertation, comprehensive exam. *Entrance requirements:* For master's and doctorate, GRE. Additional exam requirements/recommendations for international students: Required—TOEFL. *Application deadline:* For fall admission, 1/15 priority date for domestic students, 1/15 priority date for international students. Applications are processed on a rolling basis. Electronic applications accepted. *Expenses:* Tuition, state resident: full-time $4,859. Tuition, nonresident: full-time $12,000. Required fees: $589. Tuition and fees vary according to program. *Financial support:* Fellowships, research assistantships, teaching assistantships available. Financial award application deadline: 1/15.

University of Kansas, Graduate Studies Medical Center, Interdisciplinary Graduate Program in Biomedical Sciences, Department of Pharmacology, Toxicology and Therapeutics, Lawrence, KS 66045. Offers pharmacology (MS, PhD); toxicology (MS, PhD). Part-time programs available. *Faculty:* 17. *Students:* 2 full-time (0 women), 13 part-time (6 women), 6 international. Average age 28. In 2005, 5 degrees awarded. Terminal master's awarded for partial completion of doctoral program. *Degree requirements:* For master's, thesis, comprehensive exam; for doctorate, one foreign language, thesis/dissertation, comprehensive exam. *Entrance requirements:* For master's and doctorate, GRE General Test. Additional exam requirements/recommendations for international students: Required—TOEFL, TSE. *Application deadline:* For fall admission, 1/15 for domestic students. Applications are processed on a rolling basis. Application fee: $0. Electronic applications accepted. *Expenses:* Tuition, state resident: full-time $4,859. Tuition, nonresident: full-time $12,000. Required fees: $589. Tuition and fees vary according to program. *Financial support:* Fellowships, research assistantships with partial tuition reimbursements, teaching assistantships with full and partial tuition reimbursements, Federal Work-Study, institutionally sponsored loans, scholarships/grants, and traineeships available. Support available to part-time students. Financial award application deadline: 3/30; financial award applicants required to submit FAFSA. *Faculty research:* Cardiovascular pharmacology, neuropharmacology, molecular neurotoxicology, cancer chemotherapy, placental pharmacology/physiology. *Unit head:* Dr. Curtis Klaasen, Chairman, 913-588-7140, Fax: 913-588-7501, E-mail: cklaasen@kumc.edu. *Application contact:* Dorothy McGregor, Director of Admissions, 913-588-7526, Fax: 913-588-7501, E-mail: dmcgregor@kumc.edu.

University of Kentucky, Graduate School, Graduate School Programs from the College of Medicine, Program in Molecular and Biomedical Pharmacology, Lexington, KY 40506-0032. Offers pharmacology (PhD). *Faculty:* 21 full-time (5 women). *Students:* 18 full-time (10 women), 1 (woman) part-time; includes 1 minority (African American), 8 international. Average age 28. 15 applicants, 60% accepted, 6 enrolled. In 2005, 2 degrees awarded. *Median time to degree:* Of those who began their doctoral program in fall 1997, 92.8% received their degree in 8 years or less. *Degree requirements:* For doctorate, thesis/dissertation, comprehensive exam. *Entrance requirements:* For doctorate, GRE General Test, minimum undergraduate GPA of 3.0. Additional exam requirements/recommendations for international students: Required—TOEFL (minimum score 550 paper-based; 213 computer-based). *Application deadline:* For fall admission, 7/17 priority date for domestic students, 2/1 priority date for international students; for spring admission, 12/13 priority date for domestic students, 6/15 priority date for international students. Applications are processed on a rolling basis. Application fee: $40 ($55 for international students). Electronic applications accepted. *Expenses:* Tuition, state resident: full-time $6,308; part-time $331 per credit hour. Tuition, nonresident: full-time $13,968; part-time $756 per credit hour. Tuition and fees vary according to course load, degree level and program. *Financial support:* In 2005–06, 19 students received support, including 19 research assistantships with full tuition reimbursements available (averaging $22,000 per year); fellowships with full tuition reimbursements available, Federal Work-Study, scholarships/grants, traineeships, health care benefits, tuition waivers (partial), and unspecified assistantships also available. Support available to part-time students. Financial award application deadline: 3/15; financial award applicants required to submit FAFSA. *Unit head:* Dr. Robert Hadley, Director of Graduate Studies, 859-323-6656, Fax: 859-323-1981, E-mail: chadley@pop.uky.edu. *Application contact:* Dr. Brian Jackson, Senior Associate Dean, 859-257-8176, Fax: 859-323-1928.

University of Louisville, School of Medicine, Department of Pharmacology and Toxicology, Louisville, KY 40292-0001. Offers MS, PhD. *Students:* 31 full-time (18 women), 19 part-time (10 women); includes 8 minority (7 African Americans, 1 Asian American or Pacific Islander), 18 international. Average age 28. In 2005, 13 master's, 4 doctorates awarded. *Degree requirements:* For master's and doctorate, thesis/dissertation. *Entrance requirements:* For master's and doctorate, GRE General Test. *Application deadline:* For fall admission, 1/15 for domestic students. Applications are processed on a rolling basis. Application fee: $50. *Expenses:* Tuition, state resident: full-time $6,006; part-time $334 per credit hour. Tuition, nonresident: full-time $16,554; part-time $920 per credit hour. Tuition and fees vary according to course load, degree level and program. *Financial support:* Fellowships with tuition reimbursements, research assistantships with tuition reimbursements available. *Unit head:* Dr. David W. Hein, Chair, 502-852-5141, Fax: 502-852-7868, E-mail: dhein@louisville.edu. *Application contact:* Heddy R. Rubin-Teiter, Contact, 502-852-5741, Fax: 502-852-7868, E-mail: hrrubi01@gwise.louisville.edu.

University of Manitoba, Faculty of Medicine and Faculty of Graduate Studies, Graduate Programs in Medicine, Department of Pharmacology and Therapeutics, Winnipeg, MB R3T 2N2, Canada. Offers M Sc, PhD. Part-time programs available. Terminal master's awarded for partial completion of doctoral program. *Degree requirements:* For master's and doctorate, thesis/dissertation. *Entrance requirements:* For master's and doctorate, GRE. Additional exam requirements/recommendations for international students: Required—TOEFL. *Faculty research:* Clinical pharmacology; neuropharmacology; cardiac, hepatic, and renal pharmacology.

University of Maryland, School of Medicine, Graduate Program in Life Sciences, Baltimore, MD 21201. Offers biochemistry (MS, PhD); epidemiology (MS, PhD); gerontology (PhD); microbiology (PhD); molecular and cell biology (MS); molecular medicine (PhD); neuroscience (MS, PhD); pharmacology (MS); physiology (MS); rehabilitation sciences (PhD); toxicology (MS, PhD). *Faculty:* 245 full-time (52 women). *Students:* 268 full-time (165 women), 46 part-time (30 women); includes 43 minority (25 African Americans, 13 Asian Americans or Pacific Islanders, 5 Hispanic Americans), 86 international. 435 applicants, 26% accepted, 61 enrolled. In 2005, 18 master's, 46 doctorates awarded. *Median time to degree:* Of those who began their doctoral program in fall 1997, 99% received their degree in 8 years or less. *Degree requirements:* For master's, registration; for doctorate, thesis/dissertation, lab rotations, comprehensive exam, registration. *Entrance requirements:* For master's and doctorate, GRE or MCAT. Additional exam requirements/recommendations for international students: Required—TOEFL (minimum score 550 paper-based; 213 computer-based). *Application deadline:* For winter admission, 1/15 for domestic students. Applications are processed on a rolling basis. Application fee: $50. Electronic applications accepted. *Expenses:* Tuition, state resident: full-time $8,079; part-time $409 per credit hour. Tuition, nonresident: full-time $18,384; part-time $731 per credit hour. Required fees: $695; $10 per credit hour. Tuition and fees vary according to degree level and program. *Financial support:* In 2005–06, 30 fellowships with full tuition reimbursements (averaging $23,000 per year), 22 research assistantships with full tuition reimbursements (averaging $23,000 per year) were awarded; health care benefits also available. *Faculty research:* Cancer, reproduction, neuroscience, cardiovascular, immunology. *Unit head:* Dr. Margaret Merryl McCarthy, Assistant Dean for Graduate Studies, 410-706-2655, Fax: 410-706-8341, E-mail: mmcarthy@umaryland.edu.

University of Medicine and Dentistry of New Jersey, Graduate School of Biomedical Sciences, Graduate Programs in Biomedical Sciences–Newark, Department of Pharmacology and Physiology, Newark, NJ 07107. Offers PhD. *Degree requirements:* For doctorate, thesis/dissertation, qualifying exam. *Entrance requirements:* For doctorate, GRE General Test. Additional exam requirements/recommendations for international students: Required—TOEFL. *Application deadline:* For fall admission, 2/1 for domestic students. Applications are processed on a rolling basis. Application fee: $40. *Financial support:* Fellowships, research assistantships, Federal Work-Study and institutionally sponsored loans available. Financial award application deadline: 5/1. *Unit head:* Dr. Martha Nowycky, Program Director, 973-972-4391, Fax: 973-972-7950, E-mail: martha.nowycky@umdnj.edu.

University of Medicine and Dentistry of New Jersey, Graduate School of Biomedical Sciences, Graduate Programs in Biomedical Sciences–Newark, Program in Pharmacological Sciences, Newark, NJ 07107. Offers Certificate. *Application contact:* Dr. Henry E. Brezenoff, Acting Dean, 973-972-5333, Fax: 973-972-7148, E-mail: hbrezeno@umdnj.edu.

University of Miami, Graduate School, Miller School of Medicine, Graduate Programs in Medicine, Department of Molecular and Cellular Pharmacology, Coral Gables, FL 33124. Offers PhD, MD/PhD. *Faculty:* 21 full-time (7 women). *Students:* 30 full-time (18 women); includes 6 minority (3 Asian Americans or Pacific Islanders, 3 Hispanic Americans), 7 international. Average age 29. 60 applicants, 13% accepted, 5 enrolled. In 2005, 1 degree awarded. *Degree requirements:* For doctorate, thesis/dissertation, dissertation defense, laboratory rotations, qualifying exam. *Entrance requirements:* For doctorate, GRE General Test. Additional exam requirements/recommendations for international students: Required—TOEFL (minimum score 550 paper-based; 213 computer-based). *Application deadline:* For fall admission, 3/1 priority date for domestic students, 2/1 priority date for international students. Applications are processed on a rolling basis. Application fee: $50. Electronic applications accepted. *Financial support:* In 2005–06, fellowships (averaging $22,000 per year), research assistantships (averaging $22,000 per year) were awarded; institutionally sponsored loans, traineeships, and tuition waivers (full) also available. Financial award application deadline: 3/1; financial award applicants required to submit FAFSA. *Faculty research:* Membrane and cardiovascular pharmacology, muscle contraction, hormone action signal transduction, nuclear transport. *Unit head:* Dr. James D. Potter, Chairman, 305-243-5874, Fax: 305-243-6643, E-mail: jdpotter@miami.edu. *Application contact:* Dr. Kerry L. Burnstein, Director of Graduate Studies, 305-243-5732, Fax: 305-243-3420, E-mail: kburnste@miami.edu.

See Close-Up on page 1209.

University of Michigan, Medical School and Horace H. Rackham School of Graduate Studies, Program in Biomedical Sciences (PIBS), Department of Pharmacology, Ann Arbor, MI 48109. Offers PhD. *Faculty:* 23 full-time (7 women), 12 part-time/adjunct (1 woman). *Students:* 34 full-time (15 women); includes 8 minority (6 African Americans, 1 American Indian/Alaska Native, 1 Asian American or Pacific Islander), 2 international. Average age 25. 51 applicants, 33% accepted, 6 enrolled. In 2005, 9 doctorates awarded. *Median time to degree:* Of those who began their doctoral program in fall 1997, 100% received their degree in 8 years or less. *Degree requirements:* For doctorate, thesis/dissertation, oral preliminary exam, oral defense of dissertation. *Entrance requirements:* For doctorate, GRE General Test, 3 letters of recommendation. Additional exam requirements/recommendations for international students: Required—TOEFL (minimum score 560 paper-based). *Application deadline:* For fall admission, 12/31 for domestic students, 12/31 for international students. Application fee: $60 ($75 for international students). Electronic applications accepted. *Expenses:* Tuition, state resident: full-time $14,082; part-time $894 per credit hour. Tuition, nonresident: full-time $28,500; part-time $1,675 per credit hour. Required fees: $189; $189 per unit. *Financial support:* In 2005–06, 2 students received support, including 16 fellowships with full tuition reimbursements available (averaging $23,500 per year), 21 research assistantships with full tuition reimbursements available (averaging $23,500 per year); scholarships/grants, traineeships, health care benefits, and unspecified assistantships also available. Financial award application deadline: 12/31. *Faculty research:*

Signal transduction, addiction research, cancer pharmacology, drug metabolism and pharmacogenetics. Total annual research expenditures: $8.9 million. *Unit head:* Dr. Paul F. Hollenberg, Chair and Professor of Pharmacology, 734-764-8166, Fax: 734-763-5387, E-mail: phollen@umich.edu. *Application contact:* Eileen A. Ferguson, Student Services Associate, 734-764-8166, Fax: 734-763-5387, E-mail: effergie@umich.edu.

University of Minnesota, Duluth, Medical School, Program in Pharmacology, Duluth, MN 55812-2496. Offers MS, PhD. *Faculty:* 5 full-time (2 women). Terminal master's awarded for partial completion of doctoral program. *Degree requirements:* For master's, thesis, final oral exam; for doctorate, thesis/dissertation, final oral exam, oral and written preliminary exams. *Entrance requirements:* For master's and doctorate, GRE General Test. Additional exam requirements/recommendations for international students: Required—TOEFL. *Application deadline:* For fall admission, 8/1 for domestic students. Applications are processed on a rolling basis. Application fee: $40 ($50 for international students). *Financial support:* In 2005–06, research assistantships with full tuition reimbursements (averaging $18,247 per year); fellowships, institutionally sponsored loans also available. Financial award application deadline: 8/1. *Faculty research:* Drug addiction, alcohol and hypertension, neurotransmission, allergic airway disease, auditory neuroscience. Total annual research expenditures: $257,742. *Unit head:* Jean F. Regal, Associate Director of Graduate Studies, 218-726-8950, Fax: 218-726-6235, E-mail: jregal@d.umn.edu.

University of Minnesota, Twin Cities Campus, Graduate School, Department of Pharmacology, Minneapolis, MN 55455-0213. Offers MS, PhD. Terminal master's awarded for partial completion of doctoral program. *Degree requirements:* For master's and doctorate, thesis/dissertation. *Entrance requirements:* For master's and doctorate, GRE General Test. Additional exam requirements/recommendations for international students: Required—TOEFL. Electronic applications accepted. *Expenses:* Tuition, state resident: full-time $8,748; part-time $729 per credit. Tuition, nonresident: full-time $15,848; part-time $1,321 per credit. Full-time tuition and fees vary according to class time, course load, program and reciprocity agreements. *Faculty research:* Molecular pharmacology, cancer chemotherapy, neuropharmacology, biochemical pharmacology, behavioral pharmacology.

University of Mississippi, Graduate School, School of Pharmacy, Department of Pharmacology, Oxford, University, MS 38677. Offers pharmacology (MS, PhD); toxicology (PhD). In 2005, 1 degree awarded. *Degree requirements:* For master's and doctorate, thesis/dissertation. *Entrance requirements:* For master's, GRE General Test, minimum GPA of 3.0; for doctorate, GRE General Test. Additional exam requirements/recommendations for international students: Required—TOEFL. *Application deadline:* For fall admission, 4/1 for domestic students. Applications are processed on a rolling basis. Application fee: $25. *Expenses:* Tuition, state resident: full-time $4,320; part-time $240 per credit hour. Tuition, nonresident: full-time $9,744; part-time $301 per credit hour. Tuition and fees vary according to program. *Financial support:* Scholarships/grants available. Financial award application deadline: 3/1; financial award applicants required to submit FAFSA. *Faculty research:* Behavioral and biochemical pharmacology. *Unit head:* Dr. Robert C. Speth, Chairman, 662-915-7330.

University of Mississippi Medical Center, School of Graduate Studies in the Health Sciences, Department of Pharmacology and Toxicology, Jackson, MS 39216-4505. Offers pharmacology (MS, PhD); toxicology (MS, PhD). *Faculty:* 11 full-time (1 woman), 1 part-time/adjunct (0 women). *Students:* 21 full-time (6 women), 1 (woman) part-time; includes 4 African Americans, 14 international. Average age 30. 14 applicants, 29% accepted, 4 enrolled. In 2005, 4 degrees awarded. Terminal master's awarded for partial completion of doctoral program. *Degree requirements:* For master's, thesis; for doctorate, thesis/dissertation, first authored publication. *Entrance requirements:* For master's and doctorate, GRE General Test, minimum GPA of 3.0. *Application deadline:* For fall admission, 6/1 for domestic students. Applications are processed on a rolling basis. Application fee: $10. *Financial support:* In 2005–06, 23 research assistantships (averaging $16,557 per year) were awarded Financial award application deadline: 4/1. *Faculty research:* Neuropharmacology, environmental toxicology, aging, immunopharmacology, cardiovascular pharmacology. Total annual research expenditures: $5 million. *Unit head:* Dr. I. K. Ho, Dean, 601-984-1600, Fax: 601-984-1637, E-mail: iho@pharmacology.umsmed.edu. *Application contact:* Dr. Jerry M. Farley, Director, 601-984-1630, Fax: 601-984-1637, E-mail: jfarley@pharmacology.umsmed.edu.

See Close-Up on page 1211.

University of Missouri–Columbia, School of Medicine and Graduate School, Graduate Programs in Medicine, Department of Medical Pharmacology and Physiology, Columbia, MO 65211. Offers pharmacology (MS, PhD); physiology (MS, PhD). *Faculty:* 20 full-time (4 women), 1 part-time/adjunct (0 women). *Students:* 14 full-time (8 women), 10 part-time (4 women); includes 7 minority (5 African Americans, 1 Asian American or Pacific Islander, 1 Hispanic American), 3 international. In 2005, 1 master's, 4 doctorates awarded. *Degree requirements:* For master's and doctorate, thesis/dissertation. *Entrance requirements:* For master's and doctorate, GRE General Test, minimum GPA of 3.0. *Application deadline:* For fall admission, 2/1 for domestic students. Application fee: $45 ($60 for international students). *Financial support:* Fellowships, research assistantships, teaching assistantships, institutionally sponsored loans available. *Faculty research:* Endocrine and metabolic pharmacology, biochemical pharmacology, neuropharmacology, receptors and transmembrane signaling. *Unit head:* Dr. Michael Rovetto, Director of Graduate Studies, 573-882-0773.

The University of Montana, Graduate School, School of Pharmacy and Allied Health Sciences, Department of Biomedical and Pharmaceutical Sciences, Missoula, MT 59812-0002. Offers pharmaceutical sciences (MS); pharmacology (PhD); toxicology (MS, PhD). *Accreditation:* ACPE. *Degree requirements:* For master's, oral defense of thesis; for doctorate, research dissertation defense. *Entrance requirements:* For master's and doctorate, GRE General Test. Additional exam requirements/recommendations for international students: Required—TOEFL (minimum score 540 paper-based; 210 computer-based). Electronic applications accepted. *Expenses:* Tuition, state resident: part-time $267 per credit. Tuition, nonresident: part-time $665 per credit. Part-time tuition and fees vary according to course load and degree level. *Faculty research:* Cardiovascular pharmacology, medicinal chemistry, neurosciences, environmental toxicology, pharmacogenetics, cancer.

University of Nebraska Medical Center, Graduate Studies, Department of Pharmacology and Experimental Neuroscience, Omaha, NE 68198. Offers neuroscience (MS, PhD); pharmacology (MS, PhD). *Faculty:* 20 full-time, 9 part-time/adjunct. *Students:* 21 full-time (13 women), 4 international. Average age 27. 18 applicants, 22% accepted, 4 enrolled. In 2005, 1 master's, 2 doctorates awarded. Terminal master's awarded for partial completion of doctoral program. *Median time to degree:* Of those who began their doctoral program in fall 1997, 70% received their degree in 8 years or less. *Degree requirements:* For master's and doctorate, thesis/dissertation, comprehensive exam, registration. *Entrance requirements:* For master's and doctorate, GRE General Test. Additional exam requirements/recommendations for international students: Required—TOEFL (minimum score 600 paper-based; 250 computer-based). *Application deadline:* Applications are processed on a rolling basis. Application fee: $45. Electronic applications accepted. *Expenses:* Tuition, area resident: Part-time $200 per hour. Tuition, nonresident: part-time $538 per hour. Required fees: $308; $59 per term. *Financial support:* In 2005–06, 1 fellowship with full tuition reimbursement (averaging $21,000 per year), 9 research assistantships with full tuition reimbursements (averaging $21,000 per year) were awarded; career-related internships or fieldwork, institutionally sponsored loans, scholarships/grants, and unspecified assistantships also available. Support available to part-time students. *Faculty research:* Neuropharmacology, molecular pharmacology, toxicology, molecular biology, neuroscience. Total annual research expenditures: $1.8 million. *Unit head:* Dr. Jialin Zheng, Program Chair, 402-559-5656, Fax: 402-559-7495, E-mail: jzheng@unmc.edu.

The University of North Carolina at Chapel Hill, School of Medicine and Graduate School, Graduate Programs in Medicine, Department of Pharmacology, Chapel Hill, NC 27599.

Pharmacology

The University of North Carolina at Chapel Hill (continued)
Offers PhD. *Faculty:* 35 full-time (5 women), 9 part-time/adjunct (0 women). *Students:* 44 full-time (28 women); includes 11 minority (6 African Americans, 4 Asian Americans or Pacific Islanders, 1 Hispanic American). Average age 25. 82 applicants, 17% accepted, 9 enrolled. In 2005, 4 doctorates awarded. *Degree requirements:* For doctorate, thesis/dissertation, comprehensive exam, registration. *Entrance requirements:* For doctorate, GRE General Test, GRE Subject Test, minimum GPA of 3.0. *Application deadline:* For fall admission, 1/1 for domestic students. Application fee: $60. Electronic applications accepted. *Financial support:* In 2005–06, 5 fellowships with tuition reimbursements (averaging $22,000 per year), 39 research assistantships with tuition reimbursements (averaging $22,000 per year) were awarded; tuition waivers (full) and unspecified assistantships also available. *Faculty research:* Signal transduction, cell adhesion, receptors, ion channels. Total annual research expenditures: $6.7 million. *Unit head:* Dr. Gary L. Johnson, Chairman, 919-843-3107, Fax: 919-966-5640, E-mail: glj@med.unc.edu. *Application contact:* Dr. Lee M. Graves, Director of Graduate Studies, 919-966-1153, Fax: 919-966-5640, E-mail: phcograd@med.unc.edu.

University of North Dakota, School of Medicine and Graduate School, Graduate Programs in Medicine, Department of Pharmacology, Physiology, and Therapeutics, Grand Forks, ND 58202. Offers pharmacology (MS, PhD); physiology (MS, PhD). *Faculty:* 15 full-time (2 women), 1 part-time/adjunct (0 women). *Students:* 4 full-time (2 women), 9 part-time (4 women). 12 applicants, 25% accepted, 2 enrolled. In 2005, 1 degree awarded. *Degree requirements:* For master's, thesis, comprehensive exam; for doctorate, thesis/dissertation, written and oral exams. *Entrance requirements:* For master's, GRE General Test or MCAT, minimum GPA of 3.0; for doctorate, GRE General Test, minimum GPA of 3.5. Additional exam requirements/recommendations for international students: Required—TOEFL (minimum score 550 paper-based; 213 computer-based). *Application deadline:* For fall admission, 2/15 priority date for domestic students, 2/15 priority date for international students. Applications are processed on a rolling basis. Application fee: $35. Electronic applications accepted. *Financial support:* In 2005–06, 12 research assistantships with full tuition reimbursements (averaging $13,997 per year) were awarded; fellowships, teaching assistantships with full tuition reimbursements, Federal Work-Study, institutionally sponsored loans, scholarships/grants, and tuition waivers (full and partial) also available. Support available to part-time students. Financial award application deadline: 3/15; financial award applicants required to submit FAFSA. *Unit head:* Dr. Matthew Picklo, Graduate Director, 701-777-2293, Fax: 701-777-4490, E-mail: mpicklo@medicine.nodak.edu. *Application contact:* Brenda Halle, Admissions Specialist, 701-777-2947, Fax: 701-777-3619, E-mail: brendahalle@mail.und.edu.

University of North Texas Health Science Center at Fort Worth, Graduate School of Biomedical Sciences, Fort Worth, TX 76107-2699. Offers anatomy and cell biology (MS, PhD); biochemistry and molecular biology (MS, PhD); biomedical sciences (MS, PhD); biotechnology (MS); forensic genetics (MS); integrative physiology (MS, PhD); medical science (MS); microbiology and immunology (MS, PhD); pharmacology (MS, PhD); science education (MS). *Faculty:* 57 full-time (9 women), 2 part-time/adjunct (0 women). *Students:* 177 full-time (98 women), 42 part-time (33 women); includes 61 minority (15 African Americans, 3 American Indian/Alaska Native, 27 Asian Americans or Pacific Islanders, 16 Hispanic Americans), 53 international. Average age 28. 237 applicants, 62% accepted, 91 enrolled. In 2005, 37 master's, 15 doctorates awarded. Terminal master's awarded for partial completion of doctoral program. *Degree requirements:* For master's and doctorate, thesis/dissertation. *Entrance requirements:* For master's and doctorate, GRE General Test. Additional exam requirements/recommendations for international students: Required—TOEFL. *Application deadline:* For fall admission, 5/1 for domestic students. Application fee: $25 ($50 for international students). *Expenses:* Contact institution. *Financial support:* In 2005–06, 80 research assistantships (averaging $16,000 per year) were awarded; fellowships, teaching assistantships, career-related internships or fieldwork, Federal Work-Study, institutionally sponsored loans, scholarships/grants, and traineeships also available. Support available to part-time students. Financial award application deadline: 4/1; financial award applicants required to submit FAFSA. *Faculty research:* Alzheimer's disease, aging, eye diseases, cancer, cardiovascular disease. Total annual research expenditures: $21 million. *Unit head:* Dr. Thomas Yorio, Dean, 817-735-2560, Fax: 817-735-0243, E-mail: yoriot@hsc.unt.edu. *Application contact:* Carla Lee, Director of Graduate Admissions and Services, 817-735-2560, Fax: 817-735-0243, E-mail: gsbs@hsc.unt.edu.

See Close-Up on page 279.

University of Pennsylvania, School of Medicine, Biomedical Graduate Studies, Graduate Group in Pharmacological Sciences, Philadelphia, PA 19104. Offers pharmacology (PhD). *Faculty:* 78. *Students:* 81 full-time (47 women); includes 20 minority (1 African American, 1 American Indian/Alaska Native, 12 Asian Americans or Pacific Islanders, 6 Hispanic Americans), 13 international. 80 applicants, 39% accepted, 13 enrolled. In 2005, 11 doctorates awarded. *Degree requirements:* For doctorate, thesis/dissertation. *Entrance requirements:* For doctorate, GRE General Test, previous course work in physical or natural science. Additional exam requirements/recommendations for international students: Required—TOEFL. *Application deadline:* For fall admission, 12/15 priority date for domestic students, 12/1 priority date for international students. Applications are processed on a rolling basis. Application fee: $70. Electronic applications accepted. *Financial support:* In 2005–06, 76 students received support, including 17 fellowships; research assistantships, scholarships/grants, traineeships, unspecified assistantships, and departmental funds also available. *Faculty research:* Properties and regulation of receptors for biogenic amines, molecular aspects of transduction, mechanisms of biosynthesis, biological mechanisms of depression, developmental events in the nervous system. *Unit head:* Dr. Marcelo G. Kazanietz, Chair, 215-898-9736, E-mail: coord@pharm.med.upenn.edu. *Application contact:* Linda Leroy, Coordinator, 215-898-1790, Fax: 215-573-2236, E-mail: leroy@pharm.med.upenn.edu.

University of Prince Edward Island, Atlantic Veterinary College, Graduate Program in Veterinary Medicine, Charlottetown, PE C1A 4P3, Canada. Offers anatomy (M Sc, PhD); bacteriology (M Sc, PhD); clinical pharmacology (M Sc, PhD); clinical sciences (M Sc, PhD); epidemiology (M Sc, PhD), including reproduction; fish health (M Sc, PhD); food animal nutrition (M Sc, PhD); immunology (M Sc, PhD); microanatomy (M Sc, PhD); parasitology (M Sc, PhD); pathology (M Sc, PhD); pharmacology (M Sc, PhD); physiology (M Sc, PhD); toxicology (M Sc, PhD); veterinary science (M Vet Sc); virology (M Sc, PhD). Part-time programs available. *Faculty:* 76 full-time (25 women), 49 part-time/adjunct (8 women). *Students:* 54 full-time (32 women), 2 part-time. Average age 30. In 2005, 7 master's, 6 doctorates awarded. *Degree requirements:* For master's and doctorate, thesis/dissertation. *Entrance requirements:* For master's, DVM, B Sc honors degree, or equivalent; for doctorate, M Sc. *Application deadline:* Applications are processed on a rolling basis. Application fee: $50. *Expenses:* Contact institution. Tuition charges are reported in Canadian dollars. Part-time tuition and fees vary according to course level, degree level, campus/location, program and student level. *Financial support:* In 2005–06, 4 fellowships (averaging $25,000 Canadian dollars per year), 4 research assistantships (averaging $16,500 Canadian dollars per year) were awarded; career-related internships or fieldwork also available. *Faculty research:* Animal health management, infectious diseases, fin fish and shellfish health, basic biomedical sciences, ecosystem health. Total annual research expenditures: $1.2 million Canadian dollars. *Unit head:* Dr. James Bellamy, Associate Dean of Graduate Studies and Research, 902-566-0856, E-mail: bellamy@upei.ca. *Application contact:* Cheryl Gaudet, Registrar's Office, 902-566-0781, Fax: 902-566-0795, E-mail: registrar@upei.ca.

University of Puerto Rico, Medical Sciences Campus, School of Medicine, Division of Graduate Studies, Department of Pharmacology and Toxicology, San Juan, PR 00936-5067. Offers MS, PhD. *Degree requirements:* For master's, one foreign language, thesis/dissertation; for doctorate, one foreign language, thesis/dissertation, comprehensive exam. *Entrance requirements:* For master's and doctorate, GRE General Test, GRE Subject Test, interview, minimum GPA of 3.0, 3 letters of recommendation. *Expenses:* Tuition, state resident: full-time $3,600; part-time $100 per credit hour. Tuition, nonresident: full-time $4,655. Required fees: $1,734. Tuition and fees vary according to class time, degree level and program. *Faculty*

research: Cardiovascular, central nervous system, and endocrine pharmacology; anti-cancer drugs; toxicology, sodium pump.

University of Rhode Island, Graduate School, College of Pharmacy, Graduate Programs in Pharmacy, Department of Biomedical and Pharmaceutical Sciences, Kingston, RI 02881. Offers medicinal chemistry and pharmacognosy (MS, PhD); pharmaceutics and pharmacokinetics (MS, PhD); pharmacology and toxicology (MS, PhD). *Expenses:* Tuition, state resident: full-time $5,522; part-time $307 per credit. Tuition, nonresident: full-time $15,992; part-time $888 per credit. Required fees: $1,786; $73 per credit. One-time fee: $80 part-time. *Unit head:* Dr. Clinton O. Chichester, Chair, 401-874-5034.

University of Rochester, School of Medicine and Dentistry, Graduate Programs in Medicine and Dentistry, Department of Pharmacology and Physiology, Program in Pharmacology, Rochester, NY 14627-0250. Offers MS, PhD. Terminal master's awarded for partial completion of doctoral program. *Degree requirements:* For master's, thesis; for doctorate, thesis/dissertation, qualifying exam. *Entrance requirements:* For master's and doctorate, GRE General Test.

University of Saskatchewan, College of Medicine, Department of Pharmacology, Saskatoon, SK S7N 5A2, Canada. Offers M Sc, PhD. *Faculty:* 8. *Students:* 17. *Degree requirements:* For master's and doctorate, thesis/dissertation, registration. *Entrance requirements:* Additional exam requirements/recommendations for international students: Required—TOEFL. *Application deadline:* For fall admission, 7/1 for domestic students. Applications are processed on a rolling basis. Application fee: $50. *Financial support:* Fellowships, research assistantships, teaching assistantships available. Financial award application deadline: 1/31. *Faculty research:* Neuropharmacology, mechanisms of action of anticancer drugs, clinical pharmacology, cardiovascular pharmacology, toxicology: alcohol-related changes in fetal brain development. *Unit head:* Dr. V. Gopalakrishnan, Head, 306-966-6292, Fax: 306-966-6220, E-mail: gopal@sask.usask.ca. *Application contact:* Dr. J. Tuchek, Graduate Chair, 306-966-6292, Fax: 306-966-6220, E-mail: tuchek@sask.usask.ca.

University of South Alabama, College of Medicine and Graduate School, Program in Basic Medical Sciences, Specialization in Pharmacology, Mobile, AL 36688-0002. Offers PhD. *Faculty:* 20 full-time (2 women). In 2005, 1 degree awarded. *Degree requirements:* For doctorate, thesis/dissertation. *Entrance requirements:* For doctorate, GRE General Test or MCAT. *Application deadline:* For fall admission, 4/1 for domestic students. Applications are processed on a rolling basis. Application fee: $25. *Expenses:* Tuition, state resident: full-time $4,008. Tuition, nonresident: full-time $8,016. Required fees: $692. *Financial support:* Fellowships, research assistantships, institutionally sponsored loans available. Financial award application deadline: 4/1. *Faculty research:* Cardiovascular, clinical, and molecular pharmacology. *Unit head:* Dr. Mark N. Gillespie, Chair, 251-460-6497. *Application contact:* Lanette Flagge, Academic Advisor, 251-460-6153.

The University of South Dakota, School of Medicine and Health Sciences and Graduate School, Biomedical Sciences Graduate Program, Physiology and Pharmacology Group, Vermillion, SD 57069-2390. Offers MA, PhD. *Faculty:* 7 full-time (2 women). *Students:* 9 full-time (8 women), 5 international. Average age 25. 17 applicants, 41% accepted, 7 enrolled. Terminal master's awarded for partial completion of doctoral program. *Degree requirements:* For master's, thesis/dissertation, registration; for doctorate, thesis/dissertation, comprehensive exam, registration. *Entrance requirements:* For master's and doctorate, GRE General Test, minimum GPA of 3.0. Additional exam requirements/recommendations for international students: Required—TOEFL (minimum score 550 paper-based; 213 computer-based). *Application deadline:* For fall admission, 4/15 priority date for domestic students, 4/15 priority date for international students. Applications are processed on a rolling basis. Application fee: $35. *Expenses: Contact institution.* Tuition and fees vary according to course load, program and reciprocity agreements. *Financial support:* In 2005–06, 9 students received support, including 7 fellowships with partial tuition reimbursements available (averaging $20,772 per year), 4 research assistantships with partial tuition reimbursements available (averaging $10,386 per year); scholarships/grants and unspecified assistantships also available. Financial award application deadline: 4/15; financial award applicants required to submit FAFSA. *Faculty research:* Pulmonary physiology and pharmacology, drug abuse, reproduction, signal transduction, cardiovascular physiology and pharmacology. Total annual research expenditures: $637,000.

University of South Florida, College of Medicine and College of Graduate Studies, Graduate Programs in Medical Sciences, Tampa, FL 33620-9951. Offers anatomy (PhD); biochemistry and molecular biology (MS, PhD), including biochemistry and molecular biology (PhD), bioinformatics and computational biology (MS); medical microbiology and immunology (PhD); pathology (PhD); pharmacology and therapeutics (PhD), including medical sciences; physiology and biophysics (PhD). *Students:* 108 full-time (53 women), 34 part-time (26 women); includes 31 minority (9 African Americans, 9 Asian Americans or Pacific Islanders, 13 Hispanic Americans), 30 international. 117 applicants, 99% accepted, 116 enrolled. In 2005, 9 master's, 4 doctorates awarded. *Degree requirements:* For doctorate, thesis/dissertation. *Entrance requirements:* For doctorate, GRE General Test, minimum GPA of 3.0. Application fee: $30. *Expenses: Contact institution. Financial support:* Institutionally sponsored loans and scholarships/grants available. Financial award application deadline: 4/1; financial award applicants required to submit FAFSA. *Unit head:* Dr. Joseph J. Krzanowski, Associate Dean for Research and Graduate Affairs, 813-974-4181, Fax: 813-974-4317, E-mail: jkrzanow@com1.med.usf.edu.

The University of Tennessee Health Science Center, College of Graduate Health Sciences, Department of Pharmacology, Memphis, TN 38163-0002. Offers MS, PhD. *Faculty:* 19 full-time (3 women), 7 part-time/adjunct (0 women). *Students:* 4 full-time (0 women); includes 3 minority (all Asian Americans or Pacific Islanders) Average age 26. 50 applicants, 4% accepted. In 2005, 1 degree awarded. Terminal master's awarded for partial completion of doctoral program. *Degree requirements:* For master's, thesis, comprehensive exam; for doctorate, thesis/dissertation, oral and written preliminary and comprehensive exams. *Entrance requirements:* For master's, GRE General Test, minimum GPA of 3.0; for doctorate, GRE General Test, GRE Subject Test, minimum GPA of 3.0. Additional exam requirements/recommendations for international students: Required—TOEFL. *Application deadline:* For fall admission, 5/15 for domestic students. Application fee: $0. Electronic applications accepted. *Expenses:* Contact institution. *Financial support:* Fellowships, research assistantships, teaching assistantships, tuition waivers (full) available. Financial award application deadline: 2/25. *Unit head:* Dr. Burt M. Sharp, Chairman, 901-448-6000, Fax: 901-448-7206, E-mail: bsharp@utmem.edu. *Application contact:* Ida W. Mosby, Director, Enrollment Services, 901-448-5560, E-mail: imosby@utmem.edu.

The University of Texas Health Science Center at San Antonio, Graduate School of Biomedical Sciences, Department of Pharmacology, San Antonio, TX 78229-3900. Offers PhD. Part-time programs available. *Faculty:* 31 full-time (5 women). *Students:* 24 full-time (13 women), 2 part-time (both women); includes 8 minority (3 Asian Americans or Pacific Islanders, 5 Hispanic Americans), 8 international. Average age 27. 54 applicants, 17% accepted, 5 enrolled. In 2005, 3 doctorates awarded. *Degree requirements:* For doctorate, thesis/dissertation. *Entrance requirements:* For doctorate, GRE General Test, minimum GPA of 3.0. Additional exam requirements/recommendations for international students: Required—TOEFL (minimum score 560 paper-based; 220 computer-based). *Application deadline:* For fall admission, 1/15 priority date for domestic students, 1/15 priority date for international students. Applications are processed on a rolling basis. Application fee: $10. Electronic applications accepted. *Financial support:* In 2005–06, research assistantships (averaging $21,000 per year), teaching assistantships (averaging $21,000 per year) were awarded; institutionally sponsored loans, scholarships/grants, and health care benefits also available. Financial award application deadline: 3/1. *Faculty research:* Neuropharmacology, autonomic and endocrine homeostasis, aging, cancer biology. *Unit head:* Dr. Alan Frazer, Chairman, 210-567-4205, Fax: 210-567-4300, E-mail: frazer@uthscsa.edu. *Application contact:* Veronica Sprayberry, Graduate Academic Coordinator, 210-567-4220, Fax: 210-567-4300, E-mail: sprayberryv@uthscsa.edu.

Pharmacology

The University of Texas Medical Branch, Graduate School of Biomedical Sciences, Program in Pharmacology and Toxicology, Galveston, TX 77555. Offers pharmacology (MS); pharmacology/toxicology (PhD); toxicology (PhD). *Students:* 22 full-time (13 women); includes 5 minority (1 African American, 4 Hispanic Americans), 8 international. Average age 28. In 2005, 4 master's, 3 doctorates awarded. *Degree requirements:* For master's, thesis or alternative; for doctorate, thesis/dissertation. *Entrance requirements:* For master's and doctorate, GRE General Test. Additional exam requirements/recommendations for international students: Required—TOEFL (minimum score 550 paper-based; 213 computer-based). *Application deadline:* Applications are processed on a rolling basis. Application fee: $30 ($75 for international students). *Expenses:* Tuition, state resident: full-time $8,350; part-time $90 per credit hour. Tuition, nonresident: full-time $21,450; part-time $366 per credit hour. Required fees: $1,027; $11 per credit hour. $60 per term. *Financial support:* In 2005–06, fellowships (averaging $23,000 per year), research assistantships with full tuition reimbursements (averaging $23,000 per year) were awarded. Financial award applicants required to submit FAFSA. *Unit head:* Dr. Kenneth M. Johnson, Director, 409-772-9623, Fax: 409-772-9642, E-mail: kmjohnso@utmb.edu. *Application contact:* Laurie C. Shook, Coordinator II Special Programs, 409-772-9626, Fax: 409-772-9627, E-mail: lcshook@utmb.edu.

The University of Texas Southwestern Medical Center at Dallas, Southwestern Graduate School of Biomedical Sciences, Department of Pharmacology, Dallas, TX 75390.

University of the Sciences in Philadelphia, College of Graduate Studies, Program in Chemistry, Biochemistry and Pharmacognosy, Philadelphia, PA 19104-4495. Offers biochemistry (MS, PhD); chemistry (MS, PhD); medicinal chemistry (MS, PhD); pharmacognosy (MS, PhD). Part-time programs available. *Faculty:* 13 full-time (3 women). *Students:* 8 full-time (4 women), 7 part-time (3 women); includes 2 minority (both Asian Americans or Pacific Islanders), 5 international. Average age 28. In 2005, 3 master's, 4 doctorates awarded. *Degree requirements:* For master's, thesis/dissertation, qualifying exams; for doctorate, thesis/dissertation, qualifying exams, comprehensive exam. *Entrance requirements:* For master's and doctorate, GRE General Test, GRE Subject Test. Additional exam requirements/recommendations for international students: Required—TOEFL, TWE, TSE. *Application deadline:* Applications are processed on a rolling basis. Application fee: $50. *Expenses:* Tuition: Part-time $999 per credit. Tuition and fees vary according to program. *Financial support:* In 2005–06, 19 students received support, including 1 fellowship with full tuition reimbursement available (averaging $19,000 per year), 1 research assistantship with full tuition reimbursement available (averaging $19,000 per year), 12 teaching assistantships with full tuition reimbursements available (averaging $18,500 per year); institutionally sponsored loans, scholarships/grants, and tuition waivers (full) also available. Financial award application deadline: 5/1. *Faculty research:* Organic and medicinal synthesis, mass spectroscopy use in protein analysis, study of analogues of taxol, cholesteryl esters. Total annual research expenditures: $341,700. *Unit head:* , Dr. James McKee, Director, 215-596-8847, Fax: 215-596-8543, E-mail: jmckee@usip.edu. *Application contact:* Joyce D'Angelo, Administrative Assistant, 215-596-8937, E-mail: j.dangel@usip.edu.

University of the Sciences in Philadelphia, College of Graduate Studies, Program in Pharmacology and Toxicology, Philadelphia, PA 19104-4495. Offers pharmacology (MS, PhD); toxicology (MS, PhD). *Faculty:* 7 full-time (2 women), 6 part-time/adjunct (2 women). *Students:* 11 full-time (9 women), 4 part-time (1 woman); includes 2 minority (both Asian Americans or Pacific Islanders), 4 international. Average age 28. In 2005, 2 master's, 2 doctorates awarded. Terminal master's awarded for partial completion of doctoral program. *Degree requirements:* For master's, thesis/dissertation; for doctorate, thesis/dissertation, comprehensive exam. *Entrance requirements:* For master's and doctorate, GRE General Test. Additional exam requirements/recommendations for international students: Required—TOEFL, TWE, TSE. *Application deadline:* Applications are processed on a rolling basis. Application fee: $50. *Expenses:* Tuition: Part-time $999 per credit. Tuition and fees vary according to program. *Financial support:* In 2005–06, 12 students received support, including teaching assistantships with tuition reimbursements available (averaging $18,500 per year); institutionally sponsored loans and tuition waivers (partial) also available. Financial award application deadline: 5/1. *Faculty research:* Autonomic, cardiovascular, cellular, and molecular pharmacology; mechanisms of carcinogenesis; drug metabolism. Total annual research expenditures: $61,000. *Unit head:* Dr. Adeboye Adejare, Director, 215-596-8944, E-mail: a.adejar@usip.edu. *Application contact:* Joyce D'Angelo, Administrative Assistant, 215-596-8937, E-mail: j.dangel@usip.edu.

The University of Toledo, School of Graduate Studies, College of Pharmacy, Graduate Programs in Pharmacy, Program in Pharmaceutical Science, Toledo, OH 43606-3390. Offers administrative pharmacy (MSPS); industrial pharmacy (MSPS); pharmacology (MSPS). *Faculty:* 16 full-time (4 women). *Students:* 20 full-time (8 women); includes 1 minority (African American), 10 international. Average age 25. 92 applicants, 13% accepted, 11 enrolled. In 2005, 13 degrees awarded. *Degree requirements:* For master's, thesis. *Entrance requirements:* For master's, GRE General Test. Additional exam requirements/recommendations for international students: Required—TOEFL (minimum score 550 paper-based; 213 computer-based). *Application deadline:* For fall admission, 2/1 priority date for domestic students, 1/1 priority date for international students. Application fee: $45. Electronic applications accepted. *Expenses:* Tuition, state resident: full-time $6,623; part-time $308 per credit hour. Tuition, nonresident: full-time $13,232; part-time $735 per credit hour. *Financial support:* In 2005–06, 14 students received support; research assistantships with full tuition reimbursements available, teaching assistantships with full tuition reimbursements available available. *Faculty research:* Drug disposition, neuropharmacology, pharmacokinetics, product stability, pharmacy and health care administration. *Application contact:* Karen F. Papadakis, Graduate Coordinator for Pharmaceutical Sciences, 419-530-1910, Fax: 419-530-1909, E-mail: kpapada@utnet.utoledo.edu.

University of Toronto, School of Graduate Studies, Life Sciences Division, Department of Pharmacology, Toronto, ON M5S 1A1, Canada. Offers M Sc, PhD. Part-time programs available. *Degree requirements:* For master's and doctorate, thesis/dissertation. *Entrance requirements:* For master's, B Sc or equivalent; background in pharmacology, biochemistry, and physiology; minimum B+ earned in at least 4 senior level classes; for doctorate, minimum B+ average.

University of Utah, College of Pharmacy and The Graduate School, Graduate Programs in Pharmacy, Department of Pharmacology and Toxicology, Salt Lake City, UT 84112-1107. Offers MS, PhD, MD/PhD. Terminal master's awarded for partial completion of doctoral program. *Degree requirements:* For master's, thesis; for doctorate, thesis/dissertation, final exam. *Entrance requirements:* For master's, GRE; for doctorate, GRE Subject Test, BS in biology, chemistry, or pharmacy. Additional exam requirements/recommendations for international students: Required—TOEFL. *Expenses:* Tuition, state resident: full-time $2,932; part-time $369 per credit. Tuition, nonresident: full-time $10,350; part-time $1,302 per credit. Required fees: $516 per term. Tuition and fees vary according to course load and program. *Faculty research:* Neuropharmacology, neurochemistry, biochemistry, molecular pharmacology, analytical chemistry.

University of Vermont, College of Medicine and Graduate College, Graduate Programs in Medicine, Department of Pharmacology, Burlington, VT 05405. Offers MS, PhD, MD/MS, MD/PhD. *Faculty:* 12 full-time (1 woman), 1 part-time/adjunct. *Students:* 11 (6 women); includes 2 minority (1 African American, 1 Asian American or Pacific Islander) 3 international. 12 applicants, 8% accepted, 1 enrolled. In 2005, 1 doctorate awarded. *Degree requirements:* For master's and doctorate, thesis/dissertation. *Entrance requirements:* For master's and doctorate, GRE General Test, GRE Subject Test. Additional exam requirements/recommendations for international students: Required—TOEFL (minimum score 550 paper-based; 213 computer-based). *Application deadline:* For fall admission, 1/15 for domestic students. Applications are processed on a rolling basis. Application fee: $40. Electronic applications accepted. *Expenses:* Tuition, area resident: Part-time $410 per credit hour. Tuition, nonresident: part-time $1,034 per credit hour. *Financial support:* Fellowships, research assistantships, teaching assistantships available. Financial award application deadline: 3/1. *Faculty research:* Cardiovascular drugs, anticancer drugs. *Unit head:* Dr. M. Nelson, Chairperson, 802-656-2500. *Application contact:* Dr. Karen M. Lounsbury, Director of Graduate Studies, 802-656-2500.

University of Virginia, School of Medicine, Department of Pharmacology, Charlottesville, VA 22903. Offers PhD, MD/PhD. *Students:* 17 full-time (9 women), 1 part-time; includes 2 minority (1 Asian American or Pacific Islander, 1 Hispanic American), 1 international. Average age 29. In 2005, 4 degrees awarded. *Degree requirements:* For doctorate, thesis/dissertation. *Entrance requirements:* For doctorate, GRE General Test, GRE Subject Test. Additional exam requirements/recommendations for international students: Required—TOEFL. *Application deadline:* Applications are processed on a rolling basis. Application fee: $60. Electronic applications accepted. *Expenses:* Tuition, state resident: full-time $7,731. Tuition, nonresident: full-time $18,672. Required fees: $1,479. Full-time tuition and fees vary according to degree level and program. *Financial support:* Fellowships, research assistantships, teaching assistantships available. Financial award applicants required to submit FAFSA. *Unit head:* James C. Garrison, Chairman, 434-924-1919, Fax: 434-982-3878. *Application contact:* Peter C. Brunjes, Associate Dean for Graduate Programs and Research, 434-924-7184, Fax: 434-924-6737, E-mail: grad-a-s@virginia.edu.

University of Washington, School of Medicine and Graduate School, Graduate Programs in Medicine, Department of Pharmacology, Seattle, WA 98195. Offers MS, PhD. *Degree requirements:* For doctorate, thesis/dissertation. *Entrance requirements:* For doctorate, GRE General Test, minimum GPA of 3.0. *Faculty research:* Neuroscience, cell physiology, molecular biology, regulation of metabolism, signal transduction.

See Close-Up on page 1215.

University of Wisconsin–Madison, Medical School and Graduate School, Graduate Programs in Medicine, Molecular and Cellular Pharmacology Program, Madison, WI 53706-1380. Offers PhD. *Faculty:* 42 full-time (8 women). *Students:* 62 full-time (37 women); includes 7 minority (5 Asian Americans or Pacific Islanders, 2 Hispanic Americans), 19 international. Average age 26. 141 applicants, 17% accepted, 13 enrolled. In 2005, 7 doctorates awarded. *Median time to degree:* Of those who began their doctoral program in fall 1997, 100% received their degree in 8 years or less. *Degree requirements:* For doctorate, thesis/dissertation, comprehensive exam. *Entrance requirements:* For doctorate, GRE. Additional exam requirements/recommendations for international students: Required—TOEFL (minimum score 550 paper-based; 213 computer-based). *Application deadline:* For fall admission, 12/31 priority date for domestic students, 12/31 priority date for international students. Applications are processed on a rolling basis. Application fee: $45. Electronic applications accepted. *Financial support:* In 2005–06, 62 students received support, including 8 fellowships with full tuition reimbursements available (averaging $22,500 per year), 53 research assistantships with full tuition reimbursements available (averaging $22,500 per year), 1 teaching assistantship with full tuition reimbursement available (averaging $22,500 per year); scholarships/grants, traineeships, health care benefits, and unspecified assistantships also available. *Faculty research:* Protein kinases, signaling pathways, neurotransmitters, molecular recognition, receptors and transporters. *Unit head:* Dr. Richard A. Anderson, Director, 608-262-3753, Fax: 608-262-1257, E-mail: raanders@wisc.edu. *Application contact:* Lynn Louise Squire, Student Services Coordinator, 608-262-9826, Fax: 608-262-1257, E-mail: lsquire@wisc.edu.

See Close-Up on page 1217.

University of Wisconsin–Madison, School of Veterinary Medicine, Department of Animal Health and Biomedical Sciences, Program in Comparative Biosciences, Madison, WI 53706-1380. Offers anatomy (MS, PhD); biochemistry (MS, PhD); cellular and molecular biology (MS, PhD); environmental toxicology (MS, PhD); neurosciences (MS, PhD); pharmacology (MS, PhD); physiology (MS, PhD). *Degree requirements:* For doctorate, thesis/dissertation.

Vanderbilt University, Graduate School and School of Medicine, Department of Pharmacology, Nashville, TN 37240-1001. Offers PhD, MD/PhD. *Faculty:* 94 full-time, 5 part-time/adjunct. *Students:* 48 full-time (24 women); includes 5 minority (2 African Americans, 1 Asian American or Pacific Islander, 2 Hispanic Americans), 12 international. In 2005, 11 degrees awarded. *Degree requirements:* For doctorate, thesis/dissertation, preliminary, qualifying, and final exams. *Entrance requirements:* For doctorate, GRE General Test, GRE Subject Test (recommended). *Application deadline:* For fall admission, 1/15 for domestic students, 1/15 for international students. Application fee: $0. Electronic applications accepted. *Expenses:* Tuition: Part-time $1,283 per semester hour. Required fees: $2,202; $1,101 per semester. One-time fee: $30. Tuition and fees vary according to course load, program and student level. *Financial support:* Fellowships with full tuition reimbursements, research assistantships with full tuition reimbursements, Federal Work-Study, institutionally sponsored loans, traineeships, and tuition waivers (partial) available. Financial award application deadline: 1/15. *Faculty research:* Molecular pharmacology, neuropharmacology, drug disposition and toxicology, genetic mechanics, cell regulation. *Unit head:* Heidi E. Hamm, Chair, 615-322-2207, Fax: 615-936-3010, E-mail: heidi.hamm@vanderbilt.edu. *Application contact:* Joey V. Barnett, Director of Graduate Studies, 615-322-2207, Fax: 615-936-3910, E-mail: joey.barnett@vanderbilt.edu.

Virginia Commonwealth University, Medical College of Virginia-Professional Programs, School of Medicine and Graduate Programs, School of Medicine Graduate Programs, Department of Pharmacology and Toxicology, Richmond, VA 23284-9005. Offers molecular biology and genetics (PhD); neurosciences (PhD); pharmacology (PhD, CBHS); pharmacology and toxicology (MS). *Faculty:* 36 full-time (9 women). *Students:* 35 full-time (18 women), 11 part-time (all women); includes 7 minority (2 African Americans, 4 Asian Americans or Pacific Islanders, 1 Hispanic American), 5 international. 90 applicants, 33% accepted. In 2005, 6 master's, 2 doctorates awarded. Terminal master's awarded for partial completion of doctoral program. *Degree requirements:* For master's, thesis; for doctorate, thesis/dissertation, comprehensive oral and written exams. *Entrance requirements:* For master's, DAT, GRE General Test or MCAT; for doctorate, GRE General Test, MCAT, DAT. *Application deadline:* For fall admission, 4/1 for domestic students. Application fee: $50. *Expenses:* Tuition, state resident: full-time $6,268; part-time $405 per credit. Tuition, nonresident: full-time $15,904; part-time $940 per credit. Required fees: $751 per semester hour. Tuition and fees vary according to course load and program. *Financial support:* Fellowships, teaching assistantships available. *Faculty research:* Drug abuse, drug metabolism, pharmacodynamics, peptide synthesis, receptor mechanisms. *Unit head:* Dr. Billy R. Martin, Chair, 804-828-8407, Fax: 804-828-2117, E-mail: brmartin@vcu.edu. *Application contact:* Sheryol Cox, Graduate Program Coordinator, 804-828-8400, Fax: 804-828-2117, E-mail: swcox@vcu.edu.

See Close-Up on page 1219.

Wake Forest University, School of Medicine and Graduate School, Graduate Programs in Medicine, Department of Physiology and Pharmacology, Program in Pharmacology, Winston-Salem, NC 27109. Offers PhD. *Degree requirements:* For doctorate, thesis/dissertation. *Entrance requirements:* For doctorate, GRE General Test. *Faculty research:* Renal-cardiovascular system, endocrine system, neuropharmacology.

Washington State University, Graduate School, College of Pharmacy, Program in Pharmacology and Toxicology, Pullman, WA 99164. Offers MS, PhD. *Faculty:* 28 full-time (9 women), 4 part-time/adjunct (1 woman). *Students:* 17 full-time (8 women), 2 part-time (both women); includes 2 minority (1 Asian American or Pacific Islander, 1 Hispanic American), 5 international. Average age 26. 66 applicants, 17% accepted, 6 enrolled. In 2005, 1 degree awarded. *Degree requirements:* For master's and doctorate, thesis/dissertation, oral exam. *Entrance requirements:* For master's, GRE General Test, minimum GPA of 3.0, 3 letters of recommendation; for doctorate, GRE General Test, minimum GPA of 3.0. Additional exam requirements/recommendations for international students: Required—TOEFL. *Application deadline:* For fall admission, 12/1 for domestic students, 3/1 for international students; for spring admission, 9/1 for domestic students, 7/1 for international students. Applications are processed on a rolling basis. Application fee: $35. Electronic applications accepted. *Expenses:* Tuition, state resident: full-time $6,295; part-time $336 per credit. Tuition, nonresident: full-time $15,949; part-time $819 per credit. Required fees: $933. Part-time tuition and fees vary according to campus/location and program. *Financial support:* In 2005–06, 5 fellowships (averaging $6,176 per year), 10 research assistantships with full and partial tuition reimbursements (averaging

Pharmacology

Washington State University *(continued)*
$14,846 per year), 4 teaching assistantships with full and partial tuition reimbursements (averaging $14,661 per year) were awarded; Federal Work-Study, institutionally sponsored loans, tuition waivers (partial), and staff assistantships, teaching associateships also available. Financial award application deadline: 4/1; financial award applicants required to submit FAFSA. *Faculty research:* Cancer research, immunopharmacology/immunotoxicology, pharmacy, neurobiology/neuropharmacology, toxicology. Total annual research expenditures: $623,031. *Unit head:* Dr. Raymond Quock, Chair, 509-335-7598, Fax: 509-335-5902. *Application contact:* Dana Martin, Coordinator, 509-335-7598, Fax: 509-335-5902, E-mail: pharmtox@wsu.edu.

Wayne State University, Graduate School, Eugene Applebaum College of Pharmacy and Health Sciences, Department of Pharmacy, Detroit, MI 48202. Offers medicinal chemistry (MS, PhD); pharmaceutical sciences (MS, PhD); pharmaceutics (MS, PhD); pharmacology (MS, PhD); pharmacy (Pharm D). *Faculty:* 8 full-time (0 women). *Students:* 285 full-time (193 women), 18 part-time (8 women); includes 38 minority (7 African Americans, 1 American Indian/Alaska Native, 28 Asian Americans or Pacific Islanders, 2 Hispanic Americans), 88 international. Average age 26. 77 applicants, 9% accepted, 6 enrolled. In 2005, 50 first professional degrees, 3 doctorates awarded. Terminal master's awarded for partial completion of doctoral program. *Degree requirements:* For master's and doctorate, thesis/dissertation. *Entrance requirements:* For master's, GRE General Test, minimum GPA of 2.6; for doctorate, GRE General Test, minimum GPA of 3.0. Additional exam requirements/recommendations for international students: Required—TOEFL (minimum score 550 paper-based; 213 computer-based); Recommended—TWE (minimum score 6). *Application deadline:* For fall admission, 1/31 priority date for domestic students, 6/1 priority date for international students. Applications are processed on a rolling basis. Application fee: $30 ($50 for international students). Electronic applications accepted. *Expenses:* Tuition, state resident: part-time $338 per credit hour. Tuition, nonresident: part-time $746 per credit hour. Required fees: $24 per credit hour. Full-time tuition and fees vary according to program. *Financial support:* In 2005–06, 1 fellowship, 13 research assistantships with tuition reimbursements (averaging $19,079 per year) were awarded; scholarships/grants also available. Support available to part-time students. *Faculty research:* Signaling defects as basics of type 1 and type 2 diabetes, dopamine receptor and transporter inhibiorters for addition, novel nanoparticles for gene and drug delivery, role of lipocortin 1 in heavy metal mutagenesis, synthesis and anti cancer mechanisms of polyamine and histone deacetylase inhibitors. Total annual research expenditures: $1.4 million. *Unit head:* Dr. George B. Corcoran, Chair, 313-577-1737, Fax: 313-577-2033, E-mail: corcoran@wayne.edu. *Application contact:* William Lindblad, Associate Professor, 313-577-0513, E-mail: wlindbl@wayne.edu.

Wayne State University, School of Medicine and Graduate School, Graduate Programs in Medicine, Department of Pharmacology, Detroit, MI 48202. Offers MS, PhD. *Faculty:* 14 full-time. *Students:* 13 full-time (7 women), 1 part-time, 6 international. Average age 27. 16 applicants, 38% accepted, 2 enrolled. In 2005, 3 master's, 3 doctorates awarded. *Degree requirements:* For doctorate, thesis/dissertation. *Entrance requirements:* For master's and doctorate, GRE General Test. Additional exam requirements/recommendations for international students: Required—TOEFL (minimum score 550 paper-based; 213 computer-based); Recommended—TWE (minimum score 6). *Application deadline:* For fall admission, 4/30 for domestic students, 6/1 for international students. Applications are processed on a rolling basis. Application fee: $30 ($50 for international students). Electronic applications accepted. *Expenses:* Tuition, state resident: part-time $338 per credit hour. Tuition, nonresident: part-time $746 per credit hour. Required fees: $24 per credit hour. Full-time tuition and fees vary according to program. *Financial support:* In 2005–06, 13 research assistantships with tuition reimbursements (averaging $19,962 per year) were awarded; fellowships with tuition reimbursements, teaching assistantships Financial award application deadline: 4/30. *Faculty research:* Molecular and cellular biology of cancer and anti-cancer therapies; molecular and cellular biology of protein trafficking, signal transduction and aging; environmental toxicology and drug metabolism; functional cellular and in vivo imaging; neuroscience. Total annual research expenditures: $3 million. *Unit head:* Dr. Bonnie Sloane, Chair, 313-577-1580, Fax: 313-577-6739, E-mail: aa3402@wayne.edu. *Application contact:* Ray Mattingly, Associate Professor, 313-577-6022, E-mail: r.mattingly@wayne.edu.

Wright State University, School of Medicine, Program in Pharmacology and Toxicology, Dayton, OH 45435. Offers MS. *Degree requirements:* For master's, thesis optional.

Yale University, Graduate School of Arts and Sciences, Department of Pharmacology, New Haven, CT 06520. Offers PhD. *Degree requirements:* For doctorate, thesis/dissertation. *Entrance requirements:* For doctorate, GRE General Test. Expenses: Contact institution.

Yale University, School of Medicine and Graduate School of Arts and Sciences, Combined Program in Biological and Biomedical Sciences (BBS), Pharmacological Sciences and Molecular Medicine Track, New Haven, CT 06520. Offers PhD, MD/PhD. *Students:* 6 full-time. *Degree requirements:* For doctorate, thesis/dissertation. *Entrance requirements:* For doctorate, GRE General Test. Additional exam requirements/recommendations for international students: Required—TOEFL. *Application deadline:* For fall admission, 12/8 for domestic students, 12/8 for international students. Electronic applications accepted. *Financial support:* Fellowships, research assistantships available. *Unit head:* Dr. David F. Stern, Co-Director, 203-785-4832, Fax: 203-785-7467, E-mail: kathleen.fisher@yale.edu. *Application contact:* Kathy Fisher, Registrar, 203-785-4545, E-mail: kathleen.fisher@yale.edu.

Toxicology

American University, College of Arts and Sciences, Department of Chemistry, Program in Toxicology, Washington, DC 20016-8001. Offers MS, Certificate. Part-time programs available. In 2005, 5 degrees awarded. *Degree requirements:* For master's, thesis, tool of research exam, comprehensive exam. *Entrance requirements:* For master's, GRE, minimum GPA of 3.0. Additional exam requirements/recommendations for international students: Required—TOEFL (minimum score 550 paper-based; 213 computer-based). *Application deadline:* For fall admission, 2/1 for domestic students; for spring admission, 10/1 priority date for domestic students. Applications are processed on a rolling basis. Application fee: $50. *Expenses:* Tuition: Full-time $17,802; part-time $989 per credit. Required fees: $380. *Financial support:* In 2005–06, research assistantships with full tuition reimbursements (averaging $8,500 per year), teaching assistantships with full tuition reimbursements (averaging $8,500 per year) were awarded; fellowships, career-related internships or fieldwork and institutionally sponsored loans also available. Financial award application deadline: 2/1. *Faculty research:* Carbohydrate chemistry, chromatography, enzyme mechanisms and model systems, environmental analysis with monoclonal antibodies, polymers. *Unit head:* Dr. David Culver, Chair, Department of Chemistry, 202-885-2180.

A.T. Still University of Health Sciences, Kirksville College of Osteopathic Medicine, Kirksville, MO 63501. Offers biomedical sciences (MS); osteopathic medicine (DO). *Accreditation:* AOsA. *Faculty:* 57 full-time (15 women), 23 part-time/adjunct (2 women). *Students:* 680 full-time (269 women), 24 part-time (9 women); includes 108 minority (5 African Americans, 5 American Indian/Alaska Native, 86 Asian Americans or Pacific Islanders, 12 Hispanic Americans), 13 international. Average age 27. 2,600 applicants, 182 enrolled. In 2005, 150 first professional degrees awarded. *Degree requirements:* For DO, level I and 2 CE comprehensive osteopathic Medical Licensing Exam and COMLEX PE; for master's, thesis. *Entrance requirements:* For DO, MCAT, minimum undergraduate GPA of 2.5 (cumulative and science) or 90 semester hours with minimum GPA of 3.5 (cumulative and science) and minimum MCAT of 28; for master's, GRE, MCAT, or DAT, minimum undergraduate GPA of 2.5 (cumulative and science). *Application deadline:* For fall admission, 2/15 for domestic students, 2/15 for international students. Applications are processed on a rolling basis. Application fee: $60. Electronic applications accepted. *Expenses:* Tuition: Full-time $20,740. Required fees: $1,000. Tuition and fees vary according to degree level, program and student level. *Financial support:* In 2005–06, 637 students received support, including 8 fellowships with full tuition reimbursements available (averaging $9,000 per year); career-related internships or fieldwork, Federal Work-Study, institutionally sponsored loans, and scholarships/grants also available. Financial award application deadline: 5/1; financial award applicants required to submit FAFSA. *Faculty research:* Osteopathic manipulation and pneumonia in the elderly, basis of hypocholestrolemic drug-induced cataracts, membrane structure in cataracts, role of caveolin in the ocular lens, cell membrane. Total annual research expenditures: $670,095. *Unit head:* Dr. Philip C. Slocum, Vice President for Medical Affairs and Dean, 660-626-2354, Fax: 660-626-2080, E-mail: pslocum@atsu.edu. *Application contact:* Donna Sparks, Associate Director for Admissions, 660-626-2237, Fax: 660-626-2969, E-mail: admissions@atsu.edu.

Brown University, Graduate School, Division of Biology and Medicine, Program in Pathology and Laboratory Medicine, Providence, RI 02912. Offers biology (PhD); cancer biology (PhD); immunology and infection (PhD); medical science (PhD); pathobiology (Sc M); toxicology and environmental pathology (PhD). Terminal master's awarded for partial completion of doctoral program. *Degree requirements:* For master's and doctorate, thesis/dissertation, preliminary exam. *Entrance requirements:* For master's and doctorate, GRE General Test, GRE Subject Test. Additional exam requirements/recommendations for international students: Required—TOEFL. Electronic applications accepted. *Faculty research:* Environmental pathology, carcinogenesis, immunopathology, signal transduction, innate immunity.

Case Western Reserve University, School of Medicine and School of Graduate Studies, Graduate Programs in Medicine, Department of Environmental Health Sciences, Cleveland, OH 44106. Offers environmental toxicology (MS, PhD); molecular toxicology (MS, PhD). Part-time programs available. Terminal master's awarded for partial completion of doctoral program. *Degree requirements:* For master's, thesis optional; for doctorate, thesis/dissertation, qualifying exams. *Entrance requirements:* For master's and doctorate, GRE General Test, GRE Subject Test. Additional exam requirements/recommendations for international students: Required—TOEFL. *Faculty research:* Xenobiotic metabolism, microbial mutagenicity, DNA damage and repair, chromosome structure cytogenetics, carcinogenesis, photodynamic therapy.

Columbia University, College of Physicians and Surgeons and Graduate School of Arts and Sciences, Graduate School of Arts and Sciences at the College of Physicians and Surgeons, Department of Pharmacology, New York, NY 10032. Offers pharmacology (M Phil, MA, PhD); pharmacology-toxicology (M Phil, MA, PhD). Only candidates for the PhD are admitted. Terminal master's awarded for partial completion of doctoral program. *Degree requirements:* For doctorate, thesis/dissertation. *Entrance requirements:* For master's and doctorate, GRE General Test. Additional exam requirements/recommendations for international students: Required—TOEFL. *Expenses:* Tuition: Full-time $31,448. Tuition and fees vary according to course level, course load, campus/location and program. *Faculty research:* Cardiovascular pharmacology, receptor pharmacology, neuropharmacology, membrane biophysics, eicosanoids.

Cornell University, Graduate School, Graduate Fields of Agriculture and Life Sciences, Field of Environmental Toxicology, Ithaca, NY 14853-0001. Offers cellular and molecular toxicology (MS, PhD); ecotoxicology and environmental chemistry (MS, PhD); nutritional and food toxicology (MS, PhD); risk assessment, management and public policy (MS, PhD). *Faculty:* 47 full-time (10 women). *Students:* 16 full-time (9 women); includes 3 minority (1 American Indian/Alaska Native, 2 Hispanic Americans), 6 international. 24 applicants, 13% accepted, 2 enrolled. In 2005, 2 master's, 3 doctorates awarded. *Degree requirements:* For master's, thesis/dissertation; for doctorate, thesis/dissertation, comprehensive exam. *Entrance requirements:* For master's and doctorate, GRE General Test, GRE Subject Test (biology or chemistry recommended), 2 letters of recommendation. Additional exam requirements/recommendations for international students: Required—TOEFL (minimum score 600 paper-based; 250 computer-based). *Application deadline:* For fall admission, 1/15 for domestic students. Application fee: $60. Electronic applications accepted. *Financial support:* In 2005–06, 14 students received support, including 8 fellowships with full tuition reimbursements available, 6 research assistantships with full tuition reimbursements available; teaching assistantships with full tuition reimbursements available, institutionally sponsored loans, scholarships/grants, health care benefits, tuition waivers (full and partial), and unspecified assistantships also available. Financial award applicants required to submit FAFSA. *Faculty research:* Cellular and molecular toxicology, cancer toxicology, bioremediation, ecotoxicology, nutritional and food toxicology, reproductive toxicology. *Unit head:* Director of Graduate Studies, 607-255-8008, Fax: 607-755-0238. *Application contact:* Graduate Field Assistant, 607-255-8008, Fax: 607-255-0238, E-mail: envtox@cornell.edu.

Dartmouth College, Dartmouth Medical School, Program in Experimental and Molecular Medicine, Hanover, NH 03755. Offers experimental and molecular medicine (PhD); pharmcology and toxicology (PhD); phychological and brain sciences (PhD); physiology (PhD). *Degree requirements:* For doctorate, thesis/dissertation, comprehensive exam, registration. *Entrance requirements:* For doctorate, GRE General Test, 3 letters of recommendation. Additional exam requirements/recommendations for international students: Required—TOEFL. Electronic applications accepted. *Expenses:* Tuition: Full-time $31,770.

Dartmouth College, School of Arts and Sciences, Department of Pharmacology and Toxicology, Hanover, NH 03755. Offers PhD, MD/PhD. *Faculty:* 14 full-time (2 women). *Students:* 32 full-time (17 women); includes 2 minority (1 African American, 1 Hispanic American), 11 international. Average age 28. 65 applicants, 29% accepted, 10 enrolled. In 2005, 5 degrees awarded. *Degree requirements:* For doctorate, thesis/dissertation. *Entrance requirements:* For doctorate, GRE General Test, GRE Subject Test, bachelor's degree in biological, chemical, or physical science. Additional exam requirements/recommendations for international students: Required—TOEFL. *Application deadline:* For fall admission, 1/15 for domestic students. Electronic applications accepted. *Expenses:* Tuition: Full-time $31,770. *Financial support:* In 2005–06, 29 students received support, including fellowships with full tuition reimbursements available (averaging $23,250 per year), research assistantships with full tuition reimbursements available (averaging $23,250 per year); Federal Work-Study, institutionally sponsored loans, traineeships, and tuition waivers (full and partial) also available. Financial award application deadline: 4/15. *Faculty research:* Molecular biology of carcinogenesis, DNA repair and gene expression, biochemical and environmental toxicology, protein receptor ligand interactions. Total annual research expenditures: $7.1 million. *Unit head:* Dr. Ethan Dmitrovsky, Chair, 603-650-1667, Fax: 603-650-1129, E-mail: pharmacology.and.toxicology@dartmouth.edu. *Application contact:* Ruth Bedor, Coordinator of Graduate Studies, 603-650-1667, Fax: 603-650-1129, E-mail: ruth.bedor@dartmouth.edu.

See Close-Up on page 1165.

Duke University, Graduate School, Integrated Toxicology and Environmental Health Program, Durham, NC 27708. Offers PhD, Certificate. *Faculty:* 37 full-time. *Students:* 5 full-time (2 women). 15 applicants, 40% accepted, 4 enrolled. *Entrance requirements:* For doctorate, GRE General Test, GRE Subject Test (recommended). Additional exam requirements/recommendations for international students: Required—TOEFL. *Application deadline:* For fall admission, 12/31 for domestic students, 12/31 for international students. Application fee: $75. Electronic applications accepted. *Financial support:* Fellowships available. Financial award application deadline: 12/31. *Unit head:* Sidney Simon, Director, 919-684-4178.

See Close-Up on page 1167.

Duquesne University, School of Pharmacy, Graduate School of Pharmaceutical Sciences, Program in Pharmacology/Toxicology, Pittsburgh, PA 15282-0001. Offers MS, PhD. Part-time programs available. *Faculty:* 6 full-time (2 women). *Students:* 14 full-time (9 women), 4 part-time (1 woman); includes 2 minority (both African Americans), 4 international. 46 applicants, 2% accepted, 1 enrolled. In 2005, 1 degree awarded. *Degree requirements:* For master's, thesis/dissertation; for doctorate, thesis/dissertation, comprehensive exam. *Entrance requirements:* For master's and doctorate, GRE General Test. Additional exam requirements/recommendations for international students: Required—TOEFL, TSE. *Application deadline:* For fall admission, 2/1 for domestic students. Applications are processed on a rolling basis. Application fee: $50. *Financial support:* In 2005–06, 13 students received support, including 1 research assistantship with full tuition reimbursement available (averaging $12 per year), 12 teaching assistantships with full tuition reimbursements available; fellowships Financial award applicants required to submit FAFSA. *Unit head:* Dr. Paula Witt-Enderby, Head, 412-396-4346.

Florida Agricultural and Mechanical University, Division of Graduate Studies, Research, and Continuing Education, College of Pharmacy and Pharmaceutical Sciences, Graduate Programs in Pharmaceutical Sciences, Tallahassee, FL 32307-3200. Offers environmental toxicology (PhD); medicinal chemistry (MS, PhD); pharmaceutics (MS, PhD); pharmacology/toxicology (MS, PhD); pharmacy administration (MS). *Accreditation:* CEPH. *Degree requirements:* For master's and doctorate, thesis/dissertation, publishable paper, comprehensive exam. *Entrance requirements:* For master's and doctorate, GRE General Test, minimum GPA of 3.0 in last 60 hours. Additional exam requirements/recommendations for international students: Required—TOEFL. *Faculty research:* Anticancer agents, anti-inflammatory drugs, chronopharmacology, neuroendocrinology, microbiology.

Indiana University–Purdue University Indianapolis, Indiana University School of Medicine, Department of Pharmacology and Toxicology, Program in Toxicology, Indianapolis, IN 46202-2896. Offers MS, PhD, MD/PhD. *Students:* 3 full-time (2 women), 1 (woman) part-time, 1 international. Average age 26.Terminal master's awarded for partial completion of doctoral program. *Degree requirements:* For master's and doctorate, thesis/dissertation. *Entrance requirements:* For master's and doctorate, GRE General Test, GRE Subject Test, minimum GPA of 3.0. *Application deadline:* For fall admission, 1/15 for domestic students. Applications are processed on a rolling basis. Application fee: $50 ($60 for international students). *Expenses:* Tuition, state resident: full-time $5,159; part-time $215 per credit hour. Tuition, nonresident: full-time $14,890; part-time $620 per credit hour. Required fees: $614. Tuition and fees vary according to campus/location and program. *Financial support:* Fellowships with partial tuition reimbursements, research assistantships with partial tuition reimbursements, Federal Work-Study and institutionally sponsored loans available. Financial award application deadline: 1/15. *Faculty research:* Oncogenesis, liver metabolism. *Unit head:* Dr. James Klaunig, Director, 317-274-7824, Fax: 317-274-7787. *Application contact:* Victor Elharrar, Director of Graduate Studies, 317-274-1564, Fax: 317-274-7714, E-mail: velharra@iupui.edu.

Iowa State University of Science and Technology, Graduate College, College of Agriculture and College of Liberal Arts and Sciences, Department of Biochemistry, Biophysics, and Molecular Biology, Ames, IA 50011. Offers biochemistry (MS, PhD); biophysics (MS, PhD); genetics (MS, PhD); molecular, cellular, and developmental biology (MS, PhD); toxicology (MS, PhD). *Faculty:* 24 full-time, 2 part-time/adjunct. *Students:* 97 full-time (42 women), 2 part-time (1 woman); includes 2 minority (1 African American, 1 Asian American or Pacific Islander), 68 international. 28 applicants, 46% accepted, 13 enrolled. In 2005, 9 master's, 8 doctorates awarded. *Degree requirements:* For master's and doctorate, thesis/dissertation. *Entrance requirements:* For master's and doctorate, GRE General Test. Additional exam requirements/recommendations for international students: Required—TOEFL (paper score 550; computer score 213) or IELTS (score 6.5). *Application deadline:* For fall admission, 2/1 priority date for domestic students, 2/1 priority date for international students. Application fee: $30 ($70 for international students). Electronic applications accepted. *Expenses:* Tuition, state resident: full time $6,410. Tuition, nonresident: full-time $16,422. Tuition and fees vary according to program. *Financial support:* In 2005–06, 91 research assistantships with full tuition reimbursements (averaging $16,290 per year), 2 teaching assistantships with full and partial tuition reimbursements (averaging $14,455 per year) were awarded; scholarships/grants, health care benefits, and unspecified assistantships also available. *Unit head:* Dr. Alan M. Myers, Chair, 515-294-6116, E-mail: biochem@iastate.edu.

See Close-Up on page 395.

Iowa State University of Science and Technology, Graduate College, Interdisciplinary Programs, Program in Toxicology, Ames, IA 50011. Offers MS, PhD. *Students:* 19 full-time (8 women), 1 (woman) part-time; includes 3 minority (2 African Americans, 1 Hispanic American), 15 international. 11 applicants, 18% accepted, 2 enrolled. In 2005, 3 master's, 3 doctorates awarded. *Degree requirements:* For master's and doctorate, thesis/dissertation. *Entrance requirements:* For master's and doctorate, GRE General Test. Additional exam requirements/recommendations for international students: Required—TOEFL (paper score 530; computer score 197) or IELTS (score 6.0). *Application deadline:* For fall admission, 1/31 priority date for domestic students, 1/31 priority date for international students. Applications are processed on a rolling basis. Application fee: $30 ($70 for international students). Electronic applications accepted. *Expenses:* Tuition, state resident: full-time $6,410. Tuition, nonresident: full-time $16,422. Tuition and fees vary according to program. *Financial support:* In 2005–06, 15 research assistantships with full and partial tuition reimbursements (averaging $15,678 per year) were awarded; teaching assistantships with partial tuition reimbursements, scholarships/grants, health care benefits, and unspecified assistantships also available. *Unit head:* Dr. Anumantha Kanthasamy, Supervisory Committee Chair, 515-294-7697, Fax: 515-294-6669, E-mail: toxmajor@iastate.edu. *Application contact:* Linda Wild, Information Contact, 800-499-1972, Fax: 515-294-6669, E-mail: toxmajor@iastate.edu.

The Johns Hopkins University, Bloomberg School of Public Health, Department of Environmental Health Sciences, Baltimore, MD 21218-2699. Offers environmental health engineering (PhD, Sc D); environmental health sciences (MHS, Dr PH); molecular imaging (PhD, Sc D); occupational and environmental health (PhD, Sc D); occupational and environmental hygiene (MHS); physiology (PhD, Sc D); toxicological sciences (PhD, Sc D). *Faculty:* 64 full-time (19 women), 59 part-time/adjunct (25 women). *Students:* 75 full-time (50 women), 15 part-time (10 women); includes 16 minority (4 African Americans, 10 Asian Americans or Pacific Islanders, 2 Hispanic Americans), 20 international. Average age 29. 105 applicants, 57% accepted, 36 enrolled. In 2005, 11 master's, 6 doctorates awarded. *Median time to degree:* Of those who began their doctoral program in fall 1997, 75% received their degree in 8 years or less. *Degree requirements:* For master's, thesis, registration; for doctorate, thesis/dissertation, 1 year full-time residency, oral and written exams, comprehensive exam, registration. *Entrance requirements:* For master's and doctorate, GRE General Test or MCAT, 3 letters of recommendation, curriculum vitae. Additional exam requirements/recommendations for international students: Required—TOEFL (minimum score 550 paper-based; 213 computer-based). *Application deadline:* For fall admission, 1/15 priority date for domestic students, 1/15 priority date for international students. Applications are processed on a rolling basis. Application fee: $45. Electronic applications accepted. *Expenses:* Tuition: Full-time $30,960. Tuition and fees vary according to degree level and program. *Financial support:* In 2005–06, 104 students received support, including 1 fellowship (averaging $23,500 per year); Federal Work-Study, institution-

ally sponsored loans, scholarships/grants, traineeships, health care benefits, and stipends also available. Support available to part-time students. Financial award application deadline: 3/15; financial award applicants required to submit FAFSA. *Faculty research:* Respiratory and cardiovascular physiology, chemical carcinogenesis, antioxidants and drug metabolism, nuclear medicine, nuclear imaging. Total annual research expenditures: $20.5 million. *Unit head:* Dr. John Davis Groopman, Chair, 410-955-3720, Fax: 410-955-0617, E-mail: jgroopma@jhsph. edu. *Application contact:* Nina Kulacki, Senior Academic Coordinator, 410-955-2212, Fax: 410-955-0617, E-mail: nkulacki@jhsph.edu.

Long Island University, Brooklyn Campus, Arnold and Marie Schwartz College of Pharmacy and Health Sciences, Graduate Programs in Pharmacy, Division of Pharmaceutical Sciences, Brooklyn, NY 11201-8423. Offers cosmetic science (MS); industrial pharmacy (MS); pharmaceutics (PhD); pharmacology/toxicology (MS). Part-time and evening/weekend programs available. *Faculty:* 17 full-time (2 women), 3 part-time/adjunct (1 woman). *Students:* 94 full-time (40 women), 77 part-time (26 women); includes 22 minority (20 African Americans, 2 Hispanic Americans), 135 international. Average age 30. In 2005, 33 master's, 1 doctorate awarded. Terminal master's awarded for partial completion of doctoral program. *Degree requirements:* For master's, thesis optional; for doctorate, thesis/dissertation, candidacy exam. *Entrance requirements:* For master's and doctorate, minimum GPA of 3.0. *Application deadline:* Applications are processed on a rolling basis. Application fee: $30. *Financial support:* In 2005–06, 5 fellowships with full tuition reimbursements (averaging $13,500 per year), 18 teaching assistantships with full tuition reimbursements (averaging $6,000 per year) were awarded; Federal Work-Study also available. Financial award applicants required to submit FAFSA. *Unit head:* Dr. Fotios M. Plakogiannis, Director, 718-488-1101, E-mail: fotios. plakogiannis@liu.edu. *Application contact:* Edward Dettling, Director of Graduate Admissions, 718-488-1011, Fax: 718-797-2399, E-mail: admissions@brooklyn.liu.edu.

Louisiana State University and Agricultural and Mechanical College, Graduate School, School of the Coast and Environment, Department of Environmental Studies, Baton Rouge, LA 70803. Offers environmental planning and management (MS); environmental toxicology (MS). *Faculty:* 12 full-time (4 women). *Students:* 21 full-time (13 women), 11 part-time (5 women); includes 2 African Americans, 1 American Indian/Alaska Native, 3 international. Average age 29. 21 applicants, 43% accepted, 12 enrolled. In 2005, 12 degrees awarded. *Degree requirements:* For master's, thesis (for some programs). *Entrance requirements:* For master's, GRE General Test, minimum GPA of 3.0. Additional exam requirements/recommendations for international students: Required—TOEFL (minimum score 550 paper-based; 213 computer-based). *Application deadline:* For fall admission, 1/25 priority date for domestic students, 5/15 priority date for international students. Applications are processed on a rolling basis. Application fee: $25. Electronic applications accepted. *Financial support:* In 2005–06, 16 students received support, including 2 fellowships with full and partial tuition reimbursements available (averaging $11,458 per year), 12 research assistantships with full and partial tuition reimbursements available (averaging $11,036 per year); teaching assistantships with full and partial tuition reimbursements available, career-related internships or fieldwork, Federal Work-Study, institutionally sponsored loans, scholarships/grants, and unspecified assistantships also available. Support available to part-time students. Financial award applicants required to submit FAFSA. *Faculty research:* Fates and movement of pollutants, neurobiotic metabolism, application of cellular toxicity/mutagenicity testing. Total annual research expenditures: $1.2 million.

Massachusetts Institute of Technology, School of Engineering, Biological Engineering Division, Cambridge, MA 02139-4307. Offers applied biosciences (PhD, Sc D); bioengineering (PhD, Sc D); biomedical engineering (M Eng); genetic toxicology (PhD, Sc D); molecular and systems bacterial pathogenesis (PhD, Sc D); molecular and systems toxicology and pharmacology (PhD, Sc D); molecular systems toxicology (PhD, Sc D); toxicology (SM, PhD, Sc D). *Faculty:* 16 full-time (4 women). *Students:* 126 full-time (56 women); includes 25 minority (3 African Americans, 17 Asian Americans or Pacific Islanders, 5 Hispanic Americans), 39 international. Average age 26. 349 applicants, 18% accepted, 28 enrolled. In 2005, 4 master's, 12 doctorates awarded. Terminal master's awarded for partial completion of doctoral program. *Degree requirements:* For master's, thesis; for doctorate, thesis/dissertation, qualifying exams, comprehensive exam. *Entrance requirements:* For master's and doctorate, GRE General Test. Additional exam requirements/recommendations for international students: Required—TOEFL (minimum score 600 paper-based; 250 computer-based). *Application deadline:* For fall admission, 12/31 for domestic students, 12/31 for international students. Application fee: $70. Electronic applications accepted. *Expenses:* Tuition: Full-time $32,100. Required fees: $200. Part-time tuition and fees vary according to course load. *Financial support:* In 2005–06, 112 students received support, including 57 fellowships with tuition reimbursements available (averaging $27,248 per year), 64 research assistantships with tuition reimbursements available (averaging $26,559 per year), 3 teaching assistantships with tuition reimbursements available (averaging $26,180 per year); Federal Work-Study, institutionally sponsored loans, scholarships/grants, traineeships, health care benefits, and unspecified assistantships also available. Total annual research expenditures: $22.5 million. *Unit head:* Prof. Douglas A. Lauffenburger, Director, 617-252-1629, Fax: 617-258-0204, E-mail: lauffen@mit.edu. *Application contact:* Biological Engineering Academic Office, 617-253-1712, Fax: 617-258-8676, E-mail: be-acad@mit.edu.

Medical College of Wisconsin, Graduate School of Biomedical Sciences, Department of Pharmacology and Toxicology, Milwaukee, WI 53226-0509. Offers MS, PhD, MD/MS, MD/PhD. Terminal master's awarded for partial completion of doctoral program. *Degree requirements:* For master's, thesis; for doctorate, thesis/dissertation, oral and written qualifying exams, comprehensive exam, registration. *Entrance requirements:* For master's and doctorate, GRE General Test, minimum B average. Additional exam requirements/recommendations for international students: Required—TOEFL. *Faculty research:* Cardiovascular physiology and pharmacology, drugs of abuse, environmental and aquatic toxicology, central nervous system and biochemical pharmacology, signal transduction.

Michigan State University, College of Osteopathic Medicine and The Graduate School, Graduate Studies in Osteopathic Medicine and Graduate Programs in Human Medicine and Graduate Program in Veterinary Medicine, Department of Pharmacology and Toxicology, East Lansing, MI 48824. Offers pharmacology and toxicology (MS, PhD); pharmacology and toxicology-environmental toxicology (PhD). *Faculty:* 16 full-time (3 women). *Students:* 19 full-time (12 women), 1 (woman) part-time; includes 4 minority (1 American Indian/Alaska Native, 2 Asian Americans or Pacific Islanders, 1 Hispanic American), 6 international. Average age 27. 41 applicants, 20% accepted. In 2005, 1 master's, 6 doctorates awarded. *Degree requirements:* For master's, thesis; for doctorate, thesis/dissertation, oral presentation of dissertation proposal, oral exam in defense of dissertation, comprehensive exam. *Entrance requirements:* For master's, GRE General Test, minimum GPA of 3.4 in last 2 undergraduate years, 3 letters of recommendation; for doctorate, GRE General Test, minimum GPA of 3.0 in last 2 undergraduate years, 3 letters of recommendation. Additional exam requirements/recommendations for international students: Required—TOEFL (minimum score 600 paper-based; 220 computer-based). *Application deadline:* For fall admission, 12/27 for domestic students. Applications are processed on a rolling basis. Application fee: $50. Electronic applications accepted. *Expenses:* Tuition, state resident: part-time $330 per credit hour. Tuition, nonresident: part-time $685 per credit hour. Tuition and fees vary according to program. *Financial support:* In 2005–06, 12 fellowships with tuition reimbursements (averaging $9,713 per year), 8 research assistantships with tuition reimbursements (averaging $15,434 per year) were awarded; teaching assistantships, institutionally sponsored loans, scholarships/grants, and unspecified assistantships also available. Support available to part-time students. *Faculty research:* Effects of drugs and chemicals on macromolecules and actions in humans; neurological, cardiovascular, pulmonary and gastrointestinal pharmacology; environmental and analytical toxicology; endocrine pharmacology. Total annual research expenditures: $3.5 million. *Unit head:* Dr. Joseph R. Haywood, Chairperson, 517-353-7147, Fax: 517-353-8915, E-mail: haywoo12@msu.edu. *Application contact:* Diane Hummel, Graduate Secretary, 517-353-7146, Fax: 517-353-8915, E-mail: hummeld@msu.edu.

Toxicology

Michigan State University, College of Veterinary Medicine and The Graduate School, Graduate Program in Veterinary Medicine, Center for Integrative Toxicology, East Lansing, MI 48824. Offers animal science–environmental toxicology (PhD); biochemistry and molecular biology–environmental toxicology (PhD); chemistry–environmental toxicology (PhD); crop and soil sciences–environmental toxicology (PhD); entomology–environmental toxicology (PhD); environmental engineering–environmental toxicology (PhD); environmental geosciences–environmental toxicology (PhD); fisheries and wildlife–environmental toxicology (PhD); food science–environmental toxicology (PhD); forestry–environmental toxicology (PhD); microbiology–environmental toxicology (PhD); pathology–environmental toxicology (PhD); pharmacology and toxicology–environmental toxicology (PhD); zoology–environmental toxicology (PhD). *Faculty:* 39 full-time (6 women), 2 part-time/adjunct (0 women). *Students:* 23 full-time (10 women); includes 2 minority (both African Americans), 12 international. Average age 30. In 2005, 9 degrees awarded. *Entrance requirements:* Additional exam requirements/recommendations for international students: Required—TOEFL (minimum score 550 paper-based; 213 computer-based), Michigan State University ELT (85), Michigan ELAB (83). *Application deadline:* For fall admission, 12/27 for domestic students. Application fee: $50. Electronic applications accepted. *Expenses:* Tuition, state resident: part-time $330 per credit hour. Tuition, nonresident: part-time $685 per credit hour. Tuition and fees vary according to program. *Financial support:* In 2005–06, 8 fellowships with tuition reimbursements (averaging $11,614 per year), 18 research assistantships with tuition reimbursements (averaging $13,671 per year), 1 teaching assistantship with tuition reimbursement (averaging $14,823 per year) were awarded; career-related internships or fieldwork, Federal Work-Study, scholarships/grants, and unspecified assistantships also available. *Faculty research:* Environmental risk assessment, toxicogenomics, phytoremediation, storage and disposal of hazardous waste, environmental regulation. *Unit head:* Dr. Norbert Kaminski, Director, 517-353-6469, Fax: 517-355-4603, E-mail: kamis11@msu.edu. *Application contact:* Amy Swagart, Executive Secretary, 517-353-6469, Fax: 517-355-4603, E-mail: swagart@msu.edu.

Michigan State University, College of Veterinary Medicine and The Graduate School, Graduate Program in Veterinary Medicine and College of Natural Science and Graduate Programs in Human Medicine, Department of Microbiology and Molecular Genetics, East Lansing, MI 48824. Offers industrial microbiology (MS); microbiology (MS, PhD), including environmental toxicology (MS); microbiology and molecular genetics (MS, PhD); microbiology–environmental toxicology (PhD). *Faculty:* 34 full-time (10 women). *Students:* 47 full-time (28 women), 2 part-time (1 woman); includes 3 minority (2 African Americans, 1 Hispanic American), 21 international. Average age 26. 104 applicants, 13% accepted. In 2005, 3 master's, 9 doctorates awarded. *Degree requirements:* For master's, thesis (for some programs); for doctorate, thesis/dissertation, comprehensive exam. *Entrance requirements:* For master's and doctorate, GRE General Test, minimum GPA of 3.0, 3 letters of recommendation. Additional exam requirements/recommendations for international students: Required—TOEFL (minimum score 550 paper-based; 213 computer-based), Michigan State University ELT (85), Michigan ELAB (83). *Application deadline:* For fall admission, 8/27 for domestic students. Application fee: $50. Electronic applications accepted. *Expenses:* Tuition, state resident: part-time $330 per credit hour. Tuition, nonresident: part-time $685 per credit hour. Tuition and fees vary according to program. *Financial support:* In 2005–06, 12 fellowships with tuition reimbursements (averaging $6,638 per year), 29 research assistantships with tuition reimbursements (averaging $15,427 per year), 3 teaching assistantships with tuition reimbursements (averaging $14,823 per year) were awarded; scholarships/grants, traineeships, health care benefits, and unspecified assistantships also available. *Faculty research:* Microbial physiology, ecology and evolution: molecular pathogenesis and infectious diseases; genomics and genetics; cancer, differentiation, immunology and cell biology. Total annual research expenditures: $5.2 million. *Unit head:* Dr. Walter Esselman, Chairperson, 517-355-6463 Ext. 1510, Fax: 517-353-8957, E-mail: mmgchair@msu.edu. *Application contact:* Suzanne Peacock, Graduate Program Coordinator, 517-432-2288, Fax: 517-353-8957, E-mail: micgrad@msu.edu.

See Close-Up on page 887.

Michigan State University, The Graduate School, College of Agriculture and Natural Resources, Department of Animal Science, East Lansing, MI 48824. Offers animal science (MS, PhD); animal science-environmental toxicology (PhD). *Faculty:* 37 full-time (9 women). *Students:* 30 full-time (18 women), 6 part-time (4 women); includes 3 minority (1 African American, 2 Hispanic Americans), 9 international. Average age 29. 17 applicants, 18% accepted. In 2005, 5 master's, 7 doctorates awarded. *Degree requirements:* For master's, thesis or alternative, presentation of thesis/project and oral exam; for doctorate, thesis/dissertation, defense of dissertation, year of residence, comprehensive exam. *Entrance requirements:* For master's and doctorate, GRE General Test, minimum GPA of 3.0 in last 2 undergraduate years, letters of reference. Additional exam requirements/recommendations for international students: Required—TOEFL (minimum score 550 paper-based; 213 computer-based), Michigan State University ELT (85), Michigan ELAB (83). *Application deadline:* For fall admission, 12/27 for domestic students. Applications are processed on a rolling basis. Application fee: $50. Electronic applications accepted. *Expenses:* Tuition, state resident: part-time $330 per credit hour. Tuition, nonresident: part-time $685 per credit hour. Tuition and fees vary according to program. *Financial support:* In 2005–06, 5 fellowships with tuition reimbursements (averaging $10,230 per year), 19 research assistantships with tuition reimbursements (averaging $13,062 per year) were awarded; scholarships/grants and unspecified assistantships also available. *Faculty research:* Breeding and genetics, management and systems, meats and growth biology, reproductive and mammary physiology, microbiology and molecular biology. Total annual research expenditures: $5.8 million. *Unit head:* Dr. Karen I. Plaut, Chairperson, 517-355-8383, Fax: 517-353-1699, E-mail: kplaut@msu.edu. *Application contact:* Kim Dobson, Graduate Student Program Secretary, 517-353-9227, Fax: 517-353-1699, E-mail: dobsonk@msu.edu.

Michigan State University, The Graduate School, College of Agriculture and Natural Resources, Department of Crop and Soil Sciences, East Lansing, MI 48824. Offers crop and soil sciences (MS, PhD); crop and soil sciences-environmental toxicology (PhD); plant breeding and genetics-crop and soil sciences (MS, PhD). *Faculty:* 44 full-time (15 women), 13 part-time (5 women); includes 2 minority (both African Americans), 24 international. Average age 31. 30 applicants, 27% accepted. In 2005, 8 master's, 11 doctorates awarded. *Degree requirements:* For master's, thesis or alternative, oral final exam; for doctorate, thesis/dissertation, oral final exam in defense of dissertation, 1 year of residence, comprehensive exam. *Entrance requirements:* For master's, GRE General Test, minimum GPA of 3.0, bachelor's degree in crop and soil sciences or related field; for doctorate, GRE General Test, minimum GPA of 3.0, MS degree (preferred). Additional exam requirements/recommendations for international students: Required—TOEFL (minimum score 550 paper-based; 213 computer-based), Michigan State University ELT (85), Michigan ELAB (83). *Application deadline:* For fall admission, 12/27 for domestic students. Applications are processed on a rolling basis. Application fee: $50. Electronic applications accepted. *Expenses:* Tuition, state resident: part-time $330 per credit hour. Tuition, nonresident: part-time $685 per credit hour. Tuition and fees vary according to program. *Financial support:* In 2005–06, 5 fellowships with tuition reimbursements (averaging $13,440 per year), 37 research assistantships with tuition reimbursements (averaging $12,999 per year), 2 teaching assistantships with tuition reimbursements (averaging $12,470 per year) were awarded; scholarships/grants and unspecified assistantships also available. *Faculty research:* Turfgrass management, environmental toxicology, integrated pest management, water science, plant breeding and genetics. Total annual research expenditures: $8 million. *Unit head:* Dr. James J. Kells, Acting Chairperson, 517-355-0271 Ext. 1103, Fax: 517-353-5174, E-mail: kells@msu.edu. *Application contact:* Rita House, Graduate Secretary, 517-355-0271 Ext. 1111, Fax: 517-353-5174, E-mail: house@msu.edu.

Michigan State University, The Graduate School, College of Agriculture and Natural Resources and College of Natural Science, Department of Entomology, East Lansing, MI 48824. Offers entomology (MS, PhD); entomology-environmental toxicology (PhD); integrated pest management (MS). *Faculty:* 21 full-time (5 women). *Students:* 24 full-time (12 women), 3 part-time; includes 3 minority (1 African American, 2 Hispanic Americans), 4 international. Average age 30. 22 applicants, 23% accepted. In 2005, 4 master's, 2 doctorates awarded. *Degree requirements:* For master's, thesis or alternative, oral final exam; for doctorate, thesis/dissertation, oral final exam in defense of dissertation, participation in teaching program, comprehensive exam. *Entrance requirements:* For master's, GRE General Test, minimum GPA of 3.0 in last 2 undergraduate years, prior training in physical/biological sciences; for doctorate, GRE General Test, MS with thesis in entomology or related field. Additional exam requirements/recommendations for international students: Required—TOEFL (minimum score 550 paper-based; 213 computer-based), Michigan State University ELT (85), Michigan ELAB (83). *Application deadline:* For fall admission, 12/27 for domestic students. Application fee: $50. Electronic applications accepted. *Expenses:* Tuition, state resident: part-time $330 per credit hour. Tuition, nonresident: part-time $685 per credit hour. Tuition and fees vary according to program. *Financial support:* In 2005–06, 12 fellowships with tuition reimbursements (averaging $2,733 per year), 18 research assistantships with tuition reimbursements (averaging $14,266 per year), 7 teaching assistantships with tuition reimbursements (averaging $12,021 per year) were awarded; scholarships/grants and unspecified assistantships also available. *Faculty research:* Agroaquatic and forest ecology, insect physiology, agricultural entomology, integrated pest management, environmental and analytical toxicology. Total annual research expenditures: $6.8 million. *Unit head:* Dr. Richard Merritt, Chairperson, 517-355-4665, Fax: 517-355-4354, E-mail: merrittr@msu.edu. *Application contact:* Jill Kolp, Graduate Secretary, 517-355-4665, Fax: 517-355-4354, E-mail: kolpj@msu.edu.

Michigan State University, The Graduate School, College of Agriculture and Natural Resources and College of Natural Science, Department of Food Science and Human Nutrition, East Lansing, MI 48824. Offers food science (MS, PhD); food science—environmental toxicology (PhD); human nutrition (MS, PhD); human nutrition-environmental toxicology (PhD). *Faculty:* 22 full-time (8 women). *Students:* 45 full-time (30 women), 6 part-time (all women); includes 6 minority (3 African Americans, 2 Asian Americans or Pacific Islanders, 1 Hispanic American), 23 international. Average age 30. 67 applicants, 12% accepted. In 2005, 12 master's, 7 doctorates awarded. *Degree requirements:* For master's, oral examination in defense of thesis, thesis optional; for doctorate, thesis/dissertation, 1 term assistant teaching, 1 year residency, oral presentation and defense of dissertation, comprehensive exam. *Entrance requirements:* For master's, GRE General Test, minimum GPA of 3.0 in last 2 years of undergraduate course work, 3 letters of recommendation; for doctorate, GRE General Test, minimum GPA of 3.0, 3 letters of recommendation. Additional exam requirements/recommendations for international students: Required—TOEFL (minimum score 550 paper-based; 213 computer-based), Michigan State University ELT (85), Michigan ELAB (83). *Application deadline:* For fall admission, 12/27 for domestic students. Application fee: $50. Electronic applications accepted. *Expenses:* Tuition, state resident: part-time $330 per credit hour. Tuition, nonresident: part-time $685 per credit hour. Tuition and fees vary according to program. *Financial support:* In 2005–06, 11 fellowships with tuition reimbursements (averaging $8,443 per year), 30 research assistantships with tuition reimbursements (averaging $12,420 per year), 2 teaching assistantships with tuition reimbursements (averaging $13,088 per year) were awarded; scholarships/grants and unspecified assistantships also available. *Faculty research:* Food safety and toxicology, food processing and quality enhancement, biochemical nutrition, community nutrition. Total annual research expenditures: $3.2 million. *Unit head:* Dr. Gale M. Strasburg, Chairperson, 517-355-8474 Ext. 100, Fax: 517-355-8963, E-mail: stragale@msu.edu. *Application contact:* Deborah Klein, Graduate Secretary, 517-355-8474 Ext. 118, Fax: 517-353-8963, E-mail: kleinde@msu.edu.

Michigan State University, The Graduate School, College of Engineering, Department of Civil and Environmental Engineering, East Lansing, MI 48824. Offers civil engineering (MS, PhD); environmental engineering (MS, PhD); environmental engineering-environmental toxicology (PhD). Part-time programs available. *Faculty:* 18 full-time (1 woman), 1 part-time/adjunct (0 women). *Students:* 65 full-time (19 women), 12 part-time (3 women); includes 6 minority (1 African American, 4 Asian Americans or Pacific Islanders, 1 Hispanic American), 50 international. Average age 27. 124 applicants, 37% accepted. In 2005, 16 master's, 12 doctorates awarded. *Degree requirements:* For master's, thesis optional; for doctorate, thesis/dissertation, qualifying exam, oral defense of dissertation, comprehensive exam. *Entrance requirements:* For master's, GRE General Test, minimum GPA of 3.0, bachelor's degree in civil or environmental engineering or related area, 3 letters of recommendation; for doctorate, GRE General Test, minimum graduate GPA of 3.5, bachelor's or master's degree in civil or environmental engineering or related area, 3 letters of recommendation, professional writing sample. Additional exam requirements/recommendations for international students: Required—TOEFL (minimum score 570 paper-based; 230 computer-based). *Application deadline:* For fall admission, 12/27 for domestic students. Application fee: $50. Electronic applications accepted. *Expenses:* Tuition, state resident: part-time $330 per credit hour. Tuition, nonresident: part-time $685 per credit hour. Tuition and fees vary according to program. *Financial support:* In 2005–06, 22 fellowships with tuition reimbursements (averaging $5,311 per year), 35 research assistantships with tuition reimbursements (averaging $15,040 per year), 18 teaching assistantships with tuition reimbursements (averaging $14,907 per year) were awarded; scholarships/grants and unspecified assistantships also available. *Faculty research:* Environmental engineering, infrastructure, transportation, environmental toxicology and hazardous waste treatment, traffic operations and safety. Total annual research expenditures: $2.2 million. *Unit head:* Dr. Ronald S. Harichandran, Chairperson, 517-355-5107, Fax: 517-432-1827, E-mail: harichan@egr.msu.edu. *Application contact:* Laura Taylor, Graduate Program Secretary, 517-355-5107, Fax: 517-432-1827, E-mail: taylor1@egr.msu.edu.

Michigan State University, The Graduate School, College of Natural Science and Graduate Programs in Human Medicine and Graduate Studies in Osteopathic Medicine, Department of Biochemistry and Molecular Biology, East Lansing, MI 48824. Offers biochemistry and molecular biology (MS, PhD); biochemistry and molecular biology/environmental toxicology (PhD). *Faculty:* 30 full-time (5 women). *Students:* 74 full-time (31 women), 1 part-time; includes 3 minority (1 African American, 1 Asian American or Pacific Islander, 1 Hispanic American), 35 international. Average age 28. 93 applicants, 12% accepted. In 2005, 2 master's, 7 doctorates awarded. *Degree requirements:* For master's, oral defense of thesis, annual review, thesis optional; for doctorate, thesis/dissertation, teaching, annual evaluations, research proposal, comprehensive exam. *Entrance requirements:* For master's, GRE General Test, minimum GPA of 3.2 in last 2 undergraduate years, course work in biochemistry, 3 letters of recommendation from former instructors; for doctorate, GRE General Test, minimum GPA of 3.2; bachelor's or master's degree in chemistry, biochemistry or related science; course work in biochemistry; 3 letters of recommendation from former instructors. Additional exam requirements/recommendations for international students: Required—TOEFL (minimum score 550 paper-based; 213 computer-based), Michigan State University ELT (85), Michigan ELAB (83). *Application deadline:* For fall admission, 12/27 for domestic students. Application fee: $50. Electronic applications accepted. *Expenses:* Tuition, state resident: part-time $330 per credit hour. Tuition, nonresident: part-time $685 per credit hour. Tuition and fees vary according to program. *Financial support:* In 2005–06, 18 fellowships with tuition reimbursements (averaging $11,755 per year), 64 research assistantships with tuition reimbursements (averaging $15,738 per year) were awarded; scholarships/grants and unspecified assistantships also available. *Faculty research:* Biochemistry of the cell nucleus; protein structure, function, and design; plant biochemistry; biological modeling: gene expression in development and disease. Total annual research expenditures: $10.7 million. *Unit head:* Dr. S. L. Ferguson-Miller, Chairperson, 517-353-0804, Fax: 517-353-9334, E-mail: fergus20@msu.edu. *Application contact:* Jessica Hudson, Graduate Program Secretary, 517-353-0807, Fax: 517-353-9334, E-mail: jhudson@msu.edu.

See Close-Up on page 401.

Michigan State University, The Graduate School, College of Natural Science, Department of Chemistry, East Lansing, MI 48824. Offers chemical physics (PhD); chemistry (MS, PhD); chemistry-environmental toxicology (PhD); computational chemistry (MS). *Faculty:* 35 full-time (4 women). *Students:* 221 full-time (100 women), 4 part-time (2 women); includes 13 minority (4 African Americans, 1 American Indian/Alaska Native, 6 Asian Americans or Pacific

Islanders, 2 Hispanic Americans), 129 international. Average age 27. 138 applicants, 66% accepted. In 2005, 9 master's, 19 doctorates awarded. *Degree requirements:* For master's, oral defense of thesis, thesis optional; for doctorate, thesis/dissertation, oral defense of dissertation, comprehensive exam. *Entrance requirements:* For master's, GRE General Test, bachelor's degree in chemistry; course work in chemistry, physics, and calculus; 3 letters of recommendation; for doctorate, GRE General Test, minimum GPA of 3.0; bachelor's or master's degree in chemistry; coursework in chemistry, physics, and calculus; 3 letters of recommendation. Additional exam requirements/recommendations for international students: Required—TOEFL (minimum score 550 paper-based; 213 computer-based), Michigan State University ELT (85), Michigan ELAB (83). *Application deadline:* For fall admission, 12/27 for domestic students. Application fee: $50. Electronic applications accepted. *Expenses:* Tuition, state resident: part-time $330 per credit hour. Tuition, nonresident: part-time $685 per credit hour. Tuition and fees vary according to program. *Financial support:* In 2005–06, 85 fellowships with tuition reimbursements (averaging $3,132 per year), 71 research assistantships with tuition reimbursements (averaging $15,305 per year), 138 teaching assistantships with tuition reimbursements (averaging $15,041 per year) were awarded; scholarships/grants and unspecified assistantships also available. *Faculty research:* Analytical chemistry, inorganic and organic chemistry, nuclear chemistry, physical chemistry, theoretical and computational chemistry. Total annual research expenditures: $8.7 million. *Unit head:* Dr. John L. McCracken, Chairperson, 517-355-9715 Ext. 346, Fax: 517-353-1793, E-mail: chair@chemistry.msu.edu. *Application contact:* Deborah Roper, Graduate Admissions Secretary, 517-355-9715 Ext. 362, Fax: 517-353-1793, E-mail: gradoff@crm.msu.edu.

Michigan State University, The Graduate School, College of Natural Science, Department of Geological Sciences, East Lansing, MI 48824. Offers environmental geosciences (MS, PhD); environmental geosciences-environmental toxicology (PhD); geological sciences (MS, PhD). *Faculty:* 11 full-time (1 woman). *Students:* 20 full-time (12 women), 4 part-time (1 woman); includes 2 minority (both Asian Americans or Pacific Islanders), 5 international. Average age 28. 37 applicants, 59% accepted. In 2005, 8 master's, 2 doctorates awarded. *Degree requirements:* For master's, thesis for those without prior thesis work; for doctorate, thesis/dissertation, registration. *Entrance requirements:* For master's, GRE General Test, minimum GPA of 3.0, course work in geoscience, 3 letters of recommendation; for doctorate, GRE General Test, 3 letters of recommendation. Additional exam requirements/recommendations for international students: Required—TOEFL (minimum score 550 paper-based; 213 computer-based), Michigan State Univeristy ELT (85), Michigan ELAB (83). *Application deadline:* For fall admission, 12/27 for domestic students. Application fee: $50. Electronic applications accepted. *Expenses:* Tuition, state resident: part-time $330 per credit hour. Tuition, nonresident: part-time $685 per credit hour. Tuition and fees vary according to program. *Financial support:* In 2005–06, 10 fellowships with tuition reimbursements (averaging $3,479 per year), 4 research assistantships with tuition reimbursements (averaging $12,710 per year), 13 teaching assistantships with tuition reimbursements (averaging $12,997 per year) were awarded; Federal Work-Study, scholarships/grants, and unspecified assistantships also available. *Faculty research:* Water in the environment, global and biological change, crystal dynamics. Total annual research expenditures: $1.1 million. *Unit head:* Dr. Ralph E. Taggart, Chairperson, 517-355-4626, Fax: 517-353-8787, E-mail: taggart@msu.edu. *Application contact:* Information Contact, 517-355-4626, Fax: 517-353-8787, E-mail: geosci@msu.edu.

New York University, Graduate School of Arts and Science, Department of Environmental Medicine, New York, NY 10012-1019. Offers environmental health sciences (MS, PhD), including biostatistics (PhD), environmental hygiene (MS), epidemiology (PhD), ergonomics and biomechanics (PhD), exposure assessment and health effects (PhD), molecular toxicology/carcinogenesis (PhD), toxicology. Part-time programs available. *Faculty:* 26 full-time (7 women). *Students:* 42 full-time (33 women), 21 part-time (10 women); includes 11 minority (2 African Americans, 6 Asian Americans or Pacific Islanders, 3 Hispanic Americans), 21 international. Average age 32. 56 applicants, 38% accepted, 9 enrolled. In 2005, 7 master's, 6 doctorates awarded. Terminal master's awarded for partial completion of doctoral program. *Degree requirements:* For master's, thesis or alternative; for doctorate, one foreign language, thesis/dissertation, oral and written exams. *Entrance requirements:* For master's and doctorate, GRE General Test, GRE Subject Test, minimum GPA of 3.0; bachelor's degree in biological, physical, or engineering science. Additional exam requirements/recommendations for international students: Required—TOEFL. *Application deadline:* For fall admission, 11/14 for domestic students. Application fee: $80. *Financial support:* Fellowships with tuition reimbursements, teaching assistantships with tuition reimbursements, career-related internships or fieldwork, Federal Work-Study, institutionally sponsored loans, and health care benefits available. Financial award application deadline: 2/1; financial award applicants required to submit FAFSA. *Unit head:* Dr. Max Costa, Chair, 845-731-3661, Fax: 845-351-0317, E-mail: ehs@env.med.nyu.edu. *Application contact:* Dr. Jerome J Solomon, Director of Graduate Studies, 845-731-3661, Fax: 845-351-3317, E-mail: ehs@env.med.nyu.edu.

See Close-Up on page 183.

North Carolina State University, Graduate School, College of Agriculture and Life Sciences and College of Veterinary Medicine, Department of Environmental and Molecular Toxicology, Raleigh, NC 27695. Offers M Tox, MS, PhD. Terminal master's awarded for partial completion of doctoral program. *Degree requirements:* For master's, thesis (for some programs); for doctorate, thesis/dissertation. *Entrance requirements:* For master's and doctorate, GRE General Test, minimum GPA of 3.0. Electronic applications accepted. *Faculty research:* Chemical fate, carcinogenesis, developmental and endocrine toxicity, xenobiotic metabolism, signal transduction.

See Close-Up on page 1175.

Northeastern University, Bouvé College of Health Sciences Graduate School, Program in Toxicology, Boston, MA 02115-5096. Offers MS. *Entrance requirements:* For master's, GRE General Test. *Application deadline:* Applications are processed on a rolling basis. Application fee: $50. *Financial support:* Federal Work-Study available. Support available to part-time students. Financial award application deadline: 3/1; financial award applicants required to submit FAFSA. *Unit head:* Dr. Robert Schatz, Director, 617-373-3214. *Application contact:* Margaret Schnabel, Director of Graduate Admissions, 617-373-2708, Fax: 617-373-4704, E-mail: bouvegrad@neu.edu.

See Close-Up on page 1177.

Northwestern University, Northwestern University Feinberg School of Medicine and Interdepartmental Degree Programs, Integrated Graduate Programs in the Life Sciences, Chicago, IL 60611. Offers cancer biology (PhD); cell biology (PhD); developmental biology (PhD); evolutionary biology (PhD); immunology and microbial pathogenesis (PhD); molecular biology and genetics (PhD); neurobiology (PhD); pharmacology and toxicology (PhD); structural biology and biochemistry (PhD). *Degree requirements:* For doctorate, thesis/dissertation, written and oral qualifying exams, comprehensive exam. *Entrance requirements:* For doctorate, GRE General Test. Additional exam requirements/recommendations for international students: Required—TOEFL (minimum score 600 paper-based; 250 computer-based). Electronic applications accepted.

See Close-Up on page 189.

The Ohio State University, College of Veterinary Medicine and Graduate School, Program in Veterinary Medicine, Department of Veterinary Biosciences, Columbus, OH 43210. Offers anatomy and cellular biology (MS, PhD); pathobiology (MS, PhD); pharmacology (MS, PhD); toxicology (MS, PhD); veterinary physiology (MS, PhD). *Faculty research:* Microvasculature, muscle biology, neonatal lung and bone development.

Oklahoma State University Center for Health Sciences, Program in Forensic Sciences, Tulsa, OK 74107-1898. Offers forensic DNA/molecular biology (MS); forensic pathology (MS); forensic psychology (MS); forensic sciences (MFSA); forensic toxicology (MS). Part-time and evening/weekend programs available. Postbaccalaureate distance learning degree programs offered (minimal on-campus study). *Faculty:* 2 full-time (0 women), 12 part-time/adjunct (4 women). *Students:* 6 full-time (4 women), 21 part-time (17 women); includes 4 minority (all African Americans), 1 international. Average age 36. 47 applicants, 26% accepted, 11 enrolled. In 2005, 7 degrees awarded. *Degree requirements:* For master's, thesis (for some programs). *Entrance requirements:* For master's, MAT (MFSA) or GRE General Test, professional experience (MFSA). Additional exam requirements/recommendations for international students: Required—TOEFL (minimum score 600 paper-based; 250 computer-based), TWE (minimum score 5). *Application deadline:* For fall admission, 3/31 for domestic students, 3/31 for international students. Application fee: $40 ($75 for international students). *Expenses:* Tuition, state resident: full-time $16,045. Tuition, nonresident: full-time $31,265. Tuition and fees vary according to program. *Financial support:* In 2005–06, 1 student received support. Federal Work-Study available. Support available to part-time students. *Faculty research:* DNA typing, DNA polymorphism, identification through DNA, disease transmission, forensic dentistry, neurotoxicity of HIV. Total annual research expenditures: $216,000. *Unit head:* Dr. Robert T. Allen, Director, 918-561-1108, Fax: 918-561-8414. *Application contact:* Cathy Newsome, Coordinator, 918-699-8608, Fax: 918-561-8414, E-mail: newsome@osu-med.com.

Oregon State University, College of Veterinary Medicine, Program in Veterinary Science, Corvallis, OR 97331. Offers microbiology (MS); pathology (MS); toxicology (MS). Part-time programs available. *Students:* 3 full-time (2 women), 2 international. Average age 27. In 2005, 1 degree awarded. *Degree requirements:* For master's, thesis. *Entrance requirements:* For master's, minimum GPA of 3.0 in last 90 hours. Additional exam requirements/recommendations for international students: Required—TOEFL. *Application deadline:* For fall admission, 11/1 for domestic students. Application fee: $50. *Expenses:* Contact institution. *Financial support:* Research assistantships, Federal Work-Study, institutionally sponsored loans, and scholarships/grants available. Support available to part-time students. Financial award application deadline: 2/1. *Faculty research:* Calf diseases, bovine foot rot, caliciviruses, effects of toxic agents on immune systems. *Unit head:* Dr. Linda L. Blythe, Associate Dean, 541-737-2098, Fax: 541-737-4245, E-mail: linda.blythe@orst.edu.

Oregon State University, Graduate School, College of Agricultural Sciences, Department of Environmental and Molecular Toxicology, Program in Toxicology, Corvallis, OR 97331. Offers MS, PhD. *Students:* 25 full-time (16 women), 3 part-time (1 woman); includes 6 minority (1 American Indian/Alaska Native, 2 Asian Americans or Pacific Islanders, 3 Hispanic Americans), 5 international. Average age 32. In 2005, 1 master's, 1 doctorate awarded. *Degree requirements:* For master's and doctorate, thesis/dissertation. *Entrance requirements:* For master's and doctorate, GRE, bachelor's degree in chemistry or biological sciences, minimum GPA of 3.0 in last 90 hours of course work. Additional exam requirements/recommendations for international students: Required—TOEFL. *Application deadline:* For fall admission, 3/1 for domestic students. Applications are processed on a rolling basis. Application fee: $50. *Expenses:* Tuition, area resident: Part-time $301 per credit. Tuition, state resident: full-time $8,139; part-time $501 per credit. Tuition, nonresident: full-time $14,376; part-time $532 per credit. Required fees: $1,266. *Financial support:* Fellowships, research assistantships, Federal Work-Study and institutionally sponsored loans available. Support available to part-time students. Financial award application deadline: 2/1. *Faculty research:* Biochemical mechanisms for toxicology; analytical, comparative, aquatic, and food toxicology; aquaculture of salmonids; immunotoxicology; fish toxicology.

Purdue University, College of Pharmacy and Pharmacal Sciences and Graduate School, Graduate Programs in Pharmacy and Pharmacal Sciences, Department of Medicinal Chemistry and Molecular Pharmacology, West Lafayette, IN 47907. Offers analytical medicinal chemistry (PhD); computational and biophysical medicinal chemistry (PhD); medicinal and bioorganic chemistry (PhD); medicinal biochemistry and molecular biology (PhD); molecular pharmacology and toxicology (PhD); natural products and pharmacognosy (PhD); nuclear pharmacy (MS); radiopharmaceutical chemistry and nuclear pharmacy (PhD). *Faculty:* 21 full-time (2 women), 2 part-time/adjunct (0 women). *Students:* 69 full-time (39 women), 2 part-time (1 woman); includes 8 minority (2 African Americans, 2 Asian Americans or Pacific Islanders, 4 Hispanic Americans), 21 international. Average age 26. 125 applicants, 25% accepted, 15 enrolled. In 2005, 2 master's, 11 doctorates awarded. Terminal master's awarded for partial completion of doctoral program. *Degree requirements:* For master's and doctorate, thesis/dissertation. *Entrance requirements:* For master's, GRE General Test, minimum B average; BS in biology, chemistry, or pharmacy; for doctorate, GRE General Test, minimum B average; BS in biology, chemistry, or pharmacology. Additional exam requirements/recommendations for international students: Required—TOEFL. *Application deadline:* Applications are processed on a rolling basis. Application fee: $55. Electronic applications accepted. *Financial support:* Fellowships, research assistantships, teaching assistantships, traineeships available. Support available to part-time students. Financial award applicants required to submit FAFSA. *Faculty research:* Drug design and development, cancer research, drug synthesis and analysis, chemical pharmacology, environmental toxicology. *Unit head:* Dr. R. F. Borch, Graduate Head, 765-494-1403. *Application contact:* Dr. D. E. Bergstrom, Graduate Committee, 765-494-6275, E-mail: bergstrom@pharmacy.purdue.edu.

Purdue University, School of Veterinary Medicine and Graduate School, Graduate Programs in Veterinary Medicine, Department of Veterinary Pathobiology, West Lafayette, IN 47907. Offers biochemistry and molecular biology (MS, PhD); comparative epidemiology (MS, PhD); epidemiology (MS, PhD); immunology (MS, PhD); infectious diseases (MS, PhD); interdisciplinary genetics (PhD); laboratory animal medicine (MS, PhD); microbiology (MS, PhD); molecular virology (MS, PhD); parasitology (MS, PhD); pathobiology (MS, PhD); public health epidemiology (MS, PhD); toxicology (MS, PhD); veterinary anatomic pathology (MS, PhD); veterinary clinical pathology (MS, PhD); virology (MS, PhD). *Faculty:* 32 full-time (7 women). *Students:* 49 full-time (20 women), 3 part-time (1 woman); includes 2 minority (both African Americans), 31 international. Average age 35. In 2005, 3 master's, 8 doctorates awarded. Terminal master's awarded for partial completion of doctoral program. *Degree requirements:* For master's, thesis (for some programs); for doctorate, thesis/dissertation. *Entrance requirements:* For master's and doctorate, GRE General Test. Additional exam requirements/recommendations for international students: Required—TOEFL (minimum score 575 paper-based), TWE (minimum score 4). *Application deadline:* For fall admission, 8/12 for domestic students, 6/15 for international students; for spring admission, 1/12 for domestic students, 10/15 for international students. Application fee: $55. *Financial support:* Fellowships, research assistantships, teaching assistantships available. Financial award application deadline: 3/1; financial award applicants required to submit FAFSA. *Unit head:* Dr. H. Hogenesch, Head, 765-494-7543.

Queen's University at Kingston, School of Graduate Studies and Research, Faculty of Health Sciences, Department of Pharmacology and Toxicology, Kingston, ON K7L 3N6, Canada. Offers M Sc, PhD. *Degree requirements:* For master's, thesis/dissertation, registration; for doctorate, thesis/dissertation, comprehensive exam, registration. *Entrance requirements:* For master's, minimum 2nd class standing, honors bachelor of science degree (life sciences, health sciences, or equivalent); for doctorate, masters of science degree or outstanding performance in honors bachelor of science program. Additional exam requirements/recommendations for international students: Required—TOEFL (minimum score 600 paper-based; 250 computer-based). Electronic applications accepted. *Faculty research:* Biochemical toxicology, cardiovascular pharmacology and neuropharmacology.

Rutgers, The State University of New Jersey, New Brunswick/Piscataway, Graduate School, Program in Environmental Sciences, New Brunswick, NJ 08901-1281. Offers air resources (MS, PhD); aquatic biology (MS, PhD); aquatic chemistry (MS, PhD); atmospheric science (MS, PhD); chemistry and physics of aerosol and hydrosol systems (MS, PhD); environmental chemistry (MS, PhD); environmental microbiology (MS, PhD); environmental toxicology (PhD); exposure assessment (PhD); fate and effects of pollutants (MS, PhD); pollution prevention and control (MS, PhD); water and wastewater treatment (MS, PhD); water resources (MS, PhD). *Faculty:* 81 full-time, 7 part-time/adjunct. *Students:* 49 full-time (27 women), 48 part-time (19 women); includes 10 minority (3 African Americans, 6 Asian Americans or Pacific Islanders, 1 Hispanic American), 24 international. Average age 32. 79 applicants,

Toxicology

Rutgers, The State University of New Jersey, New Brunswick/Piscataway *(continued)*
41% accepted, 15 enrolled. In 2005, 8 master's, 10 doctorates awarded. Terminal master's awarded for partial completion of doctoral program. *Degree requirements:* For master's, thesis or alternative, oral final exam, comprehensive exam; for doctorate, thesis/dissertation, thesis defense, qualifying exam, comprehensive exam. *Entrance requirements:* For master's and doctorate, GRE General Test. Additional exam requirements/recommendations for international students: Required—TOEFL. *Application deadline:* For fall admission, 3/1 for domestic students; for spring admission, 11/1 for domestic students. Applications are processed on a rolling basis. Application fee: $50. Electronic applications accepted. *Expenses:* Tuition, state resident: full-time $10,440; part-time $435 per credit. Tuition, nonresident: full-time $15,520; part-time $647 per credit. Required fees: $129 per credit. Tuition and fees vary according to program. *Financial support:* In 2005–06, 10 fellowships with full tuition reimbursements (averaging $21,887 per year), 34 research assistantships with full tuition reimbursements (averaging $19,367 per year), 3 teaching assistantships with full tuition reimbursements (averaging $17,583 per year) were awarded; career-related internships or fieldwork and Federal Work-Study also available. Financial award application deadline: 1/15; financial award applicants required to submit FAFSA. *Faculty research:* Atmospheric sciences; biological waste treatment; contaminant fate and transport; exposure assessment; air, soil and water quality. Total annual research expenditures: $5.7 million. *Unit head:* John Reinfelder, Director, 732-932-8013, Fax: 732-932-8644, E-mail: reinfelder@envsci.rutgers.edu. *Application contact:* Dr. Paul J. Lioy, Graduate Admissions Committee, 732-932-0150, Fax: 732-445-0116, E-mail: plioy@eohsi.rutgers.edu.

Rutgers, The State University of New Jersey, New Brunswick/Piscataway, Graduate School, Program in Toxicology, New Brunswick, NJ 08901-1281. Offers environmental toxicology (MS, PhD); industrial-occupational toxicology (MS, PhD); nutritional toxicology (MS, PhD); pharmaceutical toxicology (MS, PhD). *Faculty:* 60 full-time. *Students:* 20 full-time (14 women), 11 part-time (7 women); includes 5 minority (4 Asian Americans or Pacific Islanders, 1 Hispanic American), 6 international. Average age 31. 17 applicants, 47% accepted, 3 enrolled. In 2005, 1 master's, 2 doctorates awarded. *Median time to degree:* Of those who began their doctoral program in fall 1997, 80% received their degree in 8 years or less. *Degree requirements:* For master's, thesis; for doctorate, thesis/dissertation, qualifying exams (written and oral), comprehensive exam, registration. *Entrance requirements:* For master's and doctorate, GRE General Test. Additional exam requirements/recommendations for international students: Required—TOEFL. *Application deadline:* For fall admission, 5/1 for domestic students, 5/1 for international students. Applications are processed on a rolling basis. Application fee: $50. Electronic applications accepted. *Expenses:* Tuition, state resident: full-time $10,440; part-time $435 per credit. Tuition, nonresident: full-time $15,520; part-time $647 per credit. Required fees: $129 per credit. Tuition and fees vary according to program. *Financial support:* In 2005–06, 9 fellowships with full tuition reimbursements (averaging $20,772 per year), 5 research assistantships with full tuition reimbursements (averaging $16,302 per year), 5 teaching assistantships with full tuition reimbursements (averaging $16,988 per year) were awarded; career-related internships or fieldwork, scholarships/grants, traineeships, and unspecified assistantships also available. Financial award application deadline: 3/1; financial award applicants required to submit FAFSA. *Faculty research:* Neurotoxicants, immunotoxicology, carcinogenesis and chemoprevention, molecular toxicology, xenobiotic metabolism. Total annual research expenditures: $10 million. *Unit head:* Dr. Kenneth R. Reuhl, Director, 732-445-6909, Fax: 732-445-0922, E-mail: reuhl@eohsi.rutgers.edu. *Application contact:* Dr. Debra L. Laskin, Chair, Admissions Committee, 732-445-5862, Fax: 732-445-0922, E-mail: laskin@eohsi.rutgers.edu.

St. John's University, College of Pharmacy and Allied Health Professions, Graduate Programs in Pharmacy, Program in Toxicology, Queens, NY 11439. Offers MS. Part-time and evening/weekend programs available. *Students:* 5 full-time (1 woman), 14 part-time (12 women); includes 9 minority (4 African Americans, 5 Asian Americans or Pacific Islanders), 3 international. Average age 27. 20 applicants, 60% accepted, 8 enrolled. In 2005, 7 degrees awarded. *Degree requirements:* For master's, residency, thesis optional. *Entrance requirements:* For master's, GRE General Test, minimum GPA of 3.0. Additional exam requirements/recommendations for international students: Required—TOEFL (minimum score 500 paper-based; 173 computer-based). *Application deadline:* For fall admission, 4/1 for domestic students, 5/1 for international students; for spring admission, 12/1 for domestic students, 11/1 for international students. Applications are processed on a rolling basis. Application fee: $40. Electronic applications accepted. *Expenses: Contact institution.* Tuition and fees vary according to program. *Financial support:* Fellowships, research assistantships, career-related internships or fieldwork and scholarships/grants available. Support available to part-time students. Financial award application deadline: 3/1; financial award applicants required to submit FAFSA. *Faculty research:* Neurotoxicology, renal toxicology, toxicology of metals, regulatory toxicology. *Unit head:* Dr. Louis Trombetta, Chair, 718-990-6025, E-mail: trombetl@stjohns.edu. *Application contact:* Matthew Whelan, Director, Office of Admissions, 718-990-2000, Fax: 718-990-2096, E-mail: admissions@stjohns.edu.

San Diego State University, Graduate and Research Affairs, College of Health and Human Services, Graduate School of Public Health, San Diego, CA 92182. Offers environmental health (MPH); epidemiology (MPH, PhD), including biostatistics (MPH); global emergency preparedness and response (MS); health behavior (PhD); health promotion (MPH); health services administration (MPH); toxicology (MS). *Accreditation:* ABET (one or more programs are accredited); ACEHSA (one or more programs are accredited); CEPH (one or more programs are accredited). Part-time programs available. *Faculty:* 29 full-time (14 women), 83 part-time/adjunct (37 women). *Students:* 254 full-time (189 women), 113 part-time (81 women); includes 128 minority (18 African Americans, 4 American Indian/Alaska Native, 53 Asian Americans or Pacific Islanders, 53 Hispanic Americans), 26 international. 469 applicants, 67% accepted, 127 enrolled. In 2005, 89 master's, 5 doctorates awarded. *Degree requirements:* For master's, thesis (for some programs), comprehensive exam (for some programs); for doctorate, thesis/dissertation. *Entrance requirements:* For master's, GMAT (health services administration MPH only), GRE General Test; for doctorate, GRE General Test. Additional exam requirements/recommendations for international students: Required—TOEFL. *Application deadline:* For fall admission, 5/1 for domestic students, 5/1 for international students; for spring admission, 11/1 for domestic students, 10/1 for international students. Applications are processed on a rolling basis. Application fee: $55. *Financial support:* Research assistantships, teaching assistantships, career-related internships or fieldwork, Federal Work-Study, and traineeships available. Financial award applicants required to submit FAFSA. *Faculty research:* Evaluation of tobacco, AIDS prevalence and prevention, mammography, infant death project, Alzheimer's in elderly Chinese. *Unit head:* Dr. Ann de Peyster, Interim Director, 619-594-6317. *Application contact:* Brenda Fass-Holmes, Coordinator, Admissions and Student Affairs, 619-594-6317, E-mail: brenda.fass-holmes@sdsu.edu.

Simon Fraser University, Graduate Studies, Faculty of Science, Department of Biological Sciences, Burnaby, BC V5A 1S6, Canada. Offers biological sciences (M Sc, PhD); environmental toxicology (Diploma), including food and drug toxicology, industrial toxicology; pest management (MPM). *Degree requirements:* For masters, doctorate, and Diploma, thesis/dissertation. *Entrance requirements:* For master's and Diploma, minimum GPA of 3.0; for doctorate, minimum GPA of 3.5. Additional exam requirements/recommendations for international students: Required—TOEFL or IELTS. Electronic applications accepted. *Faculty research:* Molecular biology, marine biology, ecology, wildlife biology, endocrinology.

State University of New York at Buffalo, Graduate School, School of Medicine and Biomedical Sciences, Graduate Programs in Medicine and Biomedical Sciences, Department of Pharmacology and Toxicology, Buffalo, NY 14260. Offers biochemical pharmacology (MS); pharmacology (MA, PhD). *Faculty:* 18 full-time (1 woman), 1 part-time/adjunct (0 women). *Students:* 13 full-time (3 women), 4 part-time; includes 3 minority (2 Asian Americans or Pacific Islanders, 1 Hispanic American), 5 international. 12 applicants, 8% accepted, 1 enrolled. In 2005, 3 doctorates awarded. Terminal master's awarded for partial completion of doctoral program. *Median time to degree:* Of those who began their doctoral program in fall 1997, 100% received their degree in 8 years or less. *Degree requirements:* For master's and doctorate, thesis/dissertation. *Entrance requirements:* For master's and doctorate, GRE General Test, 3 letters of recommendation. Additional exam requirements/recommendations for international students: Required—TOEFL. *Application deadline:* For fall admission, 2/1 priority date for domestic students, 2/1 priority date for international students. Applications are processed on a rolling basis. Application fee: $35. Electronic applications accepted. *Financial support:* In 2005–06, 14 students received support, including 2 fellowships with full tuition reimbursements available (averaging $21,224 per year), 12 research assistantships with full tuition reimbursements available (averaging $21,000 per year); teaching assistantships, Federal Work-Study, scholarships/grants, health care benefits, and unspecified assistantships also available. Financial award application deadline: 2/1; financial award applicants required to submit FAFSA. *Faculty research:* Neuropharmacology, toxicology, signal transduction, molecular pharmacology, behavioral pharmacology. Total annual research expenditures: $1.6 million. *Unit head:* Dr. Ronald P. Rubin, Chairman, 716-829-2800, Fax: 716-829-2801, E-mail: rprubin@buffalo.edu. *Application contact:* Noreen A. Harbison, Information Contact, 716-829-2800, Fax: 716-829-2801, E-mail: harbison@buffalo.edu.

Texas A&M University, College of Veterinary Medicine, Graduate Programs in Veterinary Medicine, Department of Veterinary Integrative Biosciences, College Station, TX 77843. Offers epidemiology (MS); food safety/toxicology (MS); veterinary anatomy (MS, PhD); veterinary public health (MS). *Faculty:* 13 full-time (3 women), 3 part-time/adjunct (1 woman). *Students:* 26 full-time (19 women), 4 part-time (3 women); includes 3 minority (2 Asian Americans or Pacific Islanders, 1 Hispanic American), 6 international. Average age 30. 15 applicants, 80% accepted, 11 enrolled. In 2005, 3 master's, 1 doctorate awarded. Terminal master's awarded for partial completion of doctoral program. *Degree requirements:* For master's and doctorate, thesis/dissertation, comprehensive exam, registration. *Entrance requirements:* For master's and doctorate, GRE General Test, minimum undergraduate GPA of 3.0. Additional exam requirements/recommendations for international students: Required—TOEFL. *Application deadline:* For fall admission, 7/15 priority date for domestic students, 4/1 priority date for international students; for spring admission, 10/1 priority date for domestic students, 9/15 priority date for international students. Applications are processed on a rolling basis. Application fee: $50 ($75 for international students). Electronic applications accepted. *Expenses:* Tuition, state resident: full-time $4,488; part-time $187 per credit hour. Tuition, nonresident: full-time $11,112; part-time $463 per credit hour. Required fees: $1,974. *Financial support:* In 2005–06, fellowships (averaging $18,000 per year), research assistantships (averaging $15,600 per year), teaching assistantships (averaging $15,600 per year) were awarded; institutionally sponsored loans, unspecified assistantships, and clinical associateships also available. Financial award application deadline: 7/15; financial award applicants required to submit FAFSA. *Faculty research:* Metal toxicology, reproductive biology, genetics of neural development, developmental biology, environmental toxicology. *Unit head:* Dr. Evelyn Tiffany-Castiglioni, Head, 979-845-2828, Fax: 979-847-8981, E-mail: ecastiglioni@cvm.tamu.edu. *Application contact:* Dr. Jane Welsh, Chair, Fax: 979-847-8981, E-mail: jwelsh@cum.tamu.edu.

Texas A&M University, College of Veterinary Medicine, Graduate Programs in Veterinary Medicine, Department of Veterinary Physiology and Pharmacology, College Station, TX 77843. Offers physiology and pharmacology (MS, PhD); toxicology (MS, PhD). *Faculty:* 8 full-time (0 women), 4 part-time/adjunct (2 women). *Students:* 13 full-time (7 women), 1 (woman) part-time; includes 2 minority (1 Asian American or Pacific Islander, 1 Hispanic American), 7 international. Average age 30. 16 applicants, 38% accepted, 2 enrolled. In 2005, 3 degrees awarded. *Entrance requirements:* For master's and doctorate, GRE General Test. Additional exam requirements/recommendations for international students: Required—TOEFL. Application fee: $50 ($75 for international students). *Expenses:* Tuition, state resident: full-time $4,488; part-time $187 per credit hour. Tuition, nonresident: full-time $11,112; part-time $463 per credit hour. Required fees: $1,974. *Financial support:* Fellowships, research assistantships, teaching assistantships available. Financial award application deadline: 4/1; financial award applicants required to submit FAFSA. *Faculty research:* Gamete and embryo physiology, endocrinology, equine laminitis. *Unit head:* Glen Laine, Head, 979-845-7261.

Texas A&M University, Interdisciplinary Faculty of Toxicology, College Station, TX 77843. Offers MS, PhD. Program composed of faculty members from 7 colleges and 21 departments. *Students:* Average age 26. Terminal master's awarded for partial completion of doctoral program. *Degree requirements:* For master's, thesis/dissertation, registration; for doctorate, thesis/dissertation, comprehensive exam, registration. *Entrance requirements:* For master's and doctorate, GRE General Test, minimum GPA of 3.0. Additional exam requirements/recommendations for international students: Required—TOEFL. *Application deadline:* For fall admission, 2/1 priority date for domestic students, 2/1 priority date for international students; for spring admission, 8/1 for domestic students, 8/1 for international students. Applications are processed on a rolling basis. Application fee: $50 ($75 for international students). Electronic applications accepted. *Expenses:* Tuition, state resident: full-time $4,488; part-time $187 per credit hour. Tuition, nonresident: full-time $11,112; part-time $463 per credit hour. Required fees: $1,974. *Financial support:* In 2005–06, fellowships with partial tuition reimbursements (averaging $17,500 per year), research assistantships (averaging $16,000 per year) were awarded; Federal Work-Study, institutionally sponsored loans, scholarships/grants, traineeships, health care benefits, and unspecified assistantships also available. Financial award application deadline: 3/1; financial award applicants required to submit FAFSA. *Faculty research:* Behavioral toxicology and neurotoxicology, cellular toxicology, developmental and reproductive toxicology, applied veterinary and food toxicology. *Unit head:* Dr. Timothy D. Phillips, Chair, 979-845-5529, Fax: 979-862-4929, E-mail: tphillips@cvm.tamu.edu. *Application contact:* Kimberly D. Daniel, Program Assistant, 979-845-5529, Fax: 979-862-4929, E-mail: tox@cvm.tamu.edu.

Texas A&M University System Health Science Center, Graduate School of Biomedical Sciences, Department of Medical Pharmacology and Toxicology, College Station, TX 77840. Offers PhD. *Degree requirements:* For doctorate, thesis/dissertation. *Entrance requirements:* For doctorate, GRE General Test. *Faculty research:* Medical treatment of eye disease, fetal alcohol syndrome, Alzheimer's disease, glycme receptory steroids.

Texas Southern University, Graduate School, School of Science and Technology, Program in Environmental Toxicology, Houston, TX 77004-4584. Offers MS, PhD. Part-time programs available. *Faculty:* 7 full-time/adjunct (2 women). *Students:* 23 full-time (13 women), 16 part-time (7 women); includes 30 minority (26 African Americans, 3 Asian Americans or Pacific Islanders, 1 Hispanic American), 6 international. Average age 33. 15 applicants, 100% accepted, 10 enrolled. In 2005, 2 degrees awarded. *Degree requirements:* For master's and doctorate, thesis/dissertation. *Entrance requirements:* For master's, minimum GPA of 2.75; for doctorate, GRE, minimum GPA of 2.75. *Application deadline:* For fall admission, 7/15 for domestic students; for spring admission, 11/1 for domestic students. Application fee: $35 ($75 for international students). *Expenses: Contact institution.* Tuition and fees vary according to course load and degree level. *Financial support:* In 2005–06, 5 fellowships (averaging $20,000 per year), 5 research assistantships (averaging $18,000 per year), 5 teaching assistantships (averaging $9,000 per year) were awarded; institutionally sponsored loans, scholarships/grants, tuition waivers (partial), and unspecified assistantships also available. Financial award application deadline: 5/1; financial award applicants required to submit FAFSA. *Faculty research:* Air quality, water quality, soil remediation, computer modeling. *Unit head:* Dr. Bobby L. Wilson, Director, Provost and Professor of Chemistry, 713-313-4259, Fax: 713-313-4217, E-mail: wilson_bl@tsu.edu/bilson@earthlink.net. *Application contact:* Vera McDaniels, Coordinator, 713-313-4259, Fax: 713-313-4217, E-mail: veramcd@hotmail.com.

Texas Tech University, Graduate School, College of Arts and Sciences, Department of Environmental Toxicology, Lubbock, TX 79409. Offers MS, PhD. Part-time programs available. *Faculty:* 12 full-time (1 woman). *Students:* 37 full-time (15 women), 2 part-time (1 woman); includes 3 minority (1 American Indian/Alaska Native, 2 Hispanic Americans), 25 international. Average age 30. 24 applicants, 54% accepted, 4 enrolled. In 2005, 2 master's, 6 doctorates awarded. *Degree requirements:* For master's and doctorate, thesis/dissertation. *Entrance requirements:* For master's and doctorate, GRE General Test. Additional exam requirements/recommendations for international students: Required—TOEFL (minimum score 550 paper-based; 213 computer-based). *Application deadline:* Applications are processed on a rolling basis. Application fee: $50 ($60 for international students). Electronic applications accepted. *Expenses:* Tuition, state resident: full-time $4,296. Tuition, nonresident: full-time $10,920. Required fees: $1,992. Tuition and fees vary according to program. *Financial support:* In 2005–06, 13 students received support; teaching assistantships with partial tuition reimbursements available available. Financial award application deadline: 4/15. *Faculty research:* Terrestrial and aquatic toxicology, biochemical and developmental toxicology, advanced materials and high performance computing, countermeasures to biologic and chemical threats, molecular epidemiology and modeling. *Unit head:* Dr. Ronald J. Kendall, Director and Chairman, 806-885-4567, Fax: 806-885-2132, E-mail: ron.kendall@tiehh.ttu.edu. *Application contact:* Dr. Steve Cox, Graduate Program Adviser, 806-885-4567, Fax: 806-885-2132, E-mail: stephen.cox@ttu.edu.

Université de Montréal, Faculty of Graduate Studies, Program in Toxicology and Risk Analysis, Montréal, QC H3C 3J7, Canada. Offers DESS. *Students:* 5 full-time (3 women), 17 part-time (11 women). 28 applicants, 46% accepted, 7 enrolled. In 2005, 6 degrees awarded. *Application deadline:* For fall and spring admission, 2/1. For winter admission, 11/1 for domestic students. Applications are processed on a rolling basis. Application fee: $30. Electronic applications accepted. *Unit head:* Robert Tardif, Director, 514-343-6111 Ext. 1515, Fax: 514-343-6668, E-mail: robert.tardif@umontreal.ca. *Application contact:* Micheline Dessureault, Information Contact, 514-343-2280.

University at Albany, State University of New York, School of Public Health, Department of Environmental Health and Toxicology, Albany, NY 12222-0001. Offers environmental and occupational health (MS, PhD); environmental chemistry (MS, PhD); toxicology (MS, PhD). *Students:* 24 full-time (15 women), 17 part-time (9 women). Average age 31. In 2005, 1 master's, 2 doctorates awarded. *Degree requirements:* For master's and doctorate, thesis/dissertation. *Entrance requirements:* For master's and doctorate, GRE General Test, GRE Subject Test. Additional exam requirements/recommendations for international students: Required—TOEFL (minimum score 550 paper-based; 213 computer-based). *Application deadline:* For fall admission, 1/1 for domestic students. Applications are processed on a rolling basis. Application fee: $60. Electronic applications accepted. *Financial support:* Fellowships, research assistantships available. Financial award application deadline: 2/1. *Unit head:* Dr. Laurence Kaminsky, Chair, 518-473-7553. *Application contact:* Caitlin Reid, Assistant to the Chair, E-mail: reid@wadsworth.org.

See Close-Up on page 1187.

The University of Alabama at Birmingham, Graduate Programs in Joint Health Sciences, Birmingham, AL 35294. Offers biochemistry and molecular genetics (PhD), including biochemistry; cell biology (PhD), including cell biology, cellular and molecular biology, cellular and molecular physiology, neuroscience; genetics (PhD); microbiology (PhD); neurobiology (PhD); pathology (PhD); pharmacology and toxicology (PhD), including pharmacology, toxicology; physiology and biophysics (MSBMS, PhD), including basic medical sciences (MSBMS), biophysical sciences (PhD), integrative biomedical sciences (PhD). *Students:* 439 full-time (214 women), 6 part-time (2 women); includes 74 minority (36 African Americans, 4 American Indian/Alaska Native, 28 Asian Americans or Pacific Islanders, 6 Hispanic Americans), 164 international. Average age 28. 463 applicants, 37% accepted. In 2005, 9 master's, 53 doctorates awarded. *Entrance requirements:* For master's, GRE; for doctorate, GRE, interview. *Application deadline:* Applications are processed on a rolling basis. Application fee: $35 ($60 for international students). Electronic applications accepted. *Expenses:* Tuition, state resident: part-time $170 per credit hour. Tuition, nonresident: full-time $4,612; part-time $425 per credit hour. International tuition: $10,732 full-time. Required fees: $11 per credit hour. $124 per term. Tuition and fees vary according to course load, degree level and program. *Financial support:* Fellowships, career-related internships or fieldwork available. *Unit head:* Dr. Robert R. Rich, Vice President/Dean, School of Medicine, 205-934-1111, Fax: 205-934-0333, E-mail: rrich@uab.edu.

The University of Alabama at Birmingham, Graduate Programs in Joint Health Sciences, Department of Pharmacology and Toxicology, Graduate Training Program in Toxicology, Birmingham, AL 35294. Offers PhD. *Students:* 9 full-time (4 women). Application fee: $35 ($60 for international students). *Expenses:* Tuition, state resident: part-time $170 per credit hour. Tuition, nonresident: full-time $4,612; part-time $425 per credit hour. International tuition: $10,732 full-time. Required fees: $11 per credit hour. $124 per term. Tuition and fees vary according to course load, degree level and program.

The University of Arizona, Graduate College, Graduate Interdisciplinary Programs, Tucson, AZ 85721. Offers American Indian studies (MA, PhD); applied mathematics (MS, PMS, PhD), including applied mathematics (MS, PhD), mathematical sciences (PMS); arid land resource sciences (PhD); cancer biology (PhD); comparative cultural and literary studies (MA, PhD); epidemiology (MS, PhD); genetics (MS, PhD); gerontological studies (MS, Certificate); insect science (PhD); neuroscience (PhD); nutritional sciences (MS, PhD), including dietetics (MS), epidermalogical nutrition/public health nutrition (PhD), human/clinical nutrition (PhD), molecular nutrition (PhD), nutritional biochemistry; pharmacology and toxicology (PhD); physiological sciences (PhD); planning (MS); second language acquisition and teaching (PhD). Part-time programs available. *Entrance requirements:* Additional exam requirements/recommendations for international students: Required—TOEFL.

University of Arkansas for Medical Sciences, College of Medicine and Graduate School, Graduate Programs in Medicine, Department of Pharmacology and Toxicology, Little Rock, AR 72205-7199. Offers pharmacology (MS, PhD); toxicology (MS, PhD). *Faculty:* 18 full-time (2 women), 3 part-time/adjunct (0 women). *Students:* 13 full-time, 3 part-time. *Degree requirements:* For master's and doctorate, thesis/dissertation. *Entrance requirements:* For master's and doctorate, GRE General Test. Additional exam requirements/recommendations for international students: Required—TOEFL. Application fee: $0. *Financial support:* Research assistantships, teaching assistantships available. Support available to part-time students. *Unit head:* Dr. Nancy J. Rusch, Chair, 501-686-5510. *Application contact:* Dr. Philip R. Mayeux, Graduate Coordinator, 501-686-5510, E-mail: mayeuxphilipr@exchange.uams.edu.

See Close-Up on page 1189.

University of Arkansas for Medical Sciences, College of Medicine and Graduate School, Graduate Programs in Medicine, Interdisciplinary Toxicology Program, Little Rock, AR 72205-7199. Offers MS, PhD, MD/PhD. *Faculty:* 27 full-time (3 women), 23 part-time/adjunct (3 women). *Students:* 14 full-time, 1 part-time. *Degree requirements:* For master's and doctorate, thesis/dissertation. *Entrance requirements:* For master's and doctorate, GRE General Test. Additional exam requirements/recommendations for international students: Required—TOEFL. Application fee: $0. *Financial support:* Research assistantships available. Support available to part-time students. *Unit head:* Dr. Jack A. Hinson, Director, 501-686-5766, E-mail: jahinson@uams.edu.

University of California, Davis, Graduate Studies, Graduate Group in Pharmacology and Toxicology, Davis, CA 95616. Offers MS, PhD. *Faculty:* 77 full-time. *Students:* 53 full-time (30 women); includes 16 minority (1 American Indian/Alaska Native, 12 Asian Americans or Pacific Islanders, 2 Hispanic Americans), 12 international. Average age 31. 69 applicants, 35% accepted, 13 enrolled. In 2005, 2 master's, 7 doctorates awarded. Terminal master's awarded for partial completion of doctoral program. *Median time to degree:* Of those who began their doctoral program in fall 1997, 33.3% received their degree in 8 years or less. *Degree requirements:* For master's, comprehensive exam or thesis; for doctorate, thesis/dissertation, qualifying exam. *Entrance requirements:* For master's and doctorate, GRE General Test, minimum GPA of 3.0, course work in biochemistry and/or physiology. Additional exam requirements/recommendations for international students: Required—TOEFL (minimum score 550 paper-based; 213 computer-based). *Application deadline:* For fall admission, 1/15 for

domestic students, 1/15 for international students. Application fee: $60. Electronic applications accepted. *Financial support:* In 2005–06, 51 students received support, including 9 fellowships with full and partial tuition reimbursements available (averaging $13,536 per year), 21 research assistantships with full and partial tuition reimbursements available (averaging $16,611 per year), 1 teaching assistantship with partial tuition reimbursement available (averaging $15,082 per year); career-related internships or fieldwork, Federal Work-Study, institutionally sponsored loans, scholarships/grants, tuition waivers (full and partial), and unspecified assistantships also available. Financial award application deadline: 1/15; financial award applicants required to submit FAFSA. *Faculty research:* Respiratory, neurochemical, molecular, genetic, and ecological toxicology. *Unit head:* Alan Buckpitt, Graduate Group Chair, 530-752-7874, E-mail: arbuckpitt@ucdavis.edu. *Application contact:* Judy Erwin, Graduate Administrative Assistant, 530-752-4516, Fax: 530-752-3394, E-mail: gjerwin@ucdavis.edu.

University of California, Irvine, College of Medicine, Department of Pharmacology, Graduate Program in Pharmacology and Toxicology, Irvine, CA 92697. Offers MS, PhD, MD/PhD. *Degree requirements:* For doctorate, thesis/dissertation. *Entrance requirements:* For master's, GRE, minimum GPA of 3.0; for doctorate, GRE General Test, GRE Subject Test, minimum GPA of 3.0. Additional exam requirements/recommendations for international students: Required—TOEFL (minimum score 550 paper-based; 213 computer-based). Electronic applications accepted.

University of California, Riverside, Graduate Division, Program in Environmental Toxicology, Riverside, CA 92521-0102. Offers MS, PhD. *Faculty:* 39 full-time (8 women), 1 part-time/adjunct (0 women). *Students:* 34 full-time (19 women); includes 5 minority (3 Asian Americans or Pacific Islanders, 2 Hispanic Americans), 11 international. Average age 30. In 2005, 1 master's, 2 doctorates awarded. Terminal master's awarded for partial completion of doctoral program. *Degree requirements:* For master's, thesis; for doctorate, thesis/dissertation, qualifying exams. *Entrance requirements:* For master's and doctorate, GRE General Test, minimum GPA of 3.2. Additional exam requirements/recommendations for international students: Required—TOEFL (minimum score 550 paper-based; 213 computer-based); Recommended—TSE (minimum score 50). *Application deadline:* For fall admission, 7/1 for domestic students, 2/1 for international students. For winter admission, 9/1 for domestic students; for spring admission, 12/1 for domestic students. Applications are processed on a rolling basis. Application fee: $60 ($75 for international students). Electronic applications accepted. *Expenses:* Tuition, nonresident: full-time $14,694. Required fees: $9,009. Full-time tuition and fees vary according to program. *Financial support:* In 2005–06, research assistantships with tuition reimbursements (averaging $14,000 per year), teaching assistantships with tuition reimbursements (averaging $15,000 per year) were awarded; fellowships with tuition reimbursements, career-related internships or fieldwork, Federal Work-Study, institutionally sponsored loans, scholarships/grants, health care benefits, tuition waivers (full and partial), and unspecified assistantships also available. Financial award application deadline: 3/1; financial award applicants required to submit FAFSA. *Faculty research:* Cellular/molecular toxicology, atmospheric chemistry, bioremediation, carcinogenesis, mechanism of toxicity. Total annual research expenditures: $2 million. *Unit head:* Dr. David A. Eastmond, Director, 951-827-4497, Fax: 951-827-3087, E-mail: eastmond@ucrac1.ucr.edu. *Application contact:* Deidra Kornfeld, Graduate Program Assistant, 800-735-0717, Fax: 951-827-5517, E-mail: etox@ucr.edu.

University of California, Santa Cruz, Division of Graduate Studies, Division of Physical and Biological Sciences, Environmental Toxicology Department, Santa Cruz, CA 95064. Offers MS, PhD. *Faculty:* 5 full-time (2 women). *Students:* 11 full-time (6 women); includes 1 minority (Asian American or Pacific Islander), 3 international. 18 applicants, 33% accepted, 4 enrolled. In 2005, 3 master's, 3 doctorates awarded. *Expenses:* Tuition, nonresident: full-time $14,694. Required fees: $9,437. *Unit head:* Russ Flegal, Chair, 831-459-4719. *Application contact:* Sissy Madden, Information Contact, 831-459-4719, E-mail: madden@etox.ucsc.edu.

University of Cincinnati, Division of Research and Advanced Studies, College of Medicine, Graduate Programs in Biomedical Sciences, Department of Pediatrics Developmental Biology, Program in Teratology, Cincinnati, OH 45221. Offers PhD. *Degree requirements:* For doctorate, thesis/dissertation, qualifying exam. *Entrance requirements:* For doctorate, GRE General Test, minimum GPA of 3.2. Additional exam requirements/recommendations for international students: Required—TOEFL (minimum score 520 paper-based; 190 computer-based). Electronic applications accepted. *Faculty research:* Infectious disease, endocrinology, opthalmology, immunobiology.

See Close-Up on page 1199.

University of Colorado at Denver and Health Sciences Center, School of Pharmacy, Denver, CO 80262. Offers pharmacy (Pharm D, PhD), including pharmaceutical sciences (PhD), toxicology (PhD). *Accreditation:* ACPE (one or more programs are accredited). *Students:* 501 full-time (342 women), 14 part-time (11 women); includes 155 minority (31 African Americans, 4 American Indian/Alaska Native, 81 Asian Americans or Pacific Islanders, 39 Hispanic Americans), 3 international. Average age 29. 1,137 applicants, 15% accepted, 130 enrolled. In 2005, 151 degrees awarded. *Entrance requirements:* Minimum GPA of 2.75 in pre-pharmacy course work, 3 letters of recommendation. Additional exam requirements/recommendations for international students: Required—TOEFL (minimum score 550 paper-based; 213 computer-based). *Application deadline:* For fall admission, 12/1 for domestic students. Application fee: $50. *Expenses: Contact institution.* Tuition and fees vary according to degree level and program. *Financial support:* Fellowships, research assistantships, teaching assistantships, career-related internships or fieldwork, Federal Work-Study, and institutionally sponsored loans available. Support available to part-time students. Financial award application deadline: 3/15; financial award applicants required to submit FAFSA. *Faculty research:* Antiviral clinical pharmacology, beriatric pharmacology, pharmacoepidemiology, pharmacogenomics, smoking cessation. *Unit head:* Louis Diamond, Dean, 303-315-5055. *Application contact:* Beverly Brunson, Director, 303-315-6100.

University of Colorado at Denver and Health Sciences Center, School of Pharmacy, Programs in Pharmacy, Program in Toxicology, Denver, CO 80262. Offers PhD. *Students:* 22 full-time (8 women); includes 1 minority (Asian American or Pacific Islander), 6 international. 11 applicants, 27% accepted, 3 enrolled. *Degree requirements:* For doctorate, thesis/dissertation, comprehensive exam. *Entrance requirements:* For doctorate, GRE General Test, minimum GPA of 3.0, 4 letters of recommendation. Additional exam requirements/recommendations for international students: Required—TOEFL (minimum score 550 paper-based; 213 computer-based). Application fee: $50. *Expenses:* Tuition, state resident: full-time $11,730. Tuition, nonresident: full-time $22,980. Tuition and fees vary according to degree level and program. *Unit head:* Dr. Vasilis Vasiliou, Director, 303-315-0565, Fax: 303-315-0274, E-mail: vasilis.vasiliou@uchsc.edu.

University of Connecticut, Graduate School, School of Pharmacy, Department of Pharmaceutical Sciences, Field of Pharmaceutical Sciences, Graduate Program in Pharmacology and Toxicology, Storrs, CT 06269. Offers pharmacology (MS, PhD); toxicology (MS, PhD). *Faculty:* 12 full-time (4 women). *Students:* 5 full-time (4 women). Average age 25. 40 applicants, 8% accepted, 3 enrolled. In 2005, 1 degree awarded. Terminal master's awarded for partial completion of doctoral program. *Degree requirements:* For master's, thesis/dissertation, comprehensive exam; for doctorate, thesis/dissertation. *Entrance requirements:* For master's and doctorate, GRE General Test. Additional exam requirements/recommendations for international students: Required—TOEFL (minimum score 550 paper-based; 213 computer-based). *Application deadline:* For fall admission, 2/1 priority date for domestic students, 2/1 priority date for international students; for spring admission, 11/1 for domestic students, 10/1 for international students. Applications are processed on a rolling basis. Application fee: $55. Electronic applications accepted. *Expenses:* Tuition, state resident: part-time $444 per credit hour. Tuition, nonresident: part-time $1,154 per credit hour. Tuition and fees vary according to course load. *Financial support:* In 2005–06, 1 research assistantship with full tuition reimbursement, 3 teaching assistantships with full tuition reimbursements were awarded; fellowships,

Toxicology

University of Connecticut (continued)
career-related internships or fieldwork, Federal Work-Study, scholarships/grants, traineeships, health care benefits, and unspecified assistantships also available. Financial award application deadline: 2/1; financial award applicants required to submit FAFSA. *Application contact:* Leslie Lebel, Administrative Assistant, 860-486-4066, Fax: 860-486-4998, E-mail: phrmacy8@uconnvm. uconn.edu.

See Close-Up on page 1201.

University of Florida, College of Veterinary Medicine, Graduate Program in Veterinary Medical Sciences, Gainesville, FL 32611. Offers forensic toxicology (Certificate); veterinary medical sciences (MS, PhD), including forensic toxicology (MS). Postbaccalaureate distance learning degree programs offered (no on-campus study). *Faculty:* 114. In 2005, 29 master's, 9 doctorates awarded. Terminal master's awarded for partial completion of doctoral program. *Degree requirements:* For master's and doctorate, thesis/dissertation. *Entrance requirements:* For master's and doctorate, GRE General Test, minimum GPA of 3.0. Additional exam requirements/recommendations for international students: Required—TOEFL (minimum score 550 paper-based; 213 computer-based). *Application deadline:* For fall admission, 6/1 for domestic students. Applications are processed on a rolling basis. Application fee: $30. Electronic applications accepted. *Expenses: Contact institution.* Tuition and fees vary according to program. *Financial support:* In 2005–06, 1 fellowship with partial tuition reimbursement (averaging $14,000 per year), 32 research assistantships with partial tuition reimbursements (averaging $15,623 per year), 14 teaching assistantships with partial tuition reimbursements (averaging $15,921 per year) were awarded; institutionally sponsored loans also available. *Unit head:* Dr. Charles H. Courtney, Associate Dean for Research and Graduate Studies, 352-392-4700 Ext. 5100, Fax: 352-392-8351, E-mail: courtneyc@mail.vetmed.ufl.edu. *Application contact:* Dr. Louis Archbald, Coordinator, 352-392-4700 Ext. 5641, Fax: 352-392-8351, E-mail: archbaldl@ mail.vetmed.ufl.edu.

University of Georgia, College of Pharmacy, Department of Pharmaceutical and Biomedical Sciences, Athens, GA 30602. Offers medicinal chemistry (MS, PhD); pharmaceutics (MS, PhD); pharmacology (MS, PhD); toxicology (MS, PhD). *Faculty:* 17 full-time (4 women). *Students:* 8 full-time, 3 part-time (all women); includes 3 minority (all African Americans), 3 international. 132 applicants, 11% accepted, 7 enrolled. In 2005, 1 degree awarded. *Median time to degree:* Of those who began their doctoral program in fall 1997, 100% received their degree in 8 years or less. *Degree requirements:* For master's, thesis; for doctorate, one foreign language, thesis/dissertation. *Entrance requirements:* For master's and doctorate, GRE General Test, minimum GPA of 3.0. Additional exam requirements/recommendations for international students: Required—TOEFL. *Application deadline:* For fall admission, 1/2 priority date for domestic students, 1/2 priority date for international students. Application fee: $50. Electronic applications accepted. *Financial support:* In 2005–06, fellowships with full tuition reimbursements (averaging $18,000 per year), 10 research assistantships with full tuition reimbursements (averaging $16,500 per year), 35 teaching assistantships with full tuition reimbursements (averaging $16,500 per year) were awarded; career-related internships or fieldwork, institutionally sponsored loans, tuition waivers (partial), and unspecified assistantships also available. Financial award application deadline: 1/1. *Faculty research:* Cancer and infectious diseases, drug delivery, neuropharmacology, cardiovascular pharmacology, bioanalytical chemistry, structural biology. *Unit head:* Dr. Vasu Nair, Head, 706-542-5610, Fax: 706-542-3398, E-mail: vnair@rx.uga.edu. *Application contact:* Dr. Anthony C. Capomacchia, Graduate Coordinator, 706-542-5403, Fax: 706-542-3398, E-mail: tcapomac@rx.uga.edu.

University of Georgia, College of Veterinary Medicine and Graduate School, Graduate Programs in Veterinary Medicine, Interdisciplinary Graduate Program in Toxicology, Athens, GA 30602. Offers MS, PhD. Postbaccalaureate distance learning degree programs offered (minimal on-campus study). *Students:* 7 applicants, 14% accepted, 0 enrolled. In 2005, 2 degrees awarded. *Median time to degree:* Of those who began their doctoral program in fall 1997, 100% received their degree in 8 years or less. *Degree requirements:* For master's, thesis; for doctorate, one foreign language, thesis/dissertation. *Entrance requirements:* For master's and doctorate, GRE General Test. *Application deadline:* For fall admission, 7/1 for domestic students; for spring admission, 11/15 for domestic students. Application fee: $50. Electronic applications accepted. *Financial support:* In 2005–06, fellowships with full tuition reimbursements (averaging $14,600 per year), research assistantships with full tuition reimbursements (averaging $16,000 per year) were awarded; scholarships/grants and unspecified assistantships also available. Financial award application deadline: 5/1. *Faculty research:* Neurotoxicology, cell signal modulation, biological toxins, PBPK modeling, environmental risk assessment. Total annual research expenditures: $3 million. *Unit head:* Dr. Raghubir Sharma, Graduate Coordinator, 706-542-2788, Fax: 706-542-3398, E-mail: rpsharma@vet.uga.edu.

University of Guelph, Graduate Program Services, Ontario Agricultural College, Department of Environmental Biology, Guelph, ON N1G 2W1, Canada. Offers entomology (M Sc, PhD); environmental biology and biotechnology (M Sc, PhD); environmental toxicology (M Sc, PhD); plant and forest systems (M Sc, PhD); plant pathology (M Sc, PhD). Part-time programs available. *Faculty:* 20 full-time (2 women), 21 part-time/adjunct (2 women). *Students:* 60 full-time (32 women), 13 part-time (8 women). 37 applicants, 35% accepted, 13 enrolled. In 2005, 12 master's, 7 doctorates awarded. *Median time to degree:* Of those who began their doctoral program in fall 1997, 79.5% received their degree in 8 years or less. *Degree requirements:* For master's, thesis/dissertation, registration; for doctorate, thesis/dissertation, comprehensive exam, registration. *Entrance requirements:* For master's, minimum B- average during previous 2 years of course work; for doctorate, minimum B average. Additional exam requirements/recommendations for international students: Required—TOEFL or IELTS. *Application deadline:* Applications are processed on a rolling basis. Application fee: $75. Electronic applications accepted. *Financial support:* In 2005–06, research assistantships (averaging $16,500 per year), teaching assistantships (averaging $4,606 per year) were awarded; fellowships. *Faculty research:* Entomology, environmental microbiology and biotechnology, environmental toxicology, forest ecology, plant pathology. Total annual research expenditures: $3 million. *Unit head:* Dr. M. A. Dixon, Chair, 519-824-4120 Ext. 52555, Fax: 519-837-0442, E-mail: mdixon@ uoguelph.ca. *Application contact:* Dr. H. Lee, Admissions Coordinator, 519-824-4120 Ext. 53828, Fax: 519-837-0442, E-mail: hlee@uoguelph.ca.

University of Guelph, Ontario Veterinary College and Graduate Program Services, Graduate Programs in Veterinary Sciences, Department of Biomedical Sciences, Guelph, ON N1G 2W1, Canada. Offers morphology (M Sc, DV Sc, PhD); pharmacology (M Sc, DV Sc, PhD); physiology (M Sc, DV Sc, PhD); toxicology (M Sc, DV Sc, PhD). Part-time programs available. *Faculty:* 25. *Students:* 43 (27 women). In 2005, 6 master's, 2 doctorates awarded. *Median time to degree:* Of those who began their doctoral program in fall 1997, 100% received their degree in 8 years or less. *Degree requirements:* For master's, thesis/dissertation; for doctorate, thesis/dissertation, comprehensive exam. *Entrance requirements:* For master's, honors B Sc, minimum 75% average in last 20 courses; for doctorate, M Sc with thesis from accredited institution. Additional exam requirements/recommendations for international students: Required—TOEFL (minimum score 550 paper-based; 213 computer-based). *Application deadline:* Applications are processed on a rolling basis. Application fee: $75. Electronic applications accepted. *Financial support:* Fellowships, research assistantships, teaching assistantships available. *Faculty research:* Cellular morphology; endocrine, vascular and reproductive physiology; clinical pharmacology; veterinary toxicology; developmental biology. Total annual research expenditures: $3.5 million. *Unit head:* Dr. N. Maclusky, Chair, 519-824-4120 Ext. 54904, Fax: 511-767-1450. *Application contact:* Dr. G. Kirby, Graduate Coordinator, 519-824-4120 Ext. 54948, Fax: 519-767-1450, E-mail: gkirby@uoguelph.ca.

University of Kansas, Graduate School, School of Pharmacy, Department of Pharmacology and Toxicology, Program in Pharmacology and Toxicology, Lawrence, KS 66045. Offers MS, PhD. *Students:* 12 full-time (6 women), 2 part-time (both women); includes 1 minority (Hispanic American), 6 international. Average age 28. 49 applicants, 10% accepted. In 2005, 2 degrees awarded. *Degree requirements:* For doctorate, thesis/dissertation, comprehensive exam.

Entrance requirements: For master's and doctorate, GRE. Additional exam requirements/recommendations for international students: Required—TOEFL. *Application deadline:* For fall admission, 1/15 priority date for domestic students, 1/15 priority date for international students. Applications are processed on a rolling basis. Electronic applications accepted. *Expenses:* Tuition, state resident: full-time $4,859. Tuition, nonresident: full-time $12,000. Required fees: $589. Tuition and fees vary according to program. *Financial support:* Fellowships, research assistantships, teaching assistantships available. Financial award application deadline: 1/15.

University of Kansas, Graduate Studies Medical Center, Interdisciplinary Graduate Program in Biomedical Sciences, Department of Pharmacology, Toxicology and Therapeutics, Lawrence, KS 66045. Offers pharmacology (MS, PhD); toxicology (MS, PhD). Part-time programs available. *Faculty:* 17. *Students:* 2 full-time (0 women), 13 part-time (6 women), 6 international. Average age 28. In 2005, 5 degrees awarded. Terminal master's awarded for partial completion of doctoral program. *Degree requirements:* For master's, thesis, comprehensive exam; for doctorate, one foreign language, thesis/dissertation, comprehensive exam. *Entrance requirements:* For master's and doctorate, GRE General Test. Additional exam requirements/recommendations for international students: Required—TOEFL, TSE. *Application deadline:* For fall admission, 1/15 for domestic students. Applications are processed on a rolling basis. Application fee: $0. Electronic applications accepted. *Expenses:* Tuition, state resident: full-time $4,859. Tuition, nonresident: full-time $12,000. Required fees: $589. Tuition and fees vary according to program. *Financial support:* Fellowships, research assistantships with partial tuition reimbursements, teaching assistantships with full and partial tuition reimbursements, Federal Work-Study, institutionally sponsored loans, scholarships/grants, and traineeships available. Support available to part-time students. Financial award application deadline: 3/30; financial award applicants required to submit FAFSA. *Faculty research:* Cardiovascular pharmacology, neuropharmacology, molecular neurotoxicology, cancer chemotherapy, placental pharmacology/physiology. *Unit head:* Dr. Curtis Klaasen, Chairman, 913-588-7140, Fax: 913-588-7501, E-mail: cklaasen@kumc.edu. *Application contact:* Dorothy McGregor, Director of Admissions, 913-588-7526, Fax: 913-588-7501, E-mail: dmcgregor@kumc.edu.

University of Kentucky, Graduate School, Graduate School Programs from the College of Medicine, Program in Toxicology, Lexington, KY 40506-0032. Offers MS, PhD. *Faculty:* 13 full-time (2 women), 1 part-time/adjunct (0 women). *Students:* 43 full-time (31 women), 2 part-time (both women); includes 5 minority (2 Asian Americans or Pacific Islanders, 3 Hispanic Americans), 24 international. Average age 30. 43 applicants, 72% accepted, 21 enrolled. In 2005, 2 master's, 7 doctorates awarded. Terminal master's awarded for partial completion of doctoral program. *Median time to degree:* Of those who began their doctoral program in fall 1997, 93% received their degree in 8 years or less. *Degree requirements:* For master's, thesis optional; for doctorate, thesis/dissertation, comprehensive exam. *Entrance requirements:* For master's, GRE General Test, minimum undergraduate GPA of 2.5; for doctorate, GRE General Test, minimum graduate GPA of 3.0. Additional exam requirements/recommendations for international students: Required—TOEFL (minimum score 550 paper-based; 213 computer-based). *Application deadline:* For fall admission, 7/17 priority date for domestic students, 2/1 priority date for international students; for spring admission, 12/13 priority date for domestic students, 6/15 priority date for international students. Applications are processed on a rolling basis. Application fee: $40 ($55 for international students). Electronic applications accepted. *Expenses:* Tuition, state resident: full-time $6,308; part-time $331 per credit hour. Tuition, nonresident: full-time $13,968; part-time $756 per credit hour. Tuition and fees vary according to course load, degree level and program. *Financial support:* In 2005–06, 6 fellowships with full tuition reimbursements (averaging $5,375 per year), 38 research assistantships with full tuition reimbursements (averaging $21,000 per year) were awarded; teaching assistantships with full tuition reimbursements, Federal Work-Study, institutionally sponsored loans, scholarships/grants, traineeships, health care benefits, tuition waivers (partial), and unspecified assistantships also available. Support available to part-time students. Financial award application deadline: 3/15. *Faculty research:* Chemical carcinogenesis, immunotoxicology, neurotoxicology, metabolism and disposition, gene regulation. Total annual research expenditures: $1.1 million. *Unit head:* Dr. Zhigang Wang, Director of Graduate Studies, 859-323-5784, Fax: 859-257-1059, E-mail: zwang@pop.uky.edu. *Application contact:* Dr. Brian Jackson, Senior Associate Dean, 859-257-8176, Fax: 859-323-1928.

See Close-Up on page 1207.

University of Louisville, School of Medicine, Department of Pharmacology and Toxicology, Louisville, KY 40292-0001. Offers MS, PhD. *Students:* 31 full-time (18 women), 19 part-time (10 women); includes 8 minority (7 African Americans, 1 Asian American or Pacific Islander), 18 international. Average age 28. In 2005, 13 master's, 4 doctorates awarded. *Degree requirements:* For master's and doctorate, thesis/dissertation. *Entrance requirements:* For master's and doctorate, GRE General Test. *Application deadline:* For fall admission, 1/15 for domestic students. Applications are processed on a rolling basis. Application fee: $50. *Expenses:* Tuition, state resident: full-time $6,006; part-time $334 per credit hour. Tuition, nonresident: full-time $16,554; part-time $920 per credit hour. Tuition and fees vary according to course load, degree level and program. *Financial support:* Fellowships with tuition reimbursements, research assistantships with tuition reimbursements available. *Unit head:* Dr. David W. Hein, Chair, 502-852-5141, Fax: 502-852-7868, E-mail: dhein@louisville.edu. *Application contact:* Heddy R. Rubin-Teiter, Contact, 502-852-5741, Fax: 502-852-7868, E-mail: hrrubi01@gwise. louisville.edu.

University of Maryland, Graduate School, Graduate Programs in Medicine, Department of Toxicology, Baltimore, MD 21201. Offers MS, PhD, MD/PhD. Part-time programs available. *Students:* 15 full-time (5 women), 10 part-time (9 women); includes 6 minority (2 African Americans, 3 Asian Americans or Pacific Islanders, 1 Hispanic American), 4 international. Average age 32. 19 applicants, 32% accepted, 4 enrolled. In 2005, 1 master's, 6 doctorates awarded. *Degree requirements:* For doctorate, thesis/dissertation. *Entrance requirements:* For master's and doctorate, GRE General Test, GRE Subject Test, minimum GPA of 3.0. Additional exam requirements/recommendations for international students: Required—TOEFL, TOEFL or IELTS; Recommended—IELT. *Application deadline:* For fall admission, 7/1 for domestic students, 1/15 for international students. Application fee: $50. Electronic applications accepted. *Expenses:* Tuition, state resident: full-time $8,079; part-time $409 per credit hour. Tuition, nonresident: full-time $18,384; part-time $731 per credit hour. Required fees: $695; $10 per credit hour. Tuition and fees vary according to degree level and program. *Financial support:* Fellowships, research assistantships, teaching assistantships available. Financial award application deadline: 2/15. *Unit head:* Dr. Sanford Stass, Chair, 410-328-1237, E-mail: pharris@umm.edu. *Application contact:* Dr. Katherine Squibb, Program Director, 410-706-8196, E-mail: ksquibb@umaryland.edu.

University of Maryland, School of Medicine, Graduate Program in Life Sciences, Baltimore, MD 21201. Offers biochemistry (MS, PhD); epidemiology (MS, PhD); gerontology (PhD); microbiology (PhD); molecular and cell biology (MS); molecular medicine (PhD); neuroscience (MS, PhD); pharmacology (MS); physiology (MS); rehabilitation sciences (PhD); toxicology (MS, PhD). *Faculty:* 245 full-time (52 women). *Students:* 268 full-time (165 women), 46 part-time (30 women); includes 43 minority (25 African Americans, 13 Asian Americans or Pacific Islanders, 5 Hispanic Americans), 86 international. 435 applicants, 26% accepted, 61 enrolled. In 2005, 18 master's, 46 doctorates awarded. *Median time to degree:* Of those who began their doctoral program in fall 1997, 99% received their degree in 8 years or less. *Degree requirements:* For master's, registration; for doctorate, thesis/dissertation, lab rotations, comprehensive exam, registration. *Entrance requirements:* For master's and doctorate, GRE or MCAT. Additional exam requirements/recommendations for international students: Required—TOEFL (minimum score 550 paper-based; 213 computer-based). *Application deadline:* For winter admission, 1/15 for domestic students. Applications are processed on a rolling basis. Application fee: $50. Electronic applications accepted. *Expenses:* Tuition, state resident: full-time $8,079; part-time $409 per credit hour. Tuition, nonresident: full-time $18,384; part-time $731 per credit hour. Required fees: $695; $10 per credit hour. Tuition and fees vary according to degree level and program. *Financial support:* In 2005–06, 30 fellowships with full tuition reimbursements (averaging $23,000 per year), 22 research assistantships with full tuition reimbursements (averaging $23,000 per year) were awarded; health care benefits also available. *Faculty research:* Cancer, reproduction, neuroscience, cardiovascular, immunology. *Unit head:* Dr. Margaret Merryl McCarthy, Assistant Dean for Graduate Studies, 410-706-2655, Fax: 410-706-8341, E-mail: mmcarthy@umaryland.edu.

University of Maryland Eastern Shore, Graduate Programs, Department of Natural Sciences, Program in Toxicology, Princess Anne, MD 21853-1299. Offers MS, PhD. *Expenses:* Tuition, state resident: part-time $216 per credit. Tuition, nonresident: part-time $392 per credit. Required fees: $40 per term.

University of Michigan, School of Public Health, Department of Environmental Health Sciences, Toxicology Training Program, Ann Arbor, MI 48109. Offers MPH, MS, PhD. MS and PhD offered through the Horace H. Rackham School of Graduate Studies. *Degree requirements:* For doctorate, oral defense of dissertation, preliminary exam. *Entrance requirements:* For master's and doctorate, GRE General Test. *Application deadline:* For fall admission, 2/1 for domestic students. Applications are processed on a rolling basis. Application fee: $60 ($75 for international students). Electronic applications accepted. *Expenses:* Tuition, state resident: full-time $14,082; part-time $894 per credit hour. Tuition, nonresident: full-time $28,500; part-time $1,675 per credit hour. Required fees: $189; $189 per unit. *Financial support:* Fellowships, research assistantships, teaching assistantships available. Financial award application deadline: 3/1. *Application contact:* Susan Crawford, Student Services Coordinator, E-mail: sph.ehs.inquiries@umich.edu.

University of Minnesota, Duluth, Graduate School, Program in Toxicology, Duluth, MN 55812-2496. Offers MS, PhD. *Faculty:* 11 full-time (3 women). *Students:* 8 full-time (2 women); includes 4 minority (1 American Indian/Alaska Native, 2 Asian Americans or Pacific Islanders, 1 Hispanic American). Average age 30. 9 applicants, 33% accepted, 0 enrolled. In 2005, 1 master's, 1 doctorate awarded. Terminal master's awarded for partial completion of doctoral program. *Degree requirements:* For master's, thesis; for doctorate, thesis/dissertation, oral preliminary and final exams. *Entrance requirements:* For master's, GRE General Test, minimum GPA of 3.0; for doctorate, GRE General Test. Additional exam requirements/recommendations for international students: Required—TOEFL (minimum score 550 paper-based; 213 computer-based). *Application deadline:* For fall admission, 6/15 for domestic students, 6/15 for international students; for spring admission, 10/15 for domestic students, 10/15 for international students. Applications are processed on a rolling basis. Application fee: $55 ($75 for international students). Electronic applications accepted. *Financial support:* In 2005–06, 4 students received support, including 1 fellowship with full tuition reimbursement available (averaging $10,600 per year), research assistantships with full tuition reimbursements available (averaging $18,000 per year); teaching assistantships, institutionally sponsored loans, scholarships/grants, and unspecified assistantships also available. Support available to part-time students. Financial award application deadline: 4/1. *Faculty research:* Structure activity correlations, neurotoxicity, aquatic toxicology, biochemical mechanisms, immunotoxicology. *Unit head:* Dr. Kendall B. Wallace, Director of Graduate Studies, 218-726-8899, Fax: 218-726-8014, E-mail: kwallace@d.umn.edu. *Application contact:* Luanne G. Petcoff, Office Specialist, 218-726-8892, Fax: 218-726-7559, E-mail: lpetcoff@d.umn.edu.

University of Minnesota, Twin Cities Campus, School of Public Health, Division of Environmental Health Sciences, Area in Environmental Toxicology, Minneapolis, MN 55455-0213. Offers MPH, MS, PhD. *Degree requirements:* For doctorate, thesis/dissertation. *Entrance requirements:* For master's and doctorate, GRE General Test. *Application deadline:* For fall admission, 3/1 for domestic students. Applications are processed on a rolling basis. Application fee: $55 ($75 for international students). *Expenses:* Tuition, state resident: full-time $8,748; part-time $729 per credit. Tuition, nonresident: full-time $15,848; part-time $1,321 per credit. Full-time tuition and fees vary according to class time, course load, program and reciprocity agreements. *Financial support:* Fellowships, research assistantships available. Financial award application deadline: 3/1. *Application contact:* Kathy Soupir, Major Coordinator, 612-625-0622, Fax: 612-626-4837, E-mail: soupi001@umn.edu.

University of Mississippi, Graduate School, School of Pharmacy, Department of Pharmacology, Oxford, University, MS 38677. Offers pharmacology (MS, PhD); toxicology (PhD). In 2005, 1 degree awarded. *Degree requirements:* For master's and doctorate, thesis/dissertation. *Entrance requirements:* For master's, GRE General Test, minimum GPA of 3.0; for doctorate, GRE General Test. Additional exam requirements/recommendations for international students: Required—TOEFL. *Application deadline:* For fall admission, 4/1 for domestic students. Applications are processed on a rolling basis. Application fee: $25. *Expenses:* Tuition, state resident: full-time $4,320; part-time $240 per credit hour. Tuition, nonresident: full-time $9,744; part-time $301 per credit hour. Tuition and fees vary according to program. *Financial support:* Scholarships/grants available. Financial award application deadline: 3/1; financial award applicants required to submit FAFSA. *Faculty research:* Behavioral and biochemical pharmacology. *Unit head:* Dr. Robert C. Speth, Chairman, 662-915-7330.

University of Mississippi Medical Center, School of Graduate Studies in the Health Sciences, Department of Pharmacology and Toxicology, Jackson, MS 39216-4505. Offers pharmacology (MS, PhD); toxicology (MS, PhD). *Faculty:* 11 full-time (1 woman), 1 part-time/adjunct (0 women). *Students:* 21 full-time (6 women), 1 (woman) part-time; includes 4 African Americans, 14 international. Average age 30. 14 applicants, 29% accepted, 4 enrolled. In 2005, 4 degrees awarded. Terminal master's awarded for partial completion of doctoral program. *Degree requirements:* For master's, thesis; for doctorate, thesis/dissertation, first authored publication. *Entrance requirements:* For master's and doctorate, GRE General Test, minimum GPA of 3.0. *Application deadline:* For fall admission, 6/1 for domestic students. Applications are processed on a rolling basis. Application fee: $10. *Financial support:* In 2005–06, 23 research assistantships (averaging $16,557 per year) were awarded Financial award application deadline: 4/1. *Faculty research:* Neuropharmacology, environmental toxicology, aging, immunopharmacology, cardiovascular pharmacology. Total annual research expenditures: $5 million. *Unit head:* Dr. I. K. Ho, Dean, 601-984-1600, Fax: 601-984-1637, E-mail: iho@pharmacology.umsmed.edu. *Application contact:* Dr. Jerry M. Farley, Director, 601-984-1630, Fax: 601-984-1637, E-mail: jfarley@pharmacology.umsmed.edu.

See Close-Up on page 1211.

The University of Montana, Graduate School, School of Pharmacy and Allied Health Sciences, Department of Biomedical and Pharmaceutical Sciences, Missoula, MT 59812-0002. Offers pharmaceutical sciences (MS); pharmacology (PhD); toxicology (MS, PhD). *Accreditation:* ACPE. *Degree requirements:* For master's, oral defense of thesis; for doctorate, research dissertation defense. *Entrance requirements:* For master's and doctorate, GRE General Test. Additional exam requirements/recommendations for international students: Required—TOEFL (minimum score 540 paper-based; 210 computer-based). Electronic applications accepted. *Expenses:* Tuition, state resident: part-time $267 per credit. Tuition, nonresident: part-time $665 per credit. Part-time tuition and fees vary according to course load and degree level. *Faculty research:* Cardiovascular pharmacology, medicinal chemistry, neurosciences, environmental toxicology, pharmacogenetics, cancer.

University of Nebraska–Lincoln, Graduate College, Interdepartmental Area of Toxicology, Lincoln, NE 68588. Offers MS, PhD. *Entrance requirements:* Additional exam requirements/recommendations for international students: Required—TOEFL (minimum score 550 paper-based; 213 computer-based). Electronic applications accepted.

University of Nebraska Medical Center, Graduate Studies, Graduate Program in Toxicology, Omaha, NE 68198. Offers MS, PhD. *Faculty:* 55 full-time (10 women). *Students:* 13 full-time (6 women); includes 4 minority (3 Asian Americans or Pacific Islanders, 1 Hispanic American), 4 international. Average age 31. 2 applicants, 50% accepted, 1 enrolled. In 2005, 1 degree awarded. Terminal master's awarded for partial completion of doctoral program. *Degree requirements:* For master's and doctorate, thesis/dissertation, comprehensive exam (for

some programs). *Entrance requirements:* For master's, GRE General Test, bachelor's degree in chemistry, biology, biochemistry or related area; for doctorate, GRE General Test, BS in chemistry, biology, biochemistry or related area. Additional exam requirements/recommendations for international students: Required—TOEFL (minimum score 550 paper-based; 213 computer-based). *Application deadline:* For fall admission, 6/1 for domestic students, 4/1 for international students; for spring admission, 10/1 for domestic students, 8/1 for international students. Applications are processed on a rolling basis. Application fee: $45. Electronic applications accepted. *Expenses:* Tuition, area resident: Part-time $200 per hour. Tuition, nonresident: part-time $538 per hour. Required fees: $308; $59 per term. *Financial support:* In 2005–06, 2 students received support; fellowships with full tuition reimbursements available, research assistantships, teaching assistantships, institutionally sponsored loans, traineeships, health care benefits, and tuition waivers (full) available. *Faculty research:* Mechanisms of carcinogenesis, alcohol and metal toxicity, DNA damage, human molecular genetics, agrochemicals in soil and water. *Unit head:* Dr. Ercole L. Cavalieri, Director, 402-559-7237, Fax: 402-559-8068, E-mail: uncetox@unmc.edu. *Application contact:* Sherry Cherek, Coordinator, 402-559-8924, Fax: 402-559-8068, E-mail: uncetox@unmc.edu.

University of New Mexico, School of Medicine, Biomedical Sciences Graduate Program, Albuquerque, NM 87131-5196. Offers biochemistry and molecular biology (MS, PhD); cell biology and physiology (MS, PhD); molecular genetics and microbiology (MS, PhD); neuroscience (MS, PhD); pathology (MS, PhD); toxicology (MS, PhD). Part-time programs available. Terminal master's awarded for partial completion of doctoral program. *Degree requirements:* For master's, thesis/dissertation; for doctorate, thesis/dissertation, comprehensive exam. *Entrance requirements:* For master's and doctorate, GRE General Test, minimum undergraduate GPA of 3.0. Additional exam requirements/recommendations for international students: Required—TOEFL. Electronic applications accepted. *Expenses:* Tuition, state resident: full-time $5,676. Tuition, nonresident: full-time $14,974; part-time $238 per credit hour. Required fees: $385 per term. Tuition and fees vary according to course load and program. *Faculty research:* Signal transduction, infectious disease, biology of cancer, structural biology, neuroscience.

The University of North Carolina at Chapel Hill, School of Medicine, Curriculum in Toxicology, Chapel Hill, NC 27599. Offers MS, PhD. Terminal master's awarded for partial completion of doctoral program. *Degree requirements:* For master's and doctorate, thesis/dissertation, comprehensive exam, registration. *Entrance requirements:* For doctorate, GRE General Test. Electronic applications accepted. *Faculty research:* Molecular and cellular toxicology, carcinogenesis, neurotoxicology, pulmonary toxicology, developmental toxicology.

See Close-Up on page 1213.

University of Prince Edward Island, Atlantic Veterinary College, Graduate Program in Veterinary Medicine, Charlottetown, PE C1A 4P3, Canada. Offers anatomy (M Sc, PhD); bacteriology (M Sc, PhD); clinical pharmacology (M Sc, PhD); clinical sciences (M Sc, PhD); epidemiology (M Sc, PhD), including reproduction; fish health (M Sc, PhD); food animal nutrition (M Sc, PhD); immunology (M Sc, PhD); microanatomy (M Sc, PhD); parasitology (M Sc, PhD); pathology (M Sc, PhD); pharmacology (M Sc, PhD); physiology (M Sc, PhD); toxicology (M Sc, PhD); veterinary science (M Vet Sc); virology (M Sc, PhD). Part-time programs available. *Faculty:* 76 full-time (25 women), 49 part-time/adjunct (8 women). *Students:* 54 full-time (32 women), 2 part-time. Average age 30. In 2005, 7 master's, 6 doctorates awarded. *Degree requirements:* For master's and doctorate, thesis/dissertation. *Entrance requirements:* For master's, DVM, B Sc honors degree, or equivalent; for doctorate, M Sc. *Application deadline:* Applications are processed on a rolling basis. Application fee: $50. *Expenses:* Contact institution. Tuition charges are reported in Canadian dollars. Part-time tuition and fees vary according to course level, degree level, campus/location, program and student level. *Financial support:* In 2005–06, 4 fellowships (averaging $25,000 Canadian dollars per year), 4 research assistantships (averaging $16,500 Canadian dollars per year) were awarded; career-related internships or fieldwork also available. *Faculty research:* Animal health management, infectious diseases, fin fish and shellfish health, basic biomedical sciences, ecosystem health. Total annual research expenditures: $1.2 million Canadian dollars. *Unit head:* Dr. James Bellamy, Associate Dean of Graduate Studies and Research, 902-566-0856, E-mail: bellamy@upei.ca. *Application contact:* Cheryl Gaudet, Registrar's Office, 902-566-0781, Fax: 902-566-0795, E-mail: registrar@upei.ca.

University of Puerto Rico, Medical Sciences Campus, School of Medicine, Division of Graduate Studies, Department of Pharmacology and Toxicology, San Juan, PR 00936-5067. Offers MS, PhD. *Degree requirements:* For master's, one foreign language, thesis/dissertation; for doctorate, one foreign language, thesis/dissertation, comprehensive exam. *Entrance requirements:* For master's and doctorate, GRE General Test, GRE Subject Test, interview, minimum GPA of 3.0, 3 letters of recommendation. *Expenses:* Tuition, state resident: full-time $3,600; part-time $100 per credit hour. Tuition, nonresident: full-time $4,655. Required fees: $1,734. Tuition and fees vary according to class time, degree level and program. *Faculty research:* Cardiovascular, central nervous system, and endocrine pharmacology; anti-cancer drugs; toxicology, sodium pump.

University of Rhode Island, Graduate School, College of Pharmacy, Graduate Programs in Pharmacy, Department of Biomedical and Pharmaceutical Sciences, Kingston, RI 02881. Offers medicinal chemistry and pharmacognosy (MS, PhD); pharmaceutics and pharmacokinetics (MS, PhD); pharmacology and toxicology (MS, PhD). *Expenses:* Tuition, state resident: full-time $5,522; part-time $307 per credit. Tuition, nonresident: full-time $15,992; part-time $888 per credit. Required fees: $1,786; $73 per credit. One-time fee: $80 part-time. *Unit head:* Dr. Clinton O. Chichester, Chair, 401-874-5034.

University of Rochester, School of Medicine and Dentistry, Graduate Programs in Medicine and Dentistry, Department of Environmental Medicine, Program in Toxicology, Rochester, NY 14627-0250. Offers MS, PhD. *Degree requirements:* For doctorate, thesis/dissertation, qualifying exam. *Entrance requirements:* For master's and doctorate, GRE General Test.

University of Saskatchewan, College of Graduate Studies and Research, Toxicology Centre, Saskatoon, SK S7N 5A2, Canada. Offers M Sc, PhD, Diploma. *Degree requirements:* For master's and doctorate, thesis/dissertation, registration. *Entrance requirements:* Additional exam requirements/recommendations for international students: Required—TOEFL.

University of South Alabama, Graduate School, Program in Environmental Toxicology, Mobile, AL 36688-0002. Offers MS. *Students:* 5 full-time (1 woman), 4 international. *Expenses:* Tuition, state resident: full-time $4,008. Tuition, nonresident: full-time $8,016. Required fees: $692. *Unit head:* Dr. B. Keith Harrison, Chair, 251-460-6160.

University of Southern California, School of Pharmacy and Graduate School, Graduate Programs in Pharmacy, Graduate Program in Molecular Pharmacology and Toxicology, Los Angeles, CA 90089. Offers MS, PhD, Pharm D/PhD. *Degree requirements:* For master's and doctorate, thesis/dissertation. *Entrance requirements:* For master's and doctorate, GRE General Test. *Expenses:* Tuition: Full-time $25,416; part-time $1,059 per unit. Required fees: $484; $484 per year. Tuition and fees vary according to course load and program.

The University of Texas Health Science Center at Houston, Graduate School of Biomedical Sciences, Program in Toxicology, Houston, TX 77225-0036. Offers MS, PhD, MD/PhD. *Faculty:* 18 full-time (4 women). *Students:* 3 full-time (2 women). Average age 26. 11 applicants, 64% accepted, 5 enrolled. In 2005, 2 degrees awarded. Terminal master's awarded for partial completion of doctoral program. *Degree requirements:* For master's and doctorate, thesis/dissertation. *Entrance requirements:* For master's and doctorate, GRE General Test. Additional exam requirements/recommendations for international students: Required—TOEFL, TWE. *Application deadline:* For fall admission, 1/15 for domestic students; for spring admission, 11/1 for domestic students. Applications are processed on a rolling basis. Application fee: $10. Electronic applications accepted. *Financial support:* Fellowships with full tuition reimbursements, research assistantships with full tuition reimbursements, teaching assistantships,

Toxicology

The University of Texas Health Science Center at Houston (continued)
institutionally sponsored loans, scholarships/grants, and health care benefits available. Financial award application deadline: 1/15. *Faculty research:* Apoptosis, carcinogenesis, chemotherapy, immunotoxicology, molecular and cellular mechanisms of injury. *Unit head:* Dr. David J. McConkey, Director, 713-792-8591, Fax: 713-792-8747, E-mail: dmcconke@mdanderson.org. *Application contact:* Dr. Victoria P. Knutson, Assistant Dean of Admissions, 713-500-9860, Fax: 713-500-9877, E-mail: victoria.p.knutson@uth.tmc.edu.

The University of Texas Medical Branch, Graduate School of Biomedical Sciences, Program in Pharmacology and Toxicology, Curriculum in Toxicology, Galveston, TX 77555. Offers PhD. Interdisciplinary training program. Electronic applications accepted. *Expenses:* Tuition, state resident: full-time $8,350; part-time $90 per credit hour. Tuition, nonresident: full-time $21,450; part-time $366 per credit hour. Required fees: $1,027; $11 per credit hour. $60 per term. *Faculty research:* P450, estrogen, cancer, PKC.

University of the Sciences in Philadelphia, College of Graduate Studies, Program in Pharmacology and Toxicology, Philadelphia, PA 19104-4495. Offers pharmacology (MS, PhD); toxicology (MS, PhD). *Faculty:* 7 full-time (2 women), 6 part-time/adjunct (2 women). *Students:* 11 full-time (9 women), 4 part-time (1 woman); includes 2 minority (both Asian Americans or Pacific Islanders), 4 international. Average age 28. In 2005, 2 master's, 2 doctorates awarded. Terminal master's awarded for partial completion of doctoral program. *Degree requirements:* For master's, thesis/dissertation; for doctorate, thesis/dissertation, comprehensive exam. *Entrance requirements:* For master's and doctorate, GRE General Test. Additional exam requirements/recommendations for international students: Required—TOEFL, TWE, TSE. *Application deadline:* Applications are processed on a rolling basis. Application fee: $50. *Expenses:* Tuition: Part-time $999 per credit. Tuition and fees vary according to program. *Financial support:* In 2005–06, 12 students received support, including teaching assistantships with tuition reimbursements available (averaging $18,500 per year); institutionally sponsored loans and tuition waivers (partial) also available. Financial award application deadline: 5/1. *Faculty research:* Autonomic, cardiovascular, cellular, and molecular pharmacology; mechanisms of carcinogenesis; drug metabolism. Total annual research expenditures: $61,000. *Unit head:* Dr. Adeboye Adejare, Director, 215-596-8944, E-mail: a.adejar@usip.edu. *Application contact:* Joyce D'Angelo, Administrative Assistant, 215-596-8937, E-mail: j.dangel@usip.edu.

University of Utah, College of Pharmacy and The Graduate School, Graduate Programs in Pharmacy, Department of Pharmacology and Toxicology, Salt Lake City, UT 84112-1107. Offers MS, PhD, MD/PhD. Terminal master's awarded for partial completion of doctoral program. *Degree requirements:* For master's, thesis; for doctorate, thesis/dissertation, final exam. *Entrance requirements:* For master's, GRE; for doctorate, GRE Subject Test, BS in biology, chemistry, or pharmacy. Additional exam requirements/recommendations for international students: Required—TOEFL. *Expenses:* Tuition, state resident: full-time $2,932; part-time $369 per credit. Tuition, nonresident: full-time $10,350; part-time $1,302 per credit. Required fees: $516 per term. Tuition and fees vary according to course load and program. *Faculty research:* Neuropharmacology, neurochemistry, biochemistry, molecular pharmacology, analytical chemistry.

University of Washington, Graduate School, School of Public Health and Community Medicine, Department of Environmental and Occupational Health Sciences, Seattle, WA 98195. Offers environmental and occupational health (MPH); environmental and occupational hygiene (PhD); environmental health (MS); industrial hygiene and safety (MS); occupational medicine (MPH); safety and ergonomics (MS); toxicology (MS, PhD). Part-time programs available. *Faculty:* 31 full-time (7 women), 3 part-time/adjunct (1 woman). *Students:* 69 full-time (39 women), 15 part-time (5 women); includes 16 minority (2 African Americans, 1 American Indian/Alaska Native, 9 Asian Americans or Pacific Islanders, 3 Hispanic Americans), 14 international. Average age 33. 96 applicants, 34% accepted, 23 enrolled. In 2005, 20 master's, 4 doctorates awarded. Terminal master's awarded for partial completion of doctoral program. *Median time to degree:* Of those who began their doctoral program in fall 1997, 80% received their degree in 8 years or less. *Degree requirements:* For master's, thesis (for some programs), practicum (MPH); for doctorate, thesis/dissertation, comprehensive exam, registration. *Entrance requirements:* For master's and doctorate, GRE General Test, minimum GPA of 3.0, prerequisite course work in biology, chemistry, physics, calculus. Additional exam requirements/recommendations for international students: Required—TOEFL (minimum score 580 paper-based; 237 computer-based). *Application deadline:* For fall admission, 1/15 priority date for domestic students, 1/15 priority date for international students. Application fee: $45. Electronic applications accepted. *Financial support:* In 2005–06, 68 students received support, including 16 fellowships with full tuition reimbursements available (averaging $21,060 per year), 39 research assistantships with full tuition reimbursements available (averaging $19,164 per year), 10 teaching assistantships with full tuition reimbursements available (averaging $19,164 per year); career-related internships or fieldwork, institutionally sponsored loans, traineeships, health care benefits, and unspecified assistantships also available. Financial award application deadline: 1/15. *Faculty research:* Developmental toxicology, biochemical toxicology, exposure assessment, hazardous waste, industrial chemistry. Total annual research expenditures: $16 million. *Unit head:* Dr. David Kalman, Chair, 206-543-6991, Fax: 206-616-0477, E-mail: ehadmin@u.washington.edu. *Application contact:* Rory A. Murphy, Manager, Student Services, 206-543-3199, Fax: 206-543-9616, E-mail: ehgrad@u.washington.edu.

University of Wisconsin–Madison, Graduate School, College of Agricultural and Life Sciences, Molecular and Environmental Toxicology Center, Madison, WI 53706-1380. Offers MS, PhD. *Students:* 38 full-time (25 women); includes 6 minority (1 African American, 2 American Indian/Alaska Native, 3 Asian Americans or Pacific Islanders), 6 international. Average age 28. In 2005, 1 master's, 4 doctorates awarded. *Degree requirements:* For doctorate, thesis/dissertation. *Entrance requirements:* For master's and doctorate, bachelor's degree in science-related field. Additional exam requirements/recommendations for international students: Required—TOEFL. *Application deadline:* For fall admission, 12/15 priority date for domestic students, 12/15 priority date for international students. Application fee: $45. *Financial support:* In 2005–06, 6 research assistantships with tuition reimbursements (averaging $21,500 per year) were awarded; fellowships with tuition reimbursements, traineeships, health care benefits, and unspecified assistantships also available. *Faculty research:* Toxicology cancer, genetics, cell cycle, xenobotic metabolism. *Unit head:* Dr. Jeffrey A. Johnson, Director, 608-263-4580, Fax: 608-262-5245, E-mail: uwetox@wisc.edu. *Application contact:* Eileen M. Stevens, Program Administrator, 608-263-4580, Fax: 608-262-5245, E-mail: emstevens@wisc.edu.

University of Wisconsin–Madison, School of Veterinary Medicine, Department of Animal Health and Biomedical Sciences, Program in Comparative Biosciences, Madison, WI 53706-1380. Offers anatomy (MS, PhD); biochemistry (MS, PhD); cellular and molecular biology (MS, PhD); environmental toxicology (MS, PhD); neurosciences (MS, PhD); pharmacology (MS, PhD); physiology (MS, PhD). *Degree requirements:* For doctorate, thesis/dissertation.

Utah State University, School of Graduate Studies, College of Agriculture, Program in Toxicology, Logan, UT 84322. Offers MS, PhD. *Faculty:* 14 full-time (2 women), 3 part-time/adjunct (0 women). *Students:* 1 full-time (0 women), 2 part-time (1 woman), 1 international. Average age 25. 1 applicant, 0% accepted. In 2005, 1 degree awarded. Terminal master's awarded for partial completion of doctoral program. *Degree requirements:* For master's and doctorate, thesis/dissertation. *Entrance requirements:* For master's and doctorate, GRE General Test, minimum GPA of 3.0. Additional exam requirements/recommendations for international students: Required—TOEFL. *Application deadline:* For fall admission, 6/15 for domestic students; for spring admission, 10/15 for domestic students. Applications are processed on a rolling basis. Application fee: $50 ($60 for international students). *Financial support:* In 2005–06, 1 fellowship with partial tuition reimbursement (averaging $15,000 per year), research assistantships with partial tuition reimbursements (averaging $15,000 per year) were awarded; teaching assistantships with partial tuition reimbursements, Federal Work-Study, institutionally sponsored loans, and tuition waivers (partial) also available. Support available to part-time students. *Faculty research:* Free-radical mechanisms, toxicity of iron, carcinogenesis of

natural compounds, molecular mechanisms of retinoid toxicity, aflatoxins. *Unit head:* Dr. Roger A. Coulombe, Director, 435-797-1598, Fax: 435-797-1601, E-mail: rogerc@cc.usu.edu.

Vanderbilt University, Center in Molecular Toxicology, Nashville, TN 37240-1001.

Vanderbilt University, Graduate School, Graduate Program in Toxicology and Carcinogenesis, Nashville, TN 37240-1001.

Virginia Commonwealth University, Medical College of Virginia-Professional Programs, School of Medicine and Graduate Programs, School of Medicine Graduate Programs, Department of Pharmacology and Toxicology, Richmond, VA 23284-9005. Offers molecular biology and genetics (PhD); neurosciences (PhD); pharmacology (PhD, CBHS); pharmacology and toxicology (MS). *Faculty:* 35 full-time (9 women). *Students:* 35 full-time (18 women), 11 part-time (all women); includes 7 minority (2 African Americans, 4 Asian Americans or Pacific Islanders, 1 Hispanic American), 5 international. 90 applicants, 33% accepted. In 2005, 6 master's, 2 doctorates awarded. Terminal master's awarded for partial completion of doctoral program. *Degree requirements:* For master's, thesis; for doctorate, thesis/dissertation, comprehensive oral and written exams. *Entrance requirements:* For master's, DAT, GRE General Test or MCAT; for doctorate, GRE General Test, MCAT, DAT. *Application deadline:* For fall admission, 4/1 for domestic students. Application fee: $50. *Expenses:* Tuition, state resident: full-time $6,268; part-time $405 per credit. Tuition, nonresident: full-time $15,904; part-time $940 per credit. Required fees: $751 per semester hour. Tuition and fees vary according to course load and program. *Financial support:* Fellowships, teaching assistantships available. Financial award application deadline: 1/15. *Faculty research:* Drug abuse, drug metabolism, pharmacodynamics, peptide synthesis, receptor mechanisms. *Unit head:* , Dr. Billy R. Martin, Chair, 804-828-8407, Fax: 804-828-2117, E-mail: brmartin@vcu.edu. *Application contact:* Sheryol Cox, Graduate Program Coordinator, 804-828-8400, Fax: 804-828-2117, E-mail: swcox@vcu.edu.

See Close-Up on page 1219.

Washington State University, Graduate School, College of Pharmacy, Program in Pharmacology and Toxicology, Pullman, WA 99164. Offers MS, PhD. *Faculty:* 28 full-time (8 women), 4 part-time/adjunct (1 woman). *Students:* 17 full-time (8 women), 2 part-time (both women); includes 2 minority (1 Asian American or Pacific Islander, 1 Hispanic American), 5 international. Average age 26. 66 applicants, 17% accepted, 6 enrolled. In 2005, 1 degree awarded. *Degree requirements:* For master's and doctorate, thesis/dissertation, oral exam. *Entrance requirements:* For master's, GRE General Test, minimum GPA of 3.0, 3 letters of recommendation; for doctorate, GRE General Test, minimum GPA of 3.0. Additional exam requirements/recommendations for international students: Required—TOEFL. *Application deadline:* For fall admission, 12/1 for domestic students, 3/1 for international students; for spring admission, 9/1 for domestic students, 7/1 for international students. Applications are processed on a rolling basis. Application fee: $35. Electronic applications accepted. *Expenses:* Tuition, state resident: full-time $6,295; part-time $336 per credit. Tuition, nonresident: full-time $15,949; part-time $819 per credit. Required fees: $933. Part-time tuition and fees vary according to campus/location and program. *Financial support:* In 2005–06, 5 fellowships (averaging $6,176 per year), 10 research assistantships with full and partial tuition reimbursements (averaging $14,846 per year), 4 teaching assistantships with full and partial tuition reimbursements (averaging $14,661 per year) were awarded; Federal Work-Study, institutionally sponsored loans, tuition waivers (partial), and staff assistantships, teaching associateships also available. Financial award application deadline: 4/1; financial award applicants required to submit FAFSA. *Faculty research:* Cancer research, immunopharmacology/immunotoxicology, pharmacy, neurobiology/neuropharmacology, toxicology. Total annual research expenditures: $623,031. *Unit head:* Dr. Raymond Quock, Chair, 509-335-7598, Fax: 509-335-5902. *Application contact:* Dana Martin, Coordinator, 509-335-7598, Fax: 509-335-5902, E-mail: pharmtox@wsu.edu.

Washington State University Tri-Cities, Graduate Programs, Program in Environmental Science, Richland, WA 99352-1671. Offers applied environmental science (MS); atmospheric science (MS); earth science (MS); environmental and occupational health science (MS); environmental regulatory compliance (MS); environmental science (PhD); environmental toxicology and risk assessment (MS); water resource science (MS). Part-time programs available. *Faculty:* 1 full-time (0 women), 53 part-time/adjunct. *Students:* 4 full-time (3 women), 22 part-time (10 women); includes 1 Asian American or Pacific Islander, 2 Hispanic Americans. Average age 41. 11 applicants, 55% accepted, 6 enrolled. In 2005, 1 degree awarded. *Degree requirements:* For master's, oral exam, thesis optional. *Entrance requirements:* For master's, GRE General Test, minimum GPA of 3.0, 3 letters of recommendation. Additional exam requirements/recommendations for international students: Required—TOEFL (minimum score 550 paper-based; 213 computer-based). *Application deadline:* For fall admission, 2/1 priority date for domestic students, 3/1 priority date for international students; for spring admission, 9/1 priority date for domestic students, 7/1 priority date for international students. Application fee: $35. *Expenses:* Tuition, state resident: full-time $6,295; part-time $336 per credit. Tuition, nonresident: full-time $15,949; part-time $819 per credit. Required fees: $429. Full-time tuition and fees vary according to campus/location and program. Part-time tuition and fees vary according to course load and program. *Financial support:* In 2005–06, 8 students received support, including 1 fellowship (averaging $2,200 per year); research assistantships with full and partial tuition reimbursements available, teaching assistantships with full and partial tuition reimbursements available, Federal Work-Study, scholarships/grants, health care benefits, and unspecified assistantships also available. *Faculty research:* Radiation ecology, cytogenetics. *Unit head:* Dr. Gene Schreckhise, Associate Dean/Coordinator, 509-372-7323, E-mail: gschreck@wsu.edu.

Wayne State University, Graduate School, Eugene Applebaum College of Pharmacy and Health Sciences, Department of Fundamental and Applied Sciences, Detroit, MI 48202. Offers analytical toxicology (Certificate); industrial toxicology (Post-Master's Certificate); occupational and environmental health sciences (MS); occupational health sciences (MS), including industrial hygiene, industrial toxicology, occupational medicine; occupational safety (Certificate). *Accreditation:* ABET (one or more programs are accredited). Part-time and evening/weekend programs available. *Faculty:* 15 full-time (3 women). *Students:* 8 full-time (5 women), 13 part-time (6 women); includes 7 minority (4 African Americans, 3 Asian Americans or Pacific Islanders), 5 international. Average age 33. 19 applicants, 100% accepted. In 2005, 15 master's, 4 other advanced degrees awarded. *Degree requirements:* For master's, thesis optional. *Entrance requirements:* For master's, GRE General Test, 1 year of course work in biology, mathematics, and physics; 2 years of course work in chemistry. Additional exam requirements/recommendations for international students: Required—TOEFL (minimum score 550 paper-based; 213 computer-based); Recommended—TWE (minimum score 6). *Application deadline:* For fall admission, 6/15 priority date for domestic students, 6/1 priority date for international students. Applications are processed on a rolling basis. Application fee: $30 ($50 for international students). Electronic applications accepted. *Expenses:* Tuition, state resident: part-time $338 per credit hour. Tuition, nonresident: part-time $746 per credit hour. Required fees: $24 per credit hour. Full-time tuition and fees vary according to program. *Financial support:* In 2005–06, 4 students received support; teaching assistantships, career-related internships or fieldwork and scholarships/grants available. Financial award application deadline: 7/1. *Faculty research:* Fatality assessment and control evaluation; lung toxicity of oxidants, aerosols, and tobacco smoke; modeling asthma and hypersensitivity pneumonitis; management of inflammatory lung disorders; analytical toxicology for forensic and environmental health investigations. Total annual research expenditures: $250,000. *Unit head:* Dr. Peter Frade, Chairperson, 313-577-7874, Fax: 313-577-0097, E-mail: ab8123@wayne.edu. *Application contact:* Peter Warner, Graduate Director, 313-577-1551, E-mail: aa4631@wayne.edu.

Wayne State University, Graduate School, Interdisciplinary Program in Molecular and Cellular Toxicology, Detroit, MI 48202. Offers MS, PhD. *Faculty:* 13. *Students:* 10 full-time (7 women), 7 international. Average age 28. 2 applicants, 50% accepted, 1 enrolled. In 2005, 2 degrees awarded. *Degree requirements:* For master's, thesis or alternative; for doctorate, thesis/dissertation. *Entrance requirements:* For master's, GRE General Test, GRE Subject

Test; for doctorate, GRE General Test, GRE Subject Test, minimum GPA of 3.0. Additional exam requirements/recommendations for international students: Required—TOEFL (minimum score 550 paper-based; 213 computer-based); Recommended—TWE (minimum score 6). *Application deadline:* For fall admission, 4/15 priority date for domestic students, 6/1 priority date for international students. Applications are processed on a rolling basis. Application fee: $30 ($50 for international students). Electronic applications accepted. *Expenses:* Tuition, state resident: part-time $338 per credit hour. Tuition, nonresident: part-time $746 per credit hour. Required fees: $24 per credit hour. Full-time tuition and fees vary according to program. *Financial support:* Fellowships, research assistantships, teaching assistantships, institutionally sponsored loans available. Financial award application deadline: 4/15. *Faculty research:* Molecular and cellular mechanisms of chemically-induced cell injury and death; effect of xenobiotics on cell growth, proliferation, transformation and differentiation; regulation of gene expression; cell signaling; global gene expression profiling. Total annual research expenditures: $2 million. *Unit head:* Dr. Raymond F. Novak, Director, 313-577-0100, Fax: 313-577-0082, E-mail: r.novak@wayne.edu. *Application contact:* Dharam Chopra, Professor, 313-961-2816, Fax: 313-577-0082, E-mail: d.chopra@wayne.edu.

See Close-Up on page 1221.

West Virginia University, Davis College of Agriculture, Forestry and Consumer Sciences, Interdisciplinary Program in Genetics and Developmental Biology, Morgantown, WV 26506. Offers animal breeding (MS, PhD); biochemical and molecular genetics (MS, PhD); cytogenetics (MS, PhD); descriptive embryology (MS, PhD); developmental genetics (MS); experimental morphogenesis teratology (MS); human genetics (MS, PhD); immunogenetics (MS, PhD); life cycles of animals and plants (MS, PhD); molecular aspects of development (MS, PhD); mutagenesis (MS, PhD); oncology (MS, PhD); plant genetics (MS, PhD); population and quantitative genetics (MS, PhD); regeneration (MS, PhD); teratology (PhD); toxicology (MS, PhD). *Students:* 13 full-time (7 women), 6 part-time (4 women), 11 international. Average age 28. In 2005, 2 master's, 4 doctorates awarded. *Degree requirements:* For master's, thesis/dissertation; for doctorate, thesis/dissertation, comprehensive exam. *Entrance requirements:* For master's, GRE or MCAT, minimum GPA of 2.75. Additional exam requirements/recommendations for international students: Required—TOEFL. Application fee: $45. *Expenses:* Tuition, state resident: full-time $4,582; part-time $258 per credit hour. Tuition, nonresident: full-time $13,820; part-time $741 per credit hour. *Financial support:* In 2005–06, 3 research assistantships with tuition reimbursements (averaging $9,936 per year), 4 teaching assistant-

ships with tuition reimbursements (averaging $9,936 per year) were awarded; fellowships, Federal Work-Study, institutionally sponsored loans, and tuition waivers (full and partial) also available. Financial award application deadline: 2/1; financial award applicants required to submit FAFSA. Total annual research expenditures: $1 million. *Unit head:* Dr. J. Nath, Chairman, 304-293-6023 Ext. 4333, Fax: 304-293-2960, E-mail: joginder.nath@mail.wvu.edu.

West Virginia University, School of Pharmacy, Program in Pharmaceutical and Pharmacological Sciences, Morgantown, WV 26506. Offers administrative pharmacy (PhD); behavioral pharmacy (MS, PhD); biopharmaceutics/phamacokinetics (PhD); biopharmaceutics/pharmacokinetics (MS); industrial pharmacy (MS); medicinal chemistry (MS, PhD); pharmaceutical chemistry (MS, PhD); pharmaceutics (MS, PhD); pharmacology and toxicology (MS); pharmacy (MS); pharmacy administration (MS). Part-time programs available. *Students:* 31 full-time (16 women), 6 part-time (1 woman), 22 international. Average age 27. 58 applicants, 9% accepted, 4 enrolled. In 2005, 1 master's, 5 doctorates awarded. Terminal master's awarded for partial completion of doctoral program. *Median time to degree:* Of those who began their doctoral program in fall 1997, 84% received their degree in 8 years or less. *Degree requirements:* For master's, thesis; for doctorate, one foreign language, thesis/dissertation, comprehensive exam. *Entrance requirements:* For master's and doctorate, GRE General Test, minimum GPA of 2.75. Additional exam requirements/recommendations for international students: Required—TOEFL; Recommended—TWE. *Application deadline:* For fall admission, 3/1 for domestic students. Application fee: $50. Electronic applications accepted. *Expenses: Contact institution. Financial support:* In 2005–06, 13 research assistantships with full tuition reimbursements (averaging $18,000 per year), 15 teaching assistantships with full tuition reimbursements (averaging $18,000 per year) were awarded; career-related internships or fieldwork, Federal Work-Study, institutionally sponsored loans, health care benefits, tuition waivers (full and partial), and unspecified assistantships also available. Financial award application deadline: 3/1; financial award applicants required to submit FAFSA. *Faculty research:* Pharmaceutics, medicinal chemistry, biopharmaceutics/pharmacokinetics, health outcomes research. Total annual research expenditures: $1.8 million. *Application contact:* Dr. Patrick S. Callery, Assistant Dean for Graduate Programs/Chair, 304-293-1482, Fax: 304-293-2576, E-mail: patrick.callery@hsc.wvu.edu.

Wright State University, School of Medicine, Program in Pharmacology and Toxicology, Dayton, OH 45435. Offers MS. *Degree requirements:* For master's, thesis optional.

BOSTON UNIVERSITY

School of Medicine
Department of Pharmacology and Experimental Therapeutics

Programs of Study	The graduate program at Boston University School of Medicine permits students to acquire both the broad and the specialized aspects of professional education that are necessary to a career in pharmacology. Students engage in intensive studies in the discipline itself and receive extensive exposure to the biomedical disciplines represented by other departments of the medical school. Major areas of training specialization include molecular pharmacology and neuropharmacology. Additional opportunities for interdisciplinary training are provided through interdepartmental programs in biomolecular pharmacology, neurosciences, molecular and cellular biology, human genetics, and bioinformatics. The graduate program combines course work and laboratory research experience leading to the M.A. or Ph.D. degree. Students are expected to take basic courses in biochemistry and physiology and may elect other offerings such as neurosciences and cell biology. Advanced courses designed specifically for graduate students in pharmacology include molecular neurobiology and pharmacology, laboratory techniques in modern pharmacology, advanced general pharmacology, gene regulation and pharmacology, pharmacokinetics, pharmacology of inflammation, biochemical neuropharmacology, neuropsychopharmacology, neuroendocrine pharmacology, pharmacology of learning and memory, and a seminar course on current topics in pharmacological sciences. Students are generally required to participate in laboratory rotations during their first year of study. Options for internships are available in industrial settings.
Research Facilities	Laboratories of the Department and of the School of Medicine are well equipped for the performance of pharmacological research. There are facilities for both in vitro and in vivo studies and a wide variety of techniques in molecular biology, cell biology, neurochemistry, receptor biochemistry, electrophysiology, and behavioral studies. A medical library with a computer laboratory is located on the campus at the Medical Center. Additional libraries with extensive holdings are available on the campus of Boston University. There is ready access at the Medical Center to the extensive computing facilities available on the Boston University campus. Research opportunities are enhanced by collaborations with faculty members in the Department of Medicine, Section of Molecular Medicine, and the Departments of Biomedical Engineering, Chemistry, Biochemistry, and Physiology and Biophysics. Collaborative research training is also ongoing with the Cardiovascular Institute, the Cancer Center, and the Laboratories for Mass Spectroscopy and for Transgenic Research at the Medical Center.
Financial Aid	Funds are available for full support of graduate training for students enrolled in the Ph.D. program. Students are supported with annual stipends, plus full tuition remission and health insurance. Stipends are available from federally funded research grants, the NIGMS Training Grant in Biomolecular Pharmacology, and support of the pharmaceutical industry. In addition, stipends may be supplemented by Department teaching assistantships. The Henry I. Russek Graduate Student Achievement Award of $2000 is presented annually to a student who demonstrates outstanding accomplishments. Financial aid is also available for students who are members of minority groups through supplements to NIH grants.
Cost of Study	Full-time tuition for the 2006–07 academic year is $33,330. Part-time tuition, 10 credits or fewer, is $1042 per credit.
Living and Housing Costs	The cost of living in Boston is similar to that of other major urban areas of the Northeast. There is some student housing on the Medical Center campus. Most students choose off-campus locations. The University housing office can provide information on rentals.
Student Group	A total of 41 graduate students—3 M.A., 33 Ph.D., and 5 M.D./Ph.D. candidates—were enrolled in the Department in 2005–06.
Student Outcomes	Recent graduates are pursuing careers in pharmacological research and drug development at academic medical centers and in the pharmaceutical/biotechnology industry.
Location	Boston, Cambridge, and the surrounding area provide one of the greatest concentrations of intellectual, cultural, and recreational resources in the United States. Boston University itself has superb athletic and recreational facilities as well as many student groups and activities that are open to all graduate students.
The University and The Department	Boston University is a private institution founded in 1839. Today it comprises sixteen schools and colleges. The University's history includes the graduation of the first African American woman to receive an M.D. and the first woman to receive a Ph.D. in the United States. The Boston University Medical Center comprises 500 faculty scientists and more than 600 graduate students whose research programs receive some $100 million per year in sponsored support. The Medical Center is ranked twentieth nationally in federal plus private funding. The Boston University Medical Center is located within the Biosquare Complex in Boston. Major expansion and renovation of the Department of Pharmacology and Experimental Therapeutics have occurred during the last several years and are continuing in newly acquired space. A corresponding expansion of the graduate program is also under way. The M.A. program in pharmacology is also under expansion.
Applying	Applications for admission and instructions for applying can be obtained from the Division of Graduate Medical Sciences, L-310, Boston University School of Medicine, 715 Albany Street, Boston, Massachusetts 02118; 617-638-5120; http://www.bumc.bu.edu/Dept/Home.aspx?DepartmentID=86. There is no restrictive deadline; however, applicants are urged to submit credentials by January 1. Interviews are generally conducted in February and March. Applicants should possess or be pursuing a bachelor's degree that incorporates courses in general biology, chemistry, organic chemistry, calculus, and physics. Courses in physiology, biochemistry, and statistics are recommended. Satisfactory performance on the Graduate Record Examinations or Medical College Admission Test is required.
Correspondence and Information	Dr. David H. Farb, Chairman Department of Pharmacology and Experimental Therapeutics, L-603 Boston University School of Medicine 715 Albany Street Boston, Massachusetts 02118-2526 Phone: 617-638-4300 E-mail: pharm@bu.edu Web site: http://www.bumc.bu.edu/Dept/Home.aspx?DepartmentID=65

Boston University

THE FACULTY AND THEIR RESEARCH

Carmela Abraham, Professor of Biochemistry; Ph.D., Harvard. Mechanisms of brain aging and the etiology of Alzheimer's disease.

David Atkinson, Professor of Physiology and Biophysics and Research Professor of Biochemistry; Ph.D., Council for National Academic Awards (England). Detailed structural and conformational information on the molecular organization and interactions of lipids and proteins in plasma lipoproteins, particularly high-density lipoprotein (HDL) and low-density lipoprotein (LDL).

Victoria M. Bolotina, Associate Professor of Medicine and Physiology; Ph.D., Moscow State. Ion channels and mechanisms of calcium signaling.

Richard A. Cohen, Professor of Medicine, Physiology, and Pharmacology; M.D., Johns Hopkins. Integrative molecular understanding of abnormal vascular endothelial and smooth-muscle cell function in diseased blood vessels and its contribution to abnormal vascular reactivity and atherogenesis.

Charles DeLisi, Arthur G. B. Metcalf Professor of Science and Engineering, Professor of Physics, Professor of Biomedical Engineering, and Senior Associate Provost for Biosciences; Ph.D., NYU. Development and application of computational/mathematical methods and high throughput experimental methods to analyze changes in gene and protein expression profiles of cells in response to various endogenous and exogenous signals.

Gerald Denis, Assistant Professor, Cancer Research Center; Ph.D., Berkeley. Mechanisms of transcriptional control mediated through a newly described class of coactivators and chromatin regulators called bromodomain proteins.

Howard Eichenbaum, Professor of Psychology; Ph.D., Michigan. Elucidation of the function of the hippocampus; hippocampal and cortical coding; hippocampal lesions and long-term memory; schizophrenia, dreceased function of corticolimbic NMDA receptors.

David H. Farb, Professor and Chairman of Pharmacology and Experimental Therapeutics; Ph.D., Brandeis. Pharmacology and regulation of gene expression of the GABA/benzodiazepine receptor in the vertebrate central nervous system; studies of the mechanism of action of neuromodulators and elucidation of the structure, function, and cellular dynamics of amino acid receptors in the brain.

Martin Feelisch, Professor of Medicine and Biochemistry; Ph.D., Heinrich Heine (Germany). Oxidant stress in the vascular endothelium and the role of nitric oxide in the pharmacologic treatment of cardiovascular and other chronic diseases.

Richard Fine, Professor of Biochemistry and Neurology; Ph.D., Brandeis. Cell biology and metabolism of the beta amyloid precursor protein and other proteins involved in Alzheimer's disease.

Jane Freedman, Associate Professor of Medicine and Pharmacology; Ph.D., Tufts. Molecular regulation of pathways contributing to thrombosis and vascular disease and how these factors contribute to acute coronary syndromes.

Terrell T. Gibbs, Associate Professor of Pharmacology; Ph.D., Harvard. The pharmacology of neurotransmitters and neuromodulators and mechanisms of modulation and regulation of neurotransmitter receptor function, including up-regulation, down-regulation, desensitization, and tolerance.

Mark W. Grinstaff, Associate Professor of Biomedical Engineering and Ophthalmology; Ph.D., Indiana. Designing, synthesizing, and characterizing novel biomaterials for tissue engineering and biotechnological applications.

James A. Hamilton, Professor of Physiology and Biophysics; Ph.D., Indiana. Developing novel physical approaches to address physiologically important issues; studying binding of ligands to FABP and the molecular motions of the protein in the presence and absence of ligand; new fluorescence approaches to monitor the diffusion of fatty acids into cells by measurement of intracellular pH.

Gary B. Kaplan, Professor of Psychiatry and Pharmacology; M.D., Hahnemann. Mechanisms and treatments of opioid and nicotine addictions.

John Keaney., Professor of Medicine and Pharmacology; M.D., Yale. Elucidate the role of reactive oxygen species in the modulation of nitric oxide–mediated homeostasis; examining the role of reactive oxygen species in mediating signal transduction and phenotypic changes in endothelial cells.

Conan Kornetsky, Professor of Psychiatry and Pharmacology; Ph.D., Kentucky. Research toward the determination of neuronal mechanisms involved in the behavioral effects of drugs; changes directly related to the rewarding effects associated with the action of the abused psychomotor stimulants, e.g., cocaine and opioids, e.g., heroin.

Susan E. Leeman, Professor of Pharmacology; Ph.D., Radcliffe. Studies of the two peptides, substance P (SP) and neurotensin, isolated and chemically defined in the laboratory; determining the binding domains of SP with its receptor.

Adam Lerner, Associate Professor of Medicine and Pathology; M.D., Yale. Discoveries of targets for lymphoid malignancies through analysis of apoptotic and other signaling mechanisms.

John R. Murphy, Professor of Medicine, Section of Molecular Medicine; Ph.D., Connecticut. Design, expression, and characterization of diphtheria toxin–related cytokine fusion proteins, molecular basis of diphtheria toxin gene regulation, and the molecular mechanisms by which large hydrophilic proteins are transported into eukaryotic cells.

Susan P. Perrine, Professor of Pediatrics and Pharmacology; M.D., Tufts. Development of novel molecular therapeutics for the treatment of hemoglobinopathies and thalassemias.

R. Christopher Pierce, Associate Professor of Pharmacology and Psychiatry; Ph.D., Indiana. Cellular and molecular alterations that occur in brain nuclei, such as the nucleus accumbens, following repeated treatment with psychostimulant drugs.

John A. Porco Jr., Professor of Chemistry and Pharmacology; Ph.D., Harvard. Development of new methodologies for organic synthesis and their application to synthesis of complex natural products and analogues.

Katya Ravid, Professor of Biochemistry and Medicine; Ph.D., Technion (Israel). Genetic and pharmacological control of proliferation of bone marrow and vascular cells, with a focus on platelets and smooth-muscle cells.

Karen L. Reed, Research Assistant Professor of Surgery and Pharmacology; Ph.D., Florida. Molecular and cellular characterization of proinflammatory regulators of intra-abdominal adhesion formation and inflammatory bowel disease.

Douglas Rosene, Associate Professor of Anatomy and Neurobiology; Ph.D., Rochester. Identifying the neurobiological basis of normal learning and memory and related cognitive functions in the normal brain and the disruption of these processes in various neurodegenerative diseases and forms of neurological damage.

Shelley J. Russek, Associate Professor of Pharmacology, Ph.D; Boston University. Dynamic regulation of gene expression in the nervous system for the treatment and prevention of neurological diseases.

Scott Schaus, Assistant Professor of Chemistry and Pharmacology; Ph.D., Harvard. Development of novel technologies to investigate cellular processes such as cell-cycle regulation, cell proliferation, and intracellular signaling.

G. Graham Shipley, Professor of Physiology and Biophysics; Ph.D., Nottingham (England). Structure and function of cell membranes, receptor-ligand interactions, and transmembrane-signaling mechanisms.

Jean-Jacques Soghomonian, Associate Professor of Anatomy and Neurobiology; Ph.D., Montreal. Analysis of the functional neuroanatomy of the basal ganglia and the neurobiological basis of motor control, sensorimotor integration, and learning, particularly the mechanisms of regulation of GABAergic neurons in the basal ganglia in the normal brain and in experimental models of Parkinson's disease.

Remco Spanjaard, Associate Professor of Otolaryngology and Biochemistry; Ph.D., Leiden (Netherlands). Mechanistic approaches to discovery of agents for prevention and treatment of cancers.

Joseph Z. Tsien, Professor of Pharmacology and Biomedical Engineering; Ph.D., Minnesota. Elucidation of molecular and neural mechanisms underlying normal and abnormal memory processes.

Thomas Tullius, Professor and Chairman of Chemistry; Ph.D., Stanford. Developing and applying new chemical probe methods for determining the structure of DNA and DNA-protein complexes in solution.

Carol T. Walsh, Professor of Pharmacology; Ph.D., Boston University. Gastrointestinal pharmacology; toxicology of metal compounds; pharmacokinetics.

David Waxman, Professor of Biology and Medicine; Ph.D., Harvard. Elucidation of regulatory processes for cytochorome P450 enzymes and their impact on patient responses to drugs, in particular cancer chemotherapeutic agents, at the cellular, biochemical, and molecular level.

Zhiping Weng, Associate Professor of Biomedical Engineering and Pharmacology; Ph.D., Boston University. Computational approaches to the determination of protein structure and function and the engineering of novel proteins with desired properties.

John A. White, Associate Professor of Biomedical Engineering; Ph.D., MIT. Biophysical mechanisms underlying information processing in the mammalian brain; synchronized activity important for the proper function of the hippocampal formation, a learning and memory center.

Benjamin Wolozin, Professor of Pharmacology and Neurology; M.D., Ph.D., Yeshiva (Einstein). Pathophysiology of neurodegenerative diseases.

Joyce Wong, Assistant Professor of Biomedical Engineering; Ph.D., MIT. Development of new biomaterials that interact with living cells in novel ways.

Bryan K. Yamamoto, Professor of Pharmacology; Ph.D., Syracuse. Neuropharmacology related to neurotoxicity of the amphetamines, particularly methamphetamine and MDMA (Ecstasy) as well as mechanisms related to neurodegeneration.

CORNELL UNIVERSITY

Graduate Field of Pharmacology

Program of Study

The graduate program offers an intensive course of study and research emphasizing cellular and molecular pharmacology. The program prepares each student for a productive career in biomedical research and teaching. Upon completion of the curriculum, students graduate with a Doctor of Philosophy degree. The pharmacology graduate program accepts outstanding college graduates as well as physicians and veterinarians seeking advanced training in pharmacology. The program is flexible and is designed to meet the needs of each individual student.

The first two years are usually devoted to selected courses, seminars, and laboratory rotations, providing a broad background in pharmacology. Students are encouraged to supplement their program in pharmacology by taking specialized courses in such areas as biochemistry, biophysics, chemistry, neurobiology, physiology, and toxicology. By the end of the first year, students select their major adviser and Special Committee and begin working on their thesis research project. This project is closely supervised by the major adviser, but students are also encouraged to seek advice and guidance from other faculty members involved in the pharmacology training program. Major research is in progress to elucidate basic cellular and molecular mechanisms involved in receptor and ion channel function, signal transduction, neurotransmission, and stimulus-secretion coupling. Research also emphasizes the molecular and cellular bases of diseases such as allergies, cancer, diabetes, diarrhea, epilepsy, and myasthenia gravis. In addition to the Molecular Medicine Seminar Series, students are encouraged to attend seminars in other areas, including biochemistry, biophysics, immunology, nanobiotechnology, neurobiology, physiology, and toxicology. They also participate in journal clubs and study groups, and graduate faculty members meet at least once a week with their laboratory groups to discuss research progress and developments in their specific fields of research. In addition to writing a thesis on their research, graduate students are required to pass two oral examinations. The examination for admission to Ph.D. candidacy is usually taken in the second or third year. This is a comprehensive examination that certifies that the student is eligible to undertake advanced research for a Ph.D. degree. The final examination is made up of two parts. The student first presents the thesis work in a seminar to interested students and faculty. The student then defends the thesis before the Special Committee. By this time, students have accomplished solid and original research work, with publications in major journals in their field. Previous graduates from the department have productive careers in pharmacological research and academic medicine.

Research Facilities

Many of the faculty are members of the Department of Molecular Medicine, which is housed in the well-equipped facilities in the College of Veterinary Medicine on the Ithaca campus of Cornell University.

Financial Aid

All current graduate students receive stipends, and these include a waiver of tuition. Stipend amounts vary, but all are adequate to cover normal student living expenses. Sources for these stipends include NIH training grants, predoctoral fellowships, University graduate fellowships, and faculty research grants. The College of Veterinary Medicine awards a number of assistantships on a competitive basis to students who hold the D.V.M. degree; there are also assistantships for students from minority groups.

Cost of Study

For 2006–07, tuition is expected to be about $20,800 a year, but most stipends include a tuition waiver.

Living and Housing Costs

For 2006–07, a single student needs approximately $19,360 a year in addition to the funds required for tuition. A married student without dependent children needs at least $10,000 a year more than that. University housing is available to a limited number of single students and to student families.

Student Group

Currently, there are more than 30 full-time graduate students in the pharmacology training program, half of whom are women. International students make up one third of this group. Individuals graduating from the program are prepared for productive careers in pharmacological research and academic medicine. There are approximately 19,500 students on Cornell's Ithaca campus, of whom about 6,000 are enrolled in various graduate and professional schools.

Location

Ithaca is at the tip of Cayuga Lake in the Finger Lakes region of New York State, about 250 miles northwest of New York City. The Cornell campus of 740 acres overlooks the city and the lake. The area's hills, valleys, lakes, and gorges, with numerous parks and trails, make a pleasant environment. Cornell University, Ithaca College, and the city of Ithaca offer diverse programs in art, music, theater, and sports throughout the year. Many organizations and activities oriented toward students are available in Ithaca as well as on the Cornell campus.

The University

Cornell is an internationally known university consisting of thirteen colleges. Founded in 1865, Cornell is the youngest member of the Ivy League and is New York State's land-grant institution. Cornell faculty members are leaders in their fields, and many have been honored as Nobel laureates, Pulitzer Prize winners, and members of the National Academy of Sciences.

Applying

Potential applicants are encouraged to discuss possible courses of study with individual faculty members whose research interests most closely match those of the applicant. Those who wish to apply for admission to the Graduate School should (1) hold a baccalaureate degree granted by a faculty or university of recognized standing or have completed studies equivalent to those required for a baccalaureate degree at Cornell; (2) have adequate preparation for graduate study in their chosen field; (3) be fluent in English; and (4) demonstrate evidence or promise of success in advanced study and research. In general, successful applicants have a combined score of at least 1200 on the General Test (verbal and quantitative) of the Graduate Record Examinations. Applicants from U.S. colleges and universities should be in the top third of their graduating class. Applications from members of minority groups are particularly encouraged. Applications are due by December 15.

Correspondence and Information

Graduate Education Coordinator
College of Veterinary Medicine
Cornell University
Ithaca, New York 14853-6401

Phone: 607-253-3276
Fax: 607-253-3756
E-mail: graduate_edcvm@cornell.edu
Web site: http://www.vet.cornell.edu/pharm/

Cornell University

THE FACULTY AND THEIR RESEARCH

Barbara A. Baird, Professor of Chemistry; Ph.D., Cornell. Studies examine molecular aspects of plasma membranes and receptor proteins that transmit signals during immunological cell activation, especially the receptor for immunoglobulin E. Areas of investigation include membrane structure and dynamics, lipid rafts, ligand binding to cell-surface receptors, interactions of receptor complexes during signal transduction, and development of biophysical and nanotechnology approaches to address these questions.

Richard A. Cerione, Professor of Pharmacology; Ph.D., Rutgers. Mechanisms by which signals are transmitted from cell-surface receptor proteins to biological effectors are being investigated. A variety of biochemical, structural biology–based, and genetic approaches are being used to identify and characterize novel signaling proteins that influence cell growth.

Ruth N. Collins, Assistant Professor of Molecular Medicine; Ph.D., Imperial Cancer Research Fund, London. Biochemical, cell biological, and molecular genetic techniques are being used to elucidate the mechanisms by which the *ras*-related Rab GTPase group of proteins regulate the intracellular movement of proteins and lipids and communicate with other signal transduction pathways within the cell.

Clare Fewtrell, Associate Professor of Pharmacology; D.Phil., Oxford. Cellular physiology and, in particular, the role of calcium in stimulus-secretion coupling in cells in the immune system are being studied. Biophysical, biochemical, and pharmacological approaches are being used to examine the receptor for immunoglobulin E on mast cells and the signal transduction mechanisms leading to an increase in intracellular calcium and the release of mediators of allergic reactions. Techniques include the use of fluorescent and luminescent probes to measure intracellular Ca^{2+} and membrane potential, digital video imaging of intracellular Ca^{2+} in individual cells, and single-cell amperometry.

Robert F. Gilmour Jr., Professor of Physiology; Ph.D., SUNY Upstate Medical Center. Identification of the cellular electrophysiological mechanisms for cardiac arrhythmias using ionic and dynamical computer models, voltage clamp studies of isolated myocytes, and high-resolution mapping of intact cardiac tissue. Emphasis on the application of nonlinear systems analysis techniques to the study of cellular cardiac dynamics.

Jun-Lin Guan, Professor; Ph.D., California, San Diego. Signal transduction in cell migration; FAK signaling pathways in angiogenesis and metastasis; role of putative tumor suppressor FIP200 in development and cancer.

George P. Hess, Professor of Biochemistry; Ph.D., Berkeley. The mechanisms of chemical reactions controlling signal transmission between the billions of neurons in the nervous system are being investigated. They are modified by hundreds of therapeutic agents and abused drugs and in diseases of the nervous system. New biophysical techniques are developed, including laser-pulse photolysis of caged neurotransmitters with a microsecond time resolution, and used to access new information about the reactions.

Michael I. Kotlikoff, Professor and Chair, Department of Biomedical Sciences; V.M.D., Pennsylvania; Ph.D., California, Davis. Studies focus on the molecular processes underlying excitation-contraction coupling in smooth muscle cells, seeking to understand the role of specific sarcoplasmic reticular calcium release channels and sarcolemmal ion channels. Major projects center on the molecular processes regulating calcium release from the sarcoplasmic reticulum and the development of improved methods for conditional gene inactivation and in vivo imaging.

Roy A. Levine, Associate Professor of Molecular Medicine; Ph.D., Indiana Bloomington. Understanding the molecular mechanisms by which regulators of proliferation and differentiation interact to modulate cellular behavior. The current focus is on using biochemical and molecular genetic techniques to identify the oncogenes, tumor suppressor genes, and signaling pathways that are aberrantly regulated in canine osteosarcoma and the mechanisms by which these mutant proteins and pathways alter cellular proliferation and differentiation.

Manfred Lindau, Professor of Applied and Engineering Physics; Dr.rer.nat., Berlin Technical. Investigation of the regulation and mechanisms of exocytosis in different cell types, including nerve terminals, chromaffin cells, and granulocytes. Focus is on measuring exocytosis of single vesicles to study the mechanisms of and relations between docking, fusion pore formation and expansion, and transmitter release. Methods include patch clamp capacitance measurements, amperometry, patch amperometry, electrochemical imaging, total internal reflection fluorescence microscopy, computer modeling, optical trap, molecular biology, and biochemistry.

Danny Manor, Assistant Professor; Ph.D., Yeshiva (Einstein). Understanding the signal transduction pathways that regulate normal cell growth and that are disrupted by oncogenic mutation and understanding the molecular mechanisms by which some chemo-preventative agents, such as vitamin E, offer protection from cancer.

Linda M. Nowak, Associate Professor of Pharmacology; Ph.D., Michigan. Investigations of the mechanism(s) of activation of excitatory amino acid receptor-channels and channel pore permeability are being conducted in native receptors expressed by cultured neurons and recombinant receptors expressed in *Xenopus* oocytes and mammalian cell lines. Techniques include primary neuronal and cell line culture, whole-cell and single-channel electrophysiology, hidden Markov analysis of channel kinetics, and molecular biology, including single-cell mRNA analysis.

Robert E. Oswald, Professor of Pharmacology; Ph.D., Vanderbilt. Ligand-gated ion channels (nicotinic acetylcholine receptors and glutamate receptors) and cell signaling proteins (Cdc42) are being studied. A multidisciplinary approach has combined the use of a number of techniques (radioligand binding, single-channel recording, multidimensional NMR spectroscopy, protein purification, mathematical modeling, and molecular biology) to measure functional activity and to relate the function to the structure of these proteins.

Bendicht U. Pauli, Professor; D.V.M., Ph.D., Berne (Switzerland). Studies include the molecular events of early metastasis, focusing on tumor cell/endothelial cell adhesion molecules that mediate vascular arrest of blood-borne cancer cells at preferred metastatic sites and initiate signaling cascades that promote tumor cell survival and growth at the new sites.

Mark S. Roberson, Associate Professor of Physiology; Ph.D., Nebraska. Regulatory mechanisms of reproductive hormone and growth-factor action in the anterior pituitary gland and primate and mouse placenta are studied. Cellular, molecular, and mouse genetic approaches are used to examine the molecular determinants governing fertility in mammals.

Wayne S. Schwark, Professor of Pharmacology; D.V.M., Guelph; Ph.D., Ottawa. This laboratory is concerned with pharmacokinetic/clinical pharmacological studies of domestic animals in order to provide a more rational basis for drug therapy in veterinary medicine.

Geoffrey W. G. Sharp, Professor of Pharmacology; Ph.D., Nottingham (England); D.Sc., London. Investigations are on the regulation of insulin secretion, with current emphasis on (a) the mechanisms by which G proteins mediate the effects of α-adrenergic agents, somatostatin, and galanin to inhibit exocytosis; (b) the mechanisms of augmentation of insulin release by glucose and other nutrients; and (c) the mechanism of exocytosis in pancreatic β-cells.

Holger Sondermann, Assistant Professor; Ph.D., Cologne (Germany). Deciphering the basic regulatory principles in signal transduction networks on a molecular level, focusing on growth-factor receptor signaling. In particular, studies the role of scaffolding proteins and other signal transduction regulators, using a combination of X-ray crystallography, biophysical, and cellular approaches, to elucidate how cells respond to various inputs producing distinct outputs by using a limited set of molecules.

Watt W. Webb, Professor of Applied and Engineering Physics and S. B. Eckert Professor in Engineering; Sc.D., MIT. Studies include the dynamics of biophysical processes in living cells, using modern physical optics; molecular transmembrane signaling interactions of immunological and hormone receptors; and use of digital fluorescence image processing and advanced digital microscopy techniques to track individual receptors in motion on the surfaces of living cells. Multiphoton microscopy by laser scanning probes microscopy images of biophysical processes and executes micropharmacology at the submicron scale in living cells and deep in living tissues. Fluorescence correlation spectroscopy measures molecular dynamics, molecular interactions, and diffusion in very dilute solutions and in living cells and tissues.

Gregory A. Weiland, Associate Professor of Pharmacology; Ph.D., California, San Diego. Studies of the molecular and cellular mechanisms of modulation of neurotransmitter and drug receptor function and specifically of the modulation and regulation of nicotinic acetylcholine receptor function and binding properties by cholinergic ligands, neuropeptides, and second-messenger systems. Techniques include radioligand binding, radioactive ion flux, molecular biology, electrophysiology, and cell culture.

Gary R. Whittaker, Assistant Professor of Virology; Ph.D., Leeds (England). Molecular cell biology of virus entry; receptor binding, endocytosis, and fusion, with a focus on influenza virus and coronavirus; mechanisms of action of antiviral drugs.

DARTMOUTH COLLEGE

Dartmouth Medical School
Program in Pharmacology and Toxicology

Program of Study

The Department of Pharmacology and Toxicology offers a graduate program leading to the Ph.D. degree. This program, which normally requires four to five years for completion, accommodates approximately 30 students and features close student research relationships with individual faculty members. There are opportunities for interdisciplinary research projects with faculty members in departments of Dartmouth College and Dartmouth Medical School.

All students (except those admitted with advanced standing) are expected to complete courses in biochemistry, physiology, and pharmacology in the first eighteen months. Elective courses may be taken at any time. Three separate research rotations acquaint students with technical approaches to the discipline, expose them to the investigative process, and provide an in-depth view of various laboratories. It is expected that all students choose a laboratory for their thesis research within nine months of entering the program.

The Department of Pharmacology and Toxicology at Dartmouth has well-funded research programs in the areas of cell and molecular pharmacology and toxicology, with special strengths in cancer, carcinogenesis, cellular and molecular biology of cancer, molecular basis for chemotherapy, neuropharmacology, cardiovascular pharmacology, and cellular basis for drug action. Although these areas receive particular emphasis in the graduate training program, they do not limit or exclude opportunities for training in other pharmacological subdisciplines in which the faculty has interest and expertise (details are given in The Faculty and Their Research section). Diverse postdoctoral research opportunities exist as well. The goal of the department's graduate program is to prepare individuals for research careers in academic, industrial, or government settings.

Research Facilities

Laboratory and instructional facilities of the basic science departments, including those of the Pharmacology and Toxicology faculty members, are located in the Medical School buildings (which are located on the Hanover, Lebanon, and White River Junction campuses). The department has close ties to the Norris Cotton Cancer Center, an NCI-designated comprehensive cancer center that is located at the Dartmouth-Hitchcock Medical Center and that provides state-of-the-art facilities for DNA synthesis and sequencing, protein analysis, cell imaging, cell sorting, microarray, and bioinformatics.

Financial Aid

In 2005–06, all students accepted into the program received full-tuition scholarships and a stipend award of $23,250, plus a 1-person student health insurance policy. Dartmouth is an authorized lender under the federally insured student loan program.

Cost of Study

Tuition for the 2005–06 academic year was set at $42,360, but all tuition costs are covered by scholarships for accepted students.

Living and Housing Costs

Dartmouth College assists graduate students in arranging for appropriate housing in College facilities or can provide a list of privately owned accommodations in the Hanover area. College apartments range in fees from approximately $580 to $860 per month. Rents for off-campus housing vary widely, but accommodations can often be found at rates similar to those for College housing.

Student Group

The undergraduate student body numbers approximately 5,000, and the graduate and professional school enrollment is about 1,375, including approximately 500 Ph.D. candidates in graduate programs of arts and sciences.

Location

Dartmouth College is located in Hanover, New Hampshire, a town of about 10,000 on the border of New Hampshire and Vermont. Hanover is less than 2 hours driving time from Boston and about 3 hours from Montreal. The Hanover area provides excellent opportunities for hiking, canoeing, climbing, and skiing and is located near many of northern New England's lake and ski resorts.

The Medical School and The College

Dartmouth Medical School, founded in 1797, is located on the campus of Dartmouth College. It combines modern facilities with tutorial instruction in the basic medical sciences. Dartmouth Medical School is also one of the four entities of the Dartmouth-Hitchcock Medical Center, along with the Hitchcock Clinic, the Mary Hitchcock Memorial Hospital, and the Veterans Administration Medical Center.

The cultural focus of the College is the Hopkins Center for the Performing Arts. The center sponsors a film society, many concerts, and an active drama program. In addition, all students and faculty members have access to workshops for sculpture, painting, and various crafts as well as to membership in several choral and instrumental music groups. Dartmouth also makes available to its graduate students the extensive facilities of the Dartmouth Outing Club and the Dartmouth College Athletic Council.

Applying

The basic requirements for admission are an academic degree from an accredited institution of higher education and a superior record of performance in science, engineering, pharmacy, premedicine, or another appropriate field. A Graduate Record Examinations score (or a Medical College Admission Test result) is mandatory. At least two letters of recommendation, all official transcripts, and a completed Dartmouth College application form for graduate study are required. Prospective students must submit completed application forms and all supporting materials no later than January 19, 2007, for fall entrance. Earlier application is encouraged.

Correspondence and Information

Coordinator of Graduate Recruitment
Department of Pharmacology and Toxicology
7650 Remsen, Room 522
Dartmouth Medical School
Hanover, New Hampshire 03755-3835
Phone: 603-650-1150 or
      603-650-1667
E-mail: pharmacology.and.toxicology@dartmouth.edu
Web site: http://dms.dartmouth.edu/pharmtox/

Dartmouth College

THE FACULTY AND THEIR RESEARCH

Michael Cole, Ph.D., 1978; Professor of Pharmacology and Toxicology and of Genetics and member, Norris Cotton Cancer Center. Molecular basis of cancer; oncogenic activation of transcription factors; mechanisms of chromosome-mediated transcriptional control; target genes for oncogenic pathways.

Ruth W. Craig, Ph.D., 1984; Professor of Pharmacology and Toxicology. Molecular mechanisms of induction of differentiation and reversal of the leukemic phenotype; oncogenes, tumor suppressor genes, and genes affecting the programming of differentiation and viability, such as *mcl*-1.

Joyce A. DeLeo, Ph.D., 1988; Professor of Anesthesiology and of Pharmacology and Toxicology. Neuropharmacology; neuroimmunology: mechanisms that lead to chronic pain with a focus on CNS neuroimmune responses; utilization of molecular, cellular, and in vivo behavioral pharmacological approaches; developing novel and nonaddictive therapeutics for the prevention and treatment of chronic pain.

James DiRenzo, Ph.D., 1995; Assistant Professor of Pharmacology and Toxicology. Cellular differentiation within the mammary gland; identifying the genetic and hormonal signals that underlie complex gene regulatory networks and characterizing the consequences of their subversion.

Ethan Dmitrovsky, M.D., 1980; Andrew G. Wallace Professor of Pharmacology and Toxicology and of Medicine and Chairman of Pharmacology and Toxicology. Mechanisms by which the retinoids, natural and synthetic derivatives of vitamin A, signal tumor cell differentiation and growth suppression; role of nuclear retinoid receptors in signaling tumor cell differentiation and chemoprevention.

Alan R. Eastman, Ph.D., 1975; Professor of Pharmacology and Toxicology and Associate Director for Basic Sciences, Norris Cotton Cancer Center. Targeting cancer chemotherapy selectively to tumors; mechanisms of cell-cycle arrest by DNA-damaging agents and abrogation of this arrest selectively in p53-defective tumors; defining signal transduction pathways that inhibit apoptosis in tumors and modulating these pathways to enhance drug-mediated cell killing; customizing therapy based on knowledge of the molecular phenotype of individual tumors.

Matthew J. Friedman, Ph.D., 1967, M.D., 1969; Professor of Psychiatry and of Pharmacology and Toxicology and Executive Director, Department of Veterans Affairs National Center for Posttraumatic Stress Disorder (PTSD). Psychobiological alterations associated with PTSD in humans; neurogenesis and apoptosis in brain among transgenic mice exposed to fear-potentiated startle protocols; pharmacotherapy for PTSD patients.

Alan I. Green, M.D. 1969; Raymond Sobel Professor of Psychiatry and of Pharmacology and Toxicology and Chairman of the Department of Psychiatry. Neuropharmacology: mechanisms involved in the abuse of alcohol and other substances in patients with schizophrenia and development of medications to reduce such abuse; techniques involving behavioral, molecular, and pharmacological studies in alcohol-preferring rodents and schizophrenia-model rodents as well as neuroimaging studies of brain reward circuitry in patients with schizophrenia.

Joshua W. Hamilton, Ph.D., 1985; Professor of Pharmacology and Toxicology; Adjunct Professor of Chemistry; Director, Center for Environmental Health Sciences; and Director, Toxic Metals Research Program Project. Molecular toxicology and pharmacology; molecular epidemiology; toxicogenomics and toxicoproteomics; molecular basis for effects of toxic metals and other environmental chemicals on gene expression and role in environmentally induced diseases; cellular and molecular targets for effects of arsenic and other toxic metals as endocrine disrupting chemicals.

John Hwa, M.D., 1986; Ph.D., 1996; Assistant Professor of Pharmacology and Toxicology. Molecular pharmacology and pharmacogenetics of G-protein–coupled receptors: structure-function analysis of human prostacyclin receptor polymorphisms; analysis of naturally occurring retinitis pigmentosa rhodopsin mutations.

Alexei F. Kisselev, Ph.D., 1995; Assistant Professor of Pharmacology and Toxicology and member, Norris Cotton Cancer Center. Ubiquitin-proteasome pathway; development and therapeutic application of proteasome inhibitors, especially as it relates to tumorigenesis.

Murray Korc, M.D., 1974; Joseph M. Huber Professor of Medicine, Professor of Pharmacology and Toxicology, and Chairman, Department of Medicine. Mechanisms of action of growth factors; the role of growth factors and their receptors in carcinogenesis, especially as it relates to enhanced mitogenic signaling, enhanced angiogenesis, and attenuated growth suppressive effects in pancreatic cancer.

Lionel D. Lewis, M.D., 1988; Professor of Medicine and of Pharmacology and Toxicology. Clinical of novel antineoplastic agents in phase I (early development in man), particularly the relationship between the pharmacokinetics and pharmacodynamics of these agents in "proof of concept" studies; mechanisms of toxicity of nucleoside analogues and antineoplastic agents to the mitochondrion.

Christopher H. Lowrey, M.D., 1985; Associate Professor of Medicine and of Pharmacology and Toxicology. Regulation of chromatin structure and its relationship to gene expression; gene therapy; using the human β-globin genes as a model system to understanding how domains of transcriptionally active chromatin structure are formed; designing more effective vectors for gene therapy and new pharmacologic strategies for the therapeutic manipulation of gene expression.

Kathleen A. Martin, Ph.D., 1997; Research Assistant Professor of Surgery and of Pharmacology and Toxicology. Molecular mechanisms regulating vascular smooth muscle cell proliferation and differentiation aimed at prevention/treatment of atherosclerosis and intimal hyperplasia; signaling through the mTOR (rapamycin) or PKA (prostacyclin) pathways to smooth muscle–specific transcription factors.

David W. Nierenberg, M.D., 1976; Edward Tulloh Krumm Professor of Medicine and of Pharmacology and Toxicology. Clinical pharmacology of antioxidant micronutrients and vitamins and their role in health and disease prevention; clinical pharmacology and pharmacodynamics of methotrexate and other antineoplastic agents.

Raymond P. Perez, M.D., 1985; Associate Professor of Medicine and of Pharmacology and Toxicology and Associate Director for Clinical Research, Norris Cotton Cancer Center. Clinical pharmacology and early trials of anticancer agents, particularly molecular-targeted therapies; mechanism of action and proof-of-principle for novel drugs and combinations; modulation of signal transduction and cell death, particularly focused on endogenous inhibitors of receptor kinase–mediated signaling and downstream determinants of sensitivity to apoptosis.

David J. Robbins, Ph.D., 1993; Associate Professor of Pharmacology and Toxicology. Elucidating the mechanism of action of the Hedgehog family of secreted proteins, as they pertain to cancer.

Bill D. Roebuck, Ph.D., 1975; Professor of Pharmacology and Toxicology. Modulation of toxic processes and cancer by infection; evaluation of the effects of dietary dithiolethiones on prevention of liver cancer; chemoprevention of cancer.

Yolanda Sanchez, Ph.D., 1993; Associate Professor of Pharmacology and Toxicology. Checkpoint signaling events triggered during the response to DNA damage or replication interference; how they regulate cell-cycle progression, DNA repair, and cell death; their role in the etiology and treatment of cancer.

Michael Simons, M.D., 1984; Anna Gundlach Huber Professor of Medicine (Cardiology) and of Pharmacology and Toxicology; Chief, Section of Cardiology; and Director, Angiogenesis Research Center. Mechanisms of angiogenic growth factor signaling, with emphasis on the role of heparan sulfate proteins.

Jacqueline A. Sinclair, Ph.D., 1977; Research Associate Professor of Biochemistry and of Pharmacology and Toxicology and Associate Research Career Scientist, VA Medical Center. Mechanisms by which consumption of alcoholic beverages increases liver damage from acetaminophen and arsenic decreases cytochrome P450s in the liver and the relation of such decreases to development of cancer and liver damage.

Peter R. Sinclair, Ph.D., 1970; Research Associate Professor of Biochemistry and of Pharmacology and Toxicology and Associate Career Scientist, VA Medical Center. Mechanism of hepatic disease porphyria cutanea tarda, especially the roles of alcohol, cytochrome P450s, and iron metabolism.

Stephen P. Spielberg, Ph.D., 1971, M.D., 1973; Professor of Pediatrics and of Pharmacology and Toxicology and Dean, Dartmouth Medical School. Human pharmacogenetics; pathogenesis of idiosyncratic adverse drug reactions and role of reactive electrophilic drug metabolites; pediatric and developmental pharmacology; clinical pharmacology trials in pediatric patients.

Michael J. Spinella, Ph.D., 1992; Associate Professor of Pharmacology and Toxicology. Identification of mechanistic links between stem cell pluripotency and cancer and identification of downstream genes and pathways signaling induced differentiation of human solid tumor cells, especially in response to retinoids; finding causative genes in those tumors that are cured with differentiation and cytotoxic therapy.

Michael B. Sporn, M.D., 1959; Oscar M. Cohn '34 Professor in Residence of Pharmacology and Toxicology. Chemoprevention of cancer, especially by retinoids and other ligands of the steroid receptor superfamily; peptide growth factors, especially transforming growth factor–β (TGF-β) and its mechanism of action; development of new natural products for prevention of cancer.

Craig Tomlinson, Ph.D., 1984; Assistant Professor of Medicine and of Pharmacology and Toxicology. Basis of adult diseases from in utero exposure to environmental toxicants; interaction of the aryl hydrocarbon receptor and transforming growth factor–β signaling pathways; use of high-throughput genomics as a tool to predict the outcome of gene-environment interactions.

Herbert Yohe, Ph.D., 1976; Research Associate Professor of Pharmacology and Toxicology and Research Chemist, VA Medical Center. Elucidation of sialoglycolipid structures and their function in the immune system: these critical components of the membrane, with their alteration during differentiation and oncogenesis, are directly tied to the functional status of the cell and hold potential as targets for pharmaceutical agents.

DUKE UNIVERSITY

Graduate School
Integrated Toxicology and Environmental Health Program

Program of Study

Training in toxicology and environmental health at Duke for predoctoral students and postdoctoral fellows who hold Ph.D., M.D., or D.V.M. degrees emphasizes an integrated approach to teaching and research. The interdepartmental program brings together students and faculty members who are trained in various specific disciplines to interact in the examination of toxicological and environmental health problems. Faculty members are aligned with one or more of the following subdisciplines of toxicology: genetic toxicology and epigenetics, cancer, developmental toxicology and children's health, neurotoxicology and neurological disease, environmental exposure and toxicology, and pulmonary toxicology and disease. The primary mission of this program is to give students a sound theoretical base for a career in toxicology and environmental health with sound laboratory training in experimental toxicology.

Normally, four years of study and research are required to complete the Ph.D. Preliminary examinations are administered in the second or third year by the department with which the student is affiliated. Postdoctoral training usually extends to two years of study. A two-year professional master's degree program in environmental toxicology, chemistry, and risk assessment (ETCRA) is offered by the Nicholas School of the Environment and Earth Sciences.

Research Facilities

The proximity of the University to the Duke University Medical Center provides an excellent opportunity to pursue toxicological research that is highly relevant to human health problems. Several well-funded research laboratories are available to trainees and are equipped for conducting toxicological research with state-of-the-art technology, including those at the Duke University Medical Center, the Nicholas School of the Environment and Earth Sciences, and Trinity College. Other research facilities include the Duke University Marine Laboratory in Beaufort, North Carolina; the National Institute of Environmental Health Sciences (NIEHS); CIIT Centers for Health Research; and the labs of the Environmental Protection Agency at Research Triangle Park.

The Durham campus has excellent computer facilities and a library that contains 4.7 million volumes.

Financial Aid

Outstanding students are offered admission to the training program with full fellowship support of $24,000 per year, plus tuition and fees, on the basis of their academic record, GRE scores, and letters of recommendation. Departmental aid sources and work-study funds are also available to students who can demonstrate financial need.

Cost of Study

Estimated tuition for full-time students is $28,080 per year ($14,040 per semester) in 2006–07. Students also pay a registration fee of $6900 and a health fee of $704. The program pays tuition and fees (including health insurance) for most students.

Living and Housing Costs

Estimated University housing for the 2005–06 academic year averaged $5450 per person. Costs included all utilities except telephone expenses. Rents for private housing ($400 to $900 and up per month) do not usually include the cost of utilities. Current off-campus listings are available online at http://communityhousing.duke.edu/index.php.

Student Group

The total enrollment at Duke University is 13,123. Approximately 2,400 students are currently enrolled in the Graduate School. Each year, 5–7 new students enter the toxicology program. In 2005, there were 28 trainees in the program.

Location

Durham, located midway between the Atlantic Ocean and the Appalachian Mountains, is a city of about 190,000 residents. Facilities available to graduate students include swimming pools, tennis courts, racquetball courts, a handball court, tracks, baseball fields, hiking trails, and the Duke golf course. The three cities of the Research Triangle—Durham, Chapel Hill, and Raleigh—provide a varied cultural environment. The American Dance Festival is in summer residence at Duke. There are many theater groups and chamber, symphonic, choral, and jazz music groups; arts and crafts organizations are especially active in the area. In the sciences, there is extensive interaction with the University of North Carolina at Chapel Hill, North Carolina State University at Raleigh, and the Research Triangle Park, where a number of industrial and government organizations have large research establishments. Spouses of graduate students are employed throughout Duke University, and additional employment opportunities are available in the area.

The University

Duke, a private university, dates as a corporate entity from 1924; however, its roots go back more than 150 years to Union Institute, founded in 1838. Its Graduate School has a faculty of more than 600 members. The campus is bordered by the Duke Forest, 8,500 acres of rolling, wooded land on the southwest fringe of Durham. The proximity of all of the biological science departments, including those in the Medical Center, facilitates interdepartmental cooperation and results in a wide range of course offerings, research opportunities, and seminars. Duke has a reciprocal arrangement with the University of North Carolina at Chapel Hill and with North Carolina State University at Raleigh that allows the enrollment of Duke students in courses at these nearby institutions.

Applying

Applicants who have strong undergraduate training in the biological sciences, chemistry, physics, and mathematics are favored. Scores on the General Test of the GRE are required. Students are normally admitted only in the fall semester. Applicants who wish to be considered for financial aid should submit all application materials before the end of December, including transcripts and three letters of recommendation.

Application to the program can be made in two ways. Students with a primary interest in toxicology may apply for admission directly through the Integrated Toxicology Program, indicating "Toxicology" as the primary admitting unit on the standard graduate school application, which can be found online at http://www.gradschool.duke.edu/. These students affiliate with a department, depending upon their choice of research mentor, usually in the second year.

Students with a primary interest in a departmentally based field may also apply to the Integrated Toxicology Program by indicating "Toxicology" as the secondary field on the standard graduate school application. The primary field should indicate the specific graduate department in Arts and Sciences, the School of Medicine, or the Nicholas School of the Environment and Earth Sciences. These students affiliate with a specific department in their first year. There is no difference in the eventual degree granted through either mechanism; both routes result in a Ph.D. granted by a specific department, with certification in toxicology.

Correspondence and Information

<table>
<tr><td>

Ph.D. and postdoctoral candidates:
Richard T. DiGiulio, Ph.D.
Director, Integrated Toxicology Program
Box 90328
Duke University
Durham, North Carolina 27708
Phone: 919-613-8024
E-mail: toxicology@duke.edu
Web site: http://www.duke.edu/web/toxicology

</td><td>

Master's in Environmental Health and Security:
David E. Hinton, Ph.D.
Nicholas School of the Environment and Earth   Sciences
Box 90328
Duke University
Durham, North Carolina 27708-0328
E-mail: dhinton@duke.edu

</td></tr>
</table>

Duke University

THE FACULTY AND THEIR RESEARCH

Resident Duke University faculty members, representing seven departments, are associated with both the Durham campus and the Duke University Marine Lab in Beaufort, North Carolina.

Celia Bonaventura, Ph.D., Professor of Cell Biology and Director, Marine Biomedical Center. Abnormal human hemoglobins and other respiratory proteins; marine biotechnology.

Christopher Counter, Ph.D., Assistant Professor of Pharmacology and Radiation Oncology.

Richard T. Di Giulio, Ph.D., Professor of Environmental Toxicology. Mechanisms of toxicity and adaptation in fishes; ecotoxicology.

Rodney Folz, M.D., Ph.D., Associate Professor of Medicine. The molecular and cellular impact oxidative stress has on a variety of pathophysiologic processes that occur primarily in lung, heart, and kidney.

Jonathan H. Freedman, Ph.D., Associate Professor of Environmental Toxicology. Cell and molecular toxicology (metal toxicology).

Irwin Fridovich, Ph.D., James B. Duke Professor Emeritus of Biochemistry. Mechanisms of action of enzymes; oxygen metabolism.

David Hinton, Ph.D., Professor of Environmental Toxicology. Cellular and developmental toxicity in fishes; ecotoxicology.

Randy L. Jirtle, Ph.D., Professor of Radiation Oncology and Associate Professor of Pathology. Hepatotoxicity and growth factors.

Cynthia M. Kuhn, Ph.D., Professor of Pharmacology and Cancer Biology. Developmental neuropharmacology and neuroendocrinology.

Edward D. Levin, Ph.D., Associate Professor of Psychiatry, Pharmacology, and Environment and Psychology. Cognitive effects of *Pfiesteria* to chemicals that affect the acetylcholine system; chlorpyrifos; neurobehavioral development; zebrafish neurotoxicology.

Haifan Lin, Ph.D., Associate Professor, Department of Cell Biology. Gene and drug delivery by acoustic cavitation; role of E2A during mouse oogenesis.

Elwood A. Linney, Ph.D., Professor of Microbiology and Immunology. Developmental effects in transgenic zebrafish.

Donald McDonnell, Ph.D., Professor of Pharmacology and Cancer Biology. Endocrine disruptors and developmental toxicology; mutagenesis and carcinogenesis.

David R. McClay, Ph.D., Professor of Neurobiology and Marine Science.

James McNamara Sr., M.D., Carl R. Deane Professor of Neuroscience. Neurotoxicity: cell and molecular neurobiology.

Warren H. Meck, Ph.D., Professor and Director, Undergraduate Neuroscience Program. Behavioral neuroscience.

Marie Lynn Miranda, Ph.D., Gable Associate Professor of the Practice in Environmental Ethics and Sustainable Environmental Management. Environmental economics and policy and environmental justice.

Paul L. Modrich, Ph.D., James B. Duke Professor of Biochemistry and Investigator for the Howard Hughes Medical Institute. Mismatch repair in replication fidelity, tumor biology, and drug resistance; mapping genetic mutations.

John N. Norton, D.V.M., Ph.D., Associate Professor, Department of Pathology.

Claude A. Piantadosi, M.D., Professor of Medicine. Critical-care medicine; interstitial lung diseases; sarcoidosis; pulmonary alveolar proteinosis; diving and hyperbaric medicine.

Barbara Ramsay-Shaw, Ph.D., Professor of Chemistry. Molecular mechanisms of DNA mutagenesis.

Daniel Rittschof, Ph.D., Associate Professor of Zoology. Antifouling technology; chemical and behavioral toxicology.

William H. Schlesinger, Ph.D., James B. Duke Professor of Biogeochemistry and Dean, Nicholas School of the Environment and Earth Sciences. Ecosystem analysis; global change; environmental toxicology.

Donald E. Schmechel, M.D., Associate Professor of Neurology. Alzheimer's disease; dementia; neurodegenerative disorders; toxic encephalopathy.

Sidney A. Simon, Ph.D., Professor of Neurobiology. Chemosenses as related to gustatory and trigeminal interactions.

J. H. Pate Skene, Ph.D., Associate Professor of Neurobiology. Molecular mechanisms by which the maturation of axons and their terminals leads to repression of neuronal genes and the consequences of this repression for synaptic plasticity and axon regeneration in adults.

Heather M. Stapleton, Ph.D., Assistant Professor of Environmental Sciences and Policy. Fate and biotransformation of organic contaminants in aquatic systems, with a focus on persistent organic pollutants (POPs), such as polychlorinated biphenyls (PCBs) and polybrominated diphenyl ethers (PBDEs).

Theodore A. Slotkin, Ph.D., Professor of Pharmacology and Cancer Biology and of Psychiatry. Developmental toxicology: effects of environmental contaminants, stress, and drugs on central and peripheral nerves of the fetus and neonate.

Jonathan S. Stamler, M.D., Professor of Medicine. Elucidating NO-responsive signaling pathways in mammalian systems subject to redox control; oxidative stress.

H. Scott Swartzwelder, Ph.D., Professor of Psychology–Experimental. Neurotoxicology; developmental toxicology.

Xiao-Fan Wang, Ph.D., Associate Professor of Pharmacology and Cancer Biology. Cell and molecular toxicology; cancer biology.

A. Richard Whorton, Ph.D., Associate Professor of Pharmacology and Cancer Biology. Oxidative stress.

Christina L. Williams, Ph.D., Professor and Chair, Department of Psychology–Experimental.

Adjunct Faculty

Adjunct faculty members also play an active role in the program but may not serve as primary advisers for graduate students.

Syed Ali, Ph.D. Cellular and molecular mechanisms of CNS-acting drugs and drug abuse; pesticides; organometals neurotoxicity; aging and developmental neurotoxicology. FDA.

James W. Allen, Ph.D. Environmental toxicology; heat-shock proteins in mutagenesis; cytogenetic evaluations of chemical mutagens/carcinogens and meiotic mechanisms in germ-cell mutagenesis. U.S. EPA.

Linda S. Birnbaum, Ph.D., DABT. Environmental toxicology; mechanisms of toxicity of dioxin and related compounds and developmental effects. U.S. EPA.

William K. Boyes, Ph.D. Neurophysiological toxicology; neurotoxicity of formamidine and organophosphate pesticides in the visual system. U.S. EPA.

Philip Bushnell, Ph.D. Neurotoxicity; effects of neurotoxic chemicals, including organophosphate, organic solvents, and PCBs on cognitive functions. U.S. EPA.

Byron E. Butterworth, Ph.D. Pathology; mechanisms of chemical carcinogenesis; predictive assays for carcinogenic activity, risk assessment, and public cancer policy. CIIT Centers for Health Research.

J. Donald de Bethizy, Ph.D., DABT. Pharmacology/toxicology; pharmacokinetics; physiologically based pharmacokinetic modeling; pharmacokinetic and pharmacodynamic modeling; xenobiotic metabolism. Targacept.

David C. Doolittle, Ph.D. Physiology and pharmacology; molecular biology of human lung cancer, with particular attention to genetic predispositions that make individuals more or less susceptible to this disease. R.J. Reynolds.

David C. Dorman, Ph.D. Neurotoxicology; assessment of the neurotoxicity of manganese, methanol, and other marine organisms; mechanisms of trace metal metabolism and sequestration in marine organisms. CIIT Centers for Health Research.

Timothy R. Fennell, Ph.D. Biochemistry; metabolism of toxins and carcinogens; estimation of exposure and internal dose through use of biomarkers; dosimetry and metabolism of endocrine toxicants; pharmacokinetics; NMR spectroscopy. CIIT Centers for Health Research.

Gaylia Jean Harry, Ph.D. Neurotoxicology; neural glial responses of the developing central nervous system to lead and other toxicants. NIEHS.

Gregory Kedderis, Ph.D. Biochemistry; mechanisms of genotoxicity and chemical carcinogenesis; pharmacokinetics; relationship between dose and biological effects; enzymology. CIIT Centers for Health Research.

Patrick M. Lippiello, Ph.D. Pharmacology; investigation of neuronal nicotinic receptors as novel therapeutic targets. Targacept.

Ronald P. Mason, Ph.D. Chemistry; free-radical metabolites of nitro and azo compounds. NIEHS.

Patricia McClellan-Green, Ph.D. Natural and man-made toxins in the marine environment and their effects on the metabolic activities of marine organisms.

Kevin T. Morgan, Ph.D. Experimental pathology and toxicology; upper airway function and toxicology, with emphases on regional dosimetry of inhaled gases and vapors, nasal carcinogenesis, and olfactory toxicity. GlaxoSmithKline.

Virginia Moser, Ph.D. Pharmacology/toxicology; neurobehavioral consequences of exposure to environmental chemicals. U.S. EPA.

Hernan Navarro, Ph.D. RTI International.

R. Julian Preston, Ph.D. Genetics and molecular biology; mechanism of induction of specific chromosome alterations and mutations in somatic and germ cells by chemicals and radiation. U.S. EPA.

Mark E. Stanton, Ph.D. Developmental neurotoxicology; neurobiology of learning and memory development; psychobiology. U.S. EPA.

Michael D. Waters, Ph.D. Biochemistry/genetic toxicology; chemical toxicity profiles, hazard identification, and risk assessment. U.S. EPA.

THE JOHNS HOPKINS UNIVERSITY

School of Medicine
Department of Pharmacology and Molecular Sciences

Programs of Study

The Department of Pharmacology and Molecular Sciences offers a multidisciplinary program designed to prepare highly qualified individuals for careers in biomedical research. The focus of this doctoral program is on the molecular interactions of living systems and their application to pharmacology. Within this broad framework, students are encouraged to develop individually tailored programs of study to meet their particular research interests and career objectives.

During their first year, students take courses, participate in individual laboratory research rotations, and choose a research adviser. In the following year, students complete their core courses and examinations, select their focus area of doctoral research, choose a thesis advisory committee, and submit a thesis research proposal. During the remaining two to four years, students focus on their thesis research and take elective courses, many of which are offered by other departments and divisions of the medical institutions.

Graduate studies are offered in the following areas of research: design, synthesis, screening, and biological testing of chemotherapeutic agents and lead compounds; cell-signaling mechanisms; intercellular interactions; molecular and cell biology of cancer; cancer chemoprotection; nerve-cell signaling and neuropharmacology; mass spectrometry; molecular biology and biochemistry of parasites; molecular and cell biology of herpesviruses and retroviruses; neuropharmacology and psychopharmacology; drug and drug-target imaging; drug metabolism and pharmacokinetics; and programmed cell death.

The Department also participates in the Medical Scientist Training Program for M.D./Ph.D. candidates.

Research Facilities

The Department's laboratories provide complete resources for interdisciplinary research in biological chemistry, molecular biology and genetics, virology, cell biology, and chemical biology. These include facilities for chemical synthesis; gas and liquid chromatography; infrared, ultraviolet, and nuclear magnetic resonance spectroscopy; and mass spectrometry. Containment facilities are available for biological and chemical studies. The Department houses the Middle Atlantic Mass Spectrometry Laboratory, and students have access to modern institutional facilities for peptide and oligonucleotide synthesis and sequencing; state-of-the-art microscopy; production and care of transgenic animals; interactive graphics, for use in sequence analysis and molecular modeling; and the extensive clinical facilities of the Division of Clinical Pharmacology.

Financial Aid

Students admitted to the graduate program customarily receive an award that provides a stipend of $25,200 and full payment of tuition, fees, and individual health insurance premiums. These awards are funded by training and research grants from the National Institutes of Health and other funding agencies.

Cost of Study

Tuition, $32,200 per year in 2006–07, is usually fully covered.

Living and Housing Costs

A limited number of single rooms and apartments are available on campus for approximately $400 per month. Off-campus housing is also available, and the University provides free shuttle buses for the many graduate students who live in neighborhoods surrounding the School of Medicine and the University's Homewood campus.

Student Group

There are currently more than 50 graduate students in the Department and about 500 additional Ph.D. candidates at the School of Medicine. In addition, nearly 500 medical students and more than 1,000 students in the School of Hygiene and Public Health share the campus of the Johns Hopkins medical institutions.

Student Outcomes

Graduates typically go on to postdoctoral research fellowships followed by research and teaching careers in academic institutions, government, or industry.

Location

Baltimore is a vibrant waterfront city with a wide variety of recreational, entertainment, and cultural opportunities. From its Inner Harbor attractions to its nightlife in Fells Point, from its museums and galleries to its superb sports stadiums, Baltimore provides all of the amenities of one of America's major metropolises. Among its many cultural attractions are the Baltimore Symphony Orchestra, the Baltimore Opera Company, the Peabody Conservatory, the Baltimore Museum of Art, the Walters Art Gallery, Center Stage, and the Morris Mechanic Theater. The Inner Harbor, with its shops, restaurants, and attractions (including the National Aquarium and Maryland Science Center), brings the Chesapeake Bay into the heart of Baltimore. The city also supports professional teams in major-league baseball (the Orioles), NFL football (the Ravens), and indoor soccer (the Blast). The museums, historic landmarks, and theaters in Washington, D.C., are an hour away, and the cultural offerings of Philadelphia and New York are readily accessible by train or car. Year-round opportunities for outdoor recreation are provided by the Chesapeake Bay and Atlantic Ocean, the nearby Maryland countryside, and the mountains and rivers of western Maryland, Pennsylvania, Virginia, and West Virginia.

The School and The Department

The School of Medicine was founded in 1893 as part of Johns Hopkins University, the first American institution to place a primary emphasis on graduate education. The Department, also established in 1893, has served as a model for other pharmacology departments throughout the nation.

Applying

Students are typically admitted in September, although early arrival for summer research is encouraged. Applicants should have a bachelor's degree from a qualified college or university with a major in any of the biological, chemical, or physical sciences. Entering students are expected to have completed college-level courses in biology, chemistry (inorganic, organic, and physical), calculus, and physics; a strong background in biochemistry is particularly desirable. A completed online application, scores on the Graduate Record Examinations (General Test is required; Subject Test is encouraged but not required), at least two letters of recommendation, undergraduate transcript(s), and a statement of interest must be received by the January 10 deadline.

Correspondence and Information

Dr. James T. Stivers
Director of Admissions
Department of Pharmacology and Molecular Sciences
The Johns Hopkins University School of Medicine
725 North Wolfe Street
Baltimore, Maryland 21205
Phone: 410-955-7117
E-mail: jstivers@jhmi.edu
Web site: http://www.hopkinsmedicine.org/pharmacology/

The Johns Hopkins University

THE FACULTY AND THEIR RESEARCH

Darrell R. Abernethy, Professor; M.D./Ph.D., Kansas, 1976. Role of posttranscriptional modification of drug effectors that effect responses to cardiovascular drugs; drug exposure and functional status in older individuals.

Richard F. Ambinder, Professor; M.D., 1979, Ph.D., 1989, Johns Hopkins. Virology and human cancer; antiviral therapy; antitumor therapy; lymphoma pathogenesis, prevention, and treatment; immunological approaches to virus-associated malignancies.

L. Mario Amzel, Professor; Ph.D., Buenos Aires, 1968. Structural and computational methods applied to several protein families; immunoglobulins, lectins, and other binding proteins; ATP synthase; Nudix hydrolases; redox enzymes: monooxygenases, dioxygenases, and quinone reductases.

J. Thomas August, Professor; M.D., Stanford, 1955. Immunogenetic therapy of infectious diseases and cancer; genetic vaccine chimeras with LAMP trafficking of antigens to the MHC II compartment; vaccine analysis with nonhuman primates; immunocomputational models and *ex vivo* analysis with human patient cohorts of antigen epitropes; HIV-1 and flavivirus (dengue, West Nile, and Japanese encephalitis viruses) genetic vaccines; mechanisms of antigen trafficking and immune responses.

Philip A. Cole, Professor and Director; M.D./Ph.D., Johns Hopkins, 1991. Chemical and biochemical approaches in the study of signal transduction, circadian rhythm, and gene regulation.

Robert J. Cotter, Professor; Ph.D., Johns Hopkins, 1972. Development of new analytical techniques and instrumentation for mass spectrometry; applications of mass spectrometry to the structural analysis of peptides, glycopeptides, and glycolipids.

Albena Dinkova-Kostova, Assistant Professor; Ph.D., Washington State, 1996. Protection against cancer: mechanisms and strategies; structure-activity relation of protective agents; inflammation and cancer; skin cancer prevention.

Charles W. Flexner, Professor; M.D., Johns Hopkins, 1982. Antiretroviral therapy; basic and clinical pharmacology of HIV protease inhibitors, nucleosides, and viral entry inhibitors.

Robert H. Getzenberg, Professor; Ph.D., Johns Hopkins, 1992. Cancer biomarkers; proteomic analysis of nuclear structure; field effects in cancer.

Wade Gibson, Professor; Ph.D., Chicago, 1973. Herpesvirus proteins: studies of their expression, structure, and function using genetic, biochemical, biophysical, and immunological approaches.

Marc M. Greenberg, Professor; Ph.D., Yale, 1988. Use of organic chemistry, biochemistry, and molecular biology to study DNA damage, repair, and their applications.

Carol W. Greider, Professor; Ph.D., Berkeley, 1987. Telomerase and the consequences of telomere dysfunction; chromosome stability and genomic instability in cancer.

J. Marie Hardwick, Professor; Ph.D., Kansas, 1978. Molecular mechanisms of programmed cell death in mammals, yeast, and virus infections.

Gary S. Hayward, Professor; Ph.D., Otago (New Zealand), 1972. Pathways of herpesvirus gene regulation; cell-cycle control and latency; interaction of viral immediate-early transcriptional transactivators with nuclear subdomains; chemokines and molecular piracy by KSHV; role of herpesviruses in vascular disease.

S. Diane Hayward, Professor; Ph.D., Otago (New Zealand), 1972. Epstein-Barr virus and Kaposi's sarcoma–associated herpesvirus; viral latency and tumorigenesis; viral manipulation of cell signaling pathways.

Craig W. Hendrix, Associate Professor and Director, Drug Development Unit; M.D., Georgetown, 1984. Chemoprophylaxis of HIV infection; male genital tract pharmacology; topical microbicides.

Richard L. Huganir, Professor; Ph.D., Cornell, 1982. Molecular mechanisms in the regulation of neurotransmitter receptor function and synaptic transmission and plasticity in the brain.

Thomas W. Kensler, Professor; Ph.D., MIT, 1976. Molecular mechanisms of chemical carcinogenesis; reactive oxygen and antioxidants; cancer chemoprevention.

Jun O. Liu, Professor; Ph.D., MIT, 1990. Use of drugs and natural products as probes to study intracellular signal transduction pathways; molecular mechanism of regulation of angiogenesis and cell cycle.

Ernesto T. Marques, Assistant Professor; M.D., Pernambuco (Brazil), 1993; Ph.D., Johns Hopkins, 1999. Vaccine design and immunotherapies; immunomics; immunopathogenesis of dengue and HIV.

Caren L. Freel Meyers, Assistant Professor; Ph.D., Rochester, 1999. Organic and medicinal chemistry; chemical biology; drug delivery mechanisms in bacteria; development of antibiotic prodrug strategies; study of bacterial isoprenoid biosynthesis; combinatorial biosynthesis; development of potential therapeutic agents.

William G. Nelson, Professor; M.D./Ph.D., Johns Hopkins, 1987. Contributions of inflammation to prostatic carcinogenesis; role of DNA methylation in the pathogenesis of cancer; discovery and development of small molecules targeting epigenetic gene silencing in cancer; molecular mechanisms of chemoprevention against prostate cancer..

Martin G. Pomper, Associate Professor; M.D./Ph.D., Illinois at Urbana-Champaign, 1990. In vivo cellular and molecular imaging; imaging probe synthesis and development; multimodality functional imaging of cancer and brain, with focus on radionuclide-based techniques.

Gary H. Posner, Scowe Professor; Ph.D., Harvard, 1968. Organic and medicinal chemistry aimed toward rational design and synthesis of new compounds for effective and safe chemotherapy of malaria, cancer, and psoriasis.

Jonathan D. Powell, Assistant Professor; M.D./Ph.D., Emory, 1992. A functional genomic approach to dissecting and clinically exploiting T-cell activation and tolerance.

Ronald L. Schnaar, Professor; Ph.D., Johns Hopkins, 1976. Functional glycobiology of the nervous system; axon regeneration; cell-cell recognition; inflammation.

Theresa A. Shapiro, Professor; M.D., 1976, Ph.D., 1978, Johns Hopkins. Clinical pharmacology; molecular mechanisms of antiparasitic drug action; effects of topoisomerase inhibitors on DNA of trypanosomes; structure and activity of antimalarial trioxanes.

Robert F. Siliciano, Professor; M.D./Ph.D., Johns Hopkins, 1983. HIV latency, evolution, and persistence; HIV treatment and drug resistance.

Solomon H. Snyder, Professor; M.D., Georgetown, 1962. Molecular basis of neural signal transduction.

James T. Stivers, Associate Professor; Ph.D. Johns Hopkins, 1993. Molecular mechanisms of enzymatic DNA repair; DNA topoisomerase mechanism; NMR studies of enzymes and nucleic acids; enzyme inhibition and engineering.

Paul Talalay, Professor; M.D., Yale, 1948. Molecular mechanisms in chemical and dietary protection against mutagens and carcinogens.

Craig A. Townsend, Professor; Ph.D., Yale, 1974. Organic and bioorganic chemistry: biosynthesis of natural products and biomimetic synthesis; discovery of new enzyme reactions, their mechanistic and structural characterization; molecular biology of secondary metabolism and the modification of biosynthetic systems; the role and inhibition of fatty acid synthesis in human cancer, tuberculosis, and obesity.

Jin Zhang, Assistant Professor; Ph.D., Chicago, 2000. Spatiotemporal regulation or dysregulation of protein kinases and second messengers in cell migration, energy metabolism, and cancer development; live-cell fluorescence imaging and high-throughput profiling of signaling activities for mechanistic studies and drug discovery.

Heng Zhu, Assistant Professor; Ph.D., Clemson, 1999. Protein chip and bioinformatics approaches to analyzing posttranslational modifications, signaling network, and host-pathogen interactions in humans, viruses, yeast, and bacteria.

Students accepted into the Pharmacology Graduate Training Program may also carry out thesis research with the following Oncology Center faculty members, who participate in the NCI Anti-Cancer Drug Development Training Program.

Rhoda M. Alani, Associate Professor in Oncology; M.D., Michigan, 1991. Molecular approaches to targeting melanoma and tumor angiogenesis.

Samuel R. Denmeade, Associate Professor; M.D., Columbia, 1989. Targeted therapies for prostate and breast cancer; prodrugs; protoxins; proteases; peptide libraries.

William B. Isaacs, Professor; Ph.D., Johns Hopkins, 1984. Understanding the molecular genetic events responsible for initiation and progression of prostate cancer, with particular interest in inherited susceptibility to prostate cancer.

Elizabeth M. Jaffee, Professor; M.D., New York Medical College, 1985. Dissection of the immune responses to breast and pancreas cancers; cancer immune tolerance; cancer vaccines.

Kenneth Kinzler, Professor; Ph.D., Johns Hopkins, 1988. Molecular genetics of human cancer, particularly colorectal cancer.

Paula M. Pitha-Rowe, Professor; Ph.D., Czechoslovak Academy of Sciences, 1964. Virus-cell interaction; Toll-like receptors (TLR); role of the antiviral innate immune response in viral infection and HIV-1–associated pathogenicity; novel targets of antiviral and anticellular therapy (lentiviral vectors; ubiquitination pathway).

Charles M. Rudin, Associate Professor; M.D./Ph.D., Chicago, 1993. Molecular mechanisms of apoptosis; roles of apoptosis in carcinogenesis and therapeutic resistance; novel therapeutic development in animal models of cancer.

Joel H. Shaper, Professor; Ph.D., UCLA, 1970. Molecular and cell biology of glycosyltransferases.

Bert Vogelstein, Professor; M.D., Johns Hopkins, 1974. Molecular genetics of human cancer and its implications for improved patient management.

LOYOLA UNIVERSITY CHICAGO

Stritch School of Medicine
Department of Pharmacology and Experimental Therapeutics

Program of Study

The Department of Pharmacology and Experimental Therapeutics offers a program that leads to the M.S., Ph.D., and M.D./Ph.D. degrees and trains students for research and teaching in medical schools, research centers, and pharmaceutical institutes. The program consists of formal course work in pharmacology and related fields, research, teaching experience, presentations at professional meetings, and publication. There are opportunities for a wide variety of research in autonomic, biochemical, endocrine, molecular, and neurobehavioral pharmacology; hematopharmacology; and certain areas of clinical pharmacology.

During the first two years of study, students take a core curriculum of graduate courses in molecular pharmacology, biochemistry, neuroscience, physiology, medical pharmacology, and pharmacology fundamentals. They are also encouraged to begin research during this time. In subsequent years, students take advanced courses in pharmacology and related disciplines and devote much of their time to research on their dissertation. Students choose their area of research interest in close collaboration with their dissertation adviser and members of the Ph.D. dissertation committee. Formal admission to candidacy for the Ph.D. degree is granted after a student successfully completes didactic core work and passes a qualifying examination, which is usually taken at the end of the second year. The Ph.D. degree is awarded upon successful completion of a research project culminating in a significant and original scholarly contribution presented as a doctoral dissertation. One or more published full papers and several abstracts often evolve from the dissertation work as well. For students entering with a B.Sc. degree, completion of the requirements for the M.S. degree normally requires three years; the Ph.D. degree, four to five years; and the M.D./Ph.D. degree, seven years.

The department maintains an active program of journal club presentations and research seminars in which current developments in pharmacology and related disciplines are discussed by the faculty members, students, and invited speakers.

Research Facilities

The department occupies nearly 12,800 square feet of floor space and has modern and well-equipped laboratories. Each research group has access to highly specialized and state-of-the-art facilities for electrophysiological, neuropharmacological, biochemical, immunohistochemical, neuroendocrine, molecular, cell culture, and behavioral research. Through collaborative arrangements, facilities in other departments of the Medical Center, additional research facilities in the Hines VA Hospital laboratories, and specialized University facilities, such as a computer center and a biomedical sciences library, are available to all students.

Financial Aid

The University provides doctoral candidates with basic science fellowships that carry an annual stipend of $22,000 plus full tuition remission. Dissertation fellowships are also available on a competitive basis for students in their terminal years of study. Financial support is often provided through extramural research funding.

Cost of Study

Tuition was $606 per semester-hour credit in 2005–06; tuition fees are normally covered by fellowship awards. Students are expected to take approximately 24 credits yearly.

Living and Housing Costs

With the help of the Graduate Students' Office located at the Medical Center, students generally have little difficulty finding suitable inexpensive housing close to campus.

Student Group

Approximately 525 medical and 155 basic science graduate students currently study at the Medical Center campus; there are 23 graduate students in the Department of Pharmacology and Experimental Therapeutics. The number of students in the program is kept low to achieve a close relationship between students and faculty members and to foster flexibility and individualized instruction. The interdisciplinary courses presented jointly by basic science departments in the Medical Center lead to close interaction between students in different departments.

Location

The Medical Center campus is situated in the pleasant surroundings of Maywood, a suburb about 12 miles west of downtown Chicago. An expressway and bus and subway systems provide rapid access to downtown Chicago with its full range of cultural, educational, and recreational activities. Skiing, sailing, and fishing areas are within a 2-hour drive.

The University and The Department

Loyola University Chicago is a private urban institution with four main campuses. The Medical Center campus at Maywood houses the School of Medicine. This complex, completed in 1969, also includes the Graduate School of Biomedical Sciences, the Foster McGaw Hospital and Outpatient Clinics, and the Madden Mental Health Clinic and is adjacent to the Hines VA Hospital and its research facilities. The Department of Pharmacology and Experimental Therapeutics has had graduate and postdoctoral training programs since 1948 and has awarded more than 110 Ph.D. degrees.

Applying

Applicants to the pharmacology program should request application materials from the Director of the Pharmacology Graduate Program; additional information about the program will be sent at that time. The completed application, along with official transcripts from undergraduate and graduate schools, official scores on the GRE General Test and on TOEFL (if applicable) and three letters of recommendation, should be forwarded directly to the Associate Dean of the Graduate School. The academic year begins in July; complete applications must be received by February 1. Early application is strongly advised.

Correspondence and Information

Director, Pharmacology Graduate Program
Department of Pharmacology and Experimental Therapeutics
Room 3621, Building 102
Stritch School of Medicine
Loyola University Chicago
2160 South First Avenue
Maywood, Illinois 60153
Phone: 708-216-3261
Fax: 708-216-6596
Web site: http://www.lumc.edu/pharmacology

Associate Dean
Graduate School
Room 180, Building 120
Stritch School of Medicine
Loyola University Chicago
2160 South First Avenue
Maywood, Illinois 60153
Phone: 708-216-3531
Fax: 708-216-6505
Web site: http://www.lumc.edu

Loyola University Chicago

THE FACULTY AND THEIR RESEARCH

George Battaglia, Professor of Pharmacology; Ph.D., Toronto, 1983. Biochemical psychopharmacology; pharmacology of central serotonergic systems and neurochemical mechanisms of antidepressant and therapeutic drugs affecting serotonergic pathways; impact of prenatal and prepubescent drug exposure on the development and regulation of brain serotonin systems; neurochemical effects of drugs of abuse and drug-induced neurotoxicity; regulation of receptors, receptor-effector coupling; signal transduction.

Kenneth L. Byron, Associate Professor of Pharmacology; Ph.D., Chicago, 1990. Signal transduction and regulation of intracellular calcium; mechanisms involved in the generation and regulation of action potentials in vascular smooth muscle cells; fluorescence measurements of intracellular ion concentrations, patch clamp electrophysiology, protein biochemistry, and image analysis of isolated and in situ blood vessel diameter.

Guan Chen, Associate Professor of Radiation Oncology and Pharmacology; M.D., Bengbu Medical College (China), 1982; Ph.D., Heidelberg (Germany), 1990. Cell biology and molecular pharmacology; p38 MAPK regulation of Ras signaling; interactions of nuclear receptor (ER and VDR) with Ras/MAPK pathways; mechanisms of colon cancer development and breast cancer invasion.

Neil A. Clipstone, Associate Professor of Pharmacology; Ph.D., London, 1990. Investigating the function of the Ca^{2+}-calcineurin/NFAT–signaling pathway in the regulation of cell growth and differentiation; molecular regulation of adipocyte and osteoclast differentiation and function; role of the NFATc1 transcription factor in tumorigenesis; molecular, biochemical, and genetic approaches to understanding signal transduction pathways and their effects on cell-fate decisions.

Joseph R. Davis, Professor of Pharmacology; M.D., Illinois, 1958; Ph.D., Baylor, 1961. Pharmacological and biochemical role of transmural pressure on the endothelium and smooth muscle of the isolated testicular subcapsular artery; autoregulation and effect of drugs and environmental poisons on the testicular arterial blood supply with regard to hypertension; delta-aminolevulinic acid dehydratase activity in lead poisoning.

Jawed Fareed, Professor of Pathology and Pharmacology; Ph.D., Loyola of Chicago, 1975. Pharmacology, physiology, and biochemistry of blood coagulation processes; the role of cytokines, prostaglandins, and peptides in thrombogenesis, platelet aggregation, and thrombosis; clinical studies of procoagulant and anticoagulant drugs; receptor targeting on cells as a means to develop new antithrombotic drugs; molecular markers of hemostatic activation and inhibition processes; pharmacology of recombinant antithrombotic drugs; pharmacodynamics and pharmacokinetics of antithrombotic drugs; experimental therapy for coronary restenosis; new anticoagulants for cardiovascular indications; clinical trials on new drugs; drug interactions and their pharmacodynamic consequences; pathophysiology of Alzheimer's disease and its pharmacologic modulation; pharmacologic interventions in the management of thrombotic and ischemic stroke; transplantation-induced alterations of the hemostatic process; glycosaminoglycans; malignancy-induced thrombosis and its pharmacologic modulation; proteomics and glycomics in vascular pharmacology.

Israel Hanin, Emeritus Professor of Pharmacology; Ph.D., UCLA, 1968. Animal models and biological markers in neuropsychiatry; cholinotoxins; neurotransmitter and membrane-related changes in aging and Alzheimer's disease; development of novel cognition activating (nootropic) agents; role of glycosaminoglycans in Alzheimer's disease.

Lukasz M. Konopka, Associate Professor of Psychiatry and Pharmacology; Ph.D., Loyola Chicago, 1985. From the brain to behavior: new approach to diagnosis and treatment utilizing qEEG, MRI, SPECT, pharmaco-EEG, and quantitative neuropsychology as related to interactions of classical neurotransmitters and neuropeptides in health and disease.

John M. Lee, Professor and Chairman of Pathology and Professor of Pharmacology and Neuroscience; Ph.D., 1983, M.D., 1988, Illinois. Neuropathology of neurodegenerative diseases; human neuroanatomy; receptor-effector coupling; enzyme kinetics; immunohistochemistry; protein and monoamine neurochemistry; animal models of dementia.

Stanley A. Lorens, Professor of Pharmacology; Ph.D., Chicago, 1965. Behavioral pharmacology; effects of drugs on age-related behavioral changes; histochemical and neurochemical analysis of neurotransmitter systems; glycosaminoglycans.

Adriano Marchese, Assistant Professor of Pharmacology; Ph.D., Toronto, 1998. Elucidating the molecular mechanisms mediating ubiquitin-dependent regulation of G-protein–coupled receptors; cell and molecular biology, biochemical pharmacology, and signal transduction.

Nancy A. Muma, Professor of Pharmacology; Ph.D., Louisville, 1985. Pathophysiology of Alzheimer's disease, progressive supranuclear palsy, and other neurodegenerative diseases; regulation of serotonin G-protein signaling by antidepressant and antipsychotic drugs, using cell and molecular biology approaches.

T. Celeste Napier, Professor of Pharmacology; Ph.D., Texas Tech Health Sciences Center, 1982. Characterization of neuronal systems that are destroyed in Alzheimer's disease and Parkinson's disease; drugs of abuse, with emphasis on the effects of opioids (heroin) and psychostimulants (cocaine and methamphetamine) within the basal forebrain and the neural adaptations that occur with repeated drug exposure; neuroplasticity; neuropharmacology; electrophysiology; behavior.

Tarun B. Patel, Professor and Chair of Pharmacology; Ph.D., London, 1978. Biochemical pharmacology and elucidation of molecular mechanisms involved in the regulation of adenylyl cyclases; molecular pharmacology of signaling mechanisms involved in growth factor actions and their modulation by Sprouty proteins.

Karie Scrogin, Associate Professor of Pharmacology; Ph.D., Oregon Health Sciences, 1992. Autonomic regulation of blood pressure and blood volume homeostasis in hypovolemic shock and heart failure; neuropharmacology and electrophysiology.

An aerial view of the Loyola University Chicago, Stritch Medical Center campus. In the foreground is the main Medical Center complex, housing the Stritch School of Medicine Graduate School of Biomedical Sciences, Foster G. McGaw Hospital and Outpatient Clinics, and the Stritch School of Medicine. In the background is the Hines VA Hospital.

MAYO GRADUATE SCHOOL

Graduate Program in Molecular Pharmacology and Experimental Therapeutics

Program of Study

The Department of Molecular Pharmacology and Experimental Therapeutics offers graduate training leading to the Ph.D. and M.D./Ph.D. degrees. Research training emphasizes multidisciplinary approaches—including molecular biology, biochemistry, genetics, pharmacogenomics/genetics, cell biology, clinical pharmacology, structure-based drug design, and structural biology—to enhance understanding of biology and the treatment of disease. Faculty research focuses primarily on the molecular and genetic aspects of cancer, cardiovascular diseases, and neurodegenerative diseases and the application of that understanding to the development of novel and effective therapies for those diseases. A core curriculum, advanced tutorials, and laboratory rotations provide training for advanced dissertation research. Students can complete their thesis research at campuses in Rochester, Minnesota; Jacksonville, Florida; and Scottsdale, Arizona.

Research Facilities

Faculty members of the Department of Pharmacology and Experimental Therapeutics occupy modern laboratories in the twenty-story Guggenheim Research Building in Rochester, the Birdsall and Griffin Buildings in Jacksonville, and the Johnson Research Building in Scottsdale. Investigators and students also use state-of-the-art core research facilities, including those dedicated to mass spectroscopy, DNA and protein synthesis/sequencing, NMR, optical morphology and confocal microscopy, flow cytometry, electron microscopy, and research computing. There are approximately 150 research faculty members at Mayo occupying more than 1 million square feet of laboratory space, with an annual research budget of more than $300 million.

Financial Aid

All full-time graduate students are provided with Mayo Graduate School or National Institutes of Health pharmacology predoctoral training grant research fellowships, which provide a yearly stipend ($23,600 in 2006–07) in addition to outstanding medical insurance and other benefits. Fellowship recipients are not required to work as laboratory assistants or teaching assistants. Since stipends are provided to all, students are not constrained in their choice of research advisers by funding limitations.

Cost of Study

With the exception of book purchases, all expenses are paid by the program. Students pay no tuition or ancillary fees.

Living and Housing Costs

Living costs in Rochester are comparable to those in cities of similar size within the Upper Midwest and generally lower than those in urban areas of the East and West Coasts.

Student Group

The Graduate Program in Molecular Pharmacology and Experimental Therapeutics is specifically organized to educate a small select group of students whose backgrounds include both medical and nonmedical fields of study. The low ratio of students to faculty members provides each student with an unusual opportunity for close collaboration with internationally recognized researchers.

Student Outcomes

Approximately two thirds of students enrolled in the Graduate Program in Molecular Pharmacology and Experimental Therapeutics Ph.D. programs graduate with a Ph.D. degree. Students have published an average of six publications during their thesis work. Most students pursue additional postdoctoral training and have productive careers in industry or academics.

Location

The city of Rochester combines the best of two worlds—the warmth and friendliness of a small town with the bustling commerce, entertainment, and conveniences of a metropolis. Lectures, concerts, art exhibits, and a civic theater contribute to a cosmopolitan atmosphere, unusual for a city this size. The city and its environs provide an extensive, four-season calendar of recreation. Attractions in Minneapolis and St. Paul, which are just 80 miles north of Rochester, include professional sports, theaters, museums, orchestras, shopping, and restaurants.

The Graduate School

The Mayo Graduate School is a division of Mayo Foundation, which includes Mayo Clinic, Mayo Graduate School of Medicine, Mayo Medical School, Mayo School of Health-Related Sciences, and affiliated hospitals. Mayo Foundation is accredited by the Commission on Institutions of Higher Education of the North Central Association of Colleges and Schools.

Applying

Application to the Mayo Graduate School is best accomplished by visiting the Web site (http://www.mayo.edu/mgs/index.html). General and specific course requirements, admissions deadlines, other information, and application forms for the Ph.D. and M.D./Ph.D. training programs are found at this site. Mayo Foundation is an affirmative action and equal opportunity educator and employer.

Correspondence and Information

Larry M. Karnitz, Ph.D., Graduate Program Director
Graduate Program in Molecular Pharmacology and Experimental Therapeutics
Mayo Graduate School
200 First Street, SW
Rochester, Minnesota 55905
Phone: 507-284-2747
E-mail: karnitz.larry@mayo.edu
Web site: http://www.mayo.edu/mgs/index.html

Mayo Graduate School

THE FACULTY AND THEIR RESEARCH

Michael J. Ackerman, M.D./Ph.D., Mayo, 1995. Molecular and cellular mechanisms of cardiac ion channelopathies; determinants of risk of sudden cardiac death; long QT syndrome; sudden infant death syndrome; hypertrophic cardiomyopathy. Echocardiography-guided genetic testing in hypertrophic cardiomyopathy: Septal morphological features predict the presence of myofilament mutations. *Mayo Clin. Proc.* 81:459–67, 2006 (with Binder et al.). Compendium of cardiac channel mutations in 541 consecutive unrelated patients referred for long QT syndrome genetic testing. *Heart Rhythm* 2:507–17, 2005 (with Tester et al.). Molecular diagnosis of the inherited long QT syndrome in a woman who died after near-drowning. *N. Engl. J. Med.* 341:1121–5, 1999 (with Tester et al.).

Matthew M. Ames, Ph.D., California, San Francisco, 1976. Study of anticancer agents with novel molecular mechanisms of action, with an emphasis on tumor cell–specific activation; pharmacogenetics and pharmacogenomics of genes and protein products associated with responses to anticancer agents; preclinical and clinical pharmacology of novel anticancer agents. Lack of mutations in EGFR in gastroenteropancreatic neuroendocrine tumors. *N. Engl. J. Med.* 353:209–10, 2005 (with Gilbert et al.). Primer on medical genomics part XII: Pharmacogenomics—general principles with cancer as a model. *Mayo Clin. Proc.* 79:376–84, 2004 (with Goetz et al.). Activation of the antitumor agent aminoflavone (NSC 686288) is mediated by induction of tumor cell cytochrome P450 1A1/1A2. *Mol. Pharmacol.* 62:143–53, 2002 (with Kuffel et al.).

Stephen Brimijoin, Ph.D., Harvard, 1969. Molecular neurobiology of the cholinesterases; enzymatic and gene therapy approaches to new treatments for cocaine toxicity and addiction. Gene transfer of cocaine hydrolase suppresses cardiovascular responses to cocaine in rats. *Mol. Pharmacol.* 67:204–11, 2005 (with Gao). An engineered cocaine hydrolase blunts and reverses cardiovascular responses to cocaine in rats. *J. Pharmacol. Exp. Ther.* 310:1046–52, 2004 (with Gao).

Doo-Sup Choi, Ph.D., Strasbourg (France), 1997. Genetic and genomic basis of alcoholism and drug addiction, focusing on adenosine-mediated glutamate neurotransmission in the brain using genetically engineered rodent models and genetic variants of human genes. The equilibrative nucleoside transporter type 1 modulates ethanol intoxication and preference in mice. *Nature Neurosci.* 7:855–61, 2004. Conditional rescue of protein kinase C epsilon regulates ethanol preference and hypnotic sensitivity in adult mice. *J. Neurosci.* 22:9905–11, 2002.

Alan P. Fields, Ph.D., Johns Hopkins, 1986. Genetics, molecular cell biology, pharmacology, and biochemistry of protein kinase C signaling pathways in human cancer; development of targeted therapeutics to oncogenic PKC signaling. *Cancer Res.* 66:1767–74, 2006 (with Stallings-Mann et al.). *J. Biol. Chem.* 280:31109–15, 2005 (with Regala et al.). *Cancer Res.* 65:8905–11, 2005 (with Regala et al.). *J. Cell Biol.* 164:797–802, 2004 (with Murray et al.).

Larry M. Karnitz, Ph.D., Iowa, 1989. Pharmacologic inhibition of checkpoint signaling to enhance anticancer therapy responses. *Blood* 106:318–27, 2005 (with Mesa et al.). *J. Biol. Chem.* 278:52572–7, 2004 (with Arlander). Genetic, biochemical, and molecular analysis of checkpoint signaling pathways activated by ionizing radiation and genotoxic chemotherapy agents. *J. Biol. Chem.* 278:24428–37, 2003 (with Roos-Mattjus). *J. Biol. Chem.* 278:45507–11, 2003 (with Parrialla-Castellar et al.).

Zvonimir S. Katusic, M.D., 1977, Ph.D., 1988, Belgrade. Vascular biology of nitric oxide; gene transfer to cerebral arteries; gene therapy of cerebral vasospasm. Role of endothelial NO synthase phosphorylation in cerebrovascular protective effect of recombinant erythropoietin during subarachnoid hemorrhage–induced cerebral vasospasm. *Stroke* 36:2731–7, 2005 (with Santhanam et al.). Transplantation of circulating endothelial progenitor cells restores endothelial function of denuded rabbit carotid arteries. *Stroke* 35:2378–84, 2004 (with He et al.). Human endothelial progenitor cells tolerate oxidative stress due to intrinsically high expression of manganese superoxide dismutase. *Arterioscler. Thromb. Vasc. Biol.* 24:2021–7, 2004 (with He et al.). Long-term vitamin C treatment increases vascular tetrahydrobiopterin levels and nitric oxide synthase activity. *Circ. Res.* 92(1):88–95, 2003 (with d'Uscio et al.).

Scott H. Kaufmann, M.D./Ph.D., Johns Hopkins, 1981. Regulation of anticancer drug–induced apoptosis (programmed cell death); mechanisms of resistance to antineoplastic drugs. The role of MCL-1 downregulation in the cytotoxicity of the Raf kinase inhibitor BAY 43-9006. *Oncogene* 24:6861–9, 2005 (with Yu et al.). The role of checkpoint kinase 1 in sensitivity to topoisomerase I poisons. *J. Biol. Chem.* 280:14349–55, 2005 (with Flatten et al.). Alterations in the apoptotic machinery and their potential role in anticancer drug resistance. *Oncogene* 22:7414–30, 2003 (with Vaux).

Zhenkun Lou, Ph.D., Mayo, 2001. Tumorigenesis; DNA damage checkpoints; DNA repair. MDC1 maintains genomic stability by participating in the amplification of ATM-dependent DNA damage signals. *Mol. Cell* 21:1–14, 2006. BRCA1 participates in DNA decatenation. *Nature Struct. Mol. Biol.* 12:589–93, 2005. MDC1 is coupled to activated Chk2 in mammalian DNA damage response pathways. *Nature* 421:957–61, 2003.

Michael McKinney, Ph.D., Johns Hopkins, 1982. Molecular pharmacology; neuropharmacology; molecular neuroscience; neurodegeneration; aging. Brain cholinergic vulnerability: Implications for behavior and disease. *Biochem. Pharmacol.* 70:1115–24 2005. Pontine cholinergic neurons depend on three neuroprotection systems to resist nitrosative stress. *Brain Res.* 1002:98–107, 2004 (with Williams et al.).

Cynthia T. McMurray, Ph.D., Oregon State, 1987. Cause and effects of trinucleotide expansion; mechanisms of DNA repair; pathophysiology of Huntington's disease. Mutant huntingtin impairs axonal trafficking in mammalian neurons in vivo and in vitro. *Mol. Cell. Biol.* 24:8195–209, 2004 (with Trushina et al.). The Rad 50 molecular hook: A novel structure underlying MRE11 complex functions in DNA recombination and repair. *Nature* 418:562–6, 2002 (with Hopfner et al.).

Laurence J. Miller, M.D., Thomas Jefferson, 1973. A peptide agonist acts by occupation of a monomeric G-protein–coupled receptor. Key differences in molecular complexes of the cholecystokinin receptor with structurally related peptide agonist, partial agonist, and antagonist: Focus on importance of sulfation of tyrosine in peptide position 27. *Mol. Pharmacol.* 66:545–52, 2004 (with Arlander et al.). Measurement of intermolecular distances for the natural agonist peptide docked at the cholecystokinin receptor expressed in situ using fluorescence resonance energy transfer. *Mol. Pharmacol.* 65:28–35, 2004 (with Harikumar et al.).

Nicole Murray, Ph.D., Case Western Reserve, 1995. Protein kinase C iota is required for Ras transformation and colon carcinogenesis in vivo. *J. Cell Biol.* 164:797–802, 2004 (with Fields et al.). Protein kinase C beta II and TGFbetaRII in omega-3 fatty acid–mediated inhibition of colon carcinogenesis. *J. Cell Biol.* 157:915–9, 2002.

Timothy M. Olson, M.D., Chicago, 1987. Mapping novel genes for dilated cardiomyopathy and atrial fibrillation; molecular genetic mechanisms for mechanical dysfunction and electrical instability in the heart. Sodium channel mutations and susceptibility to heart failure and atrial fibrillation. *JAMA* 293:447–54, 2005. Actin mutations in dilated cardiomyopathy, a heritable form of heart failure. *Science* 280:750–2, 1998.

Yuan-Ping Pang, Ph.D., Pittsburgh, 1990. Molecular recognition; metalloenzyme catalysis; angiogenesis; apoptosis; drug design. Three-dimensional model of a substrate-bound SARS chymotrypsin-like cysteine proteinase predicted by multiple modular dynamics simulations: catalytic efficiency regulated by substrate binding. *Proteins: Struct. Funct. Bioinf.*, in press. Rational design of alkylene-linked bis-pyridiniumaldoximes as improved acetylcholinesterase reactivators. *Chem. Biol.* 10:491–502, 2003 (with Brimijoin et al.).

Gregory A. Poland, M.D., Southern Illinois, 1980. Vaccines; vaccine-preventable diseases; vaccine immunogenetics; biodefense vaccines; vaccine development; mechanisms of immune response development to vaccines; development and testing of vaccines against biologic agents of mass destruction. Vaccines against avian influenza: A race against time. *N. Engl. J. Med.* 354:1411–3, 2006. Human leukocyte antigen haplotypes in the genetic control of immune response to measles-mumps-rubella vaccine. *J. Infect. Dis.* 193:655–63, 2006 (with Ovsyannikova et al.). Associations between measles antibody levels and the protein structure of class II human leukocyte antigens. *Hum. Immunol.* 64:696–707, 2003 (with St. Sauver et al.).

Franklyn G. Prendergast, M.B.B.S., West Indies, 1968; Ph.D., Minnesota, 1977. Fluorescence spectroscopy; protein structure and dynamics; biochemistry and bioluminescence; biophysics of the green fluorescent protein. H-bonding mediates polarization of peptide groups in folded proteins. *Protein Sci.* 12:2633–6, 2003 (with Juranic et al.). Analysis of the segmental stability of helical peptides by isotope-edited infrared spectroscopy. *Proteins* 45:81–9, 2001 (with Venyaminov et al.).

Elliott Richelson, M.D., Johns Hopkins, 1969. Basic and clinical neuropsychopharmacology. Antinociceptive, hypothermic, hypotensive, and reinforcing effects of a novel neurotensin receptor agonist, NT69L, in rhesus monkeys. *Pharmacol. Biochem. Behav.* 80:341–9, 2005 (with Fantegrossi et al.). Neurobiologic basis of nicotine addiction and psychostimulant abuse: A role for neurotensin? *Psychiatr. Clin. North Am.* 28:737–51, 2005 (with Fredrickson et al.).

Andre Terzic, M.D., Belgrade and Paris, 1985; Ph.D., Illinois, 1991. Cardioprotection and cardioregenerative medicine; stem cell therapy; genetics of cardiac disease; channelopathies; bioenergetic systems biology; nucleocytoplasmic signaling. Derivation of a cardiopoietic population from human mesenchymal stem cell yields cardiac progeny. *Nature Clin. Pract. Cardiovasc. Med.* 3:S78, 2006 (with Behfar et al.). ABCC9 mutations identified in human dilated cardiomyopathy disrupt catalytic KATP channel gating. *Nature Genet.* 36:382, 2004 (with Bienengraeber et al.). Stable benefit of embryonic stem cell therapy in myocardial infarction. *Am. J. Physiol.* 287:H471, 2004 (with Hodgson et al.). Kir6.2 is required for adaptation to stress. *Proc. Natl. Acad. Sci. U.S.A.* 99:13278, 2002. (with Zingman et al.).

Liewei Wang, M.D., FuDan Medical, 1996; Ph.D., Mayo, 2003. Understanding the genetic basis for variation in drug response and disease pathophysiology. Human thiopurine S-methyltransferase (TPMT) pharmacogenetics: Variant misfolding and aggresome formation. *Proc. Natl. Acad. Sci. U.S.A.* 102:9394–9, 2005 (with Nguyen et al.). Pharmacogenomics: Bench to bedside. *Nature Rev. Drug Discovery* 3:739–48, 2004 (with Weinshilboum). Thiopurine S-methyltransferase pharmacogenetics: Association with chaperone proteins and relation to degradation. *Pharmacogenetics* 13:555–64, 2003.

Richard M. Weinshilboum, M.D., Kansas, 1967. Molecular pharmacogenetics and pharmacogenomics. Human thiopurine S-methyltransferase (TPMT) pharmacogenetics: Variant misfolding and aggresome formation. *Proc. Natl. Acad. Sci. U.S.A.* 102:9394–9, 2005 (with Wang et al.). Pharmacogenomics: Bench to bedside. *Nature Rev. Drug Discovery* 3:739–48, 2004 (with Wang). Inheritance and drug response. *N. Engl. J. Med.* 348:529–37, 2003.

NORTH CAROLINA STATE UNIVERSITY

Department of Environmental and Molecular Toxicology

NC STATE UNIVERSITY

Programs of Study

Graduate study in toxicology at North Carolina State University (NC State) leads to the Doctor of Philosophy (Ph.D.), the Master of Science (M.S.), or the nonthesis Master of Toxicology (M.Tox.) degree. Doctoral students may elect to concentrate in molecular and cellular toxicology, environmental toxicology, or general toxicology. The Department provides a comprehensive program in course work and research training to prepare prospective toxicologists for careers in academia, government, and industry. Research in the Department spans an array of topics ranging from the molecular to population-level consequences of toxicant exposure. A common research theme in the Department involves the elucidation of toxicant-induced alterations in cell signaling and resultant changes in gene expression as it relates to toxicity at the cellular, organ, and organism level. Linkage of adverse biological endpoints to toxicant exposure is a mechanistic goal. Specific research areas include endocrine disruption, oxidative stress, cellular signaling pathways, transcriptional regulation, toxicogenomics, regulation and expression of xenobiotic metabolizing enzymes, molecular carcinogenesis, cell-cycle regulation, apoptosis, chemical exposure assessment, analytical toxicology, ecotoxicology, and risk assessment.

Students take core courses in biochemistry, general and biochemical toxicology, pharmacology, and statistics. Electives may include environmental toxicology, pathology, chemical carcinogenesis, risk assessment, and molecular genetics, among others. A weekly seminar series provides a forum for scientists and students within and outside the Department to present and discuss their research. Master's degree students take an average of two to three years to graduate. Doctoral students must pass written and oral examinations after completing the second year of study, and an original research thesis is required. Doctoral students take an average of four to five years to graduate. All students have an opportunity to interact with the Department's public outreach/extension and undergraduate programs if they desire.

Research Facilities

The Department moved into the state-of-the-art, 59,000-square-foot Toxicology Building in 2001, with excellent facilities for teaching and research. The building houses modern well-equipped laboratories for molecular, cellular, and environmental toxicology research. In addition, the building houses a small-animal facility and an aquatic facility. A multimillion-dollar genomic core facility with high throughput DNA sequencers and a full complement of microarray instrumentation is housed next door to the Toxicology Building. General facilities at NC State include numerous libraries and core facilities in electron and confocal microscopy, monoclonal antibodies, high throughput DNA sequencing, proteomics, mass spectrometry, nuclear magnetic resonance, cell and molecular imaging, and flow cytometry; the Bioinformatics Research Center; and the Center for Marine Science and Technology. Other research facilities are available in the laboratories of adjunct faculty members and collaborators at nearby institutions in Research Triangle Park, including the National Institute of Environmental Health Sciences (NIEHS), Environmental Protection Agency (EPA), CIIT Centers for Health Research, GlaxoSmithKline (GSK), the University of North Carolina at Chapel Hill, and Duke University. The aggregate resources of NC State and these other institutions have made Research Triangle Park the world's center for toxicology research.

Financial Aid

Financial assistance is available for qualified applicants through teaching and research assistantships, fellowships, and traineeships by training grants funded by National Institute of Environmental Health Science and the Environmental Protection Agency. These provide tuition, health insurance, and an annual stipend of $21,000.

Cost of Study

Tuition and fees for 2006–07 are $2651 per semester for North Carolina residents and $8675 per semester for nonresidents. Tuition is paid by the NC State Graduate Student Support Plan for all eligible graduate students supported on full teaching/research assistantships, fellowships, or traineeships.

Living and Housing Costs

North Carolina State offers housing for single and married students. There are also many off-campus apartments for rent within walking distance of the Toxicology Building, with monthly rents starting at about $450.

Student Group

The University's total enrollment is 29,640, with about one third being graduate students. The Department has 43 graduate students (32 are Ph.D. students), 13 postdoctoral fellows, and 6 visiting scientists. Students come from diverse backgrounds and from all regions of the nation and the world.

Student Outcomes

More than 200 students have received graduate degrees in toxicology at NC State. Graduates are employed almost equally among three areas: government research and regulatory agencies, academia, and private companies. Nearly all students have job offers prior to graduation. Many graduates hold prominent positions at universities, the NIH, EPA, chemical and pharmaceutical companies, and consulting firms.

Location

North Carolina State is located in Raleigh, the capital of North Carolina, which has a population of about 1 million in the extended metropolitan area. Raleigh is located approximately 2 hours from the Atlantic Ocean, 3½ hours from the Appalachian Mountains, and just 15 miles from Research Triangle Park, one of the most prominent telecommunications, environmental health, pharmaceutical, and biotechnology research centers in the United States. The Research Triangle itself is named for the triangle formed by the three major research universities located here: Duke University, the University of North Carolina at Chapel Hill, and NC State University.

The University and The Department

NC State was established in 1889 and offers bachelor's degrees in ninety-two fields of study, master's degrees in 101 fields, doctoral degrees in fifty-eight fields, and a Doctor of Veterinary Medicine degree. The University has more than 6,000 employees, including approximately 1,600 faculty and extension field faculty members.

The Department of Environmental and Molecular Toxicology began as the Interdepartmental Program in Toxicology in 1964 and became an independent department in 1988. The Department currently has 14 core research faculty members in the Toxicology Building, 14 associate faculty members located elsewhere at NC State, and 18 adjunct faculty members located at nearby research institutions. The Department offers students the cohesiveness of a well-organized department and multidisciplinary research opportunities of a University-wide program.

Applying

Prospective students should have a strong background in the biological and physical sciences. The Department requires a minimum undergraduate GPA of 3.0 on a 4.0 scale and a minimum GRE score of 1100 (combined verbal and quantitative scores). GRE subject test scores are not required. International students whose primary language is not English must submit TOEFL scores. The application deadline is January 15 for fall admission; spring admission is generally not allowed. Applications can be made online at http://www2.acs.ncsu.edu/grad/admision.htm or by contacting the Graduate School, Box 7102, North Carolina State University, Raleigh, North Carolina 27695.

Correspondence and Information

Robert C. Smart, Ph.D.
Director of Graduate Programs
Department of Environmental and Molecular Toxicology
Campus Box 7633
North Carolina State University
Raleigh, North Carolina 27695-7633
E-mail: rcsmart@unity.ncsu.edu
Web site: http://www.tox.ncsu.edu/

North Carolina State University

THE FACULTY AND THEIR RESEARCH

Core Faculty

David Buchwalter, Assistant Professor; Ph.D. Oregon State. Population level approaches for the study of environmental stressors using aquatic insects as a model organism.

W. Gregory Cope, Associate Professor and Extension Leader; Ph.D., Iowa State. Linkages among sentinel organisms and environmental/human health; effects of anthropogenic stresses on aquatic organisms, bioavailability, transport, and fate of contaminants in aquatic ecosystems.

Ernest Hodgson, William Neal Reynolds Professor; Ph.D., Oregon State. Human metabolism and metabolic interactions of public health-, agriculture- and deployment-related xenobiotics.

Christopher S. Hofelt, Teaching Assistant Professor; Ph.D., North Carolina State. Analytical methods for environmental exposure assessment; human and ecological risk assessment.

Gerald A. LeBlanc, Professor; Ph.D., South Florida. Endocrine regulation and toxicant disruption of reproduction and development; environmental sex determination and differentiation; toxicant interactions; invertebrate endocrine toxicology.

Patricia McClellan-Green, Research Assistant Professor; Ph.D., North Carolina State. Natural and synthetic toxicants in marine environment; effects on metabolic activities and reproduction of marine organisms.

Jun Ninomiya-Tsuji, Assistant Professor; Ph.D., Hiroshima (Japan). Signaling networks activated by stresses and cytokines in mammalian cells; roles of kinase cascades.

Margie Oleksiak, Assistant Professor; Ph.D., MIT (Woods Hole). Molecular effects of environmental stresses.

Randy L. Rose, Research Assistant Professor; Ph.D., LSU. Characterization of molecular basis of pesticide metabolism and resultant interactions in mammals and insects.

Damian Shea, Professor; Ph.D., Maryland. Transport and fate of chemicals in the environment; analytical toxicology; exposure assessment; bioavailability; human and ecological risk assessment.

Robert C. Smart, Professor; Ph.D., Michigan. Molecular carcinogenesis; cell-signaling pathways regulating mitotic growth, differentiation, apoptosis, and the DNA damage response network; transcription factors and gene regulation.

Julia F. Storm, Agromedicine Information Specialist; M.S.P.H., North Carolina at Chapel Hill. Rural occupational and environmental health and safety education and promotion; communication of research findings to lay audiences.

Yoshiaki Tsuji, Assistant Professor; Ph.D., Hiroshima (Japan), Gene regulation and signaling involved in iron homeostasis and detoxification in environmental and oxidative stress.

Andrew D. Wallace, Assistant Professor; Ph.D., Rochester. Toxic agent alteration of nuclear receptor signaling; mechanisms of SXR activation and gene regulation of detoxification enzymes.

Associate Faculty

Kenneth B. Adler, Professor; Ph.D., Vermont. Mechanisms of airway inflammation and mucus secretion.

Ronald E. Baynes, Associate Professor; D.V.M., Tuskegee; Ph.D., North Carolina State. Physiochemical interactions influencing dermal absorption of chemical mixtures and drug formulations; human risk assessment of veterinary drugs.

John M. Cullen, Professor; D.V.M., Pennsylvania; Ph.D., California, Davis. Hepatic injury and carcinogenesis mediated by chemical and viral mechanisms.

Hosni M. Hassan, Professor; Ph.D., California, Davis. Oxidant stress and genetic regulation of antioxidant enzymes, superoxide dismutases, and hydroperoxidases.

Jonathan M. Horowitz, Associate Professor; Ph.D., Wisconsin. Molecular mechanisms of oncogenesis and tumor suppression; transcriptional regulation; cell-cycle control, signal transduction, and development.

Michael R. Hyman, Assistant Professor; Ph.D., Bristol (England). Enzymology and physiology of bacterial degradation of environmental pollutants and recalcitrant compounds.

Scott Laster, Professor; Ph.D., Florida State. Role of tissue necrosis factor (TNF) in autoimmune diseases.

J. M. "Mac" Law, Associate Professor of Pathology; D.V.M., Ph.D., LSU; Diplomate, American College of Veterinary Pathologists (ACVP). Toxicologic and molecular pathology of small fish models for carcinogenicity testing; multistage chemical carcinogenesis; DNA adducts; environmental pathology.

Nancy A. Monteiro-Riviere, Professor; Ph.D., Purdue. Dermatotoxicology; in vitro alternative methods; drug delivery and morphology.

Elizabeth Guthrie Nichols, Assistant Professor; Ph.D. North Carolina at Chapel Hill. Chemical fate in environment, ecotoxicology, and bioremediation.

Marcelo L. Rodriguez-Puebla, Assistant Professor; Ph.D., Texas. Carcinogenesis; cell-cycle regulation in normal and neolastic proliferation.

R. Michael Roe, Professor; Ph.D., LSU. Molecular biology of esterases, epoxide hydrolases, and P450.

Philip L. Sannes, Professor; Ph.D., Ohio State. Mechanisms that control epithelial responses to growth factors and extracellular matrices in the lung.

Michael K. Stoskopf, Professor; D.V.M., Colorado State; Ph.D., Johns Hopkins. Habitat health; risk assessment for wildlife.

Adjunct Faculty

EPA: K. M. Crofton, Ph.D.: neurotoxicology. D. Dix, Ph.D.: reproductive toxicogenomics. L. E. Gray, Ph.D.: reproductive toxicology. R. J. Preston, Ph.D.: genetic risk assessment and susceptibility to cancer. M. J. Selgrade, Ph.D.: immunotoxicology. B. Veronesi, Ph.D.: neurotoxicology.

NIEHS: T. E. Eling, Ph.D.: prostaglandins and cancer. J. A. Goldstein, Ph.D.: genetic regulation of P450. K. Korach, Ph.D.: receptor biology. R. J. Langenbach, Ph.D.: molecular mechanisms of carcinogenesis. B. A. Merrick, Ph.D.: proteomics and p53 phosphorylation. D. Zeldin, Ph.D.: cytochrome P450 eicosanoid metabolism.

CIIT: W. Greenlee, Ph.D.: molecular toxicology.

Bayer Crop Science: A. Chalmers, Ph.D.: neuroreceptors and plant/insect genomics. H. Cunny, Ph.D.: regulatory toxicology, A. J. Tobia, Ph.D.: regulatory toxicology.

GSK: R. Miller, Ph.D.: hepatotoxicosis; receptor-mediated toxicosis.

NORTHEASTERN UNIVERSITY

Bouvé College of Health Sciences
Program in Pharmacology

Programs of Study	Bouvé College of Health Sciences offers a full-time pharmaceutical science training program leading to the Doctor of Philosophy degree with a concentration in pharmacology. The goal of the doctoral program is to enable the developing scientist to become an independent investigator and to provide graduates with a strong foundation in pharmaceutical science. Doctoral training normally requires four to five years for completion. The Graduate School also offers part-time and full-time Master of Science degree programs that emphasize didactic course work in pharmacology and related pharmaceutical science disciplines and that can be completed in two years. Department research strengths are in CNS, autonomic and behavioral neuropharmacology, cardiovascular pharmacology, receptor signal transduction, and biochemical toxicology. During the first two years, students take core courses in biochemistry, molecular biology, neuroscience, pharmacology, physiology, therapeutics, and toxicology. Doctoral students actively participate in departmental seminars and journal clubs, rotate in faculty research labs, and select a major thesis adviser for their doctoral research by the beginning of the second year. Ph.D. students must pass a written and oral qualifying examination and write and defend a thesis proposal at the end of the second year. Students must give written and orally defended progress reports on their thesis research until it is completed. M.S. students are encouraged to do research but are not required to complete a thesis; they are required to write a research paper.
Research Facilities	The University supports numerous centers and institutes with ties to the program, including the Barnett Institute of Chemical and Biological Analysis, the Institute for Molecular Biotechnology, the Center for Drug Discovery, the New England Inflammation and Tissue Protection Institute, the Environmental Cancer Research Program, the Center for Cardiovascular Targeting, the Center for Pharmaceutical Biotechnology and Nanomedicine, the Nano Manufacturing Research Institute, the Institute on Race and Justice, the Domestic Violence Institute, the Electronic Materials Research Institute, the Brudnick Center on Violence and Conflict, the Center for Advanced Microgravity Materials Processing, the Center for Subsurface Sensing and Imaging Systems, the Center for Urban and Regional Policy, and the Institute for Complex Scientific Software. Special University facilities include the Electron Microscopy Center, the Microfabrication Laboratory, and the Molecular Modeling Center. A high-speed data network links users and facilities on the central campus to three satellite campuses and to computing facilities around the world. Students have access to Compaq Alpha systems, public-access microcomputer labs (PC and Macintosh), a conferencing system, multimedia labs, and specialized computing equipment. Northeastern University is an Internet2 site. University libraries contain more than 985,000 books, 2.3 million microforms, 7,600 serial subscriptions, and 17,000 audiovisual materials. The libraries have licensed access to more than 13,000 electronic information sources. A central and branch library contain technologically sophisticated services, including a Web-based catalog and circulation systems and a Web portal to licensed electronic resources. The University is a member of the Boston Library Consortium and the Boston Regional Library System, giving students and faculty members access to the region's collections and information resources.
Financial Aid	The graduate schools offer aid through teaching and research assistanship awards that include tuition remission and a stipend typically averaging $7500 per semester ($22,500 per year). These assistantships require a maximum of 20 hours of work per week. Northeastern University offers need-based financial aid to graduate students through the Federal Stafford Loan, Federal Perkins Loan, and Federal Work-Study programs. In addition, the University offers a wide variety of graduate assistantships, along with minority fellowships, such as the Martin Luther King Jr. Scholarship.
Cost of Study	The cost of tuition for the 2005–06 academic year in the graduate school of Bouvé College of Health Sciences was $875 per semester hour of credit. Where applicable, special tuition charges are made for thesis, dissertation, teaching, practicums, or fieldwork. A booklet listing all fees and tuition costs is available upon request.
Living and Housing Costs	For 2005–06, on-campus room rates per semester for a single bedroom within an apartment range from $3100 to $3430. A shared bedroom in an apartment ranges from $2470 to $2890. On-campus housing for graduate students is limited and granted on a space-available basis. The latest rates for on-campus housing for graduate students may be obtained at http://www.housing.neu.edu/. An off-campus referral service is available through Housing Services by telephone (617-373-4872 or 617-373-2814) or e-mail (nucommuter@neu.edu). Although there are several board options available, graduate students typically pay $1975 per semester for ten meals per week. More information about meal plans can be obtained at http://customerservice.neu.edu/mealplan.html. An extensive public transportation system serves the greater Boston area, and there are subway and bus services convenient to the University.
Student Group	In fall 2005, 14,730 undergraduate and 4,811 graduate students were enrolled at Northeastern University, representing a wide variety of academic, professional, geographic, and cultural backgrounds. Bouvé College of Health Sciences graduate programs enrolled 1,006 students; 767 attended on a full-time basis.
Location	Boston, Massachusetts, offers a rich cultural and intellectual history and is the premiere educational center of the country, with more than thirty-five colleges in the city region. Cultural offerings—including several world-class museums, numerous art galleries, and the Boston Symphony, among others—are diverse, and the city is home to people of every race, ethnicity, political persuasion, and religion. Boston also offers world-class restaurants and a range of outdoor activities and is steeped in New England tradition.
The University	Northeastern University, located in heart of Boston, is a world leader in cooperative education and is recognized for its expert faculty and first-rate academic and research facilities. Cited for excellence for four years running by *U.S. News & World Report*, Northeastern has quickly moved up into the top-tier rankings—an impressive thirty-five spots in four years. In addition, Northeastern was named a top college in the 2006 edition of *The Princeton Review*'s annual "Best Colleges" issue. For more information, students should visit http://www.northeastern.edu.
Applying	The application deadline for the Ph.D. degree program is December 15 for students wishing to be admitted for the following fall semester. Application materials for M.S. programs must be submitted by August 1 for the following fall semester, although May is the recommended deadline for international M.S. students who will need to apply for a visa. Students whose applications have not been acted upon prior to registration or the start of classes may be permitted to register as nonmatriculated students for up to 12 semester hours of study. The following information is required with the application: official transcripts for all previous college and university work, three references, a personal statement or essay, a $50 application fee, and GRE scores (taken within the last 5 years). International students for whom English is not the primary language must submit TOEFL scores. (A minimum score of 250 computer-based or 100 Internet-based is needed).
Correspondence and Information	Molly Schnabel, Director, Graduate Admissions and Student Services Bouvé College of Health Sciences 123 Behrakis Life Sciences Center Northeastern University 360 Huntington Avenue Boston, Massachusetts 02115 Phone: 617-373-2708 Fax: 617-373-4701 E-mail: bouvegrad@neu.edu Web site: http://www.bouve.neu.edu/programs/pharmatoxms/index.php

Northeastern University

THE FACULTY AND THEIR RESEARCH

Norman R. Boisse, Associate Professor of Pharmacology and Physiology, Department of Pharmaceutical Sciences; Ph.D., Cornell, 1976. Animal models for the study of valium-like (benzodiazepine) tranquilizer withdrawal; evaluation of a GABA hypo-effectiveness hypothesis to explain tranquilizer withdrawal.

Richard C. Deth, Professor of Pharmacology, Department of Pharmaceutical Sciences; Ph.D., Miami (Florida), 1975. Involvement of D4 dopamine receptor-mediated phospholipid methylation in neurodevelopmental and neuropsychiatric disorders; role of impaired methionine synthase activity and methylcobalamin synthesis in autism.

Jonathan E. Freedman, Associate Professor of Pharmacology, Department of Pharmaceutical Sciences; Ph.D., Johns Hopkins, 1983. Signal transduction mechanisms of dopamine and other neurotransmitter receptors in the central nervous system, studied with patch-clamp electrophysiology and fluorescence microscopy.

S. John Gatley, Professor; Ph.D. Newcastle-upon-Tyne, 1975. Radiotracers and receptor imaging.

Ralph H. Loring, Associate Professor of Pharmacology, Department of Pharmaceutical Sciences; Ph.D., Cornell, 1980. Characterization of neuronal nicotinic acetylcholine receptors and excitatory amino acid receptors using biochemical, immunological, and physiological techniques; regulation of receptor expression.

Alexandros Makriyannis, Professor and Director, Center for Drug Discovery; Ph.D., Kansas, 1969. Medicinial chemistry and pharmacology of cannabinoid receptors.

Akio Ohta, Assistant Professor; Ph.D., Tohoku, 1989. Regulatory mechanism of immune and inflammatory responses.

Robert A. Schatz, Associate Professor of Pharmacology and Toxicology, Department of Pharmaceutical Sciences; Ph.D., Rhode Island, 1971. Solvent inhalation; respiratory toxicity; regulation of drug metabolizing enzymes in the respiratory tract.

Michail V. Sitkovksy, Professor of Immunophysiology, Department of Pharmaceutical Sciences, and Eleanor Black Chair of Immunophysiology and Pharmaceutical Biotechnology; Ph.D., Moscow State, 1973. Inflamation and oncology.

Barbara L. Waszczak, Professor of Pharmacology, Department of Pharmaceutical Sciences; Ph.D., Michigan, 1978. Neuropharmacology of basal ganglia motor pathways; extracellular single-unit recording and microiontophoretic techniques for evaluation of dopamine effects in substantia nigra; immunochemistry of dopamine receptor subtypes in animal models of Parkinson's disease.

Jiang Zheng, Assistant Professor of Drug Metabolism and Toxicology, Department of Pharmaceutical Sciences; Ph.D., Kansas, 1991. Metabolism and bioactivation of drugs and xenobiotic chemicals; elucidation of structures of chemically reactive metabolites; identification of cellular proteins modified by reactive metabolites.

REPRESENTATIVE PUBLICATIONS

Boisse, N. R., O. Amitay, and J. Leung. Suppression of maximal benzodiazepine withdrawal by pro-GABA-ergic drugs. I. Chlordiazeposixe, phenobarital, sodium bromide, and baclofen. *NIDA Res. Monogr.* 153(II):236, 1995.

Leung, J., **N. R. Boisse,** and O. Amitay. Sex differences in spontaneous withdrawal following acute (single dose) benzodiazepine dependence induction. *NIDA Res. Monogr.* 1153(II):237, 1995.

Zhu Q., L.-J. Qui, A. Abou-Samra, A. Shi, and **R. C. Deth.** Protein kinawe C-dependent constitutive activity of a2A/D-adrenergic receptors. *Pharmacology* 71:80–90, 2004.

Waly, M., et al. **(R. C. Deth).** P13-kinase regulates methionine synthase: Activation by IGF-1 or dopamine and inhibition by heavy metals and thimerosal. *Mol. Psychiatry* 9:358–70, 2004.

Deth, R. C. *Molecular Origins of Attention: The Dopamine-Folate Connection.* New York: Kluwer Academic Publishers, 2003.

Chen, X., H. G. Marrero, and **J. E. Freedman.** Opiod receptor modulation of a metabolically-sensitive ion channel in rat amygdala neurons. *J. Neurosci.* 21:9092–100, 2001.

Chen, C., et al. **(J. E. Freedman).** Altered gating of opiate receptor-modulated K+ channels on amygdala neurons of morphine-dependent rats. *Proc. Natl. Acad. Sci. U.S.A.* 97:14692–6, 2000.

Gatley, S. J., et al. PET imaging in clinical drug abuse research. *Curr. Pharm. Design* 11:3203–18, 2005.

Li, Z., et al. **(S. J. Gatley).** Candidate PET radioligands for cannabinoid CB1 receptors: [18F]AM5144 and related pyrazole compounds. *Nucl. Med. Biol.* 32:361–6, 2005.

Loring, R. H., V. Visalakshi, and S. Leppanen. Gene analysis of nicotine-induced upregulation of human alpha4 beta2 nicotinc receptors. *Soc. Neurosci. Abst.* 723.22, 2005.

Loring, R. H., et al. Surface expression of chimeric alpha7-GFP nicotinic receptors is cell type specific. *Soc. Neurosci. Abst.* 145.8, 2001.

Guo, J., et al. **(A. Makriyannis).** Conformational study of lipophilic ligands in phospholipid model membrand systems by solution NMR. *J. Med. Chem.* 46:4838–46, 2003.

Makriyannis, A., and A. Goutopoulos. Cannabinergics: Old and new therapeutic possibilities. In *Drug Discovery Strategies and Methods,* eds. **A. Makriyannis** and D. Biegel. New York: Marcel Dekker, Inc., 2003.

Thiel, M., et al. **(A. Ohta** and **M. V. Sitkovsky).** Oxygenation inhibits the physiological tissue-protecting mechanism and thereby exacerbates acute inflammatory lung injury. *PloS Biol.* 3:e174, 2005.

Lukashev, D., et al. **(A. Ohta** and **M. V. Sitkovsky).** Physiologic attenuation of proinflammatory transcription by the Gs protein-coupled A2A adensoine receptor in vivo. *J. Immunol.* 173:21–4, 2004.

Foy, J. W.-D., and **R. A. Schatz.** Inhibition of rat respiratory tract cytochrome P450 activity after low-level m-xylene inhalation: Role in 1-nitronaphthlene toxicity. *Inhalation Toxicol.* 16:1–8, 2004.

Vaidynathan, A., J. W.-D. Foy, and **R. A. Schatz.** Inhibition of rat respiratory-tract cytochrome P-450 isozymes following inhalation of m-xylene: Possible role of metabolites. *J. Toxicol. Environ. Health* 66:1122–43, 2003.

Sitkovsky, M. V., and **A. Ohta.** The "danger" sensors that STOP immune response: The A2 adenosine receptors? *Trends Immunol.* 26:299–304, 2005.

Sitkovsky, M. V., et al. **(A. Ohta).** Physiological control of immune response and inflammatory tissue damage by hypoxia inducible factors and adenosine A2A receptors. *Annu. Rev. Immunol.* 22:657–82, 2004.

Zahr, N. M., L. P. Martin, and **B. L. Waszczak.** Subthalamic nucleus lesions alter basal and dopamine agonist stimulated electrophysiological output from the rat basal ganglia. *Synapse* 54:119–28, 2004.

Waszczak, B. L., et al. Effects of individual and concurrent stimulation of striatal D1 and D2 dopamine receptors on electrophysiological and behavioral output from the rat basal ganglia. *J. Pharmacol. Exp. Therap.* 300:850–61, 2002.

Alvarez-Diez, T., and **J. Zheng.** 4-Ipomeanol: A mechanism-based inactivator of cytochrome P450 3A4. *Chem. Res. Toxicol.* 17:150–57, 2004.

Alvarez-Diez, T., and **J. Zheng.** Identification of glutathione conjugates derived from 4-ipomeanol metabolism in bile of rats by liquid chromatography-tandem mass spectrometry (LC-MS/MS). *Drug Metab. Disp.* 32:1345–50, 2004.

SOUTHERN ILLINOIS UNIVERSITY CARBONDALE

Department of Pharmacology
Ph.D. Program

Program of Study

The program offers the degree of Ph.D. in pharmacology. The program consists of formal course work in pharmacology and related fields, research, teaching experience, presentations at professional meetings, and publication of research. The objective of the program is to provide a thorough understanding of basic pharmacology. Students may choose from a variety of specializations when picking a research adviser and a research topic.

Students must fulfill the requirements of both the Graduate School and the Pharmacology Graduate Program to receive an advanced degree in pharmacology. Students entering the pharmacology graduate training program are required to have a strong background in physiology and biochemistry. Deficiencies, if they exist, can be remedied at the Carbondale campus, the main campus of Southern Illinois University·(SIU), before the student comes to Springfield.

During the first year in Springfield, advanced course work in pharmacology and related areas is accomplished. In addition, the student, in collaboration with the adviser, formulates a research project and selects a dissertation committee, which provides guidance for the project. After successful completion of the written comprehensive examination, taken in October of the second year in Springfield, the student submits a research proposal for dissertation work and defends it in an oral examination conducted by the Dissertation Committee. Before final admission to candidacy, tool requirements and fulfillment of residency requirements (24 credit hours) must be accomplished. After all other requirements for the degree are satisfied, the final oral exam is taken. This exam consists of a seminar that describes the dissertation work and a formal review conducted by the Dissertation Committee. Completion of a Ph.D. degree takes four to five years.

Research Facilities

The Department of Pharmacology occupies approximately 13,000 square feet of the Medical Instructional Facility of the School of Medicine, which has modern, well-equipped laboratories. Each research group has access to highly specialized, up-to-date facilities for cardiovascular, electrophysiological, neuropharmacological, molecular biological, biochemical, immunohistochemical, ultrastructural, toxicological, and behavioral research. The School of Medicine continues to render strong support in the form of state-supported, technical assistance and a complete glassware preparation facility. Through collaborative agreements, facilities in other departments of the medical school are available. This permits the students and faculty members to interact extensively with other basic and clinical research programs. Research facilities, all housed within the Medical Instructional Facility, include a computer center, an extensive biomedical sciences library, an electron microscope and image analysis facility, a biomedical communications department, a fine-instrument shop, and a laboratory animal facility under the direction of a veterinarian who provides expertise in animal housing, animal handling, and surgical procedures. The laboratory animal–care program is accredited by the American Association for Accreditation of Laboratory Animal Care. The Department's research programs have attracted many federal and private grants that have provided almost $3 million over the past five years in support of research. Some of the extramural agencies that have provided funding include the National Institutes of Health, National Science Foundation, American Heart Association, American Cancer Society, Department of Defense, American Heart Association Illinois Affiliate, Deafness Research Foundation, American Health Assistance Foundation, Alzheimer's Disease and Related Disorders Association, Illinois Department of Public Health, and Environmental Protection Agency.

Financial Aid

Students admitted into the Pharmacology Graduate Program, if qualified, may receive support for four years when entering the Ph.D. program. Financial support is given for a twelve-month period in the form of a fellowship with a competitive stipend and full tuition waiver. Financial assistance is also provided to qualified students for attendance at a national scientific meeting. Yearly renewal of this financial support is contingent upon satisfactory academic performance. The Department offers half-time graduate assistantships at $15,384 per year.

Cost of Study

In-state graduate tuition is $243 per credit hour in 2006–07. Out-of-state tuition is 2.5 times the in-state tuition rate ($607.50 per credit hour). Graduate students with at least a 25 percent appointment as a graduate assistant receive a tuition waiver. Fees vary from $441.62 (1 credit hour) to $987.30 (12 credit hours).

Living and Housing Costs

For married couples, students with families, and single graduate students, the University has 589 efficiency and one-, two-, and three-bedroom apartments that rent for $438 to $505 per month in 2006–07. Residence halls for single graduate students are also available, as are accessible residence hall rooms and apartments for students with disabilities.

Student Group

There are currently 9 students studying at the graduate level in the Department of Pharmacology.

Student Outcomes

Graduates typically pursue careers of independent research and teaching in pharmacology for academic institutions, industrial laboratories, or government research and administrative agencies.

Location

Southern Illinois University Carbondale (SIUC) is 350 miles south of Chicago and 100 miles southeast of St. Louis. Nestled in rolling hills bordered by the Ohio and Mississippi Rivers and enhanced by a mild climate, the area has state parks, national forests and wildlife refuges, and large lakes for outdoor recreation. Much of the area is a part of the 240,000-acre Shawnee National Forest. Cultural offerings include theater, opera, concerts, art exhibits, and cinema. Educational facilities available for the families of students are excellent.

The University and The Department

Southern Illinois University Carbondale is a comprehensive public university with a variety of general and professional education programs. The University offers associate, bachelor's, master's, and doctoral degrees as well as the J.D. and M.D. degree. The University is fully accredited by the North Central Association of Colleges and Schools. The Graduate School has an essential role in the development and coordination of graduate instruction and research programs. The Graduate Council has academic responsibility for determining graduate standards, recommending new graduate programs and research centers, and establishing policies to facilitate the research effort. Southern Illinois University Carbondale is a state-funded university founded in 1869. The Department of Pharmacology is part of the School of Medicine at SIU at Springfield.

Applying

Applications for admission to the graduate program may be obtained from the Director of the Pharmacology Graduate Program or directly from the Graduate School in Carbondale. Although there is no deadline for the receipt of applications, early application is advised for those applicants with outstanding qualifications to be considered for competitive fellowships. The applicant must first be admitted to the Graduate School before final acceptance into the program can be granted by the Pharmacology Graduate Program Committee. The Department of Pharmacology Graduate Program typically begins each June, but other arrangements may be possible for students already possessing related graduate degrees. A completed application and transcripts of all college and postgraduate work must be sent to the Pharmacology Graduate Program Director before consideration of admission. International students must request that a copy of their TOEFL scores be sent directly to the Director. The Graduate School requires a score of 550 or better on the TOEFL. International students must also provide official transcripts as required by the Department and the Graduate School. In addition to the above requirements of the Graduate School, all applicants must submit a brief typed statement (300–600 words) indicating why they desire to do graduate work in pharmacology. The statement should demonstrate their ability to organize and present information in proper English. It should be as specific as possible, especially if the applicant has concrete ideas about the field of research they wish to enter. Moreover, three letters of recommendation must also be received before consideration of admission. Further supporting materials that must be sent include scores on the General Test of the Graduate Record Examinations (GRE) and one Subject Test (biology or chemistry) of the GRE.

Correspondence and Information

Director, Pharmacology Graduate Program
Southern Illinois University School of Medicine
P.O. Box 19629
Springfield, Illinois 62794-9629
E-mail: ncowan@siumed.edu
Web site: http://www.siumed.edu/pharm/home.html

Southern Illinois University Carbondale

THE FACULTY AND THEIR RESEARCH

Five main areas of research are currently pursued by faculty members: neuropharmacology, cardiovascular pharmacology, molecular pharmacology, biochemical pharmacology, and toxicology. In addition to intradepartmental projects, opportunities for collaborative research with clinical investigators are available. Such efforts to date have brought fruitful results.

Department of Pharmacology Faculty
Arai C. Amy, Assistant Professor; Ph.D., Chiba (Japan), 1987. Molecular and pharmacological modulation of AMPA-type glutamate receptors and its impact on synaptic transmission and network properties.
Donald M. Caspary, Professor; Ph.D., NYU, 1973. Sensory physiology; neurophysiology; neuroanatomy; comparative physiology.
George A. Dunaway, Professor; Ph.D., Oklahoma, 1975. Regulation of energy/metabolism during diabetes; development and aging; induction of experimental ulcers in rats.
Carl L. Faingold, Professor and Chairman; Ph.D., Northwestern, 1972. Convulsive seizure mechanisms and effects of anticonvulsants; pharmacological alterations of cerebral evoked potentials.
Tony J-F. Lee, Professor; Ph.D., West Virginia, 1975. Neuromuscular transmission in cerebral blood vessels.
Louis Premkumar, Assistant Professor; Ph.D., Australian National, 1992. Molecular neurobiology; recombinant glutamate receptors; mechanism of action of drugs of abuse.
Vickram Ramkumar, Associate Professor; Ph.D., Maryland, 1992. Molecular pharmacology of adenosine receptors in cardiovascular system.
Linda A. Toth, Professor; Ph.D., Pittsburgh, 1980; D.M.V., Purdue, 1986.

Department of Pharmacology Cross-Appointed Faculty
Robert Helfert, Research Associate Professor, Department of Surgery.
Richard Katholi, Clinical Professor, Department of Medicine and Prairie Cardiovascular Consultants, Ltd.
Dean Naritoku, Associate Professor, Department of Neurology.
Leonard Rybak, Professor of Surgery, Department of Surgery.

STATE UNIVERSITY OF NEW YORK AT BUFFALO

Roswell Park Cancer Institute
Program of Molecular Pharmacology and Cancer Therapeutics

Program of Study

The Program of Molecular Pharmacology and Cancer Therapeutics offers graduate students the opportunity to study a range of topics in cancer biology and cancer therapeutics at the molecular, cellular, and biochemical levels. The program is located on the campus of Roswell Park Cancer Institute (RPCI) in downtown Buffalo, New York. It is composed of 28 faculty members and has a current enrollment of 25 predoctoral students. The program has both a rich history and distinguished alumni and constitutes the nation's largest pharmacology graduate program devoted entirely to cancer research. Students in the program have the opportunity to study basic, preclinical, and clinical research problems related to fundamental differences between normal and cancer cells. Current areas of research include the molecular mechanisms involved in cell-cycle control, differentiation, cell death, and signal transduction. In addition, development of targeted therapeutics as anticancer agents, the mode of action of currently used anticancer drugs, the interaction of anticancer agents with the host immune system, development and use of animal models of human cancer, and the mechanisms by which malignant cells become resistant to therapy are active areas of inquiry among the program faculty members.

The Ph.D. program is typically five years in length and requires 32 semester hours of course work, consisting of core requirements in pharmacology, cancer therapeutics, oncology, and biochemistry while allowing for specialization in such areas as molecular biology, cellular biology, and immunology. Courses are taught by faculty members located at both Roswell Park Cancer Institute and the State University of New York at Buffalo (SUNY) campus. All students must complete a written and oral preliminary examination and at least 32 semester hours of independent laboratory research that culminates in the successful defense of an original thesis. A formal weekly student seminar series, in which students present their research, serves to prepare students to successfully compete for future job opportunities. Degrees are awarded through the State University of New York at Buffalo, Roswell Park Graduate Division.

Research Facilities

Roswell Park Cancer Institute, the world's first research institute devoted exclusively to the study of cancer and its treatment, was founded in 1898. RPCI's 25-acre campus on the Buffalo Niagara Medical Campus includes approximately 1.5 million square feet of space distributed between clinical programs, basic and translational research, and education. The newest building on the RPCI campus, the Center for Genetics and Pharmacology, which opened in spring 2006, features a modern, open lab design and houses most program faculty members. RPCI employs more than 400 Ph.D.'s and M.D.'s and a support staff of more than 2,600. RPCI is also home to a number of other active graduate programs in the areas of cancer pathology and prevention, cellular and molecular biology, immunology, and molecular and cellular biophysics and biochemistry. RPCI maintains a sizable and modern medical library and computer center. Departmental and institutional seminar series sustain a continuous appraisal of new developments in cancer research and treatment.

Research at RPCI is facilitated by a number of NCI Comprehensive Cancer Center grant–supported core facilities in gene expression profiling, proteomics, mouse gene targeting, genomics, cancer biorepositories, and experimental tumor models, among others. RPCI has approximately 25,000 active cancer patients, offering ample opportunities for clinically related research.

RPCI is in proximity to the University at Buffalo's Center of Excellence in Bioinformatics and the Hauptman-Woodward Medical Research Institute in structural biology, providing a fertile environment for academic enrichment.

Financial Aid

In general, admission to the program qualifies students for a research stipend ($21,000 in 2006–07). Students applying prior to February 15 are also considered for the prestigious Woodburn and Presidential graduate fellowships, which offer a supplementary stipend and support for research expenses annually. In addition, the program has continuously held an NIH T32 predoctoral training grant for more than thirty-five years; this grant is typically used to provide financial support to second-year graduate students.

Cost of Study

Tuition is generally waived for students in the Program of Molecular Pharmacology and Cancer Therapeutics.

Living and Housing Costs

There is no University housing for graduate students at the Institute. Private apartments, however, are available in the University neighborhood. Typical one- and two-bedroom apartment rates are $500 to $800 per month, and the average single-family home in Buffalo sells for approximately $100,000, making housing costs in Buffalo far lower than those in comparable Eastern cities.

Student Group

At present, there are 14 women and 11 men in the program. Four to 6 new students matriculate into the program each year. Students in the program have diverse academic backgrounds and work experiences. Students holding B.A. and B.S. degrees in chemistry, biochemistry, biology, and molecular biology are represented, as are students from M.D./Ph.D. programs who are pursuing thesis research under the guidance of program faculty members.

Student Outcomes

Graduates of the program are highly competitive for research and faculty positions within universities, biomedical research centers, and the pharmaceutical industry.

Location

Buffalo is the second-largest city in New York. Its proximity to Niagara Falls, Canada, Lake Ontario, Lake Erie, and the Finger Lakes of western New York provides extensive opportunities for sailing, bicycling, golfing, swimming, rowing, fishing, canoeing, and camping in the spring, summer, and fall and skiing, skating, and hockey in the winter. The Albright-Knox Museum of Art, the Buffalo Philharmonic Orchestra, the Studio Arena Theater, and Artpark (the New York State park of visual and performing arts) sponsor numerous cultural events, and annual Shaw and Shakespeare festivals are held nearby in Canada. In addition, Buffalo fields professional teams in NFL football (Bills), NHL ice hockey (Sabres), and AAA baseball (Bisons).

The University and The Institute

The State University of New York at Buffalo is the largest single unit and has the most comprehensive graduate program of the universities that comprise the state university system. It has two major campuses and numerous affiliated units in the greater Buffalo area. These include the main campus in Amherst; the Health Sciences Center at the Main Street campus in Buffalo, which includes the Schools of Medicine, Pharmacy, Dentistry, and Nursing; and Roswell Park Cancer Institute in Buffalo. These facilities and their faculties are readily accessible to students.

The clinical and basic science research environment of Roswell Park Cancer Institute and the strong liaisons of the program with other units of the Institute and the Health Sciences Center provide excellent opportunities for advanced research and training in biochemical pharmacology. The close and frequent interactions among pharmacologists, molecular biologists, biochemists, cell biologists, and medicinal chemists working toward a common goal within the same graduate program are highly favorable for the formulation of new ideas and fruitful cooperative studies.

Applying

Formal applications are initiated online at http://www.gradmit.buffalo.edu/etw/ets/et.asp?nxappid=GRA&nxmid=getpublicapplicationsite. Applicants should submit a formal application, GRE scores, transcripts, and three letters of recommendation. Applications from women and members of ethnic minority groups are particularly encouraged.

Correspondence and Information

Dr. Adam Karpf, Director of Graduate Studies
Roswell Park Cancer Institute
Elm and Carlton Streets
Buffalo, New York 14263

Phone: 716-845-8225
Fax: 716-845-8857
E-mail: theresa.skurzewski@roswellpark.org
Web site: http://www.roswellpark.org/Site/Education/Graduate_Education/Departments/
Molecular_Pharmacology_Cancer_Therapeutics
http://www.roswellpark.org/Education/Graduate_Education/Departments

State University of New York at Buffalo

THE FACULTY AND THEIR RESEARCH

Maria Baer, Associate Professor; M.D., Johns Hopkins. Drug resistance and novel treatment approaches to leukemia.

Terry A. Beerman, Professor; Ph.D., Syracuse. Study of cellular responses to DNA damage induced by antitumor agents; study/development of novel sequence-specific DNA binding agents as inhibitors of gene expression.

Erica Berleth, Affiliate Member; Ph.D., SUNY at Buffalo. Defining differences in the apoptotic pathways between normal and cancer cells; exploitation of apoptosis-inducing anticancer therapies.

Ralph Bernacki, Professor; Ph.D., Rochester. Preclinical anticancer drug development; tumor metastasis and angiogenesis; study of cell-surface complex carbohydrates; modulation of drug resistance.

Adrian R. Black, Affiliate Member; Ph.D., Imperial College (London). Transcription factors in tumor progression; cell-cycle regulatory molecules as targets for chemotherapy.

Jennifer Black, Professor; Ph.D., Imperial College (London). Signal transduction; protein kinase C; control of cell growth/cell-cycle progression and differentiation in the intestinal epithelium; cell ultrastructure.

Michael G. Brattain, Professor and Chairman of the Department of Pharmacology and Therapeutics; Ph.D., Rutgers. Growth regulatory pathology of signal transduction by ErbB family and transforming growth factor beta.

Gokul Das, Assistant Professor; Ph.D., Baylor College of Medicine. Tumor suppressor protein p53/transcriptional regulation/genomic damage, DNA repair, and cell-cycle control–interaction between p53 and estrogen receptor signaling pathways.

Bruce J. Dolnick, Professor; Ph.D., SUNY at Buffalo. The rTS signaling pathway in growth regulation and experimental chemotherapy.

Barbara Foster, Assistant Professor; Ph.D., California, San Francisco. Prostate cancer; transgenic mouse models; prostate cancer–bone interaction; preclinical testing of novel therapeutic approaches for prevention and treatment of prostate cancer.

Allen Gao, Associate Professor; M.D., Ph.D., Texas. Identification of diagnostic markers and potential therapeutic targets for prostate cancer.

David Goodrich, Associate Professor and Chair of Graduate Studies; Ph.D., Berkeley. Tumor suppressor genes and their regulation of the cell cycle, differentiation, and cell death.

William R. Greco, Professor; Ph.D., SUNY at Buffalo. Pharmacometrics and bioinformatics.

Margot M. Ip, Professor; Ph.D., Wisconsin. Chemoprevention of breast cancer; tumor necrosis factor and its role in normal and malignant breast; stromal-epithelial interactions in normal and malignant breast.

Candace S. Johnson, Professor; Ph.D., Ohio State. Drug development of novel agents; preclinical tumor models; mechanistic studies on drugs with novel targets; steroid hormones as an antiproliferative agent and development of novel approaches in prostate cancer.

Peter M. Kanter, Assistant Professor; D.V.M., Cornell; Ph.D., SUNY at Buffalo. Preclinical toxicology; drug-DNA interactions.

Adam Karpf, Assistant Professor; Ph.D., Texas at Austin. Role of DNA methylation and DNA methyltransferase enzymes in oncogenesis, with the overall aim of exploiting this biochemical pathway as a cancer drug target.

Fengzhi Li, Assistant Professor; Ph.D., Beijing. Mechanism of anti-apoptotic protein function and regulation in cancer cells.

Athena W. Lin, Assistant Professor; Ph.D., Massachusetts Amherst. Ras/MAPK signaling; tumor suppression pathways; cell-cycle control; senescence and its relevance in carcinogenesis.

John McGuire, Associate Professor; Ph.D., Berkeley. Biochemistry of folic acid and one-carbon metabolism; biochemical pharmacology of new and established antifolates; drug resistance.

Enrico Mihich, Professor; M.D., Milan. Experimental tumor therapeutics; relationships between cancer chemotherapy and immunity; biochemical bases for selective toxicity; preclinical pharmacology and toxicology of antitumor agents.

Allan Oseroff, Professor; M.D., Yale; Ph.D., Harvard. Design and synthesis of photosensitizers and determination of structure-activity relationships; characterization of intracellular and tissue damage sites; development of combination therapies that exploit and amplify the photodamage processes; measurement of pharmacodynamics and therapeutic efficacy in in vivo model systems and in patients.

Carl Porter, Professor; Ph.D., SUNY at Buffalo. New and novel anticancer targets; biological role of polyamines in cell proliferation; drug discovery and development; mechanisms of cell-cycle and apoptosis control.

Youcef Rustum, Professor; Ph.D., SUNY at Buffalo. Clinical biochemistry and pharmacology, with emphasis on selective actions of antitumor agents.

Harry Slocum, Assistant Professor; Ph.D., SUNY at Buffalo. Cell biology of human solid tumors; cell integration and growth control; cellular heterogeneity; mechanism of action of anticancer agents.

Robert Straubinger, Associate Professor; Ph.D., California, San Francisco. Cell biology and biophysics of membranes and membrane function; use of liposomal drug carriers to improve the efficacy of cancer therapeutics.

Janice Sufrin, Associate Professor; Ph.D., Brandeis. Structure-based design/synthesis and biochemical/therapeutic effects of molecules that target S-adenosylmethionine-associated pathways, with particular emphasis on methylation, polyamine, and quorum sensing phenomena.

Donald L. Trump, Professor; M.D., Johns Hopkins. Novel therapeutic approaches to prostate and other adult solid tumors; vitamin D–based therapeutics for prevention and therapy for prostate cancer.

Michael Wong, Assistant Professor; M.D., Toronto. The interface between tumor and host, with a particular emphasis on angiogenesis, tumor matrix, and immune modulation.

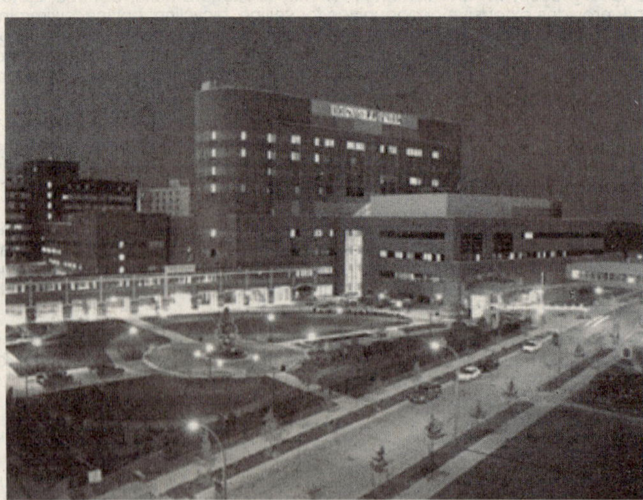

The Roswell Park Cancer Institute.

STONY BROOK UNIVERSITY, STATE UNIVERSITY OF NEW YORK

School of Medicine
Molecular and Cellular Pharmacology Graduate Program

Program of Study

The Molecular and Cellular Pharmacology Graduate Program includes faculty members of the Department of Pharmacological Sciences and other departments at Stony Brook and scientists from Brookhaven National Laboratory and Cold Spring Harbor Laboratory. Graduate training leading to the Ph.D. degree emphasizes research experience and a course of study tailored to the individual. During the first two years, students complete the elective and core course work and participate in research projects directed by program faculty members. Students select an adviser for the dissertation research during the second year and, upon completion of the qualifying examination, devote their full efforts to their research.

The goal of the program is to provide the student with the necessary knowledge and skills to pursue a career in basic research in the academic, industrial, or government sector. Students planning careers in pharmacological sciences should have a broad interdisciplinary background with strength in chemistry, biochemistry, pharmacology, physiology, and related biomedical sciences. This is accomplished through an integrated program of course work, independent study, participation in student seminars and a seminar program of visiting scientists, and extensive research experience.

Research Facilities

Faculty laboratories are equipped for all types of modern molecular and cell biological, biochemical, neurochemical, chemical, biophysical, and toxicological research. Specialized facilities are provided for tissue culture, recombinant DNA work, ultracentrifugation, scintillation and gamma spectrometry, transgenic mouse research, electron microscopy, confocal microscopy, molecular modeling, gas and high-performance liquid chromatography, nuclear magnetic resonance, X-ray crystallography, and mass spectrometry. Research activities are supported by library and core facilities, such as the proteomics center; the cell-culture and hybridoma, DNA microarray, DNA sequencing, and molecular cloning facilities; and the Division of Laboratory Animal Research, as well as by extensive computer and informatics resources. Program faculty members currently receive more than $19 million in annual research support from federal, state, and private agencies.

Financial Aid

Student support is provided by a Training Grant in Pharmacological Sciences from the National Institutes of Health, University assistantships, and faculty research grants. All students receive an annual stipend ($25,000 for the 2006–07 academic year), a full-tuition scholarship, and health insurance as a fringe benefit.

Cost of Study

Students are responsible for their books, activity fees, supplies, and their room and board.

Living and Housing Costs

A single student living on campus requires about $13,500 to $16,000 to cover normal living expenses for twelve months. Graduate students, including married couples, can be accommodated in the residence halls. On-campus apartments provide housing for nearly 1,000 graduate students and their families. Off-campus apartments and houses can be found in the vicinity of the University. Information about housing can be obtained from the University Off-Campus Housing Office or from the Health Sciences Center Office of Student Services.

Student Group

Total enrollment at the University is about 22,000, including 7,700 graduate students. The Department of Pharmacological Sciences has 36 full-time graduate students and 20 postdoctoral fellows from throughout the United States and the world.

Student Outcomes

Ph.D. graduates have been very successful in obtaining postdoctoral training in highly regarded laboratories. Approximately half of Ph.D. graduates have taken academic positions at such institutions as Stanford, University of North Carolina, Yale, Duke, Baylor, Harvard, Washington (St. Louis), University of Pennsylvania, and Wake Forest, as well as at Jackson Labs, Cold Spring Harbor Lab, and McArdle Lab at the University of Wisconsin. About half have taken positions in industry (e.g., Abbott Labs, Bristol-Myers Squibb, Novartis, Dow, R. W. Johnson, and American Cyanamid), either before or after postdoctoral training, or have furthered their education in law or public health, obtaining positions in patent law, in intellectual property law, and with the Food and Drug Administration.

Location

Stony Brook is located approximately 60 miles east of Manhattan on the wooded North Shore of Long Island, convenient to the cultural life of New York City and the recreational resources of Suffolk County's tranquil countryside and seashores. Traveling to the city is easy since the local commuter train, the Long Island Railroad, makes a stop right on the campus. Brookhaven National Laboratory and Cold Spring Harbor Laboratory are nearby. Long Island's hundreds of miles of magnificent coastline attract many swimming, boating, and fishing enthusiasts.

The University

Stony Brook University, State University of New York, is recognized as being among the nation's finest universities and has recently been ranked as one of the "100 best values in higher education" among public universities. Stony Brook offers excellent programs in a broad range of academic subjects and conducts major research and public service projects. Over the past decade, externally funded support for Stony Brook's research programs has grown faster than at any other major university in the nation. An internationally renowned faculty offers courses at the undergraduate through doctoral levels for 22,000 students through 119 undergraduate and 40 graduate departmental and interdisciplinary majors. The Health Sciences Center includes nationally recognized medical and dental schools, a veterans' nursing home, and a 520-bed hospital. Beyond its research eminence, the University is well known for its biannual film festivals and the extensive series of musical and cultural events staged each year in its performance houses, the Staller and Wang Centers.

Applying

Admission to the Graduate Program in Molecular and Cellular Pharmacology requires the following: a baccalaureate degree in an appropriate field (biology, chemistry, biochemistry, medicine, microbiology, physics, or mathematics); undergraduate transcripts with evidence of strong performance in science courses; letters of reference; scores on the General Test of the Graduate Record Examinations (a Subject Test is recommended but not required); and, in the case of international students, TOEFL scores. Previous research experience is highly desirable. Acceptance into the department's graduate program is determined by an admissions committee on the basis of the applicant's research potential and academic qualifications. Applications are considered on a rolling basis. Women, minority, and handicapped candidates are encouraged to apply.

Correspondence and Information

Molecular and Cellular Pharmacology Graduate Program
Department of Pharmacological Sciences
Health Sciences Center
Stony Brook University, State University of New York
Stony Brook, New York 11794-8651
Phone: 631-444-3057
Fax: 631-444-9749
E-mail: grad@pharm.stonybrook.edu
Web site: http://www.pharm.sunysb.edu (click on Graduate Program)

Stony Brook University, State University of New York

THE FACULTY AND THEIR RESEARCH

Daniel Bogenhagen, Professor; M.D., Stanford, 1977. Replication, transcription, and repair of mammalian mitochondrial DNA; mitochondrial proteomics.
Holly Colognato, Assistant Professor; Ph.D., Rutgers, 1999. Extracellular matrix in the brain and roles during development and during neurodegeneration.
Howard Crawford, Assistant Professor; Ph.D., Texas Southwestern Medical Center at Dallas, 1993. Pancreatic cancer.
Carlos de los Santos, Associate Professor; Ph.D., Buenos Aires, 1987. Solution structures of damaged nucleic acids and repair proteins.
Guangwei Du, Research Assistant Professor; Ph.D., Peking Union Medical College and Chinese Academy of Medical Sciences, 1999. Cellular morphogenesis and membrane trafficking.
Michael A. Frohman, Professor; M.D./Ph.D., Pennsylvania, 1985. Membrane fusion and signal transduction.
DaXiong Fu, Assistant Professor; Ph.D., Mayo, 1996. Crystallization of integral membrane proteins.
Arthur P. Grollman, Distinguished Professor; M.D., Johns Hopkins, 1959. Chemical carcinogenesis and mutagenesis.
Charles Iden, Associate Professor; Ph.D., Johns Hopkins, 1971. Biomedical applications of mass spectrometry; proteomics; characterization of DNA adducts and DNA repair mechanisms; synthesis of modified oligodeoxynucleotides.
Francis Johnson, Professor; Ph.D., Glasgow (Scotland), 1954. Synthesis of natural products; DoM reactions; antiviral agents; mechanism of action of carcinogens and mutagens; site-specific mutagenesis; DNA damage and mechanisms of action of DNA-repair enzymes.
Irwin Kurland, Associate Professor; M.D., USC, 1984; Ph.D., Vanderbilt, 1992. Regulation of insulin action and glucose metabolism.
Craig Malbon, Leading Professor; Ph.D., Case Western Reserve, 1976. Wnt-Frizzled signaling via G-proteins in development; analysis of signaling complexes.
Masaaki Moriya, Research Professor; Ph.D., Nagoya (Japan), 1981. Cellular response to DNA damage.
Sidonie A. Morrison, Associate Professor; D.Phil., Oxford, 1973. Mechanisms of infection and pathogenesis in HIV-1 disease, especially host-cell factors; restoration of immune function during antiretroviral therapy.
Jeffrey E. Pessin, Professor and Chair; Ph.D., Illinois, 1980. Insulin regulation of vesicular trafficking and signal transduction.
Basil Rigas, Professor; M.D., 1972, D.Sc., 1975, Athens. Colon cancer.
Orlando Schärer, Associate Professor; Ph.D., Harvard, 1996. Chemical biology of mammalian DNA repair.
Shinya Shibutani, Research Professor; Ph.D., Toyama (Japan), 1983. Mechanisms of translational DNA synthesis.
Ken-Ichi Takemura, Assistant Professor; Ph.D., Tokyo (Japan), 1997. Regulation of beta-catenin signaling activity.
Styliani-Anna E. Tsirka, Associate Professor and Graduate Program Director; Ph.D., Thessaloniki (Greece), 1989. Neuronal-microglial interactions in the physiology and pathology of the central nervous system.

Associated Faculty Members

James Bliska, Professor of Molecular Genetics and Microbiology; Ph.D., Berkeley, 1987. Molecular and cellular basis of bacterial–host cell interactions.
Ira Cohen, Professor of Physiology and Biophysics; M.D./Ph.D., NYU, 1974. Electrophysiology of the heart.
Grigori N. Enikolopov, Associate Professor, Cold Spring Harbor Laboratory; Ph.D., Russian Academy of Sciences, 1978. Stem cells; neurogenesis; development; signal transduction.
Marian J. Evinger, Associate Professor of Pediatrics and Neurobiology and Behavior; Ph.D., Washington (Seattle), 1978. Transcriptional regulation of PNMT gene expression; gene expression in neuronal tumors.
Robert Haltiwanger, Professor of Biochemistry and Cell Biology; Ph.D., Duke, 1986. Regulation of signal transduction by glycosylation.
Patrick Hearing, Professor of Molecular Genetics and Microbiology; Ph.D., Northwestern, 1980. Adenovirus regulation of cellular proliferation and gene expression; adenovirus vectors for human gene therapy.
A. Wali Karzai, Assistant Professor of Biochemistry and Cell Biology; Ph.D., Johns Hopkins, 1995. Biochemistry and structural biology of RNA-protein interactions; translational control of gene expression; drug discovery.
Joel Levine, Professor of Neurobiology and Behavior; Ph.D., Washington (St. Louis), 1980. Glial cells; proteoglycans and the regulation of axonal growth.
Mirjana Maletic-Savatic, Assistant Professor, Cold Spring Harbor Laboratory; M.D./Ph.D., Belgrade, 1996. Mechanisms that lead to the differentiation of neural progenitor cells.
David McKinnon, Associate Professor of Neurobiology and Behavior; Ph.D., Australian National, 1987. Molecular physiology of neurons and cardiac muscle.
W. Todd Miller, Associate Professor of Physiology and Biophysics; Ph.D., Rockefeller, 1989. Signal transduction by tyrosine kinases.
Ute M. Moll, Professor of Pathology; M.D., Ulm (Germany), 1985. Tumor suppressor gene research; mechanism of p53 inactivation.
Nicholas Nassar, Assistant Professor of Physiology and Biophysics; Ph.D., Joseph Fourier (France), 1992. Regulation of signaling proteins.
Nancy C. Reich, Professor; Ph.D., SUNY at Stony Brook, 1983. Signal transduction and gene expression induced by cytokines and viral infection.
Robert C. Rizzo, Assistant Professor of Applied Mathematics and Statistics; Ph.D., Yale, 2001. Computational research projects in cancer, HIV/AIDS, influenza, and method development.
Sami I. Said, Professor of Medicine; M.D., Cairo, 1951. Physiology and pharmacology of VIP and related neuropeptides, with special reference to their modulation of cell injury, inflammation, and cell death and their potential as therapeutic agents.
Nicole Sampson, Professor of Chemistry; Ph.D., Berkeley, 1990. Integrin receptor interactions in mammalian fertilization/enzymology of cholesterol oxidase.
Carlos Simmerling, Assistant Professor of Chemistry; Ph.D., Illinois at Chicago, 1994. Computational chemistry and structural biology; molecular dynamics of biological macromolecules.
Roy Steigbigel, Professor of Pharmacological Sciences and Molecular Genetics and Microbiology; M.D., Rochester, 1966. HIV treatment and immunoreconstitution.
Gerald H. Thomsen, Associate Professor of Biochemistry and Cell Biology; Ph.D., Rockefeller, 1988. Vertebrate embryonic development.
Peter Tonge, Professor of Chemistry; Birmingham (England), 1986. Biological chemistry and enzyme mechanisms; quantitating substrate strain in enzyme-substrate complexes using vibrational spectroscopy; rational drug design.
William Van Nostrand, Professor of Medicine; Ph.D., UCLA, 1985. Cerebrovascular pathology in Alzheimer's disease and related disorders.
Lonnie P. Wollmuth, Associate Professor of Neurobiology and Behavior; Ph.D., Washington (Seattle), 1992. Molecular mechanisms of synaptic function in the brain.
Wei-Xing Zong, Assistant Professor of Molecular Genetics and Microbiology, Ph.D., University of Medicine and Dentistry of New Jersey–Robert Wood Johnson Medical School/Rutgers, 1999. Molecular regulation of apoptotic and necrotic cell death.

Senior Advisory Program Members

Paul Adams, Professor of Neurobiology and Behavior; Ph.D., London. Thalamocortical circuitry regulating synaptic learning.
Miguel Berrios, Research Associate Professor; Ph.D., Rockefeller, 1983. Polypeptide structure of cell nucleus; cell biology of DNA damage and repair.
Moises Eisenberg, Professor; Ph.D., Caltech, 1972. Application of bioinformatics tools to study comparative gene organization.
Paul A. Fisher, Professor; M.D./Ph.D., Stanford, 1980. Structure and function of the cell nucleus; DNA metabolism and mutagenesis; human neurodegenerative diseases.
Joav M. Prives, Professor; Ph.D., McGill, 1968. Regulation of surface receptors in muscle cells.
Edward Reich, Distinguished Professor; M.D., Johns Hopkins, 1956; Ph.D., Rockefeller, 1962. Protein engineering and structure-activity analysis of plasminogen.
Thomas Rosenquist, Research Assistant Professor; Ph.D., Wisconsin–Madison, 1989. Genetic analysis of mammalian oxidative DNA damage repair.
Nisson Schechter, Professor of Psychiatry and Biochemistry; Ph.D., Western Michigan, 1971. Structure, function, and regulation of intermediate filament proteins and homeobox proteins during zebrafish neurogenesis.
Jakob H. Schmidt, Professor of Biochemistry; M.D., Munich, 1963; Ph.D., California, Riverside, 1970. Structure, function, and regulation of nicotinic acetylcholine receptors in muscle and brain.
Richard B. Setlow, Adjunct Professor; Ph.D., Yale, 1947. Macroscopic effects of tumor induction.
Charles S. Springer Jr., Professor of Chemistry and Senior Chemist, Brookhaven National Laboratory; Ph.D., Ohio State, 1967. In vivo nuclear magnetic resonance studies of biological systems.
Nora D. Volkow, Professor of Psychiatry; M.D., National (Mexico). Addiction; aging; ADHD.

THOMAS JEFFERSON UNIVERSITY

Jefferson College of Graduate Studies
Kimmel Cancer Center
Department of Biochemistry and Molecular Biology
Program in Molecular Pharmacology and Structural Biology

Programs of Study

The interdepartmental Ph.D. program provides focused training and research experience in molecular pharmacology and structural biology and in the study of drug interactions at the level of the cell and organ. It is designed to provide students with the basis for successful careers as independent scientists and scholars in either the academic or industrial sector. Students entering with a baccalaureate degree take core curriculum courses in biochemistry, molecular biology, cell biology, and genetics. Advanced courses in pharmacology and structural biology can be taken in both the first and second years, and the curriculum is flexible to accommodate the individual student's background and interests. All course work is generally completed by the end of the second year. In addition, several weekly public seminar series, frequent laboratory seminars, and invited lectureships form an integral part of the student's education. Current research interests of the faculty members include the biochemistry and molecular and cell biology of hormone-receptor interactions and signal transduction mechanisms within cells, mechanisms of membrane sorting and intracellular organization, neuropharmacology, molecular modeling and design of therapeutic agents, structural analysis of proteins and nucleic acids, the computational analysis of the structure and energetics of biological macromolecules, clinical pharmacology, and toxicology. The Graduate Program in Molecular Pharmacology and Structural Biology, along with the Programs in Biochemistry and Molecular Biology, Genetics, and Immunology and Microbial Pathogenesis, make up the Joint Graduate Programs of the Kimmel Cancer Center. Students applying to one of the Joint Graduate Programs may perform research rotations or work with faculty members in any of these programs.

Research Facilities

Research laboratories are primarily located in the Kimmel Cancer Center and the Departments of Biochemistry and Molecular Biology, Microbiology and Immunology, and Cancer Biology, which together occupy 70,000 square feet in the Bluemle Life Sciences Building and 25,000 square feet in the adjacent Jefferson Alumni Hall. In addition to extensive basic equipment and facilities, the program provides access to numerous specialized resources. These include facilities for peptide and oligonucleotide synthesis and sequencing, cell sorting by flow cytometry, protein purification and characterization, proteomics and microarray analysis, and biomolecular imaging. Also available are facilities for transgenic and knockout mouse production, extensive computer facilities, a CD spectrometer, and state-of-the-art X-ray detectors for macromolecular crystallography.

Financial Aid

Financial support is available to full-time Ph.D. students in the form of University fellowships and training grants. In 2006–07, students granted full fellowship support receive funds for payment of tuition and a stipend of $24,500. Also available to students demonstrating financial need are Title IV funds. University loan programs are also available to qualifying students.

Cost of Study

Tuition and fees for 2006–07 full-time Ph.D. students are $15,640 per year.

Living and Housing Costs

Campus housing is available at Thomas Jefferson University; reasonable alternative housing near the University in the Philadelphia area can also be found.

Student Group

Each year, approximately 15 students are admitted into the Joint Graduate Programs of the Kimmel Cancer Center. The class size in pharmacology is expected to be 3 or 4 students per year. Currently, the University enrolls about 2,500 students. The College of Graduate Studies enrolls approximately 630 students, about half of whom are women.

Student Outcomes

Recent Ph.D. graduates have accepted postdoctoral positions in prestigious academic institutions, such as Harvard, Baylor, Johns Hopkins, UC Davis, the University of Texas M. D. Anderson Cancer Center, and the National Institutes of Health, and in industry, including Bristol-Meyers Squibb and Pfizer.

Location

The 13-acre campus of Thomas Jefferson University is located in the historical downtown area of Philadelphia, within walking distance of many places of cultural interest, including concert halls, theaters, museums, art galleries, and historic sites. Numerous intercollegiate and professional athletics events take place nearby. Convenient bus and subway lines connect the University with other local universities and colleges and with several outstanding libraries. The proximity of the New Jersey shore and the Pennsylvania mountains offers year-round recreational opportunities, and New York City and Washington, D.C., are each just a few hours away.

The University

Thomas Jefferson University is an academic health center. It evolved from the Jefferson Medical College, which was founded in 1824. In addition to the medical college, the University includes the College of Graduate Studies, the College of Health Professions, the University hospital, and various affiliated hospitals and institutions. Jefferson Alumni Hall houses the TJU Fitness and Recreation Facility, study lounges, a swimming pool, a sauna, a gymnasium, and a handball/squash court.

Applying

Applications may be submitted at any time but should be received before March 1 for optimal consideration. Scores on the General Test of the Graduate Record Examinations, three letters of recommendation, and academic transcripts are required of all applicants. For those students whose native language is not English, scores on the Test of English as a Foreign Language (TOEFL) are required. Scores on an appropriate Subject Test of the Graduate Record Examinations are strongly recommended. Prospective students are encouraged to visit the Department and discuss the graduate program with members of the faculty.

Correspondence and Information

For applications:

Jessie Pervall
Director of Admissions and Recruitment
College of Graduate Studies
Thomas Jefferson University
1020 Locust Street, M-46
Philadelphia, Pennsylvania 19107-6799

Phone: 215-503-4400
Fax: 215-503-3433
E-mail: cgs-info@jefferson.edu
Web site: http://www.jefferson.edu/cgs

For program information:

Joanne Balitzky
Training Programs Coordinator
Kimmel Cancer Center
Thomas Jefferson University
233 South 10th Street, 910 BLSB
Philadelphia, Pennsylvania 19107-5541

Phone: 215-503-6687
Fax: 215-503-0622
E-mail: joanne.balitzky@jefferson.edu
Web site: http://www.jefferson.edu/jcgs/phd/mpsb/

Thomas Jefferson University

THE FACULTY AND THEIR RESEARCH

Emad S. Alnemri, Professor; Ph.D., Temple, 1991. Molecular mechanisms of programmed cell death (apoptosis); regulation of caspase activation during apoptosis and inflammation; role of mitochondria in cell death and survival.

Renato Baserga, Professor; M.D., Milan, 1949. Genetic analysis of G1 phase and control of cell proliferation; growth factors and their receptors; apoptosis; anticancer therapy.

Jeffrey L. Benovic, Professor and Program Director; Ph.D., Duke, 1986. Molecular and regulatory properties of G-protein–coupled receptors; role of receptor dysregulation in cancer and cardiovascular and neurological disorders.

Bruce M. Boman, Professor; M.D., Minnesota, 1976; Ph.D., Mayo, 1982. Mechanisms of action of the encoded product of the colon cancer susceptibility gene, adenomatous polyposis coli (APC); integration of molecular testing into diagnosis and management of colorectal cancer; colon cancer genetics.

George C. Brainard, Professor; Ph.D., Wesleyan, 1973. Photobiology, neuroendocrine, and circadian regulation; control of melatonin in humans; effects of light on physiology, behavior, and cancer progression; use of light as countermeasure in long-duration space flight.

Andrea D. Eckhart, Associate Professor; Ph.D., North Carolina at Chapel Hill, 1997. Molecular mechanisms underlying hypertension; understanding the role of regulation of 7-transmembrane spanning receptor signaling in cardiovascular disease in vivo and in vitro.

John L. Farber, Professor; M.D., California, San Francisco, 1966. Biochemical mechanisms of cell injury in ischemia; biochemical toxicology of liver cell necrosis; biochemical toxicology of activated oxygen; chemical carcinogenesis.

Barry J. Goldstein, Professor; M.D./Ph.D., Rochester, 1982. Insulin receptor signal transduction; role of protein tyrosine phosphatases in the regulation of insulin action and insulin-resistant disease states; vascular signaling of adipose secretory products, including cytokines and adiponectin.

Jan B. Hoek, Professor; Ph.D., Amsterdam, 1972. Systems biology of intracellular signal transduction networks; deregulation of cytokine and growth factor signaling in the liver associated with chronic alcohol consumption; early signaling responses during liver regeneration; bioenergetics and mitochondrial metabolism and its role in intracellular signaling and apoptosis.

Ya-Ming Hou, Professor; Ph.D., Berkeley, 1986. Genetic and biochemical studies of tRNA, including structure and function mechanism of tRNA aminoacylation, maintenance of the tRNA 3' end, maturation and processing, editing and repair, decoding on the ribosome, and development of bacterial pathogenesis in infectious diseases; targeting tRNAs as strategies against metabolic and neurodegenerative disorders.

James H. Keen, Professor; Ph.D., Cornell, 1976. Molecular mechanisms coupling signal transduction with receptor-mediated endocytosis and exocytosis; membrane transport studied by biochemical, molecular biological, and morphological approaches.

Walter J. Koch, W. W. Smith Professor of Medicine; Ph.D., Cincinnati, 1990. Molecular mechanisms of pathological cardiovascular signaling; novel gene-based therapeutics for congestive heart failure and proliferative vascular disease.

Michael Lisanti, Professor; M.D., 1991, Ph.D., 1992, Cornell. Functional role of caveolae and the caveolin genes in normal signal transduction and in a variety of pathogenic disease states, most notably the development of breast and prostate cancers.

Jose Martinez, Professor; M.D., Madrid, 1957. Role of fibrin in the differentiation of endothelial cells toward the formation of capillary tubes; role of tissue transglutaminase in the cross-linking of extracellular matrix proteins.

Andrea Morrione, Research Assistant Professor; Ph.D., Milan, 1992. Characterization of the role of adaptor protein Grb10 in regulation of IGF-I receptor ubiquitination, trafficking, and signaling; role of growth factor proepithelin and its receptor in bladder cancer.

Ulrich Rodeck, Professor; M.D., Germany, 1981. Regulation of epithelial cell survival by the epidermal growth factor receptor; coordinate regulation of cell-cycle progression and cell survival by growth factor– and adhesion-dependent signal transduction; signal transduction events controlling cell fate in the anchorage-independent state; zebrafish as a new in vivo model system for the genotoxic stress response.

Michael Root, Assistant Professor; M.D./Ph.D., Harvard, 1997. Structure and function of glycoproteins involved in viral and cell-cell membrane fusion; design of viral entry inhibitors and immunogens for vaccine development.

Charles P. Scott, Assistant Professor; Ph.D., Pennsylvania, 1997. Combinatorial drug discovery; rational drug design; chemical genomics of host-pathogen interactions and cancer.

Scott A. Waldman, Professor; Ph.D., Thomas Jefferson, 1980; M.D., Stanford, 1987. Molecular mechanisms of signal transduction, with emphasis on receptor-effector coupling and post-receptor signaling mechanisms; molecular mechanisms underlying tissue-specific transcriptional regulation; translation of molecular signaling mechanisms to novel diagnostic and therapeutic approaches to patients with cancer.

Philip B. Wedegaertner, Associate Professor; Ph.D., California, San Diego, 1991. G-protein signal transduction; molecular mechanisms and functions of covalent modifications and regulated subcellular localization.

Eric Wickstrom, Professor; Ph.D., Berkeley, 1972. Sensing, imaging, regulation, and control of oncogene expression in cells and animal models with nucleic acid derivatives; permanent bonding of drugs to implants.

John C. Williams, Assistant Professor; Ph.D., Columbia, 1995. Biochemical and biophysical studies of dynein function, regulation, and cargo recognition.

Edward P. Winter, Associate Professor; SUNY at Stony Brook, 1984. Analysis of meiotic development using molecular/genetics in yeast; signal transduction and protein kinases; transcriptional regulation.

UNIVERSITY AT ALBANY, STATE UNIVERSITY OF NEW YORK

School of Public Health
Department of Environmental Health Sciences

Programs of Study

The Department of Environmental Health Sciences at the University at Albany, State University of New York offers programs leading to the M.S. and Ph.D. degrees, preparing outstanding students for teaching and research careers in academia, public health agencies, and industry. The programs benefit from being based in the Wadsworth Center, the central laboratory complex of the New York State Department of Health. This relationship exposes the student to a range of real-world, scientifically based environmental and public health problems and to an organization with a tradition of high-quality basic research. The department offers three broad areas of specialization (tracks): environmental and occupational health, environmental chemistry, and toxicology.

Specific areas of research in the toxicology track include alterations of immune system function by environmental factors; the function and regulation of phase I and phase II enzymes in estrogen and xenobiotic metabolism in both in vivo and in vitro systems using the tools of cell biology, biochemistry, molecular biology, pharmacogenetics, toxicogenomics, and genetics; neurotoxicity of halogenated hydrocarbons and heavy metals; degradation of halogenated aromatics by anaerobic microorganisms; and detection of pathogenic organisms in water.

Research programs in the environmental chemistry track include laboratory and pilot-scale investigations of photochemical and absorption processes in the aquatic environment, development and application of analytical methods for the detection of inorganic and organic contaminants in environmental media and human tissues, atmospheric processes leading to the development of acid rain, and ozone depletion and smog.

The environmental health track focuses on such topics as environmental exposure assessment; air, soil, and water pollution; control of food and waterborne diseases; risk communication, geographic risk analysis, and pest vectors and management.

The Department of Environmental Health Sciences offers a specialized masters program, with a curriculum designed for completion in eighteen months that includes a six-month period of hands-on state-of-the-art laboratory or field experience. Students can select from pharmaco/toxicogenomics; xenobiotic and drug metabolism; environmental analytical chemistry; environmental chemistry: biomonitoring; environmental health; and environmental epidemiology/risk assessment.

Research Facilities

The program builds on the resources and state-of-the-art research equipment of the Wadsworth Center. At the Wadsworth Center, faculty and student research is supported by outstanding core facilities in biochemistry, genetics, imaging, immunology, molecular structure, mass spectrometry, and NMR, as well as national centers in biological imaging (NIH/NCRR), genetics (CDC), and nanobiotechnology (NSF). The laboratories provide scientific expertise and analytical support for numerous environmental and public health problems, including the greenhouse effect, acid rain, toxic waste, PCBs in the Hudson River, indoor air pollution, and the combustion of hazardous/municipal waste.

Financial Aid

Graduate assistantships and tuition scholarships are awarded to all incoming doctoral students. Assistantship stipends range from $20,000 to $22,000. Other forms of financial support include research assistantships from faculty research grants and merit-based University fellowships.

Cost of Study

Full-time tuition in 2005–06 was $3450 per semester for New York State residents and $5640 per semester for out-of-state and international students. University fees are approximately $500 per semester.

Living and Housing Costs

In addition to student residence halls, off-campus apartments are available at rents ranging from $400 to $600 per month.

Student Group

Of the 17,000 students at the University at Albany, 4,500 are graduate students who come from all parts of the U.S. and from more than forty other countries.

Location

Albany, the capital of the state of New York, offers a large choice of cultural attractions ranging from music, dance, and theater, to horseracing and sports. In the nearby Adirondack and Catskill Mountains and in the Berkshires, students find an unlimited variety of outdoor activities including hiking, camping, canoeing, skiing, and climbing. Other attractions include Lake George, Saratoga Springs, Tanglewood (in Massachusetts), and the Olympic Sports Complex in Lake Placid. New York City, Boston, Montreal, Vermont, and Atlantic Ocean beaches are all within a few hours' drive. Albany has an airport, bus and train services, and an extensive public transportation system.

The University

The University at Albany is the senior campus and one of four University Centers of the sixty-four-campus State University of New York (SUNY) system. The main campus is housed in a modern complex occupying a 400-acre site at the western edge of Albany.

Applying

College transcripts; Graduate Record Examinations scores on the General Test; a Subject Test in biology, physics, biochemistry, or chemistry is recommended; three references from persons familiar with the applicant's academic qualifications; a statement of the applicant's educational and professional aims; and the $60 application fee must be submitted to the Office of Graduate Admissions. International applicants must submit a minimum TOEFL score of 600. The application deadline for the Ph.D. program is January 15; the deadline for the M.S. program is April 1.

Correspondence and Information

Department of Environmental Health Sciences
Wadsworth Center
University at Albany, State University of New York
P.O. Box 509
Albany, New York 12201-0509
Phone: 518-473-7553
Fax: 518-473-8520
E-mail: ehsdept@wadsworth.org
Web site: http://www.wadsworth.org/sph/ehs

University at Albany, State University of New York

THE FACULTY AND THEIR RESEARCH

Environmental and Occupational Health

Erin M. Bell, Ph.D., North Carolina at Chapel Hill, 2000. Epidemiology of environmental exposures, specifically pesticides, cancer and adverse birth outcomes, exposure assessment for use in epidemiology studies, and immune response to environmental exposures.

Irina Birman, Ph.D., Moscow Agricultural, 1982. Drinking water quality control; source water monitoring and protection; disinfection byproducts formation.

David Carpenter, M.D., Harvard, 1964. Neurotoxicity of metals (Pb, Hg) and PCBs using electrophysiological and flow cytometric techniques.

David M. Dziewulski, Ph.D., New Hampshire, 1980. Biocorrosion and biofouling in industrial and potable water systems; metabolism and energy production of bacterial communities; environmental concentration and isolation methods.

Edward Fitzgerald, Ph.D., Yale, 1982. Environmental epidemiology; exposure assessment; human health effects; PCBs; dioxins.

Glen D. Johnson, Ph.D., Penn State, 1999. Quantifying the geographic association of human and ecological health outcomes with risk factors; small area estimation; smoothing disease maps; spatial pattern analysis and cluster detection.

Nancy K. Kim, Ph.D., Northwestern, 1969. Exposure assessments and risk assessments for environmental contaminants.

John D. Paccione, Ph.D., Rensselaer, 1993. Use of titanium dioxide and multiphase-flow reactor designs to mineralize organic compounds in potable water, recreational water, and wastewater.

Roger Sokol, Ph.D., SUNY at Albany, 1990. Biodegradation of halogenated organic contaminants; control of disinfection byproduct formation in drinking water.

Thomas Wainman, Ph.D., Rutgers, 1999. Research in assessing human exposure to environmental contaminants, utilizing GIS to examine spatial distributions of exposure sources; indoor air pollution.

Ying Wang, Ph.D., SUNY at Albany, 1989. Evaluation of health outcome data and assessment of community health concerns of populations potentially exposed to health hazards and toxic substances in the environment.

Lloyd Wilson, Ph.D., SUNY at Albany, 1995. Evaluation of the fate of environmental contaminants with a focus on understanding potential human exposures.

Environmental Chemistry

Katherine Alben, Ph.D., Yale, 1976. Chemistry of drinking water; natural and anthropogenic organic compounds in aquatic organisms, rivers, lakes, and wetlands.

Kenneth Aldous, Ph.D., Imperial College (London), 1970. Instrumental methods of analysis of environmental pollutants.

Ilham AlMahamid, Ph.D., Paris–South, 1990. Radiochemistry; migration behavior of radionuclides in the subsurface, their solubility, and complexation with organic materials; minimization of radioactive waste toxicity.

Craig M. Brown, Ph.D., Carnegie Mellon, 1997. Radiation biology; electron scattering cross section measurements in liquid water.

Liang Chu, Ph.D., Princeton, 1990. Atmospheric chemistry; kinetics and mechanisms of atmospheric reactions; reactions on ice and salt surfaces.

Kenneth Demerjian, Ph.D., Ohio State, 1973. Mechanisms of polluted and clean atmospheres.

Vincent Dutkiewicz, Ph.D., SUNY at Albany, 1977. Physical and chemical mechanisms of atmospheric sulfur and other environmentally significant pollutants.

Chia-Swee Hong, Ph.D., CUNY, 1980. Ultratrace analytical chemistry; photochemistry.

Liaquat Husain, Ph.D., Arkansas, 1968. Atmospheric transport of trace chemicals; acid rain.

Kenneth Jackson, Ph.D., Imperial College (London), 1972. Fundamental and applied ultratrace analytical chemistry.

Robert L. Jansing, Ph.D., Houston, 1975. Identification of biomarkers of human exposure to environmental chemicals.

Kurunthachalam Kannan, Ph.D., Ehime (Japan), 1994. Understanding global distribution, fate, and effects of organic pollutants in the environment.

Haider Khwaja, Ph.D., New Brunswick, 1982. Atmospheric pollutants transport studies; photochemistry.

Michael Kitto, Ph.D., Maryland, 1987. Radon in water and air; atmospheric gas and particle pollutants radioactivity.

Patrick O'Keefe, Ph.D., Oregon State, 1971. Environmental chemistry and analysis of chlorinated dioxins, pharmaceuticals, and related compounds.

Patrick J. Parsons, Ph.D., London, 1983. Trace-element analysis of biological tissues and fluids; analytical atomic spectrometry.

Thomas Semkow, Ph.D., Washington (St. Louis), 1983. Environmental radioactivity; ionizing radiation measurement, mathematical modeling.

David Spink, Ph.D., Maryland, 1983. Effects of xenobiotics on the metabolism of drugs and steroid hormones.

James Webber, Ph.D., SUNY at Albany, 1999. Hazardous microparticles in the environment; analytical electron microscopy.

Qi Zhang, Ph.D., California, Davis, 2002. Chemistry of atmospheric condensed phases (e.g., aerosol, fog, and cloud droplets); aerosol mass spectrometry; development of advanced data analysis techniques and instrumentation to characterize the sources and processes of organic particles.

Xianliang Zhou, Ph.D., Dalhousie, 1988. Deployment and development of techniques for the measurement of organic and inorganic atmospheric trace species.

Lei Zhu, Ph.D., Columbia, 1991. Tropospheric chemistry; reaction kinetics and photochemistry; homogeneous and heterogeneous reactions; atmospheric application of cavity ring-down spectroscopy; time-resolved FTIR.

Tadashi Yoshinari, Ph.D., Dalhousie, 1973. Microbial processes; sources and sinks of atmospheric trace gases.

Toxicology

Ellen Braun-Howland, Ph.D., Rensselaer, 1982. Occurrence and control of pathogens in the environment; molecular detection of environmentally-significant organisms.

Xinxin Ding, Ph.D., Michigan, 1988. Molecular toxicology; pharmacogenetics; gene regulation and biological function of cytochrome P-450 monooxygenases; transgenic/knockout mouse models of human diseases.

Christina Egan, Ph.D., Albany Medical College, 1997. Diagnostic assay development for detection of biothreat agents.

Michael Fasco, Ph.D., RPI, 1975. Molecular basis for drug interactions and tumor metastasis.

John Gierthy, Ph.D., Vermont, 1970. Development of in vitro models of cancer and assays for estrogen modulators and dioxin.

Laurence Kaminsky, Ph.D., Cape Town (South Africa), 1966. Mechanisms of action of toxifying and detoxifying systems; cytochrome P-450.

David Lawrence, Ph.D., Boston College, 1971. Cellular, molecular, and biochemical investigation of the effects of environmental factors on the immune system; regulatory interactions between the nervous and immune systems.

William Lee, Ph.D., Johns Hopkins, 1987. Cellular immunology; generation of immunologic memory; lymphocyte interactions.

Brian Pentecost, Ph.D., London, 1981. Control of steroid receptor expression including estrogen receptor and the PXR/RXR which regulates human Cytochrome P4503A4.

G-Yull Rhee, Ph.D., Cornell, 1971. Physiological ecology of algae and biodegradation of xenobiotics.

Richard Seegal, Ph.D., Georgia, 1972. Developmental neurotoxicity of PCBs and heavy metals; neuroimmune interactions and Parkinson's disease.

William Shain, Ph.D., Temple, 1972. Glial cell function in the central nervous system; nanofabricated silicon probes in the brain.

Simon Spivack, M.D., SUNY at Syracuse, 1985; M.P.H., Harvard, 1989. Molecular epidemiology in lung carcinogenesis; lung toxicology; gene regulation; individual susceptibility to lung diseases.

Robert Turesky, Ph.D., MIT, 1986. Environmental and molecular epidemiology; assessment of human exposure to toxicants; biochemical toxicology of carcinogenic heterocyclic aromatic amines.

Robert A. Waniewski, Ph.D., George Washington, 1980. Regulation of amino acid neurotransmitter synthesis; mechanisms of neurotoxicity.

Adjunct Faculty

Robert Briggs, Ph.D.; Mary O'Reilly, Ph.D.; Kevin Civerolo, Ph.D.; Brian Mayes, Ph.D.; Jay Silkworth, Ph.D.

UNIVERSITY OF ARKANSAS FOR MEDICAL SCIENCES
College of Medicine
Department of Pharmacology and Toxicology

Programs of Study

The Department of Pharmacology and Toxicology at the University of Arkansas for Medical Sciences offers separate programs of study leading to the Ph.D. in pharmacology or in interdisciplinary toxicology. Areas of research specialization include autonomic, behavioral, cardiovascular, and molecular pharmacology; immunopharmacology; pharmacokinetics and therapeutics; renal pharmacology; developmental, general, genetic, and metabolic toxicology; carcinogenesis, immunotoxicology, and neurotoxicology. The basic curriculum for both doctoral programs involves research and courses in physiology, biochemistry, biostatistics, pharmacology, and toxicology. Advanced courses suited to research are also required. Qualifying examinations are administered at the end of the second year of study, prior to admission into candidacy for the Ph.D. program. To receive the Ph.D., students must write and defend a thesis based on their research.

Research Facilities

The Department of Pharmacology and Toxicology occupies 18,000 square feet in the Biomedical Research Building. Other research facilities include the Veterans Administration Hospital and the Food and Drug Administration's National Center for Toxicological Research (NCTR) in nearby Jefferson. The interdisciplinary toxicology program is a joint effort with the FDA, and adjunct faculty members from NCTR participate in graduate student training. Students may visit the Department and University on the Web at http://www.uams.edu/.

The library at the University of Arkansas for Medical Sciences contains approximately 156,000 volumes, 1,800 current periodicals, and more than 1,000 tapes, films, and other audiovisual aids.

Financial Aid

Qualified students receive full-tuition fellowships. Current stipends are $20,000 per year plus tuition. Stipends are reviewed each year to keep up with inflation. Special scholarship awards are available for highly qualified applicants. Some students may be supported by a National Institute of Environmental Health Sciences (NIEHS) training grant.

Cost of Study

In 2004–05, tuition costs were $240 per hour for full-time in-state students and $514 per hour for out-of-state students. Students pay for a maximum of 10 hours. However, in most cases, graduate tuition is paid by fellowships.

Living and Housing Costs

In 2004–05, the University's residence hall on campus had rates varying from $306 per student per month for a double room to $600 per month for a one-bedroom apartment with a kitchenette and a private bath. Dining facilities are available in the residence hall and in the cafeteria of the University Hospital. Reasonably priced rooms, apartments, duplexes, and houses are available nearby.

Student Group

At present, there are approximately 35 students enrolled in the Department's two graduate programs. All students in the Ph.D. programs are receiving financial assistance.

Student Outcomes

Students graduating from the Ph.D. programs have taken postdoctoral and faculty positions at top universities, industries, and government organizations. These include MIT, Duke, the University of Washington, Johns Hopkins University, the University of Alabama at Birmingham, the Chemical Industry Institute of Toxicology, the National Institutes of Health, Pfizer, Monsanto Company, and Agouron Pharmaceuticals, to name but a few. Other students have taken positions as consultants and administrators in government and private industry.

Location

The population of metropolitan Little Rock, the state capital, is approximately 330,000. The Ouachita, Boston, and Ozark mountains are nearby. There are many lakes and rivers, including those in the Hot Springs resort area, within easy driving distance. The city provides numerous cultural and recreational opportunities. Adjacent to the campus are a fitness center, golf course, tennis facility, and minor-league baseball stadium. The University of Arkansas Razorbacks play football and basketball games in Little Rock.

The University

The University of Arkansas for Medical Sciences is located on a rapidly expanding campus in west-central Little Rock. In addition to the College of Medicine, the Graduate School and the Colleges of Pharmacy, Nursing, Public Health, and Health Related Professions are located on this campus. The John L. McClellan VA Medical Center adjoins the campus. Residential neighborhoods and shopping centers are located near the University. The University of Arkansas at Little Rock is located nearby.

Applying

For application forms and additional information about the graduate programs in pharmacology or interdisciplinary toxicology, prospective students should contact the appropriate program. Students who wish to enroll for the fall semester should file applications during the previous fall or winter. Usually, all positions are filled by late spring or early summer. Students are selected on the basis of their GPA, GRE General Test scores, letters of recommendation, and, in some cases, personal interviews. International applicants must have a score of at least 625 on the TOEFL. Applications are available via the Web at http://www.uams.edu/gradschool.

An undergraduate background in chemistry, biology, mathematics, or pharmacy is preferred, although exceptional students with other backgrounds are encouraged to apply. Additional information can be found on the Web at http://www.uams.edu/pharmtox.

Correspondence and Information

Dr. Philip R. Mayeux
Pharmacology Program
Department of Pharmacology and Toxicology
College of Medicine (Slot 611)
University of Arkansas for Medical Sciences
4301 West Markham Street
Little Rock, Arkansas 72205
Phone: 501-686-8895
E-mail: mayeuxphilipr@uams.edu

Dr. Jack A. Hinson
Interdisciplinary Toxicology Program
Department of Pharmacology and Toxicology
College of Medicine (Slot 638)
University of Arkansas for Medical Sciences
4301 West Markham Street
Little Rock, Arkansas 72205
Phone: 501-686-5766
E-mail: hinsonjacka@uams.edu

University of Arkansas for Medical Sciences

THE FACULTY AND THEIR RESEARCH

Core Pharmacology and Toxicology Faculty

R. Brock, Assistant Professor; Ph.D., Western Ontario, 2000. Hepatotoxicology and microvascular dynamics.
J. P. Crow, Professor; Ph.D., Auburn, 1991. Biochemical toxicology.
J. Gandy, Associate Professor; Ph.D., California, Riverside, 1985. Pesticide immunotoxicology; reproductive toxicology.
J. A. Hinson, Professor and Division Director; Ph.D., Vanderbilt, 1972. Biochemical toxicology; metabolism.
L. A. MacMillan-Crow, Associate Professor; Ph.D., Alabama at Birmingham, 1994. Renal toxicology.
P. R. Mayeux, Associate Professor; Ph.D., Tulane, 1987. Biochemical pharmacology and toxicology.
D. E. McMillan, Professor and Chairman; Ph.D., Pittsburgh, 1965. Behavioral pharmacology and toxicology.
S. M. Owens, Professor; Ph.D., North Carolina at Chapel Hill, 1981. Immunopharmacology; pharmacokinetics; drug abuse; neuroscience.
P. Palade, Professor; Ph.D., Pennsylvania, 1976. Molecular pharmacology of ion channels.
P. L. Prather, Assistant Professor; Ph.D., Georgia, 1988. Neuropharmacology; opioids; G-proteins.
M. J. Ronis, Associate Professor; Ph.D., Reading (England), 1985. Drug and toxicant metabolism.
N. J. Rusch, Professor and Chair; Ph.D., Mayo, 1983. Hypertension and ion channels.
V. Samokyszyn, Associate Professor; Ph.D., Wayne State, 1987. Biochemical toxicology.
J. R. Stimers, Associate Professor; Ph.D., USC, 1982. Cardiac electrophysiology.
G. R. Wenger, Professor and Division Director; Ph.D., West Virginia, 1971. Behavioral pharmacology and toxicology.
W. D. Wessinger, Associate Professor; Ph.D., Virginia Commonwealth, 1983. Behavioral pharmacology and toxicology.
F. Zheng, Assistant Professor; Ph.D., Texas Medical Branch, 1994. Neuropharmacology.
P. Zimniak, Professor; Ph.D., Vienna, 1976. Metabolism of endogenous toxicants.

Additional NCTR and UAMS Graduate Faculty

S. F. Ali, NCTR, Associate Professor; Ph.D., Aligarh Muslim (India), 1980. Neurochemistry.
W. T. Allaben, NCTR, Associate Professor; Ph.D., Southern Illinois at Carbondale, 1975. Chemical carcinogenesis.
T. M. Badger, Professor; Ph.D., Missouri–Columbia, 1973. Substance abuse; neuroendocrinology; neuropharmacology; nutrition.
D. A. Casciano, NCTR, Professor; Ph.D., Purdue, 1971. Chemically induced DNA damage.
C. E. Cerniglia, NCTR, Professor; Ph.D., North Carolina State, 1975. Microbial metabolism of xenobiotics.
B. Delclos, NCTR, Assistant Professor; Ph.D., Harvard, 1983. Chemical carcinogenesis.
E. K. Fifer, Assistant Professor; Ph.D., Mississippi, 1982. Medicinal chemistry; drug toxicity.
T. Flammang, NCTR, Assistant Professor; Ph.D., Arkansas Medical Sciences, 1986. Biochemical toxicology.
P. P. Fu, NCTR, Professor; Ph.D., Illinois at Chicago, 1973. Chemical carcinogenesis.
W. B. Gentry, Assistant Professor; M.D., Arkansas Medical Sciences, 1988. Pharmacokinetics; immunopharmacology; drug abuse.
D. K. Hansen, NCTR, Associate Professor; Ph.D., Indiana–Purdue at Indianapolis, 1981. Teratology.
P. C. Howard, NCTR, Associate Professor; Ph.D., Arkansas Medical Sciences, 1981. Chemical carcinogenesis.
F. F. Kadlubar, NCTR, Professor; Ph.D., Texas at Austin, 1973. Aromatic amine carcinogenesis; drug metabolism.
J. E. A. Leakey, NCTR, Associate Professor; Ph.D., Dundee (Scotland), 1976. Drug metabolism.
K. E. Light, Associate Professor; Ph.D., Indiana Bloomington, 1977. Receptor pharmacology; chronic ethanol intake.
G. Milner, NCTR, Instructor; Ph.D., Arkansas Medical Sciences, 1988. Risk assessment.
A. C. Nye, NCTR, Instructor; Ph.D., Arkansas Medical Sciences, 1986. Risk assessment.
M. G. Paule, NCTR, Professor; Ph.D., California, Davis, 1983. Behavioral pharmacology and toxicology.
R. J. S. Reis, Professor; Ph.D., Sussex (England). Molecular genetics of aging.
D. W. Roberts, NCTR, Associate Professor; Ph.D., Hahnemann, 1977. Immunotoxicology.
A. C. Scallett, NCTR, Assistant Professor; Ph.D., Wisconsin–Madison, 1981. Behavioral toxicology; neurotoxicology.
H. F. Simmons, Associate Professor; M.D., 1980, Ph.D., 1988, Arkansas Medical Sciences. Clinical toxicology.
W. Slikker, NCTR, Professor; Ph.D., California, Davis, 1978. Neurotoxicology; teratogenesis.

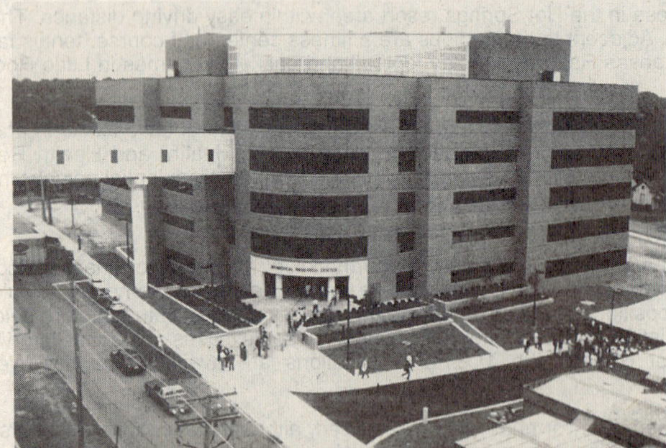

An outside view of the Biomedical Research Building on the campus of the University of Arkansas for Medical Sciences.

An inside atrium view of the Biomedical Research Building.

UNIVERSITY OF CALIFORNIA, IRVINE

School of Medicine
Department of Pharmacology

Program of Study

The Ph.D. program prepares students for careers in academia, research institutions, and the pharmaceutical industry by providing a foundation in all aspects of pharmacology, from molecular mechanisms through behavior. Specific areas of study include molecular and cellular pharmacology, neuropharmacology, psychopharmacology, and cardiovascular pharmacology. Emphasis is placed on providing an integrated understanding of receptors: their structure, location, and function; molecular aspects of drug action; receptor signaling mechanisms; structure-activity relationships and drug design; and the role of receptors and drugs in development and aging, plasticity, reinforcement and drug abuse, neural disorders, and cardiovascular physiology and disease. Students take three laboratory rotations in the first year, complete the required course work in the first two years, and then take a comprehensive written exam. Students advance to candidacy in their third year by orally defending their research proposal; they generally complete the Ph.D. degree in their fifth year.

Research Facilities

The well-equipped laboratories of the Department of Pharmacology are housed in the Med Surge II Building and the Gillespie Neurosciences Research Facility, located in the School of Medicine adjacent to the main campus. The Science Library and animal-care facility are conveniently located nearby, and a large computing facility is available on the main campus.

Financial Aid

Students who maintain satisfactory academic progress receive full financial support, i.e., tuition, fees, health insurance, and a living stipend ($25,000 for 2006–07). Sources of support include University fellowships and NIH training and research grants.

Cost of Study

The estimated cost of tuition in 2006–07 is $9669.50 for state residents and $24,630.50 for nonresidents.

Living and Housing Costs

University housing is available, as is off-campus housing in nearby communities. For rates and more information, students should visit the University housing Web site at http://www.housing.uci.edu.

Student Group

The University's total enrollment in fall 2005 was 24,434 students, of whom 4,849 were graduate students. Enrollment numbers for fall 2006 are not yet available. Pharmacology maintains approximately 35 graduate students in its program.

Student Outcomes

Recent graduates of the program are doing postdoctoral research at the University of California, Irvine (UCI); UC Los Angeles (UCLA); UC San Diego (UCSD); UC San Francisco (UCSF); Stanford; Scripps Research Institute; and the National Institute of Mental Health. Other graduates hold faculty positions (tenured or tenure track) at UCI, UCLA, UCSD, Yale, and the Universities of Colorado, Michigan, and Minnesota.

Location

UCI is located 5 miles from the Pacific Ocean and 40 miles south of Los Angeles. The surrounding hills and grazing lands give the campus an open feeling, though an estimated 2 million people live within a 20-mile radius. Concerts, repertory theater, and art galleries are available both on and off campus. Beaches, marinas, ski resorts, and mountain trails are within easy driving distance, and the mild climate permits a wide variety of year-round recreation.

The University

UCI is a young, dynamic campus that has rapidly become a full partner in the region's growing culture and economy. As a major research university, UCI offers excellent opportunities for graduate students in most traditional academic disciplines and in several interdisciplinary programs unique to the campus. Collaborations between departments in the School of Medicine and the School of Biological Sciences enhance the graduate program, providing students with access to all courses and the research expertise of all faculty members. Graduate students have access to the facilities of the entire nine-campus University of California system.

Applying

To be considered for entrance to the program in September, students should submit applications before the end of the preceding calendar year and no later than January 5. In addition to application forms provided by the Division of Research and Graduate Studies, applicants must submit official copies of all transcripts, three letters of recommendation, GRE scores from the General Test and the Subject Test in biology or chemistry, and the application fee.

Correspondence and Information

Admissions Committee
Department of Pharmacology
University of California, Irvine
Irvine, California 92697-4625

Phone: 949-824-7651
Fax: 949-824-4855
E-mail: pharm@uci.edu
Web site: http://www.ucihs.uci.edu/pharmaco/

University of California, Irvine

THE FACULTY AND THEIR RESEARCH

James D. Belluzzi, Adjunct Professor; Ph.D., Chicago, 1970. Brain substrates and pharmacology of reward; characterization and developmental modulation of nicotine and cocaine reinforcement.

Olivier Civelli, Eric and Lila Nelson Professor of Neuropharmacology; Ph.D., Zurich (Switzerland), 1979. Molecular biology of G-protein–coupled receptors; orphan receptors (receptors with unknown natural ligands); search and isolation of novel neurotransmitters and neuropeptides; determination of pharmacological and biological characteristics of novel neurotransmitters and neuropeptides, including search for their functional significance in animal models and studies on their possible involvement in human disorders.

Sue Piper Duckles, Professor; Ph.D., California, San Francisco, 1973. Pharmacology and physiology of blood vessels; unique properties of cerebral circulation; effects of gonadal steroids, including estrogen and testosterone, on vascular and mitochondrial function; impact of age on cardiovascular regulation.

Frederick J. Ehlert, Professor; Ph.D., California, Irvine, 1978. Subtypes of muscarinic receptors and their signaling mechanisms in brain, heart, smooth muscle, ocular tissue, and exocine glands; receptor–G-protein interactions; interactions between the signaling pathways of specific neurotransmitter receptors; molecular basis of drug selectivity.

Kelvin W. Gee, Professor; Ph.D., California, Davis, 1981. Molecular characterization of novel receptor sites coupled to ligand-gated ion channels; discovery of novel molecules that interact with these sites and their potential therapeutic applications.

Diana N. Krause, Adjunct Professor; Ph.D., UCLA, 1977. Cerebrovascular pharmacology; regulation of brain blood vessels by hormones, such as estrogen, testosterone, and melatonin during normal and pathological conditions.

Frances M. Leslie, Professor; Ph.D., Aberdeen (Scotland), 1977. Cellular, molecular, and behavioral characterization of actions of drugs of abuse, with particular reference to morphine and nicotine; analysis of effects of drug abuse and stress on neural plasticity, particularly during brain development.

Hans-Peter Nothacker, Assistant Adjunct Professor; Ph.D., Hamburg (Germany), 1993. Studies of orphan G-protein–coupled receptors, with special emphasis on their pharmacology, anatomical distribution, physiological functions, and development of specific agonists and antagonists.

Daniele Piomelli, Professor; Ph.D., Columbia, 1987. Various signaling pathways that involve lipids with particular interest in the endogenous cannabinoids; arachidonic acid derivatives that activate cannabinoid (marijuana) receptors in brain and other tissues; biochemical mechanisms involved in the formation and inactivation of these molecules; developing pharmacological tools to interfere with these pathways; wide range of molecular biological, pharmacological, and analytical techniques.

Ralph E. Purdy, Professor; Ph.D., UCLA, 1973. Receptor, second-messenger, and signal transduction mechanisms underlying (1) vasoconstrictor synergism between norepinephrine plus serotonin and norepinephrine plus angiotensin II and (2) microgravity-induced vascular hyporesponsiveness; cell biology of vascular smooth muscle, using techniques to address calcium influx and release from sarcoplasmic reticulum; biochemical analysis of the role of prostaglandins, endothelium-derived factors, and phosphoinositide metabolites and tyrosine kinases within vascular smooth muscle as mediators of vasoconstriction and dilation.

Paolo Sassone-Corsi, Distinguished Professor and Chair; Ph.D., Naples (Italy), 1979. Cellular and molecular pathways regulating gene expression, including regulation of transcription factors by signaling systems and chromatin remodeling; epigenetic mechanisms governing physiological functions, with particular reference to the circadian clock and relative implications in light signaling, metabolism, and cancer; epigenetic control of gene expression and cellular differentiation of male germ cells.

Larry Stein, Professor; Ph.D., Iowa, 1955. Cellular basis of reward, learning, and memory; behavioral pharmacology; drug discovery.

Qun-Yong Zhou, Associate Professor; Ph.D., Oregon Health & Science, 1992. Physiology and pharmacology of prokineticins and prokineticin receptors; circadian rhythms and neurogenesis.

Affiliated Faculty

Stephen C. Bondy, Professor; Ph.D., Birmingham (England), 1962. Attenuation of brain aging by reduction of pro-oxidant and inflammatory processes associated with senescence, using nutritional supplementation; relation of environmental exposure to metals, such as aluminum and copper, to the development and progression of age-related neurological disease, especially Alzheimer's and Parkinson's diseases.

Emiliana Borrelli, Professor; Ph.D., Naples (Italy), 1979. Molecular neurobiology; signal transduction of dopamine receptors; generation of transgenic and knockout mouse models to study human diseases involving the dopamine system and myelination; molecular basis of addiction; neurodegeneration and mechanisms of repair; behavioral studies.

William E. Bunney Jr., Professor; M.D., Pennsylvania, 1956. Clinical psychobiological and neuropsychopharmacological studies of manic-depressive illness, schizophrenia, and childhood mental illness.

Pietro R. Galassetti, Assistant Professor-in-Residence; M.D., Rome (Italy), 1986; Ph.D., Vanderbilt, 1998. Physiological and altered adaptive responses to stress (exercise, hypoglycemia, hypoxia) in healthy and dysmetabolic children and adults; non-invasive monitoring of metabolic variables through analysis of exhaled gases.

Larry Jamner, Professor and Director, the Consortium for Integrative Health Studies; Ph.D., SUNY at Stony Brook, 1985. Psychological and physiological individual differences in vulnerability to stress reactivity; development of tobacco dependence and responses to drugs, such as nicotine and opiates.

Arthur D. Lander, Professor; M.D., Ph.D., California, San Francisco, 1985. Molecular neurobiology; intracellular and extracellular signaling in the control of axon guidance and cell motility; functions of the extracellular matrix; transgenic animal approaches to neural development and function.

Ellis R. Levin, Professor in Residence; M.D., Thomas Jefferson, 1975. Molecular and cellular biology of natriuretic peptides and endothelin and their receptors in the brain and the vasculature; regulation of astrocyte growth by these peptides.

John C. Longhurst, Professor; M.D., 1973, Ph.D., 1974, California, Davis. Regulation of the cardiovascular system and the nervous systems role in this regulation.

Z. David Luo, Assistant Professor; M.D., Guangzhou Medical (China), 1984; Ph.D., SUNY at Buffalo, 1993. Molecular mechanisms of pain transduction.

John F. Marshall, Professor; Ph.D., Pennsylvania, 1973. Recovery of function after brain injury; organization of the basal ganglia motor system; neuronal circuitry of drug craving.

Rainer Reinscheid, Assistant Professor; Ph.D., Center for Molecular Neurobiology (Germany), 1993. Neurobiology of sleep, emotional behavior and stress in novel neuropeptide systems.

Michael E. Selsted, Professor; Ph.D., 1980, M.D., 1985, UCLA. Biology of host-defense processes carried out by phagocytic leukocytes and mucosal epithelium.

Marylou Solbrig, Associate Adjunct Professor; M.D., Yeshiva (Einstein), 1980. Chronic viral infections of the nervous system are used to elucidate aspects of drug abuse, movement disorders, and neurodegenerative diseases; neuropharmacologic sequelae of persistent CNS viral infections, using behavior pharmacologic, neuroanatomic, and neurochemical techniques.

Sandor Szabo, Professor in Residence; M.D., Belgrade, 1968; Ph.D., Montreal, 1973; M.P.H., Harvard, 1983. Effects of growth factors (e.g., FGF, PDGF) on ulcer healing and prevention.

UNIVERSITY OF CALIFORNIA, LOS ANGELES

School of Medicine
Department of Molecular and Medical Pharmacology

Program of Study	The Department offers a program leading to the Ph.D. degree. The objective of the program is to train students for careers in research and teaching in the university, government, or private sector. The curriculum is designed to be flexible so that students are allowed to specialize in subdisciplines of pharmacology. In the first year, students enroll in basic cellular and molecular biology courses, a graduate commentary in molecular pharmacology, and a course in pharmacological principles that includes the chemical basis of drug action. In the second year, students take a more integrated pharmacology course, enter research laboratories of their choice, and begin research. In addition, they enroll in electives designed to prepare them for their subdiscipline. The Department has a basic science and a clinical component, which is a medical imaging group, including positron emission tomography and the Crump Institute for Molecular Imaging. Thus, students are offered opportunities to use a variety of modern research tools, including imaging procedures, for their dissertation research projects. Extensive collaborative ventures with other departments at UCLA also provide numerous opportunities for interactions with other disciplines.
Research Facilities	The faculty members have well-equipped laboratories for research in various areas of pharmacology, biological imaging, nitric oxide, signal transduction, neuronal plasticity and development, virology, and immunology. The Department also houses the Stem Cells and Transgenic Mouse Facility, the Analytical Facility (gas chromatography/mass spectrometry/nuclear magnetic resonance) and the Olympic Analytical Laboratory (testing of NCAA and Olympic athletes). A full spectrum of basic and clinical experiences is available, including the exciting field of molecular imaging and positron emission tomography (PET). The Louise Darling Biomedical Library contains 425,000 volumes and computer facilities.
Financial Aid	All students are supported in the first year of study by fellowships and assistantships from the UCLA Graduate Division, the School of Medicine, or the Department. All students beyond the first year are currently supported by a combination of NIH predoctoral training grants and Department or faculty research grant assistantships. Financial support for graduate students in 2005–06 included a stipend of $25,000, registration fees, and nonresident tuition.
Cost of Study	The registration fee for the academic year 2005–06 was estimated at $8500 for California residents and $8800 for nonresidents. Nonresident tuition was $14,700 per year.
Living and Housing Costs	Privately owned apartments and married student housing units are available at varying distances from campus and at varying costs. Information about housing options can be obtained on the Web (http://www.orl.ucla.edu/) and from the University Housing Office.
Student Group	In fall 2005, the UCLA campus had an enrollment of 35,625, including 24,811 undergraduates and 10,814 graduate and professional students. The pharmacology program had 56 graduate students.
Location	UCLA is located in the foothills of the Santa Monica Mountains, only 5 miles from white-sand beaches and crashing Pacific waves and a 2-hour drive to the isolated beauty of the desert and magnificent skiing in the mountains. The campus is situated in the Westwood district, bounded by the pristine communities of Bel Air, Santa Monica, and Beverly Hills. Westwood Village, immediately adjacent to the campus, with its shops, restaurants, and theaters, has become a premier entertainment center.
The University	UCLA is one of nine campuses of the University of California, a state-supported institution with a prestigious history in advanced education. Founded in 1919, it is the second-oldest campus in the system and the largest in enrollment. The campus provides excellent recreational facilities and cultural activities. More information is available on UCLA's Web site at http://www.ucla.edu/.
Applying	Application forms for admission and financial support, as well as additional information, can be obtained from the address listed in the Correspondence and Information section. Applicants are judged on the basis of undergraduate grade point average, Graduate Record Examinations scores, three letters of recommendation, and a personal statement. Applications are accepted for fall admission only and should be submitted by December 31, although applications received after that date are considered as long as space is available.
Correspondence and Information	Graduate Information Department of Molecular and Medical Pharmacology Center for Health Sciences, 23-120, Box 951735 UCLA School of Medicine Los Angeles, California 90095-1735 Phone: 310-825-0390 Fax: 310-825-6267 E-mail: gradinfo@mednet.ucla.edu Web site: http://www.pharmacology.ucla.edu/

University of California, Los Angeles

THE FACULTY AND THEIR RESEARCH

Biological Imaging and Instrumentation

Martin Allen-Auerbach, Assistant Professor; M.D., Vienna, 1995. Nuclear medicine; clinical applications of biological and molecular imaging.

Arion F. Chatziioannou, Assistant Professor; Ph.D., UCLA, 1996. Quantitative in vivo small-animal imaging with positron emission tomography (PET) and CT.

Wei Chen, Assistant Professor; Ph.D., Harvard, 1988; M.D., Ohio State, 1992. Imaging tumor viability and proliferation using PET in brain tumor patients.

Roy Doumani, Professor; B.A., UCLA, LL.D., USC, 1962. Industry relations; entrepreneurship; biotechnology.

John C. Mazziotta, Professor; M.D., 1976, Ph.D., 1977, Georgetown. Brain mapping; functional neuroimaging with PET and fMRI.

Michael E. Phelps, Norton Simon Professor and Chair; Ph.D., Washington (St. Louis), 1970. Molecular imaging of integrated organ function in normal and disease states, from metabolism to gene expression; pharmacokinetic processes in living animals and humans using PET.

Osman Ratib, Professor; M.D., Geneva, 1981; Ph.D., UCLA, 1989. Image information systems and image management and communication; multimodality functional imaging and cardiovascular imaging.

Daniel H. Silverman, Assistant Professor; Ph.D., Harvard, 1987; M.D., Ohio State, 1992. Brain imaging of normal and disordered mood, memory, and motor function; minimally invasive cancer diagnosis, treatment planning, and therapy.

Hsian-Rong Tseng, Assistant Professor; Ph.D., Taiwan, 1998. Synthetic nanomaterials; nanoelectronic devices and biomedical sensing.

Wolfgang Weber, Associate Professor; M.D., Munich Technical, 1995. Positron emission tomography; tumor metabolism and treatment monitoring.

Hsiao-Ming Wu, Adjunct Associate Professor; Ph.D., UCLA, 1995. Biomedical physics; biological imaging and data analysis.

Lily Wu, Assistant Professor; M.D./Ph.D., UCLA, 1989. Gene-based imaging and therapy to target advanced, metastatic cancer.

Cardiovascular and Metabolic Diseases

Johannes Czernin, Associate Professor; M.D., Vienna, 1983. Oncological applications of PET.

Jamshid Maddahi, Professor; M.D., Tehran, 1975. Clinical impact of PET on health-care costs.

Srinivasa T. Reddy, Associate Professor; Ph.D., Berkeley, 1992. Atherosclerosis; regional myocardial substrate metabolism and of human aortic endothelial cells; arachidonic acid metabolism and inflammation.

Heinrich R. Schelbert, Professor; M.D./Ph.D., Würzburg, 1965. Noninvasive quantification of regional myocardial substrate metabolism and of coronary circulatory function in the human heart, using PET.

Ligia Torro, Professor; Ph.D., Centro de Investigacion y de Estudios Avanzados del IPN (Mexico), 1986. Cardiovascular biology and aging; molecular genetics and cell biology of ion channels; remodeling of cardiac and smooth muscle K channels by sex hormones.

Clinical Pharmacology

Barbara Levey, Professor; M.D., SUNY Health Science Center at Syracuse, 1961; FACP. Assistant Vice-Chancellor, Biomedical Affairs.

Drug Metabolism and Pharmacokinetics

Arthur K. Cho, Professor Emeritus; Ph.D., UCLA, 1958. Drug metabolism; pathways and their involvement in drug action.

Sung-Cheng (Henry) Huang, Professor; D.Sc., Washington (St. Louis), 1973. Tracer kinetic modeling of in vivo metabolic and pharmacological processes.

Nitric Oxide

Gautam Chaudhuri, Professor; M.D./Ph.D., London, 1975. Role of nitric oxide in reproductive processes and in cell proliferation and mechanisms by which estrogens prevent atherosclerosis.

Jon M. Fukuto, Professor; Ph.D., Berkeley, 1983. Nitrogen oxide and physiology; chemical pharmacology.

Louis J. Ignarro, Jerome J. Belzer, M.D., Distinguished Professor of Pharmacology and Nobel Laureate–Physiology of Medicine; Ph.D., Minnesota, 1966. Chemistry and pharmacology of vascular endothelium-derived substances; pharmacology and physiology of nitric oxide; mechanism of activation of guanylate cyclase and action of cyclic GMP.

Molecular Biology, Oncology, Immunology, and Virology

David B. Agus, Assistant Professor; M.D., Pennsylvania, 1992. Drug development and functional genomics in prostate cancer.

Maria G. Castro, Professor; Ph.D., National University of La Plata (Argentina), 1986. Genetic engineering of viral vectors; novel cancer therapies and preclinical cancer modeling.

Samson Chow, Associate Professor; Ph.D., Iowa, 1983. Replication and pathogenesis of human immunodeficiency virus (HIV).

Magnus Dahlbom, Professor; Ph.D., UCLA, 1987. Whole-body imaging techniques for cancer detection and evaluation of therapy.

Sanjiv Sam Gambhir, Professor; M.D., 1990, Ph.D., 1993, UCLA. Molecular imaging assays for in vivo monitoring of cellular events in living subjects; pharmacokinetic modeling; multimedia educational tools.

Thomas G. Graeber, Assistant Professor; Ph.D., Stanford, 1996. Systems biology; cancer signaling; phosphor-proteomics and computational biology.

Harvey R. Herschman, Professor and Ralph and Marjorie Crump Chair in Molecular Imaging; Ph.D., California, San Diego, 1967. Role of extracellular stimuli in eliciting changes in cell growth and differentiation as a result of induced transcription of primary response genes; imaging reporter gene expression in living animals.

Han Htun, Assistant Professor; Ph.D., Wisconsin–Madison, 1989. Molecular mechanism of steroid hormone action and the functional organization of the interphase nucleus.

Jing Huang, Assistant Professor; Ph.D., USC, 1996. Functional genomics and proteomics, chemical genetics, and bioinformatics approaches to cancer biology and neuroscience.

Daniel L. Kaufman, Professor; Ph.D., UCLA, 1988. Autoimmune disease, immunotherapeutics, and neurobiology.

Paul Krogstad, Associate Professor; M.D., Tulane, 1984. Replication, pathogenesis, and clinical treatment of human immunodeficiency virus.

Pedro R. Lowenstein, Professor; M.D., 1981, Ph.D., 1984, Buenos Aires. Gene therapy for brain diseases and neuroimmunology.

Ingo K. Mellinghoff, Assistant Professor; M.D., Munich Technical, 1995. Signal transduction in cancer; PET-probe development; molecular therapeutics.

Charles Sawyers, Professor; M.D., Johns Hopkins, 1985. Signal transduction abnormalities in cancer.

Christiaan Schiepers, Associate Professor; Ph.D., 1976, M.D., 1982, Netherlands. Metabolic imaging in oncology; noninvasive measurement of skeletal blood flow and metabolism with PET.

Desmond J. Smith, Associate Professor; Ph.D., Cambridge, 1987; B.M.B.Ch., Oxford, 1987. Functional genomics; neuropsychiatric disorders.

Ren Sun, Associate Professor; Ph.D., Yale, 1993. Molecular biology and pathogenesis of tumor-associated herpesviruses: human herpesvirus-8, Epstein-Barr virus, and murine gamma herpesvirus-68.

Yi Sun, Assistant Professor; Ph.D., Case Western Reserve, 1966. Molecular mechanisms underlying neuronal and glial cell fate specification in neural stem cells.

Jide Tian, Associate Professor; M.D., Suzhou (China), 1982. Peripheral T-cell tolerance and regulation; neuroimmunology.

Owen N. Witte, Professor; M.D., Stanford, 1976. Cell growth regulation and differentiation; function of oncogenes found in human leukemias and epithelial cancers; regulation of the immune response.

Anna M. Wu, Associate Professor; Ph.D., Yale, 1979. Biological imaging; immunotherapy of cancer.

Hong Wu, Associate Professor; M.D., Beijing, 1983; Ph.D., Harvard, 1991. Molecular genetics; blood cell formation; cancer biology.

Neurochemical Mechanisms of Drug Abuse

Don H. Catlin, Professor; M.D., Rochester, 1965. Clinical pharmacology of anabolic steroids.

Sherrel G. Howard, Associate Professor; Ph.D., UCLA, 1975. Effects of CNS stimulants and other potentially neurotoxic agents on the developing nervous system; neurotransmitter regulation of myelin formation.

William P. Melega, Associate Professor; Ph.D., UCLA, 1981. Neuroadaptations in brain after exposure to methamphetamine; gene transfer and growth factor therapies for CNS degenerative diseases.

Neuroscience and Signal Transduction

Jorge R. Barrio, Professor; Ph.D., Illinois, 1979. Development of biochemical probes for examining the chemical bases of human diseases using PET.

Bernard Kwok-Keung Fung, Professor; Ph.D., Berkeley, 1977. Molecular mechanism of signal transduction; G-protein structure and function; chaperone structure and function.

Cameron Gundersen, Professor; Ph.D., UCLA, 1977. Studies of cysteine string proteins and neurotransmitter release.

David A. Hovda, Professor; Ph.D., New Mexico, 1985. Neurochemistry; cerebral metabolism; behavior; brain injury and recovery of function.

Meisheng Jiang, Assistant Professor; Ph.D., USC, 1994. Neuroscience and signal transduction.

Harley Kornblum, Associate Professor; Ph.D., 1986, M.D., 1989, California, Irvine. Neural stem cells; brain development and repair.

Xin Liu, Assistant Professor; M.D., Beijing, 1983; Ph.D., Brandeis, 1991. Molecular genetics of nervous system development, behavior, and neurodegenerative disease.

Edythe D. London, Professor; Ph.D., Maryland, 1976. Molecular and functional imaging of neuropsychiatric diseases; neurobiology of substance abuse.

Richard W. Olsen, Professor; Ph.D., Berkeley, 1971. Structure and function of GABA-A receptors; mechanism of action of anesthetics/CNS depressant drugs; gene targeting; plasticity, epilepsy, and alcoholism.

Nagichettiar Satyamurthy, Professor; Ph.D., Madras, 1979. Synthesis of positron-emitter labeled probes for PET.

Joy A. Umbach, Associate Professor; Ph.D., UCLA, 1981. Neurotransmitter release; electrophysiology and calcium imaging.

UNIVERSITY OF CALIFORNIA, LOS ANGELES

Interdepartmental Doctoral Program in Molecular Toxicology

Program of Study	The Interdepartmental Doctoral Program in Molecular Toxicology marks a major development in health-related research and education at UCLA. The program is interdisciplinary in nature, drawing on the expertise of 19 faculty members from eight departments and four schools at UCLA. The focus of the program is on the molecular mechanisms of toxicological injury. It is the conviction of the faculty that the greatest strides in identifying, appraising, and ameliorating toxicological risk emanate from studying mechanisms. Areas of particular strength in the program include the application of genomic and proteomic technologies to toxicology, genetic susceptibility, reactive oxygen and nitrogen species, metal toxicity, radiation injury, and air pollution. Currently, it is a particularly opportune time for students to join a program such as UCLA's. There is a rapidly expanding need for individuals trained in molecular toxicology in academia, government, and the pharmaceutical and biotechnology industries. Future job prospects are therefore outstanding. The UCLA molecular toxicology program is supported by the superb resources of the UCLA biomedical research community and generous funding. The program possesses all the credentials to rank among the very finest toxicology programs in the country.

A master's degree is not a prerequisite. The ideal training for an undergraduate is to major in either chemistry or biology and to have a solid background in both of these disciplines with some quantitative training. Courses of value for toxicologists include calculus, cell biology, genetics, physiology, microbiology, molecular biology, inorganic chemistry, organic chemistry, biochemistry, and physical chemistry as well as others. Excellent students from all disciplines are considered for admission, with the opportunity to make up deficiencies during their graduate curriculum.

During their first year, students perform course work and do research rotations in the laboratories of three different professors. In their second year, they choose a research mentor with whom they do their thesis research.

Research Facilities The UCLA Medical School operates a large number of core facilities, including those for advanced research computing, biostatistics, electron microscopy, tissue culture, confocal and two-photon laser scanning microscopy, vivarium, DNA sequencing, flow cytometry, Fourier-transform infrared (FT-IR) spectrometry, gas chromatography, mass spectroscopy, X-ray crystallography, neuroimaging, nuclear magnetic resonance, spectrometry, peptide synthesis facility, statistical/biomathematical consulting, embryonic stem cell facility, transgenic mouse facility, and virology.

Financial Aid All students admitted to the Ph.D. program receive full financial support, covering an annual stipend (currently $25,500) and all tuition and fees. These funds come from a variety of sources, including the UCLA Graduate Division and a training grant from the University of California Toxic Substances Research and Teaching Program.

Cost of Study Tuition costs are covered by the program.

Living and Housing Costs Living accommodation is available both on campus and in surrounding neighborhoods. Most students live within a few miles of the campus. A new graduate student housing complex has just opened. In addition, UCLA married student housing and housing for students with children are available at below market rate.

Student Group Total enrollment at UCLA is approximately 35,000 students, including an approximate graduate enrollment in the health sciences of 4,500 students. The interdisciplinary doctoral program admits about 4 students each year.

Location UCLA is located on the west side of Los Angeles, 5 miles from the Pacific Ocean. The campus is bordered by the communities of Bel Air to the north, Westwood Hills to the west, Holmby Hills and Beverly Hills to the east, and Westwood Village to the south. Desert communities, including Palm Springs, and mountain areas, including Big Bear and Lake Arrowhead, are approximately a 1½-hour drive from the campus. UCLA is in a prime location from which to explore and enjoy all that Los Angeles and southern California have to offer.

The University and The Program UCLA is one of ten campuses in the University of California system. Since its inception in 1919 as the Southern Branch of the University of California, it has grown to become a world-renowned institution for higher education.

The Interdepartmental Doctoral Program in Molecular Toxicology was established in 2000 to provide superior education and training to future toxicologists, using UCLA's and the University of California's exceptional resources.

Applying Individuals should contact the Graduate Coordinator of the program for information. Applicants to the program are evaluated on the basis of their academic credentials, including courses taken, grades, GRE score, TOEFL score (if applicable), and interest in a career in toxicology. The program normally admits students at the beginning of the fall quarter. Those who complete their application by the preceding January 15 can be considered for admission to the following fall quarter. Prospective applicants should plan to take the GRE no later than October of the year before admission. International applicants are also required to submit their TOEFL scores.

Correspondence and Information
Sandy Valdivieso
Graduate Coordinator
Interdepartmental Doctoral Program in Molecular Toxicology
56-070 Public Health
P.O. Box 951772
Los Angeles, California 90095-1772
Phone: 310-206-1619
E-mail: jsandy@ph.ucla.edu
Web site: http://www.ph.ucla.edu/moltox

University of California, Los Angeles

THE FACULTY AND THEIR RESEARCH

Participating Faculty

Judith Berliner, Ph.D., Pathology and Laboratory Medicine. Dr. Berliner's research deals with the events mediating chronic inflammatory processes such as those that occur in atherosclerosis, cancer, and rheumatoid arthritis.

Gautam Chaudhuri, M.D., OB/GYN and Pharmacology. Dr. Chaudhuri's research focuses on the mechanism by which estradiol and nitric oxide modulate various physiological functions and the signal transduction pathways involved.

Arthur Cho, Ph.D., Molecular and Medical Pharmacology. Dr. Cho's laboratory investigates the chemical mechanisms of toxicity. He is currently examining the toxicity of quinones related to neurotransmitters and to environmental toxins.

Catherine Clarke, Ph.D., Chemistry and Biochemistry. Coenzyme Q (also known as ubiquinone or Q) is a lipid component of cellular membranes that plays an essential role in the respiratory electron transport chain. Dr. Clarke studies the biosynthesis, regulation, and function of Q in the yeast *Saccharomyces cerevisiae*.

Michael Collins, Ph.D., Environmental Health Sciences. Research in Dr. Collins' laboratory is concerned with aspects of developmental toxicology or teratology.

Curt Eckert, Ph.D., Environmental Health Sciences. Dr. Eckert's research program is concerned with elucidating the molecular basis of boron essentiality and toxicity in humans and animals.

John Froines, Ph.D., Environmental Health Sciences. Dr. Froines' research focuses on chemical mechanisms and exposure assessment related to the health effects of exposure to airborne particulate matter, evaluation of toxicokinetic factors in defining and characterizing chemical toxicity, and risk assessment, including pollution prevention.

Jon Fukuto, Ph.D., Molecular and Medical Pharmacology. Dr. Fukuto's laboratory is interested in the mechanisms of action of nitric oxide and related nitrogen oxides as cytotoxic and cytostatic agents.

Richard Gatti, Ph.D., Pathology and Laboratory Medicine. Dr. Gatti's research focuses on ataxia-telangiectasia and other chromosomal instability disorders. The lab specializes in diagnostic assays, mutation screening and identification, genetic mechanisms of disease, and functional assays for ATM and related proteins, including cellular responses to ioinizing radiation. Another project identifies therapeutic approaches to ATM based on specific groups of mutations.

Oliver Hankinson, Ph.D., Pathology and Laboratory Medicine. Dr. Hankinson's research focuses on the mechanism of carcinogenesis by polycyclic aromatic hydrocarbons (found in cigarette smoke and smog) and dioxin (a widespread pollutant) and related compounds.

Louis Ignarro, Ph.D., Molecular and Medical Pharmacology. Dr. Ignarro's research is directed toward elucidating mechanisms of regulation of nitric oxide (NO) production and cytotoxicity in macrophages, vascular cells, and tumor cells.

William McBride, Ph.D., Radiation Oncology. Dr. McBride's research focuses on degradation of proteins through the proteasome system and the inhibitory effects of exposure to radiation and other agents.

William Melega, Ph.D., Molecular and Medical Pharmacology. Dr. Melega studies the molecular mechanisms of neurodegenerative diseases and drug addiction.

Sabeeha Merchant, Ph.D., Chemistry and Biochemistry. Dr. Merchant studies the biochemistry and molecular genetics of metal metabolism.

Suzanne Paulson, Ph.D., Atmospheric Sciences. Dr. Paulson's group is developing methods to measure total nonmethane organic carbon (TNMOC) in urban and rural air.

Beate Ritz, M.D., Ph.D., Epidemiology. Dr. Ritz studies radiation and cancer, air pollution and birth outcomes, Parkinson's disease susceptibility genes, and pesticides.

Wendie Robbins, Ph.D., Nursing/Environmental Health Sciences. Dr. Robbins' research focuses on the molecular mechanisms of environmentally induced germ cell genetic damage and the development of biomarkers to detect and measure this damage in human sperm cells.

Michael Roth, M.D., Medicine. Dr. Roth's research focuses on the toxicology of inhaled substance abuse.

Robert Schiestl, Ph.D., Pathology/Environmental Health Sciences. Dr. Schiestl's work centers mostly on basic mechanisms, genetic control, and inducibility by carcinogens of homologous and illegitimate recombination, which are molecular events involved in carcinogenesis.

Joan S. Valentine, Ph.D., Chemistry and Biochemistry. Dr. Valentine's research focuses on copper-zinc superoxide dismutase and Lou Gehrig's disease and on yeast studies of oxidative stress and antioxidants.

Zuo-Feng Zhang, Ph.D., Epidemiology. Dr. Zhang's major research interest is in the molecular and genetic epidemiology of cancer as well as nutritional epidemiology.

UNIVERSITY OF CALIFORNIA, SAN DIEGO

School of Medicine
Graduate Studies in Biomedical Sciences
Department of Pharmacology

Program of Study

The School of Medicine of the University of California, San Diego (UCSD), offers multidisciplinary Ph.D. training in the pharmacological sciences. Areas of emphasis include computation and structure-guided drug design, molecular aspects of drug metabolism and disposition, cardiovascular pharmacology, and neuropharmacology. The program's strengths lie in its vigorous use of cellular and molecular approaches and its foundation in the integrative basic biomedical sciences of pharmacology, physiology, and molecular and cell biology. The Department of Pharmacology plays a major role in cross-campus programs in environmental health sciences, pharmacogenomics, molecular cardiology, and pharmacology and drug design.

A small group of faculty members with similar research interests formed the Group in Pharmacology in 1970. The current breadth of the program developed as distinguished faculty members were recruited from the School of Medicine, the Salk Institute, the Scripps Institution of Oceanography, and the Howard Hughes Medical Institute for Molecular Biology. Collaborative efforts within this diverse group of faculty members provide research opportunities that extend beyond the boundaries of traditional disciplines, where many exciting scientific advances are found.

During the first year, students complete the Biomedical Sciences core course and seminar series, more focused courses in the pharmacology track, and three laboratory rotations. Short advanced courses in statistics and scientific ethics, along with 15 units of advanced electives in specialized areas of training, are taken in this and subsequent years. Students write and orally defend a theoretical research proposal as the first part of a two-part qualifying examination given at the beginning of the second year. After choosing a dissertation adviser, they prepare for their second exam at the end of the third year. This involves obtaining preliminary data for the dissertation project and formulating and orally presenting a plan for completion of the research. Students generally receive their Ph.D. in the fifth year.

Research Facilities

The program's research laboratories are located in the Biomedical Science, Leichtag Center for Molecular Genetics, and Cell and Molecular Medicine Buildings on the Health Sciences campus in La Jolla, adjacent to the Biomedical Library. The new Skaggs School of Pharmacy and the Pharmaceutical Sciences (SSPPS) and the Departments of Biology and Chemistry also house several of the program's faculty members. Within walking distance are the San Diego Supercomputer Center, the Scripps Institution of Oceanography, the Salk Institute, and the Veterans Administration Medical Center. Additional laboratories are located in the Medical and Clinical Teaching Facilities, in the newly opened Moores Cancer Center, and at the UCSD Medical Center at Hillcrest.

Financial Aid

All students in good standing in the program receive full financial support (tuition, fees, health insurance, and a living stipend) throughout their Ph.D. training. Sources of support consist of several NIH training grants, training support from foundations, University fellowships, fellowships supported by the pharmaceutical industry, and research grants.

Cost of Study

All graduate students receive full financial support.

Living and Housing Costs

In 2005–06, limited on-campus housing was available for rents ranging from $550 per month for a single student to $1250 per month for a family apartment. A wide range of off-campus housing is available in San Diego and neighboring communities that are within a 15-minute drive of the campus and the ocean. The off-campus housing office maintains up-to-date information on rental listings.

Student Group

In 2005–06, there were 3,629 graduate and 19,763 undergraduate students at UCSD. Students interested in pharmacology matriculate in the Biomedical Sciences Graduate Program. Students in the BMS program are selected from outstanding colleges throughout the nation and abroad and include recent college graduates as well as some having substantial postcollege research experience. There are 160 students currently enrolled in the program, many of whom specialize in pharmacology. Graduates routinely obtain postdoctoral positions in excellent laboratories and move on to challenging careers in academia, research institutes, and industry. Many have risen to leadership positions in their respective fields.

Location

The 1,300-acre UCSD campus lies atop high bluffs in La Jolla, on the northern edge of San Diego, overlooking the Pacific Ocean. The intellectual activities on the campus are enriched by the proximity of the Salk Institute, the Scripps Clinic and Research Foundation, and a burgeoning group of companies engaged in biotechnological research and development. San Diego offers a year-round mild Mediterranean climate, miles of shoreline, and ready access to mountains, deserts, and Mexico. Cultural pleasures include the Old Globe Theatre and La Jolla Playhouse, the San Diego Symphony and Opera, and a wide variety of fine art galleries.

The University

In its forty-six years of existence, UCSD has progressed rapidly to become one of the premier research institutions in the United States. The University enjoys the advantage of a contiguous medical school, general campus, institution of oceanography, and supercomputer center in La Jolla. UCSD's ability to attract competitive research support is among the top six American universities. The scientific environment is full of excitement, with a growing student population and university faculty surrounded by distinguished research institutions and a burgeoning biotechnology industry.

Applying

Applications are encouraged from outstanding students who have majored in any laboratory science or in mathematics. Desirable features include a minimum GPA of 3.25 and GRE scores in the 80th percentile on the verbal, quantitative, and analytical sections of the General Test (required) and on the Subject Test in biology; in chemistry; or in biochemistry, cell and molecular biology (recommended). Undergraduate prerequisites include organic chemistry, biochemistry, physical chemistry, calculus, and biology; some exposure to cell or molecular biology and mammalian physiology is strongly advised. Prior research experience is desirable. A pre-application is required before an application is considered. Applications, considered for the fall quarter only, are due December 1; earlier submission is advantageous. Students can initiate the application process at the Web site listed in the Correspondence and Information section.

Correspondence and Information

Graduate Studies in Biomedical Sciences
School of Medicine, 0685
University of California, San Diego
9500 Gilman Drive
La Jolla, California 92093-0685
Phone: 858-534-3982
Web site: http://biomedsci.ucsd.edu

University of California, San Diego

THE FACULTY AND THEIR RESEARCH

Department Faculty

Joseph Adams, Associate Professor; Ph.D., Penn State. Mechanistic studies on protein kinases, phosphatases, and oncogenes.

Katerina Akassoglou, Assistant Professor; Ph.D., Athens. Mechanism of nervous system regeneration; fibrin signaling.

Philip E. Bourne, Professor; Ph.D., Flinders (Australia). Computational aspects of biological data representation and structure.

Joan Heller Brown, Professor and Chair; Ph.D., Yeshiva (Einstein). G protein–coupled receptors and Rho signaling pathways; cardiomyocyte growth and apoptosis; astrocytosis.

Laurence L. Brunton, Professor; Ph.D., Virginia. Cyclic nucleotide metabolism; regulatory actions of protein kinases.

Edward A. Dennis, Professor; Ph.D., Harvard. Phospholipase regulation and mechanism; prostaglandin generation.

Jack E. Dixon, Professor; Ph.D., California, Santa Barbara. Protein tyrosine phosphatases; bacterial pathogenesis.

Sylvia Evans, Professor; Ph.D., British Columbia. Biochemistry and molecular biology.

James R. Feramisco, Professor; Ph.D., California, Davis. Role of protogenes and oncogenes in cell growth and differentiation.

Tracey Handel, Professor; Ph.D., Caltech. Synthesis and characterization of polymerizeable membranes for drug delivery applications.

Vivian Y. H. Hook, Professor; Ph.D., California, San Francisco. Peptide processing and function in the cardiovascular and nervous systems.

Paul A. Insel, Professor; M.D., Michigan. Signal transduction by catecholamines, ATP, and G proteins.

Michael Karin, Professor; Ph.D., UCLA. Regulation of gene transcription; cell type specific gene expression; control of cell differentiation; signal transduction from membrane to nucleus.

Hyam L. Leffert, Professor; M.D., Yeshiva (Einstein). Growth processes of hepatocytes; role of ionic fluxes and gene expression; gene therapy.

J. Andrew McCammon, Professor; Ph.D., Harvard. Computer-aided molecular and drug design; theory of molecular recognition.

Alexandra C. Newton, Professor; Ph.D., Stanford. Signal transduction involving protein phosphorylation; protein kinase C and Akt structure and function.

Morton P. Printz, Professor; Ph.D., Pittsburgh. Molecular mechanisms of nervous system regulation of cardiovascular function; angiotensin and related neuropeptides.

Palmer Taylor, Sandra and Monroe Trout Professor; Ph.D., Wisconsin. Molecular basis for the specificity of drug-receptor interactions; regulation of enzymes and receptors involved in cholinergic neurotransmission.

Susan S. Taylor, Professor; Ph.D., Johns Hopkins. Biochemistry: structure and function of protein kinases; biophysical characterization; mutagenesis; X-ray crystallography; NMR; fluorescence probes; subcellular localization.

Roger Y. Tsien, Professor; Ph.D., Cambridge. Cell biology of intracellular signaling; molecular engineering.

Robert H. Tukey, Professor; Ph.D., Iowa. Use of recombinant DNA to characterize rabbit and human cytochromes P-450.

Tony L. Yaksh, Professor; Ph.D., Purdue. Role of spinal receptors in modulating sensory and autonomic transmission.

Adjunct Faculty

Robert Abraham, Adjunct Professor; Ph.D., Pittsburgh. Cancer chemotherapy; mTor signaling.

Tamas Bartfai, Adjunct Professor; Ph.D., Stockholm. Neuropeptides; molecular basis of cognition; inflammation and fever.

Mark R. Brann, Adjunct Associate Professor; Ph.D., Vermont. Behavioral and genetic factors of hypertension; neurotransmitter pathways; corticotropin releasing factor (CRF).

Jerold J. M. Chun, Adjunct Professor; M.D./Ph.D., Stanford. Neuronal differentiation and function; development of mammalian CNS in telencephalon; SIP and LPA signaling.

Pedro Cuatrecasas, Adjunct Professor; M.D., Washington (St. Louis). Mechanism of drug and hormone action.

Ronald M. Evans, Adjunct Professor; Ph.D., UCLA. Molecular genetics of steroid, thyroid, and retinoid receptors.

Daniel T. O'Connor, Adjunct Professor; M.D., California, Davis. Exocytosis of catecholamine storage-vesicle proteins (chromagranins) and their role in hypertension.

Melvin I. Simon, Adjunct Professor; Professor, Caltech; Ph.D., Brandeis. Molecular and cellular analysis of G-proteins; bacterial chemotaxis.

Charles F. Stevens, Adjunct Professor; M.D., Yale; Ph.D., Rockefeller. Biophysical and structural analysis of ligand-gated and voltage-sensitive ion channels.

Lynn F. Ten Eyck, Adjunct Professor; Ph.D., Princeton. Protein structure and function; hemoglobin structure; protein kinases.

Volker Vallon, Associate Adjunct Professor; M.D., Tübingen (Germany). Renal physiology; calcium channels.

J. Craig Venter, Adjunct Professor; Ph.D., California, San Diego. Prokaryotic and eukaryotic genome analysis in relation to biological function and evolution.

Gennady Verkhivker, Adjunct Professor; Ph.D., Academy of Sciences (Moscow). Structure-based drug and inhibitor design.

David S. Williams, Adjunct Professor; Ph.D., Australian National. Photoreceptor cells; retinal structure; signal transduction.

John C. Wooley, Adjunct Professor; Ph.D., Chicago. Bioinformatics; pharmacogenomics; structural genomics; computational biology.

Other Faculty Associated with the Pharmacological Sciences Training Program

Kim Barrett, Professor; Ph.D., University College (London). Role of mast cell in allergic and inflammatory diseases.

Dennis Carson, Professor; M.D., Columbia. Human autoimmune and lymphoproliferative diseases; human biochemical and molecular genetics.

Shu Chien, Professor; M.D., National Taiwan. Bioengineering and medicine.

Wolfgang H. Dillmann, Professor; M.D., Munich. Thyroid hormone effects on sarcoplasmic reticulum Ca^{2+}; ATPase function of heat shock proteins in the heart.

Mark H. Ellisman, Professor; Ph.D., Colorado at Boulder. Cellular neurobiology and ultrastructural plasticity of nervous systems.

Scott Emr, Professor; Ph.D., Harvard. Protein and lipid kinases in membrane trafficking.

Marilyn G. Farquhar, Professor; Ph.D., Berkeley. Signaling in intracellular membrane traffic; cell and molecular basis of renal function and disease.

William Fenical, Professor; Ph.D., California, Riverside. Marine natural products chemistry; synthetic and structural organic chemistry.

Lawrence S. B. Goldstein, Professor; Ph.D., Washington (Seattle). Organelle and vesicle transport and neurodegenerative disease.

Stephen B. Howell, Professor; M.D., Harvard. Pharmacology of anticancer drugs; development and testing of chemotherapeutic agents.

Martin F. Kagnoff, Professor; M.D., Harvard. Cellular and molecular immunology; lymphokines, immunoglobulin class regulation, and HLA genes.

Elizabeth Komives, Professor; Ph.D., California, San Francisco. Protein-protein interactions.

Ronald Kuczenski, Professor; Ph.D., Michigan State. Regulation and neuropharmacology of central nervous system biogenic amines.

Sanjay Nigam, Professor; M.D., Pennsylvania. Pediatrics, medicine, and cellular and molecular medicine.

Jerrold M. Olefsky, Professor; M.D., Illinois. Insulin receptors and cellular responses.

Michael G. Rosenfeld, Professor; M.D., Rochester. Development of neuroendocrine cell phenotypes; hormonal regulation of gene expression; alternate RNA splicing regulation; molecular biology of neuroendocrine gene expression; growth factor receptors.

Wylie W. Vale, Professor; Ph.D., Baylor. Isolation and characterization of hypothalamic-releasing factors; physiology and pharmacology of regulatory peptides and synthesis of potent analogues.

Ajit P. Varki, Professor; M.D., Christian Medical College, Vellore (India). Glycobiology: studies of the biochemistry and molecular biology of the oligosaccharides in development and cancer.

Tony Whyshaw-Boris, Professor; M.D., Case Western Reserve. Pharmacology; anesthesiology; sensory and autonomic spinal receptors.

Jason Yuan, Associate Professor; Ph.D., Peking Union Medical College, (Beijing). Vascular physiology and pathophysiology.

UNIVERSITY OF CINCINNATI

Graduate Program in Molecular and Developmental Biology
at the Cincinnati Children's Medical Center
Division of Teratology

Program of Study

The Graduate Program in Molecular and Developmental Biology (MDB) at the University of Cincinnati offers superb Ph.D. training. Based in the Cincinnati Children's Medical Center, one of the nation's three largest children's hospitals, the MDB program is in a unique position to offer courses and research training in the basic sciences underlying congenital and acquired diseases of children. These include fundamental mechanisms of development, organogenesis, organ dysgenesis as the basis of childhood congenital disease, infectious diseases, asthma, childhood cancer, reproductive sciences, human and molecular genetics, biomedical engineering, and neuroscience. Basic sciences are studied in a highly integrated manner with clinical research, and there are many opportunities for translational applications. Core courses in developmental biology, development and disease, molecular genetics, and cell biology provide a strong foundation for dissertation research. The combination of training in both basic science and translational research dramatically enhances the career opportunities for graduates. Typically, students in the program go on to postdoctoral positions in major academic or biotechnology laboratories. There are approximately 70 faculty members from the Children's Medical Center and other departments in the University of Cincinnati College of Medicine. They publish actively in major journals, and speak regularly at international meetings. The Children's Medical Center has outstanding research facilities and cutting-edge core technologies are available. Students take at least two laboratory rotations before deciding on a dissertation laboratory and must pass a qualifying examination in the second year for admission to candidacy.

The teratology track is offered within the development area of the program and features a core course in teratology, seminars, and journal club series in teratology and electives in toxicology and biostatistics. For those interested in neurobehavioral teratology, electives are offered in neuroscience, neuropharmacology, and psychopharmacology. The teratology track has been supported by a training grant from NIH for more than twenty-five years and is the only such program in the country. The program has been recognized as fulfilling a national need to train scientists in the study of birth defects and prenatal causes of learning disabilities. Graduates of the program are in high demand.

Research Facilities

MDB faculty members are located at Cincinnati Children's Hospital Medical Center and the University of Cincinnati College of Medicine. Research laboratories at Cincinnati Children's Hospital Medical Center occupy 500,000 square feet of research space, much of which has been constructed or renovated since 1999. Additional laboratories are located across the street in basic science departments at the University of Cincinnati College of Medicine. State-of-the-art facilities for molecular, developmental, and cell biology studies, and core facilities for creating transgenic mice (genetically modified mice), imaging, DNA analysis, genomics, proteomics, and bioinformatics facilitate student research. MDB research programs are well-supported by multiple funding agencies, including the National Institutes of Health (NIH), the American Heart Association, the American Lung Association, and the March of Dimes.

Financial Aid

Students holding baccalaureate degrees who are accepted by the program are provided with a full-tuition scholarship, health insurance, and a base stipend of $20,500 per year. Student stipends are supported by research grants from a variety of sources in addition to NIH training grants in teratology, endocrinology, and cardiac and pulmonary biology. Increased student stipends of $24,000 are available to highly qualified applicants and include the Cincinnati Children's Distinguished Research Scholar Awards, Ryan Fellowships, and externally funded predoctoral awards.

Cost of Study

All students are awarded a full-tuition scholarship, health insurance, and a stipend.

Living and Housing Costs

Furnished apartments for single and married students, maintained by the University, rented for $510 to $730 per month in 2004–05, including utilities and telephone. These apartments are limited in number, and most graduate students find less expensive apartments in adjacent areas. The cost of living in Cincinnati is relatively low.

Student Group

In 2004–05, there were 50 students in the MDB program, part of a larger graduate student population of approximately 400 graduate students in biomedical sciences at the University of Cincinnati College of Medicine.

Location

Located in the southwest corner of Ohio, Cincinnati is nestled among seven hills that overlook the Ohio River, with a population of more than 1.5 million in the greater Cincinnati area. Music, art, theater, and ballet provide an impressive array of cultural activities. Cincinnati strongly supports professional baseball (the Reds) and football (the Bengals) as well as college basketball. City and county parks provide attractive and well-maintained settings for a variety of recreational activities.

The University

Founded in 1819 as the Cincinnati College and the Medical College of Ohio, the University of Cincinnati became a municipal university in 1870; in 1977, it joined the state university system. Located on five campuses, the University consists of eighteen component colleges and divisions that provide a wide range of graduate and professional programs. It is accredited by the North Central Association of Colleges and Schools. Over the years, the University has achieved a national and international reputation in many areas of research and graduate training.

Applying

Applicants are required to take the GRE General Test and provide undergraduate transcripts (GPA score results on a 4.0 scale or equivalent). Undergraduate training in biology is required, with recommended courses in biochemistry, cell biology, molecular biology, developmental biology, and genetics. For biology majors, courses in general and organic chemistry, general physics, and algebra or calculus are recommended. For nonbiology majors, a minor in biology is recommended. Research experience is a major asset to applicants. In addition, a statement of career goals and three letters of recommendation are required. Rolling admissions are granted after February 1, and it is recommended that applicants apply before March 1 to ensure the best chance for admission. Admitted students are encouraged to enter the program by July 1 of the year in which they are admitted. Early admission requests are considered, providing all of the admission items described above are submitted before February 1. The most promising applicants are invited to visit for an interview, for which all costs and arrangements are paid by the program.

Correspondence and Information

Tim Le Cras, Ph.D.
Director of Recruiting
Division of Developmental Biology
Cincinnati Research Foundation
3333 Burnet Avenue, ML 7007
Cincinnati, Ohio 45229-3039
Phone: 513-636-4545
E-mail: mdbprog@cchmc.org
Web site: http://www.chmcc.org/dbprog/

University of Cincinnati

THE FACULTY AND THEIR RESEARCH

Division of Teratology
Kenneth Campbell, Ph.D. A role for Gsh1 in the developing striatum and olfactory bulb of Gsh2 mutant mice. *Development* 128:4769–80, 2001.
Melissa Colbert, Ph.D. Retinoic acid inhibits cardiac neural crest migration by blocking c-Jun N-terminal kinase activation. *Dev. Biol.* 232:351–61, 2001.
Chia-Yi Kuan, M.D., Ph.D. Mechanisms of programmed cell death in the developing brain. *Trends Neurosci.* 23:291–7, 2000.
Jun Ma, Ph.D. Crossing the line between activation and repression. *Trends Genet.* 21:54–9, 2005.
Jeffrey Molkentin, Ph.D. PKCα regulates cardiac contractility and propensity towards heart failure. *Nature Med.* 10:248–54, 2004.
Daniel Nebert, M.D. Clinical importance of the cytochromes P450. *Lancet* 360:1155–62, 2002.
S. Steven Potter, Ph.D. A catalogue of gene expression in the developing kidney. *Kidney Int.* 64(5):1588–604, 2003.
Charles Vorhees, Ph.D. Exposure to 3,4-methylenedioxymethamphetamine (MDMA) on postnatal days 11-20 induces reference but not working memory deficits in the Morris water maze in rats: Implications of prior learning. *Int. J. Dev. Neurosci.* 22:247–59, 2004.
Michael Williams, Ph.D. Neonatal methamphetamine administration induces region-specific long-term neuronal morphological changes in the rat hippocampus, nucleus accumbens, and parietal cortex. *Eur. J. Neurosci.* 19:23165–70, 2004.
Katherine Yutzey, Ph.D. Differential expression and function of Tbx5 and Tbx20 in cardiac development. *J. Biol. Chem.* 279:19026–34, 2004.

The following are other faculty members affiliated with the Graduate Program in Molecular and Developmental Biology.

Cancer Biology
Susan Waltz. Growth factors and receptor tyrosine kinases in tumorigenesis.
Susanne Wells. Papilloma virus and cervical cancer.
Yi Zheng. Molecular mechanisms of Rho GTPase signal transduction.

Cardiovascular Biology
D. Woodrow Benson. Genetic analysis of congenital heart malformations.
Jay Degen. Hemostatic factors and signaling systems in development and disease.
Sandra Degen. Biology of blood coagulation proteins and related growth factors.
Jeffery Molkentin. Molecular control of cardiac development and hypertrophy.
Jeffrey Robbins. Molecular basis of heart performance and hypertrophic cardiomyopathy.
Stephanie Ware. Genetics of cardiovascular development.
David Wieczorek. Contractile protein isoform function.
Katherine Yutzey. Cardiac morphogenesis and gene expression.

Developmental Biology and Stem-Cell Biology
Michael Bates. Digestive system differentiation and gene expression.
Nadean Brown. Vertebrate eye development.
Kenneth Campbell. Vertebrate forebrain development.
Daniel Choo. Mouse inner ear differentiation and patterning.
Tiffany Cook. Genes controlling color vision in *Drosophila.*
Brian Gebelein. Homeobox genes and body patterning in *Drosophila.*
Janet Heasman. Xenopus germ layer determination.
Rashmi Hegde. Structural analysis of transcription factors.
Chia-Yi (Alex) Kuan. Programmed cell death in the developing nervous system.
Richard Lang. Early development of the vertebrate eye.
James Lessard. Cell and molecular biology of muscle development and function.
Xinhua Lin. Cell-cell signaling in *Drosophila.*
Jun Ma. Transcriptional regulation and development.
S. Steven Potter. Homeobox gene regulation in mammalian development.
James Wells. Early endoderm patterning and pancreas development.
Daniel Wiginton. Development and differentiation of the intestine.
Aaron Zorn. Endoderm development and organ function.

Human Genetics, Gene Therapy, and Molecular Medicine
Bruce Aronow. Chromatin structure in T cells; leukemia and gene therapy vectors.
Mitchell Cohen. Intestinal secretion; activators of intestinal guanylate cyclase.
Timothy Cripe. Human gene therapy vectors.
Timothy Crombleholme. Fetal tissue repair and scarless healing.
Gregory Grabowski. Pathogenesis and therapy of human genetic disease.
Min-Xin Guan. Human mitochondrial genetic disease.
David Williams. Blood stem cell biology and gene therapy.

Immunobiology
Robert Colbert. Antigen processing and presentation in autoimmunity.
Hartmut Geiger. Hematopoietic stem cells: DNA integrity and aging; hematopoietic stem cell mobilization.
Gurjit Khurana Hershey. Genetics and pathogenesis of atopic disorders.
David Hildeman. Molecular biology of antigen-specific T cells.
Christopher Karp. Cytokine-mediated dysregulation of cellular immunity in human disease.
Jonathan Katz. Immunologic aspects of type 1 diabetes mellitus.
Marc Rothenberg. Eosinophil biology; chemokine receptor signaling pathways.

Neuroscience
Masato Nakafuku. Molecular and cellular control of neural stem cells.
Sarah Pixley. Olfactory signal transduction.
Nancy Ratner. Schwann cell-neuron interactions in development and disease.
David Repaske. Molecular basis of endocrine disorders.
Randy Sallee. Pharmacology and pharmacogenetics of neuropsychiatric disorders.
Michael T. Williams. Drugs of abuse, neurotoxicology, and behavioral teratology.

Pulmonary Biology
Ann Akeson. Lung vascular development.
Tim Le Cras. Newborn lung development and disease.
Ward Rice. Cell signaling in lung development and differentiation.
John Shannon. Lung developmental biology.
Timothy Weaver. Structure-function analysis of proteins critical for lung homeostasis.
Jeffrey Whitsett. Lung morphogenesis, gene regulation, and surfactant biology.

Reproductive Biology and Teratology
George Daston. Mechanisms of chemical teratogenesis.
Stuart Handwerger. Placental and decidual gene expression.
Daniel Nebert. Molecular basis of environmental toxicity and teratogenesis.
Charles Vorhees. Drug-induced developmental neurotoxicity.
Christopher Wylie. Development of the germ line.

UNIVERSITY OF CONNECTICUT

Department of Pharmaceutical Sciences
Graduate Program in Pharmacology and Toxicology
and Interdisciplinary Toxicology Training Program

Programs of Study

Programs are offered leading to the M.S. and Ph.D. degrees specializing in pharmacology or toxicology. Pharmacology/toxicology is one of three disciplines housed in the Department of Pharmaceutical Sciences. Research, considered of crucial importance in the graduate program, is active and varied and focuses on mechanistic approaches to the study of drugs and toxic agents, with an emphasis on safety evaluation, and includes biochemical and physiological mechanisms of action of therapeutic agents and toxic environmental contaminants. Students who specialize in toxicology participate in the Interdisciplinary Toxicology Program, which is administered by the Center for Biochemical Toxicology. This toxicology program also includes students and faculty members from the Departments of Molecular and Cellular Biology, Physiology and Neurobiology, Chemistry, Animal Science, and Pathobiology. Adjunct faculty members from local pharmaceutical and chemical industries also provide students with outstanding educational opportunities and on-site experience. Research interests include cardiovascular pharmacology, neuropharmacology and neurodegenerative disorders, biochemical and environmental toxicology (including hepatic, renal, and respiratory target organ toxicity), and receptor regulation of gene expression. The student's program of study and thesis work are under the direction of an advisory committee. For the Ph.D. degree, approximately 42 credits of didactic course work are required, including courses in pharmacology, toxicology, biochemistry, physiology, and statistics. A series of written and oral qualifying exams and the presentation of an acceptable dissertation based on original investigation are required.

Research Facilities

The School of Pharmacy is housed in a brand new $95-million, 221,000-square-foot building. The building is located in the center of the University Science Quadrangle, placing chemistry, engineering, pharmacy, physics, biology, and the other sciences in proximity to one another. The building houses the School of Pharmacy and the faculties from the Departments of Ecology and Evolutionary Biology and Physiology and Neurobiology as well as the Office for Animal Research Services. The north wing of the building contains six floors of research laboratories shared by pharmacy and biology. The building also contains the school's teaching, administration, and library spaces and the vivarium. The program and its faculty members are well equipped with the necessary instrumentation and equipment for conducting modern in vivo and in vitro research in pharmacology and toxicology, and interactions with other departments on campus are encouraged. On the University of Connecticut campus, students have access to extensive facilities for biomedical research, including imaging, microscopy, histopathology, and facilities for macromolecular identification and transgenic research. Available to research faculty members within the School of Pharmacy are animal facilities and audiovisual and academic libraries. The University library has more than 2.4 million volumes, with extensive holdings in biomedical sciences, as well as e-library services. There is an extensive network of terminals and microcomputers in all buildings used by the program. Additional textbooks and journals are also available at the library of the nearby University of Connecticut Health Center and through interlibrary loan programs.

Financial Aid

A limited number of teaching and research assistantships are offered through the school and are awarded on a competitive basis to qualified applicants. Stipends for first-year students typically begin at $18,000. Appointments carry remission of tuition and health benefits. A limited number of special fellowships are also available for exceptionally qualified students. Students are also eligible for support from research grants. International students must achieve a score of 260 on the TOEFL and pass a test of spoken English before a teaching assistantship can be awarded.

Cost of Study

Tuition and fees for full-time students carrying 9 or more credits were $4485 per semester in 2005–06 for in-state students and $10,515 for out-of-state students.

Living and Housing Costs

On-campus housing in graduate dormitories is available at approximately $4800 per year. Off-campus housing is also available with rents ranging from $350 to $800 per month, depending on size, location, and sharing of accommodations.

Student Group

Of a total University population of more than 28,000 students, approximately 6,000 are pursuing a graduate degree. Within the School of Pharmacy, there are about 45 graduate students, with 12 currently enrolled in the pharmacology and toxicology Ph.D. program. The demand for graduates in pharmacology and toxicology by the pharmaceutical and chemical industry and by academia remains high.

Location

The University is located on a 3,100-acre rural campus in Storrs in pastoral northeastern Connecticut. Its proximity to Hartford (30 miles), Boston (80 miles), and New York (130 miles) provides opportunities for cultural, recreational, and sporting attractions. Storrs is also near the shores of southern New England and the mountains of northern New England.

The University

Founded in 1881 as the Storrs Agricultural College, the University has grown into a major northeastern public university while retaining a rural, New England atmosphere. More than 120 buildings are housed on 3,100 acres in and around picturesque Storrs. UConn is undergoing an unprecedented transformation in infrastructure and academic stature. The University is renewing, rebuilding, and enhancing its campuses through a $2.3 billion, 20-year state investment in the University's infrastructure. State-of-the-art facilities grace every campus. A faculty of more than 1,200 serves the needs of the University's 20,000 undergraduate students.

Applying

Applicants should hold, or expect to receive, a B.S. or B.A. degree in pharmacy, biochemistry, biology, chemistry, or a related area. Ordinarily, a minimum B average or equivalent is required. GRE General Test scores, TOEFL scores (for international applicants), and three letters of reference are also required. Applicants are accepted throughout the year for admission in September or January (when space is available). Normally, requests for financial aid must be received before February 1.

Correspondence and Information

Dr. Gerald Gianutsos
Pharmacology and Toxicology Program Coordinator
University of Connecticut
69 North Eagleville Road
Unit 3092
Storrs, Connecticut 06269-3092
Phone: 860-486-4066
Fax: 860-486-5792
E-mail: gerald.gianutsos@uconn.edu

University of Connecticut

THE FACULTY AND THEIR RESEARCH

Brian J. Aneskievich, Associate Professor; Ph.D., SUNY at Stony Brook. Molecular pharmacology: regulation of epithelial growth, differentiation, malignant transformation, and gene transcription via nuclear receptors activated by retinoids, peroxisome proliferators, and xenobiotics.

Ben A. Bahr, Associate Professor and member, Center for Drug Discovery; Ph.D., California, Santa Barbara. Cell survival signaling in response to neurotoxic insults; models of Alzheimer-type pathogenesis, Parkinson's disease, stroke, and neuroprotection.

Gerald Gianutsos, Associate Professor; Ph.D., Rhode Island. Psychopharmacology: effects of antipsychotic, antidepressant, and antiparkinson drugs on CNS amines; receptor regulation; neurotoxicology.

David F. Grant, Associate Professor; Ph.D., Michigan State. Biochemical and molecular toxicology: toxicology of fatty acid epoxides, epoxide hydrolases, and pharmacogenetics.

Dennis W. Hill, Professor; Ph.D., Texas A&M. Analytical toxicology: chromatographic (TLC, GC, and HPLC) and spectrometric (UV, IR, and MS) techniques; drug and toxicant biotransformation and disposition.

Andrea K. Hubbard, Associate Professor; Ph.D., Tennessee. Immunotoxicology and pulmonary toxicity: molecular and cellular events underlying pulmonary injury in asthma or from exposure to air pollutants.

Ronald O. Langner, Professor; Ph.D., Rhode Island. Biochemical pharmacology: biochemical changes induced in blood vessels following exposure to atherogenic stimuli; development of novel methods for treating atherosclerosis.

José E. Manautou, Associate Professor; Ph.D., Purdue. Biochemical and molecular mechanisms of hepatotoxicity; role of multidrug resistance proteins in the hepatobiliary disposition of xenobiotics; changes in expression of transport proteins in response to chemical liver injury; hepatoprotective effect of peroxisome proliferators.

John B. Morris, Professor; Ph.D., Rochester. Inhalation toxicology: health effects of air pollutants, pharmacokinetic modeling; risk assessment.

Toxicology Program Faculty Affiliated with Other Departments

Ashis K. Basu, Professor of Chemistry; Ph.D., Wayne State. Molecular toxicology and chemical carcinogenesis: mechanisms of mutation and cancer induced by environmental pollutants; DNA repair; mechanistic basis of cancer chemotherapy.

Thomas T. Chen, Professor; Ph.D., Alberta. Environmental toxicology: environmental biomonitors of xenobiotic pollution; biotechnology.

Joseph F. Crivello, Professor; Ph.D., Wisconsin. Biochemical effects of aquatic pollutants.

Sylvain DeGuise, Associate Professor; Ph.D., Quebec at Montreal. Immunotoxicity; models of aquatic toxicology; toxicologic pathology.

Charles Giardina, Associate Professor of Molecular and Cell Biology; Ph.D., SUNY at Stony Brook. Gene regulation: NF-κβ–regulated genes in normal and neoplastic colonic epithelium; heat shock gene regulation; apoptosis.

Lawrence E. Hightower, Professor of Molecular and Cell Biology; Ph.D., Harvard. Environmental toxicology: development of cellular stress proteins as molecular biomarkers; development of in vitro cytotoxicity assays for fish and mammals.

Michael A. Lynes, Professor; Ph.D., North Carolina. Immunotoxicity; metals; lymphocyte activation; metallothionein; regulation of cell motility.

J. Larry Renfro, Professor of Physiology; Ph.D., Oklahoma. Comparative physiology: epithelial membrane transport of xenobiotics; ion secretion by renal epithelium in primary culture; stress-induced homeostatic responses of transport epithelia.

Lawrence K. Silbart, Professor of Animal Science and Assistant Director of the Center for Environmental Health; M.P.H., Ph.D., Michigan. Mucosal immunology and carcinogenesis; mucosal vaccine development; immunotherapeutics.

Adjunct Faculty

David E. Amacher, Research Professor and Research Advisor, Cellular Toxicology, Pfizer Central Research; Ph.D., Kent State. Hepatotoxicity; cell proliferation; hepatic enzyme induction/inhibition; bone marrow toxicity; peroxisome proliferation; in vitro toxicology methods.

Kerry T. Blanchard, Research Associate Professor and Director of Toxicology, Boehringer-Ingelheim Corporation; Ph.D., Connecticut. Gene expression profiling.

Matthew S. Bogdanffy, Research Associate Professor and Associate Director, Toxicology and Safety Assessment Boehringer-Ingelheim Corporation; Ph.D., Northeastern. Genetic toxicology, risk assessment.

Steven D. Cohen, Research Professor and Chair, Pharmaceutical Sciences, Massachusetts College of Pharmacy; D.Sc., Harvard. Biochemical and molecular toxicology: studies of xenobiotic disposition and mechanistic studies of chemically induced liver and kidney damage, with emphasis on the roles of unique molecular targets in mediating both homeostatic and detrimental cellular responses to model toxicants like acetaminophen.

Susan A. Masino, Assistant Professor of Psychology and Neuroscience, Trinity College; Ph.D., California, Irvine. Learning and memory; synaptic plasticity; excitatory responses; adenosine and synaptic modulation.

Daniel W. Rosenberg, Associate Professor of Medicine, University of Connecticut Health Center; Ph.D., Michigan. Molecular toxicology: molecular and genetic analysis of colon cancer progression; transgenic models of cancer; mechanisms of environmental chemical activation.

Raymond E. Stoll, Research Professor and Director of Toxicology and Safety Assessment, Boehringer Ingelheim Pharmaceuticals; Ph.D., Purdue. Toxicology; pathology; acute and chronic toxicity; mutagenicity; carcinogenicity; transgenic mice as alternatives to carcinogenicity bioassays.

Bruce O. Stuart, Research Professor and Senior Principal Scientist, Schering-Plough Research Institute; Ph.D., Rochester. Inhalation toxicology; radionuclides in metabolism and pharmacokinetic studies; industrial hygiene; occupational medicine; mechanisms of chronic respiratory disease and carcinogenesis; delivery systems in response to inhaled pharmaceuticals.

Cen Xu, Research Assistant Professor and Senior Research Investigator, Bristol-Myers Squibb Corporation; Ph.D., Illinois State. Peptide receptors.

UNIVERSITY OF CONNECTICUT HEALTH CENTER

Graduate Program in Cellular and Molecular Pharmacology

Program of Study

The Graduate Program in Cellular and Molecular Pharmacology at the University of Connecticut (UConn) Health Center has as its goal the production of good scientists capable of carrying out scholarly endeavors in pharmacologically related areas of work, in both academic and nonacademic environments. Graduates have found rewarding careers as researcher-educators in medical schools, research institutes, the pharmaceutical industry, and the chemical industry. The faculty of the Graduate Program in Cellular and Molecular Pharmacology has made teaching and research training of graduate students a high priority. The program faculty members are a diverse group of scientists who share an interest in signal transduction and processes of living cells, tissues, and organisms. The areas of graduate training emphasized include neuropharmacology, cardiovascular pharmacology, endocrine-reproductive pharmacology, immunopharmacology, chemotherapy and clinical mechanism of xenobiotic biotransformation. The Department of Pharmacology has a joint NIH-sponsored grant in Mechanisms of Action of Drugs and Abuse with the Center for Drug Discovery, located in the School of Pharmacology at the undergraduate campus in Storrs, Connecticut.

Research Facilities

The Department of Pharmacology comprises 16,000 square feet of laboratory space. There are 16 full-time faculty members. These faculty members, plus an additional 24 faculty members from other clinical and basic science departments, comprise the Graduate Program in Cellular and Molecular Pharmacology. In addition, there are currently 21 postdoctoral fellows and graduate students being trained in the department. The department encourages interaction and collaboration between faculty members within the department and among the various Health Center departments. The Department of Pharmacology and its faculty members are very well equipped with state-of-the-art instrumentation, including an Improvision OpenLab Imaging System with Hammamatsu ORCA-RE camera, Velocity 2.5 deconvolution and 3-D rendering software, an electron paramagnetic resonance (EPR) spectrometer, mass spectrometer (GC/MS) scanning-recording spectrofluorometers, dual-beam/split-beam recording spectrophotometers, a UV/VIS stopped flow spectrometer, high performance liquid chromatographs (HPLC), scintillation counters, gamma counters, a wide variety of centrifuges and refrigerated ultracentrifuges, oxygen and ion electrodes, fluorescence and interference contrast microscopes, single-channel and whole-cell voltage clamp electrophysiological recording equipment, cell motion detectors, bacterial and tissue culture facilities, PCR machines, and online computers in all laboratories for data handling and molecular and cell imaging. In addition, the Health Center has a number of facilities and centers for highly specialized equipment available to Health Center personnel, such as electron microscopes and X-ray diffraction equipment in the Biostructure Center and state-of-the-art imaging equipment and computers in the Imaging Center.

Financial Aid

Support for doctoral students engaged in full-time degree programs at the Health Center is provided on a competitive basis. Graduate research assistantships for 2006–07 provide a stipend of $26,000 per year, which includes a waiver of tuition/University fees for the fall and spring semesters and a student health insurance plan. While financial aid is offered competitively, the Health Center makes every possible effort to address the financial needs of all students during their period of training.

Cost of Study

For 2006–07, tuition is $3996 per semester ($7992 per year) for full-time students who are Connecticut residents and $10,368 per semester ($20,772 per year) for full-time out-of-state residents. General University fees are added to the cost of tuition for students who do not receive a tuition waiver. These costs are usually met by traineeships or research assistantships for doctoral students.

Living and Housing Costs

There is a wide range of affordable housing options in the greater Hartford area within easy commuting distance of the campus, including an extensive complex that is adjacent to the Health Center. Costs range from $600 to $800 per month for a one-bedroom unit; 2 or more students sharing an apartment usually pay less. University housing is not available at the Health Center.

Student Group

Currently, 8 students are pursuing doctoral studies with the program's faculty. The total number of graduate students at the Health Center is approximately 150, while the medical and dental schools combined currently enroll 128 students per class.

Location

The Health Center is located in the historic town of Farmington, Connecticut. Set in the beautiful New England countryside on a hill overlooking the Farmington Valley, it is close to ski areas, hiking trails, and facilities for boating, fishing, and swimming. Connecticut's capital city of Hartford, 7 miles east of Farmington, is the center of an urban region of approximately 800,000 people. The beaches of the Long Island Sound are about 50 minutes away to the south, and the beautiful Berkshires are a short drive to the northwest. New York City and Boston can be reached within 2½ hours by car. Hartford is the home of the acclaimed Hartford Stage Company, TheatreWorks, the Hartford Symphony and Chamber orchestras, two ballet companies, an opera company, the Wadsworth Atheneum (the oldest public art museum in the nation), the Mark Twain house, the Hartford Civic Center, and many other interesting cultural and recreational facilities. The area is also home to several branches of the University of Connecticut, Trinity College, and the University of Hartford, which includes the Hartt School of Music. Bradley International Airport (about 20 minutes from campus) serves the Hartford/Springfield area with frequent airline connections to major cities in this country and abroad. Frequent bus and rail service is also available from Hartford.

The Health Center

The 200-acre Health Center campus at Farmington houses a division of the University of Connecticut Graduate School, as well as the School of Medicine and Dental Medicine. The campus also includes the John Dempsey Hospital, associated clinics, and extensive medical research facilities, all in a centralized facility with more than 1 million square feet of floor space. The Health Center's newest research addition, the Academic Research Building, was opened in 1999. This impressive eleven-story structure provides 170,000 square feet of state-of-the-art laboratory space. The faculty at the center includes more than 260 full-time members. The institution has a strong commitment to graduate study within an environment that promotes social and intellectual interaction among the various educational programs. Graduate students are represented on various administrative committees concerned with curricular affairs, and the Graduate Student Organization (GSO) represents graduate students' needs and concerns to the faculty and administration, in addition to fostering social contact among graduate students in the Health Center.

Applying

Applications for admission should be submitted on standard forms obtained from the Graduate Admissions Office at the UConn Health Center or on the Web site. The application should be filed together with transcripts, three letters of recommendation, a personal statement, and recent results from the General Test of the Graduate Record Examinations. International students must take the Test of English as a Foreign Language (TOEFL) to satisfy Graduate School requirements. The deadline for completed applications and receipt of all supplemental materials is December 15. In accordance with the laws of the state of Connecticut and of the United States, the University of Connecticut Health Center does not discriminate against any person in its educational and employment activities on the grounds of race, color, creed, national origin, sex, age, or physical disability.

Correspondence and Information

Dr. Joel S. Pachter
Director, Cellular and Molecular Pharmacology Graduate Program
MC 6125
University of Connecticut Health Center
Farmington, Connecticut 06030

Phone: 860-679-3698
E-mail: pachter@nso1.uchc.edu
Web site: http://grad.uchc.edu

University of Connecticut Health Center

THE FACULTY AND THEIR RESEARCH

Alexander Amerik, Ph.D. Deubiquinating enzymes and ubiquitin homeostasis in the yeast *Saccharomyces cerevisiae.*

Stefan Brocke, M.D., Ph.D. Cellular and molecular mechanisms of brain injury in inflammatory and inflammation-associated disorders of the central nervous system.

Paul Epstein, Ph.D. Signal transduction in relation to leukemia and breast cancer; purification and cloning of cyclic nucleotide phosphodiesterases.

Maurice Feinstein, Ph.D. Phosphorylation cascades in the regulation of platelet functions.

James Hewett, Ph.D. Protein structure-function relationships and regulation of inflammatory mediator expression in the central nervous system.

Sandra Hewett, Ph.D. Cellular mechanisms underlying CNS injury.

Richard Lambrecht, Ph.D. Mechanisms controlling degradation of heme and iron homeostasis.

Eric Levine, Ph.D. Synaptic plasticity and the role of neurotrophins in brain development and learning.

Flavia O'Rourke, Ph.D. Molecular biology and function of inositol phosphate receptors.

Joel Pachter, Ph.D. Mechanisms regulating infection and inflammation of the central nervous system; use of laser-capture microdissection for gene profiling of the neurovascular unit in health and disease.

Achilles Pappano, Ph.D. Cardiac membrane receptors and regulation of ion channels.

Charlotte Ressler, Ph.D. Mechanisms of allergic reactions to drugs; toxic factors in foods.

Daniel Rosenberg, Ph.D. Molecular genetics of colorectal cancer; signaling C pathways in the development of tumors; toxicogenomics.

James Rusling, Ph.D. Use of nanofabrication and molecular self-assembly to construct functional biomolecular films on surfaces for applications in biomedical sensing and fundamental biochemical studies.

John Schenkman, Ph.D. Cytochrome P450 monooxygenase: metabolism of xenobiotics and steroids; mechanism of anticancer drug action.

Henry Smilowitz, Ph.D. Preclinical brain tumor therapy; animal models of lung injury.

Ivo Stoilov, M.D. Comparative expression profiling of mouse cytochrome P450 genes in embryonic and adult tissues.

Andrew Winokur, M.D., Ph.D. Changes in sleep physiology during depression; clinical evaluation of the efficacy and safety of drugs for the treatment of depression, bipolar disorder, anxiety disorders, and schizophrenia.

UNIVERSITY OF ILLINOIS AT CHICAGO

College of Medicine
Department of Pharmacology

Programs of Study

The Department of Pharmacology at the University of Illinois at Chicago (UIC) offers training in molecular and cellular pharmacology leading to the Ph.D. or M.D./Ph.D degrees. The department has research strengths and extensive NIH funding in the areas of cardiovascular and lung biology, cell signaling, molecular pharmacology of G proteins, immunopharmacology, molecular and cellular basis of inflammation, and neuroscience. Research in these areas is pursued using multidisciplinary methods, which emphasize intensive training at the molecular and cellular levels. The Department of Pharmacology participates in a new, integrated program, Graduate Education in Medical Sciences (GEMS), which offers interdisciplinary course work, flexibility in laboratory rotations, and choice of thesis advisers.

All first-year students are required to take GEMS courses in the first semester, which include biochemistry, cell biology and integrative physiology, molecular biology, and research methods. In the second semester, students interested in pharmacology take receptor pharmacology and cell signaling, research methods, and an elective. The second-year course requirements include medical pharmacology, scientific integrity and responsible research, and an elective chosen from molecular pharmacology of the cardiovascular system and platelets, pharmacology and biology of the vessel wall, and ion channels: structure, function, pharmacology, and pathology. Electives outside the department are also available in areas such as drug development, bioinformatics, and molecular genetics. Students initiate their research during the first year. The preliminary exam is given at the end of the second year of graduate work. Throughout their residency, students participate in seminars, colloquia, and a variety of informal discussions. The graduate student–faculty ratio is 2:1, and close tutorial relationships are emphasized.

Research Facilities

The departmental research facilities are generally located in the basic sciences research building and new state-of-the-art medicine research building. Facilities include cell and tissue culture laboratories; confocal imaging facilities; freezer, cold, and warm rooms; and radioisotope rooms. Each laboratory in the department is equipped with a complete array of sophisticated instrumentation reflecting areas of specialization in molecular pharmacology. The department has laboratories available for protein sequencing and synthesis, electron microscopy, NMR spectroscopy, and mass spectroscopy; high-speed computers; and a well-supervised animal-care unit. Students may use the extensive open-stack library at the Health Sciences Center and have online access to full-text journals.

Financial Aid

Upon admission to the program, each student is awarded financial assistance, which includes a stipend of $24,000 as a research assistant and a tuition and service fee waiver. With satisfactory progress toward completion, the student can continue to be supported throughout his or her residency. Students may qualify for other need-based financial aid by filling out the Free Application for Federal Student Aid (FAFSA) at the U.S. Department of Education's Web site.

Cost of Study

In 2004–05, fees for registration and tuition for a full program were $10,795 for Illinois residents and $25,793 for nonresidents. Tuition costs and service fees are covered for those awarded fellowships and assistantships.

Living and Housing Costs

The minimum cost of living for a graduate student in Chicago is generally offset by the stipend. Students may live in on-campus apartments, dormitories, or off-campus housing of their choice.

Student Group

Nearly 30 percent of the total enrollment at UIC is at the graduate and professional level. There are approximately 35 graduate students enrolled in the Department of Pharmacology.

Location

The University of Illinois is located on an attractive campus near the Loop of Chicago. Cultural facilities such as the Chicago Symphony Center and Art Institute are about 2 miles away. The city of Chicago, the Medical District, and neighboring universities offer a challenging scientific, cultural, and intellectual atmosphere. The downtown area is a short trip by elevated train from the Health Sciences Center campus.

The University

The University of Illinois is a major research university in the Chicago area and is one of the top fifty research universities in the United States. The campus offers bachelor's degrees in ninety-eight fields, master's degrees in seventy-nine areas, and doctorates in forty-six specializations. More than 2 million volumes are housed in the University's library and its specialized collections. The University is represented in leading academic institutions, government agencies, and industrial settings, where many graduates hold prominent positions. The environment in the University is intellectually challenging, with multiple opportunities for interchange of ideas. It has become a gathering place for visiting scientists in numerous disciplines and is one of the most active research environments in Chicago's scientific community.

Applying

The recommended application deadline is March 1 (final deadline, May 15). However, entering classes are usually filled earlier, so fall or winter application is advantageous. The deadline for international students is February 15. The General Test scores of the Graduate Record Examinations are required. Students begin their studies in the fall semester; there are no spring admissions.

Correspondence and Information

Department of Pharmacology
University of Illinois at Chicago
835 South Wolcott Avenue (M/C 868)
Chicago, Illinois 60612

Phone: 312-996-9147
E-mail: pharmacology-gs@uic.edu
Web site: http://www.uic.edu/depts/mcph

University of Illinois at Chicago

THE GRADUATE FACULTY AND THEIR RESEARCH

Athar Chishti, Professor; Ph.D., Melbourne, 1984. Function of MAGUKs (membrane associated guanylate kinase homologues); role of calpains in regulating platelet function; mechanisms of malaria parasite entry into red blood cells; cell proliferation and tumor suppression pathways.

Oscar Colamonici, Associate Professor; M.D., Uruguay, 1981. Characterization of the structure of the interferon alpha receptor (IFN R) and cytokine signaling.

Xiaoping Du, Associate Professor; M.D./Ph.D., Sydney, 1989. Thrombosis, hemostasis, and vascular biology; molecular mechanisms and molecular pharmacology of platelet and endothelial cell functions.

Ervin G. Erdös, Professor; M.D., Munich, 1950. Enzymology of peptide metabolism and its inhibitors; modes of action of the therapeutically widely used angiotensin I converting enzyme inhibitors at cellular and subcellular levels.

Tohru Kozasa, Assistant Professor; M.D./Ph.D., Tokyo, 1980. G-protein-mediated signal transduction pathways; function of RGS proteins; mechanism of regulation of Rho family GTPases (Rho, Rac, Cdc42) by G-protein-coupled receptors.

Stephen C.-T. Lam, Associate Professor; Ph.D., Toronto, 1984. Platelet adhesive functions in hemostasis and thrombosis; structure-function relationship of integrin alpha IIbβ3.

Guy C. Le Breton, Professor; Ph.D., Chicago, 1973. Fundamental mechanisms and signal transduction processes involved in platelet activation and inhibition.

Asrar B. Malik, Distinguished Professor and Head; Ph.D., Toronto, 1971. Gene regulation in endothelial cells; expression of adhesion molecules and their transcriptional regulation; regulation of endothelial transport; assembly of adherens junctions; function of caveolin-1 in endothelial cells.

Dolly Mehta, Assistant Professor; Ph.D., Delhi (India), 1989. Rho regulation of actin polymerization.

Richard Minshall, Assistant Professor; Ph.D., Illinois at Chicago, 1996. Signaling molecules that regulate caveolae-mediated endocytosis and transcytosis; delivery of therapeutic agents across the vascular barrier; role of caveolae in inflammation-induced increase in permeability.

Shigehiro Nakajima, Professor; M.D., 1955, Ph.D., 1961, Tokyo. Pharmacological and electrophysiological investigations on various receptors and channels in brain neurons, with emphasis on signal transduction mechanisms.

John P. O'Bryan, Assistant Professor; Ph.D., North Carolina at Chapel Hill, 1992. Role of scaffolding proteins in the regulation of signal transduction pathways, protein ubiquitylation, trafficking, and endocytosis.

Miodrag Radulovacki, Professor; M.D., 1959, Ph.D., 1968, Belgrade. Pharmacology of sleep and sleep apnea.

Randal A. Skidgel, Professor and Graduate Studies Director; Ph.D., California, San Diego, 1979. Role of human carboxypeptidases in activation or inactivation of peptide hormones and regulation of nitric oxide production.

Karen Snapp, Assistant Professor; Ph.D., Iowa, 1992. Leukocyte adhesion and migration; interactions mediated by the selectin family of adhesion molecules.

Chinnaswamy Tiruppathi, Associate Professor; Ph.D., Madras (India), 1977. Endothelial cell activation by thrombin: Calcium entry, transcriptional regulation of inflammatory genes, and endothelial cell injury and repair.

Tatyana Voyno-Yasenetskaya, Associate Professor; M.D./Ph.D., Institute of Pathophysiology (Moscow), 1987. G-protein regulatory mechanisms and control of cellular proliferation and cytoskeleton.

Richard D. Ye, Professor; M.D./Ph.D., Washington (St. Louis), 1988. Signaling through G-protein-coupled receptors; molecular mechanisms of leukocyte activation.

Research/Supporting Faculty

Kurt Bachmaier, Research Assistant Professor; M.D./Ph.D., Innsbruck, 1994. Molecular mechanisms in the induction of organ-specific autoimmune disease.

Viktor Brovkovych, Research Assistant Professor; Ph.D., Ukranian Academy of Science, 1988. Regulation of nitric oxide production in endothelial cells.

Peter A. Deddish, Research Assistant Professor; Ph.D., Northwestern, 1971. Mechanisms of processing of peptide hormones by converting enzyme (ACE), neutral endopeptidase 24.11 (NEP), kallikrein, and carboxypeptidase M.

Randall S. Frey, Research Assistant Professor; Ph.D., Minnesota. Signaling pathways; activation and production of oxidases and upstream signaling pathways mediating endothelial adhesiveness.

Guoquan Liu, Research Assistant Professor; Ph.D., Ohio, 1997. Signaling of G proteins and their coupled receptors.

Aleksander V. Lyubimov, Research Assistant Professor, M.D./Ph.D., Institute for Continuing Education of Physicians, St. Petersburg (Russia), 1989. Toxic effects of pharmaceuticals; cancer prevention and therapy.

Dan Predescu, Research Assistant Professor; M.D./Ph.D., Bucharest, 1988. Regulation of endothelial barrier function by transcytosis.

Sanda Predescu, Research Assistant Professor, Ph.D., California, San Diego, 1999. Molecular mechanisms of transendothelial exchange between plasma and interstitial fluid.

Jasna Saponjic, Research Assistant Professor, M.D./Ph.D., Belgrade, 1996. Central regulation of sleep breathing disorders.

Shahid Siddiqui, Research Associate Professor; Ph.D., Bombay, 1978. Functional and structural diversity of kinesin motors.

Fulong Tan, Research Assistant Professor; Ph.D., Shanghai Institute of Biochemistry, 1982. Structure-function relationship of peptidases.

Stephen M. Vogel, Research Assistant Professor; Ph.D., Virginia, 1980. Macromolecules in the regulation of lung microvascular permeability.

Jiaping Xue, Research Assistant Professor, Ph.D., Swedish University of Agricultural Sciences (Uppsala), 1994. Cell cycles.

Youyang Zhao, Research Assistant Professor, Ph.D., Shanghai Institute, 1994. Caveolin-1 regulation of G proteins in controlling endothelial cell function.

UNIVERSITY OF KENTUCKY

Graduate Center for Toxicology

Programs of Study

The Graduate Center for Toxicology (GCT) at the University of Kentucky (UK) prepares scientists for careers in molecular toxicology, carcinogenesis, immunotoxicology, neurotoxicology, and environmental toxicology. The program is interdisciplinary, providing a teaching and administrative nucleus of about 55 faculty members with research programs in toxicology. They are based in the Departments of Biochemistry, Biological Sciences, Chemistry, Medicine, Microbiology and Immunology, Neurology, Nutrition, Pharmacology, and Physiology and the Colleges of Agriculture, Medicine, and Pharmacy. Each student is guided through the program by a 4-member advisory committee, which is responsible for the student's comprehensive examination and thesis defense. Postdoctoral trainees are guided by a major adviser and may take courses and attend seminars and conferences to expand their expertise in toxicology. While the program has major strengths in the areas outlined above, other disciplines represented include analytical toxicology; behavioral toxicology; biochemical, clinical, genetic, and nutritional toxicology; environmental medicine and chemistry; and xenobiotic metabolism and equine toxicology. Noteworthy is the scientific depth available in many areas, with training at the basic, applied, and clinical levels available.

Research Facilities

The administrative offices of the GCT are in the Health Sciences Research Building, which also contains the laboratory suites of the core faculty members and is connected to the Medical Sciences wing of the College of Medicine and the College of Medicine Library. The GCT suite is within a short walking distance of the Colleges of Pharmacy and Agriculture, the Markey Cancer Center, and the Magnetic Resonance Imaging Spectroscopy Center. Students have access to a complete repertoire of analytical and synthetic chemistry techniques and equipment; environmental rooms; and hybridoma, cell-sorting, and transgenic mouse facilities.

Financial Aid

Financial aid is available on a competitive basis and includes graduate fellowships, research and teaching assistantships, minority fellowships, and stipends of $20,000 or more per year. Other support includes NIEHS predoctoral and fellowships offered through a training grant for outstanding students.

Cost of Study

Costs for Kentucky residents for 2005–06 were $4120 per semester for basic tuition plus mandatory fees and $299 per semester hour for the four-week summer session. Costs for nonresident students were $9654 per semester for basic tuition plus mandatory fees and $712 per semester hour for the four-week summer session. Out-of-state students holding fellowships or assistantships usually pay only in-state tuition and fees.

Living and Housing Costs

In 2005–06, on-campus student housing, including adequate basic furnishings, utilities, and maintenance services, cost $467 per month for an efficiency apartment, $578 for a one-bedroom unit, and $628 for a two-bedroom unit. A wide selection of furnished and unfurnished apartments is available off campus, and both city and University buses serve the campus area.

Student Group

The GCT enrollment averages 40 graduate students. The Toxicology Student Forum sponsors a fall mixer for new students, an annual summer picnic, and social events to recognize students completing their qualifying exams. Its president represents students at faculty meetings. The University's total enrollment in 2003 was approximately 34,200, including about 6,500 graduate students. The GCT encourages applications from members of minority groups, to whom financial support and fellowships are available.

Student Outcomes

Recent graduates have obtained excellent postdoctoral fellowships at major medical centers; well-compensated employment in industry and government as toxicologists, consultants in risk assessment, and scientists in contract laboratories; and positions in academia. The toxicology program was established in 1977 and has graduated more than 100 Ph.D. students.

Location

Lexington is located in the heart of the famous Bluegrass region of central Kentucky, about 85 miles from both Louisville, Kentucky, and Cincinnati, Ohio. Lexington is a progressive community with a population of approximately 265,000. The downtown area is within walking distance of the University campus.

The University

UK was established in 1865 as a land-grant institution. Close relationships are maintained with the neighboring universities and community colleges. UK employs 10,500 faculty and staff members. Opportunities for graduate study are enhanced by the presence of the University Medical Center, the Lexington VA Hospital, and various research institutes, such as the Center on Aging, situated on the main University campus.

Applying

Students interested in applying for admission to the program should have previous training in biology, biomedical sciences, chemistry, pharmacy, toxicology, or related fields. Acceptance into the graduate program in toxicology is contingent upon admission to the Graduate School of the University. Applicants must be graduates of an accredited college or university. A minimum grade point average of 3.0 on a 4.0 scale and a combined score of at least 1100 on the verbal and quantitative portions of the GRE General Test are normal requirements for admission; applicants with lesser qualifications may be admitted pending evaluation of letters of recommendation.

Correspondence and Information

Graduate Center for Toxicology
306 Health Sciences Research Building
University of Kentucky
Lexington, Kentucky 40536-0305
Phone: 859-257-3760
Fax: 859-323-1059
E-mail: gctinfo@pop.uky.edu
Web site: http://www.uky.edu

University of Kentucky

AFFILIATED FACULTY AND THEIR RESEARCH

Professors

Mary Vore, Director, Graduate Center for Toxicology; Ph.D., Vanderbilt. Biochemical toxicology.

Wesley J. Birge, Ph.D., Oregon State. Environmental toxicology.
Subbarao Bondada, Ph.D., Bombay (India). Molecular biology, immunotoxicology.
Lisa Cassis, Ph.D., West Virginia. Cardiovascular pharmacology.
Ching K. Chow, Ph.D., Illinois. Nutritional toxicology.
Donald A. Cohen, Ph.D., Cincinnati. Immunotoxicology.
Peter A. Crooks, Ph.D., Manchester (England). Medicinal chemistry.
Alan Daugherty, Ph.D., Bath (England). Cardiovascular toxicity.
Linda Dwoskin, Ph.D., Minnesota. Neurotoxicology.
James W. Flesher, Ph.D., Loyola of Chicago. Chemical carcinogenesis.
Gary Gairola, Ph.D., Illinois. Genetic and inhalation toxicology.
Howard Glauert, Ph.D., Michigan State. Nutritional toxicology and oncology.
Bernard Hennig, Ph.D., Iowa State. Nutritional biochemistry.
John Hunsaker, Ph.D., Kentucky. Forensic toxicology.
Darrell Jennings, M.D., Kentucky. Pathology.
Tae Ji, Ph.D., California, San Diego. Biological chemistry.
Davy Jones, Ph.D., California, Davis. Molecular biology.
Alan M. Kaplan, Ph.D., Purdue. Immunotoxicology.
Edward J. Kasarskis Jr., Ph.D., M.D., Wisconsin. Neurotoxicology.
Natasha Kyprianou, M.D., Leeds. Urology.
Eun Lee, Ph.D., Kyung Hee (Korea). Pathology.
Patrick McNamara, Ph.D., SUNY at Buffalo. Pharmacokinetics.
Peter R. Oeltgen, Ph.D., Loyola of Chicago. Clinical toxicology.
Vivek Rangnekar, Ph.D., Bombay (India). Apoptosis and tumor growth control.
Daret St. Clair, Ph.D., Iowa. Molecular biology and cancer.
Kevin Sarge, Ph.D., North Carolina State. Gene bookmarking.
Steven I. Shedlofsky, M.D., Michigan. Hepatotoxicology.
John Slevin, M.D., West Virginia. Neurotoxicology.
Brett Spear, Ph.D., Pennsylvania. Neurotoxicity and oxidative stress.
Thomas Tobin, D.V.M., Dublin; Ph.D., Toronto; DABT. Applied immunotoxicology.
Zhigang Wang, Ph.D., Texas. DNA repair.
Robert A. Yokel, Ph.D., Minnesota. Behavioral pharmacology; neurotoxicology.
Stephen Zimmer, Ph.D., Colorado. Molecular biology and cancer.

Associate Professors

J. Scott Bryson, Ph.D., Miami (Ohio). Murine syngeneic graft-versus-host disease.
Guo-Min Li, Ph.D., Wayne State. Mismatch repair and cancer.
Bert Lynn, Ph.D., Mississippi State. Mass spectrometry.
William Maragos, Ph.D., Michigan; M.D., Northwestern. Neurotoxicity.
Joseph McGillis, Ph.D., George Washington. Neuroimmunology; neuropeptides; neurotoxicology.
Isabel Mellon, Ph.D., Illinois. DNA repair.
Jeffrey Moscow, M.D., Dartmouth. Molecular biology of anticancer drug transport.
Dan Noonan, Ph.D., Texas. Biochemical toxicology.
Subba Palli, Ph.D., Western Ontario. Functional genomics and proteomics.
Brent Palmer, Ph.D., Florida. Endocrine disruption of lower vertebrates.
Ok-Kyong Park-Sarge, Ph.D., Illinois at Urbana-Champaign. Molecular mechanisms of steroid actions in reproductive tract.
Creed Pettigrew, M.D., Texas. Neurotoxicology.
Todd Porter, Ph.D., Illinois. Biochemical toxicology.
Peter Spielmann, Ph.D., Berkeley. Structural biology.
Hollie Swanson, Ph.D., Purdue. Biochemical toxicology.
Peter Wedlund, Ph.D., Washington (Seattle). Drug metabolism and toxicokinetics.

Assistant Professors

John D'Orazio, M.D., Miami (Florida). Hematology/oncology.
Liya Gu, Ph.D., Wayne State. Mismatch repair and cancer.
David Orren, Ph.D., North Carolina. Genetic instability and DNA repair.
Andrew Pierce, Ph.D., North Carolina at Chapel Hill. DNA repair.
Hsin-Sheng Yang, Ph.D., Arizona State. Gene expression in suppressing tumorigenesis.
Hong-Bo Zhao, M.D., Ph.D., Yichang Medical (China). Inner ear gap junctions.
Haining Zhu, Ph.D., UCLA. Proteomics of metal-ion homeostasis and neurodegenerative disease.

UNIVERSITY OF MIAMI

School of Medicine
Department of Molecular and Cellular Pharmacology

Program of Study

Molecular and cellular pharmacology is a hybrid discipline that makes use of the knowledge and techniques of physics, chemistry, and biology to study the action of drugs on living systems and, more generally, the mechanisms through which signals are recognized and transduced by cells. The goal of pharmacology is twofold: (1) to develop, study, and understand agents that may be beneficial in the treatment of disease and (2) to use drugs as experimental tools in the study of basic biological signaling processes. Because of this, the Ph.D. in pharmacology can lead to career opportunities in academic or government institutions as well as the pharmaceutical and other industries.

Students in their first year receive a good grounding in basic scientific disciplines related to pharmacology, including physiology and cell and molecular biology. They also develop short research projects in the laboratory. During their second year, students begin work on their thesis research and take a series of advanced courses in fundamental and medical aspects of pharmacology. The Department sponsors a visiting scientist program in which visitors meet graduate students and give a seminar on their research activities. In the remaining years of graduate study, most of the students' activities are centered on the thesis research project.

The Department's highly active research program has major areas of concentration in cardiovascular pharmacology/muscle contraction (e.g., structure/function relationships in the troponin complex, the role of ion channels in ventricular hypertrophy, excitation-contraction coupling in skeletal and cardiac muscle) and in neuropharmacology/neuroscience (e.g., neuronal signaling mechanisms, axon growth and synapse formation, etiology and treatment of Alzheimer's and Parkinson's disease, excitation-secretion coupling). A wide variety of technical approaches is used, including molecular genetics, protein biochemistry, immunology, patch-clamp electrophysiology, cell culture, digital video imaging, DNA microarrays, and whole-animal studies. The faculty is a mixture of senior scientists who are recognized leaders in their respective fields and more junior faculty members who have been recently trained in state-of-the-art research techniques applied to important biomedical problems. The Department's more than 40 graduate students and postdoctoral fellows contribute to the creative and stimulating scientific atmosphere.

A combined M.D./Ph.D. program is offered by the Department in conjunction with the School of Medicine; students take special courses in basic medical sciences and carry out a research dissertation in molecular and cellular pharmacology. In addition, students receive full clinical training. The pharmacology department also participates in interdepartmental/interdisciplinary programs in cellular and molecular biology, the neurosciences, and membrane biology.

Research Facilities

The Department's research laboratories, offices, and core facilities are located in 30,000 square feet of the Basic Science Building. Extensive interactions exist with the other basic science departments and with clinical departments, particularly cardiology and neurology. A variety of medical school core facilities is located in the building or nearby. The medical library has, in addition to its extensive collection of periodicals and monographs, excellent computer facilities for access to Medline, Current Contents, and more than 1,000 full-text online journals.

Financial Aid

The Department offers predoctoral stipends as well as full tuition remission to candidates who are accepted into the training program. For 2004–05, stipends were $20,772.

Cost of Study

Tuition is ordinarily waived for predoctoral students.

Living and Housing Costs

The cost of living in Miami, including housing, food, entertainment, and transportation, is about average for a major U.S. city.

Student Group

The graduate student population in the School of Medicine is approximately 350; about 5,080 graduate students are enrolled in the University.

Student Outcomes

A Ph.D. in pharmacology can open the door to a wide variety of rewarding careers. About 90 percent of the doctoral students graduating in the past ten years are in academic research and teaching positions at institutions such as Harvard, the University of Pennsylvania, the University of Florida, and the University of Texas–Southwestern. Recent graduates have also obtained jobs such as director of pharmacology at a pharmaceutical corporation and forensic scientist in a medical examiner's office.

Location

The Miami area has an international flavor and offers pleasant living conditions in a subtropical environment with numerous cultural and recreational activities. Year-round saltwater and freshwater activities are a particular feature of the area. The Everglades National Park and the Florida Keys are each an hour's drive away.

The University and The School

The University of Miami is a private, independent university composed of fourteen colleges and schools. The School of Medicine, which is located near downtown Miami, has developed into one of the country's premier research and medical institutions. The Department of Molecular and Cellular Pharmacology is housed together with the other basic science departments in the Rosenstiel Medical Science Building. The medical center complex includes Jackson Memorial Hospital and several other major hospital and research facilities.

Applying

Applications are encouraged from qualified students with a basic background in chemistry, biology, or physics. The selection of graduate students for the Ph.D. program is based upon performance and grade point average in undergraduate school, with emphasis on science subjects (an undergraduate grade level of at least B is required in order to achieve full graduate status); GRE scores; and three letters of recommendation from persons who are familiar with the scientific background of the student. Members of minority groups are especially encouraged to apply. The Department and the University of Miami are committed to increasing representation of women and members of minority groups in the biological-medical sciences. All inquiries regarding the program and requests for applications or for more information concerning the program should be sent to the contact address of the Department.

Correspondence and Information

Dr. Kerry L. Burnstein
Director of Graduate Studies
Department of Molecular and Cellular Pharmacology
University of Miami School of Medicine
P.O. Box 016189 (R189)
Miami, Florida 33101

Phone: 305-243-3419
Fax: 305-243-3420
E-mail: mcp@med.miami.edu
Web site: http://chroma.med.miami.edu/pharm/

University of Miami

THE FACULTY AND THEIR RESEARCH

James D. Potter, Ph.D., Professor and Chairman; FAHA. Molecular biological approaches to the study of the Ca^{2+} regulation of cardiac and skeletal muscle contraction by troponin and Ca^{2+}; contractile protein defects in familial hypertrophic cardiomyopathy (FHC), dilated cardiomyopathy (DCM), and a variety of skeletal muscle genetic diseases.

FHC mutations from different functional regions of troponin T result in different effects on the pH- and Ca^{2+}-sensitivity of cardiac muscle contraction. *J. Biol. Chem.* 279(15):14488–95, 2004.

Different functional properties of two troponin T mutants that cause dilated cardiomyopathy. *J. Biol. Chem.* 278(43):41670–6, 2003.

FHC-linked mutant troponin T causes stress-induced ventricular tachycardia and Ca^{2+}-dependent action potential remodeling. *Circ. Res.* 92:428–36, 2003.

Arthur L. Bassett, Ph.D., Professor. Pharmacology of ion channels in myocardial hypertrophy and the aged heart.

Cytoskeletal actin microfilaments and the transient outward potassium current in hypertrophied rat ventriculocytes. *J. Physiol.* 541(2):411–21, 2002.

Nanette Bishopric, M.D., Professor; FACC, FAHA. Genetic regulation of cardiac growth, differentiation, and apoptosis; cellular mechanisms of heart failure.

Recurrent third-trimester fetal loss and maternal mosaicism for long-QT syndrome. *Circulation* 109(24):3029–34, 2004.

Control of cardiac-specific transcription by p300/CBP through myocyte enhancer factor-2D. *J. Biol. Chem.* 276:7575–85, 2001.

John L. Bixby, Ph.D., Professor. Molecular and cellular biology of neuronal differentiation; axon guidance and regeneration.

Expression of PTPRO during mouse development suggests involvement in axonogenesis and in differentiation of NT-3 and NGF-dependent neurons. *J. Comp. Neurol.* 456:384–95, 2003.

CRYP-2/PTPRO is a neurite-inhibitory repulsive guidance cue for retinal neurons. *J. Cell Biol.* 154:867–78, 2001.

Richard J. Bookman, Ph.D., Associate Professor. Signal transduction; regulation of motility; regulation of gene expression; voltage-dependent ion channels; digital video microscopy; patch clamping; DNA microarrays.

Toxicogenomic effects of marine brevetoxins in liver and brain of mouse. *Comp. Biochem. Physiol. B: Biochem. Mol. Biol.* 136(2):173–82, 2003.

Prolonged increase in ciliary beat frequency after short-term purinergic stimulation in human airway epithelial cells. *J. Physiol.* 538(2):633–46, 2002.

Kerry L. Burnstein, Ph.D., Professor. Steroid hormone regulation of steroid receptor gene expression and prostate cancer cell proliferation.

Deregulation of the Rho GTPase, Rac1, suppresses cyclin-dependent kinase inhibitor p21(CIP1) levels in androgen-independent human prostate cancer cells. *Oncogene* 23:5513–22, 2004.

Vitamin D inhibits G1 to S progression in LNCaP prostate cancer cells through p27Kip1 stabilization and Cdk2 mislocalization to the cytoplasm. *J. Biol. Chem.* 278:46862–8, 2003.

Anthony H. Caswell, Ph.D., Professor. Proteins involved in excitation-contraction coupling in skeletal and cardiac muscle.

Membrane topography of cardiac triadin. *Arch. Biochem. Biophys.* 398:61–72, 2002.

Beatriz M. Fontoura, Ph.D., Assistant Professor. Molecular mechanisms of nuclear transport and regulation by different signaling pathways.

Sec13 shuttles between the nucleus and the cytoplasm and stably interacts with Nup96 at the nuclear pore complex. *Mol. Cell. Biol.* 23:7271–84, 2003.

Role of nucleoporin induction in releasing an mRNA nuclear export block. *Science* 295:1523–5, 2002.

The nucleoporin Nup98 associates with the intranuclear filamentous protein network of Tpr. *Proc. Natl. Acad. Sci. U.S.A.* 98:3208–13, 2001.

Sandra K. Lemmon, Ph.D., Associate Professor. Membrane traffic: roles of clathrin, actin, and phosphorylation.

The actin regulating kinase Prk1p negatively regulates Scd5p, a suppressor of clathrin deficiency, in actin organization and endocytosis. *Curr. Biol.* 13:1557–62, 2003.

Protein phosphatase-1 binding to Scd5p is important for regulation of actin organization and endocytosis in yeast. *J. Biol. Chem.* 277:48002–8, 2002.

Irene Litosch, Ph.D., Associate Professor. Regulation of phospholipase C-β G proteins, protein kinase C, and lipid messengers.

Regulation of phospholipase C-β activity by phosphatidic acid: Isoform dependence, role of protein kinase C and G protein subunits. *Biochemistry* 42:1618–23, 2003.

Phosphatidic acid modulates G protein regulation of phospholipase C-β activity in membranes. *Cell. Signal.* 14:259–63, 2002.

Regulation of phospholipase C-β activity by phosphatidic acid. *Biochemistry* 39:7736–43, 2000.

Charles W. Luetje, Ph.D., Professor and Vice Chairman. Structure, function, and pharmacology of neuronal nicotinic acetylcholine receptors.

Determinants of potency on a-conotoxin-MII, a peptide antagonist of neuronal nicotinic receptors. *Biochemistry* 43:2732–7, 2004.

Identification of residues that confer a-conotoxin PnIA sensitivity on the a3 subunit of neuronal nicotinic acetylcholine receptors. *J. Pharmacol. Exp. Ther.* 306:664–70, 2003.

Subunit dependent modulation of neuronal nicotinic receptors by zinc. *J. Neurosci.* 21:1848–56, 2001.

Lincoln T. Potter, M.D., Professor. Molecular pharmacology, biology, anatomy, and physiology of muscarinic acetylcholine receptors in relation to Alzheimer's and Parkinson's diseases.

Studies of muscarinic neurotransmission with antimuscarinic toxins. *Prog. Brain Res.* 145:121–8, 2003.

Snake toxins that bind specifically to individual subtypes of muscarinic receptors. *Life Sci.* 68:2541–7, 2001.

Vladlen Z. Slepak, Ph.D., Associate Professor. Regulation of G protein–mediated signaling in photoreceptors and neurons.

Direct binding of visual arrestin to microtubules determines the differential subcellular localization of its splice variants in rod photoreceptors. *J. Biol. Chem.*, in press.

G beta 5.RGS7 inhibits G alpha q-mediated signaling via a direct protein-protein interaction. *J. Biol. Chem.* 278(23):21307–13, 2003

Keith A. Webster, Ph.D., Professor. Redox regulation of gene expression and protein kinase signaling pathways in the heart and vascular endothelium.

Mitochondria signals initiate the activation of c-Jun N-terminal kinase by hypoxia-reoxygenation. *FASEB J.* 18:1060–70, 2004.

Therapeutic angiogenesis: A complex problem requiring a sophisticated approach. *Cardiovasc. Toxicol.* 3:283–98, 2003.

Joint Faculty

Izidore Lossos, M.D., Associate Professor (Medicine/Hematology). Pathogenesis; biology and immunology of lymphoma; gene expression profiling of lymphoma; molecular prognostic factors of lymphoma; function of recently cloned gene-HGAL; BCL6 in lymphoma; tumor immunology.

Matthias Salathe, M.D., Associate Professor (Medicine). Pulmonary medicine; asthma; COPD.

Rakesh Singal, M.D., Associate Professor (Medicine). Prostate cancer; epigenetics; transcriptional regulation; DNA methylation; chromatin structure; biomarkers.

UNIVERSITY OF MISSISSIPPI MEDICAL CENTER

Department of Pharmacology and Toxicology

Program of Study

The program in the Department of Pharmacology and Toxicology is designed primarily for the training of Ph.D. students who wish to make a professional career of independent research work, although on some occasions a master's degree is granted. Specialization in either pharmacology or toxicology is possible. Most of the first year in residence and part of the second are devoted to general course work, with some courses taken with medical and/or dental students. The preliminary examination is usually completed toward the end of the second year. All students are encouraged to become involved in research work as early as possible in their course of study. By the third year, most students have undertaken a thesis project and are well on their way toward completion of the Ph.D. Most students complete the Ph.D. within four to five years. Areas of research represented within the department are cardiovascular pharmacology, toxicology (all areas), membrane transport, autonomic pharmacology, peripheral and central neuropharmacology (especially drugs that are abused), electroneuropharmacology, and hepatic function. Students may select the particular area that appeals to them.

Research Facilities

The University of Mississippi Medical Center is a progressive medical center. Accordingly, a complete research-medical library, AALAC-accredited centralized animal facilities, and a large number and variety of academic programs exist on campus. Within the department are electrophysiological laboratories, a confocal microscope capable of measuring real-time fluorescence, a mass spectrometer facility, and extensive chemical and biochemical instrumentation; there are complete cell-culture facilities for routine biological, biochemical, and toxicological studies.

Financial Aid

Research assistantships are available for graduate students. These come from a variety of sources, including federal and state programs, and are highly competitive. In addition, whenever possible, students are supported by individual research grants.

Cost of Study

Full-time students pay a maximum of $1077 per quarter or $120 per quarter hour, and nonresidents pay $256 per quarter hour; the nonresident fee may be deferred by the state of Mississippi.

Living and Housing Costs

Apartments are available on the Medical Center campus. Demand for these apartments is high, and there is a waiting list. The cost ranges from $325 to $405 per month, including some utilities. A women's dormitory is also located on campus. Other apartments are available in the general area of the Medical Center and are competitively priced for this area of the country.

Student Group

The Department of Pharmacology and Toxicology has an average of 25 students in the program. The total graduate student population of the medical school is 217, and the average age of students is 26 to 27 years. All graduate students have completed an undergraduate degree. Students in the program represent a variety of academic backgrounds, although pharmacy, biology, and chemistry predominate. Students with training in engineering, mathematics, and physics are also encouraged to apply. There is a very diverse population composed of married, single, international, out-of-state, and in-state students.

Location

The University of Mississippi Medical Center is located in Jackson, Mississippi, the capital of the state and its largest city. The total urban population is about 428,700. Jackson has four other institutions of higher learning and offers a variety of recreational activities. Outdoor sports such as tennis, golf, soccer, boating, baseball, and swimming are particularly popular. A professional theater, an opera company, a ballet company, a symphony orchestra, and many other cultural activities are also available.

The Center

The University of Mississippi Medical Center is mainly a graduate campus. The atmosphere that prevails on the Jackson campus is one that promotes biomedical research and the training of research and health professionals. The student body of 1,673 on the campus includes medical, dental, nursing, and health-related professions students as well as Ph.D. and master's degree candidates. All of the students entering the graduate programs do so with an understanding that they are entering programs designed to train research personnel. Although some teaching opportunities for students are available, these are usually minimal. In addition to facilities for the basic science programs, the Medical Center has three hospitals on campus with 705 licensed beds.

Applying

Application for admission to the graduate program may be made at any time, although applications are usually acted upon in the spring for fall admission. Applicants with at least a B average in a recognized scientific discipline and GRE General Test verbal and quantitative scores totaling at least 1000 generally have more success in the program. An interview may be arranged after examination of the applicant's credentials. The scientific disciplines most commonly represented by successful applicants are biology, chemistry, pharmacy, engineering, physics, mathematics, and psychology.

Correspondence and Information

Jerry M. Farley Sr., Program Director
Department of Pharmacology and Toxicology
University of Mississippi Medical Center
2500 North State Street
Jackson, Mississippi 39216

Phone: 601-984-1630
Fax: 601-984-1637
E-mail: info@pharmacology.umsmed.edu
Web site: http://pharmacology.umc.edu

University of Mississippi Medical Center

THE FACULTY AND THEIR RESEARCH

Mark Austin, Associate Professor of Psychiatry and Human Behavior and Assistant Professor of Pharmacology and Toxicology; Ph.D., Washington State, 1988. Regulation of the locus coeruleus in stress and suicide; effect of cholinergic stimulation in the nucleus accumbens on locomotor behavior.

Rodney C. Baker, Professor; Ph.D., North Carolina State, 1974. Ether lipid metabolism; ethanol suppression of immune response; biotransformation of platelet activating factor.

Garth Bissette, Professor of Pharmacology and Toxicology and of Psychiatry and Human Behavior and Chief, Laboratory of Psychoneuroendocrinology; Ph.D., North Carolina, 1982. Investigation of alterations in synaptic availability of neuropeptide neurotransmitters in patients with psychosis, depression, and dementia, with the goal of identifying neuronal circuits that may be more effective therapeutic targets in these disorders.

Zhengwei Cai, Assistant Professor of Pharmacology and Toxicology and Associate Professor of Pediatrics; Ph.D., Oregon State, 1989. Developmental pharmacology and toxicology; brain injury induced by perinatal hypoxia-ischemia and maternal infection.

David B. Couch, Associate Professor; Ph.D., Virginia Commonwealth, 1975. Heritable alterations induced in somatic cells in vivo; genotoxicity of environmental agents and drugs of abuse.

Robert Cox, Associate Professor of Emergency Medicine and of Pharmacology and Toxicology; Ph.D. (chemistry), Iowa, 1980; M.D., Texas Southwestern Medical Center at Dallas, 1987. Organic solvents; chemical exposure; environmental toxins.

Roy J. Duhé, Associate Professor; Ph.D., Wisconsin, 1989. Cytokine-mediated signal transduction; redox regulation in the immune system; mechanisms regulating the catalytic activity of signal transducing enzymes such as Janus kinases (JAKs); the involvement of JAK dysregulation in oncogenesis and immunological disorders.

Jerry M. Farley, Professor and Vice Chairman; Ph.D., West Virginia, 1976. Electrophysiology, neuropharmacology; effects of xenobiotics on smooth muscle and epithelial function; drug abuse.

Elise P. Gomez-Sanchez, Professor of Medicine and Associate Professor of Pharmacology and Toxicology; D.V.M./Ph.D., Texas A&M, 1982. Hypertension; adrenal steroid synthesis and action; mineralocorticoid and glucocorticoid receptor function.

Ing K. Ho, Professor and Chairman; Ph.D., California, San Francisco, 1968. Biochemical mechanisms of neuroactive agents; neurotoxicity of environmental intoxicants.

Arthur Hume, Professor Emeritus; Ph.D., Mississippi Medical Center, 1964. Many pharmacological and toxicological effects are manifested in organ systems and are frequently related to such parameters as the distribution and metabolism of the foreign chemical compounds. Studies are under way to examine the whole-body distribution, excretion, and other actions of a number of toxins and drugs. The mechanisms of antidotes for sulfide and cyanide poisoning are also being investigated. Development of analytical toxicological methods is also of interest.

John C. Kermode, Associate Professor; Ph.D., London, 1983. Cell-surface receptors and receptor-induced signal transduction; platelets; G-protein–coupled receptors; nonreceptor protein tyrosine kinases; receptor-mediated changes in intracellular calcium.

Robert E. Kramer, Professor; Ph.D., West Virginia, 1976. Hormonal regulation of adrenal function; mechanisms of action of angiotensin, K$^+$, and ACTH; role of calcium and eicosanoids in the control of adrenal steroid secretion.

Jeffrey A. Love, Associate Professor; Ph.D., Georgetown, 1983. Neural control of pancreatic endocrine and exocrine secretion involving both intracellular recordings of synaptic transmission in pancreatic neurons and patch clamp recordings of the effects of neurotransmitters and hormones on ion channels in insulin-secreting cells.

Tangeng Ma, Assistant Professor; M.D., Hunan (China), 1973; Ph.D., Tongji (China), 1987. Biochemical mechanisms of neuroactive agents; neurotoxicity of environmental intoxicants, especially organophosphorus compounds and heavy metals.

Ian A. Paul, Assistant Professor of Pharmacology and Toxicology and Associate Professor of Psychiatry and Human Behavior; Ph.D., North Carolina at Chapel Hill, 1989. Role of excitatory amino acid neurotransmission in the pathophysiology of major depressive disorders and antidepressant therapeutics; neurobiological events mediating retrovirus-induced alteration in spontaneous and learned behaviors; neurovirological models of neurobehavioral pathology.

Soundar Regunathan, Professor of Psychiatry and Human Behavior and Associate Professor of Pharmacology and Toxicology; Ph.D., Madras (India), 1986. Agmatine and arginine decarboxylase in neuronal injury and agmatine synthesis and release during morphine exposure.

Rob Rockhold, Professor; Ph.D., Tennessee Health Sciences, 1978. Mechanism and site of action of centrally acting antihypertensive drugs; central nervous system involvement in circulatory and baroreflex homeostasis; excitatory amino acid neurotransmission in dependence upon withdrawal from and intoxication by psychomotor stimulant and opioid drugs of abuse.

James P. Shaffrey, Assistant Professor of Psychiatry and Human Behavior and of Pharmacology and Toxicology; Ph.D., Oxford, 1984. Current theory and understanding of how REM sleep plays an important biological role in the developing nervous system.

Stanley V. Smith, Assistant Professor; Ph.D., Mississippi Medical Center, 1995. Structural and conformational dynamics of proteins; ligand-binding kinetics of heme proteins; drug metabolism.

Craig Stockmeier, Professor and Director, Division of Neurobiology and Behavioral Research, Department of Psychiatry, and Associate Professor of Pharmacology and Toxicology; Ph.D., Arizona, 1983. Adaptive changes in the anatomy, pharmacology, and molecular biology of the human hippocampus in major depression.

Susan E. Wellman, Associate Professor; Ph.D., Florida State, 1986. Conformation of proteins and peptides; interactions between proteins and nucleic acids; chromatin structure.

William L. Woolverton, Professor of Pharmacology and Toxicology and of Psychiatry and Human Behavior; Ph.D., Chicago, 1977. Behavioral pharmacology of drugs of abuse, primarily intravenous self-administration and drugs as discriminative stimuli.

THE UNIVERSITY OF NORTH CAROLINA AT CHAPEL HILL

Curriculum in Toxicology

Program of Study

The Curriculum in Toxicology administers the program leading to the Ph.D. degree in toxicology. Postdoctoral training is also available to holders of appropriate doctoral degrees, such as the Ph.D. or the M.D. The Curriculum is an interdisciplinary program that attracts faculty members on the basis of research and teaching interests in toxicology. The faculty is drawn from various administrative units of the Schools of Medicine, Pharmacy, and Public Health. The faculty members in the Curriculum also include scientists who hold both adjunct faculty appointments in one of the University's administrative units and primary appointments with organizations at Research Triangle Park, including the Environmental Protection Agency and National Institute of Environmental Health Sciences.

The research interests of the faculty include most areas of toxicology, but particular emphasis is directed toward molecular and cellular toxicology, carcinogenesis, neurotoxicology, developmental toxicology, and pulmonary toxicology. The faculty generally does not conduct research in aquatic toxicology, forensic toxicology, the ecological aspects of toxicology, or invertebrate systems.

The Research Triangle area is one of the national centers for research in toxicology. The proximity of the Research Triangle Park to the University has permitted the Curriculum to utilize the expertise of various toxicologists as a part of its research and teaching activities. The participation by Research Triangle Park scientists in the training activities of the Curriculum includes direction of student research for the Ph.D. degree, membership on the research advisory and examination committees for specific students, and lecturing in their research specialties in the Curriculum's graduate courses.

Research Facilities

The research activities of the Curriculum in Toxicology are conducted in the laboratory facilities assigned to individual faculty members by the participating administrative units.

Financial Aid

Doctoral traineeships, fellowships, and research assistantships are available on a competitive basis to support students working toward the Ph.D. degree. All eligible applicants are considered for financial aid awards. Postdoctoral traineeships are also available. Funding from training grants is limited by the federal government to U.S. citizens and permanent residents. No funding is available for international students for the first two years.

Cost of Study

For the 2005–06 year, the tuition and fees were $2504 per semester for North Carolina residents and $9503 per semester for nonresidents. These charges, however, are subject to change. This program normally pays all tuition, fees, and health insurance.

Living and Housing Costs

Residences for single and married students are maintained by the University. Furnished and unfurnished apartments and houses are also available in town. An excellent, free bus service is available.

Student Group

In 2005–06, the Curriculum in Toxicology had 37 graduate students and 9 postdoctoral associates. The total enrollment of the University of North Carolina at Chapel Hill was approximately 27,000 and included 8,000 graduate students.

Student Outcomes

Of 9 recent Ph.D. graduates, 4 are employed as postdoctoral fellows in academic research laboratories, 1 is in another laboratory, and 4 are in industrial employment.

Location

Chapel Hill, with a population of approximately 50,000, is near the center of the state. The stimulating scientific and cultural climate is enhanced by the proximity of the Research Triangle Park (10 miles), Duke University (8 miles), and North Carolina State University (30 miles).

The University

The University of North Carolina was established in Chapel Hill in 1795. It has Schools of Medicine, Dentistry, Public Health, and Pharmacy. Close relationships are maintained with the neighboring universities and several of the research organizations of the nearby Research Triangle Park.

Applying

Applications for doctoral studies are considered from students who have received or expect to receive a B.S., B.A., or M.S. degree in a scientific discipline. A desirable background includes courses in biological sciences (including histology and animal physiology), chemistry (including analytical and organic chemistry), and mathematics through calculus, although all of these are not essential. A strong course in general biochemistry accelerates the student's progress. Applicants are evaluated on the basis of undergraduate (and graduate) academic performance, Graduate Record Examinations scores (on the General Test), and letters of recommendation. Students are accepted on the basis of their achievement and potential. Special circumstances, including prior research experience and publications, are considered in individual cases in the assessment of qualifications for admission. The Curriculum offers a program leading to the M.S. degree only under special circumstances. For maximum consideration for financial aid awards, applications for admission for the fall semester should be completed by January 15 and for the spring semester by September 15. The University of North Carolina at Chapel Hill is committed to the principle of equal opportunity. It is the policy of the University not to discriminate on the basis of race, sex, color, national origin, religion, or handicap with regard to its students, employees, or applicants for admission or employment.

Correspondence and Information

Curriculum in Toxicology
CB 7270
The University of North Carolina at Chapel Hill
Chapel Hill, North Carolina 27599

Phone: 919-962-8400
Fax: 919-966-6123
E-mail: bernard_weissman@med.unc.edu
Web site: http://www.med.unc.edu/toxicology/

The University of North Carolina at Chapel Hill

THE FACULTY AND THEIR RESEARCH

Professors
Louise M. Ball, Ph.D. Metabolism and genotoxicity of environmental xenobiotics.
Thomas W. Bouldin, M.D. Neuropathology; ocular pathology; neurotoxicology.
Kim Rowse Brouwer, Ph.D. Pharmacokinetics; hepatic transport; hepatobiliary disposition; biliary excretion; hepatotoxicity.
Stephen G. Chaney, Ph.D. DNA repair; platinum anticancer drugs.
Marila Cordeiro-Stone, Ph.D. DNA repair and replication in human cells; mechanisms of response to DNA damage.
Fulton T. Crews, Ph.D. Neurodegeneration and chronic drug-induced changes in brain signaling pathways.
Channing Der, Ph.D. *Ras* protein superfamily; signal transduction and oncogenesis.
Avram Gold, Ph.D. Structure-reactivity relationships in metabolism and mutagenicity of polycyclic aromatic hydrocarbons.
Milan J. Hazucha, M.D., Ph.D. Health effects of air pollutants; human studies; mechanisms of response.
David J. Holbrook, Ph.D. Biochemical toxicology; xenobiotic metabolism.
David G. Kaufman, M.D., Ph.D. DNA replication; chemical carcinogenesis.
William K. Kaufmann, Ph.D. DNA metabolism in radiation and chemical carcinogenesis.
Jean M. Lauder, Ph.D. Developmental neurobiology; developmental biology; neurotoxicology.
Terry Magnuson, Ph.D. Mammalian genetics, genomics, and development.
Richard B. Mailman, Ph.D. Parkinson's disease; CNS toxicology.
Patricia F. Maness, Ph.D. Axon guidance and signal transduction in nervous system development.
A. Leslie Morrow, Ph.D. Neurotoxicology and excitotoxicity of alcohol.
David Peden, M.D. Translational and clinical research in environmental lung disease.
Gary M. Pollack, Ph.D. Pharmacokinetics and pharmacodynamics of therapeutic and toxic agents.
Stephen M. Rappaport, Ph.D. Exposure assessment; industrial hygiene.
R. Jude Samulski, Ph.D. Development of efficient viral vectors for gene delivery into eukaryotic genes.
Aziz Sancar, M.D., Ph.D. Molecular neurobiology; DNA repair and cancer; structure and function of DNA repair enzymes; reaction mechanism of human blue-light photoreceptor.
Kathleen K. Sulik, Ph.D. Developmental toxicology; embryology.
James A. Swenberg, D.V.M., Ph.D., Director. Carcinogenesis; DNA and protein adducts; cell proliferation; risk assessment.
Bernard E. Weissman, Ph.D. Chromatin remodeling and epigenetic alterations in human cancer.
Elizabeth M. Wilson, Ph.D. Androgen receptor regulation of gene expression; environmental androgens and antiandrogens; androgen receptor regulation of prostate cancer.

Associate Professors
William B. Coleman, Ph.D. Hepatocarcinogenesis; tumor suppressor genes; biology of liver stem cells; cancer epigenetics.
Lee M. Graves, Ph.D. Protein kinases and cell signaling; regulation of cell metabolism and toxicity.
Robert C. Millikan, D.V.M., Ph.D. Cancer; genetic and molecular epidemiology.
Leena A. Nylander-French, Ph.D. Environmental and occupational exposure assessment and modeling; development of biological monitoring methods for dermal exposure.
Dale A. Ramsden, Ph.D. V(D)J recombination and DNA double-strand break repair.
Philip C. Smith, Ph.D. Toxicokinetics and xenobiotic metabolism; peptide analysis and disposition.
David W. Threadgill, Ph.D. Mammalian genetics; systems genetics; toxicogenomics; gene-environment interactions; cancer susceptibility.

Assistant Professors
Christoph H. Borchers, Ph.D. Proteomics, protein-protein, and protein-ligand interaction by mass spectrometry.
Mohanish P. Deshmukh, Ph.D. Molecular mechanisms of apoptosis in neurons and other postmitotic cells.
Ilona Jaspers, Ph.D. Cellular mechanisms of air pollutant toxicity.
Jeffrey M. Macdonald, Ph.D. Metabolomics and fluxomics, using NMR spectroscopy and imaging and tissue engineering.
Charles M. Perou, Ph.D. Characterize and classify human breast tumors into subtypes of biological and clinical importance.
Ivan Rusyn, M.D., Ph.D. Environmental genomics.

Research Associate Professor
Miroslav Stýblo, Ph.D. Metabolism and biological effects of essential and toxic metals and metalloids.

Adjunct Professors
Melvin E. Andersen, Ph.D. Pharmacokinetic and pharmacodynamic modeling of environmental compounds.
Linda S. Birnbaum, Ph.D. Chemical disposition of xenobiotics; mechanistic toxicology; dose-response and risk assessment.
John A. Cidlowski, Ph.D. Apoptosis; steroids; glucocorticoid receptors; hormone action; nucleases; gene regulation.
Daniel L. Costa, Ph.D. Cardiopulmonary and inhalation toxicology; health effects of air pollutants.
Robert B. Devlin, Ph.D. Pulmonary toxicology and molecular biology.
Steven R. Kleeberger, Ph.D. Genetic determinants of environmental lung disease.
Stephanie Padilla, Ph.D. Behavioral toxicology and neurotoxicology.
Michael D. Waters, Ph.D. Mutagenesis and carcinogenesis; toxicogenomics.

Adjunct Associate Professors
James W. Allen, Ph.D. Genetic toxicology; toxicogenomics; epigenetic mechanisms in chemical carcinogenesis.
David C. Dorman, Ph.D. Experimental neurotoxicology; nasal toxicology; pharmacokinetics.
Ronald P. Mason, Ph.D. Free-radical intermediates in the metabolism of toxic chemicals.
John M. Rogers, Ph.D. Developmental toxicology; teratology; developmental biology; embryology; nutrition.
James M. Samet, Ph.D. Inflammatory responses to pollution inhalation; cytokines and eicosanoids.
MaryJane K. Selgrade, Ph.D. Immunotoxicology.
Robert C. Sills, D.V.M., Ph.D. Molecular pathology.
Ralph J. Smialowicz, Ph.D. Immunotoxicology, with a focus on the developing immune system.
Raymond W. Tennant, Ph.D. Transgenic animals in carcinogenesis studies.
Hugh A. Tilson Jr., Ph.D. Behavioral toxicology; developmental neurotoxicology.
Kenneth R. Tindall, Ph.D. Molecular mutagenesis; somatic cell mutation; role of mutagenesis in carcinogenesis.
Douglas C. Wolf, D.V.M., Ph.D. Chemical carcinogenesis.

Adjunct Assistant Professors
Hugh A. Barton, Ph.D. Xenobiotic metabolism; pharmacokinetics; physiologically based pharmacokinetic (PBPK) and pharmacodynamic modeling.
Ronald E. Cannon, Ph.D. Cancer biology; transgenic mouse models.
Kevin M. Crofton, Ph.D. Understanding the consequences of endocrine disruption on neurodevelopment.
Michael DeVito, Ph.D. Development of models for cumulative risk to endocrine disruptors.
Suzanne Fenton, Ph.D. Environmental effects on mammary gland development and function.
Ian Gilmour, Ph.D. Experimental toxicology.
G. Jean Harry, Ph.D. Developmental neurotoxicology; molecular neurotoxicology and immunotoxicology.
Yuh-Chin T. Huang, M.D., M.P.H. Genomics and oxidative stress in vascular reactivity.
E. Sidney Hunter, Ph.D. Mechanisms of developmental toxicity; oxidative stress; embryonic stem cells in developmental toxicity.
Edward L. LeCluyse, Ph.D. Cellular and molecular mechanisms regulating liver cytochrome P450 enzyme expression.
Michael C. Madden, Ph.D. Air pollution toxicology; lung oxidative stress and inflammation.
Michael G. Narotsky, Ph.D. Developmental toxicology; pregnancy maintenance and parturition.
Nigel Walker, Ph.D. Risk assessment; receptor-mediated toxicants; environmental contaminants; mechanisms of carcinogenesis.

UNIVERSITY OF WASHINGTON

Department of Pharmacology

Program of Study	The Department of Pharmacology offers a graduate program leading to the Doctor of Philosophy degree. The program of study includes training in molecular, cellular, and other basic aspects of pharmacology. There are opportunities for thesis research in a wide variety of areas, including neuropharmacology, behavioral pharmacology, molecular pharmacology, environmental toxicology and carcinogenesis, drug disposition and metabolism, hormonal regulation, electrical excitability, metabolic regulation, ion transport, protein phosphorylation, cyclic nucleotide metabolism, regulation of gene expression, and genetic and molecular approaches to studying protein and gene function.
	During the first year of study, graduate students take courses in pharmacology, physiology, and biochemistry. Program requirements are flexible and tailored to fit students' individual needs and interests. Also during the first year, students spend a full quarter in each of three different laboratories to gain firsthand experience with different experimental approaches. A thesis adviser is chosen by the summer of the first year. During the second year, students take advanced courses in pharmacology and related disciplines and begin dissertation research. Ph.D. candidacy is granted after successful completion of a qualifying examination, taken at the end of the second year. The third and fourth years of training are spent in detailed study of an important pharmacological problem. Students design the program for the last two years in close collaboration with their dissertation advisers and committees of faculty members.
	The Department maintains an active program of research meetings and seminars in which current developments in pharmacology are discussed by the faculty, students, and invited lecturers. This program exposes students to new ideas in pharmacology by providing contact with scientists from various areas of the United States and other countries.
Research Facilities	The Department of Pharmacology is housed in the Health Sciences Center of the University of Washington School of Medicine, close to all of the other basic science and clinical departments. The Department has outstanding research facilities in each of its major research areas: molecular pharmacology, neuropharmacology, and drug metabolism. The Departmental Molecular Pharmacology Facility provides peptide microsequencing, peptide synthesis, and automated DNA sequencing services. In addition, excellent collaborative arrangements with other basic science departments provide access to many other types of research equipment and to transgenic animal facilities.
Financial Aid	The Department offers research training fellowships to promising Ph.D. students. Stipends and tuition are provided by NIH training grants, individual research grants, and University of Washington teaching assistantships. In 2006–07, students receive tuition plus an income of $25,008 per year.
Cost of Study	In 2006–07, quarterly tuition is $6396 for out-of-state students and $2713 for Washington residents. Most students receive fellowships that provide tuition and a stipend.
Living and Housing Costs	A variety of privately owned houses, apartments, and rooms can be rented close to the campus. The average Pharmacology graduate student pays between $500 and $700 per month in rent. University housing facilities are also available for students. For more information about housing, students may visit the Web site at http://www.hfs.washington.edu.
Student Group	The University of Washington is a comprehensive university enrolling 34,330 graduate and undergraduate students. This varied group includes 9,312 graduate and professional students. In addition, there are approximately 1,500 senior fellows continuing their research. At present, there are 45 graduate students and 50 postdoctoral fellows in the Department.
Student Outcomes	The graduate training program is characterized by close collegial relationships between faculty members and students, diverse research opportunities for students, and individually tailored programs of course work and research. Creative and hands-on research experience is emphasized in preparation for a career of independent research. Recent graduates have been successful in obtaining excellent positions in academic, industrial, and governmental positions involving teaching, research, and administration.
Location	Seattle offers the educational and cultural advantages of a major metropolitan center, and its setting on Puget Sound provides superb recreational opportunities. The city has an active symphony, a repertory playhouse and theaters, and it attracts outstanding performers in opera, ballet, and popular music. The Cascade and Olympic mountain ranges and three major national parks provide year-round outdoor recreation within an hour's drive.
The University	The campus is located in a residential area of Seattle on the shores of Lake Washington and Lake Union. The School of Medicine, founded in 1945, is a major center for health science education and research in the Northwest, providing research and training programs in all of the sciences basic to medicine.
Applying	Students who have emphasized either biological or physical sciences in their undergraduate careers are invited to apply online at http://depts.washington.edu/phcol/. New students enter the graduate program in mid-September. The deadline for receipt of completed applications is December 15.
Correspondence and Information	Admissions Coordinator Department of Pharmacology Health Sciences Center Box 357280 University of Washington Seattle, Washington 98195-7280 Phone: 206-685-9252 Fax: 206-685-3822 E-mail: phcoladm@u.washington.edu Web site: http://depts.washington.edu/phcol/

University of Washington

THE FACULTY AND THEIR RESEARCH

Sandra Bajjalieh, Ph.D., Associate Professor. Molecular mechanisms of neurotransmission; role of synaptic vesicle proteins and lipid modifying enzymes in regulated secretion.

Joseph A. Beavo, Ph.D., Professor. Mechanism of action of cyclic nucleotides; study of problems involving the structure, functions, and regulatory roles of cyclic nucleotide-phosphodiesterases; use of microchip arrays, crystallography, knock-out animals, RNAi, and molecular genetic techniques.

William A. Catterall, Ph.D., Professor and Chairman. Molecular basis of electrical excitability; regulation of excitability of nerve and muscle cells; mechanism of action of drugs and toxins that affect electrical excitability; molecular biology of ion channels.

Charles Chavkin, Ph.D., Professor. Neurophysiology of opioid peptides; electrophysiological and neurochemical studies of opioid receptor function, opioid receptor diversity, and regulation of receptor function.

Chris S. C. Hague, Ph.D., Assistant Professor. G-protein–coupled receptors; use of molecular and animal techniques to study the interactions of drugs and protein binding partners at adrenergic receptors; role of adrenergic receptors in the cardiovascular system.

Edwin G. Krebs, M.D., Professor Emeritus. Regulation of cellular functions by protein phosphorylation-dephosphorylation; mechanism of action of growth factor receptor protein tyrosine kinases; protein kinase structure-function relationships.

G. Stanley McKnight, Ph.D., Professor. Signal transduction pathways; use of mouse molecular genetic techniques to study the role of cAMP-dependent protein kinases in the regulation of body weight, reproduction, and drug addiction.

Randall T. Moon, Ph.D., Professor and Investigator, Howard Hughes Medical Institute. *Wnt* signal transduction pathways in vertebrates, emphasizing mechanisms of signaling and treatments for diseases that involve these pathways.

Neil M. Nathanson, Ph.D., Professor. Cell and molecular biology of neural signal transduction proteins; regulation, function, and neuroplasticity of muscarinic acetylcholine receptors, G proteins, and receptors for neuronal differentiation factors.

Nephi Stella, Ph.D., Assistant Professor of Pharmacology (joint with Psychiatry and Behavioral Sciences). Brain immunology: using cannaboid-based compounds as therapeutics for brain tumors and multiple sclerosis.

Daniel R. Storm, Ph.D., Professor. Molecular basis of memory formation; olfactory-based memory; circadian biology; neuronal signaling mechanisms; adult neurogenesis.

Bruce L. Tempel, Ph.D., Professor of Otolaryngology and Pharmacology. Molecular and behavioral genetics of hearing loss and membrane excitability studied in mice and humans.

Frank F. Vincenzi, Ph.D., Professor. Membranes and membrane transport of calcium, regulation of intracellular calcium, and the effects of free radicals, drugs, and disease on these processes; use of human red blood cells as models of normal and abnormal cellular function and abnormal function in disease; use of microcomputers in mathematical modeling, teaching, and research.

Edith Wang, Ph.D., Associate Professor. Regulation of gene transcription in human disease, with emphasis on cell-cycle control genes; mechanism of action of TAF1 acetyltransferase subunit of transcription factor IID; disruption of gene regulatory mechanisms in muscular dystrophy and neurodegeneration.

Ning Zheng, Ph.D., Assistant Professor. Structural and molecular mechanisms of ubiquitin-dependent protein degradation; DNA repair and transcription control; structure determination of protein complexes by X-ray crystallography.

Affiliated Faculty

Karol Bomsztyk, M.D., Professor of Medicine and Adjunct Associate Professor of Pharmacology. Linking signal transduction to gene expression.

David G. Cook, Ph.D., Research Assistant Professor of Medicine and Adjunct Research Assistant Professor of Pharmacology. Molecular mechanisms of Alzheimer's disease.

Mark Hamblin, Ph.D., Associate Professor of Psychiatry and Behavioral Sciences. Serotonin receptors and synaptic function.

Wim Hol, Ph.D., Professor of Biological Structure and Adjunct Professor of Pharmacology and Biochemistry. Three-dimensional structures of key proteins as starting point for design of new drugs and vaccines; protein crystallography; molecular modeling; multiprotein complexes; heat-labile enterotoxin and choleratoxin; sleeping sickness parasite enzymes; *Leishmania;* tuberculosis and malaria drug targets.

Paul Phillips, Ph.D., Assistant Professor of Psychiatry and Behavioral Sciences and Adjunct Assistant Professor of Pharmacology. Rapid dopamine neurotransmission during motivated behaviors and decision making, and its dysfunction in addictive disorders.

Gerard Schellenberg, Ph.D., Research Professor of Neurology and of Medicine. Molecular and genetic basis of aging.

Eileen L. Watson, Ph.D., Professor of Oral Biology and Adjunct Professor of Pharmacology. Mechanisms of salivary gland secretion, with emphasis on G-protein and intracellular messenger regulation; bacterial cAMP regulation and physiological function.

Zhengui Xia, Ph.D., Associate Professor of Environmental Health and Adjunct Associate Professor of Pharmacology. Mechanisms for regulation of neurogenesis and neuronal apoptosis.

UNIVERSITY OF WISCONSIN–MADISON

Molecular and Cellular Pharmacology Ph.D. Program

Program of Study	The Molecular and Cellular Pharmacology Doctoral Program is an interdisciplinary and interdepartmental program in pharmacology that broadly emphasizes study of the biochemical, molecular, and cellular approaches to investigating mechanisms of signal transduction and of the interaction of drugs and chemicals with living systems. The faculty provides expertise in such areas as phosphoinositide-generated second messengers; polyphosphoinositide regulation of cellular processes; vesicular trafficking/secretion and cell proliferation; cytoskeletal assembly; pharmacogenomics; mechanisms of gene expression; regulation of hematopoiesis and leukemogenesis; functional role of ligand-gated and voltage-dependent ion channels; glutamatergic/IP$_3$ cell signaling; mechanisms of excitation/contraction coupling in cardiac and skeletal muscle; molecular basis of neuromuscular disease; regulation of the cell cycle by second messenger–regulated and growth hormone–regulated kinases; mechanisms of action of the nuclear receptors; structure and mechanism of action of G-protein-linked receptors and the interaction of G-proteins with effectors; modulation of membrane protein complexes and enzyme-enzyme interactions; antibiotic action and resistance; and the structure and function of glutathione S-transferase. The primary objective is to train graduate students in molecular and cellular biology and the basic principles of pharmacology.

The Molecular and Cellular Pharmacology Doctoral Program offers a Ph.D. in pharmacology with great flexibility in the selection of curriculum and mentors. This allows students the opportunity to fashion their training experience to suit their individual professional goals. A program of study toward the Ph.D. degree is worked out individually with each student and emphasizes a broad base of study in biochemistry, molecular and cell biology, genetics, neuroscience, chemistry, or biological structure with an emphasis on molecular pharmacology. The research laboratories of the Molecular and Cellular Pharmacology faculty provide a rich training environment. Most faculty members also train students in neuroscience, molecular biology, biomolecular chemistry, environmental toxicology, biochemistry, or other Ph.D. programs. This mixing of students from overlapping disciplines greatly enhances the training and scientific environment. The Master of Science degree is only offered under special circumstances.

The first semesters of graduate study emphasize formal course work, and the student is encouraged to rotate through several laboratories to enhance and broaden the research experience and facilitate selection of the research mentor. After the student identifies a major professor, a Ph.D. committee of faculty advisers is formed. The Ph.D. committee advises the student and monitors his or her progress throughout the graduate training period. Each student receives training in preparation of seminars and research proposals as part of the formal training. A two-part preliminary examination consisting of a written examination (part 1) and a research proposal with an oral examination (part 2) is given between the end of the second year and the beginning of the third year. Upon satisfactory completion of this examination, the student becomes a dissertator for the Ph.D. degree. A final thesis is presented to the Ph.D. committee and an oral examination given based on the student's original research.

Research Facilities	Research is carried out in well-designed research laboratories equipped with modern instrumentation and state-of-the-art facilities. The University of Wisconsin is one of the world's largest and best research universities. Thus, the resources available to biomedical scientists at the University are vast and accessible to all students and faculty. The University of Wisconsin is world-renowned for its spirit of collegiality and scientific cooperation, which operates at all levels and is one of the major strengths of the University. Extensive library, computer, and other support facilities are available.
Financial Aid	The program offers a stipend to all of its graduate students and assists students in competing for National Science Foundation, Howard Hughes Medical Institute, and National Institutes of Health graduate fellowships. Graduate students receive a stipend of more than $22,500 per year and are exempt from nonresident fees; traineeships and fellowships offer the same stipend. Above this stipend, tuition is compensated, while a nominal fee ($13 for single students, $25 for married couples) is charged for health insurance.
Cost of Study	As noted above, students in the program are compensated for tuition and a nominal fee is charged for health insurance.
Living and Housing Costs	Madison is nationally known for its high quality of living and low cost of living. Generally, the cost of living in Madison is 20 to 50 percent lower than that on either the East or West Coast. The Madison community offers a wide variety of houses, rooms, and apartments; prices depend on location and quality. Information on privately owned housing may be obtained from the Campus Assistance Center (The Red Gym, 716 Langdon Street, Madison, Wisconsin 53706; telephone: 608-263-2400). Information on University-sponsored housing for single graduate students or those with families may be obtained from the Assignment Office, Division of University Housing (University Apartments Office, 611 Eagle Heights, Madison, Wisconsin, 53705; telephone: 608-262-3407).
Student Group	Current enrollment at the UW–Madison campus is 41,219 students, of whom 8,951 are graduate students. Most Molecular and Cellular Pharmacology faculty members are currently training between 3 and 8 graduate students in their laboratories, with an average of 5 students per laboratory. Since the faculty trains students from a number of programs on campus, the composition of students in each laboratory varies, depending upon the research interests. The diversity of students in each laboratory adds to the exchange of ideas among students with different scientific perspectives.
Location	Madison is a university city and state capital with a population of approximately 200,000 in a magnificent physical setting near four lakes. Madison has a thriving economy with many high-technology corporations. It is a city of many parks and recreational areas, including a 1,200-acre arboretum and the Vilas Park Zoo, which is located in the heart of the city. Many activities center on the University, including sports events, plays, concerts, and art exhibits. The campus has fine facilities for all indoor and outdoor sports. The surrounding countryside is truly beautiful and provides every possible outdoor recreational opportunity.
The University	The University of Wisconsin–Madison, founded in 1849, has consistently been ranked among the top universities in the country and in the world. It is characterized by a dynamic spirit of inquiry in an atmosphere of academic freedom. The University has unusual breadth and has pioneered in placing its facilities and staff at the disposal of the entire state and nation.
Applying	The Ph.D. degree must satisfy the requirements and regulations of the Graduate School. The program requires an undergraduate major in either chemistry or a biological science. Applications should be submitted by December 31. Applicants are required to take the Graduate Record Examinations, preferably in October of the year preceding admission to ensure that scores are available by December or January.
Correspondence and Information	Dr. David Wassarman Chair, Graduate Admissions for Molecular and Cellular Pharmacology Program University of Wisconsin–Madison 3770 Medical Science Center, 1300 University Avenue Madison, Wisconsin 53706 E-mail: lsquire@wisc.edu Web site: http://www.wisc.edu/molpharm

University of Wisconsin–Madison

THE FACULTY AND THEIR RESEARCH

Elaine T. Alarid, Associate Professor of Physiology (Medical School). Molecular endocrinology of estrogen action and reproduction.

Kurt J. Amann, Assistant Professor of Zoology (College of Letters and Science). Biochemical approaches to understanding the mechanisms of cell motility and shape determination.

Richard A. Anderson, Professor of Pharmacology and Director, Molecular and Cellular Pharmacology Program (Medical School). Protein and phosphoinositide kinase signal transduction; nuclear signal transduction; signaling pathway that modulates actin assembly and cell motility; cellular signal transduction mechanisms; phosphoinositide signaling; signaling to and within the nucleus.

Maureen M. Barr, Assistant Professor of Pharmaceutical Science Division (Pharmacy). Behavioral genetics and polycystic kidney disease.

Paul J. Bertics, Professor of Biomolecular Chemistry (Medical School). Characterization of the structure, signaling mechanisms, and regulation of growth factor and cytokine receptors.

Christopher A. Bradfield, Professor of Oncology (Medical School). Environmental sensors; tumor promotion dioxin toxicology; PAS proteins.

Emery H. Bresnick, Professor of Pharmacology (Medical School). Transcriptional mechanisms; hematopoiesis; stem cell biology; vascular biology.

Richard R. Burgess, Professor of Oncology (Medical School). Prokaryotic and eukaryotic RNA polymerase and protein biochemistry.

Edwin R. Chapman, Professor of Physiology (Medical School) and Investigator, Howard Hughes Medical Institute. Molecular mechanisms of exocytosis; fusion pores; receptors for Botox.

Cynthia M. Czajkowski, Professor of Physiology (Medical School). Structure and function of ligand-gated ion channels.

John M. Denu, Associate Professor of Biomolecular Chemistry (Medical School). Mechanism and biological function of reversible protein modifications involved in modulating signal transduction and chromatin dynamics.

James M. Ervasti, Professor of Physiology (Medical School). Molecular basis of muscular disease; dystrophin-glycoprotein complex.

Barry S. Ganetzky, Professor of Genetics (Medical School; College of Agricultural and Life Sciences). Genetic and molecular analysis of neuronal signaling, synaptic development, and neurodegeneration.

Michael N. Gould, Professor of Oncology (Medical School). Breast cancer, genetics, signal transduction, and experimental therapeutics.

Daniel S. Greenspan, Professor and Vice Chair for Research, Pathology and Lab Medicine (Medical School). Extracellular controls in growth-factor signaling and matrix biology.

Jeffrey D. Hardin, Professor of Zoology and Director, Biology Core Curriculum (College of Letters and Science). Epithelial migration and embryonic development.

Anna Huttenlocher, Associate Professor of Pediatrics and Pharmacology (Medical School). Cellular and molecular mechanisms that regulate cell migration; implications to tumor invasion and metastasis and inflammation.

Meyer B. Jackson, Professor of Physiology (Medical School). Ion channel mechanisms in the mammalian central nervous system; molecular studies of drug-receptor interactions in ligand-gated channels.

Colin R. Jefcoate, Professor of Pharmacology (Medical School). Differentiation of fat cells and steroidogenic cells; regulation of breast and prostate epithelia, P450 cytochromes in these cells, and the activation of polycyclic hydrocarbons.

Jeffrey A. Johnson, Associate Professor of Pharmacy (School of Medicine). Transcriptional control of antioxidant genes in neuroprotection and neurodegeneration.

Timothy J. Kamp, Associate Professor, Departments of Medicine and Physiology (Medical School). Cardiac ion channels and embryonic stem cell-derived cardiomyocytes.

Patricia J. Keely, Assistant Professor of Pharmacology (Medical School). Integrin and small GTPase signaling events in differentiation and transformation.

Scott G. Kennedy, Assistant Professor of Pharmacology (Medical School). Genetic analysis of small RNAi and RNAi in *C. elegans*.

Jonathan C. Makielski, Professor of Medicine (Cardiology) and Physiology (Medical School). Role of ion channels in cardiovascular disease.

Thomas F. J. Martin, Professor of Biochemistry (College of Agricultural and Life Sciences). Molecular mechanisms underlying stimulus-secretion coupling in neural and endocrine cells; signal transduction events in secretory granule-plasma membrane fusion.

Janet E. Mertz, Professor of Oncology (Medical School). EGFR/ErbB2/AKT/MAPK pathways in breast cancer and herpes viruses reactivation.

Shigeki Miyamoto, Associate Professor of Pharmacology (Medical School). Signaling mechanisms involved in the regulation of Rel/NF-κB transcription factors.

Deane F. Mosher, Professor of Medicine (Medical School). Biochemistry of extracellular matrix and cell adhesion and movement.

Robert A. Pearce, Professor of Anatomy, Anesthesiology, and Physiology (Medical School). Inhibitory synaptic transmission; mechanisms of anesthetic action.

Richard E. Peterson, Professor of Pharmaceutical Sciences (School of Pharmacy); Molecular and Environmental Toxicology Center (College of Agricultural and Life Sciences). Developmental and reproductive toxicity of dioxin and PCBs; Ah receptor dependent and independent mechanisms.

J. Wesley Pike, Professor of Biochemistry (College of Agricultural and Life Sciences). Transcriptional mechanisms of steroid hormone action in the skeleton.

Alan C. Rapraeger, Professor of Pathology (Medical School). Regulation of adhesion signaling in mammary carcinoma and angiogenesis by syndecan-1; heparan sulfate regulation of growth factor signaling.

Gail A. Robertson, Associate Professor of Physiology (Medical School). Molecular mechanism of K⁺ channel function and disease.

Arnold E. Ruoho, Professor and Chair of Pharmacology (Medical School). Structure and regulation of G-protein–coupled receptors, and G proteins; structure and function of vesicular monoamine transporters; structure and function of sigma receptors in regulating potassium channels and cell survival in cancer.

John P. Svaren, Assistant Professor of Comparative Biosciences (School of Veterinary Medicine. Transcriptional control of peripheral nerve myelination; role of chromatin structure in gene regulation.

Randal S. Tibbetts, Assistant Professor of Pharmacology (Medical School). Genome surveillance; DNA damage–induced signal transduction.

Jeffery W. Walker, Professor of Physiology (Medical School). Protein kinases and contractile regulation in the cardiovascular system.

David A. Wassarman, Associate Professor of Pharmacology (Medical School). Transcriptional regulation in *Drosophila*.

Jerry Yin, Professor of Genetics (Medical School). Signal transduction pathways at the synapse (atypical PKC; fragile X), retrograde to the nucleus and in the nucleus (CREB) of neurons.

VIRGINIA COMMONWEALTH UNIVERSITY

Medical College of Virginia Campus
Department of Pharmacology and Toxicology

Program of Study

The Department of Pharmacology and Toxicology of the Medical College of Virginia Campus—the Health Sciences Division of Virginia Commonwealth University—offers a program of graduate study leading to the degree of Doctor of Philosophy. The broad base offered in pharmacology and toxicology, together with basic training in physiology and biochemistry, provides the background for a successful career in academic institutions, industry, or government. Students customarily complete formal course work in physiology and biochemistry during the first year of study. Participation in research is also begun early in the first year. Students and faculty members join together in a seminar program, which includes distinguished visiting scientists from the United States and abroad. Following completion of a qualifying examination, a degree candidate is required to submit and defend a thesis embracing an original research project conducted under the guidance and supervision of an adviser. There is no foreign language requirement. The research program of the department is sufficiently broad to provide an adequate basis for entry into a wide variety of interesting areas of modern biology and medicine.

Research Facilities

In the department and associated laboratories, more than 40 faculty members carry out research in areas such as molecular and cellular pharmacology, signal transduction, alcohol and drug abuse, behavioral pharmacology, cancer pharmacology, immunotoxicology, central nervous system pharmacology, cardiovascular pharmacology, traumatic brain injury, and DNA repair mechanisms. Techniques and instrumentation include tandem mass spectrometry, voltage and patch-clamp electrophysiology, recombinant DNA techniques, flow cytometry, confocal microscopy, and other research methods. Excellent library facilities are located a block away, and electronic access to full-text journals is available for students and faculty members.

Financial Aid

All students admitted for work toward advanced degrees automatically become eligible for financial aid in the form of fellowships, scholarships, traineeships, and assistantships. Stipends are $21,000 for the academic year, and tuition and research supplies are provided.

Cost of Study

Tuition and fees in 2006–07 are $10,808 for Virginia residents and $23,117 for out-of-state students. These costs are usually covered directly or indirectly by stipends awarded to students receiving financial aid.

Living and Housing Costs

Students may live on campus in dormitories or off campus in rooming houses or apartments. University apartments for married students are located a few miles away. Information on these living facilities may be obtained from the director of housing at Box 842517.

Student Group

Virginia Commonwealth University, the largest university in the state, has a total enrollment of approximately 24,000 students. At the Medical College of Virginia Campus, more than 2,500 students undertake course work and research in medicine, dentistry, pharmacy, nursing, and various fields of graduate study. The ratio of graduate students to core faculty members in pharmacology and toxicology is about 1:1.

Location

The Medical College of Virginia is located in a historic section of downtown Richmond, near the restaurant, entertainment, and shopping areas of Shockoe Slip and the Sixth Street Marketplace. Metropolitan Richmond has a population of more than 600,000. In addition to being a historical center, Richmond offers many cultural opportunities, including museums, a symphony orchestra, ballet, opera, and theaters. Recreational areas range from the Appalachians of west-central Virginia to local rivers, the Chesapeake Bay, the Atlantic Ocean, and Washington, D.C., all within 100 miles.

The University and The Department

The Medical College of Virginia was founded in 1838 and became an independent institution in 1854. Today, it consists of the Schools of Allied Health Professions, Dentistry, Medicine, Pharmacy, and Nursing. In 1968, the Medical College of Virginia and the Richmond Professional Institute were merged to form Virginia Commonwealth University. On the academic campus, located approximately a mile away, are the School of the Arts and the Schools of Business, Community and Public Affairs, Education, Humanities and Sciences, and Social Work. The Department of Pharmacology and Toxicology belongs to the School of Medicine and has a long history of research and was the first department of the institution to offer a program of graduate studies leading to the degree of Doctor of Philosophy.

Applying

For admission to the School of Graduate Studies, applicants must have an earned at least a baccalaureate degree and be knowledgeable about the principles of biology, chemistry, mathematics, and physics. Recommended, but not required, undergraduate courses include inorganic and organic chemistry, biochemistry, quantitative analysis, physical chemistry, cell biology, physiology, mathematics through calculus, and experimental psychology. The General Test of the Graduate Record Examinations is required. Selection of applicants begins in December, and consideration of scholarship and fellowship awards for the academic year is usually completed by April 1.

Correspondence and Information

Graduate Program Director
Department of Pharmacology and Toxicology
Medical College of Virginia Campus
Virginia Commonwealth University
Box 980613
Richmond, Virginia 23298
Phone: 804-828-8400
E-mail: swcox@vcu.edu
Web site: http://www.vcu.edu/pharmtox

Virginia Commonwealth University

THE FACULTY AND THEIR RESEARCH

Galya R. Abdrakhmanova, Assistant Professor; M.D., Ph.D., Charles (Czech Republic), 2000. Pharmacology of neuronal nicotinic receptors; electrophysiology.

Mario D. Aceto, Professor; Ph.D., Connecticut, 1959. Mechanisms of action of analgesics and psychotherapeutic agents; drug dependence.

Hamid I. Akbarali, Associate Professor; Ph.D., Memorial Newfoundland (Canada) 1988. Cellular and molecular aspects of ion-channel function in gastrointestinal smooth muscle; sensory neurons; inflammation-induced modulation of ion channels; cellular signaling.

Robert L. Balster, Professor and Director of the Institute for Drug and Alcohol Studies; Ph.D., Houston, 1970. Animal models of drug dependence; behavioral pharmacology; behavioral toxicology; excitatory amino acids; inhalation studies; tobacco use; health policy.

Patrick M. Beardsley; Ph.D., Minnesota, 1982. Behavioral pharmacology; development of medications for drug dependency and other CNS disorders; abuse liability evaluation of new chemical entities.

Jill Bettinger, Assistant Professor; Ph.D., Minnesota, 1998. Neuroscience and genetics of behavior; molecular genetics of alcohol response in *C. elegans*.

Walter H. Carter Jr., Professor of Biostatistics, Pharmacology and Toxicology, and Medicine and Chairman of the Department of Biostatistics; Ph.D., Virginia Tech, 1968. Design and analysis of response surface experiments; clinical trials; toxicology.

Severn B. Churn, Associate Professor; Ph.D., Virginia Commonwealth, 1991. Neurochemical pharmacology; control of synaptic function in seizure, head trauma, and stroke.

Charles D. Cook, Assistant Professor; Ph.D., North Carolina at Chapel Hill, 2000. Behavioral pharmacology, sex/gender differences, pain, analgesia, abuse liability testing.

M. Imad Damaj, Associate Professor; Ph.D., Paris, 1991. CNS pharmacology; drugs of abuse; behavioral correlates to biochemical events; nicotine pharmacology and other drugs of abuse; pharmacology of pain and analgesia.

Andrew G. Davies, Assistant Professor; Ph.D., Melbourne (Australia), 1993. *C. elegans* behavioral genetics and neuroscience; alcohol and drugs of abuse.

Robert J. DeLorenzo, George Bliley Professor of Neurology and of Pharmacology and Toxicology; Ph.D., 1973, M.D./M.P.H., 1974, Yale. Neuroscience and molecular neurobiology; molecular basis of membrane excitability; neuropharmacology of neuroleptic drugs and biochemical bases of the effects of calcium on neuronal functions.

William L. Dewey, Professor; Ph.D., Connecticut, 1967. Mechanism of action of cannabinoids; narcotic analgesics and their antagonists; role of endogenous substances in various diseases; sudden infant death.

Thomas Eissenberg, Associate Professor; Ph.D., McMaster, 1994. Human behavioral pharmacology of nicotine, opioids, and cannabinoids; drug dependence; drug-abuse treatment; laboratory studies and clinical trials.

Earl Ellis, Professor of Pharmacology and Toxicology; Ph.D., Wake Forest, 1974. Tissue culture models of traumatic injury to brain cells; eicosanoid synthesis and actions in the nervous and cardiovascular systems; cerebral microcirculation.

David A. Gewirtz, Professor; Ph.D., CUNY, Mount Sinai, 1977. Influence of antitumor drugs and radiation on growth arrest and cell death pathways in breast cancer; senescence and recovery; radiosensitization and chemosensitization.

Steven Grant, Professor of Medicine and of Pharmacology and Toxicology, M.D., CUNY, Mount Sinai, 1973. Modulation of antileukemic drug action and apoptosis by signal transduction events.

Tai L. Guo, Assistant Professor; Ph.D., Albany Medical College, 1996. Immunotoxicology; molecular immunology; signal transduction.

Louis S. Harris, Harvey Haag Professor and Vice Chair; Ph.D., Harvard, 1958. Relationship between chemical and biochemical factors and pharmacological actions of drugs affecting the central nervous system.

Shawn E. Holt, Associate Professor; Ph.D., Texas A&M, 1994. Mechanisms of genomic instability; telomere function; telomerase action in tumor cells.

Edward J. N. Ishac, Associate Professor; Ph.D., Monash (Australia), 1983. Adrenergic receptors; signal transduction; protein kinases and JAK-STAT signaling.

Pin-Lan Li, Professor; M.D., Yichang Medical (China), 1975; Ph.D., Heidelberg, 1992. Cardiovascular and renal pharmacology; calcium signaling in the vascular smooth muscle cells and endothelial cells.

Aron H. Lichtman, Associate Professor; Ph.D., Dartmouth, 1989. Behavioral pharmacology; behavioral neuroscience; cannabinoid pharmacology and other drugs of abuse.

Billy R. Martin, Harris Professor and Chairman of Pharmacology; Director, NIDA Center on Drug Abuse Research; Ph.D., North Carolina at Chapel Hill, 1974. Characterization of receptors and second messengers mediating the action of drugs of abuse.

Everette L. May, Professor; Ph.D., Virginia, 1939. Medicinal chemistry; drug abuse; cancer chemotherapy.

Michael F. Miles, Associate Professor; Ph.D., Northwestern, 1982; M.D., Northwestern, 1983. Functional genomics and mechanisms of drug abuse; DNA arrays; ethanol/cocaine sensitization.

Richard G. Moran, Professor; Ph.D., SUNY at Buffalo, 1974. Molecular and cellular responses of cancers to drugs; protein chemistry and enzymology; proteomics; epigenetics and gene silencing; mitochondrial metabolite/drug transporters.

Prakash S. Nagarkatti, Wazeter Distinguished Professor of Pharmacology and Toxicology and Professor of Microbiology and Immunology; Ph.D., Jiwaji (India), 1980. Effect of environmental pollutants on the immune system; immunopharmacology; tumor immunology and immunotherapy; drug abuse and immunomodulation.

Katherine Nicholson, Assistant Professor; D.V.M., Georgia, 1987; Ph.D., Virginia Commonwealth, 1998. Behavioral pharmacology; excitatory amino acids; animal models of substance abuse.

Alphonse Poklis, Professor of Pathology, Affiliate Professor of Pharmacology and Toxicology, and Director, Hospital Toxicology Laboratory; Ph.D., Maryland, 1974. Analytical and forensic toxicology; drug metabolism; biological monitoring.

Lawrence F. Povirk, Professor; Ph.D., Berkeley, 1977. DNA damage and repair in cancer chemotherapy and radiotherapy; double-strand break repair and gene rearrangements; oxidative DNA damage in neurodegenerative disease.

Joseph K. Ritter, Associate Professor; Ph.D., Utah, 1987. Molecular biology of polycyclic aromatic hydrocarbon-detoxication enzymes and role in cancer susceptibility.

Susan E. Robinson, Professor of Pharmacology and Toxicology; Ph.D., Vanderbilt, 1976. Nicotine receptors; effect of prenatal exposure to drugs on developing neurotransmitters.

John A. Rosecrans, Professor; Ph.D., Rhode Island, 1963. Psychopharmacology; correlations between behavioral and biochemical effects of nicotine; drug dependency; psychoneuroimmunology.

Leslie S. Satin, Professor; Ph.D., UCLA, 1982. Biophysics of ion channels of endocrine and nerve cells; role of ion channels in pancreatic islet B cells; role of calcium ions in secretion, channel modulation, diabetes, neuronal injury, and NMDA receptors.

Stephen T. Sawyer, Professor; Ph.D., Tennessee (Knoxville), 1980. Cellular and molecular biology; erythropoietin and stem-cell factor receptors and intracellular signals controlling proliferation, apoptosis, and differentiation of hematopoietic cells.

Dana E. Selley, Assistant Professor of Pharmacology and Toxicology; Ph.D., Rochester, 1991. Biochemical neuropharmacology; mechanisms of drug efficacy, tolerance, and dependence; G-protein–mediated signal transduction.

Keith L. Shelton, Assistant Professor; Ph.D., Virginia Commonwealth, 1995. Behavioral pharmacology of abused drugs; ethanol; GABA; NMDA; animal models of drug abuse medication development.

Domenic A. Sica, Professor of Medicine and of Pharmacology and Toxicology and Chairman of Clinical Pharmacology and Hypertension; M.D., Virginia Commonwealth, 1975. Renal pharmacology; clinical pharmacology of hypertension.

Laura J. Sim-Selley, Associate Professor; Ph.D., Rochester, 1991. Functional neuroanatomy of G-protein–coupled receptors; cellular effects of psychoactive drugs and opioid and cannabinoid receptors.

Forrest L. Smith, Associate Professor; Ph.D., Texas Tech, 1989. PKA/PKC and signal transduction mechanisms in opioid tolerance/dependence; shRNA, siRNA gene expression induced changes in behaviors.

Stephen A. Varvel, Assistant Professor; Ph.D., Virginia Commonwealth, 2000. Behavioral pharmacology; learning and memory; marijuana pharmacology; endocannabinoids; neurodegenerative disorders.

Sandra P. Welch, Professor; Ph.D., Virginia Commonwealth, 1986. Cannabinoid opioid pharmacology; role of free intracellular calcium in tolerance and physical dependence to drugs of abuse; endocrine modulation of tolerance; second messenger systems.

Kimber L. White Jr., Professor of Biomedical Engineering and of Pharmacology and Toxicology; Ph.D., Virginia Commonwealth, 1981. Immunotoxicology of polycyclic aromatic hydrocarbons; chlorinated dibenzodioxins; complement.

Jenny L. Wiley, Associate Professor; Ph.D., Virginia Commonwealth, 1991. Behavioral pharmacology; cannabinoids; animal models of psychiatric disorders; developmental psychopharmacology.

WAYNE STATE UNIVERSITY

Institute of Environmental Health Sciences
Interdisciplinary Program in Molecular and Cellular Toxicology

Program of Study

The Institute of Environmental Health Sciences administers a Ph.D. graduate program in molecular and cellular toxicology that is integrated with the Interdisciplinary Biomedical Sciences (IBS) program at Wayne State University School of Medicine. The Interdisciplinary Program in Molecular and Cellular Toxicology offers a wide range of research opportunities and emphasizes investigations that probe the molecular and cellular mechanisms that underwrite environmental and metabolic disease processes. Many of the available research projects examine the effects of environmental agents on transcriptional and translational regulation of gene expression, intracellular signaling, apoptosis, oxidative stress, DNA repair, and complex mechanisms in cell growth and differentiation. The Institute of Environmental Health Sciences is enriched by its support from a nationally recognized Center in Molecular and Cellular Toxicology with Human Applications, which is funded by the National Institute of Environmental Health Sciences. Faculty members from the Institute of Environmental Health Sciences, the Wayne State University School of Medicine, the College of Pharmaceutical Sciences and Allied Health Professions, and the Barbara Ann Karmanos Cancer Institute participate in the Interdisciplinary Program in Molecular and Cellular Toxicology. The graduate program emphasizes the use of contemporary approaches, such as advanced techniques in biochemistry, cell biology, molecular biology, molecular genomics, proteomics, and bioinformatics and similar strategies to advance the understanding of fundamental biological processes as they relate to environmentally induced disease. Program requirements include didactic course work, consisting of both required and elective courses; laboratory rotations; seminar programs; and written and oral qualifying examinations. In addition, the student completing this program is required to prepare a dissertation describing the results of original research and to present an oral defense of the dissertation. The first year is course work intensive, with research rotations performed in the laboratories of two or more faculty members of the student's choice. Following the selection of a thesis adviser and thesis committee (usually at the beginning of the second year), students continue course work and perform preliminary research. Qualifying examinations for admission to Ph.D. candidacy are administered in the spring of the second year. Subsequent years are primarily research intensive in nature.

In order to prepare for emerging challenges in academics and industry, students in the program have access to research laboratories that perform innovative cell culture and molecular biology techniques such as transient and stable transfections, real-time polymerase chain reaction amplification, cellular imaging, protein-protein interaction analyses, and transgenic and knockout animal engineering. Students also have opportunities to learn how to prepare and apply recombinant plasmid-based and adenoviral constructs expressing dominant negative proteins and antisense and short interfering RNAs as molecular tools and gain valuable experience in microarray and proteomic global gene-expression studies.

Research Facilities

The Institute of Environmental Health Sciences occupies approximately 33,000 square feet of laboratory space in the Metropolitan Center for High Technology. Institute facilities include shared instrument rooms containing IBM-compatible and Macintosh computers; spectrophotometers; electrophoresis equipment; light, fluorescence, and phase-contrast microscopes; thermocyclers; flow cytometry and protein modeling workstations; a Ciphergen mass spectrometer for protein profile analysis; an Agilent microarray scanner, and bioanalyzer and bioinformatics support using a variety of software programs for performing microarray experiments. The Institute also has two darkrooms, several rooms dedicated to cell culture, an imaging and cytometry facility, a microarray/bioinformatics facility, a complete transgenic animal facility, and instrumentation for analysis of protein-protein interactions by fluorescence resonance energy transfer (FRET), surface plasmon resonance (SPR), and fluorescence polarization (FP), with additional capabilities for analysis of FRET-based biosensor signals in cells. Other available facilities include, but are not limited to, a nucleic acid sequencing core, a laser cytometer with confocal optics, a protein mass spectrometer for MALDI matrix assisted laser desorption ionization and SELDI platforms, and a Biacore instrument for quantifying macromolecular interactions. The Shiffman Medical Library maintains about 3,000 scientific periodicals and has more than 200,000 volumes of reference materials. The library also provides online access to several scientific and medical referencing databases.

Financial Aid

The graduate program in molecular and cellular toxicology provides financial support through University graduate assistantships, Interdisciplinary Biomedical Sciences fellowships, or through an NIEHS-supported T32 training grant. Fellowships are awarded to those students possessing outstanding academic credentials and demonstrable potential for achieving excellence in a research career. Assistantships and fellowships include graduate tuition in addition to the stipend. The stipend level ranges from $18,000 to $22,000. A typical stipend provides for a modest standard of living.

Cost of Study

Tuition costs for the Interdisciplinary Program in Molecular and Cellular Toxicology are generally covered by graduate assistantships, fellowships, and traineeships.

Living and Housing Costs

The University offers a variety of apartments for individuals and families. Further information and application forms are available at the University Housing Office. Off-campus rental units are also available.

Student Group

The University's many graduate programs attract a diversified student population from many states and other countries. Total enrollment at Wayne State University is approximately 34,000 students, of whom approximately 7,000 are enrolled in the Graduate School. It is anticipated that the Interdisciplinary Program in Molecular and Cellular Toxicology will admit 2 to 3 new students each year.

Location

Wayne State University is located 2 miles north of downtown in an area known as the Detroit Cultural Center, which includes the Detroit Institute of Art, Orchestra Hall, the Detroit Public Library, the Detroit Science Center, the Museum of African American History, the Detroit Historical Museum, professional theaters, and the Center for Creative Studies. Detroit is also the home of professional baseball, basketball, football, and hockey teams. The historic Henry Ford Museum at Greenfield Village in Dearborn and the Detroit Zoo are also popular attractions. Attractions in Canada are also accessible since Windsor, Ontario, is just across the Detroit River.

The University

Wayne State University is an urban university and one of three major research universities in Michigan. It has fourteen schools and colleges and offers courses in approximately 400 fields. The University's nationally ranked library system contains one of the largest computerized research collections in the country.

Applying

Applicants should possess a bachelor's degree from an accredited college and, preferably, a background in the basic sciences. An undergraduate grade point average of 3.0 or higher (on a 4.0 scale) is required as is the General Test of the Graduate Record Examinations (GRE). International students must submit their scores on the Test of English as a Foreign Language. Admission forms for the graduate school and a waiver of the application fee may be obtained from the Graduate Admissions Committee. Applications, which should be submitted by February 15, must include official transcripts from each institution attended, three letters of recommendation, a brief statement of career objectives, and GRE scores. Personal interviews may be requested. Early applications are encouraged. Notification of admission is given shortly after receipt of the completed application. Women and members of minority groups are encouraged to apply.

Correspondence and Information

Chairperson, Graduate Admissions Committee
Institute of Environmental Health Sciences
Wayne State University
2727 Second Avenue, Room 4000
Detroit, Michigan 48201-2654
Phone: 313-577-0100
Fax: 313-577-0082
E-mail: m.runge-morris@wayne.edu
Web site: http://www.iehs.wayne.edu

Wayne State University

THE FACULTY AND THEIR RESEARCH

D. Randall Armant, Associate Professor of Obstetrics and Gynecology and of Anatomy and Cell Biology; Ph.D., Virginia Tech, 1980. Mammalian embryogenesis; biochemical and genetic control of trophoblast cell adhesion; regulation of in vitro preimplantation development; embryo cryopreservation.

Dharam P. Chopra, Professor, Institute of Environmental Health Sciences; Ph.D., Newcastle (England), 1971. Oncogenes; tumor suppressor genes; growth factors; carcinogenesis; human epithelial cell culture.

Alan A. Dombkowski, Assistant Professor, Institute of Environmental Health Sciences; Ph.D., Michigan, 2000. Bioinformatics; microarray analysis; molecular modeling and protein engineering; algorithm development.

Craig N. Giroux, Associate Professor, Institute of Environmental Health Sciences; Ph.D., MIT, 1979. Systems biology and functional genomics of gene-environment interactions; genetic network analysis; oxidative stress and mechanisms of disease.

Ye-Shih Ho, Professor, Institute of Environmental Health Sciences; Ph.D., Carnegie Mellon, 1981. Transgenic models for study of lung biology and disease; regulation of gene expression in the lung in response to environmental agents.

Thomas A. Kocarek, Associate Professor, Institute of Environmental Health Sciences; Ph.D., Ohio State, 1988. Mechanisms of regulation of hepatic cytochrome P-450 gene expression.

Lawrence H. Lash, Professor of Pharmacology; Ph.D., Emory, 1985. Biochemical mechanisms of nephrotoxicity; glutathione metabolism and transport.

Xiangyi Lu, Associate Professor, Institute of Environmental Health Sciences; Ph.D., Yeshiva (Einstein), 1991. Molecular and cellular mechanisms that underlie motile cilium-mediated developmental and disease processes; environmental and epigenetic effects on cilial motility; polycystic kidney disease processes; sperm motility; asthma and cancers; *Drosophila*, cell culture, and mouse models.

Raymond R. Mattingly, Associate Professor of Pharmacology; Ph.D., Virginia, 1993. Signal transduction through Ras and heterotrimeric GTP-binding proteins.

Fred R. Miller, Professor, Barbara Ann Karmanos Cancer Institute; Ph.D., Wisconsin, 1976. Progression of preneoplastic breast disease; stromal-epithelial interactions; mechanisms of metastasis.

Raymond F. Novak, Professor and Director, Institute of Environmental Health Sciences; Ph.D., Case Western Reserve, 1973. Regulation of cytochrome P-450 expression in hepatic and extrahepatic tissues, primary hepatocytes, and cultured cells; role of intracellular signaling and ECM signaling in gene expression and cell function; microarray analysis and global gene expression profiling.

John J. Reiners Jr., Professor, Institute of Environmental Health Sciences; Ph.D., Purdue, 1977. Signal transduction processes regulating apoptosis, cell-cycle progression, and dioxin receptor function.

Barry P. Rosen, Professor and Chairman of Biochemistry and Molecular Biology; Ph.D., Connecticut, 1969. Molecular mechanisms of active transport and plasmid-mediated resistance.

Douglas M. Ruden, Associate Professor, Institute of Environmental Health Sciences; Ph.D., Harvard, 2000. Toxicogenomics of heavy metals, using *Drosophila* and mouse models; soma to germ line signaling during *Drosophila* oogenesis; epigenetic regulation of development in *Drosophila* and mammalian models.

Melissa A. Runge-Morris, Associate Professor, Institute of Environmental Health Sciences; M.D., Michigan, 1979. Molecular regulation of sulfotransferase gene expression by hormones and xenobiotics.

Bonnie F. Sloane, Professor and Chair of Pharmacology; Ph.D., Rutgers, 1976. Cancer biology; role of cysteine proteinases and their inhibitors in malignant progression.

Paul M. Stemmer, Associate Professor, Institute of Environmental Health Sciences; Ph.D., Michigan State, 1986. Ser/Thr phosphatase regulation; calmodulin-dependent processes; immunosuppressant and heavy-metal mechanisms of action.

Gan Wang, Associate Professor, Institute of Environmental Health Sciences; Ph.D., Chinese Academy of Science, 1989. DNA repair and genetic instability; transcription and gene expression regulation.

Xiaoxin Susan Xu, Assistant Professor (Research), Institute of Environmental Health Sciences; Ph.D., Connecticut, 1997. Genetic instability, DNA mismatch repair deficiency, and cancer development, using transgenic mouse models.

Section 18
Physiology

This section contains a directory of institutions offering graduate work in physiology, followed by in-depth entries submitted by institutions that chose to prepare detailed program descriptions. Additional information about programs listed in the directory but not augmented by an in-depth entry may be obtained by writing directly to the dean of a graduate school or chair of a department at the address given in the directory.

For programs offering related work, see also all other sections in this book. In Book 4, see Agricultural and Food Sciences, Chemistry, and Marine Sciences and Oceanography; in Book 5, see Agricultural Engineering and Bioengineering, Biomedical Engineering and Biotechnology, Electrical and Computer Engineering, and Mechanical Engineering and Mechanics; and in Book 6, see Optometry and Vision Sciences, Physical Education and Kinesiology, and Veterinary Medicine and Sciences.

CONTENTS

Program Directories

Announcements

Close-Ups

See also:

Cardiovascular Sciences

Albany Medical College, Graduate Programs in the Biological Sciences, Center for Cardiovascular Sciences, Albany, NY 12208-3479. Offers MS, PhD. Part-time programs available. *Faculty:* 18 full-time (3 women), 1 part-time/adjunct (0 women). *Students:* 15 full-time (13 women), 1 part-time; includes 4 minority (2 African Americans, 2 Asian Americans or Pacific Islanders). Average age 25. 5 applicants, 100% accepted, 3 enrolled. In 2005, 2 master's, 2 doctorates awarded. Terminal master's awarded for partial completion of doctoral program. *Degree requirements:* For master's, thesis; for doctorate, thesis/dissertation, oral qualifying exam, written preliminary exam, 1 published paper-peer review, comprehensive exam. *Entrance requirements:* For master's and doctorate, GRE General Test. Additional exam requirements/recommendations for international students: Required—TOEFL. *Application deadline:* For fall admission, 3/15 for domestic students. Applications are processed on a rolling basis. Application fee: $0 ($60 for international students). *Financial support:* In 2005–06, 11 research assistantships (averaging $23,000 per year) were awarded; Federal Work-Study, scholarships/grants, and tuition waivers (full) also available. Financial award applicants required to submit FAFSA. *Faculty research:* Vascular smooth muscle, endothelial cell biology, molecular and genetic bases underlying cardiac disease, reactive oxygen and nitrogen species biology, fatty acid trafficking and fatty acid mediated transcription control. *Unit head:* Dr. Peter A. Vincent, Graduate Director, 518-262-6296, Fax: 518-262-8101, E-mail: vincenp@mail.amc.edu. *Application contact:* Wendy M. Hobb, Administrative Coordinator, 518-262-8102, Fax: 518-262-8101, E-mail: hobbw@mail.amc.edu.

Baylor College of Medicine, Graduate School of Biomedical Sciences, Program in Cardiovascular Sciences, Houston, TX 77030-3498. Offers PhD, MD/PhD. *Faculty:* 30 full-time (2 women). *Students:* 23 full-time (6 women); includes 2 minority (both Asian Americans or Pacific Islanders), 16 international. Average age 29. 15 applicants, 27% accepted, 2 enrolled. In 2005, 5 degrees awarded. *Median time to degree:* Of those who began their doctoral program in fall 1997, 33% received their degree in 8 years or less. *Degree requirements:* For doctorate, thesis/dissertation, public defense. *Entrance requirements:* For doctorate, GRE General Test, GRE Subject Test (strongly recommended), minimum GPA of 3.0, strong background in biology and biochemistry. Additional exam requirements/recommendations for international students: Required—TOEFL. *Application deadline:* For fall admission, 2/1 for domestic students. Applications are processed on a rolling basis. Application fee: $30. Electronic applications accepted. *Expenses:* Tuition: Full-time $8,200. Full-time tuition and fees vary according to program. *Financial support:* In 2005–06, 9 fellowships (averaging $23,000 per year), 14 research assistantships (averaging $23,000 per year) were awarded; career-related internships or fieldwork, Federal Work-Study, institutionally sponsored loans, health care benefits, and tuition waivers (full) also available. Financial award applicants required to submit FAFSA. *Faculty research:* Cell biology of the vascular wall, cell biology of cardiac tissue, biology and models of specific cardiovascular diseases. *Unit head:* Dr. Alan Burns, 913-798-4371, Fax: 713-798-0681, E-mail: carolar@bcm.edu. *Application contact:* Carol Sentonnian, Graduate Program Administrator, 713-798-4977, Fax: 713-798-0681, E-mail: carolar@bcm.edu.

See Close-Up on page 1239.

Long Island University, C.W. Post Campus, School of Health Professions and Nursing, Department of Biomedical Sciences, Program in Cardiovascular Perfusion, Brookville, NY 11548-1300. Offers MS, Certificate. Part-time and evening/weekend programs available. Post-baccalaureate distance learning degree programs offered. *Degree requirements:* For master's, thesis.

McMaster University, Faculty of Health Sciences and School of Graduate Studies, Program in Medical Sciences, Hemostasis, Thromboembolism, and Atherosclerosis Area, Hamilton, ON L8S 4M2, Canada. Offers M Sc, PhD. *Students:* 17 full-time, 1 part-time. In 2005, 1 degree awarded. *Degree requirements:* For master's, thesis/dissertation; for doctorate, thesis/dissertation, comprehensive exam. *Entrance requirements:* For master's, honors B Sc, B+ average in related field; for doctorate, M Sc, minimum B+ average, students with proven research experience and an A average may be admitted with a B Sc degree. Additional exam requirements/recommendations for international students: Required—TOEFL (minimum score 580 paper-based; 237 computer-based). *Application deadline:* For fall admission, 9/30 for domestic students. For winter admission, 3/31 for domestic students. Applications are processed on a rolling basis. Application fee: $85. *Financial support:* Teaching assistantships available. *Unit head:* Dr. William Sheffield, Coordinator, 905-525-9140 Ext. 22701. *Application contact:* Dr. Carl Richards, Associate Dean, 905-525-9140 Ext. 22983, Fax: 905-546-1129.

Medical College of Georgia, School of Allied Health Sciences, Programs in Allied Health Sciences, Department of Respiratory Therapy, Augusta, GA 30912. Offers MS. Part-time programs available. *Faculty:* 3 full-time (1 woman). *Students:* Average age 37. In 2005, 1 degree awarded. *Degree requirements:* For master's, thesis or alternative. *Entrance requirements:* For master's, GRE General Test. Additional exam requirements/recommendations for international students: Required—TOEFL (minimum score 550 paper-based; 213 computer-based). *Application deadline:* For fall admission, 6/30 for domestic students. Applications are processed on a rolling basis. Application fee: $30. Electronic applications accepted. *Financial support:* Federal Work-Study, institutionally sponsored loans, and tuition waivers available. Support available to part-time students. Financial award application deadline: 5/31; financial award applicants required to submit FAFSA. *Unit head:* Dr. Randy Baker, Chair/Program Director, 706-721-3554, Fax: 706-721-0495, E-mail: rabaker@mail.mcg.edu.

Medical College of Georgia, School of Graduate Studies, Vascular Biology Center, Augusta, GA 30912. Offers PhD. *Faculty:* 7 full-time (3 women), 1 (woman) part-time/adjunct. *Students:* 11 full-time (7 women); includes 2 minority (1 African American, 1 Asian American or Pacific Islander), 5 international. Average age 30. *Degree requirements:* For doctorate, thesis/dissertation. *Entrance requirements:* For doctorate, GRE General Test. Additional exam requirements/recommendations for international students: Required—TOEFL. *Application deadline:* For fall admission, 6/30 for domestic students, 4/15 for international students. Application fee: $30. *Financial support:* In 2005–06, 9 research assistantships with partial tuition reimbursements (averaging $22,500 per year) were awarded; fellowships with partial tuition reimbursements, Federal Work-Study, institutionally sponsored loans, traineeships, and unspecified assistantships also available. Support available to part-time students. Financial award application deadline: 5/31. *Faculty research:* Hypertension, coronary artery disease, diabetes, angiogenesis, signal transduction. *Unit head:* Dr. John D. Catravas, Regents Professor and Director, 706-721-6338, Fax: 706-721-8545, E-mail: jcatrava@mail.mcg.edu. *Application contact:* Dr. Jennifer Pollock, Program Director, 706-721-8514, Fax: 706-721-8545, E-mail: jpollock@mail.mcg.edu.

Midwestern University, Glendale Campus, College of Health Sciences, Arizona Campus, Program in Cardiovascular Science, Glendale, AZ 85308. Offers MCVS. *Faculty:* 9 full-time (0 women), 8 part-time/adjunct (3 women). *Students:* 24 full-time (11 women), 2 part-time (both women); includes 5 minority (1 African American, 3 Asian Americans or Pacific Islanders, 1 Hispanic American), 1 international. Average age 26. 22 applicants, 86% accepted, 13 enrolled. In 2005, 10 degrees awarded. Application fee: $50. *Expenses:* Contact institution. *Unit head:* Dr. John Austin, Dean, 623-572-3616. *Application contact:* James Walters, Director of Admissions, 888-247-9277, Fax: 623-572-3229, E-mail: admissaz@midwestern.edu.

Milwaukee School of Engineering, Department of Electrical Engineering and Computer Science, Program in Perfusion, Milwaukee, WI 53202-3109. Offers MS. Part-time and evening/weekend programs available. *Faculty:* 1 full-time (0 women), 4 part-time/adjunct (1 woman). *Students:* 11 full-time (7 women). 12 applicants, 67% accepted, 6 enrolled. In 2005, 5 degrees awarded. *Degree requirements:* For master's, thesis. *Entrance requirements:* For master's, GRE General Test or GMAT, BS in an appropriate discipline, undergraduate work in human physiology or anatomy. Additional exam requirements/recommendations for international students: Required—TOEFL (minimum score 550 paper-based; 213 computer-based). Application fee: $30. *Expenses:* Tuition: Part-time $505 per credit. *Financial support:* In 2005–06, 5 students received support. Career-related internships or fieldwork available. Support available to part-time students. Financial award applicants required to submit FAFSA. *Unit head:* Dr. Ronald Gerrits, Director, 414-277-7561, Fax: 414-277-7494, E-mail: gerrits@msoe.edu. *Application contact:* Ronald T. Gaudes, Graduate Admissions, 800-332-6763, Fax: 414-277-7475, E-mail: gaudes@msoe.edu.

Northeastern University, Bouvé College of Health Sciences Graduate School, Department of Cardiopulmonary and Exercise Sciences, Program in Cardiopulmonary Science (Perfusion Technology), Boston, MA 02115-5096. Offers MS. *Students:* 15 full-time (9 women), 1 part-time. Average age 29. 12 applicants, 58% accepted. In 2005, 7 degrees awarded. *Degree requirements:* For master's, thesis optional. *Entrance requirements:* For master's, GRE General Test or MAT. *Application deadline:* Applications are processed on a rolling basis. Application fee: $50. *Financial support:* Career-related internships or fieldwork, Federal Work-Study, tuition waivers (partial), and unspecified assistantships available. Support available to part-time students. Financial award application deadline: 3/1; financial award applicants required to submit FAFSA. *Faculty research:* Compliment activation of extra corporeal circuits, application of neural networks, pathophysiology of cardiobypass. *Unit head:* Dr. Eric B. Pepin, Director, 617-373-4183, Fax: 617-373-2968, E-mail: e.pepin@neu.edu. *Application contact:* Margaret Schnabel, Director of Graduate Admissions, 617-373-2708, Fax: 617-373-4704, E-mail: bouvegrad@neu.edu.

Texas A&M University System Health Science Center, Graduate School of Biomedical Sciences, Department of Medical Physiology, College Station, TX 77840. Offers PhD. *Degree requirements:* For doctorate, thesis/dissertation. *Entrance requirements:* For doctorate, GRE General Test. *Faculty research:* Cardiovascular physiology, vascular cell and molecular biology.

Université Laval, Faculty of Medicine, Post-Professional Programs in Medical Studies, Québec, QC G1K 7P4, Canada. Offers anatomy–pathology (DESS); anesthesia–resuscitation (DESS); cardiology (DESS); care of older people (Diploma); clinical research (DESS); community health (DESS); dermatology (DESS); diagnostic radiology (DESS); emergency medicine (Diploma); family medicine (DESS); general surgery (DESS); geriatrics (DESS); hematology (DESS); internal medicine (DESS); maternal and fetal medicine (Diploma); medical biochemistry (DESS); medical microbiology and infectious diseases (DESS); medical oncology (DESS); nephrology (DESS); neurology (DESS); neurosurgery (DESS); obstetrics and gynecology (DESS); ophthalmology (DESS); orthopedic surgery (DESS); oto-rhino-laryngology (DESS); palliative medicine (Diploma); pediatrics (DESS); plastic surgery (DESS); psychiatry (DESS); pulmonary medicine (DESS); radiology–oncology (DESS); thoracic surgery (DESS); urology (DESS). *Degree requirements:* For other advanced degree, comprehensive exam. *Entrance requirements:* For degree, knowledge of French. Electronic applications accepted.

University of Calgary, Faculty of Medicine and Faculty of Graduate Studies, Department of Cardiovascular and Respiratory Sciences, Calgary, AB T2N 1N4, Canada. Offers M Sc, PhD. *Faculty:* 22 full-time (0 women). *Students:* 33 full-time (15 women). Average age 26. 25 applicants, 36% accepted, 8 enrolled. In 2005, 3 master's, 3 doctorates awarded. *Median time to degree:* Of those who began their doctoral program in fall 1997, 100% received their degree in 8 years or less. *Degree requirements:* For master's, thesis; for doctorate, thesis/dissertation, candidacy exam. *Entrance requirements:* For master's and doctorate, minimum GPA of 3.2. Additional exam requirements/recommendations for international students: Required—TOEFL (minimum score 600 paper-based; 250 computer-based). *Application deadline:* For fall admission, 6/15 for domestic students, 4/15 for international students. For winter admission, 10/15 for domestic students; for spring admission, 3/15 for domestic students. Applications are processed on a rolling basis. Application fee: $100 ($130 for international students). Electronic applications accepted. *Financial support:* In 2005–06, 33 students received support, including 12 research assistantships (averaging $4,100 per year); fellowships, teaching assistantships Financial award application deadline: 2/1. *Faculty research:* Cardiac mechanics, physiology and pharmacology; lung mechanics, physiology and pathophysiology; smooth muscle biochemistry; physiology and pharmacology. *Unit head:* Dr. S.R. Wayne Chen, Graduate Coordinator, 403-220-4235, Fax: 403-283-4841, E-mail: chenw@ucalgary.ca. *Application contact:* Kate Judycki, Graduate Program Administrator, 403-210-3937, Fax: 403-210-8109, E-mail: cvrgrad@ucalgary.ca.

University of California, San Diego, School of Medicine and Graduate Studies and Research, Molecular Pathology Program, La Jolla, CA 92093. Offers bioinformatics (PhD); cancer biology/oncology (PhD); cardiovascular sciences and disease (PhD); microbiology (PhD); molecular pathology (PhD); neurological disease (PhD); stem cell and developmental biology (PhD); structural biology/drug design (PhD). *Entrance requirements:* For doctorate, GRE General Test, GRE Subject Test. Additional exam requirements/recommendations for international students: Required—TOEFL. Electronic applications accepted.

See Close-Up on page 1119.

University of Maryland, Graduate School, Interdisciplinary Cardiovascular Training Program, Baltimore, MD 21201.

University of Medicine and Dentistry of New Jersey, School of Health Related Professions, Department of Interdisciplinary Studies, Program in Health Sciences, Newark, NJ 07107-1709. Offers cardiopulmonary sciences (PhD); clinical laboratory sciences (PhD); health sciences (MS); interdisciplinary studies (PhD); nutrition (PhD); physical therapy/movement science (PhD). *Degree requirements:* For doctorate, thesis/dissertation. *Entrance requirements:* For doctorate, interview, writing sample. Additional exam requirements/recommendations for international students: Required—TOEFL. *Application deadline:* Applications are processed on a rolling basis. Application fee: $50. *Unit head:* , Dr. Margaret L. Kildoff, Director, 973-972-8576, Fax: 973-972-5258, E-mail: scanlan@umdnj.edu.

The University of South Dakota, School of Medicine and Health Sciences and Graduate School, Biomedical Sciences Graduate Program, Cardiovascular Research Institute, Vermillion, SD 57069-2390. Offers MA, PhD. *Students:* 6 full-time (0 women). *Students:* 8 full-time (5 women), 7 international. Average age 27. 8 applicants, 50% accepted, 3 enrolled. In 2005, 1 degree awarded. Terminal master's awarded for partial completion of doctoral program. *Entrance requirements:* For master's and doctorate, GRE General Test, minimum GPA of 3.0. Additional exam requirements/recommendations for international students: Required—TOEFL (minimum score 550 paper-based; 213 computer-based). *Application deadline:* For fall admission, 4/15 priority date for domestic students, 4/15 priority date for international students. Applications are processed on a rolling basis. Application fee: $35. *Expenses:* Contact institution. Tuition and fees vary according to course load, program and reciprocity agreements. *Financial support:* In 2005–06, 8 students received support, including 7 fellowships with partial tuition reimbursements available (averaging $20,772 per year), 1 research assistantship with partial tuition reimbursement available (averaging $10,386 per year); teaching assistantships, unspecified assistantships also available. Financial award application deadline: 4/15; financial award applicants required to submit FAFSA. *Faculty research:* Cardiovascular disease. Total annual research expenditures: $1.4 million. *Unit head:* Dr. Steven B. Waller, Associate Dean of Basic Biomedical Sciences, 605-677-5157, Fax: 605-677-6381, E-mail: swaller@usd.edu. *Application contact:* Dr. Steven B. Waller, Associate Dean, Basic Biomedical Sciences, 605-677-5157, Fax: 605-677-6381, E-mail: swaller@usd.edu.

University of Virginia, College and Graduate School of Arts and Sciences, Vascular Biology Training Program, Charlottesville, VA 22903.

Molecular Physiology

Baylor College of Medicine, Graduate School of Biomedical Sciences, Department of Molecular Physiology and Biophysics, Houston, TX 77030-3498. Offers PhD, MD/PhD. *Faculty:* 33 full-time (10 women). *Students:* 17 full-time (10 women); includes 2 minority (1 African American, 1 Asian American or Pacific Islander), 10 international. Average age 29. 23 applicants, 35% accepted, 3 enrolled. In 2005, 2 degrees awarded. *Median time to degree:* Of those who began their doctoral program in fall 1997, 100% received their degree in 8 years or less. *Degree requirements:* For doctorate, thesis/dissertation, public defense. *Entrance requirements:* For doctorate, GRE General Test, GRE Subject Test (strongly recommended), minimum GPA of 3.0. Additional exam requirements/recommendations for international students: Required—TOEFL. *Application deadline:* For fall admission, 1/1 for domestic students. Application fee: $30. Electronic applications accepted. *Expenses:* Tuition: Full-time $8,200. Full-time tuition and fees vary according to program. *Financial support:* In 2005–06, 17 students received support, including fellowships (averaging $23,000 per year), research assistantships (averaging $23,000 per year); career-related internships or fieldwork, Federal Work-Study, institutionally sponsored loans, health care benefits, and tuition waivers (full) also available. Financial award applicants required to submit FAFSA. *Faculty research:* Multi-photon imaging, Magnetic Resonance Imaging (MRI), structure and function of ion channels and transport proteins, signal transduction, synaptic plasticity, cell-cycle control, reactive oxygen species, neurodegenerative diseases and cardiac development. *Unit head:* , Dr. Robia Pautler, Director, 713-798-5630, Fax: 713-798-3475. *Application contact:* Cherry McGlory, Graduate Program Administrator, 713-798-5109, Fax: 713-798-3475, E-mail: molphys@bcm.tmc.edu.

See Close-Up on page 1241.

Loyola University Chicago, Graduate School, Program in Cell and Molecular Physiology, Maywood, IL 60153. Offers MS, PhD. MS offered only to students enrolled in a first professional degree program. *Faculty:* 17 full-time (3 women). *Students:* 18 full-time (10 women), 1 part-time, 11 international. Average age 28. 15 applicants, 60% accepted, 5 enrolled. In 2005, 1 degree awarded. *Median time to degree:* Of those who began their doctoral program in fall 1997, 100% received their degree in 8 years or less. *Degree requirements:* For master's, thesis/dissertation; for doctorate, thesis/dissertation, comprehensive exam. *Entrance requirements:* For master's, GRE General Test or MCAT; for doctorate, GRE General Test. Additional exam requirements/recommendations for international students: Required—TOEFL. *Application deadline:* For fall admission, 5/15 for domestic students, 5/15 for international students. Application fee: $0. Electronic applications accepted. *Expenses:* Tuition: Full-time $11,610; part-time $645 per credit. Required fees: $55 per semester. *Financial support:* In 2005–06, 5 fellowships with tuition reimbursements (averaging $22,000 per year), 13 research assistantships with tuition reimbursements (averaging $22,000 per year) were awarded. *Faculty research:* Cardiovascular system—emphasis in neural and metabolic control of circulation, ion channels, excitation contraction coupling. *Unit head:* Dr. Donald M. Bers, Chair, 708-216-6305, Fax: 708-216-6308. *Application contact:* Dr. Ruben Mestril, Graduate Program Director, 708-327-2395, Fax: 708-216-6308, E-mail: rmestri@lumc.edu.

See Close-Up on page 1249.

Stony Brook University, State University of New York, Health Sciences Center, School of Medicine and Graduate School, Graduate Programs in Medicine, Department of Molecular Physiology and Biophysics, Stony Brook, NY 11794. Offers physiology and biophysics (PhD). *Faculty:* 17 full-time (3 women). *Students:* 21 full-time (12 women); includes 2 minority (both Asian Americans or Pacific Islanders), 8 international. Average age 29. 24 applicants, 33% accepted. In 2005, 7 degrees awarded. *Degree requirements:* For doctorate, thesis/dissertation, comprehensive exam. *Entrance requirements:* For doctorate, GRE General Test, GRE Subject Test, BS in related field, minimum GPA of 3.0. Additional exam requirements/recommendations for international students: Required—TOEFL. *Application deadline:* For fall admission, 1/15 for domestic students. Application fee: $50. *Expenses:* Tuition, area resident: Full-time $6,900; part-time $288. Tuition, state resident: Full-time $6,900. Tuition, nonresident: full-time $10,920; part-time $455. International tuition: $10,920 full-time. Required fees: $704. *Financial support:* In 2005–06, 13 research assistantships, 5 teaching assistantships were awarded; fellowships, Federal Work-Study also available. Financial award application deadline: 3/15. *Faculty research:* Cellular electrophysiology, membrane permeation and transport, metabolic endocrinology. Total annual research expenditures: $6.7 million. *Unit head:* Dr. Peter Brink, Chair, 631-444-2287, Fax: 631 444 3432. *Application contact:* Dr. Leon C. Moore, Graduate Adviser, 631-444-2287, Fax: 631-444-3432, E-mail: moore@pofvax.pnb.sunysb.edu.

See Close-Ups on pages 1259 and 473.

Tufts University, Sackler School of Graduate Biomedical Sciences, Graduate Program in Cellular and Molecular Physiology, Boston, MA 02155. Offers PhD. Applications are processed through through integrated studies program. *Faculty:* 25 full-time (7 women). *Students:* 14 full-time (8 women); includes 1 minority (Asian American or Pacific Islander), 1 international. Average age 27. In 2005, 3 degrees awarded. *Degree requirements:* For doctorate, thesis/dissertation. *Financial support:* In 2005–06, 14 students received support, including 14 research assistantships with full tuition reimbursements available (averaging $29,000 per year); scholarships/grants, health care benefits, and tuition waivers (full) also available. Financial award application deadline: 1/15. *Unit head:* Dr. Laura Liscum, Head, 617-636-6945, Fax: 617-636-0445. *Application contact:* 617-636-6767, Fax: 617-636-0375, E-mail: sackler-school@tufts.edu.

The University of Alabama at Birmingham, Graduate Programs in Joint Health Sciences, Birmingham, AL 35294. Offers biochemistry and molecular genetics (PhD), including biochemistry; cell biology (PhD), including cell biology, cellular and molecular biology, cellular and molecular physiology, neuroscience; genetics (PhD); microbiology (PhD); neurobiology (PhD); pathology (PhD); pharmacology and toxicology (PhD), including pharmacology, toxicology; physiology and biophysics (MSBMS, PhD), including basic medical sciences (MSBMS), biophysical sciences (PhD), integrative biomedical sciences (PhD). *Students:* 439 full-time (214 women), 6 part-time (2 women); includes 74 minority (36 African Americans, 4 American Indian/Alaska Native, 28 Asian Americans or Pacific Islanders, 6 Hispanic Americans), 164 international. Average age 28. 463 applicants, 37% accepted. In 2005, 9 master's, 53 doctorates awarded. *Entrance requirements:* For master's, GRE; for doctorate, GRE, interview. *Application deadline:* Applications are processed on a rolling basis. Application fee: $35 ($60 for international students). Electronic applications accepted. *Expenses:* Tuition, state resident: part-time $170 per credit hour. Tuition, nonresident: full-time $4,612; part-time $425 per credit hour. International tuition: $10,732 full-time. Required fees: $11 per credit hour. $124 per term. Tuition and fees vary according to course load, degree level and program. *Financial support:* Fellowships, career-related internships or fieldwork available. *Unit head:* Dr. Robert R. Rich, Vice President/Dean, School of Medicine, 205-934-1111, Fax: 205-934-0333, E-mail: rrich@uab.edu.

The University of Alabama at Birmingham, Graduate Programs in Joint Health Sciences, Department of Cell Biology, Graduate Program in Cellular and Molecular Physiology, Birmingham, AL 35294. Offers PhD. *Students:* 18 full-time (15 women); includes 7 minority (4 African Americans, 2 Asian Americans or Pacific Islanders, 1 Hispanic American), 12 international. 33 applicants, 21% accepted. In 2005, 8 degrees awarded. *Expenses:* Tuition, state resident: part-time $170 per credit hour. Tuition, nonresident: full-time $4,612; part-time $425 per credit hour. International tuition: $10,732 full-time. Required fees: $11 per credit hour. $124 per term. Tuition and fees vary according to course load, degree level and program. *Application contact:* Information Contact, 205-975-7145, Fax: 205-975-6748.

University of Chicago, Division of the Biological Sciences, Department of Neurobiology, Pharmacology, and Cell Physiology, Program in Cellular and Molecular Physiology, Chicago, IL 60637-1513. Offers PhD. *Faculty:* 25 full-time (5 women). *Students:* 1 (woman) full-time, 1 international. *Degree requirements:* For doctorate, thesis/dissertation, preliminary exam. *Entrance requirements:* For doctorate, GRE General Test. Additional exam requirements/recommendations for international students: Required—TOEFL. *Application deadline:* For fall admission, 12/28 priority date for domestic students, 12/28 priority date for international students. Application fee: $55. Electronic applications accepted. *Financial support:* In 2005–06, fellowships with tuition reimbursements (averaging $26,301 per year), research assistantships with tuition reimbursements (averaging $26,301 per year) were awarded; institutionally sponsored loans, scholarships/grants, traineeships, and health care benefits also available. Financial award applicants required to submit FAFSA. *Faculty research:* Molecular genetics, biochemical biological and physical approaches to cell physiology. *Unit head:* Dr. Eric Beyer, Chairman, 773-834-1498, Fax: 773-702-3774, E-mail: ebeyer@peds.bsd.uchicago.edu. *Application contact:* Diane J. Hall, Graduate Administrative Director, 773-702-6371, Fax: 773-702-1216, E-mail: d-hall@uchicago.edu.

The University of North Carolina at Chapel Hill, School of Medicine and Graduate School, Graduate Programs in Medicine, Department of Cell and Molecular Physiology, Chapel Hill, NC 27599. Offers PhD. *Faculty:* 21 full-time (6 women). *Students:* 25 full-time (18 women); includes 2 minority (both Asian Americans or Pacific Islanders), 8 international. Average age 26. 38 applicants, 13% accepted, 4 enrolled. In 2005, 7 degrees awarded. *Median time to degree:* Of those who began their doctoral program in fall 1997, 100% received their degree in 8 years or less. *Degree requirements:* For doctorate, thesis/dissertation, comprehensive exam, registration. *Entrance requirements:* For doctorate, GRE General Test. Additional exam requirements/recommendations for international students: Required—TOEFL (minimum score 550 paper-based; 213 computer-based). *Application deadline:* For fall admission, 1/1 priority date for domestic students, 1/1 priority date for international students. Applications are processed on a rolling basis. Application fee: $70. Electronic applications accepted. *Financial support:* In 2005–06, 3 students received support, including 5 fellowships with full tuition reimbursements available (averaging $22,000 per year), 20 research assistantships with full tuition reimbursements available (averaging $22,000 per year); Federal Work-Study, institutionally sponsored loans, scholarships/grants, traineeships, health care benefits, and unspecified assistantships also available. Support available to part-time students. Financial award application deadline: 3/1; financial award applicants required to submit FAFSA. *Faculty research:* Signal transduction; growth factors; cardiovascular diseases; neurobiology; hormones, receptors, and ion channels. Total annual research expenditures: $5.9 million. *Unit head:* Dr. James M. Anderson, Chair, 919-966-6411, Fax: 919-966-6927, E-mail: james_m_anderson@med.unc.edu. *Application contact:* Dr. Richard E. Cheney, Director of Graduate Studies, 919-966-0331, Fax: 919-966-6927, E-mail: cheneyr@med.unc.edu.

University of Pittsburgh, School of Medicine, Graduate Programs in Medicine, Program in Cell Biology and Molecular Physiology, Pittsburgh, PA 15260. Offers MS, PhD. *Faculty:* 29 full-time (7 women). *Students:* 20 full-time (8 women); includes 3 minority (2 African Americans, 1 Asian American or Pacific Islander), 1 international. Average age 28. 415 applicants, 22% accepted, 42 enrolled. In 2005, 1 degree awarded. *Median time to degree:* Of those who began their doctoral program in fall 1997, 95% received their degree in 8 years or less. *Degree requirements:* For doctorate, thesis/dissertation, comprehensive exam, registration. *Entrance requirements:* For doctorate, GRE General Test, GRE Subject Test, minimum QPA of 3.0. Additional exam requirements/recommendations for international students: Required—TOEFL (minimum score 600 paper-based; 250 computer-based), IELT (minimum score 7). *Application deadline:* For fall admission, 12/15 priority date for domestic students, 12/15 priority date for international students. Application fee: $40. Electronic applications accepted. *Expenses:* Tuition, state resident: full-time $13,194; part-time $537 per credit. Tuition, nonresident: full-time $25,012; part-time $1,026 per credit. Required fees: $700; $164 per term. Tuition and fees vary according to campus/location and program. *Financial support:* In 2005–06, fellowships with full tuition reimbursements (averaging $21,500 per year), research assistantships with tuition reimbursements (averaging $21,500 per year), teaching assistantships with tuition reimbursements (averaging $21,500 per year) were awarded; institutionally sponsored loans, scholarships/grants, traineeships, health care benefits, and unspecified assistantships also available. *Faculty research:* Genetic disorders of ion channels, regulation of gene expression/development, membrane traffic of proteins and lipids, reproductive biology, signal transduction in diabetes and metabolism. *Unit head:* Dr. Simon C. Watkins, Graduate Program Director, 412-648-3051, Fax: 412-648-8330, E-mail: swatkins@pitt.edu. *Application contact:* 412-648-8957, Fax: 412-648-1077, E-mail: gradstudies@medschool.pitt.edu.

University of Vermont, College of Medicine and Graduate College, Graduate Programs in Medicine, Department of Molecular Physiology and Biophysics, Burlington, VT 05405. Offers MS, PhD, MD/MS, MD/PhD. *Students:* 4 (2 women) 2 international. 16 applicants, 19% accepted, 2 enrolled. *Degree requirements:* For master's and doctorate, thesis/dissertation. *Entrance requirements:* For master's and doctorate, GRE General Test. Additional exam requirements/recommendations for international students: Required—TOEFL (minimum score 550 paper-based; 213 computer-based). *Application deadline:* For fall admission, 4/1 for domestic students. Applications are processed on a rolling basis. Application fee: $40. Electronic applications accepted. *Expenses:* Tuition, area resident: Part-time $410 per credit hour. Tuition, nonresident: part-time $1,034 per credit hour. *Financial support:* Fellowships, research assistantships, teaching assistantships available. Financial award application deadline: 3/1. *Unit head:* Dr. D. Warshaw, Chairperson, 802-656-2540. *Application contact:* Dr. C. Berger, Coordinator, 802-656-2540.

University of Virginia, School of Medicine, Department of Molecular Physiology and Biological Physics, Charlottesville, VA 22903. Offers biological and physical sciences (MS); molecular medicine and systems biology (PhD); physiology (PhD). *Students:* 32 full-time (13 women); includes 2 minority (both Asian Americans or Pacific Islanders), 6 international. Average age 28. In 2005, 21 master's, 5 doctorates awarded. *Degree requirements:* For doctorate, thesis/dissertation. *Entrance requirements:* For doctorate, GRE General Test, GRE Subject Test. Additional exam requirements/recommendations for international students: Required—TOEFL. *Application deadline:* Applications are processed on a rolling basis. Application fee: $60. Electronic applications accepted. *Expenses:* Tuition, state resident: full-time $7,731. Tuition, nonresident: full-time $18,672. Required fees: $1,479. Full-time tuition and fees vary according to degree level and program. *Financial support:* Fellowships, research assistantships, teaching assistantships available. Financial award applicants required to submit FAFSA. *Unit head:* Dr. Howard Kutchai, Interim Chair, 434-924-5108, Fax: 434-982-1616. *Application contact:* Peter C. Brunjes, Associate Dean for Graduate Programs and Research, 434-924-7184, Fax: 434-924-6737, E-mail: grad-a-s@virginia.edu.

Vanderbilt University, Graduate School and School of Medicine, Department of Molecular Physiology and Biophysics, Nashville, TN 37240-1001. Offers MS, PhD, MD/PhD. *Faculty:* 53 full-time (16 women). *Students:* 50 full-time (28 women), 1 (woman) part-time; includes 5 minority (1 African American, 1 American Indian/Alaska Native, 2 Asian Americans or Pacific Islanders, 1 Hispanic American), 10 international. In 2005, 1 master's, 6 doctorates awarded. *Degree requirements:* For doctorate, thesis/dissertation, preliminary, qualifying, and final exams. *Entrance requirements:* For doctorate, GRE General Test, GRE Subject Test (recommended). *Application deadline:* For fall admission, 1/15 for domestic students, 1/15 for international students. Application fee: $0. Electronic applications accepted. *Expenses:* Tuition: Part-time $1,283 per semester hour. Required fees: $2,202; $1,101 per semester. One-time

Molecular Physiology

Vanderbilt University (continued)
fee: $30. Tuition and fees vary according to course load, program and student level. *Financial support:* Fellowships with full tuition reimbursements, research assistantships with full tuition reimbursements, Federal Work-Study, institutionally sponsored loans, traineeships, and tuition waivers (partial) available. Financial award application deadline: 1/15. *Faculty research:* Molecular endocrinology, membrane transport biophysics, metabolic regulation, neurobiology.

Unit head: Alan D. Cherrington, Chair, 615-322-7000, Fax: 615-343-0490, E-mail: alan.cherrington@vanderbilt.edu. *Application contact:* Hussane Mchaourab, Director of Graduate Studies, 615-322-7000, Fax: 615-343-0490, E-mail: hassane.mchaourab@vanderbilt.edu.

Yale University, Graduate School of Arts and Sciences, Department of Cellular and Molecular Physiology, New Haven, CT 06520. Offers PhD. *Degree requirements:* For doctorate, thesis/dissertation. *Entrance requirements:* For doctorate, GRE General Test, GRE Subject Test.

Physiology

Albert Einstein College of Medicine, Sue Golding Graduate Division of Medical Sciences, Department of Physiology and Biophysics, Bronx, NY 10461. Offers PhD, MD/PhD. *Degree requirements:* For doctorate, thesis/dissertation. *Entrance requirements:* For doctorate, GRE General Test. Additional exam requirements/recommendations for international students: Required—TOEFL. *Faculty research:* Biophysical and biochemical basis of body function at the subcellular, cellular, organ, and whole-body level.

American University of Beirut, Graduate Programs, Faculty of Medicine, Beirut, Lebanon. Offers biochemistry (MS); human morphology (MS); microbiology and immunology (MS); neuroscience (MS); pharmacology and therapeutics (MS); physiology (MS). *Degree requirements:* For master's, one foreign language, thesis (for some programs), comprehensive exam, registration. *Entrance requirements:* For master's, GRE, letter of recommendation.

Arizona State University, Division of Graduate Studies, College of Liberal Arts and Sciences, Department of Biology, Program in Physiology, Tempe, AZ 85287. Offers MS, PhD. Terminal master's awarded for partial completion of doctoral program. *Degree requirements:* For master's, thesis; for doctorate, thesis/dissertation, oral exam. *Entrance requirements:* For master's and doctorate, GRE General Test, GRE Subject Test. Additional exam requirements/recommendations for international students: Required—TOEFL (minimum score 600 paper-based; 250 computer-based); Recommended—TSE.

Ball State University, Graduate School, College of Sciences and Humanities, Department of Physiology and Health Science, Program in Physiology, Muncie, IN 47306-1099. Offers MA, MS. *Students:* 5 full-time (2 women); includes 1 minority (African American), 1 international. Average age 22. 6 applicants, 100% accepted, 2 enrolled. In 2005, 4 degrees awarded. Application fee: $25 ($35 for international students). *Expenses:* Tuition, state resident: full-time $6,246. Tuition, nonresident: full-time $16,006. *Financial support:* Research assistantships with full tuition reimbursements, teaching assistantships with full tuition reimbursements available. Financial award application deadline: 3/1. *Unit head:* Dr. Marianna Tucker, Director, 765-285-8360, Fax: 765-285-3210.

Boston University, Sargent College of Health and Rehabilitation Sciences, Department of Health Sciences, Boston, MA 02215. Offers applied anatomy and physiology (MS, PhD); nutrition (MS). Part-time programs available. *Faculty:* 11 full-time (9 women), 6 part-time/adjunct (5 women). *Students:* 57 full-time (49 women), 5 part-time (all women); includes 7 minority (all Asian Americans or Pacific Islanders), 8 international. Average age 26. 67 applicants, 40% accepted, 20 enrolled. *Degree requirements:* For master's, thesis or alternative; for doctorate, one foreign language, thesis/dissertation. *Entrance requirements:* For master's, GRE General Test, minimum GPA of 3.0; for doctorate, GRE General Test. Additional exam requirements/recommendations for international students: Required—TOEFL (minimum score 550 paper-based). *Application deadline:* For fall admission, 3/1 for domestic students; for spring admission, 10/1 for domestic students. Applications are processed on a rolling basis. Application fee: $65. Electronic applications accepted. *Expenses:* Tuition: Full-time $31,530; part-time $985 per credit. Required fees: $316; $40 per semester. Tuition and fees vary according to course level and program. *Financial support:* In 2005–06, 20 fellowships with full tuition reimbursements, 7 research assistantships with full tuition reimbursements, 6 teaching assistantships with full tuition reimbursements were awarded; career-related internships or fieldwork, Federal Work-Study, institutionally sponsored loans, scholarships/grants, and tuition waivers (partial) also available. Support available to part-time students. Financial award application deadline: 4/15. *Faculty research:* Muscle metabolism, body acid-base balance, human performance, physical conditioning, diabetes. *Unit head:* Dr. Eileen O'Keefe, Chair, 617-353-7532, E-mail: ebokeefe@bu.edu. *Application contact:* Sharon Sankey, Director, Student Services, 617-353-2713, Fax: 617-353-7500, E-mail: ssankey@bu.edu.

Boston University, School of Medicine, Division of Graduate Medical Sciences, Department of Physiology and Biophysics, Boston, MA 02118. Offers MA, PhD, MD/PhD. Part-time programs available. *Faculty:* 24. *Students:* 26 full-time (12 women); includes 2 minority (1 Asian American or Pacific Islander, 1 Hispanic American), 20 international. Average age 27. Terminal master's awarded for partial completion of doctoral program. *Degree requirements:* For master's and doctorate, thesis/dissertation, qualifying exam. *Entrance requirements:* For master's and doctorate, GRE General Test, GRE Subject Test (strongly recommended). Additional exam requirements/recommendations for international students: Required—TOEFL. *Application deadline:* For fall admission, 1/1 for domestic students; for spring admission, 10/15 priority date for domestic students. Electronic applications accepted. *Expenses:* Tuition: Full-time $31,530; part-time $985 per credit. Required fees: $316; $40 per semester. Tuition and fees vary according to course level and program. *Financial support:* Fellowships with tuition reimbursements, research assistantships with tuition reimbursements, scholarships/grants and traineeships available. *Faculty research:* X-ray scattering, NMR spectroscopy, protein crystallography, structural electron microscopy, molecular modeling. *Unit head:* Dr. Donald M. Small, Chairman, 617-638-4001. *Application contact:* Dr. Christopher Akey, Director of Graduate Studies, 617-638-4042, Fax: 617-638-4041, E-mail: cakey@bu.edu.

See Close-Up on page 467.

Brigham Young University, Graduate Studies, College of Biological and Agricultural Sciences, Department of Physiology and Developmental Biology, Provo, UT 84602-1001. Offers neuroscience (MS, PhD); physiology and developmental biology (MS, PhD). Part-time programs available. *Faculty:* 17 full-time (0 women). *Students:* 25 full-time (11 women); includes 1 minority (Asian American or Pacific Islander), 1 international. Average age 26. 26 applicants, 38% accepted, 9 enrolled. In 2005, 5 master's, 2 doctorates awarded. Terminal master's awarded for partial completion of doctoral program. *Degree requirements:* For master's and doctorate, thesis/dissertation. *Entrance requirements:* For master's, GRE General Test, minimum GPA of 3.0 during previous 2 years; for doctorate, GRE General Test, minimum GPA of 3.0 overall. Additional exam requirements/recommendations for international students: Required—TOEFL. *Application deadline:* For fall admission, 2/1 priority date for domestic students, 2/1 priority date for international students. For winter admission, 9/10 for domestic students. Application fee: $50. Electronic applications accepted. *Financial support:* In 2005–06, 26 students received support, including 13 research assistantships with full tuition reimbursements available (averaging $15,500 per year), 13 teaching assistantships with partial tuition reimbursements available (averaging $14,900 per year); fellowships with partial tuition reimbursements available, career-related internships or fieldwork, institutionally sponsored loans, scholarships/grants, tuition waivers (full and partial), unspecified assistantships, and tuition awards also available. Financial award application deadline: 2/1. *Faculty research:* Sex differentiation of the brain, exercise physiology, developmental biology, membrane biophysics, neuroscience. Total annual research expenditures: $892,386. *Unit head:* Dr. James P. Porter, Chair, 801-422-9160, Fax: 801-422-0700, E-mail: james_porter@byu.edu. *Application contact:*

Dr. Dixon J. Woodbury, Graduate Coordinator, 801-422-7562, Fax: 801-422-0700, E-mail: dixon_woodbury@byu.edu.

See Close-Up on page 1243.

Brown University, Graduate School, Division of Biology and Medicine, Program in Molecular Pharmacology and Physiology, Providence, RI 02912. Offers MA, Sc M, PhD, MD/PhD. *Degree requirements:* For doctorate, thesis/dissertation, preliminary exam. *Entrance requirements:* For master's and doctorate, GRE General Test, GRE Subject Test. Additional exam requirements/recommendations for international students: Required—TOEFL. Electronic applications accepted. *Faculty research:* Structural biology, antiplatelet drugs, nicotinic receptor structure/function.

Case Western Reserve University, School of Medicine and School of Graduate Studies, Graduate Programs in Medicine, Department of Physiology and Biophysics, Cleveland, OH 44106. Offers cell physiology (PhD); molecular/cellular biophysics (PhD); physiology and biophysics (PhD); physiology and biotechnology (MS); systems physiology (PhD). *Faculty:* 52. *Students:* 34 full-time (15 women); includes 8 minority (3 African Americans, 5 Asian Americans or Pacific Islanders), 9 international. Average age 27. 59 applicants, 25% accepted, 7 enrolled. In 2005, 5 master's, 3 doctorates awarded. Terminal master's awarded for partial completion of doctoral program. *Degree requirements:* For master's and doctorate, thesis/dissertation. *Entrance requirements:* For master's, GRE General Test, minimum GPA of 3.28; for doctorate, GRE General Test, minimum GPA of 3.6. Additional exam requirements/recommendations for international students: Required—TOEFL. *Application deadline:* For fall admission, 3/15 for domestic students. Applications are processed on a rolling basis. Application fee: $50. Electronic applications accepted. *Financial support:* In 2005–06, fellowships with tuition reimbursements (averaging $22,000 per year), research assistantships (averaging $22,000 per year) were awarded; scholarships/grants, health care benefits, and tuition waivers (full) also available. *Faculty research:* Cardiovascular physiology, calcium metabolism, epithelial cell biology. Total annual research expenditures: $12.1 million. *Unit head:* Dr. Cathleen Carlin, Interim Chairman and Professor, 216-368-8939, Fax: 216-368-5586, E-mail: cxc39@case.edu. *Application contact:* Jean Davis, Coordinator, Graduate Training Programs, 216-368-2084, Fax: 216-368-5586, E-mail: jxd16@po.cwru.edu.

See Close-Up on page 1245.

Columbia University, College of Physicians and Surgeons and Graduate School of Arts and Sciences, Graduate School of Arts and Sciences at the College of Physicians and Surgeons, Department of Physiology and Cellular Biophysics, New York, NY 10032. Offers M Phil, MA, PhD, MD/PhD. Only candidates for the PhD are admitted. Terminal master's awarded for partial completion of doctoral program. *Degree requirements:* For doctorate, thesis/dissertation. *Entrance requirements:* For master's and doctorate, GRE General Test. Additional exam requirements/recommendations for international students: Required—TOEFL. *Expenses:* Tuition: Full-time $31,448. Tuition and fees vary according to course level, course load, campus/location and program. *Faculty research:* Membrane physiology, cellular biology, cardiovascular physiology, neurophysiology.

Cornell University, College of Veterinary Medicine, Ithaca, NY 14853-0001. Offers comparative biomedical science (PhD); immunology (MS, PhD); pharmacology (PhD); physiology (PhD); veterinary medicine (DVM); zoology (MS, PhD). *Accreditation:* AVMA. *Faculty:* 155 full-time (53 women). *Students:* 334 full-time (267 women); includes 71 minority (25 African Americans, 2 American Indian/Alaska Native, 18 Asian Americans or Pacific Islanders, 26 Hispanic Americans), 2 international. Average age 26. 871 applicants, 11% accepted, 78 enrolled. In 2005, 81 first professional degrees, 15 doctorates awarded. *Degree requirements:* For first-professional, thesis or alternative, on-site clinical training. *Entrance requirements:* GRE General Test or MCAT, undergraduate pre-medical science program, animal or veterinary experience, letter of recommendation. *Application deadline:* For fall admission, 10/1 for domestic students, 10/1 for international students. Application fee: $40. Electronic applications accepted. *Expenses:* Contact institution. *Financial support:* In 2005–06, 303 students received support, including 30 fellowships (averaging $24,854 per year), 88 research assistantships with tuition reimbursements available (averaging $24,854 per year); Federal Work-Study, institutionally sponsored loans, scholarships/grants, and unspecified assistantships also available. Financial award application deadline: 2/1; financial award applicants required to submit CSS PROFILE or FAFSA. *Faculty research:* Extensive biomedical research, comparative cancer, food safety. Total annual research expenditures: $49.3 million. *Unit head:* Dr. Donald F. Smith, Dean, 607-253-3771. *Application contact:* Jennifer A Mailey, Director of Admissions, 607-253-3700, Fax: 607-253-3709, E-mail: vet_admissions@cornell.edu.

Cornell University, Graduate School, Graduate Fields of Comparative Biomedical Sciences, Field of Physiology, Ithaca, NY 14853-0001. Offers behavioral physiology (MS, PhD); cardiovascular and respiratory physiology (MS, PhD); endocrinology (MS, PhD); environmental and comparative physiology (MS, PhD); gastrointestinal and metabolic physiology (MS, PhD); membrane and epithelial physiology (MS, PhD); molecular and cellular physiology (MS, PhD); neural and sensory physiology (MS, PhD); physiological genomics (MS, PhD); reproductive physiology (MS, PhD). *Faculty:* 57 full-time (16 women). *Students:* 15 applicants, 33% accepted, 5 enrolled. In 2005, 1 degree awarded. *Degree requirements:* For master's, thesis; for doctorate, thesis/dissertation, 1 semester of teaching experience, seminar presentation, comprehensive exam. *Entrance requirements:* For master's, GRE General Test, GRE Subject Test (biochemistry, cell and molecular biology, biology or chemistry), 2 letters of recommendation; for doctorate, GRE General Test, GRE Subject Test (biochemistry, cell and molecular biology, biology, or chemistry), 2 letters of recommendation. Additional exam requirements/recommendations for international students: Required—TOEFL (minimum score 550 paper-based; 213 computer-based). *Application deadline:* For fall admission, 12/15 for domestic students. Application fee: $60. Electronic applications accepted. *Financial support:* In 2005–06, 21 students received support, including 5 fellowships with full tuition reimbursements available, 15 research assistantships with full tuition reimbursements available, 1 teaching assistantship with full tuition reimbursement available; institutionally sponsored loans, scholarships/grants, health care benefits, tuition waivers (full and partial), and unspecified assistantships also available. Financial award applicants required to submit FAFSA. *Faculty research:* Endocrinology and reproductive physiology, cardiovascular and respiratory physiology, gastrointestinal and metabolic physiology, molecular and cellular physiology, physiological genomics. *Unit head:* Director of Graduate Studies, 607-253-3276, Fax: 607-253-3756. *Application contact:* Graduate Field Assistant, 607-253-3276, Fax: 607-253-3756, E-mail: graduate_edcvm@cornell.edu.

Cornell University, Joan and Sanford I. Weill Medical College and Graduate School of Medical Sciences, Weill Graduate School of Medical Sciences, Program in Physiology,

Biophysics and Systems Biology, New York, NY 10021-4896. Offers PhD, MD/PhD. *Faculty:* 30 full-time (8 women). *Students:* 39 full-time (19 women); includes 4 minority (1 African American, 3 Asian Americans or Pacific Islanders), 21 international. 32 applicants, 25% accepted, 4 enrolled. In 2005, 1 doctorate awarded. *Degree requirements:* For doctorate, thesis/dissertation, final exam. *Entrance requirements:* For doctorate, GRE General Test, GRE Subject Test, introductory courses in biology, inorganic and organic chemistry, physics, and mathematics. Additional exam requirements/recommendations for international students: Required—TOEFL. *Application deadline:* For fall admission, 12/15 for domestic students. Application fee: $60. *Expenses:* Tuition: Full-time $32,320. Required fees: $1,025. *Financial support:* In 2005–06, 1 fellowship was awarded; stipends also available. *Unit head:* Doris Herzlinger, Director, 212-746-6377, E-mail: daherzli@med.cornell.edu.

Dalhousie University, Faculty of Graduate Studies and Faculty of Medicine, Graduate Programs in Medicine, Department of Physiology and Biophysics, Halifax, NS B3H 4R2, Canada. Offers M Sc, PhD, MD/PhD. *Degree requirements:* For master's and doctorate, thesis/dissertation. *Entrance requirements:* For master's and doctorate, GRE Subject Test. Additional exam requirements/recommendations for international students: Required—TOEFL. *Faculty research:* Computer modeling, reproductive and endocrine physiology, cardiovascular physiology, neurophysiology, membrane biophysics.

Dartmouth College, School of Arts and Sciences, Department of Physiology, Lebanon, NH 03756. Offers PhD, MD/PhD. *Faculty:* 23 full-time (9 women). *Students:* 21 full-time (12 women); includes 2 minority (1 American Indian/Alaska Native, 1 Asian American or Pacific Islander), 3 international. Average age 29. 10 applicants, 90% accepted, 6 enrolled. In 2005, 3 degrees awarded. *Degree requirements:* For doctorate, thesis/dissertation. *Entrance requirements:* For doctorate, GRE General Test, GRE Subject Test. Additional exam requirements/recommendations for international students: Required—TOEFL. *Application deadline:* For fall admission, 1/20 for domestic students. Applications are processed on a rolling basis. Application fee: $0. *Expenses:* Tuition: Full-time $31,770. *Financial support:* In 2005–06, 19 students received support, including fellowships with full tuition reimbursements available (averaging $23,500 per year), research assistantships with full tuition reimbursements available (averaging $23,500 per year); Federal Work-Study, institutionally sponsored loans, scholarships/grants, and tuition waivers (full and partial) also available. Financial award application deadline: 4/15. *Faculty research:* Respiratory control, endocrinology of reproduction and immunology, regulation of receptors and channels, electrophysiology of membranes, renal function. Total annual research expenditures: $10.9 million. *Unit head:* Dr. Donald Bartlett, Chair, 603-650-7717, Fax: 603-650-6130, E-mail: donald.bartlet.jr@dartmouth.edu. *Application contact:* Dr. Valerie Anne Galton, Coordinator of Graduate Studies, 603-650-7717, Fax: 603-650-6130, E-mail: valerie.a.galton@dartmouth.edu.

See Close-Up on page 1247.

East Carolina University, Brody School of Medicine, Department of Physiology, Greenville, NC 27858-4353. Offers PhD. *Faculty:* 10 full-time (0 women), 4 part-time/adjunct (1 woman). *Students:* 4 full-time (1 woman), 5 part-time (4 women); includes 1 minority (African American), 3 international. Average age 28. 7 applicants, 29% accepted. In 2005, 2 degrees awarded. *Median time to degree:* Of those who began their doctoral program in fall 1997, 67% received their degree in 8 years or less. *Degree requirements:* For doctorate, thesis/dissertation, comprehensive exam, registration. *Entrance requirements:* For doctorate, GRE General Test. Additional exam requirements/recommendations for international students: Required—TOEFL. *Application deadline:* For fall admission, 6/1 for domestic students. Applications are processed on a rolling basis. Application fee: $50. *Expenses:* Tuition, state resident: full-time $2,516. Tuition, nonresident: full-time $12,832. *Financial support:* In 2005–06, fellowships with full and partial tuition reimbursements (averaging $21,500 per year) Financial award application deadline: 6/1. *Faculty research:* Cell and nerve biophysics; neurophysiology; cardiovascular, renal, endocrine, and gastrointestinal physiology; pulmonary/asthma. Total annual research expenditures: $1.3 million. *Unit head:* Dr. Robert Lust, Chairman, 252-744-2762, Fax: 252-744-3460, E-mail: lustr@ecu.edu. *Application contact:* Dr. Mike Van Scott, Graduate Director, 252-744-3654, Fax: 252-744-3460, E-mail: vanscottmi@ecu.edu.

East Tennessee State University, James H. Quillen College of Medicine, Biomedical Science Graduate Program, Johnson City, TN 37614. Offers anatomy (MS, PhD); biochemistry (MS, PhD); biophysics (MS, PhD); microbiology (MS, PhD); pharmacology (MS, PhD); physiology (MS, PhD). Part-time programs available. *Faculty:* 49 full-time (12 women), 1 (woman) part-time/adjunct. *Students:* 30 full-time (19 women), 5 part-time (4 women); includes 2 minority (1 African American, 1 Asian American or Pacific Islander), 11 international. Average age 31. 78 applicants, 13% accepted, 9 enrolled. In 2005, 1 master's, 4 doctorates awarded. Terminal master's awarded for partial completion of doctoral program. *Degree requirements:* For master's, one foreign language, thesis, comprehensive qualifying exam; for doctorate, 2 foreign languages, thesis/dissertation. *Entrance requirements:* For master's, GRE General Test, minimum GPA of 3.0, bachelor's degree in biological or related science; for doctorate, GRE General Test, GRE Subject Test. Additional exam requirements/recommendations for international students: Required—TOEFL (minimum score 550 paper-based; 213 computer-based). *Application deadline:* For fall admission, 3/15 for domestic students; for spring admission, 3/1 for domestic students. Application fee: $25 ($35 for international students). *Expenses:* Contact institution. *Financial support:* In 2005–06, 7 research assistantships with full tuition reimbursements (averaging $15,000 per year) were awarded; teaching assistantships with full tuition reimbursements, career-related internships or fieldwork, Federal Work-Study, institutionally sponsored loans, scholarships/grants, and tuition waivers (full) also available. Financial award application deadline: 7/1; financial award applicants required to submit FAFSA. Total annual research expenditures: $2.1 million. *Unit head:* Dr. Mitchell E. Robinson, Assistant Dean, Director, 423-439-4658, E-mail: robinson@etsu.edu.

Florida State University, Graduate Studies, College of Human Sciences, Department of Nutrition, Food, and Exercise Sciences, Tallahassee, FL 32306. Offers exercise science (PhD), including exercise physiology (MS, PhD), motor learning and control (MS, PhD); movement science (MS), including exercise physiology (MS, PhD), motor learning and control (MS, PhD); nutrition and food science (PhD); nutrition and food sciences (MS), including clinical nutrition, food science, nutrition and sport, nutrition science, nutrition, education and health promotion. *Faculty:* 13 full-time (10 women). *Students:* 37 full-time (24 women), 28 part-time (20 women); includes 17 minority (9 African Americans, 4 Asian Americans or Pacific Islanders, 4 Hispanic Americans), 10 international. 67 applicants, 67% accepted, 23 enrolled. In 2005, 22 master's, 2 doctorates awarded. *Degree requirements:* For master's, thesis optional; for doctorate, thesis/dissertation, registration. *Entrance requirements:* For master's and doctorate, GRE General Test, minimum GPA of 3.0. Additional exam requirements/recommendations for international students: Required—TOEFL. *Application deadline:* For fall admission, 7/1 for domestic students, 5/1 for international students; for spring admission, 11/1 for domestic students, 12/1 for international students. Application fee: $30. Electronic applications accepted. *Financial support:* In 2005–06, 43 students received support, including 3 fellowships with partial tuition reimbursements available (averaging $10,000 per year), 9 research assistantships with partial tuition reimbursements available (averaging $8,000 per year), 22 teaching assistantships with partial tuition reimbursements available (averaging $8,000 per year); career-related internships or fieldwork, Federal Work-Study, institutionally sponsored loans, scholarships/grants, and unspecified assistantships also available. Financial award application deadline: 1/15; financial award applicants required to submit FAFSA. *Faculty research:* Nutrition and exercise, vitamin A deficiency, protein biochemistry, cardiovascular responses to exercises, physiological effects of cigarette smoking related to health and wellness. *Unit head:* Dr. Bahram Arjmandi, Chair, 850-644-1828, Fax: 850-645-5000. *Application contact:* Ursula Tate, Program Assistant, 850-644-4800, Fax: 850-645-5000, E-mail: utate@mailer.fsu.edu.

Georgetown University, Graduate School of Arts and Sciences, Programs in Biomedical Sciences, Department of Physiology and Biophysics, Washington, DC 20057. Offers MS,

PhD, MD/PhD. *Degree requirements:* For doctorate, thesis/dissertation. *Entrance requirements:* For master's, GRE General Test, MCAT; for doctorate, GRE General Test. Additional exam requirements/recommendations for international students: Required—TOEFL.

Georgia Institute of Technology, Graduate Studies and Research, College of Sciences, School of Applied Physiology, Program in Prosthetics and Orthotics, Atlanta, GA 30332-0001. Offers MS.

Georgia State University, College of Arts and Sciences, Department of Biology, Program in Cell Biology and Physiology, Atlanta, GA 30303-3083. Offers MS, PhD. *Degree requirements:* For master's, thesis or alternative, exam; for doctorate, thesis/dissertation, exam. *Entrance requirements:* For master's and doctorate, GRE General Test. Additional exam requirements/recommendations for international students: Required—TOEFL. *Expenses:* Tuition, state resident: full-time $4,368; part-time $182 per semester hour. Tuition, nonresident: full-time $8,732; part-time $728 per semester hour. Required fees: $46 per semester hour.

Harvard University, Graduate School of Arts and Sciences, Department of Systems Biology, Cambridge, MA 02138. Offers PhD. *Students:* 9. 100 applicants, 12% accepted, 9 enrolled. *Degree requirements:* For doctorate, thesis/dissertation, lab rotation, qualifying examination. *Entrance requirements:* For doctorate, GRE. Additional exam requirements/recommendations for international students: Required—TOEFL. *Application deadline:* For fall admission, 12/8 priority date for domestic students, 12/8 priority date for international students. Application fee: $90. Electronic applications accepted. *Expenses:* Tuition: Full-time $28,752. Full-time tuition and fees vary according to program and student level. *Financial support:* Institutionally sponsored loans, scholarships/grants, unspecified assistantships, and all students receive a stipend and full tuition reimbursements available. *Unit head:* Judy Finkelstein, Graduate Coordinator, 617-432-5876, Fax: 617-432-5012, E-mail: pamela_silver@dfci.harvard.edu. *Application contact:* Jodi Finkelstein, Coordinator, 617-432-5202, Fax: 617-432-5012, E-mail: jodi_finkelstein@hms.harvard.edu.

See Close-Up on page 675.

Harvard University, School of Public Health, Department of Environmental Health, Boston, MA 02115-6096. Offers environmental epidemiology (SM, DPH, SD); environmental health (SM); environmental science and engineering (SM, SD); occupational health (MOH, SM, DPH, SD); physiology (SD). *Accreditation:* ABET (one or more programs are accredited); CEPH. Part-time programs available. *Degree requirements:* For doctorate, thesis/dissertation, qualifying exam. *Entrance requirements:* For master's and doctorate, GRE. Additional exam requirements/recommendations for international students: Required—TOEFL (minimum score 560 paper-based; 220 computer-based); Recommended—IELT (minimum score 7). Electronic applications accepted. *Expenses:* Tuition: Full-time $28,752. Full-time tuition and fees vary according to program and student level. *Faculty research:* Industrial hygiene and occupational safety, population genetics, indoor and outdoor air pollution, cell and molecular biology of the lungs, infectious diseases.

Howard University, Graduate School of Arts and Sciences, Department of Physiology and Biophysics, Program in Biophysics, Washington, DC 20059-0002. Offers biophysics (PhD); physiology (PhD). Part-time programs available. *Degree requirements:* For doctorate, thesis/dissertation, comprehensive exam, registration. *Entrance requirements:* For doctorate, GRE General Test, minimum B average in field. *Faculty research:* Cardiovascular physiology, pulmonary physiology, neurophysiology, endocrinology.

Howard University, Graduate School of Arts and Sciences, Department of Physiology and Biophysics, Program in Physiology, Washington, DC 20059-0002. Offers PhD. *Degree requirements:* For doctorate, thesis/dissertation, comprehensive exam. *Entrance requirements:* For doctorate, GRE General Test, minimum B average in field. *Faculty research:* Cardiovascular physiology, pulmonary physiology, neurophysiology, endocrinology.

Illinois State University, Graduate School, College of Arts and Sciences, Department of Biological Sciences, Normal, IL 61790-2200. Offers biological sciences (MS); biology (PhD); biotechnology (MS); botany (PhD); ecology (PhD); genetics (PhD); microbiology (MS); physiology (PhD); zoology (PhD). Part-time programs available. *Faculty:* 27 full-time (6 women). *Students:* 36 full-time (28 women), 25 part-time (17 women); includes 1 minority (Asian American or Pacific Islander), 23 international. 80 applicants, 21% accepted. In 2005, 14 master's, 5 doctorates awarded. *Degree requirements:* For master's, thesis or alternative; for doctorate, variable foreign language requirement, thesis/dissertation, 2 terms of residency. *Entrance requirements:* For master's, GRE General Test, minimum GPA of 2.6 in last 60 hours of course work; for doctorate, GRE General Test. *Application deadline:* Applications are processed on a rolling basis. Application fee: $30. *Expenses:* Tuition, state resident: full-time $3,060; part-time $170 per credit hour. Tuition, nonresident: full-time $6,390; part-time $355 per credit hour. Required fees: $1,411; $47 per credit hour. *Financial support:* In 2005–06, 22 research assistantships (averaging $13,909 per year), 38 teaching assistantships (averaging $12,617 per year) were awarded; Federal Work-Study, tuition waivers (full), and unspecified assistantships also available. Financial award application deadline: 4/1. *Faculty research:* CRUI: osmoregulation in euryhaline fish: physiology, ecology and molecular biology; the PRISM project: enhancing science and math education; cell structure—functions of Na pump assembly. *Unit head:* Dr. Hou Tak Takucheung, Acting Chairperson, 309-438-3669. *Application contact:* Derek A. McCracken, Graduate Adviser, 309-438-3664.

See Close-Up on page 147.

Indiana State University, School of Graduate Studies, College of Arts and Sciences, Department of Life Sciences, Terre Haute, IN 47809-1401. Offers ecology (PhD); life sciences (MS); microbiology (PhD); physiology (PhD); science education (MS); sports medicine (PhD). *Faculty:* 24 full-time (8 women), 12 part-time/adjunct (2 women). *Students:* 51 full-time (24 women), 24 part-time (10 women); includes 3 minority (all Asian Americans or Pacific Islanders), 11 international. Average age 26. 45 applicants, 73% accepted, 29 enrolled. In 2005, 9 master's, 2 doctorates awarded. *Degree requirements:* For master's, thesis (for some programs); for doctorate, thesis/dissertation, comprehensive exam. *Entrance requirements:* For master's and doctorate, GRE General Test. *Application deadline:* For fall admission, 7/1 for domestic students; for spring admission, 11/1 priority date for domestic students. Applications are processed on a rolling basis. Application fee: $35. Electronic applications accepted. *Expenses:* Tuition, state resident: full-time $6,288; part-time $262 per credit hour. Tuition, nonresident: full-time $12,504; part-time $521 per credit hour. *Financial support:* In 2005–06, 26 teaching assistantships with partial tuition reimbursements (averaging $8,005 per year) were awarded; research assistantships with partial tuition reimbursements, Federal Work-Study, institutionally sponsored loans, and tuition waivers (partial) also available. Financial award application deadline: 3/1; financial award applicants required to submit FAFSA. *Unit head:* Dr. Swapan Ghosh, Interim Chairperson, 812-237-2400.

See Close-Up on page 149.

Indiana University Bloomington, Medical Sciences Program, Bloomington, IN 47405-7000. Offers anatomy and cell biology (MA, PhD); pharmacology (MS, PhD); physiology (MA, PhD). *Students:* 11 full-time (2 women), 9 part-time (5 women); includes 5 minority (1 African American, 4 Asian Americans or Pacific Islanders), 2 international. Average age 29. *Entrance requirements:* For master's, GRE, minimum GPA of 3.0; for doctorate, GRE. Additional exam requirements/recommendations for international students: Required—TOEFL. *Application deadline:* For fall admission, 1/15 for domestic students; for spring admission, 9/1 for domestic students. Application fee: $45 ($55 for international students). *Expenses:* Tuition, state resident: full-time $5,437; part-time $227 per credit hour. Tuition, nonresident: full-time $15,836; part-time $660 per credit hour. Required fees: $821. Tuition and fees vary according to campus/location and program. *Financial support:* Fellowships available. *Unit head:* Dr. John B. Watkins, Assistant Dean/Director, 812-855-0616. *Application contact:* Kimberly Bunch, Director of Graduate Admissions, 812-855-1119, E-mail: kbunch@indiana.edu.

Physiology

Indiana University–Purdue University Indianapolis, Indiana University School of Medicine, Department of Cellular and Integrative Physiology, Indianapolis, IN 46202-2896. Offers MS, PhD, MD/PhD. *Faculty:* 12 full-time (4 women). *Students:* 37 full-time (13 women), 11 part-time (3 women); includes 5 minority (1 African American, 4 Asian Americans or Pacific Islanders), 9 international. Average age 26. Terminal master's awarded for partial completion of doctoral program. *Degree requirements:* For doctorate, thesis/dissertation. *Entrance requirements:* For master's, GRE General Test, GRE Subject Test, previous course work in biology, calculus, physical chemistry, and physics; for doctorate, GRE General Test, GRE Subject Test. *Application deadline:* For fall admission, 4/15 for domestic students. Applications are processed on a rolling basis. Application fee: $50 ($60 for international students). *Expenses:* Tuition, state resident: full-time $5,159; part-time $215 per credit hour. Tuition, nonresident: full-time $14,890; part-time $620 per credit hour. Required fees: $614. Tuition and fees vary according to campus/location and program. *Financial support:* In 2005–06, 14 students received support; fellowships with partial tuition reimbursements available, research assistantships with partial tuition reimbursements available, Federal Work-Study and institutionally sponsored loans available. *Faculty research:* Cardiovascular physiology, cell growth and development, respiratory biology, cell signaling mechanisms, cytoskeleton function. *Unit head:* Dr. Michael Sturek, Chair, 317-274-7772, Fax: 317-274-3318. *Application contact:* Dr. Fred M. Pavalko, Graduate Director, 317-274-3140, Fax: 317-274-3318, E-mail: fpavalko@iupui.edu.

The Johns Hopkins University, Bloomberg School of Public Health, Department of Environmental Health Sciences, Baltimore, MD 21218-2699. Offers environmental health engineering (PhD, Sc D); environmental health sciences (MHS, Dr PH); molecular imaging (PhD, Sc D); occupational and environmental health (PhD, Sc D); occupational and environmental hygiene (MHS); physiology (PhD, Sc D); toxicological sciences (PhD, Sc D). *Faculty:* 64 full-time (19 women), 59 part-time/adjunct (25 women). *Students:* 75 full-time (50 women), 15 part-time (10 women); includes 16 minority (4 African Americans, 10 Asian Americans or Pacific Islanders, 2 Hispanic Americans), 20 international. Average age 29. 105 applicants, 57% accepted, 36 enrolled. In 2005, 11 master's, 6 doctorates awarded. *Median time to degree:* Of those who began their doctoral program in fall 1997, 75% received their degree in 8 years or less. *Degree requirements:* For master's, thesis, registration; for doctorate, thesis/dissertation, 1 year full-time residency, oral and written exams, comprehensive exam, registration. *Entrance requirements:* For master's and doctorate, GRE General Test or MCAT, 3 letters of recommendation, curriculum vitae. Additional exam requirements/recommendations for international students: Required—TOEFL (minimum score 550 paper-based; 213 computer-based). *Application deadline:* For fall admission, 1/15 priority date for domestic students, 1/15 priority date for international students. Applications are processed on a rolling basis. Application fee: $45. Electronic applications accepted. *Expenses:* Tuition: Full-time $30,960. Tuition and fees vary according to degree level and program. *Financial support:* In 2005–06, 104 students received support, including 1 fellowship (averaging $23,500 per year); Federal Work-Study, institutionally sponsored loans, scholarships/grants, traineeships, health care benefits, and stipends also available. Support available to part-time students. Financial award application deadline: 3/15; financial award applicants required to submit FAFSA. *Faculty research:* Respiratory and cardiovascular physiology, chemical carcinogenesis, antioxidants and drug metabolism, nuclear medicine, nuclear imaging. Total annual research expenditures: $20.5 million. *Unit head:* Dr. John Davis Groopman, Chair, 410-955-3720, Fax: 410-955-0617, E-mail: jgroopma@jhsph.edu. *Application contact:* Nina Kulacki, Senior Academic Coordinator, 410-955-2212, Fax: 410-955-0617, E-mail: nkulacki@jhsph.edu.

The Johns Hopkins University, School of Medicine, Graduate Programs in Medicine, Department of Physiology, Baltimore, MD 21218-2699. Offers cellular and molecular physiology (PhD); physiology (PhD). *Faculty:* 21 full-time (6 women). *Students:* 13 full-time (8 women), 12 international. Average age 23. 46 applicants, 9% accepted, 4 enrolled. In 2005, 6 degrees awarded. *Median time to degree:* Of those who began their doctoral program in fall 1997, 100% received their degree in 8 years or less. *Degree requirements:* For doctorate, thesis/dissertation, oral and qualifying exams. *Entrance requirements:* For doctorate, GRE General Test, previous course work in biology, calculus, chemistry, and physics. Additional exam requirements/recommendations for international students: Required—TOEFL. *Application deadline:* For fall admission, 1/10 for domestic students, 1/10 for international students. Applications are processed on a rolling basis. Application fee: $75. Electronic applications accepted. *Expenses:* Tuition: Full-time $30,960. Tuition and fees vary according to degree level and program. *Financial support:* In 2005–06, 6 students received support, including 3 fellowships with full tuition reimbursements available (averaging $25,200 per year), research assistantships with full tuition reimbursements available (averaging $25,200 per year), teaching assistantships with full tuition reimbursements available (averaging $25,200 per year); scholarships/grants, traineeships, tuition waivers (full), unspecified assistantships, and stipends also available. Financial award application deadline: 1/10. *Faculty research:* Membrane biochemistry and biophysics; signal transduction; developmental genetics and physiology; physiology and biochemistry; transporters, carriers, and ion channels. *Unit head:* Dr. William B. Guggino, Director, 410-955-7166, Fax: 410-614-8331. *Application contact:* Madeline J. McLaughlin, Academic Program Coordinator, 410-955-8333, Fax: 410-614-8331, E-mail: mmclaugh@jhmi.edu.

Kansas State University, College of Veterinary Medicine and Graduate School, Graduate Programs in Veterinary Medicine, Department of Anatomy and Physiology, Manhattan, KS 66506. Offers biomedical science (MS); physiology (PhD). *Faculty:* 14 full-time (3 women), 1 part-time/adjunct (0 women). *Students:* 16 full-time (5 women), 7 part-time (4 women); includes 1 minority (Hispanic American), 12 international. Average age 30. 4 applicants, 100% accepted, 4 enrolled. In 2005, 2 degrees awarded. Terminal master's awarded for partial completion of doctoral program. *Degree requirements:* For master's, thesis; for doctorate, one foreign language, thesis/dissertation. *Entrance requirements:* Additional exam requirements/recommendations for international students: Required—TOEFL (minimum score 550 paper-based; 213 computer-based). *Application deadline:* For fall admission, 2/1 for domestic students. Applications are processed on a rolling basis. Application fee: $30 ($55 for international students). *Expenses:* Tuition, state resident: full-time $5,160; part-time $215 per credit hour. Tuition, nonresident: full-time $12,816; part-time $534 per credit hour. Required fees: $564. *Financial support:* In 2005–06, 12 research assistantships (averaging $21,325 per year) were awarded; teaching assistantships with partial tuition reimbursements, Federal Work-Study, institutionally sponsored loans, and scholarships/grants also available. Financial award application deadline: 3/1. *Faculty research:* Cardiovascular and pulmonary, immunophysiology, neuroscience, pharmacology, epithelial. Total annual research expenditures: $4 million. *Unit head:* Frank Blecha, Head, 785-532-2741, Fax: 785-532-4557, E-mail: blecha@vet.ksu.edu. *Application contact:* Chris Ross, Director, 785-532-4507, Fax: 785-432-4557, E-mail: ross@vet.ksu.edu.

Kent State University, College of Arts and Sciences, Department of Biological Sciences, Program in Physiology, Kent, OH 44242-0001. Offers MS, PhD. *Degree requirements:* For master's and doctorate, thesis/dissertation. *Entrance requirements:* For master's, GRE General Test, minimum GPA of 3.0; for doctorate, GRE General Test, minimum GPA of 3.25. Additional exam requirements/recommendations for international students: Required—TOEFL (minimum score 600 paper-based; 257 computer-based).

Kent State University, School of Biomedical Sciences and Department of Biological Sciences, Program in Physiology, Kent, OH 44242-0001. Offers MS, PhD. Offered in cooperation with Northeastern Ohio Universities College of Medicine. *Degree requirements:* For master's and doctorate, thesis/dissertation. *Entrance requirements:* For master's, GRE General Test, minimum GPA of 3.0; for doctorate, GRE General Test, minimum GPA of 3.25. Additional exam requirements/recommendations for international students: Required—TOEFL (minimum score 600 paper-based; 287 computer-based).

Loma Linda University, School of Medicine, Department of Physiology/Pharmacology, Loma Linda, CA 92350. Offers MS, PhD. Part-time programs available. *Faculty:* 28 full-time (5 women), 3 part-time/adjunct. *Students:* 8 full-time (4 women); includes 1 Hispanic American, 6 international. *Degree requirements:* For master's, thesis or alternative; for doctorate, 2 foreign languages, thesis/dissertation. *Entrance requirements:* For master's and doctorate, GRE General Test. *Application deadline:* Applications are processed on a rolling basis. Application fee: $40. *Financial support:* Tuition waivers (full and partial) available. Support available to part-time students. *Faculty research:* Drug metabolism, biochemical pharmacology, structure and function of cell membranes, neuropharmacology. *Unit head:* Dr. Cherng-ju Kim, 909-824-4564.

Louisiana State University Health Sciences Center, School of Graduate Studies in New Orleans, Department of Physiology, New Orleans, LA 70112-2223. Offers MS, PhD, MD/PhD. Terminal master's awarded for partial completion of doctoral program. *Degree requirements:* For master's and doctorate, thesis/dissertation. *Entrance requirements:* For master's and doctorate, GRE General Test. Additional exam requirements/recommendations for international students: Required—TOEFL. *Faculty research:* Host defense, lipoprotein metabolism, regulation of cardiopulmonary function, alcohol and drug abuse, cell to cell communication, cytokinesis, physiologic functions of nitric oxide.

Louisiana State University Health Sciences Center at Shreveport, Department of Molecular and Cellular Physiology, Shreveport, LA 71130-3932. Offers physiology (MS, PhD). *Faculty:* 11 full-time (2 women). *Students:* 7 full-time (2 women), 1 international. Average age 28. 3 applicants, 67% accepted. In 2005, 1 master's, 1 doctorate awarded. *Degree requirements:* For master's and doctorate, thesis/dissertation. *Entrance requirements:* For master's and doctorate, GRE General Test. Additional exam requirements/recommendations for international students: Required—TOEFL. *Application deadline:* For fall admission, 1/31 for domestic students. Application fee: $30. *Financial support:* In 2005–06, 5 students received support, including fellowships with full tuition reimbursements available (averaging $22,000 per year), research assistantships with full tuition reimbursements available (averaging $22,000 per year); institutionally sponsored loans also available. *Faculty research:* Cardiovascular, gastrointestinal, renal, and neutrophil function; cellular detoxification systems; hypoxia and mitochondria function. *Unit head:* Dr. D. Neil Grainger, Head, 318-675-6011. *Application contact:* Dr. J. Steven Alexander, Graduate Director, 318-675-4151, Fax: 318-675-4156, E-mail: jalexa@lsuhsc.edu.

See Close-Up on page 163.

Maharishi University of Management, Graduate Studies, Program in Physiology, Fairfield, IA 52557. Offers Maharishi consciousness-based health care (MS, PhD). *Degree requirements:* For doctorate, thesis/dissertation.

Marquette University, Graduate School, College of Arts and Sciences, Department of Biology, Milwaukee, WI 53201-1881. Offers cell biology (MS, PhD); developmental biology (MS, PhD); ecology (MS, PhD); endocrinology (MS, PhD); evolutionary biology (MS, PhD); genetics (MS, PhD); microbiology (MS, PhD); molecular biology (MS, PhD); muscle and exercise physiology (MS, PhD); neurobiology (MS, PhD); reproductive physiology (MS, PhD). Terminal master's awarded for partial completion of doctoral program. *Degree requirements:* For master's, thesis, 1 year of teaching experience or equivalent, comprehensive exam; for doctorate, thesis/dissertation, 1 year of teaching experience or equivalent, qualifying exam. *Entrance requirements:* For master's and doctorate, GRE General Test, GRE Subject Test. Additional exam requirements/recommendations for international students: Required—TOEFL. *Faculty research:* Microbial and invertebrate ecology, evolution of gene function, DNA methylation, DNA arrangement.

McGill University, Faculty of Graduate and Postdoctoral Studies, Faculty of Medicine, Department of Physiology, Montréal, QC H3A 2T5, Canada. Offers M Sc, PhD. *Degree requirements:* For master's and doctorate, thesis/dissertation. *Entrance requirements:* For master's and doctorate, GRE General Test, minimum GPA of 3.2. *Faculty research:* Cloning, ion channels, skeletal metabolism, chaos in biological control systems, electrophysiology.

McMaster University, Faculty of Health Sciences and School of Graduate Studies, Program in Medical Sciences, Physiology/Pharmacology Area, Hamilton, ON L8S 4M2, Canada. Offers M Sc, PhD. *Students:* 40 full-time, 4 part-time. In 2005, 8 master's, 2 doctorates awarded. *Degree requirements:* For master's, thesis/dissertation; for doctorate, thesis/dissertation, comprehensive exam. *Entrance requirements:* For master's, honors B Sc, B+ average in related field; for doctorate, M Sc, minimum B+ average, students with proven research experience and an A average may be admitted with a B Sc degree. Additional exam requirements/recommendations for international students: Required—TOEFL (minimum score 580 paper-based; 237 computer-based). *Application deadline:* For fall admission, 3/31 for domestic students. Applications are processed on a rolling basis. Application fee: $85. *Financial support:* Teaching assistantships available. *Unit head:* Dr. Jan Huizinga, Coordinator, 905-525-9140 Ext. 22590. *Application contact:* Dr. Carl Richards, Associate Dean, 905-525-9140 Ext. 22983, Fax: 905-546-1129.

Medical College of Georgia, School of Graduate Studies, Department of Physiology, Augusta, GA 30912-1500. Offers PhD. *Faculty:* 12 full-time (3 women). *Students:* 11 full-time (7 women); includes 2 minority (both African Americans), 3 international. Average age 26. In 2005, 1 degree awarded. *Degree requirements:* For doctorate, thesis/dissertation. *Entrance requirements:* For doctorate, GRE General Test. Additional exam requirements/recommendations for international students: Required—TOEFL. *Application deadline:* For fall admission, 6/30 for domestic students, 4/15 for international students. Applications are processed on a rolling basis. Application fee: $30. *Financial support:* In 2005–06, 2 students received support, including 2 fellowships with partial tuition reimbursements available (averaging $25,500 per year), 8 research assistantships with partial tuition reimbursements available (averaging $22,500 per year); institutionally sponsored loans, scholarships/grants, and traineeships also available. Support available to part-time students. Financial award application deadline: 5/31; financial award applicants required to submit FAFSA. *Faculty research:* Cardiovascular and renal physiology, hypertension, behavioral neuroscience and genetics, neurophysiology, adrenal steroid endocrinology and genetics, inflammatory mediators and cardiovascular disease. *Unit head:* Dr. Clinton Webb, Chair, 706-721-7742, Fax: 706-721-7299, E-mail: cwebb@mail.mcg.edu. *Application contact:* Dr. Michael Brands, Director, 706-721-9785, Fax: 706-721-7299, E-mail: mbrands@mail.mcg.edu.

Medical College of Wisconsin, Graduate School of Biomedical Sciences, Department of Physiology, Milwaukee, WI 53226-0509. Offers MS, PhD, MD/MS, MD/PhD. *Degree requirements:* For master's, thesis/dissertation; for doctorate, thesis/dissertation, comprehensive exam, registration. *Entrance requirements:* For master's and doctorate, GRE General Test. Additional exam requirements/recommendations for international students: Required—TOEFL. *Faculty research:* Cardiovascular, respiratory, renal, and exercise physiology; mathematical modeling; molecular and cellular biology.

See Close-Up on page 1251.

Meharry Medical College, School of Graduate Studies, Department of Anatomy and Physiology, Nashville, TN 37208-9989. Offers physiology (PhD). *Degree requirements:* For doctorate, thesis/dissertation, comprehensive exam. *Entrance requirements:* For doctorate, GRE. *Faculty research:* Neurochemistry, pain, smooth muscle tone, HP axis and peptides neural plasticity.

Michigan State University, College of Human Medicine and The Graduate School, Graduate Programs in Human Medicine and College of Natural Science and Graduate Studies in Osteopathic Medicine, Department of Physiology, East Lansing, MI 48824. Offers MS, PhD. *Faculty:* 24 full-time (7 women). *Students:* 12 full-time (7 women), 2 part-time (both women); includes 1 minority (Asian American or Pacific Islander), 8 international. Average age 30. 19 applicants, 26% accepted. In 2005, 2 master's, 1 doctorate awarded. *Degree requirements:* For master's, thesis, oral defense of thesis; for doctorate, thesis/dissertation, oral defense of dissertation, comprehensive exam. *Entrance requirements:* For master's and doctorate, GRE General Test, 3 letters of recommendation. Additional exam requirements/recommendations for international students: Required—TOEFL (minimum score 600 paper-based; 220 computer-based). *Application deadline:* For fall admission, 12/27 for domestic students. Application fee: $50. Electronic applications accepted. *Expenses:* Tuition, state resident: part-time $330 per credit hour. Tuition, nonresident: part-time $685 per credit hour. Tuition and fees vary according to program. *Financial support:* In 2005–06, 3 fellowships with tuition reimbursements (averaging $9,444 per year), 10 research assistantships with tuition reimbursements (averaging $14,545 per year), 1 teaching assistantship with tuition reimbursement (averaging $10,872 per year) were awarded; institutionally sponsored loans, scholarships/grants, and unspecified assistantships also available. Support available to part-time students. Financial award application deadline: 4/1. *Faculty research:* Cancer biology and metabolic regulation; cardiovascular, pulmonary, renal and muscle physiology; neurophysiology, neural structure, growth and development. Total annual research expenditures: $4.1 million. *Unit head:* Dr. William S. Spielman, Chairperson, 517-355-6475, Fax: 517-355-5125. *Application contact:* Angie Zell, Graduate Secretary, 517-355-6475 Ext. 1510, E-mail: pslgrad@msu.edu.

Michigan State University, College of Osteopathic Medicine and The Graduate School, Graduate Studies in Osteopathic Medicine, East Lansing, MI 48824. Offers biochemistry and molecular biology (MS, PhD); microbiology (MS); microbiology and molecular genetics (PhD); pharmacology and toxicology (MS, PhD), including pharmacology and toxicology, pharmacology and toxicology-environmental toxicology (PhD); physiology (MS, PhD). *Students:* 2 full-time (1 woman), 1 international. Average age 27. *Expenses:* Tuition, state resident: part-time $330 per credit hour. Tuition, nonresident: part-time $685 per credit hour. Tuition and fees vary according to program. *Unit head:* Dr. Veronica M. Maher, Associate Dean, Graduate Studies, 517-353-7785, Fax: 517-353-9004, E-mail: maher@msu.edu. *Application contact:* Kathie Schafer, Director of Admissions, 517-353-7740, Fax: 517-355-3296, E-mail: comadm@com.msu.edu.

New York Medical College, Graduate School of Basic Medical Sciences, Program in Physiology, Valhalla, NY 10595-1691. Offers MS, PhD, MD/PhD. Part-time and evening/weekend programs available. Terminal master's awarded for partial completion of doctoral program. *Degree requirements:* For master's, thesis/dissertation; for doctorate, thesis/dissertation, comprehensive exam. *Entrance requirements:* For master's and doctorate, GRE General Test. Additional exam requirements/recommendations for international students: Required—TOEFL. *Faculty research:* Cardiovascular physiology, renal physiology, endocrine physiology, neurophysiology, cell biology.

See Close-Up on page 1253.

New York University, School of Medicine and Graduate School of Arts and Science, Sackler Institute of Graduate Biomedical Sciences, Department of Neuroscience and Physiology, New York, NY 10012-1019. Offers neuroscience (PhD); physiology (PhD). *Degree requirements:* For doctorate, one foreign language, thesis/dissertation, qualifying exam, comprehensive exam. *Entrance requirements:* For doctorate, GRE General Test. Additional exam requirements/recommendations for international students: Required—TOEFL. *Faculty research:* Synaptic transmission, retinal physiology, signal transduction, CNS intrinsic properties, cerebellar function.

North Carolina State University, Graduate School, College of Agriculture and Life Sciences and College of Veterinary Medicine, Program in Physiology, Raleigh, NC 27695. Offers MP, MS, PhD. *Degree requirements:* For master's, thesis (for some programs); for doctorate, thesis/dissertation. *Entrance requirements:* For master's and doctorate, GRE General Test. Electronic applications accepted. *Faculty research:* Neurophysiology, gastrointestinal physiology, reproductive physiology, environmental/stress physiology, cardiovascular physiology.

Northwestern University, The Graduate School, Judd A. and Marjorie Weinberg College of Arts and Sciences, Department of Neurobiology and Physiology, Evanston, IL 60208. Offers MS. Admissions and degrees offered through The Graduate School. Part-time programs available. *Faculty:* 14 full-time (4 women), 1 part-time/adjunct (0 women). *Students:* 11 (6 women); includes 5 minority (4 Asian Americans or Pacific Islanders, 1 Hispanic American) 1 international. Average age 24. 45 applicants, 16% accepted, 4 enrolled. In 2005, 6 degrees awarded. *Degree requirements:* For master's, thesis. *Entrance requirements:* For master's, GRE General Test and MCAT (strongly recommended). Additional exam requirements/recommendations for international students: Required—TOEFL. *Application deadline:* For fall admission, 3/3 for domestic students, 3/31 for international students. Applications are processed on a rolling basis. Application fee: $60 ($75 for international students). Electronic applications accepted. *Expenses:* Contact institution. *Financial support:* In 2005–06, 5 students received support. Institutionally sponsored loans available. Financial award application deadline: 1/15; financial award applicants required to submit FAFSA. *Faculty research:* Sensory neurobiology and neuroendocrinology, reproductive biology, vision physiology and psychophysics, cell and developmental biology. Total annual research expenditures: $5.5 million. *Unit head:* David Ferster, Chair, 847-491-4137, Fax: 847-491-5211, E-mail: ferster@northwestern.edu. *Application contact:* Dr. Michael Kennedy, Associate Chair, 847-491-5521, Fax: 847-491-5211, E-mail: m-kennedy2@northwestern.edu.

Nova Scotia Agricultural College, Research and Graduate Studies, Truro, NS B2N 5E3, Canada. Offers agriculture (M Sc), including air quality, animal behavior, animal molecular genetics, animal nutrition, animal technology, aquaculture, botany, crop management, crop physiology, ecology, environmental microbiology, food science, horticulture, nutrient management, pest management, physiology, plant biotechnology, plant pathology, soil chemistry, soil fertility, waste management and composting, water quality. Part-time programs available. *Faculty:* 43 full-time (7 women), 21 part-time/adjunct (1 woman). *Students:* 50 full-time (25 women), 15 part-time (11 women); includes 7 minority (3 African Americans, 3 Asian Americans or Pacific Islanders, 1 Hispanic American). Average age 25. In 2005, 23 degrees awarded. *Degree requirements:* For master's, thesis, registration. *Entrance requirements:* For master's, honors B Sc, minimum GPA of 3.0. Additional exam requirements/recommendations for international students: Required—TOEFL (minimum score 580 paper-based; 237 computer-based), Michigan English Language Assessment Battery, IELT, Can Test, CAEL. *Application deadline:* For fall admission, 6/1 for domestic students, 4/1 for international students. For winter admission, 10/31 for domestic students; for spring admission, 2/28 for domestic students. Applications are processed on a rolling basis. Application fee: $70. *Expenses:* Tuition, state resident: part-time $2,328 per year. Tuition, nonresident: full-time $6,984; part-time $7,968 per year. International tuition: $12,624 full-time. Required fees: $481; $46 per course. Tuition and fees vary according to program and student level. *Financial support:* In 2005–06, 48 students received support, including 7 fellowships (averaging $15,000 per year), 10 research assistantships (averaging $15,000 per year), 15 teaching assistantships (averaging $900 per year); career-related internships or fieldwork, scholarships/grants, and unspecified assistantships also available. *Faculty research:* Bio-product development, organic agriculture, nutrient management, air and water quality, agricultural biotechnology. Total annual research expenditures: $4.7 million. *Unit head:* Jill L. Rogers, Manager, 902-893-6360, Fax: 902-893-3430, E-mail: jrogers@nsac.ca. *Application contact:* Marie Law, Administrative Assistant, 902-893-6502, Fax: 902-893-3430, E-mail: mlaw@nsac.ca.

The Ohio State University, College of Medicine and Public Health and Graduate School, Graduate Programs in the Basic Medical Sciences, Department of Physiology and Cell Biology, Columbus, OH 43210. Offers physiology (MS). Students need to apply through the Integrated Biomedical Science Graduate Program. *Faculty research:* Neurobiology of cell and muscle, intestinal mucosal control, autonomic neurophysiology, cardiovascular regulation, cell biology.

The Ohio State University, College of Veterinary Medicine and Graduate School, Program in Veterinary Medicine, Department of Veterinary Biosciences, Columbus, OH 43210. Offers anatomy and cellular biology (MS, PhD); pathobiology (MS, PhD); pharmacology (MS, PhD); toxicology (MS, PhD); veterinary physiology (MS, PhD). *Faculty research:* Microvasculature, muscle biology, neonatal lung and bone development.

Ohio University, Graduate Studies, College of Arts and Sciences, Department of Biological Sciences, Athens, OH 45701-2979. Offers biological sciences (MS, PhD); cell biology and physiology (MS, PhD); ecology and evolutionary biology (MS, PhD); exercise physiology and muscle biology (MS, PhD); microbiology (MS, PhD); neuroscience (MS, PhD). *Faculty:* 51 full-time (17 women), 6 part-time/adjunct (1 woman). *Students:* 88 full-time (40 women), 1 part-time; includes 1 minority (Hispanic American), 41 international. Average age 24. 50 applicants, 24% accepted, 10 enrolled. In 2005, 9 master's, 12 doctorates awarded. *Median time to degree:* Of those who began their doctoral program in fall 1997, 90% received their degree in 8 years or less. *Degree requirements:* For master's, thesis, 1 quarter of teaching experience; for doctorate, thesis/dissertation, 2 quarters of teaching experience, comprehensive exam. *Entrance requirements:* For master's and doctorate, GRE General Test. Additional exam requirements/recommendations for international students: Required—TOEFL (minimum score 620 paper-based; 260 computer-based). *Application deadline:* For fall admission, 1/15 for domestic students, 1/15 for international students. Application fee: $45. Electronic applications accepted. *Financial support:* In 2005–06, 87 students received support, including 2 fellowships with full tuition reimbursements available (averaging $15,000 per year), 10 research assistantships with full tuition reimbursements available (averaging $15,500 per year), 75 teaching assistantships with full tuition reimbursements available (averaging $15,500 per year); Federal Work-Study, institutionally sponsored loans, and tuition waivers (full) also available. Financial award application deadline: 1/15. *Faculty research:* Ecology and evolutionary biology, exercise physiology and muscle biology, neurobiology, cell biology, physiology. Total annual research expenditures: $2.8 million. *Unit head:* Dr. Ralph DiCaprio, Chair, 740-593-2290, Fax: 740-593-0300, E-mail: dicaprir@ohio.edu. *Application contact:* Dr. Donald B. Miles, Graduate Chair, 740-593-2317, Fax: 740-593-0300, E-mail: milesd@ohio.edu.

See Close-Up on page 191.

Oregon Health & Science University, School of Medicine, Graduate Programs in Medicine, Department of Physiology and Pharmacology, Portland, OR 97239-3098. Offers pharmacology (PhD); physiology (PhD). *Degree requirements:* For doctorate, thesis/dissertation. *Entrance requirements:* For doctorate, GRE General Test, MCAT. Additional exam requirements/recommendations for international students: Required—TOEFL.

The Pennsylvania State University Milton S. Hershey Medical Center, Graduate School Programs in the Biomedical Sciences, The Huck Institutes of the Life Sciences, Intercollege Graduate Program in Physiology, Hershey, PA 17033-2360. Offers MS, PhD, MD/PhD. *Students:* 16 full-time (7 women), 1 part-time; includes 2 minority (1 African American, 1 Asian American or Pacific Islander), 5 international. Average age 27.Terminal master's awarded for partial completion of doctoral program. *Degree requirements:* For master's, thesis or alternative, registration; for doctorate, thesis/dissertation, oral exam, comprehensive exam, registration. *Entrance requirements:* For master's, GRE General Test or MCAT; for doctorate, GRE General Test or MCAT, minimum GPA of 3.0. Additional exam requirements/recommendations for international students: Required—TOEFL (minimum score 500 paper-based; 213 computer-based). *Application deadline:* Applications are processed on a rolling basis. Application fee: $45. Electronic applications accepted. *Financial support:* In 2005–06, 3 fellowships with full tuition reimbursements, 14 research assistantships with full tuition reimbursements were awarded; scholarships/grants, traineeships, health care benefits, and unspecified assistantships also available. Financial award applicants required to submit FAFSA. *Faculty research:* Gene expression, diabetes obesity and insulin resistance, DNA repair and carcinogenesis, telomerase cell senescence and cancer, ion channels and cardiovascular function. *Unit head:* Dr. Leonard S. Jefferson, Chair, 717-531-8566, Fax: 717-531-7667, E-mail: physio-grad-hmc@psu.edu. *Application contact:* Beth Ditzler, Secretary, 717-531-0221, Fax: 717-531-7667, E-mail: physio-grad-hmc@psu.edu.

The Pennsylvania State University University Park Campus, Graduate School, Intercollege Graduate Programs, Intercollege Graduate Program in Physiology, State College, University Park, PA 16802-1503. Offers MS, PhD. *Students:* 17 full-time (11 women), 1 (woman) part-time; includes 1 minority (Hispanic American), 1 international. *Entrance requirements:* For master's and doctorate, GRE General Test. Application fee: $45. *Expenses:* Tuition, state resident: full-time $12,518; part-time $522 per credit. Tuition, nonresident: full-time $23,004; part-time $959 per credit. Required fees: $484. Tuition and fees vary according to course load, campus/location and program. *Unit head:* Dr. James S. Ultman, Chair, 814-865-5557, Fax: 814-865-9451, E-mail: jsu@psu.edu. *Application contact:* Dr. James S. Ultman, Chair, 814-865-5557, Fax: 814-865-9451, E-mail: jsu@psu.edu.

Purdue University, School of Veterinary Medicine and Graduate School, Graduate Programs in Veterinary Medicine, Department of Basic Medical Sciences, West Lafayette, IN 47907. Offers anatomy (MS, PhD); pharmacology (MS, PhD); physiology (MS, PhD). Part-time programs available. *Faculty:* 16 full-time (4 women), 4 part-time/adjunct (1 woman). *Students:* 24 full-time (12 women), 3 part-time (all women); includes 1 minority (Hispanic American), 18 international. Average age 27. 18 applicants, 28% accepted, 4 enrolled. In 2005, 1 master's, 3 doctorates awarded. Terminal master's awarded for partial completion of doctoral program. *Median time to degree:* Of those who began their doctoral program in fall 1997, 33% received their degree in 8 years or less. *Degree requirements:* For master's and doctorate, thesis/dissertation. *Entrance requirements:* For master's and doctorate, GRE General Test. Additional exam requirements/recommendations for international students: Required—TOEFL. *Application deadline:* For fall admission, 12/31 priority date for domestic students, 12/31 priority date for international students. Application fee: $55. Electronic applications accepted. *Financial support:* In 2005–06, 4 fellowships with partial tuition reimbursements (averaging $13,251 per year), 12 research assistantships with partial tuition reimbursements (averaging $15,012 per year), 2 teaching assistantships with partial tuition reimbursements (averaging $17,800 per year) were awarded. Financial award application deadline: 3/1; financial award applicants required to submit FAFSA. *Faculty research:* Development and regeneration, tissue injury and shock, biomedical engineering, ovarian function, bone and cartilage biology, cell and molecular biology. *Unit head:* Dr. Gordon L. Coppoc, Head, 765-494-8592, Fax: 765-494-0781, E-mail: coppoc@purdue.edu. *Application contact:* Dr. Kevin M. Hannon, Chairman, Graduate Committee, 765-494-5949, Fax: 765-494-0781, E-mail: bmsgrad@purdue.edu.

Queen's University at Kingston, School of Graduate Studies and Research, Faculty of Health Sciences, Department of Physiology, Kingston, ON K7L 3N6, Canada. Offers M Sc, PhD. *Degree requirements:* For master's, thesis/dissertation; for doctorate, thesis/dissertation, comprehensive exam. *Entrance requirements:* For master's, minimum upper B average. Additional exam requirements/recommendations for international students: Required—TOEFL. *Faculty research:* Cardiovascular and respiratory physiology, exercise, gastrointestinal physiology, neuroscience.

Rosalind Franklin University of Medicine and Science, School of Graduate and Postdoctoral Studies, Department of Physiology and Biophysics, Program in Applied Physiology, North Chicago, IL 60064-3095. Offers MS. *Entrance requirements:* For master's, GRE General Test. Additional exam requirements/recommendations for international students: Required—TOEFL, TWE. Expenses: Contact institution.

See Close-Up on page 1255.

Rosalind Franklin University of Medicine and Science, School of Graduate and Postdoctoral Studies, Department of Physiology and Biophysics, Program in Physiology, North Chicago, IL 60064-3095. Offers MS, PhD, MD/MS, MD/PhD. *Degree requirements:* For master's and doctorate, thesis/dissertation. *Entrance requirements:* For master's and doctorate, GRE General Test. Additional exam requirements/recommendations for international students: Required—TOEFL, TWE.

See Close-Up on page 1255.

Rush University, Graduate College, Department of Molecular Biophysics and Physiology, Chicago, IL 60612-3832. Offers physiology (PhD). *Faculty:* 7 full-time (2 women), 4 part-time/

Physiology

Rush University (continued)
adjunct (0 women). *Students:* 3 full-time (0 women); includes 2 minority (both Hispanic Americans) Average age 28. 25 applicants, 0% accepted, 0 enrolled. *Degree requirements:* For doctorate, thesis/dissertation. *Entrance requirements:* For doctorate, GRE General Test. Additional exam requirements/recommendations for international students: Required—TOEFL. *Application deadline:* For fall admission, 4/15 for domestic students. Applications are processed on a rolling basis. Application fee: $25. *Financial support:* In 2005–06, 3 fellowships with full tuition reimbursements (averaging $24,500 per year) were awarded; Federal Work-Study and institutionally sponsored loans also available. Support available to part-time students. Financial award application deadline: 4/15. *Faculty research:* Physiological exocytosis, raft formation and growth, voltage-gated proton channels, molecular biophysics and physiology. Total annual research expenditures: $3 million. *Unit head:* Dr. Fredric Cohen, Director, 312-942-6753, Fax: 312-942-8711, E-mail: fcohen@rush.edu. *Application contact:* Connie M. Lambert, Department Administrator, 312-563-2563, Fax: 312-563-3552, E-mail: connie_m_lambert@rush.edu.

Rutgers, The State University of New Jersey, New Brunswick/Piscataway, Graduate School, Program in Physiology and Neurobiology, New Brunswick, NJ 08901-1281. Offers PhD. *Faculty:* 76 full-time. *Students:* 4 full-time (3 women); includes 2 minority (1 African American, 1 Asian American or Pacific Islander). Average age 26. 7 applicants, 29% accepted, 2 enrolled. *Degree requirements:* For doctorate, thesis/dissertation, qualifying exam. *Entrance requirements:* For doctorate, GRE General Test, GRE Subject Test (biology or chemistry), minimum undergraduate GPA of 3.0. Additional exam requirements/recommendations for international students: Required—TOEFL. *Application deadline:* For fall admission, 3/1 for domestic students. Applications are processed on a rolling basis. Application fee: $50. *Expenses:* Tuition, state resident: full-time $10,440; part-time $435 per credit. Tuition, nonresident: full-time $15,520; part-time $647 per credit. Required fees: $129 per credit. Tuition and fees vary according to program. *Financial support:* In 2005–06, fellowships with full tuition reimbursements (averaging $17,000 per year), research assistantships with full tuition reimbursements (averaging $15,500 per year), teaching assistantships with full tuition reimbursements (averaging $16,988 per year) were awarded; tuition waivers (full) also available. Financial award application deadline: 3/1; financial award applicants required to submit FAFSA. *Faculty research:* Neuronal growth factors, neuronal gene expression, neurogenetics, circulation controls, reproduction. Total annual research expenditures: $3.4 million. *Unit head:* Jianjie Ma, Graduate Program Director, 732-235-4494, Fax: 732-235-5885, E-mail: maj2@umdnj.edu. *Application contact:* David Egger, Director of Admissions, 732-235-4522, Fax: 732-235-4029.

Saint Louis University, Graduate School and School of Medicine, Graduate Program in Biomedical Sciences and Graduate School, Department of Pharmacological and Physiological Science, St. Louis, MO 63103-2097. Offers PhD. *Faculty:* 31 full-time (13 women), 1 part-time/adjunct (0 women). *Students:* 18 full-time (12 women), 1 (woman) part-time; includes 1 minority (African American), 4 international. Average age 27. 8 applicants, 88% accepted, 7 enrolled. In 2005, 5 degrees awarded. *Degree requirements:* For doctorate, thesis/dissertation, departmental qualifying exams, comprehensive exam. *Entrance requirements:* For doctorate, GRE General Test, letters of recommendation, resumé. Additional exam requirements/recommendations for international students: Required—TOEFL (minimum score 550 paper-based; 213 computer-based). *Application deadline:* For fall admission, 7/1 for domestic students, 7/1 for international students; for spring admission, 11/1 for domestic students, 11/1 for international students. Applications are processed on a rolling basis. Application fee: $40. Electronic applications accepted. *Expenses:* Tuition: Part-time $760 per credit hour. Required fees: $55 per semester. *Financial support:* In 2005–06, 17 students received support. Traineeships and unspecified assistantships available. Financial award application deadline: 6/1; financial award applicants required to submit FAFSA. *Faculty research:* Molecular endocrinology, neuropharmacology, cardiovascular science, drug abuse, neurotransmitter and hormonal signaling mechanisms. *Unit head:* Dr. Thomas C. Westfall, Chairperson, 314-977-6400, Fax: 314-977-6411, E-mail: westfate@slu.edu. *Application contact:* Gary Behrman, Associate Dean of the Graduate School, 314-977-3827, E-mail: behrmang@slu.edu.

Salisbury University, Graduate Division, Program in Applied Health Physiology, Salisbury, MD 21801-6837. Offers MS. Part-time and evening/weekend programs available. *Faculty:* 3. *Students:* 18 full-time (13 women), 6 part-time (4 women); includes 3 minority (all African Americans) Average age 32. 11 applicants, 73% accepted, 8 enrolled. In 2005, 4 degrees awarded. *Degree requirements:* For master's, fieldwork. *Entrance requirements:* For master's, minimum GPA of 2.75, undergraduate course work in anatomy and physiology. Additional exam requirements/recommendations for international students: Required—TOEFL (minimum score 550 paper-based; 213 computer-based). *Application deadline:* For fall admission, 8/1 for domestic students; for spring admission, 1/1 for domestic students. Applications are processed on a rolling basis. Application fee: $45. *Expenses:* Tuition, state resident: part-time $249 per credit hour. Tuition, nonresident: part-time $535 per credit hour. Required fees: $47 per credit hour. *Financial support:* Applicants required to submit FAFSA. *Faculty research:* Body image and self-concept, nutrition supplements. *Unit head:* Dr. Susan Muller, Director, 410-548-5555, E-mail: smmuller@salisbury.edu. *Application contact:* Sydney Geesaman, Administrative Assistant II, 410-543-6340, Fax: 410-543-6434, E-mail: scgeesaman@salisbury.edu.

San Francisco State University, Division of Graduate Studies, College of Science and Engineering, Department of Biology, Program in Physiology and Behavioral Biology, San Francisco, CA 94132-1722. Offers MA. *Entrance requirements:* For master's, minimum GPA of 2.5 in last 60 units.

San Jose State University, Graduate Studies and Research, College of Science, Department of Biological Sciences, San Jose, CA 95192-0001. Offers biological sciences (MA, MS); molecular biology and microbiology (MS); organismal biology, conservation and ecology (MS); physiology (MS). Part-time programs available. *Students:* 44 full-time (35 women), 33 part-time (24 women); includes 39 minority (33 Asian Americans or Pacific Islanders, 6 Hispanic Americans), 12 international. Average age 31. 140 applicants, 40% accepted, 29 enrolled. In 2005, 39 degrees awarded. *Entrance requirements:* For master's, GRE. *Application deadline:* For fall admission, 6/29 for domestic students; for spring admission, 11/30 for domestic students. Applications are processed on a rolling basis. Application fee: $59. Electronic applications accepted. *Expenses:* Tuition, nonresident: part-time $339 per unit. Required fees: $1,286 per semester. Tuition and fees vary according to course load and degree level. *Financial support:* In 2005–06, 13 teaching assistantships were awarded; Federal Work-Study also available. Financial award applicants required to submit FAFSA. *Faculty research:* Systemic physiology, molecular genetics, SEM studies, toxicology, large mammal ecology. *Unit head:* Dr. Sally Veregge, Chair, 408-924-4900, Fax: 408-924-4840. *Application contact:* Dr. Howard Shellhamer, Graduate Coordinator, 408-924-4897.

Southern Illinois University Carbondale, Graduate School, Graduate Program in Medicine, Department of Physiology, Carbondale, IL 62901-4701. Offers MS, PhD. *Faculty:* 18 full-time (4 women). *Students:* In 2005, 1 degree awarded. Terminal master's awarded for partial completion of doctoral program. *Degree requirements:* For master's and doctorate, thesis/dissertation. *Entrance requirements:* For master's, GRE General Test, minimum GPA of 3.0; for doctorate, GRE General Test, minimum GPA of 3.25. Additional exam requirements/recommendations for international students: Required—TOEFL. *Application deadline:* For fall admission, 6/1 for domestic students. Applications are processed on a rolling basis. Application fee: $0. *Financial support:* In 2005–06, 18 students received support, including 3 fellowships with full tuition reimbursements available, 1 research assistantship with full tuition reimbursement available, 10 teaching assistantships with full tuition reimbursements available; institutionally sponsored loans and tuition waivers (full) also available. *Faculty research:* Hormones, neurotransmitters, cell biology, membrane protein, membranes transport. *Unit head:* Richard Steger, Chair, 618-453-1512, Fax: 618-453-1517. *Application contact:* Graduate Program Committee, 618-453-1544, Fax: 618-453-1517.

Announcement: The Department of Physiology at Southern Illinois University School of Medicine offers MS and PhD degrees in molecular, cellular, and systemic physiology. Graduates of this program have achieved highly successful careers worldwide as professors in traditional universities and in medical schools or as governmental and industrial research scientists.

See Close-Up on page 1257.

Southern Illinois University Carbondale, Graduate School, Graduate Program in Medicine, Program in Molecular, Cellular and Systemic Physiology, Carbondale, IL 62901-4701. Offers MS. *Students:* 8 full-time (4 women), 12 part-time (4 women); includes 4 minority (3 African Americans, 1 Hispanic American), 8 international. 22 applicants, 36% accepted, 5 enrolled. In 2005, 2 master's awarded. *Application contact:* Graduate Program Committee, 618-536-5513.

Stanford University, School of Medicine, Graduate Programs in Medicine, Department of Molecular and Cellular Physiology, Stanford, CA 94305-9991. Offers PhD. *Degree requirements:* For doctorate, thesis/dissertation, qualifying exams. *Entrance requirements:* For doctorate, GRE General Test, GRE Subject Test. Additional exam requirements/recommendations for international students: Required—TOEFL. Electronic applications accepted. *Faculty research:* Signal transduction, ion channels, intracellular calcium, synaptic transmission.

State University of New York at Buffalo, Graduate School, School of Medicine and Biomedical Sciences, Graduate Programs in Medicine and Biomedical Sciences, Department of Physiology and Biophysics, Buffalo, NY 14260. Offers biophysics (MS, PhD); physiology (MA, PhD). *Faculty:* 26 full-time (5 women), 1 part-time/adjunct (0 women). *Students:* 10 full-time (4 women), 2 part-time (1 woman); includes 2 minority (both Asian Americans or Pacific Islanders), 2 international. Average age 26. 9 applicants, 22% accepted, 1 enrolled. In 2005, 2 master's, 4 doctorates awarded. Terminal master's awarded for partial completion of doctoral program. *Degree requirements:* For master's, thesis, oral exam, project; for doctorate, thesis/dissertation, oral and written qualifying exam or 2 research proposals. *Entrance requirements:* For master's and doctorate, GRE General Test. Additional exam requirements/recommendations for international students: Required—TOEFL. *Application deadline:* For fall admission, 2/1 for domestic students. Applications are processed on a rolling basis. Application fee: $35. Electronic applications accepted. *Financial support:* In 2005–06, fellowships with tuition reimbursements (averaging $21,000 per year), 17 research assistantships with tuition reimbursements (averaging $21,000 per year) were awarded; Federal Work-Study, institutionally sponsored loans, health care benefits, and unspecified assistantships also available. Financial award application deadline: 2/1; financial award applicants required to submit FAFSA. *Faculty research:* Neurosciences, ion channels, cardiac physiology, renal/epithelial transport, cardiopulmonary exercise. Total annual research expenditures: $5.6 million. *Unit head:* Dr. Harold C. Strauss, Chair, 716-829-2738, Fax: 716-829-2344, E-mail: hstrauss@buffalo.edu. *Application contact:* Dr. Malcolm Slaughter, Professor, 716-829-3240, Fax: 716-829-2364, E-mail: mslaught@buffalo.edu.

State University of New York Upstate Medical University, College of Graduate Studies, Department of Neuroscience and Physiology, Syracuse, NY 13210-2334. Offers physiology (MS, PhD). Terminal master's awarded for partial completion of doctoral program. *Degree requirements:* For master's, thesis/dissertation; for doctorate, thesis/dissertation, comprehensive exam. *Entrance requirements:* For master's, GRE General Test, GRE Subject Test, interview; for doctorate, GRE General Test, interview. Additional exam requirements/recommendations for international students: Required—TOEFL.

Stony Brook University, State University of New York, Health Sciences Center, School of Medicine and Graduate School, Graduate Programs in Medicine, Department of Molecular Physiology and Biophysics, Stony Brook, NY 11794. Offers physiology and biophysics (PhD). *Faculty:* 17 full-time (3 women). *Students:* 21 full-time (12 women); includes 2 minority (both Asian Americans or Pacific Islanders), 8 international. Average age 29. 24 applicants, 33% accepted. In 2005, 7 degrees awarded. *Degree requirements:* For doctorate, thesis/dissertation, comprehensive exam. *Entrance requirements:* For doctorate, GRE General Test, GRE Subject Test, BS in related field, minimum GPA of 3.0. Additional exam requirements/recommendations for international students: Required—TOEFL. *Application deadline:* For fall admission, 1/15 for domestic students. Application fee: $50. *Expenses:* Tuition, area resident: Full-time $6,900; part-time $288. Tuition, state resident: full-time $6,900. Tuition, nonresident: full-time $10,920; part-time $455. International tuition: $10,920 full-time. Required fees: $704. *Financial support:* In 2005–06, 13 research assistantships, 5 teaching assistantships were awarded; fellowships, Federal Work-Study also available. Financial award application deadline: 3/15. *Faculty research:* Cellular electrophysiology, membrane permeation and transport, metabolic endocrinology. Total annual research expenditures: $6.7 million. *Unit head:* Dr. Peter Brink, Chair, 631-444-2287, Fax: 631-444-3432. *Application contact:* Dr. Leon C. Moore, Graduate Adviser, 631-444-2287, Fax: 631-444-3432, E-mail: moore@pofvax.pnb.sunysb.edu.

Announcement: This program is one to consider for students interested in a PhD in physiology and biophysics. Students in the program become valued members of the department. Generous stipends and waivers of tuition and fees help to ease the monetary worry as one learns, teaches, and becomes immersed in research.

See Close-Ups on pages 1259 and 473.

Temple University, Health Sciences Center, School of Medicine and Graduate School, Graduate Programs in Medicine, Department of Physiology, Philadelphia, PA 19122-6096. Offers PhD, MD/PhD. *Faculty:* 18 full-time (1 woman). *Students:* 11 full-time (4 women), 15 part-time (3 women); includes 4 minority (1 African American, 3 Asian Americans or Pacific Islanders), 7 international. 22 applicants, 64% accepted, 6 enrolled. In 2005, 2 degrees awarded. *Degree requirements:* For doctorate, thesis/dissertation, research seminars. *Entrance requirements:* For doctorate, GRE General Test, minimum GPA of 3.0. Additional exam requirements/recommendations for international students: Required—TOEFL (minimum score 575 paper-based; 230 computer-based). *Application deadline:* Applications are processed on a rolling basis. Application fee: $50. Electronic applications accepted. *Expenses:* Tuition, state resident: full-time $8,694; part-time $483 per credit. Tuition, nonresident: full-time $12,672; part-time $704 per credit. Required fees: $500; $122 per semester. Tuition and fees vary according to course level, campus/location and program. *Financial support:* Fellowships available. Financial award application deadline: 1/15; financial award applicants required to submit FAFSA. *Faculty research:* Pulmonary, microvascular, and molecular physiology; cardiac electrophysiology. *Unit head:* Dr. Ronald Tuma, Chair, 215-707-5485, Fax: 215-707-4003, E-mail: ronald.tuma@temple.edu.

Texas A&M University, College of Veterinary Medicine, Graduate Programs in Veterinary Medicine, Department of Veterinary Physiology and Pharmacology, College Station, TX 77843. Offers physiology and pharmacology (MS, PhD); toxicology (MS, PhD). *Faculty:* 8 full-time (0 women), 4 part-time/adjunct (2 women). *Students:* 13 full-time (7 women), 1 (woman) part-time; includes 2 minority (1 Asian American or Pacific Islander, 1 Hispanic American), 7 international. Average age 30. 16 applicants, 38% accepted, 2 enrolled. In 2005, 3 degrees awarded. *Entrance requirements:* For master's and doctorate, GRE General Test. Additional exam requirements/recommendations for international students: Required—TOEFL. Application fee: $50 ($75 for international students). *Expenses:* Tuition, state resident: full-time $4,488; part-time $187 per credit hour. Tuition, nonresident: full-time $11,112; part-time $463 per credit hour. Required fees: $1,974. *Financial support:* Fellowships, research assistantships, teaching assistantships available. Financial award application deadline: 4/1; financial award applicants required to submit FAFSA. *Faculty research:* Gamete and embryo physiology, endocrinology, equine laminitis. *Unit head:* Glen Laine, Head, 979-845-7261.

Texas Tech University Health Sciences Center, Graduate School of Biomedical Sciences, Department of Physiology, Lubbock, TX 79430. Offers MS, PhD, MD/PhD. Terminal master's awarded for partial completion of doctoral program. *Degree requirements:* For master's and doctorate, thesis/dissertation. *Entrance requirements:* For master's and doctorate, GRE General

Physiology

Test, minimum GPA of 3.4. Additional exam requirements/recommendations for international students: Required—TOEFL. Electronic applications accepted. *Faculty research:* Cardiovascular physiology, neurophysiology, renal physiology, respiratory physiology.

Thomas Jefferson University, Jefferson College of Graduate Studies, Program in Physiology, Philadelphia, PA 19107. Offers PhD. *Faculty:* 10 full-time (2 women). *Students:* 8 applicants, 25% accepted, 0 enrolled. *Degree requirements:* For doctorate, thesis/dissertation, comprehensive exam, registration. *Entrance requirements:* For doctorate, GRE General Test, minimum GPA of 3.2. Additional exam requirements/recommendations for international students: Required— TOEFL (minimum score 213 computer-based). *Application deadline:* For fall admission, 3/1 priority date for domestic students, 3/1 priority date for international students. Applications are processed on a rolling basis. Application fee: $50. Electronic applications accepted. *Expenses:* Tuition: Full-time $14,894; part-time $800 per credit. *Financial support:* Fellowships with full tuition reimbursements, research assistantships, Federal Work-Study, institutionally sponsored loans, scholarships/grants, and traineeships available. Support available to part-time students. Financial award application deadline: 5/1; financial award applicants required to submit FAFSA. *Faculty research:* Cardiovascular physiology, smooth muscle physiology, pathophysiology of myocardial ischemia, endothelial cell physiology, molecular biology of ion channel physiology. *Unit head:* Dr. Rosario Scalia, Program Director, 215-503-6409, Fax: 215-503-2073, E-mail: rosario.scalia@jefferson.edu. *Application contact:* Jessie F. Pervall, Director of Admissions, 215-503-0155, Fax: 215-503-9920, E-mail: jessie.pervall@jefferson.edu.

Tufts University, Sackler School of Graduate Biomedical Sciences, Integrated Studies Program, Medford, MA 02155. Offers PhD. *Students:* 12 full-time (10 women); includes 2 minority (both Asian Americans or Pacific Islanders), 4 international. Average age 26. 264 applicants, 19% accepted, 12 enrolled. *Entrance requirements:* For doctorate, GRE General Test, 3 letters of reference. Additional exam requirements/recommendations for international students: Required— TOEFL. *Application deadline:* For fall admission, 1/15 priority date for domestic students, 1/15 priority date for international students. Application fee: $65. *Financial support:* In 2005–06, 12 students received support, including 12 research assistantships (averaging $29,000 per year) Financial award application deadline: 1/15.

Tulane University, School of Medicine and Graduate School, Graduate Programs in Medicine, Department of Physiology, New Orleans, LA 70118-5669. Offers MS, PhD, MD/PhD. MS and PhD offered through the Graduate School. *Degree requirements:* For master's, one foreign language, thesis; for doctorate, 2 foreign languages, thesis/dissertation. *Entrance requirements:* For master's, GRE General Test, minimum B average in undergraduate course work; for doctorate, GRE General Test. Additional exam requirements/recommendations for international students: Required—TOEFL or TSE. Electronic applications accepted. *Faculty research:* Renal microcirculation, neurophysiology, NA+ transport, renin/angio tensin system, cell and molecular endocrinology.

Université de Montréal, Faculty of Medicine and Faculty of Graduate Studies, Graduate Programs in Medicine, Department of Physiology, Montréal, QC H3C 3J7, Canada. Offers biophysics and molecular physiology (M Sc, PhD); neurological sciences (M Sc, PhD); physiology (M Sc, PhD). *Faculty:* 27 full-time (3 women), 8 part-time/adjunct (1 woman). *Students:* 97 full-time (56 women), 1 (woman) part-time. 23 applicants, 30% accepted, 7 enrolled. In 2005, 8 master's, 1 doctorate awarded. Terminal master's awarded for partial completion of doctoral program. *Degree requirements:* For master's, thesis; for doctorate, thesis/dissertation, general exam. *Entrance requirements:* For master's and doctorate, proficiency in French, knowledge of English. *Application deadline:* For fall and spring admission, 2/1. For winter admission, 11/1 for domestic students. Application fee: $30. Electronic applications accepted. *Financial support:* Fellowships, research assistantships available. *Faculty research:* Cardiovascular, neuropeptides, membrane transport and biophysics, signaling pathways. *Unit head:* Allan Smith, Interim Director, 514-343-6347, Fax: 514-343-7072. *Application contact:* Réjean Couture, Graduate Chairman, 514-343-7060, Fax: 514-343-2111.

Université de Sherbrooke, Faculty of Medicine and Health Sciences, Graduate Programs in Medicine, Department of Physiology and Biophysics, Sherbrooke, QC J1K 2R1, Canada. Offers M Sc, PhD. Part-time programs available. *Faculty:* 8 full-time (2 women). *Students:* 20 full-time (11 women), 11 part-time (6 women). Average age 25. 8 applicants, 38% accepted, 2 enrolled. In 2005, 4 degrees awarded. Terminal master's awarded for partial completion of doctoral program. *Degree requirements:* For master's and doctorate, thesis/dissertation. *Application deadline:* For fall admission, 6/30 for domestic students. For winter admission, 10/31 for domestic students; for spring admission, 2/28 for domestic students. Application fee: $50. Electronic applications accepted. *Financial support:* Fellowships available. *Faculty research:* Electrophysiology, physiopathology of bladder, electrocardiography, ionic channels of VSM, cardiac-sarcoplasmic reticulum. *Unit head:* Dr. Nuria Basora, Director, 819-564-5307, E-mail: nuria.basora@usherbrooke.ca.

Université Laval, Faculty of Medicine, Department of Anatomy and Physiology, Québec, QC G1K 7P4, Canada. Offers M Sc, PhD. Terminal master's awarded for partial completion of doctoral program. *Degree requirements:* For master's, thesis (for some programs); for doctorate, thesis/dissertation, comprehensive exam. Electronic applications accepted.

Université Laval, Faculty of Medicine, Graduate Programs in Medicine, Programs in Physiology-Endocrinology, Québec, QC G1K 7P4, Canada. Offers M Sc, PhD. Terminal master's awarded for partial completion of doctoral program. *Degree requirements:* For master's, thesis/dissertation; for doctorate, thesis/dissertation, comprehensive exam. Electronic applications accepted.

The University of Alabama at Birmingham, Graduate Programs in Joint Health Sciences, Department of Physiology and Biophysics, Birmingham, AL 35294. Offers basic medical sciences (MSBMS); biophysical sciences (PhD); integrative biomedical sciences (PhD). *Students:* 18 full-time (8 women); includes 6 minority (3 African Americans, 1 American Indian/Alaska Native, 2 Asian Americans or Pacific Islanders), 5 international. 34 applicants, 82% accepted. In 2005, 9 degrees awarded. *Entrance requirements:* For doctorate, GRE General Test, interview, minimum GPA of 3.0. Additional exam requirements/recommendations for international students: Required—TOEFL. *Application deadline:* For fall admission, 3/1 for domestic students. Applications are processed on a rolling basis. Application fee: $35 ($60 for international students). Electronic applications accepted. *Expenses: Contact institution.* Tuition and fees vary according to course load, degree level and program. *Financial support:* Fellowships available. *Faculty research:* Standard physiology (neurological, endocrine, cardiovascular, respiratory, and renal), cell and membrane biology. *Unit head:* Dr. Dale J. Benos, Chair, 205-934-6220, Fax: 205-934-2377. *Application contact:* Director of Graduate Studies, 205-934-3969, Fax: 205-975-9028.

University of Alberta, Faculty of Graduate Studies and Research, Department of Biological Sciences, Edmonton, AB T6G 2E1, Canada. Offers environmental biology and ecology (M Sc, PhD); microbiology and biotechnology (M Sc, PhD); molecular biology and genetics (M Sc, PhD); physiology and cell biology (M Sc, PhD); plant biology (M Sc, PhD); systematics and evolution (M Sc, PhD). *Faculty:* 72 full-time (15 women), 15 part-time/adjunct (4 women). *Students:* 238 full-time (117 women), 32 part-time (15 women), 31 international. 206 applicants, 42% accepted. In 2005, 29 master's, 31 doctorates awarded. Terminal master's awarded for partial completion of doctoral program. *Degree requirements:* For master's and doctorate, thesis/dissertation, registration. *Entrance requirements:* Additional exam requirements/ recommendations for international students: Required—TOEFL. *Application deadline:* For fall admission, 3/1 for domestic students. Applications are processed on a rolling basis. Application fee: $0. Tuition and fees charges are reported in Canadian dollars. *Expenses:* Tuition, state resident: part-time $562 Canadian dollars per term. Tuition, nonresident: full-time $3,375 Canadian dollars. Required fees: $573 Canadian dollars; $84 Canadian dollars per term. *Financial support:* In 2005–06, 4 research assistantships with partial tuition reimbursements (averaging $12,000 per year), 103 teaching assistantships with partial tuition reimbursements (averaging $12,300 per year) were awarded; career-related internships or fieldwork and

scholarships/grants also available. *Unit head:* Laura Frost, Chair, 780-492-1904. *Application contact:* Dr. John P. Chang, Associate Chair for Graduate Studies, 780-492-1257, Fax: 780-492-9457, E-mail: bio.grad.coordinator@ualberta.ca.

University of Alberta, Faculty of Medicine and Dentistry and Faculty of Graduate Studies and Research, Graduate Programs in Medicine, Department of Physiology, Edmonton, AB T6G 2E1, Canada. Offers M Sc, PhD, MD/PhD. Terminal master's awarded for partial completion of doctoral program. *Degree requirements:* For master's and doctorate, thesis/dissertation. *Entrance requirements:* For master's, honors B Sc; for doctorate, M Sc. Additional exam requirements/ recommendations for international students: Required—TOEFL (minimum score 580 paper-based; 237 computer-based). Electronic applications accepted. Tuition and fees charges are reported in Canadian dollars. *Expenses:* Tuition, state resident: part-time $562 Canadian dollars per term. Tuition, nonresident: full-time $3,375 Canadian dollars. Required fees: $573 Canadian dollars; $84 Canadian dollars per term. *Faculty research:* Membrane transport, cell biology, perinatal endocrinology, neurophysiology, cardiovascular.

The University of Arizona, Graduate College, Graduate Interdisciplinary Programs, Graduate Interdisciplinary Program in Physiological Sciences, Tucson, AZ 85721. Offers PhD. *Degree requirements:* For doctorate, thesis/dissertation. *Entrance requirements:* For doctorate, GRE General Test. Additional exam requirements/recommendations for international students: Required—TOEFL. *Faculty research:* Cellular transport and signaling, receptor and messenger modulation, neural interaction and biomechanics, fluid network regulation, environmental adaptation.

University of Arkansas for Medical Sciences, College of Medicine and Graduate School, Graduate Programs in Medicine, Department of Physiology and Biophysics, Little Rock, AR 72205-7199. Offers MS, PhD, MD/PhD. *Faculty:* 16 full-time (2 women), 23 part-time/adjunct (3 women). *Students:* 27 full-time, 6 part-time. *Degree requirements:* For master's and doctorate, thesis/dissertation. *Entrance requirements:* For master's and doctorate, GRE General Test. Additional exam requirements/recommendations for international students: Required— TOEFL. Application fee: $0. *Financial support:* Research assistantships available. Support available to part-time students. *Unit head:* Dr. Michael L. Jennings, Chairman, 501-686-5123. *Application contact:* Dr. Patricia Wight, Graduate Coordinator, 501-686-5366, E-mail: wightpatricia@uams.edu.

The University of British Columbia, Faculty of Medicine and Faculty of Graduate Studies, Graduate Programs in Medicine, Department of Cellular and Physiological Sciences, Division of Physiology, Vancouver, BC V6T 1Z3, Canada. Offers M Sc, PhD. *Faculty:* 17. *Students:* 21 full-time (11 women); includes 1 American Indian/Alaska Native, 9 Asian Americans or Pacific Islanders, 2 international. Average age 26. 100 applicants, 5% accepted, 5 enrolled. In 2005, 3 master's, 4 doctorates awarded. Terminal master's awarded for partial completion of doctoral program. *Degree requirements:* For master's, thesis/dissertation; for doctorate, thesis/dissertation, comprehensive exam. *Entrance requirements:* Additional exam requirements/ recommendations for international students: Required—TOEFL (minimum score 550 paper-based; 213 computer-based), IELT (minimum score 6). *Application deadline:* For fall admission, 4/1 priority date for domestic students, 3/1 priority date for international students. Applications are processed on a rolling basis. Application fee: $90 Canadian dollars ($150 Canadian dollars for international students). Electronic applications accepted. *Financial support:* In 2005–06, 6 fellowships with tuition reimbursements (averaging $16,000 per year), 12 teaching assistantships were awarded; Federal Work-Study, institutionally sponsored loans, scholarships/grants, traineeships, tuition waivers (full), and unspecified assistantships also available. Financial award application deadline: 1/31. *Faculty research:* Neurophysiology, gastroenterology, endocrinology, cardiovascular physiology. *Application contact:* Dr. Edwin D.W. Moore, Graduate Adviser, 604-822-3424, Fax: 604-822-2316, E-mail: edmoore@interchange.ubc.ca.

University of California, Berkeley, Graduate Division, Group in Endocrinology, Berkeley, CA 94720-1500. Offers MA, PhD. *Degree requirements:* For doctorate, thesis/dissertation, oral qualifying exam. *Entrance requirements:* For master's, GRE General Test, GRE Subject Test, minimum GPA of 3.0; for doctorate, GRE General Test, GRE Subject Test, minimum GPA of 3.4. Additional exam requirements/recommendations for international students: Required— TOEFL.

University of California, Davis, Graduate Studies, Molecular, Cellular and Integrative Physiology Graduate Group, Davis, CA 95616. Offers MS, PhD. *Faculty:* 93 full-time. *Students:* 49 full-time (30 women); includes 11 minority (1 African American, 7 Asian Americans or Pacific Islanders, 3 Hispanic Americans), 7 international. Average age 29. 74 applicants, 43% accepted, 11 enrolled. In 2005, 4 master's, 2 doctorates awarded. *Median time to degree:* Of those who began their doctoral program in fall 1997, 80% received their degree in 8 years or less. *Degree requirements:* For master's, thesis (for some programs), comprehensive exam (for some programs); for doctorate, thesis/dissertation. *Entrance requirements:* For master's and doctorate, GRE General Test. Additional exam requirements/recommendations for international students: Required—TOEFL (minimum score 550 paper-based; 213 computer-based). *Application deadline:* For fall admission, 1/15 for domestic students, 1/15 for international students. Application fee: $60. Electronic applications accepted. *Financial support:* In 2005–06, 47 students received support, including 11 fellowships with full and partial tuition reimbursements available (averaging $10,540 per year), 21 research assistantships with full and partial tuition reimbursements available (averaging $17,404 per year), 7 teaching assistantships with partial tuition reimbursements available (averaging $15,082 per year); Federal Work-Study, institutionally sponsored loans, scholarships/grants, tuition waivers (full and partial), and unspecified assistantships also available. Financial award application deadline: 1/15; financial award applicants required to submit FAFSA. *Faculty research:* Systemic physiology, cellular physiology, neurophysiology, cardiovascular physiology, endocrinology. *Unit head:* James Millam, Chair, 530-752-1144, E-mail: jrmillam@ucdavis.edu. *Application contact:* Amber Carrere, Administrative Assistant, 530-752-9092, E-mail: acarrere@ucdavis.edu.

University of California, Irvine, College of Medicine and School of Biological Sciences, Department of Physiology and Biophysics, Irvine, CA 92697. Offers biological sciences (PhD). Students apply through the Graduate Program in Molecular Biology, Genetics, and Biochemistry. *Degree requirements:* For doctorate, thesis/dissertation. *Entrance requirements:* For doctorate, GRE General Test, GRE Subject Test, minimum GPA of 3.0. Additional exam requirements/ recommendations for international students: Required—TOEFL (minimum score 550 paper-based; 213 computer-based), TSE. Electronic applications accepted. *Faculty research:* Membrane physiology, exercise physiology, regulation of hormone biosynthesis and action, endocrinology, ion channels and signal transduction.

University of California, Los Angeles, Graduate Division, College of Letters and Science, Department of Physiological Science, Los Angeles, CA 90095. Offers molecular, cellular and integrative physiology (PhD); physiological science (MS, PhD). *Degree requirements:* For master's, comprehensive exam or thesis; for doctorate, thesis/dissertation, oral and written qualifying exams. *Entrance requirements:* For master's, GRE General Test, minimum GPA of 3.0; for doctorate, GRE General Test, minimum undergraduate GPA of 3.0. Electronic applications accepted. *Faculty research:* Diet and exercise in the prevention and management of degenerative diseases, neuromuscular physiology and plasticity, neural control of movement and homeostasis.

University of California, Los Angeles, School of Medicine and Graduate Division, Graduate Programs in Medicine, Department of Physiology, Los Angeles, CA 90095. Offers MS, PhD. *Degree requirements:* For doctorate, thesis/dissertation, oral and written qualifying exams. *Entrance requirements:* For master's, GRE General Test; for doctorate, GRE General Test, GRE Subject Test. *Faculty research:* Membrane physiology, cell physiology, muscle physiology, neurophysiology, cardiopulmonary physiology.

University of California, San Diego, School of Medicine and Graduate Studies and Research, Graduate Studies in Biomedical Sciences, Physiology Program, La Jolla, CA 92093. Offers PhD.

Physiology

University of California, San Diego (continued)
Degree requirements: For doctorate, thesis/dissertation, qualifying exam. *Entrance requirements:* For doctorate, GRE General Test. Additional exam requirements/recommendations for international students: Required—TOEFL. Electronic applications accepted. *Faculty research:* Cell and organ physiology, eukaryotic regulatory and molecular biology, molecular and cellular pharmacology.

University of California, San Francisco, Graduate Division, Biomedical Sciences Graduate Group, Program in Endocrinology, San Francisco, CA 94143. Offers PhD. *Degree requirements:* For doctorate, thesis/dissertation. *Entrance requirements:* For doctorate, GRE General Test.

University of California, San Francisco, Graduate Division, Biomedical Sciences Graduate Group, Program in Physiology, San Francisco, CA 94143. Offers PhD. *Degree requirements:* For doctorate, thesis/dissertation. *Entrance requirements:* For doctorate, GRE General Test, GRE Subject Test. Additional exam requirements/recommendations for international students: Required—TOEFL. *Faculty research:* Circulation and respiration, enzyme kinetics, fetal and neonatal physiology, nephrology.

University of Chicago, Division of the Biological Sciences, Department of Neurobiology, Pharmacology, and Cell Physiology, Department of Neurobiology, Pharmacology and Physiology, Chicago, IL 60637-1513. Offers cell physiology (PhD); pharmacological and physiological sciences (PhD). *Faculty:* 33 full-time (8 women). *Students:* 4 full-time (0 women); includes 1 minority (Asian American or Pacific Islander) Average age 30. In 2005, 4 degrees awarded. *Degree requirements:* For doctorate, thesis/dissertation, preliminary exam. *Entrance requirements:* For doctorate, GRE General Test. Additional exam requirements/recommendations for international students: Required—TOEFL. *Application deadline:* For fall admission, 12/28 priority date for domestic students, 12/28 priority date for international students. Application fee: $55. Electronic applications accepted. *Financial support:* In 2005–06, fellowships with tuition reimbursements (averaging $26,301 per year), research assistantships with tuition reimbursements (averaging $26,301 per year) were awarded; institutionally sponsored loans, scholarships/grants, traineeships, and health care benefits also available. Financial award applicants required to submit FAFSA. *Faculty research:* Psychopharmacology, neuropharmacology. *Unit head:* Dr. J. Murray Sherman, Chairman, 773-834-2900, Fax: 773-702-3774, E-mail: msherman@bsd.uchicago.edu. *Application contact:* Diane J. Hall, Graduate Administrative Director, 773-702-6371, Fax: 773-702-1216, E-mail: d-hall@uchicago.edu.

University of Cincinnati, Division of Research and Advanced Studies, College of Medicine, Graduate Programs in Biomedical Sciences, Department of Molecular and Cellular Physiology, Cincinnati, OH 45267. Offers physiology (PhD). *Degree requirements:* For doctorate, thesis/dissertation, publication, comprehensive exam, registration. *Entrance requirements:* For doctorate, GRE General Test, GRE Subject Test. Additional exam requirements/recommendations for international students: Required—TOEFL (minimum score 560 paper-based; 220 computer-based). Electronic applications accepted. *Faculty research:* Endocrinology, cardiovascular physiology, muscle physiology, neurophysiology, transgenic mouse physiology.

University of Colorado at Boulder, Graduate School, College of Arts and Sciences, Department of Integrative Physiology, Boulder, CO 80309. Offers MS, PhD. *Faculty:* 20 full-time (5 women). *Students:* 49 full-time (25 women), 11 part-time (4 women); includes 10 minority (1 African American, 4 Asian Americans or Pacific Islanders, 5 Hispanic Americans), 2 international. Average age 29. 27 applicants, 100% accepted. In 2005, 10 master's, 7 doctorates awarded. *Degree requirements:* For master's, thesis or alternative, comprehensive exam; for doctorate, thesis/dissertation. *Entrance requirements:* For master's, GRE General Test, minimum undergraduate GPA of 2.75. *Application deadline:* For fall admission, 1/15 priority date for domestic students, 12/15 priority date for international students. Applications are processed on a rolling basis. Application fee: $50 ($60 for international students). *Financial support:* In 2005–06, fellowships (averaging $3,442 per year), research assistantships (averaging $15,521 per year), teaching assistantships (averaging $13,190 per year) were awarded. Financial award application deadline: 3/1. *Faculty research:* Integrative or cellular kinesiology. Total annual research expenditures: $7.8 million. *Unit head:* Roger Enoka, Chair, 303-492-7232, Fax: 303-492-4009, E-mail: roger.enoka@spot.colorado.edu. *Application contact:* Marsha Cook, Graduate Program Administrator, 303-492-5362, Fax: 303-492-4009, E-mail: marsha.cook@colorado.edu.

University of Colorado at Denver and Health Sciences Center, Graduate School, Program in Biomedical Sciences, Department of Physiology and Biophysics, Denver, CO 80262. Offers PhD. In 2005, 1 degree awarded. *Degree requirements:* For doctorate, thesis/dissertation, comprehensive exam. *Entrance requirements:* For doctorate, GRE General Test, 4 letters of recommendation. Additional exam requirements/recommendations for international students: Required—TOEFL (minimum score 550 paper-based; 213 computer-based). *Application deadline:* For fall admission, 1/15 for domestic students. Application fee: $50. *Expenses:* Tuition, state resident: full-time $11,730. Tuition, nonresident: full-time $22,980. Tuition and fees vary according to degree level and program. *Financial support:* Fellowships, research assistantships, teaching assistantships, Federal Work-Study and institutionally sponsored loans available. Support available to part-time students. Financial award application deadline: 3/15; financial award applicants required to submit FAFSA. *Faculty research:* Nicotinic receptors, immunity, molecular structure, function and regulation of ion channels, calcium influx in the function of cytotoxic T lymphocytes. *Unit head:* Dr. William Betz, Chair, 303-315-8946, E-mail: bill.betz@uchsc.edu. *Application contact:* Kathy Fernandez, Information Contact, 303-315-7831, E-mail: kathy.fernandez@uchsc.edu.

University of Connecticut, Graduate School, College of Liberal Arts and Sciences, Department of Physiology and Neurobiology, Storrs, CT 06269. Offers physiology and neurobiology (MS, PhD), including comparative physiology, endocrinology, neurobiology. *Faculty:* 16 full-time (5 women). *Students:* 35 full-time (18 women), 2 part-time (1 woman); includes 5 minority (1 African American, 4 Asian Americans or Pacific Islanders), 16 international. Average age 27. 44 applicants, 32% accepted, 13 enrolled. In 2005, 5 master's, 6 doctorates awarded. Terminal master's awarded for partial completion of doctoral program. *Degree requirements:* For master's, comprehensive exam; for doctorate, thesis/dissertation. *Entrance requirements:* For master's and doctorate, GRE General Test, GRE Subject Test. Additional exam requirements/recommendations for international students: Required—TOEFL (minimum score 550 paper-based; 213 computer-based). *Application deadline:* For fall admission, 2/1 priority date for domestic students, 2/1 priority date for international students; for spring admission, 11/1 for domestic students, 10/1 for international students. Applications are processed on a rolling basis. Application fee: $55. Electronic applications accepted. *Expenses:* Tuition, state resident: part-time $444 per credit hour. Tuition, nonresident: part-time $1,154 per credit hour. Tuition and fees vary according to course load. *Financial support:* In 2005–06, 15 research assistantships with full tuition reimbursements, 14 teaching assistantships with full tuition reimbursements were awarded; fellowships, Federal Work-Study, scholarships/grants, health care benefits, and unspecified assistantships also available. Financial award application deadline: 2/1. *Unit head:* Angel L. de Blas, Head, 860-486-3285, Fax: 860-486-3303, E-mail: angel.deblas@uconn.edu. *Application contact:* J. Larry Renfro, Chairperson, 860-486-4119, Fax: 860-486-3303, E-mail: larry.renfro@uconn.edu.

See Close-Up on page 1261.

University of Delaware, College of Arts and Sciences, Department of Biological Sciences, Newark, DE 19716. Offers biotechnology (MS); cancer biology (MS, PhD); cell and extracellular matrix biology (MS, PhD); cell and systems physiology (MS, PhD); developmental biology (MS, PhD); ecology and evolution (MS, PhD); microbiology (MS, PhD); molecular biology and genetics (MS, PhD). *Faculty:* 39 full-time (11 women). *Students:* 64 full-time (46 women), 2 part-time; includes 7 minority (4 African Americans, 2 Asian Americans or Pacific Islanders, 1 Hispanic American), 18 international. Average age 26. 113 applicants, 31% accepted, 19 enrolled. In 2005, 5 master's, 4 doctorates awarded. Terminal master's awarded for partial completion of doctoral program. *Median time to degree:* Of those who began their

doctoral program in fall 1997, 100% received their degree in 8 years or less. *Degree requirements:* For master's, thesis/dissertation, preliminary exam; for doctorate, thesis/dissertation, preliminary exam, comprehensive exam. *Entrance requirements:* For master's and doctorate, GRE General Test. Additional exam requirements/recommendations for international students: Required—TOEFL (minimum score 600 paper-based; 250 computer-based); Recommended—TWE, TSE. *Application deadline:* For fall admission, 4/15 for domestic students, 1/15 for international students; for spring admission, 10/1 for domestic students. Applications are processed on a rolling basis. Application fee: $60. Electronic applications accepted. *Financial support:* In 2005–06, 26 students received support, including fellowships with full tuition reimbursements available (averaging $19,000 per year), 19 research assistantships with full tuition reimbursements available (averaging $19,000 per year), 26 teaching assistantships with full tuition reimbursements available (averaging $19,000 per year); tuition waivers (partial) also available. Financial award application deadline: 4/15. *Faculty research:* Microorganisms, bone, cancer metastasis, developmental biology, cell biology, DNA. Total annual research expenditures: $8.3 million. *Unit head:* Dr. Daniel D. Carson, Chair, 302-831-6977, Fax: 302-831-2281, E-mail: dcarson@udel.edu. *Application contact:* Dr. Melinda K. Duncan, Graduate Coordinator, 302-831-1841, Fax: 302-831-2281, E-mail: danders@udel.edu.

University of Delaware, College of Health Sciences, Department of Health, Nutrition, and Exercise Sciences, Newark, DE 19716. Offers exercise science (MS), including biomechanics, exercise physiology, motor control; health promotion (MS); human nutrition (MS). Part-time programs available. *Faculty:* 17 full-time (6 women). *Students:* 41 full-time (27 women), 26 part-time (22 women); includes 6 minority (all African Americans), 4 international. Average age 27. 73 applicants, 60% accepted, 34 enrolled. In 2005, 20 degrees awarded. *Degree requirements:* For master's, thesis, registration. *Entrance requirements:* For master's, GRE General Test, interview, minimum GPA of 3.0. Additional exam requirements/recommendations for international students: Required—TOEFL (minimum score 550 paper-based; 213 computer-based), TSE (minimum score 50). *Application deadline:* For fall admission, 3/15 priority date for domestic students, 3/15 priority date for international students; for spring admission, 10/15 priority date for domestic students, 10/15 priority date for international students. Applications are processed on a rolling basis. Application fee: $60. Electronic applications accepted. *Financial support:* In 2005–06, 10 teaching assistantships with full tuition reimbursements (averaging $11,031 per year) were awarded; fellowships with full tuition reimbursements, research assistantships with full tuition reimbursements, career-related internships or fieldwork, Federal Work-Study, scholarships/grants, and tuition waivers (full and partial) also available. Support available to part-time students. Financial award application deadline: 4/15. *Faculty research:* Sport biomechanics, rehabilitation biomechanics, vascular dynamics. Total annual research expenditures: $407,241. *Unit head:* Dr. Susan J. Hall, Chair, 302-831-2625, Fax: 302-831-4261, E-mail: sjhall@udel.edu. *Application contact:* Dr. Michelle Provost-Craig, Graduate Coordinator, 302-831-6326, Fax: 302-831-3963, E-mail: provost@udel.edu.

University of Florida, College of Medicine, Department of Physiology and Functional Genomics, Gainesville, FL 32611. Offers PhD. *Faculty:* 15 full-time (4 women). *Students:* 31 (18 women); includes 3 minority (all Asian Americans or Pacific Islanders) 8 international. In 2005, 10 degrees awarded. *Degree requirements:* For doctorate, thesis/dissertation. *Entrance requirements:* For doctorate, GRE General Test, minimum GPA of 3.0. Additional exam requirements/recommendations for international students: Required—TOEFL. *Application deadline:* For fall admission, 2/1 for domestic students. Application fee: $30. Electronic applications accepted. *Expenses:* Tuition, state resident: full-time $6,234. Tuition, nonresident: full-time $21,359. Tuition and fees vary according to program. *Financial support:* In 2005–06, 13 research assistantships with full tuition reimbursements (averaging $23,321 per year) were awarded; fellowships with full tuition reimbursements, institutionally sponsored loans and traineeships also available. *Faculty research:* Cell and general endocrinology, neuroendocrinology, neurophysiology, respiration, membrane transport and ion channels. *Unit head:* Dr. Charles Wood, Chairman, 352-392-4488, Fax: 352-846-0270, E-mail: cwood@phys.med.ufl.edu. *Application contact:* Dr. Wayne McCormack, Associate Dean of Graduate Education, 352-392-7413, Fax: 352-846-3466, E-mail: mccormac@pathology.ufl.edu.

University of Florida, College of Medicine and Graduate School, Interdisciplinary Program in Biomedical Sciences, Concentration in Physiology and Pharmacology, Gainesville, FL 32611. Offers PhD. *Faculty:* 33 full-time (9 women). *Students:* 58 full-time (16 women); includes 6 minority (all Asian Americans or Pacific Islanders) In 2005, 5 degrees awarded. *Degree requirements:* For doctorate, thesis/dissertation. *Entrance requirements:* For doctorate, GRE General Test, minimum GPA of 3.0. *Application deadline:* For fall admission, 2/15 for domestic students. Application fee: $30. Electronic applications accepted. *Expenses:* Tuition, state resident: full-time $6,234. Tuition, nonresident: full-time $21,359. Tuition and fees vary according to program. *Financial support:* In 2005–06, 24 research assistantships with full tuition reimbursements (averaging $23,320 per year) were awarded; fellowships with full tuition reimbursements, traineeships and unspecified assistantships also available. *Unit head:* Dr. Jeffrey Harrison, Director, 352-392-3227, E-mail: harrison@pharmacolog.ufl.edu. *Application contact:* Dr. Wayne McCormack, Associate Dean of Graduate Education, 352-392-7413, Fax: 352-846-3466, E-mail: mccormac@pathology.ufl.edu.

University of Florida, Graduate School, College of Health and Human Performance, Department of Applied Physiology and Kinesiology, Gainesville, FL 32611. Offers athletic training/sport medicine (MS); biomechanics (MS); clinical exercise physiology (MS); exercise physiology (MS); health and human performance (PhD); human performance (MS); motor learning/control (MS); sport and exercise psychology (MS). *Faculty:* 14 full-time (2 women), 1 part-time/adjunct (0 women). *Students:* 90 (46 women); includes 12 minority (3 African Americans, 2 Asian Americans or Pacific Islanders, 7 Hispanic Americans) 5 international. 32 applicants, 31% accepted. In 2005, 76 degrees awarded. *Degree requirements:* For doctorate, thesis/dissertation. *Entrance requirements:* For doctorate, GRE General Test. *Application deadline:* For fall admission, 6/1 for domestic students. Applications are processed on a rolling basis. Application fee: $20. Electronic applications accepted. *Expenses:* Tuition, state resident: full-time $6,234. Tuition, nonresident: full-time $21,359. Tuition and fees vary according to program. *Financial support:* In 2005–06, 9 research assistantships (averaging $14,511 per year), 33 teaching assistantships (averaging $13,181 per year) were awarded; fellowships, unspecified assistantships also available. *Unit head:* Dr. Steven Dodd, Chair, 352-392-0584 Ext. 1342, E-mail: sdodd@hhp.ufl.edu. *Application contact:* Dr. Paul A. Borsa, Coordinator, 352-392-0584 Ext. 1261, Fax: 352-392-5262, E-mail: pborsa@hhp.ufl.edu.

University of Georgia, College of Veterinary Medicine and Graduate School, Graduate Programs in Veterinary Medicine, Department of Physiology and Pharmacology, Athens, GA 30602. Offers pharmacology (MS, PhD); physiology (MS, PhD). *Faculty:* 11 full-time (3 women). *Students:* 17 full-time, 4 part-time; includes 1 minority (African American), 10 international. 22 applicants, 23% accepted, 4 enrolled. In 2005, 1 master's, 4 doctorates awarded. *Degree requirements:* For master's, thesis; for doctorate, one foreign language, thesis/dissertation. *Entrance requirements:* For master's and doctorate, GRE General Test. *Application deadline:* For fall admission, 7/1 for domestic students; for spring admission, 11/15 for domestic students. Application fee: $50. Electronic applications accepted. *Financial support:* Fellowships, research assistantships, teaching assistantships, unspecified assistantships available. *Unit head:* Dr. Thomas F. Murray, Head, 706-542-5859, Fax: 706-542-3015, E-mail: tmurray@vet.uga.edu. *Application contact:* Dr. Royal A. McGraw, Graduate Coordinator, 706-542-0661, Fax: 706-542-3015, E-mail: mcgraw@uga.edu.

University of Guelph, Ontario Veterinary College and Graduate Program Services, Graduate Programs in Veterinary Sciences, Department of Biomedical Sciences, Guelph, ON N1G 2W1, Canada. Offers morphology (M Sc, DV Sc, PhD); pharmacology (M Sc, DV Sc, PhD); physiology (M Sc, DV Sc, PhD); toxicology (M Sc, DV Sc, PhD). Part-time programs available. *Faculty:* 25. *Students:* 43 (27 women). In 2005, 6 master's, 2 doctorates awarded. *Median time to degree:* Of those who began their doctoral program in fall 1997, 100% received their degree in 8 years or less. *Degree requirements:* For master's, thesis/dissertation; for doctor-

Physiology

ate, thesis/dissertation, comprehensive exam. *Entrance requirements:* For master's, honors B Sc, minimum 75% average in last 20 courses; for doctorate, M Sc with thesis from accredited institution. Additional exam requirements/recommendations for international students: Required—TOEFL (minimum score 550 paper-based; 213 computer-based). *Application deadline:* Applications are processed on a rolling basis. Application fee: $75. Electronic applications accepted. *Financial support:* Fellowships, research assistantships, teaching assistantships available. *Faculty research:* Cellular morphology; endocrine, vascular and reproductive physiology; clinical pharmacology; veterinary toxicology; developmental biology. Total annual research expenditures: $3.5 million. *Unit head:* Dr. N. Maclusky, Chair, 519-824-4120 Ext. 54904, Fax: 511-767-1450. *Application contact:* Dr. G. Kirby, Graduate Coordinator, 519-824-4120 Ext. 54948, Fax: 519-767-1450, E-mail: gkirby@uoguelph.ca.

University of Hawaii at Manoa, John A. Burns School of Medicine and Graduate Division, Graduate Programs in Biomedical Sciences, Department of Anatomy, Biochemistry, Physiology and Reproductive Biology, Program in Physiology, Honolulu, HI 96822. Offers MS, PhD. *Faculty:* 14 full-time (1 woman), 4 part-time/adjunct (0 women). *Students:* 9 full-time (5 women), 6 part-time (2 women); includes 10 minority (9 Asian Americans or Pacific Islanders, 1 Hispanic American), 4 international. 25 applicants, 40% accepted, 5 enrolled. In 2005, 3 master's, 2 doctorates awarded. *Application deadline:* For fall admission, 3/1 for domestic students, 3/1 for international students; for spring admission, 9/1 for domestic students, 9/1 for international students. *Expenses:* Tuition, state resident: full-time $8,400; part-time $200 per credit hour. Tuition, nonresident: full-time $11,088; part-time $462 per credit hour. Tuition and fees vary according to program. *Application contact:* David Lally, 808-956-7983, Fax: 808-956-9722.

University of Illinois at Chicago, College of Medicine and Graduate College, Graduate Programs in Medicine, Department of Physiology and Biophysics, Chicago, IL 60607-7128. Offers MS, PhD. Terminal master's awarded for partial completion of doctoral program. *Degree requirements:* For master's and doctorate, thesis/dissertation. *Entrance requirements:* For master's and doctorate, GRE General Test. Additional exam requirements/recommendations for international students: Required—TOEFL. Electronic applications accepted. *Faculty research:* Neuroscience, endocrinology and reproduction, cell physiology, exercise physiology, NMR.

University of Illinois at Urbana–Champaign, Graduate College, College of Liberal Arts and Sciences, School of Integrative Biology, Program in Physiological and Molecular Plant Biology, Champaign, IL 61820. Offers PhD. *Faculty:* 14 full-time (8 women), 2 part-time (1 woman); includes 5 minority (3 Asian Americans or Pacific Islanders, 2 Hispanic Americans), 6 international. 22 applicants, 9% accepted, 0 enrolled. In 2005, 3 degrees awarded. *Application deadline:* For fall admission, 3/15 for domestic students; for spring admission, 9/15 for domestic students. Applications are processed on a rolling basis. Application fee: $50 ($60 for international students). Electronic applications accepted. *Financial support:* In 2005–06, 2 fellowships, 11 research assistantships, 9 teaching assistantships were awarded. Financial award application deadline: 2/15. *Unit head:* Hans Bohnert, Acting Director, 217-265-5475, Fax: 217-244-1224, E-mail: bohnerth@uiuc.edu. *Application contact:* Carol Hall, Secretary, 217-333-8208, Fax: 217-244-1224, E-mail: c-hall@uiuc.edu.

University of Illinois at Urbana–Champaign, Graduate College, College of Liberal Arts and Sciences, School of Molecular and Cellular Biology, Program in Molecular and Integrative Physiology, Champaign, IL 61820. Offers MS, PhD. *Faculty:* 14 full-time (3 women), 1 (woman) part-time/adjunct. *Students:* 36 full-time (16 women), 1 part-time; includes 7 minority (2 African Americans, 3 Asian Americans or Pacific Islanders, 2 Hispanic Americans), 19 international. 38 applicants, 26% accepted, 9 enrolled. In 2005, 3 master's, 6 doctorates awarded. *Degree requirements:* For doctorate, thesis/dissertation. *Entrance requirements:* For master's, GRE, minimum GPA of 3.0. *Application deadline:* Applications are processed on a rolling basis. Application fee: $50 ($60 for international students). Electronic applications accepted. *Financial support:* In 2005–06, 18 fellowships, 26 research assistantships, 21 teaching assistantships were awarded. Financial award application deadline: 2/15. *Unit head:* Philip Best, Head, 217-333-1734, Fax: 217-333-1133, E-mail: pbest@uiuc.edu. *Application contact:* Denice Wells, Administrative Secretary, 217-333-1734, Fax: 217-333-1133, E-mail: d-wells2@uiuc.edu.

See Close-Up on page 1263.

The University of Iowa, Roy J. and Lucille A. Carver College of Medicine and Graduate College, Graduate Programs in Medicine, Department of Physiology and Biophysics, Iowa City, IA 52242-1316. Offers physiology and biophysics (PhD); physiology and biophysiology (MS). *Faculty:* 15 full-time (3 women), 2 part-time/adjunct (2 women). *Students:* 20 full-time (10 women); includes 1 minority (Asian American or Pacific Islander), 7 international. Average age 25. 17 applicants, 24% accepted, 4 enrolled. Terminal master's awarded for partial completion of doctoral program. *Degree requirements:* For master's and doctorate, thesis/dissertation, teaching experience, comprehensive exam. *Entrance requirements:* For master's and doctorate, GRE General Test, minimum GPA of 3.0. *Application deadline:* For fall admission, 4/1 for domestic students, 3/1 for international students; for spring admission, 10/1 for domestic students, 9/1 for international students. Applications are processed on a rolling basis. Application fee: $60 ($80 for international students). *Expenses:* Tuition, state resident: full-time $1,882 per term. Tuition, nonresident: full-time $17,338; part-time $4,907 per term. Tuition and fees vary according to course load and program. *Financial support:* In 2005–06, 1 fellowship with full tuition reimbursement (averaging $22,000 per year), 19 research assistantships with full tuition reimbursements (averaging $22,000 per year) were awarded; traineeships also available. Financial award application deadline: 4/1. *Faculty research:* Cellular and molecular endocrinology, membrane structure and function, cardiac cell electrophysiology, regulation of gene expression, neurophysiology. *Unit head:* Kevin P. Campbell, Interim Head, 319-335-7800, Fax: 319-335-7330, E-mail: kevin-campbell@uiowa.edu. *Application contact:* Dr. Thomas Schmidt, Chairman of Graduate Admissions, 319-335-7847, Fax: 319-335-7330, E-mail: thomas-schmidt@uiowa.edu.

See Close-Up on page 1265.

University of Kansas, Graduate Studies Medical Center, Interdisciplinary Graduate Program in Biomedical Sciences, Department of Molecular and Integrative Physiology, Kansas City, KS 66160. Offers MS, PhD, MD/PhD. *Faculty:* 22. *Students:* Average age 30. In 2005, 3 degrees awarded. *Median time to degree:* Of those who began their doctoral program in fall 1997, 90% received their degree in 8 years or less. *Expenses:* Tuition, state resident: full-time $4,859. Tuition, nonresident: full-time $12,000. Required fees: $589. Tuition and fees vary according to program. *Financial support:* In 2005–06, 10 students received support; fellowships with full tuition reimbursements available, research assistantships with full tuition reimbursements available, teaching assistantships with full tuition reimbursements available, unspecified assistantships available. *Faculty research:* Neurobiology, reproduction, microcirculation and hypoxic, development, ATPASE. *Unit head:* Dr. Paul Cheney, Chairman, 913-588-7400, Fax: 913-588-7430, E-mail: pcheney@kumc.edu. *Application contact:* Dr. Thomas Imig, Director of Graduate Studies, 913-588-7407, Fax: 913-588-7430, E-mail: timig@kumc.edu.

University of Kentucky, Graduate School, Graduate School Programs from the College of Medicine, Program in Physiology, Lexington, KY 40506-0032. Offers MS, PhD. *Faculty:* 29 full-time (7 women), 1 part-time/adjunct (0 women). *Students:* 29 full-time (22 women); includes 1 minority (Asian American or Pacific Islander), 8 international. Average age 28. 18 applicants, 100% accepted, 17 enrolled. In 2005, 2 degrees awarded. *Median time to degree:* Of those who began their doctoral program in fall 1997, 77.8% received their degree in 8 years or less. *Degree requirements:* For doctorate, thesis/dissertation, comprehensive exam. *Entrance requirements:* For master's, GRE General test, minimum undergraduate GPA of 2.5; for doctorate, GRE General Test, minimum undergraduate GPA of 3.0. Additional exam requirements/recommendations for international students: Required—TOEFL (minimum score 550 paper-based; 213 computer-based). *Application deadline:* For fall admission, 7/17 priority date for domestic students, 2/1 priority date for international students; for spring admission, 12/13 priority date for domestic students, 6/15 priority date for international students. Applica-

tions are processed on a rolling basis. Application fee: $40 ($55 for international students). Electronic applications accepted. *Expenses:* Tuition, state resident: full-time $6,308; part-time $331 per credit hour. Tuition, nonresident: full-time $13,968; part-time $756 per credit hour. Tuition and fees vary according to course load, degree level and program. *Financial support:* In 2005–06, 1 fellowship with full tuition reimbursement (averaging $4,000 per year), 26 research assistantships with tuition reimbursements (averaging $21,000 per year) were awarded; teaching assistantships, Federal Work-Study, scholarships/grants, traineeships, health care benefits, tuition waivers (partial), and unspecified assistantships also available. Support available to part-time students. Financial award application deadline: 3/15. *Unit head:* Dr. Steven Estus, Director of Graduate Studies, 859-323-1412, Fax: 859-323-1070, E-mail: sestus@uky.edu. *Application contact:* Dr. Brian Jackson, Senior Associate Dean, 859-257-8176, Fax: 859-323-1928.

University of Louisville, School of Medicine, Department of Physiology and Biophysics, Louisville, KY 40292-0001. Offers MS, PhD. *Students:* 25 full-time (13 women), 10 part-time (3 women); includes 11 minority (7 African Americans, 3 Asian Americans or Pacific Islanders, 1 Hispanic American), 8 international. Average age 27. In 2005, 9 master's, 3 doctorates awarded. *Degree requirements:* For master's and doctorate, thesis/dissertation. *Entrance requirements:* For master's, GRE General Test, 36 hours of graduate course work; for doctorate, GRE General Test, minimum GPA of 3.0. *Application deadline:* For fall admission, 1/15 for domestic students. Applications are processed on a rolling basis. Application fee: $50. Electronic applications accepted. *Expenses:* Tuition, state resident: full-time $6,006; part-time $334 per credit hour. Tuition, nonresident: full-time $16,554; part-time $920 per credit hour. Tuition and fees vary according to course load, degree level and program. *Financial support:* Fellowships with full tuition reimbursements available. *Unit head:* Dr. Irving G. Joshua, Chair, 502-852-5371, Fax: 502-852-6239, E-mail: igjosh01@gwise.louisville.edu.

University of Manitoba, Faculty of Medicine and Faculty of Graduate Studies, Graduate Programs in Medicine, Department of Physiology, Winnipeg, MB R3T 2N2, Canada. Offers M Sc, PhD, MD/PhD. Terminal master's awarded for partial completion of doctoral program. *Degree requirements:* For master's and doctorate, one foreign language, thesis/dissertation. *Entrance requirements:* For master's, minimum GPA of 3.5; for doctorate, minimum GPA of 3.5, M Sc. *Faculty research:* Cardiovascular research, gene technology, cell biology, neuroscience, respiration.

University of Maryland, Graduate School, Graduate Programs in Medicine, Department of Physiology, Baltimore, MD 21201. Offers neuroscience (PhD); physiology (MS, PhD); reproductive endocrinology (PhD). *Faculty:* 28 full-time (6 women), 21 part-time/adjunct. *Students:* 28 full-time (15 women), 3 part-time (1 woman); includes 7 minority (2 African Americans, 5 Asian Americans or Pacific Islanders), 6 international. Average age 28. 41 applicants, 29% accepted, 6 enrolled. In 2005, 3 master's, 2 doctorates awarded. *Degree requirements:* For doctorate, thesis/dissertation. *Entrance requirements:* For doctorate, GRE General Test, GRE Subject Test, minimum GPA of 3.0. Additional exam requirements/recommendations for international students: Required—TOEFL, TOEFL or IELTS; Recommended—IELT. *Application deadline:* For fall admission, 7/1 for domestic students, 1/15 for international students. Application fee: $50. Electronic applications accepted. *Expenses:* Tuition, state resident: full-time $8,079; part-time $409 per credit hour. Tuition, nonresident: full-time $18,384; part-time $731 per credit hour. Required fees: $695; $10 per credit hour. Tuition and fees vary according to degree level and program. *Financial support:* Fellowships, research assistantships available. Financial award application deadline: 2/15. *Faculty research:* Membrane physiology, biophysics and morphology, central nervous system physiology, EEG analysis, information theory. *Unit head:* Dr. Mordecai P. Blaustein, Chairman, 410-706-3345, Fax: 410-706-3345, E-mail: mblaust@umaryland.edu. *Application contact:* Dr. Robert Koos, Director of Graduate Studies, 410-706-8033, Fax: 410-706-8341, E-mail: rkoos@umaryland.edu.

University of Maryland, School of Medicine, Graduate Program in Life Sciences, Baltimore, MD 21201. Offers biochemistry (MS, PhD); epidemiology (MS, PhD); gerontology (PhD); microbiology (PhD); molecular and cell biology (MS); molecular medicine (PhD); neuroscience (MS, PhD); pharmacology (MS); physiology (MS); rehabilitation sciences (PhD); toxicology (MS, PhD). *Faculty:* 245 full-time (52 women). *Students:* 268 full-time (165 women), 46 part-time (30 women); includes 43 minority (25 African Americans, 13 Asian Americans or Pacific Islanders, 5 Hispanic Americans), 86 international. 435 applicants, 26% accepted, 61 enrolled. In 2005, 18 master's, 46 doctorates awarded. *Median time to degree:* Of those who began their doctoral program in fall 1997, 99% received their degree in 8 years or less. *Degree requirements:* For master's, registration; for doctorate, thesis/dissertation, lab rotations, comprehensive exam, registration. *Entrance requirements:* For master's and doctorate, GRE or MCAT. Additional exam requirements/recommendations for international students: Required—TOEFL (minimum score 550 paper-based; 213 computer-based). *Application deadline:* For winter admission, 1/15 for domestic students. Applications are processed on a rolling basis. Application fee: $50. Electronic applications accepted. *Expenses:* Tuition, state resident: full-time $8,079; part-time $409 per credit hour. Tuition, nonresident: full-time $18,384; part-time $731 per credit hour. Required fees: $695; $10 per credit hour. Tuition and fees vary according to degree level and program. *Financial support:* In 2005–06, 30 fellowships with full tuition reimbursements (averaging $23,000 per year), 22 research assistantships with full tuition reimbursements (averaging $23,000 per year) were awarded; health care benefits also available. *Faculty research:* Cancer, reproduction, neuroscience, cardiovascular, immunology. *Unit head:* Dr. Margaret Merryl McCarthy, Assistant Dean for Graduate Studies, 410-706-2655, Fax: 410-706-8341, E-mail: mmcarthy@umaryland.edu.

University of Massachusetts Worcester, Graduate School of Biomedical Sciences, Department of Physiology, Worcester, MA 01655-0115. Offers PhD. *Faculty:* 30 full-time (5 women). *Degree requirements:* For doctorate, thesis/dissertation. *Entrance requirements:* For doctorate, GRE General Test. Additional exam requirements/recommendations for international students: Required—TOEFL (minimum score 600 paper-based; 250 computer-based). *Application deadline:* For fall admission, 12/15 for domestic students, 12/15 for international students. Applications are processed on a rolling basis. Application fee: $25 ($50 for international students). *Expenses:* Tuition, state resident: full-time $2,640. Tuition, nonresident: full-time $9,856. Required fees: $5,685. *Financial support:* In 2005–06, research assistantships with full tuition reimbursements (averaging $25,235 per year); unspecified assistantships also available. *Faculty research:* Endocrinology, regulation of cellular and tissue metabolism, electrophysiology, muscle physiology. *Unit head:* Dr. Maurice Goodman, Chair, 508-856-2101. *Application contact:* Michael Cole, Director of Admissions and Recruitment, 508-856-4779, Fax: 508-856-3659, E-mail: michael.cole@umassmed.edu.

See Close-Up on page 1267.

University of Medicine and Dentistry of New Jersey, Graduate School of Biomedical Sciences, Graduate Programs in Biomedical Sciences–Newark, Department of Pharmacology and Physiology, Newark, NJ 07107. Offers PhD. *Degree requirements:* For doctorate, thesis/dissertation, qualifying exam. *Entrance requirements:* For doctorate, GRE General Test. Additional exam requirements/recommendations for international students: Required—TOEFL. *Application deadline:* For fall admission, 2/1 for domestic students. Applications are processed on a rolling basis. Application fee: $40. *Financial support:* Fellowships, research assistantships, Federal Work-Study and institutionally sponsored loans available. Financial award application deadline: 5/1. *Unit head:* Dr. Martha Nowycky, Program Director, 973-972-4391, Fax: 973-972-7950, E-mail: martha.nowycky@umdnj.edu.

University of Medicine and Dentistry of New Jersey, Graduate School of Biomedical Sciences, Graduate Programs in Biomedical Sciences–Piscataway, Program in Physiology and Integrative Biology, Piscataway, NJ 08854-5635. Offers MS, PhD, MD/PhD. *Entrance requirements:* Additional exam requirements/recommendations for international students: Required—TOEFL. *Application deadline:* For fall admission, 1/15 for domestic students. Application fee: $40. *Unit head:* Dr. Jiange Ma, Director, 732-235-4494, Fax: 732-235-5038, E-mail: maj2@umdnj.edu.

Physiology

University of Miami, Graduate School, Miller School of Medicine, Graduate Programs in Medicine, Department of Physiology and Biophysics, Coral Gables, FL 33124. Offers PhD, MD/PhD. *Faculty:* 14 full-time (2 women). *Students:* 15 full-time (8 women); includes 3 minority (1 African American, 2 Hispanic Americans), 9 international. Average age 27. 36 applicants, 8% accepted, 3 enrolled. In 2005, 6 degrees awarded. *Degree requirements:* For doctorate, thesis/dissertation, qualifying exam. *Entrance requirements:* For doctorate, GRE General Test, minimum GPA of 3.0 in sciences. Additional exam requirements/recommendations for international students: Required—TOEFL. *Application deadline:* Applications are processed on a rolling basis. Application fee: $50. Electronic applications accepted. *Financial support:* In 2005–06, fellowships with full tuition reimbursements (averaging $22,000 per year); research assistantships, career-related internships or fieldwork and institutionally sponsored loans also available. *Faculty research:* Cell and membrane physiology, cell-to-cell communication, molecular neurobiology, neuroimmunology, neural development. *Unit head:* Dr. Karl Magleby, Chairman, 305-243-6821, Fax: 305-243-6898, E-mail: kmagleby@miami.edu. *Application contact:* Dr. David M. Landowne, Director of Graduate Studies, 305-243-6821, Fax: 305-243-5931, E-mail: dl@miami.edu.

University of Michigan, Medical School and Horace H. Rackham School of Graduate Studies, Program in Biomedical Sciences (PIBS), Department of Molecular and Integrative Physiology, Ann Arbor, MI 48109. Offers PhD. *Faculty:* 28 full-time (6 women). *Students:* 31 full-time (16 women); includes 9 minority (5 Asian Americans or Pacific Islanders, 4 Hispanic Americans). Average age 29. 21 applicants, 38% accepted. In 2005, 3 degrees awarded. *Degree requirements:* For doctorate, thesis/dissertation, oral defense of dissertation, preliminary exam. *Entrance requirements:* For doctorate, GRE General Test, 3 letters of recommendation. Additional exam requirements/recommendations for international students: Required—TOEFL, either the Michigan English Language Battery (MELAB) or the Test of English as a Foreign Language (TOEFL). *Application deadline:* For fall admission, 12/31 for domestic students. Applications are processed on a rolling basis. Application fee: $60 ($75 for international students). Electronic applications accepted. *Expenses:* Tuition, state resident: full-time $14,082; part-time $894 per credit hour. Tuition, nonresident: full-time $28,500; part-time $1,675 per credit hour. Required fees: $189; $189 per unit. *Financial support:* In 2005–06, 21 fellowships with full tuition reimbursements (averaging $23,500 per year), 10 research assistantships with full tuition reimbursements (averaging $23,500 per year) were awarded; scholarships/grants, traineeships, tuition waivers (full), and departmental funding also available. *Faculty research:* Ion transport, cardiovascular physiology, gene expression, hormone action, gastrointestinal physiology, endocrinology, muscle, signal transduction. Total annual research expenditures: $6.2 million. *Unit head:* Dr. John A. Williams, Chair, 734-936-2355, Fax: 734-936-8813, E-mail: jawillms@umich.edu. *Application contact:* Michele Boggs, Graduate Student Coordinator, 734-936-2355, Fax: 734-936-8813, E-mail: mboggs@umich.edu.

University of Minnesota, Duluth, Medical School, Graduate Program in Physiology, Duluth, MN 55812-2496. Offers MS, PhD. *Faculty:* 4 full-time (1 woman), 1 part-time/adjunct (0 women). Terminal master's awarded for partial completion of doctoral program. *Degree requirements:* For master's and doctorate, thesis/dissertation. *Entrance requirements:* For master's, GRE or MCAT; for doctorate, GRE or MCAT, 1 year of course work in each calculus, physics, and biology; 2 years of course work in chemistry; minimum GPA of 3.0 in science. Additional exam requirements/recommendations for international students: Required—TOEFL. *Application deadline:* For fall admission, 4/15 for domestic students. Applications are processed on a rolling basis. Application fee: $50 ($55 for international students). *Financial support:* In 2005–06, research assistantships with partial tuition reimbursements (averaging $16,186 per year); Federal Work-Study also available. Financial award application deadline: 4/15. *Faculty research:* Neural control of posture and locomotion, transport and metabolic phenomena in biological systems, control of organ blood flow, intracellular means of communication. Total annual research expenditures: $32,000. *Unit head:* Lorentz E. Wittmers, Associate Director of Graduate Studies, 218-726-8551, Fax: 218-726-6356, E-mail: lwittmer@d.umn.edu.

University of Minnesota, Twin Cities Campus, Medical School and Graduate School, Graduate Programs in Medicine, Department of Physiology, Minneapolis, MN 55455-0213. Offers cellular and integrative physiology (MS, PhD). Part-time programs available. Terminal master's awarded for partial completion of doctoral program. *Degree requirements:* For master's, thesis/dissertation, registration; for doctorate, thesis/dissertation, comprehensive exam, registration. *Entrance requirements:* For master's and doctorate, GRE General Test or MCAT. Electronic applications accepted. *Expenses:* Tuition, state resident: full-time $8,748; part-time $729 per credit. Tuition, nonresident: full-time $15,848; part-time $1,321 per credit. Full-time tuition and fees vary according to class time, course load, program and reciprocity agreements. *Faculty research:* Cell physiology, neuroscience, molecular biology, cardiovascular physiology, renal physiology, respiratory physiology.

University of Mississippi Medical Center, School of Graduate Studies in the Health Sciences, Department of Physiology and Biophysics, Jackson, MS 39216-4505. Offers MS, PhD, MD/PhD. *Faculty:* 24 full-time (8 women). *Students:* 5 full-time (3 women); includes 3 minority (2 African Americans, 1 Asian American or Pacific Islander). Average age 25. 9 applicants, 22% accepted, 2 enrolled. *Degree requirements:* For master's, thesis; for doctorate, thesis/dissertation, first authored publication. *Entrance requirements:* For master's and doctorate, GRE General Test, minimum GPA of 3.0. *Application deadline:* For fall admission, 8/1 for domestic students. Applications are processed on a rolling basis. Application fee: $10. *Financial support:* In 2005–06, 5 students received support, including 5 research assistantships (averaging $16,557 per year) Financial award application deadline: 4/1. *Faculty research:* Cardiovascular, renal, endocrine, and cellular neurophysiology; molecular physiology. Total annual research expenditures: $2.8 million. *Unit head:* Dr. John E. Hall, Chairman, 601-984-1801, Fax: 601-984-1817. *Application contact:* Dr. Thomas E. Lohmeier, Director, 601-984-1820, Fax: 601-984-1817, E-mail: tlohmeier@physiology.umsmed.edu.

University of Missouri–Columbia, School of Medicine and Graduate School, Graduate Programs in Medicine, Department of Medical Pharmacology and Physiology, Columbia, MO 65211. Offers pharmacology (MS, PhD); physiology (MS, PhD). *Faculty:* 20 full-time (4 women), 1 part-time/adjunct (0 women). *Students:* 14 full-time (8 women), 10 part-time (4 women); includes 7 minority (5 African Americans, 1 Asian American or Pacific Islander, 1 Hispanic American), 3 international. In 2005, 1 master's, 4 doctorates awarded. *Degree requirements:* For master's and doctorate, thesis/dissertation. *Entrance requirements:* For master's and doctorate, GRE General Test, minimum GPA of 3.0. *Application deadline:* For fall admission, 2/1 for domestic students. Application fee: $45 ($60 for international students). *Financial support:* Fellowships, research assistantships, teaching assistantships, institutionally sponsored loans available. *Faculty research:* Endocrine and metabolic pharmacology, biochemical pharmacology, neuropharmacology, receptors and transmembrane signaling. *Unit head:* Dr. Michael Rovetto, Director of Graduate Studies, 573-882-0773.

University of Missouri–St. Louis, College of Arts and Sciences, Department of Biology, St. Louis, MO 63121. Offers biology (MS, PhD), including animal behavior (MS), biochemistry (MS), biotechnology (MS), conservation biology (MS), development (MS), ecology (MS), environmental studies (PhD), evolution (MS), genetics (MS), molecular/cellular biology (MS), physiology (MS), plant systematics, population biology (MS), tropical biology (MS); biotechnology (Certificate); tropical biology and conservation (Certificate). Part-time programs available. *Faculty:* 50. *Students:* 26 full-time (15 women), 101 part-time (51 women); includes 13 minority (5 African Americans, 6 Asian Americans or Pacific Islanders, 2 Hispanic Americans), 41 international. Average age 32. In 2005, 22 master's, 2 doctorates awarded. *Degree requirements:* For master's, thesis or alternative; for doctorate, one foreign language, thesis/dissertation, 1 semester of teaching experience. *Entrance requirements:* For doctorate, GRE General Test. *Application deadline:* For spring admission, 12/1 priority date for domestic students. Applications are processed on a rolling basis. Application fee: $35 ($40 for international students). Electronic applications accepted. *Expenses:* Tuition, state resident: part-time $263 per credit hour. Tuition, nonresident: part-time $680 per credit hour. Required fees: $53 per credit hour. Tuition and fees vary according to program. *Financial support:* In 2005–06,

11 fellowships with full tuition reimbursements (averaging $30,000 per year), 15 research assistantships with full and partial tuition reimbursements (averaging $16,000 per year), 22 teaching assistantships with full and partial tuition reimbursements (averaging $16,000 per year) were awarded; career-related internships or fieldwork and Federal Work-Study also available. Support available to part-time students. Financial award application deadline: 2/1. *Faculty research:* Molecular biology, microbial genetics. *Unit head:* Zuleyma Tang-Martinez, Director of Graduate Studies, 314-516-6498, Fax: 314-516-6233, E-mail: zuleyma@umsl.edu. *Application contact:* 314-516-5458, Fax: 314-516-5310, E-mail: gradadm@umsl.edu.

University of Nebraska Medical Center, Graduate Studies, Department of Cellular and Integrative Physiology, Omaha, NE 68198. Offers physiology (MS, PhD). *Faculty:* 14 full-time (3 women). *Students:* 10 full-time (4 women), 1 (woman) part-time, 6 international. Average age 27. 9 applicants, 67% accepted, 3 enrolled. In 2005, 2 doctorates awarded. Terminal master's awarded for partial completion of doctoral program. *Degree requirements:* For master's, thesis optional; for doctorate, thesis/dissertation, comprehensive exam, registration. *Entrance requirements:* For master's, GRE General Test or MCAT, course work in biology, chemistry, math, and physics; for doctorate, GRE General Test or MCAT, course work in biology, chemistry, mathematics, and physics. Additional exam requirements/recommendations for international students: Required—TOEFL (minimum score 600 paper-based; 250 computer-based). *Application deadline:* For fall admission, 6/1 priority date for domestic students, 4/1 priority date for international students; for spring admission, 10/15 priority date for domestic students, 8/15 priority date for international students. Applications are processed on a rolling basis. Application fee: $45. Electronic applications accepted. *Expenses:* Tuition, area resident: Part-time $200 per hour. Tuition, nonresident: part-time $538 per hour. Required fees: $308; $59 per term. *Financial support:* In 2005–06, 5 students received support, including 1 fellowship with full tuition reimbursement available (averaging $21,000 per year), 2 research assistantships with full tuition reimbursements available (averaging $21,000 per year); teaching assistantships with full tuition reimbursements available, institutionally sponsored loans, traineeships, tuition waivers (full), and unspecified assistantships also available. Financial award application deadline: 2/5. *Faculty research:* Cardiovascular, renal and visual physiology, neuroscience, reproductive endocrinology. Total annual research expenditures: $3.6 million. *Unit head:* Dr. Pamela K. Carmines, Chair, Graduate Committee, 402-559-9343, Fax: 402-559-4438, E-mail: pcarmines@unmc.edu.

University of Nevada, Reno, School of Medicine and Graduate School, Graduate Programs in Medicine, Interdisciplinary Program in Cellular and Molecular Pharmacology and Physiology, Reno, NV 89557. Offers MS, PhD. *Faculty:* 8. *Students:* 13 full-time (4 women), 11 part-time (7 women); includes 1 minority (Asian American or Pacific Islander), 19 international. Average age 27. 7 applicants, 100% accepted, 7 enrolled. In 2005, 1 degree awarded. Terminal master's awarded for partial completion of doctoral program. *Degree requirements:* For master's; for doctorate, one foreign language, thesis/dissertation. *Entrance requirements:* For master's, GRE General Test, minimum GPA of 2.75; for doctorate, GRE General Test, minimum GPA of 3.0. Additional exam requirements/recommendations for international students: Required—TOEFL. *Application deadline:* For fall admission, 3/1 for domestic students; for spring admission, 11/1 for domestic students. Applications are processed on a rolling basis. Application fee: $60 ($95 for international students). *Expenses:* Tuition, area resident: Full-time $2,767; part-time $923 per semester. Tuition, state resident: full-time $5,733; part-time $1,911 per semester. Tuition, nonresident: full-time $12,679; part-time $1,911 per semester. International tuition: $13,878 full-time. Required fees: $404; $202 per term. One-time fee: $90. *Financial support:* Research assistantships, teaching assistantships available. Support available to part-time students. Financial award application deadline: 3/1. *Faculty research:* Neuropharmacology, toxicology, cardiovascular pharmacology, neuromuscular pharmacology. *Unit head:* Dr. Norman LeBlanc, Graduate Program Director, 775-784-1462.

University of New Mexico, School of Medicine, Biomedical Sciences Graduate Program, Albuquerque, NM 87131-5196. Offers biochemistry and molecular biology (MS, PhD); cell biology and physiology (MS, PhD); molecular genetics and microbiology (MS, PhD); neuroscience (MS, PhD); pathology (MS, PhD); toxicology (MS, PhD). Part-time programs available. Terminal master's awarded for partial completion of doctoral program. *Degree requirements:* For master's, thesis/dissertation; for doctorate, thesis/dissertation, comprehensive exam. *Entrance requirements:* For master's and doctorate, GRE General Test, minimum undergraduate GPA of 3.0. Additional exam requirements/recommendations for international students: Required—TOEFL. Electronic applications accepted. *Expenses:* Tuition, state resident: full-time $5,676. Tuition, nonresident: full-time $14,974; part-time $238 per credit hour. Required fees: $385 per term. Tuition and fees vary according to course load and program. *Faculty research:* Signal transduction, infectious disease, biology of cancer, structural biology, neuroscience.

University of North Dakota, School of Medicine and Graduate School, Graduate Programs in Medicine, Department of Pharmacology, Physiology, and Therapeutics, Grand Forks, ND 58202. Offers pharmacology (MS, PhD); physiology (MS, PhD). *Faculty:* 15 full-time (2 women), 1 part-time/adjunct (0 women). *Students:* 4 full-time (2 women), 9 part-time (4 women). 12 applicants, 25% accepted, 2 enrolled. In 2005, 1 degree awarded. *Degree requirements:* For master's, thesis, comprehensive exam; for doctorate, thesis/dissertation, written and oral exams. *Entrance requirements:* For master's, GRE General Test or MCAT, minimum GPA of 3.0; for doctorate, GRE General Test, minimum GPA of 3.5. Additional exam requirements/recommendations for international students: Required—TOEFL (minimum score 550 paper-based; 213 computer-based). *Application deadline:* For fall admission, 2/15 priority date for domestic students, 2/15 priority date for international students. Applications are processed on a rolling basis. Application fee: $35. Electronic applications accepted. *Financial support:* In 2005–06, 12 research assistantships with full tuition reimbursements (averaging $13,997 per year) were awarded; fellowships, teaching assistantships with full tuition reimbursements, Federal Work-Study, institutionally sponsored loans, scholarships/grants, and tuition waivers (full and partial) also available. Support available to part-time students. Financial award application deadline: 3/15; financial award applicants required to submit FAFSA. *Unit head:* Dr. Matthew Picklo, Graduate Director, 701-777-2293, Fax: 701-777-4490, E-mail: mpicklo@medicine.nodak.edu. *Application contact:* Brenda Halle, Admissions Specialist, 701-777-2947, Fax: 701-777-3619, E-mail: brendahalle@mail.und.edu.

University of North Texas Health Science Center at Fort Worth, Graduate School of Biomedical Sciences, Fort Worth, TX 76107-2699. Offers anatomy and cell biology (MS, PhD); biochemistry and molecular biology (MS, PhD); biomedical sciences (MS, PhD); biotechnology (MS); forensic genetics (MS); integrative physiology (MS, PhD); medical science (MS); microbiology and immunology (MS, PhD); pharmacology (MS, PhD); science education (MS). *Faculty:* 57 full-time (9 women), 2 part-time/adjunct (0 women). *Students:* 177 full-time (98 women), 42 part-time (33 women); includes 61 minority (15 African Americans, 3 American Indian/Alaska Native, 27 Asian Americans or Pacific Islanders, 16 Hispanic Americans), 53 international. Average age 28. 237 applicants, 62% accepted, 91 enrolled. In 2005, 37 master's, 15 doctorates awarded. Terminal master's awarded for partial completion of doctoral program. *Degree requirements:* For master's and doctorate, thesis/dissertation. *Entrance requirements:* For master's and doctorate, GRE General Test. Additional exam requirements/recommendations for international students: Required—TOEFL. *Application deadline:* For fall admission, 5/1 for domestic students. Application fee: $25 ($50 for international students). *Expenses:* Contact institution. *Financial support:* In 2005–06, 80 research assistantships (averaging $16,000 per year) were awarded; fellowships, teaching assistantships, career-related internships or fieldwork, Federal Work-Study, institutionally sponsored loans, scholarships/grants, and traineeships also available. Support available to part-time students. Financial award application deadline: 4/1; financial award applicants required to submit FAFSA. *Faculty research:* Alzheimer's disease, aging, eye diseases, cancer, cardiovascular disease. Total annual research expenditures: $21 million. *Unit head:* Dr. Thomas Yorio, Dean, 817-735-2560, Fax: 817-735-0243, E-mail: yoriot@hsc.unt.edu. *Application contact:* Carla Lee, Director of Graduate Admissions and Services, 817-735-2560, Fax: 817-735-0243, E-mail: gsbs@hsc.unt.edu.

See Close-Up on page 279.

Physiology

University of Notre Dame, Graduate School, College of Science, Department of Biological Sciences, Notre Dame, IN 46556. Offers aquatic ecology, evolution and environmental biology (MS, PhD); cellular and molecular biology (MS, PhD); genetics (MS, PhD); physiology (MS, PhD); vector biology and parasitology (MS, PhD). *Faculty:* 34 full-time (8 women), 3 part-time/adjunct (0 women). *Students:* 124 full-time (55 women); includes 11 minority (1 African American, 6 Asian Americans or Pacific Islanders, 4 Hispanic Americans), 39 international. 95 applicants, 34% accepted, 22 enrolled. In 2005, 4 master's, 11 doctorates awarded. Terminal master's awarded for partial completion of doctoral program. *Median time to degree:* Of those who began their doctoral program in fall 1997, 61% received their degree in 8 years or less. *Degree requirements:* For master's and doctorate, thesis/dissertation, comprehensive exam. *Entrance requirements:* For master's and doctorate, GRE General Test. Additional exam requirements/recommendations for international students: Required—TOEFL. *Application deadline:* For fall admission, 2/1 for domestic students; for spring admission, 11/1 for domestic students. Applications are processed on a rolling basis. Application fee: $50. Electronic applications accepted. *Financial support:* In 2005–06, 124 students received support, including 24 fellowships with full tuition reimbursements available (averaging $22,000 per year), 47 research assistantships with full tuition reimbursements available (averaging $15,250 per year), 45 teaching assistantships with full tuition reimbursements available (averaging $16,000 per year); traineeships and tuition waivers (full) also available. Financial award application deadline: 2/1. *Faculty research:* Tropical disease, molecular genetics, neurobiology, evolutionary biology, aquatic biology. Total annual research expenditures: $15.3 million. *Unit head:* Dr. Gary A. Lamberti, Director of Graduate Studies, 574-631-6552, Fax: 574-631-7413, E-mail: biology.biosadm.1@nd.edu. *Application contact:* Dr. Terrence J. Akai, Director of Graduate Admissions, 574-631-7706, Fax: 574-631-4183, E-mail: gradad@nd.edu.

See Close-Up on page 281.

University of Oklahoma Health Sciences Center, College of Medicine and Graduate College, Graduate Programs in Medicine, Department of Physiology, Oklahoma City, OK 73190. Offers MS, PhD. Part-time programs available. Terminal master's awarded for partial completion of doctoral program. *Degree requirements:* For master's, thesis (for some programs); for doctorate, thesis/dissertation. *Entrance requirements:* For master's, GRE General Test, statement of career goals and 3 letters of recommendation; for doctorate, GRE General Test, 3 letters of recommendation. Additional exam requirements/recommendations for international students: Required—TOEFL. *Faculty research:* Cardiopulmonary physiology, neurophysiology, exercise physiology, cell and molecular physiology.

University of Oregon, Graduate School, College of Arts and Sciences, Department of Human Physiology, Eugene, OR 97403. Offers MS, PhD. *Faculty:* 4 full-time (1 woman), 1 part-time/adjunct (0 women). *Students:* 45; includes 2 minority (both Asian Americans or Pacific Islanders), 9 international. 50 applicants, 82% accepted. In 2005, 20 master's, 4 doctorates awarded. *Degree requirements:* For master's, thesis optional; for doctorate, one foreign language, thesis/dissertation. *Entrance requirements:* For master's, GRE General Test, minimum GPA of 2.75 in undergraduate course work; for doctorate, GRE General Test. *Application deadline:* For fall admission, 7/18 for domestic students. Application fee: $50. *Financial support:* In 2005–06, 37 teaching assistantships were awarded; Federal Work-Study also available. *Faculty research:* Balance control, muscle fatigue, lower extremity function, knee control. *Unit head:* Dr. Gary Klug, Head, 541-346-4181. *Application contact:* Stephanie Swayne, Admissions Contact, 541-346-5430, E-mail: swayne@uoregon.edu.

University of Pennsylvania, School of Arts and Sciences, Program in Neurobiology and Physiology, Philadelphia, PA 19104. Offers PhD. *Entrance requirements:* For doctorate, GRE General Test, GRE Subject Test. Additional exam requirements/recommendations for international students: Required—TOEFL. Electronic applications accepted.

University of Pennsylvania, School of Medicine, Biomedical Graduate Studies, Graduate Group in Cell and Molecular Biology, Program in Cell Biology and Physiology, Philadelphia, PA 19104. Offers PhD, MD/PhD, VMD/PhD. *Degree requirements:* For doctorate, thesis/dissertation. *Entrance requirements:* For doctorate, GRE General Test. Additional exam requirements/recommendations for international students: Required—TOEFL. *Application deadline:* For fall admission, 12/15 priority date for domestic students, 12/1 priority date for international students. Applications are processed on a rolling basis. Application fee: $70. Electronic applications accepted. *Financial support:* Fellowships, research assistantships, scholarships/grants, traineeships, and unspecified assistantships available. *Unit head:* Dr. Morris Birnbaum, Head, 215-898-5001. *Application contact:* Emily Brady, Coordinator, 215-895-8935, Fax: 215-573-2104, E-mail: camb@mail.med.upenn.edu.

University of Prince Edward Island, Atlantic Veterinary College, Graduate Program in Veterinary Medicine, Charlottetown, PE C1A 4P3, Canada. Offers anatomy (M Sc, PhD); bacteriology (M Sc, PhD); clinical pharmacology (M Sc, PhD); clinical sciences (M Sc, PhD); epidemiology (M Sc, PhD), including reproduction; fish health (M Sc, PhD); food animal nutrition (M Sc, PhD); immunology (M Sc, PhD); microanatomy (M Sc, PhD); parasitology (M Sc, PhD); pathology (M Sc, PhD); pharmacology (M Sc, PhD); physiology (M Sc, PhD); toxicology (M Sc, PhD); veterinary science (M Vet Sc); virology (M Sc, PhD). Part-time programs available. *Faculty:* 76 full-time (25 women), 49 part-time/adjunct (8 women). *Students:* 54 full-time (32 women), 2 part-time. Average age 30. In 2005, 7 master's, 6 doctorates awarded. *Degree requirements:* For master's and doctorate, thesis/dissertation. *Entrance requirements:* For master's, DVM, B Sc honors degree, or equivalent; for doctorate, M Sc. *Application deadline:* Applications are processed on a rolling basis. Application fee: $50. *Expenses:* Contact institution. Tuition charges are reported in Canadian dollars. Part-time tuition and fees vary according to course level, degree level, campus/location, program and student level. *Financial support:* In 2005–06, 4 fellowships (averaging $25,000 Canadian dollars per year), 4 research assistantships (averaging $16,500 Canadian dollars per year) were awarded; career-related internships or fieldwork also available. *Faculty research:* Animal health management, infectious diseases, fin fish and shellfish health, basic biomedical sciences, ecosystem health. Total annual research expenditures: $1.2 million Canadian dollars. *Unit head:* Dr. James Bellamy, Associate Dean of Graduate Studies and Research, 902-566-0856, E-mail: bellamy@upei.ca. *Application contact:* Cheryl Gaudet, Registrar's Office, 902-566-0781, Fax: 902-566-0795, E-mail: registrar@upei.ca.

University of Puerto Rico, Medical Sciences Campus, School of Medicine, Division of Graduate Studies, Department of Physiology, San Juan, PR 00936-5067. Offers MS, PhD. *Faculty:* 9 full-time (3 women), 5 part-time/adjunct (0 women). *Students:* 15 full-time (10 women); all minorities (all Hispanic Americans) Average age 27. 5 applicants, 40% accepted, 2 enrolled. Terminal master's awarded for partial completion of doctoral program. *Degree requirements:* For master's, one foreign language, thesis/dissertation; for doctorate, one foreign language, thesis/dissertation, comprehensive exam. *Entrance requirements:* For master's and doctorate, GRE General Test, GRE Subject Test, interview; course work in biology, chemistry and physics; minimum GPA of 3.0; 3 letters of recommendation. *Application deadline:* For fall admission, 2/15 for domestic students, 2/15 for international students; for spring admission, 9/15 for domestic students, 9/15 for international students. Application fee: $15. *Expenses:* Tuition, state resident: full-time $3,600; part-time $100 per credit hour. Tuition, nonresident: full-time $4,655. Required fees: $1,734. Tuition and fees vary according to class time, degree level and program. *Financial support:* In 2005–06, 2 fellowships (averaging $13,800 per year), 8 research assistantships with tuition reimbursements (averaging $9,354 per year), 8 teaching assistantships (averaging $9,354 per year) were awarded; career-related internships or fieldwork and institutionally sponsored loans also available. Financial award application deadline: 4/30. *Faculty research:* Respiration, neuroendocrinology, cellular and molecular physiology, cardiovascular, exercise physiology and neurobiology. *Unit head:* Dr. Nelson Escobales, Director, 787-758-2525 Ext. 1606. *Application contact:* Dr. Carlos A. Torres-Ramos, Coordinator, 787-758-2525 Ext. 1393, Fax: 787-753-0120, E-mail: catorres@rcm.upr.edu.

University of Rochester, School of Medicine and Dentistry, Graduate Programs in Medicine and Dentistry, Department of Pharmacology and Physiology, Program in Physiology, Roches-

ter, NY 14627-0250. Offers MS, PhD. Terminal master's awarded for partial completion of doctoral program. *Degree requirements:* For master's, thesis; for doctorate, thesis/dissertation, qualifying exam. *Entrance requirements:* For master's and doctorate, GRE General Test.

University of Saskatchewan, College of Medicine, Department of Physiology, Saskatoon, SK S7N 5A2, Canada. Offers M Sc, PhD. *Faculty:* 8. *Students:* 11. *Degree requirements:* For master's and doctorate, thesis/dissertation, registration. *Entrance requirements:* Additional exam requirements/recommendations for international students: Required—TOEFL. *Application deadline:* For fall admission, 7/1 for domestic students. Applications are processed on a rolling basis. Application fee: $50. *Financial support:* Fellowships, research assistantships, teaching assistantships available. Financial award application deadline: 1/31. *Unit head:* W. Walz, Head, 306-966-6530, Fax: 306-966-8718, E-mail: walz@sask.usask.ca. *Application contact:* Dr. P. Sulakhe, Graduate Chair, 306-966-6534, Fax: 306-966-8718, E-mail: sulakhe@usask.ca.

University of Saskatchewan, Western College of Veterinary Medicine and College of Graduate Studies and Research, Graduate Programs in Veterinary Medicine, Department of Veterinary Biomedical Sciences, Saskatoon, SK S7N 5A2, Canada. Offers veterinary anatomy (M Sc); veterinary biomedical sciences (M Vet Sc); veterinary physiological sciences (M Sc, PhD). *Faculty:* 15 full-time (5 women). *Students:* 36 full-time (19 women); includes 3 minority (all African Americans) 6 applicants, 33% accepted. In 2005, 5 master's, 1 doctorate awarded. *Degree requirements:* For master's and doctorate, thesis/dissertation. *Faculty research:* Toxicology, animal reproduction, pharmacology, chloride channels, pulmonary pathobiology. *Unit head:* Dr. Barry Blakley, Head, 306-966-7349, Fax: 306-966-7376, E-mail: barry.blakley@usask.ca.

University of South Alabama, College of Medicine and Graduate School, Program in Basic Medical Sciences, Specialization in Physiology, Mobile, AL 36688-0002. Offers PhD. *Faculty:* 9 full-time (1 woman). *Degree requirements:* For doctorate, thesis/dissertation. *Entrance requirements:* For doctorate, GRE General Test or MCAT. *Application deadline:* For fall admission, 4/1 for domestic students. Applications are processed on a rolling basis. Application fee: $25. *Expenses:* Tuition, state resident: full-time $4,008. Tuition, nonresident: full-time $8,016. Required fees: $692. *Financial support:* Fellowships, research assistantships, institutionally sponsored loans available. Financial award application deadline: 4/1. *Faculty research:* Cardiovascular physiology. *Unit head:* Dr. Tom Lincoln, Chair, 251-460-7004. *Application contact:* Lanette Flagge, Academic Advisor, 251-460-6153.

The University of South Dakota, School of Medicine and Health Sciences and Graduate School, Biomedical Sciences Graduate Program, Physiology and Pharmacology Group, Vermillion, SD 57069-2390. Offers MA, PhD. *Faculty:* 7 full-time (2 women). *Students:* 9 full-time (8 women), 5 international. Average age 25. 17 applicants, 41% accepted, 7 enrolled. Terminal master's awarded for partial completion of doctoral program. *Degree requirements:* For master's, thesis/dissertation, registration; for doctorate, thesis/dissertation, comprehensive exam, registration. *Entrance requirements:* For master's and doctorate, GRE General Test, minimum GPA of 3.0. Additional exam requirements/recommendations for international students: Required—TOEFL (minimum score 550 paper-based; 213 computer-based). *Application deadline:* For fall admission, 4/15 priority date for domestic students, 4/15 priority date for international students. Applications are processed on a rolling basis. Application fee: $35. *Expenses:* Contact institution. Tuition and fees vary according to course load, program and reciprocity agreements. *Financial support:* In 2005–06, 9 students received support, including 7 fellowships with partial tuition reimbursements available (averaging $20,772 per year), 4 research assistantships with partial tuition reimbursements available (averaging $10,386 per year); scholarships/grants and unspecified assistantships also available. Financial award application deadline: 4/15; financial award applicants required to submit FAFSA. *Faculty research:* Pulmonary physiology and pharmacology, drug abuse, reproduction, signal transduction, cardiovascular physiology and pharmacology. Total annual research expenditures: $637,000.

University of Southern California, Keck School of Medicine and Graduate School, Graduate Programs in Medicine, Department of Physiology and Biophysics, Los Angeles, CA 90089. Offers MS, PhD, MD/PhD. Part-time programs available. *Faculty:* 15 full-time (3 women). *Students:* 17 full-time (11 women); includes 6 minority (1 African American, 5 Asian Americans or Pacific Islanders), 5 international. Average age 29. 3 applicants, 100% accepted, 2 enrolled. In 2005, 1 master's, 1 doctorate awarded. Terminal master's awarded for partial completion of doctoral program. *Median time to degree:* Of those who began their doctoral program in fall 1997, 100% received their degree in 8 years or less. *Degree requirements:* For master's, thesis optional; for doctorate, thesis/dissertation, comprehensive exam, registration. *Entrance requirements:* For master's and doctorate, GRE General Test, minimum GPA of 3.0. *Application deadline:* For fall admission, 2/1 priority date for domestic students, 2/1 priority date for international students; for spring admission, 10/1 priority date for domestic students, 10/1 priority date for international students. Applications are processed on a rolling basis. Application fee: $65 ($75 for international students). Electronic applications accepted. *Expenses:* Tuition: Full-time $25,416; part-time $1,059 per unit. Required fees: $484; $484 per year. Tuition and fees vary according to course load and program. *Financial support:* In 2005–06, 16 students received support, including 1 fellowship with partial tuition reimbursement available (averaging $24,876 per year), 14 research assistantships with full tuition reimbursements available (averaging $24,876 per year); teaching assistantships with full tuition reimbursements available, scholarships/grants and traineeships also available. *Faculty research:* Endocrinology, metabolism, cell transport, molecular biology, mathematical modeling. Total annual research expenditures: $6.5 million. *Unit head:* Dr. Harvey Kaslow, Director, Graduate Studies, 323-442-1244, Fax: 323-442-2283, E-mail: hrkaslow@usc.edu. *Application contact:* Elena Camarena, Graduate Coordinator, 323-442-1039, Fax: 323-442-2283, E-mail: physiol@hsc.usc.edu.

University of South Florida, College of Graduate Studies, College of Arts and Sciences, Department of Biology, Tampa, FL 33620-9951. Offers biology (PhD); botany (MS); ecology (PhD); microbiology (MS); physiology (PhD); zoology (MS). Part-time programs available. *Faculty:* 21. *Students:* 54 full-time (32 women), 26 part-time (17 women); includes 8 minority (2 African Americans, 1 American Indian/Alaska Native, 1 Asian American or Pacific Islander, 4 Hispanic Americans), 14 international. 76 applicants, 34% accepted, 13 enrolled. In 2005, 3 master's, 3 doctorates awarded. *Degree requirements:* For master's, thesis (for some programs), graduate seminar in biology; for doctorate, 2 foreign languages, thesis/dissertation, essay of research interest, comprehensive exam. *Entrance requirements:* For master's, GRE General Test, minimum undergraduate GPA of 3.0 in last 60 hours of course work; for doctorate, GRE General Test, GRE Subject Test in biology, minimum undergraduate GPA of 3.0 in last 60 hours of course work. Additional exam requirements/recommendations for international students: Required—TOEFL (minimum score 570 paper-based), TSE (minimum score 50). *Application deadline:* For fall admission, 2/1 priority date for domestic students, 3/1 priority date for international students; for spring admission, 10/1 for domestic students, 8/1 for international students. Application fee: $30. Electronic applications accepted. *Financial support:* Fellowships with full tuition reimbursements, research assistantships with full tuition reimbursements, teaching assistantships with full tuition reimbursements, Federal Work-Study and unspecified assistantships available. Financial award application deadline: 6/30. *Unit head:* Sydney Pierce, Chairperson, 813-974-3250, Fax: 813-974-3263. *Application contact:* Christine Smith, Graduate Advisor, 813-974-4747, Fax: 813-974-3263, E-mail: csmith2@chuma1.cas.usf.edu.

University of South Florida, College of Medicine and College of Graduate Studies, Graduate Programs in Medical Sciences, Tampa, FL 33620-9951. Offers anatomy (PhD); biochemistry and molecular biology (MS, PhD), including biochemistry and molecular biology (PhD), bioinformatics and computational biology (MS); medical microbiology and immunology (PhD); pathology (PhD); pharmacology and therapeutics (PhD), including medical sciences; physiology and biophysics (PhD). *Students:* 108 full-time (53 women), 34 part-time (26 women); includes 31 minority (9 African Americans, 9 Asian Americans or Pacific Islanders, 13 Hispanic

Physiology

University of South Florida *(continued)*
Americans), 30 international. 117 applicants, 99% accepted, 116 enrolled. In 2005, 9 master's, 4 doctorates awarded. *Degree requirements:* For doctorate, thesis/dissertation. *Entrance requirements:* For doctorate, GRE General Test, minimum GPA of 3.0. Application fee: $30. *Expenses: Contact institution. Financial support:* Institutionally sponsored loans and scholarships/grants available. Financial award application deadline: 4/1; financial award applicants required to submit FAFSA. *Unit head:* Dr. Joseph J. Krzanowski, Associate Dean for Research and Graduate Affairs, 813-974-4181, Fax: 813-974-4317, E-mail: jkrzanow@com1.med.usf.edu.

The University of Tennessee, Graduate School, College of Agricultural Sciences and Natural Resources, Department of Animal Science, Knoxville, TN 37996. Offers animal anatomy (PhD); breeding (MS, PhD); management (MS, PhD); nutrition (MS, PhD); physiology (MS, PhD). Part-time programs available. *Degree requirements:* For master's and doctorate, thesis/dissertation. *Entrance requirements:* For master's and doctorate, GRE General Test, minimum GPA of 2.7. Additional exam requirements/recommendations for international students: Required—TOEFL. Electronic applications accepted.

The University of Tennessee Health Science Center, College of Graduate Health Sciences, Department of Physiology, Memphis, TN 38163-0002. Offers MS, PhD. *Faculty:* 21 full-time (4 women), 2 part-time/adjunct (1 woman). *Students:* 3 full-time (2 women); all minorities (all Asian Americans or Pacific Islanders) Average age 25. 13 applicants, 23% accepted. *Degree requirements:* For master's, thesis, comprehensive exam; for doctorate, thesis/dissertation, oral and written preliminary and comprehensive exams. *Entrance requirements:* For master's, GRE General Test, minimum GPA of 3.0; for doctorate, GRE General Test, GRE Subject Test, minimum GPA of 3.0. Additional exam requirements/recommendations for international students: Required—TOEFL. *Application deadline:* For fall admission, 5/15 for domestic students. Application fee: $0. Electronic applications accepted. *Financial support:* Fellowships, research assistantships, teaching assistantships, institutionally sponsored loans and tuition waivers (full) available. Financial award application deadline: 2/25. *Unit head:* Dr. Leonard R. Johnson, Chairman, 901-448-7088, Fax: 901-448-7126, E-mail: ljohnson@utmem.edu. *Application contact:* Ida W. Mosby, Director, Enrollment Services, 901-448-5560, E-mail: imosby@utmem.edu.

The University of Texas Health Science Center at San Antonio, Graduate School of Biomedical Sciences, Department of Physiology, San Antonio, TX 78229-3900. Offers MS, PhD. Terminal master's awarded for partial completion of doctoral program. *Degree requirements:* For master's and doctorate, thesis/dissertation. *Entrance requirements:* For master's and doctorate, GRE General Test. *Faculty research:* Systemic and cellular aspects of physiology.

The University of Texas Medical Branch, Graduate School of Biomedical Sciences, Program in Cellular Physiology and Molecular Biophysics, Galveston, TX 77555. Offers MS, PhD. *Students:* 10 full-time (4 women); includes 1 minority (African American), 9 international. Average age 31. In 2005, 3 doctorates awarded. *Degree requirements:* For master's, thesis or alternative; for doctorate, thesis/dissertation. *Entrance requirements:* For master's and doctorate, GRE General Test. Additional exam requirements/recommendations for international students: Required—TOEFL (minimum score 550 paper-based; 213 computer-based). *Application deadline:* Applications are processed on a rolling basis. Application fee: $30 ($75 for international students). Electronic applications accepted. *Expenses:* Tuition, state resident: full-time $8,350; part-time $90 per credit hour. Tuition, nonresident: full-time $21,450; part-time $366 per credit hour. Required fees: $1,027; $11 per credit hour. $60 per term. *Financial support:* In 2005–06, fellowships (averaging $23,000 per year), research assistantships with full tuition reimbursements (averaging $23,000 per year) were awarded. Financial award applicants required to submit FAFSA. *Unit head:* Dr. Malcolm S. Brodwick, Director, 409-772-2973, Fax: 409-772-9382, E-mail: mbrodwic@utmb.edu. *Application contact:* Linda Spurger, Information Contact, 409-772-1942, Fax: 409-772-4687, E-mail: llspurge@utmb.edu.

The University of Toledo, School of Graduate Studies, Department of Physiology, Toledo, OH 43606-3390. Offers MS. Part-time programs available. *Degree requirements:* For master's, thesis, qualifying exam. *Entrance requirements:* For master's, GRE General Test, minimum undergraduate GPA of 3.0. *Expenses:* Tuition, state resident: full-time $6,623; part-time $308 per credit hour. Tuition, nonresident: full-time $13,232; part-time $735 per credit hour. *Faculty research:* Blood flow regulation, cell physiology, psychophysiology, contraceptive and reproductive physiology, hypertension.

University of Toronto, School of Graduate Studies, Life Sciences Division, Department of Physiology, Toronto, ON M5S 1A1, Canada. Offers M Sc, PhD. *Degree requirements:* For master's and doctorate, thesis/dissertation. *Entrance requirements:* For master's and doctorate, minimum B+ average in final year, 2 letters of reference. Additional exam requirements/recommendations for international students: Required—TOEFL (600 paper, 250 computer), MELAB (95), IELTS (8) or COPE (5).

University of Utah, School of Medicine and The Graduate School, Graduate Programs in Medicine, Department of Physiology, Salt Lake City, UT 84111-1107. Offers PhD. *Degree requirements:* For doctorate, thesis/dissertation, comprehensive qualifying exam, preliminary exam. *Entrance requirements:* For doctorate, GRE General Test, GRE Subject Test, minimum GPA of 3.0. *Expenses:* Tuition, state resident: full-time $2,932; part-time $369 per credit. Tuition, nonresident: full-time $10,350; part-time $1,302 per credit. Required fees: $516 per term. Tuition and fees vary according to course load and program. *Faculty research:* Cell neurobiology, chemosensory systems, cardiovascular and kidney physiology, endocrinology.

University of Virginia, School of Medicine, Department of Molecular Physiology and Biological Physics, Program in Physiology, Charlottesville, VA 22903. Offers PhD, MD/PhD. *Students:* 21 full-time (7 women); includes 1 minority (Asian American or Pacific Islander), 6 international. Average age 29. In 2005, 5 degrees awarded. *Entrance requirements:* Additional exam requirements/recommendations for international students: Required—TOEFL. *Application deadline:* Applications are processed on a rolling basis. Application fee: $60. Electronic applications accepted. *Expenses:* Tuition, state resident: full-time $7,731. Tuition, nonresident: full-time $18,672. Required fees: $1,479. Full-time tuition and fees vary according to degree level and program. *Financial support:* Fellowships, research assistantships, teaching assistantships available. Financial award applicants required to submit FAFSA. *Application contact:* Peter C. Brunjes, Associate Dean for Graduate Programs and Research, 434-924-7184, Fax: 434-924-6737, E-mail: grad-a-s@virginia.edu.

University of Washington, School of Medicine and Graduate School, Graduate Programs in Medicine, Department of Physiology and Biophysics, Seattle, WA 98195. Offers PhD. *Degree requirements:* For doctorate, thesis/dissertation. *Entrance requirements:* For doctorate, GRE General Test. Additional exam requirements/recommendations for international students: Required—TOEFL. *Faculty research:* Membrane and cell biophysics, neuroendocrinology, cardiovascular and respiratory physiology, systems neurophysiology and behavior, molecular physiology.

The University of Western Ontario, Faculty of Graduate Studies, Biosciences Division, Department of Physiology and Pharmacology, London, ON N6A 5B8, Canada. Offers M Sc, PhD, MD/M Sc, MD/PhD. *Degree requirements:* For master's, thesis, seminar course; for doctorate, thesis/dissertation, comprehensive exam. *Entrance requirements:* For master's, minimum B average, honors degree; for doctorate, minimum B average, honors degree, M Sc. *Faculty research:* Reproductive and endocrine physiology, neurophysiology, cardiovascular and renal physiology, cell physiology, gastrointestinal and metabolic physiology.

University of Wisconsin–La Crosse, Office of University Graduate Studies, College of Science and Health, Department of Biology, La Crosse, WI 54601-3742. Offers aquatic sciences (MS); biology (MS); cellular and molecular biology (MS); clinical microbiology (MS); microbiology (MS); nurse anesthesia (MS); physiology (MS). *Accreditation:* AANA/CANAEP. Part-time programs available. *Faculty:* 18 full-time (5 women), 1 part-time/adjunct (0 women). *Students:* 23 full-time (10 women), 52 part-time (25 women); includes 5 minority (1 American

Indian/Alaska Native, 3 Asian Americans or Pacific Islanders, 1 Hispanic American), 3 international. Average age 26. 61 applicants, 44% accepted, 23 enrolled. In 2005, 14 degrees awarded. *Degree requirements:* For master's, thesis, comprehensive exam, registration. *Entrance requirements:* For master's, GRE General Test, minimum GPA of 2.85. Additional exam requirements/recommendations for international students: Required—TOEFL (minimum score 550 paper-based; 213 computer-based). *Application deadline:* For fall admission, 3/1 for domestic students. Applications are processed on a rolling basis. Application fee: $45. Electronic applications accepted. *Expenses:* Tuition, state resident: part-time $354 per credit. Tuition, nonresident: part-time $943 per credit. Tuition and fees vary according to course load, program and reciprocity agreements. *Financial support:* In 2005–06, 10 students received support, including 4 research assistantships with partial tuition reimbursements available (averaging $10,000 per year), 10 teaching assistantships with partial tuition reimbursements available (averaging $9,600 per year); career-related internships or fieldwork, Federal Work-Study, health care benefits, and grant-funded positions also available. Support available to part-time students. Financial award application deadline: 3/15; financial award applicants required to submit FAFSA. *Faculty research:* Cell and molecular biology, physiology, environmental sciences, mycology, biomedical general. Total annual research expenditures: $700,000. *Unit head:* Dr. Tom Volk, Program Director, 608-785-6972, Fax: 608-785-6959, E-mail: volk.thom@uwlax.edu. *Application contact:* Kathryn Kiefer, Associate Director of Admissions, 608-785-8939, E-mail: admissions@uwlax.edu.

University of Wisconsin–Madison, Medical School, Endocrinology-Reproductive Physiology Program, Madison, WI 53706-1380. Offers MS, PhD. *Faculty:* 31 full-time (8 women). *Students:* 24 full-time (16 women), 1 (woman) part-time; includes 3 minority (2 African Americans, 1 Hispanic American), 6 international. Average age 30. 34 applicants, 18% accepted, 4 enrolled. In 2005, 2 master's, 3 doctorates awarded. Terminal master's awarded for partial completion of doctoral program. *Degree requirements:* For master's, thesis, oral defense of thesis, comprehensive exam, registration; for doctorate, thesis/dissertation, oral defense of dissertation, comprehensive exam, registration. *Entrance requirements:* For master's and doctorate, GRE, resumé, 3 letters of recommendation. Additional exam requirements/recommendations for international students: Required—TOEFL (minimum score 550 paper-based; 213 computer-based). *Application deadline:* For fall admission, 1/1 priority date for domestic students, 1/1 priority date for international students. Applications are processed on a rolling basis. Application fee: $45. Electronic applications accepted. *Financial support:* In 2005–06, fellowships with full tuition reimbursements (averaging $18,480 per year), 20 research assistantships with full and partial tuition reimbursements (averaging $18,480 per year), teaching assistantships with full and partial tuition reimbursements (averaging $11,262 per year) were awarded; scholarships/grants, traineeships, health care benefits, and unspecified assistantships also available. *Faculty research:* Ovarian physiology and endocrinology, fertilization and gamete biology, hormone action and cell signaling, placental function and pregnancy, embryo and fetal development. *Unit head:* Dr. Ian M. Bird, Director, 608-267-6240, Fax: 608-267-5773, E-mail: imbird@wisc.edu. *Application contact:* Tiffany Bachmann, Coordinator, 608-262-3222, Fax: 608-262-5157, E-mail: tabachmann@wisc.edu.

University of Wisconsin–Madison, Medical School and Graduate School, Graduate Programs in Medicine, Department of Physiology, Madison, WI 53706-1380. Offers PhD. *Faculty:* 22 full-time (6 women). *Students:* 28 full-time (14 women); includes 10 minority (1 African American, 8 Asian Americans or Pacific Islanders, 1 Hispanic American). Average age 22. 72 applicants, 14% accepted, 6 enrolled. In 2005, 3 degrees awarded. *Degree requirements:* For doctorate, thesis/dissertation, written exams. *Entrance requirements:* For doctorate, GRE, minimum GPA of 3.0. Additional exam requirements/recommendations for international students: Required—TOEFL (minimum score 580 paper-based; 237 computer-based). *Application deadline:* For fall admission, 1/15 for domestic students. Applications are processed on a rolling basis. Application fee: $45. Electronic applications accepted. *Financial support:* In 2005–06, fellowships with tuition reimbursements (averaging $22,500 per year), research assistantships with tuition reimbursements (averaging $22,500 per year), teaching assistantships with tuition reimbursements (averaging $22,500 per year) were awarded. *Faculty research:* Studies in molecular cellular systems, cardiovascular, neuroscience. *Unit head:* Dr. Richard L. Moss, Chair, 608-262-1939, Fax: 608-265-5072, E-mail: rlmoss@physiology.wisc.edu. *Application contact:* Sue S. Krey, Program Assistant 4, 608-262-9114, Fax: 608-265-5512, E-mail: krey@physiology.wisc.edu.

University of Wisconsin–Madison, Medical School, Training Program in Hearing, Madison, WI 53706-1380.

University of Wisconsin–Madison, School of Veterinary Medicine, Department of Animal Health and Biomedical Sciences, Program in Comparative Biosciences, Madison, WI 53706-1380. Offers anatomy (MS, PhD); biochemistry (MS, PhD); cellular and molecular biology (MS, PhD); environmental toxicology (MS, PhD); neurosciences (MS, PhD); pharmacology (MS, PhD); physiology (MS, PhD). *Degree requirements:* For doctorate, thesis/dissertation.

University of Wyoming, Graduate School, College of Arts and Sciences, Department of Zoology and Physiology, Laramie, WY 82070. Offers MS, PhD. Part-time programs available. *Faculty:* 23 full-time (4 women), 1 part-time/adjunct (0 women). *Students:* 44 full-time (16 women), 24 part-time (6 women), 6 international. 42 applicants, 29% accepted. In 2005, 13 master's awarded. *Degree requirements:* For master's, thesis/dissertation, registration; for doctorate, thesis/dissertation, comprehensive exam, registration. *Entrance requirements:* For master's, GRE General Test score of 900 minimum, minimum GPA of 3.0; for doctorate, GRE General Test (minimum 1000), minimum GPA of 3.0. Additional exam requirements/recommendations for international students: Required—TOEFL. *Application deadline:* For fall admission, 2/15 priority date for domestic students, 2/15 priority date for international students; for spring admission, 9/15 for domestic students, 9/15 for international students. Applications are processed on a rolling basis. Application fee: $50. Electronic applications accepted. *Expenses:* Tuition, state resident: full-time $3,720; part-time $155 per credit hour. Tuition, nonresident: full-time $10,704; part-time $446 per credit hour. Required fees: $666; $162 per semester. Tuition and fees vary according to course load and program. *Financial support:* In 2005–06, 1 fellowship with full tuition reimbursement (averaging $14,004 per year), 42 research assistantships with full tuition reimbursements (averaging $10,062 per year), 26 teaching assistantships with full tuition reimbursements (averaging $10,062 per year) were awarded; unspecified assistantships also available. Financial award application deadline: 3/1. *Faculty research:* Cell biology, ecology/wildlife, organismal physiology. Total annual research expenditures: $3.5 million. *Unit head:* , Dr. Graham Mitchell, Head, 307-766-4207, Fax: 307-766-5625, E-mail: mitchg@uwyo.edu. *Application contact:* E-mail: dmanore@uwyo.edu.

Virginia Commonwealth University, Graduate School, School of Allied Health Professions, Department of Physical Therapy, Program in Physiology and Physical Therapy, Richmond, VA 23284-9005. Offers PhD. *Accreditation:* APTA. *Students:* 1 full-time (0 women). 48 applicants, 96% accepted, 35 enrolled. In 2005, 1 degree awarded. *Degree requirements:* For doctorate, thesis/dissertation. *Entrance requirements:* For doctorate, GRE General Test. *Application fee:* $50. *Expenses:* Tuition, state resident: full-time $6,268; part-time $405 per credit. Tuition, nonresident: full-time $15,904; part-time $940 per credit. Required fees: $751 per semester hour. Tuition and fees vary according to course load and program. *Unit head:* Dr. Sheryl D. Finucan, Coordinator, E-mail: sfinucan@vcu.edu.

Virginia Commonwealth University, Medical College of Virginia-Professional Programs, School of Medicine and Graduate Programs, School of Medicine Graduate Programs, Department of Physiology, Richmond, VA 23284-9005. Offers neurosciences (PhD); physiology (MS, PhD, CBHS). *Faculty:* 30 full-time (3 women). *Students:* 67 full-time (36 women), 5 part-time (3 women); includes 26 minority (7 African Americans, 16 Asian Americans or Pacific Islanders, 3 Hispanic Americans), 2 international. 144 applicants, 69% accepted. In 2005, 7 master's, 5 doctorates, 18 other advanced degrees awarded. Terminal master's awarded for partial completion of doctoral program. *Degree requirements:* For master's, thesis; for doctorate, thesis/dissertation, comprehensive oral and written exams. *Entrance requirements:* For

Physiology

master's, DAT, GRE General Test, or MCAT; for doctorate, GRE General Test, MCAT, DAT. *Application deadline:* For fall admission, 2/15 for domestic students. Application fee: $50. *Expenses:* Tuition, state resident: full-time $6,268; part-time $405 per credit. Tuition, nonresident: full-time $15,904; part-time $940 per credit. Required fees: $751 per semester hour. Tuition and fees vary according to course load and program. *Financial support:* Fellowships, research assistantships, teaching assistantships, career-related internships or fieldwork and tuition waivers (full) available. *Unit head:* Dr. Margaret C. Biber, Chair, 804-828-9756, Fax: 804-828-7382, E-mail: mcbiber@vcu.edu. *Application contact:* Dr. George D. Ford, Director, 804-828-9501, Fax: 804-828-7382, E-mail: gdford@vcu.edu.

Wake Forest University, School of Medicine and Graduate School, Graduate Programs in Medicine, Department of Physiology and Pharmacology, Program in Physiology, Winston-Salem, NC 27109. Offers PhD. *Degree requirements:* For doctorate, thesis/dissertation. *Entrance requirements:* For doctorate, GRE General Test. *Faculty research:* Cardiovascular-renal physiology, endocrine physiology, neurophysiology.

Washington State University Spokane, Graduate Programs, Program in Exercise Science, Spokane, WA 99210-1495. Offers cellular physiology (MS); clinical exercise physiology (MS); clinical physiology (MS). *Faculty:* 4. *Students:* 4 full-time (2 women), 1 international. Average age 27. 10 applicants, 40% accepted, 3 enrolled. *Degree requirements:* For master's, thesis optional. *Entrance requirements:* For master's, GRE, minimum GPA of 3.0. Additional exam requirements/recommendations for international students: Required—TOEFL (minimum score 550 paper-based; 213 computer-based). *Application deadline:* For fall admission, 7/15 priority date for domestic students, 3/1 priority date for international students; for spring admission, 10/15 priority date for domestic students, 7/1 priority date for international students. Application fee: $35. *Expenses:* Tuition, state resident: full-time $6,295; part-time $336 per credit. Tuition, nonresident: full-time $15,949; part-time $819 per credit. Required fees: $593. Part-time tuition and fees vary according to campus/location and program. *Financial support:* In 2005–06, 4 students received support, including 3 research assistantships with full and partial tuition reimbursements available (averaging $9,698 per year); teaching assistantships with full and partial tuition reimbursements available, career-related internships or fieldwork, Federal Work-Study, scholarships/grants, health care benefits, and unspecified assistantships also available. *Faculty research:* Experimental exercise physiology, cellular and molecular mechanisms. *Unit head:* Dr. Sally Blank, Associate Professor/Director, 509-358-7633, E-mail: seblank@wsu.edu. *Application contact:* Erin Kincaid-McIntosh, Graduate Program Coordinator, 509-358-7626, E-mail: psh@wsu.edu.

Wayne State University, School of Medicine and Graduate School, Graduate Programs in Medicine, Department of Physiology, Detroit, MI 48202. Offers MS, PhD, MD/PhD. *Faculty:* 15 full-time (1 woman). *Students:* 7 full-time (5 women), 6 part-time (3 women); includes 1 minority (Asian American or Pacific Islander), 4 international. Average age 32. 15 applicants, 20% accepted, 2 enrolled. In 2005, 7 degrees awarded. *Degree requirements:* For master's and doctorate, thesis/dissertation. *Entrance requirements:* For master's, GRE General Test, GRE Subject Test, minimum GPA of 2.6; for doctorate, GRE General Test, GRE Subject Test, minimum GPA of 3.0. *Application deadline:* Applications are processed on a rolling basis. Application fee: $30 ($50 for international students). Electronic applications accepted. *Expenses:* Tuition, state resident: part-time $338 per credit hour. Tuition, nonresident: part-time $746 per credit hour. Required fees: $24 per credit hour. Full-time tuition and fees vary according to program. *Financial support:* In 2005–06, 25 research assistantships (averaging $18,327 per year) were awarded; fellowships, teaching assistantships Support available to part-time students. Financial award application deadline: 4/1. *Faculty research:* Regulation of brain blood flow, mechanism of hormone action, regulation of pituitary hormone secretion, regulation of cellular membranes, nano biotechnology. Total annual research expenditures: $3.7 million. *Unit head:* Joseph Dunbar, Chair, 313-577-1520, Fax: 313-577-5494, E-mail: ad4730@wayne.edu. *Application contact:* James Rillema, Professor, 313-577-1524, E-mail: ad0702@wayne.edu.

Wesleyan University, Graduate Programs, Department of Biology, Middletown, CT 06459-0260. Offers cell biology (PhD); comparative physiology (PhD); developmental biology (PhD); genetics (PhD); neurophysiology (PhD); population biology (PhD). *Faculty:* 12 full-time (3 women). *Students:* 29 full-time (15 women), 10 international. Average age 26. 131 applicants. In 2005, 2 doctorates awarded. *Degree requirements:* For doctorate, one foreign language, thesis/dissertation. *Entrance requirements:* For doctorate, GRE Subject Test. *Application deadline:* For fall admission, 2/15 for domestic students. Applications are processed on a rolling basis. Application fee: $0. *Expenses:* Tuition: Full-time $24,732. One-time fee: $20 full-time. *Financial support:* Research assistantships, teaching assistantships, stipends available. *Faculty research:* Microbial population genetics, genetic basis of evolutionary adaptation, genetic regulation of differentiation and pattern formation in *drosophila*. *Unit head:* Dr. Michael Weir, Chairman, 860-685-2402, E-mail: mweir@wesleyan.edu. *Application contact:* Marjorie Fitzgibbons, Information Contact, 860-685-2157, E-mail: mfitzgibbons@wesleyan.edu.

See Close-Up on page 323.

West Virginia University, Davis College of Agriculture, Forestry and Consumer Sciences, Division of Animal and Veterinary Sciences, Program in Animal and Veterinary Sciences, Morgantown, WV 26506. Offers breeding (MS); food sciences (MS); nutrition (MS); physiology (MS); production management (MS); reproduction (MS). Part-time programs available. *Degree requirements:* For master's, thesis, oral and written exams. *Entrance requirements:* For master's, minimum GPA of 2.5. Additional exam requirements/recommendations for international students: Required—TOEFL. Application fee: $45. *Expenses:* Tuition, state resident: full-time $4,582;

part-time $258 per credit hour. Tuition, nonresident: full-time $13,820; part-time $741 per credit hour. *Financial support:* Research assistantships, teaching assistantships, Federal Work-Study, institutionally sponsored loans, and tuition waivers (full and partial) available. Financial award application deadline: 2/1; financial award applicants required to submit FAFSA. *Faculty research:* Animal nutrition, reproductive physiology, food science. *Unit head:* Dr. Hillar Klandorf, Coordinator, 304-293-4372 Ext. 2050, Fax: 304-293-3676, E-mail: hillar.klandorf@mail.wvu.edu.

West Virginia University, Davis College of Agriculture, Forestry and Consumer Sciences, Interdisciplinary Program in Reproductive Physiology, Morgantown, WV 26506. Offers MS, PhD. Part-time programs available. *Students:* 6 full-time (5 women), 1 part-time, 1 international. Average age 27. 8 applicants, 25% accepted. In 2005, 3 degrees awarded. Terminal master's awarded for partial completion of doctoral program. *Degree requirements:* For master's, thesis/dissertation; for doctorate, thesis/dissertation, comprehensive exam. *Entrance requirements:* For master's, minimum GPA of 2.75; for doctorate, minimum GPA of 3.0. Additional exam requirements/recommendations for international students: Required—TOEFL. *Application deadline:* Applications are processed on a rolling basis. Application fee: $45. Electronic applications accepted. *Expenses:* Tuition, state resident: full-time $4,582; part-time $258 per credit hour. Tuition, nonresident: full-time $13,820; part-time $741 per credit hour. *Financial support:* In 2005–06, fellowships (averaging $12,000 per year), 4 research assistantships, 1 teaching assistantship were awarded; Federal Work-Study, institutionally sponsored loans, and tuition waivers (full and partial) also available. Financial award application deadline: 2/1; financial award applicants required to submit FAFSA. *Faculty research:* Uterine prostaglandins, luteal function, neural control of luteinizing hormone and follicle-stimulating hormone, follicular development, embryonic and fetal loss. Total annual research expenditures: $1 million. *Unit head:* Dr. E. Keith Inskeep, Chair, 304-293-2406 Ext. 4422, Fax: 304-293-2232, E-mail: einskeep@wvu.edu.

West Virginia University, School of Medicine, Graduate Programs at the Health Science Center, Biomedical Sciences Graduate Program, Program in Cellular and Integrative Physiology, Morgantown, WV 26506. *Faculty:* 36 full-time (6 women), 3 part-time/adjunct (1 woman). *Students:* 16 full-time (9 women). Average age 25. In 2005, 1 master's, 1 doctorate awarded. *Median time to degree:* Of those who began their doctoral program in fall 1997, 95% received their degree in 8 years or less. *Degree requirements:* For doctorate, thesis/dissertation, comprehensive exam. *Entrance requirements:* For doctorate, GRE General Test, minimum GPA of 3.0. Additional exam requirements/recommendations for international students: Required—TOEFL. *Application deadline:* For fall admission, 3/1 priority date for domestic students, 1/15 priority date for international students. Applications are processed on a rolling basis. Application fee: $0. Electronic applications accepted. *Financial support:* In 2005–06, research assistantships with full tuition reimbursements (averaging $20,000 per year); institutionally sponsored loans, traineeships, and health care benefits also available. *Faculty research:* Cell signaling and development of the micro vasculature, neural control of reproduction, learning and memory, airway responsiveness and remodeling. Total annual research expenditures: $2.5 million. *Unit head:* Dr. Robert L. Goodman, Chair, 304-293-1496, Fax: 304-298-3850, E-mail: bob.goodman@hsc.wvu.edu. *Application contact:* Dr. Matthew A. Boegehold, Graduate Studies Committee Chair, 304-293-5240, Fax: 304-293-3850, E-mail: matt.boegehold@hsc.wvu.edu.

See Close-Up on page 1269.

William Paterson University of New Jersey, College of Science and Health, Department of Biology, General Biology Program, Wayne, NJ 07470-8420. Offers general biology (MA); limnology and terrestrial ecology (MA); molecular biology (MA); physiology (MA). Part-time and evening/weekend programs available. *Students:* 2 full-time (1 woman), 13 part-time (9 women); includes 4 Hispanic Americans. In 2005, 4 degrees awarded. *Degree requirements:* For master's, independent study or thesis. *Entrance requirements:* For master's, GRE General Test, minimum GPA of 2.75. *Application deadline:* Applications are processed on a rolling basis. Application fee: $50. Electronic applications accepted. *Expenses:* Tuition, state resident: full-time $476. Tuition, nonresident: full-time $717. *Financial support:* Research assistantships, career-related internships or fieldwork and unspecified assistantships available. Financial award application deadline: 4/1; financial award applicants required to submit FAFSA. *Application contact:* Danielle Liautaud, Assistant Director, 973-720-3579, Fax: 973-720-2035, E-mail: liautaudd@wpunj.edu.

See Close-Up on page 331.

Wright State University, School of Graduate Studies, College of Science and Mathematics, Department of Anatomy and Physiology, Dayton, OH 45435. Offers anatomy (MS); physiology and biophysics (MS). *Degree requirements:* For master's, thesis optional. *Entrance requirements:* Additional exam requirements/recommendations for international students: Required—TOEFL. *Faculty research:* Reproductive cell biology, neurobiology of pain, neurohistochemistry.

Yale University, School of Medicine and Graduate School of Arts and Sciences, Combined Program in Biological and Biomedical Sciences (BBS), Physiology and Integrative Medical Biology Track, New Haven, CT 06520. Offers PhD, MD/PhD. *Students:* 3 full-time. *Entrance requirements:* Additional exam requirements/recommendations for international students: Required—TOEFL. *Application deadline:* For fall admission, 12/8 for domestic students, 12/8 for international students. *Unit head:* Dr. Emile Boulpaep, Director of Graduate Studies, 203-737-2215. *Application contact:* Graduate Registrar, E-mail: physiology@yale.edu.

BAYLOR COLLEGE OF MEDICINE

DeBakey Heart Center
Graduate Program in Cardiovascular Sciences

Program of Study

The DeBakey Heart Center is an independent entity within Baylor College of Medicine, and was the original National Research and Demonstration Center, funded by the National Heart, Lung, and Blood Institute. The Center has maintained its preeminence in basic cardiovascular research and offers an interdepartmental program in the biochemistry, cell biology, and physiology of the cardiovascular system, leading to the Ph.D. degree. The Center also participates in the Medical Scientist Training Program, in which students can earn the combined M.D./Ph.D. degree.

Current research covers a broad range of areas of cardiovascular physiology and biochemistry: cell biology of both cardiac and vascular tissue as well as physiologic models of cardiovascular diseases. There is extensive representation of virtually all aspects of cardiovascular research, ranging from cardiovascular development and gene transcription to cardiovascular physiology and monitoring of transgenic animals.

The exceptional perspective of this program can be found in the core course required of all students, which integrates all aspects of cardiovascular biology. Although students select a specific research area, they are also expected to be well versed in functional cardiovascular physiology. Seminar programs with student presentations are an integral part of the process.

Research Facilities

Baylor College of Medicine is part of the Texas Medical Center, which has a shared computer facility networked throughout the entire complex that is readily accessible to all students. The Texas Medical Center Library houses more than 250,000 books and journal volumes; has more than 4,000 items in its audiovisual collection, including computer software; and subscribes to more than 4,500 journals and to several major computerized information retrieval services.

Participating faculty members are from Baylor College of Medicine's Department of Medicine, Sections of Atherosclerosis and Lipoprotein, Cardiology, Cardiovascular Sciences, and Hypertension; the Departments of Anesthesiology, Cell Biology, Pathology, and Pediatrics; and Rice University's Institute of Biosciences and Bioengineering. The sections and departments are well funded, which allows for basic and extensive research and support facilities. Each section is located in either the main Baylor building or the Fondren Brown Cardiovascular Center.

Financial Aid

Currently, students receive stipends of $23,000 and full-tuition scholarships of $8200. Separate offices provide assistance to international students and to students with financial hardships.

Cost of Study

Tuition expenses and medical insurance are covered by Baylor College of Medicine. The graduate program covers the costs for the one-time matriculation fee of $25 and a $150 computer facility fee as well as all student fees. The program also pays the yearly fees (F-1 visa: $100, J-1 visa: $75) for international students.

Living and Housing Costs

A limited number of on-campus furnished apartments are available at $410 to $825 per month. Many commercial apartment complexes are located close to the Medical Center at rents ranging from $580 to $650 per month. Job opportunities for spouses are available in many institutions in the Texas Medical Center complex. The cost of living is lower in Houston than in most other major American cities.

Student Group

There are approximately 650 students enrolled at Baylor College of Medicine, of which 266 are graduate students. The low student-faculty ratio allows for close interaction between the students and faculty members.

Location

Baylor College of Medicine is located within the Texas Medical Center, a large and vigorous professional community that also includes the University of Texas Health Science Center, eight teaching hospitals, and the M. D. Anderson Hospital and Tumor Institute. The Texas Medical Center complex is situated in southwest Houston on 134 acres of land bordered on two sides by Hermann Park and on another side by Rice University. The fourth-largest city in the nation, Houston is also the home of several other colleges: Houston Baptist University, Texas Southern University, the University of Houston, and the University of St. Thomas. Students have access to all of the course offerings at these neighboring institutions; many of the DeBakey Heart Center faculty members have joint appointments as well.

Houston is an exciting cultural and metropolitan center. Ballet, opera, symphony, and theater performances are excellent and are accessible to the general population. Many fine museums and parks enhance the city life. Professional and amateur sports are very popular. The climate permits participation in a wide variety of outdoor activities, and Gulf coast beaches are a short drive from the city.

The College

Baylor College of Medicine is a private institution ranked among the best medical schools in the United States and is a nationally recognized leader in research, education, and patient care. The only private medical school in the greater Southwest, Baylor College of Medicine is an independent, nonsectarian, nonprofit corporation organized under a self-perpetuating board of trustees.

The College is part of the Texas Medical Center complex, which contains two medical schools (the Baylor College of Medicine and the University of Texas Health Science Center), and eight hospitals, with a total of more than 4,000 beds. Also included within the bounds of the Center are dental, nursing, public health, and biomedical science schools; an institution for mental sciences; institutes of religion, speech, hearing, and rehabilitation; the Houston Department of Public Health; and the Texas Medical Center Library.

Applying

Applicants must have a bachelor's degree and a strong background in biology and biochemistry. Candidates for admission must complete the application form of the Graduate School. The application should contain Graduate Record Examinations scores that are less than two years old, three letters of recommendation, and official undergraduate transcripts. One GRE Subject Test is strongly recommended. Applications are reviewed twice: once by a faculty committee of the DeBakey Heart Center and once by the Admissions Committee of the Graduate School. The program has a rolling admission policy. Applications are accepted throughout the year for beginning matriculation in August (first term). Although entrance into the program in August is preferable, the structure of the program allows students to be admitted year-round. All application materials must be sent directly to Graduate School Admissions. The application fee of $30 is waived for electronic applications. More information can be found at the department's Web site.

Correspondence and Information

For an application form:
Graduate School Admissions
Baylor College of Medicine
Houston, Texas 77030-3498
Phone: 713-798-3312

For additional information:
Dr. Alan R. Burns
Director of Graduate Studies
Graduate Program in Cardiovascular Sciences
DeBakey Heart Center
Baylor College of Medicine
Houston, Texas 77030-3498
Phone: 713-798-4977
Fax: 713-790-0681
E-mail: aburns@bcm.tmc.edu
Web site: http://www.bcm.tmc.edu/cvs/

Baylor College of Medicine

THE FACULTY AND THEIR RESEARCH

Antonio Baldini, M.D., Professor. Genetics of cardiovascular development; genetically defined mouse models of human diseases; DiGeorge syndrome.

Christie M. Ballantyne, M.D., Professor. Cell and molecular biology of adhesion molecules in atherosclerosis and arterial disease in both man and mouse; role of obesity in inflammation, biomarkers for atherosclerosis and diabetes.

Philip M. Barger, M.D., Assistant Professor. Molecular mechanisms of cardiac hypertrophy and failure; pathophysiological functions of PPARs in cardiac energy metabolism; cardiac nuclear receptor biology.

Paul F. Bray, M.D., Professor. Role of platelets and platelet risk factors in cardiovascular disease; disorders of bleeding and excessive blood clotting; role of platelets in arterial thrombosis and pharmacogenetics; signaling pathways of platelet activation; novel genes and gene products involved in platelet reactivity.

Robert M. Bryan Jr., Ph.D., Professor. Cerebral circulation; traumatic brain injury; stroke; endothelial responses; vascular smooth muscle; cerebral energy metabolism; smooth muscle–endothelial interaction; calcium regulation; potassium channels.

Alan R. Burns, Ph.D., Associate Professor. Regulation of neutrophil transendothelial migration by adhesion molecules, cytokines, and chemotactic factors; characterization of cellular and molecular events mediating the intense inflammatory response associated with myocardial ischemia.

Lawrence Chan, M.D., Professor. Cellular and molecular biology and animal models of atherosclerosis and lipoprotein metabolism; molecular genetics of RNA editing; somatic gene therapy of cardiovascular disorders.

Jing-Fei Dong, M.D., Ph.D., Associate Professor. Effects of flow shear stress on platelet functions; von Willebrand factor and its proteolysis; inflammation and thrombosis.

William Durante, Ph.D., Associate Professor. Biology of nitric oxide and carbon monoxide in the circulation; regulation of amino acid transport and metabolism by vascular cells; hemodynamic forces and vascular cell function.

Mark Entman, M.D., Professor. Muscle regulation of calcium flux by sarcoplasmic reticulum; molecular basis of cardiovascular inflammation; expression of cell adherence molecules on cardiovascular muscle cells.

Nikolaos Frangogiannis, M.D., Associate Professor. Healing of the infracted heart; role of chemokines in cardiac repair; resolution of inflammatory response in the infracted myocardium; development of fibrosis in the ischemic heart.

Margaret A. Goldstein, Ph.D., Professor. Ultrastructural changes in the Z-band lattice as a function of muscle tension, calcium levels, and change in cytoskeletal structures; computerized image processing; freeze-substitution; antibody labeling; 3-D reconstructions.

Craig Hartley, Ph.D., Professor. Design and application of implantable, intravascular, and noninvasive ultrasonic sensors; techniques for monitoring blood flow, vascular mechanics, and cardiac function in mammals from transgenic mice to humans.

Karen Hirschi, Ph.D., Associate Professor. Regulation of vascular development; vascular potential of adult stem and progenitor cells.

Dirar S. Khoury, Ph.D., Associate Professor. Cardiac electrophysiology studies of experimental animal models; methods for electrical, anatomical, and hemodynamic imaging of the heart; computer modeling of cardiac electrical activity.

Brian J. Knoll, Ph.D., Adjunct Associate Professor. G-protein–coupled receptor endocytosis and down-regulation; intracellular sorting events and control of receptor activity; ras-related GTPases in regulation of receptor trafficking; partial agonists and receptor desensitization.

Jose Lopez, M.D., Professor. Platelet adhesion to the vessel wall and in interactions of platelets with other vascular cells, such as leukocytes and endothelial cells.

Mark Majesky, Ph.D., Associate Professor. Molecular biology of vascular smooth-muscle development; cadherin-dependent cell-cell adhesion; growth factors and receptors in the vessel wall; production and alternative splicing of the matrix protein tenascin.

Douglas L. Mann, M.D., Mary and Gordon Cain Chair and Professor of Medicine and Director, Winters Center for Heart Failure Research. Cardiac injury and repair; inflammatory mediators; heart failure and translational research.

A. J. Marian, M.D., Associate Professor. Molecular genetics of cardiomyopathies; structure-function analysis of mutant sarcomeric proteins in transgenic animals and adult cardiac myocytes; identification of genetic risk factors for coronary artery disease.

Lloyd Michael, Ph.D., Professor. Myocardial ischemia; cardiac lymphatic system; platelet and leukocyte function in inflammation of the heart; experimental models of cardiovascular disease.

Joel Morrisett, Ph.D., Professor. Integrated investigation of peripheral atherosclerosis (carotid and femoral arteries) using in vivo imaging by high resolution MRI and ex vivo tissue analysis by microCT, immunohistochemistry, proteomics, and genomics to elucidate mechanisms of calcification and remodeling.

Karen Niederreither, Ph.D., Assistant Professor. Cardiovascular development; role of active vitamin A (retinoic acid) in regulation of embryonic development, cardiac growth, and cell proliferation.

Adebayo O. Oyekan, D.V.M., Ph.D., Adjunct Associate Professor. Interactions of nitric oxide, cytochrome P450–derived eicosanoids, and PPARalpha in the regulation of blood pressure and renal function.

Henry Pownall, Ph.D., Professor. Identification of the molecular basis of disorders associated with atherosclerosis, obesity, alcoholic hepatitis, and hypertriglyceridemia; dietary studies in human subjects and in animal models; cell culture and gene regulation.

Anilkumar K. Reddy, Ph.D., Assistant Professor. Development and application of noninvasive methods to measure cardiac velocities, blood flow velocities in central and peripheral vessels, arterial pulse-wave velocity, arterial wall motion, blood pressure, and aortic and arterial impedance in mice using ultrasound and other modalities.

Michael Schneider, M.D., Professor. Molecular genetics of cardiac growth and development; TGFβ receptor mutagenesis; pocket proteins and p300 as cell-cycle regulators in cardiac muscle; adenoviral gene transfer.

Robert J. Schwartz, Ph.D., Professor. Cardiovascular development and molecular gene regulation; role of tinman-like factors, serum response factor, and GATA factor in cardiac gene transcription.

C. Wayne Smith, M.D., Professor. Molecular pathway of inflammatory diseases; role of inflammatory cells; neutrophils and monocytes in reperfusion injury and their adherence to endothelium.

George Taffet, M.D., Associate Professor. Age-related changes in the heart and vasculature; physiology and related biochemistry of mouse models of human aging and disease.

Addison Taylor, M.D., Ph.D., Professor. Neuronal regulation of blood pressure; role of oxygen-free radicals and lipid-derived mediators in cardiovascular inflammatory responses.

Lubov T. Timchenko, Ph.D., Associate Professor. Molecular mechanisms of Myotonic Dystrophies types 1 and 2; regulation of myogenesis; RNA metabolism in skeletal muscle and in heart.

Jeffrey A. Towbin, M.D., Professor and Chief of Pediatric Cardiology. Molecular genetics of inherited diseases and sudden cardiac death, specifically cardiomyopathies and rhythm disturbances; genetics of congenital heart disease.

BAYLOR COLLEGE OF MEDICINE

Department of Molecular Physiology and Biophysics

Program of Study

The Department of Molecular Physiology and Biophysics at Baylor College of Medicine offers a research-oriented program of graduate study that leads to the Ph.D. degree. This program applies the most current methods of molecular and cell biology, mouse genetics, neuroscience, imaging, electrophysiology, biophysics, immunology, protein biochemistry, and pharmacology to the study of problems of physiological importance. The Department participates in Baylor College of Medicine's M.D./Ph.D. Program, whose candidates may earn both the degrees of M.D. and Ph.D.

The departmental faculty members are very interactive with ongoing research in the areas of structure and function of ion channels and transport proteins, signal transduction, synaptic plasticity, cell-cycle control, reactive oxygen species, neuronal morphology, drug and gene delivery, development of biosensors for genetic diagnosis, and small-animal in vivo neuronal tract tracing utilizing manganese enhanced magnetic resonance imaging (MEMRI). In addition, knockout and transgenic mice are used for a number of different types of animal studies, including the study of muscle and cardiovascular function, neurodegeneration, learning and memory, and cancer.

Research Facilities

The Department occupies 18,000 square feet of contiguous space with recently constructed laboratories and offices. It contains an array of modern facilities and instrumentation. There are laboratories for electrophysiology (voltage clamping, patch clamping, and planar lipid bilayer), ligand binding, fluorescence spectroscopy, protein biochemistry, automated peptide synthesis and microsequencing, fluorescence imaging, multiphoton and confocal microscopy, X-ray crystallography, cryomicroscopy, magnetic resonance imaging (MRI), echocardiography, recombinant DNA technology, and producing transgenic and genetic knockout mice. Shared facilities include cold rooms, scintillation and gamma counters, molecular modeling workstations, ultracentrifuges, a P2/P3 facility, radioisotope preparation rooms, machine and electronics shops, and darkrooms. Central computational facilities are housed within the Department, and all laboratories are connected to this facility via a local area network that also provides access to the central Baylor Computing Resources and to the Internet. The high level of collaboration among the various departments at Baylor College of Medicine and the other institutions in the Houston scientific community provides access to exceptional facilities. The library of the Texas Medical Center, across the street from Baylor College of Medicine, houses more than 333,000 books, subscribes to more than 4,000 electronic journals, and offers access to more than 400 computer databases. A Learning Resource Center, open to students 24 hours a day, provides additional computer access and a quiet area for study.

Financial Aid

Costs for Ph.D. students are normally covered by scholarships awarded by the College. Tuition scholarships are available to all students. Stipends for entering Ph.D. students are $23,000 for the 2006–07 academic year. Health insurance is provided for students; additional financial aid is also available. Requests for information regarding financial aid should be addressed to the Financial Aid Office, Baylor College of Medicine, Houston, Texas 77030.

Cost of Study

Tuition for the 2006–07 academic year is $8200. Upon acceptance into the program, tuition is covered by the Department for the first year and by student's mentor for subsequent years. Students pay a one-time matriculation fee of $25, a one-time graduation fee of $140 during the fourth year, and a student fee of $150 for the first year and $20 for subsequent years. Students on temporary visas also pay an annual international services fee of $75 for an F-1 visa or $100 for a J-1 visa.

Living and Housing Costs

One residence hall, located adjacent to Baylor College of Medicine in the Texas Medical Center, offers fully furnished rooms and apartments that are moderately priced. There are also a wide range of privately owned furnished and unfurnished apartments and houses available nearby. Housing in Houston is moderately priced compared with other major U.S. cities.

Student Group

The graduate students in the Department of Molecular Physiology and Biophysics comprise an eclectic group with varied academic backgrounds, from physics to biology. They are an integral part of the Graduate School of Biomedical Sciences, with approximately 500 students enrolled, including 100 first-year students enrolled in 2005–06. The Department also maintains a vigorous postdoctoral training program that includes research fellows and visiting scientists.

Location

As the nation's fourth-largest city, Houston is a vibrant scientific and industrial center. Its symphony orchestra, grand opera, ballet, theater companies, and art museums, many close to the Texas Medical Center, are nationally recognized. Recreational activities abound, with facilities and opportunities for a wide range of professional and amateur sports. The Texas Medical Center adjoins Hermann Park, which includes a zoo, a garden center, the Miller Outdoor Theater, the Museum of Natural Science, the Burke-Baker Planetarium, a golf course, tennis courts, jogging tracks, and picnic areas.

The College

Baylor College of Medicine, an independent, private, and fully accredited biomedical institution, is located in the heart of the Texas Medical Center, one of the largest medical centers in the world. The medical center, adjacent to residential areas, covers more than 500 acres and has forty-one institutions, 10,000 students, and 50,000 employees. Baylor is recognized as a leader in biomedical and basic science research and currently ranks tenth among the top research medical schools in the country. The graduate school is committed to excellence in graduate training. There is a high degree of interdisciplinary cooperation, not only among the faculty members in basic science areas but also with clinical investigators in the College and associated institutions in the Texas Medical Center. Ongoing research programs being carried out by productive and widely recognized investigators in both the basic sciences and the clinical faculty, coupled with the favorable faculty-student ratio, permit students to be directly involved in and contribute to significant research projects.

Applying

The school year begins in August and ends in July; it consists of five terms, each eight weeks long. Applicants must hold a bachelor's degree or a more advanced degree or must be in the final stages of a program leading to a bachelor's degree or the equivalent. Undergraduate course requirements include a year each of general biology, physics, mathematics, chemistry, and organic chemistry, all with a B average or better. The TOEFL is required of all students who have not earned a degree at a university in which the primary language is English. Official transcripts from each college or university attended, letters of evaluation from at least 3 professors, and scores from the Graduate Record Examinations (General Test and Subject Test recommended in biology, chemistry, or biochemistry, cell and molecular biology) must be mailed directly to the School rather than by the applicant. January 1 is the target date for completed applications; early application is encouraged. An electronic application can be accessed on the Web at http://www.bcm.tmc.edu/gradschool/gs-apply.html. The $30 application fee is waived for electronic applications.

Correspondence and Information

Director of Graduate Studies
Department of Molecular Physiology and Biophysics
Baylor College of Medicine
One Baylor Plaza, BCM335
Houston, Texas 77030
Phone: 713-798-5109
Fax: 713-798-3475
E-mail: molphys@bcm.tmc.edu
Web site: http://www.bcm.edu/physio

Baylor College of Medicine

THE FACULTY AND THEIR RESEARCH

Aladin M. Boriek, Associate Professor; Ph.D., Rice. Respiratory system mechanics; lung mechanics; computational models of tissue mechanics; mechanotransduction in skeletal and smooth muscles; mechanotransduction of the respiratory pump. Fiber architecture of the canine abdominal muscles. *J. Appl. Physiol.* 92(2):725–35, 2002.

Joseph Bryan, Professor; Ph.D., Pennsylvania. Molecular biology of adenosine triphosphate–sensitive potassium channels and regulation of insulin secretion. Toward linking structure with function in ATP-sensitive K+ channels. *Diabetes* 53(Suppl. 3):S104–12, 2004.

Robert Bryan, Professor; Ph.D., British Columbia. Control of cerebral circulation and mechanism of a dilator process termed endothelium-derived hyperpolarizing factor (EDHF) or endothelium-dependent hyperpolarizations. Endothelium-dependent vasodilation of cerebral arteries is altered with simulated microgravity through nitric oxide synthase and EDHF mechanisms. *J. Appl. Physiol.* 4(20), 2006.

Wah Chiu, Alvin Romansky Professor; Ph.D., Berkeley. Structural studies of macromolecules and macromolecular assemblies by electron cryomicroscopy and crystallography. Electron cryomicroscopy of membrane channels. *J. Mol. Biol.* 345:427–31, 2005. *Struct. (Camb.)* 13:363–72, 2005.

Mariella DeBiasi, Associate Professor; Ph.D., Padova (Italy). Autonomic nervous system: analysis from molecules to wake behavior; effect of stress on the mechanisms of addiction. Decreased signs of nicotine withdrawal in mice null for the beta 4 nicotinic acetylcholine receptor subunit. *J. Neurosci.* 24(5):10035–9, 2005. Co-release of dopamine and serotonin from triatal dopamine terminals. *Neuron* 46:65–74, 2005.

* Mary E. Dickinson, Assistant Professor; Ph.D., Columbia. Imaging the role of fluid mechanics in early cardiovascular development. 4-dimensional cardiac imaging of living embryos via post-acquisition synchronization of nongated slice-sequences. *J. Biomed. Optics* 054001.

Mark Entman, Professor; M.D., Duke. Muscle regulation of calcium flux by sarcoplasmic reticulum; molecular basis of cardiovascular inflammation; expression of cell adherence molecules on cardiovascular muscle cells. Extracellular matrix remodeling in canine and mouse myocardial infarcts. *Cell Tissue Res.* 2(22), 2006.

Marta Fiorotto, Associate Professor; Ph.D., Cambridge. Developmental processes specific to the perinatal period. Overexpression of des(1-3) insulin-like growth factor 1 in the mammary glands of transgenic mice delays the loss of milk production with prolonged lactation. *Biol. Reprod.* 73(6):1116–25, 2005.

* Susan L. Hamilton, Professor; Ph.D., Colorado at Denver. Skeletal muscle excitation-contraction coupling; ryanodine receptors L-type Ca^{2+} channels and diseases of skeletal muscle; skeletal muscle biology and disease, ion channels, excitation-contraction coupling. A Ca^{2+} binding domain in RyR1 that interacts with the calmodulin binding site and modulates channel activity. *Biophys. J.* (90):173–82, 2006. Heat- and anesthesia-induced malignant hyperthermia in an RyR1 knock-in mouse. *FASEB J.* 20:329–30, 2006.

Monica J. Justice, Associate Professor; Ph.D., Kansas State. Using mouse mutagenesis to analyze gene function and establish models of human disease. Mutation of l7Rn3 shows that Odz4 is required for mouse gastrulation. *Genetics* 169:285–99, 2005.

Michael Kroll, Associate Professor; M.D., Cornell. Mechanisms of von Willebrand factor–induced platelet activation; effect of physical forces on platelets; functional analyses of glycoprotein Ib/IX/V signaling. Purinergic P2Y receptor blockade inhibits shear-induced platelet phosphatidylinositol 3-kinase activation. *Mol. Pharmacol.* 63:639–45, 2003.

* Jeannette Kunz, Assistant Professor; Ph.D., Basel. Structure, function, and regulation of phosphoinositide phosphate kinases in cell polarization and cell motility. The PH domain proteins Slm1 and Slm2 are required for actin cytoskeleton organization in yeast and bind phosphatidylinositol-4,5-bisphosphate and TORC2. *Mol. Biol. Cell* E04-07-0564, 2005 (online).

Douglas L. Mann, Professor; M.D., Temple. Cardiac injury and repair; inflammatory mediators; heart failure and translational research. Molecular mechanisms underlying cardiac remodeling and decompensation. *Circulation* 109:262–8, 2004.

Sean Marrelli, Assistant Professor; Ph.D., Baylor College of Medicine. Mechanisms by which endothelium regulates vascular tone in cerebral circulation in both normal and pathological conditions, such as ischemia/reperfusion. Endothelial Ca^{2+} and endothelium-derived hyperpolarizing factor (EDHF)–mediated vasodilation. In *Microvascular Research: Biology and Pathology*, chap. 37, pp. 239–45, 2006.

Lloyd Michael, Professor; Ph.D., Ottawa. Physiological genomics and multiple model systems in myocardial biology. CCL2/monocytes chemoattractant protein-1 regulates inflammatory responses critical to healing myocardial infarcts. *Circ. Res.* 96:881–9, 2005. Critical role of engodenous thrombospondin-1 in preventing expansion of healing myocardial infarcts. *Circulation* 111:2935–42, 2005.

* Robia G. Pautler, Assistant Professor; Ph.D., Carnegie Mellon. Multidisciplinary approach combining MRI with molecular biology, genetics, computational biology, and neuroscience, with focus on understanding processes governing neurodegeneration, learning, and memory; magnetic resonance imaging: applications in mouse models of disease and novel technology development. Mouse MRI: Concepts and applications in physiology. *Physiology* 19:168–75, 2004.

* Steen E. Pedersen, Associate Professor; Ph.D., Virginia. Allosteric mechanisms of ligand-gated ion channels; functional aspects of gating, ion selectivity, and ligand binding are determined with biophysical methods. Apocalmodulin and Ca2+ calmodulin-binding sites on the CaV1.2 channel. *Biophys. J.* 85(3):1538–47, 2003.

Paul J. Pfaffinger, Associate Professor; Ph.D., Washington (Seattle). Molecular biology and biophysics of potassium channels. Structure, function, and regulation of ion channels. *Biophys. J.* 87(4):2380–96, 2004.

Florante A. Quiocho, Professor; Ph.D., Yale. X-ray crystallographic analysis of biological macromolecules; molecular recognition; structural biophysics and biology; X-ray crystallography of proteins and biological compounds. *Proc. Natl. Acad. Sci. U.S.A.* 101:15567–72, 2004. *Cell* 104:433–40, 2001.

Arun Rajan, Associate Professor; M.D., St. John's Medical (Bangalore). Stimulus-secretion coupling, physiology of pancreatic islet cells and pathophysiology of diabetes mellitus. Ion channel signal transduction in pancreatic beta cells. In *Biology of the Pancreatic Beta Cell*, ed. S. L. Howell, 1999.

Peter Saggau, Associate Professor; Ph.D., Munich. Modulation of synaptic transmission in mammalian in vitro preparations such as brain slices or neurons in primary culture, usually taken from the hippocampus, a brain structure involved in learning and memory. Presynaptic calcium modulation by metabotropic glutamate receptor activation: A differential role in acute depression of synaptic transmission and long-term depression. *J. Neurosci.*, in press.

Michael D. Schneider, Professor; M.D., Pennsylvania. Cardiac progenitor and stem cells; molecular genetics of cardiac growth and death pathways. Cardiomyocyte-specific deletion of the coxsackievirus and adenovirus receptor results in hyperplasia of the embryonic left ventricle and abnormalities of sinuatrial valves. *Circ. Res.* 98(7):923–30, 2006.

Robert Schwartz, Professor; Ph.D., Pennsylvania. Testing SRF's biological role in heart-cell commitment, differentiation, and morphogenesis; regulation of muscle gene expression. Cysteine-rich LIM-only proteins CRP1 and CRP2 are potent smooth-muscle differentiation cofactors. *Dev. Cell* 4:107–18, 2003.

Eva Sevick, Professor; Ph.D., Carnegie Mellon. Diagnostic imaging of cancer using NIR fluorescent agents. Measurements of FRET in a glucose-sensitive affinity system with frequency-domain lifetime spectroscopy. *Photochem. Photobiol.* 81(6):1386–94, 2005.

Richard Sifers, Associate Professor; Ph.D., Oklahoma. Cellular glycobiology and conformational disease. Elucidation of the molecular logic that preferentially targets misfolded alpha1-antitrypsin for intracellular degradation. *Proc. Natl. Acad. Sci. U.S.A.* 100:8229–34, 2003.

George Snipes, Associate Professor; M.D., Ph.D., Vanderbilt. Molecular pathogenesis of inherited neuropathies. Cellular mechanisms, diagnosis and rational approaches to management and therapy for Charcot-Marie-Tooth and related neuropathies. *J. Invest. Med.*, in press.

Addison Taylor, Professor; M.D., Ph.D., Missouri. Hypertension, orthostatic hypotension, syncope, and clinical pharmacology. Ethnic differences in nocturnal blood pressure decline in treated hypertensives. *Am. J. Hypertens.* 13:884–91, 2000.

Nancy Templeton, Assistant Professor; Ph.D., Wesleyan. Nonviral delivery for treatment of disease. Cell-specific cytotoxicity of human pancreatic adenocarcinoma cells using rat insulin promoter thymidine kinase directed gene therapy. *World J. Surg.* 28:826–33, 2004.

Lubov Timchenko, Associate Professor; Ph.D., Leningrad Institute of Experimental Medicine. Triplet repeat instability in heart failure molecular basis for impaired muscle differentiation in myotonic dystrophy. *Mol. Cell. Biol.* 21:6927–38, 2001.

* Xander H. Wehrens, Assistant Professor; M.D., Ph.D., Maastricht. Regulation of cardiac ion channels in normal and diseased hearts. Ryanodine receptor/calcium release channel PKA phosphorylation: A critical mediator of heart failure progression. *Proc. Natl. Acad. Sci. U.S.A.* 103:511–8, 2006.

* Gang-Yi Wu, Assistant Professor; Ph.D., Chinese Academy of Sciences. Cell signaling and activity-dependent neuronal plasticity. Calcium-calmodulin-dependent kinase II modulates Kv4.2 channel expression and upregulates neuronal A-type potassium currents. *J. Neurosci.* 24(14):3643–54, 2004.

Samuel M. Wu, Professor; Ph.D., Harvard. Retinal neurophysiology membrane biophysics and synaptic transmission; information processing in the retina. *Neuron* 43:(6)779–93, 2004. *J. Physiol.* 564(3):849–62, 2005.

* Pumin Zhang, Assistant Professor; Ph.D., Wisconsin–Madison. Molecular genetics of cell-cycle control; cell-cycle control in development and disease. *Nature Genet.* 30:31–9, 2002; *Nature* 387:151–8, 1997.

**Primary appointment is in the Department of Molecular Physiology and Biophysics.*

BRIGHAM YOUNG UNIVERSITY

College of Biology and Agriculture
Department of Physiology and Developmental Biology

Programs of Study

The Department of Physiology and Developmental Biology is located within the College of Biology and Agriculture. It offers the Master of Science and Ph.D. in physiology and developmental biology and the Master of Science and Ph.D. in neuroscience.

Each of the graduate programs within the department offer research training and classroom instruction in a wide range of areas pertaining to the disciplines of physiology and developmental biology. A biophysics research group is also part of the department.

The master's degree program in physiology and developmental biology provides students with a solid understanding of the current concepts in physiology and developmental biology. Students gain valuable research experience and techniques during the thesis research project, which teaches students the fundamentals of scientific inquiry and trains students in the latest techniques used in real world applications. Submission of the thesis to a peer-reviewed journal is encouraged but not required. In order to graduate, students must successfully complete 24 approved credit hours of research and course work, plus 6 thesis hours.

The doctoral program in physiology and developmental biology is a wide-ranging academic endeavor in physiology and developmental biology. Ph.D. candidates choose a research project that focuses in an area of either physiology or developmental biology; however, all students are expected to have a core knowledge of the key concepts relative to both disciplines. The research project includes independent inquiry and an in-depth application of the scientific method. While it is expected that each student's thesis is published in a peer-reviewed journal, it is not required. In order to graduate, students must successfully complete 36 approved credit hours of research and course work, plus 18 dissertation hours.

Research Facilities

Numerous resources are available to graduate students in this program. The equipment and facilities within the John A. Widtsoe Building and the Eyring Science Center provide students with materials and the necessary tools to conduct graduate-level research. An electron microscope laboratory, with both transmission and scanning microscopes, is available to graduate students. The University also has a DNA sequencing center in the Widtsoe Building. Areas of research include biophysics, developmental biology, genomics, exercise physiology, neuroendocrinology, and reproduction.

Students also have full access to the Harold B. Lee library, which houses more than 3 million volumes of materials, including pamphlets, journals, serials, microforms, and nonprint materials.

Financial Aid

The Department of Physiology and Developmental Biology offers teaching assistantships, research assistantships, and tuition awards. Specific endowment fund awards are also available. Further information can be obtained from the department.

Cost of Study

A substantial portion of the cost of operating the University is paid by the Church of Jesus Christ of Latter-day Saints. Students and families of students who are tithe-paying members of the Church have already made a contribution to the operation of the University. Because others have not contributed to the Church, students who are not members of the Church pay a higher rate of tuition. Tuition for full-time graduate students is $2155 per semester for Church members and $3233 per semester for nonmembers. Part-time tuition for Church members is $239 per hour and $359 per hour for nonmembers. Additional charges for lab fees, books, and campus services are added to the tuition costs. Fees are subject to change.

Living and Housing Costs

Brigham Young University (BYU) offers graduate students and students with families on-campus apartment living in two housing communities that offer one-, two-, and three-bedroom apartments, with all necessary appliances, computer lounges, and recreational areas. Rent for each month ranges from $494 to $710, depending on location and apartment size. Each apartment comes with a gas stove and oven and a family-size refrigerator. Some apartments have central air conditioning and garbage disposals. Some utilities are included in the cost of rent. Furniture is available to rent at an additional charge based on availability. Many off-campus apartments are also available at a comparable cost.

Student Group

There are approximately 20 students enrolled in the graduate programs offered by the Department of Physiology and Developmental Biology. Master's degree candidates typically finish the program in two years; doctoral degree candidates in four to five years. Because of the small size of the program, students have the opportunity to work closely with the faculty on all research projects.

Location

The campus of Brigham Young University is located in Provo, Utah, at the foot of the beautifully rugged Wasatch Range of the Rocky Mountains and bounded on the west by 23-mile-long Utah Lake. The campus is the focal point of a city of 110,000 and a valley of 331,000. Beyond it to the south and east are spectacular areas of vast sandstone canyons and monoliths, several of which are national parks. Salt Lake City is 45 miles north of the campus.

The University

Brigham Young University is sponsored by the Church of Jesus Christ of Latter-day Saints. BYU is the largest privately owned university in the country. It offers exceptional programs for the graduate student who is seeking an environment where learning experiences with dedicated scholars characterize graduate study. In a time of constantly changing human values and increased challenges for higher education, BYU holds steadfastly to a singular vision that combines reasoned and revealed learning. Along with extensive undergraduate programs, BYU offers master's and doctoral degrees in a variety of disciplines through fifty-three graduate departments.

Founded in 1875 as Brigham Young Academy, the campus has grown from one building to 500 buildings on more than 600 acres. More than 1,500 full-time faculty members instruct 33,000 students.

Applying

All graduate programs at Brigham Young University have the same general admission and entry requirements. Students wishing to enter in the fall semester should apply by the priority deadline of February 1; the final deadline is May 1. For the winter semester, the deadline is September 10. Graduate applicants must submit scores from the GRE, MCAT, or DAT. International students whose first language is not English must submit TOEFL or IELTS scores. Applicants are strongly encouraged to communicate with the Department of Physiology and Developmental Biology for further information.

Correspondence and Information

Department of Physiology and Developmental Biology
574 WIDB
Brigham Young University
Provo, Utah 84602

Phone: 801-422-2006
Fax: 801-422-0700
Web site: http://pdbio.byu.edu/home

Brigham Young University

THE FACULTY AND THEIR RESEARCH

Jeffrey Barrow, Assistant Professor; Ph.D., Utah, 1999. The role of cellular polarity in vertebrate development.

John D. Bell, Professor; Ph.D., California, San Diego, 1987. Molecular pharmacology; membrane signal transduction mechanisms of phospholipase A2; biophysics of membrane structure and function.

Michael D. Brown, Assistant Professor; Ph.D., Colorado State, 1999. Regulation of axon and dendrite extension and pathfinding during nervous system development; regeneration of the nervous system following injury; regulation of the actin cytoskeleton during cell motility and division.

David D. Busath, Professor; M.D., Utah, 1978. Molecular modeling of voltage-gated ion channels and the biophysical interactions that regulate the ion permeability of the channels.

Marc. D. Hansen, Assistant Professor; Ph.D., Stanford, 2002. Assembly and disassembly of cell-cell junctions in development and cancer metastasis.

Allan M. Judd, Associate Professor; Ph.D., West Virginia, 1981. Role of cytokines in the regulation of adrenal functions; role of arachidonate and arachidonate metabolites in anterior pituitary hormone release.

David L. Kooyman, Associate Professor; Ph.D., Ohio, 1993. Animal biotechnology, molecular genetics, and molecular biology, including the use of molecular techniques to identify the genetic component of phenotypic traits as well as the cloning of these traits for potential use in the generation of transgenic animals.

Edwin D. Lephart, Associate Professor and Director of the Neuroscience Center; Ph.D., Texas Southwestern Medical Center at Dallas, 1989. Brain aromatase and 5 alpha-reductase; endocrine disrupters: phytoestrogens, development and aging, neuroendocrine, and behavioral influences (reproductive endocrine, learning, memory, and cognition).

James P. Porter, Professor and Chair; Ph.D., California, San Francisco, 1982. Fetal programming of hypertension by maternal high-salt diet; interaction between brain and hormones in regulating the cardiovascular system.

R. Ward Rhees, Professor; Ph.D., Colorado State, 1971. Steroid hormone-brain interactions; developmental influences of hormones and prenatal stress on neuroendocrine mechanisms; biological determinants of sexual differentiation.

Robert E. Seegmiller, Professor; Ph.D., McGill, 1970. Ultrastructural, anatomical, and biochemical analyses of hereditary connective tissue disorders; analysis of developmental defects induced by drugs and environmental chemicals; epidemiological studies of factors (such as maternal age or socioeconomic status) associated with birth defects in Utah.

Roy W. Silcox, Associate Professor; Ph.D., North Carolina State, 1986.

Michael Stark, Assistant Professor; Ph.D., California, Irvine, 1998.

Sterling Sudweeks, Assistant Professor; Ph.D., Utah, 1997. How ligand-gated ion channels are modulated by gene expression.

William W. Winder, Professor; Ph.D., Brigham Young, 1971. Roles of AMP-activated protein kinase in skeletal muscle; targeting of AMP-activated protein kinase for prevention and treatment of type 2 diabetes and obesity.

Dixon J. Woodbury, Professor; Ph.D., California, Irvine, 1986. Cellular and molecular physiology, focusing on membrane biophysics, particularly vesicle/membrane fusion and its regulation by SNARE proteins.

Bruce H. Wooley, Professor; Pharm.D., USC. Pharmacology.

CASE WESTERN RESERVE UNIVERSITY

Case School of Medicine
Department of Physiology and Biophysics

Programs of Study

The Department's Ph.D. programs in cellular and molecular physiology, molecular/cellular biophysics, and systems integrated physiology are tailored to prepare students for successful careers in biomedical, pharmaceutical, and/or industrial research. The Department of Physiology and Biophysics ranks among the top ten physiology departments in the country. The programs feature individual attention from committed faculty members and are tuition-free. The training programs are designed to provide a mentored training environment that maximizes faculty-student interaction and emphasizes the use of state-of-the art experimental approaches. Prospective students should visit the University's Web sites (http://physiology.cwru.edu or http://biophysics.cwru.edu) for additional information on the Department, individual investigators, and graduate programs.

The Ph.D. program in cellular and molecular physiology embraces investigation that seeks to understand the fundamental organizational and physiological functions of the cell utilizing state-of-the-art scientific methodology and conceptual approaches. The Ph.D. program in molecular/cellular biophysics emphasizes biophysics and bioengineering concepts and technologies and seeks to develop the students' quantitative skills. The Ph.D. program in systems integrative physiology embraces the concepts of cell and molecular physiology, biochemistry, and allied sciences but seeks to understand the function of the organism at the organ system level. The M.D./Ph.D. program consists of core medical training plus advanced graduate research training, in any of the disciplines outlined above, thereby leading to a combined degree. The Ph.D. for M.D.'s program is specifically designed for individuals who already have an M.D. degree. It can be linked to research-oriented residency programs, such as the clinical investigator pathway, approved by the American Board of Internal Medicine, and other similar programs.

Research Facilities

The Department is housed in laboratory and office space on the fifth and sixth floors of the School of Medicine. These areas were specifically designed to facilitate faculty-student interaction. They are fully equipped with modern research instruments for sophisticated spectroscopic studies, video-enhanced light microscopy with image processing, molecular biology, structural biology, and extensive computer facilities. Interdisciplinary programs with other departments enable students to have access to additional specialized equipment.

Financial Aid

Ph.D. programs are tuition-free and pay a $23,000-per-year cost-of-living stipend that supports student training. Most students qualify for this support.

Cost of Study

Tuition at Case Western Reserve University Graduate School is approximately $26,000 per year.

Living and Housing Costs

Single rooms are available on campus, and a large variety of off-campus housing is available within 2 miles for married and single students. The cost of living in Cleveland is among the lowest in the United States.

Location

Case Western Reserve University is located 4 miles from downtown Cleveland in University Circle, a 500-acre area containing more than thirty educational, scientific, medical, cultural, and religious institutions. The parklike setting of University Circle contains the world-renowned Cleveland Museum of Art and Severance Hall, home of the famous Cleveland Orchestra. Metropolitan Cleveland has a population of approximately 2 million people and offers a wide array of recreational and cultural activities, including national sports events, theater, ballet, and cinema.

The University and The School

Case Western Reserve University is a private, nonprofit institution created in 1967 by the federation of the adjacent Western Reserve University (founded in 1826) and Case Institute of Technology (founded in 1880). The Case School of Medicine ranks among the top fifteen schools nationally. It is part of the dynamic and innovative consortium of biomedical research centers that include University hospitals, the MetroHealth Center, and the Cleveland Clinic Foundation. This consortium creates an exceptional array of research and learning opportunities.

Applying

Requirements for admission are an undergraduate degree with a strong background in the natural sciences from an accredited college or university, GRE General Test scores, and three letters of recommendation. Competitive candidates are invited to visit the Department to view the facility and meet the faculty. Applications should be submitted by February 1, but late applications are also considered.

Correspondence and Information

Coordinator, Graduate Degree Programs
Department of Physiology and Biophysics
Case School of Medicine
Case Western Reserve University
Cleveland, Ohio 44106-4970

Phone: 216-368-2084
Fax: 216-368-5586
E-mail: PHOL-INFO@case.edu
Web site: http://physiology.cwru.edu
       http://biophysics.cwru.edu

Case Western Reserve University

THE FACULTY AND THEIR RESEARCH

Mary Barkley, Professor; Ph.D., California, San Diego, 1964. Structure and function of reverse transcriptase.

Matthias Buck, Assistant Professor; D.Phil., Oxford, 1996. Protein structure and dynamics in transmission of signals in cells.

Marco Cabrera, Assistant Professor; Ph.D., Case Western Reserve, 1995. High-intensity exercise and energy use in children and adolescents.

Cathleen Carlin, Professor; Ph.D., North Carolina at Chapel Hill, 1979. Regulation of ErbB receptor tyrosine kinase membrane protein sorting.

Christie Cefaratti, Instructor; Ph.D., Case Western Reserve, 2000. Regulation of Mg^{2+} uptake and release; operation of Na^+/Mg^{2+} and the Ca^{2+}/Mg^{2+} exchangers.

Kim Chan, Assistant Professor; Ph.D., Arkansas, 1993. Sulfonylurea receptor role in regulating ATP-sensitive potassium channel function.

Mark Chance, Professor; Ph.D., Pennsylvania, 1986. Mass spectrometry; structural genomics; macromolecular; colon cancer; diabetes.

Margaret Chandler, Assistant Professor; Ph.D., Kent State, 1998. Myocardial energy metabolism in the pathophysiology of heart failure and diabetes.

Calvin Cotton, Associate Professor; Ph.D., North Carolina at Chapel Hill, 1984. Regulation of ion transport in cells of respiratory, gastrointestinal, and renal epithelia.

Pamela Davis, Professor; Ph.D., 1973, M.D., 1974, Duke. Structure and function of CFTR in cystic fibrosis.

Paul DiCorleto, Professor; Ph.D., Cornell, 1978. Regulation of growth factor and leukocyte adhesion molecule genes by the endothelium.

George Dubyak, Professor; Ph.D., Pennsylvania, 1979. Extracellular ATP release and metabolism in inflammation.

Mark Dunlap, Associate Professor; M.D., Tennessee Health Sciences, 1982. Nicotinic acetylcholine receptors; autonomic nervous system; heart failure; parasympathetic/sympathetic physiology; baroreceptors; reflexes; neurohumoral abnormalities.

Richard Eckert, Professor; Ph.D., Illinois at Urbana-Champaign, 1981. MAP kinase signaling regulation of gene expression; cell proliferation and apoptosis in surface epithelia.

Thomas Egelhoff, Associate Professor; Ph.D., Stanford, 1987. Signaling pathways that control actin/myosin motility during cell migration in wound healing and cancer.

Steven Fisher, Associate Professor; M.D., Pennsylvania, 1986. Cardiomyocyte apoptosis in heart development.

Joan Fox, Professor; Ph.D., McMaster, 1976; D.Sc., Southampton (England), 1983. Signaling by cell adhesion molecules.

George Gorodeski, Professor; M.D., Tel Aviv, 1973; Ph.D., Case Western Reserve, 1990. Hormonal regulation of cervical epithelial cell transport.

Edward Greenfield, Professor; Ph.D., North Carolina at Chapel Hill, 1987. Cell-cell interaction and bone turnover.

Robert Harvey, Associate Professor; Ph.D., Northwestern, 1987. Signaling mechanisms that regulate cardiac function through changes in electrical activity.

Ulrich Hopfer, Professor; M.D., Göttingen (Germany), 1966; Ph.D., Johns Hopkins, 1970. Cell biology, physiology, and pathology of epithelial transport.

Philip Howe, Professor; Ph.D., Medical College of Georgia, 1988. TGFβ and Wnt signaling regulation of cell growth and apoptosis.

Faramarz Ismail-Beigi, Professor; M.D., Johns Hopkins, 1966; Ph.D., Berkeley, 1972. Glucose transporter expression and function; control of cardiac bioenergetics.

Stephen Jones, Professor; Ph.D., Cornell, 1980. Voltage-dependent ion channels; mechanisms of channel gating, permeation, and modulation.

Jeffrey Kern, Professor; M.D., Wisconsin–Madison, 1979. Receptor tyrosine kinases, c-erbB-2, lung development, carcinogenesis, JAK-STAT, EGFR.

Michael Kinter, Assistant Professor; Ph.D., North Carolina at Chapel Hill, 1986. Molecular mechanisms of oxidant injury and oxidant defense.

John P. Kirwan, Associate Professor; Ph.D., Ball State, 1987. Insulin signaling; glucose transport; protein expression, activity, and phosphorylation; intramyocellular lipid content; exercise; low-glycemic diet; body composition.

Joseph LaManna, Professor; Ph.D., Duke, 1975. Control of adaptation to hypoxia and the pathophysiology of stroke; CNS function in cardiac arrest.

Carole Liedtke, Professor; Ph.D., Case Western Reserve, 1980. Regulation of Na-K-2Cl cotransport during fluid and electrolyte balance in epithelia.

R. Tyler Miller, Professor; M.D., Case Western Reserve, 1980. The calcium-sensing receptor and G-protein–coupled signaling.

Susanne Mohr, Assistant Professor; Ph.D., Konstanz (Germany), 1996. Role of apoptosis and inflammation in the development of diabetic retinopathy.

Virgil Muresan, Assistant Professor; Ph.D., Kansas, 1993. Microtubule motor regulation of signaling in health and disease.

Thomas Nosek, Professor; Ph.D., Ohio State, 1973. Cellular basis of muscle fatigue.

Nanduri Prabhakar, Professor; Ph.D. Baroda (India), 1980. Utilization of molecular oxygen in mammalian cells.

Andrew Resnick, Instructor; Ph.D., Alabama, 1997. Mechanosensation in epithelial cells.

Andrea Romani, Assistant Professor; M.D., Siena (Italy), 1984; Ph.D., Turin, 1990. Regulation of transmembrane Mg^{2+} transport and the role of Mg^{2+} transport in metabolism.

David Rosenbaum, Professor; M.D., Illinois at Chicago, 1983. Mechanisms of arrhythmias in myocytes.

William Schilling, Professor; Ph.D., Medical University of South Carolina, 1981. Mammalian TRP channel function; role of Ca^{2+} channels in cell death.

John Sedor, Professor; M.D., Virginia, 1978. Clinical, cellular, and genetic basis of kidney disease.

Ganes Sen, Professor; Ph.D., McMaster, 1974. Host-virus interaction.

Corey Smith, Assistant Professor; Ph.D., Colorado Health Sciences Center, 1996. Release of transmitter molecules from adrenal chromaffin cells and the sympathetic stress response.

Frank Sönnichsen, Associate Professor; Ph.D., Kiel (Germany), 1988. Nuclear magnetic resonance spectroscopy to understand protein structure and function.

Witold Surewicz, Professor; Ph.D., Lodz (Poland), 1982. Structure and function of prion proteins.

Patrick Wintrode, Assistant Professor; Ph.D., Johns Hopkins, 1997. Hydrogen exchange and mass spectrometry to study protein structure and function.

DARTMOUTH COLLEGE

Medical School
Ph.D. and M.D./Ph.D. Programs
in Molecular, Cellular and Systems Physiology

Program of Study

Research training opportunities are available in cardiovascular, cellular, comparative, endocrine, immunological, molecular, neural, renal, and respiratory physiology. Additional training may also be undertaken in theoretical aspects of biological control systems and in computer applications to biological problems. Special arrangements can be made to do combined thesis research in physiology and another field under the guidance of a faculty member in the appropriate discipline.

During the first year and a half of study, students are expected to complete the basic Medical School courses in biochemistry and physiology. Courses specifically designed for the graduate curriculum include cellular and molecular physiology and advanced courses in the various areas of physiology. During the first two years, students are introduced to research through a series of three rotations in the laboratories of individual faculty members. To qualify as a candidate for the Ph.D. degree, students take a propositional examination, in which their proposed thesis research is written up in the form of a grant application and defended before a committee consisting of three or four faculty members. Upon satisfactory completion of the qualifying examinations, students devote most of their efforts to doctoral thesis research, culminating in the Ph.D. dissertation by the end of the fourth or fifth year of graduate study.

Research Facilities

Laboratory and instructional facilities are located in the Medical School buildings at the Dartmouth Hitchcock Medical Center in Lebanon and at the Medical School buildings on the Dartmouth College campus in Hanover, a short distance away. Complete library facilities are provided by the Dana Biomedical and the Health Science medical libraries and by the extensive Dartmouth College library system. All laboratories are equipped for modern biomedical research; in addition, specialized instrumentation facilities for shared use (e.g., fluorescent flow cytometers, high-pressure liquid chromatographs, an amino acid analyzer, a patch clamp apparatus) are available. Computer facilities represent the state of the art in networked computation and provide access to complete key-served software via a campuswide network linking virtually all departments and disciplines within Dartmouth College and the Medical School. In addition, close ties to the clinical facilities at the Medical Center provide opportunities to study physiological problems of a clinical nature.

Financial Aid

All students receive a stipend, which is set at $24,000 for the 2006–07 year. Students are encouraged to apply for a variety of predoctoral fellowships awarded by private and federal institutions.

Cost of Study

The cost of tuition for 2005–06 was $44,396 per year. This cost is not paid by the students but by a variety of scholarships.

Living and Housing Costs

Dartmouth College assists graduate students in arranging for appropriate housing in College facilities or in private accommodations in the Hanover area. Unfurnished apartments with a refrigerator and stove are available for students; rents range from $565 to $740 per month plus heat and utilities. Rents for off-campus housing vary widely, but accommodations at rates comparable to those for college housing can be found. Single-student housing, usually off campus, costs $500 to $900 per month.

Student Group

Dartmouth is coeducational, with 4,000 undergraduates and 1,000 students enrolled in the professional and graduate programs, including 200 Ph.D. candidates in the sciences. About 20 students are usually enrolled in the Ph.D. Program in Molecular, Cellular and Systems Physiology, permitting a close association with faculty members and postdoctoral research associates.

Location

Dartmouth College is located in Hanover, New Hampshire, a town of about 10,000 inhabitants bordering the Connecticut River, which separates New Hampshire and Vermont. Some facilities are located in Lebanon, a short distance away. The picturesque countryside provides excellent opportunities for all types of outdoor activities, especially hiking, skiing, and water sports. Drives to Boston, Montreal, and New York City take about 2, 3, and 5 hours, respectively.

The College and The Medical School

Dartmouth was founded in 1769 as a liberal arts college, and the Medical School was established in 1797. Ph.D. programs are offered in biochemistry, biology, chemistry, engineering, mathematics and computer science, pharmacology, physics, physiology, and psychology. Advanced professional degrees are offered by the Medical School, the Thayer School of Engineering, and the Tuck School of Business Administration.

The Hopkins Center for the Performing Arts and the Hood Museum of Art serve as the cultural focus for the College community. This complex incorporates concert halls, theaters, art galleries, art studios, and crafts workshops. In addition to performances by students, the arts center sponsors a wide range of musical, dance, and theatrical productions by visiting artists.

Extensive modern sports facilities are also available to all students. Two large indoor sports arenas accommodate many sports during the snowy winter months, and the gymnasium houses two swimming pools, a weight-lifting room, and numerous squash courts.

Applying

Application forms and more detailed information about the program may be obtained from the Coordinator of Graduate Studies. The basic requirement for admission is an academic degree from an accredited institution of higher learning. Applications completed before January 20 are given priority, and students are advised to submit applications and all supporting documentation (official transcripts, GRE scores, and three letters of recommendation) as early as possible. A visit to the campus for an interview is recommended.

Correspondence and Information

Dr. Valerie Anne Galton
Director of Graduate Studies
Department of Physiology
Dartmouth Medical School
Lebanon, New Hampshire 03756
Phone: 603-650-7717
E-mail: val.galton@dartmouth.edu
Web site: http://www.dartmouth.edu/%7Ephysiol/index.html

Dartmouth College

THE FACULTY AND THEIR RESEARCH

Donald Bartlett Jr., M.D., Andrew C. Vail Professor of Physiology. Mechanisms of control and integration of breathing movements by muscles of the respiratory pump and those of the upper airway. Developmental physiology of respiratory control. *Respir. Physiol. Neurobiol.* 134:247–53, 2003. With Knuth.

Jack E. Bodwell, Ph.D., Research Associate Professor of Physiology. Molecular mechanisms of glucocorticoid hormone action and effects of endocrine disruptors. Arsenic at very low concentrations alters glucocorticoid receptor (GR)-mediated gene activation but not GR-mediated gene repression: Complex dose-response effects are closely correlated with levels of activated GR and require a functional GR DNA binding domain. *Chem. Res. Toxicol.* 17(8):1064–76, 2004. With Kingsley and Hamilton.

Robert A. Darnall, M.D., Professor of Pediatrics and Physiology. The role of medullary serotonergic neurons in Sudden Infant Death Syndrome. The effects of a GABA$_A$ agonist in the rostral ventral medulla on sleep and breathing in newborn piglets. *Sleep* 24:514–27, 2001. With Curran, Filiano, Li, and Nattie. Characterization of hypoxia-induced ventilatory depression in newborn piglets. *Exp. Physiol.* 88:509–15, 2003. With St. Jacques and St. John.

J. Andrew Daubenspeck, Ph.D., Professor of Physiology and Adjunct Professor of Biomedical Engineering. Reduced respiratory-related evoked activity in subjects with obstructive sleep apnea syndrome. *J. Appl. Physiol.* 94(2):429–38, 2003. With Akay and Leiter.

John V. Fahey, Ph.D., Research Assistant Professor of Physiology. Effect of menstrual status on antibacterial activity and secretory leukocyte protease inhibitor production by human uterine epithelial cells in culture. *J. Infect. Dis.* 185:1606–13, 2002. With Wira.

Géza Fejes-Tóth, D.M.D., Professor of Physiology. Molecular biology of membrane transporters, cell biology and development of specific renal cell types; mechanism of action of steroid hormones. Subcellular localization of mineralocorticoid receptors in living cells: Effects of receptor agonists and antagonists. *Proc. Natl. Acad. Sci. U.S.A.* 95:2973–8, 1998. With Pearce and Náray-Fejes-Tóth.

Valerie Anne Galton, Ph.D., Professor of Physiology. The roles of the iodothyronine deiodinases in the regulation of thyroid hormone action during development and in adult mammals. The studies utilize genetically altered mouse models. Targeted disruption of the type 2 selenodeiodinase gene (dio2) results in a phenotype of pituitary resistance to thyroxine. *Mol. Endocrinol.* 15:2137–48, 2001. With Schneider et al.

Alice L. Givan, Ph.D., Research Associate Professor of Physiology and Director of the Englert Cell Analysis Laboratory of the Norris Cotton Cancer Center. Development of flow-cytometric and imaging methods for the analysis of cells. Multiparameter precursor analysis of t-cell responses to antigen. *J. Immunol. Methods* 276:5–17, 2003. With Bercovici et al.

Paul M. Guyre, Ph.D., Professor of Physiology. Mechanisms of the immunosuppressive and anti-inflammatory actions of steroids in normal physiology, sepsis, and autoimmune disease. *J. Leukocyte Biol.* 76:500–8, 2004. With Pioli, Goonan, and Wardell.

Leslie P. Henderson, Ph.D., Professor of Physiology and of Biochemistry. Behavioral and physiological responses to anabolic androgenic steroids. *Neurosci. Biobehav. Rev.* 27:413–36, 2003. With Clark.

James C. Leiter, M.D., Professor of Physiology and Medicine. Heterogeneous patterns of pH regulation in glial cells in the dorsal and ventral medulla. *Am. J. Physio. Regul Integr. Comp. Physio.* 286(2):R289–302, 2004. With Erlichman, Cook, Schwab, and Budd.

Aihua Li, M.D., Research Assistant Professor of Physiology. Neural control of respiration. Central chemoreception (locations, phenotypes, and functions) and cardiorespiratory changes in awake and sleep. CO_2 dialysis in one chemoreceptor site, the RTN: Stimulus intensity and sensitivity in the awake rat. *Respir. Physiol. Neurobiol.* 133:11–22, 2002. With Nattie.

Harold L. Manning, M.D., Associate Professor of Medicine and Physiology. Respiratory sensation; respiratory control in man; ventilator-patient interactions. *J. Asthma* 38:447–60, 2001.

Robert A. Maue, Ph.D., Professor of Physiology and Biochemistry. Mechanisms underlying neuronal differentiation and dysfunction; growth factor regulation of neural morphology and ion channel expression. Adenoviral-mediated expression of functional Na+ channel beta1 subunits tagged with a yellow fluorescent protein. *J. Neurosci. Res.* 74:794–800, 2003. With Fry and Porter. Before the loss: Neuronal dysfunction in Niemann Pick Type C disease. *Biochm. Biophys. Acta,* in With Paul and Boegle.

Anikó Náray-Fejes-Tóth, M.D., Professor of Physiology. Cellular and molecular biology of steroid hormone action in epithelial cells. Molecular biology of ion transport. Hormone-regulated transepithelial Na+ transport in mammalian cortical collecting duct cells requires SGK1 (serum and glucocorticoid induced kinase-1) expression. *Am. J. Physiol.,* in press. With Helms and Fejes-Tóth.

Eugene E. Nattie, M.D., Professor of Physiology. Central chemoreceptors that sense changes in brain pH and stimulate breathing. The role of central chemoreception in the medullary raphe in the sudden infant death syndrome. Substance P-saporin lesion of neurons with NK1 receptors in one chemoreceptor site in rats decreases ventilation and chemosensitivity. CO_2. *J. Physiol.* 544(2):603–16, 2002. With Li.

Mary M. Niblock, Ph.D., Research Assistant Professor. Studying the embryonic origin and interconnections of respiratory-related brainstem neurons using a FLP recombinase-based fate mapping system in a transgenic mouse model. A subset of medullary serotonergic neurons arises from dorsal rhombencephalon corresponding to the rhombic lip. *Soc. Neurosci. Abstracts* 29. With Kinney and Dymecki.

William G. North, Ph.D., Professor of Physiology. Neuropeptides in breast cancer, in small-cell carcinoma, and in Alzheimer's disease. Gene regulation of vasopressin and vasopressin receptors in cancer. *Exper. Physiol.* 85s:27–40, 2000.

Donald L. St. Germain, M.D., Professor of Medicine and of Physiology. Importance of thyroid hormone action and metabolism during development. Complex organization and structure of sense and antisense transcripts expressed from the DIO3 gene imprinted locus. *Genomics* 83:413–24, 2004. With Hernandez, Martinez, and Croteau.

Walter M. St. John, Ph.D., Professor of Physiology. Neural control of respiration. Potential switch from eupnea to fictive gasping after blockade of glycine transmission and potassium channels. *Am. J. Physiol.* 283:R721–31, 2002. With Rybak and Paton.

Lynn A. Sheldon, Ph.D., Research Assistant Professor of Physiology. Steroid receptor–mediated gene regulation. Transcriptional regulatory mechanisms mediated by histone modifications and proteins involved in chromatin remodeling in vivo and in vitro. Steroid hormone receptor-mediated histone deacetylation and transcription at the mouse mammary tumor virus promoter. *J. Biol. Chem.* 276(35):32423–6, 2001. With Smith and Becker.

Peggy M. Simon, M.D., Associate Professor of Medicine and of Physiology. Vagal feedback in the entrainment of respiration to mechanical ventilation in sleeping humans. *J. Appl. Physiol.* 89:760–9, 2000. With Simon et al.

Bruce A. Stanton, Ph.D., Professor of Physiology. Ion channel regulation in kidney and lung. PDZ-domain interaction controls the endocytic recycling of the cystic fibrosis transmembrane conductance regulator. *J. Biol. Chem.* 277(42):40099–105, 2002. With Swiatecka-Urban et al. Increased diffusional mobility of CFTR at the plasma membrane after deletion of its C-terminal PDZ-binding motif. *J. Biol. Chem.* 279(7):5494–500, 2004. With Haggie and Verkman.

Harold M. Swartz, M.D., Ph.D., Professor of Radiology and Physiology and Adjunct Professor of Bioengineering and Chemistry. Magnetic resonance (EPR, NMR) of viable biological systems, especially measurement of pO$_2$ in vivo, redox metabolism, and free radicals. EPR studies of living animals and related model systems (in vivo EPR). In *Spin Labeling: The Next Millennium,* pp. 367–404, ed. L. J. Berlinger. New York: Plenum Publishing, 1998. With Halpern.

Charles R. Wira, Ph.D., Professor of Physiology. Physiology of reproduction; cellular and molecular actions of sex hormones and cytokines in the regulation of the mucosal immune system in the rodent and human male and female reproductive tract. The mucosal immune system in the human female reproductive tract: Influence of stage of the menstrual cycle and menopause on mucosal immunity in the uterus. In *The Endometrium,* chap. 26, pp. 371–404, eds. S. Glasser, J. Aplin, L. Giudice, and S. Tabibzadeh. London: Harwood Academic Publishers, 2002. With Fahey et al.

Hermes H. Yeh, Ph.D., Professor of Physiology and Chairman. Cellular and molecular mechanisms of neuroreceptor interactions and plasticity in the adult and developing CNS. Neuroanatomical, patch clamp electrphysiological, and molecular biological techniques are employed to reveal and investigate coordinated changes in neuronal function and gene and protein expression. Nerve growth factor rapidly increases muscarinic tone in mouse medium septum/diagonal band of Broca. *J. Neurosci.* 25:4232–42, 2005. With Wu.

LOYOLA UNIVERSITY CHICAGO

Department of Physiology
Graduate Program in Cell and Molecular Physiology

Programs of Study

The Department of Physiology offers graduate training programs with primary emphasis in cellular and molecular physiology leading to the Ph.D. and M.S. degrees. The objective of the Ph.D. program is to train independent scientific investigators through the development of their own ideas using state-of-the-art research techniques and instrumentation. The two-year M.S. program may be designed to enhance the expertise and experience of individuals interested in research career paths in academic or industrial settings, enhance the academic qualifications of individuals interested in graduate or professional programs, or provide research training for clinicians interested in academic medicine.

During the four-year Ph.D. program, students take core courses in biochemistry, neurobiology, molecular biology, organ system physiology, and cellular and molecular physiology. Students participate in four research rotations during the first year to introduce the different research lines available in the Department and help define the laboratory where they will conduct their dissertation research.

The Department has strong research programs in excitation-contraction coupling of cardiac and vascular smooth muscle, with special reference to the molecular biology, biophysics, and biochemistry of voltage-activated and ligand-activated ion channels. Cardiac electrophysiology, excitation-secretion coupling, and intracellular calcium homeostasis are also well-represented areas of research interest. Cell signaling mechanisms involved in the pathophysiology of heart failure and developing septic shock, with special reference to neural-immune interactions, are also important areas of faculty research.

Research Facilities

Each faculty member has a modern, state-of-the-art laboratory, fully equipped to meet research needs. In addition, the Department has the following shared core facilities: radioisotope (liquid scintillation and gamma counting) instrumentation, a machine and electronic shop, histology laboratory, tissue culture lab, and confocal microscopy lab.

Financial Aid

All Ph.D. graduate students in the Department receive an annual stipend of $22,000. This support is intended to allow the students to devote all of their time to completing the degree; no service is required. Graduate stipends are competitive and carry a waiver of tuition.

Cost of Study

Cost of tuition is fully waived for Ph.D. students receiving support. Tuition cost for M.S. students is $625 per credit hour, which totals $15,000 for the 24 required credits.

Living and Housing Costs

A wide variety of living arrangements can be found within the surrounding suburban communities. A director of housing assists students in locating suitable housing.

Student Group

More than 1,000 students are enrolled in degree programs at the Medical Center. The graduate student body at the Medical Center is approximately 100. In recent years, the graduate enrollment of the Department has ranged from 8 to 15 students.

Location

The Medical Center location on a 73-acre site in Maywood provides the advantages of a suburban environment as well as easy access to the attractions of the Chicago community. The Medical Center campus, adjacent to a forest preserve, includes the large Hines Veterans Hospital and numerous supporting facilities.

The University

Loyola, founded in 1870, was Chicago's first institution of higher learning. It now has four campuses, including the Medical Center. The Medical Center has become one of the top ten academic medical centers in the nation in terms of overall patient volume, trauma, cardiology, cardiac transplants, bypass surgery, neonatology, and other areas. Close research interaction exists between the Cardiovascular Institute, the Neuroscience Institute, and the Burn and Shock Trauma Institute, as well as among several clinical departments.

Applying

Students normally start their graduate training in late July when the fall semester begins. Program prerequisites include one full-year course each in physics, mathematics through calculus, and chemistry through organic chemistry. A fundamental knowledge of biology is desirable but not required. Students are required to submit scores from the GRE General Test.

Loyola University Chicago is an affirmative action agency; applications from members of minority groups and women are encouraged.

Correspondence and Information

Ruben Mestril, Ph.D.
Graduate Program Director
Department of Physiology
Stritch School of Medicine
Loyola University Chicago
2160 South First Avenue
Phone: 708-327-2395
Fax: 708-216-6308
Web site: http://www.luhs.org/depts/physio/index.html

Loyola University Chicago

THE FACULTY AND THEIR RESEARCH

Kathrin Banach, Research Assistant Professor; Ph.D., Bern (Switzerland). Differentiation of embryonic stem cells into cardiac myocytes; pacemaker activity and excitation spread in multicellular aggregates of cardiomyocytes; stem cells as a cardiac replacement tissue. Development of electrical activity in cardiac myocyte aggregates derived from mouse embryonic stem cells. *Am. J. Physiol.*, in press (with Halbach, Hescheler, and Egert). Embryonic stem cells as a model for physiological analysis of cardiovascular system. *Methods Mol. Biol.* 185:169–87, 2002 (with Hescheler, Fleischmann, and Sauer). Surface changes and conductance of human connexin37 gap junction channel. *Biophys. J.* 78(2):752–60, 2000 (with Ramanan and Brink).

Donald M. Bers, Professor and Chairman; Ph.D., UCLA. Cardiac excitation-contraction coupling. *Nature* 415:198–205, 2002. Adenoviral gene transfer of mutant phospholamban rescues contractile dysfunction in failing rabbit myocytes with relatively preserved SERCA function. *Circ. Res.* 97:252–9, 2005 (with Despa et al.). *Circ. Res.* 96:815–7, 2005 (with Ziolo, Martin, Bossuyt, and Pogwizd). Phospholemman phosphorylation mediates the β-adrenergic effects on Na/K pump function in cardiac myocytes. A mathematical treatment of integrated Ca dynamics within the ventricular myocyte. *Biophys. J.* 87:3351–71, 2004 (with Shannon, Wang, Weber, and Puglisi).

Lothar A. Blatter, Professor; M.D., Dr.med., Bern (Switzerland). Calcium regulation in excitable and nonexcitable cells of the cardiovascular system; excitation-contraction coupling; nitric oxide regulation and function; regulation of mitochondrial membrane potential and calcium; confocal microscopy; electrophysiology. Local calcium gradients during excitation-contraction coupling and alternans in atrial myocytes. *J. Physiol.* 546:19–31, 2003 (with Kockskämper et al.). Pyruvate modulates cardiac sarcoplasmic reticulum Ca^{2+} release via mitochondria-dependent and -independent mechanisms. *J. Physiol.* 550:765–83, 2003 (with Zima, Kockskämper, and Mejia-Alvarez). Nitric oxide inhibits capacitative Ca^{2+} entry and enhances endoplasmic reticulum uptake Ca^{2+} in bovine vascular endothelial cells. *J. Physiol.* 539:77–91, 2002 (with Dedkova).

Kenneth L. Byron, Associate Professor; Ph.D., Chicago. Signal transduction of vasoconstrictor hormones and regulation of ion channels in vascular smooth muscle. Signal transduction of physiological concentrations of vasopressin in A7,[5] vascular smooth muscle cells: A role for PYK2 and tyrosine phosphorylation of K^+ channels in stimulation of Ca^{2+} spiking. *J. Biol. Chem.* 277:7298–307, 2002 (with Lucchesi). Vasopressin stimulates Ca^{2+} spiking activity in A7,[5] vascular smooth muscle cells via activation of phospholipase A_2. *Circ. Res.* 78:813–20, 1996.

Leanne Cribbs, Associate Professor; Ph.D., Pittsburgh. Molecular cloning and characterization of T-type calcium channels in the cardiovascular system. Low-voltage-activated calcium channels in vascular smooth muscle: T-type channels and AVP-stimulated calcium spiking. *Am. J. Physiol. Heart Circ. Physiol.* 288:H923–5, 2005 (with Brueggemann et al.). Low-voltage-activated (T-type) calcium channels control proliferation of human pulmonary artery myocytes. *Circ. Res.* 96:864–72, 2005 (with Rodman et al.).

Samuel Cukierman, Associate Professor; M.D., Ph.D., Rio de Janeiro. Theoretical study of the structure and dynamic fluctuations of dioxolane-linked gramicidin channels. *Biophys. J.* 84:816–31, 2003 (with Yu and Pomès). The transfer of protons in water wires in proteins. *Frontiers Biosci.* 8:S1118–39, 2003 (invited review). Activation energies of proton transfer in various gramicidin channels. *Biophys. J.* 82:182–92, 2002 (with Chernyshev). On the origin of closing flickers in gramicidin channels: A new hypothesis. *Biophys. J.* 82:1329–37, 2002 (with Armstrong).

Stephen B. Jones, Professor; Ph.D., Missouri–Columbia. Neural-immune interactions, with special reference to mechanisms and consequences of sympathetic activation during injury states. Adrenergic modulation of cytokine release in bone marrow progenitor-derived macrophage following polymicrobial sepsis. *J. Neuroimmunol.*, in press (with Muthu et al.). Bone marrow norepinephrine mediates development of functionally different macrophages after thermal injury and sepsis. *Ann. Surg.* 240(1):132–41, 2004 (with Cohen et al.). Adrenergic modulation of splenic macrophage cytokine release in polymicrobial sepsis. *Am. J. Physiol. Cell Physiol.* 287:C730–6, 2004 (with Deng et al.).

Richard H. Kennedy, Professor and Senior Associate Dean of Research; Ph.D., Nebraska. Structural and functional cardiac remodeling; role of cytokines in cardiac physiology and pathophysiology. Inhibition of sarcoplasmic reticular function by chronic interleukin-6 exposure via iNOS in adult ventricular myocytes. *J. Physiol.* 566:327–40, 2005 (with Yu, Chen, and Liu). Influence of mast cells on structural and functional manifestations of radiation-induced heart disease. *Cancer Res.* 65:3100–7, 2005 (with Boerma et al.). Protective role of mast cells in homocysteine-induced cardiac remodeling. *Am. J. Physiol.* 288:H2541–5, 2005 (with Joseph et al.).

Stephen L. Lipsius, Professor; Ph.D., SUNY Downstate Medical Center. Cardiac cellular electrophysiology; atrial muscle and atrial pacemaker function; integrin-mediated regulation of ion channels; regulation of nitric oxide production in cardiomyocytes. Signaling mechanisms that mediate NO production induced by ACh exposure and withdrawal in cat atrial myocytes. *Circ. Res.* 93:1233–40, 2003 (with Dedkova, Ji, Wang, and Blatter). Laminin binding to β1-integrins selectively alters β1- and β2-adrenoceptor signaling in cat atrial myocytes. *J. Physiol.* 527:3–9, 2000 (with Wang and Samarel).

Jody L. Martin, Assistant Professor; Ph.D., California, San Diego. Signaling pathways in stress response of cardiac myocytes; role of posttranslational modifications of small heat-shock proteins in cardioprotection. Diverse mechanisms of myocardial p38-MAPK activation: Evidence for MKK-independent activation by a TAB1 associated mechanism contributing to injury during myocardial ischemia. *Circ. Res.* 93:254–61, 2003 (with Tanno et al.). Mutation of COOH-terminal lysines of a B-crystallin abrogates ischemic protection in cardiomyocytes. *Am. J. Physiol. Heart Circ. Physiol.* 283:H85–91, 2002 (with Bluhm et al.).

Rafael Mejia-Alvarez, Associate Professor; M.D., Ph.D., National of Mexico. Excitation-contraction coupling during cardiac development, from single Ca^{2+} channels to beating whole heart; techniques: Single-channel recordings, ligand-binding assays, patch clamp, confocal microscopy, and whole-heart fluorescence. Pulsed local-field fluorescence microscopy: New approach for measuring physiological signals in beating heart. *Pfluegers Arch.* 445:747–58, 2003 (with Escobar). Developmental changes of intracellular Ca^{2+} transients in beating rat hearts. *Biophys. J.* 78(1):98A, 2000 (with Escobar et al.). Unitary Ca^{2+} current through cardiac ryanodine receptor channels under quasiphysiological ionic conditions. *J. Gen. Physiol.* 113:177–86, 1999 (with Kettlun et al.). Function of cardiac ryanodine receptor from newborn rat ventricular myocytes. *Biophys. J.* 76(1):A297, 1999 (with Gómez, Escobar, and Fill).

Ruben Mestril, Associate Professor; Ph.D., Miami (Florida). Molecular and cellular studies on protective role of heat-shock proteins in cardiomyocytes. Overexpression of inducible 70-kDa heat-shock protein (HSP70i) in mouse attenuates skeletal muscle damage induced by cryolesion. *Am. J. Physiol. Cell*, in press (with Miyabara, Martin, Griffin, and Moriscot). Posttranscriptional regulation of SERCA2a: AUF1 binds to the 3'UTR but not HUR. *Am. J. Physiol. Heart* 289:H2543–50, 2005 (with Blum and Samarel). The use of transgenic mice to study cytoprotection by the stress proteins. *Methods* 35:165–9, 2005.

Gregory A. Mignery, Associate Professor; Ph.D., Texas A&M. Molecular and cellular biology of inositol 1,4,5-trisphosphate receptors; molecular cloning and structure-function characterization of ligand-gated ion channels; intracellular calcium. Isotype specific function of single $insP_3$ receptor channels. *Biophys. J.* 75:834–9, 1998 (with Ramos-Franco and Fill). Single-channel function of recombinant type-1 inositol 1,4,5-trisphosphate receptor ligand-binding domain splice variants. *Biophys. J.* 75:2783–93, 1998 (with Ramos-Franco, Caenepeel, and Fill).

Erika S. Piedras-Rentería, Assistant Professor; Ph.D., Illinois at Urbana-Champaign. Voltage-gated calcium-channel function and modulation and channelopathies; calcium-channel function and synaptic activity; modulation of calcium channels by cytoskeletal proteins; techniques: molecular biology, electrophysiology, fluorescence, and cell biology techniques. Presynaptic homeostasis at CNS nerve terminals compensates for lack of a key Ca entry pathway. *Proc. Natl. Acad. Sci. U.S.A.* 101(10):3609–14, 2004 (With Pyle et al.). Effect of expanded trinucleotide repeats in spinocerebellar ataxia type 6 (SCA6) ÿ$_{1A}$ calcium-channel function. *J. Neurosci.* 21(23):9185–93, 2001 (with Watase et al.). Ablation of P/Q-type Ca channel currents, altered synaptic transmission, and progressive action in mice lacking ÿ$_{1A}$ subunit. *Proc. Natl. Acad. Sci. U.S.A.* 96:15245–50, 1999 (with Jun et al.).

Allen M. Samarel, Professor; M.D., Harvard. Molecular and cellular biology of cardiac myocytes and vascular smooth-muscle cells; signal transduction; transcriptional and posttranscriptional regulation of muscle gene expression; molecular basis of heart failure. PYK2 regulates SERCA2 gene expression in neonatal rat ventricular myocytes. *Am. J. Physiol. Cell Physiol.* 289:C471–82, 2005 (with Heidkamp et al.). Restoration of resting sarcomere length after uniaxial static strain is regulated by PKC and FAK. *Circ. Res.* 94:642–9, 2004 (with Mansour et al.).

Charles L. Webber Jr., Professor; Ph.D., Loyola Chicago. Nonlinear dynamics of physiological systems and states; protein-folding patterns; biomathematics. Recurrence quantification analysis of nonlinear methods for the behavioral sciences. In *Tutorials in Contemporary Nonlinear Methods for the Behavioral Sciences*, chapter 2, pp. 26–94, eds. M. A. Riley and G. Van Orden, 2005 (with Zbilut et al.), http://www.nsf.gov/sbe/bcs/pac/nmbs/nmbs.pdf. Singular hydrophobicity patterns and net charge: A mesoscopic principle for protein aggregation/folding. *Physica A* 343:348–58, 2004 (with Zbilut et al). Protein aggregation/folding: Review of nonlinear analysis of proteins through recurrence quantification. *Cell. Biochem. Biophys.* 36:67–87, 2002 (with Zbilut et al.). Nonlinear signal analysis methods in elucidation of protein sequence-structure relationships. *Chem. Rev.* 102:1471–91, 2002 (with Giuliani et al.).

Robert Wurster, Professor; Ph.D., Loyola Chicago. Activation of various G-protein coupled receptors modulates Ca^{2+} channel current via PTX-sensitive and voltage-dependent pathways in rat intracardiac neurons. *J. Auton. Nerv. Syst.* 76:68–74, 1999 (with Jeong and Ikeda). Proliferation of cultured human astrocytoma cells in response to an oxidant and antioxidant. *J. Neuro-Oncol.* 44:213–21, 1999 (with Arora-Kuruganti and Lucchesi).

MEDICAL COLLEGE OF WISCONSIN

Graduate School of Biomedical Sciences
Department of Physiology

Program of Study

The Department of Physiology offers a Ph.D. program that provides unique training in molecular, cellular, functional genomics, and whole-animal physiology. This training emphasizes the integration of knowledge at all levels as well as the development of an appreciation for the relationship of this knowledge to disease processes.

Trainees are recruited internationally and selected on the basis of academic credentials, previous research experience, and commitment to a career in research and teaching. During a period of approximately five years, students complete required and elective courses and a research project, which includes the use of the techniques of molecular biology, biochemistry, physiological genomics, isolated tissue, and whole-animal investigation. Emphasis is placed on interpretation of the relationship of data obtained from subcellular systems to the normal and abnormal physiology of the whole animal, as well as on a basic understanding of genetics. Course selection is determined by the background and career objectives of the students. The student must pass qualifying exams and perform well in formal course work in order to be admitted to Ph.D. candidacy.

Early in the program, the student selects a major adviser to sponsor his or her research work. (The diverse research interests of the faculty members are listed at the end of this description, along with selected recent publications.) A favorable faculty-student ratio ensures maximal faculty contact and close, individualized supervision. In the ensuing years, formal course work is superseded by original dissertation research.

Research Facilities

Research laboratories are located primarily in the Basic Science Building Health Care Research Center and MACC Fund Cancer Research Center on the campus of the Medical Center and in the Research Service Building at the Veterans Administration Center. An area of about 30,000 square feet is allocated for research use in these three facilities, which house the Cardiovascular Research Center, the Human and Molecular Genetics Research Center, and the Bioinformatics Research Center, which are all directed by physiology faculty members. Up-to-date laboratories contain the most advanced equipment with which to investigate a wide range of physiological problems. Students also have access to other campus facilities (e.g., the National Biomedical ESR Center, the Allen-Bradley Laboratory, the Biophysics Laboratory, and the Eye Institute). Students may also engage in research programs offered jointly by the Medical College of Wisconsin and the graduate schools of Marquette University (MU), the University of Wisconsin–Milwaukee (UWM), and the Milwaukee School of Engineering.

Two animal resource centers maintain colonies of many different species of laboratory animals housed in spacious and comfortable surroundings in a total of 47,000 square feet. The Medical College Library contains more than 100,000 volumes and subscribes to about 2,000 biomedical journals. Other large science holdings are contained in the Health Sciences Library at MU and at UWM. Students have convenient access to the Medical College through several remote terminals.

Financial Aid

In 2006–07, graduate stipends of $24,000 plus a full-tuition scholarship are provided for all eligible students.

Cost of Study

Tuition for the 2006–07 academic year is $10,800 for full-time students; part-time students pay $600 per credit.

Living and Housing Costs

A variety of accommodations may be obtained in the vicinity of the Medical College of Wisconsin at relatively moderate rents.

Student Group

The Graduate School of Biomedical Sciences has 411 students; 29 are enrolled in the Department of Physiology. The medical school has 807 medical students and 690 postgraduate residents, including 28 postdoctoral research fellows. Graduate students also participate in the educational programs that are offered jointly by the Medical College and the graduate schools of Marquette University, the University of Wisconsin–Milwaukee, and the Milwaukee School of Engineering.

Location

The Medical College of Wisconsin is located on the suburban campus of the Milwaukee Regional Medical Center in Milwaukee, Wisconsin. Milwaukee is on the shore of Lake Michigan, about 90 miles north of Chicago, and has a population of 1.7 million in the metropolitan area. Cultural attractions include several repertory theaters, ballet and opera companies, and a symphony orchestra. The city also has a fine arts center, a performing arts center, museums, and a zoo. In addition, Milwaukee is the home of several major-league sports teams and several large breweries.

The College and The School

The Medical College of Wisconsin, a private institution, was the Marquette University School of Medicine until 1967. The Graduate School of Biomedical Sciences is accredited by the North Central Association of Colleges and Schools and is a member of the Council of Graduate Schools. The Graduate School grants the Ph.D. degree in cell and developmental biology, biophysics, biostatistics, physiology, pharmacology and toxicology, pathology, biochemistry, microbiology, and molecular genetics; a joint Ph.D. in functional imaging with Marquette University; an M.S. degree in epidemiology, a joint M.S. degree in health-care technologies management with Marquette University, and a joint M.S. degree in medical informatics with Milwaukee School of Engineering; and an M.A. degree in bioethics. There are also interdisciplinary programs in biomedical sciences and neuroscience. The College has 902 full-time, part-time, and visiting faculty members, more than 160 of whom hold primary appointments in the basic sciences. Administrative offices, the library, service departments, and most basic science departments are located in the Basic Medical Education facility and the Health Care Research Center, which opened in 1978. The Medical College maintains close teaching and research relationships with Marquette University, the Milwaukee School of Engineering, and the University of Wisconsin–Milwaukee, which offer biomedical programs in such areas as physical therapy, medical technology, nursing, bioengineering, dentistry, and medical informatics.

Applying

Application for admission should be made to the Graduate School of Biomedical Sciences. Admissions decisions are made only after careful consideration of students' academic backgrounds, transcripts, scores from the Graduate Record Examinations, TOEFL scores, letters of recommendation, and specific research goals. Acceptances are recommended by a Departmental admissions committee to the Dean, who makes final admission decisions.

Correspondence and Information

Director of the Graduate Program
Department of Physiology
Medical College of Wisconsin
8701 Watertown Plank Road
Milwaukee, Wisconsin 53226-0509

Medical College of Wisconsin

THE FACULTY AND THEIR RESEARCH

Dan Beard, Associate Professor; Ph.D., Washington (Seattle), 1997. Thermodynamic constraints for biomedical networks. *Am. J. Physiol.* 288:H2062–7, 2005; 288:E633–44, 2005. *Biophys. Chem.* 114:213–20, 2005.

Zeljko J. Bosnjak, Professor (also Professor of Anesthesiology); Ph.D., Medical College of Wisconsin, 1979. Anesthetics and cardiac signal transduction; cardioprotective effects of inhalation anesthetics. *Anesthesiology* 101:796–8, 2004. *J. Heart Lung Transplant.* 23:1003–7, 2004. *Anesth. Analg.* 99:1316–22, 2004.

Allen W. Cowley Jr., Professor and Chairman; Ph.D., Hahnemann, 1968. Arterial blood pressure—determinants and controllers; role of the kidney in control of body fluid volume; genetics of hypertension. *Hypertension* 45(4):766–72, 2005. *J. Appl. Physiol.* 98(5):1630–8, 2005. *Am. J. Physiol.* 288(5):F1015–22, 2005.

Gebremedhin Debebe, Associate Professor; Ph.D., Budapest, 1988. Regulation of cerebral circulation; ion channels and regulation of vascular tone; diabetes mellitus and vascular disease; cerebral signal transduction.

Hubert V. Forster, Professor; Ph.D., Wisconsin–Madison, 1969. Role of brain stem and cerebellar nuclei and carotid and intracranial chemo receptors in control of breathing; maturation of ventilatory control; hypoxia; exercise. *J. Appl. Physiol.* 97:2236–47, 2004; 97:2303–9, 2004; 97:1629–36, 2004.

Andrew S. Greene, Professor; Ph.D., Johns Hopkins, 1985. Cardiovascular physiology; microcirculation; molecular imaging; nuclear magnetic resonance imaging and spectroscopy; biotechnology development; proteomics. *Physiol. Genomics* 16(2):194–203, 2004. *J. Proteome Res.* 3(4):813–20, 2004. *Neurosci. Lett.* 335(2):95–8, 2002.

David R. Harder, Professor (also Director of Cardiovascular Research Center); Ph.D., Medical College of Wisconsin, 1976. Control of vascular smooth muscle. *J. Cerebral Blood Flow Metab.* 24:509–17, 2004. *J. Neurosci.* 23(5):1678–87, 2003. *Stroke* 33(12):2957–64, 2002.

Howard J. Jacob, Professor (also Director of Human and Molecular Genetics Center); Ph.D., Iowa, 1989. Genetics of insulin-dependent diabetes mellitus, non-insulin-dependent diabetes mellitus, hypertension, and renal failure; disease interactions. *Nature* 428(6982):493–521, 2004. *Diabetes* 53(3):735–42, 2004. *Physiol. Genomics* 6, 2004.

Elizabeth R. Jacobs, Professor (also Chief of Pulmonary Medicine); M.D., Kansas, 1977. Cell physiology; ion channels. *Am. J. Physiol. (Heart Circ. Physiol.)* 284(1):H215–24, 2003. *Respir. Physiol. Neurobiol.* 132:253–64, 2002. *J. Appl. Physiol.* 93:1391–9, 2002.

Anne E. Kwitek-Black, Associate Professor; Ph.D., Iowa, 1996. Genomic tools for study of genetic basis of diseases. *Diabetes* 54(1):259–67, 2005. *Nature* 428(6982):493–521, 2004. *Genome Res.* 14(4):750–7, 2004.

Jean-Francois Liard, Professor; M.D., 1968, Ph.D., 1973, Lausanne. Cardiovascular regulation; role of vasopressin in cardiovascular control; systemic and regional hemodynamics of experimental hypertension; use of allosteric effectors of hemoglobin. *Adv. Exp. Med. Biol.* 530:249–59, 2003. *Hepatology* 28:851–64, 1998. *Am. J. Physiol.* 266:R838–49, 1994.

Julian H. Lombard, Professor; Ph.D., Medical College of Wisconsin, 1975. Physiology of microcirculation; local regulation of blood flow in normal and pathological states; effect of high-salt diet on vascular function; microcirculation in hypertension. *Hypertension* 45(2):687–91, 2005. *Am. J. Physiol.* 288:H908–13, 2005; 287:H957–62, 2004.

David L. Mattson, Professor; Ph.D., Medical College of Wisconsin, 1990. Renal/cardiovascular physiology; renal autocrine and paracrine systems in control of blood pressure. *Acta Physiol. Scand.* 183:309–20, 2005. *Am. J. Physiol.* 287:R1478–85, 2004. *Hypertension* 44:424–8, 2004.

Meetha M. Medhora, Associate Professor; Ph.D., Jawaharlal Nehru (India), 1984. Molecular study of arachidonic acid metabolites. *Curr. Opin. Invest. Drugs* 5(9):952–6, 2004. *Rec. Res. Dev. Physiol.* 1(II):793–800, 2003. *Am. J. Physiol.* 284:H215–24, 2003.

Michael Michalkiewicz, Associate Professor; D.V.M., Ph.D., Warsaw, 1982. Neural and endocrine controls of circulation and metabolism; gene manipulations (transgenesis, rescue) in the rat. *Hypertension* 45:1–6, 2005. *Physiol. Genomics* 19:228–32, 2004. *J. Clin. Invest.* 111:1853–62, 2003.

Carol Moreno-Quinn, Assistant Professor; Ph.D., Murcia, 1999. Genomic map of cardiovascular phenotypes of hypertension in female Dahl S rats. *J. Am. Soc. Nephrol.* 16:852–6, 2005. *Physiol. Genomics* 16:178–9, 2004; 15(3):243–57, 2003.

Diane H. Munzenmaier, Assistant Professor; Ph.D., Wisconsin, 1995. Stroke-induced cerebral angiogenesis. *Stroke* 36:337–41, 2005. *Am. J. Physiol. (Heart Circ. Physiol.)* 285:H775–83, 2003. *Am. J. Physiol.* 278:1163–7, 2000.

Michael Olivier, Professor; Ph.D., Cornell, 1997. Human single nucleotide polymorphisms (SNPs) and their biology. *Mutation Res.* 573(1–2):103–10, 2005. *J. Am. Soc. Mass Spectrom.* 16:302–6, 2005. *Diabetes* 54(1):259–67, 2005.

Lawrence G. Pan, Assistant Professor; Ph.D., Medical College of Wisconsin, 1985. Respiration—regulation of ventilation and acid-base status; environmental physiology—altitude, exercise, and occupational stresses; breathing during exercise—demands, regulation, and limitations. *J. Appl. Physiol.* 96:1815–24, 2004; 94:1508–18, 2003. *Respir. Physiol.* 138:59–75, 2003.

Hershel Raff, Professor (also Professor of Medicine–Endocrine); Ph.D., Johns Hopkins, 1981. Endocrine-metabolic adaptations to hypoxia; diagnosis and differential diagnosis of Cushing's syndrome; HPA axis control in humans. *J. Appl. Physiol.* 98:981–90, 2005. *Am. J. Physiol. Endocrinol. Metab.* 288:E314–20, 2005. *J. Clin. Endocrinol. Metab.* 89:6005–9, 2004.

Richard J. Roman, Professor (also Director of Kidney Disease Center); Ph.D., Tennessee, 1977. Lipid mediators in control of epithelial transport and vascular tone; genetics of hypertension and stroke, diabetic nephropathy, and renal disease. *Am. J. Physiol.* 287:R58–68, 2004. *Diabetes* 53:735–42, 2004. *Hypertension* 43:1173–4, 2004.

Jeanne L. Seagard, Professor (also Professor of Anesthesiology); Ph.D., Medical College of Wisconsin, 1978. Central mechanisms involved in neural control of blood pressure; central integration of afferent nerve activity from barosensitive receptors, including arterial baroreceptors and cardiac mechanoreceptors, in neurons in the brain stem. *Peptides* 25:413–23, 2004. *Am. J. Physiol.* 286:H992–1000, 2004; 284:H1570–6, 2003.

David F. Stowe, Professor; Ph.D., Michigan, 1974; M.D., Wisconsin, 1983. Cardiac injury mechanisms. *Anesth. Analg.* 100:46–53, 2005. *Am. J. Physiol. (Heart Circ. Physiol.)* 287:H667–80, 2004. *Anesthesiology* 100:498–505, 2004.

Andrew K. Tryba, Assistant Professor; Ph.D., Case Western Reserve, 2000. Neural mechanisms of rhythm generation underlying respiration and epilepsy. *Curr. Opin. Neurobiol.* 14:1–10, 2004. *Neuron* 43(1):105–17, 2004. *J. Neurophysiol.* 92:2844–52, 2004.

Simon N. Twigger, Assistant Professor; Ph.D., Nottingham (England), 1994. Bioinformatics; model organism databases; comparative genomics; data mining of distributed systems.

Tetsuro Wakatsuki, Assistant Professor; Ph.D., Washington (St. Louis), 1999. Cardiovascular system and its mechanical properties and their regulation; reconstitution of tissues that mimic normal and pathological properties of hearts, blood vessels, skeletal muscles, connective tissues, and skin. *Am. J. Physiol. Cell Physiol.* 286:C8–21, 2004. *Curr. Pharm. Biotechnol.* 5:181–9, 2004. *Trends Biochem. Sci.* 29(11):609–17, 2004.

NEW YORK MEDICAL COLLEGE

School of Basic Medical Sciences
Department of Physiology

Programs of Study	The department offers programs of study and research leading to Ph.D. and M.S. degrees in physiology. The Department of Physiology is one of the leading physiology departments in the nation in both research and teaching. The program places a strong emphasis on research associated with medical physiology, so graduates are prepared for academic positions as well as positions in industry and government. A high ratio of faculty members to students contributes to close interactions in both the laboratories and the advanced classes. Researchers from clinical departments, holding joint appointments in the Department of Physiology, provide students with firsthand exposure to the potential clinical relevance of their research in physiology.
	Course offerings and research expertise of faculty members in other basic science and clinical departments further broadens the training available to doctoral candidates. There are two entry routes leading to the Ph.D. program. Students entering the Integrated Ph.D. Program (IPP) may enter the departmental program in their second year after completing core courses and lab rotations in various departments to find an appropriate research experience. Students may also enter the Ph.D. program with advanced standing after completing a master's degree program. In general, required course work is completed within the first two years, while the student also receives an introduction to hands-on research training in different laboratories. Students attend weekly seminars, which regularly feature distinguished visiting scientists. Students acquire teaching experience by making presentations in journal clubs and by participating in laboratory exercises for the medical students. Upon completing the doctoral qualifying exams at the end of the second year, candidates present their proposal for thesis research to a faculty committee for approval. Under the guidance of a thesis adviser, students concentrate on carrying out their doctoral research project and preparing a dissertation, which is defended at the conclusion of the program. A detailed description of the IPP program can be found at the Graduate School of Basic Medical Sciences of New York Medical College Web site (http://www.nymc.edu/gsbms/).
	The department also offers a program leading to the Master of Science degree in physiology. There are two tracks leading to the master's degree, involving either a research or literature review thesis. Courses are taught in the evening, so students can continue to hold outside positions while matriculating for the degree; several students currently in the master's program are employed in research laboratories. All course requirements and the master's thesis are designed to be completed within a two-year period. Many students have used the M.S. in physiology degree to improve their credentials and gain admission to medical schools or other professional studies and as a direct entry into a variety of scientific occupations.
Research Facilities	The Department of Physiology occupies one of the six wings of the Basic Sciences Building. Fourteen laboratories with modern facilities and equipment are available for performing research ranging from an integrative physiology of the cardiovascular system to cellular, subcellular, and molecular mechanisms in cardiovascular and neural sciences. All laboratories are interconnected by means of the university intranet and to other facilities by means of the Internet. The building houses a centrally located library and dining facility as well as other basic research support services.
Financial Aid	All full-time students in the Ph.D. program are awarded remission of tuition. In addition, students may qualify for fellowship stipends that are set at $22,000 per year in 2006–07. College health services for physician care and insurance coverage for hospital care are provided free of charge to Ph.D. students. Together, these benefits are equivalent to more than $30,000 per year. Neither full fellowships nor tuition waivers are available to students pursuing the M.S. degree, but students may apply for competitive scholarships after acceptance.
Cost of Study	Doctoral students are awarded remission of tuition charges and are eligible for stipend support, which should cover most expenses associated with the program.
	Tuition for students in the master's program is $645 per credit hour in 2006–07.
Living and Housing Costs	Single rooms and shared one- and two-bedroom apartments are available on campus for students in both programs by arrangement. Private rooms in nearby homes are available, and assistance in finding appropriate housing can be found in the Housing Office.
Student Group	The School of Basic Medical Sciences is one of the two graduate schools of New York Medical College. There are on average 60 doctoral students and more than 120 master's students in addition to 760 students enrolled in the medical school and more than 600 students in the Graduate School of Health Sciences. At present, the Department of Physiology typically graduates 1 to 2 Ph.D. students each year and accepts an average of 2 to 3 new doctoral students each year. There are 15 M.S. students in the department; 10 to 15 applicants are accepted into the master's program each year.
Student Outcomes	The Department of Physiology continues to be successful in placing its graduates in prestigious positions. The Ph.D. graduates have gone on to rewarding postdoctoral or professional positions in academics, government, and private industry. Master's degree students have found positions in research and teaching or used the degree for entry into other professional programs leading to more advanced degrees, including the M.D. degree.
Location	The College is located in the Westchester Medical Center, adjacent to the city of White Plains and in proximity to New York City. Students live and study in a suburban setting, yet have easy access to the varied scientific, cultural, and entertainment activities for which New York City is renowned. The nearby lakes, parks, and forests of the scenic Hudson River Valley provide outstanding recreational facilities.
The College	New York Medical College, founded in 1860, is one of the oldest and largest medical schools in the United States. Because of the later addition of the Schools of Basic Medical Sciences and Health Sciences, the College is more properly considered a health sciences university. The College has a strong commitment to scholarly activities and research as well as to the clinical training of physicians.
Applying	Admission to the Ph.D. program requires a bachelor's degree. Applicants with majors in the biological or physical sciences are preferred. In some cases, specific deficiencies in undergraduate preparation may be remedied after admission. Applicants with previous graduate training may transfer a limited number of credits to the program, subject to departmental approval. Scores on the Graduate Record Examinations General Test and the Subject Test in biology or chemistry are required. Early application for the following academic year is encouraged, especially for those requiring financial support.
	A bachelor's degree is required for admission to the master's program. The GRE tests are preferred, but the MCAT exam is also accepted. Applicants with previous graduate training may petition for transfer of a limited number of credits to the program. Students with deficiencies may be encouraged to enroll initially as a nonmatriculant in order to demonstrate their ability to qualify as an M.S. degree candidate. Applicants with previous graduate training may petition for transfer of a limited number of credits to the program. Students with deficiencies may be encouraged to enroll initially as a nonmatriculant in order to demonstrate their ability to qualify as an M.S. degree candidate.
Correspondence and Information	Dr. Carl I. Thompson Graduate Program Director Department of Physiology New York Medical College Valhalla, New York 10595 Phone: 914-594-4106 E-mail: carl_thompson@nymc.edu

New York Medical College

THE FACULTY AND THEIR RESEARCH

Professors

Gabor Kaley, Chairman; Ph.D., NYU. Microcirculation; control of blood pressure and blood flow; vascular biology and endothelial nitric oxide production; local microcirculatory control of cardiovascular function.

Francis L. Belloni, Ph.D., Michigan. Cardiovascular control; coronary blood flow; vascular and cardiac actions of adenosine.

Thomas H. Hintze, Ph.D., New York Medical College. Integrative cardiovascular functions in chronically instrumented animals, myocardial performance in exercise and heart failure; nitric oxide in cardiovascular control.

Akos Koller, M.D., Ph.D., Semmelweis (Hungary). Microcirculatory cardiovascular control; vascular endothelial cell function; blood flow in exercise; skeletal muscle vascular function; role of nitric oxide in circulation.

Christopher S. Leonard, Ph.D., NYU. Neuroscience; neuronal integration; synaptic and nonsynaptic neuromodulation; nitric oxide in the CNS: Brain cholinergic systems; neural basis of sleep and wakefulness.

Norman Levine, Ph.D., NYU. Control of fluid and electrolyte secretion in the male reproductive system.

Edward J. Messina, Ph.D., New York Medical College. Microvascular control; regulation of vascular smooth-muscle reactivity, blood flow, and blood pressure; microcirculatory control in skeletal muscle; microcirculation in diabetes.

Stanley S. Passo, Ph.D., Columbia. Neuroendocrine control of blood pressure: Vasopressin and atrial natriuretic factor.

William Ross, Ph.D., Columbia. Neuroscience; regional properties of neurons; distribution and properties of ion channels; optical methods using voltage and calcium indicator dyes; neurophysiological control of neuronal cells.

Michael S. Wolin, Ph.D., Yale. Pulmonary vascular regulation via cyclic GMP, reduced metabolites, and oxygen tension; regulation of guanylate cyclase; nitric oxide mechanisms in smooth muscle.

Associate Professors

Fabio A. Recchia, M.D., Bari (Italy); Ph.D., Torino (Italy). Control of myocardial metabolism; nitric oxide; heart failure; cardiac mechanics and efficiency; coronary circulation.

Dong Sun, M.D., Xuhou Medical College (China); Ph.D., New York Medical College. Microcirculatory control; endothelial dependent dilator function in microvessels.

Carl I. Thompson, Ph.D., Virginia. Renal and coronary hemodynamics, blood flow, and GFR control; local metabolic control of autoregulatory and renin release processes; nitric oxide production and vascular biology.

Assistant Professors

John G. Edwards, Ph.D., Iowa. Physiological control of gene transcription; regulation of transcription factors; cardiac hypertrophy; exercise biochemistry and overload alterations of the myocardial phenotype.

An Huang, M.D., Ph.D., Shanghai Second Medical (China). Microcirculatory control; endothelial dependent control of vasculature; sex differences in vascular control.

Zoltan Ungavar, MD, PhD. Semmelweis (Budapest). Molecular biology of cardiovascular aging; cytokine signaling in vascular function.

ROSALIND FRANKLIN UNIVERSITY
OF MEDICINE AND SCIENCE

Department of Physiology and Biophysics

Programs of Study

The Department of Physiology and Biophysics, through the School of Graduate and Postdoctoral Studies, offers a program leading to the M.S. and Ph.D. degrees. The Department specializes in advanced teaching of and research in membrane biophysics and transport, the mechanisms and regulation of membrane receptor function, neurophysiology, neuroendocrinology, renal function, cerebral energetics and metabolism, neuroscience, and cardiac physiology. Formal courses include topics in both basic and highly specialized aspects of physiology. The principle objectives of graduate education in physiology are to provide the student with a broad basic knowledge of the dynamic aspects of living organisms and training in the application of modern physiological research methods. Collateral formal course work is provided in other basic medical sciences by the Departments of Biochemistry, Cell Biology and Anatomy, Neuroscience, Pathology, and Cellular and Molecular Pharmacology in the Medical School. The graduate program prepares candidates for a research and teaching career in physiology. Opportunities for postdoctoral training are available. The Department also participates in the combined M.D./Ph.D. program of the University.

The faculty encourages individual consultation as well as frequent informal scientific discussion among the entire departmental staff through its seminar and journal club programs. Formal degree candidacy is granted after a student has passed the comprehensive examination, which is given after completion of the required course work. An acceptable dissertation based on original research must be presented and defended at the oral final examination. Programs of study begin in July for the entering student. A period of five years is usually required to complete the Ph.D. and two years for the M.S.

Research Facilities

Research opportunities are available in a wide range of physiological and biophysical problems. Equipment for various types of molecular, biochemical, radiochemical, electronic, graphic, electrophysiological, and cellular imaging and computer analysis is available, as well as a complete confocal and electron microscope service laboratory, a proteome center, and a multiuser X-ray crystallography facility. The University library has more than 90,000 volumes in the field of medicine, including related basic sciences, anatomy, biochemistry, physiology, pharmacology, pathology, and such medical specialties as cardiology, pediatrics, ophthalmology, radiology, and neurosciences. More than 1,000 periodicals are received annually.

Financial Aid

A limited number of fellowships were awarded with stipends of up to $21,000 per year in 2005–06, plus a full tuition waiver. Research assistantships are often available.

Cost of Study

Tuition in 2005–06 was $15,356 per year for a full course load (four quarters) or $427 per credit hour, part-time. Laboratory fees averaged $10 to $40 per quarter. The health insurance, which is offered through Density Insurance (PPO), cost $148–$331 per month for single coverage and $431–$911 per month for family coverage. Students should anticipate a 3–5 percent increase in these costs for 2006–07.

Living and Housing Costs

A variety of accommodations are available for graduate students in the area. Housing and food cost $3900 per quarter.

Student Group

There is an average total enrollment of 1,687 students in the University's four schools, the School of Graduate and Postdoctoral Studies, the School of Medicine, the College of Health Professions, and the Dr. William M. Scholl College of Podiatric Medicine.

Location

The University is located in North Chicago, adjacent to the grounds of the North Chicago Veterans Administration Medical Center, 38 miles from Chicago. The diverse activities of downtown Chicago can be reached by auto or commuter trains. In addition, the North Shore and nearby communities are rich in entertainment and cultural activities.

The Department

The Department of Physiology and Biophysics is located on the third floor of the Medical School's 375,000-square-foot Basic Medical Sciences Building. The Department has 22,000 square feet of usable space, plus teaching laboratories. The building features multidisciplinary and individual research laboratories of the most modern design and construction, with more than 16,000 square feet of space for the boarding of laboratory animals. As this is a graduate-level teaching facility, there is always ample opportunity for consultation with faculty members in different departments.

Applying

Entrance requirements include a baccalaureate degree from an accredited college. The level of academic performance should be above average (B or better) in the sciences. No undergraduate major is specified, but the following courses are recommended: general biology, qualitative and quantitative analytical chemistry, organic chemistry, physical chemistry, mathematics through calculus, and general physics. Admission depends on an evaluation of the student's undergraduate record and other credentials by the departmental faculty, the Graduate School admissions committee, and the Dean of the Graduate School. All students are admitted with probationary status until satisfactory completion of the first year of graduate study. Applicants should write to the Director of Departmental Graduate Studies or to a member of the department whose work is of interest to them as early as possible. A resume of the applicant's course work and up-to-date grades should accompany letters of inquiry. Letters of recommendation from persons familiar with the candidate's academic background are required. Applications for financial aid should be filed by March 25. Notification of acceptance or rejection is usually made by April 15. The nonrefundable application fee is $25.

Correspondence and Information

Director of Departmental Graduate Studies
Department of Physiology and Biophysics
Rosalind Franklin University of Medicine and Science
3333 Green Bay Road
North Chicago, Illinois 60064
E-mail: physiology@rosalindfranklin.edu
Web site: http://www.rosalindfranklin.edu

Rosalind Franklin University of Medicine and Science

THE FACULTY AND THEIR RESEARCH

Robert J. Bridges, Professor and Chair of Physiology and Biophysics; Ph.D., Kentucky, 1980. Physiology, biophysics, and pharmacology of epithelial ion channels.

Neil A. Bradbury, Associate Professor of Physiology and Biophysics; Ph.D., Welsh National School of Medicine, 1988. Physiology and trafficking of ion channels in health and disease.

Lisa Ebihara, Associate Professor of Physiology and Biophysics; M.D./Ph.D., Duke, 1981. Structure and function of gap-junctional proteins.

Sarah S. Garber, Associate Professor of Physiology and Biophysics; Ph.D., Brandeis, 1987. Anion channel regulation; cellular volume regulation; mechanotransduction.

Raúl J. Gazmuri, Professor of Medicine and Associate Professor of Physiology and Biophysics; M.D., Santiago (Chile), 1980; Ph.D., Finch, 1994. Role of sodium-hydrogen exchange on ischemic and reperfusion injury; apoptosis and postresuscitation myocardial dysfunction; effects of sodium-hydrogen exchange inhibition on neurological recovery after cardiac arrest.

Richard A. Hawkins, Professor of Physiology and Biophysics; Ph.D., Harvard, 1969. Cerebral energy metabolism, cerebral nutrition and transport characteristics of the blood-brain barrier.

Donghee Kim, Professor of Physiology and Biophysics; Ph.D., Michigan State, 1982. Molecular physiology of membrane ion channels.

Darryl R. Peterson, Professor of Physiology and Biophysics; Ph.D., Illinois, 1973. Blood-brain barrier transport and metabolism; regulation of brain extracellular fluid; pathophysiology of stroke; drug delivery to the brain; renal pathophysiology.

Héctor Rasgado-Flores, Associate Professor of Physiology and Biophysics; Ph.D., Center for Research and Advanced Studies; CINVESTAV (Mexico), 1984. Volume regulatory mechanisms; transport of calcium and magnesium in muscle and nerve cells; biophysics of muscle contraction.

Henry Sackin, Professor of Physiology and Biophysics; Ph.D., Yale, 1978. Electrolyte transport; mechanotransduction; volume regulation; cloned K channels; structure-function of ion channels and ion channel gating.

Sant P. Singh, Professor of Medicine and of Physiology and Biophysics; M.D., Punjab (India), 1959. Endocrinology; diabetes and alcoholism.

Janice H. Urban, Associate Professor of Physiology and Biophysics; Ph.D., Loyola of Chicago, 1987. Molecular and physiological aspects of hypothalamic neuroendocrine function.

Teaching Faculty

Timothy R. Hansen, Professor of Physiology and Biophysics; Ph.D., Michigan, 1973. Peripheral hemodynamics and hypertension; vascular smooth muscle.

Charles E. McCormack, Professor of Physiology and Biophysics; Ph.D., Wisconsin–Madison, 1963. Neural and hormonal feedback mechanisms controlling gonadotropin secretion; circadian rhythms and seasonal reproduction.

Gordon L. Pullen, Assistant Professor of Physiology and Biophysics; Ph.D., Finch, 1982. Hormone receptor regulation: membrane receptors for glucose transport and intracellular receptors for thyroid hormone.

Ernest J. Sukowski, Associate Professor of Physiology and Biophysics; Ph.D., Illinois Medical Center, 1962. Blood-brain barrier and membrane transport.

SOUTHERN ILLINOIS UNIVERSITY CARBONDALE

Department of Physiology
Ph.D. Program

Programs of Study

The Department of Physiology offers the degree of Ph.D. in molecular, cellular, and systemic physiology. This program emphasizes training in research and is designed to prepare students for research and teaching careers. These degrees provide advanced training in many subdisciplines, including mammalian and cellular physiology, molecular biology, endocrinology reproduction, neurobiology, pharmacology, and human anatomy.

Students in the Ph.D. program spend their first year taking courses and rotating through various laboratories to become familiar with the research specializations within the program. By the end of the first year, students should have selected a research mentor and begun working in their field of concentration. In the second year, students are involved in a combination of course work and research. A comprehensive written and oral preliminary examination focusing on their area of concentration is taken at the end of the second year. The remainder of their study is spent doing dissertation research, which typically takes 2 to 2½ years after passing the comprehensive examination. While all students must acquire competence in their knowledge of mammalian physiology, most research projects focus on the cellular and molecular level of physiology. Areas of research emphases within the program include molecular biology, molecular endocrinology, reproductive biology, cell physiology, and neuroscience. The course of study, as well as the dissertation research, is highly individualized to meet the students' goals. At some time during the course of study, all graduate students are required to serve as a teaching assistant.

Research Facilities

The Department of Physiology is housed in Life Sciences II and Life Sciences III on the Carbondale campus. Life Sciences II contains spacious general laboratory facilities. Life Sciences III is a modern, well-equipped research facility with graduate student offices, individual faculty laboratories, and several shared cores. Faculty laboratories and these cores contain equipment for state-of-the-art research in molecular, cellular, and systemic physiology, including electrophysiology, patch clamp studies, cell culture, reporter gene expression, HPLC, DNA sequencing and genomics, confocal and fluorescence microscopy, real-time and conventional PCR, fluorescence polarization, filmless storage phosphor autoradiography, and electrophoresis. Departmental resources also include ultracentrifuges, scintillation and gamma counters, and robotic equipment. A modern state-of-the-art vivarium for research animals is conveniently located for use by both buildings in Life Sciences III, and the Electron Microscopy and Digital Imaging Center is in an adjacent building. Research shops for making or redesigning equipment are also available nearby on campus.

Financial Aid

The Department of Physiology offers financial assistance to qualified applicants accepted by the Department. The funds that provide this assistance come from a variety of sources, including Departmental teaching assistantships and University fellowships, which are applied for directly by students, and research assistantships from grants obtained by the graduate program faculty members.

The current graduate student stipend is $14,184 for a 50 percent appointment in addition to a full tuition waiver ($6802), giving a total funding of $20,986 per year. Graduate students with at least a 25 percent appointment as a graduate assistant receive a tuition waiver.

Cost of Study

In-state graduate tuition is $243 per credit hour in 2006–07. Out-of-state tuition is 2.5 times the in-state tuition rate ($607.50 per credit hour). Graduate students with at least a 25 percent appointment as a graduate assistant receive a tuition waiver. Fees vary from $441.62 (1 credit hour) to $987.30 (12 credit hours).

Living and Housing Costs

For married couples, students with families, and single graduate students, the University has 589 efficiency and one-, two-, and three-bedroom apartments that rent for $438 to $505 per month in 2006–07. Residence halls for single graduate students are also available, as are accessible residence hall rooms and apartments for students with disabilities.

Student Group

The Department of Physiology currently has 20 students enrolled in the graduate program: 13 M.S. and 7 Ph.D. students. Five to six percent of the students are international and 50 percent are men. Annually, 3 to 5 new students are admitted. All of the full-time students (94 percent) receive financial aid.

Student Outcomes

Ph.D. graduates typically go on to a postdoctoral research appointment before moving to a permanent position as a professor at a university. Many become research scientists in government or industrial laboratories, where they often become directors.

Location

Southern Illinois University Carbondale (SIUC) is 350 miles south of Chicago and 100 miles southeast of St. Louis. Nestled in rolling hills bordered by the Ohio and Mississippi Rivers and enhanced by a mild climate, the area has state parks, national forests and wildlife refuges, and large lakes for outdoor recreation. Much of the area is a part of the 240,000-acre Shawnee National Forest. Cultural offerings include theater, opera, concerts, art exhibits, and cinema. Educational facilities available for the families of students are excellent.

The University

Southern Illinois University Carbondale is a comprehensive public university with a variety of general and professional education programs. The University offers associate, bachelor's, master's, and doctoral degrees as well as the J.D. and M.D. degrees. The University is fully accredited by the North Central Association of Colleges and Schools. The Graduate School has an essential role in the development and coordination of graduate instruction and research programs. The Graduate Council has academic responsibility for determining graduate standards, recommending new graduate programs and research centers, and establishing policies to facilitate the research effort. Southern Illinois University Carbondale is a state-funded university founded in 1869. The Department of Physiology is part of the School of Medicine.

Applying

Applications should be requested from the address given in this In-Depth Description and submitted no later than April 15 for fall admission. Early applications receive priority for funding consideration. An undergraduate degree in biology or chemistry is preferred for entering students. Admission to the master's program requires a minimum GPA of 3.0, while the Ph.D. program requires a minimum GPA of 3.25. Students requesting direct entry to the Ph.D. program from a bachelor's program should have a minimum GPA of 3.5. All applicants are required to submit a GRE General Test score. International students must also submit a TOEFL score of at least 570.

Students interested in financial assistance should request the appropriate application forms from the address given in this description. Priority for financial assistance is given on the basis of academic performance.

Correspondence and Information

Chair, Graduate Program Committee
Department of Physiology
Mail Code 6512
Life Science II 245
Southern Illinois University Carbondale
Carbondale, Illinois 62901

Phone: 618-453-1544
Fax: 618-453-1517
E-mail: physiology@som.siu.edu
Web site: http://www.siumed.edu/physiology/

Southern Illinois University Carbondale

THE FACULTY AND THEIR RESEARCH

Stuart Adler, Associate Professor; M.D./Ph.D., Duke, 1982. Regulated expression of neuroendocrine genes, especially regulation by steroid hormones and their nonsteroidal-steroidal mimics, including industrial chemicals and plant compounds.

Lydia Arbogast, Associate Professor; Ph.D., Indiana, 1988. Neuroendocrinology; hypothalamic control of prolactin secretion; cellular and molecular mechanisms involved in the regulation of hypothalamic neuronal activity.

Brent M. Bany, Assistant Professor; Ph.D., Western Ontario, 1997. Molecular biology of the uterus and conceptus-uterine interaction during implantation.

Andrzej Bartke, Professor; Ph.D., Kansas, 1965. Reproductive endocrinology; aging.

Ronald A. Browning, Professor; Ph.D., Illinois, 1971. Neuroanatomy and neurochemistry of seizures.

Varadaraj Chandrashekar, Research Professor; Ph.D., Rutgers, 1975. Reproductive endocrinology.

Michael W. Collard, Associate Professor; Ph.D., Washington State, 1987. Regulation of neuropeptide gene expression and protein processing of testicular cells; interaction of germ cell transcription factors with DNA response elements; retinoic acid and cAMP signal transduction.

Thomas C. Cox, Professor; Ph.D., Arizona State, 1979. Studies of mechanisms of ion transport across epithelial tissues, with emphasis on development of sodium transport.

James S. Ferraro, Associate Professor; Ph.D., University of Health Sciences (Chicago), 1984. Circadian rhythms.

Jodi I. Huggenvik, Associate Professor; Ph.D., Washington State, 1985. Transient transfection of mammalian cell lines with reporter plasmids and expression vectors is used to study the molecular mechanisms of transcriptional regulation by cAMP and retinoic acid.

Laura L. Murphy, Associate Professor; Ph.D., Medical College of Georgia, 1983. Neuroendocrine effects of drugs of abuse; steroidal regulation of pituitary function.

Peter Patrylo, Assistant Professor; Ph.D., University of Medicine and Dentistry of New Jersey–Robert Wood Johnson Medical School/Rutgers, 1991. Plasticity of local neuronal networks.

Michael F. Shanahan, Professor; Ph.D., Michigan, 1976. Insulin action and signal-transduction mechanisms; molecular mechanisms of glucose transport across cell membranes.

Richard W. Steger, Professor; Ph.D., Wyoming, 1974. Neuroendocrinology; seasonal reproduction; effects of age on neurochemical and endocrine function; central nervous function in diabetes mellitus.

Jena J. Steinle, Assistant Professor; Ph.D., Kansas Medical Center, 2001. Neural control of the angiogenesis in diabetes and macular degeneration; autonomic neural control of ocular blood flow and its role in disease.

STONY BROOK UNIVERSITY, STATE UNIVERSITY OF NEW YORK

School of Medicine
Department of Physiology and Biophysics

Program of Study

The Department of Physiology and Biophysics offers a program of study leading to the master of science degree and the degree of Doctor of Philosophy. Research in the department focuses on understanding biological processes at the molecular, cellular, integrative, and systems levels. Members of the department use a variety of cellular, molecular, and theoretical approaches to understand signal transduction in cells, cell-cell communication, cell metabolism, cell and systems electrophysiology and neurobiology, and cardiac preconditioning and arrhythmia prevention. The department has two tracks of study: biophysics/structural biology and cellular and systems physiology. Housed in the department is the Molecular Cardiology Institute, whose goal is to understand the molecular basis of heart disease.

Biophysics students can choose from a variety of biochemistry, biophysics, physics, and computational and structural biology courses. Physiology students can choose from a variety of physiology, neurobiology, cell biology, and biomedical engineering courses. Faculty members collaborate in both teaching and research with scientists in the biology, biomedical engineering, biochemistry, chemistry, and physics departments, and students are encouraged to do likewise. The department is formally affiliated with the Graduate Program in Molecular and Cellular Biology. Cold Spring Harbor Laboratory, Brookhaven National Laboratory, and students in the department have access to all of the courses and laboratories of this large, interdepartmental program. During their first two years, students generally rotate through three laboratories to gain research experience. After successful completion of a preliminary examination, students choose their own independent area of research under the supervision of the faculty. The requirements are flexible and can be adapted to the individual's preference and needs. Close tutorial contact between the individual student and the faculty is regarded as the most important feature of the educational program. Additional information about department requirements and programs can be obtained by writing to the Graduate Program Director.

Research Facilities

The Department of Physiology and Biophysics is well equipped with major research instrumentation for physiological, metabolic, and biochemical studies: scintillation counters for radioisotope work, ultracentrifuges, a gas-phase protein sequencer, a DNA synthesizer, and instrumentation for measuring ORD and CD, plus a wide variety of chromatographic, electrophoretic, spectrophotometric, and electronic equipment. The department has a fully equipped cell and molecular biology core facility that enables students to engage in studies involving RNA-DNA recombinant technology. A computer center containing a molecular modeling facility is also located in the department. Also available in Health Sciences Center core facilities are a peptide synthesizer, mass spectrophotometer, and laboratory for chemical synthesis of low-molecular-weight compounds. NMR instrumentation is also available through collaboration with other departments. Department faculty members are associated with a Health Sciences Center diabetes and metabolism group and have collaborative arrangements with other basic science and clinical departments.

Financial Aid

Students are normally admitted with financial support, which currently amounts to $25,000 for the twelve-month period. In addition to the stipend, there are full tuition waivers for students receiving full-time support.

Cost of Study

Full-time tuition for the 2006–07 academic year is $6900 for New York State residents and $10,920 for out-of-state residents. Miscellaneous University fees can be as high as $700 per year, including health insurance.

Living and Housing Costs

In dormitory accommodations for single students, the cost per student in a double room is $4368 per academic year. Prepaid meal plans and à la carte food service are available. Some residence halls have limited cooking facilities. There are limited accommodations for married students. Private housing is available off campus, but it is moderately expensive. Information about housing can be obtained from the University Off-Campus Housing Office or from the Health Sciences Center Office of Student Services.

Student Group

The total enrollment at the Stony Brook campus is about 16,200 students, of whom about 5,000 are graduate students enrolled in a wide range of programs. Currently, about 200 full-time students are enrolled in the various Ph.D. programs in the basic health sciences. The enrollment in the graduate Ph.D. program in physiology and biophysics is approximately 30 students, with approximately 10 students currently in the master's program.

Location

Stony Brook is located in a region of coves, beaches, and small historic villages on the North Shore of Long Island, approximately 60 miles east of New York City. The area has retained its distinctive New England flavor and couples the charm of a rural setting minutes from the water of Long Island Sound with proximity to the cultural, scientific, and industrial resources of the city. The nearby biological research facilities at Cold Spring Harbor Laboratory and Brookhaven National Laboratory provide additional advantages for the scientific community.

The University

The various departments of the medical school are located in a recently developed section of the Health Sciences Center, which is designed to foster cooperation among all the health sciences and professions in a university setting. Simultaneously, as one of the four university centers of the State University of New York, the Stony Brook campus is expanding its comprehensive graduate offerings.

Applying

Applications and information can be obtained by writing to the graduate program director, or students may apply online at https://app.applyyourself.com/?id=sunysb-gs.

The General Test of the Graduate Record Examinations is required. So that the scores are available in time for decisions, the test should be taken no later than January. The deadline for receipt of applications for admission in the fall is February 1.

Correspondence and Information

Graduate Program Director
Department of Physiology and Biophysics
Stony Brook University, State University of New York
Stony Brook, New York 11794-8661
Phone: 631-444-2299
Web site: http://www.pnb.sunysb.edu/

Stony Brook University, State University of New York

THE FACULTY AND THEIR RESEARCH

William B. Benjamin, Professor; M.D., Columbia, 1959. Endocrinology; mechanism of insulin action.
Mark Bowen, Assistant Professor; Ph.D., Illinois at Chicago, 1998. Molecular aspects of signal transduction.
Peter R. Brink, Professor and Chairman; Ph.D., Illinois, 1976. Biophysical properties of gap junction.
Roger H. Cameron, Research Assistant Professor; Ph.D., SUNY at Stony Brook, 1990. Electron microscopy; pharmacology of plasma cells secretion.
Chris Clausen, Associate Professor; Ph.D., UCLA, 1979. Electrical properties of transporting epithelia.
Ira Cohen, Leading Professor; M.D., Ph.D., NYU, 1974. Electrophysiology of the heart; ion channel expression.
Sergey Doronin, Research Assistant Professor; Ph.D., Novosibirsk Institute of Bioorganic Chemistry (Russia), 1991. Cellular biology and biochemistry of ion channel complex.
Raafat El-Maghrabi, Research Associate Professor; Ph.D., Wake Forest, 1978. Hormonal control of eukaryotic gene expression.
Junyuan Gao, Research Assistant Professor; Ph.D., SUNY at Stony Brook, 1994. Sodium potassium pump current in cardiac myocytes.
Roger A. Johnson, Professor; Ph.D., USC, 1968. Mechanism of hormone action; intracellular regulation of membrane-bound enzymes.
Varadaraj Kulandaippan, Research Assistant Professor; Ph.D., Madurai-Kamaraj (India), 1991. Lens membrane proteins and gap junctions.
Sindhu Kumari, Research Assistant Professor; Ph.D., Madurai Kamaraj (India), 1988. Biochemical and molecular characterization of gap junction channels and sodium potassium pump.
Richard T. Mathias, Professor; Ph.D., UCLA, 1975. Cardiac electrophysiology; volume regulation in lens.
Stuart McLaughlin, Professor; Ph.D., British Columbia, 1968. Biophysics of membranes.
W. Todd Miller, Professor; Ph.D., Rockefeller, 1988. Protein structure and function; molecular mechanisms of signal transduction.
Leon Moore, Professor; Ph.D., USC, 1976. Renal physiology.
Nicolas Nassar, Assistant Professor; Ph.D., Joseph Fourier (France). X-ray crystallography; signal transduction.
Irina Potopova, Research Assistant Professor; Ph.D., Novosibirsk Institute of Bioorganic Chemistry (Russia), 1990. Protein chemistry of ion channel subunits.
Barbara Rosati, Research Assistant Professor; Ph.D., Milan (Italy), 2000. Transcriptional regulation of ion channel genes in the heart.
Suzanne Scarlata, Professor; Ph.D., Illinois at Urbana-Champaign, 1984. G-protein signaling and HIV assembly.
Irene C. Solomon, Associate Professor; Ph.D., California, Davis, 1994. Reflex and central neural control of cardiovascular and respiratory function.
Ilan Spector, Associate Professor; Ph.D., Paris, 1967. Electrophysiology of nerve and muscle cell lines; ion channels; neurotoxins.
Virginijus Valiunas, Research Assistant Professor; Ph.D., Kaunas Medical (Lithuania), 1992. Gap junction; intercellular communication and cardiac electrophysiology.
William G. Van der Kloot, Distinguished Professor Emeritus; Ph.D., Harvard, 1952. Cellular neurophysiology.
Hsien Yu Wang, Research Associate Professor; Ph.D., SUNY at Stony Brook, 1989. Signal transduction and development.
Thomas W. White, Associate Professor; Ph.D., Harvard, 1984. Physiology and genetics of intercellular channels.

Faculty with Joint Appointment with the Department of Anesthesiology
James P. Dilger, Professor; Ph.D., SUNY at Stony Brook, 1980. Neuromuscular junction; ion channels in nerve membranes.
Srinivas N. Pentyala, Research Assistant Professor; Ph.D., Sri Venkateswara (India), 1989. Molecular mechanics of the action of anesthetics.
Mario Rebecchi, Research Associate Professor; Ph.D., NYU, 1984. Regulation and mechanism of phosphoinositide-specific phospholipase C.

Faculty with Joint Appointment with the Department of Biochemistry and Cell Biology
Steven O. Smith, Professor; Ph.D., Berkeley, 1985. Molecular mechanisms of signal transduction.

Faculty with Joint Appointment with the Department of Biomedical Engineering
Emilia Entcheva, Assistant Professor; Ph.D., Memphis, 1998. Cardiac cell function.
Mary Frame, Assistant Professor; Ph.D., Missouri, 1990. Microcirculation; tissue engineering; nanofabrication.
Ki H. Chon, Associate Professor; Ph.D., USC, 1993. Biomedical signal processing; identification and modeling of physiological systems and medical instrumentation.

Faculty with Joint Appointment with the Department of Medicine
Norman H. Edelman, Professor; M.D., NYU, 1961. Role of brain hypoxia in the control of breathing.
Irwin J. Kurland, Associate Professor; M.D., UCLA, 1985; Ph.D., Vanderbilt, 1992. Hepatic insulin action: role of the Pentose cycle.
Richard Z. Lin, Associate Professor; M.D., California, San Francisco, 1998. Cell signaling and cardiac dysfunction.
John Sachs, Professor; M.D., Columbia, 1960. Sodium potassium pump in the red cell.
Sami Said, Professor; M.D., Cairo, 1951. Vasoactive intestinal peptides (VIP); acute lung injury.
Gerald C. Smaldone, Professor; M.D./Ph.D., NYU, 1975. Respiratory physiology.

Faculty with Joint Appointment with Department of Molecular Genetics and Microbiology
Carol Carter, Professor; Ph.D., Yale, 1972. Assembly of the human deficiency virus (HIV).
James B. Konopka, Associate Professor; Ph.D., UCLA, 1985. G protein–coupled receptor signal transduction and yeast morphogenesis.

Faculty with Joint Appointment with the Department of Neurobiology and Behavior
John B. Cabot, Professor; Ph.D., Virginia, 1976. Central nervous system control of cardiovascular function.
David McKinnon, Associate Professor; Ph.D., Australian National, 1987. Control of ion channel expression.
Lorne Mendell, Professor; Ph.D., MIT, 1965. Physiology and modifiability of synapses in the spinal cord.

Faculty with Joint Appointment with the Department of Pharmacology
Howard Crawford, Assistant Professor; Ph.D., Texas Southwestern Medical Center at Dallas, 1993. Matrix metalloproteinases in pancreatic cancer.

Faculty with Joint Appointment with the Department of Surgery
Irvin B. Krukenkamp, Professor; M.D., Maryland, 1978. Surgical and pharmacologic preconditioning.

Faculty with Joint Appointment with the Department of Urology
Yaacov Hod, Research Associate Professor; Ph.D., Technion (Israel), 1977. Hormonal control of gene expression; control of mRNA turnover.
Sardar Ali Kahn, Professor; M.D., Bangalore (India), 1964. Urological disorders associated with erectile dysfunction.

Faculty with Adjunct Appointment with Cold Spring Harbor Laboratory, Cold Spring Harbor, New York
Scott W. Lowe, Professor; Ph.D., MIT, 1994. Molecular mechanisms of apoptosis in cancer.

Faculty with Adjunct Appointment with SUNY Old Westbury
George Stefano, Professor; Ph.D., Fordham, 1973. Opiate neurovascular immunology.

Faculty with Adjunct Appointment with the Department of Pharmacology, College of Physicians and Surgeons, Columbia University
Michael R. Rosen, Professor; M.D., SUNY Downstate Medical Center, 1964. Developmental cardiac electrophysiology.

UNIVERSITY OF CONNECTICUT

College of Liberal Arts and Sciences
Department of Physiology and Neurobiology

Programs of Study

The department offers course work and research programs leading to M.S. and Ph.D. degrees in physiology and neurobiology with concentration in areas of neurobiology, endocrinology, and comparative physiology. In addition, the Department of Molecular and Cell Biology, the Department of Ecology and Evolutionary Biology, and the Biotechnology Center provide the opportunity for students to obtain a comprehensive background in biological sciences and offer the possibility of collaborative research.

Graduate programs are designed to fit the individual student's background and scientific interests. In the first year, students take two courses on the foundations of physiology and neurobiology. Through the first two years, and occasionally into the third year of training, students select from a number of additional seminars and courses in their area of major interest and related areas. By the end of the first year, the student selects the area of dissertation research, and a committee consisting of a major adviser and 2 associate advisers is formed. Students may begin dissertation research during the first year.

Research Facilities

The Department of Physiology and Neurobiology is located primarily in the new state-of-the-art Pharmacy/Biology Building. The department houses both shared and individual laboratories for behavioral, cellular, electrophysiological, and molecular research in physiology. The department also houses the University's electron microscopy facility, which contains equipment for scanning and transmission EM as well as electron probe analysis. Departmental faculty members also utilize the Marine Research Laboratories at Noank and Avery Point, Connecticut.

Financial Aid

Several types of financial support are available to graduate students. Most students are supported either on teaching assistantships or research assistantships from faculty grants. In 2006–07, full-time assistantships (nine months) pay $18,270 for beginning graduate students, $19,226.42 for those with an M.S. or the equivalent, and $21,372 for those who have passed the Ph.D. general examination. Both half and full graduate assistantships come with a tuition waiver, and students may purchase excellent health-care coverage, heavily subsidized by the University of Connecticut. In addition, the Department of Physiology and Neurobiology awards the Edward G. Boettiger Graduate School Fellowship to an incoming Ph.D. candidate during the first year of study. Several additional research fellowships and University fellowships are awarded on a competitive basis. Many labs also provide additional funding for summer research (up to $4000).

Cost of Study

In 2006–07, tuition is $3996 per semester for legal residents of Connecticut and $10,386 per semester for nonresidents, plus fees. Tuition is prorated for students registering for fewer than 9 credits per semester. University fees are subject to change without notice. Tuition (but not the general University fee) is waived for graduate assistants.

Living and Housing Costs

Dormitory rooms are available for unmarried graduate students. University-owned and privately owned apartments are available near the campus at moderate rents. Houses and apartments for rent may also be found in the surrounding communities. In 2006–07, the fee for accommodations in graduate residence halls ranges from $1958 to $4605 per semester. Students may purchase meals à la carte, or they may opt for a meal plan of three meals a day for seven days a week while classes are in session; the cost of the plan ranges from $1773 to $1958 per semester in 2006–07.

Student Group

Approximately 16,000 undergraduates and 7,500 graduate students are enrolled at the main campus at Storrs. Eighty percent of the undergraduate students and 73 percent of the graduate students are from Connecticut. The rest of the students come from many other states and more than 100 countries. The Department of Physiology and Neurobiology has about 40 graduate students.

Location

The University is located in a scenic countryside setting of small villages, streams, and rolling hills. There is easy access by car and bus to major urban and cultural centers, including Hartford, New Haven, Boston, and New York, and to other educational institutions, such as Yale, Harvard, and MIT. Recreational opportunities include skiing, fishing, sailing, hiking, ice-skating, and athletic events. Cultural opportunities available at the Storrs campus include film series, plays, symphony and chamber music series, public lectures, and art exhibits. A small shopping center is within walking distance of the campus, and several large shopping centers are nearby.

The University

The University of Connecticut was founded in 1881 and is a state-supported institution. The 1,800-acre main campus at Storrs is the site of vigorous undergraduate and graduate programs in agriculture, liberal arts and sciences, fine arts, engineering, education, business administration, human development and family relations, physical education, pharmacy, nursing, and physical therapy. Extensive cultural and recreational programs and athletic facilities are available.

Applying

For admission to the fall semester, it is suggested that applications be submitted by January 15. To be considered for financial support, students must submit applications by March 15 for admission the following September. U.S. applicants must submit scores on the General Test of the Graduate Record Examinations and must have maintained at least a 3.0 quality point ratio (QPR) for admission as graduate students with regular status. Applications and credentials from international students must be received by March 1 for admission in the fall semester or by October 1 for the spring semester and must include TOEFL scores. The University does not discriminate in admission on the basis of race, gender, age, or national origin. Students can apply online at http://www.grad.uconn.edu/applications.html.

Correspondence and Information

Graduate Admissions Committee
Box U-3156
Department of Physiology and Neurobiology
University of Connecticut
75 North Eagleville Road
Storrs, Connecticut 06269-3156

Phone: 860-486-3304
Fax: 860-486-3303
E-mail: kathleen.kelleher@uconn.edu
Web site: http://www.pnb.uconn.edu

University of Connecticut

THE FACULTY AND THEIR RESEARCH

Lawrence E. Armstrong, Professor (primary appointment in Kinesiology); Ph.D., Ball State. Research focuses on human physiological responses to exercise, dietary intervention (i.e., caffeine, low-salt diet, glucose-electrolyte solutions, amino acid supplementation), pharmacological agents, heat tolerance, and acclimatization to heat. Laboratory measurements of local sweat production; skin blood flow; and metabolic, ventilatory, cardiovascular, fluid-electrolyte, and strength perturbations are complemented by field observations. This research includes illnesses that arise in association with exercise in hot environments.

Marie E. Cantino, Associate Professor; Ph.D., Washington (Seattle). Research in this laboratory is directed toward understanding the mechanisms of contraction in striated muscle. In particular, the lab is using electron microscopy and biochemical and mechanical assays to study the structure and organization of proteins in the contractile filaments and the mechanisms by which calcium regulates the interactions of these proteins.

William D. Chapple, Professor; Ph.D., Stanford. The interests of the laboratory center around the cellular mechanisms by which arthropods generate and control movement despite varying external forces. The model system is the hermit crab abdomen, which is used to support the animal's shell. To understand the interactions between local and global mechanisms that produce this control, identified motoneurons and interneurons are studied electrophysiologically and their properties are incorporated into a control systems model.

Thomas T. Chen, Professor (primary appointment in Molecular and Cell Biology); Ph.D., Alberta; postdoctoral study at Queen's. Structure, evolution, regulation, and molecular actions of growth hormone and insulin-like growth factor genes; regulation of foreign genes in transgenic fish; development of model transgenic fish.

Joanne Conover, Assistant Professor; Ph.D., Bath (England). Research in the laboratory focuses on neurogenesis and neuronal migration in the adult mouse brain. A combination of techniques including cell culture of neural stem cells, RT-PCR-based gene expression analysis, and examination of mouse genetic models for neurodegenerative diseases aid us in understanding neuronal proliferation, differentiation, and migration in the adult brain.

Joseph F. Crivello, Professor (joint appointment in Marine Sciences); Ph.D., Wisconsin. Research is centered in two areas. One area examines the impact of pollution on marine organisms at a biochemical and genetic level. The other area examines pollution as a selective pressure altering the genetic diversity of marine organisms.

Angel L. de Blas, Professor and Department Head; Ph.D., Indiana. Research mainly focuses on the brain receptors for the inhibitory transmitter GABA. Studies are being conducted on elucidating the molecular structure of the receptors and the molecular interactions with other proteins that determine the synaptic localization of the GABA receptors. Effects of drugs and aging on GABAergic synaptic transmission are also studied. Techniques include recombinant DNA, monoclonal antibodies, cell culture, and light, electron, and laser confocal microscopy.

Robert V. Gallo, Professor; Ph.D., Purdue. The objective of the research program is to understand the neuroendocrine mechanisms regulating luteinizing hormone release during different physiological conditions. In particular, the research examines the involvement of CNS neurotransmitters and endogenous opioid peptides in this process.

William J. Kraemer, Professor (primary appointment in Kinesiology); Ph.D., Wyoming. Research focus is directed at the neuroendocrine responses and adaptations with exercise as it relates to target tissues of muscle, bone, and immune cells. Current studies utilize receptor techniques and hormonal immuno- and bio-assays to better understand androgen, adrenal, and pituitary hormone interactions with target cells and their relationship to outcome variables of physiological function and physical performance.

Joseph J. LoTurco, Professor; Ph.D., Stanford. Research in the laboratory focuses on understanding mechanisms that direct development of the neocortex. Currently, a combination of molecular genetics, patch clamp electrophysiology, and cell culture are being used to study the mechanisms that regulate neurogenesis in the cerebral cortex.

Carl M. Maresh, Professor and Director, Human Performance Laboratory (primary appointment in Kinesiology—Department Head); Ph.D., Wyoming. Research focus is directed at the neuroendocrine, body fluid, and substrate responses and adaptations to environmental stress and training in humans. Current projects examine the efficacy of different methods of rehydration, muscle and bone adaptations to physical training in women, and exercise interventions in children at risk for obesity and diabetes.

Andrew Moiseff, Professor; Ph.D., Cornell. The laboratory is interested in the extraction and processing of sensory information by the nervous system. At present, research is concentrated on how the nervous system of the barn owl analyzes interaural time and intensity differences to obtain spatial information. Another line of research is the behavioral analysis of synchronous flashing by fireflies with an aim toward the understanding of the neural mechanisms controlling this communication behavior.

Akiko Nishiyama, Associate Professor; M.D., Nippon Medical School; Ph.D., Niigata (Japan). Research focuses on the biology of glial progenitor cells identified by the NG2 proteoglycan in normal and mutant mice and in demyelinating and excitotoxic lesions. Current studies employ immunohistochemical, tissue culture, biochemical, and molecular biological techniques to understand the mechanisms that regulate proliferation and differentiation of these glial cells and to explore their function.

Linda S. Pescatello, Associate Professor and Director, Center for Health Promotion (primary appointment in School of Allied Health); Ph.D., Connecticut. Research focus is on the interaction between the environment, neurohormones, and genetics on the exercise response in order to determine for whom exercise works best as a therapeutic modality. Current projects examine humoral, nutritional, and genetic explanations for postexercise hypotension, the exercise dose response for postexercise hypotension, and the influence of genetics on the muscle strength and hypertrophy response to resistance exercise training.

Cathy Proenza, Assistant Professor, Ph.D., Colorado. Research focuses on molecular mechanisms responsible for the intrinsic activity of myocytes in the sinoatrial node of the heart. One emphasis is on the role of hyperpolarization-activated "pacemaker" ion channels in initiating and modulating heart rate. Technical approaches include electrophysiology, fluorescent imaging, tissue culture, and biochemistry.

J. Larry Renfro, Professor; Ph.D., Oklahoma. The research program is concerned with the mechanisms and regulation of epithelial transport. Current research deals with transepithelial transport and excretion of sulfate, and phosphate and environmental pollutants in tissues isolated from the urinary systems of a variety of vertebrates, including rats, birds, and fishes. Work has concentrated on ion secretion and its regulation in primary monolayer cultures of renal epithelium and choroid plexus. The laboratory also studies transport by renal tubule brush border and basolateral membrane vesicles.

Maria E. Rubio, Assistant Professor; M.D., Ph.D., University of Alicante, Spain. Research focuses on understanding what modulates the expression and distribution of excitatory neurotransmitter receptors during development, adults and synaptic plastic events in the brain. Techniques include immunocytochemistry at light and electron microscopy level, morphometry, pharmacological, and surgical procedures in vivo and brain slices.

Randall S. Walikonis, Assistant Professor; Ph.D., Mayo. Research is directed at studying the postsynaptic signal transduction systems of excitatory synapses. The laboratory uses biochemical and molecular biological techniques to identify proteins associated with NMDA receptors and to determine their specific roles in the function of excitatory synapses. The lab also studies the role of growth factors in modifying excitatory synapses.

Steven A. Zinn, Associate Professor (primary appointment in Animal Science); Ph.D., Michigan State. The laboratory is interested in the somatotropic axis and its influence on growth and lactation. Specifically, the laboratory is investigating the influence of exogenous somatotropin on insulin-like growth factor I and the insulin-like growth factor binding proteins and their role in growth.

UNIVERSITY OF ILLINOIS AT URBANA–CHAMPAIGN

Department of Molecular and Integrative Physiology
Program in Molecular and Integrative Physiology

Programs of Study

The Department of Molecular and Integrative Physiology offers a predoctoral program in molecular and integrative physiology and cooperates with the Medical Scholars (M.D./Ph.D.) Program at the College of Medicine. The department offers an unusual breadth of training not often found in physiology departments, with particular strengths in endocrinology, reproductive biology, neuroscience, cardiac physiology, and functional genomics. The integration of graduate education with the other departments in the School of Molecular and Cellular Biology (http://www.life.uiuc.edu/mcb/) further prepares students for a successful career in academia, industry, or government.

Ph.D. candidates in physiology take two common core graduate courses with other graduate students in the School of Molecular and Cellular Biology. After completing a course in current progress in molecular and integrative physiology, additional supporting courses may be taken in biochemistry, physical chemistry, molecular biology, physiology, neurobiology, biophysics, and cell biology. A pass-fail qualifying examination is usually taken toward the end of the second year. Next, the student presents a detailed thesis research proposal at the preliminary examination. The remainder of the candidate's program is taken up by original research under the guidance of his or her thesis adviser and in close contact with other graduate students and postdoctoral collaborators in the research group of his or her choice. Most students complete their Ph.D. training and defend their thesis within four to six years.

Research Facilities

The department's research laboratories are equipped with a full range of instrumentation in molecular, cellular, and integrative physiology. Included are facilities for recombinant DNA research, tissue culture, electrophysiological measurements, and biochemical studies, as well as instrumentation for spectroscopy, microscopy, computer interfacing, data analysis, and modeling. Students and faculty members also have access to campuswide facilities for optical imaging, electron and scanning microscopy, flow cytometry, antibody production, molecular graphics computer modeling, oligonucleotide and peptide synthesis, protein sequencing and characterization, and transgenic animals. The excellent Biological Sciences Library, with extensive holdings of current research journals, is supported by the University of Illinois Library, which has more than 22 million volumes and is one of the largest public university libraries in the world.

Financial Aid

Stipends and tuition and fee waivers are given to all students who remain in good standing.

Cost of Study

In 2006–07, estimated tuition and fees are $22,240 for Illinois residents and $36,326 for nonresidents per year; however, tuition and most fees are waived for doctoral students in good standing.

Living and Housing Costs

University-owned housing is available for single and married students. Graduate dormitories cost from $4000 to $5300 per academic year. One- and two-bedroom apartments range in price from $500 to $700 per month. Rooms and apartments in the community rent at comparable or lower rates.

Student Group

Currently, 39 graduate students are enrolled in the program. Students entering the program have grade averages between A and B. Many entering students have previous experience in research, and nearly 50 percent have master's degrees from other institutions.

Location

The University is located in the center of the twin cities of Champaign and Urbana (combined population of approximately 100,000) in east-central Illinois, about 2 hours' driving time from Indianapolis, 2½ hours from Chicago, and 3 hours from St. Louis and served both by railroad and several airlines. Community life centers on the University, which provides excellent cultural opportunities in music, theater, and the arts. The Krannert Center for the Performing Arts provides outstanding programs.

The University

A land-grant institution founded in 1867, the Urbana-Champaign campus of the University of Illinois enrolls about 40,000 students. The University is well-known for its role in higher education, scientific research, and public service. It is rated one of the top ten public national universities that grant doctoral degrees. Eleven alumni and 9 faculty members have been awarded a Nobel Prize. Illinois teams participate in all major and many minor sports in Big Ten and NCAA competitions.

Applying

Important factors in the evaluation of applications are general academic performance, demonstration of interest in and commitment to scientific research, and letters of recommendation from college professors. An appropriate undergraduate preparation might include broad training in a biological science or a major in chemistry, biochemistry, physics, or engineering with exposure to general biology. The formal entrance requirement is a baccalaureate degree from an accredited college or university and a minimum GPA of 3.0 on a 4.0 scale. GRE General Test scores are required from all prospective students, and TOEFL scores are required from international applicants. The deadline for receipt of applications is January 15. Early application is encouraged, since a deadline of February 1 is in force for the department's submission of applications for fellowships and traineeships to the Graduate College. All applications to the Program in Molecular and Integrative Physiology, as well as to the Programs in Biochemistry, Cell and Developmental Biology, and Microbiology, are handled by the School of Molecular and Cellular Biology. Further details can be found at http://www.life.uiuc.edu/mcb/graduate/index.html.

Correspondence and Information

Director, Graduate Admissions
Program in Molecular and Integrative Physiology
Department of Molecular and Integrative Physiology
524 Burrill Hall, MC-114
University of Illinois at Urbana-Champaign
407 South Goodwin Avenue
Urbana, Illinois 61801
E-mail: mcbinfo@life.uiuc.edu
Web site: http://www.life.uiuc.edu/physiology/c_grad.html

University of Illinois at Urbana-Champaign

THE FACULTY AND THEIR RESEARCH

Thomas J. Anastasio, Associate Professor of Physiology and Biophysics; Ph.D., Texas. Neural mechanisms of sensorimotor integration.

Milan Bagchi, Professor of Physiology; Ph.D., Nebraska–Lincoln. Transcriptional regulation by steroid and thyroid hormone receptors.

Janice M. Bahr, Professor of Animal Science and Physiology; Ph.D., Illinois at Urbana-Champaign. Mechanisms regulating ovarian steroid biosynthesis and ovulation.

Philip M. Best, Professor of Physiology, Biophysics, and Bioengineering; Ph.D., Washington (Seattle). Physiological function and regulation of ion channels in excitable cells.

Charles L. Cox, Assistant Professor of Physiology, Pharmacology, and Biophysics; Ph.D., California, Riverside. Synaptic physiology and plasticity; neuromodulation of thalamocortical circuits.

M. Joan Dawson, Associate Professor of Physiology and Biophysics; Ph.D., Pennsylvania. In vivo imaging of metabolites using nuclear magnetic resonance.

Fred Delcomyn, Professor of Entomology and Physiology; Ph.D., Oregon. Generation of coordinated patterns of motor activity during behavior in insects.

Arthur L. DeVries, Professor of Animal Biology and Physiology; Ph.D., Stanford. Role of biological antifreeze peptides and glycopeptides in freezing avoidance in cold-water fishes.

Albert S. Feng, Professor of Physiology, Biophysics, and Bioengineering; Ph.D., Cornell. Neural basis of sound-pattern recognition.

Julia M. George, Assistant Professor of Physiology; Ph.D., Rockefeller. Role of synuclein proteins in neural functions and Parkinson's disease.

Martha U. Gillette, Professor of Cell and Developmental Biology and Biophysics (also with Physiology); Ph.D., Toronto. Cellular and molecular mechanisms in the mammalian circadian clock.

Rhanor Gillette, Professor of Physiology; Ph.D., Toronto. Mechanisms through which learning and motivation organize behavior in the sea slug *Pleurobranchaea* and in the octopus.

William T. Greenough, Professor of Psychology, Cell and Developmental Biology, and Bioengineering (also with Physiology); Ph.D., UCLA. Mechanisms whereby the developing and adult nervous system stores information.

Claudio Grosman, Assistant Professor of Physiology and Biophysics; Ph.D., Buenos Aires. Molecular physiology of neurotransmitter-gated ion channels.

Gary A. Iwamoto, Associate Professor of Kinesiology (also with Physiology); Ph.D., California, Davis. Neural control of the cardiovascular system.

Eric Jakobsson, Professor of Physiology, Biophysics, and Bioengineering; Ph.D., Dartmouth. Physics of ion permeation in membranes; coding properties of electrically excitable cells; cell homeostasis; molecular graphics.

Benita S. Katzenellenbogen, Professor of Physiology and Cell and Developmental Biology; Ph.D., Harvard. Biochemistry, molecular biology, and physiology of steroid hormone (estrogen, progesterone) receptors; mechanisms by which these proteins regulate gene expression and the growth and functioning of target cells.

Byron W. Kemper, Professor of Physiology, Pharmacology, and Cell and Developmental Biology; Ph.D., Stanford. Regulation of mammalian gene expression; intracellular targeting of proteins; structure-activity relationships.

Richard Kollmar, Assistant Professor of Physiology; Ph.D., Wisconsin–Madison. Molecular basis of hearing and balance in the zebrafish.

Kurt E. Kwast, Associate Professor of Physiology; Ph.D., Colorado at Boulder. Molecular and cellular mechanisms of oxygen sensing and signal transduction; oxygen-regulated gene expression.

Paul C. Lauterbur, Professor of Physiology, Biophysics, Chemistry, and Bioengineering; Ph.D., Pittsburgh. Development of new nuclear magnetic resonance technology, including imaging methods, spectroscopic localization, flow and perfusion imaging, microscopy, and magnetic contrast and labeling agents.

Yuqing Li, Assistant Professor of Physiology; Ph.D., Nagoya (Japan). Molecular and cellular mechanisms of synaptic plasticity during development and learning.

Esmail (Essie) Meisami, Associate Professor of Physiology; Ph.D., Berkeley. Vulnerability and growth plasticity of the developing brain; comparative quantitative functional neurocytoarchitecture; hormonal control of growth.

Ann M. Nardulli, Professor of Physiology; Ph.D., Illinois at Urbana-Champaign. Mechanisms involved in modulation of estrogen-responsive genes.

Mark E. Nelson, Professor of Physiology, Biophysics, and Bioengineering; Ph.D., Berkeley. Neural mechanisms and computations involved in the acquisition of sensory information.

Lori T. Raetzman, Assistant Professor of Physiology; Ph.D., Case Western Reserve. Exploring Notch signaling in pituitary gland development, using transgenic and knockout mouse models.

Edward J. Roy, Professor of Psychology (also with Physiology); Ph.D., Massachusetts. Neuro-oncology; neural and behavioral actions of steroid hormones.

O. David Sherwood, Professor of Physiology; Ph.D., Wisconsin. Role of the hormones relaxin and estrogen in promoting growth and development of both the cervix and mammary glands during pregnancy.

Jonathan V. Sweedler, Professor of Chemistry and Bioengineering (also with Physiology); Ph.D., Arizona. Microanalysis, capillary separations, nanoliter-volume NMR spectroscopy, neurotransmitter targeting and release, and cotransmission.

Kevin Yang Xiang, Assistant Professor of Physiology; Ph.D., Oregon Health Sciences. Beta-adrenergic receptor subtype signaling pathways and physiological functions in cardiac muscle cells.

THE UNIVERSITY OF IOWA

Department of Physiology and Biophysics

Programs of Study

The Department of Physiology and Biophysics offers a program of graduate study leading to the Doctor of Philosophy degree. The major goals of the program are to impart fundamental knowledge of physiological processes at molecular, cellular, and integrative levels and provide experience in modern research strategies and skills applicable to contemporary problems of physiology and biophysics. The principal areas of research interest in the Department are endocrinology, neurobiology, and membrane physiology and biophysics, with the unifying theme of understanding mechanisms of signal transduction involved in regulation of function at the cellular and molecular levels. Additional information is available on the Department's Web site at http://www.physiology.uiowa.edu.

The graduate program requires an average of 5.5 years for completion. During the first two years, students take required and elective courses and participate in ongoing faculty research in one or more laboratories. After satisfying course and comprehensive examination requirements, students devote their full time to thesis research. All degree candidates have experiences as classroom instructors under faculty supervision as part of their training. Students typically enter the graduate program through the Biosciences Program, giving them the opportunity to investigate several disciplines before deciding on a thesis project leading to the Ph.D. degree. Following the completion of three required research rotations in the first year, it is expected that graduate students should be able to select a research laboratory and program affiliation. Additional information is available from the Biosciences Program Web site at http://www.medicine.uiowa.edu/biosciences.

The Department of Physiology and Biophysics maintains close ties with both basic science and clinical departments of the Carver College of Medicine as well as with related departments and programs throughout the University. In addition, the Department is a major participant in the Medical Scientist Training Program (M.D./Ph.D. Program) conducted under the auspices of the Graduate College and the Carver College of Medicine (http://www.medicine.uiowa.edu/mstp/).

Research Facilities

The Department of Physiology and Biophysics occupies two floors of the Bowen Science Building, which also houses the Departments of Anatomy and Cell Biology, Biochemistry, Microbiology, and Pharmacology. In addition to specialized equipment in faculty research laboratories, the Department has multi-investigator facilities for cell culture, molecular biology, fluorescence microscopy and imaging, histology, laser densitometry, photography, and isotope analysis. The Eckstein Medical Research Building, located next to the Bowen Science Building, houses core facilities including image analysis, hybridoma production, electron microscopy, fluorescence activated cell sorting, GC/MS, and DNA analysis. The Carver Biomedical Research Building, located next to the Bowen Science Building and adjacent to the Eckstein Medical Research Building, houses facilities for research, including muscular dystrophy, pediatrics, internal medicine, and auditory regeneration.

Financial Aid

All full-time graduate students in the Ph.D. program receive financial aid in the form of stipends and tuition scholarships. The stipend for full-time graduate assistants was $22,000 in 2005–06. Qualified applicants are nominated for Presidential Graduate Fellowships, with stipends of $19,000 to $22,000, and other awards.

Cost of Study

All full-time doctoral students receive tuition scholarships.

Living and Housing Costs

In 2005–06, unfurnished University apartments for family housing rented for $400 to $600 per month, plus utilities. A variety of private rental properties are available in the immediate vicinity at reasonable rates. University residence hall accommodations are available.

Student Group

The doctoral program provides training for 15 to 20 individuals, including those in the combined M.D./Ph.D. program.

Location

The Iowa City metropolitan area is a University-oriented community with a population of more than 90,000, including 30,000 University students. Five major airlines serve the modern Eastern Iowa Airport, located 20 miles from campus. Iowa City provides a full range of commercial and professional services generally directed toward student interests.

The University

The University of Iowa is a Big Ten university and takes pride in being the first public university to admit women and men on an equal basis. The University is known for a number of additional firsts, including the design and construction of space satellites and the Iowa Writers' Workshop. It has one of the largest university-owned teaching hospitals in the United States. University enrollment includes 5,441 students in the Graduate College. A full calendar of scholarly and athletic activities is supplemented by the active programs of the Iowa Center for the Arts, a major performance center that attracts international artists from the world of music, dance, and drama.

Applying

An undergraduate grade point average of 3.0 on a 4.0 scale and a minimum combined score of 1200 on the verbal and quantitative portions of the GRE are required. An undergraduate major in one of the biological, chemical, physical, mathematical, or engineering sciences, with one or more years of course work in biology, physics, biochemistry, and calculus, is desirable. Visits to the Department are encouraged and are arranged upon request.

Correspondence and Information

Director of Graduate Studies
Department of Physiology and Biophysics
Carver College of Medicine
5-660 Bowen Science Building
The University of Iowa
Iowa City, Iowa 52242
Phone: 319-335-7803
Fax: 319-335-7330
E-mail: melissa-nies@uiowa.edu
Web site: http://www.physiology.uiowa.edu

Director, Biosciences Program
Carver College of Medicine
1186 Medical Laboratories
The University of Iowa
Iowa City, Iowa 52242-1181
Phone: 319-335-8306
     800-551-6787 (toll-free)
Fax: 319-335-7656
E-mail: biosciences@uiowa.edu
Web site: http://www.medicine.uiowa.edu/biosciences

The University of Iowa

THE FACULTY AND THEIR RESEARCH

Kevin P. Campbell, Professor and Head; Ph.D., Rochester, 1979. Cell and molecular biological studies of muscular dystrophy.

François M. Abboud, Professor (Internal Medicine); M.D., Ain Shams (Cairo), 1955. Mechanisms of activation of baroreceptors and regulation of sympathetic nerve activity.

Michael G. Anderson, Assistant Professor; Ph.D., Iowa, 1997. Genetic and molecular dissection of mechanisms contributing to ocular disease.

Nikolai O. Artemyev, Professor; Ph.D., St. Petersburg (Russia), 1988. Molecular mechanisms of sensory signal transduction; photoreceptors and visual transduction.

Michael Artman, Professor and Head (Pediatrics); M.D., Tulane, 1978. Mechanisms of excitation-contraction (EC) coupling and regulation of contractile function in the immature heart.

Mark W. Chapleau, Professor (Internal Medicine); Ph.D., LSU Medical Center, 1985. Molecular and cellular mechanisms that determine baroreceptor reflex sensitivity in normal and pathological states.

Beverly L. Davidson, Roy J. Carver Professor (Internal Medicine); Ph.D., Michigan, 1987. Neurodegenerative diseases and gene transfer to the brain.

Sarah K. England, Associate Professor; Ph.D., Medical College of Wisconsin, 1993. Molecular and electrophysiological studies of potassium channels in uterine and vascular smooth muscle.

Robert E. Fellows, Professor; M.D., McGill, 1959; Ph.D., Duke, 1969. Cellular and molecular actions of hormones and growth factors on development of neurons and glia.

Michael P. Henry, Associate Professor; Ph.D., MIT, 1995. Identification of cellular and molecular mechanisms involved in prostate cancer metastasis to bone.

Wayne A. Johnson, Professor; Ph.D., Washington (Seattle), 1985. Developmental neurobiology; sensory neuron development; sensory signal transduction.

David J. Kusner, Professor (Internal Medicine); M.D., Harvard, 1985; Ph.D., Case Western Reserve, 1994. Activation of antimicrobial defenses in macrophages, dendritic cells, and neutrophils and their relevance to infectious diseases (tuberculosis, sepsis) as well as noninfectious inflammatory disorders (atherosclerosis).

W. Scott Moye-Rowley, Professor; Ph.D., Purdue, 1986. Transcriptional control of oxidative stress tolerance; eukaryotic multidrug resistance.

Benet J. Pardini, Associate Professor (Pediatric Cardiology); Ph.D., Loyola of Chicago, 1983; PA-C. Mechanisms that control the discrete regional influence of the cardiac parasympathetic nerves and their role on cardiac function.

Robert C. Piper, Associate Professor; Ph.D., Washington (St. Louis), 1992. Molecular mechanisms of membrane traffic in the endocytic system.

Paul B. Rothman, Professor and Head (Internal Medicine); M.D., Yale, 1984. Role cytokines and cytokine signaling plays in normal lymphocyte development and leukemic transformation.

Andrew F. Russo, Professor; Ph.D., Berkeley, 1984. Molecular control of neuronal gene expression and function.

Thomas J. Schmidt, Professor; Ph.D., Cornell, 1976. Biochemistry and pharmacology of glucocorticoid receptors; regulation of gene expression by glucocorticoid and mineralocorticoid hormones.

Deborah L. Segaloff, Professor; Ph.D., Vanderbilt, 1980. Mechanism of action of the LH and FSH receptors; G-protein–coupled receptors integral to reproductive physiology.

Erwin F. Shibata, Associate Professor; Ph.D., Texas Medical Branch, 1984. Molecular and cellular cardiac electrophysiology; receptor-mediated signal transduction; ion channel regulation by caveolae/raft mechanisms.

Curt D. Sigmund, Professor; Ph.D., SUNY at Buffalo, 1987. Development of models of human cardiovascular disease using transgenic and gene targeting techniques; regulation of gene expression.

Peter M. Snyder, Professor (Internal Medicine); M.D., Iowa, 1989. Regulation of DEG/ENaC ion channels in diseases, including hypertension.

Mark A. Stamnes, Associate Professor; Ph.D., California, San Diego, 1992. Molecular mechanisms of intracellular protein transport and sorting in the mammalian Golgi apparatus.

Michael J. Welsh, Professor (Internal Medicine); M.D., Iowa, 1974. Biology of Cl^- and Na^+ channels and their dysfunction in disease such as cystic fibrosis and hypertension; gene transfer to the lung.

UNIVERSITY OF MASSACHUSETTS WORCESTER

Graduate School of Biomedical Sciences
Department of Physiology
Program in Cellular and Molecular Physiology

Program of Study

The Department of Physiology's Program in Cellular and Molecular Physiology offers students opportunities for advanced studies and independent research leading to a Ph.D. degree. In research, the faculty has particular strength in the study of physiology at the molecular, biochemical, biophysical, and cellular levels. The Department emphasizes basic research and the training of graduate students and postdoctoral fellows. The objective of the program is to provide biomedical scientists/educators with the academic background, experimental tools, and analytical skills necessary to conduct independent and innovative research and to teach modern physiology.

Incoming students should have a bachelor's degree in a biological or physical science and a strong background in mathematics (including calculus), physics, chemistry (including physical chemistry), and biology. First-year students concentrate in a core curriculum, which includes course work in biochemistry and in cellular, molecular, and systems biology. Students electing to study cellular and molecular physiology are required to take advanced courses and seminars in physiology. To the greatest extent possible, courses are arranged to suit the individual student's interests and prior background. Students complete at least two laboratory rotations prior to selecting a laboratory for thesis research, and they are expected to attend Departmental and Medical School seminars in addition to performing course and laboratory work. During the second year, students select an area of specialization and a research adviser and take the qualifying examination, the successful completion of which is required for admission to degree candidacy. The Ph.D. degree program generally requires four to six years, and the degree is awarded upon completion and successful defense of thesis research.

The Department participates actively in a schoolwide program that leads to the M.D./Ph.D. degree in six to seven years. The M.D./Ph.D. program is restricted to Massachusetts residents.

Research Facilities

The Medical School building was constructed in 1974 and has been generously endowed with laboratory, classroom, and office space. The Department of Physiology occupies 22,000 square feet in the Basic Sciences Wing of the Medical School. Facilities and equipment are the most modern available for biomedical research and include core facilities for protein sequencing, nucleic acid synthesis, electron microscopy, tissue culture, fluorescence cell sorting, and image processing. A large and well-maintained facility for the care of experimental animals and modern electronics and machine shops are available to all investigators.

Financial Aid

Full-time students were provided a graduate assistantship with an annual stipend in 2005–06 of $25,235. Tuition is waived for both graduate and research assistants. Health insurance is provided. Application for assistantships is made on the application for admission.

Cost of Study

Full-time tuition for residents of Massachusetts was estimated at $1740 per semester in 2005–06; for nonresidents, $5348 per semester. The cost of books, supplies, and fees averages $1200 per year. These figures are subject to change.

Living and Housing Costs

Apartments and rooms are available in the Worcester area. The large number of students in the area makes sharing apartments feasible at costs that fall within the budget dictated by the assistantship stipend. The Housing Office at the Medical School assists students in finding living accommodations.

Student Group

The Department has 8 graduate students enrolled in the Ph.D. program, 35 postdoctoral fellows, and a number of visiting faculty members. Current total enrollment in the Graduate Program in Biomedical Sciences is approximately 295 students. There are 200 first- and second-year medical students sharing common student facilities. Other graduate students are working at the Medical School through cooperative programs with the Worcester Consortium colleges and the University of Massachusetts at Amherst.

Student Outcomes

Most graduates pursue postdoctoral research at world-class institutions, while others acquire prestigious academic or industrial positions.

Location

Worcester, the third-largest city in New England, lies 40 miles west of Boston and 40 miles north of Providence, Rhode Island. Worcester and the surrounding towns offer living conditions varying from urban to rural and a wide range of educational and cultural activities, including those of the Worcester Art Museum and the Centrum. Clark University, the College of the Holy Cross, Worcester Polytechnic Institute, Assumption College, and Worcester State College are all located in the Worcester area.

The University and The School

The University of Massachusetts Worcester is one of five campuses of the University of Massachusetts, the other four being the main campus at Amherst and the Boston, Lowell, and Dartmouth campuses. The Medical School currently accepts 100 students to the first-year class; the Graduate Program in Biomedical Sciences accepts 50 students. The University of Massachusetts Worcester joins regional institutions and the University of Massachusetts at Amherst in training professionals in many areas of health care.

Applying

Applicants should have demonstrated superior performance in study for a bachelor's degree in a biological or physical science. Specific course requirements include a year each of biology, calculus, organic chemistry, and physics. An undergraduate course in physical chemistry is highly recommended. Applicants must take both the Graduate Record Examinations General Test. Three letters of recommendation, an application form, and payment of a fee ($25 for Massachusetts residents, $50 for nonresidents) are required.

The University of Massachusetts is an Equal Opportunity/Affirmative Action employer. The University of Massachusetts Medical School recruits minority and women candidates, considers all applicants, and, once students are accepted, ensures that they will not be discriminated against in any area.

The graduate catalog, application forms, and a bulletin entitled *Cellular and Molecular Physiology Graduate Program* may be obtained from the address below. Potential applicants may contact Dr. Jack Leonard, Graduate Director, at 508-856-6687 or visit the Departmental Web site at http://www.umassmed.edu/physiology for additional information.

Correspondence and Information

Dean
Graduate School of Biomedical Sciences S1-201
University of Massachusetts Worcester
Worcester, Massachusetts 01655

Phone: 508-856-4135
Fax: 508-856-3659
E-mail: gsbs@umassmed.edu
Web site: http://www.umassmed.edu

University of Massachusetts Worcester

THE FACULTY AND THEIR RESEARCH

Robert E. Carraway, Professor; Ph.D., Brandeis, 1972. Chemical and biological character of neurotensin-related peptides, their receptors and relationships to gastrointestinal physiology and cancer cell growth. *Prostaglandins Leukotrienes Essential Fatty Acids* 74:93–107, 2006. *Regul. Pept.* 133:105–114, 2006; 120:155–66, 2004. *Neuroscience* 126:1023–32, 2004. *Am. J. Physiol.* 287:G408–16, 2004; 281:G1413–22, 2001. *J. Pharmacol. Exp. Ther.* 309:92–101, 2004; 307:640–50, 2003. *Biochem. Pharmacol.* 66:331–42, 2003.

James E. Crandall, Research Associate Professor; Ph.D., Florida, 1980. Neuronal migration and axon guidance in the embryonic forebrain. *J. Neurobiol.* 45:195–206, 2000. *J. Comp. Neurol.* 413:218–29, 1999. *J. Neurosci.* 19:794–801, 1999.

James G. Dobson Jr., Professor; Ph.D., Virginia, 1971. Regulation of cardiac contractility and metabolism, with emphasis on catecholamine and adenosine signal transduction in the mammalian hypoxic, ischemic, and aging heart. *Am. J. Physiol.* 290:H348, 2006; 287:H1721, 2004; 285:H1471, 2003. *Life Sci.* 77:3375, 2005. *Physiol. Genomics* 15:142, 2003.

Richard A. Fenton, Associate Professor; Ph.D., Kansas, 1977. Cardiovascular physiology; role of adenosine in the regulation of cardiac function and neurotransmitter responsiveness in the ischemic and aging mammalian heart. *Am. J. Physiol.* 290:H348–56, 2006; 285:H1471–8, 2003. *Alcohol. Clin. Exp. Res.* 25:968–75, 2001. *Life Sci.* 77:3375–88, 2005.

H. Maurice Goodman, Professor and Chairman; Ph.D., Harvard, 1960. Hormonal control of metabolism in adipose tissue, with emphasis on growth hormone and insulin. *Handbook of Physiology*, vol. 2, section 7. New York: Oxford University Press, 1999. *Endocrinology* 141:513–9, 2000; 140:1219–27, 1999. *Encyclopedia of Molecular Biology*, ed. T. S. Creighton, New York: Oxford University Press, 1999.

Peter Grigg, Professor; Ph.D., SUNY Upstate Medical Center, 1969. Stretch sensitivity of mouse cutaneous afferent neurons; in vitro preparation of innervated mouse skin stretched; internal mechanical states related to responses of neurons. *J. Neurophysiol.* 87:1–11, 2003; 80:745–54, 1998. *Phys. Rev. Lett.* 82:2402–5, 1999.

Mitsuo Ikebe, Professor; Ph.D., Osaka (Japan), 1982. Molecular and cellular function of myosin super family; signaling mechanism of smooth muscle/nonmuscle contractile apparatus. *Proc. Natl. Acad. Sci. U.S.A.* 101:9630–5, 2004. *J. Cell Biol.* 165:243–54, 2004. *J. Biol. Chem.* 279:28844–54, 2004; 278:29435–41, 21352–60, 5478–87, 2003. *Nature* 415:192–5, 2002; 412:831–4, 2001. *Nature Cell Biol.* 4:302–6, 2002.

Julie A. Jonassen, Associate Professor; Ph.D., Michigan, 1980. Regulation of renal cell gene expression, signal transduction, growth and death by oxidant stressors; medical curriculum design and assessment. *Front. Biosci.* 9:797–808, 2004. *Kidney Int.* 66:1890–1900, 2004. *Nephron Exp. Nephrol.* 98:e61–4, 2004. *Crit. Rev. Eukaryotic Gene Exp.* 13:55–72, 2003. *Acad. Med.* 78:10 (Suppl) S20–23, 2003; 77:56–63, 2002; 76:529, 2001; 74:821–8, 1999. *Am. J. Nephrol.* 21(1):69–77, 2001. *Toxicol. Appl. Pharmacol.* 162:132–41, 2000. *J. Am. Soc. Nephrol.* 10:S446–51, 1999.

Daniel L. Kilpatrick, Associate Professor; Ph.D., Duke, 1980. Transcriptional regulation of neuronal differentiation, including subtype specification and internal "clock" mechanisms controlling temporal maturation; lineage-specific and stage-dependent transcriptional mechanisms regulating the differentiation of spermatogenic cells. *J. Neurosci. Methods* 149:144–53, 2005. *J. Biol. Chem.* 279:53491–7, 2004. *Mol. Cell. Biol.* 24:10681–8, 2004; 22:8478–90, 2002; 19:6048–56, 1999.

José R. Lemos, Professor; Ph.D., Wesleyan, 1979. Molecular mechanism underlying stimulus-secretion coupling; effects of drugs of abuse on synaptic transmission and regulation of hypothalamic-neurohypophysial system. *Biophys. J.* 90:2027–37, 2006. *Pflugers Arch.* 450(2):96–110, 2005; 450(6):381–9, 2005. *J. Neuroendocrinol.* 17(9):583–90, 2005. *J. Neurosci.* 24:8322–32, 2004.

Jack L. Leonard, Professor and Graduate Director, Cellular and Molecular Physiology Program; Ph.D., Berkeley, 1976. Cellular and molecular events that mediate hormone action in the central nervous system. *Brain Res.* 976(1):130–4, 2003. *Endocrinology* 143(7):2700–7, 2002. *J. Biol. Chem.* 276(38):35652–9, 2001; 276(4):2600–7, 2001; 275:31701–7, 2000; 275:25194–201, 2000. *J. Neurosci.* 20:2255–65, 2000.

Liwang Liu, Assistant Professor; M.D., Tongji Medical (China), 1985. Modulation of calcium channels in the nervous system. *Eur. Biophys. J.* 33:255–64, 2004. *Neuropharmacology* 45:281–92, 2003. *Br. J. Pharmacol.* 138:1259–70, 2003. *Proc. Natl. Acad. Sci. U.S.A.* 100:295–300, 2003. *J. Physiol.* (London) 525:391–404, 2002. *Am. J. Physiol. Cell Physiol.* 280:C1293–305, 2001.

Lawrence M. Lifshitz, Associate Professor; Ph.D., North Carolina at Chapel Hill, 1987. Computer vision and computer graphics applied to images of cells; modeling intracellular processes via reaction-diffusion equations. *J. Gen. Physiol.* 114:1–14, 1999. *J. Neurosci.* 18(1):251–65, 1998. *IEEE Trans. Med. Imaging* 1998.

Robert J. O'Connell, Professor; Ph.D., SUNY Upstate Medical Center, 1968. Olfactory neurobiology with emphasis on odor quality coding and role of pheromones in regulating reproductive behavior; functional properties of BK channels in artificial membranes. *Biophys. J.* 86:3620–33, 2004. *J. Comp. Physiol. A* 183:433–42, 1998. *Invert. Neurosci.* 3:127–35, 1997. *Physiol. Entomol.* 22:123, 1997.

William C. Okulicz, Associate Professor; Ph.D., Connecticut, 1980. Cellular and molecular regulation of hormonal response in primate endometrium. *Reprod. Biol. Endocrinology*, in press; 2:54–63, 2004. *Paediatr. Peritna. Epidemiol.* 20:79–86, 2005. *DNA Cell Biol.* 24:345–9, 2005. *Cancer Res.* 65:358–63, 2005. *Frontiers Biosci.* 8:551–8, 2003. *Biol. Reprod.* 69:1593–9, 2003; 67:1067–72, 2002.

Ann R. Rittenhouse, Assistant Professor; Ph.D., Boston University, 1984. Characterization of role of calcium channels and their modulators in nerve cell plasticity, using molecular, biochemical, and patch-clamp techniques. *Proc. Natl. Acad. Sci. U.S.A.* 100:295–300, 2003. *Am. J. Physiol.* 280:C1293–305, C1306–18, 2001. *J. Physiol.* (London) 525:391–404, 2000. *J. Gen. Physiol.* 115:277–86, 2000.

Michael J. Sanderson, Professor; Ph.D., Southampton (United Kingdom), 1981. Signal transduction mechanisms that mediate intercellular communication and regulate multicellular responses between heterologous cell types; regulation of ciliary activity of airway epithelial cells and contraction of airway smooth-muscle cells. *J. Appl. Physiol.* 95:1325–32, 2003. *J. Physiol.* 546:733–49, 2003. *J. Gen. Physiol.* 119:187–98, 2002. *Microsc. Res. Tech.* 52:289, 2001. *Mol. Cell. Biol.* II:1815, 2000. *Cell Calcium* 26:103, 1999.

Cheryl R. Scheid, Professor; Ph.D., Boston University, 1976. Cellular mechanisms underlying kidney stone disease; cellular responses to oxidant stress. *Kidney Int.* 66:189–90, 2004; 50:638–46, 2000; 57:2403–11, 2000. *Am. J. Nephrol.* 21:69–71, 2001. Oxalate-induced changes in renal epithelial cell function role in stone disease. *Front. Biosci.* 9:787–808, 2004. *Nephron Exp. Nephrol.* 98:e61–4, 2004. *Mol. Urol.* 4:371–82, 2000. Mechanisms mediating oxalate-induced alterations in renal cell functions. *Crit. Rev. Eukaryot. Gene Expr.* 13:55–72, 2003.

Joshua J. Singer, Professor; Ph.D., Harvard, 1970. Regulation of ion channels, with an emphasis on the relationship of ion channel activity to cellular biochemical and physiological processes using patch-clamp and imaging techniques. *J. Gen. Physiol.* 124:259, 2004;119:251, 2002; 114:575, 1999; 103:471, 1994. *Cell Calcium* 35:523, 2004. *Am. J. Physiol.* 284:C607, 2003; 283:C1441, 2002. *Proc. Natl. Acad. Sci.* 99:6404, 2002. *J. Physiol.* (London) 534:59, 2001; 524:3, 2000; 484:331, 1995. *FEBS Lett.* 297:24, 1992. *Science* 244:1176, 1989.

Richard A. Tuft, Associate Professor; Ph.D., Worcester Polytechnic, 1972. Development of optical/electronic instrumentation for measurement of fast changes in intracellular concentration of ions and molecules. *Biophys. J.* 90:2027–37, 2006. *Mol. Biol. Cell.* 17:1239–49, 2006. *J. Gen. Physiol.* 116:845–64, 2000. *Curr. Biol.* 9:285–91, 1999. *Science* 280:1763–6, 1998.

John V. Walsh Jr., Professor; M.D., Harvard, 1970. Molecular neurobiology; the regulation of ion channels in neurons and smooth-muscle cells studied with molecular biological, patch-clamp, and high resolution optical imaging techniques. *J. Gen. Physiol.* 103:471, 1994. *Trends Biochem. Sci.* 18:41, 1993. *Nature* 357:74, 1992.

Ronghua ZhuGe, Associate Professor; Ph.D., Iowa State, 1995. Calcium signaling and ion channel function in excitable cells. *Am. J. Physiol.* 287:C1577–88, 2004. *J. Gen. Physiol.* 120:15–27, 2002; 116:845–64, 2000; 113:215–28, 1999. *J. Physiol.* (London) 513:711–8, 1998.

Hui Zou, Assistant Professor; Ph.D., Old Dominion, 1995. Involvement of ion channels and localized calcium signaling in cellular function. *J. Gen Physiol.* 124:259–72, 2004; 114:575–88, 1999. *Cell Calcium* 35:523–33, 2004. *Proc. Natl. Acad. Sci.* 99:6404–9, 2002. *J. Physiol.* (London) 534:59–70, 2001. *J. Vasc. Res.* 37:556–67, 2000.

Affiliated Faculty

Anthony Carruthers, Professor (primary appointment in Biochemistry and Molecular Biology); Ph.D., King's College (England), 1980. Molecular mechanisms of glucose transport and glucose transport regulation. *Biochemistry* 40:15549, 2001; 39:3005, 2000.

Lawrence J. Hayward, Associate Professor (primary appointment in Neurology); M.D., Ph.D., Baylor, 1989. Neuromuscular biophysics; Na channel electrophysiology related to myotonia and periodic paralysis; motor neuron degeneration in amyotrophic lateral sclerosis (Lou Gehrig's disease). *J. Biol. Chem.* 280:29771–9, 2005; 280:17725–31, 2005; 277:15923–31, 2002. *Neurology* 52:1447–53, 1999.

J. Mark Madison, Professor (primary appointment in Medicine); M.D., Harvard, 1979. Regulation of intracellular calcium stores and the effects of cytokines on calcium in airway smooth muscle. *FASEB J.* 20:154–6, 2006. *J. Parmacol. Exp. Ther.* 313:127–33, 2005. *Am. J. Respir. Cell Mol. Biol.* 25:239–44, 2001.

David Paydarfar, Associate Professor (primary appointment in Neurology); M.D., North Carolina at Chapel Hill, 1985. Respiratory and autonomic neurophysiology. *J. Physiol.* (London) 550:287–304, 2003; 506:515–28, 1998; 483:273–88, 1995.

Adam E. Saltman, Associate Professor of Surgery and Physiology (primary appointment in Surgery); M.D., Ph.D., Columbia, 1990. Mechanism of atrial fibrillation; effects of inflammation on electrical conduction in the heart. *Intech-ISA India Reg. J. Meas. Control* 21:11–5, 2002. *J. Intech* 21:7–9, 2002. *J. Thorac. Cardiovasc. Surg.* 124:371–6, 2002. *Cardiovasc. Surg.* 10:111–5, 2002; 9:517–8, 2001. *Surg. Forum* 52:72–4, 2001.

Jeffrey S. Stoff, Professor of Medicine and Physiology (primary appointment in Medicine); M.D., Buffalo, 1968. Immunosuppressive and tolerogenic strategies in clinical transplantation; human organ allocation policy for transplantation; arachidonic acid metabolism and the kidney. Donor evaluation. In *Medical Management of the Renal Transplant Recipient*, ed. M. R. Weir. Philadelphia: Lippincott Williams & Wilkins, 2004. The report of a national conference on the wait list for kidney transplantation. *Am. J. Transpl.* 7:775–85, 2003. Primary varicella after transplantation. *Am. J. Kid. Dis.* 39(6):1310–2, 2002. Islet cell transplantation tolerance. *Transplantation* 72:43–6, 2001.

WEST VIRGINIA UNIVERSITY

Graduate Program in Cellular and Integrative Physiology

Programs of Study

The Graduate Program in Cellular and Integrative Physiology is one of seven graduate programs in West Virginia University (WVU) Schools of Medicine and Pharmacy offering interdisciplinary biomedical research training leading to the Ph.D., M.D./Ph.D., or M.S. degree. Research interests encompass the physiological and pharmacological sciences and incorporate biochemical, immunological, electrophysiological, cellular, and molecular approaches to address contemporary hypotheses. Researchers study the integrative control of cardiovascular, microvascular, endocrine, neural, and respiratory function, with an emphasis on heart, vascular, and lung diseases or processes such as aging, Alzheimer's, angiogenesis, asthma, COPD, coronary artery disease, congestive heart failure, diabetes and obesity, hypertension, and tissue edema.

Students benefit from individual attention by faculty members in a research environment that is dynamic, collaborative, and interdisciplinary. In addition to course work and laboratory research, students participate in seminars, journal clubs, and research conferences. They also attend national scientific meetings and obtain valuable speaking and teaching experience. The Ph.D. typically takes five years to complete. During Year 1, students matriculate in a common integrated core curriculum. This integrated first year allows students to build competence in key areas of contemporary science, gain exposure to the various training program options, meet potential dissertation advisers, and network scientifically and socially. In the second semester, students customize their course work by selecting from an array of program-specific electives. At the end of Year 1, students select a research adviser and can select Cellular and Integrative Physiology as their training program. Year 2 consists of advanced course work, research, teaching, and the candidacy examination. Years 3 to 5 are devoted to dissertation research. The Graduate Program in Cellular and Integrative Physiology also participates in the combined M.D./Ph.D. Scholars Program. As an M.D./Ph.D. Scholar, students take the first two years of the medical curriculum, followed typically by three years of research as required for the Ph.D. degree before returning to the M.D. program.

Research Facilities

Institutional facilities include a computer-based learning center, a centralized animal facility with a transgenic barrier, and a library housing more than 205,000 volumes and 2,400 journals. Core facilities are available for examining gene expression or genetic variation (Affymetrix platform), image analysis, confocal and electron microscopy and laser-capture microdissection, live-cell imaging, mass spectrometry, flow cytometry and high-speed cell sorting, proteomics, recombinant DNA technology, transgenic rodent biology, and functional neuroimaging (fMRI, PET/CT). Affiliated research centers include the National Institute for Occupational Safety and Health (NIOSH), Center for Advanced Imaging, Blanchette Rockefeller Neurosciences Institute, Sensory Neuroscience Research Center, and Mary Babb Randolph Cancer Center.

Financial Aid

Students in the Ph.D. or M.D./Ph.D. biomedical sciences programs receive financial support during their training, provided they remain in good academic standing and excel in research. Such support includes full tuition, health insurance, and an annual stipend of $22,000. Combined M.D./Ph.D. students also receive medical tuition waivers. M.S. students are not eligible for institutional tuition waivers or stipends but may be supported on a grant if funds are available.

Cost of Study

Students' tuition costs are covered.

Living and Housing Costs

The cost of an efficiency apartment in University-owned housing is approximately $400 per month. A limited number of University apartments are available for married students. Privately owned apartments in Morgantown cost $400 to $600 per month. In general, the cost of living is lower compared to larger cities.

Student Group

Total University enrollment is approximately 26,000 students, which includes 6,500 graduate and professional students who come from all parts of the United States and many other countries.

Location

Morgantown is a vibrant university community, with an appealing balance to life, of 80,000 residents in northern West Virginia. Located near the Pennsylvania border at the western edge of the Appalachian Mountains, abundant opportunities exist for activities such as world-class white-water rafting and kayaking, hiking and camping, mountain biking, fishing, and skiing. Morgantown has a cosmopolitan atmosphere with a range of activities usually found in much larger cities. It also enjoys proximity to major metropolitan centers: Pittsburgh is a 90-minute drive north, and Washington, D.C., is a 3-hour drive east.

The University

West Virginia University is a comprehensive, land-grant, Carnegie-designated Doctoral/Research University–Extensive public institution. The University's academic Health Sciences Center includes the Schools of Medicine, Dentistry, Nursing, and Pharmacy, all of which offer graduate degree programs. There are seven Ph.D. biomedical research training programs in the Schools of Medicine and Pharmacy, which benefit from the common, undifferentiated first year: Biochemistry and Molecular Biology, Cancer Cell Biology, Cellular and Integrative Physiology, Exercise Physiology, Immunology and Microbial Pathogenesis, Neuroscience, and Pharmaceutical and Pharmacological Sciences. Graduate faculty members in these programs are from various basic science and clinical departments throughout WVU and are members of interdisciplinary research centers in five health-related areas: cancer cell biology, cardiovascular sciences, immunopathology and microbial pathogenesis, neuroscience, and respiratory biology and lung diseases.

As a member of the Big East Conference, WVU participates in NCAA Division I sports. WVU also offers a variety of creative arts, theater, and entertainment opportunities.

Applying

Applicants must have a bachelor's degree and excellent GPA and GRE scores. Three letters of recommendation and a personal statement are required. Applicants are invited in groups of 10 for a paid, two-day visit/interview January through March. Students should visit the Web site at http://www.hsc.wvu.edu/som/resoff/gradprograms/PhD.asp for more information and an online application.

Correspondence and Information

Bernard Schreurs, Ph.D., Graduate Director
Graduate Program in Cellular and Integrative Physiology
West Virginia University
P.O. Box 9229
Morgantown, West Virginia 26506
Phone: 304-293-0497
E-mail: bschreurs@hsc.wvu.edu
Web site: http://www.hsc.wvu.edu/som/resoff/gradprograms/
PhD.asp

Office of Research and Graduate Education
Health Sciences Center
West Virginia University
P.O. Box 9104
Morgantown, West Virginia 26506
Phone: 304-293-7116
E-mail: cnoel@hsc.wvu.edu

West Virginia University

THE FACULTY AND THEIR RESEARCH

Matthew Boegehold, Professor; Ph.D., Arizona. Local and neural mechanisms controlling blood flow in the microcirculation, including role of endothelium; microvascular alterations in hypertension.

Vincent Castranova, Adjunct Professor; Ph.D., West Virginia. Pulmonary cell physiology; pulmonary inflammation; occupational lung diseases; inhalation toxicology.

Mary Davis, Professor; Ph.D., Michigan State. Toxicology, focusing on toxicity of environmental/occupational pollutants on mammalian renal function.

Richard Dey, Professor and Chair; Ph.D., Michigan State. Neuroanatomical organization and neuronal mediators in the lung.

Jeff Fedan, Adjunct Professor; Ph.D., Alabama at Birmingham. Pulmonary physiology and pharmacology; mechanisms of asthma, particularly the role of airway epithelium in airway hyperreactivity.

Mitchell Finkel, Professor; M.D., Maryland. Heart failure, prevention of cardiovascular diseases.

David Frazer, Adjunct Associate Professor; Ph.D., West Virginia. Respiration; bioengineering; lung acoustics; laser technology; animal exposure systems.

Jeffrey Frisbee, Assistant Professor; Ph.D., Guelph. Microvascular reactivity; microvessel density; microvascular dysfunction with the metabolic syndrome; regulation of tissue and organ perfusion.

Robert Goodman, Professor and Chair; Ph.D., Pittsburgh. Hypothalamic regulation of reproductive function; physiological infertility in seasonal breeders.

Leah Hammer, Research Assistant Professor; Ph.D., Monash (Australia). Gender differences and effect of aging on physiology and pathology of small blood vessels.

Pingnian He, Associate Professor; Ph.D., California, Davis. Cellular mechanisms that regulate permeability in intact microvessels; inflammation and vascular integrity.

Thomas Heming, Adjunct Professor; Ph.D. Cell biology of alveolar macrophages and host defense of the pulmonary system.

Stan Hileman, Assistant Professor; Ph.D., Kentucky. Neuronal pathways controlling food intake, nutrition, and fertility.

Chuan Hu, Research Assistant Professor; Ph.D., Columbia. Vesicle trafficking and angiogenesis.

James Lewis, Assistant Professor; Ph.D., Caltech. Mapping and exploring brain regions responsible for sound recognition and for localizing sounds in three-dimensional space; integration of auditory information with the other senses (vision, motor/touch).

Jun Liu, Assistant Professor; D.Phil., Oxford. Role of caveolae (microdomains on the cell surface involved in endocytosis, transcytosis, and signal transduction) and caveolin in angiogenesis.

Robert Mercer, Adjunct Assistant Professor; Ph.D., North Carolina at Chapel Hill. Physiology and pathophysiology of the lungs: morphological analysis of injury resulting from inhalation of airborne toxicants.

Fred Minnear, Professor; Ph.D., Oregon Health Sciences. Endothelial cell biology; barrier function of vascular endothelium; adherens junction; actin cytoskeleton.

Judy Muller-Delp, Associate Professor; Ph.D., Missouri. Effects of aging on vascular reactivity of resistance arterioles in cardiac and skeletal muscles; gender specificity; role of sex hormones and exercise in cardiovascular aging.

Eiskuke Murono, Adjunct Associate Professor; Ph.D., Rutgers. Effects of occupational chemicals in altering male and female reproductive functions, especially testicular Leydig and Sertoli, and female ovarian functions.

Jamal Mustafa, Professor; Ph.D., Lucknow (India). Cardiovascular pharmacology and physiology; blood flow regulation to the heart.

Timothy Nurkiewicz, Research Assistant Professor; Ph.D., West Virginia. Physiology and pathophysiology of the microcirculation.

Dale Porter, Research Associate; Ph.D., West Virginia. Lung toxicological response to nanoparticles and physiological and molecular regulation of toxicological responses.

Bernard Schreurs, Professor; Ph.D., Iowa. Learning and memory; synaptic plasticity; functional imaging; psychoneuroimmunology.

Anna Shvedova, Adjunct Associate Professor; Ph.D., D.Sc., Moscow. Mechanism of chronic allergic skin and lung disease caused by industrial chemicals and nanomaterials.

George Spirou, Professor; Ph.D., Florida. Neural feedback circuits mediating selective attention to sounds.

William Stauber, Professor; Ph.D., Rutgers. Muscle physiology; exercise physiology; cumulative trauma disorder; repetitive stress injuries; muscle injury and repair; muscle fibrosis and movement dysfunction.

Robert Wysolmerski, Professor; Ph.D., St. Louis. Endothelial cell biology; myosin II–based contraction; cellular tension; vasculogenesis.

Han-Gang Yu, Assistant Professor; Ph.D., Stony Brook, SUNY. Cardiac pacemaker activity; tyrosine phosphorylation of pacemaker proteins.

Section 19
Zoology

This section contains a directory of institutions offering graduate work in zoology, followed by in-depth entries submitted by institutions that chose to prepare detailed program descriptions. Additional information about programs listed in the directory but not augmented by an in-depth entry may be obtained by writing directly to the dean of a graduate school or chair of a department at the address given in the directory.

For programs offering related work, see also in this book Anatomy; Biochemistry; Biological and Biomedical Sciences; Cell, Molecular, and Structural Biology; Ecology, Environmental Biology, and Evolutionary Biology; Entomology; Genetics, Developmental Biology, and Reproductive Biology; Microbiological Sciences; Neuroscience and Neurobiology; and Physiology. In Book 4, see Agricultural and Food Sciences, Environmental Sciences and Management, and Marine Sciences and Oceanography; in Book 5, see Agricultural Engineering and Bioengineering and Ocean Engineering; and in Book 6, see Veterinary Medicine and Sciences.

CONTENTS

Program Directories

Announcements

Cross-Discipline Announcements

Close-Ups

See also:

Animal Behavior

Arizona State University, Division of Graduate Studies, College of Liberal Arts and Sciences, Department of Biology, Program in Behavior, Tempe, AZ 85287. Offers MS, PhD. Terminal master's awarded for partial completion of doctoral program. *Degree requirements:* For master's, thesis; for doctorate, thesis/dissertation, oral exam. *Entrance requirements:* For master's, GRE General Test; for doctorate, GRE General Test, GRE Subject Test. Additional exam requirements/recommendations for international students: Required—TOEFL (minimum score 600 paper-based; 250 computer-based); Recommended—TSE.

Bucknell University, Graduate Studies, College of Arts and Sciences, Department of Animal Behavior, Lewisburg, PA 17837. Offers MA, MS. Part-time programs available. *Degree requirements:* For master's, thesis. *Entrance requirements:* For master's, GRE General Test, GRE Subject Test, minimum GPA of 2.8. Additional exam requirements/recommendations for international students: Required—TOEFL.

Emory University, Graduate School of Arts and Sciences, Department of Psychology, Atlanta, GA 30322-1100. Offers clinical psychology (PhD); cognition and development (PhD); neuroscience and animal behavior (PhD). *Accreditation:* APA. *Faculty:* 32 full-time (13 women). *Students:* 71 full-time (53 women); includes 7 minority (4 African Americans, 2 Asian Americans or Pacific Islanders, 1 Hispanic American), 3 international. Average age 25. 312 applicants, 6% accepted, 11 enrolled. In 2005, 11 doctorates awarded. *Median time to degree:* Of those who began their doctoral program in fall 1997, 78% received their degree in 8 years or less. *Degree requirements:* For doctorate, thesis/dissertation, comprehensive exam, registration. *Entrance requirements:* For doctorate, GRE General Test, minimum GPA of 3.25. Additional exam requirements/recommendations for international students: Required—TOEFL. *Application deadline:* For fall admission, 1/3 for domestic students, 1/3 for international students. Application fee: $50. Electronic applications accepted. *Expenses:* Tuition: Full-time $14,400. Required fees: $217. *Financial support:* In 2005–06, 47 students received support, including 52 fellowships with full tuition reimbursements available (averaging $16,796 per year); research assistantships, teaching assistantships, career-related internships or fieldwork, Federal Work-Study, institutionally sponsored loans, scholarships/grants, and tuition waivers also available. Financial award application deadline: 4/15. *Faculty research:* Neuroscience and animal behavior; adult and child psychopathology, cognition development assessment. *Unit head:* Dr. Elaine Walker, Chair, 404-727-7438, Fax: 404-727-0372. *Application contact:* Terry Legge, Graduate Program Specialist, 404-727-7456, Fax: 404-727-0372, E-mail: tlegge@emory.edu.

Indiana University Bloomington, Graduate School, College of Arts and Sciences, Center for the Integrative Study of Animal Behavior, Bloomington, IN 47405-7000. Offers Certificate. *Expenses:* Tuition, state resident: full-time $5,437; part-time $227 per credit hour. Tuition, nonresident: full-time $15,836; part-time $660 per credit hour. Required fees: $821. Tuition and fees vary according to campus/location and program. *Unit head:* Dr. Emilia P. Martins, Director, 812-855-9663.

University of California, Davis, Graduate Studies, Graduate Group in Animal Behavior, Davis, CA 95616. Offers MS, PhD. *Faculty:* 30 full-time. *Students:* 26 full-time (19 women); includes 5 minority (1 African American, 1 Asian American or Pacific Islander, 3 Hispanic Americans). Average age 30. 69 applicants, 12% accepted, 4 enrolled. In 2005, 1 master's, 1 doctorate awarded. *Median time to degree:* Of those who began their doctoral program in fall 1997, 83.3% received their degree in 8 years or less. *Degree requirements:* For doctorate, thesis/dissertation. *Entrance requirements:* For doctorate, GRE General Test. Additional exam requirements/recommendations for international students: Required—TOEFL (minimum score 550 paper-based; 213 computer-based). *Application deadline:* For fall admission, 12/15 for domestic students, 12/15 for international students. Application fee: $60. Electronic applications accepted. *Financial support:* In 2005–06, 22 fellowships with full and partial tuition reimbursements (averaging $13,404 per year), 1 research assistantship with full and partial tuition reimbursement (averaging $11,354 per year), 5 teaching assistantships with partial tuition reimbursements (averaging $15,139 per year) were awarded; Federal Work-Study, institutionally sponsored loans, scholarships/grants, tuition waivers (full and partial), and unspecified assistantships also available. Financial award application deadline: 1/15; financial award applicants required to submit FAFSA. *Faculty research:* Wildlife behavior, conservation biology, companion animal behavior, behavioral endocrinology, animal communication. *Unit head:* Joy A. Mench, Chair, 530-752-7125, Fax: 530-752-5582, E-mail: jamench@ucdavis.edu. *Application contact:* Angelina Kuo, Graduate Staff, 530-752-2981, Fax: 530-752-8391, E-mail: abkuo@ucdavis.edu.

University of Colorado at Boulder, Graduate School, College of Arts and Sciences, Department of Ecology and Evolutionary Biology, Boulder, CO 80309. Offers animal behavior (MA); biology (MA, PhD); environmental biology (MA, PhD); evolutionary biology (MA, PhD); neurobiology (MA); population biology (MA); population genetics (PhD). *Faculty:* 26 full-time (5 women). *Students:* 44 full-time (24 women), 22 part-time (14 women); includes 16 minority (1 American Indian/Alaska Native, 3 Asian Americans or Pacific Islanders, 12 Hispanic Americans), 2 international. Average age 30. 20 applicants, 90% accepted. In 2005, 7 master's, 7 doctorates awarded. Terminal master's awarded for partial completion of doctoral program. *Degree requirements:* For master's, thesis or alternative, comprehensive exam; for doctorate, thesis/dissertation, comprehensive exam. *Entrance requirements:* For master's, GRE General Test, GRE Subject Test, minimum undergraduate GPA of 3.0; for doctorate, GRE General Test, GRE Subject Test. *Application deadline:* For fall admission, 1/2 priority date for domestic students, 12/1 priority date for international students. Application fee: $50 ($60 for international students). *Financial support:* In 2005–06, fellowships (averaging $7,844 per year), research assistantships (averaging $15,663 per year), teaching assistantships (averaging $13,829 per year)

were awarded; Federal Work-Study, institutionally sponsored loans, and tuition waivers (full) also available. Financial award application deadline: 3/1. *Faculty research:* Behavior, ecology, genetics, morphology, endocrinology. Total annual research expenditures: $5.2 million. *Unit head:* Jeffry Mitton, Chair, 303-492-0505, Fax: 303-492-8699, E-mail: mitton@colorado.edu. *Application contact:* Jill Skarstadt, Graduate Coordinator, 303-492-7654, Fax: 303-492-8699, E-mail: skarstad@colorado.edu.

University of Minnesota, Twin Cities Campus, Graduate School, College of Biological Sciences, Department of Ecology, Evolution, and Behavior, Minneapolis, MN 55455-0213. Offers ecology, evolution, and behavior (MS, PhD). Terminal master's awarded for partial completion of doctoral program. *Degree requirements:* For master's, thesis or projects; for doctorate, thesis/dissertation, comprehensive exam. *Entrance requirements:* For master's and doctorate, GRE General Test, minimum GPA of 3.0. Additional exam requirements/recommendations for international students: Required—TOEFL (minimum score 550 paper-based; 213 computer-based), MELAB. Electronic applications accepted. *Expenses:* Tuition, state resident: full-time $8,748; part-time $729 per credit. Tuition, nonresident: full-time $15,848; part-time $1,321 per credit. Full-time tuition and fees vary according to class time, course load, program and reciprocity agreements. *Faculty research:* Behavioral ecology, community ecology, community genetics, ecosystem and global change, evolution and systematics.

University of Missouri–St. Louis, College of Arts and Sciences, Department of Biology, St. Louis, MO 63121. Offers biology (MS, PhD), including animal behavior (MS), biochemistry (MS), biotechnology (MS), conservation biology (MS), development (MS), ecology (MS), environmental studies (PhD), evolution (MS), genetics (MS), molecular/cellular biology (MS), physiology (MS), plant systematics, population biology (MS), tropical biology (MS); biotechnology (Certificate); tropical biology and conservation (Certificate). Part-time programs available. *Faculty:* 50. *Students:* 26 full-time (15 women), 101 part-time (51 women); includes 13 minority (5 African Americans, 6 Asian Americans or Pacific Islanders, 2 Hispanic Americans), 41 international. Average age 32. In 2005, 22 master's, 2 doctorates awarded. *Degree requirements:* For master's, thesis or alternative; for doctorate, one foreign language, thesis/dissertation, 1 semester of teaching experience. *Entrance requirements:* For doctorate, GRE General Test. *Application deadline:* For spring admission, 12/1 priority date for domestic students. Applications are processed on a rolling basis. Application fee: $35 ($40 for international students). Electronic applications accepted. *Expenses:* Tuition, state resident: part-time $263 per credit hour. Tuition, nonresident: part-time $680 per credit hour. Required fees: $53 per credit hour. Tuition and fees vary according to program. *Financial support:* In 2005–06, 11 fellowships with full tuition reimbursements (averaging $30,000 per year), 15 research assistantships with full and partial tuition reimbursements (averaging $16,000 per year), 22 teaching assistantships with full and partial tuition reimbursements (averaging $16,000 per year) were awarded; career-related internships or fieldwork and Federal Work-Study also available. Support available to part-time students. Financial award application deadline: 2/1. *Faculty research:* Molecular biology, microbial genetics. *Unit head:* Zuleyma Tang-Martinez, Director of Graduate Studies, 314-516-6498, Fax: 314-516-6233, E-mail: zuleyma@umsl.edu. *Application contact:* 314-516-5458, Fax: 314-516-5310, E-mail: gradadm@umsl.edu.

The University of Montana, Graduate School, College of Arts and Sciences, Department of Psychology, Missoula, MT 59812-0002. Offers clinical psychology (PhD); experimental psychology (PhD), including animal behavior psychology, developmental psychology; school psychology (MA, PhD, Ed S). *Accreditation:* APA (one or more programs are accredited). *Faculty:* 19 full-time (9 women), 1 (woman) part-time/adjunct. *Students:* 42 full-time (32 women), 20 part-time (16 women); includes 13 minority (all American Indian/Alaska Native), 3 international. 139 applicants, 6% accepted, 9 enrolled. In 2005, 6 master's, 6 doctorates, 4 other advanced degrees awarded. Terminal master's awarded for partial completion of doctoral program. *Degree requirements:* For master's and doctorate, thesis/dissertation. *Entrance requirements:* For master's, doctorate, and Ed S, GRE General Test. Additional exam requirements/recommendations for international students: Required—TOEFL. *Application deadline:* For fall admission, 1/15 for domestic students. Application fee: $45. *Expenses:* Tuition, state resident: part-time $267 per credit. Tuition, nonresident: part-time $665 per credit. Part-time tuition and fees vary according to course load and degree level. *Financial support:* In 2005–06, 12 teaching assistantships with full tuition reimbursements (averaging $14,000 per year) were awarded; career-related internships or fieldwork, Federal Work-Study, scholarships/grants, traineeships, and unspecified assistantships also available. Financial award application deadline: 3/1; financial award applicants required to submit FAFSA. Total annual research expenditures: $1.8 million. *Unit head:* Dr. Nabil Haddad, Chair, 406-243-4384, Fax: 406-243-6366, E-mail: psycgrad@selway.umt.edu. *Application contact:* Graduate Secretary, 406-243-4523, Fax: 406-243-6366, E-mail: psycgrad@selway.umt.edu.

The University of Tennessee, Graduate School, College of Arts and Sciences, Department of Ecology and Evolutionary Biology, Knoxville, TN 37996. Offers behavior (MS, PhD); ecology (MS, PhD); evolutionary biology (MS, PhD). Part-time programs available. *Degree requirements:* For master's and doctorate, thesis/dissertation. *Entrance requirements:* For master's and doctorate, GRE General Test, minimum GPA of 2.7. Additional exam requirements/recommendations for international students: Required—TOEFL. Electronic applications accepted.

The University of Texas at Austin, Graduate School, College of Natural Sciences, School of Biological Sciences, Program in Ecology, Evolution and Behavior, Austin, TX 78712-1111. Offers PhD. *Entrance requirements:* For doctorate, GRE General Test. Additional exam requirements/recommendations for international students: Required—TOEFL. Electronic applications accepted.

Zoology

Auburn University, Graduate School, College of Sciences and Mathematics, Department of Biological Sciences, Auburn University, AL 36849. Offers botany (MS, PhD); microbiology (MS, PhD); zoology (MS, PhD). *Faculty:* 28 full-time (5 women). *Students:* 44 full-time (23 women), 49 part-time (27 women); includes 8 minority (5 African Americans, 3 Hispanic Americans), 17 international. 65 applicants, 51% accepted, 22 enrolled. In 2005, 11 master's, 1 doctorate awarded. *Entrance requirements:* For master's and doctorate, GRE General Test. Additional exam requirements/recommendations for international students: Required—TOEFL. *Application deadline:* For fall admission, 7/7 for domestic students; for spring admission, 11/24 for domestic students. Electronic applications accepted. *Financial support:* Research assistantships, teaching assistantships available. *Unit head:* Dr. James M Barbaree, Chair, 334-844-1647, Fax: 334-844-1645. *Application contact:* Dr. Stephen L. McFarland, Acting Dean of the Graduate School, 334-844-4700.

Clemson University, Graduate School, College of Agriculture, Forestry and Life Sciences, Department of Biological Sciences, Clemson, SC 29634. Offers biological sciences (MS, PhD); microbiology (MS, PhD); plant and environmental sciences (MS, PhD); zoology (MS, PhD). *Students:* 79 full-time (38 women), 23 part-time (13 women); includes 5 minority (2 African Americans, 1 American Indian/Alaska Native, 2 Hispanic Americans), 23 international. 61 applicants, 30% accepted, 16 enrolled. In 2005, 14 master's, 2 doctorates awarded. *Degree requirements:* For doctorate, thesis/dissertation. *Entrance requirements:* For master's

and doctorate, GRE General Test. Additional exam requirements/recommendations for international students: Required—TOEFL. *Application deadline:* Applications are processed on a rolling basis. Application fee: $50. *Financial support:* Fellowships, research assistantships, teaching assistantships available. Financial award application deadline: 3/15; financial award applicants required to submit FAFSA. Total annual research expenditures: $1.3 million. *Unit head:* Dr. Alfred Wheeler, Chair, 864-656-1415, Fax: 864-656-0435, E-mail: wheeler@clemson.edu. *Application contact:* Information Contact, 864-656-3587, Fax: 864-656-0435.

Colorado State University, Graduate School, College of Natural Sciences, Department of Biology, Fort Collins, CO 80523-0015. Offers botany (MS, PhD); zoology (MS, PhD). Part-time programs available. *Faculty:* 22 full-time (6 women). *Students:* 18 full-time (11 women), 21 part-time (11 women); includes 4 minority (2 American Indian/Alaska Native, 1 Asian American or Pacific Islander, 1 Hispanic American). Average age 28. 43 applicants, 33% accepted, 9 enrolled. In 2005, 5 master's, 4 doctorates awarded. Terminal master's awarded for partial completion of doctoral program. *Degree requirements:* For master's, thesis (for some programs), comprehensive exam; for doctorate, thesis/dissertation, comprehensive exam. *Entrance requirements:* For master's and doctorate, GRE General Test, minimum GPA of 3.0. Additional exam requirements/recommendations for international students: Required—TOEFL. *Application deadline:* For fall admission, 1/15 priority date for domestic students, 1/15 priority date for international students; for spring admission, 11/1 for domestic students, 11/1 for

Zoology

international students. Applications are processed on a rolling basis. Application fee: $50. Electronic applications accepted. *Expenses:* Tuition, state resident: full-time $3,690; part-time $205 per credit. Tuition, nonresident: full-time $14,958; part-time $831 per credit. Required fees: $1,061. *Financial support:* In 2005–06, 127 students received support, including 6 fellowships with full tuition reimbursements available (averaging $22,500 per year), 27 research assistantships with full tuition reimbursements available (averaging $13,000 per year), 94 teaching assistantships with full tuition reimbursements available (averaging $12,888 per year); career-related internships or fieldwork, Federal Work-Study, institutionally sponsored loans, and traineeships also available. Financial award application deadline: 2/15. *Faculty research:* Aquatic and terrestrial ecology, cell biology and genetics, plant/animal physiology, developmental biology, evolutionary biology. Total annual research expenditures: $3.6 million. *Unit head:* Daniel R. Bush, Chair, 970-491-7011, Fax: 970-491-0649. *Application contact:* Dorothy Ramirez, Graduate Coordinator, 970-491-1923, Fax: 970-491-0649, E-mail: dorothy.ramirez@colostate.edu.

Connecticut College, Graduate School, Department of Zoology, New London, CT 06320-4196. Offers MA, MAT. Part-time programs available. *Degree requirements:* For master's, thesis. *Entrance requirements:* For master's, GRE Subject Test or MAT. *Faculty research:* Ultra-structure, comparative physiology, marine ecology, behavioral ecology, population genetics.

Cornell University, College of Veterinary Medicine, Ithaca, NY 14853-0001. Offers comparative biomedical science (PhD); immunology (MS, PhD); pharmacology (PhD); physiology (PhD); veterinary medicine (DVM); zoology (MS, PhD). *Accreditation:* AVMA. *Faculty:* 155 full-time (53 women). *Students:* 334 full-time (267 women); includes 71 minority (25 African Americans, 2 American Indian/Alaska Native, 18 Asian Americans or Pacific Islanders, 26 Hispanic Americans), 2 international. Average age 26. 871 applicants, 11% accepted, 78 enrolled. In 2005, 81 first professional degrees, 15 doctorates awarded. *Degree requirements:* For first-professional, thesis or alternative, on-site clinical training. *Entrance requirements:* GRE General Test or MCAT, undergraduate pre-medical science program, animal or veterinary experience, letter of recommendation. *Application deadline:* For fall admission, 10/1 for domestic students, 10/1 for international students. Application fee: $40. Electronic applications accepted. *Expenses:* Contact institution. *Financial support:* In 2005–06, 303 students received support, including 30 fellowships (averaging $24,854 per year), 88 research assistantships with tuition reimbursements available (averaging $24,854 per year); Federal Work-Study, institutionally sponsored loans, scholarships/grants, and unspecified assistantships also available. Financial award application deadline: 2/1; financial award applicants required to submit CSS PROFILE or FAFSA. *Faculty research:* Extensive biomedical research, comparative cancer, food safety. Total annual research expenditures: $49.3 million. *Unit head:* Dr. Donald F. Smith, Dean, 607-253-3771. *Application contact:* Jennifer A Mailey, Director of Admissions, 607-253-3700, Fax: 607-253-3709, E-mail: vet_admissions@cornell.edu.

Cornell University, Graduate School, Graduate Fields of Agriculture and Life Sciences, Field of Zoology, Ithaca, NY 14853-0001. Offers animal cytology (MS, PhD); comparative and functional anatomy (MS, PhD); developmental biology (MS, PhD); ecology (MS, PhD); histology (MS, PhD). *Faculty:* 26 full-time (4 women). *Students:* 4 full-time (3 women), 3 international. 9 applicants, 11% accepted, 1 enrolled. *Degree requirements:* For doctorate, thesis/dissertation, 2 semesters of teaching experience, comprehensive exam. *Entrance requirements:* For doctorate, GRE General Test, GRE Subject Test (biology), 2 letters of recommendation. Additional exam requirements/recommendations for international students: Required—TOEFL (minimum score 550 paper-based; 213 computer-based). *Application deadline:* For fall admission, 2/1 for domestic students. Application fee: $60. Electronic applications accepted. *Financial support:* In 2005–06, 4 students received support, including 1 fellowship with full tuition reimbursement available, 1 research assistantship with full tuition reimbursement available, 2 teaching assistantships with full tuition reimbursements available; institutionally sponsored loans, scholarships/grants, health care benefits, tuition waivers (full and partial), and unspecified assistantships also available. Financial award applicants required to submit FAFSA. *Faculty research:* Organismal biology, functional morphology, biomechanics, comparative vertebrate anatomy, comparative invertebrate anatomy, paleontology. *Unit head:* Director of Graduate Studies, 607-253-3276, Fax: 607-253-3756. *Application contact:* Graduate Field Assistant, 607-253-3276, Fax: 607-253-3756, E-mail: graduate_edcvm@cornell.edu.

Emporia State University, School of Graduate Studies, College of Liberal Arts and Sciences, Department of Biological Sciences, Emporia, KS 66801-5087. Offers botany (MS); environmental biology (MS); general biology (MS); microbial and cellular biology (MS); zoology (MS). Part-time programs available. *Faculty:* 13 full-time (2 women), 4 part-time/adjunct (2 women). *Students:* 3 full-time (1 woman), 17 part-time (7 women), 3 international. 10 applicants, 90% accepted, 7 enrolled. In 2005, 5 degrees awarded. *Degree requirements:* For master's, comprehensive exam or thesis. *Entrance requirements:* For master's, GRE, appropriate undergraduate degree, interview, letters of reference. Additional exam requirements/recommendations for international students: Required—TOEFL (minimum score 450 paper-based; 133 computer-based). *Application deadline:* For fall admission, 8/15 for domestic students. Applications are processed on a rolling basis. Application fee: $30 ($75 for international students). Electronic applications accepted. *Expenses:* Tuition, state resident: full-time $2,890; part-time $132 per credit. Tuition, nonresident: full-time $9,258; part-time $422 per credit. Required fees: $626; $41 per credit. Tuition and fees vary according to degree level. *Financial support:* In 2005–06, 4 research assistantships with full tuition reimbursements (averaging $6,492 per year), 9 teaching assistantships with full tuition reimbursements (averaging $6,492 per year) were awarded; fellowships, career-related internships or fieldwork, Federal Work-Study, institutionally sponsored loans, health care benefits, and unspecified assistantships also available. Financial award application deadline: 3/15; financial award applicants required to submit FAFSA. *Faculty research:* Fisheries, range, and wildlife management; aquatic, plant, grassland, vertebrate, and invertebrate ecology; mammalian and plant systematics, taxonomy, and evolution; immunology, virology, and molecular biology. *Unit head:* Dr. J. Richard Schrock, Chair, 620-341-5311, Fax: 620-341-5607, E-mail: jschrock@emporia.edu. *Application contact:* Dr. Derek Zelmer, Graduate Coordinator, 620-341-5623, Fax: 620-341-5607, E-mail: dzelmer@emporia.edu.

Illinois State University, Graduate School, College of Arts and Sciences, Department of Biological Sciences, Normal, IL 61790-2200. Offers biological sciences (MS); biology (PhD); biotechnology (MS); botany (PhD); ecology (PhD); genetics (PhD); microbiology (MS); physiology (PhD); zoology (PhD). Part-time programs available. *Faculty:* 27 full-time (6 women). *Students:* 36 full-time (28 women), 25 part-time (17 women); includes 1 minority (Asian American or Pacific Islander), 23 international. 80 applicants, 21% accepted. In 2005, 14 master's, 5 doctorates awarded. *Degree requirements:* For master's, thesis or alternative; for doctorate, variable foreign language requirement, thesis/dissertation, 2 terms of residency. *Entrance requirements:* For master's, GRE General Test, minimum GPA of 2.6 in last 60 hours of course work; for doctorate, GRE General Test. *Application deadline:* Applications are processed on a rolling basis. Application fee: $30. *Expenses:* Tuition, state resident: full-time $3,060; part-time $170 per credit hour. Tuition, nonresident: full-time $6,390; part-time $355 per credit hour. Required fees: $1,411; $47 per credit hour. *Financial support:* In 2005–06, 22 research assistantships (averaging $13,909 per year), 38 teaching assistantships (averaging $12,617 per year) were awarded; Federal Work-Study, tuition waivers (full), and unspecified assistantships also available. Financial award application deadline: 4/1. *Faculty research:* CRUI: osmoregulation in euryhaline fish: physiology, ecology and molecular biology; the PRISM project: enhancing science and math education; cell structure—functions of Na pump assembly. *Unit head:* Dr. Hou Tak Takucheung, Acting Chairperson, 309-438-3669. *Application contact:* Derek A. McCracken, Graduate Adviser, 309-438-3664.

See Close-Up on page 147.

Indiana University Bloomington, Graduate School, College of Arts and Sciences, Department of Biology, Program in Zoology, Bloomington, IN 47405-7000. Offers MA, PhD. *Expenses:* Tuition, state resident: full-time $5,437; part-time $227 per credit hour. Tuition, nonresident:

full-time $15,836; part-time $660 per credit hour. Required fees: $821. Tuition and fees vary according to campus/location and program. *Application contact:* Gretchen Clearwater, Adviser for Graduate Affairs, 812-855-1861, Fax: 812-855-6705, E-mail: biograd@bio.indiana.edu.

Miami University, Graduate School, College of Arts and Sciences, Department of Zoology, Oxford, OH 45056. Offers biological sciences (MAT); zoology (MA, MS, PhD). Part-time programs available. *Degree requirements:* For master's, thesis, final exam; for doctorate, one foreign language, thesis/dissertation, final exams, comprehensive exam. *Entrance requirements:* For master's, GRE General Test, minimum undergraduate GPA of 3.0 during previous 2 years or 2.75 overall; for doctorate, GRE General Test, minimum undergraduate GPA of 2.75, 3.0 graduate. Additional exam requirements/recommendations for international students: Required—TOEFL (minimum score 550 paper-based; 213 computer-based), TWE (minimum score 4). Electronic applications accepted.

Announcement: The Department of Zoology at Miami University offers training in animal behavior, developmental biology, ecology, ecotoxicology, genetics, molecular biology, neuroscience, and organismal and ecological physiology leading to master's and PhD degrees. The 35 faculty members and 60 graduate students occupy a well-equipped building with facilities for field research located near campus. All graduate students receive annual financial assistance worth at least $25,000 in stipends and waivers in return for part-time duties.

Michigan State University, The Graduate School, College of Natural Science, Department of Zoology, East Lansing, MI 48824. Offers zoo and aquarium management (MS); zoology (MS, PhD); zoology-environmental toxicology (PhD). *Faculty:* 18 full-time (5 women). *Students:* 65 full-time (41 women), 4 part-time (2 women); includes 1 minority (Hispanic American), 9 international. Average age 28. 70 applicants, 14% accepted. In 2005, 8 master's, 10 doctorates awarded. *Degree requirements:* For master's, thesis or alternative, final oral examination; for doctorate, thesis/dissertation, comprehensive exam. *Entrance requirements:* For master's and doctorate, GRE General Test, minimum GPA of 3.0; course work in chemistry, physics, and mathematics; 3 letters of recommendation. Additional exam requirements/recommendations for international students: Required—TOEFL (minimum score 550 paper-based; 213 computer-based), SPEAK/ITAOI Test. *Application deadline:* For fall admission, 12/1 for domestic students. Application fee: $50. Electronic applications accepted. *Expenses:* Tuition, state resident: part-time $330 per credit hour. Tuition, nonresident: part-time $685 per credit hour. Tuition and fees vary according to program. *Financial support:* In 2005–06, 27 fellowships with tuition reimbursements (averaging $11,060 per year), 14 research assistantships with tuition reimbursements (averaging $13,915 per year), 29 teaching assistantships with tuition reimbursements (averaging $14,577 per year) were awarded; scholarships/grants and unspecified assistantships also available. *Faculty research:* Cellular and developmental biology, ecology and environmental toxicology, neurobiology and animal behavior, genetics, zoo and aquarium management. Total annual research expenditures: $1.9 million. *Unit head:* Dr. Fred C. Dyer, Chairperson, 517-353-9864, Fax: 517-432-2789, E-mail: fcdyer@msu.edu. *Application contact:* Lisa Craft, Graduate Secretary, 517-355-4642, Fax: 517-432-2789, E-mail: craft1@msu.edu.

Montana State University, College of Graduate Studies, College of Letters and Science, Department of Ecology, Bozeman, MT 59717. Offers biological sciences (MS, PhD); fish and wildlife biology (PhD); fish and wildlife management (MS); land rehabilitation (intercollege) (MS). Part-time programs available. *Faculty:* 15 full-time (4 women). *Students:* 4 full-time (3 women), 61 part-time (23 women); includes 1 minority (American Indian/Alaska Native), 1 international. Average age 31. 8 applicants, 75% accepted, 6 enrolled. In 2005, 9 master's, 4 doctorates awarded. *Degree requirements:* For master's, thesis (for some programs), comprehensive exam, registration; for doctorate, thesis/dissertation, comprehensive exam, registration. *Entrance requirements:* For master's and doctorate, GRE General Test. Additional exam requirements/recommendations for international students: Required—TOEFL (minimum score 550 paper-based; 213 computer-based). *Application deadline:* For fall admission, 7/15 priority date for domestic students, 5/15 priority date for international students; for spring admission, 12/1 priority date for domestic students, 10/1 priority date for international students. Applications are processed on a rolling basis. Application fee: $30. Electronic applications accepted. *Expenses:* Tuition, state resident: full-time $4,132. Tuition, nonresident: full-time $1,132. *Financial support:* In 2005–06, 3 fellowships with full tuition reimbursements (averaging $21,300 per year), 45 research assistantships with full and partial tuition reimbursements (averaging $10,830 per year), 22 teaching assistantships with full and partial tuition reimbursements (averaging $10,268 per year) were awarded; career-related internships or fieldwork, Federal Work-Study, scholarships/grants, health care benefits, tuition waivers, and unspecified assistantships also available. Financial award application deadline: 3/1; financial award applicants required to submit FAFSA. *Faculty research:* Population dynamics, landscape ecology, evolution and genetics, ecological modeling, aquatic ecosystems. Total annual research expenditures: $2.1 million. *Unit head:* Dr. David Roberts, Department Head, 406-994-4548, Fax: 406-994-3190, E-mail: droberts@montana.edu.

North Carolina State University, Graduate School, College of Agriculture and Life Sciences, Department of Zoology, Raleigh, NC 27695. Offers MS, MZS, PhD. Terminal master's awarded for partial completion of doctoral program. *Degree requirements:* For master's, thesis (for some programs), oral exam; for doctorate, thesis/dissertation, oral and written exams. *Entrance requirements:* For master's and doctorate, GRE General Test, minimum GPA of 3.0. Additional exam requirements/recommendations for international students: Required—TOEFL. Electronic applications accepted. *Faculty research:* Acquatic and terrestrial ecology, herpetology, behavioral biology, neurobiology, avian ecology.

North Dakota State University, The Graduate School, College of Science and Mathematics, Department of Biological Sciences, Fargo, ND 58105. Offers biological sciences (MS); botany (MS, PhD); cellular and molecular biology (PhD); environmental and conservation sciences (MS, PhD); genomics (MS, PhD); natural resource management (MS, PhD); zoology (MS, PhD). *Faculty:* 20. *Students:* 30 full-time (10 women), 5 part-time (1 woman); includes 1 minority (Asian American or Pacific Islander), 3 international. Average age 24. 14 applicants, 43% accepted. In 2005, 4 master's awarded. *Degree requirements:* For master's and doctorate, thesis/dissertation. *Entrance requirements:* For master's and doctorate, GRE General Test. Additional exam requirements/recommendations for international students: Required—TOEFL. *Application deadline:* For fall admission, 3/15 for domestic students; for spring admission, 10/30 priority date for domestic students. Applications are processed on a rolling basis. Application fee: $45 ($60 for international students). Electronic applications accepted. *Financial support:* In 2005–06, 3 fellowships with full tuition reimbursements (averaging $15,000 per year), 9 research assistantships with full tuition reimbursements (averaging $14,400 per year), 19 teaching assistantships with full tuition reimbursements (averaging $9,550 per year) were awarded; career-related internships or fieldwork, Federal Work-Study, institutionally sponsored loans, scholarships/grants, tuition waivers (full), and unspecified assistantships also available. Support available to part-time students. Financial award application deadline: 4/15; financial award applicants required to submit FAFSA. *Faculty research:* Comparative endocrinology, physiology, behavioral ecology, plant cell biology, aquatic biology. Total annual research expenditures: $675,000. *Unit head:* Dr. William J. Bleier, Chair, 701-231-7087, Fax: 701-231-7149, E-mail: william.bleier@ndsu.nodak.edu.

Oklahoma State University, College of Arts and Sciences, Department of Zoology, Stillwater, OK 74078. Offers conservation science (MS, PhD); zoology (MS, PhD). *Faculty:* 16 full-time (4 women), 5 part-time/adjunct (4 women). *Students:* 9 full-time (4 women), 53 part-time (26 women); includes 6 minority (3 American Indian/Alaska Native, 3 Hispanic Americans), 5 international. Average age 31. 21 applicants, 43% accepted, 7 enrolled. In 2005, 10 master's, 3 doctorates awarded. *Degree requirements:* For master's, thesis or report; for doctorate, thesis/dissertation. *Entrance requirements:* For master's and doctorate, GRE General Test, GRE Subject Test. Additional exam requirements/recommendations for international students: Required—TOEFL. *Application deadline:* For fall admission, 2/15 priority date for domestic students, 3/1 priority date for international students. Applications are processed on a rolling basis. Application fee: $40 ($75 for international students). Electronic

---

Zoology

Oklahoma State University (continued)

applications accepted. *Expenses:* Tuition, state resident: full-time $4,253; part-time $139 per credit hour. Tuition, nonresident: full-time $12,569; part-time $485 per credit hour. Required fees: $43 per credit hour. One-time fee: $20 part-time. Tuition and fees vary according to course load and program. *Financial support:* In 2005–06, 2 research assistantships (averaging $15,633 per year), 34 teaching assistantships (averaging $15,936 per year) were awarded; career-related internships or fieldwork, Federal Work-Study, scholarships/grants, health care benefits, tuition waivers (partial), and unspecified assistantships also available. Support available to part-time students. Financial award application deadline: 3/1. *Unit head:* Dr. Ron Van Den Bussche, Interim Head, 405-744-5555, E-mail: shawjh@okstate.edu.

Oregon State University, Graduate School, College of Science, Department of Zoology, Corvallis, OR 97331. Offers MA, MAIS, MS, PhD. *Faculty:* 21 full-time (7 women), 4 part-time/adjunct (2 women). *Students:* 40 full-time (23 women), 1 (woman) part-time; includes 5 minority (all Asian Americans or Pacific Islanders), 6 international. Average age 28. In 2005, 1 master's, 8 doctorates awarded. Terminal master's awarded for partial completion of doctoral program. *Degree requirements:* For doctorate, thesis/dissertation. *Entrance requirements:* For master's and doctorate, GRE General Test, GRE Subject Test, minimum GPA of 3.0 in last 90 hours. Additional exam requirements/recommendations for international students: Required—TOEFL. *Application deadline:* For fall admission, 1/15 for domestic students. Application fee: $50. *Expenses:* Tuition, area resident: Part-time $301 per credit. Tuition, state resident: full-time $8,139; part-time $501 per credit. Tuition, nonresident: full-time $14,376; part-time $532 per credit. Required fees: $1,266. *Financial support:* Fellowships, research assistantships, teaching assistantships, Federal Work-Study and institutionally sponsored loans available. Support available to part-time students. Financial award application deadline: 2/1. *Faculty research:* Cell and developmental biology, population biology and marine community ecology, behavioral physiology, comparative immunology, plant-herbivore interaction. *Unit head:* Dr. John A. Ruben, Chair, 541-737-3705, Fax: 541-737-0501. *Application contact:* Valarie Breda, Chair of Graduate Admissions, 541-737-3705, Fax: 541-737-0501, E-mail: bredav@bcc.orst.edu.

Southern Illinois University Carbondale, Graduate School, College of Science, Department of Zoology, Carbondale, IL 62901-4701. Offers MS, PhD. *Faculty:* 25 full-time (1 woman). *Students:* 9 full-time (5 women), 92 part-time (39 women); includes 4 minority (2 American Indian/Alaska Native, 2 Hispanic Americans), 14 international. Average age 25. 118 applicants, 40% accepted, 13 enrolled. In 2005, 16 master's, 4 doctorates awarded. *Degree requirements:* For master's and doctorate, thesis/dissertation. *Entrance requirements:* For master's, GRE, minimum GPA of 2.7; for doctorate, GRE, minimum GPA of 3.25. Additional exam requirements/recommendations for international students: Required—TOEFL. *Application deadline:* Applications are processed on a rolling basis. Application fee: $20. *Financial support:* In 2005–06, 49 students received support, including 5 fellowships with full tuition reimbursements available, 45 research assistantships with full tuition reimbursements available, 20 teaching assistantships with full tuition reimbursements available; Federal Work-Study, institutionally sponsored loans, and tuition waivers (full) also available. Support available to part-time students. *Faculty research:* Ecology, fisheries and wildlife, systematics, behavior, vertebrate and invertebrate biology. *Unit head:* William Muhlach, Chairperson, 618-536-2314. *Application contact:* Carey Krajewski, Director of Graduate Studies, 618-453-4132, Fax: 618-453-2806, E-mail: careyk@siu.edu.

Announcement: The Department of Zoology at SIUC provides training leading to the MS and PhD degrees. The department includes 22 graduate faculty members with research programs in biodiversity, conservation biology, wildlife ecology, fisheries biology, mined-land reclamation, terrestrial and aquatic ecology, toxicology, molecular systematics, and molecular ecology.

See Close-Up on page 1277.

Texas A&M University, College of Science, Department of Biology, Program in Zoology, College Station, TX 77843. Offers MS, PhD. *Students:* Average age 30. *Degree requirements:* For master's, thesis or alternative, registration; for doctorate, thesis/dissertation, comprehensive exam, registration. *Entrance requirements:* For master's and doctorate, GRE General Test. Additional exam requirements/recommendations for international students: Required—TOEFL. Application fee: $50 ($75 for international students). *Expenses:* Tuition, state resident: full-time $4,488; part-time $187 per credit hour. Tuition, nonresident: full-time $11,112; part-time $463 per credit hour. Required fees: $1,974. *Financial support:* Application deadline: 4/1; *Unit head:* Dr. Mark Zoran, Head, 979-845-7755.

Texas Tech University, Graduate School, College of Arts and Sciences, Department of Biological Sciences, Lubbock, TX 79409. Offers biological informatics (MS); biology (MS, PhD); microbiology (MS); zoology (MS, PhD). Part-time programs available. *Faculty:* 29 full-time (4 women). *Students:* 101 full-time (47 women), 9 part-time (2 women); includes 7 minority (2 Asian Americans or Pacific Islanders, 5 Hispanic Americans), 42 international. Average age 30. 74 applicants, 50% accepted, 16 enrolled. In 2005, 8 master's, 5 doctorates awarded. *Degree requirements:* For master's, thesis (for some programs); for doctorate, thesis/dissertation. *Entrance requirements:* For master's and doctorate, GRE General Test. Additional exam requirements/recommendations for international students: Required—TOEFL (minimum score 550 paper-based; 213 computer-based). *Application deadline:* Applications are processed on a rolling basis. Application fee: $50 ($60 for international students). Electronic applications accepted. *Expenses:* Tuition, state resident: full-time $4,296. Tuition, nonresident: full-time $10,920. Required fees: $1,992. Tuition and fees vary according to program. *Financial support:* In 2005–06, 51 students received support, including 22 research assistantships with partial tuition reimbursements available (averaging $13,849 per year), 74 teaching assistantships with partial tuition reimbursements available (averaging $13,899 per year); career-related internships or fieldwork, Federal Work-Study, and institutionally sponsored loans also available. Support available to part-time students. Financial award application deadline: 4/15; financial award applicants required to submit FAFSA. *Faculty research:* Genome organization and evolution, plant stress response, climate change on arid ecosystems, cell signaling hormone regulation. Total annual research expenditures: $3.1 million. *Unit head:* Dr. John C. Zak, Chair, 806-742-2715, Fax: 806-742-2963, E-mail: john.zak@ttu.edu. *Application contact:* Dr. Randall M. Jeter, Graduate Adviser, 806-742-2710, Fax: 806-742-2963, E-mail: randall.jeter@ttu.edu.

Uniformed Services University of the Health Sciences, School of Medicine, Programs in Biomedical Sciences, Bethesda, MD 20814-4799. Offers emerging infectious diseases (PhD); medical and clinical psychology (PhD), including clinical psychology, medical psychology; medical history (MMH); microbiology and immunology (PhD); molecular and cell biology (PhD); neuroscience (PhD); pathology (PhD), including comparative pathology, molecular pathobiology; preventive medicine and biometrics (MPH, MSPH, MTMH, Dr PH, PhD), including environmental health science (PhD), medical zoology (PhD), public health (MPH, MSPH, Dr PH), tropical medicine and hygiene (MTMH). *Faculty:* 319 full-time (100 women), 3,978 part-time/adjunct (798 women). *Students:* 135 full-time (90 women), 3 part-time (1 woman); includes 36 minority (12 African Americans, 3 American Indian/Alaska Native, 16 Asian Americans or Pacific Islanders, 5 Hispanic Americans), 9 international. Average age 28. 256 applicants, 39% accepted, 66 enrolled. In 2005, 46 master's, 13 doctorates awarded. Terminal master's awarded for partial completion of doctoral program. *Median time to degree:* Of those who began their doctoral program in fall 1997, 100% received their degree in 8 years or less. *Degree requirements:* For master's, thesis or alternative, comprehensive exam; for doctorate, thesis/dissertation, qualifying exam, comprehensive exam. *Entrance requirements:* For master's, GRE General Test; for doctorate, GRE General Test, minimum GPA of 3.0. Additional exam requirements/recommendations for international students: Required—TOEFL. *Application deadline:* For fall admission, 1/15 for domestic students. Applications are processed on a rolling basis. Application fee: $0. *Financial support:* In 2005–06, fellowships with full tuition reimbursements (averaging $23,000 per year), research assistantships with full tuition reimbursements (averaging $23,000 per year) were awarded; career-related internships or fieldwork and

tuition waivers (full) also available. *Unit head:* Dr. Eleanor S. Metcalf, Associate Dean, 301-295-1104, E-mail: emetcalf@usuhs.mil. *Application contact:* Janet M. Anastasi, Graduate Program Coordinator, 301-295-9474, Fax: 301-295-6772, E-mail: janastasi@usuhs.mil.

See Close-Up on page 237.

Uniformed Services University of the Health Sciences, School of Medicine, Programs in Biomedical Sciences, Department of Preventive Medicine and Biometrics, Program in Medical Zoology, Bethesda, MD 20814-4799. Offers PhD. *Faculty:* 59 full-time (13 women), 187 part-time/adjunct (36 women). *Students:* 2 full-time (1 woman). Average age 33. 3 applicants, 33% accepted, 1 enrolled. *Median time to degree:* Of those who began their doctoral program in fall 1997, 100% received their degree in 8 years or less. *Degree requirements:* For doctorate, thesis/dissertation, qualifying exam, comprehensive exam. *Entrance requirements:* For doctorate, GRE General Test, GRE Subject Test, minimum GPA of 3.0, U.S. citizenship. Additional exam requirements/recommendations for international students: Required—TOEFL. *Application deadline:* For fall admission, 1/15 for domestic students. Applications are processed on a rolling basis. Application fee: $0. *Financial support:* In 2005–06, fellowships with full tuition reimbursements (averaging $23,000 per year); tuition waivers (full) also available. *Faculty research:* Epidemiology, biostatistics, tropical public health, parasitology, vector biology. *Unit head:* Dr. David Cruess, Ph.D, Graduate Program Director, 301-295-3465, Fax: 301-295-1933, E-mail: dcruess@usuhs.mil. *Application contact:* Janet M. Anastasi, Graduate Program Coordinator, 301-295-9474, Fax: 301-295-6772, E-mail: janastasi@usuhs.mil.

University of Alaska Fairbanks, College of Natural Sciences and Mathematics, Department of Biology and Wildlife, Fairbanks, AK 99775-7520. Offers biological sciences (MS, PhD), including biology, botany, zoology; biology (MAT); wildlife biology (MS, PhD). Part-time programs available. *Faculty:* 29 full-time (7 women), 2 part-time/adjunct (1 woman). *Students:* 89 full-time (53 women), 24 part-time (15 women); includes 6 minority (1 African American, 4 Asian Americans or Pacific Islanders, 1 Hispanic American), 11 international. Average age 30. 74 applicants, 53% accepted, 17 enrolled. In 2005, 14 master's, 4 doctorates awarded. Terminal master's awarded for partial completion of doctoral program. *Degree requirements:* For master's and doctorate, thesis/dissertation, comprehensive exam, registration. *Entrance requirements:* For master's and doctorate, GRE General Test, GRE Subject Test. Additional exam requirements/recommendations for international students: Required—TOEFL (minimum score 550 paper-based; 213 computer-based); Recommended—TWE, TSE. *Application deadline:* For fall admission, 6/1 for domestic students, 3/1 for international students; for spring admission, 12/1 for domestic students, 9/1 for international students. Applications are processed on a rolling basis. Application fee: $50. Electronic applications accepted. *Expenses:* Tuition, state resident: full-time $4,392; part-time $244 per credit. Tuition, nonresident: full-time $8,964; part-time $498 per credit. Required fees: $800; $5 per credit. $48 per contact hour. Tuition and fees vary according to course level, course load, campus/location and reciprocity agreements. *Financial support:* In 2005–06, 36 research assistantships with tuition reimbursements (averaging $9,207 per year), 23 teaching assistantships with tuition reimbursements (averaging $5,358 per year) were awarded; fellowships with tuition reimbursements, career-related internships or fieldwork, Federal Work-Study, and scholarships/grants also available. Financial award application deadline: 2/1; financial award applicants required to submit FAFSA. *Faculty research:* Plant-herbivore interactions, plant metabolic defenses, insect manufacture of glycerol, ice nucleators, structure and functions of arctic and subarctic freshwater ecosystems. *Unit head:* Dr. Kent E. Schubegerle, Chair, 907-474-7671, Fax: 907-474-6716, E-mail: fybio@uaf.edu.

The University of British Columbia, Faculty of Graduate Studies, Faculty of Science, Department of Zoology, Vancouver, BC V6T 1Z1, Canada. Offers M Sc, PhD. *Faculty:* 49 full-time (16 women). *Students:* 112 full-time (64 women). Average age 27. 100 applicants, 25% accepted. In 2005, 14 master's, 8 doctorates awarded. *Degree requirements:* For master's, thesis, final defense; for doctorate, thesis/dissertation, comprehensive exam. final defense, comprehensive exam, registration. *Entrance requirements:* For master's and doctorate, faculty support. Additional exam requirements/recommendations for international students: Required—TOEFL. *Application deadline:* For fall admission, 3/1 for domestic students. For winter admission, 8/1 for domestic students; for spring admission, 1/1 for domestic students. Applications are processed on a rolling basis. Application fee: $90 Canadian dollars ($150 Canadian dollars for international students). Electronic applications accepted. *Financial support:* In 2005–06, fellowships (averaging $16,000 Canadian dollars per year), teaching assistantships (averaging $8,869 Canadian dollars per year) were awarded; research assistantships, institutionally sponsored loans, scholarships/grants, tuition waivers (full and partial), unspecified assistantships, and government sponsored awards also available. *Faculty research:* Cell and developmental biology; community, environmental, and population biology; comparative physiology and biochemistry; fisheries; ecology and evolutionary biology. *Unit head:* Dr. W. K. Milsom, Head, 604-822-3168, Fax: 604-822-2416, E-mail: head@zoology.ubc.ca. *Application contact:* Graduate Secretary, 604-822-5807, Fax: 604-822-2416, E-mail: gradsec@zoology.ubc.ca.

University of California, Davis, Graduate Studies, Graduate Group in Avian Sciences, Davis, CA 95616. Offers MS. *Students:* 17 full-time (9 women); includes 2 minority (1 Asian American or Pacific Islander, 1 Hispanic American), 1 international. Average age 28. 17 applicants, 53% accepted, 6 enrolled. In 2005, 4 degrees awarded. *Degree requirements:* For master's, thesis (for some programs), comprehensive exam (for some programs). *Entrance requirements:* For master's, GRE General Test, minimum GPA of 3.0. Additional exam requirements/recommendations for international students: Required—TOEFL (minimum score 550 paper-based; 213 computer-based). *Application deadline:* For fall admission, 1/15 for domestic students, 1/15 for international students. Applications are processed on a rolling basis. Application fee: $60. Electronic applications accepted. *Financial support:* In 2005–06, 14 students received support, including 6 fellowships with full and partial tuition reimbursements available (averaging $5,523 per year), 4 research assistantships with full and partial tuition reimbursements available (averaging $14,167 per year), 4 teaching assistantships with partial tuition reimbursements available (averaging $15,082 per year); career-related internships or fieldwork, Federal Work-Study, institutionally sponsored loans, scholarships/grants, tuition waivers (full and partial), and unspecified assistantships also available. Financial award application deadline: 1/15; financial award applicants required to submit FAFSA. *Faculty research:* Reproduction, nutrition, toxicology, food products, ecology of avian species. *Unit head:* Mary Delany, Graduate Chair, 530-754-9343, E-mail: medelany@ucdavis.edu. *Application contact:* Alisha L. Nork, Administrative Assistant, 530-752-2382, E-mail: alnork@ucdavis.edu.

University of Chicago, Division of the Biological Sciences, Department of Darwinian Sciences, Ecological, Integrative and Evolutionary Biology, Department of Organismal Biology and Anatomy, Chicago, IL 60637-1513. Offers functional and evolutionary biology (PhD); organismal biology and anatomy (PhD). *Faculty:* 12 full-time (1 woman), 5 part-time/adjunct (3 women). *Students:* 12 full-time (3 women), 3 international. Average age 28. In 2005, 2 doctorates awarded. *Degree requirements:* For doctorate, thesis/dissertation, registration. *Entrance requirements:* For doctorate, GRE General Test. Additional exam requirements/recommendations for international students: Required—TOEFL. *Application deadline:* For fall admission, 12/28 priority date for domestic students, 12/28 priority date for international students. Application fee: $55. Electronic applications accepted. *Financial support:* In 2005–06, 10 students received support, including fellowships with tuition reimbursements available (averaging $26,301 per year); research assistantships with tuition reimbursements available (averaging $26,301 per year); institutionally sponsored loans, scholarships/grants, traineeships, and health care benefits also available. Financial award applicants required to submit FAFSA. *Faculty research:* Ecological physiology, evolution of fossil reptiles, vertebrate paleontology. *Unit head:* Dr. Neil Shubin, Chairman, 773-702-7472, Fax: 773-702-0037, E-mail: nshubin@uchicago.edu. *Application contact:* Carolyn Johnson, Graduate Administrative Director, 773-702-9474, Fax: 773-702-4699, E-mail: csjohnso@uchicago.edu.

University of Connecticut, Graduate School, College of Liberal Arts and Sciences, Department of Ecology and Evolutionary Biology, Field of Zoology, Storrs, CT 06269. Offers MS,

Zoology

PhD. *Faculty:* 20 full-time (4 women). *Students:* 6 full-time (4 women), 2 part-time (1 woman), 1 international. Average age 31. 12 applicants, 17% accepted, 1 enrolled.Terminal master's awarded for partial completion of doctoral program. *Degree requirements:* For master's, comprehensive exam; for doctorate, thesis/dissertation. *Entrance requirements:* For master's and doctorate, GRE General Test, GRE Subject Test. Additional exam requirements/recommendations for international students: Required—TOEFL (minimum score 550 paper-based; 213 computer-based). *Application deadline:* For fall admission, 2/1 priority date for domestic students, 2/1 priority date for international students; for spring admission, 11/1 for domestic students, 10/1 for international students. Applications are processed on a rolling basis. Application fee: $55. Electronic applications accepted. *Expenses:* Tuition, state resident: part-time $444 per credit hour. Tuition, nonresident: part-time $1,154 per credit hour. Tuition and fees vary according to course load. *Financial support:* In 2005–06, 3 research assistantships, 3 teaching assistantships were awarded; fellowships, Federal Work-Study, scholarships/grants, health care benefits, and unspecified assistantships also available. Financial award application deadline: 2/1; financial award applicants required to submit FAFSA. *Application contact:* Anne St. Onje, Graduate Coordinator, 860-486-4314, Fax: 860-486-3943, E-mail: ann.st_onje@uconn.edu.

University of Florida, Graduate School, College of Liberal Arts and Sciences, Department of Zoology, Gainesville, FL 32611. Offers MS, MST, PhD. *Faculty:* 25 full-time (6 women), 1 (woman) part-time/adjunct. *Students:* 65 (32 women); includes 3 minority (all Hispanic Americans) 10 international. 111 applicants, 16% accepted. In 2005, 2 master's, 9 doctorates awarded. *Degree requirements:* For master's, thesis or alternative; for doctorate, one foreign language, thesis/dissertation. *Entrance requirements:* For master's and doctorate, GRE General Test, minimum GPA of 3.0. *Application deadline:* For fall admission, 1/14 for domestic students. Applications are processed on a rolling basis. Application fee: $20. Electronic applications accepted. *Expenses:* Tuition, state resident: full-time $6,234. Tuition, nonresident: full-time $21,359. Tuition and fees vary according to program. *Financial support:* In 2005–06, 48 students received support, including 2 research assistantships (averaging $16,080 per year), 25 teaching assistantships (averaging $16,169 per year); fellowships, unspecified assistantships also available. *Faculty research:* Behavior, ecology, evolutionary biology, comparative physiology. *Unit head:* Dr. David H. Evans, Chair, 352-392-1107, Fax: 352-392-3704, E-mail: devans@zoo.ufl.edu. *Application contact:* Dr. Craig W. Osenberg, Coordinator, 352-392-9201, Fax: 352-392-3704, E-mail: osenberg@zoo.ufl.edu.

University of Guelph, Graduate Program Services, College of Biological Science, Department of Integrative Biology, Guelph, ON N1G 2W1, Canada. Offers botany (M Sc, PhD); zoology (M Sc, PhD). Part-time programs available. *Faculty:* 37 full-time (5 women). *Students:* 100 full-time (53 women), 3 part-time (all women); includes 5 minority (all Asian Americans or Pacific Islanders), 6 international. Average age 23. 23 applicants, 83% accepted, 19 enrolled. In 2005, 26 master's, 7 doctorates awarded. Median time to degree: Of those who began their doctoral program in fall 1997, 95% received their degree in 8 years or less. *Degree requirements:* For master's, thesis, research proposal; for doctorate, thesis/dissertation, research proposal, qualifying exam. *Entrance requirements:* For master's, minimum B average during previous 2 years of course work; for doctorate, minimum A- average. Additional exam requirements/recommendations for international students: Required—TOEFL (minimum score 550 paper-based; 213 computer-based), IELT (minimum score 7). *Application deadline:* For fall admission, 7/1 for domestic students. For winter admission, 11/1 for domestic students; for spring admission, 3/1 for domestic students. Applications are processed on a rolling basis. Application fee: $75. Electronic applications accepted. *Financial support:* In 2005–06, 50 students received support, including 48 research assistantships, 78 teaching assistantships (averaging $4,606 per year); fellowships *Faculty research:* Aquatic science, environmental physiology, parasitology, wildlife biology, management. Total annual research expenditures: $3.8 million. *Unit head:* Dr. Moira Ferguson, Chair, 519-824-4120 Ext. 53598, Fax: 519-767-1656, E-mail: mmfergus@uoguelph.ca. *Application contact:* Laurie Winn, Graduate Admissions Secretary, 519-824-4320 Ext. 52730, Fax: 519-767-1656, E-mail: lwinn@uoguelph.ca.

University of Hawaii at Manoa, Graduate Division, Colleges of Arts and Sciences, College of Natural Sciences, Department of Zoology, Honolulu, HI 96822. Offers MS, PhD. *Faculty:* 38 full-time (7 women), 4 part-time/adjunct (0 women). *Students:* 85 full-time (46 women), 12 part-time (2 women); includes 14 minority (11 Asian Americans or Pacific Islanders, 3 Hispanic Americans), 14 international. Average age 30. 152 applicants, 22% accepted, 18 enrolled. In 2005, 9 master's, 3 doctorates awarded. *Degree requirements:* For doctorate, one foreign language, thesis/dissertation, seminar. *Entrance requirements:* For master's and doctorate, GRE General Test, GRE Subject Test. *Application deadline:* For fall admission, 12/15 for domestic students, 12/15 for international students. Applications are processed on a rolling basis. Application fee: $50. *Expenses:* Tuition, state resident: full-time $8,400; part-time $200 per credit hour. Tuition, nonresident: full-time $11,088; part-time $462 per credit hour. Tuition and fees vary according to program. *Financial support:* In 2005–06, 62 research assistantships (averaging $17,889 per year), 14 teaching assistantships (averaging $14,800 per year) were awarded; tuition waivers (full) also available. Financial award application deadline: 2/1. *Faculty research:* Molecular evolution, reproductive biology, animal behavior, conservation biology, avian biology. *Unit head:* Sheila Conant, Chairperson, 808-956-8617, Fax: 808-956-9812. *Application contact:* Dr. Charles Birkeland, Graduate Field Chairperson, 808-956-8956, Fax: 808-956-9812, E-mail: charlesb@hawaii.edu.

University of Illinois at Urbana–Champaign, Graduate College, College of Liberal Arts and Sciences, School of Integrative Biology, Department of Animal Biology, Champaign, IL 61820. Offers MS, PhD. *Faculty:* 9 full-time (4 women). *Students:* 28 full-time (13 women), 4 part-time (3 women); includes 5 minority (1 African American, 4 Asian Americans or Pacific Islanders), 6 international. 88 applicants, 25% accepted, 18 enrolled. In 2005, 12 master's, 2 doctorates awarded. *Degree requirements:* For doctorate, thesis/dissertation. *Application deadline:* For fall admission, 1/2 for domestic students. Applications are processed on a rolling basis. Application fee: $50 ($60 for international students). Electronic applications accepted. *Financial support:* In 2005–06, 1 fellowship, 3 research assistantships, 5 teaching assistantships were awarded; tuition waivers (full and partial) also available. Financial award application deadline: 2/15. *Unit head:* Ken Paige, Head, 217-333-7802, Fax: 217-244-4565, E-mail: k-paige@uiuc.edu. *Application contact:* Letita Cundiff, Staff Secretary, 217-333-7802, Fax: 217-244-4565, E-mail: lcundiff@life.uiuc.edu.

University of Maine, Graduate School, College of Natural Sciences, Forestry, and Agriculture, Department of Biological Sciences, Program in Zoology, Orono, ME 04469. Offers MS, PhD. *Faculty:* 15 full-time (4 women), 1 part-time/adjunct. *Students:* 10 full-time (4 women), 4 part-time (1 woman), 1 international. Average age 29. 11 applicants, 0% accepted, 0 enrolled. In 2005, 4 master's awarded. Terminal master's awarded for partial completion of doctoral program. *Degree requirements:* For master's, variable foreign language requirement, thesis; for doctorate, one foreign language, thesis/dissertation. *Entrance requirements:* For master's and doctorate, GRE General Test. Additional exam requirements/recommendations for international students: Required—TOEFL. *Application deadline:* For fall admission, 2/1 for domestic students. Applications are processed on a rolling basis. Application fee: $50. Electronic applications accepted. *Financial support:* Fellowships, research assistantships with tuition reimbursements, teaching assistantships with tuition reimbursements, career-related internships or fieldwork available. Financial award application deadline: 3/1. *Unit head:* Dr. Stelbs Tavanteis, Coordinator, 207-581-2986. *Application contact:* Scott G. Delcourt, Associate Dean of the Graduate School, 207-581-3219, Fax: 207-581-3232, E-mail: graduate@maine.edu.

University of Manitoba, Faculty of Graduate Studies, Faculty of Science, Biological Sciences Division, Department of Zoology, Winnipeg, MB R3T 2N2, Canada. Offers M Sc, PhD. *Degree requirements:* For master's and doctorate, one foreign language, thesis/dissertation.

The University of Montana, Graduate School, College of Arts and Sciences, Division of Biological Sciences, Program in Organismal Biology and Ecology, Missoula, MT 59812-0002. Offers MS, PhD. Terminal master's awarded for partial completion of doctoral program. *Degree requirements:* For master's, one foreign language, thesis; for doctorate, 2 foreign languages, thesis/dissertation. *Entrance requirements:* For master's and doctorate, GRE General Test. *Application deadline:* For fall admission, 2/1 for domestic students. Application fee: $45. *Expenses:* Tuition, state resident: part-time $267 per credit. Tuition, nonresident: part-time $665 per credit. Part-time tuition and fees vary according to course load and degree level. *Financial support:* In 2005–06, research assistantships with full tuition reimbursements (averaging $9,400 per year), teaching assistantships with full tuition reimbursements (averaging $9,400 per year) were awarded; Federal Work-Study also available. Financial award application deadline: 3/1; financial award applicants required to submit FAFSA. *Faculty research:* Conservation biology, ecology and behavior, evolutionary genetics, avian biology. *Application contact:* Janean Clark, Graduate Programs Secretary, 406-243-5222, Fax: 406-243-4184, E-mail: jmclark@selway.umt.edu.

University of New Hampshire, Graduate School, College of Life Sciences and Agriculture, Department of Zoology, Durham, NH 03824. Offers MS, PhD. Part-time programs available. *Faculty:* 25 full-time. *Students:* 18 full-time (9 women), 24 part-time (12 women); includes 2 minority (1 African American, 1 Hispanic American), 4 international. Average age 28. 23 applicants, 52% accepted, 7 enrolled. In 2005, 5 master's awarded. Terminal master's awarded for partial completion of doctoral program. *Degree requirements:* For master's, thesis; for doctorate, one foreign language, thesis/dissertation. *Entrance requirements:* For master's and doctorate, GRE General Test, GRE Subject Test. Additional exam requirements/recommendations for international students: Required—TOEFL (minimum score 550 paper-based; 213 computer-based); Recommended—TSE. *Application deadline:* For fall admission, 4/1 priority date for domestic students, 4/1 priority date for international students. For winter admission, 12/1 for domestic students. Applications are processed on a rolling basis. Application fee: $60. Electronic applications accepted. *Expenses:* Tuition, state resident: full-time $8,010; part-time $445 per credit hour. Tuition, nonresident: full-time $19,730; part-time $810 per credit hour. Required fees: $322 per semester. Tuition and fees vary according to course load and program. *Financial support:* In 2005–06, 6 fellowships, 13 research assistantships, 19 teaching assistantships were awarded; career-related internships or fieldwork, Federal Work-Study, scholarships/grants, and tuition waivers (full and partial) also available. Support available to part-time students. Financial award application deadline: 2/15. *Faculty research:* Behavior development, ecology, endocrinology, fisheries, invertebrates. *Unit head:* Dr. Jim Haney, Chairperson, 603-862-2105. *Application contact:* Diane Lavalliere, Administrative Assistant, 603-862-2100, E-mail: zoology.dept@unh.edu.

University of North Dakota, Graduate School, College of Arts and Sciences, Department of Biology, Grand Forks, ND 58202. Offers botany (MS, PhD); ecology (MS, PhD); entomology (MS, PhD); environmental biology (MS, PhD); fisheries/wildlife (MS, PhD); genetics (MS, PhD); zoology (MS, PhD). *Faculty:* 16 full-time (3 women). *Students:* 15 applicants, 13% accepted, 2 enrolled. In 2005, 5 degrees awarded. Terminal master's awarded for partial completion of doctoral program. *Degree requirements:* For master's, thesis/dissertation, final exam; for doctorate, thesis/dissertation, final exam, comprehensive exam. *Entrance requirements:* For master's, GRE General Test, GRE Subject Test, minimum GPA of 3.0; for doctorate, GRE General Test, GRE Subject Test, minimum GPA of 3.5. Additional exam requirements/recommendations for international students: Required—TOEFL (minimum score 550 paper-based; 213 computer-based). *Application deadline:* For fall admission, 10/1 for domestic students, 10/1 for international students. Application fee: $35. Electronic applications accepted. *Financial support:* In 2005–06, 8 research assistantships with full tuition reimbursements (averaging $11,375 per year), 13 teaching assistantships with full tuition reimbursements (averaging $10,813 per year) were awarded; fellowships, Federal Work-Study, institutionally sponsored loans, scholarships/grants, and tuition waivers (full and partial) also available. Support available to part-time students. Financial award application deadline: 3/15; financial award applicants required to submit FAFSA. *Faculty research:* Population biology, wildlife ecology, RNA processing, hormonal control of behavior. *Unit head:* Dr. Richard Sweitzel, Graduate Director, 701-777-4676, Fax: 701-777-2623, E-mail: richard_sweitzel@und.nodak.edu.

University of Oklahoma, Graduate College, College of Arts and Sciences, Department of Zoology, Norman, OK 73019-0390. Offers M Nat Sci, MS, PhD. *Faculty:* 45 full-time (13 women), 2 part-time/adjunct (1 woman). *Students:* 42 full-time (19 women), 7 part-time (3 women), 12 international. 17 applicants, 65% accepted, 10 enrolled. In 2005, 10 master's, 7 doctorates awarded. *Degree requirements:* For master's, thesis defense; for doctorate, dissertation defense, thesis defense. *Entrance requirements:* For master's and doctorate, GRE General Test, GRE Subject Test, 3 letters of recommendation. Additional exam requirements/recommendations for international students: Required—TOEFL, TSE. *Application deadline:* For fall admission, 1/15 for domestic students; for spring admission, 11/15 for domestic students. Applications are processed on a rolling basis. Application fee: $40 ($90 for international students). Electronic applications accepted. *Expenses:* Tuition, state resident: full-time $3,029; part-time $126 per credit hour. Tuition, nonresident: full-time $10,807; part-time $450 per credit hour. Required fees: $1,231; $44 per credit hour. Tuition and fees vary according to course load and program. *Financial support:* In 2005–06, 4 fellowships with tuition reimbursements (averaging $3,500 per year), 20 research assistantships with tuition reimbursements (averaging $13,059 per year), 30 teaching assistantships (averaging $14,130 per year) were awarded; career-related internships or fieldwork, Federal Work-Study, institutionally sponsored loans, scholarships/grants, tuition waivers (partial), and unspecified assistantships also available. Support available to part-time students. Financial award applicants required to submit FAFSA. *Faculty research:* Neuroscience, comparative physiology, evolution, ecology and behavior, cell biology and development. Total annual research expenditures: $1.6 million. *Unit head:* Dr. William J. Matthews, Academic Chair, 405-325-4821, Fax: 405-325-7560, E-mail: wmatthews@ou.edu. *Application contact:* Dr. Rosemary Knapp, Assistant Professor, Director of Graduate Studies, 405-325-4389, Fax: 405-325-6202, E-mail: rknapp@ou.edu.

Announcement: The Department of Zoology offers MS, PhD. Research interests of 34 faculty members in areas such as cell biology, genetics, physiology, neuroscience, animal behavior, evolution, systematics, and ecology. Affiliated resources include a field station, biological survey, aquatic research facility, and natural history museum. Telephone: 405-325-4821. Visit the Web site at http://www.ou.edu/cas/zoology/.

University of Puerto Rico, Medical Sciences Campus, School of Medicine, Division of Graduate Studies, Department of Microbiology and Medical Zoology, San Juan, PR 00936-5067. Offers medical zoology (MS, PhD); microbiology and medical zoology (MS, PhD). *Faculty:* 15 full-time (8 women), 2 part-time/adjunct (both women). *Students:* 32 full-time (26 women); includes 28 minority (all Hispanic Americans) Average age 23. 20 applicants, 25% accepted, 5 enrolled. In 2005, 2 master's, 1 doctorate awarded. Median time to degree: Of those who began their doctoral program in fall 1997, 50% received their degree in 8 years or less. *Degree requirements:* For master's, one foreign language, thesis/dissertation, registration; for doctorate, one foreign language, thesis/dissertation, comprehensive exam, registration. *Entrance requirements:* For master's and doctorate, GRE General Test, GRE Subject Test, interview, minimum GPA of 3.0, 3 letters of recommendation. *Application deadline:* For fall admission, 2/15 for domestic students, 2/15 for international students; for spring admission, 9/15 for domestic students, 9/15 for international students. Application fee: $15. *Expenses:* Tuition, state resident: full-time $3,600; part-time $100 per credit hour. Tuition, nonresident: full-time $4,655. Required fees: $1,734. Tuition and fees vary according to class time, degree level and program. *Financial support:* In 2005–06, 9 fellowships with full tuition reimbursements, 17 research assistantships with full tuition reimbursements, 17 teaching assistantships with full tuition reimbursements were awarded; institutionally sponsored loans and tuition waivers (full) also available. Financial award application deadline: 4/30. *Faculty research:* Molecular and general parasitology, immunology, development of viral vaccines and antiviral agents, fungal dimorphism, AIDS, pathogenesis. Total annual research expenditures: $5.1 million. *Unit head:* Dr. Guillermo J. Vázquez, Director, 787-758-2525 Ext. 1309, Fax: 787-758-

Zoology

University of Puerto Rico, Medical Sciences Campus (continued)
4808, E-mail: gvazquez@rcm.upr.edu. *Application contact:* Dr. Iraida E. Robledo, Graduate Program Coordinator, 787-758-2525 Ext. 1311, Fax: 787-758-4808, E-mail: irobledo@rcm.upr.edu.

University of South Florida, College of Graduate Studies, College of Arts and Sciences, Department of Biology, Tampa, FL 33620-9951. Offers biology (PhD); botany (MS); ecology (PhD); microbiology (MS); physiology (PhD); zoology (MS). Part-time programs available. *Faculty:* 21. *Students:* 54 full-time (32 women), 26 part-time (17 women); includes 8 minority (2 African Americans, 1 American Indian/Alaska Native, 1 Asian American or Pacific Islander, 4 Hispanic Americans), 14 international. 76 applicants, 34% accepted, 13 enrolled. In 2005, 3 master's, 3 doctorates awarded. *Degree requirements:* For master's, thesis (for some programs), graduate seminar in biology; for doctorate, 2 foreign languages, thesis/dissertation, essay of research interest, comprehensive exam. *Entrance requirements:* For master's, GRE General Test, minimum undergraduate GPA of 3.0 in last 60 hours of course work; for doctorate, GRE General Test, GRE Subject Test in biology, minimum undergraduate GPA of 3.0 in last 60 hours of course work. Additional exam requirements/recommendations for international students: Required—TOEFL (minimum score 570 paper-based), TSE (minimum score 50). *Application deadline:* For fall admission, 2/1 priority date for domestic students, 3/1 priority date for international students; for spring admission, 10/1 for domestic students, 8/1 for international students. Application fee: $30. Electronic applications accepted. *Financial support:* Fellowships with full tuition reimbursements, research assistantships with full tuition reimbursements, teaching assistantships with full tuition reimbursements, Federal Work-Study and unspecified assistantships available. Financial award application deadline: 6/30. *Unit head:* Sydney Pierce, Chairperson, 813-974-3250, Fax: 813-974-3263. *Application contact:* Christine Smith, Graduate Advisor, 813-974-4747, Fax: 813-974-3263, E-mail: csmith2@chuma1.cas.usf.edu.

University of Toronto, School of Graduate Studies, Life Sciences Division, Department of Zoology, Toronto, ON M5S 1A1, Canada. Offers M Sc, PhD. *Degree requirements:* For master's and doctorate, thesis/dissertation, thesis defense. *Entrance requirements:* For master's, minimum B average in last 2 years; knowledge of physics, chemistry, and biology.

University of Washington, Graduate School, College of Arts and Sciences, Department of Zoology, Seattle, WA 98195-1800. Offers PhD. *Degree requirements:* For doctorate, thesis/dissertation. *Entrance requirements:* For doctorate, GRE General Test, minimum GPA of 3.0. Additional exam requirements/recommendations for international students: Required—TOEFL. Electronic applications accepted. *Faculty research:* Ecology and evolution, developmental and cellular biology, physiology and ethology, mathematical biology and biomechanics.

The University of Western Ontario, Faculty of Graduate Studies, Biosciences Division, Department of Zoology, London, ON N6A 5B8, Canada. Offers M Sc, PhD. *Degree requirements:* For master's and doctorate, thesis/dissertation.

University of Wisconsin–Madison, Graduate School, College of Letters and Science, Department of Zoology, Madison, WI 53706-1380. Offers MA, MS, PhD. Part-time programs available. *Degree requirements:* For master's, thesis; for doctorate, one foreign language, thesis/dissertation. *Entrance requirements:* For master's and doctorate, GRE General Test. Additional exam requirements/recommendations for international students: Required—TOEFL. Electronic applications accepted. *Faculty research:* Developmental biology, ecology, neurobiology, aquatic ecology, animal behavior.

University of Wisconsin–Oshkosh, The School of Graduate Studies, College of Letters and Science, Department of Biology and Microbiology, Oshkosh, WI 54901. Offers biology (MS), including botany, microbiology, zoology. *Degree requirements:* For master's, thesis, comprehensive exam, registration. *Entrance requirements:* For master's, GRE General Test, minimum GPA of 3.0, BS in biology. Additional exam requirements/recommendations for international students: Required—TOEFL (minimum score 550 paper-based; 213 computer-based). Electronic applications accepted.

University of Wyoming, Graduate School, College of Arts and Sciences, Department of Zoology and Physiology, Laramie, WY 82070. Offers MS, PhD. Part-time programs available. *Faculty:* 23 full-time (4 women), 1 part-time/adjunct (0 women). *Students:* 44 full-time (16 women), 24 part-time (6 women), 6 international. 42 applicants, 29% accepted. In 2005, 13 master's awarded. *Degree requirements:* For master's, thesis/dissertation, registration; for doctorate, thesis/dissertation, comprehensive exam, registration. *Entrance requirements:* For master's, GRE General Test score of 900 minimum, minimum GPA of 3.0; for doctorate, GRE General Test (minimum 1000), minimum GPA of 3.0. Additional exam requirements/recommendations for international students: Required—TOEFL. *Application deadline:* For fall admission, 2/15 priority date for domestic students, 2/15 priority date for international students; for spring admission, 9/15 for domestic students, 9/15 for international students. Applications are processed on a rolling basis. Application fee: $50. Electronic applications accepted. *Expenses:* Tuition, state resident: full-time $3,720; part-time $155 per credit hour. Tuition,
nonresident: full-time $10,704; part-time $446 per credit hour. Required fees: $666; $162 per semester. Tuition and fees vary according to course load and program. *Financial support:* In 2005–06, 1 fellowship with full tuition reimbursement (averaging $14,004 per year), 42 research assistantships with full tuition reimbursements (averaging $10,062 per year), 26 teaching assistantships with full tuition reimbursements (averaging $10,062 per year) were awarded; unspecified assistantships also available. Financial award application deadline: 3/1. *Faculty research:* Cell biology, ecology/wildlife, organismal physiology. Total annual research expenditures: $3.5 million. *Unit head:* Dr. Graham Mitchell, Head, 307-766-4207, Fax: 307-766-5625, E-mail: mitchg@uwyo.edu. *Application contact:* E-mail: dmanore@uwyo.edu.

Virginia Polytechnic Institute and State University, Graduate School, College of Science, Department of biological Sciences, Blacksburg, VA 24061. Offers botany (MS, PhD); ecology and evolutionary biology (MS, PhD); genetics and developmental biology (MS, PhD); microbiology (MS, PhD); zoology (MS, PhD). *Faculty:* 38 full-time (9 women). *Students:* 70 full-time (30 women), 4 part-time (1 woman); includes 6 minority (2 African Americans, 1 American Indian/Alaska Native, 2 Asian Americans or Pacific Islanders, 1 Hispanic American), 12 international. Average age 27. 81 applicants, 22% accepted, 14 enrolled. In 2005, 9 master's, 8 doctorates awarded. *Entrance requirements:* For master's and doctorate, GRE General Test. Additional exam requirements/recommendations for international students: Required—TOEFL (minimum score 550 paper-based; 213 computer-based). *Application deadline:* Applications are processed on a rolling basis. Application fee: $45. Electronic applications accepted. *Expenses:* Tuition, state resident: full-time $6,558; part-time $364 per credit. Tuition, nonresident: full-time $11,296; part-time $628 per credit. Required fees: $1,419; $468 per credit. $234 per term. *Financial support:* In 2005–06, 28 research assistantships with full tuition reimbursements (averaging $16,689 per year), 37 teaching assistantships with full tuition reimbursements (averaging $13,902 per year) were awarded; career-related internships or fieldwork, Federal Work-Study, scholarships/grants, and unspecified assistantships also available. *Faculty research:* Freshwater ecology, cell cycle regulation, behavioral ecology, motor proteins. *Unit head:* Dr. Bob Jones, Chairman, 540-231-9514, Fax: 540-231-9307, E-mail: rhjones@vt.edu. *Application contact:* Sue Rasmussen, Graduate Secretary, 540-231-8929, Fax: 540-231-9307, E-mail: sueras@vt.edu.

Washington State University, Graduate School, College of Sciences, School of Biological Sciences, Department of Zoology, Pullman, WA 99164. Offers MS, PhD. *Faculty:* 31. *Students:* 25 full-time (14 women), 1 (woman) part-time; includes 1 minority (Hispanic American), 1 international. Average age 30. 29 applicants, 17% accepted, 4 enrolled. In 2005, 4 master's, 2 doctorates awarded. *Degree requirements:* For master's, thesis or alternative, oral exam; for doctorate, thesis/dissertation, oral exam, written exam. *Entrance requirements:* For master's and doctorate, GRE General Test, GRE Subject Test, minimum GPA of 3.0, 3 letters of recommendation. *Application deadline:* For fall admission, 1/15 priority date for domestic students, 3/1 priority date for international students; for spring admission, 9/15 for domestic students, 7/1 for international students. Applications are processed on a rolling basis. Application fee: $35. *Expenses:* Tuition, state resident: full-time $6,295; part-time $336 per credit. Tuition, nonresident: full-time $15,949; part-time $819 per credit. Required fees: $933. Part-time tuition and fees vary according to campus/location and program. *Financial support:* In 2005–06, 4 fellowships (averaging $4,500 per year), 3 research assistantships with full and partial tuition reimbursements (averaging $13,871 per year), 21 teaching assistantships with full and partial tuition reimbursements (averaging $14,072 per year) were awarded; Federal Work-Study, institutionally sponsored loans, and tuition waivers (partial) also available. Financial award application deadline: 4/1; financial award applicants required to submit FAFSA. *Unit head:* Dr. R. Alan Black, Associate Director, 509-335-0179, E-mail: blackra@wsu.edu. *Application contact:* E-mail: sbsgrad@wsu.edu.

Western Illinois University, School of Graduate Studies, College of Arts and Sciences, Department of Biological Sciences, Macomb, IL 61455-1390. Offers biological sciences (MS); zoo and aquarium studies (Certificate). Part-time programs available. *Students:* 43 full-time (27 women), 20 part-time (16 women); includes 6 minority (2 African Americans, 1 American Indian/Alaska Native, 1 Asian American or Pacific Islander, 2 Hispanic Americans), 3 international. Average age 26. 38 applicants, 79% accepted. In 2005, 14 master's, 6 other advanced degrees awarded. *Degree requirements:* For master's, thesis or alternative. *Entrance requirements:* Additional exam requirements/recommendations for international students: Required—TOEFL (minimum score 550 paper-based; 213 computer-based). *Application deadline:* Applications are processed on a rolling basis. Application fee: $30. Electronic applications accepted. *Expenses:* Tuition, state resident: full-time $3,599; part-time $200 per semester hour. Tuition, nonresident: full-time $7,198; part-time $400 per semester hour. Required fees: $890; $49 per semester hour. Tuition and fees vary according to campus/location. *Financial support:* In 2005–06, 21 students received support, including 7 research assistantships with full tuition reimbursements available (averaging $6,288 per year), 14 teaching assistantships (averaging $6,288 per year) Financial award applicants required to submit FAFSA. *Unit head:* Dr. Richard V. Anderson, Chairperson, 309-298-2408. *Application contact:* Dr. Barbara Baily, Director of Graduate Studies/Associate Provost, 309-298-1806, Fax: 309-298-2345, E-mail: grad-office@wiu.edu.

Cross-Discipline Announcements

Princeton University, Graduate School, Department of Ecology and Evolutionary Biology, Princeton, NJ 08544-1019.

Graduate study in the department leads to the PhD degree. Special areas of focus include evolutionary ecology; ecological and evolutionary genetics; behavioral ecology; theoretical ecology; population and community ecology; physiology; epidemiology of infectious diseases; global interactions among the biosphere, atmosphere, oceans, and climate; and conservation biology. The interests and research of the faculty range widely over these areas, so that incoming students are able to select their advisers from among several professors working in the chosen discipline. Graduate students also have excellent opportunities for combining several areas for innovative interdisciplinary work. The program is designed to develop both the breadth and depth of understanding that will enable students to respond to future advances in the field.

University of Pittsburgh, School of Arts and Sciences, Department of Biological Sciences, Program in Ecology and Evolution, Pittsburgh, PA 15260.

Department offers master's and doctoral programs in ecology and evolution. Research areas: ecological and population genetics; tropical and community ecology; evolutionary ecology; phylogeny of birds, amphibians, and reptiles; metapopulation structure; evolution of dispersal and dormancy; and plant-animal interaction, herbivory, and food webs.

SOUTHERN ILLINOIS UNIVERSITY CARBONDALE

Department of Zoology
Ph.D. Program

Program of Study

The Department of Zoology's Ph.D. program has been a major contributor to Southern Illinois University Carbondale's (SIUC's) classification by the Carnegie Foundation as a Research University with high research activity (RU/H) and is one of SIUC's top programs. Zoology faculty members have established research programs in the broad areas of environmental biology (biodiversity studies, conservation biology, wildlife ecology, fisheries biology, mined land reclamation, terrestrial and aquatic ecology, toxicology, and large-river ecology) and evolutionary genetics (population genetics, molecular systematics, molecular evolution, and conservation genetics). These programs range in scope from assessing the stability of local ecosystems and their impact on the regional economy to international collaborative efforts on basic and applied research problems. Many faculty members have ongoing collaboration with international scientists or have performed research abroad, with most of such efforts focusing on the Caribbean, South America, Australia, Africa, and Europe. There is a close affiliation between the Department of Zoology, the Fisheries and Illinois Aquaculture Center, the Cooperative Wildlife Research Laboratory, and the Centers for Ecology and Systematic Biology. Zoology faculty members maintain a strong record of external funding for research: 91 percent were grant active in 2005–06 (i.e., had submitted a proposal for, received, or were currently supported by an external grant), and 83 percent had some form of external funding during that period. The zoology Ph.D. program emphasizes original research by students under the mentorship of graduate faculty members and advanced course work in the area of specialization.

Research Facilities

The Department has well-equipped laboratories for environmental and molecular research in the Life Science II and III Buildings, the Fisheries Wetlab, and the Wildlife Annex. Zoology and related units also have extensive infrastructure for field research in terrestrial and aquatic habitats. A variety of specimen collections are also available, including mammals (1,500 specimens), birds (3,000 specimens), fish (250,000 specimens), reptiles and amphibians (10,000 specimens), parasites (1,000 specimens), insects (200,000 specimens), arachnids (125,000 specimens), and other invertebrates (200,000 specimens). Related botanical collections include vascular plants (150,000 specimens) and bryophytes (30,000 specimens). The southern Illinois region is extremely diverse and offers a natural outdoor laboratory. Many species of plants and animals are endemic to the region, which also represents the intersection of distinct biotic provinces to the north, south, east, and west. Natural areas such as the Shawnee National Forest (more than 800,000 acres), the Mississippi and Ohio Rivers, the Crab Orchard National Wildlife Refuge, numerous state parks, nature reserves, and smaller lakes and rivers provide diverse research opportunities within a short distance of campus. The Department and the College of Science maintain up-to-date computer facilities for graduate student use. SIUC's Morris Library is rated among the top 100 research libraries in the U.S. by the *Chronicle of Higher Education.*

Financial Aid

Stipends for graduate students are available in the form of fellowships, teaching assistantships, and research assistantships. Fellowship recipients are nominated by the Department and selected on the basis of academic merit in a University-wide competition each year. Teaching assistantships are assigned by the Department; research assistants are chosen by individual faculty members or programs with funding available. All assistantship assignments of quarter-time or greater carry a tuition waiver. Most active students in the Department are supported by one of these stipends.

Cost of Study

In-state graduate tuition is $243 per credit hour in 2006–07. Out-of-state tuition is 2.5 times the in-state tuition rate ($607.50 per credit hour). Graduate students with at least a 25 percent appointment as a graduate assistant receive a tuition waiver. Fees vary from $441.62 (1 credit hour) to $987.30 (12 credit hours).

Living and Housing Costs

For married couples, students with families, and single graduate students, the University has 589 efficiency and one-, two-, and three-bedroom apartments that rent for $438 to $505 per month in 2006–07. Residence halls for single graduate students are also available, as are accessible residence hall rooms and apartments for students with disabilities.

Location

SIUC is 350 miles south of Chicago and 100 miles southeast of St. Louis. Nestled in rolling hills bordered by the Ohio and Mississippi Rivers and enhanced by a mild climate, the area has state parks, national forests and wildlife refuges, and large lakes for outdoor recreation. Cultural offerings include theater, opera, concerts, art exhibits, and cinema. Educational facilities for the families of students are excellent.

The University and The Department

Southern Illinois University Carbondale is a comprehensive public university with a variety of general and professional education programs. The University offers associate, bachelor's, master's, and doctoral degrees, as well as the J.D. and M.D. degrees. The University is fully accredited by the North Central Association of Colleges and Schools. The Graduate School has an essential role in the development and coordination of graduate instruction and research programs. The Graduate Council has academic responsibility for determining graduate standards, recommending new graduate programs and research centers, and establishing policies to facilitate the research effort. SIUC is classified by the Carnegie Foundation as a Research University with high research activity. The graduate education experience in the Department is enriched by activities associated with various student organizations, such as the Zoology Club, Zoology Graduate Student Association (ZGSA), Student Chapter of the American Fisheries Society, and Student Chapter of the Wildlife Society. In conjunction with the ZGSA, the Department sponsors an excellent seminar series each semester, featuring presentations and discussions by visiting scientists from across the country.

Applying

Applications may be requested from the Department of Zoology or obtained online. Each application must include the completed Graduate School application form, the completed Departmental application form, three letters of recommendation, original transcripts of all previous academic work, and official scores for the Graduate Record Examinations. Additional materials are required for international applicants. Applications are considered for fall, spring, and summer admissions, though most new graduate students are admitted in the fall. Fellowship competitions take place in the early spring for fall admissions, so students wishing to be nominated for fellowships should have complete applications on file by December 1. Successful applicants for admission to the Ph.D. program ordinarily have an undergraduate grade point average of at least 2.8 (on a 4.0 scale).

Correspondence and Information

Director of Graduate Studies
Department of Zoology, Mailcode 6501
Southern Illinois University Carbondale
Carbondale, Illinois 62901

Phone: 618-453-4110
E-mail: jrains@zoology.siu.edu
Web site: http://www.science.siu.edu/zoology (Department)
    http://www.siu.edu/gradschl (Graduate School)

Southern Illinois University Carbondale

THE FACULTY AND THEIR RESEARCH

Frank E. Anderson, Associate Professor; Ph.D., California, Santa Cruz, 1999. Invertebrates; molecular systematics; molecular evolution.
Terence R. Anthony, Associate Professor Emeritus; M.D., 1968, Ph.D., 1975, Chicago.
Joseph A. Beatty, Associate Professor Emeritus; Ph.D., Harvard, 1969.
Ronald A. Brandon, Professor Emeritus; Ph.D., Illinois, 1962.
Brooks M. Burr, Professor; Ph.D., Illinois, 1977. Ichthyology.
Michael Eichholz, Assistant Professor; Ph.D., Alaska, 1998. Waterfowl/wetland ecology.
DuWayne C. Englert, Professor Emeritus; Ph.D., Purdue, 1964.
George A. Feldhamer, Professor; Ph.D., Oregon State, 1977. Mammalogy; wildlife ecology.
James Garvey, Associate Professor; Ph.D., Ohio State, 1997. Fisheries biology.
Richard S. Halbrook, Associate Professor; Ph.D., Virginia Tech, 1990. Wildlife toxicology.
Roy C. Heidinger, Professor Emeritus; Ph.D., Southern Illinois at Carbondale, 1970.
Edward J. Heist, Associate Professor; Ph.D., William and Mary, 1998. Population genetics; conversation genetics; fishery management.
Eric C. Hellgren, Professor; Ph.D., Virginia Tech, 1988. Wildlife ecology.
Kamal Ibrahim, Assistant Professor; Ph.D., Cambridge, 1989. Population genetics.
Anita Kelly, Assistant Professor; Ph.D., Southern Illinois at Carbondale, 1995. Fish physiology/aquaculture.
David G. King, Associate Professor; Ph.D., California, San Diego, 1975. Invertebrate neurobiology; evolution.
Christopher C. Kohler, Professor; Ph.D., Virginia Tech, 1980. Ecology: management and culture of aquatic organisms.
Carey Krajewski, Professor and Director, Graduate Studies; Ph.D., Wisconsin–Madison, 1988. Vertebrate molecular systematics.
Eugene A. LeFebvre, Associate Professor Emeritus; Ph.D., Minnesota, 1962.
William M. Lewis, Professor Emeritus; Ph.D., Iowa State, 1949.
Karen Lips, Associate Professor; Ph.D., Miami (Florida), 1995. Herpetology; conservation biology; tropical biology.
Michael J. Lydy, Associate Professor; Ph.D., Ohio State, 2001. Aquatic toxicology.
John E. McPherson Jr., Professor; Ph.D., Michigan State, 1968. Entomology; insect ecology.
William L. Muhlach, Associate Professor and Chair; Ph.D., Illinois at Chicago, 1986. Developmental biology.
John Reeve, Associate Professor; Ph.D., California, Santa Barbara, 1985. Quantitative ecology.
Eric Schauber, Assistant Professor; Ph.D., Connecticut, 2000. Wildlife ecology.
Michael W. Sears, Assistant Professor; Ph.D., Pennsylvania, 2001. Physiological ecology.
Benjamin A. Shepherd, Professor Emeritus; Ph.D., Kansas State, 1970.
Donald W. Sparling, Associate Professor; Ph.D., North Dakota, 1979. Wildlife ecology.
John B. Stahl, Associate Professor Emeritus; Ph.D., Indiana, 1958.
Richard H. Thomas, Associate Professor; Ph.D., Arizona, 1985. Molecular evolution.
George H. Waring, Professor Emeritus; Ph.D., Colorado State, 1966.
Matt R. Whiles, Associate Professor; Ph.D., Georgia, 1995. Stream ecology; freshwater invertebrates; entomology.
Gregory Whitledge, Assistant Professor; Ph.D., Missouri, 2001. Fish ecology and management.
Frank M. Wilhelm, Assistant Professor; Ph.D., Alberta, 1999. Limnology; ecology.

APPENDIXES

Institutional Changes
Since the 2006 Edition

Following is an alphabetical listing of institutions that have recently closed, moved, merged with other institutions, or changed their names or status. In the case of a name change, the former name appears first, followed by the new name.

Acupuncture & Integrative Medicine College (Berkeley, CA): name changed to Acupuncture & Integrative Medicine College, Berkeley.

Alliant International University (San Francisco, CA): name changed to Alliant International University–San Francisco Bay.

Alliant University–Fresno (Fresno, CA): name changed to Alliant International University–Fresno.

Alliant University–Los Angeles (Alhambra, CA): name changed to Alliant International University–Los Angeles.

Alliant University–San Diego (San Diego, CA): name changed to Alliant International University–San Diego.

Argosy University/Honolulu (Honolulu, HI): name changed to Argosy University/Hawai'i.

Arizona State University East (Mesa, AZ): name changed to Arizona State University at the Polytechnic Campus.

Bethany College of the Assemblies of God (Scotts Valley, CA): name changed to Bethany University.

Boston Architectural Center (Boston, MA): name changed to Boston Architectural College.

Brown Mackie College, Los Angeles Campus (Santa Monica, CA): name changed to Argosy University/Santa Monica.

California State University, Channel Islands (Camarillo, CA): name changed to California State University Channel Islands.

Center for Humanistic Studies (Farmington Hills, MI): name changed to Michigan School of Professional Psychology.

College of the Humanities and Sciences (Tempe, AZ): name changed to College of the Humanities and Sciences, Harrison Middleton University.

Des Moines University Osteopathic Medical Center (Des Moines, IA): name changed to Des Moines University.

The Graduate College of Union University (Schenectady, NY): name changed to Union Graduate College.

Indiana Institute of Technology (Fort Wayne, IN): name changed to Indiana Tech.

International College and Graduate School (Honolulu, HI): name changed to Hawai'i Theological Seminary.

Long Island University, Southampton College (Southampton, NY): name changed to Long Island University, Southampton Graduate Campus.

Luther Rice Bible College and Seminary (Lithonia, GA): name changed to Luther Rice University.

Medical University of Ohio (Toledo, OH): merged with The University of Toledo.

Midwest Theological Seminary (Wentzville, MO): name changed to Midwest University.

National College of Naturopathic Medicine (Portland, OR): name changed to National College of Natural Medicine.

New School University (New York, NY): name changed to The New School: A University.

North Shore-Long Island Jewish Graduate School of Molecular Medicine (Manhasset, NY): name changed to The Feinstein Institute for Medical Research.

The Pennsylvania State University Harrisburg Campus of the Capital College (Middletown, PA): name changed to The Pennsylvania State University Harrisburg Campus.

Reformed Theological Seminary–Washington D.C./Baltimore Campus (Bethesda, MD): name changed to Reformed Theological Seminary–Washington D.C.

Saint Martin's College (Lacey, WA): name changed to Saint Martin's University.

School of Graduate Studies in Shreveport (Shreveport, LA): name changed to Louisiana State University Health Sciences Center at Shreveport.

Southern California Bible College & Seminary (El Cajon, CA): name changed to Southern California Seminary.

Southern Christian University (Montgomery, AL): name changed to Regions University.

Tai Hsuan Foundation: College of Acupuncture and Herbal Medicine (Honolulu, HI): name changed to World Medicine Institute: College of Acupuncture and Herbal Medicine.

University of Colorado at Denver and Health Sciences Center—Health Sciences Program (Denver, CO): name changed to University of Colorado at Denver and Health Sciences Center.

The University of Lethbridge (Lethbridge, AB, Canada): name changed to University of Lethbridge.

The University of Memphis (Memphis, TN): name changed to University of Memphis.

The University of Montana–Missoula (Missoula, MT): name changed to The University of Montana.

The University of North Carolina at Wilmington (Wilmington, NC): name changed to The University of North Carolina Wilmington.

University of Phoenix–Fort Lauderdale Campus (Fort Lauderdale, FL): name changed to University of Phoenix–South Florida Campus.

University of Phoenix–Jacksonville Campus (Jacksonville, FL): name changed to University of Phoenix–North Florida Campus.

University of Phoenix–Northern California Campus (Pleasanton, CA): name changed to University of Phoenix–Bay Area Campus.

University of Phoenix–Orlando Campus (Maitland, FL): name changed to University of Phoenix–Central Florida Campus.

University of Phoenix–Sacramento Campus (Sacramento, CA): name changed to University of Phoenix–Sacramento Valley Campus.

University of Phoenix–Tampa Campus (Tampa, FL): name changed to University of Phoenix–West Florida Campus.

University of the South (Sewanee, TN): name changed to Sewanee: The University of the South.

Abbreviations Used in the Guides

The following list includes abbreviations of degree names used in the **Profiles** in the 2007 edition of the guides. Because some degrees (e.g., Doctor of Education) can be abbreviated in more than one way (e.g., D.Ed. or Ed.D.), and because the abbreviations used in the guides reflect the preferences of the individual colleges and universities, the list may include two or more abbreviations for a single degree.

Degrees

A Mus D	Doctor of Musical Arts
AC	Advanced Certificate
AD	Artist's Diploma
ADP	Artist's Diploma
Adv C	Advanced Certificate
Adv D	Advanced Diploma
Adv M	Advanced Master
AGSC	Advanced Graduate Specialist Certificate
ALM	Master of Liberal Arts
AM	Master of Arts
AMRS	Master of Arts in Religious Studies
APC	Advanced Professional Certificate
App Sc	Applied Scientist
Au D	Doctor of Audiology
B Th	Bachelor of Theology
C Phil	Certificate in Philosophy
CAES	Certificate of Advanced Educational Specialization
CAGS	Certificate of Advanced Graduate Studies
CAL	Certificate in Applied Linguistics
CALS	Certificate of Advanced Liberal Studies
CAMS	Certificate of Advanced Management Studies
CAPS	Certificate of Advanced Professional Studies
CAS	Certificate of Advanced Studies
CASPA	Certificate of Advanced Study in Public Administration
CASR	Certificate in Advanced Social Research
CATS	Certificate of Achievement in Theological Studies
CBHS	Certificate in Basic Health Sciences
CBS	Graduate Certificate in Biblical Studies
CCJA	Certificate in Criminal Justice Administration
CCMBA	Cross-Continent Master of Business Administration
CCSA	Certificate in Catholic School Administration
CE	Civil Engineer
CEM	Certificate of Environmental Management
CG	Certificate in Gerontology
CGS	Certificate of Graduate Studies
Ch E	Chemical Engineer
CITS	Certificate of Individual Theological Studies
CM	Certificate in Management
CMH	Certificate in Medical Humanities
CMM	Master of Church Ministries
CMS	Certificate in Ministerial Studies
CNM	Certificate in Nonprofit Management
CP	Certificate in Performance
CPASF	Certificate Program for Advanced Study in Finance
CPC	Certificate in Professional Counseling Certificate in Publication and Communication
CPH	Certificate in Public Health
CPM	Certificate in Public Management
CPS	Certificate of Professional Studies
CScD	Doctor of clinical Science
CSD	Certificate in Spiritual Direction
CSE	Computer Systems Engineer
CSS	Certificate of Special Studies
CTS	Certificate of Theological Studies
CURP	Certificate in Urban and Regional Planning
D Arch	Doctor of Architecture
D Ed	Doctor of Education
D Eng	Doctor of Engineering
D Engr	Doctor of Engineering
D Env	Doctor of Environment
D Env M	Doctor of Environmental Management
D Law	Doctor of Law
D Litt	Doctor of Letters
D Med Sc	Doctor of Medical Science
D Min	Doctor of Ministry
D Min PCC	Doctor of Ministry, Pastoral Care, and Counseling
D Miss	Doctor of Missiology
D Mus	Doctor of Music
D Mus A	Doctor of Musical Arts
D Phil	Doctor of Philosophy
D Ps	Doctor of Psychology
D Sc	Doctor of Science
D Sc D	Doctor of Science in Dentistry
D Th	Doctor of Theology
D Th P	Doctor of Practical Theology
DA	Doctor of Arts
DA Ed	Doctor of Arts in Education
DAOM	Doctorate in Acupuncture and Oriental Medicine
DAST	Diploma of Advanced Studies in Teaching
DBA	Doctor of Business Administration
DBS	Doctor of Buddhist Studies
DC	Doctor of Chiropractic
DCC	Doctor of Computer Science
DCD	Doctor of Communications Design
DCL	Doctor of Comparative Law
DCM	Doctor of Church Music
DCN	Doctor of Clinical Nutrition
DCS	Doctor of Computer Science
DDN	Diplôme du Droit Notarial
DDS	Doctor of Dental Surgery
DE	Doctor of Education Doctor of Engineering
DEIT	Doctor of Educational Innovation and Technology
DEM	Doctor of Educational Ministry
DEPD	Diplôme Études Spécialisées
DES	Doctor of Engineering Science
DESS	Diplôme Études Supérieures Spécialisées
DFA	Doctor of Fine Arts
DFES	Doctor of Forestry and Environmental Studies
DGP	Diploma in Graduate and Professional Studies
DH Sc	Doctor of Health Sciences
DHA	Doctor of Health Administration
DHCE	Doctor of Health Care Ethics

DHL	Doctor of Hebrew Letters
	Doctor of Hebrew Literature
DHS	Doctor of Health Service
	Doctor of Human Services
DHSc	Doctor of Health Science
DIBA	Doctor of International Business Administration
Dip CS	Diploma in Christian Studies
DIT	Doctor of Industrial Technology
DJ Ed	Doctor of Jewish Education
DJS	Doctor of Jewish Studies
DM	Doctor of Management
	Doctor of Music
DMA	Doctor of Musical Arts
DMD	Doctor of Dental Medicine
DME	Doctor of Manufacturing Management
	Doctor of Music Education
DMEd	Doctor of Music Education
DMFT	Doctor of Marital and Family Therapy
DMH	Doctor of Medical Humanities
DML	Doctor of Modern Languages
DMM	Doctor of Music Ministry
DN Sc	Doctor of Nursing Science
DNP	Doctor of Nursing Practice
	Doctor of Nursing Practice
DNS	Doctor of Nursing Science
DO	Doctor of Osteopathy
DPA	Doctor of Public Administration
DPC	Doctor of Pastoral Counseling
DPDS	Doctor of Planning and Development Studies
DPE	Doctor of Physical Education
DPH	Doctor of Public Health
DPM	Doctor of Plant Medicine
	Doctor of Podiatric Medicine
DPS	Doctor of Professional Studies
DPT	Doctor of Physical Therapy
Dr DES	Doctor of Design
Dr PH	Doctor of Public Health
Dr Sc PT	Doctor of Science in Physical Therapy
DS	Doctor of Science
DS Sc	Doctor of Social Science
DSJS	Doctor of Science in Jewish Studies
DSL	Doctor of Strategic Leadership
DSM	Doctor of Sacred Music
	Doctor of Sport Management
DSN	Doctor of Science in Nursing
DSW	Doctor of Social Work
DTL	Doctor of Talmudic Law
DV Sc	Doctor of Veterinary Science
DVM	Doctor of Veterinary Medicine
EAA	Engineer in Aeronautics and Astronautics
ECS	Engineer in Computer Science
Ed D	Doctor of Education
Ed DCT	Doctor of Education in College Teaching
Ed M	Master of Education
Ed S	Specialist in Education
Ed Sp	Specialist in Education
Ed Sp PTE	Specialist in Education in Professional Technical Education
EDM	Executive Doctorate in Management
EDSPC	Education Specialist
EE	Electrical Engineer
EJD	Executive Juris Doctor
EM	Mining Engineer
EMBA	Executive Master of Business Administration
EMCIS	Executive Master of Computer Information Systems
EMHA	Executive Master of Health Administration
EMIB	Executive Master of International Business
EMS	Executive Master of Science
EMTM	Executive Master of Technology Management
Eng	Engineer
Eng Sc D	Doctor of Engineering Science
Engr	Engineer
Ex Doc	Executive Doctor of Pharmacy
Exec Ed D	Executive Doctor of Education
Exec MBA	Executive Master of Business Administration
Exec MIM	Executive Master of International Management
Exec MPA	Executive Master of Public Administration
Exec MPH	Executive Master of Public Health
Exec MS	Executive Master of Science
GBC	Graduate Business Certificate
GCE	Graduate Certificate in Education
GDPA	Graduate Diploma in Public Administration
GDRE	Graduate Diploma in Religious Education
GEMBA	Global Executive Master of Business Administration
Geol E	Geological Engineer
GMBA	Global Master of Business Administration
GPD	Graduate Performance Diploma
GSS	Graduate Special Certificate for Students in Special Situations
IMA	Interdisciplinary Master of Arts
IMBA	International Master of Business Administration
ITMA	Master of Instructional Technology
JCD	Doctor of Canon Law
JCL	Licentiate in Canon Law
JD	Juris Doctor
JD/MAP	Juris Doctor/Master of Applied Politics
JSD	Doctor of Juridical Science
	Doctor of Jurisprudence
	Doctor of the Science of Law
JSM	Master of Science of Law
L Th	Licenciate in Theology
LL B	Bachelor of Laws
LL CM	Master of Laws in Comparative Law
LL D	Doctor of Laws
LL M	Master of Laws
LL M T	Master of Laws in Taxation
M Ac	Master of Accountancy
	Master of Accounting
	Master of Acupuncture
M Ac OM	Master of Acupuncture and Oriental Medicine
M Acc	Master of Accountancy
	Master of Accounting
M Acct	Master of Accountancy
	Master of Accounting
M Accy	Master of Accountancy
M Actg	Master of Accounting
M Acy	Master of Accountancy
M Ad	Master of Administration
M Ad Ed	Master of Adult Education
M Adm	Master of Administration
M Adm Mgt	Master of Administrative Management
M Adv	Master of Advertising

M Aero E	Master of Aerospace Engineering
M Ag	Master of Agriculture
M Ag Ed	Master of Agricultural Education
M Agr	Master of Agriculture
M Anesth Ed	Master of Anesthesiology Education
M App Comp Sc	Master of Applied Computer Sciences
M App St	Master of Applied Statistics
M Appl Stat	Master of Applied Statistics
M Aq	Master of Aquaculture
M Ar	Master of Architecture
M Arch	Master of Architecture
M Arch E	Master of Architectural Engineering
M Arch H	Master of Architectural History
M Arch UD	Master of Architecture in Urban Design
M Bio E	Master of Bioengineering
M Biomath	Master of Biomathematics
M Bus Ed	Master of Business Education
M Ch	Master of Chemistry
M Ch E	Master of Chemical Engineering
M Chem	Master of Chemistry
M Cl D	Master of Clinical Dentistry
M Cl Sc	Master of Clinical Science
M Co E	Master of Computer Engineering
M Comp E	Master of Computer Engineering
M Comp Sc	Master of Computer Science
M Coun	Master of Counseling
M Dent	Master of Dentistry
M Dent Sc	Master of Dental Sciences
M Des	Master of Design
M Des S	Master of Design Studies
M Div	Master of Divinity
M E Com	Master of Electronic Commerce
M Ec	Master of Economics
M Econ	Master of Economics
M Ed	Master of Education
M Ed T	Master of Education in Teaching
M En	Master of Engineering
M En S	Master of Environmental Sciences
M Eng	Master of Engineering
M Eng Mgt	Master of Engineering Management
M Eng Tel	Master of Engineering in Telecommunications
M Engr	Master of Engineering
M Env	Master of Environment
M Env Des	Master of Environmental Design
M Env E	Master of Environmental Engineering
M Env Sc	Master of Environmental Science
M Ext Ed	Master of Extension Education
M Fin	Master of Finance
M Fr	Master of French
M Geo E	Master of Geological Engineering
M Geoenv E	Master of Geoenvironmental Engineering
M Geog	Master of Geography
M Hum	Master of Humanities
M Hum Svcs	Master of Human Services
M Kin	Master of Kinesiology
M Land Arch	Master of Landscape Architecture
M Lit M	Master of Liturgical Music
M Litt	Master of Letters
M Man	Master of Management
M Mat SE	Master of Material Science and Engineering
M Math	Master of Mathematics
M Med Sc	Master of Medical Science
M Mgmt	Master of Management
M Mgt	Master of Management
M Min	Master of Ministries
M Mtl E	Master of Materials Engineering
M Mu	Master of Music
M Mus	Master of Music
M Mus Ed	Master of Music Education
M Nat Sci	Master of Natural Science
M Nurs	Master of Nursing
M Oc E	Master of Oceanographic Engineering
M Pharm	Master of Pharmacy
M Phil	Master of Philosophy
M Phil F	Master of Philosophical Foundations
M Pl	Master of Planning
M Pol	Master of Political Science
M Pr A	Master of Professional Accountancy
M Pr Met	Master of Professional Meteorology
M Prob S	Master of Probability and Statistics
M Prof Past	Master of Professional Pastoral
M Psych	Master of Psychology
M Pub	Master of Publishing
M Rel	Master of Religion
M Sc	Master of Science
M Sc A	Master of Science (Applied)
M Sc AHN	Master of Science in Applied Human Nutrition
M Sc BMC	Master of Science in Biomedical Communications
M Sc CS	Master of Science in Computer Science
M Sc E	Master of Science in Engineering
M Sc Eng	Master of Science in Engineering
M Sc Engr	Master of Science in Engineering
M Sc F	Master of Science in Forestry
M Sc FE	Master of Science in Forest Engineering
M Sc Geogr	Master of Science in Geography
M Sc N	Master of Science in Nursing
M Sc OT	Master of Science in Occupational Therapy
M Sc P	Master of Science in Planning
M Sc Pl	Master of Science in Planning
M Sc PT	Master of Science in Physical Therapy
M Sc T	Master of Science in Teaching
M Soc	Master of Sociology
M Sp Ed	Master of Special Education
M Stat	Master of Statistics
M Sw E	Master of Software Engineering
M Sw En	Master of Software Engineering
M Sys Sc	Master of Systems Science
M Tax	Master of Taxation
M Tech	Master of Technology
M Th	Master of Theology
M Th Past	Master of Pastoral Theology
M Tox	Master of Toxicology
M Trans E	Master of Transportation Engineering
M Vet Sc	Master of Veterinary Science
MA	Master of Administration
	Master of Arts
MA Comm	Master of Arts in Communication
MA Ed	Master of Arts in Education
MA Ed Ad	Master of Arts in Educational Administration
MA Ext	Master of Agricultural Extension
MA Islamic	Master of Arts in Islamic Studies
MA Min	Master of Arts in Ministry

MA Missions	Master of Arts in Missions
MA Past St	Master of Arts in Pastoral Studies
MA Ph	Master of Arts in Philosophy
MA Ps	Master of Arts in Psychology
MA Psych	Master of Arts in Psychology
MA Sc	Master of Applied Science
MA Th	Master of Arts in Theology
MA-R	Master of Arts (Research)
MAA	Master of Administrative Arts
	Master of Applied Anthropology
	Master of Arts in Administration
MAAA	Master of Arts in Arts Administration
MAAE	Master of Arts in Applied Economics
	Master of Arts in Art Education
MAAS	Master of Arts in Administrative Stewardship
MAAT	Master of Arts in Applied Theology
	Master of Arts in Art Therapy
MAB	Master of Agribusiness
MABC	Master of Arts in Biblical Counseling
	Master of Arts in Business Communication
MABE	Master of Arts in Bible Exposition
MABL	Master of Arts in Biblical Languages
MABM	Master of Agribusiness Management
MABS	Master of Arts in Biblical Studies
MABT	Master of Arts in Bible Teaching
MAC	Master of Accounting
	Master of Addictions Counseling
	Master of Arts in Communication
	Master of Arts in Counseling
MACAT	Master of Arts in Counseling Psychology: Art Therapy
MACC	Master of Arts in Christian Counseling
MACCM	Master of Arts in Church and Community Ministry
MACCT	Master of Accounting
MACE	Master of Arts in Christian Education
MACFM	Master of Arts in Children's and Family Ministry
MACH	Master of Arts in Church History
MACJ	Master of Arts in Criminal Justice
MACL	Master of Arts in Classroom Psychology
MACM	Master of Arts in Christian Ministries
	Master of Arts in Church Music
	Master of Arts in Counseling Ministries
MACN	Master of Arts in Counseling
MACO	Master of Arts in Counseling
MAcOM	Master of Acupuncture and Oriental Medicine
MACP	Master of Arts in Counseling Psychology
MACPC	Master of Clinical Pastoral Counseling
MACS	Master of Arts in Christian Service
MACSE	Master of Arts in Christian School Education
MACT	Master of Arts in Christian Thought
	Master of Arts in Communications and Technology
MACY	Master of Arts in Accountancy
MAD	Master in Educational Institution Administration
	Master of Art and Design
MADR	Master of Arts in Dispute Resolution
MADS	Master of Animal and Dairy Science
MAE	Master of Aerospace Engineering
	Master of Agricultural Economics
	Master of Architectural Engineering
	Master of Art Education
	Master of Arts in Economics
	Master of Arts in Education
	Master of Arts in English
	Master of Automotive Engineering
MAEd	Master of Arts Education
MAEE	Master of Agricultural and Extension Education
MAEL	Master of Arts in Educational Leadership
MAEM	Master of Arts in Educational Ministries
MAEN	Master of Arts in English
MAEP	Master of Arts in Economic Policy
MAES	Master of Arts in Environmental Sciences
MAESL	Master of Arts in English as a Second Language
MAET	Master of Arts English Teaching
MAF	Master of Arts in Finance
MAFLL	Master of Arts in Foreign Language and Literature
MAFM	Master of Accounting and Financial Management
MAFS	Master of Arts in Family Studies
MAG	Master of Applied Geography
MAGC	Master of Arts in Global Communication
MAGP	Master of Arts in Gerontological Psychology
MAGU	Master of Urban Analysis and Management
MAH	Master of Arts in Humanities
MAHA	Master of Arts in Humanitarian Assistance
	Master of Arts in Humanitarian Studies
MAHCM	Master of Arts in Health Care Mission
MAHG	Master of American History and Government
MAHL	Master of Arts in Hebrew Letters
MAHN	Master of Applied Human Nutrition
MAHS	Master of Arts in Human Services
MAIA	Master of Arts in International Administration
MAIB	Master of Arts in International Business
MAICS	Master of Arts in Intercultural Studies
MAIDM	Master of Arts in Interior Design and Merchandising
MAIPCR	Master of Arts in International Peace and Conflict Management
MAIR	Master of Arts in Industrial Relations
MAIS	Master of Accounting and Information Systems
	Master of Arts in Intercultural Studies
	Master of Arts in Interdisciplinary Studies
	Master of Arts in International Studies
MAIT	Master of Administration in Information Technology
	Master of Applied Information Technology
MAJ	Master of Arts in Journalism
MAJ Ed	Master of Arts in Jewish Education
MAJCS	Master of Arts in Jewish Communal Service
MAJE	Master of Arts in Jewish Education
MAJS	Master of Arts in Jewish Studies
MAL	Master in Agricultural Leadership
MALA	Master of Arts in Liberal Arts
MALD	Master of Arts in Law and Diplomacy
MALER	Master of Arts in Labor and Employment Relations
MALL	Master of Arts in Liberal Learning

MALM	Master of Arts in Leadership Evangelical Mobilization
MALP	Master of Arts in Language Pedagogy
MALPS	Master of Arts in Liberal and Professional Studies
MALS	Master of Arts in Liberal Studies
MALT	Master of Arts in Learning and Teaching
MAM	Master of Acquisition Management
	Master of Agriculture and Management
	Master of Applied Mathematics
	Master of Applied Mechanics
	Master of Arts in Management
	Master of Arts in Ministry
	Master of Arts Management
	Master of Avian Medicine
MAMB	Master of Applied Molecular Biology
MAMC	Master of Arts in Mass Communication
	Master of Arts in Ministry and Culture
	Master of Arts in Ministry for a Multicultural Church
MAME	Master of Arts in Missions/Evangelism
MAMFC	Master of Arts in Marriage and Family Counseling
MAMFCC	Master of Arts in Marriage, Family, and Child Counseling
MAMFT	Master of Arts in Marriage and Family Therapy
MAMM	Master of Arts in Ministry Management
MAMS	Master of Applied Mathematical Sciences
	Master of Arts in Ministerial Studies
	Master of Arts in Ministry and Spirituality
	Master of Associated Medical Sciences
MAMT	Master of Arts in Mathematics Teaching
MAN	Master of Applied Nutrition
MANM	Master of Arts in Nonprofit Management
MANT	Master of Arts in New Testament
MAO	Master of Arts in Organizational Psychology
MAOA	Master of Arts in Organizational Administration
MAOL	Master of Arts in Organizational Leadership
MAOM	Master of Acupuncture and Oriental Medicine
	Master of Arts In Organizational Management
MAOT	Master of Arts in Old Testament
MAP	Master of Applied Psychology
	Master of Arts in Planning
	Master of Public Administration
	Masters of Psychology
MAP Min	Master of Arts in Pastoral Ministry
MAPA	Master of Arts in Public Administration
MAPC	Master of Arts in Pastoral Counseling
MAPE	Master of Arts in Political Economy
MAPM	Master of Arts in Pastoral Ministry
	Master of Arts in Pastoral Music
	Master of Arts in Practical Ministry
MAPP	Master of Arts in Public Policy
MAPPS	Master of Arts in Asia Pacific Policy Studies
MAPS	Master of Arts in Pastoral Counseling/Spiritual Formation
	Master of Arts in Pastoral Studies
MAPT	Master of Practical Theology
MAPW	Master of Arts in Professional Writing
MAR	Master of Arts in Religion
Mar Eng	Marine Engineer
MARC	Master of Arts in Rehabilitation Counseling
MARE	Master of Arts in Religious Education
MARL	Master of Arts in Religious Leadership
MARS	Master of Arts in Religious Studies
MAS	Master of Accounting Science
	Master of Actuarial Science
	Master of Administrative Science
	Master of Advanced Study
	Master of Aeronautical Science
	Master of American Studies
	Master of Applied Science
	Master of Applied Statistics
	Master of Archival Studies
MASA	Master of Advanced Studies in Architecture
MASAC	Master of Arts in Substance Abuse Counseling
MASC	Master of Arts in School Counseling
MASD	Master of Arts in Spiritual Direction
MASE	Master of Arts in Special Education
MASF	Master of Arts in Spiritual Formation
MASL	Master of Arts in School Leadership
MASLA	Master of Advanced Studies in Landscape Architecture
MASM	Master of Arts in Special Ministries
	Master of Arts in Specialized Ministries
MASP	Master of Applied Social Psychology
	Master of Arts in School Psychology
MASPAA	Master of Arts in Sports and Athletic Administration
MASS	Master of Applied Social Science
	Master of Arts in Social Science
MAST	Master of Arts Science Teaching
MASW	Master of Aboriginal Social Work
MAT	Master of Arts in Teaching
	Master of Arts in Theology
	Master of Athletic Training
	Masters in Administration of Telecommunications
Mat E	Materials Engineer
MATCM	Master of Acupuncture and Traditional Chinese Medicine
MATDE	Master of Arts in Theology, Development, and Evangelism
MATE	Master of Arts for the Teaching of English
MATESL	Master of Arts in Teaching English as a Second Language
MATESOL	Master of Arts in Teaching English to Speakers of Other Languages
MATF	Master of Arts in Teaching English as a Foreign Language/Intercultural Studies
MATFL	Master of Arts in Teaching Foreign Language
MATH	Master of Arts in Therapy
MATI	Master of Administration of Information Technology
MATL	Master of Arts in Teaching of Languages
	Master of Arts in Transformational Leadership
MATM	Master of Arts in Teaching of Mathematics
MATS	Master of Arts in Theological Studies
	Master of Arts in Transforming Spirituality
MATSL	Master of Arts in Teaching a Second Language
MAUA	Master of Arts in Urban Affairs
MAUD	Master of Arts in Urban Design
MAUM	Master of Arts in Urban Ministry
MAURP	Master of Arts in Urban and Regional Planning
MAW	Master of Arts in Writing
MAWL	Master of Arts in Worship Leadership
MAWS	Master of Arts in Worship/Spirituality
MAWSHP	Master of Arts in Worship
MAYM	Master of Arts in Youth Ministry
MB	Master of Bioinformatics
MBA	Master of Business Administration

MBA-EP	Master of Business Administration–Experienced Professionals
MBA/MNO	Master of Business Administration/Master of Nonprofit Organization
MBAA	Master of Business Administration in Aviation
MBAE	Master of Biological and Agricultural Engineering Master of Biosystems and Agricultural Engineering
MBAH	Master of Business Administration in Health
MBAi	Master of Business Administration–International
MBAIM	Master of Business Administration in International Management
MBAPA	Master of Business Administration–Physician Assistant
MBATM	Master of Business in Telecommunication Management
MBC	Master of Building Construction
MBE	Master of Bilingual Education Master of Biomedical Engineering Master of Business and Engineering Master of Business Economics Master of Business Education
MBET	Master of Business, Entrepreneurship and Technology
MBIOT	Master of Biotechnology
MBIT	Master of Business Information Technology
MBMSE	Master of Business Management and Software Engineering
MBOL	Master of Business and Organizational Leadership
MBS	Master of Behavioral Science Master of Biblical Studies Master of Biological Science Master of Biomedical Sciences Master of Bioscience Master of Building Science Master of Business Studies
MBSI	Master of Business Information Science
MBT	Master of Biblical and Theological Studies Master of Biomedical Technology Master of Business Taxation
MC	Master of Communication Master of Counseling Master of Cybersecurity
MC Ed	Master of Continuing Education
MC Sc	Master of Computer Science
MCA	Master of Arts in Applied Criminology Master of Commercial Aviation
MCALL	Master of Computer-Assisted Language Learning
MCAM	Master of Computational and Applied Mathematics
MCC	Master of Computer Science
MCCS	Master of Crop and Soil Sciences
MCD	Master of Communications Disorders Master of Community Development
MCE	Master in Electronic Commerce Master of Christian Education Master of Civil Engineering Master of Construction Engineering Master of Control Engineering
MCEM	Master of Construction Engineering Management
MCH	Master of Community Health
MCHS	Master of Clinical Health Sciences
MCIS	Master of Communication and Information Studies Master of Computer and Information Science
MCIT	Master of Computer and Information Technology
MCJ	Master of Criminal Justice
MCJA	Master of Criminal Justice Administration
MCL	Master of Canon Law Master of Civil Law Master of Comparative Law
MCM	Master of Christian Ministry Master of Church Management Master of Church Ministry Master of Church Music Master of City Management Master of Communication Management Master of Community Medicine Master of Competitive Manufacturing Master of Construction Management Master of Contract Management Masters of Corporate Media
MCMS	Master of Clinical Medical Science
MCP	Master in Science Master of City Planning Master of Community Planning Master of Computer Engineering Master of Counseling Psychology
MCPD	Master of Community Planning and Development
MCR	Masters in Clinical Research
MCRP	Master of City and Regional Planning
MCRS	Master of City and Regional Studies
MCS	Master of Christian Studies Master of Clinical Science Master of Combined Sciences Master of Communication Studies Master of Computer Science
MCSE	Master of Computer Science and Engineering
MCSL	Master of Catholic School Leadership
MCSM	Master of Construction Science/Management
MCTE	Master of Career and Technology Education
MCTP	Master of Communication Technology and Policy
MCVS	Master of Cardiovascular Science
MD	Doctor of Medicine
MD/CM	Doctor of Medicine and Master of Surgery
MD/MHS	Doctor of Medicine/Master of Health Science
MDA	Master of Development Administration Master of Dietetic Administration
MDE	Master of Developmental Economics Master of Distance Education
MDR	Master of Dispute Resolution
MDS	Master of Defense Studies Master of Dental Surgery
ME	Master of Education Master of Engineering Master of Entrepreneurship Master of Evangelism
ME Sc	Master of Engineering Science
MEA	Master of Educational Administration Master of Engineering Administration
MEAP	Master of Environmental Administration and Planning
MEBT	Master in Electronic Business Technologies
MEC	Master of Electronic Commerce
MECE	Master of Electrical and Computer Engineering
Mech E	Mechanical Engineer
MED	Master of Education of the Deaf
MEDS	Master of Environmental Design Studies
MEE	Master in Education Master of Electrical Engineering Master of Environmental Engineering

MEEM	Master of Environmental Engineering and Management
MEENE	Master of Engineering in Environmental Engineering
MEEP	Master of Environmental and Energy Policy
MEERM	Master of Earth and Environmental Resource Management
MEH	Master in Humanistics Studies
MEHS	Master of Environmental Health and Safety
MEIM	Master of Entertainment Industry Management
MEL	Master of Educational Leadership
	Master of English Literature
MEM	Master of Ecosystem Management
	Master of Electricity Markets
	Master of Engineering Management
	Master of Environmental Management
	Master of Marketing
MEME	Master of Engineering in Manufacturing Engineering
	Master of Engineering in Mechanical Engineering
MEMS	Master of Engineering in Manufacturing Systems
MENVEGR	Master of Environmental Engineering
MEP	Master of Engineering Physics
	Master of Environmental Planning
MEPC	Master of Environmental Pollution Control
MEPD	Master of Education–Professional Development
MER	Master of Employment Relations
MES	Master of Education and Science
	Master of Engineering Science
	Master of Environmental Science
	Master of Environmental Studies
	Master of Environmental Systems
	Master of Special Education
MESM	Master of Environmental Science and Management
MET	Master of Education in Teaching
	Master of Educational Technology
	Master of Engineering Technology
	Master of Entertainment Technology
	Master of Environmental Toxicology
Met E	Metallurgical Engineer
METM	Master of Engineering and Technology Management
MEVE	Master of Environmental Engineering
MF	Master of Finance
	Master of Forestry
MFA	Master of Financial Administration
	Master of Fine Arts
MFAS	Master of Fisheries and Aquatic Science
MFAW	Master of Fine Arts in Writing
MFB	Master of Finance and Banking
MFC	Master of Forest Conservation
MFCC	Marriage and Family Counseling Certificate
	Marriage, Family, and Child Counseling
MFCS	Master of Family and Consumer Sciences
MFE	Master of Financial Economics
	Master of Financial Engineering
	Master of Forest Engineering
MFG	Master of Functional Genomics
MFHD	Master of Family and Human Development
MFM	Master of Financial Mathematics
MFMS	Masters in Food Microbiology and Safety
MFP	Master of Financial Planning
MFPE	Master of Food Process Engineering
MFR	Master of Forest Resources

MFRC	Master of Forest Resources and Conservation
MFS	Master of Family Studies
	Master of Financial Services
	Master of Food Science
	Master of Forensic Sciences
	Master of Forest Science
	Master of Forest Studies
	Master of French Studies
MFSA	Master of Forensic Sciences Administration
MFT	Master of Family Therapy
	Master of Food Technology
MFWB	Master of Fishery and Wildlife Biology
MFWS	Master of Fisheries and Wildlife Sciences
MFYCS	Master of Family, Youth and Community Sciences
MG	Master of Genetics
MGA	Master of Government Administration
MGD	Master of Graphic Design
MGE	Master of Gas Engineering
	Master of Geotechnical Engineering
MGH	Master of Geriatric Health
MGIS	Master of Geographic Information Science
MGP	Master of Gestion de Projet
MGS	Master of Gerontological Studies
	Master of Global Studies
MH	Master of Humanities
MH Sc	Master of Health Sciences
MHA	Master of Health Administration
	Master of Healthcare Administration
	Master of Hospital Administration
	Master of Hospitality Administration
MHCA	Master of Health Care Administration
MHCI	Master of Human-Computer Interaction
MHCL	Master of Health Care Leadership
MHE	Master of Health Education
MHE Ed	Master of Home Economics Education
MHHS	Master of Health and Human Services
MHI	Master of Health Informatics
MHIS	Master of Health Information Systems
MHK	Master of Human Kinetics
MHL	Master of Health Law
	Master of Hebrew Literature
MHM	Master of Hospitality Management
MHMS	Master of Health Management Systems
MHP	Master of Health Physics
	Master of Heritage Preservation
	Master of Historic Preservation
MHPA	Master of Heath Policy and Administration
MHPE	Master of Health Professions Education
MHR	Master of Human Resources
MHRD	Master in Human Resource Development
MHRDL	Master of Human Resource Development Leadership
MHRIM	Master of Hotel, Restaurant, and Institutional Management
MHRIR	Master of Human Resources and Industrial Relations
MHRLR	Master of Human Resources and Labor Relations
MHRM	Master of Human Resources Management
MHRTM	Master of Hotel, Restaurant, and Tourism Management
MHS	Master of Health Sciences
	Master of Health Studies
	Master of Hispanic Studies
	Master of Humanistic Studies

MHSA — Master of Health Services Administration / Master of Human Services Administration

MHSM — Master of Health Sector Management / Master of Human Services Management

MI — Master of Instruction

MI Arch — Master of Interior Architecture

MI St — Master of Information Studies

MIA — Master of Interior Architecture / Master of International Affairs

MIAA — Master of International Affairs and Administration

MIB — Master of International Business

MIBA — Master of International Business Administration

MICM — Master of International Construction Management

MID — Master of Industrial Design / Master of Industrial Distribution / Master of Interior Design / Master of International Development

MIE — Master of Industrial Engineering

MIEM — Master of Industrial Engineering and Management

MIJ — Master of International Journalism

MILR — Master of Industrial and Labor Relations

MIM — Master of Information Management / Master of International Management

MIMLAE — Master of International Management for Latin American Executives

MIMS — Master of Information Management and Systems / Master of Integrated Manufacturing Systems

MIP — Master of Infrastructure Planning / Master of Intellectual Property

MIPP — Master of International Policy and Practice / Master of International Public Policy

MIPS — Master of International Planning Studies

MIR — Master of Industrial Relations / Master of International Relations

MIS — Master of Industrial Statistics / Master of Information Science / Master of Information Systems / Master of Integrated Science / Master of Interdisciplinary Studies / Master of International Service / Master of International Studies

MISKM — Master of Information Sciences and Knowledge Management

MISM — Master of Information Systems Management

MIT — Master in Teaching / Master of Industrial Technology / Master of Information Technology / Master of Initial Teaching / Master of International Trade / Master of Internet Technology

MITA — Master of Information Technology Administration

MITE — Master of Information Technology Education

MITM — Master of International Technology Management

MITO — Master of Industrial Technology and Operations

MJ — Master of Journalism / Master of Jurisprudence

MJ Ed — Master of Jewish Education

MJA — Master of Justice Administration

MJS — Master of Judicial Studies / Master of Juridical Science

ML — Master of Latin

ML Arch — Master of Landscape Architecture

MLA — Master of Landscape Architecture / Master of Liberal Arts

MLAS — Master of Laboratory Animal Science

MLAUD — Master of Landscape Architecture in Urban Development

MLBLST — Master of Liberal Studies

MLD — Master of Leadership Development / Master of Leadership Studies

MLE — Master of Applied Linguistics and Exegesis

MLER — Master of Labor and Employment Relations

MLERE — Master of Land Economics and Real Estate

MLHR — Master of Labor and Human Resources

MLI — Master of Legal Institutions

MLI Sc — Master of Library and Information Science

MLIS — Master of Library and Information Science / Master of Library and Information Studies

MLM — Master of Library Media

MLOS — Masters in Leadership and Organizational Studies

MLRHR — Master of Labor Relations and Human Resources

MLS — Master of Legal Studies / Master of Liberal Studies / Master of Library Science / Master of Life Sciences

MLS/PMC — Master of Library Science/Graduat Certificate in Public Management

MLSP — Master of Law and Social Policy

MLT — Master of Language Technologies

MLW — Master of Studies in Law

MM — Master of Management / Master of Ministry / Master of Missiology / Master of Music

MM Ed — Master of Music Education

MM Sc — Master of Medical Science

MM St — Master of Museum Studies

MMA — Master of Marine Affairs / Master of Media Arts / Master of Musical Arts

MMAE — Master of Mechanical and Aerospace Engineering

MMAS — Master of Military Art and Science

MMB — Master of Microbial Biotechnology

MMBA — Managerial Master of Business Administration

MMC — Master of Competitive Manufacturing / Master of Mass Communications / Master of Music Conducting

MMCM — Master of Music in Church Music

MMCSS — Masters of Mathematical Computational and Statistical Sciences

MME — Master of Manufacturing Engineering / Master of Mathematics for Educators / Master of Mechanical Engineering / Master of Medical Engineering / Master of Mining Engineering / Master of Music Education / Mater of Mathematics Education

MMF — Master of Mathematical Finance

MMFT — Master of Marriage and Family Therapy

MMG — Master of Management

MMH — Master of Management in Hospitality / Master of Medical History / Master of Medical Humanities / Master of Military History

MMIS — Master of Management Information Systems

MMM — Master of Manufacturing Management / Master of Marine Management / Master of Medical Management

MMME — Master of Metallurgical and Materials Engineering

MMP	Master of Marine Policy
	Master of Music Performance
MMPA	Master of Management and Professional Accounting
MMQM	Master of Manufacturing Quality Management
MMR	Master of Marketing Research
MMRM	Master of Marine Resources Management
MMS	Master of Management Science
	Master of Marine Science
	Master of Marine Studies
	Master of Materials Science
	Master of Medical Science
	Master of Medieval Studies
	Master of Modern Studies
MMSE	Master of Manufacturing Systems Engineering
MMSM	Master of Music in Sacred Music
MMT	Master in Marketing
	Master of Music Teaching
	Master of Music Therapy
	Masters in Marketing Technology
MMus	Master of Music
MN	Master of Nursing
	Master of Nutrition
MN Sc	Master of Nursing Science
MNA	Master of Nonprofit Administration
	Master of Nurse Anesthesia
MNAS	Master of Natural and Applied Science
MNCM	Master of Network and Communications Management
MNE	Master of Network Engineering
	Master of Nuclear Engineering
MNL	Master in International Business for Latin America
MNM	Master of Nonprofit Management
MNO	Master of Nonprofit Organization
MNO/MSSA	Master of Nonprofit Organization/Master of Arts
MNPL	Master of Not-for-Profit Leadership
MNPS	Master of New Professional Studies
MNR	Master of Natural Resources
MNRES	Master of Natural Resources and Environmental Studies
MNRM	Master of Natural Resource Management
MNRS	Master of Natural Resource Stewardship
MNS	Master of Natural Science
	Master of Nursing Science
MO	Master of Oceanography
MOA	Maître d'Orthophonie et d'Audiologie
MOD	Master of Organizational Development
MOGS	Master of Oil and Gas Studies
MOH	Master of Occupational Health
MOL	Master of Organizational Leadership
MOM	Master of Manufacturing
	Master of Oriental Medicine
MOR	Master of Operations Research
MOT	Master of Occupational Therapy
MP	Master of Physiology
	Master of Planning
MP Ac	Master of Professional Accountancy
MP Acc	Master of Professional Accountancy
	Master of Professional Accounting
	Master of Public Accounting
MP Aff	Master of Public Affairs
MP Th	Master of Pastoral Theology

MPA	Master of Physician Assistant
	Master of Professional Accountancy
	Master of Professional Accounting
	Master of Public Administration
	Master of Public Affairs
MPA-URP	Master of Public Affairs and Urban and Regional Planning
MPAC	Masters in Professional Accounting
MPAD	Master of Public Administration
MPAID	Master of Public Administration and International Development
MPAP	Master of Physician Assistant Practice
	Master of Public Affairs and Politics
MPAS	Master of Physician Assistant Science
	Master of Physician Assistant Studies
	Master of Public Art Studies
MPC	Master of Pastoral Counseling
	Master of Professional Communication
MPD	Master of Product Development
	Master of Public Diplomacy
MPDS	Master of Planning and Development Studies
MPE	Master of Physical Education
MPEM	Master of Project Engineering and Management
MPH	Master of Public Health
MPHE	Master of Public Health Education
MPHTM	Master of Public Health and Tropical Medicine
MPIA	Master of Public and International Affairs
MPL	Master of Pastoral Leadership
MPM	Master of Pastoral Ministry
	Master of Pest Management
	Master of Practical Ministries
	Master of Project Management
	Master of Public Management
MPNA	Master of Public and Nonprofit Administration
MPOD	Master of Positive Organizational Development
MPP	Master of Public Policy
MPPA	Master of Public Policy Administration
	Master of Public Policy and Administration
MPPM	Master of Public and Private Management
	Master of Public Policy and Management
MPPPM	Master of Plant Protection and Pest Management
MPPUP	Master of Public Policy and Urban Planning
MPRTM	Master of Parks, Recreation, and Tourism Management
MPS	Master of Pastoral Studies
	Master of Perfusion Science
	Master of Political Science
	Master of Preservation Studies
	Master of Professional Studies
	Master of Public Service
MPSA	Master of Public Service Administration
MPSRE	Master of Professional Studies in Real Estate
MPT	Master of Pastoral Theology
	Master of Physical Therapy
MPVM	Master of Preventive Veterinary Medicine
MPW	Master of Professional Writing
	Master of Public Works
MQF	Master of Quantitative Finance
MQM	Master of Quality Management
MR	Master of Recreation
MRC	Master of Rehabilitation Counseling
MRCP	Master of Regional and City Planning
	Master of Regional and Community Planning

MRD	Master of Rural Development
MRE	Master of Religious Education
MRED	Master of Real Estate Development
MRLS	Master of Resources Law Studies
MRM	Master of Rehabilitation Medicine Master of Resources Management
MRP	Master of Regional Planning
MRRA	Master of Recreation Resources Administration
MRS	Master of Religious Studies
MRSc	Master of Rehabilitation Science
MS	Master of Science
MS Cp	Master of Science in Computer Engineering
MS Kin	Master of Science in Kinesiology
MS Acct	Master of Science in Accounting
MS Aero E	Master of Science in Aerospace Engineering
MS Ag	Master of Science in Agriculture
MS Arch	Master of Science in Architecture
MS Arch St	Master of Science in Architectural Studies
MS Bio E	Master of Science in Bioengineering Master of Science in Biomedical Engineering
MS Bm E	Master of Science in Biomedical Engineering
MS Ch E	Master of Science in Chemical Engineering
MS Chem	Master of Science in Chemistry
MS Cp E	Master of Science in Computer Engineering
MS Eco	Master of Science in Economics
MS Econ	Master of Science in Economics
MS Ed	Master of Science in Education
MS El	Master of Science in Educational Leadership and Administration
MS En E	Master of Science in Environmental Engineering
MS Eng	Master of Science in Engineering
MS Engr	Master of Science in Engineering
MS Env E	Master of Science in Environmental Engineering
MS Exp Surg	Master of Science in Experimental Surgery
MS Int A	Master of Science in International Affairs
MS Mat E	Master of Science in Materials Engineering
MS Mat SE	Master of Science in Material Science and Engineering
MS Met E	Master of Science in Metallurgical Engineering
MS Metr	Master of Science in Meteorology
MS Mgt	Master of Science in Management
MS Min	Master of Science in Mining
MS Min E	Master of Science in Mining Engineering
MS Mt E	Master of Science in Materials Engineering
MS Otal	Master of Science in Otalrynology
MS Pet E	Master of Science in Petroleum Engineering
MS Phr	Master of Science in Pharmacy
MS Phys	Master of Science in Physics
MS Phys Op	Master of Science in Physiological Optics
MS Poly	Master of Science in Polymers
MS Psy	Master of Science in Psychology
MS Pub P	Master of Science in Public Policy
MS Sc	Master of Science in Social Science
MS SEng	Master of Science in Systems Engineering
MS Sp C	Master of Science in Space Science
MS Sp Ed	Master of Science in Special Education
MS Stat	Master of Science in Statistics Master of Science in Statistics
MS Surg	Master of Science in Surgery
MS Tax	Master of Science in Taxation
MS Tc E	Master of Science in Telecommunications Engineering
MS-R	Master of Science (Research)
MSA	Master of School Administration Master of Science Administration Master of Science in Accountancy Master of Science in Accounting Master of Science in Administration Master of Science in Aeronautics Master of Science in Agriculture Master of Science in Anesthesia Master of Science in Architecture Master of Science in Aviation Master of Sports Administration
MSA Phy	Master of Science in Applied Physics
MSAA	Master of Science in Astronautics and Aeronautics
MSAC	Master of Science in Acupuncture
MSACC	Master of Science in Accounting
MSaCS	Master of Science in Applied Computer Science
MSAE	Master of Science in Aeronautical Engineering Master of Science in Aerospace Engineering Master of Science in Agricultural Engineering Master of Science in Applied Economics Master of Science in Architectural Engineering Master of Science in Art Education
MSAH	Master of Science in Allied Health
MSAM	Master of Science in Advanced Management Master of Science in Applied Mathematics
MSAOM	Master of Science in Agricultural Operations Management
MSAS	Master of Science in Administrative Studies Master of Science in Architectural Studies
MSAT	Master of Science in Advanced Technology
MSB	Master of Science in Bible Master of Science in Business
MSBA	Master of Science in Business Administration
MSBAE	Master of Science in Biological and Agricultural Engineering Master of Science in Biosystems and Agricultural Engineering
MSBC	Master of Science in Building Construction
MSBE	Master of Science in Biomedical Engineering
MSBENG	Master of Science in Bioengineering
MSBIT	Master of Science in Business Information Technology
MSBL	Master of Studies in Business Law
MSBM	Master of Sport Business Management
MSBME	Master of Science in Biomedical Engineering
MSBMS	Master of Science in Basic Medical Science
MSBS	Master of Science in Biomedical Sciences
MSC	Master of Science in Commerce Master of Science in Communication Master of Science in Computers Master of Science in Counseling Master of Science in Criminology
MSCC	Master of Science in Christian Counseling Master of Science in Community Counseling
MSCD	Master of Science in Communication Disorders Master of Science in Community Development
MSCE	Master of Science in Civil Engineering Master of Science in Clinical Epidemiology Master of Science in Computer Engineering Master of Science in Continuing Education
MSCEE	Master of Science in Civil and Environmental Engineering
MSCES	Master of Science in Computer and Engineering Sciences
MSCF	Master of Science in Computational Finance

MSChE	Master of Science in Chemical Engineering
MSCI	Master of Science in Clinical Investigation
	Master of Science in Curriculum and Instruction
MSCIS	Master of Science in Computer and Information Systems
	Master of Science in Computer Information Science
	Master of Science in Computer Information Systems
MSCIT	Master of Science in Computer Information Technology
MSCJ	Master of Science in Criminal Justice
MSCJA	Master of Science in Criminal Justice Administration
MSCM	Master of Science in Conflict Management
	Master of Science in Construction Management
MScM	Master of Science in Management
MSCP	Master of Science in Clinical Psychology
	Master of Science in Counseling Psychology
MSCPharm	Master of Science in Pharmacy
MSCRP	Master of Science in City and Regional Planning
	Master of Science in Community and Regional Planning
MSCS	Master of Science in Computer Science
	Master of Science in Construction Science
MSCSD	Master of Science in Communication Sciences and Disorders
MSCSE	Master of Science in Computer Science and Engineering
	Master of Science in Computer Systems Engineering
MSCST	Master of Science in Computer Science Technology
MSCTE	Master of Science in Career and Technical Education
MSD	Master of Science in Dentistry
	Master of Science in Design
MSDD	Master of Software Design and Development
MSDM	Master of Design Methods
MSDR	Master of Dispute Resolution
MSE	Master of Science Education
	Master of Science in Education
	Master of Science in Engineering
	Master of Science in Engineering Managment
	Master of Software Engineering
	Master of Structural Engineering
MSE Mgt	Master of Science in Engineering Management
MSECE	Master of Science in Electrical and Computer Engineering
MSED	Master of Sustainable Economic Development
MSEE	Master of Science in Electrical Engineering
	Master of Science in Environmental Engineering
MSEH	Master of Science in Environmental Health
MSEL	Master of Science in Executive Leadership
	Master of Studies in Environmental Law
MSEM	Master of Science in Engineering Management
	Master of Science in Engineering Mechanics
	Master of Science in Environmental Management
MSENE	Master of Science in Environmental Engineering
MSEO	Master of Science in Electro-Optics
MSEP	Master of Science in Economic Policy
MSES	Master of Science in Embedded Software Engineering
	Master of Science in Engineering Science
	Master of Science in Environmental Science
	Master of Science in Environmental Studies
MSESM	Master of Science in Engineering Science and Mechanics
MSET	Master of Science in Education in Educational Technology
	Master of Science in Engineering Technology
MSETM	Master of Science in Environmental Technology Management
MSEV	Master of Science in Environmental Engineering
MSEVH	Master of Science in Environmental Health and Safety
MSF	Master of Science in Finance
	Master of Science in Forestry
	Master of Social Foundations
MSFA	Master of Science in Financial Analysis
MSFAM	Master of Science in Family Studies
MSFCS	Master of Science in Family and Consumer Science
MSFE	Master of Science in Financial Engineering
	Master of Science in Financial Engineering
MSFOR	Master of Science in Forestry
MSFP	Master of Science in Financial Planning
MSFS	Master of Science in Financial Sciences
	Master of Science in Forensic Science
MSFT	Master of Science in Family Therapy
MSGC	Master of Science in Genetic Counseling
MSGL	Master of Science in Global Leadership
MSH	Master of Science in Health
	Master of Science in Hospice
MSHA	Master of Science in Health Administration
MSHCA	Master of Science in Health Care Administration
MSHCI	Master of Science in Human Computer Interaction
MSHCPM	Master of Science in Health Care Policy and Management
MSHCS	Master of Science in Human and Consumer Science
MSHE	Master of Science in Health Education
MSHES	Master of Science in Human Environmental Sciences
MSHFID	Master of Science in Human Factors in Information Design
MSHFS	Master of Science in Human Factors and Systems
MSHP	Master of Science in Health Professions
MSHR	Master of Science in Human Resources
MSHRM	Master of Science in Human Resource Management
MSHROD	Master of Science in Human Resources and Organizational Development
MSHS	Master of Science in Health Science
	Master of Science in Health Services
	Master of Science in Health Systems
MSHSA	Master of Science in Human Service Administration
MSHT	Master of Science in History of Technology
MSI	Master of Science in Instruction
MSIA	Master of Science in Industrial Administration
	Master of Science in Information Assurance and Computer Security
MSIB	Master of Science in International Business
MSIDM	Master of Science in Interior Design and Merchandising
MSIDT	Master of Science in Information Design and Technology
MSIE	Master of Science in Industrial Engineering
	Master of Science in International Economics

MSIEM	Master of Science in Information Engineering and Management
MSIM	Master of Science in Information Management
	Master of Science in Investment Management
MSIMC	Master of Science in Integrated Marketing Communications
MSIO	Master of Science of Industrial-Organizational Psychology
MSIR	Master of Science in Industrial Relations
MSIS	Master of Science in Information Science
	Master of Science in Information Systems
	Master of Science in Interdisciplinary Studies
MSIS/PhD	Master of Science in Information Systems/Doctor of Philosophy
MSISE	Master of Science in Infrastructure Systems Engineering
MSISM	Master of Science in Information Systems Management
MSISPM	Master of Science in Information Security Policy and Management
MSIST	Master of Science in Information Systems Technology
MSIT	Master of Science in Industrial Technology
	Master of Science in Information Technology
	Master of Science in Instructional Technology
MSITM	Master of Science in Information Technology Management
MSJ	Master of Science in Journalism
	Master of Science in Jurisprudence
MSJE	Master of Science in Jewish Education
MSJFP	Master of Science in Juvenile Forensic Psychology
MSJJ	Master of Science in Juvenile Justice
MSJPS	Master of Science in Justice and Public Safety
MSJS	Master of Science in Jewish Studies
MSK	Master of Science in Kinesiology
MSL	Master of School Leadership
	Master of Science in Limnology
	Master of Studies in Law
MSLA	Master of Science in Landscape Architecture
	Master of Science in Legal Administration
MSLD	Master of Science in Land Development
MSLS	Master of Science in Legal Studies
	Master of Science in Library Science
	Master of Science in Logistics Systems
MSLT	Master of Second Language Teaching
MSM	Master of Sacred Music
	Master of School Mathematics
	Master of Science in Management
MSMA	Master of Science in Marketing Analysis
MSMAE	Master of Science in Materials Engineering
MSMC	Master of Science in Marketing Communications
	Master of Science in Mass Communications
MSME	Master of Science in Mechanical Engineering
MSMFE	Master of Science in Manufacturing Engineering
MSMIS	Master of Science in Management Information Systems
MSMIT	Master of Science in Management and Information Technology
MSMM	Master of Science in Manufacturing Management
MSMO	Master of Science in Manufacturing Operations
MSMOT	Master of Science in Management of Technology
MSMS	Master of Science in Management Science

MSMSE	Master of Science in Manufacturing Systems Engineering
	Master of Science in Material Science and Engineering
	Master of Science in Mathematics and Science Education
MSMT	Master of Science in Management and Technology
	Master of Science in Medical Technology
MSN	Master of Science in Nursing
MSN-R	Master of Science in Nursing (Research)
MSNA	Master of Science in Nurse Anesthesia
MSNE	Master of Science in Nuclear Engineering
MSNM	Master of Science in Nonprofit Management
MSNS	Master of Science in Natural Science
	Master's of Science in Nutritional Science
MSOD	Master of Science in Organizational Development
MSOEE	Master of Science in Outdoor and Environmental Education
MSOES	Master of Science in Occupational Ergonomics and Safety
MSOL	Master of Science in Organizational Leadership
MSOM	Master of Science in Organization and Management
	Master of Science in Oriental Medicine
MSOR	Master of Science in Operations Research
MSOT	Master of Science in Occupational Technology
	Master of Science in Occupational Therapy
MSP	Master of Science in Pharmacy
	Master of Science in Planning
	Master of Speech Pathology
MSP Ex	Master of Science in Exercise Physiology
MSPA	Master of Science in Physician Assistant
	Master of Science in Professional Accountancy
MSPAS	Master of Science in Physician Assistant Studies
MSPC	Master of Science in Professional Communications
	Master of Science in Professional Counseling
MSPE	Master of Science in Petroleum Engineering
MSPG	Master of Science in Psychology
MSPH	Master of Science in Public Health
MSPHR	Master of Science in Pharmacy
MSPM	Master of Science in Professional Management
MSPNGE	Master of Science in Petroleum and Natural Gas Engineering
MSPS	Master of Science in Pharmaceutical Science
	Master of Science in Psychological Services
MSPT	Master of Science in Physical Therapy
MSpVM	Master of Specialized Veterinary Medicine
MSQFE	Master of Science in Quantitative Financial Economics
MSR	Master of Science in Radiology
	Master of Science in Rehabilitation Sciences
MSRA	Master of Science in Recreation Administration
MSRC	Master of Science in Resource Conservation
MSRE	Master of Science in Real Estate
	Master of Science in Religious Education
MSRED	Master of Science in Real Estate Development
MSREM	Master of Science in Real Estate Management
MSRLS	Master of Science in Recreation and Leisure Studies
MSRMP	Master of Science in Radiological Medical Physics

MSS	Master of Science in Sociology
	Master of Science in Software
	Master of Social Science
	Master of Social Services
	Master of Software Systems
	Master of Sports Science
	Master of Strategic Studies
MSSA	Master of Science in Social Administration
MSSE	Master of Science in Software Engineering
MSSEM	Master of Science in Systems and Engineering Management
MSSI	Master of Science in Strategic Intelligence
MSSL	Master of Science in Strategic Leadership
MSSLP	Master of Science in Speech-Language Pathology
MSSM	Master of Science in Sports Medicine
	Master of Science in Systems Management
MSSPA	Master of Science in Student Personnel Administration
MSSS	Master of Science in Safety Science
	Master of Science in Systems Science
MSST	Master of Science in Systems Technology
MSSW	Master of Science in Social Work
MST	Master of Science in Taxation
	Master of Science in Teaching
	Master of Science in Technology
	Master of Science in Telecommunications
	Master of Science Teaching
	Master of Science Technology
MSTC	Master of Science in Telecommunications
MSTE	Master of Science in Telecommunications Engineering
	Master of Science in Transportation Engineering
MSTIM	Master of Science in Technology and Innovation Management
MSTM	Master of Science in Technical Management
	Master of Science in Technology Management
MSTOM	Master of Science in Traditional Oriental Medicine
MSUD	Master of Science in Urban Design
MSUESM	Master of Science in Urban Environmental Systems Management
MSW	Master of Social Work
MSWE	Master of Software Engineering
MSWREE	Master of Science in Water Resources and Environmental Engineering
MSX	Master of Science in Exercise Science
MT	Master of Taxation
	Master of Teaching
	Master of Technology
	Master of Textiles
MTA	Master of Arts in Teaching
	Master of Tax Accounting
	Master of Teaching Arts
	Master of Tourism Administration
MTCM	Master of Traditional Chinese Medicine
MTD	Master of Training and Development
MTE	Master in Educational Technology
	Master of Teacher Education
MTEL	Master of Telecommunications
MTESL	Master in Teaching English as a Second Language
MTHM	Master of Tourism and Hospitality Management
MTI	Master of Information Technology
MTIM	Masters of Trust and Investment Management
MTL	Master of Talmudic Law
MTLM	Master of Transportation and Logistics Management

MTM	Master of Technology Management
	Master of Telecommunications Management
	Master of the Teaching of Mathematics
MTMH	Master of Tropical Medicine and Hygiene
MTOM	Master of Traditional Oriental Medicine
MTP	Master of Transpersonal Psychology
MTS	Master of Teaching Science
	Master of Theological Studies
MTSC	Master of Technical and Scientific Communication
MTSE	Master of Telecommunications and Software Engineering
MTT	Master in Technology Management
MTX	Master of Taxation
MUA	Master of Urban Affairs
MUD	Master of Urban Design
MUEP	Master of Urban and Environmental Planning
MUP	Master of Urban Planning
MUPDD	Master of Urban Planning, Design, and Development
MUPP	Master of Urban Planning and Policy
MUPRED	Masters of Urban Planning and Real Estate Development
MURP	Master of Urban and Regional Planning
	Master of Urban and Rural Planning
MUS	Master of Urban Studies
Mus Doc	Doctor of Music
Mus M	Master of Music
MVP	Master of Voice Pedagogy
MVPH	Master of Veterinary Public Health
MVTE	Master of Vocational-Technical Education
MWC	Master of Wildlife Conservation
MWE	Master in Welding Engineering
MWPS	Master of Wood and Paper Science
MWR	Master of Water Resources
MWS	Master of Women's Studies
MZS	Master of Zoological Science
Nav Arch	Naval Architecture
Naval E	Naval Engineer
ND	Doctor of Naturopathic Medicine
	Doctor of Nursing
NE	Nuclear Engineer
Nuc E	Nuclear Engineer
Ocean E	Ocean Engineer
OD	Doctor of Optometry
OTD	Doctor of Occupational Therapy
PBME	Professional Master of Biomedical Engineering
PD	Professional Diploma
PDD	Professional Development Degree
PE Dir	Director of Physical Education
PGC	Post-Graduate Certificate
Ph L	Licentiate of Philosophy
Pharm D	Doctor of Pharmacy
PhD	Doctor of Philosophy
PhD Otal	Doctor of Philosophy in Otalrynology
Phd Surg	Doctor of Philosophy in Surgery
PhDEE	Doctor of Philosophy in Electrical Engineering
PM Sc	Professional Master of Science
PMBA	Professional Master of Business Administration
PMC	Post Master Certificate
PMD	Post-Master's Diploma
PMS	Professional Master of Science
PPDPT	Postprofessional Doctor of Physical Therapy

PSM	Professional Master of Science
Psy D	Doctor of Psychology
Psy M	Master of Psychology
Psy S	Specialist in Psychology
Psya D	Doctor of Psychoanalysis
Re Dir	Director of Recreation
Rh D	Doctor of Rehabilitation
S Psy S	Specialist in Psychological Services
Sc D	Doctor of Science
Sc M	Master of Science
SCCT	Specialist in Community College Teaching
ScDPT	Doctor of Physical Therapy Science
SD	Doctor of Science Specialist Degree
SJD	Doctor of Juridical Science
SLPD	Doctor of Speech-Language Pathology
SLS	Specialist in Library Science
SM	Master of Science
SM Arch S	Master of Science in Architectural Studies
SM Vis S	Master of Science in Visual Studies
SMBT	Master of Science in Building Technology
SP	Specialist Degree
Sp C	Specialist in Counseling
Sp Ed	Specialist in Education
Sp Ed As	Specialist in Administrative Supervision
Sp LIS	Specialist in Library and Information Science
Sp Sch Psych	Specialist in School Psychology
SPCM	Special in Church Music
Spec	Specialist's Certificate
Spec M	Specialist in Music
SPEM	Special in Educational Ministries
SPS	School Psychology Specialist
Spt	Specialist Degree
SPTH	Special in Theology
SSP	Specialist in School Psychology
STB	Bachelor of Sacred Theology
STD	Doctor of Sacred Theology
STL	Licentiate of Sacred Theology
STM	Master of Sacred Theology
TDPT	Transitional Doctor of Physical Therapy
Th D	Doctor of Theology
Th M	Master of Theology
VMD	Doctor of Veterinary Medicine
WEMBA	Weekend Executive Master of Business Administration
WMBA	Web-based Master of Business Administration
XMA	Executive Master of Arts
XMBA	Executive Master of Business Administration

INDEXES

Close-Ups and Announcements

Directories and Subject Areas in Books 2–6

Following is an alphabetical listing of directories and subject areas in Books 2–6. Also listed are cross-references for subject area names not used in the directory structure of the guides, for example, "Arabic (*see* Near and Middle Eastern Languages)."

Accounting—Book 6
Acoustics—Book 4
Actuarial Science—Book 6
Acupuncture and Oriental Medicine—Book 6
Addictions/Substance Abuse Counseling—Book 2
Administration (*see* Arts Administration; Business Administration and Management; Educational Administration; Health Services Management and Hospital Administration; Industrial Administration; Pharmaceutical Administration; Public Administration)
Adult Education—Book 6
Adult Nursing (*see* Medical/Surgical Nursing)
Advanced Practice Nursing—Book 6
Advertising and Public Relations—Book 6
Aeronautical Engineering (*see* Aerospace/Aeronautical Engineering)
Aerospace/Aeronautical Engineering—Book 5
Aerospace Studies (*see* Aerospace/Aeronautical Engineering)
African-American Studies—Book 2
African Languages and Literatures (*see* African Studies)
African Studies—Book 2
Agribusiness (*see* Agricultural Economics and Agribusiness)
Agricultural Economics and Agribusiness—Book 2
Agricultural Education—Book 6
Agricultural Engineering—Book 5
Agricultural Sciences—Book 4
Agronomy and Soil Sciences—Book 4
Alcohol Abuse Counseling (*see* Addictions/Substance Abuse Counseling; Counselor Education)
Allied Health—Book 6
Allopathic Medicine—Book 6
American Indian/Native American Studies—Book 2
American Studies—Book 2
Analytical Chemistry—Book 4
Anatomy—Book 3
Animal Behavior—Book 3
Animal Sciences—Book 4
Anthropology—Book 2
Applied Arts and Design—Book 2
Applied Economics—Book 2
Applied History (*see* Public History)
Applied Mathematics—Book 4
Applied Mechanics (*see* Mechanics)
Applied Physics—Book 4
Applied Science and Technology—Book 5
Applied Sciences (*see* Applied Science and Technology; Engineering and Applied Sciences)
Applied Social Research—Book 2
Applied Statistics (*see* Statistics)
Aquaculture—Book 4
Arab Studies (*see* Near and Middle Eastern Studies)
Arabic (*see* Near and Middle Eastern Languages)
Archaeology—Book 2
Architectural Engineering—Book 5
Architectural History—Book 2
Architecture—Book 2
Archives Administration (*see* Public History)
Area and Cultural Studies (*see* African-American Studies; African Studies; American Indian/Native American Studies; American Studies; Asian-American Studies; Asian Studies; Canadian Studies; East European and Russian Studies; Ethnic Studies; Gender Studies; Hispanic Studies; Jewish Studies; Latin American Studies; Near and Middle Eastern Studies; Northern Studies; Western European Studies; Women's Studies)
Art Education—Book 6
Art/Fine Arts—Book 2

Art History—Book 2
Arts Administration—Book 2
Art Therapy—Book 2
Artificial Intelligence/Robotics—Book 5
Asian-American Studies—Book 2
Asian Languages—Book 2
Asian Studies—Book 2
Astronautical Engineering (*see* Aerospace/Aeronautical Engineering)
Astronomy—Book 4
Astrophysical Sciences (*see* Astrophysics; Atmospheric Sciences; Meteorology; Planetary Sciences)
Astrophysics—Book 4
Athletics Administration (*see* Exercise and Sports Science; Kinesiology and Movement Studies; Physical Education; Sports Management)
Athletic Training and Sports Medicine—Book 6
Atmospheric Sciences—Book 4
Audiology (*see* Communication Disorders)
Automotive Engineering—Book 5
Aviation—Book 5
Aviation Management—Book 6
Bacteriology—Book 3
Banking (*see* Finance and Banking)
Behavioral Genetics (*see* Biopsychology)
Behavioral Sciences (*see* Biopsychology; Neuroscience; Psychology; Zoology)
Bible Studies (*see* Religion; Theology)
Bilingual and Bicultural Education (*see* Multilingual and Multicultural Education)
Biochemical Engineering—Book 5
Biochemistry—Book 3
Bioengineering—Book 5
Bioethics—Book 6
Bioinformatics—Book 5
Biological and Biomedical Sciences—Book 3
Biological Anthropology—Book 2
Biological Chemistry (*see* Biochemistry)
Biological Engineering (*see* Bioengineering)
Biological Oceanography (*see* Marine Biology; Marine Sciences; Oceanography)
Biomathematics (*see* Biometrics)
Biomedical Engineering—Book 5
Biometrics—Book 4
Biophysics—Book 3
Biopsychology—Book 3
Biostatistics—Book 4
Biosystems Engineering—Book 5
Biotechnology—Book 5
Black Studies (*see* African-American Studies)
Botany—Book 3
Breeding (*see* Animal Sciences; Botany and Plant Biology; Genetics; Horticulture)
Broadcasting (*see* Communication; Media Studies)
Building Science—Book 2
Business Administration and Management—Book 6
Business Education—Book 6
Canadian Studies—Book 2
Cancer Biology/Oncology—Book 3
Cardiovascular Sciences—Book 3
Cell Biology—Book 3
Cellular Physiology (*see* Cell Biology; Physiology)
Celtic Languages—Book 2
Ceramic Engineering (*see* Ceramic Sciences and Engineering)
Ceramic Sciences and Engineering—Book 5
Ceramics (*see* Art/Fine Arts; Ceramic Sciences and Engineering)
Cereal Chemistry (*see* Food Science and Technology)
Chemical Engineering—Book 5
Chemical Physics—Book 4
Chemistry—Book 4
Child and Family Studies—Book 2

Child-Care Nursing (*see* Maternal/Child Nursing)
Child Development—Book 2
Child-Health Nursing (*see* Maternal/Child Nursing)
Chinese—Book 2
Chinese Studies (*see* Asian Languages; Asian Studies)
Chiropractic—Book 6
Christian Studies (*see* Missions and Missiology; Religion; Religious Education; Theology)
Cinema (*see* Film, Television, and Video Production; Media Studies)
City and Regional Planning (*see* Urban and Regional Planning)
Civil Engineering—Book 5
Classical Languages and Literatures (*see* Classics)
Classics—Book 2
Clinical Laboratory Sciences/Medical Technology—Book 6
Clinical Microbiology (*see* Medical Microbiology)
Clinical Psychology—Book 2
Clinical Research—Book 6
Clothing and Textiles—Book 2
Cognitive Sciences—Book 2
Communication—Book 2
Communication Disorders—Book 6
Communication Theory (*see* Communication)
Community Affairs (*see* Urban and Regional Planning; Urban Studies)
Community College Education—Book 6
Community Health—Book 6
Community Health Nursing—Book 6
Community Planning (*see* Architecture; Environmental Design; Urban and Regional Planning; Urban Design; Urban Studies)
Community Psychology (*see* Social Psychology)
Comparative and Interdisciplinary Arts—Book 2
Comparative Literature—Book 2
Composition (*see* Music)
Computational Biology—Book 3
Computational Sciences—Book 4
Computer Art and Design—Book 2
Computer Education—Book 6
Computer Engineering—Book 5
Computer Science—Book 5
Computing Technology (*see* Computer Science)
Condensed Matter Physics—Book 4
Conflict Resolution and Mediation/Peace Studies—Book 2
Conservation Biology—Book 3
Construction Engineering and Management—Book 5
Consumer Economics—Book 2
Continuing Education (*see* Adult Education)
Corporate and Organizational Communication—Book 2
Corrections (*see* Criminal Justice and Criminology)
Counseling (*see* Addictions/Substance Abuse Counseling; Counseling Psychology; Counselor Education; Genetic Counseling; Pastoral Ministry and Counseling; Rehabilitation Counseling)
Counseling Psychology—Book 2
Counselor Education—Book 6
Crafts (*see* Art/Fine Arts)
Creative Arts Therapies (*see* Art Therapy; Therapies—Dance, Drama, and Music)
Criminal Justice and Criminology—Book 2
Crop Sciences (*see* Agricultural Sciences; Agronomy and Soil Sciences; Botany; Plant Biology; Plant Sciences)
Cultural Studies—Book 2
Curriculum and Instruction—Book 6
Cytology (*see* Cell Biology)
Dairy Science (*see* Animal Sciences)
Dance—Book 2
Dance Therapy (*see* Therapies—Dance, Drama, and Music)
Decorative Arts—Book 2
Demography and Population Studies—Book 2
Dental and Oral Surgery (*see* Oral and Dental Sciences)
Dental Assistant Studies (*see* Dental Hygiene)
Dental Hygiene—Book 6
Dental Services (*see* Dental Hygiene)
Dentistry—Book 6
Design (*see* Applied Arts and Design; Architecture; Art/Fine Arts; Environmental Design; Graphic Design; Industrial Design; Interior Design; Textile Design; Urban Design)

Developmental Biology—Book 3
Developmental Education—Book 6
Developmental Psychology—Book 2
Dietetics (*see* Nutrition)
Diplomacy (*see* International Affairs)
Disability Studies—Book 2
Distance Education Development—Book 6
Drama/Theater Arts (*see* Theater)
Drama Therapy (*see* Therapies—Dance, Drama, and Music)
Dramatic Arts (*see* Theater)
Drawing (*see* Art/Fine Arts)
Drug Abuse Counseling (*see* Addictions/Substance Abuse Counseling; Counselor Education)
Early Childhood Education—Book 6
Earth Sciences (*see* Geosciences)
East Asian Studies (*see* Asian Studies)
East European and Russian Studies—Book 2
Ecology—Book 3
Economics—Book 2
Education—Book 6
Educational Administration—Book 6
Educational Leadership (*see* Educational Administration)
Educational Measurement and Evaluation—Book 6
Educational Media/Instructional Technology—Book 6
Educational Policy—Book 6
Educational Psychology—Book 6
Educational Theater (*see* Therapies—Dance, Drama, and Music; Theater; Education)
Education of the Blind (*see* Special Education)
Education of the Deaf (*see* Special Education)
Education of the Gifted—Book 6
Education of the Hearing Impaired (*see* Special Education)
Education of the Learning Disabled (*see* Special Education)
Education of the Mentally Retarded (*see* Special Education)
Education of the Multiply Handicapped—Book 6
Education of the Physically Handicapped (*see* Special Education)
Education of the Visually Handicapped (*see* Special Education)
Electrical Engineering—Book 5
Electronic Commerce—Book 6
Electronic Materials—Book 6
Electronics Engineering (*see* Electrical Engineering)
Elementary Education—Book 6
Embryology (*see* Developmental Biology)
Emergency Medical Services—Book 6
Endocrinology (*see* Physiology)
Energy and Power Engineering—Book 5
Energy Management and Policy—Book 5
Engineering and Applied Sciences—Book 5
Engineering and Public Affairs (*see* Management of Engineering and Technology; Technology and Public Policy)
Engineering and Public Policy (*see* Management of Engineering and Technology; Technology and Public Policy)
Engineering Design—Book 5
Engineering Management—Book 5
Engineering Mechanics (*see* Mechanics)
Engineering Metallurgy (*see* Metallurgical Engineering and Metallurgy)
Engineering Physics—Book 5
English—Book 2
English as a Second Language—Book 6
English Education—Book 6
Entomology—Book 3
Entrepreneurship—Book 6
Environmental and Occupational Health—Book 6
Environmental Biology—Book 3
Environmental Design—Book 2
Environmental Education—Book 6
Environmental Engineering—Book 5
Environmental Management and Policy—Book 4
Environmental Sciences—Book 4
Environmental Studies (*see* Environmental Management and Policy)
Epidemiology—Book 6
Ergonomics and Human Factors—Book 5
Ethics—Book 2
Ethnic Studies—Book 2

Ethnomusicology (*see* Music)
Evolutionary Biology—Book 3
Exercise and Sports Science—Book 6
Experimental Psychology—Book 2
Experimental Statistics (*see* Statistics)
Facilities Management—Book 6
Family and Consumer Sciences—Book 2
Family Studies (*see* Child and Family Studies)
Family Therapy (*see* Marriage and Family Therapy)
Filmmaking (*see* Film, Television, and Video Production)
Film Studies (*see* Film, Television, and Video Production; Media Studies)
Film, Television, and Video Production—Book 2
Film, Television, and Video Theory and Criticism—Book 2
Finance and Banking—Book 6
Financial Engineering—Book 5
Fine Arts (*see* Art/Fine Arts)
Fire Protection Engineering—Book 5
Fish, Game, and Wildlife Management—Book 4
Folklore—Book 2
Food Engineering (*see* Agricultural Engineering)
Foods (*see* Food Science and Technology; Nutrition)
Food Science and Technology—Book 4
Food Services Management (*see* Hospitality Management)
Foreign Languages (*see* specific languages)
Foreign Languages Education—Book 6
Foreign Service (*see* International Affairs)
Forensic Nursing—Book 6
Forensic Psychology—Book 2
Forensics (*see* Speech and Interpersonal Communication)
Forensic Sciences—Book 2
Forestry—Book 4
Foundations and Philosophy of Education—Book 6
French—Book 2
Game and Wildlife Management (*see* Fish, Game, and Wildlife Management)
Gas Engineering (*see* Petroleum Engineering)
Gender Studies—Book 2
General Studies (*see* Liberal Studies)
Genetic Counseling—Book 2
Genetics—Book 3
Genomic Sciences—Book 3
Geochemistry—Book 4
Geodetic Sciences—Book 4
Geographic Information Systems—Book 2
Geography—Book 2
Geological Engineering—Book 5
Geological Sciences (*see* Geology)
Geology—Book 4
Geophysical Fluid Dynamics (*see* Geophysics)
Geophysics—Book 4
Geophysics Engineering (*see* Geological Engineering)
Geosciences—Book 4
Geotechnical Engineering—Book 5
German—Book 2
Gerontological Nursing—Book 6
Gerontology—Book 2
Government (*see* Political Science)
Graphic Design—Book 2
Greek (*see* Classics)
Guidance and Counseling (*see* Counselor Education)
Hazardous Materials Management—Book 5
Health Education—Book 6
Health Informatics—Book 5
Health Physics/Radiological Health—Book 6
Health Promotion—Book 6
Health Psychology—Book 2
Health-Related Professions (*see* individual allied health professions)
Health Sciences (*see* Public Health; Community Health)
Health Services Management and Hospital Administration—Book 6
Health Services Research—Book 6
Health Systems (*see* Safety Engineering; Systems Engineering)
Hearing Sciences (*see* Communication Disorders)
Hebrew (*see* Near and Middle Eastern Languages)

Hebrew Studies (*see* Jewish Studies)
Higher Education—Book 6
Highway Engineering (*see* Transportation and Highway Engineering)
Hispanic Studies—Book 2
Histology (*see* Anatomy; Cell Biology)
Historic Preservation—Book 2
History—Book 2
History of Art (*see* Art History)
History of Medicine—Book 2
History of Science and Technology—Book 2
HIV-AIDS Nursing—Book 6
Holocaust Studies—Book 2
Home Economics (*see* Family and Consumer Sciences)
Home Economics Education—Book 6
Horticulture—Book 4
Hospice Nursing—Book 6
Hospital Administration (*see* Health Services Management and Hospital Administration)
Hospitality Administration (*see* Hospitality Management)
Hospitality Management—Book 6
Hotel Management (*see* Travel and Tourism)
Household Economics, Sciences, and Management (*see* Consumer Economics)
Human-Computer Interaction—Book 5
Human Development—Book 2
Human Ecology (*see* Family and Consumer Sciences)
Human Factors (*see* Ergonomics and Human Factors)
Human Genetics—Book 3
Humanistic Psychology (*see* Transpersonal and Humanistic Psychology)
Humanities—Book 2
Human Movement Studies (*see* Dance; Exercise and Sports Sciences; Kinesiology and Movement Studies)
Human Resources Development—Book 6
Human Resources Management—Book 6
Human Services—Book 6
Hydraulics—Book 5
Hydrogeology—Book 4
Hydrology—Book 4
Illustration—Book 2
Immunology—Book 3
Industrial Administration—Book 6
Industrial and Labor Relations—Book 2
Industrial and Manufacturing Management—Book 6
Industrial and Organizational Psychology—Book 2
Industrial Design—Book 2
Industrial Education (*see* Vocational and Technical Education)
Industrial Hygiene—Book 6
Industrial/Management Engineering—Book 5
Infectious Diseases—Book 3
Information Science—Book 5
Information Studies—Book 6
Inorganic Chemistry—Book 4
Instructional Technology (*see* Educational Media/Instructional Technology)
Insurance—Book 6
Interdisciplinary Studies—Book 2
Interior Design—Book 2
International Affairs—Book 2
International and Comparative Education—Book 6
International Business—Book 6
International Commerce (*see* International Business; International Development)
International Development—Book 2
International Economics (*see* Economics; International Affairs; International Business; International Development)
International Health—Book 6
International Service (*see* International Affairs)
International Trade (*see* International Business)
Internet and Interactive Multimedia—Book 2
Interpersonal Communication (*see* Speech and Interpersonal Communication)
Interpretation (*see* Translation and Interpretation)

Investment and Securities (*see* Business Administration and Management; Finance and Banking; Investment Management)

Investment Management—Book 6

Islamic Studies (*see* Near and Middle Eastern Studies; Religion)

Italian—Book 2

Japanese—Book 2

Japanese Studies (*see* Asian Languages; Asian Studies)

Jewelry/Metalsmithing (*see* Art/Fine Arts)

Jewish Studies—Book 2

Journalism—Book 2

Judaic Studies (*see* Jewish Studies; Religion; Religious Education)

Junior College Education (*see* Community College Education)

Kinesiology and Movement Studies—Book 6

Labor Relations (*see* Industrial and Labor Relations)

Laboratory Medicine (*see* Clinical Laboratory Sciences/Medical Technology; Immunology; Microbiology; Pathobiology; Pathology)

Landscape Architecture—Book 2

Latin (*see* Classics)

Latin American Studies—Book 2

Law—Book 6

Law Enforcement (*see* Criminal Justice and Criminology)

Legal and Justice Studies—Book 6

Leisure Studies—Book 6

Liberal Studies—Book 2

Librarianship (*see* Library Science)

Library Science—Book 6

Life Sciences (*see* Biological and Biomedical Sciences)

Limnology—Book 4

Linguistics—Book 2

Literature (*see* Classics; Comparative Literature; specific language)

Logistics—Book 6

Macromolecular Science (*see* Polymer Science and Engineering)

Management (*see* Business Administration and Management)

Management Engineering (*see* Engineering Management; Industrial/Management Engineering)

Management Information Systems—Book 6

Management of Engineering and Technology—Book 5

Management of Technology—Book 5

Management Strategy and Policy—Book 6

Manufacturing Engineering—Book 5

Marine Affairs—Book 4

Marine Biology—Book 3

Marine Engineering (*see* Civil Engineering)

Marine Geology—Book 4

Marine Sciences—Book 4

Marine Studies (*see* Marine Affairs; Marine Geology; Marine Sciences; Oceanography)

Marketing—Book 6

Marketing Research—Book 6

Marriage and Family Therapy—Book 2

Mass Communication—Book 2

Materials Engineering—Book 5

Materials Sciences—Book 5

Maternal and Child Health—Book 6

Maternal/Child Nursing—Book 6

Maternity Nursing (*see* Maternal/Child Nursing)

Mathematical and Computational Finance—Book 4

Mathematical Physics—Book 4

Mathematical Statistics (*see* Statistics)

Mathematics—Book 4

Mathematics Education—Book 6

Mechanical Engineering—Book 5

Mechanics—Book 5

Media Studies—Book 2

Medical Illustration—Book 2

Medical Informatics—Book 5

Medical Microbiology—Book 3

Medical Nursing (*see* Medical/Surgical Nursing)

Medical Physics—Book 6

Medical Sciences (*see* Biological and Biomedical Sciences)

Medical Science Training Programs (*see* Biological and Biomedical Sciences)

Medical/Surgical Nursing—Book 6

Medical Technology (*see* Clinical Laboratory Sciences/Medical Technology)

Medicinal and Pharmaceutical Chemistry—Book 6

Medicinal Chemistry (*see* Medicinal and Pharmaceutical Chemistry)

Medicine (*see* Allopathic Medicine; Naturopathic Medicine; Osteopathic Medicine; Podiatric Medicine)

Medieval and Renaissance Studies—Book 2

Metallurgical Engineering and Metallurgy—Book 5

Metallurgy (*see* Metallurgical Engineering and Metallurgy)

Metalsmithing (*see* Art/Fine Arts)

Meteorology—Book 4

Microbiology—Book 3

Middle Eastern Studies (*see* Near and Middle Eastern Studies)

Middle School Education—Book 6

Midwifery (*see* Nurse Midwifery)

Military and Defense Studies—Book 2

Mineral Economics—Book 2

Mineral/Mining Engineering—Book 5

Mineralogy—Book 4

Ministry (*see* Pastoral Ministry and Counseling; Theology)

Missions and Missiology—Book 2

Molecular Biology—Book 3

Molecular Biophysics—Book 3

Molecular Genetics—Book 3

Molecular Medicine—Book 3

Molecular Pathology—Book 3

Molecular Pharmacology—Book 3

Molecular Physiology—Book 3

Molecular Taxicology—Book 3

Motion Pictures (*see* Film, Television, and Video Production; Media Studies)

Movement Studies (*see* Dance; Exercise and Sports Science; Kinesiology and Movement Studies)

Multilingual and Multicultural Education—Book 6

Museum Education—Book 6

Museum Studies—Book 2

Music—Book 2

Music Education—Book 6

Music History (*see* Music)

Musicology (*see* Music)

Music Theory (*see* Music)

Music Therapy (*see* Therapies—Dance, Drama, and Music)

Native American Studies (*see* American Indian/Native American Studies)

Natural Resources—Book 4

Natural Resources Management (*see* Environmental Management and Policy; Natural Resources)

Naturopathic Medicine—Book 6

Near and Middle Eastern Languages—Book 2

Near and Middle Eastern Studies—Book 2

Near Environment (*see* Family and Consumer Sciences; Human Development)

Neural Sciences (*see* Biopsychology; Neuroscience)

Neurobiology—Book 3

Neuroendocrinology (*see* Biopsychology; Neuroscience; Physiology)

Neuropharmacology (*see* Biopsychology; Neuroscience; Pharmacology)

Neurophysiology (*see* Biopsychology; Neuroscience; Physiology)

Neuroscience—Book 3

Nonprofit Management—Book 6

North American Studies (*see* Northern Studies)

Northern Studies—Book 2

Nuclear Engineering—Book 5

Nuclear Medical Technology (*see* Clinical Laboratory Sciences/Medical Technology)

Nuclear Physics (*see* Physics)

Nurse Anesthesia—Book 6

Nurse Midwifery—Book 6

Nurse Practitioner Studies (*see* Advanced Practice Nursing)

Nursery School Education (*see* Early Childhood Education)

Nursing—Book 6

Nursing and Healthcare Administration—Book 6

Nursing Education—Book 6

Nutrition—Book 3

Occupational Education (*see* Vocational and Technical Education)
Occupational Health (*see* Environmental and Occupational Health; Occupational Health Nursing)
Occupational Health Nursing—Book 6
Occupational Therapy—Book 6
Ocean Engineering—Book 5
Oceanography—Book 4
Oncology—Book 3
Oncology Nursing—Book 6
Operations Research—Book 5
Optical Sciences—Book 4
Optical Technologies (*see* Optical Sciences)
Optics (*see* Applied Physics; Optical Sciences; Physics)
Optometry—Book 6
Oral and Dental Sciences—Book 6
Oral Biology (*see* Oral and Dental Sciences)
Oral Pathology (*see* Oral and Dental Sciences)
Organic Chemistry—Book 4
Organismal Biology (*see* Biological and Biomedical Sciences; Zoology)
Organizational Behavior—Book 6
Organizational Management—Book 6
Organizational Psychology (*see* Industrial and Organizational Psychology)
Organizational Studies—Book 6
Oriental Languages (*see* Asian Languages)
Oriental Medicine—Book 6
Oriental Studies (*see* Asian Studies)
Orthodontics (*see* Oral and Dental Sciences)
Osteopathic Medicine—Book 6
Painting/Drawing (*see* Art/Fine Arts)
Paleontology—Book 4
Paper and Pulp Engineering—Book 5
Paper Chemistry (*see* Chemistry)
Parasitology—Book 3
Park Management (*see* Recreation and Park Management)
Pastoral Ministry and Counseling—Book 2
Pathobiology—Book 3
Pathology—Book 3
Peace Studies (*see* Conflict Resolution and Mediation/Peace Studies)
Pediatric Nursing—Book 6
Pedodontics (*see* Oral and Dental Sciences)
Performance (*see* Music)
Performing Arts (*see* Dance; Music; Theater)
Periodontics (*see* Oral and Dental Sciences)
Personnel (*see* Human Resources Development; Human Resources Management; Organizational Behavior; Organizational Management; Organizational Studies)
Petroleum Engineering—Book 5
Pharmaceutical Administration—Book 6
Pharmaceutical Chemistry (*see* Medicinal and Pharmaceutical Chemistry)
Pharmaceutical Engineering—Book 5
Pharmaceutical Sciences—Book 6
Pharmacognosy (*see* Pharmaceutical Sciences)
Pharmacology—Book 3
Pharmacy—Book 6
Philanthropic Studies—Book 2
Philosophy—Book 2
Philosophy of Education (*see* Foundations and Philosophy of Education)
Photobiology of Cells and Organelles (*see* Botany and Plant Biology; Cell Biology)
Photography—Book 2
Photonics—Book 4
Physical Chemistry—Book 4
Physical Education—Book 6
Physical Therapy—Book 6
Physician Assistant Studies—Book 6
Physics—Book 4
Physiological Optics (*see* Physiology; Vision Sciences)
Physiology—Book 3
Planetary Sciences—Book 4
Plant Biology—Book 3
Plant Molecular Biology—Book 3

Plant Pathology—Book 3
Plant Physiology—Book 3
Plant Sciences—Book 4
Plasma Physics—Book 4
Plastics Engineering (*see* Polymer Science and Engineering)
Playwriting (*see* Theater; Writing)
Podiatric Medicine—Book 6
Policy Studies (*see* Educational Policy; Energy Management and Policy; Environmental Management and Policy; Public Policy; Strategy and Policy; Technology and Public Policy)
Political Science—Book 2
Polymer Science and Engineering—Book 5
Pomology (*see* Agricultural Sciences; Botany and Plant Biology; Horticulture; Plant Sciences)
Population Studies (*see* Demography and Population Studies)
Portuguese—Book 2
Poultry Science (*see* Animal Sciences)
Power Engineering—Book 5
Preventive Medicine (*see* Public Health; Community Health)
Printmaking (*see* Art/Fine Arts)
Product Design (*see* Environmental Design; Industrial Design)
Project Management—Book 6
Psychiatric Nursing—Book 6
Psychoanalysis and Psychotherapy—Book 2
Psychobiology (*see* Biopsychology)
Psychology—Book 2
Psychopharmacology (*see* Biopsychology; Neuroscience; Pharmacology)
Public Address (*see* Speech and Interpersonal Communication)
Public Administration—Book 2
Public Affairs—Book 2
Public Health—Book 6
Public Health Nursing (*see* Community Health Nursing)
Public History—Book 2
Public Policy—Book 2
Public Relations (*see* Advertising and Public Relations)
Publishing—Book 2
Quality Management—Book 6
Quantitative Analysis—Book 6
Radiation Biology—Book 3
Radio (*see* Media Studies)
Radiological Health (*see* Health Physics/Radiological Health)
Radiological Physics (*see* Physics)
Range Management (*see* Range Science)
Range Science—Book 4
Reading Education—Book 6
Real Estate—Book 6
Recreation and Park Management—Book 6
Recreation Therapy (*see* Recreation and Park Management)
Regional Planning (*see* Architecture; Environmental Design; Urban and Regional Planning; Urban Design; Urban Studies)
Rehabilitation Counseling—Book 2
Rehabilitation Sciences—Book 6
Rehabilitation Therapy (*see* Physical Therapy)
Reliability Engineering—Book 5
Religion—Book 2
Religious Education—Book 6
Religious Studies (*see* Religion; Theology)
Remedial Education (*see* Special Education)
Renaissance Studies (*see* Medieval and Renaissance Studies)
Reproductive Biology—Book 3
Resource Management (*see* Environmental Management and Policy)
Restaurant Administration (*see* Hospitality Management)
Rhetoric—Book 2
Robotics (*see* Artificial Intelligence/Robotics)
Romance Languages—Book 2
Romance Literatures (*see* Romance Languages)
Rural Planning and Studies—Book 2
Rural Sociology—Book 2
Russian—Book 2
Russian Studies (*see* East European and Russian Studies)
Sacred Music (*see* Music)
Safety Engineering—Book 5
Scandinavian Languages—Book 2

School Nursing—Book 6
School Psychology—Book 2
Science Education—Book 6
Sculpture (*see* Art/Fine Arts)
Secondary Education—Book 6
Security Administration (*see* Criminal Justice and Criminology)
Slavic Languages—Book 2
Slavic Studies (*see* East European and Russian Studies; Slavic Languages)
Social Psychology—Book 2
Social Sciences—Book 2
Social Sciences Education—Book 6
Social Studies Education (*see* Social Sciences Education)
Social Welfare (*see* Social Work)
Social Work—Book 6
Sociobiology (*see* Evolutionary Biology)
Sociology—Book 2
Software Engineering—Book 5
Soil Sciences and Management (*see* Agronomy and Soil Sciences)
Solid-Earth Sciences (*see* Geosciences)
Solid-State Sciences (*see* Materials Sciences)
South and Southeast Asian Studies (*see* Asian Studies)
Space Sciences (*see* Astronomy; Astrophysics; Planetary Sciences)
Spanish—Book 2
Special Education—Book 6
Speech and Interpersonal Communication—Book 2
Speech-Language Pathology (*see* Communication Disorders)
Sport Psychology—Book 2
Sports Management—Book 6
Statistics—Book 4
Strategy and Policy—Book 6
Structural Biology—Book 3
Structural Engineering—Book 5
Student Personnel Services—Book 6
Studio Art (*see* Art/Fine Arts)
Substance Abuse Counseling (*see* Addictions/Substance Abuse Counseling)
Supply Chain Management—Book 6
Surgical Nursing (*see* Medical/Surgical Nursing)
Surveying Science and Engineering—Book 5
Sustainable Development—Book 2
Systems Analysis (*see* Systems Engineering)
Systems Biology—Book 3
Systems Engineering—Book 5
Systems Management (*see* Management Information Systems)
Systems Science—Book 5
Taxation—Book 6
Teacher Education (*see* Education)
Teaching English as a Second Language (*see* English as a Second Language)
Technical Communication—Book 2
Technical Education (*see* Vocational and Technical Education)
Technical Writing—Book 2
Technology and Public Policy—Book 5
Telecommunications—Book 5

Telecommunications Management—Book 5
Television (*see* Film, Television, and Video Production; Media Studies)
Teratology (*see* Developmental Biology; Environmental and Occupational Health; Pathology)
Textile Design—Book 2
Textile Sciences and Engineering—Book 5
Textiles (*see* Clothing and Textiles; Textile Design; Textile Sciences and Engineering)
Thanatology—Book 2
Theater—Book 2
Theology—Book 2
Theoretical Biology (*see* Biological and Biomedical Sciences)
Theoretical Chemistry—Book 4
Theoretical Physics—Book 4
Theory and Criticism of Film, Television, and Video (*see* Film, Television, and Video Theory and Criticism)
Therapeutic Recreation—Book 6
Therapeutics (*see* Pharmaceutical Sciences; Pharmacology; Pharmacy)
Therapies—Dance, Drama, and Music—Book 2
Toxicology—Book 3
Transcultural Nursing—Book 6
Translation and Interpretation—Book 2
Transpersonal and Humanistic Psychology—Book 2
Transportation and Highway Engineering—Book 5
Transportation Management—Book 6
Travel and Tourism—Book 6
Tropical Medicine (*see* Parasitology)
Urban and Regional Planning—Book 2
Urban Design—Book 2
Urban Education—Book 6
Urban Studies—Book 2
Urban Systems Engineering (*see* Systems Engineering)
Veterinary Medicine—Book 6
Veterinary Sciences—Book 6
Video (*see* Film, Television, and Video Production; Media Studies)
Virology—Book 3
Vision Sciences—Book 6
Visual Arts (*see* Applied Arts and Design; Art/Fine Arts; Film, Television, and Video Production; Graphic Design; Illustration; Media Studies; Photography)
Vocational and Technical Education—Book 6
Vocational Counseling (*see* Counselor Education)
Waste Management (*see* Hazardous Materials Management)
Water Resources—Book 4
Water Resources Engineering—Book 5
Western European Studies—Book 2
Wildlife Biology (*see* Zoology)
Wildlife Management (*see* Fish, Game, and Wildlife Management)
Women's Health Nursing—Book 6
Women's Studies—Book 2
World Wide Web (*see* Internet and Interactive Multimedia)
Writing—Book 2
Zoology—Book 3

Directories and Subject Areas in This Book

NOTES

NOTES

Give Your Admissions Essay an Edge at EssayEdge.com™

FACT:

The essay is the primary tool admissions officers use to decide among hundreds or even thousands of applicants with comparable experience and academic credentials.

FACT:

More than one-third of the time an admissions officer spends on your application will be spent evaluating your essay.

Winning Essays Start at EssayEdge.com

"One of the Best Essay Services on the Internet"
—The Washington Post

"The World's Premier Application Essay Editing Service"
— The New York Times Learning Network

EssayEdge.com's Harvard-educated editors have helped more satisfied applicants create essays that get results than any other company in the world.

Visit EssayEdge.com today to learn how our quick, convenient service can help you take your admissions essay to a new level.

Use this coupon code when ordering an EssayEdge.com service and SAVE 10%

EPGRAD07

EE2006